2013 50th ACM/EDAC/IEEE Design Automation Conference

(DAC 2013)

Austin, Texas, USA
29 May – 7 June 2013

Pages 1-644

IEEE Catalog Number: CFP13DAC-PRT
ISBN: 978-1-4503-2071-9

Copyright © 2013, Association for Computing Machinery
All Rights Reserved

***This publication is a representation of what appears in the IEEE
Digital Libraries. Some format issues inherent in the e-media version may
also appear in this print version.**

IEEE Catalog Number:	CFP13DAC-PRT
ISBN 13:	978-1-4503-2071-9
ISSN:	0738-100X

Additional Copies of This Publication Are Available From:

Curran Associates, Inc
57 Morehouse Lane
Red Hook, NY 12571 USA
Phone: (845) 758-0400
Fax: (845) 758-2633
E-mail: curran@proceedings.com
Web: www.proceedings.com

TABLE OF CONTENTS

MAPPING ON MULTI/MANY-CORE SYSTEMS: SURVEY OF CURRENT AND EMERGING TRENDS 1
A. Singh, M. Shafique, A. Kumar, J. Henkel

WORKLOAD AND USER EXPERIENCE-AWARE DYNAMIC RELIABILITY MANAGEMENT IN MULTICORE PROCESSORS 11
P. Mercati, A. Bartolini, T. Rosing, L. Benini, F. Paterna

LIVENESS EVALUATION OF A CYCLO-STATIC DATAFLOW GRAPH 17
M. Benazouz, A. Munier-Kordon, T. Hujsa, B. Bodin

DOUBLE PATTERNING LITHOGRAPHY-AWARE ANALOG PLACEMENT 24
H. Chien, H. Ou, T. Chen, Y. Kuan, Y. Chang

SIMULTANEOUS ANALOG PLACEMENT AND ROUTING WITH CURRENT FLOW AND CURRENT DENSITY CONSIDERATIONS 30
H. Ou, H. Chien, Y. Chang

COUPLING-AWARE LENGTH-RATIO-MATCHING ROUTING FOR CAPACITOR ARRAYS IN ANALOG INTEGRATED CIRCUITS 36
K. Ho, H. Ou, Y. Chang, H. Tsao

DIGITAL-ASSISTED NOISE-ELIMINATING TRAINING FOR MEMRISTOR CROSSBAR-BASED ANALOG NEUROMORPHIC COMPUTING ENGINE 42
B. Liu, M. Hu, H. Li, Z. Mao, Y. Chen, T. Huang, W. Zhang

DYNAMIC BEHAVIOR OF CELL SIGNALING NETWORKS – MODEL DESIGN AND ANALYSIS AUTOMATION 48
N. Miskov-Zivanov, D. Marculescu, J. Faeder

DEFECT TOLERANCE IN NANODEVICE-BASED PROGRAMMABLE INTERCONNECTS: UTILIZATION BEYOND AVOIDANCE 54
J. Cong, B. Xiao

AN EFFICIENT AND EFFECTIVE ANALYTICAL PLACER FOR FPGAS 62
T. Lin, P. Banerjee, Y. Chang

THROUGHPUT-ORIENTED KERNEL PORTING ONTO FPGAS 68
A. Papakonstantinou, J. Cong, D. Chen, Y. Liang, W. Hwu

MEMORY PARTITIONING FOR MULTIDIMENSIONAL ARRAYS IN HIGH-LEVEL SYNTHESIS 78
Y. Wang, P. Li, P. Zhang, C. Zhang, J. Cong

BALANCING SECURITY AND UTILITY IN MEDICAL DEVICES? 86
M. Rostami, W. Burleson, F. Koushanfar, A. Juels

TOWARDS TRUSTWORTHY MEDICAL DEVICES AND BODY AREA NETWORKS 92
M. Zhang, A. Raghunathan, N. Jha

LOW-ENERGY ENCRYPTION FOR MEDICAL DEVICES: SECURITY ADDS AN EXTRA DESIGN DIMENSION 98
J. Fan, O. Reparaz, V. Rozic, I. Verbauwhede

AGING-AWARE COMPILER-DIRECTED VLIW ASSIGNMENT FOR GPGPU ARCHITECTURES 104
A. Rahimi, L. Benini, R. Gupta

EXPLOITING PROGRAM-LEVEL MASKING AND ERROR PROPAGATION FOR CONSTRAINED RELIABILITY OPTIMIZATION 110
M. Shafique, S. Rehman, P. Aceituno, J. Henkel

REGIMAP: REGISTER-AWARE APPLICATION MAPPING ON COARSE-GRAINED RECONFIGURABLE ARCHITECTURES (CGRAS) 119
M. Hamzeh, A. Shrivastava, S. Vrudhula

POLYHEDRAL MODEL BASED MAPPING OPTIMIZATION OF LOOP NESTS FOR CGRAS 129
D. Liu, S. Yin, L. Liu, S. Wei

IMPROVING ENERGY GAINS OF INEXACT DSP HARDWARE THROUGH RECIPROCATIVE ERROR COMPENSATION 137
A. Lingamneni, A. Basu, K. Palem, C. Piguet, C. Enz

EARLY PARTIAL EVALUATION IN A JIT-COMPILED, RETARGETABLE INSTRUCTION SET SIMULATOR GENERATED FROM A HIGH-LEVEL ARCHITECTURE DESCRIPTION 145
H. Wagstaff, M. Gould, B. Franke, N. Topham

XDRA: EXPLORATION AND OPTIMIZATION OF LAST LEVEL CACHE FOR ENERGY REDUCTION IN DDR DRAMS 151
S. Min, H. Javaid, S. Parameswaran

TOWARDS VARIATION-AWARE SYSTEM-LEVEL POWER ESTIMATION OF DRAMS: AN EMPIRICAL APPROACH 161
K. Chandrasekar, C. Weis, B. Akesson, N. Wehn, K. Goossens

TEASE: A SYSTEMATIC ANALYSIS FRAMEWORK FOR EARLY EVALUATION OF FINFET-BASED ADVANCED TECHNOLOGY NODES 169
A. Mallik, P. Zuber, T. Liu, B. Chava, B. Ballal, P. Bario, R. Baert, K. Croes, J. Ryckaert, M. Badaroglu, A. Mercha, D. Verkest

STITCH-AWARE ROUTING FOR MULTIPLE E-BEAM LITHOGRAPHY 175
S. Fang, I. Liu, Y. Chang

AUTOMATIC DESIGN RULE CORRECTION IN PRESENCE OF MULTIPLE GRIDS AND TRACK PATTERNS 181
N. Salodkar, S. Rajagopalan, S. Batterywala, S. Bhattacharya

MULTIPLE CHIP PLANNING FOR CHIP-INTERPOSER CODESIGN 187
Y. Ho, Y. Chang

GPU-BASED N-DETECT TRANSITION FAULT ATPG 193
K. Liao, S. Hsu, J. Li

POST-SILICON CONFORMANCE CHECKING WITH VIRTUAL PROTOTYPES 201
L. Lei, F. Xie, K. Cong

ON TESTING TIMING-SPECULATIVE CIRCUITS 207
F. Yuan, Y. Liu, W. Jone, Q. Xu

AN ATE ASSISTED DFD TECHNIQUE FOR VOLUME DIAGNOSIS OF SCAN CHAINS 213
S. Kundu, S. Chattopadhyay, I. Sengupta, R. Kapur

PREDICTING FUTURE TECHNOLOGY PERFORMANCE 219
A. Asenov, C. Alexander, C. Riddet, E. Towie

PREDICTING FUTURE PRODUCT PERFORMANCE: MODELING AND EVALUATION OF STANDARD CELLS IN FINFET TECHNOLOGIES 225
V. Kleeberger, H. Graeb, U. Schlichtmann

THE ITRS DESIGN TECHNOLOGY AND SYSTEM DRIVERS ROADMAP: PROCESS AND STATUS 231
A. Kahng

PROACTIVE CIRCUIT ALLOCATION IN MULTIPLANE NOCS 237
A. Abousamra, A. Jones, R. Melham

A HETEROGENEOUS MULTIPLE NETWORK-ON-CHIP DESIGN: AN APPLICATION-AWARE APPROACH 247
A. Mishra, O. Mutlu, C. Das

DESIGNING ENERGY-EFFICIENT NOC FOR REAL-TIME EMBEDDED SYSTEMS THROUGH SLACK OPTIMIZATION 257
J. Zhan, N. Stoimenov, J. Ouyang, L. Thiele, V. Narayanan, Y. Xie

RISO: RELAXED NETWORK-ON-CHIP ISOLATION FOR CLOUD PROCESSORS 263
H. Lu, G. Yan, Y. Han, B. Fu, X. Li

SMART HILL CLIMBING FOR AGILE DYNAMIC MAPPING IN MANY-CORE SYSTEMS 269
M. Fattah, M. Daneshtalab, P. Liljeberg, J. Plosila

HCI-TOLERANT NOC ROUTER MICROARCHITECTURE 275
D. Ancajas, J. Nickerson, K. Chakraborty, S. Roy

OPTIMIZATION OF QUANTUM CIRCUITS FOR INTERACTION DISTANCE IN LINEAR NEAREST NEIGHBOR ARCHITECTURES 285
A. Shafaei, M. Saeedi, M. Pedram

LEQA: LATENCY ESTIMATION FOR A QUANTUM ALGORITHM MAPPED TO A QUANTUM CIRCUIT FABRIC 291
M. Dousti, M. Pedram

PARETO EPSILON-DOMINANCE AND IDENTIFIABLE SOLUTIONS FOR BIOCAD MODELING 298
C. Angione, J. Costanza, P. Lio, G. Nicosia, G. Carapezza

DESIGN OF CYBERPHYSICAL DIGITAL MICROFLUIDIC BIOCHIPS UNDER COMPLETION-TIME UNCERTAINTIES IN FLUIDIC OPERATIONS 307
Y. Luo, K. Chakrabarty, T. Ho

GENE MODIFICATION IDENTIFICATION UNDER FLUX CAPACITY UNCERTAINTY 314
M. Yousofshahi, M. Orshansky, K. Lee, S. Hassoun

A FIELD-PROGRAMMABLE PIN-CONSTRAINED DIGITAL MICROFLUIDIC BIOCHIP 319
D. Grissom, P. Brisk

BDS-MAJ: A BDD-BASED LOGIC SYNTHESIS TOOL EXPLOITING MAJORITY LOGIC DECOMPOSITION 328
L. Amaru, P. Gaillardson, G. Micheli

TOWARDS OPTIMAL PERFORMANCE-AREA TRADE-OFF IN ADDERS BY SYNTHESIS OF PARALLEL PREFIX STRUCTURES 334
S. Roy, M. Choudhury, R. Puri, D. Pan

SYNTHESIS OF FEEDBACK DECODERS FOR INITIALIZED ENCODERS 342
K. Tu, J. Jiang

ON LEARNING-BASED METHODS FOR DESIGN-SPACE EXPLORATION WITH HIGH-LEVEL SYNTHESIS 348
H. Liu, L. Carloni

RUNTIME DEPENDENCY ANALYSIS FOR LOOP PIPELINING IN HIGH-LEVEL SYNTHESIS.................. 355
M. Alle, A. Morvan, S. Derrien

A HIGH-LEVEL SYNTHESIS FLOW FOR THE IMPLEMENTATION OF ITERATIVE STENCIL LOOP ALGORITHMS ON FPGA DEVICES 365
A. Nacci, V. Rana, F. Bruschi, D. Sciuto, I. Beretta, D. Atienza

CROSS-LAYER RACETRACK MEMORY DESIGN FOR ULTRA HIGH DENSITY AND LOW POWER CONSUMPTION 371
Z. Sun, W. Wu, H. Li

IMPROVING THE ENERGY EFFICIENCY OF HARDWARE-ASSISTED WATCHPOINT SYSTEMS 377
V. Karakostas, S. Tomic, O. Unsal, M. Nemirovsky, A. Cristal

LOW-POWER AREA-EFFICIENT LARGE-SCALE IP LOOKUP ENGINE BASED ON BINARY-WEIGHTED CLUSTERED NETWORKS 383
N. Onizawa, W. Gross

REAL-TIME USE-AWARE ADAPTIVE MIMO RF RECEIVER SYSTEMS FOR ENERGY EFFICIENCY UNDER BER CONSTRAINTS.. 389
D. Banerjee, S. Devarakond, S. Sen, A. Chatterjee

IMPROVING CHARGING EFFICIENCY WITH WORKLOAD SCHEDULING IN ENERGY HARVESTING EMBEDDED SYSTEMS 396
Y. Zhang, Y. Ge, Q. Qiu

CREATION OF ESL POWER MODELS FOR COMMUNICATION ARCHITECTURES USING AUTOMATIC CALIBRATION.. 404
S. Schurmans, D. Zhang, D. Auras, R. Leupers, G. Ascheid, X. Chen, L. Wang

A TRANSMISSION GATE PHYSICAL UNCLONABLE FUNCTION AND ON-CHIP VOLTAGETO-DIGITAL CONVERSION TECHNIQUE 410
R. Chakraborty, C. Larnech, D. Acharyya, J. Plusquellic

RESP: A ROBUST PHYSICAL UNCLONABLE FUNCTION RETROFITTED INTO EMBEDDED SRAM ARRAY 420
Y. Zheng, M. Hashemian, S. Bhunia

VERITRUST: VERIFICATION FOR HARDWARE TRUST.. 429
J. Zhang, F. Yuan, L. Wei, Z. Sun, Q. Xu

RASTER: RUNTIME ADAPTIVE SPATIAL/TEMPORAL ERROR RESILIENCY FOR EMBEDDED PROCESSORS... 437
T. Li, M. Shafique, J. Ambrose, S. Rehman, J. Henkel, S. Parameswaran

ABCD-L: APPROXIMATING CONTINUOUS LINEAR SYSTEMS USING BOOLEAN MODELS.................... 444
A. Karthik, J. Roychowdhury

BAYESIAN MODEL FUSION: LARGE-SCALE PERFORMANCE MODELING OF ANALOG AND MIXED-SIGNAL CIRCUITS BY REUSING EARLY-STAGE DATA 453
F. Wang, W. Zhang, S. Sun, X. Li, C. Gu

EFFICIENT MOMENT ESTIMATION WITH EXTREMELY SMALL SAMPLE SIZE VIA BAYESIAN INFERENCE FOR ANALOG/MIXED-SIGNAL VALIDATION 459
C. Gu, E. Chiprout, X. Li

VERIFICATION OF DIGITALLY-INTENSIVE ANALOG CIRCUITS VIA KERNEL RIDGE REGRESSION AND HYBRID REACHABILITY ANALYSIS 466
H. Lin, P. Li, C. Myers

MACHINE-LEARNING-BASED HOTSPOT DETECTION USING TOPOLOGICAL CLASSIFICATION AND CRITICAL FEATURE EXTRACTION 472
Y. Yu, G. Lin, I. Jiang, C. Chiang

A NOVEL FUZZY MATCHING MODEL FOR LITHOGRAPHY HOTSPOT DETECTION 478
S. Lin, J. Chen, J. Li, W. Wen, S. Chang

AN EFFICIENT LAYOUT DECOMPOSITION APPROACH FOR TRIPLE PATTERNING LITHOGRAPHY 484
J. Kuang, E. Young

E-BLOW: E-BEAM LITHOGRAPHY OVERLAPPING AWARE STENCIL PLANNING FOR MCC SYSTEM 490
B. Yu, K. Yuan, J. Gao, D. Pan

AUTOMATIC CLUSTERING OF WAFER SPATIAL SIGNATURES 497
W. Zhang, X. Li, S. Saxena, A. Strojwas, R. Rutenbar

MULTIDIMENSIONAL ANALOG TEST METRICS ESTIMATION USING EXTREME VALUE THEORY AND STATISTICAL BLOCKADE 503
S. Haralampos-G., P. Faubet, F. Mohamed, Y. Courant

HIGH-THROUGHPUT TSV TESTING AND CHARACTERIZATION FOR 3D INTEGRATION USING THERMAL MAPPING 510
K. Dev, G. Woods, S. Reda

ON EFFECTIVE AND EFFICIENT IN-FIELD TSV REPAIR FOR STACKED 3D ICS 516
L. Jiang, F. Ye, Q. Xu, K. Chakrabarty, B. Eklow

CLOUD PLATFORMS AND EMBEDDED COMPUTING – THE OPERATING SYSTEMS OF THE FUTURE 522
J. Rellermeyer, S. Lee, M. Kistler

TESSELLATION: REFACTORING THE OS AROUND EXPLICIT RESOURCE CONTAINERS WITH CONTINUOUS ADAPTATION 528
J. Colmenares, G. Eads, S. Hofmeyr, S. Bird, M. Moreto, D. Chou, B. Gluzman, E. Roman, D. Bartolini, N. Mor, K. Asanovic, J. Kubiatowicz

THE AUTONOMIC OPERATING SYSTEM RESEARCH PROJECT – ACHIEVEMENTS AND FUTURE DIRECTIONS 538
D. Bartolini, R. Cattaneo, G. Durelli, M. Maggio, M. Santambrogio, F. Sironi

ROLE OF POWER GRID IN SIDE CHANNEL ATTACK AND POWER-GRID-AWARE SECURE DESIGN 548
X. Wang, W. Yueh, D. Roy, S. Narasimhan, Y. Zheng

NUMCHECKER: DETECTING KERNEL CONTROL-FLOW MODIFYING ROOTKITS BY USING HARDWARE PERFORMANCE COUNTERS 557
X. Wang, R. Karri

HIGH-PERFORMANCE HARDWARE MONITORS TO PROTECT NETWORK PROCESSORS FROM DATA PLANE ATTACKS 564
H. Chandrikakutty, D. Unnikrishnan, R. Tessier, T. Wolf

COMPILER-BASED SIDE CHANNEL VULNERABILITY ANALYSIS AND OPTIMIZED COUNTERMEASURES APPLICATION 570
G. Agosta, A. Barenghi, M. Maggi, G. Pelosi

LIGHTING THE DARK SILICON BY EXPLOITING HETEROGENEITY ON FUTURE PROCESSORS 576
Y. Zhang, L. Peng, X. Fu, Y. Hu

SIMULTANEOUS MULTITHREADING SUPPORT IN EMBEDDED DISTRIBUTED MEMORY MPSOCS 583
R. Garibotti, L. Ost, R. Busseuil, M. Kourouma, C. Adeniyi-Jones, G. Sassatelli, M. Robert

APPLE: ADAPTIVE PERFORMANCE-PREDICTABLE LOW-ENERGY CACHES FOR RELIABLE HYBRID VOLTAGE OPERATION 590
B. Maric, J. Abella, M. Valero

AN OPTIMIZED PAGE TRANSLATION FOR MOBILE VIRTUALIZATION 598
Y. Lee, C. Hsueh

SCALABLE VECTORLESS POWER GRID CURRENT INTEGRITY VERIFICATION 604
Z. Feng

CONSTRAINT ABSTRACTION FOR VECTORLESS POWER GRID VERIFICATION 612
X. Xiong, J. Wang

THE IMPACT OF ELECTROMIGRATION IN COPPER INTERCONNECTS ON POWER GRID INTEGRITY 618
V. Mishra, S. Sapatnekar

TINYSPICE: A PARALLEL SPICE SIMULATOR ON GPU FOR MASSIVELY REPEATED SMALL CIRCUIT SIMULATIONS 624
L. Han, X. Zhao, Z. Feng

AN OPTIMAL ALGORITHM OF ADJUSTABLE DELAY BUFFER INSERTION FOR SOLVING CLOCK SKEW VARIATION PROBLEM 632
J. Kim, D. Joo, T. Kim

SMART NON-DEFAULT ROUTING FOR CLOCK POWER REDUCTION 638
A. Kahng, S. Kang, H. Lee

ROUTING CONGESTION ESTIMATION WITH REAL DESIGN CONSTRAINTS...........................645
W. Liu, Y. Wei, C. Sze, C. Alpert, Z. Li, Y. Li, N. Viswanathan

SPACER-IS-DIELECTRIC-COMPLIANT DETAILED ROUTING FOR SELF-ALIGNED DOUBLE PATTERNING LITHOGRAPHY 653
Y. Du, Q. Ma, H. Song, J. Shiely, G. Luk-Pat, A. Miloslavsky, M. Wong

21ST CENTURY DIGITAL DESIGN TOOLS...659
W. Dally, C. Malachowsky, S. Keckler

SYSTEM ARCHITECTURE AND SOFTWARE DESIGN FOR ELECTRIC VEHICLES...................665
M. Lukasiewycz, S. Steinhorst, S. Andalam, F. Sagstetter, P. Waszecki, W. Chang, M. Kauer, P. Mundhenk, S. Shanker, S. Fahmy, S. Chakraborty

MODEL-BASED DEVELOPMENT AND VERIFICATION OF CONTROL SOFTWARE FOR ELECTRIC VEHICLES 671
D. Goswami, M. Lukasiewycz, M. Kauer, S. Steinhorst, A. Masrur, S. Chakraborty, S. Ramesh

HYBRID ENERGY STORAGE SYSTEMS AND BATTERY MANAGEMENT FOR ELECTRIC VEHICLES 680
S. Park, Y. Kim, N. Chang

RELIABILITY CHALLENGES FOR ELECTRIC VEHICLES: FROM DEVICES TO ARCHITECTURE AND SYSTEMS SOFTWARE 686
G. Georakos, U. Schlichtmann, R. Schneider, S. Chakraborty

RELIABLE ON-CHIP SYSTEMS IN THE NANO-ERA: LESSONS LEARNT AND FUTURE TRENDS...695
J. Henkel, L. Bauer, N. Dutt, P. Gupta, S. Nassif, M. Shafique, M. Tahoori, N. Wehn

A LAYOUT-BASED APPROACH FOR MULTIPLE EVENT TRANSIENT ANALYSIS 705
M. Ebrahimi, H. Asadi, M. Tahoori

QUANTITATIVE EVALUATION OF SOFT ERROR INJECTION TECHNIQUES FOR ROBUST SYSTEM DESIGN 711
H. Cho, S. Mirkhani, C. Cher, J. Abraham, S. Mitra

EFFICIENTLY TOLERATING TIMING VIOLATIONS IN PIPELINED MICROPROCESSORS 721
K. Chakraborty, B. Cozzens, S. Roy, D. Ancajas

HIERARCHICAL DECODING OF DOUBLE ERROR CORRECTING CODES FOR HIGH SPEED RELIABLE MEMORIES 729
Z. Wang

POWER BENEFIT STUDY FOR ULTRA-HIGH DENSITY TRANSISTOR-LEVEL MONOLITHIC 3D ICS 736
Y. Lee, D. Limbrick, S. Lim

RAPID EXPLORATION OF PROCESSING AND DESIGN GUIDELINES TO OVERCOME CARBON NANOTUBE VARIATIONS 746
G. Hills, J. Zhang, C. Mackin, M. Shulaker, H. Wei, H. Wong, S. Mitra

MINIMUM-ENERGY STATE GUIDED PHYSICAL DESIGN FOR NANOMAGNET LOGIC...........................756
S. Liu, G. Csaba, X. Hu, E. Varga, M. Niemier, G. Bernstein, W. Porod

ULTRA LOW POWER ASSOCIATIVE COMPUTING WITH SPIN NEURONS AND RESISTIVE CROSSBAR MEMORY 763
M. Sharad, D. Fan, K. Roy

UNDERSTANDING THE TRADE-OFFS IN MULTI-LEVEL CELL RERAM MEMORY DESIGN 769
C. Xu, D. Niu, N. Muralimanohar, N. Jouppi, Y. Xie

EXPLORING TUNNEL-FET FOR ULTRA LOW POWER ANALOG APPLICATIONS: A CASE STUDY ON OPERATIONAL TRANSCONDUCTANCE AMPLIFIER 775
A. Trivedi, S. Carlo, S. Mukhopadhyay

ENERGY-OPTIMAL SRAM SUPPLY VOLTAGE SCHEDULING UNDER LIFETIME AND ERROR CONSTRAINTS 781
A. Calimera, E. Macii, M. Poncino

RELAX-AND-RETIME: A METHODOLOGY FOR ENERGY-EFFICIENT RECOVERY BASED DESIGN...787
S. Ramasubramanian, S. Venkataramani, A. Parandhaman, A. Raghunathan

POST-PLACEMENT VOLTAGE ISLAND GENERATION FOR TIMING-SPECULATIVE CIRCUITS...793
R. Ye, F. Yuan, Z. Sun, W. Jone, Q. Xu

ANALYSIS AND CHARACTERIZATION OF INHERENT APPLICATION RESILIENCE FOR APPROXIMATE COMPUTING 799
V. Chippa, S. Chakradhar, K. Roy, A. Raghunathan

DYNAMIC VOLTAGE AND FREQUENCY SCALING FOR SHARED RESOURCES IN MULTICORE PROCESSOR DESIGNS 808
X. Chen, Z. Xu, H. Kim, P. Gratz, J. Hu, M. Kishinevsky, U. Ogras, R. Ayoub

ENERGY OPTIMIZATION BY EXPLOITING EXECUTION SLACKS IN STREAMING APPLICATIONS ON MULTIPROCESSOR SYSTEMS 815
A. Singh, A. Das, A. Kumar

VERIFYING SYSTEMC USING AN INTERMEDIATE VERIFICATION LANGUAGE AND SYMBOLIC SIMULATION 822
H. Le, D. Große, V. Herdt, R. Drechsler

HANDLING DESIGN AND IMPLEMENTATION OPTIMIZATIONS IN EQUIVALENCE CHECKING FOR BEHAVIORAL SYNTHESIS 828
Z. Yang, K. Hao, S. Ray, F. Xie

A COUNTEREXAMPLE-GUIDED INTERPOLANT GENERATION ALGORITHM FOR SAT-BASED MODEL CHECKING 834
C. Wu, C. Lai, C. Huang

A ROBUST CONSTRAINT SOLVING FRAMEWORK FOR MULTIPLE CONSTRAINT SETS IN CONSTRAINED RANDOM VERIFICATION 840
B. Wu, C. Huang

SIMULATION KNOWLEDGE EXTRACTION AND REUSE IN CONSTRAINED RANDOM PROCESSOR VERIFICATION 847
W. Chen, L. Wang, J. Bhadra, M. Abadir

HARDWARE-EFFICIENT ON-CHIP GENERATION OF TIME-EXTENSIVE CONSTRAINED-RANDOM SEQUENCES FOR IN-SYSTEM VALIDATION 853
A. Kinsman, H. Ko, N. Nicolici

THE ROLE OF CASCADE, A CYCLE-BASED SIMULATION INFRASTRUCTURE, IN DESIGNING THE ANTON SPECIAL-PURPOSE SUPERCOMPUTERS 859
J. Grossman, B. Towles, J. Bank, D. Shaw

TOWARDS STRUCTURED ASICS USING POLARITY-TUNABLE SI NANOWIRE TRANSISTORS 868
P. Gaillardon, M. Marchi, L. Amaru, S. Bobba, D. Sacchetto, Y. Leblebici, G. Micheli

SACHA: THE STANFORD CARBON NANOTUBE CONTROLLED HANDSHAKING ROBOT 872
M. Shulaker, J. Rethy, G. Hills, H. Chen, G. Gielen, H. Wong, S. Mitra

ELECTRICAL ARTIFICIAL SKIN USING ULTRAFLEXIBLE ORGANIC TRANSISTOR 875
T. Sekitani, T. Sakurai, T. Yokota, T. Someya, M. Takamiya

RELAYS DO NOT LEAK – CMOS DOES 878
H. Fariborzi, F. Chen, R. Nathanael, I. Chen, L. Hutin, R. Lee, T. Liu, V. Stojanovic

SINGLE-PHOTON IMAGE SENSORS 882
E. Charbon, F. Regazzoni

NON-VOLATILE FPGAS BASED ON SPINTRONIC DEVICES 886
O. Goncalves, G. Prenat, G. Pendina, B. Dieny

A NOVEL ANALYTICAL METHOD FOR WORST CASE RESPONSE TIME ESTIMATION OF DISTRIBUTED EMBEDDED SYSTEMS 889
J. Kim, H. Oh, J. Choi, H. Ha, S. Ha

OPTIMIZATIONS FOR CONFIGURING AND MAPPING SOFTWARE PIPELINES IN MANY CORE SYSTEMS 899
J. Jahn, S. Pagani, S. Kobbe, J. Chen, J. Henkel

A SCENARIO-BASED RUN-TIME TASK MAPPING ALGORITHM FOR MPSOCS 907
W. Quan, A. Pimentel

EARLY EXPLORATION FOR PLATFORM ARCHITECTURE INSTANTIATION WITH MULTI-MODE APPLICATION PARTITIONING 913
P. Agrawal, P. Raghavan, M. Hartman, N. Sharma, L. Perre, F. Catthoor

COARX: A COPROCESSOR FOR ARX-BASED CRYPTOGRAPHIC ALGORITHMS 921
K. Shahzad, A. Khalid, Z. Rakossy, G. Paul, A. Chattopadhyay

RECONFIGURABLE PIPELINED COPROCESSOR FOR MULTI-MODE COMMUNICATION TRANSMISSION 931
L. Tiang, J. Ambrose, S. Parameswaran

ACCELERATORS FOR BIOLOGICALLY-INSPIRED ATTENTION AND RECOGNITION 939
M. Park, C. Zhang, M. Debole, S. Kestur, V. Narayanan, M. Irwin

STOCHASTIC CIRCUITS FOR REAL-TIME IMAGE-PROCESSING APPLICATIONS 945
A. Alaghi, C. Li, J. Hayes

AN EVENT-DRIVEN SIMULATION METHODOLOGY FOR INTEGRATED SWITCHING POWER SUPPLIES IN SYSTEMVERILOG 951
J. Jang, M. Park, J. Kim

A NEW TIME-STEPPING METHOD FOR CIRCUIT SIMULATION 958
G. Fang

TIME-DOMAIN SEGMENTATION BASED MASSIVELY PARALLEL SIMULATION FOR ADCS 968
Z. Ye, B. Wu, S. Han, Y. Li

A DIRECT FINITE ELEMENT SOLVER OF LINEAR COMPLEXITY FOR LARGE-SCALE 3-D CIRCUIT EXTRACTION IN MULTIPLE DIELECTRICS 974
B. Zhou, H. Liu, D. Jiao

FPGA CODE ACCELERATORS - THE COMPILER PERSPECTIVE 980
W. Najar, J. Villarreal

CAN CAD CURE CANCER? 986
S. Krishnawarny, B. Bodenmiller, D. Pe'er

LET'S PUT THE CAR IN YOUR PHONE! 988
M. Geier, M. Becker, D. Yunge, B. Dietrich, R. Schneider, D. Goswami, S. Chakraborty

THE UNDETECTABLE AND UNPROVABLE HARDWARE TROJAN HORSE 990
S. Wei, M. Potkonjak

PATH TO A TERABYTE OF ON-CHIP MEMORY FOR PETABIT PER SECOND BANDWIDTH WITH < 5WATTS OF POWER 992
S. Ghosh

RECONCILING REAL-TIME GUARANTEES AND ENERGY EFFICIENCY THROUGH UNLOCKED-CACHE PREFETCHING 994
E. Wuerges, R. Oliveira, L. Santos

INTEGRATED INSTRUCTION CACHE ANALYSIS AND LOCKING IN MULTITASKING REAL-TIME SYSTEMS 1003
H. Ding, Y. Liang, T. Mitra

PRECISE TIMING ANALYSIS FOR DIRECT-MAPPED CACHES 1013
S. Andalam, A. Girault, R. Sinha, P. Roop, J. Reineke

SSDM: SMART STACK DATA MANAGEMENT FOR SOFTWARE MANAGED MULTICORES (SMMS) 1023
J. Lu, K. Bai, A. Shrivastava

TAMING THE COMPLEXITY OF COORDINATED PLACE AND ROUTE 1031
J. Hu, M. Kim, I. Markov

ROUTABILITY-DRIVEN PLACEMENT FOR HIERARCHICAL MIXED-SIZE CIRCUIT DESIGNS 1038
M. Hsu, Y. Chen, C. Huang, T. Chen, Y. Chang

RIPPLE 2.0: HIGH QUALITY ROUTABILITY-DRIVEN PLACEMENT VIA GLOBAL ROUTER INTEGRATION 1044
X. He, T. Huang, W. Chow, J. Kuang, K. Lam, W. Cai, E. Young

OPTIMIZATION OF PLACEMENT SOLUTIONS FOR ROUTABILITY 1050
W. Liu, C. Koh, Y. Li

EXPLORATION WITH UPGRADEABLE MODELS USING STATISTICAL METHODS FOR PHYSICAL MODEL EMULATION 1059
B. Miller, F. Vahid, T. Givargis

MODULAR SYSTEM-LEVEL ARCHITECTURE FOR CONCURRENT CELL BALANCING 1065
M. Kauer, S. Naranayaswami, S. Steinhorst, M. Lukasiewycz

A METHOD TO ABSTRACT RTL IP BLOCKS INTO C++ CODE AND ENABLE HIGH-LEVEL SYNTHESIS 1075
N. Bombieri, H. Liu, F. Fummi, L. Carloni

DMR3D: DYNAMIC MEMORY RELOCATION IN 3D MULTICORE SYSTEMS 1084
D. Ancajas, K. Chakraborty, S. Roy

POWER GATING APPLIED TO MP-SOCS FOR STANDBY-MODE POWER MANAGEMENT 1093
D. Flynn

POWER MANAGEMENT AND DELIVERY FOR HIGH-PERFORMANCE MICROPROCESSORS 1098
T. Karnik, M. Pant, S. Borkar

FLEXIBLE ON-CHIP POWER DELIVERY FOR ENERGY EFFICIENT HETEROGENEOUS SYSTEMS 1101
B. Calhoun, K. Craig

POWER AND SIGNAL INTEGRITY CHALLENGES IN 3D SYSTEMS..1107
M. Corbalan, A. Keval, T. Toms, D. Lisk, R. Radojcic, M. Nowak

UNDERPOWERING NAND FLASH: PROFITS AND PERILS..1111
H. Tseng, L. Grupp, S. Swanson

NEW ERA: NEW EFFICIENT RELIABILITY-AWARE WEAR LEVELING FOR ENDURANCE ENHANCEMENT OF FLASH STORAGE DEVICES..1117
M. Yang, Y. Chang, C. Tsao, P. Huang

SAW: SYSTEM-ASSISTED WEAR LEVELING ON THE WRITE ENDURANCE OF NAND FLASH DEVICES..1123
C. Wang, W. Wong

PERFORMANCE ENHANCEMENT OF GARBAGE COLLECTION FOR FLASH STORAGE DEVICES: AN EFFICIENT VICTIM BLOCK SELECTION DESIGN..1132
C. Tsao, Y. Chang, M. Yang

DURACACHE: A DURABLE SSD CACHE USING MLC NAND FLASH..1138
R. Liu, C. Yang, C. Li, G. Chen

DISTRIBUTED STABLE STATES FOR PROCESS NETWORKS – ALGORITHM, ANALYSIS, AND EXPERIMENTS ON INTEL SCC..1144
D. Rai, L. Schor, N. Stoimenov, L. Thiele

DISTRIBUTED RUN-TIME RESOURCE MANAGEMENT FOR MALLEABLE APPLICATIONS ON MANY-CORE PLATFORMS..1154
I. Anagnostopoulos, V. Tsoutouras, A. Bartzas, D. Soudris

NETSHIP: A NETWORKED VIRTUAL PLATFORM FOR LARGE-SCALE HETEROGENEOUS DISTRIBUTED EMBEDDED SYSTEMS..1160
Y. Jung, J. Park, M. Petracca, L. Carloni

EXPLOITING JUST-ENOUGH PARALLELISM WHEN MAPPING STREAMING APPLICATIONS IN HARD REAL-TIME SYSTEMS..1170
J. Zhai, M. Bamakhrama, T. Stefanov

ON ROBUST TASK-ACCURATE PERFORMANCE ESTIMATION..1178
Y. Xu, B. Wang, R. Hasholzner, R. Rosales, J. Teich

STOCHASTIC RESPONSE-TIME GUARANTEE FOR NON-PREEMPTIVE, FIXED-PRIORITY SCHEDULING UNDER ERRORS..1184
P. Axer, R. Ernst

HADES: ARCHITECTURAL SYNTHESIS FOR HETEROGENEOUS DARK SILICON CHIP MULTI-PROCESSORS..1191
Y. Turakhia, B. Raghunathan, S. Garg, D. Marculescu

HIERARCHICAL POWER MANAGEMENT FOR ASYMMETRIC MULTI-CORE IN DARK SILICON ERA..1198
T. Muthukaruppan, M. Pricopi, V. Venkataramani, T. Mitra, S. Vishin

PEAK POWER REDUCTION AND WORKLOAD BALANCING BY SPACE-TIME MULTIPLEXING BASED DEMAND-SUPPLY MATCHING FOR 3D THOUSAND-CORE MICROPROCESSOR..1207
S. Manoj, K. Wang, H. Yu

TECHNIQUES FOR ENERGY-EFFICIENT POWER BUDGETING IN DATA CENTERS..1213
X. Zhan, S. Reda

TEMPERATURE AWARE THREAD BLOCK SCHEDULING IN GPGPUS..1220
R. Nath, R. Ayoub, T. Rosing

VAWOM: TEMPERATURE AND PROCESS VARIATION AWARE WEAROUT MANAGEMENT IN 3D MULTICORE ARCHITECTURE..1226
H. Tajik, H. Hornayoun, N. Dutt

ON THE POTENTIAL OF 3D INTEGRATION OF INDUCTIVE DC-DC CONVERTER FOR HIGH-PERFORMANCE POWER DELIVERY..1234
S. Carlo, W. Yueh, S. Mukhopadhyay

FULL-CHIP MULTIPLE TSV-TO-TSV COUPLING EXTRACTION AND OPTIMIZATION IN 3D ICS..1242
T. Song, C. Liu, Y. Peng, S. Lim

AN ACCURATE SEMI-ANALYTICAL FRAMEWORK FOR FULL-CHIP TSV-INDUCED STRESS MODELING..1249
Y. Li, D. Pan

SPEEDING UP COMPUTATION OF THE MAX/MIN OF A SET OF GAUSSIANS FOR STATISTICAL TIMING ANALYSIS AND OPTIMIZATION..1257
V. Kuruvilla, D. Sinha, J. Piaget, C. Visweswariah, N. Chandrachoodan

INTIMEFIX: A LOW-COST AND SCALABLE TECHNIQUE FOR IN-SITU TIMING ERROR MASKING IN LOGIC CIRCUITS.. 1264
F. Yuan, Q. Xu

IMPROVING PUF SECURITY WITH REGRESSION-BASED DISTILLER.. 1270
C. Yin, G. Qu

ON THE CONVERGENCE OF MAINSTREAM AND MISSION-CRITICAL MARKETS.................................. 1276
S. Girbal, M. Moreto, A Grasset, J. Abella, E. Quinones, F. Cazorla, S. Yehia

Author Index

Mapping on Multi/Many-core Systems: Survey of Current and Emerging Trends

Amit Kumar Singh[1], Muhammad Shafique[2], Akash Kumar[1], Jörg Henkel[2]

[1] Department of Electrical and Computer Engineering, National University of Singapore, Singapore

[2] Chair for Embedded Systems (CES), Karlsruhe Institute of Technology, Karlsruhe, Germany

[1]{eleaks,akash}@nus.edu.sg, [2]{muhammad.shafique,henkel}@kit.edu

ABSTRACT

The reliance on multi/many-core systems to satisfy the high performance requirement of complex embedded software applications is increasing. This necessitates the need to realize efficient mapping methodologies for such complex computing platforms. This paper provides an extensive survey and categorization of state-of-the-art mapping methodologies and highlights the emerging trends for multi/many-core systems. The methodologies aim at optimizing system's resource usage, performance, power consumption, temperature distribution and reliability for varying application models. The methodologies perform design-time and run-time optimization for static and dynamic workload scenarios, respectively. These optimizations are necessary to fulfill the end-user demands. Comparison of the methodologies based on their optimization aim has been provided. The trend followed by the methodologies and open research challenges have also been discussed.

Categories and Subject Descriptors

C.3 [**Special-purpose and application-based systems**]: Real-time systems and embedded systems

General Terms

Algorithms, Design, Performance, Reliability

Keywords

Multiprocessor Systems-on-Chip, embedded systems, application mapping

1. INTRODUCTION

The maximum operational frequency of a single-core processor has hit the roof due to power dissipation and radio frequency effects. This has forced chip manufacturers to limit the maximum frequency of the processor and shifting towards designing chips with multiple cores operating at lower frequencies [42] [10]. Moreover, the performance demands of modern complex embedded applications have increased substantially which cannot be satisfied by simply increasing the frequency of a single-core processor or by customization of the processor. Instead, there is a need of multiple processors that can cohesively communicate and provide increased parallelism. The underlying concept is to consider applications as conglomeration of many small tasks which can be efficiently distributed on multiple processors in order to execute them in parallel and thereby meeting the increased performance demands [3] [46].

Permission to make digital or hard copies of all or part of this work for personal or classroom use is granted without fee provided that copies are not made or distributed for profit or commercial advantage and that copies bear this notice and the full citation on the first page. To copy otherwise, to republish, to post on servers or to redistribute to lists, requires prior specific permission and/or a fee.

DAC'13 May 29 - June 07 2013, Austin, TX, USA.

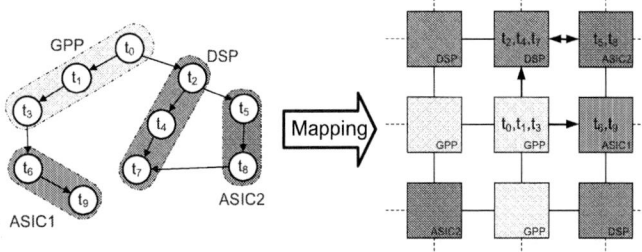

Application Task Graph Many-core Architecture

Figure 1: Application Mapping on Many-core System.

With the technological advancement and increasing performance demands, the number of cores in the same chip area has grown exponentially and different types of cores have been integrated. As nanotechnology evolves, it will become feasible to integrate thousands of cores on the same chip [10]. The large number of cores needs to employ Network-on-Chip (NoC) based interconnection infrastructure for efficiency and scalability [33] [7]. The distinct features of different types of cores can be exploited to meet the functional and non-functional requirements. This makes heterogeneous multi/many-core systems (consisting of different types of cores) a formidable computing alternative where applications witness large improvement over their homogeneous (consisting of identical cores) counterpart.

In order to map applications on multi/many-core systems, the applications need to be partitioned (parallelized) into multiple tasks that can be executed concurrently on different cores. An example of partitioned application is shown as *Application Task Graph* in Fig. 1. The application is partitioned into ten tasks $(t_0, t_1, ..., t_9)$. The partitioning job can be furnished by state-of-the-art application parallelization tools [14] [53] and manual analysis, which involves finding the tasks, adding synchronization and inter-task communication in the tasks, management of the memory hierarchy communication and checking of the parallelized code to ensure for correct functionality [59]. In case of heterogeneous platforms, a task binding process that specifies the core types on them the task can be mapped along with the cost of mapping is required [84]. The binding process analyses the implementation costs (e.g., performance, power and resource utilization) of each task on different supported core types such as general purpose processor (GPP), digital signal processor (DSP) and coarse grain re-configurable hardware.

Mapping application tasks on multi/many-core system involves assignment and ordering of the tasks and their communications onto the platform resources in view of some optimization criteria such as energy consumption and compute performance. Fig. 1 shows mapping of tasks and their communications on part of a many-core system. The communicating tasks are mapped on the same core or close to each other in order to optimize for the communication delay and energy. The optimization is necessary to satisfy performance constraints of the applications. This necessitates the need to develop efficient mapping methodologies that take application model, platform model, constraints (e.g., compute performance and power), performance model of inter-process communication (e.g., execution time and energy consumption) and estimate of the worst case execution time (WCET) of the process implementations on different cores (e.g., GPP, DSP, ASIC) as input and provide mappings that satisfy the constraints.

1.1 Mapping Problem and Challenges

Mapping and scheduling problem is similar to *Quadratic Assignment Problem*, a well-known NP-hard problem [28]. Therefore, finding optimal solution satisfying all the given constraints is very difficult and time consuming. Thus, heuristics based on the application domain knowledge need to be employed to find a nearly optimal solution.

Furthermore, the user demands (e.g., performance and power constraints) for each application need to be fulfilled. This necessitates the need to find optimal mapping solutions for each use-case[1] to be supported into the system. The optimal solutions can be explored by advance design-time analysis and then can be used at run-time. However, explosion in the number of use-cases with the increasing number of applications make the analysis unfeasible. For n applications, the analysis needs to be performed for 2^n use-cases. Additionally, such analysis cannot deal with dynamic scenarios such as run-time changing standards and addition of new applications. Run-time management is required to handle such dynamism albeit optimal mapping solutions are not found.

The application mapping problem has been identified as one of the most urgent problem to be solved for implementing embedded systems [57] [60]. This problem is being addressed by several researchers who communicate their views through various forums. A series of dedicated workshops on mapping of applications onto multi-core systems have been started to move beyond state-of-the-art. The mapping methodologies are developed by targeting specific application domain (e.g., multimedia and networking) for the most promising multi-core system.

1.2 Classification of Mapping Methodologies

There could be a number of taxonomies to classify the mapping methodologies, like target architecture based, optimization criteria based, workload based, etc. Broadly, the methodologies can be classified based on workload scenarios and other taxonomies can be included at some hierarchy in the classification as shown in Fig. 2. For static and dynamic workload scenarios, the mapping methodologies perform optimization at **design-time** and **run-time** respectively, which has led them to classify as design-time and run-time methodologies respectively. The methodologies target either **homogeneous** or **heterogeneous** multi-core systems. The run-time mapping requires a platform manager that handles mapping of tasks at run-time. In addition to mapping, the manager is also responsible for task scheduling [51], resource control, configuration control and task migration at run-time. The manager may employ **centralized management**, **distributed management** or mixture of centralized and distributed management. In centralized management, one core of the platform is used as the manager that handles the mapping process. For distributed management, the platform is divided into regions (clusters) and one core in each cluster manages the mapping process inside the cluster. The cluster managers communicate with each other through a global manager to find the best cluster for mapping an application.

Design-time mapping methodologies are suitable for static workload scenarios where a predefined set of applications

---

[1]Combination of simultaneously active applications.

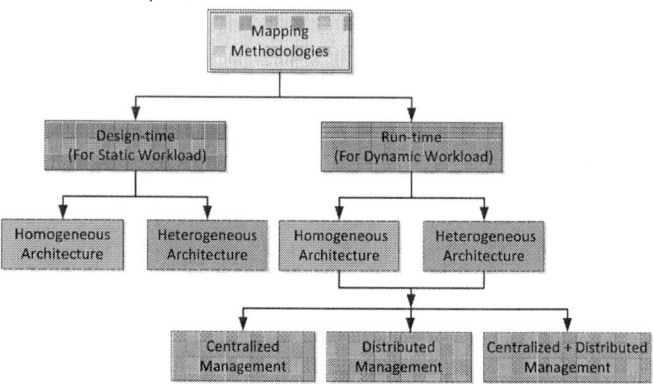

Figure 2: A Taxonomy of Mapping Methodologies.

with known computation and communication behavior and a static platform are considered. They are unable to handle dynamism in applications incurred at run-time (e.g., multimedia and networking applications). Examples of such dynamism could be adding a new application into the system at run-time. Since applications are often added to the platform at run-time (for example, downloading a Java application in a mobile-phone at run-time), workload variation takes place. We witness the need of run-time mapping methodologies to handle such dynamic workloads.

The run-time mapping methodologies face the challenge to map tasks of new applications on the platform resources to satisfy their performance requirements while keeping accurate knowledge of resource occupancy. After mapping tasks, task migration can also be used to revise placement of some of the already executing tasks if the user requirement is changed or a new application has entered into the system.

This paper performs an in-depth *survey* of the mapping methodologies reported in literature based on the earlier mentioned taxonomy. The methodologies have been *analyzed* to highlight their strengths and weaknesses. The trend followed by the methodologies over the *decade* has been observed and reported, which provides an insight for the emerging mapping methodologies. Despite being significant advancement in the development of mapping methodologies, some open issues that need to be addressed in the future are highlighted by analyzing the methodologies reported in the literature.

Paper Organization: Section 2 discusses mapping methodologies that perform design-time optimization while targeting homogeneous or heterogeneous architectures. In Section 3, run-time mapping methodologies are analyzed and elaborated. Section 4 provides the upcoming trends and open research challenges. Finally, we conclude the paper highlighting the important points of mapping methodologies in Section 5.

2. DESIGN-TIME MAPPING

Design-time mapping methodologies have a global view of the system which facilitates in making better decision for using the system resources. As optimization is performed at design-time, the methodologies can use more thorough system information to make decisions. Thus, a better quality of mapping may be achieved as compared to the run-time mapping methodologies that are restricted normally to a local view where only the neighborhood of the task mapping is considered. Most of the mapping methodologies reported in the literature fall under design-time mapping. These methodologies target either homogeneous or heterogeneous architectures.

Table 1 classifies recent works in design-time mapping and shows the target architecture and optimization goal of the mapping methodologies proposed by different authors. The target architecture (*Arch.*) is either homogeneous (*Hom.*)

Table 1: Classification of design-time mapping methodologies.

Author	Arch.	Optimization Goal
Orsila et al. [68]	Hom.	Execution time
Ruggiero et al. [74]	Hom.	Execution time
Satish et al. [76]	Hom.	Execution time
Bonfietti et al. [9]	Hom.	Mapping time & quality
Lin et al. [50]	Hom.	Throughput, Resource utilization,
Murali et al. [65]	Hom.	Energy consumption
Rhee et al. [73]	Hom.	Energy consumption
Chen et al. [16]	Hom.	Energy consumption
Hu et al. [36] [37]	Hom.	Energy consumption, Execution time
Marcon et al. [55] [56]	Hom.	Energy consumption, Execution time
Ascia et al. [5]	Hom.	Energy consumption, Execution time
Meyer et al. [62]	Hom.	Reliability
Thiele et al. [89]	Hom.	Reliability, Temperature
Thiele et al. [88]	Het.	Execution time
Choi et al. [18]	Het.	Execution time
Che et al. [15]	Het.	Execution time
Castrillon et al. [13]	Het.	Execution time
Manolache et al. [54]	Het.	Exploration time, Accuracy
Javaid et al. [41]	Het.	Exploration time, Accuracy
Wu et al. [94]	Het.	Energy consumption
Zhu et al. [100]	Het.	Reliability
Hartman et al. [30]	Het.	Reliability

or heterogeneous (*Het.*). The methodologies aim at optimizing for difference performance metrics in order to fulfill the varying user demands.

Compute Performance: Optimizing for the compute performance is of paramount importance in order to meet the timing deadlines or to minimize the time taken to finish some jobs. The compute performance may refer to total *execution time, latency, delay, period, throughput, exploration time*, etc., which are related to timing information.

Different well established search approaches have been extensively used to develop design-time mapping methodologies in order to find *optimal* or *near-optimal* placement of tasks on platform cores towards improving the compute performance. For example, Simulated Annealing (SA) is used in [68] [50], Genetic Algorithm (GA) in [18], Tabu Search in [54] and Integer Linear Programming (ILP) in [41]. Orsila et al. [68] optimize execution time and memory consumption by claiming that traditional approaches only focus on the execution time. Lin et al. [50] attempt to find mapping of tasks such that the overall system throughput is maximized. They show an improvement of 20% in the system throughput. Choi et al. [18] propose a GA-based technique for efficiently executing Synchronous Dataflow (SDF) applications on a multi-core system where each core has a limited size of scratchpad memory (SPM). Manolache et al. [54] address the problem of task mapping in the context of multi-processor applications with stochastic execution times and in the presence of constraints on the percentage of missed deadlines. Javaid et al. [41] propose a methodology consisting of ILP formulation to explore efficient mappings. *These search based approaches provide efficient mapping solutions, but they have high computational costs for large scale problems such as applications with large number of tasks.*

Different pruning strategies have been incorporated to prune the search space, thereby reducing the computational costs. Ruggiero et al. [74] combine Integer Programming with Constraint programming to speed up the executions. They target bus-based architectures that are not scalable and thus the approach enforces scalability issues. Satish et al. [76] propose a decomposition based approach to speed up constraint optimization. They optimize for the schedule length or make-span. Bonfietti et al. [9] propose an approach for throughput-maximal mapping of SDF applications. The approach speeds-up the computation and enhances the efficiency of the search by jump-starting with a high-quality bound and quickly tightening it. Thiele et al. [88] propose

a mapping framework called Distributed Operation Layer (DOL), which optimizes for computation and communication time. They integrate an analytic performance analysis strategy into DOL to alleviate the modeling and analysis of systems. Che et al. [15] consider the number of software pipeline stages to map streaming applications on SPM-based embedded multi-core system. The proposed method scales well over a wide range of cores and SPMs. Castrillon et al. [13] propose an algorithm that directly addresses mapping of tasks and their communications. The algorithm is executed repeatedly to compute mappings for real-time applications specified as Kahn Process Network (KPN). *These approaches provide mapping solutions in lesser time than the approaches performing extensive or complete search, but might miss high quality mapping solutions due to pruning of the search space.*

Energy Consumption: Optimizing for the energy consumption of modern embedded systems (e.g., mobile phones, tablets) is important as they are usually operated by stand-alone power supply like battery. The optimization needs to be performed during the system design and operation in order to increase the operational time.

Murali et al. [65] present a methodology that handles mapping of multiple use-cases while satisfying their performance constraints. The methodology shows a power savings of 54%. Rhee et al. [73] propose an ILP based approach that optimally maps cores onto mesh architecture in order to minimize energy consumption or NoC congestion. The approach achieves 81% energy savings for random benchmarks. Chen et al. [16] propose a multi-step mapping methodology where optimization is performed in different steps of the mapping process. Wu et al. [94] introduce a GA based approach that use *Dynamic Voltage Scaling (DVS)* to reduce the energy consumption by up to 51%. These methodologies show significant energy savings.

Some methodologies that perform optimization for both energy consumption and compute performance are introduced in [37], [56] and [5]. Hu et al. [37] propose a mapping methodology that reduces power consumption by decreasing the energy consumption in communication while guaranteeing the required performance. They methodology provides an energy savings of 51%. Marcon et al. [56] extend the work in [37] and propose a technique called *Communication Dependence and Computation Model (CDCM)*. In addition to communication volume as considered in [37], timing of the communication has been considered. Execution time is reduced by 98% while achieving a significant amount of energy savings. Ascia et al. [5] present a GA based approach that explore Pareto mappings efficiently and accurately while optimizing for performance and energy consumption. *The aforementioned methodologies optimize for compute performance and energy consumption, but do not take reliability of the system into account during that optimization. Therefore, the provided mapping solutions might lead to reduced lifetime of the system.*

Reliability: Design-time mapping methodologies targeting lifetime improvement of multi-core systems are proposed in [62], [89], [100] and [30]. Meyer et al. [62] propose an approach to effectively and efficiently allocate execution and storage slack in order to jointly optimize system lifetime and cost. Thiele et al. [89] propose a thermal-aware system analysis method that produces mappings with lower peak temperature of the system, leading to reliable system design. Zhu et al. [100] exploit redundancy and temperature-aware design planning to produce reliable and compact multi-core systems. Hartman et al. [30] propose a lifetime-aware task mapping methodology that produces mappings with higher lifetimes. These methodologies take preventive measures by performing reliability-aware mapping in order to reduce occurrence of faults in the systems. Such preventive measures

3

increase lifetime of systems.

2.1 Issues and Limitations of Design-time Methodologies

Most of the design-time methodologies adopt search based approaches (e.g., GA, ILP, SA) that incur high computational costs. Thus, the evaluation time might not be acceptable for large scale problems. However, they provide efficient mapping solutions for small scale systems within acceptable time. The evaluation time can be reduced by efficient pruning of the search space, but at the risk of missing the high quality mapping solutions. The reliability-aware design-time methodologies increase the system lifetime but they cannot overcome the faults incurred in the system.

Further, as the design-time methodologies find placement of tasks at design-time, they are not suitable for run-time varying workloads in the systems and run-time changing environments. Such dynamic workload scenarios require re-mapping/run-time mapping of applications. Even if these mapping methodologies are inadequate for the dynamic workload scenarios, they might be useful to find the initial task placement, or be optimized to be working at run-time.

3. RUN-TIME MAPPING

In contrast to the design-time mapping, run-time mapping needs to account for the time taken to map each task as it contributes to overall application execution time. Furthermore, the tasks are mapped one by one, unlike the static case where all the tasks are mapped at once by looking globally at the system. Therefore, typically greedy heuristic algorithms are used for efficient run-time mapping in order to optimize performance metrics such as energy consumption, communication latency, execution time, etc. The run-time mapping has several requirements, advantages and issues & research challenges for different available mapping alternatives.

Requirement: The run-time mapping caters for dynamically workload scenarios where mapping of one or more already running applications may need to be reconsidered in case of following requirements:

- Insertion of a new application into the system, which needs resources from the already executing applications.
- Modifying parameters of a running application.
- Killing a running application in order to free it's occupied resources.
- Changing performance requirements of a running application. This might need extra resources for performing extra functionality.
- When current mapping is not sufficiently optimal, it requires (re-)mapping.

Advantages: In addition to the suitability for dynamic workload scenarios, run-time mapping also offers a number of advantages. Some of them are as follows:

- *Adaptability to the available resources*: The available resources vary over time as the applications of the dynamic workload scenario enter at run-time.
- *Ability to enable unforeseeable upgrades*: It is possible to upgrade the system for new applications or changing standards that are not known at design-time, even after the delivery of the system to the end-user.
- *Ability to avoid defective parts of a SoC*: If one or more processing cores are not working properly after production of a SoC, then the defective cores can be disabled before the mapping process. Aging can lead to defective cores that are unforeseeable at design-time.

Mapping Alternatives: At run-time, mapping of new applications to be supported onto a platform can be handled either by performing all the processing at the same time, i.e. on-the-fly processing or by using previously analyzed (DSE)

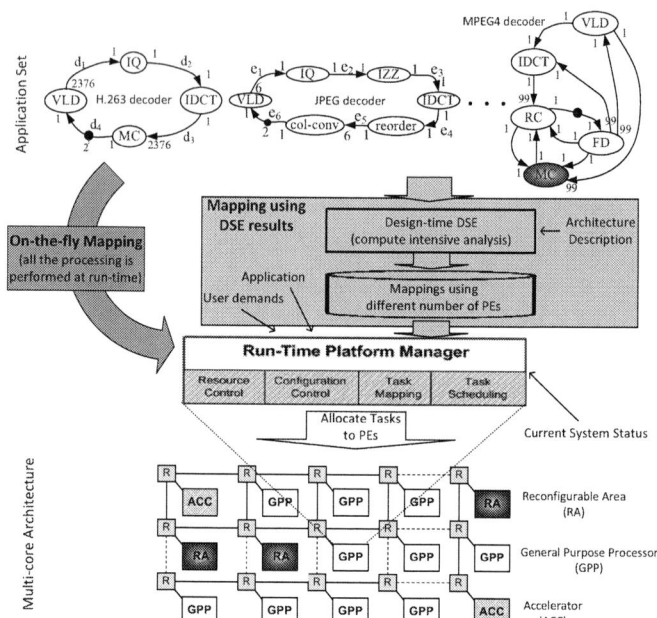

Figure 3: On-the-fly and Hybrid Mapping Flow [81].

results as shown in Fig. 3. The run-time platform manager handles the mapping of applications by taking the updated resources' status (*Current System Status*) into account.

For *on-the-fly mapping*, efficient heuristics are required to assign new arriving tasks on the platform resources. These heuristics cannot guarantee for schedulability, i.e., for strict timing deadlines due to lack of any prior analysis and limited compute power at run-time. However, these heuristics are platform independent since they do not use any platform specific analysis results computed in advance. Such heuristics lend well to map unknown applications (not available at design-time) on any platform.

For *mapping using previously analyzed (DSE) results*, the applications to be supported on a platform should be known at design-time. In such cases, light-weight heuristics are required to select the most efficient mappings for each application from the design-time (offline) analyzed mappings stored on the system (*Mappings using different number of PEs*). The selection is done subject to available system resources (extracted from *Current System Status*) and desired performance (*User demands*). The selected mapping is used to configure the platform. Compute intensive analysis is performed at design-time (*Design-time DSE*), facilitating for light-weight run-time platform manager that can configure the applications efficiently. In DSE, *application* and *architecture* description are taken as input and a number of *mappings* are produced. Such mapping methodologies have been referred to as *hybrid mapping* as they take the advantages of both design-time and run-time. The hybrid approach maps applications more efficiently than on-the-fly heuristics. However, flexibility in these approaches is limited, since all potential applications must be known in entirety at design-time and analysis results will be applicable only to the analyzed platform. Therefore, design-time analysis needs to be repeated when the application set or platform changes. Further, storing analysis results introduces additional memory overhead.

3.1 On-the-fly Mapping

Recent works on on-the-fly mapping are classified according to the proposed taxonomy and are listed in Table 2. The table reveals the target architecture (Arch.), control mechanism and optimization goal of the mapping methodologies. The methodologies target homogeneous (Hom.) or

Table 2: Classification of on-the-fly mapping methodologies

Author	Arch.	Control Manager	Optimization Goal
Hong et al. [35]	Hom.	Centr.	Execution time
Shojaei et al. [80]	Hom.	Centr.	Execution time, Solution quality
Moreira et al. [63]	Hom.	Centr.	Execution time, Resource utilization
Chou et al. [19]	Hom.	Centr.	Energy consumption, Communication cost
Mehran et al. [61]	Hom.	Centr.	Energy consumption, Mapping time
Briao et al. [11]	Hom.	Centr.	Energy consumption, Execution time
Qi et al. [72]	Hom.	Centr.	Reliability, Energy consumption
Chou et al. [20]	Hom.	Centr.	Reliability, Energy consumption, Throughput
Coskun et al. [23, 24]	Hom.	Centr.	Reliability, Temperature
Peter et al. [70]	Hom.	Distr.	Execution Time
Theocharides et al. [87]	Het.	Centr.	Execution time
Feng et al. [92]	Het.	Centr.	Execution time
Ahmed et al. [1]	Het.	Centr.	Execution time
Huang et al. [38]	Het.	Centr.	Execution time, Resource utilization
Liang et al. [17]	Het.	Centr.	Execution time, Resource utilization
Nollet et al. [67]	Het.	Centr.	Mapping time & quality
Carvalho et al. [12]	Het.	Centr.	Communication overhead
Smit et al. [84]	Het.	Centr.	Energy consumption, QoS for applications
Braak et al. [86]	Het.	Centr.	Energy consumption, Execution time
Schranzhofer et al. [78]	Het.	Centr.	Energy consumption, Execution time
Singh et al. [83]	Het.	Centr.	Energy consumption, Communication overhead
Hartman et al. [31]	Het.	Centr.	Reliability
Faruque et al. [3]	Het.	Distr.	Execution time, Mapping time, Traffic
Kobbe et al. [46]	Het.	Distr.	Execution time, Traffic
Ebi et al. [27]	Het.	HDistr.	Execution time, Temperature

heterogeneous (Het.) multi-core systems depending upon the requirement of applications. For controlling the system, a *centralized (Centr.)*, distributed (Distr.) or mix of centralized and distributed, i.e. hierarchical distributed (HDistr.) resource management strategy is used to allocate tasks on the resources at run-time. The methodologies aim at optimizing for difference performance metrics.

Compute Performance: The compute performance optimization relates to the timing optimization such as overall execution time and mapping time. Hong et al. [35] change the thread-to-processor mapping at run-time based on the workload variation in order to optimize the performance. An improvement of 29% is achieved in the overall execution time. In [80], the presented heuristic offers additional advantage to trade-off execution time versus solution quality. Moreira et al. [63] present a methodology that first assigns tasks to virtual cores (VCs) aiming to minimize total number of VCs and total bandwidth used while meeting the timing constraints. Thereafter, the VCs are mapped to real cores. Theocharides et al. [87] demonstrate a system-level bidding-based task mapping methodology that provides significant performance improvements when compared to a round robin allocation. Feng et al. [92] perform workload variation aware mapping to optimize the system performance. Ahmed et al. [1] perform adaptive resource management to maintain QoS requirement of application. Huang et al. [38] introduce self-adaptability to the run-time task allocation to achieve high system performance. The adaptability is obtained by dynamically adjusting a set of key parameters based on current resource utilization. Liang et al. [17] take the advantage of shared multi-core reconfigurable fabric to optimize the performance. Nollet et al. [67] describe a run-time task assignment heuristic for efficiently mapping the tasks in a multi-core system containing FPGA fabric tiles. With the presence of FPGA fabric tiles, the heuristic is capable of managing a configuration hierarchy and improves the task assignment success rate and quality of solutions. Carvalho et al. [12] present heuristics where tasks are mapped according to the communication requests and the load in the NoC links. Such consideration reduces the communication overhead, leading to reduced execution time.

The above mentioned methodologies to optimize compute performance use *centralized management (CM)* approach. The CM approach for large systems (many-core systems (consist of thousands of cores)) faces the following problems: *1)* single point of failure, *2)* large volume of monitoring-traffic by the CM, *3)* high computational cost to calculate mapping inside CM and *4)* bottleneck around the CM as

every core sends its status to the CM after every instance of mapping [3]. Thus, the CM becomes a hot spot. This necessitates the need of distributed management in order to reduce the monitoring traffic and computational effort.

For *distributed mapping*, the entire system is partitioned into multiple clusters. The resources within each cluster are managed by an individual cluster manager (agent) that communicates with a global platform manager in order to efficiently map an application inside the cluster. The distributed mapping methodology is better than the state-of-the-art run-time mapping methodologies using *Centralized Manager (CM)* approach when many-core systems are targeted. Peter et al. [70] present a heuristic algorithm that is distributed over the processor cores, facilitating its applicability to systems of random size. However, as each core can be considered as a cluster, resource management will become difficult due to linear increment in the number of clusters with the system size. Efficient distributed application mapping methodologies targeting large architectures such as 32×32 and 32×64 systems are presented in [3], [46] and [27]. Faruque et al. [3] consider static applications and focuses on communication, whereas Kobbe et al. [46] consider malleable applications. Ebi et al. [27] consider hierarchical distributed management that targets to trade-off the effectiveness of a centralize approach using global knowledge with the scalability of a fully distributed one.

Energy Consumption: Chou et al. [19] propose a methodology that incorporates the user behavior information in the resource allocation process; that allows system to better respond to real-time changes and adapt dynamically to user needs. This consideration saves 60% communication energy when compared to an arbitrary task allocation technique. Mehran et al. [61] present a *Dynamic Spiral Mapping (DSM)* heuristic algorithm for 2-D mesh topology where placement for a task is searched in a spiral path. The task having maximum degree (connections) is placed at the center of the mesh to facilitate closer mapping of communicating tasks, thereby reducing the communication energy. Briao et al. [11] present strategies where the system turns off idle processors and applies *Dynamic Voltage Scaling (DVS)* to processors with slack to save energy. Smit et al. [84] present an algorithm that first maps tasks needing scarce resources and then all other tasks by taking availability of the platform resources into account. The algorithm minimizes the total amount of energy consumption while providing adequate Quality of Service (QoS) for the application. Braak et al. [86] propose a run-time spatial mapping methodology that spans both the task graph and the multi-core system to find optimal mapping of tasks. Schranzhofer et al. [78] propose a polynomial-time multiple-step heuristic consisting of initial solutions followed by task re-mapping algorithms considering power constraints. Singh et al. [83] incorporate energy consumption measures and multiple tasks per core while mapping tasks according to the communication requests.

Most of the energy-aware mapping methodologies also try to optimize compute performance as shown in Table 2. Thus, they try to fulfill timing constraints while optimizing for the energy consumption. Such optimization is necessary for modern embedded systems that need to perform compute intensive operations for a long time within a limited energy budget. However, these methodologies do not take reliability of the system into account while performing different kinds of optimization.

Reliability: Some recent methodologies that perform optimization for the system reliability are introduced in Table 2. These methodologies take necessary measures to optimize reliability or cure the faults after they have been detected in the system. The detection of faults and their cure is of paramount importance for many real-time systems such as safety-critical, automotive, and avionics. Fail-

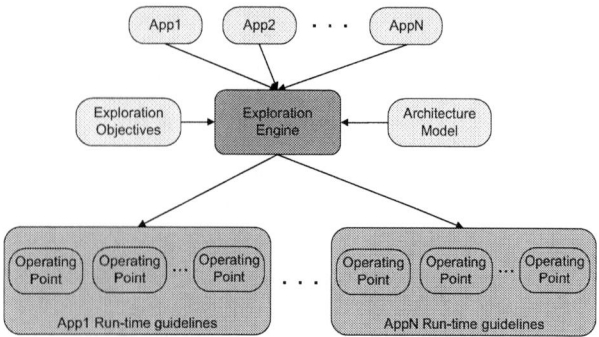

Figure 4: Design-time analysis of applications.

ing to achieve fault-tolerance in these systems may lead to catastrophic consequences. Qi et al. [72] present a technique that optimizes power while considering the system reliability. Chou et al. [20] propose a fault-tolerant application mapping methodology that optimizes system performance and energy consumption, while considering occurrence of different types of faults in the system. Coskun et al. [23, 24] target temperature aware mapping that leads to increased performance and lifetime. Hartman et al. [31] propose a run-time task mapping subsystem that mitigates faults using a wear-based heuristic. The wear-based heuristic is capable of improving system lifetime over temperature-based heuristics. The reliability consideration along with other optimization such as compute performance and energy consumption leads to a better desirable system.

At run-time, some mapping methodologies employ task migration when performance bottleneck is detected or the workload needs to be distributed more homogenously in the whole system [11] [70]. In migration, the tasks are migrated without completely stopping and restarting the already executing applications. Task migration may also be used in case user requirement is changed or a new application has entered into the system in order to revise the placement of some of the already executing tasks. Issues related to the task migration such as the cost to interrupt a given task, saving its context, transmitting all of the data to a new core and restarting the task in the new core are discussed in [66] and [8].

3.2 Based on Design-time Analysis Results

Mapping methodologies based on design-time analysis results perform compute intensive analysis at design-time and use the analyzed results at run-time. Design-time analysis strategies take application and architecture specifications as input and explore mappings with some design objectives (exploration objectives) as shown in Fig. 4. The explored mappings (operating points) provide guidelines for configuring the application at run-time, which is shown as run-time guidelines. The mappings represent trade-offs between different performance metrics. The same analysis strategies can be applied to all the applications (App1, App2, $\cdots$, AppN) one after another, which might need to be supported into the system at run-time, as shown in Fig. 4. *Exploring all the possible mappings for large application and platform size exhaustively is not feasible within a limited time. Therefore, faster analysis strategies having some design objectives are required to explore efficient mappings.*

Single Application Single Mapping Analysis: Most of the design-time analysis techniques reported in literature provide a single mapping for the application. Design-time mapping methodologies reported in Section 2 can be used to find a mapping for an application. Some other such analysis techniques are presented in [64], [2], [45] and [52].They perform exploration in view of some optimization parameters such as computational performance and energy. *The explored single mapping cannot handle dynamism in resource*

availability and performance requirement at run-time.

Single Application Multiple Mappings Analysis: Design-time analysis strategies that generate multiple mappings for the application have recently been reported in [58], [98], [4], [85], [29], [93], [43] and [71]. The generated mappings can be used to handle dynamism in resource availability and performance requirement at run-time. In [58] and [98], exploration is performed in view of optimizing for power consumption and performance in order to identify the best performance/power trade-offs. Angiolini et al. [4] optimize for the performance. Stuijk et al. [85] optimize for resource usage. Beltrame et al. [29] optimize for energy and delay. They try to minimize number of simulations required to identify the mappings providing energy/delay trade-offs. Wildermann et al. [93] present multi-objective exploration approach. Jia et al. [43] present an infrastructure called *NASA (Non Ad-hoc Search Algorithm)*, which uses different combination of search strategies to explore the mappings. Piscitelli et al. [71] propose an approach that interleaves the estimations with simulative evaluations in order to ensure that optimal mappings are explored accurately. Most of the analysis strategies use either simulation or an analytical model to evaluate mappings, where simulative evaluations are computationally costly and analytical approaches suffer from accuracy issues. In [43] and [71], simulative and analytical evaluations are combined to perform fast and accurate analysis. These strategies analyze applications one after another. To support the required applications on the system at run-time, they are mapped sequentially by using their analysis results.

Multiple Applications Multiple Mappings Analysis: There has been quite some research in multiple applications DSE. Some researchers focus on scenario based approach where multiple application mapping scenarios are explored at design-time in order to handle dynamism in the number of active applications at run-time [85], [90], [69]. A scenario contains a set of simultaneously active applications referred to as use-case [47] [65] [6]. *The scenario based approaches are not scalable as the number of scenarios increases exponentially with the number of applications, which might become intractable.*

A few strategies that perform mapping using design-time analysis results are presented in [79], [48], [97], [96], [95], [39], [81] and [82]. In [79], analysis result includes only a single mapping having minimum average power consumption. In [48], the authors target to minimize application execution time. In [97] and [96], analysis results include multiple mappings having trade-off in terms of target power consumption and performance. In [95], design-time analysis gives ideal core count and memory required for current state of the application. In [39], a set of process variation-aware schedules is analyzed at design-time. In [81] and [82], analysis results include mappings optimized from throughput point of view for homogeneous and heterogeneous platforms, respectively. The design-time analysis results have been used by run-time platform manager in order to map applications on the platform efficiently. The manager invokes run-time selection strategy to select the best mapping from the design-time analyzed mappings in order to configure the applications on the platform resources.

Reliability-aware Analysis: Reliability (fault) aware design-time analysis can be performed in order to produce solutions that can be used to cure faults incurred at run-time or to reduce the chances of system failure. Some approaches that perform such analysis and use the analysis results at run-time are presented in [25], [49], [77], [26] and [40]. In [25], design-time analysis is performed while aimed at optimizing system's life-time in terms of mean time to failure. In [49], an intensive design-time analysis for all possible failure scenarios is performed. At run-time, tasks are simply

Table 3: Comparison of various approaches for performing design-time analysis and then run-time (RT) mapping

Author	Plat.	Appl.	Maps.	RT	Optimization Goal
Mariani et al. [58]	Fix.	Hom.	Mult.	Yes	Energy consumption, Execution time
Stuijk et al. [85]	Fix.	Hom.	Mult.	No	Resource utilization
Beltrame et al. [29]	Fix.	Hom.	Mult.	No	Energy consumption, delay
Ykman et al. [97]	Fix.	Hom.	Mult.	Yes	Energy consumption, Execution time
Xue et al. [95]	Fix.	Hom.	Mult.	Yes	Resource utilization
Anup et al. [25]	Fix.	Hom.	Mult.	Yes	Reliability
Singh et al. [81]	Gen.	Hom.	Mult.	Yes	Solution quality, Execution time
Yang et al. [96]	Fix.	Het.	Mult.	Yes	Energy consumption, Execution time
Angiolini et al. [4]	Fix.	Het.	Mult.	No	Execution time
Zamora et al. [98]	Fix.	Het.	Mult.	No	Energy consumption, Execution time
Wildermann et al. [93]	Fix.	Het.	Mult.	No	Solution quality, Execution time
Schranzhofer et al. [79]	Fix.	Het.	Sing.	Yes	Energy consumption
Piscitelli et al. [71]	Fix.	Het.	Mult.	No	Solution quality, Execution time
Kwok et al. [48]	Fix.	Het.	Mult.	Yes	Execution time
Huang et al. [39]	Fix.	Het.	Mult.	Yes	Execution time
Lee et al. [49]	Fix.	Het.	Mult.	Yes	Reliability
Schor et al. [77]	Fix.	Het.	Mult.	Yes	Reliability
Derin et al. [26]	Fix.	Het.	Mult.	Yes	Reliability
Huang et al. [40]	Fix.	Het.	Mult.	Yes	Energy consumption, Reliability
Jia et al. [43]	Flex.	Het.	Mult.	No	Solution quality, Execution time
Singh et al. [82]	Gen.	Het.	Mult.	Yes	Energy Consumption, Execution time

remapped using the compile-time decisions. In [77], architectural failures are handled by allocating spare cores during design-time analysis in order to include the evaluation of all the possible failure scenarios. In [26], optimal mappings for all single-fault scenarios in the processing cores are analyzed, which are used by an online task remapping heuristic. The analysis performs optimization to minimize communication traffic and total execution time. In [40], an initial task schedule for different execution modes is generated at design-time. Then, run-time adjustment is performed at regular intervals for optimizing lifetime reliability and energy consumption.

Table 3 shows a comparison of the approaches reported in literature which consider design-time analysis and then analyzed results for run-time mapping. As can be seen, most of the existing approaches perform design-time analysis on fixed (Fix.) or flexible (Flex.) platforms and evaluate mappings that are applicable only to fixed homogeneous (Hom.), fixed heterogeneous (Het.) or a set of heterogeneous (Flex. Het.) platforms. A few approaches consider a *generic (Gen.)* platform and provide multiple mappings that are applicable to large set of platforms. Most of the analysis strategies provide multiple (Mult.) mappings. Some strategies provide support for run-time (RT) mapping that uses design-time analysis results optimized for different requirements.

3.3 Centralized vs. Distributed Management

The run-time mapping methodologies use centralized, distributed or hierarchical distributed resource management techniques. For small platforms such as 4x4 grid of cores, one core can be used as the manager that handles the mapping process. This approach is not scalable with the platform size as the monitoring traffic around the centralized manager increases which may lead to hot-spot resulting in reduced overall performance.

The distributed management caters for the large platforms such as 32x64 grid of cores. The distributed approach reduces the monitoring traffic around the Centralize Manager (CM). However, in a relatively smaller architecture, the CM approach might perform better as the distributed approach incurs additional communication overhead amongst the cluster agents without offering significant advantages. It should be noted that the cluster agents in the distributed approach behave identically to a centralized manager albeit for a small region. A detailed analysis and comparison of the centralized, distributed and hierarchical distributed approaches are provided in the next section.

4. UPCOMING TRENDS AND OPEN CHALLENGES

This section addresses some of the upcoming trends and challenges to be faced to take the mapping methodologies into the next era.

4.1 Hybrid Mapping

The reported mapping methodologies provide three alternatives: design-time mapping, on-the-fly mapping and hybrid (design-time analysis and then run-time use) mapping. Design-time techniques have pre-dominated the reported literature. However, their inability to handle dynamic workload scenarios has led to the formulation of on-the-fly mapping methodologies. On-the-fly strategies surmount the limitation of handling dynamic workloads at run-time but with the fallout of possible non-optimal mapping due to limited compute power at run-time. Recently, the issues of design-time and on-the-fly strategies have been addressed by developing hybrid mapping methodologies that attempt to incorporate the advantages of both. Hybrid strategies combine design space exploration of design-time techniques with the run-time management in order to select mapping configurations that are best suited to newly arriving applications. They involve minimum computation at run-time, facilitating for light-weight run-time manager performing efficient mapping. Our experimental results have shown that run-time mapping gets speeded up by 93% when compared to state-of-the-art on-the-fly mapping methodologies [81]. Although the advantages of hybrid strategy seem promising, it comes with its own trade-offs due to inherent pseudo-dynamic nature and inability to handle new applications without available design-space exploration. With no doubt, hybrid strategies seem to be followed in the field of mapping methodologies but due to their nascent development and lack of in-depth examination, further development of design-time and on-the-fly mapping methodologies will continue hand-in-hand with hybrid strategies.

The hybrid strategies also consider reliability-aware mapping as shown in Table 3. The strategies explore mapping alternatives at design-time for different fault scenarios incurred at run-time. Exploration by considering all the possible fault scenarios while considering different types of faults takes large time that might not be acceptable. This imposes challenge to investigate efficient exploration strategies that should overcome the exploration time bottleneck.

Modern design space exploration (DSE) strategies target optimization for multiple variables in order to satisfy several performance demands. The number of optimization variables is expected to increase with the increasing end user demands. In order to manage the challenges with increased optimization variables, an attention to prune the design space efficiently will be required. Further, heterogeneity of systems is increasing for better fulfilling the demands. This will need to be addressed with care due to the potential explosion in the number of permutations to be considered at each stage of the exploration. The exploration strategies might need to establish an upper limit on the heterogeneity in order to maintain the low complexity.

4.2 Large Scale Architectures

The technological enhancement will enable integration of hundreds and even thousands of cores [10]. Different large scale architectures have already been introduced, like Invasive Computing [32], Angstrom [34], and Intel's TeraFlop [91]. The many-core architectures impose a big challenge to manage their resources at run-time in a scalable manner. To achieve high degree of scalability in many-core architectures, resource management of large number of cores require distributed management as centralized management is not scalable with the number of cores. Some distributed management strategies are introduced in [44], [3], [46] and [27].

Kadin et al. [44] propose a Distributed Dynamic Thermal Management (D^2TM) scheme that delivers about 40% performance improvement over a standard planning scheme

7

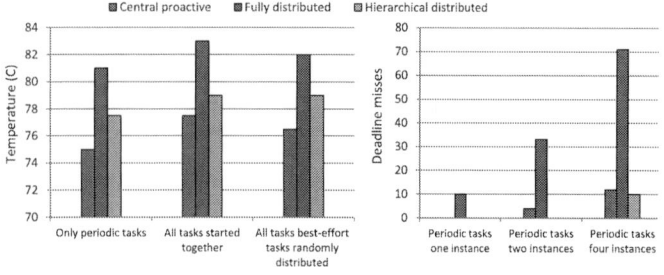

Figure 5: Peak temperatures and missed deadlines in periodic SPEC2006 tasks [27].

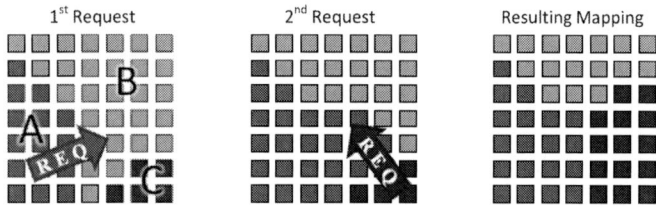

Figure 6: Three applications (A, B, and C) competing for cores in a many-core system [32].

for 16-core system without violating the temperature constraints. For the same system, an optimal central scheme delivers about 42% performance improvement. Fig. 5 shows a comparison of state-of-the-art proactive global centralized (used in [24]), fully distributed (used in [3]) and hierarchical distributed (used in [27]) approaches for peak temperature and deadline misses for applications of SPEC2006 benchmark suite. The temperature and deadline miss values for different simulations (mentioned on horizontal axis) are quoted from [27]. It can be observed that centralized proactive approach provides lower peak temperatures and hierarchical distributed approach shows minimum deadline misses. The lower peak temperatures by the centralized approaches are expected as they have a better view of the system and thus more optimization potential. However, they are limited by their scalability.

A many-core architecture to support invasive computing is introduced in [32]. The invasive many-core architecture contains large number of cores and uses distributed resource management approach [46] to achieve the required degree of scalability. In such large systems, typically many applications execute simultaneously [21] and compete for the available resources. The agent of each application executing in the system try to increase the speedup of its application by acquiring additional cores. Therefore, at run-time, each agent sends requests for cores to the nearby regions. The transfer of cores takes place if gain in the speedup is substantial over the loss for the giving application. Fig. 6 shows an example of three applications A, B, and C competing for resources. In the 1st request, application A requests additional cores and some cores are taken away from application B. In the 2nd request, application C requests more cores and the agents decide to take away some core from applications A and B. The initial mapping gets changed to a more balanced share of resources after the two optimization requests. Additionally, the mappings of applications are optimized over time as new resources might become available due to finishing of tasks of another application.

Fig. 7 shows a comparison of distributed [46] and centralized [75] resource management approaches for various system sizes. The shown results are from CES, KIT, Germany. The centralized scheme in [75] produces competitive and near-optimal schedules. Fig. 7.(a) shows average application speedup, which is computed as the total workload of all applications divided by the sum of the turnaround times of all applications. The results show that DistRM scheme per-

forms better for larger number of cores and achieves about 84% of the mapping quality of the centralized scheme. The centralized scheme always aims at the globally best solution, whereas DistRM scheme performs local changes. Fig. 7.(b) compares accumulated computational effort. In centralize scheme, all computations are performed in a single core and the computational effort increases with the system size. However, in the DistRM scheme, the effort is distributed over the system on different agents and stays constantly low. Fig. 7.(c) plots network utilization, which is calculated by multiplying the amount of messages, the average message size, and the average communication distance. The resulting communication volume required by the DistRM scheme is less for large system sizes. The messages in DistRM scheme are scattered over the NoC, which avoids communication bottlenecks that might encounter in centralized scheme. It can be observed that DistRM scheme requires lower communication volume over the centralize scheme for higher number of applications on large systems such as 32×32. Thus, DistRM scheme is more beneficial for more concurrent applications or larger systems.

Efficient exploitation of the abundant processing power of the available cores is challenging. This is one of the important problem and still needs to be investigated. The investigations need to address how applications from different domain can efficiently utilize large number of processor cores to jointly optimize compute performance, energy consumption and temperature for many-core systems.

4.3 3D Integration of Cores

Integration of multiple layers of processor cores into a single device leads to reduced area, power and signal transmission delay. These advantages make 3D multi-core architectures a potential alternative to be used in future high performance computing systems. Despite having several advantages, the 3D high integration density brings major concern in the temperature increase that causes thermal hot spots and high temperature gradients. This might lead to an unreliable system and degraded performance. Efficient thermal management of 3D architectures is challenging and requires investigation of efficient methodologies.

Thermal management techniques for 2D and 3D architectures are reported in [23, 24, 27, 89, 100] and [22, 99], respectively. However, development of efficient mapping methodologies taking thermal issues into account for 3D architectures will continue in foreseeable future. Further, 3D heterogeneous architectures will need to be considered for better fulfilling the increasing functional and non-functional demands. Heterogeneity imposes additional challenges for managing different types of cores.

Some additional challenges also need to be addressed to take the mapping methodologies into the next era. For example, development of efficient programming models for large scale and 3D architectures, efficient synchronization and control of concurrently executing tasks on such architectures and debugging of several concurrent executions if results are not as expected.

5. CONCLUSION

This paper provides a survey of the mapping methodologies targeting multi-core systems. In order to fully utilize the capabilities of multiple cores, the mapping methodologies are inevitably required to efficiently map complex applications onto them. The methodologies reported in the literature target homogeneous or heterogeneous architectures. Heterogeneous architectures provide better performance by exploiting the distinct features of the different type of cores but they are difficult to program as compared to their homogeneous counterpart. These multi-core systems employ NoC-based interconnection infrastructure for efficiency and scalability. The methodologies are classified as design-time

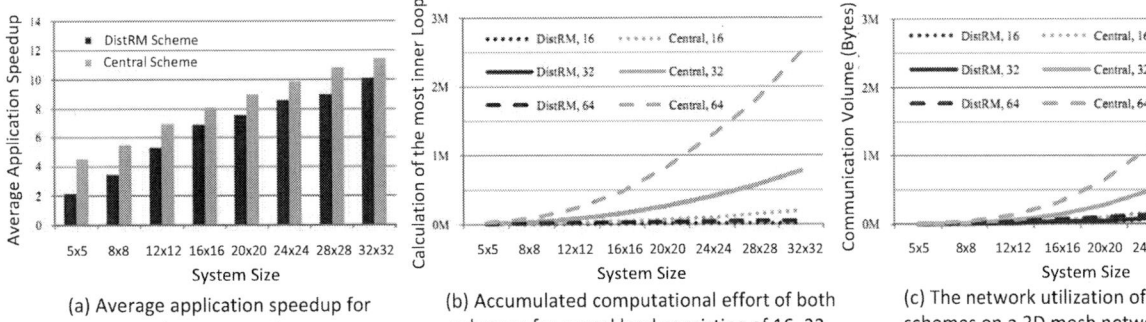

(a) Average application speedup for various system sizes

(b) Accumulated computational effort of both schemes for a workload consisting of 16, 32, and 64 applications on various system sizes

(c) The network utilization of both the schemes on a 2D mesh network for 16, 32, and 64 applications for various system sizes

Figure 7: Comparison of Distributed Resource Management (DistRM) [46] and Centralized [75] scheme.

or run-time methodologies. Their advantages and disadvantages for different type of workload scenarios are described. For dynamic workload scenarios, run-time techniques are proven to be more prevalent and useful. Additionally, they offer several other advantages over design-time techniques such as ability to enable unforeseeable upgrades, ability to avoid defective parts of a system, etc. Based on the analysis of the mapping methodologies, upcoming trends and open challenges are addressed.

6. ACKNOWLEDGMENTS

We also wish to mention that this work is partly supported by Singapore Ministry of Education Academic Research Fund Tier 1 under grant No. R-263-000-655-133 and German Research Foundation (DFG) as part of the Transregional Collaborative Research Centre "Invasive Computing" (SFB/TR 89); http://invasic.de.

7. REFERENCES

[1] W. Ahmed, M. Shafique, L. Bauer, and J. Henkel. Adaptive resource management for simultaneous multitasking in mixed-grained reconfigurable multi-core processors. In *CODES+ISSS*, pages 365–374, 2011.

[2] Y. Ahn, K. Han, G. Lee, H. Song, J. Yoo, K. Choi, and X. Feng. SoCDAL: System-on-chip design AcceLerator. *ACM Trans. Des. Autom. Electron. Syst.*, pages 1–38, 2008.

[3] M. A. Al Faruque, R. Krist, and J. Henkel. ADAM: run-time agent-based distributed application mapping for on-chip communication. In *DAC*, pages 760–765, 2008.

[4] F. Angiolini, J. Ceng, R. Leupers, F. Ferrari, C. Ferri, and L. Benini. An Integrated Open Framework for Heterogeneous MPSoC Design Space Exploration. In *DATE*, pages 1–6, 2006.

[5] G. Ascia, V. Catania, and M. Palesi. Multi-objective mapping for mesh-based noc architectures. In *CODES+ISSS*, pages 182–187, 2004.

[6] L. Benini, D. Bertozzi, and M. Milano. Resource Management Policy Handling Multiple Use-Cases in MPSoC Platforms Using Constraint Programming. In *ICLP*, pages 470–484, 2008.

[7] L. Benini and G. De Micheli. Networks on chips: a new SoC paradigm. *Computer*, (1):70–78, 2002.

[8] S. Bertozzi, A. Acquaviva, D. Bertozzi, and A. Poggiali. Supporting task migration in multi-processor systems-on-chip: a feasibility study. In *DATE*, pages 15–20, 2006.

[9] A. Bonfietti, L. Benini, M. Lombardi, and M. Milano. An efficient and complete approach for throughput-maximal sdf allocation and scheduling on multi-core platforms. In *DATE*, pages 897–902, 2010.

[10] S. Borkar. Thousand core chips: a technology perspective. In *DAC*, pages 746–749, 2007.

[11] E. W. Briáo, D. Barcelos, and F. R. Wagner. Dynamic task allocation strategies in MPSoC for soft real-time applications. In *DATE*, pages 1386–1389, 2008.

[12] E. L. d. S. Carvalho, N. L. V. Calazans, and F. G. Moraes. Dynamic task mapping for mpsocs. *IEEE Des. Test*, pages 26–35, 2010.

[13] J. Castrillon, A. Tretter, R. Leupers, and G. Ascheid. Communication-aware mapping of kpn applications onto heterogeneous mpsocs. In *DAC*, pages 1266–1271, 2012.

[14] J. Ceng et al. MAPS: an integrated framework for MPSoC application parallelization. In *DAC*, pages 754–759, 2008.

[15] W. Che and K. S. Chatha. Unrolling and retiming of stream applications onto embedded multicore processors. In *DAC*, pages 1272–1277, 2012.

[16] G. Chen, F. Li, S. Son, and M. Kandemir. Application mapping for chip multiprocessors. In *DAC*, pages 620–625, 2008.

[17] L. Chen, T. Marconi, and T. Mitra. Online scheduling for multi-core shared reconfigurable fabric. In *DATE*, pages 582–585, 2012.

[18] J. Choi, H. Oh, S. Kim, and S. Ha. Executing synchronous dataflow graphs on a spm-based multicore architecture. In *DAC*, pages 664–671, 2012.

[19] C.-L. Chou and R. Marculescu. User-aware dynamic task allocation in networks-on-chip. In *DATE*, pages 1232–1237, 2008.

[20] C.-L. Chou and R. Marculescu. Farm: Fault-aware resource management in noc-based multiprocessor platforms. In *DATE*, pages 1–6, 2011.

[21] C.-L. Chou, U. Y. Ogras, and R. Marculescu. Energy- and performance-aware incremental mapping for networks on chip with multiple voltage levels. *Trans. Comp.-Aided Des. Integ. Cir. Sys.*, pages 1866–1879, Oct. 2008.

[22] A. K. Coskun, J. L. Ayala, D. Atienza, T. S. Rosing, and Y. Leblebici. Dynamic thermal management in 3d multicore architectures. In *DATE*, pages 1410–1415, 2009.

[23] A. K. Coskun, T. S. Rosing, and K. C. Gross. Temperature management in multiprocessor socs using online learning. In *DAC*, pages 890–893, 2008.

[24] A. K. Coskun, T. v. Rosing, and K. C. Gross. Utilizing predictors for efficient thermal management in multiprocessor socs. *Trans. Comp.-Aided Des. Integ. Cir. Sys.*, pages 1503–1516, 2009.

[25] A. Das, A. Kumar, and B. Veeravalli. Reliability-Driven Task Mapping for Lifetime Extension of Networks-on-Chip Based Multiprocessor Systems. In *DATE*, 2013.

[26] O. Derin, D. Kabakci, and L. Fiorin. Online task remapping strategies for fault-tolerant Network-on-Chip multiprocessors. In *NOCS*, pages 129–136, 2011.

[27] T. Ebi, D. Kramer, W. Karl, and J. Henkel. Economic learning for thermal-aware power budgeting in many-core architectures. In *CODES+ISSS*, pages 189–196, 2011.

[28] M. R. Garey and D. S. Johnson. *Computers and Intractability; A Guide to the Theory of NP-Completeness*. 1979.

[29] B. Giovanni, L. Fossati, and D. Sciuto. Decision-theoretic design space exploration of multiprocessor platforms. *Trans. Comp.-Aided Des. Integ. Cir. Sys.*, pages 1083–1095, 2010.

[30] A. Hartman, D. Thomas, and B. Meyer. A case for lifetime-aware task mapping in embedded chip multiprocessors. In *CODES+ISSS*, pages 145–154, 2010.

[31] A. S. Hartman and D. E. Thomas. Lifetime improvement through runtime wear-based task mapping. In *CODES+ISSS*, pages 13–22, 2012.

[32] J. Henkel et al. Invasive manycore architectures. In *ASP-DAC*, pages 193–200, 2012.

[33] J. Henkel, W. Wolf, and S. Chakradhar. On-chip networks: A scalable, communication-centric embedded system design paradigm. In *VLSID*, pages 845–851, 2004.

[34] H. Hoffmann et al. Self-aware computing in the angstrom processor. In *DAC*, pages 259–264, 2012.

[35] S. Hong, S. H. K. Narayanan, M. Kandemir, and O. Özturk. Process variation aware thread mapping for chip multiprocessors. In *DATE*, pages 821–826, 2009.

[36] J. Hu and R. Marculescu. Energy-aware mapping for tile-based noc architectures under performance constraints. In *ASP-DAC*, pages 233–239, 2003.

[37] J. Hu and R. Marculescu. Energy- and performance-aware mapping for regular NoC architectures. *IEEE Trans. Comp.-Aided Des. Integ. Cir. Sys.*, (4):551–562, 2005.

[38] J. Huang, A. Raabe, C. Buckl, and A. Knoll. A workflow for runtime adaptive task allocation on heterogeneous MPSoCs. In *DATE*, pages 1–6, 2011.

[39] L. Huang and Q. Xu. Performance yield-driven task allocation and scheduling for MPSoCs under process variation. In *DAC*, pages 326–331, 2010.

[40] L. Huang, R. Ye, and Q. Xu. Customer-aware task allocation and scheduling for multi-mode MPSoCs. In *DAC*, pages 387–392, 2011.

[41] H. Javaid and S. Parameswaran. A design flow for application specific heterogeneous pipelined multiprocessor systems. In *DAC*, pages 250–253, 2009.

[42] A. Jerraya, H. Tenhunen, and W. Wolf. Guest Editors' Introduction: Multiprocessor Systems-on-Chips. *Computer*, (7):36–40, 2005.

[43] Z. J. Jia et al. NASA: A generic infrastructure for system-level MP-SoC design space exploration. In *ESTIMedia*, pages 41–50, 2010.

[44] M. Kadin, S. Reda, and A. Uht. Central vs. distributed dynamic thermal management for multi-core processors: which one is better? In *GLSVLSI*, pages 137–140, 2009.

[45] J. Keinert et al. SystemCoDesigner Üan automatic ESL synthesis approach by design space exploration and behavioral synthesis for streaming applications. *ACM Trans. Des. Autom. Electron. Syst.*, pages 1–23, 2009.

[46] S. Kobbe, L. Bauer, D. Lohmann, W. Schröder-Preikschat, and J. Henkel. Distrm: distributed resource management for on-chip many-core systems. In *CODES+ISSS*, pages 119–128, 2011.

[47] A. Kumar et al. Multiprocessor systems synthesis for multiple use-cases of multiple applications on FPGA. *ACM Trans. Des. Autom. Electron. Syst.*, pages 1–27, 2008.

[48] Y.-K. Kwok et al. A semi-static approach to mapping dynamic iterative tasks onto heterogeneous computing systems. *J. Parallel Distrib. Comput.*, 66(1):77–98, 2006.

[49] C. Lee, H. Kim, H.-w. Park, S. Kim, H. Oh, and S. Ha. A task remapping technique for reliable multi-core embedded systems. In *CODES+ISSS*, pages 307–316, 2010.

[50] L.-Y. Lin et al. Communication-driven task binding for multiprocessor with latency insensitive network-on-chip. In *ASP-DAC*, pages 39–44, 2005.

[51] C. L. Liu and J. W. Layland. Scheduling algorithms for multiprogramming in a hard-real-time environment. *J. ACM*, pages 46–61, 1973.

[52] W. Liu and other. Efficient SAT-Based Mapping and Scheduling of Homogeneous Synchronous Dataflow Graphs for Throughput Optimization. In *RTSS*, pages 492–504, 2008.

[53] A. Mallik et al. MNEMEE - An Automated Toolflow for Parallelization and Memory Management in MPSoC Platforms. In *DAC*, 2011.

[54] S. Manolache, P. Eles, and Z. Peng. Task mapping and priority assignment for soft real-time applications under deadline miss ratio constraints. *ACM Trans. Embed. Comput. Syst.*, (2):19:1–19:35, 2008.

[55] C. Marcon, A. Borin, A. Susin, L. Carro, and F. Wagner. Time and energy efficient mapping of embedded applications onto NoCs. In *ASP-DAC*, pages 33–38, 2005.

[56] C. Marcon, E. Moreno, N. Calazans, and F. Moraes. Comparison of network-on-chip mapping algorithms targeting low energy consumption. *Computers Digital Techniques, IET*, pages 471–482, 2008.

[57] R. Marculescu, U. Ogras, L.-S. Peh, N. Jerger, and Y. Hoskote. Outstanding Research Problems in NoC Design: System, Microarchitecture, and Circuit Perspectives. *IEEE TCAD*, (1):3–21, 2009.

[58] G. Mariani et al. An industrial design space exploration framework for supporting run-time resource management on multi-core systems. In *DATE*, pages 196–201, 2010.

[59] G. Martin. Overview of the mpsoc design challenge. In *DAC*, pages 274–279, 2006.

[60] P. Marwedel, J. Teich, G. Kouveli, I. Bacivarov, L. Thiele, S. Ha, C. Lee, Q. Xu, and L. Huang. Mapping of applications to MPSoCs. In *CODES+ISSS*, pages 109–118, 2011.

[61] A. Mehran, A. Khademzadeh, and S. Saeidi. DSM: A Heuristic Dynamic Spiral Mapping algorithm for network on chip. *IEICE Electronics Express*, (13):464–471, 2008.

[62] B. H. Meyer, A. S. Hartman, and D. E. Thomas. Cost-effective slack allocation for lifetime improvement in noc-based mpsocs. In *DATE*, pages 1596–1601, 2010.

[63] O. Moreira, J. J.-D. Mol, and M. Bekooij. Online resource management in a multiprocessor with a network-on-chip. In *SAC*, pages 1557–1564, 2007.

[64] O. Moreira, F. Valente, and M. Bekooij. Scheduling multiple independent hard-real-time jobs on a heterogeneous multiprocessor. In *EMSOFT*, pages 57–66, 2007.

[65] S. Murali, M. Coenen, A. Radulescu, K. Goossens, and G. De Micheli. A methodology for mapping multiple use-cases onto networks on chips. In *DATE*, pages 118–123, 2006.

[66] V. Nollet et al. Centralized Run-Time Resource Management in a Network-on-Chip Containing Reconfigurable Hardware Tiles. In *DATE*, pages 234–239, 2005.

[67] V. Nollet et al. Run-time management of a MPSoC containing FPGA fabric tiles. *IEEE Trans. Very Large Scale Integr. Syst.*, pages 24–33, 2008.

[68] H. Orsila et al. Automated memory-aware application distribution for Multi-processor System-on-Chips. *J. Syst. Archit.*, (11):795–815, 2007.

[69] G. Palermo, C. Silvano, and V. Zaccaria. Robust optimization of SoC architectures: A multi-scenario approach. In *ESTIMedia*, pages 7–12, 2008.

[70] Z. Peter et al. A Decentralised Task Mapping Approach for Homogeneous Multiprocessor Network-On-Chips. *International Journal of Reconfigurable Computing*, 2009.

[71] R. Piscitelli and A. Pimentel. Design space pruning through hybrid analysis in system-level design space exploration. In *DATE*, pages 781–786, 2012.

[72] X. Qi, D. Zhu, and H. Aydin. Global Reliability-Aware Power Management for Multiprocessor Real-Time Systems. In *ERTCSA*, pages 183–192, 2010.

[73] C.-E. Rhee, H.-Y. Jeong, and S. Ha. Many-to-Many Core-Switch Mapping in 2-D Mesh NoC Architectures. In *ICCD*, pages 438–443, 2004.

[74] M. Ruggiero et al. Communication-aware allocation and scheduling framework for stream-oriented multi-processor systems-on-chip. In *DATE*, pages 3–8, 2006.

[75] G. Sabin, M. Lang, and P. Sadayappan. Moldable parallel job scheduling using job efficiency: an iterative approach. In *JSSPP*, pages 94–114, 2007.

[76] N. Satish, K. Ravindran, and K. Keutzer. A decomposition-based constraint optimization approach for statically scheduling task graphs with communication delays to multiprocessors. In *DATE*, pages 57–62, 2007.

[77] L. Schor et al. Scenario-based design flow for mapping streaming applications onto on-chip many-core systems. In *CASES*, pages 71–80, 2012.

[78] A. Schranzhofer, J.-J. Chen, and L. Thiele. Power-Aware Mapping of Probabilistic Applications onto Heterogeneous MPSoC Platforms. In *RTAS*, pages 151–160, 2009.

[79] A. Schranzhofer, J.-J. Chen, and L. Thiele. Dynamic Power-Aware Mapping of Applications onto Heterogeneous MPSoC Platforms. *IEEE Transactions on Industrial Informatics*, (4):692–707, 2010.

[80] H. Shojaei et al. A parameterized compositional multi-dimensional multiple-choice knapsack heuristic for CMP run-time management. In *DAC*, pages 917–922, 2009.

[81] A. K. Singh, A. Kumar, and T. Srikanthan. A Hybrid Strategy for Mapping Multiple Throughput-constrained Applications on MPSoCs. In *CASES*, pages 175–184, 2011.

[82] A. K. Singh, A. Kumar, and T. Srikanthan. Accelerating throughput-aware runtime mapping for heterogeneous mpsocs. *ACM Trans. Des. Autom. Electron. Syst.*, pages 1–29, 2013.

[83] A. K. Singh, T. Srikanthan, A. Kumar, and W. Jigang. Communication-aware heuristics for run-time task mapping on NoC-based MPSoC platforms. *J. Syst. Archit.*, pages 242–255, 2010.

[84] L. Smit et al. Run-time mapping of applications to a heterogeneous reconfigurable tiled system on chip architecture. In *FPT*, pages 421–424, 2004.

[85] S. Stuijk, M. Geilen, and T. Basten. A Predictable Multiprocessor Design Flow for Streaming Applications with Dynamic Behaviour. In *DSD*, pages 548–555, 2010.

[86] T. D. ter Braak et al. Run-time spatial resource management for real-time applications on heterogeneous MPSoCs. In *DATE*, pages 357–362, 2010.

[87] T. Theocharides et al. Towards embedded runtime system level optimization for MPSoCs: on-chip task allocation. In *GLSVLSI*, pages 121–124, 2009.

[88] L. Thiele, I. Bacivarov, W. Haid, and K. Huang. Mapping Applications to Tiled Multiprocessor Embedded Systems. In *ACSD*, pages 29–40, 2007.

[89] L. Thiele, L. Schor, H. Yang, and I. Bacivarov. Thermal-aware system analysis and software synthesis for embedded multi-processors. In *DAC*, pages 268–273, 2011.

[90] P. van Stralen and A. Pimentel. Scenario-based design space exploration of MPSoCs. In *ICCD*, pages 305–312, 2010.

[91] S. Vangal et al. An 80-Tile 1.28TFLOPS Network-on-Chip in 65nm CMOS. In *ISSCC*, pages 98–589, 2007.

[92] F. Wang et al. Variation-aware task and communication mapping for mpsoc architecture. *IEEE TCAD*, (2):295–307, 2011.

[93] S. Wildermann, F. Reimann, D. Ziener, and J. Teich. Symbolic design space exploration for multi-mode reconfigurable systems. In *CODES+ISSS*, pages 129–138, 2011.

[94] D. Wu, B. M. Al-Hashimi, and P. Eles. Scheduling and Mapping of Conditional Task Graphs for the Synthesis of Low Power Embedded Systems. In *DATE*, page 10090, 2003.

[95] L. Xue, O. ozturk, F. Li, M. Kandemir, and I. Kolcu. Dynamic partitioning of processing and memory resources in embedded MPSoC architectures. In *DATE*, pages 690–695, 2006.

[96] P. Yang et al. Managing dynamic concurrent tasks in embedded real-time multimedia systems. In *ISSS*, pages 112–119, 2002.

[97] C. Ykman-Couvreur et al. Linking run-time resource management of embedded multi-core platforms with automated design-time exploration. *IET Comp. Dig. Techn.*, (2):123–135, 2011.

[98] N. H. Zamora, X. Hu, and R. Marculescu. System-level performance/power analysis for platform-based design of multimedia applications. *ACM Trans. Des. Autom. Electron. Syst.*, pages 2:1–2:29, 2007.

[99] X. Zhou, J. Yang, Y. Xu, Y. Zhang, and J. Zhao. Thermal-aware task scheduling for 3d multicore processors. *IEEE Trans. Parallel Distrib. Syst.*, pages 60–71, 2010.

[100] C. Zhu, Z. P. Gu, R. P. Dick, and L. Shang. Reliable multiprocessor system-on-chip synthesis. In *CODES+ISSS*, pages 239–244, 2007.

Workload and User Experience-Aware Dynamic Reliability Management in Multicore Processors

Pietro Mercati
UCSD
pmercati@ucsd.eng.edu

Andrea Bartolini
University of Bologna
a.bartolini@unibo.it

Francesco Paterna
UCSD
fpaterna@ucsd.eng.edu

Tajana Simunic Rosing
UCSD
tajana@ucsd.edu

Luca Benini
University of Bologna
luca.benini@unibo.it

ABSTRACT

Reliability is a major concern for nanoscale CMOS circuits. Degradation phenomena such as Electromigration, Negative Bias Temperature Instability, Time Dependent Dielectric Breakdown worsen with transistor scaling. Dynamic Reliability Management (DRM) techniques reduce reliability loss at runtime by constraining operating points, but they face the challenge of reducing user experience degradation while meeting a lifetime target. In this work we propose a sensor based hierarchical controller for multicore processor DRM, exploiting the major gap between the time scales of workload variations and reliability loss. We improve performance and user experience by locally relaxing reliability-induced operating point constraints, while meeting them over the large time windows relevant for reliability. With respect to the state-of-the-art, our solution guarantees timely execution of 100% of latency-critical applications, and have a 4% performance improvement over the whole lifetime.

1. INTRODUCTION

Technology scaling has made modern integrated circuits more susceptible to degradation phenomena such as Negative Bias Temperature Instability (NBTI), Electromigration (EM) and Time Dependent Dielectric Breakdown (TDDB) [9]. Degradation depends on many process and environmental factors, but can be controlled by managing temperature and voltage. Degradation worsens under continued stress [18], while short spikes in temperature and voltage do not affect reliability much. Aging effects are described by mathematical models in terms of Mean Time To Failure (MTTF) [15] or reliability [20].

We focus on TDDB [6, 16, 20], but our solution can apply to other phenomena as well. TDDB is a degradation phenomenon that results in a low-impedance path through transistor gate dielectric, causing high leakage current that leads to failures. With the scaling of technology, increasing the design margin and binning the chips are becoming costly strategies. runtime management techniques can overcome this by dynamically changing chip operating points [15]. Modern processors can exploit dynamic voltage and

frequency scaling (DVFS) to modify the degradation rate [18, 21]. Degradation can be estimated through voltage and temperature dependent mathematical models, but recent works also proposed devices that can directly sense degradation [13, 14] and are more accurate than temperature and voltage based estimation. In [14], TDDB-specific sensors are presented.

These elements, together with a control algorithm, are the basis for Dynamic Reliability Management (DRM). DRM has been proposed in [15, 10, 21] as a mechanism to trade off between system performance and reliability margin. DRM policies guarantee a target of reliability within the predefined lifetime of the system. However, the long time scale and the non-reversibility of degradation pose a challenge for the DRM control algorithm. To alleviate this issue, [21] proposes a predictive control for a single core CPU, based on a mathematical model.

Many systems today use multicore processors, from servers to smartphones and tablets. Embedded devices exploit multicore CPUs for data and computing intensive applications with varied requirements in terms of performance and QoS [1, 2, 8]. Android based systems exploit the mechanism of *intent* to describe and communicate to the hardware the urgency/quality of a task to be executed. Not satisfying these requirements causes significant user experience degradation. Little has been done for the reliability of multicore CPUs [5, 17, 19]. The main problem with state-of-the-art DRM solutions is that they control reliability by fixing a maximum limit of the operating conditions, disregarding the potential degradation of the user experience.

In this work we propose a novel DRM policy for multicore platforms. The proposed policy is based on a two-level controller, composed by a *Long Term Controller* and a *Short Term Controller*. The two levels operate on two different time scales, that we have called *Long Intervals*, corresponding to days that it takes for reliability to change, and *Short Intervals*, corresponding to OS scheduling ticks. Our controller monitors system reliability on a long time scale and adapts operating conditions to workload phase changes on a short time scale, with the goal of meeting a target reliability within a predefined target lifetime. Our solution uses aging sensors to improve reliability control robustness. The main novelty we introduce is the *Borrowing Strategy*, through which our solution is able to locally relax reliability-induced operating point constraints, while still meeting them over the large time windows relevant for reliability loss. This is a key feature for systems like smartphones and tablets, that emphasize user experience. We compare our policy against state-of-the-art and show that with our solution 100% of latency-critical applications meet their needs, with an overall performance improvement of 4% over the whole lifetime.

Permission to make digital or hard copies of all or part of this work for personal or classroom use is granted without fee provided that copies are not made or distributed for profit or commercial advantage and that copies bear this notice and the full citation on the first page. To copy otherwise, to republish, to post on servers or to redistribute to lists, requires prior specific permission and/or a fee.
DAC '13 May 29 - June 07 2013, Austin, TX, USA

2. RELATED WORK

Traditionally chips are designed under the assumption of worst-case conditions [10]. This approach ignores the dynamic nature of actual operating conditions, which results in an overly conservative, performance-limiting device. Srinivasan et al. [15] first introduced DRM as a technique where the processor can dynamically respond to changing application behavior to maintain its lifetime reliability target, by dynamic voltage and frequency scaling (DVFS). This was a significant enhancement over previous worst-case reliability qualification methodologies. Blome et al. [4] extended the approach to monitor and control the impact on lifetime reliability, through thread scheduling and DVFS. The authors show how to leverage the slack between the typical degradation and the worst-case one to improve performance in periods of high peak demand. Karl et.al. [10] explored DRM using a systematic model to improve performance. Both [4, 10] assume that the future workload is equal to the previous one. Therefore they are very sensitive to sudden workload variations.

Zhuo et al. [20] proposed a process variation and temperature-aware oxide reliability model, which can estimate reliability from temperature and voltage history. In [21] the authors propose a DRM framework that extends this model to periodically predict the future value of reliability at the target lifetime. Based on the difference between the predicted reliability and the target one, the controller sets a maximum operating voltage. This approach is less sensitive to workload variations compared to previous works, because it exploits a confidence-based workload estimation. Since the policy sets the maximum voltage, it is not able to guarantee speed bursts for high performance demanding tasks, causing user experience degradation. Moreover, it is entirely model-based, relying only on temperature and voltage readings, and does not use sensors to estimate aging. Therefore it has high model uncertainty.

Singh et al. [14, 13] propose oxide degradation sensors and sensor-based DRM approach. Degradation sensors are non-intrusive monitors designed to be integrated in modern CMOS circuits. The authors have designed, manufactured and tested these devices. In the rest of this paper, we refer to them as degradation sensors or aging sensors. Monitoring with these sensors helps mitigate model inaccuracy. Since reliability models are based on stress measurements, a strictly model-based DRM policy tends to be very pessimistic [14].

The presented approaches for DRM ignore the fact that aging is observable on a large time scale, and that degradation is affected by average, rather than immediate stress. Therefore, they neglect the opportunity for short performance bursts to meet quality requirements of real applications. Furthermore, none of the previous work considers multicore platforms, a key component in most computing devices today.

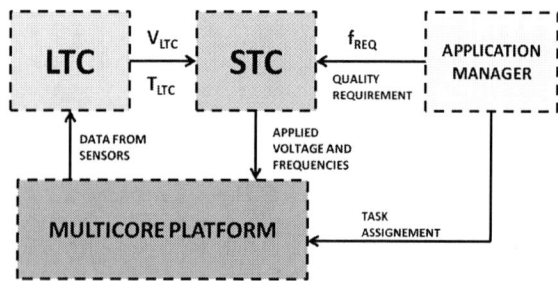

Figure 1: Two-level controller block diagram.

3. CONTROLLER ARCHITECTURE

The target platform is a homogeneous multiprocessor with N cores, with per-chip voltage setting and per-core frequency control. Each core is single threaded and has its own degradation sensors. Tasks are assigned in FIFO manner. The controller exploits voltage and frequency settings as knobs to trade off performance, while meeting the target temperature and reliability within a predefined lifetime. Figure 1 shows the basic building blocks of the proposed architecture, consisting of *Application Manager*, *Long Term Controller* and *Short Term Controller*.

Application Manager (AM): allocates tasks in FIFO manner, communicates the requested frequency f_{REQ} and the quality requirement for the task execution to the *Short Term Controller*. [1]

Long Term Controller (LTC): samples data from aging sensors at the beginning of each *Long Interval*, monitors the degradation status and calculates the average temperature and voltage. It predicts future reliability using these parameters. Since reliability loss occurs on a long time scale [15], we consider *Long Intervals* to be on the order of days. Based on the difference between predicted and target reliability, it computes a reference voltage, V_{LTC} and a reference temperature T_{LTC}, which are the inputs for the *Short Term Controller*. The constraint on reliability is met if the mean applied voltage in the *Long Interval* V_{LI} is less or equal to V_{LTC} and the temperature is below T_{LTC}.

Short Term Controller (STC): receives f_{REQ} and the quality requirement for the allocated tasks from the *Application Manager*, and V_{LTC} and T_{LTC} from the *Long Term Controller*. Based on that, it applies the *Borrowing Strategy*, adjusting voltage and frequencies at each scheduling tick given the tasks quality requirements, while keeping the mean applied voltage inside the *Long Interval* V_{LI} lower than V_{LTC}, and the temperature below T_{LTC}. The *Short Term Controller* can be coupled with state-of-the-art thermal management techniques to handle thermal emergencies [3, 12], given the thermal constraint from LTC. The operations performed by the blocks are discussed in more details in the following subsections.

3.1 Application Manager:

In modern operating systems, such as Android, the application can request a certain level of hardware and software service to provide a given QoS to the final user. Therefore, we characterize each task as either *Highly critical* (H) or *Less critical* (L) in terms of latency and user experience [2]. Executing H tasks at a frequency lower than f_{REQ} causes user experience degradation. This information allows the *Short Term Controller* to adjust its *Borrowing Strategy*. General purpose workloads for embedded devices do not contain profile information. Therefore the *Application Manager* selects $f_{REQ} = f_{MAX}$ for a running task and $f_{REQ} = f_{MIN}$ for the idle period.

3.2 Long Term Controller:

The diagram in Figure 2 shows the *Long Term Controller*. The *Long Term Controller* samples data from the aging sensors at the beginning of a new *Long Interval*, separately for each core.

The *sens2R* block estimates the reliability R_i for the i_{th} core, from the aging sensors readings S_i. R_i at time t is a number between $[0, 1]$ indicating the probability that the system will not fail before time t. It is a measure of the system degradation status [20]. For example, TDDB sensors, based on a ring oscillator whose fre-

---

[1] Our solution is orthogonal to the scheduler employed, which can be also of a different kind.

[2] Similarly, our solution could be adapted to real time systems, where the distinction between hard and soft deadline is tight.

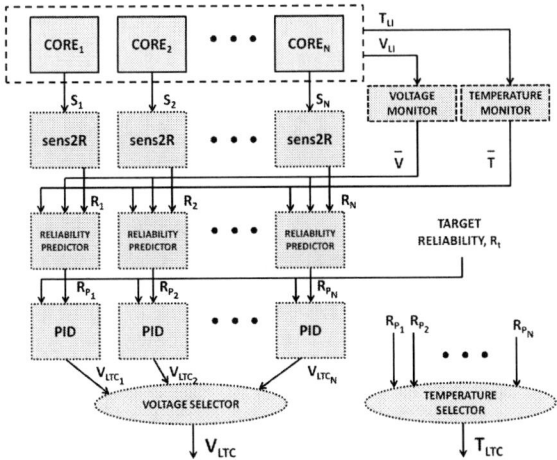

Figure 2: Long Term Controller block diagram.

quency increases as degradation takes place [13, 14], give statistically significant information on the aging status of their core [14]. In [13], authors map sensor readings for NBTI degradation to the system aging status. Based on that, they dynamically manage operating conditions to minimize NBTI degradation. In this work we assume that there exists a mapping between the output of TDDB sensors [14] and R_i and refer to papers [13, 14] for further discussion.

The *Voltage Monitor* and the *Temperature Monitor* keep the voltage and temperature history of the multicore platform by estimating the mean voltage/temperature that are going to be applied from the current time instant until the target lifetime. These values are $\bar{V}$ and $\bar{T}$, and they are calculated as an exponential moving average of the past voltages/temperatures, as:

$$
\begin{aligned}
\bar{V}_k &= \alpha_V \cdot \bar{V}_{k-1} - (1 - \alpha_V) \cdot V_{LI_{k-1}} \\
\bar{T}_k &= \alpha_T \cdot \bar{T}_{k-1} - (1 - \alpha_T) \cdot T_{LI_{k-1}}
\end{aligned}
\tag{1}
$$

Where k identifies the kth *Long Interval*, α_V and α_T are the weighting factors, $\bar{V}_{k-1}$ and $\bar{T}_{k-1}$ are the values at the previous *Long Interval*, $V_{LI_{k-1}}$ and $T_{LI_{k-1}}$ are the mean applied voltage/temperature in the previous *Long Interval*.

The *Reliability Predictor* receives R_i, $\bar{V}, \bar{T}$ and computes the predicted reliability R_{P_i} exploiting the model presented in [21]. R_{P_i} is the value of reliability that we would have at the target lifetime given a current reliability equal to R_i, and supposing that from the present time on we apply a voltage equal to $\bar{V}$ and a temperature equal to $\bar{T}$.

The *PID controller*, similarly as in [10, 21], receives R_{P_i} and the target reliability at the target lifetime R_t. Based on their difference, it calculates V_{LTC}, that asymptotically minimizes the tracking error. For the ith core at the kth *Long Interval*, we have $e_k = R_t - R_{P_k}$ and therefore:

$$
V_{LTC_k} = V_{LTC_{k-1}} +
$$
$$
+ K_P \left(e_k + K_I \sum_{j=1}^{k} (e_j \cdot \Delta_{LI}) + K_D \left(\frac{e_k - e_{k-1}}{\Delta_{LI}} \right) \right)
\tag{2}
$$

where K_P, K_I, K_D are the PID parameters and Δ_{LI} is the duration of a *Long Interval*. For example, if the system was subject to high temperature and voltage during the previous *Long Interval*, R_P will be low and the PID outputs a V_{LTC} lower than the previ-

ous one. If it was subject to low temperature and voltage, the PID outputs a V_{LTC} higher than the previous one.

The *Voltage Selector* selects the minimum V_{LTC}, outputted by the *Long Term Controller*, to guarantee the reliability of the most degraded core (the one with the lowest R_P). Only one V_{LTC} is needed[3], as the target platform only has a single voltage island. This work can be easily generalized to multiple V_{LTC}[4] Similarly, the *Temperature Selector* outputs the reference temperature T_{LTC}.

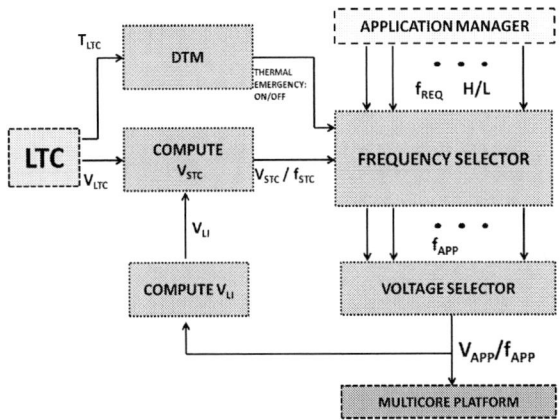

Figure 3: Short Term Controller block diagram.

3.3 Short Term Controller:

Figure 3 shows the block diagram of the *Short Term Controller*. The *Short Term Controller* receives V_{LTC} and T_{LTC} from the *Long Term Controller*, f_{REQ} and the H/L flags from the *Application Manager*. Based on these parameters, it develops the *Borrowing Strategy*, selecting the frequencies f_{APP} and the voltage V_{APP} to be applied at each *Short Interval* to meet the task quality requirements, while keeping V_{LI} less than or equal to V_{LTC}. Note that frequency selection should be coordinated with DTM. If temperature is higher than T_{LTC}, then a lower frequency is selected. Given that there has been a lot of work on DVFS for DTM, here we just focus on reliability aspects of voltage selection. The key to the *Borrowing Strategy* is the computation of the reference voltage V_{STC}:

$$
V_{STC_l} = \frac{V_{LTC} \cdot \Delta_{LI} - V_{LI_{l-1}} \cdot t_{LI}}{\Delta_{LI} - t_{LI}}
\tag{3}
$$

Where l identifies the current *Short Interval*, V_{LI_l} is the mean voltage applied from the beginning of the *Long Interval*, and t_{LI} is the time elapsed since the beginning of the *Long Interval*. For each core executing a L task, the *Short Term Controller* selects $f_{APP} = f_{STC}$, while for each core executing a H task, it selects $f_{APP} = f_{MAX}$. For a idle core, it selects $f_{APP} = f_{MIN}$. Since the system has a single voltage island, in order to execute the most performance-heavy task, the controller selects the voltage V_{APP} corresponding to the maximum f_{APP}.[5] The other cores have the same applied voltage, but run at a lower or equal frequency. The

---

[3]In case of a platform with multiple voltage islands, cores can be grouped in voltage clusters. It would then be necessary to replicate this scheme for each cluster.

[4]Both the *Reliability Predictor* and the PID could be replaced by alternative solutions. This is an area of possible future improvement.

[5]In case of multiple voltage islands, it is necessary to replicate the scheme for each of them.

K_P	K_I	K_D	Δ_{LI}	α_V	α_T
$0.73[V/y]$	$14.4[y]$	$3.6[y]$	$25[d]$	0.1	0.1

Table 1: Simulation parameters

Borrowing Strategy is based on V_{STC}. If the controller applies a voltage lower than V_{STC}, V_{STC} tends to increase, allowing the system to go faster in the next intervals. If, conversely, a H task occurs and the controller applies a voltage higher than V_{STC}, V_{STC} tends to decrease, in order to "repay the loan".

$$V_{LI_{l+1}} = \frac{V_{LI_l} \cdot t_{LI} + V_{APP_l} \cdot \Delta_{SI_l}}{t_{LI} + \Delta_{SI_l}} \qquad (4)$$

Where Δ_{SI_l} is the duration of the lth *Short Interval*. As a *Short Interval* ends, V_{STC} is updated through Equation 3 and the *Short Term Controller* performs a new frequency/voltage selection. If $V_{LTC} - V_{LI}$ at the end of a *Long Interval* is non zero, the system has either not fully exploited the available reliability margin (if positive) or it has violated the reliability constraint for the current *Long Interval* (if negative). Therefore, this difference is added to the V_{LTC} which is computed for the next *Long Interval*, so to keep track of under/over-utilization.

4. RESULTS

The target platform is composed by 4 homogeneous cores with per-chip voltage and per-core frequency settings. The voltage ranges from 0.8V to 1.4V and the frequency ranges from 223MHz to 532Mhz. Our reference is the STMicroelectronics xSTsim architecture [11]. This platform is composed by a General-purpose Processing Element (GPE) acting as *host processor* and Processing Elements (PEs) acting as *streaming engine*. The GPE is an ST231 processor and the PEs are programmable processors with a simple ISA extended with SIMD and vector mode instructions. The platform addresses the needs of data-flow dominated, highly computational intensive tasks, typical of many embedded products.

For short term simulations we test our policy with xSTsim executing Inverse Discrete Cosine Transform (IDCT) [11] on a single *Long Interval* on random frame sequences. IDCT is a representative multimedia computational kernel, used in MPEG2 and JPEG decoding. Each frame is considered as a task. The GPE acts only as a dispatcher for the PEs and performs no computation. We have modified the application so to mark each frame as either H or as L, and to control the percentages of H and L tasks. The reliability control sets the operating voltage and frequency for the PEs by following the *Borrowing Strategy*.

For long term simulations we have developed a simulation infrastructure with Matlab, following the characteristics of the described platform. With this framework we can simulate the reliability model presented in [20] over the whole system lifetime and evaluate performance. The workload is modeled as a sequence of tasks with their own requested frequency and H/L flag. The task sequences are generated to reflect different user profiles [7], by varying the percentage of idle and busy periods and the percentage of H and L tasks. Table 1 shows the value of the parameters used in our simulations. The PID gains are obtained through Ziegler-Nichols open-loop method. The system is run at V_{MAX} for the entire lifetime and its response, e.g. the reliability curve, provides the parameters for the Z-N method. Δ_{LI} is set at 25 days, for having reasonable simulation times. α_V and α_T are both 0.1. This value has been chosen among others after extensive tests of the model with different workload and temperature profiles.

4.1 Comparison With State-of-the-Art

We compare our policy against the state-of-the-art technique presented in [21]. In this work, authors present a reliability management framework which uses the model in [20] as well. The framework periodically computes the predicted reliability R_P and exploit it to set a maximum operating voltage. This work has two main limitations:

- It limits the maximum voltage, causing user experience degradation when executing H tasks. We show how our policy, thanks to the two-level controller and the *Borrowing Strategy*, overcomes this limitation by following the task quality requirements, while still meeting the target reliability.

- It does not use aging sensors, and only exploits temperature and voltage readings for calculating reliability. We show how the use of aging sensors makes the reliability control more robust with respect to model variations.

Since the policy in [21] is for single core, comparison is conducted referring to this scenario. We assume the core to have a target reliability $R_t = 0.8$ and a target lifetime of 5 years. In the following, we denote our policy as *LTST* (*Long Term - Short Term*), and the policy in [21] as *Zhuo* (from the name of the author).

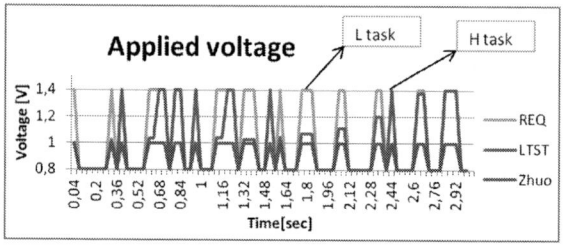

Figure 4: Voltage traces comparison on xSTsim simulator.

In Figure 4 we compare the voltage traces that we obtain with *LTST* and *Zhuo* for the execution of IDCT on a random sequence of frames in a *Long Interval*. V_{REQ} is $1.4V$ for a running task and $0.8V$ for an idle period. Executing tasks at a higher voltage allows to achieve better quality and higher performance. We assume that $V_{LTC} = 1V$. *Zhuo* is able to raise the voltage at most to $1V$. *LTST* achieves better performance for both H and L tasks. In the former case, *LTST* boosts voltage to V_{MAX}, and in the latter case, the voltage is set to V_{STC}, which is already higher than *Zhuo*'s 1V thanks to the *Borrowing Strategy*. Figure 5 shows the comparison in terms

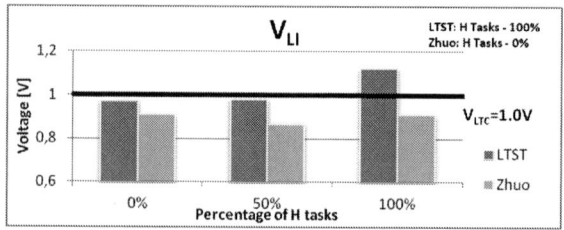

Figure 5: Mean applied voltage comparison on xSTsim simulator.

of V_{LI}. We show three cases, in which we vary the percentage of H tasks. In all of them *LTST* achieves a higher V_{LI}, and thus higher performance. In the cases of 0% and 50% of H tasks, *LTST* also keeps V_{LI} lower than V_{LTC}, respecting the reliability constraint for the current *Long Interval*. This means that *LTST* fully exploits the available reliability margin, while *Zhuo* does not. In case of 100% of *H* tasks, the V_{LI} is higher than V_{LTC}. This is not a problem, since the *Borrowing Strategy* will add the difference $V_{LI} - V_{LTC}$ to the V_{LTC} of the next *Long Intervals*. By doing this, the next *Long Interval* will be slightly penalized to recover from this violation.

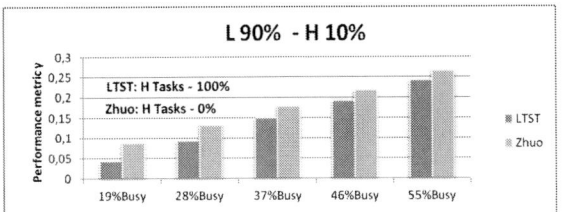

Figure 6: Performance comparison with Matlab long term simulator .

Figure 6 shows the comparison in terms of performance over the whole system lifetime, evaluated with the long term simulator. The comparison refers to different user profiles [7]. A higher percentage of *Busy* time, denotes a period of more intense user activity. For this evaluation we define a *Performance Metric* γ as:

$$\gamma = \frac{\sum_{i \in B} (f_{REQ_i} - f_{APP_i}) \cdot \Delta t_{B_i}}{\sum_{i \in B} f_{REQ_i} \cdot \Delta t_{B_i}} \cdot \frac{T_B}{T_B + T_I} \quad (5)$$

where B is the set of tasks (*Busy*), Δt_{B_i} is the duration of the task, T_B is the total *Busy* time, T_I is the total *Idle* time. Variable γ measures the frequency reduction with respect to the requested one for the executed tasks. Lower values of γ are better. Even if both solutions respect the target of reliability, Figure 6 shows that our policy presents a gain of 4% in terms of γ with respect to *Zhuo* on the entire lifetime. Moreover, our policy executes 100% of H task at their f_{REQ}, guaranteeing high quality execution to the final user. *Zhuo*, instead, slows down all the H tasks, causing significant user experience degradation.

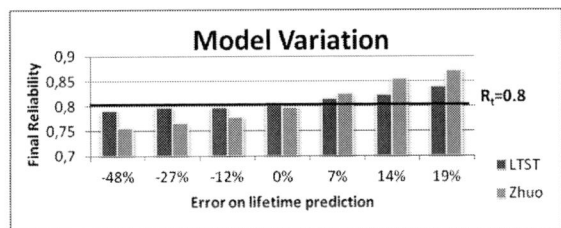

Figure 7: Robustness of sensor based approach.

To simulate the absence/presence of aging sensors, we distinguish the model that describes the real degradation M_{REAL} and the model that is adopted inside the reliability controller M_{CTRL} (through which R and R_P are computed). In case of *Zhuo*, no aging sensors are present. Therefore $M_{CTRL} \neq M_{REAL}$ and, as a consequence, the controller takes decisions based both on the wrong current reliability R and the wrong predicted reliability R_P. In *LTST* we have aging sensors that can give a better estimate of R. Therefore, only R_P is inaccurate. We define LT_{REAL} and LT_{CTRL} respectively as the lifetime obtained by controlling the system (with constant voltage and temperature) with M_{REAL} and M_{CTRL}. Therefore M_{CTRL} leads to an error on lifetime prediction, LT_{ERR}, equal to:

$$LT_{ERR} = \frac{LT_{CTRL} - LT_{REAL}}{LT_{REAL}} \quad (6)$$

LT_{ERR} measures the difference between M_{CTRL} and M_{REAL}. Figure 7 shows the comparison between *LTST* and *Zhuo* in terms of final reliability for different values of LT_{ERR}, expressed in percentages. For each case, *LTST* obtains a final reliability closer to the target with respect to *Zhuo* (final reliability is just 1% less than the target one with -48% LT_{ERR}). For this reason a sensors-based reliability control is significantly more robust.

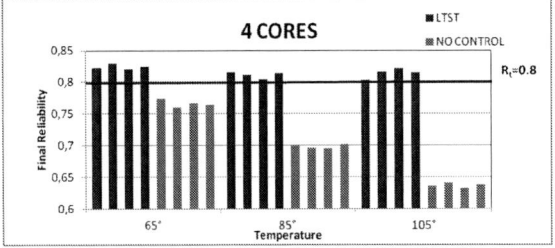

Figure 8: Robustness of the solution with temperature fluctuations

4.2 Multicore DRM:

in Figure 8 we show the benefits of adopting our solution for a platform with 4 cores, and its robustness with respect to long term temperature fluctuations. The target lifetime is 3 years, and the target reliability is 0.8. We set the nominal values of temperature T_{NOM} respectively at 65°, 85°, 105°. At each *Long Interval*, temperature is given by $T_{NOM} \pm U(-20, +20)$, where U is a uniform distribution, in order to simulate long term temperature fluctuation. For all the cases and cores, the target reliability is met and 100% of H tasks are guaranteed, whereas if control is disabled, the target reliability is violated. Figure 9 shows the reliability curve and temperature vs. time, for the 2nd core and $T_{NOM} = 85°$. The controller easily meets the target target reliability of 0.8 by carefully managing the tradeoff between better performance (higher temperatures) and reliability.

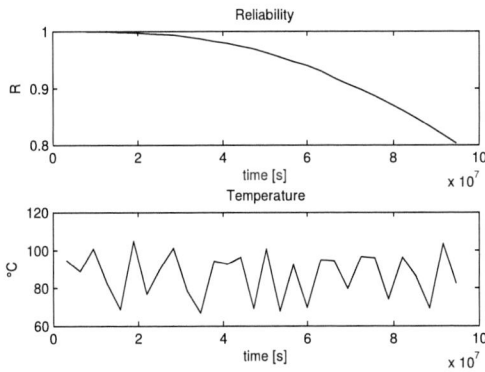

Figure 9: Reliability curve and long term temperature trace for one core.

5. CONCLUSION

In future technology scaling, many phenomena, such as TDDB, impact the system reliability. DRM techniques have been proposed to guarantee system lifetime, by constraining operating conditions, but they can cause user experience degradation. In this paper we propose a two-level controller: the first one acting on *Long Intervals* to set voltage and temperature constraints according to reliability, and the second one which sets operating conditions over *Short Intervals* to meet quality requirement, while keeping the average voltage lower or equal than V_{LTC} and the temperature below T_{LTC}. Our solution uses aging sensors to improve reliability control robustness. We compare our policy against state-of-the-art and show that with our solution, 100% of latency-critical applications meet their needs, and show an overall performance improvement of 4% over the whole lifetime.

6. ACKNOWLEDGMENTS

This work was supported, in parts, by the EU FP7 ERC Project MULTITHERMAN (GA n. 291125) and the EU FP7 Project Phidias (GA n. 318013).

7. REFERENCES

[1] Apple imovie, itunes.apple.com/en/app/-imovie/id377298193?mt=8.

[2] ionroad, http://www.ionroad.com/.

[3] A. Bartolini, M. Cacciari, A. Tilli, and L. Benini. Thermal and energy management of high-performance multicores: Distributed and self-calibrating model-predictive controller. *IEEE Transactions on Parallel and Distributed Systems*, 24(1):170–183, 2013.

[4] J. Blome, S. Feng, S. Gupta, and S. Mahlke. Self-calibrating online wearout detection. In *Microarchitecture, 2007. MICRO 2007. 40th Annual IEEE/ACM International Symposium on*.

[5] A. K. Coskun, R. Strong, D. M. Tullsen, and T. Simunic Rosing. Evaluating the impact of job scheduling and power management on processor lifetime for chip multiprocessors. In *Proceedings of the eleventh international joint conference on Measurement and modeling of computer systems*, SIGMETRICS '09, pages 169–180, New York, NY, USA, 2009. ACM.

[6] R. Degraeve, N. Pangon, B. Kaczer, T. Nigam, G. Groeseneken, and A. Naem. Temperature acceleration of oxide breakdown and its impact on ultra-thin gate oxide reliability. In *VLSI Technology, 1999. Digest of Technical Papers. 1999 Symposium on*.

[7] H. Falaki, R. Mahajan, S. Kandula, D. Lymberopoulos, R. Govindan, and D. Estrin. Diversity in smartphone usage. In *Proceedings of the 8th international conference on Mobile systems, applications, and services*, MobiSys '10, 2010.

[8] B. Gyselinckx, C. Van Hoof, J. Ryckaert, R. Yazicioglu, P. Fiorini, and V. Leonov. Human++: autonomous wireless sensors for body area networks. In *Custom Integrated Circuits Conference, 2005. Proceedings of the IEEE 2005*, pages 13 – 19, sept. 2005.

[9] C. Hu. Gate oxide scaling limits and projection. In *Electron Devices Meeting, 1996. IEDM '96., International*, pages 319 –322, dec. 1996.

[10] E. Karl, D. Blaauw, D. Sylvester, and T. Mudge. Reliability modeling and management in dynamic microprocessor-based systems. In *Design Automation Conference, 2006 43rd ACM/IEEE*.

[11] F. Paterna, A. Acquaviva, A. Caprara, F. Papariello, G. Desoli, and L. Benini. Variability-aware task allocation for energy-efficient quality of service provisioning in embedded streaming multimedia applications. *Computers, IEEE Transactions on*, 2012.

[12] S. Sharifi, R. Ayoub, and T. Rosing. Tempomp: Integrated prediction and management of temperature in heterogeneous mpsocs. In *Design, Automation Test in Europe Conference Exhibition (DATE), 2012*, pages 593 –598, march 2012.

[13] P. Singh, E. Karl, D. Blaauw, and D. Sylvester. Compact degradation sensors for monitoring nbti and oxide degradation. *Very Large Scale Integration (VLSI) Systems, IEEE Transactions on*, 2012.

[14] P. Singh, E. Karl, D. Sylvester, and D. Blaauw. Dynamic nbti management using a 45 nm multi-degradation sensor. *Circuits and Systems I: Regular Papers, IEEE Transactions on*, 58(9):2026 –2037, sept. 2011.

[15] J. Srinivasan, S. Adve, P. Bose, and J. Rivers. The case for lifetime reliability-aware microprocessors. In *Computer Architecture, 2004. Proceedings. 31st Annual International Symposium on*.

[16] J. Stathis. Physical and predictive models of ultra thin oxide reliability in cmos devices and circuits. In *Reliability Physics Symposium, 2001. Proceedings. 39th Annual. 2001 IEEE International*.

[17] S. Wang and J.-J. Chen. Thermal-aware lifetime reliability in multicore systems. In *Quality Electronic Design (ISQED), 2010 11th International Symposium on*, pages 399 –405, march 2010.

[18] E. Wu, D. Harmon, and L.-K. Han. Interrelationship of voltage and temperature dependence of oxide breakdown for ultrathin oxides. *Electron Device Letters, IEEE*, 21(7):362 –364, july 2000.

[19] M. yu Hsieh. A scalable simulation framework for evaluating thermal management techniques and the lifetime reliability of multithreaded multicore systems. In *Green Computing Conference and Workshops (IGCC), 2011 International*, pages 1 –6, july 2011.

[20] C. Zhuo, K. Chopra, D. Sylvester, and D. Blaauw. Process variation and temperature-aware full chip oxide breakdown reliability analysis. *Computer-Aided Design of Integrated Circuits and Systems, IEEE Transactions on*, 2011.

[21] C. Zhuo, D. Sylvester, and D. Blaauw. Process variation and temperature-aware reliability management. In *Design, Automation Test in Europe Conference Exhibition (DATE), 2010*.

Liveness Evaluation of a Cyclo-Static DataFlow Graph

Mohamed Benazouz
CEA, LIST,
P.C. 172, 91191
Gif-Sur-Yvette, France.
mohamed.benazouz@cea.fr

Alix Munier-Kordon
LIP6, UPMC,
Place Jussieu, 75005
Paris, France.
alix.munier@lip6.fr

Thomas Hujsa
LIP6, UPMC,
Place Jussieu, 75005
Paris, France.
thomas.hujsa@lip6.fr

Bruno Bodin
KALRAY SA,
86 Rue de Paris, 91400
Orsay, France.
bruno.bodin@kalray.eu

ABSTRACT

Cyclo-Static DataFlow Graphs (CSDFG in short) is a formalism commonly used to model parallel applications composed by actors communicating through buffers. The liveness of a CSDFG ensures that all actors can be executed infinitely often. This property is clearly fundamental for the design of embedded applications.

This paper aims to present first an original sufficient condition of liveness for a CSDFG. Two algorithms of polynomial-time for checking the liveness are then derived and compared to a symbolic execution of the graph. An original method to compute close-to-optimal buffer capacities ensuring liveness is also presented and experimentaly tested. The performance of our methods are comparable to those existing in the literature for industrial applications. However, they are far more effective on randomly generated instances, ensuring their scalability for future more complex applications and their possible implementation in a compiler.

Categories and Subject Descriptors

C.3 [**Special-purpose and application-based systems**]: Real-time and embedded systems; D.2.2 [**Software Engineering**]: Design Tools and Techniques

General Terms

Algorithms, Design, Experimentations, Theory

Keywords

Cyclo-Static Dataflow Graphs, liveness, buffer sizing

1. INTRODUCTION

Synchronous Dataflow Graphs [12] (SDFG in short) have been used for many years in the field of embedded system design such as Digital Signal Processing (DSP in short). They are used to model a large amount of applications [16, 17] and many academic results were devoted to them [9, 11, 16, 18]. However, this model is inadequate in many applications for modeling the communication between actors. Bilsen *et al.* [5] introduced Cyclo-Static Dataflow

Permission to make digital or hard copies of all or part of this work for personal or classroom use is granted without fee provided that copies are not made or distributed for profit or commercial advantage and that copies bear this notice and the full citation on the first page. To copy otherwise, to republish, to post on servers or to redistribute to lists, requires prior specific permission and/or a fee.
DAC '13, May 29 - June 07 2013, Austin, TX, USA.

Graphs which is a more accurate model to address this problem. Actors' exchanges are more detailed and this new model corresponds better to the description of applications. Analysis results are thus more pertinent. Besides, a CSDF live application may deadlock when modeled by a SDF[5].

CSDFGs were considered more recently in many areas to model data exchanges between applications processes. Those graphs are automatically extracted from a suitable description of the applications. In the field of synchronous languages, Mandel *et al.* improved the expressivity of Lustre [6] to handle processes of different rates communicating through buffers [13]. The intermediate representation of this extended language, Lucy-n, is then comparable to a CSDFG. CSDFGs are also considered to model embedded applications for its mapping on a parallel architecture. Several studies were performed in an academic context [1, 3, 19]. Another example is the dataflow compiler designed to map a CSDFG on the Massively Parallel Processor Array (MPPA in short) developed by Kalray company [10] that embeds 256 processors on a 28nm chip.

The popularity of CSDFGs comes from the fact that it is a decidable model. Its behavior is completely predictable, and its performances can be analysed. The aim of this paper is to provide an efficient method to evaluate the liveness of a CSDFG. A CSDFG is said to be live if all its actors can be executed with no deadlock. This property is clearly essential for applications. The main problem is that all algorithms developed to check it are of exponential time complexity and thus they cannot be integrated in an iterative compilation context [2, 5]. A detailed bibliography can be found in Subsection 2.2.

The main result of our study is to prove the first sufficient polynomial condition of liveness for CSDFGs. We deduce several polynomial time algorithms to ensure the liveness of a CSDFG and to compute the minimum buffer sizes. All of them were tested on industrial and academic benchmarks and compared to existing solutions.

Section 2 is dedicated to the presentation of CSDFGs and some behavioural properties. Section 3 describes the extension to CSDFGs of a polynomial transformation called normalization. It was previously introduced for SDFG by [15] and allowed to get a sufficient condition of liveness for SDFG. This sufficient condition is extended to CSDFG in Section 4 and two polynomial time algorithms for ensuring the liveness of CSDFG are deduced. Section 5 presents our experiments. Section 6 is our conclusion.

2. CYCLO-STATIC DATAFLOW GRAPHS

This section introduces Cyclo-Static Dataflow Graphs and some important basic definitions and properties. Basic notations are first introduced in Subsection 2.1. The liveness of a CSDF is introduced in Subsection 2.2 followed by a short bibliography on the methods developed to check it. Subsection 2.3 introduces the consistency of a CSDFG, that can be seen as a necessary condition on the liveness. As mentioned above, the complexity of the liveness of a CSDFG is a fundamental open problem and all the exact algorithms are of exponential time-complexity.

2.1 Notations for CSDFGs

A Cyclo-Static DataFlow Graph $\mathcal{G} = (T, A)$ is a directed graph; the set of nodes T models tasks (or actors); the set of arcs A corresponds to buffers (or channels).

2.1.1 Actors

Every actor $t \in T$ is decomposed into $\varphi(t) \in \mathbb{N} - \{0\}$ distinct phases that constitute a periodic execution sequence $t_1, \cdots, t_{\varphi(t)}$ where t_k denotes the kth phase of t for $k \in \{1, \cdots, \varphi(t)\}$. Moreover, two phases or two successive executions of an actor are supposed to not overlap.

We denote by $\langle t, n \rangle$, $n \in \mathbb{N} - \{0\}$, the nth execution of t. Similarly, for every phase $k \in \{1, \cdots, \varphi(t)\}$, $\langle t_k, n \rangle$ denotes the nth execution of the kth phase of t.

For every couple $(k, n) \in \{1, \cdots, \varphi(t)\} \times \mathbb{N} - \{0\}$, $Pred\langle t_k, n \rangle$ is the preceding execution phase of $\langle t_k, n \rangle$. More formally,

$$Pred\langle t_k, n \rangle = \begin{cases} \langle t_{k-1}, n \rangle & \text{if } k > 1 \\ \langle t_{\varphi(t)}, n-1 \rangle & \text{if } k = 1. \end{cases}$$

The execution $\langle t_{\varphi(t)}, 0 \rangle$ is fictitious and is only introduced to simplify the definition of $Pred$.

2.1.2 Buffers

Every arc $a = (t, t') \in A$ represents a buffer $b(a)$ from the actor t to the actor t'. $\forall k \in \{1, \cdots, \varphi(t)\}$, $w_a(k) \geq 0$ data are produced in $b(a)$ at the end of an execution of t_k. To enable the execution of the phase $t'_{k'}$, $\forall k' \in \{1, \cdots, \varphi(t')\}$, $v_a(k') \geq 0$ data are needed to be available in $b(a)$. They are consumed before $t'_{k'}$ starts its execution. Moreover, a buffer associated with an arc a contains initially $M_0(a)$ data (or tokens).

The cumulative number of data produced on $b(a)$ by one execution of the actor t equals $w_a \cdot \mathbb{1} = \sum_{k=1}^{\varphi(t)} w_a(k)$. Similarly, the cumulative number of data consumed from $b(a)$ by one execution of the actor t' is $v_a \cdot \mathbb{1} = \sum_{k=1}^{\varphi(t')} v_a(k)$. While it is allowed that a phase does not produce (resp. consume) data, the cumulative number of data produced (resp. consumed) must be not null, i.e. $w_a \cdot \mathbb{1} > 0$ (resp. $v_a \cdot \mathbb{1} > 0$).

For any value $n \in \mathbb{N} - \{0\}$ and $k \in \{1, \cdots, \varphi(t)\}$, let $D_a^+ \langle t_k, n \rangle$ denote the total number of tokens produced on the arc a by executions of t until the end of $\langle t_k, n \rangle$. Similarly, for any value $k' \in \{1, \cdots, \varphi(t')\}$, let $D_a^- \langle t'_{k'}, n \rangle$ denote the total number of tokens consumed from the arc a from the beginning for the execution of $\langle t'_{k'}, n \rangle$.

The total number of tokens contained in a buffer must remain non negative, that is to say any execution $\langle t'_{k'}, n' \rangle$ can be done at the completion of $\langle t_k, n \rangle$ if $M_0(a) + D_a^+ \langle t_k, n \rangle - D_a^- \langle t'_{k'}, n' \rangle \geq 0$.

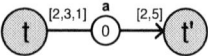

Figure 1: A buffer $b(a)$ represented by an arc $a = (t, t')$. The arc is labeled by two vectors $w_a = [2, 3, 1]$, $v_a = [2, 5]$ and by the initial number of data $M_0(a) = 0$.

Figure 1 shows a buffer $b(a)$ from t to t' that is modeled by an arc $a = (t, t')$. t (resp. t') has three (resp. two) phases i.e.,

$\varphi(t) = 3$ (resp. $\varphi(t') = 2$). The arc is labeled by vectors of production/consumption rates $w_a = [2, 3, 1]$ and $v_a = [2, 5]$. The total number of data produced in a until the completion of $\langle t_2, 2 \rangle$ is $D_a^+ \langle t_2, 2 \rangle = 6 + 5 = 11$. Similarly, the total number of tokens consumed for execution $\langle t'_1, 2 \rangle$ equals $D_a^- \langle t'_1, 2 \rangle = 7 + 2 = 9$. As $M_0(a) + D_a^+ \langle t_2, 2 \rangle - D_a^- \langle t'_1, 2 \rangle = 0 + 11 - 9 \geq 0$, $\langle t'_1, 2 \rangle$ can be executed at the completion of $\langle t_2, 2 \rangle$.

An arc $a = (t, t')$ models data-dependencies introduced by a buffer $b(a)$ between the actors t and t'. However, in most real-life embedded systems, the overall amount of memory is bounded, which implies that the capacity of the buffers cannot be considered as infinite. The bounded capacity of a buffer $b(a)$ is simply modeled by adding a feedback arc $a' = (t', t)$ with $\forall k \in \{1, \cdots, \varphi(t)\}$, $v_{a'}(k) = w_a(k)$ and $\forall k' \in \{1, \cdots, \varphi(t')\}$, $w_{a'}(k') = v_a(k')$. The initial marking $M_0(a')$ corresponds to the number of empty containers initially in $b(a)$. The capacity of the buffer $b(a)$ equals the sum $M_0(a) + M_0(a')$ (see Figure 2).

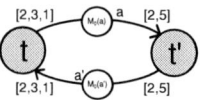

Figure 2: A bounded buffer $b(a)$ modeled by a couple of arcs $a = (t, t')$ and $a' = (t', t)$. The capacity of the buffer $b(a)$ equals $M_0(a) + M_0(a')$.

2.1.3 Particular classes of CSDFG

Synchronous DataFlow Graphs (SDFGs in short) [12] are a particular class of CSDFGs where each actor has only one phase, i.e. $\forall t \in T, \varphi(t) = 1$. Also note that any CSDFG $\mathcal{G} = (T, A)$ can be associated with an SDFG denoted by $SDFG(\mathcal{G}) = (T, A')$ where each arc $a = (t, t') \in A$ is associated with an arc $a' = (t, t')$ in A' with $w'_a = w_a \cdot \mathbb{1}$, $v'_a = v_a \cdot \mathbb{1}$ and $M_0(a') = M_0(a)$.

Homogeneous Synchronous DataFlow Graphs (HSDFGs in short) [17] are a particular class of SDFGs where $w_a = v_a = 1$ for any arc $a \in A$.

These two classes of graphs were intensively studied in the literature. Main results concerning their liveness are reviewed subsequently.

2.2 Related work on the liveness of a CSDFG

A CSDFG is said to be live if each actor can be fired infinitely often. The liveness of a CSDFG is an important basic property: finding efficient algorithms to check the liveness of such system is of great importance in an industrial context.

The complexity of the liveness of an SDFG or a CSDFG is an open problem. Up to now, all the exact algorithms for checking the liveness are of exponential time. Their main defect is that they cannot be used within a reasonable time for complex applications.

They can be grouped into two main classes: the first one is transforming an original SDFG or CSDFG into an equivalent HSDFG by replicating the actors a certain (non-polynomial) number of times [5, 17]. The liveness is then checked directly on the HSDFG, which is possible using a polynomial-time algorithm (recall that a HSDFG is live iff every circuit has at least one token). The problem is that the size of the HSDFG may be of exponential size [15], which drastically limits the efficiency of this class of methods.

Another way consists in constructing if possible a static schedule for an SDFG [8, 11] or a CSDFG [2]. If such a schedule exists, the graph is live; otherwise a deadlock is highlighted. The main drawback of this method is that the size of the static schedule may be quite large and so it cannot be computed in polynomial time.

A well-known necessary condition for the liveness of a CSDFG (or an SDFG) with bounded buffers is the consistency, as described in [12, 5]. This condition is recalled in the next subsection, and is supposed to be fulfilled by the SDFG and CSDFG considered here. A simple polynomial sufficient condition was found by [14] for *normalized* SDFG. This condition is proved to be not necessary, but allows to ensure quickly that the system is live. This paper aims to prove first that any consistent CSDFG may be normalized, and that several sufficient conditions of liveness may be obtained by expressing deadlock conditions.

2.3 Consistency of a CSDFG

Consistency is a necessary (non sufficient) condition for the existence of a valid schedule within bounded memory that was established first for SDFGs [12]. Bilsen *et al.* [5] extended this condition to CSDFG by considering the cumulative number of tokens produced/consumed by one execution of its actors. It is none other than the one proposed by Lee [12] applied to the associated SDFG of a CSDFG. This point is motivated by the fact that a CSDFG is simply obtained by refining the modeling of the data exchanges between actors of the underlying SDFG.

Let us consider the pre-post $|A| \times |T|$ matrix Γ associated with a CSDFG $\mathcal{G}$ defined by

$$
\Gamma_{at} = \begin{cases} w_a \cdot \mathbb{1} & \text{if } a = (t, t'), \ t' \in T \\ -v_a \cdot \mathbb{1} & \text{if } a = (t', t), \ t' \in T \\ 0 & \text{Otherwise.} \end{cases}
$$

The CSDFG is said to be *consistent* if the rank of Γ is $|T| - 1$.

In the following, we restrict our study to consistent strongly connected CSDFGs as not consistent graphs will either deadlock or need unbounded buffers.

3. NORMALIZATION OF A CSDFG

The *normalization* of a consistent SDFG is an operation introduced in [14] which simplifies the arc values without any influence on the data dependencies. This transformation can be simply extended to any CSDFG as long as its underlying SDFG is consistent. The normalization of an SDFG and a CSDFG are introduced and illustrated using a simple example in this section.

For any actor $t \in T$, let us denote by $\mathcal{A}^+(t) = \{a = (t, t') \in A, t' \in T\}$ the set of output arcs of t and $\mathcal{A}^-(t) = \{a = (t', t) \in A, t' \in T\}$ the set of input arcs.

An actor t is said to be *normalized* if there exists $Z_t \in \mathbb{N} - \{0\}$ such that $\forall a \in \mathcal{A}^+(t), w_a \cdot \mathbb{1} = Z_t$, and $\forall a \in \mathcal{A}^-(t), v_a \cdot \mathbb{1} = Z_t$. A CSDFG is *normalized* if all of its actors are normalized.

The *normalization* of a CSDFG consists in building an equivalent CSDFG such that all actors are normalized. The idea here is to find two vectors $Z = (Z_1, \cdots, Z_{|T|})$ and $\Delta = (\delta_1, \cdots, \delta_{|A|})$ of positive values such that

$$
\forall t \in T, \forall a \in \mathcal{A}^+(t), \delta_a \times (w_a \cdot \mathbb{1}) = Z_t
$$

and

$$
\forall t \in T, \forall a \in \mathcal{A}^-(t), \delta_a \times (v_a \cdot \mathbb{1}) = Z_t.
$$

It has been proved in [14] that every consistent SDFG can be normalized (*i.e.* the previous system has a solution). Now, if a CSDFG $\mathcal{G}$ is consistent, then so is $SDFG(\mathcal{G})$ which is normalizable; let $Z^\star$ and $\Delta^\star$ be a solution of the corresponding previous system. Vectors associated with any arc $a \in A$ and the initial markings are then replaced respectively by $\delta_a^\star \times w_a$, $\delta_a^\star \times v_a$ and $\delta_a^\star \times M_0(a)$ in order to get an equivalent normalized CSDFG. Next theorem follows:

THEOREM 1. *Let G be a strongly connected CSDFG. G is normalizable iff G is consistent.*

Let us consider as example the CSDFG pictured in left side of Figure 3. The corresponding system is:

$$
Z_1 = \delta_3 \times 5 = \delta_4 \times 5 = \delta_1 \times 2 \qquad Z_3 = \delta_2 \times 2 = \delta_3 \times 3
$$
$$
Z_2 = \delta_1 \times 3 = \delta_5 \times 5 = \delta_2 \times 5 \qquad Z_4 = \delta_4 \times 6 = \delta_5 \times 4
$$

A minimum solution is given by $\Delta^\star = (5, 3, 2, 2, 3)$ and $Z^\star = (10, 15, 6, 12)$. Initial markings obtained are $M_0(a_1) = 15$, $M_0(a_2) = 3$, $M_0(a_3) = 4$, $M_0(a_4) = 0$, $M_0(a_5) = 21$. The associated equivalent normalized CSDFG is pictured in Figure 3.

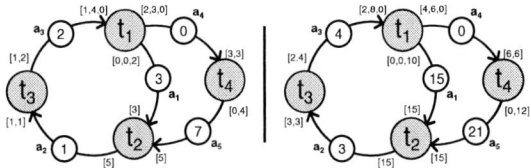

Figure 3: (Left) A CSDFG of four actors and five buffers. Initial tokens are represented by surrounded values. (Right) Its normalized version.

4. LIVENESS CHECKING ALGORITHMS

This section aims to present two original algorithms for checking a sufficient condition for the liveness of a CSDFG. The first subsection recalls that relevant values of initial markings may be limited. A first sufficient condition of liveness is then expressed, followed by a polynomial-time algorithm whose complexity depends on the total number of phases. A second condition is expressed obtained by merging the phases of the same actor in order to reduce the complexity of the corresponding algorithm. The equivalence of the two sufficient conditions of liveness is then proved as well as the fact that these conditions are not necessary. Finally, the complexity of these algorithms is expressed.

4.1 Useful tokens

It is observed that the initial markings of any buffer of an SDFG may be reduced to functions depending on the values of the arcs [14]. This result was extended to CSDFG [4, 18] as follows: let us denote for every arc $a = (t, t') \in A$,

$$
step_a = \gcd(w_a(1), \cdots, w_a(\varphi(t)), v_a(1), \cdots, v_a(\varphi(t'))),
$$

where gcd is the greatest common divisor of a given list of non negative integers. For every integer $\alpha \in \mathbb{Z}$, we also set

$$
\lfloor \alpha \rfloor^{step_a} = \left\lfloor \frac{\alpha}{step_a} \right\rfloor \cdot step_a.
$$

LEMMA 1 ([4]). *The initial marking $M_0(a)$ of any arc $a = (t, t')$ can be replaced by $\lfloor M_0(a) \rfloor^{step_a}$ without any influence on the data dependencies induced by a between the successive executions of actors t and t'.*

In the rest of this paper, it is assumed that the initial marking of any arc a is a multiple of $step_a$.

4.2 A first sufficient condition SC1

The following theorem expresses a first sufficient condition of liveness for a normalized CSDFG:

THEOREM 2. *Let $\mathcal{G}$ be a normalized CSDFG. If $\mathcal{G}$ is not live, then there exists a circuit $c = (t^1, a_1, t^2, a_2, \cdots, t^m, a_m, t^1)$ and the values $k^i \in \{1, \cdots, \varphi(t^i)\}$ for $i \in \{1, \cdots, m\}$ such that*

19

$$\sum_{i=1}^{m} M_0(a_i) \le \sum_{i=1}^{m} \left[D_{a_{i-1}}^- \langle t_{k^i}^i, 1 \rangle - D_{a_i}^+ Pred \langle t_{k^i}^i, 1 \rangle \right] - \sum_{i=1}^{m} step_{a_i}$$
with $a_0 = a_m$.

The following corollary is an immediate consequence of Theorem 2:

COROLLARY 1 (SC1). *Let $\mathcal{G}$ be a normalized CSDFG. $\mathcal{G}$ is live if for any circuit $c = (t^1, a_1, t^2, a_2, \cdots, t^m, a_m, t^1)$ of $\mathcal{G}$ and any values $k^i \in \{1, \cdots, \varphi(t^i)\}$ for $i \in \{1, \cdots, m\}$,*
$$\sum_{i=1}^{m} M_0(a_i) > \sum_{i=1}^{m} \left[D_{a_{i-1}}^- \langle t_{k^i}^i, 1 \rangle - D_{a_i}^+ Pred \langle t_{k^i}^i, 1 \rangle \right] - \sum_{i=1}^{m} step_{a_i}$$
with $a_0 = a_m$.

A polynomial-time algorithm is expressed subsequently to check this first necessary sufficient condition denoted by SC1.

4.3 Checking the liveness using condition SC1

The aim of this section is to express a polynomial time-algorithm for checking the sufficient condition of liveness SC1 on a CSDFG $\mathcal{G} = (T, A)$. For that purpose, let us consider the valued oriented graph $H_1 = (N_1, E_1)$ defined as follows:

- N_1 is the set of the phases of actors from T, *i.e.* $N_1 = \{t_k^i, i \in \{1, \cdots, |T|\}, k \in \{1, \cdots, \varphi(t^i)\}\}$.

- Any arc $a = (t^i, t^j) \in A$ corresponding to a buffer is associated with $\varphi(t^i) \times \varphi(t^j)$ arcs $u = (t_{k^i}^i, t_{k^j}^j) \in E_1$ for $k^i \in \{1, \cdots, \varphi(t^i)\}$ and $k^j \in \{1, \cdots, \varphi(t^j)\}$ valued by $W_1(u) = D_a^- \langle t_{k^j}^j, 1 \rangle - D_a^+ Pred \langle t_{k^i}^i, 1 \rangle - step_a - M_0(a)$.

The cycle-mean of a circuit $c = (t_{k^1}^1, u_1, t_{k^2}^2, \cdots, t_{k^m}^m, u_m, t_{k^1}^1)$ equals $W_1(c) = \frac{\sum_{i=1}^{m} W_1(u_i)}{m}$. Now, let $W_1^\star$ be the maximum cycle-mean value of a circuit from H_1. SC1 is equivalent to $W_1^\star < 0$. Several polynomial-time algorithms for computing $W_1^\star$ are detailed and compared experimentally in [7].

Figure 4 pictures the graph H_1 associated with the normalized CSDFG presented in Figure 3. The maximum cycle-mean value is $W_1^\star = -\frac{1}{4}$ and is reached for the circuit passing through $t_2^1, t_2^4, t_1^2, t_2^3, t_2^1$. As $W_1^\star < 0$, SC1 holds for any circuit of H_1, $\mathcal{G}$ is thus live.

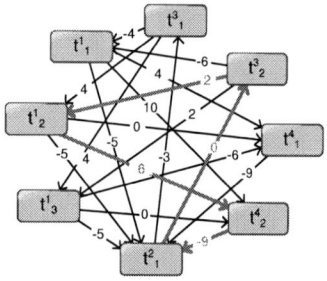

Figure 4: **Graph H_1 associated with the normalized CSDFG pictured in Figure 3. The circuit passing through t_2^1, t_2^4, t_1^2, t_2^3 and t_2^1 has a maximum cycle-mean equal to $-\frac{1}{4}$.**

The size of H_1 may grow quickly with the number of phases. Another sufficient condition may be expressed to significantly reduce the size of the underlying graph.

4.4 A second sufficient condition SC2

The number of phases for each actor may be too large to get an efficient algorithm for checking the liveness using SC1. The idea

here is to express another condition SC2 and to prove that SC1 and SC2 are equivalent.

Let us consider that a CSDFG $\mathcal{G}$ verifies condition SC2 if, for any cycle $c = (t^1, a_1, t^2, a_2, \cdots, t^m, a_m, t^1)$ of $\mathcal{G}$,
$$\sum_{i=1}^{m} M_0(a_i) > - \sum_{i=1}^{m} step_{a_i}$$
$$+ \sum_{i=1}^{m} max_{k^i \in \{1, \cdots, \varphi(t^i)\}} \left[D_{a_{i-1}}^- \langle t_{k^i}^i, 1 \rangle - D_{a_i}^+ Pred \langle t_{k^i}^i, 1 \rangle \right].$$

THEOREM 3. *Let $\mathcal{G}$ be a normalized CSDFG. Conditions SC1 and SC2 are equivalent.*

A sufficient condition of liveness follows from Corollary 1 and Theorem 3:

COROLLARY 2 (SC2). *Let $\mathcal{G}$ be a normalized CSDFG. $\mathcal{G}$ is live if for any cycle $c = (t^1, a_1, t^2, a_2, \cdots, t^m, a_m, t^1)$ of $\mathcal{G}$,*
$$\sum_{i=1}^{m} M_0(a_i) > - \sum_{i=1}^{m} step_{a_i}$$
$$+ \sum_{i=1}^{m} max_{k^i \in \{1, \cdots, \varphi(t^i)\}} \left[D_{a_{i-1}}^- \langle t_{k^i}^i, 1 \rangle - D_{a_i}^+ Pred \langle t_{k^i}^i, 1 \rangle \right].$$

Let $H_2 = (N_2, E_2)$ be a valued graph built as follows for checking SC2 on a CSDFG $\mathcal{G} = (T, A)$:

- N_2 is the set of buffers, *i.e.* $N_2 = A$.

- Each arc $e = (a, a') \in E_2$ is associated with a actor t such that $a \in \mathcal{A}^-(t)$, $a' \in \mathcal{A}^+(t)$ and is valued by
$$W_2(e) = -step_a - M_0(a)$$
$$+ max_{k \in \{1, \cdots, \varphi(t)\}} \left[D_a^- \langle t_k, 1 \rangle - D_{a'}^+ Pred \langle t_k, 1 \rangle \right].$$

As seen before, satisfying the condition SC2 is equivalent to checking that the maximum cycle-mean $W_2^\star$ of H_2 is strictly negative. The graph H_2 corresponding to the normalized CSDFG $\mathcal{G}$ pictured in Figure 3 is presented in Figure 5.

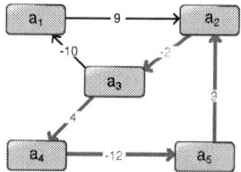

Figure 5: **Graph H_2 associated with the normalized CSDFG $\mathcal{G}$ pictured in Figure 3. A circuit of maximum mean-cycle value is highlighted.**

4.5 SC1 and SC2 are not necessary

These two conditions can be seen as a generalization of a result proved by Marchetti *et al.* [14] for SDFG expressed as follows:

THEOREM 4 ([14]). *Let $\mathcal{G}$ be a normalized SDFG. Then, $\mathcal{G}$ is live if for any cycle $c = (t^1, a_1, t^2, a_2, \cdots, t^m, a_m, t^1)$ of $\mathcal{G}$*
$$\sum_{i=1}^{m} M_0(a_i) > \sum_{i=1}^{m} (Z_i - step_{a_i}).$$

Marchetti *et al.* in [14] proved that the condition of Theorem 4 is not necessary for the liveness of an SDFG. The consequence is that conditions SC1 and SC2 are also not necessary.

4.6 Worst-case complexity of these algorithms

For any actor $t \in T$, we denote by $deg^+(t) = |\mathcal{A}^+(t)|$ the output degree of t and $deg^-(t) = |\mathcal{A}^-(t)|$ the input degree of t. We also denote by $R = (R_1, \cdots, R_n)$ the repetition vector of $\mathcal{G}$: it can be defined as the smallest positive integer vector such that, for any couple of actors $(t, t') \in T^2$, $Z_t R_t = Z_{t'} R_{t'}$ [15].

For each considered algorithm, we then define the complexity as follows.

Algorithm SE : This algorithm, developed by [2] consists of a symbolic execution. Its worst-case case complexity bounded by $\mathcal{O}(|A| \times \sum_{t \in T} R_t)$ is proved to be exponential [15].

Algorithm SC1 : It consists of checking the condition SC1 using the determination of a maximum cycle-mean in H_1. Its complexity is $\Theta(|N_1| \times |E_1|)$ corresponding to $\Theta(\sum_{t \in T} \varphi(t) \times \sum_{(t_i, t_j) \in A} \varphi(t_i) \times \varphi(t_j))$.

Algorithm SC2 : It checks SC2 using the determination of a maximum cycle-mean in H_2. Its complexity is bounded by $\Theta(|A| \times \sum_{t \in T} (deg^+(t) \times deg^-(t)))$.

These complexities can be evaluated before executing the corresponding algorithms. A simple heuristic can then choose between these three algorithms by minimizing the theoretical complexity.

5. EXPERIMENTS

This part is focused on the actual applications of our methods and their practical interest. Benchmarks are presented, followed by the experimentations of the liveness algorithms. The last subsection is dedicated to the presentation and the experimentations of an efficient algorithm to compute the minimum buffer capacity ensuring liveness.

5.1 Benchmarks

Two different benchmarks reported in Table 1 were considered to experimentally evaluate our method. The former will focus on real-life industrial applications (JPEG2000, H264, ...). The latter is generated using different generation tools: the size and the complexity of the CSDFG are higher, and can be seen as possible future instances. The first column of Table 1 reports the name of the benchmarks. The second and third ones correspond respectively to the number of actors and buffers. The last one reports the size of a schedule issued from a symbolic execution, which is exactly equal to $\sum_{t \in T} (R_t \times \varphi(t))$.

Application	Actors	Buffers	Sched. size
BlackScholes	41	40	2379
JPEG2000	240	703	29595
Echo	38	82	42003
Pdetect	58	76	4045
H264	665	3128	1471
autogen1	90	617	250992
autogen2	70	473	41331062
autogen3	154	671	308818852
autogen4	2426	2900	51301
autogen5	2767	4894	312485

Table 1: Benchmarks

In summary, the BlackScholes application is a financial tool, JPEG2000, Echo and H264 are three multimedia applications and Pdetect is an application specialized in the detection of people.

For applications that deadlock, the computation time of algorithms SC1 and SC2 will not vary, but SE will probably be faster. However, SE reaches its maximum complexity when the instances that it processes are live, that is the reason all the benchmarks we considered are live.

5.2 Experimentations on the liveness

Table 2 reports the computation times of the three algorithms (namely SE, SC1 and SC2) on our benchmark. The tests were carried out on a standard workstation based on an Intel CORE i3 processor. Framed results are those selected by our heuristic, which choose the best algorithm following the evaluation of their theorical complexity.

In most cases, the framed results correspond to the lowest computation time. We first note the complementarity between algorithms SC1 and SC2. SC2 is often faster than SC1. However, if the actors degrees are high, such as in H264, SC1 is a better choice.

In the industrial benchmarks, the computation time of the symbolic method SE remains competitive versus the SC1 and SC2 algorithms. This is not longer true for the generated one's, for which the repetition vectors are higher. This kind of instance will be of importance with the arrival of new programming tools. In this case, the computation times of SE are clearly too long to be used in an industrial context.

Application	SC1	SC2	SE[2]
BlackScholes	2.10^5/14ms	1.10^3/0ms	2.10^5/1ms
JPEG2000	8.10^6/114ms	2.10^6/18ms	4.10^7/113ms
Echo	6.10^3/1ms	1.10^4/0ms	6.10^6/95ms
Pdetect	1.10^9/2500ms	8.10^3/4ms	6.10^5/5ms
H264	3.10^7/504ms	5.10^7/936ms	9.10^6/114ms
autogen1	3.10^8/13ms	1.10^5/55ms	2.10^6/1544ms
autogen2	3.10^{10}/41ms	7.10^6/37ms	1.10^6/4min
autogen3	4.10^{11}/55ms	1.10^6/55ms	1.10^6/21min
autogen4	2.10^8/217ms	7.10^7/71ms	1.10^7/132ms
autogen5	3.10^9/787ms	5.10^7/708ms	4.10^7/1442ms

Table 2: Complexity of SC1, SC2 and SE as defined in Subsection 4.6 with their actual computation time. Surrounded results correspond to the method selected by our heuristic.

5.3 Computation of minimum buffer capacities ensuring liveness

Condition SC1 may be considered to evaluate minimum buffer capacities of a fixed CSDFG. Indeed, let us suppose that each buffer $b(a)$ associated with an arc $a = (t, t')$ is bounded. This can be modeled using a feedback arc $a' = (t', t)$ as shown in Subsection 2.1. Now, let $\mathcal{G}' = (T, A')$ be the graph obtained by adding these feedback arcs and $H_1' = (N_1', E_1)$ its corresponding oriented valued graph associated to SC1. W_1' denotes the valuation of H_1'.

The capacity of the buffer $b(a)$ equals $M_0(a) + M_0(a')$. The overall capacity of buffers of $\mathcal{G}$ is thus $\sum_{a \in A'} M_0(a)$.

Now, initial values $M_0(a)$, $a \in A'$ must be computed such that condition SC1 is fulfilled. Integer values γ_a, $a \in N_1'$ must be computed such that, for any arc $e = (u, u') \in E_1$, we get $\gamma_u - \gamma_{u'} > W_1'(e)$.

The linear system is thus:

Minimize $\sum_{a \in A'} M_0(a)$ with

$$\begin{cases} \gamma_u - \gamma_{u'} > W_1'(e), & \forall e = (u, u') \in E_2 \\ M_0(a) \in \mathbb{N}, & \forall a \in A' \\ \gamma_{t_k} \in \mathbb{R}, & \forall t \in T, \forall k \in \{1, \cdots, \varphi(t)\} \end{cases}$$

An equivalent linear program can be expressed based on SC2. Our algorithm will choose between SC1 and SC2, whichever has the lowest theoretical complexity. We use GLPK to solve a continuous version of the linear program. The buffer capacities are computed using a rounding method.

Our experimental results are compared with a greedy algorithm based on SE and inspired by [17]. Actors are executed so that the buffer sizes are minimized. Our experiments are reported in Table 3. The first column is the benchmark's names. Second and third columns report the computation time and the overall buffer size computed by SC1 or SC2 (following the theoretical evaluation of the complexity). Fourth and fifth one's report the computation time and the overall buffer size obtained using the greedy algorithm.

Application	SC1/SC2		Greedy algorithm	
	time	buf. size	time	buf. size
BlackScholes	**8 ms**	16 KB	9 ms	16 KB
JPEG2000	3089 ms	3807 KB	**2055 ms**	**3651 KB**
Echo	**5 ms**	**28 KB**	315 ms	52 KB
Pdetect	**26 ms**	3959 KB	61 ms	3959 KB
H264	4808 ms	1368 KB	**937 ms**	1368 KB
autogen1	**169 ms**	**1849 KB**	3043 ms	2009 KB
autogen2	**1704 ms**	**227 MB**	7 min	244 MB
autogen3	**2407 ms**	**1080 MB**	36 min	1296 MB
autogen4	**16605 ms**	47 KB	20522 ms	**34 KB**
autogen5	**2 min**	**1555 KB**	3 min	3069 KB

Table 3: Computed buffer sizes of the different algorithms for each CSDF.

Judging by the data in Table 3, the quality is comparable, but the running time of the greedy algorithm is much higher in more complex applications.

6. CONCLUSION

This paper presents significant advances in both fundamental and applicative point of views for evaluating the liveness of a CSDFG.

The normalization of a CSDFG is first introduced and should be used to effectively address other CSDFG problems such as the minimization of buffer considering throughput [18]. In addition, two sufficient equivalent conditions of liveness are proved. Efficient original polynomial-time algorithms for checking the liveness of a CSDFG and computing its minimal buffer sizes (ensuring liveness) are deduced.

These algorithms are the first polynomial ones to solve approximatively these two problems. They were successfully tested on industrial and academic benchmarks. The experiments highlighted that they are well suited for real-life applications and more robust than the existing methods for complex applications. Their low complexity ensures that these algorithms can safely be integrated in a compiler.

7. REFERENCES

[1] B. Akesson, S. Stuijk, A. Molnos, M. Koedam, R. Stefan, A. Andrew Nelson and Beyranvand Nejad, and K. Goossens. Virtual platforms for mixed time-criticality applications: The CoMPSoC architecture and SDF3 design flow. In *Quo Vadis, Virtual Platforms?(QVVP)*, pages 1–2, 2012.

[2] S. R. Anapalli, K. C. Chakilam, and T. W. O'Neil. Static Scheduling for Cyclo Static Data Flow Graphs. In *Parallel and Distributed Processing Techniques and Applications, PDPTA 2009*, pages 302–306. CSREA Press, 2009.

[3] M. Bamakhrama and J. Zhai. A methodology for automated design of hard-real-time embedded streaming systems. *Design, Automation & Test in Europe (DATE)*, 2012.

[4] M. Benazouz. *Buffer Sizing for Stream Processing Applications*. PhD thesis, Université P. et M. Curie, Paris, France, 2012.

[5] G. Bilsen, M. Engels, R. Lauwereins, and J. A. Peperstraete. Cycle-static data flow. *IEEE Transactions on Signal Processing*, pages 3255–3258, 1995.

[6] P. Caspi, D. Pilaud, N. Halbwachs, and J. A. Plaice. LUSTRE: a declarative language for real-time programming. In *Symposium on Principles of programming languages - POPL '87*, pages 178–188. ACM Press, 1987.

[7] A. Dasdan, S. S. Irani, and R. K. Gupta. Efficient algorithms for optimum cycle mean and optimum cost to time ratio problems. *Design Automation Conference (DAC'99)*, pages 37–42, 1999.

[8] A. Ghamarian and M. Geilen. Liveness and boundedness of synchronous data flow graphs. *Formal Methods in Computer Aided Design (FMCAD'06)*, 2006.

[9] A. Ghamarian, M. Geilen, S. Stuijk, T. Basten, B. Theelen, M. Mousavi, A. Moonen, and M. Bekooij. Throughput Analysis of Synchronous Data Flow Graphs. In *International Conference on Application of Concurrency to System Design (ACSD'06)*, pages 25–36, 2006.

[10] Kalray. Manycore processors for embedded computing. www.kalray.eu.

[11] S. F. Khasawneh, M. E. Ritcher, and T. W. O'Neil. Static Scheduling for synchronous data flow graphs. *Computers and Their Applications*, 1:38–43, 2007.

[12] E. A. Lee and D. G. Messerschmitt. Synchronous dataflow. *Proceedings of the IEEE*, 75(9):1235–1245, 1987.

[13] L. Mandel, F. Plateau, and M. Pouzet. Lucy-n: a n-synchronous extension of Lustre. *Mathematics of Program Construction*, 2010.

[14] O. Marchetti and A. Munier-Kordon. A sufficient condition for the liveness of weighted event graphs. *European Journal of Operational Research*, 197(2):532–540, Sept. 2009.

[15] O. Marchetti and A. Munier Kordon. Cyclic Scheduling for the Synthesis of Embedded Systems. In Y. Vivien and R. Frederic, editors, *Introduction to scheduling*, chapter 6, pages 135–164. Chapman and Hall/CRC Press, 2009.

[16] J. L. Pino, S. S. Bhattacharyya, and E. A. Lee. A hierarchical multiprocessor scheduling framework for synchronous dataflow graphs. Technical report, University of California, Berkeley, 1995.

[17] S. Sriram and S. Bhattacharyya. *Embedded multiprocessors: Scheduling and synchronization*. CRC, 2009.

[18] S. Stuijk, M. Geilen, and T. Basten. Throughput-Buffering Trade-Off Exploration for Cyclo-Static and Synchronous Dataflow Graphs. *IEEE Transactions on Computers*, 57(10):1331–1345, 2008.

[19] W. Thies, M. Karczmarek, and S. Amarasinghe. StreamIt: A language for streaming applications. *Compiler Construction*, pages 179–196, 2002.

APPENDIX

This appendix provides the proofs for Theorems 2 through 4.

Proof of Theorem 2

PROOF. Let us suppose that $\mathcal{G}$ is not live. Then, there exists a circuit $c = (t^1, a_1, t^2, a_2, \cdots, t^m, a_m, t^1)$, and values $n^i \in \mathbb{N} - \{0\}$ and $k^i \in \{1, \cdots, \varphi(t^i)\}$ for $i \in \{1, \cdots, m\}$ such that:

- Executions $Pred\langle t^i_{k^i}, n^i \rangle$ with $i \in \{1, \cdots, m\}$ can be performed;

- the amount of data is not sufficient to execute any execution $\langle t^i_{k^i}, n^i \rangle$, for $i \in \{1, \cdots, m\}$.

Let us first consider the arc $a_1 = (t^1, t^2)$. Since the phase $\langle t^2_{k^2}, n^2 \rangle$ cannot be executed even if $Pred\langle t^1_{k^1}, n^1 \rangle$ is, the number of tokens on a_1 must verify

$$M_0(a_1) + D^+_{a_1} Pred\langle t^1_{k^1}, n^1 \rangle - D^-_{a_1} \langle t^2_{k^2}, n^2 \rangle < 0.$$

By definition of $D^+_{a_1}$ and $D^-_{a_1}$,

$$D^+_{a_1} Pred\langle t^1_{k^1}, n^1 \rangle = D^+_{a_1} \langle t^1_{\varphi(t^1)}, n^1 - 1 \rangle + D^+_{a_1} Pred\langle t^1_{k^1}, 1 \rangle$$

and

$$D^-_{a_1} \langle t^2_{k^2}, n^2 \rangle = D^-_{a_1} \langle t^2_{\varphi(t^2)}, n^2 - 1 \rangle + D^-_{a_1} \langle t^2_{k^2}, 1 \rangle.$$

The previous inequality thus becomes

$$M_0(a_1) + D^+_{a_1} \langle t^1_{\varphi(t^1)}, n^1 - 1 \rangle + D^+_{a_1} Pred\langle t^1_{k^1}, 1 \rangle$$
$$- D^-_{a_1} \langle t^2_{\varphi(t^2)}, n^2 - 1 \rangle - D^-_{a_1} \langle t^2_{k^2}, 1 \rangle < 0.$$

Now, by Lemma 1, this sum is divisible by $step_{a_1}$, so we get

$$M_0(a_1) + D^+_{a_1} \langle t^1_{\varphi(t^1)}, n^1 - 1 \rangle + D^+_{a_1} Pred\langle t^1_{k^1}, 1 \rangle$$
$$- D^-_{a_1} \langle t^2_{\varphi(t^2)}, n^2 - 1 \rangle - D^-_{a_1} \langle t^2_{k^2}, 1 \rangle \leq -step_{a_1}.$$

Similarly, by setting $t^{m+1} = t^1$, we get for any value $i \in \{1, \cdots, m\}$,

$$M_0(a_i) + D^+_{a_i} \langle t^i_{\varphi(t^i)}, n^i - 1 \rangle + D^+_{a_i} Pred\langle t^i_{k^i}, 1 \rangle$$
$$- D^-_{a_i} \langle t^{i+1}_{\varphi(t^{i+1})}, n^{i+1} - 1 \rangle - D^-_{a_i} \langle t^{i+1}_{k^{i+1}}, 1 \rangle \leq -step_{a_i}.$$

Since $\mathcal{G}$ is normalized,
$\forall i \in \{1, \cdots, m-1\}$,

$$D^-_{a_{i-1}} \langle t^i_{\varphi(t^i)}, n^i - 1 \rangle = D^+_{a_i} \langle t^i_{\varphi(t^i)}, n^i - 1 \rangle = (n^i - 1) \cdot Z_{t^i}.$$

By summing all the previous inequalities, we then obtain that

$$\sum_{i=1}^m M_0(a_i) + \sum_{i=1}^m \left[D^+_{a_i} Pred\langle t^i_{k^i}, 1 \rangle - D^-_{a_{i-1}} \langle t^i_{k^i}, 1 \rangle \right] \leq - \sum_{i=1}^m step_{a_i}$$

which concludes the proof. $\square$

Proof of Theorem 3

PROOF. Let us suppose that $\mathcal{G}$ verifies SC2 and let $c = (t^1, a_1, t^2, a_2, \cdots, t^m, a_m, t^1)$ be a circuit of $\mathcal{G}$. Then, for any values $k^i \in \{1, \cdots, \varphi(t^i)\}$, $i \in \{1, \cdots, m\}$, we get that

$$\max_{k^i \in \{1, \cdots, \varphi(t^i)\}} \left[D^-_{a_{i-1}} \langle t^i_{k^i}, 1 \rangle - D^+_{a_i} Pred\langle t^i_{k^i}, 1 \rangle \right] \geq$$
$$D^-_{a_{i-1}} \langle t^i_{k^i}, 1 \rangle - D^+_{a_i} Pred\langle t^i_{k^i}, 1 \rangle.$$

Thus,

$$\sum_{i=1}^m M_0(a_i) > - \sum_{i=1}^m step_{a_i}$$
$$+ \sum_{i=1}^m \left[D^-_{a_{i-1}} \langle t^i_{k^i}, 1 \rangle - D^+_{a_i} Pred\langle t^i_{k^i}, 1 \rangle \right].$$

and $\mathcal{G}$ verifies SC1. Conversely, if SC1 is true, then, for any cycle $c = (t^1, a_1, t^2, a_2, \cdots, t^m, a_m, t^1)$ and any phases $k^\star_i$ of t_i maximizing $D^-_{a_{i-1}} \langle t^i_{k^i}, 1 \rangle - D^+_{a_i} Pred\langle t^i_{k^i}, 1 \rangle$, the inequality is true. SC2 is thus verified, which concludes the proof. $\square$

Proof of Theorem 4

PROOF. Any normalized SDFG $\mathcal{G} = (T, A)$ is a CSDFG for which each actor has a unique phase, thus $\forall t \in T$, $\varphi(t) = 1$. Since $\mathcal{G}$ is normalized, for any arc $a = (t, t')$, $D^-_a \langle t_1, 1 \rangle = Z_t$ and $D^+_a Pred\langle t_1, 1 \rangle = D^+_a \langle t_1, 0 \rangle = 0$. Thus, SC2 is equivalent to the condition expressed by the theorem. $\square$

Double Patterning Lithography-Aware Analog Placement *

Hsing-Chih Chang Chien[1], Hung-Chih Ou[1], Tung-Chieh Chen[3], Ta-Yu Kuan[3], and
Yao-Wen Chang[1,2]

[1]Graduate Institute of Electronics Engineering, National Taiwan University, Taipei 106, Taiwan
[2]Department of Electrical Engineering, National Taiwan University, Taipei 106, Taiwan
[3]Synopsys, Hsinchu, Taiwan

{lichee626, howard}@eda.ee.ntu.edu.tw; {donnie.chen, tayukuan}@synopsys.com; ywchang@cc.ee.ntu.edu.tw

ABSTRACT

Double patterning lithography (DPL) is one of the most promising solutions for the 28nm technology node and beyond. The main idea of DPL is to decompose the layout into two sub-patterns and manufacture the layout by two masks. In addition to traditional analog design constraints, the pre-coloring constraint should also be considered, in which patterns of critical or sensitive modules have predefined masks before layout decomposition to reduce mismatches. In this paper, we present the first work that considers DPL during analog placement and simultaneously minimizes area, wirelength, and DPL conflicts. We first propose an extended conflict graph (ECG) to represent the relation between patterns of analog modules and apply an integer linear programming (ILP) formulation to determine the orientation of each module and the color of each pattern for conflict minimization. ILP reduction schemes are proposed to further reduce the runtime. Finally, we present a three-stage flow and DPL-aware perturbations to obtain desired solutions. Experimental results show that the proposed flow can effectively and efficiently reduce area, wirelength, and DPL conflicts.

Categories and Subject Descriptors

B.7.2 [**Integrated Circuits**]: Design Aids [Placement and Routing]

General Terms

Algorithms, Performance

Keywords

Analog ICs, Physical Design, Double Patterning Lithography, Placement

1. INTRODUCTION

Double patterning lithography (DPL) is one of the most promising solutions for the 28nm technology node and beyond [3] [13]. The main idea of DPL is to decompose patterns of a layer into two sub-patterns

---

*This work was partially supported by IBM, SpringSoft, TSMC, Academia Sinica, and NSC of Taiwan under Grant No's. NSC 101-2221-E-002-191-MY3, NSC100-2221-E-002-088-MY3, NSC 99-2221-E-002-207-MY3, and NSC 99-2221-E-002-210-MY3.

Permission to make digital or hard copies of all or part of this work for personal or classroom use is granted without fee provided that copies are not made or distributed for profit or commercial advantage and that copies bear this notice and the full citation on the first page. To copy otherwise, to republish, to post on servers or to redistribute to lists, requires prior specific permission and/or a fee.
DAC '13, May 29 - June 07 2013, Austin, TX, USA.

and then use two masks to manufacture the two sub-patterns, as illustrated in Figure 1. DPL is commonly used for poly, contact, metal, and even via layers. A double patterning decomposer assigns two patterns of a layer to two different masks if their spacing is less than the specific spacing S_{min} provided by foundry. The double patterning layout decomposition problem is regarded as a two-coloring problem according to a conflict graph (CG), in which a vertex represents a pattern and an edge between two vertices represents that the spacing between the two corresponding patterns is less than S_{min}. Figure 2(a) illustrates the corresponding conflict graph for the layout in Figure 2(b).

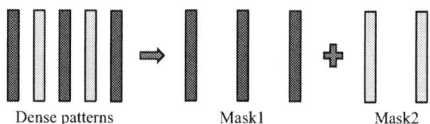

Figure 1: DPL is to decompose a dense pattern into two sub-patterns and then use mask1 and mask2 to manufacture the two sub-patterns.

A DPL conflict occurs when two patterns separated by the distance less than S_{min} belong to the same mask. To resolve DPL conflicts, a layout engineer could report the detected conflicts to designers and ask them to modify the layout so that the conflicting patterns can be relocated or removed. Nevertheless, layout modification is typically time-consuming and often very costly. Another way is to split one pattern into two sub-patterns and manufacture them by different masks to resolve the DPL conflict, which is regarded as a stitch insertion. Figure 2(c) shows an example of stitch insertion. Nevertheless, stitches might cause significant printability degradation and yield loss due to their overlay errors. Therefore, several works [16] [17] proposed simultaneous conflict and stitch minimization methods for layout decomposition. However, even with the stitches, there could be some irresolvable conflicts, called native conflicts, which cannot be resolved by stitch insertion.

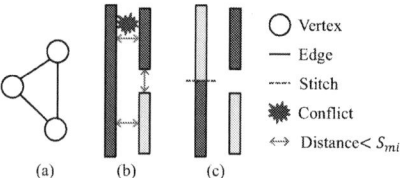

Figure 2: (a) The conflict graph corresponds to the layout on the right. (b) A DPL conflict is resulted from the two blue patterns separated by less than the specific spacing S_{min}. (c) Stitch insertion is to split one pattern into two sub-patterns, and the two sub-patterns are manufactured by different masks.

Because not every pattern can be easily decomposed into two sub-patterns, post-layout double patterning decomposition is not always feasible. Therefore, it is very important to apply DPL-aware techniques to obtain highly decomposable layout designs. Some previous works considered DPL during detailed routing to generate highly decomposable layouts. Cho et al. [6] proposed the first DPL-friendly grid-based detailed routing algorithm. Lin et al. [9] proposed an innovative conflict graph (ICG) to assist DPL-aware gridless detailed router. However, no literature is known for DPL-aware analog placement.

To handle the analog placement problem, most previous works apply topological floorplan representations such as sequence-pair [4], TCG-S [7], hierarchical B*-tree (HB*-tree) [8], O-tree [12], and corner stitching compliant B*-tree (CB-tree) [14]. Although various analog design constraints, such as symmetry, proximity, variant, fixed-boundary, minimum-separation, and boundary constraints, have been studied in previous works, DPL is not considered in these previous works. As a result, their resulting layouts may not be DPL-decomposable. Further, since the unavoidable misalignment of different masks would cause variations, analog designers usually reduce mismatches by predefining masks (colors) for the poly and contact layers of critical or sensitive modules before double patterning decomposition [2]. This *pre-coloring constraint* makes analog placement with DPL consideration much harder during design time.

The pre-coloring constraint significantly complicates DPL decomposition. Without pre-colors, DPL conflict detection is equivalent to odd cycle detection in the induced conflict graph; once the graph is two-colorable, there is no odd cycle. However, no odd cycle cannot guarantee that the layout is decomposable if a pre-coloring pattern causes a "conflict path" in the graph. Figure 3(a) shows an example of placement with pre-coloring constraints, and Figure 3(b) shows a poly layer before DPL decomposition with pre-defined color1 for critical modules. No matter what combination of colors is chosen, a conflict path always exists, as shown in Figure 3(c).

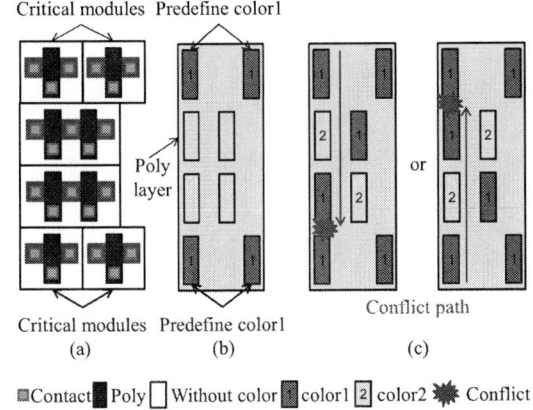

Critical modules   Predefine color1

Poly layer

or

Conflict path

Critical modules   Predefine color1
(a)   (b)   (c)

▨ Contact ■ Poly ☐ Without color ▨ color1 ② color2 ✳ Conflict

Figure 3: (a) The resulting analog placement with pre-coloring constraints. (b) A placed poly layer before DPL decomposition. (c) A conflict path exists after decomposition.

In this paper, we propose the first algorithm that solves the DPL-aware analog placement. We summarize our main contributions as follows:

- We propose an extended conflict graph (ECG) to model the global relationship among non-flipped and flipped patterns; DPL conflicts can be reduced by flipping modules to prevent area increase.

- We develop an integer linear programming (ILP) algorithm considering symmetry and pre-coloring constraints to minimize conflicts. Two speedup schemes, connected-component computation and two-edge-connected-component-computation, are proposed to reduce the runtime of ILP.

- We propose DPL-aware and module-merging perturbations to assist simulated annealing for achieving better solutions and better performance.

- A three-stage flow for DPL-aware analog placement is proposed. The three-stage flow minimizes area, wirelength, and DPL conflicts simultaneously.

The rest of this paper is organized as follows: Section 2 reviews the CB-tree representation and analog design constraints and formulates the DPL-aware analog placement. Section 3 introduces our ECG and ILP formulation. Section 4 proposes general and DPL-aware perturbations and presents our three-stage analog placement flow. Section 5 reports the experimental results. Finally, Section 6 concludes this paper.

2. PRELIMINARIES

In this section, we first give a brief review of the CB-tree [14] and then formulate the DPL-aware analog placement problem.

2.1 Review of CB-trees

For constraint-driven analog placement, CB-tree is the most effective and efficient topological representation with module adjacency information in the literature. Since we need the adjacency information to check whether there is any conflict in the placement result, we adopt the CB-tree representation in this work. A CB-tree [14] is a B*-tree integrated with modified corner stitching to augment a B*-tree for module adjacency handling. The B*-tree is an ordered binary tree representing a compacted placement [5], and corner stitching is a data structure for modeling rectangular objects with a tile plane to provide efficient geometric operations [11].

2.2 Review of Analog Constraints

For easier presentation (and also the space limit), we focus on symmetry and proximity constraints which are the two most important, intensively studied analog constraints addressed in the literature, and demonstrate how to integrate the DPL constraints with them for advanced analog placement. Though not presented in this paper, other less popular analog constraints can readily be implemented with a similar flow as ours to be presented later.

- *Symmetry constraints* are use to place some pairs of modules symmetrically along a vertical or a horizontal symmetry axis.It can reduce mismatches of sensitive modules.

- *Proximity constraints* are use to place modules at closest proximity to reduce process variation.

To handle the two most important constraints, Lin et al. proposed an HB*-tree with hierarchy and contour nodes [8]. The hierarchy nodes handle symmetry and non-symmetry modules simultaneously and guarantee feasible placement results. Since the symmetry pairs may not be rectangular, contour nodes representing top horizontal contour segments are introduced to handle rectilinear shapes of hierarchy nodes during packing. In a CB-tree, tile updating is based on the contour nodes so that it can also handle rectilinear regions.

2.3 Problem Formulation

The DPL-aware analog placement problem can be formally defined as follows:

- **The Double Patterning-Aware Analog Placement Problem:** Given a set of rectangular modules $M = \{m_k | 1 \leq k \leq |M|\}$, a set of nets $N = \{n_k | 1 \leq k \leq |N|\}$, a set of placement constraints $S = \{s_k | 1 \leq k \leq |S|\}$, and a set of pre-coloring constraints $R = \{r_k | 1 \leq k \leq |R|\}$, place all the modules in M to minimize the total area, wirelength, and DPL conflicts such that no modules overlap with each other and all the placement constraints are satisfied.

3. DPL CONFLICT HANDLING

In this section, we first discuss the techniques for resolving an analog placement with DPL conflicts. Considering pre-coloring constraints, we then present an extended conflict graph to model the global relationship among patterns/modules and derive an ILP formulation to determine the orientation of each module and the color of each pattern simultaneously to minimize DPL conflicts.

3.1 Module Flipping

To resolve an analog placement with DPL conflicts, we could simply increase the spacing between two conflicting patterns/modules. After increasing the spacing, however, we should pack the CB-tree again, and the layout area could become larger accordingly, as illustrated in Figs. 4(a) and (b). To minimize the area and the packing time, we could flip some modules to resolve the conflicts. As illustrated in Figure 4(c), we could flip the top module vertically not only to solve the conflict in

25

the contact layer but also to keep the same outline of layout. Moreover, since flipping would not change the coordinate of a module, we do not need to re-pack modules after module flipping.

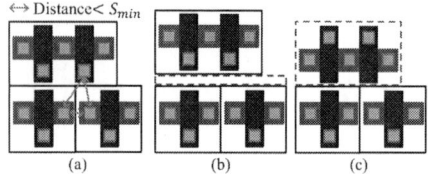

Figure 4: (a) The originally packed layout with DPL conflicts on the contact layer. (b) Conflict removal by enlarging the spacing between modules. (c) Conflict removal by flipping the top module.

3.2 Extended Conflict Graph (ECG)

In the traditional conflict graph construction, a vertex represents a pattern, and an edge is introduced if the spacing between two corresponding patterns is less than S_{min}. The traditional conflict graph considers the information from a fixed layout and cannot capture the location updates for polys and contacts with their orientation changes. Since a poly can only be placed vertically in advanced nodes, it might not be allowed to rotate a device arbitrarily. As a result, there are only two choices for module orientations: non-flipped and flipped. We use *non-flipped* and *flipped* vertices to present poly and contact patterns of non-flipped and flipped modules, respectively. For edges, since the orientations of patterns in the same module should be the same, we only construct edges from non-flipped vertices to non-flipped vertices and from flipped vertices to flipped vertices for patterns in the same module. Figure 5 illustrates a module and the corresponding vertices of poly and contact patterns.

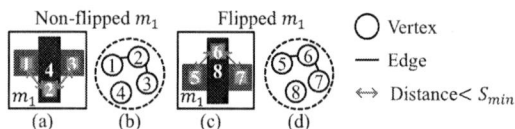

Figure 5: (a) A non-flipped module $m1$. (b) Non-flipped vertices of poly and contact patterns in $m1$. (c) A flipped module $m1$. (c) Flipped vertices of poly and contact patterns in $m1$.

To record the relation of patterns belonging to different modules, we should check the distance of all the patterns with different orientations. There are four combinations of orientations for every two modules: (1) non-flipped to non-flipped (see Figure 6(a)), (2) non-flipped to flipped (see Figure 6(b)), (3) flipped to non-flipped (see Figure 6(c)), and (4) flipped to flipped (see Figure 6(d)). We check the distances of all patterns for the four combinations and introduce an edge if the distance between two corresponding patterns is less than S_{min}. Figure 6(e) shows the ECG of module $m1$ and module $m2$.

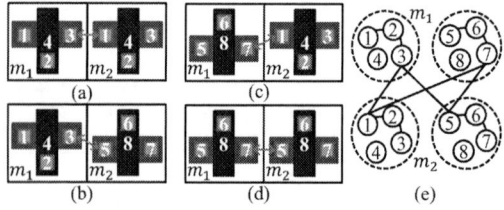

Figure 6: (a) Non-flipped $m1$ and non-flipped $m2$. (b) Non-flipped $m1$ and flipped $m2$. (c) Flipped $m1$ and non-flipped $m2$. (d) Flipped $m1$ and flipped $m2$. (e) ECG of $m1$ and $m2$.

3.3 Basic Integer Linear Programming (ILP) Formulation

After the ECG construction, we apply an ILP formulation to determine the orientation of each module and the color of each pattern in the contact and poly layers so that the number of conflicts is minimized. To reduce the mismatches of critical modules, our ILP formulation also considers symmetry and pre-coloring constraints.

The notations used in the ILP formulation are as follows:

- M: set of modules.
- M^S: set of symmetry pairs.
- V: set of vertices.
- V^n: set of non-flipped vertices.
- V^f: set of flipped vertices.
- V^1: set of vertices with pre-defined color1.
- V^2: set of vertices with pre-defined color2.
- E: set of edges.
- m_i: a module.
- m_k^S: a symmetry pair.
- p_i: a constant denoting the number of poly and contact patterns in module m_i.
- $v_{i,j}$: vertex j of module m_i.
- $e_{ij,i'j'}$: an edge connecting two vertices $v_{i,j}$ and $v_{i',j'}$.
- $r^1(v_{i,j})$: 0-1 integer variable that denotes the color of vertex $v_{i,j}$. If the color of vertex j of the module m_i is 1, $r^1(v_{i,j}) = 1$; otherwise, $r^1(v_{i,j}) = 0$.
- $r^2(v_{i,j})$: 0-1 integer variable that denotes the color of vertex $v_{i,j}$. If the color of vertex j of the module m_i is 2, $r^2(v_{i,j}) = 1$; otherwise, $r^2(v_{i,j}) = 0$.
- $c(e_{ij,i'j'})$: 0-1 integer variable that denotes the existence of a conflict between two vertices $v_{i,j}$ and $v_{i',j'}$. If the distance between the vertex j of the module m_i and the vertex j' of the module $m_{i'}$ is less than the specified spacing S_{min} and the color of them are the same, $c(e_{ij,i'j'})=1$; otherwise, $c(e_{ij,i'j'}) = 0$.

The conflict minimization problem can be formulated as follows:

$$minimize \sum_{e_{ij,i'j'} \in E} c(e_{ij,i'j'})$$

$$s.t. \quad r^1(v_{i,j}) + r^2(v_{i,j}) + r^1(v_{i,j+p_i}) + r^2(v_{i,j+p_i}) = 1,$$
$$\forall v_{i,j} \in V^n, \forall v_{i,j+p_i} \in V^f, \quad (1)$$

$$r^1(v_{i,j}) + r^2(v_{i,j}) = r^1(v_{i,j+1}) + r^2(v_{i,j+1}),$$
$$\forall v_{i,j}, v_{i,j+1} \in V^n, \quad (2)$$

$$r^1(v_{i,1}) + r^2(v_{i,1}) = r^1(v_{i',1}) + r^2(v_{i',1}),$$
$$\forall m_i, m_i' \in m_k^s, \forall m_k^s \in M^S, \quad (3)$$

$$r^1(v_{i,j}) + r^1(v_{i',j'}) \le 1 + c(e_{ij,i'j'}), \forall e_{ij,i'j'} \in E, \quad (4)$$

$$r^2(v_{i,j}) + r^2(v_{i',j'}) \le 1 + c(e_{ij,i'j'}), \forall e_{ij,i'j'} \in E, \quad (5)$$

$$r^2(v_{i,j}) = 0, \forall v_{i,j} \in V^1, \quad (6)$$

$$r^1(v_{i,j}) = 0, \forall v_{i,j} \in V^2. \quad (7)$$

The objective function is to minimize the total number of conflicts. Constraint (1) ensures that only one orientation and only one color can be chosen for each pattern. Figure 7 illustrate the corresponding patterns of the four variables in Constraint (1). If $r^1(v_{1,1})$ of the vertex $v_{1,1}$ equals one, the corresponding contact is non-flipped and colored color1. If $r^2(v_{1,1})$ equals one, the corresponding contact is non-flipped and colored color2. Since there are four patterns in the module m_1, p_1 equals four and the corresponding flipped vertex of $v_{1,1}$ is $v_{1,5}$. If $r^1(v_{1,5})$ equals one, the corresponding contact is flipped and colored color1. If $r^2(v_{1,5})$ equals one, the corresponding contact is flipped and colored color2. Constraint (2) is used to guarantee that the poly and contact patterns in the same module m_i are all non-flipped or all flipped. Constraint (3) guarantees that the modules belonging to a symmetry pair have the same orientation. Constraints (4) and (5) denote that if the distance between two vertices $v_{i,j}$ and $v_{i',j'}$ is less than S_{min} and the two vertices have the same color, a conflict occurs. Constraints (6) and (7) model the pre-coloring constraint; if a vertex has a pre-assigned color, the variable of the opposite color should be equal to zero.

THEOREM 1. *Given an ECG with $|V|$ vertices and $|E|$ edges, the number of ILP variables and the number of ILP constraints are both $O(|V| + |E|)$.*

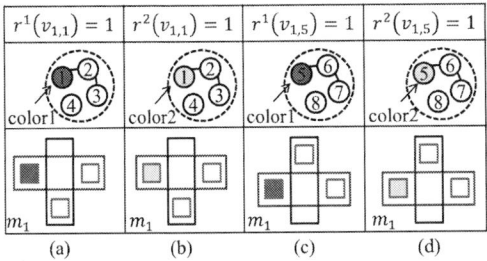

Figure 7: (a) The variable $r^1(v_{1,1}) = 1$ if the color of vertex 1 of the module m_1 is 1. (b) $r^2(v_{1,1}) = 1$ if the color of vertex 1 of the module m_1 is 2. (c) $r^1(v_{1,5}) = 1$ if the color of vertex 5 of the module m_1 is 1. (d) $r^2(v_{1,5}) = 1$ if the color of vertex 5 of the module m_1 is 2.

3.4 ILP Problem-Size Reduction

In this subsection, we propose two linear-time speedup schemes for the ILP formulation. The two schemes are based on the divide-and-conquer algorithm to divide a whole problem into subproblems.

3.4.1 Connected Component Computation

A connected component of a graph is a subgraph in which any two vertices are connected to each other with paths. Since the poly and contact patterns of the same module should be considered simultaneously, these vertices belonging to the same module are clustered before we compute the connected component. Moreover, since the modules in the same symmetry group should be with the same orientation to reduce mismatches, the vertices of modules in a symmetry pair are also clustered. Different from the traditional graph division methods, we compute the connected components for a clustered graph. Since the components are independent, the ILP formulation can be applied for each component, and the solution of the overall problem can be obtained by taking the union of all sub-solutions and maintain the optimality.

3.4.2 Two-Edge-Connected Component Computation

Two-edge-connected-component computation is to identify an edge whose removal would decompose the given graph into two components. We apply the ILP formulation to handle the two components independently. Since there is only one connection between the two components, the two sub-solutions can be merged with a simple color remapping process. For instance, we can find two components by removing an edge in Figure 8(a). After solving the two subgraphs, as shown in Figure 8(b), we recolor one of the components and merge them so that no conflict occurs. However, the conflict path may be disconnected if the pre-coloring vertices belong to different components. Figure 8(d) illustrates a graph and two pre-coloring vertices. If we remove an edge and perform the basic ILP formulation to each component, as shown in Figure 8(e), a conflict occurs after merging the components because there is a pre-coloring vertex in each component and the two components cannot be recolored. Figure 8(e) illustrates the condition of the conflict path. Therefore, we also cluster the pre-coloring vertices to handle the conflict path. After applying the ILP formulation on each two-edge-connected component, the solution of the overall problem can be obtained by merging sub-solutions.

After applying our proposed reduction schemes, the number of ILP variables and the number of ILP constraints are still $O(|V| + |E|)$. However, the schemes significantly reduce the runtime of our ILP formulation for practical applications, as will be reported in Section 5.

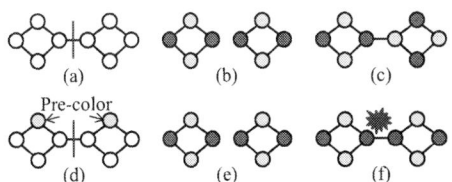

Figure 8: (a) Two-edge-connected-component computation. (b) Components are solved independently. (c) One of the components is recolored and the two components are merged. (d) Two-edge-connected-component computation with the existence of pre-coloring vertices. (e) The components are handled independently. (f) A conflict occurs after merging components.

4. THE ALGORITHM FLOW

We propose a three-stage flow to handle the simultaneous area, wirelength, and DPL conflict minimization problem. Figure 9 shows the underlying ideas of each stage. Since our DPL-aware analog placement flow is based on simulated annealing, DPL-aware perturbations are proposed to find a desired solution with fewer conflicts.

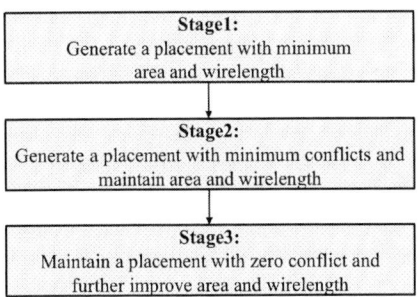

Figure 9: The overall flow of our DPL-aware analog placement.

4.1 Perturbation

We apply the following operations to perturb a CB-tree in Stage 1 and Stage 3.

- Op1: Delete a node and insert it into another place.
- Op2: Swap two nodes.
- Op3: Merge two nodes according to merging guidelines.
- OP4: Demerge two nodes.

Perturbations for a classical CB-tree include deletion/insertion, rotation, and swapping. Because polys can be placed in only one direction (say, vertically) for advanced process and a module is not allowed to be rotated to an arbitrary orientation, we only adopt the deletion/insertion and swapping operations, as in Op1 and Op2. To obtain higher performance and lower parasitic, we apply the module merging operation, which is a common technique used in analog designs. Merging modules can shorten interconnects and reduce area to achieve better solution quality. We merge modules according to two guidelines: (1) merged modules are of the same diffusion type (PMOS/NMOS), and (2) merged modules belong to the same net [10] [15]. To handle a rectilinear module during packing, we use a hierarchy node to represent a corresponding merged module and a contour node to represent a top horizontal contour segment[8]. To prevent a solution from getting stuck at a local minimum, we also propose a demerging operation to separate merged modules during the simulated annealing process.

To consider DPL effects, we further apply the following DPL-aware operations to perturb a CB-tree in Stage 2:

- Op1: Extract the node with the maximum number of conflicts (the maximum-conflict node) and insert it into another place.
- Op2: Swap the maximum-conflict node with another node.

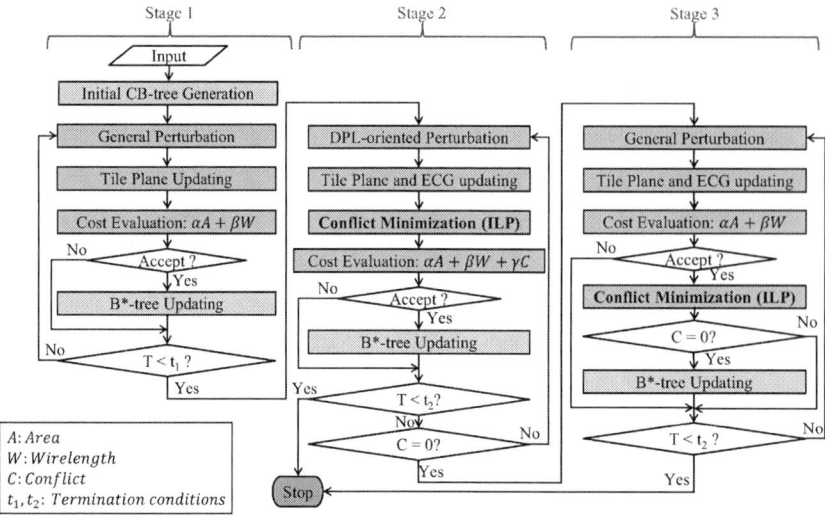

Figure 10: The detailed three-stage DPL-aware analog placement flow.

- Op3: Merge the maximum-conflict node with another node according to the merging guidelines.

- OP4: Demerge the maximum-conflict node.

Since a DPL conflict is usually caused by a module and its adjacent modules, we delete/insert or swap the node with the maximum number of conflicts to change its neighboring configuration to further reduce the conflicts. Moreover, both merging and demerging can reduce the number of conflicts because the number of contacts is changed. Figure 11 shows an example of solving a conflict with module merging, where Figure 11(a) gives the original placement and Figure 11(b) shows the placement after module merging. We also propose the demerge operation to separate a merged module with the maximum number of conflicts to change its adjacent modules.

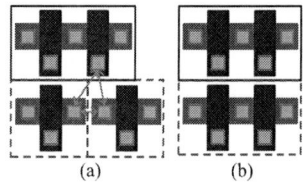

Figure 11: (a) The original placement with a conflict. (b) The placement after merging modules without conflict.

4.2 The Three-Stage Placement Flow

Figure 10 details our three-stage flow. In Stage 1, given an initial placement represented by a CB-tree, we randomly choose one perturbation and update the tile plane to handle analog constraints. The cost function ϕ_1 of the placement is defined in Equation (8), where α and β are user-specified parameters, A is the chip area and W is the half-perimeter wirelength (HPWL).

$$\phi_1 = \alpha A + \beta W. \tag{8}$$

We use the cost function ϕ_1 to evaluate the placement in Stage 1. With simulated annealing, we keep perturbing the CB-tree and record the best solution until a predefined termination condition is met. Since this stage is focused on obtaining a placement with acceptable area and wirelength and does not consider the conflict minimization, it does not apply the ILP formulation. In Stage 2, we further consider conflict minimization, while maintaining the area and wirelength quality. We perform DPL-aware perturbations to find a placement with minimum conflicts. During packing, once a node is packed, we update the tile plane and use area enumeration (in corner stitching) to check the distance between patterns of the corresponding node and patterns of

adjacent nodes for ECG updating. In particular, the time complexity of the area enumeration is only $O(1)$ because the CB-tree selects an initial searching tile (with corner stitching) near the given node. As illustrated in Figure 12, the top module is being packed and the searching area is extended by the distance S_{min} from the module boundary. After checking the distance between patterns, the ECG is updated, and then we formulate an ILP according to the updated ECG to determine the orientation of each module and the color of each pattern for conflict minimization. The cost function ϕ_2 of the placement is defined in Equation (9), where α, β, and γ are user-specified parameters and C is the number of conflicts of the placement.

$$\phi_2 = \alpha A + \beta W + \gamma C. \tag{9}$$

We use the cost function ϕ_2 to evaluate the placement in Stage 2. If the number of conflicts C is zero, we go to Stage 3. Otherwise, we keep performing DPL-aware perturbations to find solutions with fewer conflicts until a predefined termination condition is met. In Stage 3, to reduce the runtime, we first evaluate the area and wirelength after every perturbation. The ILP programs for conflict minimization can be skipped if the corresponding area and wirelength are not acceptable. Moreover, because it is desired to obtain a decomposable layout for modern design, a placement with non-zero conflict is not acceptable in this stage. Therefore, we can maintain the placement with zero conflict while further improving the area and wirelength.

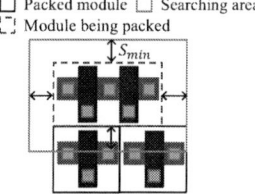

Figure 12: An example of area enumeration. The top module is being packed and the searching area is extended by the distance S_{min} from the module boundary.

Table 1: Design Statistics and Constraint Number of case1 and case2.

	PMOS	NMOS	Net	Poly	Contact	Sym.	Pre-color
Case1	8	8	11	32	230	2	8
Case2	14	16	34	92	344	2+2	8

28

Table 2: Comparison of Area, Wirelength, Conflict number and runtime between CB-tree [14] and our three-stage flow.

Circuits	[14]						Our three-stage flow with ILP reduction			
	Area (nm²)	HPWL (nm)	Conflict number		Runtime (s)		Area (nm²)	HPWL (nm)	Conflict number	Runtime (s)
			W/o ILP	W/ ILP	W/o ILP	W/ ILP				
Case1	2032050	9930	10	2	10	12	1900350	7373	0	41
Case2	3207437	19390	13	2	23	32	3014900	13155	0	197
Comp.	1.07	1.43			0.14	0.18	1.00	1.00		1.00

5. EXPERIMENTAL RESULTS

We implemented our DPL-aware analog placement flow in the C++ programming language, and all the experiments were performed on an Intel Xeon X5647 2.93GHz Linux workstation with 48GB memory. We used the CPLEX12.3 [1] library to solve the ILP problems. To evaluate our flow, we used two industrial analog circuits Case1 and Case2. The circuits were scaled down to 28 nm for the experiments. The statistics of the two scaled circuits are shown in Table 1, where "PMOS", "NMOS", and "Net" list the number of PMOSs, NMOSs, and nets, respectively. Since the number of vertices in our proposed ECG is equal to the number of poly and contact patterns, we also report the number of patterns in Table 1, where "Poly" gives the number of total poly patterns, and "Contact" gives the number of total contact patterns. Finally, "Sym." gives the number of symmetric MOSs and "Pre-color" gives the number of pre-colored poly and contact patterns.

Because there is no previous work on the DPL-aware analog placement, we compared our DPL-aware analog placement flow with the following two methods: (1) CB-tree-based placement without considering DPL and (2) CB-tree-based placement with our ILP-based conflict minimization. Table 2 compares the results, including area, half-perimeter wirelength length (HPWL), the number of DPL conflicts, and placement runtime. The column "W/o ILP" shows the results of CB-tree, which does not consider DPL; the column "W/ ILP" reports the results with our ILP formulation for conflict minimization after the CB-tree result is generated. The experimental results show that applying our ILP formulation after placement can resolve a large potion of conflicts. In particular, our flow considering DPL during placement can obtain zero conflict under reasonable runtime. With module merging (6 modules in Case1 and 8 modules in Case2), our method can achieve an average area reduction of 7% and an average wirelength reduction of 43%. Figs. 13(a) and (b) show the respective poly and contact layers of Case1 generated by our three-stage flow.

To show the efficiency of our three-stage flow, we implemented a one-stage flow, which applies DPL-aware perturbations and ILP formulation for conflict minimization in each iteration of simulated annealing. Table 3 shows the runtime comparisons between the one-stage flow and our three-stage flow, with and without applying our ILP reduction schemes. Here, "W/o Red." denotes the runtime without the reduction and "W/ Red." denotes the runtime with the reduction. The experimental results show that the runtime of the one-stage flow is more than two hours. In contrast, our three-stage flow can efficiently solve the problem because the unnecessary ILP computations can be avoided. Also, our three-stage flow can further be speeded up with the ILP reductions.

Table 3: Runtime comparison between the one-stage and our three-stage flows.

Circuits	One-stage Flow		Three-stage Flow	
	W/o Red.	W/ Red.	W/o Red.	W/ Red.
Case1	>2 hr	>2 hr	48 sec	39 sec
Case2	>2 hr	>2 hr	266 sec	197 sec
Comp.			1.33	1.00

6. CONCLUSIONS

In this paper, we have presented a DPL-aware analog placement flow to simultaneously minimize area, wirelength, and DPL conflicts. We have proposed an extended conflict graph and ILP formulations for conflict minimization. In addition, the DPL-aware perturbations and the three-stage flow have been proposed to further improve the solution quality and efficiency. Experimental results based on two industrial analog designs have shown that our flow is effective and efficient.

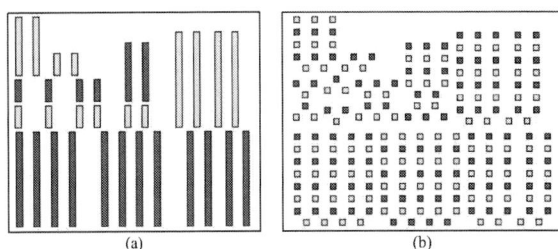

Figure 13: Layout for Case 1. (a) The poly layer of Case1. (b) The contact layer of Case1.

7. REFERENCES

[1] *IBM ILOG CPLEX Optimizer.* http://www-01.ibm.com/software/integration/optimization/cplex-optimizer/.

[2] D. Abercrombie, P. Lacour, O. El-Sewefy, A. Volkov, E. Levine, K. Arb, C. Reid, Q. Li, and P. Ghosh. Double patterning from design enablement to verification. In *Proc. of SPIE*, volume 8166, 2011.

[3] G. Bailey, A. Tritchkov, J. Park, L. Hong, V. Wiaux, E. Hendrickx, S. Verhaegen, P. Xie, and J. Versluijs. Double pattern eda solutions for 32nm hp and beyond. In *Proc. of SPIE*, volume 6521, 2007.

[4] F. Balasa and K. Lampaert. Symmetry within the sequence-pair representation in the context of placement for analog design. *IEEE Trans. on CAD*, 19(7):721–731, 2000.

[5] Y.-C. Chang, Y.-W. Chang, G.-M. Wu, and S.-W. Wu. B*-trees: a new representation for non-slicing floorplans. In *Proc. of DAC*, pages 458–463, 2000.

[6] M. Cho, Y. Ban, and D. Pan. Double patterning technology friendly detailed routing. In *Proc. of ICCAD*, pages 506–511, 2008.

[7] J.-M. Lin, G.-M. Wu, Y.-W. Chang, and J.-H. Chuang. Placement with symmetry constraints for analog layout design using TCG-S. In *Proc. of ASP-DAC*, pages 1135–1138, 2005.

[8] P.-H. Lin, Y.-W. Chang, and S.-C. Lin. Analog placement based on symmetry-island formulation. *IEEE Trans. on CAD*, 28(6):791–804, 2009.

[9] Y.-H. Lin and Y.-L. Li. Double patterning lithography aware gridless detailed routing with innovative conflict graph. In *Proc. of DAC*, pages 398–403, 2010.

[10] S. Nakatake, M. Kawakita, T. Ito, M. Kojima, K. Izumi, and T. Habasaki. Regularity-oriented analog placement with diffusion sharing and well island generation. In *Proc. of ASP-DAC*, pages 305–311, 2010.

[11] J. K. Ousterhout. Corner stitching: a data-structuring technique for vlsi layout tools. *IEEE Trans. on CAD*, 3(1):87–100, 1984.

[12] Y. Pang, F. Balasa, K. Lampaert, and C.-K. Cheng. Block placement with symmetry constraints based on the O-tree non-slicing representation. In *Proc. of DAC*, pages 464–467, 2000.

[13] J. Park, S. Hsu, D. Van Den Broeke, J. Chen, M. Dusa, R. Socha, J. Finders, B. Vleeming, A. van Oosten, P. Nikolsky, et al. Application challenges with double patterning technology (DPT) beyond 45 nm. In *Proc. of SPIE*, volume 6349, 2006.

[14] H.-F. Tsao, P.-Y. Chou, S.-L. Huang, Y.-W. Chang, M.-H. Lin, D.-P. Chen, and D. Liu. A corner stitching compliant B*-tree representation and its applications to analog placement. In *Proc. of ICCAD*, pages 507–511, 2010.

[15] L. Xiao, E. F. Y. Young, X. He, and K. P. Pun. Practical placement and routing techniques for analog circuit designs. In *Proc. of ICCAD*, pages 675–679, 2010.

[16] Y. Xu and C. Chu. A matching based decomposer for double patterning lithography. In *Proc. of ISPD*, pages 121–126, 2010.

[17] K. Yuan, J.-S. Yang, and D. Pan. Double patterning layout decomposition for simultaneous conflict and stitch minimization. *IEEE Trans. on CAD*, 29(2):185–196, 2010.

Simultaneous Analog Placement and Routing with Current Flow and Current Density Considerations *

Hung-Chih Ou[1], Hsing-Chih Chang Chien[1], and Yao-Wen Chang[1,2]

[1]Graduate Institute of Electronics Engineering, National Taiwan University, Taipei 106, Taiwan
[2]Department of Electrical Engineering, National Taiwan University, Taipei 106, Taiwan
{howard, lichee626}@eda.ee.ntu.edu.tw; ywchang@cc.ee.ntu.edu.tw

ABSTRACT

Current-flow and current-density are two major considerations for placement and routing of analog layout synthesis. The current-flow constraints are specified to the critical nets with monotonic current/signal paths to reduce parasitic impacts. The current-density constraints are usually specified on the nets with variable wire widths to avoid the IR-drop and electromigration problems. In this paper, we propose the first work to simultaneously consider current-flow and current-density constraints while placing and routing the analog circuits with minimized chip area, routed wirelength, bend numbers, via counts, and coupling noise at the same time. We first present an enhanced B*-tree representation to simultaneously model modules and interconnects for an analog circuit. Then a simultaneous placement and routing algorithm is presented to generate a layout while satisfying the current-flow and current-density constraints with minimized chip area, routed wirelength, bend numbers, via counts, and coupling noise. Experimental results show that our approach can obtain better layout results and satisfy all specified constraints while optimizing circuit performance.

Categories and Subject Descriptors

B.7.2 [**Integrated Circuits**]: Design Aids [Placement and Routing]

General Terms

Algorithms, Performance

Keywords

Physical Design, Placement, Routing, Analog ICs

1. INTRODUCTION

In modern analog and mixed-signal circuit designs, *current-flow* and *current-density* are two major considerations for placement and routing of analog layout synthesis. The current-flow constraints are specified to the critical nets in the analog circuits with monotonic current/signal paths from the power line (VDD) to the ground line (GND) to increase the current smoothness and reduce the parasitic impacts on the current/signal paths. The current-density constraints are usually specified to the nets with variable wire widths according to the amount of current flowing the nets to avoid the IR-drop and electromigration

---

*This work was partially supported by IBM, SpringSoft, TSMC, Academia Sinica, and NSC of Taiwan under Grant No's. NSC 101-2221-E-002-191-MY3, NSC100-2221-E-002-088-MY3, NSC 99-2221-E-002-207-MY3, and NSC 99-2221-E-002-210-MY3.

Permission to make digital or hard copies of all or part of this work for personal or classroom use is granted without fee provided that copies are not made or distributed for profit or commercial advantage and that copies bear this notice and the full citation on the first page. To copy otherwise, to republish, to post on servers or to redistribute to lists, requires prior specific permission and/or a fee.
DAC '13, May 29 - June 07 2013, Austin, TX, USA.

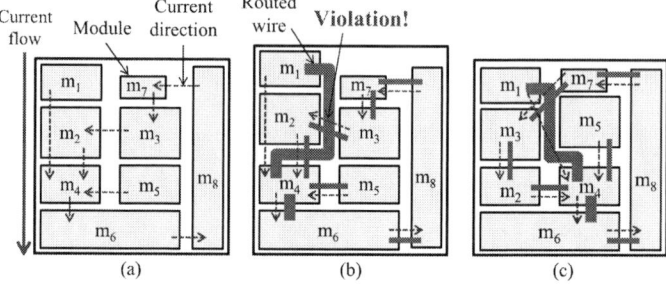

Figure 1: (a) A placement result satisfies the current-flow constraint. (b) The routing result of (a) with current-density consideration. (c) A simultaneous placement and routing result satisfies both current-flow and current-density constraints.

problems caused by high current densities. Moreover, the current-density constraints become more important to high-frequency analog circuits because the *skin effects* will confine the conducting region of a wire near the surface and further increase the current density.

Existing analog layout synthesis algorithms focus on one of the two constraints alone: current-flow constraints at the placement stage and current-density constraints at the routing stage. For the current-flow constraints, Long et al. proposed a signal-path driven partition and placement method for analog circuits [15], which simply places modules closely from left to right in each signal-path according to the pre-defined sequence to satisfy current-flow constraints. To achieve a more compact placement result, Wu et al. introduced a slicing-tree-based analog placement representation to generate multiple compact analog placements with current-path constraints [22]. For the current-density constraints, Alder et al. proposed a current driven router to calculate the required wire width of each signal net according to the current-density constraints on the fly, and route all signal nets with minimum detours in a single step [1, 2]. To further consider the electromigration problem, Lienig et al. introduced two two-stage global routing methodologies for current-driven routing to prevent electromigration in deep-sub-micron designs [10], and Lienig and Jerke further improved the routing efficiency by a current-dependent routing tree to determine the estimated routing paths [9].

To achieve better circuit performance and reduce the layout-induced parasitic impacts, it is desirable to simultaneously consider current-flow and current-density constraints at both of the placement and routing stages of analog layout synthesis. However, most of the traditional analog layout synthesis algorithms address on either placement or routing individually [3, 5, 6, 11, 13, 14, 16, 17, 18, 19, 21, 23, 24]. As a result, placement is performed followed by routing, leading to an ordered synthesis flow. This ordered analog layout synthesis flow cannot guarantee that a constraint (say, a current-flow constraint) addressed and satisfied at the placement stage will not be violated at the routing stage that considers another constraint (say, the current-density constraint). Figures 1(a) and (b) illustrate the placement and routing results considering current-flow and current-density constraints at the placement and routing stages in sequence. Since the interconnects usually cannot cross the active regions of devices [23], the current-flow constraints are violated after considering the current-density constraints at the routing stage as shown in Figure 1(b). The simultaneous placement and routing algorithm can simultaneously considers current-flow and current-density constraints and obtains a layout without any violation as shown in Figure 1(c).

Unfortunately, very few existing works, especially the topological-representation-based techniques, focus on simultaneous placement and routing of analog layout synthesis. Cohn et al. decomposed each interconnect into a two-terminal minimal-spanning tree, and incrementally reshaped or moved both modules and interconnects according to the change of the minimal-spanning trees during the annealing [7]. The later work [8] proposed a sequence-pair representation with a shaped-based formulation to simultaneously express modules and interconnects. Recently, Prieto et al. and Lin et al. worked on two-stage frameworks to consider the layout routability at the placement stage earlier [20, 12]. However, the incremental reshaping and moving may not be able to handle the multiple terminal nets, and also require some post-optimizations to further improve the routing quality. Although the sequence-pair with the shaped-based formulation can simultaneously perform placement and routing, it usually requires much time to obtain a layout and is often hard to extend the formulation to simultaneously consider current-flow and current-density constraints. Moreover, routability-driven analog placement usually uses a global router to evaluate its placement results, which might not be able to accurately capture the detailed routing results.

To simultaneously consider current-flow and current-density constraints in analog layout synthesis for better circuit performance, we propose an enhanced B*-tree representation to simultaneously model the modules and interconnects for an analog circuit while satisfying the current-flow and current-density constraints. According to the enhanced B*-tree representation, we also present a simultaneous placement and routing algorithm considering both current-flow and current-density constraints while placing and routing an analog circuit with minimized chip area, routed wirelength, bend numbers, via counts, and coupling noise at the same time.

The main contributions of this paper are summarized as follows:

- A new analog layout synthesis algorithm which simultaneously considers current-flow and current-density constraints is proposed. To the best of our knowledge, this is the first work that handles the two constraints simultaneously.

- A novel enhanced B*-tree representation for simultaneous placement and routing is presented. The enhanced B*-tree simultaneously models modules and interconnects in one representation while satisfying the current-flow and current-density constraints at the same time.

- A simultaneous placement and routing algorithm is proposed. The simultaneous placement and routing algorithm closes the gap between placement and routing and makes analog layout synthesis conform to the conventional design manner.

- Chip areas, routed wirelength, bend numbers, via counts, and coupling noise are simultaneously minimized in our simultaneous placement and routing algorithm.

- Experimental results show that our method can satisfy all the current-flow, current-density, and symmetry constraints, and achieve smaller chip area, shorter routed wirelength, and fewer vias than the previous work. The simulation results also show that our approach can achieve higher circuit performance than the previous work.

The remainder of this paper is organized as follows. Section 2 first reviews the existing hierarchical B*-tree representation and then formulates the analog layout synthesis problem with current-flow and current-density constraints. Section 3 introduces our proposed enhanced B*-tree representation. Section 4 presents our simultaneous analog placement and routing algorithm with current-flow and current-density considerations. Section 5 reports the experimental results. Finally, Section 6 concludes this paper.

2. PRELIMINARIES

In this section, we first review the hierarchical B*-tree [13] in Section 2.1, and then formally formulated the analog layout synthesis problem in Section 2.2.

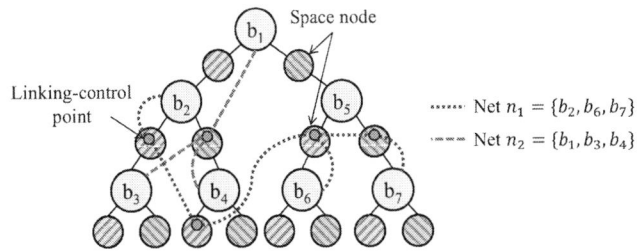

Figure 2: An example of the enhanced B*-tree representation.

2.1 Review of the Hierarchical B*-tree

The hierarchical B*-tree is an extension of B*-trees [4] that improving the ability of B*-trees to handle the symmetry constraints in analog placements. The hierarchical B*-tree is designed to represent a cluster of modules such as *symmetry islands*, *proximity groups*, or *pre-designed/existing circuits*. The function of a symmetry island and a proximity group is to connect the modules which are of the same symmetry group to each other to reduce the sensitivity to thermal gradients, process variations, or parasitic effects. The hierarchical B*-tree also inherits the benefits from the B*tree, which requires only linear line to pack modules. The following subsection reviews the hierarchical nodes for symmetry islands and proximity groups.

2.1.1 Review of the Hierarchical Nodes for Symmetry Islands and Proximity Groups

To handle symmetry constraints, the hierarchical B*-tree introduces *hierarchical nodes* and uses them in a B*-tree to model an *automatically symmetric-feasible B*-tree* (ASF-B*-tree for short) for symmetry groups. Each ASF-B*-tree consists of sets of *symmetry modules* which are symmetrical with respect to the common symmetry axis or *self-symmetry modules* whose center lines are the symmetry axis. Since the symmetry constraints cannot be violated in a symmetry group, an ASF-B*-tree records only a half of the nodes in a symmetry group, and the relative positions of the remaining nodes can be calculated directly by mirroring the nodes in ASF-B*-tree. For the self-symmetry modules, their corresponding nodes in the ASF-B*-tree should always be on the right-most branch to avoid separating the two halves of the nodes from the symmetry axis.

2.2 Problem Formulation

We formally define the analog layout synthesis problem as follows:

- **The Analog Layout Synthesis Problem:** Given a set of modules, a netlist, a set of symmetry groups, a set of current-flow constraints, a set of current-density constraints, and the design rules, place all the modules and route all the nets to optimize the chip area, routed wirelength, bend number, via count, and coupling noise such that no design rule is violated and all the current-flow and current-density constraints are satisfied.

3. THE ENHANCED B*-TREE REPRESENTATION

To simultaneously consider current-flow and current-density constraints in analog layout synthesis, we propose a novel *enhanced B*-tree representation* to facilitate the simultaneous placement and routing while satisfying the current-flow and current-density constraints. Simultaneous placement and routing is not a trivial problem because the traditional analog placement representations might not be able to model modules and interconnects in one representation simultaneously. As a result, the traditional analog layout synthesis algorithms use the half-perimeter wirelength (HPWL) to estimate wirelength and routing congestions at the placement stage, and then apply a maze- or ILP-based router to connect all the interconnects while satisfying routing constraints. Although the feedback mechanism is applied from the router to the placer, it is still difficult for the placer to fulfill all the constraints from the feedbak of the router since the interconnects usually cannot cross the active regions of devices. To remedy the deficiencies in traditional analog placement representations, the enhanced B*-tree representation is proposed to simultaneously model the modules and

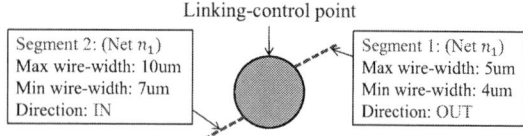

Figure 3: An example of a linking-control point. The information of each connected wire segment can be tracked in a linking-control point.

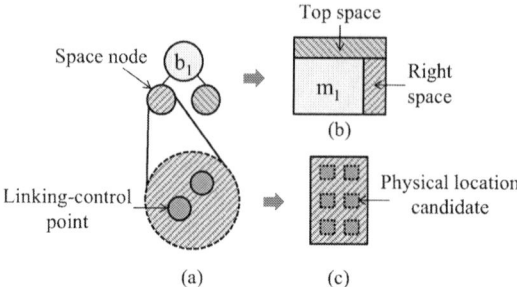

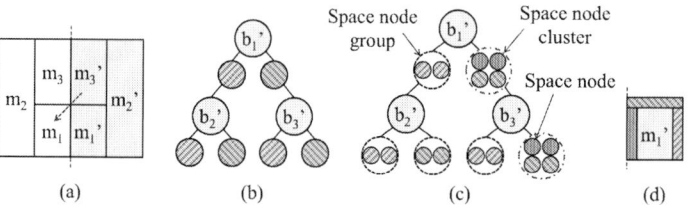

Figure 4: (a) The tree node b_1 has two space nodes on its left and right children, and each space node might contain several linking-control points. (b) The right and top space beside the module m_1. (c) The corresponding physical location candidates on the right space for the linking-control points.

Figure 5: (a) A symmetry island with one interconnect. The corresponding symmetry net is ignored in this example. (b) The corresponding ASF-B*-tree with space nodes for the left symmetry island. (c) An enhanced ASF-B*-tree with both space node group and space node cluster for the left symmetry island. (d) The module m_1' with the corresponding space node cluster.

interconnects in one topological representation while facilitating simultaneous placement and routing and satisfying the current-flow and current-density constraints at the same time.

To facilitate the modelling of interconnects and the allocation of routing resources between modules, *linking-control points* and *space nodes* are introduced in our enhanced B*-tree representation. Figure 2 shows an example of the enhanced B*-tree representation. The linking-control points are used to link the interconnects and control the routing topologies between modules, and space nodes are used to reserve and allocate routing resources for interconnects. The linking-control points appear in the middle of any interconnects and can only be placed inside the space node, and the space nodes appear on the left and right children of each tree node b_i. The space node and the linking-control point will be detailed in the following subsections.

3.1 Linking-Control Point

To control routing topologies and reserve routing resources between modules in a topological representation, the linking-control point is introduced in our enhanced B*-tree representation. The concept of the linking-control point in the enhanced B*-tree is similar to the *Steiner point* in the geometric routing space, which means that the routing topologies in an enhanced B*-tree can be changed by moving or inserting linking-control points to different space nodes to achieve better placement and routing results. A linking-control point can only connect the wire segments belonging to the same interconnect, and each linking-control point has at least two wire segments. When a linking-control point connects only two wire segments, the linking-control point can be regarded as a *physical bending point*, which can increase the degree of freedom for interconnects. Figure 3 shows an example of a linking-control point. The linking-control point also contains the information of each connected wire segment, such as the minimum/maximum wire width for current-density constraints and the current direction for current-flow constraints. This information provides an important basis for resource allocation, constraint satisfaction, and circuit optimization in the enhanced B*-tree representation.

3.2 Space Node

To reduce routing congestion and satisfy the current-flow and current-density constraints, routing resource reservation and allocation play an important role in the simultaneous placement and routing. Since the interconnects usually cannot cross the active regions of devices, the routing resource is very limited in an analog circuit. Thus, we introduce the *space node* for each tree node of the enhanced B*-tree to reserve and allocate routing resource for interconnects efficiently.

Figure 4(a) illustrates the corresponding space nodes of the tree node b_1. For each tree node in the enhanced B*-tree, there are two space nodes on the left child (left space node for short) and the right child (right space node for short). The left and right space nodes correspond to the right and top space besides the module (Figure 4(b)), respectively. Different from the previous work [12], our space node employs linking-control points to provide a substantial basis for resource reservation and allocation, which can effectively reflect the required routing resource to route the interconnects with variable wire widths and current directions, and guide the enhanced B*-tree to a better layout result.

In addition, each right or top space contains several *physical location candidates* for a linking-control point, (inside the space node) to represent its *physical location* in an enhanced B*-tree as shown in Figure 4(c). The physical location candidates can be defined according to the number of linking-control points, the required wire width, and pin locations, or can be accurately defined by analog designers for some specific patterns. With linking-control points and space nodes, the

analog layout synthesis problem can be easily solved by our enhanced B*-tree representation effectively and efficiently. The routing resource allocation and the physical location determination will be introduced in Section 4.2 and Section 4.3, respectively.

3.3 Symmetry Constraint Handling

As mentioned in Section 2.1.1, the symmetry modules can be modelled by an ASF-B*-tree to satisfy the symmetry constraints; however, the ASF-B*-tree records only a half of the nodes in a symmetry group. As a result, the space node proposed in Section 3.2 cannot be applied in the ASF-B*-tree directly since the corresponding left/top space of the other half nodes cannot be represented in the ASF-B*-tree. Figures 5(a) and (b) show an example of a symmetry-island and the corresponding ASF-B*-tree. For the interconnect in Figure 5(a), the linking-control points of that interconnect cannot explore the spaces located on the top and left of the module m_1 in the ASF-B*-tree (Figure 5(b)).

To remedy the deficiencies, we introduce *space node groups* and *space node clusters* to improve the ASF-B*-tree. Figure 5(c) shows an example of the enhanced ASF-B*-tree. The space node group in the enhanced ASF-B*-tree is formed by two space nodes to represent the right (left) and top spaces for the symmetry modules in both sides. The space node cluster has four space nodes to represent not only the right (left) and top spaces for the symmetry modules in both sides but also the space between modules, which abut on a symmetry axis (Figure 5(d)). According to the properties of the ASF-B*-tree, the space node cluster exists only on the right-most branch for the modules abutting on a symmetry axis.

4. SIMULTANEOUS PLACEMENT AND ROUTING

In this section, we propose a simultaneous placement and routing algorithm based on the enhanced B*-tree representation to si-

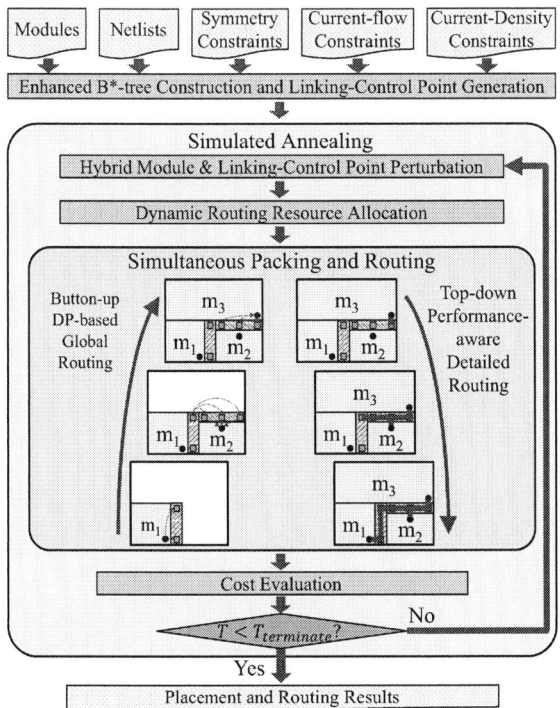

Figure 6: The simultaneous placement and routing algorithm.

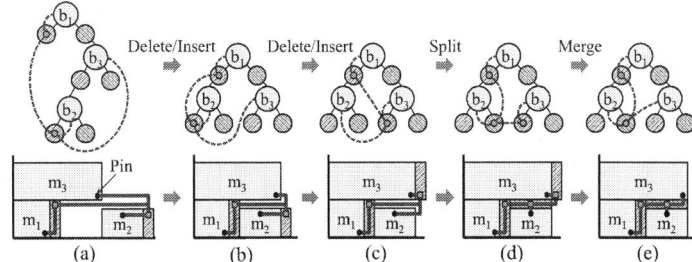

Figure 7: The enhanced B*-tree with the corresponding layout. (a) Before deleting/inserting b_2. (b) After deleting/inserting b_2. (c) After deleting/inserting the linking-control point. (d) After splitting the linking-control point. (e) After merging the linking-control points.

- *Swap*. Exchange the positions of two linking-control points.
- *Split*. Split one linking-control point into two linking-control points. The wire segments connecting to the original linking-control point will also be distributed to the two linking-control points. Figures 7(c) and (d) show an example of the change before and after the split operation.
- *Merge*. Combine two linking-control points into a new linking-control point. The wire segments belonging to the original two linking-control points will also be combined and connected to the new linking-control point. Figures 7(d) and (e) show an example of the merge operation.

4.2 Dynamic Routing Resource Reservation and Allocation

To acquire suitable routing resource and compensate the routing resource blocked by existing circuits or active regions, we propose a dynamic routing resource reservation and allocation technique to reserve and allocate appropriate routing spaces and locations according to space nodes and linking-control points after the hybrid perturbations. As mentioned in Section 3.1, the required maximum wire widths and current directions by each linking-control point can be calculated by the information of each wire segment recorded in the linking-control point. Therefore, the required total routing space by a space node can be dynamically reserved and allocated after the hybrid perturbations. Assume that R_i is the routing space required by the space node i, L_j is the maximum wire width required by the linking-control point j, and K_j is the number of wire segments connecting to the linking-control point j. For each space node i, the required routing space R_i can be defined as follows:

$$R_i = \sum_{\forall j \in i} \left(\frac{L_j \times K_j}{2} \right). \qquad (1)$$

In addition to the summation of the maximum wire width required by each linking-control point, the coupling noise induced by congested wire segments should also be considered. For this reason, we add the number of connected wire segments in Equation (1) to allocate more routing space to a space node. Since each linking-control point connects at least two wire segments, we divide the number of connected wire segments by two to avoid overestimation.

4.3 Simultaneous Packing and Routing

To place and route analog circuits with the enhanced B*-tree representation simultaneously, we propose a simultaneous packing and routing technique to generate placement and routing results at the same time. Different from the previous ordered analog layout synthesis algorithms (placement followed by routing), our enhanced B*-tree with the proposed simultaneous packing and routing technique requires only one stage to simultaneously place and route an analog circuit while optimizing its chip area, routed wirelength, bend number, via count, and coupling noise.

The simultaneous packing and routing technique consists of two phases, the bottom-up dynamic-programming-based (DP-based for

multaneously consider current-flow and current-density constraints. Figure 6 shows our proposed simultaneous placement and routing flow. After acquiring the input modules, netlist, and current-flow and current-density constraints, we first construct an enhanced B*-tree and the corresponding linking-control points. The symmetry groups, pre-designed/existing circuits, and user-specified design hierarchies are modelled as corresponding enhanced ASF-B*-trees based on the statistics of symmetry and design hierarchy constraints. This step is followed by the simulated annealing scheme to seek for a desired layout. To facilitate the simultaneous placement and routing at the tree perturbing stage, hybrid perturbations are also proposed. Then, dynamic routing resource allocation is performed to allocate suitable routing space and locations for each interconnect according to the enhanced B*-tree. After the dynamic routing resource allocation, simultaneous packing and routing is presented with a bottom-up dynamic-programming-based global routing and a top-down performance-aware detailed routing to obtain the best physical locations of modules and linking-control points while optimizing routing results and satisfying the current-flow and current-density constraints. In addition to the chip area, our simultaneous placement and routing algorithm also optimized the routed wirelength, bend number, via count, and coupling noise at the same time.

4.1 Hybrid Perturbations

To perturb both modules and interconnects in the enhanced B*-tree, the hybrid perturbations for both modules and linking-control points are proposed. For the original three perturbations of the B*-tree, *delete and insert* (delete a node and insert it to a different position), *swap* (exchange the positions of two nodes), and *rotate* (rotate a node by 90 degrees clockwise), the space nodes and the linking control points, which inside the space nodes, are regarded as a part of the tree node, and must be moved with the tree node together. Figure 7(a) and (b) show an example of the delete and insert operation for the enhanced B*-tree.

In addition to the original three perturbations, we also propose four new moves for the linking-control points to perturb the routing topologies simultaneously as follows:

- *Delete and insert*. Delete a linking-control point and insert it to another position. Figures 7(b) and (c) show an example of the change before and after the delete and insert operation.

33

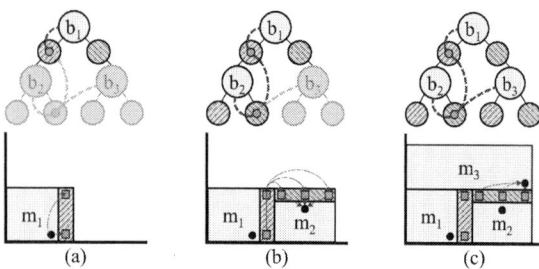

Figure 8: An example of the DP-based global routing. All the possible routing topologies and costs between two terminals are recorded by the dynamic programming as shown in the red-dotted arrows. (a) Packing the first module m_1. (b) Packing the second module m_2. (c) Packing the third module m_3.

short) global routing and the top-down performance-aware detailed routing. The bottom-up DP-based global routing is performed with an integrated packing procedure to place the modules while determining the best physical location for each linking-control point beside a module. After packing the modules, the performance-aware detailed routing is performed with the back trace of the dynamic programming to connect all the wire segments while minimizing the routed wirelength, bend number, via count, and coupling noise.

4.3.1 Dynamic-Programming-Based Global Routing

To reduce constraint violations caused by linking-control points, we propose a bottom-up dynamic-programming-based (DP-based) global routing method, which integrates the packing procedure from B*-trees, to place modules while globally exploring the best physical locations for linking-control points and satisfying the current-flow and current-density constraints.

Figure 8 illustrates a DP-based global routing procedure. Starting from the root node b_1, the coordinate of each module and its space node can be calculated by traversing a given enhanced B*-tree in a pre-order style. To determine the best physical location of a linking-control point, the dynamic programming is applied to construct a sub-problem which records all possible routing topologies with respect to all the physical location candidates of the linking-control point. As the red-dotted arrows shown in Figure 8, when the two terminals of a wire segment are connected by modules or linking-control points during packing, the dynamic programming will be activated to record all possible routing topologies and the corresponding costs between these two terminals. The current-flow constraints can be checked by examining the current directions recorded by the linking-control point in constant time during the DP-based global routing. Assume that C_{ij} is the cost to connect the wire segment i with a possible routing topology j, W_{ij} and B_{ij} are the estimated wirelength and bend numbers to connect the wire segment i with the routing topology j by A* search, and p is a penalty for unsuccessful routing or constraint violations, the cost C_{ij} can be defined as follows:

$$C_{ij} = \alpha W_{ij} + \beta B_{ij} + p. \tag{2}$$

If the A* search fails to find a feasible path or the path violates the current-flow constraint, a high penalty will be given to avoid the routing topology to be chosen by the dynamic programming.

4.3.2 Performance-Aware Detailed Routing

For better circuit performance, we propose a top-down performance-aware detailed routing after the bottom-up DP-based global routing to simultaneously trace back the sub-problems constructed by dynamic programming and detailed route the chosen wire segments. Since a routing topology might have branches at the linking-control points, we trace back the sub-problems from a leaf of an enhanced B*-tree, and follow the reverse packing order to avoid rearranging the objects. To honor the symmetry constraints and expert knowledge, the symmetry and critical nets will be traced back and routed first. During the back trace, the sub-problem which has the lowest cost and the minimum coupling effect and satisfies the current-flow and current-density constraints will be chosen and simultaneously routed.

To obtain a better layout by our enhanced B*-tree, we integrate the routing cost with the chip area to guide the simulated annealing scheme to explore more practical solutions. The total cost ϕ for the enhanced B*-tree can be defined as follows:

$$\phi = \alpha \frac{A}{\hat{A}} + \beta \frac{W}{\hat{W}} + \gamma \frac{B}{\hat{B}} + \delta \frac{V}{\hat{V}} + p, \tag{3}$$

where $\alpha, \beta, \gamma, \delta$ are user-specified weights, $A, W, B, V, \hat{A}, \hat{W}, \hat{B}, \hat{V}$ are the chip area, routed wirelength, bend number, via count, normalized chip area, normalized routed wirelength, normalized bend number, and normalized via count, respectively. p is a penalty for unsuccessful routing or constraint violations.

Table 1: Design statistics and constraint number of Comparator and Two-stage OP.

Circuits	Module	Symmetry	Current-Flow	Current-Density
Comparator	12	2	6	14
Two-stage OP	14	3	4	12

5. EXPERIMENTAL RESULTS

We implemented our simultaneous placement and routing algorithm in the C++ programming language, and implemented the algorithm in [22] in the C programming language. All experiments were performed on an Intel Xeon E5-2620 2.0GHz Linux workstation with 72GB memory. To evaluate our approach, we used two analog circuits, Comparator and Two-stage OP. These two circuits have been tape-out by the TSMC $90nm$ MSG process manually by expert analog designers. The statistics of the two circuits are shown in Table 1. We also applied the TSMC $90nm$ MSG library and used two layers to route the two circuits. After generating the layout automatically by our simultaneous placement and routing algorithm, we used Calibre nmDRC and Calibre nmLVS to check the design rules and verify the layout versus schematic.

Since there is no existing work simultaneously handling the current-flow and current-density constraints for simultaneous placement and routing, we compared our work with [22] for chip area, routed wirelength, and via count minimization. Since the current-density constraint and the routing problem are not considered in [22], we asked an expert analog designer to manually route the placement result generated by the work [22], and slightly moved the placed modules to allocate suitable area for routing without violating the current-flow/current-path and current-density constraints. As shown in Table 2, our analog layout synthesis algorithm can obtain an average chip area reduction of 7%, an average routed wirelength reduction of 12%, and an average via reduction of 8%. These results show that our enhanced B*-tree, which simultaneously models modules and interconnects, can facilitate the simultaneous placement and routing algorithm to further reduce the chip area, routed wirelength, and vias count while satisfying the current-flow and current-density constraints effectively. Since there are some current-density constraints with variable wire widths in both circuits, the modules in the placement result of [22] must be moved to allocate suitable routing area, resulting in a larger chip area and longer routed wirelength. The running time of our approach is also reported in Table 2. Note that we do not compare the running time with the work [22] since the work handles only the placement problem.

We also ran post-layout simulation to verify the quality of our automatic layout designs. We obtained the netlist with extracted parasitic RC by using Calibre xRC in the post-layout simulation, and simulated the netlist with extracted parasitic RC by using HSPICE after the RC extraction. The rise time delay, fall time delay, propagation delay, DC gain, unity gain bandwidth, and phase margin were also measured by HSPICE. Tables 3 and 4 show the simulation results of Comparator and Two-stage OP, respectively. In both results, our approach can satisfy all specified current-flow, current-density, and symmetry constraints. Since the work [22] handles only the current-flow/current-path and symmetry constraints at the placement stage, it is predictable that our approach simultaneously considering current-flow, current-density, and symmetry constraints results in much more

Table 2: Comparison of the layout results between the work [22] and our approach.

Circuits	[22]			Ours			
	Area (um^2)	Wirelength (um)	Via Number	Area (um^2)	Wirelength (um)	Via Number	Runtime (s)
Comparator	285.7	95.6	59	277.1	82.3	51	242.8
Two-stage OP	695.1	148.3	31	632.5	136.5	31	255.7
Comp.	1.07	1.12	1.08	1.00	1.00	1.00	

Table 3: Comparison of the rise time delay, fall time delay, average delay, and propagation delay between [22] and our approach.

Circuit	t_{dr} (ns)		t_{df} (ns)		t_{davg} (ns)		t_{plh} (ns)		t_{phl} (ns)	
	[22]	Ours	[22]	Ours	[22]	Ours	[22]	Ours	[22]	Ours
Comparator	1.23	1.21	1.85	1.82	1.54	1.52	0.68	0.67	0.97	0.96

Table 4: Comparison of the DC Gain, Unity Gain Bandwidth, and Phase Margin between [22] and our approach.

Circuits	DC Gain (dB)		Unity Gain Bandwidth (MHz)		Phase Margin (°)	
	[22]	Ours	[22]	Ours	[22]	Ours
Two-stage OP	40.28	40.3	17.6	17.62	53.54	53.48

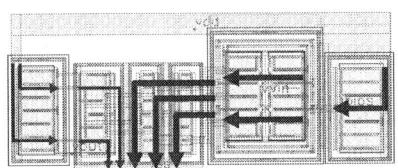

Figure 9: The layout of Comparator generated by our approach.

compact layouts, and further improves the current smoothness and reduces the parasitic impacts and IR drops on the interconnects. Moreover, our work can reduce the rise time delay, fall time delay, average delay, and propagation delay over the work [22] for Comparator. Besides, our work can also improve the DC gain and unity gain bandwidth over the work [22] for Two-stage OP. In addition, our phase margin is closest to the specification (45°). Figure 9 shows the resulting layout of Comparator generated by our approach.

6. CONCLUSIONS

In this paper, we have proposed a simultaneous placement and routing algorithm with current-flow and current-density constraints for analog circuit designs. To facilitate the simultaneous placement and routing, the enhanced B*-tree has also been proposed. Our algorithm not only minimizes the chip area but also reduces the routed wirelength, bend number, via count, and coupling noise at the same time. Experimental results have shown that our proposed algorithm is effective and efficient for analog layout synthesis.

7. ACKNOWLEDGMENTS

The authors would like to thank Mr. Jhao-Yan Liu of National Tsing Hua University for his valuable help with this paper.

8. REFERENCES

[1] T. Adler, and E. Barke, "Single Step Current Driven Routing of Multiterminal Signal Nets for Analog Applications," *Proc. of DATE*, Mar. 2000.

[2] T. Adler, H. Brocke, L. Hedrich, and F. Barke, "A Current Driven Routing and Verification Methodology for Analog Applications," *Proc. of DAC*, Jun. 2000.

[3] F. Balasa and K. Lampaert, "Symmetry Within the Sequence-Pair Representation in the Context of Placement for Analog Design," *IEEE TCAD*, vol. 19, no. 7, Jul. 2000.

[4] Y.-C. Chang, Y.-W. Chang, G.-M. Wu, and S.-W. Wu, "B*-Trees: A New Representation for Non-slicing Floorplan," *Proc. of DAC*, Jun. 2000.

[5] P.-Y. Chou, H.-C. Ou, and Y.-W. Chang, "Heterogeneous B*-trees for Analog Placement with Symmetry and Regularity Considerations," *Proc. of ICCAD*, Nov. 2011.

[6] J. M. Cohn, D. J. Garrod, R. A. Rutenbar, and L. R. Carley, "KOAN/ANAGRAM II: New Tools for Device-level Analog Placement and Routing," *IEEE JSSC*, vol. 26, pp. 330–342, Mar. 1991.

[7] J. M. Cohn, D. J. Garrod, R. A. Rutenbar, and L. R. Carley, "Techniques for Simultaneous Placement and Routing of Custom Analog Cells in KOAN/ANAGRAM II," *Proc. of ICCAD*, Nov. 1991.

[8] Y. Kubo, S. Nakatake, Y. Kajitani, and M. Kawakita, "Explicit Expression and Simultaneous Optimization of Placement and Routing for Analog IC Layouts," *Proc. of VLSID*, Jan. 2002.

[9] J. Lienig, and G. Jerke, "Current-driven Wire Planning for Electromigration Avoidance in Analog Circuits," *Proc. of ASP-DAC*, Jan. 2003.

[10] J. Lienig, G. Jerke, and T. Adler, "Electromigration Avoidance in Analog Circuits: Two Methodologies for Current-driven Routing," *Proc. of ASP-DAC*, Jan. 2002.

[11] C.-W. Lin, J.-M. Lin, C.-P. Huang, and S.-J. Chang, "Performance-driven Analog Placement Considering Boundary Constraint," *Proc. of DAC*, Jun. 2010.

[12] C.-W. Lin, C.-C. Lu, J.-M. Lin, and S.-J. Chang, "Routability-driven Placement Algorithm for Analog Integrated Circuits," *Proc. of ISPD*, Mar. 2012.

[13] P.-H. Lin, Y.-W. Chang, and S.-C. Lin, "Analog Placement Based on Symmetry-Island Formulation," *IEEE TCAD*, vol. 28, no. 6, pp. 791–804, Jun. 2009.

[14] M. P.-H. Lin, H. Zhang, M. D. F. Wong, and Y.-W. Chang, "Thermal-Driven Analog Placement Considering Device Matching," *IEEE TCAD*, vol. 30, no. 3, pp. 325–336, Mar. 2011.

[15] D. Long, X. Hong, and S. Dong, "Signal-Path Driven Partition and Placement for Analog Circuit," *Proc. of ASP-DAC*, Jan. 2006.

[16] Q. Ma, L. F. Xiao, Y.-C. Tam, and Evangeline F. Y. Young, "Simultaneous Handling of Symmetry, Common Centroid, and General Placement Constraints," *IEEE TCAD*, vol. 30, no. 1, pp. 85–95, Jan. 2011.

[17] E. Malavasi, E. Charbon, E. Felt, and A. Sangiovanni-Vincentelli, "Automation of IC Layout with Analog Constraints," *IEEE TCAD*, vol. 15, no. 8, pp. 923–942, Aug. 1996.

[18] H.-C. Ou, H.-C. Chang Chien, and Y.-W. Chang, "Non-Uniform Multilevel Analog Routing with Matching Constraints," *Proc. of DAC*, Jun. 2012.

[19] M. M. Ozdal and R. F. Hentschke, "Maze Routing Algorithms with Exact Matching Constraints for Analog and Mixed Signal Design," *Proc. of ICCAD*, Nov. 2012.

[20] J. A. Prieto, A. Rueda, J. M. Quintana and J. L. Huertas, "A Performance-Driven Placement Algorithm with Simultaneous Place&Route Optimization for Analog IC's," *Proc. of EDTC*, Mar. 1997.

[21] H.-F. Tsao, P.-Y. Chou, S.-L. Huang, Y.-W. Chang, Mark P.-H. Lin, D.-P. Chen, and D. Liu, "A Corner Stitching Compliant B*-tree Representation and Its Applications to Analog Placement," *Proc. of ICCAD*, Nov. 2011.

[22] P.-H. Wu, Mark P.-H. Lin, Y.-R. Chen, B.-S. Chou, T.-C. Chen, T.-Y. Ho, and B.-D. Liu, "Performance-driven Analog Placement Considering Monotonic Current Paths," *Proc. of ICCAD*, Nov. 2012.

[23] L. F. Xiao, Evangeline F. Y. Young, X.-Y. He, and K.P. Pun, "Practical Placement and Routing Techniques for Analog Circuit Designs," *Proc. of ICCAD*, Nov. 2010.

[24] L. Zhang, U. Kleine, and Y. Jiang, "An Automated Design Tool for Analog Layouts," *IEEE TVLSI*, vol 14, no. 8, pp. 881–894, Aug. 2006.

Coupling-Aware Length-Ratio-Matching Routing for Capacitor Arrays in Analog Integrated Circuits *

Kuan-Hsien Ho[1], Hung-Chih Ou[1], Yao-Wen Chang[1,2], and Hui-Fang Tsao[1]

[1]Graduate Institute of Electronics Engineering, National Taiwan University, Taipei 106, Taiwan
[2]Department of Electrical Engineering, National Taiwan University, Taipei 106, Taiwan
khho@eda.ee.ntu.edu.tw; howard@eda.ee.ntu.edu.tw; ywchang@cc.ee.ntu.edu.tw; sarah@eda.ee.ntu.edu.tw

ABSTRACT

Capacitance-ratio mismatch in a switched-capacitor circuit could significantly degrade circuit performance. In the nanometer era, the parasitic effects and lengths of interconnects both have significant impacts on the capacitance ratio. This paper presents the first routing work for the problem of coupling-aware length-ratio-matching routing for capacitor arrays in analog integrated circuits. The router adopts a two-stage approach of topology generation followed by detailed routing to route unit capacitors such that the coupling-aware wire length ratio can match the desired capacitance ratio. Given a length ratio, in particular, the length-ratio-matching routing problem can be handled by transforming the problem into an easier classical wirelength minimization one. Experimental results show that our algorithm can solve the addressed problem with substantially smaller costs.

Categories and Subject Descriptors

B.7.2 [**Integrated Circuits**]: Design Aids [Placement and Routing]

General Terms

Algorithms, Performance

Keywords

Physical Design, Routing, Analog ICs

1. INTRODUCTION

1.1 Interconnect Parasitic Effects

As the technology node advances into the nanometer era, the parasitic effects of interconnects have become more and more significant. The parasitic resistors and capacitors of interconnects gradually dominate the performance, signal integrity, and reliability of circuit designs [1]. Consequently, the parasitic effects of interconnects on a layout with multiple unit capacitors cannot be ignored because the effects might change the capacitance ratio. Since parasitic capacitance is highly correlated with wire length [13], if the length ratio of interconnects of each capacitor

---

*This work was partially supported by IBM, SpringSoft, TSMC, Academia Sinica, and NSC of Taiwan under Grant No's. NSC 99-2221-E-002-207-MY3, NSC 99-2221-E-002-210-MY3, NSC 100-2221-E-002-088-MY3, and NSC 101-2221-E-002-191-MY3.

Permission to make digital or hard copies of all or part of this work for personal or classroom use is granted without fee provided that copies are not made or distributed for profit or commercial advantage and that copies bear this notice and the full citation on the first page. To copy otherwise, to republish, to post on servers or to redistribute to lists, requires prior specific permission and/or a fee.
DAC'13, May 29 – June 07 2013, Austin, TX, USA.

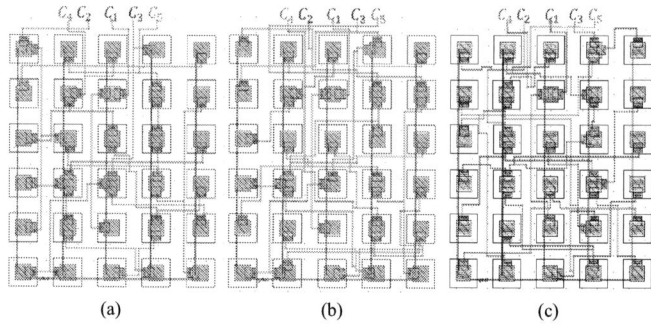

Figure 1: Example layouts generated by the TSMC $0.18\mu m$ process data. (a) Without length ratio matching. (b) With length ratio matching but no coupling considerations. (c) With both length ratio matching and coupling considerations.

can match the desired capacitance ratio, the impact of parasitic capacitance can be minimized.

We conducted an experiment to explore the influences of length-ratio mismatch and coupling on the resulting capacitance ratio. The studied case contains 30 unit capacitors of $20.28fF$ each, forming five capacitances with the capacitance ratio of $2:6:7:7:8$. We generated a layout by using the TSMC $0.18\mu m$ process data, as shown in Figure 1. Figure 1(a) shows the layout without considering length ratio matching, Figure 1(b) shows the layout with length ratio matching but no coupling considerations, and Figure 1(c) shows the layout with both length ratio matching and coupling considerations. Table 1 lists the length ratio of each layout. We assume that pads are on the top of the capacitor arrays. To evaluate the resulting capacitances, we simulated each design based on $C = I \cdot dt/dV$ using HSPICE, where I, V, C, and t denote current, voltage, capacitance, and time, respectively. Table 1 gives the resulting capacitance ratio through simulation and the deviation from the target ratio. From the results, the length and capacitance ratios are closer to the desired capacitance ratio $2:6:7:7:8$ if the length ratio matching is considered during the routing process (see Figures 1(b) and (c)). However, considering length ratio matching alone is not sufficient for modern analog circuit designs. From the simulation results of Figures 1(b) and (c) in Table 1, considering coupling effects during length ratio matching can further reduce the mismatch of the resulting capacitance ratio. In particular, the influence of parasitic capacitance is typically more significant for more advanced technology nodes.

1.2 Previous Work

Several analog routers have been proposed [2, 3, 8, 16] in the recent literature. For specific devices and interconnects, length-matching requirements are commonly imposed for analog and mixed-signal designs, where the so-called length matching refers to route two or more nets such that their wire lengths are equal. The length-matching problem has been widely studied especially for board-level routing. Ozdal *et al.* [11] proposed a general-purpose Lagrangian relaxation-based algorithm to handle the length-matching

Table 1: The simulation results of Figure 1.

		Figure 1(a)	Figure 1(b)	Figure 1(c)
	Target Cap. Ratio	2 : 6 : 7 : 7 : 8		
Wire Length	Normalized Ratio	2.00 : 6.19 : 6.55 : 6.34 : 7.43	2.00 : 5.96 : 6.88 : 6.97 : 8.03	2.00 : 5.97 : 6.82 : 7.10 : 8.11
	Target Cap. Ratio Deviation (%)	0.00 : 3.17 : 6.43 : 9.43 : 7.13	0.00 : 0.67 : 1.71 : 0.43 : 0.38	0.00 : 0.50 : 2.57 : 1.43 : 1.38
	Accumulative Deviation (%)	26.16	3.19	5.88
Resulting Cap.	Normalized Ratio	2.00 : 5.94 : 6.88 : 6.76 : 7.71	2.00 : 5.93 : 6.85 : 6.83 : 7.79	2.00 : 5.94 : 6.85 : 6.89 : 7.80
	Target Cap. Ratio Deviation (%)	0.00 : 1.00 : 1.72 : 3.43 : 3.60	0.00 : 1.40 : 2.14 : 2.43 : 2.63	0.00 : 1.00 : 2.14 : 1.57 : 2.50
	Accumulative Deviation (%)	9.75	8.60	7.21

problem. They also presented length-matching algorithms [9, 10] for the cases where all pins are aligned with each other, but the algorithms cannot be easily generalized to solve general problems. Later, Yan *et al.* [15] presented bounded sliceline grid (BSG)-based length-matching algorithms for single-layer routing. For IC-level routing, Lin *et al.* [4] proposed a matching-based placement and routing system for analog designs. Recently, Ozdal *et al.* [12] further presented a mathematical formulation that exactly models the routing matching problem.

All the aforementioned routers do not consider length ratios and may not easily be extended to handle the length-ratio-matching problem well. A possible extension of these methods, e.g., construction by correction, might route each net by constructing an obstacle-avoiding rectilinear Steiner minimal tree, followed by snaking routes to satisfy the capacitance ratio. However, it is challenging for this method to guarantee high routability with limited silicon area. Therefore, it is desirable to develop an effective routing algorithm for length-ratio consideration.

1.3 Our Contributions

To minimize the influence of length-ratio mismatch on the resulting capacitance ratio, we propose a *coupling-aware length-ratio-matching routing algorithm* to route unit capacitors such that their coupling-aware length ratio can match the desired capacitance ratio, and the capacitance-ratio mismatch can be minimized. Our major contributions are summarized as follows:

- This paper presents the first work to address the problem of length-ratio-matching routing in capacitor arrays.

- Given a length ratio, the length-ratio-matching routing problem is handled by transforming the problem into an easier classical wirelength minimization one. For the transformation, a feasible length interval is first determined, and the minimum feasible length is set as the desired length for each capacitor.

- Our simulation shows that if the coupling-aware length ratio is closer to the desired capacitance ratio, so is the resulting capacitance ratio to the desired one.

- Experimental results show that our algorithm leads to averagely a 2.00× cost reduction for test cases, compared with a reasonable heuristic. In particular, only our algorithm can solve the length-ratio-matching routing problem well for all cases.

The rest of this paper is organized as follows. Section 2 formulates the length-ratio-matching routing problem. Section 3 outlines the proposed routing algorithm. Section 4 details the routing topology generation. Section 5 presents the detailed routing. Section 6 evaluates the proposed algorithm. Finally, Section 7 concludes this paper.

2. PROBLEM FORMULATION

Consider a circuit consisting of m capacitors. Let $\mathcal{C} = \{C_1, C_2, ..., C_m\}$ be the set of m capacitors, and k_i be the number of unit capacitors in $C_i \in \mathcal{C}$ for $1 \leq i \leq m$. Thus, the circuit has the capacitance ratio, $C_1 : C_2 : ... : C_m = k_1 : k_2 : ... : k_m$. Without loss of the

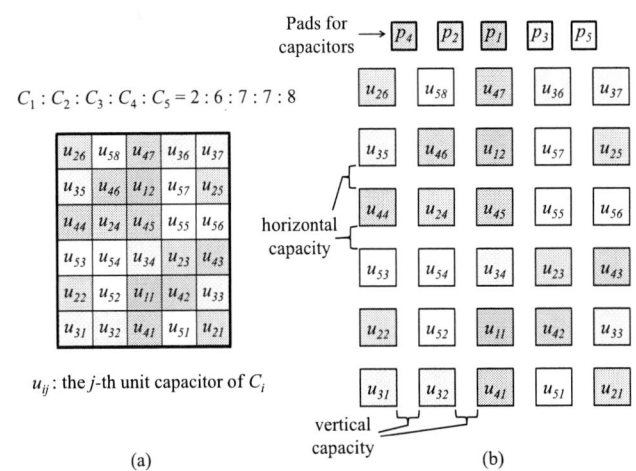

Figure 2: (a) An example capacitor placement from the previous work [6]. The number of capacitors is 5, and the capacitance ratio is $2 : 6 : 7 : 7 : 8$. (b) The capacitor placement of (a) after assigning the horizontal and vertical capacities and the location of the pad for each capacitor.

generality, we assume $k_i \leq k_j$ if $i \leq j$, where $1 \leq i, j \leq m$. Let u_{ij} be the j-th unit capacitor of $C_i \in \mathcal{C}$ for $1 \leq j \leq k_i$. Since each capacitor has a pad, we define p_i as the pad of $C_i \in \mathcal{C}$. A *capacitor net* n_i of $C_i \in \mathcal{C}$ is an interconnect that connects all of the unit capacitors of C_i and the pad of C_i. A *capacitor placement* $P = \{(x_{ij}, y_{ij}) | 1 \leq i \leq m, 1 \leq j \leq k_i\}$ is an assignment of each unit capacitor in C_i with the coordinate of u_{ij}'s bottom-left corner being (x_{ij}, y_{ij}). For two adjacent unit capacitors on the vertical (horizontal) direction, we define the *horizontal (vertical) capacity* as the number of tracks between them. Figure 2(a) shows an example capacitor placement from [6]. In this example, the number of capacitors is 5, and the capacitance ratio is $2 : 6 : 7 : 7 : 8$. Figure 2(b) illustrates the horizontal capacity, vertical capacity, and capacitor pads in this placement.

Our objective is to route all the capacitor nets such that the wire length ratio can match the desired capacitance ratio. For two capacitor nets n_i and n_j ($j > i$), ideally the length ratio equals the desired capacitance ratio:

$$L(n_i) : L(n_j) = k_i : k_j \Rightarrow \frac{L(n_i)}{k_i} = \frac{L(n_j)}{k_j} \Rightarrow L_u(n_i) = L_u(n_j),$$

where the function $L(n_i)$ and $L_u(n_i)$ denote the respective length and unit length of the capacitor net n_i (i.e., $L_u(n_i) = L(n_i)/k_i$). Therefore, if the unit lengths of all capacitor nets are the same, the length ratio exactly matches the desired capacitance ratio. As a result, we can define the *length-ratio-matching cost function* $\Phi(m)$ by the difference of unit lengths between each pair of capacitor nets as follows:

$$\Phi(m) = \sum_{\substack{i=1 \\ j>i}}^{m-1} |L_u(n_j) - L_u(n_i)|. \qquad (1)$$

With the cost definition, we formally define the addressed prob-

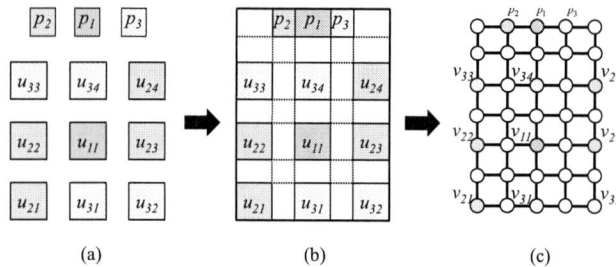

Figure 3: (a) An example capacitor placement after assigning horizontal and vertical capacities and the location of the pad for each capacitor. (b) Two kinds of tiles: colored tiles are terminal tiles and others are space tiles. (c) Routing graph modeling, where each tile is modeled as a vertex and each boundary between two adjacent tiles as an edge.

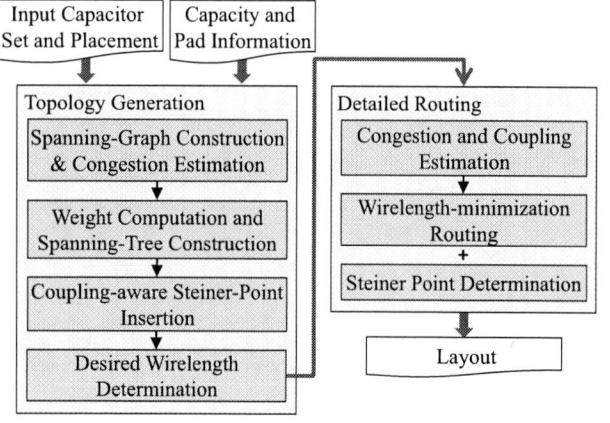

Figure 4: Our algorithm flow.

lem as follows:

- **Length-Ratio-Matching Routing Problem:** Given a capacitor set and a capacitor placement, route all of the capacitor nets such that the length-ratio-matching cost is minimized under the 100% routability guarantee and no route or via intersects any unit capacitors.

3. ROUTING MODEL AND ALGORITHM OVERVIEW

We apply a graph-search technique for routing. To do so, we first need to model the routing resource as a graph where the graph topology can represent the whole structure. For the modeling, we first partition a routing region into an array of rectangular *tiles*, each of which may accommodate assigned routing tracks in each dimension as shown in Figures 3(a) and (b). There are two kinds of tiles: *terminal tile* and *space tile*. In Figure 3(b), colored tiles are terminal tiles and others are space tiles. Then, we model a terminal tile for the unit capacitor u_{ij} as the *terminal vertex* v_{ij}, and a space tile as a *space vertex* in the routing graph, as illustrated in Figure 3(c). We further model the pad p_i as the vertex v_{p_i}, and each boundary between two adjacent tiles as an edge.

For the length-ratio-matching routing problem, we propose a two-stage algorithm: (1) topology generation, followed by (2) detailed routing. Figure 4 shows the overall flow. Topology generation determines the topologies of all capacitor nets, and each multi-pin net is decomposed into a set of 2-pin nets based on its topology. Given a length ratio, a feasible length interval is first determined, and the minimum feasible length is set as the desired length for each capacitor. Consequently, the length-ratio-matching routing problem is transformed into a classical wirelength mini-

mization problem with this process. Then, detailed routing for wirelength minimization is performed for each 2-pin net to match the coupling-aware desired length. We give more details for the two stages as follows.

The first stage, topology generation, consists of the following four steps: (1) spanning-graph construction and congestion estimation, (2) weight computation and spanning-tree construction, (3) coupling-aware Steiner-point insertion, and (4) desired wire length determination. We first construct a spanning graph for each capacitor net and estimate the congestion on the boundaries of the tiles. We then assign weights to the edges of each spanning graph and construct a spanning tree from each graph. Then, we insert Steiner points into the spanning tree to construct a flexible Steiner topology with minimized coupling effects and compute the feasible length interval for each capacitor net. With the feasible length intervals, we assign a desired length to each capacitor net and construct a fixed-length Steiner-tree topology.

The second stage, detailed routing, contains the following two steps: (1) congestion and coupling estimation, and (2) wirelength-minimization routing and Steiner-point determination. Based on the topologies generated from the first stage, we first decompose each resulting tree into a set of 2-pin nets. Then we simultaneously determine the final positions of the Steiner points to match the desired lengths and minimize routing wirelength and coupling noise.

4. TOPOLOGY GENERATION

This section details the four steps of topology generation: (1) spanning-graph construction and congestion estimation (Section 4.1), (2) weight computation and spanning-tree construction (Section 4.2), (3) coupling-aware Steiner-point insertion (Section 4.3), and (4) desired length determination (Section 4.4).

4.1 Spanning-Graph Construction and Congestion Estimation

In this step, we apply the approach proposed by [5] to construct a spanning graph for each capacitor net, where each spanning graph is an undirected connected graph on the vertex set of the corresponding capacitor net

After constructing a spanning graph, potentially congested area on the boundaries of the routing tiles can be estimated based on the spanning graph of each capacitor net. One of the most popular models for congestion estimation in digital circuit designs is the probabilistic congestion model [7, 14], which efficiently estimates routing congestion of a 2-pin net based on L-and Z-shaped pattern routes and has a good efficiency and accuracy trade-off. However, we do not use the probabilistic congestion model because the size of the routing tiles in an analog circuit design is typically much smaller than that in a digital circuit design. Instead, we propose a *propagated probability model*, which propagates probabilities tile-by-tile for a high accuracy, and use the proposed model to estimate the congestion for each edge of the spanning graph.

Our propagated probability model propagates the probability to *target-directed adjacent tiles* for each tile from the source tile t_s to a target tile t_t. A target-directed adjacent tile of a tile t_c is defined as an adjacent tile closer to t_t than t_c. Let $t[c_b, r_b]$ be the tile which locates at $[c_b, r_b]$ in the bounding box of t_s and t_t, where c_b (r_b) denotes the column (row) index. Without loss of generality, we assume that the route is from the bottom tile (t_s) to the top (t_t). (For simpler presentation, we further assume the source tile to be $t[1, 1]$, unless stated otherwise.) Let q_c be the probability of the current tile t_c, and T_{t_c} is the set of target-directed adjacent tiles of t_c. The probability that t_c propagates to its T_{t_c} equals $q_c/|T_{t_c}|$, which is defined as the estimated congestion of the boundary between each tile in T_{t_c} and t_c. For example, as shown in Figure 5, we assign probability 1 to the source tile and then propagate the probability to its target-directed adjacent

38

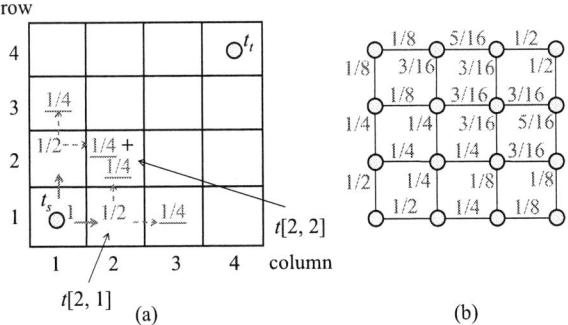

(a)

(b)

Figure 5: (a) The solid (dashed) arrow and numbers without (with) underlines shows the first (second) round of propagation. (b) The final congestion map in the routing graph.

tiles. In Figure 5(a), the solid (dashed) arrow and numbers without (with) underlines show the first (second) round of propagation. In the first round, $t[1,1]$ has two target-directed adjacent tiles, $t[1,2]$ and $t[2,1]$, to propagate, so the propagated probability is the probability of $t[1,1]$ divided by 2. Note that in the second round, the probability of $t[2,2]$ is the sum of the probabilities propagated from $t[1,2]$ and $t[2,1]$. Figure 5(b) shows the final congestion map in the routing graph, where the edge cost denotes the estimated congestion.

4.2 Weight Computation and Spanning-Tree Construction

In this step, we assign weights to all the edges in each spanning graph and then construct a spanning tree on the graph. The edge weight w_e of each edge depends on (1) wirelength weight w_{length} and (2) congestion weight w_{cong}, where the wirelength weight is the Manhattan distance of the two edge pins, and the congestion weight is the minimum value of the maximum congestions among all possible monotonic paths between t_s and t_t. The congestion weight of an edge can be computed by dynamic programming as follows. (Note that we assume the source tile to be $t[1,1]$ for easier presentation here.)

$$w_{cong}[c_b, r_b] =$$
$$\begin{cases} 0 & \text{if } c_b = 1, r_b = 1, \\ \max(w_{cong}[c_b, r_b - 1], b_V[c_b, r_b]) & \text{if } c_b = 1, r_b > 1, \\ \max(w_{cong}[c_b - 1, r_b], b_H[c_b, r_b]) & \text{if } c_b > 1, r_b = 1, \\ \min(\max(w_{cong}[c_b, r_b - 1], b_V[c_b, r_b]), \\ \quad \max(w_{cong}[c_b - 1, r_b], b_H[c_b, r_b])) & \text{if } c_b > 1, r_b > 1, \end{cases} \quad (2)$$

where $w_{cong}[c_b, r_b]$ is the minimum of the maximum congestions among all possible monotonic paths from the source tile $t[1,1]$ to $t[c_b, r_b]$, and $b_V[c_b, r_b]$ ($b_H[c_b, r_b]$) is the estimated congestion on the boundary between $t[c_b, r_b - 1]$ and $t[c_b, r_b]$ (between $t[c_b - 1, r_b]$ and $t[c_b, r_b]$). Finally, w_{cong} of the edge equals $w_{cong}[c_{t_t}, r_{t_t}]$, where c_{t_t} (r_{t_t}) denotes the column (row) index of the target tile t_t in the bounding box. With the wirelength weight and congestion weight, we define w_e as follows:

$$w_e = \begin{cases} w_{length} & \text{if } w_{cong} \le N_{track}, \\ w_{length} \times \dfrac{w_{cong}}{N_{track}} & \text{otherwise}, \end{cases}$$

where N_{track} is the minimum number of tracks between two adjacent unit capacitors in the vertical or the horizontal direction. From this definition, if the congestion is small enough, we consider only the wirelength as the edge weight; otherwise, we consider both wirelength and congestion together. After the edge weights are assigned, a minimum spanning tree can be constructed. To avoid congested areas during the construction of a minimum spanning tree, the congestion map in the routing graph needs to be updated once an edge is selected.

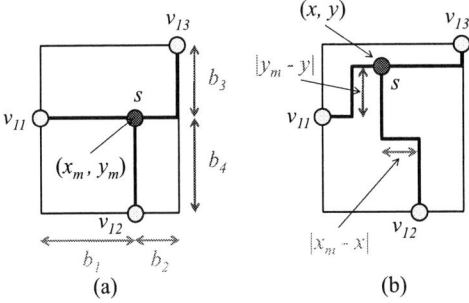

(a)

(b)

Figure 6: (a) A Steiner point s is inserted at the median point (x_m, y_m) of pins v_{11}, v_{12}, and v_{13}. (b) A Steiner point s is inserted at an arbitrary point (x, y) of pins v_{11}, v_{12}, and v_{13}.

4.3 Coupling-aware Steiner-Point Insertion

For Steiner-point insertion, we first explore some related properties as follows:

PROPERTY 1. *For a 3-pin net with the three pins not on a horizontal or vertical line, its minimum wirelength l_{min} can be obtained by inserting a Steiner point at the median point (x_m, y_m) of the three pins.*

PROPERTY 2. *Given an arbitrary Steiner point (x, y) inside the bounding box of three pins, the wirelength of the 3-pin net is $l_{min} + |x - x_m| + |y - y_m|$.*

From Properties 1 and 2, we can obtain a lower and an upper bounds of a 3-pin net wirelength by moving its Steiner point in the bounding box of the three pins. The lower bound is the wirelength directly connecting the three pins and the Steiner point at (x_m, y_m), while the upper bound is the wirelength directly connecting the pins and a Steiner point at any location inside the bounding box of the pins. Figure 6 shows an example of Steiner-point insertion, where Figures 6(a) and (b) are associated to the cases described in Properties 1 and 2, respectively. As can be seen, the lower and upper bounds of the 3-pin net wirelength are $l_{min} = (b_1 + b_2 + b_3 + b_4)$ and $l_{min} + max(b_1, b_2) + max(b_3, b_4)$, respectively. Note that we will not insert any Steiner point in a 3-pin net if the three pins are on a horizontal or vertical line because the lower and upper bounds of the 3-pin net wirelength are the same. For easier presentation, we define the *flexibility range* of a Steiner point s as the difference of the upper and lower bounds of a 3-pin net wirelength by inserting s, and define the *moving box* of s as the bounding box of the three pins.

To insert Steiner points to the proper locations, we first traverse a minimum spanning tree in a preorder sequence. Whenever a vertex is visited, we check if the vertex has a parent and a grandparent, which is the first and second ancestor of the vertex, respectively. If not, the traversal continues. Otherwise, we check if the three vertices are on a horizontal or a vertical line. If so, the traversal continues. If not, we add a Steiner point at the median of the coordinates among the three vertices in the spanning tree. Based on the aforementioned properties, the routing wirelength of the three vertices and the Steiner point is minimized. During the insertion, the tree is updated as follows. The visited vertex and its parent both become the children of the Steiner point, while the grandparent becomes the parent of the Steiner point.

However, if the moving boxes of any two Steiner points overlap, the aforementioned properties could be violated and the coupling effects need to be considered. To remove the overlaps between the moving boxes of any two Steiner points and thus reduce the coupling effects during the insertion, we propose a box-shrinking technique to relocate Steiner points and thus ensure the properties to be satisfied. Given the flexibility ranges and moving boxes of

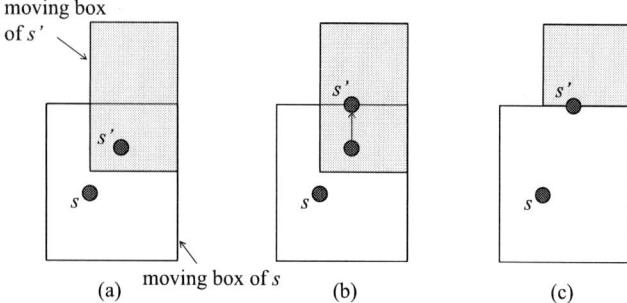

Figure 7: (a) The moving boxes of two Steiner points s and s' overlap. (b) s' is moving toward the nearest boundary of the moving box of s. (c) The moving box of s' is shrunk as the move of s'.

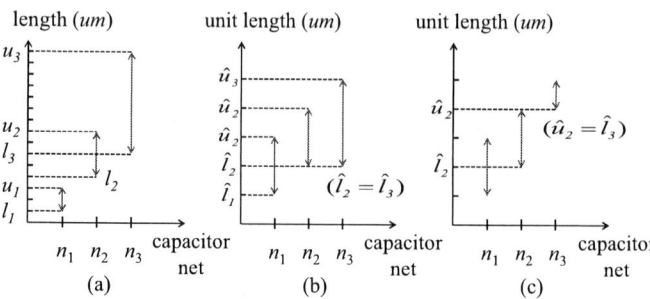

Figure 8: (a) The length intervals of a circuit with three capacitor nets. (b) The unit length intervals of the three capacitor nets of (a). This is an ideal case that a unit length, say $\hat{l}_2$, can be assigned to all capacitor nets. (c) A non-ideal case derived from (b), where at least two unit lengths, say $\hat{l}_2$ and $\hat{u}_2$, shall be selected.

two Steiner points, we first define the *shrinking cost* to relocate any Steiner point as the difference of the product of (1) the sum of the flexibility ranges and (2) the total area of the two moving boxes, before and after moving a Steiner point. Since a larger flexibility range and a larger moving box of a Steiner point imply a larger adjustable wirelength range for matching the length ratio in the remaining steps, it is desirable to minimize the shrinking cost to relocate any Steiner point. We check if there are overlapped moving boxes right after inserting a Steiner point. If the moving boxes of two Steiner points overlap, we move the Steiner point (and shrink its moving box accordingly) to the nearest boundary of the moving box of the other Steiner point that can minimize the shrinking cost the most. Figure 7 shows an example of moving a Steiner point s' and shrinking the moving box of s' accordingly.

As a result, we can obtain a length interval $[l_i, u_i]$ for each capacitor net n_i, where l_i (u_i) denotes the lower (upper) bound wirelength of n_i.

4.4 Desired Wirelength Determination

After obtaining the length intervals of all capacitor nets, our objective is to determine the *desired wirelength* d_i ($l_i \leq d_i \leq u_i$) for each capacitor net n_i to minimize Equation (1). According to Equation (1), the similarity among the unit lengths of all capacitor nets measures how good the length ratio matches the desired capacitance ratio. Therefore, we first normalize the length interval $[l_i, u_i]$ of each capacitor net n_i, to a *unit length interval* $[\hat{l}_i, \hat{u}_i]$, where we divide the length interval by k_i, and $\hat{l}_i = l_i/k_i$ and $\hat{u}_i = u_i/k_i$. The ideal case to minimize Equation (1) occurs when there exists a unit length within the unit length intervals of all capacitor nets. Then, d_i is set to k_i multiplied by the unit length. However, sometimes there does not exist a unit length within all the unit length intervals. In this case, to ensure $l_i \leq d_i \leq u_i$ for each capacitor net n_i, more than one unit length is selected to intersect all the unit length intervals. Then each capacitor net is assigned to one of the selected unit lengths to minimize Equation (1). Figure 8 illustrates the two cases. Note that the details of (1) selecting a set of unit lengths, and (2) assigning a selected unit length to each capacitor net, will be explained right after the illustration.

Figure 8(a) depicts the length intervals of a circuit with three capacitor nets, and Figure 8(b) depicts the unit length intervals of the three capacitor nets. As can been seen, Figure 8(b) is an idea case, where any one of the unit lengths within $[\hat{l}_2, \hat{u}_2]$ can be selected to minimize Equation (1). If $\hat{l}_2$ is selected, d_i is set to $\hat{l}_2 \times k_i$, for $1 \leq i \leq 3$. On the other hand, consider Figure 8(c), which is derived from Figure 8(b) by shortening the unit length interval of n_3. In this case, no unit length is within the three unit intervals. At least two unit lengths, say $\hat{l}_2$ and $\hat{u}_2$, must be selected such that $l_i \leq d_i \leq u_i$, for $1 \leq i \leq 3$. Suppose $\hat{l}_2$ is assigned to n_1, and $\hat{u}_2$ is assigned to n_2 and n_3. As a result, d_1 is set to $\hat{l}_2 \times k_1$,

and for $2 \leq i \leq 3$, d_i is set to $\hat{u}_2 \times k_i$.

Consider the problem of selecting a proper set of unit lengths to minimize Equation (1). Suppose there are m capacitor nets and there is a set of unit lengths $L_{sel} = \{\hat{l}_i, \hat{u}_i | 1 \leq i \leq m\}$ consisting of the endpoints of the unit length intervals of the m capacitor nets. Without loss of generality, suppose there is a solution with cost $\Phi(m)$ that some of the unit lengths in L_{sel} are assigned to the first $(m-1)$ capacitor nets, and a unit length not in L_{sel} is assigned to the m-th capacitor net. We observed that there must exist another solution with a lower cost $\Phi'(m)$, where the unit lengths assigned to the m capacitor nets are all in L_{sel}. Then, we have the following lemma.

LEMMA 1. $\Phi'(m) \leq \Phi(m)$.

Thus, L_{sel} is selected as the set of unit lengths to minimize Equation (1). We then have the following theorem:

THEOREM 1. *Given m capacitor nets and $L_{sel} = \{\hat{l}_i, \hat{u}_i | 1 \leq i \leq m\}$, Equation (1) can be minimized by the desired wirelength d_i of each capacitor net n_i, obtained by multiplying k_i and one of the unit lengths in L_{sel}.*

Based on the theorem, we can reduce the solution space while preserving the solution optimality. We apply the A*-search algorithm to determine the desired unit length from L_{sel}. The basic data structure is a search tree where each tree level is associated with a capacitor net, and each tree branch corresponds to a selection of the desired unit length candidates of the capacitor net. We define the cost function of a node h as the sum of (1) an accumulated cost: the total differences between determined unit lengths of capacitor nets from the root of the decision tree to h, (2) a predicted cost: the sum of minimum differences between each determined unit length and the non-determined unit length of each capacitor net from h to a leaf node of the tree, and (3) a predicted cost: the sum of minimum differences between non-determined unit lengths of the capacitor nets from node h to a leaf node of the tree.

With the unit length of each capacitor net n_i, the desired wirelength d_i can be obtained by multiplying the unit length by k_i. Then, we define the *expanding wirelength* e_i as $(d_i - l_i)$, which is the length that should be provided by moving the Steiner points in their corresponding bounding boxes. We distribute e_i of n_i to each Steiner point of n_i, where the necessary moving length M_s of each Steiner point s is set as $e_i \times (I_{max}/(u_i - l_i))$, where I_{max} is the maximum wirelength increment of moving the Steiner point located at the median inside the corresponding bounding box. I_{max} gives the flexibility of moving a Steiner point and can be computed based on the basic property of moving a Steiner point presented in Section 4.3.

With the necessary moving length of each Steiner point, we can find a candidate position for each Steiner point for the next stage,

Table 2: Statistics of the test cases modified from [6].

Test Case	#Tracks	#Cap.	Capacitance Ratio	#Unit Cap.	Matrix Size
SCF_1_4	4	5	2 : 6 : 7 : 7 : 8	30	6 × 5
SCF_1_5	5	5	2 : 6 : 7 : 7 : 8	30	6 × 5
SCF_1_6	6	5	2 : 6 : 7 : 7 : 8	30	6 × 5
SCF_1_7	7	5	2 : 6 : 7 : 7 : 8	30	6 × 5
SCF_1_8	8	5	2 : 6 : 7 : 7 : 8	30	6 × 5
SCF_2_4	4	5	1 : 2 : 2 : 10 : 17	32	8 × 4
SCF_2_5	5	5	1 : 2 : 2 : 10 : 17	32	8 × 4
SCF_2_6	6	5	1 : 2 : 2 : 10 : 17	32	8 × 4
SCF_2_7	7	5	1 : 2 : 2 : 10 : 17	32	8 × 4
SCF_2_8	8	5	1 : 2 : 2 : 10 : 17	32	8 × 4
SCF_3_4	4	4	1 : 2 : 16 : 45	64	8 × 8
SCF_3_5	5	4	1 : 2 : 16 : 45	64	8 × 8
SCF_3_6	6	4	1 : 2 : 16 : 45	64	8 × 8
SCF_3_7	7	4	1 : 2 : 16 : 45	64	8 × 8
SCF_3_8	8	4	1 : 2 : 16 : 45	64	8 × 8

Table 3: Comparisons on capacitor routing.

Test Case	PCM Method			Ours		
	Length Ratio		Time	Length Ratio		Time
	Tot. Cost	Avg. Cost	(sec)	Tot. Cost	Avg. Cost	(sec)
SCF_1_4	N/A	N/A	N/A	6.74	0.67	0.04
SCF_1_5	N/A	N/A	N/A	5.77	0.58	0.06
SCF_1_6	6.29	0.63	0.08	7.71	0.77	0.07
SCF_1_7	12.46	1.25	0.13	4.77	0.48	0.10
SCF_1_8	37.44	3.74	0.18	7.82	0.78	0.25
SCF_2_4	N/A	N/A	N/A	20.73	2.07	0.08
SCF_2_5	N/A	N/A	N/A	25.40	2.54	0.06
SCF_2_6	32.46	3.25	0.15	16.99	1.70	0.12
SCF_2_7	50.18	5.02	0.25	30.20	3.02	0.35
SCF_2_8	36.47	3.65	0.54	14.40	1.44	1.45
SCF_3_4	N/A	N/A	N/A	33.36	5.56	0.09
SCF_3_5	50.36	8.39	0.13	38.33	6.39	0.16
SCF_3_6	60.71	10.12	0.23	42.58	7.10	0.31
SCF_3_7	64.54	10.76	0.54	44.30	7.38	0.71
SCF_3_8	71.69	11.95	0.97	49.71	8.29	1.55
Comp.		2.00	0.84		1.00	1.00

detailed routing. Considering a Steiner point s in the grid-based model, we can get a diamond whose radius is equal to M_s. In order to satisfy Properties 1 and 2, we select the grids which are (1) on the diamond, (2) in the corresponding bounding box, (3) not on the location of any unit capacitor, as the candidate grids of the Steiner point. Once the Steiner point s is embedded in any candidate grid, the total wirelength can be increased by M_s.

5. DETAILED ROUTING

Based on the topologies generated from the first stage, we first decompose each resulting tree into a set of 2-pin nets, and then simultaneously determine the final positions of Steiner points and minimize routing wirelength. We assign a Steiner point to a routable candidate grid with the least congestion and coupling according to the routed nets and the routing topologies determined in the first stage. We then route all the 2-pin nets by maze routing while minimizing the bend number and coupling noise.

6. EXPERIMENTAL RESULTS

We implemented our routing algorithm in the C++ programming language and performed experiments on a Linux workstation with Intel Xeon E5620 2.4GHz CPU and 16GB memory. Since there is no previous work on the addressed problem, we implemented a reasonable heuristic for comparative studies, namely the probabilistic congestion model (PCM) method. The PCM method replaces our propagated probability congestion estimation model by the probabilistic congestion model which is widely used in previous works [7, 14]. The test cases used in our experiments were modified from the resulting capacitor placements of the work [6]. We assigned various horizontal and vertical capacities, i.e., the number of tracks, to identify the most compacted layouts. Table 2 gives the statistics of the test cases.

Table 3 shows the experimental results. The total cost denotes the total differences of unit lengths between each pair of capacitor nets, as defined in Eq. (1). The average cost is obtained by dividing the total cost by the number of different pairs of capacitor nets. As revealed in this table, our method is more effective than the PCM method. The PCM method incurs averagely a 2.0× cost over our algorithm. During the stage of topology generation, our method can generate better topologies to guide our detailed routing because our congestion model is more accurate while that for the PCM method may underestimate the routing congestion. Once the routing congestion has been underestimated, it might be difficult to achieve a desired wire length ratio because many detours might be generated for the requirement of 100% routability in subsequent detailed routing. Even worse, sometimes the requirement of 100% routability can never be met by the detailed router, see, e.g., the results with N/A to the PCM method.

7. CONCLUSION

We have presented an effective length-ratio-matching routing algorithm for the addressed problem. To the best of our knowledge, this is the first work that addresses the issue of wirelength ratios in capacitor arrays for analog circuit design. The experimental results have shown that our algorithm outperforms a reasonable heuristic by large margins. In particular, only our algorithm can solve all the test cases well. Future work includes the development of a routability-driven capacitor placement algorithm considering both capacitor placement and routing simultaneously.

8. REFERENCES

[1] L. Baldi, B. Franzini, D. Pandini, and R. Zafalon, "Design solutions for the interconnection parasitic effects in deep sub-micron technologies," *Elsevier ME*, vol. 55 (1-4), pp. 11–18, 2001.

[2] J. Cohn, D. Garrod, R. Rutenbar, and L. Carley, "KOAN/ANAGRAM II: New tools for device-level analog placement and routing," *IEEE JSSC*, vol. 26, no. 3, pp. 330–342, 1991.

[3] D. Garrod, R. Rutenbar, and L. Carley, "Automatic layout of custom analog cells in ANAGRAM," in *Proc. ICCAD*, pp. 544–547, 1988.

[4] P.-H. Lin, H.-C. Yu, T.-H. Tsai, and S.-C. Lin, "A matching-based placement and routing system for analog design," in *Proc. VLSI-DAT*, pp. 1–4, 2007.

[5] C.-W. Lin, S.-Y. Chen, C.-F. Li, Y.-W. Chang, and C.-L. Yang, "Obstacle-avoiding rectilinear Steiner tree construction based on spanning graphs," *IEEE TCAD*, vol. 27, no. 4, pp. 643–653, 2008.

[6] C.-W. Lin, J.-M. Lin, Y.-C. Chiu, C.-P. Huang, and S.-J. Chang, "Common-centroid capacitor placement considering systematic and random mismatches in analog integrated circuits," in *Proc. DAC*, pp. 528–533, 2011.

[7] J. Lou, S. Krishnamoorthy, and H. S. Sheng, "Estimating routing congestion using probabilistic analysis," *IEEE TCAD*, vol. 21, no. 1, pp. 32–41, 2002.

[8] J. Ousterhout, "Corner Stitching: A data-structuring technique for VLSI layout tools," *IEEE TCAD*, vol. 3, no. 1, pp. 87–100, 1984.

[9] M. M. Ozdal and M. D. F. Wong, "Two-layer bus routing for high-speed printed circuit boards," *ACM TODAES*, vol. 11, no. 1, pp. 213–227, January 2006.

[10] M. M. Ozdal and M. D. F. Wong, "Algorithmic study of single-layer bus routing for high-speed boards," *IEEE TCAD*, vol. 25, no. 3, pp. 490–503, 2006.

[11] M. M. Ozdal and M. D. F. Wong, "A length-matching routing algorithm for high-performance printed circuit boards," *IEEE TCAD*, vol. 25, no. 12, pp. 2784–2794, 2006.

[12] M. M. Ozdal and R.-F. Hentschke, "Exact route matching algorithms for analog and mixed signal integrated circuits," in *Proc. ICCAD*, pp. 231–238, 2009.

[13] W. Wolf, "Fabrication and Layout," in *Modern VLSI Design: A Systems Approach*, pp. 27–82, 1994.

[14] J. Xiong and L. He, "Probabilistic congestion model considering shielding for crosstalk reduction," in *Proc. ASP-DAC*, pp. 739–742, 2005.

[15] T. Yan and M. D. F. Wong, "BSG-Route: A length-matching router for general topology," in *Proc. ICCAD*, pp. 499–505, 2008.

[16] L. Zhang, U. Kleine, R. Raut, and Y. Jiang, "Aladin: A layout synthesys tool for analog integrated circuits," *Springer AICSP*, vol. 46, no. 3, pp. 215–230, 2006.

Digital-Assisted Noise-Eliminating Training for Memristor Crossbar-based Analog Neuromorphic Computing Engine*

Beiye Liu, Miao Hu, Hai Li, Zhi-Hong Mao, Yiran Chen
University of Pittsburgh
Pittsburgh, PA 15261, USA
{bel34,mih73,hal66,zhm4,yic52}@pitt.edu

Tingwen Huang
Texas A&M University
PO Box 23874, Doha, Qatar
tingwen.huang@qatar.tamu.edu

Wei Zhang
Nanyang Technological University
Singapore, 637553, SG
zhangwei@ntu.edu.sg

ABSTRACT

The invention of neuromorphic computing architecture is inspired by the working mechanism of human-brain. Memristor technology revitalized neuromorphic computing system design by efficiently executing the analog Matrix-Vector multiplication on the memristor-based crossbar (MBC) structure. However, programming the MBC to the target state can be very challenging due to the difficulty to real-time monitor the memristor state during the training. In this work, we quantitatively analyzed the sensitivity of the MBC programming to the process variations and input signal noise. We then proposed a noise-eliminating training method on top of a new crossbar structure to minimize the noise accumulation during the MBC training and improve the trained system performance, *i.e.*, the pattern recall rate. A digital-assisted initialization step for MBC training is also introduced to reduce the training failure rate as well as the training time. Experimental results show that our noise-eliminating training method can improve the pattern recall rate. For the tested patterns with 128×128 pixels our technique can reduce the MBC training time by $12.6\% \sim 14.1\%$ for the same pattern recognition rate, or improve the pattern recall rate by $18.7\% \sim 36.2\%$ for the same training time.

Categories and Subject Descriptors

C.1.3 [**Computer Systems Organization**]: PROCESSOR ARCHITECTURES—*Other Architecture Styles:Neural nets*

General Terms

Design, Algorithms, Reliability

Keywords

Memristor, Neuromorphic Computing, Pattern recognition

1. INTRODUCTION

The explosive growth of the functional variety of modern embedded systems leads to the emergence of Multiprocessor System-on-Chip (MPSoC) [7]. However, the functionalities of various processing elements on a MPSoC are usually determined at system architect and design stages. Any changes beyond the system capability may incur architecture change, circuit redesign or even new chip fabrication with high cost. The application of programmable elements, such as GPU, mitigates the redesign cost, but achieving the system reconfigurability and power efficiency simultaneously still remains as a challenge [11].

The emerging neuromorphic computing system successfully addresses this challenge by providing functionality reconfiguration as well as low power consumption. Although many neuromorphic computing algorithms have been proposed for many applications like signal processing, pattern recognition, surveillance *etc.*, limited progress was made on the VLSI realization of neuromorphic computing systems[1]. Based on the prediction of Prof. Leon Chua in 19721 [2], HP Labs discovered a memristor device. A memristor can record the historical profile of the electrical excitations applied on itself and incur corresponding resistance change. The similarity between this memristive effect and biologic synaptic have motivated many breakthroughs in the design of the neuromorphic hardware systems [4, 5]. For example, memristor-based crossbar (MBC) structure is recently introduced to improve the execution efficiency of the Matrix-Vector multiplications, which is one of the most common operations in the mathematic representation of artificial neural network (ANN) [6, 3]. However, there are three major technical challenges in such designs: 1) Due to the difficulty of real-time monitoring the memristor state, the off-line training, *e.g.*, directly programming the resistance of a single memristor to the target value is unlikely possible [4]; 2) The input vector-based iterative training methods, however, usually suffer from the long convergence time [3]; and 3) The input signal noise and process variations severely affect the training efficiency and reliability.

In this work, we quantitatively analyzed the sensitivity of the MBC programming to the process variations and input signal noise. We then proposed a noise-eliminating training method with the corresponding modified crossbar structure to minimize the noise accumulation during the MBC training and enhance the trained system performance, *i.e.*, the pattern recall rate. A digital-assisted initialization step for MBC training is also introduced to reduce the training failure rate as well as the training time. Experimental results show that our technique can significantly improve the performance and training time of neuromorhphic computing system by up to 39.35% and 23.33%, respectively.

2. PRELIMINARY

2.1 Memristor Basics

Figure 1(a) depicts an ion migration filament model of HfO_x memristors [9]. A HfO_x layer is sandwiched between two metal electrodes TE (top electrode) and BE (bottom electrode). During

---

*Permission to make digital or hard copies of part or all of this work for personal or classroom use is granted without fee provided that copies are not made or distributed for profit or commercial advantage and that copies bear this notice and the full citation on the first page. Copyrights for components of this work owned by others than ACM must be honored. Abstracting with credit is permitted. To copy otherwise, to republish, to post on servers or to redistribute to lists, requires prior specific permission and/or a fee.
DAC '13, May 29 - June 07 2013, Austin, TX, USA.

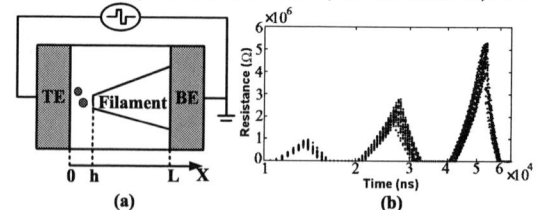

Figure 1: (a) Metal-oxide memristor [10]. (b) Resistance distributions of Multilevel Cell (MLC) memristor.

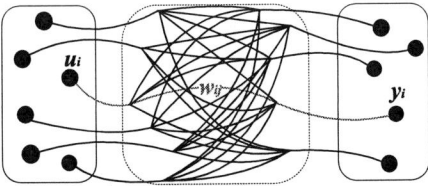

Input group of m neurons   **Synapse Net**   **Output group of n neurons**
with activity pattern Um   **with activity pattern Yn**

Figure 2: Conceptual overview of neural network [3].

the *reset* process, the memristor switches from low resistance state (LRS) to high resistance state (HRS). The oxygen ions migrate from the electrode/oxide interface and recombine with the oxygen vacancies. A partially ruptured conductive filament region with a high resistance per unit length R_{off} is formed on the left of the conductive filament region with a low resistance per unit length R_{on}, as shown in Figure 1(a). During the *set* process, the memristor switches HRS to LRS. The ruptured conductive filament region shrinks. We define L as the total thickness of the oxide layer and h as the length of the ruptured conductive filament region, respectively. The resistance of the memristor R can be calculated by [9]:

$$R = R_{off}h + R_{on}(L - h). \tag{1}$$

We note that the memristor resistance can be programmed to any arbitrary value by applying a programming current with different pulse width or magnitude. Note that the memristor resistance changes only when the applied voltage is above a threshold, *e.g.*, V_{wrth}.

2.2 MBC-based Computing

Fig. 2 depicts a conceptual overview of a neural network. Two groups of neurons are connected by a set of synapses. The output neurons collect the information from the input neurons through the synapses and process them with certain activation function. The synapses apply different weights (synaptic strengths) on the information during the transmission. In general, the relationship between the activity patterns of the input neurons **U** and the output neurons **Y** can be respectively illustrated as:

$$\mathbf{Y}_n = \mathbf{W}_{n \times m} \times \mathbf{U}_m. \tag{2}$$

Here the weight matrix $\mathbf{W}_{n \times m}$ denotes the synaptic strengths between the two neuron groups. In neuromorphic computing system, the Matrix-Vector multiplication represented in Eq. (2) is one of the most frequent operations [10]. Because of the structural similarity, reconfigurable resistive array, *e.g.*, MBC, is conceptually efficient to execute the Matrix-Vector multiplications [3]. During the operation of MBC-based computing, **U** is mimicked by the input voltage vector applied on the wordline (WL) of the MBC. Every memristor in the MBC is programmed to the resistance state representing the weight of the corresponding synapse. The current along every bitline (BL) of the MBC is collected and converted to the output voltage vector **Y** by the comparator circuit.

2.3 Training Methods of MBC

The training of MBC is defined as the process of programming the resistances of the memristors in the MBC to the value representing the connection matrix in Eq. (2). In [4], a direct open-loop training method is proposed: A programming pulse is generated and applied on the target memristor, *e.g.*, $V_{ref} + 0.5V_{wrth}$ on its WL and $V_{ref} - 0.5V_{wrth}$ on its BL, respectively. The other WLs or BLs of the MBC are set to V_{ref} and ensure that the voltage drops through all the other memristors are below $0.5V_{wrth}$. However, because of process variations, the highest resolution that the direct open-loop training can offer is only about 2-bit (or 4-level) [4], as shown in Figure 1(b). Precisely programming the memristor resis-

tance to the target value requires the real-time monitoring mechanism, which is very difficult to be implemented in MBC design.

Another type of MBC training method is also derived from the close-loop training algorithm of the weight matrix in ANN theory, *e.g.*, gradient descent training [8]. During the training process, the weight matrix W is updated iteratively until the difference between the output y and the target output $y*$ reaches the minimum. In each iteration, W is adjusted based on the gradient of the output error $|y - y*|$ as:

$$\Delta w_{ij} = \eta \cdot \left(\frac{\partial (y - y*)^2}{\partial w_{ij}} \right). \tag{3}$$

Here w_{ij} is the element in the W connecting the neuron i and j, or the resistance of the memristor at row i and column j in the MBC. Δw_{ij} is the change of w_{ij} during the iterations. η is the training rate. The choice of η is discussed in [3]. Some typical examples of gradient descent learning method include delta rule [3], back-propagation learning rule [8], *etc*. Two steps, *i.e.*, the output comparison and W programming, are needed in each training iteration. Implementing the gradient descent training algorithms in MBC design faces many technical challenges, *e.g.*, the long convergence time and the impacts of process variation and input noise. In the following sections, we will analyze these issues and propose our solutions.

3. NOISE-ELIMINATING TRAINING

3.1 Impacts of Device Variation and Noise

Process variation and signal noise are two major factors affecting the robustness of MBC-based computing and training processes. Figure 3(a) shows an example of the output comparison step in the MBC training process when a set of read voltage $V_{rd}, 0, V_{rd}/2$ is applied to the WLs of three memristors $R1 - R3$ in the same column. Here we assume the three memrsitors are all at HRS. The ideal voltage on the BL shared by these three memristors should be $V_{rd}/2$. However, the device non-uniformity and the input voltage fluctuation may cause the bias changes on the memristors. For example, if the resistance of $R1$ is larger than that of $R2$, the voltage on the BL will be below $V_{rd}/2$, as shown in Figure 3(a). Also, if the input voltages on the WL of $R1$ changes to $V_{rd} + \Delta V$, the voltage on the BL will be above $V_{rd}/2$, as shown in Figure 3(b). In both cases, the calculated difference between the current output and the target output will be different from the ideal case. Such deviation can be accumulated along with the training iterations. Together with the fluctuations of the programming voltage and the process variations, it will cause the deviation of the programmed memristor resistance from the ideal value during the programming step in the MBC training process and finally affect the computation accuracy. We use an example to illustrate the impacts of the process variation and input signal noise on the MBC training. A 64×64 MBC is implemented to realize the synapse connection of a two-layer neural network. Figure 3(d) shows the resistance difference between the ideally trained MBC (no process variation or input signal noise) and the MBCs trained with considering process variation (top row) or input signal noise (bottom row), respectively. In the evaluation of process variation's impact, the distribution of the memristor cell size in the MBC is generated randomly for every iteration with Gaussian distribution. Note that since the input noise for write will result in the variation of the MBC memristance, we consider the write input noise with precess variation together. The standard deviation of the memristance variation is assumed to be 10% ($\sigma = 0.1$), 20% ($\sigma = 0.2$), and 30% ($\sigma = 0.3$) of its nominal value. In the evaluation of the read input signal noise's impact, similarly, a random noise following Gaussian distribution is generated

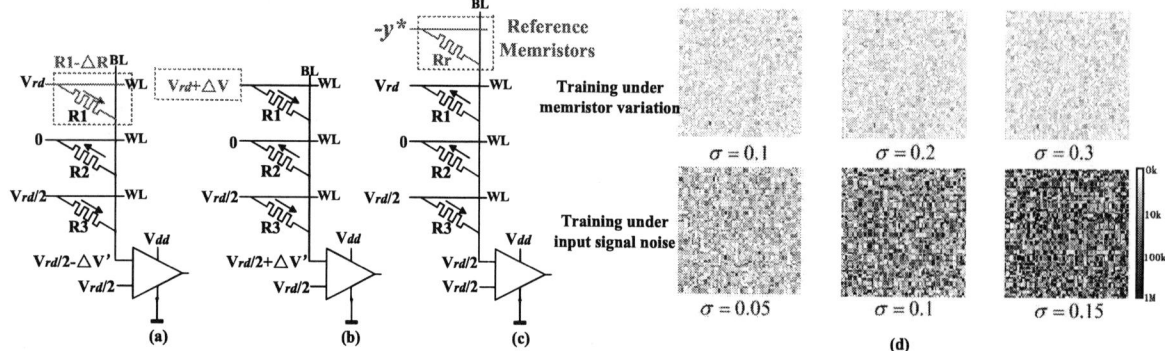

Figure 3: (a)Training under memristor variation. (b) Training with input voltage noise. (c)Redundant rows of reference memristors on top of crossbar(d)Training quality and recall performance with ideal memristor and memristor with different variations.

on the input signals of the MBC in every iteration. The standard deviation of the noise is assumed to be 10% ($\sigma = 0.05$), 20% ($\sigma = 0.1$), and 30% ($\sigma = 0.15$) of V_{rd}. The mean of the noise is zero. Gradient descent rule is applied in the training.

Our simulation shows very marginal degradation in the training robustness as the process variation increases. It is because the device variations are reflected in the difference between the current output and the target output during each iteration, and compensated by close-loop training. Similarly, write pulse noise will cause memristance change variation in each iteration, which will also be compensated by close-loop training. However, input signal noise is generated on-the-fly and accumulated during the training process, leading to a large difference from the ideal trained result.

3.2 Noise Sensitivity of MBC Training

Figure 4 illustrates how the process variations and input signal noise affect the MBC training. We assume F is the output function of the MBC, *i.e.*, comparators, which translates the output of the MBC to a digital value $\in \{1, -1\}$. The input signal noise N is added on the F before it is sent to the next iteration. Different from the conventional gradient descent training, our method tries to minimize not only the 2-norm output distance $(y - y*)^2$ but also the system's sensitivity to the noise $\frac{\partial f(\sum u_i w_{ij})}{\partial N}$ as:

$$Min_W : J = \overbrace{(y - y*)^2}^{J_1} + \overbrace{\frac{\partial f(\sum u_i w_{ij})}{\partial N}}^{J_2}. \quad (4)$$

In the above cost function J, J_1 and J_2 denote MBC output distance and the noise sensitivity, respectively. At the end of iteration t, the adjustment of the MBC in the next iteration $W_{(t+1)}$ can be derived from the current $W_{(t)}$ as:

$$W_{(t+1)} = W_{(t)} - \eta \frac{\partial J}{\partial W}, \quad (5)$$

or,

$$W_{(t+1)} = W_{(t)} - \eta \frac{\partial J_1}{\partial W} - \eta \frac{\partial^2 f(\sum u_i w_{ij})}{\partial W \partial N}. \quad (6)$$

Figure 4: Training process with noise.

The choice of training rate η is discussed in[3]. The For the second term on the right of Eq. (6), we have:

$$\frac{\partial J_1}{\partial W} = 2 \cdot (y - y*) \cdot \frac{\partial y}{\partial W} = 2 \cdot (y - y*) \cdot U. \quad (7)$$

Eq. (7) means that the variations of W (the process variation) is reflected by the output distance $(y - y*)$.

J_2 is determined by the activation function f as:

$$\frac{\partial f(\sum u_i w_{ij})}{\partial N} = \lim_{\Delta n \to 0}(f(\sum u_i w_{ij} + \Delta N) - f(\sum u_i w_{ij} - \Delta N)) \quad (8)$$
$$\approx f'(\sum u_i w_{ij}).$$

For the two popular activation functions in neuromorphic computing, *i.e.*, *sigmoid* function and *sgn* function, Eq. (8) can be expressed as:

$$f'_{sigmoid}(\sum u_i w_{ij}) = \frac{e^{\sum u_{ij} w_{ij}}}{(1 + e^{\sum u_i w_{ij}})^2} \quad (9)$$

and

$$f'_{sgn}(\sum u_i w_{ij}) = \begin{cases} \frac{1}{\Delta N} & -\Delta N < \sum u_i w_{ij} < \Delta N, \\ 0 & \text{else.} \end{cases} \quad (10)$$

respectively. In both cases, the noise sensitivity decreases when $|\sum u_i \cdot w_{ij}|$ raises, as shown in Figure 5.

3.3 Noise-Eliminating Training Scheme

Based on our observation on Eq. (9) and (10), we proposed a noise-eliminating training scheme to minimize the noise accumulation during the MBC training. Redundant rows are added on top of the memristor array to generate an offset current B that is opposite to the target output of the column y_i* during MBC training, as shown in Figure 3(c). It adds the bias $bias$ to the calculated difference between the current output and the target output of the MBC so that the $|\sum u_i \cdot w_{ij}|$ is shifted out of the sensitive region of the activation function $f(x)$ as:

$$x = \begin{cases} \sum u_i w_{ij} > bias \cdot y_i* & \text{if } y_i* = +1, \\ \sum u_i w_{ij} < bias \cdot y_i* & \text{if } y_i* = -1. \end{cases} \quad (11)$$

As shown in Figure 5, through applying $bias$, the residue of the noise in the sensitive region of the activation function is reduced and the accumulation of the noise during the training iterations is minimized. The selection of $bias$ is important in our proposed scheme: A $bias$ larger than necessary may make the training process bypass the convergence region, leading to the difficulty of convergence. If $bias$ is too small, it may not efficiently suppress the noise. A detailed evaluation on the selection of $bias$ will be given

44

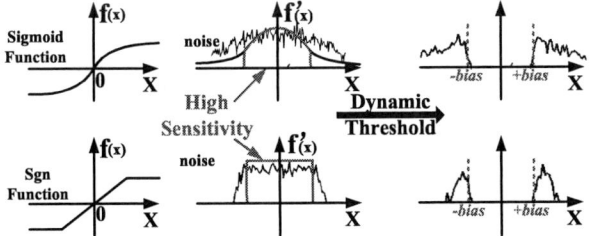

Figure 5: Noise elimination mechanism.

in Section 5.

We define *bias amplitude* a to measure the ability of the reference memristor to offset the MBC output as:

$$a = \frac{R_{on} \cdot N_{ref}}{R_{ref} \cdot N_{col}}. \tag{12}$$

Here R_{on} is the HRS of a memristor. R_{ref} is the average resistance of the reference memristors. N_{col} is number of memristors in a column. N_{ref} is the number of reference memristors in a column. During the MBC training, a training failure is defined as the unsuccessful convergence after the maximum n iterations of training. Here n is the threshold usually much more than the normal iteration number required for convergence. If a training failure happens, we will reset the reference memristors to reduce a and redo the training process until the training succeeds or $a = 0$, which indicates the training is degraded to conventional training scheme.

4. DIGITAL-ASSISTED INITIALIZATION

4.1 Basic Idea

In our noise-eliminating training scheme, the introduction of *bias* affects the convergence process of MBC training and may cause the potential convergence failure. In this section, we proposed a digital-assisted initialization step to the MBC training to reduce the training failure rate and training time.

Algorithm 1 depicts the modified MBC training process. A step called "digital-assisted training initialization" is inserted after pre-set all memristors in the MBC. As shown in Figure 6, in the initialization step, the target W, which is normally known in the algorithm, is quantized to its digital version W_D where every element is represented by a MLC data, *e.g.*, 2-bit digit. W_D is then written into the MBC by the open-loop training method, regardless the device variations. Our digital-assisted training initialization step can improve the convergence speed of MBC training by setting the initial resistance of the memristors close to the target value. The robustness of the training process is also improved as the possibility of being stuck in the local minimum reduces accordingly. Different from the open-loop training, the digital initialization does not require to program the memristor to the digitalized resistance level precisely and can tolerate the device variations. Note that the digitalization of W relies on specific training algorithms as we will show next for our approach.

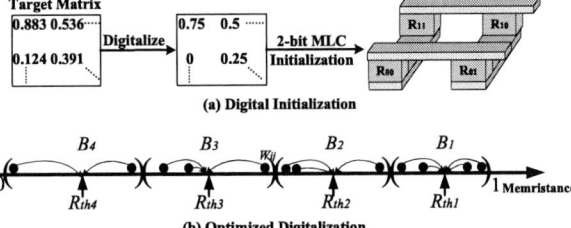

(a) Digital Initialization

(b) Optimized Digitalization

Figure 6: Digital-assisted Initialization

Algorithm 1: Digital-assisted Noise-eliminating Training

Set all memristors to LRS.;
Set input signal of reference memristors to $-y_i*$.
while $Y \neq Y*$ **do**
     $Count = 0$.;
     Digital-assisted initialization of MBC "W".;
     while $Count < n$ **do**
         $Count = Count + 1$.;
         if $Y = Y*$ **then**
             Finish!;
         else
             Upgrade W.;
     Reset reference memrisor.;

4.2 Digitalization of Weight Matrix

In the conventional MLC memory cell design, the distances between the two adjacent resistance states of the memristor must be the same to maximize the sense margin [5]. The threshold to differentiate the different MLC level is set to the cross point between the distributions of two adjacent resistance states. In MBC training , the convergence rate of the training process is conceptually determined by the distance between the target value and the initial value. Therefore, the partition method of MLC memory design does not necessarily give us the minimum distance in the digitalization of weight matrix W.

We propose a heuristic method to determine the resistance states of the memristor corresponding to the different digitalized levels of W: For a M-level digitalization, the elements of W are equally classified into M baskets B_i, $i = 1...M$ based on their values, or:

$$\forall W_{ij} \in B_k \text{ and } \forall W_{i'j'} \in B_p, k < p, \\ \text{we have } W_{ij} < W_{i'j'}. \tag{13}$$

We then find the R_{thi}, $i = 1...M$ for each basket to achieve the minimum $\Sigma |W_{ij} - R_{th1}|_{W_{ij} \in B_i}$, $i = 1,...,M$. R_{thi} is the optimal memristor resistance states for the i-level of the digitalization (shown in Figure 6(b)). Here we used 1-norm resistance distance to measure the impact of the difference between w_{ij} and $w_{D,ij}$ on the convergence rate of the MBC training. For different MBC training algorithms, other methods, *e.g.*, based on 2-norm distance or the maximum distance, may be also adopted. Considering the practical memristor programming resolution, we set $M = 4$ here. Note that this method may cause smaller MLC sensing margin, however, we do not need to read out the value of each MLC. The initialization accuracy is enough to guarantee the training quality.

5. EXPERIMENTS AND RESULTS

In this section, we evaluate the effectiveness of our proposed techniques to improve the performance, the training time of MBC-based neuromorphic computing engine for different networks and different training algorithms. For one layer iterative network, Delta Rule (DR) is adopted (Section 5.1-5.2), and for 3-layer network, Back Propagation (BP) is needed (Section 5.3). We also discuss the design tradeoffs for different technique combinations.

5.1 Noise Elimination

Figure 7 illustrates the effectiveness of the noise-eliminating training method on improving the performance of MBC-based computing engine. A hopfield network with 128 input neurons is built on a 128×128 MBC with one-layer iterative structure to remember 16 patterns. We choose conventional DR training method for comparison. In our simulation, we set the *bias amplitude a* to

45

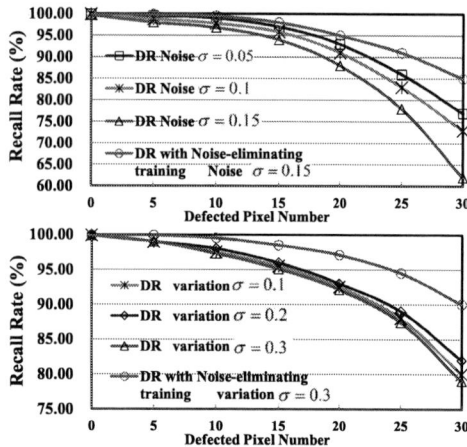

Figure 7: Effectiveness of noise-eliminating training.

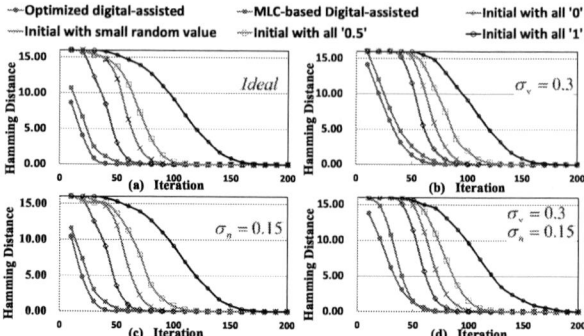

Figure 8: Comparison of the convergence rates of different initialization. (a) Ideal case. (b) Training with memristor variation $\sigma_v = 0.3$. (c) Training with input signal noise $\sigma_n = 0.15$. (d) Training with variation and noise.

0.05. Monte-Carlo simulations are conducted under different process variations and input signal noise levels to measure the success rate of image recognition. As shown in Figure 7 (a) and (b), even at the worst case of $\sigma_{variation} = 0.3$ or $\sigma_{noise} = 0.15$ at each comparison, our method still achieve the best performance. Our noise-eliminating training method significantly improves the recognition rate of the MBC-based neuromorphic computing system by 39.35% at the case of $\sigma_{noise} = 0.15$ and the defected pixel number = 30.

5.2 Digital-Assisted Initialization

Figure 8 compares the training speed of the same MBC design simulated in Section 5.1. Y-axis is the Hamming distance between the output vectors of the MBC and the target output vectors. X-axis is training iteration number. The size of training input vector set is 16 and the MBC training ends when generated output matches the target patterns. Four combinations of process variations and input signal noise levels are simulated. To exclusively measure the effects of digital-assisted initialization, noise-eliminating training is not applied in the simulations.

Among all the simulated results, initializing the states of all memristors to '1' (HRS) has the largest number of training iterations while initializing the states of all memristors to '0' (LRS) has the smallest number of training iterations among all the simulations except the ones with the digital-assisted initialization. It indicates that the majority of the target memristor states are close to '0'.

"MLC-based digital-assisted" curve denotes the results of using the digitalization method of 2-bit MLC memory design in the W initialization while "Optimized digital-assisted" curve denotes the results of using the heuristic method proposed in Section 4.2. Both of them demonstrated much lower iteration number than the other training process without the digital-assisted initialization step. Our heuristic method offers the best result among all the training methods: when both process variation and input signal noise are considered, the training iteration number of "Optimized digital-assisted" is 23.3% less than that of "initialization with '0'".

The introduction of process variation causes the deviation of the initial states of the memristors from the target states in the digital-assisted initialization step. It raises the Hamming distances of the first several iterations and increases the iteration numbers considerably, as shown in Figure 8(b) and (d).

In general, the total training time of a conventional BP training method can be calculated by:

$$T_t = (T_p \cdot n^2 + T_c \cdot n) \cdot N_{iter}. \quad (14)$$

Here n is the input size of the MBC. T_t is overall training time. T_p and T_c are the programming and comparison time consumed in each iteration. N_{iter} is the number of iterations. When the digital-assisted initialization step is applied, the initialization time T_{init} is added to the total training time. Therefore, to achieve the positive benefit, the speed up introduced by the digital-assisted initialization step must be larger than the extra initialization time. Figure 9 shows that for a MBC with the size of $n < 128$, digital-assisted initialization step does not give us any benefits on the training time reduction under the simulated conditions.

5.3 Case Study

To comprehensively evaluate effectiveness of all our proposed techniques, we implemented a three-layer feed forward neural network based on a neuromorphic computing system with multiple 512×512 MBC computing engines. BP training is used as comparison training algorithm in this case. Other simulation parameters can be found at Table 1.

Four sets of image patterns (e.g., face, animal, building and finger print) are adopted in the training nueromorphic computing systems. As shown in Figure 10, each pattern set has 8 images with a size of 128×128 pixels.

Figure 10 compares the recall success rates of the conventional back propagation (BP) training and the modified noise-eliminating

Table 1: Simulation Setup

MBC	Single MBC Size	Input Voltage Noise		Memristor Variation
	512×512	$\sigma = 0.1$		$\sigma = 0.1$
Neuromorphic Engine	Neuron Number in Input Layer	Neuron Number in Hidden Layer		Neuron Number in Output Layer
	16384 (32×MBC)	16384 (32×MBC)		3
Peripheral Circuit	G	Vref (V)	Vdd/Gnd (V)	Vrd(V)
	50	0.6	1.2/0	0.1
	Amplifier Comparator			Registers
	Programming pulse voltage(V)	Programming pulse width(ns)	Number of Comparators/MBC	Number of Registers/MBC
	2.5/-2.5	50	512	512

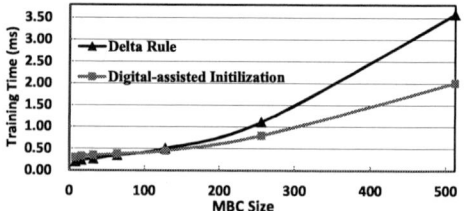

Figure 9: The impact of initialization on total training time.

46

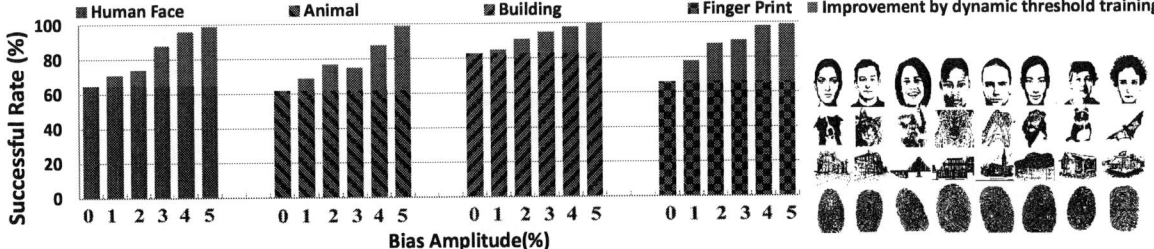

Figure 10: 3-layer network recall successful rate test of dynamic threshold training algorithm

method. Our method surpasses the conventional training method over all the simulation cases. Following the increase in the *bias amplitude*, the recall success rate improvement introduced by the noise-eliminating training method becomes more prominent.

Table 2 shows the training failure rate and the training time (without digital-assisted initialization step) under the different *bias amplitudes*. The increase in *bias amplitudes* results in the reduction of the training time for each iteration while rapidly raises the training failure rate. As aforementioned in Section 3.3, Training failure will prolong the total training time since we will redo the training with a reduced a. The overall training time T_{train} will become:

$$T_{train} = T_{t(a=a_1)} + P_{f(a=a_1)} \cdot (T_{t(a=a_2)} + P_{f(a=a_2)} \cdot (T_{t(a=a_3)} + \cdots .$$
(15)

where $T_{t(a=a_i)}$ and $P_{f(a=a_i)}$ are training time for each iteration and training failure rate for the training process with bias amplitude $a = a_i$.

Figure 11 shows the overall training time comparison between conventional back propagation (BP) training, the modified noise-eliminating training with and without the digital-assisted initialization step starting with different *bias amplitudes "a"*. Our techniques generally reduce the MBC training time by 12.6% ~ 14.1% for the same recall success rate, or improve the recall success rate by 18.7% ~ 36.2% for the same training time. Designer can pick the best combination based on the specific system requirement.

6. CONCLUSION

In this work, we proposed a noise-eliminating training method and digital-assisted initialization step to improve the training process robustness and the performance of memristor crossbar-based neuromorphic computing engine. Experimental results show that our techniques can significantly improve the recall success rate and training time by up to 18.7% ~ 36.2% and 12.6% ~ 14.1%, respectively, through suppressing the noise accumulation in the training iterations and reducing mismatch between the initial weight matrix state and the target value.

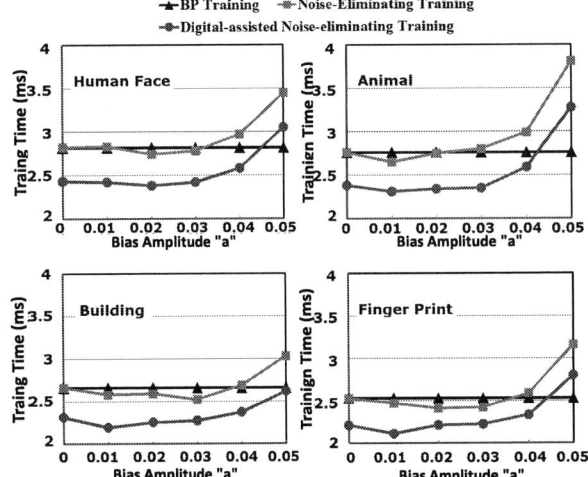

Figure 11: Comparisons of overall Training Time.

7. REFERENCES

[1] H. Chen, S. Saighi, et al. Real-Time Simulation of Biologically Realistic Stochastic Neurons in VLSI. In *IEEE Trans. on Neural Networks*, pages 1511–1517, 2010.

[2] L. Chua. Memristor-the Missing Circuit Element. In *IEEE Trans. on Circuit Theory*, pages 507–519, 1971.

[3] M. Hu et al. Hardware Realization of BSB Recall Function Using Memristor Crossbar. In *DAC*, pages 498–503, 2012.

[4] S. H. Jo. Nanoscale Memristor Device as Synapse in Neuromorphic Systems. In *Nano Letters*, pages 1297–1301, 2010.

[5] K. Kim et al. A Functional Hybrid Memristor Crossbar-Array/CMOS System for Data Storage and Neuromorphic Applications. In *Nano Letters*, pages 389–395, 2012.

[6] J. Klein et al. Hight fault tolerance in neural crossbar. In *DTIS*, pages 1–6, 2010.

[7] R. Kumar, D. Tullsen, et al. Heterogeneous chip multiprocessors. In *Computer*, pages 32–38, 2005.

[8] F. Moreno et al. Reconfigurable Hardware Architecture of a Shape Recognition System Based on Specialized Tiny Neural Networks With Online Training. In *IEEE Trans. on Industrial Electronics*, pages 3253–3263, 2009.

[9] S. Yu, Y. Wu, and P. Wong. Investigating the Switching Dynamics and Multilevel Capability of Bipolar Metal Oxide Resistive Switching Memory. In *APL*, pages 103514–3, 2011.

[10] T. Yu et al. Analog VLSI Biophysical Neurons and Synapses With Programmable Membrane Kinetics. In *IEEE Trans. on Biomedical Circuits and Systems*, pages 139–148, 2010.

[11] F. Zhuo et al. Multigrid on GPU: Tackling Power Grid Analysis on parallel SIMT platforms. In *International Conference on Computer-Aided Design (ICCAD)*, pages 647–654, 2008.

Table 2: Training Failure Rate "P_f" and Training Time "T_t"

Bias Amplitude "a"		0	0.01	0.02	0.03	0.04	0.05
Training Failure Rate "P_f" (%)	Human Face	0.61	1.9	4.2	7.3	15.4	31.5
	Animal	0.44	3.2	7.2	11.1	24.4	51.5
	Building	0.56	2.1	3.9	6.6	13.5	26.1
	Finger Print	0.39	2.7	4.5	9.1	19.4	41.7
Training Time "T_t" (ms)	Human Face	2.82	2.77	2.63	2.59	2.55	2.53
	Animal	2.76	2.57	2.56	2.40	2.28	2.25
	Building	2.66	2.5	2.48	2.34	2.31	2.28
	Finger Print	2.53	2.39	2.28	2.19	2.11	2.10

Dynamic Behavior of Cell Signaling Networks – Model Design and Analysis Automation

Natasa Miskov-Zivanov[1,2], Diana Marculescu[2], and James R. Faeder[1]

[1]Department of Computational and Systems Biology, School of Medicine, University of Pittsburgh

[2]Department of Electrical and Computer Engineering, Carnegie Mellon University

E-mail: [1]{nam66,faeder}@pitt.edu, [2]dianam@cmu.edu

ABSTRACT

Recent work has presented logical models and showed the benefits of applying logical approaches to studying the dynamics of biological networks. In this work, we develop a methodology for automating the design of such models by utilizing methods and algorithms from the field of electronic design automation. We anticipate that automated discrete model development will greatly improve the efficiency of qualitative analysis of biological networks.

Categories and Subject Descriptors

B.6.3 [Logic design]: Design Aids; I.6 Simulation and modeling; J.3 Life and medical sciences.

General Terms

Algorithms, Performance, Design, Experimentation, Theory.

Keywords

Biological networks, Logical modeling, Design automation techniques, Hardware description language, Hardware-based emulation, FPGA.

1. INTRODUCTION

Anticipation of a biological system's response to a drug or a pathogen, or prediction of tumor initiation and its temporal and spatial progression is of great interest in medicine. As cells constantly sense their environment and respond to it (Figure 1), predicting how a cell will respond to a particular stimulus is a grand challenge in cell biology.

Therefore, it is becoming increasingly important to develop models of these complex biological systems, and more specifically, models of regulatory networks that control crucial biological processes. Different computational approaches have been proposed for modeling and studying biological systems. However, the complexity of these models increases rapidly with the size of the network.

Modeling biological networks using Boolean variables and logical operators has gained increasing attention in recent years [1-6]. This approach allows for more coarse-grained yet accurate studies of the dynamics of cell signaling network and its behavior (state transitions) from initial state to steady state (or steady cycle). However, the lack of automated tools for constructing these models limits their application to large-scale networks. Even for smaller networks, the current manual

---

Permission to make digital or hard copies of all or part of this work for personal or classroom use is granted without fee provided that copies are not made or distributed for profit or commercial advantage and that copies bear this notice and the full citation on the first page. To copy otherwise, to republish, to post on servers or to redistribute to lists, requires prior specific permission and/or a fee.

DAC '13, May 29 - June 07 2013, Austin, TX, USA.

---

approach to model development is time-consuming, because it requires a trial and error process to work out the regulatory logic (*e.g.*, necessary vs. sufficient, AND vs. OR, *etc.*). We propose an automated approach to construct biological circuit models that relies on tools from the well-established field of electronic design automation (EDA).

Once the model is developed using our methodology, it is straightforward to use the model for emulating biological network behavior into Field Programmable Gate Array (FPGA) platforms. FPGAs are ideally suited to implement highly parallel architectures and they allow for efficient and accurate analysis of complex signaling and regulatory networks. The hardware-based emulation framework that has been described in [7, 8] enables orders of magnitude speedup when compared to classic simulation in silicon.

The contributions of presented work include the following:
• Allows for the development of complex logical models in which network elements may have multiple discrete values and which would otherwise be difficult (if not impossible) to develop manually;
• Allows for efficient simulation of logical models using a hardware-based emulation approach, with orders of magnitude speedup when compared to existing software-based approaches.

The rest of this paper is organized as follows. In Section 2, we review related work on developing and analyzing models of cell signaling networks. In Section 3, we present preliminaries of developing logical models for cell signaling networks. Our methodology for designing models is described in Section 4 and approaches for studying models are described in Section 5. We conclude our work with Section 6.

2. RELATED WORK

A number of methods and approaches have been used for modeling complex networks of biochemical interactions. The methods range from master equations and the Monte-Carlo method [9], ordinary differential equations (ODE) [10, 11], reaction rule-based models [12, 13], all the way to Boolean networks [1, 5, 14, 15].

Differential equations, which capture the underlying reaction kinetics in terms of rates and concentrations, are perhaps the most commonly used approach to modeling biochemical pathways and networks. Often, the information about reaction rates and parameter values is hard to determine from the existing knowledge about the system and needs to be estimated or guessed. Rule-based modeling formalisms and associated

Figure 1. The cell as a "black-box."

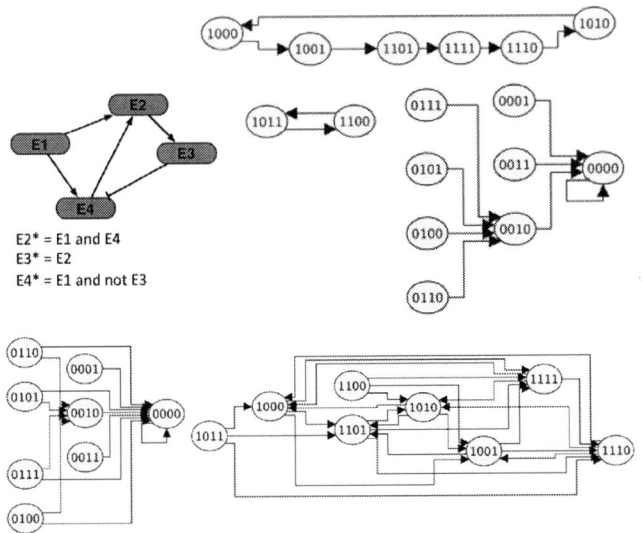

E2* = E1 and E4
E3* = E2
E4* = E1 and not E3

Figure 3. Interaction map example (top left), STG for a synchronous update scheme (top right), and STG for an asynchronous update scheme (bottom).

simulation algorithms have been developed to represent biochemical systems in terms of formal rules for biomolecular interactions [16].

Models of biological networks are usually developed manually, through literature overview, discussions with experts and by studying experimental data. There exist a number of tools to write or draw these models using formalisms mentioned above [12, 14, 17]. Rule-based modeling represents a step towards automating development of models of reaction networks. Protein regulatory networks can be assembled using 'reverse engineering' methods such as Bayesian network analysis [18] or inferred systems of differential equations. In [19], multiple linear regressions were used to develop a first order differential equation model of a gene and protein regulatory network. However, the number of reactions and biochemical species implied by rules can become enormously large, making it difficult to construct, simulate and analyze conventional models derived from rules.

An effort to automate the development of Boolean models of regulatory networks is presented in [4]. The authors in [4] present the methodology in which a tool, called CellNetOptimizer, is used to assemble Boolean models from protein signaling networks and raw data, using data management application, DataRail [20]. All possible Boolean models that are compatible with the signaling network are collected, and then applying a fitness test to each model reduces this set. The main drawback is that the complexity of this approach increases rapidly with the size of the network and the number of connections in the network.

3. LOGICAL MODELING OF BIOLOGICAL NETWORKS

In Figure 3(top left), we present an interaction map of a small system with four elements (*e.g.*, proteins). Each element has an associated discrete variable, namely **E1**, **E2**, **E3**, and **E4**. These variables can represent different aspects of protein activity and function, for example: inactive or absent (0 value) vs. active or present with different levels of activity or concentration (*e.g.*, low, medium, high can be represented using values 1, 2, 3, *etc.*). As it can be seen in Figure 3(top left), interactions between elements are represented using two types of arrows, pointed

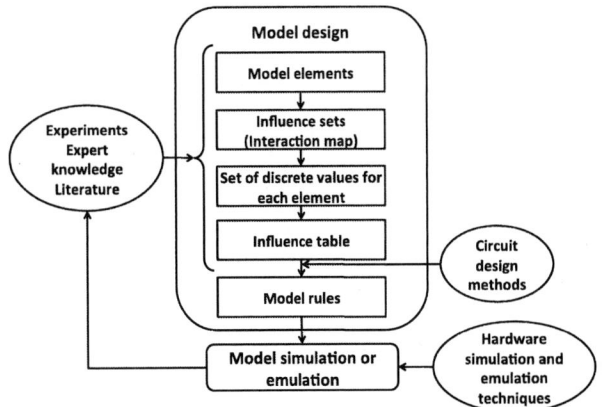

Figure 2. Model design and analysis steps.

error (positive regulation, activation, $\rightarrow$) and blunt arrow (negative regulation, inhibition, $-|$).

A model also includes a set of logic rules that more precisely define interactions, including information about whether a regulator is necessary or sufficient. In Figure 3(top left) we show rules for **E2**, **E3**, and **E4**. **E1** does not have incoming arrows and thus its value cannot change. The "*" sign in rules represents the next-state value for a variable. It can be seen from these rules that **E1** and **E4** are both necessary for activation of **E2**, while **E4** is activated only when **E1** is present and **E3** is not.

Logical models are more abstract than reaction network models, in which reaction rates and molecule counts are represented precisely. At the same time, logical models allow for studying the dynamics of a larger system, capturing external ligand and receptor interactions, internal cell signaling, and gene transcriptions inside nucleus, all within a single model. Finally, logical models can capture system behavior (state transitions) from the initial state until reaching a steady state or limit cycle.

We show in Figure 3 state transition graphs (STGs) for the example system when two different update schemes are used, synchronous and asynchronous. Synchronous update scheme assumes that all variables are updated simultaneously using update rules. State transitions are deterministic and for each state there is exactly one possible next state. State transitions resulting from synchronous update scheme applied on the example system are shown in Figure 3(top right). States in the figure represent values of vector (**E1,E2,E3,E4**). Asynchronous update scheme assumes that, in each update round, all variables (that is, those variables that are not fixed and have rules associated with them) are updated according to their logic rules, one by one, and the order of variable updates is randomly chosen in each round. The STGs at the bottom of Figure 3 represent state transitions that can occur in the example system when an asynchronous update scheme is used.

It is also possible to combine synchronous and asynchronous update schemes by grouping rules and ordering the groups, such that groups of rules are updated in order of their rank, while a random asynchronous scheme is applied to all rules within the same rank [14]. This combined update scheme is well-suited to biological signaling networks because it is able to take into account known differences in the timescale of different processes and interactions while averaging over timing differences that are not known.

4. MODEL DESIGN METHODOLOGY

In Figure 2, we show the main steps of our approach for designing and studying discrete models of cell signaling

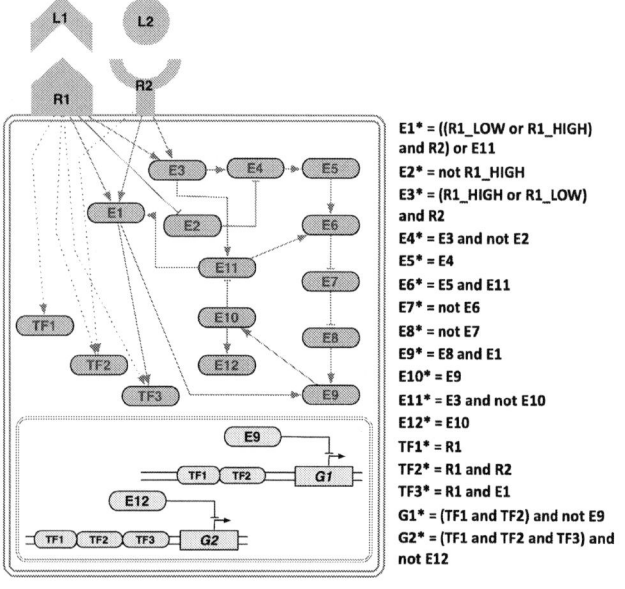

E1* = ((R1_LOW or R1_HIGH) and R2) or E11
E2* = not R1_HIGH
E3* = (R1_HIGH or R1_LOW) and R2
E4* = E3 and not E2
E5* = E4
E6* = E5 and E11
E7* = not E6
E8* = not E7
E9* = E8 and E1
E10* = E9
E11* = E3 and not E10
E12* = E10
TF1* = R1
TF2* = R1 and R2
TF3* = R1 and E1
G1* = (TF1 and TF2) and not E9
G2* = (TF1 and TF2 and TF3) and not E12

Figure 4. Example cell signaling network: interaction map, including receptors, internal signaling molecules and gene regulation (left) and logical model (right).

networks. In the following, we detail each step of our model design methodology using an example model of cell signaling network shown in Figure 4. As it can be seen from the figure, the interaction map of the network (Figure 4(left)) includes several layers of signaling:

- Receptor layer: ligands (**L1**, **L2**) bind to receptors (**R1**, **R2**) outside of the cell to initiate cell signaling;
- Cytoplasm layer: internal cell molecules (**E1-E12**) propagate signaling through events such as phosphorylation and binding, eventually activating transcription factors (**TF1**, **TF2**, **TF3**);
- Nucleus layer: transcription factors move to nucleus where they bind genes and initiate gene transcription (**G1**, **G2**).

4.1 Selection of model elements and their influence sets

The first step in model design is the identification of elements and their regulators (influence sets) through extensive literature survey and consultation with experts. Since the list of important components and interactions is rapidly evolving, this is often an iterative process. Moreover, as we outline in Figure 2, the overall model design process is expected to be iterative, with model studies providing new insights into the system that in turn allow for developing a better model.

4.2 Number of element discrete values

For each element of the network to be modeled, the number of discrete values representing different levels of element activity must be defined. Our modeling methodology is not limited to binary variables, but also allows for including multiple discrete values.

It is often the case in cell signaling that different concentration of ligand, or its affinity to binding a receptor result in different cell responses. Therefore, in our example model in Figure 4, we assume that receptor **R1** can have three different levels of activity, 0, 1, and 2. In order to model this within Boolean framework, we encode variable **R1** using two Boolean variables, high (R1_HI) and low (R1_LO):

R1 = **0** (R1_HI=0, R1_LO=0)
R1 = **1** (R1_HI=0, R1_LO=1)

R1 = **2** (R1_HI=1, R1_LO=0).

The fourth case (R1_HI=1, R1_LO=1) can be treated as a "don't care." Rules for updating variables of the model are shown on the right of the interaction map in Figure 4.

After initially assigning a single Boolean variable to all elements (except receptor **R1**) in the example model, we observed oscillations in the negative loop between elements **E9** and **E11**. In Figure 5(a), we present simulation results for two scenarios: **R1** = **1** and **R1** = **2**. The curves in Figure 5(a)(top) show how element **E11** changes from initial state until the final simulated round, averaged across 1000 simulation runs using the asynchronous update scheme. Figure 5(a)(bottom) outlines several individual simulation trajectories for element **E11**, emphasizing the effect of random asynchronous update scheme.

The oscillations observed in **E11** for R1=2 scenario occur because **E11** is assumed to be necessary for **E6** activation, which then activates **E9**, which in turn activates **E10**. Once **E10** is activated, it inhibits **E11**, preventing further activation of **E6** and activation of **E9**. Oscillations can occur in these networks as a result of negative feedback loop existing within the network. However, oscillations can also be an artifact of modeling. In case of logical modeling, oscillations can occur in negative feedback loops with odd number of inhibitions within the loop, like in the loop E6→E7—|E8—|E9→E10—|E1.

In order to avoid oscillations in the logical model, one can apply several approaches:
1. Control negative feedback loop with an element that is outside of the loop;
2. Model elements of the loop with multiple (more than two) levels of activity;
3. Change rules from necessary to sufficient (**AND** to **OR** rules) or vice versa.

The approach number 2 can be used to model damped oscillations. We will discuss the choice of update rules (approach 3) in the following sub-section. Here we discuss in more detail approach 1. We modeled **E3**, which is outside of the loop, with three levels of activity (0, 1, and 2), such that the control of the loop is different between the two scenarios. When **E3** is at the lower level (**E3**=1, scenario **R1** = 1), the inhibition of **E11** by **E10** can overcome its activation by **E3**. When **E3** is at the higher level (**E3**=2, scenario **R1** = 2), activation is stronger, and **E11** is active as long as **E3**=2. The effect of this change in levels of **E3** on average value of **E11** is shown in Figure 5(b)(top). The new rule that is used for **E11** and that captures different levels of **E3** is highlighted in Figure 5(b)(top). As can be seen from this figure, results for scenario **R1**=1 remain similar to the case where **E3** has only two levels of activity, while scenario **R1**=2 does not lead to oscillations anymore.

4.3 Update rule derivation

Once model elements, their influence sets, and levels of activity are defined, the next step is to derive element update rules. We created a framework to automate rule derivation from *influence tables*.

Influence tables are used to present the value of an element as a result of different combinations of its activators. Influence tables can be derived either from existing knowledge, or in an automated fashion from experimental time series data [19]. In the example models we are presenting here, we created the influence tables manually, from existing literature and via discussion with experts. We are currently working on developing a more formal approach for deriving these tables. In our framework, we write these tables to an input file and then use

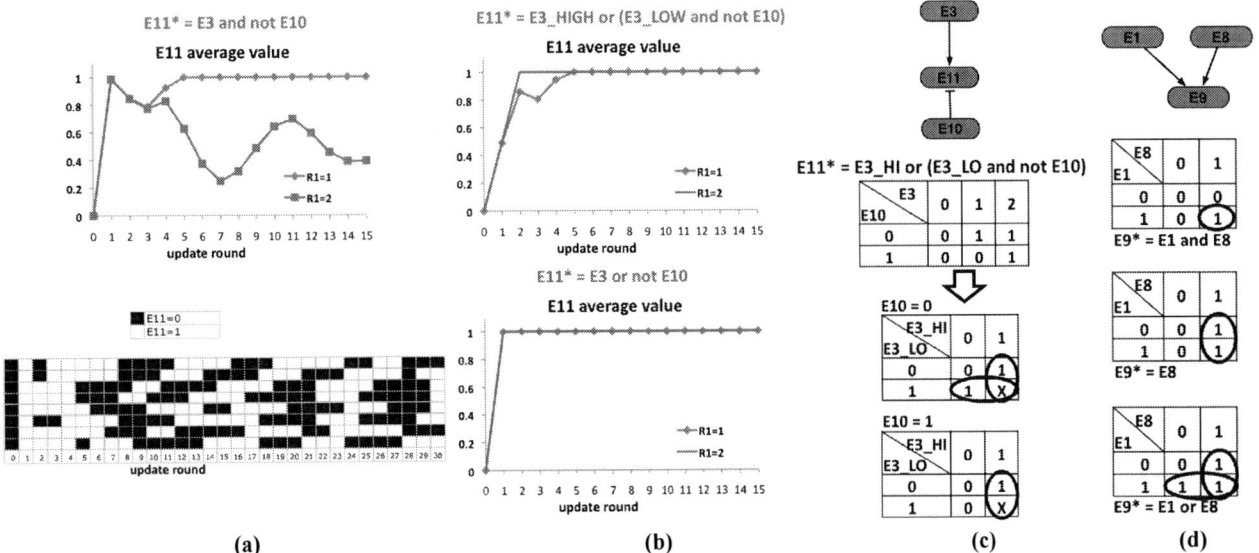

Figure 5. Limitations of representing network elements with a single Boolean variable. (a) Average trajectories for element E11 for two different scenarios using rules from **Figure 4** (top) and several individual trajectories for the "**R1=1**" scenario. (b) Average trajectories of **E11** when number of activity levels of **E3** are increased from (0,1) to (0,1,2) (top) or when a logic rule for **E11** is changed from AND to OR (bottom). (c) Rule derivation for the case from (b)(top). (d) Example of derivation of a rule that best captures regulatory relationships.

logic synthesis tools and algorithms to automate derivation of logic rules. Depending on the complexity of these tables, we use K-maps, Quine-McCluskey or Espresso [21, 22].

Since element **E11** from the model in Figure 4 has only two levels of activity (0 and 1), and is regulated by the three-level **E3** (after model modifications described above), and by the two-level **E10**, we show in Figure 5(c) how this is translated into Boolean variables. The bottom two influence tables in Figure 5(c) are same as K-maps. Although this example is simple from the perspective of logic circuits, it is important to make sure that we accurately implement biological relationships within logic rules. Furthermore, the framework that we created can handle significantly larger examples than the ones shown in Figure 5. When combinations of variables are such that they do not represent realistic biological events, we implement those cases in influence tables as "don't cares." As shown in Figure 5(c), a symbol "**x**" in the influence tables is included in cases where both E3_LO and E3_HI are equal 1, due to the fact that this combination represents a case that is not defined by modeling **E3** with three levels of activity. However, from the logic synthesis perspective, these cases can help minimize logic functions.

Figure 5(d) presents the simplest examples of influence tables, where an element has two regulators and all three of them can have only two levels of activity, 0 and 1. Element **E9** and its two regulators, **E1** and **E8** are highlighted in Figure 5(d), together with examples of deriving an update rule for **E9** from its influence table. Experiments can suggest several different regulatory relationships: (*i*) both elements **E1** and **E8** are necessary for activation of **E9** (top table), (*ii*) one of the regulators is necessary and sufficient for activation, the other one, when present, can increase the level of element **E9** (middle table) or, (*iii*) either **E1** or **E8** is sufficient for activation of **E9** (bottom table). Conditions (*i*) and (*iii*) can be accurately captured with logic rules. However, condition (*ii*) requires including multiple levels of activity when defining **E9**. As can be seen from the middle table in Figure 5(d), the logical rule can only capture activation of **E9** by regulator **E8** (necessary and sufficient regulator), but not regulation by **E1**. It can be seen

from the examples described so far that including multiple levels of activity (more than two) provides much more flexibility in modeling element relationships and system characteristics, while still being simpler to simulate than, for example, ODEs.

The examples in Figure 6 present influence tables in which both effect and its regulators have three levels of activity (0,1, and 2). Figure 6(a) shows derivation of a rule for signaling of two receptors, **R1** and **R2**, which can both be activated by binding two different types of ligands (**L1_1** or **L1_2** for **R1** and **L2_1** or **L2_2** for **R2**). In our models of biological signaling networks we often have elements with larger number of regulators, where both effect and its regulators have at least three levels of activity. In such cases, our framework uses a more efficient method, Espresso [22], to obtain logic rules. In Figure 6(b), we show an example of modeling gene regulation with two transcription factors and one inhibitor. The table shown in Figure 6(b) is slightly more complex than tables in previous examples. Thus, creating rules for the whole cell signaling network from a large number of such tables becomes impractical, and emphasizes the importance of rule design automation. Furthermore, in examples in Figure 6(a)(b), we also include statements that use integer values and capture the same relationships as influence tables. Such statements can be used in hardware description languages as an alternative to developing influence tables, something we are currently working on.

Figure 6(c) presents another example of gene regulation with three transcription factors and one inhibitor. As can be seen, increasing number of regulators by only one (going from example in Figure 6(b) to Figure 6(c)), the influence table size increases significantly. We also show in this example that, as is often the case in modeling biological systems, for some table entries there may not be an existing experimental observation to support entered values. In such cases, model design automation allows for developing alternative rules and testing which of the rules help us recapitulate best the behavior of the system. In the following section, we describe the framework that we have developed to further automate model analysis, and especially to speed up the analysis of models that have alternative rules.

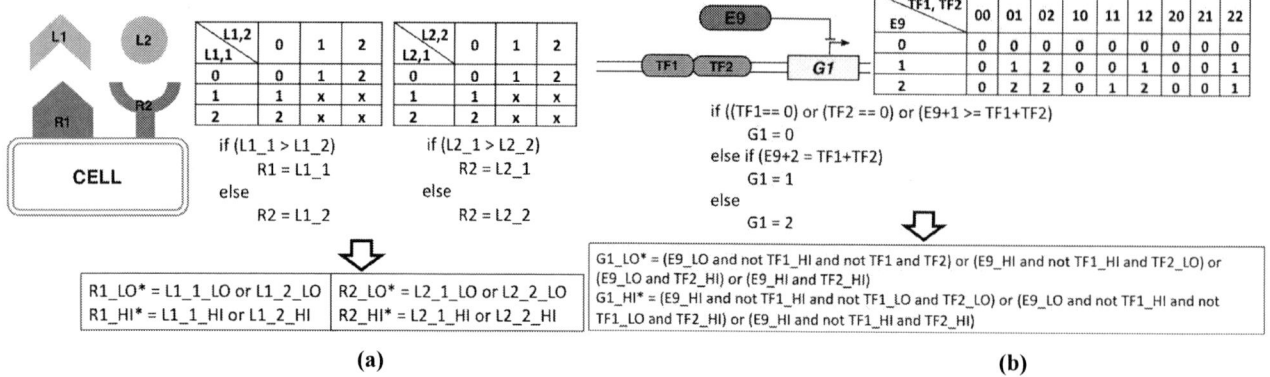

Rule 1:

G2_HI* = (TF2_LO and TF3_HI and not E12_HI and not E12_LO and TF1_LO) or (TF2_LO and TF3_HI and not E12_HI and not E12_LO and TF1_HI) or (TF2_HI and TF3_LO and not E12_HI and not E12_LO and TF1_LO) or (TF2_HI and TF3_LO and not E12_HI and not E12_LO and TF1_HI) or (TF2_HI and TF3_HI and not E12_HI and not E12_LO and TF1) or (TF2_HI and TF3_HI and not E12_HI and not E12_LO and TF1_HI)

G2_LO* = (TF2_LO and TF3_LO and not E12_HI and not E12_LO and TF1_LO) or (TF2_LO and TF3_HI and not E12_HI and TF1_LO) or (TF2_LO and TF3_HI and not E12_HI and TF1_HI) or (TF2_HI and TF3_LO and not E12_HI and TF1_LO) or (TF2_LO and not E12_HI and not E12_LO and TF1_HI) or (TF2_HI and TF3 and not E12_HI and TF1_HI) or (TF2_HI and TF3_LO and not E12_HI and TF1_HI) or (TF3_HI and not E12_HI and not E12_LO and TF1_LO) or (TF2_HI and not E12_HI and not E12_LO and TF1_LO) or (TF2_HI and TF3_HI and TF1_LO) or (TF2_HI and not E12_HI and not E12_LO and TF1_HI) or (TF2_HI and TF3_HI and TF1_HI) or (TF2_HI and TF3_HI and not E12_HI and not E12_LO)

Rule 2:

G2_HI* = (TF2_LO and TF3_HI and not E12_HI and not E12_LO and TF1_LO) or (TF2_HI and TF3_LO and not E12_HI and not E12_LO and TF1_LO) or (TF2_LO and TF3_HI and not E12_HI and not E12_LO and TF1_HI) or (TF2_HI and TF3_LO and not E12_HI and not E12_LO and TF1_HI) or (TF2_HI and TF3_HI and not E12_HI and not E12_LO and TF1_LO) or (TF2_HI and TF3_HI and not E12_HI and not E12_LO and TF1_HI)

G2_LO* = (TF2_LO and TF3_LO and not E12_HI and not E12_LO and TF1_LO) or (TF2_LO and TF3_HI and not E12_HI and TF1_LO) or (TF2_HI and TF3_LO and not E12_HI and TF1_LO) or (TF2_LO and not E12_HI and not E12_LO and TF1_HI) or (TF2_HI and TF3_HI and not E12_HI and TF1_HI) or (TF2_HI and TF3_LO and not E12_HI and not E12_LO) or (TF2_HI and not E12_HI and not E12_LO and TF1_LO) or (TF3_HI and not E12_HI and not E12_LO and TF1_HI) or (TF2_HI and not E12_HI and not E12_LO and TF1_HI) or (TF2_HI and TF3_HI and not E12_HI and not E12_LO)

E12, TF1 / TF2,TF3	00	01	02	10	11	12	20	21	22
00	0	0	0	0	0	0	0	0	0
01	0	0	0	0	0	0	0	0	0
02	0	0	1	0	0	0	0	0	0
10	0	0	1	0	0	0	0	0	0
11	0	1	1	0	0	0	0	0	0
12	0	2	2	0	1 or 0	1	0	0	0
20	0	1	1	0	0	0	0	0	0
21	1	2	2	0	1 or 0	1	0	0	1 or 0
22	1	2	2	0	1	1	0	1 or 0	1 or 0

*: Rule 1   *: Rule 2

(c)

Figure 6. Multi-valued (three-valued) model examples: (a) derivation of rules for receptors that can bind multiple ligands; (b) gene regulation with two transcription factors and one inhibitor; (c) derivation of alternative update rules for gene regulated by three transcription factors and one inhibitor.

4.4 Application of model design methodology

We have applied our model design methodology to two cell signaling networks. One network includes circuitry that controls differentiation of T cells (lymphocytes) into two different phenotypes [6]. Understanding control of T cell differentiation is critical for developing new treatments in cancer and autoimmunity. The model that we have developed for T cell differentiation has 39 elements, where several elements are modeled with three levels of activity, which leads to 45 variables overall.

The second model that we are currently working on is a model of effects of malaria infection on mosquito cells [23]. This model has 41 elements, most of which are modeled with more than two levels of activity. Many elements also have more than two regulators, making the model increasingly complex. The tool that we have developed for automating logic rule development allows us to create rules for this model, which would be otherwise time consuming and error prone if done manually.

5. MODEL ANALYSIS AUTOMATION

In this section, we describe methods for analyzing developed models. The methods presented herein emphasize again the potential that EDA techniques bring to systems biology.

5.1 Hardware design framework

In order to speedup model analysis one can use a hardware emulation approach, which can start and control multiple network simulations in parallel, as well as compare final results obtained from parallel simulations. The design of such a framework usually starts with defining larger blocks. The next step in hardware design is to define modules within each block and describe them in a hardware description language (e.g., Verilog). One issue that arises when considering the qualitative studies of biological networks in hardware is defining the time of the execution of individual interactions. FPGAs are suitable for implementing biological network models because of their concurrent processing nature. To support this, we intend to develop modeling techniques that can facilitate correct relative timing when emulating signaling networks. One way to approach the problem of implementing asynchronous network updates using synchronous hardware (that uses a clock signal) is to design a system with an embedded pseudo-random number generator, as described in [7, 8].

Details about the Verilog implementation of the hardware emulation framework and the design uploading onto an FPGA are described in [7, 8]. After all blocks of the circuit are designed in Verilog, one can use simulation to analyze the design and to view signal waveforms. As an example, ModelSim [24] can be used to provide a waveform view of simulated signals, which can be very useful for analyzing element trajectories from initial to steady state.

5.2 Simulations via multicore parallel implementation

Another approach to speedup simulation of biological networks is to use multi-core platforms that have gained a lot of attention recently due to their potential to decrease the computation runtime. We implemented a logical simulator in C under OpenCL and executed it on an 8-way multi-core platform running Ubuntu Linux on a virtual machine where only four of the eight cores are used. Similar to the BooleanNet simulations in Python, the simulation takes as input the number of nodes, the name of each node, its initial state and a list of activators and inhibitors for each node. The objective of the simulator is to allow the network to progress until a steady state is reached. Each core of the multi-core platform runs at a 2GHz clock.

In our preliminary analysis of several models, FPGA implementation is several orders of magnitude faster than Python (BooleanNet [14]) implementation, and still an order of magnitude faster than the multi-core implementation (Table I). Given that these times reflect the emulation of a single instance of the model on the FPGA and that the FPGA platform can simultaneously run several copies (up to six, as shown in [7, 8]), it is expected that the speed of hardware emulation when compared to the parallel software simulations can be 2-3 orders of magnitude larger and 4-5 orders of magnitude larger than that of sequential, Python-based simulation in BooleanNet.

6. CONCLUSION

In this work, we have presented a methodology for automating the design and analysis of logical models of biological networks. This methodology allows for efficient design of discrete (not only Boolean) models, which have been proven beneficial in studying dynamics of biological networks. Our future work will include development of tools to further automate the procedure of creating influence tables, as well as development of synthesis tools that translate a network description written in Verilog into logic rules, which can be studied using hardware emulation framework. This will allow for even more efficient construction and analysis of large-scale biological network models from which new insights can be gained.

7. ACKNOWLEDGMENTS

This work was supported by NIH Grants UL1-RR024153 and 1R01AI080799, and NSF Expeditions in Computing Grant 0926181. The authors would like to thank Solomon Sia (Electrical and Computer Engineering, Carnegie Mellon University), for his help in developing the tool that is part of the logic rule development methodology. We would also like to thank Penelope Morel (Immunology, University of Pittsburgh), Yoram Vodovotz (Surgery, University of Pittsburgh), and Shirley Luckhart (Medical Microbiology and Immunology, University of California-Davis) for beneficial discussions related to model development.

8. REFERENCES

[1]   R. Zhang, *et al.*, "Network model of survival signaling in large granular lymphocyte leukemia," *Proc Natl Acad Sci U S A*, vol. 105, pp. 16308-13, Oct 21 2008.
[2]   L. Mendoza, "A network model for the control of the differentiation process in Th cells," *Biosystems*, vol. 84, pp. 101-14, May 2006.

Table I. Hardware emulation (FPGA) speedup vs. simulations in Python (BooleanNet) or simulations in C under OpenCL on an 8-way multi-core platform (Parallel).

Model	# nodes	# rounds	BooleanNet	Parallel	FPGA
T-LGL [1]	61	15	60s	0.14	0.019s
Th diff. [2]	23	25	65s	0.16	0.022s
T cell diff. [6]	54	16	45s	-	0.016s

[3]   M. I. Davidich and S. Bornholdt, "Boolean Network Model Predicts Cell Cycle Sequence of Fission Yeast," *PLoS One*, vol. 3, pp. -, Feb 27 2008.
[4]   J. Saez-Rodriguez, *et al.*, "Discrete logic modelling as a means to link protein signalling networks with functional analysis of mammalian signal transduction," *Molecular Systems Biology*, vol. 5, pp. -, Dec 2009.
[5]   S. Bornholdt, "Boolean network models of cellular regulation: prospects and limitations," *J R Soc Interface*, vol. 5 Suppl 1, pp. S85-94, Aug 6 2008.
[6]   N. Miskov-Zivanov, *et al.*, "Boolean Modeling and Analysis of Peripheral T Cell Differentiation of Work," personal communication, 2013.
[7]   N. Miskov-Zivanov, *et al.*, "Emulation of Biological Networks in Reconfigurable Hardware," in Proc. of *ACM Conference on Bioinformatics, Computational Biology and Biomedicine (ACM-BCB)*, 2011, pp. 536-540.
[8]   N. Miskov-Zivanov, *et al.*, "Regulatory network analysis acceleration with reconfigurable hardware," *Conf Proc IEEE Eng Med Biol Soc*, vol. 2011, pp. 149-52, 2011.
[9]   D. T. Gillespie, "Exact stochastic simulation of coupled chemical reactions," *The Journal of Physical Chemistry*, vol. 81, pp. 2340-2361, 1977.
[10]   J. J. Tyson, *et al.*, "The dynamics of cell cycle regulation," *Bioessays*, vol. 24, pp. 1095-1109, Dec 2002.
[11]   M. Novak and J. J. Tyson, "A model for restriction point control of the mammalian cell cycle," *Journal of Theoretical Biology*, vol. 230, pp. 563-579, Oct 21 2004.
[12]   M. L. Blinov, *et al.*, "BioNetGen: software for rule-based modeling of signal transduction based on the interactions of molecular domains," *Bioinformatics*, vol. 20, pp. 3289-91, Nov 22 2004.
[13]   J. R. Faeder, *et al.*, "Rule-based modeling of biochemical systems with BioNetGen," *Methods Mol Biol*, vol. 500, pp. 113-67, 2009.
[14]   I. Albert, *et al.*, "Boolean network simulations for life scientists," *Source Code Biol Med*, vol. 3, p. 16, 2008.
[15]   M. Chaves, *et al.*, "Robustness and fragility of Boolean models for genetic regulatory networks," *J Theor Biol*, vol. 235, pp. 431-49, Aug 7 2005.
[16]   J. R. Faeder, *et al.*, "Investigation of early events in Fc epsilon RI-mediated signaling using a detailed mathematical model," *J Immunol*, vol. 170, pp. 3769-81, Apr 1 2003.
[17]   *The Systems Biology Markup Language (SBML)*. Available: http://sbml.org/Main_Page
[18]   K. Sachs, *et al.*, "Causal Protein-Signaling Networks Derived from Multiparameter Single-Cell Data," *Science*, vol. 308, pp. 523-529, April 22, 2005 2005.
[19]   T. S. Gardner, *et al.*, "Inferring genetic networks and identifying compound mode of action via expression profiling," *Science*, vol. 301, pp. 102-5, Jul 4 2003.
[20]   J. Saez-Rodriguez, *et al.*, "Flexible informatics for linking experimental data to mathematical models via DataRail," *Bioinformatics*, vol. 24, pp. 840-7, Mar 15 2008.
[21]   G. D. Hachtel and F. Somenzi, *Logic synthesis and verification algorithms*. Boston: Kluwer Academic Publishers, 1996.
[22]   R. K. Brayton, *Logic minimization algorithms for VLSI synthesis*. Boston: Kluwer Academic Publishers, 1984.
[23]   Y. Vodovotz, "Modeling Host-Vector-Pathogen Immuno-inflammatory Interactions in Malaria," in *Complex Systems and Computational Biology Approaches to Acute Inflammation*, ed: Springer Science+Business Media, LLC., 2013.
[24]   *ModelSim - Advanced Simulation and Debugging*. Available: http://model.com/

Defect Tolerance in Nanodevice-Based Programmable Interconnects: Utilization Beyond Avoidance

Jason Cong[1,2] and Bingjun Xiao[1]

[1]Computer Science Dept. & Electrical Engineering Dept.

[1]University of California, Los Angeles

[2]California Nano-System Institute

{cong, xiao}@cs.ucla.edu

ABSTRACT

This work focuses on defect tolerance for nanodevice-based programmable interconnects of FPGAs. First, we show that the stuck-closed defects of nanodevices have a much higher impact than the stuck-open defects. Instead of simply avoiding the stuck-closed defects, we use them by treating them as shorting constraints in the routing. We develop a scalable algorithm to perform timing-driven routing under these extra constraints. We also enhance the placement algorithm to recover logic blocks which become virtually unusable due to shorted pins. Simulation results show that at the up-to-date level of nanodevice defects (10^8–10^{11}x higher than CMOS), compared to the simple avoidance method, our approach reduces the degradation of resource usage by 87%, improves the routability by 37%, and reduce the degradation of circuit performance by 36%, at a negligible overhead of tool runtime.

1. INTRODUCTION

A number of FPGAs based on emerging nanodevices have been explored in the past few years [1–6]. The emerging nanodevices include resistive RAM (RRAM) [1], phase-change RAM (PCRAM) [2], nanoelectromechanical (NEM) relays [3,4], and molecular switches [5,6]. They can be generalized as bistable switches which can be programmed between the "on" and "off" states, as shown in Fig. 1a [7]. A single nanodevice can function as a routing switch in place

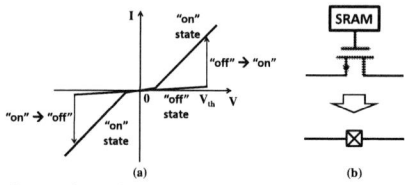

Figure 1: Illustration of nanodevices. (a) Hysteresis characteristic of a two-terminal RRAM nanodevice. (b) Function as a routing switch in place of a pass transistor and its six-transistor SRAM cell.

of a pass transistor and its six-transistor SRAM cell in conventional FPGAs, as shown in Fig. 1b. Programmable interconnects of FPGAs can therefore be built from nanodevices and have smaller footprints. In addition, these nanodevices are fabricated among metal layers and do not contribute to the footprint of CMOS transistors

Permission to make digital or hard copies of all or part of this work for personal or classroom use is granted without fee provided that copies are not made or distributed for profit or commercial advantage and that copies bear this notice and the full citation on the first page. To copy otherwise, to republish, to post on servers or to redistribute to lists, requires prior specific permission and/or a fee.
DAC'13, May 29–June 07, 2013, Austin, TX, USA.

below them. Note that conventional FPGAs prefer multiplexers to pass transistors as the basic circuits in programmable interconnects for fewer configuration bits, though signals have to pass more levels of gates in multiplexers. However nanodevices provide configuration bit storages along with signal paths. Nanodevice-based FPGAs switch back to pass transistors for higher performance [1–3,5,6]. These nanodevices are also nonvolatile devices and can save leakage power significantly. To summarize, nanodevice-based FPGAs show significant potential to save footprint, critical path delay and power consumption. For example, a NEM-relay FPGA in [3] achieves savings of 43%, 28% and 37% respectively. A RRAM-based FPGA proposed in [1] achieves savings of 80% , 56% and 39% respectively.

In nanodevice manufacturing, defects are a certainty, and reliability becomes a critical issue. The projected defect rate of nanodevices can be up to 10^{-1} which is much higher than the level of 10^{-9}–10^{-12} in CMOS systems [4,8]. A number of approaches have been proposed to improve the FPGA yield by leveraging its reconfigurable structures. Among them, we choose the component-specifc implementation [4,6–11] which works on a defect map obtained from testing as the basic frameowrk of our defect tolerance techniques (see discussion in Appendix S.1).

Defects in programmable nanodevices are manifested as losses of configurability. A defective nanodevice may be stuck at an either "on" or "off" state [7,8]. If the nanodevice is used in a logic block, it leads to a stuck-at-1 or stuck-at-0 bit. If the nanodevice is used in programmable interconnects, it leads to a stuck-closed or stuck-open switch.

Defect tolerance in logic blocks has been explored in recent years [4,8,11]. However in an FPGA chip, programmable interconnects usually occupy 2-4x more area than logic blocks [3,5,12], and are the dominant part. Tolerance of defects in programmable interconnects needs higher attention than that in logic blocks. The stuck-open switches in interconnects can be easily solved by removing the broken edges from the routing graph [6,13]. The authors of [6] showed that yield can remain nearly 100%, even at a defect rate of stuck-open switches as large as 50%. However, they ignored stuck-closed switches, which are much more challenging than stuck-open switches. In Section 2, we will show that stuck-close switches needs to remove >10x routing resources than stuck-open switches when simple defect avoidance is used. However the good thing is that a stuck-closed switch can still be used if we can guarantee that the two nodes shorted by the switch are always mapped to the same net. They can be reflected as extra shorting constraints during the routing phase [7], just like what is done for logic blocks in [4,8]. Along with huge resource savings, this method has two challenges: 1) Defects in programmable interconnects can propagate over the entire chip, and their tolerance has to be solved in a more global way with scalability taken into account; 2) Existing FPGA routing algorithms work on a directed routing graph which assumes that all the edges can programmed to be open or closed, and shorting constraints break this assumption. The second challenge pushes some researchers to switch to algorithms that can easily deal with shorting constraints but lead to poor scalability and solution quality. For example, the SAT-based method in [7] uses Boolean clauses to apply defect constraints, but has high time complexity due to the large search space, and is unable to develop a timing-drive flow based on the satisfiability solver.

The contribution of this paper are as follows. First it provides a complete defect analysis. Then this paper proposes a scalable algorithm to perform timing-driven routing under shorting constraints. We start from the negotiation-based procedure [14], the state-of-art routing algorithm of FPGAs, to maintain the circuit performance and the tool runtime. We extend the idea of the resource negotiation to balance the goals of timing and routability under shorting constraints. We also observe that a routing node will be logically inconsistent with certain nets due to shorting edges. Therefore, we add a mechanism to achieve fast pruning before routing of each net. We also develop several techniques to guide the router to map the shorting clusters to those nets with more shared paths for better utilization of routing resources while automatically balancing it with circuit performance. In addition, we found out that some logic blocks will become virtually unusable due to shorted pins found in the defect analysis. We enhance the placement algorithm to recover these logic blocks.

2. QUANTITATIVE DEFECT ANALYSIS

This section provides a complete impact analysis of both stuck-open and stuck-close defects in programmable interconnects. To the best knowledge of the authors, this is the first work that systematically evaluates the impacts of the two defect types and observes new phenomenon.

2.1 Impact on Routing

As mentioned in Section 1, defects in programmable nanodevices are manifested as losses of configurability [7, 8]. When these nanodevices are used as routing switches in programmable interconnects, the defects can be categorized into two types. The stuck-open defect indicates that the connection between two nodes cannot be used. The stuck-closed defect indicates that the two nodes on the two sides of the switch will always be shorted. Fig. 2 shows an example of the two types of defects.[1] To solve a stuck-open defect,

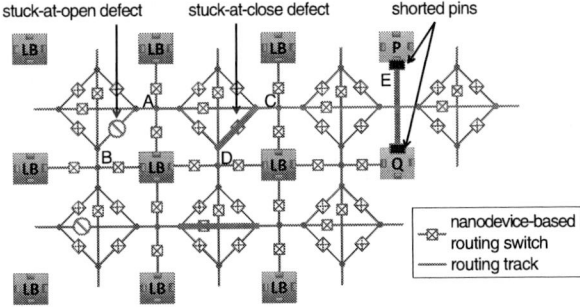

Figure 2: Illustration of stuck-open and stuck-closed defects of nanodevice-based routing switches in programmable interconnects. LB ⇒ logic block.

e.g., the switch between routing track A and B in Fig. 2, we can avoid using it during the routing process. The impact of removing a single edge from the routing graph is very limited since there are always many alternative paths between two arbitrary nodes in programmable interconnects.

However a stuck-closed defect, e.g., the switch between routing track C and D in Fig. 2, has a much higher impact. When track C and D are mapped to two different nets during routing, a logic conflict will occur due to the stuck-closed switch between the two

---
[1]For demonstration purpose, only one routing track per channel is shown in the figure. Routing buffers are also omitted. FPGAs have more complex structures and our tool works on a generalized structure.

tracks. A simple solution to guarantee logic consistence is to avoid using the two routing tracks shorted by any stuck-closed switch (as shown in Fig. 3). This is equivalent to avoidance of all the routing

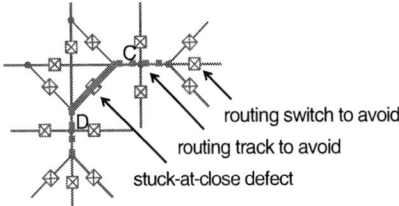

Figure 3: Solve a stuck-closed defect by simple defect avoidance. All of the 15 switches shown in this figure need to be discarded due to a single stuck-closed switch.

switches connected to the two routing tracks. In this case, all of the 15 switches shown in Fig. 3 need to be avoided. The overhead of solving a stuck-closed defect can be 15x that of solving a stuck-open defect when simple defect avoidance is used by the defect-tolerant CAD tool. To quantify the impact of a stuck-closed defect, we develop an approximate model based on probability. Let's use denotation in Table 1. Then the probability of a routing track a that

symbol	meaning
r	defect rate of stuck-closed switches
n	number of switches that a routing track is connected with

Table 1: Denotation of settings for defect impact analysis.

is connected with at least one stuck-closed switch is

$$P(a) = 1 - (1 - r)^n \approx nr \quad \text{for } r \ll 1.$$

It is also the probability of this track to be disabled due to the stuck-closed switch(es). Every routing switch is connected with two routing tracks. A switch s will be avoided if either of the two routing tracks is disabled, at probability

$$P(s) = 1 - (1 - P(a))^2 \approx 2nr - n^2 r^2 \approx 2nr \quad \text{for } r \ll 1. \quad (1)$$

This indicates that for stuck-closed defects, the effective defect rate is enlarged by $2n$ times. Depending on the structure of programmable interconnects, n could be 6–100 [1, 6, 15]. We perform simulations to verify our analytic model using a typical MCNC benchmark [16] mapped onto the RRAM-based FPGA architecture ($n = 6$) in [1] and also a heavily modified version of VPR.[2]. Fig. 4 shows an impact comparison between the stuck-open and stuck-closed defects. The tolerance level of stuck-closed defects

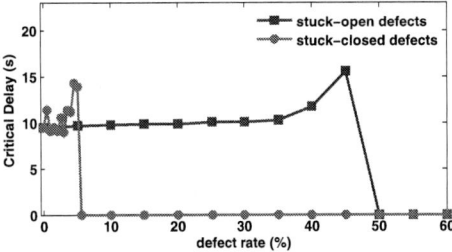

Figure 4: Impact comparison of the stuck-open and stuck-closed defects on routability. Delay going down to zero ⇒ unroutable. ~10x gap observed between the impacts of the two defect types.

is 10x lower than that of stuck-open defects when simple defect avoidance is used. Eq. (1) also leads to a dilemma. When we want to improve routability by adding more switches, i.e., by increasing n, we may not achieve desirable results due to the deteriorating impact of stuck-closed defects. To overcome these difficulties, we need to utilize stuck-closed switches by treating them as shorting constraints during routing instead of simple avoidance.

---
[2]VPR is a state-of-art FPGA CAD tool in academia [15, 17]

2.2 Impact on Placement

Another problem brought on by stuck-closed defects is shorted pins of logic blocks. As shown in Fig. 2, consecutive stuck-closed switches can form shorting paths. Some shorting paths may happen to connect pins of logic blocks together. Take the two physical logic blocks P and Q with shorted pins in Fig. 2 for example. If the two netlist logic blocks which are placed at P and Q do not share common nets in their inputs or outputs, logic inconsistency will be found during routing. Considering the large number of paths between two arbitrary pins of logic blocks, the expectation of the number of logic blocks with shorted pins is not trivial. Again we perform simulations to evaluate this impact using the same settings in Section 2.1. Fig. 5a shows that >60% logic blocks are involved with shorted pins at a stuck-closed defect rate of 5%. It indicates that though the logic inconsistency of placed logic blocks like P and Q can be solved by rejecting all the physical logic blocks with shorted pins during placement, it is not practical to reject >60% of logic blocks. Fig. 5b further shows that the number of logic blocks with shorted pins will also increase as the number of routing tracks per channel increases. This leads to another dilemma. When we

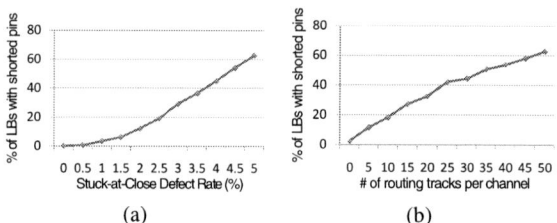

(a)   (b)

Figure 5: The number of logic blocks (LBs) with shorted pins over different stuck-closed defect rates and numbers of routing tracks in channel.

want to improve routability by adding more routing tracks, more paths will be created between pins of logic blocks, and more blocks will be involved with shorted pins. To overcome these difficulties, we need to recover the logic blocks with shorted pins via an enhanced placement algorithm.

3. DEFECT-TOLERANT ROUTING

The FPGA routing resources and their connections are represented as a directed graph $G = (V, E)$, as show in Fig. 6. The

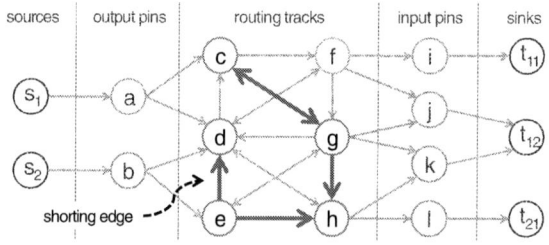

Figure 6: Example of a routing graph with shorting constraints. Bold red arrows indicate shorting edges. Nodes $\{c, d, e, g, h\}$ shorted by the shorting edges form an electrically shorted cluster (ES-cluster) as highlighted.

node set V corresponds to input/output pins of logic blocks as well as routing tracks, the edge set E to routing switches (with buffers). The edges of stuck-open defects will be removed from the graph before routing. The edges of stuck-closed defects will be marked as shorting edges, e.g., $e_s = (v_k \rightarrow v_l)$. Nodes shorted by the shorting edges will form an electrically shorted cluster (ES-cluster), e.g., $\{c, d, e, g, h\}$ in Fig. 6. Associated with each node v is a constant delay $d(v)$ and a congestion cost $c(v)$ determined by the competition among signals for v. Each net i in a netlist to be mapped onto the FPGA will place a source node s_i and multiple sink nodes t_{ij} in the graph. A routing problem is to find a routing tree $RT_i = (V_i, E_i) \subseteq G$ to connect s_i to all t_{ij}, for every i and j.

A valid routing solution requires that every node is included in only one routing tree, i.e., $\forall v \in V, c(v) \leq 1$. If the shorting constraints are applied to the routing solution, it also requires that a successor node is included in the same tree as its precedent node, i.e.,

$$\left. \begin{array}{l} \exists e_s = (v_k \rightarrow v_l) \\ v_k \in V_i \text{ of } RT_i \end{array} \right\} \Rightarrow v_l \in V_i \qquad (2)$$

This section include several technologies to enhance routing under shorting constraints. We implement them in an integrated algorithmic framework, and details can be found in Appendix S.2.

3.1 Enforcement of Shorting Constraints

The basic idea of the enforcement of shorting constraints in our tool is that we do not immediately remove the ES-cluster from the routing graph if any node in the cluster is used by a routing tree. For better solution quality, at the first few routing iterations, we allow multiple routing trees to use the nodes in the same ES-cluster and put circuit performance as the primary optimization goal. Then we gradually increase the penalty on the violation of shorting constraints to guarantee that the final routing solution complies with all the shorting constraints. The negotiation-based routing [14] balances circuit performance and resource overuse via the concept of node congestion. We extend the congestion concept so that negotiation can be performed between circuit performance and shorting constraints as well. Fig. 7 is an illustration of our method. When

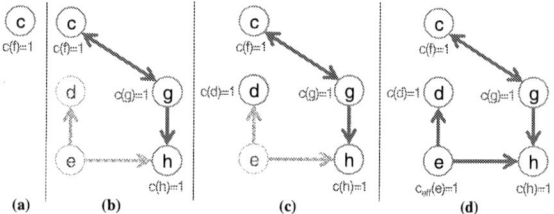

Figure 7: Our extension of the congestion concept for routing under shorting constraints. (a) Add node c to a routing tree. (b) Recursively add all the successor nodes and increase congestions. (c) Add node d to another routing tree. (d) Replace the congestion cost with the effective cost for precedent nodes, e.g., node e.

we add one node to a routing tree, e.g., node c in Fig. 7a, we recursively add all the successor nodes in the ES-cluster to the tree, e.g., nodes g and h in Fig. 7b, and increase all of their congestions by one. Other routing tress can compete for node g and h as well as node c, but with the penalty of their congestion costs. As the routing iteration continues, the router will exponentially increase the weight of the congestion costs in the node costs to eliminate any constraint violation. Other routing trees can still use node d freely since it does not violate any shorting constraint, as shown in Fig. 7c. But adding node e to other routing trees will incur a violation though its congestion has never been increased, as shown in Fig. 7d. That's because node e is a predecessor node of other used nodes in the ES-cluster. To apply shorting constraints, we replace the congestion cost with an effective cost:

$$c_{\text{eff}}(v_l) = \max\{c(v_l), c(v_{l1}), c(v_{l2}), \cdots, c(v_{ln})\} \qquad (3)$$

where v_{lk} for $1 \leq k \leq n$ are all the sink nodes in an ES-cluster of routing nodes that are reachable from v_k. Here sink nodes refer to the nodes without any outgoing shorting edges. In Fig. 7d, the qualified sink nodes for node e are node c and d. By our extension of the congestion concept, the route could utilize all the nodes in ES-clusters as much as it can while balancing with circuit performance.

3.2 Prune Invalid Solutions Before Routing

We also observe that there are nodes that will always be logically inconsistent with certain nets due to shorting edges. It takes many iterations for the router to figure out the inconsistency via the increasing congestion of these nodes. Therefore we add a mechanism to quickly prune these invalid routing solutions by analysis of

56

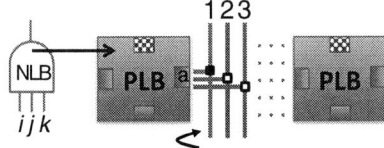

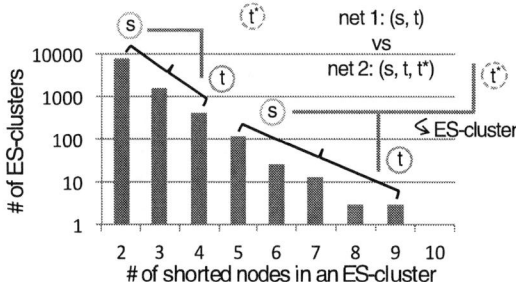

track shorted to pin: reject all nets other than {i, j, k}

Figure 8: Example of fast pruning of invalid routing solutions. Any routing solution that maps track 1 to the net in set $\overline{\{i, j, k\}}$ can be pruned ahead of time.

shorting edges. Fig. 8 shows an example. Track 1 is shorted to pin a of the PLB. It should only be used by net i, j or k since the netlist logic block (NLB) placed in the PLB contains only net i, j and k as inputs. We mark track 1 incompatible with all the nets in set $\overline{\{i, j, k\}}$. We will calculate the incompatible node set for every net before routing. We temporarily remove all the incompatible nodes from the routing graph during the routing of a net to reduce the solution space.

3.3 Smart Mapping of ES-Clusters to Nets

3.3.1 Motivation

Since the shorting constraints are new to the FPGA router, we want to develop some techniques to help the router converge at a solution that maps the ES-clusters to suitable nets for better utilization of routing resources. Fig. 9 shows a motivation example. To route a net from s_2 to t_{21}, there exist two shortest paths, one via

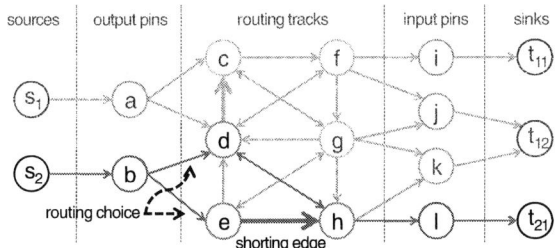

Figure 9: To route a net from s_2 to t_{21}, there exist two shortest paths—one via node d and the other via node e as highlighted. We guide the router to route via node e for better utilization of resources.

node d and the other via node e as highlighted. Conventional FPGA routers treat nodes d and e equally. However the shorting edge from node d to c leads to a waste of node c. On the other hand, the shorting edge from node e to h saves resources. It motivates us to guide the router to map ES-clusters to those nets in more shared paths.

3.3.2 Categorization of ES-Clusters

We discover that different techniques should be applied to large ES-clusters and small ES-clusters respectively. As shown in Fig. 10, while the small ES-cluster can be fully utilized by both net 1 and net 2, the large ES-cluster can be fully utilized only by net 2. This indicates that the benefits of using small ES-clusters can be judged locally during the routing of a net, since smaller ES-clusters have simple topologies and are usually fully utilized as long as they have paths shared with the net. The mapping of large ES-clusters needs to be planned globally before routing since partial utilization is a more common case, and we want to maximize the utilization ratio over more net candidates. The global and local strategies for large ES-clusters and small ES-clusters are also determined by the exponential relationship between cluster size and cluster amount, as shown in Fig. 10.

3.3.3 Global Planning of Large ES-Clusters

We formulate the global planning of large ES-clusters to nets as a search for a subset of edges in a weighted bipartite graph

Figure 10: A distribution of ES-clusters with different sizes in a defective nanodevice-based FPGA. Along with it is an example of the different potentials of small ES-clusters and large ES-clusters exposed to the same nets.

(W, C, S), with net set W, cluster set C and edge set S. The weight of an edge $s(w, c)$ refers to the distance of the shared path between an ES-cluster c and a net w (with reference to its source and sink locations). The goal is to maximize the sum of $s(w, c)$ of all the edges in the subset. The constraint is that no two edges share a common cluster. The optimal solution can be obtained by a greedy algorithm which selects the edge of an ES-cluster c_i to the net with the largest $s(w, c_i)$ among $\forall w \in W$. Proof is omitted due to page limit. Here we assume that a net can be assigned with multiple ES-clusters. In practice we find that due to the limited connectivity of the routing graph, a net is usually able to use only one ES-cluster out of all the clusters assigned to it. To eliminate the waste of clusters, we add the constraint that no two edges share a common net. Now the problem becomes the maximum weighted bipartite graph matching. The optimal solution can be obtained by using the augmenting path algorithm.

3.3.4 Runtime Mapping of Small ES-Clusters

When the router routes a net and has multiple routing node candidates to traverse towards the sink of the net, we guide the router to prefer the node which is connected to an ES-cluster with its path towards the sink, on the condition that this bias does not hurt timing. We enhance the cost function of a node candidate v into

$$\text{Cost}(v) = \text{Crit}(RT_i) \cdot d(v) + [1 - \text{Crit}(RT_i)] \cdot \text{D} \cdot c_{\text{eff}}(v)/s(v) \quad (4)$$

$s(v)$ is a factor added by us to guide the router. It is equal to one plus the distance of the path shared between the involved ES-cluster and the net towards its sink. The other parts in the formula remain unchanged. $\text{Crit}(RT_i)$ is the largest timing criticality among all the paths in the routing RT_i tree to route net i and ranges between 0–1. $d(v)$ is the delay. D is the delay normalization factor. $c_{\text{eff}}(v)$ is the effective congestion cost. In this enhanced cost function, $s(v)$ plays role only when $\text{Crit}(RT_i)$ is small and $c_{\text{eff}}(v)$ is large, i.e., applied only to an uncritical path under a tight budget of routing resources. The proposed cost function enables our use of ES-clusters to automatically balance circuit performance and routability.

4. DEFECT-TOLERANT PLACEMENT

In this section we enhance the placement algorithm to recover the logic blocks which become virtually unusable due to shorted pins. First we show that though it sounds like a good idea to place two netlist logic blocks (NLBs) with shared nets in two physical logic blocks (PLBs) with shorted pins, it is not practical. We discover that even the check of logic consistency on the placement of a group of PLBs with shorted pins cannot be divided into subproblems and therefore is hard to solve. As shown in Fig. 11, NLB pair (b,d) in the netlist has a shared input and output which match the shorted pins of PLB (B,D). NLB pair (b,c) has shared inputs which match the shorted pins of PLB (B,C). However the connection of (b,c,d) does not match the shorted pins of (B,C,D) but matches (B,E,D). In addition, it is not desirable that the placement constraint of a PLB depends on the NLB placed in another PLB—e.g. PLB B and C in Fig. 11. Here we propose a cost-effective way to decorrelate the placement of multiple logic blocks. We reduce

57

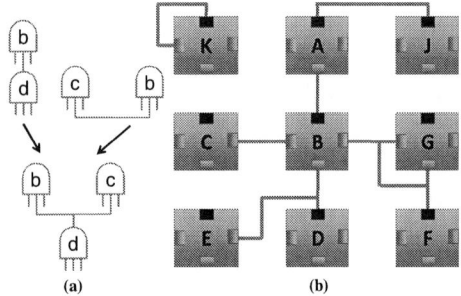

Figure 11: Example of the challenge in the placement of logic blocks with shorted pins. (a) A netlist. (b) Logic blocks with shorted pins (checked pad ⇒ output pin). Though the connections of (b,d) and (b,c) in the netlist match the shorted pins of (B,D) and (B,C) respectively, the connection of (b,c,d) does not match (B,C,D) but matches (B,E,D).

the ES-cluster size of shorted pins by disabling pins so that there is only one pin left in each cluster. All the other pins disabled in each cluster will be mapped to don't-cares for their logic blocks. Then we do not need to consider logic relationships among logic blocks and only need to focus on placement of logic blocks with different numbers of pins. This method works based on two of our observations. The first observation is that in a netlist there are many logic blocks that do not fully utilize their pins, as shown in Fig. 12. The second observation is that most ES-clusters contain only two

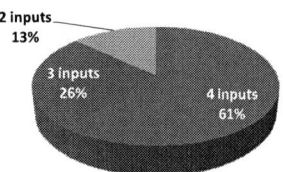

Figure 12: Distribution of logic blocks over different numbers of used inputs in a netlist after logic synthesis. Logic synthesis is performed by the Berkeley ABC tool [18] using a 4-LUT FPGA library. Pins are not fully utilized in many logic blocks.

shorted pins and only half of them need to be disabled, as shown in Fig. 13a.[3] Therefore many logic blocks remain with the full pin set, as shown in Fig. 13b. Implementation details can be found in

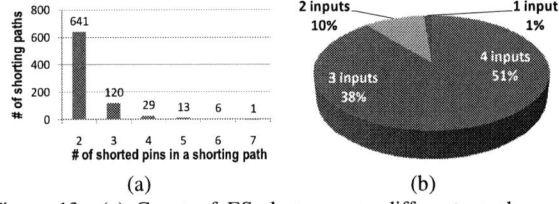

Figure 13: (a) Count of ES-clusters over different numbers of shorted pins. (b) Distribution of physical logic blocks (PLBs) over different numbers of active inputs (i.e., not disabled) after pin disabling.

Appendix S.3.2.

5. SIMULATION RESULTS

5.1 Settings

We implemented our defect tolerance methods in mrVPR [1], a modified version of the state-of-art VPR tool in FPGA CAD society [15, 17]. We chose the RRAM-based FPGA architecture

---

[3]See Section 5.1 for simulation settings.

in [1] as our experiment platform.[4] The technology node is 45nm as in [1]. The channel width is fixed when routability is evaluated. The nanodevice defect rate is set to 10% as reported in the papers published in recent years [4, 8], and stuck-open type and stuck-closed type account for this rate in equal proportion. We also provide sensitivity analysis on the defect rate in Appendix S.4. All experiments are performed on the 20 largest MCNC benchmark circuits [16] and also on two relatively large (>10k LUTs) circuits from the QUIP benchmark design set [19, 20]. Three different tool settings are used and compared: 1) conventional tool setting for defect-free circuits, 2) simple avoidance of logic blocks and routing resources affected by defects as discussed in Section 2, and 3) our method with defect utilization beyond avoidance (namely adaptive defect recovery). Note that we are unable to apply the SAT-based method in [7] to these benchmarks for comparison due to its impractical runtime for large designs. Comparisons of area and timing are made on the results of the three settings above.

5.2 Results

First we verify our defect-tolerant placement. Fig. 14 shows the area usage of benchmarks after placement. While simple defect

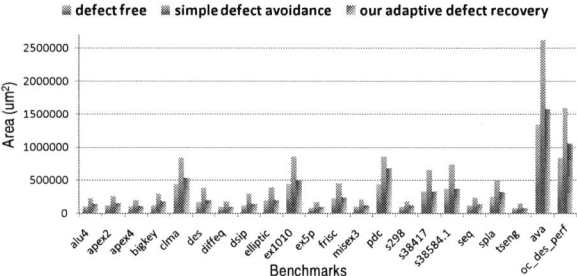

Figure 14: Area usage of benchmarks after placement.

avoidance has an average of 2.01x area compared to the defect-free case, our adaptive defect recovery has only 1.14x area. That's because we can recover most of logic blocks which become virtually unusable due to shorted pins. The result can be made even better if we improve logic synthesis to match the distribution in Fig. 12 to that in Fig. 13b.

Next, we verify our defect-tolerant routing. Fig. 15 shows the routability of benchmarks using the two defect-tolerant methods. Since simple defect avoidance wastes too many routing resources,

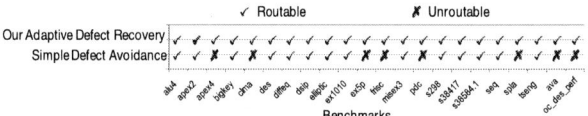

Figure 15: Routability of benchmarks using simple defect avoidance and our adaptive defect recovery.

37% of the benchmarks become unroutable. Our adaptive defect recovery can still keep 100% routability since we successfully use stuck-closed switches. We also perform experiments to justify the effect of smart mapping of ES-clusters to nets. We find that the routability is improved from 90.9% to 100%. Fig. 16 shows the critical delay of benchmarks after routing. This comparison is made on only those benchmarks routable in the case of the simple defect avoidance in Fig. 15. Simple defect avoidance shows an average of 1.43x critical delay compared to the defect-free case. That's because the tight budget of routing resources caused by simple defect avoidance leads to deviation of routing results from the optimal. Our adaptive defect recovery has only an average of 1.07x critical

---

[4]VPR does not provide opportunities to manipulate local interconnects within clustered logic blocks (CLBs). To evaluate and tolerate defects in these parts, we move these parts outside of CLBs by setting the CLB structure to a single logic block.

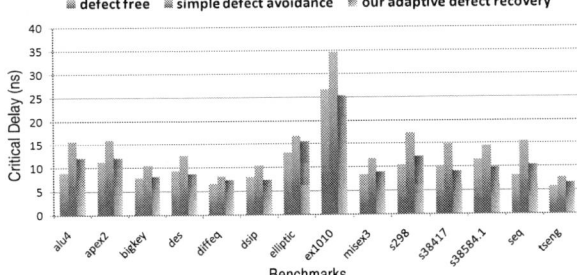

Figure 16: Critical delay of benchmarks after routing.

delay (some benchmarks show even better timing than the defect-free cases due to VPR routing noise [21]). This proves that our method balances circuit performance and routability under shorting constraints.

Fig. 17 is a comparison of runtime complexity between the SAT-based method with defect utilization in [7] and our method. The

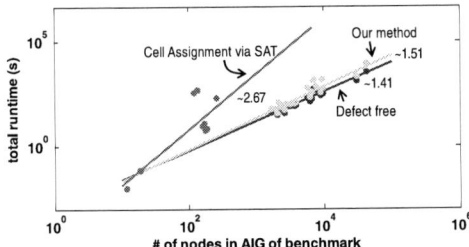

Figure 17: A comparison of runtime complexity. The scales of benchmarks are evaluated as the number of nodes in the form of an and-inverter graph (AIG) of benchmarks.

SAT-based method in [7] shows a high runtime complexity due to the large search solution. In contrast, our method shows complexity similar to the conventional CAD tool for the defect-free case.

6. CONCLUSION

This work focuses on defect tolerance for nanodevice-based programmable interconnects of FPGAs. First, we observe that the stuck-closed defects of nanodevices incur much higher impact than the stuck-open defects. Instead of simply avoiding the stuck-closed defects, we use them by treating them as shorting constraints in the routing. We develop a scalable algorithm to perform timing-driven routing under these extra constraints. We also enhance the placement algorithm to recover logic blocks which become virtually unusable due to shorted pins. Simulation results show that our method is effective for defect tolerance of nanodevice-based FPGAs.

7. ACKNOWLEDGEMENTS

This work was supported by the Center for Domain- Specific Computing (CDSC) funded by NSF "Expeditions in Computing" award 0926127 and financial contributions from Altera and Xilinx.

8. REFERENCES

[1] J. Cong and B. Xiao, "mrFPGA: A Novel FPGA Architecture with Memristor-Based Reconfiguration," in *International Symposium on Nanoscale Architectures (NANOARCH)*, Jun. 2011, pp. 1–8.

[2] P.-E. Gaillardon *et al.*, "Emerging memory technologies for reconfigurable routing in FPGA architecture," in *International Conference on Electronics, Circuits and Systems (ICECS)*, Dec. 2010, pp. 62–65.

[3] C. Chen *et al.*, "Efficient FPGAs using Nanoelectromechanical Relays," in *International Symposium on FPGAs*, 2010, pp. 273–282.

[4] R. Chakraborty *et al.*, "Low-power hybrid complementary metal-oxide-semiconductor-nano-electro-mechanical systems field programmable gate array: circuit level analysis and defect-aware mapping," *IET Computers and Digital Techniques*, vol. 3, no. 6, pp. 609–624, 2009.

[5] C. Dong *et al.*, "3-D nFPGA: A Reconfigurable Architecture for 3-D CMOS/Nanomaterial Hybrid Digital Circuits," *IEEE Transactions on Circuits and Systems I: Regular Papers*, vol. 54, no. 11, pp. 2489–2501, Nov. 2007.

[6] G. S. Snider and R. S. Williams, "Nano/CMOS architectures using a field-programmable nanowire interconnect," *Nanotechnology*, vol. 18, no. 3, p. 035204, Jan. 2007.

[7] W. N. N. Hung *et al.*, "Defect-Tolerant CMOL Cell Assignment via Satisfiability," *IEEE Sensors Journal*, vol. 8, no. 6, pp. 823–830, Jun. 2008.

[8] Y. Su and W. Rao, "Defect-Tolerant Logic Implementation onto Nanocrossbars by Exploiting Mapping and Morphing Simultaneously," in *International Conference on Computer-Aided Design (ICCAD)*, 2011, pp. 456–462.

[9] W.-J. Huang and E. J. Mccluskey, "Column-Based Precompiled Configuration Techniques for FPGA Fault Tolerance," in *International Symposium on Field-Programmable Custom Computing Machines (FCCM)*, 2001, pp. 137–146.

[10] J. Lach *et al.*, "Efficiently Supporting Fault-Tolerance in FPGAs," in *Itnernational Symposium on FPGAs*, 1998, pp. 105–115.

[11] A. Agarwal *et al.*, "Fault Tolerant Placement and Defect Reconfiguration for nano-FPGAs," in *International Conference on Computer-Aided Design (ICCAD)*, Nov. 2008, pp. 714–721.

[12] M. Lin *et al.*, "Performance Benefits of Monolithically Stacked 3-D FPGA," *IEEE Transactions on Computer-Aided Design of Integrated Circuits and Systems*, vol. 26, no. 2, pp. 681–229, Feb. 2007.

[13] R. Rubin and A. Dehon, "Choose-your-own-adventure routing," *ACM Transactions on Reconfigurable Technology and Systems*, vol. 4, no. 4, pp. 1–24, Dec. 2011.

[14] L. Mcmurchie and C. Ebeling, "PathFinder : A Negotiation-Based Performance-Driven Router for FPGAs," in *International Symposium on FPGAs*, 1995, pp. 111–117.

[15] V. Betz *et al.*, *Architecture and CAD for Deep-Submicron FPGAs*. Norwell: MA:Kluwer, 1999.

[16] S. Yang, "Logic synthesis and optimization benchmarks, version 3.0," MCNC, Tech. Rep., 1991.

[17] "VPR 5.0." [Online]. Available: http://www.eecg.utoronto.ca/vpr/

[18] "Berkeley Logic Synthesis and Verification Group, ABC: A System for Sequential Synthesis and Verification, Release 70731." [Online]. Available: http://www.eecs.berkeley.edu/~alanmi/abc/

[19] Altera, "Quartus II University Interface Program." [Online]. Available: http://www.altera.com/education/univ/research/quip/unv-quip.html

[20] J. Pistorius *et al.*, "Benchmarking Method and Designs Targeting Logic Synthesis for FPGAs," in *International Workshop on Logic and Synthesis (IWLS)*, 2007.

[21] R. Y. Rubin and A. M. DeHon, "Timing-Driven Pathfinder Pathology and Remediation: Quantifying and Reducing Delay Noise in VPR-Pathfinder," in *International Symposium on FPGAs*, 2011, pp. 173–176.

[22] Xilinx, "Xilinx: EasyPath series overview." [Online]. Available: http://www.xilinx.com/products/silicon-devices/fpga/easypath-7/index.htm

[23] Z. Hyder and J. Wawrzynek, "Defect tolerance in multiple-FPGA systems," in *International Conference on Field Programmable Logic and Applications*, vol. 153, no. 3, 2006, pp. 247–254.

[24] N. Campregher *et al.*, "Yield enhancements of design-specific FPGAs," *International Symposium on FPGAs*, pp. 93–100, 2006.

[25] F. Hatori *et al.*, "Introducing Redundancy in Field Programmable Gate Arrays," in *Custom Integrated Circuits Conference (CICC)*, 1993, pp. 7.1.1–7.1.4.

[26] A. Yu and G. Lemieux, "FPGA Defect Tolerance: Impact of Granularity," in *International Conference on Field-Programmable Technology (FPT)*, 2005, pp. 189–196.

[27] N. Mehta *et al.*, "Limit Study of Energy and Delay Benefits of Component-Specific Routing," in *International Symposium on FPGAs*, 2012, pp. 97–106.

[28] S. Kirkpatrick *et al.*, "Optimization by Simulated Annealing," *Science*, vol. 220, no. 4598, pp. 671–680, May 1983.

[29] Xilinx, "Vivado Analytical Place and Route." [Online]. Available: http://www.xilinx.com/products/design-tools/vivado/implementation/place-and-route/index.htm

Supplementary Materials

S.1 Existing Defect Tolerance Frameworks

A number of approaches have been proposed to improve the FPGA yield by leveraging its reconfigurable structures. They can be categorized into several groups, including component-specific implementation [4,6–11], design-specific testing [22–24], and adding redundancy to FPGA architecture [25,26]. Among them, component-specific implementation provides the highest level of defect tolerance at a modest area overhead. That's because it provides implementations adaptive to each defect. Defect-tolerant CAD tools are needed to configure FPGAs to work around all detected faults. Component-specific implementation proves beneficial for alleviating process variation as well, which is another main issue as feature sizes scale toward atomic limits [27]. The overhead of component-specific implementation is that the manufacturer needs to try programming every nanodevice in an FPGA chip to obtain the defect map. Then the defect-tolerant CAD flow needs to be performed for every defective chip. Given the high defect rate of nanodevices, all the approaches that target at defect tolerance in the nano era choose component-specific implementation [4, 6–11]. Our work also belongs to this group.

S.2 Implementation of Routing

Given the denotations in Table 2, Algorithm 1 shows how we implement our defect-tolerant routing into the negotiation-based routing procedure [14].

Denotations	Meanings
v, u	routing nodes
s_i	the source node of net i
t_{ij}	the jth sink of s_i
RT_i	routing tree of s_i
$e = (v_k, v_l)$	reconfigurable edge that connects v_k to v_l in routing graph
$e_s = (v_k, v_l)$	shorting edge that connects v_k to v_l in routing graph
$c(v)$	congestion of v which records how many routing trees use this node

Table 2: Denotation table for defect-tolerant routing.

Step 1 of Algorithm 1, discussed in Section 3.3.3, maps large ES-clusters to more suitable nets. The updates of congestion in Step 4, Step 15 and Step 20, discussed in Section 3.1, apply constraints to successor nodes in the routing. The calculation of node cost $\text{Cost}(u)$ in Step 11, discussed in Section 3.3.4, maps small ES-clusters to more suitable nets. The recursive addition of nodes to routing trees in Step 16, discussed in Section 3.1, applies constraints to predecessor nodes in the routing.

S.3 Implementation of Placement

S.3.1 Simulated Annealing

Simulated annealing [28] serves as the placement engine in VPR [15, 17], a state-of-art CAD tool in academia. Algorithm 2 shows how we implement our defect-tolerant placement in simulated annealing.

In step 1 of Algorithm 2, we first check hard violation for shorted pins of logic blocks against circuit rules. For example, in Fig. 11, the output pins of PLB A and J are shorted. In this case, we have to disable one of them — let's say PLB J, to avoid logic conflict.

In step 2 of Algorithm 2, we disable shorted pins in each ES-cluster to decorrelate the placement of multiple NLBs, as discussed in Section 4.

In step 3 of Algorithm 2, we need to generate an initial placement of NLBs at PLBs as the starting point of simulated annealing. For every NLB, we randomly search for a PLB with sufficient active pins. Note that the NLBs with more inputs have fewer PLB candidates. To maximize the number of NLBs that can be placed at PLBs in the given FPGA, we place the NLBs in decreasing order in terms of the number of inputs.

In step 7 of Algorithm 2, during the random swap of the two NLBs placed at two PLBs, we also need to check whether active

---

Algorithm 1: Implementation of our defect-tolerant routing in Pathfinder.

Input : source nodes s_i for each net i and their corresponding sink nodes t_{ij}
Output: $RT_i = (V_i, E_i) \in G$ for $k \to \forall i$,
s.t. $\forall e_s = (v_l \to v_k), v_k \in V_i$ of $RT_i \Rightarrow v_l \in V_i$

1   global planning of large ES-clusters;
2   **while** $\exists$ *overused resources* **do**
3   **foreach** s_i **do**
4   rip-up RT_i and $\forall v \in V_i$ of RT_i, update $c_{\text{eff}}(v)$ in eq. (3);
5   set $RT_i := s_i$;
6   **foreach** t_{ij} of s_i **do**
7   Set priority queue $PQ := RT_i$ with $\text{PathCost}(v) := \text{Crit}(RT_i) \cdot \text{delay}(v), \forall v \in RT_i$;
8   **while** t_{ij} *not found* **do**
9   pop lowest cost node v from PQ;
10   **foreach** $u := \text{fanout}(v)$ **do**
11   add u to PQ with $\text{PathCost}(u) :=$ $\text{PathCost}(v) + \text{Cost}(u)$ shown in eq. (4);
12   **end**
13   **end**
14   **foreach** *node v in path from RT_i to t_{ij}* **do**
15   update $c_{\text{eff}}(v)$ in eq. (3);
16   RecursiveAdd(v, RT_i);
17   **end**
18   **end**
19   **end**
20     $\forall v$, update historical congestion based on $c_{\text{eff}}(v)$ in eq. (3);
21   perform timing analysis and update $\text{Crit}(RT_i)$;
22   **end**

23   RecursiveAdd(v, RT_i) **begin**
24   **foreach** u *where* $\exists e_s = (v \to u)$ **do**
25   RecursiveAdd(u, RT_i);
26   **end**
27   **end**

---

pins suffice the swapped case. It is also possible that a PLB is not placed with any NLB. In this case, the check can be exempted.

S.3.2 Analytical Placement

Recent trends indicate that analytical placement is taking the place of simulated annealing as the mainstream placement method for FPGAs. For example, Xilinx released the Vivado®tool suite to replace its ISE®tool suite which was used for decades [29]. The Vivado®tool suite adopts analytical placement, and its runtime becomes 4x faster than its simulated annealing baseline. Our defect-tolerant placement can also be easily migrated into analytical placement. Enhancement is needed only in the legalization step in analytical placement. In this step, legalization is applied to each type of logic block sequentially. Each type of NLBs will be spread from unaligned locations optimized by an analytical solver for minimum cost function to nearby slots of PLBs with the same type. We can limit the spread of each NLB within those PLBs with sufficient active pins. Algorithm 3 shows how we enhance the legalization. This legalization flow will be applied to each type of logic block. Note that we call the function $\text{OriginalLegalization}(V, P)$ used in the original analytical placement. The task of this function is to spread NLBs in set V from unaligned locations to PLBs in set P. We apply this function to NLBs in set V in a decreasing order of input pins since NLBs with more pins have fewer PLB candidates. By doing so, this flow could maximize the utilization of PLBs.

S.4 Sensitivity Analysis on Defect Rates

To verify the benefits of our method over different defect rates, we also perform a sensitivity analysis. Fig. 18 shows the overall yield of all the benchmarks over multiple defect rates. Yield here is defined as the success rate of a circuit benchmark to fit into

---

Algorithm 2: Integration of our defect-tolerant placement in simulated annealing.

Input : PLB set P, NLB set V.

Denote: $|p|$:= # of inputs of a PLB $p \in P$
$|v|$:= # of inputs of a NLB $v \in V$

Output: An injective function $f : V \to P$,
s.t. $\forall v \in V$, $f(v) = p \in P$ and $|v| \le |p|$.

---

1 Check and disable PLBs with hard violation;
2 Check and disable part of shorted pins;
3 Initial random placement, s.t. $f(v) = p \Rightarrow |v| \le |p|$;
4 while *Simulated annealing continues to improve timing* **do**
5 $\quad \cdots$
6 $\quad$ Randomly select two PLBs p_1 and p_2;
7 $\quad$ Swap: $\{f^{-1}(p_1), f^{-1}(p_2)\} \to \{f^{-1}(p_2), f^{-1}(p_1)\}$, if
$\quad \begin{cases} |f^{-1}(p_i)| \le |p_j| \\ \text{or } f^{-1}(p_i) = null \end{cases}$ for $(i,j) \in \{(1,2),(2,1)\}$;
8 $\quad \cdots$
9 end

Algorithm 3: Implementation of our defect-tolerant placement in the legalization step of analytical placement.

Input : PLB set P, NLB set V.

Denote: $|p|$:= # of inputs of a PLB $p \in P$, $P_i := \{p \mid |p| = i\}$
$|v|$:= # of inputs of a NLB $v \in V$, $V_i := \{v \mid |v| = i\}$

Output: An injective function $f : V \to P$,
s.t. $\forall v \in V$, $f(v) = p \in P$ and $|v| \le |p$.

---

1 $P_c := \emptyset$;
2 $n := \max(i)$;
3 $f := null$;
4 for $i = n, n-1, n-2, \cdots, 1$ **do**
5 $\quad P_c := P_c \cup P_i$;
6 $\quad f_i : V_i \to P_c := \text{OriginalLegalization}(V_i, P_c)$;
7 $\quad$ add f_i to f;
8 end

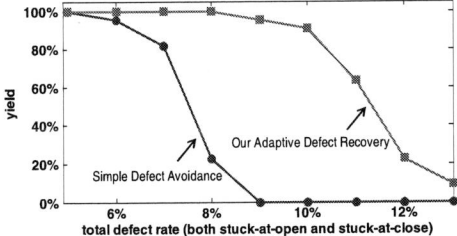

Figure 18: Yield comparison over multiple defect rates.

the logic resources in the given FPGA chip and be routable under given routing resources. Logic resource constraints are set to 1.2x of the defect-free case in this experiment.[5] Our method gains significantly higher yield over simple defect avoidance. Moreover, the improvement on yield is most significant when the defect rate is high. This indicates the capability of our adaptive defect recovery to find more successful implementations in the "difficult region."

---

[5]In practice, the maximum utilization of logic blocks on an FPGA chip is usually less than 80% for the sake of routability.

An Efficient and Effective Analytical Placer for FPGAs *

Tzu-Hen Lin[1], Pritha Banerjee[3], and Yao-Wen Chang[1,2]

[1]Graduate Institute of Electronics Engineering, National Taiwan University, Taipei, Taiwan
[2]Department of Electrical Engineering, National Taiwan University, Taipei, Taiwan
[3]Department of Computer Science and Engineering, University of Calcutta, Kolkata, West Bengal, IN
kenylin770702@eda.ee.ntu.edu.tw, banerjee.pritha74@gmail.com, ywchang@cc.ee.ntu.edu.tw

ABSTRACT

The increasing design complexity of modern circuits has made traditional FPGA placement techniques not efficient anymore. To improve the scalability, commercial FPGA placement tools have started migrating to analytical placement. In this paper, we propose the first academic multilevel timing-and-wirelength-driven analytical placement algorithm for FPGAs. Our proposed algorithm consists of (1) multilevel timing-and-wirelength-driven analytical global placement with the novel block alignment consideration, (2) partitioning-based legalization, (3) wirelength-driven block matching-based detailed placement, and (4) timing-driven simulated-annealing-based detailed placement. Experimental results show that our proposed approach can achieve 6.91× speedup on average with 7% smaller critical path delay and 1% shorter routed wirelength compared to VPR, the well-known, state-of-the-art academic simulated-annealing-based FPGA placer.

Categories and Subject Descriptors

B.7.2 [**Integrated Circuits**]: Design Aids [Placement and Routing]

General Terms

Algorithms, Performance

Keywords

Physical Design, Placement, FPGA

1. INTRODUCTION

Field programmable gate arrays (FPGAs) are programmable platforms which are increasingly competing with ASICs in the market, especially in the embedded computing environment. The main component of any island-style FPGA chip is a pre-fabricated 2D array of *configurable logic blocks* (CLB), programmable routing architecture, and input/output blocks (IOB) along the periphery of the chip. Although recent FPGA chips consist of not only the CLBs, but also other types of resources such as RAM, Multipliers, DSP blocks, and microprocessor cores, CLBs remain to be the majority resources where a *technology-mapped* netlist of designed logic

---

*This work was partially supported by IBM, SpringSoft, TSMC, Academia Sinica, and NSC of Taiwan under Grant No's. NSC 101-2221-E-002-191-MY3, NSC100-2221-E-002-088-MY3, NSC 99-2221-E-002-207-MY3 and NSC 99-2221-E-002-210-MY3.

Permission to make digital or hard copies of all or part of this work for personal or classroom use is granted without fee provided that copies are not made or distributed for profit or commercial advantage and that copies bear this notice and the full citation on the first page. To copy otherwise, to republish, to post on servers or to redistribute to lists, requires prior specific permission and/or a fee.
DAC'13, May 29 - June 07 2013, Austin, TX, USA

blocks needs to be placed. With increasing design complexity and logic capacities in the modern FPGAs, the placement of logic blocks onto the island-style FPGAs has become even more challenging. Being field programmable, FPGAs demand for very fast, scalable, place-and-route tool to map modern large scale designs of high complexity. The earlier works on placement for FPGAs, mostly simulated-annealing-based, produced high-quality placement solutions at the cost of long runtime. The min-cut partitioning-based approaches could achieve good speedup compared to simulated-annealing-based placers, however at the cost of quality loss. Modern FPGAs are finding wide spread use in diverse sets of applications, starting from embedded systems to real-time systems. Long place-and-route time in the physical design stage of FPGAs has a high impact on its time-to-market, and in particular may result in a bottleneck in its use. Thus, it is a challenge to produce a high-quality placement in significantly less runtime to cope with increasing design complexity and logic capacity of modern FPGA chips. More recently, one of the largest FPGA companies, Xilinx Inc., announced its second-generation FPGA software suite: *Vivado Design Suite* [2]. This new software suite migrates from the traditional simulated-annealing-based placements to a modern analytical technology which is expected to handle the newer high-capacity devices [3]. Therefore, now developing and applying analytical placement tools which are both fast and high-quality have become inevitable trends in FPGA design.

1.1 Previous Work

The numerous works on FPGA placement can be classified into three major categories. The first type uses *simulated-annealing-based* methods, such as the most popular, state-of-the-art academic tool VPR [4]. It uses both wirelength-driven (minimization of total half perimeter wirelength (HPWL) over all nets) and timing-driven model (minimization of critical path delay) with a tradeoff (default weight of 0.5 to each) by default. Although this method can achieve very high quality, it tends to have long runtime for large circuits.

The second type of placers use partitioning-based approach such as [15] which employs an alignment cost in the objective function for delay minimization followed by a low-temperature simulated annealing flow. Although it achieves much better speedup than VPR, it suffers from some quality loss.

Placers of the third type apply analytical approaches, which are rather fast and produce high-quality solutions in ASIC placement [7, 12]. Recently a number of analytical placement approaches such as QPF [20], CAPRI [10], StarPlace [19], and HeAP [11] have been proposed for FPGAs as well and are reported to achieve comparable quality compared to VPR. The extensive experimental results reported in [5] show that besides the analytical method, multilevel framework is the key to scalability.

However, these aforementioned analytical placers are either timing-driven or wirelength-driven instead of being both. Furthermore, except CAPRI, these analytical placers simply apply the existing ASIC analytical placement techniques and do not consider the critical path delay issue in their analytical placement framework. On the other hand, CAPRI reports desirable timing results but insufficient speedup compared to VPR.

1.2 Our Contributions

Based on the previous success of multilevel analytical placement

Table 1: Comparisons between our placer and the other timing-driven analytical FPGA placer

	CAPRI [10]	Ours
Timing Optimization	Metric embedding	Block alignment and timing net weight optimization
Multilevel Framework	No	Yes
Look-Ahead Legalization	No	Yes
Detailed Placement	Low temperature simulated annealing	Low temperature simulated annealing

Table 2: COMPARISONS BETWEEN OUR PLACER AND OTHER WIRELENGTH-DRIVEN ANALYTICAL FPGA PLACERS

	QPF [20]	FastPlace+MDP [5]	StarPlace [19]	HeAP [11]	Ours
Wirelength Model	Quadratic	Quadratic	Near-linear Start+	Quadratic	Weighted & stable LSE
Multilevel Framework	No	Yes	No	No	Yes
Look-Ahead Legalization	No	No	No	Yes	Yes
Detailed Placement	Low temperature simulated annealing	Interval bipartite block matching	No	Greedy random block swapping	Window-based bipartite block matching

[5], in this paper we propose the first multilevel analytical placer which considers both wirelength and critical path delay in a more efficient and practical way to place a technology mapped netlist onto a two-dimensional array of CLBs on an FPGA chip. Our main contributions are summarized as follows.

- We propose the first academic multilevel and both timing-driven and wirelength-driven (timing-and-wirelength-driven) analytical placement algorithm for FPGAs. Our proposed algorithm consists of four stages: (1) multilevel timing-and-wirelength-driven analytical global placement with block alignment consideration, (2) partitioning-based legalization, (3) wirelength-driven block matching-based detailed placement, and (4) timing-driven simulated-annealing-based detailed placement.

- We propose a novel differentiable block alignment function for analytical FPGA placement which considers the variable-length pre-fabricated wire segments.

Table 1 summarizes the main features of our proposed timing-driven analytical placement stage against the existing timing-driven analytical placer CAPRI [10]. Furthermore, Table 2 summarizes the main features of our proposed analytical placement against the three existing wirelength-driven analytical FPGA placers: QPF [20], FastPlace+MDP [5], StarPlace [19], and HeAP [11] for clarity. Specifically, our proposed method can achieve 6.91× speedup on average with 7% smaller critical path delay and 1% shorter total routed wirelength compared to VPR.

The remainder of the paper is organized as follows. Section 2 gives the problem formulation of FPGA placement and an overview of the analytical placement flow. Section 3 states our proposed algorithm, and Section 4 shows the experimental results. Finally, we conclude in Section 5.

2. PRELIMINARIES

In this section, we introduce the problem formulation of FPGA placement and the analytical placement framework.

2.1 Problem Formulation

The FPGA placement problem can be formulated as a hypergraph $H = (V, E)$ placement problem. Let vertices $V = \{v_1, v_2, ..., v_{n^2}\}$ represent n^2 logic blocks, and hyperedges $E = \{e_1, e_2, ..., e_m\}$ represent nets. Let x_i and y_i be the x and y coordinates of the center of block v_i, respectively. Given a 2D array of a placement region with size $n \times n$, the placement problem is to determine the optimal coordinate position (x_i, y_i) of each logic block v_i minimizing the total wirelength and the critical path delay.

The analytical placement problem is usually solved in three steps: (1) global placement, (2) legalization, and (3) detailed placement. Global placement finds the best positions for logic blocks minimizing a target cost metric, such as the half-perimeter wirelength (HPWL), ignoring the overlaps among logic blocks. Then, legalization removes all block overlaps. Also in FPGA placement, blocks are placed at the regular integer coordinate positions of a 2D array during legalization. Finally, detailed placement refines the FPGA placement solution.

2.2 Analytical Placement Framework

The placement region is first divided into uniform non-overlapping bin grids in order to spread the blocks. Then, the global placement problem is formulated as a constrained minimization problem [18] as follows:

$$
\begin{aligned}
\min \quad & W(\mathbf{x}, \mathbf{y}) \\
\text{s.t.} \quad & D_b(\mathbf{x}, \mathbf{y}) \leq M_b, \text{ for each bin b,}
\end{aligned}
\tag{1}
$$

where $W(\mathbf{x}, \mathbf{y})$ and $D_b(\mathbf{x}, \mathbf{y})$ are the wirelength and density function respectively. The wirelength function is the total half-perimeter wirelength (HPWL) over all nets. The density function is the total area of movable blocks in bin b and M_b is the maximum allowable area of movable blocks in bin b. Thus, $M_b = w_b h_b$, where w_b and h_b are the width and height of the bin respectively. The density function assists spreading the blocks as they tend to concentrate at the center of the chip for smaller wirelength during HPWL minimization.

For the above constrained minimization problem, we can convert it into an unconstrained optimization problem by introducing a penalty multiplier λ. As a result, we solve a sequence of unconstrained nonlinear optimization problems of the form

$$
\min \quad W(\mathbf{x}, \mathbf{y}) + \lambda \sum_b (D_b(\mathbf{x}, \mathbf{y}) - M_b)^2
\tag{2}
$$

with increasing λs. The solution of the previous problem is used as the initial solution for the next one. We solve the unconstrained problem in Equation (2) by the conjugate gradient (CG) method. To apply gradient search, the optimized function must be differentiable everywhere on the function. Therefore, we need to find differentiable wirelength and density functions to be used for the above formulation.

The HPWL ($W(\mathbf{x}, \mathbf{y})$) is typically defined as follows:

$$
W(\mathbf{x}, \mathbf{y}) = \sum_{\text{net } e} (\max_{v_i, v_j \in e} |x_i - x_j| + \max_{v_i, v_j \in e} |y_i - y_j|).
\tag{3}
$$

Since $W(\mathbf{x}, \mathbf{y})$ is not differentiable everywhere, a smooth approximation function, the *log-sum-exp* (LSE) wirelength model

$$
\begin{aligned}
\hat{W}(\mathbf{x}, \mathbf{y}) = \gamma \sum_{e \in E} (&\ln \sum_{v_i \in e} \exp(\frac{x_i}{\gamma}) + \ln \sum_{v_i \in e} \exp(\frac{-x_i}{\gamma}) + \\
&\ln \sum_{v_i \in e} \exp(\frac{y_i}{\gamma}) + \ln \sum_{v_i \in e} \exp(\frac{-y_i}{\gamma})),
\end{aligned}
\tag{4}
$$

proposed in[18] is used in our framework. The parameter γ in Equation (4) represents the smoothing parameter and a small value of γ approximates the LSE model to HPWL. This model is reported to achieve the best result in the literature [6] for ASIC placement.

Next, the density function is expressed as follows:

$$
D_b(\mathbf{x}, \mathbf{y}) = \sum_{v \in V} P_x(b, v) P_y(b, v),
\tag{5}
$$

where $P_x(b, v)$ and $P_y(b, v)$ denote the overlaps between block v and bin b along the x and y directions respectively. Since the density $D_b(\mathbf{x}, \mathbf{y})$ is neither smooth nor differentiable, bell-shaped function $\hat{D}_b(\mathbf{x}, \mathbf{y})$ is employed as in [6, 7, 12].

3. THE PROPOSED ALGORITHM

In this section we propose a new four-stage optimization algorithm for FPGA placement. Our proposed algorithm consists of four stages: (1) analytical global placement including an improved wirlength approximation, the critical path delay optimization by block alignment and net weighting in a multilevel framework, (2) a simple, yet fast enough partitioning-based look-ahead legalization, (3) wirelength-driven detailed placement by window-based block matching, and (4) timing-driven simulated-annealing-based detailed placement. The input to the global placement is the locations of IO blocks on the periphery of the chip and the netlist of logic blocks. The global placement generates a high-quality placement solution optimizing both the critical path delay and wirelength and is followed by legalization and detailed placement in order to legalize and refine the result of the analytical global placement in both wirelength and critical path delay. Figure 1 summarizes the flow of our proposed algorithm.

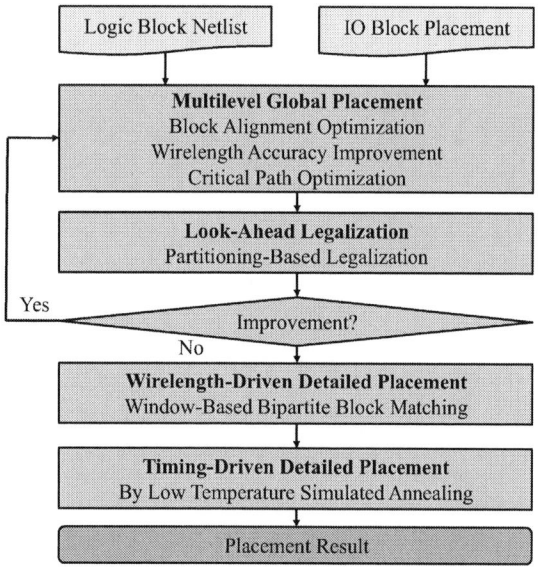

Figure 1: Flow of the proposed algorithm.

3.1 FPGA Global Placement

Our global placement is based on the aforementioned analytical technique, which includes the following methods: (1) multilevel framework, (2) block alignment optimization, (3) wirelength approximation improvement, and (4) critical path optimization.

3.1.1 Multilevel Framework

Solving only the unconstrained optimization problem is not sufficient for scalability when the number of blocks becomes large [5]. To enhance the scalability, we apply the V-cycle multilevel framework similar to that applied in [7]. The V-cycle multilevel framework consists of two main stages: bottom-up coarsening (clustering) followed by top-down uncoarsening (declustering). During the coarsening stage, logic blocks are clustered using the first-choice clustering algorithm in the order of highest to lowest connectivity. The coarsening phase runs till the number of blocks in a given cluster is less than a user-defined threshold. The clustered netlists at the root of the clustering hierarchy are then placed by solving an unconstrained nonlinear objective function. The clusters are declustered to the next lower level of hierarchy and are placed by solving the unconstrained optimization problem as before. The process continues till it reaches the lowest level in the clustering hierarchy.

As an FPGA consists of identical CLBs arranged in a 2D array, a CLB has unit width and height, and thus the value of M_b in Equation (1) is set to 1 for each bin b at the finest level of this multilevel framework. This way, when the density value approaches 1, the placement of blocks also approaches to a legal FPGA place-

ment. We use a timing weight (will be described in section 3.1.4) for critical path delay optimization in the objective function. As we observed that the placement tends to converge much slower on introducing the timing weight in the beginning of the global placement, we use the timing weight only during the finest level of global placement. Each time our placer will obtain the static timing analysis information as in VPR and compute the timing weight of each net. During the finest level of placement, we also use a partitioning-based *look-ahead legalization* step (will be described in section 3.2) to assist optimization to converge faster.

3.1.2 Block Alignment Optimization

FPGA designs have *variable-length pre-fabricated wire segments* which makes them very different from ASICs during routing. A wire segment that spans N CLBs is called a *legnth-N line*; a wire that can span the whole FPGA dimension is called a *long-line*. In particular, the delay of a net in FPGA designs is determined by the number of switch blocks which connect wire segments on routed nets. As switch block delay is much larger than wire delay, more usage of switch blocks in the routed net significantly increases the net delay. Because the number of switch blocks on a net is equal to the number of wire segments on the net minus one, the delay of a net is commonly estimated by the number of wire segments. Besides, many timing-driven FPGA routers tend to minimize the number of wire segments according to the placement result given to them. Therefore, different placement results may result in different net delay. Figures 2(a) and (b) show two kinds of placement examples. The placement in Figure 2(a) results in the same routed wirelength compared to Figure 2(b). However, the placement in Figure 2(a) results in larger delay since its blocks are routed by 3 length-1 wire segments while the blocks in Figure 2(b) are routed by 1 length-3 wire segment (long-line in this example).

If we can align blocks either in horizontal or vertical direction, the delay of routed wires can be reduced [15]. Therefore to minimize the wire delay, only considering wirelength is not sufficient. To remedy this insufficiency, we propose a novel *alignment force* in the unconstrained minimization problem which aligns blocks either in horizontal or in vertical direction. Also, the alignment cost function is smooth and defined as the summation of the product of both horizontal and vertical span of each net as follows:

$$A(\mathbf{x}, \mathbf{y}) = \sum_{\text{net } e} (\max_{v_i, v_j \in e} |x_i - x_j| \cdot \max_{v_i, v_j \in e} |y_i - y_j|). \quad (6)$$

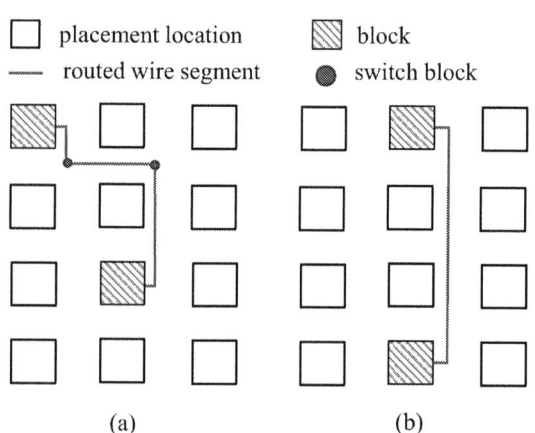

Figure 2: The effect of alignment consideration. (a) Placement without considering alignment. The routed wirelength = 3 and the number of wire segments = 3. (b) Placement considering alignment. The routed wirelength = 3 and the number of wire segments = 1. Smaller number of wire segments stands for smaller delay.

This $A(\mathbf{x}, \mathbf{y})$ is minimized to 0 when either of the two maximum terms becomes 0, which means that blocks are aligned in either horizontal or vertical direction. Therefore, our problem here can be formulated as a constrained optimization problem:

$$\begin{aligned} \min \quad & W(\mathbf{x}, \mathbf{y}) + \mu A(\mathbf{x}, \mathbf{y}) \\ \text{s.t.} \quad & D_b(\mathbf{x}, \mathbf{y}) \leq M_b, \text{ for each bin } b, \end{aligned} \tag{7}$$

where μ is the weight for the alignment cost. To relax the constraint in Equation (7) to objective and solve it by CG method, this alignment cost function has to be differentiable. Therefore, we propose the differentiable alignment cost $\hat{A}(\mathbf{x}, \mathbf{y})$ as follows:

$$\hat{A}(\mathbf{x}, \mathbf{y}) = \gamma \sum_{e \in E} ((\ln \sum_{v_i \in e} \exp(\frac{x_i}{\gamma}) + \ln \sum_{v_i \in e} \exp(\frac{-x_i}{\gamma})) \cdot$$

$$(\ln \sum_{v_i \in e} \exp(\frac{y_i}{\gamma}) + \ln \sum_{v_i \in e} \exp(\frac{-y_i}{\gamma}))). \tag{8}$$

The maximum function in Equation (6) can be approximated by the LSE wirelength model. With this smoothed alignment cost, we can solve Equation (7) by transforming it into an unconstrained problem as follows:

$$\min \quad \hat{W}(\mathbf{x}, \mathbf{y}) + \mu \hat{A}(\mathbf{x}, \mathbf{y}) + \lambda \sum_b (\hat{D}_b(\mathbf{x}, \mathbf{y}) - M_b)^2, \tag{9}$$

Furthermore, since Equation (8) shares the same log-sum-exponential terms with Equation (4), computing this alignment cost will not take much extra time.

3.1.3 Wirelength Accuracy Enhancement

To enhance the accuracy of wirelength approximation for FPGA placement, we adopt two techniques: use of (1) stable LSE, and (2) net-weighting in the wirelength model of Equation (4).

The LSE approximation in wirelength model is a good approximation theoretically. It approaches the exact HPWL when the smoothing parameter γ in Equation (4) approaches zero. However, numeric overflow of the exponential term may occur due to the small γ on implementation, causing numerical instability. To overcome this instability, we apply stable log-sum-exp model (SLSE) proposed in [9] as follows:

$$\gamma \sum_{e \in E} (\frac{X_{max_e}}{\gamma} + \ln \sum_{v_i \in e} \exp(\frac{x_i - X_{max_e}}{\gamma}) +$$

$$\ln \sum_{v_i \in e} \exp(\frac{X_{min_e} - x_i}{\gamma}) - \frac{X_{min_e}}{\gamma} +$$

$$\frac{Y_{max_e}}{\gamma} + \ln \sum_{v_i \in e} \exp(\frac{y_i - Y_{max_e}}{\gamma}) +$$

$$\ln \sum_{v_i \in e} \exp(\frac{Y_{min_e} - y_i}{\gamma}) - \frac{Y_{min_e}}{\gamma}), \tag{10}$$

where $X_{max_e}(X_{min_e})$ is a constant equal to the maximum (minimum) value among the x positions of all blocks connected to net e, and $Y_{max_e}(Y_{min_e})$ the maximum (minimum) among the y positions of all blocks connected to net e. The key is the introduction of the terms $X_{max_e}(X_{min_e})$ and $Y_{max_e}(Y_{min_e})$ so that the exponent of the exponential term will always be less or equal to zero. Thus, the exponential term will never overflow resulting in greater accuracy.

The HPWL metric is a widely used wirelength approximation method. It estimates the wirelength of a net as half the perimeter of the bounding box that encloses all the fanouts of the net to be connected. However, it is very likely to underestimate the wirelength of high-fanout nets since it only looks at the bounding box and ignores the wirelength within the bounding box. Figures 3(a) and (b) illustrate how the HPWL approximation underestimates wirelength of a high-fanout net against a 3-fanout net. The LSE model which is a smooth approximation to HPWL works perfectly for ASIC designs since the percentage of high-fanout nets is small enough to be ignored. However, in circuits for FPGA placement, the number of high-fanout nets are usually much more [5]. In order to compensate for the fact that the LSE model underestimates the wirelength required for nets with more than three fanouts, we

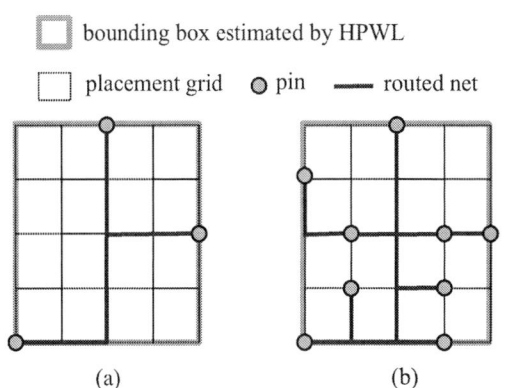

Figure 3: Inaccuracy of HPWL estimation in high-fanout nets. Assume that each placement grid has unit size. (a) A 3-fanout net, HPWL = 8, wirelength = 8. (b) A 9-fanout net, HPWL = 8, wirelength = 14.

use the compensating factor $w_f(e)$ as defined in [4, 8] as the *net weight* in Equation (10). Thus $w_f(e)$ is set to 1 for nets up to three fanouts. Then, it slowly increases to 2.79 for nets with 50 fanouts [8], and for nets with more than 50 fanouts, it linearly increases as follows [4]:

$$w_f(e) = 2.79 + 0.02616(f(e) - 50), \tag{11}$$

where $f(e)$ is the number of fanouts in net e. For nets with more than 85 fanouts, it quadratically increases as follows:

$$w_f(e) = 2.7933 + 0.011f(e) - 0.0000018f(e)^2. \tag{12}$$

Therefore, our weighted SLSE model has a multiplicative term $w_f(e)$ associated to Equation (10). This weighted SLSE model is used to remedy the inaccuracy of $W(\mathbf{x}, \mathbf{y})$ in the nonlinear objective function described in Equation (7) during the global placement stage for FPGAs.

3.1.4 Critical Path Optimization

Besides minimizing the wirelength and delay of each net, minimizing the critical path delay is also crucial for an FPGA circuit. VPR [17] minimizes the critical path delay by evaluating the timing cost as the summation of the timing cost of all connections. Here are the details of how VPR computes the timing cost of each connection of node i and node j (denoted as $connection(i, j)$):

First of all, the arrival time for each node i is defined as:

$$T_{arrival}(i) = \max_{j \in fanin(i)} T_{arrival}(j) + Delay(i, j), \tag{13}$$

where $delay(i, j)$ is the delay of $connection(i, j)$. Similarly, the required time for each node i is defined as:

$$T_{required}(i) = \max_{j \in fanout(i)} T_{required}(j) - Delay(i, j), \tag{14}$$

Then, the slack of a $connection(i, j)$ can be computed as:

$$Slack(i, j) = T_{required}(j) - T_{arrival}(i) - Delay(i, j). \tag{15}$$

Also, the criticality of $connection(i, j)$ is defined as:

$$Criticality(i, j) = 1 - \frac{Slack(i, j)}{D_{max}}, \tag{16}$$

where D_{max} is the maximum arrival time of all nodes. Finally, the timing cost of $connection(i, j)$ as follows:

$$Timing\_Cost(i, j) = Delay(i, j) \cdot Criticality(i, j)^{\alpha}, \tag{17}$$

where α is a user-specified constant.

Instead of summing up all the timing costs of all connections as VPR, we only sum up the timing costs of connections within a single net. Therefore, we define the timing weight of a net e as:

$$w_t(e) = \sum_{i, j \in e} Timing\_Cost(i, j). \tag{18}$$

Then we apply the timing cost of each net as a timing weight of the net wirelength in our analytical placement framework. Therefore, our wirelength model becomes:

$$\gamma \sum_{e \in E} w_t(e) \cdot w_f(e) \left(\frac{X_{max_e}}{\gamma} + \ln \sum_{v_i \in e} \exp\left(\frac{x_i - X_{max_e}}{\gamma}\right) + \right.$$

$$\ln \sum_{v_i \in e} \exp\left(\frac{X_{min_e} - x_i}{\gamma}\right) - \frac{X_{min_e}}{\gamma} +$$

$$\frac{Y_{max_e}}{\gamma} + \ln \sum_{v_i \in e} \exp\left(\frac{y_i - Y_{max_e}}{\gamma}\right) +$$

$$\left. \ln \sum_{v_i \in e} \exp\left(\frac{Y_{min_e} - y_i}{\gamma}\right) - \frac{Y_{min_e}}{\gamma} \right). \quad (19)$$

3.2 Legalization

In FPGAs, the logic blocks need to be mapped to integer coordinate positions on a regular two-dimensional grid. We use a fast partitioning-based approach similar to [13] to obtain a feasible placement either for all the logic blocks. The placement area is recursively cut either vertically or horizontally alternately, pushing the logic blocks to its nearest partition with respect to their x or y positions respectively, depending on the direction of the cut. The process continues until there is exactly one logic block in each partition.

Unlike ASIC designs, logic blocks on an FPGA can only be placed at regular integer coordinate positions on the chip. However, blocks are not guaranteed to be at such positions after the global placement, and this may cause a significant displacement of logic blocks before and after the legalization step. This in turn may generate large differences between the wirelength estimation before and after the legalization. Thus, we apply the legalization step during each iterations in the finest level of global placement stage. This look-ahead legalization not only improves the accuracy of wirelength estimation but also assists in faster convergence with improved quality of placement results.

3.3 Wirelength-Driven Detailed Placement

Since our legalization method is simple, some other better local combinations of block positions may be missed. Therefore, we apply a refinement method similar to that proposed in [7] to further refine the wirelength of the placement solution: a window-based bipartite block matching. This method is reported to be fast and effective among modern detailed placers.

The *block matching* approach uses *bipartite matching* on a bipartite graph having a small group of independent logic blocks (blocks with no common nets) in a window as one set of vertices and the possible new locations as another set of vertices. The cost of an edge (i, j) is given as the difference in HPWL if a block i is placed at location j. Since bipartite matching is used on a very small number of vertices each time, it is very fast.

3.4 Timing-Driven Detailed Placement

The timing-and-wirelength-driven analytical global placement generates excellent placement solutions having good timing and wirelength results efficiently. Beyond that, we can still refine the critical path delay for better results. As a result, we employed default VPR [17] which is both wirelength and timing-driven with weight 0.5 using a low-temperature simulated annealing to the high-quality placement obtained by our proposed analytical global placer. The temperature we set depends on the size of the circuits and is much lower than that of VPR. The low-temperature simulated annealing generates a very high-quality solution converging much faster than a typical simulated-annealing flow while achieving better critical path delay and total wirelength after routing.

4. EXPERIMENTAL RESULTS

To evaluate the performance of our proposed algorithm, we conducted experiments on the 20 MCNC benchmark circuits with 4-input LUT, cluster size 10 clustered with T-VPack [16] most widely used by the academic researchers for FPGA placement. Also, to test the scalability of our multilevel framework, we also generated two larger circuits ourselves. The number of CLBs, IOBs, and the number of nets for each of the circuits are summarized in Table 3

in order of their increasing sizes. Our proposed algorithm was implemented using the C++ programming language, and all the experiments were performed on the same Linux workstation with an Intel Xeon 2.13 GHz CPU and 16 GB memory. Since the VPR tool has become a golden state-of-the-art benchmark for comparison in all previous works on FPGA placement, we compared the quality of our placement results with *default* VPR placer (VPR 5.0.2) [1, 14, 17]. Note that *default* VPR uses a weighted objective function to optimize the wirelength and the critical path delay with an equal weight of 0.5 to each objective, thereby trading off between two objectives. Also, to see the effectiveness of our block alignment optimization, we also compared with our algorithm without block alignment. For fair comparisons, all placers were executed for the same FPGA architecture and the same pre-specified IOB locations. In all cases, we prohibited the movements of the IOBs during the placement iterations. To evaluate the quality of placement solutions, we routed all the placement results by the VPR router. Furthermore, we used the Xilinx' Virtex-5 routing architecture with length-2 lines, length-5 lines and long-lines to place all the benchmark circuits.

Table 4 summarizes the comparison results of critical path delay and total routed wirelength reported by the VPR router, placement runtime, and minimum number of tracks required for routing. In terms of critical path delay, our proposed placer achieves 7% smaller than VPR placer and 5% smaller critical path delay on average. For total routed wirelength, our proposed placer achieves 1% shorter on average than both VPR placer and our placer without block alignment. As for runtime, our proposed placer achieves $6.91\times$ speedup on average compared to VPR and is slightly faster than our placer without block alignment. Moreover, our proposed placer achieves 8% and 1% smaller minimum number of tracks required for routing on average than VPR and our placer without block alignment, respectively. To sum up, from the experimental results we can see that our placer can achieve much more speedup and even better quality than VPR. Moreover, the results also demonstrate the effectiveness of our block alignment algorithm. This shows the efficiency and effectiveness of our placement algorithm.

Figures 4(a) and (b) show the circuit *spla* after global placement and after the final place-and-route with critical paths shown respectively.

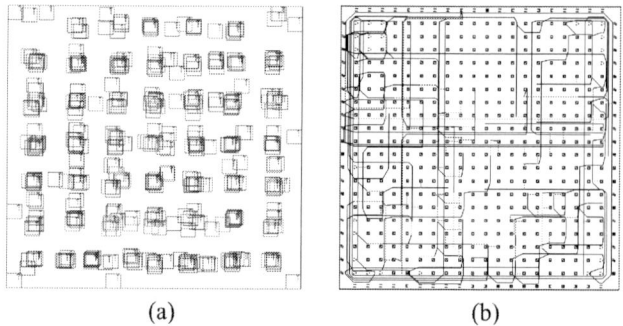

(a)   (b)

Figure 4: Placement of *spla* by our proposed method: (a) After global placement. (b) After the final place-and-route with critical paths plotted in VPR GUI.

5. CONCLUSIONS

Traditional simulated-annealing-based FPGA placers which can achieve very high-quality have suffered from scalability problem for the modern high-complexity designs. In the past, analytical placers were not expected to achieve competitive quality against simulated-annealing-based placers. This paper has proposed an efficient yet high-quality FPGA placer which is the first academic multilevel timing-and-wirelength-driven analytical placer for FPGAs. The proposed placer consists of four stages: (1) multilevel timing-and-wirelength-driven analytical global placement with block alignment consideration, (2) partitioning-based legalization, (3) wirelength-driven block matching-based detailed placement, and (4) timing-driven simulated-annealing-based detailed placement. We com-

Table 4: Comparisons of critical path delay (in nanoseconds), routed wirelength, placement runtime (in seconds), and minimum number of tracks required for routing with VPR and our placer without block alignment optimization.

Circuits	Critical Path Delay (ns)			Routed Wirelength			Placement Runtime (s)			Number of Tracks Required		
	VPR	Ours w/o Align	Ours	VPR	Ours w/o Align	Ours	VPR	Ours w/o Align	Ours	VPR	Ours w/o Align	Ours
tseng	5.74	5.61	5.61	8949	9855	9855	10.97	1.05	1.04	60	52	52
ex5p	5.29	7.41	5.34	11238	13005	11370	7.19	0.8	0.83	138	106	152
apex4	8.42	6.95	6.26	14192	13982	13397	6.05	2	1.76	102	108	108
dsip	5.75	4.35	4.35	22125	20097	20097	22.19	2.85	2.91	44	34	34
misex3	6.2	4.96	6.53	12738	12363	12967	6.66	1.55	1.88	92	92	86
diffeq	6.07	5.34	5.99	9923	10091	10055	9.83	1.33	1.26	52	58	52
alu4	5.87	7.33	5.99	11202	11972	10892	5.92	1.75	1.9	74	74	68
des	7.35	5.8	7.56	32740	27319	29376	35.83	5.16	5.12	60	48	36
bigkey	4.88	3.74	3.74	23235	19422	19422	43.73	2.95	2.97	48	46	46
seq	5.72	5.9	5.9	18006	18347	18347	12.14	1.13	1.44	100	100	100
apex2	6.91	9.12	7.6	19451	19418	19268	10.65	2.34	2.3	100	100	100
s298	11.67	11.64	9.7	11564	12069	11763	6.65	1.83	2.01	52	52	52
frisc	10.19	10.37	10.32	38711	40009	36477	31.97	3.71	3.83	114	112	116
elliptic	11.31	9.64	9.28	37690	38776	40039	38.05	5.12	4.87	102	98	100
spla	10.46	8.16	7.93	44054	43382	44521	28.55	5.6	5.42	122	124	130
pdc	10.91	10.67	9.45	68394	68783	65853	37.07	6.87	6.78	156	154	158
ex1010	8.82	10.34	8.8	49603	51567	51459	36.49	7.16	6.54	122	124	126
s38417	6.48	6.86	6.96	44957	45080	47546	50.27	7.61	7.72	74	76	74
s38584.1	6.06	6.44	6.38	51831	54903	53329	68.45	7.4	8.04	74	70	76
clma	12.84	11.75	11.75	99145	105399	105399	86.76	8.65	8.76	130	140	140
large1	15.7	15.01	13.2	982842	1053637	1014466	4850.17	714.09	739.75	146	144	146
large2	11.72	16.21	15.95	523549	528408	543093	4016.01	770.39	762.13	138	76	74
Normalized	1.07	1.05	1.00	1.01	1.01	1.00	6.91	1.01	1.00	1.08	1.01	1.00

Table 3: Statistics of the benchmark circuits.

Circuits	#CLBs	#Nets	#IOBs	Circuits	#CLBs	#Nets	#IOBs
tseng	107	808	174	s298	194	751	10
ex5p	109	748	71	frisc	356	1983	136
apex4	132	826	28	elliptic	363	2171	245
dsip	137	661	426	spla	374	1996	62
misex3	142	816	28	pdc	463	2436	56
diffeq	151	977	103	ex1010	480	2996	20
alu4	153	738	22	s38417	642	4033	135
des	160	1161	501	s38584.1	645	4126	342
bigkey	171	929	426	clma	842	5164	144
seq	176	1033	76	large1	8420	51631	1431
apex2	190	1180	41	large2	9700	37501	451

pared our placer with the state-of-the-art simulated-annealing-based placer VPR. The experimental results have shown that unlike earlier works, our placer has established the successful use of analytical approaches for fast, high-quality, and practical FPGA placement.

Acknowledgements

We thank the authors of [1, 4, 14, 17] for their open source VPR package.

6. REFERENCES

[1] VPR 5.0.2. http://www.eecg.toronto.edu/vpr/.

[2] Xilinx inc. *White Paper: Vivado Design Suite*, June 2012.

[3] B. Bailey. Second generation for FPGA software. *EE Times*, Apr 2012.

[4] V. Betz and J. Rose. VPR: A new packing, placement and routing tool for FPGA research. In *Proc. Int. Conf. Field Programmable Logic and Applications*, pages 213–222. Springer-Verlag, 1997.

[5] H. Bian, A. C. Ling, A. Choong, and J. Zhu. Towards scalable placement for FPGAs. In *Proc. ACM/SIGDA Int. Symp. Field Programmable Gate Arrays*, pages 147–156, 2010.

[6] T. Chan, J. Cong, and K. Sze. Multilevel generalized force-directed method for circuit placement. In *Proc. ACM Int. Symp. Physical Design*, pages 185–192. ACM, 2005.

[7] T.-C. Chen, Z.-W. Jiang, T.-C. Hsu, H.-C. Chen, and Y.-W. Chang. NTUplace3: An analytical placer for large-scale mixed-size designs with preplaced blocks and density constraints.

IEEE Trans. on Computer-Aided Design of Integrated Circuits and Systems, 27(7), 2008.

[8] C.-L. E. Cheng. RISA: Accurate and efficient placement routability modeling. In *Proc. IEEE/ACM Int. Conf. Computer-Aided Design*, pages 690–695, 1994.

[9] N. Funatsu and Y. Takashima. Overlap-aware analytical placement based on Stable-LSE. In *Proc. of Synthesis and System Integration of Mixed Information Technologies*, pages 318–323, 2009.

[10] P. Gopalakrishnan, X. Li, and L. Pileggi. Architecture-aware FPGA placement using metric embedding. In *Proc. ACM/IEEE Design Automation Conf.*, pages 460–465, 2006.

[11] M. Gort and J. H. Anderson. Analytical placement for heterogeneous FPGAs. In *Proc. Int. Conf. Field Programmable Logic and Applications*, pages 143–150, 2012.

[12] A. B. Kahng and Q. Wang. Implementation and extensibility of an analytic placer. *IEEE Trans. on Computer-Aided Design of Integrated Circuits and Systems*, 24(5), May 2005.

[13] M. Kleinhans, G. Sigl, F. M. Johannes, and K. J. Antreich. GORDIAN: VLSI placement by quadratic programming and slicing optimization. *IEEE Trans. on Computer-Aided Design of Integrated Circuits and Systems*, 10(3), 1991.

[14] J. Luu, I. Kuon, P. Jamieson, T. Campbell, A. Ye, W. M. Fang, K. Kent, and J. Rose. Vpr 5.0: Fpga cad and architecture exploration tools with single-driver routing, heterogeneity and process scaling. *ACM Transactions on Reconfigurable Technology and Systems (TRETS)*, 4(4):32, 2011.

[15] P. Maidee, C. Ababei, and K. Bazargan. Timing-driven partitioning-based placement for island style FPGAs. *IEEE Trans. on Computer-Aided Design of Integrated Circuits and Systems*, 24(3):395–406, 2005.

[16] A. Marquardt, V. Betz, and J. Rose. Using cluster-based logic blocks and timing-driven packing to improve FPGA speed and density. In *Proc. ACM/SIGDA Int. Symp. Field Programmable Gate Arrays*, pages 37–46, 1999.

[17] A. Marquardt, V. Betz, and J. Rose. Timing-driven placement for FPGAs. In *Proc. ACM/SIGDA Int. Symp. Field Programmable Gate Arrays*, pages 203–213, 2000.

[18] W. C. Naylor, R. Donelly, and L. Sha. US patent 6,301,693: Non-linear optimization system and method for wire length and delay optimization for an automatic electric circuit placer. 2001.

[19] M. Xu, G. Gréwal, and S. Areibi. Starplace: A new analytic method for FPGA placement. *Integration*, 44(3):192–204, 2011.

[20] Y. Xu and M. Khalid. QPF: Efficient quadratic placement for FPGAs. In *Proc. Int. Conf. Field Programmable Logic and Applications*, pages 555 – 558, Aug 2005.

Throughput-Oriented Kernel Porting onto FPGAs

Alexandros
Papakonstantinou
ECE Department
University of Illinois
Urbana-Champaign, IL, USA
apapako2@illinois.edu

Deming Chen
ECE Department
University of Illinois
Urbana-Champaign, IL, USA
dchen@illinois.edu

Wen-Mei Hwu
ECE Department
University of Illinois
Urbana-Champaign, IL, USA
w-hwu@illinois.edu

Jason Cong
CS Department
University of California
Los Angeles, California, USA
cong@cs.ucla.edu

Yun Liang
EECS School
Peking University
Beijing, China
ericlyun@pku.edu.cn

ABSTRACT

Reconfigurable devices are often employed in heterogeneous systems due to their low power and parallel processing advantages. An important usability requirement is the support of a homogeneous programming interface. Nevertheless, homogeneous programming interfaces do not eliminate the need for code tweaking to enable efficient mapping of the computation across heterogeneous architectures. In this work we propose a code optimization framework which analyzes and restructures CUDA kernels that are optimized for GPU devices in order to facilitate synthesis of high-throughput custom accelerators on FPGAs. The proposed framework enables efficient performance porting without manual code tweaking or annotation by the user. A hierarchical region graph in tandem with code motions and graph coloring of array variables is employed to restructure the kernel for high throughput execution on FPGAs.

1. INTRODUCTION

Tighter integration of latency oriented CPUs with throughput oriented compute architectures with massive parallelism and low power characteristics is becoming common in many compute domains (e.g. mobile, high-performance, compute clusters, etc) [1, 17, 16]. Programming efficiency is a prerequisite for leveraging the benefits of heterogeneous systems. The introduction of parallel programming models and semantics such as CUDA [15], OpenCL [2] and OpenACC [3] addresses the need for programming heterogeneous processors through a homogeneous programming interface. Homogeneous programming models facilitate functionality porting but often necessitate device-specific code tweaking to achieve performance porting.

In this work we propose a throughput oriented performance porting (TOPP) framework that leverages code restructuring techniques to enable automatic performance porting of CUDA kernels onto FPGAs. CUDA offers explicit control over (i) data memory spaces, (ii) computation distribution across cores, and (iii) thread synchronization. Hence, CUDA kernels designed for the GPU architecture may not map efficiently on reconfigurable devices. The TOPP framework pro-

posed in this work, leverages the hierarchical region graph (HRG) representation to efficiently analyse and restructure the kernel code. Restructuring entails a wide range of transformations including code motions, synchronization elimination (through array renaming), data communication elimination (through rematerialization), and idle thread elimination (through control flow fusion and loop interchange). As data handling plays a critical role in the performance of massively parallel CUDA kernels, the proposed flow employs advanced dataflow and symbolic analysis techniques to efficiently manage data. Graph coloring in tandem with throughput estimation techniques is used to optimize kernel data structure allocation and utilization of on-chip memories. Through orchestration of different code transformation and optimization techniques, the TOPP framework generates C code which is fed to high-level synthesis (HLS) to generate high-throughput custom accelerators on the reconfigurable architecture. Our experimental study shows that the proposed flow improves FPGA execution performance by more than 4X without manual code tweaking from the user.

The main contributions of this work are summarized below:

- Introduction of the hierarchical region graph representation of CUDA kernels.
- Implementation of an automated performance porting flow from CUDA to FPGAs.
- Description of efficient throughput metrics for throughput oriented kernel restructuring.
- Experimental evaluation of the performance porting capability of the TOPP framework.

In the next Section we provide further background information on CUDA-to-FPGA flows and introduce the HRG representation. Section 3 offers an overview of the TOPP framework which is complemented by algorithms and other implementation details in the Appendices. Finally, Section 4 contains the experimental evaluation of TOPP followed by conclusion in Section 5.

2. MOTIVATION AND BACKGROUND

CUDA employs a SIMT (single instruction, multiple threads) parallel programming interface which efficiently expresses multiple fine-grained threads executing as groups of cooperative thread arrays (CTA). The GPU architecture comprises high-throughput compute cores grouped in Streaming Multiprocessors (SMs). Computation is distributed across SMs at CTA granularity [15]. A carefully crafted interconnect scheme between SMs and off-chip memory facilitates high-bandwidth data accesses at low latency overhead.

Permission to make digital or hard copies of all or part of this work for personal or classroom use is granted without fee provided that copies are not made or distributed for profit or commercial advantage and that copies bear this notice and the full citation on the first page. To copy otherwise, to republish, to post on servers or to redistribute to lists, requires prior specific permission and/or a fee.
DAC '13, May 29 - June 07 2013, Austin, TX, USA.

Listing 1: CUDA code for DWT kernel

```
1   for ( tid =0; tid <bdim ; tid ++){
2       shr [ tid ] = id [ idata ];
3       __syncthreads ();
4       data0 = shr [2* tid ];
5       __syncthreads ();
6       od [ tid_global ] = data0 *SQ2;
7       shr [ tid ] = data0 *SQ2;
8       __syncthreads ();
9       numThr = bdim >> 1;
10      int  d0 = tid * 2;
11      for (int  i =1; i<lev ;++ i){
12          if ( tid < numThr ){
13              c0 = id0 +( id0 >>LNB );
14              od [ gpos ] = shr [ c0 ]* SQ2;
15              shr [ c0 ] = shr [ c0 ]* SQ2;
16              numThr = numThr >>1;
17              id0 = id0 <<1; }
18      __syncthreads ();    } }
```

Listing 2: C code for DWT kernel

```
1   for ( tid =0; tid <bdim ; tid ++){
2       shr [ tid ] = id [ idata ]; }
3   for ( tid =0; tid <bdim ; tid ++){
4       d0 [ tid ] = shr [2* tid ];}
5   for ( tid =0; tid <bdim ; tid ++){
6       od [ tid_glob ] = d0 [ tid ]* SQ2;
7       shr [ tid ] = d0 [ tid ]* SQ2;}
8   for ( tid =0; tid <bdim ; tid ++){
9       numThr = bdim >> 1;
10      id0 [ tid ] = tid * 2;}
11      for (int  i =1; i<lev ;++ i){
12          for ( tid =0; tid <bdim ; tid ++){
13              if ( tid < numThr ){
14                  c0 = id0 [ tid ]+( id0 [ tid ]>>LNB );
15                  od [ gpos ] = shr [ c0 ]* SQ2;
16                  shr [ c0 ] = shr [ c0 ]* SQ2;
17                  numThr = numThr >>1;
18                  id0 [ tid ] = id0 [ tid ]<<1;  }}}
```

Reconfigurable devices, on the other hand, offer grids of fine-grained compute and storage resources that can be synthesized into parallel processing custom cores (CCs) at different granularities. FPGAs offer the benefit of application-driven compute customization at the cost of area overhead for reconfigurability. HLS flows enhance FPGA design efficiency by facilitating fast and easy design at higher abstraction. Achieving high-throughput implementations on FPGA requires cautious allocation and coordination of the available compute and storage resources. This depends heavily on parallelism expression and organization in the input code of HLS design flows. The TOPP framework combines advanced code transformations to enable high-throughput designs in CUDA-to-RTL HLS flows.

2.1 SIMT-to-C Compilation

Previous works have described SIMT-to-C (S2C) compilation flows porting kernels onto multicore CPUs [23] and FPGAs [20, 22, 19, 13]. A common characteristic of these S2C flows is the expression of threads as loops over the CTA thread ID (tID), hereafter referred to as tID-loop. Thread synchronization is enforced through loop fission (e.g. tID-loop in line 1 of List. 1 is split into 5 loops in List. 2), loop interchange (e.g. loops in lines 11, 12 in List. 2) and variable privatization transformations (e.g. d0 in line 4 of List. 2). Thread-loop unrolling in tandem with vector loads/stores may be used to exploit the CUDA thread parallelism in the kernel.

Kernel decomposition into computation (COMP) and communication (COMM) tasks has been proposed in [20, 22, 21]. Communication tasks comprise data transfers to/from off-chip memory. Task decomposition is critical in optimizing CTA execution latency on the reconfigurable fabric. Aggregating off-chip memory accesses across CTA threads within COMM tasks facilitates efficient off-chip memory bandwidth through data transfer bursts. Decomposition may also benefit COMP task latency by eliminating data fetch latency through data prefething. The kernel decomposition proposed in [20, 22, 21] is based on user-injected annotations that assist the compiler in identifying COMP and COMM tasks.

This work employs the kernel decomposition philosophy but eliminates the need for user-injected annotations. The proposed flow leverages sophisticated analysis and transformation techniques to identify and re-organize kernel tasks so as to optimize execution throughput on the FPGA architecture.

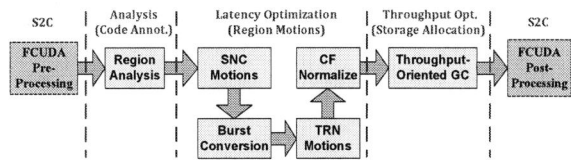

Figure 1: TOPP framework integrated in FCUDA

We will use the Nvidia SDK kernel for discreet wavelet transforms (DWT) as a running example to motivate the importance of throughput-oriented performance porting (TOPP). The DWT kernel (List. 1) contains thread-dependent control flow (line 12), thread synchronization directives (line 3, 5, 8, 18), and intermingled computation and communication regions/statements (lines 6,14) which render manual task annotation cumbersome. Moreover, code restructuring may be required to eliminate kernel fragmentation into fine grained COMP/COMM tasks; e.g. tsk(6), tsk(7), tsk(9,10), etc. in List. 1, where tsk($x[, y]$) denotes the task contained within line(s) x (to y). Fine-grained tasks can negatively impact performance through (i) overhead of implicit thread synchronization across tasks and (ii) increased storage overhead due to variable privatization for variables referenced across tasks (e.g. data0 is referenced in tsk(4) and tsk(6) and hence it is privatized with respect to tid).

The proposed framework leverages rigorous analysis and transformations in tandem with throughput estimation techniques to maximize execution throughput on FPGA. With regard to previous works, the proposed flow elevates the importance of data communication and storage in execution throughput and tries to balance task latency optimization with memory resource allocation for each task in order to maximize kernel execution throughput. Dataflow, value range and symbolic expression analysis in tandem with efficient code motions, and memory allocation optimizations are applied in a phased approach depicted in Figure 1. TOPP is integrated in the FCUDA flow [20, 21] and comprises three major phases: (i) kernel analysis (code analysis and annotation), (ii) task latency optimization (code restructuring) and (iii) throughput optimization (efficient storage allocation). A hierarchical region graph (HRG) representation of the kernel is built during the analysis phase and it is used throughout the subsequent transformation stages.

2.2 Hierarchical Region Graph (HRG)

Hierarchical task graphs (HTGs) have been previously proposed for code representation as a means of extracting parallelism in compilers [10] and HLS flows [12]. In these works the HTG is generated from a sequential low-level 3-address representation of the application and incorporates control and data dependence information along with control-flow hierarchy. The HRG, on the other hand, is generated from high-level SIMT code and summarizes the computation, communication and synchronization characteristics of the application along with data and control flow dependence. The HRG represents the kernel as a tree graph $G_{HRG} = (V, E)$, where each leaf vertex, $v_l \in V$, represents a code region of type t_r and each internal vertex, $v_h \in V$, represents a control-flow (CF) structure of type t_f. Fig. 2 depicts the HRG for the DWT kernel. Region type, t_r, specifies whether the leaf node corresponds to a compute (CMP), communication (TRN), or synchronization (SNC) region. Control-flow type, t_f, identifies the dependence of the CF condition expression from thread ID (tID). Double-rimmed (purple) nodes represent tID-variant (TVAR) CF whereas single-rimmed (green) nodes represent tID-invariant (TiVAR) CF. HRG edges, E, include a set of hierarchy edges, E_H (non-dashed edges connecting nodes into a tree, which denote CF structure), and a set of data dependence edges, E_D (dashed edges). Control flow dependencies

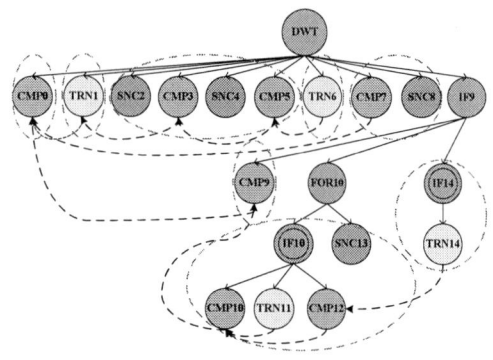

Figure 2: HRG for DWT kernel

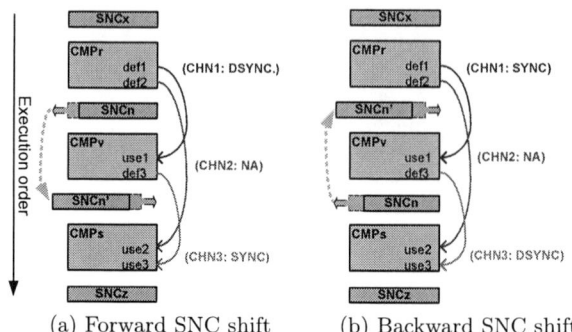

(a) Forward SNC shift   (b) Backward SNC shift

Figure 3: SNC region motions

can be extracted by a depth-first traversal of the HRG tree (non-dashed edges), i.e. child nodes are ordered in control-flow order.

The HRG summarizes the kernel region organization and enables easy and throughput-driven kernel decomposition into COMP and COMM tasks through depth-first traversals (DFT) of the HRG tree. Moreover, it facilitates efficient feasibility and cost/gain analysis of different code transformations. Hence, transformations may be evaluated on the HRG representation before being applied on the SIMT code. Sequences of transformations implemented on the HRG can be applied to source code by translating the resulting HRG tree to C or CUDA code. The algorithm for CUDA-to-HRG translation is described in Appendix A.

3. TOPP FRAMEWORK OVERVIEW

TOPP has been implemented in the FCUDA flow [20, 21], as an interleaved sequence of analysis and transformation phases (Fig. 1). An overview of the integration of TOPP in FCUDA is provided in Appendix B. TOPP analysis and transformation phases are discussed in the following subsections.

3.1 Analysis Phase

Region analysis, (Fig. 1), identifies the CMP, TRN and SNC regions of the kernel and annotates each statement with region and thread-variance (TVAR) information. The annotated information is used in the generation of the HRG. The analysis process is carried out as a sequence of six steps: ($A1$) Identify global memory accesses, ($A2$) Normalize multi-type statements, ($A3$) Build Def-Use chains [18], ($A4$) Find tID-variant (TVAR) statements, ($A5$) Annotate TVAR statements, and ($A6$) Build the kernel HRG. Initially global memory variables are identified and all global memory references are collected in step $A1$. Global memory variables include CUDA __constant__ variables and C-pointer parameters of the kernel procedure, as well as all of their alias definitions through pointer arithmetic. During step $A2$, kernel statements are scanned for multi-type statements, i.e. statements entailing both CMP and TRN operations. Each such statement is converted into separate single-typed CMP and TRN statements. Subsequently, dataflow analysis is used to build Def-Use chains (step $A3$) which facilitate tID-variant (TVAR) variable and statement identification during step $A4$ and tagging during step $A5$ (a statement is tagged as TVAR, if it contains the definition of a TVAR variable). Finally the HRG is constructed in step $A6$ (Alg. 1) based on the analysis information annotated on the kernel statements.

3.2 Latency Optimization Phase

The latency optimization phase in TOPP comprises different region motion stages which aim to eliminate the execution latency overhead resulting from excessive (i) CMP and TRN interleaving, and (ii) synchronization directives. Hence, the

goal of the transformations applied in this phase is to reduce CTA execution latency through region reorganization so as to enable the creation of coarser COMP and COMM tasks. As an example we can use the organization of regions in the initial DWT HRG (Figure 2) which can be arranged into eight tasks (marked with red dashed circles). Since each task is outlined in a separate task procedure in the FCUDA flow, task boundaries represent implicit synchronization points (ISPs) imposing synchronization overhead and bounding ILP extraction space at the thread level. Moreover, multiple fine-grained tasks result in extra TVAR variables with ISP-crossing lifetimes. This is dealt in FCUDA with variable privatization along the tID dimension, leading to higher BRAM resource usage (e.g. variables d0 and id0 in List. 1 are privatized after task decomposition in List. 2). The TOPP framework considers the impact of privatization on BRAM allocation and employs region motions and merging to reduce ISP count.

The HRG in tandem with the annotated Def-Use chain information plays a critical role in region motion feasibility analysis and cost/gain estimation during this optimization phase. Each Def-Use chain that crosses multiple regions is characterized as either *thread shared chain* (TSC) or *thread private chain* (TPC). TSCs refer to chains corresponding to __shared__ or global variables, where explicit synchronization between the definition region and the use region may be required (e.g. chain corresponding to def and use of __shared__ variable shr in lines 2 and 4, respectively of List. 1; represented with dependence edge between TRN1 and CMP3 regions in Figure 2). TPCs, on the other hand, correspond to variables that host values read by the thread that wrote them (i.e. same def and use thread per value) which are not affected by CTA synchronization dependence-wise. Nonetheless, synchronization might affect the storage allocation of TPC variables as discussed earlier. Hence, TSCs affect the feasibility of region motions, whereas TPCs affect the cost/gain estimation analysis of region motions. There are three possible effects that region motions may have on Def-Use chains: (i) Desynchronization (DSYNC), (ii) Synchronization (SYNC), or (iii) Not affected (NA). Desynchronization happens in the case that the explicit or implicit synchronization points between source and sink regions of a chain are removed. For example, chain CHN1 comprised of def1 and use1, in Fig. 3(a), is desynchronized when SNCn region is shifted below CMPv becoming SNCn'. Correspondingly, chain CHN3 between def3 and use3 is synchronized for the same motion of SNCn, whereas CHN2 is not affected by this region motion. Determining which case a region motion corresponds to, is based on the partial ordering enforced by the region identifiers (rIDs) of the involved regions.

The feasibility of a region motion with respect to a TSC is determined by the motion effect on the chain (i.e. DSYNC, SYNC or NA) in combination with the value of its dependence distance vector [5]. Specifically, in case of SYNC or NA motion effects on the TSC, feasibility is positive regardless of the

70

dependence vector distance (e.g. CHN2 and CHN3 in Figure 3(a)). However, in case of DSYNC motion effect on the TSC, the dependence distance vector needs to be examined in order to determine feasibility. We leverage the work in [11] and extend it by applying dependence distance vectors in determining region motion feasibility. Specifically, the authors of [11] show that it is feasible to remove implicit synchronization points (ISPs) between the source and sink of a Def-Use chain as long as one of the following rules holds with respect to the chain's distance dependence vector v:

- $v[0] == 0$
- $v[0] < 0 \wedge v[1 : (|v| - 1)] == 0$
- $v[0] == v[i] : i \in [1 : (|v| - 1)] \wedge v[1 : i] == 0$

where $v[0]$ corresponds to the index of the tID-loop and $v[0] < 0$ denotes an inter-thread data dependence. For the purpose of determining the feasibility of a region motion we also apply this test to *explicit synchronization points* (ESPs). Distance vectors are evaluated leveraging symbolic analysis ([18]) in combination with range analysis ([6, 9]) and array dependence analysis ([18, 5]). If none of the conditions can be proven, feasibility is not confirmed and the corresponding region motion is rejected. In each of the TRN motions and SNC motions stages, cost-function based evelation is used to quantify the benefit of a motion with regard to the following factors:

- De-synchronized TPCs gain
- Synchronized TPCs cost
- Explicit synchronization point (ESP) elimination gain
- Implicit synchronization point (ISP) overhead cost

Appendix C discusses in further detail the transformation stages in the latency optimization phase of TOPP.

3.3 Througput Optimization Phase

During this phase TOPP leverages throughput estimation techniques along with resource information to guide kernel restructuring. Hence, the optimization goal is shifted toward CTA grid execution throughput (vs. CTA execution latency, previously), taking into account the available resource on the target device.

3.3.1 Throughput Factors and Metrics

Throughput of system configuration C with N custom cores (CCs), TP_C, can be expressed as: $TP_C = \frac{EP_N}{cp}$, where EP_N represents the cumulative CTA execution progress across all CCs completed per clock period, cp. For the purpose of throughput-oriented kernel restructuring we leverage the clock period selection feature offered by the HLS engine used in our flow. That is, the generated RTL is pipelined according to the selected clock period, and operation cycle latencies are adjusted accordingly. We have created cycle latency tables (CLT_{cp}) by characterizing operation cycle latencies for different clock periods (cp). These tables are used in TOPP to estimate cycle latency and throughput for a chosen clock period. Hence, the CTA execution throughput metric can be expressed in terms of cycle latencies as: $TP_C = N_{CC} \div (CL_{COMP} + CL_{COMM})$, where configuration C has N_{CC} cores with compute and communication task cycle latencies of CL_{COMP} and CL_{COMM}, respectively. The number of cores, N_{CC}, is estimated for the selected FPGA device based on (i) the number of arrays required per CTA by configuration C and (ii) resource allocation feedback provided from the HLS engine. Latencies CL_{COMP} and CL_{COMM} are calculated as the sums of the sequential CMP and TRN region latencies per CTA in configuration C, respectively: $CL_{COMP} = \sum_i CL_{CMP_i}$, and $CL_{COMM} = \sum_j CL_{TRN_j}$. Concurrent tasks are represented by the latency of the longer task (the HLS engine schedules tasks in a bulk synchronous

way; tasks may either start concurrently, if not dependent, or sequentially, otherwise.) Cycle latency CL_{CMP_i} of compute region CMP_i, is estimated by determining the task's critical execution path. Def-Use chains are used for identifying the critical execution path, while operation cycle latencies are referenced from CLT_{cp} tables. Cycle latency estimate, CL_{TRN_j}, of data transfer region TRN_j is affected by two main factors: (i) the on-chip memory bandwidth and (ii) the off-chip memory bandwidth. The former is estimated based on the on-chip SRAM memory port bandwidth (BW_S), the execution frequency and the read/write data volume. The latter depends on the off-chip DDR memory system peak bandwidth, (BW_D), provided by the user, the extent of static coalescing achieved by the burst conversion stage in the latency optimization phase and the read/write data volume of the task. The final COMM task latency is calculated as $CL_{TRN_j} = max(CL_{SM_j}, CL_{DM_j})$, where CL_{SM_j} corresponds to the on-chip memory access latency and CL_{DM_j} corresponds to the off-chip memory access latency. As described above, CL_{SM_j} is mainly dependent on the architecture of the chosen configuration, C, while CL_{DM_j} is mainly constrained by the value of BW_D provided by the user.

3.3.2 Throughput-Driven Graph Coloring

Graph coloring is often used in compilers for the allocation of registers to program variables and temporary values [8, 7], due to its ability to lead to efficient solutions. Registers represent the most scarce but efficient storage resource at the topmost level of memory hierarchy and thus good register allocation is critical to performance. The SIMT programming model used in FCUDA offers visibility of different memory address spaces with different memory attributes. The goal of the throughput-driven graph coloring (TDGC) transformation in TOPP is to enhance the allocation of kernel arrays onto FPGA Block-RAM (BRAM) memories, considering both kernel characteristics and resource availability. The proposed TDGC algorithm leverages the throughput metrics described in Section 3.3.1 to optimize performance through efficient (i) allocation of arrays onto BRAMs and (ii) off-chip data transfer scheduling.

TDGC entails three main steps: ($GC1$) Array coloring, ($GC2$) Throughput estimation, and ($GC3$) Data communication task (COMM) rescheduling. The three steps may be iterated until no more throughput improving rescheduling alternatives are available. In most cases, the number of iterations is small (not exceeding 3). Initially, candidate arrays for allocation are identified (step $GC1$) and an interference graph, G_I, is generated (Fig. 12(a)). Vertices in G_I correspond to array lifetimes, whereas edges represent overlapping array lifetimes in the kernel. The interference graph, G_I, is colored using a modified R-coloring [18] algorithm (R represents the number of BRAMs per CTA). Coloring determines a BRAM allocation configuration which is used in step $GC2$ to estimate throughput using the metric discussed in 3.3.1. The number of instantiated CCs, N_{CC}, is determined based on the BRAM allocation selected in step $GC1$ and resource estimation feedback from the HLS with respect to other type of resources. If BRAM turns out to be the throughput limiting resource (i.e. it constrains N_{CC}), we employ COMM task rescheduling in step $GC3$ as a means to reduce BRAM requirements. Specifically, G_I nodes are characterized based on their interference degree, L_{ID}, and their *idle lifetime intervals* (ILI), L_{II}; we define as idle the intervals of an array lifetime that correspond to HRG regions where the array is not accessed. Subsequently, nodes are sorted with respect to *lifetime scatter*: $L_S = L_{ID} * \frac{L_{II}}{L_T}$, where L_T represents the total lifetime interval. Nodes are examined in decreasing L_S order with regard to the feasibility of reducing their ILI (and subsequently their interference degree) through TRN region motions and the benefit of such motions in the interference degree of the G_I graph. If a node fulfilling these requirements is found, the HRG is modified and the TDGC steps reiterated until no further candidate nodes are available. At each iteration of the TDGC steps, the TP_C of the new configuration is estimated (step $GC2$) and the TRN region motion is committed only for configurations with higher TP_C (See Appendix D for further details in TDGC and R-coloring).

4. EXPERIMENTAL EVALUATION

TOPP framework is implemented within the FCUDA flow and its analysis phase essentially replaces the (manual) annotation task (Fig. 7). Moreover, the latency and throughput optimization phases of TOPP apply performance oriented code restructuring prior to compiling the SIMT code into explicitly parallel C code for the HLS engine. The HLS engine integrated in the flow is Vivado-HLS [4], which is the successor of AutoPilot [24] used in [20, 22]. The CUDA kernels used in our experiments have been selected from the Nvidia SDK [15] suite. Our experimental evaluation is centered around exposing the effect of the employed TOPP transformations on performance. Specifically, in the next section we measure the performance impact from the individual latency oriented transformations on execution. Subsequently, the effectivenes of the metric used to guide throughput optimization is tested in Section 4.2. Finally, we evaluate the total kernel execution speedup achieved by integrating TOPP into FCUDA, in Section 4.3.

4.1 Latency Optimization Evaluation

First we evaluate the effect of the transformations applied during the latency optimization phase (Figure 1) on CTA compute task latency. Specifically, we measure the effect of each individual transformation on latency by disabling the TDGC transformation in the throughput optimization phase of TOPP and enabling only the desired latency transformations in the latency optimization phase. Initially we enable only the SNC motions (SM) transfomation and measure the relative speedup of the compute latency over FCUDA. We gradually enable the other transformations of the latency optimization phase in the order executed within TOPP (1) and compare the cummulative speedups achieved over the original FCUDA flow in [20] (Fig. 4). The speedup achieved by each set of enabled latency transformations depends on the kernel and code structure characteristics. Kernels that either contain long dataflow paths (e.g. FWT2) or more convoluted control flow paths (e.g DWT) offer more opportunities for optimization. We observe that TRN motions (TM) can have significant impact in the compute latency (e.g. FWT2 and DWT). This is due to enabling the generation of coarser COMP tasks by shifting interleaved TRN regions. It is interesting to observe that burst conversion (BC) results in good speedups for some kernels (e.g. FWT1 and FWT2), even though COMM task latency is not considered in Fig. 4. This is mainly due to the address calculation simplification from consolidating the

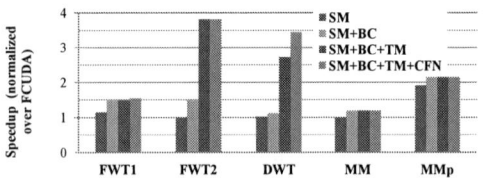

Figure 4: Execution speedup over FCUDA from cummulative application of latency transformations: (i) SNC region motions (SM), (ii) burst conversion (BC), (iii) TRN region motions (TM), and (iv) control flow normalization (CFN).

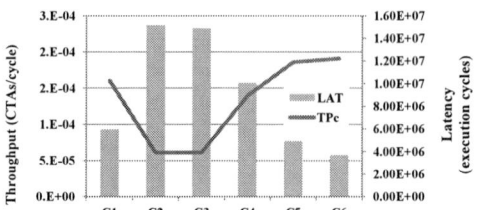

Figure 5: Effectiveness of TP_c metric (Left axis shows TPc value, right axis shows execution latency).

memory address computation from all the threads into burst address computations at the CTA level. On the other hand, SNC region motions do not seem to affect compute latency in a considerable way. However, they enable elimination of excessive variable privatization during FCUDA postprocessing, optimizing BRAM resource per CC, and hence throughput. Finally, note that the MMp kernel is an optimized version of the MM kernel derived through loop pipelining during HLS. The MMp speedup values are normalized over FCUDA latency of MM kernel. This shows that TOPP and HLS optimizations can be applied cummulatively.

4.2 Throughput Metric Evaluation

Here we measure the correlation of the throughput estimation metric to the actual execution latency. For this purpose we use the DWT kernel that has served as a running example throughout the previous sections. Specifically, intermediate configurations Ci, of DWT during compilation through the TOPP stages are extracted and fed to FCUDA postprocessing stage to collect execution results. TP_C is calculated for each configuration and it is depicted with execution latency results in Fig. 5. The gray bars correspond to execution latency, whereas the blue line corresponds to calculated TP_C values. We can observe the inverse correlation between the two performance metrics. This shows the effectiveness of the throughput metric in guiding the selection of high-performance configurations during the throughput latency phase.

4.3 TOPP vs. FCUDA

This evaluation measures the total kernel execution speedup achieved with TOPP over FCUDA [20]. Here, all transformations are enabled in both latency and throughput optimization phases of TOPP. In order to evaluate the performance effect of the code transformations in TOPP, we target the same execution frequency for all the kernels. Thus, we eliminate the fuzziness induced by the effect of synthesis and place-and-route optimizations on different RTL structures. Instead, we synthesize all the kernels at 200MHz, but run them at 100MHz to ensure that routing will not affect our evaluation (note that overconstraining the clock period during synthesis is a common practice in industry, in order to absorb the frequency hit from routing delays). In terms of memory interface and bandwidth we model in our evaluation a similar memory interface as the one used in the [14] hybrid computer, where the compute-acceleration FPGA leverages the high-speed serial tranceivers to transfer data to off-chip memory controllers that support high-banwidth DDR memory accesses.

Figure 6 depicts the speedup of the TOPP-compiled kernels against the FCUDA-compiled ones. The FP-SX50 and FP-SX95 bars use floating point kernels and target SX50T and SX95T Virtex-5 devices, respectively. The third bar (INT-SX50) uses integer kernels and targets device SX50T. Each bar is normalized against the execution latency of FCUDA for the same device and kernel. We can observe that the speedup achieved on the bigger SX95T device is slightly lower than the SX50T (even though in absolute terms latency on SX95T is lower from latency on SX50T). The main reason for this trend

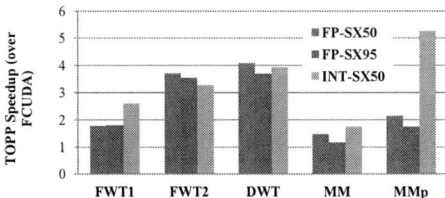

Figure 6: TOPP execution speedup over FCUDA

is due to the fact that the compute/memory resource capacity of SX95T is 1.8X higher then SX50T resource capacity, but the off-chip bandwidth of SX95T is only 50% higher than the off-chip bandwidth in SX50T thus limiting the speedup that can be achieved by the TOPP transformations. With regard to speedup of the integer kernels, this is similar to speedup for floating point kernels in most cases. FWT1 and MMp stand out for different reasons; FWT1 optimizes away integer multipliers for powers of two, while MMp expoits loop pipelining more efficiently with integer operations (Note: the MMp kernel speedup is here, also, normalized against the FCUDA-compiled latency of MM kernel.)

Finally, comparing the speedup corresponding to bars FP-SX50 with the compute latency results in Section 4.1, we can observe that the performance advantage of the TOPP flow is further improved. This is partially due to the better allocation of BRAMs achieved by the TDGC stage and partially due to more efficient exploitation of the off-chip memory bandwidth (i.e. transfers can be more efficiently disentangled from compute and converted to bursts).

5. CONCLUSIONS

In this paper we present the throughput-oriented parallelism synthesis (TOPP) framework which aims to provide throughput-oriented performance porting of CUDA kernels onto FPGAs. The techniques applied in this work could potentially be employed in other application programming interfaces with similar SIMT programming semantics that target heterogeneous compute systems (e.g. OpenCL [2]). Our experimental evaluation demonstrates the effectiveness of performance porting achieved through orchestration of advanced analysis with latency and throughput optimizations in the TOPP framework.

As computing is moving toward massively parallel processing for *big data* applications, it is critical to increase the abstraction level of optimization and transformation techniques. Representing and leveraging application algorithms at a higher level is crucial for delivering high throughput and high performance in massively-parallel compute domains. In this work, we have dealt with the issue of raising the abstraction level in the field of high-level synthesis of parallel custom processing cores. We have developed efficient throughput estimation and optimization techniques that improve performance beyond thread latency by dealing with conflicting performance factors at the CTA level and managing the compute and storage resources accordingly.

6. ACKNOWLEDGMENTS

This work is partially supported by the Gigascale Systems Research Center (GSRC) and Intel Corporation. We also thank Steven Burns, Mustafa Ozdal, Kanupriya Gulati and Taemin Kim of Intel and Kyle Rupnow of ADSC (Illinois Center in Singapore) for their helpful comments and discussions.

7. REFERENCES

[1] AMD Fusion family of APUs: Enabling a superior, immersive PC experience. White Paper. http://sites.amd.com/us/Documents/48423B_fusion_whitepaper_WEB.pdf, Mar. 2010.

[2] The OpenCL specification. http://www.khronos.org/registry/cl/specs/opencl-1.1.pdf, Sept. 2010.

[3] The OpenACC application programming interface. http://www.openacc.org/sites/default/files/OpenACC.1.0_0.pdf, Nov. 2011.

[4] Vivado design suite user guide: High-level synthesis. UG902(v2012.2). http://www.xilinx.com/support/documentation/sw_manuals/xilinx2012_2/ug902-vivado-high-level-synthesis.pdf, July 2012.

[5] R. Allen and K. Kennedy. *Optimizing compilers for modern architectures.* Morgan Kaufmann, first edition, 2002.

[6] W. Blume and R. Eigenmann. The range test: A dependence test for symbolic, non-linear expression. In *Proc. ACM/IEEE Conf. on Supercomputing (SC'94)*, Nov. 1994.

[7] P. Briggs, K. D. Cooper, and L. Torczon. Improvements to graph coloring register allocation. *ACM Transactions on Prog. Languages and Systems*, 16(3):428–455, May 1994.

[8] G. Chaitin. Register allocation and spilling via graph coloring. *ACM SIGPLAN Notices - Best of PLDI 1979-1999*, 39(4):66–74, Apr. 2004.

[9] C. Dave, H. Bae, S. J. Min, S. Lee, R. Eigenmann, and S. Midkiff. Cetus: A source-to-source compiler infrastructure for multicores. *IEEE Computer*, 42(12):36–42, Dec. 2009.

[10] M. Girkar and C. Polychronopoulos. Extracting task-level parallelism. *ACM Transactions on Prog. Languages and Systems*, 17(4):600–634, 1995.

[11] Z. Guo, E. Z. Zhang, and X. Shen. Correctly treating synchronizations in compiling fine-grained spmd-threaded programs for cpu. In *Proc. ACM Int'l Conference on Parallel Architectures and Compilation Techniques (PACT'11)*, Sept. 2011.

[12] S. Gupta, R. Gupta, and N. Dutt. Coordinated parallelizing compiler optimizations and high-level synthesis. *ACM Transactions on Design Automation of Electronic Systems*, 9(4):441–470, 2004.

[13] S. Gurumani, K. Rupnow, Y. Liang, H. Cholakkail, and D. Chen. High level synthesis of multiple dependent CUDA kernels for FPGAs. In *Proc. IEEE/ACM Asia and South Pacific Design Automation Conference*, Jan. 2013.

[14] The Convey HC-1: The world's first hybrid core computer. Datasheet. http://www.conveycomputer.com/Resources/HC-1\%20Data\%20Sheet.pdf, 2009.

[15] CUDA: Parallel programming and computing platform. http://www.nvidia.com/object/cuda_home_new.html, 2012.

[16] Zynq-7000 all programmable SoC. http://www.xilinx.com/products/silicon-devices/soc/zynq-7000/index.htm, 2012.

[17] Tegra super processors. http://www.nvidia.com/object/tegra-4-processor.html, 2013.

[18] S. Muchnick. *Advanced compiler design and implementation.* Morgan Kaufmann, first edition, 1997.

[19] M. Owaida, N. Bellas, K. Daloukas, and C. Antonopoulos. Synthesis of platform architectures from opencl programs. In *Proc. IEEE Symposium on Field-Programmable Custom Computing Machines (FCCM'11)*, May 2011.

[20] A. Papakonstantinou, K. Gururaj, J. Stratton, D. Chen, J. Cong, and W. Hwu. FCUDA: enabling efficient compilation of cuda kernels onto FPGAs. In *Proc. IEEE Symposium on Application Specific Processors*, June 2009.

[21] A. Papakonstantinou, K. Gururaj, J. Stratton, D. Chen, J. Cong, and W. Hwu. Efficient compilation of CUDA kernels for high-performance computing on FPGAs. *ACM Transactions in Embedded Computing Systems*, Vol. 13, 2014.

[22] A. Papakonstantinou, Y. Liang, J. Stratton, K. Gururaj, D. Chen, W. Hwu, and J. Cong. Multilevel granularity parallelism synthesis on FPGAs. In *Proc. IEEE Int'l Symposium on Field-Programmable Custom Computing Machines*, May 2011.

[23] J. Stratton, V. Grover, J. Marathe, B. Aarts, M. Murphy, Z. Hu, and W. Hwu. Efficient compilation of fine-grained SPMD-threaded programs for multicore cpus. In *Proc. ACM Int'l Symposium on Code Generation and Optimization (CGO'10)*, Feb. 2010.

[24] Z. Y. Zhang, F. W. Jiang, G. Han, C. Yang, and J. Cong. Autopilot: A platform-based ESL synthesis system. In P. Coussy and A. Moraviec, editors, *High-Level Synthesis: From Algorithm to Digital Circuit*, chapter 6, pages 99–112. Springer, 2008.

APPENDIX

A. HRG GENERATION

As described in Section 3.1 the HRG is generated during step $A6$ in the analysis phase. Having identified the global memory accesses in step $A1$ and split the multitype statements into single-type ones in step $A2$, a depth-first traversal (DFT) is carried out on the kernel AST (abstract syntax tree) representation used in the FCUDA compiler [9]. DFT is implemented by the recursive **hrgGen()** procedure (Alg. 1) which classifies each kernel statement as TRN, CMP, SNC or CF. Non CF statements are collected into a list ($sLst$) which is used to build HRG leaf nodes from statements of same type (**sTyp()**) and same AST level (lvl). Each region is assigned a region ID (rID) which helps maintain partial ordering of the HRG nodes. CF statements form HRG internal nodes by themselves and get assigned the smallest rID of their child leaf nodes (Alg. 1). Region IDs infer execution ordering and facilitate region motion feasibility checks during latency and throughput optimization phases of TOPP. The generated HRG is also structured as an AST and each node contains pointers to the code statements summarized by the HRG node. Hence, it is easy to reconstruct a new kernel AST from an optimized HRG. Finally, a similar traversal of the HRG tree is used to group HRG nodes into tasks that satisfy two rules, (i) a task may contain control-flow (CF) nodes as long as every child node of a contained CF node is also included in the task, and (ii) a task may contain nodes across different control-flow hierarchy levels as long as the corresponding CF nodes are also included in the task (e.g. grouping nodes CMP12 and SNC13 in Fig. 2 within the same task is only allowed if nodes IF10, CMP10 and TRN11 are also included in the task). Note that HRG leaf nodes are grouped into tasks based on their type; COMM tasks comprise only TRN nodes whereas COMP tasks may include CMP and SNC nodes.

B. FCUDA FLOW DETAILS

FCUDA (Fig. 7(a)) provides the underlying basis flow on which TOPP is built to provide throughput-oriented kernel restructuring. The proposed flow (Fig. 7(b)) leverages FCUDA utilities during preprocessing and postprocessing stages for conditioning the input CUDA code and translating the TOPP output into parallel C code, respectively (Fig. 1). Moreover, integration of TOPP in FCUDA removes the burden of manual annotation from the user. FCUDA annotations consist of lightweight pragma directives that provide user guidelines for the transformations and optimizations applied in the flow. The main types of annotations are COMPUTE and TRANSFER directives which guide decomposition of the kernel into COMP and COMM tasks. Other types of annotation include SYNC and BLOCK directives which guide the synchronization of tasks and the logical layout of threads in CTAs. The analysis phase in TOPP in combination with the generated HRG representation and the throughput estimation metric eliminate the need for user annotations and help increase the exploited optimization opprotunities.

Preprocessing utilities entail kernel procedure identification and code conditioning through a sequence of transformations: (i) declaration normalization (i.e. hoist declarations out of executable kernel regions), (ii) return normalization (i.e. convert multiple return points into a single return point), (iii) procedures inlining (i.e. inline non-library procedure calls within the kernel procedure) and (iv) unsupported code identification and assertions (i.e. check and flag unsupported code structures). Note that callee inlining in the current implementation facilitates easier kernel restructuring in subsequent processing stages, but is not required for most transformations. Unsupported code structures include unstructured

---

Algorithm 1: HRG generation

```
/* hrgGen(hAST,level,cID,sLst):  Generate HRG
through code depth first traversal (DFT)      */
```

Input: Abstract syntax tree of code hierarchy $hAST$
Input: Level in kernel lvl, region ID pID
Output: region ID rID
Output: region statement list: $sLst$

```
1   cID ← pID                        // Update current ID
2   cTyp ← -1                        // Invalidate current type
3   while S ← next(hAST) do          // Get next statement
4       switch typ ← sTyp(S) do      // get type of S
5           case TRN                 // S type is TRN
6               if typ ≠ ctyp then   // Different region
7                   n ← hrgNod(sLst) // New HRG node
8                   pn ← getNod(hAST) // parent node
9                   addChld(n, pn)   // Link n to pn
10                  empty(sLst)      // Clean sLst
11                  cID ← cID + 1    // Update region ID
12              annot(S, TRN)        // Annotate S
13              annot(S, cID)        // Annotate S
14              annot(S, lvl)        // Annotate S
15              push(sLst,S)         // Push S into sLst list
16              cTyp ← typ           // Update type
17          case CMP                 // S type is CMP
18              ...                  // Similar to TRN case
19          case SNC                 // S type is SNC
20              ...                  // Similar to TRN case
21          case CF                  // Control flow
22              n ← hrgNod(sLst)     // New HRG node
23              pn ← getNod(hAST)    // Get parent node
24              addChld(n, pn)       // Link n to pn
25              empty(sLst)          // Clean sLst
26              cAST ← getAST(S)     // get AST of S
27              annot(S, cID)        // Annotate S with cID
                // Recurse for next AST level
28              cID ← hrgGen(cAST,lvl+1,cID,sLst)
29              n ← hrgNod(S)        // New HRG node
30              addChld(n, pn)       // Link n to pn
31      return cID
```

---

control flow (e.g. goto statements) and other CUDA features not currently supported by the compiler (e.g. texture memory variables). Postprocessing entails (i) variable privatization (i.e. variables referenced across different HRG regions with thread-variant data), (ii) kernel task outlining (i.e. extracting the TOPP-generated COMP and COMM tasks into separate procedures), (iii) intra-CTA parallelism extraction (i.e. tID-loop unrolling and on-chip memory banking), and (iv) inter-CTA parallelism extraction (i.e. replication of task calls, through CTA-loop unrolling). Preprocessing and postprocessing transformations, as well as annotation directives are discussed in further detail in [20, 23, 21].

C. LATENCY PHASE TRANSFORMATIONS

The transformations applied in the latency phase of TOPP aim to facilitate the generation of coarser tasks through region motions in the kernel HRG. Each region is assigned a region ID, rID, which is updated after every region motion to ensure that partial ordering with respect to rID reflects execution ordering. Reflection of execution ordering in the rID value in tandem with Def-Use chains facilitates easy motion feasibility testing. HRG nodes corresponding to CF structures are initially assigned the same rID as their first (in DFT order) leaf node. The space of rIDs in the HRG may be sparse due to transformations that result in region merging or elimination. A region motion can be encoded with respect to the initial rID, i and the final rID, j, as mot(i,j). Reflecting the execution ordering in the rID field during region motion mot(i,j) may require updates in the rID of regions with rID = k, where $i < k \leq j$, if $i < j$ or $j \leq k < i$, if $i > j$.

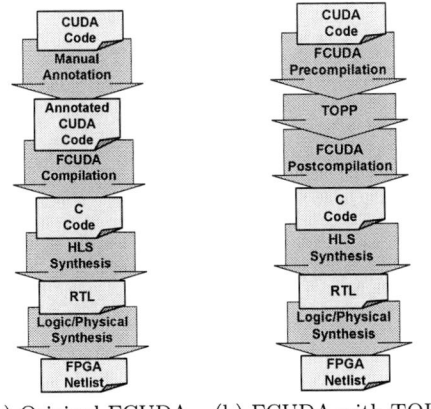

(a) Original FCUDA   (b) FCUDA with TOPP

Figure 7: Integration of TOPP in FCUDA

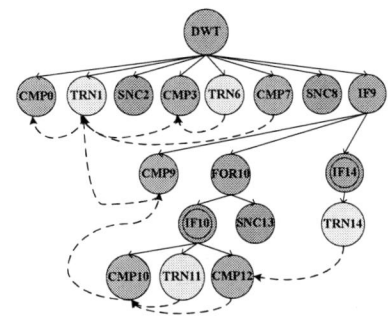

Figure 8: DWT HRG after SNC motions (SM)

C.1 SNC Region Motions (SM)

This stage identifies feasible SNC motions that facilitate region merging. The only type of SNC region motions considered are those with destinations within the same HRG level. SM stage involves four main steps: ($SM1$) Collect all SNC regions, ($SM2$) Get feasible destinations, ($SM3$) Estimate motion cost/gain, and ($SM4$) Perform motion. Initially SNC regions are collected (step $SM1$) and ordered with respect to their region identifier, rID. For each synchronization region, SNCn (with rID(SNCn) = n), feasible destination candidates are identified in step $SM2$. Candidate destination locations are explored in two sweeps of the corresponding HRG level: a forward and a backward sweep (Fig. 3) starting from the initial location of SNCn in the HRG. During a sweep, the candidate destinations are sequentially evaluated until (i) another SNC region is encountered, (ii) a destination is assesed as non-feasible or (iii) no candidates are left. Feasibility is tested as described in Section 3.2. TSCs with source rID, $r : r < n$ and sink rID, $v : v > n$ do not break feasibility if the destination rID, n', satisfies the following condition: $n' > v$, for forward sweeps (Fig. 3(a)), or $n' < v$, for backward sweeps (Fig. 3(b)). A destination location is selected based on the evaluation of all feasible destinations. Application of SM on the DWT HRG (Fig. 2) shifts SNC4 immediately after SNC2 region, effectively resulting in its elimination (Figure 8).

C.2 Burst Conversion (BC)

This latency optimization stage analyzes the address computation patterns of global memory accesses and determines the feasibility for coalescing memory accesses across threads into burst transfers. Moreover, this transformation offers two additional benefits: (i) reduces address computation overhead (base address calculation shared by all threads) and (ii) facilitates region consolidation in the HRG (address computation combined with data transfer in one region) which may reveal

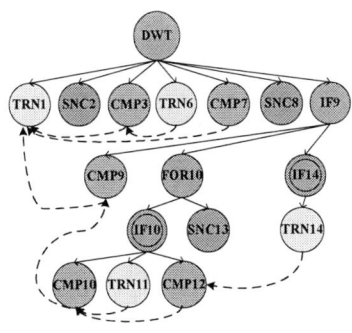

Figure 9: DWT HRG after burst conversion (BC)

new optimization opportunities. Region consolidation from burst conversion results in the elimination of CMP0 region in our DWT running example (Fig. 9).

Burst conversion involves four main steps: ($BC1$) Identify all CMP regions involved in address calculation, ($BC2$) Analyze coalesced accesses with respect to tID, ($BC3$) Analyze address coalescing with respect to TiVAR loop indexes in the kernel, and ($BC4$) Perform HRG restructuring. Initially, the Def-Use chains computed earlier in the flow are leveraged to identify statements containing address computation (step $BC1$). Subsequently, symbolic analysis and value range analysis is used to determine whether the range of computed addresses per CTA is coalesced with respect to tID. In particular, forward substitution is used to derive the address calculation expression, E_A. Then, the tID variant (TVAR) analysis performed during region analysis stage is used to decompose the expression into a TVAR part, E_{TVAR}, and a tID invariant part, E_{TiVAR}: $E_A = E_{TVAR} + E_{TiVAR}$. Symbolic and range analyses are used to examine the E_{TVAR} expression and determine whether memory accesses are coalesced in piecewise ranges, (s_i, e_i), of the tID domain. If such piecewise domain ranges can be identified, their maximum range value is returned; otherwise, a negative value is returned. Subsequently, a similar analysis of the address calculation expressions is carried out to identify coalescing opportunities across piecewise ranges of non tID-loops (step $BC3$). Any additional piecewise ranges found are used to extend the tID piecewise ranges identified previously (step $BC2$). Finally, during the last step of this stage, the address computation analysis results are utilized to perform any required HRG modifications. In the case of statically identified coalesced address ranges, individual thread accesses are converted into `memcpy` calls where E_{TiVAR} serves as the source/destination address and the size of the piecewise address range, (s_i, e_i), as the transfer length. `memcpy` calls are subsequently transformed into DMA-based bursts by the HLS engine. In the case that no address ranges are returned by static analysis, address computations are kept within CMP regions and computed addresses are stored for use by the corresponding TRN regions.

C.3 TRN Region Motions (TM)

This transformation stage shifts TRN regions to more profitable locations within their current HRG level. In particular, *TRN-Read* (TRN-R) regions (i.e. off-chip reads) are shifted toward the beginning of the HRG level, whereas *TRN-Write* (TRN-W) regions (i.e. off-chip writes) are shifted toward the end of the HRG level. This transformation aims to enable coarsening of CMP regions into bigger regions with more opportunity for ILP extraction and resource sharing. For example, Fig. 10 depicts the DWT HRG after TM transformation, where TRN6 is shifted to the right of the level and CMP3 and CMP7 are merged into CMP3. TRN motions involve four main steps: ($TM1$) Collect all TRN regions in two lists representing TRN-R and TRN-W regions, respectively, ($TM2$) Get feasible shift destinations, ($TM3$) Estimate motion cost/gain,

75

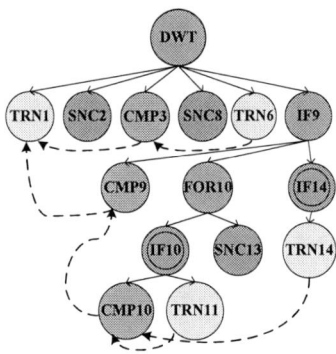

Figure 10: DWT HRG after TRN motion (TM)

Listing 3: Unnormalized CF with CMP and TRN

```
1   // tID := threadIdx.x
2   tidx=(blockIdx.x*blockDim.x);
3   for(tID=0; tID<blockDim.x; tID++) // tID-loop
4     for (pos=tidx+tID; pos<N; pos+=numThreads){// TVAR
5       locA1 = (locA0 * (locB * rcpN));  // CMP
6       d_A[pos] = loc1A;      }          // TRN
```

and (*TM4*) Perform TRN motion.

Initially, TRN regions are sorted with regard to their region ID into two lists corresponding to TRN-R and TRN-W transfers (step *TM1*). Regions in the TRN-R (TRN-W) list are processed in increasing (decreasing) rID order to find candidate destination locations (step *TM2*) within their current HRG level. In the case of TRN-R (TRN-W) regions with rID = r, candidate destinations include earlier (later) points in the HRG level with rID = z. The TRN motion, $mot(r,z)$, is feasible unless there is a region with rID = q that bears true dependence to the TRN-R (TRN-W) region and its rID satisfies the expression $z \le q < r$ ($r < q \le z$). The candidate destination locations are examined for feasibility in increasing order of: $|r - z|$; if a nonfeasible destination is identified, any remaining candidates are dumped from the candidate list. Finally, a destination location for each considered TRN region is selected based on the candidate destination evaluation (step *TM3*) and the motion is applied (step *TM4*).

C.4 Control Flow Normalization (CFN)

This stage handles control-flow (CF) structures that use TVAR expressions as conditions. TVAR CF structures containing CMP and TRN regions need special handling in order to expose the implicit synchronization points (ISPs) between compute and communication tasks. Exposing the ISPs is critical in exploiting data transfer coalescing across neighboring threads in TRN regions as well as exposing the data-level compute parallelism in CMP regions. Exposing the ISPs requires interchanging the TVAR CF with the tID-loop which expresses the CTA threads. List. 3 depicts a TVAR loop (line 4) within the tID-loop (line 3) which contains a CMP (line 5) and a TRN statement (line 6). The CF normalization converts the TVAR loop into a TiVAR loop (line 6 in List. 4) preceded by initialization of the induction variable (lines 3-4) of the original TVAR loop. Thus, the ISPs between regions are exposed through tID-loops wrapped around each region (lines 7, 10). Note that variables in List. 4 are in SIMT notation. Postprocessing stage in FCUDA determines whether they should be privatized (storage redundancy) or reimplemented (compute redundancy). Variable `pos`, for example, would become an array of size `blockDim.x` in the case of privatization, whereas reimplementation would result in the code shown in List. 5. Figure 11 depicts the resulting HRG representation of the DWT after CF normalization: *IF10* node is split into *IF10* and *IF11* nodes

Listing 4: CF normalization using privatization

```
1    // tID := threadIdx.x;
2    tid=(blockIdx.x*blockDim.x);
3    for (tID=0; tID<blockDim.x; tID++)
4      pos = tidx+tID;
5    cfCond=true;
6    while(cfCond){
7      for(tID=0;tID<blockDim.x;tID++)//tID-loop
8        if (pos<N)
9          locA1 = (locA0 * (locB * rcpN));
10     for(tID=0; tID<blockDim.x;tID++)//tID-loop
11       if (pos<N)
12         d_A[pos] = loc1A;
13       cfCond = false;
14     for(tID=0;tID<blockDim.x;tID++)//tID-loop
15       if (pos<N) {
16         pos += numThreads;
17         cfCond |= (pos<N);       }}
```

Listing 5: CF normalization using reimplementation

```
1    // tID := threadIdx.x;
2    tidx=(blockIdx.x*blockDim.x);
3    pos = tidx;
4    cfCond=true;
5    while(cfCond) {
6      for(tID=0;tID<blockDim.x;tID++)//tIDX-loop
7        if ((pos+tID)<N)
8          locA1 = (locA0 * (locB * rcpN));
9      for(tID=0;tID<blockDim.x;tID++)//tIDX-loop
10       if ((pos+tID)<N)
11         d_A[pos] = loc1A;
12     cfCond = false;
13     pos += numThreads;
14     cfCond |= (pos<N); }
```

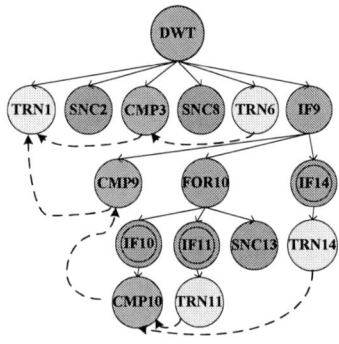

Figure 11: DWT HRG after CF normalization (CFN)

D. TDGC ALGORITHM DETAILS

Alg. 2 provides an overview of the TDGC flow described in Section 3.3.2. The three steps of the TDGC flow (*GC1*, *GC2*, *GC3*) are distinguished in the algorithm comments. During step *GC3* a recursive call to `tdgc` procedure with the rescheduled HRG is made (if node n can be shifted.) If the rescheduled HRG does not provide a higher throughput (TPc), the rescheduled HRG is discarded and the next candidate node n is evaluated for rescheduling. *R*-coloring is discussed in the next section.

D.1 R-Coloring Algorithm

As mentioned in Section 3.3.2 an interference graph, G_I, is generated based on the analysis of the array lifetimes with respect to the regions in the HRG. The interference graph is colored using a modified *R*-coloring [18] algorithm which dynamically determines the value of R, i.e. the number of allocated BRAMs. Note that the interference graph represents the lifetime interferences of arrays per CTA; these interferences affect the BRAM resource requirement per CC. For a total BRAM count of N_B at the system level, there is a tradeoff between the number of instantiated CCs, N_{CC}, and the number of BRAMs, R, allocated per CC: $R = \lfloor \frac{N_B}{N_{CC}} \rfloor$. Unlike traditional graph coloring implementations where the number of colors (resource units), R, is a fixed constraint, the number of allocated BRAMs per CC, in TDGC, can range across a set of values that fulfill the previous constraint on R. In other words, by modifying the CTA region schedule we can generate different HRG configurations that have different

Algorithm 2: TDGC Flow overview

Input: HRG: Gin, Throughput: TPci
Output: New HRG: $Gout$

1   $arrs \leftarrow$ getArr(Gin)   // Collect array variables
     // GC1 step
2   $Gi \leftarrow$ bldInterf($arrs, Gin$)   // Build interference
                                        // graph
3   rColor(Gi)   // do R-coloring
     // GC2 step
4   $TPco \leftarrow$ getTput(Gin)   // Estimate throughput
5   **if** $TPco < TPci$ **then**   // Previous TPc is better
6   **return** *0*
7   **if** tpcLim($TPco$) $== BRAM$ **then**   // Is BRAM
                                   // throughput limiter?
       // GC3 step
8   calcILI(Gi)   // Calculate ILI in Gi
9     $nods \leftarrow$ calcLs(Gi)   // Calculate scatter,
                                  // (Ls) in Gi and
                                  // sort nodes wrt Ls in nods
10   **foreach** $n \in nods$ **do**
         // Check dependency constraints for move
11   **if** canMov(n, Gin) **then**
12        $Gout \leftarrow$ movNod(n, Gin)   // Build new HRG
13        $Gout \leftarrow$ tdgc($Gout, TPco$)   // Try TDGC on Gout
14   **if** $Gout \neq 0$ **then**   // if success
15   break   // Do not check more nodes
16   **else**
17        $Gout \leftarrow Gin$ // Move failed
18 **return** $Gout$

Algorithm 3: Graph coloring of the interference graph

```
/* tdgc(Gin):  Throughput Driven Graph Coloring
*/
```
Input: Uncolored interference graph G_I
Output: Colored interference graph G_I'

1   $nodes \leftarrow$ nods(G_I)
2   $R \leftarrow 1$ // initialize max R
3   **while** $nodes \neq \emptyset$ **do**   // Node pushing
4   sort($nodes$)   // Sort nodes wrt interf. degree
5     $n \leftarrow$ getNod($nodes$)   // Get first node
6     $d \leftarrow$ minDegree(n)   // Get interference degree
7     $R \leftarrow$ max(R, $(d+1)$)   // Update degree
8   push(n, $stack$)   // Push to stack and prune graph
9   **while** $stack \neq \emptyset$ **do**   // Node poping
10    $n \leftarrow$ pop($stack$)   // Pop node from stack and
                                 // add back to graph
11   getMinColor(n, R)   // Allocate min color id$\leq$R
                                 // not used by neighbors of n

The coloring process comprises an initial *node pushing* phase, during which, nodes are pruned in increasing order of interference degree from G_I and pushed into a stack. (This resembles traditional R-coloring with fixed R, where nodes with degree less than R are pruned first based on the observation that a graph with a node of degree less than R is R-colorable if and only if the graph without that node is R-colorable.) During each node pruning, R is adjusted as depicted in line 7 of Alg. 3 and the interference degrees of its neighboring nodes are decremented. Once all of the nodes are pushed into the stack, they are popped back into the graph in reverse order and assigned a color (Alg. 3). The assigned color for each popped node is the minimum color number that has not been assigned to any of the previously popped neighboring nodes. At the end of *node popping* all the graph nodes are going to be colored with at most R_m colors, where R_m is the maximum value of R used during the node pushing phase of coloring.

Figure 12(b) depicts the updated interference graph for DWT kernel after the throughput optimization phase. The new interference graph entails lower BRAM pressure and coloring results in the allocation of two BRAMs (compared to three BRAMs for the initial interference graph in Figure 12(a)).

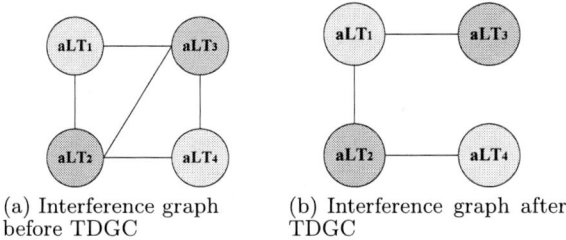

(a) Interference graph before TDGC

(b) Interference graph after TDGC

Figure 12: DWT interference graphs

resource requirements and latencies, hence different N_{CC} and throughput. The goal of TDGC is to identify the value of R (and the corresponding feasible HRG configuration) that maximizes system execution throughput under a given resource constraint. Hence, our graph coloring initially sets R to one and dynamically adjusts its value during the first phase of graph coloring.

Memory Partitioning for Multidimensional Arrays in High-level Synthesis

Yuxin Wang,[1] Peng Li,[1] Peng Zhang,[2] Chen Zhang,[1] Jason Cong[1,2,3]

[1] Center for Energy-Efficient Computing and Applications, Computer Science Department, Peking University, China
[2] Computer Science Department, University of California, Los Angeles, USA
[3] UCLA/PKU Joint Research Institute in Science and Engineering
{ayerwang, peng.li, chen.ceca}@pku.edu.cn, {pengzh, cong }@cs.ucla.edu

ABSTRACT

Memory partitioning is widely adopted to efficiently increase the memory bandwidth by using multiple memory banks and reducing data access conflict. Previous methods for memory partitioning mainly focused on one-dimensional arrays. As a consequence, designers must flatten a multidimensional array to fit those methodologies. In this work we propose an automatic memory partitioning scheme for multidimensional arrays based on linear transformation to provide high data throughput of on-chip memories for the loop pipelining in high-level synthesis. An optimal solution based on Ehrhart points counting is presented, and a heuristic solution based on memory padding is proposed to achieve a near optimal solution with a small logic overhead. Compared to the previous one-dimensional partitioning work, the experimental results show that our approach saves up to 21% of block RAMs, 19% in slices, and 46% in DSPs.

Categories and Subject Descriptors

B.5.2 [**Hardware**]: Design Aids–*automatic synthesis*

General Terms

Algorithms, Performance, Design

Keywords

High-Level Synthesis, Memory Partitioning, Memory Padding

1. INTRODUCTION

To balance the requirements of high performance, low power and short time-to-market, field programmable gate array (FPGA) devices have gained a growing market against ASICs and general-purpose processors over the past two decades. Recently, FPGAs have also been used as general computing platforms as alternatives to CPUs and GPUs. Although FPGAs provide plenty computational units for parallelization, how to supply those units with the required high-speed data streams is a major challenge.

This is especially true after loop unrolling and pipelining, when multiple data elements from the same array are often required simultaneously in a single clock cycle. Typical on-chip block RAMs (BRAMs) in FPGAs have two access ports. A straightforward solution is to duplicate the array into multiple

Permission to make digital or hard copies of all or part of this work for personal or classroom use is granted without fee provided that copies are not made or distributed for profit or commercial advantage and that copies bear this notice and the full citation on the first page. To copy otherwise, to republish, to post on servers or to redistribute to lists, requires prior specific permission and/or a fee.
DAC'13, May 29 - June 07 2013, Austin, TX, USA.

copies [13]. Although the duplication approach can support simultaneous read operations, it may have significant area and power overhead and introduce memory consistency problem. A better approach is to partition the original array into multiple memory banks. Each bank holds a portion of the original data and serves a limited number of memory requests.

Memory partitioning has been studied in the distributed computing domain for decades [8, 15, 16], where data elements are partitioned into different processors to reduce communication among the processors. While some of the partitioning algorithms in distributed computing can be directly applied to high-level synthesis, the freedom of creating memory banks tailored to the target application can lead to more efficient memory partitioning algorithms for high-level synthesis [19, 3, 6, 20, 12]. In [19], different fields of a single *structure* are partitioned into multiple memory banks for data parallelism based on profiling results. In [3], a single array is decomposed into disjoint memory banks for storage minimization purposes through accurate lifetime analysis using a polyhedral model. The purpose of the memory partitioning algorithm presented in this paper is to improve system performance by assigning memory accesses to disjoint memory banks and providing simultaneous conflict-free memory accesses [6, 20, 12], which is orthogonal to the problem in [3]. In [6], an automated memory partition algorithm is proposed to support multiple simultaneous affine memory references to the same array. The algorithm can be extended to efficiently support memory references with modulo operations (common after data reuse using scratchpad memory) with limited memory paddings [20]. In [12], memory accesses in different loop iterations can be partitioned into different memory banks and scheduled into the same cycle to minimize the number of required memory banks.

However, previous memory partitioning algorithms are designed for one-dimensional arrays, while many designs for FPGAs are often specified by nested loops with multidimensional arrays—such as image, video, and scientific computing applications. In previous works, a multidimensional array is first flatted into a single-dimensional array before memory partitioning. However, memory addresses after array flattening are dependent on the array size. For different array sizes, different partitioning schemes are generated, many of which are suboptimal. In this paper we focus on providing an effective and efficient memory partition algorithm for multidimensional arrays based on linear transformation.

The main contributions of this work are described as follows:

1) A linear-transformation-based multidimensional memory partition algorithm is proposed to generate the smallest memory bank numbers regardless of the size of input array.
2) An optimal inner-bank offset generation scheme is proposed based on point counting in polytopes.

3) A heuristic solution based on memory padding is proposed to achieve a near-optimal inner-bank offset generation with a comparative small logic overhead and storage overhead.

The remainder of this paper is organized as follows: Section 2 provides a motivational example for the multidimensional memory partitioning; Section 3 formulates the problem, and Section 4 describes the detailed solution; Section 5 analyzes the experimental results, and is followed by conclusions in Section 6.

2. MOTIVATIONAL EXAMPLE

Our motivational example, as shown in Fig. 1(a), is from a loop kernel of the 2D denoise algorithm, which is a key application in medical image processing [2]. The kernel has five accesses to the array A in the inner loop. Fig. 1(c) shows the access pattern of the inner loop iteration and the partition based on linear transformation, where x_0 is the lower-dimension index and x_1 denotes the index in higher dimension. The light points in Fig. 1(c) represent the data elements in the array with the dark points representing the elements accessed in a single loop iteration. We assume that the physical memory has one read port—only one data element can be read from a physical memory in each clock cycle. To improve the processing throughput of the loop kernel, we need to pipeline the execution of successive inner loop iterations, which means that multiple accesses to the same array will happen in one clock cycle. If array elements are not properly allocated in multiple physical memory banks, memory conflicts will occur and pipeline performance will be impacted.

Previous memory partitioning solutions mainly focus on 1-D arrays, as in [6]. It flattens the array first, as shown in Fig. 1(b), and then partitions the flattened array. In order to fully pipeline the loop, five elements of data are required in each clock cycle. Thus the minimum number of memory banks for a non-conflict partitioning is five. However a cyclic partition with five banks can not satisfy the non-conflict constraint according to the code in Fig. 1(b). Take iteration (i, j)=(1, 1) for example, the second reference (A[64*j+i-1]=A[64]) and the forth reference (A[64*j+i+64]=A[129]) will access the same bank (64%5 = 129%5). Using the approach in [6] on the flattened array, we can prove that at least six banks are required.

```
int A[64][64];
for j= 1 to 62
  for i = 1 to 62
    b[j][i] = f(A[j][i], A[j][i-1], A[j-1][i ], A[j+1][i], A[j][i+1]);
    //accesses to down, up, left, and right
```
(a)

```
int A[4096];
for j= 1 to 62
  for i = 1 to 62
    b[j][i] = f(A[64*j+i], A[64*j+i-1], A[64*j+i-64], A[64*j+i+64],
    A[64*j+i +1]);
```
(b)

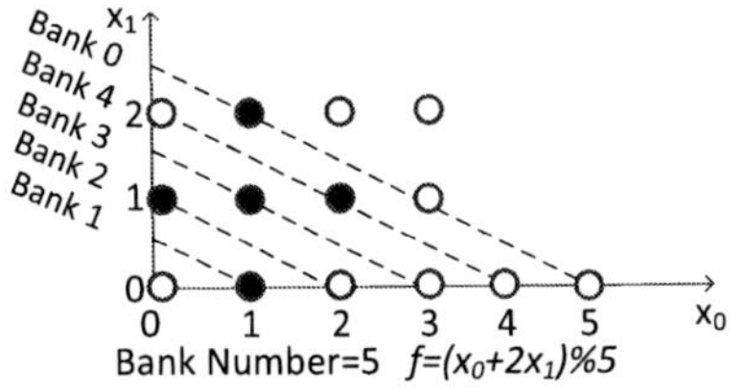

Bank Number=5 $f=(x_0+2x_1)\%5$

(c)

Fig. 1 Denoise: (a) original loop kernel, (b) loop kernel with flattened array, (c) multidimensional partitioning based on linear transformation

In fact, using the linear transformation based multidimensional partitioning method proposed in this work, the original code (Fig. 1(a)) can be fully pipelined with five memory banks. As illustrated in Fig. 1(c), the data elements on the same dotted line will be partitioned into the same memory bank, e.g., the data A[0][2] and A[1][0] are in the same bank. Whereas, the five data elements accessed in one inner-loop iteration are mapped into five different banks; i.e., in iteration (i, j)=(1, 1), the second reference (A[1][0]) and the forth reference (A[2][1]) are no longer in the same bank. Based on the linear transformation method we proposed, the code in Fig. 1(a) is partitioned with a linear transformation $f = (x_0 + 2x_1)\%5$ (as shown in Fig. 1(c)). We will describe the detailed partitioning algorithm in Sections 3 and 4.

3. PROBLEM FORMULATION

In this paper we will describe how we partition several multidimensional memory references in a multidimensional loop nest to separate memory banks to enable loop pipelining with simultaneous memory accesses. For simplicity, loop initiation interval (II) and physical memory port number are both assumed to be 1 in this paper. Algorithms and formulations can be extended for any constant loop initiation interval and physical memory port number by scheduling and mapping the accesses onto different time intervals and physical memory ports (as presented in [6]).

DEFINITION 1 (ITERATION DOMAIN [10]) Given a l-level loop nest with the iteration variables $i_0, i_1, ..., i_{l-1}$ from outermost to innermost loop, the iteration vector is a vector of iteration variables, $\vec{i} = (i_0, i_1, ..., i_{l-1})^T$. The *iteration domain D* is a set of all iteration vectors in the loop bounds.

DEFINITION 2 (AFFINE MEMORY REFERENCE) Given a d-dimensional array, a d-dimensional *affine memory reference* to the array is a set of linear combinations of iteration vectors and a constant:

$$R = \begin{pmatrix} a_{0,0} & \cdots & a_{0,l} \\ \vdots & \ddots & \vdots \\ a_{d-1,0} & \cdots & a_{d-1,l} \end{pmatrix} * (i_0, i_1, ..., i_{l-1}, 1)^T$$

where $a_{k,j} \in \mathbb{Z}$ is the coefficient of the j-th iteration vector in the k-th dimension.

DEFINITION 3 (DATA DOMAIN) Given a loop with m affine memory references $R_0, R_1, ..., R_{m-1}$ on the same array, the *data domain M* of the array is defined as a set of all memory elements accessed by any memory reference in any loop iteration. Assuming the memory element accessed by memory reference R_j in iteration $\vec{i}$ is represented as $R_j(\vec{i})$, then

$$M = \bigcup_{\vec{i} \in D; 0 \le j < m} R_j(\vec{i})$$

DEFINITION 4 (MEMORY PARTITION) A *memory partition* of an array with data domain M is described as a pair of mapping functions $(f(d), g(d))$, $\forall d \in M$, where $f(d)$ is the bank number that d is mapped to, and $g(d)$ is the corresponding inner bank offset. Also $f(d) \ge 0$, and $g(d) \ge 0$.

After memory partitioning, a data element in the original array is allocated on a new memory bank with a new array offset (inner bank offset). The validation of the partitioning is interpreted as two distinct data elements mapped onto either different memory banks or the same bank with different inner bank offsets. A valid memory partition of an array with data domain M is described as $\forall d_1, d_2 \in M$,

$$d_1 \ne d_2 \Leftrightarrow (f(d_1), g(d_1)) \ne (f(d_2), g(d_2))$$

where $(f(d_1), g(d_1)) \neq (f(d_2), g(d_2))$ means

$$f(d_1) \neq f(d_2) \qquad \text{or} \qquad f(d_1) = f(d_2), g(d_1) \neq g(d_2)$$

An access conflict between two memory references R_j and R_k $(0 \leq j < k < m)$ means that $\exists \vec{i} \in D, R_j(\vec{i}), R_k(\vec{i}) \in M$

$$f(R_j(\vec{i})) = f(R_k(\vec{i}))$$

This access conflict constraint is under the assumption that each physical memory only has one port. With the preceding definitions and formulations, we use Problem 1 defined below to formulate the multidimensional memory partitioning problem. Eqn. (1) defines the optimality of memory partitioning, as our main objective is to minimize the memory bank number. Eqn. (2) is responsible for the validity of the partitioning. Eqn. (3) ensures no conflict access in any iteration, which is required for fully-pipelined loops.

PROBLEM 1. (BANK NUMBER MINIMIZATION). Given a loop with m affine memory references $R_0, R_1, ..., R_{m-1}$ on the same array, find the optimal memory partition f, such that:

Minimize $bank\_num = \max_{0 \leq i < m}\{f(R_i)\}$	(1)
Subject to $\forall d_1, d_2 \in M, (f(d_1), g(d_1)) \neq (f(d_2), g(d_2))$	(2)
$\forall \vec{i} \in D, 0 \leq j < k < m, f(R_j(\vec{i})) \neq f(R_k(\vec{i}))$	(3)

The storage overhead minimization problem is formulated as Problem 2 under the same valid partition and non-conflict constraints as Problem 1.

PROBLEM 2 (STORAGE MINIMIZATION). Given a loop with m affine memory references $R_0, R_1, ..., R_{m-1}$ on the same array, a memory partition number N, find the inner bank offset function g and check globally for consistency such that:

Minimize $storage = \sum_{n=0}^{N-1} G_n$
Subject to $0 \leq n < N, G_n = \max_{0 \leq i < m, f(R_i) = n} g(R_i)$
$0 \leq i < m, f(R_i) \leq N$
$\forall d_1, d_2 \in M, (f(d_1), g(d_1)) \neq (f(d_2), g(d_2))$
$\forall \vec{i} \in D, 0 \leq j < k < m, f(R_j(\vec{i})) \neq f(R_k(\vec{i}))$

4. PARTITIONING ALGORITHM

In this paper, we propose a Linear Transformation Based (LTB) memory partitioning algorithm. The algorithm is general enough to cover the solutions from previous array flattening based approaches. We only consider cyclic partitioning strategy in this work. Other partitioning schemes (as block and block-cyclic) can be applied based on this solution.

A d-dimensional memory index $\vec{x} = (x_0, x_1, ..., x_{d-1})^T$ is first transformed by $\vec{x} \rightarrow \vec{\alpha} * \vec{x}$, where $\vec{\alpha} = (\alpha_0, \alpha_1, ..., \alpha_{d-1})$, $\alpha_i \in \mathbb{Z}$. According to the properties of cyclic partitioning, the bank mapping function f is described as

$$bank: f(\vec{x}) = (\vec{\alpha} * \vec{x}) \% N.$$

From a geometrical point of view, $\vec{\alpha} * \vec{x} = $ c represents a series of hyperplanes in the data domain, where $c \in \mathbb{Z}$, and $f(\vec{x})$ assigns the hyperplanes to different banks according to the value of $c\%N$. The traditional array flattening approach is just a special case of LTB when $\vec{\alpha}$ is decided by the dimensional width, as shown in Example 1.

EXAMPLE 1 (Flattening Partition) Supposing that the dimensional width of the target array from low dimension to high dimension is $w_0,...,w_{d-1}$, the traditional approach will first flatten the reference

into one dimension. Then the array is cyclically partitioned, using modulo and division operations to generate the bank number and inner bank offset. The bank mapping function f and inner bank offset function g are described as below.

$$f(\vec{x}) = (x_{d-1} * \prod_{k=0}^{d-2} w_k + x_{d-2} * \prod_{k=0}^{d-3} w_k +$$
$$... + x_1 * w_0 + x_0)\%N$$

$$g(\vec{x}) = (x_{d-1} * \prod_{k=0}^{d-2} w_k + x_{d-2} * \prod_{k=0}^{d-3} w_k +$$
$$... + x_1 * w_0 + x_0)/N$$

We can see that the flattening partition is just a special case in LTB method with the coefficient $\vec{\alpha}$ equal to $(1, w_0, w_1, ..., \prod_{k=0}^{d-3} w_k, \prod_{k=0}^{d-2} w_k)$.

4.1 Bank Mapping

Extending the constraint provided by work in [6], we build our own non-conflict constraint for d-dimensional array references as Theorem 1. It offers a sufficient condition for the conflict-free accesses regulated by Eqn. (3). With the constraint, we can find the candidate linear transformation vectors that meets the requirement. Assuming that there are two d-dimensional array references as

$$R_0 = \begin{pmatrix} a_{0,0} & \cdots & a_{0,l} \\ \vdots & \ddots & \vdots \\ a_{d-1,0} & \cdots & a_{d-1,l} \end{pmatrix} * (i_0, i_1, ..., i_{l-1}, 1)^T \text{ and}$$

$$R_1 = \begin{pmatrix} b_{0,0} & \cdots & b_{0,l} \\ \vdots & \ddots & \vdots \\ b_{d-1,0} & \cdots & b_{d-1,l} \end{pmatrix} * (i_0, i_1, ..., i_{l-1}, 1)^T,$$

the bank mapping for R_0 and R_1 with a linear transformation vector $\vec{\alpha} = (\alpha_0, \alpha_1, ..., \alpha_{d-1})$ is

$$f(R_0) = (\vec{\alpha} * R_0) \% N \qquad \text{and} \qquad f(R_1) = (\vec{\alpha} * R_1) \% N.$$

THEOREM 1. *Assuming that a d-dimensional array is accessed by two references R_0 and R_1 in an l-level loop nest, the array is cyclically partitioned into N banks with a linear transformation vector $\vec{\alpha}$ and a bank mapping function f so that the simultaneous accesses are not in conflict in the iteration domain, if*

$$gcd(\vec{\alpha} * \Delta_0^T, \vec{\alpha} * \Delta_1^T, ..., \vec{\alpha} * \Delta_{l-1}^T, N) \nmid \vec{\alpha} * \Delta_l^T \qquad (4)$$

where

$$\Delta_k = (\Delta_{0,k}, \Delta_{1,k}, ... \Delta_{d-1,k}), \Delta_l = (-\Delta_{0,l}, -\Delta_{1,l}, ... -\Delta_{d-1,l}),$$
$$\Delta_{j,k} = a_{j,k} - b_{j,k}, \forall 0 \leq k < l, \ 0 \leq j < d,$$

The detailed proof is in Appendix.

EXAMPLE 2. For a two dimensional array A[64][64] with two array references A[j][i], and A[j+1][i+1] in the inner loop iteration, the linear transformation vector $(\alpha_0, \alpha_1) = (1,2)$ and N=2 meets the non-conflict constraint according to gcd(0,0,2)=2 $\nmid$ (1+2).

The candidate $\vec{\alpha}$ can be generated by exhaustive enumeration. We can use some constraints to reduce the searching space. First, it is obvious that $gcd(\alpha_0, \alpha_1, ..., \alpha_{d-1}) = 1$. Second, the optimal partition number is the number of the references m. For this target N, the searching space for the $\vec{\alpha}$ is N^d ($\forall 0 \leq j < d, 0 \leq \alpha_j < N$). If $\vec{\alpha}$ is a candidate, for $\forall k \in \mathbb{Z}$ and $\forall 0 \leq j < d, \alpha_{j\_k} = k * N + \alpha_j$, $\vec{\alpha}_k = (\alpha_{0\_k}, \alpha_{1\_k}, ..., \alpha_{d-1\_k})$ also meets the constraint. In addition, the theorem can be easily extended to multiple references by detecting the conflict between each pair of references.

4.2 Constructing Inner Bank Offset Functions

Using techniques in Section 4.1, the candidate linear transformation vectors can be generated. In this section, we will specify how to calculate the inner bank offset for a given linear transformation vector. The principle is to keep the validation of the partitioning, which is that two different data can't be mapped to the same physical location. Our goal is to optimize Problem 2, for with different mapping functions, some physical locations may be mapped without any data so that an extra storage overhead is induced. Two approaches are introduced in this section.

4.2.1 Optimal Approach

An optimal approach to generate the inner bank offset is to scan the data in sequence. Since all of the data elements on the same hyperplane set $((\alpha * x)\%N = c)$ are in the same bank, scanning the data along the hyperplane set in sequence and use the sequence number as the inner bank offset can generate a valid memory partition without any extra storage overhead. The problem can be converted by integer point counting in a polytope using Ehrhart polynomial [9]. Two polytopes (a base polytope and an offset polytope) are formulated for a given point $x' = (x'_0, x'_1, ..., x'_{d-1})$. Then the sum of the point number in the two polytopes is used as the inner bank offset for the point. We illustrate this process in Example 3. The detailed formulation and theory of integer point counting using Ehrhart polynomial is given in Appendix.

EXAMPLE 3. Given a candidate vector $\vec{\alpha}=(1,2)$, the hyperplanes are described as $x_0 + 2x_1 = c$. For a given point $x' = (3,1)$, the two polytopes are formed as in Fig. 2, in which the base polytope contains the hyperplanes with $c < 5$, and the offset polytope is on $c=5$. According to the theory in [9], the point numbers in the two polytopes are the functions of c and x'_0 separately. By using the Ehrhart tool in Polylib [21], we get the Ehrhart polynomials for each polytope as L_{base} and L_{offset}.

$$L_{base}(c) = \frac{1}{20} \times c^2 + \left(-\frac{1}{10} \times c \right) + [0, \frac{1}{20}, 0, -\frac{3}{20}, -\frac{2}{5}\frac{1}{4}, -\frac{1}{5}\frac{1}{4}, -\frac{2}{5}, -\frac{3}{20}]_c$$

$$L_{offset}(x'_0) = \frac{1}{2} \times x'_0 + [1, \frac{1}{2}]_{x'_0},$$

where $[u_0, u_1, ... u_{p-1}]_c = u_l$ when $c\%p = l, p \in Z$.

When $x'=(x'_0, x'_1) = (3,1)$, c=5,

$L_{base}(5)=\frac{25}{20} - \frac{5}{10} + \frac{1}{4} = 1$, $L_{offset} = 2$,

$g(3,1) = L_{base} + L_{offset} = 3$

Using Ehrhart's point-counting method, we have the optimal solution to Problem 2, but we find that the area required for computing the optimal g fuctions can be very large. In Example 2, four multiplications and two tables (generating the constants in the end of the polynomials) are used. Although we can get the optimal solution to both Problem 1 and Problem 2, compared to the straightforward array-flattening method, the complex address generation in this method makes it not worth doing. As a result a trade-off between practicality and optimality is considered using a heuristic approach presented next.

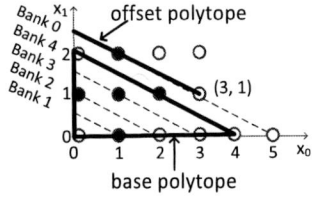

Bank Number=5 f=(x₀+2x₁)%5

Fig. 2 An example of Ehrhart's point-counting

4.2.2 Heuristic Approach

Our heuristic method to efficiently find a linear transformation vector with a comparative simple inner bank offset function g is to do memory padding in the data domain. As stated before, a linear transformation vector $\vec{\alpha}$ for flattening partition is

$$\vec{\alpha}=(1, w_0, w_1, ..., \prod_{k=0}^{d-3} w_k , \prod_{k=0}^{d-2} w_k).$$

It may lead to suboptimal partitioning as we depicted in the motivational example. Our memory padding method finds the coefficient vector $\vec{\alpha}$ with the validity guaranteed based on this given vector. Firstly, a padding vector $\vec{q}=(q_0, q_1, ..., q_{d-1})$ is introduced, in which q_k represents the increase of size in k-th dimension. For a sub-domain formed by dimension j and dimension k $(0 \leq j < k < d)$ with a given bank number N and bank linear transformation vector $\vec{\alpha} = (...\alpha_j, ..., \alpha_k ...)$, a padding size q_k, a valid partition should satisfy

$$\alpha_k(w_k + q_k) \bmod N = \alpha_j \bmod N.$$

It is equal to Eqn. (5).

$$N | \alpha_k(w_k + q_k) - \alpha_j \qquad (5)$$

The new linear transformation vector $\vec{\alpha}$ is

$$\vec{\alpha}=(1, wp_0, wp_1, ..., \prod_{k=0}^{d-3} wp_k , \prod_{k=0}^{d-2} wp_k),$$

where $0 \leq k < d, wp_k = w_k + q_k$.

The geometric meaning of the memory padding is that as each hyperplane only has one data element with the vector based on array flattening, the address is actually generated by scanning along a certain dimension. With a certain bank number, the allocation of the banks needs to be continuous between the last data element in the previous line and the first data element in the next line so that the partitioning validity is met. Fig. 3 shows an example for our method. To meet the validity of the partitioning with an optimal bank number five, dimension x_0 is increased by 2 (size increased from 64 to 66).

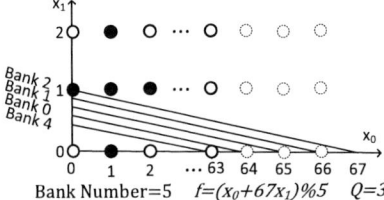

Bank Number=5 f=(x₀+67x₁)%5   Q=3

Fig. 3 An example for memory padding

The above discussion is based on a fixed dimension scanning order. But in fact the value of the element in $\vec{\alpha}$ implies the scanning order. For example, $\alpha_k = 1$ implies the scanning starts from dimension k. Thus when we change the value of $\vec{\alpha}$ in a range, the scanning order of the array and the total padding size will be changed. Through this, we minimize the extra storage overhead induced by memory padding. The padding size on each dimension is bounded within N, as each dimension is cyclically partitioned according to bank mapping function f. Eqn. (5) could be simplified as $N|w_k + q_k$, then we have

$$q_k = N * \left\lceil \frac{w_k}{N} \right\rceil - w_k \qquad (6)$$

The maximum padding size on a d-dimensional array with N partition banks is calculated as

$$(N-1) * \frac{\prod_{k=0}^{d-1} w_k}{w_0} + (N-1) * \frac{\prod_{k=0}^{d-1} w_k}{w_1} +, ..., +(N-1) * \frac{\prod_{k=0}^{d-1} w_k}{w_{d-2}}$$

4.3 Overall Flow

This section describes the overall flow while using memory padding based heuristic method. As the interplay between the padding size and the bank number, we give our flow to find the tradeoff between the optimal partition and extra storage overhead. The lower bound for the bank number N is the reference number m in the inner loop iteration. First, we fix the bank number N ($N \geq m$). Second, we find the possible padding $\vec{q} = (q_0, q_1, ..., q_{d-1})$ under various array dimension orders with linear transformation coefficient $\vec{\alpha} = (\alpha_0, \alpha_1, ..., \alpha_{d-1})$. And we'll get the best candidate vector $\vec{\alpha}_p$ with the total padding size minimized. Then we check whether $\vec{\alpha}_p$ satisfy the conflict-free constraint in Eqn. (4) with the bank number N. The detailed LTB algorithm is described as follows.

Step 1: Give the partition bank number $N = m$.
Step 2: Find every possible $\vec{\alpha}$ ($\forall 0 \leq j < d, 0 \leq \alpha_j < N$) with a padding vector $\vec{q}$ according to Eqn. (6). Queue all the $\vec{\alpha}$ by the increase of the total padding size. Find $\vec{\alpha}_p$ with the minimum padding size.
Step 3: Check if $\vec{\alpha}_p$ meets the conflict-free requirement according to Eqn. (4). If $\vec{\alpha}_p$ cannot meet the requirement, find the next solution in the queue and recheck the conflict-free constraint.
Step 4: If there is no solution for N, $N=N+1$, go back to Step 2.

The complexity of searching for an array dimension order is $\prod_{k=1}^{d} k$. And according to Eqn. (6), we can actually calculate the padding size based on a given dimension order. This flow is capable to find a solution for both Problem 1 and Problem 2. It is optimal in Problem 1, and it provides a near optimal solution to Problem 2 with a bounded maximum extra storage overhead and a low complexity. Our experiments prove that in some cases the padding method can find an optimal solution and in other cases the gap between it with optimality is small.

5. EXPERIMENTAL RESULTS

5.1 Experiment Setup

The automatic multidimensional memory-partitioning flow is implemented in C based on the open source compiler infrastructure ROSE [14]. ROSE is a flexible translator supporting source-to-source code transformation. We use Vivado from Xilinx [17] as the high-level synthesis tool. The RTL output is implemented by Xilinx ISE 13.1 [18] on the target FPGA platform Xilinx Virtex-6. The implementation flow is illustrated in Fig. 4.

The high-level abstraction is parsed into the flow with the partition directives and constraints, such as target II. After memory partitioning analysis and source-to-source code transformation, the transformed code is synthesized by the high-level synthesis tool and followed by logic synthesis.

Six loop kernels are selected from the real applications as the benchmarks. As we focus on the effects brought by different access patterns, several of the benchmarks are the loop kernels from the same application with different access patterns. DENOISE_1 and DENOISE_2 are from the Rician-denoise algorithm [11] in medical image applications. DENOISE_1 is the original access pattern which accesses five data elements in the inner-loop iteration. DENOISE_2 is the access pattern by unrolling the loop in DENOISE_1 by 2. MOTION_LV and MOTION_C are the different loop kernels of motion compensation from official H.264 decoder JM 14.0 [4]. MOTION_LH is the motion compensation for luma samples in the video frame in the vertical direction, and MOTION_C is the interpolation for the chroma components. BICUBIC_INTER [1] is from bicubic interpolation process. And SOBEL [16] is from Sobel edge detection algorithm. (The detailed access patterns of the benchmarks are illustrated in Appendix)

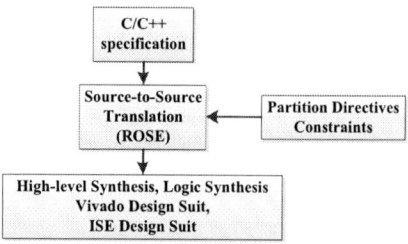

Fig. 4 Implementation flow

5.2 Experimental Results

The detailed experimental results are shown in Table 1, Table 2 and Table 3. We compared the experimental results for the state-of-art 1-dimensional partition algorithm with flattened arrays and our proposed linear transformation based algorithm (LTB). Table 1 shows the percentage of extra storage overhead when applying different memory size on DENOISE_1. The results after source-to-source transformation are shown in Table 2. And Table 3 shows the results after synthesizing. The algorithm from [6] for the flattening memory partitioning is re-implemented for comparison. As shown in Table 2, we list the *original II* of the pipelined loop and the *target II*. Our target throughput is *II=1*. The partitioning in both of the methods can meet the throughput requirement. The bank number for achieving the target throughput by using the flattening method and LTB are represented in the next two columns, followed by the essential *padding size* after applying LTB. Physical resource usage (block RAMs, slices, and DSPs) and timing information are reported by Xilinx ISE, and power estimation is given by Xilinx XPower Analyzer. The block RAMs are dual-port in the Xilinx Virtex-6.

Table 1 represents the percentage of padding size compared to the original array size (also the optimal solution). 140 different array sizes are applied in the experiments. And we found that the padding size is related tiny. In Table 2, we can see that our proposed LTB method improves the partitioning bank number on all of the six benchmarks. And five benchmarks have extra padding by using our LTB approach (DENOISE_1, DENOISE_2, MOTION_C, BICUBIC_INTER, and SOBEL). As each piece of the partition is relatively not too large and can fit in a BRAM, the padding size is totally negliable. However, if the arrays in the benchmarks are originally large, memory padding may introduce extra memory overhead; this means that more block RAMs are required.

Table 3 represents the use of logical units on FPGA. The utilization of block RAMS, Slices and DSPs are very related to the bank number. The average BRAMs improvement after using LTB is up to 21%. In the benchmarks DENOISE_2 and MOTION_C, the reduction of DSPs is up to 96% and 100%. In these cases, the partitioning number is reduced to a power of 2, which can be implemented as data shifting rather than using DSPs for the dividers. Although the use of the physical resources in DENOISE_1, DENOISE_2, BICUBIC_INTER, and SOBEL is reduced, the power estimation increases in these benchmarks, especially in SOBEL (about 48.85%). Based on our analysis of the transformed code, LTB uses more logic to implement the address generation for the array indices due to the extra padding size (It introduces an extra multiplication in each index). We could optimize it with some common address generation strategies (as the scheme proposed in [12]). However, in the benchmark SOBEL, as the flattening method uses 25% more block RAMs in this benchmark and the critical path is much longer than the one in LTB, the target CP (5ns) cannot be met.

In all, there is an average 21% reduction in BRAMs, 19% reduction in slices, 46% reduction in DSPs, and 14.69% more overhead in power. The CP has a small increase of 0.6% on average.

Table 1 Storage overhead of padding method

Array Size(# of data)	Padding Rate
<1000	0.0706
1000~5000	0.0281
5000~10000	0.0161
10000~20000	0.0116
>20000	0.0098

Table 2 High-level partitioning results

	Original II	Target II	Bank (Flatten)	Bank (LTB)	Padding size
DENOISE_1	5	1	6	5	64
DENOISE_2	8	1	10	8	128
MOTION_C	4	1	6	4	64
MOTION_LV	6	1	7	6	0
BICUBIC_INTER	4	1	6	5	64
SOBEL	9	1	12	9	64

Table 3 Synthesis experimental results

		Block RAM	Slice	DSP	CP (ns)	Power
DENOISE_1	Flatten	6	531	8	3.826	537
	LTB	5	441	8	4.451	685
	comp.(%)	-16.7	-16.9	0	16.3	27.5
DENOISE_2	Flatten	10	1114	75	4.995	1097
	LTB	8	767	3	4.563	1367
	comp.(%)	-20	-31.1	-96	-8.6	24.6
MOTION_C	Flatten	6	515	4	4.215	670
	LTB	4	255	0	4.068	484
	comp.(%)	-33.3	-50.5	-100	-3.5	-27.8
MOTION_ LV	Flatten	7	627	9	4.143	1263
	LTB	6	601	9	3.846	1026
	comp.(%)	-14.3	-4.1	0	-7.2	-16.15
BICUBIC_ INTER	Flatten	6	456	4	3.870	512
	LTB	5	441	4	4.451	672
	comp.(%)	-16.7	-3.3	0	15	31.25
SOBEL	Flatten	12	1302	105	5.222	1441
	LTB	9	1195	15	4.808	2145
	comp.(%)	-25	-8.2	-85.7	-7.9	48.85
AVERAGE(%)		-21	-19	-46	0.6	14.69

6. CONCLUSIONS

Memory partitioning is a crucial technology to enable data-level parallelism in FPGA designs. In this work we propose an automatic memory-partitioning method for multidimensional arrays. Linear transformation on the multidimensional array indices is introduced to extend the design space for the possible optimal solution. An optimal solution based on Ehrhart points counting and a heuristic solution based on memory padding are proposed. Experimental results demonstrate that compared with the state-of-art partitioning algorithm, our proposed algorithm can reduce the number of block RAMs by 21%.

7. ACKNOWLEDGMENTS

This work was supported in part by the National High Technology Research and Development Program of China 2012AA010902, RFDP 20110001110099 and 20110001120132,

and NSFC 61103028. We would like to thank UCLA/PKU Joint Research Institute in Science and Engineering (JRI) and the support from Xilinx.

8. REFERENCES

[1] Bicubic interpolation http://www.mpi-hd.mpg.de/astrophysik/HEA/internal/Numerical_Recipes/f3-6.pdf

[2] Center for Domain-Specific Computing http://www.cdsc.ucla.edu/

[3] F.Balasa, H.Zhu, I.I.Lucian, "Computation of Storage Requirements for Multi-Dimensional Signal Processing Applications," Signal Processing Systemsm," in *IEEE Trans. Very Large Scale Integration Systems (TVLSI)*,VOL.15, No.4,2007.

[4] JM Software, H.264/AVC Software Coordination, http://iphome.hhi.de/suehring/tml/

[5] J. Cong, P. Zhang and Y. Zou, "Optimizing Memory Hierarchy Allocation with Loop Transformations for High-Level Synthesis", *Proceedings of the 49th Annual Design Automation Conference (DAC 2012)*, pp. 1233-1238, 2012.

[6] J. Cong, W. Jiang, B. Liu, and Y. Zou, "Automatic Memory Partitioning and Scheduling for Throughput and Power Optimization," in *ACM Trans. on Design Automation of Electronic Systems (TODAES)*, 2011, Vol. 16 Issue 2, Article 15

[7] L. T. Yang,Y. Pan, et al, *High performance scientific and engineering computing: hardware/software support*, Springer, 2003

[8] M. Gupta, "Automatic Data Partitioning on Distributed Memory Multicomputers," 1992.

[9] P. Clauss, V. Loechner, "Parametric Parametric Analysis of Polyhedral Iteration Spaces," in *Journal of VLSI signal processing systems for signal, image and video technology*, Volume 19, Issue 2, pp 179-194, 1998.

[10] P. Feautrier, "Some efficient solutions for the affine scheduling problem, part I, one dimensional time," in *International Journal of Parallel Processing*, 21(6), December 1992

[11] P. Getreuer, "tvreg: Variational imaging methods for denoising, deconvolution, inpainting, and segmentation," online available: http://code.google.com/p/cdsc-image-processing-pipeline/downloads/list

[12] P. Li, Y. Wang, P. Zhang, G. Luo, T. Wang, and J. Cong, "Memory Partitioning and Scheduling Co-optimization in Behavioral Synthesis", in *Inter. Conf. on Computer-Aided Design (ICCAD)*, 2012, pp. 488-495.

[13] Q. Liu, T. Todman, W. Luk, "Combining Optimizations in Automated Low Power Design," in *Proc.of Design, Automation and Test Europe(DATE)*, 2010, pp. 1791-1796.

[14] ROSE compiler infrastructure, http://rosecompiler.org/

[15] S. Chatterjee, et al, "Generating Local Addresses and Communication Sets for Data-parallel Programs," *Journal of Parallel and Distributed Computing*,1995.

[16] S. Verdoolaege, H. Nikolov, and T. Stefanov, "pn: A Tool for Improved Derivation of Process Networks," *EURASIP Journal on Embedded Systems*, vol. 2007, pp. 1-13, 2007.

[17] Vivado High-Level Synthesis , http://www.xilinx.com/products/design-tools/vivado/integration/esl-design/hls/index.htm

[18] Xilinx ISE Design Suite, http://www.xilinx.com/

[19] Y. Ben-Asher, N. Rotem, "Automatic Memory Partitioning: Increasing Memory Parallelism via Data Structure Partitioning," in *Proc. of the 8th Int. Conf. on Hardware/Software Codesign and System Synthesis (CODES+ISSS)*, 2010, pp. 155-162.

[20] Y. Wang, P. Zhang, X. Cheng, and J. Cong, "An Integrated and Automated Memory Optimization Flow for FPGA Behavioral Synthesis," in *Asia and South Pacific Design Automation Conf. (ASP-DAC)*, 2012, pp. 257-262.

[21] Polylib, http://www.irisa.fr/polylib/

Appendix

1. The proof of Theorem 1

Assuming that there are two d-dimensional array references in the iteration domain as

$$R_0 = \begin{pmatrix} a_{0,0} & \cdots & a_{0,l} \\ \vdots & \ddots & \vdots \\ a_{d-1,0} & \cdots & a_{d-1,l} \end{pmatrix} * (i_0, i_1, \ldots, i_{l-1}, 1)^T \quad \text{and}$$

$$R_1 = \begin{pmatrix} b_{0,0} & \cdots & b_{0,l} \\ \vdots & \ddots & \vdots \\ b_{d-1,0} & \cdots & b_{d-1,l} \end{pmatrix} * (i_0, i_1, \ldots, i_{l-1}, 1)^T.$$

The bank number mapping functions with a linear transformation vector $\vec{\alpha} = (\alpha_0, \alpha_1, \ldots, \alpha_{d-1})$ are

$$f(R_0) = (\vec{\alpha} * R_0) \% N \quad \text{and} \quad f(R_1) = (\vec{\alpha} * R_1) \% N$$

THEOREM 1. *Assuming that a d-dimensional array is accessed by two references R_0 and R_1 in an l-level loop nest, the array is cyclically partitioned into N banks with a linear transformation vector $\vec{\alpha}$ and a bank mapping function f so that the simultaneous accesses are not in conflict in the iteration domain, if*

$$gcd(\vec{\alpha} * \Delta_0{}^T, \vec{\alpha} * \Delta_1{}^T, \ldots, \vec{\alpha} * \Delta_{l-1}{}^T, N) \nmid \vec{\alpha} * \Delta_l{}^T$$

where

$\Delta_k = (\Delta_{0,k}, \Delta_{1,k}, \ldots \Delta_{d-1,k}), \Delta_l = (-\Delta_{0,l}, -\Delta_{1,l}, \ldots -\Delta_{d-1,l}),$

$\Delta_{j,k} = a_{j,k} - b_{j,k}, \forall 0 \le k < l, \ 0 \le j < d$

Proof

The converse-negative proposition of theorem is proved as:

$$. \exists \vec{i} \text{ s.t. } f(R_0) = f(R_1)$$

$$\Leftrightarrow \vec{\alpha} * R_0 \equiv \vec{\alpha} * R_1 \bmod N$$

$$\Leftrightarrow \exists \vec{i}, k \text{ s.t.} \vec{\alpha} * \begin{pmatrix} \Delta_{0,0} & \cdots & \Delta a_{0,l-1} \\ \vdots & \ddots & \vdots \\ \Delta_{d-1,0} & \cdots & \Delta_{d-1,l-1} \end{pmatrix} * (i_0, i_1, \ldots, i_{l-1})^T$$
$$+ kN = - \vec{\alpha} * (\Delta_{0,l}, \Delta_{1,l}, \ldots \Delta_{d-1,l})^T$$

$$\Leftrightarrow \gcd(\vec{\alpha} * (\Delta_{0,0}, \Delta_{1,0}, \ldots \Delta_{d-1,0})^T, \vec{\alpha} * (\Delta_{0,1}, \Delta_{1,1}, \ldots \Delta_{d-1,1})^T,$$
$$\ldots, \vec{\alpha} * (\Delta_{0,l-1}, \Delta_{1,l-1}, \ldots \Delta_{d-1,l-1})^T, N)|$$
$$\vec{\alpha} * (\Delta_{0,l}, \Delta_{1,l}, \ldots \Delta_{d-1,l})^T$$

$$\Leftrightarrow gcd(\vec{\alpha} * \Delta_0{}^T, \vec{\alpha} * \Delta_1{}^T, \ldots, \vec{\alpha} * \Delta_{l-1}{}^T, N)| \vec{\alpha} * \Delta_l{}^T$$

where

$\Delta_k = (\Delta_{0,k}, \Delta_{1,k}, \ldots \Delta_{d-1,k}), \Delta_l = (-\Delta_{0,l}, -\Delta_{1,l}, \ldots -\Delta_{d-1,l}),$

$\Delta_{j,k} = a_{j,k} - b_{j,k}, \forall 0 \le k < l, \ 0 \le j < d$

2. Ehrhart's Points-Counting Theory

The following definitions and theorems are referenced from [9], as supplemental materials to section 4.2.1 to help understand the optimal approach.

Let Q denote the set of rational numbers and Z the set of integers. A convex polyhedron is defined by a finite set of linear inequalities:

$$P = \{ x \in Q^d \mid A \cdot x \le b \},$$

where A is a rational matrix and b a rational vector.

Definition 1 (homothetic-bordered system [9]). Let H_N, $N = (n_1, n_2, \ldots, n_q)$, be a system defined by constraints of the form $\sum a_i x_i = \sum b_j n_j + c$, $\sum a_i x_i < \sum b_j n_j + c$, $\sum a_i x_i \le \sum b_j n_j + c$, where the a_i's, the b_i's and the c's are given integers, the x_i's are free variables and the n_i's are positive integral parameters.

Such a system is homothetic-bordered if and only if the polytope it defines has vertices whose coordinates are affine combinations of the parameters.

Counting the number of integer points is based on the decomposition of a parametric polytope into several homothetic-bordered systems, associated with validity domains.

Example

$$\begin{aligned} P_{n_1} &= \{x \mid x \ge 0, x \le n, 2x \le n+6\}, \\ P_{n_2} &= \{x \mid x \ge 0, n \ge 0, 2x \le n+6\}, \\ P_{n_3} &= \{x \mid 0 \le x \le n\} \end{aligned}$$

P_{n_2} and P_{n_3} are homothetic-bordered system and P_{n_1} is not homothethic-bordered system.

$$P_{n_1} = \cup \begin{cases} P_{n_2}, & \text{when } 0 \le n \le 6 \\ P_{n_3}, & \text{when } 6 \le n \end{cases}$$

Definition 2 (periodic number [9]). A one-dimensional periodic number $u(n) = [u_1, u_2, \ldots, u_p]n$ is equal to the item whose rank is equal to $n \bmod p$, p is called the period of $u(n)$.

$$u(n) = \begin{cases} u_1 & if n = 1 (mod\ p), \\ u_2 & if n = 2 (mod\ p), \\ \quad \cdots \\ u_p & if n = 0 (mod\ p). \end{cases}$$

Example

$$\begin{aligned} (-1)^n &= [-1, 1]_n \\ (-1)^{n-m} &= \begin{bmatrix} 1 & -1 \\ -1 & 1 \end{bmatrix}_{(n,m)} \end{aligned}$$

Definition 3 (denominator [9]). The denominator of a rational point is the lowest common multiple of the denominators of its coordinates. The denominator of a rational polyhedron is the least common multiple of the denominators of its vertices.

Theorem 1 (Ehrhart's fundamental theorem [9]). *The enumerator j_n of any homothetic-bordered k polyhedron P_n is a polynomial in n of degree k if P_n is integral; and it is a pseudo-polynomial in n of degree k whose pseudo-period is the denominator of P_n if P_n is rational.*

EXAMPLE Bank mapping function: $f = (x_0 + 2x_1)\%5$, $0 \le x_0, x_1 \le 64$, for $(x_0, x_1) = (32,15)$, find the inner bank address.

There are two polytopes: base polytope and offset polytope

- The base polytope is

$$\begin{cases} x_0 + 2 + 5k = c \\ 0 \le x_0, x_1 \le 64 \\ \quad k \ge 1 \\ \quad c \ge 0 \end{cases}$$

There are four Ehrhart polynomials for the base polytope. For different domain of d, they are:

Domain1: c -197 >= 0

Ehrhart Polynomial: $L_{base}(c) = 845$

Domain2: c -133 >= 0 and - c + 197 >= 0

Ehrhart Polynomial:

$$L_{base}(c) = -\frac{1}{20} \times c^2 + \frac{98}{5} \times c +$$

$$[-1076, -\frac{21511}{20}, -1076, -\frac{21507}{20},$$

$$-\frac{5378}{5}, -\frac{4303}{4}, -\frac{5379}{5}, -\frac{4303}{4}, -\frac{5378}{5}, -\frac{21507}{20}]$$

Domain3: c -69 >= 0 and - c + 133 >= 0

Ehrhart Polynomial: $L_{base}(c) = \frac{13}{2} \times c + [-218, -\frac{435}{2}]$

Domain4: - c + 69 >= 0 and c -5 >= 0

Ehrhart Polynomial:

$$L_{base}(c) = \frac{1}{20} \times c^2 + \left(-\frac{1}{10} \times c\right)$$

$$+[0, \frac{1}{20}, 0, -\frac{3}{20}, -\frac{2}{5}, \frac{1}{4}, -\frac{1}{5}, \frac{1}{4}, -\frac{2}{5}, -\frac{3}{20}]$$

- The offset polytope is

$$\begin{cases} x_0 + 2x_1 = x_0' + 2x_1' \\ 0 \le x_0 \le x_0' \\ 0 \le x_0', \ x_1' \le 64 \end{cases}$$

$$L_{offset}(x_0') = \frac{1}{2} \times x_0' + [1, \frac{1}{2}]$$

$(x_0, x_1) = (32,15)$, c=62,

$g(32,15) = L_{base}(62) + L_{offset} = 186 + 17 = 203$

3. Detailed descriptions of the benchmarks

The detailed description of the benchmarks is listed in Table 2. DENOISE_1 and DENOISE_2 are from the Rician-denoise algorithm [11] from medical image applications, and their access patterns are shown in Fig. 5(a), Fig. 5(b). DENOISE_1 and DENOISE_2 are the original access patterns in the application. DENOISE_2 is the access pattern by unrolling DENOISE_1 by 2. MOTION_LV and MOTION_C are the different loop kernels of motion compensation from official H.264 decoder JM 14.0 [4]. MOTION_C is the interpolation for the chroma components. MOTION_LV is the motion compensation for the luma samples in the video frame in the vertical direction. Their access patterns are shown in Fig. 5(c) and Fig. 5(d). BICUBIC_INTER [1] is from bicubic interpolation process. And SOBEL [16] is from Sobel edge detection algorithm. The access patterns of them are illustrated in Fig. 5(e) and Fig. 5(f).

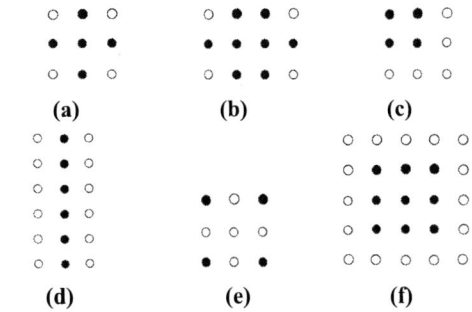

Fig. 5 The access patterns of the benchmarks: (a) DENOISE_1 (b) DENOISE_2 (c) MOTION_C (d) MOTION_LV (e) BICUBIC_INTER (f) SOBEL

Table 4 Benchmark Description

	Benchmark description
DENOISE_1	2D Rician-denoise, as Fig. 5(a)
DENOISE_2	2D Rician-denoise, with loop titling, as Fig. 5(b)
MOTION_C	H.264 motion compensation for chroma samples, as Fig. 5(c)
MOTION_LV	H.264 Motion compensation for luma samples in horizontal direction, as Fig. 5(d)
BICUBIC_INTER	Bicubic interpolation, as Fig. 5(e)
SOBEL	2D Sobel edge detection algorithm, as Fig. 5(f)

Balancing Security and Utility in Medical Devices?

Masoud Rostami
Rice University
Houston, TX
masoud@rice.edu

Wayne Burleson
University of Massachusetts
Amherst, MA
Burleson@ecs.umass.edu

Ari Juels
RSA Laboratories
Cambridge, MA
ari.juels@rsa.com

Farinaz Koushanfar
Rice University
Houston, TX
farinaz@rice.edu

ABSTRACT

Implantable Medical Devices (IMDs) are being embedded increasingly often in patients' bodies to monitor and help treat medical conditions. To facilitate monitoring and control, IMDs are often equipped with wireless interfaces. While convenient, wireless connectivity raises the risk of malicious access to an IMD that can potentially infringe patients' privacy and even endanger their lives.

Thus, while ease of access to IMDs can be vital for timely medical intervention, too much ease is dangerous. Obvious approaches, such as passwords and certificates, are unworkable at large scale given the lack of central authorities and frequent emergencies in medical settings. Additionally, IMDs are heavily constrained in their power consumption and computational capabilities. Designing access-control mechanisms for IMDs that can meet the many constraints of real-world deployment is an important research challenge.

In this paper, we review proposed approaches to the access-control problem for IMDs, including the problem of secure pairing (and key distribution) between an IMD and another device, such as a programmer. (We also treat related technologies, such as body-area networks.) We describe some limitations of well-conceived proposals and reveal security weaknesses in two proposed cryptographic pairing schemes. Our intention is to stimulate yet more inventive and rigorous research in the intriguing and challenging areas of IMD security and medical-device security in general.

Categories and Subject Descriptors

J.3 [**Computer Applications**]: Life and Medical Sciences - Medical information systems; C.3 [**Computer Systems Organization**]: Special-Purpose and Application-Based System, Real-time and embedded systems

General Terms

Security, Design, Usability

Permission to make digital or hard copies of all or part of this work for personal or classroom use is granted without fee provided that copies are not made or distributed for profit or commercial advantage and that copies bear this notice and the full citation on the first page. To copy otherwise, to republish, to post on servers or to redistribute to lists, requires prior specific permission and/or a fee.
DAC 13, May 29 - June 07 2013, Austin, TX, USA.

Keywords

Implantable Medical Devices, IMD Security

1. INTRODUCTION

Implantable Medical Devices (IMDs) are increasingly being embedded into patients to monitor medical conditions and to apply a range of therapies, from medication infusion to cardiac pacing and neurostimulation [1]. State-of-the-art IMDs often contain electronic components capable of computation, storage, and wireless communication, and monitor patient conditions in order to adjust their therapeutic regimens. Over the past few decades, IMDs have greatly improved patient care, quality of life, and life expectancy. Next-generation IMDs will provide even more benefit, as they improve existing therapies and enable new ones.

Wireless interfaces contribute significantly to the utility and successful deployment of IMDs, as they enable convenient and non-invasive control, monitoring, and maintenance of the IMD using a control device typically known as a "programmer." A drawback to such wireless access, though, is its inherently open nature, which raises IMDs susceptibility to over-the-air adversarial threats ranging from eavesdropping to unauthorized access and control. IMD manufacturers are subject to strict safety and reliability requirements, and the safety of IMDs, including the problem of unexpected failures, has been a subject of ongoing research [2]. IMDs are not subject, however, to similar standards around *logical security* and *access control*. In many cases, the approach has been protection of IMDs through security-by-obscurity: The main barrier to unauthorized access is no more than secret and proprietary design.

Researchers have consequently demonstrated a range of practical attacks, some executed remotely over the air, permitting unauthorized access to IMDs such as cardiac defibrillators and insulin pumps [3,4]. These attacks enable an adversary to eavesdrop on IMD communications and in some cases emulate a programmer and modify the therapies applied by IMDs, potentially threatening patient privacy and even patients' health and lives. While there are no documented examples of such attacks "in the wild," the need for better secured access control in IMDs (and robust logical security more generally) is urgent and clear.

One major challenge in designing good access-control and other security mechanisms for IMD is their severely constrained resources. IMDs, like other portable electronics, have limited available battery energy. For IMDs, the situation is particularly problematic, as battery replacement usually entails invasive surgery and removal/replacement of the IMD or its components. Additionally, the desire for minimally invasive implantation favors small form-factors for IMDs, further limiting battery size. Remote power delivery and energy scavenging, although promising, are currently available in only a very limited set of applications.

Three ongoing trends suggest that energy challenges will persist for IMDs. First, the devices are getting increasingly complex and power-hungry due to demand for new, sophisticated therapeutic and monitoring functionality. Power requirements are even outstripping the benefits of Moore's Law and low-power design techniques, as with smart-phones. Second, IMDs are collecting ever more data as new sensors are added to monitor patient health. Transmitting sensor data from an IMD involves wireless communication, which is power intensive. Third, well designed security protocols, including authentication and code verification require the use of cryptography, and cryptographic primitives are notoriously computation- and power-intensive.

A second major challenge in securing IMDs is the tension between the demands of reliable access on the one hand, and protection against access by an adversary or unauthorized entity on the other. In an emergency, medical personnel may need to monitor or reprogram a patient's IMD immediately and thus access it with as little impediment as possible. But an IMD that can be accessed too easily may be vulnerable to eavesdropping on its data transmissions or tampering with its operation.

The conflicting requirements of security, reliability, and usability in IMDs have given rise to an important and vigorous line of research. Halperin et al. [5] first discussed the security and privacy challenges arising from the resource constraints and inflexibility of existing IMD designs, and highlighted fundamental tensions among privacy, security, safety, and utility in IMDs. Increased networking of embedded devices and the emergence of pervasive health care technologies have also motivated closely related security and privacy research for general sensor networks and body-sensor networks (e.g., [6, 7]), healthcare information technology (e.g., [8]), and patient health data (e.g., [9]).

In this paper, we briefly survey the problem of enabling authorized access to IMDs by programmers. We use the terms IMD and Programmer generically in this paper, but much of the literature we explore treats or is also relevant to other types of medical devices, such as body-area networks (BANs)—arrays of medical / physiological devices that may or many not be implanted.

Security model and goals

The security goals for a device architecture naturally depend upon the participating trustworthy entities and the motives, access, capabilities, and resources of a potential adversary. IMDs typically communicate with a Programmer and potentially with other IMDs or outside-the-body medical devices. The adversary of main concern in these settings is one that acts remotely, over the air, against the IMD's network. Given the open nature of wireless networks, such an adversary may be "active," meaning that it has complete control of the network. Such an adversary can replay, modify, forge, drop, and jam message within the network at will. We can (at least in some cases) assume, however, that one side of the communication, the IMD, is not directly accessible to the adversary, as it is implanted in the body.

Again, a main objective is to allow a Programmer to gain logical access to an IMD while an adversary can't feasibly do so. An obvious approach would be to authenticate an entity communicating with an IMD using a predetermined key or password. The main obstacle to this approach is a fundamental challenge of cryptography: Key distribution. It's impractical to ensure that all valid programmers and/or medical personnel (in medical settings around the world) have valid keys but an adversary can't feasibly gain access

to one.

A somewhat more flexible approach to key distribution is the use of public-key cryptography. The TLS protocol, which is ubiquitous on the Internet, relies upon the distribution of public keys to servers. A global public-key infrastructure (PKI) permits designated authorities to certify these public keys, ensuring a binding of the public key to a suitable server identity (domain name). Building a PKI for all medical programmers worldwide—and adequately securing all of their private keys—would be a formidable and probably impractical effort. Recent breakdowns in the trustworthiness of certificate issuance for the Internet, e.g., [10], warn in general of the challenges of constructing sound PKIs.

Recent research on access control and authentication for IMDs and BANs has mainly focused on approaches in which Programmer authorization is determined as a function of *physical access or proximity*. As we explain, there are a number ways to determine whether a Programmer is in suitable proximity to an IMD. The problem often boils down, however, to one of *key-agreement* or *pairing*. Ideally, only an authorized Programmer should be able to establish a (secret) cryptographic key with an IMD; this key enables the establishment of a secure (confidential and integrity-protected) channel between the two devices.

Thus we focus here mainly on proposed approaches to IMD access control through key agreement, which may occur directly between a Programmer and IMD or may be mediated by an additional trusted device carried by a patient. While there are several sound and well-conceived proposed approaches, none in our view provides a fully satisfactory balance of utility and security. Achieving rigorous security guarantees can also be quite challenging: We present attacks against two such proposed protocols, IMD-Guard [11] and OPFKA [12]. Thus, the challenge of good IMD access-control remains an important open research problem.

Organization: The remainder of the paper is organized as follows. We review several IMD key-agreement schemes in Section 2. Distance bounding, jamming, and shielding approaches are discussed in Section 3. In Section 4, we analyze and present attacks against two proposed protocols. We discuss the prospect of using new hardware architectures and device technologies to secure IMDs in Section 5. Section 6 concludes the paper.

2. KEY-AGREEMENT SCHEMES

As we have noted, use of pre-distribution of secret or public keys among IMDs and programmers presents unworkable key-distribution and certification challenges. Proposed access-control protocols for IMDs thus generally avoid reliance on pre-established relationships between IMDs and programmers and instead generate keys on the fly. In this section, we review two general methods proposed in earlier work: (1) Transmitting a secret key using the human body and (2) Key generation using physiological values.

2.1 Key distribution by intra-body signaling

One idea for sharing a secret key between an IMD and Programmer is to generate the key in the IMD and transmit it to the Programmer through the human body itself. This approach requires that the Programmer be in close enough proximity to receive the key; generally, it may make physical contact with a patient. The critical security assumption is that an adversary must operate at a distance from the patient too great to intercept a key; the minimum required distance for such assurance depends on the specific scheme. We now briefly review three proposed intra-body carriers of IMD secret keys: acoustic, electric, and electromagnetic signals.

Acoustic broadcasting. Halperin et al. [3] proposed a scheme in which an implanted piezo device generates a random key and emits it acoustically. The method results in a rather fast key agreement, requiring only 400ms to transmit a 128-bit key. A serious drawback, however, is the requirement for special implantation of the piezo device. This implantation must be at a depth of 1 cm or so from the skin, ruling out incorporation into deep-body IMDs, such as Implantable Cardioverter Defibrillators (ICDs). The electronic circuits that produce acoustic signals can be shielded with a Faraday cage against electromagnetic interception. An adversary can potentially resort to eavesdropping on acoustic emanations, however, to attack the system. Acoustic eavesdropping of this kind is not well studied and merits further investigation.

Electric and Electromagnetic broadcasting. Zimmerman [13,14] proposed transmitting information through the human body using a pico-amp electric current, in effect using the body as a low-frequency ($\approx$ 1MHz) electrical carrier. Chang et al. [15] discussed securing body area networks (BANs) by distributing a secret key using electrical currents below the action potential of human cells. They used empirical data to analyze the characteristics of the human body as a communication medium. They estimated 0.469 bits per hour as a lower bound on the bandwidth achievable with their proposed method. This is unacceptably slow, of course, for practical IMD key establishment.

In general, key distribution by intra-body communication has the potential to combine strong security against eavesdropping at a distance with minimal power consumption. The actual resistance of such methods to eavesdropping has not been well studied, however. These methods also require approval from government regulators that, to the best of our knowledge, has not yet been granted in the United States even for trial use.

2.2 Key generation using physiological values

The idea of extracting secret keys from physiological values (PVs) to secure IMDs was first suggested in [16]. PVs such as Electrocardiograph (ECG) and Electroencephalography (EEG) signals are suitable candidates for key generation because they provide continuous sources of true randomness. In other words, these PVs may be viewed as entropy sources inside the human body that constantly generate and broadcast (unpredictable) random bits. The randomness of PVs has been documented in an extensive body of medical literature [17–19].

Due to its availability throughout the body and its ease of measurement, the most frequently proposed PV for securing IMDs has been the ECG signal, the electrical signal associated with the activity of the heart.

A number of challenges need to be addressed to achieve practical and secure use of PVs in key agreement. One obstacle is that PV readings are highly sensitive to probe locations on the body and to environmental conditions. Chang et al. [15] assert that if a transformation of the full ECG signal is used for authentication, as suggested in [20, 21], then the PV readings may be so noisy, given a poorly placed probe, that they can be decoded as effectively at a distance by an adversary as by a Programmer with physical contact. The full ECG signal cannot be consistently decoded because the shape of the associated waveform is subject to distortion. Time intervals between specific waveform features, however, can in fact be reliably measured from nearly anywhere in the body. One such feature is the prominent R-Peak of the ECG signal: The time between two R-peaks, which is equal to the heartbeat duration, is essentially invariant to the positioning of probes on the body.

It has been shown that if the heartbeat duration is appropriately quantized, some of its least significant bits are truly random [22–25]. These random bits may differ across probe points on

the body due to measurement noise, limiting their naïve use as key bits shared between an IMD and Programmer. To address the challenge of noisy key sources, several methods have been advanced in previous work. For example, Xu et al. [11] proposed the IMD-Guard protocol to securely pair an IMD with an external device using noisy ECG data to construct a key. IMDGuard looks to establish a persistent, cryptographic-strength key under non-emergency conditions. Unfortunately, the IMDGuard pairing protocol lacks a rigorous security analysis; Section 4 describes a man-in-the-middle attack that reduces its effective key length and hence its claimed level of security.

Another possible approach for securely extracting a cryptographic key from noisy PV readings is to use "fuzzy" cryptographic primitives, e.g., [26, 27]. Some proposed scheme use the "fuzzy vault" construction in [27] or variants thereof to authenticate devices in a body-area network [20, 21, 28]. Recently, Hu et al. [12] proposed a variant algorithm for PV-based key-agreement, called OPFKA, that is designed to reduce the storage costs associated with fuzzy vaults. OPFKA, however, lacks a rigorous security analysis and, as we explain in Section 4, has notable security weaknesses. The design of a reliable PV-based IMD key-agreement protocol with rigorously analyzed security properties, low power consumption, and a minimal hardware footprint remains a significant open research problem.

3. SECURITY USING DISTANCE BOUNDING OR JAMMING

Another approach to establishing a secure channel between an IMD and a Programmer (or other external device) is to make use of a trusted device to intermediate access to the IMD. This trusted device can be external to a patient's body, and thus well resourced. It can shield the IMD from unauthorized attempts at access and even potentially jam malicious ones.

The idea of blocking inappropriate access to an IMD was first proposed in [29] via a device called a Communication Cloaker. The idea saw a follow-up exploration by Gollakota et al. [30]. Their proposed device, called a *shield*, is worn near the body and used to authenticate / mediate Programmer (or other) communications with the IMD. Helpfully, a shield doesn't require modification of existing IMDs. It protects communications with the IMD using a full duplex radio device acting as a jammer-cum-receiver. It simultaneously listens to and jams IMD messages as appropriate, as well as unauthorized Programmer commands.

Shen et al. [31] have recently proposed a smart jamming technique in which the shield jams the communication channel intermittently; a trusted Programmer knows in advance the intervals in which the channel is clear, and can thus communicate with the IMD. With this solution, the patient has the option of keeping the shield active even during Programmer communication with IMDs.

These approaches have the advantage of being compatible with legacy IMD, so they can be applied seamlessly to the currently deployed devices. A drawback, however, is that jamming, when employed to counteract attacks as in [11, 30], can disrupt the communications of other RF devices and violate laws regarding radio interference.

A promising related approach, by Rassmussen et al. [32], uses ultrasound-based distance bounding to authenticate Programmer access to an IMD, achieving an access policy of proximity similar to those in Section 2 that use intra-body key transmission or key establishment using PVs. Their system requires RF shielding, however, amplifying the engineering complexity of an IMD. It also relies on RF communication. For some IMDs, e.g., brain implants, RF antennas are of prohibitive length, and alternatives, e.g., infrared, are preferred. Distance-bounding protocols' security models have also historically proven fragile (see, e.g., [33, 34]). Finally, this approach uses ultrasound transmission, which usually requires more power than RF transmission.

4. CASE STUDIES: SECURITY FLAWS

We now describe security weaknesses in two proposed protocols for authenticated key-agreement between devices in a body-area network. One is the setup protocol for the IMDGuard system [11], which pairs a protective device called a "Guardian" with an IMD. The other, OPFKA (Ordered-Physiological-Feature-based Key Agreement), is a generic body-area network pairing protocol [12]. Both IMDGuard and OPFKA rely on ECG measurements as a common source of entropy to establish shared secret keys. While terminology differs across papers, and some involve devices in BANs, we continue to refer to devices generically as the Programmer and IMD.

4.1 Attack on IMDGuard

We briefly describe the IMDGuard scheme for key agreement between a Programmer (in IMDGuard, the Guardian) and IMD. We then show a simple man-in-the-middle attack that reduces the effective key length from 129 bits to 86 bits.

IMDGuard key-agreement protocol The Programmer and IMD each measure ECG data in a succession of four-bit blocks. Let $\alpha = (\alpha_1, \alpha_2, \alpha_3, \alpha_4)$ and $\beta = (\beta_1, \beta_2, \beta_3, \beta_4)$ denote respective measurements of one such block by the IMD and Programmer. As these readings are noisy, IMDGuard includes the following noise-reducing "reconciliation" scheme for extraction of key material.

In Round 1, the Programmer and IMD exchange parity bits: The IMD sends $\alpha_1 \oplus \alpha_2 \oplus \alpha_3 \oplus \alpha_4$, while the Programmer sends $\beta_1 \oplus \beta_2 \oplus \beta_3 \oplus \beta_4$, where $\oplus$ denotes XOR. If the parity bits agree, the two devices accumulate the first three bits as key material. (They discard a bit to compensate for the one bit leaked by parity-symbol disclosure.) Once 43 blocks pass the parity check, the two sides each possess 129 bits of key material. They hash their respective key material to generate check values h_α and h_β, which they then exchange. The Programmer compares these check values and sends an `accept` message if $h_\alpha = h_\beta$, and a `reject` message otherwise.

Round 2 takes place if (and only if) the Programmer determines that $h_\alpha \neq h_\beta$ (`reject`). In Round 2, the IMD transmits $\alpha_3 \oplus \alpha_4$ and the Programmer, $\beta_3 \oplus \beta_4$. If these two bits agree, the IMD retains α_2 and α_3 as key material, while the Programmer retains β_2 and β_3. I.e., the first bit of each block is discarded to compensate for parity-bit leakage. Further blocks are read and reconciled as needed. When enough bits have accumulated, check values are again compared. (The authors assert a Round 3 is never required.)

Man-in-the middle attack A man-in-the-middle adversary Adv can do the following. Adv allows the IMD and Programmer to proceed normally with parity-bit exchange in Round 1. Suppose that $h_\alpha = h_\beta$ (as happens with high probability). Adv makes two message substitutions at the end of the round: (1) Adv substitutes a random value for the check value h_α transmitted by the IMD, causing the Programmer to send a Round-1 `reject` message and proceed to Round 2 and (2) Adv substitutes an `accept` message for the Programmer's `reject` message, causing the IMD to terminate the protocol with the key established in Round 1.

The Programmer thus proceeds with Round 2. It transmits a second parity bit ($\beta_3 \oplus \beta_4$) for each block from Round 1; at the same time, Adv simulates Round-2 parity-bit transmissions by the IMD.

Adv intercepts Programmer parity-bit transmissions to recover an additional bit of information for each block. (For a given block α, the IMD uses $(\alpha_1, \alpha_2, \alpha_3)$ as key bits. Adv learns $\alpha_1 \oplus \alpha_2$.)

While the resulting effective key length of 86 bits is an infeasible target for brute-force attack today, this attack demonstrates a serious design weakness in IMDGuard.

4.2 Attack on OPFKA

In OPFKA, the IMD and Programmer each perform local ECG readings on a human subject over the same interval of time. They translate these readings into a temporally ordered sequence of "features," short (e.g., 12-bit) values. The two devices exploit overlap in their respective feature sequences to construct a shared secret key κ, much as with IMDGuard.

OPFKA adopts a different approach than IMDGuard, however, to specify this overlap. In OPFKA, the Programmer transmits its features obscured with spurious *chaff* values to the IMD in what is called a *coffer*. The IMD indicates to the Programmer those feature values in the coffer that lie in its own feature sequence. Each device, then, can determine the intersection of their two respective feature sequences which is used to construct the shared key κ.

Here is a more detailed specification of the protocol. For simplicity, we assume 12-bit features, one option in OPFKA. We omit protocol parameters and messages not germane to our analysis. For clarity, we also change some of the original notation for OPFKA.

1. **Feature reading:** Each device reads a sequence of N features. (In OPFKA, $N = 30$.) Let $\tilde{F}^{imd} = \{\tilde{f}_0^{imd}, \tilde{f}_1^{imd}, \ldots \tilde{f}_{N-1}^{imd}\}$ be the IMD's features, in temporal order, and $\tilde{F}^{pro} = \{\tilde{f}_0^{pro}, \tilde{f}_1^{pro}, \ldots \tilde{f}_{N-1}^{pro}\}$, the Programmer's.

2. **"Hashing":** Feature values in $\tilde{F}^{imd}$ and $\tilde{F}^{pro}$ are mapped into 20-bit feature values via a "hash" function $H : \{0,1\}^{12} \to \{0,1\}^{20}$. Let $F^{imd} = \{f_0^{imd}, f_1^{imd}, \ldots f_{N-1}^{imd}\}$ be the resulting set of feature values for the IMD and F^{pro} similarly for the Programmer.

3. **Coffer transmission:** The Programmer randomly selects M chaff features $F' = \{f_0', f_1', \ldots, f_{M-1}'\}$, where $f_i' \in_R \{0,1\}^{20} - F^{pro}$. It randomly permutes elements in $C = F \bigcup F'$ and sends the resulting *coffer* C to the IMD.

4. **Coffer opening:** Starting with an empty set J, for each element $f_j^{imd} \in F^{imd}$, the IMD adds j to J if $f_j^{imd} \in C$. The result is an ordered set $J = \{j_0, \ldots, j_{n-1}\}$ of feature positions in F^{imd}. Opening is considered successful if $n \geq q$ for some predetermined parameter q.

5. **Key computation:** The IMD computes $\kappa = h(f_{i_0}^{imd} \| f_{i_1}^{imd}, \ldots, \| f_{i_n}^{imd})$ for a hash function h. The IMD sends (J, m, μ) to the Programmer, where $\mu = MAC_\kappa[m]$ for a message m (whose details are unimportant here).

Attack on small hash range. OPFKA has a security weakness resulting from the use of hashing in step 2. If the IMD selects in step 4. ("Coffer opening") a feature that is in $C \bigcup F^{imd}$, but not in F^{pro}, then the Programmer cannot then compute κ, and the protocol fails. To reduce the rate of such failures, the authors intend for step 2. ("Hashing") to expand the range of feature values in C.

But application of a "hash" function H *does not* expand the possible range of feature values *for a fixed domain D*. Let $D = \{0,1\}^{12}$ be the set of possible values for a 12-bit feature $\tilde{f}$. The hash of $\tilde{f}$ is computed as $H(s, \tilde{f})$, for pre-agreed salt s (a random nonce). Let $R = \{H(s, \tilde{f})\}_{\tilde{f} \in D}$ denote the range of $H(s, \cdot)$ over D. Then it is easy to see that $|R| \leq |D| = 2^{12}$, as $H(s, \cdot)$ is a deterministic function. (In fact, given the 12-bit domain and 20-bit range of H in OPFKA, with high probability over s, $|R| < |D|$.)

Thus the vast majority of chaff values in C will be invalid feature values lying outside R. Let $\hat{R} = C \cap R$ denote the set of values in the coffer that are valid feature values. (Note that $F_P \subseteq \hat{R}$.) The probability that a randomly selected chaff value $f' \in \{0,1\}^{20} - F^{pro}$ lies in $\hat{R}$ is bounded above by $|R|/(2^{20} - N) < 0.004$.

By excluding invalid chaff values (those not in $\hat{R}$), an adversary can greatly constrain its search space in a brute-force attack against the key κ, as shown in Algorithm 1. (Here, Π_n denotes the set of permutations over $\mathbb{Z}_n$ and $\pi \in \Pi_n$ is a permutation $\pi : \mathbb{Z}_n \leftrightarrow \mathbb{Z}_n$.)

Algorithm 1 Key search algorithm for OPFKA

```
Inputs: J, m, μ, C, R̂
Output: Key κ
```

for all $< f_0, f_1, \ldots, f_n > \in \hat{R}^n$ **do**
    for all $\pi \in \Pi_n$ **do**
        $\kappa' \leftarrow h(f_{\pi(0)} \parallel f_{\pi(1)} \cdots \parallel f_{\pi(n-1)})$;
        if $MAC_{\kappa'}[m] = \mu$ **then** output κ'; halt
        end if
    end for
end for

For example, for one proposed parameterization for OPFKA ($M = 1000$, $N = 30$, and $q = 12$) a key-strength equivalent of 120 bits is claimed in [12]—well beyond feasibility for a brute-force attack. With probability about 63%, though, there will be at most 4 valid chaff values in $\hat{R}$. In this case, assuming $n = q$, the maximum running time of Algorithm 1 will be $\binom{4+30}{12} \times 12! \approx 2^{58}$—equivalent to breaking a 58-bit key, and requiring vastly less effort than the claimed 120-bit strength of OPFKA. Cracking a 58-bit key is within the realm of feasibility, as shown by successful cracking of a 64-bit (RC5) key in 2002 [35].

Remark: Distinct salt values might be used in hashing for different positions. This might seem more secure, but isn't, as F^{pro} would no longer in general contain valid feature values for all positions.

Adaptive attack. For large M, such as the proposed parameter $M = 5000$, OPFKA is vulnerable also to an adaptive attack in which an adversary simulates a Programmer to extract the key κ from the IMD. Due to lack of space, our description is brief.

Adv constructs a coffer C as follows. R is partitioned into (arbitrary) equal sized (size 2^{11}) sets R_0 and R_1. A coffer C is constructed that includes R_0 and a subset $R_1' \subseteq R_1$ (to be specified); other features in C are selected to be invalid (drawn from $\{0,1\}^{20} - R$). With high probability, F^{imd} will include $n \geq q$ feature values in R_0. Therefore the IMD will respond to transmission of C with a set of indices J for any choice of R_1'.

Now, in an initial transmission, Adv sets $R_1' = R_1$. With high probability, F^{imd} will include at least one feature value in R_1' with index j. By recursing on halves of R_1' and observing whether $j \in J$ in the IMD's response, Adv can perform a binary search and learn f_j^{imd} with $log_2|D| = 12$ transmissions. By choosing different initial partitions (R_0, R_1), Adv can learn q feature values in F^{imd} and successfully impersonate a legitimate Programmer.

Variant attacks are possible with smaller M and with parallelization to search for multiple IMD feature values simultaneously.

The attack here assumes an ability to query the IMD fairly rapidly, and arises in part because OPFKA includes no throttling or back-off mechanism. A simple countermeasure is for the IMD, after a failed key-agreement with a Programmer, to refuse connections until it collects a fresh set of IPIs. Of course, this raises the risk of denial-of-service attacks and delays due to protocol failures.

5. TOWARD STRONGER AUTHENTICATION TECHNOLOGIES

Perhaps the most significant design constraint on pairing protocols for IMDs are computational power and bandwidth—and thus battery power and energy. With more resources would come a richer design space for signal processing algorithms and also for cryptographic protocols, including various secure two-party computation schemes capable of handling noisy key material, e.g., [36].

Given a specific primitive (e.g., AES) and IMD platform, implementation optimization can save only a fairly limited amount of energy. Thus, enhancing the security capabilities and/or limited battery lifetime of an IMD significantly would require a major change in underlying hardware or energy supply technology. This section discusses three active research areas that promise to extend IMD security capabilities and deployment lifetimes.

Memory technology IMDs typically contain multiple sensors on whose output they perform intensive data processing, storage, and retrieval. This trend, and the energy cost of existing memory devices, argue a need for improved storage technologies. For instance, novel nonvolatile memory technologies with zero leakage power and low read/write energy costs could drastically reduce an IMD's energy consumption; several ongoing research and development efforts, including those in Phase Change Memory (PCM) and Spin-Transfer Torque Random-Access Memory (STT RAM), aim to realize such efficient nonvolatile memory structures [37].

Energy storage technology Active IMDs often obtain power from conventional electro-chemical batteries. Such batteries suffer from slow charge cycles, limited lifetimes, limited power density, and slow rates of improvement in energy capacity. Newer energy-supply technologies, such as nano-scale super-capacitors and fuel cells, with improved cost, size, energy/power storage density, and recycling ability are becoming available [38]. Development, integration, and operation of such novel energy supply devices for IMDs pose an interesting and significant research challenge.

Energy scavenging technology Instead of using an attached energy source internal to the patient's body, IMDs can harvest energy from external sources such as the patient's physical movement, ambient heat, light, radio, or vibrations [39]. There are at least two sets of challenges in the use of energy-scavenging solutions for IMDs. First, the small energy output of typical energy scavenging devices is insufficient for most IMD applications. Second, the inherent uncertainty in harvesting energy from external sources conflicts with IMDs' strict safety and reliability requirements. Nevertheless, development of new energy harvesting, storage, and transfer methods is an active research area which could in time potentially bring significant improvement in energy availability for IMDs.

6. CONCLUSIONS

Presently available IMDs can be wirelessly accessed under loose security policies, allowing attackers to endanger the health and privacy of patients. This paper addressed the problem of authenticating IMDs to external Programmers. Particularly, we reviewed the problem of secure pairing and key distribution between an IMD and Programmers. Several of the currently available proposals to secure IMDs were analyzed. We presented attacks against two such protocols, namely IMDGuard [11] and OPFKA [12]. Securing IMDs remains to be a challenging open research problem which calls for the development of new and innovative solutions.

7. ACKNOWLEDGMENT

This research was supported in part by an Office of Naval Research grant to the ACES lab at Rice University (ONR R16480).

8. REFERENCES

[1] W. Burleson, S. S. Clark, B. Ransford, and K. Fu, "Design challenges for secure implantable medical devices," in *Proceedings of Design Automation Conference*, pp. 12–17, 2012.

[2] W. Maisel, "Safety issues involving medical devices," *Journal of the American Medical Association*, vol. 294, pp. 955–958, Aug. 2005.

[3] D. Halperin, T. Heydt-Benjamin, B. Ransford, S. Clark, B. Defend, W. Morgan, K. Fu, T. Kohno, and W. Maisel, "Pacemakers and implantable cardiac defibrillators: Software radio attacks and zero-power defenses," in *IEEE Symp. on Security and Privacy (S& P)*, pp. 129–142, 2008.

[4] C. Li, A. Raghunathan, and N. K. Jha, "Hijacking an insulin pump: Security attacks and defenses for a diabetes therapy system," in *IEEE Int. Conf. on e-Health Networking Applications and Services*, pp. 150–156, 2011.

[5] D. Halperin, T. Kohno, T. Heydt-Benjamin, K. Fu, and W. Maisel, "Security and privacy for implantable medical devices," *IEEE Pervasive Computing*, vol. 7, pp. 30–39, Jan.-Mar. 2008.

[6] F. Stajano and R. Anderson, "The resurrecting duckling: Security issues for ad-hoc wireless networks," in *Int. Workshop of Security Protocols*, pp. 172–194, 1999.

[7] S. Warren, J. Lebak, J. Yao, J. Creekmore, A. Milenkovic, and E. Jovanov, "Interoperability and security in wireless body area network infrastructures," in *IEEE Engineering in Medicine and Biology Society*, pp. 3837–3840, 2005.

[8] M. Meingast, T. Roosta, and S. Sastry, "Security and privacy issues with health care information technology," in *IEEE Engineering in Medicine and Biology Society*, pp. 5453–5458, 2006.

[9] F. Hu, Q. Hao, M. Lukowiak, Q. Sun, K. Wilhelm, S. Radziszowski, and Y. Wu, "Trustworthy data collection from implantable medical devices via high-speed security implementation based on IEEE 1363," *IEEE Trans. on Info. Tech. in Biomedicine*, vol. 14, pp. 1397–1404, Nov. 2010.

[10] K. Zetter, "DigiNotar files for bankruptcy in wake of devastating hack," *Wired*, 20 Sept. 2011.

[11] F. Xu, Z. Qin, C. Tan, B. Wang, and Q. Li, "IMDGuard: Securing implantable medical devices with the external wearable guardian," in *Proc. of IEEE INFOCOM*, pp. 1862–1870, 2011.

[12] C. Hu, X. Cheng, F. Zhangand, D. Wuand, X. Liao, and D. Chen, "OPFKA: Secure and efficient ordered-physiological-feature-based key agreement for wireless body area networks," in *Proc. of IEEE INFOCOM, To Appear*, 2013.

[13] T. G. Zimmerman, "Personal area networks: near-field intrabody communication," *IBM systems Journal*, vol. 35, pp. 609–617, 1996.

[14] T. G. Zimmerman, J. R. Smith, J. A. Paradiso, D. Allport, and N. Gershenfeld, "Applying electric field sensing to human-computer interfaces," in *Proceedings of the SIGCHI conference on Human factors in computing systems*, pp. 280–287, 1995.

[15] S.-Y. Chang, Y.-C. Hu, H. Anderson, T. Fu, and E. Y. Huang, "Body area network security: robust key establishment using human body channel," in *Proceedings of the USENIX conference on Health Security and Privacy*, pp. 5–5, 2012.

[16] S. Cherukuri, K. Venkatasubramanian, and S. Gupta, "Biosec: a biometric based approach for securing communication in wireless networks of biosensors implanted in the human body," in *Parallel Processing Workshop*, pp. 432–439, 2003.

[17] A. L. Goldberger, D. R. Rigney, and B. J. West, "Chaos and fractals in human physiology," *Scientific American*, vol. 262, pp. 42–49, 1990.

[18] M. Signorini, F. Marchetti, and S. Cerutti, "Applying nonlinear noise reduction in the analysis of heart rate variability," *Engineering in Medicine and Biology Magazine, IEEE*, vol. 20, no. 2, pp. 59–68, 2001.

[19] R. Yulmetyev, P. Hänggi, and F. Gafarov, "Quantification of heart rate variability by discrete nonstationary non-Markov stochastic processes," *Physical Review E*, vol. 65, no. 4, p. 046107, 2002.

[20] K. K. Venkatasubramanian, A. Banerjee, and S. K. S. Gupta, "PSKA: usable and secure key agreement scheme for body area networks," *IEEE Trans. on Information Technology in Biomedicine*, vol. 14, no. 1, pp. 60–68, 2010.

[21] K. K. Venkatasubramanian, A. Banerjee, and S. Gupta, "Plethysmogram-based secure inter-sensor communication in body area networks," in *IEEE Military Communications Conference*, pp. 1–7, 2008.

[22] C. Poon, Y. Zhang, and S. Bao, "A novel biometrics method to secure wireless body area sensor networks for telemedicine and m-health," *IEEE Communications Magazine*, vol. 44, no. 4, pp. 73–81, 2006.

[23] S. Bao, C. Poon, Y. Zhang, and L. Shen, "Using the timing information of heartbeats as an entity identifier to secure body sensor network," *IEEE Trans. on Info. Tech. in Biomedicine*, vol. 12, no. 6, pp. 772–779, 2008.

[24] K. Venkatasubramanian and S. Gupta, "Physiological value-based efficient usable security solutions for body sensor networks," *ACM Trans. Sensor Networks*, vol. 6, pp. 31:1–31:36, July 2010.

[25] K. Cho and D. Lee, "Biometric based secure communications without pre-deployed key for biosensor implanted in body sensor networks," in *Information Security Applications*, pp. 203–218, 2012.

[26] Y. Dodis, L. Reyzin, and A. Smith, "Fuzzy extractors: How to generate strong keys from biometrics and other noisy data," *SIAM Journal on Computing*, vol. 38, no. 1, pp. 97–139, 2008.

[27] A. Juels and M. Sudan, "A fuzzy vault scheme," *Designs, Codes and Cryptography*, vol. 38, no. 2, pp. 237–257, 2006.

[28] X. Liao, "Body area network security: A fuzzy attribute-based signcryption scheme," *IEEE Journal on Selected Areas in Communications, To Appear*, 2013.

[29] T. Denning, K. Fu, and T. Kohno, "Absence makes the heart grow fonder: New directions for implantable medical device security," in *USENIX HotSec*, 2008.

[30] S. Gollakota, H. Hassanieh, B. Ransford, D. Katabi, and K. Fu, "They can hear your heartbeats: non-invasive security for implantable medical devices," in *ACM SIGCOMM*, pp. 2–13, 2011.

[31] W. Shen, P. Ning, X. He, and H. Dai, "Ally friendly jamming: How to jam your enemy and maintain your own wireless connectivity at the same time," in *IEEE Symp. on Security and Privacy*, 2013.

[32] K. Rasmussen, C. Castelluccia, T. Heydt-Benjamin, and S. Capkun, "Proximity-based access control for implantable medical devices," in *Proc. of Computer and communications security*, pp. 410–419, 2009.

[33] C. Cremers, K. Rasmussen, B. Schmidt, and S. Capkun, "Distance hijacking attacks on distance bounding protocols," in *IEEE Symp. on Security and Privacy*, pp. 113–127, 2012.

[34] M. Poturalski, M. Flury, P. Papadimitratos, J.-P. Hubaux, and J.-Y. L. Boudec, "Distance bounding with ieee 802.15.4a: Attacks and countermeasures," *IEEE Trans. on Wireless Comms.*, vol. 10, no. 4, pp. 1334–1344, 2011.

[35] distributed.net, "Project RC5." http://www.distributed.net/RC5, Referenced 2012.

[36] B. Kanukurthi and L. Reyzin, "Key agreement from close secrets over unsecured channels," in *Eurocrypt*, pp. 206–223, 2009.

[37] G. Burr, B. Kurdi, J. Scott, C. Lam, K. Gopalakrishnan, and R. Shenoy, "Overview of candidate device technologies for storage-class memory," *IBM Journal of Research and Development*, vol. 52, no. 4.5, pp. 449–464, 2008.

[38] F. Koushanfar and A. Mirhoseini, "Hybrid heterogeneous energy supply networks," in *IEEE Int. Symp. on Circuits and Systems*, 2011.

[39] J. Paradiso and T. Starner, "Energy scavenging for mobile and wireless electronics," *IEEE Pervasive Computing*, vol. 4, no. 1, pp. 18–27, 2005.

Towards Trustworthy Medical Devices and Body Area Networks

Meng Zhang
Department of Electrical
Engineering
Princeton University
Princeton, NJ 08544
mengz@princeton.edu

Anand Raghunathan
School of Electrical and
Computer Engineering
Purdue University
West Lafayette, IN 47907
raghunathan@purdue.edu

Niraj K. Jha
Department of Electrical
Engineering
Princeton University
Princeton, NJ 08544
jha@princeton.edu

ABSTRACT

Implantable and wearable medical devices (IWMDs) are commonly used for diagnosing, monitoring, and treating various medical conditions. A general trend in IWMDs is towards increased functional complexity, software programmability, and connectivity to body area networks (BANs). However, as medical devices become more "intelligent," they also become less trustworthy – less reliable and more vulnerable to malicious attacks. Various shortcomings – hardware failures, software errors, wireless attacks, malware and software exploits, and side-channel attacks – could undermine the trustworthiness of IWMDs and BANs. The trustworthiness of IWMDs must be addressed aggressively and proactively due to the potential for catastrophic consequences. While some recent efforts address the defense of IWMDs against specific security attacks, a holistic strategy that considers all concerns and types of threats is required. This paper discusses trustworthiness concerns in IWMDs and BANs through a comprehensive identification and analysis of potential threats and, for each threat, provides a discussion of the merits and inadequacies of current solutions.

Categories and Subject Descriptors

J.3 [**Computer Applications**]: Life and Medical Sciences—*Medical information systems*; C.3 [**Computer Systems Organization**]: Special-Purpose and Application-Based Systems—*Real-time and embedded systems*

General Terms

Security, Reliability, Design

1. INTRODUCTION

Recent years have witnessed an explosion of activity in the development and use of implantable and wearable medical devices (IWMDs) for a variety of diagnostic, monitoring, and therapeutic applications. Advances in electronics promise to revolutionize the capabilities of IWMDs, leading to new generations of devices with increased functionality, programmability, and connectivity to body area networks (BANs). IWMDs are also increasingly being connected to personal computers and smartphones, and to web or cloud based medical data repositories, to provide patients with complete personal healthcare systems (PHSs). Figs. 1(a) and 1(b) present some examples of IWMDs. Fig. 1(c) presents a generic architecture for a PHS, which consists of four parts:

Permission to make digital or hard copies of all or part of this work for personal or classroom use is granted without fee provided that copies are not made or distributed for profit or commercial advantage and that copies bear this notice and the full citation on the first page. To copy otherwise, to republish, to post on servers or to redistribute to lists, requires prior specific permission and/or a fee. DAC '13, May 29 - June 07 2013, Austin, TX, USA.

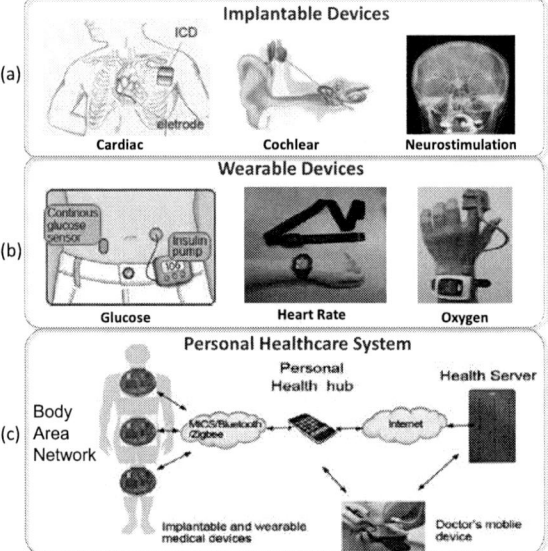

Figure 1: (a) Implantable and (b) wearable medical devices (adopted from http://www.wikipedia.org/), and (c) integration of IWMDs into a PHS.

medical sensors/actuators/controllers, a health hub such as the patient's smartphone or PC, a remote health server, and the doctor's smartphone or PC. Within the BAN, the IWMDs may communicate with each other or with the hub using short-distance communication technology, such as Bluetooth, ZigBee, and Medical Implant Communications Service [39].

The above advances in IWMDs, however, also open them up to the possibility of malicious attacks. Recent demonstrations of successful attacks on IWMDs, such as cardiac pacemakers [17] and insulin pumps [25, 35], have already placed them squarely in the focus of attackers. Any concerns regarding trustworthiness in IWMDs must be addressed aggressively and proactively due to the potential for catastrophic consequences. Unfortunately, IWMDs come with extreme size and power constraints, making it infeasible to simply borrow reliability or security solutions from the general-purpose computing arena. Therefore, this is an area that demands immediate attention of the information security and embedded systems research communities, medical device manufacturers, and regulatory bodies.

This paper aims to provide a comprehensive analysis and categorization of the threats posed to the trustworthiness of IWMDs and BANs, and promising approaches to address them. Table 1 summarizes major PHS trustworthiness requirements. Fig. 2 classifies the threats to the trustworthiness of PHSs, and various approaches to address them. The

Table 1: PHS Trustworthiness Requirements

Requirements	Description
Reliability	– IWMDs should function correctly even under extreme environmental conditions.
Confidentiality	– Information transmitted within the BAN should be secured from unauthorized access. – Patient data stored on IWMD, health hub, or remote server should be kept confidential.
Integrity	– Information transmitted within the BAN should be authentic and complete. – Patient data stored on IWMD, health hub, or remote server should not be altered.
Availability	– The BAN should be available even during jamming and denial-of-service attacks. – Patient data stored on IWMD, health hub, or remote server should be readily retrievable.
Privacy	– Using a PHS or carrying a device should not disclose patient condition.

table on the left lists IWMDs and other PHS components, and their relevant functional characteristics. The table in the middle lists the threats that lead to trustworthiness concerns. The table on the right includes possible countermeasures against the various threats. We discuss these challenges and prospective solutions in detail in the rest of the paper.

2. TRUST CHALLENGES/SOLUTIONS

Each of the following subsections analyzes one type of threat faced by a PHS. Suitable countermeasures as well as their merits and inadequacies are also discussed. Some of these countermeasures are widely used, or have even been included in standards, whereas others are based on recent research.

2.1 Hardware Failures/Software Errors

Hardware failures can be caused by undetected manufacturing defects, wear-and-tear faults due to electromigration, hot carrier injection, dielectric breakdown, *etc.*, as well as transient errors induced by a complex physical environment (*e.g.*, noise, power disturbance, extreme temperature, vibration, electromagnetic interference, *etc.*). Studies have shown that electromagnetic interference may cause temporary or permanent malfunction in pacemakers and implantable cardioverter defibrillators (ICDs) [22]. The critical nature of their functionality and the fact that they are in close contact with human organs leave little tolerance for hardware failure. A glitch in the operation of a cellphone may go unnoticed, whereas a glitch in a pacemaker can be life-threatening.

Software trustworthiness of IWMDs is just as important as hardware reliability. Many medical devices are essentially embedded computing systems with significant software complexity. Unfortunately, designing bug-free software is difficult, especially in complex devices that might be used in unanticipated contexts. More than a fourth of the recalls of defective medical devices during the first half of 2010 were likely caused by software defects [38]. Currently, there are no widely accepted techniques in use for the development or verification of software in medical devices [9]. Verification often consists of testing the device with specific test cases, with little assurance regarding completeness of the tests or the properties that they verify. It has been argued that using open-source software is more secure and reliable for medical applications, as it enables continuous and broad peer review that identifies and eliminates software errors [38]. However, understandably, medical device manufacturers are reluctant to adopt this approach.

Solutions. Next, we discuss two widely-used techniques for enhancing hardware/software reliability, which may also be applied to the components of a PHS.

2.1.1 Fault-tolerant Design

Although manufacturing-time test typically identifies a large number of hardware defects, attaining complete fault or defect coverage may not be feasible. By performing concurrent detection, diagnosis, and correction of fault effects, fault-tolerant designs enable a system to continue operating properly in the event of faults in its components [32]. Fault tolerance can also be extended to cope with software errors

caused by design inadequacies [36].

In general, fault tolerance requires some form of redundancy, either in time, hardware, or information. Hence, it incurs performance degradation and/or hardware overhead. For example, triple modular redundancy (TMR) [28], a well-known fault tolerance scheme that employs three copies of a module and uses a majority voter to determine the final output, has more than three times the cost of the original circuit. Unlike other safety-critical systems (*e.g.*, space electronics), IWMDs are subject to very tight resource constraints. Therefore, there is a great need for cost-effective fault-tolerant design techniques that can be applied to safety-critical IWMDs.

safety-critical IWMDs.

2.1.2 Formal Verification

Formal methods have been suggested as a means to design and develop reliable medical systems [9, 13, 21]. Formal methods are mathematical techniques for the specification, development, and verification of software and hardware systems. The specifications used in formal methods are well-formed statements in a mathematical logic, and formal verification consists of rigorous deductions in that logic. Therefore, formal methods provide a means to symbolically examine the entire state space of a system and establish that a correctness or safety property is true for all possible inputs.

Formal verification may be used to ensure that the software running on the devices is free of vulnerabilities, such as buffer overflows. However, this is far from sufficient to ensure that the medical device would operate in a trustworthy manner. Two key challenges must be addressed in order to truly leverage the power of formal verification in the context of medical devices [24]. First, current software verification tools target specifications written in high-level programming languages, and are not suitable for the highly platform-specific and low-level programs that are written for medical devices. These programs interact with hardware peripherals, such as medical sensors and actuators, in addition to timers, ADCs, UARTs, *etc.* In addition, they often adopt a highly interrupt-driven software architecture. It is necessary to verify the operation of these programs with sufficient semantics of the hardware platform that they execute on, while avoiding the state space explosion that results from excessive detail in modeling the hardware. Second, properties need to be verified at the interfaces of the IWMDs with the real world. In other words, rather than merely verifying that a program running on a pacemaker is free of buffer overflows, it is also important to verify whether any execution path in the pacemaker program leads to the cardiac pacing signal not being generated within the specified time window.

2.2 Radio Attacks

A common IWMD design pitfall is relying on proprietary protocols for secrecy [12]. Since cryptographic protection is typically not employed, the wireless channels between PHS components [*e.g.*, the link between medical devices in Fig. 1(c)], are highly prone to attacks.

A successful attack on an ICD [17] demonstrates how the ICD design, which involves wireless communication with an external programmer, can be exploited by an attacker. By deciphering and replaying packets, the attacker can launch radio attacks, with consequences ranging from disclosure of private data to alteration of device settings. In the worst case, the attacker can maliciously reconfigure the ICD to harm a patient by inaction (failure to deliver treatment when necessary) or by delivering an electrical stimulus when the heart is beating normally.

Using a similar approach, a successful attack is implemented on a glucose monitoring and insulin delivery system [25]. By mimicking the remote control, the attacker can configure the insulin pump to disable or change the intended therapy, stop the insulin injection, or inject a much higher dose than necessary. By mimicking the glucose meter, the attacker can send bogus data to the insulin pump, causing the patient to incorrectly adjust insulin delivery.

Another such man-in-the-middle attack is demonstrated on a Bluetooth-enabled pulse oximeter system in [34]. With

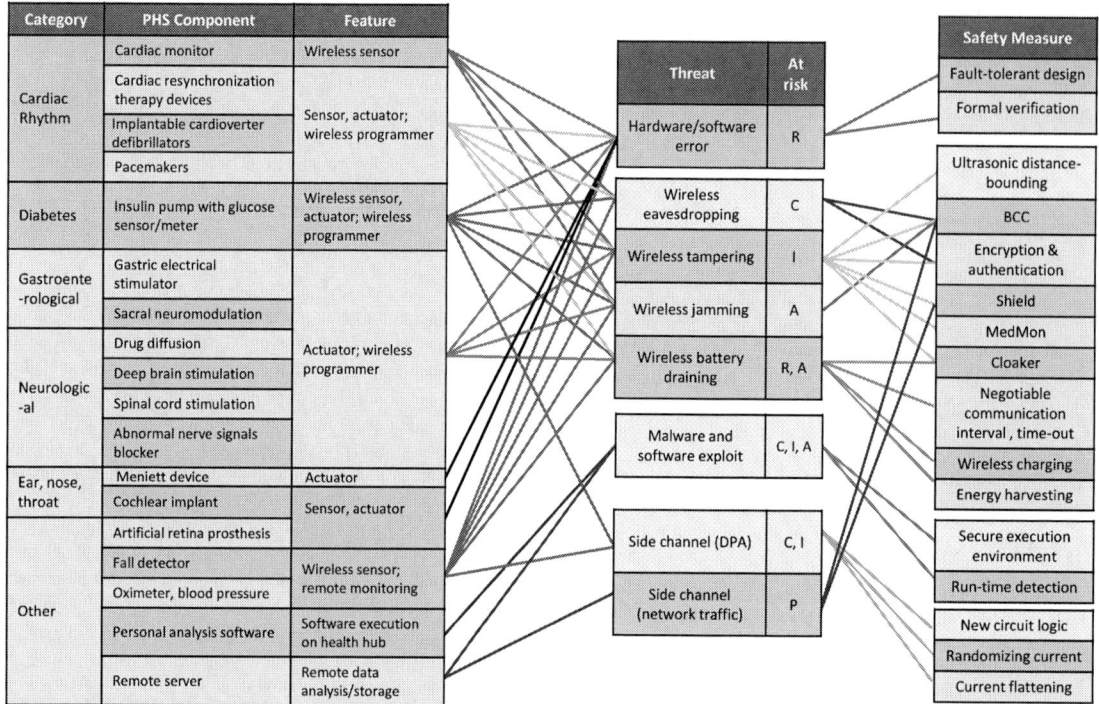

Figure 2: IWMDs, threats to trustworthiness, and countermeasures. Reliability, confidentiality, integrity, availability, and privacy are abbreviated to R, C, I, A, P, respectively.

the assumption that the PIN used in standard Bluetooth pairing is known, the attack shows that these wearable devices can be made to communicate with an unauthenticated intermediary equipped with a Bluetooth-enabled laptop.

Finally, with the knowledge of the communication protocol, denial-of-service attacks that aim to drain the battery of an implantable medical device (IMD) may be launched through the wireless channel. If the device responds to each incoming communication request from attackers, its battery may simply die and need to be surgically replaced. In addition, an attacker could also generate a large amount of noise to jam normal communication if he simply knows the approximate frequency of transmission.

Solutions. Next, we discuss several methods to detect, defend against, or mitigate radio attacks.

2.2.1 Close-range Communication

Limiting the communication range is a simple and intuitive way of limiting radio attacks. A radio frequency identification (RFID)-based channel between IWMDs and external controllers is often proposed in this context [20]. However, an attacker with a strong enough transmitter and a high-gain antenna can attack the wireless channel even if the channel is only for RFID-based communication. For an RFID channel, the attacker can access the IWMD from up to 10 m away [18]. A better alternative may be near-field communication (NFC), an extension of RFID, which is gaining increasing attention, especially due to its integration on mobile phones. The typical working distance for NFC is up to 20 cm. However, there is no guarantee that an attacker with a high-gain antenna cannot read the signal from outside the intended range, *e.g.*, from 1 m away [19].

Another technology that can help limit the communication range is body-coupled communication (BCC). In contrast to conventional wireless communication, BCC uses the human body as the transmission medium. The communication range is limited to the proximity of the human body [11]. Experimental results presented in [25] show a promising attenuation in signal strength at even small distances when comparing the BCC channel to a conventional

wireless channel. BCC works at low frequencies (ranging from 10 kHz to 10 MHz) and can only achieve very low data rates; however, this is not a problem for most medical devices.

In addition to physical communication layers that are designed to be inherently short-range, measures can be taken to enforce close-range communication. An access control scheme based on ultrasonic distance-bounding is introduced in [37]. In this scheme, an IWMD grants access to its resources to only those devices that are close enough. However, limiting the communication range is only effective when radio attacks are launched from beyond a certain distance. It is quite possible that an attacker can approach within a small distance of the patient and even make physical contact without raising suspicion (*e.g.*, in a crowded subway station).

2.2.2 Cryptography

Cryptography is a well-proven approach to secure the wireless communication channel and prevent unauthorized access. The high energy and implementation costs of *asymmetric* cryptography currently preclude its use in many IWMDs, making device authentication and shared key generation a challenge. The use of symmetric ciphers, while possible, may still significantly shorten device battery time. It is well known that compression techniques can be used before encryption to reduce encryption as well as transmission cost. In this context, compressive sensing [15] is particularly well suited since it can be realized with a very low computational and energy footprint that is compatible with the energy constraint of IWMDs. An evaluation of encompression (compressive sensing [15] + encryption + integrity checking) shows that an energy reduction of up to 78% can be achieved with a reasonable compression ratio of 6-10× [48].

Cryptographic protection is more effective when the secret keys shared by IWMDs and the hub can be renewed periodically, as in the 802.11 WiFi standards [31], since fixed pre-shared keys are prone to attacks. Furthermore, the secret keys should be updated automatically, with minimal user intervention. Works described in [8, 29] focus on extracting

94

shared secret keys from unsecured wireless channels by using physical channel characteristics and by exploiting the directional symmetry of wireless links. The most commonly used metric in these works is the received signal strength indicator (RSSI), a measure of signal power in logarithmic units. However, generation of randomness relies on relative movement between the communicating peers, or the environment being dynamic [8]. In the case where both IWMDs and the hub are mounted on the body at fixed positions, the measured RSSI traces may not provide sufficient randomness. In addition, rapid fluctuations in RSSI due to blockage by clothes or sudden movement may cause asymmetry in signals received at the two ends of the communication, resulting in failure to establish shared keys. IMDGuard [46] introduces an alternative cryptographic scheme for medical devices that utilizes the patient's physiological signals (such as the electrocardiogram) for key extraction.

Conventional cryptographic methods are also not directly applicable to IWMDs due to their unique usage models, which may require key distribution to legitimate parties outside the BAN. For example, encryption prevents medical professionals from accessing a patient's medical device in an emergency. As a possible solution, a master key (common to all instances of a specific model or manufacturer) may be preloaded in the device that the ambulance staff can request from the manufacturer or patient's doctor in emergencies. However, this scheme is not very secure as attackers can discover the secret key of a particular model through side-channel attacks (Section 2.4) or by hacking into the doctor's computer. Another straightforward solution is to ask patients to carry cards or bracelets imprinted with the secret keys of their devices. To prevent the imprints from being lost or damaged, the keys could be printed into the patient's skin using ultraviolet-ink micropigmentation – "tattoos" that only become visible under ultraviolet light [40]. In summary, with the widespread use of cryptography for information security, there is great interest in applying it to medical devices and BANs. However, challenges related to the energy cost of cryptography, shared key generation, and key distribution need to be addressed.

2.2.3 External Devices

To preserve IMD battery power, verification of incoming requests can be offloaded to a trusted external device, which can be easily recharged. A wearable device, called Communication Cloaker [14], is proposed to mediate communications between the IMD and pre-authorized parties and cause the IMD to ignore incoming communications from all unauthorized programmers. In emergency situations, the medical staff can remove the Cloaker in order to access the IMD. Since the burden of computation is offloaded to the external device, this approach can protect the IMD against battery-draining attacks.

"Shield" [16] is another proposal for an external device. It works as a relay between an IMD and an external programmer. It receives and jams messages from the IMD at the same time, so that others cannot receive them. It then uses an encrypted communication channel to relay the IMD's message to the programmer. The shield also protects the IMD from unauthorized incoming commands by jamming all messages sent directly to the IMD. All commands must be encrypted and sent to the shield first, which then relays legitimate commands to the IMD. Therefore, the shield does not require any change in commercial IMDs, but requires changes in all programmers. Since the messages from the IMD are jammed and the communication between the programmer and the shield is encrypted, the confidentiality of IMD messages is protected. However, when the shield sends the programmer's commands to the IMD, confidentiality may be compromised. The shield is specifically designed to protect IMD-to-programmer communications, but is not applicable when two medical devices (e.g., glucose meter and insulin pump) need to communicate, neither of which can be modified to support encryption.

MedMon [49] is an external monitor that snoops on all the radio frequency (RF) wireless communications to/from IWMDs and uses anomaly detection to identify potentially

malicious transactions. Anomalies are detected by analyzing the physical characteristics of the transmitted signal (e.g., RSSI), the time of arrival (TOA), the differential time of arrival (DTOA), and the angle of arrival (AOA), or application layer information, i.e., the semantics of data contained in the packets. Upon detection of a potentially malicious communication, MedMon takes appropriate response actions, which could range from passive (notifying the user) to active (jamming the packets). By acting like a firewall, the monitor protects the BAN against integrity attacks, which are arguably the most dangerous type of attacks. It does not protect confidentiality and privacy, nor can it protect the BAN against jamming. However, it provides protection against battery-draining attacks, as frequent transmissions may be classified as abnormal and jammed by MedMon and never reach the IWMD. A key benefit of MedMon is that it is applicable to existing IWMDs and programmers without the need for any hardware or software modifications.

2.2.4 Alleviating the Battery Constraint

Compared to wearable medical devices whose battery can be readily recharged or replaced, battery-draining attacks pose much greater threat to IMDs, such as electroencephalogram implants and pacemakers, since replacing the battery usually requires surgery. Zero-power defenses (security at no cost to the battery) have been proposed for such devices, in which the induced RF energy is harvested for notification, authentication, and key exchange [17]. In addition, efforts are being undertaken to design BAN protocols to mitigate this problem. For example, the IEEE 802.15.6 BAN standard allows a node and a hub to negotiate their communication intervals by encoding them in authenticated messages. The node thus will not wake up to receive any messages outside the negotiated time intervals [4].

An ideal defense against battery-draining attacks on IMDs is to relax battery constraints. One solution is to make the implant wirelessly rechargeable [26]. Another is to harness kinetic energy from the human body [30]. These technologies are still in the research phase and must go through rigorous testing and evaluation before commercial use.

2.3 Malware and Vulnerability Exploits

Various forms of malware, including viruses, worms, Trojans, keyloggers, and rootkits, have emerged and attackers keep evolving and adapting them to new classes of computing platforms. With the increasing programmability and connectivity of PHS components, it is just a matter of time before the first appearance of malware that specifically targets PHSs.

Furthermore, since software is inherently complex, abstract, and intangible, software vulnerabilities are inevitable and difficult to detect. With some knowledge of system software, attackers can exploit the software vulnerabilities to steal private information, tamper with medical data, and even change device settings.

While BANs are subject to unique threat models and attacks, as described in previous sections, software attacks will continue to be a commonly utilized approach for compromising their security, due to the relative ease and low cost of launching such attacks. In this context, the "weakest link" of a BAN, i.e., the component that exposes the largest attack surface and is the most accessible to software attacks, is the health hub, which is commonly a smartphone that executes the medical applications (for logging of health data, display of data to the user, and communication with remote medical professionals and health information services). As reflected by the rapid proliferation of "application" marketplaces for mobile devices, users are likely to use their smartphones to execute untrusted and potentially vulnerable applications as well. Smartphone platforms, such as Android and iOS, have been breached by mobile malware [2,5]. In the extreme case, the operating system (OS) on the hub may itself be compromised, making it trivial to subvert the medical applications that execute under its full control. Thus, it becomes essential to provide a secure execution environment for the medical applications in the face of other untrusted applications and also an untrusted OS.

Solutions. The following techniques can be used to defend medical software/data against malware, vulnerability attacks, and malicious OS.

2.3.1 Secure Execution Environment

While it may not be feasible to secure all applications from a compromised OS, it is possible to achieve a secure execution environment that provides isolation for selected, security-critical applications. The isolation may be based on physical separation (*e.g.*, IBM's secure co-processor [41]) or logical separation, in which both the sensitive and untrusted code are run on the same processor, but are isolated either using an additional layer of software or through additional hardware support, such as ARM TrustZone [1].

A secure execution environment based on logical separation for medical applications is illustrated in Fig. 3. It is based on two key technologies: secure virtualization and trusted computing. Security-critical medical applications can be protected in a separate virtual machine (VM), which we refer to as the medical VM. The medical VM is a restricted environment in which only medical applications and the supporting software libraries are executed, isolated from the other applications running on the system. Trusted computing [45] is a set of standards that is widely gaining popularity in general-purpose computing systems. Trusted computing requires a "root of trust" in the system for tamper-proof storage and attestation, which is typically realized by adding a separate tamper-proof hardware component called the Trusted Platform Module (TPM) to the system. In size-constrained and resource-constrained platforms, such as smartphones, it is currently not common to see hardware TPMs. In such cases, the use of a software TPM based on software emulation of TPM functions within an isolated execution environment has been demonstrated [7].

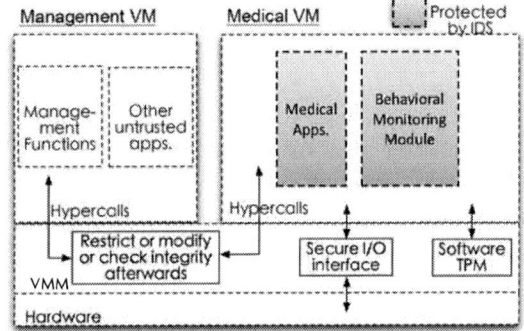

Figure 3: Secure execution environment.

In addition to logical separation, data confidentiality and integrity can also be achieved by physically separated and secured data storage on the health hub. For example, Plug-n-Trust [43] is a plug-in smart card that provides a trusted computing environment and keeps medical data safe. Assuming the data sent by the medical sensors are encrypted, they remain encrypted while stored on the hub, and are only decrypted within the smart card. Application programming interfaces are provided by the card to allow data modification by medical applications.

A more aggressive approach would be to completely separate health-related applications from untrusted applications and OS by making the hub an independent device. One such wrist-worn device, called Amulet [42], is dedicated to communications with IWMDs. It occasionally communicates with the smartphone in order to connect with health servers. In addition to physical separation from potential software attacks, another strong argument in favor of a dedicated hub is that IWMDs must be able to operate continuously and securely without relying on smartphones or other non-wearable personal computing devices, which can easily be lost, stolen, or run out of power.

2.3.2 Run-time Monitoring

Isolating medical applications from other software does not protect against vulnerabilities in the medical applica-

tions themselves. Vulnerabilities are commonly introduced into software as artifacts of the software development process. Intrusion detection techniques based on dynamic binary instrumentation (DBI) have been extensively investigated [6]. An application is first tested by running it against a large input set in a virtualized test environment (*e.g.*, on a server in the cloud), wherein DBI is used to apply extensive security policies. If it passes the evaluation in the test environment, its execution is characterized by generating behavioral models, which can be seen as a database of safe behaviors. On the target platform (*e.g.*, smartphone), lightweight run-time monitoring is used to restrict the application's behavior to the pre-generated behavioral model. As much of the workload is shifted from the target platform to the test environment, the performance penalty can be greatly reduced compared to traditional monitored execution and intrusion detection techniques.

Even though the monitoring work is minimized, the delay overhead for fine-grained monitoring can still be significant. To overcome this problem, the embedded processor can be augmented with a hardware monitoring engine that observes the processor's dynamic execution trace, checks whether the execution trace falls within the allowed program behavior, and flags any deviations from the expected behavior to trigger appropriate responses [10].

2.4 Side-channel Attacks

Side-channel attacks exploit information leaked through physical channels, such as power consumption, execution time, electromagnetic emission, *etc.* For example, the Intel Health Guide system [3] is equipped with integrated cameras, allowing online health sessions and video consultations through the Internet. Even if the communications are encrypted, the network traffic flow may leak some private information. The schedule of health sessions and video calls, for example, could be deduced from monitoring the network traffic flow.

Various forms of side-channel attacks have been extensively applied to break cryptographic systems, including timing analysis, fault analysis, and differential power analysis (DPA). DPA employs statistical analysis of measured power consumption traces, which are correlated with the data processed by the device [23]. DPA attacks can extract secret keys from extremely noisy signals and are hence very difficult to guard against. Although no DPA attacks on medical devices have been reported, it is not hard to envision such a scenario. Suppose a heart-rate monitor uses a symmetric cipher (such as AES) with a built-in secret key to encrypt the measured heart rates before sending them to the hub. If an attacker gains access to the heart-rate monitor, the secret key can be recovered through DPA by feeding the device with known data, measuring the corresponding current consumption, and analyzing the measured current traces. If a common key is used for many or all units of the same model, the attacker could publicize the revealed secret key and thus make the cryptographic protection ineffective for a large number of devices.

Solutions. Next, we introduce some proposed countermeasures against DPA attacks.

As the reason for the vulnerability of classical CMOS logic circuits to DPA attacks lies in the imbalance of charging and discharging behavior between 0-to-1 and 1-to-0 transitions, novel logic styles with data-independent power consumption have been proposed as circuit-level solutions to reduce the dependence of power dissipation on input patterns [44]. Other system-level countermeasures either try to suppress the differential signal used in the DPA attacks or randomize the power profile [27,33,47]. Unfortunately, in most of these methods, DPA resistance still comes at the expense of large area and power overheads, which are not compatible with resource-constrained IWMDs.

3. CONCLUSION

A general trend in IWMDs is towards increased functional complexity, software programmability, and wireless network connectivity. An undesirable, yet inevitable, side-effect of these trends is that IWMDs and BANs are increasingly vulnerable to security attacks. Trustworthiness concerns are likely to become a hindrance to the ubiquitous deployment of IWMDs and BANs. We analyzed various aspects of threats faced by them and discussed suitable solutions for each threat. Given the critical functions IWMDs perform, these issues should be addressed aggressively and proactively by medical electronics manufacturers together with the information security and embedded systems research communities.

4. ACKNOWLEDGMENTS

This work was supported by NSF under Grant no. CNS-0914787 and CNS-1219570. The authors would like to thank Prof. Susmita Sur-Kolay for her valuable feedback.

5. REFERENCES

[1] ARM trustzone technology overview. http://www.arm.com/products/CPUs/arch-trustzone.htm.
[2] Infected apps found in Google's Android market, 2011. http://www.securitynewsdaily.com.
[3] Intel Health Guide, 2011. http://www.careinnovations.com/products/ healthguide.
[4] An overview of IEEE 802.15.6 standard, 2011. arxiv.org/pdf/1102.4106.
[5] iPhone malware paradigm, 2012. http://www.crosstalkonline.org.
[6] N. Aaraj, A. Raghunathan, and N. K. Jha. Virtualization-based framework for malware defense. In *Proc. Conf. Detection of Intrusions and Malware & Vulnerability Assessment*, pages 64–87, Jul. 2008.
[7] N. Aaraj, A. Raghunathan, and N. K. Jha. Analysis and design of a hardware/software trusted platform module for embedded systems. *ACM Trans. Embed. Comput. Syst.*, 8:8:1–8:31, Jan. 2009.
[8] S. T. Ali, V. Sivaraman, and D. Ostry. Zero reconciliation secret key generation for body-worn health monitoring devices. In *Proc. ACM Conf. Security and Privacy in Wireless and Mobile Networks*, pages 39–50, 2012.
[9] D. Arney, R. Jetley, P. Jones, I. Lee, and O. Sokolsky. Formal methods based development of a PCA infusion pump reference model: Generic infusion pump (GIP) project. In *Proc. Joint Workshop High Confidence Medical Devices, Software, and Systems and Medical Device Plug-and-Play Interoperability*, pages 23–33, Jun. 2007.
[10] D. Arora, S. Ravi, A. Raghunathan, and N. K. Jha. Secure embedded processing through hardware-assisted run-time monitoring. In *Proc. Design, Automation and Test in Europe Conf.*, volume 1, pages 178–183, Mar. 2005.
[11] H. Baldus, S. Corroy, A. Fazzi, K. Klabunde, and T. Schenk. Human-centric connectivity enabled by body-coupled communications. *IEEE Comm. Mag.*, 47:172–178, Jun. 2009.
[12] W. Burleson, S. S. Clark, B. Ransford, and K. Fu. Design challenges for secure implantable medical devices. In *Proc. Design Automation Conf.*, pages 12–17, 2012.
[13] L. Cordeiro, B. Fischer, H. Chen, and J. Marques-Silva. Semiformal verification of embedded software in medical devices considering stringent hardware constraints. In *Proc. IEEE Int. Conf. Embedded Software and Systems*, pages 396–403, May 2009.
[14] T. Denning, K. Fu, and T. Kohno. Absence makes the heart grow fonder: New directions for implantable medical device security. In *Proc. Conf. Hot Topics in Security*, pages 1–7, Jul. 2008.
[15] D. Donoho. Compressed sensing. *IEEE Trans. Information Theory*, 52(4):1289–1306, Apr. 2006.
[16] S. Gollakota, H. Hassanieh, B. Ransford, D. Katabi, and K. Fu. They can hear your heartbeats: Non-invasive security for implantable medical devices. In *Proc. ACM Conf. Special Interest Group on Data Communication*, Aug. 2011.
[17] D. Halperin, T. S. Heydt-Benjamin, B. Ransford, S. S. Clark, B. Defend, W. Morgan, K. Fu, T. Kohno, and W. H. Maisel. Pacemakers and implantable cardiac defibrillators: Software radio attacks and zero-power defenses. In *Proc. IEEE Symp. Security and Privacy*, pages 129–142, May 2008.
[18] G. P. Hancke and S. C. Centre. Eavesdropping attacks on high-frequency RFID tokens. In *Proc. Workshop RFID Security*, pages 100–113, Jul. 2008.
[19] E. Haselsteiner and K. Breitfuss. Security in near field communication. In *Proc. Workshop RFID Security*, pages 3–13, Jul. 2006.
[20] C. Israel and S. Barold. Pacemaker systems as implantable cardiac rhythm monitors. *American J. Cardiology*, 88(4):442–445, Aug. 2001.
[21] R. Jetley, S. P. Iyer, P. L. Jones, and W. Spees. A formal approach to pre-market review for medical device software. In *Proc. Int. Conf. Computer Software and Applications*, pages 169–177, Sep. 2006.
[22] C. Jilek, S. Tzeis, T. Reents, H.-L. Estner, S. Fichtner, S. Ammar, J. Wu, G. Hessling, I. Deisenhofer, and C. Kolb.

Safety of implantable pacemakers and cardioverter defibrillators in the magnetic field of a novel remote magnetic navigation system. *J. Cardiovascular Electrophysiology*, 21(10):1136–1141, Oct. 2010.
[23] P. Kocher, J. Jaffe, and B. Jun. Differential power analysis. In *Proc. Int. Cryptology Conf.*, pages 388–397, Aug. 1999.
[24] C. Li, A. Raghunathan, and N. K. Jha. Improving the trustworthiness of medical device software with formal verification methods. under review.
[25] C. Li, A. Raghunathan, and N. K. Jha. Hijacking an insulin pump: Security attacks and defenses for a diabetes therapy system. In *Proc. IEEE Int. Conf. e-Health Networking, Applications and Services*, Jun. 2011.
[26] P. Li and R. Bashirullah. A wireless power interface for rechargeable battery operated medical implants. *IEEE Trans. Circuits and Systems*, 54(10):912–916, Oct. 2007.
[27] P.-C. Liu, H.-C. Chang, and C.-Y. Lee. A low overhead DPA countermeasure circuit based on ring oscillators. *IEEE Trans. Circuits and Systems*, 57:546–550, Jul. 2010.
[28] R. E. Lyons and W. Vanderkulk. The use of triple-modular redundancy to improve computer reliability. *IBM J. Res. Dev.*, 6:200–209, Apr. 1962.
[29] S. Mathur, W. Trappe, N. Mandayam, C. Ye, and A. Reznik. Radio-telepathy: Extracting a secret key from an unauthenticated wireless channel. In *Proc. ACM Int. Conf. Mobile Computing and Networking*, pages 128–139, 2008.
[30] P. D. Mitcheson, E. M. Yeatman, G. K. Rao, A. S. Holmes, and T. C. Green. Energy harvesting from human and machine motion for wireless electronic devices. *Proc. IEEE*, 96(9):1457–1486, Sep. 2008.
[31] T. Moore. IEEE 802.11-01/610r02: 802.1.x and 802.11 key interactions. Technical report, Microsoft Research, 2001.
[32] W. R. Moore. A review of fault-tolerant techniques for the enhancement of integrated circuit yield. *Proc. IEEE*, 74(5):684–698, May 1986.
[33] R. Muresan and S. Gregori. Protection circuit against differential power analysis attacks for smart cards. *IEEE Trans. Comput.*, 57:1540–1549, Nov. 2008.
[34] V. Pournaghshband, M. Sarrafzadeh, and P. Reiher. Securing legacy mobile medical devices. In *Proc. Int. Conf. Wireless Mobile Communication and Healthcare*, 2012.
[35] J. Radcliffe. Hacking medical devices for fun and insulin: Breaking the human SCADA system. In *Proc. Black Hat Technical Security Conf.*, July-Aug. 2011.
[36] B. Randell. System structure for software fault tolerance. *SIGPLAN Not.*, 10:437–449, Apr. 1975.
[37] K. B. Rasmussen, C. Castelluccia, T. S. Heydt-Benjamin, and S. Capkun. Proximity-based access control for implantable medical devices. In *Proc. ACM Conf. Computer and Communications Security*, pages 410–419, Nov. 2009.
[38] K. Sandler, L. Ohrstrom, L. Moy, and R. McVay. Killed by code: Software transparency in implantable medical devices, 2010. http://www.softwarefreedom.org.
[39] H. S. Savci, A. Sula, Z. Wang, N. S. Dogan, and E. Arvas. MICS transceivers: Regulatory standards and applications. *Proc. IEEE*, pages 179–182, Apr. 2005.
[40] S. Schechter. Security that is meant to be skin deep: Using ultraviolet micropigmentation to store emergency-access keys for implantable medical devices. Technical Report MSR-TR-2010-33, Microsoft Research, Apr. 2010.
[41] S. W. Smith and S. Weingart. Building a high-performance, programmable secure coprocessor. *Comput. Netw.*, 31:831–860, Apr. 1999.
[42] J. Sorber, M. Shin, R. Peterson, C. Cornelius, S. Mare, A. Prasad, Z. Marois, E. Smithayer, and D. Kotz. An amulet for trustworthy wearable mHealth. In *Proc. Workshop Mobile Computing Systems Applications*, pages 7:1–7:6, 2012.
[43] J. M. Sorber, M. Shin, R. Peterson, and D. Kotz. Plug-n-Trust: Practical trusted sensing for mHealth. In *Proc. Int. Conf. Mobile Systems, Applications, and Services*, pages 309–322, 2012.
[44] K. Tiri and I. Verbauwhede. Charge recycling sense amplifier based logic: Securing low power security ICs against DPA. In *Proc. European Solid-State Circuits Conf.*, pages 179–182, 2004.
[45] "Trusted Computing Group". https://www.trustedcomputinggroup.org.
[46] F. Xu, Z. Qin, C. C. Tan, B. Wang, and Q. Li. IMDGuard: Securing implantable medical devices with the external wearable guardian. In *Proc. IEEE Int. Conf. Computer Communications*, pages 1862–1870, Apr. 2011.
[47] M. Zhang and N. K. Jha. FinFET-based power management for improved DPA resistance with low overhead. *ACM J. Emerg. Technol. Comput. Syst.*, 7(3):10:1–10:16, Aug. 2011.
[48] M. Zhang, M. M. Kermani, A. Raghunathan, and N. K. Jha. Energy-efficient and secure sensor data transmission using encompression. In *Proc. Int. Conf. VLSI Design*, Jan. 2013.
[49] M. Zhang, A. Raghunathan, and N. K. Jha. MedMon: Securing medical devices through wireless monitoring and anomaly detection. *accepted for publication in IEEE Trans. Biomedical Circuits and Systems*.

Low-Energy Encryption for Medical Devices: Security Adds an Extra Design Dimension

Junfeng Fan, Oscar Reparaz, Vladimir Rožić, Ingrid Verbauwhede
KU Leuven - COSIC and IMINDS
Kasteelpark Arenberg 10
Leuven, Belgium
{junfeng.fan,oscar.reparaz,vladimir.rozic,ingrid.verbauwhede}@esat.kuleuven.be

ABSTRACT

Smart medical devices will only be smart if they also include technology to provide security and privacy. In practice this means the inclusion of cryptographic algorithms of sufficient cryptographic strength. For battery operated devices or for passively powered devices, these cryptographic algorithms need highly efficient, low power, low energy realizations. Moreover, unique to cryptographic implementations is that they also need protection against physical tampering either active or passive. This means that countermeasures need to be included during the design process.

Similar to design for low energy, design for physical protection needs to be addressed at *all design abstraction levels*. Differently, while skipping one optimization step in a design for low energy or low power, merely reduces the battery life time, skipping a countermeasure, means opening the door for a possible attack. Designing for security requires a thorough threat analysis and a balanced selection of countermeasures.

This paper will discuss the different abstraction layers and design methods applied to obtain low power/low energy and at the same time side-channel and fault attack resistant cryptographic implementations. To provide a variety of security features, including location privacy, it is clear that medical devices need public key cryptography (PKC). It will be illustrated with the design of a low energy elliptic curve based public key programmable co-processor. It only needs $5.1\mu J$ of energy in a 0.13 μm CMOS technology for one point multiplication and includes a selected set of countermeasures against physical attacks.

1. INTRODUCTION

It is widely known that medical data needs the highest protection against disclosure and against tampering [2]. Indeed medical devices and medical data have a long lifespan. For instance, the battery of a pacemaker will last for 5 to 15 years before it is replaced. In addition, pacemakers and other medical (implantable) devices, wireless sensors, RFID

tags, and others have become more sophisticated over the years. Instead of only issuing a fixed electrical pulse, they are now tuned to the patient. Similar arguments can be made for body sensors based on BAN, WAN or RFID technology: they pick up vital signs and transmit them to a wearable collector of data, e.g. the patient's cellular phone. The longevity of medical data explains the need for security levels that last for many years , since the attackers only get stronger over time due to Mooer's law. Unfortunately, longer key length translates in a larger computational load.

On top, the devices itself are not protected inside computer rooms or behind walls. Therefore, physical attacks, active or passive, are possible. For instance, pacemakers can be remotely updated or tuned. This wireless link can be eavesdropped, or it can be used to interfere with the readings or settings of the pacemaker. Wireless tags which are used to monitor the health status, give a patient much more freedom of movement and allow medical staff to monitor a patient without being bedridden. However, this can also be used to track patients and therefore location privacy is an important concern.

The goal of this paper is to provide insight on how to combine efficiency (in terms of energy or power consumption) with high security levels. We claim that this can only be reached by considering all design abstract levels. In this sense, design for security is similar to design for low power or low energy. It is also different, as a designer has to decide which is the right abstraction level to address particular attacks.

To illustrate this, the paper is organized as follows. We first describe some typical scenario's and the associated security analysis in section 2. A security analysis is used to select the type of cryptographic algorithms and protocols required for the application. In section 3, we discuss the design abstraction levels, which we call the security pyramid. Algorithms and protocols to address the security requirements are addressed in section 4. The architecture level is discussed in section 5 and the circuit level in section 6. Finally, the security evaluation is discussed in section 7.

2. SECURITY ANALYSIS

In a typical scenario, we assume that a patient wears several medical devices and sensors, some of them are worn on the body such as a hearing aid, or an insulin pump, others are implanted, such as a pacemaker or a brain monitoring or stimulation device [2].

In this typical scenario, these sensors and actuators communicate over a wireless channel: this could be a BAN

Permission to make digital or hard copies of all or part of this work for personal or classroom use is granted without fee provided that copies are not made or distributed for profit or commercial advantage and that copies bear this notice and the full citation on the first page. To copy otherwise, to republish, to post on servers or to redistribute to lists, requires prior specific permission and/or a fee.
DAC '13, May 29 - June 07 2013, Austin, TX, USA.

(body-area-network), RFID, Zigbee or similar. One example is the Human++ project of IMEC [15]. In this scenario a device, a cellular phone or similar, will serve as the local mini server, meaning that it collects the data and controls the network. It is assumed that this mini server is energy rich compared to the sensors and actuators. In a typical use case, the sensors will transmit patient data, e.g. his or her vital signs to the mini-server. Confidentiality, source authentication as well as integrity of this data are important. This is needed in both directions, i.e. mutual authentication is needed.

Recently, privacy received a lot of attention, more specifically location privacy or resistance to tracking. While early protocols did not make a distinction in types of privacy [9], more recent papers aim at providing strong privacy levels [14].

A careful choice of the protocols is only half of the security analysis. It is a necessary but not sufficient requirement to attain the desired security goals such as confidentiality, authentication and integrity.

For the system to be secure, we also require the implementation of the protocols be secure against an adversary that may have physical or short-distance access to the medical device. Techniques belonging to the broad class of side-channel attacks, such as Power Analysis [7, 8] have been used to extract keys from embedded devices by only monitoring the execution time of a cryptographic computation or the power consumption of the device (even in a remote contact-less fashion with specialized antennas that pick up the electromagnetic emanations of the chip).

In general, side-channel attacks exploit additional information that is available during the cryptographic computation aside from the input and output values, in contrast to classical cryptanalytical attacks. It is clear that should an attacker extract cryptographic keys from an embedded medical device, the security of the protocol is compromised.

Hence, the implementation of a medical embedded device should provide some degree of protection against physical attacks, such as tampering or side-channel attacks. In the next sections, we address the design procedure, best practices and decisions that are made in different abstraction levels when designing an exemplary crypto-processor for efficiency, low power, low energy and security.

3. DESIGN METHODS FOR LOW POWER AND SECURITY

Over the years, many design methods for low power and/or low energy have been developed. These design methods are situated at different abstraction levels. It is generally accepted and shown in practice that optimizations at higher abstraction levels have a bigger impact than those at lower abstraction levels.

An early paper showing the importance of transformations for low power is the paper by Chandrakasan et al. [3]. It shows that techniques like pipelining and parallelism can be used to reduce the power consumption. Others showed the importance of transformations or techniques at system level, e.g. the memory transformations introduced in [13]. At circuit level, also many techniques have been introduced: well known examples are gated clock strategies, reduced swing strategies or the introduction of power domains.

We distinguish the following abstraction levels for our pur-

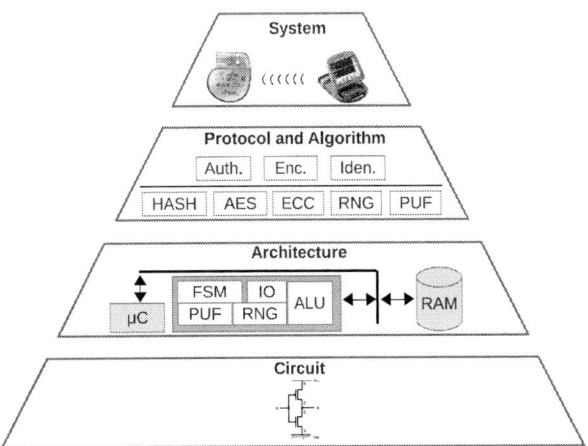

Figure 1: Security pyramid.

pose [21], as shown in Figure 3. The top level is the application or 'system' level. The selection of the protocol and the associated cryptographic algorithms has a huge influence on the final power or energy consumption of the embedded medical device. Next, is the algorithm level. For secret key and certainly for public-key algorithms there is a wide selection of algorithms and implementation strategies. The next level is the architecture level, where a digital platform for the implementation of the protocol and algorithm is selected. In most embedded applications that require extra physical protection, a HW/SW co-design platform is chosen. Typically there is an embedded micro-controller with programmable co-processors that support cryptographic algorithms. This is typically the platform of choice for applications like RFID tags, smart cards, portable devices and also medical devices. The design is eventually mapped to circuit-level implementation. In order to resist side-channel analysis, circuit-level countermeasures are crucial.

4. PROTOCOL AND ALGORITHM

Protocols are designed based on a variety of cryptographic primitives, such as hash functions, symmetric key ciphers, public key ciphers. It can also include non-algorithmic primitives, such as Random Number Generators (RNG), secure storage, or Physically Unclonable Functions (PUFs). Traditionally, protocol designers consider the security of a protocol the first design priority, with minimizing the computational complexity the second. Since implanted medical devices has strict power and energy budget, protocol designers need to consider many factors besides security. We identify a few of them below.

- *Security properties.* The security properties of a protocol should be clearly specified. In case of a pacemaker, mutual authentication is required to prevent impersonation. In order to protect the privacy of the patient, sensitive data should also be encrypted. Note that a modification on the ciphertext may also lead to a corrupted therapy that endanger the patient's life. Therefore, data authentication is also required. As such, the communication protocol between the pacemaker and the server should at least include mutual authentication, data authentication and encryption.

- *Location privacy* is a separate concern. It protects the user against tracking. Location based services are offered as part of many phone applications. In this case, accepting or denying it, should be the users choice. In medical applications, e.g. tracking as a means of protecting older people, should be strictly limited to medical personal. Location privacy heavily relies on public key based protocols.

- *The asymmetry between the parties* in a protocol, for instance a tag and a reader should also be considered in the protocol design. Protocols should be designed such that the heaviest computation load is for the reader (or other energy rich device) while the load for a tag or a sensor is minimized. This reduces the computation load. Other options are specific for the interaction of light-weight internet of things devices and are based on threshold cryptography [18].

- *Implementation size.* Implanted medical devices have a strict budget for the cryptographic modules: silicon area for hardware implementations or code size for software implementations. Close interaction with implementation people is needed: e.g. protocol designers tend to believe that hash functions are very cheap in hardware, thus should be used in light-weight protocols. For the most recent generation of hash functions, this is no longer true. The smallest SHA-1 implementation [12] uses 5527 gates, while an ECC core uses about 12k gates [10].

- *Energy usage reduction.* Protocols can be improved in at least three ways to reduce the energy usage on the device. Firstly, the computation on the device should be reduced as much as possible. Secondly, the communication should be minimized since wireless communication is power-hungry. Thirdly, the protocol should be designed to minimize energy consumption due to *useless* computations. Consider the mutual authentication between a pacemaker and a server, server authentication should be performed before other operations. As such, the protocol session stops immediately on the device when the server authentication fails.

Protocols based on secret key algorithms, like AES, are often cheaper in computation cost but not necessarily in communication cost. Secret key algorithms have also the problem of key distribution and management. Several exercises to evaluate the computation versus communication cost of secret-key versus public-key based security protocols have been made: the conclusions depend on the cryptographic algorithm, the digital platform and the wireless distance over which the communication occurs [5, 4].

Vaudenay [20] showed that public key algorithms are needed in order to provide strong privacy. However, not all PKC-based protocols achieves strong privacy. For example, tags using the Schnorr identification protocol [17] can be easily traced. We use the identification protocol by Peeters and Hermans [14] as an example. The protocol is shown in Figure 4, and it achieves *wide-forward-insider* privacy [1].

---

[1] *Wide-forward-insider* privacy is a widely used privacy notion in security analysis of private RFID identification protocols. It covers most practical use cases of private RFID identification. For a detailed definition, see [14].

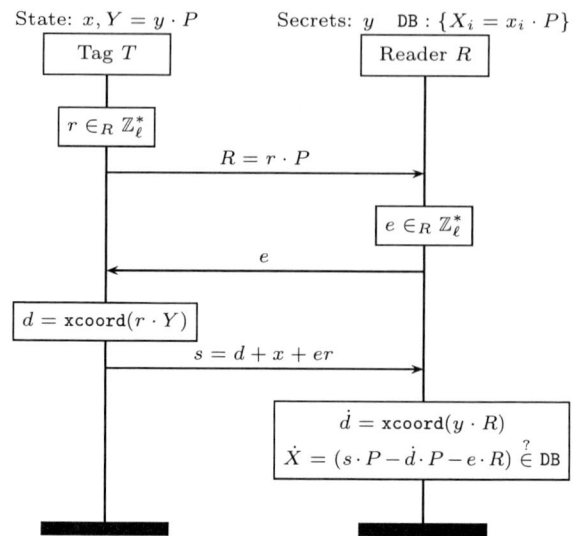

Figure 2: Peeters-Hermans identification protocol [14].

Elliptic curve cryptography is a favorable choice due to its relatively small key size and high security level compared to RSA. An elliptic curve over a finite field $\mathbb{F}_{2^m}$ is defined using the simplified Weierstrass function:

$$y^2 + xy = x^3 + ax^2 + b, \tag{1}$$

where $a, b \in \mathbb{F}_{2^m}$ are curve parameters. All the points $P(x, y)$ on the curve together with a point at infinity form an Abelian group $E(\mathbb{F}_{2^m})$. We can also add two points and obtain another point on the curve. Given an integer k, $k <$ Order(P), we can define $Q = k \cdot P = P + P + ... + P$ (k times). The computation of $k \cdot P$ is called elliptic curve point multiplication (ECPM). The security of ECC is based on the so-called Elliptic Curve Discrete Logarithm Problem (ECDLP), namely, given Q and P to find k such that $Q = k \cdot P$. ECDLP is believed to be computationally infeasible.

As shown in Figure 4 the main operation on the tag is two point multiplications (namely, $r \cdot P$ and $r \cdot Y$), and one modular multiplication (namely, er). All the operations should be protected against side-channel attacks and fault attacks. The challenge is how to securely perform these operations and at the same time meet the area and power budget.

The first step is to select curve parameters, which largely determines the security level and implementation size. Our ECC chip uses a Koblitz curve [1] defined over $\mathbb{F}_{2^{163}}$, which provides 80-bit security, equivalent to 1024-bit RSA. Besides, multiplication in binary extension fields it is carry-free. As a result, the multiplier is smaller and faster than integer multipliers.

The point multiplication algorithm directly determines the performance, the size of temporary storage, the performance and also its resistance against side-channel attacks. Our ECC chip uses the Montgomery powering ladder (MPL) for ECPMs. Note that MPL also allows us to use only the x coordinate to represent a point. One coordinate requires 163 bits of memory. Our ECC chip uses six 163-bit registers for the whole point multiplication. On the contrary, the best known algorithm for ECPM over a prime field uses

100

Algorithm 1 Point Multiplication using MPL

Require: An EC $y^2 + xy = x^3 + ax^2 + b$, a point $P = (x, y)$,
a t-bit integer k, $k = (1, k_{t-2}, ..., k_0)$, $k_i \in \{0, 1\}$
Ensure: $R = kP$
$\quad R \leftarrow (xr, r)$ //projective coordinate randomizaton
$\quad Q \leftarrow 2 \cdot P$
$\quad$**for** $i = t - 2$ downto 0 **do**
$\quad\quad$**if** $k = 1$ **then**
$\quad\quad\quad R \leftarrow R + Q, Q \leftarrow 2 \cdot Q$
$\quad\quad$**else**
$\quad\quad\quad Q \leftarrow Q + R, R \leftarrow 2 \cdot R$
$\quad\quad$**end if**
$\quad$**end for**
$\quad R \leftarrow RecoverY(P, R)$
$\quad$Return R

---

8 registers excluding a and b [6]. The MPL algorithm is also resistant against Timing and Simple Power Analysis Attacks. In order to prevent Differential Power Analysis, we use randomized projective coordinates. More details about the countermeasures are discussed in the following section.

5. ARCHITECTURE LEVEL

At this level, we still have the same design dimensions (area, speed, power, energy, security) but estimations become more accurate. Optimizing the design on one dimension may lead to deterioration on the others. Here we describe several methods for algorithm level optimization.

- *Identify the root of trust.* Adding countermeasures leads to larger area or longer running time. Therefore, it is important to partition the design into a *secure* zone and an insecure zone. The secure zone operates on the sensitive data, this part should be protected using state-of-the-art countermeasures. The insecure zone contains the non-critical parts of the systems such as parts of the algorithm that don't depend on the secret information: this part can be implemented using a standard design flow. As long as the insecure zone is not compromised, the security of the system as a whole remains. One elegant solution is using a secure co-processor for the critical parts of the algorithm and an ordinary processor for everything else. The secure co-processor can then be strengthened by applying the countermeasures at the circuit level, or even using a full-custom approach.

- *Architecture-level security evaluation.* A crypto co-processor usually includes both hardware and software. The hardware part helps to achieve a high energy efficiency, while the software part provides flexibility. Sensitive data should appear only on the internal databus, and should not be available through the instruction set. So, no strange combination of instructions should release the key or the private date. For example, a procedure that reads the secret key from the memory and sends it to the output should not be programmable with the given instructions. Moreover, countermeasures against side-channel attacks need to be included. At a minimum, to prevent timing attacks, all instructions should execute with a constant number of cycles.

- *Area-power-security trade-off.* Although global optimization seems to be difficult, local optimization is possible for the trade-off between area, execution time and power consumption. For instance, in our ECC co-processor, a digit-serial multiplier for $\mathbb{F}_{2^{163}}$ is used. The choice of the digit-size determines the power needed for the computation, as well as the latency and area [16]. By using a digit serial multiplication with a 163×4 modular multiplier we achieve the optimal area-energy product within the given latency constraints. Moreover, the execution time is independent of the key length.

6. CIRCUIT LEVEL

If the basic design building blocks, the logic gates, are not designed to support security, the problem propagates to higher levels of abstraction and compromises the security of the complete system. Ignoring the problems at the circuit level, can make countermeasures at protocol, algorithm or architecture level irrelevant.

Most digital integrated circuits use standard CMOS logic due to its compact area, low power and availability of a standard cell design methodology. However, CMOS circuits have one fundamental security weakness. During the $0 \rightarrow 1$ transition at the output, a CMOS gate consumes power from the source which is not the case for $0 \rightarrow 0$, $1 \rightarrow 1$ or $1 \rightarrow 0$ transitions. This asymmetry is what enables the attacker to develop a power consumption model and, by comparing the model prediction with the actual measurements, extract the secret information.

Sense amplifier based logic (SABL) consumes the same amount of energy regardless of the data being processed which is achieved by using complementary outputs and dynamic operation. In order to have a meaningful improvement, this countermeasure has to be accompanied by a balanced layout of dual signal wires. Alternative style, Wave Dynamic Differential Logic (WDDL) operates using the same principle, and is compatible with regular synthesis, and place and route tools [19]. Side-channel resistant logic styles are the most efficient countermeasures to prevent power analysis, however they come with high area and power cost.

Making the power consumption data independent seems to be the most promising approach so far. Even when no dedicated logic style is used, there are several tricks that can be used to reduce the information leakage in combination with standard cell based design. These tricks do not provide the same level of protection as using specialized logic styles do, but they still increase the attack effort in practice.

- *Balance critical signals* to reduce the risk of SPA. Critical signals are typically control signals driving the multiplexers. These control signals usually connect to many multiplexers (164 in the presented ECC co-processor) as well as to a complex network of long wires and signal repeaters. Due to this high capacitive load, signal transitions will cause a noticeable pattern in the power trace. E.g. Figure 6 illustrates how multiplexer control signals can depend on the value of the key bit k. These signals have to be encoded in such way that the corresponding hamming differences are constant, otherwise the unbalance will reflect in the power trace. Regular layout structure and identical routing of these signals will make this countermeasure more effective.

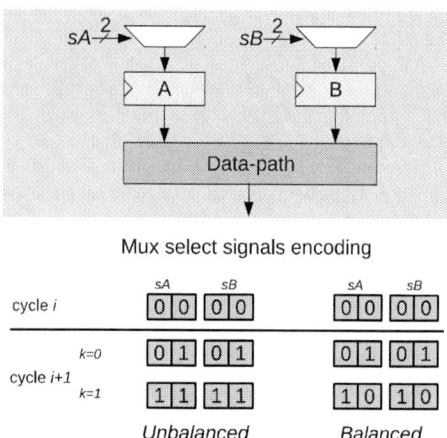

Figure 3: Register updating scheme and multiplexer encoding.

- *Avoid data-dependent clock-gating.* Clock gating may be a tempting solution to reduce dynamic power, however, in some cases, overly aggressive clock gating can introduce security risks. If different registers are enabled depending on the secret key different parts of the clock tree will be activated. The corresponding difference in power consumption will result in a clearly visible pattern in the power trace, thereby enabling an SPA. The mere fact that a different set of registers is gated, can be linked to a particular instruction sequence and directly or indirectly to the key.

- *Isolate the inputs to the data-paths.* When register outputs are connected to the data-path, updating the register value will cause spurious signal transitions inside the data-path. This will increase the power consumption but it will also compromise the security since the power consumption is correlated with the data loaded to the register. The solution is to set data-path inputs to a fixed value when it is not used. This can usually be implemented using AND gates and enable signals.

- *Avoid glitches.* This is a good practice for low power design since unwanted glitches result in higher power consumption. Even when the number of $0 \rightarrow 1$ transitions is balanced at the higher abstraction levels, glitches that appear in the data-path can cause data-dependent power consumption thereby enabling an attack. Please note that dynamic differential logic (such as SABL and WDDL) provide inherent protection against glitching. Other circuit styles were broken based on glitches [11].

The presented circuit-level optimization techniques reduce the risk of side-channel attacks and increase the security of the system. Some of these design practices align well with the power reduction techniques, while others are in clear contradiction.

A prototype chip for the ECC co-processor, based on the architecture presented in [10] is fabricated using UMC $0.13\mu m$ process. At the operating frequency of $847.5 kHz$ and core voltage $Vdd = 1V$, the processor consumes $50.4\mu W$ and uses only $5.1\mu J$ for one point-multiplication. At this

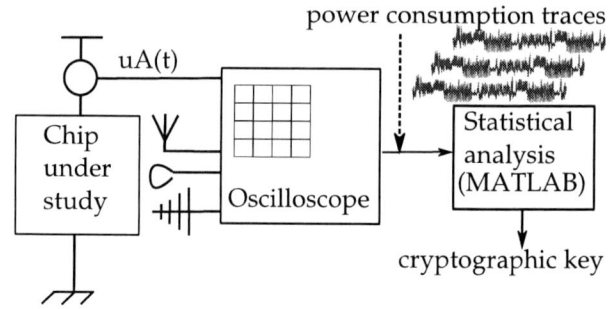

Figure 4: Typical workflow for side-channel attacks

frequency, the throughput is 9.8 point multiplications per second.

7. SECURITY EVALUATION

A security evaluation typically starts with a white-box evaluation of a prototype chip and system. In a white-box evaluation, the attacker has complete access to the inner working details of the chip, including a precise description of the countermeasures implemented, and is generally regarded as a worst-case evaluation. A real attacker, later-on in a practical setting, will have less information available. The countermeasures used in the prototype co-processor were evaluated in a worst-case lab setting as Figure 4 depicts. The setup allows high sampling resolution of the instantaneous power consumption of the device.

Timing attacks exploit the timing variance with different inputs to provide some information about the key [7]. The prototype co-processor is intrinsically resistant to timing attacks. This is due to the fact that the computation time of a point multiplication is the same for different key values. This is achieved by careful optimizations on two abstraction levels. At the algorithm level, the Montgomery powering ladder requires the same number of iterations, while at architecture level, it is ensured that each iteration uses a constant number of clock cycles.

Moreover, since the same operations are executed in the same order in every invocation of the scalar multiplication routine (regardless of the value of the key), the device is mostly secure against attacks that inspect the power consumption signature of the device, a.k.a Simple Power Analysis (SPA) attacks. We identified a complex attack that could extract the key since a small source of SPA leakage was detected in our white-box evaluation. However, in order for the attacker to exploit it, he has to perform a complex profiling phase with an identical device that is under his total control, which is outside of the scope of our initial requirements. One of the causes of this SPA leakage might be that, although at the layout level the design was carefully balanced, slight unbalances are still present in the layout.

Differential Power Analysis (DPA) [8] is a statistical technique used to recover the key of a cryptographic chip from its instantaneous power consumption, provided that the power consumption is related to the intermediate data processed. Informally, DPA recovers the key in a divide-and-conquer fashion by comparing the measured power consumption with several hypothesized power consumptions, one for each subkey hypothesis. It is generally expected that the similarity between the measured power consumption and the predicted power consumption will be high only for the correct key hypothesis. Note that the power consumption can be picked up remotely from the electromagnetic emission of the chip by using specific-purpose antennas.

To prevent DPA, the chip randomizes the internal points representation by using a random Z coordinate in each execution. Since the intermediate values cannot be predicted, DPA attacks cannot be mounted.

We empirically confirmed that DPA attacks are correctly thwarted by using randomized projective coordinates. When the countermeasure is disabled, a DPA attack succeeds with as low as 200 traces. When the countermeasure is enabled, but the randomness is known, the attack also succeeds. This scenario is only possible in a white-box evaluation and does not correspond to the normal operation of the chip. In the normal operation, the randomness is generated by the chip and kept secret to the adversary. The fact that the attack works in this lab setting provides confidence on the soundness of the attack. When the countermeasure is enabled, and the randomness is unknown, the attack does not succeed. Even 20000 traces are not enough to reveal a single key bit, using the same DPA attack.

8. CONCLUSIONS

Making a device secure adds an extra design dimension. Indeed, for the design of medical devices, a trade-off between security, power and energy needs to be made. We have described the security traps on each abstraction level and presented the corresponding design guidelines. This is illustrated with the design of a light-weight co-processor for elliptic curve cryptography.

9. ACKNOWLEDGMENTS

Oscar Reparaz is funded by an FWO fellowship. In addition, this work is supported in part by the Flemish Government through FWO G.0550.12N and G.0130.13, the Hercules Foundation AKUL/11/19, and by the European Commission through the ICT program under contract FP7-ICT-2011-284833 PUFFIN.

10. REFERENCES

[1] FIPS PUB 186-3, Digital Signature Standard (DSS).

[2] W. P. Burleson, S. S. Clark, B. Ransford, and K. Fu. Design Challenges for Secure Implantable Medical Devices. In *DAC 2012*, June 2012. Invited paper.

[3] A. Chandrakasan, M. Potkonjak, R. Mehra, J. Rabaey, and R. Brodersen. Optimizing Power Using Transformations. *IEEE TCAD*, 14(1):12–31, 1995.

[4] G. de Meulenaer, F. Gosset, F.-X. Standaert, and O. Pereira. On the Energy Cost of Communications and Cryptography in Wireless Sensor Networks. In *(extended version), SecPriWiMob 2008*, pages 580–585, 2008.

[5] A. Hodjat and I. Verbauwhede. The Energy Cost of Secrets in ad-hoc Networks. In *Proc. IEEE Circuits and Systems Workshop on Wireless Communications and Networking*, page 4, 2002.

[6] M. Hutter, M. Joye, and Y. Sierra. Memory-Constrained Implementations of Elliptic Curve Cryptography in Co-Z Coordinate Representation. In *AFRICACRYPT*, pages 170–187, 2011.

[7] P. C. Kocher. Timing Attacks on Implementations of Diffie-Hellman, RSA, DSS, and Other Systems. In *Advances in Cryptology - CRYPTO*, pages 104–113, 1996.

[8] P. C. Kocher, J. Jaffe, and B. Jun. Differential Power Analysis. In *Advances in Cryptology - CRYPTO*, pages 388–397, 1999.

[9] Y. Lee, L. Batina, and I. Verbauwhede. EC-RAC (ECDLP Based Randomized Access Control): Provably Secure RFID Authentication Protocol. In *IEEE International Conference on RFID*, pages 97–104, 2008.

[10] Y. K. Lee, K. Sakiyama, L. Batina, and I. Verbauwhede. Elliptic-Curve-Based Security Processor for RFID. *IEEE Trans. Computers*, 57(11):1514–1527, 2008.

[11] S. Mangard, T. Popp and B. M. Gammel. Side-Channel Leakage of Masked CMOS Gates. In *CT-RSA*, pages 351–365, 2005.

[12] M. O'Neill. Low-cost SHA-1 Hash Function Architecture for RFID Tags. In *Workshop on RFID Security - RFIDSec*, pages 41–51, 2008.

[13] P. R. Panda, F. Catthoor, N. D. Dutt, K. Danckaert, E. Brockmeyer, C. Kulkarni, A. Vandecappelle, and P. G. Kjeldsberg. Data and Memory Optimization Techniques for Embedded Systems. *ACM Trans. Design Autom. Electr. Syst.*, 6(2):149–206, 2001.

[14] R. Peeters and J. Hermans. Wide Strong Private RFID Identification Based on Zero-Knowledge. *IACR Cryptology ePrint Archive*, 2012:389, 2012.

[15] V. Pop, R. de Francisco, H. Pflug, J. Santana, H. Visser, R. J. M. Vullers, H. de Groot, and B. Gyselinckx. Human++: Wireless Autonomous Sensor Technology for Body Area Networks. In *ASP-DAC 2011*, pages 561–566, 2011.

[16] K. Sakiyama, L. Batina, B. Preneel, and I. Verbauwhede. Multicore Curve-Based Cryptoprocessor with Reconfigurable Modular Arithmetic Logic Units over $GF(2^n)$. *IEEE Trans. Computers*, 56(9):1269–1282, 2007.

[17] C.-P. Schnorr. Efficient Identification and Signatures for Smart Cards. In G. Brassard, editor, *Advances in Cryptology - CRYPTO'89, LNCS*, volume 435, pages 239–252. Springer-Verlag, 1989.

[18] K. Simoens, R. Peeters, and B. Preneel. Increased Resilience in Threshold Cryptography: Sharing a Secret with Devices That Cannot Store Shares. In *Pairing 2010*, volume 6487 of *LNCS*, pages 116–135, 2010.

[19] K. Tiri and I. Verbauwhede. A Digital Design Flow for Secure Integrated Circuits. *IEEE Transactions on Computer-Aided Design of Integrated Circuits and Systems*, 25(7):1197–1208, 2006.

[20] S. Vaudenay. On Privacy Models for RFID. In *ASIACRYPT 2007*, volume 4833 of *LNCS*, pages 68–87, 2007.

[21] I. Verbauwhede and P. Schaumont. Design Methods for Security and Trust. In *DATE 2007*, pages 1–6, NICE, FR, 2007. IEEE.

Aging-Aware Compiler-Directed VLIW Assignment for GPGPU Architectures

Abbas Rahimi
CSE, UC San Diego
La Jolla, CA 92093, USA
abbas@cs.ucsd.edu

Luca Benini
DEIS, University of Bologna
40136 Bologna, Italy
luca.benini@unibo.it

Rajesh K. Gupta
CSE, UC San Diego
La Jolla, CA 92093, USA
gupta@cs.ucsd.edu

ABSTRACT

Negative bias temperature instability (NBTI) adversely affects the reliability of a processor by introducing new delay-induced faults. However, the effect of these delay variations is not uniformly spread across functional units and instructions: some are affected more (hence less reliable) than others. This paper proposes a NBTI-aware compiler-directed very long instruction word (VLIW) assignment scheme that uniformly distributes the stress of instructions with the aim of minimizing aging of GPGPU architecture without any performance penalty. The proposed solution is an entirely software technique based on static workload characterization and online execution with NBTI monitoring that equalizes the expected lifetime of each processing element by regenerating aging-aware *healthy* kernels that respond to the specific health state of GPGPU. We demonstrate our approach on AMD Evergreen architecture where iso-throughput executions of the *healthy* kernels reduce NBTI-induced voltage threshold shift up to 49% (11%) compared to naïve kernel executions, with (without) architectural support for power-gating. The kernel adaption flow takes average of 13 millisecond on a typical host machine thus making it suitable for practical implementation.

Keywords: NBTI, GPGPU, Aging-aware Compilation, VLIW, Adaptive Kernel, Dynamic Binary Optimizer.

1. INTRODUCTION

Variability across manufactured parts and aging over time are emerging challenges in IC chips [1]. Among various aging mechanisms, the generation of interface traps under NBTI in PMOS transistors has become a critical reliability issue in determining the lifetime of CMOS devices [2]. NBTI effects can be significant: its impact on circuit delay is about 15 percent on a 65nm technology node [3] and it gets worse in sub-65nm nodes [4]. This imposes an excessive guardband over circuit lifetime causing performance loss and increased costs.

When a PMOS transistor is negatively biased ($V_{gs} = -V_{dd}$), the dissociation of $Si-H$ bonds along the silicon oxide interface, causes the generation of interface traps, while removal of the bias ($V_{gs} = 0$) causes a reduction in the number of interface traps due to annealing [1]–[5]. The rate of generation of these traps is accelerated by temperature, and the time of applied stress. The threshold voltage (V_{th}) of the PMOS transistors increases as more traps form, reducing the drive current, which in turn slows down the rising propagation delay of logic gates over time. Thus, the NBTI-induced performance degradation strongly depends on the amount

of time during which a PMOS transistor is *stressed*, that is, when a logic '0' is applied to the gate. The increase in V_{th} is a logarithmic function of the corresponding stress time [6], which is distributed non-uniformly across a logic circuit, leading to 2−5× difference in the degradation rate of V_{th} [7]. When the stress condition is *relaxed*, aging can be recovered partially, and the V_{th} decreases toward the nominal value [7], [8].

Non-uniform stress caused by non-uniform workload is a major concern for general purpose graphical processing units (GPGPUs) [9] with up to 512 CUDA cores [10], or 320 five-way VLIW processors [11]. To ensure necessary observability for non-uniform aging degradation, *in situ* NBTI and oxide degradation sensors with digital outputs have been proposed and validated on silicon [12]. These sensors enable high-volume data collection to guide dynamic management schemes and warn of impending device failure. Using NBTI sensors, adaptive guardbanding has been proposed earlier to reduce the otherwise conservative guardbands due to better than worst-case operating conditions [13]. For controllability, power-gating is known as an effective technique to mitigate NBTI-induced aging [14], since PMOS stress is removed during periods of power-gating. In this context, Paterna *et. al.* [15] propose a dynamic workload allocation to mitigate aging-induced unbalanced cores lifetimes by means of core activity duty cycling on a multi-core platform.

1.1 Contributions

This paper makes following contributions:

I. We propose an online adaptive reallocation strategy to mitigate NBTI-induced performance degradation in GPGPU machines. This is accomplished through a NBTI-aware compiler that uses a dynamic binary optimizer. During dynamic recompilation, the binary is optimized by customizing the kernel's code with respect to specific health state of GPGPU. This technique leverages a compiler-directed scheme that uniformly distributes the stress of instructions throughout various VLIW resource slots, results in a *healthy* code generation that keeps the underlying GPGPU hardware *healthy*. Section 3 and 4 describe NBTI model and GPGPU.

II. We propose a fully software solution that uses static (offline) workload characterization and online availability of NBTI sensors. The dynamic binary optimizer correlates the device stress time with instructions distribution, and equalizes the expected lifetime of each processing element without any architectural modification. Section 5 covers this technique in detail.

III. In Section 6, we demonstrate our approach on AMD Evergreen GPGPU architecture and its tool-chain to adapt kernels to the health state of GPGPU. The throughput of our *healthy* kernel execution is the same as naïve kernel execution (iso-throughput). In comparison with the naïve kernels, our *healthy* kernels execution achieves a maximum 49% reduction in NBTI-induced V_{th} shift over five years if GPGPU supports power-gating during idle states. Power-gating is intrinsically protective against NBTI by

---

Permission to make digital or hard copies of part or all of this work for personal or classroom use is granted without fee provided that copies are not made or distributed for profit or commercial advantage and that copies bear this notice and the full citation on the first page. Copyrights for components of this work owned by others than ACM must be honored. Abstracting with credit is permitted. To copy otherwise, to republish, to post on servers or to redistribute to lists, requires prior specific permission and/or a fee.

DAC '13, May 29 - June 07 2013, Austin, TX, USA

providing sleep states that spare gates from stress that produces NBTI effects. In the absence of power-gating, our uniform self-healing NOP execution technique mitigates the V_{th} shift by 11%. On average, the total execution time of the entire adaptation process is 13 millisecond on an Intel i5 CPU 2.67GHz.

2. RELATED WORK

Various techniques [15]–[17] have been proposed to slow down the aging of traditional coarse-grained multi-core architectures. These techniques range from selective frequency scaling to manage the aging process, dynamic control of the usage of processing units through shutdown that together seeks to equalize the level of aging seen across the cores. A brief review of important contributions follows.

Selective Speed Scaling: Chip-wide voltage scaling has been applied to switch the processor from a slow-aging mode to a high-speed mode [16] selectively over its lifetime. This affects performance and to combat the performance loss, Bubblewrap [17] supports multiple modes based on [16], for instance, by running the slow cores at a higher supply voltage for a shorter service life until they entirely wear-out and are discarded. For fine-grain many-core architectures, this technique loses effectiveness because after the early lifetime, the difference between the adaptive voltage and the over-designed supply voltage is small [18].

Selective Shutdown: To combat the impact of within-die core-to-core frequency variations on GPGPU throughput, two techniques are proposed in [19]: (i) disabling the slowest cores, and (ii) running each core at its maximum frequency independently. Both of these solutions impose a non-negligible performance penalty: the first directly diminishes the throughput of a cluster, and the second imposes extra latency for synchronization of cores with different frequencies. Further, these techniques only consider the effects of static process variation, and do not cover aging of GPGPUs which is dynamic in nature.

At a finer architectural granularity, Colt [20] equalizes the duty cycle ratio and the usage frequency of the functional units in a microprocessor. To mitigate aging effects, it uses a number of measures such as complement mode execution, cache set rotation, and operand identifier swapping schemes. These measures are intrusive and fairly complicated: the complement mode is applied to the whole data path, control path, and storage hierarchy. In a similar vein, a linear programming scheme is employed to find a new instruction to replace the processor's default NOP instruction for minimizing the NBTI effects [21]. This approach also requires architectural supports and pipeline modification. Wearout-aware compiler-directed register assignment techniques have been proposed in [22] that attempt to distribute the stress-induced wearout throughout the register file. Another aging-aware assignment of registers has been proposed to balance the duty cycle ratio of the internal bits in register files [23]. Even though [22], [23] do not impose architectural overheads and modification, their compiler strategies are limited to the utilization of the register file.

NBTI-aware power-gating [14] exploits the sleep state where a circuit is intrinsically immune to aging. Caliman *et al.* [24] propose static and dynamic strategies to compensate the aging effects on the sleep transistors. Here, the benefit of power-gating is strongly dependent on the fraction of time that a circuit spends in sleep mode. In practice, high power-gating factors are accompanied by significant performance degradation. As an alternative, in Section 5.3, we show how a VLIW machine can instead arrange instructions to utilize the power-gating factor without any performance penalty.

3. DEVICE-LEVEL NBTI MODEL

We briefly review the dynamic NBTI model for its use in compiler optimizations. In NBTI, the PMOS transistor undergoes alternate *stress* ($V_{gs} = -V_{dd}$) and *recovery* ($V_{gs} = 0$) periods, derived from the Reaction-Diffusion theory [7], [8]. When logic input '0' is applied to the gate of a PMOS transistor ($V_{gs} = -V_{dd}$), the presence of holes in the channel causes $Si-H$ bonds to break. The resulting H diffuses away, leaving positive traps ($Si+$) in the interface, which increase voltage threshold by $\Delta V_{th\text{-}stress}$:

$$\Delta V_{th\text{-}stress} = \left(K_v\sqrt{t_{stress}} + \sqrt[2n]{\Delta V_{th\text{-}t0}}\right)^{2n} \tag{1}$$

where t_{stress} is the time that PMOS is under stress; K_v has dependence on electrical field, temperature (T), and V_{dd}; n is the time exponent parameter, and for H_2 diffusion is 1/6; and $\Delta V_{th\text{-}t0}$ is the initial V_{th} variation of PMOS at time zero due to process variation caused by random dopant fluctuations. When logic input '1' is applied to the gate ($V_{gs} = 0$), the transistor turns off, and hydrogen atoms diffuse back, eliminating some of the traps in a recovery phase that can recover part of the V_{th} shift:

$$\Delta V_{th\text{-}recov} = \Delta V_{th\text{-}stress}\left(1 - \frac{2\xi_1 t_e + \sqrt{\xi_2 C\, t_{recov}}}{(1+\delta)t_{ox} + \sqrt{Ct}}\right) \tag{2}$$

where t_{recov} is the time under recovery; t_{ox} is the oxide thickness; t_e is the effective oxide thickness; t is the total time; C has temperature dependence; ζ_1, ζ_2, δ are constants in [7]. Duty cycle (α) is the ratio of the time spent in stress to the period of one stress-recovery cycle. ΔV_{th} has been shown to be a monotonically increasing function of higher duty cycle (α), t, V_{dd}, T [25]. The NBTI-induced V_{th} shift is also a function of process-dependent parameters, and relatively insensitive to the switching frequency (f) when it is above 100Hz [8]. The duty cycle (α) can be directly tuned by the software to reduce or eliminate the NBTI-induced effects.

If a transistor has a larger threshold voltage than expected, its transconductance is smaller, it has a lower drive current and increased delay during a transition. The transistor switching delay can be approximately expressed as the alpha-power law:

$$\tau \propto \frac{V_{dd}L}{\mu(V_{dd}-V_{th})^{\alpha'}} \tag{3}$$

where μ is the mobility of carriers; $\alpha' \approx 1.3$ is the velocity saturation index; and L is the channel length. Hence, the delay variation $\Delta\tau/\tau$ can be derived as follows:

$$\frac{\Delta\tau}{\tau} = \frac{\Delta L}{L} + \frac{\Delta\mu}{\mu} + \frac{\alpha'}{V_{dd}-V_{th}}\Delta V_{th} \tag{4}$$

considering only the effect of ΔV_{th} shift, and neglecting other terms, the delay degradation $\Delta\tau$ is given by

$$\Delta\tau = \frac{\alpha'\Delta V_{th}}{V_{dd}-V_{th\text{-}t0}}\tau_0 \tag{5}$$

where $V_{th\text{-}t0}$ is the original transistor threshold voltage (at the life of time t_0), and τ_0 is its corresponding delay before degradation. There might be several ΔV_{th} of different PMOSs in a circuit, thus we consider the largest one to calculate the worst case delay degradation. In our analysis, we set all the internal node states to a '0' during stress mode to determine the worst case circuit degradation that limits the lifetime of a chip. Although this is a conservative assumption and during runtime there exists no such input vector that makes the internal nodes all 0s; this assumption is only used to calculate the maximum possible degradation and the potential of NBTI mitigation technique. Section 5 describes how an online calibrator regulates overestimates and underestimates of degradation due to the complex input patterns and inaccurate estimations.

4. GPGPU ARCHITECTURE

We focus on the Evergreen family of AMD GPGPUs (a.k.a. Radeon HD 5000 series), designed to target not only graphics applications but also general-purpose data-intensive applications. The Radeon HD 5870 GPGPU compute device consists of 20 Compute Units (CUs), a global front-end ultra-thread dispatcher, and a crossbar to connect the global memory to the L1-caches [11]. Every CU has access to a global memory, implemented as a hierarchy of private 8KB L1-caches, and 4 shared 512KB L2-caches. Each CU contains a set of 16 Stream Cores (SCs) that have access to a shared 32KB local data storage. Within a CU, a shared instruction fetch unit provides the same machine instruction for all SCs to execute in a SIMD fashion. Finally, each SC contains five Processing Elements (PEs), labeled X, Y, Z, W, and T constituting an ALU engine to execute Evergreen machine instructions in a vector-like fashion. The SC has also a general-purpose registers file to support private memory. The block diagram of architecture is shown in Figure 1.a.

Every SC is a five-way VLIW processor capable of issuing up to five floating point scalar operations from a single very long instruction word consists primarily of five slots ($slot_X$, $slot_Y$, $slot_Z$, $slot_W$, $slot_T$). Each slot is related to its corresponding PE. Four PEs (X, Y, Z, W) can perform up to four single-precision operations separately and perform two double-precision operations together, while the remaining one (T) has a special function unit for transcendental operations. In each cycle, VLIW slots supply a bundle of data-independent instructions to be assigned to the related PEs for simultaneous execution. In an N-way VLIW processor, up to N data-independent instructions, available on N slots, can be assigned to the corresponding PEs and be executed simultaneously. Typically, this is not done in practice because the compiler may fail to find sufficient Instruction-Level Parallelism (ILP) to generate complete VLIW instructions. On average, if M out of N slots are filled during an execution, we call the achieved packing ratio is M/N. The actual performance of a program running on a VLIW processor largely depends on the packing ratio.

4.1 GPGPU Workload Distribution

In this subsection, we analyze the workload distribution on the Radeon HD GPGPUs architecture, where there are many PEs to carry out computations. As it is mentioned in Section 3, NBTI-induced degradation strongly depends on the resource utilization, which depends on the execution characteristics of the workload. Thus, it is essential to analyze how often the PEs are exercised during the runtime execution of the workload. To this end, we first monitor the utilization of various CUs (inter-CU), and then the utilization of PEs within a CU (intra-CU).

To examine the inter-CU workload variation, the total number of executed instructions by each CU is collected during a kernel execution as per a methodology described in Section 6. Figure 1.b shows that the CUs execute almost equal number of instructions, and there is a negligible workload variation among them. We have configured six compute devices with different number of CUs, {2, 4,..., 64}, to finely examine the effect of the workload variation on a variety of GPGPU architecture (The latest Radeon HD 5000 series, HD 5970, has 40 CUs featuring 4.3 billion transistors in 40nm). During DCT kernel execution, the workload variation between CUs ranges from 0% to 0.26% depends to the number of physical CUs on the computation device. The DCT input kernel parameters are fixed for all configured compute devices, thus they carry out the same amount of workload— note that the total number of executed instructions per CU is inversely proportional to the number of available CUs on the compute device. Execution of all kernels listed in Section 6 confirms that the inter-CU workload variation is less than 3%, when running on the device with 20 CUs (HD 5870). This nearly uniform inter-CU workload distribution is accomplished by load balancing and uniform resource arbitration algorithms of the ultra-thread dispatcher.

Next, we examine the workload distribution among the PEs. Figure 1.c shows the percentage of the executed instructions of ALU engine by various PEs during execution of different kernels. ALU engine in this paper refers to four PEs (PE_X, PE_Y, PE_Z, PE_W) which are identical in their functions [26]; they differ only in the vector elements to which they write their result at the end of the VLIW. As shown, the instructions are not uniformly distributed among PEs. For instance, the PE_X executes roughly half of the ALU engine instructions (50.7%) during Rdn kernel execution, while only about one quarter of the ALU engine instructions (27.1%) are executed by PE_X during SF kernel execution. Execution of all kernels listed in Section 6 shows that seven kernels execute more than 40% of the ALU engine instructions only on PE_X. This non-uniform workload variation causes non-uniform aging among PEs, and exhausts some PEs more than others and shortening their lifetime. Unfortunately, this non-uniformity happens within all CUs since their workload is highly correlated together, therefore no PE throughout the entire compute device is immune from this unbalanced utilization.

Thus, root cause of non-uniform aging among PEs is the frequent and non-uniform execution of VLIW slots. For example, higher utilization of PE_X implies that $slot_X$ of VLIW is occupied more frequently than the other slots. This substantiates that the compiler does not uniformly assign the independent instructions to various VLIW slots, mainly because the compiler only employs optimizations for increasing the packing ratio through finding more ILP to fully pack the VLIW slots. The VLIW processors are designed to give the compiler tight control over program execution; however,

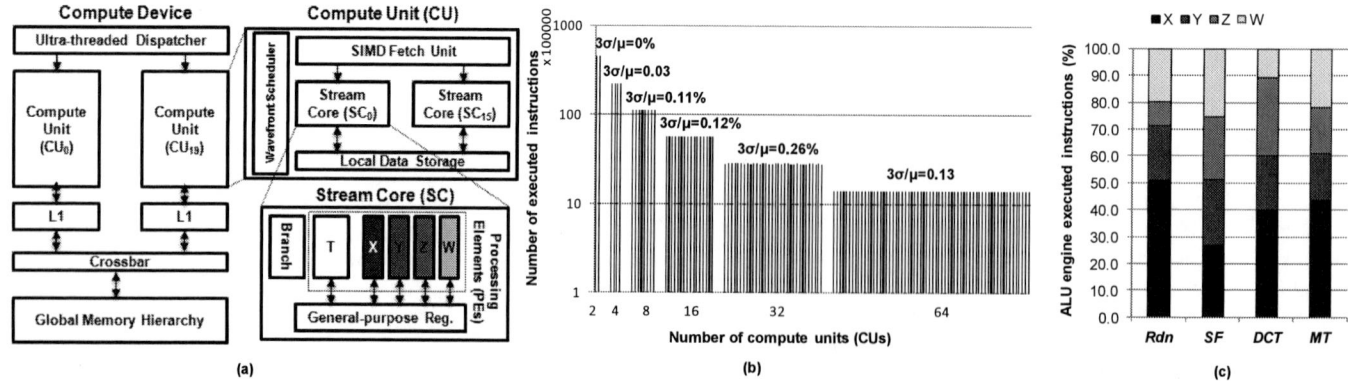

Figure 1.(a) Block diagram of the Radeon HD 5870 architecture. (b) Inter-CU workload variations for six configured compute devices. (c) Inter-PE ALU instructions distribution for various naïve kernels in the HD 5870 compute device (#CUs = 20).

106

the flexibility afforded by such compilers, for instance to tune the order of instructions packing, is rarely used towards reliability improvement.

5. AGING-AWARE COMPILATION
The key idea of an aging-aware compilation is to assign independent instructions uniformly to all slots: idling a *fatigued* PE and reassigning its instructions to a *young* PE through swapping the corresponding slots during the VLIW bundle code generation. This basically exposes the inherent idleness in VLIW slots and guides its distribution that does matter for aging. Thus, the job of dynamic binary optimizer, for *K-independent* instructions, is to find *K-young* slots, representing *K-young* PEs, among all available *N* slots, and then assign instructions to those slots. Therefore, the generated code is a *"healthy"* code that balances workload distribution through various slots maximizing the life time of all PEs. In this section, we describe how these statistics can be obtained from silicon, and how compiler can predict and thus control the non-uniform aging. The adaptation flow is illustrated in Figure 2 through four steps: 1) reading aging sensors; 2) kernel disassembler, static code analysis, and calibration of predictions; 3) uniform slot assignment; 4) *healthy* code generation.

5.1 Observability: Aging Sensors
The compiler needs to access the current aging data (ΔV_{th}) of PEs to be able to adapt the code accordingly. The ΔV_{th} is caused by the temporal degradation due to NBTI and/or the intrinsic process variation, thus PEs even during early life of a chip might have different aging. Employing the compact per-PE NBTI sensors [12] which provide ΔV_{th} measurement with 3σ accuracy of 1.23 mV for a wide range of temperature, enables large scale data-collection across all PEs. The performance degradation of every PE can be reliably reported by a per-PE NBTI sensor, thanks to the small overhead of these sensors. Test chips efficiently consider multiple sensors banks containing up to total 256 NBTI sensors (in 45nm), hence the power overhead of laying out thousands of sensors would only be a few hundreds of μW at maximum, which is a small fraction of power relative to a PE [12]. The sensors support digital frequency outputs that are accessed through memory-mapped I/O by the dynamic binary optimizer in arbitrary epochs of the post-silicon measurement.

5.2 Prediction: Wearout Estimation Module
As described, the dynamic binary optimizer accesses to the ΔV_{th} of various PEs, and evaluates their current performance ($\tau_{\{X,...,W\}}[t]$) using Equations (3)–(5). In addition to the current aging data, the compiler needs to have an estimate regarding the impact of future workload stress on the various PEs. This is accomplished by wearout estimation module shown in Figure 2. Since every naïve kernel binary can be considered as the future workload, code analysis techniques are required to predict the future workload in presence of branches. A just-in-time disassembler disassembles the desired naïve kernel binary to a device-dependent assembly code in which the assignment of instructions to the various slots (corresponding PEs) are explicitly defined, and thus observable by the dynamic binary optimizer. Then, a static code analysis technique is applied that estimates the percentage of instructions that will be carried out on every PE in a static sense. It extracts the future stress profile, and thus the utilization of various PEs using the device-dependent assembly code. Then, the static code analysis technique predicts the future ΔV_{th} shift of PEs (Pred-$\Delta V_{th-\{X,...,W\}}[t+1]$).

If the predicted ΔV_{th} of a PE is overestimated or underestimated, mainly due to the static analysis of the branch conditions of the

kernel's assembly code, a linear calibration module fits the predicted ΔV_{th} shift to the observed ΔV_{th} shift, in the next adaptation period. For every PE, e.g. PE_X, the linear calibration module uses the *simple linear regression* with an explanatory variable (Pred-$\Delta V_{th-X}[t+1]$), and a dependent variable ($\Delta V_{th-X}[t+1]$). The *simple linear regression* fits a straight line through the set of *m* points (each kernel execution) in such a way that makes the sum of squared residuals of the model as small as possible. The model is developed during online measurement by observing the actual ΔV_{th} shift reported by NBTI sensors ($\Delta V_{th-X}[t]$) after each kernel execution. Therefore, the linear calibration for every PE determines the curve that best describes the relationship between expected and observed sets of ΔV_{th} data; it projects the future ΔV_{th} of PEs ($\Delta V_{th-\{X,...,W\}}[t+1]$) by minimizing the sums of the squares of deviation between observed and expected values. Finally, $\Delta V_{th-\{X,...,W\}}[t+1]$ is used to calculated the future NBTI-induced performance degradation ($\Delta \tau_{\{X,...,W\}}[t+1]$).

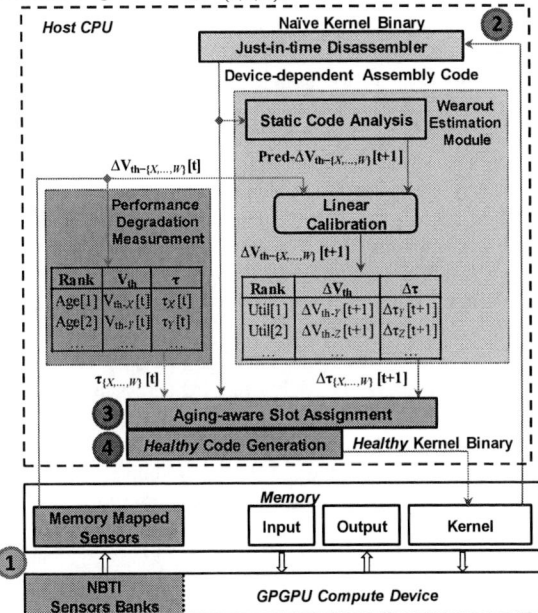

Figure 2. Aging-aware kernel adaptation flow.

5.3 Controllability: Uniform Slot Assignment
Thus far, we have described how the dynamic binary optimizer evaluates the current performance degradation (aging) of every PE ($\tau_{\{X,...,W\}}[t]$), and their future performance degradation ($\Delta \tau_{\{X,...,W\}}[t+1]$) due to the naïve kernel execution. Then, the compiler uses that information to perform code transformations with the goal of improving reliability, without any penalty in the throughput of code execution (maintaining the same ILP). To minimize stresses, the compiler sorts the predicted performance degradation of the slots increasingly and the aging of the slots decreasingly, and then applies a permutation to assign fewer/more instructions to higher/lower stressed slots. This algorithm for every period of adaptation [t] is shown below:

Degrad$_{[1, 2, 3, 4]}$ = Rank_degradation_increasingly ($\Delta \tau_{(X,Y,Z,W)}[t+1]$)

Age$_{[1, 2, 3, 4]}$ = Rank_aging_decreasingly ($\tau_{(X,Y,Z,W)}[t]$)

For *i* =*1* to *4*

    Reallocate (slot (Age[*i*]) ← slot (Degrad[*i*]))

where slot(Degrad[*1*]) is the slot that will have the minimum number of instructions during the future execution of the kernel, and slot(Age[*1*]) is the slot that its corresponding PE has the highest aging. To take into account both initial and temporal deg-

107

radations, our algorithm considers the highest aging value across the same type of PE since the lifetime of the chip is limited by the most aged component. Moreover, there is no means in the assembly code to distinguish the same type of PEs spread out among all CUs, unless the hardware architectural scheduler provides support. As a result of the slot reallocation, the minimum/maximum number of instructions is assigned to the highest/lowest stressed slot for the future kernel execution, thus uniforming the lifetime of PEs.

Execution of all examined kernels shows that the average packing ratio is 0.3 which means there is a large fraction of empty slots in which PEs can be relaxed during kernels execution. Evergreen ISA states that when a slot is empty, i.e. no instruction is specified for that slot in a VLIW bundle, the corresponding PE implicitly execute a NOP instruction [26]. Overall, our solution slips the *pre-assigned* instructions from high stressed slot, thus they will have more NOP instructions to execute instead of the stress-full instructions. This reduces their total stress time and effectively decreases α and thus ΔV_{th}. We can assume that during a NOP execution the PE is power-gated as it invalidates the written result in the corresponding vector elements at the end of NOP execution [26]. The feasibility of single-cycle power-gating is validated by Intel through a fine-grained power-gating for a 45nm SIMD tile [27]. Nevertheless, even in the absence of power-gating, the NOP instruction execution is self-healing that can reduce the stress time of the PE adequately. Moreover, the NOP instruction itself can be designed to highly minimize the NBTI effect [21]. We compare the benefit of a GPGPU architecture with and without power-gating for our approach in Section 6.

Among the available software knobs to mitigate NBTI, our algorithm aims to equalize the duty cycle (α) across all the slots. Another knob is the input pattern which is impractical to predict both in the complex workloads and circuits, thus our wearout estimation module relies on the online NBTI-induced measurement feedback through the linear calibration module for better adaptation. The proposed compiler-directed reliability approach superposes on top of all optimization performed by naïve compiler and does not incur any performance penalty, since it only reallocates the VLIW slots (slips the scheduled instructions from one slot to another) within the same scheduling and order determined by the naïve compiler. In other words, this dynamic binary optimizer guarantees the iso-throughput execution of the *healthy* kernel. It also runs fully in parallel with GPGPU on a host CPU, thus there will be no penalty for GPGPU kernel execution if dynamic compilation of one kernel can be overlapped with the execution of another kernel.

6. EXPERIMENTAL RESULTS

Our methodology is based on AMD Accelerated Parallel Processing (APP) software ecosystem suitable for stream applications written in OpenCL. The stream kernels are compiled into GPGPU device-specific binaries using the OpenCL compiler tool-chain which uses a standard off-the-shelf compiler front-end (g++), as well as the low-level virtual machine framework with extensions for OpenCL as the back-end. We have implemented our dynamic binary optimizer tool using C++ leveraging AMD Compute Abstraction Layer (CAL) APIs. CAL provides a runtime device driver library that supports code generation, kernel loading and execution, and allows applications to interact with the stream cores at the lowest-level. Multi2Sim [28] cycle-accurate simulation framework − a CPU-GPU model for heterogeneous computing targeting Evergreen ISA − is modified to collect the ALU engines

statistics. We have also equipped the simulator with the NBTI sensors where our tool has access to them; in a GPGPU chip those digitally-output memory-mapped sensors can be accessed by the device management part of CAL.

The following naïve binaries of AMD APP SDK 2.5 [29] kernels are run on the simulator: Reduction (*Rdn*), Binary Search (*BSe*), Haar1D (*DH1D*), Bitonic Sort (*BSo*), Fast Walsh Transform (*FWT*), Floyd Warshall (*FW*), Binomial Option (*BO*), Discrete Cosine Transform (*DCT*), Matrix Transpose/Multiplication (*MT/M*), Sobel Filter (*SF*), Uniform Random Noise Generator (*URNG*). Before invoking the kernel, our adaptation flow is triggered: the assembly code of the kernel using CAL APIs runtime library (*aticalrt*) in conjunction with NBTI sensors data is passed to the wearout estimation module, and a new code is generated that adapts the binary to the specific health state of GPGPU. In our experiments, to keep track of aging, this flow of adaptation is also run periodically in parallel on a host CPU every hour so as to impose negligible overhead.

We consider cycle-by-cycle architectural NBTI analysis [8] in the 65nm PTM technology with V_{gs}=1.2V, T=300K, and the stress statistics of the kernels execution obtained from the simulator; it is common to assume that all PMOS in a circuit degrade by the same amount [16], [17], and [18]. Figure 3.a shows the NBTI-induced V_{th} degradation when executing a *healthy Rdn* kernel compared to the naïve execution at time zero, and after one year. For this experiment, we consider a HD 5870 which is not affected by the process variability (initial inter-PE ΔV_{th}=0mV), and without power-gating support. As shown in Figure 3.a, at time 0, all PEs have the equal V_{th} since there was no stress, but after one year execution of naïve *Rdn*, PE_X has a maximum V_{th} of 435mV, because of executing 50.7% of the total ALU engine instructions (see Figure 1.c). However, the *healthy Rdn* kernel execution eliminates this non-uniformity by adapting itself every hour, and thus results in 14mV lower V_{th} shift after one year (for all PEs, V_{th}=421mV).

We also evaluate the effectiveness of the proposed approach when executing the *healthy Rdn* kernel on a process variability-affected HD 5870 (initial inter-PE ΔV_{th}=10mV) and without power-gating support compared to the naïve execution. Figure 3.b shows the V_{th} shift over time due to the naïve kernel execution, and at the end of 360hr, there is an 8mV V_{th} variation among PEs which limits the lifetime of PE_X (V_{th-X}=413mV). On the other hand, Figure 3.c shows that adapting the kernel periodically leads to a uniform V_{th} shift among all PEs (V_{th} variation is ~0.6mV), and the maximum V_{th} shift is 406mV at the end of 360hr − with power-gating support it further reduces to 402mV.

Indeed, the benefit of our technique is further pronounced for a larger time scale. Figure 4 shows the reduction in ΔV_{th} over five years execution of *healthy* kernels with and without power-gating support of GPGPU architecture. In comparison with the naïve execution of kernels, GPGPU with power-gating achieves a maximum 49% reduction in ΔV_{th}, while without power-gating the self-healing NOP execution provides a maximum of 11% reduction in ΔV_{th}. Since during power-gating the circuits are in the sleep state their aging mechanism are recovered quickly as derived in [14]. In average, compared to the naïve kernels, the execution of *healthy* kernels reduces ΔV_{th} by 34% and 6% in the presence and absence of power-gating supports respectively. Furthermore, the impact of our technique is higher if we consider the local temperature reduction due to idleness and power-gating.

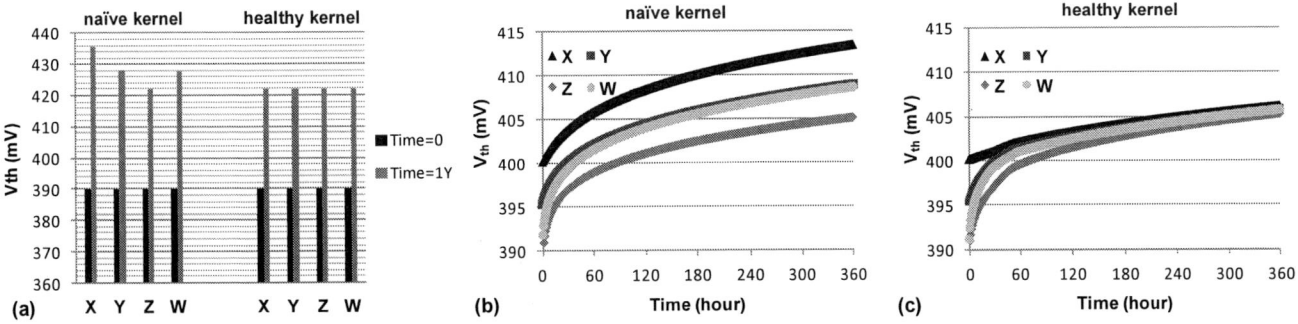

Figure 3. V_{th} shift for *Rdn* kernel: (a) NBTI-induced for 1 year; (b) Process variation and NBTI-induced for 360 hours.

The total execution time of the proposed adaptation flow is measured. Figure 5 shows the average execution time of the entire process, starting from disassembler up to the *healthy* code generation. It also shows the fastest and slowest execution we measure, as error bars. More than 95% of execution time is spent through the kernel disassembly using online CAL APIs, so the assembly code can be cached for faster iterations in future adaptation. The uniform slot assignment algorithm always runs below 2K cycles for all kernels, and the static code analysis is done between 220K-900K cycles depend to the size of kernel. Overall, the total execution time is bounded by 35 millisecond, and on average 13 millisecond on a host machine with an Intel i5 CPU 2.67GHz.

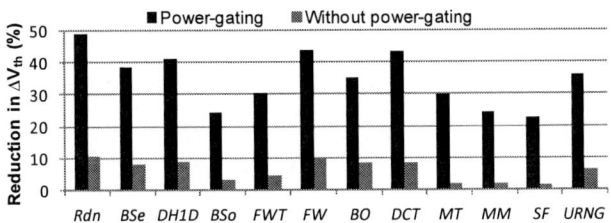

Figure 4. Reduction in ΔV_{th} due to the *healthy* kernels execution compared to naïve kernels for 5 years.

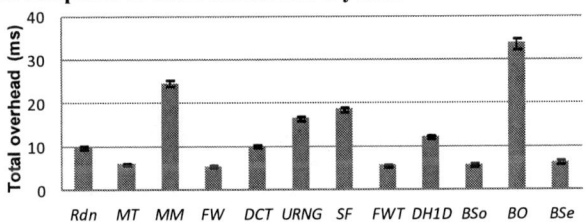

Figure 5. Total execution time of adaptation process.

7. CONCLUSION

Although the workload distribution among Compute Units (CUs) of GPGPU is nearly uniform, its Processing Elements (PEs) suffer from non-uniform VLIW distribution. To mitigate the impacts on lifetime uncertainty and unbalancing among the PEs, an online adaptive VLIW reallocation strategy is proposed that leverages a compiler-directed scheme to uniformly distribute the stress of instructions throughout various VLIW slots. This technique periodically regenerates *healthy* codes that heal over GPGPU aging. Compared to the naïve kernels, the execution of *healthy* kernels not only imposes 0% throughput penalty but also reduces ΔV_{th}: up to 49%(11%) and on average 34%(6%) in presence(absence) of architectural power-gating supports. On average, the total execution time of the adaption process is 13 millisecond.

Ongoing work is focused on generalizing the proposed approach on memory subsystems and variety of architectures.

8. ACKNOWLEDGMENTS

This work was supported by the NSF Variability Expedition under award n. 1029783, ERC-AdG MultiTherman GA n. 291125, and FP7 Virtical GA n. 288574.

9. REFERENCES

[1] P. Gupta, et al., "Underdesigned and Opportunistic Computing in Presence of Hardware Variability," IEEE Trans. on CAD of Integrated Circuits and Systems, pp. 489-499, 2012.

[2] G. Chen, et al., "Dynamic NBTI of p-MOS transistors and its impact on MOSFET scaling," IEEE Electron Device Letters, pp. 734–736, Dec. 2002.

[3] K. Bernstein, et al., "High-performance CMOS variability in the 65-nm regime and beyond," IBM Journal of Research and Development , pp.433–449, 2006.

[4] G. Chen, et al., "Dynamic NBTI of PMOS transistors and its impact on device lifetime," Proc. IEEE Reliability Physics Symposium, pp. 196-202, 2003.

[5] S. Chakravarthi, et al., "A Comprehensive Framework for Predictive Modeling of Negative Bias Temperature Instability," Proc. IEEE *Reliability Physics Symposium*, April 2004.

[6] S. V. Kumar, et al., "An analytical model for negative bias temperature instability," Proc. ACM/IEEE *ICCAD*, pp. 493–496, 2006.

[7] W. Wang, et al., "The Impact of NBTI Effect on Combinational Circuit: Modeling, Simulation, and Analysis," IEEE Trans. on VLSI Systems, Feb. 2010.

[8] S. Bhardwaj, et al., "Predictive modeling of the NBTI effect for reliable design," Proc. IEEE *CICC*, pp. 189–192, 2006.

[9] J.T. Adriaens, et al., "The case for GPGPU spatial multitasking," Proc. IEEE *HPCA*, 2012.

[10] J. Nickolls, et al., "The GPU Computing Era," IEEE Micro, March-April 2010.

[11] AMD Corporation. ATI Radeon HD 5870 Graphics.

[12] P. Singh, et al., "Dynamic NBTI Management Using a 45 nm Multi Degradation Sensor" IEEE Trans. on Circuits and Systems, pp.2026-2037, Sept. 2011.

[13] A. Rahimi, et al., "Hierarchically Focused Guardbanding: An Adaptive Approach to Mitigate PVT Variations and Aging," Proc. ACM/IEEE DATE, 2013.

[14] A. Calimera, et al., "NBTI-aware power gating for concurrent leakage and aging optimization," Proc. ACM/IEEE *ISLPED*, pp. 127–132, 2009.

[15] F. Paterna, et al., "Adaptive Idleness Distribution for Non-Uniform Aging Tolerance in MultiProcessor Systems-on-Chip," Proc. ACM/IEEE *DATE*, 2009.

[16] A. Tiwari and J. Torrellas, "Facelift: Hiding and slowing down aging in multi-cores," Proc. ACM/IEEE *MICRO*, pp. 129–140, 2008.

[17] U. Karpuzcu, et al., "The bubblewrap many-core: popping cores for sequential acceleration," Proc. ACM/IEEE *MICRO*, pp. 447–458, 2009.

[18] T. Chan, et al., "On the efficacy of NBTI mitigation techniques," Proc. ACM/IEEE *DATE*, 2011.

[19] J. Lee, et al., "Analyzing throughput of GPGPUs exploiting within-die core-to-core frequency variation," Proc. IEEE *ISPASS*, pp.237–246, 2011.

[20] E. Gunadi, et al., "Combating aging with the colt duty cycle equalizer," Proc. IEEE/ACM *MICRO*, pp. 103–114, 2010.

[21] F. Firouzi, et al., "NBTI Mitigation by Optimized NOP Assignment and Insertion," Proc. IEEE/ACM *DATE*, pp. 218–223, 2012.

[22] F. Ahmed, et al., "Wearout-aware compiler-directed register assignment for embedded systems," Proc. IEEE *ISQED*, pp.33–40, 2012.

[23] S. Wang, et al., "Low Power Aging-Aware Register File Design by Duty Cycle Balancing," Proc. IEEE/ACM *DATE*, pp. 546−549, 2012.

[24] A. Calimera, et al., "Design Techniques for NBTI-Tolerant Power-Gating Architectures," IEEE Transactions on Circuits and Systems II, April 2012.

[25] W. Wang, et al., "An efficient method to identify critical gates under circuit aging," Proc. IEEE/ACM *ICCAD*, pp.735-740, 2007.

[26] AMD Evergreen Family Instruction Set Architecture, 2011.

[27] H. Kaul, et al., "A 300 mV 494GOPS/W Reconfigurable Dual-Supply 4-Way SIMD Vector Processing Accelerator in 45 nm CMOS," IEEE Journal of Solid-State Circuits, Vol.45, No.1, pp.95–102, Jan. 2010.

[28] Multi2Sim [Online]. Available: http://www.multi2sim.org/

[29] AMD APP SDK 2.5 [online]. Available: www.amd.com/stream/

Exploiting Program-Level Masking and Error Propagation for Constrained Reliability Optimization

Muhammad Shafique, Semeen Rehman, Pau Vilimelis Aceituno, Jörg Henkel

Chair for Embedded Systems (CES), Karlsruhe Institute of Technology (KIT), Germany

{muhammad.shafique, henkel}@kit.edu; semeen.rehman@student.kit.edu

Abstract—Since embedded systems design involves stringent design constraints, designing a system for reliability requires optimization under tolerable overhead constraints. This paper presents a novel reliability-driven compilation scheme for software program reliability optimization under tolerable overhead constraints. Our scheme exploits program-level error masking and propagation properties to perform reliability-driven prioritization of instructions and selective protection during compilation. To enable this, we develop statistical models for estimating error masking and propagation probabilities. Our scheme provides significant improvement in reliability efficiency (avg. 30%-60%) compared to state-of-the-art program-level protection schemes.

I. INTRODUCTION AND RELATED WORK

Reliability has become a major design objective for on-chip systems due to advanced technology scaling [1]. Several hardware level (TMR, pipeline protection, etc. [3]) and software level (SWIFT-R [6], EDDI [7], in-register duplication [8], etc.) schemes have emerged. These schemes principally rely on full-scale duplication either at logic or instruction level, thus they incur significant overhead. Hence, trends have been set for adaptive and selective protection, i.e. applying reliability to more susceptible logic blocks [4][5] or program code [12] to obtain cost-efficient reliable designs. Hardware-level adaptive techniques target at reducing the power overhead by activating and deactivating the redundant hardware while monitoring the architectural vulnerability [2][4][5] or instruction vulnerability [19][30]. However, these schemes are still prohibitively expensive in terms of area and power for embedded computing. The area overhead may be reduced by using, for instance, reliable ultra-reduced instruction set co-processors to detect permanent faults [20][28].

In order to complement/alleviate hardware level schemes or to provide reliability in cases where hardware redundancy is prohibitive, *reliability-driven compilation with selective instruction protection* has emerged as a promising layer to improve the overall system reliability. Principally, these selective instruction protection schemes [11]-[17] identify so-called *critical instructions* that are more sensitive with respect to correct program execution *either* for a user-defined output data range *or* in order to avoid program crashes. The scheme of [11] performs fault injection to identify an instruction vulnerability factor (IVF) and protect instructions for IVF greater than a user-defined threshold. The scheme in [12] first partitions the program output in user-defined tolerable and non-tolerable sets for multimedia applications. Afterwards, it identifies critical instructions through static analysis of instruction dependencies in the program data flow graph. The scheme of [13] also targets multimedia applications and performs a static analysis on the program to identify instructions that affect the control flow, loop counters, etc. The schemes of [14][15] exploit inherent resilience of multimedia applications to reduce power consumption. The schemes of [16][17][29] perform transformations during the compilation or dynamic code compilation to reduce the probability of errors for improved soft error resilience. In summary, the above-mentioned state-

of-the-art schemes suffer from three main limitations:

1) They mainly target multimedia applications (which have a high degree of intrinsic error tolerance) but they do not efficiently protect non-multimedia applications.
2) They account only for instruction dependencies and ignore the probability of error masking at instruction level.
3) They protect all instructions that ultimately affect the program output but they do not provide measure for the relative importance of instructions w.r.t. reliability. Therefore, these schemes are hardly applicable in cases where constraints are applied to guide reliability.

A. Motivational Case Study on Program Reliability Analysis

Our reliability analysis experiments in Fig. 1 show that different instructions have varying vulnerabilities[1] to errors due to hardware-level faults. Even different programs have varying vulnerability distributions. This is due to their varying instruction profiles and vulnerable time they dwell in different pipeline components. Fig. 1 shows that the instruction vulnerabilities in "SusanC" are relatively higher compared to that in "ADPCM". "SusanC" has a relatively sparse distribution. Variations in instruction vulnerabilities and data/control flow properties of "SusanC" and "ADPCM" also hint towards their varying degree of instruction-level error masking and error propagation. This leads to a noticeable difference in program output errors of both applications: i.e. 5% output errors in "SusanC" vs. 45% in "ADPCM" at 5faults/MCycle (using Monte Carlo simulations; see details of experimental setup in Supplementary Section **S3**). This shows that despite low instruction vulnerabilities in "ADPCM", there is a significantly high propagation of unmasked errors to different execution paths and program outputs that leads to a higher susceptibility to program errors.

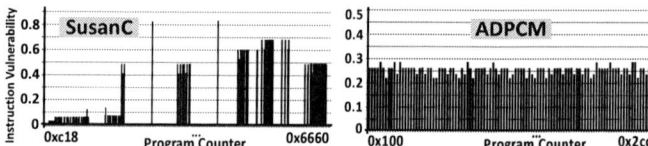

Fig. 1 Distribution of Instruction Vulnerability Index [10] for Different Instructions in "SusanC" and "ADPCM"

Summary and conclusion from observations: The above analysis illustrates that ignoring inherent error masking properties and only accounting for instruction vulnerability or instruction dependencies may lead to *inefficient protection*. Exploiting error masking and propagation properties provide a strong potential for efficient program reliability optimization under performance constraints. Furthermore, *not all* parts of a given program may require the same level of protection due to their diverse data/control flow properties. The error propagation and masking properties may be leveraged to limit the growing overhead of reliability optimization, while utilizing the tolerable overhead for protecting the sensitive parts of a program.

B. Problem and Research Challenges

The problem of constrained program reliability optimization poses the following **key research challenges**:

1) Choosing a set of instructions for program level protection while accounting for program level error masking and propagation properties under a given performance overhead budget.

Permission to make digital or hard copies of all or part of this work for personal or classroom use is granted without fee provided that copies are not made or distributed for profit or commercial advantage and that copies bear this notice and the full citation on the first page. To copy otherwise, to republish, to post on servers or to redistribute to lists, requires prior specific permission and/or a fee.

DAC '13, May 29 - June 07 2013, Austin, TX, USA.

---

[1] Computed using the model of [10], see Section **S2** for further details.

2) To enable constrained program optimization, it is important to accurately model and estimate the error masking and propagation probabilities at instruction level, such that instructions can be prioritized w.r.t. their impact on reliability.

C. Our Novel Contributions and Concept Overview

To address the above-mentioned challenges, a novel scheme for constrained program reliability optimization is proposed that employs:

1) **A Selective Instruction Protection Algorithm (Section IV)** that chooses a set of instructions for program-level protection considering their reliability-wise higher/lower impact on the program output, (i.e. jointly accounting for instruction vulnerability, masking probabilities, and error propagation probabilities) under a given performance overhead budget. To enable this, our scheme requires:

2) **Modeling and Estimation of Instruction-Level Error Masking Index (Section III.A)** that jointly accounts for masking effects due to program data/control flow properties and microarchitectural effects.

3) **Modeling and Estimation of Instruction-Level Error Propagation Index (Section III.B)** as a joint function of non-masking probability of all the successor instructions and all the possible execution paths along with their path execution probabilities.

Fig. 2 illustrates an overview of our novel scheme that is integrated in a reliability-aware compiler.

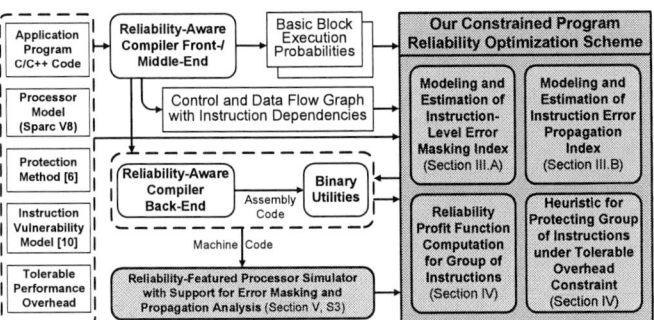

Fig. 2 Our Constrained Program Reliability Optimization Scheme (in highlighted boxes) Integrated into a Reliability-Aware Compiler

II. SYSTEM MODELS

Application Model: A program function $\mathbf{F}$ is a set of basic blocks $\mathbf{B}=\{B_1,...,B_m\}$ each having a control flow probability P_{CF}. $B_i=\{\mathbf{I}_i, f_i, l_i\}$. f_i and l_i denote the basic block number of executions and execution time, respectively. $\mathbf{I}_i$ denotes the set of instructions in the basic block, such that the complete function can be represented as a graph $\mathbf{G=(V,E)}$. $\mathbf{V}$ is the set of instruction nodes, s.t. $\mathbf{V}=\{I_1,...,I_n\}$. L_G defines a set of leaf nodes of $\mathbf{G}$. Each instruction node is given as a tuple $I=(l,o,S,P,\xi,IMI,\phi)$. l and o denote the instruction latency and set of operands, respectively. S and P denote the set of successor and predecessor instructions, respectively. ξ, IMI, and ϕ denote the instruction error propagation index, instruction masking index, and instruction vulnerability index, respectively. For estimation of the instruction vulnerability index ϕ, we employ state-of-the-art technique of [10] (see details in supplementary Section **S2**). $\mathbf{E}$ is the set of edges that denote dependencies between instructions, s.t. $\mathbf{E}=\{e_{xy}|(I_x,I_y)\in\mathbf{V}\}$. The weight of the edge represents the latency of moving from instruction I_x to I_y. We define the term *instruction path* p in the graph G as a sub-graph with a sequential set of instructions, such that each instruction in p has exactly one predecessor and successor, i.e. $\forall i \in p | i \notin Leafs \ |i.S| = 1 \wedge |i.P| = 1$. Note that an instruction may appear in different paths, as it has several successors. We define the term *execution path* ep in the control flow as a set of instructions that execute with a conditional probability of $P_{CF}(ep|I)$, given an instruction I executes.

Processor and Fault Model: in-order RISC single core processor with multiple *pipeline* stages subjected to "transient faults" with single or multiple bit upsets.

III. MODELING AND ESTIMATION OF PROGRAM-LEVEL ERROR MASKING AND ERROR PROPAGATION PROPERTIES

In the following, we model the error masking and error propagation properties at instruction level and identify different parameters that affect error masking/propagation in software programs. These models jointly provide a measure of severity of an error at a given instruction, thus can be used to prioritize instructions for constrained reliability optimization. Fig. 3 provides an overview and flow of steps to estimate the error masking probabilities and the error propagation index (for the ease of reader, we have provided the complete set of parameters in Table I, Supplementary Section **S1.D**).

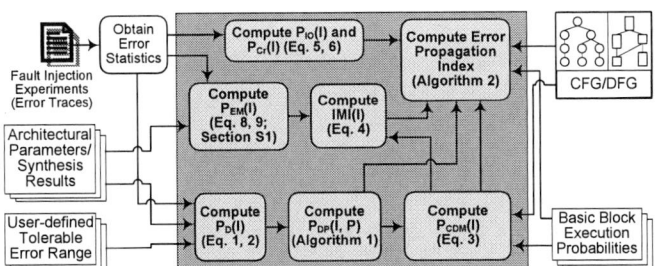

Fig. 3 Flow of Steps to Compute the Instruction-Level Masking Probabilities and Error Propagation Index

A. Program-Level Error Masking

Definition: we quantify program-level error masking as the *Instruction Masking Index* (IMI(I)) of an instruction I, which is defined as the total probability of an error at instruction I being masked until the last instruction of all of its instruction paths p (i.e. leaf nodes), such that the output of p is correct.

Parameter Identification: The masking of an error at an instruction I can occur due to the following factors that constitute the parameters of the IMI(I).

1) *Error masking due to data flow properties* ($P_{DP}(I,p)$): The error at instruction I may be masked due to the successor instruction in path p depending upon two factors: (a) instruction type, and (b) value of the operand variables.

2) *Error masking due to the changing control flow properties* ($P_{CDM}(I)$), where a highly probable execution path may exhibit masking instructions that block the propagation of the error to the relevant program output. It may also happen that a highly probable execution path does not even use the erroneous value from the preceding basic block. Example: Fig. 4(a) illustrates an example of control flow graph showing different basic blocks and the control flow probabilities. The error in basic block B_1 may be blocked in B_3 due to the "&" and/or "or" instructions (see Fig. 4(b)). However, if the control flow follows B_2, the error will propagate to B_4 and ultimately to the visible program output. Note that B_3 has a higher probability of execution compared to B_2. Fig. 4(b) shows that the error masking occurs only in case of the "&" and "or" instructions due to the value of the other operand, while "+" does not mask the error.

3) *Masking in pipeline components during the execution of instruction I* ($P_{EM}(I)$) due to microarchitecture-level logical masking effects, i.e. the error within a pipeline component (combinatorial logic) is not visible at the output latch as the error propagation through different logic elements/gates is blocked due to subsequent logic element(s) and the output of the pipeline component remains correct (see an example in Section **S1.A**).

From software program's perspective the most important challenge is to estimate the masking probabilities $P_{DP}(I,p)$ and $P_{CDM}(I)$ that represent the novel contributions of this paper and are discussed in the following. $P_{EM}(I)$ is estimated using fault injection in different pipeline components (see details in Section **S1.B** and **S3**).

111

Estimation of $P_{DP}(I,p)$: For each instruction I, the masking probability depending upon the data flow can be modeled through Eq. 1.

$$P_D(I) = \sum_{x \in I.O.Bits} P_D(x,I) \times P_e(x) \qquad (1)$$

O is the set of operands of instruction I with a set of Bits. $P_D(x,I)$ is the probability of masking of each bit depending upon the instruction type (as discussed in Fig. 4(b)). $P_e(x)$ is the error probability of each bit, which can be simplified to $1/N_{Bits}$ by assuming the same error probability for all bits, where N_{Bits} is the bit-width of operand registers. In case, the user specifies a tolerable range th for the error in the output value ($log_2(th)$ provides the number of bits), Eq. 1 can be modified as Eq. 2.

$$P_D(I) = \sum_{x \in I.O.Bits \setminus log_2(th)} P_D(x,I) \times P_e(x) + log_2(th)/N_{Bits} \qquad (2)$$

In a given instruction path p, the error masking depends upon the individual instructions as well as the combination of consecutive instructions. Note, the scheme of [21] does not account for combined masking effects of consecutive instructions, thus may provide inaccurate estimates for the masking probability. Depending upon their masking behavior, instructions can be classified into two categories:

Type A: instructions like "&" and "or" with a variable value assuming a random bit masking; the theoretical masking probability for such instructions is given as ½;

Type B: instructions like "shift" left or right by a constant value; where the masking bits can be inferred from the source code; these instructions affect the computation of masking probabilities of the predecessor instruction. In case the joint masking effects of consecutive instructions are ignored, the total masking probabilities are under-estimated (see an example in Section **S1.C**).

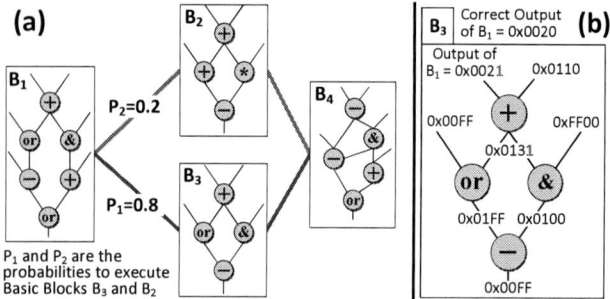

Fig. 4 An Example Control & Data Flow Graph showing the Error Masking due to Successor Instructions and Changing Control Flow

Algorithm 1 illustrates the pseudo-code for computing the error masking probability $P_{DP}(I,p)$ for instruction I of path p in an iterative manner starting from the leaf nodes. First, for all instructions in G, $P_D(I)$ is computed using Eqs. 1-2 (line 1). Afterwards, $P_{DP}(I,p)$ is initialized for all the leaf nodes with their corresponding $P_D(I)$, as they represent the last nodes of the graph (line 2). The list of ready nodes (i.e. instructions for which $P_{DP}(I,p)$ can be computed) L is initialized with the predecessors of leaf nodes (lines 3). The while loop (lines 4-19) to compute the $P_{DP}(I,p)$ iterates till the list is empty. For each iteration of the loop, first, the set of all possible instruction paths Paths is generated from every instruction I in L until the leaf nodes L_G (line 6). Afterwards, for all possible paths of every instruction, i.e. all possible combinations of (I,p), the number of consecutive instructions of *type B* N_B is computed (lines 9-12) and the masking probabilities $P_{DP}(I,p)$ are computed (lines 13-16). If there are 2 or more consecutive instructions of *type B*, their cumulative masking probabilities are considered to compute $P_{DP}(I,p)$ (lines 14-16); otherwise, the independent masking probabilities are considered (line 13). After computing $P_{DP}(I,p)$, the instruction I is removed from the list and its predecessors are added to the list (line 18). The loop of instructions is iterated until the list is empty (lines 18, 19). Fig. 9 in Section **S1.C** illustrates step-by-step Algorithm 1 with the help of an example instruction graph.

Algorithm 1: Computing the Error Masking Probability $P_{DP}(I, p)$

INPUT: Instruction flow graph G=(V,E), Set of leaf nodes L_G, Set of predecessors and successors for each instruction (P, S).
OUTPUT : $P_{DP}(I, p)$ – Masking probabilities due to data flow for each instruction I for path p.
BEGIN
1. $\forall I \in G$ $P_D(I) \leftarrow compute P_D(I)$;   *//Eq. 1; Eq. 2*
2. $\forall I \in L_G(G)$ $P_{DP}(I,p) \leftarrow P_D(I)$;
3. List $L()$; $\forall x \in L_G.P$ $L.add(x)$; *// List of ready nodes*
4. *while*(! $L.isEmpty()$){
5. $\forall I \in L${
6. $I.Paths \leftarrow generatePaths(I)$; *// generate all instruction paths*
7. $\forall p \in I.Paths${
8. $N_B \leftarrow 0$; $p' \leftarrow p$;
9. $\forall x \in p'${
10.   *if*($x = typeB$){ $N_B \leftarrow N_B+1$; $p'.remove(x)$; }
11.   *else* { $N_B \leftarrow 0$; $p'.remove(x)$; };
12.   }
13.   *if*($N_B \leq 1$) $P_{DP}(I,p) \leftarrow P_D(I)+(1-P_D(I)) \times P_D(I.S))$;
14.   *else*{ *// add masking probabilities of* N_B *consecutive successor instrs.*
15. $P_D'(I) \leftarrow \sum_{x=1 \ to \ (1+N_B)} P_D(x)$;
16. $P_{DP}(I,p) \leftarrow P_D'(I)+(1-P_D'(I)) \times P_D(I.S))$; }
17.   } *// end loop of paths*
18. $L.remove(I)$; $\forall ip \in I.P$ $L.add(ip)$;   }
19. } *// end While loop when the List of nodes is empty.*
20. **END**

Estimation of $P_{CDM}(I)$: For each instruction I, the error masking probability due to both data and control flow $P_{CDM}(I)$ can be modeled as the weighted masking probability due to data flow $P_{DP}(s,p)$ for all the *successor* instructions s of instruction I and in all the corresponding instruction paths s.Paths while considering the execution path probabilities $P_{CF}(ep|I)$; see Eq. 3.

$$P_{CDM}(I) = \sum_{\forall ep|I \in ep} \left(P_{CF}(ep|I) \times \prod_{\forall s \in I.S; \forall p \in s.Paths} P_{DP}(s,p) \right) \qquad (3)$$

The execution path probabilities $P_{CF}(ep|I)$ are estimated using the GCC framework with option "*-fguess-branch-prob*".

Estimation of Instruction Masking Index ($IMI(I)$): we model the *Instruction Masking Index* ($IMI(I)$) using Eq. 4. $P_{CDM}(I)$ and $IMI(I)$ are computed using the *breadth-first search starting from the leaf nodes* and explores G in a bottom-up fashion.

$$IMI(I) = P_{EM}(I) + (1-P_{EM}(I)) \times P_{CDM}(I) \qquad (4)$$

Non-masked errors may propagate to subsequent instructions in different execution paths, thus corrupting the correct program output. Error propagation to multiple execution paths denote a high susceptibility of program errors. Therefore, we *leverage the error masking probabilities to model the program-level error propagation to quantify the severity of the unmasked errors.*

B. Program-Level Error-Propagation

Definition: we quantify the program-level error propagation effects as the *Instruction Error Propagation Index* $\xi(I)$, which is defined as the product of the non-masking probability (i.e. the probability that an error is *visible* at the program output) of all the *successor* instructions of a given instruction I for all possible instruction paths.

Modeling and Estimation of Instruction Error Propagation Index ($\xi(I)$): Algorithm 2 shows the procedure for estimating ξ for each instruction in graph G. First the error propagation index for all the leaf nodes (outputs) is initialized with 1, as the errors in the leaf nodes are considered as propagated to the next stages of the program execution (line 1). Afterwards, a list C is created that contains the traversed instructions for which ξ is computed (line 2). The algorithm uses a FIFO based queue Q to store the instructions for which ξ can be computed considering all successors are completed (line 2). Initially the predecessors of the leaf nodes are inserted in Q (lines 3). $\xi(I)$ is computed for all the instructions whose successors are in the C list (lines 4-14). Otherwise, the instruction I is inserted back into the queue Q

Algorithm 2: Computing the Instruction Error Propagation Index

INPUT: Instruction flow graph $G=(V,E)$, Set of leaf nodes L_G, masking probabilities due to data $P_{DP}(I,p)$, control flow probabilities $P_{CF}(p|I)$.

OUTPUT: $\xi(I)$ – Set of error propagation index for each instruction

BEGIN

1. $\forall I \in L_G(G) \quad \xi(I) \leftarrow 1;$
2. *List* $C(L_G); \qquad$ *Queue* $Q();$
3. $\forall I \in L_G \quad \{ \quad \forall i \in I.P \quad Q.Enqueue(i); \quad \}$
4. *while*$(!Q.isEmpty())\{$
5. $\quad I \leftarrow Q.Dequeue();$
6. $\quad if (\forall s \in I.S \quad s \in C)\{$
7. $\quad\quad \xi \leftarrow 0;$
8. $\quad\quad \forall s \in I.S\{$
9. $\quad\quad\quad if (IMI(I)=0) \quad \xi \leftarrow \xi + \xi(s) \times P_{Execution}(s|I);$
10. $\quad\quad\quad else \quad \xi \leftarrow \xi + \sum_{\forall p \in s.Paths} ((1 - P_{DP}(s,p)) \times P_{CF}(p|I) \times \xi(L_G(p)));$
11. $\quad\quad \} \; // \; end \; loop \; of \; successors$
12. $\quad\quad \xi(I) \leftarrow \xi \times (P_{IO}(I)/(P_{IO}(I)+P_{Cr}(I))); \quad C.add(I); \quad \forall i \in I.P \; Q.Enqueue(i);$
13. $\quad \} \; else \; Q.Enqueue(I);$
14. $\} \quad // \; end \; while \; loop \; when \; the \; Queue \; of \; instructions \; is \; empty$
15. **END**

(line 13). For $\xi(I)$ computation for an instruction I, all of its successors instructions $I.S$ are considered (line 6-11). In case, a successor instruction s is a non-masking instruction, its $\xi(s)$ is directly used in the ξ computation along with its corresponding execution probability $P_{Execution}(s|I)$ (line 9). Otherwise, ξ is computed as the weighted sum of non-masking probability $(1-P_{DP}(s,p))$ multiplied with ξ of the leaf nodes of the path p, for all the successors $I.S$ and their instruction paths $s.Paths$ (line 10). Note that the weight is determined using the path probability $P_{CF}(p|I)$. An error in an instruction may lead to an "incorrect output" or "program failure" (like crash, halt, abort). Therefore, $\xi(I)$ accounts for the probability of crash $P_{Cr}(I)$ (see Eq. 5) and probability of incorrect output $P_{IO}(I)$ (see Eq. 6); line 12. After computing the $\xi(I)$ for instruction I, all of its predecessors $I.P$ are added into the queue Q (line 12).

$$P_{Cr}(I) = \sum_{b \in COB(I)} P_{eOP}(b,I) + (1-IMI(I)) \times \sum_{b \in CAB(I)} P_{eAd}(b,I); \quad (5)$$

$$P_{IO}(I) = 1 - IMI(I) - P_{Cr}(I) + IMI(I) \times P_{Cr}(I); \quad (6)$$

$COB(I)$ is the number of critical opcode bits that lead to a "non-decodable instruction" error. $P_{eOP}(b,I)$ and $P_{eAd}(b,I)$ are the error probabilities in the opcode and address bits. $CAB(I)$ is the number of critical address bits that lead to a "memory segmentation" error due to an access to the invalid or restricted memory region.

The above-discussed error propagation index ($\xi(I)$) captures the error masking and propagation properties of the software program, which are crucial for prioritizing the instructions in basic blocks with respect to reliability. Therefore, our instruction-level error masking and propagation models ($IMI(I)$ and $\xi(I)$) *enable constrained software reliability optimization on unreliable hardware* that facilitates tradeoff between performance loss and reliability improvement.

IV. OUR CONSTRAINED INSTRUCTION PROTECTION SCHEME

We propose a heuristic that selectively protects instructions or group of consecutive instructions in different execution paths in a given function. For this, our scheme leverages the error propagation index ξ, and instruction vulnerabilities ϕ to compute the instruction reliability profit.

Reliability Profit Function (RPF) for Choosing Instructions for Protection: Our instruction protection heuristic employs a *reliability profit function* (RPF, Eq. 7) which is defined as the accumulated *reliability efficiency* of a group of instructions g, such that the reliability efficiency is given as the product of error propagation index ξ and instruction vulnerability ϕ divided by the protection overhead ω. The overhead of the instruction group g is computed using Eq. 7; csi is the set of consecutive instructions and ci is the checking instruction inserted only at the end of g or at the point of multiple outputs.

$$RPF = (\xi(g) \times \phi(g)) / \omega_g; \quad s.t., \quad \omega_g = \sum_{\forall csi \in g} t_{csi} + \sum_{\forall ci \in g} t_{ci} \quad (7)$$

Example: Fig. 5 shows an excerpt from an example instruction graph with 13 instructions and the effect of different parameters in the optimization goal on the efficiency of selective instruction protection under a tolerable overhead budget of 20 cycles. The table in Fig. 5 provides ϕ, ξ, and ω of each instruction I.

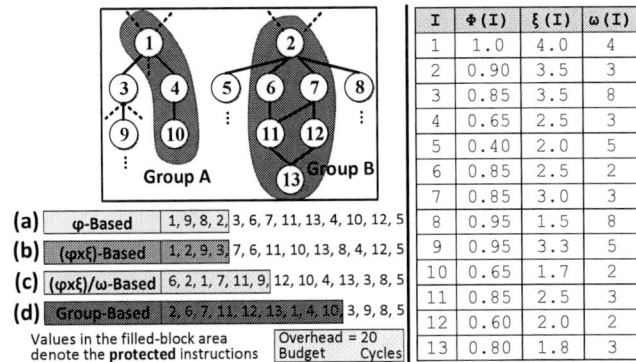

I	Φ(I)	ξ(I)	ω(I)
1	1.0	4.0	4
2	0.90	3.5	3
3	0.85	3.5	8
4	0.65	2.5	3
5	0.40	2.0	5
6	0.85	2.5	2
7	0.85	3.0	3
8	0.95	1.5	8
9	0.95	3.3	5
10	0.65	1.7	2
11	0.85	2.5	3
12	0.60	2.0	2
13	0.80	1.8	3

(a) φ-Based — 1, 9, 8, 2, | 3, 6, 7, 11, 13, 4, 10, 12, 5
(b) (φ×ξ)-Based — 1, 2, 9, 3, | 7, 6, 11, 10, 13, 8, 4, 12, 5
(c) (φ×ξ)/ω-Based — 6, 2, 1, 7, 11, 9, | 12, 10, 4, 13, 3, 8, 5
(d) Group-Based — 2, 6, 7, 11, 12, 13, 1, 4, 10, | 3, 9, 8, 5

Values in the filled-block area denote the **protected** instructions. Overhead = 20 Budget Cycles

Fig. 5 An Example Showing the Effect of Different Parameters on the Reliability Efficiency of the Selective Instruction Protection

a) *φ-based Selection:* Fig. 5(a) illustrates that when considering only instruction vulnerability ϕ, only four instructions are chosen for protection. There may be cases, where an instruction's vulnerability to error is quite high, but the probability that this error will be masked until the visible output is also high.

b) *(φ×ξ)-based Selection:* Fig. 5(b) illustrates that when jointly accounting for ϕ and ξ, instruction 3 is chosen instead of instruction 8. However, still a total of 4 instructions are protected due to the high protection overhead of instruction 3 and 9. In such scenarios, it might be beneficial to choose several instructions with a slightly lower $(\phi \times \xi)$ profit and low protection overhead, rather than protecting only few instructions with high protection overhead. Note, depending upon the instruction types and protection scheme, the protection overhead may vary significantly for different instructions.

c) *Instruction Reliability Efficiency based Selection:* Fig. 5(c) shows that 6 instructions are protected, while the total reliability efficiency is 0.854 (computed using Eq. 7 and values of table in Fig. 5), which is 46% and 29% better compared to the case (a) having a reliability efficiency of 0.585 and case (b) having a reliability efficiency of 0.663, respectively. Note, the protection overhead depends upon the protection scheme. For instance, in case of simple software level error recovery like SWIFT-R [6], the voting instructions are inserted at the store instructions or leaf nodes of the groups. Therefore, protection overhead may be curtailed by computing the group reliability efficiency, i.e. cumulative reliability efficiency for a group of consecutive instructions.

d) *Group Reliability Efficiency based Selection:* Fig. 5(d) shows the marked regions in the graph as groups of protected instructions. Note, using group reliability efficiency, 9 instructions are protected, while the overall reliability efficiency is 0.966, which is 13% better compared to that of the case (c).

Constrained Protection Algorithm: Algorithm 3 shows the pseudo-code of our selective instruction protection heuristic. The *inputs* are: unprotected function F, user-provided tolerable performance overhead budget β_F in cycles, set of instruction vulnerabilities and error propagation indexes (ϕ, ξ), and a user-defined reliability method R. The *output* is the protected function F'. The *optimization goal* is to choose individual or groups of instructions while maximizing the total reliability profit function. Our scheme operates on the instruction graph G. First, the reliability profit function for each instruction is individually computed and inserted into a list that is then sorted in descending order (lines 1-3). The *while* loop iterates until the list is

113

empty or the budget β_F is exhausted. Since the generation of all groups of all instructions leads to a significant complexity, the heuristic starts with protecting individual instructions (lines 5-7) and incrementally builds groups of instructions for protection considering the predecessor and successors of the protected instructions (lines 8-17). For each predecessor and successor instruction of the protected instruction I, all possible instruction groups GI are generated (line 9) and their combined protection overhead is computed (lines 10-16). If an instruction i in the group g has more successors and a successor s do not belong to the group g, a *check* instruction is inserted at instruction i (lines 12-15). Afterwards, the reliability profit function is computed for all the groups (line 16). The list of instruction is re-sorted such that the instructions of group g_{Best} with the highest reliability profit RPF_{Best} (Eq. 7) appears first in the list (line 17), which is later evaluated for protection in the subsequent iteration of the loop (lines 6-7). For each protected instruction or group of instructions, the overhead budget β_F is updated accordingly (line 7).

Note, in this work, we assume that the control flow is protected using standard techniques like basic block signature [9]. This work employs SWIFT-R [6] as a protection scheme. However, our proposed models for error propagation and masking, and our scheme for constrained program reliability optimization is equally beneficial for selective applicability of other protection schemes and orthogonal to improvements in such program-level recovery schemes.

V. RESULTS AND EVALUATION

A. Experimental Setup and Tool Flow

The proposed error masking and propagation models along with our constrained program reliability optimization scheme are integrated in a GCC-based compiler. For evaluation, we have employed a reliability-featured ISS with an integrated fault injection engine. Numerous fault injection campaigns have been carried out at three different fault rates (1, 5, 10 faults/MCycles). Parameters to the fault injection module are particle strike rate (obtained from the flux calculator [23]), coordinates of the location, processor layout and frequency (in this work, a Leon-II processor [24] @ 100MHz is used), etc. An extensive analysis of errors, vulnerabilities, and reliability efficiency has been performed at the software program layer. Results of fault injection campaigns, vulnerability analysis, and program data/control flow are used to build and verify the models for error masking and propagation in MATLAB. Section **S3** provides details on the tool flow, experimental setup, and fault injection configurations.

For evaluation, we use different applications: "ADPCM", "SUSAN" from MiBench [25], and kernels from a complex "H.264 video encoder" [18], namely "SAD" for motion estimation and coding mode decision; and "DCT" for transformation. The selected applications vary significantly in terms of their instruction vulnerabilities, error masking index, and error propagation properties due to distinct control and data flow. All the applications are compiled with –O3 option.

B. Comparison to State-of-the-Art

We compare our constrained program reliability optimization scheme with state-of-the-art program-level protection schemes [12][11][6] (Fig. 7 b, c, d) and the baseline unprotected case (Fig. 7d) for varying number of tolerable performance overhead cases for different applications. Fig. 7 illustrates the reliability efficiency improvement of our scheme over state-of-the-art.

Fairness of Comparison: for fairness, we have provided the same fault scenarios, same compiler options, thus same application binaries, control and data flow graph, same input data, and same instruction protection method (i.e. instruction level TMR). The results solely represent the effect due to difference in the protection cost function and selection scheme.

Overall Comparison with all State-of-the-Art Schemes: First, we discuss experimental observations that are common in all comparisons. For "Susan" application, our scheme obtains a very high reliability

---

Algorithm 3: Pseudo-Code of Our Selective Instruction Protection

INPUT: Unprotected Function F from the software program as G= (V,E), User-provided tolerable performance overhead in cycles β_F; Set of instruction vulnerabilities ϕ, Set of error propagation indexes for all instructions ξ, User-provided program reliability method R (like SWIFT-R [6]), etc.

OUTPUT: F' – Function with selective instruction protection

BEGIN
1. *List L;*
2. $\forall I \in G \quad RPF(I) \leftarrow (\xi(I) \times \phi(I)) / \omega(I);$
3. *Sort*$(L, RPF, DescendingOrder);$
4. *while*$(!L.Empty() || (\beta_F > 0))\{$
5. $I \leftarrow L.Pull();$
6.   *if*$(\omega(I) \le \beta_F)\{$
7.   *Protect*$(I); \quad \beta_F \leftarrow \beta_F - \omega(I);$
8. $\forall i \in (I.S \vee I.P)\{$
9. $GI = generateGroups(I,i); // groups of consecutive instructions$
10. $\forall g \in GI\{ // compute overhead of instruction groups$
11. $g' \leftarrow g;$
12. $\forall i \in g\{$
13. $if((|i.S| > I) \& (\exists_{s \in i}.inGroup(s,g') = False)) \, setCheckInstrPt(i,g');$
14. $setCheckInstructionPt(Leaf(g'), g'); \quad \}$
15. $\omega(g') \leftarrow getOverhead(g', R); // see Eq. 7$
16. $\} \; RPF(i) \leftarrow (\xi(i) \times \phi(i)) / \omega(i,g');$
17. $\} \; Sort(L, RPF, DescendingOrder);$
18. $\}$
19. $\} // end While loop of Budget finished or all instructions protected$
END

---

improvement (avg. 30%-60%) over all comparison partners because "Susan" exhibits instructions with high error propagation index having varying distribution (see Fig. 6) and at the same time high IMI and high vulnerability. Additional results and discussion on instruction masking and error propagation is given in Supplementary Section **S4**. Due to the joint consideration of error propagation/masking probabilities and vulnerability in the cost function (Eq. 7), our scheme stays superior compared to all schemes and achieves a reliability efficiency improvement of up to 60%-99% (avg. 30%-60%). Fig. 7 shows that for 50% tolerable overhead, the reliability of "Susan" reaches close to 100%, because all the important non-masking instructions are protected within this budget, while masking instructions are left un-protected as errors during these instructions do not affect the correct program output. This illustrates the benefit of our scheme since it accounts for error masking and propagation probabilities in the protection cost function.

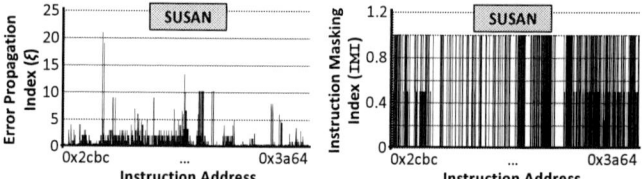

Fig. 6 Distribution of Error Masking and Propagation Indexes

The improvements are also high in case of "ADPCM" for overhead cases of 30% and higher (avg. 30%-40% reliability efficiency improvement). Below 20% the efficiency is low, as important instructions require more overhead than the tolerable overhead due to their high execution frequency. The improvements in "SAD" are also noticeable (up to 45% and avg. 10%-30% reliability efficiency improvement). However, the improvements for "DCT" are relatively low, due to limited masking probability of instructions and dependency on the earlier instructions of the algorithm (avg. 8%-10% reliability efficiency improvement). For low overhead cases, in several cases (like in "DCT", "SAD", and rarely in "ADPCM"), the savings of our scheme are below 10%. This is because of the fact that important instructions (that typically occur in loops) have many executions and their required protection overhead cannot be fulfilled under a cap of 5% or 10% tolerable overhead.

Now, we discuss specific observations for different comparison cases.

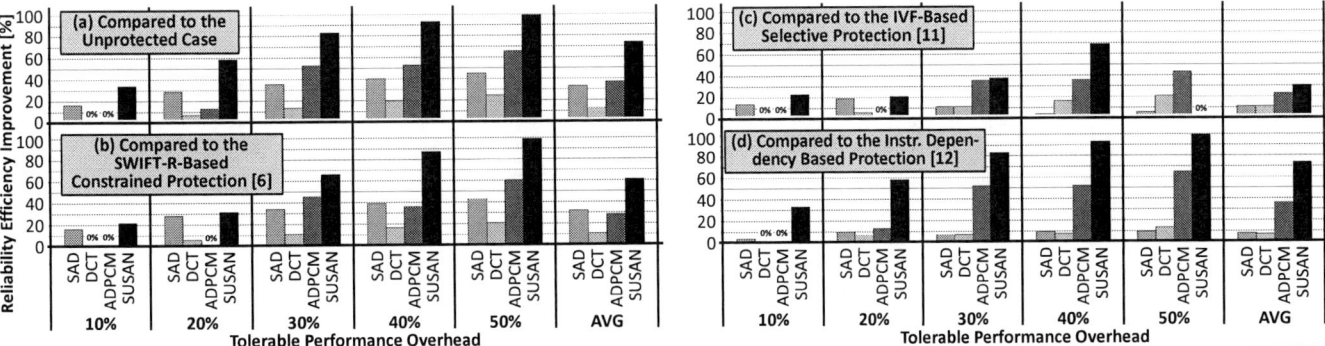

Fig. 7 Comparing the Reliability Efficiency Improvement of our Scheme over (a) Unprotected Case; and 3 State-of-the-Art (b) SWIFT-R [6] under Constraint; (c) IVF-Based Selective Protection [11]; and (d) Instruction Dependency based Selective Protection [12]

Comparing to the Unprotected Case: Comparison with the unprotected case shows the best possible reliability efficiency improvement of our scheme. Fig. 7a shows that our scheme provides up to 25%-99% and average 30%-70% improvement in the reliability efficiency compared to the unprotected case for different applications.

Comparing to the Scheme of SWIFT-R [6]: SWIFT-R [6] is the most prominent program-level instruction protection scheme that employs TMR with majority voting for protecting all instructions. Compared to original SWIFT-R (which is unconstrained to performance overhead), our scheme achieves >3x better reliability efficiency, since the overhead of SWIFT-R is >5-6x. For fairness of comparison, we have modified the SWIFT-R implementation towards constrained optimization, such that the overhead constraint is used to determine the number of instructions that can be protected. Afterwards, the instructions are selected for protection in a sequential manner, i.e. first execute, first protect. Fig. 7b shows that our scheme provides up to 20%-97% and average 10%-60% improvement in the reliability efficiency compared to the constrained SWIFT-R variant.

Comparing to the IVF-based Selective Protection Scheme [11]: this scheme computes the instruction vulnerability and protects the instruction with the highest vulnerability factor first. However, this scheme does not account for the error propagation properties and instruction dependencies. Therefore, this scheme works well only for the cases where vulnerability is dominant and error propagation is very low with smooth distribution. In contrast, our scheme accounts for both vulnerability and error propagation. As a result, our scheme provides up to 20%-70% (avg. 10%-30%) improvement in reliability efficiency, compared to the scheme of [11].

Comparing to the Instruction Dependency based Selective Protection Scheme [12]: this scheme prioritizes and protects the instructions with the highest number of dependent instructions. However, this scheme ignores the error masking probabilities (which is an important parameter to be considered in the error propagation) and instruction vulnerabilities. As a result, in some cases (like "Susan") the scheme of [12] provides significantly less protection. However, in cases where error propagation is crucial and dominant over vulnerability (i.e. "DCT"), this scheme provides good reliability. In contrast, our scheme provides high reliability efficiency in all cases, as it jointly accounts for error propagation and masking probabilities, vulnerabilities, and overhead of different instructions individually or jointly in a group. Our scheme thereby achieves up to 12%-99% (avg. 7.5%-80%) improved reliability efficiency compared to the scheme of [12], see Fig. 7d.

VI. CONCLUSIONS

We presented a constrained program reliability optimization scheme that exploits the program level error masking and propagation along with instruction vulnerabilities to selectively protect critical parts of the program for a given tolerable performance overhead. Our scheme is integrated in a reliability-driven compiler and exploits the data and control flow properties to estimate the error masking and propagation indexes. Our scheme provides significant improvement in reliability

efficiency (avg. 30%-60%) compared to state-of-the-art protection schemes. Due to the novel enhancements in Section III and IV, state-of-the-art selective program protection techniques cannot reach the level of program reliability efficiency of our scheme. Our proposed models enable a whole new range of program-level or program-guided hardware-level constrained reliability optimizations.

ACKNOWLEDGMENT

This work is supported in parts by the German Research Foundation (DFG) as part of the priority program "Dependable Embedded Systems" (SPP 1500 - spp1500.itec.kit.edu).

REFERENCES

[1] J. Henkel et al., "Reliable On-Chip Systems in the Nano-Era: Lessons Learnt and Future Trends", IEEE DAC, 2013.

[2] S. S. Mukherjee, C. Weaver, J. Emer, S.K. Reinhardt, T.Austin, "A systematic methodology to compute the architectural vulnerability factors for a high-performance microprocessor", MICRO, pp. 29-40, 2003.

[3] D. Ernst et al., "Razor: circuit-level correction of timing errors for low-power operation," IEEE MICRO, vol. 24, no. 3, pp. 10-20, 2004.

[4] R. Vadlamani et al.,"Multicore soft error rate stabilization using adaptive dual modular redundancy", IEEE DATE, pp. 27-32, 2010.

[5] M. D. Powell, A. Biswas, S. Gupta, S. S. Mukherjee, "Architectural core salvaging in a multi-core processor for hard-error tolerance", International Symposium on Computer architecture (ISCA), pp. 93-104, 2009.

[6] G. A. Reis, J. Chang, D. I. August, "Automatic instruction-level software only recovery", IEEE MICRO, pp. 36–47, 2007.

[7] N. Oh et al., "Error detection by duplicated instructions in super-scalar processors", IEEE Transaction on Reliability, 51-1, pp. 63-75, 2002.

[8] J. Hu et al., "In-Register Duplication: Exploiting Narrow-Width Value for Improving Register File Reliability," DSN, pp. 281-290, 2006.

[9] E. Borin , C. Wang, Y. Wu, G. Araujo, "Software-Based Transparent and Comprehensive Control-Flow Error Detection", CGO, pp. 333-345, 2006.

[10] S. Rehman et al., "Reliable software for unreliable hardware: Embedded code generation aiming at reliability", Codess+ISSS, pp. 237-246, 2011.

[11] D. Borodin et al., "Protected Redundancy Overhead Reduction Using Instruction Vulnerability Factor," IEEE CF, pp. 319-326, 2010.

[12] J. Cong, K. Gururaj, "Assuring Application-Level Correctness Against Soft Errors", ICCAD, pp. 150-157,2011.

[13] A. Sundaram et al., "Efficient fault tolerance in multi-media applications through selective instruction replication", WREFT, pp. 339-346, 2008.

[14] M. Shafique et al., "Power-Efficient Error-Resiliency for H.264/AVC Context-Adaptive Variable Length Coding", DATE, pp. 697-702, 2012.

[15] M. A. Makhzan et al., "A low power JPEG2000 encoder with iterative and fault tolerant error concealment", IEEE TVLSI, vol. 17, no. 6, pp. 827-837, 2009.

[16] S. Rehman, M. Shafique, J. Henkel, "Instruction Scheduling for Reliability-Aware Compilation", IEEE DAC, pp. 1288-1296, 2012.

[17] J. Lee et al., "Dynamic Code Duplication with Vulnerability Awareness for Soft Error Detection on VLIW Architectures", ACM TACO, vol. 9, no. 4, article 48, 2013.

[18] M. Shafique, L. Bauer, J. Henkel, "Optimizing the H.264/AVC Video Encoder Application Structure for Reconfigurable and Application-Specific Platforms", JSPS, vol. 60, no. 2, pp. 183-210, 2010.

[19] T. Li et al., "CSER: HW/SW Configurable Soft-Error Resiliency for Application Specific Instruction-Set Processors", IEEE DATE, 2013.

[20] A. Rajendiran et al., "Reliable computing with ultra-reduced instruction set co-processors", IEEE DAC, pp. 697-702, 2012.

[21] S. Rehman et al., "Leveraging Variable Function Resilience for Selective Software Reliability on Unreliable Hardware", IEEE DATE, 2013.

ADDITIONAL REFERENCES

[22] S. Z. Shazli, M. B. Tahoori, "Obtaining Microprocessor Vulnerability Factor Using Formal Methods", DFTVS, 2008.

[23] Flux calculator: www.seutest.com/cgi-bin/FluxCalculator.cgi.

[24] J. Gaisler, "A portable and fault-tolerant microprocessor based on the SPARC v8 architecture", DSN, pp. 409-415, 2002.

[25] MiBench (http://www.eecs.umich.edu/mibench/).

[26] IBM® XIV®: http://publib.boulder.ibm.com/infocenter/ibmxiv/r2/index.jsp.

[27] AMD PhenomTM II Processor Product Data Sheet 2010.

[28] S. Ananthanarayan, S. Garg, H. D. Patel, "Low Cost Permanent Fault Detection Using Ultra-Reduced Instruction Set Co-Processors", IEEE DATE, 2013.

[29] S. Rehman et al., "RAISE: Reliability Aware Instruction SchEduling for Unreliable Hardware", IEEE ASP-DAC, pp.671-676, 2012.

[30] T. Li et al., "RASTER: Runtime Adaptive Spatial/Temporal Error Resiliency for Embedded Processors", IEEE DAC, 2013.

SUPPLEMENTARY MATERIAL

S1. Details on Modeling Program-Level Error Masking

A. Example –Error Masking in Pipeline Components

Fig. 8 illustrates an example of an instruction executing through different pipeline stages/components (PC) and corresponding masking probabilities $P_{EM}(I,PC)$, s.t., $PC=\{F,D,E,M,W\}$. An example of microarchitecture-level logical masking for an adder circuit (which is a part of the ALU) is shown in Fig. 8, where different error cases are denoted by red dots. It is shown in Fig. 8 (case-2) that an error at the "AND" gate is blocked by the subsequent "OR" gate. However, the error at the "XOR" gate is propagated to the output (Fig. 8, case-1). Note that only "OR" and "AND" gates may mask the error with a theoretical probability of 0.5, while any "XOR" gate does not mask an error. Moreover, the logical masking properties of a pipeline component depend upon its microarchitecture. Therefore, the logical masking properties of a carry-lookahead adder are different from a ripple-carry adder.

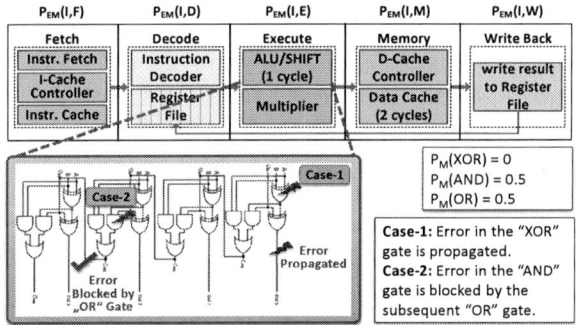

Fig. 8 Different Pipeline Stages Exhibit Distinct Masking During the Instruction Execution due to Combinatorial Logic

B. Parameter Estimation of Masking Probability due to Microarchitectural Effects ($P_{EM}(I)$)

The masking probability $P_{EM}(I)$ during the execution of an instruction I can be estimated as the cumulative probability of masking in different pipeline components PC it uses; see Eq. 8. Each pipeline component c can be seen as a set of connected logic elements LE (logic gates, latches, etc.) each having a certain masking probability $P_{EM}(c,l,t)$ and error probability $P_e(c,l,t)$. T is the total time spent in the pipeline in terms of cycles.

$$P_{EM}(I)=\sum_{\forall c \in PC; \forall l \in c.LE; \forall t \in T}\left(P_{EM}(c,l,t) \times P_e(c,l,t)\right) \quad (8)$$

We can simplify Eq. 8 by assuming similar error probability for each logic element of a given pipeline stage; see Eq. 9.

$$P_{EM}(I)=\sum_{\forall c \in PC; \forall t \in T}\left(P_{EM}(c,t)/(N_{LE}(c) \times t(c))\right) \quad (9)$$

$N_{LE}(c)$ is the number of logic elements in each pipeline component c obtained from the hardware synthesis; $t(c)$ is the time spent in each pipeline component, and $P_{EM}(c,t)$ is the logical masking of each component. Several methods have been researched by the hardware-level community to estimate $P_{EM}(c,t)$, ranging from analytical/statistical approaches (like EPP [22]) to fault injection. In this paper, we employ a fault injection based technique to estimate $P_{EM}(c,t)$. See Section **S3** for the details of experimental and fault injection setup.

C. Example –Effect of Joint Masking Probabilities of Consecutive Instructions

Fig. 9 illustrates an example showing the computation of error masking probabilities of different instruction in a given instruction graph, while showing the effect of consecutive instructions of type B on the total masking probability. It is shown in Fig. 9 that for instruction "1", the total masking probability is equal to "0.803" when considering the masking effects of consecutive instructions of type B (i.e. instructions "1" and "2"). If the masking effects of consecutive instructions are ignored, the total masking probability of instruction "1" is equal to "0.770", i.e. a difference of "0.033" compared to the earlier case. This shows that ignoring the effects of consecutive instructions may lead to an under-estimation of masking probabilities. Therefore, accurate estimation of instruction-level error masking needs to account for the joint error masking effects due to the combination of consecutive successor instructions of *type B* in the path p.

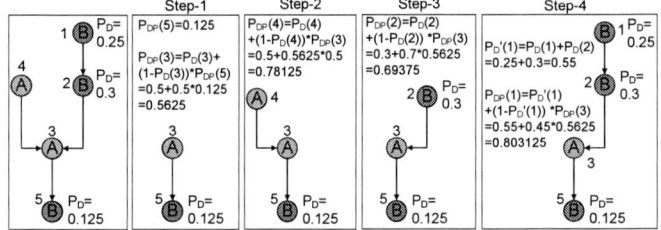

Fig. 9 An Example showing the Computation of Error Masking Probabilities illustrating the Effect of Consecutive Instructions of type B in the Path on the Total Masking Probability

D. Table of Model Parameters, Variables, and Notations

Table I provides the description of various model parameters, variables, and notations used in this paper. The table provides a summary, which is helpful in cross-referring from different algorithms and equations, in order to increase the readability and ease-of-understanding of the reader.

Table I: Description of Different Model Parameters and Variables

Parameter	Description
F	Program Function
I	Instruction
IMI(I)	Instruction masking index: The probability that an error in instruction I will not be seen at the output (any output)
p	Path of dependent instructions, such that each instruction has exactly one successor and one predecessor. Note that an instruction may appear in different paths, as it has several successors.
$P_{DP}(I,p)$	Error masking probability due to data flow properties for an instruction I in path p.
$P_{CDM}(I)$	Error masking probability for instruction I due to control and data flow properties. For instance, if an instruction is executed only in half of the executions, then its errors will be masked with a probability of 0.5.
$P_{EM}(I)$	Probability of masking in pipeline due to microarchitectural logical masking. An error appearing in an instruction may be masked due to the logic gates during the instruction execution.
$P_D(I)$	Probability of instruction I masking an error on the input due to data flow (operation type and operands).

O	Operands of an instruction
Bits	Bits of the operands of an instruction
$P_D(x,I)$	Probability that instruction I will mask an error in bit x
$P_e(x)$	Probability of an error in bit x
N_{Bits}	Number of operand bits
th	Threshold for tolerable output error, i.e. acceptable output value in presence of faults. Even if it is not exactly the correct output result, in some applications we may still have an acceptable result. The threshold is the accepted deviation from the correct result.
G	Graph of a program function given as a set of vertices and edges, where vertex denotes an instruction and edge denotes the dependency
L	List of nodes for which $P_{DP}(I,p)$ is computed
L_G	Set of dependencies from a node to the leaf (outputs)
N_B	Consecutive masking instructions of type B
$P_{DP}(s,p)$	Look $P_{DP}(I,p)$
$P_{CF}(ep|I)$	Probability of execution path ep if the instruction I is executed.
$\xi(I)$	Error propagation index of instruction I. It is a measure of how many outputs are likely to be affected by an error at I.
C	List of nodes for which $\xi(I)$ is computed
I.S	Successors of instruction I
s	A particular successor instruction
I.P	Predecessors of instruction I
$P_{Execution}(s|I)$	Probability of executing s if we know that I is executed
s.Paths	Dependency paths from s to leaf nodes
$P_{CF}(p|I)$	Probability of executing the a set of instructions that contains path p knowing that instruction I is executed
$P_{Cr}(I)$	Probability that an error in instruction I will generate a crash
$P_{IO}(I)$	Probability that an error in instruction I will generate an incorrect output
COB(I)	Number of critical opcode bits in I. Critical opcode bits means those bits that if wrong will generate a crash
$P_{eOP}(b,I)$	Probability of error in bit b of opcode of instruction I
CAB(I)	Number of critical address bits. An error in one of those will generate a crash when the address is accessed
$P_{eAd}(b,I)$	Probability of error in bit b of address used in instruction I
$\phi(I)$	Instruction Vulnerability Index for instruction I
$\omega(I)$	Overhead needed to protect instruction I
ω	Overhead to protect the whole function
p_{Best}	Group of instructions with the highest reliability profit function value
RPF_{Best}	The reliability profit function value corresponding to the best selected group of instructions
csi	Consecutive successor instructions
ci	Checking instructions added at the end of a set of protected instructions to verify the results
β_F	Tolerable performance overhead for protecting a function F
F'	Protected Function
R	Protection Method

S2. Overview of Instruction Vulnerability Estimation [10]

The instruction vulnerability index (ϕ) is a program-level reliability model that quantifies the program reliability at instruction [10]. The instruction vulnerability index is bridges the gap between hardware and software by quantifying the effects of hardware-level faults at the software program level. It defines the vulnerability of an instruction as the accumulated sum of vulnerabilities when an instruction executes in different pipeline components PC={F,D,E,M,W} each having a logic area A_C and probability of fault $P_E(C)$. The instruction vulnerability in a pipeline stage can be quantified as the product of *spatial*

vulnerability (i.e. vulnerable bits/logic area of each pipeline stage during the instruction execution) and *temporal vulnerability* (i.e. vulnerable residency period in different pipeline stages); see Eq. 10.

$$\phi(I)=\sum_{\forall C \in PC}\big(\phi(C,I)\times A_C \times P_E(C)\big)\Big/\sum_{\forall c \in PC}A_C \qquad (10)$$

The instruction vulnerability index is estimated by executing the application binary on a given processor architecture for a given input data and analyzing the vulnerable periods and bits in different pipeline stages. Note, that the vulnerable bits and period vary significantly depending upon the data and control flow properties. This fact is evident from the instruction vulnerability distributed shown in Fig. 1(a) in Section I.A.

S3. Details on the Experimental Setup and Fault Injection

Fig. 10 shows the tool flow and experimental setup. The masking and error propagation models are developed in MATLAB and the input for parameter evaluation is provided from the program reliability analysis which is performed on a reliability-featured processor simulator. The input to the modeling is the program error distribution, data and control flow graph, and application program execution traces. The developed models are then implemented in our reliability-driven compiler for estimation of error masking and propagation probabilities. These models are also integrated the reliability analysis module of the processor simulator for evaluation of error masking and propagation properties of application programs.

The program is compiled on a GCC-based compiler extended with program reliability features. In this paper, we have proposed error masking and propagation models along with a constrained program reliability protection scheme. These models and protection scheme are integrated in the compiler. The error masking and propagation operate on the data and control flow graph. The instructions are protected using TMR [6]. We have also implemented and integrated various state-of-the-art protection techniques for comparison purposes.

The compiled program is simulated and evaluated for reliability on a reliability-featured processor simulator. Our simulator features a basic processor instruction set simulator (a Leon-II based on SPARC-V8 architecture is simulated), an integrated fault generation and inject module, and an enhanced program reliability analysis module.

Fault Generation and Injection Module: the fault generation module is used to generate various fault scenarios, such that same scenarios can be provided to all the comparison partners and the experiments can be repeatable. Numerous fault scenarios are generated considering the following fault parameters.

1) *Particle Flux Rate:* The particle flux rate is determined by the neutron flux calculator [23], which takes city coordinates and altitude (at which the device is operating) as an input and provides the flux rate in particle strikes per mm^2 per sec. An example input is 59° 55' N, 10° 45' O, altitude ranging from 1- 20km (covering various use cases from terrestrial to the aerial).

2) *Processor Layout/Area:* This parameter specifies the size of the complete target device. The value of this parameter is typically in mm^2 size or logic gates and technology parameters from where the absolute area in mm^2 can be obtained, which is important to determine the surface area exposed to particle strikes.

3) *Vulnerable/Unprotected Fault Location:* A list of processor components is specified, where the faults are injected. This list has option to specify protected or unprotected components, which is important to accelerate the fault injection simulations by simply skipping the fault injection in the protected parts. An example list of processor resources is: register file, program counter, instruction word, instruction memory, data memory, etc.

4) *Fault Probability:* This parameter specifies the probability that a neutron strike on a processor component results in a architecturally visible error.

5) *Fault Distribution:* The fault distribution model describes how faults in different processor components will be injected, for instance randomly or following a certain correlation model.

6) *Fault Model:* This parameter specify the distribution of faults in terms of number of even upsets, i.e. single even upsets or multiple even upsets. This parameter may also partly cover fault sources other than soft errors. In this

7) *Processor Frequency:* This parameter specifies the operating frequency of the processor and it is used to convert the fault rate from per-sec unit to the per-MCycle unit.

A fault rate (in faults/MCycles) is computed using the neutron flux, city coordinates, fault probability, processor area/layout, and the processor frequency. In our experiments, we used various a Leon-II embedded processor from Gaisler [24] and obtained three different faults rates (1, 5, 10 faults/MCycles) for different altitude levels. In this paper, we employ single bit flip faults. In this work, we assume that caches are protected following the common practice in in the prominent industrial and research projects by AMD [27] and IBM [26].

After generating various fault scenarios, the integrated configurable fault injection module injects the faults during the application execution on the processor instruction set simulator. For instance, in case of the instruction decoder or instruction word, corrupted operation or wrong source/destination addresses of processor is modeled by making one/multiple fields of the instruction word corrupted, consequently leading to a wrong opcode or wrong operand. The program output is categorized in "Correct Output", "Incorrect Output", and "Program Failure", where the program failures are further characterized depending upon the type of fault, like memory segmentation fault, non-decodable instruction, etc. For each simulation run, the reliability analysis is performed, where instruction vulnerabilities, masked and propagated errors are monitored and analyzed. Note that the fault injection does not contribute in an undesirable consequence such as changing performance counters. The fault injection operates in parallel to the application execution while preserving the processor state in backup data structures.

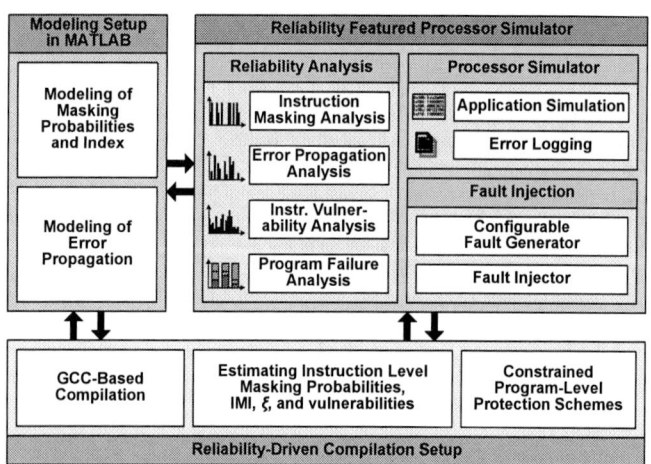

Fig. 10 Our Experimental Setup with Reliability-Featured Processor Simulator, Fault Injection Engine, Reliability-Aware Compiler, and Reliability Analysis Flow

S4. Additional Results on Program-Level Masking and Error Propagation

Fig. 11 shows the distribution of error propagation index (ξ) and instruction masking index (IMI) for 3 additional applications "ADPCM", "DCT", and "SAD" (see similar plots for "Susan" in Section V.B). the distribution illustrates summary from a selected execution run out of many. Horizontal axis shows the instruction address in the execution sequence.

Fig. 11 shows that in "ADPCM" and "DCT", several instructions have zero value for ξ. This denotes that these instructions do not have any dependent instructions or the dependent instructions mask the errors completely. Such masking dependent instructions can be identified by comparing the plots of ξ and IMI corresponding to the same instruction address. These instructions are relatively less important for protection compared to those instructions which exhibit high error propagation index ξ.

In "ADPCM" and "SAD", the ξ value is much lower compared to the ξ plot of "DCT". Therefore, in cases of "ADPCM" and "SAD", the protection is dominated by the instruction vulnerability index. In case of "DCT", the ξ value for many instructions is low, and the number decreases exponentially as the value grows the same way. That is due to the butterfly form of the instruction dependencies. This illustrates that it is important to protect earlier executing instructions compared to the later ones. In case of "SAD", the ξ value is low and the distribution is homogeneous. This is due to the fact that there are many parallel instruction paths with similar dependency structure in the data flow graph. Note that in case of "SAD" the loops are completely unrolled, therefore, the ξ plot is very dense. In case of "SAD", the protection decision is primarily dominated by the value of the instruction vulnerability index.

Fig. 11 also shows the plots for IMI. It is noteworthy that most of the IMI values are close to 0, 0.5, and 1. This primarily reflects three important cases:

1) **Case-1:** an instruction is either not important (IMI=1, EPI=0) and therefore its output will not matter;

2) **Case-2:** IMI value close to 0.5 indicates the cases of comparison or logical instructions, where the error probability to become visible is controlled by the 50% probability of the operation type.

3) **Case-3:** IMI value close to 1 indicates the cases where a dependency path directly leads to an error without any intermediate masking, e.g., a sequence of arithmetic instructions leading to an output value as in case of "SAD".

Note that the case-2 is dominant in "SAD" due to comparison operations of the absolute operation. However, case-1 is dominant in "DCT" due to dependent arithmetic instructions.

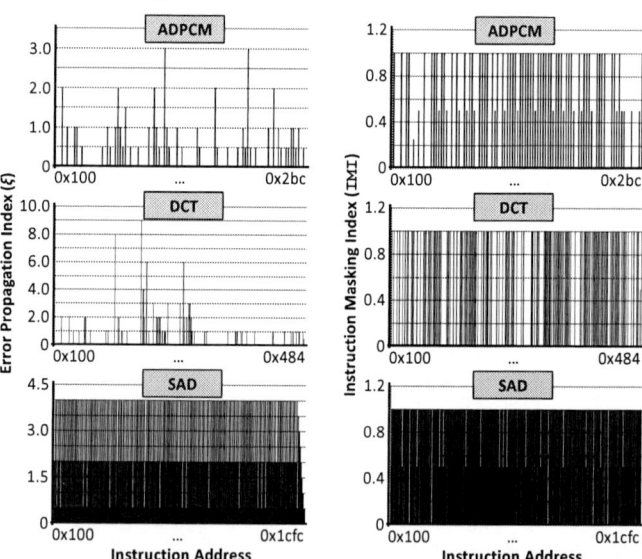

Fig. 11 Distribution of Error Masking and Propagation Indexes

118

REGIMap: Register-Aware Application Mapping on Coarse-Grained Reconfigurable Architectures (CGRAs)

Mahdi Hamzeh, Aviral Shrivastava, and Sarma Vrudhula
School of Computing, Informatics, and Decision Systems Engineering
Arizona State University, Tempe, AZ
{mahdi, aviral.shrivastava, vrudhula}@asu.edu

ABSTRACT

Coarse-Grained Reconfigurable Architectures (CGRAs) are an extremely attractive platform when both performance and power efficiency are paramount. Although the power-efficiency of CGRAs can be very high, their performance critically hinges upon the capabilities of the compiler. This is because a CGRA compiler has to perform explicit pipelining, scheduling, placement, and routing of operations. Existing CGRA compilers struggle with two main problems: 1) effectively utilizing the local register files in the PEs, and 2) high compilation times. This paper significantly improves the state-of-the-art in CGRA compilers by first creating a precise and general formulation of the problem of loop mapping on CGRAs, considering the local registers, and from the insights gained from the problem formulation, distilling an efficient and constructive heuristic solution. We show that the mapping problem, once characterized, can be reduced to the problem of finding maximal weighted clique in the product graph of the time-extended CGRA and the data dependence graph of the kernel. The heuristic we've developed results in average of 1.89 X better performance than the state-of-the-art methods when applied to several kernels from multimedia and SPEC2006 benchmarks. A unique feature of our heuristic is that it learns from failed attempts and constructively changes the schedule to achieve better mappings at lower compilation times.

1. INTRODUCTION

The *holy grail* of computer hardware and software design across all market segments, including battery powered hand-held devices, tablets and laptops, desktop PCs, and high performance servers, is to simultaneously improve performance and power-efficiency (performance-per-watt). Multicores solve this problem to some extent, but accelerators are needed to improve power-efficiency to levels much higher than what multicores can provide. Although special purpose or function specific hardware accelerators (e.g. for FFT) can be very power efficient, they are expensive, not programmable, and therefore limited in usage. Graphics Processing Units (GPUs) are becoming very popular; although programmable, they are limited to accelerating only "parallel loops." Field Programmable Gate

Permission to make digital or hard copies of all or part of this work for personal or classroom use is granted without fee provided that copies are not made or distributed for profit or commercial advantage and that copies bear this notice and the full citation on the first page. To copy otherwise, to republish, to post on servers or to redistribute to lists, requires prior specific permission and/or a fee.
DAC'13, May 29 - June 07, 2013, Austin, TX, USA.

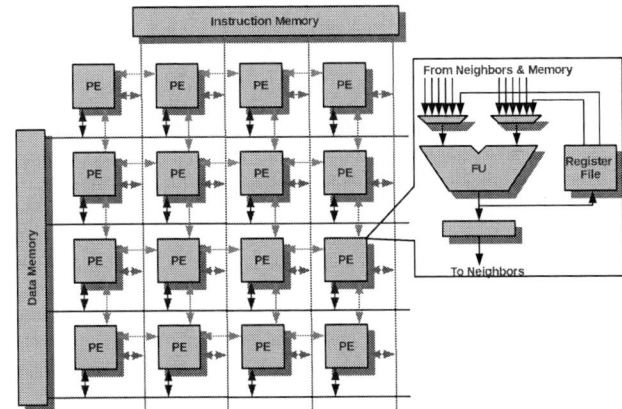

Figure 1: A 4 × 4 CGRA. PEs are connected in a 2-D mesh. Each PE is an ALU plus a local register file.

Arrays are general-purpose, but due to their fine grain reconfigurability, have poor power-efficiency. In this field of accelerators, Coarse-Grain Reconfigurable Architectures or CGRAs are a very attractive platform [24]. A CGRA is simply a two dimensional mesh of PEs, with each PE equipped with an ALU and a small register file (Figure 1). The PEs are connected to neighboring PEs, and the output of a PE is accessible to its neighbors in the next cycle. In addition, a common data bus from the data memory provides data to all the PEs in a row. It is referred to as coarse grained reconfigurable because each PE can be programmed to execute different instructions at the cycle level granularity.

CGRAs are much more power-efficient, with power efficiencies close to hardware accelerators. They are also programmble and easier to program than FPGAs due to their coarse-level of reconfigurability. Finally, they are more general purpose accelerators than GPUs, as CGRAs can accelerate even non-parallel loops[1].

Attracted by the promise of CGRAs, more than a dozen CGRAs including XPP [2], PADDI [4], PipeRench [9], KressArray [13], Morphosys [18], MATRIX [20], and REMARC [21] were proposed over the past decade. In particular, the ADRES CGRA has been shown to achieve power efficiency of 60 GOps/W in 32 nm CMOS technology [3]. However, achieving the promised power-efficiency critically hinges on the compiler technology, and a CGRA compiler is much more complex than a regular compiler, since it has to perform code analysis to extract parallelism, schedule operations in time, bind the operations to PEs, route the data dependencies between the PEs, and perform explicit software pipelining. One major limitation of existing CGRA compilers is their inability to

---
[1]Non-parallel loops cannot be efficiently accelerated by GPUs; they can only be accelerated to a theoretical extent, depending on the inter-iteration loop dependencies. However, CGRAs permit acceleration of such loops.

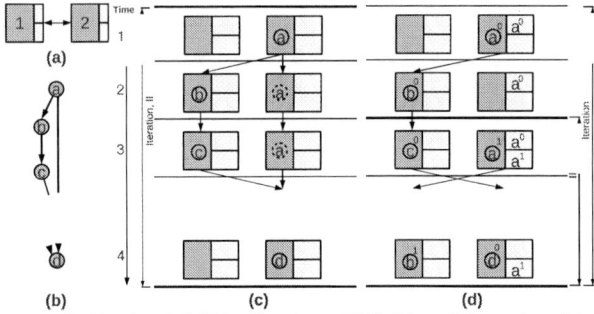

Figure 2: (a) a 2 × 1 CGRA, (b) an input DFG, (c) a valid mapping of the given DFG (*b*) on (*a*) with iteration latency =II= 4, (d) another mapping for the given DFG with iteration latency= 4 and II = 2. Lower II is achieved because two iterations of the loops are executed simultaneously which becomes possible because internal registers of PE_2 are used to route data from PE_2 at cycle 1 to PE_2 at cycle 4.

use register files to improve performance. Most existing compilers simply do not use register files, and transfer the operands among the PEs through a computational path with a PE. We know of only two schemes [6, 22] that generate mapping that use the registers in the CGRA. However, they suffer from the twin problems of i) poor acceleration of loops, and ii) high compilation times.

Towards improving the compiler technology for CGRAs, this paper makes several contributions:

1. A precise formulation for the CGRA mapping problem while using register files: We formally define the problem of register aware application mapping on CGRAs. In contrast to the previous ad-hoc problem definitions, our problem formulation is quite general and allows for various flexibilities in constructing mappings, including recomputation, and sharing of routing paths with dependencies.

2. A novel formulation for integrated operation placement and register allocation: We show that the problem of simultaneous placement and register allocation can be formulated as one of finding a *constrained maximal clique* on the product of two graphs: the data flow graph (DFG) of the computation and the graph representing a time-extended CGRA.

3. An effective heuristic for mapping loops to CGRAs: From the insights gained from the problem definition, we derive an effective heuristic to partition the mapping problem into sub-problems: (1) scheduling and (2) integrated placement, and register allocation. Our method, iteratively and constructively, solves these problems until a mapping can be found.

Experimental results on multimedia applications loops and some SPEC2006 benchmark kernels demonstrate the effectiveness of our heuristic. Our approach improves the performance of computationally bounded loops on average by 1.89 X, while reducing the compilation time by 56X than the existing state-of-the art technique [6].

2. BACKGROUND AND RELATED WORK

More than a dozen CGRA architectures have been designed till now [12]. These CGRAs were primarily targetted for embedded systems to perform signal processing in a power-efficient manner, and therefore programmed manually. However, as we envision the use CGRAs as more general purpose accelerators, compiler technology to automatically map parts of applications onto CGRAs is needed. Recognizing this, much of the research on CGRAs in this century has focused on advancing compilation for CGRAs [1, 7, 8, 11, 14, 17, 22, 25].

Since applications spend most of the time on loops [23], existing compilation techniques (as well as this paper) focus on the problem of mapping the innermost loops on a CGRA. Figure 2(a) shows a CGRA with 2 PEs, and Figure 2(b) shows the data flow graph of a simple loop. Figure 2(c) shows one valid mapping of the data flow graph on the CGRA. In this graph, the CGRA is extended in time to

4 cycles. In cycle 1, operation a is performed on PE_2. Operations b is executed on PE_1 in cycle 2, receiving the value of operation a from the output register of PE_2 (see Figure 1. Note each PE has an output register). Similarly operation c and d are executed on PE_1 PE_2 at times 3 and 4, respectively. To enable this computation, the result of operation a must remain in the output register for 3 cycles. This is also called *routing* of the dependency from a to d. The schedule length of this mapping is 4 cycles, and the II (Initiation Interval) of this mapping is also 4 cycles. II means the difference in time between the initiation of successive iterations of the loop. Since performance is inversely proportional to II, the goal of mapping is to minimize II (rather than schedule length).

Note that the mapping in figure 2(c) does not use registers in the PE. It routes the dependencies through PEs (actually the output register in the PE). As shown in the figure, each PE has 2 registers. Figure 2(d) shows the mapping using the registers in the PEs. In the mapping, after a is executed on PE_2 at cycle 1, its result is stored in one of its registers until d is executed by PE_2 at time 3. The result of a is also made available via its output register to PE_1, which executes b at cycle 2. Although the schedule length of this mapping is 4, the II is reduced to 2. This is because the next iteration of the loop can start at cycle 3. This is shown in the figure by operations a^1, b^1 being mapped in cycles 3 and 4. All operations are present in cycles 3 and 4, and this kernel can execute repeatedly, giving an II of 2.

This simple example shows how using register files can result in higher performance of loops on CGRAs, but most of the existing compiler techniques for CGRAs do not exploit register files well. We know of only two techniques that do use register files to obtain better mappings. However, both the existing techniques are "exploratory" mapping techniques. EMS [22] allocates registers during the scheduling and placement of operations on CGRA. The method places the input DFG, arc by arc, onto the CGRA. An arc can be placed, if the nodes on the arc can be placed, and the dependencies can be routed. If at any point this cannot be done, II is increased, and this mapping is tried all over again. The method described in [6], called DRESC, expands the time-extended CGRA graph to explicitly include registers as nodes (one node per register file). The method uses simulated annealing to find a mapping. Operations are randomly moved to decrease the number of overused resources. Once all resources are used only once, the mapping is completed. If no mapping can be found, II is increased and mapping is tried again. No control strategy, e.g. the temperature schedule is derived for the algorithm.

The drawbacks of the existing techniques, namely poor mapping and high compilation times are due to their exploratory nature. The basic strategy employed is to find two mappings successively: first, a mapping of the operations in the DFG to the PEs of the time-extended CGRA, followed by a mapping of arcs in the DFG to paths between the corresponding PEs in the time-extended CGRA. Even though the mappings allow paths to start and end in PEs, with the intermediate nodes being allowed to be PEs or registers, the approaches is restrictive as they do not permit recomputation, in which one operation of the DFG can be mapped to multiple PEs in the CGRA [11].

In this paper, we will present a general formulation of the problem of mapping a kernel on the CGRA while using its registers to minimize II. The formulation enables us to partition this problem into a scheduling problem and an integrated placement and register allocation problem. We reduce the placement to one of finding a *constrained maximal clique* in the product graph of the time-extended CGRA and the input DFG. Then we develop an efficient and constructive heuristic to map loop kernels onto the CGRA. An

120

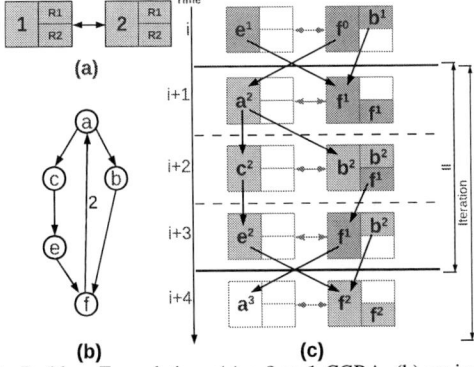

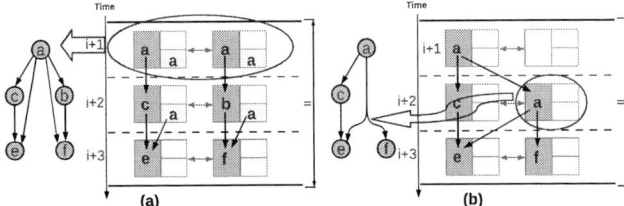

Figure 3: Problem Formulation: (a) a 2×1 CGRA, (b) an input DFG, (c) Mapping of the DFG to a CGRA. II of this mapping is 3 but to better visualize the overlap of loop execution, operations from iteration 1 and 3 are also shown.

important aspect of our heuristic is its ability to learn from failures, which in turn results in better performance, at lower compilation times.

3. PROBLEM DEFINITION

Let $I = (V_I, E_I)$ be the input DFG representing a loop. Given a DFG and a CGRA, we first determine the lower bound on II by extracting resource minimum II ($ResMII$) and recurrence minimum II ($RecMII$) [23], denoted as MII. Given an MII, we make a time extended resource graph denoted by $R_{II} = (V_R, E_R)$ which is constructed by replicating the nodes in the CGRA (PE and registers), II times, representing cycles 0 through $II-1$. For every pair (u, v) of adjacent nodes in the CGRA, there is an arc from (replication of) u at time t to (replication of) v at time $t + 1$. Note that every node in the CGRA is adjacent to itself. This time-extended resource graph is the same as the MRRG graph used in [6].

Figure 3 (a) shows a 2×1 CGRA, and (b) shows input DFG. First the minimum II of this is calculated as 3. The CGRA graph is then unrolled 3 times, and a time-extended CGRA resource graph is constructed. We do not show that graph in the figure, since it has too many arcs, and does not help in visualization. Figure 3 (c) shows the mapping of the input DFG on this time-extended resource graph. The time-extended CGRA must be extended only 3 times (and therefore 3 rows), but we add an extra row at the top and bottom for clarity. The following terminology is needed to simplify the statement of the problem.

Definition: A node v in a time extended resource graph R_{II} is said to be *associated with* an operation i in the DFG if v is a PE executing operation i at t or v is a register in a PE that stores the result of operation i and is available at time t. For instance, in Figure 3(c) PE_2^{i+1} (PE_2 at times $i + 1$) is associated with operation f^1 (operation f of first iteration), PE_2^{i+2} is associated with operation b^2, and PE_2^{i+3} is associated with operation f^1. The first register of PE_2^{i+1} is not associated with any operation, but the first register of PE_2^{i+2} is associated with operation b^2, and the first register of PE_2^{i+3} is associated with b^2.

Now, given a DFG $I = (V_I, E_I)$ and a CGRA, the problem is to construct a time extended resource graph $R_{II} = (V_R, E_R)$ of minimum extension for which:

1. there exists a surjective mapping $M : V_R^* \rightarrow V_I$, where $V_R^* \subseteq V_R$, and

2. for every arc in $(i, j) \in E_I$, the following property holds: *for each node $r_n \in R_{II}$ associated with j, there is a path $P = (r_1, \ldots, r_\ell, \ldots r_n)$ such that r_1 is a PE associated with i, r_2 through $r_{\ell-1}$ are associated with i, the rest are associated with j, and r_ℓ is a PE.*

Figure 4: Problem Formulation: Our problem definition allows both routing and recomputation by allowing multiple PEs to map to the same operation. (a) If the PEs associated with same operations connected through a path, then it is routing, and if there is no path between them, then it is recomputing. operation a in this mapping is recomputed because it is associated with PE_1 and PE_2 at time i+1 and there is no path between them. (b) Our problem definition allows for path sharing.

The succinct statement of the problem makes it difficult to understand. First note that the mapping is counter-intuitive: it is from the time-extended resource graph to the DFG and not vice-versa, which is typically easier to understand. Second, the mapping is from subsets of the time-extended resource graph to an operation in the DFG. In the simplest case, the subset can be a single PE. Consider mapping presented in Figure 3. For example, PE_1^{i+1} is mapped to a^2. In a more general case, several resources may be mapped to an operation. For example, PE_2^{i+2}, the first register of PE_2^{i+2} and PE_2^{i+3} are mapped to operation b^2. The mapping for operation f^1 is also from a subset of resources (PE_2^{i+1}, and PE_2^{i+3}, and the second register of PE_2^{i+1} and PE_2^{i+2}). Third thing to note is that the mapping is surjective. This simply implies that every operation in the DFG is included in the mapping.

The second condition in the definition essentially says that if there is a dependency between two operations, then there must be a path between the subsets to which the two operations are mapped. Thus, if we pick any resource r_n which is associated with the destination operation, then a path must exist from r_1 to r_n through some r_ℓ, such that r_1 is a PE associated with the source operation (i.e., source operation is executed here), and r_ℓ is a PE associated with the destination operation (i.e., destination operation is executed here). The path between r_1 to r_ℓ can be formed of arbitrary combination of PEs and registers all associated with source operation (routing). Also, the path after r_ℓ up to r_n can also be composed of arbitrary combination of PEs and registers all associated with destination (routing).

An example of the simple case is the arc between a^2 and c^2 in Figure 3(c). For this arc, only PE_1^1 is mapped to a^2, and PE_1^2 is mapped to c^2. The path PE_1^1, PE_1^2 meets the criterion, where $r_1 = PE_1^1$, and $r_\ell = PE_1^2$. A more complicated case is the arc between a^2 and b^2. PE_1^1 is mapped to a^2, but the subset consisting of PE_2^{i+2}, and the first register of PE_2^{i+2} and PE_2^{i+3}, is mapped to b^2. The definition requires that from any resource which is mapped to b^2, there must exist a path that satisfies the second criterion. Let's take r_n to be the first register of PE_2^{i+3}. The the path (PE_1^{i+1}, PE_2^{i+2}, first register of PE_2^{i+2}, and first register of PE_2^{i+3}) satisfies the criterion, with $r_1 = PE_1^{i+1}$, and $r_\ell = PE_2^{i+2}$.

This problem definition is quite general, and allows for routing, recomputation [11], and path sharing between multiple dependencies of the same operation [5]. Routing is allowed by second condition of the definition, according to which, for each dependency, a path of resources must exist, connecting the source and destination of the dependency. Most problem definitions do not allow recomputation, since they require an operation to be mapped to one PE. As shown in Figure 4 (a), our problem definition allows for recomputation by allowing multiple PEs to be mapped to one operation. Hamzeh et al. [11] show that recomputation can result in better mappings that result in up to 2X better performance. Finally, our problem definition also allows for sharing the paths by multiple dependencies of an operation. As Figure 4 (b) shows, if an operation has two dependents, then the resources that the dependencies use can be shared between the paths. Chen et al. [5] demonstrate that path sharing can improve resource utilization by more than 50%. Although path sharing can also be done in the methods of DRESC[6] and EMS[22], it is not explicit aspect of the solution method.

4. OVERVIEW OF OUR APPROACH

The general CGRA mapping problem is NP-complete [11]. Even so, another major challenge is that size of time-extended resource graph of the CGRA can grow very large, even for small problem sizes. For a $m \times m$ CGRA, with r registers in each PE, if we intend to map a DFG with an $II = i$, the time-extended resource graph will have $N = m \times m \times r \times i$ nodes[2]. The problem then is to find a $n + 1$-partition[3] of this graph, where n is the number of operations in the DFG. The number of possible solutions to this problem is a Sterling number [10] ($\{\frac{N}{n}\}$), which is of the order of $O(n^N)$ even when the input DFG is fixed (no extra operations to be added to DFG). Due to the explosive growth of the number of partitions of time-extended resource graph, greedy approaches that attempt to find a mapping by incrementally exploring the search space are computationally very slow and result in poor solutions, i.e. high values of II. The method developed here is a constructive solution that avoids the explosive growth of the search space.

We partition the mapping problem into two sub-problems: **1) scheduling**, and **2) integrated placement and register allocation** (referred to as placement). The routing is implicitly accomplished as part of scheduling and placement. This separation of the problem into scheduling and placement results in a significant reduction of the search space. Since the operations of the scheduled DFG are bound to be placed at matching time slot in the time-extended resource graph, the number of ways to partition is $\{\frac{m}{n'}\}^i$, where n' is the maximum number of operations scheduled at any time (width of the graph). A more important advantage of breaking this problem into scheduling and placement is that *instead of explicitly enumerating the registers as nodes, we put the number of registers required as arcs weights*. This further reduces the search space by a factor of r, and makes the problem more scalable. Although the partitioning into two subproblems is in general, not optimal, we can further improve the results by performing placement and scheduling in a loop. *If a placement is not possible, then we learn from the failure, identify which operations could not be placed, and then constructively change the schedule of DFG and place them at higher priority in the next round.*

5. REGIMAP

Let $D = (V_D, E_D)$ be the scheduled graph. When operations are scheduled and II is extracted, the $R_{II} = (V_R, E_R)$ can be constructed. R_{II} is time-extended PE graph (without registers).

Step 1: Construct a compatibility graph between D and R_{II}:
The compatibility graph $P = (V_P, E_P)$ is a subgraph of the product of the DFG and the R_{II}. It is a directed graph. Its vertices, which are a resource-operation pair, represent the possible mappings of operations to PEs. Its edges (ignoring the directions) indicate *compatibility* between the two corresponding mappings. Scheduling reduces the size of compatibility graph because some resource-operation pairs become incompatible, i.e. the resource is not available at the time the operation is scheduled. For instance, the number of vertices in the product graph is 16, whereas the number of

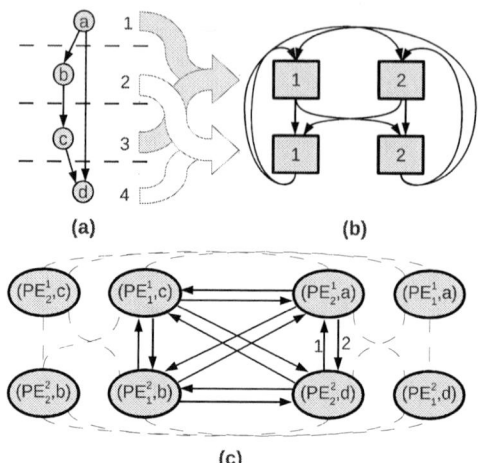

Figure 5: (a) scheduled DFG, D (b) time extended PE graph, R_2 (c) Compatibility graph P between D and R_2.

vertices in the compatibility graph shown in Figure 5 is only 8. An edge between two mapping means that both can co-exist in a solution. Details of the construction of the compatibility graph is given in Appendix A.

Step 2: Assign weights to arcs of compatibility graph: The weight of an arc (directed edge) denotes the number of registers required from the time the source node (a mapping) is executed to the time the destination node is executed. Note this is asymmetric. For instance in Figure 5, the arc $[(PE_2^1, a), (PE_2^1, d)]$ has weight 2, which means that from the time a is executed on PE_2 to the time that d is executed on PE_2, two registers are required in PE_2 because the $II = 2$(see Figure 2). The arc $[(PE_2^1, d), (PE_2^1, a)]$ has a weight of 1, because when d is executed, one register can be released (see Figure 2). The process of arc weight assignment in the compatibility graph is described in detail in Appendix B.

Step 3: Find maximal clique in the compatibility graph: The weight of a node is the sum of the weights of its outgoing arcs. We reduce the problem of finding a placement of D onto R_{II} to the problem of finding the largest clique whose total node weight is less than the number of available registers. By construction, the maximal clique in the graph can be no larger than $|V_D|$. Therefore, if we find a maximal clique of size $|V_D|$, in which the sum of the node weights is less than the number of registers in each PE, we have a mapping. The algorithm to find maximal clique in the compatibility graph is described in Appendix section D.

A mapping can fail because either a node in the clique does not satisfy the register constraint, or the maximal clique is smaller than $|V_D|$. In the first case, we search for another clique. If no such clique exists, or we find that an operation is not present in the clique, we reschedule that node and proceed to find a new mapping. After we run out of rescheduling options, we reduce the width of the input DFG (max. no. of operations at any cycle), and try again. If the MII increases as a result of thinning the graph, we increase II. Details of our algorithm are described in Appendix E.

6. EXPERIMENTAL RESULTS

We have modified backend GCC and integrated REGIMap right before register allocation. Loops are selected among digital signal processing applications and spec2006 benchmark suite. We compare REGIMap with register-aware DRESC [6]. It has been shown in [22] that DRESC [19] without register allocation [6] maps loops at lower II than [22], and [23].

We conduct experiments on CGRAs with different number of PEs and registers. We assume all registers are rotating registers. In our experiments, CGRA has enough memory to hold the instructions as well as variable in the loops. Since CGRA executes loops,

---

[2]$m \times m$ PEs each with r registers is replicated i times.
[3]the $(n + 1)^{th}$ partition maps to no operation

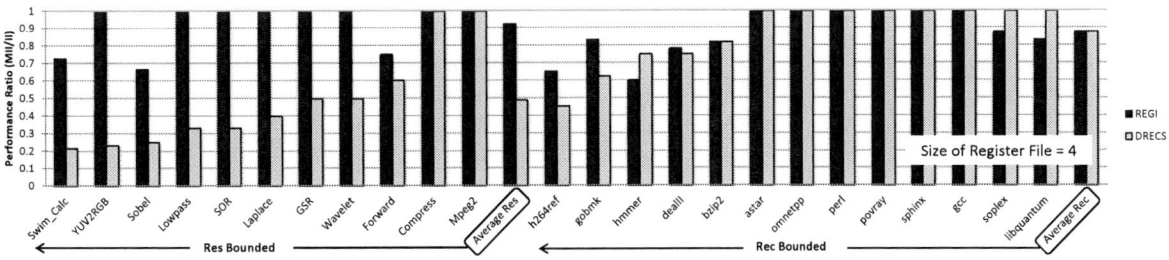

Figure 6: The performance of mapping loops to a 4×4 CGRA with 4 registers using REGIMap vs. DRESC. REGIMap compiles *Res-bounded* loops at on average 1.89 times lower *II* than DRESC.

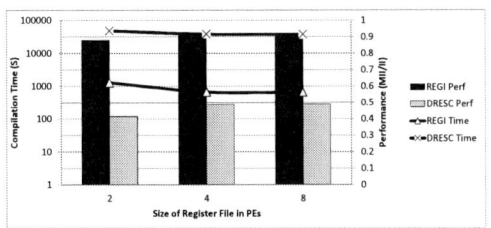

Figure 7: Compilation time using REGIMap vs. DRESC for mapping loops on 4×4 CGRA when the size of register file varies from 2 to 8.

and instructions within the kernel repeat every *II* cycles, a highly optimized multi-way cache can provide such memory requirement. PEs are connected in a 2D mesh-like network similar to Figure 1. We also assume that all PEs have access to the data memory but data bus is shared among PEs in a row; thus, only one PE of a row has access to the data memory at each cycle. For load and store operations, two instructions should be executed, the first instruction generates the address and the second one generates/loads data. For memory operations data to be loaded/stored is read/generated at the same row where the address bus has been asserted. The CGRA is homogeneous and PEs are capable of performing logical and fixed-point operations. All operations have latency of 1 cycle.

6.1 1.89 Times Better Performance for Resource bounded Loops

In modulo scheduling, MII is theoretically the lower bound of II for scheduled loops. Our performance metric to evaluate mappings is MII divided by the achieved II at each loop. We have divided the loops into two groups, in the first group, MII is primarily limited by ResMII, referred to as *res-bounded*. In this type of loops, placement is more challenging because the number of resources at this type of loops is relatively close to the number of operations at loop. In the second group, MII is limited by recurrence cycles (RecMII) in the DFG, referred to as *rec-bounded*. Therefore, resources are under-utilized and there are more options for each operation to be placed at. In Figure 6, the performance comparison of different applications compiled for a 4×4 CGRA with 4 registers using REGIMap and DRESC is presented.

Our evaluation reveals that on average the performance of compiled loops using REGIMap is 1.89 times higher that using DRESC technique for *res-bounded* loops. Placement in these applications is more complicated and time consuming than *rec-bounded* loops. However, the average performance of compiled loops using both mapping techniques is relatively close for *rec-bounded* loop. More details about the performance of compiled loops at different CGRA and register file sizes is given in Appendix G.

6.2 56 Times Lower Compilation Time for Resource bounded Loops

The compilation time of loops on REGIMap is much lower than compilation time of them on DRESC. DRESC is a simulated annealing (SA) based technique which spends a lot of time to move

operations at time and resource dimensions. For *res-bounded* loops, REGIMap compiles them on average 37 times faster than DRESC when the size of register file of PEs is 2. The ratio of compilation time of REGIMap to DRESC increases to 56 times faster when the size of register file of PEs increases to 4 and 8. For *rec-bounded* loops, the compilation time using REGIMap is about 6 times faster than DRESC when the size of register file of PEs is 2. REGIMap becomes 8 times faster than DRESC when the size of register file increases to 4 or 8. For *rec-bounded* loops, placement is relatively easy because there are many free (not utilized) resources available in the resource graph. Those resources can be utilized for routing data between operations making placement easier. Thus, DRESC faces with less number of failed attempts which significantly decreases its running time. Both techniques compile loops faster when the size of register file increases. It is primarily because both methods face with less failure at register allocation. More details about compilation time is presented in Appendix G.

6.3 Learning from Failure Improves Performance and Reduces Compilation Time

More detailed evaluation reveals the effectiveness of operation rescheduling after placement in REGIMap. When a mapping attempt at a given II fails, most existing mapping techniques increase II. It is basically because if they schedule the DFG again and initiates a new placement attempt, they will end up with another failure. The reason is that almost all scheduling techniques use a justifiable *static* policy for ordering the nodes. Operations are to be selected by that order for scheduling. Because this ordering is static, for a DFG, the scheduling of the graph does not change unless the structure of the graph changes or the nodes are ordered differently.

The first heuristic is to change the order of the nodes. REGIMap changes this order and consequently schedule the nodes so that the next schedule is different from the previous one. Therefore, it exploits both time and resource dimensions in a guided manner.

The second heuristic REGIMap applies is virtually reducing the number of available resources at CGRA. Therefore, after many placement failure, when this heuristic is applied, the schedule of DFG further changes. Therefore, REGIMap constructively changes the schedule or the DFG to find a mapping.

Our experiments show that the rescheduling heuristic is more effective on *res-bounded* loops. To demonstrate rescheduling effectiveness, we have disabled rescheduling from REGIMap. About 90% of loops are compiled at higher *II*s than when compiled with rescheduling enabled. For *rec-bounded* loops, this heuristic becomes less effective and about 30% of loops are compiled with higher *II*s when it is disabled. Our further investigation reveals that most of those 30% of loops has close ResMII and RecMII.

6.4 Scalability

Figure 7 shows the average performance of compiling resource-bounded loops for a 4×4 CGRA using REGIMap and DRESC versus different sizes of register file of PEs. The left Y axis is the

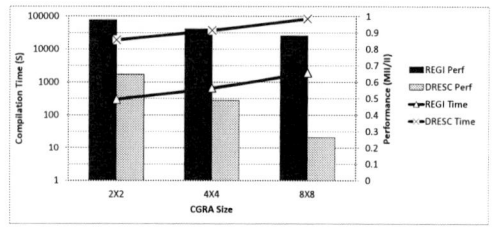

Figure 8: Compilation time using REGIMap vs. DRESC for mapping loops on 4×4 CGRA when the size of register file varies from 2 to 8.

compilation time (logarithmic) and the left one is the performance. The compilation time of loops using REGIMap and DRESC relatively decreases by increasing the size of register files. The performance of those mapping also increases when more registers are available. It is primarily because when the number of register increases, both technique can utilize more registers to compile loops. However, the performance of compiling loops using REGIMap is significantly higher (1.5X to 1.9X) compared to DRESC.

In Figure 8, the performance of compiling resource-bounded loops using two mapping techniques for different CGRA sizes when the size of register file of PEs is two is presented. The left Y axis is the compilation time (logarithmic) and the left one is the performance. The performance of loops using both techniques decreases when the size of CGRA increases. However, the performance of those mapping using DRESC dramatically decreases while the compilation time increases. The compilation time of those loops using REGIMap increases as well because the search space is proportional to the CGRA size too.

6.5 Power-Efficiency Estimation

In this section, we discuss power efficiency improvement when compute intensive loops are offloaded to the CGRA. The synthesis information presented in [3] shows that ADRES CGRA at 312 MHz consumes only 81 mW. REGIMap compiles *res-bounded* loops and achieves average of 10.75 instruction per cycle (IPC). Further details about IPC of compiled loops is presented in Appendix G. Therefore, it can be estimated that CGRA would approximately execute 3.3 GOps per second. This implies only 24 pW per instruction for CGRA and compared to 12 nW for an Intel Core2 processor [16]. Therefore, by offloading highly parallel regions in the code, instructions can be executed at approximately 500 X less energy. Considering maximum of 2 instruction per cycle for Intel Core2 processor and clock frequency of 2.6 GHz, 5.2 G instructions can be executed per second. Assuming each instruction is equivalent to one operation, we conclude that the power efficiency can approximately improve by 250 X when those loops are offloaded.

7. SUMMARY

Coarse-Grained Reconfigurable Architectures (CGRAs) are extremely attractive platform when both performance and power efficiency are paramount. However, the achievable performance and power efficiency of CGRAs critically hinges upon compiler capabilities. One of the main challenges in CGRA compilers is to efficiently utilize registers which is specially difficult due to their distributed nature. This paper make three contributions: i) we formulate the problem of mapping loops on CGRAs while efficiently using registers, ii) we present a unified and precise formulation of the problem of simultaneous placement and register allocation, and iii) an efficient and effective heuristic solution, REGIMap is distilled from our problem formulation. REGIMap leads to an average of 1.89 X better performance on several kernels from multimedia and SPEC2006 benchmarks suits when compared to the existing state-of-the-art compilation technique at lower compilation time.

8. ACKNOWLEDGMENTS

This work was partially supported by funding from National Science Foundation grants CSR-EHS 0509540, CCF-0916652, CCF 1055094 (CAREER), NSF I/UCRC for Embedded Systems (IIP-0856090), Center for Embedded Systems grant DWS-0086; Science Foundation Arizona (SFAz) grant SRG 0211-07, Raytheon, and by the Stardust Foundation.

9. REFERENCES

[1] BANSAL, N., GUPTA, S., DUTT, N., NICOLAU, A., AND GUPTA, R. Network topology exploration of mesh-based coarse-grain reconfigurable architectures. In *Proc. DATE* (2004), pp. 474–479.

[2] BECKER, J., AND VORBACH, M. Architecture, memory and interface technology integration of an industrial/ academic configurable system-on-chip (csoc). In *Proc. ISVLSI* (2003), pp. 107–112.

[3] BOUWENS, F., BEREKOVIC, M., SUTTER, B. D., AND GAYDADJIEV, G. Architecture enhancements for the adres coarse-grained reconfigurable array. In *Proc. HiPEAC* (2008), pp. 66–81.

[4] CHEN, D. C. *Programmable arithmetic devices for high speed digital signal processing.* PhD thesis, University of California, Berkeley, 1992.

[5] CHEN, L., AND MITRA, T. Graph minor approach for application mapping on cgras. In *Proc. FPT* (2012).

[6] DE SUTTER, B., COENE, P., VANDER AA, T., AND MEI, B. Placement-and-routing-based register allocation for coarse-grained reconfigurable arrays. In *Proc. LCTES* (2008), pp. 151–160.

[7] DIMITROULAKOS, G., GALANIS, M., AND GOUTIS, C. Exploring the design space of an optimized compiler approach for mesh-like coarse-grained reconfigurable architectures. In *Proc. IPDPS* (2006), pp. 113–122.

[8] FRIEDMAN, S., CARROLL, A., VAN ESSEN, B., YLVISAKER, B., EBELING, C., AND HAUCK, S. Spr: an architecture-adaptive cgra mapping tool. In *Proc. FPGA* (2009), pp. 191–200.

[9] GOLDSTEIN, S., SCHMIT, H., MOE, M., BUDIU, M., CADAMBI, S., TAYLOR, R., AND LAUFER, R. Piperench: a coprocessor for streaming multimedia acceleration. In *Proc. ISCA* (1999), pp. 28 –39.

[10] GRAHAM, R. L., KNUTH, D. E., AND PATASHNIK, O. *Concrete Mathematics: A Foundation for Computer Science*, 2nd ed. Addison-Wesley Longman Publishing Co., Inc., Boston, MA, USA, 1994.

[11] HAMZEH, M., SHRIVASTAVA, A., AND VRUDHULA, S. Epimap: using epimorphism to map applications on cgras. In *Proc. DAC* (2012), pp. 1284–1291.

[12] HARTENSTEIN, R. A decade of reconfigurable computing: a visionary retrospective. In *Proc. DATE* (2001), pp. 642–649.

[13] HARTENSTEIN, R., AND KRESS, R. A datapath synthesis system for the reconfigurable datapath architecture. In *Proc. ASP-DAC* (1995), pp. 479 –484.

[14] HATANAKA, A., AND BAGHERZADEH, N. A modulo scheduling algorithm for a coarse-grain reconfigurable array template. In *Proc. IPDPS* (2007), pp. 1–8.

[15] HUFF, R. A. Lifetime-sensitive modulo scheduling. In *Proc. PLDI* (1993), pp. 258–267.

[16] KEJARIWAL, A., VEIDENBAUM, A., NICOLAU, A., TIAN, X., GIRKAR, M., SAITO, H., AND BANERJEE, U. Comparative architectural characterization of spec cpu2000 and cpu2006 benchmarks on the intel core 2 duo processor. In *Proc. SAMOS* (july 2008), pp. 132 –141.

[17] LEE, J.-E., CHOI, K., AND DUTT, N. D. Compilation approach for coarse-grained reconfigurable architectures. *IEEE Design and Test of Computers 20*, 1 (2003), 26–33.

[18] LEE, M.-H., SINGH, H., LU, G., BAGHERZADEH, N., KURDAHI, F. J., FILHO, E. M. C., AND ALVES, V. C. Design and implementation of the morphosys reconfigurable computingprocessor. *J. VLSI Signal Process. Syst. 24* (2000), 147–164.

[19] MEI, B., VERNALDE, S., VERKEST, D., DE MAN, H., AND LAUWEREINS, R. Dresc: a retargetable compiler for coarse-grained reconfigurable architectures. In *Proc. IEEE FPT* (dec. 2002), pp. 166 – 173.

[20] MIRSKY, E., AND DEHON, A. Matrix: a reconfigurable computing architecture with configurable instruction distribution and deployable resources. In *Proc. FPGAs for Custom Computing Machines* (1996), pp. 157 –166.

[21] MIYAMORI, T., AND OLUKOTUN, K. Remarc: Reconfigurable multimedia array coprocessor. *IEICE Trans. on Information and Systems* (1998), 389–397.

[22] PARK, H., FAN, K., MAHLKE, S. A., OH, T., KIM, H., AND KIM, H.-S. Edge-centric modulo scheduling for coarse-grained reconfigurable architectures. In *Proc. PACT* (2008), pp. 166–176.

[23] RAU, B. R. Iterative modulo scheduling: an algorithm for software pipelining loops. In *Proc. MICRO* (1994), pp. 63–74.

[24] TAYLOR, M. B. Is dark silicon useful?: harnessing the four horsemen of the coming dark silicon apocalypse. In *Proc. DAC* (2012), pp. 1131–1136.

[25] YOON, J., SHRIVASTAVA, A., PARK, S., AHN, M., AND PAEK, Y. A graph drawing based spatial mapping algorithm for coarse-grained reconfigurable architectures. *IEEE Trans. on VLSI Systems 17*, 11 (2009), 1565–1578.

APPENDIX

A. COMPATIBILITY GRAPH CONSTRUCTION

Construction of Compatibility Graph $P = (V_P, E_P)$ is completed in two steps. In the first step, the set of nodes in this graph is formed. Then directed unweighed arcs between nodes are created.

A.1 Nodes

First we define the Cartesian product of set of nodes in graph R_{II} and D. Every element in this set represents a pair of an operation and a resource in R_{II}. Basically a pair of nodes represents potential mapping of an operation to a resource. Since operations at graph D are scheduled, we can easily reject a large number of potential mapping of operations represented by some elements in CP. For instance, consider $u_i = (v_i^R, v_i^D) \in CP$. This potential mapping of operation v_i^D on v_i^R can be rejected under the following conditions. If the ALU of v_i^R does not support operation of v_i^D, then the mapping of this of v_i^D to v_i^R is not feasible. Similarly, if v_i^R is not present at the cycle when v_i^D is scheduled, then u_i is not a potential solution. Based on observation above, we create the set of nodes in P.

Let $CP = \{V_R \times V_D\}$ (Cartesian product). The set of nodes in P is a subset of CP. Thus, each node $u_i = (v_i^R, v_i^D) \in CP$ represents a pair of a resource[4] at R_{II} and an operation. We define v_i^R and v_i^D compatible if resource v_i^R supports (its ALU) the boolean function of operation v_i^D and it is present at the same cycle (in R_{II}) that the operation is scheduled. In Figure 2(c), let $PE(t, n)$ represents PE_n at cycle t in the resource graph. Assume that operation a is scheduled at cycle 1. Then pair $(PE(1, 2), a)$ is a compatible pair because $PE(1, 2)$ is present at the cycle 1. None of pairs $(PE(2, 1), a)$ nor $(PE(2, 2), a)$ are compatible because they are not present at time 1.

A.2 Binary Arcs

Next, REGIMap constructs the set of arcs in P. In this step, we define a reflexive and symmetric relation, called compatibility, between pairs of nodes in V_P. Note that each element in V_P represents a potential mapping of a node to a resource. compatibility between elements $u_i = (v_i^R, v_i^D)$ and $u_j = (v_j^R, v_j^D)$ in V_P essentially implies that both mapping can co-exist in the solution. More precisely, when v_i^D is mapped to v_i^R, does it restrict v_j^D from being mapped to v_j^R. If there is an arc from u_i to u_j, then if u_i is in a solution mapping, u_j can also be in that solution.

Two nodes u_i and u_j where $i \neq j$ are incompatible if one of the following conditions holds:

1. u_i and u_j represent the same operation ($v_i^D = v_j^D$).

2. u_i and u_j represent the same resource ($v_i^R = v_j^R$).

3. There is an arc between the operation u_i represents to the one u_j represents, but there is no communication from resources u_i represents to u_j represents.

There are two considerations for the 3rd case. v_i^D and v_j^D can be scheduled apart (more than one cycle apart) and there is no arc in R_{II} to connect resources apart more than one cycle. For this case, REGIMap tags u_i and u_j incompatible if they represent different PEs in the CGRA.

---

[4] We do not use the term PE to avoid confusion between physical PE or replicated PEs in R_{II}

The second problems arises when there is an inter-iteration data dependency between operations u_i and u_j represent. Similar to the previous case, REGIMap tags them incompatible, if they represent different PEs in the CGRA.

For example, consider the following nodes $(PE(1, 2), a)$ and $(PE(2, 1), b)$ in Figure 2(d). These nodes are compatible because there is an arc from a to b in DFG as well as an arc between $PE(1, 2)$ and $PE(2, 1)$. However, $(PE(1, 2), a)$ and $(PE(1, 1), a)$ are not compatible because they both represent operation a. On the other hand, $(PE(3, 2), a)$ and $(PE(4, 2), d)$ are compatible[5] because both nodes represent the same physical PE at CGRA where a and d are scheduled 3 cycles apart.

When there is a symmetric compatibility between u_i and u_j, two directed arcs between them are to be formed. Next, REGIMap looks for a pair of nodes representing two operations with intra-iteration data dependency scheduled more than one cycle apart. Next, weigh of arcs needs to be assigned.

B. ARC WEIGHT ASSIGNMENT TO THE COMPATIBILITY GRAPH

The weight of arcs represent the number of registers required to establish a path between a pair of mappings. Let $T : V_D \to N$ be the schedule function. Consider $u_i = (v_i^R, v_i^D)$ and $u_j = (v_j^R, v_j^D)$ where there is an arc between v_i^D and v_j^D and they are not scheduled in two consecutive cycles. The weight of arc between u_i and u_j is to be increases by:

$$R = \left\lceil \frac{T(v_j^D) - T(v_i^D)}{II} \right\rceil \tag{1}$$

When these operations are scheduled less than II cycles, one register is sufficient to carry-out data dependency between them. However, when they are scheduled more than II cycles apart, this data dependency must be carried-out across multiple iterations of the loop. Thus, when a new iteration starts, when v_i^D of the next iteration of the loop is executed, if the previous output has not been consumed yet, the output of this operation must be stored in a different register. Otherwise, it overwrites a register that has not been read yet by the consumer operations.

Note that a register is allocated when v_i^D is executed on v_i^R until v_j^D is executed on v_j^R at all resources representing same PE as v_i^R represents. Therefore, the weight of arcs from all u_k representing those resources to v_i^R must be increased by R. At $T(v_j^D)$, a register is to be released at v_j^R until a new data is produced by v_i^D at the next iteration of the loop. Thus the arc from nodes representing resources between v_j^R and v_i^R to v_i^R must be increased by $R - 1$.

In Figure 2(d), operations a and d are scheduled more than one cycle apart. The weight of arc from $(PE(3, 2), a)$ to $(PE(4, 2), d)$ should to be increased by 2 [6]. However, the weight of arc from $(PE(4, 2), d)$ to $(PE(3, 2), a)$ is only increased by 1 because one register is to be released when d is executed.

Operations with inter-iteration data dependencies are to be considered next. Let $u_i = (v_i^R, v_i^D)$ and $u_j = (v_j^R, v_j^D)$ represent two operations with inter-iteration data dependency. Let $e_{(v_i^D, v_j^D)}$ be the inter-iteration distance of these operations. The weight of arc

---

[5] $II = 2$

[6] $\lceil \frac{4-1}{2} \rceil = 2$

between u_i and u_j is to be increases by:

$$R = e_{(v_i^D, v_j^D)} - \begin{cases} \left| \frac{T(v_j^D) - T(v_i^D)}{II} \right| - 1 & \text{if } T(v_j^D) \geq T(v_i^D) \\ \left| \frac{T(v_j^D) - T(v_i^D)}{II} \right| & otherwise \end{cases}$$

(2)

When v_j^D is scheduled after v_i^D and $T(v_j^D) - T(v_i^D) > II$, because of the repetitive execution of the loop, the inter-iteration distance between them decreases because they belong to different iteration of the loop at execution. On the contrary, when v_i^D is scheduled after v_j^D, by increasing the schedule distance, their inter-iteration distance increases.

Similar to the previous case, at all resources between v_i^R to v_j^R, R registers should be available to establish a path between v_i^R to v_j^R. Therefore, the weight of arcs from all nodes u_k representing above-mentioned resources to u_i must be increased by R. However, the weight of arcs from nodes representing the resources between v_j^R to v_i^R that represent same PE of v_i^R to u_i must be increase by $R - 1$. This step also considers operations with self inter-iteration data dependency. Note that the weight of arcs between two node in P is to be update only if those nodes are compatible.

C. PROBLEM REDUCED TO NODE TOTAL WEIGHT CONSTRAINED MAXIMUM CLIQUE

THEOREM C.1. *In graph P, there is a maximum clique C (of size $|V_D|$) such that the sum of the weight of outgoing arc is less than the size of register file,i.e. $\forall u_i \in V_C : S_{u_i} < N_R$ where $S_{u_i} = \sum_{\forall (u_i, u_j) \in E_C} \{e_{(u_i),(u_j)}\}$ and where N_R is the size of register file of PEs, iff there is a placement of operation so that the maximum of N_R registers are used at each PE.*

D. FINDING MAXIMAL CLIQUE ALGORITHM

REGIMap conducts a greedy strategy to find a clique in P. Although P is a directed graph, the arcs between nodes are formed symmetrically though the weight of arcs can be different. Thus, finding a clique in P is similar to finding a clique in an undirected graph.

To find a clique, REGIMap starts with an empty clique. To maximize the clique size, REGIMap finds a node that is connected to all nodes present at clique. In the case of many, REGIMap selects the one with the maximum number of arcs to the nodes outside the clique. This heuristic is applied to increase the chance of maximizing the clique size. When the size of a clique cannot be further increased, if finds a nodes outside the clique connected to all nodes inside the clique expect one. In that case, REGIMap replaces this new node with the old one and tries to further increase the size of the clique. When this condition is checked for all nodes at P REGIMap stores this clique in a list.

REGIMap later finds the intersect of pairs of cliques. This intersect is the next initial clique to be maximized. At the end, REGIMap returns the maximum clique it finds after the above steps. At every step to increase a clique size, when a nodes is selected, the sum of the weights of outgoing arcs from the selected node to all nodes in clique is checked to be less than N_R. If a node violates this condition, it cannot be added to the clique. Note that during intersection of cliques, there is no need to verify the number of allocated registers.

Once the size of the clique increases to $|V_D|$, the placement is complete. A clique at the end of this step, represents the placement of operations and the sum of weight of outgoing arcs of a node, determines how many registers are used at all PEs. If the size of maximum clique is less than $|V_D|$, some nodes in D could not be placed, thus REGIMap, change those nodes at the next step.

E. REGIMAP ALGORITHM

REGIMap initially extracts the minimum II using technique in [15]. Then operations are scheduled with the goal of minimizing II. In the next step, REGIMap constructs a time extended resource graph. In this graph, only PEs are present. Afterwards, a compatibility graph P is to be generated from scheduled DFG and R_{II}. When P is constructed, a maximum clique $C = (V_C, E_C)$ in graph P where the sum of weight of outgoing arcs at all nodes is less than the register file size must be found. The mapping is completed when $|V_C| = |V_D|$. If REGIMap fails to find such a clique, it reschedules operations not present in the clique and tries again until a mapping is found. The REGIMap algorithm is presented in Algorithm 1.

Algorithm 1: REGIMap(Input D, Input $CGRA$)

begin
  $MII \leftarrow$ DetermineMII($D, |V_D|$);
  $S \leftarrow |V_C|; D_s \leftarrow D$;
  while *true* **do**
    $N \leftarrow \infty$;
    while *true* **do**
      $D_S, II \leftarrow$ Schedule(D_S, S);
      if $II > MII$ **then**
        $MII \leftarrow MII + 1$;
        $S \leftarrow |V_C|; D_S \leftarrow D$;
        break;
      $R_{II} \leftarrow$ Construct_Resource_Graph(C, MII);
      $P \leftarrow$ Construct_Compatibility_Graph(D_S, R_{II});
      $C \leftarrow$ Weight_Constrained_Max_Clique(P);
      if $|V_C| = |V_{D_s}|$ **then**
        Return C;
      else
        if $|V_{D_S}| - |V_C| > N$ **then**
          $S \leftarrow S - 1; D_S \leftarrow D$;
          break;
        else
          $D_S \leftarrow$ Re-Schedule($V_{D_S} - V_C$);
          $N \leftarrow |V_{D_S} - V_C|$;

Re-Scheduling: If a placement fails to place and allocate registers for some operations, the set of unplaced operations are rescheduled. REGIMap reschedules those operations to one cycle earlier than their current schedule cycle, or insert extra routing nodes if failed operation requires register for placement (relaxing routing problem). If the schedule cycle of an operation is decreased to a cycle lower than least feasible one (determined by scheduling), or adding extra nodes increases lower bound II, then the DFG is to be rescheduled with a new heuristic. At this step, the number of available PEs is set to be $N - 1$ where N was the number of available PEs at the previous scheduling attempt. N is initially set to the number of PEs. Please note that decreasing the number of available resources can increase II at scheduling. In such a case, REGIMap

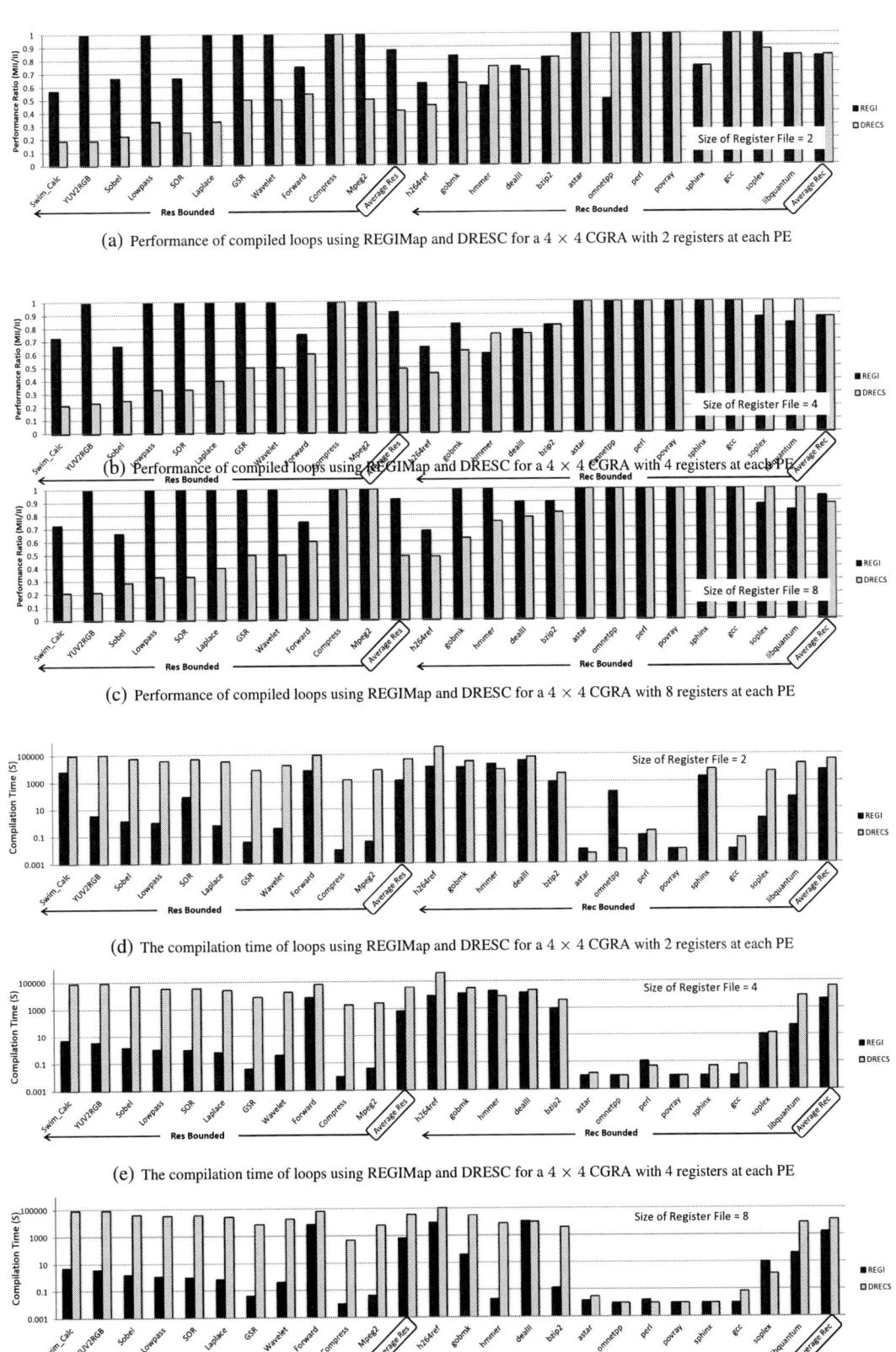

(a) Performance of compiled loops using REGIMap and DRESC for a 4×4 CGRA with 2 registers at each PE

(b) Performance of compiled loops using REGIMap and DRESC for a 4×4 CGRA with 4 registers at each PE

(c) Performance of compiled loops using REGIMap and DRESC for a 4×4 CGRA with 8 registers at each PE

(d) The compilation time of loops using REGIMap and DRESC for a 4×4 CGRA with 2 registers at each PE

(e) The compilation time of loops using REGIMap and DRESC for a 4×4 CGRA with 4 registers at each PE

(f) The compilation time of loops using REGIMap and DRESC for a 4×4 CGRA with 8 registers at each PE

Figure 9: The performance of compiling and compilation time of loops using REGIMap versus DRESC for different size of register file at PEs.

increases MII by one and reset the number of available resources to be the number of PEs. When MII increases, REGIMap proceeds with a new scheduling and placement attempt.

F. HOW IS THE PROBLEM OF REGISTER ALLOCATION ON CGRA DIFFERENT THAN ON VLIW ARCHITECTURES?

Register allocation in the problem of mapping application to VLIW architectures is essentially different from CGRAs. In a VLIW architectures, the register file is central providing connection to all functional units at a high bandwidth. Therefore, when dependent operations mapped to separate resources, data dependency between them can readily satisfied using central register file. In CGRA, however, the register files are distributed along with functional units or PEs. When two dependent operation are mapped to separate resources, data should explicitly routed between them. Therefore, communication cost is proportional to the distance of resources where dependent operation are mapped. This implies a tight dependency between placement and register allocation in CGRA application mapping while it is not the case for VLIW architectures where communication cost is approximately constant.

If registers are to be allocated after scheduling and placement in CGRAs, in the case of a failure in register allocation, the cost of inserting spill code is to reconstruct an entire new mapping. This cost is primarily due to the modulo scheduling where the execution is repetitive. Insertion of spill code is a typical technique utilized in compilers for VLIW processor. Therefore, a different policy for register allocation is required in the area of CGRA application mapping.

G. MORE RESULTS

Application	REGIMap IPC
Swim_Calc	10.52
YUV2RGB	12.33
Sobel	9.67
Lowpass	13.5
SOR	12.5
Laplace	11.5
GSR	12
Wavelet	10
Forward	7.25
Compress	9
Mpeg2	10
Average Res- Bounded	10.75

Table 1: Instruction per cycle (IPC) of mapping of res-bounded loops of CGRA using REGIMap.

Polyhedral Model based Mapping Optimization of Loop Nests for CGRAs[*]

Dajiang Liu, Shouyi Yin, Leibo Liu, Shaojun Wei
Institute of Microelectronics
Tsinghua University, Beijing 100084, China
liudj09@mails.tsinghua.com,{yinsy,liulb,wsj}@tsinghua.edu.cn

ABSTRACT

The coarse-grained reconfigurable architecture (CGRA) is a promising platform that provides both high performance and high power-efficiency. The compute-intensive portions of an application (e.g. loops) are often mapped onto CGRA for acceleration. To optimize the mapping of loop nests to CGRA, this paper makes two contributions: i) Establishing a precise CGRA performance model and formulating the loop nests mapping as a nonlinear optimization problem based on polyhedral model, ii) Extracting an efficient heuristic loop transformation and mapping algorithm (PolyMAP) to improve mapping performance. Experiment results on most kernels of the PolyBench and real-life applications show that our proposed approach can improve the performance of the kernels by 21% on average, as compared to one of the best existing mapping algorithm, EPIMap. The runtime complexity of PolyMAP is also acceptable.

Categories and Subject Descriptors

C.3 [**Special-purpose and application-based systems**]: Real-time and embedded systems; D.3.4 [**Processors**]: Code generation, Compilers, Optimization

General Terms

Algorithms, Design, Performance

1. INTRODUCTION

The CGRA is becoming a promising platform that provides the potential for high performance, high energy efficiency, high flexibility and low cost. A CGRA is typically constituted of a host controller and a 2-D mesh processing element array (PEA), which is shown in Figure 1. The host

---

[*]This work was supported in part by the NNSF of China grant (No.61274131), the International S&T Cooperation Project of China grant (No. 2012DFA11170), the Tsinghua Indigenous Research Project (No.20111080997) and the China National High Technologies Research Program (No. 2012AA012701).

Permission to make digital or hard copies of all or part of this work for personal or classroom use is granted without fee provided that copies are not made or distributed for profit or commercial advantage and that copies bear this notice and the full citation on the first page. To copy otherwise, to republish, to post on servers or to redistribute to lists, requires prior specific permission and/or a fee.
DAC '13, May 29 - June 07 2013, Austin, TX, USA.

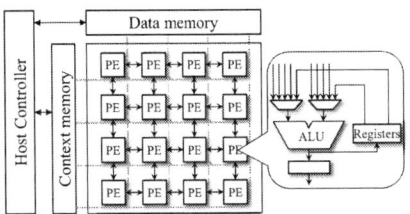

Figure 1: General Architecture of CGRA

controller controls the executions of the whole system. And each PE includes an ALU and a register file. The size of PEA can vary from 4 PEs arranged in a row up to 64 PEs arranged in an 8×8 grid. The functionality of PE could be configured to be different word-level operations of fixed-point numbers. The routing style of PEs also has great variety, such as connections between two PEs from neighbor rows, connections between each PE and its four or eight neighbors, buses connecting each node to other nodes in the same row or column. The degree of register file sharing also ranges from individual register files at each PE, to multiple register files each shared by a subset of PEs, to a single central register file shared by several or all PEs. On one hand these hardware characters bring CGRA with the potentials of high performance and high power-efficiency, but on the other hand the special features make the design of CGRA compiler is much more complex than a regular compiler.

Since the compute-intensive portions of an application (e.g. loops) are often mapped onto CGRA for acceleration, the most important task of CGRA compiler is automatically and efficiently mapping loop nests onto PEA. The mapping of loop nests could be partitioned into two subproblems, namely, loop transformation, and loop body placement and routing (*P&R*). The *P&R* of the loop body is to decide which PE to perform an operation and how to route data between PEs which has been well studied in the past several decades [1], while loop transformation is still the most challenging problem. To efficiently map loops on CGRAs, several loop transformation techniques have previously been proposed. Loop unrolling [2][3] is a common technique to generate a mapping scheme with greater parallelism. It unrolls a loop and transforms the unrolled loop into a data flow graph (DFG). Then the DFG is mapped onto CGRA. Modulo scheduling is another widely used technique, which improves the parallelism of CGRA by overlapping the execution of different iterations of a loop. In the modular scheduling works [4][5], initial interval (II) is usually used as a performance metric to guide the loop transformation. Recently, loop affine transformation is applied to CGRA. For example, in [6], loop affine transformation is used to optimize the PE utilization rate and communication cost between PEA and controller.

However, all the previous works have two major drawbacks. First, the CGRA's architectural features are not well considered in loop transformation. The major architectural difference between CGRA and general purpose processor (GPP) is the reconfiguration mechanism of CGRA. The CGRA needs additional clock cycles to finish reconfiguration, while GPP has no such feature. As a result, the traditional performance metrics are not accurate and effective for CGRA. For example, in [4] and [5], II is used as performance metric as traditional software pipelining, which cannot reflect reconfiguration cost of CGRA. In [6], although PE utilization and communication cost are considered, the reconfiguration cost is also missed. Therefore, a new performance metric which reflects the overall CGRA's architectural features (in other words, the performance influencing factors, such as reconfiguration, communication between controller and PEA, and 2-D parallelism) must be designed for loop transformation. Second, most previous works handle only the innermost loop of a loop nests and the parallelism is not explored adequately. For example, both loop unrolling and modulo scheduling usually deal with the innermost loop. Due to the 2-D topology of CGRA, the innermost 2-level loops should be taken into account to explore the parallelism.

Towards improving the whole performance of CGRA, this paper makes the following two important contributions:

1. **CGRA performance model establishment:** By analyzing the operation mechanism of a CGRA, we establish an analytic total execution time (TET) model that takes all the performance influencing factors of CGRA into an overall consideration based on polyhedral model[7]. Taking this TET model as an optimization target, we formulate the mapping of loop nests on CGRA as a constrained optimization problem.

2. **PolyMAP approach formation:** Based on genetic algorithm [8], we extract an efficient heuristic for the above optimization problem and further form a loop transformation and mapping approach (PolyMAP) to improve the performance of CGRA.

The rest of the paper is organized as follows: In section 2, we give the basic idea of our work and why a unified CGRA TET metric is needed. Then, section 3 describes the formulation of the optimization problem based on polyhedral model. Section 4 gives an efficient solution of the optimization problem and the whole mapping process. Then, section 5 gives the experimental results that demonstrate the effectiveness of our optimization works. At last, we conclude in section 6.

2. BASIC IDEA

Since the ultimate goal of loop mapping is to minimize the total execution time of the entire loop, the most rational way is to use the total execution time (TET) as performance metric to guide loop mapping. Due to the architectural features, the TET of CGRA is affected by the reconfiguration of PEA, data communication between the host controller

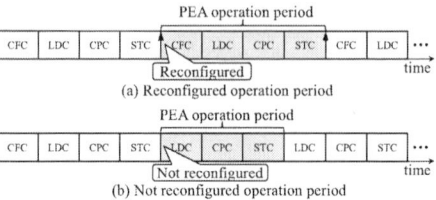

Figure 2: Operation mechanism of a CGRA

and PEA, and computation of PEA. Moreover these factors are correlated. Therefore simply taking one or two factors (e.g. II and communication cost) as optimization targets may lead to poor mapping performance. The only possible way is to build a unified analytical model of TET.

The operation mechanism of a CGRA is shown in Figure 2, where CFC, LDC, CPC and STC denotes reconfiguration cycles, data loading cycles, PEA computing cycles and data storing cycles, respectively. In the process of executing a task, CGRA first reads configuration context from context memory to configure each PE. Then, input data is loaded from data memory for preparation of computing. Next, the PEA performs the computing of current parts of the task. Finally, the result data of this PEA operation is stored in data memory. Again and again like this, the whole task is executed, as shown in Figure 2(a). At some time, if the functionality of an PEA operation is the same as the previous one, configuration cycles are not needed in this PEA operation, as the case in Figure 2(b). Considering that most programs consist of a number of consecutive PEA operations $p = [1, P]$ with different characteristics, CGRA performance can be defined in terms of the operating frequency f, reconfiguration cycles, data communication cycles (CMC) (data loading and data storing) and PEA computing cycles as follow:

$$TET = \frac{1}{f} \sum_{p \in P} (CFC_p + LDC_p + CPC_p + STC_p) \quad (1)$$

As depicted in Figure 2(b), the CFC_p could be passed over in some PEA operations. Thus, we give a 0-1 variate x_p to distinguish the reconfigured and not reconfigured PEA operations. In addition, the reconfiguration time of PEA is a constant (denoted as CFC) in practice. As a result, the total configuration cycles of a task can be represented as follow:

$$\sum_{p \in P} CFC_p = \sum_{p \in P} x_p \cdot CFC \quad (2)$$

As shown in Figure 2, the communication cycles (CMC_p) in a PEA operation includes data loading cycles (LDC_p) and data storing cycles (STC_p):

$$\sum_{p \in P} CMC_p = \sum_{p \in P} LDC_p + STC_p \quad (3)$$

The computing cycle has a close relation to the PEA route style (RS) and the data dependence length (L_e^p) of input applications. So the computing cycles of a PEA operation could be represented as a function of RS and L_e^p:

$$\sum_{p \in P} CPC_p = \sum_{p \in P} Func(RS, L_e^p) \quad (4)$$

From the above discussion, the CGRA TET could be represented by variables x_p, LDC_p, STC_p and L_e^p. And all these variables are closely related to the form of loop transformations. For example, the transformed loop body would lead to different times of reconfiguration and have different data dependence lengths compared to the original one. Moreover the data volume that need to be transferred between loop instances would be also different. Therefore, there should be an optimal loop transformation that generates optimal variable x_p, LDC_p, STC_p and L_e^p leading to minimal TET. Accordingly, we need to formulate TET as a function of loop transformation. Then by minimizing TET, the optimal loop transformation and mapping could be achieved.

3. PROBLEM FORMULATION

To formulate TET and the optimization problem, we leverage the polyhedral model and affine transformation theory.

3.1 Polyhedral Model based Loop Transformation for CGRAs

The polyhedron model[7] provides a powerful abstraction to the transformation of loop nests by viewing every iteration of each statement as an integer point in a well-defined space called polyhedron. Then, affine transformations for optimization are performed on the such representation. The transformations finally reflect in the generated code with improved execution performance. It is feasible for Static Control Part (SCoP) of a program, where the loop bounds, *if* the conditions and array subscripts are made of affine expressions involving only outer loop iterations, integer constants and integer literals. The detailed information about the polyhedral model is presented in Appendix A. In this paper, we just consider the mapping of innermost two loops of multi-level loop nests. As shown in Figure 3(b), the original iteration domain of an input loop nest is a rectangle, and it would be affine transformed(i.e., finding two hyperplanes Θ and Π) into a new iteration domain depicted as a parallelogram in Figure 3(e). At last, the new iteration domain will be tiled into small tiles mapped onto PEAs.

The input loop nests are divided into two parts: small loop body (the number of operators is less than the number of PEs in a PEA) and big loop body (the number of operators is more than the number of PEs in a PEA), which undergo different processes. For small loop body loops, we combine one or more loop bodies into a big body matching the size of PEA. For the big loop body loops, we would first distribute [9] them into small loops and handle them one by one. In practice, the size of PEA could be big enough (e.g., 8×8) to hold loop bodies in most cases. In addition, our optimization is applied to perfectly nested loops. For imperfectly nested loops, the embedding approach proposed in [10] could convert the imperfectly nested loops into perfectly nested loops. As a result, our proposed approach could handle most cases of computation-intensive applications.

3.2 PEA Resource Tile

Let $L(M, N, W_{lb}, L_{lb})$ be the input 2-D nested loops, where M, N indicate the bounds of outer and inner loop and W_{lb}, L_{lb} indicate the size of the loop body, as shown in Figure 3(a). As the loop body is small, the $W_{lb} \times L_{lb}$ loop body size could be determined simply by an efficient $P\&R$ algorithm[1]. Let W_{pea}, L_{pea} be the size of 2-D PEA, shown in Figure 3(c). We define a PEA Resource Tile (PRT) to indicate the maximal size of iterations a PEA could hold, as shown in Figure 3(d). The size of PRT is $\eta \times \zeta$, which could be simply obtained by

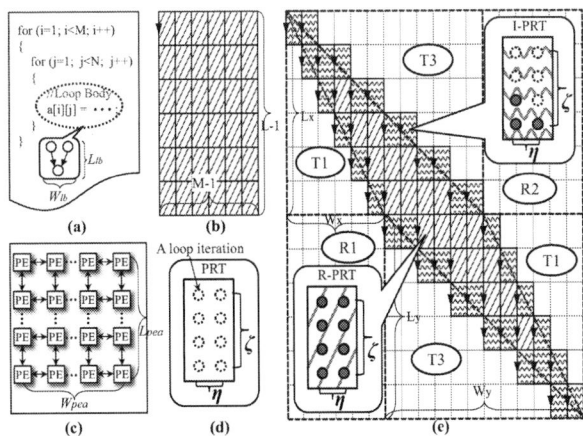

Figure 3: (a)A loop nest; (b)Original Iteration domain covered by PRTs; (c)PEA size; (d)PRT size; (e)Transformed iteration domain covered by PRTs

PEA size (W_{pea}, L_{pea}) and loop body size as follow:

$$\eta = \lfloor \frac{W_{pea}}{W_{lb}} \rfloor, \zeta = \lfloor \frac{L_{pea}}{L_{lb}} \rfloor \qquad (5)$$

Let the affine transformation coefficients of hyperplane Θ and hyperplane Π are (c_1, c_2) and (d_1, d_2). In order to guarantee the transformed iteration domain is compact, we give the unimodular transformation constraints:

$$|c_1 d_2 - c_2 d_2| = 1 \qquad (6)$$

Then, the transformed parallelogram (Figure 3(e)) would be covered by PRTs one by one. Every PRT on the parallelogram indicates a PEA operation of CGRA. As depicted in Figure 3(e), the slanting-line rectangle indicates a Regular PRT (R-PRT) that is full of iterations and the wave-line rectangle indicates an Irregular PRT (I-PRT) that is at the edge of the parallelogram and not full of iterations. The total number of PRT (T-PRT) is the sum of the R-PRT and I-PRT. We use $N_{rp}, N_{ip}, N_{tp}, N_{row}$ to indicates the number of R-PRT, I-PRT, T-PRT and PRT rows, respectively. In Appendix B, we give the derivative process of the four parameters.

3.3 Determination of Optimization Target

In this subsection, we will further consummate the performance model as an optimize target with the number of PRT and the coefficients of loop transformation based on polyhedral model.

3.3.1 Configuration Cycles

After the original loop nest is transformed and tiled with PRT, we now could determine configuration cycles. As shown in Figure 3(e), the loop nest would be executed from top to bottom, from left to right one PEA operation by one PEA operation and a PEA operation needs to be reconfigured if its functionality is changed from the previous PEA operation. The switch between two nearby I-PRTs needs a reconfiguration (indicated by an arrow in Figure 3(e)) and the switch between an R-PRT and an I-PRT also needs a reconfiguration (also indicated by an arrow) because the functionality of the two adjacent PRTs is different due to the different number of iterations they hold. On the other hand, the switch between two R-PRTs does not need any reconfigurations. As a result, the reconfiguration times could be

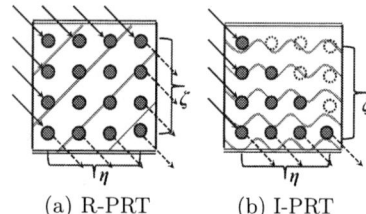

(a) R-PRT	(b) I-PRT

Figure 4: Communication volume of PRTs

roughly represented as follows:

$$N_{cfg} = N_{ip} + N_{row} \tag{7}$$

Then we put the result of Equation 7 into Equation 2, we obtained the whole reconfiguration cycles of the input loop task:

$$\sum_{p \in P} CFC_p = (N_{ip} + N_{row}) \cdot CFC \tag{8}$$

3.3.2 Communication Cycles

Let $\vec{i_s^e}$ and $\vec{i_t^e}$ be the iteration index of the source node and target node of dependence e in a loop nest. The set of dependence (E) could be calculated by the dependence polyhedron as described in Appendix A. As the work in [7], we take the number of hyperplanes the dependence e traverses as a communication cost function δ. Then, the communication cost of hyperplane Θ and Π could be represented as follows:

$$\delta_e(\Theta) = \left(\Theta(\vec{i_t^e}) - \Theta(\vec{i_s^e}) \right), \quad \delta_e(\Pi) = \left(\Pi(\vec{i_t^e}) - \Pi(\vec{i_s^e}) \right) \tag{9}$$

We first consider the communication volume of the R-PRT, which holds $\eta \times \zeta$ loop iterations. As shown in Figure 4(a), the communication volume of one R-PRT could be represented as follows:

$$2 \sum_{e \in E} (\eta \cdot \delta_e(\Theta) + \zeta \cdot \delta_e(\Pi) - \delta_e(\Theta)\delta_e(\Pi)) \tag{10}$$

The communication volume of I-PRT (Figure 4(a)) is different because some PEs are not occupied. Thus, the communication volume of I-PRT is a part of R-PRT and we give a scale factor $\beta \in (0,1)$ to indicate the difference. Therefore, the communication volume of I-PRT could be represented as follow:

$$2\beta \sum_{e \in E} (\eta \cdot \delta_e(\Theta) + \zeta \cdot \delta_e(\Pi) - \delta_e(\Theta)\delta_e(\Pi)) \tag{11}$$

Actually, once the loop nest is transformed and tiled, β could be determined. For instance, as depicted in Figure 4(a), the communication volume of the R-PRT is 14 (7 inputs and 7 outputs) and the communication volume of I-PRT (Figure 4(b))is 8 (4 inputs and 4 outputs). As a result, the value of β in this transformation is $8/14 \approx 0.57$.

Considering Equation 3, the total number of communication cycles of the loop task is obtained:

$$\sum_{p \in P} (CMC_p) = 2N_{rp} \sum_{e \in E} (\eta \cdot \delta_e(\Theta) + \zeta \cdot \delta_e(\Pi) - \delta_e(\Theta)\delta_e(\Pi)) \\ + 2N_{ip}\beta \sum_{e \in E} (\eta \cdot \delta_e(\Theta) + \zeta \cdot \delta_e(\Pi) - \delta_e(\Theta)\delta_e(\Pi)) \tag{12}$$

3.3.3 Computing Cycles

As analyzed in Equation 4, the computing cycles (CPC) has a close relationship with the RS and L_e^p. We also define the length of dependence as the number of hyperplanes a dependence e transverse from the source node to target node.

So the modified computing cycles could be represented as follow:

$$\sum_{p \in P} (CPC_p) = N_{tp} Func \left(RS, \Theta(\vec{i_t^e}) - \Theta(\vec{i_s^e}), \Pi(\vec{i_t^e}) - \Pi(\vec{i_s^e}) \right) \tag{13}$$

There are mainly three interconnection topologies of CGRA[11], Mesh, Mesh_plus and Morphosys. In the Mesh style CGRAs, a PE connects to its 4 neighbor PEs, $Func(\cdot)$ could be represented as $Max_{e \in E}(L_e^p) - 1$ as dependence more than 1 would be passed by registers. In the Mesh_plus CGRAs, $Func(\cdot)$ could be represented as $Max_{e \in E}(L_e^p) - 2$ as dependence more than 2 would be passed by registers. In the Morphosys CGRAs, $Func(\cdot)$ could be represented as $Max_{e \in E}(L_e^p) - 4$ as dependence more than 4 would be passed by registers.

From all the discussion above, now we could give the integral optimization target base on affine transformation:

$$TET = (N_{ip} + N_{row}) \cdot CFG \\ = 2N_{rp} \sum_{e \in E} (\eta \cdot \delta_e(\Theta) + \zeta \cdot \delta_e(\Pi) - \delta_e(\Theta)\delta_e(\Pi)) \\ + 2N_{ip}\beta \sum_{e \in E} (\eta \cdot \delta_e(\Theta) + \zeta \cdot \delta_e(\Pi) - \delta_e(\Theta)\delta_e(\Pi)) \\ + N_{tp} Func \left(RS, \Theta(\vec{i_t^e}) - \Theta(\vec{i_s^e}), \Pi(\vec{i_t^e}) - \Pi(\vec{i_s^e}) \right) \tag{14}$$

3.4 Determination of Optimization Problem

Based on the preparation above, now we could establish the optimization problem as follow:

$$Minimize \quad TET$$
$$Subject\ to \quad \Theta(\vec{i_t^e}) - \Theta(\vec{i_s^e}) \geq 0, \quad e \in E \tag{15a}$$
$$\Pi(\vec{i_t^e}) - \Pi(\vec{i_s^e}) \geq 0, \quad e \in E \tag{15b}$$
$$|c_2 d_1 - c_1 d_2| = 1 \tag{15c}$$
$$c_1, c_2, d_1, d_2 \in \mathbb{Z} \tag{15d}$$

Where Equation 15a and Equation 15b give the constraints of legal affine transformation, which is the same as the work in [7]. Equation 15c gives the constraint of tightness of transformed space. Based on the performance metric, we could find the optimal coefficients (c_1, c_2) and (d_1, d_2) to perform affine transformation.

4. EFFICIENT SOLUTION AND POLYMAP

The calculation of TET in Equation 14 is non-convex and an enumerate-based approach is needed. Instead of enumerating all legal transformation coefficients in a brute-force way, we used the features of constraints to greatly reduce the computation complexity of finding optimal solutions for large design space. We note that the condition 15c is an equation of the 4 variables c_1, c_2, d_1 and d_2 and we could reduce the dimensionality of the problem to 3. Thus, the variable d_2 could be expressed by the combination of c_1, c_2 and d_1. As we only consider the innermost two loops of loop nests, the design space is not quite big in fact. Thus, we adopt Genetic Algorithm[8] to solve the problem to guarantee the accuracy of the results. In order to crossover the variables, a N bits binary B_N is used to indicate a variable, where the MSB indicates the sign bit and the rest indicates the value. Thus, a solution of the problem (c_1, c_2, d_1) could be represented by a $3N$ bits binary (chromosome). We use the inverse of TET as the fitness function and the definition domain (DD) of variables is formed by inequality 15a, 15b and 15d.

Algorithm 1 LoopTrans($M = 4, G = 32, N = 4, P_c = 0.65, P_m = 0.003$)

1: $M \leftarrow$ PopulationNumber;
2: $G \leftarrow$ MaxGenerationNumber;
3: $N \leftarrow$ VariableWidth;
4: $i \leftarrow$ FirstGenerationNumber;
5: $P_i^j \leftarrow$ Random(DD), $j \in (1, M)$;
6: **repeat**
7:    Fitness(P_i^j) $\leftarrow 1/\text{TET}(P_i^j)$;
8:    $P_i^j \leftarrow$ RouletteWheel(P_i^j,Fitness(P_i^j));
9:    **for** $j \leftarrow 1$ to M/2
10:     **if** (Random(0,1) $> P_c$) **then**
11:      **repeat**
12:       $CrossBit \in \leftarrow$ Random($1, 3N - 1$);
13:       $P_i^j, P_i^{j+M/2} \leftarrow$ Crossover($P_i^j, P_i^{j+M/2}, CrossBit$);
14:      **until** $P_i^j, P_i^{j+M/2} \in DD$
15:    **for** $j \leftarrow 1$ to M
16:     **if** (Random(0,1) $> P_m$) **then**
17:      **repeat**
18:       $P_i^j \leftarrow$ Mutation(P_i^j);
19:      **until** $P_i^j \in DD$
20:    **for** $j \leftarrow 1$ to M
21:     $P_{i+1}^j \leftarrow P_i^j$, $j \in (1, M)$;
22:**until** i > G

In **Algorithm 1**, i) we first initialize the population number ($M = 4$), the maximum generation number ($G = 32$) and the bit width ($N = 4$) of the 4 variables (c_1, c_2, d_1 and d_2). ii) Then, we randomly generate 4 different solutions $P_i^j, j \in (0, 4)$ in DD. iii) Next, we calculate the fitness function (inverse of TET) of every solution P_i^j. iv) Then roulette wheel scheme[8] is used for selection population. v) Next, we perform one point crossover to generate new solutions in DD with the probability P_c. vi) Next, we execute one point mutation for every solutions with the probability P_m. vii) Then, we get the next generation population from step v) and vi). viii) Repeat the previous 5 step until exceeding the maximal generation number. At last, we get optimized transformation coefficients c_1, c_2, d_1 and d_2; Obviously, the time complexity of this Genetic Algorithm based algorithm is $O(G \cdot M \cdot N)$.

Based on **Algorithm 1**, the complete loop mapping scheme, named PolyMAP, is designed, as shown in Figure 5. We take loop kernels in high-level specifications like C/C++ as input and preprocess the kernels for easy management (e.g., branch removal and imperfect loops conversion). Then, we compare the loop body size with PEA size to decide if loop distribution is needed. Next, we analyze the polyhedral model intermediate represent (IR) with dependence by LLVM-Polly framework[12]. With the CGRA architecture description, TET optimization problem is established and transformation is performed with the optimal coefficients. At last, CGRA contexts are generated from the new polyhedral IR.

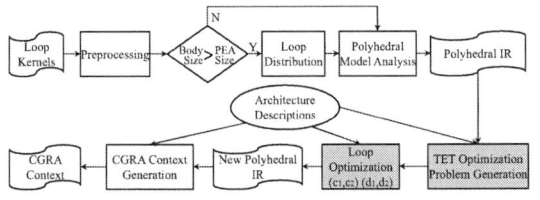

Figure 5: The Processing Flow of PolyMAP

5. EXPERIMENT RESULTS

We have taken some loops from Polybench2.0[13] and media applications H.264 decoder for experiments.

The target CGRA architecture is a 8×8 homogeneous PEA and every PE could perform fixed-point and logic operations, which is an instance of ADRES [11] template. Each PE has a access to the data memory and the interconnection topology is Morphosys. Each PE has its own local RF consisting of 5 rotating registers with 5 read and 5 write ports, which could be accessed by the local ALU and its 4 neighbor PEs. The configuration cycles of one PEA operation is 2 cycles if needed.

To evaluate the effectiveness of our approach (PolyMAP) we select two of the latest presented approaches as references. The first reference is EPIMap[5]. EPIMap used recomputation for resource limitation and obtained optimized II for a single-level loop. The second reference is communication minimized optimization of nested loop based on polyhedra model[6](labeled PolyCOM), where the communication of PEA is taken as an optimization metric. All the evaluations were taken on an Intel Dual-Core CPU machine running at 1.9GHz with 2GB memory. Mapping results were verified with cycle accurate simulator.

5.1 The Accuracy of Performance Model

Our optimization approach is based on the deduced performance model in Equation 14. Thus, the accuracy of the performance model is very important to the results of the optimization. To evaluate the accuracy of our performance model, we use PolyMAP to mapping loops onto CGRA and give a report of execution time obtained by performance model. Then, the experimental results is also obtained by running configurations on the target CGRA. Figure 6(a) give the comparison between performance model and actual experiments results. We can see the execution time obtained by performance model is accurate enough compared with the actual experimental result. Consequently, our mapping optimization based on the performance model is well-founded.

5.2 The Execution Time Comparison

The performance comparison of three different mapping techniques is shown in Figure 6(b).

Comparison with PolyCom. Although both PolyCom and PolyMap use polyhedra model as the loop optimization tools to perform transformation, PolyMap always outperforms PolyCom by more than 36% on average. The poor performance of PolyCom is due to the improper performance metric. Figure 7 is used to give further explanation. The goal of PolyCom is to minimize the communication between host controller and PEA. Indeed, PolyCom works quit well for communication minimization. As shown in Figure 7, the communication cycles (t_{com}) of PolyCom on kernel adi and gemm are less than that of EPIMap and PolyMap. However, the communication cycles is only part of total execution time. The excessive optimization of communication cycles results in exacerbation of reconfiguration cost and computation cost. For instance, the configuration cycles (t_{cfg}) of PolyCom are more than that of EPIMap and PolyMAP, as shown in Figure 7(a). On the other hand, PolyMAP uses the unified metric, TET, which can achieve global optimial result. As a result, the overall performance of PolyMap exceeds that of PolyCom.

Comparison with EPIMap. As shown is Figure 6(b), PolyMap performs better than EPIMap by more than 21% on average. EPIMap works well on the mapping of resource limited CGRA and could achieve the minimized II on most

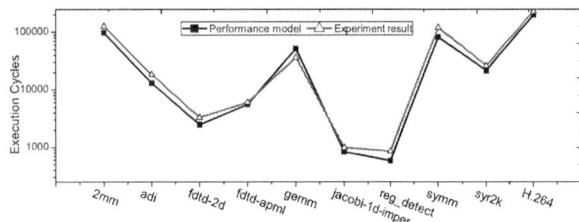

(a) The accuracy of the proposed performance model

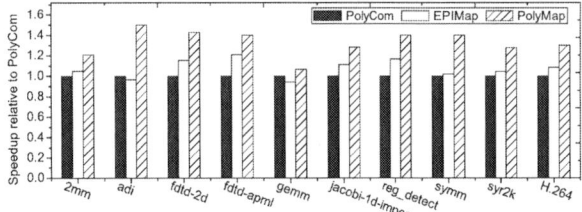

(b) Speedup of PolyMap compared to PolyCom and EPImap

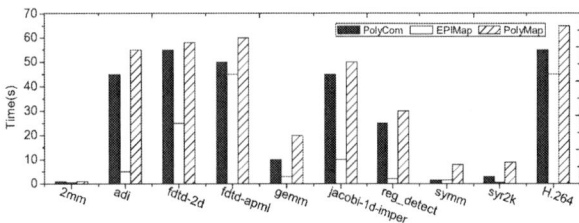

(c) Compilation time of PolyMap compared to PolyCom and EPImap

Figure 6: Experiments of PolyMap compared to PolyCom and EPImap

cases. However, take II as a metric can not fully reflect the performance of CGRA since CGRA has a special operation mechanism. When we take other performance influencing factors into account(e.g., reconfiguration), II minimized mapping may not be the mapping that could achieve highest performance. For instance, the configuration cycles t_{cfg} of EPIMap are more than that of PolyMap in Figure 7. In addition, EPIMap only maps the innermost loop of a nested loop and the communication overhead of the outer loop is neglected. In contrast, PolyMAP deals with the innermost 2-level loops which leads to greater parallelism.

5.3 The Compilation Time

Figure 6(c) demonstrated the compilation time of PolyMAP for different kernels, as compared to PolyCom and EPIMap. The compilation time of PolyMAP varies from several milliseconds to dozens of seconds and more than that of PolyCom and EPIMap on average. Actually, the number of dependence is an important factor influencing the compilation time. For instance, *adi,dftd-2d,fdtd-apml* and *jacobi-1d-imper* have data dependence more than two and they all have a longer compilation time in PolyCom and PolyMAP. As a result, PolyMAP achieves better average performance at the cost of compilation time to find the optimal coefficients to perform loop transformation. However, the actual compilation time is acceptable.

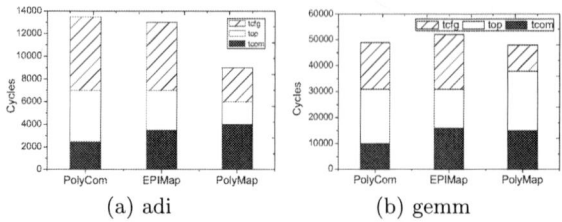

    (a) adi   (b) gemm

Figure 7: The time component parts of the three approaches on kernel adi and gemm

6. CONCLUSIONS

Many compute-intensive loops are often mapped on to CGRAs for acceleration and there is not a unified metric to perform loop transformations. To this end, we first build a CGRA performance model considering synthetically performance influencing factors and formulate an optimization problem for coefficients to perform affine transformations based polyhedral model. Then, we extract an efficient heuristic to reduce the searching space. At last, the effectiveness of our proposed algorithm is demonstrated with Poly-Bench and real-life applications.

7. REFERENCES

[1] W. Böhm, J. Hammes, B. Draper, M. Chawathe, C. Ross, R. Rinker, and W. Najjar. Mapping a single assignment programming language to reconfigurable systems. *The Journal of Supercomputing*, 21(2):117–130, 2002.

[2] O.S. Dragomir and K. Bertels. Extending loop unrolling and shifting for reconfigurable architectures. *Architectures and Compilers for Embedded Systems (ACES), Edegem, Belgium*, pages 61–64, 2010.

[3] S. Yin, C. Yin, L. Liu, M. Zhu, and S. Wei. Configuration context reduction for coarse-grained reconfigurable architecture. *IEICE TRANSACTIONS on Information and Systems*, 95(2):335–344, 2012.

[4] H. Park, K. Fan, S.A. Mahlke, T. Oh, H. Kim, and H. Kim. Edge-centric modulo scheduling for coarse-grained reconfigurable architectures. In *Proceedings of the 17th international conference on Parallel architectures and compilation techniques*, pages 166–176. ACM, 2008.

[5] M. Hamzeh, A. Shrivastava, and S. Vrudhula. Epimap: Using epimorphism to map applications on cgras. In *Design Automation Conference (DAC), 2012 49th ACM/EDAC/IEEE*, pages 1280–1287. IEEE, 2012.

[6] D. Liu, S. Yin, C. Yin, L. Liu, and S. Wei. Mapping optimization of affine loop nests for reconfigurable computing architecture. *IEICE transactions on electronics*, E95-D(12):1284–1290, 2012.

[7] U. Bondhugula, A. Hartono, J. Ramanujam, and P. Sadayappan. A practical and fully automatic polyhedral program optimization system. In *ACM SIGPLAN PLDI*, 2008.

[8] D.E. Goldberg. Genetic algorithms in search, optimization, and machine learning. 1989.

[9] O.S. Dragomir and K. Bertels. Loop distribution for k-loops on reconfigurable architectures. In *Design, Automation Test in Europe Conference Exhibition (DATE), 2011*, pages 1–6. IEEE, 2011.

[10] N. Ahmed, N. Mateev, and K. Pingali. Synthesizing transformations for locality enhancement of imperfectly-nested loop nests. *International Journal of Parallel Programming*, 29(5):493–544, 2001.

[11] F. Bouwens, M. Berekovic, B. De Sutter, and G. Gaydadjiev. Architecture enhancements for the adres coarse-grained reconfigurable array. *High Performance Embedded Architectures and Compilers*, pages 66–81, 2008.

[12] T. Grosser. *Enabling polyhedral optimizations in llvm*. PhD thesis, Masterąfs thesis, University of Passau, 2011.

[13] LN Pouchet. Polybench: The polyhedral benchmark suite (2011). *http://www-roc.inria.fr/ pouchet/software/polybench*.

APPENDIX

A. POLYHEDRAL MODEL

Polyhedra Model is based on four main concepts: the iteration domain, the access function, dependence polyhedra and affine transformation. A program part that can be represented using the polyhedron model is called a Static Control Part or SCoP for short.

The iteration domain is defined by a system of affine inequalities, $\mathcal{D}_S(\vec{i}_S) \geq \vec{0}$, derived from the bounds of loops surrounding statement S. Using matrix to present the inequalities, the iteration space polytope is presented as:

$$\mathcal{D}_S \cdot \begin{pmatrix} \vec{i}_S \\ \vec{g}_S \\ 1 \end{pmatrix} \geq \vec{0} \tag{16}$$

where $\mathcal{D}_S$ is a matrix of n affine constraints on the execution of statement S. $\vec{i}_S$ is the iteration vector of statement S and $\vec{g}_S$ is a vector of global parameters.

Each reference in a statement is also affine functions of loop indices and global parameters, which could also be represented using matricesčž

$$\mathcal{F}_{kAS}(\vec{i}_S) = \mathcal{F}_{kAS} \cdot \begin{pmatrix} \vec{i}_S \\ \vec{g}_S \\ 1 \end{pmatrix} \tag{17}$$

Where $\mathcal{F}_{kAS}$ is a matrix representing an affine mapping form the iteration space of statement S to the data space of array S .

The Polyhedral Dependence Graph (PDG) is a directed multi-graph with each vertex representing a statement, and an edge, $e \in E$, from node S_i to S_j representing a polyhedral dependence from a dynamic instance of S_i to one of S_j: it is characterized by a polyhedron,Pe, called the dependence polyhedron that captures the exact dependence information corresponding to e. The dependence polyhedron is in the sum of the dimensionality of the source and target statement's polyhedra. The h-transformation [7]h_e maps the target iteration vector $\vec{i}_t$ to the source iteration vector$\vec{i}_s$ that is the last access the same index of a array. So the dependence polyhedra can be represented as:

$$\mathcal{P}_e \equiv \left(\begin{array}{c|c} \mathcal{D}_s & \\ & \mathcal{D}_t \\ \hline & h_e \end{array} \right) \begin{pmatrix} \vec{i}_s \\ \vec{i}_t \\ \vec{g} \\ 1 \end{pmatrix} \begin{pmatrix} \geq & \vec{0} \\ \geq & \vec{0} \\ = & \vec{0} \end{pmatrix} \tag{18}$$

The affine transformation of a statement S is defined as an affine mapping that maps an instance of S in the original program to an instance in the transformed program. The transform function of a statement S is given by:

$$\Phi(\vec{i}_S) = \mathcal{T}_S \cdot \begin{pmatrix} \vec{i}_S \\ \vec{g}_S \\ 1 \end{pmatrix} \tag{19}$$

Where $\mathcal{T}$ is a row vector and the affine transformation is a one-dimensional mapping, which can be also called an affine hyperplane. When mapping loop nests on a 2-D PEA, we use II to denote row related hyperplane, and Θ to denote the column related hyperplane.

B. THE NUMBER OF PRTS

THEOREM 1. *Let W and L be the width and length of a right-angled triangle T. c_1, d_1 are mutually prime integers satisfying equation $W/L = c_1/d_1$. T is covered by minimal number of squares S with sides of 1. The S well within T is called S_{rp}. The S with portions in T and the left portions out of T is called S_{ip}. Let N_{rp}, N_{ip} and N_{tp} be the number*

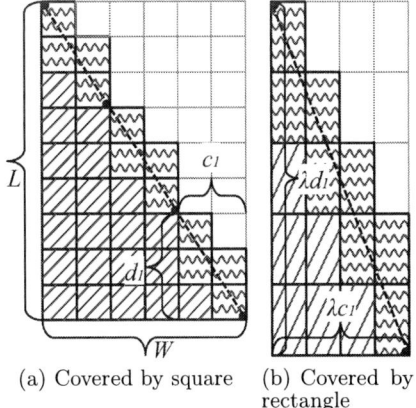

(a) Covered by square   (b) Covered by rectangle

Figure 8: Triangle covered by PRTs

of $S_{rp}s, S_{ip}s$ and all Ss. Then

$$N_{ip} = \frac{W}{c_1}(c_1 + d_1 - 1)$$

$$N_{rp} = \frac{1}{2}\left(WL - \frac{W}{c_1}(c_1 + d_1 - 1) \right)$$

$$N_{tp} = \frac{1}{2}\left(WL + \frac{W}{c_1}(c_1 + d_1 - 1) \right)$$

PROOF. As shown in Figure 8(a), we first consider the situation of smallest repeat triangle (SRT) with two mutually prime square edges c_1 and d_1. We note that the slanting edge crossed one edge of the two squares at the end of the slanting edge and crossed two edge of the squares inside the slanting edge. Thus, the number of squares that the slanting edge crossed is

$$\frac{2(c_1 - 1 + d_1 - 1) + 2}{2} = c_1 + d_1 - 1$$

Now we could extend to the situation of a right-angled triangle without two mutually prime square edges. With the condition $W/L = c_1/d_1$, N_{ip} could be obviously obtained

$$N_{ip} = \frac{W}{c_1}(c_1 + d_1 - 1)$$

The the number of the Ss covering the whole triangle T is the sum of area of T and half of N_{ip}, we obtained

$$N_{tp} = \frac{1}{2}\left(WL + \frac{W}{c_1}(c_1 + d_1 - 1) \right)$$

At last, the number of S_{rp} could be calculated by N_{ip} and N_{tp}

$$N_{rp} = N_{tp} - N_{ip} = \frac{1}{2}\left(WL - \frac{W}{c_1}(c_1 + d_1 - 1) \right)$$

□

THEOREM 2. *Let W and L be the width and length of a right-angled triangle T. c_1, d_1 are mutually prime integers satisfying equation $W/L = c_1/d_1$. T is covered by minimal number of rectangle R with width of η and length of ζ. The R well within T is called R_{rp}. The R with portions in T and the left portions out of T is called R_{ip}. Let N_{rp}, N_{ip} and N_{tp} be the number of $R_{rp}s, R_{ip}s$ and all Rs. Let $lcm(a, b)$ be the* *least common multiple of two integers a and b. Then*

$$N_{ip} = \lceil \frac{W}{\lambda c_1} \rceil \left(\frac{\lambda c_1}{\eta} + \frac{\lambda d_1}{\zeta} - 1 \right)$$

$$N_{rp} = \frac{1}{2} \left(\frac{WL}{\eta \zeta} - \lceil \frac{W}{\lambda c_1} \rceil \left(\frac{\lambda c_1}{\eta} + \frac{\lambda d_1}{\zeta} - 1 \right) \right)$$

$$N_{tp} = \frac{1}{2} \left(\frac{WL}{\eta \zeta} + \lceil \frac{W}{\lambda c_1} \rceil \left(\frac{\lambda c_1}{\eta} + \frac{\lambda d_1}{\zeta} - 1 \right) \right)$$

PROOF. As shown in Figure 8(b), we first determine the size of smallest repeat triangle (SRT). As the side of R is no longer 1, the sides of SRT are no longer c_1 and d_1. Actually, the sides of SRT are multiples of c_1 and d_1 and the multiply could be represented as

$$\lambda = lcm \left(\frac{lcm(c_1, \eta)}{c_1}, \frac{lcm(d_1, \zeta)}{d_1} \right)$$

Now we could consider the situation of SRT $(\lambda c_1 \times \lambda d_1)$, which could be normalized as situation in Theorem 1 by dividing the size of SRT by the size of R. As a result

$$N_{ip} = \left\lceil \frac{W}{\lambda c_1} \right\rceil \left(\frac{\lambda c_1}{\eta} + \frac{\lambda d_1}{\zeta} - 1 \right)$$

$$N_{rp} = \frac{1}{2} \left(\frac{WL}{\eta \zeta} - \left\lceil \frac{W}{\lambda c_1} \right\rceil \left(\frac{\lambda c_1}{\eta} + \frac{\lambda d_1}{\zeta} - 1 \right) \right)$$

$$N_{tp} = \frac{1}{2} \left(\frac{WL}{\eta \zeta} + \left\lceil \frac{W}{\lambda c_1} \right\rceil \left(\frac{\lambda c_1}{\eta} + \frac{\lambda d_1}{\zeta} - 1 \right) \right)$$

□

THEOREM 3. *Let $M - 1$ and $N - 1$ be the width and length of original rectangle O. (c_1, c_2) and (d_1, d_2) be the coefficients two affine transformations Φ and Π satisfying $|c_2 d_1 - c_1 d_2| = 1$. O is transformed by Φ and Π into a new parallelogram P shown in Figure 3(e). P is covered by minimal number of rectangle PRT with width of η and length of ζ. The PRT well within P is called $R - PRT$. The PRT with portions in P and the left portions out of P is called $I - PRT$. Let N_{rp}, N_{ip}, N_{tp} and N_{row} be the number of $R - PRT$s, $I - PRT$s, all PRTs and the rows of PRT. Then*

$$N_{ip} = 2\frac{M-1}{\lambda} \left(\frac{\lambda|c_1|}{\eta} + \frac{\lambda|d_1|}{\zeta} - 1 \right) + 2\frac{N-1}{\lambda'} \left(\frac{\lambda'|c_2|}{\eta} + \frac{\lambda'|d_2|}{\zeta} - 1 \right)$$

$$N_{tp} = \lceil \frac{M-1}{\eta} \rceil \lceil \frac{N-1}{\zeta} \rceil$$
$$+ \frac{M-1}{\lambda} \left(\frac{\lambda|c_1|}{\eta} + \frac{\lambda|d_1|}{\zeta} - 1 \right) + \frac{N-1}{\lambda'} \left(\frac{\lambda'|c_2|}{\eta} + \frac{\lambda'|d_2|}{\zeta} - 1 \right)$$

$$N_{rp} = \lceil \frac{M-1}{\eta} \rceil \lceil \frac{N-1}{\zeta} \rceil$$
$$- \frac{M-1}{\lambda} \left(\frac{\lambda|c_1|}{\eta} + \frac{\lambda|d_1|}{\zeta} - 1 \right) - \frac{N-1}{\lambda'} \left(\frac{\lambda'|c_2|}{\eta} + \frac{\lambda'|d_2|}{\zeta} - 1 \right)$$

$$N_{row} = \left\lceil \frac{|d_1|(M-1) + |d_2|(N-1)}{\zeta} \right\rceil$$

where,

$$\lambda = \begin{cases} \frac{\zeta}{|d_1|}, & c_1 = 0 \\ \frac{\eta}{|c_1|}, & d_1 = 0 \\ lcm(\frac{lcm(|c_1|, \eta)}{|c_1|}, \frac{lcm(|d_1|, \zeta)}{|d_1|}), & else \end{cases}$$

$$\lambda' = \begin{cases} \frac{\zeta}{|d_2|}, & c_2 = 0 \\ \frac{\eta}{|c_2|}, & d_2 = 0 \\ lcm(\frac{lcm(|c_2|, \eta)}{|c_2|}, \frac{lcm(|d_2|, \zeta)}{|d_2|}), & else \end{cases}$$

PROOF. Without loss of generality, we first discuss the case: $c_1 c_2 d_1 d_2 \neq 0$ and $|c_2 d_1| > |c_1 d_2|$. As shown in Figure 3(e), the size parameter L_x, W_x, L_y and W_y could be obviously represented as

$$W_x = |c_1|(M - 1)$$
$$L_x = |d_1|(M - 1)$$
$$W_y = |c_2|(N - 1)$$
$$L_y = |d_2|(N - 1)$$

Actually, the N_{pr} could be calculated by the number of $R - PRT$s in T_1, T_2, T_3, T_4 and the number of PRTs in R_1 and R_2. With Theorem 2, the number of the $R - PRT$s in the 6 mentiond parts is

$$\left(\frac{W_x L_x}{\eta \zeta} - \lceil \frac{W_x}{\lambda c_1} \rceil \left(\frac{\lambda c_1}{\eta} + \frac{\lambda d_1}{\zeta} - 1 \right) \right)$$
$$+ \left(\frac{W_y L_y}{\eta \zeta} - \lceil \frac{W_y}{\lambda' c_2} \rceil \left(\frac{\lambda' c_2}{\eta} + \frac{\lambda' d_2}{\zeta} - 1 \right) \right) + 2\frac{W_x L_y}{\eta \zeta}$$

The N_{tp} could be calculated by taking $R - PRT$s from the big rectangle $((W_x + W_y) \times (L_x + L_y))$

$$N_{tp} = \frac{(W_x + W_y)(L_x + L_y)}{\eta \zeta} - 2\frac{W_x L_y}{\eta \zeta}$$
$$- \left(\frac{W_x L_x}{\eta \zeta} - \lceil \frac{W_x}{\lambda c_1} \rceil \left(\frac{\lambda c_1}{\eta} + \frac{\lambda d_1}{\zeta} - 1 \right) \right)$$
$$- \left(\frac{W_y L_y}{\eta \zeta} - \lceil \frac{W_y}{\lambda' c_2} \rceil \left(\frac{\lambda' c_2}{\eta} + \frac{\lambda' d_2}{\zeta} - 1 \right) \right)$$
$$= \frac{W_y L_x - W_x L_y}{\eta \zeta} + \lceil \frac{W_x}{\lambda c_1} \rceil \left(\frac{\lambda c_1}{\eta} + \frac{\lambda d_1}{\zeta} - 1 \right)$$
$$+ \lceil \frac{W_y}{\lambda' c_2} \rceil \left(\frac{\lambda' c_2}{\eta} + \frac{\lambda' d_2}{\zeta} - 1 \right)$$
$$= \lceil \frac{M-1}{\eta} \rceil \lceil \frac{N-1}{\zeta} \rceil$$
$$+ \frac{M-1}{\lambda} \left(\frac{\lambda|c_1|}{\eta} + \frac{\lambda|d_1|}{\zeta} - 1 \right) + \frac{N-1}{\lambda'} \left(\frac{\lambda'|c_2|}{\eta} + \frac{\lambda'|d_2|}{\zeta} - 1 \right)$$

As shown in Figure 3(e), N_{ip} could be obtained by the $I - PRT$s in the 4 triangles, T_1, T_2, T_3 and T_4. Using Theorem 2, we obtained

$$N_{ip} = 2 \lceil \frac{W_x}{\lambda c_1} \rceil \left(\frac{\lambda c_1}{\eta} + \frac{\lambda d_1}{\zeta} - 1 \right)$$
$$+ 2 \lceil \frac{W_y}{\lambda' c_2} \rceil \left(\frac{\lambda' c_2}{\eta} + \frac{\lambda' d_2}{\zeta} - 1 \right)$$
$$= 2\frac{M-1}{\lambda} \left(\frac{\lambda|c_1|}{\eta} + \frac{\lambda|d_1|}{\zeta} - 1 \right) + 2\frac{N-1}{\lambda'} \left(\frac{\lambda'|c_2|}{\eta} + \frac{\lambda'|d_2|}{\zeta} - 1 \right)$$

Naturally, N_{rp} could be calculated by N_{ip} and N_{tp}

$$N_{rp} = N_{tp} - N_{ip} = \lceil \frac{M-1}{\eta} \rceil \lceil \frac{N-1}{\zeta} \rceil$$
$$- \frac{M-1}{\lambda} \left(\frac{\lambda|c_1|}{\eta} + \frac{\lambda|d_1|}{\zeta} - 1 \right) - \frac{N-1}{\lambda'} \left(\frac{\lambda'|c_2|}{\eta} + \frac{\lambda'|d_2|}{\zeta} - 1 \right)$$

Finally, the number of rows of PRTs could be calculated by

$$N_{row} = \frac{L_x + L_y}{\zeta} = \frac{|d_1|(M-1) + |d_2|(N-1)}{\zeta}$$

In the case of $c_1 c_2 d_1 d_2 = 0$, we redefine the parameter λ and λ' as follow

$$\lambda = \begin{cases} \frac{\zeta}{|d_1|}, & c_1 = 0 \\ \frac{\eta}{|c_1|}, & d_1 = 0 \\ lcm(\frac{lcm(|c_1|, \eta)}{|c_1|}, \frac{lcm(|d_1|, \zeta)}{|d_1|}), & else \end{cases}$$

$$\lambda' = \begin{cases} \frac{\zeta}{|d_2|}, & c_2 = 0 \\ \frac{\eta}{|c_2|}, & d_2 = 0 \\ lcm(\frac{lcm(|c_2|, \eta)}{|c_2|}, \frac{lcm(|d_2|, \zeta)}{|d_2|}), & else \end{cases}$$

With the newly defined λ and λ', the result of Theorem 3 could also apply to the case of $c_1 c_2 d_1 d_2 = 0$. □

Improving Energy Gains of *Inexact* DSP Hardware Through *Reciprocative Error Compensation*

Avinash Lingamneni
Dept. of ECE, Rice University
Houston, TX 77005, USA
avinash.l@rice.edu

Arindam Basu
School of EEE
Nanyang Technological University
50 Nanyang Ave, Singapore 639798

Christian Enz
Wireless & Integrated Systems Division, CSEM SA
Neuchatel, CH-2002
Switzerland

Krishna V Palem
Dept. of CS, Rice University
Houston, TX 77005, USA
palem@rice.edu

Christian Piguet
Wireless & Integrated Systems Division, CSEM SA
Neuchatel, CH-2002
Switzerland

ABSTRACT

We present a *zero hardware-overhead* design approach called *reciprocative error compensation*(REC) that significantly enhances the energy-accuracy trade-off gains in *inexact* signal processing datapaths by using a two-pronged approach: (a) deliberately redesigning the basic arithmetic blocks to effectively compensate for each other's (expected) error through *inexact logic minimization*, and (b) "reshaping" the response waveforms of the systems being designed to further reduce any residual error. We apply REC to several DSP primitives such as the FFT and FIR filter blocks, and show that this approach delivers 2-3 orders of magnitude lower (expected) error and more than an order of magnitude lesser Signal-to-Noise Ratio (SNR) loss (in dB) over the previously proposed inexact design techniques, while yielding similar energy gains. Post-layout comparisons in the 65nm process technology show that our REC approach achieves upto 73% energy savings (with corresponding delay and area savings of upto 16% and 62% respectively) when compared to an existing *exact* DSP implementation while trading a relatively small loss in SNR of less than 1.5 dB.

1. INTRODUCTION AND BACKGROUND

A large class of emerging applications, in particular embedded, multimedia, DSP systems and Recognition, Mining and Synthesis(RMS) workloads, can tolerate varying amounts of error, yet remain potentially useful. This is primarily attributed to the fact that the information produced by these systems is consumed by our senses that possess "cognitive filling" capabilities, and the absence of a unique and well-defined "golden" result in the end. Hence, in these systems, error (either caused probabilistically due to inherent variations/perturbations [12] or introduced ex-

plicitly [13]) has been viewed as a commodity that can be traded for lowering of hardware costs such as the energy consumed, delay and/or area.

However, the existing work in inexact design had a few major shortcomings: *(a)* the study of efficient ways to compose the building blocks to realize a system so as to maximize the hardware savings remains largely unaddressed. Straightforward approaches would more often than not lead to prohibitive accumulation of error as we combine atomic blocks to realize complex systems, *(b)* algorithmically well-founded approaches to navigate the combinatorially explosive design space of these inexact logic functions was found lacking, and *(c)* taking advantage of the knowledge of the underlying inexactness to redesign the DSP system or more specifically, the algorithm it implements, was seldom done. Overcoming these shortcomings has the potential for achieving more gains for a given error when compared to naive designs and approaches.

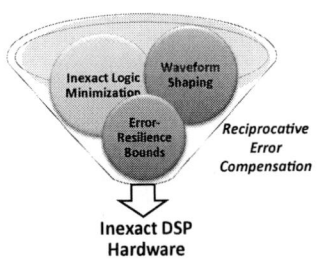

Figure 1: A framework for inexact DSP systems through *Reciprocative Error Compensation*.

In this paper, we present an approach called *reciprocative error compensation* (shown in Figure 1) to address and overcome these shortcomings. The REC approach provides for efficient datapath realizations by advocating a redesign of the basic building blocks through careful selection of "compensation buddies" for individual computational blocks, thereby enabling an effective compensation effect to overcome the first shortcoming of the existing works. Considering the second shortcoming, we tap into the substantial research that has been done in the domain of logic synthesis and optimization over the last couple of decades and demonstrate that traditional logic synthesis principles that were largely inap-

---

Permission to make digital or hard copies of all or part of this work for personal or classroom use is granted without fee provided that copies are not made or distributed for profit or commercial advantage and that copies bear this notice and the full citation on the first page. To copy otherwise, to republish, to post on servers or to redistribute to lists, requires prior specific permission and/or a fee.
DAC'13, May 29 - June 07 2013, Austin, TX, USA.

plicable in the context of traditional datapath design can in fact be extremely well-suited to guide the design space exploration of the inexact datapaths. Using this as a basis, we then propose a system-level technique termed *waveform shaping* technique that effectively reshapes the frequency (and phase) response curves of the DSP datapaths we are designing, to further reduce the error to address the third shortcoming. As the waveform shaping is applied at design time, and only involves modifying the coefficients of the DSP blocks to be stored in the memories, it does not incur any hardware overheads. Thus, succinctly, REC consists of: *(i)* inexact logic minimization through *compensation buddy identification* step followed by *(ii)* a system-level DSP error compensation through *waveform shaping* step. We demonstrate the efficacy of our approach by applying it to widely-used signal processing circuits—a pipelined complex Fast Fourier Transform (FFT), and a Finite Impulse Response (FIR) filter.

The rest of the paper is organized as follows: we present a brief review of the previously published inexact design techniques in Section 2. We then start developing our novel REC technique in Section 3, which includes a mathematical analysis of the drawbacks of existing approaches, compensation buddy identification of a node(s), leading to a general algorithm for finding them in a given DSP datapath graph. In Section 4, we provide details of the second step of REC— to provide for a system-level compensation in DSP subsystems through *waveform shaping*. In Section 5, we describe the experimental framework and DSP architectures used for demonstrating the efficacy of the REC framework along with the results and analysis. We conclude the paper and identify some possible directions for future work in Section 6.

2. RELATED WORK

A plethora of papers, as outlined in [10] and [11], have started taking advantage of the principle of trading accuracy of the hardware designs in exchange for significant resource savings in a wide variety of error-resilient applications. This was originally advocated in the works of Palem [18, 19]. Most of the early work focused on utilizing overscaling of physical parameters (such as supply voltage) to control and vary the energy-accuracy tradeoffs. These approaches such as voltage overscaling suffered from inefficient amortization of the hardware overheads in the targeted embedded systems [17, 1]. However, in the recent years, there has been a shift of focus to innovations at the architectural-level [1, 4, 17, 20, 9] that provided zero hardware overhead implementations. These techniques advocated reduction in the logic density by either pruning/deletion of a component with a low *significance*—determined by contribution to the output being computed—[1, 3], or by transformations to "similar" logic that consumes lesser energy [15, 5, 2]. While, by definition, pruning allowed the definitive inclusion or exclusion of a collection of gates and thus, offered a coarse-grained outcome, transformation of logical structures to "similar" lower-cost counterparts afforded more fine-grained control over the energy-accuracy tradeoffs due to a much larger associated design space. The later techniques, termed *inexact logic minimization*, involved logic-level manipulation of boolean function [15, 5, 2] using the notion of intentional *bit flips* or *cube-swaps* (implies an intentional forcing of the output from $0 \rightarrow 1$ or $1 \rightarrow 0$ for certain input vectors to assist in further logic minimization). The value of these architectural-layer approaches has been demonstrated through specific arithmetic structures such as adders or multipliers [16, 20, 5, 1,

17, 2]. However, efficient techniques for realizing inexact circuits of larger scale, such as signal processing primitives using these smaller nodes as building blocks, has not been explored. An important part of the problem is that most of these previously proposed techniques do not scale efficiently as we combine simpler systems, such as adders and multipliers, to realize more complex designs. Furthermore, most of these existing techniques either provide a unidirectional design space [1, 17] leading to accumulative error or have a narrow and restrictive design space [16, 20]. On the other hand, the inexact logic minimization techniques provided a richer design space, but overlooked the inherent compensation ability that could be gleaned from these logic transformations and relied on *absolute* or *worst case* statistical measures as the guiding principles that more often than not lead to pessimistic under-designed inexact systems.

In this paper, we focus on building upon and improving these known techniques through a scalable REC approach, resulting in efficient design of large-scale datapath networks from the DSP domain. In particular, we leverage the empirical evidence that inexact logic minimization yields error distributions that are (approximately) gaussian and therefore, minimizing the mean value of these error distributions (when variance is kept constant) will improve the SNR in the context of DSP hardware.

3. *INEXACT LOGIC MINIMIZATION* THROUGH RECIPROCATIVE ERROR COMPENSATION

3.1 Drawbacks of Existing Techniques for Systems Composed of Inexact Blocks

To demonstrate the rapid degradation in the quality (accuracy) of the results using the existing techniques as the scale of the composition increases, we present a mathematical analysis of a linear network of adders—present in the DSP datapaths—as shown in Figure 2 in this subsection, and empirical results for more complex networks with multipliers in the subsequent subsection.

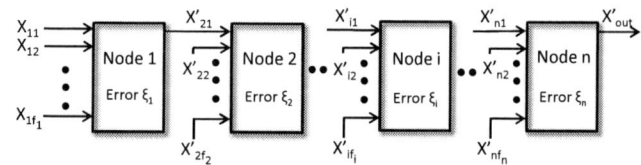

Figure 2: A typical composition of datapath elements.

Let there be n nodes in this path with a node at level i having inputs (fan-in of) f_i and an error represented by a random variable ξ_i with a probability distribution function $g(\xi_i)$. The *expected value* of ξ_i is denoted by $\mu_i = E(\xi_i) = \sum \xi_i \cdot g(\xi_i)$ with a variance of $\sigma_i^2 = \sum (\xi_i - \mu_i)^2 \cdot g(\xi_i)$.

Let the error at the primary output of the network be denoted by $\hat{\xi}$ with its expected value given by

$$\hat{\mu} = \mu_n + f_n \cdot \mu_{n-1} + \cdots + (\prod_{j=2}^{n} f_j) \cdot \mu_1 \qquad (1)$$

For the ease of exposition, let us consider a simple case where all the nodes have the same error properties, i.e., $g(\xi_i) = g(\xi_j)$ *for all* nodes $1 \leq i \leq j \leq n$ with a mean μ, and with a uniform number of inputs (fan-in of) M for ($M \in \mathbb{N}$,

$M > 1$), i.e. $f_i = M$ *for all* nodes $1 \leq i \leq n$, then equation (1) reduces to

$$\hat{\mu} = \mu \cdot \left(\sum_{i=0}^{n-1} M^i \right) = \mu \cdot \frac{M^n - 1}{M - 1} \qquad (2)$$

showing an exponential increase in the expected error value with increasing number of nodes.

3.2 Nullifying *Expected* Error Through *Compensation Buddies*

To avoid such an exponential error increase, through our REC approach, we will find a *compensation buddy* for each inexact node in the circuit. Formally:

DEFINITION 1. Given a node i with its associated error represented by a random variable with mean μ_i, it is said to be a *compensation buddy* for the set of its input (fanin) node(s) Φ iff

$$\mu_i = -\sum_{j \in \Phi} \mu_j.$$

Using this definition of compensation buddies in Equation (1), we get

$$
\begin{aligned}
\hat{\mu} &= -f_n \cdot \mu_{n-1} + f_n \cdot \mu_{n-1} + \cdots \\
&\quad + \left(\prod_{j=3}^{n} f_j \right) \cdot (-f_2 \cdot \mu_1) + \left(\prod_{j=2}^{n} f_j \right) \cdot \mu_1 \\
&= 0 \qquad (3)
\end{aligned}
$$

Without loss of generality, we can assume that n is even and the case that it is odd, we can handle by inserting an *exact* node n. Hence, it has been shown that the proposed REC technique theoretically achieves a zero expected error value using compensation buddies wherever such buddies exist for the design of interest . However, while such a theoretical limit is possible, the practicality of this approach relies on the ability to find appropriate compensation buddies which we will address the next subsection. An illustrative example highlighting the distinction in the error profiles of the conventional and REC-based inexact logic minimization techniques is shown in Section 8.1 of the appendix.

3.3 Finding the *Compensation Buddy*

To understand approaches to identifying compensation buddies, let us consider a circuit that computes a completely-specified boolean function $\mathcal{F} : B^n \to B^m$ that maps n-input boolean vector $\mathbf{x} = <x_1, x_2, \ldots x_n>$ to an m-output boolean vector $\mathbf{y} = <y_1, y_2, \ldots y_m>$ with an associated hardware cost $\mathbf{C}_{\mathcal{F}}$. The goal of inexact logic minimization is to find a Boolean function $\mathcal{F}' : B^{n'} \to B^{m'}$ where $n' \leq n$ and $m' \leq m$, such that its cost $\mathbf{C}_{\mathcal{F}'}$ is minimal subject to

$$\sum_{\forall \vec{i} \in \mathbf{I}} \frac{|\mathcal{F}(\vec{i}) - \mathcal{F}'(\vec{i})|}{T} \leq \mathcal{E}r_{th}$$

where $\mathbf{I}$ is a set of test vectors of size $T >> 0$ to this circuit and $\mathcal{E}r_{th}$ represents the tolerable error (here denoted by the average error) in the given circuit. For an output $y_i \in \mathbf{y}$, we define the *on-set* and *off-set* of $\mathcal{F}$ as $\mathbf{x}^{\mathbf{ON}} \subseteq B^n$ such that $\mathcal{F}(\mathbf{x}^{\mathbf{ON}}) = 1$ and $\mathbf{x}^{\mathbf{OFF}} \subseteq B^n$ such that $\mathcal{F}(\mathbf{x}^{\mathbf{OFF}}) = \mathbf{0}$ respectively. Inexact logic minimization relies on *swapping* the cubes between the $\mathbf{x}^{\mathbf{ON}}$ and $\mathbf{x}^{\mathbf{OFF}}$ sets so as to expand the cube(s) size leading to a decrease in the logic complexity—calculated by the number of literal reduction or basic gate equivalents. This is the key difference between

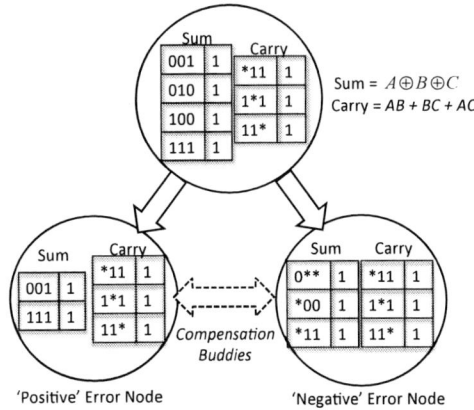

Figure 3: Example of an *exact* node (1-bit full adder) and its corresponding inexact *compensation buddies*.

traditional logic optimization techniques which reduce redundant logic to a minimal logic form while maintaining the functional equivalence, and *inexact logic minimization which reduces a minimal logic even further by relaxing the requirement of functional equivalence between the minimized and the original boolean formulae.* By convention, we denote the inexact nodes having expected or average output value less than their corresponding correct nodes as *positive error nodes* and if not, we designate them as *negative error nodes*. For those familiar with the usage of hypercube representation for logic minimization, essentially, the dominance of $\mathbf{x}^{\mathbf{ON}} \to \mathbf{x}^{\mathbf{OFF}}$ cube-swaps leads to positive error nodes while the dominance of $\mathbf{x}^{\mathbf{OFF}} \to \mathbf{x}^{\mathbf{ON}}$ cube-swaps provides negative error nodes. An example of this convention on a 1-bit full adder node is shown in Figure 3.

3.4 A General Algorithm for Design Space Exploration of REC-based *Inexact Logic Minimization*

A full-fledged design space exploration using inexact logic minimization is computationally intractable due to a combinatorial explosion of the solution space for a given accuracy constraint. As an example, for an n-input m-output function, in the worst case (such as the XOR-dominated functions common in most datapath circuits), we get $\sum_{i=0}^{k/2} C(2^k, i)$ possible logic functions to evaluate for each output, where $k = 2^n$ is the cardinality of the set of input vector combinations. This can lead to a doubly exponential complexity bound of $\Theta(2^{m \cdot 2^n})$ in the number of input variables—a very large number indeed!. However, in this paper, by restricting ourselves to XOR-dominated arithmetic circuits (adders and multipliers) and their use in building symmetrically-structured DSP blocks (FFTs and FIR filters), the solution space is quite dense with elements having similar energy-error tradeoff gains. Hence, rather than having to do a full search to find a global minimum, we employ greedy significance-guided heuristics (for example, alternating positive- and negative-error configurations) combined with randomized optimization schemes (for example, branch and bound, simulated annealing or stochastic gradient descent) and are able to achieve solutions of good quality. As in past work [13, 4], we use an output-significance driven assignment for the datapath circuits, i.e., nodes feeding outputs with higher (binary) significance are assigned higher significance values. A pseudo-code of the proposed algorithm along with a brief

139

description of its important steps is provided in Algorithm 1 in the Appendix Section 8.2.

3.5 Notes on Extending the Reach of Logic Synthesis

Historically, datapath circuits did not lend themselves favorably to traditional logic synthesis techniques given the dominance of XOR gates that did not exhibit the *adjacency* properties (thereby, offering no possibilities of cube expansion and hence, logic reduction). Therefore, most of the prior approaches to designing the best datapath circuits involved custom design as opposed to automatic logic synthesis guided approaches. However, we have made initial strides towards extending the reach of logic synthesis through inexact design to include datapath design. For doing so, we will rely on the foundational principles from Quine-McCluskey's work [14] in identifying *adjacency* of the minterms as an important criterion to decide the sequence of inexact logic minimizations. For example, to identify the candidate cubes for *cube swapping* between the $\mathbf{x^{ON}}$ and $\mathbf{x^{OFF}}$ sets in step IV of Algorithm 1, we *greedily* select only those cubes that are adjacent to *atleast* one cube in the original sets satisfying the following necessary and sufficient conditions:

Necessary and Sufficient Conditions for *Cube-Swapping*

> (i) The hamming weight of the swapped cube and the original cube should differ by 1.
> (ii) The difference between decimal value of the swapped cube index and original cube index should be a power of 2.
> (iii) If the hamming weight of one cube is greater than the other, then its corresponding decimal index should be greater as well.

These conditions guarantee that the swapped cubes are indeed adjacent to atleast one other cube in the original subsets, thereby ensuring the reduction in logic complexity by increasing cube cover, thereby, reducing the total size of the literals needed to represent these cubes).

3.6 Results and Analysis for Datapath Networks

In this subsection, we focus our efforts on quantifying the gains in error reduction of the proposed REC approach for inexact logic minimization over conventional techniques under similar energy, delay and area constraints. We use two quality metrics for comparison: expected error and SNR(dB) loss, the latter calculated as the absolute value of the signal-to-noise ratio difference between that computed by exact circuits and that by its inexact relaxation. We do this through empirical comparisons using various datapath elements (adders and multipliers) in a pipelined 64-256 point radix-2^2 FFT network [8]. We employ two types of inexact configurations of this FFT network: (i) inexact multipliers only, and (ii) both inexact adders and multipliers. For the adders, a combination of parallel prefix (Kogge-Stone) and ripple carry adders were used while for the multiplier networks, we used the standard truncated array multipliers with variable correction [6]. We used 16-bit uniformly random input vectors in the range [-1, 1) generated by Matlab and incorporated operand scaling at each node as appropriate to maintain the constant bit-width. As evident from the results in Table 1, the proposed REC technique reduces the mean error by two to three orders of magnitude, and degradation in SNR by one order of magnitude when compared to existing techniques under iso-energy conditions.

Table 1: Average error, variance, and SNR Loss results at the outputs of FFT network using existing inexact designs (denoted by Conv.) and the proposed REC technique (denoted by Prop.)

Inexact Multipliers in FFT Blocks						
	Average error		Variance		SNR(dB) Loss	
Path Depth	Conv.	Prop.	Conv.	Prop.	Conv.	Prop.
3	0.0125	0.0023	0.165	0.076	0.66	0.055
6	0.036	0.0004	0.22	0.11	1.11	0.09
9	0.051	0.0003	0.037	0.016	4.45	0.39
Inexact Adders & Multipliers in FFT Blocks						
	Average error		Variance		SNR(dB) Loss	
Path Depth	Conv.	Prop.	Conv.	Prop.	Conv.	Prop.
3	0.014	0.0048	0.033	0.017	0.74	0.06
6	0.036	0.0004	0.041	0.025	1.25	0.13
11	0.042	0.0015	0.043	0.019	7.78	1.09

4. *WAVEFORM SHAPING* TO ENCHANCE RECIPROCATIVE ERROR COMPENSATION

The second contribution of this paper involves further enhancing the REC approach by utilizing of the knowledge of the underlying properties of the algorithm to provide for system-level compensation. We demonstrate this in the context of DSP systems whose system level model is shown in Figure 4(a). As observed in this figure, a typical DSP system model consists of two key components: a computational datapath and a coefficient memory (for storing the system parameters). The existing inexact techniques modify the computational datapath but use pre-computed parameters (coefficients) that were optimized for an exact DSP algorithm. However, if we could alter the DSP system parameters based on the knowledge of the underlying inexact datapath elements, we would potentially minimize the overall error in the system.

In this paper, we achieve this by casting the DSP networks into equivalent neuronal models and employing a variation of the gradient descent based backpropagation algorithm called the *Normalized-Least-Mean-Sqare* (NLMS) algorithm [7] as shown in Figure 4(b). In DSP systems, this is referred to as waveform shaping of the frequency and phase response characteristics. As this approach only involves modifying the coefficients of the DSP blocks that will be stored in the memories, it will be applied during design time and hence, does not incur any hardware overheads. For equivalence between the neuronal models and the actual DSP hardware implementations, refer to Section 8.3 of the Appendix. In the following subsection, we provide a neuronal model for the DSP blocks, and show that it can be solved efficiently using the NLMS algorithm to modify the inexact DSP system parameters.

4.1 Neuronal Model for DSP Blocks

The behavior of an exact DSP block that can be modeled as a linear neuron as follows:

$$\Im : \{\vec{x}(i), \vec{t}(i)\}; \quad i = 1, 2, 3 \ldots k$$

where, $\vec{x}(i) = [x_1(i), x_2(i), \ldots x_n]^T$ is an n-dimensional in-

140

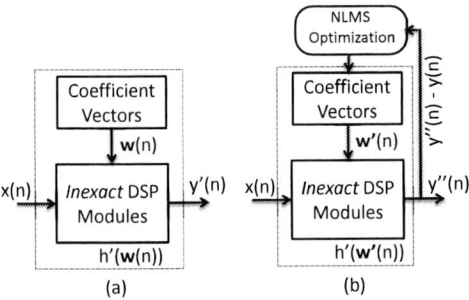

Figure 4: A DSP system level model for (a) conventional inexact circuits (b) proposed REC-based inexact circuits.

put vector and $\vec{t}(i) = [t_1(i), t_2(i), \ldots t_k]^T$ is the desired k-dimensional output vector at time (or for sample) i. Similar to a *supervised* machine learning algorithm, we have two phases—a feedforward computation that propagates the input vectors across the inexact blocks to the outputs using a initial set of weights obtained from the *exact* DSP block and an error propagation mode that *backpropagates* the error values from the output nodes to iteratively adjust the weights for the inexact block to minimize the targeted error metric.

To elaborate, in the first phase, we compute the output of the inexact system as follows:

$$y(i) = \sum_{k=1}^{n} w_k \cdot x_k(i) = \vec{x}^T(i) \cdot \vec{w}$$

where, $\vec{w} = [w_1, w_2, \ldots w_n]$ is an n-dimensional coefficient vector. We compute the error of the *inexact* neuron as

$$e(i) = t(i) - y(i) = t(n) - \vec{x}^T(n) \cdot \vec{w} \quad (4)$$

Since, the targeted DSP blocks – FIR/FFT primitives– can be modeled as a single-layer linear neurons (refer to supplemental section 8.3 for details), we employ the NLMS algorithm [7]. Without loss of generality, we limit ourselves to presenting the NLMS algorithm on a single neuron in this section, but we can use this algorithm to iteratively optimize multiple neurons in a single layer (e.g. FFT neuronal model shown in the supplemental section 8.3). Hence, the NLMS algorithm solves the constrained optimization problem of minimizing the squared Euclidean norm of the change in the weight vector and can be described as follows:

$$\text{Minimize} \quad \|\Delta \vec{w}(n)\|^2 = \|\vec{w}(n+1) - \vec{w}(n)\|^2$$
$$\text{subject to :} \quad \vec{x}^T \cdot \vec{w}(n+1) = t(n)$$

We can find a solution for this optimization problem using Lagrange multipliers [7] leading to weight update function of:

$$\vec{w}(n+1) = w(n) + \mu \frac{e(n)\vec{x}(n)}{\vec{x}^T(n)\vec{x}(n)} \quad (5)$$

We refer the reader to the pseudocode provided in Algorithm 2 in Section 8.4 of the appendix for more details.

5. APPLICATION TO DSP DATAPATHS

5.1 Implementation Framework

The desired DSP blocks have been described in VHDL and synthesized using the Cadence RTL compiler using industry-standard 65nm (low power, low leakage) technology libraries.

The Place and Route of the synthesized designs has been done using Cadence SoC Encounter and the post-layout spice netlists (with parasitics) have been extracted and simulated using the Mentor Graphics ADiT fast-spice simulator to calculate the power and delay values accurately. This framework is tied into in-house C and Matlab based simulators that have been extended to be compatible with Algorithms 1 and 2 to determine the desired error metrics.

5.2 Hardware Implementation of DSP Blocks Synthesized with REC

For the hardware implementation, we consider a pipelined 256-point radix-2^2 complex FFT [8] that is capable of processing 16-bit complex data. for demonstrating the effectiveness of the proposed REC approach. We opt for the single-path delay feedback (SDF) architecture with a continuous sampling of inputs. Additionally, we have also implemented a data broadcast structure based 30-tap low-pass FIR filter with 16-bit inputs.

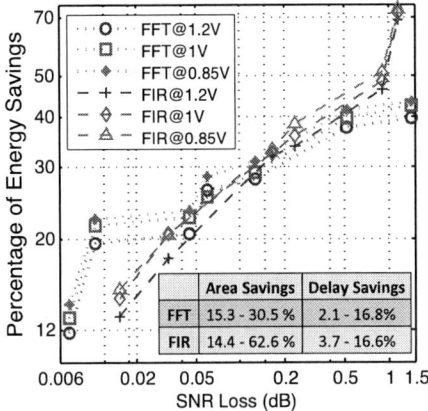

Figure 5: Energy Savings vs SNR Loss (dB) plot for the REC-based inexact logic minimization technique on a pipelined 256-point radix-2^2 complex FFT and 30-tap FIR filter with the conventional *(exact)* circuits as a comparison baseline.

In Figure 5, we show the plots of the energy savings of the proposed REC-based inexact logic minimization approach as a function of SNR loss with reference to the conventional "exact" implementation of the 256-point FFT and 30-tap FIR filter. As evident from this figure, the proposed approach achieves between 12%–73% energy savings (a multiplicative factor of 1.14–3.7), with corresponding delay and area savings between 2%–16% and 15%–62% respectively, when compared to the *exact* implementations with an associated loss in the SNR of 0.006 dB upto 1.5 dB. Another interesting and crucial aspect of the proposed technique is the ability to maintain these savings over the conventional exact implementations even at scaled supply voltages. Owing to the adoption of an aggressive pipelined strategy, the delay savings from the proposed approach are not as significant as the energy and area savings. This is because our randomized approach synthesized datapath blocks with non-uniform levels of logic minimization, wherein the presence of even a single "not-so-minimized" block in a pipeline stage would inflate the critical path delay.

Turning to the idea of optimizing the application parameters, the waveform shaping technique further enhances the scope of our approach in the context of DSP applications.

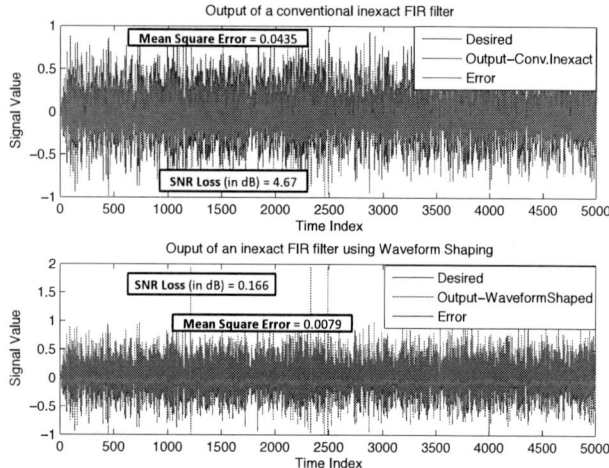

Figure 6: The output of a conventional inexact and a waveform-shaped inexact band-pass FIR filters for 5000 samples of white noise input.

The results of the waveform shaping when applied to a inexact 30-tap band-pass FIR filter—using two different sets of 5000 white noise samples for training and testing—are shown in Figure 6. The proposed waveform shaping technique is able to effectively compensate for most of any residual error from inexact logic minimization in the DSP blocks. To provide the reader some intuition as to how the "shaped" system response curves look, we have included the frequency and phase responses in Figure 10 of the Appendix.

6. CONCLUSION AND FUTURE WORK

We have presented a novel *zero-hardware overhead* design technique called *reciprocative error compensation*(REC) to enable inexact logic synthesis that scales to the level of subsystems from the DSP domain. To the best of our knowledge, this is the first time that feasibility of inexact systems is being demonstrated at this scale. Our technical contributions are two-tiered: (a) synthesis of basic building blocks using inexact logic minimization to compensate for each others' error through REC, and (b) system-level DSP error compensation methodology through *waveform shaping*. Through the application of the proposed techniques on a pipelined Fast Fourier Transform (FFT) architecture, we show that 2-3 orders of magnitude lower (expected) error and more than an order of magnitude lower Signal-to-Noise Ratio (SNR) loss (in dB) can be achieved, over the previously proposed inexact design techniques under *iso-energy* conditions. The gains over conventional *exact* FFT and FIR filter implementations include upto 73% (or a multiplicative factor of 1.14–3.7) energy savings at a minimal SNR loss of less than 1.5 dB. The corresponding delay and area savings are upto 16% and 62% respectively. Future work includes developing efficient CAD algorithms and optimization frameworks for further enhancing the gains of our proposed framework and application of this framework to a broader set of applications such as those embodying RMS workloads.

7. REFERENCES

[1] A. Lingamneni et al. Energy parsimonious circuit design through probabilistic pruning. *in proc. of DATE*, pages 764–769, Mar 2011.

[2] A. Lingamneni et al. Parsimonious circuit design for error-tolerant applications through probabilistic logic minimization. *in the proc. of the PATMOS*, pages 204–213, 2011.

[3] A. Lingamneni et al. Algorithmic methodologies for ultra-efficient inexact architectures for sustaining technology scaling. *in proc. of ACM Intl. Conference on Computing Frontiers*, pages 3–12, May 2012.

[4] A. Lingamneni et al. Synthesizing parsimonious inexact circuits through probabilistic design techniques. *in proc. of ACM Transactions on Embedded Computing Systems*, 2013.

[5] D Shin et al. Approximate logic synthesis for error tolerant applications. *in the proc. of DATE*, pages 957 – 960, 2010.

[6] E.J King et al. Data-dependent truncation scheme for parallel multipliers. *Asilomar Conf. on Signals, Systems & Computers*, 1997.

[7] S. Haykin. *Adaptive Filter Theory*. Prentice-Hall, Inc., Englewood Cliffs, NJ., 2002.

[8] S. He and M. Torkelson. A new approach to pipeline FFT processor. *in proc. of Parallel Processing Symposium*, (766-770), 1996.

[9] J. Huang et al. A methodology for energy-quality tradeoff using imprecise hardware. *in the 49th DAC*, pages 504–509, 2012.

[10] K. Palem et al. What to do about the end of moore's law, probably! *in the 49th DAC*, pages 924–929, 2012.

[11] K. Palem et al. Ten years of building broken chips: The physics and engineering of inexact computing. *in proc. of ACM Transactions on Embedded Computing Systems*, 2013.

[12] K.V. Palem et al. Sustaining moore's law in embedded computing through probabilistic and approximate design: retrospects and prospects. In *in proc. of CASES*, pages 1–10, 2009.

[13] L.N.B. Chakrapani et al. Highly energy and performance efficient embedded computing through approximately correct arithmetic: A mathematical foundation and preliminary experimental validation. In *proc. of IEEE/ACM CASES*, pages 187–196, 2008.

[14] E. McCLUSKEY Jr. Minimization of boolean functions. *Bell System Technical Journal*, 1956.

[15] M.R. Choudhury et al. Approximate logic circuits for low overhead, non-intrusive concurrent error detection. *in the proc. of DATE*, pages 903 – 908, Mar 2008.

[16] N. Zhu et al. Design of low-power high-speed truncation-error-tolerant adder and its application in digital signal processing. *IEEE Trans. Very Large Scale Integration Sys.*, 18(8):1225–1229, 2010.

[17] P. Kulkarni et al. Trading accuracy for power with an underdesigned multiplier architecture. *in the proc. of Intl Conf on VLSI Design*, (346-351), 2011.

[18] K. V. Palem. Energy aware algorithm design via probabilistic computing: From algorithms and models to Moore's law and novel (semiconductor) devices. In *proc. of CASES*, pages 113 – 116, 2003.

[19] K. V. Palem. Energy aware computing through probabilistic switching: A study of limits. *IEEE Transactions on Computers*, 54(9):1123–1137, 2005.

[20] V. Gupta et al. Impact: imprecise adders for low-power approximate computing. *in proc. of ISLPED*, (409-414), 2011.

8. SUPPLEMENTAL MATERIAL

8.1 An Illustrative Example of Conventional and REC-based Error Profiles

As a simple example, consider Figure 7(a) where we show the average error in uniformly random 2000 data samples in the range [-1, 1) by using 14 conventional inexact full adder nodes in a simple 16-bit ripple carry adder. A unidirectional positive error profile is evident from the figure that would lead to accumulation of error when used in larger structures. On the other hand, if these inexact nodes can be designed to have error profile similar to Figure 7(b) (which has a 20X lower average error) by using the concept of compensation buddies, error accumulation can be prevented even in larger datapath networks.

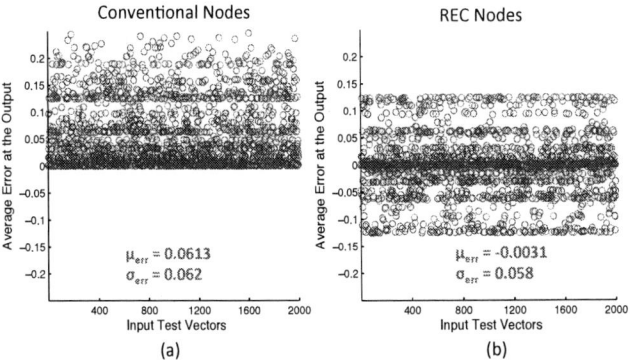

Figure 7: Graph showing the average error in Ripple carry adders using conventional and REC based full adder nodes. Uniformly random 16-bit inputs between [-1, 1) were used.

8.2 Pseudo-code of the REC-based Inexact Logic Minimization Algorithm

In this section, a pseudo-code of the proposed algorithm for REC-based inexact logic minimization along with a brief description of its important steps is provided in Algorithm 1.

8.3 Some Notes on the Neuronal Models for the DSP Blocks

As described in Section 4, an FIR filter can be modeled as a single neuron (a perceptron) with a linear activation function as shown in Figure 8. While an FIR filter can be trivially modeled as a single multi-input neuron – not far-fetched from it circuit implementation, the neuronal model of an FFT is slightly non-intuitive as it is very different from the typical hardware implementations, for example, the pipelined radix-2^2 implementation that was chosen for this paper. Hence, we would like to make a clear distinction between the actual hardware implementation of the FFT [8] and the neuronal model of the FFT shown in Figure 9 that is only used for applying the proposed idea of waveform shaping.

Given the independent and iterative application of the NLMS optimization to determine the weights, the coefficient sharing that was feasible with the *exact* FFT twiddle factors is no longer possible as each of the weights can iteratively converge to a different value. But it should noted that the area overhead of adding additional rows to the memory bank storing the coefficients is pretty negligible compared to the area gains that the proposed FFT achieves. If area

---

Algorithm 1 Pseudo-code for applying REC-based inexact logic minimization to a circuit represented as a Directed Acyclic Graph $\mathcal{G}$

//Inputs: Circuit Graph $\mathcal{G}$, Primary Output Significance Array $\mathbf{S}_o$, Error threshold $\mathcal{E}r_{th}$
//Outputs: Circuit Graph after REC approach $\mathcal{G}_{REC}$

// Step I : Graph with node clustering. Nodes with fanin and fanout between 3 to 10 are clustered together.
$\mathcal{G}_c \leftarrow$ **ClusterNode**$(\mathcal{G})$;

// Step II : Identify a set of all unique nodes types in a clustered circuit graph
$\mathbf{N}_u \leftarrow$ **Uniquify** $(\mathcal{G}_c)$

// Step III : Use sensitivity analysis w.r.t the output Significance Array to determine the significance of each node cluster
$\mathbf{S} \leftarrow$ **ComputeSignificance** $(\mathcal{G}_c, \mathbf{N}_u, S_o)$;

// Step IV: // Local Exploration – Compute the design space (both positive and negative configurations) of each unique node cluster using greedy approach with incremental (upto 'L') cube swaps, or until the node is reduced to a wire.
for $j \in \{pos, neg\}$ **do** *//Positive- or negative-error*
    for all $i \leftarrow 1$ to $|\mathbf{N}_u|$ **do**
        $\mathbf{ErCfg}_{j,i} \leftarrow \{\phi\}$
        while $(n_i \neq wire)$ **do** *//foreach $n_i \in \mathbf{N}_u$*
            $\mathbf{ErCfg}_{j,i} \leftarrow$ GreedyCubeSwap(n_i, j, L)
    end for
end for

// Step V: Global Exploration – Compute the positive- or negative-error configuration assignment for each node cluster across dataflow graph $\mathcal{G}_c$.
$\text{Cost}(\mathcal{G}_{REC}) = \text{Cost}(\mathcal{G})$
$(\mathbf{N}_{pos}, \mathbf{N}_{neg}) \leftarrow$ **BipartiteParition**$(\mathcal{G}_s)$
//DSP graphs can be partitioned into bipartite subsets due to their symmetry
while *iteration* < *bound* **repeat**
    $\mathcal{G}' \leftarrow$ RandomErrCfg $(\mathbf{N}_{pos}, \mathbf{ErCfg}_{pos}, \mathcal{S})$
    s.t. if $s_k > s_l$, then $ErCfg_k < ErCfg_l, \forall\ k, l \in \mathbf{N}_u$
    $\mathcal{G}' \leftarrow$ CompensationBuddy $(\mathbf{N}_{neg}, \mathbf{ErCfg}_{neg})$
    //Apply the definition of compensation buddy to determine the error configuration of negative error nodes
    if $\text{Cost}(\mathcal{G}') < \text{Cost}(\mathcal{G}_{min})$ && $\mathcal{E}r(\mathcal{G}') < \mathcal{E}r_{th}$
        $\mathcal{G}_{REC} \leftarrow \mathcal{G}'$
end while
return $\mathcal{G}_{REC}$

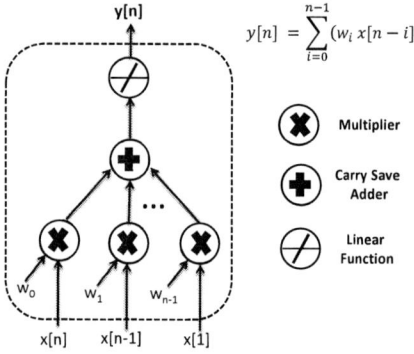

Figure 8: A linear neuronal model representing an FIR circuit.

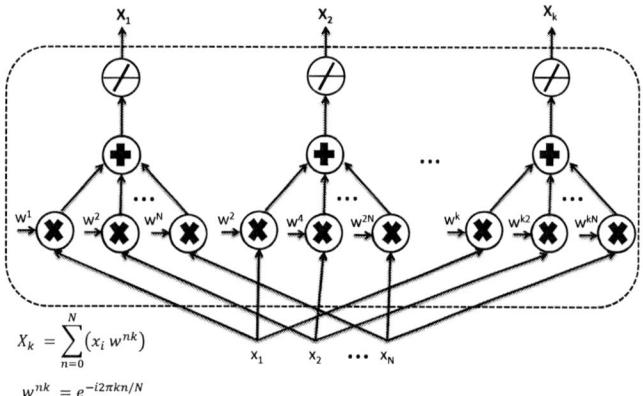

$$X_k = \sum_{n=0}^{N} \left(x_i \, w^{nk} \right)$$

$$w^{nk} = e^{-i2\pi kn/N}$$

Figure 9: A linear neuronal model representing an FFT circuit.

is still a primary concern in any design, we can still apply the NLMS algorithm and store the average values of the modified twiddle factors as a possible option.

8.4 Pseudo-code for the REC-based Waveform Shaping Algorithm

In this section, a pseudo-code of the proposed algorithm for REC-based waveform shaping along with a brief description of its important steps is provided in Algorithm 2.

Algorithm 2: NLMS-based Coefficient Update

Training Vectors: Input Vectors $= \vec{x}(n)$
                        Number of Neurons $= k$
                        Exact circuit weight vectors $= \vec{w}_j(n)$
                        Exact circuit response $= t_j(n)$
                        for $j = 1, 2, \ldots k$.
User Selected Parameters: μ_j
Initial Conditions: $\vec{w}(0) = \vec{w}_{exact}$
NLMS Algorithm:
    for n = 1, 2, 3,... loop
        for j = 1, 2, ..., k loop
            $e_j(n) = t_j(n) - \vec{w}_j \cdot \vec{x}^T(n)$
            $\vec{w}_j(n+1) = w_j(n) + \mu_j \frac{e_j(n)\vec{x}(n)}{\vec{x}^T(n)\vec{x}(n)}$
        end for
    end for

8.5 Response Plots for the Waveform-shaped a FIR Filter

In this section, we provide the frequency and phase responses curves of a 30-tap bandpass FIR filter used in Figure 6 before and after the waveform-shaping technique.

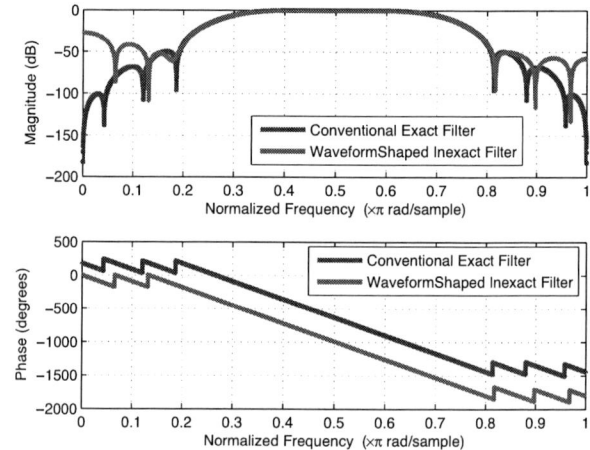

Figure 10: The frequency and phase responses of a conventional exact and a waveform-shaped inexact FIR filters.

144

Early Partial Evaluation in a JIT-compiled, Retargetable Instruction Set Simulator Generated from a High-Level Architecture Description

Harry Wagstaff   Miles Gould   Björn Franke   Nigel Topham

Institute for Computing Systems Architecture
School of Informatics
University of Edinburgh
h.wagstaff@sms.ed.ac.uk, mgould1@inf.ed.ac.uk, bfranke@inf.ed.ac.uk, npt@inf.ed.ac.uk

ABSTRACT

Modern processor design tools integrate in their workflows generators for instruction set simulators (ISS) from architecture descriptions. Whilst these generated simulators are useful for design evaluation and software development, they suffer from poor performance. We present an ultra-fast JIT-compiled ISS generated from an ARCHC description. We also introduce a novel partial evaluation optimisation, which further improves JIT compilation time and code quality. This results in a simulation rate of 510MIPS for an ARM target across 45 EEMBC and SPEC benchmarks. On average, our ISS is 1.7 times faster than SIMIT-ARM, one of the fastest ISS generated from an architecture description.

1. INTRODUCTION

Instruction set simulators (ISS) are indispensable tools for hardware architects and software developers alike. Consequently, modern processor design suites such as *Synopsys Processor Designer* [13] or *Target IP Designer* [14] integrate ISS generator tools, which synthesise ISS from high-level architecture descriptions. However, these machine generated simulators are typically slower than their hand-coded counterparts targeting only a single architecture for which they have been optimised [10].

Currently, the fastest available ISS are based on dynamic binary translation (DBT) and use a parallel just-in-time (JIT) compiler [4, 11] for the translation of regions of target machine code to the host system's ISA, possibly interleaving a detailed performance model of the simulated processor for cycle-accurate pipeline modelling [5]. Unfortunately, these fast DBT based ISS are – for performance reasons – hand-coded and, thus not easily retargetable. Naïve approaches to generating JIT DBT simulators from high-level architectural descriptions suffer from two problems: poor quality of the JIT compiled code and excessive JIT compilation times. The common root of these two problems lies in the complex be-

haviours of target machine instructions containing different possible execution paths dependent on, for example, processor state.

In this paper we present a novel approach to generating ultra-fast JIT DBT simulators from high-level ARCHC based architecture descriptions [1]. We apply a partial evaluation optimisation [8], which eliminates dead execution paths from complex instruction behaviours, early on in the JIT compilation process as part of the intermediate representation (IR) generation stage. This not only relieves the underlying low-level JIT compiler from performing this work, which possibly utilises expensive analyses, but aids better code generation. In turn, this results in faster JIT compilation and, at the same time, less – and less complex – code is emitted by the JIT compiler due to compile-time specialisation.

The **key idea** is to simplify instruction behaviours as soon as possible in the high-level JIT code generator (see Figure 2), rather than deferring optimisation of the generated code to the low-level JIT compiler (see Figure 1). Through the use of early partial evaluation at IR generation time we ensure that instruction behaviours that are impossible in a specific context are not emitted, and thus are never presented to the low-level JIT compiler. Rather than emitting complex instruction behaviour patterns and relying on the low-level JIT compiler to eliminate dead code later, we spend a little more effort in the high-level IR generator. In return, we gain significantly higher performance in, and achieve better code quality from, the low-level JIT compiler, resulting in greater overall simulation speed.

We have evaluated our ISS generation approach using an ARM V5 architecture model written in ARCHC against 45 EEMBC and SPEC CPU2006 benchmarks. On a standard x86 simulation host we demonstrate an average simulation rate of 191 MIPS for our baseline ISS using a naïve generation scheme (see Figure 1), and 510 MIPS after enabling early partial evaluation (see Figure 2). Using the same ARM V5 target and the same simulation host machine (see Table 1) this is approximately 21 times faster than the original ARCHC simulator (24 MIPS) [1], 78 times faster than FAC-SIM (6.5 MIPS) [9], 1.7 times faster than SIMIT-ARM (300 MIPS) [11], which relies heavily on manual instruction specialisation as part of the modelling and retargeting process, and only 24.5% slower than QEMU-ARM [2], which has been hand-tuned for the target, requires significant low-level retargeting effort and unlike our ISS sacrifices instruction observability for performance.

Permission to make digital or hard copies of all or part of this work for personal or classroom use is granted without fee provided that copies are not made or distributed for profit or commercial advantage and that copies bear this notice and the full citation on the first page. To copy otherwise, to republish, to post on servers or to redistribute to lists, requires prior specific permission and/or a fee.
DAC '13, May 29 - June 07 2013, Austin, TX, USA.

Figure 1: Naïve JIT implementation: entire instruction implementation functions are passed to the LLVM optimiser, resulting in long optimisation and back-end compilation times.

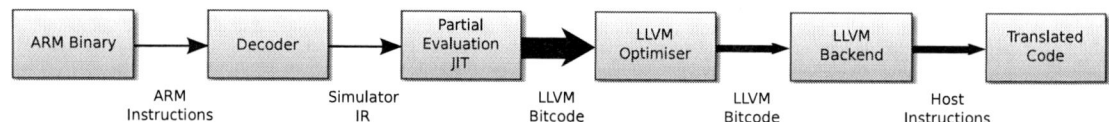

Figure 2: Partial Evaluation JIT: only required instruction implementation code is passed to the LLVM optimiser, resulting in faster optimisation and compilation and better code quality.

1.1 Contributions

Among the contributions of this paper are:

1. A methodology for fast *and* retargetable instruction set simulation in a JIT-compiled ISS generated from an ARCHC architecture description,

2. the introduction of a high-level JIT code optimisation based on *early partial evaluation*, and

3. an extensive evaluation of the proposed scheme against industry standard EEMBC and SPEC benchmarks and an ARM V5 architecture model allowing for a direct performance comparison against other retargetable ISS such as SIMIT-ARM, ARCHC, FACSIM, and QEMU-ARM.

1.2 Motivating Example

Consider the snippet of ARM assembly in Figure 3. This shows a comparison, followed by a conditional **add**, such that the execution of the **add** is dependent on the result of the comparison. Listing 2 shows an example implementation of such an instruction in an instruction set simulator. The implementation is not specialised to the condition code, or the arguments to the instruction (other than the PC). Clearly, there are several statements here which could be simplified or specialised:

1. The call to condition_passed (line 1) could be replaced with a direct calculation of the result of the instruction condition.

2. The call to ROR (line 3) could be replaced with its result, since both of its operands are fields within the instruction.

3. The call to read_register() (line 4) can be replaced with the actual PC value, since the PC does not change for an individual instruction.

4. The inst.S == true path (lines 7–13) can be removed since this particular instruction is not flag setting.

In a naïve JIT, these statements may be simplified at optimisation time (assuming optimisations are applied). However, this requires that IR is first emitted, then analysed, and then pruned, which can be costly. Standard optimisations are also frequently not strong enough to fully take advantage of early-evaluable expressions.

In contrast, our early partial evaluation scheme already evaluates statically evaluable expressions (lines 1, 3 and 4) at IR generation time and avoids emitting dead code such as the flag setting code in lines 7–13 of Listing 2. This means, we produce less IR for the code generator to process and, hence, we speed up the JIT compiler.

For the example in Figure 1 our partial evaluation scheme is able to fully compute the result of the example **add** instruction at compile-time (see Listing 3). In comparison, the naïve scheme requires much more expensive '-O3' LLVM optimisations to achieve a similar effect. The naïve scheme generates more LLVM IR instructions, which are then processed by the optimiser, resulting in a long latency of the performance-critical JIT compiler.

2. METHODOLOGY

2.1 Simulator Generation

Our simulator generation system is based on there being a distinct boundary between the processor implementation and the simulator framework (see Figure 4). Processor models are written using a variant of ARCHC [1], and then processed into a set of C++ source files implementing instruction decoding, an interpreter, a JIT module, disassembly, etc. This system is sufficiently general and flexible to support a variety of ISAs (so far we have models for ARM V5, ARM V7A, ARM V7M, POWERPC, INTEL 8086 and MOS 6502) and is modular enough that different implementations of interpreters, JIT systems, etc can be swapped in and out.

The baseline JIT system precompiles each instruction implementation to LLVM bitcode functions. When an instruction is JIT compiled, a call to the appropriate function is generated (with the instruction fields as parameters) and inlined.

2.2 Early Partial Evaluation

The early partial evaluation JIT is generated completely from a processor ISA description. The implementation of each instruction is written in a C-like language, and compiled into an SSA form consisting of variable reads and writes, register and memory reads and writes, unary and binary operations, some intrinsics, and control flow. The SSA form does not have Φ-nodes, since Φ-analysis is unnecessary at this stage and can be done later by the LLVM back-end. Function calls and subroutines are supported, and all func-

Listing 1: ARM V5 Assembly snippet

```
1   ...
2   @ compare the values in r0 and r1 and set flags
3   cmp r0, r1
4   @ if the zero flag is clear, add 0x3fc to the PC
5   addne r0, pc, #0x3fc
6   ...
```

Listing 2: Simplified high-level ARCHC description of the behaviour of an add instruction in ARM V5

```
1   if(condition_passed(inst.cond))
2   {
3    uint32 imm = ROR(inst.imm, inst.rot);
4    uint32 rn_val = read_register(inst.rn);
5    uint32 rd_val = rn_val + imm;
6    write_register(inst.rd, rd_val);
7    if(inst.S)
8    {
9     write_flag(N, !!(rd_val & 0x80000000));
10    write_flag(Z, rd_val == 0);
11    write_flag(C, carry_from(rn_val + imm));
12    write_flag(V, ovrflw_from(rn_val + imm));
13   }
14  }
```

Listing 3: LLVM code for the ARM V5 code snippet in Listing 1 after partial evaluation

```
1    ...
2    ; %43 = r0, %44 = r1
3    %45 = sub %43, %44
4    ; %46 and %47 calculate and store C flag
5    ; calculate the Z flag
6    %48 = icmp eq %45, 0
7    ...
8    ; check the 'ne' predicate
9    %57 = icmp ne i32 %48, 0
10   br i1 %57, label %60, label %61
11
12  ; <label>:60
13   ; Don't need to do any "add" at runtime.
14   ; PC and immediate operand are fixed,'
15   ; JIT compiler can evaluate this statically.
16
17   ; store the calculated value (pc + 0x3fc = 1040)
18   ; in r0 (%38 contains the address of r0)
19   store i32 1040, i32* %38, align 4
20   br label %61
21
22  ; <label>:61
23   ; next instruction
24   ...
25   ...
```

Figure 3: Motivating example: ARM assembly snippet containing an add operation in Listing 1, semantic action with conditional behaviours for the add instruction in Listing 2, and resulting LLVM output in Listing 3.

Algorithm 1 Computing 'fixedness' of variable modifying statements in an instruction implementation.

> **function** INSNIMPLFIXEDNESS(*action*)
>   **for all** $b \leftarrow BB \in action$ **do**
> $b.dyn\_in \leftarrow []$
> $b.dyn\_out \leftarrow []$
> $b.ctrlflow \leftarrow invalid$
> $b.mark\_variable\_accesses\_as\_fixed()$
> $wl \leftarrow [action.entry\_block]$
>   **while** wl is not empty **do**
> $b \leftarrow wl.pop\_front()$
> $result \leftarrow$ BBFIXEDNESS(b)
>   **if** $result = False$ **then**
> $wl.insert(b.successors)$
> **function** BBFIXEDNESS(*block*)
>   **for all** $p \leftarrow BB \in block.predecessors$ **do**
>   **if** $p.ctrlflow = dynamic \lor \neg p.final\_stmt.is\_fixed$ **then**
> $block.ctrlflow \leftarrow dynamic$
> $block.dyn\_in \leftarrow block.dyn\_in \cup p.dyn\_out$
> $dyn\_now \leftarrow block.dyn\_in$
>   **for all** $s \leftarrow Statement \in block.statements$ **do**
>   **if** s writes a dynamic value to a variable v **then**
> $dynamic\_now \leftarrow v$
>   **if** s reads a variable in dyn_now **then**
>   mark s as dynamic
> $block.dyn\_out \leftarrow dyn\_now$
>   **if** $block.ctrlflow$ changed $\| dyn\_now \neq dyn\_in$ **then**
>   **return** $False$
>   **return** $True$

---

tion calls are inlined immediately. Although this results in a larger JIT code module (and precludes the use of recursive functions), it allows for more straightforward analysis and implementation.

The partial evaluation scheme itself is based on computing whether or not each SSA statement is 'fixed' (i.e., relies only on JIT-time information such as instruction fields and constant values) or 'dynamic' (i.e., relies on information loaded from registers or memory). For most statements, this is trivial: a binary operation is fixed if both of its operands are fixed, a memory or register write is fixed if its address and value are fixed etc. However, computing the fixedness of variable reads (and thus statements which depend on those variable reads) requires slightly more careful analysis.

The algorithm for doing this analysis can be seen in Algorithm 1. The analysis is done on each instruction implementation. Each basic block in each implementation has three variables: a list of variables which are 'dynamic' at the entry point of the block, a list of variables dynamic at the exit of the block, and a variable stating whether or not control flow into this block is fixed or dynamic. We start by emptying the lists and considering the control flow 'invalid' (not yet computed).

We then process each block in turn, starting with the entry block of the instruction. If the block processing algorithm reports that the state of the block has changed (either its control flow or dynamic IN/OUT lists) then we add its successors to the work list. Once the work list becomes empty the fixedness computation is complete. A single block may be processed multiple times if it is in the body of a loop.

For each basic block we first compute whether or not control flow into the block is fixed or dynamic. Control flow is dynamic if any conditional branch or switch in any ancestor of the block depends on a dynamic value. Otherwise, control flow into the block is fixed. We also compute the dynamic IN variables as the union of the dynamic OUT variables of the block's predecessors.

We then loop over each of the statements in the block in program order. For each statement, if the statement writes

147

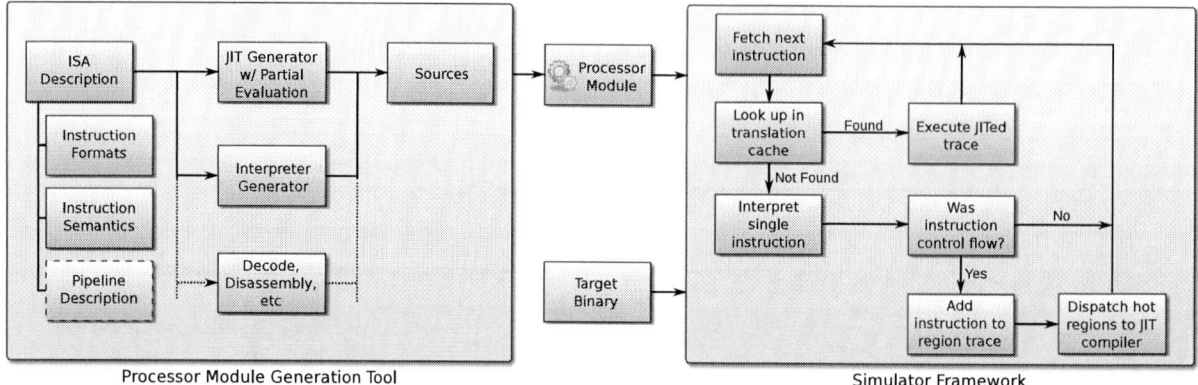

Figure 4: Generation of a processor module from a high-level architecture description and its use within the main simulation loop of a retargetable Iss. Early partial evaluation is added to the JIT compiler at generation time and performed during program simulation when traced regions are translated to native code.

dynamic values to any variables, we add those variables to the dynamic OUT list. For variable reads, if the read variable is dynamic at this point, we mark the read as dynamic.

2.3 JIT Compiler Generation

When generating the JIT compiler itself, we again work on an instruction type by instruction type basis. For each instruction type we generate a function which implements the fixed portions of the instruction directly, and generates LLVM instructions for the dynamic portions. So, fixed control and data flow in the instruction becomes C++ control and data flow, and dynamic control and data flow becomes LLVM control and data flow.

Since a variable may be fixed at one point during execution but dynamic later on, we do further analysis to determine points at which variables must be 'spilled' into the dynamic context. This is only strictly necessary when a variable is sometimes dynamic on entry to a block (e.g. if we write a dynamic value to it in the 'then' portion of an 'if' statement but write a static value in the 'else' portion). If a dynamic statement reads the value of a fixed variable we write the value directly into the output LLVM statement rather than 'spilling' the variable.

As static control flow is executed completely at JIT time, multiple SSA blocks may become a single LLVM block. This produces one of the main improvements in code generation speed in the form of 'on-the-fly' dead code elimination, meaning that there is much less code for LLVM to generate and optimise. Blocks which have dynamic control flow are only emitted 'on-demand' (i.e., if either a static or dynamic control flow statement which has the block as a target is encountered) which also helps to reduce the amount of code passed into LLVM.

3. EMPIRICAL EVALUATION

We have implemented the presented early partial evaluation technique in our Iss framework and evaluated it against the full EEMBC 1.1 and SPEC CINT2006 benchmark suites. We have compiled the benchmarks with the ARM v5 port of the GCC 4.5.2 compiler. For each of the benchmarks we have measured the time from start of the simulator to completion, with and without early partial evaluation, and used

Simulation Host	
CPU	Intel Xeon X5650
Frequency	2.67GHz
# Cores	12
Target Platform	
ISA	ARM v5
Compiler	GCC 4.5.2
Executable Format	Linux ELF
Simulator Configuration	
# JIT threads	4
JIT compiler	LLVM 2.9
JIT optimisation flags	-O3
System calls	Emulated

Table 1: Simulation host, target platform and Iss.

these times to calculate absolute simulation rates (in MIPS) and speedups. For comparison, we have executed the same binaries of the benchmarks on SIMIT-ARM v3 [11] Iss using an equivalent configuration and performed the same timing measurements. Details of the simulation host, the target platform and Iss configuration are summarised in Table 1.

3.1 Key Performance Results

Our main results showing relative speedups resulting from our early partial evaluation scheme over the parallel, JIT-compiled SIMIT-ARM v3 Iss for both short-running EEMBC and long-running SPEC CPU2006 benchmarks are presented in Figure 5.

For all but two of the EEMBC benchmarks our partial evaluation Iss significantly outperforms SIMIT-ARM v3. On individual benchmarks (*cacheb01* and *idctrn01*) speedups of up to nearly 4 can be observed, with a geometric mean of 1.86 over all EEMBC benchmarks. This indicates that for these short-running benchmarks, where JIT compilation times contribute for a relatively larger amount of the overall execution time, partial evaluation has a large positive effect.

Although the performance of our JIT system is very good, there are a few benchmarks in the EEMBC suite in which it does not perform as well as SIMIT-ARM, namely the *rotate01* and *bitmnp01* benchmarks. On closer examination, these benchmarks are extremely heavy in control flow. The *rotate01* benchmark, on which SIMIT-ARM performed the

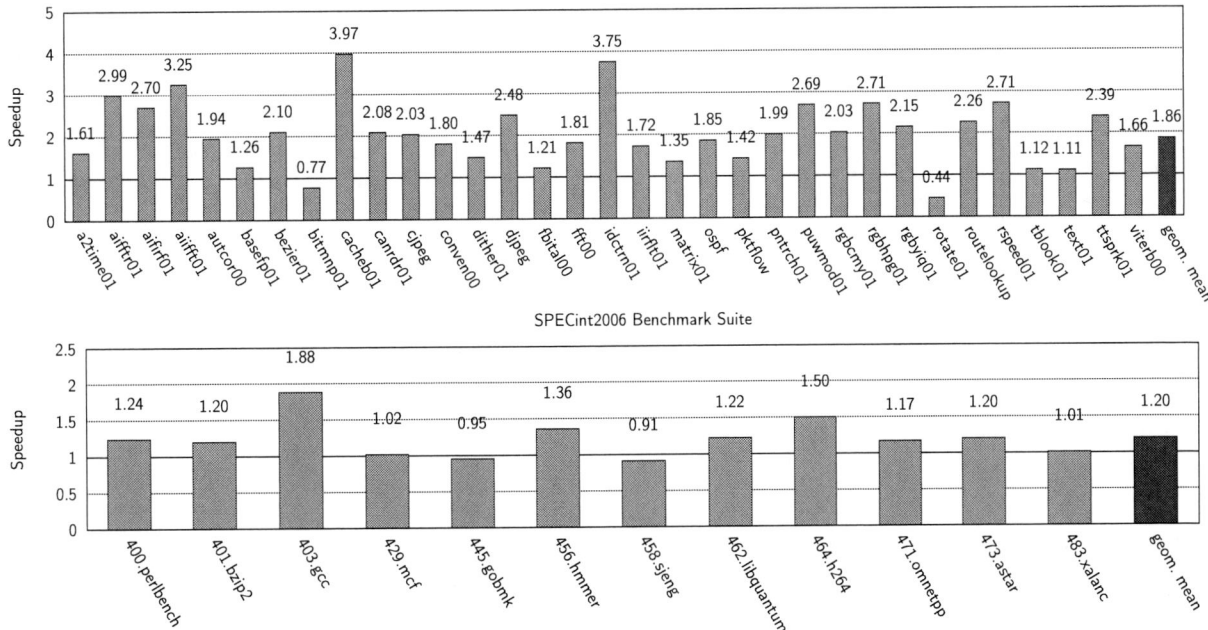

Figure 5: **Performance improvements due to early partial evaluation in our** Iss **for short-running** Eembc **and long-running** Spec Cpu2006 **benchmarks. Speedups are calculated over the** Adl-**retargetable** Simit-Arm v3 Iss.

best, is over 35% branch instructions (not including more exotic branches such as loads-to-PC or pop instructions). On the other end of the scale, the two benchmarks on which our system performed the best (*cacheb01* and *idctrn01*) are particularly light in control flow, with relative branches making up around 4% of the dynamic instruction count. With this in mind, it seems that the reason for the difference in performance between our Iss and Simit-Arm on these benchmarks is due to Gcc, which is used as a Jit compiler in Simit-Arm, outperforming Llvm, which is used as a Jit compiler in our Iss, at optimising control flow.

On the Spec benchmark suite the performance of our system is generally on par with, or better than, that of Simit-Arm. We perform particularly well on the *gcc* and *h264* benchmarks. This is due to the highly phase-oriented nature of the applications, where execution moves through various distinct sections of the program. We perform better than Simit-Arm in these situations due to our superior code generation speed. The Spec suite contains benchmarks which generally run for longer periods of time than the Eembc suite, thus code quality is much more important for these benchmarks (perhaps with the exception of *gcc*). For these long-running benchmarks our partial evaluation approach delivers improved code quality, resulting in an average speedup of 1.2 across all Spec benchmarks. We suffer minor performance losses on two benchmarks (*gobmk* and *sjeng*), where the Gcc compiler used by Simit-Arm does a better job of handling bit manipulation operations than the Llvm optimiser in our Iss.

3.2 Code Size Analysis

One of the main benefits of using our partial evaluation scheme is the extreme code size reduction benefit. Figure

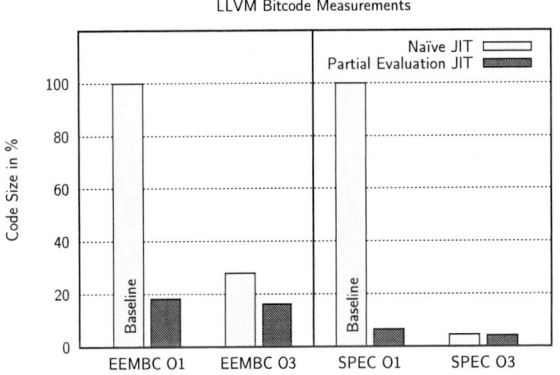

Figure 6: Llvm **bitcode size reduction obtained with partial evaluation in combination with standard** Llvm **optimisations at levels '-O1' and '-O3', respectively, for** Eembc **and** Spec.

6 shows the reduction in Llvm bitcode emitted at standard '-O1' and '-O3' optimisation levels when using our scheme compared to a naïve scheme. An 80% reduction in code size is obtained when optimising to '-O1' (i.e., the partial evaluation scheme emits only 20% as many instructions as the naïve scheme) on the Eembc suite, and an even larger reduction is seen for Spec. A smaller but still significant reduction in code size is observed when optimising to '-O3'. This produces a significant reduction in compilation time which can be exploited to improve Jit warmup speed (i.e., translate code faster) or to more aggressively optimise translations.

149

This is particularly pronounced on the *gcc* benchmark in the SPEC suite. When using the naïve JIT, around 155 million LLVM instructions are generated in total when using '-O1' optimisations. When using the partial evaluation JIT, this drops to just 4 million.

4. RELATED WORK

Retargetable ISS have seen an increased interest in recent years. SIMIT-ARM [11] is the ARM V5 port of such an ISS, where retargeting is achieved through provision of a high-level ISA model from which C language simulator modules are generated. SIMIT-ARM uses parallel and distributed JIT compilation for dynamic code generation, which makes it one of the fastest available ISS today. Still, instruction specialisation is left to the user to be performed manually as part of the retargeting and modelling process. In [7] a simulation framework is presented, which is based on a structural architecture description language (ADL) and that uses the LLVM open-source compiler infrastructure to dynamically translate instruction sequences of the simulated architecture into machine instructions of the host machine. The code generation scheme in [7] is relatively simple and operates at basic block level and relies entirely on LLVM standard optimisations. In contrast, our simulator operates on trace regions and adds partial evaluation on top of standard code optimisations, thus providing higher performance without manual user intervention. Another popular, retargetable ISS is QEMU [2]. Retargeting of QEMU involves rewriting blocks of target code using low-level *tiny code generator* (TCG) operations, which are a machine-independent intermediate notation. Subsequently this notation is being compiled for the host's architecture by TCG subject to optional optimisation passes. Unlike our simulator and also SIMIT-ARM, which employ high-level ISA descriptions, TCG requires that there be dedicated low-level TCG code written to support each target instruction.

An interesting approach is presented in [3]. It aims at generating an ISS from a pseudo-formal document such as a datasheet. However, this approach still requires lots of manual adaptation. In [12] specialisation of instruction behaviours in a generated ISS is discussed.

The open-source ARCHC tool-suite [1] contains an ISS generator, which is based on the same architecture description language as we are using. The ARCHC simulation methodology, however, is very different in that they translate the architecture model to a set of SYSTEMC classes for processor modelling, whereas we generate a JIT based ISS. As a result our simulator is more than one order of magnitude faster than the original ARCHC simulator.

Commercially available, retargetable ISS are included in the SYNOPSYS PROCESSOR DESIGNER [13] and TARGET IP DESIGNER [14] tool suites. These packages are aimed at ASIP design support and unlike the simulator presented here offer less support for complex features typically found in general-purpose embedded processors. We explicitly consider user and kernel mode operation, interrupt handling, memory management and make provisions for efficiently handling self-modifying code.

Partial evaluation [8] is a widely-used program optimisation based on specialisation, which precomputes effects of static input at compile time, thus reducing the complexity and runtime of the generated code. In JIT compilation partial evaluation has the potential to both reduce compilation time and increase code quality. This has been demonstrated, for example, in the PYPY Python JIT compiler [6].

5. SUMMARY AND CONCLUSION

In this paper we have presented an early code optimisation based on partial evaluation, which can be applied in JIT-compiled ISS generated from high-level architecture descriptions. We have demonstrated by implementation that early partial evaluation reduces the overhead of JIT compilation and improves the quality of the generated code, hence contributing to increased overall performance of the ISS. For an ARM V5 architecture model evaluated against the EEMBC and SPEC CPU2006 benchmarks our ISS delivers an average simulation rate of 510 MIPS, outperforming the state-of-the-art, ADL-retargetable SIMIT-ARM ISS by as much as 297%.

6. REFERENCES

[1] R. Azevedo, S. Rigo, M. Bartholomeu, G. Araujo, C. Araujo, and E. Barros. The ArchC architecture description language and tools. *Int. J. Parallel Program.*, 33(5):453–484, Oct. 2005.

[2] F. Bellard. QEMU, a fast and portable dynamic translator. In *Proceedings of the Annual Conference on USENIX*, ATEC '05, pages 41–41, Berkeley, CA, USA, 2005. USENIX Association.

[3] F. Blanqui, C. Helmstetter, V. Joloboff, J.-F. Monin, and X. Shi. Designing a CPU model: from a pseudo-formal document to fast code. *CoRR*, abs/1109.4351, 2011.

[4] I. Böhm, T. J. Edler von Koch, S. C. Kyle, B. Franke, and N. Topham. Generalized just-in-time trace compilation using a parallel task farm in a dynamic binary translator. In *Proceedings of the 32nd ACM SIGPLAN Conference on Programming Language Design and Implementation*, PLDI '11, pages 74–85, New York, NY, USA, 2011. ACM.

[5] I. Böhm, B. Franke, and N. Topham. Cycle-accurate performance modelling in an ultra-fast just-in-time dynamic binary translation instruction set simulator. In *2010 International Conference on Embedded Computer Systems (SAMOS)*, pages 1–10, July 2010.

[6] C. F. Bolz, A. Cuni, M. Fijałkowski, M. Leuschel, S. Pedroni, and A. Rigo. Allocation removal by partial evaluation in a tracing JIT. In *Proceedings of the 20th ACM SIGPLAN Workshop on Partial Evaluation and Program Manipulation*, PEPM '11, pages 43–52, New York, NY, USA, 2011. ACM.

[7] F. Brandner, A. Fellnhofer, A. Krall, and D. Riegler. Fast and accurate simulation using the LLVM compiler framework. In *1st Workshop on Rapid Simulation and Performance Evaluation: Methods and Tools (RAPIDO)*, January 2009.

[8] C. Consel and O. Danvy. Tutorial notes on partial evaluation. In *Proceedings of the 20th ACM SIGPLAN-SIGACT Symposium on Principles of Programming Languages*, POPL '93, pages 493–501, New York, NY, USA, 1993. ACM.

[9] J. Lee, J. Kim, C. Jang, S. Kim, B. Egger, K. Kim, and S. Han. FaCSim: a fast and cycle-accurate architecture simulator for embedded systems. In *Proceedings of the 2008 ACM SIGPLAN-SIGBED Conference on Languages, Compilers, and Tools for Embedded Systems*, LCTES '08, pages 89–100, New York, NY, USA, 2008. ACM.

[10] R. Leupers, J. Elste, and B. Landwehr. Generation of interpretive and compiled instruction set simulators. In *Proceedings of the Asia and South Pacific Design Automation Conference (ASP-DAC '99)*, pages 339 –342 vol.1, jan 1999.

[11] W. Qin and S. Malik. Flexible and formal modeling of microprocessors with application to retargetable simulation. In *Proceedings of the Xonference on Design, Automation and Test in Europe - Volume 1*, DATE '03, pages 10556–, Washington, DC, USA, 2003. IEEE Computer Society.

[12] J. Song, H. Hao, C. Helmstetter, and V. Joloboff. Generation of executable representation for processor simulation with dynamic translation. In *2008 International Conference on Computer Science and Software Engineering*, volume 4, pages 106–109, dec. 2008.

[13] Synopsys. Processor Designer. www.synopsys.com.

[14] Target Compiler Technologies. IP Designer. www.retarget.com.

XDRA: Exploration and Optimization of Last-Level Cache for Energy Reduction in DDR DRAMs

Su Myat Min   Haris Javaid   Sri Parameswaran

School of Computer Science and Engineering, University of New South Wales, Sydney, Australia

{sumyatmins, harisj, sridevan}@cse.unsw.edu.au

ABSTRACT

Embedded systems with high energy consumption often exploit the idleness of DDR-DRAM to reduce their energy consumption by putting the DRAM into deepest low-power mode (self-refresh power down mode) during idle periods. DDR-DRAM idle periods heavily depend on the last-level cache. Exhaustive search using processor-memory simulators can take several months. This paper for first time proposes a fast framework called XDRA, which allows the exploration of last-level cache configurations to improve DDR-DRAM energy efficiency.

XDRA combines a processor-memory simulator, a cache simulator and novel analysis techniques to produce a Kriging based estimator which predicts the energy savings for differing cache configurations for a given main memory size and application. Errors for the estimator were less than 4.4% on average for 11 applications from mediabench and SPEC2000 suite and two DRAM sizes (Micron DDR3-DRAM 256MB and 4GB). Cache configurations selected by XDRA were on average 3.6× and 4× more energy efficient (cache and DRAM energy) than a common cache configuration. Optimal cache configurations were selected by XDRA 20 times out of 22. The two suboptimal configurations were at most 3.9% from their optimal counterparts. XDRA took a few days for the exploration of 330 cache configurations compared to several hundred days of cycle-accurate simulations, saving at least 85% of exploration time.

1. INTRODUCTION

Better energy efficiency increases reliability and improves battery life of embedded systems. Main memories in embedded systems consume up to 80% of total system power (excluding I/O power) [1] (also illustrated by our experiments reported in Figure 11, Appendix C). Thus, reducing energy consumption of main memory can have a great impact on the overall energy efficiency of the system.

Figure 1, drawn to scale, shows the current drawn (which in turn governs the power) by a modern DDR3 DRAM memory. The activate current (which consists of activation and precharge currents) is shown with a (red) dashed line, whereas the read and write current is shown in a (blue) dotted line. The former is used to activate DRAM, whereas the latter is used to perform read/write operations. Much like other DDR DRAMs, DDR3 has to be refreshed periodically, which consumes the refresh current shown in a (purple) dashed-dotted line. The (green) solid line shows the background current, which is continuous. However, depending upon the memory mode, the amount of background current varies. Figure 1 shows this background current for different low-power modes. These modes are used when DDR3 DRAM is not performing any read/write operations, and hence is idle. According to our experiments, it is the background power (due to background current) that consumes 90% of DRAM power and about 70% of total system power (see Figure 11, Appendix C). In this paper, we focus on the

Permission to make digital or hard copies of all or part of this work for personal or classroom use is granted without fee provided that copies are not made or distributed for profit or commercial advantage and that copies bear this notice and the full citation on the first page. To copy otherwise, to republish, to post on servers or to redistribute to lists, requires prior specific permission and/or a fee.
DAC'13 May 29 - June 07, 2013, Austin, TX, USA.

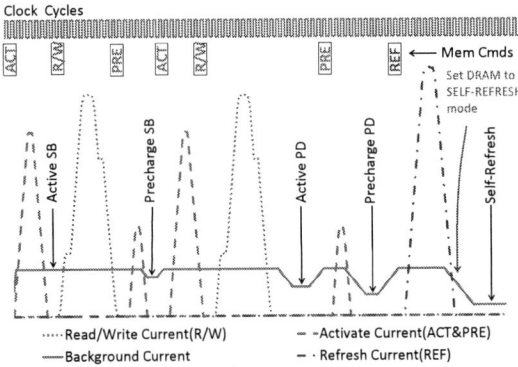

Figure 1: Different currents drawn by a DDR3 DRAM, adapted from Micron [2].

use of self-refresh Power Down (PD) mode to reduce this background power consumption, because it is the deepest low-power mode, and consumes approximately only a tenth of the background power of the shallowest low-power mode, Active StandBy (SB) mode (see Table 2, Appendix B).

Transition to any of the low-power modes comes at the cost of a wakeup latency, which is reported in Table 2 (Appendix B) for different low-power modes of DDR3 DRAM. Efficient exploitation of a low-power mode requires DRAM idle periods to be long enough to amortize the performance penalty due to the wakeup latency. High wakeup latency of self-refresh PD mode (highest amongst all the low-power modes) has typically hampered its use, and only two works [3, 4] are known to date which employ this mode. The authors of [3] proposed history based predictor to forecast durations of DRAM idle periods, and a memory controller policy for selective use of precharge PD and self-refresh PD modes. They focused on maximal exploitation of DRAM idle periods, without attempting to prolong those periods for complete use of self-refresh PD mode. On the other hand, Amin et al. [4] proposed a cache replacement policy to skew cache-memory traffic in such a way that idle periods for some of the DRAM ranks (a DRAM consists of multiple ranks, where a rank is the smallest granularity at which low-power modes are applied) are prolonged to enable the use of self-refresh PD mode for those ranks. Such a policy, as shown in [4], is effective, but requires the use of a completely new cache policy, and thus significant hardware modification.

In this paper, we also focus on prolonging of DRAM idle periods; however, for the first time, we propose to do so by the use of a suitable last-level cache configuration (cache size, line size and associativity), which is obtained by exploration of the last-level cache design space. The selected cache configuration is the one which maximally reduces total energy consumption of the last-level cache and DRAM. Our proposal is motivated by the fact that many modern embedded processors such as ARM, Xtensa and Microblaze allow configuration of the cache so that one can choose the cache configuration to be implemented. Additionally, since embedded systems execute an application or a class of applications repeatedly, one can tune cache parameters for better energy efficiency. In fact, the authors of [5] also advocated the importance of last-level cache exploration for a given application; however,

151

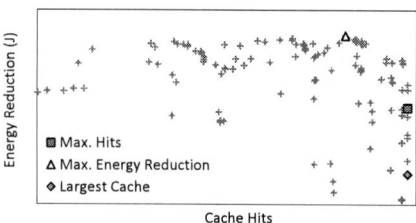

Figure 2: Total cache and DRAM energy reduction for different L2 (last-level) cache configurations.

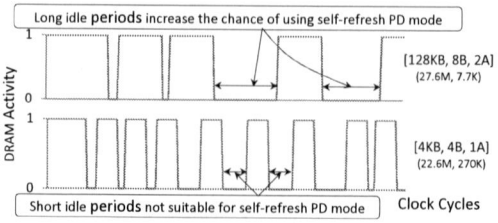

Figure 3: Effect of two distinct L2 (last-level) cache configurations on DRAM idle periods. PD stands for PowerDown.

they focused on energy reduction of only the cache rather than both the cache and DRAM.

The design space of last-level cache can be explored either using full-system cycle-accurate simulations (including processor, caches and memory) or cache simulators. Cycle-accurate simulations are exorbitantly slow [6], especially when multi-level caches and DRAM are involved. Hence, they are not feasible for exploration of hundreds of cache configurations, which are typical of an embedded system design [6, 7]. Fast cache simulation approaches [7, 8], an alternative option, use memory trace of an application to output the number of hits and misses for all the cache configurations. Calculation of DRAM energy consumption is a complex process [2, 9], and using only the number of last-level cache hits/misses or cache size is not sufficient. Figure 2 plots total energy reduction of cache and DRAM against the number of hits for different L2 cache configurations in a uniprocessor system with two levels of cache, executing mpeg2enc application compared to a system without second level cache. Cache configurations with maximum energy reduction and maximum number of hits are marked along with the largest cache configuration. It can be seen that the MaxER configuration is different from both MaxHits and the largest cache. When cache area footprint is compared, MaxER (0.052 mm^2) is significantly smaller than MaxHits (3.22 mm^2) and the largest cache (6.21 mm^2). This plot illustrates the fact that cache hits or cache sizes alone are not sufficient for selection of a suitable cache configuration. Hence, *the challenge is to quantify the effect of last-level cache configurations on DRAM energy consumption with minimal number of cycle-accurate simulations.* Therefore, in this paper, we propose a novel DRAM energy reduction estimator to quickly predict energy reductions for differing cache configurations, and a novel framework around the estimator for quick exploration and selection of a suitable last-level cache configuration.

Novel Contributions. In particular, we make the following contributions:

- A novel DRAM energy reduction estimator. The estimator uses five parameters: self-refresh PD cycles, self-refresh PD count, row buffer conflicts, refresh count and DRAM traffic to predict energy reduction of DRAM for a given last-level cache configuration. The estimator is based upon a kriging model that is derived through a small number of cycle-accurate simulations. Our estimator has an average error of less than 4.4% when tested on several applications from mediabench and SPEC2000 applications, and 256MB and 4GB DRAMs.
- XDRA framework. Our framework integrates processor-memory and cache simulators with our estimator with the help of novel analysis techniques. These techniques do not require cycle-accurate simulations of all the cache configurations for computation of the estimator parameters, and thus enable fast exploration of last-level cache configurations. When used to select cache configurations with maximal reduction in total energy of last-level cache and DRAM, XDRA found 20 out of 22 times the same configurations as from cycle-accurate simulations. The suboptimal configurations were off by a maximum of 3.9% from their optimal counterparts.

Motivational Example. To illustrate how differing cache configurations can affect DRAM idle periods and improve its energy efficiency, let us consider a JPEG encoder application running on a uniprocessor system with L1 and L2 cache and a DRAM memory. Figure 3 reports DRAM activity for several thousand cycles, where a value of 1 means the DRAM is accessed and a value of 0 means that it is idle. Two L2

cache configurations are used: (1) a 4KB, 4B line size, direct-mapped cache, denoted as [4KB, 4B, 1A] and shown in lower plot, and (2) a 128KB, 8B line size and 2-way associative cache, denoted as [128KB, 8B, 2A] and shown in the upper plot. The values inside the parentheses report DRAM idle cycles and the number of idle periods. For instance, in [128KB, 8B, 2A] L2 cache system, DRAM was idle for 27.6 million cycles, distributed across 7,700 idle periods. The [128KB, 8B, 2A] L2 cache increased DRAM idle cycles by 22% which was expected due to its larger size. The number of idle periods has reduced from 270,000 to only 7,700, and this reduction will be advantageous in reducing DRAM energy consumption because: (1) idle periods will be longer enabling DRAM's transition to self-refresh PD mode; (2) DRAM can remain in self-refresh PD mode for longer periods, and (3) fewer wakeups from self-refresh PD mode mean less performance penalty. The plot for [128KB, 8B, 2A] cache corroborates our belief, where several short idle periods of [4KB, 4B, 1A] cache that are not suitable for self-refresh PD mode, have mostly been consolidated into two longer idle periods which in fact are long enough to transition DRAM into self-refresh PD mode. In this example, [128KB, 8B, 2A] cache reduced 38% more DRAM energy consumption than [4KB, 4B, 1A] cache. This example shows the importance of exploring last-level cache, and choosing a suitable configuration in the first place can tremendously reduce energy consumption of DRAM.

2. RELATED WORK

Recent research has focused on reducing DRAM energy consumption, consisting of memory architectures and controller policies, cache replacement policies, OS-level and compiler-directed optimizations.

Delaluz et al. [10] predicted memory idle periods for cacheless systems to control the use of low-power modes. Fan et al. [11] extended their work for multi-level cache hierarchy through an analytical model of Rambus DRAM. The work in [12] proposed an adaptive history based scheduler with memory commands' throttling to increase memory idle periods. Liu et al. [13] introduced a temperature and power management policy by buffering DRAM write operations to improve page hit rate, resulting in reduction of DRAM active power. Prefetching of data from DRAM into an intermediate buffer has been explored [14, 15] with the focus on reducing DRAM active power only. The authors of [16] proposed an architectural technique where a DRAM rank is divided into multiple mini-ranks to reduce the number of active devices for a memory access. In contrast to above memory architectures and controller policies, Amin et al. [4] proposed a cache replacement policy to skew cache-memory traffic to prolong DRAM idle periods for use of self-refresh PD mode. Recently, the authors of [3] proposed history based predictor to forecast durations of DRAM idle periods for selective use of precharge PD and self-refresh PD modes. OS-level approaches [17, 18], on the other hand, manage memory traffic at kernel layer through page migration, power-aware page allocation and similar policies. Compiler-directed approaches [19, 20] statically analyze application code to detect memory idle periods in addition to data access patterns to optimally place both code and data in DRAM.

Unlike all the above works, our work differs in its very proposal of exploring the last-level cache and the selection of a suitable configuration for maximal improvement in combined energy efficiency of the cache and DRAM. Therefore, our proposal is orthogonal to aforementioned works, and can be applied in conjunction with them.

3. PROBLEM STATEMENT

We target a uniprocessor system with multi-level cache hierarchy and

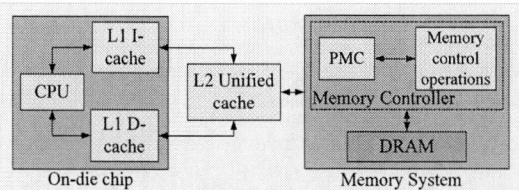

Figure 4: An example of target system.

DDR3 DRAM. An example system is shown in Figure 4, where the processor has on-chip separate L1 instruction and data caches, which are connected to a unified off-chip L2 cache. The L2 cache is interfaced to the DRAM memory through a memory controller. Here, L2 is the last-level cache and we use this system as an example throughout the paper in addition to its use in our experiments.

DDR3 DRAM contains an internal mode controller which automatically sets it into one of the low-power modes whenever it is idle. Power Mode Controller (PMC) is another module that is typically embedded into memory controller firmware to control the use of low-power modes [15]. In our work, PMC transitions DRAM into the deepest low-power mode, self-refresh PD, if it is idle for at least threshold number of clock cycles as used in [12, 4]. This is done to avoid significant degradation of performance that may arise from greedy use of self-refresh PD mode for every idle period. We use 30 cycles as the self-refresh PD threshold (PDthreshold) which was obtained through experimentation [12, 4].

Given an application, a uniprocessor system with unified last-level cache and DRAM, a PDthreshold and last-level cache configurations (size, line size and associativity), our goal is to determine, for all the last-level cache configurations, the reduction in combined energy consumption of the last-level cache and DRAM compared to the system without last-level cache. Using the aforementioned energy reductions, we aim to select the cache configuration with maximal energy reduction with/without a constraint on last-level cache area footprint.

4. DRAM ENERGY REDUCTION ESTIMATOR

Estimation of DRAM energy consumption is a complex process [2, 9] due to the involvement of different DRAM states and their transitions during the execution of an application. In this paper, we build a DRAM energy reduction estimator based upon the concept of kriging models [21]. Note that CACTI [22] tool does not consider detailed DRAM architecture, and hence is not suitable for estimation of DRAM energy consumption [9].

Estimator Parameters. The parameters of an estimator are a key choice as they directly influence the accuracy, and hence usefulness of an estimator. One should only include the most influential parameters because other parameters with little influence increase an estimator's complexity without any useful benefit. We conducted several experiments with differing last-level cache configurations to record DRAM energy reduction, and computed correlation to obtain the significance of the following most common parameters (some of which are explained in more detail later) on the amount of energy reduction:

- *Application parameters:* total memory requests (TMR).
- *Cache parameters:* size (CS), line size (CLS), associativity (CA), hits (CH) and misses (CM).
- *DRAM parameters:* accesses (DAs), non-accesses (DNAs), memory traffic (MT), self-refresh PD cycles (PDcycles), number of times self-refresh PD mode is used (PDCnt), row buffer conflicts (RBConf) and number of times DRAM is refreshed (RefCnt).

Figure 5 depicts the average value of correlation coefficients for all the parameters over 2 DRAM sizes (256MB and 4GB) and 11 applications (from mediabench and SPEC2000 benchmarks). A value of +1 signifies a perfect positive correlation (that is, increase in parameter value will increase energy reduction), whereas a value of -1 signifies a perfect negative correlation (that is, increase in parameter value will decrease energy reduction). A value close to zero means no correlation. It can be seen from the figure that PDcycles, PDCnt, RBConf, RefCnt and MT are the most significant parameters (marked with rectangles) as their correlation values are either greater than 0.8 or less than -0.8. An intuitive reasoning for these parameters is as follows:

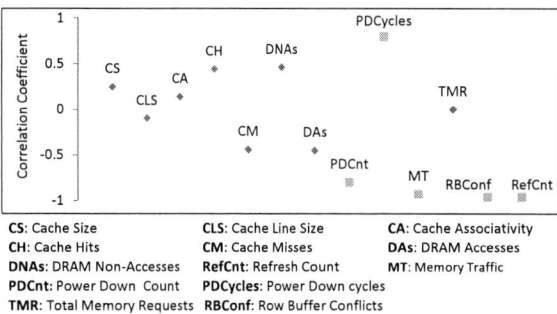

CS: Cache Size   CLS: Cache Line Size   CA: Cache Associativity
CH: Cache Hits   CM: Cache Misses   DAs: DRAM Accesses
DNAs: DRAM Non-Accesses RefCnt: Refresh Count   MT: Memory Traffic
PDCnt: Power Down Count PDCycles: Power Down cycles
TMR: Total Memory Requests RBConf: Row Buffer Conflicts

Figure 5: Correlation coefficients of most common parameters, averaged over 2 DRAM sizes and 11 applications.

- *PDcycles:* the total number of cycles for which DRAM is in PD mode during the execution of an application. This parameter depends on the total number of DRAM idle cycles and the PDthreshold. More PDcycles mean DRAM remains in PD mode for longer periods, providing more energy reduction.
- *PDCnt:* the total number of times DRAM is transitioned to self-refresh PD mode. More transitions to self-refresh PD mode mean more overhead, which results in less energy reduction.
- *RBConf:* In DRAM, data is brought to a row buffer before it can be accessed [2]. A row buffer conflict occurs when the requested data is not present in the row buffer, resulting in its reloading (precharging and/or activation), which increases DRAM energy consumption.
- *RefCnt:* DRAM must be refreshed periodically to retain its contents. Refreshing DRAM consumes refresh power which in turn increases DRAM energy consumption.
- *MT:* The product of last-level cache line size and total number of DRAM accesses measures the amount of traffic going to DRAM during the execution of an application. More traffic to DRAM means that it will be active for longer time, increasing its energy consumption.

Kriging Model. Kriging models take into account the spatial correlation between the current design point, x_i, and an initial set of design points (training set), x, to estimate the output at x_i. We chose kriging model because it allows various polynomial functions and correlation functions, and does not restrict to just linear regression models. Additionally, kriging models can capture complex interactions between parameters due to the use of spatial correlations, and thus perform better than linear regression models [23]. Our experiments reveal similar results, which are detailed in Section 6. For details of kriging model, readers are referred to Section A.1.

5. XDRA FRAMEWORK

Our framework, XDRA, to quickly explore last-level cache in the context of DRAM energy reduction is shown in Figure 6. XDRA integrates our estimator with a cycle-accurate processor-memory simulator, a cache simulator, and employs our novel analysis techniques to quickly compute the parameters used in the estimator. The following paragraphs explain core components of XDRA in more detail.

Last-level Cache Idle (LCI) Profile Generation. An application is simulated in a cycle-accurate processor-memory simulator *without last-level cache and PMC's power down mechanism* to record two entities at the second-last-level cache and memory interface: (1) No Last-level Cache (NoLC) memory trace, and (2) LCI profile. NoLC trace will contain only those memory requests that will be missed in lower level caches. For instance, in Figure 4 without L2 cache, the memory trace captured at the L1–memory interface will only contain L1 cache misses. *An LCI period refers to an application's execution period that does not access DRAM.* For instance, at L1–memory interface without L2 cache in Figure 4, LCI periods will be the execution periods with no memory requests from the application (consecutive non-load and non-store instructions) and execution periods with memory requests that will hit in L1 caches. Hence, L2 cache will be idle during LCI periods. Figure 12 (Appendix C) illustrates such an LCI period. *An LCI profile captures all LCI periods in clock cycles from the execution of*

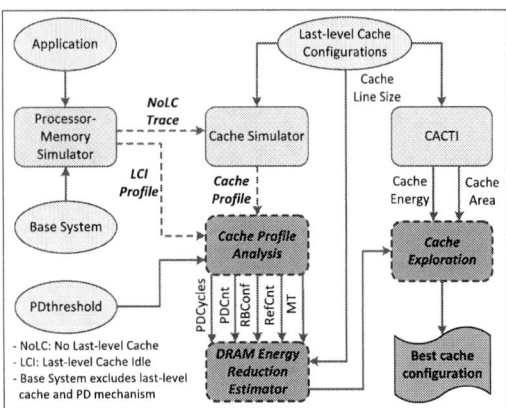

Figure 6: XDRA Framework. Dotted-lined rectangles and broken arrows show our novel contributions.

an application. For an in-order processor, which is typical of embedded processors, an LCI profile of the application will not change across different last-level cache configurations because the processor pipeline is stalled during each memory request and lower level caches remain unchanged. Note that DRAM will be idle during all LCI periods and hence these periods will contribute to DRAM idle periods. The DRAM energy consumption from this simulation is used as the reference point for calculation of energy reductions for all the last-level cache configurations. This is the only step in XDRA where an application is cycle-accurately simulated (excluding a few simulations required to train the estimator).

Cache Simulation. In this step, we feed NoLC trace to a cache simulator to generate cache profile for each of the last-level cache configuration. *A cache profile reports whether DRAM will be accessed or not for each memory request in NoLC trace.* A memory request in NoLC trace is considered DRAM access or non-access in the cache profile depending upon the cache policy (such as write-back, write-allocate, etc.) and whether the request hits or misses the cache. For more details, readers are referred to [24].

Cache Profile Analysis. This step analyzes the cache profile of each last-level cache configuration to compute the five parameters that are used in our estimator. *A DRAM Idle (DI) profile captures the duration of each DI period from a given cache profile and LCI profile of the application where a DI period refers to consecutive cycles of DRAM idleness.* First, we insert LCI periods from LCI profile between appropriate DRAM accesses and non-accesses in the cache profile. Then, all DI periods are marked where an idle period consists of consecutive DRAM non-accesses and LCI periods. PDcycles are computed by summing all DI periods, whereas PDCnt is equal to the number of DI periods that are greater than PDthreshold. Note that this step allows a designer to apply any threshold that he/she deems suitable. Readers are referred to Section A.2 for an example.

The number of row buffer conflicts, RBConf, depends upon the DRAM address mapping, and row buffer policy (such as open-page, closed-page, etc.). A row buffer conflict occurs when the requested data is not present in the row buffer, which requires its loading from DRAM through the process of activation. We calculate the location (rank, bank and row ids [2]) of each DRAM access using the DRAM address mapping, which is known a priori. Afterwards, if the rank and bank ids of two successive DRAM accesses are the same, but the row ids are different, then we record a row buffer conflict. Depending upon the row buffer policy, row buffer may be empty (closed) between DRAM accesses to the same bank, which are accounted for using the information from the row buffer policy. Note that the aforementioned method does not capture all the row buffer conflicts because DRAM needs to be refreshed periodically to retain its contents. During the refresh process, the contents of the row buffer are lost, which results in a row buffer conflict for the next DRAM access. Thus, the inaccuracies in RBConf depend upon RefCnt, the number of times DRAM is refreshed during the execution of an application.

The parameter RefCnt depends upon the time for which DRAM is active. Accurate computation of DRAM's active time is not possible

because DRAM accesses take differing cycles depending upon DRAM state [25], which is unknown unless cycle-accurate simulation is performed. Therefore, we estimate DRAM active time as [DRAM accesses × fixed DRAM access latency], where the fixed latency is an average of latencies of all DRAM accesses from cycle-accurate simulation performed during the 'LCI profile generation' step. This DRAM active time is then divided by refresh frequency of DRAM, which is known a priori (for example, 64 ms for DDR3 from Micron [2]), to get an estimate of RefCnt. Since both RBConf and RefCnt are estimated, they might cause inaccuracies in the estimator; however, our results (see Section 6) show that such inaccuracies are small.

The parameter MT measures the total traffic going to DRAM during the execution of an application. MT is calculated in bytes by multiplying the number of DRAM accesses by last-level cache line size. In this way, we analyze the cache profile of each last-level cache configuration to calculate the values of five parameters for use in our estimator, without the need for cycle-accurate simulations of those configurations.

Cache Exploration. At the final step, DRAM energy reduction is computed for each last-level cache configuration using the PDcycles, PDCnt, RBConf, RefCnt and MT in our estimator. The predicted energy reductions are adjusted by subtracting the energy consumptions of respective cache configurations which are obtained from CACTI [22]. Once net DRAM energy reductions for all last-level cache configurations are available, the cache configuration with maximum reduction is selected. Here, the decision to search all the cache configurations for global optima or to use a heuristic search is left to designer. We chose to use the former approach in this paper.

6. EXPERIMENTS & RESULTS

We use the target system of Figure 4 with LRU and write-back policies for L2 cache. The target system is implemented using Tensilica's LX4 processor [26] with 2KB L1 instruction and 1 KB L1 data caches, both direct-mapped with 4B line size. We used -125E DDR3 DRAM (1600 Million Transfers per second) from Micron [2] to implement DRAM memory. We created two target systems with 256MB (1 rank) and 4GB (4 ranks) memories to observe the applicability of our estimator. These DRAMs used an interface width of 4B, open-page row buffer policy and internal refresh time of 64 ms. A threshold of 30 cycles and a wakeup latency of 64 cycles was used for self-refresh Power Down (PD) mode.

We used the cycle-accurate simulator from [25][1] and tool from [7] as processor-memory simulator and cache simulator in XDRA. In addition, CACTI 6.5 [22] was configured for a given 90nm technology to obtain energy consumption and area of L2 cache configurations. The L2 cache design space consisted of 330 configurations, constituted by changing cache size from 4KB to 4MB, line size from 4B to 128B and associativity from 1 to 16. These configurations are typically explored in an embedded system's design [6, 7]. For evaluation purposes, we used adpcmenc/dec, jpegenc/dec, g721enc/dec, mpeg2enc/dec applications from mediabench, and memory-bound vpr, gzip and bzip2 applications from SPEC2000 [28]. For SPEC2000 applications, first 500 million memory requests were used [29]. All experiments were conducted on an Intel Xeon 64 core machine with 256GB RAM.

For selection of design points in the training set, we use a well-known design of experiments technique, Latin Hypercube Sampling (LHS). LHS is a statistical technique that evenly distributes multiple points in the design space so that all segments of the design space's dimensions are represented. The minimum requirement of LHS for the number of design points is summation of ranges of all the design space dimensions, which in our case is 22 (11 cache sizes, 6 line sizes and 5 associativities). Additionally, we used 8 corner design points [30] from the design space to create a training set with 30 design points. The training set was cycle-accurately simulated to capture actual energy reductions and values of the parameters in the kriging model. We created kriging models (separately per application) using the DACE toolbox [31] for MATLAB, where constant and linear polynomials were used for the regression function $f(x)$, while linear, gaussian, exponential, spherical and spline functions were used as the correlation function $z(x)$. Note that training of a kriging model takes a few seconds, and

---

[1]Their cycle-accurate simulator integrated DRAMSim [27] with LX4's simulator to create a detailed processor-memory simulator.

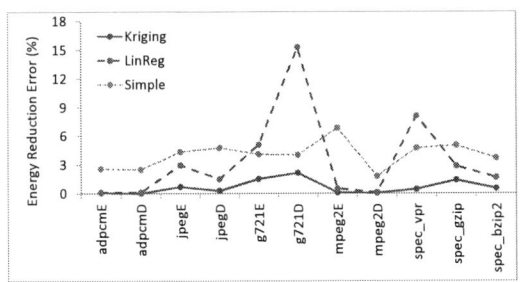

Figure 8: Average error in estimated energy reduction for 256MB DRAM.

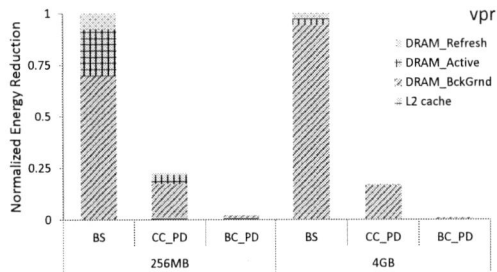

Figure 9: Normalized DRAM energy consumption breakdown of vpr for different L2 caches and DRAM sizes.

hence several different functions can be evaluated to best fit the training set.

For rigorous evaluation of our estimator, we created two other estimators: (1) LinReg, based on a linear regression model, and (2) Simple, based on a constant energy per DRAM access model. Details of these models are in Section A.3.

Estimator Accuracy. We evaluated the accuracy of the three estimators by calculating average errors using the estimated energy reductions and the actual energy reductions from cycle-accurate simulations for all the cache configurations. Figure 8 reports the average errors when 256MB DRAM is used for all the applications. Our estimator had a worst average error of 2.2% across all the applications. On the other hand, LinReg and Simple had maximum average errors of 15.3% and 6.8% respectively. For 4GB DRAM, the errors are 3.4%, 26.2% and 99.4% for kriging, LinReg and Simple estimators, which are illustrated in Figure 14 (Appendix C). It is noteworthy that the average error of our estimator is consistently low compared to significant variations for LinReg and Simple, which means better applicability for our estimator.

To further stress the accuracy of the estimators, we performed constrained optimization on the estimated cache design spaces. We searched for a cache configuration with maximum energy reduction under a constraint on area footprint of the cache, which is typical of embedded systems. For each application, we applied several constraints, ranging from largest to smallest cache area. The cache configurations selected by the estimators were compared to corresponding cache configurations from actual design space (obtained using the same constraints) to record minimum and maximum errors in energy reduction over all the constraints. The results for 256MB DRAM are illustrated in Figure 7 for all the applications. Again, our estimator performs consistently well with maximum error less than 4.4% over all the applications. LinReg and Simple estimators had significant variations in the selection of cache configurations, with errors up to 74.1% and 92.3% respectively, which renders them unreliable compared to our estimator. Results for 4GB DRAM are reported in Figures 13 (Appendix C), where kriging, LinReg and Simple estimators had errors up to 3.3%, 67.9% and 7% respectively. These results show that our estimator is better at consistently modelling DRAM energy reductions with differing last-level cache configurations, and thus the rest of this section reports results for our estimator only.

Cache Selection with XDRA. For each application, we used XDRA to explore L2 cache to select the cache configuration with maximum DRAM energy reduction. Table 3 (Appendix C) reports the L2 cache configurations selected by XDRA with their area footprints. Out of 22 cache configurations (two configurations for two different DRAM sizes per application), 20 times XDRA found the same configuration as from cycle-accurate processor-memory simulations. The suboptimal configurations had a maximum increase of just 3.9% in total energy

consumption of the cache and DRAM.

Once the best cache configuration was known from XDRA, for comparison purposes, we simulated the following three systems in processor-memory simulator to obtain their actual DRAM energy reduction:

- **BS:** Base system without L2 cache and self-refresh PD mechanism, but with PD mechanism of DDR3 internal mode controller.
- **CC_PD:** System with a reasonable Common Cache configuration (64KB, 64B line size, direct-mapped L2 cache, which has been reported in [32]) and PD mechanism.
- **BC_PD:** System with Best Cache configuration and PD mechanism.

The energy reduction comparison of the above three systems for one of the applications with 256MB and 4GB DRAMs is discussed here (results for other applications are similar and are reported in Figures 15– 24, Appendix C). Figure 9 reports the normalized energy consumption breakdown for 'vpr' application from SPEC2000. Both CC_PD and BC_PD systems significantly reduce the energy consumption of DRAM; however, BC_PD is more energy efficient than CC_PD, by a factor of 12× and 24× for 256MB and 4GB DRAMs respectively. These results show that a suitable cache configuration with self-refresh PD reduces both active and background power of DRAM (although reduction in background power is more significant than active power) because (1) most of the memory requests are serviced by the cache, which reduces DRAM activity, and (2) DRAM can be put into self-refresh PD mode more often. In summary, BC_PD system from XDRA reduced on average 3.6× and 4× more cache and DRAM energy compared to CC_PD for 256MB and 4GB DRAMs respectively. These results indicate the usefulness of XDRA in selecting a suitable cache configuration, and use of an arbitrary cache configuration always is not the most energy efficient choice.

The performance penalty due to wakeup latency of self-refresh PD mode is measured by comparing the execution times of BC_PD with a similar system but without self-refresh PD mechanism. In our experiments, a maximum penalty of 2% was observed. This result illustrates the fact that a suitable cache not only increases DRAM idle periods but also consolidates them into longer periods to make them suitable for self-refresh PD mode and to reduce the number of DRAM wakeups, hence reducing the overall performance penalty. It is important to note that none of the BC_PD systems incurred any performance penalty compared to BS system because the reduction in execution time due to cache hits amortized the wakeup latency of self-refresh PD mode. The above results point to the fact that a suitable last-level cache with self-refresh PD mode can tremendously increase DRAM energy efficiency with (1) marginal performance penalty compared to a similar system but without PD mechanism, and (2) performance improvement compared to a similar system but without any cache.

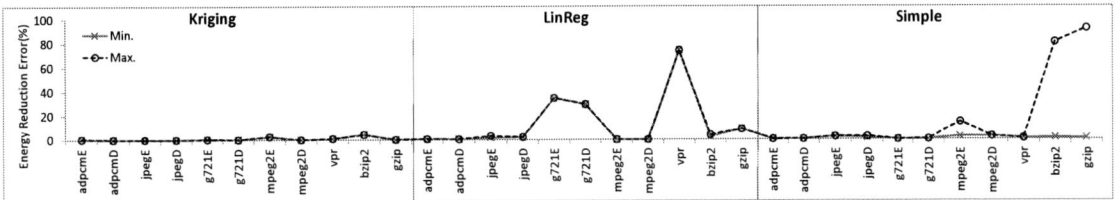

Figure 7: Error in energy reduction from cache configurations selected under differing area constraints for 256MB DRAM.

155

App.	Cycle-Accurate Simulator	XDRA			
		TS	LCI	CPA	Total
adpcmE	3.2h	17m	45s	6m	23m
adpcmD	2.1h	12m	33s	6m	17m
jpegE	21.4h	2h	5m	23m	2.4h
jpegD	7.9h	43m	2m	9m	55m
g721E	13.8d	1d	1h	7h	1.5d
g721D	14.6d	1d	1h	7h	1.7d
mpeg2E	155.2d	14d	12h	1d	15.5d
mpeg2D	41.7d	4d	4h	13h	4.5d
vpr	103.7d	9d	9h	2d	11.3d
bzip2	143.9d	13d	12h	2d	15.1d
gzip	124.1d	11d	10h	1d	13.4d

Table 1: Time comparison of cycle-accurate processor-memory simulator and XDRA for 256MB DRAM.

Table 1 reports the time taken by cycle-accurate processor-memory simulator and XDRA for exploration of 330 L2 cache configurations with 256MB DRAM. The total time for XDRA has been broken down into: time to train our estimator (TS); time to generate LCI profile and cache simulation (LCI); and, time for cache profile analysis and cache exploration (CPA). XDRA reduces exploration time from several hundred days to a few days, resulting in savings of at least 85% of simulation time, which enables quick exploration of last-level cache. Note that the total time of XDRA with LinReg estimator is the same as our estimator because of similar training set. However, if Simple estimator is used in XDRA, then its time will be equal to LCI time only, making it faster than our estimator. However, as shown before, Simple estimator had average errors as high as 99.4% and does not perform consistently across various applications, limiting its practical use. Results for 4GB DRAM reveal similar savings, which are reported in Table 4 (Appendix C).

7. ADVANTAGES & LIMITATIONS

XDRA features several advantages. XDRA (1) is fast as it only uses one cycle-accurate simulation per application (excluding a few simulations to train the estimator), (2) integrates processor-memory and cache simulators with analysis techniques to quickly compute parameters for DRAM energy reduction estimator and (3) uses the estimator for fast exploration of last-level cache configurations. Although several cycle-accurate simulations are required to build the estimator, the number of simulations is less than 10% (30 out of 330, see Section 6) of the whole design space.

XDRA is very flexible as any processor-memory and cache simulators can be used given the profiles described earlier can be produced, whcih requires minor modifications to existing simulators. Finally, a designer has the flexibility to use the cache policies, PDthreshold, DRAM address mapping, row buffer policies, etc. of his/her system under design, although we did not test with all the possible combinations of such options.

XDRA can also be used to find best last-level cache configuration for a class of applications. In this case, the trace from combined execution of the applications should be captured and input to cache simulation. A truly representative trace from multiple applications' execution might not be possible due to indeterminism; however, this is a different problem and not the focus of our work. XDRA will explore last-level cache configurations to find the best one for a given trace irrespective of whether that trace captures execution of single or multiple applications.

XDRA as such cannot explore last-level cache in a multi-processor system because the LCI periods will be different across different last-level cache configurations (unlike a uniprocessor system where LCI periods are the same across different last-level configurations) due to inter-processor dependencies and cache coherency. In future, we will look into extending XDRA for multi-processor systems.

8. CONCLUSION

This paper for the first time proposes an estimator to predict the energy reductions when last level caches are used and the main memory power consumption is reduced by aggressively switching the memory to self-refresh power mode whenever the memory is idle. The estimator

is Kriging based and a complex system containing a processor-memory simulator, a cache simulator and novel analysis techniques. The predictor is accurate to within 4.4% on average for 11 applications from mediabench and SPEC2000 suite and two DRAM sizes. We used the estimated energy reductions to find suitable energy configurations, and were able to do so in 20 of the 22 cases. In the two cases which were non-optimal the errors were within 4% of optimal values. XDRA is comparatively fast, and reduces 85% of the time taken by complete simulation.

9. REFERENCES

[1] F. Catthoor, E. d. Greef, and S. Suytack, *Custom Memory Management Methodology: Exploration of Memory Organisation for Embedded Multimedia System Design*. Norwell, MA, USA: Kluwer Academic Publishers, 1998.

[2] I. Micron, "Micron ddr3." http://www.micron.com/products/dram/ddr3/.

[3] G. Thomas, K. Chandrasekar, B. Akesson, B. Juurlink, and K. Goossens, "A predictor-based power-saving policy for dram memories," in *Proc. 15th Euromicro Conference on Digital System Design*, (Izmir, Turkey), September 2012.

[4] A. M. Amin and Z. A. Chishti, "Rank-aware cache replacement and write buffering to improve dram energy efficiency," in *Proceedings of the 16th ACM/IEEE international symposium on Low power electronics and design*, ISLPED '10, 2010.

[5] K. Swaminathan, E. Kultursay, V. Saripalli, V. Narayanan, and M. Kandemir, "Design space exploration of workload-specific last-level caches," in *Proceedings of International symposium on Low power electronics and design*, ISLPED, 2012.

[6] L. Benini, A. Macii, and M. Poncino, "Energy-aware design of embedded memories: A survey of technologies, architectures, and optimization techniques," *ACM Trans. Embed. Comput. Syst.*, vol. 2, pp. 5–32, February 2003.

[7] M. Haque, J. Peddersen, A. Janapsatya, and S. Parameswaran, "Dew: A fast level 1 cache simulation approach for embedded processors with fifo replacement policy," in *Design, Automation Test in Europe Conference Exhibition (DATE)*, 2010.

[8] X. Li, T. Mitra, H. S. Negi, and A. Roychoudhury, "Design space exploration of caches using compressed traces," in *In Proceedings of the 18th annual international conference on Supercomputing*, pp. 116–125, ACM Press, 2004.

[9] K. Chandrasekar, B. Akesson, and K. Goossens, "Improved power modeling of ddr sdrams," in *DSD*, pp. 99–108, 2011.

[10] V. Delaluz, M. Kandemir, N. Vijaykrishnan, A. Sivasubramaniam, and M. J. Irwin, "Dram energy management using software and hardware directed power mode control," in *Proceedings of the 7th International Symposium on High-Performance Computer Architecture*, HPCA '01, pp. 159–, IEEE Computer Society, 2001.

[11] X. Fan, C. Ellis, and A. Lebeck, "Memory controller policies for dram power management," in *Proceedings of the 2001 international symposium on Low power electronics and design*, ISLPED '01, (New York, NY, USA), pp. 129–134, ACM, 2001.

[12] I. Hur and C. Lin, "A comprehensive approach to dram power management," in *HPCA*, pp. 305–316, 2008.

[13] S. Liu, S. Ogrenci Memik, Y. Zhang, and G. Memik, "An approach for adaptive dram temperature and power management," in *Proceedings of the 22nd annual international conference on Supercomputing*, ICS '08, pp. 63–72, ACM, 2008.

[14] J. Lin, H. Zheng, Z. Zhu, Z. Zhang, and H. David, "Dram-level prefetching for fully-buffered dimm: Design, performance and power saving," in *Performance Analysis of Systems Software, IEEE International Symposium on*, pp. 94–104, 2007.

[15] J. Trajkovic, A. V. Veidenbaum, and A. Kejariwal, "Improving sdram access energy efficiency for low-power embedded systems," *ACM Trans. Embed. Comput. Syst.*, vol. 7, pp. 24:1–24:21, May 2008.

[16] H. Zheng, J. Lin, Z. Zhang, E. Gorbatov, H. David, and Z. Zhu, "Mini-rank: Adaptive dram architecture for improving memory power efficiency," *Microarchitecture, IEEE/ACM International Symposium on*, vol. 0, pp. 210–221, 2008.

[17] H. Huang, K. G. Shin, C. Lefurgy, and T. Keller, "Improving energy efficiency by making dram less randomly accessed," in *Proceedings of the 2005 international symposium on Low power electronics and design*, ISLPED '05, 2005.

[18] M. Lee, E. Seo, J. Lee, and J. soo Kim, "Pabc: Power-aware buffer cache management for low power consumption," *IEEE Transactions on Computers*, vol. 56, 2007.

[19] C.-G. Lyuh and T. Kim, "Memory access scheduling and binding considering energy minimization in multi-bank memory systems," in *Proceedings of the 41st annual Design Automation Conference*, DAC '04, 2004.

[20] G. Chen, F. Li, and M. Kandemir, "Compiler-directed channel allocation for saving power in on-chip networks," *SIGPLAN Not.*, vol. 41, pp. 194–205, January 2006.

[21] T. J. Santner, W. B., and N. W., *The Design and Analysis of Computer Experiments*. Springer-Verlag, 2003.

[22] HP, "Cacti 6.5." http://www.hpl.hp.com/research/cacti/.

[23] G. Mariani, A. Brankovic, G. Palermo, J. Jovic, V. Zaccaria, and C. Silvano, "A correlation-based design space exploration methodology for multi-processor systems-on-chip," in *Proceedings of the 47th Design Automation Conference*, DAC '10, 2010.

[24] J. L. Hennessy and D. A. Patterson, *Computer Architecture, Fourth Edition: A Quantitative Approach*. San Francisco, CA, USA: Morgan Kaufmann Publishers Inc., 2006.

[25] S. Min, J. Peddersen, and S. Parameswaran, "Realizing cycle accurate processor memory simulation via interface abstraction," in *VLSI Design (VLSI Design), 2011 24th International Conference on*, pp. 141–146, jan. 2011.

[26] Tensilica, Inc., "Xtensa Configurable Processors." http://www.tensilica.com.

[27] D. Wang, B. Ganesh, N. Tuaycharoen, K. Baynes, A. Jaleel, and B. L. Jacob, "Dramsim: a memory system simulator," *SIGARCH Computer Architecture News*, vol. 33, no. 4, pp. 100–107, 2005.

[28] A. Jaleel, "Memory characterization of workloads using instrumentation-driven simulation." http://www.jaleels.org/ajaleel/workload/SPECanalysis.pdf.

[29] S. Sair and M. Charney, "Memory Behavior of the SPEC2000 Bechmark Suite," tech. rep., IBM T.J. Watson Research Center, Oct 2000.

[30] R. Baysal, B. Nelson, and J. Staum, "Response surface methodology for simulating hedging and trading strategies," in *Simulation Conference, 2008*, dec. 2008.

[31] "Kriging toolbox for matlab." http://www2.imm.dtu.dk/ hbni/dace/.

[32] A. Gordon-Ross, F. Vahid, and N. Dutt, "Fast configurable-cache tuning with a unified second-level cache," *Very Large Scale Integration (VLSI) Systems, IEEE Transactions on*, vol. 17, pp. 80–91, jan. 2009.

[33] D. R. Jones, M. Schonlau, and W. J. Welch, "Efficient global optimization of expensive black-box functions," *J. of Global Optimization*, Dec. 1998.

APPENDIX

A. SUPPLEMENTARY TEXT

A.1 Kriging Model

Kriging models take into account the spatial correlation between the current design point, x_i, and an initial set of design points (training set), x, to estimate the output at x_i. More formally, a kriging model combines a global model with local trends in the form of:

$$y(x_i) = f(x) \quad z(x_i)$$

where $y(x)$ is to be estimated, $f(x)$ is a known approximation function, and $z(x)$ is a stochastic process with mean zero, variance σ^2, and non-zero covariance. While $f(x)$ globally approximates the design space, $z(x)$ creates local deviations which are interpolated by the kriging model with the use of spatial correlations. In other words, regression function $f(x)$ captures global impact of the parameters, whereas correlation function $z(x)$ captures local impact of the parameters. In many cases, $f(x)$ is taken as either a constant or a linear polynomial in x's parameters [31]. Additionally, many correlation functions such as exponential, gaussian, etc. are available to be used as $z(x)$, where the correlation of two points depends upon the weighted distance between them. For example, the widely used exponential correlation function is:

$$corr(x_i, x_j) = \prod_{h=1}^{p} e^{-\theta_h |x_i(h) - x_j(h)|^{s_h}}$$

where p is the total number of parameters in a design point, and θ_h and s_h are unknowns that govern the correlation distance weights. θ_h represents the significance of parameter h, whereas s_h represents the smoothness of function in the direction of parameter h. Given a training set, the coefficients of $f(x)$ and weights of $z(x)$ are estimated using the maximum likelihood technique [33]. Once these are known, the final model is written as:

$$y(x_i) = f(x) \quad r(x)' R^{-1}(y - f(x))$$

where $r(x)$ is the correlation vector between x_i and x, R is the correlation matrix representing correlation between all the pairs of design points in x, and y contains output values at the design points in the training set x.

A.2 Calculation of PDcycles and PDCnt

Figure 10 shows an example of how PDcycles and PDCnt are derived from a cache profile and LCI profile. The cache profile reports DRAM accesses and non-accesses while the LCI profile reports L2 cache idle periods. First, each LCI period from LCI profile is inserted after appropriate Memory Request (MR) in the cache profile. For example, second LCI period occurs after MR3 and hence is inserted between MR3 and MR4 in the DI profile. After merging of these two profiles, the DI profile shows all the DI periods marked in dotted-lined rectangles. The cycles in each DI period are calculated using a 4 cycle DRAM non-access latency (L2 cache hit latency). For example, for the first DI period, there are 20 and 110 cycles from two LCI periods and 16 cycles from 4 DRAM non-accesses, totalling to 146 cycles. Finally, the DI

profile is converted to PDcycles by applying the PDthreshold, that is, the initial threshold number of cycles in each DI period will not contribute to PDcycles. Note that the last DI period will be filtered since its duration is only 28 cycles. For this example, DRAM will remain in self-refresh PD mode for 156 cycles, whereas PDCnt will be 2.

A.3 LinReg and Simple Estimators

The first estimator is based on linear regression where the same parameters as of kriging estimator are used. We refer to it as LinReg estimator, and is written as:

$$ER = \beta_0 PDCycles \quad \beta_1 PDCnt \quad \beta_2 RBConf$$
$$\beta_3 RefCnt \quad \beta_4 MT$$

The coefficients of LinReg estimator were computed using the same training set as used for kriging estimator.

The Simple estimator is written as:

$$ER = [DA_{wc} - DA_c] \times [DRAMEnergy_{wc}/DA_{wc}]$$

where DA_{wc} and DA_c are the number of DRAM accesses without last-level cache and with a given last-level cache configuration respectively. $DRAMEnergy_{wc}$ is the energy consumption of DRAM in a system without last-level cache and is computed from cycle-accurate simulation of such a system. The first factor estimates the number of DRAM accesses reduced by a given cache configuration, which is then multiplied by a constant energy consumption per DRAM access (second factor) to compute overall DRAM energy reduction.

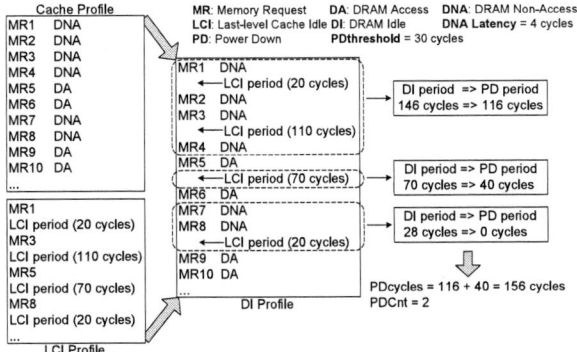

Figure 10: An example of calculating PDcycles and PDCnt.

B. SUPPLEMENTARY TABLES

Low-power Mode	Current (mA)	Wakeup Latency (clock cycles)
Active SB	57	0
Precharge SB	55	3
Active PD	35	4
Precharge PD	12	13
Self-refresh PD	6	64

Table 2: Power modes of Micron DDR3 DRAM. SB and PD stand for StandBy and PowerDown respectively.

Apps.	256MB DRAM	4GB DRAM
adpcmenc	[4KB, 16B, 16A] (0.03)	[16KB, 32B, 8A] (0.05)
adpcmdec	[4KB, 16B, 16A] (0.03)	[16KB, 32B, 4A] (0.05)
jpegenc	[256KB, 128B, 1A] (0.5)	[256KB, 128B, 1A] (0.5)
jpegdec	[256KB, 128B, 1A] (0.5)	[64KB, 16B, 16A] (0.31)
g721enc	[8KB, 16B, 16A] (0.05)	[32KB, 4B, 1A] (0.5)
g721dec	[8KB, 64B, 8A] (0.03)	[8KB, 16B, 16A] (0.05)
mpeg2enc	[32KB, 128B, 8A] (0.05)	[1MB, 128B, 8A] (1.5)
mpeg2dec	[16KB, 16B, 16A] (0.08)	[512KB, 128B, 1A] (0.86)
spec_vpr	[64KB, 16B, 16A] (0.31)	[256KB, 128B, 8A] (0.5)
spec_bzip2	[512KB, 128B, 4A] (0.9)	[1MB, 128B, 2A] (1.5)
spec_gzip	[256KB, 128B, 8A] (0.52)	[512KB, 32B, 4A] (1.6)

Table 3: L2 cache configurations with maximum DRAM energy reduction (BC_PD) from XDRA for different DRAM sizes. The numbers in parentheses are area footprint in mm^2.

Note that in Table 3, the area footprint of CC_PD configuration is 0.145 mm^2.

App.	Cycle-Accurate Simulator	XDRA			
		TS	LCI	CPA	Total
adpcmE	4.9h	27m	1m	6m	32m
adpcmD	3.5h	19m	47s	6m	25m
jpegE	1.3d	3h	6m	22m	3.3h
jpegD	12.8h	1h	3m	8m	1.3h
g721E	19.5d	2d	2h	17h	2.1d
g721D	14.4d	1d	1h	19h	1.7d
mpeg2E	183.6d	17d	14h	3d	18.6d
mpeg2D	46.9d	4d	4h	1d	4.9d
vpr	129.8d	12d	11h	2d	14.1d
bzip2	192d	18d	15h	22h	19.3d
gzip	295.7d	27d	23h	2d	29.5d

Table 4: Time comparison of cycle-accurate processor-memory simulator and XDRA for 4GB DRAM.

C. SUPPLEMENTARY FIGURES

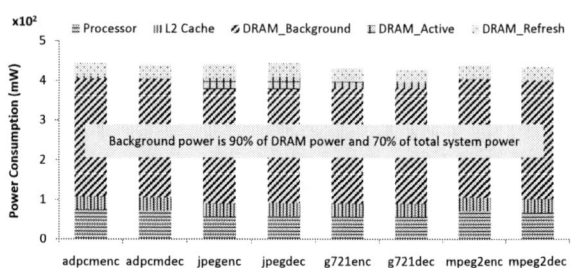

Figure 11: Power consumption breakdown in a uniprocessor system with on-chip L1 cache, off-chip L2 cache and DRAM memory.

Note that power consumption breakdown for applications from SPEC2000 benchmark are not reported in Figure 11. This is because we used first 500 million memory requests for SPEC2000 applications, and stopped the cycle-accurate simulation afterwards. Stopping a simulation in Tensilica's XTensa MultiProcessor (XTMP) environment does not generate a partial energy consumption report for the processor and caches.

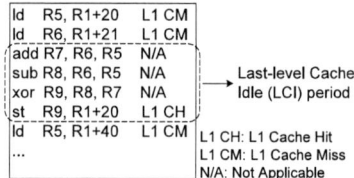

Figure 12: An example of LCI period for target system of Figure 4 (L2 is the last-level cache).

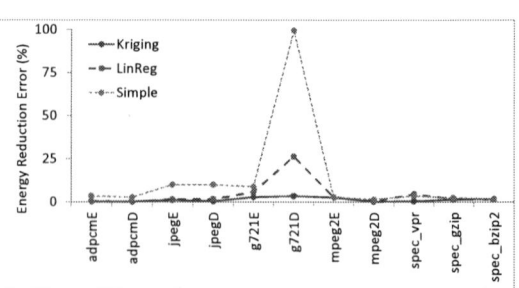

Figure 14: Average error in estimated energy reduction for 4GB DRAM.

158

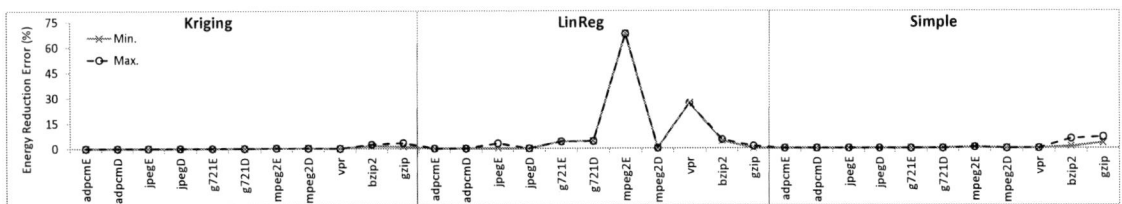

Figure 13: Error in energy reduction from cache configurations selected under differing area constraints for 4GB DRAM.

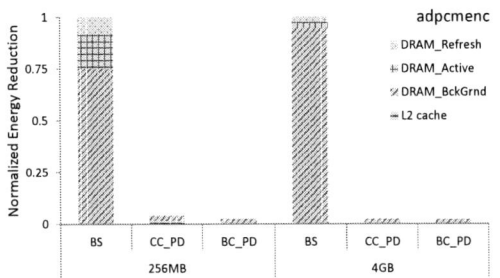

Figure 15: Normalized DRAM energy consumption breakdown of adpcmenc for different L2 caches and DRAM sizes.

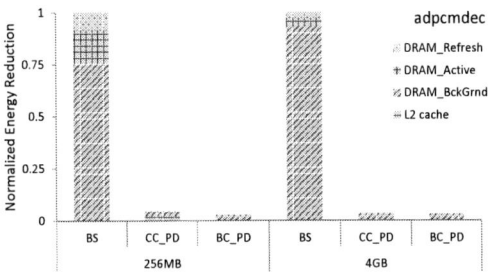

Figure 16: Normalized DRAM energy consumption breakdown of adpcmdec for different L2 caches and DRAM sizes.

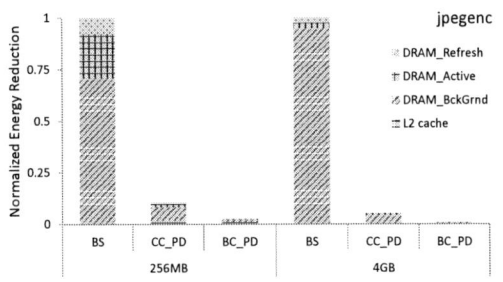

Figure 17: Normalized DRAM energy consumption breakdown of jpegenc for different L2 caches and DRAM sizes.

Figure 18: Normalized DRAM energy consumption breakdown of jpegdec for different L2 caches and DRAM sizes.

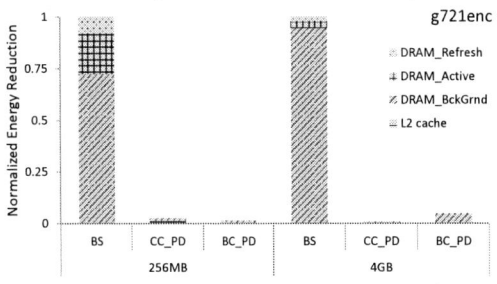

Figure 19: Normalized DRAM energy consumption breakdown of g721enc for different L2 caches and DRAM sizes.

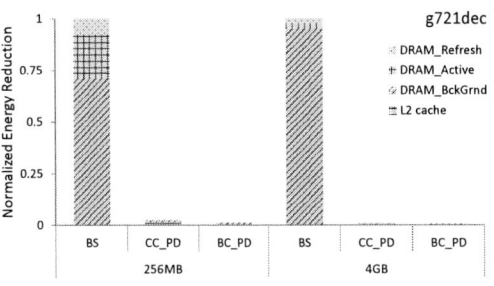

Figure 20: Normalized DRAM energy consumption breakdown of g721dec for different L2 caches and DRAM sizes.

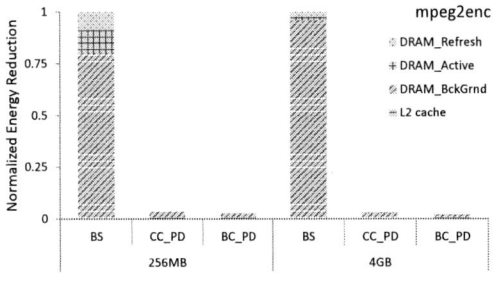

Figure 21: Normalized DRAM energy consumption breakdown of mpeg2enc for different L2 caches and DRAM sizes.

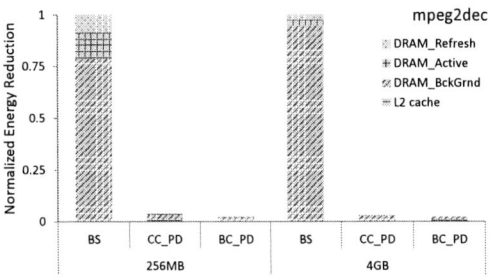

Figure 22: Normalized DRAM energy consumption breakdown of mpeg2dec for different L2 caches and DRAM sizes.

159

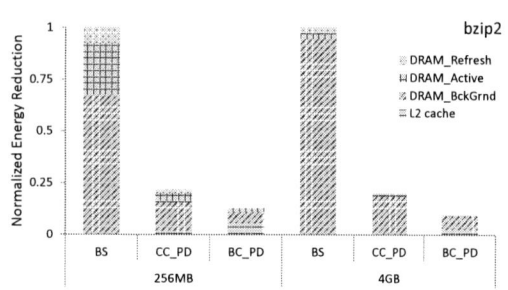

Figure 23: Normalized DRAM energy consumption breakdown of bzip2 for different L2 caches and DRAM sizes.

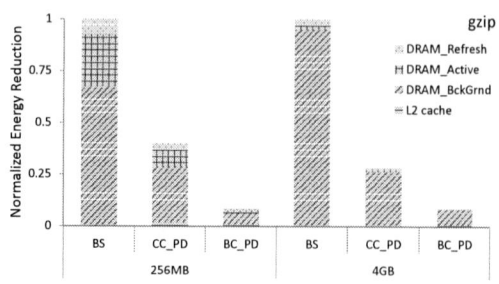

Figure 24: Normalized DRAM energy consumption breakdown of gzip for different L2 caches and DRAM sizes.

Towards Variation-Aware System-Level Power Estimation of DRAMs: An Empirical Approach

Karthik Chandrasekar[1], Christian Weis[2], Benny Akesson[3], Norbert Wehn[2], Kees Goossens[4]

[1]Computer Engineering, TU Delft, The Netherlands
[2]Microelectronic Systems Design, TU Kaiserslautern, Germany
[3]CISTER-ISEP Research Centre, Polytechnic Institute of Porto, Portugal
[4]Electronic Systems Group, TU Eindhoven, The Netherlands

ABSTRACT

DRAM vendors provide pessimistic current measures in memory datasheets to account for worst-case impact of process variations and to improve their production yield, leading to unrealistic power consumption estimates. In this paper, we first demonstrate the possible effects of process variations on DRAM performance and power consumption by performing Monte-Carlo simulations on a detailed DRAM cross-section. We then propose a methodology to empirically determine the actual impact for any given DRAM memory by assessing its performance characteristics during the DRAM calibration phase at system boot-time, thereby enabling its optimal use at run-time. We further employ our analysis on Micron's 2Gb DDR3-1600-x16 memory and show considerable over-estimation in the datasheet measures and the energy estimates (up to 28%), by using realistic current measures for a set of MediaBench applications.

1. INTRODUCTION

DRAM memories account for a significant share of any system's power and energy consumption, be it battery-driven mobile devices [1] or high-performance computing servers [2]. With system power and energy budgets getting tighter, it becomes absolutely essential to employ highly accurate power models and energy estimates for every component in the system, including processors and DRAMs. Unfortunately, with the impact of process variations [3, 4] on power consumption scaling significantly at technologies below 90nm, existing power models are becoming less and less accurate, while worst-case power estimates are just too pessimistic to use. Hence, it has become imperative to estimate the expected impact of process variations on power consumption, for all system components, for an accurate system power analysis.

In the case of processors, many solutions have been proposed, both by vendors and academia that estimate [5, 6] and even help mitigate [7,8], the expected performance and power impact. However, when it comes to DRAMs, vendors merely sort the memories into discrete speed-bins and furnish one set of worst-case current measures per speed-bin in datasheets, leading to over-estimation of DRAM power consumption. With DRAM memories becoming increasingly prominent in a system's power/energy profile, employing such worst-case datasheet measures leads to unrealistic over-dimensioning of the system. This calls for variation-aware DRAM power-estimation methodologies that address the pessimism in the datasheets and improve the accuracy of the power models and energy estimates.

Permission to make digital or hard copies of all or part of this work for personal or classroom use is granted without fee provided that copies are not made or distributed for profit or commercial advantage and that copies bear this notice and the full citation on the first page. To copy otherwise, to republish, to post on servers or to redistribute to lists, requires prior specific permission and/or a fee.

Figure 1 shows the impact of process-variation observed by a memory vendor in the production analysis of a lot of 11,000 DDR3 1Gb memories with 533MHz frequency and x8 width, manufactured at 70nm, in batch U6PN8XBS-13G3.

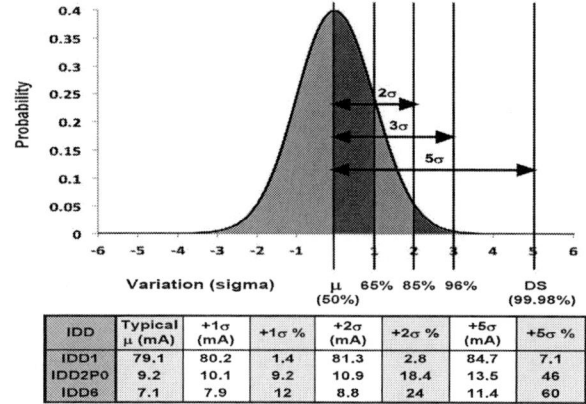

IDD	Typical μ (mA)	$+1\sigma$ (mA)	$+1\sigma$ %	$+2\sigma$ (mA)	$+2\sigma$ %	$+5\sigma$ (mA)	$+5\sigma$ %
IDD1	79.1	80.2	1.4	81.3	2.8	84.7	7.1
IDD2P0	9.2	10.1	9.2	10.9	18.4	13.5	46
IDD6	7.1	7.9	12	8.8	24	11.4	60

Figure 1: Distribution of Current Consumption

This data shows very large difference between the datasheet current measures (DS) and the typical (μ) current values (by a factor of 5σ) of up to 46% and 60% for the low-power modes (power-down: I_{DD2P0} and self-refresh: I_{DD6}) and up to 7% for the activate-read-precharge (I_{DD1}) current [9]. With DRAM memories now being manufactured at technologies below 50nm, these current variations are only expected to worsen, and so is the accuracy of the power models employing the datasheet measures. *Unfortunately, such current distributions are not provided for all DRAMs, and only worst-case measures are given in vendor datasheets* [10].

Intel in [11,12] and others in [13,14] observed similar power variation in DRAM memories and suggested different techniques to work around this problem [14–17]. However, there are no known realistic models or studies that estimate the actual impact of variation on power consumption of a DRAM memory, impairing the usage of these proposed solutions. Existing power models [18–22] choose to ignore the impact of variations on power consumption due to the lack variation data, which reflects poorly on their accuracy.

In this paper, we intend to provide an insight into the possible effects of process variations on DRAM power consumption to help improve the accuracy of DRAM power models and to enable the optimal use of DRAMs at run-time. The three important contributions of this work are: (1) We demonstrate the impact of process variations by performing Monte-Carlo simulations on a detailed DRAM cross-section modeled in NGSPICE [24]. (2) We propose a methodology to empirically determine this impact for any given DRAM memory, by assessing its actual performance characteristics during the DRAM calibration phase [9] at system boot-time. (3) We extend the Monte-Carlo analysis to examine the impact of DRAM architecture parameters, such as capacity, width and frequency, on variations and current estimates.

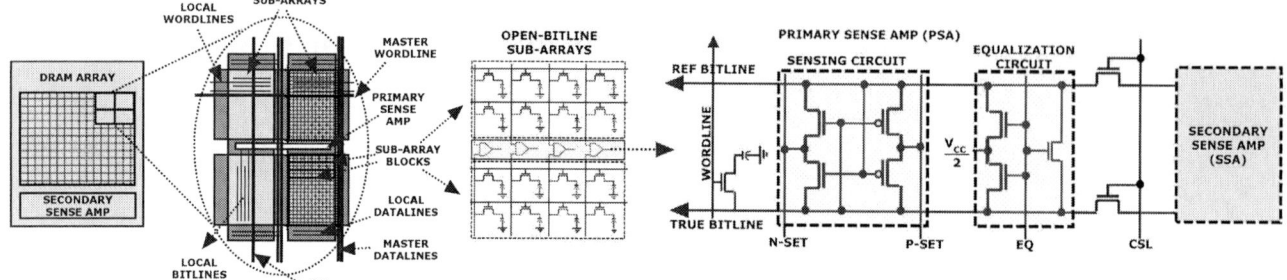

Figure 2: DRAM Cross-Section Model

Using these mechanisms, we derive possible current distributions for DRAM memories of any configuration, and also determine the actual performance measures and realistic current measures for a given DRAM memory using the characterization step at system boot-time. The derived performance measures can be used to improve the performance of the given DRAM memory and the realistic current measures can be employed in place of the worst-case datasheet values to obtain variation-aware DRAM power and energy estimates. We evaluate our proposed solution by deriving the current distributions for Micron's 2Gb DDR3-1600-x16 memories [10] based on the the Monte-Carlo analysis, and employing them with a system-level power model [20,23] to show significant differences between typical and worst-case datasheet current measures and the corresponding energy estimates (up to 28%) for four MediaBench applications [33].

2. RELATED WORK

When it comes to studying the impact of process variation on power consumption in DRAMs, Intel observed performance degradation and power variation in DRAM memories in [11,12] and suggested performance throttling to maintain an average power budget assuming datasheet estimates in [15], as a work-around to this problem. M. Gottscho et al. in [13] also observed variations of around 15% in power consumption across several 1GB DIMMs from the same vendor and around 20% across different vendors, although they did not establish the causes for the observed extent of variations. L. Bathen et al. in [16,17] employed these observations by [13], and suggested memory mapping and partitioning solutions to exploit this variability. S. Desai et al. in [14] on the other hand, performed Monte-Carlo analysis on a single DRAM cell and basic circuit components and together with interconnect delay models estimated the variation impact for an entire DRAM memory. They further proposed using adaptive body biasing to improve the yield of DRAMs. Although the variation estimates may be acceptable for the basic circuit components, such an extrapolation to an entire DRAM is at best a coarse approximation. Unfortunately, there are no known realistic models or studies that provide acceptable estimates of the expected impact of process-variations on DRAM power consumption and no variation data is made available by DRAM vendors, undermining the applicability of the solutions suggested in [14–17]. In this work, we derive realistic estimates of the impact of variations on DRAM currents to enable use of such solutions.

When it comes to DRAM power estimation, many power models have been proposed. Among the circuit-level models, CACTI 5 [22] was proposed for embedded DRAMs, Rambus presented a circuit-level open-source DRAM power model in [18] and Weis et al. employed a SPICE based model in [19] for 3D-stacked DRAMs. At the system-level, Micron presented a datasheet-based power model in [21] and Chandrasekar et al., proposed a transaction-based power model in [20] that also uses datasheet measures. Unfortunately, none of these power models consider the impact of process-variations on power consumption in DRAMs, due to lack of variation analysis and data. In this work, we provide possible distributions of the current measures for different DRAM operations, which can be employed with the datasheet-based power models, such as [20, 21, 23], to obtain more realistic DRAM power and energy estimates.

3. DRAM MODELING AND ANALYSIS

In this section, we first describe the baseline DRAM cross-section model to be used for our NGSPICE simulations. In these simulations, we observe the timings for different DRAM operations and verify the functional correctness of our design. We then perform Monte-Carlo analysis [27] on this cross-section to derive the variation-impact in DRAMs.

3.1 Baseline DRAM Cross-Section Model

The basic DRAM cell is modeled as a transistor-capacitor (1T1C) pair and stores a single bit of data in the capacitor as a charge. As shown in Figure 2, the transistor is controlled by a local wordline (lwl) at its gate, which connects the capacitor to the local bitline (lbl) when turned on (activated). Before reading the data from the memory cell, the bitlines in the memory array are precharged (set to halfway voltage level) using an equalization circuit. When connected, the cell capacitors change the precharged (PRE) voltage levels on the bitlines very slightly. Hence, a set of primary sense amplifiers (PSA) (or row buffer) distributed across memory sub-arrays are used to detect the minute changes and pull the active bitline voltage all the way to logic level 0 or 1. Once the bitline voltage is amplified, it also recharges the capacitors as long as the transistors remain on. The primary sense amplifiers hold the data till all column accesses to the same row are completed or till a precharge is issued. In our architecture, we used the open bitline array structure and hence differential sense amplifiers (in PSA), which use a reference bitline from a neighboring inactive array segment to detect the minute difference in active bitline voltage. When the Read (RD) command is issued, the data/charge is read out using column select lines (CSLs) from the row buffer (PSA). The data is then switched via master datalines from the PSA to the secondary sense amplifiers (SSA), which connects to the I/O buffers. Once finished, the wordlines can be switched off, safely restoring the charge in the memory cells, before starting to precharge (PRE) the bitlines again.

The memory arrays are organized in a hierarchical structure of memory sub-arrays for efficient wiring. A memory sub-array consists of 256k cells connecting up to 512 cells per local bitline and per local wordline. 16 memory sub-arrays connect to one master wordline forming 4Mb blocks. 16 master wordlines and 16 column select lines (CSLs) connect the 256 memory sub-arrays to form 64Mb memory array macros. The row and column decoders and the master wordline drivers are placed per memory array. The N-Set and P-Set control signal drivers used for activating the primary sense amplifiers are shared between a set of sub-arrays in the memory array. The voltage regulators and charge pumps are shared between the different banks.

162

3.2 DRAM Cross-Section SPICE Simulations

In our NGSPICE [24] modeling of the DRAM cross-section, we employed BSIM [27] model cards built on Low Power Predictive Technology models (LP-PTM) [28], since there are no openly available technology libraries specific to DRAMs. As a result, the LP-PTM devices had to be adapted appropriately, to ensure functional and timing correctness of the simulated DRAM cross-section.

We modeled the memory cell architecture (of $6F^2$ area), the equalization circuit, the wordline driver, and the sense amplifier using the designs suggested in [18], [25] and [26]. The baseline DRAM configuration targets a 1Gb DDR3-1066 (533MHz) x8 memory with core timings of 7-7-7 cc (refer Section 4.1) at 45nm. We chose 45nm, since it is the common technology node employed by vendors for DDR3 memories including Samsung, Micron and Hynix. To verify our DRAM cross-section, we present the timings and voltages of the different signals [26] corresponding to basic DRAM operations (ACT-RD-PRE) in Figure 3.

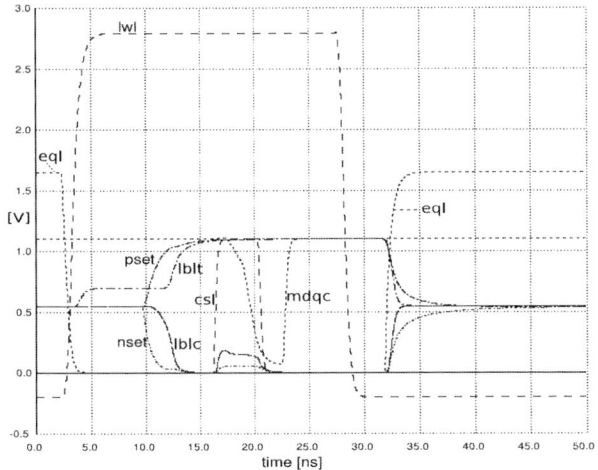

Figure 3: Behavior of DRAM Cross-Section

As depicted in the figure, first the equalization circuit (eql) forces both the true (active) bitline (lblt) and the complementary (reference) bitline (lblc) to the same reference voltage (0.55V). This is followed by the local wordline (lwl) going high to begin the activation process that switches the relevant transistors on and connects the cell capacitors to the corresponding local bitlines. Simultaneously, the equalization circuit (eql) de-activates to enable sensing of the change in bitline voltage due to the charge transfer. As the wordline high reaches the required voltage of 2.8V at around 5ns, the pre-sensing phase begins to create a minimum voltage difference (around 200mV) between the reference (lblc) and active (lblt) bitlines at the PSA. This is followed by the activation of the sensing circuit by N-Set and P-Set control signals, which drives the active bitline (lblt) to logic level 1 (the charge stored in the cell corresponds to 1 here) and the reference bitline (lblc) to 0 at around 15ns. Both the lblt and pset signals are driven to the core voltage of 1.1V, while lblc and nset signals are driven to 0V. This is followed by the read operation depicted by the rising column select line (csl) voltage at 18ns. Following this, the charge detected at all the PSAs in a memory sub-array are transferred via their respective local datalines to a master dataline, which is reflected by the current drawn from the NMOS components of the PSA (lblc) and the gradual drop in voltage level of master dataline complement signal (depicted by mdqc). Once the mdqc drops by around 200mV (in relation to the core voltage), the data is sensed at the SSA at around 20ns. Once the read operation finishes, (data received by SSA) the mdqc (master dataline complement) is precharged back

to its reference voltage (1.1V) at around 24ns and the local wordline is switched off at around 28ns. After a short delay to close the transistor and avoid destroying the charge in the cell, the sensing circuit in PSA is deactivated and the bitline equalization re-starts at around 33ns. This precharges both the local bitlines back to reference voltage levels, finishing at 50ns, as expected for a DDR3-1066 memory [9].

Besides the basic ACT-RD-PRE operations depicted in this figure, we also modeled the write and refresh operations in a similar manner and observed accurate functionality and timing [26], thus, verifying our modeling of the DRAM cross-section. We employ this DRAM cross-section to perform Monte-Carlo analysis to observe the impact of variation on delay and power consumption in the next section.

3.3 Baseline Monte-Carlo Analysis

In this section, we present the results from Monte-Carlo analysis on our verified 1Gb DDR3-1066 x8 DRAM cross-section, described in Section 3.2. Towards this, we vary global device parameters such as channel length, channel mobility, and oxide thickness and the local device threshold voltage (V_{th}) (primarily the variations in line edge roughness (LER) [31]), besides the interconnect parameters including wire width and wire thickness, within pre-defined variation ranges. We obtained the variability ranges (scaling metric (σ) in the corresponding Gaussian distributions) for these parameters from the ITRS technology requirements on Design for Manufacturability [29] and Modeling and Simulation [30] and the variation models of transistors [31,32]. We also introduce spatial-correlations in the variations among neighboring transistors, due to expected similarity in the parametric variations. Using these variability values, we performed Monte-Carlo runs on 1000 circuit instances reflecting the variations in all the device and interconnect parameters. From our observations, the variation in the device V_{th} parameter had the biggest impact on the circuit delay and current consumption [31], since it is also directly influenced by the variations in the global device parameters. As expected, the active (dynamic) DRAM currents and frequency increased linearly, while the leakage currents increased exponentially against the variations in the V_{th} parameter [34,35]. Hence, we analyzed the variations in leakage currents on the natural logarithmic scale [34] to obtain the σ values of their distributions corresponding to those of the V_{th} parameter.

The variations in the local and global device parameters at 45nm based on [29–32], as used in our simulations are presented in Table 1. These measures correspond to the variability introduced in the device parameters per σ change in their Gaussian distributions. In the table, the σ% value gives the relative variation to the nominal values (μ) obtained from the PTM models [28], while the σ values correspond to the absolute values of variation.

Table 1: Transistor Process Parameter Variations

Tech nm	Mobility σ (%)	V_{th} (LER) σ (V)	Length σ (m)	T_{ox} σ (%)
45	8.2	3.0e-9/$\sqrt[2]{(w \times l)}$	45e-9×0.03	1.67

Amongst the different characteristic DRAM currents, we identify the dynamic (active and background) currents as: I_{DD0}, I_{DD1}, I_{DD2N}, I_{DD3N}, I_{DD4R}, I_{DD4W}, and I_{DD5}, and the static (leakage) currents as: I_{DD2P0} and I_{DD6} (when the clock is disabled). The different DRAM currents are described in detail in Section A1 and in [9,10].

In Table 2, we show the impact of process-variation on the different currents for a baseline 1Gb DDR3-1066 (533MHz) x8 DRAM memory. In this table, we present the nominal measures along with the 1σ, 2σ and 5σ estimates in the different I_{DD} currents. We also provide σ% value to get relative variation for the different current measures.

Table 2: Variation Impact on Current Measures

Current	μ mA	$\sigma\%$	$+1\sigma$ mA	$+2\sigma$ mA	$+5\sigma$ mA
I_{DD0}	98.4	2.37	100.7	103.1	110.6
I_{DD1}	104.3	2.32	106.7	109.1	116.9
I_{DD2N}	37.7	4.77	39.5	41.4	47.6
I_{DD3N}	41.5	5.71	43.8	46.3	54.7
I_{DD4R}	118.1	2.96	121.6	125.2	136.6
I_{DD4W}	123.4	2.75	126.7	130.2	141.2
I_{DD5}	146.1	2.15	149.6	153.1	164.2
I_{DD2P0}	8.41	13.69	9.56	10.9	15.9
I_{DD6}	8.04	14.02	9.17	10.5	15.5

As can be noticed from the table, the $+5\sigma$ estimate is significantly higher than the nominal (μ) values for the different I_{DD} currents. In Sections A2 and A5, we present the impact of variation on the timing and power consumption corresponding to $\pm 1\sigma$ variations in device and interconnect parameters, as observed from the 1000 Monte-Carlo runs.

4. DRAM MEMORY CHARACTERIZATION

In this section, we relate a DRAM's actual performance and current measures to the impact of process variations. Towards this, we first begin by reviewing the process of speed-binning in DRAMs in Section 4.1. We then propose a methodology to determine optimal functional measures for the performance parameters and conservative estimates for the current measures of a particular DRAM memory during the calibration phase at system boot-time, in Section 4.2.

4.1 Variation and DRAM Speed-Binning

DRAM memories manufactured in a particular generation are down-binned into predefined speed-bins based on their minimum guaranteed frequency and memories within these speed-bins are classified according to their core-timings [9]. Table 3 presents the speed-bins and the core-timings in clock cycles (cc) used to classify Micron's DDR3 memories [10].

Table 3: Micron Speed-Bins and Core-Timings

Speed Bin	Freq (MHz)	Fast Core (cc) n_{CL}-n_{RCD}-n_{RP}	Slow Core (cc) n_{CL}-n_{RCD}-n_{RP}
800	400	5-5-5(12.5ns)	6-6-6(15ns)
1066	533	7-7-7(13.125ns)	8-8-8(15ns)
1333	666	9-9-9(13.5ns)	10-10-10(15ns)
1600	800	10-10-10(12.5ns)	11-11-11(13.75ns)

In Table 3 in the DDR3-800 speed-bin, the memories are guaranteed to work at the lower bound (F_{LB}) of 400MHz. All memories that fall short of the lower bound of the next speed-bin (and upper bound F_{UB} of the current speed-bin: $\leq$532MHz), are down-binned into the 400MHz speed-bin, ignoring the fact that they can operate at higher frequencies. Besides this frequency-sorted speed-binning, 3 core-timings (in cc) are used to define a DRAM memory within a speed-bin [9]: (1) n_{CL} - RD to Data Latency, (2) n_{RCD} - ACT to RD/WR Latency and (3) n_{RP} - PRE Latency (see Section A3 for details). However, only two sets of core-timings (fast and slow) are used to sub-categorize the memories within a speed-bin. In the case of DDR3-800, the fast memories have core-timings of 5-5-5 cc and slow memories have core-timings of 6-6-6 cc. Hence, memories that may achieve core-timings of 5-5-6 cc are conservatively categorized among the slow 6-6-6 memories, further ignoring their individual core-timings.

When it comes to providing current measures for these memories, only one set of worst-case measures per speed-bin are provided in datasheets [10], thus ignoring the actual performance characteristics of the DRAM memories completely. Determining the actual functional F_{MAX} and core-timings is important to derive the actual performance and current measures of a particular DRAM memory and we present a methodology to obtain the same in Section 4.2.

4.2 Conservative Memory Characterization

As discussed in Section 4.1, when reporting the worst-case current measures in datasheets, DRAM vendors consider only the memories operating near the upper frequency bound with the fastest core timings of their speed-bins. However, since slower memories draw less current than the faster ones, their individual F_{MAX} and core-timings should be determined and used to identify the actual current measures.

In this section, we propose a methodology to determine the frequency and core-timings of a particular DRAM memory and to relate them to the corresponding current measures from the distributions derived in Table 2. Towards this, we derive a performance metric, *Functional Speed* (FS), defined as the product of the sum of the memory's core-timings (CL+RCD+RP) and the corresponding clock period ($1/F_{MAX}$), to represent both these performance parameters. (Lower the FS, faster the memory and higher the performance.) The goal of this proposed methodology is two-fold: (1) to derive the fastest overall DRAM functional speed (Max_FS or lowest common FS) based on core timings and F_{MAX}, at which the entire DRAM can function, to improve the DRAM's performance and (2) to derive the fastest individual bank functional speed (Min_FS or lowest individual bank FS) at which any individual DRAM bank may function, to conservatively identify the actual worst-case current measures, by relating this FS to the current distributions obtained in Table 2. The relation between the delays and currents is also shown in Section A5.

In Algorithm 1, we present this methodology to derive the overall DRAM and individual bank functional speeds. We propose to employ this algorithm once during the memory's calibration phase [9] at system boot-time. Currently, this calibration phase in DRAMs is employed for timing synchronization and skew corrections in DRAM signals, to enable proper DRAM functionality. We propose to merely add a performance assessment step to this phase, to obtain realistic performance and current measures for use at run-time.

Algorithm 1 Frequency and Variation Estimation

Require: var_check (F_{LB}, F_{UB})
1: {Comment: $\sum$CT = [n_{CL} + n_{RCD} + n_{RP}]}
2: # Define: CT_Min[] = {5,5,5} and CT_Max[] = {8,8,8}
3: # Define: F_σ = (F_{UB} - F_{LB})/12 {Here: F_σ=11MHz}
4: # Define: Banks = 8
5: CT[i] = CT_Max[i] {Initialized}
6: **for** $i = 0 \to 2$ **do**
7:     {Comment: Representing CL, RCD and RP}
8:     **for** $j = 0 \to 2$ **do**
9:         {Comment: Representing CT range 8cc to 5cc}
10:         CT[i] = CT[i] - j - 1
11:         **for** $k = 0 \to Banks$ **do**
12:             {Comment: Iterating over all 8 banks}
13:             **for** $f = F_{LB} \to F_{UB}$ **do**
14:                 {Comment: Checking all frequency levels}
15:                 **if** CT_check (k, f, CT*) == True **then**
16:                     FS_Bank[i][j][k] = $\sum$CT $\times$ 1/f
17:                     {Comment: Store corresponding f and CT*}
18:                     f = f + F_σ
19:                 **else**
20:                     Break;
21:                 **end if**
22:             **end for**{f}
23:             {Comment: Stores least FS for bank k for set CT*}
24:         **end for**{k}
25:     **end for**{j}
26: **end for**{i}
27: Min_FS = min (FS_Bank[*][*][*])
28: {Comment: Return corresponding f and CT*}
29: Max_FS = max (FS_Bank[*][*][*])
30: {Comment: Return corresponding f and CT*}

In this algorithm, we begin by identifying the fastest and slowest core-timings (in clock cycles) in a speed-bin (say DDR3-800) at the upper frequency bound of this speed-bin (532MHz). We then propose to start with the slowest set of memories [8-8-8] (15ns at 532MHz) (in Step 5) and reduce one core-timing parameter (say n_{CL}) by 1cc (Step 10), while maintaining the others at 8cc and increasing the memory frequency in steps along the 13 frequency values in steps of F_σ (here 11MHz between 400MHz and 532MHz, given by Step 13). With these new core-timing settings, we propose to execute a core-timings check (CT_check), which is based on JEDEC's I_{DD1} Measurement-Loop test [9] (described in Section A3) over all the 8 banks in the memory (Steps 11 and 15). This CT_check comprehensively checks the activation, reading and precharging operations on a given bank in different rows, which tests all the important DRAM timings [9]. If the test completes without any errors, the frequency is increased by another step F_σ (Step 18). If not, the last explored working frequency gives the lowest FS value for that bank for the selected set of core-timings (Step 16). We store this lowest FS value and the corresponding F_{MAX} and core-timings for reference. Steps 13 to 22 are repeated for all the banks and the lowest FS values are obtained for all the banks with the selected set of core-timings. Now we reduce the considered n_{CL} parameter further by 1cc and the tests are repeated with the new set of core-timings and the corresponding lowest FS values are noted, till the minimum functional n_{CL} value is reached. The same procedure is then employed with the other core-timing parameters (n_{RCD} and n_{RP}), assuming the fastest n_{CL} value at which the memory continued to work. All the corresponding lowest FS values for the different banks and set of core-timings are stored. Finally, the lowest FS value obtained for any of the DRAM banks (Min_FS) is used to conservatively identify the actual worst-case current measures of the memory and the maximum of the lowest FS values supported by all banks of the memory (Max_FS) and the corresponding core-timings and F_{max} are used to identify the new performance parameters.

Using the current distributions derived in Table 2, and the distribution of the functional speeds observed in Algorithm 1, in Figure 4, we overlap the two distributions to obtain the complete performance-power-variation relation in the ±6σ form. Here, the FS range for DDR3-800 is identified between 28.2ns (fastest memories {∑CT=15}) at 532 MHz and 45ns (slowest memories {∑CT=18}) at 400 MHz. The datasheet current measures are identified at +5σ position at 29.6ns (fastest) and the datasheet performance is identified at -6σ position at 45ns (slowest). An example of the new performance parameters is highlighted at Max_FS - 1σ position (fastest overall DRAM FS) at 41ns and an example of realistic current measures is highlighted at Min_FS + 1σ position (fastest individual bank FS) at 34ns.

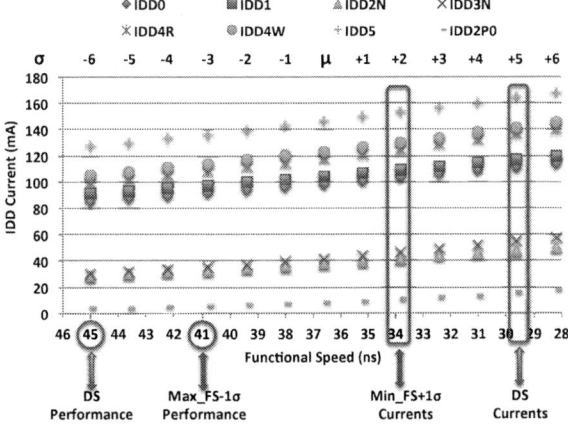

Figure 4: FS Vs. Current Consumption

From this analysis, we derive new conservative performance parameters and realistic current measures for a DRAM, for its optimal run-time usage and power management.

5. RESULTS AND ANALYSIS

In this section, we apply our variation study on different DRAM system configurations. Towards this, we first repeat the Monte-Carlo Analysis with modified system parameters, such as capacity, frequency and data-bus width in Section 5.1. We then apply the results on a 2Gb DDR3-1600 (800MHz) x16 memory from Micron in Section 5.2 and present the μ and $\sigma\%$ estimates for current measures for this memory. Finally, in Section 5.3, we employ these results on a set of MediaBench applications [33], and show up to 28% difference in energy estimates by using more realistic $\mu+2\sigma$ current estimates instead of the datasheet (DS) measures.

5.1 System Parameters Impact on Variation

DRAM vendors sort the memories by three system parameters: frequency, capacity and width. In this experiment, we repeat the Monte-Carlo simulations to analyze the impact on μ and $\sigma\%$ for the different currents when these system parameters change. Our baseline configuration targeted 1Gb DDR3-533MHz x8 memories. In this experiment, we alter the system parameters individually to simulate different configurations. Accordingly, we change: (1) the frequency to 800MHz and simulate a 1Gb-800MHz-x8 memory, (2) the capacity to 2Gb and simulate a 2Gb-533MHz-x8 memory, and (3) the data-width to x16 to simulate a 1Gb-533MHz-x16 memory and observe the impact on μ and $\sigma\%$ in Table 4.

Table 4: System Parameters Vs. Current Measures

	Baseline		Freq		Capacity		Width	
Config	1Gb-533-x8		1Gb-800-x8		2Gb-533-x8		1Gb-533-x16	
I_{DD} Type	μ mA	σ %	μ mA	σ %	μ mA	σ %	μ mA	σ %
I_{DD0}	98.4	2.4	112	2.6	99.3	2.5	98.4	2.37
I_{DD1}	104	2.3	118	2.2	105	2.5	113	2.22
I_{DD2N}	37.7	4.8	42.7	4.5	46.5	6.1	37.7	4.77
I_{DD3N}	41.5	5.7	56.7	4.5	49.9	5.3	41.5	5.71
I_{DD4R}	118	2.9	153	3.5	127	3.3	208	3.14
I_{DD4W}	123	2.7	159	4.1	132	3.7	213	2.6
I_{DD5}	146	2.1	161	2.4	184	2.2	146	2.15
I_{DD2P0}	8.4	13.7	8.4	13.7	16.6	16.1	8.4	13.7
I_{DD6}	8	14	8	14	13.7	19.9	8	14

As shown in the results, when increasing the frequency from 533MHz to 800MHz, all currents except the leakage currents scale up linearly due to their dependency on the clock. When increasing the memory density, all currents scale up linearly due to the doubling of the number of memory cells and primary sense amplifiers. However, when the data-width is doubled, while retaining the same page-size (1KB), only the currents reflecting data transfer, viz., I_{DD1}, I_{DD4R} and I_{DD4W} are affected, since only the number of data bits accessed during the column accesses increases.

5.2 Reverse Engineering Datasheet Values

In Section 5.1, we presented the impact of three system parameters on DRAM currents. However, since more than one parameter can be different between two DRAM memories, to estimate this impact, one should combine the influence of each of the concerned system parameters, from the observations in Table 4. We present the impact of all possible combinations in the appendix Section A4, since these are merely derived from the results in Table 4.

When applying this analysis on a 2Gb DDR3-800MHz x16 memory from Micron [10], all the three system parameters change at once. Accordingly, we estimate the possible current distributions in Table 5. As observed from the results in Table 5, the nominal estimates for the active currents are up to 30% lower (for I_{DD3N}) than the datasheet (DS) measures, while those for the leakage currents are up to 86% lower (for I_{DD6}). These large differences in the current measures highlight the pessimism in the datasheets.

165

Table 5: Datasheet Values Vs. Nominal-Case

Current	DS mA	μ (mA)	μ vs DS %	2σ mA	2σ vs DS %
IDD0	110	98	-12.2	102.8	-6.96
IDD1	125	112.2	-11.4	117.3	-6.53
IDD2N	42	33.5	-25.4	36.9	-13.8
IDD3N	45	34.6	-30.1	38.7	-16.1
IDD4R	270	232.2	-16.2	247.3	-9.15
IDD4W	280	246.7	-13.5	260	-7.67
IDD5	215	193.6	-11.1	202.1	-6.34
IDD2P0	12	6.62	-81.1	8.77	-36.7
IDD6	12	6.45	-85.9	8.67	-38.4

5.3 Variation Impact on Application Energy

In these experiments, we employed four randomly selected
MediaBench applications [33] including: (1) Ray Tracing,
(2) EPIC Encoder, (3) JPEG Encoder, and (4) GSM Decoder. These applications were independently executed on
the SimpleScalar simulator [36] with a 16KB L1 D-cache,
16KB L1 I-cache, 128KB L2 cache and 64-byte cache line
configuration. We filtered out the L2 cache misses meant
for the DRAM and forwarded them through a DRAM controller [37], which generated the memory commands. We
also employed the power-down mode conservatively [38] during the idle periods. We compare the energy estimates,
when employing the nominal (μ), datasheet (DS) and realistic $\mu+2\sigma$ I_{DD} measures from Table 5, since this covers
more than 85% of the memories in one generation. We used
the I_{DD} measures with the DRAMPower tool [20, 23], to
estimate DRAM energy consumption, depicted in Figure 5.

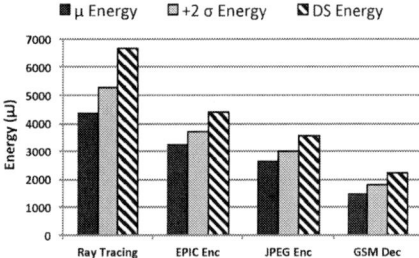

Figure 5: Application Energy using μ and 2σ vs. DS

As can be noticed, the energy consumption when using
$+2\sigma$ current estimates is up to 28% lower for the Ray tracing application, compared to using the datasheet estimates.
This difference increases to 58%, if the nominal (μ) measures are employed. Similarly, considerable differences are
observed for other applications as well, highlighting the significance of variation-aware power and energy estimation.

6. CONCLUSION

In this paper, we demonstrated the effects of process variations on DRAM performance and power consumption. Towards this, we defined a detailed circuit-level DRAM cross-
section in NGSPICE and performed Monte-Carlo analysis
to derive the impact on DRAM performance and current
measures. We also presented a methodology that assesses
the performance characteristics of a given DRAM at system
boot-time and conservatively identifies new performance parameters (in terms of functional speeds, core-timings and
F_{max}) and realistic current measures, for use at run-time.
We further extended the Monte-Carlo analysis to review
the impact of system parameters on current consumption
and applied the same on a Micron DDR3 memory, showing significant pessimism in the datasheet measures. In a
nutshell, the contributions of this work can be employed to
improve DRAM performance and obtain variation-aware realistic and accurate power consumption estimates.

Acknowledgments

This work was partially supported by HiPEAC collaboration
grant and projects EU FP7 288008 T-CREST and 288248
Flextiles, Catrene CA104 COBRA, DFG SPP 1500, PT
FCT, NL STW 10346 NEST and Artemis 100202 RECOMP.

7. REFERENCES

[1] C.H.Berkel, *Multi-core for mobile phones*, In Proc. DATE 2009.
[2] L.Minas et al., *Energy Efficiency for Information Technology: How to Reduce Power Consumption in Servers and Data Centers*, Intel Press, 2009.
[3] Y.Cao et al., *Mapping Statistical Process Variations Toward Circuit Performance Variability: An Analytical Modeling Approach*, In Proc. DAC 2005.
[4] S.Bhardwaj et al., *Modeling of intra-die process variations for accurate analysis and optimization of nano-scale circuits*, In Proc. DAC 2006.
[5] S.Borkar et al., *Parameter variations and impact on circuits and microarchitecture*, In Proc. DAC 2003.
[6] K.Bowman et al., *Impact of Die-to-Die and Within-Die Parameter Variations on the Throughput Distribution of Multi-Core Processors*, In Proc. ISLPED 2007.
[7] X.Liang et al., *Mitigating the Impact of Process Variations on Processor Register Files and Execution Units*, In MICRO 2006.
[8] J.Tschanz et al., *Adaptive Frequency and Biasing Techniques for Tolerance to Dynamic Temperature-Voltage Variations and Aging*, In Proc. ISSCC 2007.
[9] JEDEC SST Assn., *DDR3 Specification*, JESD79-3E, 2010.
[10] Micron Tech. Inc., *2Gb: X4, X8, X16 DDR3 Datasheet*, 2010.
[11] Intel, *Memory 3-sigma Power Analysis Methodology*.
[12] S.Y.Ji et al., *An Empirical Study of Performance and Power Scaling of Low Voltage DDR3*, In Proc. EPEPS 2010.
[13] M.Gottscho et al., *Power Variability in Contemporary DRAMs*, IEEE Embd. Sys. Letters, Vol. 4, No. 2, 2012.
[14] S.Desai et al., *Process Variation Aware DRAM Design Using Block Based Adaptive Body Biasing Algorithm*, In ISQED 2012.
[15] H.David et al,. *RAPL: memory power estimation and capping*, In Proc. ISLPED 2010.
[16] L.Bathen et al., *Vamv: Variability-aware memory virtualization*, In Proc. DATE 2012.
[17] L.Bathen et al., *ViPZonE: OS-Level Memory Variability-Driven Physical Address Zoning for Energy Savings*, In Proc. CODES+ISSS 2012.
[18] T.Vogelsang, Rambus Inc., *Understanding the Energy Consumption of Dynamic Random Access Memories*, In Proc. MICRO 2010.
[19] C.Weis et al., *Design Space Exploration of 3D-stacked DRAMs*, In Proc. DATE 2011.
[20] K.Chandrasekar et al., *Improved Power Modeling of DDR SDRAMs*, In Proc. DSD 2011.
[21] Micron Technology Inc., *TN-41-01: Calculating Memory System Power for DDR3*, Tech Report, 2007.
[22] S.Thoziyoor et al., *CACTI 5.1*, HP Labs, 2008.
[23] K.Chandrasekar et al., *DRAMPower: Open-source DRAM power & energy estimation tool, www.es.ele.tue.nl/drampower*
[24] NGSPICE, *http://ngspice.sourceforge.net/*
[25] T. Schloesser et al., *A $6F^2$ Buried Wordline DRAM Cell for 40nm and Beyond*, In Proc. IEDM 2008.
[26] B.Jacob et al., *Memory Systems: Cache, DRAM, Disk*, Morgan Kaufmann Publishers, 2008.
[27] BSIM4.6, *http://www-device.eecs.berkeley.edu/bsim/*
[28] Predictive Technology Model, *http://ptm.asu.edu/*
[29] ITRS, *Design*, Report and Tables, 2011.
[30] ITRS, *Modeling & Simulation*, Report and Tables, 2011.
[31] Y. Ye et al., *Statistical Modeling and Simulation of Threshold Variation Under Random Dopant Fluctuations and Line-Edge Roughness*, IEEE Trans. VLSI Sys. vol.19, no.6, 2011.
[32] C.Lin et al., *Compact Statistical Compact Modeling of Variations in Nano MOSFETs*, In Proc. VLSI-TSA 2008.
[33] C.Lee et al., *MediaBench: a tool for evaluating and synthesizing multimedia and communications systems*, In Proc. MICRO 1997.
[34] A.Keshavarzi et al., *Leakage and Process Variation Effects in Current Testing on Future CMOS Circuits*, IEEE Design & Test Comp., vol.19, no.5, 2002. ASP-DAC 2002.
[35] S.Herbert et al., *Variation-aware dynamic voltage/frequency scaling*, In Proc. HPCA 2009.
[36] T.Austin et al., *SimpleScalar: An infrastructure for computer system modeling*, IEEE Computer, vol.35, no.2, 2002.
[37] B.Akesson et al., *Architectures and Modeling of Predictable Memory Controllers for Improved System Integration*, In Proc. DATE 2011.
[38] K.Chandrasekar et al., *Run-Time Power-Down Strategies for Real-Time SDRAM Memory Controllers*, In Proc. DAC 2012.

Appendix

A1: DRAM Currents and Power Consumption

In this section, we describe the different DRAM currents, when and how they are measured, and the state of the banks and changes to the DRAM settings, when they are measured. These currents are also described in detail in [9].

(1) I_{DD0} (**One Bank Active-Precharge Current**): Measured across ACT and PRE commands to one bank. Other banks are retained in the precharged state.

(2) I_{DD1} (**One Bank Active-Read-Precharge Current**): Measured across ACT, RD and PRE commands to one bank, while other banks are retained in the precharged state. This measurement is performed twice, targeting two different memory locations and toggling of all data bits.

(3) I_{DD2N} (**Precharge Standby Current**): Measured when all banks are closed (precharged state).

(4) I_{DD2P0} (**Precharge Power-Down Current - Slow Exit**): Measured during power-down mode, with CKE (Clock Enable) Low and the DLL locked (slow-exit), while the external clock is On and all banks are closed (precharged).

(5) I_{DD3N} (**Active Standby Current**): Measured when all banks are open (active state).

(6) I_{DD4R} (**Burst Read Current**): Measured during Read (RD) operation, assuming seamless read data burst with all data bits toggling between bursts and all banks open, with the RD commands cycling through all the banks.

(7) I_{DD4W} (**Burst Write Current**): Measured during Write (WR) operation, assuming seamless write data burst with all data bits toggling between bursts and all banks open, with the WR commands cycling through all the banks and the ODT (On Die Termination) stable at HIGH.

(8) I_{DD5} (**Refresh Current**): Measured during Refresh (REF) operation, with REF commands issued every nRFC.

(9) I_{DD6} (**Self Refresh Current**): Measured during self-refresh mode, with CKE Low and the DLL off, while the external clock is Off and all banks are closed (precharged).

A2: Monte Carlo on DRAM Cross-Section

In this section, we present the impact of process-variation on the timing behavior of the DRAM cross-section presented in Section 3.2 by performing using Monte-Carlo analysis on the same, considering $\pm 1\sigma$ variations in the device and interconnect parameters. In Figure 6, we present the effects on the local wordline activation (lwl), and the sensing of the true (lblt) and complementary (lblc) bitlines by the PSA.

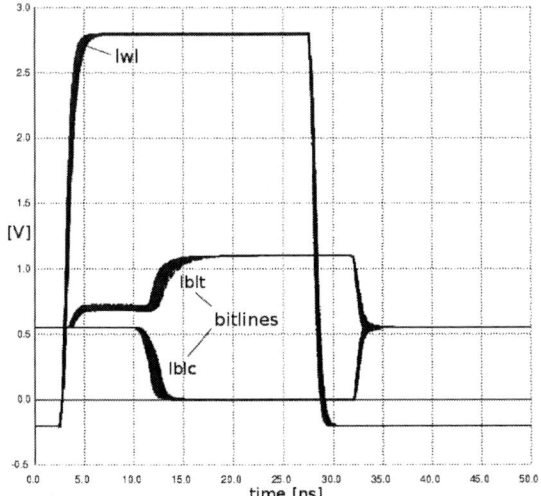

Figure 6: Variation Impact on Bitline and Wordline

As can be observed from the figure, there is a significant impact on the timings of the operations associated with the wordline and bitlines. For instance, the local word line reaches its required potential (upon activation) of 2.8V at around 6ns instead of 5ns, which was the case for the baseline configuration without any variation (shown in Figure 3). Similarly, the bitlines reach their potential (upon sensing by the PSA) at around 17ns, compared to around 15ns in the baseline configuration (Figure 3). The variations in the bitlines and wordline impact the activation latency given by the core-timing parameter n_{RCD}, thereby impacting both the DRAM frequency (delay) and power consumption.

A3: Core-Timings Check

In this section, we present the Core-Timings Check function in Algorithm 2, which is an adaptation of the I_{DD1} Measurement Loop test proposed by JEDEC for DRAMs in [9]. We begin by first providing background information on the three core-timings of a DRAM memory.

(1) The n_{CL} parameter corresponds to the minimum CAS latency, which is the delay between the Read command and the availability of the first bit of output data.

(2) The n_{RCD} parameter corresponds to ACT to RD/WR latency, which defines when an RD/WR can be issued after the ACT has been issued to assure completion of activation.

(3) The n_{RP} parameter defined the precharge (PRE) latency, which defines the time required by the precharge operation to completely precharge the local bitlines.

Another important timing constraint to review is the n_{RAS} timing constraint, which gives the minimum delay between ACT and PRE commands to the same bank, thus encompassing both the n_{RCD} and n_{CL} core-timings [9].

The original I_{DD1} test on which the CT_check function is based, is employed by memory vendors to measure the worst-case estimates for I_{DD1} current. Interestingly, this test performs Activation, Read and Precharge operations that employ the three core-timing parameters viz., n_{CL}, n_{RCD} and n_{RP}, which form the core of DRAM performance assessment methodology proposed in this work. Hence, we selected this I_{DD1} test as a part of our methodology, by adapting it to check for functional accuracy of the memory, when the three core-timing parameters are modified by Algorithm 1. We do not use this test for current measurements, as this requires expensive current measurement hardware, which is generally available only with DRAM vendors.

Algorithm 2 Core-Timings Check

Require: CT_check $(k, l, CT[])$
1: {Initializing, Bank, Frequency, Data and Address Offsets}
2: Bank $= k$; Set_Freq $= l$;
3: Set_CL $=$ CT[0]; Set_RCD $=$ CT[1]; Set_RP $=$ CT[2]
4: Data_0 $=$ 0xAAAAAAAA; Addr_0_Offset $=$ 0x0000
5: Data_1 $=$ 0x55555555; Addr_1_Offset $=$ 0x000F
6: {Comment: I_{DD1} Test Phase}
7: **for** $i = 0 \rightarrow 1$ **do**
8:     {Comment: Representing two sets of data and addresses}
9:     Issue: ACT, Addr[i]
10:     wait(RCD);
11:     Issue: RD, Addr[i]
12:     wait(RAS-RCD);
13:     Recv: Recv_Data[i]
14:     **if** Recv_Data[i] == Data[i] **then**
15:         check $=$ TRUE
16:         Issue: PRE, Addr[i]
17:         wait(RP);
18:     **else**
19:         check $=$ FALSE
20:         Issue: PRE, Addr[i]
21:         wait(RP);
22:         Break;
23:     **end if**
24: **end for**{i}
25: **return** check

The inputs to Algorithm 2 include selected core-timings, operating frequency and target bank provided by Step 16 in Algorithm 1. We begin by initializing these settings (Steps 2 and 3) and selecting two unique data values with all bits toggling and two address offsets within the target bank, as required by the JEDEC I_{DD1} test loop (Steps 4 and 5). Before the test phase commences (in Step 7), the two data values are written at the corresponding addresses.

In the I_{DD1} test phase, two sets of ACT-RD-PRE operations are performed, with each targeting a different row in the same memory bank and all the data-bits toggling across the two accesses. Note that since the n_{RAS} parameter covers the period between ACT and PRE, it encompasses both the n_{RCD} and n_{CL} core-timings within itself. As a result, the total latency for one set of ACT-RD-PRE operations performed in this test is given by $n_{RAS}+n_{RP}$. The test begins by issuing an ACT to the address of the first specified transaction and waiting n_{RCD} clock cycles for it to complete (Steps 9 and 10). This is followed by issuing the READ command to the same address and waiting $n_{RAS} - n_{RCD}$ clock cycles (representing the n_{CL} core-timing parameter and the complete data transfer period) to read out the data from the corresponding address (Steps 11 to 13). Once the data is received, we adapted the I_{DD1} test to merely check this data against the expected value in Step 14, to verify the correct functioning of the memory for the ACT and RD operations. If this test passes, a precharge is issued (Step 16) and after waiting for n_{RP} cycles for the completion of the precharge operation (Step 17), the second transaction starts. If the test fails, the test issues a precharge and waits for its completion (Steps 20 and 21) returns a FALSE to the CT_check call in Algorithm 1. If both set of ACT-RD-PRE tests pass, a TRUE is returned instead.

To enable this performance assessment, the core-timings and frequency are set using the Mode register settings [9] and frequency scaling is performed using existing support in standard DRAM memory controllers.

A4: Combination of System Parameters

Using the analysis presented in Table 4 in Section 5.1, it is now possible to estimate the impact of the three system parameters viz., capacity (C), frequency (F) and width (W), on DRAM current consumption. However, when a combination of system parameters differ between two DRAM memories, the influence all the concerned system parameters must be taken into account. This can be derived directly from the results in Table 4 by adding the corresponding impact (μ and $\sigma\%$) for one parameter at a time, considering the most influential parameter first (determined by % change in μ). In this section, we present the corresponding impact of the combinations in Table 6, by extrapolating the results presented in Section 5.1, using the same system parameter values. Similar extrapolation was performed to derive the current distributions in Section 6 for the Micron memory. The baseline values for μ and $\sigma\%$ are presented in Table 2.

Table 6: Multi-Parameter Impact on Currents

	F&C		F&W		C&W		F&C&W	
Config	2Gb-800-x8		1Gb-800-x16		2Gb-533-x16		2Gb-800-x16	
I_{DD} Type	μ mA	σ %	μ mA	σ %	μ mA	σ %	μ mA	σ %
I_{DD0}	113	2.5	112	2.4	99.3	2.4	113	2.4
I_{DD1}	119	2.4	127	2.3	115	2.3	128	2.3
I_{DD2N}	48.6	6.3	42.7	5.0	43.7	5.1	48.6	5.1
I_{DD3N}	52.4	5.5	56.7	5.9	47.2	6.0	52.4	6.0
I_{DD4R}	159	3.4	244	3.2	214	3.2	250	3.2
I_{DD4W}	163	3.8	250	2.7	217	2.7	253	2.7
I_{DD5}	194	2.3	161	2.2	179	2.2	194	2.2
I_{DD2P0}	14.1	18.3	8.4	15.5	14.1	15.9	14.1	16.2
I_{DD6}	11.2	22.6	8.0	15.9	11.2	16.8	11.2	17.2

A5: Impact on Power and Delay

In this section, we present the impact of $\pm 1\sigma$ variations in device and interconnect parameters on basic memory operations including activation, read, precharge and power-down.

The impact of variation on I_{DD1} active power and the corresponding operation latency ($t_{RCD}+t_{Data}$) is depicted in Figure 7. This represents activation, read and precharge operations in a particular memory row, with t_{RCD} being the activation period and t_{Data} being the latency to read the data out. Similarly, the impact on leakage power is plotted against the t_{RCD} delay in Figure 8.

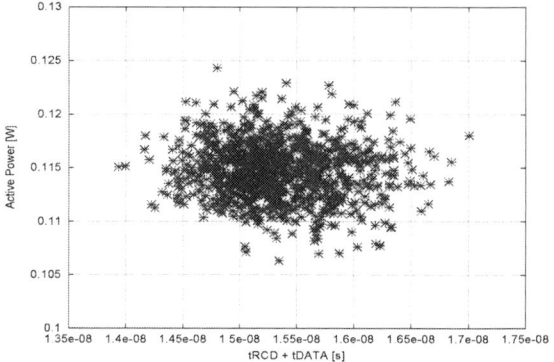

Figure 7: Impact on ACT-RD-PRE Operations

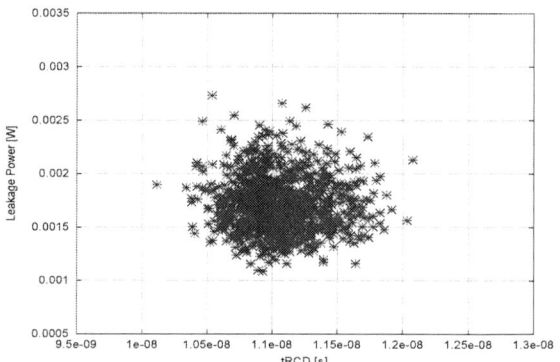

Figure 8: Impact on Leakage current

In Figure 9, we present a Q-Q (quantile) plot comparing the distributions observed in active and leakage currents (power) and the delay measures (t_{RCD}, t_{Data}) corresponding to $\pm 1\sigma$ variations. The linearity in the four measures shows a Gaussian distribution in the variation, as expected.

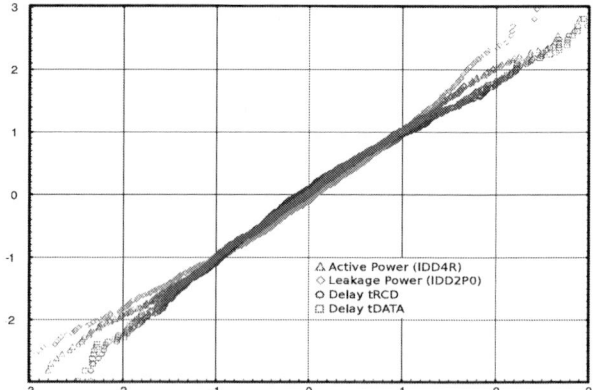

Figure 9: Impact on Currents and Timing

These results show the impact of device and interconnect variations on the delay and power consumption of DRAM memories, highlighting the significance of this work.

TEASE: A Systematic Analysis Framework for Early Evaluation of FinFET-based Advanced Technology Nodes

Arindam Mallik, Paul Zuber, Tsung-Te Liu, Bharani Chava, Bhavana Ballal, Pablo Royer Del Bario, Rogier Baert, Kris Croes, Julien Ryckaert, Mustafa Badaroglu, Abdelkarim Mercha, Diederik Verkest

IMEC, Belgium; Arindam.Mallik@imec.be

ABSTRACT

This paper proposes TEASE (Technology Exploration and Analysis for SoC-level Evaluation), a framework to systematically analyze and evaluate system design in finFET-based technology node. The proposed framework combines both lithography and electrical constraints of a particular technology node to optimize the standard cell library performance. Growing complexity of logic design at nodes below 20nm causes to adopt a design style that can embrace the simplicity required to enable manufacturing, along with a process technology that can be finely tuned to the desired performance constraints. Additionally, the introduction of finFET based devices poses a new challenge for the designers to come up with an efficient standard cell template. The proposed framework can be used to detect the technology constraints that act as the bottleneck for the enablement of design at these advanced nodes. Results presented in this paper show by optimizing these bottlenecks we can improve the performance of a standard cell library significantly. Furthermore, adapting to such an analysis framework at an early stage of technology development helps to take the design constraints into the decision loop for realization of technology research into real products.

Categories and Subject Descriptors

B.7.1 [INTEGRATED CIRCUITS]: Types and Design Styles – *advanced technologies, standard cells.*

General Terms

Performance, Design.

Keywords

FinFET technology, standard cell architecture, SoC evaluation.

1. INTRODUCTION

The traditional advancement of semiconductor industry is governed by the process engineering community. Governed by Moore's law and the goal to reduce the cost of a single processor, the primary objective has been to put an increasing number of devices onto a single wafer. While the increasing complexity had forced the costs of technology higher [1], the benefits of technology scaling have been successfully maintained by ensuring sufficiently high product yields.

Permission to make digital or hard copies of all or part of this work for personal or classroom use is granted without fee provided that copies are not made or distributed for profit or commercial advantage and that copies bear this notice and the full citation on the first page. To copy otherwise, to republish, to post on servers or to redistribute to lists, requires prior specific permission and/or a fee.
DAC '13, May 29 - June 07 2013, Austin, TX, USA.

Unfortunately, the sub-90 nm technology nodes have experienced the increasing constraint of decreasing yield landscape. In particular, the inability of process engineers to guarantee the necessary process window across the entire design space has led to a sharp increase in the number of systematic layout geometry induced faults. In parallel, process variability has amplified and has been established as a major cause for parametric yield losses. As a result, maintaining silicon yield has been the major bottleneck for successful technology scaling in the future as the cost benefit outlined by Moore rapidly eroded.

The design rules have played the role of a bridge between the design and manufacturing community. Technology scaling has resulted in an exponential increase of design rules to maintain an industry-acceptable yield at a particular node. With increasing complexity of the advanced nodes, the process community has exposed the details of the manufacturing process to the design community. However, the designer has been poorly adept in making the proper manufacturing trade-offs at the current abstraction level. This has convinced industrial community to involve the designer from the start of the development of a technology node. By looking at the design constraints and the electrical properties of the future generation node at a system level, a realistic estimate of advantages of technology node transition can be perceived by the design community.

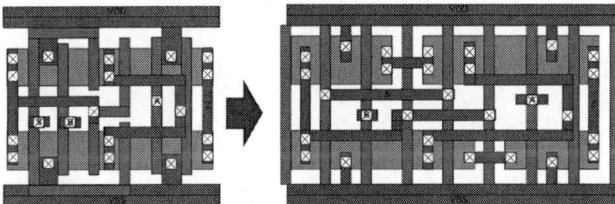

Figure 1. Transition in Standard Cell Design

The current state of the industry makes it clear that further increases in the complexity at the design and process level can no longer be economically sustained. The layout illustrated in Figure 1 shows the change in layout styles for advanced nodes. Advanced process technologies and lithography constraints have imposed strong restrictions on uniform poly-pitch, poly-jogging as shown in the figure. A design style that can embrace the simplicity required to enable manufacturing, along with a process technology that can be finely tuned to the desired design constraints, must become the industry adopted solution for future technology nodes. Consequently, bringing the design perspective and the manufacturing constraints under a single umbrella will be an efficient and intelligent way of developing a particular technology node. Specifically, the shift to the electrical and litho-aware restricted design rules will allow the future technology to be more cost-effective and optimized for performance at the same time.

The recent evolution to Bulk finFETs (BFF) [1] just paved the way to the transformations that CMOS technology will undergo in the next decade. Performance scaling will not occur just by feature size reduction but by appropriately blending a plethora of technological features, such as standard cells based on finFETs (FEOL), Middle-end of Line (MOL) interconnects, multiple-patterned BEOL. The constraints brought by innovations at these features need to be considered when architecting a system. This evolution signs the era of versatile scaling in which performance scaling is not exclusively determined by technology features but rather by a symbiotic combination of technologies conditioned by system requirements.

Based on these observations, we have proposed an analytical framework. It uses an abstraction of the cell architecture and few representative cells to detect the bottleneck design constraints for the future technology node. This enables economical trade-offs of the advanced technology nodes at a very early stage. **The main contribution of this paper is the foundation of a single framework named as TEASE (Technology Exploration and Analysis for SoC-level Evaluation) that combines both technology and electrical constraints of a particular technology node to optimize the performance of a standard cell library.** The future success of the semiconductor industry will depend on its willingness to accept such a paradigm shift, which aims to reduce process and design complexity while maintaining the cost scaling trends. Henceforth, there is an increasing need for such a framework that helps to evaluate the feasibility of future technology node and the corresponding design automation tools that support these systematic optimization methodologies.

In the next section, we present overall flow of the framework for the system design and analysis at advanced technology node. The different technology constraints are discussed in section 3. In Section 4, the experimental results based on the proposed standard cell templates and different optimizations are reported. Section 5 reports the related works followed by a conclusion in section 6.

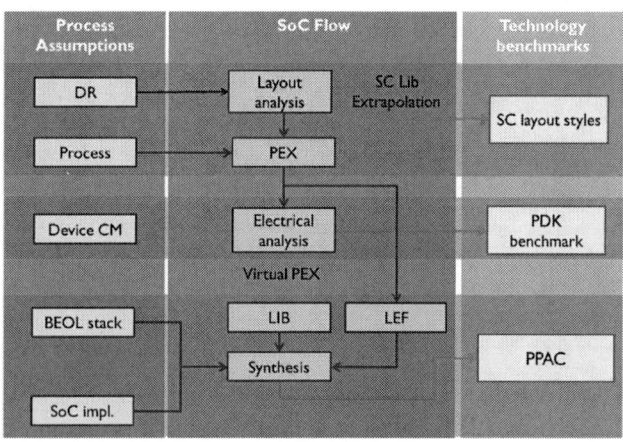

Figure 2. TEASE Framework

2. TEASE FRAMEWORK

The TEASE framework dataflow is described in Figure 2. The main goal of the framework is to establish a platform to **identify** the bottlenecks of designing in an advanced technology node, **propose** standard cell level optimizations to solve such design issues and **evaluate** the benefit of such optimizations in a holistic way at both standard cell and SoC level. Naturally, it consists of two major analysis blocks, namely Standard Cell analysis and SoC level evaluation for standard cell library.

2.1 Standard Cell Analysis

Traditionally, the design rules for a particular technology are determined based on the manufacturing constraints as detected by the process engineers and the lithography experts. Once the process engineers have decided on the number of layers to be used in a particular technology, they come up with a preliminary set of design rules that will help the designer to build a standard cell on that particular node. In the current framework, we have taken these initial design rules as the starting point of the whole flow. The process assumptions used in the particular logic node is dependent on the major innovations introduced at different process module and the standard cell architecture. The cell architecture we have used for our analysis is depicted in Figure 3. As shown in the figure, the logic cell consists of the following critical layers: Active(finFET); Wells; Poly-Lines; Poly-Cut; Intermediate Metallization1 (IM1); Intermediate Metallization2 (IM2); Via-0 (IM2-M1 contact); Metal1 (M1); Via-1 (M1-M2 contact); Metal2 (M2). Figure 3 shows a 3-dimensional setup of these critical layers in a particular cell. As evident from the figure, we have used intermediate metallization in two different layers (IM1 and IM2), for the connection between source to drain or power-rail to active connections. Due to stability problems, the IM1 does not run over the STI and we use IM2 for that purpose.

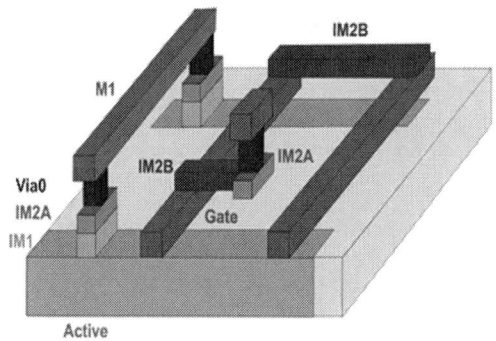

Figure 3. Cell architecture model

Based on the layers available and their physical connectivity, we construct the architecture shown in Figure 3 to design standard cells. This standard cell architecture is in line with established technology [4]. As shown in Figure 2, a layout analysis is performed by constructing a few representative cells for a particular architecture. We propose it as the baseline standard cell architecture under the context of finFET based standard cell design. As described in the section 4, we will use this template using the **TEASE** framework to detect constraints for technology scaling, propose innovations to come up with alternative architectures to remove such bottlenecks and evaluate them systematically using the method discussed next.

2.2 Scorecard Evaluation

To evaluate different cell templates for standard cell, we introduced the scorecard based approach that looks into the different parameters from patterning, electrical and SoC compatibility perspective. Table 1 details the justification for different parameters in the scorecard template. This is a systematic way to evaluate different standard cell template at an early stage of decision making. Each of the parameters can be quantified if an evaluation framework is in place. Otherwise, they can be subjective and colour-coded from green (best), yellow (medium), or red (worst) based on level of difficulty. For all the proposed cell templates we used the scorecards to evaluate their merit in an all-round way.

Table 1: Scorecard based evaluation parameters.

Parameter	Effect
Patterning scorecard	
Cell area	Direct consequence on scaling. Target across technology node should be 50% decrease
2D pattern occurrence	More complex patterning induces constraint on litho. Determines cost of manufacturing
Contact strategy	In-line contacts or staggered
Tip-to-Tip spacing	Primary constraint for shrinking cell width as metal CDU scaling is 0.85
Minimum Metal area	Determines how densely metal segments can be placed
SoC compatibility scorecard	
Cell abutment	Vertical symmetry of cells needs to be maintained. Additional complexity due to presence of Local interconnects in the boundary region
Port accessibility	Helps in P&R, improves cell density
Dummy Poly Penalty	Additional dummy poly penalty due to fin processing or SADP constraint amounts to significant area penalty
Well boundary symmetry	Placement of cells horizontally becomes easier
Electrical scorecard	
Power Rail Width	Thicker power rail helps to reduce IR Drop
Dynamic Power	Direct effect on power consumption of the system
Leakage power	Determines static power of the system
Speed	Needs to follow the scaling trend

2.3 SoC level evaluation

In this section, we verify the main features of the standard cell styles on an IP-block level. The method follows industry standard design flows. It consists of characterizing the standard cell libraries after layout, LVS, and PEX. Next, instances of verilog netlists of representative designs (DesignWare components and OpenCores designs) in register level transfer (RTL) file formats are synthesized using the previously generated libraries, and industry-standard synthesis tools. Power, performance, and area analysis is performed using Design Compiler and different clock period targets. In addition, a place-route step with varying fill rates is performed to assess the routability of the cell styles, to deliver a more accurate estimation of the SoC area. All steps use industry standard tools, such as Assura, Liberate, Design Compiler and Encounter.

3. TECHNOLOGY CONSTRAINTS

At advanced nodes, process engineers have innovated multiple ways to boost the performance while enabling area shrinking. To keep performance at pace, devices are evolving from a planar to 3D known as Bulk finFET. To contact devices, Middle of Line inter-metals are introduced, self-aligned, to limit resistance and keep parasitics contained [3]. The tight interconnect pitches force the lower and intermediate level BEOL to double patterning such as Litho-Etch-Litho-Etch (LELE) or Self-aligned Double patterning (SADP) and self-aligned Vias. However, each of these innovations result in severe bottlenecks from a circuit design perspective. This section will provide the details for three such constraints, and propose alternative standard cell architectures to overcome the bottlenecks.

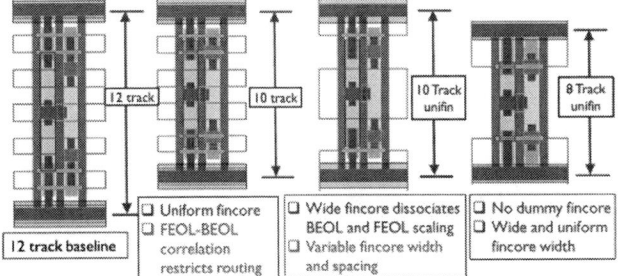

Figure 4: Various alternatives for FEOL optimization.

3.1.1 FEOL constraints and optimization

As shown in Figure 4, the tight correlation between finpitch and metalpitch (same pitch, i.e. 45nm) induces design rule restrictions in standard cells. Such a correlation in addition to uniform fin-pitch along the cell height results in a cell with 12-track height as shown in Figure 4. The cell height can be reduced using either of the three ways as demonstrated in Figure 4 As shown in the figure, the 8track **one-fin** cell template looked like the most preferred solution in terms of area.

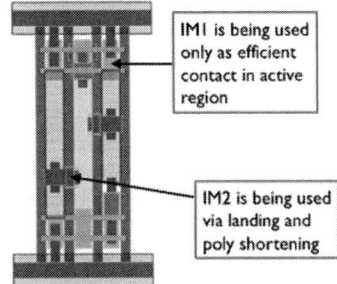

Figure 5: MOL constraints for baseline cell architecture.

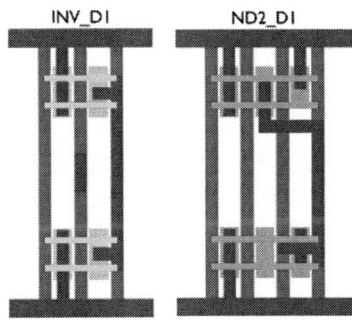

Figure 6: Representative NAND cell with MOL optimization.

3.1.2 MOL constraints and optimization

As shown in Figure 5, the baseline standard cell architecture does not utilize the local interconnect layers fully. The IM1 layer has been used as a connecting layer for active fins and better landing for the contacts in active region. As for the IM2 layers, they are mostly used for via landing and gate shortening. We propose a MOL optimized cell architecture where IM2 layers have been aggressively used as a routing layer. It assumes that crossing or running on top of an active gate using IM2 won't short-circuit the IM2 and poly layer(presence of nitride layer in between). Furthermore, instead of using two IM layers, we propose the use of single IM cell architecture with shallow (horizontal) and deep (vertical) trenches. This reduces the number of lithography steps involved in the process and thus helps to reduce wafer cost. This MOL optimization reduces the height of the cells from 12 tracks to 10 tracks.

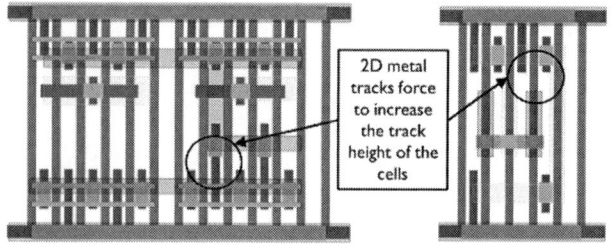

Figure 7: BEOL constraints for baseline standard cell

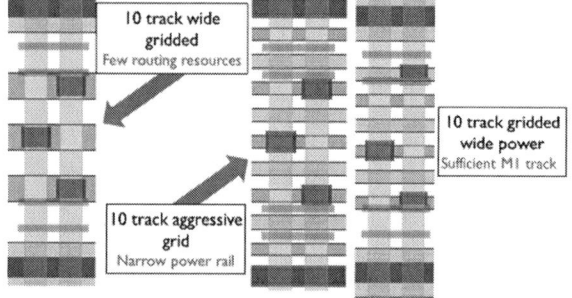

Figure 8: Various gridded standard cell architecture for BEOL optimization.

3.1.3 BEOL constraints and optimization

In the baseline standard cell template, the Metal-1 layer is assumed to be two-dimensional. The aggressive pitch of this layer constrains the design due to tip-to-line and tip-to-tip constraints in both directions (Figure 7). It forced the designers to design standard cells of height 12 tracks. We propose a BEOL optimized cell architecture with unidirectional metal. The use of unidirectional metal introduces a grid involving Metal 1 and 2 within the standard cell. Figure 8 demonstrates a few variations of the gridded cell architecture based on the amount of aggressiveness. We ended up choosing a cell template with wide power rails and aggressive grid in at a safe distance from the wide metal lines. It offers the convenience of gridded metals without losing performance due to thin power rails

4. EXPERIMENTAL RESULTS

To evaluate the validity of the analysis framework, we applied the proposed methodology in exploring a prototype standard cell library with BFF devices at 14nm technology node. The device characteristics and cell architecture used in the simulation framework is as described in Section 2. The transistors models used in the experimentation have been verified against state-of-the-art publications reporting the expected behavior of these devices at 28nm technology and beyond [5].

4.1 Process Assumptions

Any library development should start with the basic process assumptions. We used a poly pitch of 62nm and a metal pitch of 45nm to analyze a representative 14nm technology node. It is in line with the 0.7x scaling assumption at 14nm from 20nm industry-standard technology node. Please note, Metal-1 Critical Dimension Unit (CDU) used in the standard cells is half of the metal pitch.

Once we have fixed the process assumption, the next step is to develop a set of prototype standard cell libraries to evaluate the proposed methodology. We start with a track budget height of 12 Metal-tracks for the library development for the baseline standard cell architecture. For early evaluation, we developed the most frequently used combinational standard cells in any representative

SoC, namely Inverter, NAND, and NOR gates. The use of such a limited number of cells definitely has an effect on the accuracy of benefit of absolute numbers in area or performance changes. However, as shown in Section 4.4 fidelity report, such limited number of cells can provide a valuable insight in terms of relative benefits from each of the proposed cell architecture at an early stage of technology node definition.

	FEOL	MOL	BEOL
Patterning			
Cell area	-33%	-20%	-20%
2D pattern occurrence	More frequent	Distributed	None
Contact strategy	Same as before	Less dense	Dense
Tip-to-Tip spacing	Same as before	Severe	Unidirectional
Minimum Metal area	Same as before	Less dense	Less dense
SoC Compatibility			
Cell abutment	Same as before	Same as before	Same as before
Port accessibility	Difficult	Better	Through M3
Dummy Poly Penalty	Same as before	Same as before	Same as before
Well boundary Symmetry	Same as before	Same as before	Same as before
Electrical			
Power Rail Width	-33%	Same as before	Same as before
Dynamic Power	-22%	-6%	4%
Leakage Power	-50%	Same as before	Same as before
Performance	-44%	6%	-8%

Figure 9. Scorecard based comparison

4.2 Cell level analysis

The scorecard evaluation technique is employed to evaluate the benefits of each of the optimizations proposed above. We used representative cells namely INV, ND2 and NR2 to look at the benefits offered by each of the optimizations individually. As recorded in Figure 9, the FEOL optimized 8track **onefin** cells offer the best area benefit of 33% at the cost of significant loss of performance 44%. This can be attributed to the use of reduced number of active fins in the proposed architecture. The MOL optimization based on **singleIM** cells manages to reduce the cell area by 20% and furthermore reduce the 2D pattern occurrence due to distribution of routing layers in both metal-1 and IM2. The reduced cell height and the use of IM2 as a routing layer help to reduce cell parasitic and result in a performance increase of 6%. The BEOL optimization in **gridded** cells manages to reduce 20% area for the standard cells but at the cost of 8% performance loss. This can be attributed to the use of additional metal-2 layer for connecting to the gates in gridded metal connections.

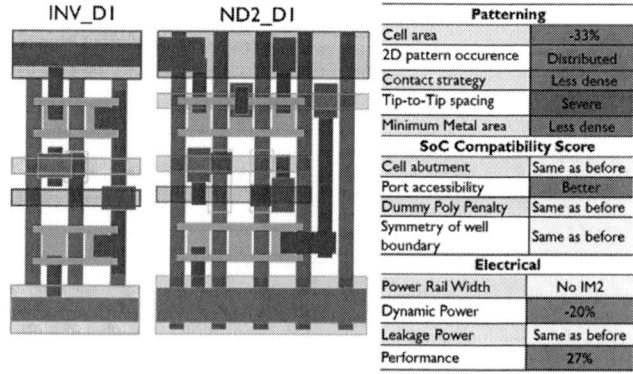

INV_D1	ND2_D1	**Patterning**	
		Cell area	-33%
		2D pattern occurence	Distributed
		Contact strategy	Less dense
		Tip-to-Tip spacing	Severe
		Minimum Metal area	Less dense
		SoC Compatibility Score	
		Cell abutment	Same as before
		Port accessibility	Better
		Dummy Poly Penalty	Same as before
		Symmetry of well boundary	Same as before
		Electrical	
		Power Rail Width	No IM2
		Dynamic Power	-20%
		Leakage Power	Same as before
		Performance	27%

Figure 10: Optimized cell architecture and scorecard.

4.3 An optimized standard cell template

Based on the evaluation scores of each of the proposed optimization, we tried to come up with an **optimized** cell architecture that reflects the combined benefit of each of the proposed solution. The significant performance loss of **onefin** cell

172

template disqualifies it to be used in the combined cell optimization. We utilized the aggressive routing of IM2 layers from the **singleIM** cell template with the additional constraint that IM2 can route horizontally only while crossing a dummy gate. Furthermore, we adapt to the structure of unidirectional metal from the **gridded** cell template while using only Metal-1 inside the standard cell. The resultant **optimized** cell architecture produces a standard cell of 8track height as shown in Figure 10. As demonstrated in the Figure 10, the particular cell template can reduce the cell area by 33%. Furthermore, small cell height and use of IM2 as routing layers help to reduce the cell parasitic significantly and therefore improve the performance of the standard cells by 27% on average. The results show that such a standard cell template is the most promising candidate for an extended standard cell library development.

4.4 SoC level evaluation

Complexity Reduction: Cell design usually comprises the design of several hundreds of cells, and is a task that is difficult to automate, the more so in an environment of premature design rules and process assumptions at an early stage of technology development. We addressed this problem by generating a few cells, and then directly running the synthesis and analysis without the full library. In the following, we assess the fidelity of the results generated by this method, requiring a few test runs in existing libraries.

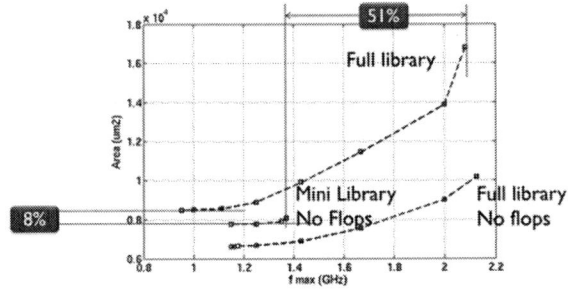

Figure 11: Fidelity test.

For the test runs we use a design A, synthesize it targeting a library from foundry 1 for different timing constraints. The resulting area-delay trade-off can be seen in the Figure 11, curve "*Full library*". We then repeat this step, first taking out the flipflops from the library (curve "*Full Library No Flops*") and then taking out all cells but those for which we have layout for (curve "*Mini Library No Flops*"). Comparing the full library versus the mini library design space shows that area-wise, the impact of removing the flipflops (and setting their area to zero) is higher than the area penalty of a reduced cell type and drive set to select from. Timing-wise, there is about a 50% penalty in maximum frequency by reducing the cell selection.

fmax	40L	40G	28M	Area	40L	40G	28M
Cordic	42%	46%	51%	Cordic	11%	9%	8%
Mult64x64	n/a	41%	48%	Mult64x64	n/a	-1%	-2%
Mult32x32	45%	52%	49%	Mult32x32	0%	-1%	0%

Figure 12: Fidelity results.

Fidelity can be verified if we repeat the experiment on different libraries, and achieve similar results. The fidelity checking results are shown in Figure 12. The numbers show that the timing accuracy is between 41% and 51% depending on library, and largely independent of the design. The area inaccuracy is between 8% and 11% for the Cordic design and about -1% for a multiplier.

It can be concluded that fidelity can be reached by synthesizing a design with a drastically reduced cell library if a cross-check on the inaccuracy done on a similar, existing library is performed.

Using the 14nm standard cell variants previously described and the method above, we generated SOC level Power-Perforamce-Area(PPA) figures. The goal was to verify which cell layout style is likely to meet at all scaling targets without designing a full library in each style. In the following, performance and power are shown as tradeoff with pre-placement area figures:

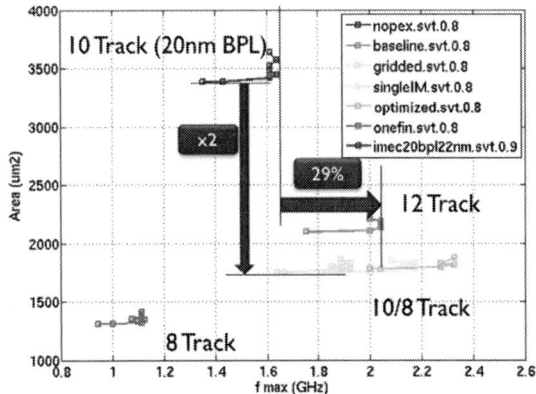

Figure 13: Performance versus area results.

Figure 13 shows the synthesis results with the 14nm libraries in comparison to the in-house 20 bulk planar (BPL) library (using the same cell set). It can be seen that the styles *gridded*, *simgleIM*, *optimized* and *onefin* are able to meet the area scaling target of 50% as opposed to the *baseline* library. However, only the *optimized* and the *singleIM* are able to meet the scaling target of 30% improvement in timing.

The dual-fin solutions are significantly faster (90-110%) than the one-fin solution. Differences in maximum SOC frequency among different dual-fin styles are up to 25% (15% at constant area). *Onefin* library should be used for designs with very low speed requirement and the *optimized* standard cell library for very high speed, and *singleIM* in between. The *gridded* and *baseline* libraries do not have any Pareto-optimal point.

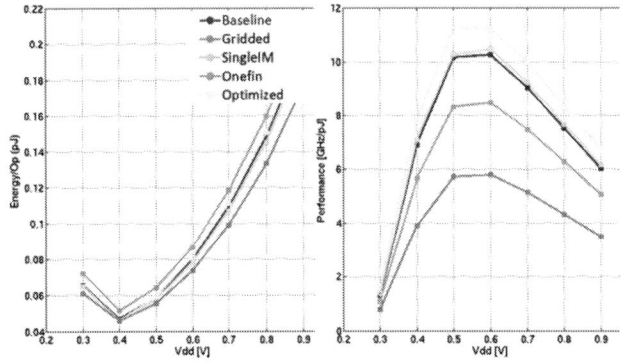

Figure 14: Energy/op & performance efficiency (1/EDP)

The results on dynamic energy are very similar to that of timing, with the exception that only *optimized* and *onefin* are Pareto-optimal in an energy-area tradeoff. The advantage of scaling to 14nm is about 50% energy saving. In terms of leakage, ideally there is no difference between 20nm and 14nm since the technologies are targeted for constant I_{off}. For all the variations, there is still a 30% reduction in leakage for the 14nm version. There is no significant difference between all the dual-fin

173

solutions, and the *onefin* library achieves about ½ of the leakage of dual-fin versions. The leakage depends on the number of gates in the design and, thus on the synthesis constraints set. It is intuitive that the *onefin* library has the lowest energy per operation, and the *optimized* library has the best energy-delay product (EDP). It is illustrated in Figure 14 for different voltages (no area/performance targets were set in this specific comparison).

Apart from the direct impact of the raw cell area, there is an impact of the layout style on routability. This is reflected in the cell density of a design. The impact of cell parasitic on gate count and size was not further investigated in the context of this paper. We used an industry standard place-and-route flow, and completed the physical implementation by sweeping the area fill factor. Minimum effective area is achieved before the big jump in number of DRC errors due to scarce routing resources (Figure 15).

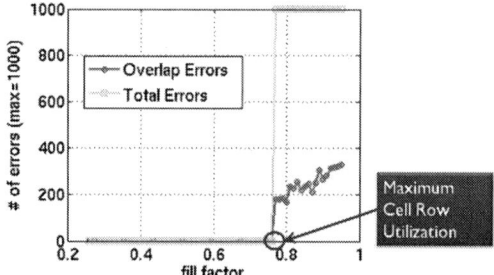

Figure 15: Deriving the maximum possible fill rate

Running this exercise on all standard cell styles resulted in the final area components summarized in Figure 16. The *gridded* solution has swapped rank with *baseline* but both of them were dominated by all other solutions. The use of unidirectional metal has a direct impact on routing overhead as shown in *gridded* and *optimized* cell libraries.

To summarize the results, a full library designed in *optimized* style (extensive use of IM2) will meet all system PPA constraints. A cost-efficient alternative could be a MOL variant (*singleIM*) with slight performance penalty. On the other hand, ultra low power/low area/low performance implementations will benefit greatly from the *onefin* library.

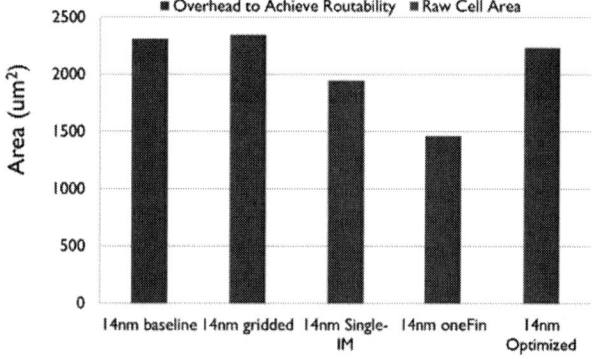

Figure 16: Place and Route Overheads

5. RELATED WORK

Growing complexity of design rules beyond 32 nm has generated a major bottleneck to continue silicon scaling with the established two year node-to-node cycle. As a result, a closer interaction between design and technology research has become a necessity in advanced technology[6]. Liebmann et. al [7] presented a

collaborative framework that facilitates rigorous design-technology co-optimization. Tejas et. al [8] proposed a regular design methodology that exploited the regularity of unidirectional layers. However, their framework is limited to the constraints due to lithography. Our framework contains circuit level analysis to ensure that both lithography and system constraints are met. Vaidyanathan et. al [9] proposed a framework to create design-efficient, and cost-effective standard cell layouts. However, their framework is not compatible with the traditional standard cell placement, power-rail design techniques. Whereas, TEASE is based on an industry-compliant system-design flow that does not require any change in the EDA tool assumptions.

6. CONCLUSIONS

This paper proposed TEASE framework to combine both lithography and electrical constraints of a FinFET based technology node for optimizing system performance and chip area. A scorecard-based evaluation for standard cell development has been introduced. This analysis detected the bottleneck for the enablement of design at FinFET and local interconnect based technology for a given cell architecture. By optimizing the bottlenecks design rules, an optimized standard cell architecture have been introduced. The prototype cell libraries were characterized at a SoC level using the proposed framework. Experimental results show an area improvement of 20% and reduced energy consumption up to 50% for optimized standard cell library. Enabling such improvement would require aggressive process technology and lithography techniques as suggested in the paper. To summarize, the TEASE framework can serve as a valuable tool for early evaluation of technology assumptions made for future technology nodes.

7. ACKNOWLEDGMENTS

Our sincere thanks to Keizo Hiraga from Sony Corporation for his valuable feedback. Work done in this paper has been supported by the European FP7 funded project ParaDIME.

8. REFERENCES

[1] A. Mallik et al., "The need for EUV lithography at advanced technology for sustainable wafer cost", SPIE 2013

[2] D. Perlmutter et al., "Sustainability in Silicon and Systems Development", ISSCC 2012 paper 1.4

[3] M. Stucchi et al., "Impact of advanced patterning options, 193nm and EUV on local interconnect performance", IEEE International Interconnect Technology Conf – IITC, 2012

[4] Chipworks, "Intel's 22-nm Trigate Transistors Exposed," http://chipworksrealchips.blogspot.in/2012/04/intels-22-nm-trigate-transistors.html

[5] T. Yamashita et al., "Sub-25nm FinFET with Advanced Fin Formation and Short Channel Effect Engineering", VLSI 2011

[6] G. Northrop, "Design technology co-optimization in technology definition for 22nm and beyond," VLSI Technology (VLSIT), 2011

[7] L. Liebmann et al., "Simplify to survive: prescriptive layouts ensure profitable scaling to 32nm and beyond", SPIE, 2009

[8] T. Jhaveri et al., "Co-Optimization of Circuits, Layout and Lithography for Predictive Technology Scaling Beyond Gratings", ICCAD, 2010

[9] K. Vaidyanathan et. al, "Design and manufacturability tradeoffs in unidirectional and bidirectional standard cell layouts in 14 nm node", SPIE, 2012

Stitch-Aware Routing for Multiple E-Beam Lithography*

Shao-Yun Fang[1], Iou-Jen Liu, and Yao-Wen Chang[1,2]

[1]Graduate Institute of Electronics Engineering, National Taiwan University, Taipei 106, Taiwan
[2]Department of Electrical Engineering, National Taiwan University, Taipei 106, Taiwan
{yuko703, yrliu}@eda.ee.ntu.edu.tw; ywchang@cc.ee.ntu.edu.tw

ABSTRACT

Multiple e-beam lithography (MEBL) is one of the most promising next generation lithography (NGL) technologies for high volume manufacturing, which improves the most critical issue of conventional single e-beam lithography, throughput, by simultaneously using thousands or millions of e-beams. For parallel writing in MEBL, a layout is split into stripes and patterns are cut by stripe boundaries, which are defined as *stitching lines*. Critical patterns cut by stitching lines could suffer from severe pattern distortion or even yield loss. Therefore, considering the positions of stitching lines and avoiding stitching line-induced bad patterns are required during layout design. In this paper, we propose the first work of stitch-aware routing framework for MEBL based on a two-pass bottom-up multilevel router. We first identify three types of stitching line-induced bad patterns which should not exist in an MEBL-friendly routing solution. Then, stitch-aware routing algorithms are respectively developed for global routing, layer/track assignment and detailed routing. Experimental results show that our stitch-aware routing framework can effectively reduce stitching line-induced bad patterns and thus may not only improve the manufacturability but also facilitate the development of MEBL.

Categories and Subject Descriptors

B.7.2 [**Integrated Circuits**]: Design Aids

General Terms

Algorithms, Design, Performance

Keywords

Multiple electron beam lithography, stitch, routing, manufacturability

1. INTRODUCTION

E-beam lithography (EBL) is one of the most expected Next Generation Lithography (NGL) technologies for overcoming the manufacturing limitations of conventional optical lithography. However, the relatively low throughput due to the maskless direct write process constrains EBL from high volume manufacturing. Thus, EBL was only applied to few applications such as photomask fabrication [15]. In recent years, the concept of multiple e-beam lithography (MEBL) has been proposed, which utilizes massively parallel exposure with thousands or even millions of beams to dramatically improve the throughput. Also, several innovative MEBL systems have been under development and have

---

*This work was partially supported by IBM, SpringSoft, TSMC, Academia Sinica, and NSC of Taiwan under Grant No's. 101-2221-E-002-191-MY3, NSC100-2221-E-002-088-MY3, NSC 99-2221-E-002-207-MY3, and NSC 99-2221-E-002-210-MY3.

Permission to make digital or hard copies of all or part of this work for personal or classroom use is granted without fee provided that copies are not made or distributed for profit or commercial advantage and that copies bear this notice and the full citation on the first page. To copy otherwise, to republish, to post on servers or to redistribute to lists, requires prior specific permission and/or a fee.
DAC'13, May 29 - June 07, 2013, Austin, TX, USA.

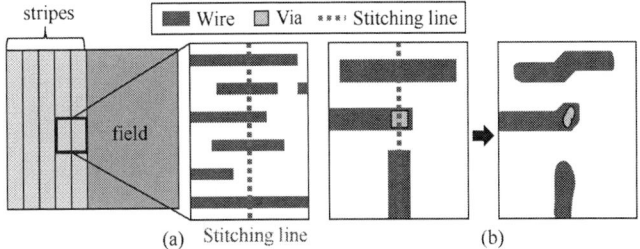

Figure 1: Layout division and overlay error in MEBL. (a) A layout is split into stripes and the stripe boundaries are defined as the stitching lines. (b) Features cut by stitching lines suffer from different degrees of pattern distortion.

shown very promising lithography performance and cost effectiveness [11, 13, 14, 17].

Due to the deflection limitation of each beam and parallel writing strategies in MEBL, a layout (a main field) is split into stripes (subfields) as shown in Figure 1(a), and we define the stripe boundaries as the *stitching lines*. Since patterns in different stripes are written by different beams or in different writing passes, a pattern cut by a stitching line suffers from overlay error between two beams or two writing passes [7, 16]. Note that the overlay error could cause different impacts on different types of patterns cut by stitching lines. As illustrated in Figure 1(b), a horizontal wire can be patterned well even if an overlay error exists. On the other hand, some patterns with critical dimension, such as vias or vertical wires, can have severe pattern distortion and electrical variation due to the overlay error. Therefore, it is desirable to consider stitching lines for MEBL-friendly layout designs to enhance manufacturability. However, to the best of our knowledge, no previous work has addressed the stitching line-induced printability problems during physical design for MEBL.

In current semiconductor manufactruing, metal layers become one of the most critical parts with respect to reliability, manufacturability, and circuit performance, and thus routing plays a crucial role in the VLSI design flow. In MEBL, routing without considering stitching lines may cause stitching line-induced bad patterns. As shown in Figure 2(a), without stitching line consideration, a via is cut by the stitching line on the wire A, and a part of the wire B is vertically routed on the stitching line. Another undesired pattern occurs on the wire C, which is a short wire segment cut by the stitching line with a landing via. We define this type of patterns as *short polygons*. Short polygons may also cause severe manufacturing defects, which will be explained in Section 2. Figure 2(b) shows a better routing result, where no stitching line-induced bad pattern is produced. Avoiding vias cut by stitching lines and avoiding wires vertically routed on stitching lines are not difficult. For example, removing routing tracks vertically overlapped with stitching lines can prevent wires from vertically routing on stitching lines. However, avoiding the generation of short polygons is not trivial. In fact, considering this type of bad patterns could significantly increase the design complexity.

In this paper, we propose the first work of stitch-aware routing for MEBL, which minimizes the number of stitching line-induced bad patterns during routing. The framework is based on a two-

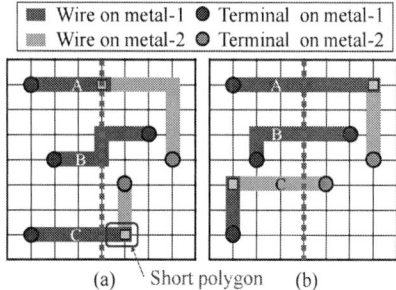

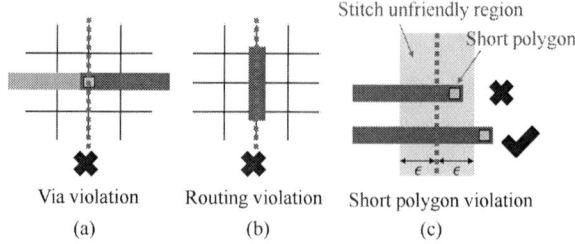

Figure 2: Routing with and w/o stitching consideration. (a) Stitching line-induced bad patterns are generated without considering stitching lines during routing. (b) A better routing solution derived from a stitch-aware router.

Figure 4: Three routing constraints for stitch-aware routing. (a) Via constraint. (b) Vertically routing constraint. (c) Short polygon constraint.

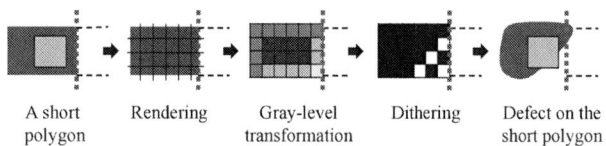

Figure 3: The rasterization process of a short polygon. A severe defect occurs due to dithering with error diffusion.

pass bottom-up multilevel router (similar to [3] for double via optimization). We first identify three types of stitching line-induced bad patterns and establish three corresponding stitch-aware routing constraints. Then, the stitch-aware routing algorithms are respectively proposed in each routing stage: global routing, layer/track assignment and detailed routing. Experimental results show that our algorithms can effectively avoid the generation of stitching line-induced bad patterns for MEBL.

The rest of this paper is organized as follows: Section 2 introduces the three routing constraints. In Section 3, the stitch-aware routing algorithms in global routing, layer/track assignment, and detailed routing are respectively presented. Section 4 reports our experimental results. Finally, we conclude our work in Section 5.

2. STITCH-AWARE ROUTING CONSTRAINTS

As mentioned in Section 1, patterns cut by stitching lines suffer from overlay errors between two different beams or two different writing passes. Although these pattern segmentations are inevitable during circuit design, stitching lines should avoid cutting critical patterns to reduce severe pattern distortion or even yield loss. For example, as mentioned in Section 1, vias should not be cut by stitching lines and wires should not vertically route on stitching lines.

Another type of stitching line-induced bad patterns, *short polygons*, is due to the data preparation flow in MEBL. Because of the maskless lithography process, *rasterization* is required to transform a layout into a pixel-based black/white bitmap, and thus patterns can be exposed on a wafer by controlling each independent beam to be "on" or "off" [7, 10]. Rasterization consists of two major steps: (1) *rendering* followed by (2) *dithering* with error diffusion. In rendering, a layout is sliced into grids, and patterns are converted into pixel-based gray-level data with intensity proportional to the pattern coverage in each pixel. Then, in dithering, the resulting gray-level bitmap is transformed into a black/white bitmap. The error of each pixel due to dithering is not neglected but diffused to its neighboring unprocessed pixels.

A short polygon may cause a severe defect after the rasterization process. Figure 3 shows an example. The short polygon cut by the stitching line undergoes rendering and dithering during the data preparation flow. Due to the error diffusion process,

the short polygon has irregular pixels on the bottom-right corner. These problematic pixels could account for a large percentage of the pixels associated with the short polygon and thus can result in serious pattern distortion after e-beam exposure. Then, the misalignment between the polygon and the via becomes a circuit defect or causes unacceptable electrical variation. Therefore, short polygons with landing vias should be avoided in a routing solution for better MEBL control.

Hence, given a set of stitching lines, we define the following three routing constraints:

- Via constraint: vias cannot be cut by stitching lines (see Figure 4(a)).

- Vertical routing constraint: wires cannot vertically route on stitching lines (see Figure 4(b)).

- Short polygon constraint: vias should not land on short polygons. As illustrated in Figure 4(c), we define the area within the distance ϵ from a stitching line as the *stitch unfriendly region* of the stitching line. A horizontal wire has a short polygon violation if it satisfies the following two conditions: (1) The wire is cut by a stitching line. (2) At least a line end of the wire lies in the corresponding stitch unfriendly region with a landing via. Therefore, in Figure 4(c), the upper wire has a short polygon violation, and the lower wire is a preferred routing instance without any violation.

Note that in our routing framework, the via constraint and the vertical routing constraint are hard constraints that a routing solution must always satisfy, and the short polygon constraint is a soft constraint that our routing framework should optimize.

3. STITCH-AWARE ROUTING FRAMEWORK

In this section, stitch-aware global routing, layer/track assignment, and detailed routing are presented in the following subsections.

3.1 Stitch-Aware Global Routing

In the global routing stage, a routing plane is first divided into global tiles and transformed into a routing graph, in which a vertex represents a global tile and each pair of adjacent global tiles is connected by an edge, as shown in Figure 5(a). Then, nets sequentially find their global routing paths on the graph with minimized routing costs. The routing cost of a routing path is usually computed according to the routing congestion on the path.

Resource estimation in global routing for MEBL is quite different from conventional routing problems due to the existence of stitching lines. For example, in Figure 5(b), the capacity of each boundary (the maximum number of wires that can pass through the boundary) of the global tile is originally six without considering stitching lines. However, the capacities of the top boundary and the bottom boundary are reduced by one since no wire can route on the track occupied by the stitching line due to the vertical routing constraint. Furthermore, it is undesirable that many line ends of vertical segments lie in the same tile. As shown in Figure 5(b), only two vertical tracks are not in stitch unfriendly regions. If there are three vertical segments whose line ends lie in the tile, at least one line end will lie in the stitch unfriendly region, and the line end may cause a short polygon violation on the connected horizontal wire, as the segment C in Figure 5(b).

To consider both of the situations, in a global routing graph, each edge is assigned an edge capacity indicating the maximum number of wires that can pass through the tile boundary without overflow, and each vertex is also assigned a vertex capacity denoting the number of tracks not in stitch unfriendly regions. Then, the cost of an edge e_i ($\psi_e(i)$) and the cost of a vertex v_j ($\psi_v(j)$) are respectively defined as follows:

$$\psi_e(i) = 2^{d_e(i)/c_e(i)} - 1, \quad (1)$$

$$\psi_v(j) = 2^{d_v(j)/c_v(j)} - 1, \quad (2)$$

where $c_e(i)$ is the capacity of e_i, $c_v(j)$ is the capacity of v_j, $d_e(i)$ is the demand of e_i, which is the number of segments that have routed on e_i, and $d_v(j)$ is the demand of v_j, which is the number of line ends that have lain on v_j. Thus, the routing cost of a global routing path is the summation of the vertex costs and the edge costs in the path.

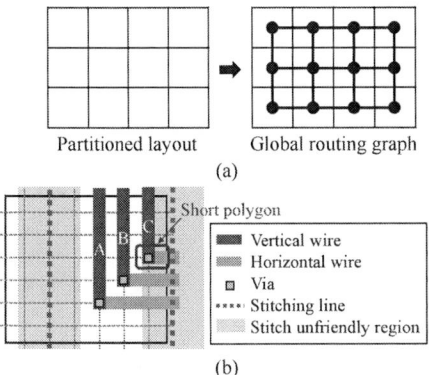

Partitioned layout → Global routing graph

(a)

(b)

Figure 5: Global routing model and routing resource estimation for MEBL. (a) A layout is divided into global tiles and transformed into a graph model. (b) The routing resource is reduced due to the stitching lines.

3.2 Stitch-Aware Layer Assignment

Layer/track assignment has been proven as an effective intermediate stage between global routing and detailed routing [1, 6] for improving the routing quality of high complexity designs. In addition, many manufacturability issues can be optimized during layer/track assignment, such as crosstalk, antenna effect, and wire density uniformity [4, 9, 12, 18]. In our work, stitch-aware layer and track assignment algorithms are also proposed for optimizing stitching line-induced bad patterns.

In layer assignment, we assign the vertical (horizontal) segments in a column (row) panel to different vertical (horizontal) routing layers. A column (row) panel is defined as a column (row) of global tiles in a global routing graph. The conventional objective in layer assignment is to uniformly distribute segments in a panel. However, as mentioned in Section 3.1, line ends of segments should be also scattered to different layers to avoid generating short polygons.

To solve the stitch-aware layer assignment problem, we first construct a segment conflict graph for each panel, in which a vertex v_i represents a segment i and an edge connecting two vertices if the two segments intersect in some tiles. For a column panel, we set an edge weight $w(v_i, v_j)$ for each edge (v_i, v_j) as follows:

$$w(v_i, v_j) = D_{segment}(v_i, v_j) + D_{end}(v_i, v_j), \quad (3)$$

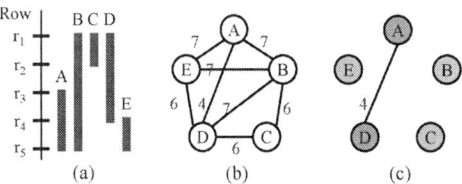

Figure 6: Layer assignment considering segment and line-end uniformities. (a) A set of segments in a vertical panel. (b) The corresponding segment conflict graph. (c) A layer assignment solution by solving the maximum-cut k-coloring problem.

where $D_{segment}(v_i, v_j)$ is the maximum segment density in the rows where the segment i and the segment j are overlapped, and $D_{end}(v_i, v_j)$ is the maximum line-end density in the rows where the line ends of i and j are overlapped. (Note that we simply remove the second item in Equation (3) for row panels since we consider line-end densities only in column panels.) Figure 6(b) shows a conflict graph for the segments in a column panel shown in Figure 6(a). To uniformly distribute segments and line ends to k layers, the layer assignment problem can be solved by finding a maximum-cut k-coloring solution [5] of the segment conflict graph, which is equivalent to finding a k-coloring solution with the minimum total edge weight [5]. Figure 6(c) shows a three-coloring solution with the minimum total edge weight of the segment conflict graph.

Since the maximum-cut k-coloring problem is NP-complete [5], previous work has proposed a heuristic approach that first constructs a maximum spanning tree on a conflict graph and then solves the k-coloring problem on the tree. Note that a tree is always k-colorable when $k \geq 2$. This heuristic can solve the maximum-cut k-coloring problem well as k equals two; however, as k is greater than two, solving the maximum-cut k-coloring problem with the maximum spanning tree approach may degrade the solution quality since more edges can be simultaneously considered as more colors are available. As illustrated in Figures 7(a) and (b), if three vertical layers are available, after constructing a maximum spanning tree and three-coloring the tree according to the tree level of each vertex, a layer assignment solution is generated with total edge weight equal to 13.

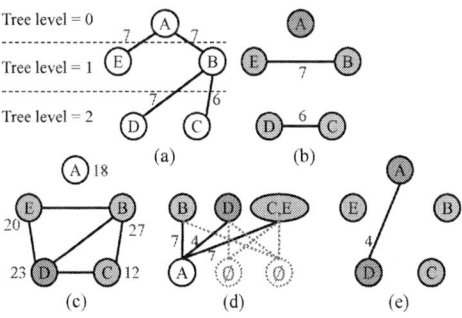

Figure 7: Heuristics for solving the maximum-cut k-coloring problem. (a)(b) The maximum spanning tree approach. (c)(d)(e) Our algorithm that can generate a better solution.

In this work, we propose another heuristic algorithm to get better solutions. We first compute the vertex weight for each vertex by summing the weights of the incident edges. Then, we find a set of k-colorable vertices with the maximum total vertex weight. Although this problem is NP-complete in general graphs, it can be solved in polynomial time for segment conflict graphs, which are interval graphs, by using a minimum cost flow algorithm [2]. As shown in Figure 7(c), $V_1 = \{v_B, v_C, v_D, v_E\}$ is a

177

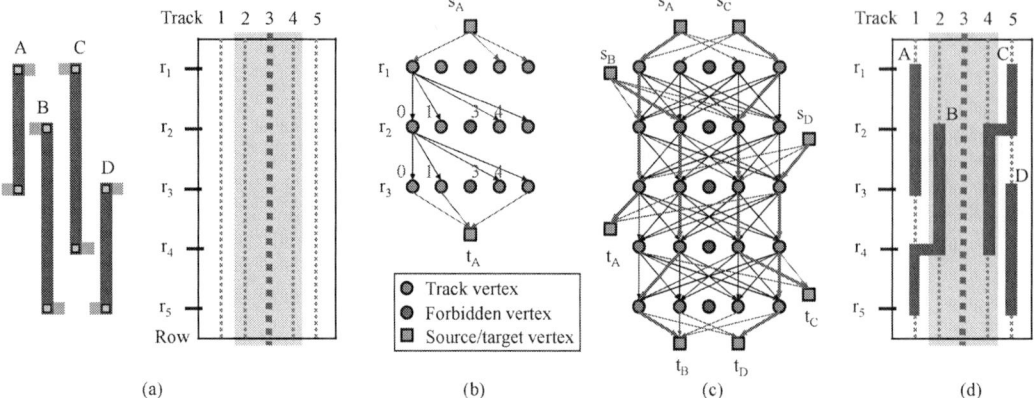

Figure 8: The ILP-based track assignment approach. (a) A track assignment instance. (b) The multi-commodity flow model of the segment A. (c) The multi-commodity flow model of all segments and the solution derived from the ILP formulation. (d) The corresponding track assignment solution.

three-colorable vertex set in the segment conflict graph with the maximum total vertex weight, and $\{v_B\}$, $\{v_D\}$ and $\{v_C, v_E\}$ are the three-coloring groups of V_1. The algorithm then finds the next k-colorable vertex set with the maximum total vertex weight on the remaining graph. To merge the coloring groups of the two vertex sets, a perfect bipartite matching algorithm is applied to minimize the total conflict edge weight. As illustrated in Figure 7(d), two pseudo coloring groups $\emptyset$ are first created since only the vertex v_A remains, and thus the three-coloring groups of the second vertex set V_2 are $\{v_A\}$, $\emptyset$ and $\emptyset$. To combine the coloring groups of V_1 and V_2, a complete bipartite graph is constructed and the edge weights are set as the total conflict edge weight between two groups. By solving the minimum weight perfect bipartite matching problem, coloring groups are merged with the minimum conflict edge weight. The above process is performed iteratively until no vertex is left. Figure 7(e) shows the layer assignment result with smaller total edge weight equal to 4.

3.3 Short Polygon-Avoid Track Assignment

In track assignment, segments of the same layer in a panel are assigned exact track numbers, which is a crucial stage for short polygon avoidance. A desired track assignment solution which can avoid short polygon generation is a track assignment without *bad ends*. A bad end is a line end of a vertical wire segment lying in the stitch unfriendly region of a stitching line, and the connected horizontal wire is cut by the stitching line. For example, the lower end of the wire segment C in Figure 5(b) is a bad end. To derive a track assignment solution without bad ends, an ILP-based algorithm and a graph-based algorithm are proposed, which are detailed in the following subsections. Note that the short polygon-avoiding track assignment algorithms are only applied to column panels. Segments in row panels can be assigned by using conventional track assignment algorithms.

3.3.1 ILP-Based Approach

First, the short polygon-avoiding track assignment problem can be intuitively transformed into a multi-commodity flow model, which is a directed graph $G = (V, E)$. Figure 8(a) shows a track assignment instance. To find an exact track number for the segment A, for example, the multi-commodity flow graph model is constructed as shown in Figure 8(b), where a track vertex represents a track in a global tile, a forbidden vertex is a track occupied by a stitching line, and the source vertex s_A and the target vertex t_A are the top end and the bottom end of the segment A. The source edges connect s_A to the track vertices of the tile where the top end of A lies. Similarly, the target edges connect the track vertices of the tile where the bottom end of A lies to t_A. A

source/target edge is removed if the line end becomes a bad end on the corresponding track. For example, s_A causes a bad end if it starts on the second track, and thus the second source edge is removed from the graph. Also, track vertices of adjacent tiles are connected with track edges, and the edge weight of a track edge is set to be the difference of the track numbers of the two track vertices to minimize wirelength and the number of routing bends. The whole multi-commodity flow graph model of the four segments is shown in Figure 8(c). Then, we find a track assignment solution with an ILP formulation. The notation used in our ILP formulation is listed as follows:

- K: a set of segments in a track assignment problem.

- $s(k)$: the source vertex of the segment k.

- $t(k)$: the target vertex of the segment k.

- V_{track} a set of track vertices.

- $w(u, v)$: the weight of the directed edge (u, v).

- $f_k(u, v)$: 0-1 integer variable that denotes if the segment k is routed through the directed edge (u, v).

- $\mathcal{C}$: a set of crossed edge pairs.

Based on the notations, the short polygon-avoiding layer assignment problem can be formulated as follows:

$$minimize \quad \sum_{(u,v)\in E}\left(w(u,v)\times\sum_{k\in K}f_k(u,v)\right)$$

$$subject\ to \quad \sum_{(s(k),v)\in E}f_k(s(k),v)=1,\forall k\in K, \quad (4)$$

$$\sum_{(u,t(k))\in E}f_k(u,t(k))=1,\forall k\in K, \quad (5)$$

$$\sum_{u\in V}f_k(u,v)=\sum_{w\in V}f_k(v,w),\forall k\in K,\forall v\in V_{track}, \quad (6)$$

$$\sum_{u\in V}\sum_{k\in K}f_k(u,v)\le 1,\forall v\in V_{track}, \quad (7)$$

$$\sum_{k\in K}f_k(u_1,v_1)+\sum_{k\in K}f_k(u_2,v_2)\le 1,\forall((u_1,v_1),(u_2,v_2))\in\mathcal{C}. \quad (8)$$

The objective of the ILP formulation is to minimize the total edge weight of a flow solution such that the wirelength and

178

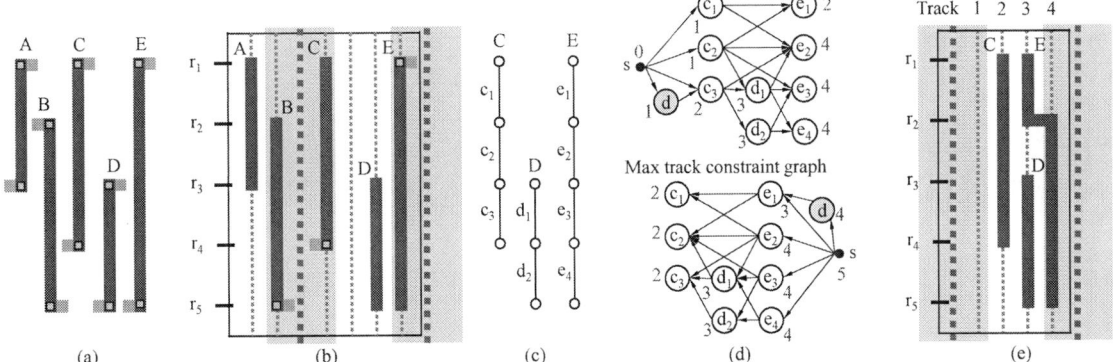

Figure 9: The graph-based track assignment approach. (a) A track assignment instance. (b) A segment order is first determined. (c) The segments C, D and E between two stitching lines are simultaneously considered and are divided into intervals. (d) The feasible track assignment solution space of each interval is computed by using the minimum and maximum track assignment constraint graphs (c) The final track assignment solution of C, D and E.

the number of wire bends can be minimized. Constraint (4) and Constraint (5) ensure that each segment can find a unique path from its source vertex to the target vertex. Constraint (6) is used to guarantee that the number of paths flowing into a node equals that draining from the node. Constraint (7) guarantees that a track in a tile is occupied by at most one segment. Finally, Constraint (8) prevents segments from crossing with each other.

THEOREM 1. *In the above ILP formulation, the number of variables is $O(T^2 R)$ and the number of constraints is $O(TR|K| + T^4 R)$, where T is the number of tracks in a column panel and R is the number of rows in the global routing graph*

Note that the complexity is dominated by Constraints (6) and (8).

Using "doglegs" to avoid short polygon generation is one of the advantage of the ILP-based approach. However, since the short polygon-avoiding track assignment process is performed for every panel in all vertical layers, the runtime of iteratively solving the ILP formulation may be prohibitively long as the chip size increases. Therefore, we propose another graph-based track assignment heuristic, which can efficiently utilize doglegs for short polygon avoidance.

3.3.2 Graph-Based Approach

The graph-based short polygon-avoiding track assignment algorithm first determines the segment order in a panel, and then tries to resolve bad ends with doglegs by using a graph-based algorithm.

The approach starts from assigning longer segments next to stitching lines. As shown in Figures 9(a) and (b), the segments B, C and E are placed adjacent to stitching lines. Longer segments have larger flexibility to avoid short polygon generation by applying doglegs. Then, some bad ends of those longer segments will be generated if the segments do not change their track numbers. For example, the bottom end of B, the bottom end of C, and the top end of E are currently bad ends. After that, segments not overlapped with the bad ends are assigned next to those longer segments such that the bad ends can be easily resolved with doglegs. Therefore, as illustrated in Figure 9(b), the segment A is assigned next to B and the segment D is assigned next to E. For the remaining segments having less impact on bad ends, the track numbers are arbitrarily assigned.

After determining the segment order, doglegs are used to resolve bad ends. A set of segments between two stitching lines are considered at a time. Each segment is first divided into intervals according to global tiles, as the segments C, D, and E shown

in Figure 9(c). Then, two constraint graphs are constructed to record the geometry relationship among these intervals. As illustrated in Figure 9(d), the first one is the *minimum track constraint graph*, where a vertex represents an interval and a directed edge (v_i, v_j) indicates that the two intervals are overlapped in the x-direction and the interval i is left to the interval j. A dummy vertex d is created and connected to a vertex v_i if the interval i should not be assigned to the leftmost track. For example, the interval c_3 has a bad end if it is assigned to the leftmost track between the two stitching lines, and thus a dummy vertex is created and connected to v_{c_3} through the edge (d, v_{c_3}) in the minimum track constraint graph. After creating a source vertex s and connecting s to the vertices of the leftmost intervals and the dummy vertices, a longest path algorithm is applied to compute the minimum track number m of each interval, which indicates the leftmost feasible track number. For the *maximum track constraint graph*, the construction is almost the same except that an edge (v_i, v_j) connects the vertex of the interval i right to the vertex of the interval j and a dummy vertex connected to a vertex v_i if the interval i should not be assigned to the rightmost track. A similar algorithm is applied to compute the maximum track number M of each interval. As shown in Figure 9(d), the two numbers of a vertex in the minimum and maximum track constraint graphs give a feasible solution space $[m, M]$ of track assignment for each corresponding interval.

Finally, we sequentially determine the track numbers from the leftmost segment to the rightmost segment between the two stitching lines according to their feasible solution spaces. The wirelength and the number of bends of each segment are greedily optimized during track assignment. For example, all intervals of the segment C are assigned to the second track for wirelength and bend optimizations, and the final assignment solution is shown in Figure 9(e).

3.4 Stitch-Aware Detailed Routing

The final stage of our routing framework is stitch-aware detailed routing, which finds pin-to-segment and segment-to-segment detailed routes with a conventional A*-search routing algorithm [8]. To satisfy the via and vertical routing constraints, wires passing through stitching lines can only route in the x-direction (perpendicular to stitching lines). For minimizing the number of generated short polygons, we give a larger routing cost if a wire in a stitch unfriendly region routing in the z-direction, and thus a detailed path with the minimum number of line ends lying in stitch unfriendly regions will be found.

Table 1: Experimental results that compare the effectiveness and efficiency among different track assignment algorithms.

Circuit	w/o stitch consideration				ILP-based Approach				Graph-based Approach			
	Rout. (%)	#VV	#SP	CPU (s)	Rout. (%)	#VV	#SP	CPU (s)	Rout. (%)	#VV	#SP	CPU (s)
Struct	99.90	584	3282	2	100.00	597	23	505	100.00	597	15	2
Primary1	99.80	212	1431	1	100.00	215	1	8909	100.00	226	0	1
Primary2	98.63	643	5268	6	100.00	667	161	74780	100.00	688	76	6
S5378	96.32	660	584	1	99.10	643	37	3904	99.23	645	9	1
S9234	98.67	312	452	1	99.90	297	30	2983	99.90	294	20	1
S13207	97.43	45	1212	2	99.87	58	92	12767	100.00	58	62	2
S15850	97.14	43	1524	2	99.75	63	124	14912	99.76	62	90	2
S38417	97.89	24	3499	2	NA	NA	NA	> 100000	99.73	41	182	5
S38584	97.71	2163	4020	10	NA	NA	NA	> 100000	99.32	2037	153	11
Dma	96.00	1748	5242	5	99.40	1764	117	3084	99.49	1792	46	6
Dsp1	97.08	2326	5613	4	99.82	2434	59	4743	99.81	2364	17	7
Dsp2	96.82	2367	5770	4	99.74	2389	77	4133	99.75	2402	16	5
Risc1	96.09	3149	8947	11	99.56	3188	161	9883	99.62	3200	44	11
Risc2	96.64	3199	8801	8	99.67	3233	111	8881	99.66	3248	42	9
Comp.	1.000	1.000	1.000	1.0	1.023	1.070	0.034	4159.5	1.022	1.110	0.022	1.1

4. EXPERIMENTAL RESULTS

Our algorithm was implemented in the C++ programming language on a 2.93 GHz Linux workstation with 48 GB memory. The minimum cost flow problem and the minimum weight bipartite matching problem in layer assignment were solved by adopting the LEDA package [20], and the ILP formulation in track assignment was solved by using the CPLEX12.3 library [19]. In our routing framework, the number of stitching lines was set to be that of global tiles in a row, and the stitching lines were uniformly distributed in a layout. In addition, the tracks adjacent to stitching lines fell into stitch unfriendly regions.

We show the effectiveness of avoiding short polygon generation by applying three different track assignment approaches: (1) track assignment without considering stitching lines, (2) track assignment by solving the ILP formulation, and (3) track assignment by applying the graph-based algorithm. Note that all the three approaches use the same stitch-aware algorithms in other routing stages. Two suits of benchmarks were used, the MCNC benchmarks and the Faraday benchmarks. The experimental results are shown in Table 1, where "Rout." gives the routability, "#VV" reports the number of via violations, "#SP" shows the number of short polygon violations, and "CPU" lists the runtime in second. Due to the fixed pin positions of nets, the three approaches have similar numbers of via violations. On the other hand, the results show that by considering stitching lines, both the ILP-based approach and the graph-based approach slightly improve the routability. It is because not considering stitching lines can cause lots of vertical routing violations. Since the routing constraint is a hard constraint a routing solution must satisfy, the violated wires are ripped-up and will be rerouted in the detailed routing stage, which may cause lots of failed nets. Also, both the approaches can effectively reduce the number of short polygon violations by more than 96%. However, the ILP-based approach is too time-consuming to generate a routing solution in a reasonable runtime, and thus the graph-based approach is more appropriate for large scale routing instances. Note that although the ILP formulation can find a track assignment solution without any short polygon if such a solution exists, the ILP-based approach generates more short polygon violations than the graph-based approach. It is because once the ILP formulation fails to find a solution for a track instance, we directly route the segments in the detailed routing stage, which may cause more short polygons. In contrast, if the graph-based approach fails to find a legal solution, we can simply remove the rightmost segment and keep other segments connected.

5. CONCLUSIONS

In this paper, we have proposed the first work of stitch-aware routing framework for MEBL. We first identify three types of stitching line-induced bad patterns which could cause severe pattern distortion, electrical variation, or even yield loss. Then, we provide solutions to avoid generating these bad patterns during each routing stage. Experimental results show that our algorithms can efficiently and effectively reduce the number of short polygons. To remove the via violations due to the fixed pin positions of nets, it is also desirable to develop stitch-aware algorithms at the placement stage, which is our future work to further improve the manufacturability and facilitate the development of MEBL.

6. REFERENCES

[1] Batterywala et al., "Track assignment: a desirable intermediate step betweeen global routing and detailed routing," Proc. ICCAD, pp. 59–66, 2002.

[2] Carlisle and Lioyd, "On the k-coloring of intervals," DAM, vol. 59, no. 3, pp. 225–235, 1995.

[3] Chen et al., "Full-chip routing considering double-via insertion," IEEE TCAD, vol. 27, no. 5, pp. 844–857, 2008.

[4] Chen et al., "A novel wire-density-driven full-chip routing system for CMP variation control," IEEE TCAD, vol. 28, no. 2, pp. 193–206, 2009.

[5] Cho et al., "Fast approximation algorithms on maxcut, k-coloring, and k-color ordering for VLSI applications." IEEE TC, vol. 47, no. 11, pp. 1253–1266, 1998.

[6] Cong et al., "DUNE–a mulitilayer gridless routing system," IEEE TCAD, vol. 20, no. 5, pp. 633–647, 2011.

[7] Hakkennes et al., "Demonstration of real time pattern correction for high throughput maskless lithography," Proc. SPIE, vol. 7970, pp. 79701A, 2011.

[8] Hart et al., "A formal basis for the heuristic determination of minimum cost paths," IEEE SSC, vol. 4, no. 2, pp. 100-107, 1968.

[9] Ho et al., "Crosstalk- and performance-driven multilevel full-chip routing," IEEE TCAD, vol. 24, no. 6, pp. 869–878, 2005.

[10] Hung et al., "Bottlenecks in data preparation flow for multi-beam direct write," Proc. SPIE, vol. 8166, pp. 81662C, 2011.

[11] Klein et al., "PML2: the maskless multibeam solution for the 22nm node and beyond," Proc. SPIE, vol. 7271, pp. 72710N, 2009.

[12] Lee and Wang, "Simultaneous antenna avoidance and via optimization in layer assignment of multi-layer global routing," Proc. ICCAD, pp. 312–318, 2010.

[13] Lin, "Future of multiple-e-beam direct-write systems," Proc. SPIE, vol. 8323, pp. 832302, 2012.

[14] McChord et al., "REBL: design progress toward 16 nm half-pitch maskless projection electron beam lithography," Proc. SPIE, vol. 8323, pp. 832311

[15] Rizvi, "Handbook of photomask manufacturing technology," Taylor & Francis, 2005.

[16] Ronse, "E-beam maskless lithography: prosoects and challenges," Proc. SPIE, vol. 7637, 2010.

[17] Wieland et al., "MAPPER: high throughput maskless lithography," Proc. SPIE, vol. 7637, pp. 76370F, 2010.

[18] Wu et al., "Antenna avoidance in layer assignemnt," IEEE TCAD, vol. 25, no. 4, pp. 734–738, 2006.

[19] IBM ILOG CPLEX Optimizer. http://www-01.ibm.com/software/integration/optimization/cplex-optimizer/

[20] The LEDA package. http://www.algorithmic-solutions.com/leda

Automatic Design Rule Correction in Presence of Multiple Grids and Track Patterns

Nitin Salodkar
Synopsys India Pvt. Ltd.,
RMZ Infinity, Benniganahallli,
Bangalore, India - 560016
salodkar@synopsys.com

Subramanian Rajagopalan
Synopsys India Pvt. Ltd.,
RMZ Infinity, Benniganahallli,
Bangalore, India - 560016
rsubbu@synopsys.com

Sambuddha Bhattacharya
Synopsys India Pvt. Ltd.,
RMZ Infinity, Benniganahallli,
Bangalore, India - 560016
sbb@synopsys.com

Shabbir Batterywala
Synopsys India Pvt. Ltd.,
RMZ Infinity, Benniganahallli,
Bangalore, India - 560016
battery@synopsys.com

ABSTRACT

Traditionally, automatic design rule correction (DRC) problem is modeled as a Linear Program (LP) with design rules as difference constraints under minimum perturbation objective. This yields Totally Uni-Modular (TUM) constraint matrices thereby guaranteeing integral grid-compliant solutions with LP solvers. However, advanced technology nodes introduce per-layer grids or discrete tracks that result into non-TUM constraint matrices for the DRC problem. Consequently, LP solvers do not guarantee integral solutions. In this work, we propose a novel formulation using an 'unrolling' technique. Our formulation guarantees TUM constraint matrices and hence integral multiple grid/track compliant solutions. We demonstrate its efficacy on layouts at advanced nodes.

Keywords

Layout automation, difference constraints, linear program, integer linear program, total uni-modularity

1. INTRODUCTION

Advanced processes employ complex lithography [1]. While this enables continued scaling, this has also led to an explosion in the number of design rules. Design rules today comprise complex tables, forbidden regions and depend on the context of surrounding geometry. Increased number of rules impose a challenge in designing violation free layouts. This is especially important in the context of manual layout design for custom blocks, digital standard cells and analog subsystems. The complexity of such layout designs arises

Permission to make digital or hard copies of all or part of this work for personal or classroom use is granted without fee provided that copies are not made or distributed for profit or commercial advantage and that copies bear this notice and the full citation on the first page. To copy otherwise, to republish, to post on servers or to redistribute to lists, requires prior specific permission and/or a fee.
DAC '13 May 29 - June 07 2013, Austin, TX, USA

from simultaneous consideration of multiple context dependent rules, and the tedium of checking and modifying layouts iteratively. Automatic design rule correction engines are an important aid in generating violation free layouts.

Automatic design rule correction (DRC) problem, hereafter called layout legalization, can be modeled as modified compaction problem [2]. The idea is to perform line sweeps on layout geometry to capture applicable design rules, and impose these as constraints to remove violations. These constraints are generally linear(-ized) and can be modeled through either constraint graphs or linear programs (LP) [3]. The work in [4] proposed a technique for design rule correction while minimizing perturbation of layout edges. Minimum perturbation is particularly effective in the presence of contextual design rules that are often approximated with a collection of simpler rules since it avoids inadvertent introduction of context dependent violations.

Advanced process nodes bring new challenges to layout legalization. These processes employ restrictive design rules (RDR) where all shapes in some layers (especially polysilicon) have the same orientation, the same width, and the same pitch [5]. The work in [6] explores how these RDRs are used to improve manufacturability. Unlike a single manufacturing grid in older processes, these RDRs impose layer specific grids. Thus, layouts now need to obey multiple grid constraints. Furthermore, as the work in [1] pointed out, at 22nm these grids need not be uniform for a layer. Rather, different grid pitches could be defined for the same layer at different locations in the design. To distinguish these from uniform grid for a layer, we hereafter refer them as *tracks*. For such layers, centerlines of layout shapes need to be incident on grids/tracks. Similarly, via centers need to be at the intersection of the grids/tracks of their corresponding bottom and top layers.

Figure 1 illustrates uniform and non-uniform grids and the desired legal layouts. These per layer grids and tracks pose additional challenges to layout automation. Specifically, in the context of layout legalization, constraints pertaining to multiple grids usually necessitate usage of expensive integer linear program (ILP) solvers. Non-uniform grids or tracks complicate it further. The work in [7] and [8] discuss tech-

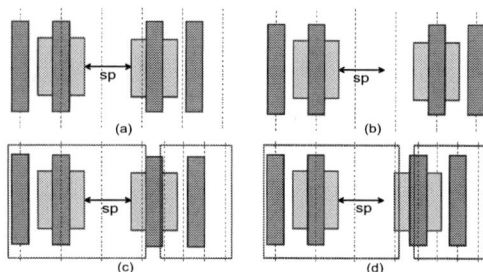

Figure 1: (a) Layout with violations for uniform grid (b) Layout legalized to uniform grid; (c) Layout with violations for region-based non-uniform grids (d) Legalized layout - shapes jump across to different grids.

niques to address grid compliance in the context of layout migration. While both layout migration and layout legalization are closely related, the proposed techniques have some shortcomings. Firstly, both [7] and [8] focus only on uniform per layer grids. Non-uniform grids or tracks pose additional challenges which are not addressed. The work in [7] uses two step solution framework where positions for grid variables are determined in the first step. The techniques in [7], [8] use forward and reverse runs of longest path to estimate a valid grid compliant position. As shown later in this work, this may miss minimum perturbation solution points.

In this work, we solve the multi-grid/track based layout legalization problem with minimum perturbation using regular LP solvers. An LP solver is guaranteed to yield integral solution for totally uni-modular (TUM) constraint matrices [3]. We exploit the constraint structure and incorporate a variable *unrolling* technique to obtain a TUM constraint matrix. This formulation ensures multi-grid/track compliant integral solution with regular LP solvers. Specifically, the contributions of this work are: (a) the variable unrolling based formulation that guarantees multi-grid/track compliance with regular LP solvers, (b) a minimum perturbation solution ensured through a layout-aware Bellman-Ford algorithm.

The paper is organized as follows. Section 2 provides a brief primer on traditional layout legalization, highlights the importance of integrality of solution and identifies the challenges due to multiple grids and track patterns. Section 3 outlines the unrolling technique for converting the constraints into a set of implications; these can be expressed as difference constraints on 0-1 variables. Section 4 discusses the important aspects and practical considerations for the legalization framework. Section 5 lists experimental observations and Section 6 concludes the paper.

2. BACKGROUND AND MOTIVATION

In this section, we describe the traditional formulation of the DRC correction as a constrained optimization problem. We then discuss the new challenges due to advanced processes and point out the shortcomings of the traditional formulation in presence of these.

2.1 Traditional Formulation

An automatic DRC correction tool reads in an input layout and fractures its mask layer geometry into rectangles. Each rectangle edge is considered as a positional variable.

Each rectangle, thus, has four positional variables associated with it, two each in horizontal and vertical directions. All applicable design rules are modeled as as linear(-ized) constraints. Constraints are of the difference form $x_i - x_j \geq val$, where x_i and x_j are the positional variables corresponding to two rectangle edges, and val corresponds to the required distance between the variables to obey the particular rule. All layer and layer-pair constraints, constraints with upper bounds, constraints for imposing alignments, etc. can be modeled as difference constraints [2]. These constraints are then solved with a minimum perturbation objective function to generate the legalized (DRC correct) layout. The problem is modeled as:

$$\text{minimize} \quad \sum_{i=1}^{N} |x_i - x_i^0| \quad \text{subject to :} \quad \mathbf{Ax} \geq \mathbf{b}, \quad (1)$$

where $\mathbf{x}$ is a column vector consisting of the variables corresponding to rectangle edges, $\mathbf{x}^0$ are the rectangle edge positions in input layout. The matrix $\mathbf{A}$ is the coefficient matrix for the constraints, where each row $\mathbf{A}_i$ represents a single constraint. The vector $\mathbf{b}$ represents the rule values. The term $|x_i - x_i^0|$ models minimum perturbation on the input layout. The objective function minimizes total perturbation across all layout edges. Heng et al [4] introduced a technique to linearize the minimum perturbation objective and solve the resulting problem as a standard LP.

Ensuring Integrality: Note that if the constraints are only of difference type, then the matrix has a TUM structure and LP yields integral solutions [3]. However, the layout also needs to obey the manufacturing grid. If the grid is g then replacing variables $x_i = g.y_i$ and solving the scaled down problem with LP still yields integral solutions for $\mathbf{y}$. Therefore the solution $\mathbf{x}$ is manufacturing grid compliant.

2.2 Challenges in Advanced Processes

As mentioned earlier, at advanced technology nodes, each mask layer can have its own grid which can even be non-uniform. This breaks the mathematical paradigm of traditional formulation for layout legalization.

Multiple Grids: Let the variable x_i correspond to a rectangle in layer l_i for which the applicable uniform grid is g_i. For grid compliant solution we should have $x_i = g_i.y_i$ where y_i needs to take integral value. Substituting this into the difference constraints we get $g_i.y_i - g_j.y_j \geq b_{ij}$. The objective function becomes, $\sum_{i=1}^{N} |g_i.y_i - x_i^0|$. The corresponding constraints are of the form $\mathbf{Ay} \geq \mathbf{b}$, but the matrix $\mathbf{A}$ is no longer in TUM form. The resulting solution ($\mathbf{y}$) by LP solver is no longer guaranteed to be integral [3].

Non-uniform Grids or Tracks: In a more general scenario, multiple region-based grids can be defined for the same mask layer. In such cases, the positional variables x_i can take discrete, possibly non-uniformly separated values. This can not be modeled as above and warrants a more general formulation. This leads us to the next section which presents our variable unrolling based formulation.

3. FORMULATION WITH VARIABLE UN-ROLLING

In this section, we present a new formulation for the legalization problem with multiple non-uniform grids. In our formulation, we ensure that all constraints have at most two variables per inequality (TVPI). LP problems with this structure have been studied well in operation research. We incorporate a *variable unrolling* based solution technique of Hochbaum-Naor [9] for TVPI systems. To the best of our knowledge, application of this technique for automatic design rule correction has not been reported in EDA. Our formulation retains the TUM structure of the constraint matrix. Therefore LP solvers yield integral solutions compliant with multiple non-uniform grids. First, we discuss the concept of variable unrolling. Then we illustrate how constraints and the minimum perturbation objective are modeled in this framework.

3.1 Unrolled Variables

Let the variable x_i take values in the range $[L_i, \ldots, U_i]$. Let the set $\mathcal{X}_i = \{v_i^1, \ldots, v_i^{|\mathcal{X}_i|}\}$ with $v_i^1 = L_i$ and $v_i^{|\mathcal{X}_i|} = U_i$ represent the enumeration or *unrolling* of the discrete and possibly non-uniformly separated values for variable x_i. For now, we assume that computing the range and unrolling is possible. We describe an approach for achieving this in Section 4.1.

Associate with each value $v_i^p \in \mathcal{X}_i$, a binary variable b_i^p. Express variable x_i in terms of its associated binary variables:

$$x_i = v_i^1 + \sum_{p=1}^{|\mathcal{X}_i|} b_i^p \delta_i^p, \qquad (2)$$

where $\delta_i^p = v_i^p - v_i^{p-1}, \delta_i^1 = 0$, is difference between successive track locations. If the track locations are uniform then $\delta_i^p = \delta, \forall p$. If the variable x_i takes a value v_i^p then the boolean variables corresponding to $p = 1, \ldots, P$ all take value 1, and rest of the variables corresponding to $p = P+1, \ldots, |\mathcal{X}_i|$ take value 0. Note that $b_i^p = 1 \implies b_i^{p-1} = 1$, i.e.,

$$b_i^{p-1} - b_i^p \geq 0. \qquad (3)$$

3.2 Modeling Constraints

Consider a typical constraint:

$$x_i - x_j \geq b_{ij}, \ x_i \in \mathcal{X}_i, \ x_j \in \mathcal{X}_j, \text{i.e.,}$$
$$x_i \geq b_{ij} + x_j.$$

Recall that the set $\mathcal{X}_i = \{v_i^1, \ldots, v_i^{|\mathcal{X}_i|}\}$ represents the unrolled discrete values corresponding to x_i. And likewise there exists a set for x_j. If $x_j \geq v_j^p$, then $x_i \geq b_{ij} + v_j^p$. Let $b_{ij} + v_j^p = v_i^m$ for some $m \in 1, \ldots, |\mathcal{X}_i|$. Then we have, $x_j \geq v_j^p \implies x_i \geq v_i^m$. In terms of binary variables this translates into, $b_j^p = 1 \implies b_i^m = 1$, i.e., $b_i^m \geq b_j^p$, i.e.,

$$b_i^m - b_j^p \geq 0. \qquad (4)$$

Observe that the unrolled binary variables incorporate the discrete and possibly non-uniform nature of grids. Note that all the variable implications, i.e., $b_i^{p-1} - b_i^p \geq 0$ and constraint implications, i.e., $b_i^m - b_j^p \geq 0$ are difference constraints. The constraint matrix $\mathbf{A}$ with such constraints has the TUM form and produces integral solutions with regular LP solvers.

Figure 2 shows a simple example for a layout with gridding requirements and its corresponding set of implications. The nodes v_1^1 (300) and v_1^2 (400) represent the pair of track points for variable x_1 and likewise v_2^1 (600) and v_2^2 (800) for x_2. It is easy to see that v_1^1 implies v_2^2 in order to satisfy the separation constraint of 300. Also, choosing v_2^2 implies v_2^1; the implications for the other nodes follow likewise.

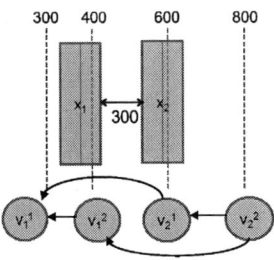

Figure 2: A layout with grid requirement and corresponding constraint graph

3.3 Modeling Minimum Perturbation Objective

The objective function in (1) can be written as,

$$\text{minimize} \quad \sum_{i=1}^{N} |x_i - x_i^0| \qquad (5)$$

Since we have, $x_i = v_i^1 + \sum_{p=1}^{|\mathcal{X}_i|} b_i^p \delta_i^p$, x_i^0 can be expressed as, $x_i^0 = v_i^1 + \sum_{p=1}^{C_i} b_i^p \delta_i^p$, where C_i is the index of the input layout location $v_i^{C_i}$ of variable x_i. The objective function for the term x_i can be expressed as,

$$|x_i - x_i^0| = \left| v_i^1 + \sum_{p=1}^{|\mathcal{X}_i|} b_i^p . \delta_i^p - v_i^1 - \sum_{p=1}^{C_i} b_i^p \delta_1^p \right|. \qquad (6)$$

The second summation in RHS has $b_i^p = 1$, for $p = 1..C_i$. The equation becomes,

$$|x_i - x_i^0| = \left| \sum_{p=1}^{C_i} (b_i^p - 1)\delta_i^p + \sum_{p=C_i+1}^{|\mathcal{X}_i|} b_i^p \delta_i^p \right|. \qquad (7)$$

If first term in RHS is negative then second term is zero. If first term in RHS is zero then second term is positive. Hence absolute operation ($|.|$) in the RHS can be taken away. The equation simplifies as,

$$|x_i - x_i^0| = \left(\sum_{p=1}^{C_i} (1 - b_i^p)\delta_i^p + \sum_{p=C_i+1}^{|\mathcal{X}_i|} b_i^p \delta_i^p \right)$$

$$|x_i - x_i^0| = \left(-\sum_{p=1}^{C_i} b_i^p \delta_i^p + \sum_{p=C_i+1}^{|\mathcal{X}_i|} b_i^p \delta_i^p \right). \qquad (8)$$

Note that δ_i^p being constant for all values of p, this is a linear objective function involving the b_i^p variables.

4. TRACK COMPLIANT DRC FIXING

Legalization involves generation of constraints corresponding to design rules, ensuring that the set of constraints is feasible, and then solving the set of constraints under the minimum perturbation objective. Our legalization framework draws from previous work on constraint generation and resolution of infeasible constraints [2][8]. In this section we focus on some practical considerations for our formulation. First, we discuss how we obtain the ranges for unrolling the variables. However, unrolling leads to solving a problem with much larger number of variables. We then describe

183

several techniques that mitigate the effect of larger problem size.

4.1 Range Computation

Here we describe how to generate the ranges for the variables. Our algorithm is based on the well known Bellman-Ford algorithm (BF) [10]. Note that determining longest paths by running BF on the constraint graph results into layout compaction [2]. This idea can be employed to generate bounds for the variables as well. Indeed, in [8], the authors generate bounds for the variables using two runs of BF – a forward and a reverse longest path run. The forward run generates the lower bound for variables (while achieving compaction to the left) whereas the reverse run generates the upper bound (while achieving compaction to the right).

However, if the bounds are generated in this manner, certain points in the solution space where minimum perturbation can be achieved are missed out. We demonstrate this using a four node example depicted in Figure 3. The nodes x_1, x_2, x_3 and x_4 have layout locations 0, 50, 50 and 100 respectively. For simplicity, assume that these nodes do not have any track restrictions, i.e., they can take all positions on the layout. The constraints between the variables be the following: $x_2 - x_1 \geq 40$, $x_3 - x_1 \geq 40$, $x_4 - x_2 \geq 60$, and $x_4 - x_3 \geq 60$. The bounds generated by the forward and reverse BF runs are listed in columns 3 and 4 (labeled BF LB and BF UB) in Table 1.

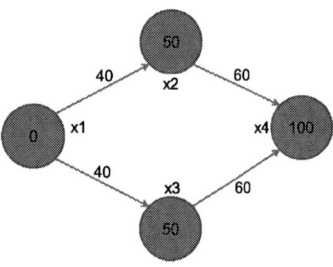

Figure 3: Example graph with layout locations noted inside nodes and constraints on edges

Var	Lyt loc	BF LB	BF UB	LABF LB	LABF UB
x_1	0	0	0	-10	0
x_2	50	40	40	40	50
x_3	50	40	40	40	50
x_4	100	100	100	100	110

Table 1: Ranges generated using BF and LABF

It is easy to see that in this case, the minimum perturbation solution is 0, 50, 50, 110 with a total perturbation of 10 units. However, it is not possible to achieve this minimum perturbation using the ranges in columns labeled BF LB and BF UB in Table 1 while satisfying constraints. This is because probable minimum perturbation solution points 50, 50, and 110 are missed out in ranges of variables x_2, x_3, and x_4 respectively.

To address this problem, we suggest a modification to BF called as the Layout-aware Bellman-Ford Algorithm (LABF). The essence of this algorithm is that a variable is perturbed

only if dictated by a constraint. To achieve this, unlike BF, the variables are initialized with their initial layout locations (see line 6 in Algorithm 1). In a forward run of LABF, a variable is updated only when a constraint pushes it to the right. Note that unlike BF, forward LABF run (coupled with track aware rounding achieved through $\lceil t \cdot \rceil^t$ operator in line 15) generates the upper bound. The reverse LABF (with graph edges and weight signs inverted) generates the lower bound. The bounds generated by the forward and reverse LABF runs are listed in columns labeled LABF LB and LABF UB in Table 1. Note that the lower bounds as well as upper bounds serve as minimum perturbation solutions with a perturbation of 10 units.

Algorithm 1 LayoutAwareBellmanFord (list vertices, list edges, vertex source)

1: {Step 1. Initialization}
2: **for** vertex v in vertices **do**
3:     **if** v is source **then**
4:         $v.bound \leftarrow 0$
5:     **else**
6:         $v.bound \leftarrow initial\_layout\_location$
7:     **end if**
8: **end for**
9: {Step 2. Relax edges repeatedly}
10: **for** i = 1 to size(vertices) - 1 **do**
11:     **for** each edge uv in edges **do**
12:         $u \leftarrow uv.source$
13:         $v \leftarrow uv.dest$
14:         **if** $u.bound + uv.constraint > v.bound$ **then**
15:             $v.bound \leftarrow \lceil {}^t u.bound + uv.constraint \rceil^t$
16:         **end if**
17:     **end for**
18: **end for**

All constraint infeasibilities are resolved prior to running LABF using techniques in [8]. Given a set of feasible constraints, LABF should not have node updates beyond the maximum number of iterations. In this context, we note the following properties for LABF.

LEMMA 1. *LABF does not incorrectly detect positive cycles.*

LEMMA 2. *Ranges generated using LABF include all points in the solution space where minimum perturbation can be achieved.*

For justifying Lemma 1 and Lemma 2, we consider the following two scenarios:

- Constraints are tight over a path, i.e., the layout locations are such that variables must move from their initial positions to satisfy constraints.

- Constraints are not tight over a path, i.e., there exists at least one variable such that it need not be moved from its initial position since all constraints to it are already satisfied in the initial layout. Let us call such a variable a 'slack' variable.

Lemma 1: The constraint graph in both LABF and BF algorithms is the same. Only initialization of nodes is different in LABF from BF. We now argue that this initialization does not trigger additional relaxations in LABF compared to BF. In the first scenario, the relaxations are same as BF and initialization has no bearing on relaxations. In the second scenario, by initializing variables to their initial layout values instead of $-\infty$, we prevent certain updates for slack variables. Hence number of relaxations in LABF is less than or equal to that in BF. If BF relaxations terminate LABF relaxations terminate as well. Consequently LABF does not detect cycles incorrectly.

Lemma 2: The minimum perturbation objective function has the following property: if there are multiple ways in which constraints can be satisfied, choose the one with minimum total perturbation. We argue how LABF satisfies this property in the two scenarios described above. In the first scenario, minimum perturbation objective has no bearing on the variable movement since it is purely constraint dictated. In the second scenario, note that constraints can be satisfied in different ways. For simplicity, consider a path with exactly one slack variable v. Let us divide the path into two parts both of which have tight constraints, with the second path beginning at v. Now, in both paths, to satisfy constraints, variables must move in a constraint dictated manner. Moreover, v remains at its initial location (precisely as done in LABF). This case can be easily generalized to recursively handle cases where there is more than one slack variable in a path. By moving variables in a constraint dictated manner and not moving variables unnecessarily, LABF is able to generate ranges that encompass all minimum perturbation solution points.

The complexity of LABF is same as that of the rounding schedule in [8]. While LABF runs in polynomial time, the number of unrolled points that it can generate can be large; exponential in the worst case. However, in practical layouts, often the ranges are not large because of the limited space in which variables can move. This results into manageable number of unrolled points. Moreover, we also discuss methods that limit the number of points generated. Given the variable unrolling, since we have a TUM constraint matrix and a linear objective function, the time complexity of the solver is the time complexity of LP which is polynomial in the number of variables.

4.2 Reducing Unrolled Variables

Once the track compliant bounds are available, one can enumerate the feasible track locations within that range. These locations for a variable x_i form the set $\mathcal{X}_i$. The ranges and intermediate points, if determined naively, may lead to a large number of unrolled variables. However, subsequent steps can limit that.

4.2.1 Selective Unrolling

Here we describe an approach for generating feasible points selectively or *selective unrolling* instead of unrolling all track points within a range. The idea is to generate a point each time an update is performed on a variable. This can be achieved by maintaining a track point set $\mathcal{X}_i$ for variable x_i and inserting the newly generated bound (on line 15 in Algorithm 1) during variable update into this set. In this way, one can generate points as dictated by constraints.

4.2.2 Value Propagation

In a practical layout, certain objects are fixed or pinned at pre-determined locations, hence, the upper bound and lower bound are equal to the current layout location. These variables have a single binary variable associated with them. Moreover, the value of this variable is equal to one. Since position of the pinned variable is already known, this information is propagated through a breadth first search [10] to variables related through constraints. Thus, the set of values that the related variables can take is appropriately shrunk.

5. EXPERIMENTAL RESULTS

We first demonstrate our solution in practice. Figure 4 (a) depicts part of a layout with a bus structure. The paths and vias in this layout are track objects and need to be on the horizontal or vertical tracks. Figure 4 (b) depicts the same layout post automatic DRC correction run. Notice that the track objects have moved to associated tracks.

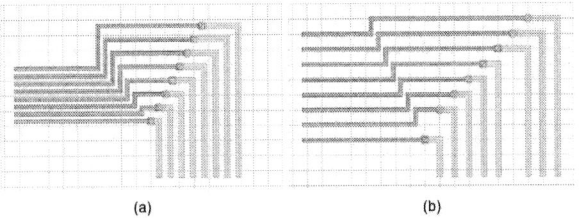

(a)   (b)

Figure 4: (a) Initial layout with objects not on track (b) Corrected layout with objects on track

In order to demonstrate the efficacy of our algorithm we first discuss an Integer Linear Programming (ILP) formulation for the DRC problem. This approach views the DRC problem with non-uniform tracks as a selection problem.

5.1 Formulation as a Selection Problem

Associate with each value v_i^p that a variable x_i can take, a binary variable b_i^p. A variable x_i can then be expressed in terms of the new binary variables as:

$$x_i = \sum_{p=1}^{|\mathcal{X}_i|} b_i^p v_i^p. \tag{9}$$

Moreover to ensure that b_i's are binary variables and since x_i can take only one value at any point of time, we have:

$$b_i^p \geq 0, \quad b_i^p \leq 1, \quad \sum_{p=1}^{|\mathcal{X}_i|} b_i = 1. \tag{10}$$

A constraint of the form $x_i - x_j \geq c_{ij}$ can be expressed as:

$$\sum_{p=1}^{|\mathcal{X}_i|} b_i^p v_i^p - \sum_{p=1}^{|\mathcal{X}_j|} b_j^p v_j^p \geq c_{ij}. \tag{11}$$

Moreover, the minimum perturbation objective function can be expressed as:

$$\text{obj} = \sum_{i}^{N} |\sum_{p=1}^{|\mathcal{X}_i|} b_i^p v_i^p - x_i^0|. \tag{12}$$

The problem is to minimize the objective in (12) subject to constraints of the form (9), (10) and (11). Note that we need to ensure that b_i's are binary i.e., they are either 0 or 1. This makes it a binary (or Integer) programming problem with

185

linear constraints and objective function. We call this the 'Selection' formulation. Integer Linear Programming Problem (ILP) solvers can be used to solve this problem. Running the ILP solver on designs with few thousand binary variables and constraints on a 3 Ghz, 8 core machine with 16 GB RAM took more than 2 hours. This prohibited the use of Selection algorithm for solving this problem.

We now present experimental results for our algorithm. These experiments have been performed on selected layout regions on various designs. In Table 2, we present the details

Expt	#lyt obj	#Edg var	#Cnstr
1	87	94	592
2	322	284	2250
3	102	404	655
4	642	544	4658
5	202	804	1205
6	404	1604	4272
7	1027	4104	6668

Table 2: Details of experiments

of the conducted experiments. Performance of the proposed algorithm without optimizations is presented in Table 3.

Expt	#Bin Var	#Bin Cnstr	#Slvr time (s)
1	392	4987	0.162
2	600	10190	1.010
3	1999	4864	0.273
4	2790	8945	1.211
5	7734	15701	6.472
6	14050	36351	24.606
7	4104	10261	0.538

Table 3: Performance of the proposed algorithm without optimizations

In Table 4, we present the performance results of our algorithm after incorporating optimizations like selective unrolling and value propagation. It can be seen that these optimizations greatly improve the performance of solver.

Expt	#Bin Var	#Bin Cnstr	#Slvr time (s)	Impr
1	42	452	0.001	162x
2	64	916	0.005	202x
3	304	759	0.002	136.5x
4	114	2314	0.025	48.4x
5	604	1409	0.006	1078.6x
6	1504	5476	0.128	192.2x
7	4104	10261	0.507	1.06x

Table 4: Performance of the proposed algorithm with optimizations

6. CONCLUSIONS AND FUTURE WORK

We have studied the problem of DRC fixing in presence of multiple layer-specific non-uniform grids. This breaks the usual TUM nature of the constraint matrix and precludes

application of traditional methods. We made an important observation that even for multiple non-uniform grids, the constraints have at most two variables per inequality (TVPI). We used this property to incorporate a variable unrolling based formulation for solving linear optimization problems with TVPI and integrality requirements. Using the variable unrolling technique, we modeled constraints and minimum perturbation objective for solving the DRC fixing problem. Furthermore, the formulation needs realistic bounds for each variable. For this we proposed a layout-aware modification on the Bellman-Ford algorithm. The generated bounds are shown to contain minimum perturbation solution points. We also presented additional steps to reduce the unrolled problem size.

Future work in this area would involve additional techniques to further reduce the unrolled problem size. The formulation also needs to be extended to handle hierarchical layouts in presence of non-uniform layer grids.

7. REFERENCES

[1] D. Abercrombie, P. Elakkumanan, and L. W. Liebmann, "Restrictive design rules and their impact on 22nm design & physical verification," *Proc. of Electronic Design Process Symposium*, Apr. 2009.

[2] D. Boyer, "Symbolic layout compaction review," *Proc. IEEE/ACM Design Automation Conf.*, pp. 383–389, Jun. 1988.

[3] L. Luenberger, *Linear and Nonlinear Programming*. Reading, MA, USA: Addison-Wesley, second ed., 1984.

[4] F.-L. Heng, Z. Chen, and G. Tellez, "A VLSI artwork legalization technique based on a new criterion of minimum layout perturbation," *Proc. Int. Symp. Physical Design*, pp. 116–121, Apr. 1997.

[5] L. W. Liebmann, A. Barish, Z. Baum, H. Bonges, S. Bukofsky, C. Fonseca, S. Halle, G. Northrop, S. Runyon, and L. Sigal, "High-performance circuit design for the ret-enabled 65nm technology node," *Proc. of SPIE Design and Process Integration for Microelectronics Manufacturing II*, pp. 20–29, 2004.

[6] M. Lavin, F.-L. Heng, and G. Northrop, "Backend cad flows for 'restrictive design rules'," *Proc. Int. Conf. Computer Aided Design*, pp. 739–746, Nov. 2004.

[7] X. Yuan, K. McCullen, F.-L. Heng, R. Walker, J. Hibbeler, R. Allen, and R. Narayan, "Technology migration technique for designs with strong RET-driven layout restrictions," *Proc. Int. Symp. Physical Design*, pp. 175–182, Apr. 2005.

[8] X. Tang and X. Yuan, "Technology migration techniques for simplified layouts with restrictive design rules," *Proc. Int. Conf. Computer Aided Design*, pp. 655–660, Nov. 2006.

[9] D. S. Hochbaum and J. Naor, "Simple and fast algorithms for linear and integer programs with two variables per inequality," *SIAM Journal of Computing*, vol. 23, pp. 1179–1192, Dec. 1994.

[10] T. H. Cormen, C. E. Leiserson, and R. L. Rivest, *Introduction to Algorithms*. Cambridge, MA, USA: MIT Press, 1990.

Multiple Chip Planning for Chip-Interposer Codesign *

Yuan-Kai Ho[1] and Yao-Wen Chang[1,2,3]

[1]Graduate Institute of Electronics Engineering, National Taiwan University, Taipei 106, Taiwan
[2]Department of Electrical Engineering, National Taiwan University, Taipei 106, Taiwan
[3]Research Center for Information Technology Innovation, Academia Sinica, Taipei 115, Taiwan
yuankai@eda.ee.ntu.edu.tw; ywchang@cc.ee.ntu.edu.tw

ABSTRACT

An interposer-based three-dimensional integrated circuit, which introduces a silicon interposer as an interface between chips and a package, is one of the most promising integration technologies for modern and next-generation circuit designs. Inter-chip connections can be routed on the interposer by chip-scale wires to enhance design quality. However, its design complexity increases dramatically due to the extra interposer interface. Consequently, it is desirable to simultaneously consider the co-design of the interposer and multiple chips mounted on it. This paper addresses the first work of chip-interposer codesign to place multiple chips on an interposer to reduce inter-chip wirelength. For this problem, we propose a new hierarchical B*-tree to simultaneously place multiple chips, macros, and I/O Buffers. An approach based on bipartite matching is then proposed to concurrently assign signals from I/O buffers to micro bumps. Experimental results show that our approach is effective and efficient for the codesign problem.

Categories and Subject Descriptors

B.7.2 [**Integrated Circuits**]: Design Aids

General Terms

Algorithms, Design

Keywords

Physical Design, 2.5D-IC, Interposer, Codesign

1. INTRODUCTION

As technology advances, interposer-based three-dimensional integrated circuits (interposer-based 3D ICs, and also known as 2.5D ICs) become one of the most promising solutions for enhancing system performance, decreasing power consumption, and supporting heterogeneous integration [14, 22]. Figure 1

---

*This work was partially supported by IBM, SpringSoft, TSMC, Academia Sinica, and NSC of Taiwan under Grant No's. NSC 101-2221-E-002-191-MY3, NSC100-2221-E-002-088-MY3, NSC 99-2221-E-002-207-MY3, and NSC 99-2221-E-002-210-MY3.

Permission to make digital or hard copies of all or part of this work for personal or classroom use is granted without fee provided that copies are not made or distributed for profit or commercial advantage and that copies bear this notice and the full citation on the first page. To copy otherwise, to republish, to post on servers or to redistribute to lists, requires prior specific permission and/or a fee.
DAC'13, May 29 - June 07 2013, Austin, TX, USA

shows the structure of an interposer-based 3D IC, which introduces a silicon interposer as an interface between chips and a package. Multiple chips can be mounted on the interposer, and inter-chip nets are routed on the redistribution layers (RDLs) of the interposer by chip-scale wires. In addition, different from stacked 3D ICs [11, 19] which use through-silicon vias (TSVs) as vertical interconnects among different layers, the interposer-based 3D ICs contain TSVs only in their interposer. As a result, the fabrication cost and design limitation/complexity with TSVs can be reduced, and many industrial products adopt the structure of the interposer-based 3D ICs, such as Xilinx Virtex-7 2000T [4, 12].

In a conventional design flow, chips are often designed independently, then placed on a silicon interposer, and finally routed with inter-chip connections on the RDLs of the interposer. This conventional flow might incur interposer-unfriendly micro bump assignments, and thus requires considerable extra efforts for inter-chip routing, causing a bottleneck of the time-to-market. To meet the micro bump assignments, furthermore, the inter-chip routing often incurs longer total wirelength and consumes more routing layers, dramatically degrading circuit performance and increasing manufacturing cost.

To improve the inter-chip routing quality, it is desirable to simultaneously consider a silicon interposer and multiple chips mounted on it. Figure 2 illustrates an example for the importance of chip-interposer codesign. In Figure 2(a), each of the four chips is designed independently and then placed on an interposer, resulting in a design with complex and hard-to-route inter-chip connections. The interposer, which mounts these chips, might need to be customized specifically for the design with more RDLs and thus significantly higher manufacturing cost. In contrast, Figure 2(b) shows another chip planning result considering the chip-interposer codesign. The inter-chip connections of this result are shorter and simpler than that in Figure 2(a). As a result, these inter-chip connections would

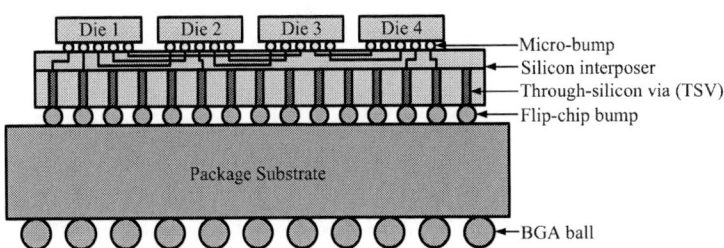

Figure 1: The structure of an interposer-based 3D IC. The structure has a silicon interposer between chips and a package. The interposer provides chip-scale connections to chips.

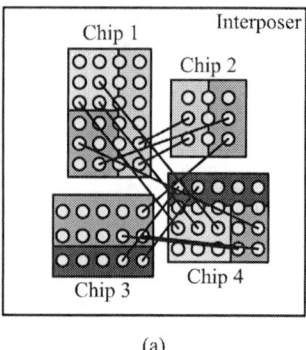 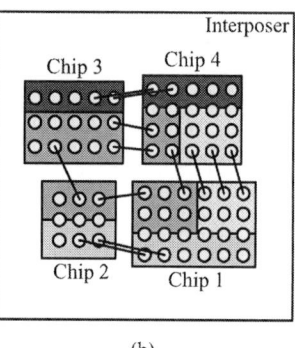

(a)   (b)

Figure 2: (a) Multiple-chip planning without the chip-interposer codesign consideration. (b) Multiple-chip planning with the chip-interposer codesign consideration. In (a), the inter-chip connections are more complex and hard-to-route. In (b), the inter-chip connections are simpler and shorter.

consume fewer layers and vias in a standard interposer and thus improve its signal quality and reduce the manufacturing cost.

Many previous works have addressed various codesign problems as follows: (1) chip-package codesign [5, 6, 13, 17, 18, 21, 23], (2) package-board codesign [8, 16], and (3) chip-package-board codesign [7, 15, 20]. However, no previous work is focused on silicon interposers, key components of interposer-based 3D ICs. For chip-package codesign, Fang and Chang [5] proposed a network-flow based routing algorithm to route free-assignment signals from block ports to bump balls. Xiong et al. [23] defined some constraints for chip-package codesign and proposed an effective multi-step algorithm to solve a constraint-driven I/O placement problem. In addition, some previous works focused on the signal skew between chips and packages [6, 13, 18, 21]. For package-board codesign, Fang et al. [8] used Delaunay triangulation and Voronoi diagrams to create triangular tile models and then route signal nets with the models. The work [16] presented a pin-out assignment method to consider the interactions between packages and boards. For the chip-package-board codesign problem, Park [20] proposed a die abstract model and a board-driven methodology to reduce the overall design cycle and cost. Lee and Chang [15] proposed a Λ-shaped codesign flow to route signals among boards, packages, and chips. To consider differential pairs, an integer linear programming (ILP) based routing algorithm was proposed in [7]. Although these previous works have addressed many codesign issues, the inter-chip connection optimization for chip-interposer codesign has not been considered.

In this paper, we address the chip-interposer codesign problem and propose an effective multiple chip planning algorithm to minimize inter-chip wirelength for interposer-based 3D IC designs. We first propose a new hierarchical B*-tree (HB*-tree, for short) to simultaneously place multiple chips, macros, and I/O Buffers. We also propose a cost function to evaluate an HB*-tree for simulated annealing (SA) based optimization. Finally, an approach based on bipartite matching [3] is presented to assign signals from I/O buffers to micro bumps. Experimental results show that our approach is effective and efficient for the chip-interposer codesign problem.

We summarize our contributions as follows.

- This paper presents the first work to address the problem of chip-interposer codesign. To the best of our knowledge, no previous codesign work considers silicon interposers,

key components of novel interposer-based 3D ICs.

- A new HB*-tree is proposed to solve the addressed problem effectively in reasonable running time. The HB*-tree can be used to simultaneously place multiple chips, macros, and I/O Buffers.

- A bipartite-matching-based algorithm [3] is proposed to find the connections between I/O buffers and micro bumps.

The rest of this paper is organized as follows. Section 2 gives the formulation of the chip-interposer codesign problem and then reviews the B*-tree representation [1] on which our work is based. Section 3 proposes our planning algorithm for interposer-based 3D IC designs. Section 4 reports our experimental results, and Section 5 gives the conclusion.

2. PRELIMINARIES

In this section, we first formulate the chip-interposer codesign problem in Section 2.1. Then, we review the B*-tree representation [1] in Section 2.2 because our proposed algorithm is based on this representation.

2.1 Problem Formulation

We first give some notations used in this paper:

- $C = \{c_1, c_2, \ldots, c_{|C|}\}$ is the set of chips mounted on a silicon interposer.

- $M_i = \{m_1^i, m_2^i, \ldots, m_{|M_i|}^i\}$ is the set of macros in the chip c_i.

- $B_i = \{b_1^i, b_2^i, \ldots, b_{|B_i|}^i\}$ is the set of I/O buffers in the chip c_i.

- $U_i = \{u_1^i, u_2^i, \ldots, u_{|U_i|}^i\}$ is the set of micro bumps in the chip c_i.

- $M = M_1 \cup M_2 \cup \ldots M_{|C|}$ is the set of all macros.

- $B = B_1 \cup B_2 \cup \ldots B_{|C|}$ is the set of all I/O buffers.

- $U = U_1 \cup U_2 \cup \ldots U_{|C|}$ is the set of all micro bumps.

The predefined wire connections in a silicon-interposer-based design can be classified as follows: (1) intra-chip connections and (2) inter-chip connections. Intra-chip connections define the connections of macros in M_i inside a single chip C_i, $1 \leq i \leq |C|$, while inter-chip connections define the wire connections among different chips through I/O buffers. Note that we need to connect these buffers to suitable micro bumps in our proposed algorithm to accomplish the inter-chip connections because these connections are routed on a silicon interposer. Figure 3 illustrates an instance of a silicon-interposer-based design. Since the chips in C are area-I/O flip-chip designs, the I/O buffers in B can be placed in the whole area of a chip.

The problem of multiple chip planning in an interposer-based design can be defined as follows:

- **Chip-Interposer Codesign Problem:** Given C, M, B, U, and a silicon interposer, place macros and I/O buffers within corresponding chips without any overlaps, and find the locations of the chips on the interposer, so that no constraints are violated, and the total wirelength is minimized.

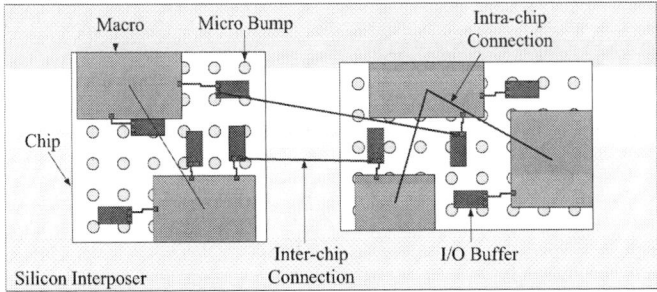

Figure 3: An illustration of a silicon-interposer-based design. Each chip is an area-I/O flip-chip. The I/O buffers can be placed in the whole area of a chip.

2.2 Review of B*-tree

Our proposed algorithm is based on the B*-tree representation presented in [1] to tackle the chip-interposer codesign problem. Therefore, we shall give a brief review of the B*-tree representation.

Given a compacted placement, where each module cannot move bottom and left anymore, we can construct a B*-tree to represent the compacted placement. A B*-tree is an ordered binary tree, and each node of the B*-tree represents a module of a compacted placement. The root of a B*-tree corresponds to the module on the bottom-left corner. In addition, a B*-tree has the geometric relationship between two nodes n_i and n_j as follows: (1) if n_j is the left child of the node n_i, the corresponding module of n_j is the lowest adjacent module on the right side of the corresponding module of n_i, and (2) if n_j is the right child of the node n_i, the corresponding module of n_j is the first module above the corresponding module of n_i with the same x-coordinate. As a result, we can derive the x-coordinate of each module in a compacted placement by traversing the corresponding B*-tree. To derive the y-coordinate of each module, the contour data structure in [9] is adopted. The contour structure uses a doubly linked list to record the contour line segments. The time complexity of computing the y-coordinate of a module is in amortized constant time, and the total packing time is linear to the number of nodes of a B*-tree.

Figure 4 gives an example of a compacted placement and its corresponding B*-tree. The node n_i corresponds to the module m_i, $0 \le i \le 9$. We can traverse the B*-tree to derive a compacted placement. For example, since the node n_0 is the root, the module m_0 is on the bottom-left corner. The left child of n_0 is n_1, and the right child of n_0 is n_3. Therefore, m_1 is on the right side and adjacent to m_0, and m_3 is above m_0 with the same x-coordinate.

3. OUR PROPOSED ALGORITHM

In this section, we present our planning algorithm for chip-interposer codesign. Figure 5 shows the overall flow of the algorithm. We first create a hierarchical B*-tree to represent multiple chips mounted on a silicon interposer and their own macros and I/O buffers. Simulated annealing (SA) based optimization is used to find the solution. We iteratively perturb a hierarchical B*-tree under some constraints and evaluate the corresponding placement by a specific cost function until the final result is good enough or no better solution can be found. Since we derive a compacted placement of chips after the SA-based optimization, we separate the chips to feasible locations on the silicon interposer. In addition, since the inter-chip connections among these chips are to connect micro bumps on the silicon interposer, we

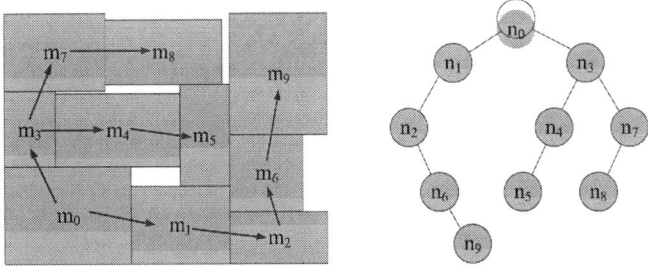

Figure 4: A compacted placement and the corresponding B*-tree [1]. For a node n_i and its corresponding module m_i, the left child of the node n_i represents the lowest adjacent module on the right of the corresponding module m_i; the right child of the node n_i represents the first module above the corresponding module m_i with the same x-coordinate.

find feasible micro bumps and connect them to I/O buffers. See the following sections for the detailed algorithms.

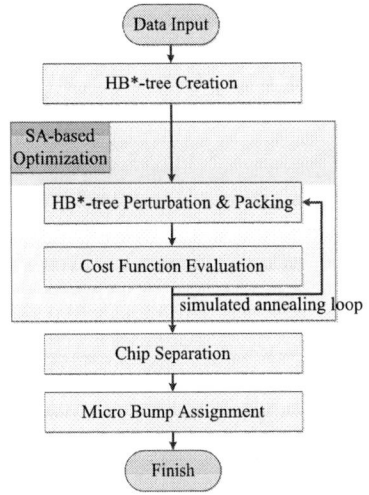

Figure 5: Proposed algorithm flow for chip-interposer codesign.

3.1 HB*-tree Construction

Given a set C of chips, a set M of macros, and a set B of I/O buffers, we can construct the corresponding nodes in our hierarchical B*-tree (HB*-tree for short). The nodes of an HB*-tree can be categorized as follows:

- $N_c = \{n_1^c, n_2^c, \ldots, n_{|N_c|}^c\}$ is the set of chip nodes, corresponding to the chips in C.
- $N_{m_i} = \{n_1^{m_i}, n_2^{m_i}, \ldots, n_{|N_{m_i}|}^{m_i}\}$ is the set of macro nodes, corresponding to the macros in M_i.
- $N_{b_i} = \{n_1^{b_i}, n_2^{b_i}, \ldots, n_{|N_{b_i}|}^{b_i}\}$ is the set of buffer nodes, corresponding to the I/O buffers in B_i.
- $N_m = N_{m_1} \cup N_{m_2} \cup \ldots N_{m_{|C|}}$ is the set of all macro nodes.
- $N_b = N_{b_1} \cup N_{b_2} \cup \ldots N_{b_{|C|}}$ is the set of all buffer nodes.

To reduce the wirelength between I/O buffers and macros, we cluster a macro and its connected I/O buffers into a group. For this group, we introduce more notations as follows:

- $N_{g_i} = \{n_1^{g_i}, n_2^{g_i}, \ldots, n_{|N_{g_i}|}^{g_i}\}$ is the set of group nodes, where $n_j^{g_i}, 1 \le j \le |N_{g_i}|$, represents a macro m_j^i and its connected I/O buffers in the chip c_i.

- $N_g = N_{g_1} \cup N_{g_2} \cup \ldots N_{g_{|N_c|}}$ is the set of all group nodes.

- $N = N_c \cup N_m \cup N_b \cup N_g$ is the set of all nodes in an HB*-tree.

Figure 6 shows an example of a group node. As shown in Figure 6(a), the macro m_1^1 connects to six I/O buffers. We cluster the macro and its connected I/O buffers into a group to reduce the wirelength among them. The corresponding nodes in an HB*-tree are shown in Figure 6(b). The group node $n_1^{g_1}$ includes a sub-B*-tree to represent the placement in Figure 6(a).

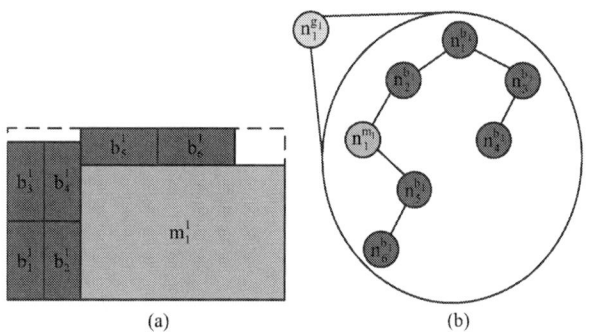

Figure 6: (a) A macro and its connected I/O buffers. (b) The corresponding nodes in an HB*-tree. A group node is used in our algorithm to represent a sub-B*-tree.

Now we introduce the construction of an initial HB*-tree. An HB*-tree is a three-level hierarchical B*-tree: (1) In the first level, which is the interposer level, we consider all chip nodes in N_c and randomly connect them to form a binary tree. There are one subtree composed of chip nodes in the HB*-tree. The nodes in the first level decide the placement of multiple chips on a silicon interposer. (2) In the second level, which is the chip level, for each chip node n_i^c, we consider all group nodes in N_{g_i} and randomly connect them to form a sub-B*-tree included in the chip node n_i^c. Obviously, there are $|N_c|$ subtrees composed of group nodes in the HB*-tree. The nodes in the second level determine the placement of each chip. (3) In the third level, which is the group level, for each group node in $n_j^{g_i}$, we randomly connect the macro m_j^i and its connected I/O buffers to form a sub-B*-tree included in the group node $n_j^{g_i}$. The number of subtrees composed of macro nodes and I/O buffer nodes in the HB*-tree is $\sum_{i=1}^{|N_c|} |N_{g_i}|$. The nodes in the third level determine the placement of each group.

Figure 7 shows the structure of an HB*-tree. The tree composed of the chip nodes n_1^c, n_2^c, n_3^c, and n_4^c in the interposer level can represent the placement of multiple chips on a silicon interposer. The tree composed of the group nodes $n_1^{g_3}$, $n_2^{g_3}$, $n_3^{g_3}$, $n_4^{g_3}$, and $n_5^{g_3}$ is in the chip level and represents the placement of the chip node n_3^c. The tree composed of the macro node $n_5^{m_3}$ and buffer nodes $n_1^{b_3}$, $n_2^{b_3}$, and $n_3^{b_3}$, is in the group level and represents the placement of the group node $n_5^{g_3}$.

3.2 HB*-tree Perturbation and Packing

To perturb an HB*-tree in simulated annealing, we apply the following operations.

- Op1: Rotate a chip, a macro, an I/O buffer, or a macro-buffer group.

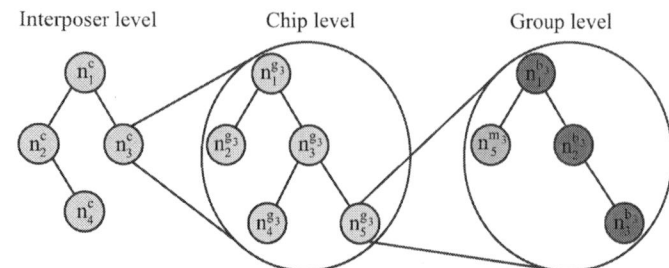

Figure 7: An illustration of an HB*-tree. The HB*-tree has three levels: a tree in the interposer level composed of chip nodes, a tree in the chip level composed of group nodes, and a tree in the group level composed of macro nodes and buffer nodes.

- Op2: Move a node of a subtree to another place of the same subtree.

- Op3: Swap two nodes within a subtree.

For Op1, we rotate a chip, a macro, an I/O buffer, or a macro-buffer group for a tree node. For Op2, we only allow a node of a subtree to be moved to another place of the subtree due to the special structure of an HB*-tree. For example, moving a chip node to the left child of another chip node is legal, but moving a macro node from one chip to another chip is illegal. Similarly, for Op3, we only allow swapping two nodes within one subtree.

We pack an HB*-tree from the bottom level to the top level. That is, we first pack each subtree in the third level and then pack that in the second level. Finally, each subtree in the first level is packed. The subtree packing is the same as the B*-tree packing by applying the depth first search (DFS) procedure.

3.3 Cost Function Evaluation

To evaluate the quality of a placement solution during simulated annealing, we define the cost function $\Phi(P)$ of a placement P as follows:

$$\Phi(P) = A + \alpha W_1 + \beta W_2 + \gamma \sum_{i=0}^{|C|} (R_i - R_i^*)^2, \qquad (1)$$

where α, β, and γ are user-specified weighting parameters, A is the total area of chips, W_1 is the total wirelength of inter-chip connections, W_2 is the total wirelength of intra-chip connections, R_0 is the current aspect ratio of the placement of all chips, R_0^* is the pre-defined aspect ratio of the placement of all chips on an interposer. R_i, $1 \le i \le |C|$, is the current aspect ratio of the chip i, and R_i^*, $1 \le i \le |C|$, is the pre-defined aspect ratio of the chip i. Note that if the chip i is rotated, we adjust R_i^* for the rotation.

To prevent a placement solution from out of the boundaries of chips, the aspect ratio penalty, $\sum_{i=0}^{|C|}(R_i - R_i^*)^2$, is adopted in the cost function. As suggested in [2], we use the square of the aspect ratio difference and impose a huge weight for the aspect ratio penalty to guarantee that our placement solution can place all macros and I/O buffers into the corresponding chips, and also place all chips on an interposer. Then, we shall decrease the weight of the aspect ratio penalty to concentrate more on the wirelength optimization.

Since we have not established the connections between I/O buffers and micro bumps yet, we shall use the Manhattan distance between two I/O buffers as the wirelength of inter-chip connections in the cost function. We will assign signals from I/O buffers to micro bumps in Section 3.5.

3.4 Chip Separation

Since we derive a compacted placement of chips after the SA-based optimization, a partition-based approach is proposed to spread the chips in the whole feasible region of a silicon interposer. In our approach, we vertically (horizontally) partition the placement of chips into the left sub-placement and right sub-placement (top sub-placement and bottom sub-placement). Similarly, the feasible region of a silicon interposer can be partitioned in the same way. If a chip is in the right sub-placement (top sub-placement), we shift it with distance d along the x-axis (y-axis), where d is the distance between the sub-placement and the sub-region of the interposer. The partitioning is iteratively performed until the sub-placement size is sufficiently small, or each sub-placement has only one chip. Finally, we put the separated placement at the center of the feasible region of the silicon interposer.

Figure 8 illustrates our chip separation approach. As shown in Figure 8(a), we vertically partition the placement of chips and the feasible region of an interposer. The distance between the right sub-placement and the right sub-region is Δx. Therefore, the chips in the right sub-placement are shifted with Δx along the x-axis as shown in Figure 8(b). Then, we horizontally partition each sub-placement and sub-region and derive the distances Δy_1 and Δy_2. The chips can further be separated as shown in Figure 8(c). Finally, we put the separated placement at the center of the feasible region as shown in Figure 8(d).

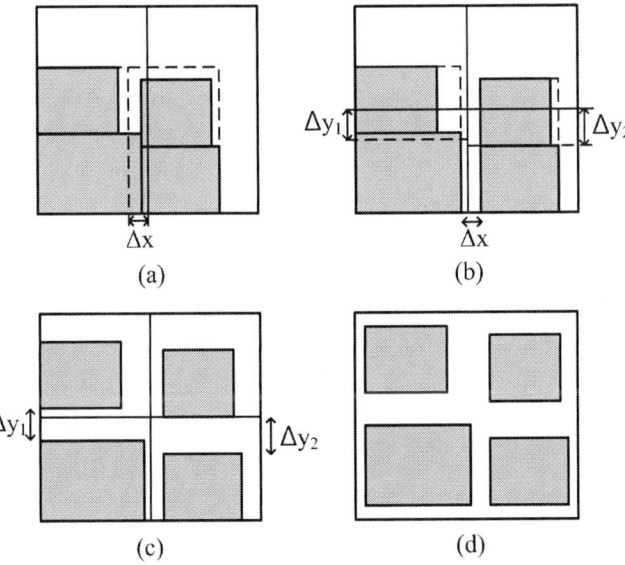

Figure 8: An example of our chip separation approach. (a) The placement and the whole feasible region are partitioned vertically, and the chips in the right part of the placement are shifted with distance Δx. (b) The placement and the feasible region are further partitioned horizontally, and the distances Δy_1 and Δy_2 are derived. (c) The top-left chip is shifted with distance Δy_1, and the top-right chip is shifted with distance Δy_2. (d) The separated placement in (c) is placed at the center of the feasible region.

3.5 Micro Bump Assignment

The given inter-chip connections describe the wire connections among multiple chips through their I/O buffers. However, the multiple chips on a silicon interposer are connected each other by micro bumps. Therefore, we need to find the connec-

tions between I/O buffers and micro bumps. The micro bump assignment problem can be modeled as follows: Given a set B of I/O buffers and a set U of micro bumps, the objective is to assign signals from I/O buffers to micro bumps so that no constraint is violated, and the total wirelength is minimized.

For the assignment problem, we propose an algorithm based on bipartite matching [3] to establish the connections between I/O buffers and micro bumps. A bipartite graph $G = (V, E)$ is first constructed, where V is $B \cup U$, B is the set of I/O buffers, U is the set of micro bumps, and E is the set of edges between B and U. For any pair, $b_x^i \in B$ and $u_y^j \in U$, we create an edge between the pair if they are in the same chip, that is, $i = j$. In contrast, there is no edge between the pair if $i \neq j$. We define $e(b_x^i, u_y^i)$ to be the edge between b_x^i and u_y^i. Figure 9 shows an example for constructing a bipartite graph. The I/O buffers and the micro bumps are in the two chips c_1 and c_2. There is an edge between an I/O buffer and a micro bump if the buffer and the bump are in the same chip. For example, there is an edge between b_1^2 and u_1^2 because they are in the same chip c_2, and there is no edge between b_2^2 and u_1^1 because they are in the different chips.

The cost function Ψ of the edge $e(b_x^i, u_y^i)$ is defined as follows:

$$\Psi(e(b_x^i, u_y^i)) = \begin{cases} D(b_x^i, u_y^i) + \delta D(u_y^i, b_z^j), & \text{if } b_x^i \text{ connected with } b_z^j \\ D(b_x^i, u_y^i), & \text{otherwise,} \end{cases} \quad (2)$$

where δ is a user-specified parameter, $D(b_x^i, u_y^i)$ is the distance between the I/O buffer b_x^i and the micro bump u_y^i, and $D(u_y^i, b_z^j)$ is the distance between the bump u_y^i and the I/O buffer b_z^j. If the buffer b_x^i in chip i connects with the buffer b_z^j in chip j, $i \neq j$, we want to select a bump so that the sum of the distance between the buffer b_x^i and the bump, and the distance between the bump and the buffer b_z^j is minimized. If the buffer b_x^i has no pre-defined connection with another buffer, we only consider the distance between the buffer b_x^i and its connected micro bump.

After constructing the bipartite graph G, we apply a bipartite matching algorithm to match the I/O buffers with the micro bumps based on the proposed cost function.

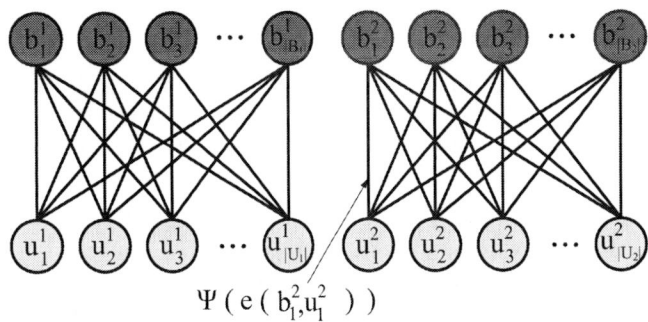

Figure 9: Bipartite graph construction. An edge between an I/O buffer and a micro bump is introduced if the buffer and the bump are in the same chip.

4. EXPERIMENTAL RESULTS

We implemented our algorithm in the C++ programming language and performed experiments on an Intel Xeon 2.93GHz Linux workstation with 48GB memory.

To validate the effectiveness of our approach, we tested our approach on six industrial testcases. We slightly modified these testcases to suit the interposer-based 3D IC structures. Table 1 shows the statistics of these testcases. Column "#Chips" represents the number of chips in a silicon interposer. Columns "#Macros" and "#Buffers" give the total number of macros and I/O buffers in all chips, respectively. The number of inter-chip connections is listed in Column "#Inter-chip nets".

Table 1: The statistics of the benchmark circuits.

Circuits	#Chips	#Macros	#Buffers	#Inter-chip nets
case1	2	12	50	15
case2	2	24	336	55
case3	2	46	640	127
case4	2	56	768	179
case5	4	86	1194	304
case6	4	92	1280	514

We compared our approach with a conventional design flow because no existing work has addressed the chip-interposer codesign problem. We say that the design flow is a single-chip-driven method, which optimizes a chip at a time without considering the other chips. For the single-chip-driven method, the macros and the I/O buffers of each chip are placed first, and then signals are sequentially assigned from I/O buffers to their closest micro bumps. Finally, all chips are put in a silicon interposer. Table 2 shows our experimental results. Column "Wirelength" lists the total wirelength of inter-chip nets and intra-chip nets; column "CPU times" gives the total running time of each testcase. In addition, since the single-chip-driven method is to individually place chips, we sum up the running time of each chip as the total running time of the method.

Table 2: Comparison of wirelength and CPU times for the single-chip-driven method and our method.

Circuits	Single-chip-driven		Ours	
	Wirelength	CPU times (s)	Wirelength	CPU times (s)
case1	92980	6	84826	9
case2	513204	32	482508	43
case3	993020	172	855264	210
case4	1686745	216	1409126	295
case5	3869600	340	3149400	476
case6	6616723	358	5154129	542
	1.17	0.72	1.00	1.00

Compared with the single-chip-driven method, our proposed algorithm can reduce the total wirelength for all testcases in reasonable running time. This is because our proposed algorithm can simultaneously place chips, macros, and I/O buffers of an interposer to consider the inter-chip connections. By observing the comparison, we find that we get higher improvements for larger cases with bigger chip sizes or larger chip numbers. The reason is that these larger cases have many possible locations of I/O buffers and complex inter-chip connections; considering the inter-chip connections in these cases can significantly improve the total wirelength. In addition to the inter-chip connections, we also consider the very complicated intra-chip connections of each chip in our algorithm, where the locations of buffers and macros might be restricted by the intra-chip connections.

5. CONCLUSIONS

In this paper, we have addressed a chip-interposer codesign problem for interposer-based 3D IC designs. For this problem, we have proposed a multiple chip planning algorithm. To consider the interaction between chips and an interposer, we have designed a three-level hierarchical B*-tree to simultaneously place multiple chips, macros, and I/O Buffers. We also have proposed a cost function to evaluate an HB*-tree for SA-based optimization. Further, to assign signals from I/O buffers to micro bumps, we have proposed a bipartite-matching-based algorithm that can find the connections between I/O buffers and micro bumps. Experimental results have shown that our approach is effective and efficient for the chip-interposer codesign problem.

6. REFERENCES

[1] Y.-C. Chang, Y.-W. Chang, G.-M. Wu, and S.-W. Wu, "B*-trees: A new representation for non-slicing floorplans," in *Proc. of DAC*, pp. 458–463, 2000.

[2] T.-C. Chen and Y.-W. Chang, "Modern floorplanning based on fast simulated annealing," in *Proc. of ISPD*, pp. 104–112, 2005.

[3] T. H. Cormen, C. E. Leiserson, R. L. Rivest, and C. Stein. *Introduction to Algorithms*. The MIT Press, 2009.

[4] P. Dorsey, "Xilinx Stacked Silicon Interconnect Technology Delivers Breakthrough FPGA Capacity, Bandwidth, and Power Efficiency", Xilinx White paper: Virtex-7 FPGAs, 2010

[5] J.-W. Fang and Y.-W. Chang, "Area-I/O flip-chip routing for chip-package co-design," in *Proc. of ICCAD*, pp. 518–522, 2008.

[6] J.-W. Fang and Y.-W. Chang, "Area-I/O flip-chip routing for chip-package co-design considering signal skews," *IEEE TCAD*, vol. 29, no. 5, pp. 711–721, 2010.

[7] J.-W. Fang, K.-H. Ho, and Y.-W. Chang, "Routing for chip-package-board co-design considering differential pairs," in *Proc. of ICCAD*, pp. 512–517, 2008.

[8] J.-W. Fang, M. D.-F. Wong, and Y.-W. Chang, "Flip-chip routing with unified area-I/O pad assignments for package-board co-design," in *Proc. of DAC*, pp. 336–339, 2009.

[9] P.-N. Guo, C.-K. Cheng, and T. Yoshimura, "An O-tree representation of non-slicing floorplan and its applications," in *Proc. of DAC*, pp. 268–273, 1999.

[10] S. Kirkpatrick, C. D. Gelatt, and M. P. Vecchi, "Optimization by Simulated Annealing," *Science*, vol. 220, no. 4598, pp. 671–680, 1983.

[11] D. H. Kim, K. Athikulwongse, and S. K. Lim, "A study of Through-Silicon-Via impact on the 3D stacked IC layout," in *Proc. of ICCAD*, pp. 674–680, 2009.

[12] N. Kim, D. Wu, D. Kim, Arif Rahman, and P. Wu, "Interposer design optimization for high frequency signal transmission in passive and active interposer using through silicon via (TSV)," in *Proc. of ECTC*, pp. 1160–1167, 2011.

[13] M.-F. Lai and H.-M. Chen, "An implementation of performance-driven block and I/O placement for chip-package codesign," in *Proc. of ISQED*, pp. 604–607, 2008.

[14] J. H. Lau, Y. S. Chan, and R. S. W. Lee, "3D IC integration with TSV interposers for high-performance applications," *Chip Scale Review*, pp. 26–29, September/October 2010.

[15] H.-C. Lee and Y.-W. Chang, "A chip-package-board co-design methodology," in *Proc. of DAC*, pp. 1082–1087, 2012.

[16] R.-J. Lee and H.-M. Chen, "Fast flip-chip pin-out designation respin for package-board codesign," *IEEE TVLSI*, vol. 17, no. 8, pp. 1087–1098, 2009.

[17] R.-J. Lee and H.-M. Chen, "Row-based area-array I/O design planning in concurrent chip-package design flow" in *Proc. of ASPDAC*, pp. 837–842, 2011

[18] K.-S. Lin, H.-W. Hsu, R.-J. Lee, and H.-M. Chen, "Area-I/O RDL routing for chip-package codesign considering regional assignment," in *Proc. of EDAPSS*, pp. 1–4, 2010.

[19] X.-D. Liu, Y.-F. Zhang, G. Yeap, and X. Zeng, "An integrated algorithm for 3D-IC TSV assignment," in *Proc. of DAC*, pp. 652–657, 2011.

[20] John F. Park, "Board driven I/O planning & optimization," in *Proc. of ICCAD*, pp. 395–397, 2010.

[21] C.-Y. Peng, W.-C. Chao, Y.-W. Chang, and J.-H. Wang, "Simultaneous block and I/O buffer floorplanning for flip-chip design" in *Proc. of ASPDAC*, pp. 213–218, 2006.

[22] M. Sunohara, T. Tokunaga, T. Kurihara, and M. Higashi, "Silicon interposer with TSVs (Through Silicon Vias) and fine multilayer wiring," in *Proc. of ECTC*, pp. 847–852, 2008.

[23] J. Xiong, Y.-C. Wong, E. Sarto, and L. He, "Constraint driven I/O planning and placement for chip-package co-design," in *Proc. of ASPDAC*, pp. 207–212, 2006.

GPU-Based N-Detect Transition Fault ATPG

Kuan-Yu Liao, Sheng-Chang Hsu, and James Chien-Mo Li
Lab. of Dependable Systems (LaDS), Department of Electrical Engineering
National Taiwan University
Taipei, Taiwan

Abstract—This is a massively parallel ATPG that explores device-level, block-level and word-level parallelism in GPU. Eight-detect transition fault ATPG experiments on large benchmark circuits show that our technique achieved 5.6 and 1.6 times speedup compared with a single-core and 8-core CPU commercial tool, respectively. Test patterns selected from our test set are about the same length and quality as those selected from commercial N-detect ATPG. To the best of our knowledge, this is the first proposed GPU-based ATPG algorithm.

Keywords- GPU, test generation, *N*-detect, parallel.

1. Introduction

N-detect test sets have been shown to be more effective than single-detect test sets [1]. In an N-detect test set, each fault is detected by at least N different patterns. Test sets for detecting small delay defects (SDDs) are gaining importance in nanometer designs [2]. Test pattern selection from N-detect test sets has been shown to be an effective alternative to timing-aware ATPG for detecting SDDs [3]. However, N-detect transition fault ATPG still suffer from long test generation time. In our experiment, it took more than 10 days for a CPU-based single-core commercial tool to generate an 8-detect transition fault test set for a 1.1M gate circuit.

Parallel programming is a popular technique to speed up ATPG [4]. Parallel ATPG algorithms on CPU can be divided into three categories according to different partitioning schemes: fault partitioning [5], search space partitioning [6], and circuit partitioning [7]. The fault partitioning approach dynamically or statically partitions the fault list into multiple partitions and each of which is handled by a different core. The search space partitioning approach partitions the solution space into different portions, each of which is searched by different cores. The circuit partitioning approach divides the circuit into multiple subcircuits, each of which is taken care of by a core. Recent research showed that two to five times of speedup has been achieved using fault partitioning approach on eight-core CPU [5].

In addition to multi-core CPU, graphics processing unit (GPU) has become a popular general-purpose computing platform for various EDA algorithms. Modern GPUs contain several stream multiprocessors, each of which executes a *thread block* (or a *block* for short). Many researchers proposed GPU-based logic simulation and fault simulation techniques in recent years. Based on the partitioning schemes, simulation techniques can be classified into three categories: fault partitioning [8], pattern partitioning [9], and circuit partitioning [10]. Different partitions are assigned to different blocks for parallelization. All approaches above can be easily combined with word-level parallel techniques, such as well-known parallel fault simulation [11] and parallel pattern single fault propagation (PPSFP) [12], which are widely used on CPU. GPU accelerated Boolean SAT solver has been proposed in [13]. By transforming a circuit into conjunctive normal form representation, SAT solvers can be used to generate test patterns [14]. However, SAT-based ATPGs suffer from long runtime due to loss of circuit structure information. So far there is no efficient GPU-based ATPG algorithm available yet.

This paper presents a fast GPU-based N-detect transition fault ATPG, which is an extension to a CPU-based parallel ATPG algorithm, SWK [15]. All three components of this ATPG system are parallelized on GPU: test generation, logic simulation, and fault simulation. In test generation, three levels of parallelism are implemented: *device*-level, *block*-level and *word*-level partitioning. The partitioning techniques in each level are listed below:

1) Device-level: fault partitioning.
2) Block-level: fault and circuit partitioning.
3) Word-level: fault and search space partitioning.

Device-level fault partitioning statically partitions faults into different fault lists, each of which is assigned to a different device. Block-level fault partitioning assigns different target faults to different blocks, whereas block-level circuit partitioning assigns different logic levels to different blocks. World-level fault partitioning assigns target faults to different bits in a word, whereas word-level search space partitioning assigns different branches of the decision tree to different bits. That means, the proposed algorithm converts decision making into bitwise logic operation so different branches of the decision tree can be explored at the same time. Suppose the number of devices is v, the number of blocks is t, and the size of a word is w. Our ATPG generates $v \times t \times w$ patterns concurrently. For example, if $v=2$, $t=64$, and $w=32$, then 4,096 patterns can be generated simultaneously. This is a massively parallel ATPG that cannot be achieved on traditional CPU architecture.

In logic and fault simulation, two-level parallelism is implemented: block-level circuit partitioning and word-level pattern partitioning. Block-level circuit partition assigns different logic levels to different blocks, whereas word-level pattern partitioning assigns different patterns to different bits.

Overall, our GPU-based ATPG is 5.9 and 1.6 times faster than a single-core and 8-core CPU-based commercial ATPG in generating 8-detect transition fault patterns, respectively. The delay test quality (DTC) of the test patterns selected from our generated test set is similar to those selected from 8-core commercial ATPG generated test set. To our best knowledge, this is the first GPU-based ATPG algorithm proposed.

The structure of this paper is as follows. Section 2 gives the background of GPU architecture. Section 3 describes the details of our proposed algorithm. Section 4 shows the experimental results and section 5 concludes this paper. Detailed example and memory analysis can be found in the supplement.

2. GPU Architecture

GPU has become a highly parallel general computing device which supports up to tens of thousands of threads running concurrently. Our ATPG is implemented using *compute unified device architecture* (CUDA) on NVIDIA's graphic card. A GPU *device* contains hundreds of *stream processors*. A task running on GPU is a *kernel*, which specifies how each thread is executed. To manage the threads running on GPUs, CUDA provides a three-level architecture: *grid, block*, and *thread*. On GPU, all threads of the same tasks reside in the same grid. Each grid consists of multiple blocks and each block consists of multiple threads.

The memory architecture of the GPU can be divided into three levels: grid-level, block-level, and thread-level. Grid-level memories, *global, constant*, and *texture*, can be accessed by all threads. Constant and texture memories are read-only with cache, whereas global memory is readable and writable without cache. Block-level memory is *shared memory*, which can be accessed by all threads within the same block. Thread-level memory is *registers*, which can only be accessed by each thread.

2.1 Memory Allocation on GPU

Figure 1 shows memory allocation of our ATPG on GPU. Netlist is allocated in the grid-level texture memory since it is read-only and accessed by all threads. For each gate in the netlist (such as *G0*), information such as gate type and fanin/fanout are stored in a slot. Because 128 gates in a logic level are accessed simultaneously, empty slots are filled with dummies. Fault list is allocated in the global memory to store the location and the number of detection of each fault. Circuit signals (seven words per gate, see Section 3.1.1) are stored in global memory. Each of *t* blocks keeps its own circuit signals so that block-level parallel test generation can access these signals independently. Temporary values during run time of test generation and fault simulation are stored in the shared memory of each block.

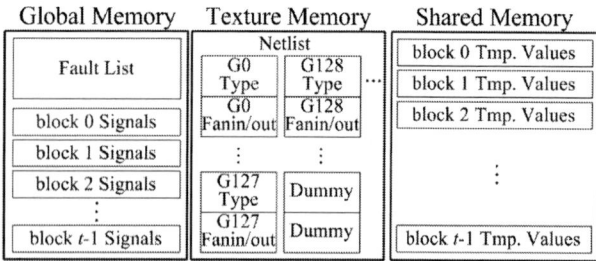

Figure 1. Memory Allocation on GPU

2.2 Memory Access Optimization

Memory access plays a crucial role in the performance of GPU programs. Three techniques have been implemented to optimize global memory access latency, access efficiency, and communication with CPU main memory.

1) The netlist information, such as gate types and connections, is allocated in texture memory to reduce global memory access latency. Texture memory has its own cache which can reduce access latency by one hundred times if data is cached.

2) *Coalesced access* is a NVIDIA GPU feature to maximize global memory access efficiency [16]. Coalesced access requires data access to be aligned to 128-byte *cache lines*— that means data addresses must start at multiples of 128. To utilize coalesced access, signal values are packed into 128-bytes and aligned to cache lines.

3) *Zero-copy* [16] technique is implemented to transfer generated patterns back to main memory to hide the communication overhead. A block of main memory is allocated as zero-copy memory. After each fault simulation, the patterns are *"zero-copied"* back to main memory while GPU continues to generate patterns for undetected faults at the same time.

For more details about memory access optimization, please refer to the supplement section.

3. GPU-Based ATPG Algorithm

Figure 2 shows the overall flow of our ATPG system. In the preprocess stage, the netlist is parsed, levelized and then duplicated into two time frames. After the preprocess stage, the amount of memory needed for storing netlist and fault list is allocated on the GPU. The *test generation kernel* is then invoked by CPU to generate test patterns for undetected faults. After test generation kernel, the *fault simulation kernel* is invoked to simulate the test patterns. The number of detection of each fault is recorded. A fault is dropped if the number of detection is larger than or equal to *N*. The generated test patterns are zero-copied to CPU main memory after each simulation. At the same time, test generation continues to generate tests for remaining faults. After all faults have been tried, test selection is performed based on DTC to obtain a high quality compact test set.

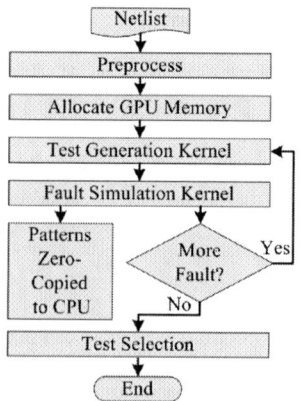

Figure 2. Proposed ATPG Overall Flow

3.1 Test Generation Kernel

The test generation kernel consists of three levels of parallelism: word-level parallelism for a block, block-level parallelism for a kernel, and device-level parallelism for a circuit under test (CUT). Each block executes a word-level parallel test generation algorithm and each device handles a statically partitioned fault list. The following subsections describe each level of parallelism in detail with a simple example. Please refer to [15] or the supplement for a full example.

3.1.1 Word-level Parallel Test Generation for a Block

The word-level parallel test generation algorithm extends a previous work, bitwise parallel SWK ATPG on CPU [15], to handle N-detect transition faults test generation on GPU. This PODEM-based [17] algorithm generates test patterns by repeatedly backtracing objectives to inputs and propagating fault effects to outputs. The main concept of this algorithm is converting backtrace and propagation into bitwise logic operation so different branches of the decision tree can be explored at the same time.

In this algorithm, a signal Y is represented by seven words of w bits: $Y^0, Y^1, Y^d, Y^{\bar{d}}, Y^{b_0}, Y^{b_1}$, and Y^p. Each bit in a word represents an individual *clone*, which performs an independent search. The meaning of the first four words is as follows.

$Y^0 =1$: signal Y is zero (good 0/faulty 0).

$Y^1 =1$: signal Y is one (good 1/faulty 1).

$Y^d =1$: signal Y is d (good 1/faulty 0).

$Y^{\bar{d}} =1$: signal Y is $\bar{d}$ (good 0 /faulty 1).

For a given clone, the above four values are mutually exclusive so at most one of them equals one at a time. If all of them are zero, Y is unknown. When Y is unknown, the other three words indicate whether Y is on the *propagation path* or the *objective path*. The former is a path that faulty effect (d or $\bar{d}$) will potentially propagate to reach an output. The latter is a path that the objective backtrace follows to reach an input. For each clone,

$Y^p =1$: signal Y is on a propagation path.

$Y^{b_0} =1$: signal Y is on an objective 0 backtrace path.

$Y^{b_1} =1$: signal Y is on an objective 1 backtrace path.

Again, these three values are mutually exclusive so at most one of them equals one at a time.

Figure 3 shows the test generation flow of each block in our GPU-based test generation for N-detect transition delay faults. Given a set of faults F, word size w, and the target number of detection N, each fault in F is handled by N clones. For example, if $w = 32$, $F = \{f_1, f_2, f_3, f_4\}$, and $N = 8$, then the first eight clones generate patterns for f_1, and the following eight clones generate patterns for f_2, and so on. For transition faults test generation, the two initial objectives are fault-free values at the fault site in two time frames. A backtrace is then performed from the fault site to the inputs. After inputs are assigned, fault effect propagation is performed from inputs to the outputs. The test generation of a clone is *successful* if a d or $\bar{d}$ has reached any output. This process ends if test generations for all w clones are successful or time limit has reached.

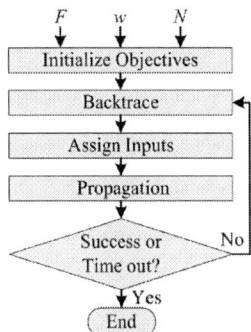

Figure 3. Test Generation Kernel Flow (For A Single Block)

Figure 4 to Figure 7 illustrate an example to generate a test pattern for a single transition fault. This is a two-time frame circuit, each of which consists of three AND gates, three inputs (E, F, G), and one flip-flop (H). LV stands for logic level. Suppose that the target fault is $G2$ slow-to-fall. We use a two-bit word ($w=2$) to represent two clones: clone #1 and clone #2. Since the fault is slow-to-fall, the initial objectives in time frame one and two are one and zero, respectively. They are denoted as $\{b_1, b_1\}$ and $\{b_0, b_0\}$ in Figure 4.

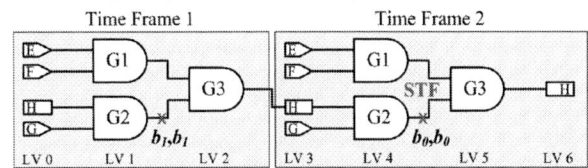

Figure 4. Test Generation Example (Initialize Objectives)

Figure 5 shows the first backtrace after Figure 4. Backtracing $G2$ in time frame one is an *implication backtrace* because both gate inputs must be one to justify the output objective b_1. Backtracing $G2$ in time frame two requires a decision because either $H=0$ or $G=0$ justifies the output objective b_0. An *objective split* is performed to assign two clones with different objectives. In this example, clone #1 backtraces b_0 on input H and clone #2 backtraces b_0 on input G.

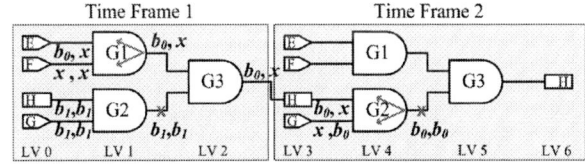

Figure 5. Test Generation Example (First Backtrace)

Figure 6 shows the propagation after Figure 5. The fault is excited for both clones so $G2$ in time frame two is $\{\bar{d}, \bar{d}\}$ in the figure. Output H is $\{p, p\}$ in the figure. The symbol 'p' indicates that output H is on the propagation path.

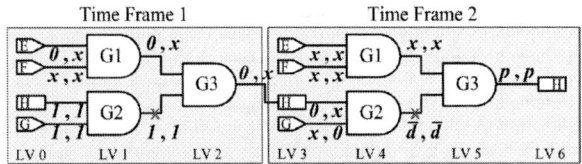

Figure 6. Test Generation Example (First Propagation)

Figure 7 shows the second backtrace followed by another propagation. To propagate the faulty effect to output, both clones of *E* and *F* in time frame two are set to one. Two test patterns *EFG*={(0x1, 11x), (xx1, 110)} and *H*={1, 1} are successfully generated.

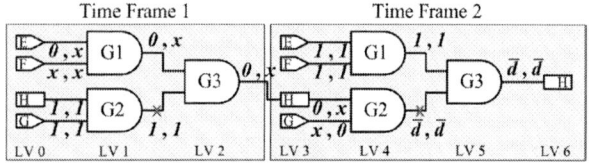

Figure 7. Test Generation Example (Second Backtrace and then Propagation)

3.1.2 Block-Level Parallel Test Generation for a Kernel

Figure 8 illustrates the block-level parallelism of *t* blocks generating test patterns simultaneously. Each block handles a logic level and each thread in a block handles a gate. Each *x* represents a target fault. The i_{th} thread in the block is shown as th_i in the figure. For example in Figure 8, block 0 handles logic level 4. In block 0, th_0 and th_1 handle the upper and the lower gate in logic level 4, respectively. The number of threads needed in a block is equal to the maximum number of gates in a logic level. If the number of gates in a logic level is larger than the maximum threads allowed in a block, multiple blocks are assigned to the logic level. Each block has a copy of signals and performs test generation for different faults independently. As shown in Figure 8, block 0 is performing backtrace and block 1 is performing propagation. All blocks are synchronized after all threads are finished.

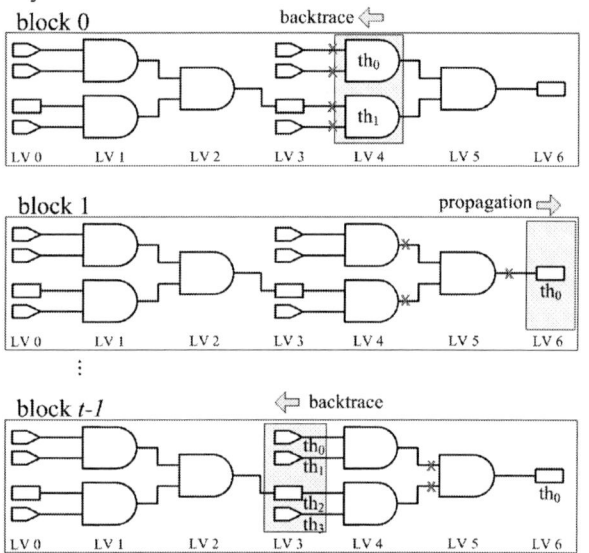

Figure 8. Test Generation Block-Level Parallelism

Two variables, *current_level* and *current_status*, are stored in the shared memory of each block. *current_level* which represents the current logic level the block is executing. *current_status* indicates the four status of the block (namely *initialize*, *backtrace*, *propagation*, and *finished*). *current_level* and *current_status* are updated by the last thread in a block. The *current_level* is incremented every time when the block is propagating signals and is decremented when the block is backtracing signals. The *current_status* is set to *backtrace*

when a block propagates to outputs and is set to *propagation* when a block backtraces to inputs. If all clones have at least one *d* or $\bar{d}$ at any output, *current_status* is set to *finished*. Take Figure 8 for example, th_1 of block 0 updates *current_level* to 3, th_3 of block *t*-1 updates *current_level* to 2, and th_0 of block 1 updates *current_status* to *backtrace*.

The test generation kernel terminates when all blocks are finished or execution time reaches a user defined limit. To balance the load between different blocks, all blocks are working on faults in the same logic level at a time.

3.1.3 Device-Level Parallel Test Generation for a CUT

All transition faults are statically partitioned into groups, each of which is handled by a device. To balance the load between devices, faults within the same logic level are evenly distributed into different groups.

3.4 Fault Simulation Kernel

The fault simulation kernel simulates the patterns generated by the test generation kernel. There are two levels of parallel fault simulation. Word-level parallelism is implemented in a way similar to conventional PPSFP on CPU [12]. This section only explains the block-level parallelism. The fault simulation kernel reuses memory space allocated for the test generation kernel, so no additional memory is required. Figure 9 shows an example of the block-level parallel fault simulation kernel. In this example, seven blocks are created to simulate a pattern set generated by a block in the test generation kernel, so a total of 7*t* blocks are created to perform fault simulation for all patterns generated by *t* blocks. As in test generation, a block handles a logic level and each thread in a block handles a gate in the logic level. However, in fault simulation, the logic level assigned to a block is fixed. Each block simulates gates in its logic level and passes the results to the next block.

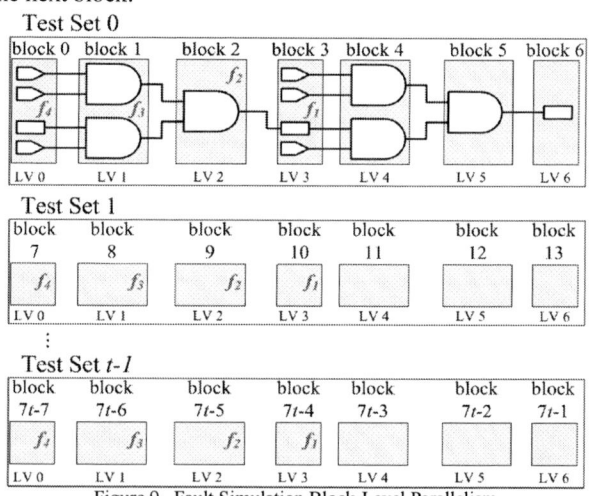

Figure 9. Fault Simulation Block-Level Parallelism

The fault simulation consists of three phases. 1) Patterns in each test set are randomly filled. Parallel pattern logic simulation is then performed. 2) PPSFP is performed for all undetected faults. 3) Last logic level blocks check for *d* or $\bar{d}$ and update the detection number for the target fault.

196

Faults are shifted from low logic level to high logic level in a pipelined fashion. Take Figure 9 for example, suppose there are four faults to be simulated, f_1, f_2, f_3, and f_4. The level 0 blocks (block 0, block 7, ..., and block 7t-7) first simulate f_1. After each block synchronization, faults are shifted to the next block to the right. Figure 9 shows the status after three block synchronizations. The fault simulation kernel ends when logic level 6 blocks finished simulating f_4.

4. Experimental Results

To validate the proposed GPU-based ATPG, experiments are performed on large ISCAS'89, ITC'99, and IWLS'05 benchmark circuits and a one-million gate benchmark, RSA, a cipher circuit designed by us. Results are compared with a commercial ATPG tool with multi CPU core support. The hardware specifications are summarized in Table II. The GTX590 graphic card contains two devices.

Table II. Platform Specification

	CPU	GPU
Model	i7-2600	GTX590
#Cores	8 Processors	16 Stream Multiprocessors 512 Stream Processors
Clock Speed	3.4GHz	1.22GHz
Memory	16GB	1.5GB

A single-core CPU-based version of our SWK algorithm is implemented to compare with our proposed GPU-based version. Table III shows the runtime (RT), test length (TL), and fault coverage (FC) of our CPU-based version and GPU-based version using one device. The last row shows the averaged results normalized with respect to those of the CPU-based version. Overall, the GPU-based version has 6.7 times of speedup compared with CPU-based version.

Table III. Comparison with our CPU Version

Circuits	Our CPU [13]			Our 1 Device		
	RT(s)	TL	FC(%)	RT(s)	TL	FC(%)
s38417	101	1,591	98.55	21	1,602	98.58
s38584	187	3,046	92.14	22	3,138	92.25
b17	3,982	17,983	81.57	571	19,873	81.68
b18	27,903	69,750	78.62	3,687	71,691	78.65
b19	76,455	138,734	77.61	10,657	139,566	77.68
Normalize	1.00	1.00	1.00	0.15	1.04	1.00

Table IV shows the results of 8-detect transition fault test generation of a commercial tool compared with proposed GPU-based ATPG using two devices. In our test generation kernel, 64 blocks are used except leon3mp and RSA, which only uses 16 blocks due to insufficient global memory. The first two columns show benchmark names and primitive gate counts. Single-core, 4-core, and 8-core commercial ATPGs are used to

generate test patterns. No significant speedup is observed beyond 8-core parallel ATPG. The last row shows the averaged result normalized to that of the single-core commercial ATPG. Commercial tool results for RSA are not finished at the time of submission. We stopped the single-core commercial tool before it is finished and recorded its results in the table. The normalized numbers excludes the RSA case. Compared with CPU-based single-core commercial ATPG, our 2-device GPU-based ATPG achieved 5.9 times of speedup with 41% test length overhead. Compared to 8-core CPU-based parallel ATPG, our ATPG is 1.6 times faster.

To evaluate the quality of our test sets, we use DTC [18] as test quality metrics. DTC calculates the effectiveness of a test set in detecting transition faults through long paths, and is used to evaluate the quality of our test patterns. DTC is defined as

$$DTC = \frac{\sum_{f \in F} W_f}{M} \quad \text{(eq. 1)}$$

$$W_f = \frac{D_f^a}{D_f^s} \quad \text{(eq. 2)}$$

where D_f^s is the structural longest path through fault f, and D_f^a is the actual longest path through fault f sensitized by the test set. F is the set of transition delay faults and M is the number of total faults in F. The upper bound of delay test coverage is equal to the transition delay fault test coverage. The larger the DTC, the better the test set is in detecting transition delay faults through long paths. Table V shows the DTC of our ATPG and 8-core commercial tool generated test sets. The last row shows the average results normalized with respect to those of commercial 8-core ATPG. The DTC of our generated test sets is higher than those of commercial ATPG but our test length is longer. The data for leon3mp and RSA is not shown because the evaluation program is not finished at the time of submission.

Table V. DTC Comparison

Circuits	Commercial 8 Cores	Our 1 Device	Our 2 Devices
s38417	92.45	93.89	93.92
s38584	84.48	85.70	85.75
b17	57.86	58.24	58.87
b18	60.84	62.61	63.57
b19	60.32	62.15	64.66
Normalize	1.00	1.02	1.03

Test selection is applied to show that a compact test set selected form our generated tests remains high DTC coverage. Test selection is performed by [19]. Commercial 8-core ATPG and our 2-device ATPG are compared in this experiment.

Table IV. Comparison with Commercial CPU-Based ATPG

Circuit	#Primitive Gates	Commercial 1 Core			Commercial 4 Cores			Commercial 8 Cores			Our 2 Devices		
		RT(s)	TL	FC(%)	RT(s)	TL	FC(%)	RT(s)	TL	FC(%)	RT(s)	TL	FC(%)
s38417	24K	55	1,288	98.48	15	1,280	98.69	12	1,345	98.71	13	1,682	98.71
s38584	21K	86	2,715	92.36	31	2,820	92.44	18	2,884	92.46	13	3,584	92.46
b17	42K	2,614	13,572	75.61	884	16,623	81.03	693	17,424	81.76	372	20,670	81.80
b18	116K	16,237	54,041	74.08	6,397	60,711	77.76	5,419	63,909	78.55	1,989	75,053	79.62
b19	234K	41,189	108,840	73.30	14,698	123,031	77.06	12,106	127,461	77.55	6,797	142,471	78.73
leon3mp	978K	326,139	168,585	94.16	150,807	208,285	97.32	106,533	229,351	98.53	68,003	268,338	98.61
RSA	1,103K	1,429,670	145,425	95.45	705,584	171,281	97.12	526,774	183,243	97.78	387,249	197,939	98.13
Normalize	-	1.00	1.00	1.00	0.38	1.13	1.03	0.29	1.19	1.04	0.18	1.40	1.04

Table VI shows the DTC and test length of the patterns selected. The last row shows the averaged results normalized to that of commercial 8-core ATPG. The results show that the DTC of the test patterns selected from our ATPG is slightly higher than that of selected from the commercial 8-core ATPG with 2% test length overhead only. Benchmarks leon3mp and RSA failed test selection due to insufficient CPU memory.

Table VI. Test Selection Comparison

Circuits	Commercial 8 Cores		Ours 2 Devices	
	TL	DTC(%)	TL	DTC(%)
s38417	356	90.15	411	90.22
s38584	498	83.22	547	83.27
b17	2,793	54.40	2,878	56.70
b18	9,252	58.34	9,261	58.89
b19	18,503	57.97	19,728	60.44
Normalize	1.00	1.00	1.07	1.02

This paragraph analyzes the scalability of our ATPG. In our current architecture, the memory consumption of a block in the test generation kernel is proportional to netlist size. In Figure 10, the diamond line shows the maximum number of blocks allowed with respect to different netlist size on our GPU device (1.5GB global memory). The number of blocks used in experiment is rounded down to multiples of 16 since there are 16 stream multiprocessors on our device.

The upper dashed line shows the upper bound of blocks due to the stream multiprocessors. In our experiments, using more than 64 blocks has no additional speedup so at most 64 blocks are used. The lower dashed line shows the minimum number of blocks needed to make full use of all 16 stream multiprocessors. According to Figure 10, our proposed GPU-based ATPG is scalable to approximately 1.2M gate design while maintaining 16 blocks parallel test generation, given 1.5GB global memory. When the size of design increases beyond 1.2M gates, the number of blocks decreases below 16.

The GPU hardware is improving every year in terms of memory capacity, number of cores, and clock speed. Our tool can be easily scaled to larger designs with the hardware improvement on GPU.

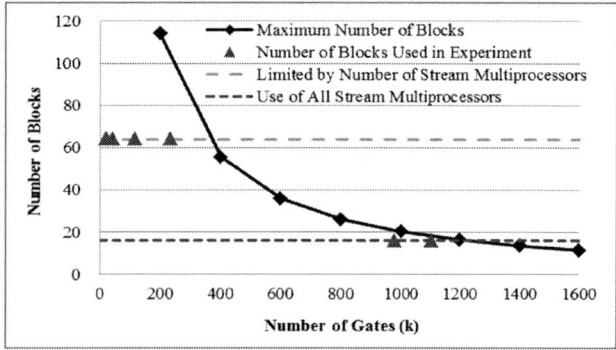

Figure 10. Number of Blocks to Number of Gates

5. Conclusion

A GPU-based massively parallel ATPG algorithm to generate high quality N-detect transition delay fault test patterns is proposed. Three-level parallelism is exploited in the test generation kernel. Experimental results show that our GPU-based ATPG achieves 5.6 times speedup over single-core CPU-based commercial ATPG. The DTC of our generated test patterns is similar to those of 8-core CPU-based commercial ATPG. To the best of our knowledge, this is the first proposed GPU-based ATPG algorithm.

References

[1] C.-W. Tseng and E. J. McCluskey, "Multiple-Output Propagation Transition Fault Test," *Proc. Int'l Test Conf.*, pp. 358-366, 2001.

[2] N. Ahmed, M. Tehranipoor, and V. Jayaram, "Timing-Based Delay Test for Screening Small Delay Defects," *Proc. IEEE Design Automation Conf.*, pp. 320–325, 2006.

[3] H. Lee, S. Natarajan, S. Patil, and I. Pomeranz, "Selecting High-quality Delay Tests for Manufacturing Test and Debug," *Proc. IEEE Int'l Symp. on Defect and Fault Tolerance in VLSI Systems Defect Fault Tolerance*, pp. 59-70, 2006.

[4] R. H. Klenke, R. D. Williams, and J. H. Aylor, "Parallel-Processing Techniques for Automatic Test Pattern Generation," *IEEE Computer*, Vol.25, No.1, pp.71-84, 1992.

[5] X. Cai, P. Wohl, J. A. Waicukauski, and P. Notiyath, "Highly Efficient Parallel ATPG Based on Shared Memory," *Proc. Int'l Test Conf.*, pp. 1-7, 2010.

[6] S. Patil and P. Banerjee, "A Parallel Branch and Bound Algorithm for Test Generation," *IEEE Trans. Computer-aided Design*, Vol. 9, No.3, pp. 313-322, 1990.

[7] S. P. Smith, B. Underwood, and M. R. Mercer, "An Analysis of Several Approaches to Circuit Partitioning for Parallel Logic Simulation," *Proc. IEEE Int'l Conf. on Comput. Design*, pp. 664-667, 1987.

[8] M. Li and M. S. Hsiao, "FSimGP^2: An Efficient Fault Simulator with GPGPU," *Proc. IEEE Asian Test Symp.*, pp. 15-20, 2010.

[9] K. Gulati and S. P. Khatri, "Towards Acceleration of Fault Simulation using Graphics Processing Units," *Proc. IEEE Design Automation Conf.*, pp. 822-827, 2008.

[10] M. A. Kochte, M. Schaal, H.-J. Wunderlich, and C. G. Zoellin, "Efficient Fault Simulation on Many-Core Processors," *Proc. IEEE Design Automation Conf.*, pp. 380-385, 2010.

[11] S. Sesuh and D. N. Freeman, "On Improved Diagnosis Program," *IEEE Trans. Electron. Comput.*, EC-14(1), pp. 76-79, 1965.

[12] J. A. Waicukauski, E. B. Eichelberger, D. O. Forlenza, E. Lindbloom, and T. McCarthy, "Fault Simulation for Structured VLSI," *Proc. VLSI Syst. Des.*, 6(12), pp. 20-32, 1985.

[13] S. Bechers, G. De Samblanx, F. De Smedt, T. Goedeme, L. Struyf, and J. Vennekens, "Parallel Hybrid SAT Solving Using OpenCL," *Proc. Int'l Conf. on Applied Computing*, pp. 435-441, 2011.

[14] T. Larrabee, "Test Pattern Generation Using Boolean Satisfiability," *IEEE Trans. Comput.-Aided Design Intergr. Circuits Syst.*, vol. 11, Issue 1, pp. 4-15, 1992.

[15] K.-Y. Liao, C.-Y. Chang, and J. C.-M. Li, "A Parallel Test Pattern Generation Algorithm to Meet Multiple Quality Objectives," *IEEE Trans. Comput.-Aided Design Intergr. Circuits Syst.*, Vol. 30, Issue 11, pp. 1767-1772, 2011.

[16] nVidia (2012). *CUDA C Best Practices Guide* (Version 4.1) [Online]. Available: http://www.nvidia.com

[17] P. Goel, "An Implicit Enumeration Algorithm to Generate Tests for Combinational Logic Circuits", *IEEE Trans. Computers*, Vol. C-30, Issue 3, pp. 215-222, 1981.

[18] S. Mitra, E. Volkerink, E.J. McCluskey, and S. Eichenberger, "Delay Defect Screening using Process Monitor Structures," *Proc. IEEE VLSI Test Symp.*, pp. 43-52, 2004.

[19] C.-Y. Chang, "Compact Test Pattern Selection for Small Delay Defect," M. S. thesis, Dept. Elect. Eng., National Taiwan Univ., 2011.

Supplement

S1 SWK Test Generation Equations [13]

Here we give the detailed SWK test generation equations using the same example in Section 3. Each subsection explains a block in the test generation kernel flow (Figure 3).

S1.1 Initialize Objectives

Figure 11 illustrates the initial objectives of test generation for the target fault Y slow-to-fall. Y_{first} and Y_{second} represent the signal in time frame one and time frame two, respectively. The status of the objective generation is stored in a w-bit OBJ flag vector. $OBJ_k = 1$ means that the k_{th} clone has an objective generated; otherwise, the k_{th} clone is still waiting for an objective to be generated. The initial objective generation is shown in the following equations.

$$Y_{first}^{b_1} = Y_{first}^{x} \qquad \text{(eq. 3a)}$$

$$Y_{second}^{b_0} = Y_{second}^{x} \qquad \text{(eq. 3b)}$$

$$OBJ = Y_{second}^{x} \qquad \text{(eq. 3c)}$$

Equation 3a and 3b generate the objective values in time frame one and time frame two, respectively. Equation 3c sets the OBJ flag to prevent another objective from being generated again.

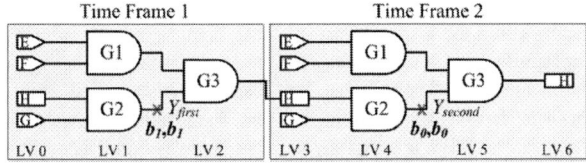

Figure 11. Initial Objectives

S1.2 Backtrace

After the initial objectives are set, backtrace is performed from outputs to inputs. The backtrace function can be divided into two *phases*: the o-*backtrace* and the p-*backtrace*. The former searches for input assignments to achieve the objectives, whereas the latter searches for paths to propagate the fault effect. In the example, since the target fault has not been excited yet, we show the equations for performing o-backtrace here. Please refer to S1.4 for p-backtrace after the fault is excited. Please note that every clone is independent so each clone can be in different backtrace phase at the same time. The following equations implement the o-backtrace of an AND gate with inputs AB and output C,

$$A^{b_1} = A^x(B^d + B^{\bar{d}})C^p\overline{OBJ} + A^x(B^0 + B^{\bar{d}})'C^{b_1} \qquad \text{(eq. 4a)}$$

$$A^{b_0} = A^x(B^1 + B^d + B^{\bar{d}})C^{b_0} + A^p B^p C^{b_0} S \qquad \text{(eq. 4b)}$$

$$B^{b_1} = B^x(A^d + A^{\bar{d}})C^p\overline{OBJ} + B^x(A^0 + A^{\bar{d}})'C^{b_1} \qquad \text{(eq. 4c)}$$

$$B^{b_0} = B^x(A^1 + A^d + A^{\bar{d}})C^{b_0} + B^p A^p C^{b_0} \overline{S} \qquad \text{(eq. 4d)}$$

where a prime sign indicates a complement.

S is a random *split vector* of w bits, half of which are ones and the other half are zeros. Three sample split vectors and their complements are listed below.

$S_1 = \{0000...01111...1\}$ $\overline{S_1} = \{1111...10000...0\}$

$S_2 = \{1010101010...10\}$ $\overline{S_2} = \{0101010101...01\}$

$S_3 = \{11001100...1100\}$ $\overline{S_3} = \{00110011...0011\}$

where S_1 and $\overline{S_1}$ is a pair of complement split vectors. Figure 12 shows the backtrace from fault site to inputs. Take $G2$ in the second time frame for example, the objective paths are split into different clones by applying equation 4. The upper input of $G2$ has objective b_0 in clone #1 whereas the lower input has objective b_0 in clone #2. In the example, the objectives are split into different clones at $G2$ in time frame two, $G1$ in time frame one, and $G3$ in time frame one. The objectives are implied instead of split at $G2$ in time frame one.

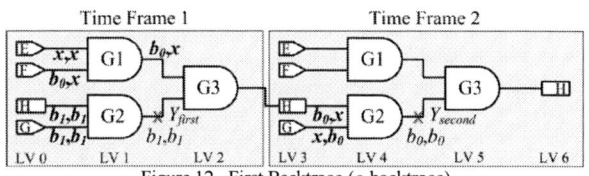

Figure 12. First Backtrace (o-backtrace)

S1.3 Assign Inputs

Inputs are assigned to their backtrace values when the test generation kernel backtraces to inputs. For example in Figure 12, two test cubes {X011, XX11} are assigned to inputs E, F, G, H in time frame one and {XXX, XX0} are assigned to inputs E, F, and G in time frame two.

S1.4 Propagation

After inputs are assigned, propagation is performed from inputs to outputs. A d or $\bar{d}$ is generated when the fault site is controlled to its fault-free value. d-generation can be easily performed by inserting a logic operation at the fault site. For Y slow-to-fall fault in Figure 11, the corresponding d-generation can be simply implemented by:

$$Y_{second}^{\bar{d}} = Y_{second}^{0} \qquad \text{(eq. 5)}$$

A p is generated when one of the gate inputs is d or $\bar{d}$ while the other input is unknown. A p could be propagated from gate input to gate output when three conditions are met: 1) one of the inputs is p, and 2) the other input is non-controlling value, and 3) the gate output is unknown. For a two-input AND gate, the *p-generation* and *p-propagation* can be written as the following bitwise logic operation.

$$C^p = A^x(B^d + B^{\bar{d}}) + B^x(A^d + A^{\bar{d}}) + \overline{A^0} B^p C^x + A^p \overline{B^0} C^x \qquad \text{(eq. 6)}$$

where first two terms represent p-generation, while the last two terms take care of p-propagation.

In Figure 13, {p, p} is generated at the output of $G3$ in the second time frame because the upper input of $G3$ is {x, x} and the lower input of $G3$ is {$\bar{d}$, $\bar{d}$}. p-propagation does not appear in the example but is shown in equation 6.

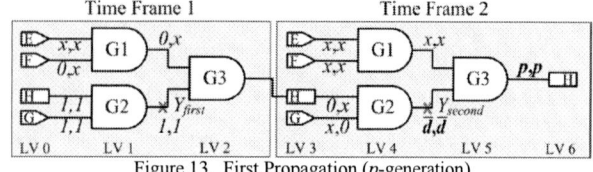

Figure 13. First Propagation (*p*-generation)

S1.5 Success or Time Out

After each propagation, d or $\bar{d}$ is checked at the outputs. The test generation backtraces again from outputs to inputs if no d or $\bar{d}$ has reached outputs. The following equations implement the *p*-backtrace of an AND gate with inputs AB and output C,

$$A^p = A^p(B^1 + \overline{B^p})C^p \cdot \overline{OBJ} + A^p B^p C^p S \cdot \overline{OBJ} \qquad \text{(eq. 7a)}$$

$$B^p = B^p(A^1 + \overline{A^p})C^p \cdot \overline{OBJ} + A^p B^p C^p \overline{S} \cdot \overline{OBJ} \qquad \text{(eq. 7b)}$$

After the fault is excited, the goal is to propagate d or $\bar{d}$ to an output. An objective is generated at the gate input when two conditions are met: 1) no objective has been generated so far, and 2) the gate is a *d*-frontier. Take an AND gate with inputs AB and output C for example, equation 8a generates an objective when the gate is a *d*-frontier ($B = d$ or $\bar{d}$, and $A = x$) and there is no existing objective. Equation 8b set the control flag to prevent another objective from being generated again for the same clone.

$$A^{b_1} = A^x(B^d + B^{\bar{d}})C^p \overline{OBJ} \qquad \text{(eq. 8a)}$$

$$OBJ = A^{b_1} + OBJ \qquad \text{(eq. 8b)}$$

Figure 14 shows the example of the second backtrace. The objectives $\{b_1, b_1\}$ are generated at the upper input of *G3* in the second time frame. Then we perform *o*-backtrace so both inputs of *G1* in the second time frame are assigned to one.

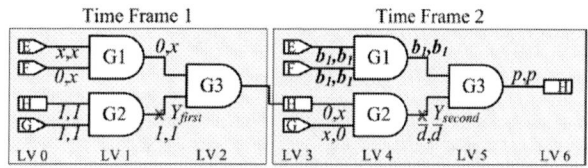

Figure 14. Second Backtrace (*o*-backtrace)

After inputs are assigned, the fault effect (d or $\bar{d}$) is propagated from a gate input to its gate output when two conditions are true: 1) at least one of the input is d or $\bar{d}$, and 2) none of the gate input holds the controlling value. For an example of an AND gate with inputs AB and output C, the following logic operations implement *d-propagation*.

$$C^1 = A^1 B^1 \qquad \text{(eq. 9a)}$$

$$C^0 = A^0 + B^0 + A^d B^{\bar{d}} + A^{\bar{d}} B^d \qquad \text{(eq. 9b)}$$

$$C^d = A^1 B^d + A^d B^1 + A^d B^d \qquad \text{(eq. 9c)}$$

$$C^{\bar{d}} = A^1 B^{\bar{d}} + A^{\bar{d}} B^1 + A^{\bar{d}} B^{\bar{d}} \qquad \text{(eq. 9d)}$$

Figure 15 shows an example of fault effect propagated through *G3* in the second time frame to the output. At this point, both two clones have successfully propagated the fault

effect to the output. Two test cubes $EFG = \{(X01,11X), (XX1, 110)\}$ and $H = \{1, 1\}$ are successfully generated. Here we only show test generation equations for AND gates. Similar equations can be derived for other gate types.

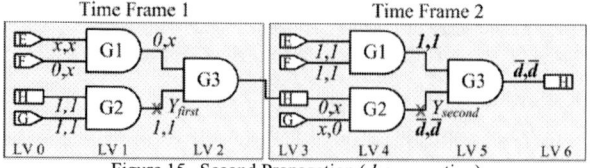

Figure 15. Second Propagation (*d*-propagation)

S2 GPU Memory Access Optimization Analysis

We analyze the three memory access optimization techniques in section 2. Table VII shows the type, level, latency, and size of different available memories on our NVIDIA GTX590 GPU.

Table VII. NVIDIA GTX590 GPU Memory

Type	Level	Latency cycles (cached)	Size	Accessibility
Global	Grid	400-600	1.5GB	R/W
Texture	Grid	400-600 (4-6)	1.5GB	Read-only
Constant	Grid	400-600 (4-6)	64KB	Read-only
Shared	Block	4-6	48KB	R/W
Register	Thread	immediate	32KB	R/W

Figure 16 shows the runtime after applying each optimizing technique to benchmark circuit b19 using single device. Technique A is allocating netlist in texture memory. Technique B is coalesced access to global memory plus technique A. Technique C is using zero-copy to transfer patterns from global memory to CPU main memory plus technique B. By comparing techniques A, B, and C with the version without optimization, runtime is reduced by 7%, 25%, and 32%, respectively.

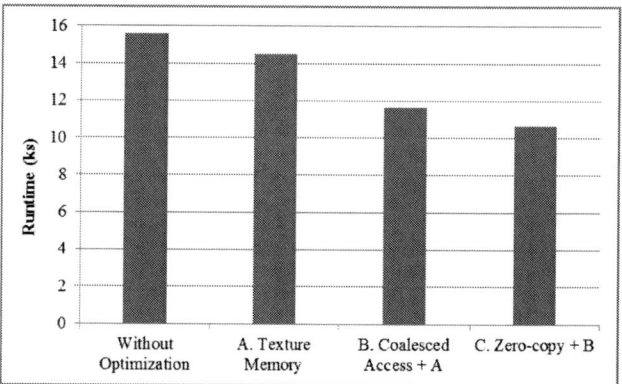

Figure 16. Runtime of Memory Access Optimization Techniques (b19).

Post-silicon Conformance Checking with Virtual Prototypes

Li Lei
Dept. of Computer Science
Portland State University
leil@cs.pdx.edu

Fei Xie
Dept. of Computer Science
Portland State University
xie@cs.pdx.edu

Kai Cong
Dept. of Computer Science
Portland State University
congkai@cs.pdx.edu

ABSTRACT

Virtual prototypes are increasingly used in device/driver co-development and co-validation to enable early driver development and reduce product time-to-market. However, drivers developed over virtual prototypes often do not work readily on silicon devices, since silicon devices often do not conform to virtual prototypes. Therefore, it is important to detect the inconsistences between silicon devices and virtual prototypes.

We present an approach to post-silicon conformance checking of a hardware device with its virtual prototype, i.e., a virtual device. The conformance between the silicon and virtual devices is defined over their interface states. This approach symbolically executes the virtual device with the same driver request sequence to the silicon device, and checks if the interface states of the silicon and virtual devices are consistent. Inconsistencies detected indicate potential errors in either the silicon device or the virtual device. We have evaluated our approach on three network adapters and their virtual devices, and found 15 inconsistencies exposing 15 real bugs in total from the silicon and virtual devices. The results demonstrate that our approach is useful and efficient in facilitating device/driver co-validation at the post-silicon stage.

Categories and Subject Descriptors

B.6.2 [**Logic Design**]: Reliability and Testing—*Error-checking*

General Terms

Design, Reliability, Verification

Keywords

Post-silicon validation, conformance checking, virtual prototypes

1. INTRODUCTION

Virtual prototyping has emerged as a promising technique for device/driver co-development. A notable example is how Intel utilizes virtual devices to enable early driver development for their new generation, 40 Gigabit Ethernet (40GbE) adapter, before the FPGA prototype is available [13].

Permission to make digital or hard copies of all or part of this work for personal or classroom use is granted without fee provided that copies are not made or distributed for profit or commercial advantage and that copies bear this notice and the full citation on the first page. To copy otherwise, to republish, to post on servers or to redistribute to lists, requires prior specific permission and/or a fee.
DAC 2013 May 29 - June 07 2013, Austin, TX, USA

The use of virtual prototypes, i.e., virtual devices, has potential to shorten device/driver development cycles. Nevertheless to achieve this benefit, a key challenge has to be addressed. As silicon devices often do not conform to virtual devices, drivers developed over virtual devices often do not work readily on silicon devices due to either silicon device bugs or driver bugs hidden on virtual devices. Unfortunately, troubleshooting these bugs today heavily depends on ad-hoc and time-consuming system testing/debugging at the post-silicon stage. By detecting the inconsistencies between the virtual and silicon devices, conformance checking provides a systematic and efficient way to (1) expose the virtual or silicon device errors; (2) reveal the causes of driver bugs hidden on the virtual device.

We present a novel approach to post-silicon conformance checking of a hardware device with its virtual device and discovery of their inconsistencies. The conformance of a silicon device and its virtual device is defined between their interface states (cf. Section 3.1). This approach symbolically executes a virtual device with the same driver request sequence to its silicon device, and checks if the interface states of the silicon and virtual devices are consistent. There are three major steps: (1) recording the driver requests to the silicon device; (2) symbolically executing the virtual device by taking the recorded request sequence; (3) checking if the silicon and virtual device interface states are consistent after executing the virtual device on each driver request.

Post-silicon validation often suffers from the limited observability of hardware, a critical problem that our approach needs to address. A silicon device has internal states that are difficult to observe. Moreover, the outside environment inputs to the silicon device are hard to capture. We use symbolic execution to address this problem. Instead of observing the internal states and environment inputs, our approach models them using variables with symbolic values when simulating the silicon device behaviors on the virtual device. This way symbolic execution covers their possible values.

We implement our approach using the virtual devices from the QEMU virtual machine [2] as virtual prototypes, and discover 15 inconsistencies behind which there are 15 real bugs in the silicon and virtual devices. These bugs can cause severe problems, e.g., system crashes. Detecting these bugs can significantly facilitate hardware/software co-validation at the post-silicon stage.

The rest of this paper is organized as follows. Section 2 introduces related background. Section 3 presents our approach. Section 4 shows how we implement the approach. Section 5 reports our experiment results. Section 6 presents the related work. Section 7 concludes and discusses the future work.

2. BACKGROUND

In this section, we introduce three related concepts: QEMU virtual devices which we adopt as our virtual prototypes, non-deterministic interleaving which we utilize in generating execution harnesses for virtual devices, and symbolic execution with which we replay driver requests on virtual devices.

2.1 QEMU and Virtual Devices

QEMU [2] is a virtual machine that can emulate different processor architectures, such as x86, SPARC, and ARM. It also emulates virtual devices for different peripheral devices, e.g., network adapters and mass storage devices. Such virtual devices are widely used for device driver developments.

A QEMU virtual device is a software component integrated into QEMU. We illustrate the virtual device concept with the Intel e1000 network adapter, a PCI (Peripheral Component Interconnect) device. As shown in Figure 1, the e1000 virtual device has the following major components:

- PCI device state, as defined by E1000State, which keeps track of the state of the PCI device;

- PCI device module functions, which simulate the basic functionalities of the PCI device. As Figure 1 shows, function e1000_mmio_writel simulates how the e1000 device responds to the driver write request, and e1000_receive simulates how the e1000 device receives network packets and notifies QEMU via interrupts.

```
typedef struct E1000State_st{
    PCIDevice dev;
    NICState *nic;
    NICConf conf;
    uint32_t mac_reg[0x8000];
    uint16_t phy_reg[0x20];
    uint16_t eeprom_data[64];
    ... ...
}E1000_state;

static void
e1000_mmio_writel(void *opaque,
        target_phys_addr_t addr, uint32_t val)
{
    ... ...
    if (index < NWRITEOPS && macreg_writeops[index])
    {
        macreg_writeops[index](s, index, val);
    }
    ... ...
}

static ssize_t
e1000_receive(VLANClientState *nc,
const uint8_t *buf, size_t size)
{
    ... ...
    //Fire an interrupt after receiving packets
    set_ics(s, 0, n);
}
```

Figure 1: Excerpts from the e1000 QEMU virtual device.

2.2 Non-deterministic Interleaving

Non-deterministic interleaving [11] is a transaction-level modeling technique for hardware concurrency. Hardware devices are concurrent in nature. For example, a network adapter processes driver requests and receives data concurrently. To model this concurrency using non-deterministic interleaving, there are three steps: (1) identify the concurrent modules (e.g., processing driver requests, receiving data, etc.) of the target hardware device; (2) specify the modules using separate C functions, which we refer to as module functions; and (3) non-deterministically invoke these module functions in a loop. When the loop is executed multiple times, these module functions are executed in a non-deterministic sequence. The possible effects of hardware concurrency can be captured by the set of hardware states after non-deterministic many executions of the loop. For example, a network adapter concurrently processing a driver request and receiving a packet can be captured by either processing a driver request followed by receiving a packet, or invoking these two modules in the reverse order. Since a QEMU virtual device already implements the first and second steps, we only need to complete the third step before executing it symbolically (cf. Section 4.2).

2.3 Symbolic Execution

Symbolic execution [10] executes a program with symbolic values as inputs instead of concrete ones and represents the values of program variables as symbolic expressions. Consequently, the outputs computed by the program are expressed as functions of input symbolic values. The symbolic state of a program includes the symbolic values of program variables, a path condition, and a program counter. The path condition is a Boolean expression over the symbolic inputs; it accumulates constraints which the inputs must satisfy for the symbolic execution to follow the particular associated path. The program counter points to the next statement to be executed. A symbolic execution tree captures the paths explored by the symbolic execution of a program: the nodes represent the symbolic program states and the arcs represent the state transitions.

3. CONFORMANCE CHECKING

This section presents the basic workflow of our conformance checking framework. As illustrated in Figure 2, the framework has two major components: a trace recorder and a conformance checker. The trace recorder records the driver request sequence to

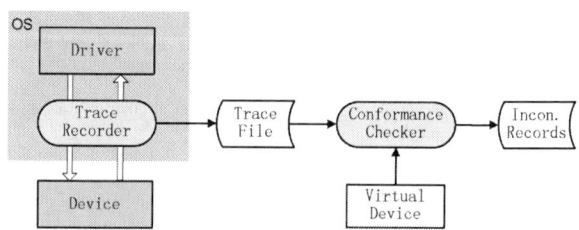

Figure 2: Workflow of conformance checking

the silicon device. The conformance checker replays the sequence on the virtual device and checks the conformance. The discovered inconsistencies are recorded. An inconsistency record contains the inconsistent registers, the driver request causing the inconsistency, and the virtual device execution trace under the driver request.

3.1 Preliminaries

Before discussing the details of this workflow, we first introduce our notion of conformance, which is defined between the states of the silicon and virtual devices. The state of the silicon device is determined by the values of its interface and internal registers. The interface registers of the silicon device are observable while the internal registers are generally not observable and are sometimes even unknown. The virtual device is a model of the silicon device. It models interface registers of the silicon device with a set R_I of corresponding variables and defines a set R_N of variables to capture device internal behaviors. However, the variables in R_N often have no correspondence with the internal registers of the silicon device. We define a virtual device state as follows.

DEFINITION 1. *A* **virtual device state** *is denoted as* $V = \{V_I, V_N\}$ *where* V_I *is the device interface state, i.e., the assignments to variables in* R_I *and* V_N *is the device internal state, i.e., the assignments to variables in* R_N.

We represent the silicon device state with the same sets of variables: R_I and R_N. The variables in R_I are assigned values observed from the corresponding interface registers of the device. The variables in R_N are assigned symbolic values with no constraints since the device internal is not observable.

DEFINITION 2. *A* **silicon device state** *is denoted as* $S = \{S_I, S_N\}$ *where* S_I *is the assignments to variables in* R_I *and* S_N *is the symbolic assignments to variables in* R_N.

A **concrete device state** is a device state whose state variable values are all concrete. A **symbolic device state** is a device state some of whose state variable values are symbolic and there can also be constraints on these symbolic values. A symbolic device state can be viewed as a set of concrete states. In our approach, we treat both V and S as symbolic states, which can be viewed as two set of concrete device states, denoted as $set(V)$ and $set(S)$ respectively. Based on this generalization, we give a conformance definition between a silicon device state and a virtual device state, as described in Definition 3.

DEFINITION 3. *A silicon device state* S *and a virtual device state* V *conform to each other if* $set(S) \cap set(V) \neq \emptyset$.

To compute $set(S) \cap set(V)$, we denotes the device state variables as $var_1, var_2, ..., var_n$ and the values of the state variables of S as $Val(var_1)_S, Val(var_2)_S, ..., Val(var_n)_S$. We construct the expression of S as $Expr(S)$: $(var_1 == Val(var_1)_S) \wedge (var_2 == Val(var_2)_S) \wedge ... \wedge (var_n == Val(var_n)_S)$. Similarly, assume the constraints of V as $Cont(V)$, the expression of V, $Expr(V)$, is $(var_1 == Val(var_1)_V) \wedge (var_2 == Val(var_2)_V) \wedge ... \wedge (var_n == Val(var_n)_V)) \wedge Cont(V)$. Given $Expr(S)$ and $Expr(V)$, $set(S) \cap set(V) \neq \emptyset$ if and only if $Expr(S) \wedge Expr(V)$ is satisfiable.

3.2 Trace Recorder

The trace recorder captures: (1) each driver request issued to the silicon device; (2) the silicon device interface state before each driver request is issued. A sequence of such state-request pairs captured on the silicon device can be viewed as a **device trace**. We define such a device trace as $T = \langle S_{I_0}, D_0 \rangle, \langle S_{I_1}, D_1 \rangle, ..., \langle S_{I_n}, D_n \rangle$, where the pair $\langle S_{I_k}, D_k \rangle$ $(0 \leq k \leq n)$ represents a driver request D_k to the current silicon device interface state S_{I_k}.

3.3 Conformance Checking Algorithm

The conformance checker replays T on the virtual device using symbolic execution. To compare the virtual and silicon device states, given S_{I_k} of T, the conformance checker constructs its silicon device state S_k (cf. Section 3.1). In this way, the conformance

checker converts T to a new device trace $T' = \langle S_0, D_0 \rangle, \langle S_1, D_1 \rangle, ..., \langle S_n, D_n \rangle$, where $S_k (0 \leq k \leq n)$ is a silicon device state derived from S_{I_k}. The conformance checking algorithm works as follows:

1. Initialize the virtual device state V_0 to be S_0 from T' and set $k = 0$.

2. Take the next driver request D_k of T' and symbolically execute the virtual device from V_k on D_k. Symbolic execution may produce a set G of virtual device states.

3. Check the conformance between G and S_{k+1} (see below for details). If not conforming, report an inconsistency; otherwise continue checking.

4. Set the virtual device state V_{k+1} to be the silicon device state S_{k+1}; Increment k and go to step 2.

5. The conformance checker terminates when it finishes the last driver request of T'.

As discussed above, symbolic execution of a virtual device may produce a set of virtual device states $G = \{V_i | 0 \leq i \leq n\}$. The next silicon device state under D_k is denoted as S_{k+1}. We define the conformance between G and S_{k+1} as follows.

DEFINITION 4. *Given* G *and* S_{k+1}, *the virtual device and the silicon device conform to each other at* D_k *if* $\exists V_i \in G$ *where* $0 \leq i \leq n$, $set(S_{k+1}) \cap set(V_i) \neq \emptyset$.

3.4 Discussions

Our conformance definition is essentially the conformance between the interface states of the silicon and virtual devices since the internal variables of S have unconstrained symbolic values. Therefore, our algorithm may not detect internal state non-conformance. Moreover, to reduce symbolic execution complexities, we synchronize the virtual device state to the silicon device state after each drive request (Step 4). This may miss inconsistencies that only surface after several driver requests. (How to check for such inconsistencies will be discussed in a future paper.) Under this conformance definition, our approach is sound theoretically as symbolic execution explores all possible interface states of the virtual device. Nevertheless in practice, for practicality and efficiency, our approach may introduce false negatives due to optimizations of symbolic execution (cf. Section 4).

4. IMPLEMENTATION

4.1 Selective Capturing

The trace recorder captures values of the interface registers of the silicon device. However, it is difficult to capture all interface registers since a device often has a large range of interface registers. For example, Intel e1000 network adapter, a PCI device, has 128KB of interface registers. Capturing all these registers will heavily degrade the system performance. To address this problem, we propose a method, namely selective capturing, which captures a smaller set of important registers rather than the complete set.

To decide which registers to capture, we statically analyze the virtual device: symbolically execute the virtual device by using symbolic inputs and record the registers accessed in execution. As the registers can be accessed by using symbolic addresses, which may lead to an unnecessarily large range of registers to record. Therefore, we only record the registers accessed by concrete addresses. This may miss certain registers. As a supplement, we

203

allow the user to specify which registers they want to capture. Selective capturing does not affect the soundness of our approach although it may miss inconsistencies. As selective capturing has a critical impact on conformance checking results, we will develop a systematic method that balances overhead and effectiveness in future work.

4.2 Harness Generation for Virtual Devices

A QEMU virtual device is not a stand-alone program, which is executed as part of the QEMU virtual machine. Therefore, we need an execution harness for symbolically executing the virtual device. We generate an execution harness based on the concepts of non-deterministic interleaving and symbolic inputs.

- *Non-deterministic interleaving.* As Section 2.2 illustrates, to capture the hardware concurrency, it requires non-deterministic many executions of a loop where the module functions are invoked non-deterministically. We define such a loop as the **main loop** of the execution harness. The condition of the main loop is a non-deterministic choice and module functions are invoked non-deterministically in the main loop.

- *Symbolic inputs.* As outside environment inputs are not captured from the silicon device, we assign symbolic values to these input variables so that symbolic execution can cover the possible inputs from the outside environment.

Example. We illustrate harness generation using the e1000 network adapter. Figure 3 shows an excerpt from the harness we generate for the e1000 virtual device. There are two module functions: (1) Access_Register; (2) e1000_receive. The function Access_Register models how the device responds to a driver request, e.g., writing to or reading from a register. The function e1000_receive models how the device receives packets from the network, which takes several input parameters. We call the function dcc_make_symbolic to assign symbolic values to the input variables. The function choice() implements a non-deterministic choice which returns a symbolic value. In the main loop, the two module functions are invoked non-deterministically.

```
    ... ...

    dcc_make_symbolic(buff, BUFF_SIZE, "buff");
    dcc_make_symbolic(size, sizeof(uint32_t), "size");

    //Non-deterministic many executions
    while(choice()){

        //Non-deterministic Interleaving
        switch (choice()) {

            // Respond to write/read registers
            case 0:    Access_Register();  break;

            //Receive packets
            case 1:    e1000_receive(nc, buff, size);  break;

            // Do nothing
            default:    break;
        }
        ... ...
```

Figure 3: Excerpts of execution harness of e1000 virtual device

4.3 Termination of Symbolic Execution

Symbolic execution might not terminate when it encounters a loop without a statically known number of iterations, e.g., the main loop in the execution harness. We refer to such a loop as an unbounded loop. To address this issue, we set constant bounds for all such loops in the virtual device. We leverage runtime behaviors of the virtual device in the QEMU virtual machine to decide the loop bound for each unbounded loop. The method contains three steps:

1. Statically analyze the virtual device: symbolically executing the virtual device using symbolic inputs, to identify the unbounded loops.

2. When the virtual device is running within the QEMU virtual machine, for each unbounded loop identified by static analysis, we record the largest number of iterations that the loop has been executed. If we encounter an unbounded loop while replaying the silicon device trace, we use its recorded maximum number of iterations as its bound.

3. As a supplement, we allow the user to adjust the loop bound for a specific loop. For example, if using a large bound induces high time and memory costs or even path explosions, the user may lower the bound.

Remarks. Loop bounding may lead to false negatives since it potentially reduces the virtual device behaviors. However, we argue that the false negative ratio is low due to two reasons. First, static analysis shows that for most unbounded loops, increasing the numbers of loop iterations does not affect the virtual device interface state. Therefore, the conformance checking result will not be affected most of the time. Second, the loop bounds cover most virtual device behaviors if the runtime test cases for identifying loop bounds have a high coverage of the virtual device (herein we use the code coverage metrics such as statement coverage). Moreover, a discovered false negative may be eliminated thereafter by the user incrementing the loop bounds. However, since setting the bounds too large may lead to high time and memory costs and even path explosions, sometimes false negatives cannot be completely eliminated. Therefore, the user may need to search for a "sweet spot" to achieve minimum false negatives with reasonable symbolic execution costs. Our evaluation results give more details (cf. Section 5).

4.4 Implementation Details

We implement our approach on Linux. The trace recorder is implemented as a Linux kernel library. A standard Linux device driver always calls Linux kernel functions to access its device. For instance, a driver calls function writel to write a long integer to a device register. We hook these kernel functions. As a result, the trace recorder is invoked to record the driver requests when the driver calls these functions to issue requests.

We construct our conformance checker using the symbolic execution engine KLEE [4]. We modify KLEE in two aspects. First, we set the loop bounds during symbolic execution. Second, we realize our own module for conformance checking.

5. EVALUATION

This section evaluates our approach from two aspects: usefulness and efficiency. Regarding usefulness, we present the inconsistencies and real bugs we discovered in three network adapters and their QEMU virtual devices. Furthermore, we evaluate our framework in terms of time usages and memory usages, demonstrating that our approach is efficient.

5.1 Experiment Setup

All experiments were conducted on a workstation with a dual-core Intel Pentium D Processor at 3.20 GHz and 4GB of RAM, running Linux with kernel version 2.6.35. The devices evaluated are three types of widely used network adapters. Information of these devices and their virtual devices are summarized in Table 1. It also shows the size of the registers we selectively capture in each network adapter. The virtual device size is measured in Lines of Code (LOC). Intel e1000 and Intel eepro100 virtual devices are included in QEMU 0.15.1 source code. Broadcom bcm 5751 virtual device is newly created following the QEMU 0.15.1 interface.

Table 1: Summary of Devices for Case Studies

Devices	Virtual Device Size (LOC)	Selective Captured Size (Bytes)
Intel e1000 Gigabit NIC	2099	1224
Broadcom bcm5751 Gigabit NIC	4519	412
Intel eepro100 Megabit NIC	2178	74

5.2 Inconsistencies and Bugs

We discovered 15 inconsistencies between the three network adapters and their virtual devices under test: 7 in e1000, 6 in bcm5751, and 2 in eepro100. By analyzing the inconsistency records generated by the conformance checker, we also discovered 13 bugs from the virtual devices, and 2 bugs from the silicon devices. As the result shows, most of these inconsistencies are caused by the bugs of the virtual devices. This is because on one hand the silicon devices are stable products which have gone through extensive testing and bug-fixing procedures; on the other hand, their virtual devices are not heavily tested through any rigorous testing procedures. However, these virtual device bugs are still possible to appear in silicon prototypes at the early stage of hardware development, since these bugs are common violations of hardware designs. We believe that if this approach is conducted at the post-silicon testing stage before devices are released, it can also discover many inconsistencies caused by the bugs of silicon devices/prototypes.

5.2.1 Types of device bugs

We summarized the bugs which cause the inconsistencies. As shown in Table 2, there are 7 types of device bugs we discovered by analyzing the inconsistencies. VD indicates the virtual device bugs while SD indicates the silicon device bugs. Most of these bugs are very common violations of hardware designs. For example, firing interrupts too many times and failing to fire interrupts are both common defects in hardware devices. We discuss the silicon device bugs and the virtual device bugs respectively.

- *Silicon device bugs.* The bugs of the first type are silicon device bugs. The device updates the register specified as reserved in the device specification. This bug can be serious since it may cause unnecessary device behaviors, expose additional device information, and consume extra power.

- *Virtual device bugs.* The bugs of second to fourth types are all related to interrupts. The bugs of the fifth type and sixth type can cause the driver to read incorrect values. These bugs often cause serious driver and system errors or even crashes, and similar silicon device errors have been reported [9].

5.2.2 Consequences of inconsistencies

These inconsistencies can have serious consequences. Here we use an inconsistency found in Intel e1000 as an example. In this

Table 2: Types of Bugs in Virtual Devices and Silicon Devices

No.	Bug Type	Num.	Devices
1	Update reserved register bits which not allowed	2	SD
2	Generate unnecessary interrupts	2	VD
3	Fail to generate necessary interrupts	1	VD
4	Fail to clear the interrupt when the driver requests	1	VD
5	Fail to update registers when necessary	4	VD
6	Write incorrect values to registers	4	VD
7	Incorrect data types used for modeling device states	1	VD

scenario, the device driver writes certain values to register MDIC to transfer data into the internal module of the device. After the data transfer finishes, according to the value of a specific bit in register MDIC, the device determines whether to fire an interrupt.

```
static void
set_mdic(E1000State *s, int index, uint32_t val) {
    ... ...
    s->mac_reg[MDIC] = val | E1000_MDIC_READY;
    set_ics(s, 0, E1000_ICR_MDAC);
}
```

Figure 4: Excerpt of e1000 virtual device

However, Figure 4 shows how the virtual device responds under such the scenario by invoking the function set_MDIC. In this function, no matter what is the value of register MDIC, the virtual device always generates an interrupt by invoking the interrupt function set_ics. Due to this feature, the driver developed on the virtual device may always expect an interrupt after the device finishes transferring data. However, the silicon device does not always generate an interrupt to notify the driver when the data transfer is completed. Therefore, if the driver is not well written, it will treat no interrupt as an incorrect data transfer in the silicon device, and report an exception by mistake. The driver's normal work flow will be disrupted on the silicon device. By detecting such an inconsistency, our tool helps users easily figure out why the driver does not work properly with the silicon device. This case illustrates how our approach can help post-silicon device/driver co-debugging.

5.3 Efficiency

We evaluate the efficiency of our approach, in terms of time usages, memory usages, and false negative ratios. We issue four kinds of test cases to the network adapters to collect device traces. These test cases are all common usages of network adapters as shown in Table 3. "NIC test-suite" contains a family of typical test cases on network interface controllers (NIC), which manipulate a NIC in different ways, e.g., sending UDP packets and setting MTU size.

Table 3: Summary of Test Cases

Test Cases	Description
Reset Network Interface	Bring down and then bring up the network interface
Ping	Ping another network interface
Transfer files	Copy large files with total size 3.2 GB
NIC test-suite	A set of typical test cases on NIC

5.3.1 Time and memory usages

We evaluate the time and memory usages of conformance checking. Table 4 shows the results. The "Time Usage" column shows the average time usages for the conformance checker processing each driver request of the device trace collected under the test cases. We also recorded the maximum values of memory usages. Consider that our approach is an offline checking approach, the time usage is acceptable and the memory usage is low.

Table 4: Time/Memory Usages and False Negatives

Devices	Test Cases	Time Usage (sec)	Memory Usage (MB)	Inconsistency (Discovered /Verified)
e1000	Reset NIC	0.24	212.60	8/8
	Ping	2.92	300.00	8/8
	Transfer files	3.11	308.14	**12/9**
	NIC test-suite	3.06	288.23	11/11
bcm5751	Reset NIC	0.19	166.51	9/9
	Ping	2.88	255.16	8/8
	Transfer files	2.87	251.02	**8/6**
	NIC test-suite	2.33	218.65	7/7
eepro100	Reset NIC	0.26	207.73	4/4
	Ping	2.10	220.15	2/2
	Transfer files	2.45	236.77	2/2
	NIC test-suite	2.31	226.84	4/4

5.3.2 *False negative ratios*

To assess the number of false negatives introduced by our optimizations, we verified all the inconsistencies discovered. In the "Inconsistency" column of Table 4, we show the numbers of discovered inconsistencies and verified inconsistencies.

Most of the inconsistencies are verified. We encountered false negatives in the traces of transferring files on e1000 and bcm5751 (marked as bold). Both virtual devices have only one unbounded loop whose number of iterations affects the virtual device interface state. The number of iterations of the loop depends on the total size of packets received by the silicon device between two consecutive driver requests. In the virtual device, one iteration of the loop would receive a fixed number of packets. Therefore, one iteration of the loop captures the silicon device behaviors when the network traffic is modest. Occasionally when the network traffic is heavy, it requires executing the loop more than once. Therefore, our setting the bound to one produces false negatives. Nevertheless, as we adjust the bound by incrementing it to two, all previously encountered false negatives are eliminated while the time and memory costs remain modest. This demonstrates that (1) our approach has a low false negative ratio; (2) The supplementary loop bounding method is effective in eliminating false negatives.

6. RELATED WORK

Recently formal methods have been increasingly used for facilitating post-silicon validation. Some of these work focus on improving observability and traceability of hardware at the post-silicon stage. A notable work is "backspace" [6], it uses SAT-solving techniques to provide an execution trace to a crashed post-silicon state, thus facilitating off-line debugging. Several approaches [3, 8, 12] integrate formal specifications into post-silicon checking of hardware by observing its execution trace. In [14], hardware monitors are introduced to ameliorate observability requirements on silicon. It uses pre-silicon RTL models to construct hardware monitors.

Symbolic execution is widely used for software testing. SAGE [7], KLEE [4], and S2E[5] use symbolic execution to test software systems that intensively interact with environments. Other tools [16, 15, 1] also employ symbolic execution to generate test cases for testing software programs.

7. CONCLUSIONS AND FUTURE WORK

We have presented an approach to conformance checking of a hardware device with its virtual prototype. Preliminary evaluation shows that our approach is useful and efficient. In three network adapters, we discover 15 inconsistencies caused by 15 bugs in the silicon and virtual devices while incurring low memory and time usages. This demonstrates our approach's major potential in facilitating hardware/software co-validation at the post-silicon stage. In future work, we plan to use virtual prototypes to estimate silicon hardware functional coverage and to validate hardware at runtime.

8. ACKNOWLEDGMENT

This research received financial support from National Science Foundation (Grant #: 0916968). A pending patent filed on this research by Portland State University has been licensed to Virtual Device Technologies (VDTech) where Fei Xie is a partner.

9. REFERENCES

[1] M. Baluda, P. Braione, G. Denaro, and M. Pezzè. Structural coverage of feasible code. In *Proc. of AST, 2010*.

[2] F. Bellard. QEMU, a fast and portable dynamic translator. In *Proc. of ATEC, 2005*.

[3] M. Boule, J. Chenard, and Z. Zilic. Adding Debug Enhancements to Assertion Checkers for Hardware Emulation and Silicon Debug. In *Proc. of ICCD, 2006*.

[4] C. Cadar, D. Dunbar, and D. R. Engler. KLEE: Unassisted and Automatic Generation of High-Coverage Tests for Complex Systems Programs. In *Proc. of OSDI, 2010*.

[5] V. Chipounov, V. Kuznetsov, and G. Candea. S2E: a platform for in-vivo multi-path analysis of software systems. In *Proc. of ASPLOS, 2011*.

[6] F. M. De Paula, M. Gort, A. J. Hu, S. Wilton, and J. Yang. BackSpace: Formal Analysis for Post-Silicon Debug.

[7] P. Godefroid, M. Y. Levin, and D. A. Molnar. SAGE: Whitebox Fuzzing for Security Testing. *ACM Queue*, 10(1), 2012.

[8] A. J. Hu, J. Casas, and J. Yang. Efficient Generation of Monitor Circuits for GSTE Assertion Graphs. In *Proc. of ICCAD, 2003*.

[9] A. Kadav, M. J. Renzelmann, and M. M. Swift. Tolerating hardware device failures in software. In *Proc. of SOSP, 2009*.

[10] J. C. King. Symbolic execution and program testing. *Commun. ACM*, 19:385–394, July 1976.

[11] J. Li, F. Xie, T. Ball, V. Levin, and C. McGarvey. Formalizing hardware/software interface specifications. In *Proc. of ASE, 2011*.

[12] J. A. M. Nacif, F. M. de Paula, H. Foster, C. J. N. C. Jr., and A. O. Fernandes. The chip is ready. am i done? on-chip verification using assertion processors. In *VLSI-SOC, 2003*.

[13] S. Nelson and P. Waskiewicz. Virtualization: Writing (and testing) device drivers without hardware. In *Proc. of Linux Plumbers Conference*, 2011.

[14] S. Ray and W. A. Hunt, Jr. Connecting Pre-silicon and Post-silicon Verification. In A. Biere and C. Pixley, editors, *Proc. of FMCAD, 2009*.

[15] K. Sen, D. Marinov, and G. Agha. CUTE: a concolic unit testing engine for C. In *Proc. of ESEC/FSE, 2005*.

[16] W. Visser, C. S. Păsăreanu, and S. Khurshid. Test input generation with java pathfinder. In *Proc. of ISSTA, 2004*.

On Testing Timing-Speculative Circuits

Feng Yuan[†], Yannan Liu[†], Wen-Ben Jone[‡] and Qiang Xu[†]

[†]CUhk REliable Computing Laboratory (CURE)
Department of Computer Science & Engineering
The Chinese University of Hong Kong, Shatin, N.T., Hong Kong
Email: {fyuan,ynliu,qxu}@cse.cuhk.edu.hk

[‡]School of Electronics and Computing System
University of Cincinnati, USA
Email: jonewb@ucmail.uc.edu

ABSTRACT

By allowing the occurrence of infrequent timing errors and correcting them online, circuit-level timing speculation is one of the most promising variation-tolerant design techniques. How to effectively test timing-speculative circuits, however, has not been addressed in the literature. This is a challenging problem because conventional scan techniques cannot provide sufficient controllability and observability for such circuits. In this paper, we propose novel techniques to achieve high fault coverage for timing-speculative circuits without incurring high design-for-testability cost. Experimental results on various benchmark circuits demonstrate the effectiveness of the proposed solution.

1. INTRODUCTION

With aggressive technology scaling, the timing behavior of integrated circuits (ICs) is increasingly sensitive to process, voltage, and temperature (PVT) variations and aging effects [1, 2]. Conventional circuit design and optimization techniques guarantee that all circuit paths meet their timing requirements in all conditions to ensure error-free computing [3]. Such worst-case design methodologies, however, inevitably lead to pessimistic designs because a large design guardband needs to be incorporated to prevent any timing failure.

To address the above problem, a "better-than-worst-case (BTWC) design" methodology that allows reliability to be traded off against power and performance was proposed, which is able to dramatically improve the energy-efficiency of computation [4]. The basic idea behind BTWC design methodology is that, since circuit non-idealities mainly manifest themselves as infrequent timing errors on critical paths of the circuit (if sufficient design guardband is not incorporated) [5], we can over-clock the chip and/or reduce the supply voltage of the chip to a point where timing errors occur, and achieve resilient computation (instead of error-free computation) by performing timing error detection and correction. This approach is generally referred to as *timing speculation*, which has attracted lots of interests from both academia and industry.

To enable timing speculation, a number of timing speculators (e.g., [6, 7, 12, 13]) were presented in the literature, which capture timing errors occurred at flip-flops driven by critical paths (referred to as *suspicious FFs*) based on double sampling or late transition detection. With timing error detection capability, a circuit can react to each error quickly and recover from it by rolling back to a known-good pre-error

---

Permission to make digital or hard copies of all or part of this work for personal or classroom use is granted without fee provided that copies are not made or distributed for profit or commercial advantage and that copies bear this notice and the full citation on the first page. To copy otherwise, to republish, to post on servers or to redistribute to lists, requires prior specific permission and/or a fee.

---

state. By doing so, the circuit energy-efficiency can be significantly improved. Recently, Intel [7] has demonstrated in their test chip that a timing-speculative microprocessor is able to achieve more than 30% throughput gain when compared to a conventional microprocessor design under the same supply voltage. The above benefits have motivated a large amount of recent research efforts on design and optimization techniques for timing-speculative circuits (e.g., [8–11]).

While great potential for timing-speculative circuits was shown with promising test chip measurement results, there are still many challenges to overcome before they can be fully commercialized. One of the key challenges is how to conduct effective and efficient manufacturing test for timing-speculative circuits, which is fundamentally different from conventional circuit testing that targets zero timing failure. The main differences lie in the following aspects:

- Conventional VLSI testing regards a chip to be defective if it cannot pass at-speed delay tests, while timing-speculative circuits are inherently tolerant to timing errors. Therefore, the pass/fail criteria for timing-speculative circuits need to be re-examined.

- To detect timing errors, a certain period right after the clock edge (namely *detection window*) is used to monitor late transitions on suspicious FFs. Any occurrence of transitions in this detection window is regarded as a timing error. To distinguish late transitions on critical paths from those early arrivals on short paths so as to guarantee the correctness of error detection, the propagation delay on each short path driving any suspicious FF must be larger than the detection window, denoted as *min-delay constraint*. It is therefore essential to identify those defective chips that violate the above min-delay constraint during manufacturing test of timing-speculative circuits.

- the newly-introduced circuitries for timing speculation, including each timing speculator itself and the system-level error signal collection logic, need to be fully tested.

Due to the above, it is essential to develop new test methodologies for timing-speculative circuits, which are addressed in this paper. To the best of our knowledge, this is the first work that considers timing-speculative circuit testing. The contributions of this paper include the following:

- We conduct fault analysis and identify new types of faults that need to be considered for timing-speculative circuits.

- We introduce new test flow and design-for-test (DfT) structures for timing-speculative circuits.

- We present novel test solutions to achieve high fault coverage for timing speculators and timing error collection logic without incurring high DfT cost. The associated test pattern generation procedure is largely compatible with conventional ATPG techniques.

The remainder of this paper is organized as follows. Section 2 presents the preliminaries of this work on timing-speculative circuit designs. In Section 3, we conduct fault analysis for timing-speculative circuits. The proposed test methodologies and the corresponding test generation techniques are detailed in Section 4 and Section 5, respectively. Experimental results on various benchmark circuits are next presented in Section 6. Finally, Section 7 concludes this paper.

2. PRELIMINARIES

Various timing-speculative designs have been presented in the literature (e.g., [6, 7]). While they differ in some aspects (e.g., the internal structure of timing speculators), the basic design principle is quite similar. In this work, we present the proposed test methodology based on the recent Intel timing-speculative design shown in [7], and we briefly discuss it in this section.

2.1 Timing Speculator

The timing speculator design in [7] is depicted in Fig. 1(a), which consists of a latch, a shadow master-slave FF (MSFF) and a *XOR* gate. The latch operates as a datapath state element for normal computation, while the MSFF samples the input value again for timing error detection. The timing diagram depicted in Fig. 1(b) demonstrates the operation of the timing speculator. Since the latch is on during high clock phase while the MSFF captures data at clock rising edge, normally, the input value is stable before clock rising edge and both the latch and MSFF hold the same value. When a late transition occurs, the value captured by the MSFF is different from the one just captured by the latch, and the *XOR* gate sets the error signal accordingly. Consequently, with such a timing speculator design, the high clock phase is the detection window for timing errors.

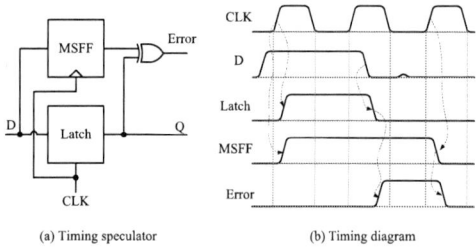

(a) Timing speculator   (b) Timing diagram

Figure 1: Timing speculator in [7].

2.2 Timing-Speculative Pipeline Design

Fig. 2 presents the overall timing-speculative design in [7], which is a three-stage pipelined microprocessor with timing speculation capability. It can be observed that all the pipeline flip-flops (FFs) are replaced with timing speculators, and error signals from timing speculators in the same stage are grouped with an *OR*-gate tree, whose output is captured by a *Final Error FF*. The three pipeline error signals are propagated to the input buffer controller in one cycle, which determines the appropriate instructions to replay based on the three pipeline error signals. They are also pipelined to the output buffer controller to invalidate erroneous data. In a microprocessor, the instruction replay circuits could leverage existing circuit used to recover from a branch miss-prediction. If a timing error occurs, the input buffer signals the *clock divider* to halve the clock frequency to ensure correct operation during replay, meanwhile, to maintain a constant high clock phase to avoid min-delay constraint violation (i.e., the detection window size remains unchanged).

2.3 Clock Divider and Duty-Cycle Controller

The clock divider and duty-cycle control circuits and the corresponding conceptual timing diagram are presented in Fig. 3. A clock generator with a differential pulse-splitter creates differential inputs CLKIN and CLKIN#. The Half_Fre input is controlled by the input buffer shown in Fig. 2. The CLK output is distributed throughout the chip.

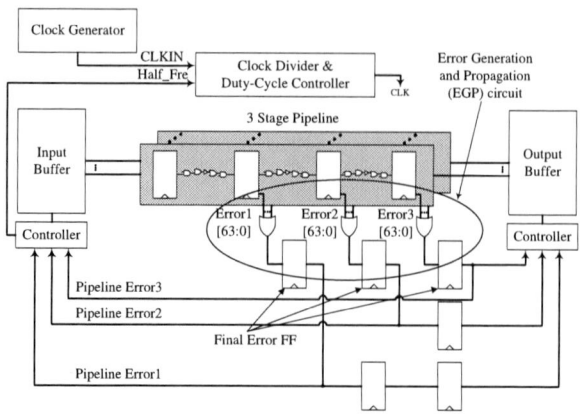

Figure 2: Timing-speculative microprocessor design in [7].

CLKIN and CLKIN# are inputs to a differential amplifier that generates an intermediate clock signal. This intermediate clock signal and the output of the falling edge-triggered MSFF in Fig. 3(a) are inputs to a logic-AND gate to produce the clock divider output (CLK0). When the Half_Fre input is logic '0' (i.e., no timing error occurs, and the circuit is running in normal functional mode), the output of the falling edge-triggered MSFF remains logic '1', and thus CLK0 and CLKIN have the same frequency. When the Half_Fre input is asserted (i.e. timing error is detected, and the circuit is running in recovery mode), the output of the falling edge-triggered MSFF toggles every other cycle, enabling the clock divider circuit to skip every other high phase of CLKIN as illustrated in Fig. 3(b). The duty-cycle control is performed with a logical-AND of CLK0 and a delayed CLK0# (i.e., inversion of CLK0) with CLK as the output, and the delayed CLK0# determines the length of the CLK high phase. With this duty-cycle control circuit, the CLK high phase delay remains constant value (T_H) at both normal and recovery modes, which is essential to ensure min-delay constraint is not violated. It is worth noting that the CLK high phase T_H is tunable with the reconfigurable duty-cycle control circuit, controlled by scan bits (see Fig. 3(a)). Considering we can use on-chip PLL to control the clock cycle time T_{cycle}, T_L is tunable as well. We leverage these tunable units in our proposed test methodologies (discussed later).

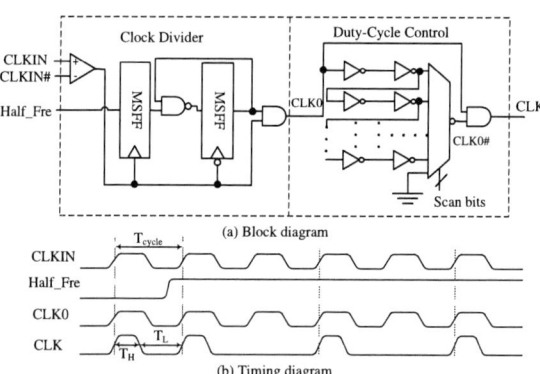

(a) Block diagram

(b) Timing diagram

Figure 3: Illustration for Clock Divider & Duty-Cycle Controller Block Proposed in [7].

3. FAULT ANALYSIS FOR TIMING-SPECULATIVE CIRCUITS

In a timing-speculative circuit, for a regular FF that is not driven by critical paths, the following timing constraints apply:

$$T_{max\_r} < T_{cycle} - T_{setup,clk\_r}$$
$$T_{min\_r} > T_{hold,clk\_r}, \tag{1}$$

where, T_{cycle} is the clock cycle time, T_{max_r} and T_{min_r} are the maximum path delay and the minimum path delay driving the regular FF,

T_{setup,clk_r} and T_{hold,clk_r} are the setup time constraint and the hold time constraint based on the clock rising edge, respectively.

For a suspicious FF, since there is an extra timing error detection window that is the high clock phase, denoted as T_H (similarly, we denote the low clock phase as $T_L = T_{cycle} - T_H$), the maximum path delay constraint T_{max_s} is as follows:

$$T_{max\_s} < T_{cycle} + T_H - T_{setup,clk\_f}, \qquad (2)$$

where T_{setup,clk_f} is setup time based on the falling clock edge. As discussed earlier, in order not to treat early transitions from short paths as timing errors, we have the minimum path delay constraint T_{min_s} for a suspicious FF as

$$T_{min\_s} > T_H + T_{hold,clk\_f}, \qquad (3)$$

and T_{hold,clk_f} is the hold time based on the falling clock edge.

Next, let us consider error generation and propagation (EGP) circuit path (i.e., from a timing speculator to the Final Error FF, see Fig. 2). In the worst case, the timing error may appear at the end of detection window, thus the maximum error propagation delay T_{max_e} should satisfy

$$T_{max\_e} < T_{cycle} - T_{setup,clk\_r} - T_H + T_{setup,clk\_f}. \qquad (4)$$

It can be observed from the above equation that the worst case error propagation time is reduced roughly by T_H.

With conventional delay testing, we usually conduct at-speed test of critical paths with functional clock only. However, for timing-speculative circuits, we need to conduct the following new types of path delay faults: (i) long path delay fault for suspicious FFs according to Equation 2; (ii) short path delay fault for suspicious FFs according to Equation 3; (iii) EGP circuit path delay fault according to Equation 4. Moreover, we need to target those static faults (e.g., stuck-at faults) that affect the logic function of the EGP circuit.

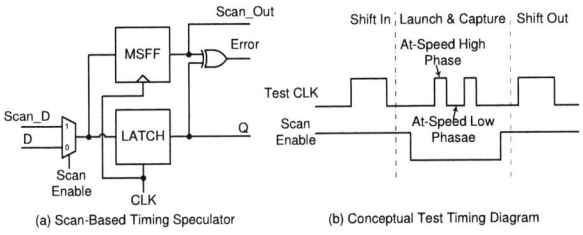

(a) Scan-Based Timing Speculator   (b) Conceptual Test Timing Diagram

Figure 4: Scan Architecture for Timing-Speculative Circuits

4. PROPOSED TEST METHODOLOGY FOR TIMING-SPECULATIVE CIRCUITS

In this section, we first present the DfT structure for timing-speculated circuits. Following that, we present the proposed test methodologies.

4.1 DfT for Timing-Speculative Circuits

In order to control and observe the values of timing speculators, we need to be able to scan them. Since latches are not capable of performing shift operations properly, we propose to construct scan chains only using the shadow MSFF of each timing speculator, as shown in Fig. 4(a). By doing so, test stimuli can be shifted in to setup the logic values of both the MSFF and the latch in a timing speculator while test responses captured by the shadow MSFF will be shifted out. To perform at-speed delay test, we use functional clock during the capture phase, as shown in Fig. 4(b). Moreover, we leverage the clock divider and duty-cycle controller in timing-speculative designs (see Section 2.3) to manipulate test clocks, whenever necessary. That is, we are able to control T_H, T_L and hence T_{cycle} during testing.

The above scan design, however, is not sufficient to test the EGP circuit due to the difficulty in error signal generation. To be specific, asserting the error signal of a particular timing speculator requires careful timing control so that the values latched in the MSFF and the latch are different, which is quite difficult to achieve. It is even harder to

control the timing of a late signal transition at the end of T_H (see Equation 4), which is required to tolerate dynamic variations. A straightforward method to resolve this issue would be adding a dedicated falling edge-triggered shadow flip-flop to generate an artificial *error signal* directly when testing the EGP circuits. However, this method incurs high DfT cost and also complicates the design of timing speculators. Instead of doing so, we propose *non-intrusive* test solutions to address this problem, as discussed in the following subsection.

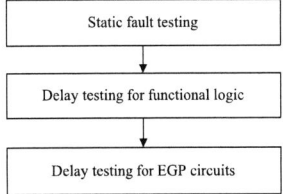

Figure 5: The overall flow for testing timing-speculative circuits

4.2 Proposed Test Methodologies

According to the timing-related fault analysis discussed in Section 3, for a timing-speculative circuit, we need to test long paths driving suspicious FFs, long paths driving regular FFs, short paths driving suspicious FFs, as well as EGP circuit testing.

The test flow is presented in Fig. 5. First, we conduct static fault testing (e.g., stuck-at faults) to ensure that circuit does not contain logic errors, including static faults in EGP circuits (discussed later). Next, we perform the above-mentioned delay tests. To test a short path driving a specific suspicious FF, we apply its test patterns and observe the corresponding error signal result at the end of the timing error detection window T_H. To ensure that the error can be correctly captured, we fix T_H as its functional value $T_{H\_functional}$ and prolong T_{cycle} to remove the impact of possible EGP circuit delay fault (which is addressed later in the test flow). The test results are then captured into Final Error FFs and scanned out for observation. Long path delay test for regular FFs can be performed with conventional delay testing methods using functional clock cycle $T_{cycle\_functional}$. Finally, to test a long path driving a suspicious FF, we configure the test clock cycle to be $T_{cycle} = 2T_{H\_functional} + T_{L\_functional}$ (due to timing speculation), and then apply its test patterns and observe the test response. Any error identified during the above at-speed delay testing process means the circuit is defective and the chip under test must be abandoned.

As shown in Fig. 5, the last step in the test flow is to conduct delay testing for EGP circuits. Without adding dedicated DfT circuits for such faults, we discuss how to achieve high fault coverage by novel test application techniques after introducing our test strategies for static faults in EGP circuits.

4.2.1 Static Fault Testing for EGP Circuits

In order to test static faults in EGP circuits, we intentionally generate and capture each error signal with a low-speed test clock, as illustrated in Fig. 6. This figure shows the scenario where a $0 \rightarrow 1$ transition propagates through two consecutive scan cells. It can be observed that, starting from Scan_Out0, the transition takes delay D_{sc} to arrive at Scan_D1 because of the scan chain delay. Since the test clock is low-speed, we can assume T_H is much larger than D_{sc}, and thus logic '1' will be captured by Latch1 first. MSFF1 will latch the transition at the next clock's rising edge, different logic values occur on Latch1 and MSFF1, and Error1 is set to logic '1'. Then, we set Scan_Enable to logic '0', so that the circuit is applied in functional mode and Error1 will be captured by the corresponding Final Error FF.

Test pattern generation is quite straightforward. Based on the static fault model, we only need to control the proper transition position in the scan data to align with the targeted suspicious FF position in the scan chain. Nevertheless, one thing needs to be pay attention to is that two timing speculators belonging to the same error group cannot be tested by the same pattern, because their error signals will mask each other with the *OR*-gate tree used for error grouping and propagation.

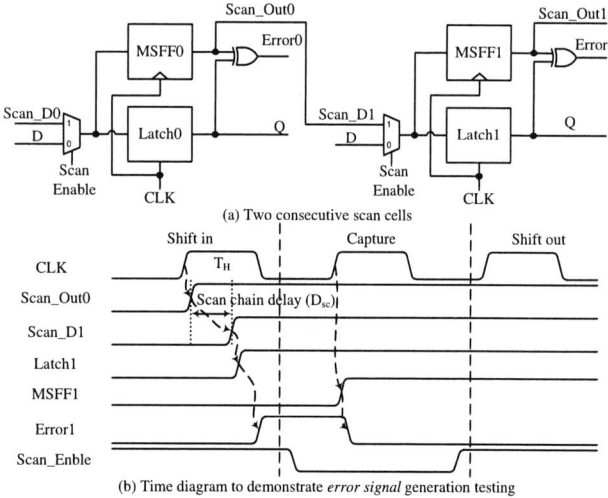

(a) Two consecutive scan cells

(b) Time diagram to demonstrate *error signal* generation testing

Figure 6: The proposed error signal generation method.

4.2.2 Delay Testing for EGP Circuits with Long Path Sensitization

To apply traditional long path delay patterns to detect delay faults in EGP circuits, the basic principle is to purposely generate a timing error on a particular long path, and make it propagate through the OR tree under the at-speed clock condition. Due to design guardband and process variation, however, there is no guarantee for the occurrence of timing errors, since whether the targeted path delay is larger than functional clock cycle ($T_{cycle\_functional}$) is unknown for us. Therefore, both the logic value captured by the MSFF of a targeted suspicious FF and its corresponding Final Error FF state have to be observed to make the pass/fail decision, and the criteria are listed in Table 1. It is important to emphasize that, the logic value stored in the MSFF of the suspicious FF appears in the second at-speed test clock cycle, while the error signal can only be latched into its Final Error FF in the next clock cycle. Therefore, to observe both of them, we have to apply the same test patterns twice: one is to capture the MSFF state with two at-speed clock cycles, while the other is to capture the Final Error FF state with three at-speed clock cycles.

Case ID	Captured Value vs Golden Value	Final Error FF State	Result
A	Same	0	Unknown
B	Same	1	Fail
C	Different	0	Fail
D	Different	1	Pass

Table 1: Pass/Fail criteria for EGP circuit delay testing with long path sensitization.

As shown in the table, there are four possible test results. Case A means no timing error occurs, and hence EDP circuit is not tested. Case B presents the scenario that the targeted path delay is smaller than one clock cycle while an error signal is still generated. Such kind of result may occur when static faults exists in EGP circuit or relevant short paths violating the min-delay constraint are sensitized (due to incomplete testing with previous test steps in our test flow). Therefore, it is a test fail. Case C represents the situation where a timing error occurs but its error signal is not captured because of a delay fault in the EGP circuit, and thus it is a failed test. Case D is a passed test, and it implies that the timing error is correctly captured under the at-speed test clock.

Due to the existence of Case A, delay testing for EGP circuits with long path sensitization cannot guarantee high fault coverage[1]. We therefore propose another method to test such faults using short path sensitization, as discussed in the following.

---

[1]Note that, tightening the clock period with smaller T_{cycle} may facilitate to set the error signal, but it is not trustworthy due to possible glitches on long paths.

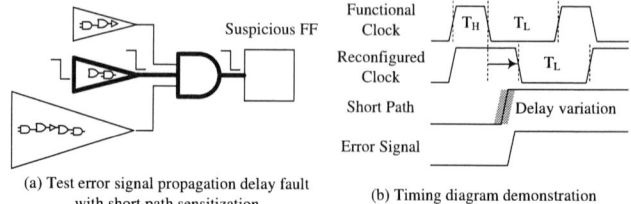

(a) Test error signal propagation delay fault with short path sensitization

(b) Timing diagram demonstration

Figure 7: EGP circuit delay testing with short path sensitization

4.2.3 Delay Testing for EGP Circuits with Short Path Sensitization

In circuit normal function mode, all short paths have to obey the min-delay constraint, i.e., $T_{H\_functional}$ must be smaller than all short path delays in order to guarantee that transitions on the short paths are not accidentally treated as timing errors. By making T_H longer than $T_{H\_functional}$ with the help of the clock divider and duty-cycle controller, however, we can purposely select a short path to drive each suspicious FF to generate an "early transition", thus generating the error signal, as shown in Fig. 7. We use this method for delay testing of EGP circuits.

Due to process variation, the estimated short path delay based on timing analysis is inherently inaccurate and it is very difficult to setup test clock configurations for error signal generation. We use Fig. 8(a) for illustration. Cases A, B and C depict three possible situations after test clock reconfiguration, wherein the light line denotes a short path delay and the dark line depicts an error signal propagation delay. As shown in Case A, the error signal generated by a timing speculator will be captured by its corresponding Final Error FF, if we correctly estimate the short path delay and no delay fault occurs on the error signal propagation path, the circuit passes the test. However, if no error signal is captured by the Final Error FF, there are two possibilities: (i) there exists a delay fault on the error signal propagation path (e.g., Case B); (ii) the short path fails to generate an error signal (e.g., Case C due to process variation). Without distinguishing these two cases, we cannot make a correct test decision.

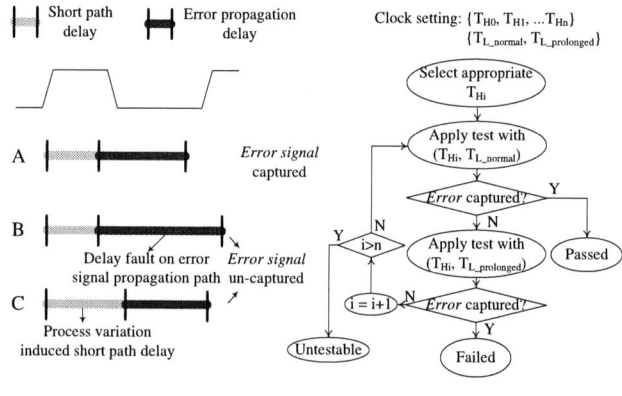

(a) Ambiguous test result

(b) Process variation aware test flow

Figure 8: Short path variation-aware test methodology

One observation to solve this problem is that, for Case C, once T_H is fixed we can never capture the error signal even if a prolonged T_L clock setting is applied, while we can still capture the error signal by prolonging T_L for Case B. Based on this observation, we propose a process-variation-aware test flow to distinguish the above two cases, as shown in Fig. 8(b). Here, we assume there exists a set of clock configuration $\{T_{H0}, T_{H1}, ..., T_{Hn}\}$ and $\{T_{L\_functional}, T_{L\_prolonged}\}$, where T_{Hi} is sorted in an ascending order, $T_{L\_functional}$ is simply the functional T_L, and $T_{L\_prolonged}$ is a much larger value than $T_{L\_functional}$. Considering a single suspicious FF, we first select an appropriate T_{Hi} based on the static timing analysis result of a sensitized short path, and apply at-speed delay testing under the clock setting ($T_{Hi}, T_{L\_functional}$). If an error signal is captured by the Final Error FF, the delay test passes successfully (Case A). Otherwise, we reconfigure the clock as ($T_{Hi}, T_{L\_prolonged}$) and

210

apply the same test again. A delay fault on the propagation path can be detected if the error signal is captured (Case B). If the error signal is not captured, we first check whether we can further prolong T_H with another configuration. If yes, we sweep to $T_{H(i+1)}$ and repeat the test flow again, otherwise the suspicious FF is untestable even with short path sensitization.

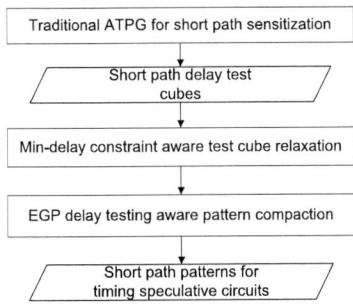

Figure 9: Test pattern generation flow

5. TEST PATTEN GENERATION

In this section, we introduce test pattern generation algorithm for short path sensitization, and the generated patterns are used for EGP delay testing and short path min-delay constraint checking.

As discussed earlier, testing EGP circuits with short path sensitization usually require multiple test clock configurations. To reduce the associated testing time cost, we propose several techniques to enhance test parallelism (i.e., test multiple faults concurrently whenever possible), and the test generation flow is shown in Fig. 9. We first acquire test cubes for short paths with conventional ATPG tool. Next, we propose a novel test cube relaxation technique by taking advantage of the min-delay constraint checking requirement. By doing so, we are able to have a more compacted test pattern set, and our test compaction algorithms are described afterwards.

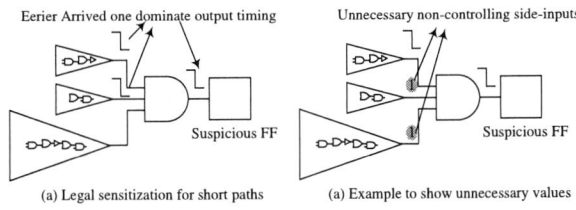

(a) Legal sensitization for short paths   (a) Example to show unnecessary values

Figure 10: Example to motivate short path test cube relaxation

5.1 Test Cube Relaxation

As the objective of short path testing is to check min-delay constraint violations, we try to use one pattern to sensitize multiple short paths simultaneously. An example is shown in Fig. 10(a). As can be observed for this example circuit, the falling transition at the *AND* gate's output will be dominated by the earliest arrived transition since logic '0' is the controlling value of each *AND* gate. Consequently, the delay of the shortest path determines the transition time and it does not matter which short path generates the earliest transition. With conventional ATPG, however, the side-inputs are set to be non-controlling values to test each individual path, as shown in Fig. 10(b). This is not necessary and we use this feature for test cube relaxation and we use the algorithm proposed in [14] to achieve the above objective.

5.2 Test Pattern Compaction

By merging compatible test cubes into a single test pattern, both test pattern count and testing time can be reduced. However, as the testing time is mainly determined by the number of test clock configurations for EGP circuit delay testing, it is essential to take this into consideration when conducting test pattern compaction. Let us consider the following example with two short paths, e.g., path A with 0.25ns propagation delay and path B with 1.25ns propagation delay. According to

benchmark	SFF (#)	SFF (#) detectable using long path	SFF(#) detectable using short path	
			One config.	Multiple config.
s38417	469	36.113	340.16	469
s38584	424	3.5616	348.99	424
ethernet	21	18.00	3.73	21
wb_conmax	86	15.1618	62.245	86
des_perf	161	160.678	115.095	161

Table 2: Results for EGP circuit coverage with different sensitization methods.

the test flow in Fig. 8(b), we need to set different clock configurations to test these two paths, due to the large gap between their propagation delays. In this case, even though these two paths can be sensitized simultaneously by the same pattern, we still need to apply this pattern twice with our test application strategy. From this perspective, it is preferable to sensitize multiple paths with close delay values simultaneously by a single test pattern, so that they have a high chance to share the same clock configuration when sweeping the T_H value.

Based on the above observation, we propose to divide the sensitized short paths into several groups according to their propagation delay values (obtained with static timing analysis), and two test cubes can be compacted into one pattern if their sensitized paths belong to the same short path delay group. We denote the above constraint as *delay group constraint* during test compaction. In addition, as we need to observe the error signal for making test pass/fail decision, all short paths whose driven suspicious FFs belong to the same error group (see Fig. 2) cannot be compacted into one test pattern either, because otherwise error signals may cancel each other. We name this rule as the *error group constraint* during test compaction. With the above two constraints, we use the algorithm in [15] for test pattern compaction.

6. EXPERIMENTAL RESULTS

6.1 Experimental Setup

To evaluate the effectiveness of our proposed timing-speculative circuit testing technique, we conduct experiments on two large ISCAS'89 benchmark circuits, *s38417* and *s38584*, as well as three large IWLS benchmark circuits, *wb_conmax*, *ethernet* and *des_perf*, which are the largest benchmark circuits available to the public domain.

In the experimental flow, we first synthesize benchmark circuits with commercial tools to obtain the optimized circuit netlist. Next, based on timing analysis results reported with Synopsys PrimeTime, we set 70% of the circuit's longest path delay as the threshold to differentiate regular FFs and suspicious FFs. That is, for those FFs whose driving path delays are larger than this threshold, they are replaced with timing speculators. These timing speculators are randomly grouped for error propagation, and each group holds up to 32 error signals. After that, we manually insert *OR* gate trees and Final Error FFs to obtain the timing-speculative designs. Finally, path delay patterns generated with Synopsys TetraMax are converted to timing-speculative circuit test patterns with our proposed pattern generation algorithms. To evaluate the impact of process variation on our proposed solution, we conduct Monte Carlo simulation to generate 10,000 circuits for each benchmark under 10% gate delay variation based on Gaussian distribution.

6.2 Results and Analysis

In the first experiment, we show the EGP circuit delay fault coverages achieved by long path sensitization and short path sensitization. The experiments are conducted with 10,000 circuits under process variation, and results are shown in Table 2. For each benchmark circuit, the total number of suspicious FFs is shown in Column 2. Columns 3-5 present the average number of detectable circuit paths following suspicious FFs in the EGP circuit, with long path sensitization, short path sensitization without test clock reconfigurations, and short path sensitization with test clock reconfigurations, respectively. It can be seen from the table, the proposed method using short paths to sensitize EGP circuits with test clock reconfigurations is able to achieve 100% coverage for EGP circuits, while the other two methods cannot achieve this objective.

benchmark	Short path (#)	Ori.	Compacted pattern count			
			Ori. (DG & EG)		Relaxed (DG & EG)	
			#	INC	#	INC
s38417	3445	126	197	56.35%	129	-34.51%
s38584	2521	111	152	36.94%	116	-23.68%
ethernet	33	12	23	91.67%	21	-8.70%
wb_conma	841	21	40	90.48%	30	-0.25%
des_perf	300	3	10	233.33%	10	0.00%
Avg.				101.75%		-18.38%

Table 3: Test pattern count results for EGP circuit testing

benchmark	Compacted pattern count			Testing time ($\times 10^3$ clock cycles)		
	EG	DG & EG	INC	EG	DG & EG	INC
s38417	86	129	50.00%	1449.916	985.6	-32.02%
s38584	78	116	48.72%	936.48	774.435	-17.30%
ethernet	21	21	0.00%	125.425	114.88	-8.41%
wb_conmax	19	30	57.89%	398.86	274.215	-31.25%
des_perf	10	10	0.00%	59.405	57.805	-2.69%
Avg.			31.32%			-18.34%

Table 4: Testing time results for EGP circuit testing

In the next experiment, we show the test pattern count for EGP circuits with our proposed solution, as shown in Table 3. Column 2 presents the number of sensitized short paths for each benchmark circuit. Without considering both the error group constraint (*EG*) and delay group constraint (*DG*), we conduct test compaction on the raw patterns generated by the ATPG tool directly, denoted as *Ori.*. When taking both *DG* and *EG* constraints into consideration for test compaction, we can see from Columns 4 and 5 that the pattern count increases 101% on average. With the proposed min-delay constraint aware test cube relaxation technique, we are able to mitigate the negative effect caused by these constraints. The average saving is about 18% . In particular, when the number of original test patterns is high, the proposed method is able to achieve better savings, e.g., for s38417 and s38584.

Table 4 evaluates the testing time with the proposed methodology that intentionally sensitizes paths with close delay values. Because the number of test patterns is determined by the real circuit short path delays, this experiment is also conducted with 10,000 circuits under process variation. Two test sets are compacted from the relaxed patterns, where one of them only considers the *EG* constraint and the other one considers both *EG* and *DG* constraints. Despite that the pattern count increases 31% on average (Column 4) after considering *DG* constraint, on average we can reduce testing time by 18% (Column 7).

In the last experiment, we plot the relationship between EGP test coverage with the number of performed test clock reconfigurations, as shown in Fig. 11. For all benchmark circuits, the trend is that the coverage increases significantly in the first few configurations and quickly saturates to a stable value. To be specific, all benchmarks can achieve more than 90% suspicious FF coverage after 5 test clock configurations. To achieve 100% coverage for the EGP circuits, however, the average number of configurations is 5 for all the benchmarks, and the best case is benchmark des_perf using 2 configurations and the worst case is benchmark wb_conmax using 10 configurations.

7. CONCLUSION

Without developing efficient and effective test methods for timing-speculative circuits, it is impossible to push forward for volume production. The major difficulty in dealing with timing-speculative circuit testing comes from the fact that the timing behavior of such circuits is non-deterministic. In this paper, for the first time, novel test solutions were proposed to address the above problem. With the proposed test flow and the novel test pattern generation techniques, we are able to achieve high fault coverage for timing-speculative circuits without incurring high DfT cost.

8. ACKNOWLEDGEMENT

This work was supported in part by the Hong Kong SAR Research Grants Council under General Research Fund No. CUHK418111 and No. CUHK418812.

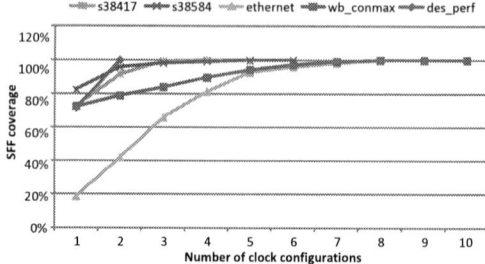

Figure 11: EGP circuit coverage vs. the number of test clock configurations

9. REFERENCES

[1] S. Borkar, *et al.*, Parameter variations and impact on circuits and microarchitecture. In *Proc. ACM/IEEE Design Automation Conference (DAC)*, pp. 338–342, 2003.

[2] D. Frank, R. Puri, and D. Toma, Design and CAD challenges in 45nm CMOS and beyond. In *Proc. ACM/IEEE International Conference on Computer-Aided Design (ICCAD)*, pp. 329–333, 2006.

[3] S. Borkar, Designing reliable systems from unreliable components: the challenges of transistor variability and degradation. In *IEEE Micro*, vol. 25, no. 6, pp. 10–16, 2005.

[4] T. Austin, V. Bertacco, D. Blaauw, and T. Mudge, Opportunities and challenges for better than worst-case design. In *Proc. IEEE/ACM Asia South Pacific Design Automation Conference (ASP-DAC)*, pp. 2–7, 2005.

[5] S. R. Sarangi, *et al.*, VARIUS: A model of process variation and resulting timing Errors for microarchitects. In *IEEE Transactions on Semiconductor Manufacturing*, vol. 21, no. 1, pp. 3–13, February 2008.

[6] D. Ernst, *et al.*, Razor: a low-power pipeline based on circuit-level timing speculation. In *Proc. IEEE/ACM International Symposium on Microarchitecture (MICRO)*, pp. 7-18, 2003.

[7] K. Bowman, *et al.*, Energy-Efficient and Metastability-Immune Resilient Circuits for Dynamic Variation Tolerance. In *IEEE Journal of Solid-State Circuits*, pp. 49-63, 2009.

[8] B. Greskamp, *et al.*, Blueshift: Designing processors for timing speculation from the ground up. *Proc. IEEE International Symposium on High-Performance Computer Architecture (HPCA)*, pp. 213-224, 2009.

[9] A. B. Kahng, *et al.*, Slack redistribution for graceful degradation under voltage overscaling. *Proc. IEEE/ACM Asia South Pacific Design Automation Conference (ASP-DAC)*, pp. 825-831, 2010.

[10] R. Ye, F. Yuan and Q. Xu. Online clock skew tuning for timing speculation. *Proc. ACM/IEEE International Conference on Computer-Aided Design (ICCAD)*, pp. 442–447, 2011.

[11] Y. Liu, *et al.*, On logic synthesis for timing speculation. *Proc. ACM/IEEE International Conference on Computer-Aided Design (ICCAD)*, pp. 591–596, 2012.

[12] M. Favalli and C. Metra, Sensing circuit for on-line detection of delay faults. In *IEEE Transactions on VLSI Systems*, pp. 130-133, 1996.

[13] Y. Tsiatouhas *et al.*,A sense amplifer based circuit for concurrent detection of soft and timing errors in CMOS ICs. In *IEEE International On-Line Testing Symposium (IOLTS)*, pp. 12-16, 2003.

[14] A. EI-Maleh and A. AI-Suwaiyan, An efficient test relaxation technique for combinational & full-scan sequential circuits. In *IEEE VLSI Test Symposium (VTS)*, pp. 53-59, 2002.

[15] R.K. Roy, J.H. Patel and J.A. Abraham, Test Compaction for sequential circuits. In *IEEE Transaction on Computer-Aided Design of Integrated Circuits and Systems*, pp. 206-267, 1992.

[16] The International Technology Roadmap for Semiconductors: 2011 Edition - Design. In *Semiconductor Industry Association*, (http://public.itrs.net), San Jose, CA, 2011.

[17] E. S. Fetzer, Using adaptive circuits to mitigate process variations in a microprocessor design. In *IEEE Design & Test*, vol. 23, no. 6, pp. 476-483, Nov. - Dec. 2006.

[18] V. Chickermane *et al.*, A power-aware test methodology for multi-supply multi-voltage designs. In *International Test Conference*, paper 9.1, 2008.

[19] X. Kavousianos *et al.*, Test schedule optimization for multicore SoC: handling dynamic voltage scaling and multiple voltage islands. In *IEEE Transactions on Computer-Aided Design*, vol. 31, no. 11, pp. 1754-1766, Nov. 2012.

[20] N. B. Z. Ali *et al.*, Dynamic voltage scaling aware delay fault testing. In *Proc. of 11th European Test Sym.*, pp. 15-20, 2006.

[21] S. Khursheed *et al.*, Gate-sizing-based single test for bridging defects in multivoltage designs. In *IEEE Transactions on Computer-Aided Design*, vol. 27, no. 2, pp. 1117-1127, June 2010.

An ATE Assisted DFD Technique for Volume Diagnosis of Scan Chains[*]

Subhadip Kundu[1], Santanu Chattopadhyay[1], Indranil Sengupta[1] and Rohit Kapur[2]

Indian Institute of Technology Kharagpur[1]; Synopsys Inc. USA[2]

(subhadip,isg)@iitkgp.ac.in[1], santanu@ece.iitkgp.ernet.in[1], Rohit.Kapur@synopsys.com[2]

ABSTRACT

Volume Diagnosis is extremely important to ramp up the yield during the IC manufacturing process. Limited observability due to test response compaction negatively affects the diagnosis procedure. Hence, in a compaction environment, it is important to implement Design For Diagnosis (DFD) methodology to restore diagnostic resolution. In this paper, a novel DFD technique which makes the faulty chains to behave as good chains during loading, has been proposed. As a result, the errors introduced in the responses, must occur during unloading of the scan chains. Diagnosis can then be performed by directly comparing the actual and expected responses without any fault simulation - leading to significant reduction in time. Results on benchmark circuits show that the average number of suspected cells for single chain failure is 1.27 (ideal value being 1) and the time taken for diagnosis is in the order of milli-seconds.

Categories and Subject Descriptors

B.7.2 [**INTEGRATED CIRCUITS**]: Design Aids

General Terms

Diagnosis

Keywords

Scan Chain Diagnosis, Design For Diagnosis, Automatic Test Equipment

1. INTRODUCTION

When a logic circuit fails a test, diagnosis is the process of narrowing down the possible locations of defect. Diagnosis helps to reduce the product debug time and thus reduces the time to market and product cost. By reducing the candidate locations down to possibly only a few, subsequent physical

---

[*]This work is partially supported by "Synopsys CAD Laboratory Projects", sponsored by Synopsys Inc., India.

Permission to make digital or hard copies of all or part of this work for personal or classroom use is granted without fee provided that copies are not made or distributed for profit or commercial advantage and that copies bear this notice and the full citation on the first page. To copy otherwise, to republish, to post on servers or to redistribute to lists, requires prior specific permission and/or a fee.
DAC '13, May 29 - June 07 2013, Austin, TX, USA.

failure analysis becomes faster and easier when searching for the root causes of failure.

A failure can occur in a circuit due to the defects present in the logic circuit or in the scan chains [24] [23]. While many defects reside in the logic part of a chip, defects in scan chains are becoming more and more common. Scan chains are the most important Design for Test (DFT) mechanisms used in today's VLSI industry. Thus, it is very important to test the integrity of scan chains. Scan chain failures are the cause for a substantial proportion of failing chips. As 30%-50% of logic gates of a typical chip impact the operation of scan chains [13], it is very likely that scan chain operations will be impacted by random and/or systematic defects. Therefore, scan chain failure diagnosis is important for effective scan-based testing.

Scan chain diagnosis starts with a flush test application [18]. A flush pattern, consists of shift-in and shift-out operations without pulsing capture clocks. The scan cells between the scan chain input and the input to a particular scan cell are known as the upstream cells of that scan cell. The scan cells between the scan cell output and the scan chain's output terminal are called the downstream cells of that scan cell. Table 1 shows an example of identifying faulty chains and modeling chain defects by flush patterns. Suppose a scan chain with 12 scan cells is loaded with flush pattern 001100110011. The second column gives the unloaded faulty values for each type of permanent fault given in column 1. By using this table, the fault model to be used for diagnosis can be identified.

Table 1: Scan chain fault models and their effects. (Fault-free unloaded values are 001100110011)

Fault Type	Unloaded values with one permanent fault
Stuck-at 0	000000000000
Stuck-at 1	111111111111
Slow-to-rise	00100010001X
Slow-to-fall	01110111011X
Fast-to-rise	X01110111011
Fast-to-fall	X00100010001

Thoush by using flush patterns, the fault type and the faulty chain can be identified easily, it is insufficient for pinpointing the index of a failing flip-flop. So, different techniques have been proposed in the literature to diagnose scan chains effectively. The techniques can be classified into three categories: (a) Tester based, (b) Hardware assisted, and (c) Software based methods. Each of these methods have been discussed in details in Section II.

In a different context, encoding of test stimuli and test response compaction [24] [22] are the widely used techniques to reduce the test data volume. Compaction of test responses negatively impacts fault diagnosis due to reduced observability. Methods to improve fault diagnosis for circuits using test response compaction often use: (i) bypass of compaction circuits, and (ii) additional tests [13] [14] [5]. Bypassing compaction circuitry requires additional on-chip circuits and increased test data volume. Using additional tests to improve diagnosis can be done in two ways. One is to augment production tests. However, since this approach increases test application time it is typically not used. The other approach is to use production tests first to detect defects and then use additional tests for diagnosis purpose. However, multiple test sessions increase test cost.

Today, in all compression schemes masking logic [2] [25] is available to isolate chains on the compactor side. This masking was put in to handle X's but has also been used in improving diagnosability [20]. However, there is no DFD method used in compression on the decompresser side. In this work, a decompresser side masking technique is proposed to help in diagnosis of scan chains. The proposed method adds an AND-OR network in between the decompresser and the scan chains. For each chain, one set of AND-OR gate is required. The basic idea of the approach is that the test stimuli, loaded to the scan chains, can be controlled using the AND-OR network. Let there be a faulty chain with stuck-at-1 fault (the type of the fault can be easily found using flush test). Now, the AND-OR network can be controlled to change the values to be loaded to that chain only to all 1's. Since, there will not be any conflict while loading the faulty chain, the change in the failing bits in the responses should be caused during the unloading process. Next, a bound calculation process (similar to [6]) is used to narrow down the suspected list of cells. Since, the loaded pattern is known, the bound can be calculated by observing only the responses. So, no further fault injections and simulations are required (though a true value simulation with the modified pattern is required) which are the main bottleneck of the software based scan diagnosis procedures. Thus, the diagnosis time can be reduced to a great extent with no loss of diagnostic resolution, making it suitable for volume diagnosis. The proposed method assumes multiple chain environment which is state-of-the-art industry technique.

The method has the following advantages:

- The proposed decompresser side masking method dynamically changes the test stimuli in such a way that the faulty chains behave as good chains during scan loading.

- The proposed method is independent of the decompresser/compactor design.

- Time taken to diagnose a chain is negligible which is the main requirement for volume diagnosis.

- Production test sets are enough for diagnosis. No extra patterns are required.

- The proposed method can handle simultaneous failures at multiple chains. All the failed chains can be successfully diagnosed with no additional overhead.

Rest of the paper is organized as follows: In Section II, literature on scan chain diagnosis has been reviewed. The details of the proposed architecture has been described in Section III. An illustration with an example of the working of the proposed method has been given on Section IV. In Section V, experimental results have been presented. Finally, conclusion and future works have been given in Section VI.

2. PREVIOUS WORKS

Tester-based diagnosis techniques [3] [17] use tester to control scan chain shift operations and physical failure analysis (PFA) equipment to observe defective responses at different locations in order to identify failing scan cells. These techniques normally provide good diagnosis resolution. However, they require expensive, time consuming, and often destructive sample preparation.

Hardware-assisted methods use special scan chain and scan cell designs to facilitate diagnosis. In [16], authors have proposed to connect each scan cell's output to a scan cell in another scan chain so that its value could be observed from the other scan chain (partner chain) in diagnostic mode. In [4], authors have proposed to insert XOR gates between scan cells to enhance chain diagnosis. The proposed scheme will always identify the fault closest to the scan output if there are multiple faults. The scheme makes a trade-off between the number of XOR gates added and the diagnostic resolution. In [15], the authors have proposed to add simple circuitry to a scan flip-flop to enable its scan-out port to be either set or reset. Tekumulla and Lee [21] have proposed a partitioning of scan chains into segments and bypassing segments that contain hold-time violations. When a hold-time violation is located on a scan chain segment, the flip-flop in that segment is bypassed and new test patterns are created. In [26] a special circuit has been proposed to flip, set, or reset scan cells to identify the defective ones. After shifting in a chain pattern, the circuit can invert, set, or reset each flipflop's state. The faulty cell is located via the observed unloaded value. However, since these methods typically require extra hardware overhead, they are not acceptable in many products. In addition, defects could occur in the extra control hardware, which make diagnosis more complicated.

Software-based techniques use diagnosis algorithms to identify faulty scan cells. Compared to hardware-based techniques, software-based techniques do not need modification of the conventional scan design and are more widely adopted in industry. The inject-and-evaluate paradigm, commonly used for logic diagnosis, can also be applied to scan chain diagnosis with a specific fault model [8]. This technique is generally categorized as model based technique. In [6], authors have proposed an algorithm that identifies an upper bound (UB) and a lower bound (LB) of scan cells within a chain to locate a faulty cell. A jump simulation technique has been proposed in [12] to diagnose a single chain fault. For each failing pattern, a simulator performs multiple simulations to quickly determine the UB and LB. After the range is finalized, a detailed simulator performs parallel pattern simulation for every fault in the final range. In [7] an effect-cause method has been proposed using dynamic learning. This method is based on several learning rules, which analyze the circuit, patterns, and mismatched bits. It backtraces the logic cones to determine which cells should be simulated in the next iteration. As a result, only a few cells need to be simulated to find suspects instead of simulating every scan cell within a range. In [9], a method for multiple fault diagnosis has been proposed. The work based on (1) double candidate range calculation, (2) dynamic learning, and (3) two dimensional space linear search and can successfully identify the dominant fault pair in a chain.

Another set of software-based approaches, known as signal profiling based approach or data-driven chain diagnosis approach, are proposed in [11]. These methods use special patterns for chain diagnosis. These patterns could be either functional test patterns that start from an initial state, or scan patterns that starts with all 0's or all 1's. The main objective of such patterns is to avoid (or minimize) any faulty value introduced during loading of scan chains. Therefore all (or most) of the failing bits are caused in the process of unloading scan chains. Then diagnosis can be performed by monitoring from which scan cell the signal probability has been significantly changed. The algorithms select patterns to randomize signal probability of scan cells before unloading. The main disadvantages of the data-driven approaches are: (1) Manufacturing ATPG scan patterns cannot be used for diagnosis. This is because the faulty values during scan chain loading procedures could be propagated to faulty chain itself and will compromise signal profiling results. (2) It cannot be applied to circuit with embedded compression logic without using bypass mode. In [1] [10], an adaptive signal profiling algorithm using manufacturing ATPG patterns have been proposed.

With this background, in the next section, we have proposed a DFD architecture which will assist in diagnosis of scan chains.

3. PROPOSED DFD ARCHITECTURE

The proposed DFD architecture inserts an "AND" and an "OR" gate in between the on-chip decompresser and every scan chain. The basic structure is shown in Figure 1.

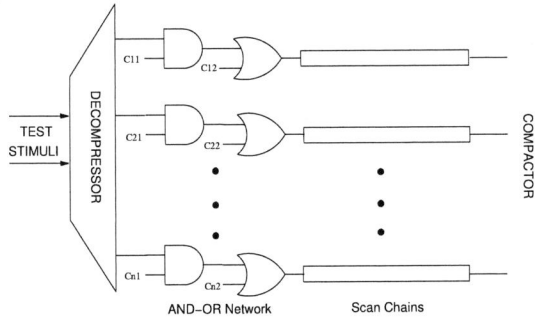

Figure 1: Proposed DFD Architecture

Test patterns are stored in a compressed format in the ATE. The compressed stimuli is decompressed using an on-chip decompresser. The decompressed test patterns are then loaded into the scan chain. In the proposed architecture, the test patterns go through an AND-OR network before being loaded into the scan chains.

The AND-OR network is controlled using pairs of control signals ($C11$, $C12$) to ($Cn1$ and $Cn2$). For n scan chains, $2n$ such controls signals are required. These control signals are generated by ATE considering the flush test results for each chain. For i^{th} chain, if the chain is fault free, the control signal $Ci1$ is set to 1 and $Ci2$ is set to 0. This will allow the normal test pattern bits to be loaded in chain i. If the chain has a stuck-at 0 fault, the values of $Ci1$ and $Ci2$ are both set to 0. This will change the test pattern bits to be shifted to chain i to all 0's. Since, there is a stuck-at 0 fault present in the chain, there will not be any conflict while loading the chain. So, the faulty chain will behave as a good chain while loading and all-zero pattern will be loaded into the chain. Similarly, the control signal values, which will make the ith

faulty chain to behave as a good chain during loading for different fault types, are given in Table 2.

Table 2: Different control values for chain i

Fault Type	$Ci1$	$Ci2$
Stuck-at 0	0	0
Stuck-at 1	X	1
Slow-to-rise	0	0
Slow-to-fall	X	1
Fast-to-rise	X	1
Fast-to-fall	0	0
Fault Free	1	0

By controlling the AND-OR network, we can make sure that either all 1's or all 0's are being loaded into the faulty chain(s). The fault free chains are loaded with the corresponding fault free bits of the test patterns. Now, after the capture clock, the responses of the modified patterns are stored back into the scan chains. During unloading, the response bits which are upstream cells of the faulty cell will be modified. Since, the loaded pattern is known, by comparing the observed response with the modified response using the bound calculation algorithm given in [6], the faulty cell can be easily located.

The $2n$ control signals are generated by ATE and applied to the AND-OR network for the whole duration of testing. These extra channels will increase the cost of the ATE. To reduce this extra channels for ATE, we have added two flip-flops (FFs) for each chain to hold the control signals during testing. These FFs are connected as a scan chain which is known as the set up chain. The integrity of the set up chain can be tested similarly using flush test. After the flush test, once the faulty chains and fault types are known, the set up chain is initialized via a single ATE channel (This process is similar to the loading of a chain). Once, the FFs are loaded with corresponding control signals, normal testing mode can be resumed. The setup chain configuration is shown in Figure 2.

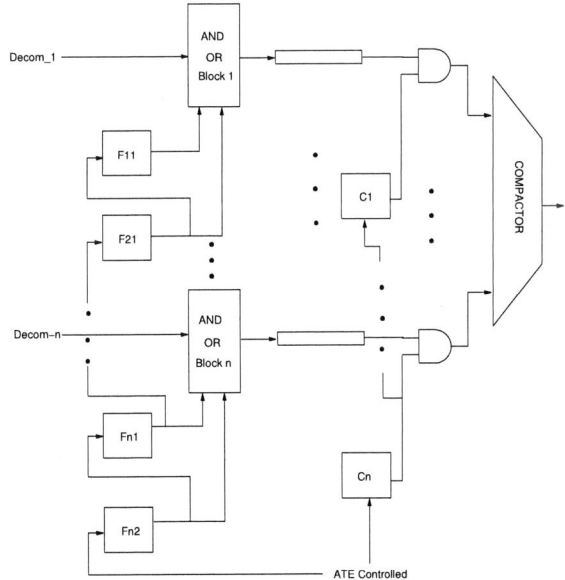

Figure 2: Proposed Setup Chain Configuration

In Figure 2, $F11$, $F12$, $\cdots$, $Fn1$, $Fn2$ are the FFs used to hold the control values. The FFs $Fi1$ and $Fi2$ hold the control values for the ith chain, that is, $Ci1$ and $Ci2$ respectively. The AND-OR block receives the decompresser output and depending on the control values, modified/unchanged values are loaded into the scan chain.

In the compactor side, an equivalent setup chain with AND gates to block or pass individual chains have been assumed. The FFs are named from $C1$ to Cn. As already discussed, these AND gates are generally used to block X's but they can also aid in diagnosis.

3.1 Hardware Overhead

The major problem with any DFD technique is the extra hardware overhead of the design. Generally, all the approaches proposed in literature, are based on modification of scan cells. Since, the number of scan cell in a large design can be very high, the hardware overhead of the approaches are also very high. The major advantage of the proposed approach is that the approach requires additional hardware for scan chains (not for scan cells). As the number of chains in a large design is much smaller compared to the scan cells (typically by a factor of 100), the hardware overhead is almost negligible compared to other approaches.

The proposed approach requires two FFs per chain to hold the control values and an AND and an OR gate to change the test pattern bits (if required) to be loaded into a scan chain. The AND-OR block can also be implemented with a NAND-NOR block or with a 3:1 Multiplexer. The multiplexer inputs will be - decompresser output, 1, and 0.

The number of FFs required for the proposed approach can also be reduced. Consider that a scan chain may be loaded with one of these three values: (i) the original test pattern bits, (ii) all 1's, and (3) all 0's. So, if there are n chains, there can be 3^n possible ways the chains can be loaded. Let there be k FFs to hold the control signals. Clearly, the inequality $2^k \geq 3^n$ should hold. Therefore, $k \geq n \log 3/2$. However, it will require a more complex selection network.

Assuming the more traditional way to implement the proposed method, two FFs and two gates are required for each chain. Let there be n_g number of gates in the combinational part of a circuit. Also, let there be n_s scan FFs distributed over k chains. Assuming a Master-Slave DFF requires 10 gates, and a scan FF requires 14 gates (4 for the mux), the additional gate overhead of the proposed approach over the scan design is given by:

$$\%Overhead = \frac{(2*10+2)*k}{n_g + 14*n_s} \qquad (1)$$

For a moderate circuit with one million gates and 10000 FFs distributed in 100 chains, the gate overhead is only 0.2%, which is negligible.

There will also be a test time overhead caused by the loading of the set up chain with the controlling values. Assuming that the set up chain can operate on the same scan clock as the normal chains, $2*k$ scan cycles are required to initialize the set up chain. Since, the contents of the set up chains are not changing, the overall test time will not be affected that much by the initialization of the setup chain.

4. AN EXAMPLE WITH BOUND CALCULATION ALGORITHM

Consider a circuit with 25 scan flops, equally distributed in 5 chains. The chains are numbered as 0 to 4. Let the

scan cells in a chain be numbered from 4 down to 0 (where scan cell 4 is connected to scan in and cell 0 is connected to scan out). Let chain 2 has stuck-at-0 fault at cell 2. Now, using flush test, faulty chain and the type of the fault can be identified. Based on the result, ATE can initialize the control flip flops to the appropriate values. The original and the modified test patterns are shown in Table 3.

Table 3: Control signals with loaded pattern

Chain	$Ci1$	$Ci2$	Original Values (to be loaded)	Modified values (Actually loaded)
0	1	0	01010	01010
1	1	0	11011	11011
2	0	0	01110	00000
3	1	0	11011	11011
4	1	0	01011	01011

The proposed AND-OR network passes the original test bits to be loaded to all the fault free chains. It modifies and loads all 0's to the faulty chain 2. As there is no conflict while loading the test pattern, we know with certainty the modified pattern shifted in. Now, let the expected output for the modified pattern from chain 2 be: 01010. If the observed output after unloading is 00010, the mismatch of the output value is only due to the errors introduced during unloading process. Now, comparing the two responses, we can conclude that the faulty cell will be among the upstream cells of cell 1 and the downstream cells of cell 4 (including cell 4). This way, by applying the other patterns, the faulty cell location can be pinpointed. The bound calculation method though similar to [6], is more accurate (and fast) because definite values (either all 1's or 0's) are loaded in the failed chain instead of don't cares [6]. Also, unlike the simulation based approaches, the proposed approach does not require any fault simulation to narrow down the suspected cell list. Thus, it reduces the time to diagnose to a great extend. The bound calculation process can be run on a remote computer which is connected to ATE mainframe. So, when the pattern application process reaches its end, the computer can return a fault hypothesis. Thus, the proposed approach can be used to diagnose scan chains in volumes. The flow of the proposed approach is given in Figure 3.

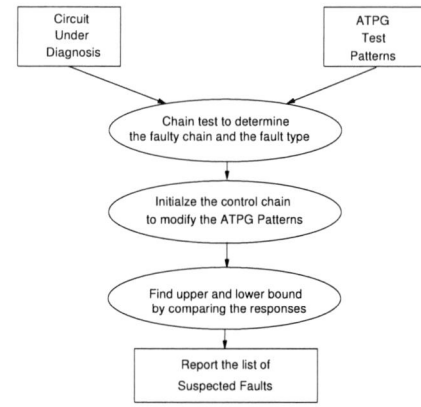

Figure 3: Flow Chart of the Proposed Approach

The proposed approach only shifts either all 1's or all 0's to the faulty chains. The contents of the fault free chains remain same as the pattern applied. Hence, it is expected to have different output values in the faulty chains for different patterns which will help in diagnosis. Results obtained while conducting the experiment validates this point.

Recently, it has been found that failures on multiple scan chains are being observed much more frequently. In [9], it has been reported that 32% of 529 units with scan chain failures contained failure at multiple chains. The proposed approach can handle this scenario easily. The control flip-flops of the faulty chains will be loaded with appropriate values. The chains can be diagnosed in a similar way as a single chain diagnosis.

5. EXPERIMENTAL RESULTS

We have implemented the proposed approach in C and obtained results on ISCAS'89 and ITC'99 benchmark circuits. We have assumed a full scan version of the circuit. The scan cells are distributed over multiple chains. The test patterns are generated using Synopsys TetraMAX [19] tool. In Table 4, the details of each circuit is given. The second column notifies the compaction ratio (CR) considered in this approach. A $30X$ compaction implies that 30 scan chains are observed through a single compactor output. In the next column, the % gate overhead (HWO) for each circuit is given. The hardware overhead is calculated using Eqn. 1. As it can be seen, the hardware overhead is almost negligible.

Table 4: Details of the circuits

Circuits	#CR	% HWO	#Test Patterns
s5378	9X	2.04	124
s9234	12X	1.68	147
s13207	15X	1.02	273
s15850	15X	1.04	131
s38417	30X	0.8	100
s38584	30X	0.92	146
b13	6X	6.52	31
b14	10X	0.89	737
b15	25X	1.97	457
b20	30X	1.33	697
b21	30X	1.31	711

In Table 5, the average size of suspected list obtained by the proposed method for different circuits is given. The list contains the scan cells suspected to be faulty. The results for both compactor bypass mode and with compacted response mode are given. In compactor bypass mode, contents of all the scan chains for all the patterns are observed by bypassing the compactor. So, for any faulty chain, all the patterns can be used to diagnose that chain. Naturally, the diagnostic resolution in this case will be higher.

However, diagnosis with compacted responses is preferred for Volume diagnosis. The problem with compacted responses is that the bound calculation algorithm cannot be directly applied to them. The masking circuitry at the compactor side can be used in this case. The set up chain at the compactor side can be initialized to pass one faulty chain and mask all other chains. So, the compacted response will only consist of the unloaded values of failed chain and the bound calculation algorithm can be used to narrow down the suspected lists. In case of multiple scan chain failures, the patterns are distributed equally such that each failed chain is observed for some patterns. For example, s38417 has 100 test patterns and let there be four faulty chains. For each faulty chain, 25 different patterns are there which pass that chain and mask all other chains. As the number of patterns used for diagnosis of each chain reduces with increase in the number of failed chains, the average size of suspected list becomes higher than in the bypass mode.

Average suspected list indicates the difference between lower and upper bound (including the lower bound) given by

the bound calculation algorithm. It is calculated as: (Upper Bound - Lower bound) + 1. Under the columns 1 chain, 2 chains, 3 chains, and 4 chains, we indicate the number of faulty chains present in the circuit simultaneously. For all the cases, the fault is present within the given bound. The results are shown with the assumption that each chain contains at most one fault, but multiple chains can be faulty simultaneously. Faults in different chains may be off different types also.

The faulty chains are loaded with either all 1's or all 0's and the fault free chains are loaded with the actual test bits. Now, when more chains are faulty, the correlation between the successive loaded patterns become higher. So, there might not be enough information in the unloaded responses to narrow down the suspected list further. Thus, the average number of suspected faults are generally increased when more chains become faulty.

For each case, we have repeated the experiment 100 times and the average results are given. The length of average suspected lists are very close to 1 which implies that the proposed method can successfully pin point the faulty cell even in presence of multiple scan chain failures.

The CPU time taken by the proposed approach is shown in the last column of Table 5. The simulations are run on Intel I5 processor with 3 GB ram memory. The CPU time for diagnosis is in the order of milli-seconds.

5.1 Discussions

The proposed approach has several advantages like low hardware overhead, generic design, excellent diagnosis resolution and also time for diagnosis is very small. It has few limitations as well. The test patterns are dynamically modified. So, the responses stored in the ATE, are off less importance. We do not need to store the responses of the modified patterns back to ATE. We can perform a true value simulation of the modified patterns in the host computer and obtain the responses. We only need the observed responses which can be obtained in one of the two possible ways: (a) Store the observed responses into ATE for the entire test set and then download the ATE log and perform diagnosis off-line, or (b) After applying one pattern, download the observed response from ATE to the host computer where diagnosis can be performed based on that response. The second approach is on-line and requires continuous communication between ATE and the host computer.

The other limitation is due to multiple faults of different polarity in a chain. When multiple faults of same polarity occurs, the proposed approach is always able to diagnose the fault which is closer to the scan-out. Different polarity faults can also be diagnosed with some fault simulations. However, the main feature of the proposed work, scan chain behaves as a good chain while loading, does not hold. We can make the scan chain to behave as a good chain by dynamically changing the control values. It will increase the total test time and also requires continuous communication between ATE and host computer.

6. CONCLUSION

In this paper, we have proposed a novel decompresser side scan chain masking technique to diagnose scan chains. The proposed approach has negligible hardware overhead and can successfully narrow down the suspected list of faults. The proposed approach does not require any additional test patterns. The time for diagnosis is in the order of milli-

Table 5: Average Suspected List for different circuits obtained by the proposed method

Circuit	Bypassing Compactor				With Compactor				CPU Time in
	Avg. Suspect List for Fault in				Avg. Suspect List for Fault in				
	1 chain	2 chains	3 chains	4 chains	1 chain	2 chains	3 chains	4 chains	millie-seconds
s5378	1.00	1.00	1.00	1.00	1.00	1.00	1.00	1.00	0.003
s9234	1.05	1.13	1.19	1.28	1.05	1.14	1.21	1.29	0.005
s13207	1.05	1.15	1.22	1.30	1.05	1.17	1.26	1.34	0.018
s15850	1.01	1.09	1.14	1.19	1.01	1.10	1.14	1.21	0.010
s38417	1.17	1.14	1.29	1.32	1.17	1.14	1.29	1.34	0.044
s38584	1.06	1.04	1.06	1.07	1.06	1.04	1.07	1.08	0.063
b13	1.32	1.74	2.31	2.97	1.32	1.74	2.37	3.10	0.001
b14	1.01	1.07	1.29	1.63	1.01	1.08	1.44	1.91	0.084
b15	1.54	1.54	1.49	1.65	1.54	1.55	1.49	1.71	0.045
b20	1.86	2.27	2.28	2.55	1.86	2.30	2.66	2.95	0.177
b21	1.91	2.41	2.41	2.82	1.91	2.91	2.71	3.26	0.184
Avg.	**1.27**	**1.42**	**1.52**	**1.71**	**1.27**	**1.47**	**1.6**	**1.84**	

seconds and thus the method is appropriate for volume diagnosis. Currently, we are working on to extend the approach towards multiple fault diagnosis in a single chain.

7. REFERENCES

[1] W.-T. Cheng and Y. Huang. Enhance Profiling-Based Scan Chain Diagnosis by Pattern Masking. In *Proceedings of Asian Test Symp.*, pages 255–260, 2010.

[2] V. Chickermane, B. Foutz, and B. Keller. Channel masking synthesis for efficient on-chip test compression. In *Proceedings of Int. Test Conf.*, pages 452–461, oct. 2004.

[3] K. De and A. Gunda. Failure Analysis for Full-Scan Circuits. In *Proceedings of Int. Test Conf.*, pages 636–645, 1995.

[4] S. Edirisooriya and G. Edirisooriya. Diagnosis of scan path failures. In *Proceedings of 13th IEEE VLSI Test Symp.*, pages 250–255, apr-may 1995.

[5] R. Guo, Y. H., and W.-T. Cheng. A complete test set to diagnose scan chain failures. In *Proceedings of Int. Test Conf.*, pages 1–10, oct. 2007.

[6] R. Guo and S. Venkataraman. A technique for fault diagnosis of defects in scan chains. In *Proceedings of Int. Test Conf.*, pages 268 –277, 2001.

[7] Y. Huang. Dynamic learning based scan chain diagnosis. In *Proceedings of Design, Automation Test in Europe Conf. Exhibition*, pages 1–6, april 2007.

[8] Y. Huang, W. Cheng, C.-J. Hsieh, H. Y. Tseng, A. Huang, and Y.-T. Hung. Efficient Diagnosis for Multiple Intermittent Scan Chain Hold-Time Faults. In *Proceedings of Asian Test Symp.*, pages 44–49, nov. 2003.

[9] Y. Huang, W.-T. Cheng, and R. Guo. Diagnose Multiple Stuck-at Scan Chain Faults. In *Proceedings of 13th European Test Symp.*, pages 105–110, may 2008.

[10] Y. Huang, W.-T. Cheng, R. Guo, T.-P. Tai, F.-M. Kuo, and Y.-S. Chen. Scan chain diagnosis by adaptive signal profiling with manufacturing atpg patterns. In *Proceedings of Asian Test Symp.*, pages 35–40, 2009.

[11] J-S.Yang and S-Y.Huang. Quick Scan Chain Diagnosis Using Signal Profiling. In *Proceedings of the 2005 International Conference on Computer Design*, pages 157–160, 2005.

[12] Y. Kao, W.-S. Chuang, and J. C.-M. Li. Jump Simulation: A Technique for Fast and Precise Scan Chain Fault Diagnosis. In *Proceedings of Int. Test Conf.*, pages 1–9, oct. 2006.

[13] S. Kundu. Diagnosing scan chain faults. *IEEE Trans. on Very Large Scale Integration (VLSI) Systems*, 2(4):512–516, dec 1994.

[14] J.-M. Li. Diagnosis of multiple hold-time and setup-time faults in scan chains. *IEEE Trans. on Computers*, 54(11):1467–1472, nov. 2005.

[15] S. Narayanan and A. Das. An efficient scheme to diagnose scan chains. In *Proceedings of Int. Test Conf.*, pages 704–713, nov 1997.

[16] J. Schafer, F. Policastri, and R. McNulty. Partner SRLs for improved shift register diagnostics. In *Proceedings of 10th IEEE VLSI Test Symp.*, pages 198–201, april 1992.

[17] P. Song, F. Stellari, A. Weger, and T. Xia. A Novel Scan Chain Diagnostics Technique Based on Light Emission from Leakage Current. In *Proceedings of Int. Test Conf.*, pages 140–147, 2004.

[18] K. Stanley. High-accuracy flush-and-scan software diagnostic. *IEEE Design Test of Computers*, 18(6):56–62, nov/dec 2001.

[19] Synopsys. Tetramax atpg guide, 2006.

[20] X. Tang, R. Guo, W.-T. Cheng, and S. Reddy. Improving compressed test pattern generation for multiple scan chain failure diagnosis. In *Proceedings of Design, Automation Test in Europe Conf. Exhibition,*, pages 1000–1005, april 2009.

[21] R. C. Tekumulla and D. Lee. On identifying and bypassing faulty scan segments. In *Proceedings of North Atlantic Test Workshop*, pages 134–143, 2007.

[22] N. Touba. Survey of Test Vector Compression Techniques. *IEEE Design Test of Computers*, 23(4):294–303, april 2006.

[23] J. Waicukauski and E. Lindbloom. Failure diagnosis of structured VLSI. *IEEE Design Test of Computers*, 6(4):49–60, aug 1989.

[24] L. Wang, C. Wu, and X. Wen. *VLSI Test Principles and Architectures: Design for Testability*. 1st Edition, Elsevier, 2006.

[25] P. Wohl, J. Waicukauski, and S. Ramnath. Fully X-tolerant combinational scan compression. In *Proceedings of Int. Test Conf.*, pages 1–10, oct. 2007.

[26] W. Yuejian. Diagnosis of scan chain failures. In *Proceedings of IEEE International Symp. on Defect and Fault Tolerance in VLSI Systems*, pages 217–222, nov 1998.

Predicting Future Technology Performance

Asen Asenov and Craig Alexander
Gold Standard Simulations
The Rankine Building, Oakfield Avenue
Glasgow G12 8LT
+44 (0)141 330 4790
a.asenov@goldstandardsimulations.com
c.alexander@goldstandardsimulations.com

Craig Riddet and Ewan Towie
Device Modelling Group
University of Glasgow, Oakfield Avenue
Glasgow G12 8LT
+44 (0)141 330 4790
craig.riddet@glasgow.ac.uk
ewan.towie@glasgow.ac.uk

ABSTRACT

In this paper we highlight the important role of full-scale 3D Ensemble Monte Carlo (EMC) transport simulations in the performance analysis of contemporary and future decananometer MOSFETs. Considering both electron and hole transport in alternative device structures and materials we demonstrate that conventional drift diffusion (DD) simulations using standard mobility models fail to capture the non-equilibrium transport effects present in these devices, limiting their effectiveness in terms of performing predictive simulation of Si based FinFETs. We clearly demonstrate the capabilities and the power of EMC in evaluating the scaling potential and performance of FinFETs and quantum well transistors employing high mobility materials and the impact that additional scattering sources has on their performance.

Keywords

MOSFET, FinFET, Silicon, Germanium, InGaAs, Monte Carlo, Drift Diffusion.

1. INTRODUCTION

Low performance, intolerable levels of random dopant induced statistical variability, and corresponding increased leakage and SRAM yield and reliability problems has shifted the attention away from bulk MOSFETs that have been the workhorse of the semiconductor industry for decades. Fully depleted SOI and multi-gate MOSFETs, both with superior electrostatic integrity and tolerance to low channel doping, are the competing successors [11]. In particular, Intel's adoption of FinFET technology at the 22nm CMOS technology node [12] has invigorated the interest in their optimization and scaling.

In addition to this, the use of high-mobility channel materials such as III-Vs and germanium (Ge) as an alternative to conventional silicon (Si) is a promising technology option that has gained recognition in the 2011 edition of the International Roadmap for Semiconductors (ITRS) [13]. In this regard, TCAD based device simulation, analysis and optimization of such new transistor architectures plays an important role.

Permission to make digital or hard copies of all or part of this work for personal or classroom use is granted without fee provided that copies are not made or distributed for profit or commercial advantage and that copies bear this notice and the full citation on the first page. To copy otherwise, to republish, to post on servers or to redistribute to lists, requires prior specific permission and/or a fee.
DAC '13, May 29 - June 07 2013, Austin, TX, USA.

However, the drift diffusion (DD) approach, commonly used in commercial TCAD tools, can not deliver predictive simulations of these advanced devices with decananometer channel lengths due to non-equilibrium, quasiballistic transport [25]. This is further exacerbated by the use of strain and channel orientation to enhance transport and drive current.

Fully quantum simulations, such as the non-equilibrium Green's functions (NEGF), are more suited to simulating future devices in terms of properly capturing the complicated transport, but are computationally demanding and therefore there use has been limited to low dimensional transport or to the simulation of small structures such as nanowire transistors [8]. This limits their applicability to practical transistors design, especially when full 3D transport is required as is the case for the FinFET structure.

In contrast, ensemble Monte Carlo (EMC) device simulations are capable of resolving non-equilibrium transport effects in addition to the impact of orientation and strain on the bandstructure, transport and device performance [14], [23]. This provides the means to reliably investigate device behaviour early in the design stage making EMC more predictive than DD while less computationally intensive than NEGF.

The EMC approach can be extended by the use of complicated bandstructure models, quantum corrections, degenerate statistics and a full suite of scattering mechanisms allowing scaled devices using alternative channel materials and architectures to be accurately assessed.

In this paper we describe the full 3D EMC simulation module that is part of the GSS 3D device simulator GARAND [2] which is capable of accurately characterizing the transport in advanced device structures employing conventional and high mobility channel materials. The models at the heart of this approach are described in Section 2. The capabilities are than demonstrated in Section 3 by the simulation of FinFET and implant free quantum well (IFQW) transistors in comparison to DD simulation to highlight the predictive power of EMC.

2. SIMULATOR DESCRIPTION

The 3D EMC module of GARAND is self-consistently coupled to the solution of Poisson's equation coupling the carrier transport to the field and is essential to accurately describe transport and transistor performance, particularly at high fields [19].

2.1 Bandstructure Model

For electron transport a multi-band analytical ellipsoidal, non-parabolic description is employed, while for hole transport a full 6-band k•p approach is used that includes the effects of spin orbit coupling [6], [10]. The energy surfaces for Si and Ge calculated using this approach are shown in Figure 1. Both methods allow the impact of crystallographic orientation and applied strain to be easily accounted for.

Carrier statistics can be evaluated using either a non-degenerate Maxwell-Boltzmann or a fully degenerate Fermi-Dirac model, the latter of which includes the Pauli-Exclusion principle and is vital for the simulation of III-V materials with low density of states.

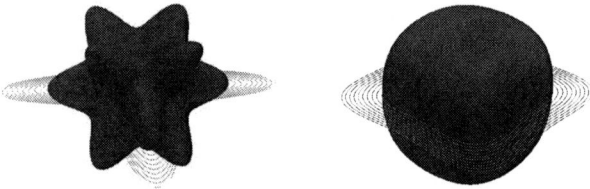

Figure 1 Energy surfaces of the heavy hole valence band at 50meV for Si (left) and Ge (right). Energy contours on the (001) surface are shown out to 1eV.

2.2 Quantum Corrections

The continued scaling of CMOS devices makes the inclusion of quantum effects absolutely essential for accurate predictive simulations [25]. Numerous approaches have been proposed to include quantum effects in EMC, ranging from a crude modification of the workfunction and oxide capacitance [5], to the effective potential method [7] and up to self-consistent coupling to the solution of the Schrodinger equation. All of these approaches have strengths and weaknesses in terms of accuracy, stability and efficiency and their impact on the other models used within the simulator.

In the EMC module of GARAND, an approach based on the Density Gradient formalism [1] is used that captures the essential quantum mechanical effects while having little effect on the computational efficiency of the simulator as a whole [19].

Here, the total driving force applied to the particles is based on the sum of classical potential from the solution of Poisson's equation and a quantum correction term [4]:

$$F_q = -\nabla \left(\psi_{cl} + \psi_{qc} \right)$$

The quantum correction term is taken from an initial DD simulation as the difference between the classical and quantum potentials and is stored at the beginning of the EMC simulation. The Density Gradient solver is calibrated to 1D Poisson-Schrodinger simulations in order to match the carrier distribution normal to the gate.

2.3 Scattering Mechanisms

A full range of scattering mechanisms is included within the EMC module for both electron and hole transport.

Acoustic phonon scattering (IAP) is modeled using an inelastic approach [14], [20] that includes a full dispersion [17]. For n-channel simulations this is an intravalley process, while for hole transport interband transitions are included, which is necessary due to the degenerate heavy and light hole valence bands. Inelastic optical phonon scattering is modeled without dispersion [14], with intra- and inter-valley transitions included. Both polar (IPOP) and non-polar (INPOP) modes are included and are used where applicable. The phonon scattering parameters are carefully calibrated to match measured low field mobility and velocity-field characteristics in undoped samples.

Ionized impurity scattering (II) is treated using Ridley's Third Body Exclusion [21] with an empirical correction to match experimental data applied for each material and carrier type.

Due to the use of quantum corrections, the use of a specular/diffusive interface roughness scattering (IR) is not appropriate [16], [22]. Therefore we follow Ando's approach [18] for this mechanism, where a conventional scattering rate is used. Alloy scattering (AL) is included using the approach discussed in [9], with the alloy potential calibrated to match experimentally observed mobilities. This is applied to alloy materials such as InGaAs, InAlAs and SiGe.

When Fermi-Dirac statistics is used, a suitable modification is made to the rates of the inelastic processes only [24].

3. RESULTS AND DISCUSSION
3.1 Si FinFET

The simulated Si FinFET is a 20nm gate length SOI FinFET, illustrated in Figure 2. The fin width and fin height are 10 nm and 25 nm respectively. The equivalent oxide thickness of the high-k/metal gate stack is 0.8 nm. The top of the conducting fin is insulated from the gate by a thicker layer of silicon nitride. As these devices were designed for SRAM applications the simulations are carried at the worst temperature corner of 358K. The same structure is employed for n- and p-channel simulations, and IAP, INPOP and II scattering is included, with IR scattering neglected to give an indication of the peak achievable performance assuming an ideal interface. A substrate orientation of (001) is used, with a channel of $\langle 100 \rangle$ for the n-channel and $\langle 110 \rangle$ for the p-channel transistors, giving sidewall surfaces of (001) and (110) respectively.

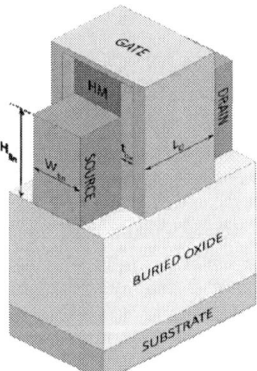

Figure 2 Device structure of the FinFET simulated in this study.

3.1.1 Comparison with DD

MC simulation yields a greater on-current as a result of nonequilibrium transport leading to a greater injection velocity compared to the corresponding DD velocity that cannot exceed the default silicon saturation velocity. Figure 3 shows the average carrier velocity profiles, from source to drain, from both EMC and DD simulation at V_{DS}=0.90V for the 20nm n-channel Si FinFET. At the virtual source (marked by the dashed line in the figure) the difference in velocity between EMC simulations and the DD simulations with default mobility is significant. While the EMC injection velocity exceeds 1.5×10^7cms^{-1}, the DD velocity is below 1.0×10^7cms^{-1}. By increasing the value of the saturation velocity in the DD field dependent mobility model, it should be possible to better reproduce the magnitude of the injection velocity within DD simulations and hence to match the EMC on-current.

220

Using the EMC simulated transfer characteristics as a target the DD mobility model was calibrated. First the low field mobility and its vertical field dependence was adjusted to reproduce the low field EMC simulated I_D-V_G characteristics. Then the saturation velocity was adjusted to match the EMC simulated on-current at high drain bias. The procedure was iterated until self-consistent values of the low field mobility and the saturation velocity were obtained.

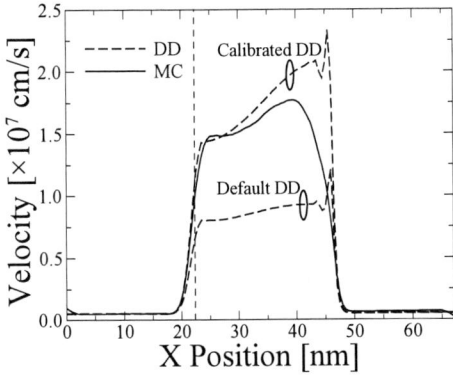

Figure 3 Electron velocity from source to drain from EMC (solid lines) and DD (dashed lines) in the n-channel FinFET at V_D=V_G=0.9V. DD results are shown using default mobility model parameters and after calibration to EMC results.

Figure 4 shows the calibrated transfer characteristics compared with the target from EMC simulation. Excellent agreement is achieved for the drain current at both low and high drain bias. DD simulation results using default mobility parameters are also shown for comparison. The sub-threshold slope determined by the device electrostatics is largely unaffected by the calibration maintaining a good agreement between DD and EMC.

This highlights the benefits of using EMC as a predictive simulation tool over DD, which cannot reliably estimate the on-current performance of a given device without calibration of mobility models to either measured transistors or to EMC simulations.

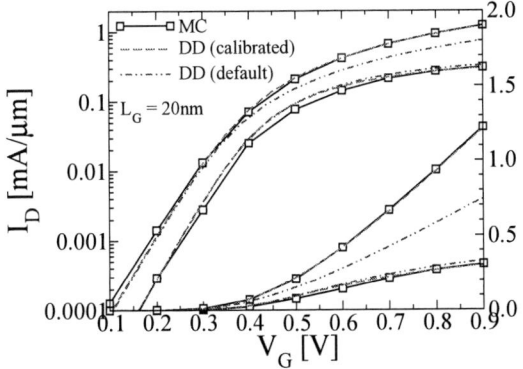

Figure 4 I_D-V_G characteristics for 20nm gate length n-channel FinFET from DD and EMC simulation. DD results are shown using default mobility model parameters and after calibration.

A similar procedure was followed for the p-channel FinFET, at the same operating conditions. Since no strain is used in this simulations, as can be seen from Figure 5 the underestimation of the carrier velocity at the source is less severe when using the default mobility model parameters at high drain bias compared to the n-channel case. However, as shown in Figure 6 careful calibration is required for properly predictive simulations.

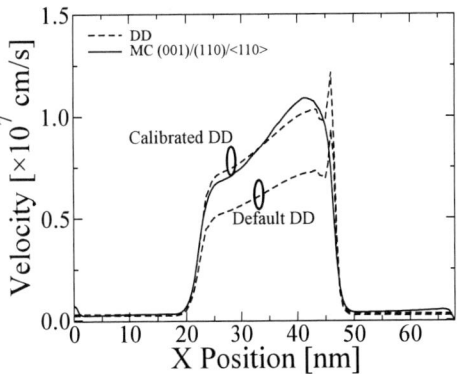

Figure 5 Hole velocity from source to drain from EMC (solid lines) and DD (dashed lines) simulation in the p-channel FinFET at V_D=V_G=0.9V. DD results are shown using default mobility model parameters and after calibration to EMC results.

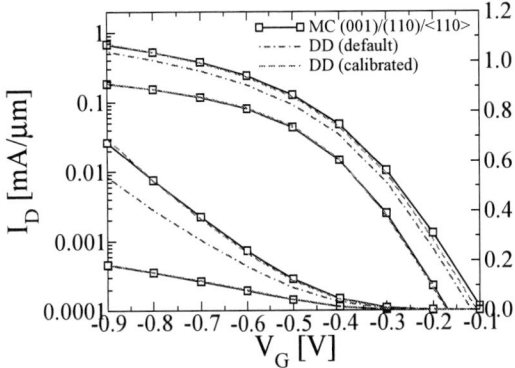

Figure 6 I_D-V_G characteristics for 20nm gate length p-channel FinFET from DD and EMC simulation. DD results are shown using default mobility model parameters and after calibration.

3.1.2 Impact of Crystallographic Orientation

A further complication arises when the impact of surface and channel orientation is taken into account for the p-channel transistor. In Figure 7 DD simulations have been calibrated to match EMC simulations with two different surface/channel orientations at low drain bias using only the low field mobility. As the EMC simulations demonstrate, at high drain bias the DD simulations firstly underestimate the performance and secondly fail to capture the change in the preferential orientation as carriers access different areas of the bandstructure. These effects are not captured in the DD simulations due to the absence of non-equilibrium effects in this simulation model.

This is a further limitation of DD as a method for predicting device performance in future technology options, and further demonstrates the benefits of using EMC as these effects are captured via the calculated bandstructure.

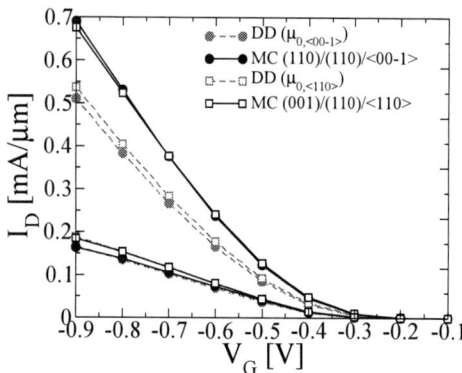

Figure 7 I_D-V_G characteristics for the p-channel FinFET showing two different substrate/sidewall/channel orientations. In both cases the low field mobility is calibrated to match the low drain bias characteristic from EMC.

3.2 IFQW

The structure of the 15nm gate length Implant Free Quantum Well (IFQW) transistors is illustrated in Figure 8 targeting the 10nm technology generation and has been designed following the ITRS guidelines [13]. For the n-channel case, the channel is $In_{0.53}Ga_{0.47}As$ with an $In_{0.52}Al_{0.48}As$ substrate, while the p-channel device employs Ge for the channel with a Si substrate. The transistor utilizes a 3.75nm thick QW channel with epitaxial in-situ doped raised source and drain regions. The source and drain regions are doped to 9.1×10^{19}cm^{-3}, the channel doping is 1.82×10^{17}cm^{-3} and the substrate 3.65×10^{18}cm^{-3}. A common gate oxide of Al_2O_3 is used with an EOT=0.51nm, and the lateral spacers are Si_3N_4 with a width ranging from 1 to 5nm. The IFQW transistor also includes the diffusion of dopants from the source/drain regions into the channel layer [15], which is referred to as sub-diffusion.

For the n-channel simulations, IAP, INPOP, IPOP and II scattering are included, while for the p-channel simulation the IAP, INPOP and II mechanisms are employed.

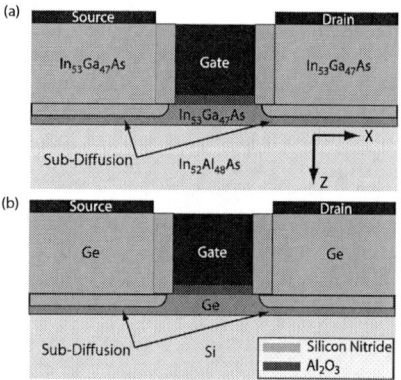

Figure 8 Structure of the IFQW device showing the (a) n-channel InGaAs and (b) p-channel Ge transistors.

3.2.1 Spacer Scaling

The impact of the width of the lateral Si_3N_4 spacer between the raised source/drain regions and the gate is known to be a critical factor in defining device performance [3], and here EMC simulations are used to demonstrate its impact on the on-current in both the n- and p-channel transistors. In both cases $I_{OFF} = 0.1\mu A/\mu m$ and $|V_D| = 1V$.

The impact of the spacer scaling on the nIFQW transistor is demonstrated in Figure 9, with an overall increase of 35% in the drive current as the spacer is scaled from 5nm to 1nm. It is also clear that the difference between the 2nm and 1nm cases is small compared to the other incremental changes, though SS and DIBL increase leading to the conclusion that the 2nm spacer is optimal.

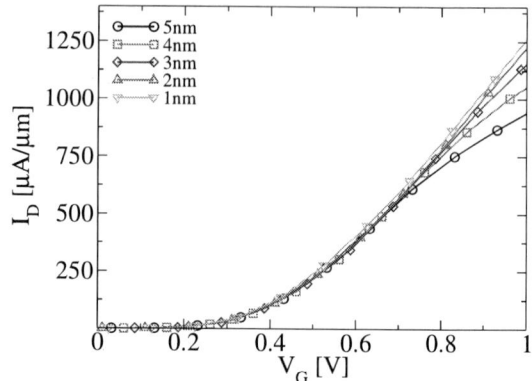

Figure 9 Transfer characteristics for $I_{OFF}=0.1\mu A/\mu m$ at $|V_D|=1V$ against lateral spacer width for the nIFQW device.

For the p-channel transistor (Figure 10), the impact of the spacer scaling is greater, with an increase of 118% as the spacer is scaled from 5nm down to 1nm. The stronger impact of degeneracy and the differences in confinement in the III-V material are the causes of this difference.

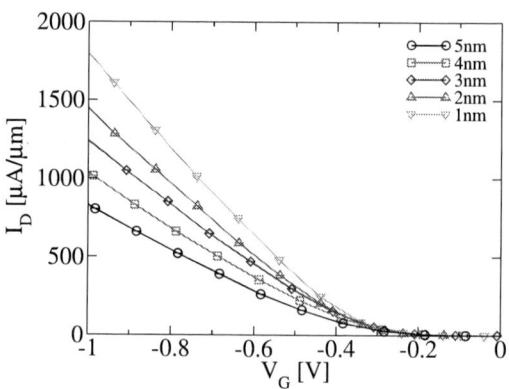

Figure 10 Transfer characteristics for $I_{OFF}=0.1\mu A/\mu m$ at $|V_D|=1V$ against lateral spacer width for the pIFQW device.

In both cases the use of sub- diffusion doping reduces the impact of the barrier between the contact regions and the channel by introducing doping into the channel layer and under the lateral spacer, improving the overall on- current as well as reducing the impact of variations in the lateral spacer thickness on device performance.

3.2.2 Comparison with DD

The transfer characteristics for the n- and p-channel IFQW devices are shown in Figure 11 and Figure 12 respectively for a device with a 2nm lateral spacer. In both cases the DD simulations use default mobility models, and hence significantly underestimate the drive current. Indeed, the current for the high drain bias DD simulations is less than half that predicted by the EMC simulations.

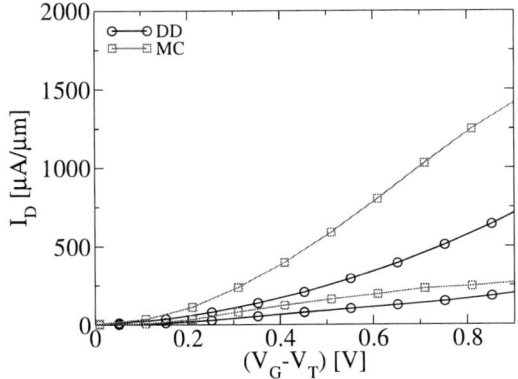

Figure 11 Comparison of the transfer characteristics at V_D=0.05V and V_D=1V between EMC and DD for the nIFQW device.

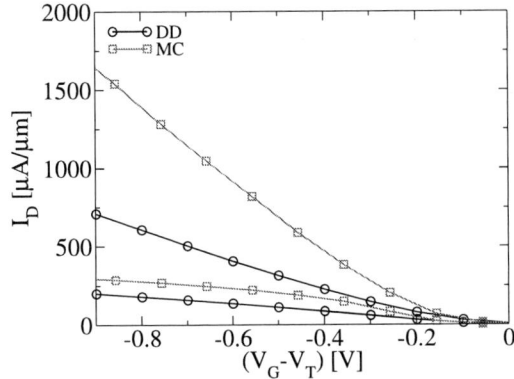

Figure 12 Comparison of the transfer characteristics at V_D=0.05V and V_D=1V between EMC and DD for the pIFQW device.

3.2.3 *Impact of Surface Roughness on IFQW Performance*

Further simulations of the IFQW devices with IR scattering applied to the interface between the Al_2O_3 gate dielectric and the channel have been carried out for devices with a range of lateral spacer widths.

Figure 13 shows the I_D-V_G characteristics for the n- and pIFQW transistors with and without this additional scattering mechanism. While the impact is greater in the nIFQW device, in both cases the impact increases as the spacer width reduces, though the scaling of the spacer width still results in an improvement in device performance.

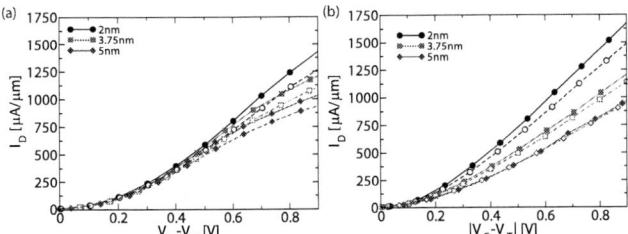

Figure 13 I_D-V_G characteristics of (a) the nIFQW and (b) the pIFQW transistor without (solid symbols) and with (hollow symbols) IR scattering at V_D=1V.

In the n-channel case, the high electron velocity dictates the drive current and the additional scattering from IR reduces the channel velocity, and in turn reduces the drive current. For the pIFQW

device the larger inversion density is responsible for the drive current, and again the additional scattering influences this, in particular at the source end of the channel leading to the reduction in the drive current when IR scattering is introduced.

3.3 III-V FinFET

To compare the two device concepts introduced in the previous sections, and to consider the impact of using a high mobility channel material in place of Si, a 15nm gate length FinFET shown in Figure 14 using $In_{0.53}Ga_{0.47}As$ as a channel material has been simulated using the 3D EMC module. These simulations employ scattering from IAP, INPOP, IPOP and II.

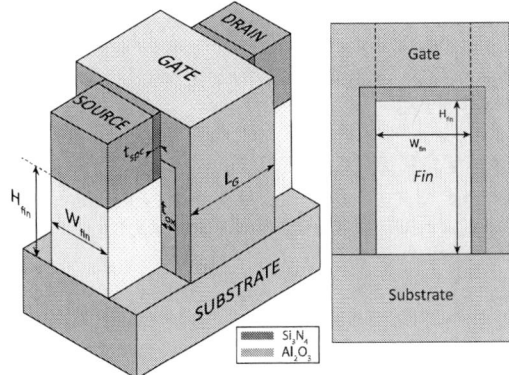

Figure 14 Device structure of the III-V n-type FinFET device showing a perspective and mid-gate cross-section.

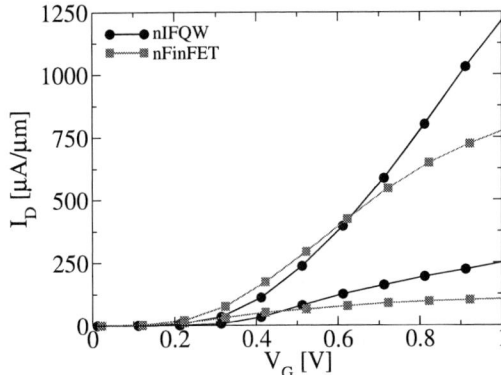

Figure 15 I_D-V_G characteristics of the III-V nMOSFETs for I_{OFF}= 0.1μA/μm.

The I_D-V_G characteristics for the two devices are presented in Figure 15 and show the relationship of drive current per unit width between device architectures. The SS is vastly improved from 88mV/decade in the IFQW to 68mV/dec in the FinFET, and DIBL improves from 85mV/V to 29mV/V.

Due to the low density of states of InGaAs, the larger electron density in the FinFET channel increases the impact of degeneracy and forces the electrons into the heavier effective mass L-valleys impacting on the channel velocity. Coupled with increased access resistance from the raised source/drain regions results in a reduced drive current per unit width. Though the increased effective gate area of this device leads to an increase in the drive current (the effective channel width of the IFQW device is 15nm, while for the FinFET it is 60nm). With further improvements to the source/drain design, combined with the better electrostatics suggests that the FinFET remains a promising candidate for this generation of device.

223

4. CONCLUSIONS

In this paper we have clearly demonstrated that accurate 3D EMC simulations are needed in order to reliably predict the performance of contemporary decananometer scale MOSFETs. DD simulations fail to accurately predict device performance using default mobility models, but via careful calibration to EMC transport simulations can reproduce the predicted performance. However, this calibration will not necessarily hold for a given device using, for instance, an alternative surface/channel orientation. We have also shown the usefulness of 3D EMC in the analysis of alternative architectures and channel materials, where non-equilibrium effects can be stronger due to the lower carrier mass. As these effects are well represented by the EMC simulation model, the impact of scaling and scattering on performance can be fully evaluated, giving a full indication of performance potential of a given device.

5. REFERENCES

[1] Ancona, M. G., and Iafrate, G. J. 1989. Quantum correction to the equation of state of an electron gas in a semiconductor. *Phys. Rev. B* 39 (May 1989) 9536–9540.

[2] Asenov A., Brown, A. R., Roy G., Cheng, B., Alexander C., Riddet C., Kovac U., Martinez A., Seoane N. and Roy S. 2009. Simulation of statistical variability in nano-CMOS transistors using drift-diffusion, Monte Carlo and non-equilibrium Green's function techniques. *J.Comp El.* 8 (November 2009) 349–373.

[3] Benbakhti, B., Kalna, K., Chan, K., Towie, E., Hellings, G., Eneman, G., Meyer, K. D., Meuris, M., and Asenov, A. 2011. Design and analysis of the $In_{0.53}Ga_{0.47}As$ implant-free quantum-well device structure. *Microelectronic Engineering.* 88 (April 2011) 358–361.

[4] Brown, A. R., Watling, J. R., Roy, G., Riddet, C., Alexander, C., Kovac, U., Martinez, A., and Asenov A. 2010. Use of density gradient quantum corrections in the simulation of statistical variability in MOSFETs. *J. Comp. El.* 9, 3-4 187–196.

[5] Bufler, F. M., Hude, R., and Erlebach, A. 2006. On a simple and accurate quantum correction for Monte Carlo simulation. *J. Comp. El.* 5 467–469.

[6] Dijkstra, J. E. and Wenckebach, W. T. 1997. Hole transport in strained Si. *J. Appl. Phys.* 81 (February 1997) 1259–1263.

[7] Ferry, D. K., Akis, R., and Vasileska, D. 2000. Quantum Effects in MOSFETs: Use of an Effective Potential in 3D Monte Carlo Simulation of Ultra-Short Channel Devices. In *IEDM Tech. Dig.* (December 2000) 287–290.

[8] Georgiev, V. P., Towie, E. and Asenov A. 2013 Impact of Precisely Positioned Dopants on the Performance of an Ultimate Silicon Nanowire Transistor: A Full Three-Dimensional NEGF Simulation Study. *IEEE Trans. El. Dev.* 60 (March 2013) 965-971.

[9] Harrison, J. W. and Hauser, J. R. 1976. Alloy scattering in ternary III–V compounds. *Phys. Rev. B, Condens. Matter.* 13, 12 (June 1976) 5347–5350.

[10] Hinckley, J. M. and Singh, J. 1994. Monte Carlo studies of ohmic hole mobility in silicon and germanium: Examination

of the optical phonon deformation potential. *J. Appl. Phys.* 76 (October 1994) 4192–4200.

[11] Hu, C. 2011. New sub-20nm transistors – why and how. In *Proc. Design Automation Conference (DAC)* 2011, pp. 460-463

[12] Intel 22nm 3-D tri-gate transistor technology [online] http://newsroom.intel.com/DOC-2032

[13] ITRS, International Roadmap for Semiconductors [online]. http://www.itrs.net/Links/2011/ITRS/

[14] Jacoboni C. and Lugli P. 1989, The Monte Carlo method for semiconductor devices, Springer-Verlag Wien New York.

[15] Mitard, J., *et al.* 2011 1mA/um-ION Strained SiGe45 Raised and Embedded S/D in *2011 Symposium on VLSI Technology* 134–135.

[16] Palestri, P., Eminente, S., Esseni, D., Fiegna, C., Sangiorgi, E., and Selmi, L. 2005. An improved semi-classical Monte-Carlo approach for nano-scale MOSFET simulation. *Solid-State Elec.* 49 727–732.

[17] Pop, E., Dutton, R. W., and Goodson, K. E. 2004 Analytic band Monte Carlo model for electron transport in Si including acoustic and optical phonon dispersion. *J. Appl. Phys.* 96 (November 2004) 4998–5005.

[18] Ramey, S. M. and Ferry, D. K. 2003. Implementation of Surface Roughness Scattering in Monte Carlo Modeling of Thin SOI MOSFETs Using the Effective Potential. *IEEE Trans. Nanotech.* 2 (June 2003) 110–114.

[19] Riddet, C., Alexander, C., Brown, A. R., Roy S., and Asenov A. 2011. Simulation of "Ab Initio" Quantum Confinement Scattering in UTB MOSFETs Using Three-Dimensional Ensemble Monte Carlo. *IEEE Trans. El. Dev.* 58 (March 2011) 600–608.

[20] Riddet, C., Watling, J. R., Chan, K., Parker, E. H. C., Whall, T. E., Leadley, D. R., and Asenov, A. 2012. Hole Mobility in Germanium as a Function of Substrate and Channel Orientation, Strain, Doping, and Temperature. *IEEE Trans. El. Dev.* 59 (July 2012) 1878–1884.

[21] de Roer, T. G. V. and Widdershoven, F. P. 1986. Ionized Impurity scattering in Monte Carlo calculations. *J. Appl. Phys.* 59 813–815.

[22] Sangiorgi, E. and Pinto, M. R. 1992. A Semi-Empirical Model of Surface Scattering for Monte Carlo Simulation of Silicon n-MOSFET's. *IEEE Trans. El. Dev.*, 39 (February 1992) 356–361.

[23] Sangiorgi, E., Palestri, P., Esseni, D., Fiegna, C. and Selmi, L. 2008. The Monte Carlo approach to transport modeling in deca-nanometer MOSFETs. *Solid-State Elec.* 52 1414–1423.

[24] Ungersboeck, E. and Kosina, H. 2005. The Effect of Degeneracy on Electron Transport in Strained Silicon Inversion Layers, in *SISPAD.* 311–314.

[25] Vasileska, D., Khan, H. R., Ahmed, S. S., Ringhofer, C. and Heitzinger, C. 2005. Quantum and Coulomb Effects in Nanodevices. *Int. J. Nanoscience.* 4, 3 305–361.

Predicting Future Product Performance: Modeling and Evaluation of Standard Cells in FinFET Technologies

Veit B. Kleeberger, Helmut Graeb, Ulf Schlichtmann
Institute for Electronic Design Automation, Technische Universität München, Munich, Germany
kleeberger@tum.de, graeb@tum.de, ulf.schlichtmann@tum.de

ABSTRACT

With continued scaling of CMOS technology it becomes increasingly difficult to maintain reliable circuits. Early predictive technology and design exploration help to understand major effects of variability sources and their impact on circuit performances. With each new technology basic circuit blocks have to be redesigned to appropriately evaluate the impact of technology scaling. Therefore, this paper presents an approach which is able to find the optimal sizing of basic circuit blocks considering process variation. We utilize this approach to predict the impact of scaling in FinFET technologies and the influence of process variations in future technology nodes.

Categories and Subject Descriptors

B.7.1 [**Hardware**]: Integrated Circuits—*Types and Design Styles, Advanced Technologies*; B.8.2 [**Hardware**]: Performance and Reliability—*Performance Analysis and Design Aids*

General Terms

Algorithms, Performance, Reliability

Keywords

FinFET, predictive modeling, discrete sizing, standard cells, process variations, NBTI

1. INTRODUCTION

With ongoing technology scaling conventional planar CMOS devices suffer from increasing susceptibility to numerous variation sources. Enhanced sensitivity of circuit performances, such as delay or leakage, to process and environmental variations makes these devices increasingly unreliable [11]. One answer to this problem is the creation of new device structures, such as 3D FinFETs. These devices, already starting to be used in upcoming 22nm SRAM architectures [7], promise to solve some of these problems, e.g. by tolerating low channel doping.

Permission to make digital or hard copies of all or part of this work for personal or classroom use is granted without fee provided that copies are not made or distributed for profit or commercial advantage and that copies bear this notice and the full citation on the first page. To copy otherwise, to republish, to post on servers or to redistribute to lists, requires prior specific permission and/or a fee.
DAC'13, May 29 – June 07 2013, Austin, TX, USA

While FinFET transistors promise to solve some of the scaling related problems they also introduce other issues. Due to their 3D structure their manufacturing process gets more complicated and also new variability sources, such as fin edge roughness, are introduced.

Recently, there has been some effort to create predictive transistor models for FinFETs to allow early technology exploration. Such models are either created based on scaling experiences from productive technologies [16] or directly based on TCAD simulations [3,17]. Generally these models are used to predict transistor performances such as threshold voltage, I_{on} or I_{off}. When it comes to predicting circuit performance often just the smallest devices, i.e., performances of SRAM cells, are studied.

For larger circuit blocks, such as combinational or sequential cells, it becomes increasingly difficult to predict their performance in a new technology as this often requires a redesign of the whole cell to accurately account for technology aspects. This comprises sizing of the transistors of a circuit block such that given performance specifications are met under worst-case manufacturing and operating conditions. Usually sizing is solved as a continuous optimization problem with subsequent rounding, accepting that the solution is suboptimal. However, in modern technologies using FinFETs sizing is only done by using multiple fins or fingers. This suggests to introduce proper discrete sizing techniques.

2. PREDICTIVE CIRCUIT MODELING

In the following we will give some definitions, then we will present a discrete sizing approach for proper predictive modeling, followed by an extension to consider manufacturing and operating tolerances.

2.1 Parameters, Performance, Simulation

Sizing is based on evaluating the performances $\mathbf{f}$ in dependence of circuit parameters $\mathbf{x}$ by means of numerical simulation:

$$\boldsymbol{\varphi} : \mathbf{x} \in \mathcal{R}^{n_x} \to \mathbf{f} \in \mathcal{R}^{n_f} \qquad (1)$$

Three types of parameters can be distinguished:

$$\mathbf{x} = \begin{cases} \mathbf{s} \sim \mathcal{N}(\mathbf{d}_0, \mathbf{C}) & \text{statistical parameters} \\ \mathbf{o} \in T_o = [\mathbf{o}_L, \mathbf{o}_O] & \text{operational parameters} \\ \mathbf{d} & \text{design parameters} \end{cases} \qquad (2)$$

Statistical parameters are for instance transistor model parameters. They reflect the manufacturing process variations with various types of statistical distributions, which are transformed into a normal distribution with mean value vector

$\mathbf{d}_0$ and covariance matrix $\mathbf{C}$ without loss of generality. Operational parameters are for instance temperature or supply voltage. They reflect the varying operating conditions, which are defined by lower and/or upper bounds $\mathbf{o}_L, \mathbf{o}_U$ of ranges within which the product is guaranteed to work. Design parameters are for instance transistor widths and lengths and are subject to sizing/optimization. We distinguish two types of design parameters:

$$\mathbf{d} = \begin{cases} \mathbf{d}_c \in \mathcal{R}^{n_{dc}} & \text{continuous} \\ \mathbf{d}_d \in \mathcal{D} = \{ \quad \mathbf{d}^{(j)}, j = 1, \ldots, n_D, & \text{ordered..} \\ \quad \forall_i d_i^{(j)} \le d_{i+1}^{(j)} \} & \text{..discrete} \end{cases} \quad (3)$$

Please note that we assume that the discrete design parameters are ordered real numbers, e.g. of an equidistant or nonlinear or arbitrary grid, which yields a discrete value domain $\mathcal{D}$. Please note also that we assume that simulation (1) is possible for the relaxed domain $\tilde{\mathcal{D}}$ of continuous values that can be obtained by interpolation among several points of $\mathcal{D}$:

$$\tilde{\mathcal{D}} = \{\mathbf{d}_d \in \mathcal{R}^{n_{dd}} | \quad \mathbf{d}_d = \alpha \cdot \mathbf{d}_i + (1 - \alpha) \cdot \mathbf{d}_j, \quad (4)$$
$$\mathbf{d}_{i,j} \in \mathcal{D}, \quad 0 \le \alpha \le 1\}$$

A sizing method that is suitable when simulation is only possible on the discrete value domain $\mathcal{D}$ is presented in [12].

In Sec. 2.2, we will first describe nominal sizing, which does not consider distributions or tolerance intervals of statistical and operational parameters. In Sec. 2.3, we will extend the method to include statistical distributions and tolerance intervals.

2.2 Automatic Discrete Nominal Sizing

The optimization objective $c(\mathbf{d})$ is designed as a weighted (η_i), normalized ($\frac{\cdot}{\|f_{U,i}\|}$), one-sided ($\max(0, .)$) least-squares approach for given performance specifications ($f_{U,i}$):

$$c(\mathbf{d}) = \sum_{i=1}^{n_f} \left(\eta_i \cdot \max \left(0, \frac{f_i(\mathbf{d}) - f_{U,i}}{\|f_{U,i}\|} \right) \right)^2 \quad (5)$$

Here we have assumed upper performance bounds, lower performance bounds can be considered analogously. Eq. (5) is particularly practicable for circuit sizing. The performance normalization makes different objectives with different orders of magnitude and different physical units comparable. Least-squares is a favorable cost function type, which is enhanced by allowing performances to overfulfill the required spec ($\max(0, .)$). The moment a performance objective is satisfying its specification, it is no longer contributing to the cost function. This leads to a termination of the optimization process when all spec bounds are fulfilled, no matter how much. The cost function can be further tuned by setting different weights among the performance objectives. Please note that after termination of the sizing process, the specs can be tightened to search for even better product quality.

The optimization task now is formulated as minimization of the cost function (5) subject to sizing constraints:

$$\min c(\mathbf{d}) \text{ subject to } \boldsymbol{\varrho}(\mathbf{d}) \ge \mathbf{0} \quad (6)$$

This represents a special case of a discrete optimization problem with ordered discrete parameter values (3).

We have developed a problem-specific solution approach, which is suitable for both digital and analog circuits [13]. This approach is based on a combination of Branch&Bound and Feasible Sequential Quadratic Programming (FSQP). Algorithm 1 sketches the basic Branch&Bound approach in

---

Algorithm 1: Branch&Bound+ $(\mathbf{d}_{inc}, \tilde{\mathcal{D}})$

Data: current solution $\mathbf{d}_{inc}$, relaxed domain $\tilde{\mathcal{D}}$
1 solve (6) on relaxed domain $\tilde{\mathcal{D}}$ to get $\mathbf{d}'$
   // Pruning:
2 **if** $\mathbf{d}'$ *not feasible or worse, i.e.,* $c(\mathbf{d}') \ge c(\mathbf{d}_{inc})$ **then**
3 | return
4 **end**
5 **if** $\mathbf{d}'$ *fits discrete grid, i.e.,* $\mathbf{d}'_d \in \mathcal{D}$ **then**
6 | $\mathbf{d}_{inc} = \mathbf{d}'$, return
7 **end**
   // Model-Based intermediate Branch&Bound:
8 Branch&Bound on quadratic performance model from FSQP solution in line 1 to get $\mathbf{d}''$
9 **if** $\mathbf{d}''$ **then**
10 | $\mathbf{d}_{inc} = \mathbf{d}''$, return
11 **end**
   // Branching:
12 choose parameter d_i, partition relaxed domain at two neighbor grid points of solution d'_i:
13 $\tilde{\mathcal{D}}_L = \left\{ \mathbf{d} \in \tilde{\mathcal{D}} \,|\, d_i \le \lfloor d'_i \rfloor \right\}, \tilde{\mathcal{D}}_U = \left\{ \mathbf{d} \in \tilde{\mathcal{D}} \,|\, d_i \ge \lceil d'_i \rceil \right\}$
14 Branch&Bound+ $(\mathbf{d}_{inc}, \tilde{\mathcal{D}}_L)$
15 Branch&Bound+ $(\mathbf{d}_{inc}, \tilde{\mathcal{D}}_U)$

---

recursive form. Its main idea is to relax the discrete domain to a continuous domain $\tilde{\mathcal{D}}$ and optimize on this domain (line 1). The resulting solution is taken to open two new search branches at a certain parameter such that in a lower branch the domain may maximally reach the lower neighbor grid point and in an upper branch the domain may minimally reach the upper neighbor grid point (lines 12-15). In that way, and by recursively calling the Branch&Bound with the respective new branches, the final solution will be forced to a discrete grid point. The algorithm keeps an incumbent solution and returns if no further improvement is possible in the respective branch (line 2) or if a better point is found that fits into the grid (line 5).

A reasonable choice to solve (6) in line 1 is an FSQP approach. This requires numerical simulation for performance evaluation. The results of FSQP provide a quadratic performance model, which is much faster than simulation. This property is exploited by inserting a Branch&Bound step with the quadratic model after line 7. From experiments it resulted that this model-based Branch &Bound in line 8 very often gives a solution that satisfies the given specs, which provides a significant reduction in branching and hence CPU times. The corresponding extension is given in lines 8 to 11 of Algorithm 1.

2.3 Automatic Discrete (Parametric) Yield Optimization

In [5], the duality between worst-case analysis and yield analysis has been described in detail. It will be used here to solve the yield optimization problem by a corresponding worst-case optimization problem. Towards that, a certain minimum yield requirement is translated into a tolerance region T_s of statistical parameters:

$$Y' = \int_{-\infty}^{\beta_w} \frac{1}{\sqrt{2\pi}} \exp \left(-\frac{\beta^2}{2} \right) d\beta \quad (7)$$

$$T_s = \left\{ \mathbf{s} \,\middle|\, (\mathbf{s} - \mathbf{s}_0) \cdot \mathbf{C} \cdot (\mathbf{s} - \mathbf{s}_0)^T \le \beta_w^2 \right\} \quad (8)$$

Then, a worst-case analysis is set up, which computes the worst performance values $f_{w,i}$ that are obtained for the given tolerance regions T_s (8) and T_o (2):

$$\max_{\mathbf{s} \in T_s, \mathbf{o} \in T_o} f_i, i = 1, \ldots, n_f$$

$$\longrightarrow \mathbf{s}_{w,i}, \mathbf{o}_{w,i}, f_{w,i}(\mathbf{d}) = f_i(\mathbf{s}_w, \mathbf{o}_w, \mathbf{d}), i = 1, \ldots, n_f(9)$$

Here we have again assumed upper performance bounds, which refer to worst-case values in the direction of increasing performance. Lower performance bounds can be considered analogously. Equation (9) can be solved, e.g., by an advanced worst-case analysis tool [10]. Please note that every performance gets its own worst-case parameter set, and that the respective worst-case performance value can be guaranteed under any operating condition with a β_w-sigma robustness (e.g., 3 sigma = 99.9% yield) [5]. Therefore, we replace the performance simulation in the cost function (5) with a worst-case analysis (9) to obtain a worst-case cost function:

$$f_i(\mathbf{d}) \to f_{w,i}(\mathbf{d}) \Rightarrow c(\mathbf{d}) \to c_w(\mathbf{d}) \qquad (10)$$

Equation (10) is then used in (6) and Algorithm 1 to optimize the worst-case cost for a given yield requirement. As this process implies a large number of circuit simulation, a relaxed worst-case analysis can be used that performs only one iteration step before switching back to the Branch&Bound loop [14]. This has been applied in the circuit examples provided in this paper.

3. EXPERIMENTAL RESULTS

3.1 Modeling Assumptions

We base our study on the freely available predictive transistor models from http://ptm.asu.edu [16, 18]. Table 1 summarizes key parameters from some these models.

Parameter	Unit	PTM20	PTM14	PTM7
L_g	nm	24	18	11
T_{fin}	nm	15	10	7
H_{fin}	nm	28	23	18
T_{ox}	nm	1.4	1.3	1.15
V_{DD}	V	0.9	0.8	0.7
N_A	$10^{18}/cm^{-3}$	0.5	0.05	0.001

Table 1: Key parameters of used FinFET transistor models [16]

To model process variation we consider the following variation parameters. To account for global process variability we model channel length L_g, fin thickness T_{fin}, and fin height H_{fin} by a Gaussian distribution. We assume for the standard deviation of each of these parameters $3\sigma = 10\%$ of their respective nominal values [6, 9]. Similar we model global oxide thickness variation as Gaussian distribution with $3\sigma = 5\%$ of its nominal value [6]. Additionally we assume local variability in L_g and T_{fin} due to line-edge roughness (LER) and we model them by a 3σ truncated normal distribution. The corresponding values for LER are taken from ITRS [6]. We assume that fin height and oxide thickness are not subject to local variability as they depend on film thickness and not on lithography [9].

We model Random Dopant Fluctuations (RDF) by assuming the number of dopants in the channel to be Poisson distributed. The standard deviation due to RDF is then given by:

$$\sigma_{N_A} = \sqrt{\frac{N_A}{L_g \cdot T_{fin} \cdot H_{fin}}} \qquad (11)$$

Metal gates are used in newer technologies to overcome limitations from using polysilicon gate electrodes. These metal gates introduce another variability source due to the random orientation of the metal grains. Metal gate granularity results in a variation of the gate work function Φ_g. We assume in the following that the gate work function is Gaussian distributed with $\sigma = 20$ meV [2, 4, 15]. Please note, that this assumes that the number of metal grains stays constant, i.e., the grain size is reduced with technology scaling.

To evaluate the impact of *Bias Temperature Instability* (BTI) we use the following degradation equation [1]:

$$\Delta V_{th} = A(V, T) \sum_{i=1}^{2} \alpha_i \log\left(1 + t_{stress}/\tau_{c,i}\right) \qquad (12)$$

$$A(V, T) = K \cdot \exp\left(\frac{-E_0}{k_B T}\right) \cdot \exp\left(\frac{B \cdot V}{t_{ox} \cdot k_B T}\right) \qquad (13)$$

Fig. 1 shows this aging model together with all its fitting parameters. Although this model is intended for a planar CMOS process we assume that it is also valid for FinFET transistors. Comparisons from [8] indicate that this will approximately capture the influence of aging or will at least provide a lower bound to it.

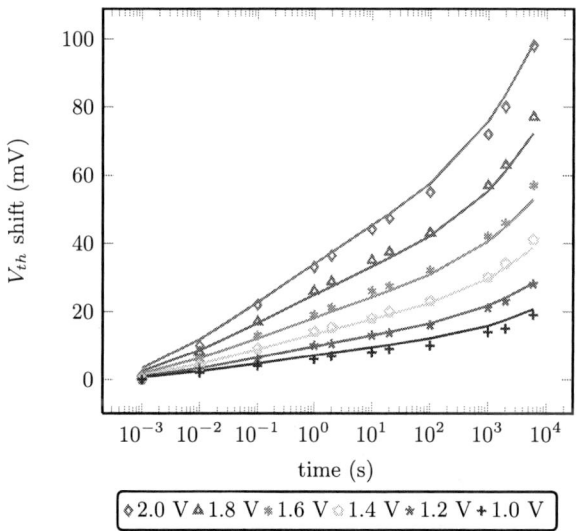

(a) Model and measurements for 65nm (T=125°C, various V_{DD})

$\tau_{c1} = 1$ ms	$\tau_{c2} = 1$ ks	$\alpha_1 = 1 - \alpha_2 = 0.3$
$K = 420$ mV	$E_0 = 0.190$ eV	$B = 0.08$ eVnm/V

(b) Model Parameters

Figure 1: NBTI Degradation Model

3.2 Technology Characteristics

We start our study by nominal sizing of the inverter circuit shown in Fig. 2 for different planar and FinFet technology nodes. This circuit represents an inverter driving four times a copy of its own. Thus, we assume that all p-channel transistors (i.e., M1, M3, ...) are equally sized and that all n-channel transistors (i.e., M2, M4, ...) are equally sized.

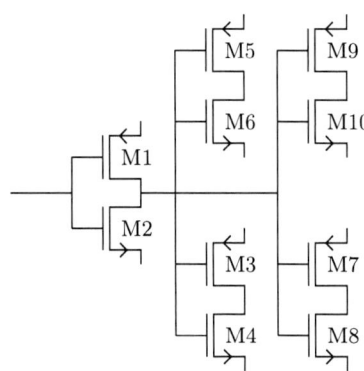

Figure 2: FO4 Circuit

We target as sizing objective an approximately equal rise and fall delay as well as an equal rise and fall slope at the output of the first inverter built from M1 and M2. We model this target using the following constraints:

$$-C \leq \frac{\text{delay}_{\text{rising}} - \text{delay}_{\text{falling}}}{\text{delay}_{\text{rising}} + \text{delay}_{\text{falling}}} \leq C \qquad (14)$$

$$-C \leq \frac{\text{slope}_{\text{rising}} - \text{slope}_{\text{falling}}}{\text{slope}_{\text{rising}} + \text{slope}_{\text{falling}}} \leq C \qquad (15)$$

As fins are discrete by nature exactly equal rise and fall delays and slope are unfeasible. Using equations (14) and (15) we allow them to be different by a given percentage C. We repeat this sizing process several times using the measured output slope after one optimization run as new stimuli at the circuit input until we get approximately equal input and output slopes.

We size this circuit for different technology nodes and extract characteristic performances such as P/N sizing ratio, output slope, delay and equivalent input capacitance of the inverter (Fig. 3).

We can see from the P/N ratio that while we have to choose different sizings for the P- and N-transistors in a planar technology, the number of fins remains the same for the P- and N-transistor. This points out that both transistor types behave approximately equal in this setup making different sizing to maintain equal rise and fall unnecessary. Although the inverter delay initially scales better for FinFETs, we see that the scaling trend quickly reverts back to the initial scaling trend from planar CMOS. The slope scales strongly for FinFETs, while it stays constant for conventional MOSFETs. The input capacitance scales similar compared to conventional MOSFET but is larger for the same channel length. Both, the scaling of slope and capacitance, can be very well explained by the three-dimensional gate controlling the channel.

3.3 Influence of Process Variations

In this section we want to study the influence of process variations in FinFET technologies. We apply the PTM14 model, which has a nominal channel length of 18nm. As characteristic representatives we choose a NAND and a NOR cell, which are shown in Fig. 4 together with their obtained sizings using the algorithm from Sec. 2.

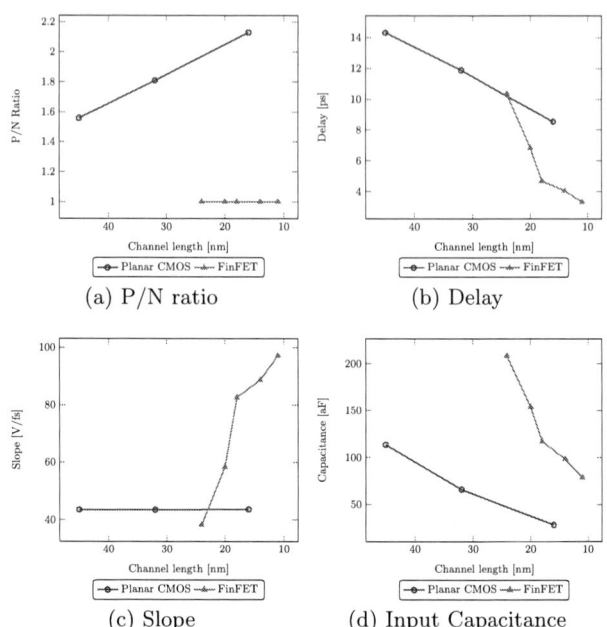

Figure 3: Scaling of inverter performances

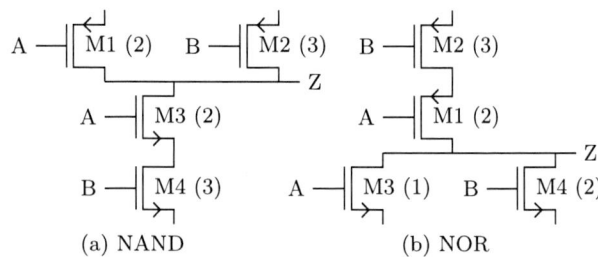

Figure 4: NAND and NOR cells with chosen fin numbers in brackets

To evaluate these cells we run a Monte-Carlo simulation with 1000 samples and measure delay, slope, and leakage power for every input combination. The corresponding Q-Q plots are shown in Fig. 5 and Fig. 6.

We see that delay is very well standard normal distributed as we expect it from previous technologies. For the slope inverse (i.e., switching time) this holds also for some timing arcs, but especially for the timing arcs where the whole transistor stack is involved (e.g., $B_r \rightarrow Z_f$ for the NAND) this property does not hold. For leakage power we test for a lognormal distribution in Fig. 6. Leakage variability fits very well a lognormal distribution for most samples with some deviations at the tail of the distribution.

Additionally we conduct a worst case analysis as described in Sec. 2.3. Based on the gradient pointing to the worst-case direction we compute the contribution of the different variability sources as well as the single transistors compared to global variability (Fig. 7 and 8).

Both cells show metal gate granularity as a major variability source for delay and leakage in our setup. Besides, fin thickness also contributes considerably, especially for leakage. It can also be seen that the contribution of random dopant

228

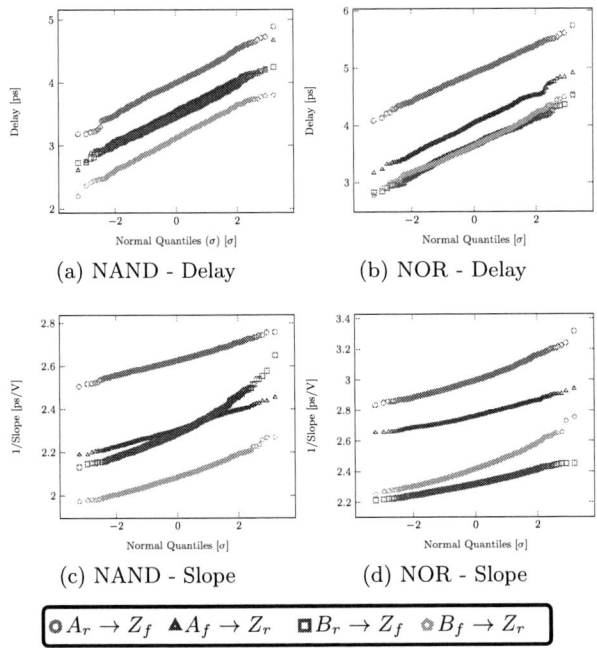

(a) NAND - Delay   (b) NOR - Delay

(c) NAND - Slope   (d) NOR - Slope

$$\odot A_r \to Z_f \quad \blacktriangle A_f \to Z_r \quad \square B_r \to Z_f \quad \oplus B_f \to Z_r$$

Figure 5: Q-Q Plot – Delay & Slope

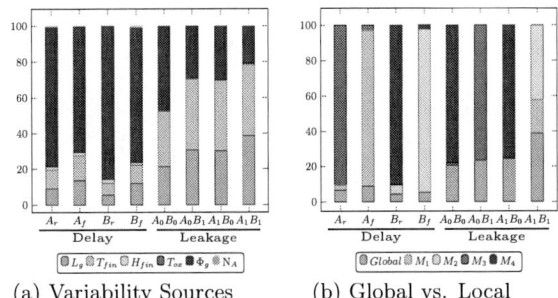

(a) Variability Sources   (b) Global vs. Local

Figure 7: Delay and Leakage Variability – NAND

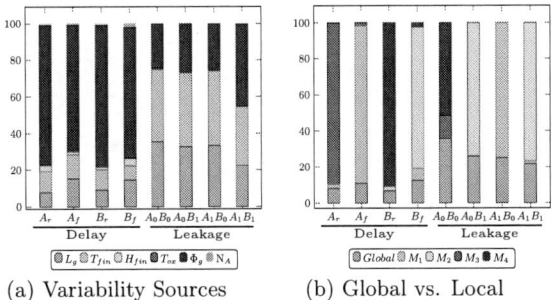

(a) Variability Sources   (b) Global vs. Local

Figure 8: Delay and Leakage Variability – NOR

fluctuations is well suppressed due to the low channel doping. Additionally we can see that most variability is caused by local variations of the transistors where again metal gate granularity dominates, while the remaining local variation is caused by line edge roughness effects.

3.4 Scaling Trends

To evaluate the scaling trend in FinFET design we size a mid-performance inverter and evaluate its behavior under different specifications, which are listed in Table 2. While specification 0 evaluates the inverter under nominal conditions we allow in each subsequent specification one additional variation source: Process variation in spec 1, temperature variation in spec 2, V_{DD} variation in spec 3, and aging in spec 4. For BTI aging we assume a signal probability of 0.5, which means for 10 years lifetime the transistors are for 5 years under stress. For simplicity we do not consider recovery

here.

Table 2: Specifications for performance evaluation

Spec-Nr.	Yield	T	V_{DD}	Lifetime
0	Nominal	27° C	Nominal	Fresh
1	3σ	27° C	Nominal	Fresh
2	3σ	$0 - 60°$ C	Nominal	Fresh
3	3σ	$0 - 60°$ C	$\pm 10\%$	Fresh
4	3σ	$0 - 60°$ C	$\pm 10\%$	10 years

Figure 9 shows the change in delay over technology nodes, while Fig. 10 shows the change in leakage.

The relative delay change caused by process variations increases only slightly. Spec 1 causes about 15% delay increase at 24nm channel length, while it causes about 20% delay increase at 11nm channel length. Compared to this increase temperature variability has a negligible influence in our specifications. Voltage droop causes an increase comparable to process variations. Aging has a more extreme impact (Fig. 9a). As we assumed no improvement of NBTI induced threshold voltage degradation over process technology scaling this effect has a severe impact on delay.

Leakage shows process variation as the major variability source for the 3σ-design due to its lognormal distribution, followed by temperature and supply voltage.

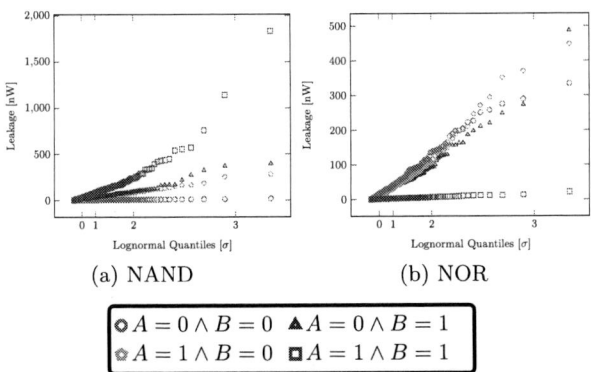

(a) NAND   (b) NOR

$$\odot A = 0 \wedge B = 0 \quad \blacktriangle A = 0 \wedge B = 1$$
$$\oplus A = 1 \wedge B = 0 \quad \square A = 1 \wedge B = 1$$

Figure 6: Q-Q Plot – Leakage

4. CONCLUSION

We presented an automatic discrete sizing method suitable for the design of basic building circuit blocks in FinFET technologies considering variations. This approach was utilized to evaluate scaling trends in digital logic based on FinFET transistors. We see metal gate granularity becoming a major

229

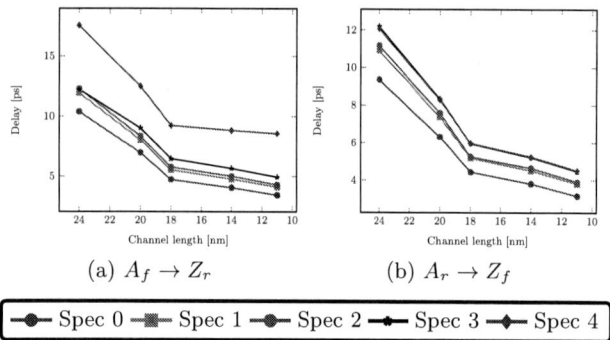

(a) $A_f \to Z_r$ (b) $A_r \to Z_f$

```
—●— Spec 0  —※— Spec 1  —●— Spec 2  —▲— Spec 3  —◆— Spec 4
```

Figure 9: Scaling Trends for Delay

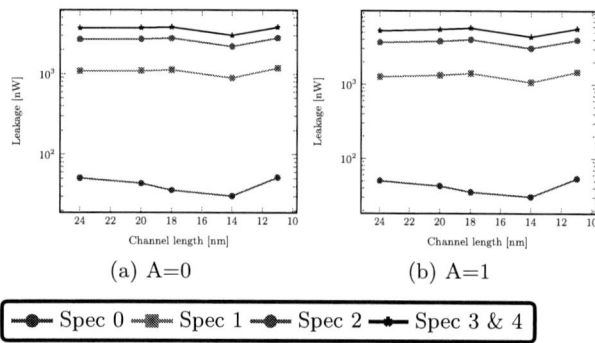

(a) A=0   (b) A=1

```
—●— Spec 0  —※— Spec 1  —●— Spec 2  —▲— Spec 3 & 4
```

Figure 10: Scaling Trends for Leakage

variability source in future technology nodes. Also the impact of aging compared to other variations sources seems to increase. We propose therefore further studies of this effect based on degradation equations which accurately consider FinFET specifics.

ACKNOWLEDGMENTS

The authors would like to thank Dr. Christoph Werner from TUM, Institute for Technical Electronics for help with the NBTI degradation model. The authors would like to thank Dr. Sani Nassif from IBM Austin Research Lab for inspiring discussions, ongoing cooperation and important contributions.

This work was supported in parts by the German Research Foundation (DFG) as part of the priority program "Dependable Embedded Systems" (SPP 1500 – http://spp1500.itec.kit.edu).

5. REFERENCES

[1] M. Barke, V. B. Kleeberger, C. Werner, D. Schmitt-Landsiedel, and U. Schlichtmann. Analysis of Aging Mitigation Techniques for Digital Circuits Considering Recovery Effects. In *edaWorkshop*, May 2013.

[2] A. R. Brown, N. M. Idris, J. R. Watling, and A. Asenov. Impact of metal gate granularity on threshold voltage variability: A full-scale three-dimensional statistical simulation study. *IEEE Electron Device Letters*, 31(11):1199–1201, 2010.

[3] B. Cheng, X. Wang, A. Brown, C. Millar, A. Asenov, J. Kuang, and S. Nassif. Statistical TCAD Based PDK

Development for a FinFET Technology at 14nm Technology node. In *IEEE International Conference on Simulation of Semiconductor Processes and Devices (SISPAD)*, 2012.

[4] H. Dadgour, K. Endo, V. De, and K. Banerjee. Modeling and analysis of grain-orientation effects in emerging metal-gate devices and implications for SRAM reliability. In *IEEE International Electron Devices Meeting*, 2008.

[5] H. Graeb. *Analog Design Centering and Sizing*. Springer, 2007.

[6] International Technology Roadmap for Semiconductors. http://www.itrs.net, 2012.

[7] E. Karl, Y. Wang, Y.-G. Ng, Z. Guo, F. Hamzaoglu, M. Meterelliyoz, J. Keane, U. Bhattacharya, K. Zhang, K. Mistry, and M. Bohr. A 4.6 GHz 162 Mb SRAM Design in 22 nm Tri-Gate CMOS Technology With Integrated Read and Write Assist Circuitry. *IEEE Journal of Solid-State Circuits*, 48(1):150 –158, Jan. 2013.

[8] S. Kim, K. Han, B. Choi, S. Kong, J. Lee, and J. Lee. Negative bias temperature instability of bulk fin field effect transistor. *Japanese Journal of Applied Physics: Part 1 – Regular Papers, Short Notes, and Review Papers*, 45(3A):1467, 2006.

[9] D. D. Lu, C.-H. Lin, A. M. Niknejad, and C. Hu. Compact Modeling of Variation in FinFET SRAM Cells. *IEEE Design & Test of Computers*, 27(2):44–50, 2010.

[10] MunEDA. *WiCkeD Tool Suite*, Jan. 2013.

[11] S. R. Nassif, V. B. Kleeberger, and U. Schlichtmann. Goldilocks Failures: not too soft, not too hard. In *IEEE International Reliability Physics Symposium (IRPS)*, 2012.

[12] M. Pehl. *Discrete Sizing of Analog Integrated Circuits*. PhD thesis, Technische Universitaet Muenchen, Nov. 2012.

[13] M. Pehl and H. Graeb. An SQP and Branch-and-Bound Based Approach for Discrete Sizing of Analog Circuits. In E. Tlelo-Cuautle, editor, *Advances in Analog Circuits*, chapter 13, pages 297–316. InTech, Feb. 2011.

[14] M. Pehl and H. Graeb. Tolerance Design of Analog Circuits Using a Branch-and-Bound Based Approach. *Journal of Circuits, Systems, and Computers*, 21(8), Dec. 2012.

[15] S. H. Rasouli, K. Endo, and K. Banerjee. Variability analysis of FinFET-based devices and circuits considering electrical confinement and width quantization. In *IEEE International Conference on Computer-Aided Design (ICCAD)*, 2009.

[16] S. Sinha, G. Yeric, V. Chandra, B. Cline, and Y. Cao. Exploring sub-20nm FinFET design with predictive technology models. In *ACM/IEEE Design Automation Conference (DAC)*, 2012.

[17] X. Wang, A. R. Brown, B. Cheng, and A. Asenov. Statistical variability and reliability in nanoscale FinFETs. In *IEEE International Electron Devices Meeting (IEDM)*, 2011.

[18] W. Zhao and Y. Cao. New generation of predictive technology model for sub-45 nm early design exploration. *IEEE Transactions on Electron Devices*, 53(11):2816–2823, 2006.

The ITRS Design Technology and System Drivers Roadmap: Process and Status

Andrew B. Kahng
CSE and ECE Depts., Univ. of California at San Diego
abk@ucsd.edu

ABSTRACT

The Design technology working group (TWG) is one of 16 working groups in the *International Technology Roadmap for Semiconductors* (ITRS) effort. It is responsible for the ITRS' Design Chapter, which roadmaps design technology requirements and potential solutions for elements of the semiconductor supply chain that are produced by the electronic design automation (EDA) industry. The Design TWG is also responsible for the ITRS' System Drivers Chapter, which roadmaps the key product classes that drive the leading-edge requirements for process and design technologies. Through these activities, the Design TWG sets a number of fundamental parameters in the overall ITRS: layout density, die size, maximum on-chip clock frequency, total chip power, SOC and MPU architecture models, etc. This paper reviews the process by which the Design TWG evolves its roadmap content, and some of the key modeling and roadmapping questions that the semiconductor and EDA industries will face in the near term.

1. INTRODUCTION

As noted in [13], technology roadmaps seek "precompetitive" specifications of future technical requirements and challenges. Potential solutions are identified, investigated, pruned, productized, standardized, and delivered to the marketplace – in a synchronized, timely, and cost-effective manner – to ensure a continued stream of technology benefits. The *International Technology Roadmap for Semiconductors* (ITRS) [22] is one of the most successful roadmapping efforts ever: well over 1000 scientists and engineers worldwide collaborate to synchronize a wide range of industries and technologies (automated test equipment, assembly and packaging, photomask, electronic design automation (EDA), lithography, interconnect, device, etc.) so that the "Moore's Law" semiconductor value proposition can continue. The broad scope of the ITRS is essential, e.g., the roadmap for design technology must comprehend (i) lithography and restricted design rules; (ii) die stacking and 3D integration; (iii) device and interconnect electrical performance, variability and robustness; (iv) ATE, BIST and BISR overheads and production costs; (v) product-level trajectories for RF blocks, IO bandwidth and processing capability; and many other futures. The ITRS's 15-year horizon reflects the lead times needed to identify and develop production-worthy technologies.

All technology roadmaps struggle with the tension between "roadmapping" and "extrapolation". An uncalibrated roadmap lacks credibility. On the other hand, unthinking extrapolation from historical data risks "driving by the rear-view mirror", and can result in absurd projections at the 15-year horizon. Meaningful roadmapping of technology requirements and potential solutions requires at least the following elements.

- *Metrics.* What cannot be measured cannot be tracked or improved. EDA tools heuristically address large-scale, NP-hard optimizations, and design quality is strongly determined by flow and methodology ("it's the magician, not the wand"). Thus, it is challenging to identify metrics that capture the

progress of design technology.

- *Understanding of contexts and needs for technology.* Contexts ranging from process technology to market forces affect the need for technology. For example, the trajectory of mobile consumer SOC products has driven rapid innovation in low-power design techniques spanning embedded memory design, power and clock gating, dynamic voltage scaling, etc. At the same time, these low-power design techniques must acknowledge process and material attributes such as discreteness of FinFET device widths starting at the 16nm foundry node, or increasingly dominant reliability and aging mechanisms.

- *Holistic selection of potential solutions.* Technology roadmapping must holistically model and predict impacts of potential technology solutions, at many levels. For example, solutions to a "power crisis" in IC design may come from manufacturing technologists (e.g., process innovation to reduce Vth variation), device and circuit technologists (introduction of FinFET and resistive RAM), and system designers (heterogeneous multi-core SOC architectures) – as well as design and test technologists (asynchronous design flow, on-chip variability monitoring and adaptivity, etc.). All potential solutions cost money to develop and deploy. Thus, as discussed in [11], a mindset of "shared red bricks" in the semiconductor technology roadmap is critical to achieve proper allocation of R&D resources.[1]

The ITRS Design Technology Working Group. The Design technology working group (TWG) is one of 16 TWGs in the ITRS. With over 50 industry and academic contributors from all five regional semiconductor industry associations (USA, EU, Japan, Taiwan, Korea), the Design TWG is responsible for the ITRS Design Chapter, which roadmaps design technology requirements and potential solutions relevant to the EDA industry, and the ITRS System Drivers Chapter, which roadmaps the key product classes that drive leading-edge requirements for process and design technologies.

Figure 1 shows how the Design and System Drivers chapters have consistently evolved over the past decade. First, the Design Chapter gives a *quantified* Design Technology roadmap with metrics, potential solutions, and mappings from requirements to potential solutions. This matches the structure and metrics-oriented "look and feel" of other ITRS chapters. Second, an increasingly comprehensive set of System Drivers has been developed that maintain alignment to key segments of the semiconductor industry. Each update to the System Drivers (e.g., the acknowledgment of a hard platform power limit in the MPU roadmap, starting in 2007) has ripple effects across Overall Roadmap Technology Characteristics (ORTCs) such as layout density, transistor count, die size, chip power and frequency – as well as fundamental technology metrics owned by other technology working groups. These interactions are conceptually depicted in Figure 2.[2] The System Drivers also enable

---

[1] In ITRS parlance, a "red brick" is a technology requirement that has no known solution (the term stems from the coloring convention in ITRS technology requirement tables). For example, to solve the problem of poor interconnect RC scaling, are R&D dollars best invested in new dielectric materials, new interconnect and barrier materials, better overlay control, more accurate signal integrity analyses in EDA tools, scalable many-core GALS architectures, or ...? Or, to solve the problem of exploding (and widening) modes and corners in signoff, should variation be reduced in the process itself, or should statistical signoffs be adopted, or should "signoff at typical" be adopted in combination with adaptivity [3], or ...?

[2] In over 17 years of NTRS and ITRS roadmap participation, I have witnessed a steady rise in the prominence of "design" within the ITRS. Originally highly process-centric, the roadmap now increasingly relies on "design-based equivalent scaling" [24] and

---

Permission to make digital or hard copies of all or part of this work for personal or classroom use is granted without fee provided that copies are not made or distributed for profit or commercial advantage and that copies bear this notice and the full citation on the first page. To copy otherwise, to republish, to post on servers or to redistribute to lists, requires prior specific permission and/or a fee.
DAC'13, May 29 - June 07 2013, Austin, TX, USA.

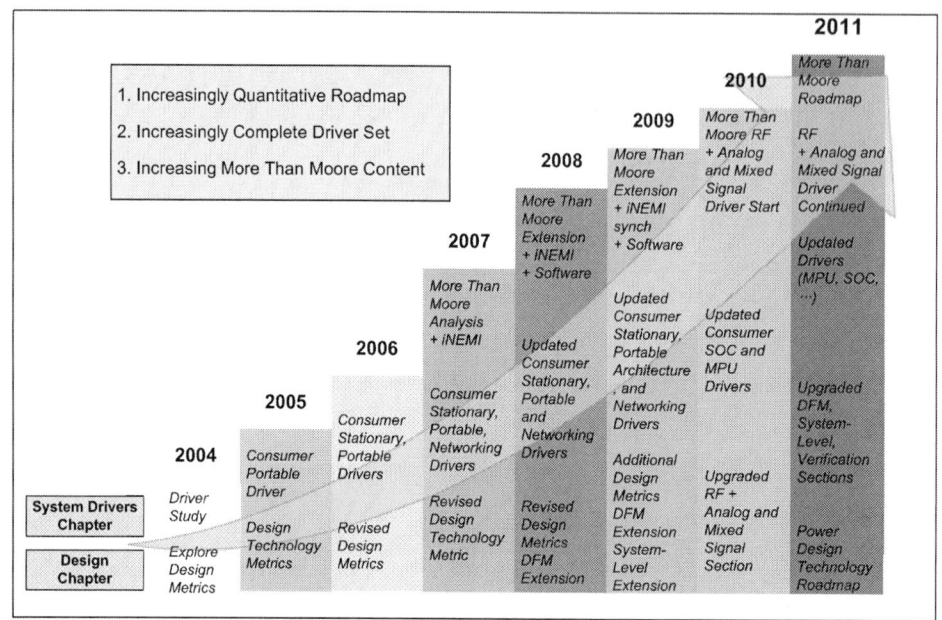

Figure 1: Roadmap from ITRS System Drivers and Design chapters. [Source: ITRS Design ITWG 2011 Public Conference presentation, December 2011, Songdo, Korea.]

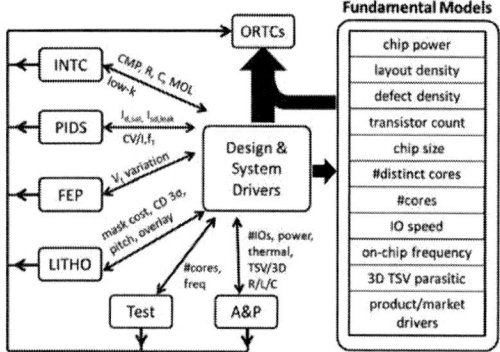

Figure 2: Increasingly central role of Design TWG in ITRS roadmap definition.

stronger alignment (cf. "More Than Moore") between the ITRS's chip-level roadmap and system product-level roadmaps such as iN-EMI [21].

Organization of This Paper. The remainder of this paper is organized as follows. Section 2 outlines the process and over-arching objectives that guide the evolution of Design and System Drivers content. Several examples then give the "flavor" of how the roadmap evolves. Two aspects of the System Drivers Chapter are the System Driver model evolution, which is discussed in Section 3, and the "A-factor" approach that underlies projection of density scaling in the ITRS, which is discussed in Section 4. Two aspects of the Design Chapter are the low-power design technology roadmap, which is discussed in Section 5, and the evolution of Design for Manufacturability (Variability, Reliability) content, which is discussed in Section 6. Section 7 concludes with some thoughts on modeling and roadmapping issues that the semiconductor and EDA industries will face in the near term.

2. DESIGN TWG GOALS AND PROCESS

Like every other technology working group in the ITRS, the Design TWG places the interests of its industry and R&D community – i.e., EDA and VLSI CAD – first and foremost. In ITRS cross-TWG interactions, the Design TWG must respond to questions such as "How much variability can designers tolerate?" (Lithography TWG) or "What is the J_{max} limit for on-chip global interconnects?" (Interconnect TWG) or "What tradeoff between leakage and drive currents is best for mobile SOCs?" (Process Integration, Devices and Structures (PIDS) TWG). The roadmap for DFT is

jointly owned with the Test TWG. The roadmap for off-chip IO bandwidth is jointly owned with the Test TWG and the Assembly and Packaging (A&P) TWG. And the roadmap for 3D/TSV based integration is jointly owned with a number of other TWGs, notably A&P, Test, Interconnect and Front-End Processing (FEP). All of these interactions entail asynchronous, off-line dialogues year-round with designers, EDA technologists and researchers so that perspectives from IC design, and from IC design automation, are correctly represented.

ITRS challenges and technology requirements directly inform the research priorities and funding allocations of a number of government funding agencies and industry consortia worldwide, and the phrase "According to the ITRS, ..." is often given as motivation in academic research papers. Thus, Design TWG activities often include advocacy for the importance of EDA technology and academic research. Furthermore, "key messages" in the Design Chapter can seed future trends in academic research and research funding. Three examples of such advocacy and messaging are as follows.

- *The Design Cost Model.* Although tremendous product differentiation comes from design and design technology, EDA industry revenues, and levels of R&D investment and academic research funding, have been stagnant. With this in mind, quantifying the *value of design technology* has been one of the high-level goals for the Design TWG within the ITRS effort. Since 2001, the Design Chapter has included a highly influential Design Cost model [14] [12] that now encompasses both hardware and software development costs (salary and overhead of engineers, EDA tool cost per seat, interoperability costs, etc.). The cost model quantifies the impact of design technology innovation and resulting productivity improvements. For example, the hardware design costs for a consumer portable SOC design in 2011 are estimated at $25.7M, versus $7708M had design technology innovations between 1993 and 2009 not occurred.

- *Key Messages.* Over the years, the Design TWG has formulated specific key messages within the ITRS. Since 2001, an overarching message has been that "cost of design is the greatest obstacle to continuation of semiconductor roadmap". In the 1998-2001 time frame, the Design TWG also advocated a "Living ITRS" mindset wherein all technology roadmap projections and models could be implemented on a common platform, to enable interoperability and cross-checking for consistency.[3] More specific messages have also been given over the years. For example, in 2009 the Design Chapter's

---

"More Than Moore" to deliver scaling of semiconductor product value in the face of non-ideal performance, power, density and variability scaling.

---

[3]The GTX (MARCO GSRC Technology Extrapolation) package [4] for some years provided a realization of this goal, but is no longer maintained.

key messages were that (i) software and system-level design productivity are critical to the roadmap of semiconductor value; (ii) design reliability roadmapping was a necessary addition to the roadmap; (iii) system-level design techniques would ultimately be crucial to managing power; and (iv) design technology innovations must keep on schedule through the end of the roadmap in order to contain design costs. New messages in 2011 and 2012 included (i) roadmapping focus at the design-manufacturing interface has evolved from "manufacturability" to a more general "variability", which now entails an even broader question of how systems will maintain reliability and be resilient; (ii) design technology innovations must keep on schedule through the end of the roadmap in order to contain power; and (iii) the importance of cross-TWG interactions is continually growing, whether for More Than Moore, 3D, Beyond CMOS, or even the basic device and lithography roadmaps.

- *Grand Challenges.* The ITRS Executive Summary calls out a subset of each working group's "difficult challenges", and categorizes these as either "Enhancing Performance" or "Cost-Effective Manufacturing", and as either near-term (within the next seven years) or long-term (between eight and 15 years out). In the 2005-2011 ITRS editions, power management, design productivity, and DFM were consistently listed as near-term grand challenges for design. The roadmap noted that power management challenges would need to be addressed across multiple levels, especially system, design, and process technology. Moreover, to maintain design quality in advanced process nodes, design implementation productivity must improve to the same degree that design complexity is scaled – with improvement of design productivity and IP reuse being key considerations. Long term challenges have evolved from management of leakage power consumption in the 2005-2009 roadmaps to design of concurrent software and design for reliability and resilience in the 2011 roadmap.

The Design TWG operates in a distributed manner, with each major Design Chapter section or System Driver model maintained by a distinct subteam. Different geographies tend to assume natural responsibilities for content, e.g., European contributors have responsibility for the AMS/RF content, and Japanese contributors have responsibility for the SOC system driver models. New content is constantly developed according to identified gaps in roadmap coverage, e.g., Design Chapter updates in 2009 and 2011 include (i) a 3D/TSV design technology section, (ii) a hardware-related software development cost component for Design Cost model, and (iii) a low-power design technology roadmap. Following ITRS convention, the U.S. TWG co-chairs coordinate worldwide efforts and serve as the editors for all published content.[4]

3. KEY SYSTEM DRIVER MODELS

As noted above, the System Drivers Chapter models and projects key semiconductor product classes that create the need for continued semiconductor innovation [5–7]. The 2011 System Drivers Chapter identifies three microprocessor (MPU) drivers (high-performance (HP), cost-performance (CP) and power-connectivity-cost (PCC)) and three System-On-Chip (SOC) drivers (consumer portable (CP), consumer stationary (CS) and networking (NW)).[5] Each driver should provide impetus for specific technology objectives, e.g., the SOC-CP driver drives lower leakage (or standby) power consumption, given the severe battery life requirement of mobile devices. For each MPU and SOC system driver, the ITRS roadmaps scaling of parameters such as number of cores, number of SRAM and logic transistors, layout density, frequency and power.

---

[4] Resources and dedicated bandwidth in support of the ITRS have not yet recovered from the 2008-2009 economic downturn. All suggestions, participation in ITRS meetings, and other contributions are always welcome; interested individuals should contact the Design TWG co-chairs, Dr. Andrew B. Kahng (abk@ucsd.edu) and Dr. Juan-Antonio Carballo (jantonio@ieee.org).

[5] MPU-HP are server products, e.g., Intel Xeon and AMD Opteron. MPU-CP are desktop products, e.g., Intel Core i7 and AMD Phenom. MPU-PCC are handheld and micro-server products, e.g., Intel Atom and Marvell Armada. SOC-CP are handheld products, e.g., Qualcomm Snapdragon and Samsung Exynos. SOC-CS are products for game consoles, e.g., IBM Cell BE and WonderMedia (Via) WM series. SOC-NW are multi-core network processors, e.g., Broadcom XLP864 and Calxeda ECX-1000.

MPU Driver Modeling

The ITRS MPU driver model has for many years scaled the number of logic transistors and the number of SRAM transistors by $2\times$ per technology node. Since dimensions shrink by $0.7\times$ per node, and nominal layout density therefore doubles, this simple scaling model allows die size to remain constant across technology nodes.

MPU Die Size. The 2009 MPU model update [10] set a constant die area of $260mm^2$ for MPU-HP and $140mm^2$ for MPU-CP. The model for logic density ($D_{tr,logic}$) is

$$D_{tr,logic} = \frac{N_{tr,nand2}}{O_{logic} \cdot U_{logic}} \quad (1)$$

where $N_{tr,nand2}$ (number of transistors in a NAND2 gate) is four, O_{logic} (logic overhead due to design integration) is 2.0 (i.e., 100% area overhead for whitespace), and U_{logic} (the area of a unit NAND2 gate) is calculated using the "A-factor" described below in Section 4. The model for SRAM density ($D_{tr,SRAM}$) [10] is

$$D_{tr,SRAM} = \frac{N_{tr,bitcell}}{O_{SRAM} \cdot U_{SRAM}} \quad (2)$$

where $N_{tr,bitcell}$ is the number of transistors in a SRAM bitcell, O_{SRAM} (overhead due to peripheral circuits) is assumed to be 1.6 (i.e., 60% area overhead), and U_{SRAM} (the area of a unit SRAM bitcell) is calculated using another A-factor, also described in Section 4. While the 2009 MPU model remains accurate with respect to number of cores, or total number of transistors, die areas of recent server MPU products have grown rapidly, reaching $\sim 530mm^2$ in the 2012-2013 time frame. Moreover, the simple model of cores + SRAM does not acknowledge the growth of "uncore" elements (memory controllers, IO controllers, GPU cores, on-chip networking, etc.) in MPU products. These considerations make it likely that the 2013 ITRS edition will see substantial revision of the MPU-HP model with respect to both A-factors and architecture.

MPU Frequency. Figure 3 overlays historical changes in the ITRS maximum on-chip frequency roadmap with product data from the Stanford CPUDB [18]. The 2001 System Drivers Chapter observed that rapid MPU frequency increases up to that time had been enabled by reduction in the number of fanout-of-four (FO4) delays per clock period. That is, microarchitecture (aggressive pipelining, with fewer stages of logic per pipeline stage) had been used to increase frequency at a faster rate than the intrinsic growth of device switching speed. At that time (2001), a basic limit of 12 FO4 delays (in which useful computation could be performed during a clock cycle) was being reached, and so the roadmap was modified to improve frequency only as device speeds improved (17%/year improvement in CV/I metric, in the PIDS roadmap).

In 2007, a market-driven platform power limit of 130W per die was acknowledged, and the MPU frequency roadmap was revised to increase by just 8% per year to meet this power limit.[6] The slowing of frequency enabled the PIDS device roadmap to also slow the CV/I improvement to 13%/year, which eased the challenge of managing leakage currents. Subsequently, during the 2009-2011 roadmapping cycle, device technologists found that even the 13%/year CV/I improvement was incompatible with leakage current requirements; hence, the likely scenario for 2013 and beyond is for 4%/year frequency increase in MPU products (still with design-based equivalent scaling in the form of switching factor reductions), along with some limited "headroom" of 8%/year improvement in the device CV/I metric.

System Driver Futures

During the 12 years since the System Drivers Chapter was introduced, many structural changes have occurred in the marketplace. As these shifts occur, the set of system drivers, and their intrinsic models, are subject to change.

- The SOC-CS driver was introduced at a time when the IBM Cell BE was highly visible in the game console market. Today, game consoles are primarily driven by high-end CPU-GPU fusion products such as AMD A10-5800K, which is essentially an instance of the MPU-CP driver. Thus, the need for an SOC-CS driver may be obsolete.

---

[6] With this 8%/year frequency growth, a "magic, design-based" 5% reduction per year in the chip's switching activity factor had to be added into the MPU model, to keep MPU power flat. Although actual product frequencies were already visibly flattening, it was felt that a model with 0% frequency increase would stall device and circuit innovation needed by other semiconductor products.

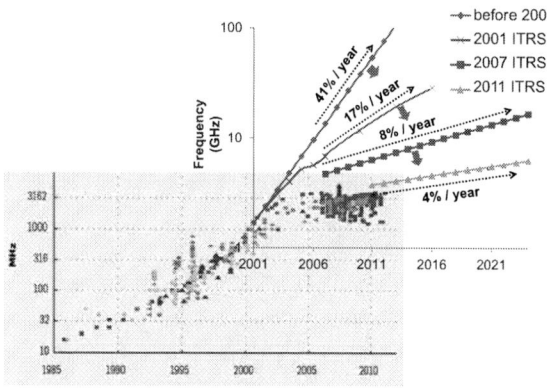

Figure 3: Frequency scaling roadmap.

- While the MPU-CP driver originally reflected the desktop PC market and the "shrink" version of the MPU-HP "lead" processor, today we see that desktop processors simply use the same architecture as either server- or handheld-class products. Accordingly, MPU-CP may also be considered for removal as a separate driver.

- If SOC-CS and MPU-CP drivers are both removed from the roadmap, the key remaining drivers will be MPU-HP and SOC-CP. SOC-CP, which reflects the handheld market, is rapidly becoming more general-purpose and integrates GPU IPs such as Mali, PowerVR, etc. The architecture and scaling models for SOC-CP may be considered for change.

- It may also be noted that the MPU-PCC driver is evolving toward the micro-server market and away from the handheld market. Thus, the roadmap for MPU-PCC may need to change as well.

- Even as the MPU-CP and SOC-CS drivers become less important to the technology roadmap, new system drivers may arise from automotive, defense, medical and energy management applications, aligning with recent More Than Moore foci.

4. LAYOUT DENSITY A-FACTORS

In the ITRS System Drivers Chapter and Overall Roadmap Technology Characteristics, *A-factors* enable the modeling of unit cell areas of SRAM and standard-cell logic circuit fabrics, in terms of the M1 half-pitch, F. SRAM layout density is mainly determined by Mx pitches and poly pitch in a bulk technology. With FinFET devices, the fin pitch (P_{fin}) becomes the dominant factor for SRAM layout. On the other hand, the density of *standard cells* is mainly decided by the cell height (in M2 tracks) and the poly pitch. Since the 2009 ITRS, the A-factor for a 6T SRAM bitcell has been $60F^2$, and the A-factor for a 2-input NAND gate has been $175F^2$ [10]. These values are based on various ratios between, e.g., poly, M1, and M2 layer pitches (design rules) as summarized in the left half of Table 1, as well as on the canonical layouts shown in Figures 4(b) and 5(b) [10].

As the industry moves to double-patterning, FinFETs with discrete gate widths, and "middle of line" (MOL) layers to enable local access to transistors, the fundamental A-factor scaling models will likely require significant revisions. For example, in future NAND2 cell layouts, M1 may no longer be the most congested metal layer, so M2 pitch (P_{M2}) may shrink to be the same as M1 pitch (P_{M1}). Furthermore, with emerging FinFET (multi-gate) devices, fin pitch (P_{fin}) cannot be arbitrarily small, and gate width is in quanta of fins. Based on these considerations, the A-factor of the bulk NAND2 layout may evolve to $144F^2$ (Figure 4(b)), i.e., $W_{cell} = 3P_{poly}$, $H_{cell} = 8P_{M2}$, and hence $A_{Bulk,NAND2} = W_{cell} \times H_{cell} = 144F^2$. The area of the FinFET NAND2 layout may be set to $162F^2$ (Figure 4(a)), i.e., $W_{cell} = 3P_{poly}$, $H_{cell} = 9P_{M2}$, and hence $A_{FinFET,NAND2} = W_{cell} \times H_{cell} = 162F^2$.[7] Industry colleagues have observed that contacted poly pitch (CPP)

appears more difficult to scale than Mx (local metal) pitch. In other words, Mx pitch seems to be scaling at a rate faster than $0.7\times$ per node, while CPP scales at a rate slower than $0.7\times$ per node, even as the product of the two pitches achieves $0.5\times$ area scaling. Such a trend, if continued, may eventually change the A-factor modeling and A-factor values.

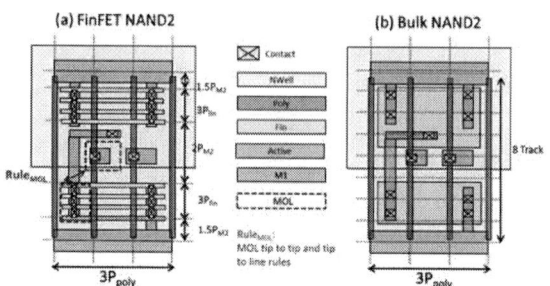

Figure 4: Layout of NAND2 cells for (a) FinFET and (b) bulk.

Table 1: Pitch conversions used in the A-factor models.

Layer	Pitch/M1 (2009)	Pitch/M1 (revised)
F	0.50	0.50
P_{M1}	1.00	1.00
P_{poly}	1.50	1.50
P_{M2}	1.25	1.00
P_{fin}	N/A	0.75
P/G track width	N/A	1.50

Figures 5(a) and (b) respectively give canonical layouts for FinFET and bulk 6T SRAM bitcells. Each layout uses two poly channels, so the bitcell height is $2P_{poly}$. The width of the bitcell depends on (i) the distances between bitline to wordline on each end; (ii) transistor separations (P to N); and (iii) distance between n-active (N to N); these parameters differ for FinFET and bulk. The width of the 6T bulk SRAM is $5P_{M1}$ from Figure 5(b), and the A-factor of the bulk 6T SRAM is therefore $60F^2$ [10].

Derivation of an A-factor for a FinFET-based 6T SRAM bitcell must consider two main issues. First, a pitch conversion between P_{fin}, P_{M1} and P_{poly} must be determined; industry experts suggest $P_{fin} = 0.75 \times P_{M1}$. Second, the β ratio (i.e., the ratio of fin counts between the PU and PD transistors in FinFET SRAM bitcell) is critical for read stability [9, 16], and affects fin counts and layout. For example, using a β ratio of 2 along with the pitch ratios in Table 1 would set the width of routing regions of bitlines to $2 \times 0.75P_{fin}$, PD NMOS to $2 \times P_{fin}$, P/N channel separation to $2 \times 1.5P_{fin}$, and PU PMOS to $1 \times P_{fin}$. The A-factor of the FinFET 6T SRAM would then be calculated as $67F^2$. (Note that the area overhead can be less in 8T FinFET bitcells compared to 6T FinFET bitcells, since additional read transistors in 8T bitcells provide read margin protection. By assuming β = 1.0 and the layout in Figure 6, the A-factor of 8T FinFET SRAM would be calculated as $72F^2$.)

5. LOW-POWER DESIGN

In response to power and energy being identified as *the* grand challenge for the semiconductor roadmap, the Design TWG in 2011 added a Low-Power Design technology roadmap to the Design Chapter. The low-power design roadmap contains a mix of future solutions spanning electrical, functional and software realms [13]. Projected low-power design innovations include (i) frequency islands and near-threshold computing at the circuit level; (ii) heterogeneous parallel processing, many core software development tools, and hardware/software co-partitioning at the architecture level; and (iii) power-aware software and software virtual prototyping at the software level. Figure 8 shows that with low-power innovations the SOC-CP driver dissipates 3.5W (with 48.8M logic gates) in 2011. Low-power design innovations will help limit the power to 8.22W when the number of logic gates grows by more than 40x to 1995.5M in 2026.

---

[7]The A-factor calculation for FinFET NAND2 may be based on the following assumptions: (i) the heights of P/G rails, which scale poorly in recent nodes due to current delivery and electromigration reliability reasons, are assumed to be $1.5P_{M2}$; (ii) pullup fin count is the same as pulldown fin count; and (iii) cell width of $3P_{poly}$ is still valid for the FinFET-based design. Based on these assumptions and

the pitch conversions given in the second column of Table 1, width of the cell is $3P_{Poly}$, and height of the cell is $1.5P_{M2} + 3P_{fin} + 2P_{M2} + 3P_{fin} + 1.5P_{M2} = 9.5P_{M2}$, if $P_{fin} = 0.75P_{M1}$. These calculations suggest that the track number should be more than eight. Recently, GlobalFoundries and ARM have implemented a 14nm FinFET library with 9-track cells [20]. In light of this, the preceding discussion has assumed a track height of nine.

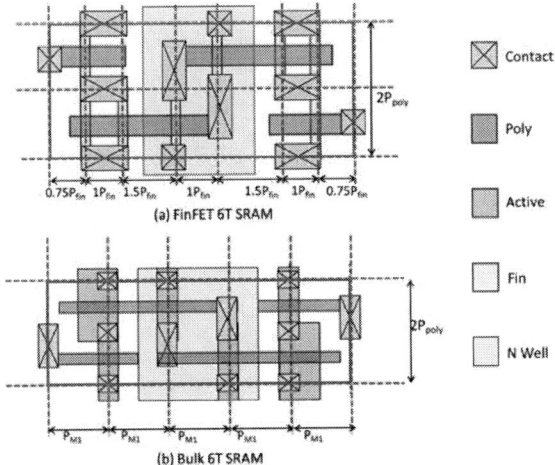

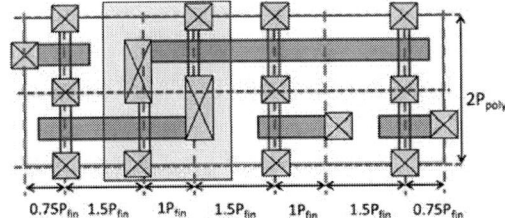

Figure 5: Layout of 6T SRAMs for (a) FinFET and (b) bulk.

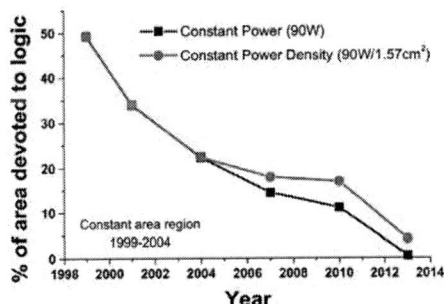

Figure 6: Layout of 8T FinFET SRAM.

Figure 8 shows that even if future low-power innovations are developed and deployed according to the low-power design roadmap, power of mobile SOC-CP designs will keep increasing. This is unacceptable in the mobile context; indeed, the SOC-CP driver has a flat power consumption requirement of ∼2W through the end of the roadmap. This is not a new story: Figure 7 from the 2001 Design Chapter predicts that percentage of logic that can be turned on reduces steadily to 2%-6% around 2012, i.e., what researchers have recently termed "dark silicon" [8], [17]. The inability to manage power limits the amount of (switched) logic content in an SOC, which in turn limits product value.

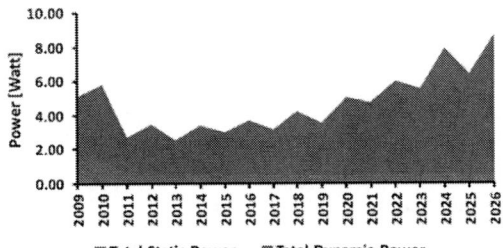

Figure 8: Impact of low-power design innovations on SOC-CP power consumption. [Source: 2011 System Drivers Chapter.]

duced the design for manufacturing (DFM) section in 2005 to discuss DFM requirements and the corresponding solutions. DFM requirements can be broadly classified as (1) *fundamental economic limitations*, and (2) *variability and lithography limitations*. Requirements due to economic limitations focus on mask cost, which is a key limiter for SOC innovations coming from small companies and emerging-market entities. Requirements due to variability and lithography limitations include quantified bounds on the variability of supply voltage, threshold voltage, critical dimension, circuit performance and circuit power consumption.

Since variability can cause circuits to exhibit faulty behavior, the DFM section in the 2009 Design Chapter adds projections for circuit-level impacts of variability, focusing on three canonical CMOS logic circuits which are the key components of a digital CMOS design, i.e., (i) SRAM bitcell for storage[8]; (ii) latch for circuit synchronization[9]; and (iii) inverter for logic functions. Failure probabilities for the three canonical circuits in future high-performance technology nodes are obtained by simulating their behavior under the influence of manufacturing process variability. The simulations use *Predictive Technology Model* (PTM) [23] with variability estimates down to 12nm node.

Revised DFM discussion in the 2011 ITRS observes that SRAM failure rate has already become a significant problem in the current technology node. Furthermore, although the latch has a lower failure rate compared to the SRAM, this circuit, too, is predicted to be problematic by the 20nm foundry node. The 2011 analysis also shows that enlarging circuits (i.e., reverse scaling) can be moderately effective in controlling the impact of variability. Other analyses show that failure rate can be reduced by more than an order of magnitude when supply voltage is increased from 90% to 120% of its nominal value, i.e., there is a clear engineering tradeoff between power and robustness.

Over the eight-year history of the Design Chapter's DFM section, potential DFM solutions have been divided into three categories, (i) solutions that address fundamental economic limitations; (ii) solutions that address the impact of variability; and (iii) solutions that address the impact of lithography limitations. Among these, early solutions that directly handle variability (e.g., in timing analysis) have emerged as predicted. The embedding of statistical methods throughout the design flow has been slower than initially forecast, but is still viewed as inevitable. DFM techniques that directly model and simulate lithographic non-idealities are becoming more popular, but will take longer to become qualified in production flows as a consequence of their tighter link to manufacturing models.

7. CONCLUDING THOUGHTS

The Design Chapter in the ITRS has for well over a decade defined technology requirements and design challenges for the EDA industry and the VLSI CAD research community. Design technology roadmaps for DFM, low-power design, 3D/TSV integration, More Than Moore, etc. are continually added to maintain relevance of the roadmap. Recent Design Cost and Low-Power Design models highlight the challenges of design productivity, software design cost, and power management in future SOC and MPU designs. At the same time, the System Drivers Chapter has provided models for key market drivers as well as basic chip parameters (layout density,

Figure 7: Dark Silicon projection. [Source: 2001 Design Chapter.]

In 2012, new additions to the low-power design roadmap include (i) approximate computing (variable-accuracy computing, e.g., flexibly from 64b to 16b); (ii) 4D computing (reconfiguration of circuits on the fly); and (iii) adaptivity (recapturing overdesign due to wearout and variation margins, etc.). To manage power to extreme limits, future low-power innovations must also improve the accuracy of power modeling and estimation. Chips are becoming heterogeneous systems (complex entities with multi-processor software environments) with unpredictable behavior and performance (more of a chip is turned off at any given moment, i.e., dark silicon). In this context, accurate estimation of chip power becomes very difficult.

6. DFM, VARIABILITY, RESILIENCE

Increasing process variability, mask cost, data size and lithography hardware limitations pose significant design challenges across different abstraction levels. The ITRS Design Chapter first intro-

---

[8] An SRAM bitcell is considered to be faulty when the SRAM is unable to store the correct logic value during a write operation or when the it fails to preserve the stored logic value during a read operation.

[9] A latch or an inverter is considered to be faulty when its signal delay (e.g., clock-to-output delay for latch) is 10 times the nominal value.

clock frequency, power dissipation, etc.) that bind the ITRS together via the Overall Roadmap Technology Characteristics. The MPU driver model has evolved frequency and power attributes in response to disappearing microarchitectural knobs, emergence of power limits, and challenges of device leakage; further changes (adding uncore elements, evolution of MPU-PCC for micro-server, updated die area modeling) are likely in the near future. The past decade has also seen increased reliance on "design-based equivalent scaling" (e.g., methods for activity factor reduction without compromising throughput or performance) to continue the semiconductor value proposition, and rapidly growing involvement in cross-TWG issues ranging from variability limits to device requirements.

The future of design technology roadmapping, and of the Design TWG's work in the ITRS, will be affected by a variety of technical, business and cultural factors.

- Past foundations of the ITRS seem increasingly shaky. For example, A-factors may no longer be constant across multiple technology nodes. Mx and poly pitches (i.e., horizontal vs. vertical densities) may scale at different rates. The fundamental assumption of 2× density scaling per node may be already long past; whether the industry can flourish with, e.g., 1.4× density scaling per node is an open question.

- Tremendous uncertainty with respect to patterning technology (e.g., timing of EUV, directed self-assembly), cost models (e.g., triple- and quadruple-patterning), device and interconnect structures and properties (tunnel FETs, resistive RAMs, drive vs. leakage currents), and high-value applications all present challenges to the roadmapping of design technology requirements.

- Fewer resources are available for ITRS activity even as the scope of the roadmap widens (MEMS, More Than Moore, new storage and switch elements, 3D integration) and the difficulty of the roadmapping task increases. Greater automation is needed to check consistency and impacts of proposed roadmap changes, a la the "Living ITRS" efforts of a decade ago [4].

- An oligopolistic EDA industry, along with continued consolidation and disaggregation in the semiconductor industry, as well as unwillingness to share competitive (as opposed to pre-competitive) data,[10] means that leading companies more frequently "opt out" of roadmap participation. There is a risk of a "vicious cycle" of decreased roadmap participation and decreased roadmap value.

- Communication across supplier industries, across the design-manufacturing interface, and across academia-industry boundaries is increasingly needed to optimize technology investments and maximize the returns from the roadmapping process. As the industry faces an explosion of post-CMOS, post-optical technology options, it seems appropriate to at least revisit the concept of "shared red bricks".

Against this backdrop, there is some good news: Members of the design, EDA and research communities are willing to find common cause in the design technology roadmap. At the 2009 and 2010 EDA Roadmap Workshops [19], representatives from leading EDA companies, semiconductor companies, and research consortia commenced a dialogue to analyze needs and status of EDA roadmapping.[11] Other discussions sought new mechanisms by which more of the community could contribute to the design technology roadmap. And the really good news for EDA and VLSI CAD: If anything

---

[10] It is suboptimal for students at UCSD to "predict" designs and cell libraries that industry has already developed, or for students at Purdue to develop ab initio models for device structures that again have already been developed. Yet, these are the mechanisms by which core material and data is generated in the ITRS today.

[11] The 2009 workshop addressed such questions as "What would make an EDA roadmap more useful?", "Which EDA areas lack most in roadmap efforts?", and "Which EDA areas are behind what the roadmaps say?" The 2010 workshop then identified gaps in the EDA roadmap (system-level executable specification, design-space exploration and pathfinding, EDA scaling requirements in light of evolving computing platforms, power-driven design, and design for resilience), reached agreement on the nature of EDA, and identified challenges in filling in the EDA roadmap gaps (incremental design flows, new design for cost methodologies, and an expanded scope of EDA moving to system-level design).

---

remains essential to the future of Moore's Law scaling, it will be design technology, and design-based equivalent scaling.

Acknowledgments

Dr. Juan-Antonio Carballo has co-chaired the U.S. and International Design TWGs with me for the past decade, and has been particularly influential in the conception of the System Drivers Chapter as well as iNEMI and More Than Moore interactions. Dr. Kwangok Jeong developed and maintained the MPU, power, frequency and A-factor models during the critical years of 2007-2011, which saw many Design-PIDS interactions regarding roadmap for device power vs. performance. This paper would not exist without the help of UCSD Ph.D. students Tuck-Boon Chan, Siddhartha Nath, Wei-Ting Jonas Chan, and Ilgweon Kang. Many participants in the ITRS Design and System Drivers efforts, and in the overall ITRS effort, have contributed valuable insights and perspectives over the years. I also thank Dr. Sani Nassif (who has for years driven the DFM section of the Design Chapter) for organizing the special session which led to the writing of this paper.

8. REFERENCES

[1] C. Auth, C. Allen, A. Blattner, D. Bergstrom et al., "A 22nm High Performance and Low-Power CMOS Technology Featuring Fully-Depleted Tri-Gate Transistors, Self-Aligned Contacts and High Density MIM Capacitors", *Proc. Symposium on VLSI Technology*, 2012, pp. 131-132.

[2] V. S. Basker, T. Standaert, H. Kawasaki, C.-C. Yeh et al., "A 0.063 μm^2 FinFET SRAM Cell Demonstration with Conventional Lithography Using a Novel Integration Scheme with Aggressively Scaled Fin and Gate Pitch", *Proc. IEDM*, 2010, pp. 19-20.

[3] T.-B. Chan and A. B. Kahng, "Tunable Sensors for Process-Aware Voltage Scaling", *Proc. ICCAD*, 2012, pp. 7-14.

[4] A. E. Caldwell, Y. Cao, A. B. Kahng, F. Koushanfar, H. Lu, I. L. Markov, M. R. Oliver, D. Stroobandt and D. Sylvester, "GTX: The MARCO GSRC Technology Exploration System", *Proc. DAC*, 2000, pp. 693-698.

[5] J.-A. Carballo and A. B. Kahng, "ITRS Chapters: Design and System Drivers", *Future Fab International* (36) (2011), pp. 45-48.

[6] J.-A. Carballo and A. B. Kahng, "ITRS Chapters: Design and System Drivers", *Future Fab International* (40) (2012), pp. 54-59.

[7] J.-A. Carballo and A. B. Kahng, "ITRS Chapters: Design and System Drivers", *Future Fab International* (44) (2013), pp. 52-56.

[8] H. Esmaeilzadeh, E. Blem, R. S. Amant, K. Sankaralingam and D. Burger, "Dark Silicon and The End of Multicore Scaling", *Proc. ISCA*, 2011, pp. 365-376.

[9] Z. Guo, S. Balasubramanian, R. Zlatanovici, T.-J. King and B. Nikolic, "FinFET-Based SRAM Design", *Proc. ISLPED*, 2005, pp. 2-7.

[10] K. Jeong and A. B. Kahng, "A Power-Constrained MPU Roadmap for the International Technology Roadmap for Semiconductors (ITRS)", *Proc. ISOCC*, 2009, pp. 49-52.

[11] A. B. Kahng, "The Road Ahead: Shared Red Bricks", *IEEE Design and Test of Computers*, 19(2) (2002), pp. 70-71.

[12] A. B. Kahng, "The Road Ahead: The cost of design", *IEEE Design and Test*, 19(4) (2002), pp. 136-137.

[13] A. B. Kahng, "The Road Ahead: Roadmapping Power", *IEEE Design and Test of Computers*, 28(5) (2011), pp. 104-106.

[14] A. B. Kahng and G. Smith, "A New Design Cost Model for the 2001 ITRS", *Proc. ISQED*, 2002, pp. 190-193.

[15] H. Kawasaki, M. Khater, M. Guillorn, N. Fuller et al., "Demonstration of Highly Scaled FinFET SRAM Cells with High-k Metal Gate and Investigation of Characteristic Variability for the 32 nm Node and Beyond", *Proc. IEDM*, 2008, pp. 1-4.

[16] D. Lekshmanan, A. Bansal and K. Roy, "FinFET SRAM: Optimizing Silicon Fin Thickness and Fin Ratio to Improve Stability at Iso Area", *Proc. CICC*, 2007, pp. 623-626.

[17] G. Venkatesh, J. Sampson, N. Goulding, S. Garcia, V. Bryksin, J. Lugo-Martinez, S. Swanson and M. B. Taylor, "Conservation Cores: Reducing the Energy of Mature Computations", *Proc. ASPLOS*, 2010, pp. 205-218.

[18] CPUDB. *http://cpudb.stanford.edu/*

[19] EDA Roadmap Workshop at DAC 2010. *http://vlsicad.ucsd.edu/EDARoadmapWorkshop/*

[20] "GlobalFoundries Details 14nm-XM FinFET Technology Performance, Power and Area Efficiency with a Dual-Core Cortex-A9 Processor Implementation". *http://www.globalfoundries.com/newsroom/2013/20130205-ARM.aspx*

[21] iNEMI. *http://www.inemi.org*

[22] ITRS Edition Reports. *http://public.itrs.net/reports.html*

[23] Predictive Technology Model. *http://ptm.asu.edu*

[24] Design-Based "Equivalent Scaling" to the Rescue of Moore's Law. *http://vlsicad.ucsd.edu/Presentations/talk/UCI-Colloquium-121031-v7-distributed.pdf*

Proactive Circuit Allocation in Multiplane NoCs*

Ahmed Abousamra
University of Pittsburgh
Department of Computer
Science
Pittsburgh, PA, USA
abousamra@cs.pitt.edu

Alex K. Jones
University of Pittsburgh
Department of Electrical and
Computer Engineering
Pittsburgh, PA, USA
akjones@ece.pitt.edu

Rami Melhem
University of Pittsburgh
Department of Computer
Science
Pittsburgh, PA, USA
melhem@cs.pitt.edu

ABSTRACT

This work explores a method for efficient pre-allocation of circuits in network-on-chip (NoC) to reduce communication latency and improve performance. Circuit pre-allocation eliminates the time cost of circuit establishment by using request messages to reserve the circuits for their anticipated reply messages. Requests reserve circuits in a priority order rather than for a particular time slot, avoiding delays or blocking even if the newly requested circuits conflict with previously reserved ones. Benchmark simulations show speedup in execution time of up to 16%, with an average of 8% for communication sensitive benchmarks, over a leading proposal in pre-configuring circuits.

1. INTRODUCTION

The scaling of semiconductor technology enables packing many cores on a single chip. For example, Intel integrates 48 cores on a single chip [8], and Tilera produces chips with up to a 100 cores on a single chip [19]. Communication between the chip components is carried out by the NoC. The communication latency has a significant effect on chip multiprocessor (CMP) performance; this effect continues to increase as the technology scales down enabling even more cores on a single chip [13].

In general purpose chips, the NoC must support full connectivity between the chip components to accommodate general programming/application. Full connectivity can be achieved through packet-switching, in which packets are examined at each NoC router and appropriately routed to their destinations. However, the overhead of making routing decisions increases communication latency. Conversely, circuit-switching reduces communication latency since packets avoid the routing overhead by traveling on pre-configured paths to their destinations. Unfortunately, limited resources allow only a subset of all possible circuits to simultaneously exist. Thus, hybrid NoCs have been proposed to support

---

*This work is supported, in part, by NSF award CCF-1064976.

Permission to make digital or hard copies of all or part of this work for personal or classroom use is granted without fee provided that copies are not made or distributed for profit or commercial advantage and that copies bear this notice and the full citation on the first page. To copy otherwise, to republish, to post on servers or to redistribute to lists, requires prior specific permission and/or a fee.
DAC'13, May 29 - June 07 2013, Austin, TX, USA.

---

full connectivity through packet switching, while speeding up communication whenever possible by configuring paths on which packets bypass routers' pipelines, which is often referred to as circuit-switching in the NoC literature [11, 9, 1, 2, 14] and in the rest of this paper.

Establishing a circuit between two nodes incurs time overhead since a control message must first be sent to reserve or configure the circuit. Such overhead can decrease the benefit of circuit switching unless it is reduced or eliminated. Different methods have been proposed for reducing this overhead:

(1) Amortizing the overhead over many circuit re-uses [9]; circuits are configured on-demand and kept in place until they are removed to allow the establishment of conflicting circuits. Without circuit re-use, the overhead can become too expensive.

(2) Time-Division-Multiplexing (TDM) [16, 6, 7]: removes the circuit setup overhead by enforcing a static schedule for realizing circuits. Although the NoC design is simplified, the NoC may suffer from underutilization and may not be suitable for general purpose CMPs, where communication requirements may not be known a priori.

(3) Hiding circuit setup overhead: Flit reservation flow control [15] performs accurate time-based reservations of the buffers and ports of the routers that a flit will pass through, but requires a dedicated faster plane to carry the reservation flits that are sent ahead of the corresponding message flits for which reservations are made. Li et al. [12] also propose time-based circuit reservations. As soon as a clean data request is received at a node, a circuit reservation is injected into the NoC to reserve the circuit for the data message, optimistically assuming that the request will hit in the cache and assuming a fixed cache latency. However, the proposal is conceptual and missing the necessary details to handle uncertainty in such time-based reservations. Déjà Vu Switching [1] amortizes circuit setup overhead for data messages carrying requested cache lines through early sending of circuit reservations once a cache hit is detected. Instead of time-based reservations, circuit reservations are queued and realized on a first-come-first-serve (FCFS) basis, which enables reservations to proceed unblocked to their destinations. While the early sending of FCFS reservations helps hide the circuit configuration overhead, in a CMP optimized for performance this advantage may be insufficient. If the on-chip cache is sufficiently fast, circuit reservations may not have a large enough lead time on their corresponding data messages, particularly in higher diameter NoCs.

In this work we explore proactive circuit allocation to *completely remove* the overhead of circuit configuration for anticipated reply messages without making the assumption that circuit reservations travel on a plane that is clocked at a speed which is higher than that of the data plane [15]. In our proposal data request messages reserve the circuits for their anticipated reply data messages. In this setting accurate time-based reservations as in [15] are impractical, since at the time that a request is reserving a circuit, we cannot be certain of the actual time at which the reply message will be injected in the NoC as other network traffic may cause unforeseen delays. Moreover, simple FCFS reservations [1] can under-utilize the NoC by delaying the realization of circuits for data messages that have already arrived, as we explain later. Rather, our proposal combines the ideas of both *queued* and *time-based* circuit reservations; reservations are still queued but instead of an FCFS ordering for realizing circuits, reservations are ordered based on *estimates* of circuit utilization times.

The remainder of the paper is organized as follows. Section 2 describes the proposed circuit pre-allocation scheme. Section 3 discusses handling the cases when circuit pre-allocation is not possible. Section 4 describes the simulation environment and results. Related work is presented in Section 5. Finally, Section 6 concludes the paper.

2. PROACTIVE CIRCUIT ALLOCATION

In this section we describe the proposed proactive circuit allocation scheme. We start by describing the network architecture, then how data requests reserve circuits, and finally how circuits are realized.

2.1 Network Architecture

The interconnect is composed of two planes[1] organized in a regular two dimensional mesh topology, where every router is connected with its four neighboring routers via bidirectional point-to-point links and with a single processor tile via the local port. One plane is packet-switched while the other is circuit-switched. Control and coherency messages such as data access requests (e.g. read and exclusive requests), invalidation messages, and acknowledgments travel on the packet-switched plane, which is referred to as the *control plane*. Data messages carrying cache lines, whether replies to data requests or write-back messages of modified cache lines, travel on the circuit switched plane, which is referred to as the *data plane*.

Data request messages travel on the control plane making circuit reservations at the corresponding data plane routers for their anticipated data reply messages, while data plane routers inform their corresponding control plane routers of space availability in the circuit reservation buffers.

2.2 Reserving Circuits

The purpose of circuit pre-allocation by request messages is to remove the circuit configuration overhead. To be able to reserve circuits for their replies, a request and its reply should travel the same path *but in opposite directions*; hence we refer to the circuits reserved by requests as *reverse or backward circuits*. To avoid delaying request messages if they attempt to reserve previously reserved ports, routers support storing and realizing multiple reverse circuit reservations. However, the order of realizing *reverse circuits* cannot be FCFS since it can poorly utilize the interconnect resources

---

[1]The interconnect may be composed of more than two planes but in this work we assume it is composed of two.

as it may delay the realization of circuits even when their data messages are ready.

For example, in Fig. 1 the data request Req_A is traveling to a far node, R_N, and reserves a circuit, C_A, at routers R_1 and R_2, for its anticipated reply. On the other hand, Req_B is traveling to a near node, R_2, and reserves a circuit C_B also at routers R_1 and R_2 immediately after Req_A. In this example, Req_B arrives at R_2 much earlier than the time at which Req_A arrives at R_N. Assuming both requests hit in the cache, $Reply_B$, the reply to Req_B, becomes ready much earlier than $Reply_A$, the reply to Req_A. However, with FCFS ordering, circuit C_A would be realized before C_B, thus delaying the ready message $Reply_B$. Conversely, if circuits are realized based on their expected utilization times, C_B would be realized before C_A, and $Reply_B$ would not suffer unnecessary delay. The proposed circuit pre-allocation improves the circuits realization order using approximate predictions of the arrival times of reply messages as described next.

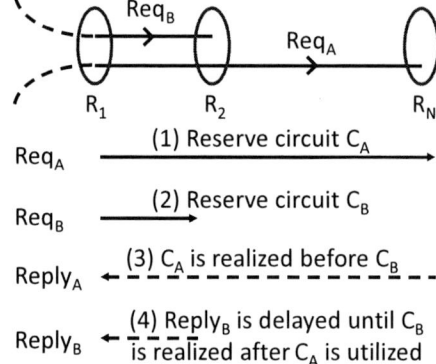

Figure 1: Example showing that realizing reverse circuits in a FCFS order can result in poor utilization of the NoC resources (See Section 2.2)

Approximate Time-Based Circuit Reservation.

Consider the following example. Router R_1 sends a data request to R_N and this request has to traverse 10 routers on the path to R_N. Assume that a hop takes 3 cycles on the packet-switched control plane. Assume that the request will hit in the cache and that it takes 5 cycles to read the cache line. On the circuit-switched data plane communication latency is 1 cycle per hop. Thus, assuming the request and reply face no delays, the minimum duration of the round-trip since sending the request and until receiving the first flit of the reply is: the request travel time + cache processing time of the request + the reply travel time = 3x10 + 5 + 1x10 = 45 cycles. Assume that R_1 sends the request at cycle 100. Then the request reserves the circuit at R_1 with expected utilization cycle = 145, and on the next router, R_2, the request reserves the circuit with expected utilization cycle = 144 and so on, until it reaches R_N and reserves the circuit with expected utilization cycle = 136. Essentially, the request carries the estimate, c, of the cycle number at which the circuit is expected to be utilized at the next router, R, where the circuit will be reserved. After the circuit reservation is successfully added to R, the request's carried estimate is decreased by one to become $c = c - 1$, and the request message advances to the next router on the path to the request's destination.

In the example, the expected circuit utilization cycle is based on the minimum time for the round-trip that starts with injecting the request and ends with receiving the reply message. Unfortunately, the three components that make up the round-trip time: request travel time, request processing time, and reply travel time, will not always take the minimum time, nor can they be precisely determined. The travel time of the request and reply messages may be affected by other traffic in the NoC. Similarly, the processing time may vary depending on whether the cache can process the request immediately, whether the request hits or misses in the cache, the cache may forward the request to the requested cache line's owner, or the cache may even reply with a negative acknowledgment indicating that the request should be retried.

Since determining the round-trip precisely is not possible, the next best thing is to *estimate* how long a round-trip would take, and include with the circuit's reservation at each router the circuit's estimated utilization cycle at that router. Routers would then realize circuits in ascending order of their estimated circuit utilization cycles, which need not exactly coincide with the actual cycles that the reply messages traverse the routers as long as the traversal order is preserved.

An intuitive way to estimate the round-trip time from R_1 to R_N is to assume it is similar to the observed round-trip time when R_1 last sent a request to R_N. However, large variability in request processing times can adversely affect the round-trip estimation. Better estimates can be derived by averaging or using the median of previously observed round trip times[2]. Next we describe how reservations are ordered and realized.

2.3 Realizing Reserved Circuits

When a circuit is reserved at a router, the expected utilization cycle (EUC) of the circuit is included in the reservation. Each port – whether input or output – has a separate reservation buffer (RB) to store its circuit reservations. Rather than realizing circuits in the order they were added to the reservation buffers, routers realize circuits in ascending order of their expected utilization cycles. Specifically, each port maintains a pointer, p_{min} to the reservation, res_{min}, having the earliest EUC. When a new reservation, res_{new} is added to RB, its expected utilization cycle, EUC_{new}, is compared to EUC_{min}, the EUC of res_{min}, and p_{min} is updated if necessary (Fig. 2). Circuits are realized by matching the reservations pointed to by p_{min} pointers in each of the RBs of the input and output ports, as the following example demonstrates.

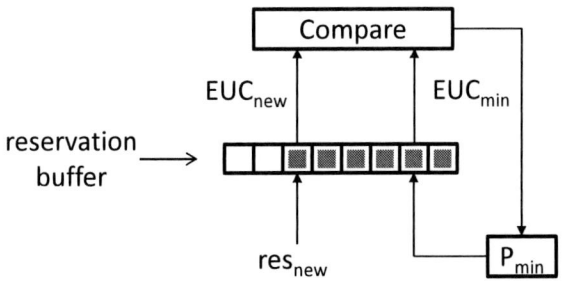

Figure 2: Checking if the new reverse reservation has the earliest EUC among existing reservations.

Consider for example that at some router two data re-

---

[2]Estimating round-trip times is discussed in Section A in the appendix.

quest messages, r_1 and r_2, reserve the crossbar connections: west-east (i.e., west output port and east input port) and south-east, respectively, such that the EUC of r_1's reservation is earlier than that of r_2's. Assume both reservations become the ones with the earliest EUCs in the RBs of the west and south input ports (Fig. 3). Because the EUC of r_1's reservation is earlier than that of r_2's, the east output port realizes r_1's reservation before r_2's, i.e., the west-east crossbar connection gets realized before the south-east.

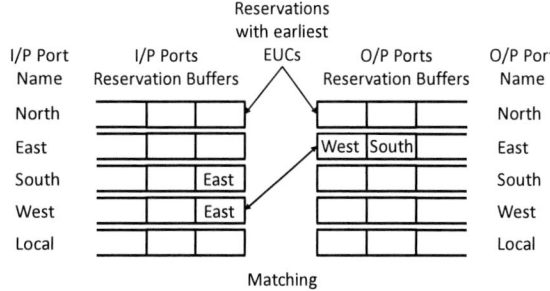

Figure 3: Example: Realizing circuit reservations in ascending order of their EUCs. The west-east connection is realized before the south-east connection.

Once a circuit's connection is realized at a router, the connection remains active until the tail flit of the message traveling on the circuit traverses the crossbar, at which time the input and output ports of the connection become free to participate in realizing subsequent circuit reservations. Correct routing requires that each node injects data messages in the data plane in ascending order of the their circuit reservations' EUCs. Further, the EUCs of any two circuit reservations ensure a consistent realization order of the circuits in all the ports they share on their paths[3].

Note that since EUCs are only estimates that may not coincide with the cycles at which packets traverse routers, and since there is always the chance that a new reservation having an earlier EUC than all reservations in a port's RB may be added[4], circuits are realized only after a packet is incoming to an input port, which can be detected through a look-ahead signal: each output port matched during switch allocation signals its corresponding input port on the next router that a packet is incoming. Once a circuit is realized, its input and output ports update their p_{min} pointers to point to the next reservation with the earliest EUC.

3. HANDLING CASES WHEN CIRCUIT PRE-ALLOCATION IS NOT POSSIBLE

There are cases when circuit pre-allocation is not possible. For example, write-back messages sent upon evicting a dirty cache line are not preceded by a request, hence there are no pre-allocated circuits for such messages. Additionally, data request messages may not always reserve circuits. For example, when sending a data request if there is not a good[5] estimate for when the reply data message will arrive at the requester, it may be better not to pre-allocate a circuit.

---

[3]Section D in the appendix discusses ensuring consistent ordering of realizing circuits.

[4]Section B in the appendix discusses ensuring correct routing even when a new circuit with an early EUC is reserved, while Section C discusses deadlock prevention.

[5]Section A in the appendix discusses the quality of round-trip estimates.

There are also cases when a circuit is partially or completely reserved but should be removed. For example, when a request misses in the cache, the requested cache line is fetched from the off-chip memory, which takes a relatively long time. If this request's circuit is kept until the line is fetched, it can delay the realization of other circuits, which hurts performance; instead a message should be dispatched in place of the data reply message to utilize and remove the circuit. Another example is a reservation conflict[4], which – although rare – may occur while reserving a new circuit. If not handled, a reservation conflict can cause miss-routing of already in-flight data messages; thus the partial reservation of the new circuit need to be removed. In all the above cases the data messages still need to be sent, and because we design the data plane to be circuit-switched, we choose to fall back to FCFS [1] to reserve *forward* circuits. In this section we explain how the reverse and forward reservations are simultaneously supported in the NoC.

There are two main distinctions between reverse and FCFS [1] reservations: the direction of reserving the circuit and the order of circuit realization. These distinctions require the reverse and FCFS reservations be separated and require that the packets traveling on these two types of reserved circuits be separated as well. I.e., each port maintains future reverse and FCFS reservations in separate buffers, and two virtual channels (VCs) are required on the data plane, one for packets traveling on reverse circuits and the other for packets traveling on FCFS circuits. Configuring the crossbar of a data plane router is based on the result of matching either: reverse reservations having the earliest EUCs in the RBs of the input and output ports (Section 2.3), or the heads of FCFS reservation queues of input and output ports – since the FCFS reservations are already queued in their order of realization. To improve the quality of matching, in each cycle separate matching of the reverse and FCFS reservations is carried out with priority given to the decisions of one of them based on a particular arbitration policy such as round robin. Fig. 4 shows the architecture of the control and data plane routers which support both kinds of reservations. The top router depicts the control plane router which is packet-switched, and communicates to the data plane router reverse and FCFS circuit reservations made by data request and FCFS circuit reservation messages, respectively. The bottom router depicts the data plane router connected to the control plane router at the same node. It is circuit-switched and has reservation buffers for both reverse and FCFS circuits, and has two VC flit buffers at each input port, one for the packets traveling on reverse circuits and one for packets traveling on FCFS circuits.

4. EVALUATION

We evaluate the proposed proactive circuit allocation (PRO) scheme through simulations of benchmarks from the SPLASH-2 [20], PARSEC [3], and Specjbb [18] suites using the functional simulator Simics [17]. We assume a 16-core CMP with 3 GHz UltraSPARC III in-order cores with instruction issue width of three. Each core has private 16 KB L1 data and instruction caches with an access latency of one cycle. The CMP has a distributed shared L2 with 1MB per core. Cache lines are 64 bytes, and each is composed of eight 8-byte words. Cache coherency is maintained with the MESI protocol. Similar to [1], a stalled instruction waiting for an L1 miss to be satisfied is able to execute once the critical

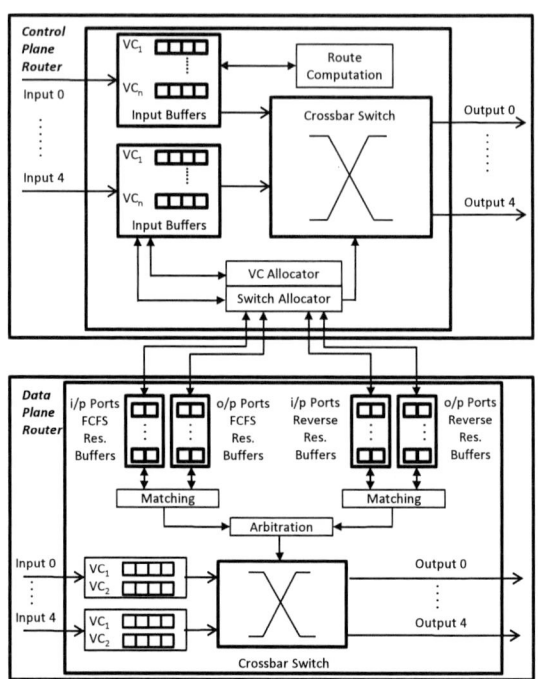

Figure 4: Diagrams of the control and data plane's routers with support for both FCFS and reverse reservations.

word is received, which is sent as the first word in the data reply packet. We assume the cache is optimized for fast access. From Cacti [4], at 3 GHz and 32nm technology the access cycles of the L2 tag and data arrays are two and four cycles, respectively, for a 1MB L2 per tile partitioned into two banks. The NoC's topology is a 2D mesh.

We evaluate a CMP with the PRO NoC against CMPs with: (1) a purely packet-switched NoC (PKT), (2) the state-of-the-art FCFS circuit reservations NoC [1], and (3) a zero-overhead *Ideal* NoC. Each of the evaluated NoCs is composed of two planes: a control plane that carries control and cache coherency messages, and a data plane that carries data messages. The control plane is packet-switched in all four NoCs, while the data plane is only packet-switched in the PKT NoC and circuit-switched in the other three NoCs. In the data plane of the *Ideal* NoC all possible circuits are assumed to simultaneously exist, such that all the circuit-switched flits experience only one-cycle per hop without suffering any network delays due to contention. Below, we describe the configuration of the simulated NoCs.

Packet-Switching and Message Sizes We simulate a three cycle pipeline for packet-switched routers. In general, messages on the control plane are one flit long, while messages on the data plane are five flits long. For PRO, *data request* messages may be composed of either one or two flits. If the request will reserve a circuit for its reply, the request message is composed of two flits due to the additional space required to carry the circuit's *EUC*; otherwise it is composed of one flit.

Virtual Channels The control plane has four virtual channels (VCs). Control plane routers have a FIFO buffer for two packets per VC per input port. The data plane of PKT, FCFS, and the Ideal NoCs, each has only one channel for data messages, while the data plane of the proposed PRO NoC has two VCs, one for the messages traveling on reverse

240

circuits and one for the messages traveling on FCFS circuits. The routers of the data plane have a FIFO buffer for two data packets per input port. In the case of PRO, the FIFO buffer of each VC can hold one data packet.

Circuit Reservation Buffers In the PRO NoC, each router port has two circuit reservation buffers, one for the reverse and one for the FCFS reservations. The buffers can hold 12 reverse reservations and 5 FCFS reservations, per port. In the FCFS NoC, each port has only one buffer for FCFS reservations; we set its size to 17, the total number of reservations a port on the PRO NoC can store.

Estimating Round-Trip Time At the requesting node, we estimate the round-trip time by computing the median of the last observed three round-trip times for the request message's destination. However, large estimates tend to be inaccurate which hurt performance[6]. To reduce such inaccurate estimates, we allow a data request message to reserve a circuit only if the estimate is at most X times the minimum round-trip time. After experimenting with the design space, we chose to set $X = 2$.

4.1 Performance Evaluation

We simulate the parallel section of each benchmark. First, we compare the average latency of satisfying an L1 miss that hits in the L2, or simply the average L2 hit latency, which is essentially the average round-trip time for sending a request that hits in the L2 and receiving its reply. Fig. 5 shows the average L2 hit latency of the three CMPs: (1) with the FCFS NoC; (2) with the PRO NoC; and (3) with the Ideal NoC. The results displayed in all the figures are relative to the CMP with the purely packet-switched NoC (PKT). With the FCFS NoC there is only a modest improvement in the L2 hit latency, while with the PRO NoC we see a significant improvement for almost all the benchmarks except for a couple of benchmarks (the contiguous version of *LU* and *Water Spatial* did not benefit from the PRO NoC).

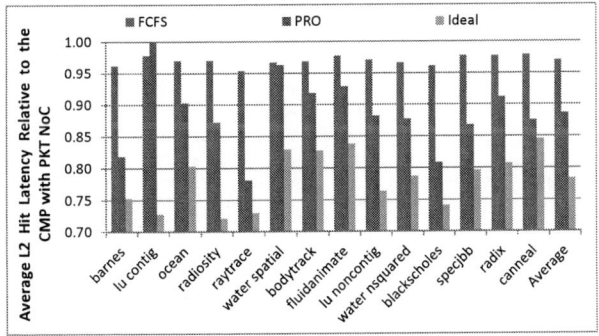

Figure 5: Average L2 hit latency normalized to the purely packet-switched system (the Y-axis starts at 0.7).

Since the execution time of each benchmark may not be sensitive to the communication latency over the NoC, we examine the execution time speedup achievable with the Ideal NoC (Fig. 6) and classify the benchmarks into two groups: *communication sensitive* with a speedup of at least 4% and *communication insensitive* with a speedup of less than 4%. Based on this classification, we compare the execution time speedup achievable with the FCFS and PRO NoCs

---

[6]Section A in the appendix discusses estimating round-trip times.

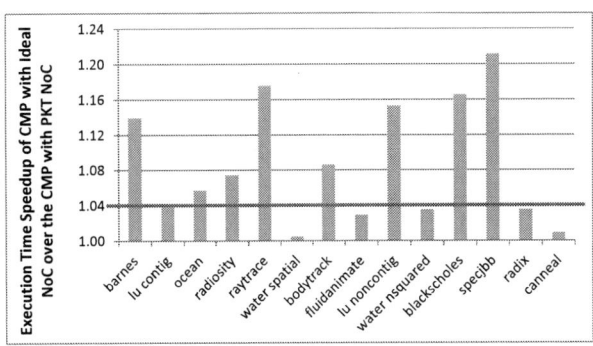

Figure 6: Identification of communication sensitive benchmarks by examining the execution time speedup using the Ideal NoC (the Y-axis starts at 1.0)

in Fig. 7. The speedups of the *communication sensitive* benchmarks are displayed on the right side of the chart. The system with *FCFS* achieves an average speedup of *only 2%* over the system with PKT. The system with *PRO* achieves up to 16% speedup (Raytrace and Specjbb), with an average of 8% over the system with *FCFS*, and an average of 10% over the system with PKT. On the left side of the chart the speedups of the *communication insensitive* benchmarks are displayed. With FCFS there is almost no speedup, while with PRO there is a nominal speedup (2%, on average).

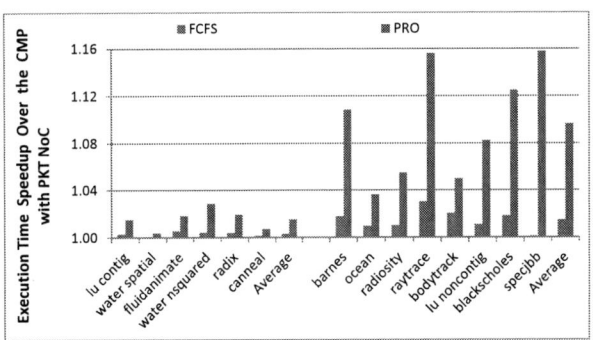

Figure 7: Execution time speedup of CMPs with the FCFS and PRO NoCs (the Y-axis starts at 1.0). Communication sensitive benchmarks are displayed on the right of the chart.

Fig. 8 shows how much of the potential execution time speedup achievable with the Ideal NoC that the systems with the FCFS and PRO NoCs achieve. The CMP with FCFS gains only between 1% to 24%, with an average of 12%, compared with the ideal case, while the CMP with PRO gains much more; between 40% and 89%, with an average of 68%.

4.1.1 FCFS Circuits as a Fallback

As mentioned in Section 3, there are situations that require releasing reverse circuit reservations. In Fig. 9 we examine the percentage of released circuits relative to the number of circuit reservations. We find that the majority of circuits are released due to processing times that exceed a threshold (for example, upon a cache miss to the off-chip memory), which can reach more than 25% for several applications. The percentage of circuits released due to potential

241

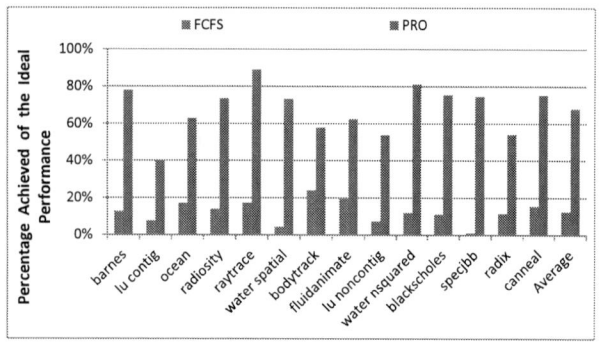

Figure 8: Percentage achieved of the performance of the CMP with the ideal NoC.

deadlocks and reservation conflicts represent a very small percentage of less than 3% and 2% of circuit reservations, respectively.

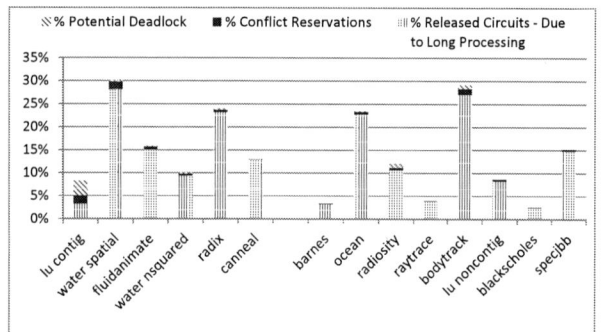

Figure 9: Percentage of released circuits relative to the number of requests performing circuit reservations.

When reverse circuits are released, forward (FCFS) circuits are used. Additionally, forward circuits are used when data requests do not reserve circuits due to round-trip estimates that exceed the stated threshold (2 times the minimum round trip time) and for write-back messages of modified cache lines. Sending messages to release circuits can increase the traffic volume, however, we find that this increase is small. Specifically, assuming flit sizes of 6- and 16-bytes on the control and data planes, respectively, we find that the percentage increase in traffic volume in the PRO NoC compared to the FCFS NoC is 2%, on average, for the communication sensitive benchmarks.

5. RELATED WORK

Some of the related work [9, 16, 6, 7, 15, 12, 1] has already been described in the introduction. Other related work includes: Duato et al. [5] propose a router architecture with multiple switches for concurrently supporting wormhole and circuit switching in the interconnection. Kumar et al. [11] propose express virtual channels to improve communication latency in 2D mesh NoCs through static and dynamic circuit establishment. Ahn and Kim [2] and Park et al. [14] suggest router designs that exploit temporal locality in the traffic to allow flits to bypass the router pipeline. Kumar et al. [10] suggest a design in which routers announce the availability of buffers and VCs to other routers such that packets may quickly traverse routers on less congested routes. These proposals support both packet and circuit switching

on the same plane and speedup communication by using local router information to let packets bypass router pipelines whenever possible, which is orthogonal to our work and can be incorporated in the control plane in our design.

6. CONCLUSION

Circuit-switching is effective in speeding up communication when the overhead of setting up circuits is reduced or amortized with re-use of circuits. This work proposes a proactive scheme for circuit allocation to completely hide the circuit setup overhead for reply messages by having the request messages reserve the circuits for their anticipated replies. Reserving circuits by requests requires using time-based reservations to avoid holding NoC resources unnecessarily idle which under-utilizes the NoC. However, variability in network traffic conditions and request processing times make it impossible to use accurate time-based reservations. Hence, we use approximate time-based reservations by estimating the round-trip time from the time when a request is sent and until its reply is received. We demonstrate the benefit of our design through simulations of parallel benchmarks. For a CMP with a fast on-chip cache, while the state-of-the-art circuit-switched NoC using FCFS circuit reservations achieves execution time speedup of up to 3% and an average of about 2% over a purely packet-switched NoC, our proposed scheme enables execution time speedup of up to 16% and an average of about 10% over the purely packet-switched NoC; outperforming the state-of-the-art FCFS reservations scheme by up to 16% and an average of 8%.

7. REFERENCES

[1] A. K. Abousamra, R. G. Melhem, and A. K. Jones. Deja vu switching for multiplane nocs. *Networks-on-Chip, International Symposium on*, 0:11–18, 2012.

[2] M. Ahn and E. J. Kim. Pseudo-circuit: Accelerating communication for on-chip interconnection networks. In *Proceedings of the 2010 43rd Annual IEEE/ACM International Symposium on Microarchitecture*, MICRO '43, pages 399–408, Washington, DC, USA, 2010. IEEE Computer Society.

[3] C. Bienia, S. Kumar, J. P. Singh, and K. Li. The parsec benchmark suite: Characterization and architectural implications. In *Proceedings of the 17th International Conference on Parallel Architectures and Compilation Techniques*, October 2008.

[4] "CACTI". http://quid.hpl.hp.com:9081/cacti/.

[5] J. Duato, P. López, F. Silla, and S. Yalamanchili. A high performance router architecture for interconnection networks. In *ICPP, Vol. 1*, pages 61–68, 1996.

[6] K. Goossens and A. Hansson. The aethereal network on chip after ten years: Goals, evolution, lessons, and future. In *Design Automation Conference (DAC), 2010 47th ACM/IEEE*, pages 306 –311, june 2010.

[7] A. Hansson, M. Subburaman, and K. Goossens. Aelite: A flit-synchronous network on chip with composable and predictable services. In *Design, Automation Test in Europe Conference Exhibition, 2009. DATE '09.*, pages 250 –255, april 2009.

[8] J. Howard et al. A 48-core ia-32 message-passing processor with dvfs in 45nm cmos. In *Solid-State*

Circuits Conference Digest of Technical Papers (ISSCC), 2010 IEEE International, pages 108 –109, feb. 2010.

[9] N. D. E. Jerger, L.-S. Peh, and M. H. Lipasti. Circuit-switched coherence. In *NOCS*, pages 193–202, 2008.

[10] A. Kumar, L.-S. Peh, and N. K. Jha. Token flow control. In *Proceedings of the 41st annual IEEE/ACM International Symposium on Microarchitecture*, MICRO 41, pages 342–353, Washington, DC, USA, 2008. IEEE Computer Society.

[11] A. Kumar, L.-S. Peh, P. Kundu, and N. K. Jha. Express virtual channels: towards the ideal interconnection fabric. In *ISCA*, pages 150–161, 2007.

[12] Z. Li, C. Zhu, L. Shang, R. P. Dick, and Y. Sun. Transaction-aware network-on-chip resource reservation. *Computer Architecture Letters*, 7(2):53–56, 2008.

[13] J. D. Owens, W. J. Dally, R. Ho, D. N. Jayasimha, S. W. Keckler, and L.-S. Peh. Research challenges for on-chip interconnection networks. *IEEE Micro*, 27(5):96–108, 2007.

[14] D. Park, R. Das, C. Nicopoulos, J. Kim, N. Vijaykrishnan, R. Iyer, and C. R. Das. Design of a dynamic priority-based fast path architecture for on-chip interconnects. In *Proceedings of the 15th Annual IEEE Symposium on High-Performance Interconnects*, HOTI '07, pages 15–20, Washington, DC, USA, 2007. IEEE Computer Society.

[15] L.-S. Peh and W. J. Dally. Flit-reservation flow control. In *HPCA*, pages 73–84, 2000.

[16] M. Schoeberl, F. Brandner, J. Sparso, and E. Kasapaki. A statically scheduled time-division-multiplexed network-on-chip for real-time systems. *Networks-on-Chip, International Symposium on*, 0:152–160, 2012.

[17] "Simics". http://www.windriver.com/products/simics/.

[18] SPEC. Spec benchmarks. http://www.spec.org/.

[19] "Tilera". http://www.tilera.com/products/processors/.

[20] S. C. Woo, M. Ohara, E. Torrie, J. P. Singh, and A. Gupta. The splash-2 programs: Characterization and methodological considerations. In *ISCA*, pages 24–36, 1995.

APPENDIX

A. IMPROVING QUALITY OF ESTIMATION

Inaccuracy in estimating circuit utilization times may hurt resource utilization and interconnect performance. Specifically, if an estimation is too optimistic assuming that a circuit, C_i, would be utilized much sooner than actually occurs, a message traveling on another circuit sharing a sub-path with C_i but scheduled later than C_i may be delayed until C_i is utilized. Conversely, if an estimation is too pessimistic assuming C_i would be utilized much later than what actually happens, the message traveling on C_i may suffer delays if circuits sharing sub-paths with C_i *but* having earlier EUCs are reserved; as these circuits would be realized before C_i on the shared sub-paths even though C_i arrives first.

Obtaining an accurate EUC can be reduced to determining a good mechanism for estimating round-trip times for

satisfying requests. The request and reply travel times depend on network conditions, while the request processing time depends on the status of the requested line in the cache, which can cause great variability in the request processing time. For example, a request that hits in the cache takes much less time to send the data reply than if the request misses and the line has to be retrieved from the off-chip memory. Large variability in request processing times can greatly affect the accuracy of EUCs. Therefore, we restrict estimates to the cases of short request processing times, *which we assume is the typical case for an efficient cache design.*

When the request requires long processing due to the memory system (e.g., a cache miss), a *release circuit* message is immediately dispatched in place of the data reply message to release the circuit reservation. In this case, another method (e.g., traditional packet switching) can be used to send the data reply message. In this work a FCFS reservation is used for reserving the reply's circuit when the reply is ready (Supporting FCFS circuits as a fallback for reverse circuits is discussed in Section 3). Focusing on replies with short processing times reduces the variability in round-trip times primarily induced by the memory system, and makes it dependent mainly on network conditions.

An intuitive estimate of the round-trip time from node A to node B utilizes previously observed round-trip times. Alternative methods exhibit different trade-offs between quality of estimation and hardware resources. For example, each node may keep a *per hop* estimate that is the average of: the current per hop estimate and the last observed per hop latency (the last observed round-trip time to any destination normalized per hop). Similarly, a node may keep a running average *per destination*, or even more information.

However, when high traffic load causes the estimated round-trip times to be large, inaccuracy of the EUC can become amplified. In such cases, the benefit of early circuit reservation is often outweighed by the potentially poor resource preallocation due to the inaccuracy in the round-trip time estimation. Therefore, we propose capping estimates by a factor of the minimum round-trip time, such that an estimate greater than the cap value does not reserve a circuit for the reply. For example, if the zero-load round-trip time from A to B is 30 cycles and the maximum cap factor is two, then for any estimate greater than 60, A's request does not reserve the circuit for the reply. In these cases we choose to fallback to FCFS for reserving the reply's circuit.

A.1 Round-trip estimation methods

In this section we compare three different methods for estimating round-trip times: (1) MedianOf3: the requesting node estimates the round-trip time as the median of the last three observed round-trip times to the destination. (2) DestinationAvg: each node maintains a running average of the round-trip latency per destination and uses these averages as the estimates for the round-trip times. (3) HopAvg: a requesting node maintains a running average of the round-trip latency normalized per hop for all messages returning to the requesting node, and uses it to estimate the round-trip latency to any destination. Fig. 10 compares the execution time speedup of our proposed scheme using each of the three methods for the communication sensitive benchmarks. We observed little differences between three estimation methods, except in the case of Specjbb where the MedianOf3 greatly out performs the other two.

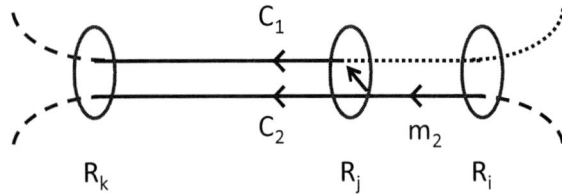

Figure 12: Circuit C_1 is scheduled before C_2, but the right part of C_1 in the dotted line is not yet reserved. Message m_2 starts traversing the shared sub-path between C_1 and C_2 before C_1 is completely reserved on it. If no corrective measure is taken, m_2 would wrongly travel on C_1 instead of remaining on C_2.

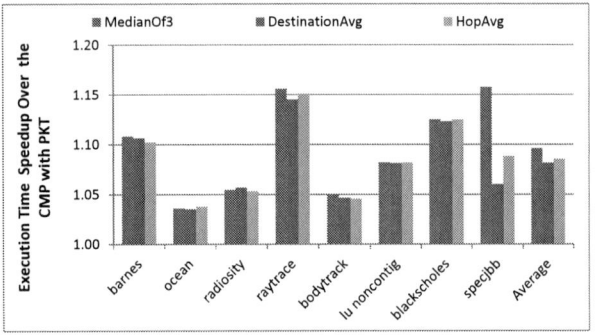

Figure 10: Comparing the execution time speedup with different round-trip times estimation methods.

B. ENSURING CORRECT ROUTING ON RESERVED CIRCUITS

There are two conditions to ensure that each message travels on the right circuit from source to destination. The first is that each node injects the data plane messages in the same order in which the reserved circuits at the local input port will be realized. The second is maintaining a consistent order of realizing any two circuits that share routers relative to each other in all the shared routers. In other words, for any two circuits, C_1 and C_2, that share a sub-path, p, either C_1 is realized before C_2 in all the ports on p, or C_2 is realized before C_1 (see Fig. 11). The later condition ensures that a message does not jump from one circuit to another.

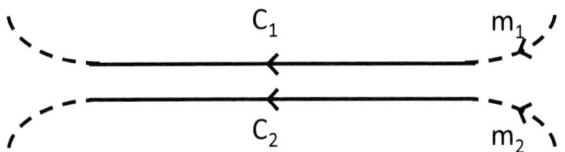

Figure 11: The solid line represents the shared sub-path between circuits C_1 and C_2. C_1 is scheduled before C_2, thus m_1 crosses the shared sub-path before m_2

Infrequently, a circuit request may arrive that includes a shared sub-path with a circuit already in flight, but with the new request having an earlier EUC. Consider Fig. 12. The two circuits C_1 and C_2 share the sub-path, p, which starts at router R_i and ends at R_k. Let m_2 be the message traveling on C_2 and assume that C_1's EUC is earlier than C_2's. If C_1 is reserved at all the routers on p before m_2 starts traversing p, then m_2 cannot be mistakenly routed on C_1, since the situation would be similar to the one in Fig. 11; m_2 will be held in R_i until the message m_1 traveling on C_1 traverses the sub-path, p.

In contrast, if m_2 starts traversing p while C_1 is only reserved at some but not all of p's routers, then at the first router $R_j \in p$ where m_2 meets the reservation of C_1 (remember that circuits are reserved backwards, from destination to source), C_1 would be realized instead of C_2, thus miss-routing m_2 on C_1. The above miss-routing problem occurred due to a *reservation conflict* between two circuits sharing a sub-path. Before explaining how to detect and handle such a conflict we need some notation.

Assuming a two dimensional mesh topology, each router, R_i, on the data plane has five ports. Each port, π, where $\pi \in \{$Local, North, West, South, East$\}$, has: an input flit buffer, IFB^i_π, for storing the flits of incoming messages; an input reservation buffer, $RB^i_{in,\pi}$, for storing the reservations of circuits at the π input port; and an output reservation buffer, $RB^i_{out,\pi}$, for storing the reservations of circuits at the π output port. Note that below we use *input* and *output* ports from the perspective of circuits, i.e., input and output ports, respectively, of data plane routers.

B.1 Detecting Reservation Conflicts

To detect reservation conflicts, we first recognize that only local information is available at the router where a new reservation is being added; a router does not know when messages are going to arrive at its input ports. Thus, detection can only occur at the "point" where a message meets a reservation, for example, in Fig. 12, m_2 meets C_1's reservation at R_j. Second, we precisely define the *meeting point*: we choose the *meeting point* to be the router's output port that is on the path of the new circuit reservation.

Consider again the example in Fig. 12. The reservation conflict between C_1 and C_2 occurred because at some router, R_j, along the shared sub-path, p, circuit C_2 was being utilized (i.e., m_2 was traveling on C_2) when the new reservation of C_1, which has an earlier EUC than C_2, was added to the RB of R_j's output port that is shared by both C_1 and C_2. Hence, a reservation conflict is detected if the EUC of the new reservation is earlier than the EUC of the *last realized circuit* at the output port. Note that the last realized circuit may be currently being utilized by a message traversing the port, or it may have been utilized and removed.

Consider the example in Fig. 13. Circuit C_2 passes through the two consecutive routers R_j and R_{j+1}, where R_j precedes R_{j+1} on C_2's path, and message m_2 is traveling on C_2. The data request Req_1 is reserving a new circuit, C_1, which shares the routers R_j and R_{j+1} with C_2. In particular, C_1 and C_2 share $RB^j_{out,West}$ and its corresponding $RB^{j+1}_{in,East}$. Req_1 has arrived at R_j, which indicates C_1's reservation was successfully added to $RB^{j+1}_{in,East}$. Before reserving C_1 at R_j, the conflict detection mechanism compares C_1's EUC with that of the last realized circuit at C_1's required output port (i.e., the west output port). In the example, the detection mechanism compares C_1's EUC with C_2's EUC. If C_1 has a later EUC, then no conflict is detected, but if C_1 has an earlier EUC, then a conflict is detected.

When a circuit is reserved at a router the reservation is either successfully added to the RBs of both of the circuit's

244

required input and output ports or to neither of them. The detection mechanism determines whether adding the new reservation to the output port's RB presents a reservation conflict. The reservation succeeds at the router if no reservation conflict is detected.

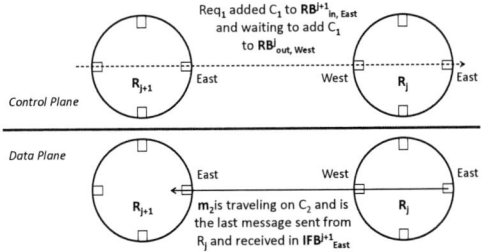

Figure 13: Detecting conflict when reserving an output port: Circuit C_1 failed to be added to $RB^j_{out,West}$ due to a reservation conflict with the last realized circuit, C_2.

B.2 Handling Reservation Conflicts

A reservation conflict indicates there is a *potential* of miss-routing a message on the new circuit. To be safe, the partially reserved new circuit is removed and the request that was reserving this new circuit is allowed to proceed but without reserving the remainder of the circuit path[7]. Furthermore, a simple miss-routing test is performed to eliminate the potential chance that a message gets miss-routed on the new circuit before its removal.

Consider again the example in Fig. 13, upon detecting the conflict at R_j, the request reserving the new circuit, C_1, is allowed to proceed but without reserving the remainder of C_1. Further, R_j signals $RB^{j+1}_{in,East}$, the last RB where C_1 was reserved, to remove C_1's reservation. A reservation is removed by injecting a one-flit *remove conflicting circuit* message to travel on the already reserved part of C_1 to utilize and remove it. To simplify the process of identifying which reservation should be removed, each input port may receive a one bit signal that means: remove the last added circuit to the input port's RB – which in this example is $RB^{j+1}_{in,East}$. Correct operation requires that R_{j+1} does not deliver another circuit reserving request to R_j until R_j indicates that it is safe.

B.3 Detecting and Handling Miss-Routing

Recall that a reservation conflict indicates a potential of miss-routing a message on the incorrectly partially reserved new circuit. In practice, miss-routing is very rare (on average, reservation conflicts represented less than 2% of circuit reservations; see simulation results in Section 4). However, miss-routing should be detected and corrected.

Consider the example in Fig. 13. If m_2 was traversing C_2 at R_j when the reservation conflict was detected, it is possible that C_1 gets realized at R_{j+1} before $RB^{j+1}_{in,East}$ receives the signal to remove C_1's reservation, which would result in miss-routing m_2 on C_1.

To prevent miss-routing, R_{j+1}'s east input port should check that m_2 is traveling on the correct circuit. In general,

---

[7]The request message has a bit that indicates whether the request performs circuit reservation; this bit is reset upon detecting a reservation conflict.

at any router, R_i, each input port, π, performs a simple

check on the destination node of the next message in IFB^i_π to make sure it matches the destination node of the currently realized circuit at the π input port. To retain one cycle per hop latency on the data plane, the message's destination node and currently realized circuit matching is performed in parallel with the head flit traversing the switch to the next router.

To demonstrate how to correct miss-routed packets, again consider the case in Fig. 13. Assume that the check indicates that m_2 is being miss-routed on C_1. Therefore, R_{j+1} stops sending m_2, and does not remove m_2's head flit from IFB^{j+1}_{East}. Each router's input port receives a *data valid* signal, which indicates whether a flit is being received during the current cycle. The result of checking m_2's destination node is logically ANDED with the *data valid* signal of the next input port on C_1's path. Because the check failed, the *data valid* signal would be cleared causing the next input port to ignore m_2's head flit.

Further, we know that $RB^{j+1}_{in,East}$ will receive a signal from R_j to remove C_1. However, if miss-routing is detected before receiving the signal, the one-flit *remove conflicting circuit* message can be sent at the next cycle instead of waiting for the signal from R_j. With this optimization, $RB^{j+1}_{in,East}$ would have to ignore the next *remove circuit* signal that R_j sends.

After sending the *remove conflicting circuit* message, normal operation of R_{j+1} resumes, which includes: $RB^{j+1}_{in,East}$ finding the next reservation with the earliest EUC (C_2 in the example), realizing that circuit, and sending the buffered message on the circuit (sending m_2 on C_2).

C. DEADLOCK

In the proposed scheme, each of the control and data planes can be designed to avoid deadlock. We assume a 2D mesh topology where the control plane uses X-Y routing and the data plane uses Y-X routing. The routers of the data plane have two different kinds of buffers: flit buffers for storing messages (or packets), and reservation buffers for storing circuit reservations. These two types of buffers have a dependence relationship. On the data plane, messages travel on circuits, which require space in the circuit reservation buffers. Similarly, new reservations require free space in the RBs. RB space becomes available only when messages are able to advance so that circuit reservations are utilized and removed from the RBs. Because circuits are reserved backwards; from destination to source, a circular dependency may develop causing potential deadlock in the NoC, as in the following scenario:

In Fig. 14, a data request is attempting to reserve a new circuit, C_1. Unfortunately, when the request arrives at router R_a, there is no free space in the RB of C_1's required input port, $RB^a_{in,North}$. If the request waits, free space may become available allowing C_1 to be reserved and allowing the request to advance to its destination. Free space becomes available only if the next message, m, in IFB^a_{North} is able to exit R_a, thus making room for C_1's reservation. However, m may be blocked and unable to advance due to a full buffer at the input port of the next router on m's path. Let m_2 be the message at the head of the chain of blocked messages and assume that m_2 is stopped at router R_z and is traveling on circuit C_2. A circular dependency occurs if m_2 is unable to move because C_2 cannot be realized at R_z before the new circuit being reserved, C_1, is consumed at the same router,

245

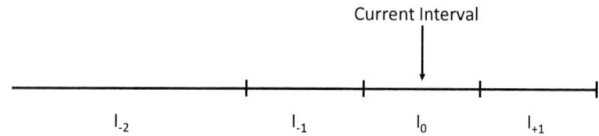

Figure 15: Tracked time intervals. All reservations falling in I_{-2} are maintained in ascending order in the reservation buffers.

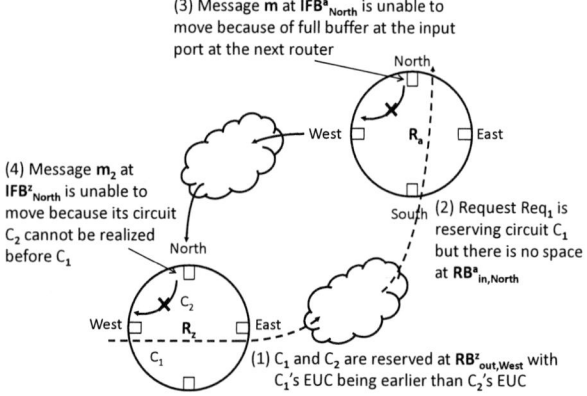

Figure 14: A circular dependency that causes deadlock.

R_z. An example of this might be if C_1 and C_2 share the west output port at R_z, and C_1 has an earlier EUC than C_2. We can detect that a deadlock *may have developed* if a request is unable to reserve a circuit due to unavailability of RB space and this situation persists for a specified number of cycles (i.e., a timeout mechanism).

Resolving this potential deadlock is similar to handling a reservation conflict (Section B.2): the router at which the request is unable to make the reservation (R_a in Fig. 14) signals the last input port's RB at which C_1 was successfully reserved to mark C_1 for removal, and the request is allowed to proceed to its destination *without* reserving C_1, thus breaking the deadlock. Note that C_1's reservation in the signaled RB is not necessarily the one with the earliest EUC. Consequently, C_1's reservation may not be released immediately; rather it is *marked for removal* so that when it becomes the earliest one in the RB, a *remove conflicting reservation* message is injected to consume the partially reserved C_1.

D.   IMPLEMENTATION ISSUES

To demonstrate hardware implementation feasibility, we discuss the representation of EUC and a scheme to keep track of the current cycle number, as well as breaking ties between reservations that have equal EUCs.

D.1   EUC Representation

To minimize the number of bits for representing EUC, we treat time as being composed of consecutive time intervals of equal lengths, with a counter, CLOCK, recording the cycle number in the current interval. EUC is a cycle number which is relative to either: the current (I_0), previous (I_{-1}), or the next time interval (I_{+1}) – thus, two bits are sufficient to represent an interval. At the end of the current interval, I_0,

CLOCK is reset to 0 and the intervals of EUCs are shifted, such that EUCs in I_i are now considered to be in I_{i-1}, where $i \in \{+1, 0, -1\}$. For example, assume that the length of the time interval is 1024 cycles and assume that a router, R_a, on the data plane has a reservation, Res_k, with EUC = 1020 in I_0. Also, assume that when I_0 ends, some router, R_b, on the control plane has a request, Req_l, carrying an EUC of 26 in I_{+1}. When I_0 ends, CLOCK is reset to 0, Res_k's EUC in R_a becomes 1020 in I_{-1}, and the EUC carried by Req_l becomes 26 in I_0.

To handle the case that a circuit reservation may age to be in a time interval older that I_{-1}, we consider the time before I_{-1} as one infinite interval, I_{-2} (See Fig. 15). Reservations in I_{-2} are realized before the reservations in other time intervals. Reservations that age and become in I_{-2} are kept in the sequential order of their realization while their EUCs are discarded. I.e., if at an RB one or more circuit reservations age and become in I_{-2}, these reservations are ordered relative to each other using their EUCs, and then added after any reservations that are already in I_{-2}.

Because EUCs for reservations held in I_{-2} are not retained, it is necessary to guarantee that no data request can insert a new reservation in I_{-2}. The first step to achieve this guarantee is choosing an appropriate length, T, of the time intervals. Let M be the maximum acceptable round-trip time (in cycles) between any two nodes. By choosing T to be at least M cycles, no request can insert a reservation in an interval beyond I_{+1} in the future, and choosing T to be at least $2M$ cycles reduces the probability that a request will attempt to insert a reservation in I_{-2}. To eliminate this probability, a request should stop reserving a circuit if the reservation will be in I_{-2}, as follows.

A request's carried EUC continues to be decremented by one cycle per hop as the request advances to its destination. When a request is sent, its initial carried EUC can be in either I_{+1} or I_0. Thus, if the current time interval ends and the request's carried EUC becomes in I_{-2} due to a severly delayed resevation packet, this indicates that the request's carried EUC is now very inaccurate. In such a case, the request's partially reserved circuit should be removed while allowing the request to proceed without reserving the remainder of the circuit. The partial circuit is removed in the same way a circuit is removed when a potential deadlock is detected (Section C).

D.2   Breaking Ties

It may happen that two different requests reserve two circuits with equal EUCs across the two circuits' shared ports. We need to guarantee a consistent ordering of realizing these two circuits on their shared sub-path. To enforce a total ordering, we associate two pieces of information – besides the EUC – with a circuit's reservation: (1) the number of the circuit's destination node, $d_{node} \in \{d_0, ..., d_{N-1}\}$, where N is the number of nodes in the network; and (2) the id, r_{id}, of the outstanding request at d_{node} that reserved the circuit, such that $r_{id} \in \{0, ..., s\}$, where a node can have at most s outstanding requests. If two circuits C_1 and C_2 have equal EUCs, then we attempt to break the tie by comparing their destination nodes (there is a total ordering of destination nodes), and if they share the same destination, we break the tie by comparing their request ids. Tie breaking can be simplified to having to compare only destination nodes when EUCs are equal while enforcing that a requesting node does not issue two or more requests with identical EUCs.

246

A Heterogeneous Multiple Network-On-Chip Design:
An Application-Aware Approach

Asit K. Mishra
Intel Corporation
Hillsboro, OR 97124, USA
asit.k.mishra@intel.com

Onur Mutlu
Carnegie Mellon University
Pittsburgh, PA 15213, USA
onur@cmu.edu

Chita R. Das
The Pennsylvania State University
University Park, PA 16802, USA
das@cse.psu.edu

ABSTRACT

Current network-on-chip designs in chip-multiprocessors are agnostic to application requirements and hence are provisioned for the general case, leading to wasted energy and performance. We observe that applications can generally be classified as either *network bandwidth-sensitive* or *latency-sensitive*. We propose the use of two separate networks on chip, where one network is optimized for bandwidth and the other for latency, and the steering of applications to the appropriate network. We further observe that not all bandwidth (latency) sensitive applications are equally sensitive to network bandwidth (latency). Hence, within each network, we prioritize packets based on the relative sensitivity of the applications they belong to. We introduce two metrics, *network episode height and length*, as proxies to estimate bandwidth and latency sensitivity, to classify and rank applications. Our evaluations show that the resulting heterogeneous two-network design can provide significant energy savings and performance improvements across a variety of workloads compared to a single one-size-fits-all single network and homogeneous multiple networks.

Categories and Subject Descriptors

C.1.2 [**Computer Systems Organization**]: Multiprocessors; Interconnection architectures

Keywords

Heterogeneity, On-chip Networks, QoS, Packet Scheduling

1. INTRODUCTION

Network-on-Chips (NoCs) are envisioned to be a scalable communication substrate for building multicore systems, which are expected to execute a large number of different applications and threads concurrently to maximize system performance. A NoC is a critical shared resource among these concurrently-executing applications, significantly affecting each application's performance, system performance, and energy efficiency. Traditionally, NoCs have been designed in a monolithic, one-size-fits-all manner, agnostic to the needs of different access patterns and application characteristics. Two common solutions are to design a single NoC for (1) the common-case, or average-case, application behavior or (2) the near-worst case application behavior, by over-provisioning the design as much as possible to maximize network bandwidth and to minimize network latency. However, applications have widely different demands from the network, e.g., some require low latency, some high bandwidth, some both, and some neither. As a result, both design choices are suboptimal in either performance or energy efficiency. The "average-case" network design cannot provide good performance for applications that require more than

Permission to make digital or hard copies of all or part of this work for personal or classroom use is granted without fee provided that copies are not made or distributed for profit or commercial advantage and that copies bear this notice and the full citation on the first page. To copy otherwise, to republish, to post on servers or to redistribute to lists, requires prior specific permission and/or a fee.
DAC '13, May 29 - June 07 2013, Austin, TX, USA.

the supported bandwidth or that benefit from lower latency. Both network designs, especially the "over-provisioned" design, are power- and energy-inefficient for applications that do not need high bandwidth or low latency. Hence, monolithic, one-size-fits-all NoC designs are suboptimal from performance and energy standpoints.

Ideally, we would like a NoC design that can provide just the right amount of bandwidth and latency for an application such that the application's performance requirements are satisfied (or its performance maximized), while the system's energy consumption is minimized. This can be achieved by dedicating each application its own NoC that is dynamically customized for the application's bandwidth and latency requirements. Unfortunately, such a design would not only be very costly in terms of die area, but also requires innovations to dynamically change the network bandwidth and latency across a wide range. Instead, if we can categorize applications into a *small* number of classes based on similarity in resource requirements, and design multiple networks that can efficiently execute each class of applications, then we can potentially have a cost-efficient network design that can adapt itself to application requirements.

Building upon this insight and drawing inspiration from the embedded and ASIC designs where a single network is customized for *a single* application, this paper proposes a new approach to designing an on-chip interconnect that can satisfy the *diverse* performance requirements of general-purpose applications in an energy-efficient manner. We observe that applications can be divided into two general classes in terms of their requirements from the network: bandwidth-sensitive and latency-sensitive. Two different NoC designs, each of which is customized for high bandwidth or low latency, can, respectively, satisfy requirements of the two classes in a more power-efficient manner than a monolithic single network. We, therefore, propose designing two separate heterogeneous networks on a chip, dynamically monitoring executing applications' bandwidth and latency sensitivity, and steering/injecting network packets of each application to the appropriate network based on whether the application is deemed to be bandwidth-sensitive or latency-sensitive. We show that such a heterogeneous design can achieve better performance and energy efficiency than current average-case one-size-fits-all NoC designs.

To this end, based on extensive application profiling, we first show that a high-bandwidth, low-frequency network is best suited for bandwidth-sensitive applications and a low-latency, high-frequency network is best for latency-sensitive applications. Next, to steer packets into a particular network, we identify a packet's sensitivity to network latency or bandwidth. For this, we propose a new packet classification scheme that is based on an application's intrinsic network requirements. We introduce two new metrics, *network episode length* and *height*, to dynamically identify the communication requirements (latency and bandwidth sensitivity) of applications. Observing that not all applications are equally sensitive to latency or bandwidth, we propose a fine-grained prioritization mechanism for applications within the bandwidth and latency optimized networks. Thus, our mechanism consists of first *dynamically classifying* an application as latency or bandwidth sensitive, then *steering* it into the appropriate network and, finally within each network *prioritizing* each application's packets based on its relative potential to improve overall system performance and reduce energy.

Our evaluations on a 64-core 2D mesh architecture, considering 9 design alternatives with 36 diverse applications, show that our heterogeneous two-network NoC design consisting of a 64b link-width latency-

247

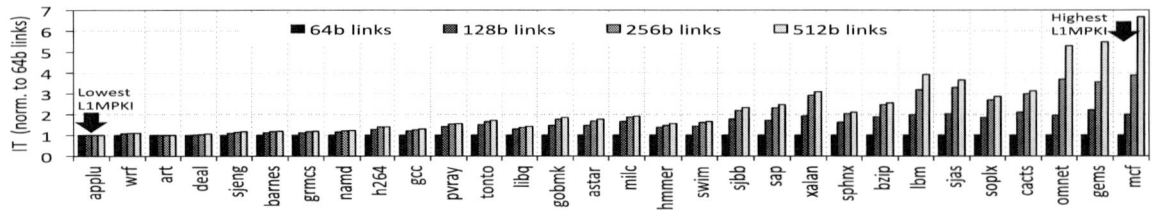

Figure 1: Instruction throughput (IT) scaling of applications with increase in network bandwidth.

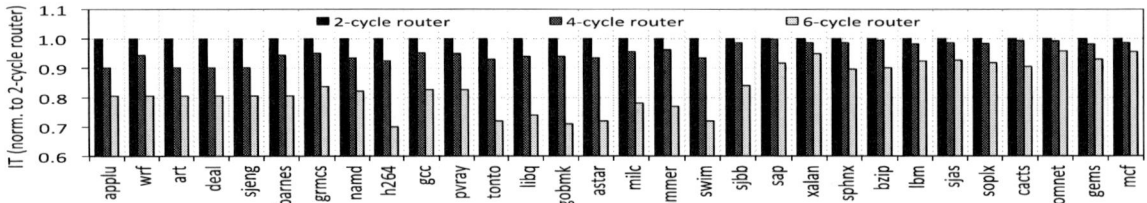

Figure 2: Instruction throughput (IT) scaling of applications with increase in router latency.

optimized network and a 256b link-width bandwidth-optimized network, provides 5%/3% weighted/instruction throughput improvement and 31% energy reduction over an iso-resource (320b link-width) monolithic network design. When compared to a baseline 256b link-width monolithic network, our proposed design provides 18%/12% weighted/ instruction throughput improvement and 16% energy reduction.

2. COMMUNICATION CHARACTERIZATION

We provide observations that highlight the intrinsic heterogeneity in network demand across applications. These observations form the motivation for an application-aware NoC design. We start by looking at two *fundamental* parameters: network channel bandwidth and latency.

Impact of channel bandwidth on performance scaling: Channel or link bandwidth is a critical design parameter that affects network latency, throughput and energy/power. To study the sensitivity of an application to variation in link bandwidth, we use a 64-core chip-multiprocessor (CMP) on an 8x8 mesh network, where both cores and network clocked at the same frequency, and run a copy of the same application on all nodes on the network. Table 1 shows the system configuration. We chose applications from commercial, SPEC CPU2006, SPLASH and SPEC OMP suites. We analyze scenarios where we double the bandwidth starting with 64-bit links up to 512-bit links. Figure 1 shows the results of this analysis for 30 of the 36 applications in our benchmark suite (6 non-network-sensitive applications are omitted to reduce clutter in the plots). In this figure, the applications are shown on the X-axis in order of their increasing L1MPKI (L1 misses per 1000 instructions). The Y-axis shows the average instruction throughput when normalized to the instruction throughput of the 64b network.

Observations from this analysis are: (1) Of the 30 applications shown, performance of 12 applications (the rightmost 12 in the figure after swim) scales with increase in channel bandwidth. For these applications, an 8X increase in bandwidth results in at least a 2X increase in performance. We call these applications *bandwidth-sensitive* applications. (2) The other 18 applications (all applications to the left of and including swim), show very little to no performance improvement with increase in network bandwidth. (3) Even for bandwidth-sensitive applications, not all applications' performance scales equally with increase in bandwidth. For example, while omnet, gems and mcf show more than 5X performance improvement for 8X bandwidth increase, applications like xalan, soplex and cacts show only 3X improvement for the same bandwidth increase. (4) L1MPKI is not necessarily a good predictor of bandwidth-sensitivity of applications. Intuitively, applications that have high L1MPKI would inject more packets into the network, and hence would benefit more from a higher-bandwidth network. But this intuition does not hold entirely true. For instance, bzip, despite having a higher L1MPKI than xalan, is less performance-sensitive to bandwidth than xalan. Thus, we need a better metric to identify bandwidth-sensitive applications.

Impact of network latency on performance scaling: Next, we analyze the impact of network/router latency on the instruction throughput of these

applications. For this, we add an extra pipeline latency of 2 and 4 cycles to each router (in the form of dummy pipeline stages) on top of the baseline router's 2-cycle latency. The cores and the network are clocked at the same frequency for this analysis. Increasing the pipeline stages at each router *mimics* additional contention in the routers when compared to the baseline network. Figure 2 shows the results for this analysis, where the channel bandwidth is 128b (although the observation from this analysis holds true for other channel bandwidths as well).

Our observations are the following: (1) Bandwidth-sensitive applications (the rightmost 12 applications) are not very responsive to increase in network/router latency. On average, for a 3X increase in per-hop latency, there is only 7% degradation in application performance (instruction throughput) for these applications, i.e., an extra 4 cycle latency per router is tolerated by these applications. (2) On the other hand, for all applications to the left of and including swim, there is about 25% performance degradation when the router latency increases from 2-cycles to 6-cycles. We call these *latency-sensitive* applications. (3) L1MPKI is *not* a perfect indicator of latency-sensitivity (hmmer, despite having a higher L1MPKI than h264, does not show proportional performance improvement with reduction in router latency).

Application-level implications on network design: The above analysis suggests that a single monolithic network is not the best option for various application demands. Therefore, an alternative approach to designing an on-chip interconnect is to have multiple networks, each of which is specialized for common application requirements, and dynamically steer requests of each application to the network that matches the application's requirements. Based on Figures 1 and 2, a wide and low-frequency network is suitable for bandwidth-sensitive applications, while a narrow and high-frequency network suitable for latency-sensitive ones. To improve the performance of the latency-sensitive applications, a network architect can reduce the router pipeline latency from 2-cycles (our baseline) to a single cycle, while keeping the frequency constant, or increase the network frequency (to reduce network latency). Although there are proposals that advocate for single-cycle routers [10, 13, 12, 6], such designs often involve speculation, which increases complexity and can be ineffective at high or adverse load conditions, and require relatively sophisticated arbiters that are not necessarily energy efficient. Hence, while single-cycle routers are feasible, in this paper, we use frequency as a knob to reduce the network latency. According to our analysis, increasing the frequency of the network from 1 GHz to 3 GHz (3 times the core frequency) leads to less than 1.5% increase in *energy* for latency-sensitive applications (results for energy with frequency scaling are omitted for brevity).

Designing latency- and bandwidth-customized networks is the first step in achieving customization in the network. We also need a runtime mechanism to classify applications into one of the two categories: *latency or bandwidth sensitive*. In addition, since not all applications are equally sensitive to bandwidth or latency, we would like a mechanism that ranks the applications within each category in a more fine-grained manner within the latency- and bandwidth-customized networks. The next section discusses how we perform application classification.

3. DYNAMIC CLASSIFICATION OF APPLICATIONS

The goal of dynamically identifying an application's sensitivity to latency or bandwidth is to enable the local network interface (NI)

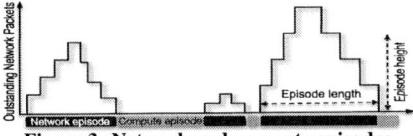

Figure 3: Network and compute episodes.

at each router to steer packets into a network that has been optimized for either latency or bandwidth. We propose two new metrics, called *network episode length* and *episode height*, that effectively capture the network latency and bandwidth demands of an application.

Episode length and height: During its life cycle, an application alternates between two kinds of episodes (shown in Figure 3): (1) *network episode*, where the application has at least one packet (to L2 cache or to DRAM) in the network, and (2) *compute episode*, where there are no outstanding cache/memory requests by the thread. During the network phase, there may be multiple outstanding packets from the application in the network owing to various techniques that exploit memory-level parallelism (MLP) [5, 14]. During this network phase, the core is likely stalled, waiting for L2 and memory requests to be serviced. Because of this, the instruction throughput of the core is low. During the compute episode, however, the instruction throughput is high [9]. We characterize a network episode by its length and height. *Length* is the number of cycles the episode lasts starting from when the first packet is injected into the network until there are no more outstanding packets belonging to that episode. *Height* is the average number of packets (L1 misses) injected by the application during the network episode. To compute height, the core hosting the application keeps track of the number of outstanding L1 misses (when there is at least 1 L1 miss) in the re-order buffer on a per-cycle basis.

A short episode height suggests that the application has low MLP. In other words, the latencies of the requests are not overlapped, and as a result the application's progress is sensitive to network latency. The application also does not require significant bandwidth because it has a small number of outstanding requests at any given time. On the other hand, a tall episode height suggests that the application has a large number of requests in the network, and the network latency of the packets are overlapped. This indicates that the application likely needs significant bandwidth from the network to make progress while its progress is less sensitive to network latency. Hence, we use the network episode height of an application as the main indicator of latency or bandwidth sensitivity of an application.

A short episode length (on average) suggests that the application is less network intensive. Contrast this with an application that has a long episode length, i.e., more network intensive. An equal amount of network delay would slow down the former application more than it does the latter application. As a result, the former application with a short average episode length is likely to be more sensitive to network latency, and therefore would likely benefit from being prioritized over another application with a longer average episode length. Note that this observation is similar to Kim et al.'s for memory controllers [9], which showed that prioritizing an application with shorter memory episodes over a one with longer memory episodes is likely to improve performance.

We compute running averages of the episode height and length to keep track of these metrics at runtime. We quantize episode height as *tall*, *medium* or *short* and episode length as *long*, *medium* and *short*. This allows us to perform fine-grained dynamic application (or, application phase) classification based on episode length and height. Figures 4 and 5 show these metrics for 30 applications in our benchmark suite. Note that, these figures show the *average* metrics for an entire application and that there are intra-application latency/bandwidth sensitive phases that our dynamic scheme (Section 4) captures. Based on Figures 1 and 2, we classify all applications whose average episode length and height are shorter than sjbb's episode length and height, respectively, to be short in length and height (shaded black in the figures). Applications whose average episode height is larger than sjbb's episode height but lower than 7 (empirically chosen) are classified as having medium (shaded blue in the figures) and the remaining as having tall episode heights (shaded with hatches in Figure 4). Empirically, a cut-off of 10K cycles is chosen to classify applications as having medium vs. long episode length.

Classification and ranking of applications: Figure 6 (left) shows the classification of applications based on their episode height and length. The figure also shows the bandwidth-sensitive and latency-sensitive applications based on such a classification. We use this fine-grained classification to steer applications to the two networks as well as rank applications for prioritization within a network.

Episode height, as a general principle, is used to identify the bandwidth/latency sensitivity of applications. Applications with tall height are considered bandwidth sensitive (due to high MLP, which leads to high bandwidth demand as well as high latency tolerance), and steered to the bandwidth-optimized network. Applications with low height are latency sensitive (due to low MLP, which leads to low latency tolerance for each packet) and are steered to the latency-optimized network. Figure 6 shows the resulting classification of applications. Note that applications whose episode height is medium but episode length is short are classified as latency-sensitive as they have neither short nor tall episode heights, but demand less from the network due to short episodes.

Episode length, for the most part, is used to determine the *ranking of applications within each network*. The general principle is to prioritize applications with shorter network episode length over others since delaying such applications in the network has much more of an effect on their performance (slowdown) than delaying applications with long episode lengths (since the latter set of applications are already slow due to long network episodes anyway). Figure 6 (right) shows the resulting ranking of the applications in their respective networks based on relative episode length and height.

4. DESIGN DETAILS

Since we use a canonical 2D network in our study, instead of discussing the standard router and network designs, we focus on the design aspects for supporting our classification and prioritization schemes in this section.
Computing episode characteristics: To filter out short-term fluctuations in episode height/length, and adapt our techniques to handle long-term traffic characteristics, we use running averages of the metrics. On every L1 miss, the NI computes the running average of episode height/length. To compute episode height, the outstanding L1 miss count is obtained from the miss-status handling registers (MSHRs). Counting the number of cycles (using an M-bit counter) the L1 MSHRs are occupied gives the information to compute episode length. This M-bit counter is reset every batching interval, B. We use the notion of batching [3, 15] to prevent starvation due to ranking of applications in each network, as done in [3, 15, 2] (more information on this is in Section S.3).

When the NI of a router receives a packet, it: (1) updates the episode height/length metric for the current application phase and (2) decides which network this packet is to be steered to, based on the classification scheme (Section 3). Thus, the episode metrics are computed per phase of an application. All packets belonging to a particular phase are steered to the network optimized for either latency or bandwidth. Note that each application's rank and network sensitivity are decided at runtime (although the classification analysis we provided in Section 3 was for the whole application). This helps our scheme to capture within-application variation in latency and bandwidth sensitivity. No central coordination is required in our technique to decide a uniform central ranking across all the applications in the system, which was needed in past works that ranked applications for prioritization [3, 2]. Once a packet's rank is decided, it is consistently prioritized with that rank across the entire network until it reaches its destination.

The NI tags the transmitted packet with its rank (2-bits) and its batch-id (3-bits). At each router, the priority bits in the header-flit are utilized by the priority arbiters in a router to allocate the virtual channels (VCs) and the switch. To prevent priority inversion due to VCs in routers, where a packet belonging to an older batch or higher rank is queued behind a lower ranked packet, we use atomic buffers [16].
Customized network design choices: As mentioned earlier, we opt for a high-frequency but low link-width network for the latency-sensitive applications and a high-bandwidth network operating at the same frequency as the cores for the bandwidth-sensitive applications. We use a 2-stage baseline router and increase the router frequency up to 3 times for the latency-sensitive network. High-frequency routers can be designed by a combination of both micro-architecture and circuit optimizations as shown

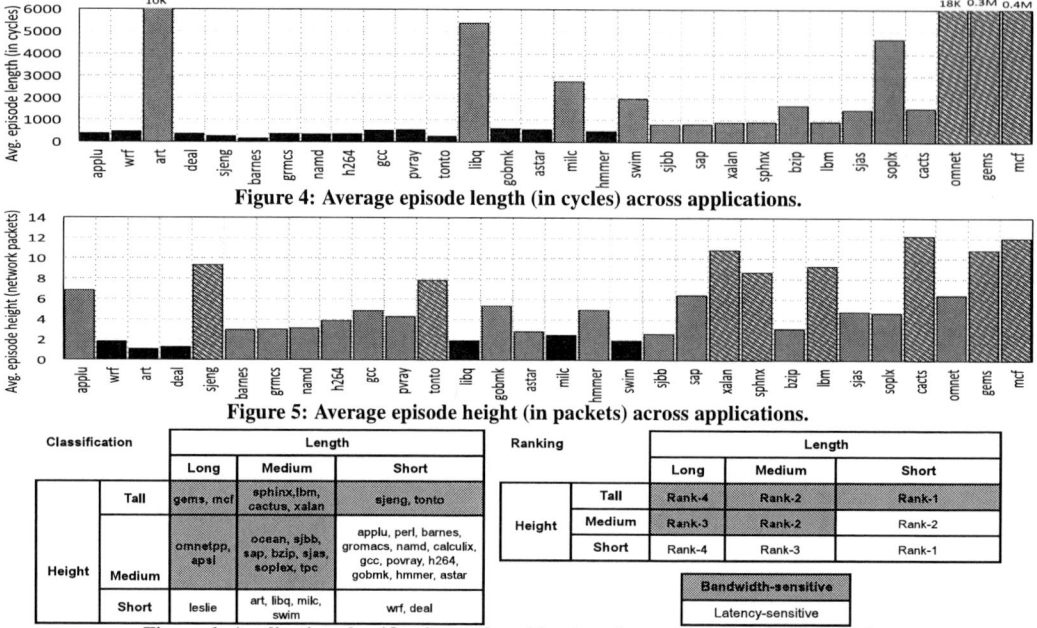

Figure 4: Average episode length (in cycles) across applications.

Figure 5: Average episode height (in packets) across applications.

Figure 6: Application classification and ranking based on episode length and height.

by previous works [10, 11] as well as industrial prototypes [18, 7]. In canonical router designs, the arbitration and crossbar traversal stages are the bottleneck stages in terms of critical path [11]. In our design, since the latency-optimized network has only 3 VCs per physical channel and a narrow link width (64b), our analysis (based on synthesis of the RTL of VC and switch allocation and crossbar stages) shows that it is feasible to clock the routers in this network at a higher frequency.

Network bandwidth depends on link-width and frequency. A designer could think of increasing the frequency of the bandwidth-customized network to increase the total network bandwidth. However, increasing the frequency of the wider 256b network would adversely affect the power of this network. Hence, in our network designs, we only increase the frequency of the narrow 64b link-width network whose power envelope is 43% lower than the wider network. Increasing the frequency of the latency-customized network also increases this network's bandwidth, but since we steer only latency-sensitive applications (which are agnostic to bandwidth increase) into this network, the performance improvement of these applications is primarily due to network latency reduction.

The design space for optimizing network latency or bandwidth is large, and not possible to cover in this paper. Although we articulate frequency as a knob to improve latency in the latency-optimized network, a designer could use different topologies, flow-control, and arbitration mechanisms as various knobs to improve latency. The motivation of this paper is to demonstrate that heterogeneous multiple networks (customized for latency and bandwidth separately) provides a better design than a monolithic network or homogeneous multiple networks. The 64b 3X-frequency network for the latency-sensitive applications and 256b 1X-frequency network for the bandwidth-sensitive applications are only two design points to demonstrate the concept.

5. EVALUATION METHODOLOGY

Design scenarios: Starting with a monolithic network, we show the benefits of having two networks, each customized for either bandwidth or latency. We also show the benefits of our scheme compared to an iso-resource single network (with the same total bandwidth). Following are the nine design scenarios we evaluate on our experimental platform:
❶ **1N-128**: This design has a single homogeneous 128b link network. We assume this to be our starting point. Starting with this monolithic network, we increase its bandwidth to create a bandwidth-optimized network, and reduce its bandwidth (and increase its frequency) to design a latency-optimized network. ❷ **1N-256**: This configuration has a single homogeneous network with 256b links. We chose this as our *baseline* net-

work. Starting with this network, we first design a homogeneous multiple network design (where each network has equal bandwidth, 2N-128x128) and then customize one network for latency-sensitive applications and the other network for bandwidth-sensitive applications. ❸ **2N-128x128**: This design has two parallel networks, each with 128b link width. The buffer resources in each network is half that of the 1N-128 network and each of the networks operate at the same frequency as the cores. Packets are steered into each network with a probability of 0.5, i.e., there is load balancing across the networks. ❹ **1N-512**: This design has a single network with 512b link width. We call this a *high-bandwidth* configuration and analyze it to see how our proposal fares compared to a very high bandwidth network. ❺ **2N-64x256-ST**: In this design, there are two parallel networks, one with 64b link width and the other with 256b link width. The buffering resources in each network is half that of a single network, so that the total buffering resources are constant across this design and a design that has a single network. Further, in this configuration, the bandwidth-sensitive packets are steered (hence, the annotation **ST**) into the 256b network and the latency-sensitive packets are steered into the 64b network. Each network in this configuration is clocked at the frequency of the cores. ❻ **2N-64x256-ST+RK(no FS)**: This design is same as the previous design except that the network also prioritizes applications based on their ranks (hence, the annotation **RK**) at every cycle in a router. ❼ **2N-64x256-ST+RK(FS)**: This design is same as the previous configuration except that the 64b network is clocked at 3X the frequency of cores. The 256b network is still clocked at the core frequency. This configuration is analyzed to see the benefits of frequency scaling (hence, the annotation **FS**) the latency-optimized network. ❽ **1N-320(no FS)**: In this design, there is a single network with 320b (=64b+256b) links. The network operates at the frequency of the core. This configuration is iso-bandwidth with all our 64x256 networks and is analyzed to see the benefits of our proposal over an equivalent configuration. ❾ **1N-320(FS)**: This design is similar to the above design, except that the network is now clocked at 3X the core frequency. This design is analyzed to see the effectiveness of our scheme over a scheme that is iso-resource as well as over-clocked to help latency-sensitive applications.

Experimental setup: Our proposals are evaluated on an instruction-trace-driven, cycle-level x86 CMP simulator. Table 1 provides the configuration of our baseline, which contains 64 cores in a 2D, 8x8 mesh NoC. The network connects the cores, shared L2 cache banks, and memory controllers (these all stay constant across all evaluated designs). A data packet consists of 1024b (=cache line size) and is decomposed into flits depending upon the link width in each design. Since wiring resources on die are abundant [1, 19], when simulating parallel networks, we assume the

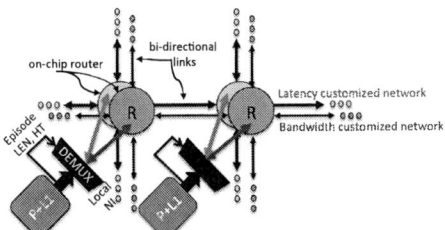

Figure 7: Schematic of the proposed CMP.

Table 1: Baseline core, cache, memory and network configuration

Core
128-entry instruction window, 2 INT/FP operations and 1 LD/ST per cycle
Caches and Main Memory
L1 Caches: 32 KB per-core (private), 4-way set associative, 128B block size, 2-cycle latency, write-back, 32 MSHRs
L2 Caches: 1MB per bank, shared, 16-way set associative, 128B block size, 3-cycle bank latency, 32 MSHRs
Main Memory: 4GB; up to 16 outstanding requests per processor, 320 cycle access
Network and Router
Network Router: 2-stage wormhole switched, virtual channel flow control, 6 VC's per port, 5 flit buffer depth, 1 flit/address packet, X-Y routing
Network Topology: 8x8 mesh, each node has a router, processor, private L1 cache and shared L2 cache bank (all nodes), 4 memory controllers (1 at each corner node), 256b bi-directional links (= data packet's flit width)

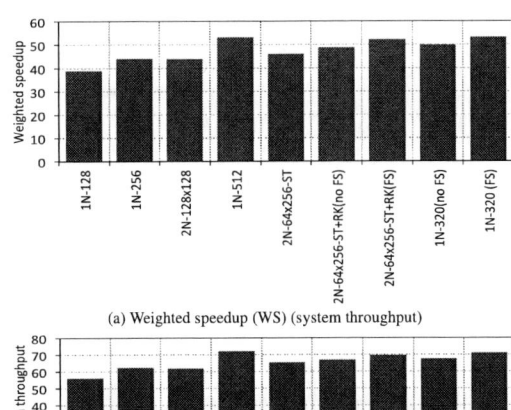

(a) Weighted speedup (WS) (system throughput)

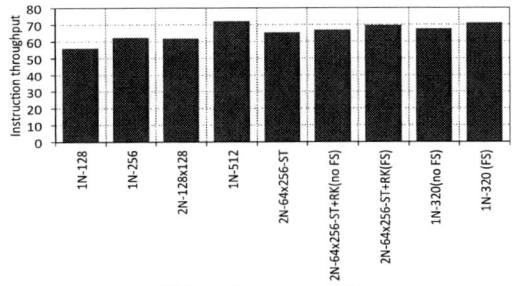

(b) Instruction throughput (IT)

Figure 8: Performance comparison of various network designs, averaged across 25 multiprogrammed workload mixes.

networks to be implemented in the same 2D substrate as the cores. A high-level schematic of the proposed interconnection network is shown in Figure 7. The dynamic and leakage energy numbers for the network were extracted using Orion 2.0 [8] and Rock Creek router data [7], and incorporated into our simulator for detailed network energy analysis at 32nm technology node. Based on Orion 2.0 and Rock Creek layout estimates, the area of two networks (router and links) consisting of 256b and 64b links is only 1% larger than the iso-resource network with 320b links (2.4X larger area when compared to the 128b link network), and the power envelope of these two networks is 20% lower than the iso-area network (2.3X higher power when compared to the 128b link network). The counter configurations used in our techniques are: (1) counter size for number of cycles in a network episode, M=14bits (2) batching interval, B=16,000 cycles.

Application setup: We use a diverse set of multiprogrammed application workloads comprising scientific, commercial, and desktop benchmarks. We use the SPEC CPU2006 benchmarks, applications from SPLASH-2 and SPEC-OMP benchmark suites, and four commercial workload traces (sap, tpcc, sjbb, sjas) totaling 36 applications. All our experiments analyze multiprogrammed workloads, where each core runs a separate application. We simulate at least 640 Million instructions across 64 processors (minimum 10 Million instructions per core; a core keeps exerting pressure even after 10M instructions; on average each run simulates 3.5 Billion instructions in the system). All our results are aggregated across *25 workload combinations*. In each of these workload combinations, 50% (32) of the applications are latency-sensitive and 50% (32) of the applications are bandwidth-sensitive. This provides a good mix of bandwidth/latency-sensitive applications, which is likely to be a common mix for future multicore systems. Within each of these two categories, applications are *randomly picked* to form the workload.

Evaluation metrics: Our primary performance evaluation metrics are instruction throughput [4] and weighted speedup [17]. Instruction throughput is defined to be the sum total of the number of instructions committed per cycle (IPC) in the entire CMP. The weighted speedup metric sums up the speedup (inverse of slowdown) experienced by each application in a workload, compared to its standalone run in the same configuration, and represents *system throughput* [4].

6. ANALYSIS OF RESULTS

Performance comparison: Figure 8 shows the performance comparison across the network designs. The following observations are in order:

❶ Two 128b networks (2N-128x128) provide similar performance (both system and instruction throughput) as a bandwidth-equivalent single monolithic network with 256b links (1N-256). This is in spite of the increase in packet serialization in the networks. The primary reason for this per-

formance improvement is reduction in congestion in each network (each network now sees 50% fewer packets than a monolithic wider network). The total bandwidth in the 2N-128x128 design is the same as the 1N-256 design and hence bandwidth-sensitive applications' performance is not affected. On the other hand, the performance of latency-sensitive applications is improved because of the load balancing (reduced congestion is each network), and thus, the degradation in performance due to serialization latency increase is compensated by performance improvement due to reduced congestion.

❷ Bandwidth- and latency-optimized parallel networks operating at the same frequency as the processor along with steering of packets based on their bandwidth/latency sensitivity (2N-64x256-ST) provide 4.3%/5% system/instruction throughput improvement, over the baseline (1N-256) design. By providing bandwidth-sensitive applications more bandwidth (than 1N-128) and reducing the congestion compared to a monolithic network, the performance of both bandwidth- and latency-sensitive applications is improved. Prioritizing and ranking packets based on their criticality after steering them into a network (2N-64x256-ST+RK(no FS)) provides an additional 6%/3% improvement in system/instruction throughput over the 2N-64x256-ST design. This is because, our ranking scheme prioritizes the relatively network-sensitive applications in each network, and ensures no starvation using batching.

❸ Frequency scaling the latency-optimized network along with steering and ranking the applications (2N-64x256-ST+RK(FS)) provides the maximum performance improvement among our proposals: 18%/12% system/instruction throughput improvement over the baseline network. With frequency scaling, the latency-optimized network is clocked at a higher frequency, accelerating the latency-sensitive packets and this brings the additional benefits in performance.

❹ Frequency scaling and steering along with ranking of applications (2N-64x256-ST+RK(FS)) is better than an iso-resource network (1N-320(no FS)) by 5%/3% in weighted/instruction throughput. 2N-64x256-ST+RK(FS) design is within 2.0%/2.2% system/instruction throughput of the high-frequency iso-resource network with frequency increased by 3X (1N-320(FS)). High frequency of the 1N-320(FS) network helps latency-sensitive applications and high bandwidth of the same network (compared to 256b links) helps bandwidth-sensitive applications. But, as we will show shortly, the energy consumption of such a wide network is higher than our proposal.

❺ Our proposed 2N-64x256-ST+RK(FS) design's system performance is within 1.8% of a very high bandwidth network (1N-512). A high bandwidth network helps bandwidth-sensitive applications, but provides little

251

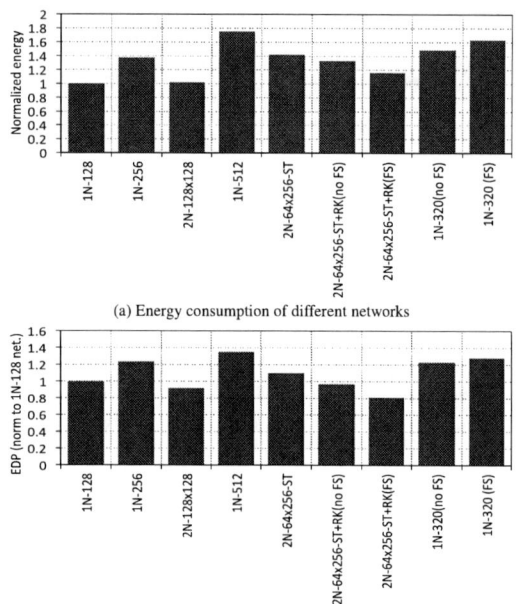

(a) Energy consumption of different networks

(b) Energy-delay product (EDP) of networks

Figure 9: Energy and EDP comparison of various network designs (all results normalized to the 1N-128 network).

benefit for other applications. Additionally, as will be shown next, a wide-link network's energy consumption is very high (about 75% higher than a 128b link-width network). Hence, although our proposed network provides similar performance as a very high bandwidth network, it does so at a lower energy envelope.

Energy and EDP comparison: Increasing the link width decreases the serialization (and zero-load) latency and hence, end-to-end latency is reduced. However, increasing the link width also affects router crossbar power quadratically. Figure 9 shows the energy and energy-delay product (EDP) of the applications across the 9 designs. We find that:

❶ The average energy consumption of a 256b link network (1N-256) is 38% higher than a 128b link network (1N-128). However, the two 128b network design (2N-128x128) has similar energy consumption as a single 128b link monolithic network. The energy reduction going from one network to two networks comes primarily from reduction in network latency (by reducing the congestion in each network). In fact, we observed that the energy consumption of two parallel networks, each with link width $N/2$, is always lower than a single network with link width N. When compared to the 1N-128 network, the 2N-128x128 network has the same buffering resources but double the number of routers and links, which leads to higher power. However, the 2N-128x128 has better performance and less contention, which leads to lower energy. These two effects mostly cancel out, leading to similar energy consumption for 2N-128x128 and 1N-128.

❷ The average energy consumption of a high bandwidth (512b links) network (1N-512) is 75% and 26% higher than a 128b and 256b link network respectively. When link width increases, although serialization latency reduces, the crossbar power starts to dominate the energy component.

❸ Our proposed design with two heterogeneous networks and fine-grained application prioritization in each network (2N-64x256-ST+RK(FS)) consumes 16% lower energy than the baseline 1N-256 network. This is 39% lower energy when compared to a high-bandwidth network (1N-512) and 31% lower energy than an iso-resource network which is frequency scaled (1N-320(FS)). Overall, our proposed design consumes lower energy than an iso-resource 320b link network, and the 2N-64x256-ST network. Note that using fine-grained prioritization within each network provides 6.7% energy reduction over not doing so.

❹ On the EDP metric, our proposed design (2N-64x256-ST+RK(FS)) is 35% better than the baseline (1N-256). Our scheme reduces network latency significantly, which lowers the delay component in the EDP metric. Even without frequency scaling, the 2N-64x256-ST+RK(no FS) design has 22% lower EDP than the baseline. Our proposed design always has lower EDP than a high-bandwidth network (1N-512), the iso-resource

320b link network, or the iso-resource 2N-64x256-ST network.

Summary: Overall, compared to an iso-resource 320b link monolithic network operating at 1X frequency, we find that the combination of a 64b network operating at 3X frequency and a 256b network operating at 1X frequency (along with our steering and prioritization algorithms) provides better performance, energy and EDP.

7. CONCLUSIONS

We proposed an application-driven approach for designing high-performance and energy-efficient heterogeneous on-chip networks (NoCs). The goal of our design is to cater to the applications' requirements more effectively and more efficiently than an application-agnostic monolithic NoC design. The main idea is to have one network customized for low latency and another customized for high bandwidth, and steer latency-sensitive applications to the former and the bandwidth-sensitive applications to the latter. Within each network, applications that are more likely to improve system throughput are prioritized over others. We find that network episode height and length are simple-to-measure metrics to classify applications in terms of bandwidth- and latency-sensitivity. Our results show that the proposed heterogeneous two-network design outperforms a single monolithic network or two homogeneous networks. We conclude that our application-driven methodology for designing heterogeneous NoCs provides high system and application performance at reduced energy, and hope that the proposed dynamic application classification scheme can provide a simple framework for designing heterogeneous on-chip networks.

References

[1] J. Balfour et al. "Design Tradeoffs for Tiled CMP On-Chip Networks". In *ICS*. 2006.

[2] R. Das et al. "Aergia: Exploiting Packet Latency Slack in On-Chip Networks". In *ISCA*. 2010.

[3] R. Das et al. "Application-Aware Prioritization Mechanisms for On-Chip Networks". In *MICRO*. 2010.

[4] S. Eyerman et al. "System-Level Performance Metrics for Multiprogram Workloads". In *IEEE Micro* (2008).

[5] A. Glew. "MLP Yes! ILP No!" In *ASPLOS WACI*. 1998.

[6] M. Hayenga et al. "The NoX router". In *MICRO*. 2011.

[7] J. Howard et al. "A 48-Core IA-32 Processor in 45 nm CMOS Using On-Die Message-Passing and DVFS for Performance and Power Scaling". In *J. Solid-State Circuits* 46.1 (2011).

[8] A. B. Kahng et al. "ORION 2.0: A Fast and Accurate NoC Power and Area Model for Early-Stage Design Space Exploration". In *DATE*. 2009.

[9] Y. Kim et al. "ATLAS: A Scalable and High-Performance Scheduling Algorithm for Multiple Memory Controllers". In *HPCA*. 2010.

[10] A. Kumar et al. "A 4.6Tbits/s 3.6GHz Single-cycle NoC Router with a Novel Switch Allocator in 65nm CMOS". In *ICCD*. 2007.

[11] A. K. Mishra et al. "A Case for Dynamic Frequency Tuning in On-Chip Networks". In *MICRO*. 2010.

[12] T. Moscibroda et al. "A Case for Bufferless Routing in On-Chip Networks". In *ISCA*. 2009.

[13] R. Mullins et al. "Low-Latency Virtual-Channel Routers for On-Chip Networks". In *ISCA*. 2004.

[14] O. Mutlu et al. "Efficient Runahead Execution: Power-Efficient Memory Latency Tolerance". In *IEEE Micro* (2006).

[15] O. Mutlu et al. "Parallelism-Aware Batch Scheduling: Enhancing both Performance and Fairness of Shared DRAM Systems". In *ISCA*. 2008.

[16] C. A. Nicopoulos et al. "ViChaR: A Dynamic Virtual Channel Regulator for Network-on-Chip Routers". In *MICRO*. 2006.

[17] A. Snavely et al. "Symbiotic Jobscheduling for a Simultaneous Multithreaded Processor". In *ASPLOS*. 2000.

[18] S. Vangal et al. "An 80-Tile 1.28TFLOPS Network-on-Chip in 65nm CMOS". In *ISSCC*. 2007.

[19] D. Wentzlaff et al. "On-Chip Interconnection Architecture of the Tile Processor". In *IEEE Micro* (2007).

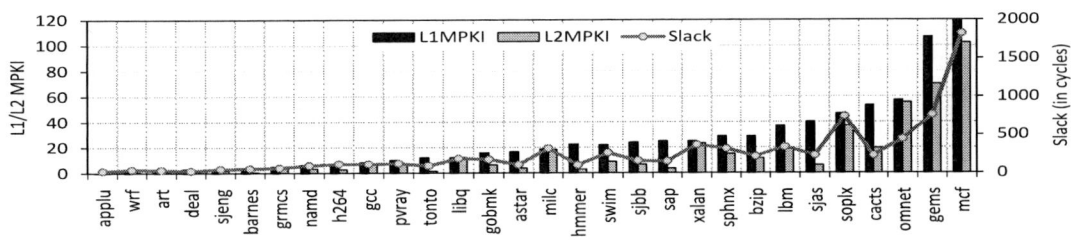

Figure 10: Average L1MPKI, L2MPKI and slack of applications.

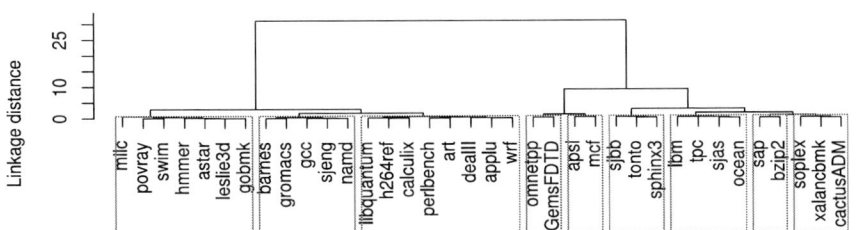

Figure 11: Hierarchical clustering of applications. Applications on the left branch of the root are latency-sensitive, those on the right branch are bandwidth-sensitive.

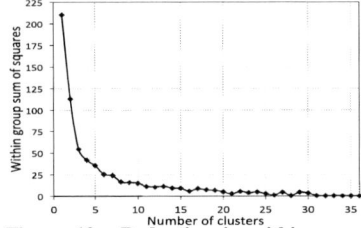

Figure 12: Reduction in within-group sum of squares with increase in number of clusters.

S. SUPPLEMENTAL

In this supplemental section, we present further analyses of our proposed metrics and how they compare to previously proposed criticality metrics, explain more the rationale for our application classification into sub-classes, discuss subtle but important design aspects and how we addressed them, provide more discussion on application characteristics, and present how our scheme compares to closely related previous works.

S.1 Network Episodes vs. Other Metrics

In Section 3, we use the notion of network episodes for classifying applications into latency-sensitive or bandwidth-sensitive categories. We propose two new metrics, network episode height and length for the classification. We contrast these proposed metrics to two recently proposed heuristics, L1MPKI [8] and slack [7], which were proposed to estimate a packet's criticality in the network.

Private cache misses per instruction (MPI): This metric captures an application's network intensity. If the network intensity is low, the application likely (but not always) has low MLP and hence its request are likely to be latency-sensitive as opposed to bandwidth-sensitive. Figure 10 shows the L1MPKI and L2 MPKI of 30 applications. We find that MPI (or MPKI) can help in identifying latency-sensitive applications from bandwidth-sensitive ones. In Figure 10, all applications to the left of sjbb have a lower MPKI than sjbb's MPKI. Since these applications are latency-sensitive, empirically we can think of having a threshold in MPKI (equal to sjbb's MPKI) to classify applications as bandwidth- or latency-sensitive. However, as mentioned earlier, this metric is not as effective in distinguishing between applications *within* the latency-sensitive class or bandwidth-sensitive class (as we will demonstrate in Section S.5.4).

Packet slack: Slack was investigated by Das et al. [7] to identify a packet's criticality in the network. It quantifies how long a packet can be delayed without affecting an application's performance. We measured an instruction's slack from when it enters the re-order buffer (ROB) to when the instruction actually becomes the oldest in the ROB and is ready to commit. Figure 10 shows how average slack varies across applications. Intuitively, the slack of an L1-miss instruction directly translates to the packet's criticality in the network: an application that has a larger average slack is more tolerant to network delay versus an application that has a smaller slack. We find that, although slack is very useful in prioritizing individual packets' criticality, it has low correlation in identifying increase in performance with increase in bandwidth or frequency for a given application.

S.2 Rationale for Application Classification

In Section 3, we classify applications into two gross classes, latency- and bandwidth-sensitive, and nine sub-classes. The sub-classes aid our mechanism in fine-grained prioritization within the networks to further improve performance and reduce energy. We take two decisions in our

application-level classification: (1) choosing sjbb's episode length and height as a threshold to distinguish between applications, and (2) choosing 9 smaller sub-classes after classifying the applications as bandwidth- or latency-sensitive. Here, we present more empirical results that had led us to these decisions.

Figure 11 shows the results of the *hierarchical clustering* of all the applications in our benchmark suite. Hierarchical clustering incrementally groups objects that are similar, i.e., objects that are close to each other in terms of some distance metric [13]. The input to the clustering algorithm consists of the improvement in IPC with bandwidth scaling (from 64b to 512b) and improvement in IPC with router latency scaling (from 1X to 3X), i.e., values from Figures 1 and 2. Our purpose is to observe whether a clustering algorithm perceives noticeable difference between applications' performance with frequency and bandwidth scaling. To perform clustering, we tried using various *linkage distance* metrics (linkage metric determines the affinity between sets of data-points as a function of the pairwise distances between them), such as Euclidean distance, Pearson correlation, and average distance between the objects, and found similar clustering results with all. In all cases, the clustering was consistent with that shown in Figure 11 (shown for Euclidean distance). Although the eventual hierarchical cluster memberships are different from that shown in our classification matrix in Figure 6, the broader classification of how hierarchical clustering groups applications into bandwidth- and latency-sensitive clusters (from the root, the left dendrogram represents latency-sensitive applications and the right dendrogram represents bandwidth-sensitive applications) matches exactly with our classification scheme, which is based on episode height and length – with the exception of sjeng. The reason for sjeng's misclassification is that its performance does not scale with bandwidth and hence, hierarchical clustering classifies it as a latency-sensitive application. However, sjeng's episode height is tall and length is short, on average, meaning it is very bursty during a small interval of time (as such, it has high MLP). Because of this, we classify it as the highest ranking application in the bandwidth-optimized network.

Why 9 sub-classes? To answer this question, we measure the total within-group sum-of-squares (WG-SS) of the clusters resulting from hierarchical clustering. Figure 12 shows this metric as the number of clusters increases. The total WG-SS is a measure of the total dispersion between individual clusters and is regarded as a metric to decide the optimal number of clusters from a hierarchical or K-means clustering algorithm [18, 27]. When all clustering objects are grouped into one cluster, the total WG-SS is maximized, whereas, if each object is classified as a separate object, WG-SS is minimized (=0). Figure 12 suggests that 8 or 9 clusters have similar WG-SS and, 8 or 9 clusters reduce the total WG-SS by 13X compared to a single cluster. Based on this, we choose 9 classes for our application classification and, hence, quantize episode height and length into three classes each.

S.3 More Design Details

We present two important design decisions we made while architecting our scheme: 1) how to handle starvation, 2) how to handle packet reordering.

Handling starvation: Prioritizing highly-ranked packets in a network may lead to starvation of other packets. To prevent starvation, we combine our application-aware prioritization with a "batching mechanism" proposed by Das et al. in [8, 7]. Each packet is part of a batch, and packets belonging to older batches are prioritized over packets from younger batches. Only if two packets belong to the same batch, they are prioritized based on their applications' rank order that is based on episode height/length. A batch also provides a convenient granularity in which the ranking of the applications is enforced.

To support batching, each node keeps a local copy of a batch-ID (BID) register containing the current (injection) batch number and maximum supported batch-ID register containing the maximum number of batching priority levels (L). BID is incremented every B cycles, and thus, BID values across all nodes are the same. Due to batch-ID wrap-around, a router cannot simply prioritize packets with lower batch-IDs over others with higher batch-IDs, and we use schemes suggested in [8, 7] to handle relative priorities inside a router. In all our experiments we used $L=8$ as the number of batching levels.

Handling packet reordering: Due to deterministic routing in the baseline design, for every pair of source and destination in the network, there is always a single path from the source to the destination. Multiple packets on this path do not get re-ordered. When using our scheme with multiple networks, since there are multiple routes between a source and destination pair (via the two different networks), there is a chance of packet reordering in the network. The effect arising out of reordering could be handled either by the software or application layer or by the network itself.

In our case, we use packet sequence numbers to handle reordering. For each destination, a source router maintains sequence numbers (8 bit) which every outgoing packet is tagged with. These sequence numbers are reset at the end of every batch. With this, whenever a cache controller or memory controller receives a request packet to be serviced (in the attached cache bank or memory), the controller inspects the sequence number of the packet. If the sequence number of the packet is *not* the next in sequence to the last serviced packet, then the controller does not service the packet and buffers it in the NIC queues. With this scheme, even if a node (cache or memory controller) receives packets out-of-order, the packets are always serviced in order. Note that reordering in the network could also be introduced due to the use of adaptive routing or deflection routing [10, 11]. Our scheme could leverage any mechanism that handles packet reordering in such systems.

S.4 Application Characteristics

Table 2 characterizes our application suite. The reported parameters are for the applications running alone on the baseline system without any interference. The table shows application characteristics based on network load intensity (high/low), episode height (tall/medium/short), episode length (long/medium/short) and the fraction of execution time spent in network episodes. It is this fraction of execution time spent in the network that is reduced by our scheme which leads to improvement in performance. Energy efficiency benefits come from 1) steering each application to a network that is more appropriately provisioned for the application's demand, 2) prioritizing applications appropriately such that contention that degrades both energy and performance reduces.

S.5 More Results and Analyses

S.5.1 Reply packets from L2 cache (DRAM) to L1 cache (L2 cache)

In all of our earlier evaluations, we route the L2 cache (DRAM) replies to the L1 cache (L2 cache) in either the 64b or the 256b network depending on where the request packet traversed the network: if the request packet was bandwidth-sensitive, the matching reply is sent on the 256b network and vice-versa.

Reply packets are L1/L2 cache line sized packets (1024b). Transmitting them over the 64b network increases their serialization latency. However, the 64b network is relatively less congested than the 256b network (because of lower injection ratio of latency-sensitive applications). Since

Table 2: Application characteristics when run on the baseline (Load: High/Low depending on network injection rate, Episode height: Tall/Medium/Short, Episode length: Long/Medium/Short, Net. fraction: Fraction of execution time spent in network episodes)

#	Benchmark	Load	Episode height	Episode length	Net. fraction
1	applu	Low	Medium	Short	8.2%
2	wrf	Low	Short	Short	9.4%
3	perlbench	Low	Medium	Short	8.8%
4	art	Low	Short	Medium	82.3%
5	dealII (deal)	Low	Short	Short	27.9%
6	sjeng	Low	Tall	Short	28.4%
7	barnes	Low	Medium	Short	72.5%
8	gromacs (grmcs)	Low	Medium	Short	48.6%
9	namd	Low	Medium	Short	51.6%
10	h264ref (h264)	Low	Medium	Short	61.5%
11	calculix	Low	Medium	Short	48.2%
12	gcc	Low	Medium	Short	47.6%
13	povray (pvray)	Low	Medium	Short	59.6%
14	tonto	Low	Tall	Short	53.0%
15	libquantum (libq)	Low	Short	Medium	99.0%
16	gobmk	Low	Medium	Short	64.9%
17	astar	Low	Medium	Short	82.8%
18	milc	Low	Short	Medium	88.2%
19	ocean	Low	Medium	Medium	90.1%
20	hmmer	Low	Medium	Short	66.1%
21	swim	Low	Short	Medium	41.0%
22	sjbb	High	Medium	Medium	87.3%
23	sap	High	Medium	Medium	88.9%
24	xalancbmk (xalan)	High	Tall	Medium	89.9%
25	sphinx3 (sphnx)	High	Tall	Medium	83.9%
26	bzip2 (bzip)	High	Medium	Medium	84.9%
27	lbm	High	Tall	Medium	81.1%
28	sjas	High	Medium	Medium	89.5%
29	soplex (soplx)	High	Medium	Medium	81.2%
30	tpc	High	Medium	Medium	86.8%
31	cactusADM (cacts)	High	Tall	Medium	82.3%
32	leslie3d	High	Short	Long	99.7%
33	omnetpp	High	Medium	Long	92.6%
34	GemsFDTD	High	Tall	Long	97.3%
35	apsi	High	Medium	Long	95.2%
36	mcf	High	Tall	Long	99.2%

the 64b network is clocked at 3X frequency, the network latency in this network is also lower. Our analysis shows that transmitting *all* the reply packets in the 256b network increases the system/instruction throughput by an additional 1.6%/2.4% and reduces energy consumption by an additional 4% when compared to the baseline network. Also, since coherence packets are usually latency-sensitive, our design can route them in the latency-optimized network.

S.5.2 Intra-application latency/bandwidth sensitivity

Figure 13 shows the percentage of packets in an application *dynamically steered* into each network via our mechanism. The data for this figure is collected by averaging 25 workload combinations consisting of 50% latency and 50% bandwidth-sensitive applications (the same workloads used for Figures 8 and 9). The results highlight the dynamic nature of our scheme, where each application phase is classified as either latency or bandwidth-sensitive and all packets belonging to a phase are steered into either of the networks. The figure depicts the intra-application variance: at one extreme is gems which has 99% bandwidth-sensitive packets and on the other extreme is wrf which has 98% latency-sensitive packets. All other applications in our benchmark suite lie within these two extremes.

S.5.3 Sensitivity to fraction of bandwidth- and latency-sensitive applications in the workload

In all results shown so far, each workload is a multiprogrammed mix of applications with equal percentage of latency- and bandwidth-sensitive applications. To analyze the sensitivity of our scheme across more diverse multiprogrammed application mixes, we vary the fraction of bandwidth-sensitive applications in a workload from 0% to 100%. Figure 14 shows the results of this analysis (the results are normalized to the 128b monolithic network). We find that our proposal, in general, has higher system/instruction throughput across the entire spectrum of workload mixes. However, as expected, the benefits are small (4%/9% system/instruction

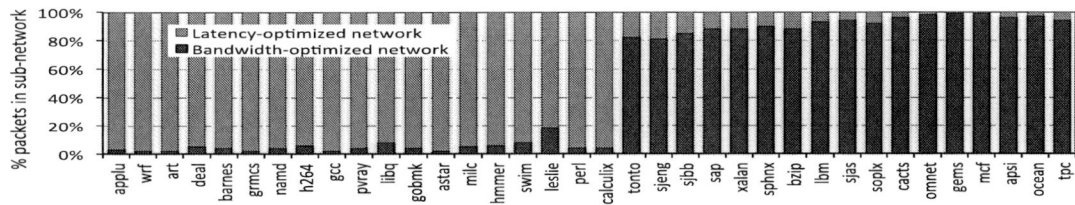

Figure 13: Fraction of packets dynamically steered into latency- vs. bandwidth-optimized networks.

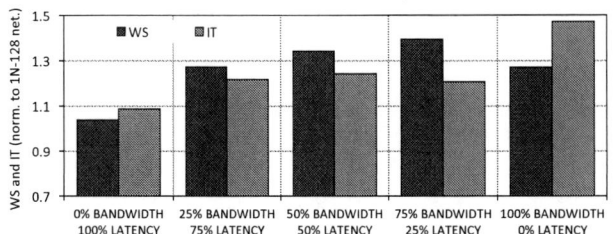

Figure 14: Performance improvement versus the proportion of bandwidth- and latency-sensitive applications in each workload.

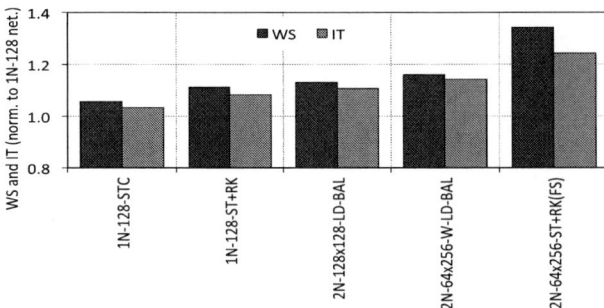

Figure 15: Weighted speedup (WS) and instruction throughput (IT) versus state-of-the art designs [8, 5] (normalized to 1N-128).

throughput improvement over 1N-128 design) when the system has 100% latency-sensitive applications. When the application mix is skewed (i.e., the system has *only* bandwidth *or* latency-sensitive applications), we weighted-load-balanced the two networks (as described in Section S.5.4). As such, with 100% latency-sensitive applications in the workload mix, benefits arise only due to load distribution and reduction in congestion, rather than customization of the network to application requirements (which is the purpose of heterogeneity). We conclude that, even though our design provides performance (and energy) benefits when the application mix is homogeneous in terms of latency- and bandwidth-sensitivity, our heterogeneous network design provides the largest performance and energy improvements when the application mix is heterogeneous.

S.5.4 Comparison with prior works

Das et al. [8] proposed a ranking framework, called stall time criticality (STC), which is based on the dynamic identification of the criticality of a packet in the network. The authors use the L1MPKI metric to estimate the criticality of an application's packets and use a ranking framework that ranks applications with lower L1MPKI over applications with higher L1MPKI. In their work, a *central* decision logic periodically gathers information from each node, determines a global application ranking and communicates this information to each node. Each node then prioritizes packets 1) belonging to the oldest batch over others and 2) in their application rank order within a batch. Since our proposal also prioritizes applications in the network using a rank order, we compare our scheme with STC. When comparing with STC for a single network design, we utilize a 2-level ranking scheme for our technique. The first-level ranking prioritizes latency-sensitive applications over bandwidth-sensitive applications. Within the latency- and bandwidth-sensitive application groups, we use episode width and height to rank the applications (Section 3).

Balfour and Dally [5] showed the effectiveness of load balancing traffic *equally* over two parallel networks. In this work, each of the networks

is a concentrated mesh with similar bandwidth. With detailed layout/area analysis, the authors found that a second network has no impact on the chip area since the additional routers can reside in areas initially allocated for wider channels in the first network. As we also propose parallel networks (although our results in Section 6 show heterogeneous networks are better than homogeneous), we compare our scheme with a similar load-balancing scheme proposed by Balfour and Dally [5].

Figure 15 shows the results, where we compare the performance of our schemes with the two prior proposals mentioned above. All numbers in these plots are normalized to that of a 128b-link network with no prioritization (the link width used in the STC work [8]). The STC schemes are annotated as **-STC** and the load-balancing schemes are annotated as **-LD-BAL** with a given network design in the figure.

The performance improvement of STC is 6%/3% system/instruction in a 128b link monolithic network. Our 2-level ranking scheme shows 11%/8% system/instruction throughput improvement over the 1N-128 design. As shown earlier, L1MPKI, which STC uses, is not a very strong metric to decide the fine-grained latency/bandwidth sensitivity of applications. L1MPKI metric does not take into account the *time* factor, which is taken into account by network *episode* height and length metrics. We found that taking time into account provides a better indication of 1) an application's average MLP (bandwidth demand), 2) how important prioritizing the application will be in relative terms to other applications. Both of these factors are inherently is affected by the notion of *time*. Averaging *episode height* over time, and measuring *episode length* provides better estimators for these two factors than L1MPKI, which leads to the performance benefit with our schemes compared to STC.

In Figure 15, 2N-128x128-LD-BAL design is the one proposed by Balfour and Dally in [5]: *equal* load balancing across two parallel networks. Since we propose heterogeneous networks, when load balancing between two networks in comparison to [5], we steer packets in the weighted-ratio of $\frac{256}{256+64}$ and $\frac{64}{256+64}$ between the 256b and the 64b network. This scheme is annotated as **-W-LD-BAL**. We consider weighted-ratio load balancing, because our evaluations show that, steering packets with *equal* probability into each network leads to more congestion in the 64b-link network and under-utilizes the 256b-link network, leading to suboptimal performance. Overall, we find that our proposal (2N-64x256-ST+RK(FS)) has an additional 18%/10% system/instruction throughput improvement over the weighted load-balancing scheme (2N-64x256-W-LD-BAL). The load balancing scheme is oblivious to application characteristics. With this scheme, a latency-sensitive packet is steered into the bandwidth-optimized network with a probability of 0.8 and a bandwidth-sensitive packet is steered into the latency-optimized network with a probability of 0.2. As a result, in both cases performance either degrades or does not improve. We conclude that, with heterogeneous networks, application-unaware load balancing (either equal or weighted-ratio) is suboptimal and that intelligently steering packets based on the applications' latency versus bandwidth sensitivity can lead to significant performance benefits.

S.6 Related Work

To our knowledge, this is the first proposal for a heterogeneous on-chip network where one network is customized for latency-sensitive workloads and the other for bandwidth-sensitive ones. Our paper is also the first to use the metrics network episode height and length to characterize the latency- and bandwidth-sensitivity of applications. Our proposal is related to other proposals of multiple networks, heterogeneous networks, and request prioritization, which we review below.

Multiple networks: We have already compared our scheme with Balfour and Dally's proposal for homogeneous multiple networks [5], and showed that our scheme is more effective than a (weighted) load balancing scheme. Other prior works that have proposed multiple networks for NoCs include TRIPS [24], RAW [26], Tilera [29], and IBM cell [14, 2]. The motivation for including multiple networks in all these designs is entirely different from ours. In TRIPS, multiple networks are used to connect operand networks. RAW has two static networks (routes specified at compile time) and two dynamic networks (one for trusted and other for untrusted clients). Cell's EIB has a set of four unidirectional concentric rings (arranged in groups of four and interleaved with ground and power shields) primarily to reduce coupling noises. DASH multiprocessor [19] had multiple networks (request and reply meshes) to eliminate request-reply deadlocks. Tilera's iMesh network consists of five separate networks to handle different packet sizes: memory access, streaming packet transfers, user data, cache misses, and interprocess communication. In contrast to all these multiple-network designs, our proposal customizes each network to cater to a different class of applications.

Heterogeneity in networks: HeteroNoC [20], polymorphic NoCs [15] and Kilo-NOC [12] have explored heterogeneity in NoCs from router micro-architecture, topology and QoS perspectives, respectively. HeteroNoC constructs heterogeneous networks by using two types of routers and is agnostic to application properties. Polymorphic NoCs provide per-application network reconfiguration (that needs to be done before the application is run) and incurs a high area overhead to provide performance benefits. Kilo-NOC uses two kinds of routers, QoS-enabled and not QoS-enabled, to provide low cost, scalable and energy-efficient QoS guarantees in a network. Prior work has also investigated co-designing the NoC with caches [6, 4], asymmetric cores [20] and memory controllers [1, 20]. In particular, works in [6, 4] have examined heterogeneous wires with varying width, latency and energy, and proposed mapping coherence messages with differing latency and bandwidth characteristics onto the different wires. Similar to ours, Volos et al. [28] proposes two asymmetric networks, one customized for coherence and short messages and the other for cache block reply packets. Most of these past works have investigated heterogeneity or customization in the network based on micro-architectural (long or short packets) or hardware characteristics (coherence). Our approach is different because it provides an *application-aware* design for heterogeneous networks – as such, we expect our techniques can potentially be combined with these other proposals for heterogeneity.

Request Prioritization: There has been extensive research on prioritizing memory accesses based on their importance to overall application performance and system throughput/fairness [8, 7, 17, 22, 16, 23, 9, 21, 3, 25]. We already presented comparisons of our proposal to [8]. Closely related to our work is Thread Cluster Memory Scheduling (TCM) by Kim et al. [17]. In this work, the authors propose mechanisms to group threads with similar memory access behavior into either latency-sensitive or bandwidth-sensitive clusters and prioritize memory accesses (to DRAM) of the latency-sensitive cluster over those of the bandwidth-sensitive cluster at the memory controller. Our work is related to TCM in the sense that we also exploit latency and bandwidth sensitivity of applications to improve system performance. However, our proposal is different and complementary since it is in the context of NoCs, uses new metrics for classification of applications, and exploits heterogeneous networks.

The concept of compute and non-compute episodes has been used in ATLAS [16]. ATLAS defines *memory* episode length to be the duration for which an application awaits at least one memory request (i.e., L2 miss). ATLAS exploits the notion of memory episodes to prioritize threads with the least attained memory service times. In contrast, we use the concept of *network* episode length, which is the duration for which an application awaits at least one L1 miss. We use network episode characteristics to classify applications into categories. Further, we introduce the notion of network episode *height* to measure an application's MLP, which no previous work has done.

Acknowledgments

We would like to thank the anonymous reviewers, Rachata Ausavarungnirun, and Kevin Kai-Wei Chang for their feedback. This research is supported in part by NSF grants #1213052, #1152479, #1147388, #1139023, #1017882, #0963839, #0811687, #147397, #1212962, #0953246 and grants from Intel. Onur Mutlu is in part supported by an Intel Early Career Faculty Honor Program Award.

References

[1] D. Abts et al. "Achieving Predictable Performance Through Better Memory Controller Placement in Many-Core CMPs". In *ISCA*. 2009.

[2] T. W. Ainsworth et al. "Characterizing the Cell EIB On-Chip Network". In *IEEE Micro* (2007).

[3] R. Ausavarungnirun et al. "Staged Memory Scheduling: Achieving High Performance and Scalability in Heterogeneous Systems". In *ISCA*. 2012.

[4] R. Balasubramanian et al. "Microarchitectural Wire Management for Performance and Power in Partitioned Architectures". In *HPCA*. 2005.

[5] J. Balfour et al. "Design Tradeoffs for Tiled CMP On-Chip Networks". In *ICS*. 2006.

[6] L. Cheng et al. "Interconnect-Aware Coherence Protocols for Chip Multiprocessors". In *ISCA*. 2006.

[7] R. Das et al. "Aergia: Exploiting Packet Latency Slack in On-Chip Networks". In *ISCA*. 2010.

[8] R. Das et al. "Application-Aware Prioritization Mechanisms for On-Chip Networks". In *MICRO*. 2010.

[9] E. Ebrahimi et al. "Parallel Application Memory Scheduling". In *MICRO*. 2011.

[10] C. Fallin et al. "CHIPPER: A Low-Complexity Bufferless Deflection Router". In *HPCA*. 2011.

[11] C. Fallin et al. "MinBD: Minimally-Buffered Deflection Routing for Energy-Efficient Interconnect". In *NOCS*. 2012.

[12] B. Grot et al. "Kilo-NOC: a heterogeneous network-on-chip architecture for scalability and service guarantees". In *ISCA*. 2011.

[13] T. Hastie et al. *The Elements of Statistical Learning (2nd edition)*. Springer-Verlag, 2008.

[14] J. A. Kahle et al. "Introduction to the Cell Multiprocessor". In *IBM J. of Research and Development* (2005).

[15] M. M. Kim et al. "Polymorphic On-Chip Networks". In *ISCA*. 2008.

[16] Y. Kim et al. "ATLAS: A Scalable and High-Performance Scheduling Algorithm for Multiple Memory Controllers". In *HPCA*. 2010.

[17] Y. Kim et al. "Thread Cluster Memory Scheduling: Exploiting Differences in Memory Access Behavior". In *MICRO*. 2010.

[18] W. J. Krzanowski et al. "A Criterion for Determining the Number of Groups in a Data Set Using Sum-of-Squares Clustering". In *Biometrics* 44.1 (1988).

[19] D. Lenoski et al. "The Stanford Dash multiprocessor". In *IEEE Computer* (1992).

[20] A. K. Mishra et al. "A Case for Heterogeneous On-Chip Interconnects for CMPs". In *ISCA*. 2011.

[21] S. P. Muralidhara et al. "Reducing Memory Interference in Multicore Systems via Application-Aware Memory Channel Partitioning". In *MICRO*. 2011.

[22] O. Mutlu et al. "Parallelism-Aware Batch Scheduling: Enhancing both Performance and Fairness of Shared DRAM Systems". In *ISCA*. 2008.

[23] O. Mutlu et al. "Stall-Time Fair Memory Access Scheduling for Chip Multiprocessors". In *MICRO*. 2007.

[24] K. Sankaralingam et al. "Exploiting ILP, TLP, and DLP with The Polymorphous TRIPS Architecture". In *ISCA*. 2003.

[25] L. Subramanian et al. "MISE: Providing Performance Predictability and Improving Fairness in Shared Main Memory Systems". In *HPCA*. 2013.

[26] M. B. Taylor et al. "The Raw Microprocessor: A Computational Fabric for Software Circuits and General Purpose Programs". In *IEEE Micro* (2002).

[27] R. Tibshirani et al. "Estimating the Number of Clusters in a Data Set via the Gap Statistic". In *Journal of the Royal Statistical Society*. 63.2 (2001).

[28] S. Volos et al. "CCNoC: Specializing On-Chip Interconnects for Energy Efficiency in Cache-Coherent Servers". In *NOCS*. 2012.

[29] D. Wentzlaff et al. "On-Chip Interconnection Architecture of the Tile Processor". In *IEEE Micro* (2007).

Designing Energy-Efficient NoC for Real-Time Embedded Systems through Slack Optimization

Jia Zhan[*], Nikolay Stoimenov[†], Jin Ouyang[‡], Lothar Thiele[†],
Vijaykrishnan Narayanan[*], Yuan Xie[*,§]
[*]The Pennsylvania State University, {juz145,vijay,yuanxie}@cse.psu.edu
[†]Computer Engineering and Networks Laboratory, ETH Zurich, {stoimenov,thiele}@tik.ee.ethz.ch
[‡]NVIDIA, jouyang@nvidia.com
[§]AMD Research, yuan.xie@amd.com

ABSTRACT

Hard real-time embedded systems impose a strict latency requirement on interconnection subsystems. In the case of network-on-chip (NoC), this means each packet of a traffic stream has to be delivered within a time interval. In addition, with the increasing complexity of NoC, it consumes a significant portion of total chip power, which boosts the power footprint of such chips. In this work, we propose a methodology to minimize the energy consumption of NoC without violating the pre-specified latency deadlines of real-time applications. First, we develop a formal approach based on network calculus to obtain the worst-case delay bound of all packets, from which we derive a safe estimate of the number of cycles that a packet can be further delayed in the network without violating its deadline—the *worst-case slack*. With this information, we then develop an optimization algorithm that trades the slacks for lower NoC energy. Our algorithm recognizes the distribution of slacks for different traffic streams, and assigns different voltages and frequencies to different routers to achieve NoC energy-efficiency, while meeting the deadlines for all packets.

Categories and Subject Descriptors

C.2 [**Computer-Communication Networks**]: Network Architecture and Design

General Terms

Algorithms, Design

Keywords

Network-on-Chip, Network calculus, Voltage-frequency scaling

1 Introduction

Contemporary embedded systems and SoCs feature an increasing number of processing elements (PE) and other components, a sign that interconnection will play a more vital role in these chips. Network-on-chips (NoC) is a promising design paradigm for future many-core chips as found by many previous researches [1, 7]. However, the fundamental challenge of using NoCs in many-core embedded systems is that these systems often have

---

This work is in part supported by NSF CCF-0903432, CNS-0905365, and SRC grant.

Permission to make digital or hard copies of all or part of this work for personal or classroom use is granted without fee provided that copies are not made or distributed for profit or commercial advantage and that copies bear this notice and the full citation on the first page. To copy otherwise, to republish, to post on servers or to redistribute to lists, requires prior specific permission and/or a fee.
DAC '13, May 29 - June 07 2013, Austin, TX, USA

---

very limited resources and stringent processing latency requirements, which places very different constraints than general-purpose processors on NoC design. There are two major differences between embedded systems and general-purpose processors: 1) General-purpose processors are often designed to achieve a high aggregate throughput, and therefore the NoCs for them are allocated sufficient resources to sustain the peak performance. In contrast, embedded systems are designed to provide *just enough* performance to accommodate specific tasks. Thrift is a virtue in designing NoC for those systems, in order for power and area reduction. 2) General-purpose processors care about the overall progress of all tasks running on all cores. In contrast, embedded systems often provide certain guarantees for individual tasks' progress. In the so-called *hard real-time embedded systems*, to provide certain quality-of-service (QoS), each task has an associated maximum allowed communication delay. Reflected on NoC, each network packet needs to be delivered to the destination before a *deadline*; otherwise the corresponding task may not be able to deliver the required quality-of-service, and even causes catastrophic outcomes.

One way to address the conflicting requirements of energy and latency is to leverage the inherent heterogeneity in NoC traffics, and use voltage frequency scaling (VFS) to improve the energy-efficiency of NoC. A lot of previous work [15,19,21] have adopted DVFS to reduce the energy consumption of NoC while still providing high throughput. Heterogeneity can also be utilized to improve the efficiency of NoC. Das *et al.* [9] was the first to propose the idea of **network slack**, which refers to the number of cycles that a packet can be delayed in the network without affecting execution time. In their work, packets with smaller slacks (those more likely to impact execution time) are prioritized. This slack-based approach improves the throughput of all running tasks. However, the above researches are still focused on designing NoC for general-purpose processors, and aimed at improving the overall throughput. For example, the estimated slack proposed by Das *et al.* [9] does not consider precise deadlines on individual packets and only serves as a hint in assigning priorities to packets. There is no guarantee that a packet will arrive in time before it is needed. Therefore, these approaches cannot be applied to NoC in embedded systems where violating deadlines could be disastrous.

Unlike previous work, we focus on improving NoC energy-efficiency in hard real-time embedded systems by leveraging the heterogeneity in NoC traffics. We propose a design methodology that provides *just enough power* to NoC in order to meet the latency requirements (deadlines) of all traffic streams. Inspired by Das *et al.*'s work [9], we first calculate the worst-case *slacks* for packets of different streams. Then an energy optimization algorithm is proposed that leverages the slacks to allocate differentiated resources (energy) to different portions of the network. Different from their work, the slack calculation must be precise and conservative in order to guarantee the timing-correctness of real-time systems. To solve this problem, we adopt network calculus [3, 4] to predict the worst-case latency of different packets, from which the worst-case slacks can be obtained. Leveraging these slacks, we can progressively reduce the voltages and frequencies of individual routers in the network to reduce energy consumption while still meeting all deadlines. To summarize, the contributions of this work are:

- We develop a formal method based on network calculus to obtain the worst-case slacks of packets in the NoC for hard real-time embedded systems. We improve over previous work [18] by taking virtual channels and heterogeneous router frequencies into consideration.

- We propose an effective algorithm that trades slacks for energy-efficiency of NoC, and thus minimizes the total communication energy while still maintaining timing-correctness. This algorithm adjusts energy and performance by applying voltage and frequency scaling (VFS) to individual routers in the network.

Finally, the voltage-frequency assignments computed using the proposed approach are static, which fits into a design category where a large number of real-time embedded systems fall into. While not considered here, the framework devised in this work can be applied in dynamically reconfigured NoCs by periodically performing voltage/frequency scaling based on run-time network states.

2 Worst-Case Delay Analysis

2.1 Router Architecture

Most of the state-of-the-art NoC researches assume a baseline wormhole router that achieves high energy-efficiency [7]. To provide guaranteed services in NoC, researchers extended the baseline router architecture to either pre-allocate switching time slots for critical packets [10, 11], or preserve a virtual channel for each traffic stream [2, 20]. Pre-allocating switching time slots eliminates run-time contentions altogether, while preserving virtual channels only prevents head-of-line blocking and still needs proper arbitration schemes for performance guarantee.

In this paper, we assume a baseline wormhole router architecture with five router stages (1. **BW**: Buffer Write; 2. **RC**: Routing Computation; 3. **VA**: Virtual Channel Allocation; 4. **SA**: Switch Allocation; 5. **ST**: Switch Traversal). Recently NoC router architectures with fewer stages were also proposed. However, changing the number of router stages affects only the initial latency in our analysis (refer to details below), and therefore our approach can be applied to router architectures with fewer pipeline stages. In addition, we assume that each traffic stream uses a dedicated virtual channel throughout the network, which is in line with the designs proposed by [2, 20]. We do not opt for pre-allocating time-slots for each packet [10, 11], because this approach eliminates the flexibility of scaling voltage and frequency to reduce energy consumption.

In this section, the detailed analytic models for the two types of router pipeline stalls are presented, and the worst-case packet delay bound is derived from these models. Table 1 summarizes symbols used in our modeling and analysis.

2.2 Principles of Network Calculus

Network calculus [3] is a theory of deterministic queuing systems for communication networks. In particular, this approach is based on three important concepts:

Arrival Curve: If $A[s, t]$ denotes the number of packets (here we define a packet as a fixed-length basic unit in network traffics; variable-length packets can be viewed as a sequence of fixed-length packets) that arrive in the time interval $[s, t]$, then we say the flow A is constrained by an arrival curve α if and only if for all $s < t$:

$$A[s, t] \leq \alpha(t - s) \tag{1}$$

Table 1: Symbols used for modeling and analysis

Symbols	Description
α	arrival curve
β	service curve
β^{R_i}	overall service curve of router R_i
$\beta^{R_i'}$	ideal service curve of router R_i without back-pressure
$A[s, t]$	the number of packets that arrive during $[s, t)$
$C[s, t]$	the number of packets that can be processed during $[s, t)$
d_{worst}	worst-case packet delay
s_i	number of slots assigned to flow i in the scheduling model
B	VC buffer size
η	router frequency scaling factor
D_i	deadline constraint for flow i
$\otimes$	min-plus convolution e.g. $a \otimes b = \min_{0 \leq s \leq t}\{a(s) + b(t-s)\}$
$\wedge$	infimum e.g. $a \wedge b = min\{a, b\}$
δ_T	burst delay function $\delta_T = +\infty$ if $t > T$, else 0
$\gamma_{r,b}$	affine arrival curve $\gamma_{r,b}(t) = rt + b$ if $t > 0$, else 0
$\beta_{\lambda,T}$	rate-latency function $\lambda[t - T]^+ = \lambda(t - T)$ if $t > T$, else 0
$\overline{f}$	sub-additive closure $\overline{f} = \delta_0 \wedge f \wedge (f \otimes f) \wedge (f \otimes f \otimes f) \wedge \ldots$

Service Curve: If $C[s, t]$ denotes the number of packets that can be processed by a router or a whole network over the time interval $[s, t)$, and C is bounded by a service curve β if and only if for all $s < t$:

$$C[s, t] \geq \beta(t - s) \tag{2}$$

Delay Bound: Assume a packet stream, constrained by an arrival curve α, traverse a system that offers a service curve β. Then the worst-case packet delay d_{worst} can be bounded as:

$$d_{worst} \leq \sup_{t \geq 0}\{\inf\{\tau \geq 0 : \alpha(t) \leq \beta(t + \tau)\}\} \tag{3}$$

An example is shown in Figs. 1a and 1b for a single router, where we show an affine arrival curve $\gamma_{r,b}$, defined by: $\gamma_{r,b}(t) = rt + b$ for $t > 0$, and $\gamma_{r,b} = 0$ otherwise, and a rate-latency service curve $\beta_{\lambda,T}$, defined by: $\beta_{\lambda,T}(t) = \lambda[t - T]^+ = \lambda(t - T)$ for $t > T$, and $\beta_{\lambda,T} = 0$ otherwise. The arrival curve $\gamma_{r,b}$ implies that the source can send at most b packets at once, but no more than r packets/cycle in the long run, while the service curve $\beta_{\lambda,T}$ implies a pipeline delay T for a packet to traverse a router and an average service rate of λ packets/cycle. As shown in Fig. 1b, the worst-case delay bound d is the maximum horizontal distance between arrival curve and service curve.

When extended to multiple interconnected components, as shown in Fig. 1c, the end-to-end packet delay becomes more unpredictable. Fig. 1d is the worst-case delay analysis for one flow f_2. As we can see, arrival curve α remains as the given injection pattern, while service curve is now the concatenation of all the routers that f_2 traverses from source to destination, which can be calculated through server concatenation [3]. For instance, the concatenation of two routers with service curve β^{R_1} and β^{R_2} is:

$$\beta^{R_{\{1,2\}}} = \beta^{R_1} \otimes \beta^{R_2} = \min_{0 \leq s \leq t}\{\beta^{R_1}(s) + \beta^{R_2}(t - s)\} \tag{4}$$

So far we only consider the case where routers with infinite buffers provide service to a single flow. In reality, the router is designed with a finite buffer size that exerts back-pressure, and many flows may share routers in NoC experiencing reduced service quality. Both factors introduce additional stalls (flow-control stall and switch-contention stall). In the rest of this section, we consider the additional stalls resulted from back-pressure and resource sharing in the worst-case delay analysis.

2.3 Flow-Control Stall

With credit-based flow control [8], the upstream router keeps a count of the number of free buffers in each virtual channel downstream. No packets will be forwarded if their intended

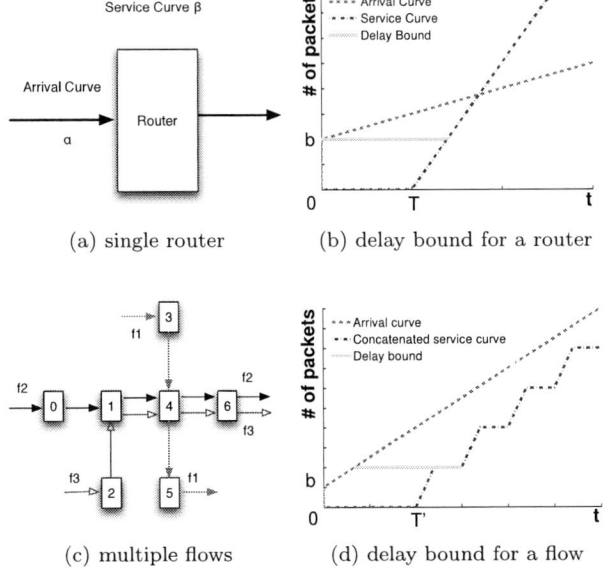

(a) single router   (b) delay bound for a router

(c) multiple flows   (d) delay bound for a flow

Figure 1: Delay bound from network calculus

buffers are full, until the downstream buffer forwards a packet and sends a credit back to the upstream router. Here we adapt Chang *et al.*'s work [5] to derive the worst-case latency bound under the back-pressure of credit-based flow control. For simplicity, we consider two adjacent routers R_1 and R_2 in our demonstration, and the results can be easily applied to the case when more routers are involved. Fig. 2 shows a graphical view of the two-router case.

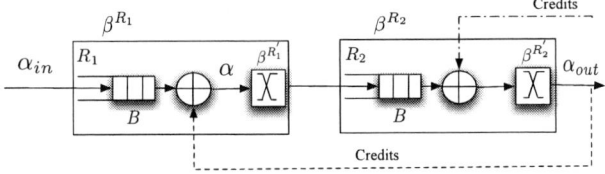

Figure 2: Analysis of credit-based flow control

Let α_{in} be the generic input process to router R_1 while α be the effective input process to the internal crossbar of the router, which is the outcome of both α_{in} and back-pressure. α_{out} is the output process of router R_2. Suppose the overall service curve β^{R_2} of R_2 seen by R_1 is known, and the ideal service curve (without back-pressure) of R_1 is $\beta^{R_1'}$ (provided by the crossbar), then according to [5], the overall service curve β^{R_1} of R_1 considering back-pressure is given as:

$$\beta^{R_1} = \beta^{R_1'} \otimes \overline{(I_B \otimes (\beta^{R_1'} \otimes \beta^{R_2}))} \tag{5}$$

where B is the buffer size, and I_B is defined as $I_B(t) = \infty$ for $t > 0$ and $I_B(0) = B$. The horizontal bar is the operation for sub-additive closure. In this way, we can derive the service curve of each router and then recursively concatenate them based on Equation (4) from destination to source to get the concatenated service curve for all routers along a flow's path.

2.4 Switch-Contention Stall

Packet stall can happen at switch allocation stage, when all front packets in different virtual channels compete for the same crossbar input or output port. Here we model a generic switch arbiter that allocates time slots to different input ports according to their priorities. In Fig. 3a, we show an example where two flows arrive at a router and compete for the same output link to illustrate service curves experienced by each flow.

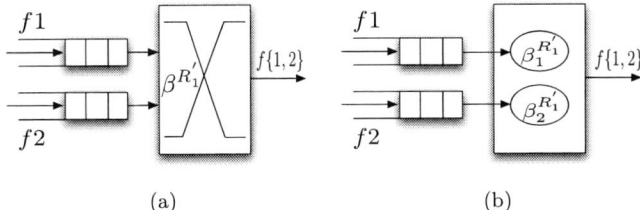

Figure 3: Analysis of switch contention

The length of individual slot $s_i(i = 1, 2)$ assigned for flow f_i is proportional to the relevant priorities of incoming flows and should only take values that are multiples of a cycle length, and the corresponding total cycle length is $s = \sum_i s_i$. Assume the full ideal service curve of the router is β^{R_1}, then the partial ideal service curve for f_i, as shown in Fig. 3b, is proportional to the slot distribution:

$$\beta_i^{R_1'} = \frac{s_i}{s} \beta^{R_1'} \otimes \delta_{s-s_i} \tag{6}$$

where δ_{s-s_i} is the delay bound when stream f_i just missed its slots in the worst case and has to wait for the next round.

As for stream f_i, (5) can be re-written as:

$$\beta_i^{R_1} = \beta_i^{R_1'} \otimes \overline{(I_B \otimes (\beta_i^{R_1'} \otimes \beta_i^{R_2}))} \tag{7}$$

Then by plugging Equation (6) into (7) we can derive the allocated service curve for a specific stream at each router it traversed, and the concatenated service curve according to (4). Finally, the worst-case delay bound can be obtained by applying the principles of network calculus. In this paper, we assume a ***round-robin*** arbiter for every router in the network, which implies that *the priorities of each flow are proportional to their arrival rates to a specific router*. While not considered here, such results can be obtained in a similar way for other scheduling policies like Fixed Priority (FP), Rate Monotonic (RM), and Earliest Deadline First (EDF).

3 Slack Optimization for Saving Energy

Applying the methodology in the previous section, we are able to bound the worst-case packet delay for individual application streams, hence obtaining *worst-case slack*—the time interval between the delay bound and its pre-specified transmission deadline, which indicates the number of cycles that a packet can be further delayed in the network. Based on this idea, we propose an energy-aware voltage and frequency scaling approach to optimize network energy-efficiency under deadline constraints.

3.1 Frequency and Voltage Scaling

The existence of packet slack implies that we can still achieve the required performance while lowering the operating frequency of some routers instead of making them run at a homogeneous high speed. The supply voltage, in the meantime, can be reduced together with the frequency to reduce energy.

Voltage-frequency islands (VFI) [14] have been adopted for achieving fine-grain system-level power management. A fine granularity partitioning could assume that each module in the design belongs to a different island [16] for best flexibility, or find the optimum partitioning via island merging [17] for energy savings. Here we are doing static voltage-frequency assignment and model these routers to be able to run at its own voltage and frequency. For completeness, we also explore the energy gain of operating all routers at homogeneous voltage and frequency.

In order for the network routers to operate at different frequencies, they should communicate in a Globally Asynchronous

259

Locally Synchronous (GALS) mechanism. To address synchronization latency, we adopt the fast synchronizer proposed by Dally $et\ al.$ [6] which adds only half cycle of synchronization delay that can be well absorbed in the buffer write stage.

Note that if we assign a lower frequency to a router, its service curve is affected accordingly. Specifically, the packet service rate will decrease and the time it takes to traverse a router will increase. For example, for a router R_k with rate-latency service curve $\beta_{\lambda,T}^{R_k}(t) = \lambda[t-T]^+$ as discussed in Section II, when its operating frequency is scaled by a factor η, its new service curve is modeled as:

$$\beta_{\eta\lambda,T/\eta}^{R_k}(t) = \eta\lambda[t - T/\eta]^+ \qquad (8)$$

Then the worst-case packet delay should be updated to check if there is remaining slack time for further optimization. Our energy optimization algorithm is described below in detail.

3.2 Energy-Aware Heuristic Search Algorithm

There are two straightforward ways to trade slack for energy savings. One is to simply scale down all the routers simultaneously by the same factor, which will keep them running at homogeneous frequency and voltage. The other is through exhaustive search to find out the optimum assignment. However, the former approach is not flexible enough for adjustment, because the degree to which the network speed can be reduced is limited by the packet flow with minimum slack. The latter is time-consuming and not scalable for a large network with multiple voltage-frequency levels.

Therefore, we propose an energy-aware heuristic search (EHS) algorithm, which can find an efficient solution leveraging network heterogeneity and avoiding exhaustive search at the same time. Specifically, we abstract a NoC energy model and integrate it into our worst-case delay analysis framework to automatically generate the frequency-voltage assignments.

3.2.1 Energy Models

The set of nodes in the network is denoted by $T = \{0, 1...N-1\}$. The supply voltage-frequency pairs of each node $i \in T$ are given by (V_i, f_i). Then the sum of dynamic and static energy consumption associated with node i is:

$$E(V_i, f_i) = E_d(V_i, f_i) + E_s(V_i, f_i) \qquad (9)$$

The dynamic energy part can be calculated through:

$$E_d(V_i, f_i) = M_i * E_p(V_i, f_i) \qquad (10)$$

where M_i is the total number of packets that traverse node i during execution, and $E_p(V_i, f_i)$ is the energy consumption when a packet traverses node i:

$$E_p(V_i, f_i) = E_{buffer} + E_{switch} + E_{link} \qquad (11)$$

where E_{buffer}, E_{switch}, and E_{link} represent the energy dissipated at input buffers, switch and link and are found experimentally using ORION 2.0 [13].

The static energy $E_s(V_i, f_i)$ part is defined as:

$$E_s(V_i, f_i) = P_s(V_i, f_i) * t \qquad (12)$$

where t is the system execution time, while $P_s(V_i, f_i)$ is static power for node i and can be obtained as:

$$P_s(V_i, f_i) = I_{static}^i * V_i \qquad (13)$$

where I_{static}^i is the leakage current for node i and can also be extracted from ORION 2.0 [13].

Thus, combining Equation (9), (10) and (12), the total NoC energy consumption for an application can be expressed as:

$$E = \sum_{i=0}^{N-1} (M_i * E_p(V_i, f_i) + P_s(V_i, f_i) * t) \qquad (14)$$

3.2.2 Algorithm Description

If node i is scaled from the current voltage-frequency level (V_i^k, f_i^k) to the next lower level (V_i^{k+1}, f_i^{k+1}), then energy reduction can be expressed as:

$$\Delta E_i = \sum_{i=1}^{N} (M_i * (E_p(V_i^k, f_i^k) - E_p(V_i^{k+1}, f_i^{k+1}))$$
$$+ P_s(V_i^k, f_i^k) * t^k - P_s(V_i^{k+1}, f_i^{k+1}) * t^{k+1}) \qquad (15)$$

Under deterministic routing, the only undetermined variable is the system execution time t^k. Assume the set of application streams is denoted by $S = \{s_1...s_m\}$. The number of packets injected by s_j is M_{s_j} with average injection rate r_{s_j}. Then t^k can be approximated from the slowest stream:

$$t = \max_{s_j \in S}\left\{M_{s_j}/r_{s_j}\right\} \qquad (16)$$

This is because the end-to-end packet delay is negligible compared to t, especially when the packet number is large.

At the same time, the service curve β^{R_i} is modified based on Equation (8) if scaling i and the new stream delay d_{s_j} is calculated via the worst-case delay analysis in section II. The accumulated slack cost after scaling is represented as:

$$\Delta d_i = \sum_{s_j \in S} \Delta d_{s_j} \qquad (17)$$

where Δd_{s_j} is the reduced slack for stream s_j.

Our heuristic search algorithm uses $\Delta d_i/\Delta E_i$ as a measure of the slack cost and the energy gain if we adjust the voltage-frequency level of a router i. The pseudo-code below outlines this algorithm. Specifically, the algorithm iterates through all routers in the network. At each iteration, it generates a list containing the slack costs and the energy gains for all routers in the network, picks the router i with lowest $\Delta d_i/\Delta E_i$ without causing deadline violations, and steps down the router's voltage and frequency. Finally, the algorithm terminates when there is no router in the network of which the voltage and the frequency can be further reduced without causing timing violations.

Algorithm 1: Energy-aware heuristic search algorithm

Result: Frequency-voltage assignment for all nodes
Initialize: Flag = 0;
while *Flag == 0* **do**
    Flag = 1;
    for $i \leftarrow 0$ to $N-1$ **do**
        if f_i *is at its lowest level* **then**
            continue;
        else
            /* worst-case delay analysis   */
            Calculate $d_{s_j} (\forall s_j \in S)$, Δd_i and ΔE_i if scaling down f_i by one level, and insert them into a list L as one element ;
        end
    end
    Sort L in ascending order of $\Delta d_i/\Delta E_i$, associated with the original index i;
    for *each entry in the list L* **do**
        if $d_{s_j} < D_{s_j} (\forall s_j \in S)$ **then**
            Scale down f_i by one level; Flag = 0; **break**;
        end
    end
end

For an N-node NoC with k voltage-frequency levels for each node, the algorithm complexity of our EHS algorithm is $(k-1)*N^2logN$, compared to k^N for exhaustive search.

4 Experiments

4.1 Experimental Setup

We implemented a cycle-accurate network simulator based on the booksim 2.0 simulator [12], with dynamic and leakage power numbers extracted from ORION 2.0 [13] .

We also analyze the timing behavior of some video applications and characterize the arrival curves for packet streams. Table 2 shows the simulator and benchmark configurations.

Table 2: Simulation Parameters

Baseline Network Configuration			
Topology	2D mesh	Phit width	128bits
Size	4×4=16	Frequency	2GHz
VC #	3	Voltage	1.5v
Buffer depth	4 flits	Routing	dimension-order
Video Application Configuration			
MJPEG		PiP (HR)	PiP (LR)
Frame	352×240	Frame	704×576
Period	90,000	PE service	226.57
Throughput	307.2KB	Macroblock #	1584
JPEG size	8Kb	Rate: 25 frames/s	Rate: 12.5 frames/s
Deadline Constraint (cycle)			
$D_1 = 50$		$D_2 = 95$	$D_3 = 50$

Motion-JPEG (MJPEG) decoder: The MJPEG decoder is a video codec in which each video frame is compressed as a JPEG image. The video of 352×240 pixels is spilt into JPEG image size of 8 Kb. The maximum throughput is 307.2 KB per invocation with a period of 90,000 cycles.

Picture-in-picture (PiP): We use two sets of video clips: Regular clips with moderate to high motion content and clips displaying still images. These two sets characterize the two streams high-resolution (HR) and low-resolution (LR). Incoming streams have the same frame resolution of 704×576 pixels but will be down-scaled for LR, and each frame consists of 1584 macroblocks. Frames are read at a constant rate of 25 frames/s for HR and 12.5 frames/s for LR. The service offered by a processing element is 226.57 macroblocks/ms.

A JPEG image or a macroblock is treated as a packet, and we derive arrival curves for the three packet streams:

$$\text{MJPEG stream } f_1: \ \alpha_1(t) = 0.218t + 3.0 \quad (18a)$$

$$\text{PiP HR stream } f_2: \ \alpha_2(t) = 0.175t + 13.109 \quad (18b)$$

$$\text{PiP LR stream } f_3: \ \alpha_3(t) = 0.086t + 4.37 \quad (18c)$$

And the baseline service curve is shown in (19) for a generic wormhole NoC router that can process one packet per cycle with a total pipeline length of five cycles.

$$\text{Baseline router: } \beta(t) = [1.0 \times t - 5]^+ \quad (19)$$

As a case study, we consider that the three application streams are mapped in a 4×4 mesh network shown in Fig. 4a with deterministic routing. Fig. 4b shows the resource sharing, including feedback loops in stream f_1 as an example. A detailed network configuration is shown in Table 2.

4.2 Experimental Results

We use the methodology in Section II to analyze the worst-case latency in our case study, and results are shown in Fig. 5.

At the same time, we run simulation to get the maximum packet latency for individual streams. A comparison between the calculated worst-case delay bound (solid lines) and the simulated maximum packet latency (dashed lines) when varying virtual channel buffer size B is shown in Fig. 6a. We can see that the calculated delay bounds are fairly tight.

Apart from the case study with three applications streams, we duplicate each of the three sample streams to generate more streams and map them on the NoC platform to form different traffic scenarios. The whole process is conducted randomly.

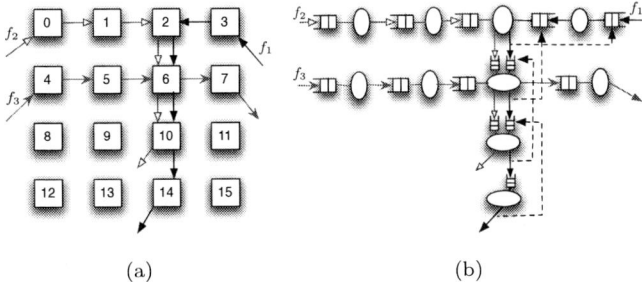

(a) (b)

Figure 4: Case study: Three video streams

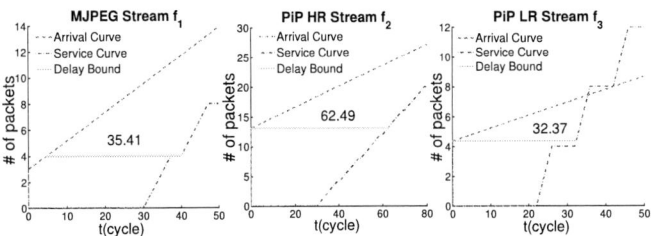

Figure 5: Delay bounds for three flows

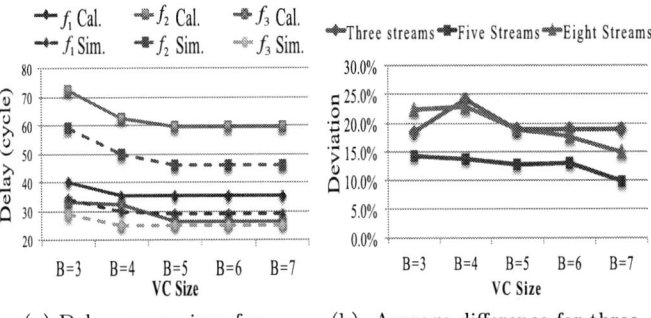

(a) Delay comparison for three streams

(b) Average difference for three, five and eight streams

Figure 6: Calculated (*Cal.*) vs simulated (*Sim.*) delay bound

Here we consider running another five streams (Two MJPEG, two PiP HR, one PiP LR) and eight streams (Three MJPEG, three PiP HR, two PiP LR). Similarly, we do worst-case delay analysis to derive the delay bound and run simulation to get the maximum packet latency. Due to space limitation, we only show the differences between calculated delay bounds and simulated results as the ratio with respect to simulated results, when buffer size varies from 3 to 7, as shown in Fig. 6b.

We find that the average difference is 17.2% while the ratio is generally decreasing as VC size increases. This is because with a small VC size, the effect of back-pressure is more salient and the results from our analytical model is more pessimistic. With VC size increased, the effect of back-pressure is smoothed out, resulting in a tighter delay estimate that converges with the simulation results.

Furthermore, we perform the proposed EHS algorithm to reduce the energy of routers, assuming that three discrete frequencies are available (1.0, 1.5 and 2.0 GHz), and the minimum required voltage is 0.8, 1.2 and 1.5 volts in 45nm CMOS, respectively. As a comparison, we also evaluate homogeneous scaling (*Homo*) in which case the frequency-voltage level of the routers in the whole network is scaled together. Results of the worst-case delay bounds and the normalized total network energy consumption are shown in Fig. 7, where we refer to the

jth duplicate of the ith sample stream as f_i^j.

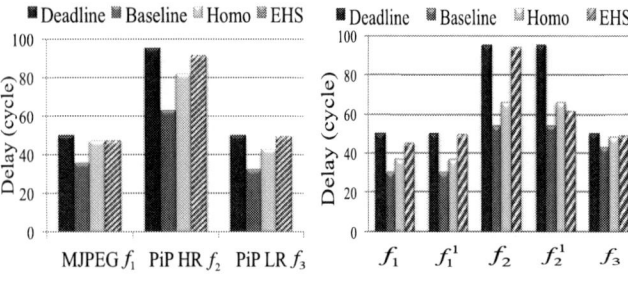

(a) Delay bounds for three streams

(b) Delay bounds for five streams

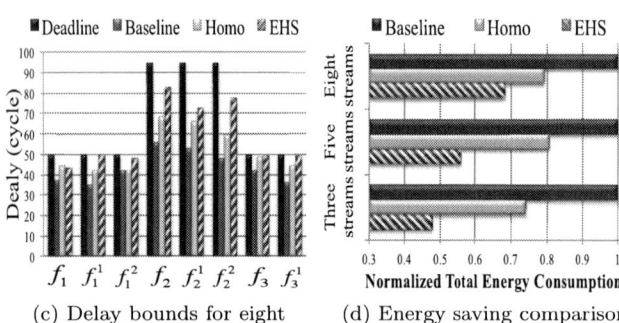

(c) Delay bounds for eight streams

(d) Energy saving comparison

Figure 7: Worst-case delay bound after frequency scaling and energy saving comparison for three, five and eight streams

As we can see from Figs. 7a to 7c, there is no deadline miss using any of these scaling mechanisms. We define *slack utilization* as the amount of slack scavenged by the algorithm to save energy, divided by the amount of initial slack under baseline configuration. Our proposed *EHS* algorithm has effectively exploited individual stream slack, making the completion time (delay) close to the deadline with a slack utilization of 80.7% on average. In contrast, *Homo* only has a slack utilization of 53.9% on average.

As for energy optimization, shown by Fig. 7d, our proposed *EHS* mechanism significantly outperforms *Homo*. On average, *EHS* achieves 42.7% energy reduction, while *Homo* saves 22.0%. These results confirm the effectiveness of our energy optimization algorithm. *EHS* can efficiently utilize the available slack of individual application streams for energy optimization, while for *Homo* case, it cannot exploit slack in fine granularity and therefore leads to relatively poor energy savings.

In addition, for sensitivity evaluation, we apply the proposed EHS algorithm with different slack ratios available. Slack ratio is the amount of slack over baseline worst-case latency. As shown in Fig. 8, Our EHS algorithm works well under different slack ratios, saving energy by 23.1%, 31.1%, 38.4%, 52.0%, and 59.7%, respectively.

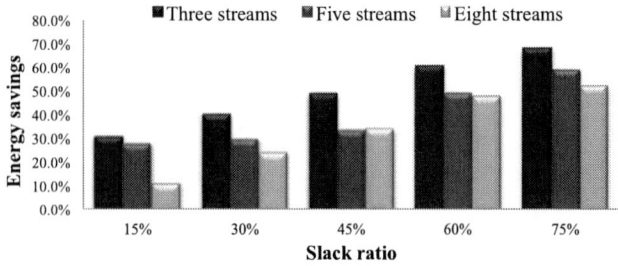

Figure 8: Energy savings when slack varies

5 Conclusion

In this paper, a formal analysis based on network calculus is adopted to obtain the worst-case slacks of packets in the NoC for hard real-time embedded systems, and used to trade slacks for energy savings by applying different voltages and frequencies to individual routers. Experimental results show that our worst-case delay analysis can derive a upper bound for packet latency, and our energy-aware heuristic search algorithm can effectively find the frequency-voltage assignment that can reduce network energy significantly under variable slack ratios.

6 References

[1] L. Benini and G. D. Micheli. Networks on chips: A new SoC paradigm. *IEEE Computer*, 35(1):70–78, 2002.

[2] T. Bjerregaard and J. Sparso. A router architecture for connection-oriented service guarantees in the MANGO clockless network-on-chip. In *DATE*, pages 1226–1231, 2005.

[3] J.-Y. L. Boudec and P. Thiran, editors. *Network Calculus: A Theory of Deterministic Queuing Systems for the Internet*. Lecture Notes in Computer Science. Springer-Verlag, Berlin, Heidelberg, 2001.

[4] S. Chakraborty, S. Kunzli, and L. Thiele. A general framework for analysing system properties in platform-based embedded system designs. In *DATE*, pages 190–195, Munich, Germany, 2003.

[5] C.-S. Chang, editor. *Performance Guarantees in Communication networks*, pages 78–83. Springer-Verlag, London, UK, 2000.

[6] W. Dally and S. Tell. The even/odd synchronizer: A fast, all-digital, periodic synchronizer. In *ASYNC*, pages 75–84, 2010.

[7] W. J. Dally and B. Towles. Route packets, not wires: On-chip interconnection networks. In *DAC*, pages 684–689, 2001.

[8] W. J. Dally and B. Towles, editors. *Principles and Pracitices of Interconnection Networks*, pages 245–247. Morgan Kaufmann Publishers Inc., San Francisco, CA, USA, 2003.

[9] R. Das, O. Mutlu, T. Moscibroda, and C. R. Das. Aergia: exploiting packet latency slack in on-chip networks. In *ISCA*, pages 106–116, 2010.

[10] K. Goossens, J. Dielissen, and A. Radulescu. Æthereal network on chip: concepts, architectures, and implementations. *Design & Test of Computers, IEEE*, 22(5):414–421, 2005.

[11] K. Goossens and A. Hansson. The Æthereal network on chip after ten years: Goals, evolution, lessons, and future. In *DAC*, pages 306–311, 2010.

[12] N. Jiang, G. Michelogiannakis, D. Becker, B. Towles, and W. J. Dally. *BookSim 2.0 User's Guide*. Standford University, 2010.

[13] A. B. Kahng, B. Li, L.-S. Peh, and K. Samadi. ORION 2.0: A power-area simulator for interconnection networks. *IEEE Trans. on VLSI*, pages 191–196, 2012.

[14] D. Lackey, P. Zuchowski, T. Bednar, D. Stout, S. Gould, and J. Cohn. Managing power and performance for system-on-chip designs using voltage islands. In *ICCAD*, pages 195–202, 2002.

[15] A. K. Mishra, R. Das, S. Eachempati, R. Iyer, N. Vijaykrishnan, and C. R. Das. A case for dynamic frequency tuning in on-chip networks. In *MICRO*, pages 292–303, 2009.

[16] K. Niyogi and D. Marculescu. Speed and voltage selection for GALS systems based on voltage/frequency islands. In *ASPDAC*, pages 292 – 297, 2005.

[17] U. Ogras, R. Marculescu, D. Marculescu, and E. G. Jung. Design and management of voltage-frequency island partitioned networks-on-chip. *IEEE Trans. on VLSI*, 17(3):330–341, 2009.

[18] Y. Qian, Z. Lu, and W. Dou. Analysis of worst-case delay bounds for best-effort communication in wormhole networks on chip. In *NOCS*, pages 44–53, 2009.

[19] L. Shang, L.-S. Peh, and N. K. Jha. Dynamic voltage scaling with links for power optimization of interconnection networks. In *HPCA*, pages 91–102, 2003.

[20] A. Sharifi, H. Zhao, and M. Kandemir. Feedback control for providing QoS in NoC based multicores. In *DATE*, pages 1384–1389, 2010.

[21] P. Zhou, J. Yin, A. Zhai, and S. S. Sapatnekar. NoC frequency scaling with flexible-pipeline routers. In *ISLPED*, pages 403–408, 2011.

RISO: Relaxed Network-on-Chip Isolation for Cloud Processors

Hang Lu[†‡], Guihai Yan[†], Yinhe Han[†‡], Binzhang Fu[†] and Xiaowei Li[†‡]

[†]State Key Laboratory of Computer Architecture, Institute of Computing Technology,
Chinese Academy of Sciences, Beijing, China
[‡]University of Chinese Academy of Sciences, Beijing, China
{luhang, guihai_yan, yinhes, fubinzhang, lxw}@ict.ac.cn

ABSTRACT

Cloud service providers use workload consolidation technique in many-core cloud processors to optimize system utilization and augment performance for ever extending scale-out workloads. Performance isolation usually has to be enforced for the consolidated workloads sharing the same many-core resources. Networks-on-chip (NoC) serves as a major shared resource, also needs to be isolated to avoid violating performance isolation. Prior work uses strict network isolation to fulfill performance isolation. However, strict network isolation either results in low consolidation density, or complex routing mechanisms which indicates prohibitive high hardware cost and large latency. In view of this limitation, we propose a novel NoC isolation strategy for many-core cloud processors, called *relaxed isolation (RISO)*. It permits underutilized links to be shared by multiple applications, at the same time keeps the aggregated traffic in check to enforce performance isolation. The experimental results show that the consolidation density is improved more than 12% in comparison with previous strict isolation scheme, meanwhile reducing network latency by 38.4% on average.

Categories and Subject Descriptors

C.1.2 [**Multiple Data Stream Architectures (Multiprocessors)**]: Interconnection architectures; C.1.4 [**Parallel Architectures**]: Distributed architectures; D.4.1 [**Process Management**]: Concurrency, Threads

General Terms

Performance, Design, Algorithms

Keywords

Networks-on-Chip, Cloud Computing, Cloud Processors, Workload Consolidation, Performance Isolation, Relaxed Isolation

1. INTRODUCTION

Cloud computing has emerged as a fundamental platform to deploy increasing on-line services like web searching, social networks and so forth. To tackle expanded power and space consumption brought by the growing cloud infrastructure, the processor with tens even hundreds of cores is believed to play a critical role in the coming cloud computing era, or simply called many-core "cloud processors". Intel's 48-core "Single-chip Cloud Computer" [1] and Tilera's "TILE-Gx

Permission to make digital or hard copies of all or part of this work for personal or classroom use is granted without fee provided that copies are not made or distributed for profit or commercial advantage and that copies bear this notice and the full citation on the first page. To copy otherwise, to republish, to post on servers or to redistribute to lists, requires prior specific permission and/or a fee.
DAC '13, May 29 - June 07, 2013, Austin, Texas, USA.

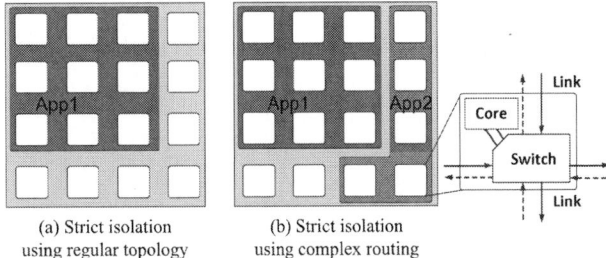

(a) Strict isolation using regular topology (b) Strict isolation using complex routing

Figure 1: Two traditional schemes of strict isolation

3000 Series" [2] are two representatives. The cloud processor chips can accommodate abundant parallelism for concurrent **scale-out workloads**, which demand various amounts of computational resources and will be scheduled back and forth by the data center software [3][4]. For such cloud computing platform, one key optimization is to reduce the TCO (total cost of ownership) to avoid low hardware utilization by aggressive "workload consolidation" technique [5][6]. In such computing paradigm, **performance isolation** between scale-out workloads has to be enforced to provide controllable QoS and priority-based services which are critical for cloud computing service providers, such as Google and Amazon.

Performance isolation imposes two orthogonal requirements: 1) Computation isolation, i.e. the computing cores, and the associate cache, memory bandwidth etc., should not be preempted by other workloads in uninterruptable computing sessions [6][7]. 2) Communication isolation, or network isolation, i.e. the on-chip network (NoC) traffic of different workloads should not block each other because the performance of many parallel applications are highly sensitive to network latency [8]. Our work focuses on communication isolation, the same important but less extensively studied topic than computation isolation.

Communication isolation strategies involve making tradeoffs between the regularity of network topology and complexity of routing mechanism. Usually, a regular topology, e.g. rectangular-shaped network, has more efficient routing algorithm, but lower server consolidation density, hence less hardware utilization. By contrast, enabling the network isolation to support more flexible topologies, thereby achieving high consolidation density, will inevitably complicate the routing mechanism. This tradeoff can be further explained with the following example.

Figure 1(a) shows the principle of regular-shaped isolation. The 16-core processor accommodates one application (App1) mapped into a rectangular-shaped region. However, the consolidation density is poor to meet the regular shape requirement: there are seven routers which have to be idle although the incoming application (App2) only needs a five-router network. To relax this limitation, a viable solution is to enable the network isolation to support irregular shapes through implementing flexible routing mechanisms such as Up*/Down* [9] or table-based routing [10], as Figure 1(b) shows. However, the cost of employing those complex routing can hardly

263

be justified for cloud processors given the prohibitive TCO and sporadic performance variations.

Previous work fails to resolve the above contradiction between consolidation density and complexity of routing mechanisms. The reason is that those schemes follow the concept of "**Strict Isolation**", i.e. resorting to strict physical isolation to achieve performance isolation. However, enforcing such strict rule is quite conservative and often leads to overdesign. The on-chip routers and links are often heavily underutilized, especially those on the application region boundaries. By judiciously sharing some routers and links, we can still achieve performance isolation without nailing down to strict isolation. This paper thereby proposes the concept of "**Relaxed Isolation (RISO)**" for many-core cloud processors. Following this concept, we can achieve the consolidation density of schemes supporting irregular topology, but employ the same efficient routing mechanism as in schemes using regular topology. In particular, this paper makes the following contributions:

• *We propose relaxed NoC isolation scheme enforcing workload performance isolation.* We find the traditional strict isolation is highly conservative for performance isolation, which impairs the consolidation density. RISO does not resort to physical isolation, but permits conditional network resource sharing without violating performance isolation.

• *We propose an application mapping algorithm to maximally exploit the potential of RISO.* This algorithm is fully with respect to performance isolation by preventing overlaid traffic exceeding a safe threshold. Also, irregular-shaped regions, which are wasted in previous work to compromise with routing complexity, are also taking into consideration to further improve consolidation density.

The rest of this paper is organized as follows. Section II describes the motivation of RISO by further specifying the limitation of traditional strict isolation schemes. Section III presents the key algorithms to implement RISO. Section IV shows experiment setup and results, followed by related work in Section V and conclusion in Section VI.

2. MOTIVATION

2.1 "Strict Isolation" Degrading Consolidation Density

Improving consolidation density is an effective approach to reduce the TCO of data centers for cloud service providers. In the future, the effectiveness of consolidating multiple workloads into a single many-core cloud processor is just like today's sever consolidation across racks in data centers [5]. However, enforcing performance isolation would be more challenging in cloud processors because the on-chip network is much less flexible than its traditional off-chip counterpart. In particular, the routing mechanism for large-scale many-core system usually follows dimensional order routing (DOR) which is table-less and tailored for on-chip networks [11]. DOR imposes network topology constraints for every workload. By the doctrine of "strict isolation", the consolidation density will be degraded. For example, Figure 2(a) shows dynamic application mapping over time. In this example, App1, App2, ⋯, and App5 have already been mapped into the system at that time. The remaining free cores constitute a contiguous but irregular region. Supposing a ten-core workload, App6, is waiting to be served. However, the DOR topology constraint renders this application fail to be mapped because a regular rectangle shape cannot be found under "strict isolation", although there are still 16 free cores which is more than App6 required. Furthermore, the block of App6 probably prevents the subsequent applications, for example App7, from execution, which further degrades the consolidation density.

2.2 Breaking Strict Isolation — RISO

We propose *Relaxed Isolation (RISO)* to tackle the limitation of *strict isolation*. We find that using strict isolation in on-chip networks to ensure communication isolation is highly

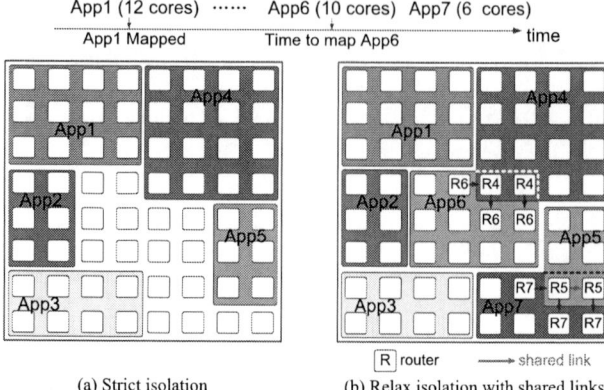

(a) Strict isolation   (b) Relax isolation with shared links

Figure 2: Application mapping on strict isolation and RISO

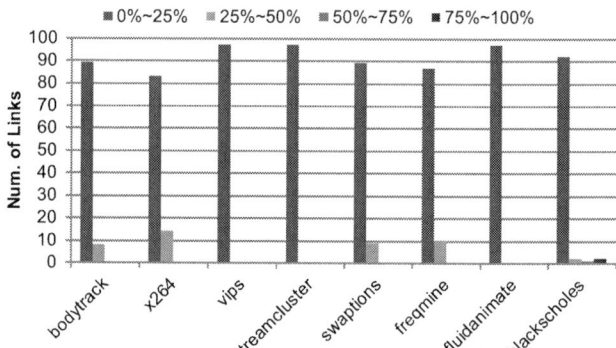

Figure 3: Link utilization distribution

conservative. Since our ultimate goal is performance isolation, the App6 can be mapped into the irregular regions as long as the aggregated traffic on the "overlapped" routers and links won't degrade the latency of each other. As Figure 2(b) shows, by permitting the router and link sharing — RISO, both App6 and subsequent App7 can be served without delay, hence improving the consolidation density.

The rationale behind RISO is to exploit under-utilized routers and links. We find that low NoC utilization is not uncommon in reality, which sufficiently justifies the concept of RISO. The utilization (U) of a link within N link cycles is calculated by Eq.(1) [12].

$$U = \frac{\sum_{i=0}^{N} \alpha(i)}{N} \qquad \alpha(i) = \begin{cases} 0 & \text{Link is } idle \text{ at cycle } i; \\ 1 & \text{Otherwise.} \end{cases} \quad (1)$$

We survey the link utilization of a set of workloads. The result is shown in Figure 3, represented by "histgram". The link utilization is divided into four ranks: 0~25%, 25~50%, 50~75%, 75~100%. The result shows that the lowest rank, 0%~25%, dominates in all applications. Very few links can reach up to the second rank, without mention the third and forth ranks. Hence, such severely low utilization should provide a unique opportunity for RISO.

RISO won't violate communication isolation as long as the aggregated traffic is kept below a certain threshold, called "congestion point" in this paper, as Figure 4 shows. We find that network latency starts to increase only when the link utilization increases beyond the congestion point. This trend applies to various traffic patterns. Experimental study shows the congestion point is application-independent and at the link utilization around 65%, which agrees with prior work [13]. Given that link utilization is usually much less than 25% in reality, we can safely conclude that the minority of link sharing in RISO won't cause obvious latency increase,

264

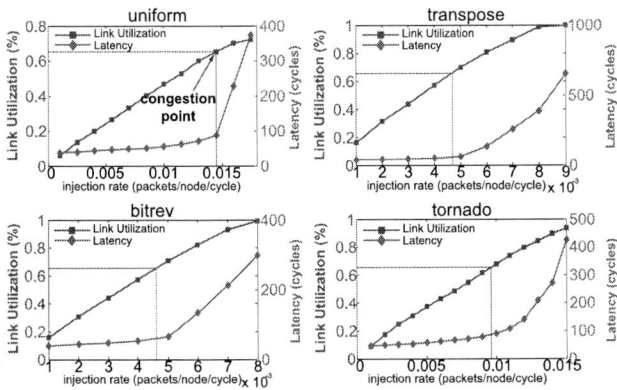

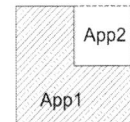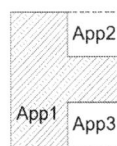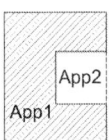

Figure 5: RISO supported shapes

Figure 4: Link utilization and latency under various traffic patterns

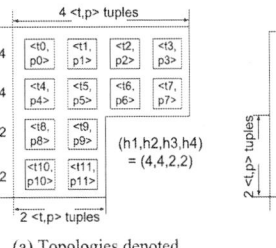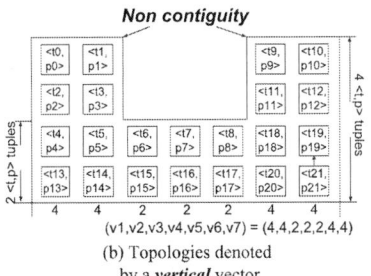

(a) Topologies denoted by a *horizontal* vector

(b) Topologies denoted by a *vertical* vector

Figure 6: Shapes represented by horizontal (H) or vertical vector (V), and thread organization

therefore keeping the performance isolation intact.

In this paper, we confine our scope to many-core cloud processors connected with "Mesh" networks, which has been proved with superior scalability. But the concept of relaxed isolation is applicable to other types of on-chip networks.

3. APPLICATION MAPPING ALGORITHM BASED ON RISO

Mapping application into a many-core system is to assign the application a specified number of physical cores whose network is organized to the routing-allowed topology, meanwhile without violating the criteria of performance isolation. The basic mapping procedure can be divided into two steps: 1) Search for the physical core groups (regions) which meet the topology requirement under the condition of relaxed NoC isolation, instead of strict isolation; 2) Verify the performance isolation criteria by checking for the utilization of shared links carrying the overlapped traffic. The following will delve into the details of the proposed mapping algorithm.

3.1 Topology Representation

For some applications with intensive intra-application communications, the performance and power consumption can be topology-specific [14][15]. We therefore assume each application to be mapped has a *preferred topology*, which serves as an input to our mapping algorithm. An application's preferred topology incorporates two unique characteristics: *physical shape* and *threads organization*. The proposed algorithm first searches for the candidate regions in NoC that can accommodate the preferred physical shape. To maximize the consolidation density, the mapping algorithm should be capable to handle not only regular shape, i.e. rectangle, but also various irregular shapes ignored in previous works [16]. We abstract those irregular shapes into the following three basic types: "L", "⊢", "⊏", as Figure 5 shows, with various rotations and mirrors.

To represent the physical shapes in Figure 5, we define *horizontal vector*, $H[h_1, h_2, h_3, \cdots, h_n]$, and *vertical vector*, $V[v_1, v_2, v_3, \cdots, v_n]$, where h_n and v_n is the number of cores in the nth row and column respectively in the preferred topology. H applies to shapes that exhibit complete contiguity in horizontal dimension. For instance, Figure 6(a) shows a "L" shape: the 1st and 2nd row each requires 4 cores; the 3rd and 4th row each requires 2 cores and every row is contiguous; hence this shape is represented by the horizontal vector $H[4, 4, 2, 2]$. However, for the shape that is non-contiguous horizontally but contiguous vertically, we use a vertical vector as Figure. 6(b) shows. Its corresponding vertical vector is $V[4, 4, 2, 2, 2, 4, 4]$. The other irregular shapes both H and V cannot describe are beyond the scope of this paper.

For the multiple threads of an incoming application, we use the tuple $\langle t_i, p_j \rangle$ to indicate each thread and its corresponding position in the physical shape. Specifically, parameter t_i

and p_j means that the ith thread is located at the jth position, as Figure 6 shows. We use a tuple set $S = \{\langle t_i, p_j \rangle\}$ to represent all threads of an application and their positions in the physical shape.

3.2 Problem Formulation

Based on the above specification, the problem can be formulated as follows:
- Given: 1) the NoC topology, $T(F, B)$, where F and B indicates the set of *free* and *busy* cores, respectively; 2) the traffic matrix, $M_{running}$, whose elements denote the historical communication volume between thread pairs of various applications; 3) the preferred topology, denoted by H or V and tuple set S; 4) the link utilization threshold $U_{congest}$ under which performance isolation can be enforced.
- Determine: 1) the mapping of every $< t_i, p_j >$ tuple from S to F, $< t_i, p_j >: S \rightarrow F$. After mapping, relevant position of threads remains the same as in preferred topology; 2) the shared link set L, and link utilization U of every shared link $l \in L$;
- With respect to the constraint: $\forall l \in L$, $U_l < U_{congest}$.

3.3 Algorithm in Detail

Given the inputs and constraints, the mapping algorithm solves the problem in two steps: 1)Topology searching, represented in Algorithm 1; 2) Performance verification, described in Algorithm 2.

Step 1: Topology Searching: The algorithm will firstly search the NoC for proper candidate shapes to serve the incoming application. If the number of free cores in F is fewer than that the application requires (number of $< t_i, p_j >$ tuples in S), the searching process returns directly with a failure. Otherwise, it tries to find the candidate shapes specified by the H or V. The detail of topology searching is described in Algorithm 1. Line 3 through line 13 are responsible for searching the target shape denoted by H (horizontal vector in this example). The algorithm starts searching T column by column (line 5) to satisfy every element in H. If it finds a busy node (line 8), the algorithm starts searching from another node in F. As long as every element in H is satisfied, the target topology is found (line 4 to 11). Otherwise, if the algorithm has traversed all nodes in set F but still does not find shape identical to H (line 13), the algorithm returns with a failure. Algorithm 1 aims at shapes indicated by the horizontal vector, and for the shapes indicated by vertical vector (V), the overall process is the same, except that it searches T row by row to satisfy every element in V (line 5). Note that Algorithm 1 is not limited to searching 'L' shape, and to further boost consolidation density, it is also applicable to

265

Algorithm 1 Topology Searching

Input: Requested shape: H or V; NoC topology: $T(F, B)$;
Output: *Found* or *NotFound*
1: **if** H **then**
2:    $h_{len} = \text{length}(H[h_1, h_2, \ldots, h_i, \ldots, h_n])$;
3:    **for** each $Node(row, col) \in F$ **do**
4:      **for** $i = 0; i < h_{len}; i + +$ **do**
5:        **if** $\{[Node(row + i, col), Node(row + i, col + h_i)]\} \subseteq F$ **then**
6:          continue;
7:        **else**
8:          break; //there is a $Node \in B$, start from another $Node \in F$
9:        **end if**
10:      **end for**
11:      return *Found*; //the shape denoted by H is found in T
12:    **end for**
13:    return *Notfound*; //no shape is identical to H in T
14: **end if**

Algorithm 2 Performance Verification

Input: the traffic matrix, $M_{running}$, shared link set, L; time interval, *time*, link utilization threshold, $U_{congest}$;
Output: The validity, *validity*;
1: int $sum = 0$;
2: **for** each shared link $l(< (x_{from}, y_{from}), (x_{to}, y_{to}) >) \in L$ **do**
3:    **for** each $node(a, b) \in lower\_set$ of l **do**
4:      **for** each $node(c, d) \in upper\_set$ of l **do**
5:        $sum += R_{(a,b)(c,d)}$; //sum up the values in traffic matrix
6:      **end for**
7:    **end for**
8:    $U = \frac{sum}{time}$;
9:    **if** $U > U_{congest}$ **then**
10:      $validity = false$; //congestion will happen after mapping
11:    **end if**
12: **end for**
13: $validity = true$; //U of every shared link is lower than $U_{congest}$, performance isolation is ensured

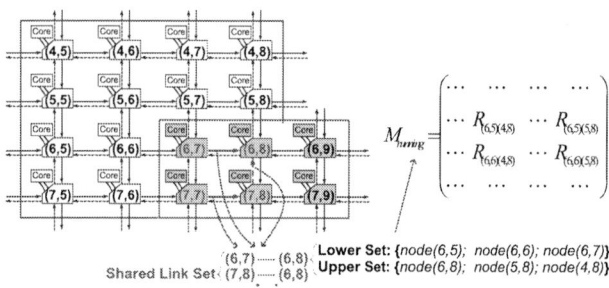

Figure 7: Shared link set identification

other shapes shown in Figure 5.

Step 2: Performance Verification: After successfully finding the target topology, we need to verify the criteria of performance isolation, represented by the constraint that the aggregated link utilization of every shared link U_l will not exceed $U_{congest}$. Firstly, we need to figure out the shared link set L. As an example, Figure 7 shows the scenario after mapping topology in Figure 6(a). Every node has a coordinate marked by the horizontal and vertical location in NoC. A link is denoted by the coordinates of associate nodes as $< (x_{from}, y_{from}), (x_{to}, y_{to}) >$. For example in Figure 7, there are three shared links in DOR routing mechanism. The links can be presented as $< (6, 7), (6, 8) >, < (7, 7), (7, 8) >$ and $< (7, 8), (6, 8) >$, respectively.

Secondly, to calculate the aggregated traffic on a shared link, we divide the associate nodes of the shared link into two sets: *lower_set* and *upper_set*, defined by Eq. 2. In DOR, a shared link will only carry the traffic generated from nodes in *lower_set* and terminated at nodes in *upper_set*. By looking up into the traffic matrix $M_{running}$, we can get the aggregated traffic volume on the shared link.

$$\begin{cases} lower\_set : \{Node(x,y) \mid x = x_{from} \&\& y \leq y_{from}\} \\ upper\_set : \{Node(x,y) \mid x \leq x_{to} \&\& y = y_{to}\} \end{cases} \quad (2)$$

Algorithm 2 shows in detail the verification process. Line 3 through line 7 determines the aggregated traffic volume on each shared link. U is calculated based on Eq. 1 in line 8. Line 10 indicates that if the U of any shared link exceeds $U_{congest}$, this candidate topology is not valid and must search another candidate from Step 1. It is valid only if all shared links are satisfied, as line 13 describes.

Note that our algorithm takes the "first-fit" principle [16]. That is if multiple topology candidates can pass the Step 2, we take the first one.

3.4 Traffic Prediction

As many prior NoC flow management solutions, RISO relies on accurate traffic prediction. At each scheduling interval, the operating system, based on the traffic history, predicts the traffic distribution and thereby identifies the

sharable links to implement RISO. Most prior work uses liner predictor to fulfill this purpose. We find that although liner predictor is capable for gradually-changed traffic patterns, it's highly unreliable to cope with bursty traffic patterns which, if fail to predict, can jeopardize the performance isolation. Therefore, we take a "conservative" approach in traffic prediction: for bursty traffic, since the traffic volume changes sharply, we just exclude the associated links from sharing. This may slightly degrade the consolidation density, but the performance isolation is well guaranteed. Specifically, we modified last value predictor (LVP) [17] to handle both bursty and non-bursty scenarios. The prediction function is implemented as Eq. 3.

$$T_{prediction} = \begin{cases} h_2 & \mid \frac{(h_2 - h_1)}{h_1} \mid < 10\%; \\ +\infty & Otherwise. \end{cases} \quad (3)$$

The predictor stores two most recent traffic volumes as h_1, h_2 for every source-destination pair. If the two values differ sharply, we exclude the associated links from sharable links by assigning a $+\infty$ to the final prediction value; otherwise we still follow the LVP. We find that setting the bar to 10% works well.

4. EVALUATION

4.1 Experimental Setup

4.1.1 Metric for Consolidation Density

We use *system utilization* (U_{system}) [16] as a metric for consolidation density evaluation. For a N-node system during T period of time, U_{system} is defined by Eq. 4.

$$U_{system} = \frac{\sum_{i=1}^{N} T_i}{N \times T} \quad (4)$$

where T_i is the busy time of node i over T period of time. A high system utilization means high consolidation density.

System utilization depends on the "*load*" condition [16] which is defined by Eq. 5.

$$Load = \frac{R \times S}{N \times I} \quad (5)$$

where R is average requested resources (i.e. cores in this paper) of all applications; I is average inter-arrival time between consecutive applications; S is average application running time. *Load* below 1 means application arrival rate is lower than departure rate; otherwise the system will be overloaded and improving system utilization will be critical. The values of these parameters used in the experiment are listed in Table 1.

4.1.2 Performance Simulation Setup

We modified Booksim2.0 [18] to evaluate performance after using the proposed application mapping algorithm. The

Table 1: Parameters of system utilization evaluation

Parameter	Value
Topology	Mesh (32*32 and 16*16)
Scheduling mechanism	FCFS
R	64
S	2000 (in cycles)
distribution of requested Num. of cores	uniform
Num. of consolidated workloads	10000 (per experiment)
load range	[0.1~1.6], step by 0.1

baseline NoC topology is a 8x8 mesh. The router is configured with a two-stage pipeline plus one cycle for link traversal. We use two virtual channels and each has an eight flit buffer. The congestion threshold ($U_{congest}$) is set at 65%, in accordance with the result shown in Figure 4.

4.1.3 Workloads

For consolidation density evaluation, we map 10000 consolidated workloads at each *load* condition. We log the U_{system} in real time. This measurement is repeated ten times at each *load*. The final U_{system} result is the average of the ten measurements.

For performance evaluation, we run application traces [8] and each application acts as a workload. Application traces are obtained from GEMS [19], a full system simulator. The detailed configuration is shown in Table 2.

Table 2: Full system simulator configuration

Parameter	Value
Cloud Processor	16 in-order cores
Coherence Protocol	MOESI
L1 I/D Cache	32KB (2-way associative)
L1 Cache Access Latency	1 cycle
Private L2 Cache	256KB (4-way associative)
L2 Cache Access Latency	8 cycles
Main Memory Latency	90 cycles

4.1.4 Baselines Compared

We compare our scheme with two previous representative schemes: the first scheme employs efficient routing but at the expense of lower consolidation density [20], denoted by "Regularity-oriented" scheme. Clearly, "Regularity-oriented" scheme ideally enforces performance isolation. The second scheme takes the opposite, i.e. emphasizing the density, but paying for more complex routing mechanisms [16], denoted by "Density-oriented" scheme.

In the experiment we will show that our scheme, "RISO", can also preserve performance isolation as "Regularity-oriented" scheme provides, and meanwhile achieves the consolidation density of "Density-oriented" scheme, but with much better performance.

4.2 Result 1: Consolidation Density

Figure 8 shows the system utilization from underload to overload scenarios. The result shows that *RISO* improves the system utilization by up to 12% higher than *Regularity-oriented* scheme in the overload condition. Surprisingly, RISO performs almost equally well to *Density-oriented* scheme (within 0.1%). Even though RISO cannot exploit all irregular regions due to the link utilization constraint, but it can deal with some unique regions such as "⊏" which cannot be supported in "Density-oriented" scheme, which makes our scheme match density-oriented scheme in consolidation density.

Moreover, our scheme uses DOR as the underlying routing algorithm which is more efficient and cost-effective regarding network performance, as the following results show.

4.3 Result 2: Performance

4.3.1 Performance Isolation Analysis

Regularity-oriented schemes are regarded for ideal performance isolation by enforcing strict physical isolation, as described in Section II. Hence, we compared our relaxed isolation scheme to the regularity-oriented scheme to verify the capability of performance isolation. We run application traces

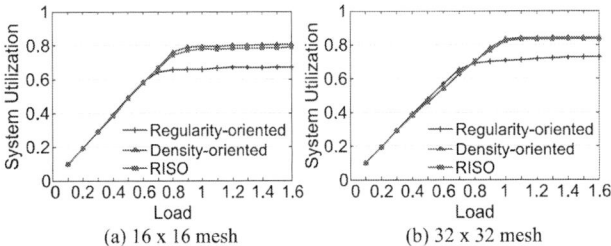

Figure 8: System utilization for 16x16 and 32x32 mesh

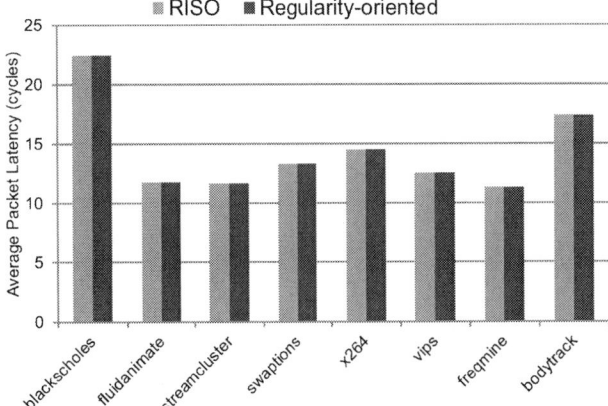

Figure 9: Performance comparison between routing-oriented scheme and RISO

under RISO and regularity-oriented condition respectively. The results are shown in Figure 9. RISO won't degrade the latency compared to the regularity-oriented scheme. This is because we put a hard constraint for each shared link, as Section III describes. Therefore, we can safely conclude that RISO preserves performance isolation, but with higher consolidation density as shown in Figure 8.

4.3.2 Network Latency Analysis

Density-oriented scheme is notorious for traffic latency, although it can enforce performance isolation and provide high consolidation density. Hence, we show how much latency can be improved with RISO. We run various combinations of two application traces under relaxed and density-oriented isolation respectively. Note that RISO uses efficient DOR, while density-oriented scheme uses the most favorable Up*/Down* [9] routing mechanism.

Figure 10 shows the comparison results for 25 groups of consolidated applications. Clearly, our scheme wins for all. The latency, if using density-oriented scheme, will degrade over 100% for some groups such as blackscholes_bodytrack (114%) and swaptions_bodytrack (101%), which are usually the mix of computation intensive and memory intensive applications. Such combination usually renders more under-utilized links which RISO can take advantage of while density-oriented scheme cannot. The average latency improvement is 38.4% for 25 groups of consolidated applications.

5. RELATED WORK

The concept of RISO is similar to the topology virtualization techniques proposed in [21] and [22], in which NoC is reconfigured and the software always observes a logically regular-shaped topology when faults are happening. However, our work targets the mapping of the arbitrary-shaped topology required by an application into the appropriate region to improve consolidation density.

Quality of service (QoS) in NoC level is introduced to fairly allocate on-chip resources according to specific service poli-

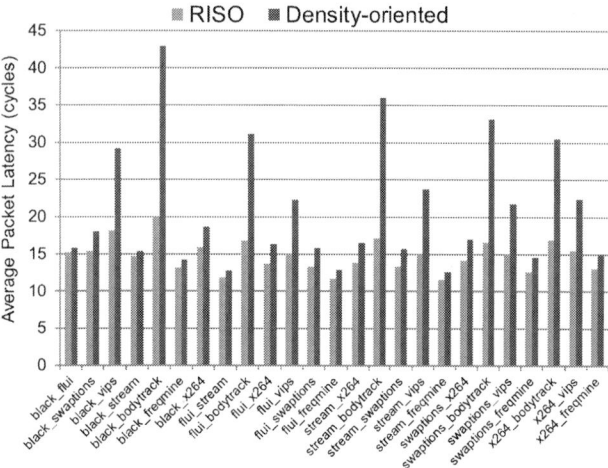

Figure 10: Latency comparison between density-oriented scheme and RISO

cies. For instance, Brand et al [13] uses link utilization as the congestion measure, and proposed that the bandwidth and latency will be guaranteed for best-effort communication service if the contribution to link utilization by different workloads stays below a certain threshold. In [23], virtual point-to-point links are used to ensure reliable communication if the threads of an application are distributed into disjoint NoC regions. Our work is orthogonal to them.

Performance isolation technique is highly required to achieve controllable QoS in NoC level, and is firstly introduced in [24]. It clarifies the basic items needed to solve in this area. Particularly, the tradeoff between regularity of topology and complexity of routing is the most important in that it relates directly to the network performance and power consumption.

Some proposed techniques follow the strict NoC isolation strategy using rectangular shapes for performance isolation, such as [20]. These methods are restrained by the maximum number of consolidated workloads, which will degrade the consolidation density. Unlike those regular shaped performance isolation methods, Solheim et al. [16] proposed an irregular-shaped isolation based on complex routing mechanism. This method also follows strict isolation between workloads and to some extent improves consolidation density compared with rectangle based isolation. However, its routing mechanism is less efficient and exhibits substantial degradation with respect to network performance. To the best of our knowledge, the proposed relaxed isolation scheme is the first work that simultaneously takes the advantage of regularity of topology and efficient routing mechanisms in network isolation.

6. CONCLUSION

This paper proposes relaxed isolation (RISO) strategy to enforce performance isolation constraint in workload consolidation. Unlike the traditional strict isolation strategy such as regularity-oriented and density-oriented approach, RISO allows under utilized links to be shared by multiple applications, as long as the aggregated contribution to link utilization is lower than the congestion threshold. Compared with regularity-oriented approach, RISO supports more flexible topologies and can greatly improve consolidation density. RISO does not complicate the routing mechanism required by density-oriented approach; it also uses efficient and cost-effective DOR routing mechanism and hence yields higher network performance. In other words, RISO strikes better tradeoff between regularity of topology and complexity of routing algorithm in implementing network isolation. We therefore believe that RISO is a promising scheme for workload consolidation in many-core cloud processors.

7. ACKNOWLEDGEMENT

This work is supported in part by National Basic Research Program of China (973) under grant No. 2011CB302503, in part by NSFC under grant No. (61202056, 61100016, 61076037, 60921002). Correspondence should be addressed to Prof. Yinhe Han. We also thank Prof. Ninghui Sun and Prof. Lixin Zhang for their support.

8. REFERENCES

[1] Intel Single-chip Cloud Computer, http://www.intel.com /content/www/us/en/research/intel-labs-single-chip-clo ud-computer.html.

[2] TILE-Gx 3000 Series Cloud Processor, http://www.tiler a.com/products/processors/TILE-Gx-3000.

[3] M. Ferdman et al, "Clearing the clouds: a study of emerging scale-out workloads on modern hardware," in *ASPLOS 2012*, pp. 37–48.

[4] P. Lotfi-Kamran et al, "Scale-out processors," in *ISCA 2012*, pp. 500–511.

[5] Amazon Elastic Cloud Computing, http://aws.amazon. com/ec2/.

[6] M. Marty and M. Hill, "Virtual hierarchies to support server consolidation," in *ISCA 2007*, pp. 46–56.

[7] S. Ma, N. Jerger, and Z. Wang, "Dbar: an efficient routing algorithm to support multiple concurrent applications in networks-on-chip," in *ISCA 2011*, pp. 413–424.

[8] C. Bienia et al, "The parsec benchmark suite: characterization and architectural implications," in *PACT 2008*, pp. 72–81.

[9] M. Schroeder, "Autonet: A high-speed, self-configuring local area network using point-to-point links," tech. rep., 1990.

[10] A. Mejia et al, "Segment-based routing: an efficient fault-tolerant routing algorithm for meshes and tori," in *IPDPS 2006*, pp. 10–19.

[11] S. Bell et al, "Tile64 - processor: A 64-core soc with mesh interconnect," in *ISSCC 2008*, pp. 88–99.

[12] S. Li, L. Peh, and N. Jha, "Dynamic voltage scaling with links for power optimization of interconnection networks," in *HPCA 2003*, pp. 91–102.

[13] van den Brand et al, "Congestion-controlled best-effort communication for networks-on-chip." in *DATE 2007*, pp. 1–6.

[14] M. Kandemir, O. Ozturk, and S. Muralidhara, "Dynamic thread and data mapping for noc based cmps," in *DAC 2009*, pp. 852–857.

[15] B. Fu, Y. Han, and J. Ma, "An abacus turn model for time/space-efficient reconfigurable routing," in *ISCA 2011*, pp. 259–270.

[16] A. Solheim et al, "Routing-contained virtualization based on up*/down* forwarding," in *HiPC 2007*, pp. 500–513.

[17] Y. Huang et al, "Ntpt: On the end-to-end traffic prediction in the on-chip networks," in *DAC 2010*, pp. 449–452.

[18] Booksim2.0, https://nocs.stanford.edu/.

[19] GEMS, http://research.cs.wisc.edu/gems/publications. html.

[20] V. Gupta and A. Jayendran, "A flexible processor allocation strategy for mesh connected parallel systems," in *ICPP 1996*, pp. 166–173.

[21] L. Zhang and Y. Han et al, "Defect tolerance in homogeneous manycore processors using core-level redundancy with unified topology," in *DATE 2008*, pp. 891–896.

[22] L. Zhang and Y. Han et al, "On topology reconfiguration for defect-tolerant noc-based homogeneous manycore systems," in *TVLSI*, pp. 1173–1186, 2009.

[23] M. Asadinia et al, "Supporting non-contiguous processor allocation in mesh-based cmps using virtual point-to-point links," in *DATE 2011*, pp. 1–6.

[24] J. Flich et al, "On the potential of noc virtualization for multicore chips," in *CISIS 2008*, pp. 801–807.

Smart Hill Climbing for Agile Dynamic Mapping in Many-Core Systems

Mohammad Fattah, Masoud Daneshtalab, Pasi Liljeberg, Juha Plosila

Department of Information Technology, University of Turku, Turku, Finland

{mofana, masdan, pakrli, juplos}@utu.fi

ABSTRACT

Stochastic hill climbing algorithm is adapted to rapidly find the appropriate start node in the application mapping of network-based many-core systems. Due to highly dynamic and unpredictable workload of such systems, an agile run-time task allocation scheme is required. The scheme is desired to map the tasks of an incoming application at run-time onto an optimum contiguous area of the available nodes. Contiguous and un-fragmented area mapping is to settle the communicating tasks in close proximity. Hence, the power dissipation, the congestion between different applications, and the latency of the system will be significantly reduced. To find an optimum region, we first propose an approximate model that quickly estimates the available area around a given node. Then the stochastic hill climbing algorithm is used as a search heuristic to find a node that has the required number of available nodes around it. Presented agile climber takes the steps using an adapted version of hill climbing algorithm named $\underline{S}$mart $\underline{Hi}$ll $\underline{C}$limbing, SHiC, which takes the run-time status of the system into account. Finally, the application mapping is performed starting from the selected first node. Experiments show significant gain in the mapping contiguousness which results in better network latency and power dissipation, compared to state-of-the-art works.

Categories and Subject Descriptors: D.4.7 [**Operating Systems**]: Organization and Design – *Real-time systems and embedded systems.*

General Terms: Algorithms, Management, Performance, Design.

Keywords: On-chip many-core systems, Application mapping, Task allocation, AI algorithms, Hill climbing.

1. INTRODUCTION

The Future Multi-Processor Systems-on-Chip (MPSoCs) are likely to have tens or hundreds of resources connected together. Networks on chip [1] (NoCs) have emerged as a promising solution for communication infrastructure of such systems. NoCs provide a regular platform for connecting the system resources and makes the communication architecture scalable and flexible compared to traditional bus or hierarchical bus architectures.

Many-core systems will feature an extremely dynamic workload where an unpredictable sequence of different applications enter and leave the system at run-time. In order to handle the featured dynamic nature, a run-time system manager is required to efficiently map an incoming application onto the system resources [2]–[4]. Applications are modeled as a set of communicating

---

Permission to make digital or hard copies of all or part of this work for personal or classroom use is granted without fee provided that copies are not made or distributed for profit or commercial advantage and that copies bear this notice and the full citation on the first page. To copy otherwise, to republish, to post on servers or to redistribute to lists, requires prior specific permission and/or a fee.

DAC '13, May 29 - June 07 2013, Austin, TX, USA.

---

tasks, and the mapping function of the central manager (CM) of the system decides on the appropriate node for each task.

The system performance is significantly influenced by the utilized mapping approach. For example, assume a dispersed application mapping where tasks of an application are mapped onto a distant and fragmented set of nodes. This increases the average hop count within tasks of an application and places tasks of different applications between each other. Accordingly, the power dissipation and congestion probability (latency) of the network will dramatically increase. On the other hand, consider a convex and contiguous application mapping where tasks of an application are placed on relatively close nodes without fragmentation. This will reduce the average distance between mapped tasks and isolates the communication of different applications from each other which decreases the congestion probability between them.

To have an efficient mapping, it is desired to place an application on a near convex and contiguous set of nodes where the remaining set of available nodes is also kept contiguous. Finding a convex region of nodes is a polynomial, $O(n^3)$, problem [5]. However, considering more aspects such as near-convexity condition, remaining nodes contiguity, etc. turns it into clustering problems [6] with deterministic algorithms of NP-hard complexity. Indeed, these are not tolerable time complexities regarding the growing size and dynamic nature of the systems.

In this work, we propose an algorithm with significantly decreased complexity in order to find the appropriate area for a given application. The task allocation is then efficiently carried out through our *CoNA* method [4]. Briefly stated, *CoNA* starts from a *first node* and attempts to map the application tasks onto a set of contiguous nodes around it. Thus, the job of finding the appropriate area is to select a *first node* which has the required number of contiguous and near convex available nodes around it and leads to least fragmentation of remaining nodes. For this, we first propose an approximate model to estimate the number of the available (contiguous and near convex) nodes around any given node. Then, hill climbing search heuristic is adapted in order to find the optimum *first node* rapidly among all the available nodes. The proposed Smart Hill Climbing, *SHiC*, is equipped with a level of intelligence in which the steps are not taken fully stochastic. The proposed near optimal (non-deterministic) solution tackles the problem in $O(\sqrt{n})$, n is the given network size, in the best case and $O(n^2)$ in the worst case; which is an impractical case.

The rest of the paper is organized as follows: In Section 2 we present related works and motivate the impact of the optimum area (*first node*) selection. Section 3 formally defines the mapping problem and different metrics to evaluate the mapping results. Section 4 describes our *SHiC* heuristic in the *first node* selection. The simulation details along with the experimental results are presented and discussed in Section 5. Finally, Section 6 concludes the paper.

2. RELATED WORK AND MOTIVATION

There are several works dealing with dynamic management of workload in multi- and many-core systems. In this Section we explore different *first node* or generally the mapping area selection methods used by different works. We also motivate the importance of the selection method by presenting several examples.

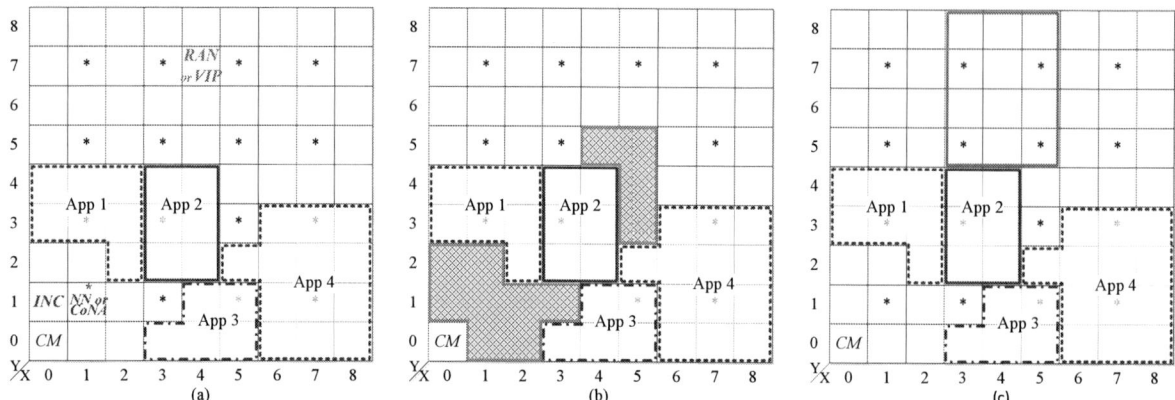

Figure 1. Possible scenarios resulting in the area dispersion. (a) Initial configuration of the system and the *first node* for the next application in different methods. (b) Mapped area (green hatched area) dispersion in *NN*, *INC* and *CoNA* methods. (c) Contiguous mapping of *DistRM*, *STDM* and *VIP*-supported approaches, however they lead to fragmentation of remaining nodes.

Carvalho et al. [2], [7] present different heuristics, such as Nearest Neighbor (*NN*), Best Neighbor (*BN*), etc. In all of their heuristics, a clustering mechanism for the *first node* selection is considered. A set of sparse nodes (cluster nodes) are assumed to select the *first node* of the mapping algorithm among them; i.e. a mapping solution can exist only when a free cluster node exists. This is to assure some amount of available nodes around the chosen *first node*.

Chou et al. [3], in their incremental (*INC*) approach, break down the mapping problem into two steps: the region selection, and the task allocation. In the region selection step, they start from the closest node to the CM and include it in the region. Then, they iteratively add nodes to the selected region trying to keep both the selected region and remaining nodes contiguous. Afterward, in the task allocation step, application tasks are mapped inside the selected region.

As an advanced approach, CoNA [4] selects the closest node to the CM with all its four neighbors available. Thus a minimum number of available nodes are assured. Then, task graph is traversed in breath-first order and tasks are mapped onto the neighboring of their parents in which a smaller square is formed.

There are lots of possible scenarios in which *NN*, *INC* and CoNA methods would easily result in area fragmentation. Figure 1 (a) and (b) depict one of those scenarios. Notice that the cluster nodes of the *NN* algorithm are indicated by asterisks. The current status of the system is shown in Figure 1 (a), where four applications are running on the system after entrance and exit of several applications. As indicated in the Figure 1 (a), the selected *first node* for the next application will be nodes (1, 1) or (0, 1) for CoNA and *NN* or *INC* algorithms, respectively. Now, an application with 12 tasks enters the system while there are only 8 contiguous nodes available regarding the selected *first node*. Accordingly, a fragmented area of nodes (the hatched area in Figure 1 (b)) will be chosen for the application. Consequently, as can be seen, the communication paths between dispersed nodes will be stretched and can be congested by communications of other existing applications (1 to 3).

As a decentralized mapping algorithm for tree-structured applications, Weichslgartner et al. [8] suggest three different methods for the *first node* selection. Two of them need prior knowledge about all incoming applications, while the one which is more dynamic, the *farthest-away* algorithm, leads to inefficient results as it does not consider the size of applications.

Both the decentralized *DistRM* [9] and *STDM* [10] approaches, start mapping from a random node. The randomly chosen node in *DistRM* starts exploring its neighboring nodes as well as distant nodes in the system to find the node with enough surrounding resources to start allocation. To this end, *DistRM* utilizes a negotiation algorithm in cooperation with a proposed distributed directory service. On the other hand, the randomly chosen node in *STDM* tries to map the incoming application around it. In case of

failure a new random node is chosen until several times. Both methods impose huge negotiation traffic on the network in order to compensate the applied randomness.

Asadinia et al. [11], in their VIP-supported approach, find all contiguous regions of free nodes, and select the smallest region with the size greater than or equal to the application size. Then, they start mapping from one of the selected region nodes with the maximum number of neighbors. This is to guarantee the contiguity of the mapped region and select a *first node* with maximum available nodes around it.

As another possible scenario, let us assume that in the system status of Figure 1 (a), the node (4, 7) is chosen by *VIP* as it is one of the nodes with maximum number of neighbors, *DistRM* or *STDM* after several random node generations. The selected *first node* has enough resources around it for the proposed application with 12 tasks and, as shown in Figure 1 (c), a contiguous set of nodes are allocated to the application tasks. However, the remaining set of available nodes is dispersed and the future applications may suffer from the area fragmentation of the available nodes.

Note that optimum *first node* selection is not only a function of the current state of the system but also the size of the application. For instance, *CoNA*, *NN* and *INC* approaches could have worked optimally, in the described scenario, if the size of the application had been less than or equal to 8. Moreover, the area selection method not only has to be agile but also must avoid random trial and error approaches. This is to minimize time and communication penalties imposed onto the system.

In this work, we propose an algorithm for optimum *first node* selection in order to achieve an agile and efficient mapping of applications. It uses the general knowledge of currently running applications and imposes no additional traffic on the network.

3. DEFINITIONS

In the following, we present formal definitions of an application, the NoC architecture, and the mapping problem. In order to reduce the problem size and simplify the analysis, we consider a homogenous mesh-based NoC in our definitions and experiments. Moreover, we define several evaluation metrics as assessment tools to compare different algorithms. The metrics weigh the resulted mapping of an application. Later in Section 5, we evaluate effect of our *first node* selection method, *SHiC*, on the network performance using these metrics along with extracted results of the network simulation.

3.1 Problem Definition

Mapping algorithms try to allocate system resources, connected together through an on-chip network, to tasks of a requested application in an optimal way.

Each application in the system is represented by a directed graph denoted as a task graph $A_p = TG(T, E)$. Each vertex $t_i \in T$

represents one task of the application A_p, while the edge $e_{i,j} \in E$ stands for a communication between the source task t_i, and the destination task t_j. Task graph of an example application with 6 tasks is shown in Figure 2. The amount of data transferred from a task t_i to t_j of edge $e_{i,j}$ is indicated on the edge as $w_{i,j}$.

An architecture graph $AG(N, L)$ describes the communication infrastructure, which is a simple $M \times M$ 2D-mesh NoC with the XY routing (Figure 3 (a)). The AG contains a set of nodes $n_{x,y} \in N$, connected together through communication links $l_k \in L$. Each node $n_{x,y}$ contains a 5-port router $r_{x,y}$ connected to the local processing element $pe_{x,y}$ by its local port.

Mapping of an application onto the NoC-based multi-core system is defined as a one-to-one mapping function from the set of application tasks T, to the set of NoC nodes N:

$$map: T \rightarrow N, s.t. \ map(t_i) = n_{x,y}; \forall t_i \in T, \exists n_{x,y} \in N \qquad (1)$$

Based on the definition, a mapping function is started if and only if there are enough available nodes to map onto them. Figure 3 (a) illustrates a possible mapping of the application in Figure 2, onto the described NoC platform. For simplicity, we denote a node where a task t_i is mapped onto as nt_i and the packet corresponding to the edge $e_{i,j}$ as $pck_{i,j}$; i.e. the packet sent from nt_i to nt_j. The set of running applications on the system is also denoted by $APPS$ which changes at run-time due to the system's dynamic nature.

Note that the cardinality of a set shows the number of elements in that set, while the cardinality of a number means the absolute value of that number; e.g. $|APPS|$ means number of running applications, while $|-4|$ equals the number 4.

3.2 Evaluation Metrics

3.2.1 Average Weighted Manhattan Distance

The dissipated energy related to a packet delivery is a function of both the packet size, and the path the packet traverses [12]. As a metric to evaluate the power consumption of a mapped application, *Average Weighted Manhattan Distance (AWMD)* is the sum product of MD and corresponding weight of communicating nodes, averaged by the total communication volume of the application:

$$AWMD_{map(A_p)} = \frac{\sum_{\forall e_{i,j} \in E} w_{i,j} \times MD(nt_i, nt_j)}{\sum w_{i,j}} \qquad (2)$$

In other words, the $AWMD$ of a mapped application determines the number of hops that each bit of the application data traverses in the network. The $AWMD$ value in Figure 3 (a), for instance, is $63/35 = 1.8$, which means each bit will dissipate 1.8 times of energy unit despite the networking overheads.

3.2.2 Mapped Region Dispersion

The network performance is highly correlated to the network congestion [13], [14]. Congestion increases the network latency dramatically [15] and also increases the network dynamic power consumption considerably [16]. This is why there are plenty of works aiming to diminish the network congestion in different aspects, like routing [16], [17] .

Two types of congestion can be defined from the perspective of dynamic application mapping: external and internal congestions. External congestion occurs when a network channel is contented

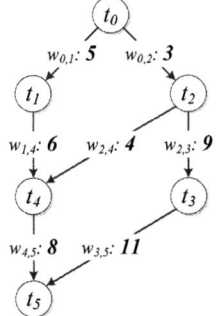

Figure 3. An application with 6 tasks and 7 edges.

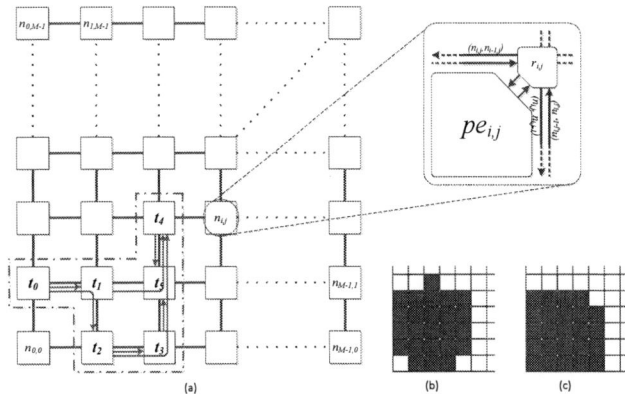

Figure 2. (a) NoC-based platform, with an application mapped onto it (the highlighted region.). (b) An area of 24 nodes with minimum *MRD* value; and (c) the same number of nodes with the least *NMRD* value.

by the packets of different applications; while the internal congestion is related to the packets of the same application. Hence, the internal congestion probability is related to the utilized mapping algorithm [4], and thus it is out of this paper scope.

As aforementioned, to decrease the external congestion probability, the mapped area of an application should be as convex as possible and minimally fragmented. Several works, e.g. [3], [4], considered the average pairwise MD of the allocated nodes as a metric to assess the *Mapped Region Dispersion*:

$$MRD_{map(A_p)} = \frac{\sum_{t_i, t_j \in T} MD(nt_i, nt_j)}{\binom{|T|}{2}} \qquad (3)$$

Distant node allocation will increase the MRD for the obtained mapping; i.e. more external congestion probability. The increased congestion probability is originated from communication of tasks of different applications that are mapped among each other.

The area with the smallest MRD is almost circular [18]. Regarding the mesh topology of the network, however, a circular region will generate irregularity in remaining available nodes and more area fragmentation in long term. This is shown in Figure 3 (b) for allocation of 24 nodes with the least MRD value. On the other hand, a rectangular allocation forms regular regions, decreases applications overlap and thus isolates their communications. Thus, the best mapped area would be square (shown in Figure 3 (c)) as it is the rectangle with the smallest MRD. It can be shown that the MRD of a square with $|T|$ nodes will be:

$$MRD_{SQ(|T|)} = \frac{2 \times \sqrt{|T|}}{3} \qquad (4)$$

Thus the *Normalized MRD* metric is defined which assesses the squareness of the mapped region independent of the size of the application:

$$NMRD_{map(A_p)} = 1 + \frac{|MRD_{map(Ap)} - MRD_{SQ(|T|)}|}{MRD_{SQ(|T|)}} \qquad (5)$$

The *NMRD* value of 1 means a squared area. *NMRD* increases as the mapped area is getting more fragmented and less similar to a square shape. For example, the *NMRD* of the dispersed mapped area in Figure 1(b) is 1.86 while it is 1.01 in the rectangle region of Figure 1(c). Moreover, as the MRD of the first area is almost twice that of the second one (4.3 vs. 2.3); the $AWMD$ will get almost doubled using the same mapping strategy. Thus, dispersed allocation not only increases the congestion probability but also results in more power dissipation of the network. This shows the importance of this work in finding the optimum area for an application.

4. SMART FIRST NODE SELECTION

Based on the motivational examples and analysis presented in previous sections, the optimum area for an application mapping is a (i) contiguous and (ii) square region of available nodes. CoNA [4] attempts to form a contiguous area around its *first node* while favoring square shapes. Given that, appropriate selection of the mapping *first node* is the important step towards the contiguity of the allocated area. Accordingly, the appropriate *first node* is the one with the required number of available nodes in a square shape around it while the minimum fragmentation of remaining nodes occurs after the mapping.

In this section, we first define our approximate model which estimates the number of available nodes in a square shape around a given node. Then the proposed model is utilized in the adapted hill climbing search hubristic to find the appropriate *first node* in an agile and smart manner.

4.1 Square Factor

The *square factor* of a given node, $SF(n_{i,j})$, is the estimated number of contiguous, almost square-shaped, available nodes around that node. Hence, the appropriate *first node* for mapping of an application would be the node with the *SF* equal to the application size.

Each running (already mapped) application in the system is modeled as a rectangle defined by its *left-down* and *right-up* corner nodes. Regarding the modeled rectangle of a running application, there might be (1) some nodes within the rectangle which do not belong to the application and/or (2) some nodes of application which are excluded from the rectangle. The rectangle of each application is modeled such that minimizing the number of these two types of nodes. The rectangle models of running applications 1 to 3 of Figure 1 are shown in Figure 4 (a). For instance, the rectangle of the application 1 excludes one node ($n_{2,2}$) of the application and the rectangle of the application 3 includes one node ($n_{3,1}$) which is not part of the application. However, they are the best fit rectangles according to the included and excluded nodes.

To calculate the $SF(n_{i,j})$, we first find the largest square centered on $n_{i,j}$, $SQ_{max} = (n_{i,j}, r_{max})$, where it fits within the mesh limits and has no conflict with other rectangles (running applications) of the system. This is shown in Figure 4 (b) for the node $n_{7,1}$ which is the *first node* of the application 4. In addition to the SQ_{max} area, there might be also some more nodes beyond the square borders not belonging to system rectangles, as marked with asterisk in Figure 4(b). These nodes are counted in order to prevent available nodes from being isolated while keeping the mapped area close to the square shape. The square factor of a given node $n_{i,j}$ is calculated by summing up the area of the SQ_{max} with the available nodes beyond the square borders. For instance, the $SF(n_{7,1})$ will be the square area, 9, summed up with marked nodes, 5, which is 14.

During the *SF* calculation of a node, the algorithm also calculates a direction, called open direction (*openDir*), which indicates one of the eight neighbors of the node estimated to have a larger *SF*. The *openDir* is towards that side of SQ_{max} where there is the maximum number of nodes beyond it. In the example shown in Figure 4 (b), there are 3 asterisked nodes beyond the up side of the SQ_{max} versus 1 in the left side (corners are not counted in). Thus the *openDir* would be upward, meaning the upper neighbor is probably holding a larger *square factor*. When there is more than one side with the maximum number of nodes beyond them, the *openDir* is then towards their vector addition. An openDir of value zero means no specific direction is predicted to result in a larger *square factor*. The climber function of the hill climbing heuristic uses the computed open direction to provide smartness in taking the steps.

Note that there can be available nodes inside an application rectangle; e.g. $n_{3,1}$ is inside the rectangle of the application 3. In such cases the *square factor* is evaluated by counting the available neighbors of the node, and the open direction is toward the side exiting from the rectangle. For instance, $SF(n_{3,1}) = 3$ because two of the node's neighbors are available ($n_{2,0}$ and $n_{2,1}$); the open direction is leftward to exit from the rectangle of the application. Moreover, the system *CM* is always assumed as a rectangle of one node.

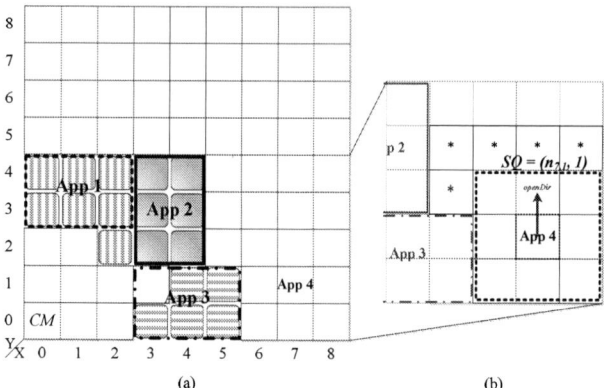

Figure 4. (a) Applications are modeled as rectangles in our approach. (b) *Square factor* calculation of the node $n_{7,1}$ before application 4 being mapped.

As the algorithm for the *SF* calculation explores the distance between the given node and all rectangles (running applications, *APPS*) of the system to find the SQ_{max}, the *SF* calculation has a linear time complexity of O (|*APPS*|).

4.2 SHiC: Smart Hill Climbing

To find the appropriate *first node*, one can calculate the *SF*, $O(|APPS|)$, for all $M \times M$ nodes of the system, and select the one with the best *SF* regarding the application size (exhaustive search). However, this will take $O(M^2|APPS|)$ time which is not a tolerable complexity for the future many-core systems with hundreds of nodes. Instead, *SHiC* starts from a randomly selected node and walks smartly through the network nodes by means of the defined open direction to reach the optimum node. This significantly reduces the amount of traversed nodes, resulting in an agile mapping algorithm. The pseudo code of our proposed algorithm is shown in Figure 5.

The original hill climbing heuristic starts from a node and moves in the direction of the increasing value (of the *SF*) to the uphill (max. *SF*) [19]. However, the appropriate *first node* is not the one on the uphill but the one in a specific height; i.e. the node with the *SF* equal to the application size. Hence, *SHiC* moves in the direction of the optimum *SF* instead of the maximum one. *SHiC* looks for the node with the optimum *SF* according to a preference

---

Inputs: Size (|T|) of the requested application A_p, current set of running applications APPS.

Output: chosen first node: n_{fn}.

(1)   **repeat** $2+\sqrt{|APPS|}$ *times*
(2) $n_{cur} \leftarrow$ *choose a random available node*;
(3) $iter \leftarrow 0$;
(4) $yawAngle \leftarrow 0$;
(5)   **while** ($iter < (M/2)$ **AND** $SF(n_{cur}) \neq |T|$)
(6)   **if** $openDir(n_{cur}) = (0,0)$ **then**
(7) $moveDir \leftarrow$ a random direction;
(8)   **else**
(9) $moveDir \leftarrow (SF(n_{cur}) < |T|)$? $openDir(n_{cur})$: $-openDir(n_{cur})$;
(10)   divert $moveDir$ by ($yawAngle$);
(11)   **end if**
(12) $n_{next} \leftarrow$ move in $moveDir$;
(13)   **if** n_{next} is available **AND** preferred to n_{cur} **then**
(14) $n_{cur} \leftarrow n_{next}$;
(15) $yawAngle \leftarrow 0$;
(16)   **else**
(17) $yawAngle \leftarrow$ a random in opposite side ;
(18)   **end if**
(19)   **end while**
(20)   **if** n_{cur} is preferred to n_{fn} **then**
(21) $n_{fn} \leftarrow n_{cur}$;
(22)   **end if**
(23) **end repeat**

Figure 5. Smart Hill Climbing, *SHiC*, pseudocode.

function. Standing on a node n_{cur}, a step to the next node n_{next} is called to be a preferred step when:

$$
\begin{aligned}
&(SF(n_{cur}) < |T| \text{ AND } SF(n_{next}) > SF(n_{cur})) \text{ OR} \\
&(SF(n_{next}) \geq |T| \text{ AND } SF(n_{next}) < SF(n_{cur}))
\end{aligned} \quad (6)
$$

By means of the defined preference function, *SHiC* first looks for the node with the smallest *SF* value which is larger than or equal to the application size. Otherwise, the node with the largest *SF* value is preferred. Note that, when there are two nodes with equal *SF*, the one closer to mesh corner is preferred to decrease the incurred defragmentation of remaining nodes.

The original hill climbing heuristic examines all possible neighbors to find the best choice, which will increase the execution time significantly. As an alternative, stochastic (first-choice) hill climbing generates random moves until one is generated that is preferred to the current one [19]. Moreover, *SHiC* uses the calculated *openDir* to impart smartness to the steps it takes (lines 6 to 11). Standing on the current node, the climber first moves according to the *openDir*. If the taken step is not preferred (line 17), the climber yaws and chooses a random direction on the opposite side of the yaw angle for the next move. The climber takes M/2 steps (line 5) as it is the required number of steps to start from a node in the mesh corner and reach the center node of the mesh.

On the other hand, the hill climbing heuristic might get stuck on local optimums. For example, the climber might stuck to the node $n_{1,1}$ of the Figure 1, while looking for a node with the *SF* equal to 12. As a solution, we use the random restart approach in which the proposed stochastic hill climbing approach is executed several times starting from different randomly chosen nodes [19]. The algorithm is repeated $(2+\sqrt{|APPS|})$ times (line 1) as the number of the fragmented regions in the system is related to the number of running applications.

Finally, the best found node is passed to the mapping algorithm as the *first node*. The *SHiC* outer loop is executed $O(\sqrt{|APPS|})$ times, the *while* loop is executed $O(M)$ times and $SF(n_{next})$ is calculated in each iteration in $O(|APPS|)$. Thus, the time complexity of the *SHiC* will be $O(M \times |APPS|^{3/2})$. This is significantly faster than the exhaustive search, and gets equal when $|APPS|$ is equal to the mesh size; i.e. one application per node which is an impractical case.

5. RESULTS AND ANALYSIS

In this section, we assess the impact of the *SHiC* method on improving the mapping results. Several set of applications with 4 to 35 tasks are generated using TGG [20] where the communication volumes ($w_{i,j}$) are randomly distributed between 2 to 16 flits of data. Experiments are performed on our in-house cycle-accurate SystemC many-core platform which utilizes a pruned version of Noxim [21], as its communication architecture. Different mapping and *first node* selection methods are evaluated over the network size varying from 8×8 to 20×20 nodes. These sizes are according to the current industry trends [22], [23] as well as the future many-core systems.

A random sequence of applications is entered into the scheduler FIFO according to the desired rate, λ. The sequence is kept fixed in all experiments for the sake of fair comparison. Applications are scheduled based on First Come First Serve (FCFS) policy and the maximum possible scheduling rate is called λ_{full}. An allocation request for the scheduled application is sent to the *CM* of the platform residing in the node $n_{0,0}$. The *first node* is selected using our *SHiC* method as well as other approaches. Then the

application is mapped according to the desired mapping algorithm and tasks are allocated to the system nodes regarding the mapping result. Nodes emulate the behavior of their allocated task and inform the *CM* upon the task termination. Hence, the *CM* is kept updated about the status of the nodes at run-time without using any monitoring facilities. In order to have a holistic view of the results and enable real case comparisons, each set of experiments are performed over 10 million cycles where hundreds of applications enter and leave the system.

5.1 Square Factor Accuracy and SHiC Success

In our first study, we assess the accuracy of the *square factor* and *SHiC* success in finding the optimum node. We run several experiments while different combinations of *first node* selection approaches and mapping algorithms are set. In order to assess the *square factor* accuracy, in one of the experiments we perform exhaustive search to select the node with the optimum *SF* as the *first node*. The network size is set to 16×16 and applications enter the system with $0.8\lambda_{full}$ rate. Table I shows the average latency of the network (L_{avg}), the percentage of packets delivered by external congestion (*Ext. Cong.*), and normalized values of *NMRD* and *AWMD* metrics for different experiments.

As can be seen, the best results belong to exhaustive search on the *SF* values. This shows the accuracy of the proposed model. Results demonstrate that *SHiC* has been successful in finding the optimum node, as the dispersion (*NMRD*) and power (*AWMD*) metrics show only 4% of increase. We remind the significant effect of area dispersion on the external congestion and network latency. This highlights the importance of the contiguous and convex mapping towards performance improvement of the system. Moreover, *SHiC* approach significantly enhances the performance of the system under the same utilized mapping algorithm; e.g. 50% reduction in the power consumption of *INC* mapping. This proves the effectiveness of the *first node* selection method despite the utilized mapping algorithm. Note that, the extracted power values follow the same trend as *AWMD*; thus omitted for brevity.

When the λ increases towards fully utilized state (λ_{full}), it is less possible to find a well-shaped region with required number of nodes. As the second evaluation, we study the λ effect on the system performance. Figure 6 shows *AWMD* and *NMRD* values of *CoNA/SHiC* case as well as their ratio to the *CoNA/NN* case, indicated by R(*AWMD*) and R(*NMRD*). Other approaches follow almost the same trend. As can be seen both the dispersion and power dissipation of applications increase versus increase of λ, while *SHiC* keeps its preeminence over other approaches and outperforms by 10 to 30 percent.

5.2 Scalability Evaluation

The *SHiC* performance in different network sizes is evaluated as

Table 1. Extracted Results for Different Methods

Mapping / first node	L_{avg}	Ext. Cong.	NMRD	AWMD
CoNA / INC	42	7.71	1.26	1.29
CoNA / NN	43	7.65	1.20	1.23
CoNA / CoNA	41	5.63	1.09	1.06
CoNA / SHiC	40	3.96	1.00	1.00
CoNA / Exh. SF	39	3.55	0.97	0.96
INC / INC	52	11.51	1.27	1.49
INC / SHiC	42	4.02	1.02	1.04

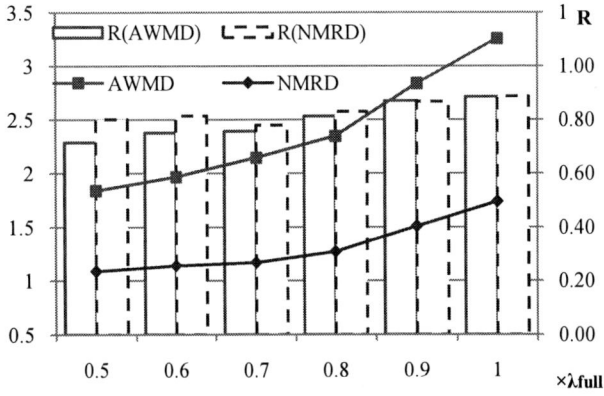

Figure 6. *AWMD* and *NMRD* values for *SHiC* approach in different λ values, along with their ratio to *NN*, valued in the right axis.

273

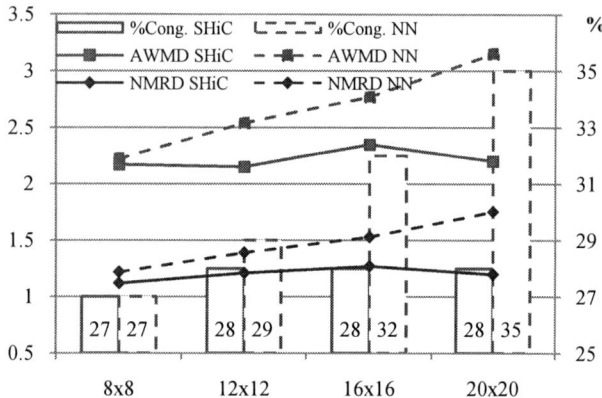

Figure 7. *AWMD, NMRD and % of congested packets in CoNA/SHiC and NN/NN cases over different network sizes. The congestion values are according to the right axis.*

our third study. Different experiments are done over various network sizes while the λ is kept $0.8\lambda_{full}$. Figure 7 shows *AWMD, NMRD* and percentage of congestion in different configurations for both the *CoNA/SHiC* and the *NN/NN* cases. As can be seen *SHiC* scales well as the network size increases and keeps the system performance at the same level. While the system performance in other approaches drops. Thus, the *SHiC* approach is suitable for both the current network sizes and future many-core systems.

6. CONCLUSION

In this paper, we proposed an agile scheme, *SHiC*, to find the optimum *first node* for run-time application mapping of many-core systems. The adapted stochastic hill climbing algorithm finds a node that has the required number of available nodes around it. This algorithm utilized an approximate model which quickly estimates the available area around a given node. The provided open direction aided the climbing algorithm to reach the optimum node faster by taking smart steps.

Simulation results over different network sizes and system utilizations showed significant improvement influenced by *SHiC* in different network parameters such as congestion and power dissipation. More precisely, results emphasized the significant impact of convex mapping on congestion reduction of the network.

As future work a distributed version of the proposed work collaborating with the utilized mapping algorithm is planned. Moreover, a more precise model of mapped applications is desired in order to increase the achieved contiguity of the mapped applications.

7. REFERENCES

[1] W. J. Dally and B. Towles, "Route packets, not wires: on-chip interconnection networks," in *Design Automation Conference, 2001. Proceedings*, 2001, pp. 684– 689.

[2] E. Carvalho, N. Calazans, and F. Moraes, "Heuristics for Dynamic Task Mapping in NoC-based Heterogeneous MPSoCs," in *18th IEEE/IFIP International Workshop on Rapid System Prototyping, 2007. RSP 2007*, 2007, pp. 34–40.

[3] C.-L. Chou, U. Y. Ogras, and R. Marculescu, "Energy- and Performance-Aware Incremental Mapping for Networks on Chip With Multiple Voltage Levels," *IEEE Transactions on Computer-Aided Design of Integrated Circuits and Systems*, vol. 27, no. 10, pp. 1866–1879, Oct. 2008.

[4] M. Fattah, M. Ramirez, M. Daneshtalab, P. Liljeberg, and J. Plosila, "CoNA: Dynamic application mapping for congestion reduction in many-core systems," in *2012 IEEE 30th International Conference on Computer Design (ICCD)*, 2012, pp. 364 –370.

[5] D. P. Dobkin, H. Edelsbrunner, and M. H. Overmars, "Searching for empty convex polygons," in *Proceedings of the fourth annual symposium on Computational geometry*, New York, NY, USA, 1988, pp. 224–228.

[6] S. Fortunato, "Community detection in graphs," *arXiv:0906.0612*, Jun. 2009.

[7] E. L. d. S. Carvalho, N. L. . Calazans, and F. G. Moraes, "Dynamic Task Mapping for MPSoCs," *IEEE Design & Test of Computers*, vol. 27, no. 5, pp. 26–35, Oct. 2010.

[8] A. Weichslgartner, S. Wildermann, and J. Teich, "Dynamic decentralized mapping of tree-structured applications on NoC architectures," in *2011 Fifth IEEE/ACM International Symposium on Networks on Chip (NoCS)*, 2011, pp. 201–208.

[9] S. Kobbe, et al., "DistRM: distributed resource management for on-chip many-core systems," in *Proceedings of the seventh IEEE/ACM/IFIP international conference on Hardware/software codesign and system synthesis*, New York, NY, USA, 2011, pp. 119–128.

[10] M. Hosseinabady and J. L. Nunez-Yanez, "Run-time stochastic task mapping on a large scale network-on-chip with dynamically reconfigurable tiles," *IET Computers Digital Techniques*, vol. 6, no. 1, pp. 1 –11, Jan. 2012.

[11] M. Asadinia, M. Modarressi, A. Tavakkol, and H. Sarbazi-Azad, "Supporting non-contiguous processor allocation in mesh-based CMPs using virtual point-to-point links," in *Design, Automation & Test in Europe Conference & Exhibition (DATE), 2011*, 2011, pp. 1–6.

[12] T. T. Ye, G. D. Micheli, and L. Benini, "Analysis of power consumption on switch fabrics in network routers," in *Proceedings of the 39th annual Design Automation Conference*, New York, NY, USA, 2002, pp. 524–529.

[13] C.-L. Chou and R. Marculescu, "Contention-aware application mapping for Network-on-Chip communication architectures," in *IEEE International Conference on Computer Design, 2008. ICCD 2008*, 2008, pp. 164–169.

[14] C.-Q. Yang and A. V. Reddy, "A taxonomy for congestion control algorithms in packet switching networks," *IEEE Network*, vol. 9, no. 4, pp. 34–45, Aug. 1995.

[15] J. W. Brand, C. Ciordas, K. Goossens, and T. Basten, "Congestion-Controlled Best-Effort Communication for Networks-on-Chip," in *Design, Automation & Test in Europe Conference & Exhibition, 2007. DATE '07*, 2007, pp. 1–6.

[16] S. Ma, N. Enright Jerger, and Z. Wang, "DBAR: an efficient routing algorithm to support multiple concurrent applications in networks-on-chip," in *Proceedings of the 38th annual international symposium on Computer architecture*, New York, NY, USA, 2011, pp. 413–424.

[17] M. Ebrahimi, et al., "HARAQ: Congestion-Aware Learning Model for Highly Adaptive Routing Algorithm in On-Chip Networks," in *2012 Sixth IEEE/ACM International Symposium on Networks on Chip (NoCS)*, 2012, pp. 19 –26.

[18] C. M. Bender, M. A. Bender, E. D. Demaine, and S. P. Fekete, "What is the optimal shape of a city?," *J. Phys. A: Math. Gen.*, vol. 37, no. 1, p. 147, Jan. 2004.

[19] S. Russell and P. Norvig, *Artificial Intelligence: A Modern Approach*, 3rd ed. Prentice Hall, 2009.

[20] *Task graph generator (TGG)*. [Online]. Available at: http://sourceforge.net/projects/taskgraphgen/.

[21] *Noxim: the NoC Simulator*. [Online]. Available at: http://noxim.sourceforge.net/.

[22] J. Howard, et al., "A 48-Core IA-32 message-passing processor with DVFS in 45nm CMOS," in *Solid-State Circuits Conference Digest of Technical Papers (ISSCC), 2010 IEEE International*, 2010, pp. 108 –109.

[23] Tilera Corporation, "Tile-GX Processor Family," 2011.

HCI-Tolerant NoC Router Microarchitecture

Dean Michael Ancajas   James McCabe Nickerson   Koushik Chakraborty   Sanghamitra Roy

USU BRIDGE LAB, Electrical and Computer Engineering, Utah State University
{dbancajas, jmnickerson}@gmail.com {koushik.chakraborty, sanghamitra.roy}@usu.edu

ABSTRACT

The trend towards massive parallel computing has necessitated the need for an On-Chip communication framework that can scale well with the increasing number of cores. At the same time, technology scaling has made transistors susceptible to a multitude of reliability issues (NBTI, HCI, TDDB). In this work, we propose an HCI-Tolerant microarchitecture for an NoC Router by manipulating the switching activity around the circuit. We find that most of the switching activity (the primary cause of HCI degradation) are only concentrated in a few parts of the circuit, severely degrading some portions more than others. Our techniques increase the lifetime of an NoC router by balancing this switching activity. Compared to an NoC without any reliability techniques, our best schemes improve the switching activity distribution, clock cycle degradation, system performance and energy delay product per flit by 19%, 26%, 11% and 17%, respectively, on an average.

1. INTRODUCTION

In the forthcoming era of many-core computing, fueled by the tremendous growth in on-chip resources from technology scaling, Network-on-Chip (NoC) architectures have emerged as the design of choice for on-chip communication. On the other hand, rapid technology scaling has severely undermined the device level reliability, forcing the chip designers to critically consider long term sustainability in system design. While a large body of recent works targets on-chip computing resources (processing cores), many-core systems must consider reliability and sustainability of NoCs. Various aging mechanisms such as *Negative Bias Temperature Instability (NBTI)*, *Hot Carrier Injection (HCI)*, *Time Dependent Dielectric Breakdown (TDDB)*, and *Electromigration* play a major role in degrading performance characteristics of NoCs over time. Such a performance degradation can have a massive system level impact in NoCs, and may ultimately shorten the chip lifetime prematurely [2, 3].

To extend the period of fault-free execution, few recent works have addressed aging challenges in NoCs by mitigating NBTI or Electromigration. For example, Fu et al. propose

---

Permission to make digital or hard copies of all or part of this work for personal or classroom use is granted without fee provided that copies are not made or distributed for profit or commercial advantage and that copies bear this notice and the full citation on the first page. To copy otherwise, to republish, to post on servers or to redistribute to lists, requires prior specific permission and/or a fee.
DAC '13, May 29 - June 07 2013, Austin, TX, USA.

---

techniques to mitigate NBTI aging in NoCs by balancing the duty cycle in the Virtual Allocator circuits [10]. Bhardwaj et al. propose aging-aware adaptive routing to throttle NBTI and Electromigration degradation [3]. NBTI is a critical but recoverable device aging mechanism. In contrast, HCI is an unrecoverable aging phenomena [15], which affects the components due to their dependence on switching activity [17]. Due to aggressive transistor scaling, the thinner gate dielectric in CMOS transistors increases the probability of HCI degradation. In fact, HCI can account for a major component of aging in a 10-year product lifetime [19]. To the best of our knowledge, none of the existing works consider HCI aging in the NoC architecture.

In this work, we perform a holistic cross-layer analysis of HCI degradation in the NoC router microarchitecture. We focus on the crossbar structure of the router microarchitecture, due to its profound significance in dictating the router frequency [16]. Combining application level traffic profile with bit level logic analysis, we find that the crossbar structure is highly vulnerable to HCI aging. Due to the data communication patterns in many-core applications, we observe that a majority of gate-level switching activities are restricted to a small portion of the entire crossbar circuit topology, resulting in a large HCI degradation. To throttle HCI aging in the crossbar, we propose a series of low-overhead techniques that evenly distribute the switching activity in the crossbar, without affecting the architecture level routing latency and bandwidth.

We make the following contributions in this paper.

- We develop a cross-layer framework for HCI aging analysis of an NoC router. Our framework combines application traces, RTL gate-level simulation of a crossbar circuit, logic analysis, and HSPICE simulation of HCI degradation effect (Sections 3.2 and S1).

- We analyze the switching activity of the crossbar, a major circuit in an NoC router using real-world applications and find that only a small group of gates account for most of the switching activity. On an average for PARSEC benchmarks, only 25% of the gates account for more than 75% of the switching activity, severely damaging some gates while leaving others unscathed (Section 3).

- We propose four schemes using low overhead techniques to evenly distribute the switching activity and minimize HCI degradation (Section 4). Our four schemes are: *Bit Cruising* that distributes the high activity bits around the channel; *Distributed Cycle Mode* that exploits idle cycles in the NoC; *Crossbar Lane Switching* that manipulates the port in the crossbar by utilizing the virtual channels; and a combination of Bit-Cruising and Crossbar Lane Switching.

- We present a holistic evaluation of our proposals spanning full-system simulation down to RTL and gate-level HSPICE simulation (Section 6). Our best schemes improve the switching activity distribution by up to 31% (ave: 19%). We also see a maximum of 30% (ave: 26%) improvement in the clock cycle degradation, while the system performance degradation is reduced by up to 17.6% (ave: 11%) compared to the baseline scheme. The Energy Delay Product per Flit is improved by up to 27% (ave: 17%).

2. BACKGROUND

In this section, we introduce our HCI model that correlates threshold voltage degradation with the time spent by a transistor in stress.

HCI occurs when a carrier overcomes the potential barrier between silicon and the gate oxide and leaves the channel. A portion of the carriers (hole/electrons) that leave the channel are deposited into forbidden regions in the transistor such as the gate oxide. Throughout a transistor's lifetime, these deposited carriers change the conductive properties of the transistor and ultimately lead to degradation of the threshold voltage (V_{th}), drain saturation current (I_{on}) and transconductance (Δg_m).

The HCI effect on the transistor parameters described above can be modeled as a power-law with respect to the stress time (t) [6,22]. We only discuss the V_{th} model as the one for I_{on} is similar. The model for Δg_m can be seen in [6].

$$\Delta V_{th} = A \cdot t^n \tag{1}$$

where A and n are technology dependent parameters. Parameter n has been widely accepted as ~ 0.5 over a wide range of processes [1]. Parameter t is the time the transistor is under stress, while A is derived as:

$$A = \frac{q}{C_{ox}} K \sqrt{C_{ox}(V_{GS} - V_{th})} \cdot e^{\frac{E_{ox}}{E_0}} e^{\frac{\varphi_{it}}{q\lambda E_m}} \tag{2}$$

All relevant parameters in Equation 2 can be obtained from Wenping et al. [23]. The stress time of a transistor is derived from the transition density and the pertinent transitions, since not all inputs that cause switching have a significant contribution to HCI aging [13]. We give a brief background of how we estimate pertinent transitions in our framework in Section S1.1.

3. MOTIVATION

In this section, we motivate the need for HCI-aware design of components in an NoC router. We first discuss major reliability concerns in the datapath of an NoC router. We then explain our framework for holistic HCI aging analysis of the NoC crossbar. Lastly, we discuss our results, demonstrating the need for HCI-aware techniques in the design of resilient NoC routers.

3.1 HCI Degradation in the NoC Crossbar

Massively parallel programs running in the many-core use the NoC as an interconnect fabric due to scalability demands. Processors communicate with each other through messages sent as packets in the NoC. Since on-chip wiring

Circuit	Logic Depth	# of gates
Crossbar Switch	4	5760
64-bit ALU	46	4728
Address Generator	43	491
Issue Queue Logic	33	189

Table 1: Logic Depth of Various Modules

is abundant, a lot of these packets that were previously sent over narrow off-chip buses now cannot fully utilize the whole channel bandwidth available. Coupled with the fact that most data sent through the network are narrow width [8], this trend leads to uneven sensitization of transistors, eventually causing unbalanced HCI degradation across the channel.

The crossbar switch is at the heart of the communication infrastructure in an NoC router[1], largely dictating the cycle time [16]. There are three critical reliability issues in an NoC crossbar. First, the gate level activity in a crossbar is only concentrated in a very few bits of the channel width, due to the bit patterns being sent. This asymmetry causes unbalanced HCI degradation. Second, since most upper bit transistors do not switch and only maintain their values, they can undergo NBTI degradation. Third, since the crossbar is a wide circuit with a shallow logic depth (Table 1), minor delay variations caused by both HCI and NBTI will have a profound effect on its overall critical path delay.

3.2 Aging Analysis Framework for the NoC Crossbar

Figure 1 shows the methodology we employ in assessing HCI degradation in the crossbar circuit. Our cross-layer approach comprises system level simulation of 16-thread parallel programs and their gate-level HCI degradation in a crossbar circuit. Since HCI depends on switching activity, we acquire the switching activity of each gate by capturing cycle-by-cycle actual data values traversing the crossbar. We then evaluate its overall degradation effect for each transistor in the circuit using our model discussed in Section 2. However, using real-world applications to assess gate-level degradation is a computationally intensive task. As such, we have adopted several important steps to efficiently avoid long simulation times, while still providing a holistic analysis of HCI aging effect.

First, we pick multiple sample points in different phases of execution of the program. The sample points are chosen according to traffic intensity in the NoC. Each sample phase contains about 1 million flits. Second, we run our simulation setup (Section 5) and take the traces of data traffic at the specified points. Third, we feed these data traces to an Open Source RTL Verilog model of a 16-core NoC and gather cycle-by-cycle inputs in the crossbar circuit. Lastly, we use our novel HCI Aging Analyzer Framework (Section S1) to analyze degradation in the circuit.

3.3 Results

3.3.1 Logic Depth Analysis

Table 1 shows the results for the logic depth analysis we perform on major circuits from NoC and processor systems. We analyzed the crossbar switch from an NoC, the Arithmetic and Logic Unit (ALU), the memory address generator

---

[1]We provide an NoC primer on Section S3.

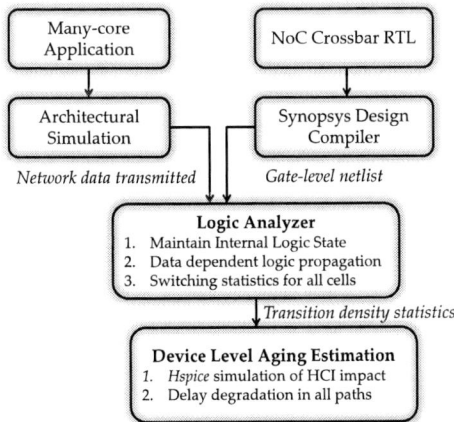

Figure 1: HCI Aging Analysis Framework

and the issue queue selector of a Fabscalar core [7]. Among all these modules, the crossbar has the shallowest logic depth that can be 10× lower than the other circuits. This characteristic makes it more susceptible to aging as there is little chance that a different signal path can hide the delay incurred by degraded transistors. Thus, we need to implement efficient aging mitigation techniques in the crossbar data path.

3.3.2 HCI Degradation Results

Figure 2 shows the switching activity data in the crossbar circuit. The x-axis shows the percentage of gates while the y-axis shows its accumulated switching activity as a percentage of the total activity. Ideally, a 1:1 ratio between the percentile gates and the switching activity is optimal for HCI aging (i.e. a straight line with a 45° slope). However, it can be seen that on an average, only 25% of the gates account for 75% of the total switching activity. This large asymmetry leads to unbalanced HCI degradation between different parts of the circuit and can accelerate failure of NoCs before their rated lifetime.

We also show the corresponding clock cycle degradation of an NoC router (22nm, 7 years) in Figure 3 as a result of the unbalanced HCI degradation. From this data, *swaptions* experiences the most clock cycle degradation at 10.51%, while *canneal* has the least at 8.99%. We can also verify this trend from Figure 2, where swaptions (left most curve) has the most concentrated switching activity among all programs.

Both the results above show that the inherent imbalance in switching activity caused by data patterns sent over the network causes non-uniform HCI aging in the crossbar circuit. This asymmetrical aging causes some path delays to increase disproportionately and will eventually lead to premature router failure. In the succeeding sections, we will discuss our proposed designs that primarily shift the switching activity from one part of a circuit to another in order to slow down HCI degradation and balance aging impact.

4. DESIGN OVERVIEW

In this section, we discuss our proposed techniques for mitigating HCI effect in the router crossbar. Our techniques aim to balance HCI degradation by distributing the switching activity. We explore four techniques in the router microarchitecture: *Bit Cruising (BC)*; *Distributed Cycle Mode (DCM)*; *Crossbar Lane Switching (CLS)*; and *BCCLS* that is a combination of schemes BC and CLS. Apart from DCM,

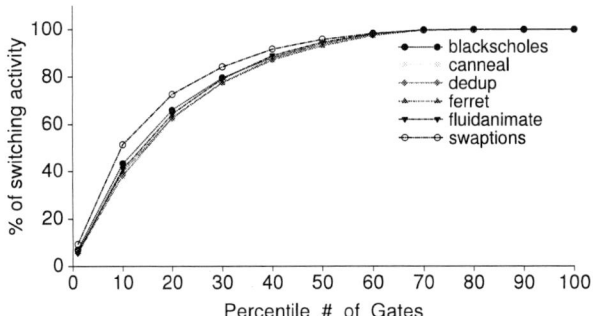

Figure 2: Cumulative Distribution Function of the Switching Activity vs Gate Count

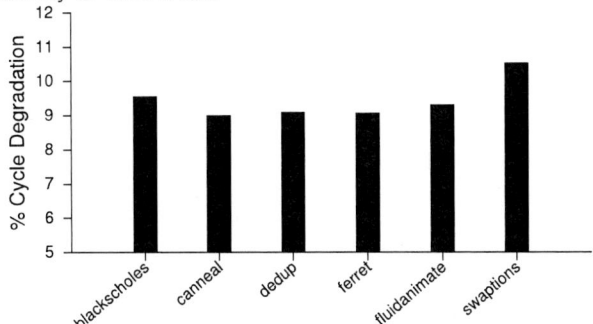

Figure 3: Clock Cycle Degradation of a 22nm NoC Router due to HCI after 7 years

all of our schemes involve minimal modifications at the front-end of the router and do not affect the critical path of the pipeline (crossbar traversal).

4.1 Bit Cruising (BC)

Bit Cruising swaps the different portions of the data being transmitted in the crossbar. This technique is largely motivated by two properties of the programs. First, most data traversing the NoC do not occupy the full channel width of the network because most data in the cache line are aggregated at the lower bits. In some cases, all data bits are actually zero. Recent works have also exploited this characteristic by compressing flits and sending only those that have important data [8]. Second, control requests, while being sent as a single flit also do not store information in the most significant portions of the channel as the routing information can fit in the first few bytes of the whole channel. In our setup, the control flit only utilizes 25% of the channel width, leaving the remaining 75% constant. Together, these two characteristics radically lower the switching activity in certain bits while emphasizing others.

To prevent this asymmetry in HCI degradation, the data being sent across the network must be such that the switching activity across the channel is distributed. By passing different data values each time a gate is used, it will balance the switching activity and hence also uniformly degrade all gates. This is the primary working philosophy of Bit Cruising, where highly changing bits are being *cruised* around the channel. The Bit Cruiser circuit is situated in the Network Interface (NI) and **does not add any overhead in the critical path of the pipeline** of an NoC. We explain in detail the functionality and the circuit implementation of the BC circuit in Section S2.

4.2 Distributed Cycle Mode (DCM)

The Distributed Cycle Mode aims to balance out degradation of transistors by latching an input value in the crossbar during idle times such that unswitched transistors in previous cycles will transition and experience equivalent aging. As such, it does not relieve any HCI aging compared to our other schemes but can be beneficial as equally aged transistors have smaller leakage power. The DCM mode can also be coupled with NBTI recovery schemes such as [21]. We explain the DCM mode in more detail in Section S4.

4.3 Crossbar Lane Switching (CLS)

Our two previous techniques focused on distributing the switching activity across an entire channel of an input port to balance HCI degradation. However, another asymmetrical degradation also occurs in the crossbar lanes that are immune to techniques applied in the channel level.

This type of asymmetric degradation arises when some input-output pairs are used more than others. We demonstrate this occurrence with an example in Figure 4 where there are two paths (p0 and p1) that both use the same *East* output port. For instance, if path p0 is used more than p1, then the transistors along the path p0 will be sensitized more and hence, experience more HCI degradation.

Our third technique, CLS, is also situated at the front-end of the router pipeline and aims to balance the usage of the crossbar lanes[2]. In the canonical router model, an input port directly forwards flits to the output ports by establishing a physical connection between the two via the crossbar switch. As such, flits coming from the same input port will always use the same crossbar lane to connect to different output ports. However, the introduction of Input Buffers (IB) and Virtual Channels (VC) in modern router architectures decouples this one-to-one association because the flits are first stored in the IB before being transmitted to the output ports. With trivial modifications in the VC allocator and the Route Calculation part of the pipeline, we can control which crossbar lane an input port will utilize at any given time.

This new allocation and routing policy will now cause the crossbar circuit to use a different path and activation circuit, but still send the same data as if it were coming from the original input port. Thus, we preserve the correctness of the flit and the route. Similar to the Bit Cruising technique's *cruise setting*, CLS will need a *knob* input to indicate the new mapping between input ports and crossbar lanes. We expand on this and explain the required overheads in implementing CLS in Section S5.

4.4 Bit Cruising and Crossbar Lane Switching (BCCLS)

Our last technique is a combination of the BC and CLS schemes. BCCLS combines both the benefit of switching distribution inside a channel (BC scheme) and the distribution of activity across many channels (CLS scheme). The implementation of BCCLS comes naturally because both BC and CLS tackle different portions of the router circuit. BC reshuffles the data sent through the network while CLS effectively changes the port a flit is coming from by modifying the VC allocation and route calculation.

5. METHODOLOGY

---
[2]a lane is the path taken by an input port to the output port

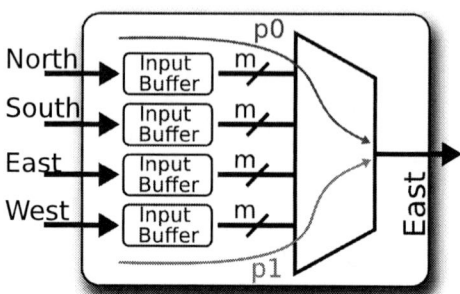

Figure 4: East Section of A Crossbar Switch. CLS works on the **inter**-lane[3] (by changing the path of the data) level while BC works only on the **intra**-lane level (by changing the bit ordering within a path).

In this section, we discuss our simulation infrastructure that combines multiple tools across different abstraction layers. Our methodology can be broadly classified into three categories: Architectural Setup, RTL and Switching Activity Simulation and HCI degradation analysis using SPICE.

5.1 Architectural Setup

Our simulation setup is composed of a 16-node mesh system arranged in a 4×4 grid. Each node in the system is composed of 1 processor, 1 L1 Cache and a slice of a system-shared L2 cache. Each router in the system has seven sets of input and output ports including the ones for the processor and caches. The flit size is configured at 16-bytes (128 bits). A single control request fits in a single flit while data flits needed to transfer a 64-byte cache line are sent in five (4 data + 1 control) consecutive flits. Each processor's L1 and L2 cache sizes are 64kB and 512kB, respectively.

5.2 RTL and Switching Activity Simulation

The first step in obtaining an accurate switching activity is to produce real-word data vectors from standard benchmark programs as inputs to the RTL circuits. We use the PARSEC [4] benchmark suite (large inputs) running on gem5 [5] to collect data traces. We collect data traces for the four center most routers in a 16-node mesh.

After the traces are taken, we implement a trace feeder through a Verilog VPI based functional verification framework called Teal [18]. This module allows us to easily obtain cycle-by-cycle values in any sub-module of the router such as the crossbar.

5.3 HCI Degradation Analysis

Using the outputs from the previous step, our logic analysis tool is then used to obtain the transition densities of each transistor (Figure 1). We post-process all the results in our HAAF (Section S1) to calculate V_{th} degradation and simulate them in HSPICE to obtain clock cycle degradation data for all paths and for different benchmarks. In all our analysis, we use the 22nm [24] technology and an aging period of 7 years.

6. RESULTS

In this section we present the effectiveness of our schemes across different metrics.

6.1 Comparative Schemes and Evaluation Metrics

We compare the following five schemes:

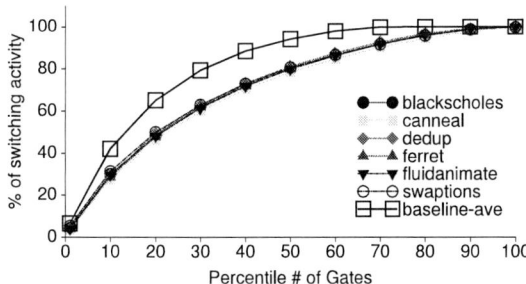

Figure 6: CDF for BCCLS scheme.

- **BASE**: Baseline configuration where the system is unmodified.
- **BC**: Bit Cruising scheme, the channel is divided into four segments and a bit cruiser circuit is placed between the NI and the router.
- **DCM**: Distributed Cycle Mode technique presented in Section S4.
- **CLS**: Crossbar Lane Switching scheme discussed in Section 4.3.
- **BCCLS**: Combination of BC and CLS schemes.

We evaluate all these schemes in terms of switching activity distribution through Cumulative Distribution Function (CDF) plots, clock frequency degradation, Energy-Delay Product Per Flit (EDPPF) and System Performance. The circuits used to facilitate all these schemes (except DCM) are added in the front-end of the pipeline without affecting the actual crossbar circuit, and as such do not incur any additional timing overhead in the critical path[3].

6.2 Switching Activity Distribution

We show the CDF plots (Figures 5 and 6) of the switching activity distribution of all schemes. The average of the baseline scheme is superimposed in each figure. All of our schemes outperform the baseline by having a lower value (y-axis) at any percentile point. Hence, it is evident that our schemes achieved their aim of distributing the switching activity. At an evaluation point of 20 percentile, our best performing scheme (BCCLS in Fig. 6) shows 31% less switching activity compared to the baseline.

6.3 Clock Cycle Degradation

Figure 7(a) shows the cycle degradation for the NoC router at the end of a 7 year aging period using the ASU 22nm predictive technology model [24] operating at 1 Ghz. On an average, the base scheme degrades the clock cycle by 9.4%. Our schemes improve this degradation by 20.6%, 0%, 12% and 25.5% for BC, DCM, CLS and BCCLS, respectively. Combining both BC and CLS schemes results in the least amount of clock cycle degradation while DCM provides no improvement from the baseline. As HCI is an unrecoverable degradation [15], any damage done during normal operation cannot be rectified. The difference between DCM and all other schemes is that it is reactive while the others are proactive (preventing aging beforehand). However, DCM improves other aspects of the circuit such as the EDPPF which will be discussed next.

6.4 Energy Delay Product Per Flit (EDPPF)

---

[3]DCM's cycle degradation is taken without the timing of the additional multiplexers as we want to show the timing degradation in the crossbar circuit only across all schemes.

We show in Figure 7(b) the EDPPF of all schemes. The base scheme is shown as a line at 100%. Most schemes have lower EDPPF compared to the baseline except for some outliers. For the BC scheme, *dedup* and *ferret* have larger EDPPFs while for CLS, *swaptions* has a slightly larger EDPPF than the baseline. Upon further investigation, although BC has helped achieve less degradation and a more distributed switching activity, its dynamic switching activity for benchmarks *dedup* and *ferret* are actually 63% and 30% more compared to the average of all other programs. This unusual activity increase is due to the workload-dependent bit patterns being sent across the network. For *swaptions*, the switching activity for the benchmark is unusually high in all schemes except for BC.

Even though DCM does not provide any improvement in the clock cycle, it provides consistent reduction in EDPPF. This reduction is because optimally aged transistors have higher threshold voltages and will have lesser leakage power. Leakage power cannot be ignored in small technologies such as the one we are using (22nm). On an average, DCM improves the EDPPF by 18% compared to the baseline.

6.5 System Performance

Figure 7(c) shows the overall system performance impact of all schemes relative to the baseline. DCM shows no improvement because it has the same clock degradation as the baseline. On an average, performance degradation is reduced by 9.3%, 8% and 11% for BC, CLS and BCCLS schemes. Maximum is 17.6% for the BCCLS scheme running *ferret*. Overall, the system performance improvement is less than the clock cycle degradation improvement due to the sublinear dependence of clock frequency and performance.

7. RELATED WORK

The aggressive scaling in CMOS technology has made reliability a primary design constraint in modern computing systems. While there has been a wide scope of studies tackling different reliability issues (NBTI, TDDB, HCI) in processing elements [11, 21], there is only a limited number of works which address wear-out mitigation in the on-chip communication infrastructure of such systems. Bhardwaj et al. implemented a dynamic routing algorithm to equalize NBTI and electromigration aging across the on-chip network [3]. Fu et al. created new virtual channel allocation and routing algorithms in order to improve process variation and NBTI effects in key components of the router [10]. Park et al., Fick et al. and Kim et al. explored fault tolerant NoC architectures by decoupling modules and having redundancies in order to recover from intermittent errors in the network or provide graceful degradation [9] [14] [20].

Most of the studies mentioned above focus on recovering from intermittent errors or minimizing NBTI effect on storage elements by balancing the duty cycle. On the contrary, our work focuses on HCI, an unrecoverable aging phenomena that affects combinational components. HCI mitigation presents a different set of challenges because of its dependence on the switching activity of transistors, as opposed to NBTI which depends only on the input bias. To the best of our knowledge, our study is the first work to tackle HCI in an NoC router microarchitecture.

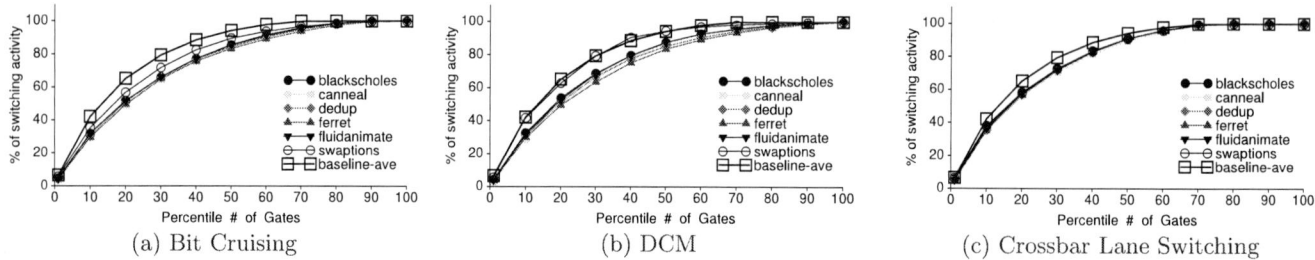

(a) Bit Cruising   (b) DCM   (c) Crossbar Lane Switching

Figure 5: Cumulative Distribution Graph of BC, DCM and CLS schemes. Solid line in each graph is the baseline average. (Lower is better.)

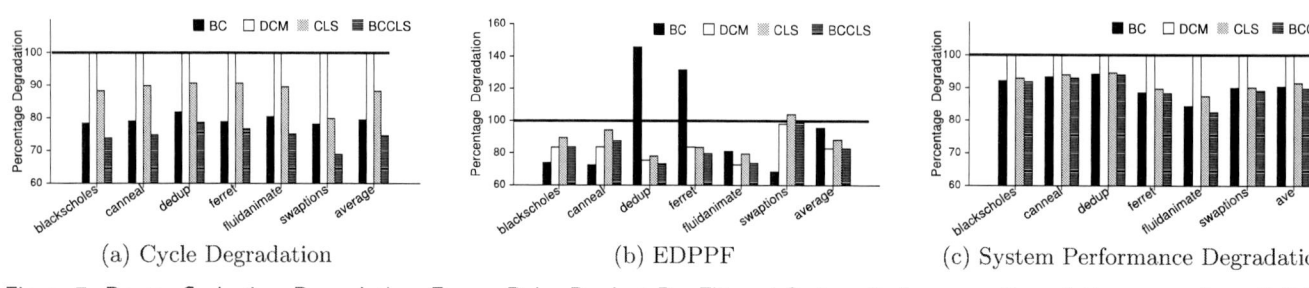

(a) Cycle Degradation   (b) EDPPF   (c) System Performance Degradation

Figure 7: Router Cycle time Degradation, Energy Delay Product Per Flit and System Performance Degradation comparison. Solid line indicates baseline. (Lower is better).

8. CONCLUSION

In this paper, we find out that Network On Chip architectures running parallel programs produce communication patterns that lead to unbalanced HCI degradation through asymmetrical gate switching activity. We exploit this property and present four novel proposals in HCI mitigation in the crossbar circuit, a major component in an NoC router which dictates the operating frequency of the network. Overall, our schemes distribute the switching activity, improve the clock cycle degradation, energy delay product per flit and system performance.

Acknowledgments

This work was supported in part by National Science Foundation grants CNS-1117425 and CAREER-1253024, and donation from the Micron Foundation.

9. REFERENCES

[1] BERTOLINI, C. AND OTHERS Relation between HCI-induced performance degradation and applications in a RISC processor. *Proc. of OLTS* (2012), 67–72.

[2] BHARDWAJ, K. AND OTHERS An MILP Based Aging Aware Routing Algorithm for NoCs. In *Proc. of DATE* (2012), pp. 326–331.

[3] BHARDWAJ, K. AND OTHERS Towards Graceful Aging Degradation in NoCs Through an Adaptive Routing Algorithm. In *Proc. of DAC* (2012), pp. 382–391.

[4] BIENIA, C. AND OTHERS The PARSEC benchmark suite: characterization and architectural implications. In *PACT* (2008), pp. 72–81.

[5] BINKERT, N. AND OTHERS The gem5 simulator. *SIGARCH Comput. Archit. News 39*, 2 (Aug. 2011), 1–7.

[6] BRAVAIX, A. AND OTHERS Hot-Carrier acceleration factors for low power management in DC-AC stressed 40nm NMOS node at high temperature. In *Proc. of RPS* (2009), pp. 531 –548.

[7] CHOUDHARY, N. K. AND OTHERS FabScalar: composing synthesizable RTL designs of arbitrary cores within a canonical superscalar template. In *Proc. of ISCA* (2011), pp. 11–22.

[8] DAS, R. AND OTHERS Performance and power optimization through data compression in Network-on-Chip architectures. In *HPCA* (2008), pp. 215–225.

[9] FICK, D. AND OTHERS Vicis: a reliable network for unreliable silicon. In *Proc. of DAC* (2009), pp. 812–817.

[10] FU, X. AND OTHERS Architecting reliable multi-core network-on-chip for small scale processing technology. In *Proc. of DSN* (2010), pp. 111–120.

[11] GUPTA, S. AND OTHERS StageNetSlice: a reconfigurable microarchitecture building block for resilient CMP systems. In *Proc. of CASES* (2008), pp. 1–10.

[12] HESTNESS, J. AND OTHERS Netrace: dependency-driven trace-based network-on-chip simulation. In *Proc. of WNOCA* (2010), pp. 31–36.

[13] KAMAL, M. AND OTHERS An Efficient Reliability Simulation Flow for Evaluating the Hot Carrier Injection Effect in CMOS VLSI Circuits. In *ICCD* (2012), pp. 352–357.

[14] KIM, J. AND OTHERS A Gracefully Degrading and Energy-Efficient Modular Router Architecture for On-Chip Networks. In *Proc. of ISCA* (2006), pp. 4–15.

[15] KUFLUOGLU, H. *Mosfet Degradation due to NBTI and HCI and its Implications for Reliability-Aware VLSI Design.* PhD thesis, Purdue University, 2007.

[16] KUNDU, P. On-Die Interconnects for Next Generation CMPs. In *Proc. of WOCIN* (2006).

[17] LORENZ, D. AND OTHERS Aging analysis at gate and macro cell level. In *Proc. of ICCAD* (2010), pp. 77–84.

[18] MINTS, M., AND EKENDAHL, R. *Hardware Verification with C++: A Practitioners Handbook*, vol. 1. Springer, 2006.

[19] NIGAM, T. AND OTHERS Accurate product lifetime predictions based on device-level measurements. In *Proc. of RPS* (2009), pp. 634 –639.

[20] PARK, D. AND OTHERS Exploring Fault-Tolerant Network-on-Chip Architectures. In *Proc. of DSN* (2006), pp. 93–104.

[21] SIDDIQUA, T., AND GURUMURTHI, S. Enhancing NBTI Recovery in SRAM Arrays Through Recovery Boosting. *IEEE Trans. on VLSI Systems. 20*, 4 (2012), 616–629.

[22] TAKEDA, E., AND SUZUKI, N. An empirical model for device degradation due to hot-carrier injection. *Electron Device Letters 4*, 4 (1983), 111 – 113.

[23] WANG, W. AND OTHERS Compact Modeling and Simulation of Circuit Reliability for 65-nm CMOS Technology. *IEEE Trans. on Device and Materials Reliability* (2007), 509 –517.

[24] ZHAO, W., AND CAO, Y. *Predictive Technology Model.* http://ptm.asu.edu/.

Supplemental Materials

S1. HCI AGING ANALYZER FRAMEWORK (HAAF)

In this section, we discuss our Aging Analyzer Framework used to evaluate HCI degradation of all gates in a circuit. We first give an overview of pertinent transitions of a gate and then discuss our simulation framework.

S1.1 Pertinent Transitions

HCI affects a transistor during a switching activity. However, for a reliability evaluation of a VLSI circuit consisting of thousands of transistors operating for years (typically 7-10), accurate HCI degradation analysis using HSPICE takes too long. As such, it has been determined by [13] that only certain type of transitions in a logic gate generate interface traps in its transistors. Hence, we only calculate the HCI impact of these transitions, allowing for a practical simulation time. We list the pertinent transitions of the gates we used in our design (INV, NAND, NOR) in Table 2, the transitions indicated in the second column and third column induce HCI degradation for NMOS and PMOS transistors, respectively. We simulate all these transitions and evaluate their HCI aging impact on the logic gates. Only transitions that affect the transistor near the output node are counted as they contribute the most to HCI [13].

GATE	NMOS	PMOS
INV	$\uparrow$A	$\downarrow$A
NOR	$(\uparrow$A,$\overline{B})$ $(\overline{A},\uparrow$B$)$	$(\downarrow$A,B$)$
NAND	$(\uparrow$A,B$)$	$(A,\downarrow$B$)$ $(\downarrow$A,B$)$

Table 2: Pertinent Transitions of Various Gates

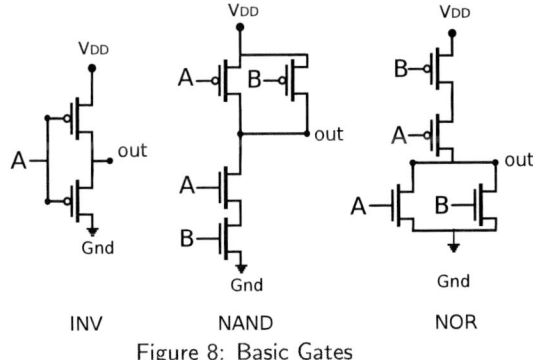

INV   NAND   NOR

Figure 8: Basic Gates

S1.2 Aging Framework

In the gate level, HCI degradation is manifested during transistor switching. We developed a tool to examine the possible HCI impact on all gates of a circuit through extensive logic analysis. Our HAAF tool works by taking the input in every clock cycle and propagating the logic in a domino fashion until it reaches the output. During the course of this propagation, some gates will switch while others will not. We record all these transitions in all clock cycles and use them to calculate the transition density of the gate. Note that we post-process all the transition events to deter-

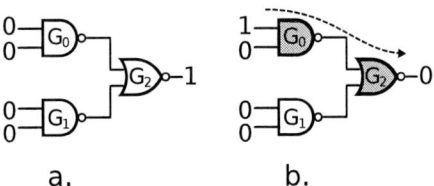

a.   b.

Figure 9: HCI Analysis

mine if they are pertinent transitions before calculating the transition densities.

Figure 9a shows a detailed example of how this analysis is done on a circuit with three gates indicated as G_0, G_1, G_2, with initial states as shown[4]. Figure 9b shows a new set of inputs being fed and denotes the specific gates that will change (highlighted in gray). G_0 and G_2 changes in this cycle while G_1 does not. We calculate the transition density (TD_g) of a gate g as follows:

$$TD_g = \frac{\sum_{n=1}^{x} S_{gn}}{x} \quad (3)$$

where x is the total number of cycles simulated and $S_{gn} = 1$ if gate g made a pertinent transition at cycle n (0 otherwise). We then use the transition density to calculate the new V_{th} using our model in Equation 1. A new propagation delay t_g is then obtained for each gate g using HSPICE simulation. Note that we simulate t_g for all gates and not just the ones in the critical path because the critical path can change depending on the extent of degradation in different parts of the circuit. Finally, we calculate the new propagation delay (T_P) of the whole circuit as:

$$T_P = max(X_0, X_1, ..., X_Y) \quad (4)$$

$$X_y = \sum_{g=1}^{H_y} t_g \quad (5)$$

where Y is the set of all paths in the circuit and X_y is the total propagation delay of path y. H_y is the set of all gates in path y.

The process discussed above forms the bulk of our evaluation framework and although it is very computationally intensive, its thoroughness allows us to accurately evaluate the benefits of our architectural techniques at a circuit-level accuracy.

S2. CIRCUIT IMPLEMENTATION OF BC

We discuss the implementation and overhead of the Bit Cruiser circuit as a continuation of the discussion in Section 4.1.

Figure 10 shows the Bit Cruiser circuit that is responsible for cruising the bits around the channel. Bit Cruising can be implemented at different granularities. However, in this work we use a granularity of four (i.e. the whole channel is segmented into four equal parts) because the most lower quarter of the channel bits have the most activity based on our input traces. The Bit Cruiser circuit takes in as inputs

---

[4]initial states are a result of a previous execution

281

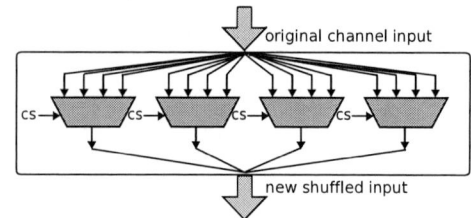

Figure 10: Bit Cruiser Circuit

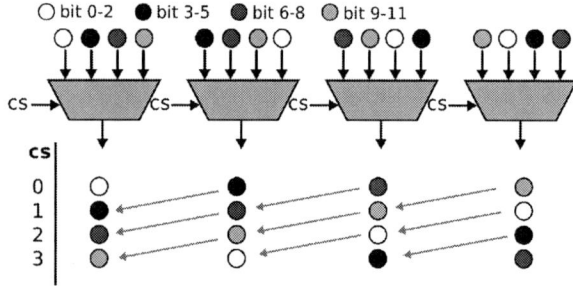

Figure 11: Time Lapse Example of a Bit Cruiser Circuit for a 12-bit channel. Signal cs is the cruise setting.

the *channel bits* and a 2-bit *cruise setting*. The *cruise setting* is then used as an input to a 4-to-1 multiplexer in order to reshuffle the bits as desired.

We show in Figure 11 an example of the effect of bit cruising on channel bits. The *cs* signal in the figure represents the cruise setting. In this example, we assume that in each clock cycle, *cs* is increased by one[5]. The shaded circles are used to indicate one segment of a channel. In the figure, when *cs* is equal to zero, the BC circuit output is the same as the channel input (i.e. or when there is no BC circuit at all). When *cs*=1 the lowermost segment is transferred to the uppermost and the second lowermost segment is shifted to the former's place (direction indicated by an arrow). All other segments follow in unison.

S2.1 Overhead of BC

The circuit in Figure 10 will be placed in the Network Interface right before sending the flit to the router of the source node. Since the bits being sent through the network are now jumbled, the router front end must be able to appropriately identify the header flit bits in order to route the circuit correctly. To this end, we introduce a Routing Information Extraction (RIE) circuit. The RIE circuit extracts the appropriate bits from the shuffled channel bits and places it in a Routing Information Register (RIR), which will be accessed by the Routing Calculation module in the succeeding pipeline stages.

Figure 12 shows the implementation of the RIE circuit. Every time a flit arrives and is about to be written in the virtual channel, the RIE circuit (using the cruise setting information) will determine if the flit is a head flit. If it is, the routing information is latched into the RIR. The RC module in the next pipeline stage will then use the contents of the RIR to route the flit in the corresponding VC. Since there is only going to be one packet in each virtual channel, the overhead for the RIR is minimal. We next calculate its overhead in a typical modern NoC configuration.

In modern NoCs, the network width is often large as on-chip wiring is abundant and bandwidth is also important.

---

[5]Note that the cruise setting can also be altered for larger time granularities.

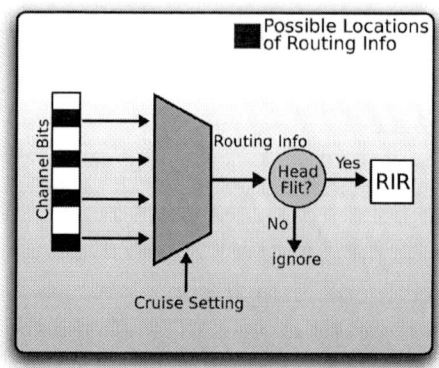

Figure 12: Routing Information Extraction Circuit

For a 128-bit flit width in an 8×8 network with an algorithmic routing algorithm that uses the number of hops in the x and y directions, the RIR will only require 3 bits in each direction for a total of 6 bits. If there are 5 flits in each buffer, then the overhead is 0.9%. For deeper flit buffers, the overhead further goes down.

S3. NOC AND CROSSBAR CIRCUIT PRIMER

In a network on chip, the crossbar circuit acts as the heart of communication of all inputs and outputs in the router. Hence, it's reliability is of extreme importance in the functioning of the whole NoC. Moreover, the critical latency delay of a crossbar circuit dictates the clock frequency of the whole NoC [16]. Thus, any minor deviations in its critical path could corrupt the transmitted data and will lead to total failure in the router and the NoC itself. This makes the crossbar circuit a good candidate as a case study for our HCI mitigation techniques.

Packets in NoCs are sent from one node to another by hopping through intermediate nodes. Figure 13 shows a 16-node mesh where node 0 sends a packet to node 8. Before reaching node 8, the packet will traverse through nodes 1,4 and 7. The router in each node is responsible for the correct transmission of a packet to its adjacent nodes.

Figure 14 shows a simplified model of a typical router. The end goal of a router is to transfer flits from an input port to a specified output port. In each cycle, the router will receive multiple requests to route, an allocator circuit will then determine an optimal connection of input and output ports. In the succeeding cycle, the allocator signals the crossbar and the flits traverse the switch, and are sent to the next adjacent router. This process continues in each node until the flit reaches the destination node.

As the crossbar is at the focal point of packet transmission in an NoC, we choose the crossbar circuit as a case study for our HCI mitigation techniques, our techniques could be easily adapted to work on other parts of an NoC.

S4. DISTRIBUTION CYCLE MODE (DCM)

The Distribution Cycle Mode technique utilizes idle cycles to balance out the HCI degradation of transistors. Most real-world applications spend a considerable time waiting for information from the NoC. Moreover, cache coherence requests are self-throttling or that succeeding requests are not sent unless a reply is received [12]. As such the crossbar spends most time (average of 85% in our setup) sending no data through the crossbar. This presents us with tremen-

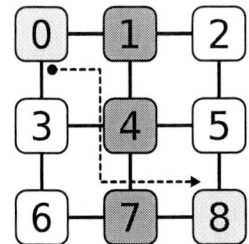

Figure 13: 3x3 Mesh NoC. Node 0 sending a packet to node 8.

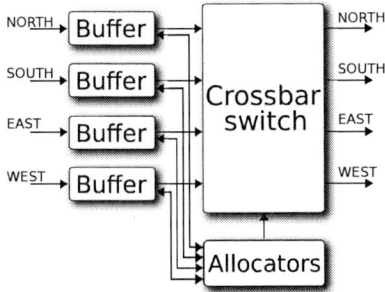

Figure 14: Simple NoC Router

dous opportunities in using reliability techniques without incurring a latency overhead. Our second technique, DCM, introduces a new mode of operation for the crossbar circuit during idle phases of execution. In DC mode, the crossbar uses optimized inputs to achieve a near-uniform switching activity across the data channel.

To explain the idea behind DCM, we introduce a simple example of a 2-bit NOR-gate in DC mode. Figure 15 shows a 2-bit NOR gate in standard CMOS executing optimized inputs. For each set of inputs, the transition number is indicated along with an arrow in the circuit that indicates the corresponding path that the output signal has taken. For instance, transition 0: inputs A=0 and B=0, switches both the PMOS transistors while transition 1 (A=0, B=1) only switches NMOS B. The key to successfully implementing the DC Mode is to balance out the number of switching transitions that each transistor makes.

In this example, executing transitions 0-2 once gives each transistor a balanced switching count. Note that there are four possible inputs for the NOR gate but we only need three in order to get a balanced HCI degradation. Being able to accurately determine the needed inputs, rather than exhaustively using all possible, is the key to optimal DCM operation.

S4.1 Implementing DCM in the Crossbar

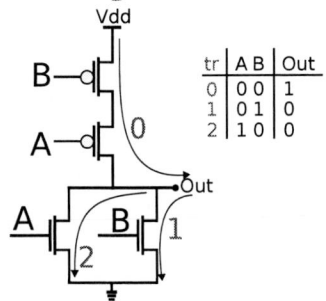

tr	A	B	Out
0	0	0	1
1	0	1	0
2	1	0	0

Figure 15: Two-bit NOR gate showing the different transitions with respect to inputs.

Applying DCM to a big circuit such as a router crossbar poses some major challenges because optimal HCI degradation is only achieved when the inputs are carefully constructed to balance the switching activity. However, despite the enormity of the crossbar, its regular structure allows us to analyze a small subset of the circuit and use our results to optimize the whole component.

There are three key requirements to seamlessly applying DCM while maintaining the correct and unobstructed execution of the NoC Router. We outline them here and discuss each one in detail. They are:

1. **Idle time identification** - To engage the crossbar in the Distributed Cycle Mode, idle cycles must be correctly identified or else the correct value that is supposed to be transferred during the *switch traversal* stage of an NoC is going to be overwritten. This overwriting can corrupt a running program.

2. **Identification of optimal inputs** - The optimal inputs to the crossbar circuit are derived using an offline analysis similar to the one discussed in the previous subsection. This is a one time effort that can be used throughout the lifetime of the NoC router.

3. **Feeding mechanism of customized inputs** - The crossbar must have an option of using the inputs provided by the analysis above in order distribute HCI aging in all of its transistors.

In a typical router in an NoC, the crossbar switch has multiple lanes to handle simultaneous demands of multiple inputs to multiple outputs (NORTH, SOUTH, EAST, WEST). As such, when no input port is scheduled to transfer a data to a specific output port in a particular time, that output port (or lane) is considered idle. Thus, correctly identifying the idle cycles of a crossbar depends mostly on the output of the scheduling algorithm of the switch.

The main mechanism to identify idle cycles is already present in any Switch Allocator (SA) implementation as it outputs a schedule of the switches every clock cycle. Figure 16 shows an NoC router along with the supplementary logic and components to identify idle cycles and implement DCM. Aside from the main DCM module that serves as the control unit for DCM operation, a lookup table and an additional multiplexer is added for the purpose of storing the optimized values and to have the ability to load them when desired, respectively.

In each clock cycle, the SA takes in as input the requests of different virtual channels and input ports and gives the permission to specific input ports to use the output ports in the next cycle. If there are no contention of requests, all requests could be permitted to traverse in the crossbar the next cycle. However, if there is, it is resolved based on a scheduling priority. In other cases though, there simply are not enough requests to keep the switch/crossbar fully utilized. When this happens, our DCM module immediately senses this and queries the lookup table and instructs the multiplexer to load an HCI aging-optimized value in the subsequent cycle.

S5. CROSSBAR LANE SWITCHING

In this section, we elaborate in more details the implementation of the Crossbar Lane Switching scheme discussed in Section 4.3. We first discuss the baseline implementation

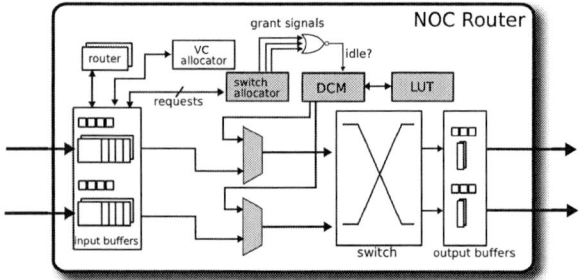

Figure 16: Modified NoC Router to Accommodate DCM operation.

of a modern NoC Router and then explain our modifications in order to implement CLS.

Figure 17 shows a logical diagram for a traditional Virtual Channel (VC) flow NoC Router with two input ports and two virtual channels per input port. The virtual channels are used to handle multiple concurrent streams per input port, each waiting for its turn to use the crossbar switch, hence improving the overall bandwidth of the network. In our example, the north input port can only utilize VCs 1 and 2, while the south utilizes 3 and 4. In each clock cycle, all VCs request usage of the crossbar for the succeeding cycle. The switch allocator will then determine a winner and subsequently connect the virtual channel to the desired output buffers.

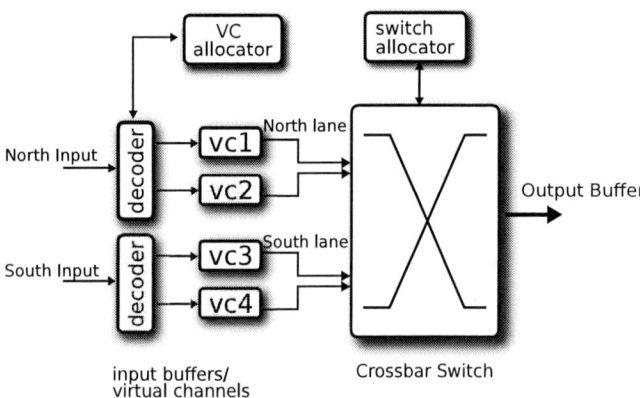

Figure 17: Baseline implementation of an NoC router showing the virtual channels, input ports and the crossbar switch. Output Ports and Output Virtual Channels are not shown.

As we have discussed in Section 4.3, the lanes of the crossbar can undergo uneven degradation when certain input and

output pairs are used more. CLS aims to balance this degradation by evenly distributing the paths taken by a flit. Figure 18 shows the necessary modifications on the NoC Router to be able to implement CLS. Also, the VC allocator must be able to assign any incoming flit to any virtual channel (additional lines in the decoder)[6]. As the virtual channels are implemented as SRAM arrays [SR1] similar to a register file in a processor, there will be no additional logic needed to access the different virtual channels. The only extra logic needed will be for the VC allocator to distribute

---

[6]Note that there are many possible implementations of the Input Buffers. Our overhead is analyzed with respect to an open-source RTL implementation of a modern NoC router [SR1].

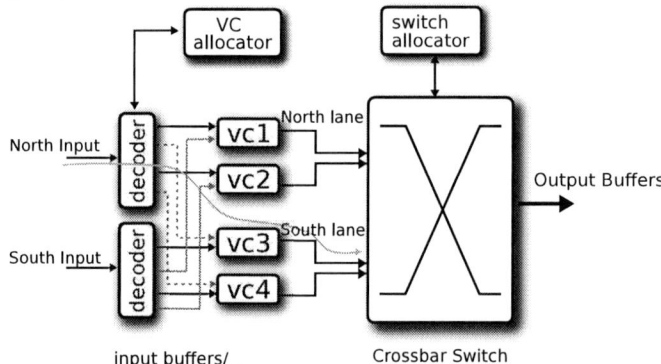

Figure 18: CLS Implementation. VC Allocator Can Assign Incoming Flits to Any Virtual Channel.

the flits across the many virtual channels which can be accomplished by a simple counter circuit which is added to the offset of the decoding stage. The Route Calculation (RC) stage will automatically determine the route of the flit since the routing information is stored in the head flit. The RC will then send the SA the appropriate commands, preserving the correctness of the flit and its route.

The light blue line in Fig. 18 shows the path taken for a flit arriving at the North input and traversing the South lane of the crossbar. This is made possible by storing the flit in virtual channels 3 or 4 and then informing the SA to use the same channel as input to the crossbar. In summary, an incoming flit uses the same input port, a different virtual channel and crossbar lane, and the same output port.

S6. REFERENCES

[SR1] DANIEL BECKER AND WILLIAM DALLY Open Source NoC Router RTL *https://nocs.stanford.edu/cgi-bin/trac.cgi/wiki/Resources/Router.*

Optimization of Quantum Circuits for Interaction Distance in Linear Nearest Neighbor Architectures

Alireza Shafaei   Mehdi Saeedi   Massoud Pedram

Department of Electrical Engineering
University of Southern California
Los Angeles, CA 90089
{shafaeib,msaeedi,pedram}@usc.edu

ABSTRACT

Optimization of the interaction distance between qubits to map a quantum circuit into one-dimensional quantum architectures is addressed. The problem is formulated as the Minimum Linear Arrangement (MINLA) problem. To achieve this, an interaction graph is constructed for a given circuit, and multiple instances of the MINLA problem for selected subcircuits of the initial circuit are formulated and solved. In addition, a lookahead technique is applied to improve the cost of the proposed solution which examines different subcircuit candidates. Experiments on quantum circuits for quantum Fourier transform and reversible benchmarks show the effectiveness of the approach.

Categories and Subject Descriptors

B.6.3 [Logic Design]: Design Aids—*Automatic synthesis*

General Terms

Algorithms, Design

Keywords

Logic synthesis, quantum circuits, interaction distance, quantum architectures.

1. INTRODUCTION

Current technologies for quantum computing often need gates that involve geometrically *adjacent* qubits. The architecture of a quantum computing system can be described by a simple connected graph $G = (V, E)$ where vertices V represent qubits and edges E represent adjacent qubit pairs where gates can be applied on [1]. Accordingly, a complete graph expresses the absence of any *constraints*. Quantum algorithms usually consider no interaction constraint between qubits. However, physical implementation may impose additional geometrical constraints. Therefore, the developed quantum algorithms or quantum circuits should be modified to consider the effect of various technological limitations.

Permission to make digital or hard copies of all or part of this work for personal or classroom use is granted without fee provided that copies are not made or distributed for profit or commercial advantage and that copies bear this notice and the full citation on the first page. To copy otherwise, to republish, to post on servers or to redistribute to lists, requires prior specific permission and/or a fee.
DAC '13, May 29 - June 07 2013, Austin, TX, USA.

Quantum computation technologies arrange qubits of a physical layout in a one (1D), two (2D), or three (3D) dimensional architecture.[1] The *Linear Nearest Neighbor* (LNN) architecture corresponds to a graph where an edge exists between only neighboring vertices in a line. *Two-dimensional square lattices* (2DSL) corresponds to a graph on a Manhattan grid with four neighboring qubits. The *three-dimensional square lattices* (3DSL) model is a set of stacked 2D lattices with six neighboring qubits. Generally, 3DSL is less restrictive. However, it can suffer from the difficulty of controlling 3D qubits. Several quantum computing systems of trapped ions [2] and liquid NMR [3] have been designed based on the interactions in a line. 2DSL proposals include arrays of trapped ions [2] and Josephson junctions [4]. The architecture in [5] is based on the 3DSL model.

Exploring an efficient realization of a given quantum algorithm or quantum circuit for a restricted architecture — the focus of this work — has been followed by different researchers during the recent years. Physical implementation of the quantum Fourier transformation (QFT) [6,7], Shor's factorization algorithm [8–10], quantum addition [11], quantum error correction [12], and general reversible circuits [13] for the LNN/2DSL architectures have been explored in the past. Worst-case synthesis cost of a general/Boolean unitary matrix under the nearest neighbor restriction has been discussed in [14–17]. In [18,19] heuristic methods for converting an arbitrary quantum circuit to its equivalent circuit on the LNN architectures have been proposed.

In this work, we model the problem of improving *locality*, i.e., reducing *interaction distance*, of a given quantum circuit by graph theory. Precisely, we use the *minimum linear arrangement* (MINLA) problem in graph theory to find optimized *local quantum computation*, in terms of the total synthesis cost or latency, in architectures with qubits arranged in a line. The rest of this paper is organized as follows. In Section 2, basic concepts are introduced. Prior work is discussed in Section 3. Section 4 describes the proposed approaches for locality improvement of quantum circuits. Experimental results are given in Section 5 and finally Section 6 concludes the paper. We also discuss how the proposed (MINLA)-based techniques can be generalized for 2D quantum architectures.

2. BASIC CONCEPTS

---
[1]Quantum technologies proposed mainly for quantum communication, such as photon-based model, is not considered here.

In the following two subsections, we briefly discuss related concepts in quantum circuits and quantum architectures.

2.1 Quantum gates and circuits

A quantum bit, *qubit*, can be considered as a mathematical object which represents a quantum state with two basic states $|0\rangle$ and $|1\rangle$. In addition to the selected basis, a qubit can get any linear combination of its basic states. A quantum system which contains n qubits is often called a *quantum register* of size n. An n-qubit *quantum gate* performs a specific $2^n \times 2^n$ unitary operation on selected n qubits. The unitary matrix implemented by several gates acting on different qubits independently can be calculated by the tensor product of their matrices. Two or more quantum gates can be cascaded to construct a *quantum circuit*. For a set of k gates $g_1, g_2, \cdots, g_k$ cascaded in a quantum circuit C in sequence, the matrix of C can be calculated as $M_k M_{k-1} \cdots M_1$ where M_i is the matrix of the i-th gate ($1 \leq i \leq k$). Given any unitary U over m qubits $|x_1 x_2 \cdots x_m\rangle$, a controlled-$U$ gate with k control qubits $|y_1 y_2 \cdots y_k\rangle$ may be defined as an $(m+k)$-qubit gate that applies U on $|x_1 x_2 \cdots x_m\rangle$ iff $|y_1 y_2 \cdots y_k\rangle = |11 \cdots 1\rangle$. For example, CNOT is the controlled-NOT with a single control, Toffoli is a NOT gate with two controls. A multiple-control Toffoli gate C^kNOT is a NOT gate with k controls. In circuit diagrams, $\bullet$ is used for conditioning on the qubit being set to value one. A SWAP gate maps $|ab\rangle$ into $|ba\rangle$. We use $\times$ on qubits of a SWAP gate in circuit diagrams. More information is in [20].

2.2 Physical layout

In a particular realistic physical layout, e.g., in an ion-trap quantum architecture [21], each qubit has a specific *physical* location at each time step. To apply a 2-qubit gate in a quantum computing system which uses *mobile* qubits, both qubits should be available at one location.[2] This is done by moving qubits from one physical location to another during the computation. Some quantum architectures provide a "MOVE" operator which moves one qubit at a time. Other ones provide a "SWAP" operator which exchanges the location of two adjacent qubits at one time step. We will discuss these cases later in the paper. If all gates use local (adjacent) qubits, physical implementation can be done with no further effort; otherwise *additional* movements are necessary.

For a given quantum circuit, an initial "trivial" qubit ordering $1, 2, \cdots, n$ is usually assumed independent of the circuit. This qubit ordering reflects the physical locations of qubits in the physical layout, i.e., qubit $\#i$ is assigned to the i-th physical location ($1 \leq i \leq n$). However, for a given quantum circuit this arrangement may not be the best in terms of the resulting *latency*.[3]

To improve the total latency, or equivalently the number of 2-qubit local gates, following [18], one can *globally* reorder qubits to change the initial physical locations as

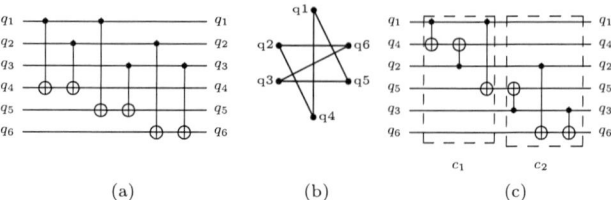

Figure 1: (a) A sample circuit, and its interaction graph (b). (c) Circuit in (a) after applying global reordering.

$l_1, l_2, \cdots, l_n$ for $1 \leq l_i \leq n$. Global reordering consumes no additional gates. To illustrate, consider a sample circuit in Figure 1(a) which is the X error syndrome using Shor-EC method for $[[9, 1, 3]]$ code [23]. The resulting circuit after applying global reordering is shown in Figure 1(c). As can be seen, the original circuit has 6 non-local gates, where four of them are local in the new circuit. Since local gates use adjacent qubits, there is no need to move the involved qubits for a gate application. This reduces the circuit latency. The problem is that even after global reordering, some gates may remain (or be) non-local. In this case, one needs to add additional local operations, MOVE or SWAP, to *move* or *permute* qubits such that the qubits that are involved in the original non-local gates will be local afterward. This is called *local reordering* [18]. Note that after a local qubit reordering, qubit locations may be changed and working with the remaining gates may need additional orderings. This is done by applying extra SWAP gates.

3. PRIOR WORK

For quantum architectures which support SWAP operation, a straightforward method to overcome the interaction constraints is to insert local SWAP gates in front of a non-local gate to permute lines (qubits) and move the involved lines toward each other. This should be followed by adding SWAP gates after the computation to recover the initial qubit ordering. For specific quantum circuits, one can explore more efficient implementations [6–9, 11, 12]. Additionally, one may try to use local gates "during" a general synthesis [13] instead of trying to reduce SWAP gates by a post-process approach. Although this seems interesting, considering locality besides other important metrics during the synthesis can complicate the overall process significantly. On another side, several researchers considered the overall impact of the interaction constraints on their developed circuits/constructions instead of working with actual circuits. In this case, they may "prove" that total cost may increase by a *constant* factor (e.g., 10 in [15] and < 2 in [14, 16]).

To work with arbitrary circuits, the authors in [18] developed exact and heuristic post-synthesis methods to reduce the number of SWAP gates. The exact method, which is limited to small circuits, was used in a peephole optimization approach. Along with 3 templates, the authors suggested two reordering strategies, global and local, where in global a qubit with the highest interaction impact is placed at the middle line continuingly until no further improvement can be achieved. In local, the algorithm inserts SWAP gates only before a non-local gate, and the new ordering is used for the remaining gates.

In [19], the authors showed that a bubble sort generates

---

[2]Quantum technologies with constantly moving phenomena (e.g., photons) use "flying" qubits. In this case, gates are fixed and qubits are affected upon flowing through the gates. In quantum technologies with physical locations for qubits, gates are applied in fixed locations and "mobile" qubits may travel between locations. More detail is in [22].

[3]After arranging qubits, gates should be applied. This may need to move qubits from one physical location to another. Latency is defined as the total number of time steps required to perform all gates. This time includes the time required to move qubits between gate operations, and the time required to apply all gates.

286

the minimum number of SWAPs required to construct an arbitrary permutation of qubits for each gate. They additionally showed that in an n-qubit circuit, for two qubits of the i-th gate positioned at locations q_1^i and q_2^i only qubits placed between q_1^i and q_2^i should be considered instead of working with all qubits (i.e., $|q_1^i - q_2^i|!$ permutations instead of $n!$ permutations). Note that finding the *best* local orderings for all 2-qubit gates needs considering all $|q_1^i - q_2^i|!$ permutations for all gates at the same time (i.e., $|q_1^1 - q_2^1|! \times |q_1^2 - q_2^2|! \cdots$. To avoid this huge exponential search, authors worked with at most w consecutive gates.

The authors of [24] considered circuits that perform "specified" operations *spanning* n wires with focus on *depth*. They showed that *rotation* of n wires with local gates can be done in depth $n + 5$, *reversing* n wires with local gates is possible with depth $2n + 2$, *swapping* across n wires by local gates can be done in depth $n + 7$ for even n and in depth $n + 8$ for odd n with size $6n - 9$. More information is in [24].

4. THE PROPOSED METHOD

Basically, a quantum computer technology, e.g., architectures based on ions, may support the MOVE operation to transform a qubit from one physical location to another. A physical location may also be shared by several qubits at the same time. In this case, to make a local two-qubit gate, one needs to change the locations of far qubits, by applying a sequence of single-qubit MOVE operations. Note that there should be enough room to hold the moving qubit(s) in intermediate and final physical locations. On the other hand, a quantum architecture may not provide the MOVE operation e.g., in architectures based on superconductors. For this case, one needs to apply SWAP gates which physically change the locations of both involved qubits. For quantum architectures with 1D interaction distance, the MOVE operation and a physical location for multiple qubits are not usual, and the previous approaches discussed in Section 3 worked with SWAP gates. The same limitations exist for some 2D quantum architectures too. For other 2D quantum architectures, the MOVE operation and a multi-qubit physical location exist. In the following sections, we propose our methods to make gates local in 1D quantum architectures. Potential problems to extend our approach for 2D quantum architectures are outlined in Section 6.

The MINLA problem is defined for a weighted graph $G = (V, E)$. The goal is to arrange the vertices V of G on an integer line by a one-to-one function $f : V \to [1 \cdots |V|]$ to minimize $\sum_{\{u,v\} \in E} w_{\{u,v\}} |f(u) - f(v)|$ where $w_{\{u,v\}}$ is the weight of the edge between nodes u and v. This problem can be considered as a *label assignment* of the given graph G. The MINLA problem is NP-hard in general. However, polynomial time algorithms to compute exact solutions for some particular graphs are known. In addition, some approximation algorithms have been proposed in the past. More information can be found in e.g., [25]. In this paper, the degree of a vertex v in graph G is represented as $\deg(v)$. The maximum degree of a graph G, denoted by $\Delta(G)$, is the maximum degree of its vertices.

4.1 Label assignment for qubit reordering

Consider a given circuit $\mathcal{C}$ with n qubits $q_1, q_2, \cdots, q_n$ and m 2-qubit gates $g_1, g_2, \cdots, g_m$.[4] Working on $\mathcal{C}$, we construct

---

[4] We ignore single-qubit gates for locality improvement since they

a weighted graph G, *called interaction graph*, with n vertices $v_1, v_2, \cdots, v_n$ corresponding to qubits in $\mathcal{C}$. Additionally, we add one edge with weight $w_{i,j}$ between nodes v_i and v_j if there are $w_{i,j}$ 2-qubit gates between qubits q_i and q_j in the circuit. If $w_{i,j} = 0$, we omit the edge and for $w_{i,j} = 1$, we omit the weight. Figure 1(b) illustrates the interaction graph for the circuit in Figure 1(a). For a gate g_i with qubits i and j (and $i \neq j$), interaction distance (ID) is defined as $|i - j - 1|$. A gate with ID$= 0$ is a local gate. Total interaction distance of a circuit is a summation over the interaction distances of its gates. Accordingly, minimizing the interaction distance in $\mathcal{C}$ is equivalent to solving the MINLA problem for G. Figure 1(c) shows the circuit in Figure 1(a) after applying the MINLA algorithm where the initial ID $= 12$ in (a) is improved to ID $= 4$ in (b). While applying the MINLA problem on the whole circuit can improve the total interaction distance, some gates may remain non-local, see Figure 1(c). Next, we show how the MINLA problem can be used to make all gates local.

4.2 Subcircuits with consecutive gates

Consider a set of w consecutive 2-qubit gates $A = \{g_1, g_2, \cdots, g_w\}$ for an n-qubit LNN architecture. Assume that the gate g_i works on qubits q_1^i and q_2^i (and $q_1^i \neq q_2^i$). There are many different qubit arrangements. For an interaction graph G, we may have:

- $\Delta(G) = 0$. This is a trivial case with no gate.

- $\Delta(G) = 1$. In this case, all gates use distinct qubits. Accordingly, one can find a qubit (re)ordering when for each gate g_i, the qubits q_1^i and q_2^i are adjacent. To achieve this, group q_1^i and q_2^i as a new qubit for all gates in A for a total of $n - w$ qubit groups — each group can include either one qubit (if the qubit is not used by any gates in A) or two qubits (if exactly one gate in A uses the two qubits). There are $2 \times (n - w)!$ qubit orderings to make all gates in A local. No SWAP gate is required in this case.

- $\Delta(G) = 2$ and there is no cycle in G. For this case, there is a "staircase" construction where all gates are local. Assume there are k_0 vertices with $\deg(v) = 0$, k_1 vertices with $\deg(v) = 1$, and k_2 vertices with $\deg(v) = 2$ — equivalently k_0 qubits with no interaction, etc. For $k_0 > 0$ or $k_1 > 2$ (k_1 is even) the interaction graph is unconnected and the number of connected components is $k_0 + k_1/2$. To construct a local circuit, place all qubits (equivalently vertices in G) with $\deg(v) = 0$ close to each other, and remove such vertices from G. For a vertex with $\deg(v) = 1$, place the related qubit at the next available physical location, and remove the vertex and its edge from G. Continue the same approach for the "new" vertex (created after removing the previous vertex) with $\deg(v) = 1$. Apply this approach for all connected components in G. After all, group qubits which belong to one connected component. This leads to $k_0 + k_1/2$ groups in total. Exchanging all physical locations of one group

---

have no effect on circuit locality. Another reason is that single-qubit gates can be absorbed into surrounding two-qubit gates. The resulting circuit is called a "skeleton" circuit in [7]. Throughout the paper, we simply use circuit to mean a skeleton circuit.

with the ones for another group, in $(k_0 + k_1/2)!$ total ways, has no effect on total interaction distance. Overall, no SWAP gate is required in this case.

- In any other non-trivial cases with $\Delta(G) \geq 3$ or $\Delta(G) = 2$ and with cycle(s) in G, at least one SWAP gate is required. Assume that we divide all 2-qubit gates in $\mathcal{C}$ into k sets with only local gates, each set with at most s SWAP gates. For $s = 0$, all sets belong to either case 1 or case 2. Otherwise, one needs to add SWAPs, called **intra-set** SWAP gates. If it is not possible to make the current w gates local with s SWAP gates, decrement w to $w - 1$ and recheck. If not, re-decrement and proceed. It can be verified that at least for $w = 2$, we can find a local subcircuit with no SWAP gate, i.e., $s = 0$. A larger s limit leads to a larger w.

Inter-set SWAP gates. For a circuit $\mathcal{C}$ with m gates and n qubits, assume that one finds $w_1, w_2, \cdots, w_k$ consecutive gates, $w_1 + w_2 + \cdots + w_k = m$, such that all w_i gates in set i need at most s SWAPs to be local. Assume all w_i gates in the i-th set work on $n_i \leq n$ qubits. Each set i with w_i gates needs a new qubit reordering for the involved n_i qubits. Accordingly, to have a local circuit $\mathcal{C}'$ for $\mathcal{C}$, one needs to add SWAP gates between sets i and $i + 1$ for $1 \leq i \leq w - 1$ to change the qubit ordering in set i to the one in set $i + 1$. These SWAP gates are called **inter-set** SWAP gates. Working with an unlimited s leads to no inter-set SWAP gates. Figure 2(a) illustrates the concept. Note that the methods in [18,19] are special cases of the method discussed above in the sense that they assume m sets for a circuit with m non-local gates, no intra-set SWAPs, and then apply inter-set SWAPs to locally construct adjacent gates.

Label assignment for local reordering. To find the best possible ordering for each set, we use the MINLA problem. This is done by constructing an interaction graph G for gates in each set (vs. the whole circuit) and applying the MINLA problem for each set accordingly. The solution to the MINLA problem may not be unique, in this case the one that leads to the minimum number of inter-set SWAP gates is preferred. In case of a tie, one is selected randomly. To select sets, one can try $w = 3$ consecutive gates starting from gate i in the circuit, and then apply the MINLA problem. If a relabeling and at most s SWAP gates are sufficient to make the gates local, increment w and redo. Otherwise, use $w - 1$ and restart with $i + w$. Note that the exact solution of the MINLA problem for some particular graphs can be found in a polynomial time.[5] Hence, one may be able to find optimal MINLA solutions in several sets.

Intra-set SWAP & inter-set SWAP minimization. To minimize the number of SWAP gates one should particularly consider the value of s for each set. Consider k sets indexed from 1 to k each of which with w_i gates and working on n_i qubits. Assign s_i as the maximum number of intra-set SWAP gates for set i. The value of s_i affects the final qubit ordering of set i as well as its following sets, and also the total number of sets. Figure 2 illustrates this effect with one example. Accordingly, s_i values and the total number of sets should be carefully determined. On the other hand, after distinguishing sets and appropriate qubit orderings for each set, we apply [19, Theorem 1] to find the minimum

---

[5]Trees, rectangular and square meshes and hypercubes are examples of graphs that can be solved in polynomial time with optimal MINLA solution [25].

number of inter-set SWAP gates. This method is based on simulating the bubble sort algorithm.

4.3 SWAP minimization with lookahead

Consider an n-qubit subcircuit $\mathcal{C}$ with W gates divided into a set of k smaller subcircuits $c_1, c_2, \cdots, c_k$ each of which with w_i gates ($\Sigma_{i=1}^{i=k} w_i = W$). Applying the MINLA problem on set i leads to a qubit ordering $\mathcal{O}_i$ for this set. Given that the ordering $\mathcal{O}_i$ for set i has no effect on its successive qubit orderings for sets $i + 1, \cdots, k$ (i.e., solutions of the MINLA problem in the successive sets), all k MINLA problems can be theoretically solved in parallel. To find a local subcircuit c_i' for c_i after considering the effect of $\mathcal{O}_i$, we apply the approach in [18]. Additionally, the number of inter-set SWAP gates is determined by considering qubit orderings $\mathcal{O}_i$ and $\mathcal{O}_{i+1}$ ($1 \leq i \leq k - 1$) and the method in [19, Theorem 1].

Altogether, one can find the number of intra-set SWAP gates for all sets and the number of inter-set SWAPs accordingly to determine the total number of SWAP gates for the subcircuit $\mathcal{C}$. However, w_i values are not pre-determined and are subject to an optimization (Figure 2). To improve, one can move the last gate of set i to set $i + 1$ (or the first gate to set $i - 1$) and update the number of intra-set and inter-set SWAP gates accordingly as discussed. This should be followed by selecting the best sets to minimize the total number of SWAP gates for $\mathcal{C}$. To Achieve this, we apply a lookahead. Starting from the first gate of subcircuit $\mathcal{C}$, one needs to determine initial set borders. We begin from the gate at position $i = 0$ and include gates at positions $1 \cdots j$ until the added gate at position j leads to an interaction graph G with $\Delta(G) > 2$, or creation of a cycle in the graph. Then, we close the current set without the j-th gate, and restart the same approach from the gate at position j and a new set. Continuing this process leads to an *initial set configuration*. Then, we increment/decreamnet the number of gates at each set by $\min(L, w_i)$ for a lookahead of size L, and construct a search tree to obtain a minimized number of SWAPs for $\mathcal{C}$.

5. EXPERIMENTAL RESULTS

We implemented the proposed optimization method in C++ and all experiments were done on an Intel Core i7-3770 machine with 16GB memory. To achieve this, the program initially extracts an interaction graph from a given circuit. Then, it determines the number of sets and set borders in such a way that each set can be implemented locally with no SWAP gate. This step is followed by running several instances of the MINLA algorithm for each set. Finally, the algorithm inserts SWAP gats between adjacent sets as discussed. To reduce the number of SWAPs, for each set the ordering which leads to the minimum number of inter-set SWAP gates is selected.

To evaluate the proposed interaction distance optimiza-

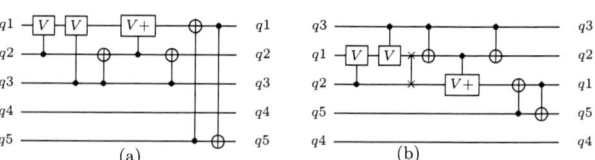

Figure 4: The result of applying the proposed method on the 4gt11_84 benchmark. (a) Non-local circuit, (b) local circuit.

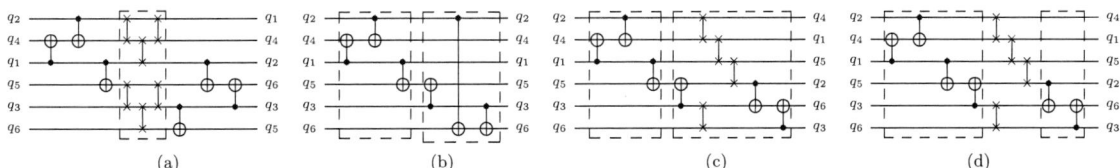

(a)   (b)   (c)   (d)

Figure 2: (a) Circuit in Figure 1(c) with only local gates. In this circuit, our algorithm was applied for subcircuits c_1 and c_2 in Figure 1(c), and 6 inter-set SWAP gates are added and $s_1 = s_2 = 0$. (b) The value of s_i in each set affects the total number of SWAP gates. Circuit in (b) has the same functionality in (a) but with $s_1 = s_2 = 4$ (redrawn in (c) with SWAP gates). The second subcircuit in (b) uses intra-set SWAP gates as shown in (c). No inter-set SWAP gate is used in (c). Note that in (c) one may consider two sets with four and two gates, shown in (d). In this case, $s_1 = s_2 = 0$, and four inter-set SWAP gates are applied. Comparing circuits in (d) and (a) further reveals the effect of selecting appropriate subcircuits on total cost.

Table 1: The synthesis results (# of SWAPs) for benchmarks in [26] as well as for quantum Fourier transform circuits after applying the method in [18] and ours. Runtime results (all in second) for [18] vary from ≈ 0 for small circuits to 1300 for large circuits. On average, the results in [18] are improved by 28%. S, m/M/A/T, and % represent # of sets, minimum/maximum/average/total numbers of SWAP gates in each set, and improvement.

Circuit	n	[18]	S	Ours (m,M,A,T)	Time	%	Circuit	n	[18]	S	Ours (m,M,A,T)	Time	%
3_17_13	3	6	5	(1,1,0.8,4)	0.007	33	hwb8_118	8	24541	6205	(1,9,2.3,14361)	389.1	41
4_49_17	4	20	10	(1,2,1.2,12)	0.006	40	hwb9_123	9	36837	7344	(1,14,2.9,21166)	1200	43
4gt10-v1_81	5	30	14	(1,3,1.4,20)	0.013	33	mod5adder_128	6	85	31	(1,4,1.6,51)	0.064	40
4gt11_84	5	3	2	(2,2,1,1)	0.01	67	mod8-10_177	5	77	43	(1,3,1.8,72)	0.02	6
4gt12-v1_89	5	35	23	(1,3,1.5,35)	0.021	0	rd32-v0_67	4	2	3	(1,1,0.6,2)	0.006	0
4gt13-v1_93	5	11	6	(1,2,1,6)	0.11	45	rd53_135	7	76	29	(1,7,2.3,66)	0.097	13
4gt4-v0_80	5	34	16	(1,5,2.2,34)	0.035	0	rd73_140	10	62	22	(1,8,2.5,56)	12.472	10
4gt5_75	5	17	9	(1,2,1.3,12)	0.015	29	sym9_148	10	5480	1736	(1,8,1.9,3415)	833.33	38
4mod5-v1_23	5	16	8	(1,2,1.2,9)	0.015	44	sys6-v0_144	10	62	21	(1,7,2.8,59)	9.814	5
4mod7-v0_95	5	28	15	(1,2,1.4,21)	0.012	25	urf1_149	9	60235	19952	(1,11,2.2,,44072)	896.1	27
aj-e11_165	4	39	23	(1,4,1.5,36)	0.018	8	urf2_152	8	25502	8652	(1,9,2.0,17670)	61.14	31
alu-v4_36	5	23	10	(1,5,1.8,18)	0.017	22	urf5_158	9	52440	17705	(1,9,2.2,39309)	1191.7	25
decod24-v3_46	4	4	3	(1,2,1,3)	0.01	25	QFT5	5	12	3	(3,3,2,6)	0.008	50
ham7_104	7	84	32	(1,7,2.1,68)	0.082	19	QFT6	6	22	4	(2,7,3,12)	0.053	45
hwb4_52	4	14	8	(1,2,1.2,10)	0.005	29	QFT7	7	39	5	(3,10,5.2,26)	0.253	33
hwb5_55	5	79	42	(1,5,1.5,63)	0.037	20	QFT8	8	60	5	(5,13,6.6,33)	1.77	45
hwb6_58	6	136	54	(1,6,2.1,118)	0.049	13	QFT9	9	87	6	(4,16,9,54)	22.332	38
hwb7_62	7	3660	961	(1,8,2.2,2128)	11.99	42	QFT10	10	123	7	(2,18,10,70)	4.261	43

tion method, we compared our results with those obtained by applying the method in [18] for reversible benchmarks in [26] as well as for the quantum Fourier transform circuits. For all cases, the number of SWAP gates added by each method to construct a local circuit was compared. In [18], quantum costs before and after the optimization were reported where SWAP gate was considered as a unit-cost gate (see [18, Table 3]). Accordingly, the number of SWAP gates is the difference of quantum cost before and after the optimization. We limited the runtime to 30 minutes and reported those cases that our algorithm leads to a solution. Table 1 reports the results.

For each circuit in Table 1, besides the number of SWAP gates we reported the number of sets and the minimum, maximum and average numbers of SWAP gates in each set. Note that we limited the algorithm to use $s = 0$ in each set. Accordingly, no intra-set SWAP gate is used and all SWAPs are the result of applying inter-set SWAP insertion method. We also limited the algorithm to use a lookahead of $depth = 1$ to reduce runtime. Increasing the lookahead depth improves the results with the penalty of runtime. As can be seen in Table 1, the average number of SWAP gates required to transform a local ordering from one set to another set is small, and our algorithm leads to a considerable reduction in the number of SWAP gates — 28%, on average and up to 60%. Figure 3 and Figure 4 illustrate the results of applying the proposed method on two benchmarks. In these circuits, all SWAP gates are inserted between sets.

6. CONCLUSION AND FUTURE WORK

In this paper, the interaction distance constraint in quantum architectures with 1D interactions was addressed. We modeled the interactions between gate qubits in a given circuit by an interaction graph, and used the well-known Minimum Linear Arrangement (MINLA) problem to find qubit reordering to improve circuit locality. The proposed approach divides a given circuit into several subcircuits and uses the MINLA instances within each subcircuit to find qubit locations for qubits involved in each subcircuit. Next, local SWAP gates are inserted inside each subcircuits to make the remaining non-local gates local. Finally, additional SWAP gates are inserted between subcircuits to transform one qubit ordering to another to keep circuit functionally unchanged. The proposed approach applies a lookahead to determine subcircuit borders and to minimize the total number of SWAP gates. Given that the MINLA is NP-hard, the problem of minimizing the number of necessary SWAP gates to run an arbitrary quantum circuits on LNN architectures is NP-hard. This addresses the **conjecture** in [19, page 25].

In addition to considering circuit depth and improving runtime to handle larger circuits, a major step in future direction is to consider the interaction distance constraint in 2D quantum architectures. This path has been followed for specific circuits in the past e.g., [10]. However, the case of general circuits needs attention.

- For architectures which do not support MOVE and use one physical location per qubit, we can extend MINLA to the 2-dimensional grid arrangement problem [27]. New methods are required to insert intra-set SWAPs and to determine initial qubit locations.

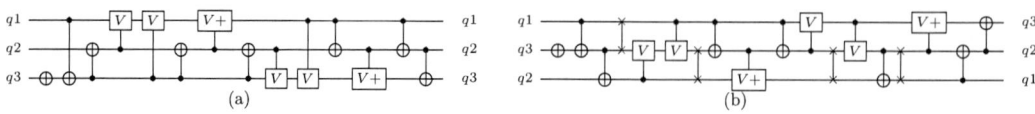

Figure 3: The result of applying the proposed method on the 3_17_13 benchmark. (a) Non-local circuit, (b) local circuit.

- For quantum architectures which support MOVE and may hold several qubits in one physical location our ideas should be revised. In this case, one can construct a graph for physical locations (vs. qubits) and use one node for qubits with the same location. This can be followed by a 2-dimensional grid arrangement problem to determine a qubit ordering. Besides the challenges stated above, the algorithm should handle the maximum size of intermediate and final physical locations when one qubit is passing from one physical location. Method in [28] is a related approach.

7. REFERENCES

[1] D. Cheung, D. Maslov, and S. Severini. Translation techniques between quantum circuit architectures. *Workshop on Quant. Inf. Proc.*, Dec 2007.

[2] H. Häffner et al. Scalable multiparticle entanglement of trapped ions. *Nature*, 438:643–646, Dec 2005.

[3] M. Laforest et al. Using error correction to determine the noise model. *Phys. Rev. A*, 75(1):133–137, 2007.

[4] B. Douçot, L. B. Ioffe, and J. Vidal. Discrete non-Abelian gauge theories in Josephson-junction arrays and quantum computation. *Phys. Rev. B*, 69(21):214501, Jun 2004.

[5] C. A. Pérez-Delgado, M. Mosca, P. Cappellaro, D. G. Cory. Single spin measurement using cellular automata techniques. *Phys. Rev. Lett.*, 97(10):100501, 2006.

[6] Y. Takahashi, N. Kunihiro, and K. Ohta. The quantum Fourier transform on a linear nearest neighbor architecture. *Quant. Inf. Comput.*, 7:383–391, 2007.

[7] D. Maslov. Linear depth stabilizer and quantum Fourier transformation circuits with no auxiliary qubits in finite neighbor quantum architectures. *Phys. Rev. A*, 76, 2007.

[8] A. G. Fowler, S. J. Devitt, and L. Hollenberg. Implementation of Shor's algorithm on a linear nearest neighbour qubit array. *Quant. Inf. Comput.*, 4:237–245, 2004.

[9] S. A. Kutin. Shor's algorithm on a nearest-neighbor machine. *Asian Conf. on Quant. Inf. Sci.*, 2007.

[10] P. Pham, K. Svore. A 2D nearest-neighbor quantum architecture for factoring. *arXiv:1207.6655*, 2012.

[11] B.-S. Choi and R. Van Meter. On the effect of quantum interaction distance on quantum addition circuits. *J. Emerg. Technol. Comput. Sys.*, 7(3):11:1–11:17, August 2011.

[12] A. G. Fowler, C. D. Hill, L. Hollenberg. Quantum error correction on linear nearest neighbor qubit arrays. *Phys. Rev. A*, 69:042314.1–042314.4, 2004.

[13] M. Arabzadeh, M. Saheb Zamani, M. Sedighi, and M. Saeedi. Depth-optimized reversible circuit synthesis. *Quant. Inf. Proc.*, 2012.

[14] M. Möttönen, J. J. Vartiainen. Decompositions of general quantum gates. *Ch. 7 in Trends in Quantum Computing Research, NOVA Publishers*, 2006.

[15] V. V. Shende, S. S. Bullock, and I. L. Markov. Synthesis of quantum-logic circuits. *IEEE Trans. CAD*, 25(6):1000–1010, June 2006.

[16] M. Saeedi, M. Arabzadeh, M. Saheb Zamani, and M. Sedighi. Block-based quantum-logic synthesis. *Quant. Inf. Comput.*, 11(3-4):0262–0277, 2011.

[17] M. Saeedi, M. Saheb Zamani, M. Sedighi, and Z. Sasanian. Reversible circuit synthesis using a cycle-based approach. *J. Emerg. Technol. Comput. Sys.*, 6(4):13:1–13:26, 2010.

[18] M. Saeedi, R. Wille, and R. Drechsler. Synthesis of quantum circuits for linear nearest neighbor architectures. *Quant. Inf. Proc.*, 10(3):355–377, 2011.

[19] Y. Hirata, M. Nakanishi, S. Yamashita, and Y. Nakashima. An efficient conversion of quantum circuits to a linear nearest neighbor architecture. *Quant. Inf. Comput.*, 11(1–2):0142–0166, 2011.

[20] M. Saeedi and I. L. Markov. Synthesis and optimization of reversible circuits - a survey. *ACM Computing Surveys, to appear, arXiv:1110.2574*, 2012.

[21] D. Kielpinski, C. Monroe, and D. J. Wineland. Architecture for a large-scale ion-trap quantum computers. *Nature*, 417:709–711, Jun 2002.

[22] R. Van Meter and M. Oskin. Architectural implications of quantum computing technologies. *J. Emerg. Technol. Comput. Sys.*, 2(1):31–63, 2006.

[23] M. Nielsen and I. Chuang. *Quantum Computation and Quantum Information*. Cambridge Univ. Press, 2000.

[24] S. Kutin, D. Moulton, and L. Smithline. Computation at a distance. *Chicago J. of Theor. Comput. Sci.*, 2007.

[25] J. Petit. Experiments on the minimum linear arrangement problem. *J. Exp. Algorithmics*, 8, 2003.

[26] R. Wille et al. RevLib: An online resource for reversible functions and reversible circuits. *Int'l Symp. on Multiple-Valued Logic*, pages 220–225, May 2008.

[27] M. Oswald, G. Reinelt, and S. Wiesberg. Exact solution of the 2-dimensional grid arrangement problem. *Discrete Optimization*, 9(3):189–199, 2012.

[28] D. Maslov, S. M. Falconer, M. Mosca. Quantum circuit placement. *IEEE Trans. CAD*, 27(4):752–763, 2008.

Acknowledgments

This research was supported by the Intelligence Advanced Research Projects Activity (IARPA) via Department of Interior National Business Center contract number D11PC20165. The U.S. Government is authorized to reproduce and distribute reprints for Governmental purposes notwithstanding any copyright annotation thereon. The views and conclusions contained herein are those of the authors and should not be interpreted as necessarily representing the official policies or endorsements, either expressed or implied, of IARPA, DoI/NBC, or the U.S. Government.

LEQA: Latency Estimation for a Quantum Algorithm Mapped to a Quantum Circuit Fabric

Mohammad Javad Dousti and Massoud Pedram
Department of Electrical Engineering, University of Southern California, Los Angeles, CA 90089, U.S.A.
{dousti, pedram}@usc.edu

ABSTRACT

This paper presents *LEQA*, a fast latency estimation tool for evaluating the performance of a quantum algorithm mapped to a quantum fabric. The actual quantum algorithm latency can be computed by performing detailed scheduling, placement and routing of the quantum instructions and qubits in a quantum operation dependency graph on a quantum circuit fabric. This is, however, a very expensive proposition that requires large amounts of processing time. Instead, LEQA, which is based on computing the neighborhood population counts of qubits, can produce estimates of the circuit latency with good accuracy (i.e., an average of less than 3% error) with up to two orders of magnitude speedup for mid-size benchmarks. This speedup is expected to increase superlinearly as a function of circuit size (operation count).

Categories and Subject Descriptors

B.7.2 [**Integrated Circuits**]: Design Aids – *Simulation, Placement and routing.*

General Terms

Algorithms, Performance, Design.

Keywords

Quantum computing, latency estimation, algorithm, quantum fabric, CAD tool.

1. INTRODUCTION

To accurately calculate the latency (total execution time) of a software program, one needs to simulate or run it on a *specific* processor. Changing the processor architecture including the size of cache memories or internal buffers can affect the latency dramatically. A number of approaches have been proposed to estimate program latency without performing time consuming simulations [1][2]. Researchers in the area of quantum computing face the same issue for estimating the latency of a quantum algorithm, programmed in a high-level quantum programming language such as QPL. In this field, the problem is even harder because the size of quantum programs for real-size problems is so huge that the simulation time is much more time consuming than that for classical programs [3].[1]

Devising a new quantum algorithm is a challenging task because of the complex structure of today's quantum computers and the non-intuitive

---

[1] By simulation, we only mean tracing the execution of quantum operations. Simulation of a quantum program and calculating the results cannot be performed efficiently on classical computers even for mid-size problems.

Permission to make digital or hard copies of all or part of this work for personal or classroom use is granted without fee provided that copies are not made or distributed for profit or commercial advantage and that copies bear this notice and the full citation on the first page. To copy otherwise, to republish, to post on servers or to redistribute to lists, requires prior specific permission and/or a fee.

principles (i.e. quantum physics) they are built upon. Currently, quantum algorithms are designed and evaluated by asymptotic runtime analysis, i.e. big O notation [4]. Unfortunately, in many cases, the asymptotic analysis is too coarse-grained to be of practical use to algorithm developers. Another problem is that quantum computers built using the current technology are only capable of executing toy-size programs, so they cannot be used to experimentally determine the latency of a quantum program. Hence devising a fast, yet accurate, method for estimating the latency of a program is necessary. This method would enable quantum algorithm designers to evaluate their new algorithms and *learn* efficient ways of *coding their quantum algorithms* by quickly comparing the latency of different software coding techniques. Moreover, this method allows designers of *quantum error correction codes* (QECC) to investigate the effect of different error correction codes on the latency of quantum programs.

Latency is an important factor for QECC designers since quantum computers allow only a limited amount of time for running a quantum program without using error correction. QECC has a high impact on the latency. At the same time, one needs to know the latency of a quantum program to know how much error correction it needs. So there is a complex inter-dependency between the quantum algorithm and its latency on one hand and the QECC used on the other hand.

In this paper, we present a procedural method to accurately and quickly estimate the latency of a quantum program. A tool called *quantum algorithm latency estimator* (LEQA) is developed based on this method. To the best of our knowledge, no research has been conducted on this topic before.

The rest of this paper is organized as follows. Section 2 uses the prior art (such as [5] and [6]) to describe a (somewhat novel) design flow for compiling a quantum algorithm and mapping it to primitive quantum structures on a 2-D plane. Section 3 explains the estimation method used for the latency calculation. A procedural method is presented for estimating the average routing latency for the CNOT gates. This section introduces a new parameter called d_{uncong}, which is the average routing latency of a qubit in an average-size presence zone when the routing channels are not congested. Estimation of this parameter is explained followed by the detailed description of LEQA (a prototype software implementation of the proposed method). Section 4 presents experimental results while Section 5 concludes the paper.

2. A QUANTUM DESIGN FLOW

A typical quantum circuit fabric consists of an infinite 2-D array of identical primitive structures (called *quantum templates* in this paper), each structure containing some sites for generating/initializing qubits, measuring them, performing operations on one or two qubits, and channels for moving qubits or swapping their information. Unfortunately, dealing with this primitive template array is very cumbersome and unwieldy. So in practice another 2-D array of super-templates (which we call *tiles*) is built. Each tile comprises a number of primitive templates. Instead of mapping a quantum circuit directly to the quantum fabric, quantum circuit is mapped to this tiled architecture (see below). A quantum logic synthesis tool (surveyed in reference [7]) generates a reversible quantum circuit. Every qubit in the output circuit is called a *logical qubit*, which is subsequently encoded into several *physical qubits* to detect and correct potential errors.

To prevent the propagation of errors in the quantum circuit, the (reversible) logic gates in the synthesized circuit (which are typically NOT, CNOT, and Toffoli gates [8]) must be converted into *Fault-Tolerant* (FT) quantum operations. A possible universal (but redundant) set of FT quantum operations includes CNOT, H (Hadamard), T ($\pi/4$ rotation), $T^{\dagger}$ ($-\pi/4$ rotation), S (phase), X, Y and Z gates. Note that these gates are all one and two-qubit gates. Implementation of these FT quantum operations depend on the picked error correction method. Note that the set {CNOT, H, T} constitutes a universal basis for quantum circuit realization–the other operations are included to enable more logical simplification in the process of converting the logic synthesis output to the FT quantum operation realization. Each quantum fabric is natively capable of performing a universal set of one and two-qubit instructions (also called native quantum instructions). This set differs among various quantum fabrics. Each FT quantum operation can be implemented by using a composition of these native quantum instructions.

The transformation from logical gates (results of the quantum logic synthesis) to the FT quantum operations and from the FT quantum operations to the native quantum instructions can be called *quantum FT synthesis* and *quantum fabric synthesis*, respectively. Quantum FT and quantum fabric synthesis are outside of the scope of this paper. Each of these FT quantum operations performs a desired function on one or two logical qubits as input producing one or two logical qubits as output; each of the input qubits is encoded with some number of physical qubits. The output qubits will also be coded. Moreover, each of these FT quantum operations requires syndrome extraction circuitry following the quantum gate in order to detect and correct errors (up to a certain limit) that may have been introduced by the quantum operation. Based on the adopted encoding scheme, implementation of each of the aforementioned FT quantum operations may require hundreds to tens of thousands of native quantum instructions in a given quantum fabric.

Various works (e.g. [9] and [10]) have suggested using the *tiled quantum architecture* (TQA), composed of a regular two-dimensional array of *Universal Logic Blocks* (ULBs) to avoid dealing with this complexity. Notice that each ULB in TQA is capable of performing any FT quantum operations. ULBs are separated by the routing channels, which are needed to move logical qubits (or information about these qubits) from some source ULBs to a target ULB in the TQA. A pictorial representation of the TQA is shown in Figure 1. The quantum structures placed at the junctions of routing channels may be thought out as *quantum crossbars* (possibly with some qubit purification capability [11]). Routing channels and quantum crossbars are also built from quantum templates.

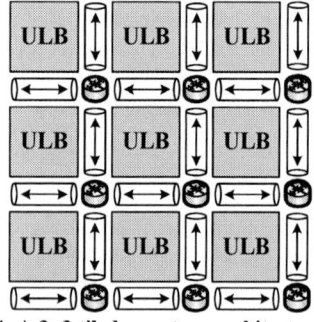

Figure 1. A 3×3 tiled quantum architecture (TQA)

A ULB is analogous to a *Configurable Logic Block* (CLB) in an FPGA device, in that it can implement any of a set of target functions. Moreover, the same ULB (as identified by its unique row and column indices in the ULB array) can be configured to perform different FT quantum operations at different times as needed. This is analogous to an on-the-fly-reconfigurable CLB. After appropriate high-level

transformations, a quantum algorithm may be represented as a *quantum operation dependency graph* (QODG), in which nodes represent FT quantum operations and edges capture data dependencies. A one-qubit operation is represented by a node with one edge entering it and one edge leaving it. On the other hand, a two-qubit operation is shown using a node with two edges entering it and two edges leaving it. One edge is called *control edge* while the other is called *target edge*. It is assumed that the order of gates does not change after the synthesis step. If two edges in the QODG come from one node and go to another node, the edges are combined in order to keep the graph *simple*. Also, due to the no-cloning theorem, a fan-out in the circuit is forbidden. A *start node* is added which connects to the first-level nodes in order to satisfy the initial dependencies. Also an *end node* is added where all the last-level nodes are connected to it. These two extra nodes simplify the problem formulation. A sample synthesized quantum circuit and the QODG constructed from it is presented in Figure 2.

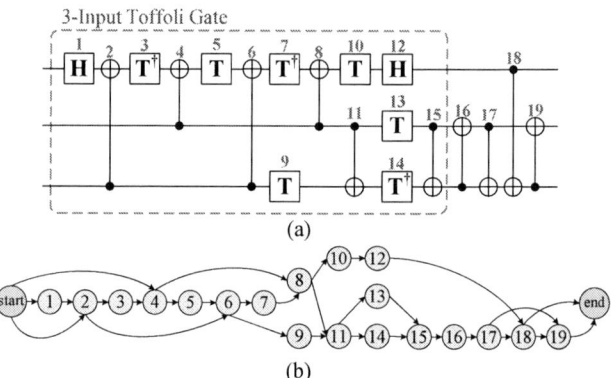

Figure 2. (a) The synthesized *ham3* circuit [12] for size 3 Hamming optimal coding. Note that this circuit only contains FT gates. (b) A QODG constructed from the circuit shown in (a). Numbers are added to relate each node to its corresponding operation in the circuit.

Based on the target quantum fabric and error threshold, a particular quantum coding is selected, and subsequently, a high-level tool maps the QODG into a TQA, where each ULB (tile) in this architecture can implement any operation in a fault-tolerant way. The latency of the quantum algorithm mapped to the TQA can be calculated as the length of the longest path (critical path) in the mapped QODG, where the length of a path in the QODG is the summation of latencies of operations located on the path plus routing latencies of their qubit operands. Note that the critical path of the mapped QODG may not be the same as the critical path of the original QODG because the latter does not contain routing latencies of logical qubits. These latencies change the scheduling slacks and hence may change the critical path of the entire graph.

Mapping a QODG to a TQA comprises of three intertwined steps: scheduling, placement, and routing. These steps depend on each other. For example, the result of placement and routing can increase the routing latency of a logical qubit and hence the qubit may fail to meet the timing requirements of the scheduling. As a result, the operation should be deferred by one or more scheduling steps. The quantum mapping problem, similar to the corresponding problem in the traditional VLSI area, is a hard problem. Hence, several heuristics have been proposed in the literature for solving it near-optimally [9][10][13][14]. Unfortunately, these heuristics are still very slow. They produce the mapping solution with the details of every qubit movement on the TQA. Since quantum computers are still not mature enough to handle large-scale problems [15], detailed information that a quantum mapper produces is excessive and not very useful. Hence, we introduce a model to quickly estimate the latency of a quantum algorithm as explained next.

3. ESTIMATING LATENCY OF A QUANTUM ALGORITHM

The latency of a quantum algorithm may be calculated as follows:

$$D = N_{CNOT}^{critical}\left(d_{CNOT} + L_{CNOT}^{avg}\right) + \sum_{g \in O} N_g^{critical}\left(d_g + L_g^{avg}\right) \quad (1)$$

where O is the set of one-qubit FT operations (such as H, T, S, etc.); $N_{CNOT}^{critical}$ and $N_g^{critical}$ are the number of CNOTs and operations of type g (one-qubit FT operations) on the critical path; d_{CNOT} and d_g determine the delay of CNOT and the operation of type g respectively; L_{CNOT}^{avg} and L_g^{avg} capture the average routing latency for CNOT and the operation of type g. Note that the equation treats one and two-qubit operations differently. The only two-qubit FT operation is CNOT while there are different one-qubit operations. $N_{CNOT}^{critical}$ and $N_g^{critical}$ can be determined by calculating the critical path of the circuit. As explained earlier, considering the critical path of the original QODG instead of the critical path of the mapped QODG introduces some errors to the estimation model. So the values of L_{CNOT}^{avg} and L_g^{avg} will be added to the operation delays in the QODG in order to determine the critical path more accurately. d_{CNOT} and d_g depend on the underlying fabric technology, the error correction and the control techniques used. These parameters are the output of a ULB fabric designer tool which has a very low runtime execution (in the order of at most a few minutes) and produces exact results which can be used for any algorithms. Hence, values of these parameters for all types of FT operations are assumed to be given. L_g^{avg} can be estimated empirically since routing of a qubit is not too complex. A one-qubit operation can be done in the ULB where the qubit currently resides or in the nearest free ULB if the current location is also occupied by another qubit. Value of L_g^{avg} is set to $2 \times T_{move}$ where T_{move} is a physical parameter which captures the time that a logical qubit needs to move from any ULBs, channels, or quantum crossbars to another ULB, channel or quantum crossbar in its neighborhood. This empirical result shows that on average each qubit needs to move to its nearest ULB for a one-qubit operation. The main challenge is to estimate L_{CNOT}^{avg} which is more interesting as it represents the average traveling (routing) time of two logical qubits from their source locations to the target ULB (i.e., the ULB where the two qubits will interact). This value accounts for the traffic congestion in the routing channels. In this paper, a procedural method for estimating L_{CNOT}^{avg} is suggested. Knowing this value and estimating the critical path, one can calculate the latency of a quantum program using Equation (1).

3.1 Estimating the Average Routing Latency for CNOT

The wire length estimation problem in the traditional VLSI area [16] approximates the average total wire length among all of the connected standard logic cells before performing the time-consuming cell placement and routing steps. Our problem is similar to the aforesaid, but in fact it is more complex. This is mainly because L_{CNOT}^{avg} also depends on the scheduling of a QODG. More precisely, mapping of a QODG to a quantum fabric consists of three steps: scheduling, placement and routing. These steps are interrelated, and none can be optimally solved without solving the others. Placement and routing affect the result of scheduling (which in turn affects the timing slacks in the QODG.) Placement cannot be done optimally without considering the effect of routing and channel congestions. Also note that in the placement problem, one should assign both the logical qubits and the logical operations to ULBs. Compared to the VLSI placement, this problem has (*dynamically*) *moveable cells* since the qubits move during the execution of a program. Also two or more operations may be assigned to a ULB as long as they are scheduled to be done in different time slots. Moreover, the size of QODG for real-size problems is generally far larger than any standard VLSI gate-level

netlists [10]. So, this estimation problem is more complex than the traditional VLSI counterpart.

For each logical qubit, a hypothetical presence zone is assumed in which the qubit performs most of its interactions. This zone also shows the area where the other qubits that interact with the qubit in question are located at some point in time. These zones are located in different places of the TQA fabric. They can overlap with each other. An overlap resembles congestion since it is possible that more than one qubit pass the overlapping area at the same time. Figure 3 depicts an illustration of five presence zones placed randomly on a fabric showing the interaction among five qubits. The overlapping area among zones 3, 4, and 5 is the most congested area.

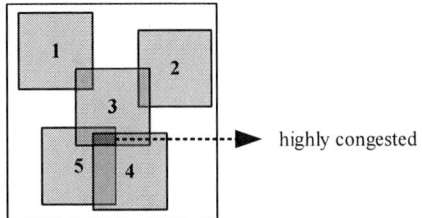

Figure 3 Five presence zones placed randomly on the fabric

Since the result of the placement is not known a priori, the zones are assumed to be placed randomly (uniformly and independently) on the fabric. L_{CNOT}^{avg} can be estimated using Equation (2).

$$L_{CNOT}^{avg} \approx \frac{\sum_{q=1}^{Q} E[S_q] \times d_q}{\sum_{q=1}^{Q} E[S_q]} \quad (2)$$

$$\sum_{q=0}^{Q} E[S_q] = A \quad (3)$$

where Q is the total number of logical qubits; $E[S_q]$ is the expected surface of the fabric covered by exactly q overlapping presence zones; d_q is the average routing latency of a qubit in an average-size presence zone when all the routing channels are occupied by q qubits; and A is the area of the fabric and it is equal to the total number of ULBs assuming that each ULB is a 1×1 square. Equation (3) shows a constraint on $E[S_q]$. Note that the summation index in the constraint starts at 0 instead of 1 since some parts of the fabric may not be covered by any presence zones. Since calculating the latency for the unoccupied surface is meaningless, $E[S_q]/\sum_{q=1}^{Q} E[S_q]$ is used in Equation (2) as the normalized value for $E[S_q]$. To calculate $E[S_q]$, Equation (4) is used:

$$E[S_q] = \binom{Q}{q} \sum_{x=1}^{a} \sum_{y=1}^{b} (P_{x,y})^q (1 - P_{x,y})^{Q-q} \quad (4)$$

where $P_{x,y}$ is the probability that the ULB at position (x,y) on the fabric is being covered by a qubit's presence zone randomly placed on the fabric; a and b are width and length of the fabric. (Remember that a fabric is modeled as a grid of $a \times b$ square-shape ULBs and $a \times b = A$.) The coefficient is the number of ways to choose q presence zones from the total presence zones (i.e. Q which equals to the total number of logical qubits). The two summations add the probability that each of the ULBs on the fabric is covered by exactly q presence zones. The overall equation calculates the expected surface of the fabric covered by exactly q presence zones. Calculating this equation Q times and using it in Equation (2) is time consuming. Hence, only the first 20 terms are calculated in practice. Simulation results show that this choice does not dramatically affect the accuracy of the estimation while it substantially improves the runtime of LEQA.

Equation (5) calculates $P_{x,y}$. (In the nominator, two min{.} functions are multiplied. Note that they are written in two lines.)

$$P_{x,y} = \frac{\left(\begin{array}{l}\min\{x, a-x+1, \lceil\sqrt{B}\rceil, a-\lceil\sqrt{B}\rceil+1\} \times \\ \min\{y, b-y+1, \lceil\sqrt{B}\rceil, b-\lceil\sqrt{B}\rceil+1\}\end{array}\right)}{(a-\lceil\sqrt{B}\rceil+1)(b-\lceil\sqrt{B}\rceil+1)} \quad (5)$$

where B is the average area of presence zones. Figure 4 depicts how Equation (5) is derived. The nominator counts the number of ways that placing a presence zone of size ($\lceil\sqrt{B}\rceil \times \lceil\sqrt{B}\rceil$) on a ($a \times b$) fabric covers the ULB located at position (x,y). Min{.} functions are used to account for the boundary situations. The denominator counts the number of ways a ($\lceil\sqrt{B}\rceil \times \lceil\sqrt{B}\rceil$) presence zone can be placed on a ($a \times b$) fabric.

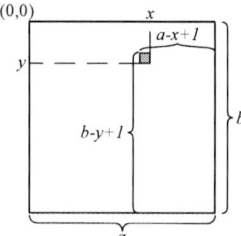

Figure 4. Calculation of $P_{x,y}$

To estimate B, the average area of presence zones, a new graph called *interaction intensity graph* IIG(V,E) is built as follows. Nodes of this graph are logical qubits which are denoted by n_i. An edge e_{ij} is added between nodes n_i and n_j if these two qubits interact with each other. Weight of this edge, which is denoted by $w(e_{ij})$, is equal to the number of two-qubit operations between n_i and n_j. Note that edges are not directed, so e_{ij} and e_{ji} refer to the same edge. Clearly, IIG(V,E) has no self-loops since no edges are added for one-qubit operations. M_i is defined as the number of neighbors of node n_i in the IIG(V,E). It is equal to $\deg(n_i)$ which is the degree of node n_i in the IIG(V,E). We model the area of the presence zone associated with n_i, which is denoted by B_i, as follows:

$$B_i = \sqrt{M_i + 1} \times \sqrt{M_i + 1} \quad (6)$$

Addition of one to the term M_i accounts for the qubit n_i itself. (There are $M_i + 1$ qubits in the presence zone.) Qubit n_i travels inside this zone and interacts with M_i qubits. It visits $\sum_{\forall n_j \in adj(n_i)} w(e_{ij})$ number of ULBs which may not be necessarily unique during the program execution. The average size of presence zones, B, can be calculated by using a weighted average over the size of the presence zone of all logical qubits:

$$B = \frac{\sum_{i=1}^{Q}\left(\sum_{\forall n_j \in adj(n_i)} w(e_{ij})\right) \times B_i}{\sum_{i=1}^{Q} \sum_{\forall n_j \in adj(n_i)} w(e_{ij})} \quad (7)$$

$\sum_{\forall n_j \in adj(n_i)} w(e_{ij})$ sums over the weights of all adjacent edges of the node n_i in IIG(V,E). It increases the weight of the term B_i if the qubit n_i is involved in more two-qubit operations.

To calculate d_q, which was used in Equation (2), the following equation is used:

$$d_q = \begin{cases} d_{uncong}, & q \leq N_c \\ \dfrac{(1+q)d_{uncong}}{N_c}, & otherwise \end{cases} \quad (8)$$

where N_c is the capacity of routing channels and d_{uncong} is the average routing latency of a qubit for interacting with another qubits in an average-size presence zone when all the routing channels are *uncongested*. A channel is considered as *uncongested* if the number of qubits inhibiting the channel is less than or equal to N_c. In this case, qubits can pass through the channel with the minimum delay (i.e. d_{uncong}). If q is greater than N_c, the channel is called *congested* and qubits will form a pipeline for passing through it (hence, they will be delayed depending on their position in the pipeline). We capture this increase in the routing latency by modeling the routing channels as an M/M/1/∞ queue. Figure 5 shows a pictorial view of this model. Green blocks show logical qubits that are currently using the channel. Red blocks show the qubits waiting to get access to the channel. We assume that the arrival rate of qubits has Poisson distribution with parameter λ since the inter-arrival time of qubits are independent and memory-less. Hence, a Poisson distribution can model it very well. The service rate is assumed to have an exponential distribution with parameter μ. This assumption is made to simplify the calculations. Experimental results show that this simple model performs well in practice.

Knowing that the average routing latency for each qubit under the service is d_{uncong}, μ can be calculated as N_c/d_{uncong}. Moreover, the average length of the queue, l_{queue}^{avg}, is q which is the number of qubits in the queue.

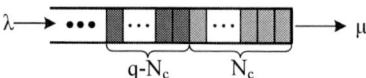

Figure 5. An M/M/1/∞ queue model for routing channels

Based on the queuing theory [17], l_{queue}^{avg}, (i.e., the average length of the queue) can be calculated as in Equation (9). Exploiting this equation and knowing the value of l_{queue}^{avg}, λ can be calculated as shown in Equation (10).

$$l_{queue}^{avg} = \frac{\lambda}{\mu - \lambda} = \frac{\lambda}{\dfrac{N_c}{d_{uncong}} - \lambda} \quad (9)$$

$$q = \frac{\lambda}{\dfrac{N_c}{d_{uncong}} - \lambda} \rightarrow \lambda = \frac{qN_c}{(1+q)d_{uncong}} \quad (10)$$

Now the values of the arrival rate (λ) and the average queue length (l_{queue}^{avg}) are known. With these values, Little's formula [17] gives the average waiting (service) time in the queue (W_{avg}):

$$q = \frac{qN_c}{(1+q)d_{uncong}} \times W_{avg} \rightarrow W_{avg} = \frac{(1+q)d_{uncong}}{N_c} \quad (11)$$

This is the expression used in Equation (8). Estimation of d_{uncong} is not a trivial task and explained in the next section.

3.2 Estimating d_{uncong}

To estimate d_{uncong}, a new parameter called $d_{uncong,i}$ is defined. This variable represents the average routing latency of the qubit n_i in an average-size presence zone when the routing channels are not congested. A weighted average over all $d_{uncong,i}$ values, similar to the Equation (7), gives an estimation of d_{uncong}:

$$d_{uncong} = \frac{\sum_{i=1}^{Q}\left(\sum_{\forall n_j \in adj(n_i)} w(e_{ij})\right) \times d_{uncong,i}}{\sum_{i=1}^{Q} \sum_{\forall n_j \in adj(n_i)} w(e_{ij})} \quad (12)$$

One way to estimate $d_{uncong,i}$ is to randomly place $M_i + 1$ qubits in the presence zone of the qubit n_i and calculate the expected length of the shortest Hamiltonian path ($E[l_{ham,i}]$) which goes through these qubits. These qubits can be placed anywhere in the presence zone, even they can be placed at the same location. This captures the fact that two qubit can travel to the same ULB for interaction. The reason for selecting Hamiltonian path is that according to the assumption, in a presence zone, only one qubit interacts with others, so it has to travel to M_i locations (not necessarily unique) and interact with M_i unique qubits. Interactions among other qubits are considered in their own presence zone calculation. A shortcoming of the aforementioned approach is that the problem of calculating the expected shortest Hamiltonian path is NP-hard [18]. Hence, the exact calculation of $E[l_{ham,i}]$ is infeasible for a *quick* estimation method. An upper bound and a lower bound for the expected path length of *traveling salesman problem* (TSP) are presented in reference [19]. It assumes that ($M_i + 1$) ≫ 1 points are randomly distributed in a 1×1 square. Equation

294

(13) presents a lower bound and equation (14) shows an upper bound for the expected path length of TSP.

$$\text{lower bound: } 0.708\sqrt{M_i + 1} + 0.551 \qquad (13)$$

$$\text{upper bound: } 0.718\sqrt{M_i + 1} + 0.731 \qquad (14)$$

Taking the average of the upper bound and the lower bound gives a good estimation for the expected path length of TSP. In our problem, the square length is $\sqrt{B_i}$ times greater so the result should be multiplied by $\sqrt{B_i}$. Moreover, since TSP solution is a tour, the result should also be multiplied by $(M_i - 1)/M_i$ to give the Hamiltonian path length which has one edge less than the tour. Equation (15) shows the resultant estimation for $E[l_{ham,i}]$.

$$E[l_{ham,i}] \approx \sqrt{B_i} \times \left(0.713\sqrt{M_i + 1} + 0.641\right) \times \frac{M_i - 1}{M_i} \qquad (15)$$

By knowing the value of $E[l_{ham,i}]$, $d_{uncong,i}$ can be calculated as follows:

$$d_{uncong,i} = \frac{E[l_{ham,i}]}{v \times M_i} \qquad (16)$$

where v is a parameter depending on the physical characteristics of the fabric technology mostly the speed of moving a logical qubit through the channels. This parameter also can be used for tuning the LEQA with different quantum mappers. M_i is presented in the denominator to give the average routing latency for an operation.

3.3 LEQA Algorithm and Its Performance

Algorithm 1 shows the implementation of LEQA based on the presented procedural method. Note that QODG is an input of the algorithm. One can easily construct it from a synthesized quantum circuit as shown in Figure 2. Size of the fabric is another input. This value can be changed to find the optimal size for the fabric which results in the minimum delay. The other inputs are physical parameters. The runtime complexity of the algorithm may be summarized as follows:

$$O\left(|V_{QODG}| + |E_{QODG}| + Q.A.\log Q\right) \qquad (17)$$

More details on the analytical analysis to derive this time complexity are presented in the Supplemental Material section.

4. EXPERIMENTAL RESULTS

4.1 Simulation Setup

LEQA is implemented in Java. For the baseline, a quantum scheduling, placement, and routing tool (called *QSPR*) [20] was used. QSPR was minimally modified to work on the tile-based architecture of Figure 1. Table 1 lists the physical parameters of the TQA used for simulations. QSPR was also used to calculate the delay of performing FT operations on an ion-trap circuit fabric (left table). The [[7,1,3]] Steane code was used as the encoding and error correction scheme. Hence, delays of the T and $T^\dagger$ gates (d_T and $d_{T\dagger}$) which are non-transversal in this coding, are higher than the others. These numbers can be adjusted based on any underlying technologies and does not limit the functionality of LEQA to a specific quantum realization technique. In the right table, the specifications of a TQA are presented.

Table 1. List of physical parameters of the TQA

Parameter	Value	Parameter	Value
d_H	5440μs	N_c	5
$d_T, d_{T\dagger}$	10940μs	v	0.001
d_X, d_y, d_z	5240μs	A=a × b	3600 = 60 × 60
d_{CNOT}	4930μs	T_{move}	100μs

Benchmarks are taken from reference [12] and synthesized using the fault-tolerant gate library. The simple method presented in reference [4] is used to decompose n-input Toffoli and n-input Fredking gates (n>3) to several 3-input Toffoli and Fredking gates. Note that this method adds ancillary qubits to the circuit. Also no ancillary sharing is performed among the decomposed gates. The resultant 3-input Fredkin gates are replaced by three 3-input Toffoli gates. Finally, 3-input

---

Algorithm 1: LEQA

Inputs: QODG quantum operation dependency graph, a, b width and length of the fabric, d_{CNOT} and d_g delays of logical gates, N_c the capacity of routing channels, v speed of a logical qubit through the routing channels, Q number of logical qubits

Outputs: D estimated latency of the input program

1 Make IIG(V,E) from the given QODG
2 Let $M_i = \deg(n_i)$ for every n_i and calculate B_i from Eq (6).
3 Calculate B from Eq (7).
4 **For** ($i = 1$ to Q)
5   Calculate $E[l_{ham,i}]$ using Eq (15).
6   Calculate $d_{uncong,i}$ using Eq (16).
7 **End**
8 Calculate d_{uncong} from Eq (12).
9 **For** (x= 1 to a)
10   **For** (y= 1 to b)
11   Calculate $P_{x,y}$ using Eq (5).
12   **End**
13 **End**
14 **For** ($q = 1$ to Q)
15   Calculate d_q from Eq (8).
16   Calculate E$[S_q]$ from Eq (4).
17 **End**
18 Calculate L_{CNOT}^{avg} from the approximation given in Eq (2)
19 Update the $QODG$ based on the value of L_{CNOT}^{avg} and empirical value for L_g^{avg} and then calculate $N_{CNOT}^{critical}$ and $N_g^{critical}$ for all operations types
20 Calculate D using the estimation given in Eq (1).
21 **Return** D

Toffoli gates are decomposed to a set of fault-tolerant gates using the method presented in reference [21] and shown in Figure 2.

LEQA and QSPR share the same parsers for parsing the inputs, the TQA specification, and physical parameters. A PC with Intel Pentium Dual-Core E5500 CPU clocked at 2.80GHz with 4GB RAM running Windows 7 and Java Development Kit (JDK) 7 is used for simulations.

4.2 Simulation Results

Table 2 shows the comparison between the actual delay computed by QSPR and the estimated delay calculated by LEQA. As can be seen, the average estimation error is equal to 2.11% while the maximum error is below 9%.

Table 3 lists the information about the benchmarks as well, i.e. the qubit count and operation count. The benchmarks are sorted based on the operation count. Also Table 3 compares the runtime of LEQA and QSPR. Evidently, when the operation count grows, LEQA performs better. In the largest benchmark, which its netlist file size is more than 12MB, LEQA performs more than two orders of magnitude faster than QSPR. This trend shows that as the size of netlist grows, LEQA beats QSPR in terms of speed and still gives accurate results.

As an interesting case, consider the last two benchmarks, i.e. gf2^128mult and gf2^256mult. The operation count of the latter benchmark is almost 4 times of the former one. By comparing the runtime of LEQA and QSPR for these two benchmarks, it can be seen that runtime of LEQA is increased by a factor of 3 while the runtime of QSPR is increased by a factor of 4.5. This further depicts the scalability of LEQA compared to QSPR.

LEQA achieves 114X speedup over QSPR for the largest benchmark, i.e. gf2^256mult. This factor increases for larger benchmarks. Precisely, QSPR runtime scales super linearly with operation count in the circuit (with degree of 1.5) whereas LEQA runtime depends only linearly on this count (see Equation (17)). Reference [10] reports that Shor algorithm for a 1024-bit integer has 1.35×10^{15} physical operations. Using two-level [[7,1,3]] Steane code, each logical operation results in about 10^5 physical operations. So this algorithm has almost 1.35×10^{10} logical operations. Using extrapolation, QSPR would compute the latency in ~2 years whereas LEQA needs only 16.5 hours!! Moreover, multiple QSPR runs are needed to select minimum overhead QECC design.

Table 2. Comparison between the actual latency computed by QSPR and the estimated latency calculated by LEQA

Benchmark	Actual Delay (sec)	Estimated Delay (sec)	Absolute Error (%)
8bitadder	1.617E+00	1.667E+00	3.10
gf2^16mult	4.460E+00	4.524E+00	1.45
hwb15ps	1.940E+01	1.993E+01	2.76
hwb16ps	1.852E+01	1.903E+01	2.76
gf2^18mult	5.085E+00	5.109E+00	0.46
gf2^19mult	5.393E+00	5.407E+00	0.25
gf2^20mult	5.654E+00	5.660E+00	0.11
ham15	2.518E+01	2.530E+01	0.51
hwb20ps	3.026E+01	3.106E+01	2.66
hwb50ps	1.236E+02	1.274E+02	3.10
gf2^50mult	1.474E+01	1.495E+01	1.44
mod1048576adder	2.027E+02	1.958E+02	3.38
gf2^64mult	1.904E+01	1.935E+01	1.64
hwb100ps	3.427E+02	3.402E+02	0.72
gf2^100mult	3.015E+01	2.998E+01	0.57
hwb200ps	9.638E+02	8.839E+02	8.29
gf2^128mult	3.886E+01	3.838E+01	1.24
gf2^256mult	7.936E+01	7.654E+01	3.55

Table 3. Information about benchmark circuits and comparison between the runtime of QSPR and LEQA

Benchmark	Qubit Count	Operation Count	QSPR Runtime (sec)	LEQA Runtime (sec)	Speedup (X)
8bitadder	24	822	0.9	0.115	8.2
gf2^16mult	48	3,885	3.0	0.289	10.3
hwb15ps	47	3,885	2.7	0.256	10.7
hwb16ps	55	3,811	2.9	0.250	11.5
gf2^18mult	54	4,911	3.5	0.276	12.6
gf2^19mult	57	5,469	3.7	0.259	14.2
gf2^20mult	60	6,019	5.1	0.301	17.1
ham15	146	5,308	4.3	0.257	16.6
hwb20ps	83	6,395	3.8	0.272	13.9
hwb50ps	370	25,370	11.8	0.450	26.3
gf2^50mult	150	37,647	16.9	0.398	42.5
mod1048576adder	1,180	37,070	20.2	0.382	52.8
gf2^64mult	192	61,629	29.4	0.461	63.8
hwb100ps	1,106	67,735	26.7	0.575	46.4
gf2^100mult	300	150,297	65.2	0.859	76.0
hwb200ps	3,145	175,490	66.7	0.915	72.9
gf2^128mult	384	246,141	106.0	1.381	78.3
gf2^256mult	768	983,805	524.8	4.576	114.7

5. CONCLUSION

This paper presented *LEQA*—a fast latency estimation tool for evaluating the latency of a quantum algorithm mapped to a tiled quantum architecture. It uses a procedural method to calculate the latency of an algorithm based on computing the neighborhood population counts of qubits. Simulation results showed that in mid-size circuits, LEQA is two orders of magnitude faster than the modern quantum mapper that performs detailed scheduling, placement and routing of the quantum instructions and qubits in a quantum operation dependency graph to a quantum fabric. This speedup is expected to increase superlinearly as a function of circuit size (operation count). Moreover, LEQA could produce quick estimates of the circuit latency with sufficient accuracy i.e., an average of 2.11% error.

6. ACKNOWLEDGEMENT

This research was supported by the Intelligence Advanced Research Projects Activity (IARPA) via Department of Interior National Business Center contract number D11PC20165. The U.S. Government

is authorized to reproduce and distribute reprints for Governmental purposes notwithstanding any copyright annotation thereon. The views and conclusions contained herein are those of the authors and should not be interpreted as necessarily representing the official policies or endorsements, either expressed or implied, of IARPA, DoI/NBC, or the U.S. Government.

7. REFERENCES

[1] C. Cascaval and D. A. Padua, "Estimating cache misses and locality using stack distances," in *Proceedings of the 17th International Conference on Supercomputing*, New York, NY, USA, 2003, pp. 150–159.

[2] J. R. Bammi, E. Harcourt, W. Kruitzer, L. Lavagno, and M. T. Lazarescu, "Software performance estimation strategies in a system-level design tool," in *Proceedings of the 8th International Workshop on Hardware/Software Codesign*, 2000, pp. 82 –86.

[3] D. D. Thaker, T. S. Metodi, and F. T. Chong, "A Realizable Distributed Ion-Trap Quantum Computer," in *Proceedings of the 13th International Conference on High Performance Computing*, Bangalore, India, 2006, vol. 4297, pp. 111–122.

[4] M. A. Nielsen and I. L. Chuang, *Quantum Computation and Quantum Information*. Cambridge University Press, 2010.

[5] K. M. Svore, A. V. Aho, A. W. Cross, I. Chuang, and I. L. Markov, "A Layered Software Architecture for Quantum Computing Design Tools," *Computer*, vol. 39, no. 1, pp. 74–83, 2006.

[6] N. C. Jones, R. Van Meter, A. G. Fowler, P. L. McMahon, J. Kim, T. D. Ladd, and Y. Yamamoto, "Layered Architecture for Quantum Computing," *Phys. Rev. X*, vol. 2, no. 3, p. 031007, Jul. 2012.

[7] M. Saeedi and I. L. Markov, "Synthesis and Optimization of Reversible Circuits - A Survey," *arXiv:1110.2574*, Oct. 2011.

[8] V. V. Shende, A. K. Prasad, I. L. Markov, and J. P. Hayes, "Synthesis of reversible logic circuits," *IEEE Transactions on Computer-Aided Design of Integrated Circuits and Systems*, vol. 22, no. 6, pp. 710 – 722, Jun. 2003.

[9] T. S. Metodi, D. D. Thaker, and A. W. Cross, "A Quantum Logic Array Microarchitecture: Scalable Quantum Data Movement and Computation," in *Proceedings of the 38th International Symposium on Microarchitecture*, Washington, DC, USA, 2005, pp. 305–318.

[10] M. G. Whitney, N. Isailovic, Y. Patel, and J. Kubiatowicz, "A Fault Tolerant, Area Efficient Architecture for Shor's Factoring Algorithm," in *Proceedings of the 36th International Symposium on Computer Architecture*, New York, NY, USA, 2009, pp. 383–394.

[11] J. I. Cirac, A. K. Ekert, and C. Macchiavello, "Optimal Purification of Single Qubits," *Phys. Rev. Lett.*, vol. 82, no. 21, pp. 4344–4347, May 1999.

[12] "Reversible Benchmarks." [Online]. Available: http://webhome.cs.uvic.ca/~dmaslov/. [Accessed: 26-Nov-2012].

[13] D. D. Thaker, T. S. Metodi, A. W. Cross, I. L. Chuang, and F. T. Chong, "Quantum Memory Hierarchies: Efficient Designs to Match Available Parallelism in Quantum Computing," *Proceedings of the 33rd International Symposium on Computer Architecture*, vol. 34, no. 2, pp. 378–390, May 2006.

[14] L. Kreger-Stickles and M. Oskin, "Microcoded Architectures for Ion-Tap Quantum Computers," in *Proceedings of the 35th International Symposium on Computer Architecture*, Washington, DC, USA, 2008, pp. 165–176.

[15] T. D. Ladd, F. Jelezko, R. Laflamme, Y. Nakamura, C. Monroe, and J. L. O'Brien, "Quantum computers," *Nature*, vol. 464, no. 7285, pp. 45–53, Mar. 2010.

[16] D. Stroobandt, *A Priori Wire Length Estimates for Digital Design*, 1st ed. Springer, 2001.

[17] S. M. Ross, *Introduction to Probability Models*, 10th ed. Academic Press, 2009.

[18] D. S. Johnson, L. A. McGeoch, and E. E. Rothberg, "Asymptotic experimental analysis for the Held-Karp traveling salesman bound," in *Proceedings of the 7th ACM-SIAM Symposium on Discrete Algorithms*, Philadelphia, PA, USA, 1996, pp. 341–350.

[19] "Travelling salesman problem," *Wikipedia, the free encyclopedia*. 25-Nov-2012.

[20] M. J. Dousti and M. Pedram, "Minimizing the latency of quantum circuits during mapping to the ion-trap circuit fabric," in *Proceedings of Design Automation and Test in Europe*, 2012, pp. 840–843.

[21] V. V. Shende and I. L. Markov, "On the CNOT-cost of TOFFOLI gates," *Quantum Information & Computation*, vol. 9, no. 5, pp. 461–486, May 2009.

Supplemental Material

1. PERFORMANCE ANALYSIS OF LEQA

The number of nodes in a QODG is equal to the number of operations in the circuit plus two (because of the dummy *start* and *end* nodes) and designated as $|V_{QODG}|$. The number of edges in this graph is also shown by $|E_{QODG}|$. Knowing these parameters, the runtime complexity of each line (or set of lines) in Algorithm 1 can be calculated as follows:

Line 1: Making of graph IIG(V,E) needs a traversal of QODG which takes $\mathcal{O}(|V_{QODG}| + |E_{QODG}|)$.

Line 2: Calculation of M_i and B_i can be done in $\mathcal{O}(Q)$.

Line 3: Calculating the weights need to sum over all edges in the IIG(V,E) which has at most $\mathcal{O}(|V_{QODG}|)$ edges. Calculating the summation over weighted B_is takes $\mathcal{O}(Q)$. Overall this line takes $\mathcal{O}(|V_{QODG}| + Q)$ to be done.

Lines 4-7: Calculation of $E[l_{ham,i}]$ and $d_{uncong,i}$ can be done in constant time and hence the for-loop takes $\mathcal{O}(Q)$ to complete.

Line 8: Same as line 3, it takes $\mathcal{O}(|V_{QODG}| + Q)$. One can reuse the calculated weights in line 3 to reduce the calculation time to $\mathcal{O}(Q)$.

Lines 9-13: The nested for-loops iterate $A(= a \times b)$ times in total. In each iteration, the value of P_{ab} is calculated in constant time. So it takes $\mathcal{O}(A)$ time to complete.

Lines 14-17: The for-loop iterates Q times and in each iteration, line 15 takes $\mathcal{O}(1)$ whereas line 16 takes $\mathcal{O}(A.\log Q)$. $\mathcal{O}(A)$ is the result of the double summation over the area and $\mathcal{O}(\log Q)$ is the time needed to

calculate $(P_{x,y})^q$ and $(1 - P_{x,y})^{Q-q}$. The value of $\binom{Q}{q}$ can be calculated in constant time using the following recursive formula:

$$f(Q, 0) = 1$$
$$f(Q, q) = f(Q, q-1) \times \frac{Q - q + 1}{q}, \qquad 0 < q \le Q \tag{18}$$

Overall these lines take $\mathcal{O}(Q.A.\log Q)$ for completion. As explained in the paper, only the first 20 values for $E[S_q]$ is calculated in practice, i.e. for $q = 1$ to 20. Hence, in action LEQA performs much faster than $\mathcal{O}(Q.A.\log Q)$.

Line 18: The calculation takes $\mathcal{O}(Q)$.

Line 19: Updating the delay of all instructions takes $\mathcal{O}(|V_{QODG}|)$. Calculation of the critical path in a directed acyclic graph (DAG) takes $\mathcal{O}(|V_{QODG}| + |E_{QODG}|)$ (Chapter 24 of the reference [1] explains an algorithm with this time complexity). Deriving the values of $N_g^{critical}$ and $N_{CNOT}^{critical}$ can be done by traversing the critical path which has the length $\mathcal{O}(|V_{QODG}|)$ in the worst case.

Line 20: Calculation of D can be done in constant time.

So, the overall runtime of the algorithm may be summarized as follows:

$$\mathcal{O}(|V_{QODG}| + |E_{QODG}| + Q.A.\log Q) \tag{19}$$

2. REFERENCE

[1] T. H. Cormen, C. E. Leiserson, R. L. Rivest, and C. Stein, *Introduction to Algorithms*, 3rd ed. The MIT Press, 2009.

Pareto epsilon-Dominance and Identifiable Solutions for BioCAD Modeling

Claudio Angione
Computer Laboratory,
University of Cambridge, UK
claudio.angione@cl.cam.ac.uk

Jole Costanza
Dept of Maths & CS,
University of Catania, Italy
costanza@dmi.unict.it

Giovanni Carapezza
Dept of Maths & CS,
University of Catania, Italy
carapezza@dmi.unict.it

Pietro Lió
Computer Laboratory,
University of Cambridge, UK
pietro.lio@cl.cam.ac.uk

Giuseppe Nicosia
Dept of Maths & CS,
University of Catania, Italy
nicosia@dmi.unict.it

ABSTRACT

We propose a framework to design metabolic pathways in which many objectives are optimized simultaneously. This allows to characterize the energy signature in models of algal and mitochondrial metabolism. The optimal design and assessment of the model is achieved through a multi-objective optimization technique driven by epsilon-dominance and identifiability analysis. A faster convergence process with robust candidate solutions is permitted by a relaxed Pareto dominance, regulating the granularity of the approximation of the Pareto front. Our framework is also suitable for black-box analysis, enabling to investigate and optimize any biological pathway modeled with ODEs, DAEs, FBA and GPR.

1. INTRODUCTION

What is needed to carry out a meaningful and comprehensive analysis of a biological model? In this work, we associate the Pareto optimal principle, the ϵ-dominance analysis and the identifiability analysis to perform an in silico analysis for biological pathways. Our framework designs robust metabolic networks able to perform specific tasks, i.e., biological functions. The design is focused on the levels of genes, reactions, enzymes and metabolites. Here, we focus on the ϵ-analysis and on identifiability analysis. This is the first time that identifiability analysis is used in a design automation framework. We test our framework on the mitochondrial network to analyze its energetic yield. In details, we investigate three biological circuits mathematically described by different methods, each of which represents the mitochondrial metabolism.

Mitochondria are organelles of eukaryotic cells and play a key role in the cell. Mitochondria are responsible for the energy productivity: they are the energy source of the cell, since they synthesize adenosine triphosphate (ATP), the chem-

ical energy in the cell. Second, the mitochondrion is the site of carbohydrates metabolism, fatty acid oxidation and urea cycle. Mitochondria are also essential for several other processes, including the regulation of calcium homeostasis and other inorganic ions, cellular differentiation, cell death (apoptosis), as well as the control of the cell cycle and cell growth. Mitochondria have been also detected as responsible for several human diseases, including mitochondrial disorders, cardiac dysfunction, and type 2 diabetes [1].

In the recent work by Bazil et al. [2], 73 algebraic differential equations are implemented to model the mitochondrial bioenergetics, including 34 biochemical reactions. We perform an in silico analysis in order to find the metabolites that are important for optimizing the energetic productivity, i.e., for maximizing ATP and NADH productions in the mitochondrial matrix space. We conduct five different studies with five different matrix calcium concentrations. As above introduced, mitochondria regulate calcium homeostasis, that is strictly linked to ATP and NADH productions. The model includes kinetic parameters useful to mime regulatory effects such as activation of enzymes by protein kinases.

In our work we take into account also a mitochondrial model by Smith et al. [3] realized with flux balance analysis (FBA) [4]. Here, the system is described considering a steady state for all the metabolites involved in the network. The FBA mitochondrial model is composed of 423 reactions (including transformation reactions, transport reactions between compartments, and between internal and external environment) and 228 metabolites. The computational time to solve the problem with FBA is highly reduced. Finally, the third model we take into account represents the metabolic network of the alga *Chlamydomonas reinhardtii* [5], which is modeled through FBA and includes also the mitochondrial compartment.

We believe that an effective Biocad tool should include the following analyses: evolutionary many-objective optimization, robustness analysis, ϵ-dominance analysis, sensitivity and identifiability analyses.

Permission to make digital or hard copies of all or part of this work for personal or classroom use is granted without fee provided that copies are not made or distributed for profit or commercial advantage and that copies bear this notice and the full citation on the first page. To copy otherwise, to republish, to post on servers or to redistribute to lists, requires prior specific permission and/or a fee.
DAC '13, May 29 - June 07 2013, Austin, TX, USA.

2. CONCURRENT OPTIMIZATION OF MULTIPLE OBJECTIVES

We exploit the concept of many-objective Pareto optimality to maximize (or minimize) two or more desired metabolite productions or concentrations in a model, thus obtaining new artificial strains that are optimal in many parameters or variables concurrently. A multi-objective optimization is needed when the system performs multiple tasks, and a given phenotype cannot be optimal at all of them (e.g., when two tasks are in contrast with each other).

Let us assume to have r objective functions $f_1, ..., f_r$ to optimize. The problem of multi-objective optimization can be formalized as $\max_{x} (f_1(x), f_2(x), ..., f_r(x))^{\mathsf{T}}$, where x is the variable in the search space. Without loss of generality, we have assumed that all the functions have to be maximized; indeed, minimizing a function f_i can be thought of as maximizing $-f_i$.

The solution of a multi-objective problem is a set of points called Pareto optimal solutions or *Pareto front*. A point y^* in the solution space is said to be Pareto optimal if there does not exist a point y such that $f(y)$ dominates $f(y^*)$, i.e. $f_i(y) > f_i(y^*), \forall i = 1, ..., r$, where f is the vector of r objective functions to optimize in the objective space. We use the Pareto-front concept to find the set of designs that represent the best trade-off between two or more requirements. The Pareto front is the set of all the phenotypes that remain after eliminating all the feasible phenotypes dominated on all tasks [6].

3. BIOLOGICAL DESIGN 1: FBA ALGAL MITOCHONDRIA

In this first case, we take into account the flux balance analysis algal model [5] of *Chlamydomonas reinhardtii*, which contains reactions related to the mitochondrial functions. By using a multi-objective optimization algorithm based on the Non-dominated Sorting Genetic Algorithm II, also known as NSGA-II [7], we maximize ATP and NADH productions. This approach overcomes the state-of-the-art design and optimization methods in biological networks [8]. In order to measure ATP and NADH productions, we add two reactions to represent the transport of ATP and NADH from the matrix to the external environment. In this way, we can calculate their rate in the FBA framework. The aim is to find the optimal genetic strategies for increasing the algal bioenergy yield. Therefore, the decision variables are the genes and, in particular, their presence or not in the metabolic network. The gene knockout strategies are represented as a binary vector y. Hence, the combinatorial optimization problem consists of finding the optimal string of bits y^* that represents the optimal genetic strategy. We consider two cases in which the maximum number of knockouts allowed is equal to 10 and 50 respectively.

We look also for the optimal environment that increases the algal bioenergy yield. In this case, the decision variables are the 44 input fluxes. We search for the best values of uptake rate fluxes, up to 1000 $mmolh^{-1} gDW^{-1}$. The results of the optimization carried out on the algal metabolism are shown in Figure 1. We show how optimal environment conditions, obtained by changing the input fluxes, reach higher values

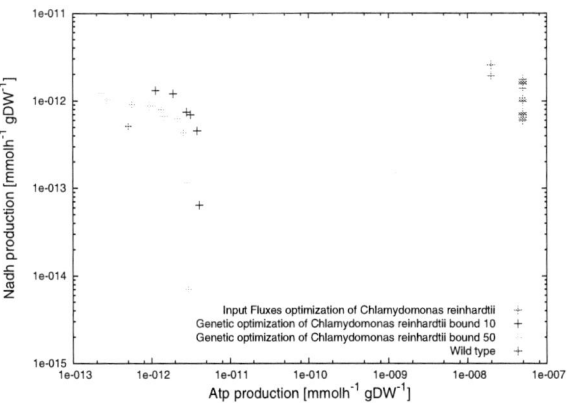

Figure 1: Comparison of the Pareto fronts obtained for the algal metabolism. The blue point represents the starting point. The green points represent the genetic strategies found with a bound of 50 knockouts allowed, whereas the black points represent the genetic strategies found with a bound of 10 knockouts allowed. The red points are the ATP and NADH corresponding to the input fluxes strategies.

of ATP and NADH production, while the genetic strategies remain near the starting point.

4. BIOLOGICAL DESIGN 2: FBA MITOCHONDRIAL MODEL

The FBA mitochondrial model [3] contains more reactions and metabolites than the differential algebraic equations (DAEs) mitochondrial model. As in the previous case study, by using a multi-objective optimization algorithm we maximize ATP and NADH productions. The aim is to find the optimal environment for mitochondria so as to increase its bioenergy yield. The decision variables are the 73 input fluxes. We search for the best values of uptake rate flux, which can assume a maximum value of 1000 $mmolh^{-1} gDW^{-1}$. The optimization algorithm finds only one Pareto-optimal solution, showed in red in Figure 2. Therefore, we conduct another experiment, maximizing ATP production and simultaneously minimizing NADH production. After 1000 generations of the optimization algorithm, as shown in Figure 2 we observe that ATP production grows more rapidly when NADH is consumed. Moreover, the first generations find Pareto fronts constituted by many non-dominated points. Instead, in the last generations (after the 900th generation), the algorithm finds only two points, which represent our final results.

We set the input fluxes of the mitochondrial model as described in the original work [3]. In this condition, the ATP production is equal to 139.4264 $mmolh^{-1} gDW^{-1}$, while NADH is totally consumed, i.e., the production is equal to 0. At the end of the optimization process, we find that the first non-dominated solution reaches NADH = -140.5484 and ATP = 971.0874 $mmolh^{-1} gDW^{-1}$, and the second one reaches NADH = -139.5166 and ATP = 971.1778 $mmolh^{-1} gDW^{-1}$. By comparing the initial state with the optimal state, we remark that ATP and NADH change when the

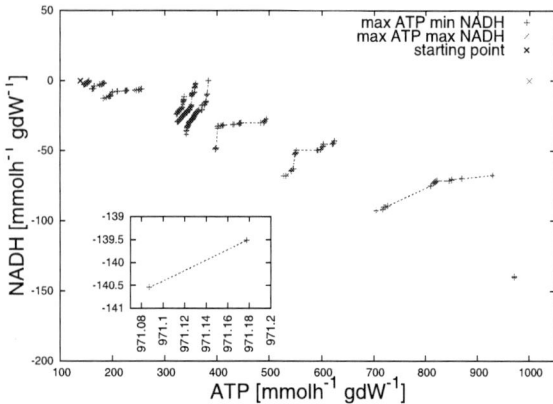

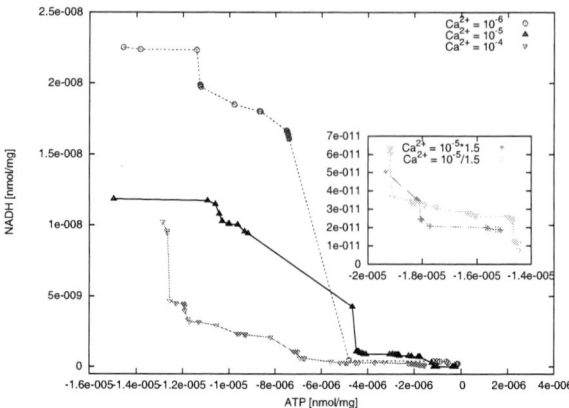

Figure 2: Effect of the genetic algorithm on the Pareto front when optimizing ATP and NADH in the FBA mitochondrial model. This evolution has been carried out with 100 individuals and halted at the 1000th generation. The Pareto front evolves from the starting point until the generation 1000. We optimized the uptake rate fluxes (73 exchange fluxes) in order to increase the energy state of the cellular organelle. The Pareto optimal points, reported in the inset, are the two non-dominated points with respect to all the other points.

Figure 3: ATP and NADH production maximization in the mitochondrial DAEs model [2]. We optimized the concentration of 55 metabolites and in particular the initial conditions to solve the differential equations system. We simulated five differential states based on the concentration of calcium in the matrix.

uptake rates linked to (R)-3-hydroxybutanoate, isocitrate, alpha-D-glucose, citrate and oxygen increase. The oxygen is the most sensitive uptake rate flux, and changes from 19.8 $mmolh^{-1}$ gDW^{-1} to 143.17 $mmolh^{-1}$ gDW^{-1}. In this case, the optimization does not consider the limitation of substrates (as glucose or oxygen) in the biological environment. They are free to enter and exit from the system. This is an asymptotic analysis for investigating the potentiality of mitochondria. To give a more in-depth interpretation of our optimization, we select a minimal set of decision variables, which are the following twelve fluxes: oxygen, arginine, lysine, proline, aspartate, alpha-D-Glucose, (R)-3-Hydroxybutanoate, isoleucine, valine, hexadecanoic acid, (S)-Lactate, HCO3-. Moreover, we change the maximum uptake rate allowed for each variable, that is increased of 33% with respect to the nominal value from the original work [3]. In this condition, NADH production does not increase, ATP increases and, by excluding all the solutions where NADH is negative, we find ATP= 185.4299 $mmolh^{-1}$ gDW^{-1}. Also in this analysis, the most sensitive uptake rate is the oxygen flux.

5. BIOLOGICAL DESIGN 3: DAEs MITOCHONDRIAL MODEL

Through NSGA-II we optimize multiple energy-related objectives. The model we adopt here consists of 73 differential-algebraic equations (DAEs) to model the mitochondrial bioenergetics [2]. As in the previous case studies, we maximize the bioenergy yield, and in particular the concentration of ATP and NADH in the mitochondrial matrix. The variable space is defined as the space of feasible initial concentrations of metabolites. Before the optimization, at the fully oxidized state we obtain NADH = $1.5987 \cdot 10^{-10}$ nmol/mg

(formation) and ATP = -0.0014 nmol/mg (consumption). After the optimization, we obtain the Pareto-optimal points shown in black in Figure 3.

Finally, we analyze more thoroughly two particular Pareto-optimal solutions, i.e., the point with maximum ATP synthesis (and lower NADH formation) and the point with maximum NADH formation (and lower ATP synthesis). Maintaining the calcium concentration in the matrix constant at 10^{-5} nmol/mg (Pareto front in black), the first solution provides NADH = $6.17 \cdot 10^{-15}$ nmol/mg and ATP = 2027.34 nmol/mg, with over-production of SUC_{mtx}, $SCoA_{mtx}$, Co-ASH_{mtx}, H^+_{mtx} and ATP_{ims} (ims=intermembrane space, mtx=matrix) and under-production of $ISOC_{mtx}$, aKG_{mtx}, MAL_{mtx}, CIT_{ims}, $ISOC_{ims}$, aKG_{ims}, SUC_{ims}, MAL_{ims} and GLU_{cyt}, ASP_{cyt} (cyt= cytosolic space). The second solution provides NADH = $6.07 \cdot 10^{-6}$ nmol/mg and ATP = -3734.6 nmol/mg (consumption), over-producing the following metabolites: H^+_{mtx}, $ISOC_{mtx}$, SUC_{mtx} and ATP_{ims}, whereas CIT_{mtx}, MAL_{ims} and AMP_{ims}, PYR_{ims}, $GLU_{ims,cyt}$ and aKG_{ims} are totally consumed.

If the matrix calcium content is increased from 10^{-5} to 10^{-4} nmol/mg, the ATP synthesis and NADH formation decrease (see Figure 3, red signs). If Ca^{2+} decreases to 10^{-6} nmol/mg, ATP synthesis remains constant, but NADH formation increases (see Figure 3, blue circles). Small perturbations in Ca^{2+} (to $1.5 \cdot 10^{-5}$ nmol/mg and to $1.5/10^{-5}$ nmol/mg) reduce NADH formation, but improve ATP synthesis in the matrix (see Figure 3, green and purple signs). This experiment can demonstrate that a perturbation in mitochondrial Ca^{2+} homeostasis has major implications for cell function at the level of ATP synthesis and NADH generation. The results are summarized in Table 1.

6. ϵ-DOMINANCE ANALYSIS

Here we introduce the concept of ϵ-dominance (inspired by Laumanns et al. [9]) to have more insights into the under-

State	ATP	NADH
before optimization	-2.2929e-005	-1.2904e-012
Trade-off solution		
$Ca^{2+} = 10^{-5}$	-2.2174e-007	8.9483e-012
$Ca^{2+} = 10^{-4}$	-1.5470e-006	7.3746e-011
$Ca^{2+} = 10^{-6}$	-1.2562e-007	2.1660e-010
$Ca^{2+} = 1.5 \cdot 10^{-5}$	-1.5107e-005	1.8484e-011
$Ca^{2+} = 10^{-5}/1.5$	-1.4455e-005	7.6942e-012
min ATP max NADH		
$Ca^{2+} = 10^{-5}$	-1.4965e-005	1.1825e-008
$Ca^{2+} = 10^{-4}$	-1.2873e-005	1.0283e-008
$Ca^{2+} = 10^{-6}$	-1.4562e-005	2.2719e-008
$Ca^{2+} = 1.5 \cdot 10^{-5}$	-1.9341e-005	6.8836e-011
$Ca^{2+} = 10^{-5}/1.5$	-1.9175e-005	7.1143e-011

Table 1: Results obtained from the multi-objective optimization in the DAEs mitochondrial model [2]. The trade-off solution is equal to the point that maximizes ATP and minimizes NADH.

standing of the information carried by the Pareto front. This technique improves the diversity of the solutions and the convergence of the optimization algorithm. Once the Pareto-optimal solutions have been obtained, we consider all the dominated and non-dominated solutions of all the generations, and we seek solutions that may have been discarded because they are dominated by a small amount ϵ that, for our purposes, can be considered negligible. In other words we apply a "relaxed" condition of dominance, thus building a new set of solutions.

After the optimization, we perform the ϵ-dominance analysis to search accurately near the edge of the Pareto-optimal region. That is, we use a condition of approximated dominance to perform a post-processing analysis in order to calculate an approximated Pareto front. Once the optimization routine has been carried out, all the sampled points are revisited and a new set of solutions is built, called "ϵ-non-dominated" set. Formally, let f be the array of the objective functions, and suppose that all the objective functions are positive and must be maximized. Let $\epsilon > 0$ be the tolerance of our relaxed condition. We seek all points (solutions) w belonging to the set $\{w : f_i(w) - \epsilon \geq f_i(u), \ \forall \ i = 1, ..., r\}$, where f is the vector of the r objective functions and u represents all the other points sampled. This set will contain both the new "ϵ-non-dominated" solutions and the old non-dominated ones.

In Figure 4 we report the results obtained when maximizing ATP and NADH in the model of the algal metabolism of *C. reinhardtii* [5] with 10 knockouts allowed at most. In this experiment, knockouts are considered as Boolean decision variables. The figure shows the Pareto front and the points given by the ϵ-dominance analysis. In fact, if we consider the non-relaxed condition of dominance, interesting solutions may be discarded although dominated by a small amount. In Figure 5 we report the ϵ-dominance analysis performed on the same algal metabolism when maximizing the fluxes of ATP and NADH. In this experiment, we consider the fluxes as real-valued decision variables.

7. IDENTIFIABILITY ANALYSIS

In this section we augment the optimization and sensitivity analysis with the identifiability analysis, which allows to find functional relations involving components of the sys-

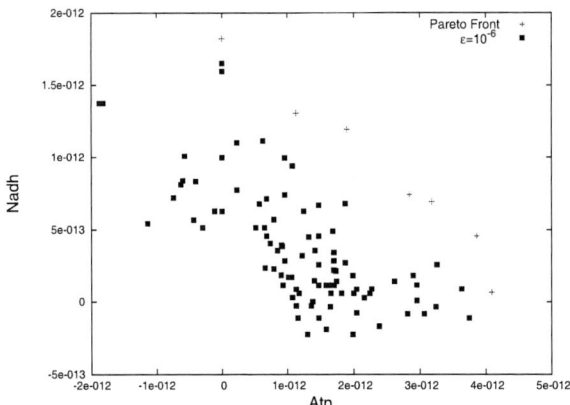

Figure 4: ϵ-dominance analysis in the algal mitochondria model for the maximization of ATP and NADH, with the knockouts as Boolean decision variables.

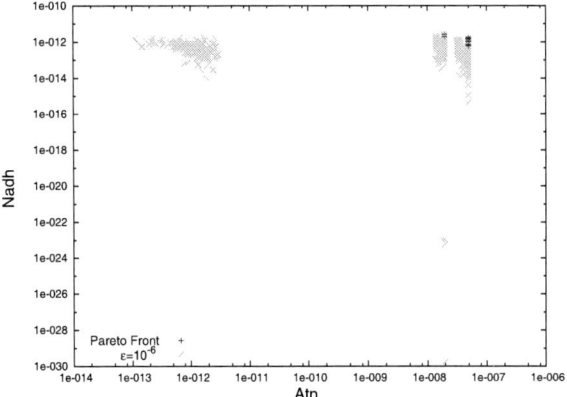

Figure 5: ϵ-dominance analysis in algal mitochondria model for the maximization of ATP and NADH, with the fluxes as real-valued decision variables.

tem. By considering the sensitivity together with the Pareto fronts from the multi-objective optimization, one can obtain a detailed investigation and comparison of the models investigated. Our general approach could be easily tailored for other organisms.

Biological processes are usually modeled with components (e.g. parameters, variables) determined by measuring data and fitting to experiments. A *non-identifiable* component is a part of the system for which no unique solution exists. There are two different kinds of non-identifiability: (i) the *structural non-identifiability* occurs when some components are functionally related and therefore they cannot be determined unambiguously; (ii) the *practical non-identifiability* occurs when it is not possible to estimate precisely the component, due to low amount or quality of data available. The identifiability analysis (IA) detects structural non-identifiable components of a model by fitting it repeatedly to experimental data and by analyzing the estimates of each component.

Formally, let $K = [v_1, ..., v_m] \in \mathbb{R}^{n \times m}$ be the matrix of the n values for the m decision variables $\{x_1, ..., x_m\}$, where each column $v_i \in \mathbb{R}^n$ contains the n estimates for the ith variable. Let us suppose that the variables are related by unknown linear or non-linear functional relations. The true transformations that linearize these relations are denoted by α and β_j, namely $\alpha(x_i) = \sum_{j \neq i}^m \beta_j(x_j) + \xi$, where ξ represents a Gaussian noise. The ACE algorithm [10] estimates the optimal transformations $\hat{\alpha}(x_i)$ and $\hat{\beta}_j(x_j)$, $j \neq i$, such that $\hat{\alpha}(x_i) = \sum_{j \neq i}^m \hat{\beta}_j(x_j)$, where x_i is the response and all the other variables are the predictors.

Here we consider the output of the multi-objective optimization carried out on the FBA model of the mitochondrion by Smith et al. [3] when maximizing ATP and NADH. The mitochondrial FBA model is composed of 423 reactions: 73 reactions represent the "input fluxes", i.e., the transport reactions from the external environment into the mitochondrion; 135 reactions represent the matrix reactions, i.e. all the reactions that take place in the matrix compartment, such as the reactions of the Krebs cycle and beta-oxidation. The other reactions take place in the inner membrane space. We computed 1000 states of flux balance in mitochondria, adding constrains in the fumarate flux, in particular we act at level of the reaction Fumarate + H$_2$O -> (S)-Malate. We constrain the fumarate flux, by imposing the knockout condition (i.e., forcing the flux of the reaction to be zero) and increasing the flux by 0.014 mmolh^{-1} gDW^{-1}, until 13.986 mmolh^{-1} gDW^{-1}. Therefore, we obtain a 423 $\times$ 1000 matrix V containing all the fluxes in the model corresponding to fixed fluxes of fumarate. Remarkably, in healthy conditions and when the objective function in FBA is the ATP production, we observe a fumarate flux equal to: 6.9721 mmolh^{-1} gDW^{-1}. As described by Smith et al. [3], in the Fumarase deficiency conditions (about 2-3 mmolh^{-1} gDW^{-1}) the ATP production is reduced until 75% of the maximum value.

We apply the IA to the 135 matrix reactions in the 1000 mitochondrial conditions obtained with 1000 different values of fumarate. In other words, we take into account 135 decision variables of the model, namely the fluxes of the reactions in the matrix. We use the method proposed by Hengl et al. [11] in order to detect structural identifiability consisting of functional relations between decision variables. The functional relations are inferred using the alternating conditional expectation algorithm (ACE) [10]. The process of repeating estimates in the matrix K is replaced by taking into account all the points given as output by the FBA run in different fumarate conditions. In other words, a single fitting sequence K is obtained by considering the entire matrix V. Thus, the problem of identifiability analysis is mapped onto the problem of detecting groups of the functionally related fluxes that are part of the matrix V. Specifically, the connection between the identifiability analysis and a constraint structure stems from the fact that a non-identifiable constraint involving decision variables causes them to be functionally related. In our case, the constraint is detected through 1000 estimates of all the 135 variables (fluxes). Each estimate corresponds to a non-dominated point of the FBA output fluxes obtained to maximize the ATP and NADH productions. We adopt the Mean Optimal Transformation Approach (MOTA) [11], by fixing at 5 the maximal number of parameters allowed to enclose a functional relation. The

results are shown in Table S1 in Appendix.

The "groups" column indicates the functional relations between variables. For instance, R01361MM and R01978MM are functionally related. In other words, the response variable x_{47} is strongly related to the predictors x_{62}. Conversely, the flux R01801MM (x_{55}) does not have any functional relation with any other flux in the matrix of the mitochondrion. The r^2 column indicates how much variance of the response can be explained by the predictors. A high amount of variance of the response that can be explained by the predictors indicates a large effect of the fixation of the predictors on the standard deviations of the response. The $cv(x) = std(x)/mean(x)$ helps distinguish practical identifiable from non-identifiable parameters [11]. In case of practical non-identifiability, the choice of the parameter to fix depends on the experiments and on the reference values found in the literature. In the algorithm, all the variables are once considered as response variable, thus a functional relation involving k variables is tested k times. However, for real models and real data it is unlikely that the same functional group is detected all the k times. Specifically, a variable is detected in a group depending on the contribution strength of a predictor to the response.

In Figure S3 in Appendix we show the functional relations between the reaction R01361MM (R-3-Hydroxybutanoate+NAD$^+$ -> Acetoacetate + NADH + H$^+$) and R01978MM (S-3-Hydroxy-3-methylglutaryl-CoA + CoA -> Acetyl-CoA + H$_2$O + Acetoacetyl-CoA). This relation has been detected by the identifiability analysis applied to both reactions, and thus marked by a double asterisk in Table S1, which indicates a strong relation (proved to be linear by the plot in Figure S3). Figure 6 shows that the optimal transformation β found for both reactions are similar to each other. This indicates the structural non identifiability of both variables. Since their cv is high in Table S1, we can infer they are also practically non-identifiable.

The interdependent fluxes, which are non-identifiable, may be fixed at an arbitrary value in order to improve identifiability. This would not affect the model's dynamical properties, as the variables functionally related to the fixed variable change accordingly. Specifically, the identifiability analysis reveals whether a component of a model can be uniquely determinable or not, thus providing a more realistic picture of what can be inferred from a model.

8. CONCLUSIONS

In this work we have analyzed how the genetic and the energy-converting pathways of mitochondria can be explored using multi-objective optimization, sensitivity, identifiability and ϵ-dominance. These techniques are framed in a unique pipeline applicable to a variety of conditions and organisms. In a related work [12], we have also used the robustness analysis as part of the same framework to augment the multi-objective optimization. As well as optimizing simultaneously two or more outputs of a model, the pipeline can also provide interesting insights into clusters of chemical reaction networks, which are often found in the cell and reflect the presence of different pathways with different responses to external or internal perturbations.

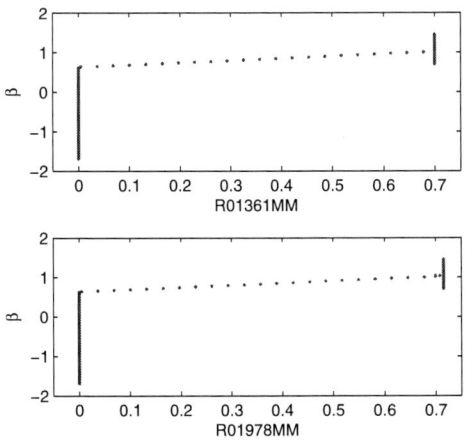

Figure 6: Optimal transformations β (y axis) found for the two fluxes R01361MM and R01978MM (x axis) [mmolh^{-1} gDW^{-1}]. This plot proves that there is a strong relation between these two fluxes, with slightly different and noisier behavior in the neighborhood of 0 and 0.7.

The sensitivity analysis has detected the variables playing the major role on the output of the model. Specifically, the PoSA algorithm assesses the sensitivity of each pathway, rather than focusing only on a single reaction or component of the model. Most importantly, PoSA is able to handle Boolean inputs [12]. Conversely, the identifiability analysis has detected the functional groups of variables. The elements of a functional group are variables functionally related with each other, which cannot be determined unambiguously. In our work, the identifiability analysis is applied to the input variables of each model, considering only the values in the variable space that correspond to the optimal points in the objective space. The ϵ-dominance analysis is performed to investigate the neighborhood of the suitable genetic designs. To our knowledge, this is the first time that the ϵ-*dominance analysis* is exploited in biological pathways and, in general, in a design framework.

The interplay between these techniques in our general-purpose framework can be exploited not merely to reach the optimal configuration for a model, but also to conduct tentative analyses on the variables and components of any model including ODEs, DAEs, FBA and GPR mappings. Striking applications of our pipeline could be in the field of metabolic engineering of a whole biological network. In particular, we propose a general and automated way to optimize biological models and assess their optimal solutions. In a network of organisms interacting with each other, our approach can be used for analyzing all the network even if it is mapped to different submodels. The results given by the framework applied to each submodel can be easily integrated, since they are output of the same pipeline, thus allowing a convergence of different modeling techniques [13].

In each optimization procedure we have considered the single organelle, while in the cells there are usually many compart-

ments, each of which contains an organelle. Compartments, also referred to as submodules, may differ for their activity depending on their location in the cell. In a module of interacting organelles, most of the reactions involve more than one compartment. Indeed, any kind of circuit can be split into submodules to increase its efficiency. An appropriate approach would therefore be to build a Pareto front where each objective belongs to a different compartment in [14], linking compartments with a set of Delay Differential Equations (DDEs) to account for events that depend on the state of the system at an earlier time (e.g., diffusion processes or maturation events). In this way, we could envisage our framework in a larger common pipeline to investigate not merely biological circuits, but also human body monitoring techniques, biosensors design, as well as a possible integration with microelectronics, e.g. CMOS biomicrosystems.

Acknowledgments. C.A. and P.L. acknowledge funding from FP7-Health-F5-2012 under grant agreement 305280 (MI-MOmics).

9. REFERENCES

[1] U. Sengupta, S. Ukil, N. Dimitrova, and S. Agrawal. Expression-based network biology identifies alteration in key regulatory pathways of type 2 diabetes and associated risk/complications. *PloS one*, 4(12):e8100, 2009.

[2] J.N. Bazil, G.T. Buzzard, and A.E. Rundell. Modeling mitochondrial bioenergetics with integrated volume dynamics. *PLoS computational biology*, 6(1):e1000632, 2010.

[3] A.C. Smith and A.J. Robinson. A metabolic model of the mitochondrion and its use in modelling diseases of the tricarboxylic acid cycle. *BMC systems biology*, 5(1):102, 2011.

[4] J. D. Orth, I. Thiele, and B.O. Palsson. What is flux balance analysis? *Nature Biotechnology*, 28(3):245–248, 2010.

[5] R.L. Chang, L. Ghamsari, A. Manichaikul, E.F.Y. Hom, S. Balaji, W. Fu, Y. Shen, T. Hao, B.O. Palsson, and K. Salehi-Ashtiani. Metabolic network reconstruction of chlamydomonas offers insight into light-driven algal metabolism. *Molecular systems biology*, 7(1):518, 2011.

[6] O. Shoval, H. Sheftel, G. Shinar, Y. Hart, O. Ramote, A. Mayo, E. Dekel, K. Kavanagh, and U. Alon. Evolutionary trade-offs, pareto optimality, and the geometry of phenotype space. *Science*, 2012.

[7] K. Deb, A. Pratap, S. Agarwal, and T. Meyarivan. A fast and elitist multiobjective genetic algorithm: Nsga-ii. *IEEE Transactions on Evolutionary Computation*, 6(2):182–197, 2002.

[8] R. Umeton, G. Stracquadanio, A. Sorathiya, P. Liò, A. Papini, and G. Nicosia. Design of robust metabolic pathways. In *Proceedings of the 48th Design Automation Conference*, pages 747–752. ACM, 2011.

[9] M. Laumanns, L. Thiele, K. Deb, and E. Zitzler. Combining convergence and diversity in evolutionary multiobjective optimization. *Evol. Comput.*, 10(3):263–282, September 2002.

[10] L. Breiman and J.H. Friedman. Estimating optimal transformations for multiple regression and correlation. *Journal of the American Statistical Association*, 80(391):580–598, 1985.

[11] S. Hengl, C. Kreutz, J. Timmer, and T. Maiwald. Data-based identifiability analysis of non-linear dynamical models. *Bioinformatics*, 23(19):2612–2618, 2007.

[12] J. Costanza, G. Carapezza, C. Angione, P. Liò, and G. Nicosia. Robust design of microbial strains. *Bioinformatics*, 28(23):3097–3104, 2012.

[13] M.E. Dumas. Metabolome 2.0: quantitative genetics and network biology of metabolic phenotypes. *Molecular BioSystems*, 2012.

[14] C. Angione, G. Carapezza, J. Costanza, P. Liò, and G. Nicosia. Rational design of organelle compartments in cells. *EMBnet. journal*, 18(B):p. 20, 2012.

[15] M.D. Morris. Factorial sampling plans for preliminary computational experiments. *Technometrics*, 33(2):161–174, 1991.

APPENDIX

10. SENSITIVITY ANALYSIS

In this section, we implement the Species-oriented Sensitivity Analysis (SoSA) and the Pathway-oriented Sensitivity Analysis (PoSA) methods, inspiring by Morris [15], to perturb a system with the aim of finding the parameters that mainly affect its behavior. We adopt this methods to investigate the light-driven algal metabolism of *Chlamydomonas reinhardtii* [5] and the FBA model of the mitochondrion by Smith and Robinson [3].

We calculate the distribution of elementary effects [15] for each parameter of the model. For a combinatorial problem (such as PoSA), we define the "elementary effect" for the input b_s as

$$EE_s = \frac{\left[f(b_1, b_2, \ldots, b_{s-1}, \tilde{b}_s, b_{s+1}, \ldots, b_p) - f(\tilde{y}) \right]}{\Delta_s},$$

where $\tilde{b}_s$ is the mutation on the input b_s, and consists of the *switch* of bits chosen randomly in b_s: if a bit is equal to 0 (or 1), the permutation turns it in 1 (or 0). The output $f(y)$ considered in our analysis is the array of all the fluxes in the network. $\tilde{y}$ is the mutation carried on the inputs defined in the binary region of interest $\Omega = \{0, 1\}^L$. Instead, in SoSA the inputs/parameters analyzed are real-valued variables. For FBA models we consider the uptake rates. The elementary effect for the fluxes v_{exj}, $j = 1, \ldots, n$ (where n is the number of uptake rate fluxes) is

$$EE_j = \frac{\left[f(v_{ex1}, \ldots, v_{exj} + \Delta, \ldots, v_{exn}) - f(v_{ex}) \right]}{\Delta}.$$

For each input, an estimation of the distribution of the elementary effects is calculated by using N trials.

The estimation of the mean μ^* and standard deviation σ^* will be used as indicator of which inputs should be considered important. A large (absolute) central tendency for EE indicates an input with an important overall influence on the output. A large spread indicates an input whose influence is highly dependent on the values of the inputs [15]. In Figure S1 we show the results of the sensitivity analysis on the FBA mitochondrial model. We consider the upper bounds of the 73 exchange fluxes as inputs, and the sum of the fluxes obtained after the simulation as output. The plot shows that the oxygen uptake rate is the most sensitive input, followed by HCO^{3-}, Serine, Aspartate, Glutamate and Valine. The other inputs are also relevant, while the few inputs that are not influent on the output are octanoic acid, FAD, glutathione, sulfate, urea, biomass and fumarate.

For the light-driven algal metabolism, we detect the most sensitive inputs using the Pathway oriented Sensitivity Analysis (PoSA) [12]. PoSA enables us to rank the genetic manipulations according to their influence on the output of the model. Each input of the model is represented through a set of binary variables. The results are presented in Figure S2. The figure highlights that the pathway related to the transport and the mitochondrial compartment is the most sensitive. Other sensitive pathways are the transport and chloroplast, pyruvate metabolism, pyrimidine metabolism, transport glyoxysome and glycerolipid metabolism. This result underlines that the mitochondrion plays a key role in

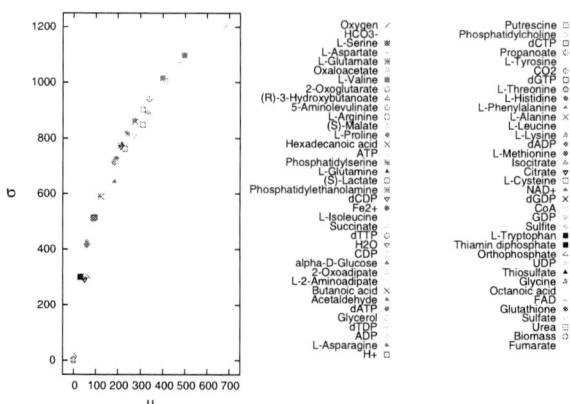

Figure S1: Sensitivity analysis on the mitochondrial FBA model. The plot shows the mean and the standard deviation of the elementary effects computed through the Morris' method applied to the upper bounds of the exchange reaction fluxes.

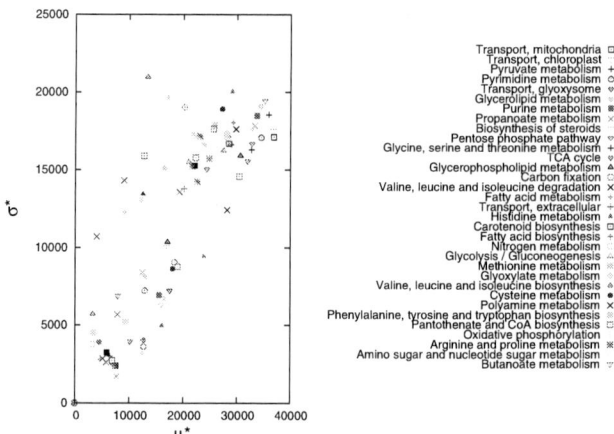

Figure S2: PoSA algorithm applied to the algal metabolism of *C. reinhardtii*. In the key, only the most sensitive pathways have been reported.

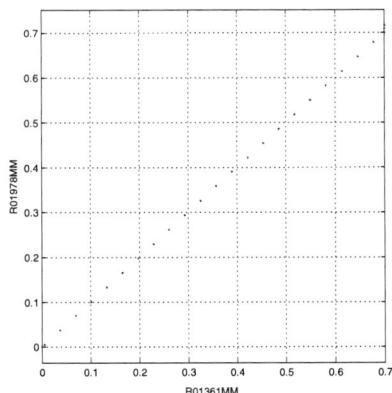

Figure S3: Functional relation among the two fluxes R01361MM and R01978MM [mmolh^{-1} gDW^{-1}] in the matrix, thus highlighting the structural non-identifiability of these variables. This group has been detected for both R01361MM and R01978MM, indicating a strong relation (two asterisks in the IA in Table S1). This plot shows that the relation detected with the IA is linear, although it shows a noisier behavior in the neighborhood of 0.7, as confirmed by Figure 6.

the algal metabolism.

Variable	Flux	Flux groups	r^2	cv
x_1	$R00004MM$	$x_1,x_2,x_{12},x_{27},x_{72},x_{82}*$	0.999	2.231
x_2	$R00014MM$	$x_2,x_{21},x_{37},x_{58},x_{72},x_{82}*$	0.999	0.381
x_3	$R00081MM$	$x_3,x_{64}*$	1.000	0.590
x_4	$R00086MM$	$x_4,x_{13},x_{21},x_{45},x_{65}*$	1.000	0.575
x_5	$R00127MM$	$x_5,x_{38},x_{98}*$	1.000	1.373
x_6	$R00157MM$	$x_6,x_{20},x_{61},x_{78}*$	1.000	1.372
x_7	$R00205MM$	$x_7,x_{19},x_{117},x_{128}*$	0.999	1.003
x_8	$R00238MM$	$x_8,x_{27},x_{37},x_{61},x_{78}*$	1.000	1.018
x_9	$R00243MM$	x_9*	1.000	5.512
x_{10}	$R00245MM$	$x_{10},x_{16},x_{57},x_{61},x_{98}*$	1.000	0.645
x_{11}	$R00256MM$	$x_{11},x_{118},x_{128}*$	0.999	1.411
x_{12}	$R00258MM$	$x_8,x_{12},x_{58}*$	0.998	0.430
x_{13}	$R00275MM$	$x_{13},x_{16},x_{57},x_{65},x_{80},x_{134}*$	1.000	0.557
x_{14}	$R00330MM$	$x_{14},x_{21},x_{78},x_{109},x_{113}*$	0.996	4.213
x_{15}	$R00342MM$	$x_{15},x_{27},x_{52},x_{66},x_{134}*$	1.000	0.588
x_{16}	$R00351MM$	$x_{16},x_{27},x_{57},x_{66},x_{134}*$	1.000	0.705
x_{17}	$R00355MM$	$x_2,x_{17},x_{72},x_{82}*$	0.999	0.382
x_{18}	$R00371MM$	$x_{18},x_{19},x_{43}*$	0.999	1.003
x_{19}	$R00388MM$	$x_{12},x_{19},x_{21},x_{106},x_{112},x_{128}*$	1.000	0.498
x_{20}	$R00430MM$	$x_{14},x_{20},x_{52},x_{66},x_{85}*$	0.988	4.075
x_{21}	$R00432MM$	$x_{21},x_{37},x_{52},x_{57},x_{73}*$	1.000	0.706
x_{22}	$R00512MM$	$x_{22},x_{129}**$	0.999	2.855
x_{23}	$R00551MM$	$x_{16},x_{23},x_{57}*$	1.000	0.378
x_{24}	$R00572MM$	$x_{19},x_{24},x_{33},x_{98},x_{112},x_{128}*$	0.998	2.165
x_{25}	$R00667MM$	$x_{25},x_{27},x_{61},x_{109},x_{113}*$	0.999	0.378
x_{26}	$R00705MM$	$x_{26},x_{31}**$	1.000	31.623
x_{27}	$R00709MM$	$x_{15},x_{21},x_{27},x_{45},x_{52},x_{70}*$	1.000	0.705
x_{28}	$R00713MM$	$x_{12},x_{28},x_{88},x_{91},x_{114},x_{116}*$	0.998	2.429
x_{29}	$R00716MM$	$x_{29},x_{69}**$	1.000	0.982
x_{30}	$R00740MM$	$x_{26},x_{30}*$	0.999	31.623
x_{31}	$R00830MM$	x_{26},x_{31}	n.a.	n.a.
x_{32}	$R00833MM$	$x_{32},x_{61},x_{78}*$	0.999	0.414
x_{33}	$R00851MM$	$x_{24},x_{33}*$	0.999	2.855
x_{34}	$R00927MM$	$x_{34},x_{81},x_{93}*$	1.000	1.289
x_{35}	$R00941MM$	$x_{35},x_{36}**$	1.000	1.000
x_{36}	$R00945MM$	$x_{35},x_{36}**$	1.000	1.000
x_{37}	$R01082MM$	$x_{16},x_{21},x_{37},x_{52},x_{57}*$	1.000	0.725
x_{38}	$R01175MM$	$x_{38},x_{58},x_{82},x_{104},x_{108},x_{112}*$	1.000	0.865
x_{39}	$R01177MM$	$x_{39},x_{86},x_{90},x_{111},x_{116}*$	1.000	0.976
x_{40}	$R01214MM$	$x_{31},x_{40}*$	1.000	0.378
x_{41}	$R01218MM$	$x_{15},x_{41},x_{58},x_{106},x_{108}*$	0.999	1.333
x_{42}	$R01253MM$	$x_{39},x_{42},x_{91},x_{98},x_{105}*$	1.000	0.378
x_{43}	$R01279MM$	$x_{33},x_{43},x_{87},x_{98},x_{106},x_{118}*$	1.000	0.864
x_{44}	$R01280MM$	$x_{44},x_{103}*$	1.000	1.731
x_{45}	$R01325MM$	$x_{16},x_{25},x_{27},x_{45},x_{57},x_{66}*$	1.000	0.705
x_{46}	$R01360MM$	$x_8,x_{46},x_{91},x_{107},x_{114}*$	1.000	1.248
x_{47}	$R01361MM$	$x_{47},x_{62}**$	1.000	1.248
x_{48}	$R01624MM$	$x_{48},x_{97}*$	1.000	1.731
x_{49}	$R01626MM$	$x_{44},x_{49},x_{102}*$	1.000	1.731
x_{50}	$R01648MM$	$x_8,x_{15},x_{50},x_{66}*$	0.994	3.206
x_{51}	$R01655MM$	$x_{35},x_{51}*$	1.000	1.000
x_{52}	$R01700MM$	$x_{15},x_{16},x_{45},x_{52},x_{70},x_{134}*$	1.000	0.707
x_{53}	$R01706MM$	$x_{53},x_{103}*$	1.000	1.731
x_{54}	$R01799MM$	$x_{24},x_{54}*$	0.999	2.855
x_{55}	$R01801MM$	$x_{55}*$	0.998	25.993
x_{56}	$R01859MM$	$x_{31},x_{56}*$	1.000	1.289
x_{57}	$R01900MM$	$x_{15},x_{16},x_{27},x_{57},x_{66}*$	1.000	0.705
x_{58}	$R01923MM$	$x_{58},x_{72},x_{88},x_{91},x_{98},x_{109}*$	1.000	0.864
x_{59}	$R01939MM$	$x_{59},x_{85},x_{128}*$	0.999	0.982
x_{60}	$R01940MM$	$x_{38},x_{52},x_{60},x_{66}*$	1.000	0.934
x_{61}	$R01975MM$	$x_{16},x_{61},x_{73},x_{74},x_{78},x_{134}*$	1.000	0.938
x_{62}	$R01978MM$	$x_{47},x_{62}**$	1.000	1.248
x_{63}	$R02030MM$	$x_8,x_{22},x_{63},x_{78},x_{129}*$	0.998	2.855
x_{64}	$R02161MM$	$x_{27},x_{37},x_{45},x_{64},x_{134}$	-1.396	0.590
x_{65}	$R02163MM$	$x_4,x_{13},x_{15},x_{16},x_{65}*$	1.000	0.557
x_{66}	$R02164MM$	$x_{15},x_{27},x_{45},x_{52},x_{66},x_{70}*$	1.000	0.706
x_{67}	$R02199MM$	$x_{67},x_{132}*$	1.000	1.289
x_{68}	$R02241MM$	$x_{33},x_{68}*$	0.999	2.855
x_{69}	$R02313MM$	$x_{29},x_{69}**$	1.000	0.982
x_{70}	$R02487MM$	$x_{70},x_{85}*$	0.998	0.934
x_{71}	$R02529MM$	$x_{18},x_{71}*$	1.000	1.003
x_{72}	$R02569MM$	$x_2,x_{27},x_{58},x_{72},x_{73},x_{82}*$	1.000	0.381
x_{73}	$R02570MM$	$x_{16},x_{21},x_{27},x_{45},x_{73}*$	1.000	0.707
x_{74}	$R02571MM$	$x_{74},x_{78},x_{85},x_{89},x_{118}*$	0.999	0.934
x_{75}	$R02661MM$	$x_{31},x_{75}*$	1.000	0.378
x_{76}	$R02662MM$	$x_{31},x_{76}*$	1.000	0.378
x_{77}	$R02765MM$	$x_{58},x_{77},x_{98},x_{104},x_{108}*$	0.999	1.289
x_{78}	$R03026MM$	$x_8,x_{61},x_{70},x_{78},x_{113},x_{116}*$	1.000	0.938
x_{79}	$R03102MM$	$x_{79}*$	0.998	0.982
x_{80}	$R03172MM$	$x_{38},x_{80},x_{98}*$	0.999	1.289

Variable	Flux	Flux groups	r^2	cv
x_{81}	$R03174MM$	$x_{81}, x_{93}*$	1.000	1.289
x_{82}	$R03270MM$	$x_2, x_{72}, x_{82}*$	1.000	0.381
x_{83}	$R03314MM$	$x_{58}, x_{83}, x_{88}, x_{111}, x_{112}*$	0.998	0.378
x_{84}	$R03381MM$	$x_{84}*$	0.997	0.378
x_{85}	$R03777MM$	$x_{85}, x_{87}, x_{89}, x_{117}, x_{118}*$	1.000	0.864
x_{86}	$R03778MM$	$x_{39}, x_{86}, x_{90}, x_{106}, x_{116}*$	1.000	0.976
x_{87}	$R03857MM$	$x_{43}, x_{85}, x_{87}, x_{89}, x_{118}**$	1.000	0.864
x_{88}	$R03858MM$	$x_{39}, x_{70}, x_{86}, x_{88}, x_{91}, x_{111}*$	1.000	0.976
x_{89}	$R03990MM$	$x_{58}, x_{89}, x_{110}, x_{112}, x_{115}*$	1.000	0.864
x_{90}	$R03991MM$	$x_{39}, x_{70}, x_{90}, x_{105}, x_{109}, x_{113}*$	1.000	0.976
x_{91}	$R04170MM$	$x_{39}, x_{70}, x_{88}, x_{91}, x_{107}, x_{114}*$	1.000	0.976
x_{92}	$R04203MM$	$x_{92}, x_{132}*$	1.000	1.289
x_{93}	$R04204MM$	$x_{26}, x_{31}, x_{93}*$	1.000	1.289
x_{94}	$R04224MM$	$x_{31}, x_{94}*$	1.000	0.378
x_{95}	$R04355MM$	$x_{49}, x_{95}*$	1.000	1.731
x_{96}	$R04428MM$	$x_{96}, x_{99}**$	1.000	1.731
x_{97}	$R04430MM$	$x_{97}, x_{122}*$	1.000	1.731
x_{98}	$R04433MM$	$x_2, x_4, x_{13}, x_{58}, x_{65}, x_{98}*$	1.000	0.858
x_{99}	$R04533MM$	$x_{96}, x_{99}**$	1.000	1.731
x_{100}	$R04536MM$	$x_{100}, x_{120}**$	1.000	1.731
x_{101}	$R04537MM$	$x_{101}, x_{123}**$	1.000	1.731
x_{102}	$R04543MM$	$x_{102}, x_{125}*$	1.000	1.731
x_{103}	$R04544MM$	$x_{103}, x_{125}**$	1.000	1.731
x_{104}	$R04737MM$	$x_{104}, x_{106}, x_{108}, x_{110}, x_{115}*$	1.000	0.976
x_{105}	$R04738MM$	$x_{88}, x_{90}, x_{105}, x_{107}, x_{116}*$	1.000	0.976
x_{106}	$R04739MM$	$x_{104}, x_{106}, x_{107}, x_{112}, x_{113}*$	1.000	0.976
x_{107}	$R04740MM$	$x_{88}, x_{90}, x_{91}, x_{107}, x_{116}*$	1.000	0.976
x_{108}	$R04741MM$	$x_{104}, x_{108}, x_{110}, x_{111}, x_{115}*$	1.000	0.976
x_{109}	$R04742MM$	$x_{39}, x_{90}, x_{91}, x_{109}, x_{113}*$	1.000	0.976
x_{110}	$R04743MM$	$x_{86}, x_{108}, x_{110}, x_{112}, x_{115}*$	1.000	0.976
x_{111}	$R04744MM$	$x_{90}, x_{107}, x_{111}, x_{113}, x_{116}*$	1.000	0.976
x_{112}	$R04745MM$	$x_{108}, x_{109}, x_{110}, x_{112}, x_{115}*$	1.000	0.976
x_{113}	$R04746MM$	$x_{86}, x_{108}, x_{109}, x_{113}, x_{116}*$	1.000	0.976
x_{114}	$R04747MM$	$x_{90}, x_{105}, x_{113}, x_{114}, x_{115}*$	1.000	0.976
x_{115}	$R04748MM$	$x_{106}, x_{108}, x_{110}, x_{112}, x_{115}*$	1.000	0.976
x_{116}	$R04749MM$	$x_{86}, x_{109}, x_{111}, x_{113}, x_{116}*$	1.000	0.976
x_{117}	$R04751MM$	$x_{43}, x_{85}, x_{89}, x_{110}, x_{117}*$	1.000	0.864
x_{118}	$R04754MM$	$x_{43}, x_{85}, x_{87}, x_{89}, x_{118}**$	1.000	0.864
x_{119}	$R04952MM$	$x_{119}*$	0.998	1.731
x_{120}	$R04953MM$	$x_{100}, x_{120}**$	1.000	1.731
x_{121}	$R04954MM$	$x_{121}, x_{122}**$	1.000	1.731
x_{122}	$R04956MM$	$x_{121}, x_{122}**$	1.000	1.731
x_{123}	$R04959MM$	$x_{101}, x_{123}**$	1.000	1.731
x_{124}	$R04968MM$	$x_{17}, x_{124}*$	0.998	1.731
x_{125}	$R04970MM$	$x_{103}, x_{125}**$	1.000	1.731
x_{126}	$R05064MM$	$x_{126}*$	1.000	0.378
x_{127}	$R05066MM$	$x_{127}*$	1.000	0.378
x_{128}	$R07162MM$	$x_{12}, x_{19}, x_{27}, x_{109}, x_{111}, x_{128}*$	1.000	0.498
x_{129}	$R07390MM$	$x_{22}, x_{129}**$	0.999	2.855
x_{130}	$R07599MM$	$x_{31}, x_{130}*$	1.000	0.378
x_{131}	$R07600MM$	$x_{31}, x_{131}*$	1.000	0.378
x_{132}	$R07603MM$	$x_{132}, x_{133}**$	1.000	1.289
x_{133}	$R07604MM$	$x_{132}, x_{133}**$	1.000	1.289
x_{134}	$R07618MM$	$x_{16}, x_{27}, x_{45}, x_{66}, x_{70}, x_{134}*$	1.000	0.591
x_{135}	$R08157MM$	$x_{31}, x_{135}*$	1.000	0.378

Table S1: Identifiability analysis applied to the mitochondrial FBA model. We take into account the matrix V obtained as output of the FBA applied to maximize ATP and NADH in different fumarate conditions. The 135 matrix fluxes are grouped according to functional relations. The full name of the fluxes can be found in [3]. The r^2 column indicates how much variance of the response can be explained by the predictors. A high ratio $cv(x) = std(x)/mean(x)$ suggests that the data are scattered, and therefore there may be practical non-identifiability. "n.a." stands for "not available", since those fluxes do not play any role in the 2-objective maximization of ATP and NADH. The asterisk denotes the cases such that $r^2 > 0.9$ and $cv > 0.1$. Another asterisk is added when the same functional group is detected more than one time with different variables as response, indicating a strong interdependence.

Design of Cyberphysical Digital Microfluidic Biochips under Completion-Time Uncertainties in Fluidic Operations[*]

Yan Luo[†], Krishnendu Chakrabarty[†], and Tsung-Yi Ho[‡]

[†]Electrical & Computer Engineering Department, Duke University, Durham, NC 27708, USA
[‡]Computer Science and Information Engineering Department, National Cheng Kung University, Tainan, Taiwan
E-mail: {yan.luo, krish}@duke.edu; tyho@csie.ncku.edu.tw

ABSTRACT

Cyberphysical digital microfluidics enables the integration of fluid-handling operations, reaction-outcome detection, and software-based control in a biochip. However, synthesis algorithms and biochip design methods proposed in the literature are oblivious to completion-time uncertainties in fluidic operations, and they do not meet the requirements of cyberphysical integration in digital microfluidics. We present an operation-interdependency-aware synthesis method that uses frequency scaling and is responsive to uncertainties that are inherent in the completion times of fluidic operations such as mixing and thermal cycling. Using this design approach, we can carry out dynamic on-line decision making for the execution of fluidic operations in response to detector feedback. We use three common laboratorial protocols to demonstrate that, compared to uncertainty-oblivious biochip design, the proposed dynamic decision making approach is more effective in satisfying realistic physical constraints. As a result, it decreases the likelihood of erroneous reaction outcomes, and it leads to reduced time-to-results, less repetition of reaction steps, and less wastage of precious samples and reagents.

Categories and Subject Descriptors

B.2.2 [**Hardware**]: Performance Analysis and Design Aids

General Terms

Algorithms, Performance, Design.

Keywords

Digital microfluidics, electrowetting-on-dielectric, lab-on-chip.

1. INTRODUCTION

Digital microfluidic biochips have emerged in recent years as a promising platform for implementing laboratory procedures in biochemistry [1–5]. A digital microfluidic biochip consists of an array of electrodes that can be applied actuation voltages under

---

[*]The work of Y. Luo and K. Chakrabarty was supported in part by the US National Science Foundation under grants CCF-0914895 and CNS-1135853. The work of T.-Y. Ho was supported in part by the Taiwan National Science Council under grant no. NSC 101-2220-E-006-016 and 101-2628-E-006-018-MY3 and the Ministry of Education, Taiwan, R.O.C. under the NCKU Aim for the Top University Project Promoting Academic Excellence & Developing World Class Research Centers.

Permission to make digital or hard copies of all or part of this work for personal or classroom use is granted without fee provided that copies are not made or distributed for profit or commercial advantage and that copies bear this notice and the full citation on the first page. To copy otherwise, to republish, to post on servers or to redistribute to lists, requires prior specific permission and/or a fee.
DAC 2013, May 29 - June 07 2013, Austin, Texas, USA.

clock control through a set of control pins. Biochemical assays, such as the dilution of samples and reagents [3], crystallization of protein molecules [4], on-chip chemistry for DNA sequencing [1], multiplexed real-time polymerase chain reaction (PCR) [2], protein crystallization for drug discovery [5], and glucose measurement for blood serum [3], have been successfully implemented on such biochips.

The precision of fluidic operations is vital for the accuracy of analytical bioassays. For example, in the quantitative measurement for glucose concentration in blood [3], accurate measurements cannot be obtained if the mixing time for blood sample and enzymatic reagent is not controlled precisely. In order to determine appropriate parameter settings and increase the precision of on-chip operations, bioassays need to be thoroughly characterized [6] [7] [8]. Fluidic operations must be repeatedly executed and monitored to obtain statistically significant results [6]. Based on these results, a module library that defines the execution time for each type of operation is derived, and this library is used as the guideline for the execution of on-chip operations.

However, even after careful characterization, the problem of inaccuracy remains due to the inherent variability and randomness of biological/chemical processes [6–11]. The bioassay yield, defined as the percentage of bioassay instances that terminate with outcomes within calibrated ranges and within the predetermined time, remains low when a large number of fluidic operations are involved. An analysis of bioassay yield is presented in Part A of the appendix. In addition to the problem of compounding inaccuracy for multi-step assays, the characterization procedure suffers from other drawbacks. Since characterization requires repeated execution of each operation, it is a time-consuming process and leads to the wastage of sample and reagent droplets.

In order to overcome the drawbacks associated with characterization, biochips integrated with sensing systems, i.e., cyberphysical microfluidic biochips, are being developed [3] [7] [12] [13]. The feedback provided by the sensing system enables real-time concentration checking, error detection, and error correction for fluidic operations [3]. Therefore, essential operations such as droplet dispensing and mixing can be precisely implemented on cyberphysical microfluidic biochips without the need for characterizing a bioassay or specifying a module library [3] [12].

However, today's synthesis algorithms for mapping biochemistry protocols to the chip rely on characterization procedures for bioassays [14–17]. Hence the advantages of cyberphysical integration are not fully exploited, and precious samples/reagents and time are wasted during characterization. In addition, current design methods for cyberphysical biochips suffer from following three limitations:

1. Prior work is oblivious to variability and uncertainty in biochemical processes. The competition-time of fluidic operations in practical applications may be different from the time defined in a module library, therefore the accuracy of fluidic operations cannot be guaranteed.

2. Based on the feedback from sensors, the control software in previous work needs to perform on-line computation with high complexity (details are presented in Part B of the appendix). The additional computation time may lead to the

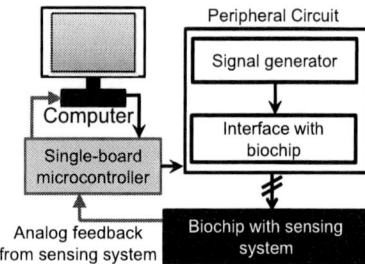

Figure 1: The schematic of a cyberphysical digital microfluidic system [15].

interruption of operations and affect time-to-result [15].

3. On-line computation in prior work does not generate the information needed for droplet transportation, which is essential for the execution of a bioassay.

To overcome the above drawbacks, we propose a new design method that facilities on-line decision making using cyberphysical microfluidic biochips. The key contributions of this paper are as follows:

1. We propose the design of microfluidic biochips using multiple clock frequencies. The execution time of the bioassay can be reduced without any additional degradation of electrodes or hardware cost.

2. We propose an "operation-interdependence-aware" synthesis algorithm — the first on-chip biochemistry synthesis procedure that does not use the module library as a design guideline. Using this algorithm, the characterization process can be eliminated.

3. We propose a design approach that considers completion-time uncertainties for fluidic operations, hence the accuracy of fluidic operations is improved.

4. We describe an on-line computation approach with low complexity, hence the response time of the system is negligible.

The remainder of this paper is organized as follows. Section 2 describes cyberphysical integration in a digital microfluidic biochip. Section 3 presents the design of microfluidic biochips with multiple clock frequencies. Section 4 introduces the framework of operation-dependency-aware synthesis. Simulation results for three widely used bioassays are presented in Section 5. Section 6 concludes the paper.

2. CYBERPHYSICAL INTEGRATION

An example of closed-loop integration in cyberphysical microfluidics is shown in Figure 1 [15]. The biochip and the control software are coupled together by the peripheral circuit (which includes the signal generator and biochip's input interface) and the sensing system. The peripheral circuit converts actuation values derived by the control software to electrical signals that are applied to electrodes on the biochip, and the sets of voltages and frequencies for the electrical signals are both programmable [18]. The sensing system monitors the status of droplets (attributes such as color, concentration, volume, diameter, and position), and sends the feedback information to the control software. In this way, the software can control fluidic operations based on feedback signals.

For example, in the measurement of glucose in blood, serum samples need to be well-mixed with an enzymatic reagent [3]. During the mixing procedure, the status of the droplet is monitored by an image sensor. The extent to which mixing has been completed can be quantified by analyzing images for a droplet. Therefore, the control software will force the biochip to continue mixing until the feedback information shows that the droplets are sufficiently well-mixed. Hence the mixing operation can be precisely controlled without knowing the precise mixer execution

time in advance, and the on-chip measurement results for glucose concentration can be as precise as the results derived by a traditional bench-top analyzer used by a laboratory technician [3].

Such cyberphysical biochips enable automatic error recovery for bioassays [14] [16]. The on-chip sensing system measures the concentration of intermediate droplets and compares the results with pre-determined expected values. Droplets that are identified as "unqualified droplets" are discarded, and the biochip re-executes operations for generating new droplets to replace the discarded ones.

3. BIOCHIPS WITH MULTIPLE CLOCK FREQUENCIES

Experimental results published in the literature demonstrate that the degradation of an electrode is directly proportional to the number of times that it is switched on and off [18], i.e., with the same sequence of electrode actuation vectors, electrodes will degrade more quickly under higher clock frequency. On the other hand, an increase in the clock frequency can reduce the execution time of some fluid-handling operations [8]. Hence, in order to ensure the reliability of electrodes on the biochip, and at the same time complete the bioassay under timing constraints, it is important to choose an appropriate clock frequency.

Fluid-handling operations are divided into two categories: frequency-sensitive operations and frequency-insensitive operations. The completion time of droplet transportation and dispensing is determined by clock frequency, because the droplet will be moved from one unit cell to another adjacent unit cell in each clock cycle; hence the rate at which a droplet is transported or dispensed is proportional to the clock frequency. Note that, if the transportation or dispensing path for a droplet consists of P electrodes, then the number of clock cycles required to move or dispense the droplet is also P. This number of clock cycles only relates to the length of the transportation path, and it is independent of the clock frequency. If we increase the electrode switching frequency for droplet transposition and dispensing, the time needed for these operations can be reduced without any additional degradation of electrodes. Hence, we conclude that by increasing the clock frequency, the transportation and dispensing time of droplets can be accelerated without affecting chip reliability.

The execution times of mixing and dilution operations cannot be reduced significantly by increasing the clock frequency. For example, experimental results for droplet mixing show that, at a frequency of 8 Hz, the time spent on the mixing operation is 12 seconds; when the frequency is increased to 16 Hz, the mixing time decreases to 11 seconds [8]. Hence the mixing time only decreases 8.3% while the rate of degradation of electrodes increases 100%. Hence we conclude that increasing the clock frequency for dilution/mixing operation will adversely affect the lifetime of the biochip without any significant reduction in the time needed for these steps of a bioassay.

In order to minimize the time required to complete a bioassay with least impact on chip reliability, it is desirable to run different categories of operations at different clock frequencies. Hence we propose to schedule transportation/dispensing operations and dilution/mixing operations at different time segments. The time segment to implement droplet transportation is defined as the "transportation phase" (T phase), and the segment to implement dilution/mixing operations is defined as the "dilution/mixing phase" (D/M phase); see Figure 2(a). For each phase, only transportation operations or the dilution/mixing operations are carried out on the chip.

Assume that we have already determined the set of dilution/mixing operations to be implemented at each D/M phase before the execution of a bioassay. At run-time, the biochip operates under clock frequency f_T in the T phase. Output droplets of previous steps and droplets dispensed from reservoirs are moved to the modules where the subsequent dilution/mixing operations are to be carried out. After all the droplets arrive at their destination modules, the biochip enters the D/M phase. The dilution/mixing operations that are scheduled in the same phase start together, and they are carried out under clock frequency $f_{D/M}$. When the

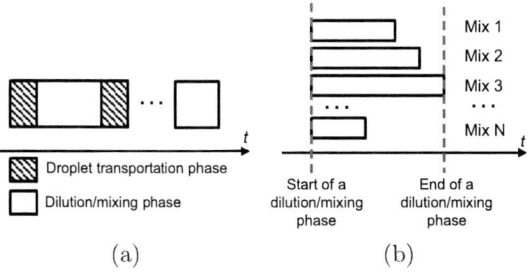

(a)   (b)

Figure 2: (a) Droplet transportation and dilution/mixing operations are scheduled in different phases; (b) start and end time of a D/M phase.

feedback from sensors indicates that all the dilution/mixing operations have already been completed, the D/M phase ends and the biochip enters the next T phase, as shown in Figure 2(b).

In this way, based on sensor feedback, the biochip "switches" between the T phase and the D/M phase with different clock frequencies. Since the output frequency of the signal generator is controlled by software, this biochip design using multiple clock frequencies can be implemented by the hardware setup shown in Figure 1 without any extra cost. Recent work in a different context, viz. to understand the reliability impact of multiple frequencies, has demonstrated the feasibility of such a hardware setup [18]. We use this same setup here for cyberphysical adaptation. In the D/M phase, the biochip operates at a nominal frequency (for example, $f_{D/M} = 1$ Hz), while in the T phase, higher-frequency signals (for example, $f_T = 16$ Hz) are applied to the electrodes. Hence, the time spent on dispensing and transporting droplets can be reduced significantly without any additional degradation of electrodes or any additional cost in hardware.

Since the time spent on each dilution/mixing operation is determined by sensor feedback rather than a pre-determined module library, we can derive a "semi-deterministic" design for biochips with the consideration of timing uncertainties. The design includes the synthesis result which is introduced next in Section 4.

4. OPERATION-DEPENDENCY-AWARE SYNTHESIS

In this section, we describe how the synthesis results can be derived in the presence of completion-time uncertainties of fluidic operations. Using the proposed algorithm, we can obtain the synthesis results that include: (i) the module placement for each operation; (ii) the set of operations to be implemented in each D/M phase. The exact start and end time of operations is determined on the basis of feedback from sensors during bioassay run-time, hence they are not included in the results derived by the proposed synthesis algorithm. Therefore, the synthesis results that we derive are semi-deterministic.

The proposed synthesis algorithm focuses on the interdependency among dilution/mixing operations that are given by the sequencing graph of a bioassay. A sequencing graph is an abstract description for a bioassay; each node in it represents a fluidic operation, and each edge represents the interdependency for a pair of operations [11]. For any two operations O_a and O_b, if the output droplet of O_a is the input of O_b, then there is an edge from O_a to O_b. The sequencing graph in this method is reduced by deleting all nodes that represent dispensing operations.

A sequencing graph for a bioassay has two important properties: (i) it is a directed acyclic graph, because there is no infinite loop or a repeated step under identical conditions in an assay protocol; (ii) the numbers of input droplets and output droplets for each operation are at most 2. Therefore, the in-degree and out-degree of each node in the sequencing graph are at most 2.

The pseudocode for the operation-interdependency-aware synthesis approach is shown in Figure 3. In the following parts, we first introduce the synthesis algorithm for sequencing graphs with the structure of directed-trees, and then introduce the steps for synthesizing a bioassay in the general case.

1: Partition a sequencing graph G into directed-trees $\{T_1, T_2, ..., T_n\}$;
2: **for** each $T_i \in \{T_1, T_2, ..., T_n\}$ **do**
3:     Start from leaf nodes, determine relative positions of dilution/mixing modules on a level-by-level basis;
4:     Determine schedules of operations;
5:     Package synthesis result for the entire directed tree as a "macro-module" M_{T_i};
6: **end for**
7: Sort the directed trees based on operation interdependencies;
8: Derive new sequencing graph which consists of macro-modules for directed trees;
9: Merge synthesis results of directed trees based on their orders.

Figure 3: Pseudocode for operation-interdependency-aware synthesis.

4.1 Synthesis method for sequencing graphs with directed tree structure

First we study bioassays whose sequencing graphs are directed trees. Examples are shown in Figure 4(a) and Figure 4(b). The node without children nodes in Figure 4(a) and the node without parent nodes in Figure 4(b) are defined as the root nodes of the directed trees. The difference between these two sequencing graphs is that, all edges in Figure 4(a) are directed towards the root of the tree, while in Figure 4(b), all edges are directed away from the root. After switching the directions of all of the edges, the directed tree in Figure 4(b) can be analyzed in the same way as the one in Figure 4(a). Thus, we use the structure with all edges directed towards the root of the tree (shown in Figure 4(a)) to analyze the proposed synthesis procedure.

The leaf nodes are defined as the lowest-level nodes of the tree, and the operations represented by these leaf nodes are defined as the lowest-level operations of the bioassay. In the synthesis procedure, starting from the lowest-level operations, we can determine the schedule and module placement for the operations on a level-by-level basis. Consider the following example. Operations 1, 2, and 6 are three mixing operations shown in Figure 4(a). Here 1 and 2 are the lowest-level operations and their outputs are the inputs of 6, hence 1 and 2 must be completed before 6 starts. Based on the interdependency relationship, operations 1 and 2 are implemented in the same D/M phase, while the operation 6 is schedule to be implemented in the next D/M phase, as shown in Figure 4(c). If we write the set of operations to be implemented at the i-th D/M phase as S_i, then the schedules of operations 1, 2, and 6 can be written as: $\{1, 2\} \subseteq S_1$ and $\{6\} \subseteq S_2$.

Next, mixing operations 1 and 2 are mapped to two mixers on the biochip. The sizes of available mixers on a biochip can be found in [8] [11]. Since the mixing operation 6's inputs are the outputs of 1 and 2, we can refer to the region that overlaps with the modules for 1 and 2 as the "execution region" of operation 6, as shown in Figure 4(d). In this way, the outputs of mixing operations 1 and 2 can be fed directly into 6 without any module-to-module transportation, and the module placement and schedule for operation 1, 2, and 6 are obtained. After operation 6 is finished, the region will be assigned for operation 8.

Next, operations 1, 2, 6, and 8 are packaged as a "macro-module" M_{S_L}; similarly, operations 3, 4, 7, and 9 are packaged as a "macro-module" M_{S_R} as shown in Figure 4(e). The schedule for operations $1 - 4$ and $6 - 9$ are: $\{1, 2, 3, 4\} \subseteq S_1$, $\{6, 7\} \subseteq S_2$, and $\{8, 9\} \subseteq S_3$.

Similarly, the placement of modules for M_{S_L}, M_{S_R} and N_R in Figure 4(e) can be determined based on their interdependency. The module for M_{S_L} will be placed "beside" the region occupied by M_{S_R}, and after the completion of M_{S_L} and M_{S_R}, N_R will be mapped to the same region that is assigned to M_{S_L} and M_{S_R}. In this way the schedule and resource assignment results for all the operations shown in Figure 4(a) can be derived level by level.

Suppose we arbitrarily pick a node N_R from the directed-tree structure shown in Figure 4(a), and write the set of operations on the left and right sub-trees of N_R as S_L and S_R, respectively. Then the synthesis results derived from the proposed operation-interdependency-aware synthesis algorithm have the following two characteristics:

309

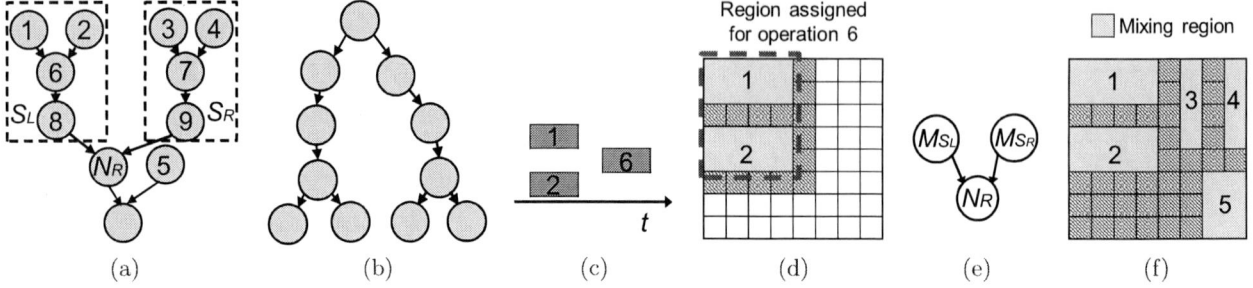

Figure 4: Sequencing graph with tree structure (a) all edges are directed towards the root of the tree; and (b) all edges are directed away from the root; (c) scheduling results for mixing operations 1, 2, and 6; (d) module placement for 1, 2 and 6; (e) packaged macro-modules M_{S_L} and M_{S_R}; (f) a feasible solution for module placements of operations $1-5$ on an 8×8 array.

Characteristic 1: Operations in S_L and S_R do not share any on-chip resources, i.e., they are executed in two separate regions of the biochip.

Characteristic 2: Suppose operations in S_L and S_R are carried out on the sets of electrodes which are written as E_{S_L} and E_{S_R}, respectively, and the operation N_R is carried out on the set of electrodes E_{N_R}, then we have: $E_{N_R} \subseteq (E_{S_L} \cup E_{S_R})$.

Based on Characteristic 2, we conclude that the resource bound to an operation (i.e., the module placement) is determined in turn by its predecessor operations. Hence, for a sequencing graph with a directed-tree structure, the resources for other operations can be determined easily when resources assigned to operations represented by leaf nodes are known. For example, the synthesis result for the sequencing graph shown in Figure 4(a) can be determined as follows. According to the proposed resource binding steps, the modules for operations 1 and 2 will be placed beside each other, and the modules for operations 3 and 4 will be placed beside each other. An example of a feasible result for the placement of the modules on an 8×8 array is shown in Figure 4(f). Based on the module placement for operations 1-5, module placement for other succeeding operations can be derived. For example, the regions where operations 1 and 2 are implemented will be assigned to their successor operations. In this way, the module placement of operations in the directed-tree structure are determined. The schedules of operations in the directed-tree can also be determined based on their input/output interdependency.

4.2 Synthesis method for sequencing graphs in general cases

For sequencing graphs that are not directed trees (i.e., there are nodes whose out-degrees are greater than 1, as shown in Figure 5(a)), we can derive their synthesis results using the following steps:

1. Graph partitioning: determine the set of nodes $\{N_{n_1}, N_{n_2}, ..., N_{n_k}\}$ whose out-degrees are greater than 1, then remove all the edges directed away from these nodes. Then the graph is partitioned into multiple directed trees $\{T_1, T_2, ..., T_n\}$, and each node in $\{N_{n_1}, N_{n_2}, ..., N_{n_k}\}$ becomes the root node in a directed tree.

2. Synthesis for directed trees: apply operation-interdependency-aware synthesis to each directed tree, and derive the corresponding synthesis result.

3. Sorting of directed trees: for any pair of trees T_A and T_B, suppose there exists an operation $O_{T_{A_1}} \in T_A$ and an operation $O_{T_{B_1}} \in T_B$, such that the output of $O_{T_{A_1}}$ is the input of $O_{T_{B_1}}$. Then we express the relationship between trees T_A and T_B as $T_A < T_B$. Any two elements T_x and T_y in $\{T_1, T_2, ..., T_n\}$ may stand in any of three mutually exclusive relationships to each other: $T_x < T_y$, or $T_x > T_y$, or $T_x = T_y$ (neither of the other two). The conflicting case (i.e., $T_x < T_y \cap T_x > T_y$) will never occur. The proof can be found in Part C of the appendix.

4. Merge the synthesis results: the synthesis result of $\{T_1, T_2, ..., T_n\}$ are merged together based on their order. If

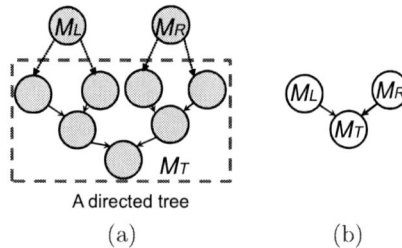

Figure 5: (a) Partitioning of a sequencing graph; (b) the interdependency of macro-module M_T, module M_L and M_R.

$T_x < T_y$ then operations in T_x must be implemented before T_y; if $T_x > T_y$ then operations in T_x must be implemented after T_y; if $T_x = T_y$ then operations in T_x and T_y are implemented in parallel.

From the above steps, the synthesis results can be derived for general sequencing graphs. An example is shown in Figure 5(a). The graph is divided into a directed tree T and the nodes that represent operation M_L and M_R. The synthesis result for T is packaged as a "macro-module" M_T. Note that M_L, M_R and the macro-module of M_T together form another tree, as shown in Figure 5(b). The module placement can be derived in a similar way as the sequencing graph shown in Figure 4(e).

The operation-interdependency-aware algorithm described above can derive the semi-deterministic synthesis result for a bioassay. The synthesis result assigns each dilution/mixing operation to a specific D/M phase, while completion-times of these operations are determined on-line during the execution of the bioassay. The robust module placement thus derived is independent of the execution time of each operation. When timing uncertainties of dilution/mixing operations exist, the module placement of a bioassay remains unchanged.

5. SIMULATION RESULTS

In this section, we first present the simulation results for three widely used laboratory protocols, namely exponential dilution of a protein sample, interpolation dilution of a protein sample, and polymerase chain reaction (PCR), respectively [5] [11]. The sequencing graph and the detailed description of the protocol for these bioassays can be found in [5] [11]. We compare our results with prior work on biochip synthesis in [11], and a recently published cyberphysical software-based recovery method based on a greedy algorithm [14] [15]. As with all prior work in this area, both these baseline methods are oblivious of timing uncertainties in fluidic operations.

In order to compare the uncertainty-oblivious baseline methods with the proposed design method and study the synthesis results in the presence of timing uncertainty, the baseline methods are adjusted using three methods: (i) adding additional time to module library specifications; (ii) running dynamic resynthesis when a timing overshoot is detected; (iii) adding interruption whenever there is a timing overshoot.

5.1 Yield and completion time of bioassays

In Part A of the appendix, we compute the probability P_{success} of successful implementation. We show that P_{success} decreases exponentially with an increase in the number of dilution/mixing operations. The mixing/dilution operation times are modeled as Gaussian random variables, each with a mean and standard deviation. Next, we compute numerical values of P_{success} for the three test-case bioassays.

Let T_i be the time spent on operation O_i when a static synthesis method is used based only on a module library, and P_i be the probability that O_i is successfully completed before the next operation is started. Based on the characteristics of the Gaussian distribution, if we set the time spent on operation O_i as $T_i = \mu_i + \sigma_i$, then $P_i = 0.84$; if we set $T_i = \mu_i + 2\sigma_i$, then $P = 0.98$. Table 1 lists the probabilities that bioassays are successfully completed (i.e., with no unfinished operations) when no cyberphysical adaptation is used. When we conservatively set $T_i = \mu_i + 2\sigma_i$ (and increase the operation times considerably) and obtain $P_i = 0.98$, the probability for successfully implementing the exponential mixing of protein bioassay is still low, only 0.34. We can further calculate that, in order to improve the yield of exponential dilution bioassay to above 0.90, we need to set $T_i = \mu_i + 4\sigma_i$.

We therefore conclude that the bioassay execution on conventional biochip platforms without a sensing system, closed-loop control, and uncertainly-aware synthesis, will lead to unacceptably low bioassay yield and low confidence in reaction outcomes.

From Table 1, we also find that P_{success} can be increased by increasing T_i. However, on the other hand, increasing T_i will increase the reaction completion time, and increase the risks of excessive heating and evaporation of droplets [19]. Hence T_i should be set to be within in a reasonable range. Due to the lack of sufficient data thus far from real experiments on fabricated chips, we do not consider here the additional probability of bioassay failure caused by excessive heating and evaporation. In this way, we do not quantify this important shortcoming of the baseline methods that we use for comparison.

Next we compare bioassay completion times of the proposed operation-interdependency-aware synthesis algorithm with the parallel recombinative simulated annealing (PRSA)-based synthesis algorithm [11]. Here we use the module library defined in [20], and run simulations by considering timing uncertainty. We assume that for each operation, its average execution time μ_i is the time defined in the library, and σ_i is $0.1\mu_i$. For example, for a dilution operation executed on a 2×4 array, $\mu_i = 5$ seconds, and $\sigma_i = 0.5$ second.

In order to increase the yield for the baseline design, when we run the PRSA-based algorithm, we add extra execution time ΔT_i for each operation. In the simulations, we assume that the PCR bioassay, which has relatively few operations, is executed on an 8×8 array. Exponential dilution and interpolation dilution are executed on a 10×10 array. The simulation results with ΔT_i set to σ_i, $2\sigma_i$, and $4\sigma_i$ are shown in Table 2. We find that the completion times of the bioassays increase with ΔT_i, and by applying the proposed method, we can reduce the time-to-results.

As shown in Table 2, for the exponential dilution bioassay, the completion time derived by PRSA increases only slightly when the execution time for each dilution/mixing operation increases. This is because in this bioassay, most of the execution time is spent on dispensing operations, whose execution time is known to have low variability.

5.2 Number of droplets consumed

For the synthesis algorithm proposed in [14] [15], bioassays need to be characterized before they are executed on the biochip. In the characterization procedure, each operation needs to be executed at least three times [12]. The comparison of the number of droplets consumed for each bioassay in [14] [15] [16] and the proposed design is shown in Figure 6. Note that the prior methods consume the same number of droplets each. We find that the number of droplets is greatly reduced in the proposed design

Table 1: Probabilities of bioassays being successfully implemented without unfinished operations.

Bioassay	No. of dilution/ mixing operations	P_{success}, T_i listed		
		μ_i	$\mu_i + \sigma_i$	$\mu_i + 2\sigma_i$
PCR	7	0.01	0.30	0.85
Interpolating dilution	35	~ 0	~ 0	0.45
Exponential dilution	47	~ 0	~ 0	0.34

Table 2: Comparison of bioassay completion time derived by PRSA-based algorithm [11] and proposed method.

	Completion time (s)				
	PRSA-based algorithm [11], ΔT_i listed				Proposed method
Bioassay	0	σ_i	$2\sigma_i$	$4\sigma_i$	
PCR	26	28	31	35	25
Interpolation dilution	177	186	199	220	154
Exponential dilution	239	245	251	258	172

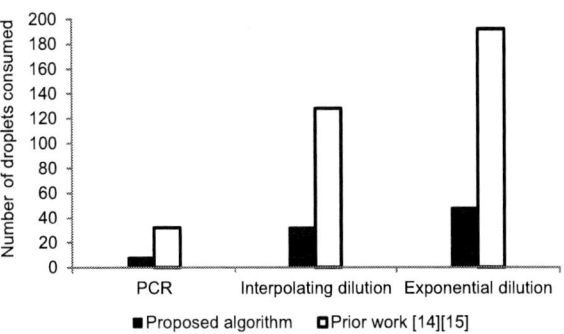

Figure 6: Comparison between the number of droplets consumed in the biochips of [14] [15] (droplet consumption is the same for both these methods) and the proposed method.

based on a cyberphysical platform and uncertainly-aware synthesis.

5.3 Number of operations interrupted under uncertainty

Suppose we run the synthesis results derived by the PRSA-based algorithm on a cyberphysical biochip. If the execution time of an operation O_{longer} is longer than the time defined by the module library, then the controller can stop all other operations until O_{longer}. In this way, time-consuming on-line resynthesis can be avoided. However, the executions of other operations have to be interrupted.

Based on the Gaussian assumption in Section 5.1, the probability that an operation's execution time is longer than the time defined in the module library is 0.5 if we do not add extra execution time for each operation. We simulate the bioassay and count the number of operations that must be interrupted. During the execution of bioassay, an operation may be interrupted multiple times; however, we count such cases as one interrupted operation. Hence the number of interruptions reported for the baseline method is less than the actual number of interruptions experienced.

The total number of operations, and the number of operations that are interrupted for the three bioassays, are listed in Table 3. We note that nearly all the operations are interrupted for the synthesis results derived using the PRSA-based algorithm. The interruptions that occur during the execution of the bioassay may influence the quality of output droplets [14]. On the other hand,

Table 3: Comparisons for the number of operation interrupted between PRSA-based algorithm [11] and the proposed method.

Bioassay	Total no. of operations	No. of operations interrupted [11]	No. of operations interrupted (proposed method)
Exponential dilution	103	97	0
Interpolating dilution	71	65	0
PCR	15	3	0

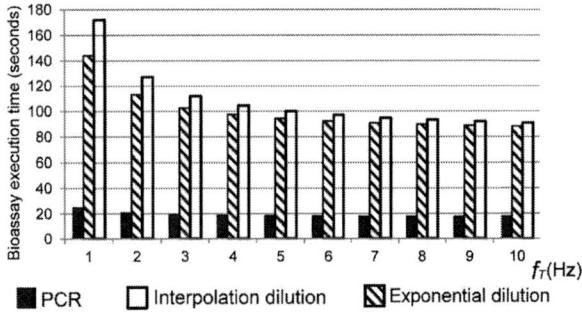

Figure 7: The relationship between execution time of bioassays and f_T.

in the proposed algorithm, no operations are interrupted and the assay proceeds in an unimpeded manner.

5.4 Completion time with multiple clock frequencies

For the biochip with multiple clock frequencies, if we increase the clock frequency for the T phase, then the execution time for the bioassay can be reduced. The relationship between execution time of bioassays and the clock frequency of the T phase is shown in Figure 7.

By increasing f_T from 1 Hz to 10 Hz, the execution time for exponential dilution can be reduced from 172 seconds to 91 seconds. The completion time for interpolation dilution and PCR are also listed in Figure 7. When we increase the clock frequency of the T phase, the execution time of PCR does not decrease significantly because the time consumed by droplet transportation is relatively less.

6. CONCLUSION

We have shown how cyberphysical integration in digital microfluidics can be used to carry out on-chip bioassays despite the timing uncertainties inherent in fluidic operations such as mixing, dilution, and thermal cycling. We have presented an operation-interdependency-aware synthesis method that is responsive to such uncertainties. The proposed design approach facilitates dynamic on-line decision making for the execution of fluidic operations in response to detector feedback. We have also incorporated the use of multiple clock frequencies to accelerate the time-to-response without any adverse impact on electrode reliability. We have used three common laboratorial protocols to demonstrate that, compared to uncertainty-oblivious biochip synthesis, the proposed dynamic synthesis approach decreases the likelihood of erroneous reaction outcomes, and it leads to reduced time-to-result, less repetition of reaction steps, and less wastage of precious samples and reagents. This work is therefore a significant step forward towards fully automated on-chip biochemistry with reliable assay outcomes and low cost.

7. REFERENCES

[1] R. Fair, "Digital microfluidics: Is a true lab-on-a-chip possible?", *Microfluidics and Nanofluidics*, vol. 3, pp. 245-281, 2007.

[2] Z. Hua et al., "Mutiplexed real-time polymerase chain reaction on a digital microfluidic platform", *Anal. Chem.*, vol. 82, pp. 2310-2316, 2010.

[3] B. Hadwen et. al, "Programmable large area digital microfluidic array with integrated droplet sensing for bioassays", *Lab on a Chip*, pp. 3305-3313, 2012.

[4] T. Xu, K. Chakrabarty, and V. Pamula, "Defect-tolerant design and optimization of a digital microfluidic biochip for protein crystallization", *IEEE Transactions on Computer-Aided Design of Integrated Circuits and Systems*, vol. 29, Issue 4, pp. 552-565, 2010.

[5] H. Ren, V. Srinivasan, and R. Fair, "Design and testing of an interpolating mixing architecture for electrowetting-based droplet-on-chip chemical dilution", *Int. Conf. on Solid-State Sensors, Actuators and Microsystems*, pp. 619-622, 2003.

[6] S. Park, P. Wijethunga, H. Moon H, and B. Han, "On-chip characterization of cryoprotective agent mixtures using an EWOD-based digital microfluidic device", *Lab on a Chip*, pp. 2212-2221, 2011.

[7] M. Schertzer, *Characterization of the motion and mixing of droplets in electrowetting on dielectric devices*, PhD thesis, University of Toronto, Toronto, CA, 2010.

[8] P. Paik, V. Pamula, and R. Fair, "Rapid droplet mixers for digital microfluidic systems", *Lab on a Chip*, vol. 3, pp. 253-259, 2003.

[9] M. Iyengar and M. McGuire, "Imprecise and qualitative probability in systems biology", *International Conference on Systems Biology*, 2007.

[10] O. Levenspiel, *Chemical Reaction Engineering*, New York: Wiley, 1999.

[11] K. Chakrabarty and F. Su, *Digital Microfluidic Biochips: Synthesis, Testing, and Reconfiguration Techniques*, Boca Raton, FL: CRC Press, 2006.

[12] J. Gong, *Portable Digital Microfluidic System: Direct Referencing EWOD Devices and Operating Control Board*, PhD thesis, UCLA, 2007.

[13] M. Schertzer et al., "Using capacitance measurements in EWOD devices to identify fluid composition and control droplet mixing", *Sensors and Actuators B: Chemical*, volume 145, pp. 340-347, 2010.

[14] Y. Luo, K. Chakrabarty, and T.-Y. Ho, "Error recovery in cyberphysical digital-microfluidic biochips", *IEEE Transactions on Computer-Aided Design of Integrated Circuits and Systems*, vol. 32, pp. 59-72, Issue 1, 2013.

[15] Y. Luo, K. Chakrabarty, and T.-Y. Ho, "Dictionary-based error recovery in cyberphysical digital-microfluidic biochips", *Proc. IEEE/ACM International Conference on Computer-Aided Design*, pp. 369-376, 2012.

[16] M. Alistar, P. Pop, and J. Madsen, "Online synthesis for error recovery in digital microfluidic biochips with operation variability", *Symposium on Design, Test, Integration and Packaging of MEMS/MOEMS*, pp. 53-58, 2012.

[17] D. Grissom and P. Brisk, "Fast online synthesis of generally programmable digital microfluidic biochips", *Proc. CODES+ISSS*, pp. 413-422, 2012.

[18] L. Huang, B. Koo, and C.-J. Kim, "Evaluation of anodic Ta2O5 as the dielectric layer for EWOD devices", *IEEE Micro Electro Mechanical Systems*, pp. 428-431, 2012.

[19] V. Kumar and N. Sharma, "SU-8 as hydrophobic and dielectric thin film in electrowetting-on-dielectric based microfluidics device", *Journal of Nanotechnology*, pp. 1-6, 2012.

[20] F. Su and K. Chakrabarty, "Unified high-level synthesis and module placement for defect-tolerant microfluidic biochips", *Proc. IEEE/ACM Design Automation Conference*, pp. 825-830, 2005.

APPENDICES

A. YIELD ESTIMATION FOR BIOCHIPS WITH NO FEEDBACK-BASED ADAPTATION

As discussed in Section 1, the execution time of a fluidic operations needs to be considered as a random variable rather than a known constant. If no feedback-based control is used and the synthesis of the biochip is based on a module library with constant operation times, there is a high likelihood that many operations will not be completed within the allocated time. The outputs of these "unfinished operations" will be unqualified droplets, and the outcome of the bioassay will be unacceptable.

Next we compute the probability that a bioassay will fail due to the occurrence of unfinished operations. Assume that the execution time for each operation is Gaussian random variable, and for any two operations in a bioassay, their execution times are independent of each other. Suppose that for any operation O_i, its completion time T_i has mean μ_i and variance σ_i^2.

If we define the time spent on operation O_i as $T_i = \mu_i + \sigma_i$, and the real execution time T_i^* does not exceed T_i, then O_i can be executed correctly; otherwise O_i will be deemed to have failed. A bioassay terminates with an acceptable outcome only if all the operations are executed correctly. Let P_i be the probability that a dilution/mixing operation O_i is executed correctly, and let the number of dilution/mixing operation in a bioassay be N_b. Then the probability P_{success} that the bioassay is implemented successfully is given by: $P_{\text{success}} = \prod\limits_{i=1}^{N_b} P_i = P^{N_b}$ (if $P_i = P$ for all i).

Therefore, the probability of successful implementation P_{success} decreases exponentially with the number of dilution/mixing operation. Hence for a realistic bioassay with a large number of fluidic operations, the compound yield will be unacceptably low. A quantitative analysis of bioassay yield was presented in Section 5.1.

B. INSTABILITY OF PRIOR SYNTHESIS METHODS

The synthesis results derived in [11] [14] [16] [17] do not provided an appropriate baseline for comparison with this paper. Dynamic adaption based on these synthesis results and sensor feedback suffer from high computational complexity. To search for a feasible solution during resynthesis, the control software not only needs to consider the timing constraints of operations, but it also needs to ensure that there is no conflict in resource assignment.

None of the synthesis procedures of [11] [14] [16] [17] consider the timing overshoot that can arise in fluid-handling operations. Hence in their synthesis results, two independent operations may be assigned to use the same resource at different time segments. An example is shown in Figure 8. Consider two operations O_A and O_B that are independent from each other; their module placements are shown in Figure 8(a). As shown in Figure 8(b), these two operations are originally scheduled to be executed during different time segments, hence the overlap of their corresponding modules will not lead to resource conflicts. However, if the start and stop time of O_A are delayed due to a previous operation, and O_A cannot be completed before the start of O_B, then the overlap of their corresponding modules will lead to resource conflicts. Hence the synthesis result for O_B also needs to be changed.

From the above discussion, we find the synthesis results in [11] [14] [16] [17] are not robust when timing uncertainties of operations exist. In each resynthesis procedure, the control software must adjust the schedule and module placement for a series of operations. These cumbersome on-line computation steps require high CPU time. The resulting response time will interrupt the execution of the bioassay. Hence in order to run bioassays on cyberphysical biochips with reduced computational complexity and short response times, a new synthesis algorithm needs to be developed to derive robust results in the presence of timing variabilities for fluidic operations.

In this paper, the baseline methods were designed by adding ad-

ditional time to module library specifications, adding interruption whenever there is a timing overshoot, and dynamic resynthesis.

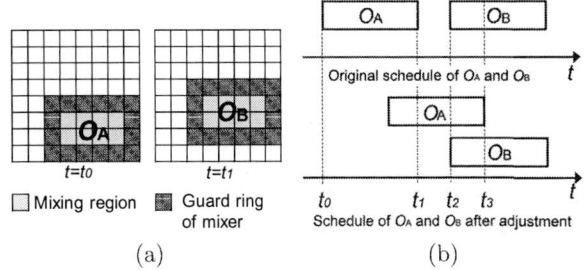

(a)   (b)

Figure 8: (a) Module placement for operations O_A and O_B; (b) scheduling result for operations O_A and O_B.

C. PROOF FOR THE NONEXISTENCE OF CONFLICT RELATIONSHIPS BETWEEN TWO TREES

In this part, we present the proof for the following conclusion (used in Section 4):

LEMMA 1. *For any two elements T_x and T_y in the set of directed tree $\{T_1, T_2, ...T_n\}$ partitioned from a sequencing graph, the conflict relationship $T_x < T_y \cap T_x > T_y$ never occurs.*

Proof: This lemma can be proved by *reductio ad absurdum*. First we assume that these two directed tree have the relationships $T_x < T_y$ and $T_x > T_y$. Then as shown in Figure 9, we can find two operation $O_{x_1} \in T_x$ and $O_{y_1} \in T_y$, such as the output of O_{x_1} is the input of O_{y_1}; we can also find two operation $O_{x_2} \in T_x$ and $O_{y_2} \in T_y$, such as the output of O_{y_2} is the input of O_{x_2}.

Next we can prove that O_{x_1} is the root node for T_x. As introduced in Section 4, we first find out all the nodes $\{N_{n_1}, N_{n_2}, ..., N_{n_k}\}$ whose out-degrees are more than 1. Then remove all the edges directed away from these nodes. Hence in the trees we derived from the partitioning procedure, the out-degrees of all the nodes except but root node is equal to 1. The out-degree of the root node is equal to 0. Assume O_{x_1} is not the root of T_x. Then in T_x, there must be an edge from O_{x_1} to another node in T_x. As shown in Figure 9, there is another edge from node O_{x_1} to O_{y_1}. Hence the out-degree of O_{x_1} in the sequence graph is equal to 2.

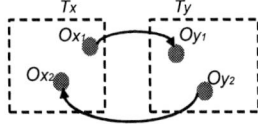

Figure 9: Assume two directed trees T_x and T_y have the relationships that $T_x < T_y$ and $T_x > T_y$.

Based on the step of graph partitioning, O_{x_1} whose out-degree is equal to 2, must be a root node of a directed tree derived. Hence we have reached the conclusion that O_{x_1} is the root node for T_x. Similarly, we can reach the conclusion that O_{y_2} is the root node for T_y. Then based on the characteristic of directed tree, there must exist a directed path from the node O_{x_2} toward the root node O_{x_1} in T_x, and there must exist a directed path from the node O_{y_1} towards the root node O_{y_2} in T_y. Since there are the edges from O_{x_1} to O_{y_1}, and the edge from O_{y_2} to O_{x_2}, we get a directed circle that connects the nodes O_{x_1}, O_{x_2}, O_{y_1} and O_{y_2} in the sequencing graph. However, as introduced in Section 4, a sequencing graph for bioassay will not have a directed cycle.

Hence we have proved that, the conflict relationships between T_x and T_y will not exist. This completes the proof of the lemma. □

Gene Modification Identification Under Flux Capacity Uncertainty

Mona Yousofshahi[1], Michael Orshansky[2], Kyongbum Lee[3], and Soha Hassoun[1]

[1] Department of Computer Science, Tufts University

[2] Department of Electrical and Computer Engineering, University of Texas at Austin

[3] Department of Chemical and Biological Engineering, Tufts University

Abstract

Re-engineering cellular behavior promises to advance the production of commercially significant biomolecules and to enhance cellular function for many applications. To achieve a desired cellular objective, it is necessary to identify within a metabolic network a set of reactions whose fluxes should be changed using gene modifications. We develop a computational method, CCOpt, to optimize the selection of an intervention set that consists of gene up/down-regulation using uncertainty-aware chance-constrained optimization. In contrast to deterministic approaches where constraints are met with 100% certainty, constraints in CCOpt are probabilistically met at a user-specified confidence level. We investigate the application of CCOpt to two case studies that utilize the Chinese Hamster Ovary (CHO) cell metabolism. Our results demonstrate that CCOpt is capable of identifying optimal intervention sets without the run-time cost of a sampling based (Monte Carlo) approach.

1. Introduction

Biological cells have been engineered to produce various commercially significant biomolecules ranging from therapeutic antibodies to biofuels. Engineered cells have been used as therapeutic agents, e.g. to accelerate wound healing or regenerate damaged or lost tissue function, and play an implicit role in the efficacy of drugs and other molecular therapeutics designed to mitigate a harmful (e.g. uncontrolled growth) or beneficial (e.g. detoxification) process in the cell. Systematic engineering of cell behavior is just beginning to benefit from computational approaches that provide analysis and optimization capabilities.

To enable cellular engineering to fully exploit computational advances, computational models and methods must account for uncertainties inherent to biological systems: stochastic fluctuations of molecular processes, incomplete knowledge, and the imprecision of engineering interventions. Addressing uncertainty is a challenging issue that has become increasingly important not only for engineering biological systems, but also human-made systems such as electronic systems. Indeed, the past decade has witnessed a paradigm shift in the design of electronics, where circuits are now designed to maximize tolerance to manufacturing and operational variations or to include tuning circuitry for post-manufacturing re-calibration.

As biological engineering efforts progress from proof-of-principle to scaled-up manufacturing, computational methods to effectively address biological and engineering uncertainties at the design stage will become increasingly important in ensuring the identification of the most robustly optimal gene modifications.

Several computational methods have been recently developed to identify optimal gene modifications to achieve a desired metabolic engineering objective. Here, a gene modification refers to changing the enzyme expression level, and thus the activity, for a particular network reaction. The problem of identifying optimal genetic modifications translates to identifying reactions whose fluxes should be modified to achieve a metabolic objective. The problem can be expressed in terms of operating state variables (flux associated with each reaction), and control (decision) variables such as the presence or absence of gene modifications. The optimal design "tunes" the control variables such that the solution maximizes a desired engineering objective while satisfying several constraints reflecting physicochemical considerations and experimental observations. Notable computational methods include OptReg [1], OptKnock [2], and GDLS [3]. None of the techniques however address uncertainty. Due to biological variability, stochastic effects associated with gene expression, and imprecision in genetic engineering implementation, it is questionable that enzyme levels can be precisely tuned to exactly match the target values calculated using computational design tools. More likely, the target enzyme levels, and thus the corresponding reaction flux capacities, can only be achieved with some uncertainty.

The uncertainty in achieving the targeted enzyme activities suggests that the enzyme levels, and hence the corresponding flux carrying capacities, could be considered statistical distributions rather than fixed value parameters. In this statistical interpretation, a flux constraint in a conventional deterministic optimization problem represents the most conservative point in the flux capacity distribution, since a deterministic problem enforces all constraints with zero uncertainty. Such deterministic approaches [1-3], might lead to choosing an intervention set that could fail to achieve in practice the best possible yield. Repeatedly applying deterministic approaches to problem instances obtained using Monte Carlo sampling of flux capacities could identify some intervention sets. However, such an approach is computationally expensive. An alternate approach is to use chance-constrained programming (CCP). CCP selects an optimal solution with a user-defined degree of probabilistic confidence in meeting constraints. Chance-constrained programming was first introduced to solve the problem of temporal planning when uncertainty is present [4]. Since then, CCP has been utilized in numerous applications such as circuit sizing [5], soil conservation [6], ground water management [7], and energy management [8].

Permission to make digital or hard copies of all or part of this work for personal or classroom use is granted without fee provided that copies are not made or distributed for profit or commercial advantage and that copies bear this notice and the full citation on the first page. To copy otherwise, to republish, to post on servers or to redistribute to lists, requires prior specific permission and/or a fee.

DAC '13, May 29 - June 07 2013, Austin, TX, USA.

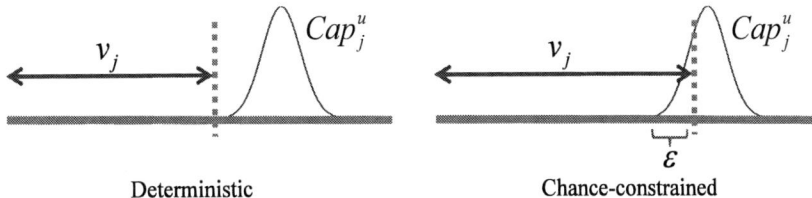

Figure 1: Deterministic vs. chance-constrained upper-bound constraint interpretation. The red dotted lines represent the upper-bounds for reaction j. The arrows show the flux ranges. In the deterministic approach, v_j is strictly less than any value in the upper-bound distribution. The chance-constrained approach allows a violation of the upper-bound distribution by ε.

We present in this paper a mathematical formulation to identify gene modification while addressing uncertainty due to engineering interventions. We investigate how the results of our CCP optimization (CCOpt) method compare with those obtained using Monte Carlo optimization. We apply the optimization methods to two test cases that describe the Chinese Hamster Ovary (CHO) metabolism.

1. Background

A cellular process is represented using a biochemical network consisting of a set of reactions and compounds. A network is at steady state if the total rate of producing each of its internal compounds is equal to the total rate of consuming the compound. A network may be able to operate at more than one such steady state. The steady state flux is a quantitative measure that characterizes the degrees of engagements of the reactions in a network. One of the most widely used techniques for flux calculation is Flux Balance Analysis (FBA) [9]. FBA assumes that the cell is at a metabolic steady state, which dictates that fluxes leading to the formation of a metabolite are balanced by the fluxes leading to the degradation of the metabolite. There are typically more reaction fluxes in the system than balance equations that constrain the magnitude and direction of these fluxes. Consequently, the system of equations is typically underdetermined. Using linear programming, FBA finds flux values for a particular cellular objective, such as maximizing the production of a metabolite, minimizing nutrient uptake, or maximizing biomass production (cellular growth). Flux Variability Analysis (FVA) is a related technique useful for computing steady-state flux ranges [10]. Each reaction flux is separately minimized and maximized, subject to similar balance constraints as FBA, to calculate the maximal allowable range supported by the cells operating at steady state. Measurements on metabolite uptake or release, when available, can further restrict such ranges. In the present study, we use the steady state ranges calculated using FVA as upper and lower bounds on the reaction fluxes, which are the operating state variables for optimization.

2. Formulations
2.1 Chance-Constrained Optimization (CCOpt)
In this section, we investigate a way of capturing uncertainty in parameter values when formulating an optimization problem with the objective of maximizing the production of a desired metabolite through gene up/down-regulating operations. The objective function can be formulated as:

$$maximize \ v_{target} \qquad (1)$$

where v_{target} is the production rate of a desired metabolite. The objective function in equation (1) is subject to certain restrictions on reaction fluxes and decision variables. These constraints include:

- The flux for each reaction is limited within a possible range of zero and an upper bound (inequality (4)). The upper bound is an uncertain parameter and will be discussed in detail.
- The modified cell needs to remain viable which can be guaranteed with a minimal growth rate constraint.
- The rate of production of each intracellular metabolite is equal to its rate of consumption.
- The number of allowed interventions is limited by an upper bound.
- The gene manipulations are exclusive, i.e. a reaction can be either up- or down-regulated in a solution, but not both.
- The uncertain parameters in this formulation are the regulated flux capacities which are defined as the flux upper bounds and are set by the expression levels of the corresponding genes.

Figure 1 illustrates the deterministic and chance-constrained interpretation of an uncertain upper-bound constraint on the flux of reaction j. In a deterministic interpretation, the value of flux v_j is enforced to be strictly less than all the values in the upper-bound distribution Cap_j^u. In the probabilistic interpretation, there is a nonzero probability that flux v_j will be equal to or larger than some of the values in the distribution Cap_j^u. In the case of CCP, the constraint is relaxed by introducing a parameter ε which reflects the confidence level for the probability that the solution satisfies the constraint:

$$\text{Prob}\{v_j < Cap_j^u\} \ge 1 - \varepsilon \qquad (2)$$

To generalize the previous inequality to also consider the effects of up/down-regulating an enzyme (or the expression of the gene that encodes the enzyme), we introduce two sets of binary decision variables y_j^u and y_j^d. A value of 1 indicates that the corresponding enzyme is up/down-regulated, whereas a value of 0 indicates the corresponding enzyme expression is unchanged.

$$\text{Prob}\{v_j \le SSU_j + y_j^u(Cap_j^u - SSU_j) + y_j^d(Cap_j^d - SSU_j)\} \ge 1 - \varepsilon \qquad (3)$$

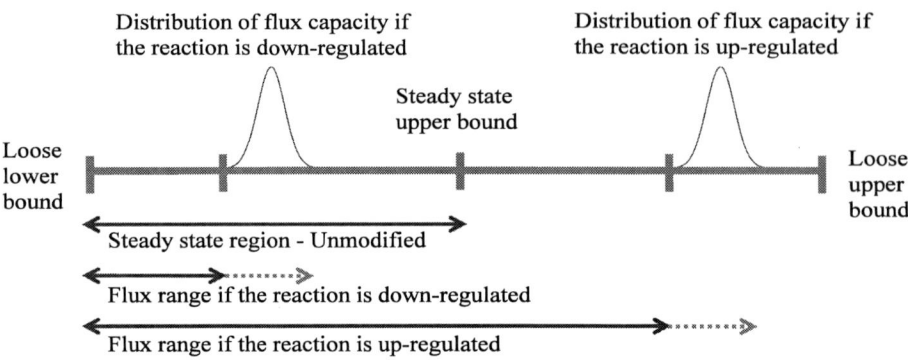

Figure 2. Maximum production rates and intervention sets obtained by CCOpt for the small CHO cell model (a), and for the large CHO model (b). The blue circles represent the maximum production rates obtained by CCOpt with $\varepsilon = 0.25$. The reactions involved in each intervention set are denoted above each data point.

When there is no modification in the enzyme level, the flux for is reaction j is less than the upper reference level (SSU_j). When the reaction is up-regulated ($y_j^u = 1$), v_j can go above the reference state upper bound up to Cap_j^u. In the case of down-regulation, the upper bound for v_j, decreases to Cap_j^d. This probabilistic constraint is illustrated in Figure 2.

Various approaches have been developed to solve CCP problems. One method to solve CCP is to convert the probabilistic constraints into their deterministic equivalents at their specified confidence level ε. This approach works if the chance constraints are linear, independent and assume the form of random right-hand side as described in [8]. Our formulation meets all of these conditions; therefore, the chance constraints can be converted into their deterministic equivalents. Using the inverse of the cumulative distribution functions (CDF) for Cap_j^u and Cap_j^d, inequality (3) can be reformulated as:

$$v_j < SSU_j + y_j^u\left(F_{j,u}^{-1}(\varepsilon) - SSU_j\right) \\ + y_j^d\left(F_{j,d}^{-1}(\varepsilon) - SSU_j\right) \quad (4)$$

where $F_{j,u}^{-1}$ and $F_{j,d}^{-1}$ denote the inverse CDFs of Cap_j^u and Cap_j^d respectively, which can be numerically calculated if needed.

3. Results

3.1 Test Cases

We use two test cases based on models for the metabolism of Chinese Hamster Ovary (CHO) cells to assess the benefits of our chance-constrained optimization technique. The first test case, which is a small-scale model describing CHO cell metabolism in fed-batch culture [11], comprises 24 metabolites and 46 reactions. The optimization objective is to maximize the production of a therapeutic antibody. The second test case, which is a significantly larger model, represents a reduced version [12] of a genome scale model [13]. We augmented the model from [12] by adding a biomass reaction, as the test case objective was to maximize cell growth. The modified model comprises 270 metabolites and 423 reactions.

Traditionally, gene modification has been modeled as a deterministic event leading to a fold-change in the level of the corresponding enzyme, and hence a fold-change in the flux

capacity of the reaction catalyzed by the enzyme. Here, we model enzyme level modification as an uncertain event using a probability distribution. We assume a normal distribution $N_u(\mu, \sigma^2)$ [14] with an average fold-change of $\mu = 6$ following gene up-regulation and a standard deviation of $\sigma = 1.3$. The average fold-change value reflects experimental data reported in [15]. The standard deviation is chosen such that $\mu - 3\sigma > 1$, which ensures that the flux capacity after up-regulating the enzyme level is higher than the unmodified state. A decrease in enzyme level, and hence reaction flux capacity, is modeled by a normal distribution $N_d(\mu, \sigma^2)$ with an average fold-change of $\mu = 0.5$ and a spread of $\sigma = 0.17$.

Based on the probabilistic interpretation of fold-changes in enzyme levels, we also estimate the resulting reaction flux capacities as probability distributions. A fold-change in enzyme level is assumed to directly correlate with a fold-change in the maximal reaction velocity ($v_{j,max}$). Therefore, flux capacity distributions were calculated by multiplying the enzyme fold-change distributions with $v_{j,max}$.

3.2 Identifying Intervention Sets

Figure 3(a) shows the maximum antibody production rate and the intervention sets obtained by CCOpt using the small CHO cell. The intervention sets are shown above their corresponding flux values. An empty set indicates that no interventions were identified. Reaction 17 represents the antibody synthesis reaction, which upon up-regulation, allows more antibody production. Reactions 13 and 25 are in series with each other with a common goal of synthesizing cysteine. As reported in [16], one of the rate-limiting steps along the antibody production process in CHO cells is the folding and assembly of polypeptides in endoplasmic reticulum where cysteine residues enter the assembly process. Up-regulating reaction 4, increases the production of α-ketoglutarate which is an intermediate of the tricarboxylic acid cycle. A recent study showed that supplementing the culture medium with α-ketoglutarate has a positive effect on growth and antibody production of mouse hybridoma cell lines [17].

The intervention sets obtained by applying CCOpt to the large CHO cell models are shown in Figure 3(b). Sets A, B, C and D represent reaction sets {89, 256, 276}, {89, 90, 109, 110, 122, 209, 256, 276, 359, 363}, {86, 89, 90, 109, 110, 122, 167, 168, 209, 256, 276, 359, 363}, and {89, 109, 110, 256, 276, 363}, respectively. As for the small CHO cell, the maximal predicted

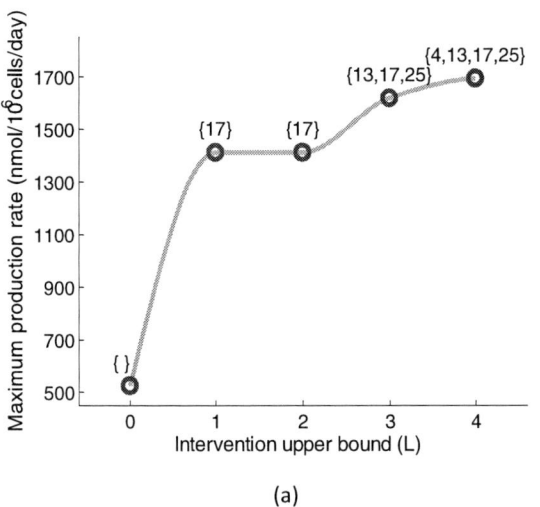
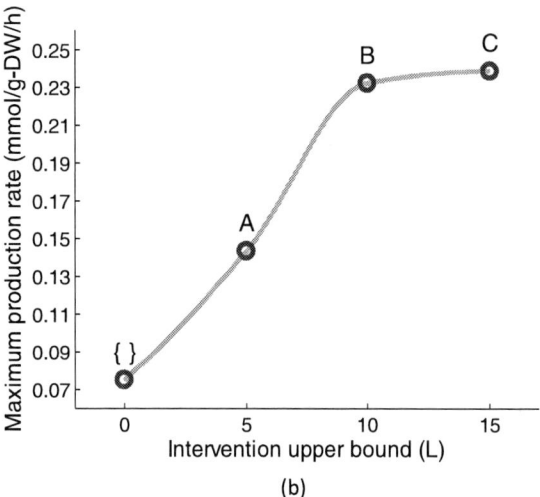

(a)   (b)

Figure 3. Maximum production rates and intervention sets obtained by CCOpt for the small CHO cell model (a), and for the large CHO model (b). The blue circles represent the maximum production rates obtained by CCOpt with $\varepsilon = 0.25$. The reactions involved in each intervention set are denoted above each data point.

target flux increases with the number of allowed interventions. A general trend is that the smaller sets of interventions are largely subsets of the larger sets of interventions. The intervention set represented by point A ($L = 5$) in Figure 3(b) includes hexokinase, nucleoside-diphosphate kinase and the biomass reaction. Set B ($L = 10$) adds glucose phosphate isomerase, UDP-glucose pyrophosphorylase, glycogen synthase, malic enzyme, phosphatidylinositol synthase, ribulose-5-phosphate-3-epimerase and ribose 5-phosphate isomerase. The selection of these enzymes in set B for up-regulation is consistent with experimental observations that CHO cell cultures in log-phase growth exhibit large fluxes through glycolysis (from glucose to lactate), anaplerosis (from pyruvate to oxaloacetate and from glutamate to a-ketoglutarate), and the malate cycle [18].

3.3 Comparison against Monte Carlo Sampling

To further understand the quality of solutions obtained using CCOpt, we compare the maximum production rate with those obtained using Monte Carlo sampling. We use Monte Carlo methods to randomly select a flux capacity bound from the flux capacity distribution and then determine an intervention set. Using this approach, the repeated solutions eventually identify multiple intervention sets with some frequencies. We iterated the Monte Carlo simulation 1,000,000 times.

The results for the small CHO cell are shown in Figure 4(a). When the number of allowed interventions is 1 or 2 ($L = 1, 2$), Monte-Carlo based optimization generates only a single intervention set that is identical to the result of CCOpt. For $L = 3$, Monte Carlo identifies two sets of interventions. The intervention set identified with the greater frequency exactly matches the set identified by CCOpt. Interestingly, this intervention set also corresponds to the highest target production rate among all intervention sets comprising three reactions

($L = 3$). For $L = 4$, the trend is the same as for $L = 3$; that is, among the different solutions generated by Monte Carlo sampling, the intervention set with the highest frequency is the same as the set identified by CCOpt.

The CCOpt results for the large CHO cell are also compared to solutions generated using the Monte Carlo approach. As was the case for the smaller CHO cell model, the intervention sets identified by CCOpt (shown in red) represent the most frequently generated Monte Carlo solutions (Figure 4(b)). CCOpt solutions A, B and C occur with 100%, 98% and 97% frequency, respectively. These solutions also have the highest maximum fluxes compared to other possible intervention sets.

Together, these results indicate that the chance-constrained optimization approach, CCOpt, is as capable of identifying the best intervention sets *without* the run-time cost of multiple optimizations used in the Monte Carlo sampling approach.

4. Conclusion

The presented work addresses an efficient computational framework to identify an optimal set of gene modifications that can be used to maximize cellular production of a particular metabolite. The novelty is in explicitly accounting for likely variations in flux capacities due to engineering modifications. Our algorithm, CCOpt, is based on chance-constrained programming where constraints are probabilistically met at a user-specified confidence level. Evaluation of the approach for two test cases demonstrates that CCOpt consistently finds the solution most frequently found when using Monte Carlo sampling, but at a fraction of the computational cost. The CCOpt formulation is the first work to incorporate uncertainty when computing gene modifications. It can be extended to capture other types of uncertainties, such as biological variability in measured data and cell transfection efficiency, making CCOpt an effective technique for probabilistic strain optimization.

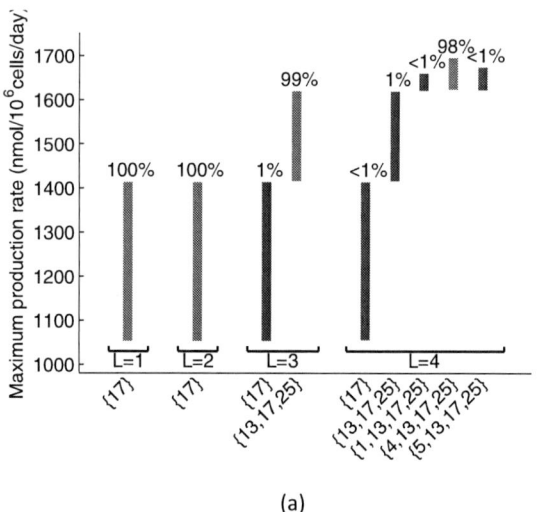 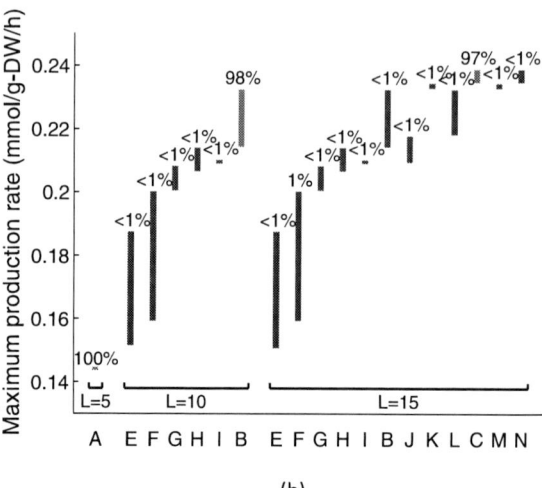

(a)   (b)

Figure 4: (a) The results of the Monte Carlo Sampling approach with the objective of maximizing antibody in the small CHO cell. The x-axis represents the obtained intervention sets for each *L*. The percentage values at the top of each bar represent the frequency at which the corresponding intervention set was selected among the pool of generated results with the same upper bound *L*. The intervention sets matching those of CCOpt are colored in red. (b)The results of Monte Carlo Sampling approach for the large CHO cell.

5. Acknowledgement

This work was supported by the National Science Foundation under Grant no. 0829899 and the Wittich Family Foundation.

6. References

[1] P. Pharkya and C. D. Maranas, "An optimization framework for identifying reaction activation/inhibition or elimination candidates for overproduction in microbial systems," *Metab. Eng.,* vol. 8, pp. 1-13, 1, 2006.

[2] A. P. Burgard, P. Pharkya and C. D. Maranas, "Optknock: A bilevel programming framework for identifying gene knockout strategies for microbial strain optimization," *Biotechnol. Bioeng.,* vol. 84, pp. 647-657, 2003.

[3] D. S. Lun, G. Rockwell, N. J. Guido, M. Baym, J. A. Kelner, B. Berger, J. E. Galagan and G. M. Church, "Large-scale identification of genetic design strategies using local search," *Mol Syst Biol,* vol. 5, 08/18, 2009.

[4] A. Charnes and W. W. Cooper, "Chance-Constrained Programming," *Management Science,* vol. 6, pp. pp. 73-79, Oct., 1959.

[5] M. Mani and M. Orshansky, "A new statistical optimization algorithm for gate sizing," *Computer Design: VLSI in Computers and Processors, 2004. ICCD 2004. Proceedings. IEEE International Conference On,* pp. 272-277, 2004.

[6] M. Zhu, D. B. Taylor, S. C. Sarin and R. Kramer, "Chance Constrained Programming Models for Risk-Based Economic and Policy Analysis of Soil Conservation," *Agric. Resour. Econ. Rev.,* vol. 23, 1994.

[7] A. M. Yeou-Koung Tung, "Groundwater Management by Chance-Constrained Model," *J. Water Resource Planning Manage.,* vol. 112, pp. 1, 1986.

[8] W. v. Ackooij, R. Zorgati, R. Henrion and A. Möller, "Chance constrained programming and its applications to energy management, stochastic optimization," in *Stochastic Optimization—Seeing the Optimal for the Uncertain,* pp. 291-320, 2011.

[9] A. Varma and B. O. Palsson, "Metabolic Flux Balancing: Basic Concepts, Scientific and Practical Use," *Nat Biotech,* vol. 12, pp. 994-998, 1994.

[10] R. Mahadevan and C. H. Schilling, "The effects of alternate optimal solutions in constraint-based genome-scale metabolic models," *Metab. Eng.,* vol. 5, pp. 264-276, 10, 2003.

[11] R. P. Nolan and K. Lee, "Dynamic model of CHO cell metabolism," *Metab. Eng.,* vol. 13, pp. 108-124, 1, 2011.

[12] L. Quek, S. Dietmair, J. O. Krömer and L. K. Nielsen, "Metabolic flux analysis in mammalian cell culture," *Metab. Eng.,* vol. 12, pp. 161-171, 3, 2010.

[13] L. QUEK and L. K. NIELSEN, "On the reconstruction of the Mus musculus genome-scale metabolic network model," *Genome Informatics,* vol. 21, pp. 253, 2008.

[14] X. Deng, J. Xu, J. Hui and C. Wang, "Probability fold change: A robust computational approach for identifying differentially expressed gene lists," *Comput. Methods Programs Biomed.,* vol. 93, pp. 124-139, 2, 2009.

[15] H. C. Wang, Y. H. Ko, H. J. Mersmann, C. L. Chen and S. T. Ding, "The expression of genes related to adipocyte differentiation in pigs," *Journal of Animal Science,* vol. 84, pp. 1059-1066, May 01, 2006.

[16] S. L. Davies and D. C. James, "Engineering Mammalian Cells for Recombinant Monoclonal Antibody Production," vol. 6, pp. 153-173, 2009.

[17] S. Nilsang, A. Kumar and S. Rakshit, "Effect of α-Ketoglutarate on Monoclonal Antibody Production of Hybridoma Cell Lines in Serum-Free and Serum-Containing Medium," *Appl. Biochem. Biotechnol.,* vol. 151, pp. 489-501, 12/01, 2008.

[18] W. S. Ahn and M. R. Antoniewicz, "Metabolic flux analysis of CHO cells at growth and non-growth phases using isotopic tracers and mass spectrometry," *Metab. Eng.,* vol. 13, pp. 598-609, 9, 2011.

A Field-Programmable Pin-Constrained Digital Microfluidic Biochip

Daniel Grissom, Philip Brisk
Department of Computer Science and Engineering
University of California, Riverside
{grissomd, philip}@cs.ucr.edu

ABSTRACT

As digital microfluidic biochips (DMFBs) have matured over the last decade, efforts have been made to 1.) reduce the cost, and 2.) produce general-purpose chips. While work done to generalize DMFBs typically depends on the flexibility of individually controlled electrodes, such devices have high wiring complexity, which requires costly multi-layer printed circuit boards (PCBs). In contrast, pin-constrained DMFBs reduce the wiring complexity, but reduce the flexibility of droplet coordination. We present a field-programmable pin-constrained DMFB that leverages the cost-savings of pin-constrained designs, but is general-purpose, rather than assay-specific. We show that with just a few more pins than the state-of-the-art pin-constrained designs, we can execute arbitrary assays almost as fast as the most recent general-purpose DMFB designs.

Categories and Subject Descriptors

B.7.2 [**Integrated Circuits**]: Design Aids; J.3 [**Life and Medical Sciences**]: Biology and Genetics, Health

General Terms

Algorithms, Design, Performance.

Keywords

Digital Microfluidic Biochip (DMFB), Laboratory-on-Chip (LoC), Pin-Constrained, Field-Programmable.

1. INTRODUCTION

This paper presents a design for a field-programmable, pin-constrained digital microfluidic biochip (DMFB). Just as a field-programmable gate array (FPGA) can be programmed by an end-user in the "field," a field programmable pin-constrained DMFB can be programmed to execute any *assay* (biochemical protocol) after it has been designed and manufactured. In contrast, prior pin-constrained DMFBs have been assay-specific [9][17].

Direct-addressing DMFBs provide independent control over each electrode; these devices are costly because the large number of control inputs and high wiring complexity increases the number of printed circuit board (PCB) layers. Pin-constrained DMFBs, in contrast, have fewer control inputs and low wiring complexity, but lack flexibility. This paper introduces *the first* pin-constrained DMFB with sufficient flexibility to enable field-programmability.

Permission to make digital or hard copies of all or part of this work for personal or classroom use is granted without fee provided that copies are not made or distributed for profit or commercial advantage and that copies bear this notice and the full citation on the first page. To copy otherwise, to republish, to post on servers or to redistribute to lists, requires prior specific permission and/or a fee.
DAC '13, May 29 – June 07 2013, Austin, TX, USA.

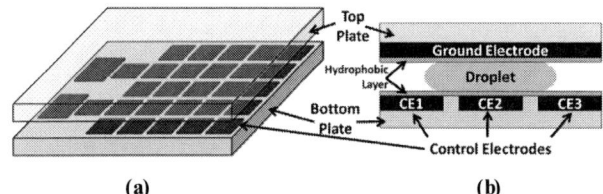

(a)   **(b)**

Figure 1. (a) Planar array of electrodes; (b) Cross-sectional view of electrode array.

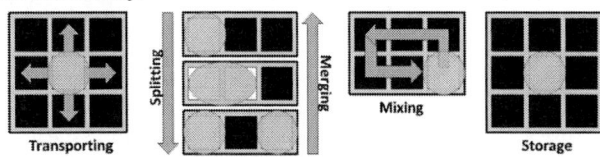

Figure 2. Basic microfluidic operations form the building blocks for assays to be executed on an array of electrodes.

1.1 DMFB Technology Overview

1.1.1 Background: Physical Droplet Manipulation

DMFBs execute assays by manipulating nanoliter-sized droplets of fluid. DMFBs are typically based on a phenomenon known as electrowetting [11]. An electrowetting-based DMFB, as seen in **Figure 1**, consists of a top and bottom plate coated with a hydrophobic layer. The bottom plate has an array of droplet-sized control electrodes, while the top plate has a single conducting electrode that spans the entire array of control electrodes. Each droplet is sandwiched between the bottom and top plates and will hold its place if its underlying electrode remains activated.

In **Figure 1(b)**, a droplet overlaps neighboring electrodes; if a neighboring electrode is activated, the droplet will begin to flow toward the newly activated electrodes. Thus, if CE3 is activated and CE2 is simultaneously deactivated, the entire droplet will move to cover CE3. As seen in **Figure 2**, with the proper sequence of electrode activations, several basic microfluidic operations can be performed. Sensor-based detection operations execute by moving a droplet to a detector (placed above an electrode) and storing the droplet there. Dispense and output operations are performed by I/O reservoirs on the perimeter.

If a droplet is not centered over or adjacent to any activated electrodes, it will *drift* across the DMFB in an undetermined and unpredictable manner.

1.1.2 Background: High-level Assay Synthesis

Figure 3 illustrates the process of synthesizing an assay onto a DMFB. A directed acyclic graph (DAG) represents the assay; each node represents a microfluidic operation (i.e., dispense, output, split, mix/merge, detect), while the edges represent dependencies and order of operations (e.g., in **Figure 3**, *M1* cannot be executed until *I1* and *I2* are complete).

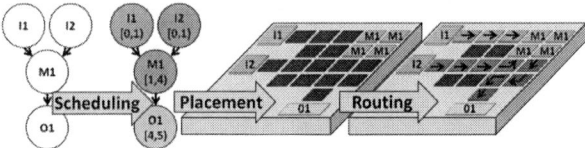

Figure 3. A microfluidic assay is represented in the form of a DAG; its operations are then *scheduled* and *placed* onto the DMFB array; droplets are then *routed* between operation locations.

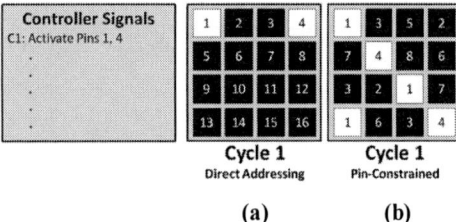

Figure 4. Activating a pin on a (a) direct-addressing DMFB activates (white) exactly 1 electrode per pin; (b) a pin on a pin-constrained DMFB activates 1+ electrodes per pin, depending on the pin layout.

The DAG is scheduled such that each operation has a specific start and stop time-step; a *time-step* is the basic scheduling unit for operations and usually lasts for 1s or 2s. Next, the placer selects a specific I/O port or group of cells, called a *module*, to perform each assay operation. Lastly, the router computes droplet pathways between all operations at the start of each time-step.

1.1.3 Background: Low-Level Pin Mapping

The output of the compiler is a list of pins to active each cycle; a *cycle* is the time it takes to move a droplet from one electrode to the next. In **Figure 4**, a "dry" controller (e.g., a PC) sends signals to activate pins during each cycle on the "wet" DMFB.

In a direct-addressing DMFB, each pin is electrically tied to a single electrode such that an $m \times n$ array of electrodes has $m \times n$ external pins driven by the dry controller. As each electrode is individually controllable, direct addressing allows for maximal flexibility in coordinating droplet-movement. Unfortunately, the complexity of routing $m \times n$ pins underneath a $m \times n$ electrode array is complicated and requires increasingly more PCB layers as the array increases in size, leading to expensive products [17].

Pin-constrained DMFBs connect each control pin to multiple electrodes (**Figure 4(b)**) to reduce the number of wires routed underneath the electrode array; this reduces the number of PCB layers, which, in turn, reduces cost. For a pin-constrained DMFB, activating a control pin will active multiple electrodes, as shown in **Figure 4(b)**. This complicates assay compilation, as independent control over individual electrodes no longer exists. Thus far, pin-assignment has been proposed to reduce the cost of assay-specific DMFBs [9][17], but generalized pin-constrained DMFBs that execute arbitrary assays have not yet been realized.

1.2 Contribution

The contribution of this paper is a pin assignment scheme that facilitates all basic microfluidic operations at pre-determined locations in a pin-constrained DMFB; the resulting DMFB is therefore field-programmable, rather than assay-specific. A high-level synthesis flow targeting this device establishes automatic compilation. Experiments demonstrate that field-programmability is achieved with a handful of additional control pins compared to state-of-the-art assay-specific pin-constrained DMFBs, and that the performance overhead incurred at the cost of field-programmability is marginal.

Figure 5. Pin diagram for a 12×15 field-programmable, pin-constrained DMFB which can accommodate 4 mix modules and 6 split/store/detect (SSD) modules. Routing and mixing pins are shared; the interference region does not contain actual electrodes. Holding and I/O electrodes are independently wired to single control pins for flexibility and programmability.

2. RELATED WORK

Griffith and Akella [1] and Grissom and Brisk [2][3] impose virtual topologies on top of direct-addressing DMFBs; a virtual topology is a mesh-like network of streets and rotaries that perform droplet transport, and dedicated reaction chambers that perform all other assay operations. This limits flexibility, but simplifies certain aspects of dynamic recompilation in response to operation variability and errors. The field-programmable pin-constrained DMFB employs a physical topology/architecture in order to facilitate general-purpose assay execution.

Xu and Chakrabarty [17] introduced a multi-functional pin-constrained DMFB that can execute a pre-defined set of assays; however, it is not field-programmable. Luo and Chakrabarty [9] introduced a pin-assignment algorithm that ensures that two droplets can move independently on a pin-constrained DMFB without conflicting; their algorithm reduced the number of pins required to realize the multi-functional architecture. To the best of our knowledge, their approach is not field-programmable when more than two droplets move concurrently.

Other works have since come and optimized various aspects of the pin-constrained problem. One such recent work is presented by Huang and Ho and combines the droplet routing and pin-count reduction problems together [6]. This is different from the typical approach which starts with droplet routes that have been computed on a direct-addressing array and then attempts to reduce pins. Their solution uses sequential global routing and incremental integer linear programming (ILP) stages to compute solutions. Zhao and Chakrabarty also offer an ILP and heuristic solution to the droplet routing and pin-count co-optimization problem [19].

Lin and Chang present work which focuses on the problem of droplet cross-contamination on pin-constrained devices [8]; they describe scalable algorithms which allow wash droplets to safely clean up contaminated areas on pin-constrained devices. Finally, S. Roy describes an orientation strategy to connect and wire external pins to electrodes on multi-chip devices executing identical assays in lockstep [12].

3. PIN-CONSTRAINED ASSIGNMENT

The field-programmable pin-constrained DMFB employs a pin assignment scheme that enables all of the basic assay operations (**Figure 2**) to execute in a conflict-free manner. **Figure 5** shows a

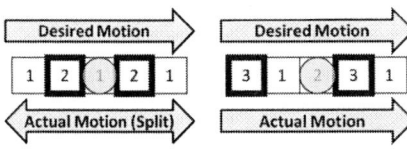

Figure 6. At least 3 repeatable pins are needed to move a droplet along a straight path without causing the droplet to split. Electrodes with bold borders indicate electrodes being activated next cycle.

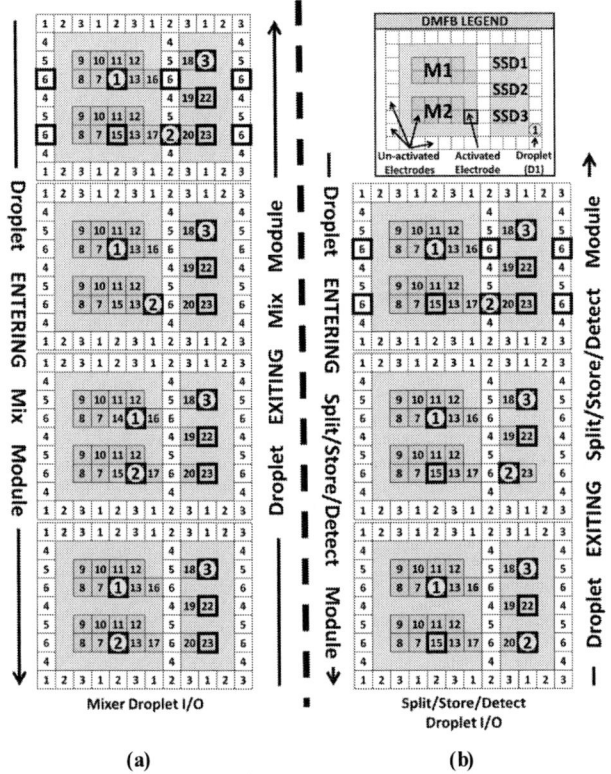

Figure 7. Pin-activation sequence showing how a single droplet (D2) can enter/exit (a) mix modules and (b) split/store/detect modules. Sequences are designed to allow a droplet to enter/exit any module without adversely affecting droplets (D1, D3) in other modules.

12×15 field-programmable, pin-constrained DMFB. The topology/architecture reserves specific DMFB regions for assay operations and others for routing. The topology contains a vertical column of mixing modules on the left (blue/orange electrodes, 7-17) and a vertical column of different modules on the right (orange electrodes, 28-33) that perform splitting, storage, and detection (which requires an external detector affixed above the module); we call these modules *SSD modules*.

White electrodes labeled 1-6 surround the two columns of modules and define the droplet routing regions. I/O reservoirs can be placed anywhere along the perimeter of the chip. The gray electrodes labeled 18-27 indicate pins that allow droplets to enter and exit each module. An *interference region* (pink) surrounds each module to isolate droplets within the module from droplets in the routing region or adjacent modules. These regions are not functional, and do not contain electrodes.

The layout is designed for operation concurrency and scalability. Since routing times are so much shorter than operation times [16], we devote more pins to modules (as opposed to the routing region) to allow more operations to be executed simultaneously at

any time-step. The architecture can also be lengthened or shortened in the vertical dimension to produce a DMFB with any desired number of modules.

3.1 DMFB Operations and Synchronization

The following sub-sections briefly describe how the basic microfluidic operations are performed on our field-programmable, pin-constrained DMFB.

3.1.1 Droplet Transport

Figure 6 shows that at least 3 pins are required to successfully transport a droplet along a straight path; this is called a 3-phase transport bus [13]. In **Figure 5**, pins 1-3 control two horizontal transport buses; pins 4-6 construct three vertical transport buses. This pin-constrained virtual architecture facilitates droplet transfer between horizontal and vertical transport busses, and routable paths exist between all modules and I/O reservoirs on the chip's perimeter. Chips of arbitrary vertical height can be instantiated without remapping the transport electrodes. The mix and SSD module hold electrodes remain active during routing to ensure that droplets within the modules do not drift.

Droplets are routed one at a time because the 3-phase transport busses do not provide a sufficient number of unique pins to sufficiently hold droplets in the routing area while another droplet enters/exits a module. Additional cells could be added to the bus to increase routing parallelism; however, given that routing times (milliseconds) are much smaller than operations (seconds), they are typically considered negligible and are often ignored [16]. See supplemental **Section S2** for more details on sequential routing.

3.1.2 Droplet Dispensing and Outputting

I/O reservoirs are placed on the DMFB perimeter. Each I/O reservoir has an individually controlled electrode that leads a droplet to the edge of the array; these are left off the diagram in **Figure 5** because they are common to all DMFB designs.

3.1.3 Merging/Mixing

Figure 7(a) illustrates a droplet (D2) entering and exiting a mixing module (M2) without conflicting with droplets in other modules (D1, D3). At the top, D2 has reached the routing electrode adjacent to the mixing module (M2) it will enter; D1 is stored in mixing module M1 and D3 is stored in SSD module SSD1. All SSD module electrodes are activated (pins 21-23) to hold all stored droplets in place during mixing module I/O. Activating pin 17 (M2's I/O cell) moves droplet D2 to a position adjacent to M2. Activating pin 13 draws D2 into M2, while transporting D1 to an adjacent cell within M1. Next, all droplet hold cells (pins 14 and 15) move D1 and D2 to identical positions within M1 and M2 respectively. **Figure 7(a)** also shows that the electrode sequence is simply reversed to facilitate a droplet leaving a mixing module.

Before mixing, two droplets must first merge (i.e., collide into each other). A nearly-identical electrode sequence as the one seen in **Figure 7(a)** handles this case. For the interested reader, we demonstrate this in **Figure S1** in the supplementary section. Once M1 and M2 each contain a merged droplet, they can perform mixing operations concurrently by activating cells 7-13 in sequence, followed by 14 and 15 together. This permits both droplets to complete one clockwise cycle in the mixing modules. Mixing pauses by holding droplets on their hold cells whenever another mixing module I/O operation occurs.

3.1.4 Storage, Detection, and Splitting

SSD modules perform storage and detection (if equipped with an external detector). Both operations require a droplet to enter an

321

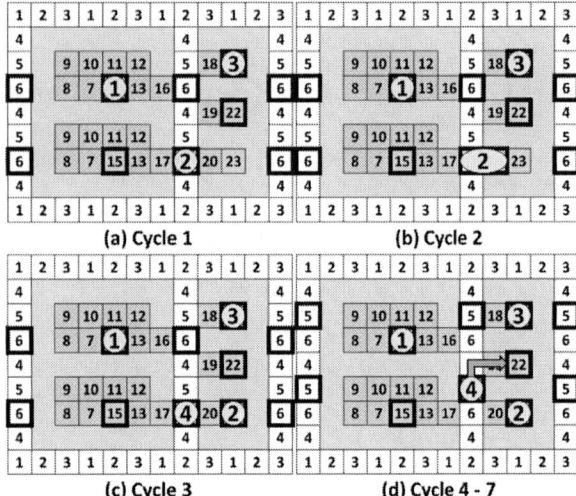

(a) Cycle 1   **(b) Cycle 2**

(c) Cycle 3   **(d) Cycle 4 - 7**

Figure 8. Pin-activation sequence for splitting a droplet (*D2*) and storing in split/store/detect (SSD) modules. Sequences are designed to allow a droplet to split and store without adversely affecting droplets (*D1, D3*) in other modules. NOTE: Legend same as Figure 7.

SSD module and remain in place. **Figure 7(b)** illustrates the process by which a droplet enters/exits an SSD module (*SSD3*) without affecting other droplets in other modules. All SSD hold electrodes are kept on, except for *SSD3*'s, which allows droplet *D2* to enter. *SSD3*'s I/O electrode is activated, followed by its hold electrode, to complete the entrance. This sequence is reversed to facilitate droplets exiting SSD modules.

Figure 8(a)-(c) illustrates droplet splitting. The initial position of droplet *D2*, which will be split, is on a vertical transport bus next to an SSD module's I/O cell (a). The cell on the transport bus is activated throughout the split. The I/O cell is then activated, which stretches *D2* to cover both cells (b). Next, the SSD module's hold cell is activated, and the I/O cell is deactivated; this splits *D2* into two separate droplets: *D2*, on the hold cell, and *D4*, in the transport bus. If storage is required for *D4*, then it must be routed to an available SSD module, as shown in **Figure 8(d)**.

4. FIELD-PROGRAMMABLE SYNTHESIS

This section describes the synthesis flow (**Figure 3**) that maps an assay to the field-programmable pin-constrained DMFB.

4.1 Scheduling

List scheduling [3][16] is a fast, greedy, single-path scheduling algorithm. List scheduling targeting the field-programmable, pin-constrained DMFB differs from prior implementations in several respects. The most important difference is that prior list schedulers use one generic module type for all assay operations, rather than distinguishing between mixing and SSD modules.

As shown in **Figure 8(d)**, split modules may require two SSD modules if both droplets that are produced must be stored. As shown in **Figure 9**, the split node is converted into an instantaneous split followed by two storage operations.

The scheduler reserves one SSD module to address routing deadlocks, as explained later in **Section 4.3**. Thus, in **Figure 5**, only 5 of the 6 SSD modules are available for storage and detection. Prior list schedulers may transport droplets between modules for storage for a variety of reasons [3][10]. Since only SSD modules perform storage and each stores at most one droplet, there is no motivation to transport droplets between SSD modules during storage; thus, a stored droplet remains in a single SSD

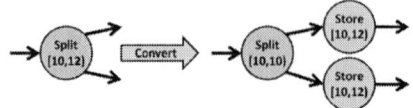

Figure 9. Split operations are converted to a split and two stores for synthesis.

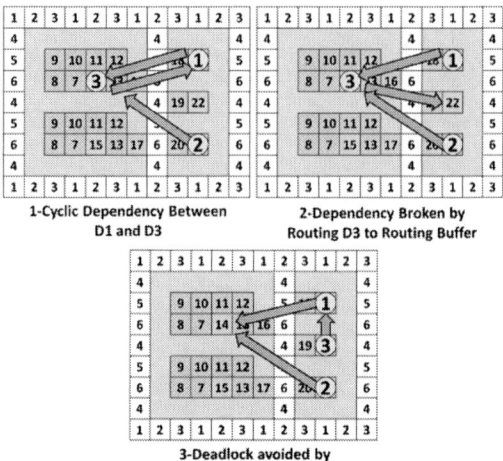

1-Cyclic Dependency Between D1 and D3

2-Dependency Broken by Routing D3 to Routing Buffer

3-Deadlock avoided by routing D1 before D3

Figure 10. Cyclic routing dependencies can be broken by first routing a droplet in the cycle to the routing buffer module (one of the SSD modules). Arrows indicate that the droplet at the tail end is about to travel to the module at the head end. NOTE: Legend same as Figure 7.

module for the entirety of its storage lifetime. This may reduce the number of droplets that must be routed in certain cases.

4.2 Placement/Binding

Similar to Grissom and Brisk [2][3], we reduce placement to a binding problem, which is solved using the left-edge algorithm [7]. One minor difference between our binder and others is that we do not bind split operations since they yield two immediate storage nodes (**Figure 9**). Instead, we simply bind the two storage children directly. In the interest of space, we direct the reader to other references for a more-complete description and psuedocode of the left-edge binder [2][3].

4.3 Routing

4.3.1 Route Computation

A routing *sub-problem* refers to the set of droplets that must be routed just before each time-step begins in the schedule. Before each time-step (operation) begins, droplets are routed sequentially, one at-a-time. Given the topology in **Figure 5**, three different types of routes must be computed: input reservoir to module, module to module, and module to output reservoir.

To route a droplet from an input reservoir to a module, it suffices to compute the shortest distance from the input reservoir to the electrode adjacent to the target module's I/O electrode. The main question is to determine whether the clockwise or counter-clockwise path is shorter. Once the droplet arrives, the appropriate module input sequence is applied, as discussed in **Section 3.1**.

Module-to-module routing uses the vertical column in the center of the DMFB. The router applies the output sequence to extract the droplet from the source module, routes the droplet north or south as appropriate, and applies the input sequence to deliver the droplet to its target module.

322

Table 1. Experimental results comparing the direct-addressing DMFB (DA) [3] with our field-programmable, pin-constrained DMFB (FP).

Benchmarks	Array Dim.		# Electrodes Used		# Pins		Routing Time (s)		Operations Time (s)		Total Time (s)	
	DA	FP	DA	FP	DA	FP	DA	FP	DA	FP	DA	FP
PCR	15x19	12x21	285	153	285	43	0.7	2.1	11	11	11.7	13.1
In-Vitro 1	15x19	12x21	285	153	285	43	0.7	2.6	14	14	14.7	16.6
In-Vitro 2	15x19	12x21	285	153	285	43	1.2	3.8	18	18	19.2	21.8
In-Vitro 3	15x19	12x21	285	153	285	43	1.9	6.2	22	18	23.9	24.2
In-Vitro 4	15x19	12x21	285	153	285	43	1.8	8.8	24	19	25.8	27.8
In-Vitro 5	15x19	12x21	285	153	285	43	2.9	11.6	32	25	34.9	36.6
Protein Split 1	15x19	12x21	285	153	285	43	1.8	2.9	71	71	72.8	73.9
Protein Split 2	15x19	12x21	285	153	285	43	6.2	6.1	106	106	112.2	112.1
Protein Split 3	15x19	12x21	285	153	285	43	13.9	13.5	176	176	189.9	189.5
Protein Split 4	15x19	12x21	285	153	285	43	32.9	29.3	316	316	348.9	345.3
Protein Split 5	15x19	12x25	285	177	285	49	63.6	61.4	670	596	733.6	657.4
Protein Split 6	15x25	12x29	375	203	375	55	161.2	127.4	1156	1156	1317.2	1283.4
Protein Split 7	15x25	12x31	375	239	375	63	290.3	260.6	2353	2276	2643.3	2536.6
Avg. Normalized Improvement: (> 1 is improvement)			1.82		6.53		0.68		1.07		0.98	

Table 2. We present results from Xu's [17] and Luo's [9] pin-constrained designs for chips which can run PCR, In-Vitro 1, Protein Split 3 and a multi-functional chip which can run all three.

Benchmark	Array Dim.	# Electrodes Used	# Pins		Total Time (s)	
			Xu	Luo	Xu	Luo
PCR	15x15	62	14	22	20	30
In-Vitro 1	15x15	59	25	21	73	90
Protein Split 3	15x15	54	26	20	150	170
Multi-Function	15x15	81	37	17	150	170

Table 3. We demonstrate the three benchmark assays from Xu [17] and Luo [9] on our field-programmable, pin-constrained DMFB design of various sizes.

Array Dim.	#Module (Mix/SSD)	# Electrodes Used	# Pins	Total Time(s)		
				PCR	In-Vitro 1	Protein Split 3
12x9	2/3	62	23	18.59	19.00	-
12x12	3/4	89	27	15.89	17.26	-
12x15	4/6	111	33	12.88	16.56	-
12x18	5/7	133	39	13.00	16.60	189.65
12x21	6/9	153	43	13.08	16.60	189.53

To route a droplet from a module to an output reservoir, the output sequence is applied to extract the droplet from the source module; then the droplet is routed either clockwise or counter-clockwise along the shortest path to the output reservoir.

4.3.2 Droplet Dependencies and Deadlock

Special care must be taken to prevent droplet dependencies from turning into deadlock. Routing deadlock occurs when one or more droplets are waiting for resources to become available that will never become free. This can occur when a droplet dependency cycle occurs, as seen in **Figure 10**. *D1* is in *SSD1* and waiting for droplet *D3* to leave *M1*, while droplet *D3* is in *M1* and waiting for droplet *D1* to leave *SSD1*. To break the cycle, we pick *D3* to first route itself to an empty SSD module (*SSD2*), as shown in step 2 of **Figure 10**. The scheduler always keeps one SSD module unallocated, as it cannot predict routing dependencies a-priori.

As seen in step 3 of **Figure 10**, although the cyclic dependency is broken between droplets *D1* and *D3*, deadlock can still occur if the sequential droplet routing order is chosen poorly. Now, droplet *D3* travels to *SSD1*, while droplets *D1* and *D2* travel from *SSD1* and *SSD3*, respectively, to *M1*. Droplet *D2* can be routed at any time, because no droplets will travel to *SSD3* (its source) and no droplets remain at *M1* (its destination) that must first move. If the router tries to route droplet *D3* before routing *D1*, then deadlock will occur because *SSD1* is not yet free to receive new droplets. If there is no such dependency check, droplet contamination will occur. We describe a general algorithm to eliminate droplet dependencies and provide the details and pseudocode in supplemental **Section S3**.

5. EXPERIMENTAL RESULTS

We implemented our field-programmable, pin-constrained DMFB in C++; we compare with Grissom and Brisk's fast online synthesis framework [3], which is publicly available online [5]. All tests were run using a 2.8GHz Intel Core i7 CPU and 4GB RAM on a 64-bit version of Windows 7.

5.1 Comparison to General DMFB

We first compare our implementation to the most recent generally programmable direct-addressing DMFB design [3]. We run a set of 13 assays based on the PCR [15], in-vitro diagnostics [14][15] and protein-split benchmarks [4]. **Table 1** shows the number of seconds spent both routing and executing assay operations; the total time is the sum of the two. Results are also given for the number of usable electrodes (i.e., tied to a control pin) and number of external control pins for the DMFB size used. For Protein Split 5-7, the array dimensions had to be increased to execute the assay for one or both of the DMFBs.

Our DMFB has longer routing times for the first 7 benchmarks because of sequential routing; however, it actually has shorter routing times for Protein Split 2-7 because additional routes are not generated between storage nodes, as described in **Section 4.1**. Modules in the direct-addressing DMFB [3] can store up to two droplets at any time. To utilize as many resources as possible, droplets stored alone in separate modules will consolidate in order to free up more modules to do useful work; routing these droplets adds to the routing time, and therefore the total time as well.

The bottom row of **Table 1** shows the average improvement of our field-programmable DMFB compared to the direct-addressing DMFB. We calculated this metric by computing the improvement of FP over DA (baseline) for each benchmark and then averaging these values over the entire set of benchmarks. Any value over 1 means FP is an improvement. Notice that, although FP's average routing time is 32% slower, its average operation time is 7% faster. On average, the field-programmable pin-constrained DMFB only suffers an average 2% slowdown in total execution time, while reducing the pin count by 6-7×.

5.2 Comparison To Pin-Constrained DMFBs

Table 2 presents results for two prior pin-constrained assay-specific and multi-functional DMFB architectures [9][17]; assay-specific architectures were generated for PCR, In-vitro 1, and Protein Split 3 assays, while the multi-functional chip can perform all three assays. Many differences exist between these designs, most notably that they are assay specific while ours is field-programmable, and that they use linear array mixing modules, which have longer latencies than the 4×2 mixers used here. Thus, the schedules are different, and it is unclear if their reported results include droplet routing times, as their primary objective was to reduce the cost of their pin-constrained DMFBs by reducing the pin-count. **Table 2** is reproduced from Ref. [9].

Table 3 reports the performance and pin-count for PCR, In-Vitro 1, and Protein Split 3 for field programmable pin-constrained DMFBs of varying sizes. For PCR and In-Vitro 1, execution times decrease as the DMFB size (and thus the number of available modules) increase, saturating at 12×15. For larger DMFBs, performance degrades slightly due to longer routing times.

The Protein Split 3 assay requires 6 droplets to be stored at several instances during the assay; thus, the 12×18 array with 7 SSD modules (6 available to the scheduler) is the smallest compatible device. The total execution time remains steady, regardless of resources (we also tested on a 12×81 DMFB with abundant resources) at 189s. In this case, the total execution time is not limited by resource availability, but by the 7s droplet dispense times [15]. It is unclear what droplet dispense times were assumed in prior work [9][17]; reducing the dispense times to 2s instead of 7s reduces the assay execution time to approximately 100s.

In general, the field-programmable pin-constrained DMFBs require more pins than the assay-specific or multi-functional pin-constrained DMFBs reported in **Table 2**; this is to be expected because our device is optimized for field-programmability, while their devices are optimized for reduced pin-count.

Luo and Chakrabarty's pin assignment scheme [9] theoretically provides some flexibility, as two droplets are guaranteed to be able to move without interfering with one another; however, they did not provide details on how synthesis was performed, so it is difficult to provide a direct comparison.

In contrast, the different versions of our field-programmable pin-constrained DMFB reported in **Table 3** are the same generic 2-column architecture, but with a different number of resources. As seen in **Table 1**, if we pick dimensions of reasonable size (12×21), we can run all three assays in **Table 2**, as well as others, due to the field-programmable nature of our design.

6. CONCLUSION

This paper has introduced the first field-programmable pin-constrained DMFB that can execute arbitrary assays; prior pin-constrained DMFBs have all been assay-specific or multi-functional, but not field-programmable. To program the device, we describe modifications to a synthesis flow for DMFBs, which address architectural issues that are specific to our design.

Compared to field-programmable, direct-addressing DMFBs, the field-programmable, pin-constrained DMFB offered comparable or improved performance, while reducing the pin-count by 6-7x. Compared to assay-specific, pin-constrained DMFBs, the field-programmable device offered better performance and a comparable pin-count for the PCR and In-vitro 1 benchmarks, but degraded performance and a 2x higher pin-count for Protein Split 3. Compared to the multi-functional, pin-constrained DMFB, the field-programmable pin-constrained DMFB required 1.44x more pins for a comparably sized array (12x18 vs. 15x15). Thus, field-programmable does come at a price in terms of pin-count and, sometimes, assay execution time, compared to state-of-the-art assay-specific and multi-functional pin-constrained DMFBs; however, the flexibility provided is unmatched by prior DMFBs and offers a significant advancement in terms of programmability.

7. ACKNOWLEDGEMENTS

This work was supported in part by NSF Grant CNS-1035603. Daniel Grissom was supported by an NSF Graduate Research Fellowship. Any opinions, findings, and conclusions or recommendations expressed in this material are those of the authors and do not necessarily reflect those of the NSF.

8. REFERENCES

[1] E. Griffith and S. Akella. Performance characterization of a reconfigurable planar-array digital microfluidic system. IEEE Trans. Comput.-Aided Design Integr. Circuits Syst, 25(2), Feb. 2006.

[2] D. Grissom and P. Brisk. A high-performance online assay interpreter for digital microfluidic biochips. In Proc. of GLSVLSI, pages 103-106, Salt Lake City, UT, USA, May 3-4, 2012.

[3] D. Grissom and P. Brisk. Fast online synthesis of generally programmable digital microfluidic biochips. In Proc. of CODES+ISSS, pages 413-422, Tampere, Finland, Oct. 7-12, 2012.

[4] D. Grissom and P. Brisk. Path scheduling on digital microfluidic biochips. In Proc. of DAC, pages 26-35, San Francisco, CA, USA, Jun. 3-7, 2012.

[5] D. Grissom, et al. A digital microfluidic biochip synthesis framework. In Proc. of VLSI-SoC, Santa Cruz, CA, Oct. 7-10,2012.

[6] T-W. Huang and T-Y. Ho. A two-stage integer linear programming-based droplet routing algorithm for pin-constrained digital microfluidic biochips. IEEE Trans. Comput.-Aided Design Integr. Circuits Syst, 30(2), Feb. 2011.

[7] F. J. Kurdahi and A. C. Parker, "REAL: a program for REgister ALlocation. In Proc. of DAC, pages 210-215, Miami, FL, USA, 1987.

[8] C. C-Y. Lin and Y-W. Chang. Cross-contamination aware design methodology for pin-constrained digital microfluidic biochips. IEEE Trans. Comput.-Aided Design Integr. Circuits Syst, 30(6), Jun. 2006.

[9] Y. Luo and K. Chakrabarty. Design of pin-constrained general-purpose digital microfluidic biochips. In Proc. of DAC, pages 18-25, San Francisco, CA, USA, Jun. 3-7, 2012.

[10] K. O'Neal, D. Grissom, and P. Brisk. Force-directed list scheduling for digital microfluidic biochips. In Proc. of VLSI-SoC, Santa Cruz, CA, USA, Oct. 7-10, 2012.

[11] M. G. Pollack, A.D. Shenderov, and R. B. Fair. Electrowetting-based actuation of droplets for integrated microfluidics. Lab on a Chip, 2:96-101, 2002.

[12] S. Roy, D. Mitra, B. B. Bhattacharya, K. Chakrabarty. Congestion-aware layout design for high-throughput digital microfluidic biochips. ACM Journal on Emerging Technologies in Computing Systems, 8(3): article #17, Aug. 2012.

[13] V. Srinivasan, V. Pamula, and R. Fair. An integrated digital microfluidic lab-on-a-chip for clinical diagnostics on human physiological fluids. Lab on a Chip, 4:310-315, 2004.

[14] F. Su and K. Chakrabarty. Architectural-level synthesis of digital microfluidics-based biochips. In Proc. of ICCAD, pages 223-228, San Jose, CA, USA, Nov. 7-11, 2004.

[15] F. Su and K. Chakrabarty. "Benchmarks" for digital microfluidic biochip design and synthesis. Duke University, Department of Electrical and Computer Engineering, 2006. http://www.ee.duke.edu/~fs/Benchmark.pdf

[16] F. Su and K. Chakrabarty. High-level synthesis of digital microfluidic biochips. ACM Journal on Emerging Technologies in Computing Systems, 3(4): article #16, Jan., 2008.

[17] T. Xu and K. Chakrabarty. Broadcast electrode-addressing for pin-constrained multi-functional digital microfluidic biochips. In Proc. of DAC, pages 173-178, Anaheim, CA, USA, Jun. 8-13, 2008.

[18] P-H. Yuh, C-L. Yang, and Y-W. Chang. BioRoute: a network-flow-based routing algorithm for the synthesis of digital microfluidic biochips. IEEE Trans. Comput.-Aided Des. Integr. Circuits Syst., 27(11):1928-1941, Nov. 2008.

[19] Y. Zhao and K. Chakrabarty. Simultaneous optimization of droplet routing and control-pin mapping to electrodes in digital microfluidic biochips. IEEE Trans. Comput.-Aided Design Integr. Circuits Syst, 31(2), Feb. 2012.

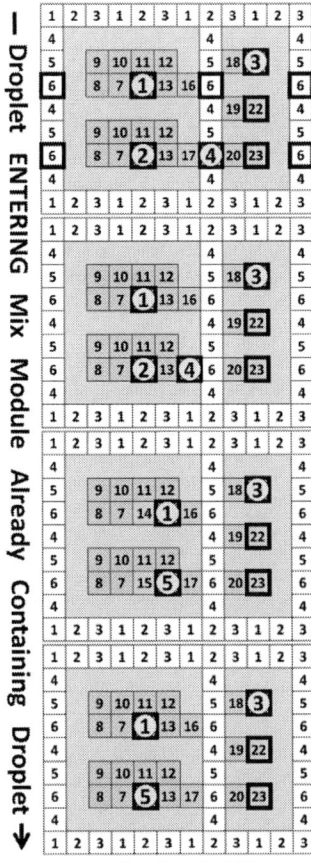

Mixer Droplet I/O

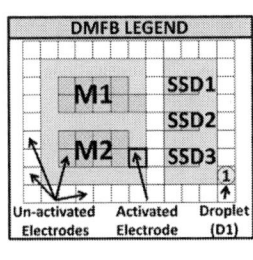

DMFB LEGEND

M1   SSD1
           SSD2
M2   SSD3

Un-activated   Activated   Droplet
Electrodes   Electrode   (D1)

Figure S1. Shows the electrode/pin activation sequence (from top to bottom) that causes *D4* to merge with *D2* (in *M2*) to become *D5* (twice the volume) and re-sync with any other droplets in mix modules (i.e., *D1* in *M1*).

S1. MERGING TWO DROPLETS

Figure S1 shows how a droplet (*D4*) merges with an existing droplet (*D2*) in *M2* to become *D5*. Once merged, the new droplet (*D5*, with twice the volume) is synced with *D1* back to the mixers' hold locations (the bottom image in **Figure S1**). If *D1* has already been merged, then mixing can begin; if not, the general process in **Figure S1** must be repeated to merge a new droplet with *D1* before the merged droplets in *M1* and *M2* can be mixed.

S2. SEQUENTIAL ROUTING

In this section, we provide more details to explain why sequential routing was chosen. As mentioned in **Section 3**, operation times are on the order of seconds, while droplet transport times are on the order of milliseconds. A typical time-step is 1 or 2 seconds [16], while a typical droplet actuation is at most 10ms (100Hz) [18]. With this in mind, we chose the 3-phase bus approach because, although it restricts droplet routing parallelism, it simplifies the general-purpose nature of the device.

Consider **Figure S2** which shows our field-programmable, pin-constrained design with two different numbers of modules and module sizes. Notice that, despite the central vertical bus ending with pin 4 or pin 5, a clean transition can be made between buses because all of the pins adjacent to the intersection are guaranteed to be unique [9]. Thus, the same algorithms can be used to map assays to field-programmable, pin-constrained arrays of various

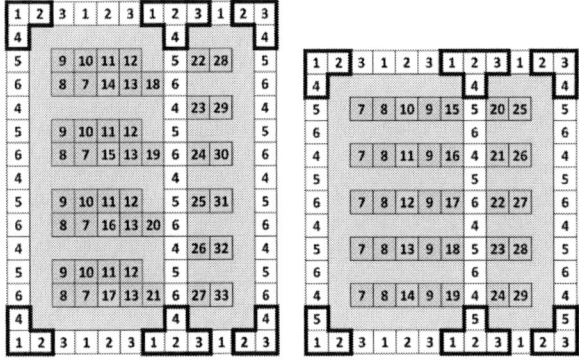

Figure S2. Shows that the number or size of modules can be changed and the 3-phase bus can be repeated, regardless of array size, without causing pin-conflict at the vertical-horizontal bus intersections (bold borders).

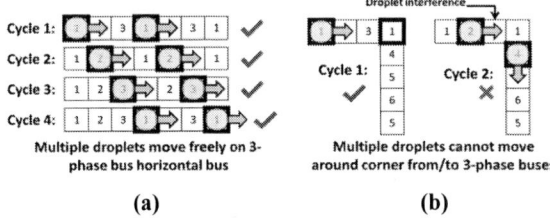

(a)   **(b)**

Figure S3. Shows that moving two droplets concurrently is (a) feasible when moving in a straight path, but (b) not always possible when moving around a bend because droplet interference can occur.

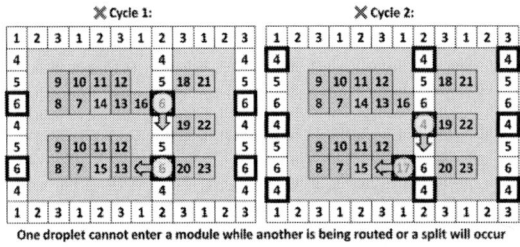

One droplet cannot enter a module while another is being routed or a split will occur

Figure S4. Shows that multiple droplets moving through the vertical bus will result in an unintentional split when one tries to enter a module.

sizes, given that they keep the same general form. This is important because it would allow an end-user to design an assay and then go purchase the cheapest compatible pin-constrained DMFB; here, compatibility means that there are sufficient resources available (meaning mixing and SSD modules) and that the SSD modules have appropriate detectors.

As seen in **Figure S3(a)**, it is always possible to move multiple droplets along a straight path on the 3-phase bus because there is sufficient space between repeating pin numbers. However, **Figure S3(b)** shows that droplet interference can occur when moving around a corner. In cycle two, if the next two pins are activated (pin 3 and pin 5), the droplets will most likely merge. It would be possible to hold pin 2 in cycle 3 such that the top droplet would stall and avoid the droplet interference in cycle 3. However, consider that droplets will only be making this transition when traveling to or from an I/O reservoir.

Most of the opportunities to parallelize routing occur when routing between modules. In light of this, consider **Figure S4**, which shows multiple droplets in the central vertical routing bus.

325

```
1    Given sequence graph G = (V, E)
2    int timeStep = 0;
3    Repeat {
4      graph d = ∅; // Dependencies
5      for (∀v ∈ V: v.startTime = timeStep)
6        for (∀p ∈ v.parents)
7          d.add(p.location, v.location);
8      end for
9
10     list cc  = ∅; // Connected Components
11     list scc  = ∅; // Strongly Connected Components
12     cc = findAllConnectedComponents(d);
13     scc = findStronglyConnectedComponents(cc);
14     resolveDependencies(scc);
15     reverseTopologicalSort(cc); // scc ⊆ cc
16
17     for (∀c ∈ cc)
18       for (∀o ∈ c: o.startTime = timeStep)
19         for(∀op ∈ o.parents)
20           routeSrcToDest(op.location, o.location);
21
22     timeStep++;
23   } until (time ≥ max (v.startTime, ∀v ∈ V)
```

Figure S5. Psuedocode for route computation.

For the lower droplet to enter the lower mixing module (*M2*), the DMFB must activate pin 17, while simultaneously deactivating pin 6. This is possible, but notice that the top droplet requires pin 4 to be activated to continue downward on its path. Activating this pin will cause two adjacent electrodes to be activated near the lower droplet, which will result in a split. Moving the top droplet up, down or keeping it stationary will require pins 5, 4, or 6 to be activated in cycle 2, respectively, which will each cause the bottom droplet to split. If pins 4-6 are not activated, then the top droplet will drift and the assay will not execute correctly.

Rather than deal with these complications, we chose to route droplets one-at-a-time instead, as the impact on total assay execution time is minimal.

S3. ROUTING ALGORITHM

This section elaborates on the routing process discussed in **Section 4.3**; **Figure S5** presents pseudocode. The router receives a scheduled and placed DAG $G = (V, E)$, where vertices represent operations and edges represent droplets that must be transferred between operations. Each vertex has a *location*, which indicates the module or I/O reservoir where the corresponding operation will take place.

Each vertex in V is scheduled to begin at a certain time-step, as computed by the scheduler. A time-step typically lasts one or two seconds and represents the time when operations are processed by their respective modules or I/O reservoirs. When a new time-step begins, then a new operation may start. This requires droplets to be routed to the module that will execute the operation. Thus, we start at time-step 0 (*Line 2*) and repeat the routing process for each time-step until the last scheduled operation begins (*Lines 3-23*); each iteration handles one routing sub-problem (time-step).

First, a graph of dependencies (*d*) is created based on the location of each node that is relevant to the current time-step (*Lines 4-8*). An edge (D_x, D_y) in the dependency graph means that droplet D_x will be routed to droplet D_y's current location, so D_y must be routed first. As seen in *Line 7*, dependencies are added to the graph based on the *location* field because droplets are being routed from the parents' location to the newly-executing node's location.

The next step is to decompose *d* into its connected components (*Line 12*), which can be computed using a simple recursive multi-directional, depth-first search [S1]. Connected components are processed on-by-one. To simplify further discussion, we will assume that *d* is composed of a single connected component.

Routing is simple if *d* is acyclic. Since the algorithm routes droplets one-at-a-time, edge (D_x, D_y) indicates that D_y must be routed before D_x; otherwise, D_x would merge inadvertently with D_y upon completing its route. A legal routing solution for the sub-problem can be achieved by routing the droplets one-by-one in reverse topological order [S2]. *Lines 10-20* in **Figure S5** solve the more complicated cyclic case, which is described next; in the simple acyclic case, *Lines 11, 13,* and *14* are unnecessary.

If *d* is cyclic, routing becomes more complicated, as a cycle means that no droplet can complete its route without inadvertently merging with a droplet waiting at its destination. This problem is solved by temporarily allocating DMFB resources for storage.

The first step is to compute *strongly connected components (SCCs)* (*Line 13*) from the connected components using Gabow's path-based, depth-first search [S3]. One minor modification is that we only need to identify the SCCs that contain more than one node, as single-node SCCs do not have cyclic droplet dependencies.

Once the SCCs that represent cycles are identified, the cycles must be resolved (*Line 14*). As demonstrated in **Figure 10**, the router randomly selects a droplet D_y from the SCC and routes it to an empty SSD module for temporary storage, which breaks the dependency cycle. The dependency graph *d* is then modified to account for the relocated droplet's new location: each edge of the form (D_x, D_y) is removed from *d* as D_x is now free to move to its destination, since D_y has moved out of the way.

The scheduler always leaves at least one SSD module free so that there is room to break one cycle in the SCC. If the SCC contains multiple intersecting cycles, then any other free SSD or mixing module could be used for temporary storage. This process repeats until *d* becomes acyclic. Once *d* becomes acyclic, a legal routing solution can be found, as previously discussed.

One optimization that can reduce the extra storage requirement (not shown in **Figure S5**) is to break SCCs one-by-one. Droplets corresponding to vertices with no predecessors in *d* are routed immediately, and the corresponding vertex is removed from *d*. Then, an SCC is chosen that satisfies the following property: for every vertex D_x belonging to the SCC and each outgoing edge (D_x, D_y), D_y also belongs to the SCC. Breaking all of the cycles in this particular SCC will ensure that at least one vertex in the updated graph *d* will have no successors.

The advantage of the second approach is that it reduces the need for temporary storage resources. As an example, suppose that *d* has two SCCs, scc_1 and scc_2, and that each requires one additional storage resource to resolve. Under the first approach, two storage resources must be allocated in order to convert *d* to an acyclic graph before the droplets can be routed. Under the second scheme, all of the droplets in scc_1 will be routed before all of the droplets in scc_2, or vice-versa. Therefore, both SCCs can use the same storage resource, so just one available module suffices. In general, if *d* contains *k* SCCs, and scc_i requires m_i storage modules, then the first scheme requires $M_1 = m_1 + m_2 + ... m_k$ modules for storage, whereas, the second requires $M_2 = max\{m_1, m_2, ..., m_k\}$ modules.

That being said, we did not encounter a single droplet dependency cycle in any of the 25 benchmarks seen in **Table 1** and **Table 3**; the largest assay, Protein Split 7, contains 2556 nodes. Although a droplet dependency problem can still occur in theory, it seems unnecessary, from a practical standpoint, to devote a large number of resources to resolving droplet dependency cycles, even for large assays.

S4. SUPPLEMENTAL REFERENCES

[S1] J. Hopcroft and R. Tarjan. Algorithm 447: efficient algorithms for graph manipulation. Communications of the ACM, 16(6):372-378, Jun 1973.

[S2] A. Kahn. Topological sorting of large networks. Communications of the ACM, 5(11):558-562, Nov 1962.

[S3] H. Gabow. Path-based depth-first search for strong and biconnected components. Information Processing Letters, 74(3-4):107-114, May 2000.

BDS-MAJ: A BDD-based Logic Synthesis Tool Exploiting Majority Logic Decomposition

Luca Amarú, Pierre-Emmanuel Gaillardon, Giovanni De Micheli

Integrated Systems Laboratory (LSI), EPFL, Switzerland.

Abstract—Despite the impressive advance of logic synthesis during the past decades, a general methodology capable of efficiently synthesizing both control and datapath logic is still missing. Indeed, while synthesis techniques for random control logic (AND/OR-intensive) are well established, no dominant method for automated synthesis of datapath logic (XOR/MAJ-intensive) has yet emerged. Recently, *Binary Decision Diagrams* (BDDs) have been adopted to create an optimization system, named BDS, that supports integrated synthesis of both AND/OR- and XOR-intensive functions through functional logic decomposition on the BDD structure. However, it does not support direct decomposition and manipulation of majority logic which, instead, is widely used in datapath circuits. In this paper, we present the first BDD-based majority logic decomposition method and a logic decomposition system, BDS-MAJ, that enables efficient logic synthesis for both random control and datapath circuits. Experimental results show that logic synthesis based on BDS-MAJ produces CMOS circuits having on average 28.8% and 26.4% less area and, at the same time, 12.8% and 20.9% smaller delay with respect to academic ABC and BDS synthesis tools. Compared to commercial Synopsys *Design Compiler* synthesis tool, BDS-MAJ reduces on average the circuit area by 6.0% and decreases the delay by 7.8%.

Categories and Subject Descriptors

B.6.3 [Design Aids]: Automatic Synthesis, Optimization

General Terms

Algorithms, Design, Performance, Theory.

Keywords

Majority Logic, Decomposition, BDD, Logic Synthesis.

I. INTRODUCTION

Virtually all digital integrated circuits are realized using logic synthesis techniques [1]. Whereas most circuits contain datapath and control functions, current logic synthesis tools are better at synthesizing control logic as compared to datapaths, especially when arithmetic functions are involved. Indeed, original logic synthesis techniques [2]–[5], which are the basis for current commercial tools, exploited algorithms using AND/OR representations, while arithmetic/datapath circuits are rich in XOR and MAJORITY functions. A major aim for today's synthesis tools is to handle properly both random control and datapath logic to fully and efficiently automatize ASIC designs.

A step toward this direction is enabled by the use of *Binary Decision Diagrams* (BDDs) [6]–[8] as logic representation structures, because they are typically compact for a wide class of functions, including AND/OR- and XOR/MAJ-intensive functions. To exploit this opportunity, BDDs are considered in [9]–[14]. In these works, BDDs support automated synthesis through efficient logic decomposition. In particular, a BDD-based decomposition theory is proposed in [10] to support various logic structures, *e.g.*, AND, OR, XOR and MUX. Based on this theory, a practical synthesis tool named BDS is described in [10] and refined in [11]. BDS advantageously synthesizes both AND/OR- and XOR-intensive functions thanks to an unified methodology. However, BDS still does not manipulate majority logic losing further optimization opportunities in datapath circuits.

Permission to make digital or hard copies of all or part of this work for personal or classroom use is granted without fee provided that copies are not made or distributed for profit or commercial advantage and that copies bear this notice and the full citation on the first page. To copy otherwise, to republish, to post on servers or to redistribute to lists, requires prior specific permission and/or a fee.

DAC 13, May 29 - June 07 2013, Austin, TX, USA.

In this paper, we aim to extend the capability of current logic synthesis methods by focusing on majority decomposition which is useful for both for datapath (XOR/MAJ-intensive) and control (AND/OR-intensive) logic. To this end, we present the first majority logic decomposition method based on BDD. We integrate the proposed majority decomposition technique in the current state-of-art BDD-based decomposition tool, BDS-PGA [11], in order to create a new complete decomposition tool, BDS-MAJ. Large MCNC and custom datapath benchmarks are used to evaluate BDS-MAJ. Thanks to its runtime efficient algorithms, BDS-MAJ decomposes the largest benchmarks in few seconds. BDS-MAJ produces decomposed networks having on average 29.1% fewer nodes as compared to BDS-PGA. To exploit the new decomposition feature, we employ BDS-MAJ in a standard *optimization-mapping* synthesis flow. Experimental results over MCNC and custom datapath benchmarks show that BDS-MAJ outperforms academic ABC and BDS synthesis tools by 28.8% and 26.4% in less area, respectively, and by 12.8% and 20.9% in smaller delay, respectively. When compared to commercial Synopsys *Design Compiler*, BDS-MAJ produces, on average, circuit with 6.0% less area and 7.8% smaller delay.

The remainder of this paper is organized as follows. Section II provides a background on BDD-based logic decomposition. In Section III, the new majority logic decomposition method is presented. Then, in Section IV, the implementation of the BDS-MAJ decomposition system is discussed. Experimental results for BDS-MAJ are presented and compared with state-of-art commercial and academic synthesis tools in Section V. We conclude the paper in Section VI.

II. BACKGROUND AND MOTIVATION

This section presents relevant background about *Binary Decision Diagrams* (BDDs) and related logic decomposition. Notations and definitions for BDDs are also introduced.

A. Binary Decision Diagrams

BDDs are logic representation structures that were first introduced by Lee [6] and Akers [7]. The notions of ordering and reduction of BDDs were introduced by Bryant in [8], where it was shown that, with these restrictions, BDDs are a canonical logic representation form. Canonical reduced and ordered BDDs are often compact and easy to manipulate, and are therefore widely used in EDA and other fields. We assume that the reader is familiar with basic concepts of Boolean algebra and BDDs (for a review see [1], [10]). We review hereafter only the basic notation used in the rest of the paper.

B. Notation

A BDD is a *Direct Acyclic Graph* (DAG) representing a Boolean function. A BDD is uniquely identified by its *root*, the set of *internal nodes*, the set of *edges* and the *1/0-sink nodes*.

Each internal node in a BDD is labeled by a Boolean variable v and has two out-edges labeled 0 and 1. Each internal node represents the Shannon's expansion with respect to its variable v and the 1- and 0-edges connect to positive and negative Shannon's cofactors, respectively.

Edges are characterized by a regular/complemented attribute: complemented edges indicate to invert the function pointed by that edge.

We refer hereafter to BDDs as to *canonical reduced and ordered* BDDs [8], that are BDDs where (i) each input variable is encountered at most once in each root to sink path and in the same order on all such paths, (ii) each internal node represent a distinct logic function and (iii) only 0-edges can be complemented.

C. BDD-based Logic Decomposition

BDDs are exploited to achieve efficient logic function decomposition [9]–[14] thanks to the notable characteristics of the BDD structure. Special classes of nodes, defined as dominators, are used in [10], [11], [14] to guide the decomposition process directly on the BDD structure. Dominator nodes allow us to uniquely identify substructures in the BDD that are corresponding to specific decomposition types. In [14], 0- and 1-dominators are introduced to support disjoint AND/OR decompositions. Similarly, in [10], x-dominators are defined to support disjoint XNOR decomposition. Generalized 0-, 1- and x-dominators are also introduced in [10] to achieve general non-disjoint decompositions. The decomposition system BDS [10] is based on dominator nodes driven decomposition. BDS exhibits better decomposition results for XOR/XNOR-intensive circuits as compared to traditional AND/OR optimization techniques while maintaining good results quality also for random control logic [10]. A successive version of BDS has been proposed in [11], named BDS- PGA, incorporating further decomposition schemes that generate area-minimal logic networks. Despite BDS and its evolutions are efficient to decompose a great variety of logic functions, they are still missing the opportunity to identify majority decomposition structures, that are widely used in datapath circuits as well as in some random control applications. Note that early attempts to achieve majority logic decomposition are already reported in the 60's [15], but, due to their intractable complexity, failed to gain momentum later in automated logic synthesis. We address, in this paper, the unique opportunity led by efficient majority logic decomposition employed in a contemporary synthesis flow.

III. MAJORITY LOGIC DECOMPOSITION

In this section, we present our novel majority logic decomposition theory. First, the existence of majority decompositions for general logic functions is studied. Then, the search for majority dominator nodes on BDDs is introduced, and, subsequently, the actual majority decomposition construction, optimization and selection phases are described. Finally, the computational complexity of the proposed decomposition method is evaluated.

A. Majority Decomposition

The aim of majority logic decomposition is to express a Boolean function F in the form $Maj(F_a, F_b, F_c)$.

Theorem 3.1: (Existence of majority decomposition) Given a Boolean function F, there always exists a decomposition for F in the form $F = Maj(F_a, F_b, F_c)$.

Proof: (By construction) We consider, in the context of the proof, to operate at truth table level to build functions F_a, F_b and F_c. The truth table for the original function F has $2^{|supp(F)|}$ rows, each one indicating a distinct input combination and the corresponding logic value assumed by F. Now, we add in the truth table for F, other additional three columns indicating the logic values assumed by F_a, F_b and F_c for the same input combinations considered in F. For each of these input combinations (row of the truth table) we impose that two functions over three (the choice of the couple is free for each row) among F_a, F_b and F_c must assume the same value of F while the remaining one is free to be set to any logic value. In this way, $Maj(F_a, F_b, F_c) = F$ is respected for each row, and therefore for all the possible input combinations. ∎

Thus, as a consequence of the symmetry in the majority operator, there exist many possible majority decomposition structures for a function F. In order to reduce the search space for majority decompositios, it is useful to first determine one of the three functions and then apply majority construction methods for the remaining two.

The choice of the first function, F_a, is a crucial point that determines the quality of the resulting majority decomposition. BDDs offer the opportunity to efficiently identify candidates for F_a through the use of a special class of nodes, *majority dominators*, defined in the next subsection.

Our majority decomposition method is presented in Algorithm 1. First, candidates for the function F_a are searched on the BDD (α)

Algorithm 1 Majority decomposition method

INPUT: Boolean function F
OUTPUT: $F = Maj(F_a, F_b, F_c)$
FUNCTION: MajDecomp(F)

Build a BDD for F
m-dom-list ←Search for *non-trivial* m-dominators   (α)
for all nodes v in m-dom-list **do**
   F_a is the function rooted at node v (α)
   $F_b = ITE(F_a \oplus F, F, F_{F_a})$ (β)
   $F_c = ITE(F_a \oplus F, F, F_{F_a'})$ (β)
   while (improvement)&&(iterations<limit) **do**
      for all couples (X, Y) among F_a,F_b and F_c **do**
         $F_x = X \oplus Y$ (γ)
         XOR-decompose F_x in M and K (balance)   (γ)
         $X_{opt} = ITE(F_x, K, X)$ (γ)
         $Y_{opt} = ITE(F_x, M, Y)$ (γ)
      end for
      evaluate improvement
      increase iteration count
   end while
   if isbest(current decomposition)==1 **then**
      best decomposition ← current decomposition   (ω)
   else
      keep previous best decomposition   (ω)
   end if
end for

---

using majority dominators. Then, an initial solution for the majority decomposition $F = Maj(F_a, F_b, F_c)$ is determined (β). Successively, the initial solution is optimized via an iterative procedure (γ). Finally, the best decomposition over all the F_a candidates is selected (ω). The four major phases of Algorithm 1 are detailed in the following subsections.

B. Majority Dominator on a BDD

BDDs allow us to identify advantageous logic decompositions. Dominator nodes have been reported in [10] to support AND, OR, XOR, XNOR and MUX decompositions. Similarly to this approach, we search for nodes whose characteristics lead to an efficient majority decomposition, (α)-phase in Algorithm 1. We call such nodes m-dominators. A m-dominator node is the root for the candidate function F_a. Following Theorem 3.1, every node in the BDD can be a valid m-dominator but some of them lead to a non-advantageous decomposition. For this reason, we introduce some characteristics to select *non-trivial* m-dominators leading to potential advantageous majority decompositions. We refer to a *non-trivial* m-dominator as to an internal BDD node that:

(i) it is not a simple x-, 0- or 1- dominator.

(ii) has more than one non complemented 0-incoming edges and 1-incoming edges,

Condition (i) avoids simple x-, 0- and 1- dominators [10] since they uniquely indicate XNOR, OR and AND disjoint decompositions. The intuition behind condition (ii) is that the function F_a in $Maj(F_a, F_b, F_c)$ must be reached for all the input combinations corresponding to $Maj(F_a, 0, 1)$ and $Maj(F_a, 1, 0)$ and therefore is most likely a highly connected node (high fan-in) in the BDD. Fig. 1 depicts the BDD for a simple $F = ab + bc + ac$ and highlights its non-trivial m-dominator.

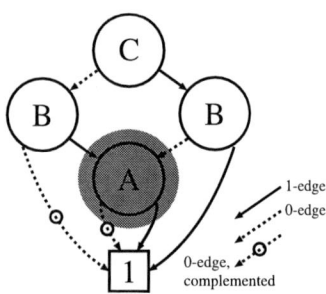

Fig. 1: BDD for the function $ab + bc + ac$, *non-trivial* m-dominator highlighted in red.

C. Majority Decomposition Construction

Once a candidate F_a is identified, the majority decomposition is constructed computing two F_b and F_c functions, (β)-phase in Algorithm 1. These two functions, F_b and F_c, must respect $F = Maj(F_a, F_b, F_c)$ for all the possible input combinations. In particular, for the input combinations such that F and F_a have different values, the functions F_b and F_c must assume the same logic value as F to guarantee $Maj(F_a, F_b, F_c) = F$. On the other hand, for the remaining input combinations, F and F_a have the same value and only one function between F_b and F_c must assume the same value of F to guarantee $Maj(F_a, F_b, F_c) = F$, the remaining one is free to assume any logic value. This concept is formalized in the following theorem.

Theorem 3.2: (Majority logic construction given F_a) Given a function F and a candidate function F_a, a majority decomposition $F = Maj(F_a, F_b, F_c)$ is valid *if and only if*:

$$\begin{cases} F_b = ITE(F_a \oplus F, F, H) \\ F_c = ITE(F_a \oplus F, F, W) \end{cases} \quad (1)$$

where ITE is the *if-then-else* logic operator and H, W are logic functions satisfying the equation:

$$(H \overline{\oplus} F) + (W \overline{\oplus} F) = 1 \quad (2)$$

Proof: Let S be the set of all the possible input combinations for F. Let $S_{(F \neq F_a)}$ be the subset of S such that $F_a \neq F$ (i.e., $F_a = \overline{F}$). Let $S_{(F=F_a)}$ be the subset of S such that $F_a = F$. Note that $S_{(F=F_a)} \cap S_{(F \neq F_a)} = \emptyset$ and $S_{(F=F_a)} \cup S_{(F \neq F_a)} = S$. We need to prove the theorem valid for all the input combinations (S). To this end, we divide the rest of the proof in two parts, first for the input combinations in $S_{(F \neq F_a)}$ and successively for $S_{(F=F_a)}$.

(i) Input combinations from $S_{(F \neq F_a)}$. For these inputs, the ITE operators in Equation 1 always return the *then* part, since $F \oplus F_a$ is always true when $F \neq F_a$. Consequently, $F_b = F_c = F$ and $Maj(F_a, F_b, F_c) = Maj(\overline{F}, F, F) = F$ which is the only valid majority decomposition for $S_{(F \neq F_a)}$ input set. Indeed, note that if F_b or F_c assume any other value than F, the decomposition $Maj(F_a, F_b, F_c)$ will be equal to $\overline{F}$ and not anymore to F.

(ii) Input combinations from $S_{(F=F_a)}$. For these inputs, the ITE operators in Equation 1 always return the *else* part. Therefore, $F_b = H$, $F_c = W$ and $Maj(F_a, F_b, F_c) = Maj(F, H, W)$. Since Equation 2 imposes that, for every input combinations, at least one function between H and W is equal to F, the majority decomposition always have at least two terms equal to F, which is a sufficient condition to say $Maj(F, H, W) = F$. To show that this is the only valid decomposition, consider to use the complement of Equation 2. In this case, both H and W are not equal to F but to its complement $\overline{F}$, therefore the decomposition $Maj(F, H, W) = Maj(F, \overline{F}, \overline{F})$ will be equal to $\overline{F}$ and not anymore to F. ∎

Following Theorem 3.2, the choice of H and W functions respecting Equation 2 is the only freedom left to design a majority decomposition for F, given F_a.

Consequently, the choice of H and W is a key point to obtain minimum-sized F_b and F_c functions. A trivial solution is $H = F$ and $W = "Don't\ Care"$. Obviously, this is an inefficient solution since F_b reduces to the original F itself.

Unfortunately, exact-methods to find the best H and W optimizing a given metric and respecting Equation 2 are intractable. For this reason, we propose to use as *initial seed* the two following functions:

$$\begin{cases} H = F_{F_a} \\ W = F_{F_a'} \end{cases} \quad (3)$$

where the expression X_Y stands for the generalized cofactor of function X w.r.t. function Y. The generalized cofactor can be efficiently computed using BDD algorithms such as *restrict* [17] and *constraint* [18]. We prove that the proposed *initial seed* lead to a valid majority decomposition in the following theorem.

Theorem 3.3: (H and W initial seed) The H and W functions of Equation 3 respect the condition in Equation 2.

Proof: $(H \overline{\oplus} F) + (W \overline{\oplus} F)$ condition from Equation 2 reduces to $(F_{F_a} \overline{\oplus} F) + (F_{F_a'} \overline{\oplus} F)$. Expanding F into $F_{F_a} F_a + F_{F_a'} F_a'$ inside the formula, we get $F_{F_a} F_{F_a'} + F_{F_a}' + F_{F_a'}' + F_{F_a} F_a' + F_{F_a'} F_a$ that can be further simplified in $F_{F_a} F_{F_a'} + (F_{F_a} F_{F_a'})' + F_{F_a} F_a' + F_{F_a'} F_a$ which is indeed a tautology. Equation 2 is therefore respected. ∎

Using the initial seeds for H and W functions, a *non-trivial* majority decomposition can be constructed starting from the original function F and the candidate F_a function.

Example (Majority decomposition construction): $F = ab + bc + ac$. F_a is a as highlighted by Fig. 1. $H = F_{F_a} = b + c$. $W = F_{F_a'} = bc$. Applying the ITE operator we get $F_b = b + c$ and $F_c = bc$. $Maj(F_a, F_b, F_c) = Maj(a, b + c, bc) = ab + bc + ac$ is valid.

As evidenced by the previous example, the H and W formula in Equation 3 may not highlight the most convenient F_b and F_c functions. In order to further optimize the couple of functions (F_b, F_c), but also (F_a, F_c) and (F_a, F_b), we propose a cyclic optimization procedure.

D. Majority Decomposition Optimization

Given a majority decomposition $F = Maj(F_a, F_b, F_c)$, it is possible to minimize F_a, F_b and F_c, while maintaining the decomposition validity. This is done during (γ)-phase in Algorithm 1, exploiting the majority operator functionality. Indeed, for each possible input combination, if a pair of functions (X,Y) among $(F_a, F_b$ and $F_c)$ assume the same value, this is the output value of the majority operator, while if the logic values of (X,Y) are opposite, the majority operator is uniquely determined by the remaining function. In the latter case, the actual values of (X,Y) are not important, it is only needed that X≠Y. This opens up the possibility to restructure and balance the pair of functions (X,Y) in order to reduce the complexity of the current majority decomposition. The majority balancing concept is illustrated by Fig. 2 and formalized in the following theorem.

Theorem 3.4: (Majority decomposition balancing) Given a majority decomposition $F = Maj(F_a, F_b, F_c)$, any pair of functions from $(F_a, F_b$ and $F_c)$, say for example (F_b, F_c), can be restructured as:

$$\begin{cases} F_{b-res} = ITE(F_b \oplus F_c, K, F_b) \\ F_{c-res} = ITE(F_b \oplus F_c, M, F_c) \end{cases} \quad (4)$$

where K and M are logic functions satisfying the equation:

$$(M \oplus K) = (F_b \oplus F_c) \quad (5)$$

330

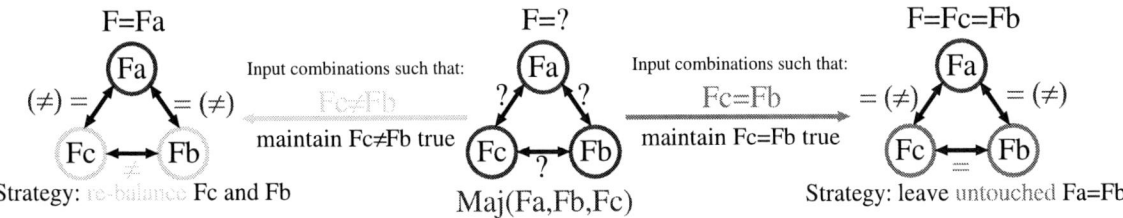

Fig. 2: Majority decomposition balancing. For the input combinations such that $F_b = F_c$, the original function becomes $F = F_b = F_c$ and the values of F_b, F_c cannot be touched, without involving F_a. Instead, for the input combinations such that $F_b \neq F_c$, the original function becomes $F = F_a$ and the actual values of F_b, F_c do not matter provided that $F_b \neq F_c$ remains valid. It is then possible to restructure F_b, F_c.

Proof: Similarly to the proof of Theorem 3.2, we define S as the set of all the possible input combinations for F, $S_{(F_b = F_c)}$ the subset of S such that $F_b = F_c$ and $S_{(F_b \neq F_c)}$ the subset of S such that $F_b \neq F_c$. We divide the proof in two parts, first for the input combinations in $S_{(F_b = F_c)}$ and successively for $S_{(F_b \neq F_c)}$.

(i) Input combinations from $S_{(F_b = F_c)}$. For these inputs, the ITE operators in Equation 4 always return the *else* part. Consequently, $F_{b-res} = F_b$ and $F_{c-res} = F_c$ and $Maj(F_a, F_b, F_c)$ remains valid.

(ii) Input combinations from $S_{(F_b \neq F_c)}$. For these inputs, the ITE operators in Equation 4 always return the *then* part. Consequently, $F_{b-res} = K$ and $F_{c-res} = M$ and $Maj(F_a, F_b, F_c) = Maj(F_a, K, M)$. Note that for the considered inputs in $S_{(F_b \neq F_c)}$, $Maj(F_a, F_b, F_c) = F_a$ since F_b and F_c assumes opposite logic values. Indeed, $Maj(F_a, K, M) = Maj(F_a, K, \overline{K})$ due to Equation 5 evaluated for the same inputs in $S_{(F_b \neq F_c)}$. Finally, $Maj(F_a, K, M) = Maj(F_a, K, \overline{K}) = F_a$ remains a valid majority decomposition. ∎

The effectiveness of the decomposition balancing operation depends on the way K and M functions are chosen. BDD-based XOR decomposition methods in [10] offer an efficient opportunity to compute balanced M and K functions starting from $(F_b \oplus F_c)$, and therefore respecting Equation 5. In Algorithm 1, we employ such core techniques to obtain M and K functions. Then, the decomposition balancing/optimization operation is iterated over all the possible functions pairs and till there is a complexity reduction, or the maximum number of iterations is reached.

We present the majority optimization method applied to the previous example.

Example (Majority decomposition balancing): $F = ab + bc + ac$. $F_a = a$, $F_b = b + c$ and $F_c = bc$ from the previous example. $(F_b \oplus F_c) = ((b + c) \oplus (bc)) = (b \oplus c)$. The XOR-decomposition of $(b \oplus c)$ leads to $K = b$, $M = c$. The ITE operators of Equation 4 finally achieve $F_b = b$ and $F_c = c$. No more optimization is needed: $F_a = a$, $F_b = b$ and $F_c = c$, $Maj(a, b, c) = ab + bc + ac$.

E. Majority Decomposition Selection

Algorithm 1 produces a number of different majority decompositions $F = Maj(F_a, F_b, F_c)$ equal to the number of *non-trivial m*-dominators, corresponding to F_a candidates, in the BDD for F. In order to evaluate the best majority decomposition among the others, ($\boldsymbol{\omega}$)-phase in Algorithm 1, some metric is needed. We use as a first metric the size ($|F|$) of the decomposed functions: a majority decomposition 1 is superior to another majority decomposition 2 if $(|F_{a1}| + |F_{b1}| + |F_{c1}|) < (|F_{a2}| + |F_{b2}| + |F_{c2}|)$. However, this condition alone does not permit to evaluate the specific balance between each decomposed function. We employ as additional condition taking in account this property $(k|F_{a1}| \leq |F_{a2}|)\&(k|F_{b1}| \leq |F_{b2}|)\&(k|F_{c1}| \leq |F_{c2}|)$, where $\&$ stands for the logical AND operator and k is a sizing factor determined heuristically.

F. Majority Decomposition Computational Complexity

In order to estimate the computational complexity of the proposed decomposition method, we denote by N the number of nodes in the BDD for the original function F. We consider in this paper a fully BDD-based implementation of the majority decomposition method. Thanks to the efficiency of BDDs manipulation algorithms and to the BDDs representation canonicity [8], the ITE and 2-operands Boolean operators are always guaranteed to produce minimized BDDs, for a given variable order, despite having redundancy in the inputs. The most expensive operation in Algorithm 1 is the ITE operator that, as employed there, has a computational complexity of $O(N^3)$ [19]. Instead, any Boolean operator of 2 arguments [8] has a computational complexity of $O(N^2)$. The cyclic majority optimization loop is limited to take a prefixed maximum number of iterations. The number of *non-trivial m*-dominators is in general $O(N)$ but can be adjusted on the fly specifying tighter selection constraints about the fan-in of m-dominators. The overall majority decomposition algorithm has $O(N^4)$ computational complexity. Note that, in practice, the runtime is much less than $O(N^4)$. Indeed, (i) the ITE and *2-operand* Boolean operators have a typical runtime performance close to the size of the resulting function ($|F|$) [19] and (ii) the number of *non-trivial m*-dominators can be made remarkably small with tight selection constraints.

IV. BDS-MAJ SYSTEM IMPLEMENTATION

This section introduces a complete logic optimization system, BDS-MAJ. BDS-MAJ integrates the majority decomposition method proposed in Section III with the *BDD Decomposition System* (BDS) presented in [10], [11]. The synthesis flow for BDS-MAJ is shown in Fig. 3. The *network partitioning* and *factoring trees optimization* phases are maintained the same as in BDS. We refer the reader to [10] for a detailed description of these phases. The BDD-based decomposition engine from [10] is here adapted to support MAJ decomposition in addition to XOR, AND, OR and MUX decompositions that are currently supported in BDS [10], [11]. A concise description of the network partitioning phase is provided in the next subsection. Then, details for the BDD-decomposition engine are given. Finally, factoring trees optimization is briefly reviewed.

A. Network Partitioning

Since the manipulation of a global BDD may be impractical for large logic circuits [20], in BDS [10] a preprocessing of the input Boolean network is proposed. It consists of a partial collapsing of the input circuit into a set of supernodes. Each of these super nodes is then efficiently represented as a local BDD. The actual partial collapsing method is implemented in [10] using an evolution of the *eliminate* procedure described in [21]. In BDS-MAJ, we maintain untouched the network partitioning phase from BDS [10].

B. BDD Decomposition Engine

The BDD-decomposition engine in [10] takes in input each BDD produced by the network partitioning phase. As a first step, it performs variable reordering to compact the size of the input BDD. Then,

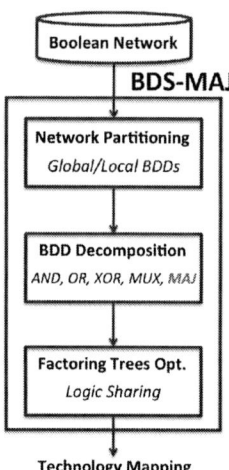

Fig. 3: BDS-MAJ synthesis flow and its main phases.

it starts a search for efficient BDD decompositions. Dominator nodes are used to guide the decomposition process. Simple dominator nodes (0-, 1- and x- dominators) are first considered since they indicate advantageous disjoint decompositions. If no simple dominator is found, the search continues for general dominators that enable non-disjoint decompositions. As a last resort, if no dominator nodes are found, the BDD is decomposed by cofactoring with respect to the top variable (MUX).

We embed our majority decomposition method on the top of the dominator nodes search. Even though that the proposed method obtains general non-disjoint decomposition, a complex radix-3 decomposition (MAJ) is potentially much more advantageous than the traditional radix-2 decompositions (XOR, AND, OR).

In BDS-MAJ, if no m-dominators are found or if the obtained majority decomposition is considered not advantageous for the global decomposition, the standard dominator nodes search in [10] is continued. In order to evaluate if the majority decomposition is useful, we use a similar metric to the one defined in Section III-E where the right hand of the equations is substituted with the size of the original BDD to be decomposed. We distinguish this metric as *global* majority selection while we refer to the one described in Section III-E as *local* majority selection. For the *global* majority selection, the sizing factor k is set to 1.6 by extensive simulations. In a similar way, for the *local* majority selection, the sizing factor k is set to 1.5. Note that the number of iterations in the majority decomposition cyclic optimization is set to 5.

C. Factoring Trees Optimization

In BDS [10], the result of the decomposition is stored in a *factoring tree*. Logic sharing between *factoring trees* is applied in order to further optimize the synthesis result. The bottom up construction of factoring trees (corresponding to the top-down decomposition of BDDs) enables efficient on-line logic sharing detection during the decomposition process. Moreover, the canonicity of BDDs simplifies the actual sharing detection task. In BDS-MAJ, the *factoring trees* optimization procedure of [10] is maintained.

V. EXPERIMENTAL RESULTS

In this section, we evaluate the advantage of the proposed majority logic decomposition method at both logic optimization and logic synthesis levels. First, BDS-MAJ is employed to decompose large logic functions, comprising random control and datapath logic, taken both from the MCNC suite and *ad hoc* large HDL descriptions. Then, the decomposed circuits are mapped onto a simple standard cell library, characterized for CMOS 22nm technology node [22]. Logic optimization and synthesis based on BDS-MAJ are compared to academic BDS-PGA [11], ABC [16], and commercial Synopsys *Design Compiler* (DC) tools, fed with the same standard cell library.

A. Logic Optimization

We present here experimental methods and results for logic optimization performed by BDS-MAJ.

1) Methods: BDS-MAJ is compared to BDS-PGA [11] in terms of node count in the decomposed network. The benchmarks are taken both from the MCNC suite and custom HDL descriptions. HDL descriptions are converted in blif format using a HDL-to-blif translator. Default execution options are used for BDS-PGA and maintained in BDS-MAJ. Local and global majority selection sizing factors are kept the same as in Section IV.

2) Results: Table I summarizes experimental results for logic circuit decomposition. The average node count of BDS-MAJ is 29.1% smaller than BDS-PGA, highlighting the superior decomposition power enabled by the use of majority logic. Majority logic nodes account for 9.8% of the total node count in BDS-MAJ. This result evidences that even a small fraction of majority nodes advantageously restructures the logic function producing consistently more compact logic circuits compared to ordinary techniques. The runtime of BDS-MAJ is almost the same as the one of BDS-PGA, only a slight 4.6% average runtime increase is reported.

B. Logic Synthesis

Experimental methods and results for BDS-MAJ based logic synthesis are presented hereafter.

1) Methods: We evaluate the advantage of BDS-MAJ employed in a traditional *optimization-mapping* synthesis flow. To this end, a standard cell library consisting of MAJ-3, XOR-2, XNOR-2, NAND-2, NOR-2 and INV logic gates is characterized for CMOS 22nm technology [22]. Technology mapping after BDS-MAJ logic optimization is performed in two steps. First, MAJ, XOR and XNOR nodes are directly assigned to logic cells in order to preserve such highlighted functions, otherwise potentially hidden by standard technology mappers. Then, the rest of the logic circuit is mapped using ABC [16] mapper. BDS-MAJ synthesis flow is compared to academic ABC, BDS and commercial Synopsys *Design Compiler* (DC) synthesis tools. Defaults and options for ABC, BDS and DC flows are:

- ABC: ABC *resyn2* optimization script and ABC mapper.
- BDS: BDS logic optimization and ABC mapper.
- DC: Synopsys Design Compiler *compile -area effort high*.

2) Results: Table II summarizes experimental results for logic synthesis using BDS-MAJ. The average area of circuits synthesized by BDS-MAJ is 26.4% and 28.8% smaller than BDS and ABC, respectively. The average delay is 20.9% and 12.8% smaller than BDS and ABC, respectively. Considering the commercial DC flow, BDS-MAJ produces logic circuits that have on average 6.0% less area and 7.8% smaller delay.

3) Discussion: BDS-MAJ based logic synthesis exhibits promising results. The advantage enabled by the majority decomposition method leads to faster and smaller circuits compared to state-of-art synthesis tools. Indeed, the majority decomposition is a radix-3 decomposition that naturally leads to more compact circuits compared to traditional radix-2 decomposition structures. The efficient runtime of the decomposition techniques employed in BDS-MAJ highlight the interest of its use in logic synthesis for real-life applications. On a standard workstation (2.2 GHz Intel dual-core processors and 4 GB of RAM), BDS-MAJ took, on average, only 1.4 ms per gate count of the final circuit, to run the optimization procedure.

VI. CONCLUSIONS

We presented in this paper the first BDD-based majority logic decomposition technique enabling unprecedented logic synthesis opportunities for both datapath and random control logic. We integrated the proposed majority decomposition method with the state-of-art BDD-based decomposition engine, BDS-PGA, to form a complete

TABLE I: Decomposition Results: BDS-MAJ *vs.* BDS-PGA

	BDS-MAJ							BDS-PGA						
	Node number						Seconds	Node number						Seconds
Benchmarks	AND	OR	XOR	XNOR	MAJ	Total	Runtime	AND	OR	XOR	XNOR	MAJ	Total	Runtime
MCNC Benchmarks														
alu2	45	99	4	10	13	171	0.9	71	129	7	13	0	220	0.4
C6288	369	378	66	320	139	1272	0.6	711	764	65	355	0	1895	0.6
C1355	14	44	14	80	31	183	0.1	46	26	46	66	0	184	0.3
dalu	126	408	80	21	133	768	1.4	463	895	25	62	0	1445	2.3
apex6	253	289	9	10	16	577	0.4	243	437	7	7	0	694	0.3
vda	65	203	0	0	22	290	0.2	24	392	0	0	0	416	0.3
f51m	18	24	1	10	4	57	0.1	26	41	1	7	0	75	0.1
misex3	337	704	0	1	21	1063	1.0	377	860	2	2	0	1241	0.9
seq	331	1175	0	0	55	1561	6.7	1159	1471	1	2	0	2633	5.6
bigkey	400	1494	64	87	194	2239	2.8	1058	1834	4	31	0	2927	4.0
HDL Benchmarks														
SQRT 32 bit	162	289	60	158	142	811	0.5	254	471	74	132	0	931	0.4
Wallace 16 bit	208	189	178	302	158	1035	0.6	491	785	169	259	0	1704	0.4
CLA 64 bit	179	208	41	53	167	648	0.1	320	481	35	47	0	883	0.2
Rev (1/X) 19 bit	1223	2109	401	1265	599	5597	13.4	2263	4199	383	1121	0	7966	11.2
Div 18 bit	705	1598	255	422	188	3168	7.1	1290	2918	136	308	0	4652	6.4
MAC 16 bit	322	487	177	541	160	1687	0.5	532	891	187	365	0	1975	1.4
4-Op ADD 16 bit	30	32	10	86	52	210	0.1	87	89	9	85	0	270	0.1
Average	281.6	572.5	80.0	198.0	123.0	**1255.1**	**2.1**	553.8	981.3	67.7	168.3	0.0	1771.2	2.0

TABLE II: Logic Synthesis, CMOS 22nm Technology Node

	BDS-MAJ			BDS-PGA			ABC			Design Compiler		
Benchmark	A. (μm^2)	G.C.	D. (ns)	A. (μm^2)	G.C.	D. (ns)	A. (μm^2)	G.C.	D. (ns)	A. (μm^2)	G.C.	D. (ns)
MCNC Benchmarks												
alu2	34.16	238	0.34	40.81	295	0.40	66.50	503	0.41	50.54	373	0.57
C6288	348.78	1422	0.98	360.78	1441	1.11	355.18	1350	1.08	355.11	1453	1.26
C1355	55.23	188	0.30	56.42	200	0.33	60.69	213	0.29	55.44	190	0.31
dalu	111.30	825	0.40	244.09	1731	0.47	171.36	1292	0.44	103.74	743	0.41
apex6	94.85	811	0.25	106.40	813	0.30	100.73	733	0.26	96.04	745	0.31
vda	71.26	567	0.24	114.24	893	0.20	133.56	1035	0.20	70.98	564	0.25
f51m	13.23	78	0.15	13.86	88	0.19	26.18	199	0.17	17.85	135	0.22
misex3	186.90	1440	0.30	236.25	1825	0.28	225.12	1753	0.28	185.01	1424	0.36
seq	266.35	2086	0.33	541.17	4167	0.27	488.32	3678	0.26	304.15	2325	0.30
bigkey	428.29	3512	0.24	528.22	4121	0.22	713.79	5692	0.22	434.49	3526	0.22
HDL Benchmarks												
SQRT 32 bit	205.22	920	3.22	236.81	1029	4.17	226.31	1058	3.66	211.40	990	3.44
Wallace 16 bit	291.89	1455	0.65	385.49	1995	0.88	413.56	2118	0.77	319.41	1541	0.69
CLA 64 bit	145.32	1455	0.65	170.17	1160	1.08	181.44	1126	0.76	161.07	1114	0.67
Rev (1/X) 19 bit	1044.26	5339	3.09	1506.96	7425	4.56	1545.67	8175	4.26	1160.60	5432	3.14
Div 18 bit	702.03	4255	8.54	957.53	6403	10.24	931.35	6302	9.52	734.02	4948	9.22
MAC 16 bit	365.22	1492	0.67	449.33	2150	0.95	491.12	2560	0.72	383.67	1431	0.70
4-Op ADD 16 bit	59.93	171	0.40	65.17	221	0.51	86.18	391	0.50	63.63	201	0.44
Average	**260.25**	**1510.41**	**1.22**	353.75	2115.12	1.54	365.71	2245.76	1.40	276.89	1596.18	1.32

logic decomposition tool, BDS-MAJ. BDS-MAJ produces decomposed circuits having 29.1% less nodes on average compared to BDS-PGA. This advantage traduces to smaller and faster logic circuits when BDS-MAJ is employed in a traditional *optimization-mapping* synthesis flow. Experimental results show that BDS-MAJ produces on average CMOS circuits having 28.8% and 26.4% smaller area and, at the same time, 12.8% and 20.9% smaller delay with respect to academic ABC and BDS synthesis tools. Compared to commercial Synopsys *Design Compiler*, BDS-MAJ produces on average circuits with 6.0% less area and 7.8% smaller delay.

ACKNOWLEDGEMENTS

This research was supported by ERC-2009-AdG-246810.

REFERENCES

[1] G. De Micheli, *Synthesis and Optimization of Digital Circuits*, McGraw-Hill, New York, 1994.

[2] R.L. Rudell, A. Sangiovanni-Vincentelli, *Multiple-valued minimization for PLA optimization*, IEEE Trans. CAD, Vol. 6, Iss. 5, pp. 727-750, 1987.

[3] R.K. Brayton, *et al.*, *MIS: A Multiple-Level Logic Optimization System*, IEEE Trans. CAD, vol. 6, pp. 1062-1081, Nov.1987.

[4] E. Sentovich, *et al.*, *SIS: A System for Sequential Circuit Synthesis*, ERL, Dept. EECS, Univ. California, Berkeley, UCB/ERL M92/41, 1992.

[5] R.K. Brayton, C. Mc Mullen, *The Decomposition and Factorization of boolean expressions*, Proc. ISCAS 1982.

[6] C.Y. Lee, *Representation of Switching Circuits by Binary-Decision Programs*, Bell Systems Technical Journal, 1959.

[7] S.B. Akers, *Binary Decision Diagrams*, IEEE Trans. Comp., C-27(6):509-516, June 1978.

[8] R.E. Bryant, *Graph-based algorithms for Boolean function manipulation*, IEEE Trans. Comput., C-35: 677-691, 1986.

[9] V. Bertacco, M. Damiani, *The Disjunctive Decomposition of Logic Functions*, Proc. ICCAD, 1997

[10] C. Yang and M. Ciesielski, *BDS: A BDD-Based Logic Optimization System*, IEEE Trans. CAD, vol. 21, pp. 866-876, July 2002.

[11] N. Vemuri, P. Kalla and R. Tessier, *BDD-based Logic Synthesis for LUT-based FPGAs*, ACM Trans. TODAES, Vol.7, pp. 501-525, Oct. 2002.

[12] T. Bengtsson, A. Martinelli, E. Dubrova *A BDD-Based Fast Heuristic Algorithm for Disjoint Decomposition*, Proc. ASP-DAC 2003.

[13] S. Plaza, V. Bertacco, *STACCATO: Disjoint Support Decompositions from BDDs through Symbolic Kernels*, Proc. ASP-DAC 2005.

[14] K. Karplus, *Using if-then-else DAGs for multi-level logic minimization*, Univ. California, Santa Cruz, UCSC-CRL-88-29, 1988.

[15] Y. Tohma, *Decompositions of Logical Functions Using Majority Decision Elements*, IEEE Trans. Electronic Computers, pp. 698-705, 1964.

[16] ABC Logic Synthesis Tool [Online]. Available: http://www.eecs.berkeley. edu/alanmi/abc/

[17] O. Coudert, J.C. Madre, *A unified framework for the formal verification of sequential circuits*, Proc. ICCAD, 1990

[18] O. Coudert, C. Berthet, J.C. Madre, *Verification of sequential machines using boolean functional vectors*, Proc. International Workshop on Applied Formal Methods for Correct VLSI Design, 1989.

[19] K.S. Brace, R.L. Rudell, R.E. Bryant, *Efficient implementation of a BDD package*, Proc. DAC, 1990.

[20] R.E. Bryant, *On the Complexity of VLSI Implementations and Graph Representations of Boolean Functions with Application to Integer Multiplication*, IEEE Trans. on Computers, vol. 40, no. 2, p. 205, Feb. 1991.

[21] R. Chaudry *et al.*, *Area-oriented synthesis for PTL*, Proc. ICCD 1998.

[22] Predictive Technology Model (PTM), http://ptm.asu.edu/

Towards Optimal Performance-area Trade-off in Adders by Synthesis of Parallel Prefix Structures

Subhendu Roy[‡], Mihir Choudhury[†], Ruchir Puri[†], David Z. Pan[‡]

[‡]Department of Electrical and Computer Engineering, University of Texas at Austin, USA
[†] IBM T. J. Watson Research Center, Yorktown Heights, USA
subhendu@utexas.edu, {choudhury,ruchir}@us.ibm.com, dpan@ece.utexas.edu

ABSTRACT

This paper proposes an efficient algorithm to synthesize prefix graph structures that yield adders with the best performance-area trade-off. For designing a parallel prefix adder of a given bit-width, our approach generates prefix graph structures to optimize an objective function such as size of prefix graph subject to constraints like bit-wise output logic level. Besides having the best performance-area trade-off our approach, unlike existing techniques, can (i) handle more complex constraints such as maximum node fanout or wire-length that impact the performance/area of a design and (ii) generate several feasible solutions that minimize the objective function. Generating several optimal solutions provides the option to choose adder designs that mitigate constraints such as wire congestion or power consumption that are difficult to model as constraints during logic synthesis. Experimental results demonstrate that our approach improves performance by 3% and area by 9% over even a 64-bit full custom designed adder implemented in an industrial high-performance design.

Categories and Subject Descriptors

B.2.m [**Hardware, Arithmetic and Logic Structure**]: Miscellaneous;

General Terms

Algorithms, Design, Performance

Keywords

Logic synthesis, Parallel prefix adder, Bottom-up approach

1. INTRODUCTION

Datapath logic constitutes a significant portion of a general purpose microprocessor and frequently occurs on the timing-critical paths in high-performance designs. Arithmetic components, such as adders, multipliers, shifters are the basic building blocks in datapath logic and hence, to a great extent dictate the performance of the entire chip. Binary addition is one of the most fundamental and widely used arithmetic operations in microprocessors. Today, adders are designed in 2 ways - either manually through full custom design or in an automated manner using synthesis tools. In a custom adder design methodology, a designer has to manually choose between regular adder structures such as Kogge-Stone [1], Sklansky [2], Brent-Kung [3] and tune physical design parameters such as placement, gate sizing, buffer optimization to maximize performance under power constraints for the target technology [4][5]. Hence, custom adder design methodology is expensive, takes a long time to converge to a satisfactory design, and is inflexible to late design changes.

In contrast, automated synthesis approach is productive and flexible to late design changes but traditionally has lagged behind in performance as compared to custom designs. Therefore, the prevalent design approach for high-performance datapath logic continues to be custom design. In the past, several algorithms have been proposed to generate parallel prefix adders targeting minimization of the size of the prefix graph (s) under given bit-width (N) and logic level (L) constraints. Snir [6] has given a theoretical bound of s for $L \geq 2\log_2 N - 2$ with uniform input profile. [7] presents a recursive construction of parallel prefix graphs to obtain a trade-off between size and level, but it could not achieve the bound provided by [6]. Other existing algorithms like a greedy depth-decreasing heuristic [8], dynamic programming based approaches ([9], [10]) or non-heuristic optimization [11] could achieve this bound for some cases but yield a non-optimal result as logic level constraints are reduced (for e.g. to $\log_2 N$) – which is more relevant for high performance adders. The most recent approach [9], that uses dynamic programming (DP) on a restricted search space to generate a seed prefix graph followed by an area-heuristic to further reduce the size of the seed prefix graph, is also the most effective in minimizing the size of the prefix graphs. However, the quality of the area-heuristic solution depends on the selection of seed solution from DP, which is not unique. Furthermore, this algorithm cannot handle fanout/wire-length constraints on nodes in the prefix graph or arrival/required time constraints on individual input/output bits that impact the performance, area, and power consumption of the adder after physical design. In [12], an exhaustive approach is attempted to explore the optimal arithmetic-circuit architectures through selective factorization, but it is very limited in terms of scalability.

To tackle these issues, this paper proposes an efficient al-

Permission to make digital or hard copies of all or part of this work for personal or classroom use is granted without fee provided that copies are not made or distributed for profit or commercial advantage and that copies bear this notice and the full citation on the first page. To copy otherwise, to republish, to post on servers or to redistribute to lists, requires prior specific permission and/or a fee.
DAC '13 May 29 - June 07 2013, Austin, Tx, USA

gorithm to generate prefix graphs for synthesizing adders with the best performance-area trade-off. In this approach, prefix graph structures are constructed in bottom-up fashion by exhaustively generating all possible $n+1$ bit prefix graphs from n bit prefix graphs. For scalability to large adders up to 128 bits, our approach proposes a novel compact data structure for manipulating prefix graphs, efficient memory management techniques like lazy copy for storing several prefix graph solutions, and search space reduction strategies like level-restriction, dynamic size pruning, repeatability pruning for targeting prefix graph structures relevant for achieving the best performance-area trade-off. Compared to existing algorithms our approach has the following advantages:

1. It is more effective than all existing algorithms in minimizing the size of the prefix graph for given bit-width N and bitwise input/output logic level constraints.

2. It provides greater opportunity for improving performance of the adder because the algorithm can handle fanout/wire-length constraints on nodes in the prefix graph and arrival/required time constraints on individual input/output bits.

3. It generates many candidate prefix graph structures for a given set of constraints, which can also be evaluated for placement and wiring congestion to yield efficient physical and routing implementation.

The rest of the paper is organized as follows. Section 2 describes binary addition as a prefix graph problem. Section 3 presents our algorithm for generating prefix graph structures. Section 4 presents the results of this approach with a conclusion in section 5.

2. PRELIMINARIES

Given an ordered n inputs x_0, x_1, ..., x_{n-1} (where x_{n-1} is the MSB and x_0 is the LSB) and an associative operation o, prefix computation of n outputs is defined as follows:

$$y_i = x_i \ o \ x_{i-1} \ o...o \ x_0 \quad \forall i \in [0, n-1] \quad (1)$$

where i-th output depends on all previous inputs x_j ($j \leq i$). A prefix graph of width n is a directed acyclic graph (with n inputs/outputs) whose nodes correspond to the associative operation "o" in the prefix computation and there exists an edge from node v_i to node v_j if v_i is an operand of v_j. Fig. 1 represents a prefix graph for 6 bit. In this example, we can write y_5 as

$$
\begin{aligned}
y_5 = i_1 \ o \ y_3 &= (x_5 \ o \ x_4) \ o \ (i_0 \ o \ y_1) \\
&= (x_5 \ o \ x_4) \ o \ ((x_3 \ o \ x_2) \ o \ (x_1 \ o \ x_0)) \quad (2)
\end{aligned}
$$

Next, we will explain this prefix graph in the context of binary addition.

Binary addition problem is defined as follows: given n bit augend $A = a_{n-1}....a_1 a_0$ and n bit addend $B = b_{n-1}....b_1 b_0$, compute the sum $S = s_{n-1}....s_1 s_0$ and carry out $C_{out} = c_{n-1}$, where $s_i = a_i \oplus b_i \oplus c_{i-1}$ and $c_i = a_i b_i + a_i c_{i-1} + b_i c_{i-1}$.

With bitwise (group) generate function g (G) and propagate function p (P), n bit binary addition can be mapped to a prefix computation problem as follows:

- Pre-processing: Bitwise g, p generation

$$g_i = a_i.b_i \text{ and } p_i = a_i \oplus b_i \quad (3)$$

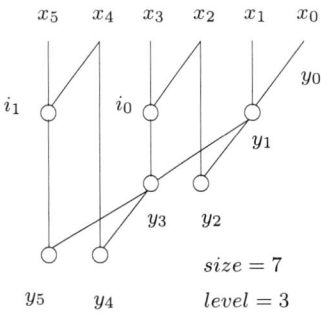

Figure 1: Prefix Graph Representation

- Prefix-processing: The concept of generate/propagate is extended to multiple bits and $G_{[i:j]}$, $P_{[i:j]}$ ($i \geq j$) are defined as

$$P_{[i:j]} = \begin{cases} p_i & \text{if } i = j \\ P_{[i:k]}.P_{[k-1:j]} & otherwise \end{cases}$$

$$G_{[i:j]} = \begin{cases} g_i & \text{if } i = j \\ G_{[i:k]} + P_{[i:k]}.G_{[k-1:j]} & otherwise \end{cases} \quad (4)$$

The computation for (G, P) is expressed in terms of associative operation o as:

$$
\begin{aligned}
(G,P)_{[i:j]} &= (G,P)_{[i:k]} \ o \ (G,P)_{[k-1:j]} \quad (5) \\
&= (G_{[i:k]} + P_{[i:k]}.G_{[k-1:j]}, P_{[i:k]}.P_{[k-1:j]})
\end{aligned}
$$

- Post-processing: Sum generation

$$s_i = p_i \oplus c_{i-1} \text{ and } c_i = G_{[i:0]} \quad (6)$$

Among the three components of binary addition problem, both pre-processing and post-processing parts are fixed structures. However, o being an associative operator, provides the flexibility of grouping the sequence of operations in prefix processing part and executing them in parallel. So the structure of the prefix graph determines the extent of parallelism.

At the technology independent level, size of the prefix graphs (# of prefix nodes) gives the area measure and the logic levels of the nodes estimate roughly the timing. It is important to note that the actual timing depends on other parameters as well like fan-out distribution and size of the prefix graph. Smaller sizes of prefix graph offer better flexibility during post-synthesis gate sizing.

3. OUR APPROACH

This section describes a compact data structure for storing and manipulating a prefix graph, efficient memory management strategies for storing several prefix graph solutions, and pruning strategies to scale our approach up to 128 bit adders. Due to the associative nature of the prefix operation o, each output bit (m) can be constructed by combining the previous input bits 0, 1 ... m in any way keeping their relative orders intact and the number of possible ways is $catalan(m)$, where $catalan(m) = \frac{1}{m+1}\binom{2m}{m}$. Let G_n denotes the set of all possible prefix graphs with bit-width n. Then size of G_n grows exponentially with n and is given by $catalan(n-1) * catalan(n-2) *.... catalan(0)$. For example, $|G_8| = 332972640$, $|G_{12}| = 2.29 * 10^{24}$. As the search space is huge, we require compact data structure, efficient memory management and search space reduction techniques to scale this approach.

3.1 Compact Notation and Data Structure

We represent the prefix graph by a sequence of indices. Each prefix node is represented by an index, which is the most significant bit (MSB) of the node. Fig.2 illustrates the compact notation, where the sequence is determined in topological order, and in addition, precedence is given to higher significant bits in the sequence of indices. For instance, in Fig.2 (right side), indices $\{3,1\}$ and $\{3,2\}$ occur at first and second topological levels respectively. With only topological ordering, 4 possible sequences are possible - 3132, 3123, 1332, 1323. Since 3 is given precedence over 1 and 2 at the first and second topological levels respectively, the only possible sequence here is 3132. On the other hand, we can construct a prefix graph by traversing the sequence of indices from left to right in the following way: for each index i in the sequence, we add a node p which is derived from 2 nodes – the most recent node r with index i (or input bit i) and the node just before p in the sequence (or the input bit $LSB(r) - 1$). For example, in the sequence '3132' in Fig.2, the node for first 3 is constructed from input bits 3 and 2, where as that for second 3 is constructed from the node for first 3 and the node (with index 1) just before it.

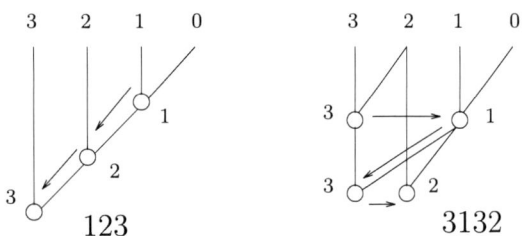

Figure 2: Compact Notation for a prefix graph

Apart from storing the *index*, we also need to track the *LSB*, *level*, *fanout* for each node in the prefix graph. We store all this information using a single integer for each node, and represent a prefix graph by a list/sequence of integers. Since we want to explore adders up to 128 bits and provision a carry-in as the 129th bit, we reserve 8 bits ($\lceil \log_2(129) \rceil$) for *index*, *level*, *fanout* and *LSB*. Thus, all information for a node can be stored in a single integer as shown in Fig.3.

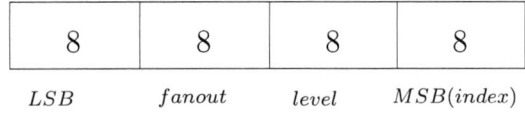

Figure 3: Bit Slicing

This compact data structure helps in reducing memory usage and runtime (due to faster copy/delete operation for a prefix node) as compared to using a structure to store *index*, *LSB*, *level*, and *fanout* as individual integers.

3.2 Exhaustive Bottom-up Enumeration

We start from a prefix graph of 2 bits (represented by a single index sequence '1') and construct the prefix graph structures for higher bits in an inductive way, i.e. given all possible prefix graphs (G_n) for n bit, we construct all possible prefix graphs (G_{n+1}) of $n + 1$ bit. The process of generating such graphs of $n+1$ bit from an element of G_n by inserting n at appropriate positions is a recursive procedure. Fig.4 explains this for an element ('12') of G_3 with the help of a recursion tree.

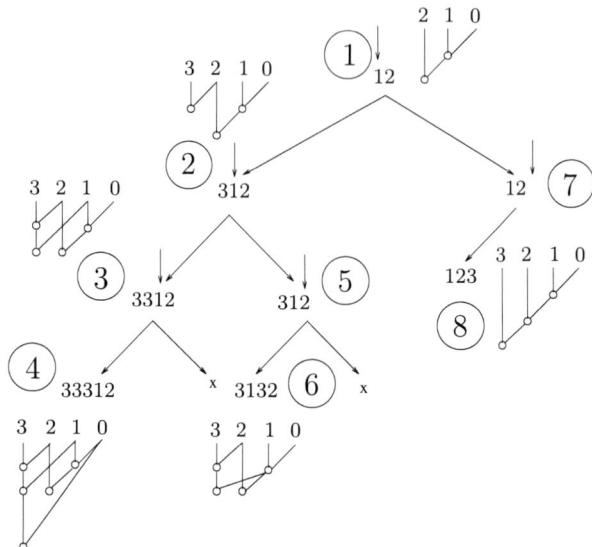

Figure 4: An illustrative example

At the beginning of this recursive procedure (RP), we have a sequence '12' (node 1) with an arrow on '1'. The arrow points to the index before which 3 can be inserted. At any stage, there are two options, either insert 3 and call RP, or move the arrow to a suitable position and then call RP. This position is found by iterating the list/sequence in forward direction until $searchIndex$ $(= LSB(RecentNode(3)) - 1)$ is found, where $RecentNode(i)$ signifies the most recent node with index i in the sequence. The left subtree denotes the first option and the right subtree indicates the second option. So the procedure either inserts '3' at the beginning of '12' and goes to node 2 or it goes to node 7 by moving the arrow to the appropriate position. We can see that, $searchIndex = LSB(RecentNode(3)) - 1 = 3 - 1 = 2$ for this case. Similarly, for node 2, the $searchIndex$ has become $2 - 1 = 1$, and so this procedure either inserts '3' (in node 3) or shifts the pointer after '1' (in node 5). The traversal is done in pre-order and this recursion is continued till $LSB(RecentNode(3))$ becomes '0' or alternatively, a 4 bit prefix graph is constructed. The right subtree of a node is not traversed if a prefix graph for 4 bits has been constructed at the left child of the node. For example, we do not traverse the right subtree of node 3 and node 5.

Algorithm 1 illustrates the steps of this exhaustive enumeration technique. The algorithm preserves the uniqueness of the solutions by inserting the indices at appropriate positions. In the 'buildRecursive' procedure, $nodeList$ is an STL list (*insert* and *erase* operations are thus $O(1)$ operations), $recentNode$ is passed as a parameter which is used to find $searchIndex$ and to track if a solution has been generated. $currIter$ is the iterator corresponding to $\downarrow$ in Fig.4. The return value of the procedure is true, when $nodeList$ is a solution of G_{n+1}, thereby indicating that the right subtree of parent of $nodeList$ does not require traversal.

3.3 Efficient Recursion Implementation

The key step of Algorithm 1 is the recursive procedure as explained in Fig.4. In a pre-order traversal of typical recursion tree implementation, when we move from root node to

Algorithm 1 Exhaustive Bottom-up Enumeration

1: //Given G_n construct G_{n+1}..
2: **for all** $g \in G_n$ **do**
3: buildRecursive(g, $null$, $g.begin$, n);
4: **end for**
5: **Procedure** buildRecursive($nodeList$, $recentNode$, $currIter$, $index$)
6: **if** $recentNode \neq null$ **and** $LSB(recentNode) = 0$ **then**
7: save solution $nodeList$ in G_{n+1};
8:   **return** true;
9: **end if**
10: $searchIndex \leftarrow LSB(recentNode) - 1$;
11: $newIter \leftarrow nodeList.insert(currIter, index)$;
12: $newNode \leftarrow$ value at $newIter$;
13: $flag \leftarrow$ buildRecursive($nodeList$, $newNode$, $currIter$, $index$);
14: **if** $flag = true$ **then**
15:   **return** false;
16: **end if**
17: $nodeList.erase(newIter)$;
18: **repeat**
19: $node \leftarrow$ value at $currIter$;
20: $currIter \leftarrow currIter + 1$;
21: **until** $MSB(node) \neq searchIndex$ **and** $currIter \neq nodeList.end$
22: buildRecursive($nodeList$, $recentNode$, $currIter$, $index$);
23: **end Procedure**

its left subtree, a copy of the root node is stored to traverse the right subtree at later stage. In our approach, we copy the sequence only when we get a valid prefix graph, otherwise keep on modifying the sequence. As for example, we do not store the sequences ('312', '3312') in Fig.4, i.e. when we move to the left subtree of a node in the recursion tree, we insert the index and delete it while coming back to the node in the pre-order traversal, and store only the leaf nodes. This notion of late copy is motivated by a concept in object-oriented-programming, known as lazy copy or copy-on-write [13] which is a combination of deep copy and shallow copy. In lazy-copy, when an object is copied initially, a shallow copy (fast) is used and then deep copy (slow) is performed when it is absolutely necessary (for example, modifying a shared object). Lazy copy helps to significantly reduce run time by replacing list copy and delete operations with list entry insertion and deletion operations at a given position (iterator) which, being an $O(1)$ operation, does not impact the runtime. For the simple example shown in Fig. 4, an implementation without lazy copy needs 5 list copy and 2 list delete operations whereas an implementation with lazy copy only needs 3 list copy operations and no list delete operations. The benefits of lazy copy increase exponentially with bit-width.

3.4 Search Space Reduction

As the size of the solution space of all prefix graphs is huge, it is not feasible to generate all possible prefix graphs. Many prefix graphs are also not relevant because they do not have a good performance-area trade-off. We are interested only in generating candidate solutions to optimize performance (prefix graphs with minimum logic levels) and area (prefix graphs with minimum number of prefix nodes). Hence, the following search space reduction techniques are employed to scale this approach.

Level Pruning: The performance of an adder depends directly on the number of logic levels of the prefix graph. Our approach intends to minimize the number of prefix nodes with given bit-width and logic level (L) constraints. In Algorithm 1, we keep track of the levels of each prefix node and solutions are discarded if the level of the inserted node (or index) becomes greater than L. This work focusses on synthesizing adders with maximum performance and hence, constrains the level at each output bit to the smallest possible value, i.e., bit m is constrained to be at level $\lceil \log_2 m \rceil$.

Dynamic Size Pruning: As discussed in section 3.2, we construct the set G_{n+1} from G_n. While doing this, we prune the solution space based on size (# of prefix nodes) of elements in G_n. Let s_{min} be the size of the minimum sized prefix graph(s) of G_n. Then we prune the solutions (g) for which $size(g) > s_{min} + \Delta$. For example, suppose the sizes of the solutions in $G_n = [9 \quad 10 \quad 11]$ and $\Delta = 2$. To construct G_{n+1}, we select the graphs of G_n in increasing order of sizes and build the elements of G_{n+1}. Let the graphs with sizes $X_1 = [12 \quad 13 \quad 14 \quad 15]$, $X_2 = [11 \quad 14]$ and $X_3 = [13 \quad 16]$ be respectively constructed from the graphs of sizes 9, 10, 11 in G_n. In this case, the minimum size solution is the solution with size 11 and so the sizes of the solutions stored in $G_{n+1} = [[12 \quad 13], [11], [13]]$. This pruning is done to choose the potential elements of G_{n+1}, which can give minimum size solution for the higher bits. The selection of Δ is critical to reduce the search space and we found empirically that $\Delta = 3$ is sufficient to get minimum size solutions for $\log_2 N$ level till 128 bit. But any kind of restriction (like fanout) on the graph structure requires higher Δ to achieve feasible solutions. In that case, we store a fixed number of solutions of G_n for each size s ($s_{min} \leq s \leq s_{min} + \Delta$), which allows higher Δ without increasing memory usage too much.

However, pruning the superfluous solutions after constructing the whole set G_{n+1} can cause peak memory overshoot. So we employ the strategy "Delete as early as possible", i.e. we generate solutions on the basis of current minimum size $s_{min}^{current}$. Let us take the same example to illustrate this. In X_1, $s_{min}^{current} = 12$ and so we do not construct the graph with size 15, as $15 > 12 + 2$. Similarly, when we get the solution with size 11 in X_2, we delete the graph with size 14 from X_1 and do not construct the graph with size 14 in X_2 and 16 in X_3. Indeed, whenever the size of the list/sequence in algorithm 1 exceeds $s_{min}^{current}$ by $\Delta + 1$, the flow is returned from RP. Apart from reducing the peak memory usage, this dynamic pruning of solutions helps in improving run time by reducing copy/delete operations.

Repeatability Pruning: The sequence (in our notation) denoting a prefix graph can have consecutive indices. We denote the maximum number of consecutive indices in a sequence by R. For example, '33312' in Fig.4 has 3 consecutive 3's in the sequence so $R = 3$. We have observed that $R = 1$ does not degrade the solution quality, but significantly reduces the search space at an early stage. For example, in Fig.5, '3132' is a better solution than '33312' both in terms of logic level and size. Algorithm 1 is modified to track repeatability and prune solutions with $R > 1$.

Prefix Structure Restriction: This is a special restriction in prefix graph structure for 2^n bit adders with n logic levels. For example, if we need to construct an 8 bit adder with logic level 3, the only possible way to realize the MSB using the same notation as Eqn.(2) is given by

$$y_7 = ((x_7 \; o \; x_6) \; o \; (x_5 \; o \; x_4)) \; o((x_3 \; o \; x_2) \; o \; (x_1 \; o \; x_0)) \quad (7)$$

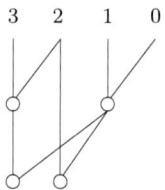 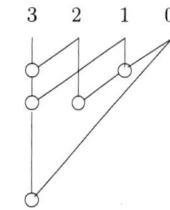

sequence: 3132   sequence: 33312
level = 2, size = 4   level = 3, size = 5

Figure 5: 3132 is better prefix structure than 33312

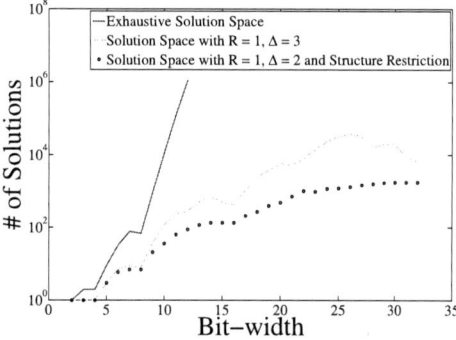

Figure 6: Search Space Reduction for each output bit m at level $\lceil log_2 m \rceil$

So 7 nodes or alternatively $(2^n - 1)$ prefix nodes are fixed for the 2^n bit adder with n level. We impose this restriction in our implementation for generating the sequence of indices, which helps in improving the run time significantly.

Fig.6 plots the number of solutions (each output bit m being at level $\lceil log_2 m \rceil$) with bit-width for 3 cases, first the exhaustive solution space which grows exponentially with bit-width, next the solution space with $\Delta = 3$, $R = 1$ and without any structure restriction and finally that with structure restriction and $\Delta = 2$, $R = 1$. We have observed that the third case is able to generate the same first 1786 minimum size solutions for 32 bit as that of second case which reinforces that the prefix structure restriction can help in achieving same solution quality with less search space exploration, thereby reducing runtime.

4. RESULTS

We have implemented our approach in C++ and executed on a linux machine with 12 GB RAM and 2.8 GHz CPU. First, we present our results at the logic synthesis (technology independent) level. As the dynamic programming based area-heuristic approach presented in [9] has achieved better results compared to the other existing techniques, we have implemented this approach as well to compare with our experimental results. Table 1 presents the comparison of number of prefix nodes for adders with different bit-width (N) with $log_2 N$ logic level constraint for all output bits. In this case, the input profile is uniform, i.e. the arrival times of all input bits are assumed to be the same. Results for non-uniform profile for 32 bit adder are shown in table 2. We can see that our approach outperforms [9] in both cases. The runtime of our approach for generating 128 bit prefix graphs with level constraint of 7 is 25 seconds, which is acceptable for any logic synthesis tool.

As mentioned earlier, the existing approaches ([9], [10], [11] etc) are not flexible in restricting parameters like fan-

Table 1: Prefix Graph size for $log_2 N$ level

Bit-width	Our Approach	Area Heuristic [9]
16	31	31
24	45	46
32	74	74
48	105	106
64	167	169
128	364	375

Table 2: Prefix Graph size for non-uniform input profile in a 32 bit adder

Profile	Our Approach	Area Heuristic [9]
A	55	56
B	55	58
C	56	60
D	54	59
E	53	59
F	55	59
G	53	57

out, which is a critical parameter to optimize post-synthesis design performance. Usually, electrical violations at high-fanout points are mitigated by buffer-insertion and gate-sizing, but at the cost of performance. Hence, for high-performance designs, Kogge-Stone [1] is the most effective adder structure. An important property of this structure is that maximum fan-out (MFO) of a n bit adder is less than $log_2 n$ (without any buffer insertion) and the fan-out for prefix nodes at logic level $log_2 n - 1$ is 2. Table 3 shows that, even with a fan-out restriction of 2 for *all* prefix nodes, the prefix graph generated by our approach has fewer prefix nodes than the prefix graph for a Kogge-Stone adder.

We have integrated our approach to a placement driven synthesis [14] tool and run the tool on the minimum size solutions of 8,16,32,64 bit adders. A cutting-edge technology node is used for technology mapping. We present the various metrics like area, worst negative slack (WNS), wire-length, figure of merit (FOM) after placement in Table 4 for the solution having best WNS. FOM signifies the sum of the total negative slacks at the timing end-points. Both wirelength and area are unitless. Area is reported as the number of icells and wirelength as the number of tracks. An icell has a constant area based on pitch. Our approach is compared against regular adders like Brent-Kung (BK), Kogge-Stone (KS) adders, adders generated by Dynamic Programming (DP) [9], and 64 bit full custom adder (CT). It is to be noted that we have prevented V_{th}-swapping in the placement tool, so leakage power would be proportional to area.

Fig.8 represents the plot of area versus WNS for the solutions provided by our approach along with those provided by other methods. We can draw a pareto curve with the solution points obtained using our approach, which gives the option to select the individual points on the pareto curve based on area/power budget. We see that the solution points of the other methods are above and/or to the right of this curve, which indicates that we can always get some solution on the pareto front, which is better in terms of performance and/or area than each of the other methods. For a 16 bit adder, the total number of pareto-optimal points is 4 and the single point $p1$ provides better solution than DP, KS

Table 3: Comparison with Kogge-Stone Adder

Bit-width	Our Approach (MFO = 2)	Our Approach (MFO = $log_2 N$)	Kogge-Stone
8	14	13	17
16	42	35	49
32	114	89	129
64	290	238	321
128	706	631	769

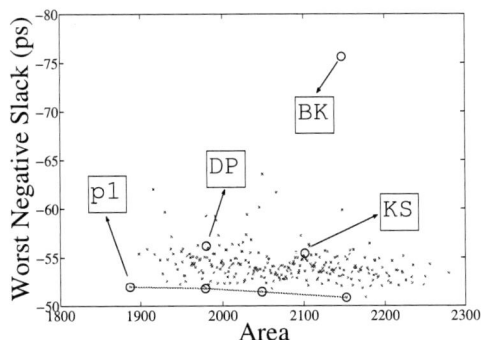 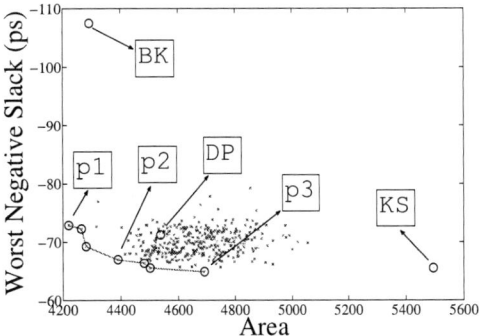

Figure 8: Area vs. Worst Negative Slack plot for 16 and 32 bit adders

Table 4: Post Placement Comparison

n	Method	Area	Worst Slack (ps)	Wire Length	FOM (ps)
8	Brent-Kung	828	-71.7	3996	-527
	Kogge-Stone	1146	-48.9	5889	-391
	Dyn. Prog.	853	-47.4	3761	-371
	Our Approach	871	-43.4	3804	-351
16	Brent-Kung	2147	-75.7	12712	-1156
	Kogge-Stone	2101	-55.5	13604	-878
	Dyn. Prog.	1980	-56.2	9776	-852
	Our Approach	2152	-50.7	11102	-812
32	Brent-Kung	4292	-107.5	26397	-3072
	Kogge-Stone	5495	-65.5	39474	-2082
	Dyn. Prog.	4538	-71.3	25784	-2096
	Our Approach	4692	-64.9	24683	-2074
64	Brent-Kung	9832	-120.3	59402	-6931
	Kogge-Stone	13389	-84.5	120600	-5181
	Dyn. Prog.	10718	-88.9	66249	-5334
	Custom	10905	-89.1	71054	-5709
	Our Approach	10048	-83.8	60450	-5230

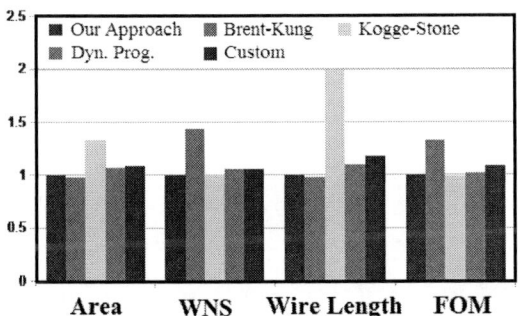

Figure 7: 64 bit adder after placement.

and BK. For a 32 bit adder, the points $p1$, $p2$, $p3$ are better solutions than BK, DP, KS respectively.

Fig.7 compares these metrics for single solution (with best WNS) of 64 bit adder with other approaches. Our approach improves performance by 19% with 2% higher area over a Brent-Kung adder, improves performance and area by 0.4% and 33%, respectively over a Kogge-Stone adder, improves performance and area by 3% and 6.7%, respectively over Dynamic Programming [9], and improves performance and area by 3.2% and 8.5% over a full custom adder design. Note that the performance improvement was computed based on the actual critical path delay value and not the worst negative slack. Our approach also improves wire-length and FOM over both Kogge-Stone and full custom adder design.

5. CONCLUSION AND FUTURE WORK

In this paper, a highly efficient parallel prefix graph generation driven high performance adder synthesis technique is presented. The complexity of parallel prefix graph generation problem for adders is exponential in the number of bits. We presented efficient pruning strategies and implementation techniques to scale this approach up to 128 bit adders. The results, both at the technology-independent level and after physical synthesis (post placement) show that this approach significantly improves over existing techniques by yielding better quality of results in terms of both timing and wire length for high performance adders in state of the art microprocessor designs. The proposed approach improves over even the manually designed custom adders yielding, up to 3% better delay and 9% better area. As our approach can generate multiple prefix graph structures for given constraints, it provides a framework for further exploration to identify structures that can account for practical design issues like wire congestion and power consumption.

6. REFERENCES

[1] P. M. Kogge and H. S. Stone. A parallel algorithm for the efficient solution of a general class of recurrence equations. *IEEE Trans. Computers*, pages 786–793, 1973.

[2] J. Sklansky. Conditional sum addition logic. *IRE Trans. on Electronic Computers*, pages 226–231, 1960.

[3] R. P. Brent and H. T. Kung. A regular layout for parallel adders. *IEEE Trans. Computers*, pages 260–264, 1982.

[4] C. Zhou et al. 64-bit prefix adders: Power-efficient topologies and design solutions. *Custom Integrated Circuit Conference*, pages 179–182, 2009.

[5] J. Liu et al. Optimum prefix adders in a comprehensive area, timing and power design space. *ASPDAC*, 2007.

[6] M. Snir. Depth-size trade-offs for parallel prefix computation. *Journal of Algorithms*, pages 185–201, 1986.

[7] R. E. Ladner and M. J. Fischer. Parallel prefix computation. *Journal of ACM*, pages 831–838, 1980.

[8] J. P. Fishburn. A depth decreasing heuristic for combinational logic; or how to convert a ripple-carry adder into a carry-lookahead adder or antything in-between. *DAC*, pages 361–364, 1990.

[9] T. Matsunaga and Y. Matsunaga. Area minimization algorithm for parallel prefix adders under bitwise delay constraints. *Great Lakes Symposium on VLSI*, pages 435–440, 2007.

[10] J. Liu et al. An algorithmic approach for generic parallel adders. *International Conference on Computer Aided Design*, pages 734–740, 2003.

[11] R. Zimmermann. Non-heuristic optimization and synthesis of parallel prefix adders. *International Workshop on Logic and Architecture Synthesis*, pages 123–132, 1996.

[12] A. K. Verma and P. Ienne. Towards the automatic exploration of arithmetic-circuit architectures. *DAC*, pages 445–450, 2006.

[13] H. Sutter. *More Exceptional C++*. Addison Wesley, 2002.

[14] H. Ren et al. Sensitivity guided net weighting for placement driven synthesis. In *International Symposium on Physical Design*, pages 10–17, 2004.

APPENDIX

A. NON-UNIFORM INPUT PROFILE

In Table 2, we have compared the result for non-uniform input profile. The required time of arrival for all output bits are set to 9 and the input arrival levels have been randomly generated between 0-4. Table 5 presents those arrival levels of each input bit for all profiles.

Table 5: Input Arrival Times for Table 2

Bit	A	B	C	D	E	F	G
0	1	2	1	2	1	2	2
1	2	1	3	3	2	3	1
2	1	3	2	1	1	3	1
3	3	2	3	1	1	1	2
4	4	1	2	2	1	2	1
5	2	0	1	3	3	1	2
6	1	4	3	2	2	1	1
7	3	3	2	1	3	1	2
8	1	2	1	4	4	1	1
9	2	1	3	3	2	2	3
10	1	3	4	2	3	2	2
11	0	2	2	1	1	2	2
12	3	2	1	3	2	3	2
13	2	1	2	2	3	2	2
14	1	4	4	1	2	3	2
15	4	2	1	1	1	2	1
16	2	1	3	2	2	1	3
17	2	2	1	3	2	1	1
18	1	3	2	2	1	4	1
19	0	1	1	1	2	1	2
20	1	4	2	3	1	2	3
21	3	0	1	1	1	1	4
22	4	2	1	4	2	1	4
23	1	1	2	1	2	2	4
24	2	2	3	2	2	1	2
25	2	1	2	3	2	2	1
26	1	3	4	2	1	1	4
27	3	2	1	3	3	2	2
28	1	1	3	1	2	1	4
29	2	3	2	2	4	1	1
30	2	1	2	1	1	2	2
31	3	2	1	2	2	1	3

Table 6 compares our approach with [9] and [11] for correlated input profile, like late higher words or monotonically increasing inputs, appeared in [11].

Table 6: Comparison on Zimmermann's examples

DATA	Our Approach	Area [9] Heuristic	Zimmermann [11]
A	49	49	50
B	59	61	61
C	56	56	56
D	63	64	63
E	50	55	55
F	73	73	73
G	56	58	59
H	78	79	78
I	68	68	68

B. IMPACT OF MFO ON POST PLACEMENT RESULT

In Fig.9, the worst negative slack (WNS) is plotted against the size of the prefix graph for 16 bit adders. We can see that the prefix graphs of higher node count and smaller maximum fan-out (MFO) are better for timing.

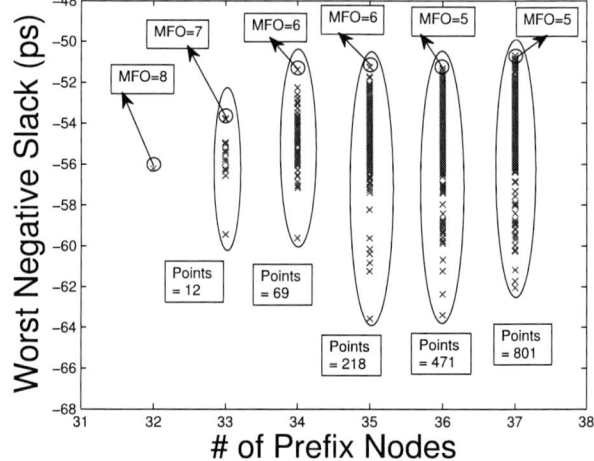

Figure 9: # of Prefix Nodes vs. WNS for 16 bit adder

C. PREFIX GRAPHS

Bit Width = 16, Size = 42
Max. Level = 4, Max. Fanout = 2

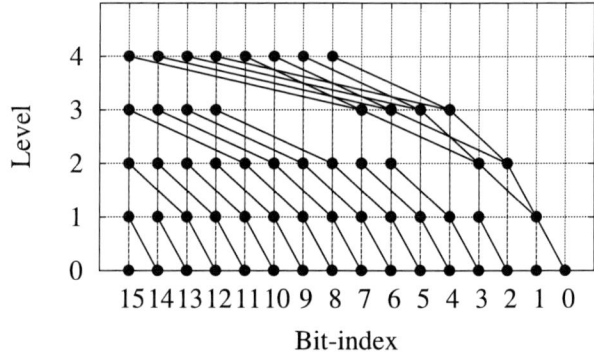

Figure 10: Size of a 16 bit prefix graph with level 4 and fanout 2 generated by our approach is less than that of Kogge Stone by 7

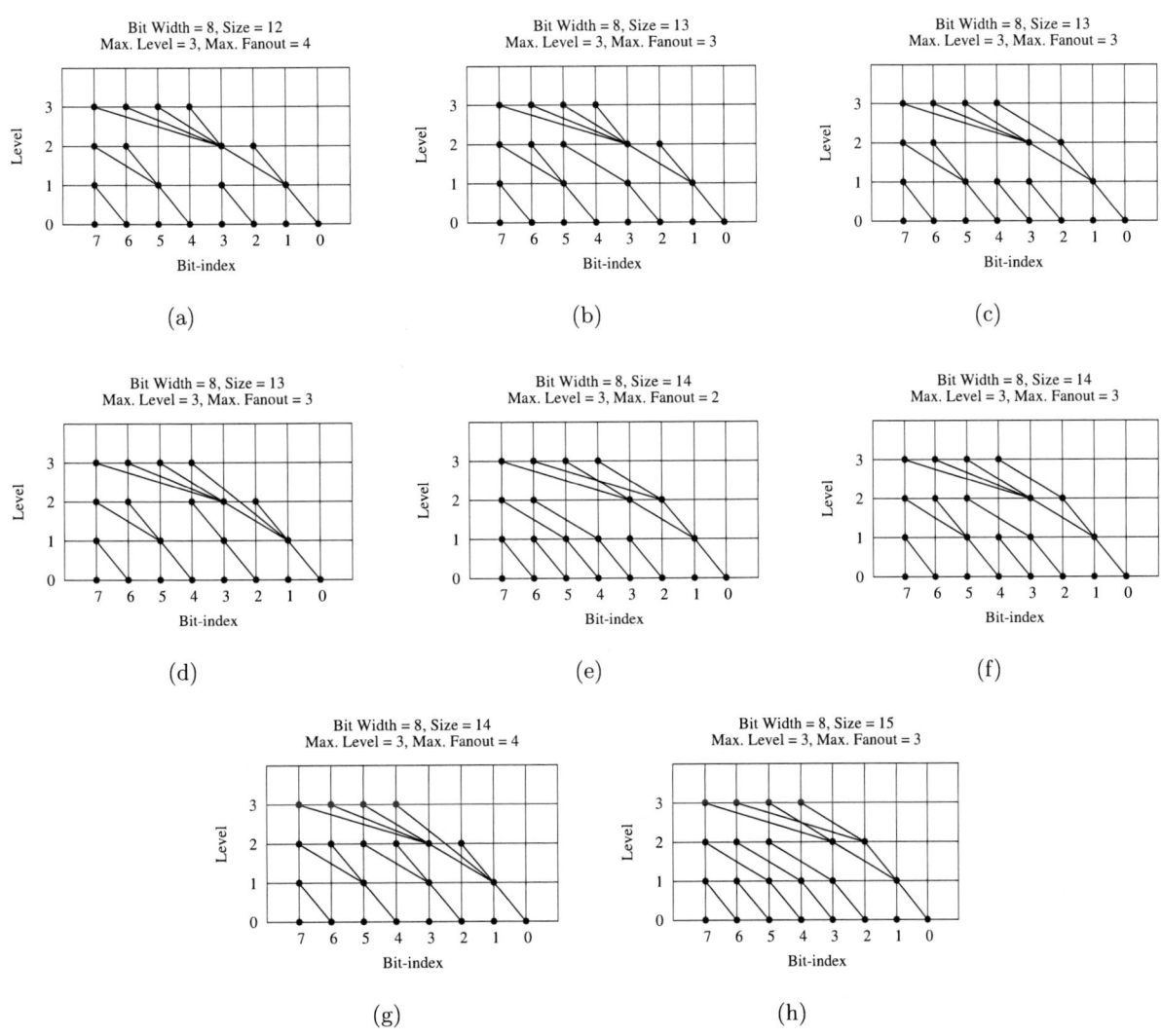

Figure 11: 8 bit prefix graphs with level 3

Synthesis of Feedback Decoders for Initialized Encoders

Kuan-Hua Tu and Jie-Hong R. Jiang

Department of Electrical Engineering / Graduate Institute of Electronics Engineering
National Taiwan University, Taipei 10617, Taiwan

ABSTRACT

Encoding and decoding are common practice in data processing. Designing encoder and decoder circuitry manually can be error prone and time consuming. Although great progress has been made on automating decoder synthesis from its encoder specification, prior specification was limited to an uninitialized encoder only, whose decoder in turn cannot depend on the entire execution history of the encoder. Prior decoder existence condition is unnecessarily stringent as encoders are often initialized to some specific starting states. This paper shows how decoders of initialized encoders can be practically synthesized. Experimental results demonstrate effective decoder synthesis of initialized encoders, beyond existing methods' capabilities.

Categories and Subject Descriptors

B.6.3 [**Logic Design**]: Design Aids—*automatic synthesis*

General Terms

Algorithms, logic synthesis, verification

Keywords

Craig interpolation, decoder, encoder, finite-state transition system, satisfiability solving

1. INTRODUCTION

Data processing is pervasive in computation and communication, and relies on an encoding and decoding scheme for effective and robust data manipulation. An encoder transforms some input data to encoded data, whereas a decoder recovers the original input data from the encoded data (possibly being modified by the underlying communication channel). Given an encoder specification (and possibly channel characteristics), its decoder design may be non-trivial and hard to formally verify. Automating decoder synthesis eases the design and verification tasks.

There have been prior efforts on decoder synthesis. Shen et al. [11] proposed a bounded decoder synthesis method with no termination guarantee. Complete synthesis methods were later established independently by Shen et al. [12] and Liu et al. [7, 8]. Particularly, Liu et al. [7, 8] exploited incremental satisfiability (SAT) solving [5] and Craig interpolation [3, 9, 6] techniques for efficient computation.

All prior methods, however, can only synthesize a special class of decoders, whose decoding process must depend on a

Permission to make digital or hard copies of all or part of this work for personal or classroom use is granted without fee provided that copies are not made or distributed for profit or commercial advantage and that copies bear this notice and the full citation on the first page. To copy otherwise, to republish, to post on servers or to redistribute to lists, requires prior specific permission and/or a fee.
DAC 2013, May 29 - June 07, 2013, Austin, Texas, USA.

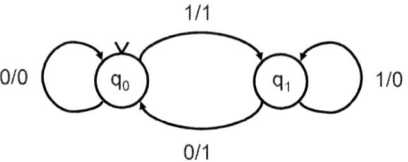

Figure 1: A 0-1 alternation detector with a designated initial state q_0.

bounded execution history of their encoders. Therefore the considered decoder must be in a pipelined form without sequential feedbacks. Not all decoders, however, can be in such simplicity. Many encoders in real applications inevitably require their decoding to depend on the entire histories of output sequences starting from initial states. In essence, sequential feedbacks are a necessity for such decoders. Figure 1, modified from [7, 8], shows one such encoder example, whose decoder exists with respect to any initial state but does not exist if no initial state is specified. The decoding must depend on the entire input history, which is unbounded and accounts for the necessity of sequential feedbacks. Moreover, decoders with sequential feedbacks are potentially more compact than those without feedbacks especially when the number of pipeline stages is large for feedback-free decoders. On the other hand, prior work required encoders in their normal operation to be in non-dangling states (a state is called *dangling* if it cannot be reached by any states, including itself, or can be reached only by dangling states), and permitted not all the input data being recovered. However in certain applications such data loss is disadvantageous, if not unacceptable.

This paper investigates the fundamental question how to synthesize decoders without restricting to a bounded execution history of their encoders and without admitting any data loss. The main results include 1) a sound and complete decoder existence checking algorithm, 2) practical decoder existence checking techniques using incremental SAT solving and property directed reachability analysis [2, 4], 3) a decoder synthesis method based on Craig interpolation. Experimental results demonstrate the effectiveness of decoder existence checking and synthesis, while the obtained decoders, unlike those derived previously, can depend on an unbounded encoder execution history and recover data without any loss.

This paper is organized as follows. Section 2 provides some preliminaries. After the decoding problem is stated in Section 3, our new results on decoder existence checking and decoder synthesis are presented in Sections 4 and 5, respectively. Experimental results are shown in Section 6, and finally concluding remarks are given in Section 7.

2. PRELIMINARIES

In the sequel the cardinality of a set S is denoted as $|S|$. The set of truth valuations of a vector $\vec{x} = (x_1, \ldots, x_k)$ of Boolean variables is denoted $[\![\vec{x}]\!]$, for instance, $[\![(x_1, x_2)]\!] =$

$\{(0,0),(0,1),(1,0),(1,1)\}$. Symbols $\neg$, $\wedge$, $\vee$, and $\Rightarrow$ stand for logical connectives negation, conjunction, disjunction, and implication, respectively.

2.1 SAT Solving and Craig Interpolation

We assume the reader's familiarity with Boolean satisfiability (SAT) solving [10, 5] and the circuit to conjunctive normal form (CNF) formula conversion [13]. A more detailed exposition can be found in [6].

For decoder synthesis, the following theorem is useful.

THEOREM 1 (CRAIG INTERPOLATION THEOREM). *[3]*
Given two Boolean formulas ϕ_A and ϕ_B, if $\phi_A \wedge \phi_B$ is unsatisfiable, then there exists a Boolean formula ψ_A, called the interpolant of ϕ_A with respect to ϕ_B, referring only to the common variables of ϕ_A and ϕ_B such that $\phi_A \Rightarrow \psi_A$ and $\psi_A \Rightarrow \neg\phi_B$.

The interpolant ψ_A can be constructed in linear time from a refutation proof of $\phi_A \wedge \phi_B$ produced by a SAT solver [9].

2.2 State Transition Systems

A *state transition system* consists of a state transition relation $T(\vec{x},\vec{s},\vec{y},\vec{s}')$ and a set $I(\vec{s})$ of initial states, where $\vec{s}$, $\vec{s}'$, $\vec{x}$, and $\vec{y}$ are referred to as the current-state variables, next-state variables, input variables, and output variables, respectively. (In the sequel, state sets are represented with characteristic functions. We shall not distinguish between a characteristic function and the set that it represents.) For a deterministic system as we shall assume for an encoder, the transition relation $T(\vec{x},\vec{s},\vec{y},\vec{s}')$ can be alternatively treated as the transition function $T : [\![\vec{x}]\!] \times [\![\vec{s}]\!] \to [\![\vec{y}]\!] \times [\![\vec{s}']\!]$.

A time-frame expansion of the state transition system $T(\vec{x},\vec{s},\vec{y},\vec{s}')$ is the time unrolling of T into multiple time-indexed copies, denoted $T^t = T(\vec{x}^t,\vec{s}^t,\vec{y}^t,\vec{s}^{t+1})$, the transition relation at time t. In contrast, in the sequel $T^* = T(\vec{x^*},\vec{s^*},\vec{y^*},\vec{s^*}')$ denotes a renamed copy of T with variables $\vec{x}$, $\vec{s}$, $\vec{y}$, and $\vec{s}'$ of T substituted with fresh new variables $\vec{x^*}$, $\vec{s^*}$, $\vec{y^*}$, and $\vec{s^*}'$, respectively.

3. PROBLEM STATEMENT

Given a state transition system with transition relation $T(\vec{x},\vec{s},\vec{y},\vec{s}')$ and initial states $I(\vec{s})$, as an encoder it transforms an input sequence to an output sequence. The decoder synthesis problem asks whether a decoder exists that recovers the original input sequence from the observed encoded sequence, and furthermore how to synthesize it if it exists.

A realistic decoder should satisfy the following two properties. First, the decoder must have a finite amount of memory elements for practical implementation. Second, the decoding must be an online process as the input and output sequences can be indefinitely extended without a pre-specified length upper bound. That is, the decoder must causally recover a prefix of the original input sequence on-the-fly based on a so-far observed encoded sequence. Unlike prior work [11, 12, 7, 8], the encoder under our consideration can have normal operation under non-dangling states, and the synthesized decoder can recover the original input sequence starting from the very first input (though with some time delay).

4. DECODER EXISTENCE CHECKING

4.1 Bounded Decoder Existence Checking

Given an encoder starting in some known state at time $t = 0$, the following proposition states the necessary and sufficient condition that the original input to the encoder at time $t = 0$ can be uniquely determined by the encoded sequence of length p generated by the encoder.

PROPOSITION 1. *Given an encoder with transition relation $T(\vec{x},\vec{s},\vec{y},\vec{s}')$ and initial states $I(\vec{s})$, let formula $\varphi_{M(p)}$ be*

$$\bigwedge_{t=0}^{p-1}\left(T^t \wedge T^{*t} \wedge (\vec{y}^t = \vec{y^*}^t)\right) \wedge (\vec{x}^0 \neq \vec{x^*}^0) \wedge (\vec{s}^0 = \vec{s^*}^0) \quad (1)$$

where predicate "$=$" asserts the bit-wise equivalence of its two argument variable vectors and "$\neq$" asserts the corresponding negation. Then the original input $\vec{i}^0 \in [\![\vec{x}^0]\!]$ can be uniquely determined by observing the encoded outputs $\vec{o}^0,\dots,\vec{o}^{p-1} \in [\![\vec{y}^0]\!] \times \dots \times [\![\vec{y}^{p-1}]\!]$ of length $p \geq 1$ if and only if the formula

$$\varphi_{M(p)} \wedge I(\vec{s}^0) \quad (2)$$

is unsatisfiable.

In the sequel, we shall call Formula (1), namely $\varphi_{M(p)}$, the miter formula.

To ensure that the original input $\vec{x}^t$ at any time $t \geq 0$ can be uniquely determined by observing an encoded output sequence of length p starting from a known state at time t, Proposition 1 has to be strengthened by relaxing the state variable constraint of $\vec{s}^0$ from the initial states to any states reachable from the initial states.

THEOREM 2. *For a given encoder with (deterministic) transition relation $T(\vec{x},\vec{s},\vec{y},\vec{s}')$ and initial states $I(\vec{s})$, let $R(\vec{s})$ be the set of reachable states. Then the original input $\vec{x}^t \in [\![\vec{x}^t]\!]$ at any time $t \geq 0$ can be uniquely determined by observing the encoded outputs $\vec{o}^t,\dots,\vec{o}^{t+p-1} \in [\![\vec{y}^t]\!] \times \dots \times [\![\vec{y}^{t+p-1}]\!]$ of length $p \geq 1$ if and only if the formula*

$$\varphi_{M(p)} \wedge R(\vec{s}^0) \quad (3)$$

is unsatisfiable.

PROOF. ($\Longrightarrow$) For the sake of contradiction, assume $\varphi_{M(p)} \wedge R(\vec{s}^0)$ is satisfiable. Then there exists some reachable state $\vec{q} \in [\![\vec{s}]\!]$ such that $\varphi_{M(p)} \wedge (\vec{s}^0 = \vec{q})$ is satisfiable. By Proposition 1, we know the original input $\vec{i}^0$ with current state $\vec{q}$ cannot be uniquely determined by the encoded outputs $\vec{o}^0,\dots,\vec{o}^{p-1}$. (Note that $t = 0$ is a relative reference time point.) Since $\vec{q}$ is a reachable state, it follows that not every original input at any time can be uniquely determined from an encoded output sequence of length p.

($\Longleftarrow$) If Formula (3) is unsatisfiable, no state $\vec{q} \in [\![\vec{s}]\!]$ reachable from I satisfies $\varphi_{M(p)} \wedge (\vec{s}^0 = \vec{q})$. For $\vec{q}$ being an initial state, the unsatisfiability of $\varphi_{M(p)} \wedge (\vec{s}^0 = \vec{q})$ implies the original input $\vec{i}^0 \in [\![\vec{x}^0]\!]$ can be uniquely determined from the encoded output sequence $\vec{o}^0,\dots,\vec{o}^{p-1}$ by Proposition 1. Thereby the next state $\vec{q}'$ of $\vec{q}$ can be uniquely determined from the deterministic transition relation T under current state $\vec{q}$ and current input $\vec{i}^0$. Again since $\vec{q}'$ is a reachable state, $\varphi_{M(p)} \wedge (\vec{s}^0 = \vec{q}')$ must be unsatisfiable, and the original input under current state $\vec{q}'$ can be uniquely determined from the encoded output sequence $\vec{o}^1,\dots,\vec{o}^p$ and so is its next state. Repeating this argument, we know that the original input $\vec{i}^t$ at any time t can be determined from the encoded output sequence $\vec{o}^t,\dots,\vec{o}^{t+p-1}$. ■

Notice that the above decoder existence condition, Formula (3), is with respect to some pre-specified length bound p. The decoder non-existence at $p = n$ does not exclude the decoder existence at a larger $p = n+k$ for $k \geq 1$ however. To determine if there exists no decoder for arbitrary $p \geq 1$, additional constraints need to be imposed to make the checking finitary and complete as we show below.

4.2 Unbounded Decoder Existence Checking

The following proposition provides a necessary and sufficient condition for determining decoder existence without referring to a pre-specified length bound.

PROPOSITION 2. *Given an encoder with transition relation $T(\vec{s}, \vec{x}, \vec{s}', \vec{y})$ and initial states $I(\vec{s})$, its decoder does not exist if and only if the encoder starting from some reachable state $\vec{q} \in [\![\vec{s}^t]\!]$ at time t produces the same infinite encoded output sequence $\vec{o}^t, \vec{o}^{t+1}, \ldots \in [\![\vec{y}^t]\!] \times [\![\vec{y}^{t+1}]\!] \times \cdots$ under two input sequences $\vec{i_1}^t, \vec{i_1}^{t+1}, \ldots$ and $\vec{i_2}^t, \vec{i_2}^{t+1}, \ldots \in [\![\vec{x}^t]\!] \times [\![\vec{x}^{t+1}]\!] \times \cdots$ with $\vec{i_1}^t \neq \vec{i_2}^t$ at time t.*

The following theorem provides a computational means to demonstrate the non-existence of decoders, that is, Formula (1) is satisfiable for any arbitrary $p \geq 1$.

THEOREM 3. *The decoder of a transition system $T(\vec{x}, \vec{s}, \vec{y}, \vec{s}')$ with initial states $I(\vec{s})$ does not exist if and only if the formula*

$$\varphi_{M(p)} \wedge R(\vec{s}^0) \wedge \bigvee_{i=0}^{p-1} \bigvee_{j=i+1}^{p} \left((\vec{s}^i = \vec{s}^j) \wedge (\vec{s^*}^i = \vec{s^*}^j) \right), \quad (4)$$

is satisfiable for some p, where R characterizes the set of states reachable from I under T.

PROOF. ($\Longrightarrow$) If there exists no decoder, then there exist two distinct inputs $\vec{i}^0 \in [\![\vec{x}^0]\!]$ and $\vec{i^*}^0 \in [\![\vec{x^*}^0]\!]$ at time $t = 0$ that are consistent (in terms of input-output traces confined by T) yielding the same infinite encoded output sequence. Let the corresponding two state traces be $\vec{q}^0, \vec{q}^1, \ldots$ and $\vec{q^*}^0, \vec{q^*}^1, \ldots$, respectively, with $\vec{q}^0 = \vec{q^*}^0$. Because of the finite state space of the encoder, there must be a state pair $(\vec{q}^k, \vec{q^*}^k)$ at time k in the two traces repeating itself at some other time $k + l$, that is, $(\vec{q}^{k+l}, \vec{q^*}^{k+l}) = (\vec{q}^k, \vec{q^*}^k)$. Hence Formula (4) is satisfiable.

($\Longleftarrow$) Assume Formula (4) is satisfiable under some p. Then there must exist two infinite input sequences with two distinct current inputs and the same current reachable state that result in the same infinite encoded output sequence. By Proposition 2, the decoder does not exist. ∎

PROPOSITION 3. *If Formula (4) is satisfiable under $p = 1, 2, \ldots, |R|^2$, where $|R|$ denotes the cardinality of the reachable state set R of the encoder, then Formula (4) remains satisfiable for arbitrary $p > |R|^2$.*

Suppose the decoder existence condition of Formula (4) is checked by incrementing p starting from 1 until decoder existence or non-existence is concluded. Then the looping sub-formula of Formula (4), i.e.,

$$\bigvee_{i=0}^{p-1} \bigvee_{j=i+1}^{p} \left((\vec{s}^i = \vec{s}^j) \wedge (\vec{s^*}^i = \vec{s^*}^j) \right). \quad (5)$$

can be simplified to

$$\varphi_{L(p)} = \bigvee_{i=0}^{p-1} \left((\vec{s}^i = \vec{s}^p) \wedge (\vec{s^*}^i = \vec{s^*}^p) \right). \quad (6)$$

This simplification is possible due to the fact that the looping sub-constraints of Formula (5) corresponding to $j < p$ have been checked and shown unsatisfiable in conjuncting with $\varphi_{M(p-1)} \wedge R(\vec{s}^0)$. Since $\varphi_{M(p)} \Rightarrow \varphi_{M(p-1)}$, these looping sub-constraints of $j < p$ play no role contributing to the satisfiability of Formula (4) and can be removed. That is, Formula (4) can be simplified to

$$\varphi_{M(p)} \wedge R(\vec{s}^0) \wedge \varphi_{L(p)} \quad (7)$$

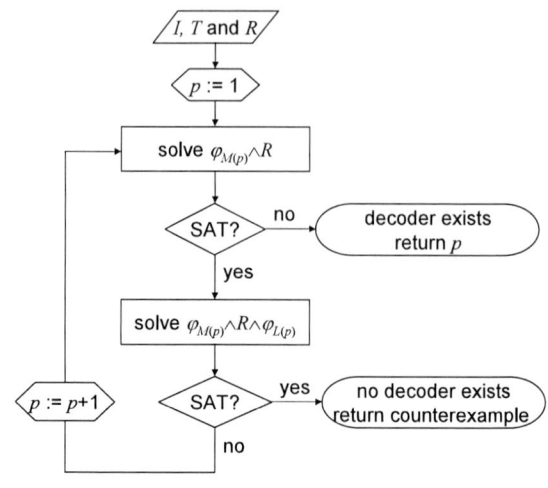

Figure 2: Flow of decoder existence checking with a priori reachability knowledge.

in incremental satisfiability solving.

The flow of decoder existence checking, with the reachable state set R given, is shown in Figure 2. In the procedure, p starts from 1 and is increased by 1 until either $\varphi_{M(p)} \wedge R$ is unsatisfiable (a decoder exists) or $\varphi_{M(p)} \wedge R \wedge \varphi_{L(p)}$ is satisfiable (no decoder exists). By Proposition 3, the procedure is guaranteed to terminate at some $p \leq |R|^2$.

4.3 Decoder Existence Checking without A Priori Reachability Knowledge

The above discussion assumes the reachable state set R is given. Exact state reachability analysis, however, is often too expensive to be practically computed. Fortunately, recent advances in SAT-based unbounded model checking (UMC), in particular the interpolation method [9] and the property directed reachability method [2, 4], allow efficient computation of over-approximated reachable state sets. Given a state transition system T, initial state set I, and final state set F as input, an UMC algorithm, denoted $UMC(I, T, F)$, returns either an input trace as an evidence in the case of F reachable from I, or an over-approximated state set $R^\dagger$ satisfying

$$\forall \vec{s}.\ I(\vec{s}) \Rightarrow R^\dagger(\vec{s}), \quad (8)$$

$$\forall \vec{x}, \vec{s}, \vec{y}, \vec{s}'.\ R^\dagger(\vec{s}) \wedge T(\vec{x}, \vec{s}, \vec{y}, \vec{s}') \Rightarrow R^\dagger(\vec{s}'), \text{ and} \quad (9)$$

$$\forall \vec{s}.\ R^\dagger(\vec{s}) \Rightarrow \neg F(\vec{s}) \quad (10)$$

in the case of F not reachable from I. It is immediate that $R \Rightarrow R^\dagger$ and thus $R^\dagger$ over-approximates R.

An UMC algorithm can be exploited as a black-box tool in decoder existence checking as follows. Instead of checking directly whether Formula (7) is satisfiable, we check if any state $\vec{q} \in [\![\vec{s}]\!]$ satisfying $\varphi_{F(p)}(\vec{s}) =$

$$\exists \vec{x}^0, \ldots, \vec{x}^{p-1}, \vec{y}^0, \ldots, \vec{y}^{p-1}, \vec{s}^1, \ldots, \vec{s}^p.\varphi_{M(p)} \wedge \varphi_{L(p)} \quad (11)$$

can be reached from the initial states I. (Notice that in Formula (11) the original free variables $\vec{s}^0$ are renamed to $\vec{s}$ to avoid confusion as the time index 0 of $\vec{s}^0$ is merely a relative reference time point rather than the absolute initial time point.) That is, Formula (11) is treated as the final state set F. Notice that no quantifier elimination needs to be performed on these existentially quantified variables as they are simply free variables during SAT-based UMC computation.

Intuitively $\varphi_{F(p)}$ characterizes the set of bad states (starting from them, state pair recurrence happens within p steps

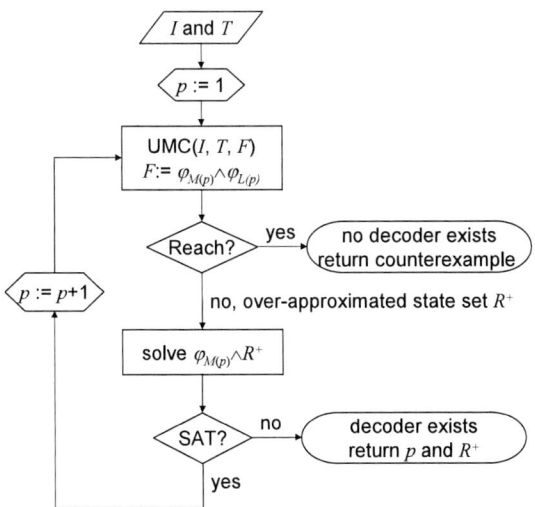

Figure 3: Flow of decoder existence checking without *a priori* reachability knowledge.

in the state trace, i.e., $\varphi_{L(p)}$ is satisfied), whose inputs cannot be uniquely determined by observing output sequences of finite lengths. So if any state of $\varphi_{F(p)}$ is reachable from I, then no decoder exists and an UMC algorithm returns an input trace that accounts for the non-existence. Otherwise, an over-approximated reachable state set $R^\dagger$ is returned by an UMC algorithm. It allows the replacement of Formula (3) with

$$\varphi_{M(p)} \wedge R^\dagger(\vec{s}^0). \tag{12}$$

If Formula (12) is unsatisfiable, then decoder exists and can be further synthesized (as to be detailed in Section 5). Otherwise the length parameter p is incremented for a new iteration of computation. The overall flow of the computation is summarized in Figure 3.

The correctness of the above computation flow is based on the following theorems.

THEOREM 4. *For a given encoder with transition relation $T(\vec{x}, \vec{s}, \vec{y}, \vec{s}')$ and initial states I, its decoder does not exist if and only if there is some state $\vec{q} \in [\![\vec{s}^0]\!]$ that satisfies*

$$\varphi_{M(p)} \wedge \varphi_{L(p)} \tag{13}$$

for some $p \geq 1$ and is reachable from I under transition relation T.

PROOF. ($\Longrightarrow$) If there exists no decoder, formula $\varphi_{M(p)} \wedge (\vec{s}^0 = \vec{q})$ is satisfiable for any reachable state $\vec{q}$ and any p. Moreover, because the cardinality of reachable state set is finite, a state pair $(\vec{q}_1, \vec{q}_2) \in [\![\vec{s}]\!] \times [\![\vec{s}^\star]\!]$ in the state trace satisfying $\varphi_{M(p)}$ must recur at some p. Hence $\varphi_{L(p)}$ will eventually be satisfiable, and so will Formula (13).

($\Longleftarrow$) If Formula (13) is satisfiable, then there exists a reachable state whose input cannot be uniquely determined by observing output sequences of length up to p. Since a state pair recurs within the state trace as asserted by $\varphi_{L(p)}$, appending the time-frames corresponding to the recurrent state trace of length n to the miter formula $\varphi_{M(p)}$ makes $\varphi_{M(p+n)} \wedge \varphi_{L(p+n)}$ remain satisfiable. By continuing appending time-frames, formula $\varphi_{M(p+kn)} \wedge \varphi_{L(p+kn)}$ remains satisfiable for arbitrary $k \geq 0$. Therefore no decoder exists. ∎

THEOREM 5. *For a given encoder with transition relation $T(\vec{x}, \vec{s}, \vec{y}, \vec{s}')$ and initial states I, let $R^\dagger$ be an over-approximated*

set of reachable states. Then the original input $\vec{i}^t \in [\![\vec{x}^t]\!]$ at any time $t \geq 0$ can be uniquely determined by observing the encoded outputs $\vec{o}^t, \ldots, \vec{o}^{t+p-1} \in [\![\vec{y}^t]\!] \times \cdots \times [\![\vec{y}^{t+p-1}]\!]$ of length $p \geq 1$ if Formula (12) is unsatisfiable.

PROOF. Since $R \Rightarrow R^\dagger$, the unsatisfiability of Formula (12) implies that of Formula (3). By Theorem 2, the statement holds. However the converse is not true since Formula (12) can be satisfiable under the unreachable states $R^\dagger \wedge \neg R$ while $\vec{i}^t$ can still be uniquely determined. ∎

Due to the reachability over-approximation, it is possible that the computation flow of Figure 3 terminates at some $p = n$ that is much larger than that of Figure 2. Nevertheless it is possible to overcome this deficiency with a modified computation flow of Figure 3 by replacing the SAT solving $\varphi_{M(p)} \wedge R^\dagger$ with the computation UMC$(I, T, \varphi_{M(p)})$. If the final states $F = \varphi_{M(p)}$ are unreachable from I, then a decoder exists. Otherwise, p is incremented for the next computation iteration. (Surely the price to pay is to perform the more expensive UMC computation rather than SAT solving.)

THEOREM 6. *The decoder existence checking procedure of Figure 3 with the SAT solving $\varphi_{M(p)} \wedge R^\dagger$ replaced by the computation UMC$(I, T, \varphi_{M(p)})$ (such that a decoder exists if final states $F = \varphi_{M(p)}$ are reachable from I, and p is incremented for next iteration otherwise) terminates at $p = n$ for some $n \geq 1$ same as the procedure of Figure 2.*

PROOF. Assume the procedure of Figure 2 terminates at iteration $p = n$ under decoder existence. Then $\varphi_{M(n)} \wedge R(\vec{s}^0)$ is unsatisfiable, for R the exact reachable state set. That is, any state $\vec{q} \in [\![\vec{s}]\!]$ cannot satisfy both $R(\vec{q})$ and $\varphi_{M(n)} \wedge (\vec{s}^0 = \vec{q})$. Hence any state in the final states $F = \varphi_{M(n)}$ is not in the reachable state set R. Because the final states $F = \varphi_{M(n)}$ are unreachable from I, UMC$(I, T, \varphi_{M(p)})$ should return unreachability when $p = n$ and thus the modified procedure terminates at the same iteration n.

On the other hand, assume the procedure of Figure 2 terminates at iteration $p = n$ under decoder non-existence. Then $\varphi_{M(n)} \wedge R \wedge \varphi_{L(n)}$ is satisfiable. Since UMC is complete in proving reachability, UMC$(I, T, \varphi_{M(n)} \wedge \varphi_{L(n)})$ must establish the reachability of $\varphi_{M(n)} \wedge \varphi_{L(n)}$ from I, and thus the modified procedure terminates at iteration n. ∎

5. DECODER SYNTHESIS

By the unsatisfiability of $\varphi_{M(p)} \wedge R^\dagger(\vec{s}^0)$ (which includes $\varphi_{M(p)} \wedge R(\vec{s}^0)$ as a special case), a decoder can be synthesized from the corresponding resolution refutation by Craig interpolation. The decoding function f_i corresponding to every output bit $x_i^0 \in \vec{x}^0$ of the decoder is synthesized one at a time. The actual length p_i ($1 \leq p_i \leq p$) of the observation window for f_i is usually smaller than p. Let $\varphi_{M_i(p_i)}$ be

$$\bigwedge_{t=0}^{p-1} \left(T^t \wedge T^{\star t} \right) \wedge \bigwedge_{t=0}^{p_i-1} \left(\vec{y}^t = \vec{y}^{\star t} \right) \wedge (x_i^0 \neq x_i^{\star 0}) \wedge (\vec{s}^0 = \vec{s}^{\star 0}). \tag{14}$$

Then $\varphi_{M_i(p_i)} \wedge R^\dagger(\vec{s}^0)$ must remain unsatisfiable and f_i can be obtained as follows, similar to the synthesis approach proposed in [6].

THEOREM 7. *Given a transition system $T(\vec{x}, \vec{s}, \vec{y}, \vec{s}')$ and its (over-approximated) reachable state set $R^\dagger$, if $\varphi_{M_i(p_i)} \wedge R^\dagger(\vec{s}^0)$ is unsatisfiable, then the interpolant ψ_{i_A} of ϕ_{i_A} with*

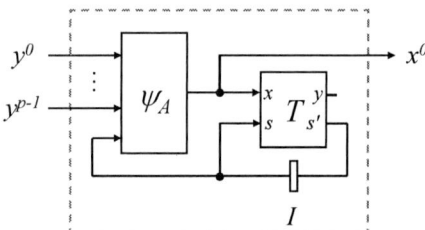

Figure 4: Block diagram of synthesized decoder.

respect to ϕ_{i_B} is a valid decoding function for $x_i^0 \in \vec{x}^0$, where

$$\phi_{i_A} : \bigwedge_{t=0}^{p-1} T^t \wedge x_i^0, \ and \tag{15}$$

$$\phi_{i_B} : \bigwedge_{t=0}^{p-1} T^{*t} \wedge \bigwedge_{t=0}^{p_i-1} \left(\vec{y}^t = \vec{y}^{*\,t} \right) \wedge \neg x_i^{*\,0} \wedge (\vec{s}^0 = \vec{s}^{*\,0}) \wedge R^\dagger (\vec{s}^0). \tag{16}$$

PROOF. The common variables of ϕ_{i_A} and ϕ_{i_B} are $\vec{s}^0$ and $\vec{y}^0, \ldots, \vec{y}^{p_i-1}$. By Theorem 1, the interpolant ψ_{i_A} refers only to these common variables. Furthermore its onset (respectively offset) contains the assignments to the common variables that satisfy ϕ_{i_A} (respectively ϕ_{i_B}). Because x_i^0 is asserted to be true in ϕ_{i_A} and $x_i^{*\,0}$ is asserted to be false in ϕ_{i_B}, the onset (respectively offset) of the interpolant contains the assignments to the common variables that determine x_i^0 to true ($x_i^{*\,0}$ to false). So the interpolant validly implements the decoding function of variable x_i^0. ∎

Let $\vec{\psi}_A$ be the collection of obtained decoding functions. Then the decoder can be constructed by composing $\vec{\psi}_A$ and the encoder transition function T as the circuit shown in Figure 4. Note that the decoder should start its operation after its first p inputs $\vec{o}^0, \ldots, \vec{o}^{p-1} \in \llbracket \vec{y}^0 \rrbracket \times \cdots \times \llbracket \vec{y}^{p-1} \rrbracket$ are ready. So it produces its first output at the pth clock cycle.

THEOREM 8. *For a given encoder with transition relation $T(\vec{x}, \vec{s}, \vec{y}, \vec{s}')$ and initial states I, its decoder has transition relation $T_d(\vec{x}_d, \vec{s}_d, \vec{y}_d, \vec{s}_d\,')$ equal to*

$$\left(\vec{y}_d = \vec{\psi}_A(\vec{x}_d, \vec{s}_d) \right) \wedge \left(\exists \vec{y}.T(\vec{y}_d, \vec{s}_d, \vec{y}, \vec{s}_d\,') \right), \tag{17}$$

with $\vec{x}_d = (\vec{y}^0, \ldots, \vec{y}^{p-1})$, and has initial states $I_d(\vec{s}_d)$ equal to $I(\vec{s}_d)$.

PROOF. By Theorem 7, $\vec{\psi}_A(\vec{x}_d, \vec{s}_d)$ for $\vec{x}_d = (\vec{y}^0, \ldots, \vec{y}^{p-1})$ and $\vec{s}_d = \vec{s}^0$ form the decoding functions of variables $\vec{x}^0 = \vec{y}_d$. The output function $\llbracket \vec{x}_d \rrbracket \times \llbracket \vec{s}_d \rrbracket \to \llbracket \vec{y}_d \rrbracket$ of the decoder can be represented as relation $\vec{y}_d = \vec{\psi}_A(\vec{x}_d, \vec{s}_d)$.

In addition, because the decoding functions $\vec{\psi}_A$ need the output and state information of the encoder to reconstruct the original input sequence while the decoder only receives the encoded output sequence from the encoder, the decoder needs to compute the state information by itself. Embedding the transition relation of the encoder inside allows the decoder to keep track of the state transition in synchronization with the encoder. Notice that the output of the embedded encoder transition relation is of no use in the decoder and can be removed through logic minimization. The overall transition relation of the decoder is generated as the conjunction of relations $\vec{y}_d = \vec{\psi}_A(\vec{x}_d, \vec{s}_d)$ and $\exists \vec{y}.T(\vec{y}_d, \vec{s}_d, \vec{y}, \vec{s}_d\,')$. ∎

Notice that the above discussion imposes no constraint on the cardinality of the initial state set, i.e., $|I| \geq 1$. Although

Table 1: Benchmark Statistics.

circuit	#gate/#level	#reg	#input
PCIE	265/24	22	14
XGXS	177/15	15	9
Scrambler	535/9	58	65
T2Ethernet	1094/18	48	8
CC13	63/8	13	1
CC14	66/8	14	1
CC3	12/4	3	1
CC4	24/4	4	1
AD	5/2	1	1
Huff-alphabet	297/18	9	5
Huff-jpeg	1528/27	12	8
Huff-ran8	2553/32	13	8
Huff-ran9	5155/32	14	9
Huff-skew5	538/21	10	5
Huff-skew6	1092/23	12	6
LFSR-12-6-4	12/6	12	1
LFSR-26-6-2	12/6	26	1
LFSR-32-7-6	12/6	32	1
SBC-Add(4,4)	171/23	16	4
C2670	717/21	0	233
s38417	9219/31	1636	28
s444	155/13	21	3
s5378	1343/17	164	35
s6669	2263/80	239	83

there are multiple initialization choices when $|I| > 1$, the start state of the encoder and that of the decoder should match. Otherwise the decoder may not correctly recover the expected original input sequence of the encoder.

6. EXPERIMENTAL RESULTS

The proposed method, named DECOSY-I, was programmed in the C language and implemented within the ABC system [1], where command pdr [4] was used for the UMC reachability computation. The experiments were conducted on a Linux machine with Xeon 2.53GHz CPU and 48GB RAM.

The considered benchmark circuits are listed in Table 1, where the numbers of gates, logic levels, registers, and inputs are shown. Circuits XGXS, Scrambler, PCIE, and T2Ethernet are from [11]; the CC series circuits are convolutional code encoders from [7, 8]; circuit AD is the encoder of Figure 1; the Huff series circuits are Huffman code encoders; the LFSR series circuits correspond to linear feedback shift registers; the SBC circuit is a sliding block code encoder; others are from the MCNC and ISCAS benchmark suites. In the following experiments, the obtained decoder circuits were optimized in ABC under the script "strash; scleanup; dsd; strash; dc2; dc2; dch; map", where technology mapping was performed using the *mcnc.genlib* library.

Tables 2, 3, and 4 show the statistics of DECOSY [7, 8] assuming uninitialized encoder operation and our new DECOSY-I assuming initialized encoder operation. It should be emphasized that the two approaches cannot be directly compared as their decoding problems are fundamentally different. The circuits shown in Table 2 have decoders under both initialization assumptions; those in Table 3 have decoders only under the initialized assumption; those in Table 4 have no decoders at all.

Table 2 lists the window size of observing output sequences in Columns 2 and 6, the numbers of decoder inputs/registers in Columns 3 and 7, decoder area/delay in Columns 4 and 8, and CPU time (including decoder generation time plus script optimization time in parentheses) in Columns 5 and 9. As shown, the runtimes are comparable except for circuits CC-13 and CC-14, where DECOSY timed out at 3600 seconds (with a window size of 13 upon timeout). On the other hand, the decoders generated by our new DECOSY-I are usually slightly larger than those by DECOSY since we require the transition

346

Table 2: Comparison on Decoders Synthesized under Different Encoder Assumptions.

circuit	Decosy (for uninitialized encoder)				Decosy-I (for initialized encoder)			
	window size	#in/#reg	area/delay	time (s)	window size	#in/#reg	area/delay	time (s)
PCIE	4	11/0	147/5.2	0.13 (+0.10)	3	10/0	168/5.6	0.31 (+0.10)
XGXS	3	11/0	294/7.7	0.04 (+0.05)	2	10/1	468/10.2	0.18 (+0.09)
Scrambler	3	65/64	640/3.8	0.54 (+0.07)	1	64/58	727/3.8	1.26 (+0.09)
T2Ethernet	6	11/0	388/9.9	4.96 (+0.08)	5	10/0	423/10.2	0.94 (+0.04)
CC3	3	4/0	10/3.8	0.00 (+0.09)	2	1/2	12/1.9	0.01 (+0.00)
CC4	3	6/0	15/6.9	0.00 (+0.00)	2	1/3	18/3.8	0.01 (+0.00)
CC13	13	–	–	> 3600	2	1/12	52/5.7	0.09 (+0.02)
CC14	13	–	–	> 3600	2	1/13	86/5.5	0.02 (+0.02)

Table 3: Comparison on Decoder (Non-)Existence Checking under Different Encoder Assumptions.

circuit	Decosy (for uninitialized encoder)			Decosy-I (for initialized encoder)				
	window size	exist?	time (s)	window size	exist?	#in/#reg	area/delay	time (s)
AD	1	no	0.00	1	yes	1/1	10/1.9	0.01 (+0.00)
Huff-alphabet	5	no	0.02	10	yes	10/9	261/9.6	0.52 (+0.04)
Huff-jpeg	5	no	0.44	16	yes	16/12	1058/16.5	19.16 (+0.20)
Huff-ran8	7	no	1.57	19	yes	19/13	1995/14.3	72.07 (+0.28)
Huff-ran9	7	no	2.37	19	yes	19/14	4110/16.5	467.96 (+0.76)
Huff-skew5	3	no	0.00	31	yes	31/10	395/25.9	7.17 (+0.05)
Huff-skew6	3	no	0.02	63	yes	63/12	826/51.5	289.21 (+0.12)
LFSR-12-6-4	31	no	11.63	1	yes	1/12	36/4.0	0.01 (+0.05)
LFSR-26-6-2	55	no	> 3600	1	yes	1/26	57/4.0	0.01 (+0.02)
LFSR-32-7-6	15	no	0.14	1	yes	1/32	66/3.8	0.01 (+0.01)
SBC-Add(4,4)	5	no	0.29	1	yes	7/16	48584§/22.4§	50.65 (+12.25§)

Table 4: Comparison on Decoder Non-Existence Checking under Different Encoder Assumptions.

circuit	Decosy (for uninitialized encoder)		Decosy-I (for initialized encoder)	
	window size	time (s)	window size	time (s)
C2670	1	0.00	1	0.03
s38417	3	0.15	2	346.88
s444	1	0.00	1	0.01
s5378	1	0.01	1	1.00
s6669	1	0.02	1	91.59

function of the encoder to be embedded as part of the decoder although our decoders require smaller observation windows. Note that, unlike our obtained decoder, the decoder generated by Decosy assumes steady state operation and may not recover the entire original input sequence.

Table 3 reveals that, under different initialization assumptions, decoder existence checking may exhibit very different characteristics. For circuits LFSR-12 and LFSR-26, for example, Decosy required large window sizes to conclude decoder non-existence whereas Decosy-I efficiently synthesized their decoders with window sizes 1. (Decosy timed out on LFSR-26 with a window size of 55.) Another special case is SBC-Add(4,4), where logic optimization failed in dsd of our ABC synthesis script. The reported decoder area/delay and script runtime exclude dsd transformation and are marked with "§" in superscript.

Table 4 suggests that the runtime of Decosy-I is proportional to circuit and window sizes (particularly dominated by pdr computation) while Decosy concludes decoder non-existence within 0.15 seconds for all the circuits.

7. CONCLUSIONS

A sound and complete approach has been proposed to synthesizing decoders for initialized encoders. An obtained decoder can depend on an unbounded history of encoder execution (which should be distinguished from a bounded observation window), and recover the original input sequence without any prefix loss. This approach exceeds the capabilities of prior methods, and, as justified by experimental results, achieves computational efficiency by SAT-based approximative reachability analysis and interpolation-based synthesis.

Acknowledgments

This work was supported in part by the National Science Council under grants NSC 99-2221-E-002-214-MY3, 99-2923-E-002-005-MY3, and 101-2923-E-002-015-MY2.

8. REFERENCES

[1] Berkeley Logic Synthesis and Verification Group. *ABC: A system for sequential synthesis and verification.* http://www.eecs.berkeley.edu/~alanmi/abc/

[2] A. R. Bradley. SAT-based model checking without unrolling. In *Proc. Int'l Conf. on Verification, Model Checking, and Abstract Interpretation (VMCAI)*, pp. 70-87, 2011.

[3] W. Craig. Three uses of the Herbrand-Gentzen theorem in relating model theory and proof theory. *J. Symbolic Logic*, 22(3):269-285, 1957.

[4] N. Eén, A. Mishchenko, and R. Brayton. Efficient implementation of property-directed reachability. In *Proc. Int'l Conf. on Formal Methods in Computer Aided Design (FMCAD)*, pp. 125-134, 2011.

[5] N. Eén and N. Sörensson. An extensible SAT-solver. In *Proc. Int'l Conf. on Theory and Applications of Satisfiability Testing (SAT)*, pp. 502-518, 2003.

[6] J.-H. R. Jiang, C.-C. Lee, A. Mishchenko, and C.-Y. Huang. To SAT or not to SAT: Scalable exploration of functional dependency. *IEEE Trans. on Computers*, 59(4):457-467, April 2010.

[7] H.-Y. Liu, Y.-C. Chou, C.-H. Lin, and J.-H. R. Jiang. Towards completely automatic decoder synthesis. In *Proc. Int'l Conf. on Computer-Aided Design (ICCAD)*, pp. 389-395, 2011.

[8] H.-Y. Liu, Y.-C. Chou, C.-H. Lin, and J.-H. R. Jiang. Automatic decoder synthesis: methods and case studies. *IEEE Trans. on Computer-Aided Design of Integrated Circuits and Systems*, 31(9): 1319-1331, September 2012.

[9] K. McMillan. Interpolation and SAT-based model checking. In *Proc. Int'l Conf. on Computer Aided Verification (CAV)*, pp. 1-13, 2003.

[10] M. Moskewicz, C. Madigan, L. Zhang, and S. Malik. Chaff: Engineering an efficient SAT solver. In *Proc. Design Automation Conference (DAC)*, pp. 530-535, 2001.

[11] S. Shen, Y. Qin, K. Wang, L. Xiao, J. Zhang, and S. Li. Synthesizing complementary circuits automatically. *IEEE Trans. on Computer-Aided Design of Integrated Circuits and Systems*, 29(8):1191-1202, August 2010.

[12] S. Shen, Y. Qin, J. Zhang, and S. Li. A halting algorithm to determine the existence of the decoder. In *IEEE Trans. on Computer-Aided Design of Integrated Circuits and Systems*, 30(10): 1556-1563, October 2011.

[13] G. Tseitin. On the complexity of derivation in propositional calculus. *Studies in Constructive Mathematics and Mathematical Logic*, pp. 466-483, 1970.

On Learning-Based Methods for Design-Space Exploration with High-Level Synthesis

Hung-Yi Liu and Luca P. Carloni

Department of Computer Science, Columbia University, New York, NY, USA

{hungyi, luca}@cs.columbia.edu

ABSTRACT

This paper makes several contributions to address the challenge of supervising HLS tools for design space exploration (DSE). We present a study on the application of learning-based methods for the DSE problem, and propose a learning model for HLS that is superior to the best models described in the literature. In order to speedup the convergence of the DSE process, we leverage *transductive experimental design*, a technique that we introduce for the first time to the CAD community. Finally, we consider a practical variant of the DSE problem, and present a solution based on *randomized selection* with strong theory guarantee.

Categories and Subject Descriptors

B.6.3 [**Design Aids**]: Automatic Synthesis

General Terms

Algorithms, Design, Performance

Keywords

System-Level Design, High-Level Synthesis.

1. INTRODUCTION

It has been a longtime dream that Electronic-System-Level (ESL) design can be automatically synthesized from high-level specifications (e.g. C/C++ or SystemC) to optimized low-level implementations (e.g. RTL or gate-level netlists), a methodology known as High-Level Synthesis (HLS). After many years of endeavor and evolution [9], modern HLS tools can now also take micro-architecture choices as input constraints. By elaborating different sets of constraints, HLS tools allow designers to evaluate multiple implementation alternatives, a process known as Design Space Exploration (DSE). DSE with HLS is already a major leap from DSE with Logic Synthesis, since the latter starts from design specifications given at a lower-level of abstraction using Verilog or VHDL. These hardware-description languages make it more difficult and time-consuming for designers to specify many substantially different micro-architectures.

The industrial adoption of HLS tools is, however, still at an evaluation stage [9]. One of the major bottlenecks is that DSE with HLS still requires substantial efforts for setting the micro-architecture constraints, whose cardinality in general grows exponentially with the size of real design. Another bottleneck is the long runtime of HLS tools: in fact, a simple Discrete Fourier Transform (DFT) design in SystemC may still require an hour of CPU time for one HLS

Permission to make digital or hard copies of all or part of this work for personal or classroom use is granted without fee provided that copies are not made or distributed for profit or commercial advantage and that copies bear this notice and the full citation on the first page. To copy otherwise, to republish, to post on servers or to redistribute to lists, requires prior specific permission and/or a fee.

DAC '13, May 29 - June 07 2013, Austin, TX, USA.

run to complete on a modern computer. Combined, these two challenges are a major roadblock towards the realization of the dream of automatic ESL design.

We make several contributions to address these critical challenges:

- We approach the DSE problem from a machine-learning viewpoint. We propose *Random Forest* [2], a learning model for HLS that is superior to the best known models.

- We present new methods to apply a state-of-the-art DSE framework, which has been proposed for processor design [8, 11, 15] and IP-block macro generators [17], to the context of HLS. Our methods consist of more accurate learning models particularly tailored for HLS. Moreover, for the *first* time in the CAD community, we introduce Transductive Experimental Design (TED) [16], which can judiciously sample *representative* and *hard-to-predict* micro-architecture choices, and use them for training the learning models.

- For complex high-level specifications, we point out a major scalability issue of the existing DSE framework [8, 11, 15, 17], and propose novel and scalable algorithms, which are inspired by recent advancement of machine-learning theory.

2. RELATED WORK

Due to the large solution space, general DSE algorithms rely on local-search techniques, e.g., Genetic Algorithms [7] or Simulating Annealing [13]. When applied with CAD tools, however, these algorithms require at every step actual simulation/synthesis to acquire solution qualities, thereby still suffering from long simulation/synthesis runtime.

Learning-based methods were proposed to guide the DSE process by predicting solution qualities before running actual simulation/synthesis [1, 8, 10, 11, 17]. Compared with local-search techniques, learning-based methods can yield better solution quality as well as require shorter simulation/synthesis runtime. Among these methods, Markov Decision Process was adopted in [1], but this approach may not scale well as it intrinsically traverses an exponentially-growing state space. Alternatively, an efficient framework based on *iterative refinement* was first presented in [11]. Within this framework, two very recent papers [8, 17] independently reported that Gaussian Process [17], a.k.a. Kriging [8], was the most promising learning model, superior to Artificial Neural Network [10, 11] and other simple models. These results were obtained from DSE with *processor simulators* [8] or *IP generators* [17].

In the context of DSE with HLS, learning-based methods were just adopted recently [3, 15]. These works still rely on either local-search techniques [3] or common learning models [3, 15]. In contrast, our work proposes a novel model particularly for HLS, and presents an effective model-training scheme. Moreover, with theory guarantee, our DSE methods are scalable for complex ASIC design.

Table 1: Some Common HLS Knobs with Settings

Knob	Setting
Loop Manipulation	Breaking, Unrolling, Pipelining
State Insertion	Adding State Registers
Array Implementation	Registers, Embedded Memory
Function Inlining	Yes or No

3. PROBLEM FORMULATION

We refer to the directives used in a HLS script for determining micro-architecture choices as *knobs*. Table 1 lists some common knobs and their settings available in modern HLS tools. Depending on the knob types, the choices of each knob setting can vary. For example, while the choices of function inlining is binary (either to inline or not to), loop manipulation offer many alternative choices. Let us denote by c the max number of choices of a knob setting.

Given a design specified in high-level languages, e.g. C or SystemC, HLS tools generate an optimized RTL by taking a set of knob settings as input constraints. For a given design specification, the number of places where HLS knobs can be applied can vary and is generally large. For instance, even a small SystemC design may have many functions, many (nested) loops, many arrays, and so on. Let p denote the number of knob-applicable places. Then the total combination of knob settings grows exponentially ($O(c^p)$).

The huge knob-setting space makes DSE with HLS very different from DSE with processor simulators or IP generators. In the latter cases, p can be assumed a small constant in practice. For example, for processor simulators, only a limited set of parameters needs to be determined, e.g., the number of cores, the size of L1/L2 cache, etc., and similarly for IP generators, for which the parameters are the I/O size, algorithms, etc. Instead, p is typically a large constant for DSE with HLS, which therefore becomes an even more challenging problem. Since to exhaustively explore the set of of all possible RTL designs that can be obtained with HLS is unfeasible, the goal of DSE with HLS is to derive an approximation of the set of Pareto-optimal designs.

We consider the *DSE with HLS* problem as follows.

PROBLEM 1. *Given a high-level design specification and HLS tool with a budget of b runs, find the best approximate Pareto-optimal set of RTL designs without exceeding b.*

Note that we consider a multi-objective DSE problem. Although throughout this paper we focus on 2-objective cases for simplicity, all our discussions and findings are generally applicable to higher dimensional cases.

In order to measure the quality of an approximate set of Pareto-optimal designs, we utilize the metric of *average distance from reference set* (ADRS) [11]. Consider a two-dimensional (area $\mathcal{A}$ vs. effective-latency $\mathcal{T}$) design space[1]. For both objectives, the smaller the objective, the better the RTL implementation. Given a reference Pareto set $\Pi = \{\pi_1, \pi_2, \dots \mid \pi_i = (a, t), a \in \mathcal{A}, t \in \mathcal{T}\}$ and an approximate Pareto set $\Lambda = \{\lambda_1, \lambda_2, \dots \mid \lambda_j = (a, t), a \in \mathcal{A}, t \in \mathcal{T}\}$,

$$\text{ADRS}(\Pi, \Lambda) = \frac{1}{|\Pi|} \sum_{\pi \in \Pi} \min_{\lambda \in \Lambda} \delta(\pi, \lambda),$$

where

$$\delta(\pi = (a_\pi, t_\pi), \lambda = (a_\lambda, t_\lambda)) = \max\{0, \frac{a_\lambda - a_\pi}{a_\pi}, \frac{t_\lambda - t_\pi}{t_\pi}\}.$$

Note that the lower $\text{ADRS}(\Pi, \Lambda)$, the closer the approximate set Λ to the reference set Π.

---

[1] We define the *effective latency* as the product of the clock period and the clock cycle count.

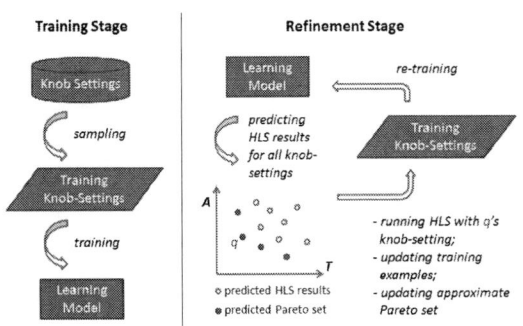

Fig. 1: The iterative-refinement DSE framework [8, 11, 15, 17].

Algorithm 1 Iterative-Refinement Framework

Input: HLS tool H, HLS budget b, input design D
Output: Approximate Pareto set Λ
1: Let $\mathcal{K}$ be the knob-setting space of D
2: Let $\tilde{H}$ be a learning model
3: Let $\tilde{\mathcal{K}} \subset \mathcal{K}$ be a training set
4: Let $S = \phi$ be the set of all HLS results
5: /** Training Stage **/
6: synthesize all $\tilde{\mathbf{k}} \in \tilde{\mathcal{K}}$; add the results to S
7: remove $\tilde{\mathcal{K}}$ from $\mathcal{K}$
8: train $\tilde{H}$ by $(\tilde{\mathcal{K}}, S)$
9: /** Refinement Stage **/
10: $\Lambda \leftarrow$ current Pareto approximation of S
11: **for** $i = |S| + 1 \rightarrow b$ **do**
12: $\quad \tilde{\mathcal{Q}} \leftarrow$ predicted HLS results $\forall \mathbf{k} \in \mathcal{K}$ by $\tilde{H}$
13: $\quad$ pick one $q \in \tilde{\mathcal{Q}}$ for HLS; add q's result to S
14: $\quad$ move q's knob setting from $\mathcal{K}$ to $\tilde{\mathcal{K}}$
15: $\quad$ retrain $\tilde{H}$ by $(\tilde{\mathcal{K}}, S)$
16: $\quad \Lambda \leftarrow$ current Pareto approximation of S
17: **end for**

---

4. PRELIMINARY

We start with a case study of DFT designed in SystemC. The DFT design implements an iterative-FFT algorithm [5] with a synthesizable fix-point library using 45nm technology. The total number of knob-setting combinations for the DFT design is 360,448. In order to establish a ground truth for the analysis of Pareto optimality, we synthesized a restricted set of 242 knob settings, which cover 11 loop manipulations, 2 function-inlining choices, and 11 feasible clock periods. The 242 HLS runs took an aggregated 84-hour CPU time with a commercial HLS tool. Throughout this paper, we refer to the 242 knob settings as the knob-setting space of the DFT.

Next we review a state-of-the-art DSE framework [8, 11, 15, 17], which is also illustrated in Fig. 1. The main idea of the framework is twofold: *(i)* to approximate the HLS tool H by a learning model $\tilde{H}$, and *(ii)* to use the fast $\tilde{H}$ for predicting the quality-of-result space, instead of invoking the time-consuming H. Algorithm 1 describes the major steps of the framework. Initially, the framework trains $\tilde{H}$ by spending some HLS runs (the *training* stage in Lines 6–8), and then finds an approximate Pareto set by iteratively refining $\tilde{H}$ and the current Pareto approximation (the *refinement* stage in Lines 9–16).

5. ALGORITHMS

Clearly, the effectiveness of the iterative-refinement framework relies on the accuracy of $\tilde{H}$. Intuitively, guided by a highly-accurate $\tilde{H}$, the refinement stage could search in the right space. However,

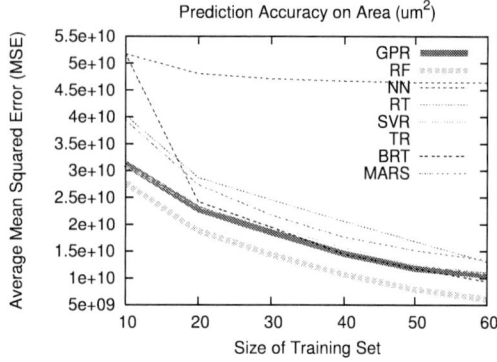

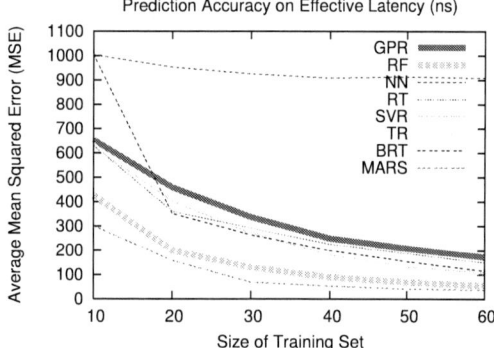

Fig. 2: Learning-model accuracy for predicting DFT area (top) and effective latency (bottom). The training sets are randomly sampled.

spending too many HLS runs in the training stage while aiming for highly-accurate $\tilde{H}$, may not necessarily guarantee the final Pareto optimality, since the total number of HLS runs is limited to a given budget b. We approach this dilemma by suggesting a more accurate learning model for HLS in Section 5.1 and a more effective training scheme that is beneficial for general models in Section 5.2. Moreover, in Sections 5.3 and 5.4, we identify a scalability issue of the framework, and propose novel algorithms. All our algorithmic findings are inspired by recent machine-learning theory.

5.1 Learning Models for HLS

We examined eight advanced learning models for predicting the DFT's area and effective latency. These models include Gaussian Process Regression (GPR), Random Forest (RF), Neural Network (NN), Regression Tree (RT), Support Vector Regression (SVR), Transductive Regression (TR), Boosted Regression Tree (BRT), and Multivariate Adaptive Regression Splines (MARS)[2]. Note that we examined five of these models (namely RF, SVR, TR, BRT, and MARS) for the first time in the DSE literature.

Fig. 2 shows the prediction accuracy of these models that were trained on randomly-sampled training sets[3]. The results suggest that RF is consistently more accurate than GPR, which was previously reported as the best model for processor simulators and IP generators. The adoption of RF for HLS instead of GPR brings the following benefits.

---

[2] All these models are publicly available in separate R packages [12], except TR [6], which we implemented in R by ourselves. Notice that we finely tuned all these models via *model selection* [14] to achieve the best accuracy.

[3] Throughout this paper, all the results involving randomized procedures are the average results obtained from 100 trials.

Algorithm 2 Sequential TED

Input: Set $\mathcal{K}$ of n knob settings, training-set size m
Output: Training set $\tilde{\mathcal{K}}$
1: $\mathbf{F} \leftarrow \mathbf{F}_{\mathbf{k},\mathbf{k}}$
2: $\tilde{\mathcal{K}} \leftarrow \phi$
3: **for** $i = 1 \rightarrow m$ **do**
4:     select $\mathbf{k}_i \in \mathcal{K}$ with the largest $||\mathbf{F}_{\mathbf{k}_i}||^2/(f(\mathbf{k}_i, \mathbf{k}_i) + \mu)$, where $\mathbf{F}_{\mathbf{k}_i}$ and $f(\mathbf{k}_i, \mathbf{k}_i)$ are $\mathbf{k}_i$'s corresponding column and diagonal entry in current $\mathbf{F}$
5:     add $\mathbf{k}_i$ to $\tilde{\mathcal{K}}$,
6:     $denom \leftarrow f(\mathbf{k}_i, \mathbf{k}_i) + \mu$
7:     $(\mathbf{F})_{jk} \leftarrow (\mathbf{F})_{jk} - \frac{(\mathbf{F})_{ji}(\mathbf{F})_{ki}}{denom}, \forall 1 \leq j, k \leq n$
8: **end for**

- HLS knobs which provide *binary* choices are common, e.g., function inlining for a function call, state insertion for a certain edge in the control/data flow graph, array implementation in either registers or memory, and others. In this regard, GPR assumes that every knob variable follows a Gaussian distribution, which is obviously not proper for the binary-valued knobs. Instead, the tree-based RF model can easily handle binary decisions by introducing a node with two branches separating the two decisions.

- RF is an ensemble model consisting of multiple regression trees [2]. Given a training set, the set is internally and randomly partitioned for training individual trees. Then, the final prediction is made by collective vote from the individual trees. The two steps combined are capable of minimizing both the generalization error and prediction variance. A recent study in the Machine Learning literature also shows the superior accuracy of RF, especially for *high dimensional* data [4].

- We observe that the CPU time required for training and prediction with RF is around 50% less than that with GPR. Moreover, the internal partitioning-then-training scheme by nature makes RF suitable for running on multi-core machines. From an implementation viewpoint, this is another advantage of adopting RF.

Given $\tilde{H}$ being either GPR or RF, we conclude this section by discussing how to select a best knob-setting for next HLS in the refinement stage, i.e., Line 13 in Algorithm 1. **For GPR**, the prediction of a design objective consists of a mean and a variance. The mean represents the predicted value, and the variance suggests the uncertainty of GPR about the prediction. Therefore, as suggested in [17], a *predicted* Pareto set $\tilde{\Lambda}$ is first extracted from the predicted objective space $\tilde{\mathcal{Q}}$ (i.e., Line 12 in Algorithm 1), and then the element $\in \tilde{\Lambda}$ with the max variance (uncertainty) across all objectives is selected for the next HLS. On the other hand, **for RF**, since the prediction uncertainty is minimized by RF's collective-vote scheme, we just randomly pick one element $\in \tilde{\Lambda}$ for next HLS.

5.2 Transductive Experimental Design (TED)

In the prior work [8, 17], the training set $\tilde{\mathcal{K}}$ is randomly sampled from $\mathcal{K}$ (Line 3 in Algorithm 1). Alternatively, we introduce *transductive experimental design* (TED) [16], that aims for selecting *representative* as well as *hard-to-predict* $\tilde{\mathcal{K}}$, in order to effectively train the learning model for predicting $\mathcal{K}$. *Note that TED assumes no priori knowledge about the learning model and should therefore be beneficial for any model.*

Assume that overall we have n knob settings ($|\mathcal{K}| = n$), from which we want to select a training set $\tilde{\mathcal{K}}$ such that $|\tilde{\mathcal{K}}| = m$. In

350

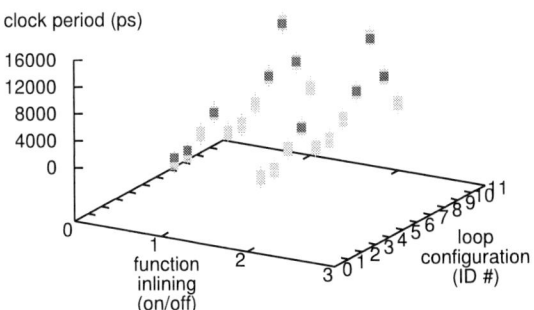

DFT Knob-Setting Space

Fig. 3: Training-set sampling by Transductive Experimental Design (TED). See Section 5.2 for details.

order to minimize the prediction error $H(\mathbf{k}) - \tilde{H}(\mathbf{k})$ for all $\mathbf{k} \in \mathcal{K}$, TED is shown to be equivalent to the following problem.

$$\max_{\tilde{\mathcal{K}}} \quad T\left[\mathcal{K}\tilde{\mathcal{K}}^\top(\tilde{\mathcal{K}}\tilde{\mathcal{K}}^\top + \mu\mathbf{I})^{-1}\tilde{\mathcal{K}}\mathcal{K}^\top\right] \\ \text{s.t.} \quad \tilde{\mathcal{K}} \subset \mathcal{K}, |\tilde{\mathcal{K}}| = m, \tag{1}$$

where $T[\,]$ is a matrix trace and $\mu > 0$ given. The solution to the problem can be interpreted as follows. It tends to find *representative* data samples $\tilde{\mathcal{K}}$ that span a linear space to retain most of the information of $\mathcal{K}$ [16].

Equation 1 corresponds to TED sampling for *linear* regression. For *non-linear* regression, TED can be extended by a kernel function

$$f(\mathbf{u}, \mathbf{v}) = \theta(\mathbf{u}) \cdot \theta(\mathbf{v}),$$

where $\mathbf{u}, \mathbf{v} \in \mathcal{R}^d, \theta : \mathcal{R}^d \to \mathcal{F}$ is a function mapping from the knob-setting space to a feature space. In this paper, we use Gaussian kernel as in [16]:

$$f(\mathbf{u}, \mathbf{v}) = e^{\frac{||\mathbf{u}-\mathbf{v}||^2}{2\sigma^2}},$$

where σ is a given non-zero constant. Then, kernelized TED can be expressed as

$$\max_{\tilde{\mathcal{K}}} \quad T\left[\mathbf{F}_{\mathbf{k}\tilde{\mathbf{k}}}(\mathbf{F}_{\tilde{\mathbf{k}}\tilde{\mathbf{k}}} + \mu\mathbf{I})^{-1}\mathbf{F}_{\tilde{\mathbf{k}}\mathbf{k}}\right] \\ \text{s.t.} \quad \tilde{\mathcal{K}} \subset \mathcal{K}, |\tilde{\mathcal{K}}| = m, \tag{2}$$

where matrix entries $(\mathbf{F}_{\mathbf{k}\tilde{\mathbf{k}}})_{ij} = f(\mathbf{k}_i, \tilde{\mathbf{k}}_j), (\mathbf{F}_{\tilde{\mathbf{k}}\tilde{\mathbf{k}}})_{ij} = f(\tilde{\mathbf{k}}_i, \tilde{\mathbf{k}}_j)$, $(\mathbf{F}_{\tilde{\mathbf{k}}\mathbf{k}})_{ij} = f(\tilde{\mathbf{k}}_i, \mathbf{k}_j)$, vectors $\mathbf{k}_i, \mathbf{k}_j \in \mathcal{K}$, vectors $\tilde{\mathbf{k}}_i, \tilde{\mathbf{k}}_j \in \tilde{\mathcal{K}}$.

Unfortunately, both TED problems (Equations 1 and 2) are proven to be NP-hard [16]. Therefore, we apply an efficient greedy algorithm, *sequential TED* [16], to solve Equation 2. The algorithm (given as Algorithm 2) can be interpreted as follows. Once a best $\mathbf{k}_i \in \mathcal{K}$ is selected (Line 4), the kernel matrix $\mathbf{F}$ is updated (Line 7) to represent the residual knob settings, so that the next selection would be picked among those under-represented by previously selected settings. In other words, the algorithm tends to select knob settings that cannot be well represented by selected ones, i.e., to favor potentially *hard-to-predict* knob settings if not being selected as training examples.

Fig. 3 illustrates the TED sampling results of the DFT knob-setting space. We follow the suggestion in [16] to set $\mu = 0.1$ for Algorithm 2 and $\sigma = 0.1$ for Gaussian kernel. As a result, a training set of 10 knob settings (red squares) is selected from the 242 candidates (green pluses). We can see that TED indeed selects *representative* samples by distributing its selections in the whole space, without having dense clusters. Besides, TED distributes 6 out of the 10 selections to the subspace where the loop-

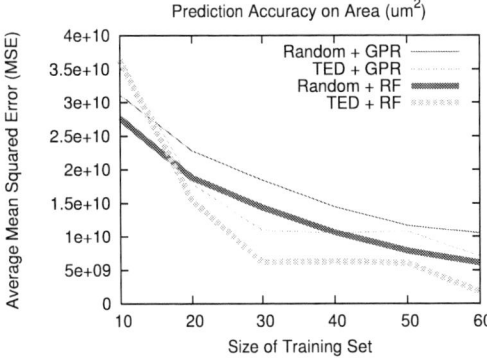

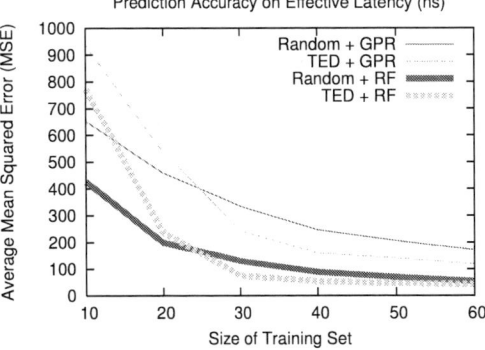

Fig. 4: Learning-model accuracy for predicting DFT area (top) and effective latency (bottom). "Random" and "TED" indicates training-set sampling algorithms. "GPR" and "RF" are learning models.

configuration ID number is greater than 6, a subspace where candidates are sparser than its counterpart. The loop configurations are generated by varying one loop-manipulation knob for consecutive configurations. Hence, the knob settings (green pluses) that are close to each other are very likely to produce similar HLS results. Based on this observation, TED considers the samples in the sparse subspace *hard-to-predict*, thereby selecting more samples there.

Fig. 4 shows the prediction accuracy of the combination of random/TED sampling with GPR/RF models. Starting from the sample size of 20 knob-settings for predicting area and 30 knob-settings for predicting effective latency, TED indeed reduces the prediction error for GPR and RF, respectively (see the dashed vs. solid lines). On the other hand, for very small training-set size, random sampling is helpful to escape from local optimal where the sequential-TED algorithm could be trapped. In general, TED is very effective for sampling good training sets for any learning models. Finally, we remark that with the aid of TED sampling, the prediction error of RF is still consistently lower than that of GPR (see the dashed lines in Fig. 4).

5.3 Randomized Selection

The size of knob-setting space $\mathcal{K}$ grows exponentially as explained in Section 3. Therefore, even if a very fast learning model $\tilde{H}$ is used, the exhaustive search in $\mathcal{K}$ (Line 12 in Algorithm 1) makes the iterative-refinement framework not scalable. From now on, we refer to Problem 1 with very large $\mathcal{K}$ as the **Extreme DSE-with-HLS Problem**, and refer to Problem 1 with tractable $\mathcal{K}$ as the **Basic DSE-with-HLS Problem**.

To conquer the extreme DSE-with-HLS problem, we introduce the following theorem on *randomized selection* [14].

Table 2: Comparison of DSE Methods

Method	Reference	Learning Model	Training-Set Sampling	Next-HLS Selection
state-of-the-art	[17]	GPR	random sampling	exhaustive search
basic	Section 5.1	RF	random sampling	exhaustive search
basic-ST	Section 5.2	RF	sequential TED	exhaustive search
extreme	Section 5.3	RF	random sampling	randomized selection
extreme-RT	Section 5.4	RF	randomized TED	randomized selection

---

Algorithm 3 Randomized TED

Input: Set $\mathcal{K}$ of n knob settings, training-set size m
Output: Training set $\tilde{\mathcal{K}}$
1: $\tilde{\mathcal{K}} \leftarrow \phi$
2: **for** $i = 1 \rightarrow m$ **do**
3:   $M = \{\mathbf{m}_1, \ldots, \mathbf{m}_{N_{rted}}\} \leftarrow$ a random subset of $\mathcal{K}$
4:   $\tilde{M} = \tilde{M} \cup \tilde{\mathcal{K}}$; $\mathbf{F} \leftarrow \mathbf{F}_{\mathbf{m},\mathbf{m}}$
5:   $\forall \mathbf{m}_i \in \tilde{\mathcal{K}}$, update $\mathbf{F}$ as Lines 6–7 in Algorithm 2
6:   select $\mathbf{m}_i \in \tilde{M}$ as Line 4 in Algorithm 2; add $\mathbf{m}_i$ to $\tilde{\mathcal{K}}$
7: **end for**

THEOREM 1 (**Ranks on Random Subsets**). *Let $M = \{x_1, \ldots, x_\alpha\} \subset \mathcal{R}$, and let $\tilde{M} \subset M$ be a random subset of size β. Then the probability that $\max \tilde{M}$ is greater than γ elements of M is at least $1 - \left(\frac{\gamma}{\alpha}\right)^\beta$.*

According to the theorem, if we draw a random subset $\tilde{M}$ of size 59 ($\beta = 59$), then $\max \tilde{M}$ would be greater than 95% ($\frac{\gamma}{\alpha} = 95\%$) elements of M with at least $1 - 5\% = 95\%$ confidence, since $\left(\frac{\gamma}{\alpha}\right)^\beta = 0.95^{59} < 5\%$. Note that the sample size 59 is a *constant* for any large-sized M to achieve a 95% confidence in the 95% percentile range.

Based on Theorem 1, we propose a simple modification for selecting the next HLS in Algorithm 1 (Lines 12–13): draw a random subset $\tilde{M} \subset \mathcal{K}$ of size N_{next}, and then pick the $q \in \tilde{M}$ with the smallest

$$\tilde{H}_A(q) + \tilde{H}_T(q), \qquad (3)$$

where $\tilde{H}_A(q)$ and $\tilde{H}_T(q)$ are the predicted area and effective latency of q, respectively. Clearly, Theorem 1 also applies to selecting a minimum element. Consequently, we pick the best $q^* \in \tilde{M}$ for HLS, based on the cost function defined as Equation 3, which predicts the quality-of-result of any $q \in \tilde{M}$. Note that to set $N_{next} = 59$ should suffice to achieve a very good approximation even for a very large $\mathcal{K}$, as inferred from Theorem 1.

5.4 Randomized TED

Consider again the prohibitively large size of $\mathcal{K}$ of the extreme DSE-with-HLS problem. This makes even the training-set sampling by TED not scalable, since an $O(|\mathcal{K}|^2)$ matrix computation is required in the sequential-TED algorithm. To address this scalability issue, inspired again by Theorem 1, we propose the *randomized TED* algorithm (as Algorithm 3). The main idea is that in each iteration we draw a random subset $\tilde{M} \subset \mathcal{K}$ of size N_{rted}, and add previously selected samples to $\tilde{M}$. $\tilde{M}$ is now treated as the $\mathcal{K}$ in the sequential-TED algorithm. We update the kernel matrix as if the previously selected samples are selected again in this iteration. Then we follow the same criterion as in the sequential-TED algorithm to select the best residual $\mathbf{m} \in \tilde{M}$, and we iterate until m samples are selected.

Overall, our algorithm utilizes the randomized-selection scheme to reduce the computational cost, while still preserving the principle of TED. Note that since we want to approximate the best m elements, as opposed to the best one only, we should expect the constant N_{rted} to be larger than 59 in order to achieve an equivalent approximation as discussed in Section 5.3.

Table 3: Number of HLS Runs to Find the Exact Pareto Set

Training-Set Size	10	20	30	40	50	60
state-of-the-art	120	NA	NA	NA	NA	NA
basic	115	113	NA	NA	NA	NA
basic-ST	112	91	100	100	109	119

6. EXPERIMENTAL RESULTS

We continue our DFT case study to compare the four DSE methods that we presented in the previous section among them and with respect to the existing method proposed in [17], which we denote as state-of-the-art. Table 2 summarizes the five methods. We call basic our method that utilizes RF as the learning model, random sampling for selecting training knob-settings, and exhaustive search for selecting the next HLS knob-setting. We call basic-ST the variant of basic that utilizes sequential TED for selecting training knob-settings. Then, we refer to our method that utilizes RF as the learning model, random sampling for selecting training knob-settings, and randomized selection for selecting the next HLS knob-setting, as extreme. And finally we call extreme-RT the variant of extreme that utilizes randomized TED for selecting training knob-settings.

For each DSE method, we prepared training sets of size $m \in \{10, 20, \ldots, 60\}$. For each value of m, we set a HLS budget $\in \{m+10, m+20, \ldots, 120\}$ and we tracked the average ADRS that each method can achieve, with the exact Pareto set as a reference set. The HLS budget was capped at 120, i.e., less than 50% of DFT's knob-setting space. The results of all the experiments are collected in Table 6, which is placed in the Appendix for space reasons. Note that across all training-set sizes, each method should have $11 + 10 + 9 + 8 + 7 + 6 = 51$ ADRS records.

Results for the basic DSE-with-HLS problem. For the basic problem, we compare three methods: basic, basic-ST, and state-of-the-art.

Table 3 summarizes the total number of HLS runs that each method requires to find the exact Pareto set (i.e., ADRS = 0). We can see that basic-ST, which utilizes TED to sample training sets, is the only method that can find the exact Pareto set within 120 HLS runs for any training-set size. Besides, for any training-set size, both basic and basic-ST, which adopt the RF learning model, outperform state-of-the-art, which adopts GPR instead. These results confirm that a more accurate learning model with a more effective training-set sampling scheme (in our case, RF with TED) indeed facilitates the convergence of the iterative-refinement framework. Also note that the best training-set size happens at 20, followed by 30 and 40. This result suggests that basic-ST only requires small training sets to achieve its best performance, because TED can judiciously select those *representative* and *hard-to-predict* knob-settings as training examples.

As explained in Section 3, in real applications of DSE with HLS, it is infeasible to search for the exact Pareto set. Therefore, we now examine the performance of those three methods given different HLS budgets. For HLS budget $b \in \{20, 30, \ldots, 120\}$, Table 4 summarizes the best DSE method, its average ADRS, and the training-set size it requires. We can observe that for smaller budgets ($20 \leq b \leq 50$), basic (abbreviated as bs in Table 4) can yield the

352

Table 4: `state-of-the-art` [17] vs. `basic (bs)` vs. `basic-ST (bs-ST)` given HLS budget b

Budget b	20	30	40	50	60	70	80	90	100	110	120
Best Method	bs	bs	bs	bs	bs-ST	bs-ST	bs-ST	bs-ST	bs-ST	bs-ST	bs-ST
Average ADRS (%)	21.54	9.01	3.58	1.74	0.98	0.21	0.11	0.01	0.00	0.00	0.00
Training-Set Size	10	20	20	20	30	30	30	20–30	20–40	20–50	>=10

Table 5: `extreme (ex)` vs. `extreme-RT (ex-RT)` given HLS budget b

Budget b	20	30	40	50	60	70	80	90	100	110	120
Best Method	ex-RT	ex-RT	ex-RT	ex-RT	ex-RT	tie	ex	ex	ex-RT	ex-RT	tie
Average ADRS (%)	19.12	9.23	5.82	3.52	2.00	1.09	0.58	0.17	0.10	0.04	0.01
Training-Set Size	10	20	20	20	20	30/50	40	40	40	40	40

minimum average-ADRS, requiring a training-set of size around 20. For greater budgets ($60 \leq b \leq 120$), `basic-ST` (abbreviated as `bs-ST`) stands out to achieve the minimum average-ADRS, requiring a training-set of size around 30. In general, `basic-ST` starts with a higher ADRS because TED sampling does not favor the knob-settings that can result in Pareto-optimal RTLs, whereas they could be sampled at random instead. However, `basic-ST` can approach the exact Pareto front faster, due to the more accurate learning model resulting from the aid of TED. These two reasons combined explain why `basic` performs better for smaller HLS budgets, whereas `basic-ST` is better for greater budgets. Overall, both our methods, `basic` and `basic-ST`, outperform `state-of-the-art`.
Results for the extreme DSE-with-HLS problem. For the extreme problem, we focus on two methods: `extreme` and `extreme-RT`.

First we compare `extreme` with `basic` to see how effective the approximation using randomized-selection can be. For `extreme`, we set $N_{next} = 59$ for drawing a random subset of size 59 to select the next knob-setting for HLS. As expected, most of the ADRS achieved by `extreme` are higher (worse) than those achieved by `basic`, but we find that the difference is small: for training-set size $10 \leq m \leq 30$, the max difference is less than 3.0%, and for $40 \leq m \leq 60$, the max difference is even less than 0.8%. These small differences show that the approximation using randomized-selection can be very effective. Moreover, since the ADRS differences are marginal, we observed that 41 out of 51 (80%) ADRSs achieved by `extreme` are also lower (better) than those achieved by `state-of-the-art`. For $N_{next} \in \{59, 79, 99\}$, we see no significant difference on ADRS: in fact, 46 out of 51 (90%) ADRS differences are less than 0.9%. Therefore, we set $N_{next} = 59$ by default for `extreme`.

Next, we show the results of `extreme-RT`. For `extreme-RT`, we fix its $N_{next} = 59$ and set $N_{rted} \in \{59, 79, 99\}$ for drawing a random subset of size N_{rted} to select a training knob-setting by the randomized-TED algorithm. Compared with `extreme`, if `extreme-RT`'s $N_{rted} = 59$, we observe that only 17 out of 51 (33%) ADRSs achieved by `extreme-RT` are lower (better). If we increase N_{rted} to 79, then 30 out of 51 (59%) ADRSs are better. The improvement saturates at $N_{rted} = 99$, where 31 out of 51 (61%) ADRSs are better. As a result, the randomized-TED algorithm works best with $N_{rted} = 99$. Therefore, we set $N_{rted} = 99$ by default for `extreme-RT`.

Finally, for the extreme problem with different HLS budgets, Table 5 summarizes the best method, its average ADRS, and the training-set size it requires. We see that `extreme-RT` (abbreviated as `ex-RT`) almost outperforms `extreme` (abbreviated as `ex`) for every budget b. This result is different from the `basic` vs. `basic-ST` result (see Table 4), where `basic` (`basic-ST`) works better for smaller (greater) budgets. This is because the random-subset drawing used in the randomized-TED algorithm can alleviate the high initial ADRS otherwise caused by the sequential-TED algorithm. Overall, we find `extreme-RT` very effective for addressing the ex-

treme problem, because our randomized-TED algorithm successfully avoids the exhaustive search in the full knob-setting space while still selecting *representative* and *hard-to-predict* knob-settings as training examples.

7. CONCLUSIONS

We have presented novel learning-based methods for DSE with HLS. Our methods based on the Random-Forest learning model, transductive experimental design, and randomized selection, can effectively find an approximate Pareto set of RTL designs.

Acknowledgments. This work is partially supported by an ONR Young Investigator Award, the National Science Foundation (Award #1219001), and the DARPA Perfect program.

8. REFERENCES

[1] G. Beltrame, L. Fossati, and D. Sciuto. Decision-theoretic design space exploration of multiprocessor platforms. *IEEE TCAD*, 29(7):1083–1095, July 2010.

[2] L. Breiman. Random forests. *Mach. Learn.*, 45(1):5–32, Oct. 2001.

[3] B. Carrion Schafer and K. Wakabayashi. Machine learning predictive modelling high-level synthesis design space exploration. *Computers Digital Techniques, IET*, 6(3):153–159, May 2012.

[4] R. Caruana, N. Karampatziakis, and A. Yessenalina. An empirical evaluation of supervised learning in high dimensions. In *Proc. of the 25th Intl. Conf. on Machine learning*, pages 96–103, 2008.

[5] T. H. Cormen, C. E. Leiserson, R. L. Rivest, and C. Stein. *Introduction to Algorithms*. The MIT Press, 3rd edition, 2009.

[6] C. Cortes and M. Mohri. On transductive regression. In *Advances in Neural Information Processing Systems (NIPS)*, pages 305–312, 2006.

[7] K. Deb, A. Pratap, S. Agarwal, and T. Meyarivan. A fast and elitist multiobjective genetic algorithm: Nsga-ii. *IEEE Trans. on Evolutionary Computation*, 6(2):182–197, Apr. 2002.

[8] G. Mariani, G. Palermo, V. Zaccaria, and C. Silvano. Oscar: An optimization methodology exploiting spatial correlation in multicore design spaces. *IEEE TCAD*, 31(5):740–753, May 2012.

[9] G. Martin and G. Smith. High-level synthesis: Past, present, and future. *IEEE Design Test of Computers*, 26(4):18–25, Aug. 2009.

[10] B. Ozisikyilmaz, G. Memik, and A. Choudhary. Efficient system design space exploration using machine learning techniques. In *Proc. of DAC*, pages 966–969, 2008.

[11] G. Palermo, C. Silvano, and V. Zaccaria. Respir: A response surface-based pareto iterative refinement for application-specific design space exploration. *IEEE TCAD*, 28(12):1816–1829, Dec. 2009.

[12] R Core Team. *R: A Language and Environment for Statistical Computing*. R Foundation for Statistical Computing, Vienna, Austria, 2012.

[13] B. C. Schafer, T. Takenaka, and K. Wakabayashi. Adaptive simulated annealer for high level synthesis design space exploration. In *Proc. of VLSI-DAT*, pages 106–109, Apr. 2009.

[14] B. Scholkopf and A. J. Smola. *Learning with Kernels: Support Vector Machines, Regularization, Optimization, and Beyond*. MIT Press, 2001.

[15] S. Xydis, G. Palermo, V. Zaccaria, and C. Silvano. A meta-model assisted coprocessor synthesis framework for compiler/architecture parameters customization. In *Proc. of DATE*, 2013.

[16] K. Yu, J. Bi, and V. Tresp. Active learning via transductive experimental design. In *Proc. of the 23rd Intl. Conf. on Machine learning*, pages 1081–1088, 2006.

[17] M. Zuluaga, A. Krause, P. Milder, and M. Püschel. "Smart" design space sampling to predict Pareto-optimal solutions. In *Proc. of LCTES*, pages 119–128, 2012.

Table 6: Average ADRS achieved by different DSE methods given HLS budget b. The parenthesized number after `extreme` denotes N_{next} used for selecting next HLS knob-setting. The parenthesized number after `extreme-RT` denotes N_{rted} used in the randomized TED algorithm. For `extreme-RT`, N_{next} is fixed to 59.

Budget b	20	30	40	50	60	70	80	90	100	110	120
Method	Average ADRS (%)										
Training-Set Size = 10											
state-of-the-art [17]	27.37	20.53	15.39	9.36	5.50	2.88	1.56	1.11	0.84	0.20	0.00
basic	21.54	14.95	7.13	3.85	2.23	1.45	1.05	0.50	0.17	0.02	0.00
basic-ST	165.79	98.43	34.63	11.37	6.88	3.82	2.13	0.93	0.24	0.01	0.00
extreme (59)	21.09	16.52	9.49	4.71	3.27	2.54	2.17	1.88	1.49	0.39	0.08
extreme (79)	22.02	13.11	6.79	5.22	3.41	2.70	2.30	1.97	1.27	0.36	0.07
extreme (99)	25.13	18.48	9.51	5.48	3.45	2.42	2.12	1.83	1.46	0.45	0.07
extreme-RT (59)	25.20	18.65	10.08	5.55	3.51	2.24	1.68	1.37	0.82	0.15	0.03
extreme-RT (79)	16.10	9.69	7.74	4.54	3.14	2.63	2.17	1.62	0.80	0.13	0.02
extreme-RT (99)	19.12	11.77	8.15	4.98	3.23	2.20	1.72	1.40	1.08	0.26	0.06
Training-Set Size = 20											
state-of-the-art [17]	NA	12.95	9.98	8.23	6.90	4.38	2.16	0.59	0.24	0.15	0.15
basic	NA	9.01	3.58	1.74	0.99	0.50	0.34	0.18	0.07	0.03	0.00
basic-ST	NA	15.25	12.39	7.47	3.23	0.97	0.31	0.01	0.00	0.00	0.00
extreme (59)	NA	10.31	6.57	4.02	2.49	1.64	1.22	0.94	0.70	0.24	0.04
extreme (79)	NA	11.50	6.83	4.23	2.25	1.47	1.12	0.95	0.68	0.30	0.09
extreme (99)	NA	11.01	7.20	4.82	2.82	1.86	1.55	1.41	0.97	0.37	0.12
extreme-RT (59)	NA	10.97	7.86	5.28	3.15	2.14	1.77	1.41	0.85	0.26	0.04
extreme-RT (79)	NA	9.45	6.09	3.71	1.56	0.95	0.75	0.70	0.64	0.25	0.03
extreme-RT (99)	NA	9.23	5.82	3.52	2.00	1.49	1.34	1.30	0.64	0.25	0.02
Training-Set Size = 30											
state-of-the-art [17]	NA	NA	10.43	6.98	4.06	2.88	1.88	1.10	0.28	0.15	0.03
basic	NA	NA	5.01	2.24	1.05	0.48	0.32	0.16	0.08	0.04	0.01
basic-ST	NA	NA	5.62	2.30	0.98	0.21	0.11	0.01	0.00	0.00	0.00
extreme (59)	NA	NA	7.20	3.72	2.13	1.09	0.64	0.48	0.27	0.21	0.07
extreme (79)	NA	NA	6.46	3.92	2.01	1.20	0.68	0.48	0.25	0.18	0.08
extreme (99)	NA	NA	7.58	4.55	2.23	1.35	0.72	0.39	0.23	0.17	0.15
extreme-RT (59)	NA	NA	8.15	4.30	1.96	1.05	0.65	0.51	0.40	0.21	0.03
extreme-RT (79)	NA	NA	6.95	4.21	2.36	1.31	0.84	0.70	0.64	0.41	0.20
extreme-RT (99)	NA	NA	7.02	3.89	2.26	1.32	0.91	0.62	0.45	0.39	0.03
Training-Set Size = 40											
state-of-the-art [17]	NA	NA	NA	8.16	5.33	3.02	1.43	1.28	1.07	0.33	0.18
basic	NA	NA	NA	4.81	1.86	1.07	0.48	0.27	0.10	0.03	0.01
basic-ST	NA	NA	NA	3.12	1.11	0.29	0.17	0.16	0.00	0.00	0.00
extreme (59)	NA	NA	NA	5.37	2.64	1.32	0.58	0.17	0.12	0.05	0.01
extreme (79)	NA	NA	NA	4.86	2.58	1.33	0.78	0.35	0.15	0.05	0.01
extreme (99)	NA	NA	NA	5.53	2.52	1.12	0.62	0.28	0.10	0.06	0.01
extreme-RT (59)	NA	NA	NA	5.50	2.85	1.78	1.09	0.52	0.40	0.24	0.05
extreme-RT (79)	NA	NA	NA	5.70	2.95	1.64	1.04	0.53	0.42	0.27	0.06
extreme-RT (99)	NA	NA	NA	4.90	2.95	1.78	0.96	0.19	0.10	0.04	0.01
Training-Set Size = 50											
state-of-the-art [17]	NA	NA	NA	NA	6.33	3.75	2.12	0.88	0.63	0.39	0.30
basic	NA	NA	NA	NA	3.33	1.31	0.69	0.42	0.25	0.09	0.03
basic-ST	NA	NA	NA	NA	2.22	0.56	0.29	0.19	0.14	0.00	0.00
extreme (59)	NA	NA	NA	NA	2.71	1.30	0.90	0.43	0.19	0.11	0.03
extreme (79)	NA	NA	NA	NA	3.36	1.64	0.89	0.43	0.17	0.09	0.03
extreme (99)	NA	NA	NA	NA	3.18	1.62	0.87	0.46	0.17	0.08	0.03
extreme-RT (59)	NA	NA	NA	NA	3.65	1.97	1.22	0.87	0.55	0.42	0.26
extreme-RT (79)	NA	NA	NA	NA	3.19	1.49	0.82	0.39	0.16	0.07	0.04
extreme-RT (99)	NA	NA	NA	NA	2.40	1.09	0.74	0.56	0.34	0.27	0.19
Training-Set Size = 60											
state-of-the-art [17]	NA	NA	NA	NA	NA	4.84	2.47	1.38	0.71	0.46	0.23
basic	NA	NA	NA	NA	NA	2.77	0.85	0.55	0.38	0.20	0.09
basic-ST	NA	NA	NA	NA	NA	2.38	0.79	0.32	0.19	0.09	0.00
extreme (59)	NA	NA	NA	NA	NA	2.41	1.21	0.73	0.41	0.18	0.09
extreme (79)	NA	NA	NA	NA	NA	2.62	1.16	0.66	0.37	0.13	0.07
extreme (99)	NA	NA	NA	NA	NA	2.19	1.10	0.65	0.39	0.14	0.08
extreme-RT (59)	NA	NA	NA	NA	NA	1.97	0.90	0.61	0.39	0.14	0.05
extreme-RT (79)	NA	NA	NA	NA	NA	1.93	0.91	0.62	0.39	0.11	0.06
extreme-RT (99)	NA	NA	NA	NA	NA	1.78	0.94	0.61	0.37	0.14	0.07

Runtime Dependency Analysis for Loop Pipelining in High-Level Synthesis

Mythri Alle
IRISA/University of Rennes 1
mythri.alle@irisa.fr

Antoine Morvan
INRIA/ENS cachan
antoine.morvan@inria.fr

Steven Derrien
IRISA/University of Rennes 1
steven.derrien@irisa.fr

ABSTRACT

Research on High-Level Synthesis has mainly focused on applications with statically determinable characteristics and current tools often perform poorly in presence of data-dependent memory accesses. The reason is that they rely on conservative static scheduling strategies, which lead to inefficient implementations. In this work, we propose to address this issue by leveraging well-known techniques used in superscalar processors to perform runtime memory disambiguation. Our approach, implemented as a source-to-source transformation at the C level, demonstrates significant performance improvements for a moderate increase in area while retaining portability among HLS tools.

1. INTRODUCTION

High-Level Synthesis (HLS) enables the derivation of custom hardware from high-level algorithmic specification (in C, C++, SystemC, etc.) There exists several robust and mature HLS tools [1, 2] used as production tools by world-class chip vendor companies. Even though these tools provide impressive improvement in productivity, there is a large gap between "accepted" codes and "efficiently handled" codes. They hence rely on the designer to deeply restructure the program source code and to use sophisticated compiler directives (usually in the form of #pragmas) to drive the synthesis flow.

Many of the techniques used in HLS borrow from earlier research results on optimizing compiler back-end for DSP or VLIW processors. In such a compiler, instructions are statically scheduled. Such an approach prevents scheduling optimization opportunities for HLS tools. This choice in design flow has not yet been questioned and this is easy to explain: the application domain targeted by HLS mostly consists of kernels for which the aforementioned approaches perform relatively well.

However, as HLS usage becomes widespread, it is likely that user expectations and target application domains will expand beyond their current needs. For example, current generation HLS tools perform poorly when trying to schedule computations that have data-dependent memory ac-

cesses. For these kernels, it is generally not possible to assert at compile time that two array/memory accesses will *never* alias[1], even if they very rarely (or even never) do in practice. In such situations, HLS tools fallback to a conservative (worst case static) scheduling, which guarantees correctness but is inefficient.

While this problem may currently be perceived as a *corner case* by HLS tool providers, we believe that improving the support for such dynamic behavior is important to broaden the use of HLS to more application domains. For example, recent research work advocated the use of FPGA for accelerating data-analytic applications [3] and sparse linear algebra operations [4]. Both of these domains contain data-dependent memory accesses. For such application domains, providing efficient high-level design tools is a key issue and current HLS tools are clearly not ready for that.

In this work, we study how dynamic scheduling techniques can be used to improve the efficiency of HLS tools in the presence of complex data-dependent memory accesses. More precisely, we make the following two contributions:

- We propose a technique based on *runtime memory disambiguation* to improve the efficiency of loop pipelining in the presence of data-dependent memory dependencies.

- We implement our approach as a semi-automatic (i.e., user driven) source-to-source transformation, enabling portability among HLS tools.

We validate our approach on a set of representative kernels using two leading-edge HLS tools, with FPGA as target technology. Our results show that our technique can lead to significant improvement in throughput at the price of moderate area overhead.

The remainder of the article is organized as follows. Section 2 describes the problem we address in this work. Section 3 describes our technique and its implementation. Section 4 provides experimental validation of our approach. Conclusion and future direction are discussed in Section 5. We provide detailed algorithms and results in Appendix A and Appendix B respectively. Appendix C presents a survey of related work. Current limitations of our approach are discussed in Appendix D.

2. PROBLEM STATEMENT

Loop pipelining is a key optimization in High-Level Synthesis tools. It builds on the software pipelining technique

---

Permission to make digital or hard copies of all or part of this work for personal or classroom use is granted without fee provided that copies are not made or distributed for profit or commercial advantage and that copies bear this notice and the full citation on the first page. To copy otherwise, to republish, to post on servers or to redistribute to lists, requires prior specific permission and/or a fee.
DAC '13, May 29-Jun 07 2013, Austin, TX, USA

---

[1]That is, they will never access the same memory location.

proposed by Lam et al. back in 1988 [5]. In this technique, parallelism across loop iteration is exploited by initiating the next iteration of the loop before the completion of the current iteration. This allows instructions from several iterations to overlap and hence improve throughput. The throughput achieved is limited by the delay between initiations of two successive iterations. This delay is called as the initiation interval(II). Typically II is determined both by hardware constraints and data dependencies in the application. However, since HLS tools are not constrained by a predefine micro-architecture, data dependencies alone determine the minimum II that can be achieved. Thus, the efficiency of loop pipelining in the context of HLS is dependent on the accuracy of dependency analysis provided by these tools. Most HLS tools employ basic (hence conservative) dependency analysis techniques, which limit the applicability of loop pipelining.

For example, consider the loops shown in Figure 1. Loop L1 contains a non-uniform, non-affine dependency that is difficult to analyze. Hence, HLS tools are forced to make a conservative assumption that the write in the current iteration is consumed in the next iteration. This prevents the tool from pipelining this loop. To cope with this limitation, HLS tool vendors provide additional user directives (in the form of pragmas). These directives allow users to bypass the tool dependency analysis and force the compiler to generate a pipelined schedule. For example, asserting that the minimum number of iterations between dependent memory operations is 5 allows the HLS tool to overlap five iterations of the loop. This enables the tool to pipeline the loop.

```
1    #pragma pipeline, max_latency=5
2  L1: for (int i=1; i<N; i++) {
3        // reuse distance >=5 when N>1
4        z[i*N+2] = foobar(z[i-1]);
5      }
6  L2: for (int i=0; i<N; i++) {
7        addr = lookup[i];
8        x[addr] = foo(x[addr]);
9      }
```

Figure 1: Two types of dependences.

However, in many cases it is impossible to determine dependencies at compile-time. Such cases typically occur when data-dependent array accesses are involved in the loop body. This is the case of the loop L2 in Figure 1. For such loops, the only solution to extract additional parallelism is to perform a dependency analysis at runtime, by using so-called *runtime memory disambiguation*.

2.1 Related Work

In this section, we discuss an earlier work that employed runtime memory disambiguation in the context of HLS. Interested readers are referred to a more detailed discussion of related work in Appendix C. Ravi et al. [6] proposed the use of runtime memory disambiguation to obtain higher throughput implementations for applications containing memory accesses that cannot be disambiguated at compile time. The basic idea behind their technique is to generate two different schedules and to choose one of them at runtime. One schedule assumes that there are no dependencies between memory operations and hence provides more opportunities to exploit parallelism. The other schedule is conservative and assumes there is a dependency. Verification operations are introduced to check for a dependency at runtime. Depending on the outcome of these operations, appropriate schedule is chosen at runtime. The verification operations check for a dependency violation between

the current iteration and the previous iteration. This allows only two loop iterations to overlap. We extend this idea by allowing multiple iterations to overlap to achieve a higher throughput. When multiple iterations overlap, it is necessary to add multiple runtime checks, one for each pending memory operation in the pipeline. We discuss this in detail in section 3. Further, the earlier technique adds new states to the FSM to incorporate different schedules and verification operations. This could potentially lead to complex FSM, which may affect the clock frequency. We borrow the technique employed in superscalar processor by introducing necessary control to stall the pipeline. Though our technique also affects clock frequency, we expect it to have a smaller impact. In the following, we discuss two issues that affect efficiency of our approach.

2.2 Customizing Disambiguation Hardware

In the context of HLS, implementing runtime memory disambiguation has much fewer drawbacks than in a complex super-scalar processor. In HLS, the architecture is customized for a given application kernel, hence it is possible to analytically measure the benefit of dynamic disambiguation for this kernel (or part if it) before deciding to use it.

For example, if the HLS compiler front-end fails to analyze a memory dependence and if profiling pass shows that the probability for an actual manifestation of the dependency during the loop execution is low, using this technique is likely to be beneficial. The main challenge is to customize the memory disambiguation hardware to have "good enough" accuracy, while minimizing its hardware footprint and critical path length.

2.3 Working at the Kernel C Source Level

Our approach should ideally be implemented in a commercial HLS tools. However, these tools are closed source compiler infrastructures and cannot be modified/extended by third parties. Although it would be possible to use an open-source HLS tool, it turns out that they lack the level of robustness that we expect to be able to make our technique work properly.

One solution is to implement this approach as a source-to-source transformation operating directly at the C level. Source-to-source (S2S) compilers are well known in the parallel computing community [7, 8], but are rarely used in the context of HLS [9, 10].

In a S2S-HLS flow, the HLS tool is used as a back-end in charge of low-level scheduling/binding stages. This permits the S2S compiler to be vendor independent, and eases the evaluation of the approach on several different HLS backends. It however brings additional challenges, we cannot have access or control the low-level scheduling/binding decisions performed at the back-end level.

3. PROPOSED APPROACH

We propose to transform the loop structure in the initial C specification to include additional control logic for memory dependency hazard detection. This control logic is in charge of stalling the pipelined schedule when necessary. *Note that our technique does not pipeline the loop; it transforms the loop to allow pipelining of the loop.* This section provides a description of our technique and its implementation as an automatic source-to-source transformation.

3.1 An Illustrative Example

We first illustrate our approach using a simple example

shown in Figure 2. In this kernel, which computes an image histogram, the array `hist[]` holds the frequency of occurrence of pixel intensity values in `pixel[][]`. Array `hist[]` is indexed by the value of the pixel at hand and its corresponding occurrence count is incremented.

```
uint i,j;
uchar val;
uchar pixel[W][H];
uint hist[256];
// histogram building loop
for (i=0; i<W; i++) {
  for (j=0; j<H; j++) {
    val = pixel[i][j];
    hist[val] = hist[val]+1;
  }
}
```

Figure 2: Histogram kernel and dependency graph

The innermost loop body contains a potential RAW (Read After Write) loop carried dependency over array `hist[]`. Since the access pattern of array `hist[]` depends on the data (`pixel[i][j]`) that is being processed, static dependency analysis will fail at disambiguating memory references for `hist[]`. The loop pipeline scheduler will hence have to consider this loop carried dependency when looking for a pipelined schedule. Assuming one cycle latency for the array read operation and one cycle latency for the addition, we will obtain a lower bound for $II = 3$ as illustrated in Figure 3.

Figure 3: Original pipeline of the histogram kernel

However, since successive pixels within an image have a relatively low probability of having *exactly* the same value, this dependency will not manifest itself very often, making the statically pipelined schedule inefficient. The use of runtime disambiguation techniques brings the possibility of implementing a tighter ($II = 1 + \epsilon$) pipeline schedule, at the price of additional runtime checks to ensure the schedule correctness. These runtime checks perform runtime address disambiguation and consists checking if two pending memory reference addresses alias.

In the context of loop pipelining, there can be multiple instances of the same instruction in-flight in the pipeline. It is therefore necessary to check for a potential dependency violation for each of the pending memory operations.

We use a shift register to store the addresses accessed by write operations, on which other read operations could be dependent. In the following, we refer to such set of write operations as *dynamic writes*[2], and similarly all read operations that may alias as *dynamic reads*. Note that the memory operations that can be statically analyzed and disambiguated are not considered as *dynamic read/write*. All dynamic reads are checked against the addresses stored in the shift register to detect any aliases, i.e. dependence violations at runtime. The entire loop body, including the

[2]Note that though we say writes, these operations could be memory reads in the case of WAR dependencies as explained later in Section 3.3.

increment of the loop index, is guarded by the result of this check. If we detect an alias at runtime, no new computation is issued (this can be seen as dynamically introducing a *bubble* in the pipeline). In this approach, a distinct shift register is needed for every dynamic write instruction. Similarly, alias detection hardware is needed for related pairs of *dynamic reads/writes* (This issue is further discussed in Section 3.5).

To help the reader understand our approach, we show in Figure 4 the transformed code corresponding to the histogram kernel. An array `shift_reg[]` is introduced to store the addresses of pending dynamic write operations in the pipeline. Alias detection is performed by the loop labeled L2, which we completely unroll (using a `#pragma`) to enable parallel execution of the address comparisons (this is only possible if the `shift_reg[]` is mapped to registers). Both the depth of the shift register and the iteration count of the loop depend on the latency of the pipeline.

```
      #pragma II=1, ignore_dependency hist
      for(i=0; i<H; i++) {
L1:     while (j<W) {
            val = pixel[i][j];
            // dependency violation detection logic
            #pragma UNROLL
L2:         for(k=LATENCY-1, stall=0; k>0; k--) {
                stall = stall | (shift_reg[k-1] == val);
                shift_reg[k] = shift_reg[k-1];
            }
            shift_reg[0] = stall?(-1):val;
            if (!stall ) {
                hist[val] = hist[val] + 1;
                j = j + 1;
            }
        }
      }
```

Figure 4: Transformed code for histogram kernel.

The results of all the address comparisons are *or*ed to generate a `stall` signal. The value of this signal indicates if an immediate execution of the current loop iteration induces a memory dependency violation. If there is a violation, no new operation is issued until the pending conflicting memory operation has been completed. This is ensured by guarding all the instructions in the loop body using the `stall` signal as predicate. It is important to note that the loop increment is also guarded by this condition.

In this transformed loop, since the stall logic detects any possible violation and stalls the execution, the *while* loop (labeled L1) can now be safely pipelined, by forcing the HLS tool to ignore the target memory dependency. An illustration of the pipelined execution is provided in Figure 5. In the following sections, we discuss various aspects of this technique in detail.

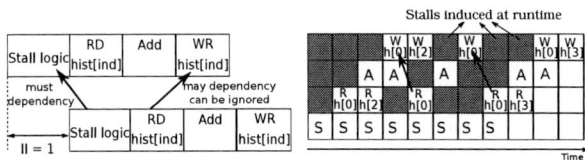

Figure 5: Pipeline of the transformed histogram kernel.

3.2 Identifying Ambiguous Memory Accesses

The first step in transforming the loop is to identify memory accesses that need to be handled by this approach. In

357

this work, we restrict the use of the technique to arrays mapped to on-chip memory banks, where no inter-array alias is allowed (inter-array alias is usually not supported by HLS tools).

The detailed algorithm to identify ambiguous memory accesses pairs is provided in Appendix A (algorithm 1). This algorithm analyzes RAW, WAR and WAW dependencies. In the case of a RAW or WAW dependency we check that a memory read (resp. memory write) does not alias a pending write operation (we hence need to keep track of write addresses). In the case of WAR dependencies, we check issuing writes against pending reads (hence keeping track of the read address).

However, it is to be noted that not all dependencies require dynamic memory disambiguation. In the following, we explain which dependencies are relevant for our approach.

1. Fully-static dependences consist of memory accesses that can be statically analyzed and the compiler is able to prove the absence of alias. Obviously, these need not be considered by our approach.

2. Partially-static dependences consist of accesses that can be statically analyzed but for which the compiler identifies a subset of the loop iteration domain where alias happens (see Appendix C.1). Some of these references may be good candidates for our technique which could be more area efficient than fully static solutions [10].

3. Fully-dynamic dependences consist of all pairs of memory accesses that cannot be statically analyzed and for which designer cannot assert that alias will never occur. An example of such a dependency is shown in our earlier example in Figure 2.

3.3 Extracting Address Expressions

Once we have computed potentially aliasing memory accesses, we must identify their index/address expressions. In the case of multi-dimensional arrays, index expressions for a given dimension are ignored for disambiguation when they are found to be the same in both elements of the pair.

Once identified, the set of index expressions that require disambiguation must be *sliced*[3] out of the loop body where these indices are originally computed. An example of a program and the *slice* corresponding to the computation of the index val is shown in Figure 6.

```
for ( i = 2; i < n; i++ ) {        int slice(...) {
  inc = 10;                          if ( i == X[i] )
  if ( i == X[i] ) {                   val = Y[i];
    val = Y[i];                      else
    inc = inc + 30;                    val = Y[i] + 10;
  } else {                           return val;
    val = Y[i] + 10;               }
  }
  C[i] = foo(C[val]) + inc;
}
```

Figure 6: Example showing the program slice of index expression

This step is very important as this *slice* will be moved out of the initial body and will become part of the pipeline stalling control logic. Note that it is possible only if the execution of the slice is side-effect free and if the slice does not involve ambiguous memory accesses (these limitations are discussed in Appendix D).

[3]Program slicing [11, 12] is a well studied program transformation that we don't detail here.

3.4 Simplifying Disambiguation Logic

Since we are deriving a custom micro-architecture tailored to a specific kernel, we also have the opportunity to customize the disambiguation logic to minimize its hardware footprint. Instead of storing a complete pending address in the shift register it is possible to use a hash of its value, to store (and compare) a smaller number of bits. This kind of technique is used in super-scalar processors to reduce the hardware cost of the disambiguation engine [13, 14].

This saving in area comes at a cost: using fewer bits to encode the address will lead to false positives in the alias detection, leading to increases in number of (unnecessary) stall cycles.

However, because we are designing application specific hardware, we can tailor the choice of the hashing function to the kernel at hand (such a customization can be driven by memory access execution traces of the kernel). One can for example search for the sub-set of bits in the address that, when used as hash, minimize the numbers of conflicts[4]. More sophisticated hashing functions used for super-scalar processors and based on Bloom filters [13] or non-singular binary matrix multiplication [15] could also be considered. However, their larger hardware footprint makes them unpractical for our approach.

Depending on the hashing function employed, not all bits in the address are required to check if the pipeline has to be stalled. This in turn can shorten the combinatorial path of alias detection logic, as we may start detecting a possible alias even before all address bits have been computed.

3.5 Reducing Shift Register Area Cost

As mentioned earlier, we introduce a shift register to hold addresses (either for a read or for a write) for all in-flight memory operations involved in an ambiguous dependency pair. One obvious simplification is to make sure a same address sequence is not mapped to two different shift registers.

It is also possible to take advantage of memory operations occurring in mutually exclusive execution paths inside the loop body. In this case, the same shift-register and alias detection logic can be used for both memory operations. Finding the minimum number of such registers is a classical resource conflict allocation problem that can be modeled (and solved) as a graph coloring problem, where non-mutually exclusive registers are considered as conflicting nodes in the graph. Our current implementation however uses a simple greedy strategy detailed in Algorithm 2.

In addition to the number of shift register components, we also need to determine their individual depths. These depths depend on the number of stages separating ambiguous access pairs in the loop pipeline. In our running example in Figure 2, the number of entries in the shift register depends on the schedule of the write operation in the pipeline and its corresponding read. In this example, the write is scheduled at stage 3, while the read is scheduled at stage 1. This hence requires a two-entry shift register.

Since we are implementing this technique as a front-end optimization, such a detailed knowledge about the pipelined schedule may not be available. In our current implementation, we assume the worst-case latency, which is the latency of the pipeline itself. We assume that this information is provided by the user through a compiler directive. This could be improved and automated as most HLS tools provide a script based interface to access cycle accurate schedule in-

[4]Our work on automating the selection of a hashing function is still on-going.

formation.

4. IMPLEMENTATION & RESULTS

In this section, we first briefly describe the implementation of our technique in an existing compiler infrastructure and then present some experimental results

4.1 Implementation in an S2S Compiler

The technique discussed in this paper was implemented as a part of GeCoS[5], a source-to-source compiler framework targeted at High-Level-Synthesis. To implement this transformation we leverage the compiler high-level intermediate representation that offers an SSA based representation, while preserving the program structure. Our SSA model is able to deal with both scalars and array accesses (using *may/must* dependency information) as we assume no inter-array alias and do not support pointers.

For these experiments, we consider that all the memory read/write pairs that do not alias have been flagged either by the compiler or by user directives. The remaining ambiguous access pairs are then considered for runtime dependency analysis. In our current prototype, the choice of the hashing function is limited to simple function where the hash is a concatenation of the n least significant bits from index expression of each dimension of the array.

4.2 Chosen Benchmark

There are currently very few publicly available benchmark suites for HLS tools. The CHStone[16] seems the only one containing kernels with loops and array accesses. However it only contains examples that are well supported by most HLS tools (regular access patterns, no complex loop carried dependencies, etc.). It is therefore irrelevant in our case.

To validate our approach we hence chose a set of kernels with memory dependencies for which our approach can be applied. The first set of kernels exposes *partially-static* memory dependencies. They are the same as those used by Morvan et al. [10] in their work and correspond to loop nest that underwent a loop coalescing transformation. This set includes BBFIR, Matrix-mult, Floyd-Warshall and Jacobi-2D kernels. For the sake of completeness, we compared our results with their approach (which provides a compile-time schedule correction mechanism). This comparison is discussed in Table 1 of Appendix B.

The second set of kernels contains *fully-dynamic* dependencies. It consists of Knapsack, histogram and tree-traversal kernels. For these loops, performance improvements provided by our technique are data-set dependent. Rather than trying to characterize realistic workloads (which is inherently difficult since these kernels can be used in a wide variety of applications), we provide performance improvement by using synthetic benchmarks with two different probabilities (10% and 50% chances) for a memory dependency to manifest during the kernel execution.

4.3 Experimental Results

We used two leading edge commercial High-Level Synthesis tools, which we are not allowed to name due to licensing issues. The first one, that we will name LEC-HLS, is considered as being one of the most efficient and robust HLS tool currently on the market. The second one is targeted for a FPGA technology vendor X and is now sold as a standard component of their FPGA design flow. We will call this second tool FPGA-HLS.

---
[5]http://gecos.gforge.inria.fr/doku.php

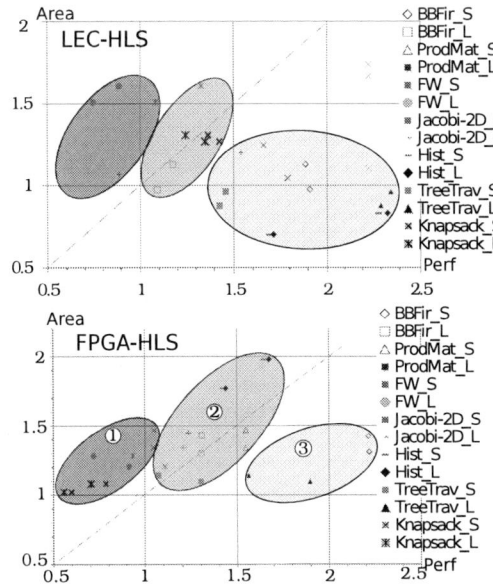

Figure 7: Area/performance trade-off for LEC-HLS and FPGA-HLS

Each kernel was transformed into several versions, each one with a different hashing strategy. The goal is to explore the trade-off between the alias detection accuracy (minimizing false positive dependencies) and area and clock speed overhead of the detection logic. We synthesized both the original and transformed kernels for Virtex-6 and Stratix-IV FPGAs and compared their area, frequency, clock cycles count (#cycles) and overall performance (combining both frequency and #cycles). Due to page limit, the complete results are provided in Appendix. We also discuss the impact of increasing pipeline depth on area and performance in Appendix B.

We provide a graphical summary of the results in Figure 7. We normalized the area and execution time of each transformed kernel by scaling them w.r.t the area and execution of their corresponding original kernel. For few kernels, inner-most loop did not contain any dependencies and hence it was possible to pipeline the loop without employing our technique. For such kernels, the comparison shown in the figure is with that of the pipelined version. We refer to this implementation as pipeline-1D. For such kernels, our technique is applied on the coalesced loop [10], which allows pipelining of outer levels of the loopnest in addition to the inner-most loop.

Performance is shown for two versions of each benchmark kernel (the L-suffix stands for large iteration count, the S suffix stands for small iteration count). Area is provided for full address alias detection (e.g. no hashing) and for a hashing based on the 4 least significant bits of the address. On average, we observe an increase of 18% overhead in area and a reduction of 74% in execution time for LEC-HSL tool and 29% and 25% for FPGA-HLS tool over pipeline-1D implementation.

To quantify the benefit of our approach, we compared our results against a theoretic/hypothetical optimization which, given an extra area of x% would decrease execution time by the same amount. This reference is represented in Figure 7 as dashed diagonal lines. Additionally we represent in the figure three regions named ①,② and ③, which correspond to different cases of performance/area trade-offs.

- Region ① contains all transformed kernels for which our optimization is counter-productive. Those include (in both HLS tools) Floyd-Warshall and Jacobi-2D (for large iteration counts). They correspond to situations where the number of false dependence eliminated using our technique is limited and where the degradation in clock frequency due to the alias detection logic is high.

- Region ② contains all transformed kernels where our optimization helps improving overall performance, but for a considerable area overhead. These include Matrix-mult and knapsack kernels. In both these kernels the inner-loop is parallel and can be pipelined without using our technique. Hence, the scope for performance improvement is limited.

- Region ③ contains all transformed kernels for which our optimization significantly improves the efficiency of the accelerator at a minor area overhead. These include tree-traversal and BBFIR in both tools. In LEC-HLS tool, histogram also belongs to this region. In FPGA-HLS tool, though we could achieve performance improvement for histogram kernel, the area overhead is also significant. So this belongs to Region ② in FPGA-HLS tool. area overhead.

Interestingly, kernels in regions ② and ③ are different from one tool to another. However, it seems that the LEC-HLS better supports our technique. Nevertheless the results show that the approach is beneficial to kernels where data-dependent memory accesses prevent pipelining (for ex: tree traversal, histogram). It also performs relatively well in less favorable cases where inner loop can be pipelined without our technique (for ex: BBFIR, knapsack).

5. CONCLUSION

In this paper, we have proposed an original technique for improving the ability of current HLS to deal with loops involving dynamic memory dependencies. Our approach was implemented in a source-to-source compiler and shows significant performance improvements.

We consider our work as the first step toward more aggressive dynamic scheduling technique which could borrow from approaches used in high-performance processors. We believe that there exist many interesting challenges that could benefit from advanced source-to-source transformations mixing aggressive static analysis features with customized speculative micro-architectures.

Acknowledgments

This work was funded in part by French ANR Compa and INRIA-STMicroelectronics Nano2012-S2SHLS projects.

6. REFERENCES

[1] M. Graphics, "Catapult-C Synthesis." http://www.mentor.com.

[2] Xilinx corp., *Xilinx Vivado Design Suite User Guide : High-Level Synthesis*, ug902 (v2012.2) ed., 2012.

[3] B. Betkaoui, D. Thomas, W. Luk, and N. Przulj, "A Framework for FPGA Acceleration of Large Graph Problems: Graphlet Counting Case Study," in *International Conference on Field-Programmable Technology*, pp. 1 –8, December 2011.

[4] J. Johnson, T. Chagnon, P. Vachranukunkiet, P. Nagvajara, and C. Nwankpa, "Sparse LU Decomposition using FPGA," in *International Workshop on State-of-the-Art in Scientific and Parallel Computing*, 2008.

[5] M. S. Lam, "Software Pipelining: An Effective Scheduling Technique for VLIW Machines," in *Proceedings of the ACM SIGPLAN conference on Programming Language design and Implementation*, PLDI '88, pp. 318–328, 1988.

[6] S. Ravi, G. Lakshminarayana, and N. K. Jha, "Removal of Memory Access Bottlenecks for Scheduling Control-flow Intensive Behavioral Descriptions," in *Proceedings of the IEEE/ACM international conference on Computer-aided design*, pp. 577–584, 1998.

[7] R. Keryell, C. Ancourt, F. Coelho, B. Creusillet, F. Irigoin, and P. Jouvelot, "PIPS: A Framework for Building Interprocedural Compilers, Parallelizers and Optimizers," Technical Report 289, CRI, École des mines de Paris, Apr. 1996.

[8] C. Dave, H. Bae, S.-J. Min, S. Lee, R. Eigenmann, and S. Midkiff, "Cetus: A Source-to-Source Compiler Infrastructure for Multicores," *IEEE Computer*, vol. 42, no. 12, pp. 36–42, 2009.

[9] J. M. Cardoso, J. Teixeira, J. C. Alves, R. Nobre, P. C. Diniz, J. G. Coutinho, and W. Luk, "Specifying Compiler Strategies for FPGA-based Systems," *IEEE Symposium on Field-Programmable Custom Computing Machines*, vol. 0, pp. 192–199, 2012.

[10] A. Morvan, S. Derrien, and P. Quinton, "Efficient Nested Loop Pipelining in High Level Synthesis using Polyhedral Bubble Insertion," in *International Conference on Field-Programmable Technology*, pp. 1–10, IEEE, Dec. 2011.

[11] M. Weiser, "Program Slicing," in *Proceedings of International Conference on Software engineering*, ICSE '81, (Piscataway, NJ, USA), pp. 439–449, IEEE Press, 1981.

[12] F. Tip, "A Survey of Program Slicing Techniques.," technical report, CWI, The Netherlands, 1994.

[13] S. Sethumadhavan, R. Desikan, D. Burger, C. R. Moore, and S. W. Keckler, "Scalable Hardware Memory Disambiguation for High ILP Processors," in *Proceedings of IEEE/ACM International Symposium on Microarchitecture*, pp. 399–, IEEE, 2003.

[14] L. Baugh and C. Zilles, "Decomposing the Load-store queue by Function for Power Reduction and Scalability," *IBM Journal on Reearch and Development*, vol. 50, pp. 287–297, Mar. 2006.

[15] D. M. Gallagher, W. Y. Chen, S. A. Mahlke, J. C. Gyllenhaal, and W. W. Hwu, "Dynamic Memory Disambiguation using the Memory Conflict Buffer," *ACM SIGOPS Operating Systems Review*, vol. 28, pp. 183–193, Dec. 1994.

[16] Y. Hara, H. Tomiyama, S. Honda, and H. Takada, "Proposal and Quantitative Analysis of the CHStone Benchmark Program Suite for Practical C-based High-level Synthesis.," *Journal of Information Processing*, vol. 17, pp. 242–254, 2009.

APPENDIX

A. DETAILED ALGORITHMS

In this section, we provide formal specification of the algorithms. Algorithm 1 is used to identify ambiguous memory operations. The input for this algorithm is a dependency graph that includes anti, output and true dependencies.

Algorithm 1 Identify the dependent memory operations

Require: DG dependency graph that includes anti and output dependencies, Δ the latency of the pipeline
  procedure IDENTIFYDEPENDENTPAIRS(DG, Δ)
    for all $d \in DG$ **do**
      if $depDist(d) \geq \Delta$ over entire iteration space **then**
        continue;
      end if
      if d holds over whole iteration space **then**
        continue;
      end if
      $src \leftarrow source(d)$
      $tgt \leftarrow target(d)$
      $depPairs \leftarrow depPairs \cup (src, tgt)$
    end for
  end procedure

Another optimization we briefly mentioned in section 3.5 is to reuse shift registers across memory operations that are on mutually exclusive paths. To motivate the need for such an optimization, consider a case shown in Figure 8.

```
1  for ( int i = 0; i < N; i++ ) {
2      for ( int j = 0; j < W; j++ ) {
3          if ( j > w[i] ) {
4  lab1:      T[i][j] = f(T[i][a[i]]);
5             output[i] = f(T[i][w[i]]);
6          } else {
7  lab2:      T[i][b[i]] = g(T[i-1][b[i]]);
8          }
9  lab3:  output[i] = k(T[i][j]);
10     }
11 }
```

Figure 8: An example kernel

Using a simple scheme, we would require three shift registers, one per each memory operation. However, shift registers can be shared between two write operations (at labels `lab1` and `lab2`) since they are on mutually exclusive paths. Algorithm 2 describes the method to assign shift registers to memory operations.

B. DETAILED RESULTS

In this section, we summarize the experimental results. Table 1 summarizes the hardware area (in terms of LUTs, Flipflops and DSPs) and the frequency of each of the implementations obtained using LEC-HLS tool. Table 5 summarizes the area and frequency of the implementations obtained using FPGA-HLS tool. The performance comparison of these implementations is summarized in tables Table 3 and Table 4 for LEC-HLS and FPGA-HLS tools respectively. We compare the results obtained using LEC-HLS tool with the results of approach proposed by Morvan et al.[MDQ11].

Another interesting aspect to study is the impact of pipeline depth on the efficiency of the our technique. Increasing pipeline depth leads to an increase in the number of entries in the shift register and the comparison logic. The secondary effect on area is due to the increase in number

---

Algorithm 2 Assigning shift registers for dependent memory operations

---

  procedure ASSIGNSHIFTREG($depPairs, latency$)   ▷
  If latency is not available for all operations, we assume a worst case latency of Δ
    $regMap \leftarrow$ MergeDepPairs($depPairs$);
    for all $reg \in regMap$ **do**
      $maxLat \leftarrow$ maxLatency($regMap[reg], latency$)
      $numEntries[reg] \leftarrow maxLat$
    end for
  end procedure
  procedure MERGEDEPPAIRS($depPairs$)
    $workingSet \leftarrow depPairs$
    while $workingSet \neq \emptyset$ **do**
      $mergedSet \leftarrow \emptyset$
      for all $pair \in workingSet$ **do**
        $tgt \leftarrow target(pair)$
        if tgt is mutually exclusive to $mergedSet$ **then**
          $mergedSet \leftarrow mergedSet \cup pair$
        end if
      end for
      $tgtOps \leftarrow tgtOps - mergedSet$
      $regMap[reg_i] \leftarrow mergedSet$
      $i \leftarrow i + 1$
    end while
    $return regMap$
  end procedure
  procedure MAXLATENCY($regPairs, latency$)
    $maxLat \leftarrow -1$
    for all $pair \in regPairs$ **do**
      $src \leftarrow source(pair)$
      if $latency[src] \geq maxLat$ **then**
        $maxLat \leftarrow latency[src]$
      end if
    end for
    $return(maxLat)$
  end procedure

Application	Version	Hardware characteristics			
		ALUT	REG	DSP	Freq (MHz)
BBFIR	original	553	152	4	185
	[MDQ11]	649	241	4	241
	FullAddr	534	175	4	173
	4bit Hash	469	144	4	175
ProdMat	original	489	215	4	272
	[MDQ11]	559	226	4	231
	FA-16bit	499	230	4	230
	8-bit	466	215	4	225
	4bit Hash	295	276	4	242
FloydWarshal	original	383	74	0	271
	[MDQ11]	859	87	0	210
	FA-16bit	569	206	0	160
	8-bit	504	172	0	173
	4bit Hash	475	160	0	173
Jacobi-2D	original	1012	845	8	164
	[MDQ11]	1417	975	8	172
	FA-16bit	1143	825	8	143
	8-bit	1065	827	8	150
	4bit Hash	929	754	8	143
Knapsack	original	372	84	0	282
	FA-8bit	429	167	0	336
	4bit Hash	423	155		370
Histogram	original	159	71	1	372
	FA-8bit	126	65	1	318
	4bit Hash	112	49	1	320
TreeTraversal	original	370	123	0	295
	FA-8bit	307	167	0	227
	4bit Hash	273	159	0	222

Table 1: Hardware characteristics of different implementations using LEC-HLS tool

of bits used in the hash function. As the pipeline depth increases, the number of false positives would also increase.

In order to maintain the same rate of false positives, we have to increase the number of bits used for hashing the address of memory operations. The number of additional bits required depends on the actual data pattern. Further, increase in the number of pipeline stages also increases the penalty we pay in the case of an actual dependency. Table 2 shows the results we obtained for an example kernel by increasing the number of stages from 5 to 22. The example contains a series of 32 bit multiplications guarded by conditions based on the output of previous multiplication. We obtain implementations with various pipeline stages by controlling the required clock frequency. To measure the impact on performance, we assume uniform probability for a memory dependence to manifest at runtime. The table shows the incremental performance/area overhead over the implementation with fewer stages.

Pipeline Depth	HashSize (in bits)	ALUT	REG	Freq (in MHz)	% inc Area	% inc perf.
5	8	746	193	31.3	-	-
8	10	913	345	57	33.9	20.88
13	12	1073	604	99.1	33.3	7.3
22	13	1074	1113	148.6	30.4	2.49

Table 2: Impact of pipeline stages on Area and Performance

Application	Version	Hardware characteristics			
		ALUT	REG	DSP	Freq (MHz)
BBFIR	original	216	290	3	185.6
	FA-8bit	383	339	3	186
	4bit Hash	346	311	3	186
Prodmat	original	231	304	3	185.7
	FA-16bit	422	444	3	186
	8bit Hash	396	388	3	186
	4bit Hash	356	360	3	185.8
FloydWarshall	original	233	155	0	311.5
	FA-16bit	428	214	0	208.5
	8bit Hash	331	166	0	214.1
	4bit Hash	307	160	0	275.9
Jacobi-2D	original	385	307	4	182
	FA-16bit	892	517	4	162
	8bit Hash	578	422	4	165.2
	4bit Hash	522	404	4	167.5
Knapsack	original	340	201	0	342.9
	FA-8bit	378	206	0	235.9
	4bit Hash	367	185	0	188.9
Histogram	original	62	29	1	528.5
	FA-8bit	107	73	1	325.6
	4bit Hash	104	57	1	380
TreeTraversal	original	188	83	0	411
	FA-8bit	192	117	0	264.9
	4bit Hash	192	105	0	321.2

Table 5: Hardware characteristics of different implementations using FPGA-HLS tool

C. DISCUSSION ON RELATED WORK

Runtime dependence analysis (and also dependency analysis) is a widely studied topic and is studied in various contexts. In the following, we summarize some of the important research work.

C.1 Compiler Based Dependency Analysis

Dependency information is important for parallelizing compilers. Hence, static dependency analysis[6] has been (and is still) a very widely studied topic. Earlier works have focused on regular (i.e. affine) access patterns such as those found in scientific codes. The developed approaches range from fast but very conservative, to exact ones [Fea91, Pug91]. More recent works focus on extending the scope of these analyses to more irregular computation patterns [CBF95, OR12].

Dependency information obtained from such static analysis is used by compilers when performing optimizations. In the context of loop pipelining, the dependency information is an important factor that determines the efficiency of the pipeline schedule. In many cases, a loop-carried dependency will hold only over a sub-set of the iteration space and sophisticated analysis can compute this information. This information can be used to perform an index set splitting transformation, which isolates the iteration subset that can be pipelined (or parallelized) [GFL00]. Another approach, specifically targeted at HLS corrects the loop schedule by inserting wait-states at compile time. This approach was proposed by Morvan et al. [MDQ11] for correcting dependency violations occurring after a loop coalescing transformation.

C.2 Software Runtime Dependency Analysis

The idea of performing software based runtime dependency analysis was proposed as early as 1989 by Nicolau [Nic89]. The basic technique is to introduce checks in the source code to disambiguate memory references at runtime.

---

[6]The book of Allen and Kennedy [KA02] offers a good survey of the topic.

This allows compiler to aggressively schedule operations assuming memory references do not alias on one path. Huang et al. [HH94] present a similar technique for architectures that support conditional execution. They employ predicates to guard the statements instead of explicit branches.

Salami et al. [SCAV02] extend this technique to disambiguate an entire loop instead of disambiguating at the level of iterations. This was proposed in the context of multimedia applications. Another work by Rus et al. [RRH03] uses a representation called *RT_LMAD*, *Run-Time Linear Memory Access Descriptor* to summarize the memory references in the program. This representation is used to disambiguate memory references at the level of loops efficiently. Both of these approaches aim at detecting loops containing *DOALL* parallelism.

C.3 Hardware Runtime Dependency Analysis

The idea of performing memory dependence analysis at the micro-architectural level dates back to early 80s. Smith [Smi84] proposed Decoupled Access Execute architectures to perform "an associative compare of each newly issued load address with all the addresses in the Write Address Queue" to ensure *store forwarding* for aliasing load/store pairs.

This kind of mechanism was later extensively used in Load-Store Queues (LSQ) of wide issue out-of-order superscalar processor architectures to handle dependence violations due to speculative execution of load/store operations. Such super-scalar processors have very deep execution pipelines (31 stages for the Pentium D). This mechanism is therefore quite costly in terms of transistor count. For such architectures, the challenge is to provide a scalable mechanism to deal with hundreds of in flight instructions [BZ06, SDB+03]. In the context of embedded hardware platforms, the energy consumed by such an approach also needs to be taken into account. In most of the embedded platforms even when the processor pipeline depth and issue width remain limited, the performance improvement provided by this mechanism rarely outweighs its area and

Application	Size	orig	[MDQ11]	FA-16bit	8bitHash	4bitHash
BBFIR	256×8	22413.3	9279	-	11895.24	11751.18
	1024×32	221790.66	148050	-	189456.84	203727.09
ProdMat	4^3	1182.72	294	295.1	301.92	280.16
	128^3	15457812.48	9078597	9101657	9311372.64	8640282.72
FloydWarshall	16^3	31454.08	22816	25750	23813.6	29131.2
	128^3	10760494.08	11650905	13107350	12132844.24	14453814.8
Jacobi-2D	30×16	91816.2	44000	62539.5	59587.02	103431.03
	30×256	14382664.2	11262604	13751755.5	13264369.02	24248289.03
Knapsack	128×16	1034.48	-	-	454.672	612.5
	1024×64	8238.32	-	-	3549.456	4812.5
Histogram	128	8223.54	-	-	5646.96	5782.5
	1024	231898.42	-	-	99143.44	101425.5
TreeTraversal	128×4	9065.94	-	-	6670.62	6375.24
	1024×16	246500.82	-	-	199221.66	183598.92

Table 3: Execution time (in ns) for different implementations synthesized using LEC-HLS tool

Application	Size	orig	FA-16bit	8bitHash	4bitHash
BBFIR	256×8	24914.1	-	11162.6	11156.4
	1024×32	230067.6	-	176408.6	176552.0
ProdMat	4^3	882.8	569.6	569.6	570.2
	128^3	11818828.6	11270127.0	11270127.092	11282710.04
FloydWarshall	16^3	17976	19749.9	19231.06	16217.4
	128^3	7046592	10058046.5	9800961.6	7692168.3
Jacobi-2D	30×16	63828.9	52830.9	51795.5	53055.3
	30×256	11255393.7	12143992.5	11907988.2	11852163.3
Knapsack	128×16	728.4	-	441.6	512.6
	1024×64	5814.1	-	3468.3	4046.0
Histogram	128	4673.7	-	4354.0	3595.5
	1024	127063.4	-	81187.4	66969.9
TreeTraversal	128×4	7467.8	-	9512.3	12489.6
	1024×16	203049.8	-	284335.1	359979.2

Table 4: Execution time (in ns) for different implementations synthesized using FPGA-HLS tool

energy overheads.

Interestingly, some mixed static/dynamic approaches have also been proposed. In the work of Gallagher et al. [GCM+94] the hardware support for runtime dependency analysis is only activated for load/store pairs that are flagged by the compiler as being possibly dependent and use compiler generated repair code in case of a dependency violation. This technique was later extended by Mahadevan et al. [MNJH00] in the context of modulo scheduled loops targeting the EPIC IA64 architecture.

C.4 Runtime Scheduling in HLS

High-Level Synthesis frameworks focus on kernels that exhibit limited data-dependent behavior[7] inside loop kernels. For this reason, most of HLS tools have been relying on static scheduling techniques.

Several research work [KW02, RB94, GSK+01, VCG05] have addressed issues related to runtime scheduling (in which we include speculative techniques) in order to support efficient scheduling (in both latency and area) over possibly complex control-flow execution paths. However these contributions do not employ runtime dependency analysis techniques and assume that memory disambiguation is performed at compile time (hence conservatively).

More recently, Thielman et al. [THK11] studied the automatic generation of speculative application specific microarchitecture from high-level specifications. While similar in its goal, their approach differs from ours in two ways: (1) They focus on value prediction techniques to reduce impact of external memory accesses. However, they do not consider runtime hardware memory disambiguation techniques. (2) They focus on a specific architectural model (the PreCORE machine); whereas our approach aims at a seamless integra-

tion into existing HLS flows.

D. LIMITATIONS OF THE APPROACH

In this section, we discuss the limitation of the approach both from the point of efficiency and from the applicability of the technique.

D.1 Performance Bottleneck

It is important to understand how our stall logic mechanism is implemented to realize its limitations.

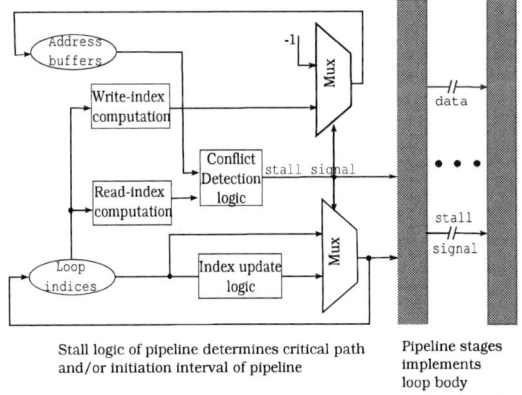

Stall logic of pipeline determines critical path and/or initiation interval of pipeline

Pipeline stages implements loop body

Figure 9: Block diagram to illustrate the technique

Figure 9 shows the organization of our pipeline stall logic. It comprises two critical loops that limit the initiation interval of the pipeline. One dependency is through the increment of the loop index variable and the second is the update to the shift register. Both these operations are conditional and are control dependent on the stall signal. Thus, the

---

[7]When it does, the data dependent behavior does not impact parallelization opportunities.

computation of `stall` signal is on the critical path. Further, update to the shift register requires computation of index expressions of memory operations that are being analyzed by our technique. This computation may become critical depending on the complexity of the index computation.

For example, consider a hypothetical example shown in Figure 10. To obtain accurate information about the write accesses to array `T` we need to evaluate the condition `j > w[i] * w[i]`. Evaluating such complex conditions may stretch the initiation interval and/or increase the clock width thereby degrading the performance obtained.

```
1  for ( int i = 0; i < N; i++ ) {
2      for ( int j = 0; j < W; j++ ) {
3          if ( j > (w[i] * w[i]) ) {
4              T[i][j] = max(T[i-1][j],T[i-1][j-w[i]] + v[i
                  ])
5          } else {
6              T[i][j-1] = T[i-1][j];
7          }
8      }
9  }
```

Figure 10: A hypothetical example to illustrate the critical path of our approach

One way to work around this problem is to compromise on accuracy and record both writes without evaluating the condition. Note that, in such cases, we need to provide two different shift registers, one for each write operation. In such cases, the optimization discussed in Section 3.5 (of merging shift registers on mutually exclusive paths) cannot be performed. Another way to improve performance is to use speculation to reduce the critical path by allowing the pipeline to proceed as long as there is no change to the global state. We are still looking at the details of this technique.

D.2 Scope of the Approach

Our approach cannot find a schedule with an $II = 1$ for applications that contain dynamic memory references in the program *slice* (refer section 3.3) of the index expression. To be able to pipeline such a loop we need to use speculation. To illustrate such a case, consider the example in Figure 11. In this example, the `index1`, which is used to access data from array `Y`, is based on a read to the same array `Y`. The reads and writes to the array `Y` are data dependent and cannot be statically disambiguated. Hence, we cannot read value from `Y` before computing the stall logic. However, we require a read to the array for computing the stall logic. This causes a circular dependency and can only be broken using speculation.

```
1  for ( int i = 0; i < N; i++ ) {
2      index0 = X[i];
3      index1 = foo(Y[index0]);
4      Y[i] = bar(Y[index1]);
5  }
```

Figure 11: Chained ambiguous memory access pairs

E. References

[BZ06] L. Baugh and C. Zilles. Decomposing the Load-store queue by Function for Power Reduction and Scalability. *IBM Journal on Reearch and Development*, 50(2/3):287–297, March 2006.

[CBF95] Jean-François Collard, Denis Barthou, and Paul Feautrier. Fuzzy Array Dataflow Analysis. *SIGPLAN Notices*, 30(8):92–101, 1995.

[Fea91] Paul. Feautrier. Dataflow Analysis of Array and Scalar References. *International Journal of Parallel Programming*, 1991.

[GCM+94] David M. Gallagher, William Y. Chen, Scott A. Mahlke, John C. Gyllenhaal, and Wenmei W. Hwu. Dynamic Memory Disambiguation using the Memory Conflict Buffer. *ACM SIGOPS Operating Systems Review*, 28(5):183–193, December 1994.

[GFL00] Martin Griebl, Paul Feautrier, and Christian Lengauer. Index Set Splitting. *International Journal of Parallel Programming*, 28(6):607–631, December 2000.

[GSK+01] Sumit Gupta, Nick Savoiu, Sunwoo Kim, Nikil Dutt, Rajesh Gupta, and Alex Nicolau. Speculation Techniques for High Level Synthesis of Control Intensive Designs. In *Proceedings of the 38th conference on Design automation*, pages 269–272. ACM Press, June 2001.

[HH94] AS Huang and S Httang. Speculative Disambiguation: A Compilation Technique for Dynamic Memory Disambiguation. *SIGARCH Computer Architecture News*, pages 200–210, 1994.

[KA02] Ken Kennedy and John R. Allen. *Optimizing Compilers for Modern Architectures: a Dependence-based Approach*. Morgan Kaufmann Publishers Inc., San Francisco, CA, USA, 2002.

[KW02] Apostolos A. Kountouris and Christophe Wolinski. Efficient Scheduling of Conditional Behaviors for High-level Synthesis. *ACM Transactions on Design Automation of Electronic Systems*, 7(3):380–412, July 2002.

[MDQ11] Antoine Morvan, Steven Derrien, and Patrice Quinton. Efficient Nested Loop Pipelining in High Level Synthesis using Polyhedral Bubble Insertion. In *International Conference on Field-Programmable Technology*, pages 1–10. IEEE, December 2011.

[MNJH00] Uma Mahadevan, Kevin Nomura, Roy Dz-ching Ju, and Rick Hank. Applying Data Speculation in Modulo Scheduled Loops. In *Proceedings of the International Conference on Parallel Architectures and Compilation Techniques*, PACT '00, pages 169–. IEEE Computer Society, 2000.

[Nic89] Alexandru Nicolau. Run-Time Disambiguation: Coping with Statically Unpredictable Dependencies. *IEEE Transactions On Computers*, 38(5):663–678, May 1989.

[OR12] Cosmin E. Oancea and Lawrence Rauchwerger. Logical Inference Techniques for Loop Parallelization. In *Proceedings of the 33rd ACM SIGPLAN conference on Programming Language Design and Implementation*, volume 47, page 509. ACM Press, June 2012.

[Pug91] William Pugh. The Omega Test: a Fast and Practical Integer Programming Algorithm for Dependence Analysis. In *Proceedings of the 1991 ACM/IEEE conference on Supercomputing - Supercomputing '91*, pages 4–13. ACM Press, August 1991.

[RB94] Ivan Radivojevic and Forrest Brewer. Incorporating Speculative Execution in Exact Control-dependent Scheduling. In *Proceedings of the Design Automation Conference*, DAC '94, pages 479–484, 1994.

[RRH03] Silvius Rus, Lawrence Rauchwerger, and Jay Hoeflinger. Hybrid Analysis: Static & Dynamic Memory Reference Analysis. *International Journal of Parallel Programming*, 31(4):251–283, August 2003.

[SCAV02] Esther Salamí, Jesús Corbal, Carlos Álvarez, and Mateo Valero. Cost Effective Memory Disambiguation for Multimedia Codes. In *Proceedings of the 2002 international conference on Compilers, architecture, and synthesis for embedded systems*, CASES '02, pages 117–126, New York, NY, USA, 2002. ACM.

[SDB+03] Simha Sethumadhavan, Rajagopalan Desikan, Doug Burger, Charles R. Moore, and Stephen W. Keckler. Scalable Hardware Memory Disambiguation for High ILP Processors. In *Proceedings of IEEE/ACM International Symposium on Microarchitecture*, pages 399–. IEEE, 2003.

[Smi84] James E. Smith. Decoupled access/execute Computer Architectures. *ACM Transaction on Computer Systems*, 2(4):289–308, November 1984.

[THK11] Benjamin Thielmann, Jens Huthmann, and Andreas Koch. Precore-A Token-Based Speculation Architecture for High-Level Language to Hardware Compilation. *International conference on Field Programmable Logic*, pages 123–129, September 2011.

[VCG05] Girish Venkataramani, Tiberiu Chelcea, and Seth Copen SC Goldstein. HLS Support for Unconstrained Memory Accesses. In *IEEE International Workshop on Logic Synthesis (IWLS)*, Lake Arrowhead, CA, 2005.

A High-Level Synthesis Flow for the Implementation of Iterative Stencil Loop Algorithms on FPGA Devices

Alessandro Antonio Nacci,
Vincenzo Rana,
Francesco Bruschi,
Donatella Sciuto
Politecnico di Milano
Dipartimento di Elettronica e Informazione (DEI)
Milan, Italy
{nacci, rana, bruschi, sciuto}@elet.polimi.it

Ivan Beretta,
David Atienza
École Polytechnique Fédérale de Lausanne
Embedded Systems Laboratory (ESL)
Lausanne, Switzerland
{ivan.beretta, david.atienza}@epfl.ch

ABSTRACT

The automatic generation of hardware implementations for a given algorithm is generally a difficult task, especially when data dependencies span across multiple iterations such as in iterative stencil loops (ISLs). In this paper, we introduce an automatic design flow to extract parallelism from an ISL algorithm and perform a design space exploration to identify its best FPGA hardware implementation, in terms of both area and throughput. Experimental results show that the proposed methodology generates hardware designs whose performance is comparable to the one of manually-optimized solutions, and orders of magnitude higher than the implementations generated by commercial high-level synthesis tools.

Categories and Subject Descriptors

B.5.2 [**Design Aids**]: *Automatic synthesis*; F.1.2 [**Modes of Computation**]: Process Management—*Parallelism and concurrency*

General Terms

Design, Algorithms, Languages

Keywords

High Level Synthesis, Iterative stencil loops, Symbolic Execution, Performance and Area Estimation

1. INTRODUCTION

A large number of interesting algorithms for scientific computation and multidimensional signals processing, come in the form of iterative applications of a given transformation t. That is, starting from a signal f (a *frame*), the overall transformation T is defined as the repeated application of t:

$$f_1 = t(f), f_2 = t(f_1), ..., f_n = t(f_{n-1}) = T(f)$$

Typically, the desired $T(f)$ is a fixed point of the single step transformation $t : t(T(f)) = T(f)$. In this case, the ideal output of the

Permission to make digital or hard copies of all or part of this work for personal or classroom use is granted without fee provided that copies are not made or distributed for profit or commercial advantage and that copies bear this notice and the full citation on the first page. To copy otherwise, to republish, to post on servers or to redistribute to lists, requires prior specific permission and/or a fee. DAC '13, May 29 - June 07 2013, Austin, TX, USA.

process is the fixed point to which the transformation converges starting from the initial frame. This class of algorithms is known in the literature as *iterative stencil loops* (ISLs) [6], and it has been analysed within the compiler community to find good implementations targeted to CPUs [6] and GPUs [7].

The design of dedicated hardware circuits for ISL algorithms, on the other hand, still presents unsolved challenges, and no automatic design flow can guarantee high performance implementations, mainly because of their complex data dependencies. The typical state-of-the-art approach for the implementation of generic iterative algorithms (such as ISLs) on FPGAs consists in employing two frame buffers [1] [2] [3], A and B, and a logic to compute t. The initial frame is loaded in one of the buffers, and then the following iteration is computed and stored in the other buffer (f (in A) $\xrightarrow{t} f_1$ (in B) $\xrightarrow{t} f_2$ (in A), ...). The procedure continues until the desired number of transformations has been performed. This architecture shows a substantial shortcoming: the area and on-chip memory required by the transformation logic is proportional to the frame size, making it too costly in real-world conditions.

In this work, we propose a high level synthesis (HLS) methodology that specifically targets the class of ISL algorithms, as well as scientific and multimedia algorithms that show similar data dependencies between subsequent iterations. The proposed methodology stems from this observation: most of these algorithms feature a peculiar form of *spatial locality*, since the value of each element p at iteration $i + 1$ (p_{i+1}) depends only on a small number of elements in the neighbourhood of p at iteration i (p_i). By exploiting this, the proposed synthesis methodology automatically generates custom hardware modules that work on a portion of the frame, and that output a subset of the intermediate results used by the subsequent iterations. Suppose we want to compute a single element p of the final resulting matrix, obtained after a number n of iterations, and let us call it p_n. The value of p_n depends on a set $P_{n-1} = \{p_{n-1}^1, ... p_{n-1}^m\}$ of elements computed at iteration $n - 1$ and, by propagating these data dependency relations back to the starting input frame, we obtain the domain of the function that computes p_n. Since these algorithms are typically *translation invariant*, such function is uniquely determined by the number of levels we want to traverse, and we call it a *cone* of *depth n*, such as the one shown in Figure 1. We can generalise this concept considering cones that compute a set P_n of elements of the n-th iteration: in this broader definition, a cone is also characterised by P_n, which we call *window*.

It is intuitive (but we are showing it formally in Section 3.1) that it is possible to perform the desired processing by repeatedly applying a cone to portions of the input matrix. This approach leads to hardware implementations whose on-chip memory requirements

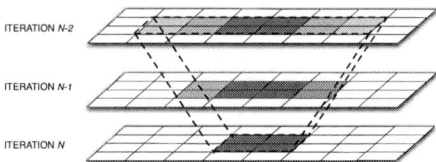

Figure 1: A cone of depth 2 and window size 4

are independent from the frame size. Let us call *cone architecture* a set of cones that perform the processing. The parameters that define a cone are the *cone window size* (let us consider only square windows, for the sake of illustration), the *cone depth*, and the *number of cones* simultaneously present in hardware. Finding the best cone architecture that satisfies a specified throughput constraint (such as a frame rate lower bound) is a challenging problem, as there is a large number of trade-offs at stake that make the space search complex. Moreover, the estimation of the area and the throughput of each cone architecture may require a very long synthesis time, which can reach the order of dozens of hours for realistic values of the window size and the cone depth.

In this paper, we face the previously described problems by presenting an automatic High Level Synthesis (HLS) flow that, given a processing algorithm, generates a set of hardware implementations that are Pareto-optimal with respect to cost and performance (measured as throughput). In particular, the flow: 1) takes a high level description (C language) of the algorithm as input, 2) systematically analyses data dependencies between elements of subsequent iterations, 3) generates synthesizable VHDL descriptions of all the *cones* that are best suited to custom fine-grained programmable devices (e.g., FPGAs), 4) efficiently explores the cone architecture solution space with a minimum number of synthesis runs.

2. ITERATIVE STENCIL LOOPS

Iterative stencil loops are designed to iteratively apply the same core operation (the *stencil*) on an *n*-dimensional matrix . The number of iterations can either be known in advance (as, for instance, in an iterative convolution filter [13], where the amount blur corresponds to a number of filtering steps), or potentially unbounded (as in fixed point algorithms, where one would ideally iterate until an equilibrium is reached). Without loss of generality, we assume that the number of iterations is known *a priori*, and hence we formally define a generic ISL algorithm with the following pseudo-code:

Algorithm 1 Generic ISL Algorithm
---
for i in { 1..N (number of iterations)} **do**
    for p in f_{i+1} **do**
        p=$t_p(f_i)$ (t_p is the function that computes only the element p of the target frame)
    end for
end for

Many algorithms for multimedia processing follow the pattern presented in Algorithm 1, such as the ones presented in [13, 14, 15, 16, 18] and [3], as well as algorithms for scientific computation, such as convolution and the *Jacobi* iterative algorithm to solve linear eigenvalue problems [17].

If we analyse the typical structure of the elementary function t_p, we find that it shows two interesting properties: 1) the set of elements required to compute an element at the iteration $i+1$ is a small subset of the frame f_i produced at the *i*-th iteration, and these elements are close to element p that has to be computed, and 2) given two target elements that are separated by a translation, the

corresponding dependency schemas have the same shape, but they are translated by the same distance as the target element. In this paper, we will refer to property 1) with the term "*domain narrowness*", and to property 2) with "*translational invariance*", which can be observed in all the algorithms we referenced in this section.

2.1 State-of-the-Art Implementations

From a computational point of view, ISL algorithms have traditionally been a challenging problem for the designers, mainly because of the complexity of their data dependencies. In the literature, the problem of designing efficient implementations for this class of algorithms has been addressed for both CPUs ([5], [6]) and GPGPUs ([7], [8]): on such architectures, the main problems that have been faced are the memory organization and the data transfers.

On the other hand, in the context of configurable devices (e.g., FPGAs), no complete design flow has been proposed so far, at best of our knowledge. In fact, the existing approaches for the hardware implementation of ISL algorithms either apply generic and ineffective optimizations, or impose very strict and limiting constraints. For instance, the work in [9] proposes a methodology to generate a hardware pipeline that spans across multiple iterations, but it is limited to only one floating-point operation per iteration, and no design space exploration is possible as the depth of the pipeline is uniquely determined. Conversely, generic HLS tools such as Xilinx Vivado [25] or Synopsys Synphony C Compiler [24] are able to handle any instance of ISL algorithms, but they perform a set of predefined and *general purpose* array and loop optimizations (unrolling, merging, flattening, pipelining, array partitioning, etc.) on the input algorithm, which is described in C. Since these frameworks do not take into account the peculiarities of the specific algorithm, the performance of the FPGA implementations they generate are generally unsatisfying for ISL algorithms, especially when compared to manually optimized implementations (as shown in Section 4).

Given the lack of support for the automatic generation of hardware designs for ISLs, many *ad-hoc* implementations have been proposed for specific ISL algorithms. For example, [4] proposes an optimized implementation for non-iterative 2D convolutions, and [19] provides an efficient hardware approach for the Chambolle algorithm [18]. However, since these solutions are manually tailored for a specific algorithm, they lack of generality and reusability, and the effort required to adapt one of these solutions to a different problem (if possible) is generally not negligible, even if the algorithms are structurally similar.

The high level synthesis flow proposed in this paper fills the lack of automation for the implementation of ISL algorithms on FPGAs. The flow is based on the abstract methodology presented in [20], and it addresses issues that are different from the ones faced by the existing approaches targeting CPUs or GPGPUs. In fact, the hardware structure (e.g., number and kinds of cores) of the latter architectures is fixed and cannot be modified or adapted to the characteristics of the input algorithm. On FPGAs, on the other hand, the definition of an efficient hardware architecture is a crucial aspect, which must be specifically handled by the synthesis tool.

2.2 Design Challenges for ISLs

In this subsection, we provide an overview of the main design challenges that arise while implementing a generic ISL algorithm (like the one shown in Algorithm 1) on a custom hardware device, such as an FPGA. When performing two subsequent iterations, the intermediate results produced by the first one have to be stored (typically in an on-chip memory for performance reasons), since they will be the input to the second iteration. The straightforward way to implement this iterative structure on a custom hardware platform uses a temporary buffer to store the intermediate data ([1] [2] [3]). However, if the dependencies of the particular algorithm are not taken into account, *it is necessary to completely compute an intermediate frame (f_i) before continuing with the following one (f_{i+1})*. In this case, if the on-chip memory is not large enough to hold a frame, it is necessary to transfer part of the intermediate results to the off-chip memory, and to get them back on-chip as soon as the next iteration starts. In this context, a memory/performance conflict arises. On the one hand, in order to keep high hardware performance, it is necessary to employ an on-chip memory large enough to hold all the intermediate results, but this typically requires several MBs of memory, which leads to expensive and power-consuming solutions. On the other hand, if the on-chip memory size is limited (only a few kBs for most of the devices used in the multimedia field), the performance is bound by the memory transfers that take place between the off-chip and the on-chip memories at each iteration.

The way we propose to handle this conflict exploits the structure of the dependencies in the algorithms that have to be implemented. In particular, by taking advantage of the *domain narrowness* (see Section 2), it is possible to design a new class of architectures, where a small portion of the input is processed *through all the iterations* by modules that we call *cones*. In this way, all the intermediate results can be stored in the on-chip memory and a transfer between on- and off-chip memories is only necessary when the logic starts to process a new portion of the initial frame. For instance, let us consider the case of the Chambolle algorithm [18] for optical flow estimation. At the best of our knowledge, an implementation of the algorithm that could sustain a real time frame rate was never proposed until [19], where the authors introduced an architecture that avoided computing a whole frame at a time, thus solving the memory/performance conflict. However, [19] proposes only a specific architecture, which was designed by hand and by studying the peculiarities of the algorithm, which requires a considerable effort. In this paper, we start from the architecture in [19] and, using the theoretical considerations shown in [20], we extend this approach by creating a high level synthesis flow that automatically analyzes the dependencies among iterations, and generates a set of Pareto-optimal implementations with respect to area and throughput.

3. THE PROPOSED HLS FLOW

The solutions space in which we seek a Pareto-optimal set of hardware implementations is composed by instantiations of the structural template we propose in Section 3.1, which allows to achieve very high performance even with modest on-chip memory requirements. The design flow that we use to generate and explore the design space is shown in Figure 2, and it consists of two main phases, which are described in the following sections: (1) analysis of the data dependency of the algorithm; (2) estimation of performance and area requirements for each architecture and design space exploration. As described in Section 3.2, the dependency analysis is performed with a novel combination of *symbolic execution* and register reuse. Then, in Section 3.3 we propose an original method to estimate the area usage of a generic cone architecture, starting from a high level representation of its structure, that provides a realistic estimation even with a low number of synthesis runs.

3.1 Architecture Template

If the input algorithm features *domain narrowness* (defined in Section 2), then it is possible to build a computational structure that is different from the straightforward one-whole-frame-at-a-time approach. To this end, in the proposed approach, data dependencies are extracted automatically by using the *symbolic execution* described in the following section, which makes it possible to express

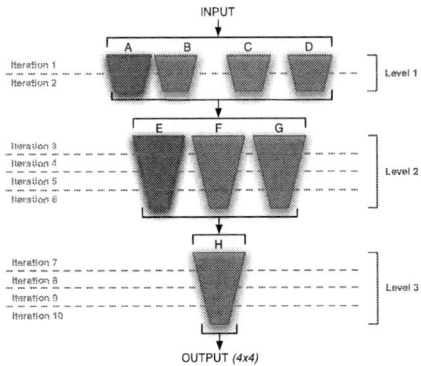

Figure 3: An example of the *cone*-based architectural template

the result of the $(i+m)$-th iteration as a function of (part of) the elements computed at the i-th iteration. As a consequence, given the data available from the i-th iteration, instead of trying to compute the whole f_{i+1}, we can focus on a subset of the matrix elements and directly compute the results of a generic m-th iteration (with $m \geq 1$), thus obtaining a subset of f_{i+m}. In accordance to what anticipated in the introduction, we refer to the core that performs such multi-iteration computation as a *cone* of *depth m*.

We define an *architectural template* by combining multiple levels of cones of different depths, which are able to compute the result of multiple iterations of the elementary transformation t. The proposed template (an instance of which is shown in Figure 3) works as follows: a small subset (window) of the input data (stored in the off-chip memory) is transferred to the on-chip memory to feed the cones of the first level of the architecture (A, B, C ad D in Figure 3). The output of each level is then used as input for the subsequent level, until all the necessary iterations are performed. The output of the last level (Level 3 in Figure 3) is finally sent back to the off-chip memory and the whole process starts over on a different window of the input data, until all the matrices has been computed.

The number and the depth of the cones in the actual architecture has to be tailored to the algorithm to be implemented, since the dependencies can significantly vary from algorithm to algorithm. Thus, multiple *instances* of the template may exist, and each one is uniquely characterized by: (1) the size of the output window of each *cone*; (2) the number of levels in which the computation is divided or, equivalently, the number of iterations that are performed at once by each cone.

Figure 3 shows an instance of the template with an output window of 4×4 elements and 3 levels of computation: the first one involves 2 iterations, while the other two levels involve 4 iterations each. It is worth noting that, since the amount of data exchanged between two levels x and $x+1$ (the output of level x is the input of level $x+1$) only depends on the size of the output of level $x+1$ and on the number of iterations considered by the two levels of computation, the parameters previously introduced suffice to completely specify any architecture. The only requirement for an instance to be feasible is that, if cones of different depths are required, at least one cone of each depth must be implemented on the device. For instance, the instance in Figure 3 is feasible if the available resources are sufficient to fit cones A and E because, in this case, the first level can be implemented by sequentially executing cone A four times (in order to cover B, C and D as well), and cone E four times (3 executions are required for level 2, and one for level 3). Many instances are generally feasible, and the same instance may be implemented in different ways by instantiating different numbers of cores of different depths, according to the resources availability.

367

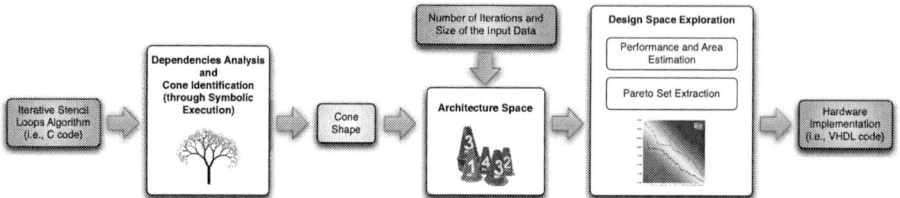

Figure 2: Schema of the proposed synthesis flow

As a consequence, multiple different trade-offs between area usage and achievable throughput (the more cones, the better) need to be evaluated, which is an aspect we address in Section 3.3.

3.2 Dependencies Analysis

In order to generate the cones, it is necessary to express the value of an element $p \in f_{i+m}$ as a function of a set of elements of the frame produced at the i-th iteration (i.e., f_i). This functional relation is often computed by hand (such as in [19]) but, when different numbers of levels need to be evaluated during the design space exploration, an efficient and automatic way to determine the equations for all the $m = 1, ..., N$ is required.

In the solution proposed here, the algorithm analysis is automated by running an optimized *symbolic execution* on a C description of the input algorithm. Symbolic execution is a well-known technique [21] that has traditionally been employed for testing purposes, which consists in executing the algorithm by propagating symbolic expressions rather than the actual values of the variables. Thus, after executing the algorithm from iteration i to $i + m$, the output is not the numeric value of f_{i+m}, but a set of equations that relate each element of f_{i+m} to a subset of elements of f_i.

The main problem that arises while performing symbolic execution in the general case is the exponential growth of the number of symbols included in the expressions, that makes it impractical for complex algorithms. In the proposed flow, we overcome this issue by exploiting the properties defined in Section 2, which enables an efficient symbolic execution for the targeted class of algorithms. Firstly, it is not necessary to find an equation for *all* the elements of f_{i+m}: if *translation invariance* holds, the dependencies of the elements in the frame only differ by a translation, which allows tracking only one element in order to get the desired expressions for the whole f_{i+m}. Secondly, data dependencies between two consecutive iterations i and $i+1$ are the same for each value of $i \in \{1, ..., N-1\}$. As a consequence, it suffices to perform symbolic execution for just one iteration to find the relation between f_{i+1} and f_i, which in turn can be used as a building block to compute the dependencies between any pair of f_{i+m} and f_i during the VHDL generation.

The equations returned by the symbolic execution are exploited to automatically generate a synthesizable VHDL description of the cones. During the equations-to-VHDL translation, the exponential explosion of the number of symbols is avoided by enforcing data reuse. In fact, a large number of operations on the same elements is repeated multiple times to satisfy the data dependencies, as shown in the example in Figure 4. As we mentioned above, this redundancy is not detected by the symbolic execution itself, which would instead introduce a large number of repeated symbols and operations in the equations. In our flow, we handle it by unrolling the dependencies between f_{i+m} and f_i through m iterations and, for each operation between two elements, we store the result in a register: whenever the operation appears more than once, the register is reused. This generates a slim VHDL code with a high degree of resource reuse, which can later handled by any synthesis tool for

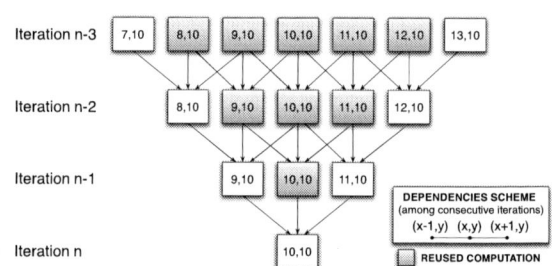

Figure 4: Example of the data-reuse technique

FPGAs that also performs area optimization.

3.3 Performance and Area Estimation

In order to determine the Pareto-optimal architectures to implement a given algorithm, it is necessary to know the cost and the throughput of each architecture of the solution space. Unfortunately, a synthesis is generally needed to know the area required by a cone, which is directly related to the number of cones deployable on the target system, and thus to the performance of the whole architecture. An obvious way to determine area and performance would be to synthesize all the cones of every window size and depth but, for typical problem sizes, the synthesis may take days of CPU time, making a complete design space exploration unfeasible.

As a consequence, we hereby propose a novel methodology to quickly estimate the area requirements of a cone architecture. The proposed evaluation only requires a very small number (as low as two) of circuit syntheses, and its accuracy is related to the number of syntheses that the designer is willing to perform (the higher the number, the more accurate the estimation). Describing the area requirements in an analytical form presents several challenges, the main one arising from the non-linear growth of the area with respect to the number of cones in the architecture, which is due to the optimization and the logic reuse performed by the synthesis tool. However, we observed that the trend of the area occupation follows the growth of the number of registers allocated into the cones. We captured the observed trend with the following relation:

$$A_i^{est} = A_{i-1}^{est} + (Reg_i - Reg_{i-1}) \cdot Size_{reg} \cdot \alpha \qquad (1)$$

Where A_i^{est} is the estimated area requirement for an architecture whose cones have an output window of size i. Reg_i is the number of registers in a cone with an output window of size i, and this quantity is already known when the VHDL description of the algorithm is generated and data reuse is enforced. $Size_{reg}$ represents the average size of a register on the target architecture. Finally, the α correction factor takes into account the degree of logic reuse performed by the synthesis tool, which can be experimentally evaluated by interpolating two or more initial syntheses (if a higher accuracy is needed, more initial synthesis need to be performed). The proposed estimation proves to be very effective in practice (see Section 4).

The estimation of the throughput follows the traditional approach, i.e., summing the delays of the operations included in each cone, and counting the number of cones that can run in parallel. This information is immediately available after the VHDL generation, when the kind and the number of operations are analyzed.

The precise estimation of the area of each cone makes it possible in turn to simply evaluate the latency (and thus the throughput) of any solution. Once the architectures space is completely characterized (thus, the area and throughput of each possible implementation has been estimated) the flow is finally able to extract the Pareto set by means of an exhaustive search that typically requires the evaluation of a few hundreds of solutions.

4. EXPERIMENTAL RESULTS

We applied the proposed flow on different case studies, of which we discuss the most significant two for the sake of illustration: an iterative gaussian filter [13] and the Chambolle algorithm [18]. The two algorithms are characterized by data dependencies of different complexities. The aim of the shown experiments is the validation of the proposed area estimation model, as well as the performance of the final architecture on two different ISL algortimhs.

4.1 Iterative Gaussian Filter (IGF)

The first case study considered is the *blur* effect, obtained by convolving an image f with a Gaussian kernel G. A common approach to implement gaussian convolutions with large kernels is to use an iterative gaussian filter (IGF) with a smaller kernel [11]. We exploit this property to formulate the filter as an iterative convolution of the frame f with a small kernel g.

The proposed flow performed the dependencies analysis, and then the area estimation. To verify the precision of our technique for the latter phase, we performed most of the syntheses, and compared them with our estimations: the results are presented in Figure 5 with respect to the output window size and to the number of iterations involved in the optimization. The maximum estimation error is 6.58%, and the average error is 2.93%, hence the proposed model provides a very accurate evaluation without requiring a full synthesis. Let us now analyze the Pareto set of optimal cone architectures. Figure 6 shows the resulting Pareto curve, with respect to performance (in this case, the time to process a single fame) and area requirements (i.e., the number of slices on a FPGA), for the convolution of a 1024x768 image. The set of Pareto solutions is reported into the zoomed window.

If the design is targeted to a specific FPGA device, and hence the amount of resources is know in advance, the synthesis tool uses all the available area to maximize the throughput, thus obtaining the results shown in Figure 7. This chart shows the degree of the throughput variation on a Xilinx Virtex-6 XC6VLX760 FPGA when the size of the output window is varied. It can be observed that the cores that lead to best performances are those whose depth is a divider of the number of overall iterations (in the example, 10 iteration are best performed with cores of depth 1, 2 and 5). The reason why cores of depth 3 and 4 achieve worse performance is that they are not dividers of 10, hence it is necessary to allocate an additional specific core (of depth 1 and 2, respectively) in order to implement the remaining iterations, thus making the exploitation of the available area suboptimal. Even by considering a single cone depth, the trend reported in Figure 7 is not monotone because, although larger cones typically lead to better throughputs, it may happen that smaller cones allow to better fit the device area.

A comparison between our cone-based solutions and the ones presented in the literature show a significant speed-up when the amount of resources is comparable. For instance, [16] presents a

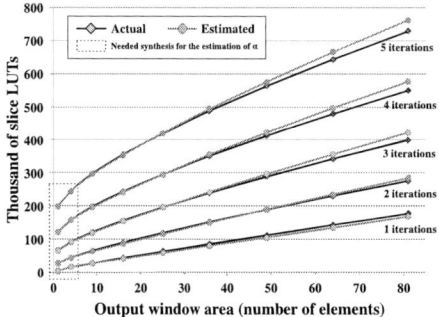

Figure 5: IGF area estimation

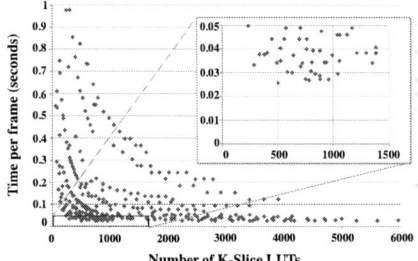

Figure 6: IGF Pareto curve (image size: 1024x768)

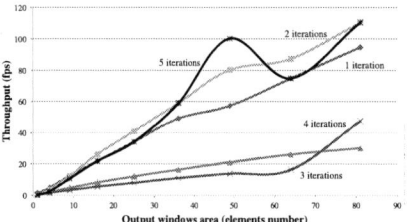

Figure 7: IGF throughput (image size: 1024x768)

20-iterations convolution with a 3x3 kernel working on a Xilinx Virtex-II Pro at 13.5fps with 1024x768 images, and at less than 5fps with Full-HD images, while our architecture achieves, on the same FPGA, up to 35fps on Full-HD images. With a modern FPGA such as a Virtex 6, our architecture reaches 110fps on 1024x768 images.

4.2 Chambolle Algorithm

Chambolle ([18], [19]) is an algorithm for total variation minimization, which is used in such fields as de-noising, zooming, and optical flow computation. As for the gaussian filter, we estimated the areas of each possible cone architecture for Chambolle, and we compared them to the actual synthesis results. Figure 8 reports the results, which are again very accurate, as the maximum area estimation error we observed is 6.36%, and the average one is 2.19%.

Starting from the area estimation, we computed the Pareto curve, which is illustrated in Figure 9. When a specific FPGA is targeted, the behavior of the throughput is similar to the one discussed for the iterative filter. In this example, it can be observed that the best solution in terms of throughput is not the one with the largest output window (9×9), but rather the solution with 8×8 cones, since in this case 2 instances of the cone can be deployed simultaneously on the device (see Figure 10). The performance of the cone-based architectures detected by our flow are competitive with respect to state-of-the-art implementations. For example, the architectures in [3], [22] and [23]), are unable to reach the real-time threshold (i.e., 30fps) even on small images because of their intrinsic absence of

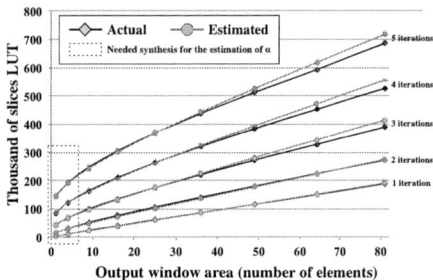

Figure 8: Chambolle area estimation

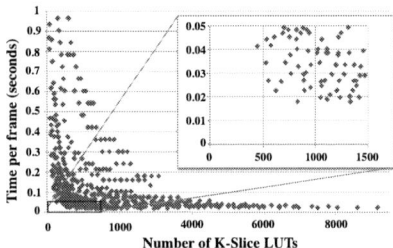

Figure 9: Chambolle Pareto curve (image size: 1024x768)

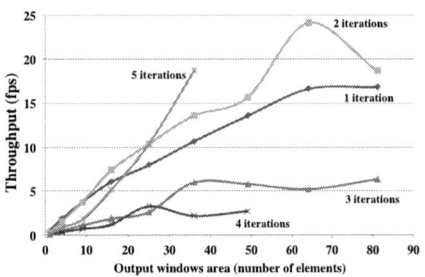

Figure 10: Chambolle throughput (image size: 1024x768)

parallelism. The architecture in [19], based on a similar locality exploitation, was designed by hand in several months of work, and it reaches 38fps on 1024x768 images and 99fps on 512x512 resolutions. With our flow we *automatically* obtained comparable results: 24fps on 1024x768 images and 72fps on 512x512 images.

4.3 Evaluation of Commercial HLS Tools

As a conclusion, we synthesized the two aforementioned case studies with two market-leading HLS tools: *Synphony C Complier* by Synopsis [24], and *Vivado HLS* by Xilinx [25]. Both the tools are able to perform a set of optimizations starting from a C code (even though Vivado HLS offers more choices than Synphony C Compiler), including loop unrolling, merging, flattening, pipelining, array partitioning, and many others. However, even for the simple iterative gaussian filter, the tools show their limitations in finding an efficient solution. For instance, by combining the possible loop manipulations and the array partitioning of Vivado HLS, the best implementation found by the tool has a throughput of only 0.14fps on a 1024x768 image. When loop merging is enabled, a solution cannot be found because of the data dependencies between subsequent iterations (which is peculiar of ISL algorithms). Conversely, when pipelining and loop flattening are employed, the execution cannot be completed because of memory shortage (an out-of-memory exception is generated even on a powerful Intel i7 with 16GB of RAM), thus showing poor scalability on ISL algorithms.

5. CONCLUDING REMARKS

In this paper, we considered the problematic implementation of Iterative Stencil Loop algorithms on custom hardware platform (such as FPGAs) from the high level synthesis perspective. Starting from the characterization of the family of algorithms, we proposed a hardware architecture template that overcomes the problem of storing the intermediate results by exploiting the dependencies between subsequent iterations. We also designed an automatic synthesis flow that produces a set of Pareto-optimal solutions with respect to area and throughput, starting from the C description of the input algorithm. In addition to the advantage of automatic synthesis, experimental results showed that the performance of the solutions found by the proposed flow are comparable to (and, in some cases, significantly better than) state-of-the-art manual implementations.

6. REFERENCES

[1] D. Crookes and K. Benkrid, "FPGA implementation of image component labelling", in Reconfigurable Technology: FPGAs for Computing and Applications, SPIE vol 3844, 17- 23 (1999)

[2] K. Benkrid, S. Sukhsawas, D. Crookes, and A. Benkrid, "An FPGA-based image connected component labeller", in Field-Programmable Logic and Applications. Springer Berlin, 1012- 1015 (2003)

[3] T. Pock et al., "A duality based algorithm for TV-L1 optical-flow image registration," in Proc. of *MICCAI*, 2007, pp. 511–518.

[4] J. Fowers at al., "A performance and energy comparison of FPGAs, GPUs, and multicores for sliding-window applications", *in Proc. of FPGA '12*, pp. 47-56

[5] M. Christen , "PATUS: A Code Generation and Autotuning Framework for Parallel Iterative Stencil Computations on Modern Microarchitectures", *IPDPS 2011*, pp. 676-687

[6] Z. Li and Y. Song, "Automatic tiling of iterative stencil loops", *ACM Trans. Program. Lang. Syst.* 26, Nov. 2004, pp. 975-1028.

[7] J. Meng and K. Skadron, "Performance modeling and automatic ghost zone optimization for iterative stencil loops on GPUs.", *ICS*, 2009, pp. 256-265.

[8] J. Meng and K. Skadron, "A Performance Study for Iterative Stencil Loops on GPUs with Ghost Zone Optimizations", International Journal of Parallel Programming, 2011, 39, pp. 115-142.

[9] C. Alias et al., "Automatic generation of FPGA-specific pipelined accelerators," in Proc. of *ARC*, 2011, pp. 53–66.

[10] D. V. Rao et al. , "Implementation and evaluation of image processing algorithms on reconfigurable architecture using c-based hardware descriptive languages," *JATIT*, vol. 1, pp. 9–34, 2006.

[11] Y. Park et al., "A new method of illumination normalization for robust face recognition," in *Progress in Pattern Recognition, Image Analysis and Applications*, Springer, 2006, vol. 4225, pp. 38–47.

[12] S. L. Park, "Retinex method based on cmsb-plane for variable lighting face recognition," in Proc. of *ICALIP*, 2008, pp. 499 –503.

[13] E. Jamro et al., "Convolution operation implemented in FPGA structures for real-time image processing," in Proc. of *ISPA*, 2001, pp. 417–422.

[14] C. Charoensak and F. Sattar, "A single-chip FPGA design for real-time ica-based blind source separation algorithm," in Proc. of *ISCAS*, 2005, pp. 5822–5825, vol. 6.

[15] K. Mohammad and S. Agaian, "Efficient FPGA implementation of convolution," in Proc. of *SMC*, 2009, pp. 3478 –3483.

[16] B. Cope, "Implementation of 2D Convolution on FPGA, GPU and CPU," Master's thesis, Department of Electrical & Electronic Engineering, Imperial College London, 2006.

[17] L. Gerard et al., "A Jacobi-Davidson Iteration Method for Linear Eigenvalue Problems," SIAM Review , Vol. 42, No. 2, 2000, pp. 267-293.

[18] A. Chambolle, "An algorithm for total variation minimization and applications," *Journal of Mathematical Imaging and Vision*, vol. 20, pp. 89–97, 2004.

[19] A. Akin et al., "A high-performance parallel implementation of the Chambolle algorithm," in Proc. of *DATE*, 2011, pp.1,6.

[20] V. Rana et al., "Design Methods for Parallel Hardware Implementation of Multimedia Iterative Algorithms," *IEEE Design & Test of Computers*, 2012.

[21] J. C. King, "Symbolic execution and program testing," *Commun. ACM*, vol. 19, no. 7, pp. 385–394, 1976.

[22] C. Zach et al., "A duality based approach for realtime TV-L1 optical flow," *DAGM conference on Pattern recognition*, 2007, pp. 214–223.

[23] A. Weishaupt et al., "Tracking and Structure from Motion," Master's thesis, EPFL, 2010.

[24] Synopsys, "Synphony C Compiler," 2012.

[25] Xilinx Inc., "Vivado Design Suite User Guide, High-Level Synthesis," UG902, 2012.

Cross-Layer Racetrack Memory Design for Ultra High Density and Low Power Consumption

*Zhenyu Sun, †Wenqing Wu, and *Hai (Helen) Li
email: *{zhs25, hal66}@pitt.edu †wenqingw@qualcomm.com
*Swanson School of Engineering, University of Pittsburgh, Pittsburgh, PA, USA
†Qualcomm Incorporated, San Diego, CA, USA

ABSTRACT

The racetrack memory technology utilizes magnetic domains along a nanoscopic wire to obtain ultra-high data storage density. The recent success in the planar racetrack nanowire promised its fabrication feasibility and future scalability, bringing more design challenges and opportunities. In this paper, we initialize the optimization of racetrack memory embracing design considerations across multiple layers, including cell design, array structure, architecture organization, and data management. Our evaluation shows that racetrack memory based cache can achieve 6.4× area reduction, 25% performance enhancement, and 62% energy saving, compared to STT-RAM cache design. The benefit over SRAM technology is even more significant.

Categories and Subject Descriptors

B.3.2 [**Memory Structures**]: Design Styles—*Cache memories*

General Terms

Design

Keywords

Racetrack, Cross-layer design

1. INTRODUCTION

In the modern chip multiprocessor designs, large on-chip caches and main memories have become the dominant contributors to the overall power and thermal budget. The worries about the continuous scaling of SRAM and DRAM triggered the investment in *spin-transfer torque RAM* (STT-RAM) with higher density, competitive access speed, and more controllable power consumption under the scaled technology [1][2]. Very recently, ***Everspin began shipping working samples of 64MB STT-RAM*** [3], announcing the commercialization era after many years of joint effort from both academia and industry [4][5].

However, restricted by the theoretical limit of $9F^2$, further shrinking memory cell size and hence improving performance and power consumption in STT-RAM is difficult [6]. To offer a "faster-than-Moore's law" scaling path, a team led by Dr. Parkin in IBM proposed *racetrack* memory that uses a spin-coherent electric current to move magnetic domains along a nanoscopic permalloy wire for data storage [7]. The cell area is expected to be as small as $2F^2$. Moreover, the continuous progress in device physics [8][9][10] and the recent successes in fabrication process [11][12][13] promise the

Permission to make digital or hard copies of all or part of this work for personal or classroom use is granted without fee provided that copies are not made or distributed for profit or commercial advantage and that copies bear this notice and the full citation on the first page. To copy otherwise, to republish, to post on servers or to redistribute to lists, requires prior specific permission and/or a fee.
DAC'13, May 29 - June 07 2013, Austin, TX, USA

feasibility of racetrack memory. Very recently, *TapeCache*, an early stage of estimation to utilize racetrack memory as data cache, was presented [14]. It showed 2.3× higher density, 1.4× power reduction and similar system performance compared to STT-RAM in *last-level cache* (LLC). However, the potential of the racetrack memory has not been fully explored.

Compared with array-style random access memory, including STT-RAM, integrating tape-style racetrack memory faces several unique design challenges: (1) To effectively utilize the stripe structure, new circuit layouts and optimizations are necessary. (2) The stripe-based memory structures require new logical abstractions of memories. (3) Moving from random access (i.e., wordlines and bitlines) to sharing one access device (i.e., writing/reading requires shifting) requires careful design and scheduling of data access.

In this work, we comprehensively consider these design requirements across different abstraction layers. An ultra-dense on-chip racetrack memory was proposed enabling significant enhancement, in system performance and energy saving: compared to STT-RAM cache design, the racetrack memory based last-level cache can achieve 6.4× area reduction, 25% performance enhancement, and 62% energy saving. The benefit over SRAM technology is even more significant.

The primary design considerations and contributions include:

(1) **The efficient cell and array designs** are presented that eliminate the area constraint of the access transistor size and enable the uniform access ports for read and write operations.

(2) **An optimized racetrack architecture** is presented after carefully exploring sub-array and architecture configuration. The according physical-to-logic mapping scheme and hardware design are offered.

(3) **An application-driven data management policy** which tends to allocates the access-intensive data blocks close to access ports of racetracks was proposed to minimize racetrack shifting and the induced overhead.

(4) **The impact of racetrack cell size** under the proposed architecture was evaluated by comparing two racetrack geometrical dimensions. The benefits and potential design challenges are discussed.

In the rest of the paper, Section 2 describes the fundamental of racetrack memory. We then present the proposed cell and array designs, architecture, and data management policy in Section 3, 4, and 5, respectively. The simulation results are reported and analyzed in Section 5. At last, Section 6 concludes the paper.

2. BACKGROUND

The racetrack memory comprises an array of magnetic stripes, namely, *racetracks* (RTs), arranged vertically [7] or horizontally on a silicon chip [12]. Figure 1 illustrates a horizontal RT structure that will be discussed in this work. It consists of many magnetic domains separated by ultra-narrow domain walls. Each domain has its own magnetization direction. Similar as STT-RAM, the binary

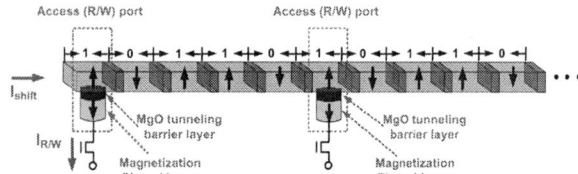

Figure 1: A racetrack in horizontal structure.

values can be represented by the magnetization direction of every domain. And several domains share one access port for read and write operations. A select device together with an *magnetic tunneling junction* MTJ sensor is built at an access port. The motion of magnetic domain walls in a RT can be controlled by applying short current pulse I_{shift} on the head or tail of the RT. To access a domain, we need a two-step operation: first *shift* it to an access port and then *read* or *write* the domain by applying an appropriate current (I_R or I_W).

The device engineering of RT memory can be classified into two types – in-plane and perpendicular – according to the anisotropy direction of its magnetic layer. The *perpendicular magnetic anisotropy* (PMA) RT memory can provide a higher energy barrier even when the volume of domain cell is very small, enabling the continuous scalability for racetrack memory [15].

3. CELL AND ARRAY DESIGNS

The accelerated storage density improvement introduced by the advanced RT memory enlarges the area gap between small memory element and relatively large NMOS select transistor. Figure 2(a) and (b) illustrate the schematic of a column of RT memory and the corresponding layout. The table in the figure summarizes the used components and symbols. The layout design shows that a RT (the blue strip) consumes only a small portion of the space above access transistors. The area highlighted in the gray shadow however is wasted.

A straightforward way to improve RT layout area efficiency is decreasing access transistor size. However, the driving current provided by small transistor might not be sufficient to switch magnetic domains. We can use these access points (named as R-ports) for only read operations. One or a few write-only ports (W-ports) associated with large access transistors are still necessary for a RT. The fewer W-ports result in longer RT shifting in write operations and a larger overhead of domain cells. This is how macro cell was

designed in *TapeCache* [14]. Without re-engineering the RT cell design, it is difficult to increase array area efficiency and read/write (R/W) accessibility at the same time. Here, we propose new RT cell and array designs to achieve better optimization.

Memory cell design: Figure 2(c) depicts the proposed layout of a column of RT memory. Multiple RTs are arranged side by side to cover the whole space above select transistors and their access points to the select transistors are placed in a diagonal manner. In this design, the number of RTs per column is determined by the widths of RT and selection transistor. For example, Figure 2(c) assumes 4 RTs per columns, resulting in $4\times$ memory density compared to the baseline layout. Note that the four transistors share one source-line (SL), but each of them is connected to only one RT and hence one bit-line (BL).

The proposed RT layout design can maximize the utilization of the space above CMOS layer. The size of the access transistor size, as far as it is large enough for write operations, is not the limiting design factor. In fact, the selection of access transistor size becomes more flexible and can be used to facilitate architecture optimization.

Memory array design: Figure 3 shows the circuit schematic of a basic RT memory array, which supports the following three basic operations:

- *Shift:* Shifting a RT up (su) or down (sd) is realized by a bi-directional shifting current (I_{shift}), which is controlled by signal 'su+', 'su−', 'sd+' and 'sd−'.
- *Write:* The write current (I_W) in a write-'1' (or write-'0') operation is provided by enabling 'wr1+' and 'wr1−' (or 'wr0+' and 'wr0−').
- *Read:* 'rd' and 'wr1−' are turned on so that a small read current I_R can be supplied to the target cell. The voltage generated on its BL will be delivered to a sense amplifier (not included in the figure) for data detection.

Note that extra magnetic domains, or *RT-overhead*, shall be added at both ends of a RT. The RT-overhead provides the extra space to store the bits shifted out of the original data portion during accesses and prevent data loss.

In summary, our proposed memory cell and array designs significantly improve the area efficiency of RT memory, leading to a unprecedentedly high density. *The design is the first one enabling read and write operations at every access port without producing side effect on the area efficiency.*

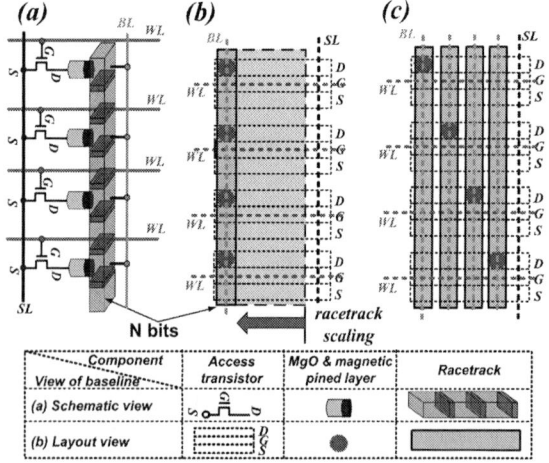

Figure 2: (a,b) Schematic and layout of the baseline RT design, (c) The proposed RT layout.

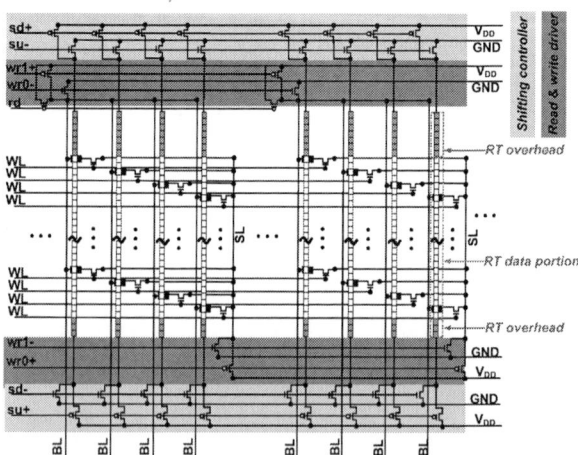

Figure 3: The circuit schematic of a RT memory array.

372

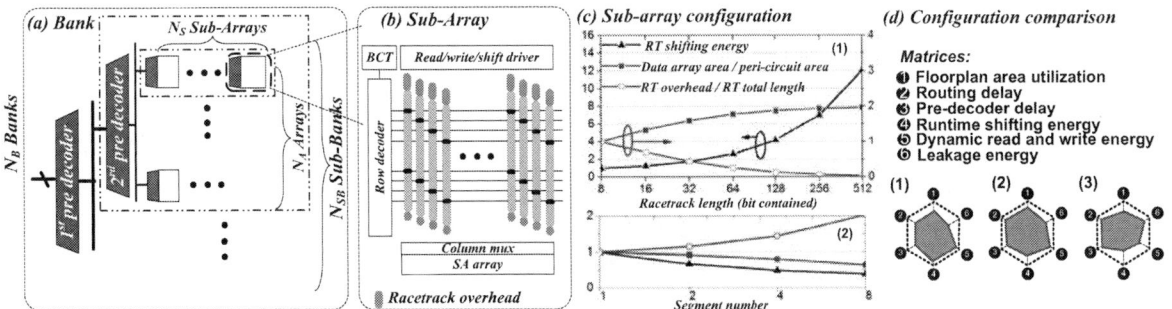

Figure 4: The RT memory architecture design exploration.

4. ARCHITECTURE EXPLORATION

The efficiency of a RT architecture is related to the basic array configuration, architectural organization, and physical-to-logical mapping. In this section, we comprehensively investigate the impacts of these design considerations in the floorplan utilization, performance optimization, and energy consumption based on the proposed cell and array design.

Figure 4(a) illustrates the proposed *hierarchical and dense architecture* for RT (**HDART**). The proposed HDART maintains the same I/O interface as current memory hierarchy to ease the technology adoption. Within the architecture, however, the bank organization could be flexible. For example, an entire cache architecture can be physically partitioned into N_B banks, each of which has its own I/O ports to support concurrent transactions.

Sub-array configuration: As the smallest component in architecture construction, a sub-array in Figure 4(b) can significantly affect the overall performance of the entire architecture. The sub-array based on the memory array structure in Figure 3 shall be carefully configured according to design requirements. Three parameters are used when evaluating various sub-array configurations: (a) *the RT shifting energy*, (b) *the sub-array area efficiency* defined as the ratio of the data array area and the peripheral circuit area, and (c) *the RT overhead ratio* which is the ratio between the RT-overhead and the total length of RT.

RT length is an important design parameter related to RT shift energy. A longer RT produces higher runtime shifting energy, while a shorter RT degrades the sub-array area efficiency and has the higher RT-overhead ratio. We can also divide a long RT into several segments. Each segment needs its own shifting controller and the entire RT has one shared read/write driver. In general, more segments indicates lower shifting energy, but lower sub-array area efficiency and higher RT overhead. Figure 4(c) compares the different sub-array configurations in normalized scale.

Architecture exploration: Based on the basic array design, the memory architecture configuration, including banks, sub-banks, arrays etc., can be adjusted to satisfy the different design specifications including the criteria of performance, energy, and area constraint. We evaluated and compared the 4MB RT LLC designs based on different basic array configurations. Here, three typical basic array configurations were selected, representing the designs with (1) more segments in one RT, (2) one medium-length RT (e.g., 64 bits long), and (3) one long RT, respectively.

The detail comparison in terms of area efficiency, performance, and energy consumption at the LLC architectural level were conducted from six design matrices and shown by the hexagon graphs in Figure 4(d). The dotted border of a hexagon map indicates the optimal expectation. All these design matrices are relevant with each other and determined by the both of the architecture and sub-array configuration. In the work, we are primarily interested in a RT-based cache design with high performance and low power consumption. We selected configuration (2) in Figure 4(d) for following evaluations, due to its lower access delay and energy.

Physical-to-logic mapping: The physical-to-logic mapping is orthogonal to the memory architecture design but shall be optimized based on the specific design requirement. Figure 5 gives an example of mapping on the top of the proposed HDART architecture, in which a sub-array contains 64 groups of RTs, each of which has 4 RTs. And each RT has 8 access points. The magnetic domains connected to a single R/W port in the physical design corresponds to the same bit number of cache blocks within the same set from different ways. For instance, as illustrated in the figure, Bit 0 (b0) of 32 cache blocks belonging to all the 32 ways (w) within Set 0 (s0) are all mapped to the first group of RTs in sub-array 0 of array 0. The RT shifting during an access can controlled by a physical-login mapping unit, e.g., *LUT* in Figure 5. The shifting distance shall be determined by the block's way number and the current track position stored in a track status register T_{reg}.

Because all the same bits from the different ways are within the same array and controlled by a single access port, such a design is in favor of data block reordering among different ways, the data management method that shall be discussed in Section 5. Nevertheless, unlike way reordering that can be easily implemented with tag mechanism, set reordering requires set re-mapping table which results in extra area, delay and energy overhead. Thus, we select

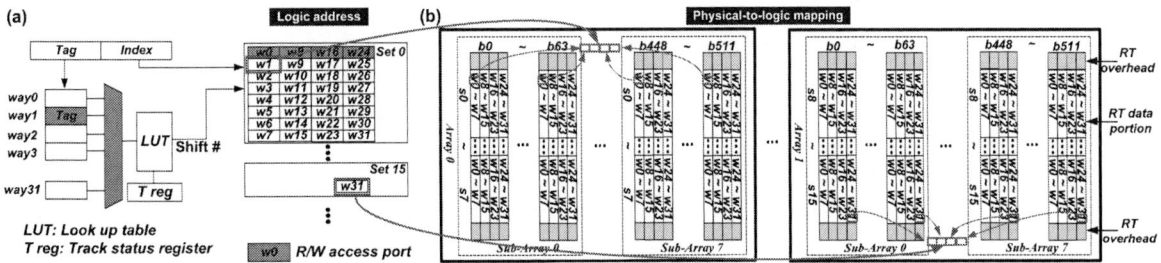

Figure 5: A physical to logic address mapping scheme for RT memory.

373

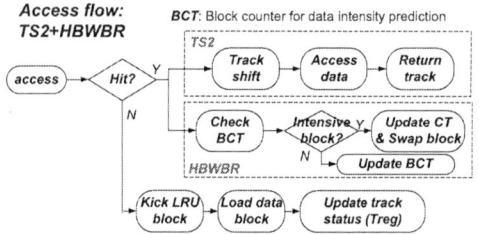

Access flow: TS2+HBWBR

Figure 6: The RT memory access flow with TS2 and HBWBR.

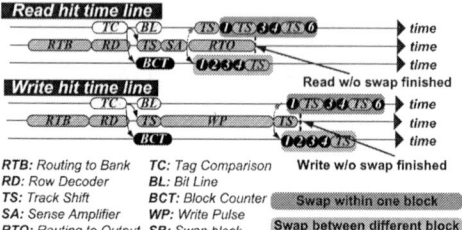

RTB: Routing to Bank   TC: Tag Comparison   **Read w/o swap finished**
RD: Row Decoder   BL: Bit Line
TS: Track Shift   BCT: Block Counter   ▓ Swap within one block
SA: Sense Amplifier   WP: Write Pulse
RTO: Routing to Output   SB: Swap block   ▓ Swap between different block

Figure 7: Cache access timing.

way reordering for the data management method.

Hardware design complexity: Besides the RT memory itself, the design complexity of other related hardware need to be considered. For example, though tag array contributes only 5% of total area in SRAM LLC design, it could become the major bottle in RT memory resulting from the unbalanced scaling trends of data storage and tag array. Here, utilizing STT-RAM for tag design can alleviate the impact, which is adopted in our design. Several components are introduced in the proposed RT memory, including a track status register (T_{reg}) to store RT position, a look-up-table (*LUT*) to assist physical-to-logic mapping, and a block counter (*BCT*) indicating the data access intensity for data management in Section 5.

5. DATA MANAGEMENT POLICY

In a traditional random access memory, every storage element has its own access path. In contrast, many magnetic domains in a RT memory share one R/W port. During an access, a domain need to be shifted to R/W port, inducing extra overhead in access latency and energy consumption. We propose two *track shifting policies*:

- *TS1:* After an access is completed, a RT stays where it is.
- *TS2:* A RT returns to its original position after each access. *TS1* benefits when cache accesses show strong spacial locality, but generates frequent RT shiftings when randomly distributed cache accesses dominate. Resetting RT in *TS2* potentially increases the frequency of RT shifting. But the data management in *TS2* is easier because of the fixed relation of the memory cells to their R/W ports.

Considering that the cache accesses in many applications are unevenly distributed and only a small portion of cache blocks are frequently accessed [16], we propose a RT data management policy, named as *hardware based way block reorder* (HBWBR), to alleviate the shifting overhead. By tracing the data access pattern, HBWBR can identify the cache blocks with intensive accesses and then place/swap them to the physical locations onto the R/W ports.

The access flow and timing: Figure 6 depicts the HDART access flow when applying TS2 and HBWBR together. The corresponding access timing diagram is shown in Figure 7. Note that track shifting (TS2) and data management (HBWBR) are executed simultaneously and are independent to each other. Thus, the access to block counter for data intensity prediction (BCT) does not introduce any extra latency overhead. An access hit in HDART triggers the examination of BCT. If the access block is predicted to be ac-

cess intensive, we swap it with the one at R/W port. Otherwise, the BCT is updated by increasing 1. During a cache miss, the least-recently used (LRU) policy is adopted.

The timing flow and the critical path in HDART accesses is similar to STT-RAM cache access, except extra RT shifting delay shall be included in every data access. The track reset delay can be hidden with the routing to output (RTO) delay during read. A data swap induces a relative big delay overhead, but it occurs much less frequently with regard to total access number. The detail explanation of timing components are listed at the bottom of Figure 7 for reference. And the timing component parameters can be found in Table 1. The effectiveness of HBWBR relies on the efficiency of both the intensive block prediction and the data swap.

Intensive block prediction: For design simplicity, a counter-based scheme is used for cache access intensity prediction. In the design, each data block is associated with a *block counter* (BCT). When a cache hit occurs, the corresponding counter of the data block increments by one. All the counters are self-decremented periodically till it becomes to '0'. A data block is considered as an access-intensive block once its counter exceeds the predefined threshold.

Block swap: In Section 4, we propose to map the same bits from all the ways within one set into one array. Therefore, the data on some ways sitting right on the R/W ports can be accessed without any track shifting. We name these ways as 'fast way'. Moreover, since data exchanges occurs within the same array in such a design, the way-based block swapping is convenient and energy efficient. Therefore, we propose to swap an access-intensive block with those on fast ways. When a block outside fast way is regarded as access-intensive block, it swaps with the data block in one of the four fast ways with the smallest access number.

As shown in Figure 5, each set consists of four fast ways on four RTs. Accordingly, the data swap could occurs within the same track or between two different tracks. The different latency and energy overhead caused by these two types of data swap operations will be included in system evaluations in Section 6.

Easy integration with HDART: *HBWBR* is an efficient data management for RT memory that has variable access latencies. In the previous RT memory design, a RT macro cell has only one R/W port but multiple R-ports [14]. The data management in such a design requires to distinguish the read- and write-intensive data

Table 1: Design parameters for different cache types

Cache parameters	Area					Timing (Timing components are illustrated in Figure7(a))							
	Cell size	Total area	Data	Tag	Periph	RTB	RD	BL	SA	TC	RTO	WP	TS
SRAM	$125\ F^2$	$9.09\ mm^2$	93.43%	4.67%	1.90%	**2.03**ns	$0.19ns$	$0.12ns$	$0.2ns$	$0.5ns$	**2.10**ns	–	–
STT	$32\ F^2$	$2.51\ mm^2$	86.28%	11.44%	2.28%	$0.56ns$	$0.16ns$	$0.10ns$	$0.2ns$	$0.3ns$	$0.58ns$	$10ns$	–
OP-STT	$28\ F^2$	$2.24\ mm^2$	84.55%	12.80%	2.65%	$0.52ns$	$0.15ns$	$0.10ns$	$0.2ns$	$0.3ns$	$0.53ns$	$5ns$	–
Baseline RT	$16\ F^2$	$1.41\ mm^2$	76.91%	20.37%	2.72%	$0.31ns$	$0.14ns$	$0.07ns$	$0.2ns$	$0.3ns$	$0.32ns$	$5ns$	$0.5ns$/shift
HDART ($4\ F^2$)	$4\ F^2$	$0.599\ mm^2$	45.37%	48.08%	6.55%	**0.13**ns	$0.13ns$	$0.03ns$	$0.2ns$	$0.3ns$	**0.13**ns	$10ns$	$0.5ns$/shift
HDART ($1\ F^2$)	$1\ F^2$	$0.390\ mm^2$	17.42%	73.84%	8.74%	**0.08**ns	$0.1ns$	$0.01ns$	$0.2ns$	$0.3ns$	**0.08**ns	$5ns$	$0.5ns$/shift

Note: One shift operation means shift the whole racetrack memory cell up or down with the unit distance of one domain cell bit. The shifting current can be tuned by sizing the shifting controller transistor. The racetrack shifting (domain wall motion) velocity is determined by the shifting current density. With carefully tuned current, it takes one cycle (0.5ns) to shift the racetrack cell for one unit of distance.

Table 2: Energy components of diff. memory technologies.

	Write eng	Read eng	Leakage pwr	Shift Eng
SRAM	$0.35nJ$	$0.42nJ$	$4100mW$	–
STT	$1.92nJ$	$0.36nJ$	$130mW$	–
OP-STT	$1.52nJ$	$0.34nJ$	$120mW$	–
Baseline RT	$1.07nJ$	$0.22nJ$	$83mW$	–
HDART($4F^2$)	$0.57nJ$	$0.074nJ$	$46mW$	$0.62nJ$/shift
HDART($1F^2$)	$0.46nJ$	$0.037nJ$	$33mW$	$0.31nJ$/shift

Note: The write, read and shift energy is for one cache block.

Table 3: Processor configuration

Processors	8 cores, 2GHz, 4 threads/CPU core, 1-way issue
SRAM L1 Cache	Local, 16KB I/D, 2-way, 64B line, 2-cycle, write-back
LLC Cache	Shared, 4MB, 4 banks, 32-way, 64B line, write-back, 1 read/write port, 4 write buffers.
Main Memory	4GB, 400-cycle latency.

blocks and allocate them to different ports. Especially, the limited W-port number constrains the optimization space. So it is less adaptable for efficient data management. In contrast, HDART supports both read and write at every access ports, easing the design complexity and enhancing the efficiency of data management. The data management can be naturally integrated on the HDART.

6. SIMULATION SETUP

Cache design parameters: We evaluated and compare 4MB LLC design by using different memory technologies, including SRAM, STT-RAM, and RT memory. The cache configuration is set as $N_B = 4$, $N_{SB} = 8$ and $N_S = 8$ (refer Figure 3(a)). The cache latency and energy parameters were obtained based on SPICE simulation and the modified NVsim [17]. The domain wall shifting energy was calculated from micro-magnetic simulations. These parameters are summarized in Table 1 and Table 2.

Evaluation platform: We performed the evaluations on an 8-core UltraSPARC T1 processor by adopting various memory technologies as 4MB LLC. Table 3 summarizes the process configurations. The cache model of Simics toolset [18] was modified according to the different memory requirements. The multi-threaded benchmarks from Parsec Benchmark Suite [19] were adopted in simulations. For each benchmark, we warm up the cache for 200 million instructions, and then execute 500 million instructions. The following baseline memory technologies were selected for comprehensive comparison. They are: **SRAM, STT-RAM, OP-STT** [2], and **Baseline RT** (according to layout in Figure 2(b)).

We evaluated the impact of the RT technologies under the proposed HDART by comparing two RT geometrical dimensions: a moderate domain size of $4F^2$ to reflect current device engineering and an aggressive racetrack design with domain size of $1F^2$, corresponding to the racetrack with a width of $1F$. After applying different tracking shifting and data management policies proposed in the work, totally six RT memory configurations were examined. They are: $4F^2$+TS1, $4F^2$+TS2, $4F^2$+TS2+HBWBR, $1F^2$+TS1, $1F^2$+TS2, and $1F^2$+TS2+HBWBR.

7. SIMULATION RESULTS

Comparison to baseline memories: To demonstrate the potential of RT memory, we first compare the HDART with baseline memory technologies. Figure 8(a) shows the performance results represented by the normalized *instruction per cycle* (IPC). HDART with *TS1* achieves $10\% \sim 15\%$ IPC enhancement over SRAM and OP-STT. The shorter routing latency for both read and write operations in HDART dominate the IPC performance improvement, though the extra delay caused by track shifting slightly offsets the benefit. Figure 8(b) shows the overall energy of different configurations.

The overall energy of HDART is $40\times$ smaller than that of SRAM cache. Compared whti the most advanced OP-STT, HDART+TS1 can achieve an average 19% of energy saving. The detail energy breakdowns of three most energy efficient memory technologies are shown in Figure 8(c). *Baseline RT* has more R/W ports and hence consumes $\sim 2\times$ less track shifting energy than HDART. However, HDART's overall energy consumption is 18% less than Baseline RT. The saving comes from less leakage and dynamic energies due to its higher density.

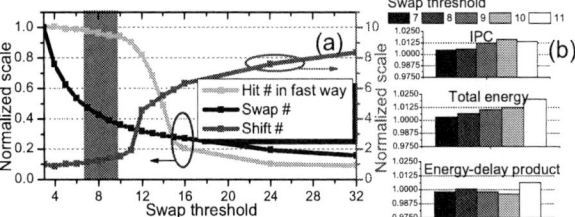

Figure 9: Swap threshold selection.

The swap threshold selection: directly determines the effectiveness of data intensity prediction and the frequency of runtime data block swaps. Figure 9(a) summarizes the average statistical data of all benchmarks including the hit number on fast ways, the shift number, and the swap number when changing the swap threshold from 3 to 32. When the threshold exceeds 11, the hit number in fast way decreases dramatically, resulting in a significant increasing of racetrack shift number. Moreover, the swap number reduces quickly before the threshold approaches 11 and becomes flat at a large threshold. According to Figure 9(b), we set the swap threshold as 10 for the best energy-delay product.

HBWBR effectiveness: The above comparison to baseline memories, HDART is only equipped with *TS1*. We evaluate the effectiveness of the *HBWBR* in this section. Figure 8(c) show that a big portion of HDART energy comes from track shifting. Therefore, reducing the track shifting is necessary to improve energy consumption. Figure 10(a) shows the trend of shift numbers at the beginning of simulations for different benchmarks. With the assistant of *HBWBR*, he shift number decreases with time . Note that the different benchmarks apply the different observation windows in the figure for their different reduction rate of shift number. Figure 10(b) summarizes the total shift reduction in HDART+*TS2* before and after applying *HBWBR*. On ovarge, *HBWBR* helps remove 60% of track shift, indicating 60% shifting energy reduction.

Figure 11 shows the IPC performance comparison of HDART under different policies. And the detail energy comparison can be found in Figure 12. *TS2* alone suffers from more track shift operations, resulting in 2.5% performance degradation than *TS1*. *HBWBR* effectively reduces shifting overhead and improves IPC 7.5%

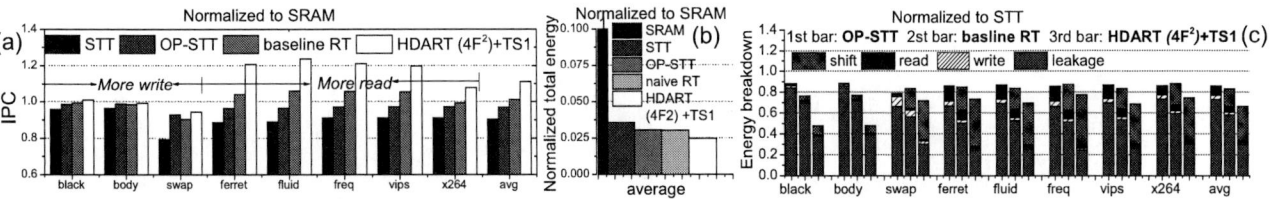

Figure 8: Comparison with baseline memory technologies (a) IPC performance (b) Total energy (c)Energy breakdown.

375

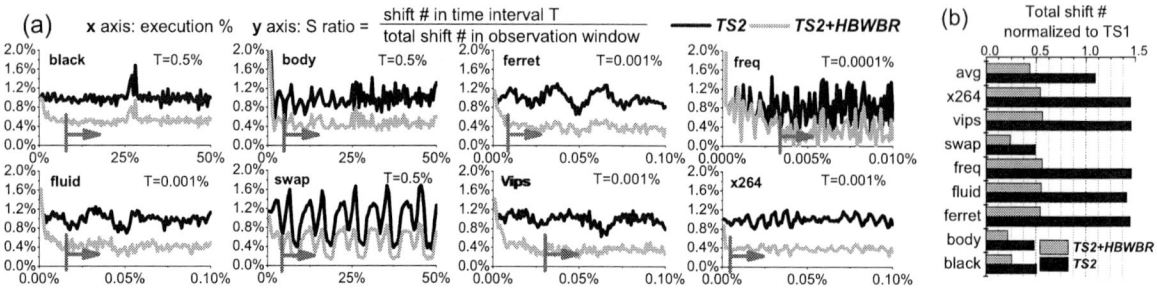

Figure 10: Shifting reduction by *HBWBR*.

on average. In summary, the HDART design $4F^2+TS2+HBWBR$ achieves $4.2\times$ area reduction, 20% performance enhancement, and 49% energy saving, compared to STT-RAM cache. Compared to SRAM, the performance is improved by 13% and the energy consumption is reduced by $40\times$.

Impact of RT memory cell sizes: Smaller racetrack domain cell results in an even more compact LLC, leading to faster access and less energy (including both dynamic and leakage) consumption. The possible side effect could be more shifting numbers, because one R/W port will be shared by more magnetic domains. But, smaller domain dimension could lower per shift energy so that the side effect can be partially compensated. We compared the HDART designs built with a moderate domain size of $4F^2$ and an aggressive racetrack design with domain size of $1F^2$. The performance and energy comparison can be found in Figure 11 and Figure 12. The results show that designs of $1F^2$ can achieve 3% and 11% performance and energy saving compared to designs of $4F^2$.

8. CONCLUSION

In this work, we performed a comprehensive exploration and design enhancement for *Racetrack* memory across multiple layers. We initialize the design exploration with a novel layout approach which enables an all R/W ports memory array structure. A flexible hardware architecture (HDART) is proposed based on the memory cell and array design. A data management scheme that can be naturally integrated onto the proposed HDART, further improves the efficiency for the RT based LLC. The RT based HDART with data management can achieve $6.4\times$ area reduction, 25% performance enhancement, and 62% energy saving, compared to STT-RAM cache design. The improvement obtained from the proposed HDART is much higher than *TapeCache*.

9. REFERENCES

[1] G. Sun and et al, "A Novel Architecture of the 3D Stacked MRAM L2 Cache for CMPs," in *Proc. of HPCA*, 2009, pp. 239–249.

[2] C. Smullen and et al., "Relaxing Non-Volatility for Fast and Energy-Efficient STT-RAM Caches," in *Proc. of HPCA*, 2011, pp. 50–61.

[3] "Everspin Throws First ST-MRAM Chips Down, Launches Commercial Spin-Torque Memory Era," in

http://www.engadget.com/2012/11/14/everspin-throws-first-st-mram-chips-down/, 2012.

[4] R. Nebashi and et al, "A 90nm 12ns 32Mb 2T1MTJ MRAM," in *Conf. of ISSCC*, 2009, pp. 462–463.

[5] Y. Chen and et al, "Combined Magnetic-and-Circuit-Level Enhancements for the Nondestructive Self-Reference Scheme of STT-RAM," in *Proc. of ISLPED*, 2010, pp. 1–6.

[6] "Itrs," http://www.itrs.net.

[7] S. Parkin, "Racetrack Memory: A Storage Class Memory Based on Current Controlled Magnetic Domain Wall Motion," in *DRC*, 2009, pp. 3–6.

[8] T. Koyama and et al, "Observation of the Intrinsic Pinning of a Magnetic Domain Wall in a Ferromagnetic Nanowire," *Nat. Mat.*, vol. 10, no. 3, pp. 194–197, 2011.

[9] X. Jiang and et al, "Discrete Domain Wall Positioning due to Pinning in Current Driven Motion along Nanowires," *Nano Lett.*, vol. 11, no. 1, pp. 96–100, 2010.

[10] L. Thomas, R. Moriya, C. Rettner, and S. Parkin, "Dynamics of magnetic domain walls under their own inertia," *Science*, vol. 330, no. 6012, pp. 1810–1813, 2010.

[11] S. Fukami and et al, "Low-current perpendicular domain wall motion cell for scalable high-speed mram," in *Symp. on VLSI Technology*, 2009, pp. 230–231.

[12] A. Annunziata and et al, "Racetrack Memory Cell Array with Integrated Magnetic Tunnel Junction Readout," in *Symp. on IEDM*, 2011, pp. 24–3.

[13] R. Nebashi and et al, "A Content Addressable Memory using Magnetic Domain Wall Motion Cells," in *IEEE Symp. on VLSI Circuits*, 2011, pp. 300–301.

[14] R. Venkatesan, , and et al, "TapeCache: a High Density, Energy Efficient Cache based on Domain Wall Memory," in *Proc. of ISLPED*, 2012, pp. 185–190.

[15] A. Brataas and et al., "Current-induced torques in magnetic materials," *Nat. Mat.*, vol. 11, no. 5, pp. 372–381, 2012.

[16] Z. Sun. and et al., "Multi Retention Level STT-RAM Cache Designs with a Dynamic Refresh Scheme," in *Proc. of Micro*, 2011, pp. 329–338.

[17] NVSim., "http://www.rioshering.com/nvsimwiki/index.php."

[18] Simics., "http://www.windriver.com/products/simics/."

[19] Parsec., "http://parsec.cs.princeton.edu/index.htm."

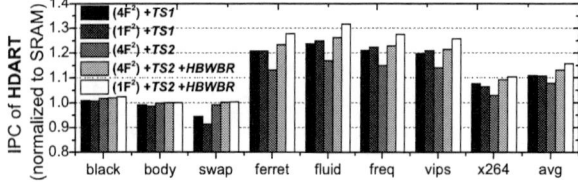

Figure 11: Performance enhancement by different policy for HDART (Normalized to $(4F^2+TS1)$).

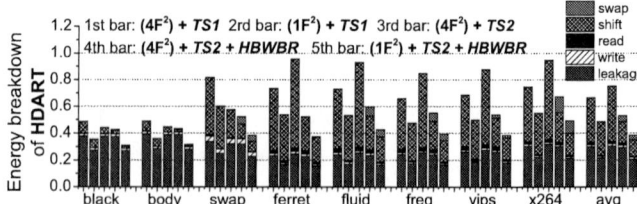

Figure 12: Energy breakdown by different policy for HDART (Normalized to STT).

Improving the Energy Efficiency of Hardware-Assisted Watchpoint Systems

Vasileios Karakostas[1,2], Sasa Tomic[1], Osman Unsal[1], Mario Nemirovsky[3], Adrian Cristal[1,4]

[1]Barcelona Supercomputing Center
[2]Universitat Politecnica de Catalunya
[3]ICREA Senior Research Professor at Barcelona Supercomputing Center
[4]Spanish National Research Council (IIIA-CSIC)
{first.lastname}@bsc.es

ABSTRACT

Hardware-assisted watchpoint systems enhance the execution of numerous dynamic software techniques, such as memory protection, module isolation, deterministic execution, and data race detection. In this paper, we show that previous hardware proposals may introduce significant energy overheads, and propose WatchPoint Filtering (WPF), a novel filtering mechanism that eliminates unnecessary watchpoint checks. We evaluate WPF on two state-of-the-art proposals for hardware-assisted watchpoints using two common memory checkers. WPF eliminates 83% of the watchpoint checks (up to 99.7%) and reduces 57% of the dynamic energy overhead (up to 78%) on average, without introducing additional performance execution overhead.

Categories and Subject Descriptors

B.3.2 [**Memory Structures**]: Design Styles—*cache memories*; C.0 [**General**]: hardware/software interfaces; D.2.0 [**Software Engineering**]: General—*protection mechanisms*

General Terms

Design, Performance

Keywords

Watchpoints, Metadata cache, TLB, Filtering, Optimization

1. INTRODUCTION

Writing bug-free code is a difficult task that presumes dedicating a significant amount of time to testing and debugging. However, the demand for higher productivity and for meeting tight release deadlines often results in insufficiently tested software. For instance, a very common type of bugs are memory-related. These bugs often pass development tests and manifest themselves under obscure conditions only after release. To bridge this gap, runtime systems should provide *always-on* and *low-overhead* support for analysis tools that improve the quality of executed code and increase the reliability of the system.

Permission to make digital or hard copies of all or part of this work for personal or classroom use is granted without fee provided that copies are not made or distributed for profit or commercial advantage and that copies bear this notice and the full citation on the first page. To copy otherwise, to republish, to post on servers or to redistribute to lists, requires prior specific permission and/or a fee.
DAC '13, May 29 - June 07 2013, Austin, TX, USA.

Tools based on dynamic binary instrumentation [1, 5, 6, 8, 12] provide the desired functionality but they impose significant performance degradation (e.g. more than 30× slowdown in Valgrind's MemCheck [12]). The root cause is the instrumentation overhead: all memory accesses are checked in software. Furthermore, commodity processors lack hardware support to accelerate such memory checks. The Translation Lookaside Buffer (TLB) checks access rights at coarse page-level granularity, while debug registers are limited to watching only a few memory locations.

To overcome these limitations, hardware-assisted watchpoint systems [4, 11, 14, 15, 17, 19, 21] have been proposed to enhance analysis tools in production runs. The hardware support typically includes a metadata-cache that accelerates watchpoint checks. Although these proposals seem to have good performance, they often make unneeded watchpoint checks and introduce significant dynamic energy overhead. We find that a high performance scheme that accesses the metadata-cache in *every* memory operation introduces up to 15.8% dynamic energy overhead - only due to the metadata-cache - with respect to the dynamic energy consumed in the private per-core caches. Taking into account the increasing importance of energy consumption in future processors, vendors are expected to employ hardware support for unlimited watchpoints only if they can provide both minimal energy and performance execution overheads.

In this paper, we propose WatchPoint Filtering (WPF), a novel mechanism that filters unnecessary watchpoint checks. Our proposal leverages the existing hardware in most commodity processors – the Virtual Memory mechanism. WPF repurposes an unused bit in the TLB entry, to mark whether the page has any defined watchpoints. This way, WPF manages to eliminate unnecessary watchpoint checks, reduces the number of accesses to the metadata-cache and improves the energy efficiency and performance of the system. Being orthogonal to the previous proposals for watchpoint support, WPF can be applied to almost any proposed implementation.

The main contributions of this paper are:

- We identify a gap between energy efficiency and performance in the state-of-the-art hardware-assisted watchpoint systems, and propose a filtering mechanism that eliminates unnecessary watchpoint checks.

- We demonstrate how WPF integrates with two previously proposed hardware watchpoint mechanisms and show that applying WPF requires only minor modifications to the original proposals.

- We evaluate the performance of WPF with two memory checkers. WPF eliminates on average 83% of the metadata checks and reduces 57% of the dynamic energy overhead, without introducing additional performance overhead.

In Section 2 we discuss the shortcomings of the hardware-assisted watchpoint systems. We describe the WPF mechanism and demonstrate how WPF can be applied to two state-of-the-art proposals in Section 3. We evaluate the efficiency of WPF in Section 4. In Section 5 we discuss related work and, finally, in Section 6 we conclude our study.

2. BACKGROUND & MOTIVATION

In this section we provide information about watchpoint systems, and we elaborate on the motivation for our work.

2.1 Watchpoints

Watchpoints allow the detection or the prevention of accesses to certain memory locations in user-level [4, 11, 15, 17, 21]. A memory watchpoint is defined by: (i) a memory range that needs to be watched, (ii) the desired access rights, and (iii) an exception handler. An access violation invokes a lightweight user-level handler. This handler may report an error, start a debugger, or perform a healing action to automatically recover from the access violation. All watchpoint information (the metadata) is stored in data structures managed by the software runtime.

Watchpoints can enhance a multitude of analysis tools to operate efficiently. For example, Venkataramani et al. [17] explain how watchpoints can prevent common memory bugs from manifesting. Shriraman et al. [15] apply watchpoints to enhance multi-module software engineering in Apache. Greathouse et al. [4] show how watchpoints can accelerate dynamic dataflow analysis, deterministic execution, and data race detection tools among others.

To accelerate the *always-on* execution of such analysis tools, watchpoint systems employ additional hardware support that mitigates the overheads of checking memory accesses in software. The hardware support typically includes a *metadata-cache* [4, 15, 17, 19, 21]. In case the hardware detects a watchpoint violation or a metadata-cache miss, the software runtime takes control of the execution.

2.2 Motivation

There are two main categories of hardware-assisted watchpoint systems. The first category targets high-performance; the processor accesses the metadata-cache in parallel with the L1 cache on *every* memory access [4, 17, 19, 21]. MemTracker [17] is a typical representative of this category. We find that a MemTracker-like approach introduces up to 15.8% of dynamic energy overhead only due to the metadata-cache with respect to the dynamic energy consumed in the private per-core caches (i.e. TLB, L1 and L2 cache) during normal execution (Section 4.2).

The second category targets energy efficiency; a representative example of this approach is Sentry [15]. The metadata-cache is consulted in the L1 miss path trading off performance for energy reduction. Indeed, the Sentry-like approach increases performance overhead around 2× compared to MemTracker [15]. Furthermore, we find that in such an approach, the total dynamic energy overhead can be up to 13.6% due to the high interaction of the metadata-cache with the L2 cache (Section 4.2).

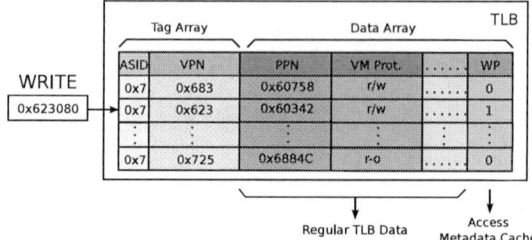

Figure 1: TLB filtering of memory-watchpoint checks. The WP-bit marks whether there are watchpoints in the corresponding page. The WP-bit is set and cleared by user-level instructions.

Therefore, we identify a gap between energy efficiency and high performance in the proposed hardware-assisted watchpoint systems. In this paper we narrow the gap with a filtering mechanism that orthogonally improves the energy efficiency in both categories without affecting the performance.

3. WATCHPOINT FILTERING

In this section we describe the WatchPoint Filtering (WPF) mechanism, we demonstrate how WPF integrates into two state-of-the-art hardware-assisted watchpoint systems, and we explain the implementation details of WPF.

3.1 The Idea

WPF reuses the Translation Lookaside Buffer (TLB) and assigns novel functionality to a currently unused (reserved) bit in the TLB entry. We name it as WatchPoint (WP) bit and it indicates the existence of watchpoints in the corresponding page (Figure 1). In case the WP-bit is set, the page contains defined watchpoints that should be checked in the metadata-cache. Otherwise, there are no watchpoints set and no further checks are performed.

WPF improves the energy efficiency in several ways. First, WPF filters out metadata checks for pages that do not have defined watchpoints. This reduces the dynamic energy spent in the metadata-cache. Second, WPF eliminates the pollution of the metadata-cache with non-restrictive watchpoints, i.e. useless entries that do not enforce any restriction in memory accesses. This results in higher metadata hit-ratio and enables the employment of smaller metadata-caches without affecting the performance. Third, by improving the metadata hit-ratio, WPF also filters out the transitions to the software runtime, reducing further the execution overhead and the cache interference due to these routines. Finally, by reusing the existing hardware of TLB (extending it with simple logic), WPF increases neither the latency, nor the static power dissipation, or the area of the TLB.

WPF avoids unnecessary checks in the metadata-cache leveraging the observation that watchpoints are typically set on a limited part of the address space. The amount of the reduced checks depends on how the analysis tool utilizes the watchpoints. Since WPF is an unintrusive optimization mechanism, the resulting system achieves lower dynamic power dissipation and better overall efficiency. Being orthogonal to previous mechanisms, WPF can be applied to almost any existing watchpoint implementation.

3.2 Integrating WPF

WPF with MemTracker-like approach. MemTracker [11] is the main representative of high performance hardware-assisted watchpoint designs. In MemTracker, the metadata-

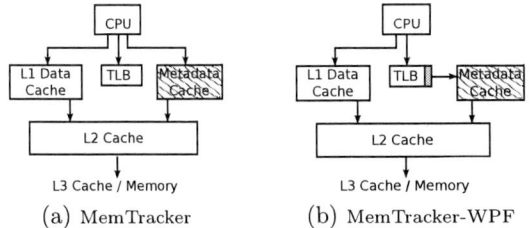

(a) MemTracker   (b) MemTracker-WPF

Figure 2: MemTracker architecture with and without WPF. The hardware support for watchpoints is hatched, while WPF is shaded in gray.

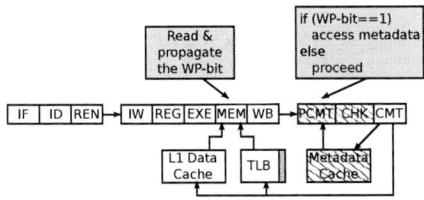

Figure 3: Integration of WPF into the processor pipeline of Mem-Tracker. The processor fetches the WP-bit in the MEM stage of the pipeline when the TLB is accessed. In the pre-commit stage (PCMT), the WP-bit is checked. According to its value, the metadata-cache is accessed. Finally, the memory access is checked in the CHK stage.

cache is accessed in parallel with the L1 cache on every memory operation (Figure 2a). Although the metadata-cache typically has a small size (around 4KB) it can introduce a significant amount of dynamic energy consumption. In Section 4 we show that such an approach introduces up to 15.8% overhead of dynamic energy consumption with respect to the per-core cache structures (i.e. TLB, L1 and L2 cache).

In MemTracker with WPF, the watchpoint checks take place only if the WP-bit is set (Figure 2b). MemTracker checks for watchpoint violations in the last stage of the execution pipeline, just before an instruction commits. To integrate WPF with a MemTracker-like system, no additional pipeline stages are required (Figure 3). In this way, WPF reduces the number of accesses to the metadata-cache and improves the energy efficiency of MemTracker.

WPF with Sentry-like approach. Sentry [15] targets energy efficiency and caches watchpoints in a metadata-cache placed on the L1 miss path. The metadata-cache reuses the L1 cache coherence states to elide checks on L1 hits. For example, if a cache-line is in shared state, it can be read directly without consulting the metadata-cache. If the watchpoint access rights are down-graded, the corresponding L1 cache lines are invalidated. The next memory reference to this cache line causes a miss in L1 and a check in the metadata-cache.

Figure 4 shows the design of Sentry and its enhanced version with WPF. Although Sentry is able to reduce the metadata-cache accesses compared to a performance-aggressive design (such as MemTracker), WPF can improve the dynamic energy and performance overheads even further.

Discussion. When we refer to MemTracker and Sentry in this paper, we focus on when the metadata-cache is accessed: (i) in parallel with the L1 cache (MemTracker), or (ii) in the L1 miss path (Sentry). The rest of their key mechanisms (e.g. the programmable state machine in MemTracker or the protection domain support in Sentry) are independent of our proposal. However, in Section 5 we explain how WPF

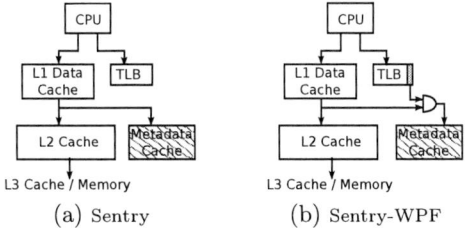

(a) Sentry   (b) Sentry-WPF

Figure 4: Sentry architecture with and without WPF. The metadata-cache is accessed only when a memory reference misses in L1 and the WP-bit for the corresponding page is set.

compares to their optimization techniques for reducing the number of accesses to the metadata-cache.

3.3 Implementation Details

WP-bit in memory. The software runtime holds various data structures that associate watchpoints with access rights and exception handlers. Thus, the software runtime is slightly extended to hold and control the WP-bit information (one bit per virtual page). In case of a TLB miss, the WP-bit is set and the software runtime is invoked to update it lazily without affecting the TLB-miss critical path.

Updating the WP-bit. Every time the programmer modifies a watchpoint, the software runtime serves the requested modification by updating its data structures and the metadata-cache. This is when the software runtime updates the corresponding WP-bit through a new unprivileged instruction added to the Instruction Set Architecture, which modifies the WP-bit from the user-level code. Security issues are prevented using the same mechanisms that disable arbitrary code to update watchpoints (e.g. state permission in MemTracker, or protection domains in Sentry).

Multi-core configurations. In multi-core systems, each processor has its own private TLB. To maintain the WP-bit coherent across the TLBs, we should invalidate the relevant entries. This could happen using the classic TLB shootdown approach, where the initiator core uses inter-processor interrupts to notify the rest of the cores to update the WP-bit of their entry. Note that the shootdown algorithm takes place immediately only when the check of the watchpoints is enforced (WP-bit goes from 0 to 1) so that no watchpoint checks are missed. The best approach is to build WPF on top of hardware-coherent TLBs, as proposed in [13] and [18]. This allows better performance execution by eliminating the need for inter-processor interrupts.

Large pages. Modern operating systems provide support for large pages [2, 10]. The use of large pages somewhat reduces the benefits of WPF, since the WP-bit represents a larger memory range. However, we show that WPF provides decent advantages even with large pages in Section 4.2.

4. EVALUATION

In this section we evaluate how WPF improves the dynamic energy consumption and the performance of hardware-assisted watchpoint systems.

4.1 Simulation Methodology

Watchpoint Tools and Benchmarks. We implement two memory checkers. The first is the *return-address checker*; it protects the control flow of an application by disallowing writes to the return addresses of functions (stored in

379

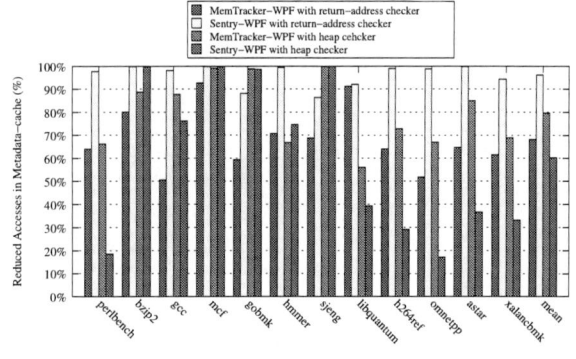

(a) SPECint2006

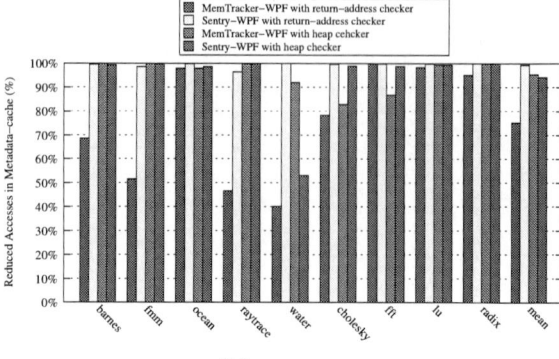

(b) SPLASH2

Figure 5: Percentage of reduced checks in the metadata-cache due to WPF.

Feature	Description
Processor(s)	8 cores, x86 in-order
L1 Cache	32 KB, 64B cache line, 4-way assoc., private, writeback, 2 cycles latency
L2 Cache	256 KB, 64B cache line, 8-way assoc., private, writeback, 8 cycles latency
L3 Cache	8 MB, 64B cache line, 16-way assoc., shared, writeback, 16 cycles latency
Data TLB	256 entries, 4-way assoc., 4 KB page, accessed in parallel with L1
main memory	4GB, 200 cycles latency
MemTracker [17]	4KB, 4-way assoc., in parallel with L1
Sentry [15]	4KB, 4-way assoc., on L1 miss path

Table 1: Simulator Configuration.

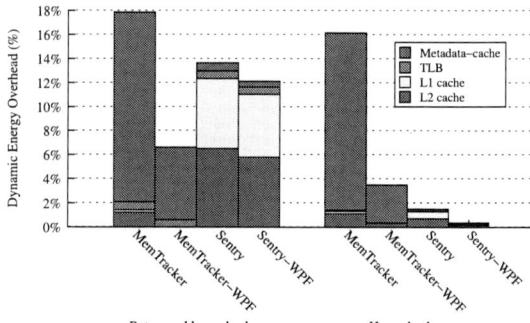

Figure 6: Breakdown of dynamic energy overhead. The overhead is estimated with respect to the per-core caches (i.e. TLB, L1 and L2 cache). WPF reduces significantly the dynamic energy consumption in both the metadata-cache and the cache hierarchy.

the stack) with watchpoints. The second is the *heap checker* that sets a no-access watchpoint in each memory deallocation to prevent dangling pointer usage. In a memory allocation, the checker removes possible watchpoints so that the allocated memory can be used, and prevents from out-of-bounds accesses by setting no-access watchpoints before and after the allocated area. On top of these memory checkers, we run the SPECint2006 benchmarks [7] with the test inputs and the Splash2 [20] benchmarks with the default inputs.

Simulation Infrastructure. To evaluate the efficiency of WPF, we use Pin [8] to implement the two checkers (by capturing the events that trigger watchpoint updates according to the checker) and to simulate an x86 chip multiprocessor. The simulated system consists of simple x86 in-order cores with IPC=1 except on memory accesses. The TLB and the data cache hierarchy are modeled in detail. Caches are inclusive and are kept coherent through a MESI directory-based protocol (Table 1). To evaluate the energy savings of WPF, we use CACTI 6.5 [9] with 32nm technology. For TLB, L1 caches, L2 caches, and the MemTracker's metadata-cache we use high-performance transistors ("itrs-hp"), while for the metadata-cache of Sentry we use low-power transistor technology ("itrs-lop").

4.2 Evaluation Results

We first evaluate how efficiently WPF filters unnecessary checks in the metadata-cache, and what improvements WPF provides in dynamic energy consumption. We then show how WPF affects the system's performance and increases the hit-ratio of the metadata-cache. Finally, we make a sensitivity analysis of the parameters that WPF depends on.

Reducing Metadata Checks. Figure 5 shows the percentage of reduced watchpoint checks that WPF achieves for

MemTracker and Sentry, using the return-address checker and the heap checker.

WPF eliminates 79% of metadata checks for MemTracker and 87% for Sentry on average. The efficiency of WPF with the evaluated checkers depends on various characteristics of the applications, such as the percentage of stack/heap accesses, the frequency of calling functions and the frequency of memory allocations/deallocations. For example, MemTracker-WPF with the return-address checker shows less benefits for applications that call functions frequently, such as gcc, omnetpp and xalancbmk. However, even for these applications, WPF eliminates more than 50% of the metadata checks. Similarly, Sentry-WPF with the heap checker exhibits smaller improvements for the applications that stress the memory allocator, but still eliminates more than 61% of metadata checks on average for this checker.

Reducing Dynamic Energy. Figure 6 shows the dynamic energy overhead spent in the per-core private caches (i.e. TLB, L1, L2 and metadata-cache) with respect to that of the normal execution without watchpoint support and checkers. The figure shows important findings for both original schemes and the potential benefits of WPF.

Regarding the original schemes, we find that MemTracker introduces up to 15.8% of dynamic energy overhead spent only in the metadata-cache. In Sentry the energy overhead of the metadata-cache is significantly lower than in MemTracker, but the total dynamic energy overhead can be up to 13.6% due to the high interaction with the L2 cache. Hence, we conclude that the hardware-assisted watchpoint systems may introduce significant energy overheads.

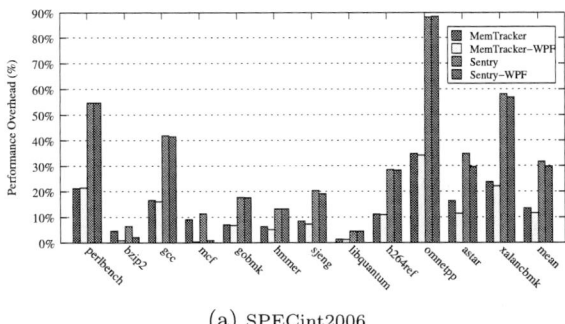

(a) SPECint2006

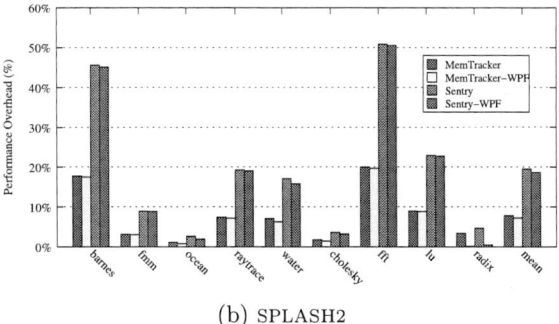

(b) SPLASH2

Figure 7: Performance overhead for the return-address checker. The overheads for the heap checker are less than 2% on average across all configurations. WPF achieves slight performance improvement by reducing the pressure in the metadata-cache.

WPF reduces by 71% and by 44% the total dynamic energy overhead of MemTracker and Sentry respectively. While WPF reduces the dynamic energy spent in the metadata checks, it cannot eliminate the metadata updates. However, the checks are more frequent than the updates, and therefore WPF brings significant improvements in the dynamic energy overhead of the metadata-cache itself. Besides this, WPF also achieves to reduce the watchpoint-induced overheads in the rest of the memory hierarchy by improving the hit-ratio of the metadata-cache. Better hit-ratio translates to less memory accesses and less interference with the application's data in the cache hierarchy because less metadata misses have to be resolved.

Figure 6 shows also that MemTracker-WPF can be more energy efficient than Sentry-WPF (6.6% vs. 13.7%) with tools that update watchpoints frequently (return address checker), and less energy efficient (3.5% vs. 0.4%) with tools that update watchpoints less often (heap checker). This tradeoff should be considered in the implementation of hardware-assisted watchpoint systems.

Improving Performance. Among the two evaluated checkers, the return-address checker demands more frequent watchpoint updates and introduces higher overheads. Figure 7 compares the performance overhead of MemTracker and Sentry with and without the WPF mechanism running the return-address checker.

The results show that WPF achieves a slight performance improvement for MemTracker and Sentry. The main sources of overhead for the original schemes are: (i) the software runtime updates, and (ii) the implications of accessing the metadata-cache in every memory operation (in every L1 miss for Sentry), i.e. increased pressure in the metadata-cache, unnecessary resolutions of metadata-cache misses and interference with the data in the memory hierarchy. WPF can reduce only the second source of overhead. On average, WPF reduces the performance overhead from 6.1% to 5% for MemTracker, and from 13.6% to 12.5% for Sentry.

Increasing Metadata Hit-Ratio. WPF eliminates the pollution of the metadata-cache with non restrictive watchpoint entries by not accessing the metadata-cache on every memory operation. Figure 8 shows the hit-ratio of the metadata-cache with various sizes, from 4KB (default) to 256B. The results show that WPF increases the hit-ratio of the metadata-cache for each size configuration. Moreover, WPF maintains high hit-ratio as the metadata-cache size reduces - for MemTracker the hit-ratio remains practically the same. This way, WPF can further improve the energy efficiency of the watchpoint system by employing a smaller

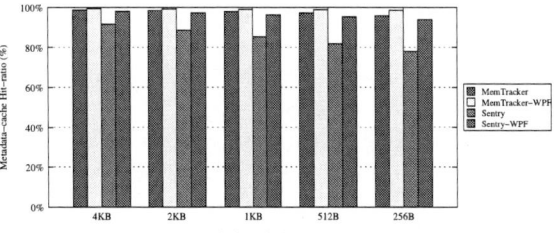

Figure 8: Sensitivity analysis for the metadata-cache hit-ratio. WPF can further improve the energy efficiency by employing a smaller metadata-cache, while maintaining similar performance.

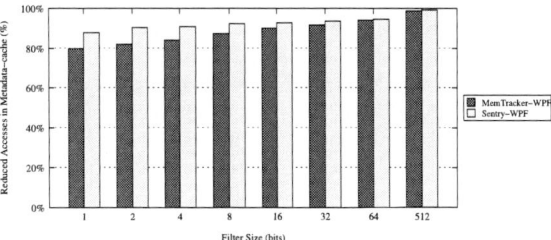

Figure 9: Sensitivity analysis for the filter size. The proposed WPF mechanism with 1 bit per TLB entry is sufficient regarding the tradeoff of performance against area and power.

metadata-cache, while maintaining similar performance.

Sensitivity Analysis for Filter Size. The efficiency of WPF comes from its ability to filter out unnecessary metadata checks by repurposing a single unused bit in each TLB entry. We evaluate WPF with various sizes of filters, where the filter applies as a mask to the corresponding TLB entry.

Figure 9 shows the results of the sensitivity analysis for the filter size ranging from 1 to 512 bits which represents an ideal per-word filter. As the filter size increases, the percentage of reduced accesses to the metadata-cache also increases. We found no significant differences for 1 to 8-bits filters, however for filters that are larger than 16-bits, the improvement becomes significant, while an ideal filter achieves more than 98% reduced metadata accesses. Considering the tradeoff of performance against area and power, we conclude that WPF with 1 bit per TLB entry is sufficient.

Sensitivity Analysis for Page size. WPF's improvements also depend on the page size. Figure 10 shows the results for various page sizes from 4KB to 4MB. Regarding the return-address checker, WPF is not significantly affected by the page size since most of the watchpoints are concentrated in a small region of the memory space. Even with 4MB page size, WPF reduces more than 66% of the metadata checks for MemTracker and 93% for Sentry. On the other hand, the

381

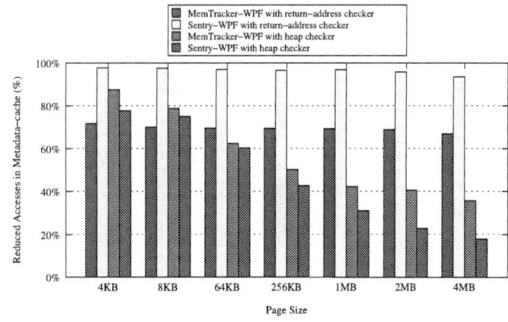

Figure 10: Sensitivity analysis for page size. Even with large pages, WPF eliminates a significant amount of metadata checks.

efficiency of WPF with the heap checker reduces as the page size increases. However, WPF still eliminates the metadata checks by 36% for MemTracker and 18% for Sentry.

5. RELATED WORK

Several proposals reduce metadata checks using dedicated registers. iWatcher [21] introduces a register to completely turn on/off the watchpoint system. Mondrian [19] uses side-car registers (SRs) to cache the metadata for an arbitrary range of memory addresses. MemTracker [17] can execute multiple checkers concurrently. To minimize overheads, Mem-Tracker uses an event mask register that masks out unused checkers. WPF acts complementary to the event mask register without completely turning off a checker.

HeapMon [14] adds a bit per cache line to indicate that a line access should be checked in software. Chen *et al.* proposed idempotent filters (IF) to eliminate redundant metadata checks in Log-Based Architectures [3]. WPF acts complementary to IFs for deciding whether the metadata check that skipped IF need to be performed or not.

SoftSig [16] relies on the programmer to mark code regions whose memory accesses should be checked in order to reduce the accesses to the signature file. WPF depends on the distribution of watchpoints across the address space. Thus, WPF improves SoftSig in an orthogonal way.

Greathouse *et al.* [4] accelerate watchpoint support introducing two different on-chip caches: a bitmap and a range cache. Incorporating WPF in [4] is straightforward, since the range cache contains the necessary information for updating the WP-bit.

Sentry [15] utilizes the F-bit that has some relevance to the WP-bit in the sense that they both reduce accesses to the metadata-cache. However, the F-bit is stored in the Page Table Entry (PTE). Updating the F-bit requires a slow call to the *operating system*; the authors claim that an optimized low-overhead system call takes around 300 cycles in modern microarchitectures. In addition, they do not further evaluate the benefits of the F-bit. In this paper, we propose and quantify the benefits of having *lightweight user-space* control of the WP-bit, we explore the WPF mechanism as a universal optimization method applicable to a general hardware watchpoint mechanism, and we show that WPF enables high-performance hardware-assisted watchpoint systems being more energy efficient than those that target solely energy efficiency under tools that update watchpoints frequently.

6. CONCLUSIONS

In this paper we presented WPF, a filtering mechanism for hardware-assisted watchpoint systems. Using only one bit

per TLB entry, we showed that WPF eliminates up to 99.7% of the watchpoint checks, and reduces up to 78% of the dynamic energy overhead. WPF introduces nearly no area or static power dissipation overhead and can orthogonally enhance almost any existing watchpoint implementation.

7. ACKNOWLEDGMENTS

The authors would like to thank Srdjan Stipic, Nehir Sonmez, Adria Armejach and Daniel Nemirovsky for their feedback. This work was partially supported by the cooperation agreement between the Barcelona Supercomputing Center and Microsoft Research, by the Ministry of Science and Technology of Spain and the European Union (FEDER funds) under contracts TIN2007-60625 and TIN2008-02055-E, and by the European Network of Excellence on High Performance Embedded Architecture and Compilation (HiPEAC).

8. REFERENCES

[1] D. Bruening. *Efficient, transparent, and comprehensive runtime code manipulation.* PhD thesis, MIT, 2004.

[2] K. Chen et al. Improving enterprise database performance on intel itanium architecture. In *Ottawa Linux Symposium*, 2003.

[3] S. Chen, M. Kozuch, T. Strigkos, B. Falsafi, et al. Flexible hardware acceleration for instruction-grain program monitoring. ISCA, 2008.

[4] J.L. Greathouse, H. Xin, Y. Luo, and T. Austin. A case for unlimited watchpoints. ASPLOS, 2012.

[5] S. Hangal and M. Lam. Tracking down software bugs using automatic anomaly detection. ICSE, 2002.

[6] R. Hastings and B. Joyce. Purify: Fast detection of memory leaks and access errors. USENIX, 1991.

[7] J. Henning. Spec cpu2006 benchmark descriptions. *SIGARCH Comput. Archit. News*, 34:1–17, Sept. 2006.

[8] C. Luk, R. Cohn, R. Muth, H. Patil, et al. Pin: building customized program analysis tools with dynamic instrumentation. PLDI, 2005.

[9] N. Muralimanohar et al. Architecting efficient interconnects for large caches with cacti 6.0. *IEEE Micro*, 28:69–79, January 2008.

[10] J. Navarro et al. Practical, transparent operating system support for superpages. OSDI, 2002.

[11] N. Neelakantam and C. Zilles. Ufo: A general-purpose user-mode memory protection technique for application use. Technical report, UIUC, 2007.

[12] N. Nethercote et al. Valgrind: a framework for heavyweight dynamic binary instrumentation. PLDI, 2007.

[13] B. Romanescu, A. Lebeck, D. Sorin, and A. Bracy. Unified instruction/translation/data (unitd) coherence: One protocol to rule them all. HPCA, 2010.

[14] R. Shetty et al. Heapmon: A helper-thread approach to programmable, automatic, and low-overhead memory bug detection. *Ibm Journal of Research and Development*, 50:261–276, 2006.

[15] A. Shriraman and S. Dwarkadas. Sentry: light-weight auxiliary memory access control. ISCA, 2010.

[16] J. Tuck, W. Ahn, L. Ceze, and J. Torrellas. Softsig: software-exposed hardware signatures for code analysis and optimization. ASPLOS, 2008.

[17] G. Venkataramani et al. Memtracker: Efficient and programmable support for memory access monitoring and debugging. HPCA, 2007.

[18] C. Villavieja et al. Didi: Mitigating the performance impact of tlb shootdowns using a shared tlb directory. PACT, 2011.

[19] E. Witchel, J. Cates, and K. Asanovic. Mondrian memory protection. *Sigplan Notices*, 37:304–316, 2002.

[20] S. Woo et al. The splash-2 programs: characterization and methodological considerations. ISCA, 1995.

[21] P. Zhou, F. Qin, et al. iwatcher: Efficient architectural support for software debugging. ISCA, 2004.

Low-Power Area-Efficient Large-Scale IP Lookup Engine Based on Binary-Weighted Clustered Networks

Naoya Onizawa and Warren J. Gross
Dept. of Electrical and Computer Engineering, McGill University Montreal, QC, Canada H3A 2A7
naoya.onizawa@mail.mcgill.ca,warren.gross@mcgill.ca

ABSTRACT

We propose a novel architecture for low-power area-efficient large-scale IP lookup engines. The proposed architecture greatly increases memory efficiency by storing associations between IP addresses and their output rules instead of storing these data themselves. The rules can be determined by simple hardware using a few associations read from SRAMs, eliminating a power-hungry search of input addresses in TCAMs. The proposed hardware that stores 100,000 144-bit entries is evaluated under TSMC 65nm CMOS technology. The dynamic power dissipation and the area of the proposed hardware are 4.6% and 30.6% of a traditional TCAM, respectively while maintaining comparable throughput.

Categories and Subject Descriptors

B.3.2 [**Memory Structures**]: Design Styles

General Terms

Algorithm, Design, Performance

Keywords

Associative memory, TCAM, neural network

1. INTRODUCTION

As the use of applications that require high bandwidth increases, Internet routers have become key hardware in the Internet backbone. Data transmission rates in the routers are large, such as in OC-768 (40Gb/s). For these requirements, Internet routers require fast IP-lookup operations utilizing hundred thousands of entries or more. Each router forwards packets toward their final destinations based on longest prefix matching (LPM) that determines the closest location to the final destination among several candidates. The length of a packet is up to 32 bits for IPv4 and 144 bits for IPv6. The packets contain binary strings and wildcards.

The hardware of the LPM has been designed based on the use of ternary content-addressable memory (TCAM) [1, 2,

3, 4], trie-based schemes [5, 6], and hash-based schemes [7, 8]. TCAMs compare a search word with all entries stored in TCAM cells in parallel and realize high-speed lookup operations. In contrast, the large area of the cell (16 transistors vs. 6 transistors in a static random access memory (SRAM) cell) and the brute-force searching cause large power dissipation and inefficient hardware architecture for large forwarding tables. Trie-based schemes store prefixes and locations based on a binary-tree structure that is created based on portions of stored IP addresses. The searching is performed by traversing the tree until an LPM is found. The hardware can be designed using SRAMs instead of TCAMs, which potentially lowers power dissipation. However, deep trees require multi-step lookups. Hash-based schemes use one or more hash tables to store prefixes. The benefit of hashing is scalability as table size is increased with length-independent searching speed. The hash-based schemes have a possibility of collisions that requires post-processing to decide on only one output and requires reading many hash tables for each length of stored strings.

In this paper, a low-power large-scale IP lookup engine based on clustered neural networks (CNNs) is proposed. The CNN is a neural network recently presented in [9] and stores data using only binary weighted connections between several clusters. Compared with a classical Hopfield Neural Network (HNN) [10], the CNN needs less complex functions while learning (storing) larger number of messages. The hardware implementation of the CNN has also been reported in [11]. For the IP lookup engine, the proposed architecture based on novel algorithms is designed to store three values ("0", "1", "don't care") by extending the original CNN that can store only binary data. Unlike TCAMs that store IP addresses themselves, the proposed hardware stores the associations between IP addresses and output rules, increasing memory efficiency. The output rule can be determined by simple hardware using a few associations read from SRAMs, which greatly reduces the search power dissipation compared with that of a TCAM that requires a brute-force search. As both IP addresses and their rules can be stored as associations in the proposed IP lookup engine, an additional SRAM that stores rules in the conventional TCAM-based IP lookup engine is not required.

The rest of this paper is organized as follows. Section 2 reviews an IP lookup scheme based on a traditional TCAM. Section 3 describes the proposed IP lookup scheme based on CNNs. Section 4 presents the hardware implementation. Section 5 evaluates the performance and compares to the related work. Section 6 concludes the paper.

Permission to make digital or hard copies of all or part of this work for personal or classroom use is granted without fee provided that copies are not made or distributed for profit or commercial advantage and that copies bear this notice and the full citation on the first page. To copy otherwise, to republish, to post on servers or to redistribute to lists, requires prior specific permission and/or a fee.
DAC '13, May 29 - June 07 2013, Austin, TX, USA.

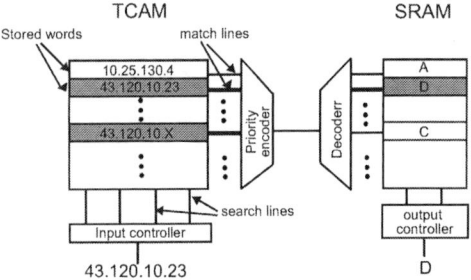

Figure 1: IP lookup using TCAM and SRAM. IP addresses and output rules themselves are stored.

2. REVIEW OF IP LOOKUP SCHEME

Fig. 1 shows an IP lookup scheme using a TCAM and a SRAM [12]. TCAMs contain large number of entries from hundreds to several hundred thousands. Each entry contains binary address information and a wildcard (X) in TCAMs, while only binary information is stored in binary CAMs. The size of the entry is several dozen to hundreds (e.g. 128, 144 bits for IPv6 [13]). An input address is broadcast onto all entries through search lines and one or more entries are matched. The priority encoder finds the longest prefix match among these matched entries and determines a matched location that is an address of a SRAM contains rules. Since the matched location in the TCAM corresponds to an address of the SRAM, the corresponding rule is read. For example in Fig. 1, an input address is 42.120.10.23 and matches two entries: 42.120.10.23 and 42.120.10.X. Although two matched locations are activated, the matched location corresponding to 42.120.10.23 is selected. Finally, a "rule D" is read from the SRAM. TCAMs perform high-speed matching based on one clock cycle in small number of entries: however there are some drawbacks when the number of entries is large, such as network routing. Since the search lines are connected to all entries, large search-line buffers are required, causing large power dissipation. The power dissipation of the search lines is the main portion of the all power dissipation. In terms of area, the number of transistors in a TCAM cell is 16, while it is 6 in a SRAM cell, causing area inefficient hardware implementation.

3. IP LOOKUP BASED ON CLUSTERED NEURAL NETWORK

3.1 IP lookup scheme without wildcards

The proposed IP lookup engine is designed by extending the CNN [9]. Fig. 2 shows an IP lookup scheme without wildcards. The IP lookup scheme contains both functions of a TCAM and a SRAM by storing associations between IP addresses and their rules. There are c input clusters and c' output clusters. Each input cluster consists of l neurons and each output cluster consists of l' neurons. In the example shown in Fig. 2, $c=4$ and $l=16$ are set in the input cluster, while $c'=1$ and $l'=16$ are set in the output cluster. The input address has 16 bits and the output rule has 4 bits.

3.1.1 Learning process

In the learning process, IP addresses and their corresponding rules are stored. Suppose that a k-th learning address m_k is composed of c-sub messages $m_{k1}...m_{kc}$ and a k-th learn-

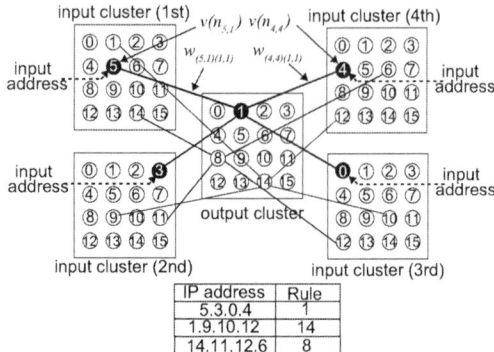

IP address	Rule
5.3.0.4	1
1.9.10.12	14
14.11.12.6	8

Figure 2: IP lookup without wildcards based on CNN. The IP lookup scheme contains both functions of TCAM and SRAM.

ing rule m'_k is composed of c'-sub messages $m'_{k1}...m'_{kc'}$. The length of the address is $c*log_2 l$ bits and that of the rule is $c'*log_2 l'$ bits. The address is partitioned into c-sub messages $m_{kj}(1 \leq j \leq c)$, whose size is $\kappa = log_2 l$ bits. Each sub message is converted into a l-bit one-hot signal, which activates the corresponding neuron in the input cluster. The rule is also partitioned into c'-sub messages $m_{k'j'}(1 \leq j' \leq c')$, whose size is $\kappa' = log_2 l'$ bits. Each sub message is converted into a l'-bit one-hot signal, which activates the corresponding neuron in the output cluster. During the learning process of M messages $m_1 ... m_M$ that include input addresses and rules, the corresponding patterns (activated neurons) $C(m)$ are learned. Depending on these patterns, connections $w_{(i_1,j_1)(i_2,j_2)}$ between i_1-th neuron of j_1-th input cluster and i_2-th neuron of j_2-th output cluster are stored in the following:

$$w_{(i_1,j_1)(i_2,j_2)} = \begin{cases} 1, & \text{if} \begin{cases} \exists m \in \{m_1...m_M\} \\ \text{and } C(m)_{j_1} = i_1 \\ \text{and } C(m)_{j_2} = i_2 \end{cases} \\ 0, & \text{otherwise} \end{cases} \quad (1)$$

This process is called **"Global learning"**. Suppose that M messages are uniformly distributed. An expected density d defined as memory utilization for **"Global learning"** is given by the following equation:

$$d = 1 - \left(1 - \frac{1}{ll'}\right)^M. \quad (2)$$

For example, $d=0.3$ means that stored bits of "1" is 30% of the whole bits in the memory.

3.1.2 Retrieving process

In the retrieving process, an output neuron (rule) is retrieved using a $c*\kappa$-bit input messages. The input message m_{in} is partitioned into c-sub messages $m_{inj}(1 \leq j \leq c)$, which are converted into l-bit one-hot signals. In each input cluster, one neuron that corresponds to the l-bit one-hot signal is activated. The value of each neuron $v(n_{i_1,j_1})(i_1 < l, j_1 \leq c)$ in the input cluster is given by:

$$v(n_{i_1,j_1}) = \begin{cases} 1, & \text{if } m_{inj_1} = i_1 \\ 0, & \text{otherwise} \end{cases} \quad (3)$$

This process is called **"Local decoding"**. Then, values of neurons $v(n'_{i_2,j_2})$ in the output cluster, where $n'_{i_2,j_2}(i_2 < l', j_2 \leq c')$ is the j_2-th neuron of the i_2-th input cluster, are

384

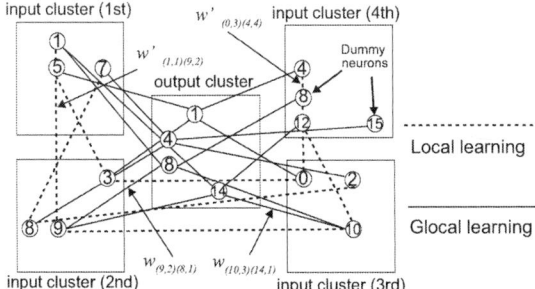

Figure 3: Learning process in IP lookup scheme with wildcards based on CNN.

Table 1: Learned messages in Figs. 3 and 4.

IP address	IP address w/ dummy	Rule
5.3.0.4	5.3.0.4	1
1.9.10.12	1.9.10.12	14
1.9.10.X	1.9.10.8	8
7.8.2.X	7.8.2.15	4

updated using the following equation:

$$v(n'_{i_2,j_2}) = \sum_{i_1=0}^{l-1} \sum_{j_1=1}^{c} w_{(i_1,j_1)(i_2,j_2)} v(n_{i_1,j_1}). \quad (4)$$

In each output neuron, the maximum value $v_{Max_{j_2}}$ is found and then output neurons are activated using the following equations:

$$v_{Max_{j_2}} = \max v(n'_{i_2,j_2}) \quad (5)$$

$$v(n'_{i_2,j_2}) = \begin{cases} 1, & \text{if } v(n'_{i_2,j_2}) = v_{max_{j_2}} \\ 0, & \text{otherwise} \end{cases} \quad (6)$$

These (4)-(6) processes are called **"Global decoding"**. In the example shown in Fig. 2, the input message is 5.3.0.4. After the process (2), the value of the 1-th neuron in the output cluster is 4 and becomes the maximum value. Hence, the rule "1" is retrieved.

3.2 IP lookup scheme with wildcards

3.2.1 Learning

As an IP routing table contains wildcards (X), the IP lookup scheme is extended to store wildcards in the table. Fig. 3 shows a learning process in the proposed IP lookup scheme with wildcards. There are two different connections: $w'_{(i,j)(i',j')}$ and $w_{(i_1,j_1)(i_2,j_2)}$. $w'_{(i,j)(i',j')}$ represents connections between activated neurons in the input clusters, while $w_{(i_1,j_1)(i_2,j_2)}$ represents connections between activated neurons in the input and output clusters. In the learning process for $w'_{(i,j)(i',j')}(i, i' < l : j, j' \leq c)$ shown in Fig.3, an address learned is partitioned into c-sub messages whose size is $\kappa = log_2 l$ bits. If the sub message is binary information, it is converted into a l-bit one-hot signal, which activates the corresponding neuron in the input cluster. If it is a wildcard, no neuron is activated in the input cluster. The output neurons are activated to store connections using the same algorithm in the IP lookup scheme without wildcards. Depending on corresponding patterns (activated neurons) $C(m)$, connections $w'_{(i,j)(i',j')}$ between i-th neuron of j-th input cluster and i'-th neuron of j'-th input cluster are stored in the fol-

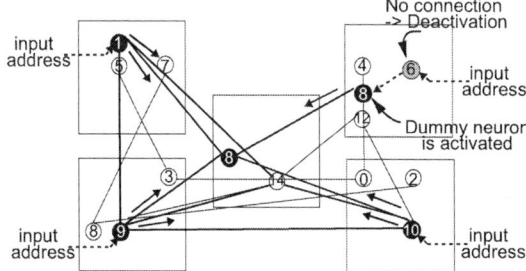

Figure 4: Retrieving process in IP lookup scheme with wildcards based on CNN.

lowing:

$$w'_{(i,j)(i',j')} = \begin{cases} 1, & \text{if } \begin{cases} \exists m \in \{m_1 ... m_M\} \\ \text{and } j+1 = j' \\ \text{and } C(m)_j = i \\ \text{and } C(m)_{j'} = i' \end{cases} \\ 0, & \text{otherwise} \end{cases} \quad (7)$$

This process is called **"Local learning"**. In the example shown in Fig. 3, there are two connections for a stored address is 7.8.2.X because $C(m)_1 = 7$, $C(m)_2 = 8$, $C(m)_3 = 2$ $C(m)_4 = $ X.

In the learning process for connections between the input and the output clusters $w_{(i_1,j_1)(i_2,j_2)}(i_1 < l : i_2 < l' : j_1 \leq c : j_2 \leq c')$ shown in Fig. 3, dummy neurons are activated when learned messages include wildcards. Suppose that the last half of learned messages has wildcards randomly. If a sub message is a wildcard, the wildcard is replaced by dummy information that is binary one. The dummy information is determined by a function using the first half of learned messages. In the paper, the function is realized by XORing two sub messages of the first half of learned messages. The k-th learning sub-messages m_{kj} that contain wildcards (X) are replaced as md_{kj} by the following equation:

$$md_{kj} = \begin{cases} m_{k(j-\frac{c}{2})} \oplus m_{k(j-\frac{c}{2}+1)(\mod \ \frac{c}{2})}, & \text{if } \begin{cases} m_{kj} = X \\ \text{and } j > \frac{c}{2} \end{cases} \\ m_{kj}, & \text{otherwise} \end{cases}$$
$$(8)$$

In the example shown in Fig. 3, a learned message 1.9.10.X has a wildcard. The wildcard is replaced by XORing $9 \oplus 1$ and becomes 8. After making dummy neurons, process (1) is performed by using md_k instead of m_k in order to make connections between the input and output clusters and then these connections are stored.

3.2.2 Retrieving

In the retrieving process, the proposed IP lookup scheme checks whether input messages are stored or not using stored connections. If an activated neuron corresponding to an input sub-message m_{inj_1} in the j_1-th input cluster doesn't have a connection to an activated neuron in the precedent input cluster, m_{inj_1} can be decided as "non stored messages". Hence, the activated neuron is de-activated and alternatively a dummy neuron is activated as the following two rules. First, the number of connections con_{j_1} in the j_1-th input cluster is given by:

$$con_{j_1} = \sum_{j=1}^{j_1} w_{(m_{in(j-1),j-1})(m_{inj,j})}. \quad (9)$$

385

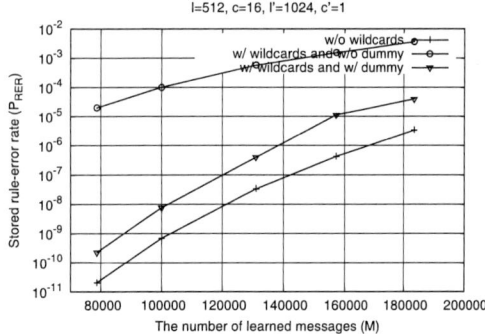

Figure 5: P_{RER} vs. M. **The IP lookup scheme can store 100,000 144-bit IP addresses at negligible low probability of P_{RER} ($<10^{-8}$).**

If con_{j_1} is less than $(j_1 - 1)$, the input sub-messages are not stored. Then, the input sub-messages m_{inj_1} are replaced as ms_{inj_1} by dummy information as follows:

$$ ms_{inj_1} = \begin{cases} \begin{aligned} & m_{in(j_1 - \frac{c}{2})} \\ & \oplus\, m_{in(j_1 - \frac{c}{2}+1)(\text{mod }\frac{c}{2})}, \end{aligned} & \text{if } \begin{cases} con_{j_1} < j_1 - 1 \\ \text{and } j_1 > \frac{c}{2} \end{cases} \\ m_{inj_1}, & \text{otherwise} \end{cases} $$
(10)

These (9),(10) processes are called **"Input selection"**. Then, $v(n_{i_1,j_1})$ is obtained based on equation (3) by using ms_{inj_1} instead of m_{inj}. In the example shown in Fig. 4, the input address is 1.9.10.6. Since the activated neuron "6" in the 4th input cluster doesn't have a connection to the activated neuron "10" in the 3rd input cluster, the activated neuron is de-activated. Instead, a dummy neuron "4" in the 4th input cluster is activated.

The subsequent process is **"Global decoding"** based on (4)-(6). As the first-half input clusters have only binary information, the summation of $w_{(i_1,j_1)(i_2,j_2)}v(n_{i_1,j_1})$ from the first-half ones must be the half of the number of input clusters, if output neurons are a candidated rule. In the IP lookup scheme with wildcards, (4) is modified by the following three equations:

$$ ena_{i_2,j_2} = \sum_{i_1=0}^{l-1} \sum_{j_1=1}^{c/2} w_{(i_1,j_1)(i_2,j_2)} v(n_{i_1,j_1}). \quad (11) $$

$$ sum_{i_2,j_2} = \sum_{i_1=0}^{l-1} \sum_{j_1=\frac{c}{2}+1}^{c} w_{(i_1,j_1)(i_2,j_2)} v(n_{i_1,j_1}). \quad (12) $$

$$ v(n'_{i_2 j_2}) = \begin{cases} sum_{i_2,j_2}, & \text{if } ena_{i_2,j_2} = \frac{c}{2} \\ 0, & \text{otherwise} \end{cases} \quad (13) $$

After these processes, (5) and (6) are performed.

Fig. 5 shows stored rurle-error rates (P_{RER}) vs. the number of learned messages (M) in the IP lookup schemes ($l = 512, c = 16, l' = 1024, c' = 1$). In the proposed schemes, there might be ambiguity that more than two neurons (rules) are retirieved in the stored rules because they might have the maximum value after **"Global decoding"**. P_{RER} is defined as the probability of the ambiguity. In the figure, the proposed IP lookup schemes store 144-bit ($c * log_2 l$) addresses (messages) for IPv6 [13]. P_{RER} are evaluated by simulations. Unlike IPv4, since packet traces for the long

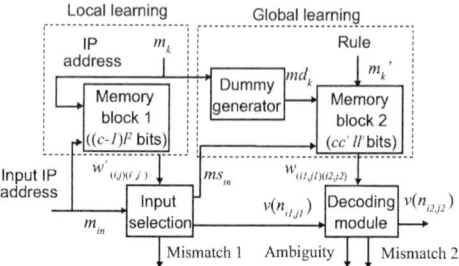

Figure 6: **The proposed IP lookup engine.**

prefix table are not available to the public and the prefix table is still small [14], we choose addresses randomly for the evaluation. The stored addresses are uniformly distributed. Random-length wildcards appear in the last half of addresses (72 bits). If the range of addresses that have wildcards is changed, the prefix length can be changed.

P_{RER} is strongly dependent of M. In the proposed IP lookup scheme with wildcards, if a dummy neuron is the same neuron that already stored, the both outputs (rules) might be retrieved, which slightly increases P_{RER} compared with that of the IP lookup scheme without wildcards. Adding dummy neurons are very effective to lower P_{RER} that is about five orders of magnitude reduction of that without dummy at M=78,643. The proposed IP lookup scheme with wildcards can store 100,000 144-bit IP addresses at negligible low probability ($P_{RER} < 10^{-8}$).

P_{RER} is also simulated when the learned messages are correlated. The word length in the IP lookup scheme with wildcards is 64 bits ($l = 256, c = 8, l' = 1024, c' = 1$). The first 8 bits of the learned messages are selected from 64 fixed patterns of 256 (2^8). The rest of the learned messages are uniformly distributed. Random-length wildcards appear in the last half of addresses (32 bits). At M=10,000, P_{RER} using the correlated patterns is $5.30 * 10^{-7}$ while P_{RER} using uniformly distributed patterns is $1.69 * 10^{-7}$.

Unlearned input messages are detected as "mismatch" using two steps. In the first step, the number of local connections $con_{c/2}$ in (9) is checked in **"Input selection"**. If it is not the same as $(c/2) - 1$, an unlearned input message can be detected as "mismatch" because all local connections related to the input message are not stored ("Mismatch 1"). If an unlearned input message is detected as "match" by other stored connections, the number of global connections ena_{i_2,j_2} is checked at **"Global decoding"**. If all ena_{i_2,j_2} are not the same as $(c/2)$, an unlearned input message can be detected as "mismatch" because all global connections between the input message and all output rules are not stored ("Mismatch 2").

4. HARDWARE IMPLEMENTATION

Fig. 6 shows an overall structure of the proposed IP lookup engine with wildcards. The learning process is "**Local learning**" in (7) and "**Global learning**" in (8) and (1). The retrieving process is that: 1) each input-sub message is replaced by a dummy message if it is not stored ("**Input selection**") in (9) and (10), 2) connections between c activated neurons in the input clusters are read from a memory block 1 ("**Local decoding**") in (3), and 3) an output neuron is retrieved based on connections between neurons of the input and the output clusters in a memory block 2 in (11)-(13), (5), and (6) ("**Global decoding**").

386

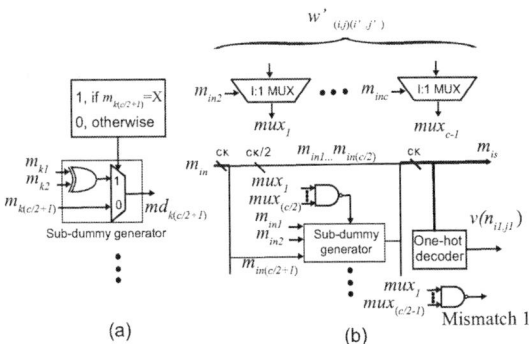

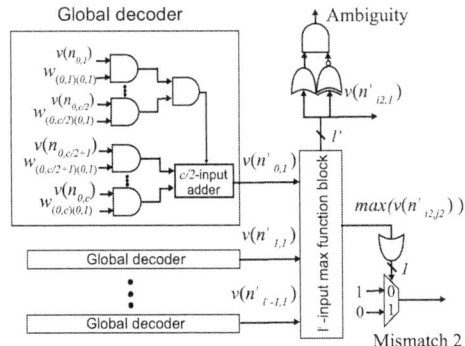

Figure 7: Circuit diagrams: (a) dummy generator, and (b) Input selection block.

There are $((c\text{-}1)l^2)$-bit SRAMs for "**Local learning**" and $c'll'$-bit SRAMs for "**Global learning**". In "**Local learning**", a sub-input k-th learning address m_{kj} and the subsequent one $m_{kj'}$ are converted into one-hot l-bit signals at row and column decoders, respectively. Then, a $w_{(i,j)(i',j')}$ is stored in the memory block 1 if both are not a wildcard. In "**Global learning**", the last half of an input address is replaced by dummy information using a dummy generator that includes $\frac{c}{2}$ sub-dummy generators if each sub-input address is a wildcard shown in Fig. 7 (a). The sub-dummy generator contains l 2-input XOR gates and multiplexors. Using the sub-input address with dummy information md_{kj_1} and the corresponding rule m'_{kj_2}, a $w_{(i_1,j_1)(i_2,j_2)}$ is stored in the memory block 2.

In the retrieving process, an input address m_{in} is partitioned into c-sub messages and $((c\text{-}1)l)$ connections $w_{(i,j)(i',j')}$ are read from the memory block 1. $w_{(m_{inj},j)(m_{in(j+1)},j+1)}$ in (9) are selected in a multiplexor of an input-selection module shown in Fig. 7 (b). Then, the last half of the input address is replaced by dummy information if these corresponding connections are not found. The output ms_{in} that contains the first half of the input address and the generated last half of input address is sent to the memory block 2. The one-hot decoder transforms ms_{in} to $v(n_{i_1,j_1})$. In the memory block 2, $(c'l')$ connections $(w_{(i_1,j_1)(i_2,j_2)})$ are read by ms_{in} and are sent to a decoding module shown in Fig. 8. The decoding module contains $(c'l')$ global decoder, c' max-function blocks and c' ambiguity checkers, where c' is set to 1 in Fig. 8. In the global decoder, $\frac{c}{2}$ 2-input AND gates and $\frac{c}{2}$-input AND gate generate an enable signal to a $\frac{c}{2}$-input adder, where these circuits are corresponding to (11)-(13). There are a l'-input max-function block that decides an activated neuron $v(n_{i_2,j_2})=1$. The ambiguity checker checks that two neurons are activated simultaneously.

5. EVALUATION

We designed the proposed IP lookup engine with wildcards based on TSMC 65nm CMOS technology. The parameters of the hardware are determined by simulations shown in Fig. 5 as l=512, c=16, l'=1024 and c'=1. The IP lookup engine stores 100,000 144 entries with P_{RER} is less than 10^{-8}. 15 SRAMs (256kb) are designed for the local learning block and 32 SRAMs (256kb) are designed for the global learning block. For the purpose of comparison, a reference TCAM is also designed. The TCAM also stores the same

Figure 8: Decoding module

Table 2: Performance comparisons.

	Reference TCAM	Proposed
Throughput [Gbps]	52.0	48.3
Dynamic power [W]	3.03	0.14
Static power [W]	0.24	0.09
Energy metric [fJ/bit/search]	0.584	0.028
Memory [Mb]	14.4 (TCAM)	11.75 (SRAM)
Equivalent size (SRAM) [Mb]	38.4	11.75
Number of transistors	256M	77M

entries as the proposed IP lookup engine. A TCAM cell is designed using a NAND-type cell that is composed of 16 transistors [12]. Each entry has 144 TCAM cells and is designed based on a hierarchy design style for high-speed matching operations. The TCAM is divided into 20 sub-TCAMs that have 5,000 entries for a power reduction of search lines. A priority encoder attached to the TCAM has 100,000 inputs and 17-bit outputs.

Table 2 shows performance comparisons between the reference TCAM and the proposed IP lookup engine by HSPICE simulation. For learning (storing) process, the proposed IP lookup engine takes 100,000 clock cycles to store 100,000 messages (entries) as well as the conventional TCAM. For retrieving (matching) process, the proposed engine takes two clock cycles. The worst-case delay is 1.44 ns in the latter part, where the delay of the max function block is 60 % of the whole delay. Throughput is defined as $length/worst\text{-}case\ delay$. The dynamic power dissipation of the proposed circuit is just 4.6 % of that of the reference TCAM, because the proposed circuit doesn't require search lines whose power dissipation is a large portion of that of the TCAM. In terms of area, the TCAM cell contains 16 transistors while the SRAM cell contains 6 transistors. Hence, the required memory size of the proposed circuit is 11.75 Mb that is 30.6 % of the equivalent area of the reference TCAM. The memory area is 96 % of the total area in the proposed circuit. A lookup speed of the proposed IP lookup engine is 201 Gb/s that is over 40Gb/s (OC-768), while a packet is considered to be 75 bytes.

The performance of the proposed IP lookup engine depends on the number of clusters and neurons. Under the same learned messages, large number of neurons and small number of clusters achieve low P_{RER} while requiring large memory sizes and computation blocks. In contrast, small number of neurons and large number of clusters reduce memory sizes and computation blocks while having high P_{RER}. Hence, depending on applications that can afford P_{RER}, the

Table 3: Performance comparisons with related works.

	TCAM [1]	DTCAM [2]	IPCAM [3]	eDRAM [4]	Trie [6]	Hash [7]	Proposed
Length [bits]	512	144	32 (128)[a]	23	63	32	144
Number of entries	21,504	32,768	65,536 (1.38M)[b]	16,000,000	318,043	524,288	100,000
Throughput [Gbps]	76.8	20.6	32	4.6	12.6	6.4	48.3
Power dissipation [W]	12.26	2.0	7.33	0.6	-	5.5	0.14
Energy metric [fJ/bit/search]	5.53	2.96	0.159	0.007	-	1.64	0.028
Memory [Mb]	10.5	4.5	2	432	31.19	60	11.75
Equivalent size (SRAM) [Mb]	28	-	7.33	-	-	-	-
Equivalent size (SRAM) for 144-bit length [Mb]	7.88	4.5	32.99	10^{39}		-	11.75
Technology [nm]	130	130	65	40	(FPGA)	130	65

[a]This method can be extended to 128 bits for IPv6.
[b]1 IPCAM word is approximately equivalent to 22 TCAM words.

performance can be determined by changing the number of neurons and clusters.

Table 3 shows performance comparisons with related works. The design in [1] is a straightforward implementation using TCAMs. In [2], eDRAMs are used to reduce the size of TCAM cells for low power dissipation; however they tend to be complex process. In [3], several entries are shared using special-purpose CAM cells to reduce the number of entries required. In [4], the prefix match is realized by reading candidates from eDRAMs and hence the energy metric is very small. However, as the memory size is $O(2^n)$ where n is the word length, an unacceptably large memory of 10^{39} Mb is required for long words (e.g. 144 bits). A trie-based method in [6] (PC trie-4) realizes a memory-efficient IP lookup engine using a prefix-compressed trie and also uses a hash function to reduce memory accesses to off-chip memory in order to achieve high throughput. A hash-based method in [7] reduces power dissipation using a collision-free hash function compared with the TCAM in [1].

6. CONCLUSIONS

We have proposed the low-power large-scale lookup engine based on binary-weighted clustered neural networks (CNN). By extending the CNN, the proposed IP lookup engine efficiently stores IP addresses and their outputs (rules) as connections in SRAMs, reducing the memory requirement and eliminating power-hungry TCAMs. The retrieving (matching) process is performed in parallel, realizing high-throughput IP lookup engines. Under TSMC 65nm CMOS technology by HSPICE simulation, the proposed IP lookup engine reduces the energy dissipation and the area to 4.6 % and 30.6 % of those of a traditional TCAM, respectively. Compared with other related works, the proposed IP lookup engine realizes one of the lowest energy metric while dealing with large number of long entries. In future work, we will apply the extended CNN to many applications that currently use TCAMs. In addition, we will consider eliminating errors related to P_{RER}.

7. ACKNOWLEDGMENTS

The authors would like to thank Vincent Gripon and Hooman Jarollahi for their helpful discussions.

8. REFERENCES

[1] B. Gamache et. al., "A fast ternary CAM design for IP networking applications," in *Proc. 12th IEEE ICCCN*,
Oct. 2003, pp. 434 – 439.

[2] H. Noda et. al., "A cost-efficient high-performance dynamic TCAM with pipelined hierarchical searching and shift redundancy architecture," *IEEE JSSC*, vol. 40, no. 1, pp. 245 – 253, Jan. 2005.

[3] S. Maurya et. al., "A dynamic longest prefix matching content addressable memory for IP routing," *IEEE TVLSI*, vol. 19, no. 6, pp. 963 –972, June 2011.

[4] Y. Kuroda et. al., "A 200Msps, 0.6W eDRAM-based search engine applying full-route capacity dedicated FIB application," in *Proc. CICC*, 1/4, Sept. 2012.

[5] W. Eatherton et al., "Tree bitmap: hardware/software IP lookups with incremental updates," *SIGCOMM Comput. Commun. Rev.*, vol. 34, no. 2, pp. 97–122, Apr. 2004. [Online]. Available: http://doi.acm.org/10.1145/997150.997160

[6] M. Bando et. al., "Flashtrie: Beyond 100-Gb/s IP route lookup using hash-based prefix-compressed trie," *IEEE/ACM Transactions on Networking*, vol. 20, no. 4, pp. 1262 –1275, Aug. 2012.

[7] J. Hasan et. al., "Chisel: A storage-efficient, collision-free hash-based network processing architecture," in *Proc. 33rd ISCA*, 203/215, June 2006.

[8] S. Dharmapurikar et al., "Longest prefix matching using bloom filters," *IEEE/ACM Trans. Networking*, vol. 14, no. 2, pp. 397 – 409, April 2006.

[9] V. Gripon and C. Berrou, "Sparse neural networks with large learning diversity," *IEEE Trans. Neural Networks*, vol. 22, no. 7, pp. 1087 –1096, July 2011.

[10] J. J. Hopfield, "Neural networks and physical systems with emergent collective computational abilities," in *Proc. the National Academy of Sciences*, vol. 79, no. 8, pp.2554-2558, Apr 1982.

[11] H. Jarollahi et. al., "Architecture and implementation of an associative memory using sparse clustered networks," in *Proc.ISCAS*, 2901/2904, May 2012.

[12] K. Pagiamtzis and A. Sheikholeslami, "Content-addressable memory (CAM) circuits and architectures: a tutorial and survey," *IEEE JSSC*, vol. 41, no. 3, pp. 712–727, Mar. 2006.

[13] H. Po-Tsang and H. Wei, "A 65 nm 0.165 fJ/bit/search 256 × 144 TCAM macro design for IPv6 lookup tables," *IEEE JSSC*, vol. 46, no. 2, pp. 507 –519, Feb. 2011.

[14] "Border gateway protocol (BGP)," http://bgp.potaroo.net.

Real-Time Use-Aware Adaptive MIMO RF Receiver Systems for Energy Efficiency under BER Constraints

Debashis Banerjee, Shyam Devarakond, Shreyas Sen* & Abhijit Chatterjee

Georgia Institute of Technology, Atlanta, USA, *Intel Corporation, USA

debashis.banerjee@gatech.edu, chat@ece.gatech.edu

ABSTRACT

Modern MIMO RF transceiver systems are designed to operate reliably under diverse channel conditions leading to incorporation of significant performance margins in RF transceiver systems. In general, across dynamically varying channel conditions, the fidelity of the RF front end devices can be traded-off against power consumption without compromising system-level BER limits. In this work such a real-time performance vs. power consumption modulation of RF front-end devices in MIMO systems is demonstrated. Through a multi-dimensional optimization technique, power-optimal configuration of the front-end for varying channel conditions are created. Additionally multiple low-power operating modes for the MIMO system are proposed depending on the performance metric (data rate or energy-per-bit) that need to be optimized for different applications.

Categories and Subject Descriptors

B.4.1 [**Data Communications Devices**]: Data Communications Devices- Transmitters/Receivers.

General Terms

Performance, Design, Algorithms

Keywords

MIMO, Low Power, Adaptive, Diversity, Spatial Multiplexing, Receiver. Radio-Frequency, OFDM.

1. INTRODUCTION

In today's harsh radio propagation environments providing high bit rates (for data-hungry applications) and extended range of mobile signal coverage has become increasingly challenging. Higher data rates demand higher power or larger bandwidth, both of which are at a premium in modern systems. Since the cell size determines the number of base stations required to cover a target population or area, the range of coverage is intricately tied to the economics of network deployment. A key technology that enables current systems to meet the stringent demands of the latest communication standards is multiple input multiple output (MIMO) communication.

MIMO systems can transmit/receive multiple parallel streams of data within the same channel bandwidth. Thus MIMO implementation results in an effective increase in data-rate (Spatial Multiplexing(SM)) or extended range of operation (Spatial Diversity (SD)) over Non Line of Sight (NLOS) faded

Permission to make digital or hard copies of all or part of this work for personal or classroom use is granted without fee provided that copies are not made or distributed for profit or commercial advantage and that copies bear this notice and the full citation on the first page. To copy otherwise, to republish, to post on servers or to redistribute to lists, requires prior specific permission and/or a fee.

DAC '13, May 29 - June 07 2013, Austin, TX, USA.

noisy channels. The benefits of using MIMO as opposed to single input, single output (SISO) systems have been illustrated in prior studies[1].

A parallel thrust in modern research is in the development of ultra-low power communication systems. The use of portable battery-powered multimedia devices has forced the issue of low power wireless communication and multi-media system design. Thus, while MIMO systems provide higher data bandwidth and increased range of operation, due to increased number of transceiver chains they must also be designed for low energy consumption across diverse operating conditions. Current standards allow for support of up to 8 chains on a single chip demanding high performance requirements from on-chip battery. This problem can be addressed at various levels of design abstraction: protocol/algorithmic level, system (module) level and circuit (transistor) level.

Prior work in this domain end has been reported in [2-6]. In [2] the power consumption has been optimized at the transmitter for several Power Amplifiers (PA) topologies through intelligent power allocation schemes. [3-4] discuss low-power baseband techniques for MIMO transceivers. In [5] the system is optimized to maintain a constant throughput across various channel conditions by switching between SISO, single input multiple output (SIMO) and MIMO configurations. [6] optimizes the power consumption of the system across channel conditions by switching between antenna modes and encoding schemes. Thus previous work actually optimizes power by switching between modulation schemes/MIMO modes and number of transmit/receive chains depending on the channel conditions. Hence, all of the radio frequency (RF) front-ends discussed are static in nature, power being saved by switching the number of chains.

Prior literature has demonstrated how real-time modulation of a SISO transmitter and receiver front-end [7-8][13] can be performed to save power under dynamically changing channel conditions. *However unlike SISO systems, the presence of multiple-modes of operation in MIMO RF systems results in different non-monotonically varying performance margins across different points of operation of a MIMO transceiver system.* As illustrated in [9] the maximum benefit can be derived from a MIMO system if it is used in Spatial Multiplexing (SM) mode when the wireless channel is good and in Spatial Diversity (SD) mode when the channel is bad. Depending on wireless channel conditions, current MIMO systems switch across different operational modes and signal modulation rates in real time to maximize the wireless data transmission rate for any given channel condition. However, fine-grained co-modulation of the RF circuit level performance along with the operational modes and signal modulation rates for MIMO systems has not been fully exploited by prior research. In this work, it is demonstrated that a finer power-performance trade-off can be achieved if vertically integrated operational mode vs. signal modulation rate vs. circuit level reconfigurations are performed. Moreover the amount of power that can be saved *is inherently dependent on the data-*

throughput requirements of the end-user application. Certain end-user applications have constraints of operating at the maximum data rate possible. While in other applications data rate can be compromised to achieve optimal energy-per-bit. Thus multiple low power modes could exist, each optimized for a different application (data rate vs energy-per-bit). In summary, the paper demonstrates:

a) *Seamless environment-adaptive low-power operation of a MIMO receiver as it switches between several modulation techniques and MIMO modes over richly faded Non-Line-of-Sight (NLOS) channels*

b) *Two different low-power operation modes: a Data Priority (DP) mode and an Energy Priority (EP) mode catering to two different data reception needs. The 2 operation modes illustrate the trade-off in energy and throughput for different applications and result in a "use-aware" environment-adaptive MIMO RF receiver front-end.*

The rest of the paper is organized as follows: section 2 describes MIMO system concepts, section 3 introduces low-power methodology, section 4 details system simulation setup, section 5 describes optimization algorithm and section 6 presents simulation results followed by Hardware demonstration of proposed concept and references. Additional figures and graphs are presented in Supplement to illustrate and support the paper.

2. KEY MIMO SYSTEM CONCEPTS

2.1 System and Feedback Description

Figure 1 shows a 2x2 MIMO system including the RF transmitter, MIMO channel and a detailed view of the proposed channel-adaptive RF receiver. The MIMO RF transmitter processes a bit stream by encoding, modulating it and transmits across two spatially separated antennas.

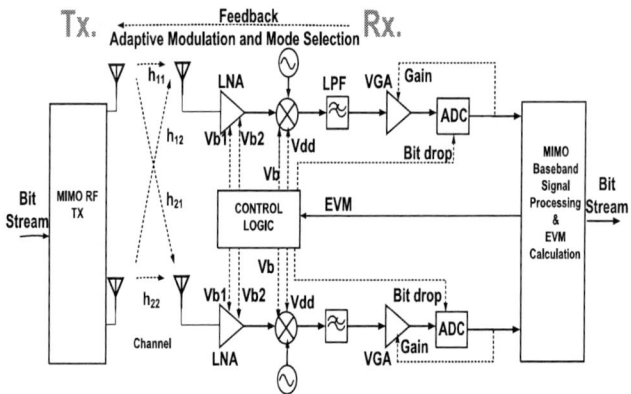

Figure 1: Adaptive MIMO Receiver RF Front End

The receiver consists of 2 parallel and symmetric front-end chains interfacing to the 2 antennas. Each chain consists of an low noise amplifier (LNA), a mixer, variable gain amplifier (VGA) and an ADC. The digitized output from each chain is fed to the baseband unit that processes the data to decode the transmitted bit stream. As shown in Figure 1 for the purpose of low-power adaptive operation a feedback loop within the receiver itself has been introduced. The feedback loop consists of a control block that depending upon the current error vector magnitude (EVM) of the received signal controls the built-in tuning knobs of several modules (LNA, Mixer, ADC etc.). This allows us to modulate the performance of the receiver in real-time. The functioning of the feedback loop and the description of the elements are covered in Section 3 and 4.

2.2 MIMO Modes and Performance Margins

One of the primary reasons for the commercial success of MIMO systems is that they provide the flexibility to operate in both SM and SD Mode. In SM Mode, MIMO systems transmit two parallel data streams over the two antennas thereby doubling the data rate over the same frequency bandwidth. Minimum mean square error (MMSE) decoding [1] within the receiver is used to extract the two data streams. SM is only possible when the received signal SNR is very high. When the wireless channel degrades, SM operation fails to meet system level performance specifications for values of the SNR metric below a predetermined level. Consequently, the MIMO system switches to SD mode of operation (Figure 2) which can support lower SNR values at correspondingly lower data throughput rates. In SD mode, using the Alamouti algorithm [10] the effective data rate is reduced to that of a SISO system. However using multiple spatially dispersed antennas the receiver effectively combines the signal power across multiple paths leading to an increase in signal strength (diversity gain).

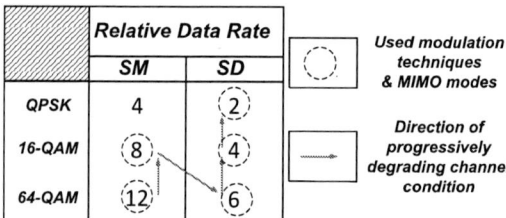

Figure 2: MIMO Data Rates and Switching Order

Due to this degree of redundancy the effective SNR is increased and the relevant fade margin required is reduced allowing the MIMO system to operate more reliably than a corresponding SISO system (larger range of operation/lower bit error rate). During the operation of the receiver in each of these modes the higher protocol layers adaptively change the modulation technique depending on the channel SNR so as to always transmit at the maximum data rate possible.

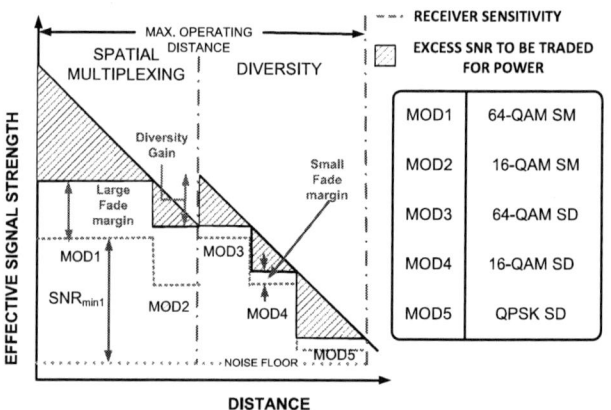

Figure 3: MIMO Operation and Performance Margins

Figure 3 shows the effective received signal power versus distance plots, annotated with noise floor, fade margin and SNR margin required for each mode and modulation technique used. As evident the signal strength reduces with distance. For the best received SNR the system operates at the highest data rate possible and remains in this mode till the signal strength is just sufficient to support the fade margin and the minimum required SNR to maintain an acceptable BER (SNR_{min}) for this mode of operation. Thereafter it switches to a mode having a lower data rate and hence a lower SNR_{min}. The received signal power at which the system switches from the ith mode to the next lower data rate is given by:

$$P_{switch,i} = Noise\ Floor + SNR_{min,i} + (Fade\ Margin)_i \quad (1)$$

Fade margin is the margin kept in the link budget ensure a certain probability of signal reception. While switching between modes the system follows the order shown in Figure 2. However, as shown in the diagram, even after adaptive modulation a significant signal margin (hatched region) is available which may be exploited to reduce power consumption. It can be seen in Figure 3 that at the point of switching from SM to SD mode the effective signal strength is increased by the diversity gain and the fading margin is reduced due to the fact that signals are combined in SD mode. This increases the SNR of the system and provides further headroom for trading-off power versus performance in the MIMO architecture. By carefully tracking the channel conditions we propose to achieve Zero-SNR-margin operation at ultra-low power levels.

3. MIMO LOW-POWER OPERATION

In order to trade-off power for performance it is necessary to determine the threshold of acceptable operation during the real-time operation of the device. The chosen system-level metric for performance is Bit Error Rate (BER). However BER calculation requires transmission of a large number of symbols and is slow for practical purposes. A metric which has good correlation with BER is Error Vector Magnitude (EVM). It has been shown in [7] that a limit on the BER for a particular system and modulation rate translates to a threshold on the corresponding EVM. Prior work in [8] shows that for an acceptable maximum BER threshold of $5x10^{-4}$ the EVM thresholds for 64-QAM, 16-QAM and QPSK are 5%, 14% and 35% respectively. Moreover, EVM requires significantly less time to determine from received symbols and is thus more suitable as a real-time metric for system-level performance.

For minimum power operation, the EVM of received constellations is monitored in real-time and performance such as gain, noise figure and linearity of the MIMO receiver is dynamically traded off for power consumption using the feedback control of Figure 1. This feedback control is implemented in a way such that a constant EVM value corresponding to a prescribed BER is always maintained. For a given operating mode and signal modulation rate, as the channel improves, the noise figure (NF) and linearity performance metrics of the MIMO front end is degraded and vice versa.

In order to always operate at this EVM threshold, the specifications of the front-end must be adaptively varied with time. To achieve this, the RF front-end modules are designed to have tunable specifications such as Gain, IIP3, and ADC resolution. First, the entire gamut of channel conditions is partitioned into a finite number (say m) of representative channels. Then the MIMO system is characterized to find the combination of tuning knob settings for each of the m channels such that, the EVM is maintained at the threshold (EVM_{th}) of acceptable operation while the system-level power consumption is minimized. These combinations are stored in an $m \times n$ table (where n is the number of tuning knobs) as shown in Figure 13 (Supplementary Section 10). The channels are ordered such that the first row represents the *best* channel (and hence minimum power consumption) and the last row represents the *worst* case channel (maximum power consumption). Since this optimization is a one-time exercise performed during the design phase the

complexity or time involved for this procedure is not a major concern. The determination of the best tuning knob values that minimizes the power consumption across the range of selected channel conditions while maintaining the received symbol EVM threshold value, is performed using a constrained optimization algorithm.

The table so generated is then stored on-chip (see Figure 13). During real time operation the EVM of the system is constantly monitored and a control algorithm running on the baseband processor continuously attempts to maintain the EVM at EVM_{th} by changing the tuning knob settings. If for a specific tuning knob setting of the MIMO front end, the monitored EVM value increases the feedback loop attempts to bring it below EVM_{th} by moving the pointer to the following row of the table (having higher power consumption) and vice-versa, saving power across dynamically changing channel conditions.

The controller is implemented as a bang-bang controller where the sign change in EVM controls the direction of movement along the look up table. If the system needs to save power more aggressively, the guard-band needs to be small and hence the number of channels large (i.e. channel spacing small). Thus there exists a trade-off between the number of channels stored (memory used in the baseband) and the granularity in power savings.

However depending on the constraints on data throughput the low-power operation can be further optimized. We define two different low-power modes as explained below:

1) Data Priority (DP) mode: In certain applications the data rate is of prime importance. For example in a video or voice call the degradation in data rate may cause a drop in call quality and may even lead to dropped calls. Hence in this mode the mandate is to:

(a) Ensure that the system operates always at the highest possible data rate for the given channel condition.

(b) Excess SNR is traded off for lower power consumption in the front-end while ensuring condition (a) is satisfied.

This mode would lead to the highest possible throughput. However the energy per bit of the data may not be optimal across all possible channel conditions.

2) Energy Priority (EP) mode: In contrast with DP mode, there are certain applications where data rate is not of prime importance. These applications include background downloads of updates on mobile devices and synchronization of data across cloud platform devices. In these applications since the restriction on data rate is relaxed the same can be optimized to obtain lowest energy per bit of the RF front-end. This ensures that the battery lifetime is extended for the longest possible time. This is called the Energy priority mode. Thus in this mode the goal is to minimize:

$$Energy\ per\ bit = \frac{Energy\ Consumption}{No.\ of\ bits\ transferred} \quad (2)$$

Now differentiating both numerator and denominator with respect to time we get:

$$Energy\ per\ bit = \frac{Power\ Consumption}{Bit\ Rate} \quad (3)$$

Bit-rate again is a product of bits per symbol (subcarrier), Multiplexing factor, number of Orthogonal Frequency Division Multiplexing (OFDM) subcarriers & frequency of OFDM symbol generation. *Assuming* a constant OFDM symbol frequency and a constant number of OFDM subcarriers, it effectively means that we need to minimize the proxy metric(M):

$$M = \frac{Power}{Bits\ per\ Symbol\ *\ Multiplexing\ factor} \quad (4)$$

Here bits per symbol is $\log_2(x_i)$ where $x_i = 64,16$ or 4 depending on whether the modulation used is 64-QAM,16-QAM or OPSK. The multiplexing factor is 2 for SM mode and 1 for SD mode. In static receivers the numerator of M is constant and the denominator changes according to the mode of operation. Thus in a static receiver M is minimum when the bit rate is the highest. Thus for any channel the goal is to maximize the data rate However, this is not true in our dynamically varying receiver. Thereby for static receivers both EP and DP modes are one and the same. However the ability to trade-off excess SNR for power implies that the numerator is a function of the channel conditions. Thus in this paper, in order to achieve lowest Energy per bit the metric M has been be minimized in the EP mode.

4. SYSTEM SIMULATION SETUP

4.1 Transmitter

The system considered here is a 2x2 WiMAX MIMO OFDM transceiver operating at 2.4 GHz. In this paper, we have restricted ourselves to 64-QAM, 16-QAM and QPSK modulation schemes to demonstrate the low-power operation of the receiver. The transmitter is modeled as transmitting with a fixed power level throughout the course of operation. As shown in Figure 1 there is a feedback information path from the receiver to the transmitter which enables adaptive modulation and adaptive MIMO modes. Each of the paths (h_{ij}) is assumed to have a power of 30 dBm at the transmitter.

4.2 Receiver

The receiver incorporates 2 chains, each with an LNA, Mixer and ADC. The output of ADC feeds to the baseband unit.

1) LNA: [11] discusses an LNA in which the Gain-NF and the linearity can be tuned independent or orthogonal to each other. This LNA is used as a tunable front-end block in the receiver chain.

2) Mixer: A tunable mixer with 2 tuning knobs (V_b and V_{DD}) is used. The two bias voltages can be used to trade-off Gain-NF versus Power. It has a maximum gain of 11.84 dB and an IIP3 of 21 dBm. The maximum power consumption is 33.75 mW,

3) ADC: An 10-bit ADC is implemented with a capability to reduce power by reducing resolution through bit-dropping. The power is scales linearly with the number of bits dropped linearly. The maximum power consumption for the ADC is 40mW.

4.3 Channel

The channel is modeled to have attenuation due to Path Loss (PL) and rapidly changing fading characteristics. The fading on each of the four propagation paths is independent of each other and is assumed to follow a Rayleigh distribution. A key assumption here is that fading does not change significantly over the duration of a packet. Additionally the PL changes at a much slower rate than fading. The system is designed to track the attenuation due to Path Loss whereas EVM is calculated over a sufficiently long time to always get an average effect due to fading. Each of the *h* parameters(shown in Figure 1) follows a similarly distributed (but not temporally identical) fading channel.

5. OPTIMIZATION

The problem of finding a tuning knob combination for a given channel condition for near-zero EVM margin operation is a constrained global optimization problem. One approach to solve

such an optimization problem is to use a multistart constrained optimization[12]. In this approach multiple points (say k) in the n-dimensional search space are generated which comply with the constraints of the problem (EVM<EVM_th for this case). Each of these points is then used as a starting point by a local optimizer to converge to a local basin of attraction. Thus at the end of the process a list of minima is obtained, each minima corresponding to a different starting point. If sufficient number of starting points is selected then it can be stated with a high confidence that the minimum of the local minima so obtained is the global minima. This paper utilizes this approach to solve for the minimum power point. The above discussion is summarized in the form of a flowchart in Figure 14(shown in Supplement section 10.2).

The optimization is carried out in the off-line characterization phase of the device. Since it is not done in real-time, the time to convergence is not a critical criterion to evaluate the effectiveness of the algorithm. Moreover the algorithm must work with discretely spaced values of the tuning knob variables (since in practice only discrete states of the bias, supply and bit drop values is implementable). Considering these requirements a gradient search algorithm is chosen as the local optimizer for the algorithm. The multi-start technique serves to increase the confidence of finding the global optimum to a large extent. The starting points are chosen at random (while satisfying the EVM constraint). The nominal operating knob combination (corresponding to the highest power consumption) is always included as a starting point. The convergence of power for a particular channel condition with 3 starting points is shown in Figure 4.

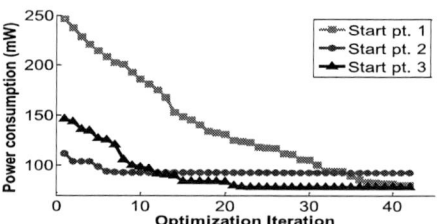

Figure 4: Convergence of multistart optimization

6. RESULTS

In order to demonstrate channel adaptive low-power operation we must identify the EVM margins that inherently exist in the MIMO RF front-end. Figure 5 shows the working of a static MIMO RF system across varying channel conditions. For a given MIMO mode (say SM 16-QAM) as PL increases, the SNR degrades resulting in a reduced EVM margin. This continues until the EVM is equal to EVM_th for that MIMO mode. If the PL increases any further, the system must switch to a lower data rate (SD 64-QAM in this case) to operate within the maximum BER constraint. Let us denote the PL at which a system switches from i^{th} mode to the $(i+1)^{th}$ mode to be PL_i. The maximum EVM margin for a particular mode occurs at the switching point between the previous (higher data rate mode) and the particular mode considered. This is because switching from a higher data rate to a lower data rate opens up new SNR headroom through:

a. reduction in SNR_{min} requirement across modulation rates and hence an increase in EVM threshold (EVM_{th})
b. an increase in effective SNR (when switching from multiplexing to diversity).

Thus the maximum power savings can be achieved at this point of transition PL_i. Figure 5 shows the variation of EVM as Path Loss increases from 90dB to 150dB. The graph also shows the EVM_th for different modulation techniques. The difference between the EVM at any PL and the EVM_{th} for the modulation technique used at that PL gives the EVM margin for that PL. As shown in Figure 5, the system switches between different modes and modulation techniques in the sequence illustrated in Figure 2. The EVM margin at a particular PL is a key indicator of how much power can be saved at that particular PL. We next analyze and present the results for the two modes of low-power operation.

1) Data Priority (DP) Low-Power Mode: In this mode the system operates at the highest data rate possible at any particular channel condition. Hence the EVM margins illustrated in Figure 5 are traded-off for power using a multi-dimensional constrained optimization algorithm discussed in Section 5. Figure 6 shows the optimized power consumption versus the PL for the MIMO RF front-end. It can be seen from Figure 6 that the maximum power consumption takes place at those values of PL where EVM is nearly equal to EVM_{th}. That is maximum power consumption occurs when the channel is just good enough to sustain a certain communication mode and thus the block level specifications of the receiver components cannot be degraded any further. The channel-adaptive receiver leads to 3x power savings over a static-receiver.

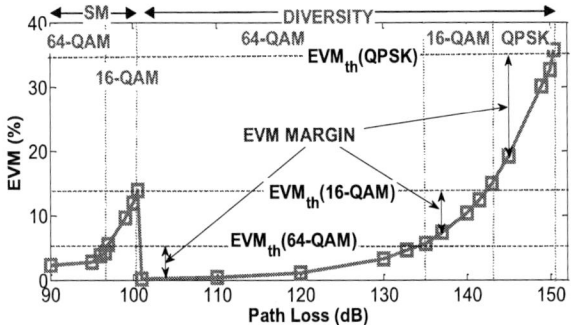

Figure 5: EVM values for a nominal receiver over different Path Loss

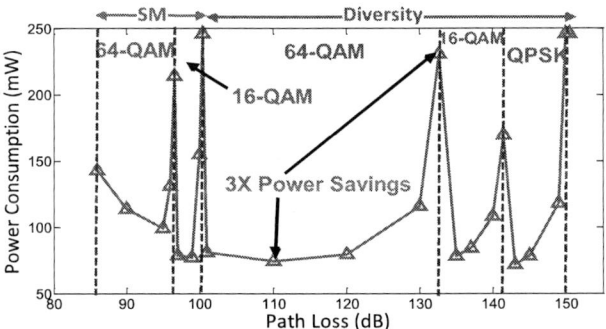

Figure 6: MIMO Receiver Power consumption in DP mode

2) Energy Priority (EP) Low-Power Mode: In Energy Priority mode the restriction to operate at the highest data rate is relaxed. Let P_{kj} be the optimized power consumption for the kth communication mode for the jth channel condition. Then for the jth channel the mode of operation should be such that the optimum proxy metric M is achieved, i.e.

$$M_{opt,j} = \min(\{P_{kj}/(b_k * m_k)\}) \qquad (5)$$

Here b_k is the bits per subcarrier in an OFDM system and m_k is the multiplexing factor (1 for SD and 2 for SM). However P_{kj} is a very complicated non-linear function of channel conditions and the power versus tuning knob profile of the individual blocks in the receiver. Moreover power consumption and bit rate do not scale linearly. Another approach is to plot the proxy metric for all

communication modes over the entire gamut of channel conditions. This is shown in Figure 7. The power consumption is optimized for the MIMO front-end for all modes of operation over the set of discrete PL conditions spanning the entire gamut of channels. Thereafter the corresponding values of M are plotted. The value of M for a operational mode (say SD 64-QAM, i.e. point "A" in Figure 7) peaks sharply near the boundary of its operation (PL=135 dB) i.e. where EVM $=EVM_{th}$. Near these points the M values for this mode of operation are higher than the next lower mode (i.e. SD 16-QAM). Thus although it is possible to transmit at the higher data rate near this boundary, in order to have the lowest energy per bit it is required to switch to a lower data rate. In summary, EP mode is achieved by following the curve having the trajectory of lowest M value at any point of time (in Figure 7) at any particular path loss.

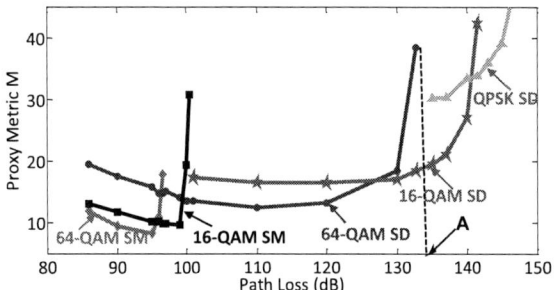

Figure 7: Proxy metric (M) over channel conditions for different MIMO mode

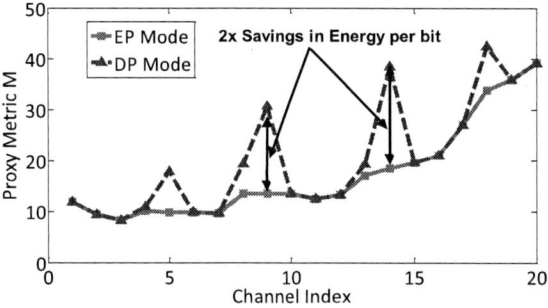

Figure 8: Relative Energy per bit comparison between EP and DP mode

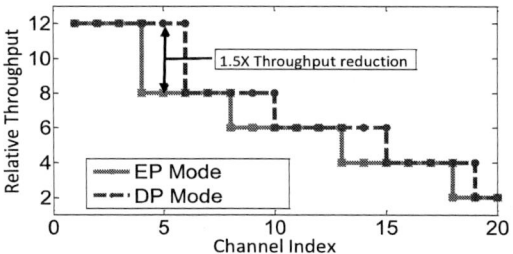

Figure 9: Relative throughput comparison between EP and DP mode

Figure 8 shows the plot of M versus the quantized channel index for both EP and DP modes of operation. Each of the discrete PL points, for which the front-end is optimized, is a channel condition and can be denoted by a channel index. The highest channel index corresponds to the maximum PL and hence the worst case channel. Similarly, the lowest Channel index corresponds to the lowest PL and hence the best case channel. EP mode can save upto 2x energy per bit over the DP mode. The penalty is paid in terms of data throughput. The relative throughput of both modes

393

versus the channel index is shown in Figure 9. It is seen that for certain channels EP mode can be 1.5x times slower than the DP mode.

7. HARDWARE IMPLEMENTATION

In order to prove the feasibility of the proposed concept a hardware demonstration consisting of a 2x2 MIMO system operating in spatial multiplexing and diversity modes at 2 GHz is setup. In the setup, a tunable RF MIMO front-end is setup as shown in Figure 10. Each front-end chain consists of an up-conversion mixer (MAX2039) (part of the transmitter) and the RF receiver front end with LNA (RF2370) and down-conversion mixer (ADL5801). Each of the LNA and the down-conversion mixer has tunable bias and supply knobs (a total of 4 knobs per RF chain) to trade-off performance versus power. An OFDM modulated random data stream is generated in MATLAB and sent across the RF chain through a DAC(NI PXI-5412). The output of the RF chain is acquired by an ADC card (NI PXI-5105). A 2x2 channel is implemented in software after the signal is acquired by the digitizer. Fifteen different channel conditions are created through attenuation and noise addition. Channels are ordered from best to worst, with channel 1 being the best and channel 15 the worst.

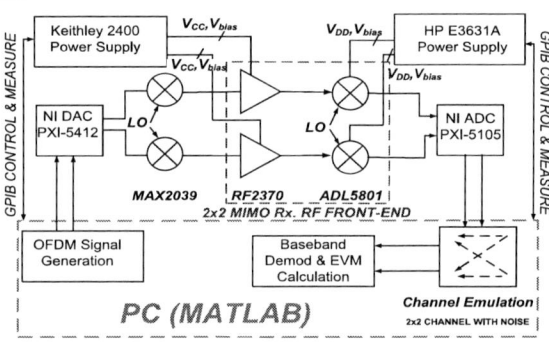

Figure 10: Hardware setup

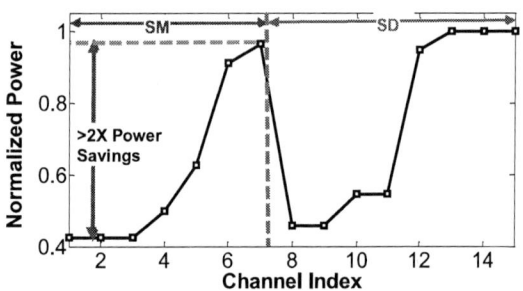

Figure 11: DP mode demonstrated in hardware setup

In the setup the performance of the system in terms of power consumption and EVM is recorded across all possible combinations of the 4 tuning knobs for 15 different channel conditions (enumerated by channel index). This is done for both spatial diversity and spatial multiplexing mode for QPSK modulation. The variation of power consumption across two (out the four) knob combinations is shown in Figure 16. Subsequently, assuming an EVM threshold of 33%, the tuning knob combination with lowest power consumption is selected for each channel. The relative power consumption in an adaptive RF front-end across 15 channels is shown in Figure 11. From this graph it can be seen that substantial savings (up to 2X) in power by real-time tracking

This work was funded by the United States National Science Foundation under Contract CCR - 0916270 and in part by MARCO under Contract GSRC/FCRP 2009-DT-2049.

of the channel conditions can be obtained. The operation in EP mode is shown in Figure 12. EP mode demonstrates additional energy per bit savings (upto 10%) over DP mode operation. This hardware setup thus demonstrates the efficacy of the proposed low-power technique for MIMO RF front-end.

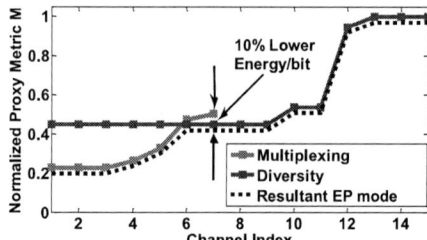

Figure 12: EP mode demonstrated in hardware setup

8. CONCLUSION

This work demonstrates a channel-adaptive MIMO RF front-end which operates at the threshold of acceptable operation across diverse channel conditions. Multiple low-power modes are proposed to address the throughput and energy-per-bit needs of different end-user applications. The proposed methodology is proved through system level software simulations and hardware demonstration. The concepts proposed are highly relevant for ultra-low power wireless devices utilizing MIMO RF front-ends. Future work could include investigations to make the system tolerant to process variations.

9. REFERENCES

[1] Gesbert, D.; Shafi, M.; Da-shan Shiu; Smith, P.J.; Naguib, A.; , "From theory to practice: an overview of MIMO space-time coded wireless systems," IEEE JSAC , vol.21, Apr 2003

[2] An He et. al. , "Power Consumption Minimization for MIMO Systems — A Cognitive Radio Approach," *IEEE JSAC*, vol.29, February 2011

[3] Lei Wang; Shanbhag, N.R.; , "Low-power MIMO signal processing," *Very Large Scale Integration (VLSI) Systems, IEEE Transactions on* , vol.11, no.3, June 2003

[4] Kim, E.P.; Shanbhag, N.R.; , "An energy-efficient multiple-input multiple-output (MIMO) detector architecture," *Signal Processing Systems (SiPS), 2011 IEEE Workshop on*, Oct. 2011

[5] Hongseok Kim et.al. , "Energy-efficient adaptive MIMO systems leveraging dynamic spare capacity," *Information Sciences and Systems, 2008. CISS 2008. 42nd Annual Conference on* , March 2008

[6] Shuguang Cui; Goldsmith, A.J.; Bahai, A.; , "Energy-efficiency of MIMO and cooperative MIMO techniques in sensor networks," *Selected Areas in Communications, IEEE Journal on* , vol.22, no.6, pp. 1089- 1098, Aug. 2004

[7] Senguttuvan, R.; Sen, S.; Chatterjee, A.; , "Multidimensional Adaptive Power Management for Low-Power Operation of Wireless Devices," IEEE TCAS II, Sept. 2008

[8] Sen, S. et. al.., "Pro-VIZOR: Process Tunable Virtually Zero Margin Low Power Adaptive RF for Wireless Systems", DAC 2008.

[9] Heath, R.W.; Paulraj, A.J.; , "Switching between diversity and multiplexing in MIMO systems," *Communications, IEEE Transactions on* , vol.53, no.6, pp. 962-968, June2005

[10] Alamouti, S.M.; , "A simple transmit diversity technique for wireless communications ," *Selected Areas in Communications, IEEE Journal on* , vol.16, no.8, pp.1451-1458, Oct 1998

[11] Shreyas Sen et. al. "A Power-Scalable Channel-Adaptive Wireless Receiver Based on Built-In Orthogonally Tunable LNA." IEEE Trans. on Circuits and Systems 59-I(5): 946-957 (2012)

[12] Rafael Martí, Mauricio G.C. Resende, Celso C. Ribeiro, "Multi-start methods for combinatorial optimization", European Journal of Operational Research,

[13] Banerjee, D;Sen, S; Banerjee, A; Chatterjee, A;, "Low-power adaptive RF system design using real-time fuzzy noise-distortion control." ISLPED 2012. ACM, New York, NY, USA, 249-254.

10. SUPPLEMENT

10.1 Adaptive MIMO Operation

The schematic shown in Figure 13 describes the two phases of Adaptive MIMO operation. It consists of an off-line design phase optimization combined with real-time EVM based channel adaptation. This scheme has been discussed in detail in Section 3 of this paper.

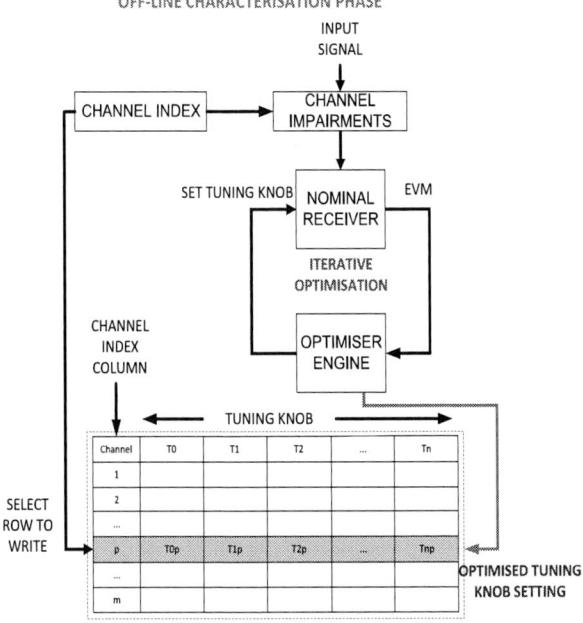

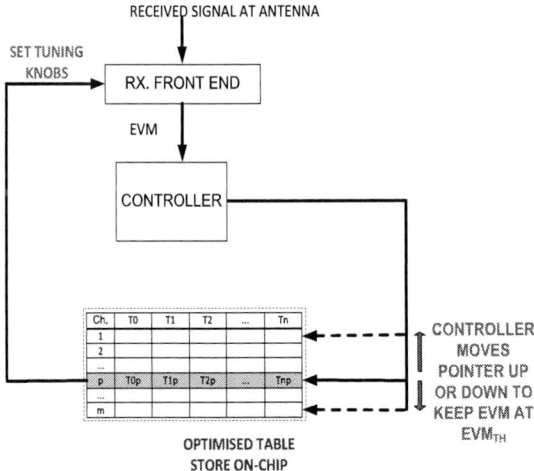

Figure 13: Overview of Adaptive MIMO operation

10.2 Optimization Algorithm

Figure 14 shows the multistart optimization algorithm described in Section 5. In this algorithm k points in the n-dimensional search space are selected. The nominal operating point is always included in this k element set. Then starting from each of these k points a local optimizer converges to a minima. The best solution of the k solutions thus obtained can be stated to be the global optimum with a high degree of confidence.

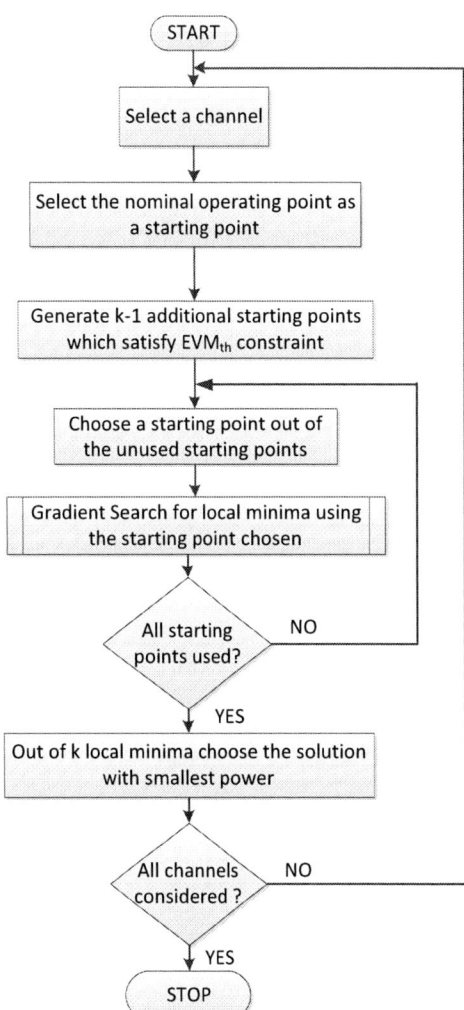

Figure 14: Optimization Algorithm

10.3 Hardware Power Profile

Figure 15 shows the power versus bias knob profile for each chain of the front-end in the hardware experiment. The other 2 knobs i.e. the supply voltages of the LNA and Mixer have been kept at nominal voltages to get the surface shown in this figure. Performance of the front-end degrades as bias voltages are lowered. Hence there is a clear power-performance trade-off in the front-end chain thus constructed.

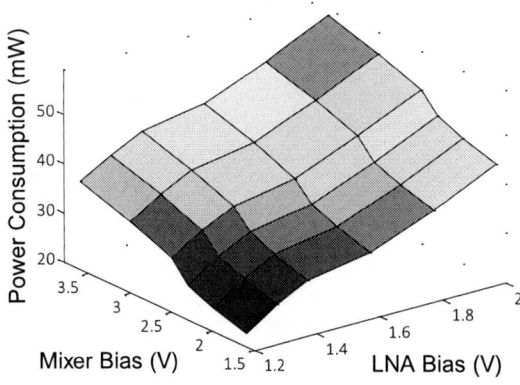

Figure 15: Power consumption as a function of 2 tuning knobs with the supply knobs kept at nominal

Improving Charging Efficiency with Workload Scheduling in Energy Harvesting Embedded Systems

Yukan Zhang, Yang Ge and Qinru Qiu

Department of Electrical Engineering and Computer Science, Syracuse University
Syracuse, New York, 13244, USA
{yzhan158, yage, qiqiu}@syr.edu

ABSTRACT

In energy harvesting embedded systems, if the harvested power is sufficient for the workload, extra power will be stored in the electrical energy storage (EES) bank. How much energy can be stored is affected by many factors including the efficiency of the energy harvesting module, the input/output voltage of the DC-DC converters, the status of the EES elements, and the characteristics of the workload. This paper investigates the impact of workload scheduling of the embedded system on the storage efficiency of the EES bank. We first provide an approximated but accurate power consumption model of the DC-DC converter. Based on this model, we analytically prove that an optimal workload schedule is to always execute high power task first. Experimental results confirm that proposed scheduling strategy outperforms all other possible scheduling and increases the amount of stored energy by up to 10.41% in average.

Categories and Subject Descriptors

C.3 [**Special Purpose and Application Specific Systems**]: Real-time and embedded systems

General Terms

Algorithms, Design.

Keywords

Energy harvesting embedded system, electrical energy storage, scheduling

1. INTRODUCTION

According to a recent survey, longer battery life is more important to user satisfactory than any other features [1] for mobile embedded systems such as smartphones and tablet computers. One promising technique to extend the battery lifetime of these embedded systems is to integrate an environmental energy harvesting module as a supplement energy source. Many research works have proposed to power the embedded system directly from the harvested energy in order to reduce the battery charging and discharging overhead [2][6]. When the harvested energy is abundant, extra energy will be stored into the battery, or in a more general term, *electrical energy storage* (EES) element. When the harvesting rate is low, the embedded system will be powered by the EES element as usual.

The energy harvesting embedded system is a complex platform consisting of many components that work collaboratively. And the energy efficiency of the system can be improved in many aspects and at different levels of abstraction. From the perspective of embedded system optimization, the scheduling and operating frequency/voltage of the embedded processor should be carefully selected to satisfy the performance requirement as well as power/energy constraint. For example, the authors in [6] proposed an energy-harvesting aware dynamic voltage frequency selection (EA-DVFS) algorithm, which adjusts the speed of task execution based on the current energy reserve in the system. In [7], the energy savings is further improved by exploring task slacks.

From the perspective of energy harvesting optimization, the operating conditions of the harvesting modules should be set carefully to generate more power. For example, in order to draw maximum amount of power from a photovoltaic (PV) array, various maximum power point tracking (MPPT) methods [8] have been proposed that dynamically adjust the output current to match the output impedance. In a recent work [9], the authors proposed a technique to improve the PV cells' efficiency under partial shading conditions.

The DC-DC converters are also important components in the system and their energy efficiency has drawn lots of attention in recent works. In [10], the authors proposed an analytic power model for DC-DC converters and based on this model, the authors found that the overall energy consumption of the embedded application and the converter is a convex function of the processor's supply voltage. Thus the optimal operating voltage can be solved by minimizing a convex function. The DC-DC converter's power efficiency is also essential to the energy transfer between different components in the system. For example, [3][11][12] try to adjust the voltage of *charge transfer interconnect* (*CTI*) and active EES banks to reduce the power wasted on DC-DC converters. And [4] proposed to reconfigure the parallel and serial connection of the EES bank to match the input and output voltage of the DC-DC converter to reduce the DC-DC converter power consumption.

In this work, we investigate the impact of task scheduling on the efficiency of an energy harvesting embedded system. More specifically, we consider the scenario where the harvested energy is more than enough to power the embedded system and the extra power can be used to charge an EES bank. Apparently, how much power is consumed by the embedded system directly affects the amount of remaining power that goes into the EES bank. For most EES modules (e.g. supercapacitors), their terminal voltage has a monotonically increasing relation with the amount of energy stored. Different input power will increase energy storage at different rate and hence causes different terminal voltage. This affects the efficiency of the DC-DC converter, which matches the voltages between the EES and other system components. As a result, not all input energy can be stored in the EES and the efficiency of the EES charging process is affected. Our goal is to find an optimal schedule of the embedded workload such that the most energy can be stored into the EES bank.

To facilitate the development of the scheduling algorithm, we first present a simplified yet accurate approximation to model the power consumption of a DC-DC converter. Then we consider the most basic scheduling problem that has only two tasks with equally short execution time and different power consumptions under the scenario of fixed energy harvesting rate. Using the approximated power model, we analytically prove that, in this special scenario, executing the high power task first enables more energy to be stored in EES. We then generalize this result to multiple tasks with arbitrary lengths under general energy harvesting scenarios.

The following summarizes the technical contributions of this work:

- Although there have been many publications on battery aware task scheduling, most of them consider the non-linear property of battery when it is in discharge mode [13]. To the best of our knowledge, this is the first work that optimizes the task scheduling to increase the efficiency of EES charging process.

- Many of the existing works on EES efficiency optimization either ignore the impact of DC-DC converters ([6] and [7]) or minimize the energy waste on DC-DC using dynamic voltage selection with an assumption of fixed task scheduling [10]. This is the first work that investigates the impact of task scheduling on the energy efficiency with the consideration of the DC-DC converters.

- Our scheduling algorithm is rigorously proved to be optimal. The scheduling algorithm has very low complexity as it only has to sort the tasks based on their power consumptions.

The rest of this paper is organized as follows. The energy harvesting embedded system model and the problem definition are presented in section 2. The proposed scheduling algorithm is described in section 3. Experimental results and discussions are presented in section 4. We conclude our paper in section 5.

2. SYSTEM MODEL

In this section, we present the energy harvesting embedded system model, the DC-DC converter power consumption model and our problem definition.

2.1 Energy Harvesting Embedded System Model

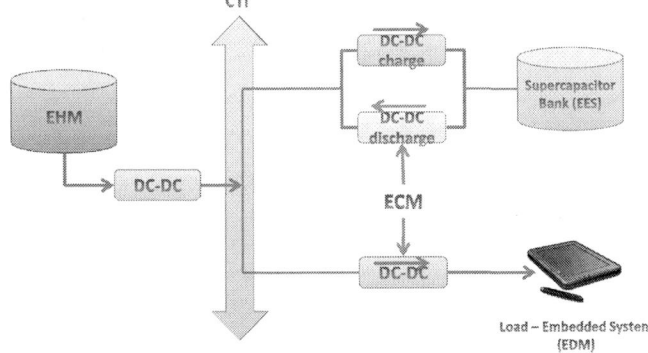

Figure 1. Energy harvesting embedded system architecture

Figure 1 shows the block diagram of the architecture of the energy harvesting embedded system considered in this paper. The system consists of the following components, an *energy harvesting module* (*EHM*), several heterogeneous *electrical energy storage* (*EES*) banks (only one bank is shown in the figure) and the

embedded systems, i.e. the *energy dissipation module* (*EDM*). All these components are connected together through *Charge Transfer Interconnect* (*CTI*) and DC-DC converters. We refer those DC-DC converters as *energy conversion modules* (*ECM*). The DC-DC converters connected to EHM, EDM and ESS are called ECM_{EH}, ECM_{ED}, and ECM_{ES} respectively. The ECM_{ES} can further be divided into two function units, ECM_{ES_charge} and $ECM_{ES\_discharge}$, which involve in ESS charge and discharge processes.

Energy is harvested from the environment by EHM and distributed to other components. Any type of EHM can be integrated in our system. When the harvested power is more than sufficient for the embedded workload, the extra power will be stored in the EES bank for future use. In order to minimize the power wasted on the DC-DC converter, we assume only one EES bank is turned on at any time to accept the extra power. The efficiency of DC-DC converter increases when the difference between its input and output voltage reduces [4], by setting the voltage of the Charge Transfer Interconnect (V_{cti}) to be equal to the V_{dd} of the embedded processor, we can minimize the power wasted on ECM_{ED}. The very low V_{dd} level used by today's deep submicron technology will very likely keep V_{cti} lower than the terminal voltage of the EES bank, and consequently makes ECM_{ES} operate in the boost mode. Therefore, in this paper, we confine our discussion to the scenario where ECM_{ES} operates in boost mode.

2.2 The DC-DC Converter Power Model under Boost Mode

DC-DC converters are placed between system components and the charge transfer interconnect for voltage and current regulation. In this paper, we assume that the converters are uni-directional switching buck-boost converters as shown in Figure 2 ([4]). When the input voltage V_{in} is greater than the output voltage V_{out}, the DC-DC converter operates at buck mode. Otherwise the DC-DC converter operates at boost mode. The power consumption P_{dcdc} of the DC-DC converter strongly depends on V_{in}, V_{out} and the output current I_{out}. It consists of three components: conduction loss P_{cdct}, switching loss P_{sw} and controller loss P_{ctrl}.

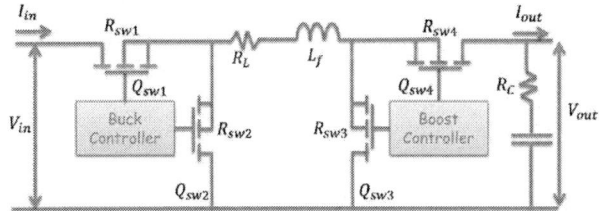

Figure 2. Energy harvesting embedded system architecture

As mentioned earlier, we focus our discussion in the scenario where the DC-DC converter operates at boost mode. In boost mode, these three power consumption components can be expressed as the following ([3]):

$$P_{cdct} = (I_{out}^2/D^2) \cdot (R_L + (1-D) \cdot R_{sw3} + D \cdot R_{sw4} + R_{sw1} + D \cdot (1-D) \cdot R_C) + (\Delta I)^2/12 \cdot (R_L + (1-D) \cdot R_{sw3} + D \cdot R_{sw4} + R_{sw1} + D \cdot R_C)$$

$$P_{sw} = V_{out} \cdot f_s \cdot (Q_{sw3} + Q_{sw4}), P_{ctrl} = V_{in} \cdot I_{ctrl}$$

where $D = V_{in}/V_{out}$ is the duty ratio and $\Delta I = V_{out} \cdot (1-D)/(L_f \cdot f_s)$ is the maximum current ripple. f_s is the switching frequency; I_{ctrl} is the current flowing into the controller; R_L and R_C are the equivalent series resistance (ESR) of inductor L and capacitor C, respectively; $R_{sw1...4}$ and $Q_{sw1...4}$ are the turn-on resistances and gate charges of the four switches in Figure 2 respectively. As we could see from the above equations, P_{dcdc} is a complex non-linear function of V_{in}, V_{out} and I_{out}. To facilitate the development of an efficient task scheduling algorithm, it is desirable to derive a simple yet accurate approximation to model power consumption of the DC-DC converter. We first divide the P_{cdct} to two parts:

$$P_{cdct1} = \frac{I_{out}^2}{D^2} \cdot (R_L + (1-D) \cdot R_{sw3} + D \cdot R_{sw4} + R_{sw1} + D \\ \cdot (1-D) \cdot R_C)$$

$$P_{cdct2} = \frac{(\Delta I)^2}{12} \cdot (R_L + (1-D) \cdot R_{sw3} + D \cdot R_{sw4} + R_{sw1} + D \\ \cdot R_C)$$

It has been shown in [5] that the resistance R_{sw3} and R_{sw4} have about the same value, so the above equations can be reduced to

$$P_{cdct1} = \frac{I_{out}^2}{D^2} \cdot (R_L + R_{sw4} + R_{sw1} + D \cdot (1-D) \cdot R_C)$$

$$P_{cdct2} = \frac{(\Delta I)^2}{12} \cdot (R_L + R_{sw4} + R_{sw1} + D \cdot R_C)$$

Because the DC-DC converter operates at boost mode, the possible region of V_{out} is $[V_{in}, +\infty)$. When $V_{in} = V_{out}$, $P_{cdct2} = 0$; when $V_{out} \rightarrow +\infty$, $P_{cdct1} \rightarrow +\infty$ and P_{cdct2} is a finite number. In both cases, P_{cdct2} is negligible compare to P_{cdct1}. Therefore, we eliminate P_{cdct2} in the approximation model. Furthermore, because $D = 1$ when $V_{in} = V_{out}$ and $D = 0$ when $V_{out} = +\infty$, we also drop the term $D(1-D) \cdot R_C$ in the approximation model since its value is below $0.25R_C$ which is much smaller than the other terms in the function. The final approximation model is as following:

$$P_{dcdc} = P_{cdct1} + P_{sw} + P_{ctrl} \\ = \left(\frac{I_{out} V_{out}}{V_{in}} \right)^2 \cdot (R_L + R_{sw4} + R_{sw1}) \\ + V_{out} \cdot f_s \cdot (Q_{sw3} + Q_{sw4}) + V_{in} \cdot I_{ctrl}$$

To validate our approximation model, we fix the V_{out} at 5V and vary V_{in} from 1V to 5V and show the actual power consumption and the approximated power consumption of DC-DC converter in Figure 3. The red, blue and green lines are for $I_{out} = 1A$, $2A$ and $3A$ respectively. We can see in the figure that the approximation model closely tracks the original model.

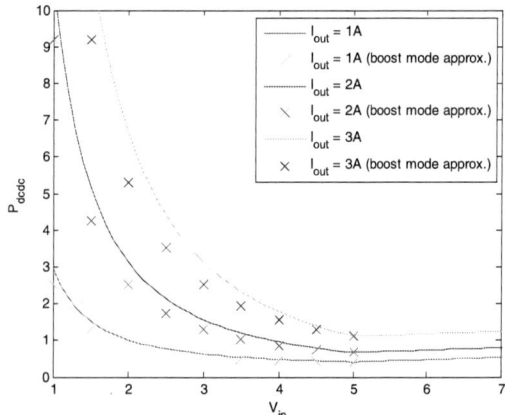

Figure 3. Approximation power model under boost mode

2.3 Problem Definition

We define our scheduling problem as following. Given an energy harvesting embedded system, which is used to process n periodic tasks $T_1, T_2, \ldots, T_n$. The set of tasks are immediately available at the beginning of a new period. Their durations are $t_1, t_2, \ldots, t_n$. Tasks are independent, non-preemptive, and the processor works in single task mode. The power consumption of task T_i is denoted as P_i. In this work, we do not consider dynamic voltage and frequency scaling of the embedded processor; therefore, the power consumption of a task is assumed to be fixed during the runtime. We assume that the energy harvesting rate is higher than the power consumption of the EDM, therefore the EES bank works in charge mode. As we explained in section 2.1, the DC-DC converter (ECM$_{ES}$) connecting the EES to the CTI operates in boost mode. Our goal is to find the optimal task schedule such that by the time when all tasks are completed the most energy can be stored into the EES bank.

3. Task Scheduling for Efficient EES Charging

We refer to the different charging characteristics as *charging phases* of the EES bank. When the harvesting rate is given, the identification of charging phases is solely determined by task execution. Given a set of tasks and an energy harvesting rate, executing the task with higher power consumption is equivalent to charging the EES bank using a lower input power, and vice versa. Therefore, the task scheduling can also be viewed as charge phase scheduling. In this work, we use the name *highest power workload first* (HPWF) to refer to the scheme that gives the top priority to execute the workload with the highest power consumption, and use the name *lowest power workload first* (LPWF) to refer to the scheme that gives the top priority to execute the workload with the lowest power consumption. We also use names *highest power charge first* (HPCF) and *lowest power charge first* (LPCF) to refer to the scheduling scheme that generates a sequence of EES charging phases whose input power follows descending or ascending order respectively. It is obvious that when the harvesting rate is fixed, LPWF is equivalent to HPCF and HPWF is equivalent to LPCF.

In this section, we first discuss the optimal charging phase scheduling and show that LPCF is superior to HPCF. This is proved first with the assumption that there are only two charging phases and both have the same durations that are extremely short. Then we extend this conclusion to scenarios with multiple charging phases that are equally short. Finally, we remove all constraint and generalize this conclusion to multiple charging phases with arbitrary duration. Based on this result, we further prove that HPWF is superior to LPWF.

3.1 Scheduling of Two Charging Phases with Very Short Duration

In this subsection, we consider two extremely short charging phases τ_1 and τ_2 with duration $\Delta t \rightarrow 0$, and their charging powers has the relation $P_{lo} < P_{hi}$. We denote the initial energy and voltage of the bank as (E_0, V_0). With the LPCF scheme, the bank is charged with P_{lo} in the first phase and then charged with P_{hi} in the second phase. The stored energy and voltage of the bank at the end of phase 1 and phase 2 are denoted as (E_1^{LPCF}, V_1^{LPCF}) and (E_2^{LPCF}, V_2^{LPCF}) respectively. Similarly, with the HPCF scheme, the bank is charged with P_2 in the first phase and then is charged with P_1 in the second phase. The bank energy and voltage at the end of phase 1 and phase 2 are denoted as (E_1^{HPCF}, V_1^{HPCF}) and (E_2^{HPCF}, V_2^{HPCF}) respectively.

Lemma 1. Given an EES with initial energy E_0 and initial voltage V_0. Charge this EES using two charging phases that has equal duration Δt that is extremely small. Denote the charging powers in the first and second phases as P_A and P_B respectively, and the bank voltage at the end of first and second phases as V_A and V_B respectively. At the end of second phase, the amount of energy stored in EES is:

$$E_B = E_0 + \frac{\sqrt{\Delta(P_A, V_0)} - V_{in}^2}{2\alpha} \Delta t + \frac{\sqrt{\Delta(P_B, V_A)} - V_{in}^2}{2\alpha} \Delta t,$$

where $\Delta(P, V) = V^4 + 4\alpha(P - \beta V - \gamma V)V^2$, and α, β, and γ are positive constant parameters for the given DC-DC converter.

Proof: Based on the DC-DC power approximation model, when in boost mode, P_{dcdc} can be written as

$$P_{dcdc} = \alpha \left(\frac{P_{out}}{V_{in}}\right)^2 + \beta V_{in} + \gamma V_{out} \qquad (1)$$

Substitute P_{dcdc} with $P_{in} - P_{out}$ in the above equation and rearrange the equation, we obtain

$$\alpha P_{out}^2 + V_{in}^2 P_{out} + (\beta V_{in} + \gamma V_{out} - P_{in})V_{in}^2 = 0 \qquad (2)$$

Although P_{out} can have two solutions, because $P_{out} > 0$, the only valid solution is

$$P_{out} = \frac{\left(\sqrt{V_{in}^4 + 4\alpha(P_{in} - \beta V_{in} - \gamma V_{out})V_{in}^2} - V_{in}^2\right)}{2\alpha} \qquad (3)$$

Based on the definition of $\Delta(P, V)$, we have

$$P_{out} = \frac{\sqrt{\Delta(P_{in}, V_{out})} - V_{in}^2}{2\alpha}$$

The initial state of the bank is (E_0, V_0). After being charged using P_A for Δt, the state of the bank becomes (E_A, V_A). Because Δt is very small, we assume the output voltage of the DC-DC converter is the same during this period. Also because P_{in} and V_{in} are constant values during the same charging phase, the output power is also a constant value and is denoted as $P_{out}(P_A, V_0)$. Then we have

$$E_A = E_0 + P_{out}(P_A, V_0) \cdot \Delta t$$
$$= E_0 + \frac{\sqrt{\Delta(P_A, V_0)} - V_{in}^2}{2\alpha} \Delta t \qquad (4)$$

For the same reason, in the second charging phase, we charge the bank using P_B for Δt and we have

$$E_B = E_A + P_{out}(P_B, V_A) \cdot \Delta t = E_A + \frac{\sqrt{\Delta(P_B, V_A)} - V_{in}^2}{2\alpha} \Delta t$$

$$= E_0 + \frac{\sqrt{\Delta(P_A, V_0)} - V_{in}^2}{2\alpha} \Delta t + \frac{\sqrt{\Delta(P_B, V_A)} - V_{in}^2}{2\alpha} \Delta t \qquad \square$$

Based on Lemma 1, we have the following theorem. The proof of the theorem is provided in the supplemental section.

Theorem 1. Given that both charging phases have equal durations which are approaching to 0, the system follows LPCF scheme has more stored energy at the end of phase 2 than the system follows HPCF scheme, i.e. $E_2^{LPCF} > E_2^{HPCF}$.

3.2 Scheduling of Multiple Charging Phases with Equally Short Durations

In this subsection, we consider the scheduling problem for arbitrary number of extremely short charging phases $\tau_1, \tau_2, ..., \tau_n$ with duration $\Delta t \to 0$, and their charging powers are $P_1, ..., P_n$. Before giving the theorem, we first give a lemma.

Lemma 2. Given two identical EES bank B_1 and B_2 with initial energy $E_1 < E_2$. After charging them using the same power P for the same duration T, the energy in B_1 is less than or equal to the energy in B_2, i.e. $E_1' \leq E_2'$.

Proof: Assume it takes time t to charge B_1 from E_1 to E_2 using P. Then after time $T - t$, bank B_2 reaches E_1', which equals the final energy of B_1 after time T. Continue to charge B_2 for time t, we have $E_2' \geq E_1'$.

Theorem 2. Given n charging phases $\tau_1, \tau_2, ..., \tau_n$, with duration $\Delta t \to 0$, scheduling them based on the ascending order of their power maximizes the amount of energy stored in EES. Such scheduling policy is referred as LPCF (i.e. Lowest Power charging First).

Proof: We will prove this theorem using induction and contradiction. We know from Theorem 1 that the statement is true when $n = 2$. Assume LPCF scheme is the optimal scheduling policy for any i charging phases as long as $i \leq N$, we are going to prove LPCF scheme is the best for $N + 1$ charging phases.

Assume the energy optimal scheduling for the $N + 1$ charging phases are $S_{opt} = \{\tau_1, \tau_2, ..., \tau_{N+1}\}$, and their power are not in the ascending order. This means that there are two tasks τ_i and τ_{i+1} such that $P_i > P_{i+1}$. Then we construct a new schedule by switching phases τ_i and τ_{i+1} and get $S' = \{\tau_1, ..., \tau_{i+1}, \tau_i, ..., \tau_{N+1}\}$. Note that after charging phase τ_{i-1}, the energy stored by both S_{opt} and S' are the same, because the first $i - 1$ charging phases are the same for the two schedules. For the next two charging phases, schedule $\{\tau_i, \tau_{i+1}\}$ is worse than schedule $\{\tau_{i+1}, \tau_i\}$ because $P_{i+1} < P_i$. Therefore, at the end of $(i + 1)$th charging phase, schedule S' stores more energy than the schedule S_{opt}. Because the remaining $N - i - 1$ charging phases are the same for both schedules, based on Lemma 2, S' stores more energy than S_{opt} at the end of all charging phases. This contradicts the assumption that S_{opt} is the energy optimal scheduling. Therefore, for $N + 1$ charging phases, the optimal scheduling is still the LPCF scheme.

From the above discussions, we have seen that for arbitrary number of very short charging phases, the LPCF scheme is the most energy efficient scheduling.

3.3 Scheduling of Arbitrary Charging Phases

In this subsection, we consider the scheduling problem for multiple charging phases with arbitrary duration. We claim that the LPCF scheme is still the best scheduling. This can be proved by dividing charging phases into very small slices such that $t_i = N_i \Delta t$, where t_i is the duration of the ith charging phase. We consider each slice as a sub-phase. All sub-phases belonging to the ith phase have the same charging power P_i.

Based on the discussion of previous sections, to reach the highest energy efficiency, these sub-phases should be arranged based on LPCF scheme. All sub-phases having the lowest charging power will be scheduled first, followed by the sub-phases having the second lowest charging power. This is equivalent as scheduling the charging phases from low power to high power, in another word, to execute tasks with highest power consumptions first.

3.4 Task Scheduling to Improve Charging Efficiency

It is easy to know that when the energy harvest rate is fixed, task scheduling and charging phase scheduling have direct correspondence. HPWF is the optimal task scheduling because it leads to LPCF, which has been proved to be optimal in previous sections. In the next we will show that HPWF is the optimal even if the energy harvesting rate is time varying.

We first consider the case when the harvesting power is monotonically increasing or decreasing. When the harvesting power is monotonically increasing, the highest power workload first (HPWF) scheduling will produce a sequence of charging phases with monotonically increasing input power. So it is equivalent to the LPCF, which is optimal as proved.

In the next, we will show that the HPWF scheduling is still optimal when the harvesting rate is monotonically decreasing. Similar to section 3.1, we first consider two very short tasks t_{hi} and t_{lo}, with duration $\tau \to 0$ and power consumption R_{hi} and R_{lo}. Without loss of generality, we assume $R_{hi} > R_{lo}$. The harvesting power during time period $[0, \tau]$ and $[\tau, 2\tau]$ is denoted as Q_1 and Q_2 respectively. We assume harvesting power is more than enough to power either one of the two tasks, i.e. $\min(Q_1, Q_2) > \max(R_{hi}, R_{lo})$. We focus our discussion to the case where the harvesting energy is decreasing, i.e. $Q_1 > Q_2$.

Based on the relations between R_{hi}, R_{lo} and Q_1, Q_2, we have two possible cases:

Case1: $Q_1 - R_{hi} > Q_2 - R_{lo}, Q_1 - R_{lo} > Q_2 - R_{hi}$.

Case 2: $Q_1 - R_{hi} < Q_2 - R_{lo}, Q_1 - R_{lo} > Q_2 - R_{hi}$.

We denote $P_1 = Q_1 - R_{hi}$, $P_2 = Q_2 - R_{lo}$ and $P_1' = Q_1 - R_{lo}, P_2' = Q_2 - R_{hi}$. It is easy to see that if we execute task t_{hi} during the period $[0, \tau]$ and task t_{lo} during the period $[\tau, 2\tau]$, we are also charging the EES with input power P_1 during the period $[0, \tau]$ and input power P_2 during the period $[\tau, 2\tau]$. On the other hand, if we execute t_{lo} followed by task t_{hi}, we are charging the EES with P_1' for the duration $[0, \tau]$ and P_2' for the duration $[\tau, 2\tau]$. Furthermore, $P_1 + P_2 = P_1' + P_2' = P$.

We will first prove that, under case 1, HPWF is better than LPWF, i.e. charging EES with (P_1, P_2) is better than charging with (P_1', P_2').

Based on the lemma 1, using HPWF, the final energy stored in EES bank can be calculated as:

$$E_2 = E_0 + \frac{\sqrt{\Delta(P_1, V_0)} - V_{in}^2}{2\alpha}\Delta t + \frac{\sqrt{\Delta(P_2, V_1)} - V_{in}^2}{2\alpha}\Delta t$$

We are interested in finding the derivative of E_1' against P_1, i.e. dE_1'/dP_1. Based on the definition of $\Delta(P_1, V_0)$ and given that $P_2 = P - P_1$, and E_0, V_{in} do not depend on P_1, the derivative is:

$$\frac{dE_1'}{dP_1} = \frac{1}{\sqrt{\Delta(P_1, V_0)}}\Delta t - \frac{1 + \frac{\gamma}{CV_0\sqrt{\Delta(P_1, V_0)}}}{\sqrt{\Delta(P_2, V_1)}}\Delta t$$

Based on the definition of $\sqrt{\Delta(P_1, V_0)}$, we have $\sqrt{\Delta(P_1, V_0)} > \sqrt{\Delta(P_2, V_1)}$ because $P_1 > P_2$. Since $\gamma > 0$, we also have $1 + \frac{\gamma}{CV_0\sqrt{\Delta(P_1, V_0)}} > 1$. Together, we have $dE_1'/dP_1 < 0$, when $P_1 > P_2$. This means E_1' is a decreasing function against P_1, if $P_1 > P_2$. Because $P_1' > P_1$, charging with (P_1, P_2) is better than charging with (P_1', P_2'). So for case 1, HPWF is better than LPWF.

For case 2, we note that, first, charging with (P_1, P_2) is better than (P_2, P_1), because (P_1, P_2) is LPCF. Then based on the discussion for case 1, we know (P_2, P_1) is better than (P_1', P_2'), because $P_2 < P_1'$. Therefore HPWF is still better for case 2.

We have proved that HPWF scheduling is optimal for two very short tasks under monotonically decreasing harvesting power. This result can be extended to any number of tasks with arbitrary duration by using the same techniques in section 3.2 and 3.3.

Because HPWF is the optimal scheduling policy for scenarios with either monotonously increasing or decreasing energy harvesting rate, we can conclude that no matter how the harvesting changes, the HPWF scheduling scheme could always store the most energy into the EES banks.

4. EXPERIMENTAL RESULTS

To demonstrate the effectiveness of the proposed algorithm, we implement a C++ simulator to model the energy harvesting embedded system. We assume the system has one customized supercapacitor [14] with 40F capacitance and 15V rated voltage as the EES element. This configuration is similar to the one provided in [15]. We obtain the parameters of the DC-DC power converter model from [5]. These parameters are obtained from the datasheets of the real devices. We also assume the V_{dd} of the embedded system is 1.0V and the V_{cti} is also operated at 1.0V to match V_{dd}. The initial bank terminal voltage is also set to 1.0V.

4.1 Scheduling for Two Charging Phases

In the first set of experiments, we examine the impact of the scheduling for two charging phases with different charging power. Because there are only two charging phases, only two possible schedules are available: the LPCF and the HPCF scheme.

We first set the input power of one charging phase to be 0.5W and sweep the other from 0.2W to 1.0W. Note that the input power here is the extra harvested power after supplying the embedded system. We skip the case when both charging phases are 0.5W. We also set the duration of both charging phases to be 30 min. This duration is similar to the military radio application [11]. Table 1 Shows the energy stored in the EES element for the two scheduling schemes. As we can see, the LPCF scheme always performs better than the HPCF scheme as expected, and the difference could be up to 24.6%.

Table 1 Comparison of energy stored by HPCF &LPCF

Power (W)	0.2	0.3	0.4	0.6	0.7	0.8	0.9	1.0
HPCF (J)	419.9	419.9	495.8	711.4	789.5	868.3	946.8	1069.2
LPCF (J)	442.9	496.9	562.5	773.3	910.5	1046.2	1179.9	1311.4
Impr. (%)	5.46%	18.34%	13.45%	8.70%	15.33%	20.49%	24.61%	22.65%

As the input power increases in one of the charging phases and remains fixed in the other, the total energy stored in the EES element generally increases for both HPCF and LPCF. However, we also note that for the HPCF scheme, when the incoming power in one of the charging phase changes from 0.2W to 0.3W, the stored energy remains the same and all input power is consumed by the DC-DC converter. In fact, there is a minimum input power that is required to charge the EES bank. It can be estimated by setting $P_{in} = P_{dcdc}$ and $P_{out} = 0$ in our approximation model given by Equation (1). The minimum required input power is a function of V_{in} and V_{out}:

$$P_{in} = \beta V_{in} + \gamma V_{out}$$

which means we need more input power in order to break even of the consumption in the DC-DC converter when the terminal voltage of EES bank (i.e. V_{out}) increases. This can be viewed as an intuitive explanation of why LPCF is always better than the HPCF scheme. With the HPCF scheme, high power charging phase will first raise the terminal voltage of the EES bank to a high level, which reduces the opportunity for the bank to be charged in the future.

In the next experiment, we sweep the charging power from 0.2W to 1.0W with a step of 0.1W for both phases. The duration of each phase is still kept at 30 min. Figure 4 shows the improvement of LPCF over HPCF for all possible combinations of charging phases. We could see that LPCF performs better than HPCF in all cases. And the average improvement is 10.41%. When the duration of the charge phase is set to 20min and 10min, the average improvement is 9.14% and 6.86% respectively.

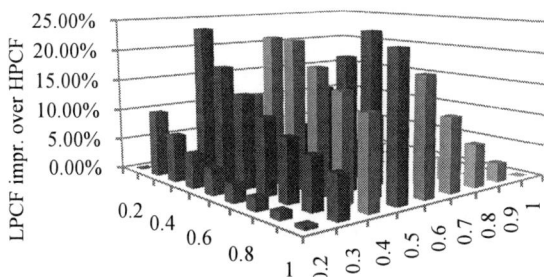

Figure 4. Improvement of LPCF over HPCF when sweep power from 0.2w to 1w

4.2 Scheduling Results for Multiple Tasks

In the second set of experiments, we examined the impact of task scheduling for multiple tasks on the energy stored in the EES bank. Various system configurations are tested. The number of tasks in each configurations are set to 6, 8 and 10; the task durations varies from 100s, 300s to 600s; and the harvesting power varies from 1.2W, 1.4W to 1.6W. The power consumption of each task is uniformly distributed between [0.2W, 1.0W]. We compare our LPCF scheme with HPCF scheme and the average results of 400 random generated schedules.

Table 2. HPCF, LPCF and Random for Multiple Tasks

		1.2 W			1.4 W			1.6 W		
	Sec	100	300	600	100	300	600	100	300	600
8 tasks	HPCF (J)	384.3	835.1	1364.5	514.4	1174.0	1935.4	644.2	1545.0	2600.1
	LPCF (J)	405.7	939.6	1598.6	535.7	1294.2	2238.7	663.6	1652.1	2901.3
	avg (J)	393.3	879.6	1458.4	524.0	1227.7	2065.1	653.0	1593.3	2735.6
	Im. H	5.58%	12.5%	17.2%	4.15%	10.2%	15.7%	3.01%	6.94%	11.6%
	Im. A	3.15%	6.83%	9.61%	2.23%	5.42%	8.41%	1.62%	3.69%	6.06%
6 tasks	HPCF (J)	389.3	807.1	1262.8	488.5	1086.9	1783.2	585.4	1364.1	2307.4
	LPCF (J)	400.2	873.0	1450.9	498.4	1145.4	1951.3	594.6	1417.1	2459.4
	avg (J)	394.7	839.0	1354.4	493.4	1115.4	1865.1	589.9	1390.0	2381.8
	Im. H	2.81%	8.17%	14.9%	2.03%	5.39%	9.43%	1.56%	3.89%	6.59%
	Im. A	1.42%	4.05%	7.12%	1.03%	2.69%	4.62%	0.79%	1.95%	3.26%
10 tasks	HPCF (J)	493.8	1154.6	1921.9	656.7	1592.5	2677.4	816.4	2049.8	3536.0
	LPCF (J)	519.8	1275.2	2208.9	679.9	1716.3	3012.5	837.6	2160.9	3840.4
	avg (J)	507.3	1215.3	2065.4	668.8	1657.0	2852.5	827.5	2107.9	3696.8
	Im. H	5.27%	10.5%	14.9%	3.53%	7.78%	12.5%	2.59%	5.42%	8.61%
	Im. A	2.47%	4.93%	6.94%	1.66%	3.58%	5.61%	1.22%	2.51%	3.88%

Table 2 reports the amount of stored energy for EES under LPCF, HPCF and the average random scheduling for all test cases. It also reports the relative improvement of LPCF over HPCF and random scheduling. We could see from this table that the LPCF scheme consistently outperforms the HPCF as well as the random scheduling policy in all test cases. And the improvement can be up to 17.15% over the HPCF scheme and 9.61% over the random scheduling. The HPCF always stores the least energy, which is even less than the worst cases in the random schedules. We observe that in some settings where the phase duration is short, the terminal voltage of the EES bank stays relatively low all the time. These cases are in general less sensitive to the scheduling order, so the difference between the three scheduling schemes is small. This explains the trend that the results for cases with 600s phase duration are generally better than the cases with 100s phase duration. For the similar reason, when the harvesting power increases, the system becomes less sensitive to the charging phase order. On the other hand, some settings are more sensitive to the scheduling. For example when the harvesting power is 1.2W and duration is 600s. In these cases, inappropriately charging the EES bank with high power first will raise the terminal voltage of EES bank and prevent it to be further charged at low power phase.

5. CONCLUSIONS

In this paper, we investigate the effects of workload scheduling on the efficiency of the EES charge process in an energy harvesting embedded system. We found that low power first scheme always performs better than the high power first scheme. It is proved using an approximated but accurate power model of the DC-DC converter. Experimental results show that the LPCF outperforms the HPCF by up to 10.41% for two charging phases. For multiple tasks, LPCF outperforms HPCF by up to17.15% and outperforms the random scheduling scheme by up to 9.61%.

6. REFERENCES

[1] K. Kumar, Y. Lu, "Cloud Computing for Mobile Users: Can Offloading Computation Save Energy?," in *IEEE Computer*, Apr. 2010.

[2] C. Moser, L. Thiele, D. Brunelli and L. Benini, "Adaptive Power Management in Energy Harvesting Systems," in *Proc. Design, Automation & Test in Europe Conference*, Apr. 2007.

[3] Q. Xie, Y. Wang, Y. Kim, N. Chang and M. Pedram, "Charge allocation for hybrid electrical energy storage systems," in *Proc. of the International Conference on Hardware/Software Codesign and System Synthesis*, Oct. 2011.

[4] Y. Kim, S. Park, Y. Wang, Q. Xie, N. Chang, M. Poncino, and M. Pedram, "Balanced reconfiguration of storage banks in a hybrid electrical energy storage system," in *Proc. of Int'l Conference on Computer Aided Design*, Nov. 2011.

[5] D. Shin, Y. Wang, N. Chang, and M. Pedram, "Battery-supercapacitor hybrid system for high-rate pulsed load applications," in *Proc. of Design Automation and Test in Europe*, Mar. 2011.

[6] S. Liu, Q. Qiu, Q. Wu, "Energy Aware Dynamic Voltage andFrequency Selection for Real-Time Systems with Energy Harvesting", in *Proc. of Design Automation and Test in Europe*, Mar. 2008.

[7] S. Liu, Q. Wu and Qinru Qiu, "An Adaptive Scheduling and Voltage/Frequency SelectionAlgorithm for Real-time Energy Harvesting Systems", in *Proc. Of Design Automation Conference*, Jul. 2009.

[8] D. Hohm and M. Ropp, "Comparative study of maximum power point tracking algorithms using an experimental, programmable, maximum power point tracking test bed," in *IEEE PSC*, 2000.

[9] X. Lin, Y. Wang, S. Yue, D. Shin, N. Chang and M. Pedram, "Near-optimal, dynamic module reconfiguration in a photovoltaic system to combat partial shading effects." in *Proc. of Design Automation Conference*, Jun. 2012.

[10] Y. Choi, N. Chang and T. Kim, "DC–DC Converter-Aware Power Management for Low-Power Embedded Systems," in *IEEE Trans. Computer-Aided Design of Integrated Circuits and Systems*, vol.26, no.8, Aug. 2007.

[11] Q. Xie, Y. Wang, M. Pedram, Y. Kim, D. Shin and N. Chang, "Charge replacement in hybrid electrical energy storage systems," in *Proc. of Asia and South Pacific Design Automation Conference*, Jan. 2012.

[12] Y. Wang, Y. Kim, Q. Xie, N. Chang, and M. Pedram, "Charge migration efficiency optimization in hybrid electrical energy storage (HEES) systems," in *Proc. of Int'l Symp. Low Power Electronics and Design*, Aug. 2011.

[13] P. Stanley-Marbell and D. Marculescu, "Dynamic fault-tolerance and metrics for battery powered, failure-prone systems," in *Proc. of Int'l Conference on Computer Aided Design*, Nov. 2003.

[14] http://www.tecategroup.com/ultracapacitors-supercapacitors/custom-modules.php

[15] http://www.tecategroup.com/capacitors/datasheets/powerburst/PBD.pdf

7. SUPPLEMENTAL MATERIAL

7.1 Proof of Theorem 1

Theorem 1. Given that both charging phases have equal durations which are approaching to 0, the LPCF scheme gives more stored energy than the HPCF scheme, i.e. $E_2^{LPCF} > E_2^{HPCF}$.

Proof:

For the LPCF scheme, we charge the bank using P_{lo} followed by P_{hi}. Both charging phases lasts for Δt. Based on Lemma 1 the final energy stored in the EES is:

$$E_2^{LPCF} = E_1 + P_{out}(P_{hi}, V_1^{LPCF}) \cdot \Delta t$$

$$= E_1 + \frac{\sqrt{\Delta(P_{hi}, V_1^{LPCF})} - V_{in}^2}{2\alpha} \Delta t$$

$$= E_0 + \frac{\sqrt{\Delta(P_{lo}, V_0)} - V_{in}^2}{2\alpha} \Delta t + \frac{\sqrt{\Delta(P_{hi}, V_1^{LPCF})} - V_{in}^2}{2\alpha} \Delta t$$

,where V_1^{LPCF} is the bank voltage at the end of phase 1 for the system using LPCF scheme.

Similarity, with the HPCF scheme, the final energy in EES is:

$$E_2^{HPCF} = E_0 + \frac{\sqrt{\Delta(P_{hi}, V_0)} - V_{in}^2}{2\alpha} \Delta t + \frac{\sqrt{\Delta(P_{lo}, V_1^{HPCF})} - V_{in}^2}{2\alpha} \Delta t$$

After eliminating the common terms we can see that, in order to prove $E_2^{LPCF} > E_2^{HPCF}$, we only need to prove the following inequality:

$$\sqrt{\Delta(P_{lo}, V_0)} + \sqrt{\Delta(P_{hi}, V_1^{LPCF})}$$
$$> \sqrt{\Delta(P_{hi}, V_0)} + \sqrt{\Delta(P_{lo}, V_1^{HPCF})} \quad (5)$$

Before proving (5), let us first find the relation between V_0 and V_1^{LPCF} or V_1^{HPCF}. For simplicity, we use supercapacitor as the EES bank in our proof. (The same discussion can be applied to any EES bank as long as its terminal voltage is a monotonically increasing function of its energy.)

Let P denote the input power, V_0 and V_1 denote the initial bank voltage and the bank voltage after first charging phase. The energy of a supercapacitor can be represented as $E = \frac{1}{2}CV^2$. We used this to replace E in (4), and have

$$\frac{1}{2}CV_1^2 = \frac{1}{2}CV_0^2 + \frac{\sqrt{\Delta(P, V_0)} - V_{in}^2}{2\alpha} \Delta t$$

or equivalently

$$C(V_1 - V_0)(V_1 + V_0) = \frac{\sqrt{\Delta(P, V_0)} - V_{in}^2}{\alpha} \Delta t$$

Because $\Delta t \to 0$, we could assume that $V_1 \approx V_0$, and $V_1 + V_0 \approx 2V_0$. Therefore, we have

$$V_1 = V_0 + \frac{\sqrt{\Delta(P, V_0)} - V_{in}^2}{2\alpha C V_0} \Delta t.$$

Therefore, we have:

$$V_1^{LPCF} = V_0 + \frac{\sqrt{\Delta(P_{lo}, V_0)} - V_{in}^2}{2\alpha C V_0} \Delta t \text{ and } V_1^{HPCF} = V_0 + \frac{\sqrt{\Delta(P_{hi}, V_0)} - V_{in}^2}{2\alpha C V_0} \Delta t$$

We expand the V_1^{LPCF} and V_1^{HPCF} in $\Delta(P_{hi}, V_1^{LPCF})$ and $\Delta(P_{lo}, V_1^{HPCF})$:

$$\Delta\left(P_{hi}, V_1^{LPCF}\right) = V_{in}^4 + 4\alpha\left(P_{hi} - \beta V_{in} - \gamma(V_0 + \frac{\sqrt{\Delta(P_{lo}, V_0)} - V_{in}^2}{2C\alpha V_0} \Delta t)\right) V_{in}^2$$

$$= \Delta(P_{hi}, V_0) - \frac{2\gamma(\sqrt{\Delta(P_{lo}, V_0)} - V_{in}^2)V_{in}^2}{CV_0} \Delta t$$

$$\Delta\left(P_{lo}, V_1^{HPCF}\right) = V_{in}^4 + 4\alpha\left(P_{lo} - \beta V_{in} - \gamma(V_0 + \frac{\sqrt{\Delta(P_{hi}, V_0)} - V_{in}^2}{2C\alpha V_0} \Delta t)\right) V_{in}^2$$

$$= \Delta(P_{lo}, V_0) - \frac{2\gamma(\sqrt{\Delta(P_{hi}, V_0)} - V_{in}^2)V_{in}^2}{CV_0} \Delta t$$

We plug the above two equations into equation (5), the left side becomes:

$$\sqrt{\Delta(P_{lo}, V_0)} + \sqrt{\Delta(P_{hi}, V_1^{LPCF})} =$$
$$\sqrt{\Delta(P_{lo}, V_0)} + \sqrt{\Delta(P_{hi}, V_0) - \frac{2\gamma(\sqrt{\Delta(P_{lo}, V_0)} - V_{in}^2)V_{in}^2}{CV_0} \Delta t} \quad (6)$$

While the right side becomes:

$$\sqrt{\Delta(P_{hi}, V_0)} + \sqrt{\Delta(P_{lo}, V_1^{HPCF})} =$$
$$\sqrt{\Delta(P_{hi}, V_0)} + \sqrt{\Delta(P_{lo}, V_0) - \frac{2\gamma(\sqrt{\Delta(P_{hi}, V_0)} - V_{in}^2)V_{in}^2}{CV_0} \Delta t} \quad (7)$$

Then we take square on both (6) and (7), and denote $\Delta(P, V_0)$ as $f(P)$, we have

$$f(P_{lo}) + f(P_{hi}) - \frac{2\gamma(\sqrt{f(P_{lo})} - V_{in}^2)V_{in}^2 \Delta t}{CV_0}$$
$$+ 2\sqrt{f(P_{lo})f(P_{hi}) - \frac{2f(P_{lo})\gamma(\sqrt{f(P_{lo})} - V_{in}^2)V_{in}^2 \Delta t}{CV_0}}$$

And

$$f(P_{lo}) + f(P_{hi}) - \frac{2\gamma(\sqrt{f(P_{hi})} - V_{in}^2)V_{in}^2 \Delta t}{CV_0}$$
$$+ 2\sqrt{f(P_{lo})f(P_{hi}) - \frac{2f(P_{hi})\gamma(\sqrt{f(P_{hi})} - V_{in}^2)V_{in}^2 \Delta t}{CV_0}}$$

Eliminate the common terms in the above two equations, to prove (5), it is sufficient to prove the two inequalities below

$$-\frac{2\gamma(\sqrt{f(P_{lo})} - V_{in}^2)V_{in}^2 \Delta t}{CV_0} > -\frac{2\gamma(\sqrt{f(P_{hi})} - V_{in}^2)V_{in}^2 \Delta t}{CV_0}$$

$$-\frac{2f(P_{lo})\gamma(\sqrt{f(P_{lo})} - V_{in}^2)V_{in}^2 \Delta t}{CV_0}$$
$$> -\frac{2f(P_{hi})\gamma(\sqrt{f(P_{hi})} - V_{in}^2)V_{in}^2 \Delta t}{CV_0}$$

Here we need to state two properties of $f(P)$.

(1) Because $\frac{df(P)}{dP} = 4\alpha > 0$, $f(P)$ is an increasing function of P. So $f(P_{hi}) > f(P_{lo})$.

(2) Based on Equation (3), we have

$$\frac{\sqrt{f(P_{lo})} - V_{in}^2}{2\alpha} = \frac{\sqrt{\Delta(P_{lo}, V_0)} - V_{in}^2}{2\alpha} = P_{out} > 0$$

Therefore, $\sqrt{f(P_{hi})} > \sqrt{f(P_{lo})} > V_{in}^2$. And consequently

$$f(P_{lo})\left(\sqrt{f(P_{lo})} - V_{in}^2\right) < f(P_{hi})\left(\sqrt{f(P_{hi})} - V_{in}^2\right).$$

Given these two properties, it is not difficult to see that the two inequalities holds for $P_{hi} > P_{lo}$.

Combining the discussions above, we proved Theorem 1.

Creation of ESL Power Models for Communication Architectures using Automatic Calibration

Stefan Schürmans, Diandian Zhang, Dominik Auras, Rainer Leupers, Gerd Ascheid
Institute for Communication Technologies and Embedded Systems
RWTH Aachen University, Germany
{schuerma,zhang,auras,leupers,ascheid}@ice.rwth-aachen.de

Xiaotao Chen, Lun Wang
Huawei Technologies Co., Ltd.
Bridgewater, NJ, USA / Plano, TX, USA
{xiaotaochen,lun.wang}@huawei.com

ABSTRACT

Power consumption is an important factor in chip design. The fundamental design decisions drawn during early design space exploration at electronic system level (ESL) have a large impact on the power consumption. This requires to estimate power already at ESL, which is usually not possible using standard ESL component libraries due to missing power models. This work proposes a methodology that allows extension of ESL models with a power model and to automatically calibrate it to match a power trace obtained by gate-level simulation or measurements. Two case studies show that the methodology is suitable even for complex communication architectures.

Categories and Subject Descriptors

B.8.2 [**Performance and Reliability**]: Performance Analysis and Design Aids; I.6.5 [**Simulation and Modeling**]: Model Development—*Modeling methodologies*

General Terms

Design, Experimentation, Performance

Keywords

Electronic System Level, Power Estimation, Power Model

1. INTRODUCTION

Nowadays, the computational performance of embedded devices is increasing steadily. This increase is enabled by the advances in silicon technology according to Moore's Law [13], which allows to pack an exponentially increasing number of transistors onto a single chip. In recent years, this development has led to Multi-Processor Systems on Chip (MPSoCs), which integrate several processors together with complex communication architectures and large

---

Permission to make digital or hard copies of all or part of this work for personal or classroom use is granted without fee provided that copies are not made or distributed for profit or commercial advantage and that copies bear this notice and the full citation on the first page. To copy otherwise, to republish, to post on servers or to redistribute to lists, requires prior specific permission and/or a fee.
DAC '13, May 29 - June 07 2013, Austin, TX, USA.

---

memories on a single piece of silicon. The large number of transistors causes high power consumption, which heats up the chip, drains the battery in mobile devices and has thus become a major design concern.

Exploration of the design options for a future MPSoC usually starts at electronic system level (ESL) using SystemC [6] in order to allow for easy adaptions and fast simulations. Common building blocks are either supplied as black box intellectual property (IP) components by commercial vendors like Synopsys [4] or are available as open source like SoCLib [3]. However, those allow only to simulate functionality and timing, but not power consumption. Unfortunately, the fundamental design decisions drawn at ESL have a large impact on the power consumption, which becomes visible only in later design stages when it is very difficult to revise those decisions.

Therefore, it is highly desired to obtain first power consumption estimates during the ESL simulations, even if those are not as accurate as power simulations at later design stages. This work proposes a methodology to extend ESL models with a power model. The state contained in the internals of the ESL model is made available by simple manual instrumentation in the form of state traces, which are used as input to a parametrized power model. Its parameters, called *power state factors*, can be obtained automatically by a process called *calibration*, which takes a power trace and corresponding state traces as input. The output of the calibration process are the power state factors resulting in the best match of the power estimate with the provided power trace. Afterwards, the calibrated power model can be used to estimate the power consumption in different scenarios.

The key differentiator of the proposed methodology is to *simultaneously support the following features*:

- Creation of power models for existing ESL models is **fast**, supported by **automatic tools** and requires only little manual work.
- Both **source-based** models and **black box IP components** are supported.
- Identification of **different power consumption phases** is possible, as the result of the power estimation includes a power trace over time.
- Power models can be created for different types of components, even for **communication architectures**.

- Calibration of power models needs **only a power trace** of a suitable scenario and no details about the hardware structure or the technology library.
- The **ESL simulation speed only drops slightly** when adding power models.

This paper is organized as follows: After discussing related work in section 2, the proposed methodology and the new power model will be introduced in section 3. The calibration process is described in section 4. Section 5 introduces our case studies for ESL power estimation and presents the results. The paper is concluded in section 6.

2. RELATED WORK

Commercial power estimation tools like Synopsys Prime-Time [5] typically operate at post-synthesis or post-layout gate-level and can thus be only used at late stages in MPSoC design. Additionally, the long simulation time is prohibitive for simulation of many options and thus does not allow for a thorough design space exploration.

To alleviate this problem, researchers have raised the abstraction level of power estimation. Early system-level power estimation approaches like [8] were still tailored towards a fixed platform and included manually created power models based on switching activity of RTL signals. Recent tools like Docea Aceplorer [2] have eliminated those limitations and can be used for a wide range of systems. The tool separates the functional description from the power behavior by managing user-defined, high-level power models in parallel to the existing models. This allows to keep the existing design flow for timed functional ESL simulation unchanged.

A comprehensive approach from academia is the ESL framework for rapid prototyping presented in [9]. It supports non-functional system properties like power consumption estimated by pluggable external models for the different system components. Starting from separate application and platform descriptions, it constructs a timing and power aware virtual prototype for analysis of different options in the design space. High flexibility is achieved by only defining interfaces and relying on external tools for creation of the actual timing and power models of the components.

Wattch [7] and SimplePower [18] are popular approaches for creating the required power models for processors. They perform simulations on the architecture level and use the state of the signals in every cycle to drive power models that have been carefully parametrized. The abstraction level can be raised further by running an instruction set simulator and estimating the power consumption just from the executed instructions according to [16]. However, due to inter-instruction effects, a lot of different training instruction sequences have to be analyzed in order to obtain a power model of suitable accuracy.

Functional Level Power Analysis (FLPA, [11], [12]) does not abstract the architecture completely, but avoids the need for simulating all architectural details during power estimation. The power model contains the power consumptions of the different functional units of a processor. Those have been obtained by running training instruction sequences activating only some of the different units and then applying curve fitting in order to obtain a value for each unit. During simulation, the power consumption is derived from the utilization of the units.

All of the approaches discussed so far are processor-centric and will only work for simple buses when applied in a system context. Because modern MPSoCs with a large number of processing elements require complex communication architectures, the power consumption of Networks on Chip (NoCs) is estimated in [14]. A significant increase in accuracy is achieved by using a rate-based model instead of a volume-based one as in earlier works. However, this limits the approach to NoCs and makes it not applicable to other communication architectures like crossbars.

A generalization of FLPA is used to create power models for different types of components in [15]. The code of timed functional models used in common ESL simulations is instrumented to output information about their states to power models. Those power models are created using several manually created training instruction sequences and linear regression. The methodology presented in this paper is also based on the regression approach. In contrast, it does not rely on hand-crafted inputs as it calibrates the power model automatically from a power trace of a known scenario.

The approach from [15] has been extended in [17] to also support black box IP components, i.e. models without access to the source code. As the IP block internals cannot be observed by extension of the code, their inputs and outputs are observed by newly created *estimators*. First, those estimators make the actions on the ports available for power estimation. Second, they can also contain state machines to track the internal state of the IP block, which is not visible from outside the model. Our approach adopts the estimator concept, but generalizes it to be suitable also for complex communication architectures instead of only processors and simple buses. Nevertheless, it does not require manual work during calibration of the power model.

PowerDepot [10] instruments ESL models to export some signals from the ESL simulation for driving a power model. Those signals can be exported directly by an extended ESL model or by a *monitor* observing the ports of an IP block, which is the same concept as the estimator from [17]. The power models are created in a multi-step process from the cells of the standard library and the netlist of the components. This process selects a few hardware signals as *key signals*, which the power model will expect as input. The results obtained by estimation of an H.264 application look very promising, but it is not clear from [10] if this application has also been involved in the characterization of the power model. A major advantage compared to the works discussed before is its capability to output a power trace from the ESL simulation instead of only an average at its end. However, the power model dictates which key signals have to be provided by the ESL model. This limits the approach to models in which information about the state of those signals is available. The methodology presented in this paper does not have this limitation and is still able to generate a power trace at ESL.

3. POWER EXTENSION OF ESL MODELS

For ESL design space exploration, the focus is rather on low setup effort and high simulation speed than on full accuracy. Therefore, the methodology presented in this paper allows to create power models for existing ESL models by little manual instrumentation work and automatic calibration. Instrumentation is used to capture the internal states of the ESL models and to make them available for power estimation. The calibration process creates the parameters of the power model from a reference power trace. Although this work focuses on communication architectures, the methodology is general enough to be applicable to all types of models.

3.1 Information at ESL

ESL models are usually SystemC models written in transaction level modeling (TLM) style, i.e. their internals are mainly modeled using member function calls and member variables. The models can be functional models or pure performance models, which implement only the timing and use dummy data on their ports. However, even a performance model has to implement some control functionality for tracking its state, which is needed for correct timing behavior. In general, ESL models do not contain detailed information about the implementation structure of the component or the type of circuit connected to their outputs, like the length of the wires or the load the component has to drive.

In actual hardware, power consumption is caused by leakage and switching. Leakage is present whenever a circuit is powered and does not depend on its activity. Switching depends on the clock signal for registers and on data signals changing their value along combinational paths. Thus, switching power is mainly influenced by clock gating and activity in different parts of the circuit, which are both dependent on controls signals, i.e. the state of the component.

The actual data signals of the data path also influence switching as different sequences of bits pass through them. However, the control signals have a larger impact on switching than the data bits, because the data bits usually exhibit a nearly fixed switching rate and the control bits determine if data bits are passing through a certain submodule.

In total, the major part of the power consumption depends on the control information, which is available in typical ESL models.

3.2 State Tracing

The state of an ESL model can be hidden within the internal data structures. Manual instrumentation is needed to make the state available for power estimation as depicted in Figure 1, because it is not easily possible to detect the member variables and functions that provide the relevant information using automatic methods. The instrumentation should be simple and have a low overhead at simulation run time. The work presented in this paper uses a singleton class *StateTracker*. It allows to register state traces in the constructor of the model (Listing 1, lines 6-8) and then log state changes at any location in the module code using a single line.

ESL models can contain two types of states: *natural states* and *events*. Natural states, like the activity of a sub-block or the number of pending requests, are usually available in member variables and can be traced by recording the updates (Listing 1, line 14). Events, like data arriving or a register being updated are usually modeled as a function call. As the event models a single-cycle action in hardware, the corresponding state trace has to change from 0 to 1 when the function is called and back to 0 in the next cycle. This can also be accomplished with a single line of code (Listing 1, line 12).

ESL simulation models obtained from commercial ven-

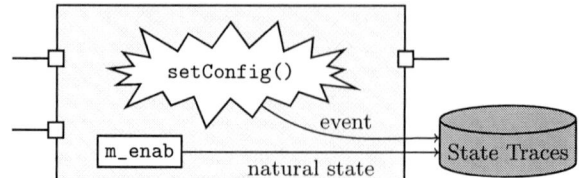

Figure 1: State tracing for ESL module

```
1   class MyModule: public sc_module
2   {
3     MyModule(const sc_module name &name):
4       sc_module(name)
5     {
6       m_tr = StateTracker::get();
7       m_state = m_tr.create(name() + ".enab", 0);
8       m_event = m_tr.create(name() + ".cfg", 0);
9     }
10    void setConfig(bool enab)
11    {
12      m_tr.event(m_event, 1, m_cycle_time, 0);
13      m_enab = enab;
14      m_tr.update(m_state, m_enab ? 1 : 0);
15    }
16  };
```

Listing 1: Instrumentation for tracing a natural state and an event using StateTracker

dors are often delivered as black box IP components. Those blocks can be used in arbitrary ESL simulations, but it is not possible to add state tracing to them, as the source code is not accessible and cannot be instrumented. For those blocks, the approach presented in [17] is chosen. As shown in Figure 2, the activity on all ports is monitored by a small ESL block that just forwards its inputs to its outputs, but additionally traces the states.

If the internals of the IP model involve state information that cannot be tracked at the single ports, a *power state machine* (PSM) is added next to the IP block. This state machine is fed with the observed information and will keep track of the IP block state. A state trace is created for each state of the PSM. The trace is set to 1 while the PSM is in this state and to 0 otherwise.

3.3 Power Model

A linear power model with a constant part is used in the proposed methodology.

Let $\mathbf{s}_i \in \mathbb{N}^T$ for $i \in \{2, \dots, N\}$ be the state traces provided by an instrumented ESL model during a simulation of T cycles of duration t_{cyc}. Further, let $\mathbf{s}_1 = \mathbf{1}$ be an artificial state trace which is always 1, modeling the constant part of the power consumption, which is caused by leakage, the clock network, etc. For each state trace $\mathbf{s}_i$, a so-called *power state factor* $f_i \in \mathbb{R}$ is defined, which will be determined during the calibration of the power model.

Figure 2: State tracing for IP component: observing ports, PSM to track internal state

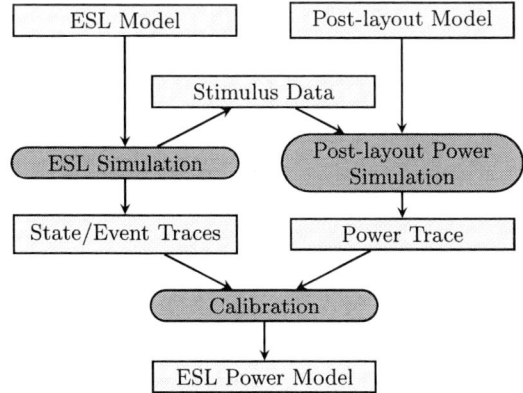

Figure 3: Work flow for ESL power model calibration using post-layout power simulation

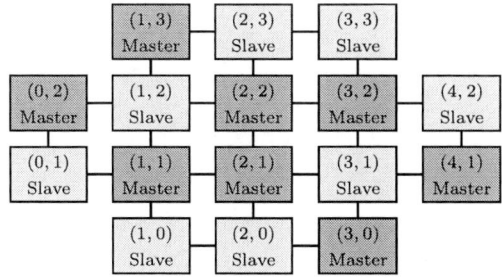

Figure 4: 2D Mesh NoC based system

The estimated ESL power trace $\mathbf{P}_{est}$ of the model is then the weighted sum of the state traces. The energy E_{est} and the average power $\overline{P_{est}}$ can be calculated from it:

$$\mathbf{P}_{est} := \sum_{i=1}^{N} f_i \mathbf{s}_i \quad E_{est} := t_{cyc} \sum_{j=1}^{T} P_{est,j} \quad \overline{P_{est}} := \frac{1}{T} \sum_{j=1}^{T} P_{est,j}$$

4. AUTOMATIC CALIBRATION

The remaining task in order to use the power model together with an existing ESL model is to determine values for the power state factors f_i. Because the ESL model is abstract, it is not possible to derive the power state factors directly from it. Instead, some information about the power consumption of the component is needed for this purpose.

The methodology presented here relies on a power trace recorded in a reference scenario and uses this for *calibration*. It does not matter if the power trace has been obtained by power simulation at RTL, post-synthesis or post-layout level, by measurements of actual hardware or by any other means. The only requirement is that the reference scenario is known and can also be simulated at ESL in order to obtain the corresponding state traces.

Without loss of generality, the remainder of this section describes calibration to data obtained from post-layout simulations.

4.1 Obtaining a Power Trace using Post-layout Gate-Level Simulation

If an implementation of the component is available, post-layout power simulation can be used to generate the power trace for calibration of the ESL power model. This allows to achieve a high accuracy also for large components, in which wires and the clock network have a big impact on power consumption.

The work flow is depicted in Figure 3. First, the ESL simulation is run using the non-power extended ESL model to obtain both the state traces and the traces of the data on the ports of the ESL model. The port data are used as stimulus data for a post-layout gate-level simulation, whose outputs are processed by a time-based power simulation in order to obtain a cycle-accurate power trace.

4.2 Calculation of Power State Factors

Because the power model is linear, the minimization of the mean square error is a natural approach for calculation of

power state factors f_i that provide a best fit of the estimated power $\mathbf{P}_{est}$ to the reference power trace $\mathbf{P}_{ref}$ for given state traces $\mathbf{S} = (\mathbf{s}_1 \; \ldots \; \mathbf{s}_N)$:

$$\mathbf{f} := (\mathbf{S}^\top \mathbf{S})^{-1} \cdot \mathbf{S}^\top \mathbf{P}_{ref}$$

Due to redundancies in the recorded state traces, the matrix $\mathbf{S}^\top \mathbf{S}$ might be singular or unstable (i.e. almost singular), making the inversion impossible or imprecise, respectively. In order to avoid this, some of the state traces have to be excluded. The trace selection is started with just the constant state trace $\mathbf{s}_1$ and then iteratively selects further state traces as long as the matrix does not become singular. If there is some information about the relevance priorities of the state traces for power consumption, it is possible to sort the state traces $\mathbf{s}_2, \ldots, \mathbf{s}_N$ in order of descending relevance before starting the selection.

The calibration will deliver values of the power state factors for selected traces. All values for excluded traces are set to zero. The power model presented in section 3.3 can then be used to estimate the power consumption of other scenarios at ESL. This allows to avoid running the slow power simulations at lower level.

5. CASE STUDIES

The communication architecture is one of the parts of an MPSoC for which power estimation is most difficult, because it contains not only combinational logic and registers like processors or peripherals, but typically also long wires consuming high amounts of switching power. Therefore, two different communication architectures with considerable complexity have been selected as case studies for the ESL power estimation methodology presented here.

As communication architectures cannot be operated properly without connecting them to subsystems, 8 master subsystems and 8 slave subsystems have been modeled on ESL, without including them in the power estimation. The master subsystems consist of processors and local memories while the slave subsystems contain only memories.

5.1 Network on Chip

A Network on Chip (NoC) is used as the first case study. It is organized as a semi-regular, 2-dimensional mesh of size 5x4, as shown in Figure 4. Each node contains either a master or a slave subsystem, an NoC router and a network interface connecting the subsystem to the router. Each router has up to 5 ports of width 128 bit with 4 virtual channels (VC), each containing a buffer for 8 flits. The packets are limited to 4 kB of data, contain a 4 bit priority value and are transmitted in flits of 128 bit using wormhole routing according to an adapted x/y-routing scheme. For post-layout simulations used as reference, the NoC has been implemented

407

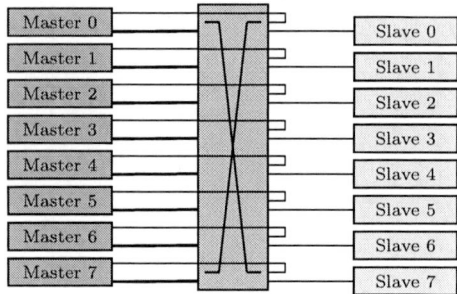

Figure 5: AXI based system

scenario	# active subsys.	traffic per subsys. [MB/s]		
		min.	average	max.
A	16	332	1078	3318
B	16	1296	1296	1296
C	7	55	1345	3988

Table 1: Traffic scenarios: synthetic (A, B), basestation of mobile communication services (C)

calibration scenario	estimation scenario		
	A	B	C
A	212.17 mW 0.00 %	299.91 mW −4.48 %	168.20 mW −1.05 %
B	222.38 mW 4.81 %	313.96 mW 0.00 %	131.45 mW −21.03 %
C	173.73 mW −18.12 %	296.63 mW −5.52 %	166.45 mW 0.00 %
post-layout	212.17 mW	313.96 mW	166.45 mW

Table 2: ESL power estimation results for NoC

calibration scenario	estimation scenario		
	A	B	C
A	59.35 mW 0.00 %	90.48 mW −3.92 %	63.54 mW −2.44 %
B	55.39 mW −6.67 %	94.17 mW 0.00 %	71.41 mW 9.64 %
C	50.80 mW −14.41 %	89.85 mW −4.59 %	65.13 mW 0.00 %
post-layout	59.35 mW	94.17 mW	65.13 mW

Table 3: ESL power estimation results for AXI

at RTL and a layout has been created using a proprietary 65 nm standard-cell library, Synopsys Design Compiler [5] for synthesis and Cadence Encounter [1] for place & route.

The ESL models of the NoC router and the network interface have been modeled in SystemC from scratch. After timing verification against the RTL models, state tracing has been added. The number of buffered flits and the number of pending requests to routing, switch allocation and VC allocation are traced as natural states. Event tracing is used for flits entering and leaving the buffer and for performing routing, switch allocation and VC allocation. The manual one-time instrumentation took approximately two hours.

5.2 AXI Crossbar

An AXI crossbar, which is commercially available in Synopsys DesignWare [4], has been chosen as a second case study. It is configured to have a fully registered data path of 128 bit width and to provide 8 master and 16 slave ports. Figure 5 presents a diagram of the AXI system. Each master subsystem is connected to a master and a slave port, which is being used for remote access to the local memory. Each slave subsystem is connected to a slave port. The layout for AXI has been created using the same 65 nm library and the same tools as the NoC.

At ESL, a black box IP model from Synopsys [5] is used to model the AXI. Because the internals of the IP block are not accessible for state tracing, the approach depicted in Figure 2 has been used. Sending or receiving an address, a data word or a ready flag on a master or slave port is traced as an event. The necessary implementation work was completed within a single work day.

5.3 Results

For evaluation of the ESL power estimation methodology, three periodic traffic scenarios have been simulated for 3 periods of 1.58 ms. The simulations were done on both the NoC and the AXI communication architecture at ESL and post-layout gate-level using a clock frequency of 316 MHz. The ESL simulations have been performed with Synopsys tools [5]: Platform Architect at ESL and VCS/PrimeTime at gate-level. For each case, the ESL power model has been calibrated and was then used for ESL power estimation for

all of the scenarios.

The power estimates $\overline{P_{est}}$ were compared to the post-layout power consumption $\overline{P_{ref}}$ and the relative error e was calculated:

$$e := \frac{\overline{P_{est}} - \overline{P_{ref}}}{\overline{P_{ref}}} \cdot 100 \%$$

The results for the NoC are shown in Table 2 and for AXI in Table 3. It can be seen that the overall maximum error is 21 %. This accuracy allows to **substitute gate-level power estimation with much faster ESL power estimation** during design space exploration. Thus, either more configurations can be simulated or the exploration is finished earlier.

When comparing the accuracies achieved by different calibration scenarios, it shows that scenario A leads to power models resulting in estimation errors below 5 %, which is remarkable considering the large difference of abstraction levels between post-layout gate-level and ESL. The reason is that this scenario is suited very well for the calibration process. First, it uses all subcomponents of the communication architectures. This allows to calibrate the power state factors for all of them. Second, the traffic produced by the subsystems is different enough to avoid a high correlation between the state traces. This allows the regression during calibration to work well.

The post-layout and ESL power traces are plotted in Figure 6. To facilitate readability, their resolution has been reduced to 5 k cycles by averaging. The plots show that the proposed methodology is able to **predict the phases of different power consumption** correctly. This is extremely beneficial for design space exploration, as the designer is enabled to identify which computations cause the highest power consumption and focus on those during optimization.

To investigate the tradeoff between power estimation accuracy and simulation speed, the execution times of the post-layout power simulations and the ESL simulations with and without power estimation have been measured. The numbers in Table 4 show that the ESL simulations are only slightly slower for AXI when including ESL power estima-

408

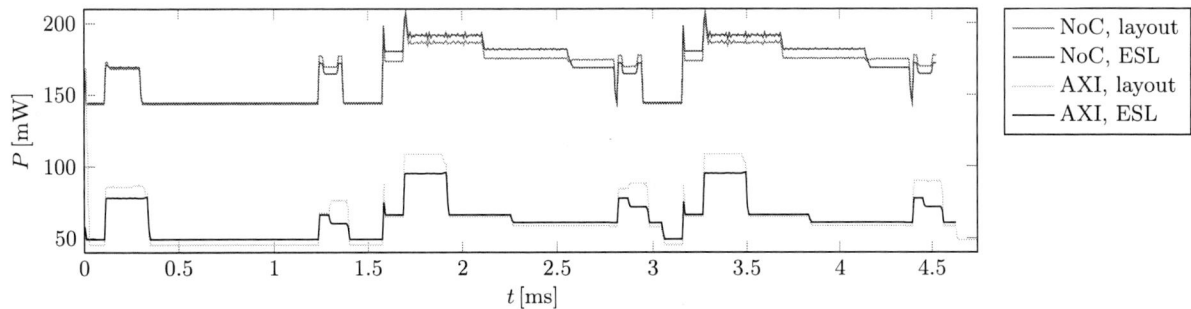

Figure 6: Power traces of scenario C (calibration of ESL model using scenario A)

system & scenario		post-layout power	ESL sim.	ESL & power	speed-up vs. layout	overhead vs. ESL sim.
NoC	A	45 h	101 s	274 s	591	2.71
	B	91 h	152 s	479 s	684	3.15
	C	34 h	39 s	107 s	1144	2.74
AXI	A	38 h	166 s	196 s	698	1.18
	B	75 h	167 s	195 s	1385	1.17
	C	27 h	108 s	124 s	784	1.15

Table 4: Speedup of ESL power estimation vs. post-layout power estimation, and overhead of ESL power estimation vs. ESL simulation

tion. The slowdown of about factor 3 for NoC is mainly caused by disk I/O for writing the traces for each NoC component. Compared to time-based post-layout power simulation a speedup of factor 880 is achieved on average.

6. CONCLUSIONS

The presented methodology for ESL power estimation extends available ESL models with power models, which are automatically calibrated to match power traces obtained by low-level power simulation or measurements. This enables inclusion of the important design criterion of power consumption in early design space exploration and thus development of more power-efficient systems. Two case studies have shown the applicability of the methodology to complex communication architectures. Depending on the calibration scenario, the accuracy ranges between 5 % and 21 % compared to post-layout power simulations while achieving a gain of multiple orders of magnitude in simulation time.

Currently, the quality of the generated power model depends on the calibration scenario. This requires a subsequent validation of the model. Further, the proposed methodology has only been applied to communication architectures yet, although it was created with applicability to all kinds of models in mind. An evaluation for other components like processors, accelerators and peripherals will be performed in the future.

7. REFERENCES

[1] Cadence digital implementation. [Online] http://www.cadence.com/products/di/ (accessed 11/2012).

[2] Docea Aceplorer. [Online] http://www.doceapower.com/products-services/aceplorer.html (accessed 11/2012).

[3] SoClib. [Online] http://www.soclib.fr (accessed 11/2012).

[4] Synopsys IP. [Online] http://synopsys.com/IP (accessed 11/2012).

[5] Synopsys tools. [Online] http://synopsys.com/Tools (accessed 11/2012).

[6] SystemC. [Online] http://www.accellera.org/downloads/standards/systemc (accessed 11/2012).

[7] D. Brooks, V. Tiwari, and M. Martonosi. Wattch: A framework for architectural-level power analysis and optimizations. In *Proceedings of the 27th Annual International Symposium on Computer Architecture*, ISCA '00, New York, NY, USA, 2000. ACM.

[8] W. Fornaciari, P. Gubian, D. Sciuto, and C. Silvano. Power estimation of embedded systems: A hardware/software codesign approach. In *Very Large Scale Integration (VLSI) Systems, IEEE Transactions on*, volume 6, Jun. 1998.

[9] K. Gruttner, K. Hylla, S. Rosinger, and W. Nebel. Towards an ESL framework for timing and power aware rapid prototyping of HW/SW systems. In *Specification Design Languages (FDL 2010), 2010 Forum on*, Sep. 2010.

[10] C.-W. Hsu, J.-L. Liao, S.-C. Fang, C.-C. Weng, S.-Y. Huang, W.-T. Hsieh, and J.-C. Yeh. Power depot: Integrating IP-based power modeling with ESL power analysis for multicore SoC designs. In *Proceedings of the 48th Design and Automation Conference*, ACM, New York, NY 10121, Jun. 2011. ACM.

[11] N. Julien, J. Laurent, E. Senn, and E. Martin. Power consumption modeling and characterization of the TI C6201. *Micro, IEEE*, 23(5), Sep. 2003.

[12] J. Laurent, N. Julien, E. Senn, and E. Martin. Functional level power analysis: An efficient approach for modeling the power consumption of complex processors. In *Proceedings of the Conference on Design, Automation and Test in Europe*, DATE '04, Washington, DC, USA, 2004. IEEE Computer Society.

[13] G. E. Moore. Cramming more components onto integrated circuits. *Electronics*, 38(8), Apr. 1965.

[14] L. Ost, G. Guindani, F. Moraes, L. Indrusiak, and S. Määttä. Exploring NoC-based MPSoC design space with power estimation models. *IEEE Design and Test*, 28, Mar. 2011.

[15] S. K. Rethinagiri, R. ben Atitallah, and J.-L. Dekeyser. A system level power consumption estimation for MPSoC. In *2011 International Symposium on System on Chip*. IEEE, Nov. 2011.

[16] V. Tiwari, S. Malik, and A. Wolfe. Power analysis of embedded software: A first step towards software power minimization. *Very Large Scale Integration Systems, IEEE Transactions on*, 2(4), Dec. 1994.

[17] C. Trabelsi, R. Ben Atitallah, S. Meftali, J.-L. Dekeyser, and A. Jemai. A model-driven approach for hybrid power estimation in embedded systems design. *EURASIP Journal on Embedded Systems*, 2011.

[18] W. Ye, N. Vijaykrishnan, M. Kandemir, and M. J. Irwin. The design and use of SimplePower: A cycle-accurate energy estimation tool. In *Proceedings of the 37th Annual Design Automation Conference*, DAC '00, New York, NY, USA, 2000. ACM.

A Transmission Gate Physical Unclonable Function and On-Chip Voltage-to-Digital Conversion Technique

Raj Chakraborty, Charles Lamech
Intel Corp.
raj.k.chakraborty@intel.com,
charles.d.lamech@intel.com

Dhruva Acharyya
AdvanTest Inc.
Dhruva.Acharyya@advantest.com

Jim Plusquellic
University of New Mexico
jimp@ece.unm.edu

ABSTRACT

A physical unclonable function (PUF) is an embedded integrated circuit (IC) structure that is designed to leverage naturally occurring variations to produce a random bitstring. In this paper, we evaluate a PUF which leverages resistance variations which occur in transmission gates (TGs) of ICs. We also investigate a novel on-chip technique for converting the voltage drops produced by TGs into a digital code, i.e., a voltage-to-digital converter (VDC). The analysis is carried out on data measured from chips subjected to temperature variations over the range of -40°C to +85°C and voltage variations of +/- 10% of the nominal supply voltage. The TG PUF and VDC produce high quality bitstrings that perform exceptionally well under statistical metrics including stability, randomness and uniqueness.

Categories and Subject Descriptors

K.6.5 [**Management of Computing and Information Systems**]: Security and Protection -- *Authentication.*

General Terms

Security

Keywords

Hardware security, unique identifier, process variations

1. INTRODUCTION

Physical Unclonable Functions (PUFs) continue to gain momentum as an alternative to embedding 'secrets' using fuses and non-volatile memory on ICs. PUFs derive secrets from variations that occur in the physical parameters of the on-chip wires and transistors. These variations are unique to each chip and, depending on the parameter, can be leveraged to produce large numbers of random bits. PUFs can produce repeatedly random bitstrings on the fly, and therefore eliminate the need for a specialized non-volatile on-chip memory to store them.

A PUF produces a bitstring by applying a set of "challenges" to specialized circuit primitives and measuring the corresponding "responses". The challenges are typically 'digital' and therefore can be generated on-chip using a pseudo-random number generator such as a linear feedback shift register (LFSR). The challenges

Permission to make digital or hard copies of all or part of this work for personal or classroom use is granted without fee provided that copies are not made or distributed for profit or commercial advantage and that copies bear this notice and the full citation on the first page. To copy otherwise, to republish, to post on servers or to redistribute to lists, requires prior specific permission and/or a fee. DAC 2013, May 29 - June 07 2013, Austin, TX, USA.

are used to configure one or more PUF circuit primitives prior to the application of a stimulus. The stimulus elicits an analog response from the PUF primitives, which is measured and digitized by other components of the PUF circuit. The digitized responses are then compared in a variety of combinations to produce a digital bitstring.

The PUF response is analog in nature, e.g., it can be a voltage drop or the propagation delay of a signal through the PUF primitive. The analog nature of the underlying random variable make the PUF sensitive to environmental variations such as temperature and power supply noise. Several important applications of a PUF require that they produce the same bitstring for a fixed challenge. Therefore, PUF architectures must be both random and resilient to noise sources.

In this paper, we investigate a PUF primitive that leverages resistance variations that occur in transmission gates (TGs). Hardware experiments are carried out on a set of chips at 9 temperature-voltage (**TV**) corners, using all combinations of the temperatures -40°C, 25°C and 85°C and voltages 1.08 V, 1.2 V and 1.32 V. A novel embedded test structure called a **voltage-to-digital converter (VDC)** is also evaluated under these environmental conditions. The VDC is used to digitize the voltage drops produced by the TG PUF.

Beyond these novel aspects of this work, we also investigate several noise resilient bit-flip avoidance schemes, that are designed to increase the probability that the bitstring can be reproduced under varying environmental conditions. The first technique derives a threshold from a chip's digitized voltage drop distribution profile that is used to decide whether a given comparison generates a **strong** bit or a **weak** bit. A second triple-module-redundancy (TMR-based) scheme is proposed for fixed length bitstrings that further improves bit-flip resilience. Although these techniques discard a significant fraction of bits, they provide several significant advantages. The public (helper) data associated with these methods reveals nothing about the secret bitstrings that they encode. Second, for applications where the PUF responses are made public, the difficulty of model building is significantly increased (assuming the public data is obfuscated) because bitstrings are constructed using only a subset of all possible voltage pairings. These techniques are investigated on data obtained from 63 copies of a test chip fabricated in a 90 nm technology.

2. BACKGROUND

Random bitstrings form the basis for encryption, identification, authentication and feature activation in hardware security. The introduction of the PUF as a mechanism to generate random bitstrings began in [3], although their use for chip identifiers began a couple years earlier [2]. Since their introduction, there have been many proposed architectures that are promising for PUF imple-

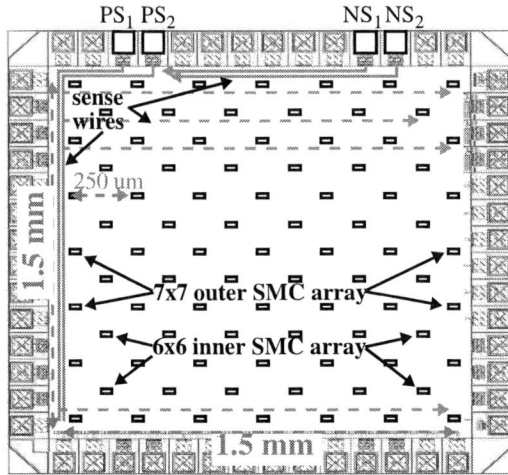

Fig. 1. Block diagram of 90 nm chips with 85 embedded stimulus-measure circuits (SMCs).

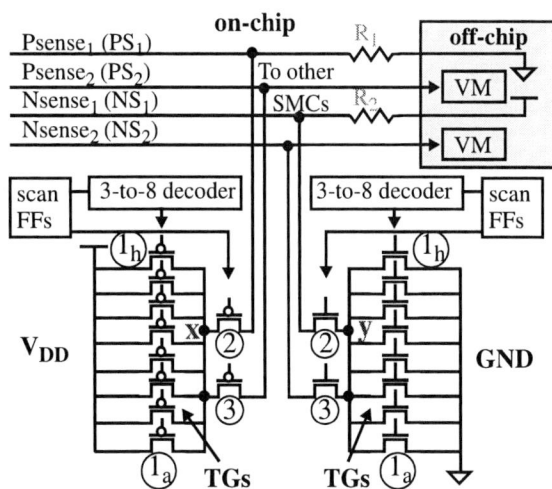

Fig. 2. SMC schematic in 90 nm chips.

mentations, including those that leverage variations in transistor threshold voltages [2], in speckle patterns [3], in delay chains and ROs [4-7], in SRAMs [8], in metal resistance [9][10], in sensors [11], and many others. The TG PUF proposed in this research is also based on resistance variations as in [10]. However, this paper for the first time investigates the reproducibility of the bitstrings across 9 industrial range TV corners after digitization using an on-chip VDC.

3. EXPERIMENT SETUP
3.1 TG Array, TGVs and TGVDs

Fig. 1 gives a block diagram of the 90 nm test chip architecture. The chip padframe consists of 56 I/Os, and surrounds a chip area of approx. 1.5 mm x 1.5 mm. Four PADs labeled PS_1, PS_2, NS_1 and NS_2 refer to *voltage sense* connections, the 'P' version for sensing voltages near V_{DD} and the 'N' version for voltages near GND. These four terminals wire onto the chip and connect to 85 copies of a *Stimulus/Measure circuit* (SMC). The SMCs are distributed across the entire chip (see small rectangles) as two arrays, a 7x7 outer array and a 6x6 inner array. Although not shown, a controlling scan chain connects serially to each of the SMCs.

The schematic diagram of the SMC is shown in Fig. 2. A set of 20 'pseudo' pass gates (hereafter referred to as transmissions gates or **TGs**) serve as both the PUF primitives and voltage sensing elements. Eight of the TGs, labeled I_a through I_h, connect to the V_{DD} grid, as shown on the left side of Fig. 2, while the other eight connect to the GND grid. Two additional TGs, labeled as *2* and *3*, connect to the drains of the I_{a-h} TGs. Separate scan FFs control their connection to the chip-wide wires that route to the P/NS_x pins shown in Fig. 1. The PS_1 and NS_1 sense wires are connected off-chip to GND and V_{DD}, resp., to create the stimulus condition described below. PS_2 and NS_2 are routed to off-chip Agilent 34401A voltmeters (VMs).

A voltage drop measurement is carried out by enabling three TGs, both of those labeled *2* and *3* and one from the group I_a through I_h. For example, using the PFET TGs, enabling TGs I_a and *2* create a short between the V_{DD} grid on-chip and a GND node off-chip. The voltage falls across the two TGs as well as the PS_1 wire. The voltage on the node **x** between TG I_a and *2* can be

sensed with TG 3^1. The on-resistances of the TGs (and the resistance of the PS_1 wire) determine how much of the V_{DD} voltage falls across each of TG I_a and *2*. Random variations in the on-resistances of the TGs I_a through I_h (referred to subsequently as the **stack**) produce different voltage drops as each is enabled. We refer to the voltages at node **x** as **TGVs**.

The component of the TGV that falls across the sense wires represents a bias because the length of the sense wires is different for each SMC in the array. The bias is eliminated by creating TGV differences (**TGVDs**) using the 8 TGVs measured within each SMC, separately for NFETs and PFETs. The TGVDs are obtained by subtracting pairs of TGV values. With 8 TGVs, a total of 8*7/2 = 28 TGVDs can be created in each stack. The total number of TGVDs obtained per chip is 2,380 for each of the PFETs and NFETs, obtained as 85 SMCs * 28 TGVDs/SMC.

The NFET and PFET TGVDs, in turn, can be compared under all combinations to produce bitstrings of length 2,380*2,379/2 * 2 = **5,662,020 bits**. The NFET and PFET TGVDs cannot be compared with each other primarily because of channel width differences (PFETs are 2.5x wider than the NFETs) and mobility variations with doping (NFET variations are larger than PFET variations). As a consequence, PFET voltage variations are only about half as large as the NFET variations.

In our experiments, the order in which the comparisons are made is randomized using *srand(seed)* and *rand()* from the C programming library. This operation is easily implemented on chip using an LFSR and a seed.

3.2 Voltage-to-Digital Converter (VDC)

In addition to analyzing the TG voltage drops directly, we also analyze a digital representation of them that is produced by an on-chip VDC, similar to designs described in [12]. The architecture of the VDC is shown in Fig. 3. The VDC is designed to 'pulse shrink' a negative input pulse as it propagates down a current-starved inverter chain. As the pulse moves down the inverter chain, it activates a corresponding set of latches to record the passage of the pulse, where activation is defined as storing a '1'. A

---

1. Only a negligible amount of current flows through TG *3* to the voltmeter so the voltage on node **x** is nearly identical to that at the voltmeter.

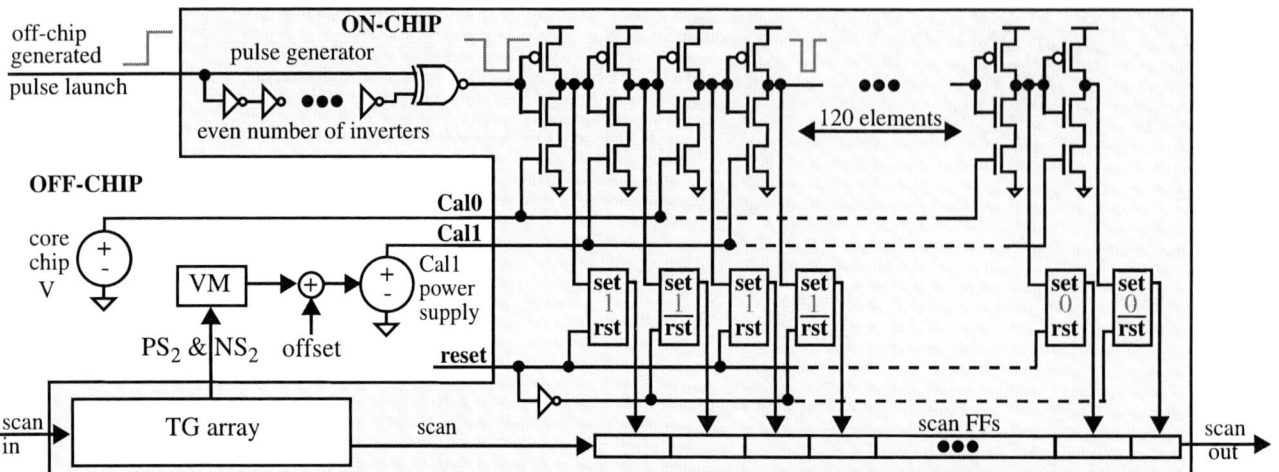

Fig. 3. Voltage-to-Digital Converter (VDC). On the left side is off-chip instrumentation that measures a voltage from the TG array, adds an offset and programs a power supply to drive the Cal1 input of the VDC.

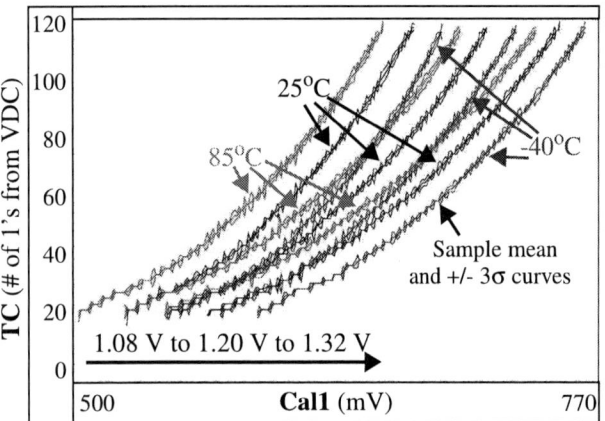

Fig. 4. VDC Cal1 vs. thermometer code (TC) curves across 9 TV corners on one chip.

thermometer code, i.e., a sequence of '1's followed by a sequence of '0's, represents the digitized voltage.

The voltage-to-digital conversion is accomplished by introducing a fixed-width (constant) input pulse, which is generated by the pulse generator shown on the left side of the Fig. 3. Two analog voltages, labeled Cal0 (which is held constant) and Cal1 (the voltage to be digitized) connect to a set of NFET transistors in the inverter chain, with Cal0 connecting to NFETs in even numbered inverters and Cal1 to the NFETs in odd numbered inverters. The propagation speed of the two edges associated with the pulse are controlled separately by these voltages. The pulse will eventually die out at some point along the inverter chain when the trailing edge of the pulse 'catches up' to the leading edge. This is ensured by fixing Cal0 at a voltage higher than Cal1. A digital representation of the Cal1 voltage can then be obtained by counting the number of '1's in the latches.

In order to enable this type of pulse shrinking behavior, Cal1 needs to be set to a value between 500 mV and 800 mV. The voltage-divider (series) arrangement of the identically-sized TGs shown in Fig. 2 should provide voltages at the midpoint of the supply voltage, e.g., approx. 600 mV. This is not the case, however, for two reasons; 1) a portion of the voltage falls across the NS_1 and PS_1 sense wires resistances labeled R_1 and R_2 in Fig. 2, and 2) the

series-connected transistors in the shorting path, e.g., 1_a and 2 in Fig. 2, operate in linear mode and saturation modes, resp. (See Section S4 for details.) As a consequence, the range of the TGVs observed in our experiments at node **x** in Fig. 2 for PFETs is between 950 mV to 1050 mV, and at node **y** for NFETs is 150 mV to 250 mV. In order to move Cal1 into the 600 mV range, an **offset** voltage is added (subtracted) to the voltages measured by the VM as shown in Fig. 3 for NFETs (PFETs). This offset voltage is computed using a calibration process described below.

The calibration process is needed because the required offset voltage changes as a function of TV conditions. The curves in Fig. 4 depict the behavior of the VDC over the 9 TV corners for one chip. The graph plots Cal1 on the x-axis against the number of '1' bits in the thermometer code, referred to as **TC**, on the y-axis. The mean and 3σ curves are superimposed. The average 3σ, computed using the individual 3σ in each curve, is less than 1 for all curves. The small non-linearity in the curves does not degrade the statistical properties of the bitstrings, as shown below. The sensitivity of the VDC is approx. 1 TC bit per millivolt change in Cal1. The TGVs for a typical chip vary over the range of 40 to 60 mVs so less than half of the 120 bit range of the VDC is used in our experiments.

Although the VDC remains stable across the TV corners, the shift of the curves along the x-axis causes overflow in the VDC; a situation where the pulse propagates through all 120 delay chain elements. A calibration process is carried out that tunes the 'offset' at each TV corner, and effectively eliminates the adverse effects of the curve shift. The calibration process tests a distributed set of 9 TGs, e.g., of the 680 NFET TGs, and uses binary search to find an offset voltage that produces a 'target' TC, separately for each of the 9 tests. We set the target TCs for NFET and PFET TGVs to 65 and 85, resp. These targets worked well to prevent overflow in all of the 1,360 TG measurements, across all TVs and chips used in our experiments. The **median offset** from the 9 calibration tests is used as the offset during the subsequent data collection process. This calibration procedure only *approximates* the best offset, but does not need to be precise because the goal is only to prevent overflow in the VDC. A more detailed explanation of the process is given in Section S1.

We plan to integrate the instrumentation used to measure the TGVs, to add an offset and to control the Cal1 voltage, as shown

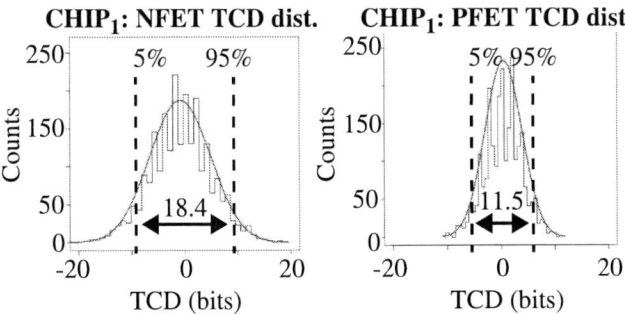

Fig. 5. Enrollment NFET (left) and PFET (right) TCD distributions with 2,380 components from one chip, with inter-percentile ranges delineated.

on the left side of Fig. 3, in the next version of the chip. The Cal1 offset voltages can be derived using a resistor-ladder network [13], and added to the TG voltage using a voltage subtractor/adder circuit [14]. The offset only needs to be accurate to approx. 5 mVs, which significantly reduces the area overhead of the ladder network. With the availability of these on-chip components, a state machine can be easily designed to carry out the calibration process described above.

3.3 Data Collection Process

The calibration process is used to select an offset voltage, separately for the PFET and NFET elements on each of the 63 chips. Each of the 680 components are then enabled, one at a time, and the corresponding TGV is measured using the VM as shown in Fig. 3. The Cal1 power supply is programmed with this TGV plus the offset and 11 TC samples are collected from the VDC. This process is repeated for both the NFET and PFET components. The mean value of the 11 samples is used to compute a 'difference' value, synonymous to the TGVDs described above. We use the term **TCD** to refer to these thermometer code differences in the remainder of this paper.

3.4 Overhead

Each SMC occupies an area of approx. 500 um^2, so the total area occupied by the array of 85 SMCs is approx. $42,500 \text{ um}^2$. If the SMCs are placed adjacent to each other (instead of being distributed as in Fig. 1), the array would occupy a 206 um x 206 um region. The VDC occupies an area of 136 um x 60 um. The area of the digital components, i.e., the LFSR and bit generation engine, is estimated at 300 um x 300 um. On-chip memory requirements for the array of 680 NFET and PFET TGs is approx. 2,380 bytes.

3.5 Thresholding Technique

As discussed above, TCDs are computed by subtracting TCs within the same SMC as a means of eliminating the voltage bias introduced by the sense wires. Computing differences also has the benefit of significantly increasing the number of bits that can be produced from each chip. For example 2,380 TCDs are produced from the 680 NFET TCs.

Using difference values, however, has two main drawbacks. First, subtracting two TCs reduces the signal-to-noise ratio because the noise from two separate measurements is combined in the difference. More importantly, TCDs 're-use' the base entropy of the array, which is defined by the 1,360 NFET and PFET TCs for each chip. Therefore, re-use makes model building attacks possible in cases where the bitstring is made public.

We propose a thresholding technique as a means of dealing

with model-building attacks and preventing information leakage in the public helper data. Our thresholding technique discards TCD comparisons that are susceptible to producing bit flips in the bitstring. Bit flips occur when the relative ordering of a pair of TCDs defined during enrollment reverse order during regeneration. This is much more likely to occur for pairs of TCDs that are similar in magnitude. We show in our experimental results that it is possible to define a threshold that filters all TCD pairings that introduce bit flips during regeneration at one or more of the TV corners. The threshold is derived using the distribution characteristics of TCDs obtained during **enrollment**, which is carried out in our experiments at 25°C and 1.20V.

Fig. 5 shows the TCD enrollment distributions for NFETs and PFETs from one of our chips. It is clear from the spread of the distributions that the NFET TCDs have more variation than the PFET TCDs as discussed in Section 3.1. The objective is to derive a threshold from these distributions that serves three primary goals: 1) avoids bit flips under different TV conditions in the subsequent bit generation phase, 2) preserves as many strong bits as possible for each chip and 3) makes the number of strong bits as consistent as possible across chips, i.e., scales with the range of variation that occurs on each chip. We define **strong bits** as those generated by TCD comparisons where the differences in the TCDs exceeds the threshold.

In our experiments, we found the limits defined by the two vertical lines labeled 5% and 95% in Fig. 5 achieve these goals. These limits capture the spread of the distribution while ignoring the outliers on the tails of the distributions, which, when included, introduce large variations in the number of strong bits preserved across the chip population, i.e., they degrade criteria 3 above. We then multiply the 2 **inter-percentile ranges** defined as the distances between these limits by 2 **scaling factors**, one for NFETs and one for PFETs, to define the 2 TCD **thresholds** for the chip.

Figs. 6(a) and (b) provide an illustration of the thresholding process applied using TCD data from one of the chips. The graphs plot bit number along the x-axis against the **differences** of the TCDs being compared. Only the first 390 strong bits are shown. The horizontal lines at 9.7 and -9.7 delineate the threshold boundaries for the NFET TCDs, which are derived from Fig. 5 using a scaling factor of 0.53.

Fig 6(a) shows those TCD differences which produce strong bits during enrollment. In addition to generating the secret bitstring, a **thresholding bitstring** is also constructed during enrollment which indicates which comparisons produce strong bits and which produce **weak bits**. The thresholding bitstring is recorded in public data storage, and using techniques such as run-length encoding, is proportional in size to the secret bitstring (see Section S3). This type of public data reveals nothing about the secret bitstring, and represents the helper data for our PUF.

Fig. 6(b) superimposes the TCD difference data points generated under the remaining 8 TV corner experiments, which represent the regeneration scenarios in our experiments. The thresholding bitstring is consulted to ensure regeneration uses the same comparisons as enrollment[1]. The data points associated with the regenerations appear above and below the enrollment data points. Only those that move toward 0 line are problematic how-

---

1. The thresholding process is implemented only during enrollment, and is disabled during regeneration.

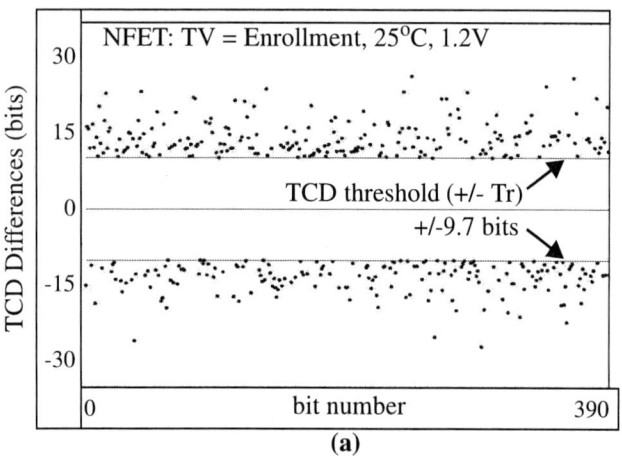
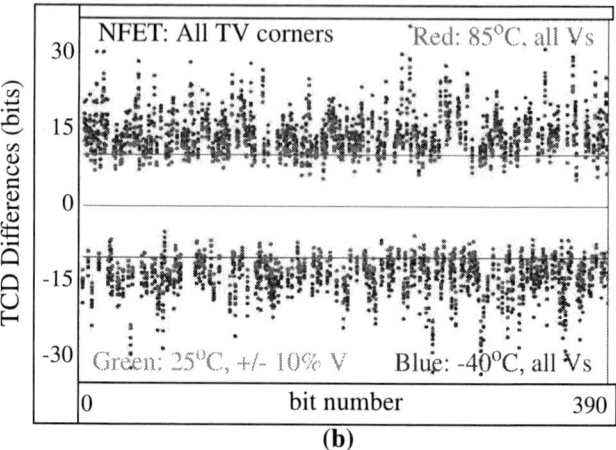

Fig. 6. Threshold method showing the first 390 strong bit comparisons during (a) enrollment and (b) regeneration across 8 TV corners.

ever. Although none occur in these plots, points that move over the 0 line from above or below indicate the relative ordering has changed in the TCD pairing. A bit flip will occur during regeneration if this condition is met.

The TCD differences plotted in the figure span a larger range than the TCDs used to compute the inter-percentile range from Fig. 5 because the TCDs themselves are both positive and negative. Despite their larger range, only about 21% of the 2,831,010 possible comparisons, i.e., approx. 595,000 bits, survive the thresholding for NFETs. A similar analysis using the TGVDs shows approx. 33% surviving the thresholding, which suggests that the digitization process adds to the noise. This is even more dramatic in the PFET analysis, where approx. 7% of the TCDs survive and approx. 36% of the TGVDs survive. The smaller variation in the PFET TCDs reduces the signal-to-noise for the VDC even further. However, the 832,343 TCD-based bits for this chip that survive are reproducible across the TV corners and exhibit excellent statistical characteristics as we show below.

3.6 Fixed Length Bitstrings and TMR

In actual applications, only a fixed number of bits are needed. With encryption, the values vary between 128 to 1024 bits, depending on the encryption algorithm. The large number of bits available from the PUF can be beneficial, however, by allowing a distinct set of fixed-length secret keys to be generated over time during successive enrollments.

A second possible usage scenario leverages this large pool of strong bits to further increase the resiliency to bit flip failures, i.e., beyond that provided by thresholding. We propose a bitstring replication method that mimics a popular scheme used in fault tolerance called triple-module-redundancy or TMR. In this technique, a fixed length, e.g., 1,024-bit, bitstring is generated as described above. TMR is then applied to generate two more copies of the bitstring. The two copies are generated by parsing the strong bit sequence until a match is found to each bit in the first bitstring. During regeneration, a majority voting scheme is applied to each of the columns in the three identically regenerated bitstrings as a means of avoiding single bit flip failures. In other words, the final bitstring is constructed by using the majority of the 3 column bits as the final bit for each bit position, i.e., a '1' is assigned in the final bitstring when 2 or more of the 3 bits in the column are '1', and a '0' otherwise. An illustrative example is given in Section S2.

A PUF that is able to generate strong bit sequences that are

locally random (a quality measured by the NIST tests [1] presented in the Section 4) ensures that a match occurs for each bit during the generation of the two copies every 2 bits on average. Under these conditions, it follows that a TMR-based bitstring, and its public data, consumes on average 5 times more strong bits than a non-TMR-based bitstring. The benefit, on the other hand, is a significant decrease in the 'probability of failure', i.e., the likelihood of a bit flip occurring during regeneration, as we show in Section 4. Moreover, this scheme offers flexibility by allowing a trade-off between tolerance to bit flips and public data size.

4. EXPERIMENTAL RESULTS

In this section, we evaluate the several important statistical properties of the TGVD and TCD-derived bitstrings including randomness, uniqueness and probability of bit flips, e.g., failures to regenerate the bitstring under different environmental conditions. As discussed in Section 3.2, the process of digitizing the voltages using the VDC adds noise and reduces the number of corresponding strong bits. The penalty of the digitization process is evaluated by carrying out the same analysis using the TGVDs directly, and serves to illustrate the best that can be achieved in the absence of digitization noise.

Fig. 7a) gives the inter-chip hamming distance (HD) distribution using the TGVDs while Fig. 7b) shows the distribution using TCDs. The graphs plot HD along the x-axis against the number of instances on the y-axis[1]. With 63 chips, the total number of instances is 63*62/2 = 1,953. The distributions are 'fitted' with Gaussian curves to illustrate the level of conformity they exhibit to this distribution.

Since HDs must be computed across bitstrings of equal length, it was necessary to truncate the bitstrings used in Fig. 7 to the length obtained for the chip with the fewest number of strong bits. Truncation reduced the lengths to 1,901,845 for the TGVD analysis and 725,230 for the TCD analysis, which are approx. 33.6% and 12.8%, resp., of the maximum possible length, i.e., 5,662,020 bits. The chip with the longest bitstring, in comparison, uses 35.6% of the maximum for the TGVD analysis and 15.0% for the TCD analysis. The term **truncated bitstrings** is used to refer to the shorter, equal-length bitstrings.

---

1. HD is computed by counting the number of bits that are different in the bitstrings from two chips.

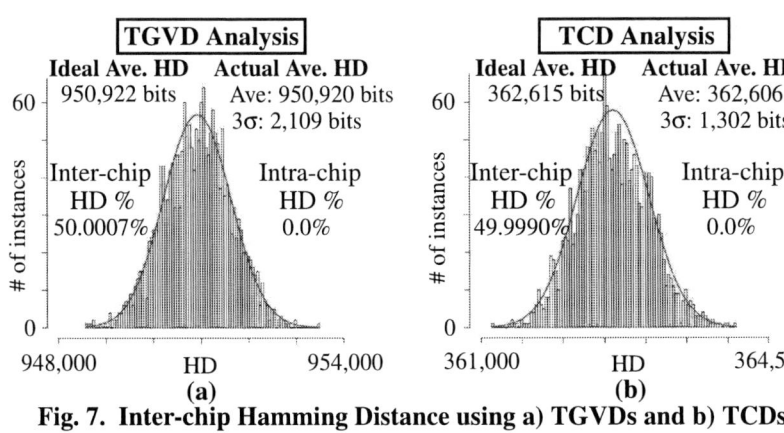

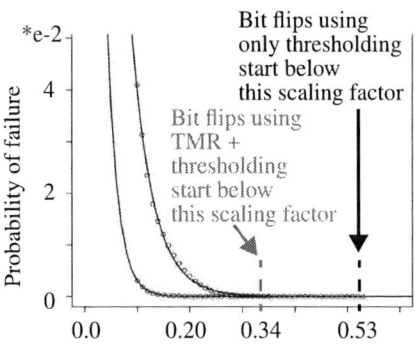

Fig. 7. Inter-chip Hamming Distance using a) TGVDs and b) TCDs.

Fig. 8. NFET TCD scaling factor (x-axis) vs. probability of failure (y-axis).

The actual average inter-chip HDs listed in Fig. 7 are nearly equal to the ideal value of 50%. In contrast, the average inter-chip HDs for the bitstrings of length 5,662,020, i.e., those with the weak bits included (not shown), is 48.4% and 48.5% for TGVD and TCD, resp., so removing the weak bits improves the inter-chip HDs. The 3σ values shown in the figure are derived from the Gaussian curves and represent the spread of the distributions (where smaller is better). These values are small relative to the length of the truncated bitstrings, e.g., they are only 0.11% and 0.18% of the lengths for the TGVD and TCD analysis, resp.

The scaling factors are set to 0.42 (NFET) and 0.39 (PFET) for the TGVD analysis and 0.53 (NFET) and 0.78 (PFET) for the TCD analysis. These values were derived by analyzing the bitstrings across all 9 TV corners and tuning the values until no bit flips occurred (Section S2 discusses how this can be done in practice). Therefore, the intra-chip HD is 0.0% as shown in Fig 7 for both analyses. However, the underlying noise levels can be measured by disabling the thresholding technique, yielding intra-chip HDs of 5.11% and 8.68% for the TGVD and TCD analyses, resp. The increase in the TDC intra-chip HD over that given for TGVD reflects the noise added by the VDC digitization process.

We applied the NIST statistical tests [1] to the truncated bitstrings of the 63 chips at a significance level of 0.01 (the default). The TGVD and TCD bitstrings **pass all tests**, with no fewer than 60 passing chips per test (the number required by NIST for the test to be considered 'passed'). Moreover, all tests passed the **Pvalue-of-the-Pvalues** metric.

Fixed-length bitstrings were also created using the TMR-based scheme proposed in Section 3.6. In our experiments, we were able to create, on average, 381 1024-bit TMR-based bitstrings per chip using TGVD data, and 156 on average using TCD data. Although not shown, the statistical test results are similar to those discussed above for the longer bitstrings.

As discussed in Section 3.6, the TMR scheme improves resiliency to bit flips over the thresholding scheme alone. The curves shown in Fig. 8 illustrate the improvement. The scaling factor used for NFETs (the PFET scaling factor is also changed proportionally) is plotted along the x-axis against the probability of failure on the y-axis. The probability of failure is computed at each scaling factor value by dividing the number of bit flips that occur in all 63 chips by the total number of strong bits produced. The curve on the left is the result obtained using the TMR + thresholding technique, while the curve on the right uses only thresholding. Both curves are exponential in shape (see Section S2 for curve fits and further analysis). However, from the positions of the curves, it is clear that the TMR scheme requires a lower scaling factor, 0.34 vs. 0.53, before any bit flips occur. Using 0.53 as the scaling factor, the probability of failure is 1.1e-6 with thresholding but improves significantly to 1.5e-12 after adding TMR.

5. CONCLUSIONS

A transmission gate (TG) PUF and on-chip voltage-to-digital conversion circuit are evaluated on 63 copies of a 90 nm chip, at 9 temperature-voltage corners. Thresholding and triple-module-redundancy techniques are proposed as a means of avoiding bit flips. Results from statistical tests confirm that cryptographic quality bitstrings are obtained using either the TG voltages or their digitized representations. The proposed bit flip avoidance schemes allow the user to trade-off the probability of failure with helper data overhead for applications requiring bitstring regeneration.

6. REFERENCES

[1] NIST: Computer Security Division, Statistical Tests, http://csrc.nist.gov/groups/ST/toolkit/rng/stats_tests.html

[2] K. Lofstrom, *et al.*, "IC Identification Circuits using Device Mismatch," *SSCC*, 2000, pp. 372-373.

[3] R. S. Pappu, *et al.*, "Physical One-Way Functions," *Science*, 297(6), 2002, pp. 2026-2030.

[4] B. Gassend, *et al.*, "Controlled Physical Random Functions," *Conference on Computer Security Applications*, 2002.

[5] M. Majzoobi, *et al.*, "Lightweight Secure PUFs", *ICCAD*, 2008.

[6] G. Qu and C. Yin, "Temperature-Aware Cooperative Ring Oscillator PUF", *HOST*, 2009, pp. 36-42.

[7] A. Maiti and P.Schaumont, "Improving the Quality of a Physical Unclonable Function using Configurable Ring Oscillators", *FPLA*, 2009. pp. 703-707.

[8] J. Guajardo, *et al.*, "Physical Unclonable Functions and Public Key Crypto for FPGA IP Protection," *FPLA*, 2007, 189-195.

[9] R. Helinski, *et al.*, "Physical Unclonable Function Defined Using Power Distribution System Equivalent Resistance Variations", *DAC*, 2009, pp. 676-681.

[10] J. Ju, R. Chakraborty, R. Rad, J. Plusquellic, "Bit String Analysis of Physical Unclonable Functions based on Resistance Variations in Metals and Transistors", *HOST*, 2012, pp. 13-20.

[11] K. Rosenfeld, *et al.*, "Sensor Physical Unclonable Functions", *HOST*, 2010, pp. 112-117.

[12] L. Guansheng, Y.M. Tousi, A. Hassibi and E. Afshari, "Delay-Line-Based Analog-to-Digital Converters,", Trans. on *CAS II*, Volume: 56, Issue: 6, 2009, pp. 464-468.

[13] Dan O'Sullivan & Tom Igoe, "Physical Computing: Sensing and Controlling the Physical World with Computers," Thomson Course Technology Publishers, 2004, pp 388-391.

[14] R. Fried and C. C. Enz, "Simple and Accurate Voltage Adder/ Subtractor," *Electronics Letters*, vol. 33, 1997, pp. 944-945.

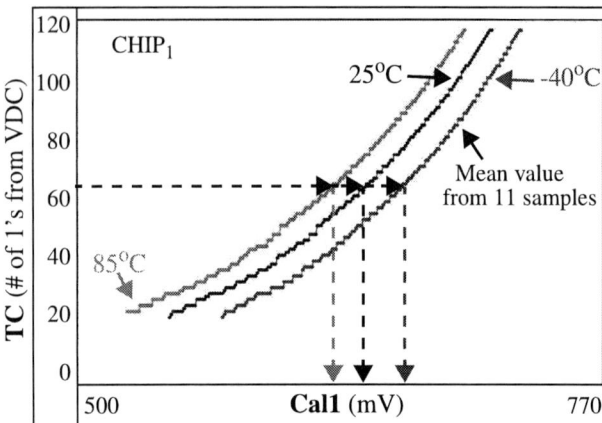

Fig. 9. VDC calibration curves at 85, 25, -40°C and 1.2V illustrating offset calculation process.

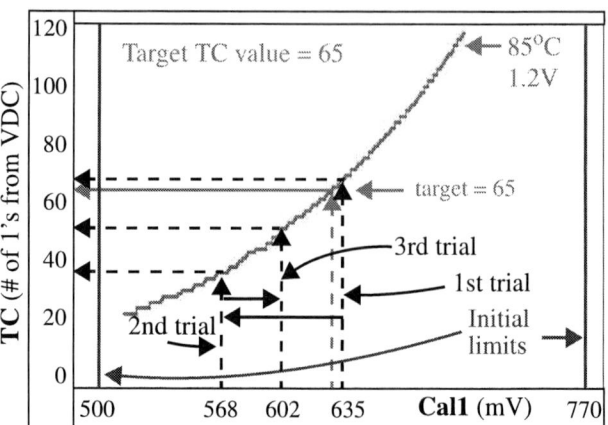

Fig. 10. Illustration of the binary search process used during calibration at 85°C, 1.2V.

SUPPLEMENTARY MATERAL
S1 VDC Calibration Process

The calibration process described in Section 3.2 is further illustrated using the Cal1 vs. TC curves shown in Fig. 9. As indicated earlier, calibration is carried out before enrollment and regeneration, and its objective is to find an appropriate Cal1 voltage offset that prevents overflow in the VDC for any of the TGVs that will be measured during bit generation. We determined that testing a subset of 9 TGs during calibration is sufficient to obtain a good predictor for offset voltage that prevents overflow.

The goal of calibration is to select an offset voltage such that the TG-under-test produces the same TC value independent of the TV corner. This objective is illustrated in Fig. 9 with the horizontal dashed line at TC = 65. The 3 curves shown represent the mean values produced by the VDC on CHIP$_1$ as the Cal1 voltage is swept across a range of values (similar to the process described in Section 3.2 in reference to Fig. 4) at 3 different temperatures. The different positions of the dashed vertical lines from each curve make it clear that the offset voltage needs to change in order to maintain a value of 65 in the VDC. Note that the TGV itself measured from the TG-under-test will also change as a function of temperature. This situation is handled by using the TGVs directly in the calibration process (as opposed to using a special voltage source).

Calibration is carried out by enabling each of a select, distributed group of TGs, one at a time, and performing a binary search. The search process varies the Cal1 voltage offset until the TG-under-test produces a specific TC value. The process is illustrated in Fig. 10 using the 85°C Cal1-TC curve from Fig. 9. The initial limits are set to 500 mV and 770 mV. The 1st trial selects the midpoint between these limits, i.e., 635 mV. Note this midpoint voltage is the sum of the TGV and the offset voltage that is being tuned in the search. The 1st trial produces a TC of approx. 68, which is larger than the target. Therefore, the next trial uses 635 mV as the upper limit and the new midpoint voltage becomes 568. The 2nd trial produces a TC of 35, so 568 is used as the lower limit for the new midpoint. The process continues until an offset is found that produces a TC of 65. The binary search process is repeated using 9 TGs as a means of obtaining a value that best approximates the average behavior. The median value from the 9 calibration tests is used as the final offset, which is added to all subsequent TGVs measured at this TV corner.

S2 Thresholding & TMR-based scheme

The thresholding and TMR-based schemes are described in Sections 3.5 and 3.6. This section of the Supplementary Material is designed to clarify this process with an example. The thresholding scheme shares characteristics with the shielding function proposed in [15] but is simpler because it is based entirely on strong bits, referred to as 'robust' bits in the reference. This fact changes the nature of the public data and eliminates information leakage that, although unlikely, is possible with shielding functions.

Fig. 11 illustrates the proposed thresholding and TMR-based scheme using data from a hypothetical chip. The x-axis plots a sequence of comparisons that would be used to generate a bitstring, while the y-axis plots the differences between the pairings of TCDs. Each difference reflects the relative ordering of the two TCDs, e.g., positive difference values indicate that the first TCD is larger than the second. For strong bits, the TCD difference data points must lie above or below the thresholds, labeled '+Tr' and '-Tr' in the figure. This condition, when met, is recorded using a '1' in the thresholding bitstring shown below the data points. Weak bits, on the other hand fall within the thresholds and are indicated with a '0'. The bold (and blue) '0's indicate strong bits that are skipped under the TMR scheme described below.

As discussed in Section 3.6, the TMR-based method constructs 3 identical bitstrings during enrollment as shown along the bottom of Fig. 11. The left-most bitstring labeled 'Secret BS' is generated from the first 4 strong bits encountered as the sequence of data points is parsed from left to right. The second bitstring labeled 'Redundant BS$_1$' is produced from the next sequence of data points but has the additional constraint that each of its bits must match those in the first bitstring. During its construction, it may happen in the continued left-to-right parsing of the data points that a strong bit is encountered that does not match the corresponding position in the 'Secret BS'. In the example, this occurs at the position indicated by the left-most bold '0' in the thresholding bitstring. Here, we encountered a strong bit with a value of '0'. But the 'Secret BS' requires the first bit to be a '1', so this strong bit is skipped. This process continues until redundant bitstrings BS$_1$ and BS$_2$ bitstrings are constructed.

The number of strong bits required to generate a secret bitstring of length 4 is approx 5x or 20. From the example, this is evaluated by counting the number of '1's and bolded '0's in the thresholding bitstring, which is given as 19. The benefit of creating these redun-

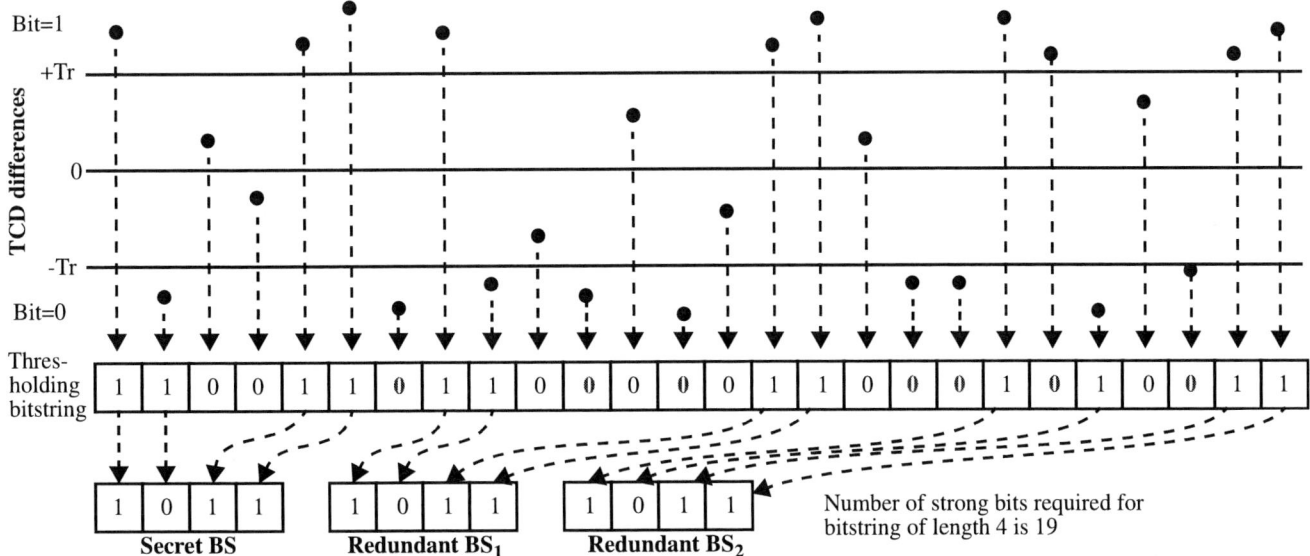

Fig. 11. Secret bitstring generation example using the proposed thresholding and TMR-based method.

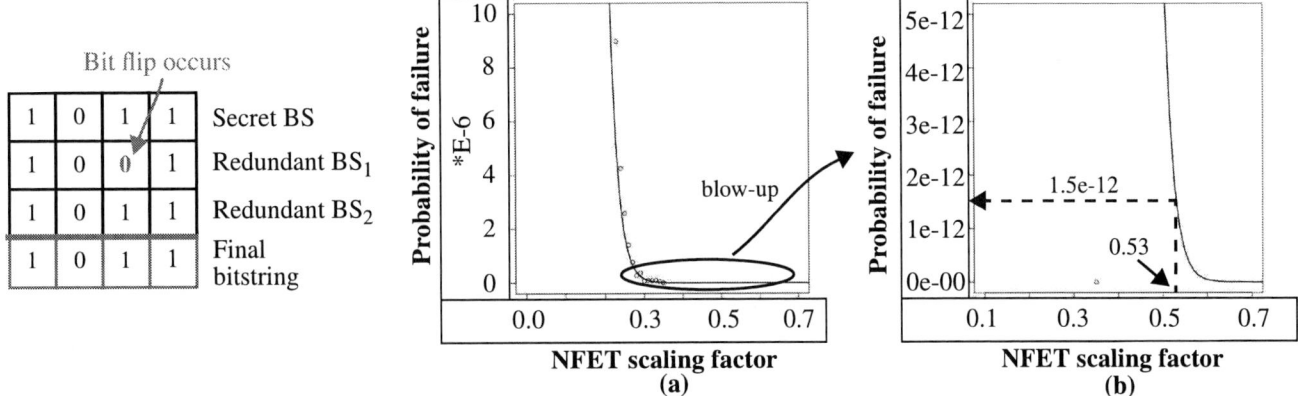

Fig. 12. Bit flip avoidance illustration using example from Fig. 11.

Fig. 13. a) TMR probability of error curve and b) blow-up of the designated region. The discrete curve is fitted with a superimposed exponential function.

dant bitstrings is the improved tolerance that they provide to bit flips. For example, during regeneration, the three bitstrings are again produced, but this time using the thresholding bitstring to determine which TCDs to compare.

In scenarios where the threshold is set too low, it is possible that a strong data point used in enrollment is displaced across both the threshold and the '0' line because of different TV conditions in regeneration, causing a bit flip. However, with TMR, a bit flip can be avoided if no more than 1 bit flip occurs in a single column of the matrix of bits created from the 3 bitstrings. For example, the first 3 rows of the matrix of bits in Fig. 12 is constructed during regeneration in a similar way to those shown in Fig. 11 for enrollment. The bottom row represents the final secret bitstring and is constructed by using a **majority vote** scheme (in the spirit of TMR). The bit flip shown in the third column has no effect on the final bitstring because the other two bits in that column are '1', and under the rule of majority voting, the final secret bit is therefore defined as '1'[1].

In Section 4, the probability of failure using thresholding alone

---

1. TRM can be extended to include 5, 7, etc. copies of the bitstring to further enhance bit flip resiliency.

and in combination with TMR was discussed, with the latter improving significantly on the former, from 1.1e-6 to 1.5e-12. These values were obtained by fitting the discrete-valued curves produced from repeatedly running the analysis at different scaling factors with exponential functions. Fig. 13(a) shows the data for the TMR + thresholding curve in Fig. 8 with the fitted exponential curve. The exponential is clearly a good fit to the data points. Fig. 13(b) shows a blow-up of the region around the NFET scaling factor of 0.53 from which the estimate of 1.5e-12 was derived.

S3 Run-Length Encoding of Public Data

The size of the public (helper) data under the thresholding and TMR-based schemes can be reduced using compression techniques such as run-length encoding. The benefit of run-length encoding is its simplicity. Fig. 14 shows an example of a thresholding bitstring with 26 bits. The long strings of '0's can be run-length encoded by simply counting them and replacing the '0' sequence with a field which represents the number of '0's in each sequence. In the example, the run-length encoded bitstring uses 19 bits instead of 26. The longer the sequences of '0's, the more efficient the scheme becomes. The best choice for the field width depends on the nature of the public data, i.e., the average length of the '0' strings.

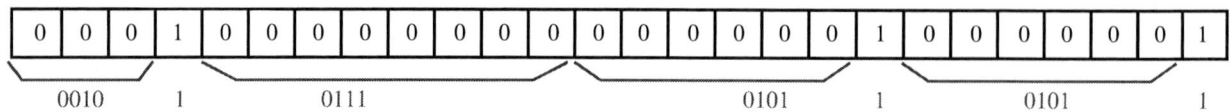

Fig. 14. Examples of run-length encoding as a compression technique to reduce public data size. Original public data string has 26 bits. Run-length encoded using a field width of 4 yields 19 bits.

The public data for the TCD analysis from Section 4 indicates that approx. 14% of the bits survive the thresholding, and even fewer, approx. 8.4%, are marked with '1's in the public data when TMR is added. The public data is therefore expected to contain strings of 0's with average lengths of approx. 11 under thresholding + TMR. Therefore, a field width between 3 and 4 (which allows counting up to 8 and 16, resp.) should be optimal. We found that a field width of 5 is best and yields a 42% reduction on average to the size of the original public data string. We plan to explore other compression techniques in future work.

In addition to compression, obfuscation is required for the thresholding bitstring when the PUF usage scenario involves authentication. This is true because the 'secret' bitstring is not kept on chip as it is for encryption but rather is also made public. With both bitstrings available, an adversary can reverse engineer the relative ordering of the TCDs. In order to prevent this, we propose to obfuscate a portion of the thresholding bitstring as follows. During enrollment, the first n strong bits, e.g., 128, are used as a key to encrypt the thresholding bitstring, excluding those public data bits that correspond to the encryption key itself. These bits do not need to be encrypted because the key is never made public.

S4 Underlying Stability Characteristics of the TG PUF

As discussed in Section 3.5, bit flips introduced by TV variations represent the primary threat to the TG-PUF's reliability. Here, we investigate the underlying mechanism that cause TGVs to vary as a function of temperature and voltage. Although the main focus of this paper has been on the TCDs, instability in the TGVs is the primary component of the instability observed in the TCDs and is therefore the focus of our analysis.

Fig. 15 shows a portion of the NFET stack shown on the right side of Fig. 2 for two arbitrary SMCs, SMC_X and SMC_Y. As discussed earlier, TGVDs are created to eliminate the sense wire bias in which the TGVs from two distinct tests in the same SMC are subtracted. The figure includes only 2 NFETs (of the 8) from the stack as an illustration of this operation. In each of the two tests, two transistors are enabled, e.g., TG_{1a} and TG_{1b}, which establishes a current path labeled I_{1ab} from the off-chip power supply, V_{DD}, through the sense wire resistor (R_{w1}) and the two TGs to the on-chip GND grid. The voltage between the two transistors labeled TGV_{1ab} is measured off-chip using a voltmeter (VM) by enabling a third transistor TG_{1S}. A second voltage drop, TGV_{1ac} (not shown), is obtained in similar fashion by enabling TG_{1a} and TG_{1c} within SMC_X. The TGVD is defined as $TGVD_1 = TGV_{1ab} -$ TGV_{1ac}. The exact same process is carried out within SMC_Y to obtain $TGVD_2$.

In order to understand how the TGVD change as a function of TV variations, we need to first determine the modes of operation of the two transistors in the shorting path. The shorting path defines a voltage divider network with, e.g., $R_{w1} + R_{1a}$ (the sense wire resistance and **TG_{1a} R_{on} resistance**) as one element and R_{1b}

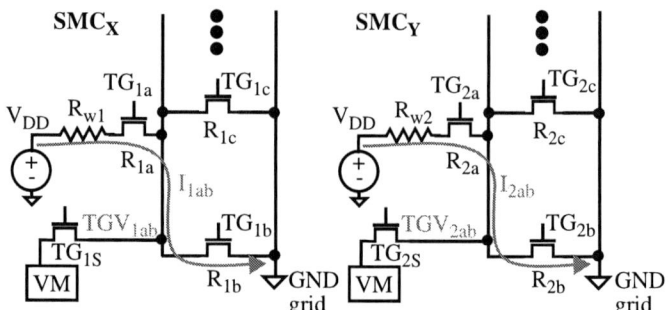

Fig. 15. Example and analysis of TV variations and its impact on TGVDs.

$$TGVD_1 = V_{DD}\left[\frac{1}{1+\dfrac{R_{1a}(1+Y_{1a})}{R_{1b}(1+X_{1b})}} - \frac{1}{1+\dfrac{R_{1a}{}'(1+Y_{1a}{}')}{R_{1c}(1+X_{1c})}}\right] \quad \textbf{Eq. 1.}$$

$$TGVD_2 = V_{DD}\left[\frac{1}{1+\dfrac{R_{2a}(1+Y_{2a})}{R_{2b}(1+X_{2b})}} - \frac{1}{1+\dfrac{R_{2a}{}'(1+Y_{2a}{}')}{R_{2c}(1+X_{2c})}}\right] \quad \textbf{Eq. 2.}$$

(the **TG_{1b} R_{on} resistance**) as the second element. The R_w's vary from approx. 100 Ohms (upper left-most SMC in Fig. 1) to 1.5 KOhms (lower right-most SMC). Unfortunately, there is no way to measure the R_w's by themselves (the values above are obtained from the layout geometries and the design manual's resistance/ square values) so they are lumped together with the transistor resistances R_{1a} and R_{2a} for the purposes of this analysis.

Eqs. 1 and 2 are the defining equations for the 2 TGVDs. Each equation incorporates two voltage divider network equations, one for each of the 2 TGVs. The R_{on}'s are obtained by dividing, e.g., $(V_{DD}\text{-}TGV_{1a})$ and TGV_{1a} by the current I_{1ab} measured using the off-chip power supply. The two voltage divider subexpressions will be referred to as the **1st term** and the **2nd term** subsequently within which the R_{on} ratios at 25°C, 1.2V (the enrollment corner) are referenced. Note that R_{1a} and R_{2a} from the 1st terms are designated as R_{1a}' and R_{2a}' in the 2nd terms because these resistances are a function of the drain-to-source voltage (V_{DS}), which are different in the two ratios as discussed below. The X and Y terms are defined as the percentage changes in the R_{on} of the associated transistors at a specific TV corner with respect to the R_{on} measured at enrollment.

The magnitude of the R_{on}'s are determined primarily by the mode of operation of the two transistors. NFET transistors whose sources are connected to the on-chip GND grid, e.g., TG_{1b}, TG_{1c}, TG_{2b} and TG_{2c} operate in the linear region. This is true because the V_{DS} for these transistors are in the range of 200 mV while V_{GS} is equal to V_{DD}, e.g., 1.2 V. The design manual specifies that threshold voltages are > 300 mV in this 90 nm technology. There-

418

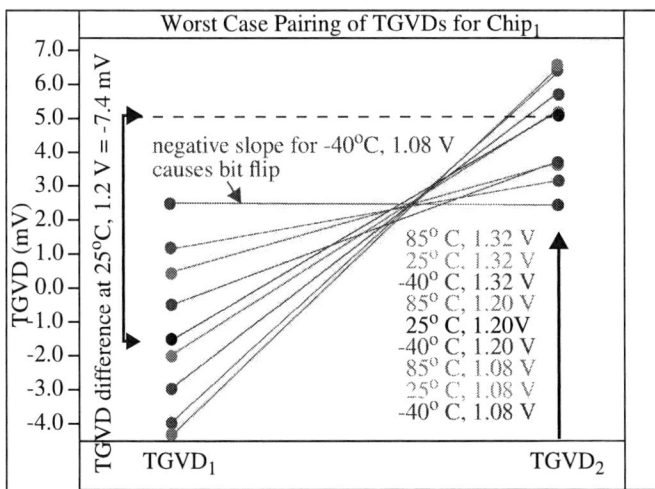

Fig. 16. 'Worst-case' TGVD comparison for Chip₁, where worst case is defined to be the comparison that has the largest enrollment TGVD difference and a bit flip.

fore, $V_{DS} < V_{GS} - V_t$ indicating the operating mode is linear.

NFET transistors TG_{1a} and TG_{2a} on the other hand operate in saturation. This is true because the voltage drops across the R_w's are less than 300 mV (typical currents for I_{1ab} and I_{2ab} are approx. 180 uAs). Therefore, with V_{DD} at 1.2 V, V_{GS} is approx. (1.2 - 0.2) = 1.0 V while V_{DS} is, in the worse case, (0.9 - 0.2) = 0.7. Moreover, threshold voltages increase when the V_{SB} (source-to-substrate) is greater than 0, a condition that holds true for these NFET transistors. Therefore, $V_{DS} > V_{GS} - V_t$ indicating the operating mode is saturation.

The resistances given in Eqs. 1 and 2 will change as a function of TV conditions. If the percentage change in all R_{on}'s are identical, i.e., all X and Y are the same, then TV variations would not increase the number of bit flips that occur over the number introduced by measurement noise alone. This is not the case, however. Therefore, the R_{on}'s and the corresponding X and Y percentage change values from the equations must vary at different rates across the TV corners.

This characteristic of the NFET resistances is demonstrated using data from a special, worst-case, pairing of TGVDs. In particular, we analyze the pairing (from the 5,662,020 pairings described in Section 4) from Chip₁ that possesses the largest difference in the TGVDs at enrollment AND has a bit flip. This pairing defines the minimum threshold (see Section 3.5) that can be used to avoid bit flips across the 9 TV corners.

Fig. 16 shows the behavior of the two TGVDs used in this pairing. The 9 data points for each TGVD, one for each TV corner, are plotted as a vertical sequence under each TGVD labeled on the x-axis. Each of the points from TGVD₁ is line-connected with the point in TGVD₂ corresponding to the same TV experiment. If the sign of the difference TGVD₁ - TGVD₂ remains the same, then the set of lines would all have positive or all have negative slopes. Instead, they cross over and depict a near complete reversal in order. For example, the ordering from top-to-bottom of the points for TGVD₁ is opposite to the legend's ordering, which lists the TV corners in descending order according to voltage and then temperature, while the points for TGVD₂ are consistent with it. Note that

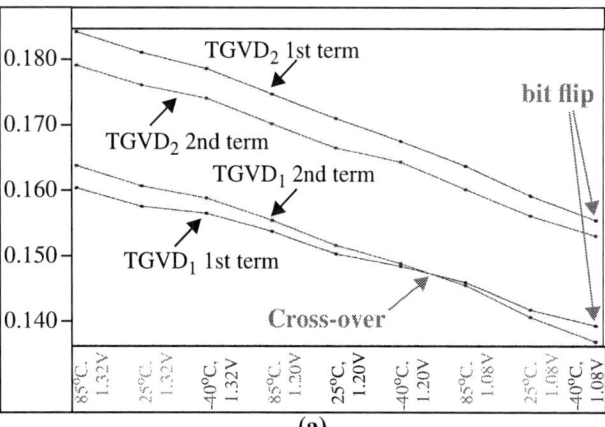

(a)

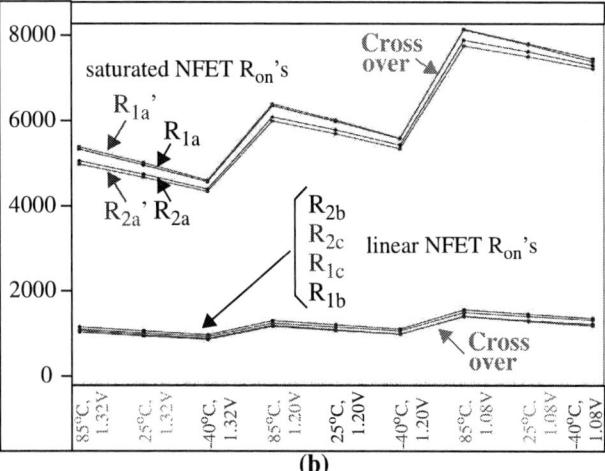

(b)

Fig. 17. Behavior of the a) 1st and 2nd terms and b) the individual R_{on}'s from Eqs. 1 and 2 across the TV corners for TGVD₁ and TGVD₂ given in Fig. 16.

the slope of the line associated with the -40°C, 1.08 V is negative while the others are positive. This condition reflects a bit flip, i.e., TGVD₁ > TGVD₂ at this TV corner while TGVD₁ < TGVD₂ in the others.

The behavior of the 1st and 2nd terms in Eqs. 1 and 2 as a function of TV corners are shown in Fig. 17(a), which plots the two terms for each TGVD as separate curves. Each curve consists of 9 points (one for each TV corner). Interestingly, all 4 terms decrease monotonically as TV decrease, which illustrate the self-compensation property of the NFET pair. Unfortunately, the rate at which the terms decrease, which is reflected in slope of the curves, is not constant. The larger difference in the slopes between the 1st and 2nd terms for TGVD₁ cause the curves to cross over and eventually introduce a bit flip at the -40°C, 1.08V corner. The curves in Fig. 17(b) plot the behavior of the individual R_{on}'s within the ratios of Eqs. 1 and 2. Although the R_{on}'s vary significantly with TV, especially for the saturated NFET R_{on}'s shown along the top of the figure, the corresponding changes in the R_{on}'s of the linear NFETs compensate for most, but not all, of the variations. In particular, the R_{on}'s for all 1.08V TV corners cross over for TGVD₁.

S5 Supplementary Material References

[15] B. Skoric, P. Tuyls, W. Ophey, "Robust Key Extraction from Physical Uncloneable Functions", Chapter in Applied Cryptography and Network Security, 2005.

RESP: A Robust Physical Unclonable Function Retrofitted into Embedded SRAM Array

Yu Zheng, Maryam S. Hashemian and Swarup Bhunia
Case Western Reserve University, Department of EECS, Cleveland, Ohio, 44106
{yu.zheng3, mxh460, skb21}@case.edu

ABSTRACT

Physical Unclonable Functions (PUFs) have emerged as an attractive primitive to address diverse hardware security issues in Integrated Circuits (ICs). A majority of existing PUFs rely on a dedicated circuit structure for generating chip-specific signatures, which often imposes concerns due to area/power overhead and extra design efforts. Furthermore, existing PUF-based signature generation cannot be employed to authenticate chips already in the market. In this paper, we propose RESP, a novel PUF structure realized in embedded SRAM array, a prevalent component in processors and system-on-chips (SOCs), with virtually no design modification. RESP leverages on voltage-depend memory access failures (during write) to produce large volume of high-quality challenge-response pairs. Since many modern ICs integrate SRAM array of varying size with isolated power grid, RESP can be easily retrofitted into these chips. Circuit-level simulation of 1000 chips using realistic process variation model shows high uniqueness of 49.2% average inter-die Hamming distance and good reproducibility of 2.88% intra-die Hamming distance under temperature $< 85°C$. The device aging effect, e.g. bias temperature instability (BTI), results in only 4.95% estimated unstable bits for ten-year usage.

Categories and Subject Descriptors

K.6.5 [**Security and Protection**]: Authentication

General Terms

Design, Security

Keywords

Hardware security, PUF, SRAM, Signature, BTI

1. INTRODUCTION

In recent years, Physical Unclonable Functions (PUFs) have been widely investigated as a security primitive of integrated circuits (ICs) in variety of applications such as Intellectual Property (IP) counter-plagiarism, chip authentication and embedded system security. PUFs have obvious advantages over traditional digital-key storage in a non-volatile memory (NVM). First, PUFs avoid the high cost of building tamper-resistant NVM system, since any invasive attack

Permission to make digital or hard copies of all or part of this work for personal or classroom use is granted without fee provided that copies are not made or distributed for profit or commercial advantage and that copies bear this notice and the full citation on the first page. To copy otherwise, to republish, to post on servers or to redistribute to lists, requires prior specific permission and/or a fee.
DAC'13, May 29 - June 07 2013, Austin, TX, USA.

may alter internal behavior of an IC leading to incorrect signatures [1]. Moreover, a PUF can produce a large amount of challenge-response pairs that are random and usually difficult to predict, which overcome the limitation of insufficient number (usually only one) of digital-key storage.

PUFs transform the inherent random variations in a manufacturing process (e.g. threshold voltage (V_{th}), channel length (L)) to variations in circuit-level parameters for random digital-key generation. A majority of existing PUFs require dedicated circuit structures [4–6]. Apart from the substantial cost in silicon area, their integration into a system-on-chip (SOC) design needs extra effort on the placement, routing and verification. On the other hand, a separate class of relatively few PUF implementations generates signature from existing on-chip structures, such as PUFs that exploit random mismatch in inner node voltages of memory elements (e.g. SRAM or Flip-Flops) [7–10]. This class of PUFs, however, often requires considerable modifications of the original design. For example, the PUF in [7] adds four extra transistors into each 6-T SRAM cell as twisted NOR gates for initializing the inner voltages, and a programmable word line duty cycle controller is inserted into the SRAM array in [9]. Although the PUF in [8] requires no such modification, the residual charge in the SRAM cell severely impacts the power-up randomness of signature, thus compromising the quality of signature. The intrinsic PUF uses the power-up state of flip-flops in FPGA, however, it requires altering the bit configuration procedure to retain the values and read it out [10]. Moreover, a common disadvantage of these PUFs is small challenge-response space and Shannon entropy. The SRAM cells only generate a signature with the entropy of 1 bit/cell in the best case.

In this paper, we propose *RESP*, **R**etrofitted **E**mbedded **SRAM P**UF. Unlike existing PUF structures, which implement a PUF either through insertion of a dedicated PUF IP in a design [4–6] or through design modifications of on-chip structures [7–10], RESP utilizes voltage scaling induced access failures in SRAM array to generate large set of robust signatures. It leverages on the fact that modern ICs usually adopt separate power delivery network (PDN) for the functional blocks and embedded memory [13]. Signature generation in RESP can be accomplished for practically any IC in the market or large volume of legacy ICs with pins to externally control the supply voltage of embedded SRAM. Furthermore, the idea of RESP can be applied during any chip design process by creating separate voltage (VDD) island for SRAM in a die to integrate PUF into SRAM array.

RESP exploits the fact that for a set of SRAM cells in an array, under scaled supply voltage, write access failure occurs only in specific cells in the set depending on device-level process variations. After an initial value is written to this set at the scaled supply, the content in the SRAM cells can be read out to create a random signature for a chip. Fig. 1 illustrates this approach for a typical two-dimensional SRAM array. We first initialize the cells C1,

C2 and C3 as '1' to all the chips. Then write an initial value, say all '0' into them under reduced VDD. Next, we read out the values at nominal voltage to generate chip-specific random signatures. Large choice of initial values and voltage levels enables generation of large volume of random signatures from a chip.

We have presented detailed description of the signature generation process. To improve the robustness of signature with temporal variations (e.g. temperature, aging effect), we propose an iterative procedure that satisfies a preset write failure probability by adjusting VDD. We have studied the effectiveness of RESP with extensive simulation using realistic process variation model (for 45nm process) with 15% inter- and 10% intra-die standard deviations on V_{th}. The variations on L and other device parameters are represented as additional contribution to V_{th} variation [14]. We observe that RESP can achieve high uniqueness (49.2% average inter-die Hamming distance) and reproducibility (2.88% unstable bits below 85 °C). We also show that the entropy can be up to 6 bits/cell. The aging-induced V_{th} increase of 17 mV in PMOS due to negative bias temperature instability (NBTI) effect with ten-year usage results in only 4.95% flipped bits. Furthermore, we show that multiple measures can enhance the robustness of signature under simultaneous switching noise (SSN) of VDD.

The remainder of the paper is organized as follows. Section 3 presents the basic structure of an SRAM array and its failure modes. Section 4 introduces the RESP methodology. Simulation results on uniqueness, robustness and aging effect are shown in Section 5. Section 6 analyzes the signature stability under VDD variations. The discussions are in Section 7. Section 8 concludes and provides future directions.

2. RELATED WORK ON PUF

A majority of PUFs rely on dedicated circuit structures. Optical PUF depends on laser speckle fluctuation of coherent radiation to disordered media [2]. In the coating PUF, a sensor array on the top metal layer measures the unit capacitance of coating of random dielectric particles for signature [3]. However, both of them resort to the special material or equipment. To produce signature, the delay-based PUFs compare the delay values of two similar components under process variations [4] [5]. The butterfly PUF employs two cross-coupled latches to construct a cell [6]. Besides the spatial correlation among components that compromises entropy, the dedicated circuits incur large area overhead.

Another class of PUFs consider on-chip resources to reduce area overhead. In [11], the standard FLASH interface measures the distribution of transistor threshold voltage as signature of each FLASH chip. The intrinsic PUF collects the random power-up content of flip-flops as signature in FPGA by modifying the configuration procedure [10]. The symmetric structure of SRAM cell become unbalanced under process variations that enables generating random content [7] [8]. The PUF in [7] adds another four transistors into each 6T SRAM cell. In [8], the entropy of each cell may be greatly reduced due to residual charge in the cross-coupled inverters. *MECCA* PUF incorporates the word-line duration into the challenge that induces write failure in SRAM [9]. *ScanPUF* obtains the signature by exploiting variations in scan path delay [12]. These PUFs, however, do not completely eliminate design modifications and/or compromise quality of signature.

3. BACKGROUND AND PRELIMINARIES

3.1 SRAM Array Structure for RESP

The SRAM organization in Fig. 1 is considered for the

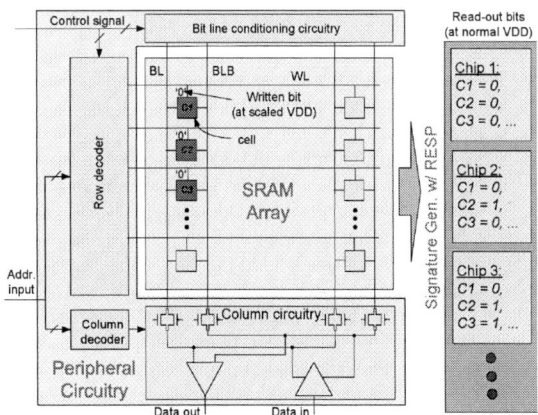

Figure 1: Conventional SRAM architecture.

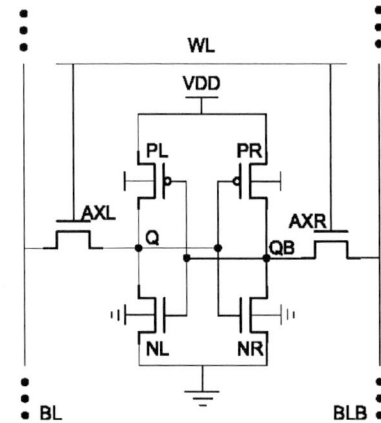

Figure 2: Structure of a typical 6-T SRAM cell [14].

proposed PUF due to its wide-spread use as embedded memory. It is comprised of a 2-D cell array and the peripheral components. The cell array is the aggregation of 6-T cells as shown in Fig. 2, while the periphery includes row/column decoder, bitline conditioning and column circuitry.

The read and write of SRAM are divided into the pre-charge and evaluation phase. In the pre-charge phase, BL and BLB in Fig. 2 are both charged to high level ('1') and will be floating high at the beginning of evaluation phase. Then the word line (WL) is raised to move the transistor AXL and AXR to 'ON' state, connecting BL (or BLB) to node Q (or QB). For reading '1' (Q='1' and QB='0'), since NR is stronger than AXR and QB remains below the trip point of inverter PL-NL (readability), BLB is pulled down to a lower voltage (e.g. 0.1VDD) which is sensed by the sense amplifier. During writing '0' into a cell storing '1', since PL is weaker than AXL (writability), Q is pulled lower than the trip point of inverter PR-NR.

The peripheral and cell array either share the same VDD or have distinct ones to improve writability or standby power [15] [16]. In the proposed approach, we consider single power supply, since the double power supply is equivalent to making the two voltages equal.

3.2 SRAM Failure Modes

Under process variations, an SRAM array experiences the following four types of parametric failures [14].

1. Read Failure: It occurs if the data stored in SRAM cells flip during read. For a cell storing '1', it flips if the voltage on QB rises to a value higher than the trip point of the inverter PL-NL.

2. Write Failure: It occurs if a cell cannot be flipped while writing a different value due to insufficient WL duration or weak writability.

3. Access Failure: If the voltage difference between BL and BLB at the time of sense amplifier firing is lower than the required offset voltage, it leads to incorrect sensing of stored content.

4. Hold Failure: The destruction of cell content at a lower VDD is known as hold-stability failure. For a cell storing '1', higher trip point of PR-NR makes it easier to flip thus requiring higher retention voltage.

3.3 Choice of Failure Mode

Appendix B shows the write and access failures are more sensitive to VDD scaling. The six transistors in each SRAM cell can determine the occurrence of write failure, while the access failure additionally relies on sense amplifier that influences the probability of sensing '0' (or '1') for a column. For example, if the sense amplifier is 'strong' under process variation, it can correctly sense small voltage difference between BL and BLB. Hence, the bits in the column are highly likely to be read out correctly providing little or no random information. Hence, we focus on write failure in RESP, although the approach can be extended to other mode of failures.

3.4 Retrofitting PUF into SRAM

Definition 1: A PUF is *retrofitted* into a hardware component, if the component is treated as a black box and no modification (e.g. structure, in/out ports) is incorporated into it. The extra hardware resources outside the component used to support retrofitting a PUF are referred to as retrofitting overhead.

Under nominal voltage and temperature, the failure rate of SRAM cells is expected to be rather low (e.g. typically less than 10^{-8} [14]) with little random information. In RESP, we deliberately reduce the VDD to induce failures during memory access for write. It allows us to retrofit a PUF into memory. Moreover, unlike most PUFs integrated into SOC die as a separate circuit, RESP facilitates obtaining responses to challenge vectors by using existing on-chip resources for memory read/write operations.

Next, we discuss the impact of VDD scaling on peripheral circuits as in Fig. 1 that provides the control signals (e.g. WL) for synchronous and asynchronous SRAM. RESP may not be effective in asynchronous SRAM due to timing errors (e.g. WL rising up in the evaluation phase) caused by the large and unbalanced skewing of control signals at scaled VDD. For synchronous SRAMs, the clock period and duty cycle influence the pre-charge and evaluation phase. However, for synchronous SRAM, when path delay increases beyond clock period due to VDD scaling, timing violation may occur for some cells during write (Appendix A).

3.5 Voltage Scaling Mechanism in RESP

RESP can be realized through off-chip or on-chip control of memory voltage. If a manufactured chip reserves a separate pin for the VDD of embedded SRAM, we can control the supply voltage of SRAM outside the chip using a supply tuning block, which incurs no on-chip design modification. In this case, we need test equipment for external control of supply voltage. It would make signature generation using RESP possible for many legacy chips with separate PDN and dedicated supply pin for the embedded memory. On the other hand, during design process of a new chip, RESP can be realized by deliberately incorporating separate supply pin and PDN for the memory as well as an on-die memory voltage tuning circuit to facilitate signature generation. Since,

majority of modern SOC designers typically incorporate separate PDN for embedded memory, the design overhead for RESP is expected to be low, as discussed in Section 7.1.

4. METHODOLOGY FOR IMPLEMENTING RESP

4.1 Voltage Scaling Driven by Write-failure Prob.

Write duration T_{WR} of writing '0' (similarly writing '1'), which is the interval that node Q is discharged to the voltage of trip point V_{TRR} of PR-NR inverter, can be expressed as:

$$T_{WR} \doteq \begin{cases} |\int_{V_{TRR}}^{VDD} \frac{C_Q(V_Q)dV_Q}{I_{dsPL}(V_Q)-I_{dsAXL}(V_Q)}| & if\ V_{WRL} < V_{TRR} \\ \infty & otherwise \end{cases}$$

$$(1)$$

where C_Q is the net capacitance at Q, and V_{WRL} is the voltage on Q during write, which is determined by the strength of AXL and PL. In the CMOS processes, T_{WR} is affected by both inter-die and intra-die V_{th} variations. The inter-die V_{th} variation is shard by the transistors on the same die and varies from die to die. The intra-die V_{th} is a random shift on each transistor within a die and the standard deviation $\sigma_{th,intra}$ of a gate with L and W is modeled as [14]:

$$\sigma_{th,intra} = \sigma_{th0,intra}\sqrt{\frac{W_{min}L_{min}}{WL}} \qquad (2)$$

where $\sigma_{th0,intra}$ is the standard deviation of intra-die V_{th} for a gate of minimum width W_{min} and length L_{min} in a specific CMOS process. Fig. 3(a) shows the distribution of T_{WR} in 200 chips (128 memory cells for each) following the Gaussian-distributed V_{th} variation ($\sigma_{th,inter} = 15\%$ and $\sigma_{th0,inter} = 10\%$) with $VDD = 1.00$ V. The probability density function (PDF) of T_{WR} with a long tail is modeled as a noncentral F function [14]. The write failure probability P_{WF} is

$$P_{WF}(t_{wp}) = Pr\{T_{WR} > t_{wp}\} = 1 - \Phi(t_{wp}) \qquad (3)$$

where $\Phi(\cdot)$ is the cumulative density function (CDF) of T_{WF} and t_{wp} is the WL duration. We assume $t_{wp} = 0.7$ ns as the minimum WL duration in Fig. 3(a). Fig. 3(b) illustrates the samples of T_{WR} in the fast and slow chips for the scaled voltage $VDD = 0.70$ V. All the SRAM cells in those chips simultaneously experience (or avoid) write failure for $t_{wp} = 0.7$ ns, which implies no random information. Hence, the distinct inter-die V_{th} deviations lead to different P_{WF} of the chips under identical VDD. We specify the VDD around $P_{WF} = 0.5$ at temperature 25 °C without aging effect. Fig. 4(a) shows that P_{WF} increases to 0.90 on average, when the temperature rises to 85°C at the nominal VDD. In Fig. 4(b), if V_{th} of PMOS is increased by 0.20 V due to the aging effect such as NBTI, P_{WF} becomes 0.65 on average. As P_{WF} reflects the number of zeros in the signature, each chip needs its own VDD to achieve a specific P_{WF}. The choice of VDD also should consider temperature fluctuation and aging effect.

4.2 RESP Architecture and Procedure

The challenges of retrofitted PUF include the cell locations, pre-stored values ('0' or '1') and VDD tracking the process and temporal variations, implying more challenge-response pairs than other SRAM PUFs using power-up state.

Fig. 5(a) explains the architecture of RESP, including a programmable reference generator and a voltage regulator, which is similar to the VDD tuning structure in [17]. Assume the reference generator outputs the lower voltage set $[V_1, V_2, ..., V_n]$ satisfying $V_1 > V_2 > ... > V_n$. The voltage we need is output of an n:1 multiplexor and stabilized by a regulator. Note Fig. 5(a) shows only the functional blocks for

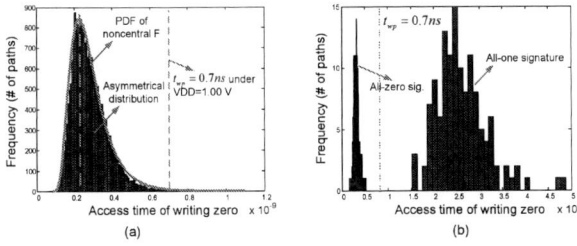

Figure 3: Histogram of T_{WR} for writing '0': (a) $VDD = 1.00$ V; and (b) fastest and slowest chips for $VDD = 0.70$ V.

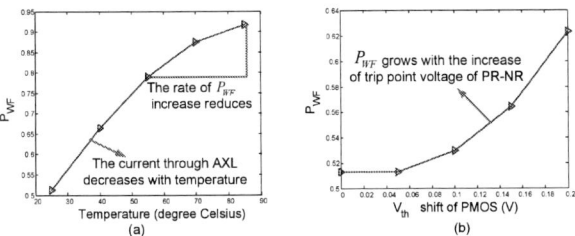

Figure 4: P_{WF} variation with: (a) temperature; and (b) NBTI-induced aging effect.

RESP that can be implemented either on the PCB (off-chip scheme) or on the die (on-chip scheme). Fig.5(b) illustrates the two steps of RESP: (1) preparation, (2) iterative VDD adjustment to find the acceptable signature for each chip. The preparation specifies the address of SRAM cell to generate signature with $P_{WF} = p_{wf}$, as well as the acceptable error Δp between p_{wf} and p'_{wf}. Note p_{wf} is a preset value of P_{WF} in RESP and each chip can have its own p_{wf}; p'_{wf} is the measured value of P_{WF} for the chip in each iteration. Δp is determined by the resolution of VDD. With higher resolution, we can set smaller Δp. Assume at least one voltage among $[V_1, V_2, ..., V_n]$ satisfies $|p_{wf} - p'_{wf}| \leq \Delta p$ found by iteratively tuning VDD. In each iteration, write '0'(or '1') into the SRAM cells successfully at the normal supply voltage $VDD = V_{norm}$. Then write '1'(or '0') into the same address under $VDD = V_i \leq V_{norm}$, where i is initialized as $\lfloor n/2 \rfloor$. Read out the contents with $P_{WF} = p'_{wf}$. If $|p'_{wf} - p_{wf}| \leq \Delta p$, accept it as a signature and stop; if $p'_{wf} - p_{wf} > \Delta p$ (or $p'_{wf} - p_{wf} < -\Delta p$), $i \leftarrow i - 1$ (or $i \leftarrow i + 1$) and repeat a new iteration. The procedure is amenable for automation. In addition, signature generation can be performed at run time using the idle SRAM blocks.

p_{wf} helps us to obtain unique and robust signatures from a chip. Appendix E shows that, if an attacker knows part of the signature, revealing p_{wf} may reduce the min-entropy of unknown part. To address it, the system can randomly choose a p_{wf} among multiple candidates when the signature is needed.

5. SIMULATION RESULTS AND ANALYSIS

RESP is evaluated for uniqueness and robustness of 128-bit signature for $m = 1000$ 32KB SRAM prototypes in the Monte-Carlo Hspice simulation using PTM 45nm CMOS Process, with V_{th} variation of $\sigma_{th,inter} = 15\%$ and $\sigma_{th0,intra} = 10\%$ [18]. VDD is chosen between 0.67 V and 0.87 V with the tuning step 0.01 V. VDD is assumed ideal and the noise impact is discussed in Section 6.

5.1 Uniqueness Analysis

For authentication, the signatures need to be as different as possible from each other. We use the Hamming distance (HD) to evaluate the signature distance quantitatively. The uniqueness of signature is evaluated among the chips for a

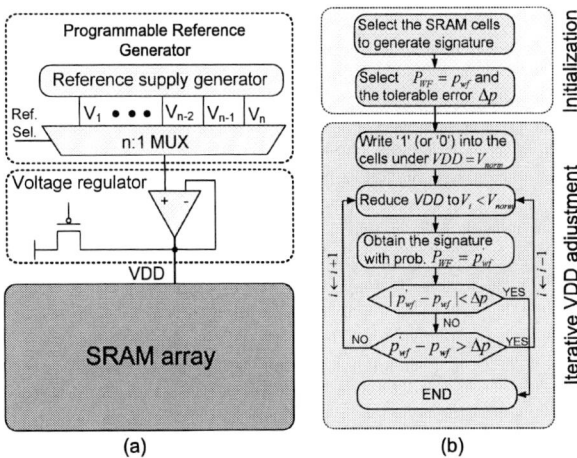

Figure 5: (a) Hardware architecture of RESP; and (b) Signature generation procedure.

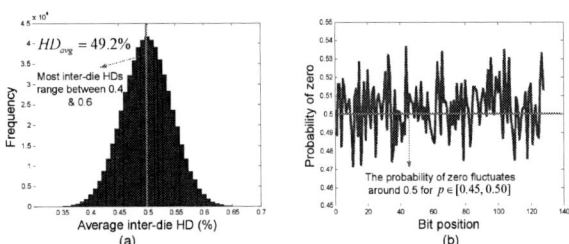

Figure 6: For $p_{wf} \in [0.40, 0.60]$, (a) inter-die HD of 1000 chips; and (b) probability of zero in each bit position.

specific range of p_{wf}. We select an arbitrary value of p_{wf} within $[0.40, 0.60]$). Fig. 6(a) shows about 50% (64 bits) inter-die HD, i.e. one-bit entropy for each SRAM cell. The probability of zero for each bit in Fig. 6(b) is around 0.5, which is skewed from 0.5 to 0.4 and 0.6, respectively for $[0.50, 0.70]$ and $[0.30, 0.50]$, as shown in Fig. 7. Assuming $HD_{i,j}$ is the HD between chip i and chip j, the average HD for m chips, denoted by HD_{avg}, is calculated as [20]

$$HD_{avg} = \frac{2}{m \cdot (m-1)} \sum_{i=1}^{m-1} \sum_{j=i+1}^{m} HD_{i,j} \qquad (4)$$

When $p_{wf} \in [0.40, 0.60]$, HD_{avg} is 49.2% near the theoretical value 50% ($128 \times 0.5 = 64$ bits). For $[0.50, 0.70]$ and $[0.30, 0.50]$, HD_{avg} is reduced to 46.70% and 47.32% respectively. Since HD_{avg} is not affected severely as the chips choose different p_{wf}, the accuracy of tuning VDD can be equivalently reduced. The Shannon entropy is dependent on VDD selection, which is up to 6 bits/cell as described in Appendix C.

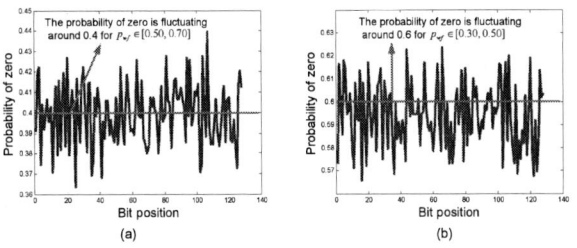

Figure 7: Probability of zero on each bit position for: (a) $p_{wf} \in [0.50, 0.70]$; and (b) $p_{wf} \in [0.30, 0.50]$.

423

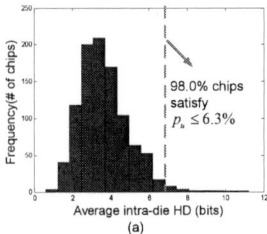
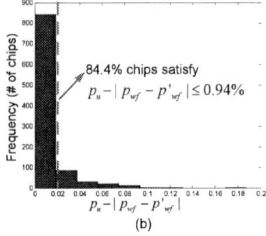

Figure 8: Distribution of: (a) p_u under temperature variation; and (b) $p_u - |p_{wf} - p'_{wf}|$ for $m = 1000$ chips.

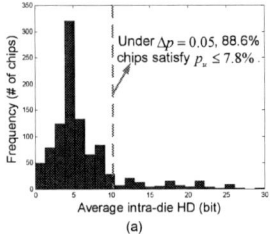
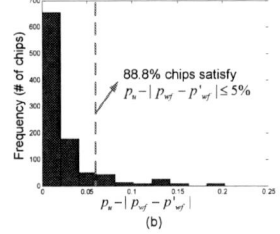

Figure 9: Distribution of: (a) p_u under NBTI; and (b) $p_u - |p_{wf} - p'_{wf}|$ for $m = 1000$ chips.

5.2 Robustness under Temperature Variation

The robustness measures the signature reproducibility under non-ideal environment such as temperature variation, which results in unstable bits. Considering precise value of $P_{WF} = p_{wf}$ may be difficult to achieve due to environmental fluctuations and limited VDD resolution, we accept an error of $|p_{wf} - p'_{wf}| \leq \Delta p$ to stop tuning VDD in Fig. 5(b). Under this condition, the overall ratio of flipped bits is p_u for a given temperature. Therefore, we can derive $p_u \geq |p_{wf} - p'_{wf}|$ for each chip. When the temperature rises from 25 °C to 85 °C at the interval of 15 °C, p_u is nearly 2.88% on average for $\Delta p = 0.05$. In particular, Fig.8(a) shows 95.4% chips with $p_u \leq 6.3\%$. $p_u - |p_{wf} - p'_{wf}|$ can correspond to the flipped bits under the unlimited VDD resolution yielding $p_{wf} = p'_{wf}$. Fig. 8(b) shows the histogram of 1000 chips for $p_{wf} \in [0.40, 0.60]$ and $\Delta p = 0.05$. More than 84.4% chips satisfy $p_u - |p_{wf} - p'_{wf}| \leq 0.94\%$. It means 84.4% chips would have only 0.94% unstable bits under $p_{wf} = p'_{wf}$. Hence, the non-zero $|p_{wf} - p'_{wf}|$ may contribute significantly to p_u. The unstable bits can be tolerated by the post-processing procedures [21].

5.3 Aging Effect

In the nano regime, the V_{th} degradation of PMOS due to NBTI plays a major role in temporal performance degradation of circuits. If V_{th} of PMOS PL and PR is increased by 17.3 mV after a simulated ten-year usage [19], p_u is 4.95% on average for $\Delta p = 0.05$ with 25 °C. Furthermore, Fig. 9(a) shows that 88.6% chips achieve $p_u \leq 7.8\%$. In Fig. 9(b), the histogram of 1000 chips shows that 88.8% chips satisfy $p_u - |p_{wf} - p'_{wf}| \leq 5\%$. We can estimate the proportion of unstable bits caused by NBTI in the design cycle and choose more powerful error-correcting codes to tolerate them in the fuzzy extractor [21].

6. STABILITY UNDER SUPPLY FLUCTUATION

6.1 Signature with Supply Fluctuation

Simultaneous switching noise (SSN) from current variation of the switching logic generates the voltage spikes on the supply [22]. Therefore, the VDD across the selected SRAM cells may deviate from the desired value. In the

Monte-Carlo simulation, we specify inter-die V_{th} deviation percentage as low ($\pm 5\%$), medium ($\pm 15\%$) and high ($\pm 25\%$) respectively, and set $\sigma_{th0,intra}$ as 10% and 20% to obtain the unstable percents in Table 1 for $p_{wf} \in [0.4, 0.6]$.

Table 1: Unstable % of signature bits under different process and voltage variations

Inter-die corner, $\sigma_{th0,intra}$	Supply shift (mV)				
	± 2	± 4	± 6	± 8	± 10
low, 10%	2.85%	5.76%	8.58%	11.46%	14.09%
low, 20%	1.57%	3.09%	4.71%	6.36%	7.93%
medium, 10%	2.82%	5.54%	8.23%	10.86%	13.40%
medium, 20%	1.63%	3.08%	4.49%	6.16%	7.62%
high, 10%	2.74%	5.53%	8.20%	10.88%	13.46%
high, 20%	1.55%	3.05%	4.78%	6.24%	7.56%

In Table 1, the number of unstable bits grows with the fluctuation of VDD. Up to 10 mV, the worst case is 14.09% flipped bits in a signature on the low inter-die variation corner. However, the number of unstable bits decrease with the increasing intra-die variation. For example, for 5% inter-die deviation, when $\sigma_{th0,intra}$ is increased to 20% from 10%, the unstable percent becomes 5.79%. It is reasonable, since the larger $\sigma_{th0,intra}$ can expect more chances to have the cells of large $|T_{WR} - t_{wp}|$. With the same $\sigma_{th0,intra}$, the inter-die shift makes negligible impact on the number of unstable bits, because VDD is tuned to track the inter-die corner for p_{wf}. The aggressive scaling of CMOS process leads to increase in intra-die variation (e.g. random dopant fluctuation) resulting in better tolerance of supply fluctuation.

6.2 Stable Signature Extraction through Multiple Measurements

Considering the current variation at each clock cycle, the fluctuation of VDD denoted by $\Delta V(t)$, can be simplified as a stochastic process within $[V_r - \Delta V_1, V_r + \Delta V_1]$ $(\Delta V_1 > 0)$, where V_r is the desired value. Two assumptions are made about $\Delta V(t)$:

1. It is the ergodicity process that $\Delta V(t_0)$, $\Delta V(t_1)$,..., can traverse all values in $[V_r - \Delta V_1, V_r + \Delta V_1]$.

2. $\Delta V(t_i)$, $i = 0, 1, ...$ is statistically independent and follows the identical distribution.

Further, we model $\Delta V(t)$ as [22]

$$\Delta V(t) = \Delta V_1 sin(\omega t + \theta) \qquad (5)$$

where ω is the oscillation frequency and θ is the initial phase. Assume the range $[V_r - \Delta V_e, V_r + \Delta V_e]$ yields the acceptable intra-die HD as Table. 1. Based on Appendix E, within N times signature measurements, the probability P_{hit} that the VDD is sampled in $[V_r - \Delta V_e, V_r + \Delta V_e]$ is

$$P_{hit} = 1 - (1 - (2/\pi)arcsin(\Delta V_e/\Delta V_1))^N \qquad (6)$$

P_{hit} relies on N and the ratio $\Delta V_e/\Delta V_1$ as shown in Fig. 10. For ΔV_e takes up to half of V_1 (small noise or high fluctuation tolerance), $N = 10$ achieves the high probability (approximately 1) to sample VDD within $[V_r - \Delta V_e, V_r + \Delta V_e]$. When $\Delta V_e/\Delta V_1 = 0.1$, N becomes 50 for $P_{hit} \approx 0.9$. Hence, the high supply noise needs a large N. Correspondingly, the multiple measurements under the same VDD need be incorporated into the procedure of RESP in Fig.5(b).

7. DISCUSSIONS

7.1 Hardware Overhead Analysis

If a chip provides a pin to control the VDD of embedded SRAM array from outside, no modification is carried out and the on-chip retrofitting overhead is zero. Or if the embedded SRAM has the VDD self-tuning structure as [17],

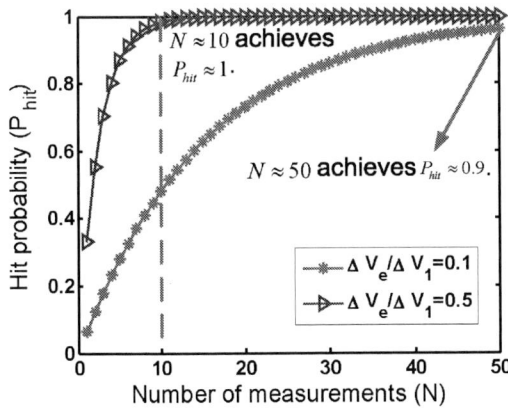

Figure 10: P_{hit} **for different** $\Delta V_e / \Delta V_1$.

RESP can be realized in it directly, resulting in virtually zero retrofitting overhead when no more components are needed to increase the number of references voltages. In cases of SOCs equipped with multiple voltage islands, we can use the on-chip voltage regulators during design process to minimize overhead. Hence, the retrofitting overhead would be primarily due to a multiplexor to choose specific voltage level. Moreover, such an approach can also be employed to reduce the SRAM power through dynamic voltage fluctuation as in [17].

7.2 Resolution of Supply Voltage Adjustment

In RESP, VDD resolution (i.e. the steps at which the VDD is adjusted), represents a trade-off between number of signatures (due to varying challenge vectors) and robustness of signatures. In our simulation, we have observed that the voltage step of 0.01 V provides acceptable value of robustness. Increasing voltage levels compromises reproducibility of signature under temporal voltage variations. However, the multiple-measurement approach as described in Section 6 can be used to achieve finer voltage steps leading to increased signature space.

8. CONCLUSION AND FUTURE WORK

We have presented RESP, a methodology to retrofit PUF into embedded SRAM array, a prevalent on-chip structure, by exploiting voltage scaling induced random write failures in SRAM cells. Using extensive simulations under realistic process variations, we have shown that it can generate large set of unclonable signatures. These signatures provide high entropy (since vast set of challenge-response pairs can be generated with selection of cell locations and scaled supply voltages), good reproducibility, and high stability under temporal fluctuations in voltage/temperature as well as aging effects. Since embedded SRAM is an integral component of modern processors and most SOCs, and often a memory core is associated with isolated supply grid/pin, RESP can be retrofitted to existed chips not designed with PUFs. Furthermore, RESP may provide the advantage of virtually zero hardware and design overhead for these chips.

Future work may exploit other forms of memory failures. In particular, hold or data retention failure can be used to generate random signatures. In addition, one can also target application of RESP into field programmable gate array (FPGA) devices.

9. REFERENCES

[1] J. R. Anderson, G. M. Kuhn, "Low cost attacks on tamper resistant devices," *IWSP*, 1997, pp. 125-136.

[2] R. Pappu, "Physical one-way functions," PhD thesis, Massachusetts Institute of Technology, 2001.

[3] P. Tuyls, G. Schrijen, B. Skoric, "Read-proof hardware from protective coatings," *CHES*, 2006, pp. 369-381.

[4] D. Lim, *et al*, "Extracting secret keys from integrated circuits," *IEEE Trans. VLSI Syst.*, vol. 13, no. 10, pp. 1200-1205, 2005.

[5] G. E. Suh, S. Devadas, "Physical unclonable functions for device authentication and secret key generation," *DAC*, 2007, pp. 9-14.

[6] S. Kumar, *et al*, "The butterfly PUF: protecting IP on every FPGA," *HOST*, 2003, pp. 2809-2825.

[7] Y. Su, J. Holleman and B. Otis, "A 1.6J/bit 96% stable chip ID generating circuit using process variation," *ISSCC*, 2007, pp. 406-611.

[8] D. E. Holcomb, W. P. Burleson and K. Fu, "Power-up SRAM state as an identifying fingerprint and source of true random numbers," *IEEE Trans. on Computers*, vol. 58, no. 9, pp. 1198-1210, Sep., 2009.

[9] A. R. Krishna, *et al*, "MECCA: a robust low-overhead PUF using embedded memory array," *CHES*, 2011, pp. 407-420.

[10] R. Maes, P. Tuyls and I. Verbauwhede, "Instrinsic PUFs from flip-flops on reconfigurable devices," *WISSec*, 2008.

[11] Y. Wang, *et al*, "Flash memory for ubiquitous hardware security functions: true random number generation and device fingerprints," *SP*, 2012, pp. 33-47.

[12] Y. Zheng, A. Krishna and S. Bhunia, "ScanPUF: robust ultralow-overhead PUF using scan chain," *ASP-DAC*, 2013.

[13] D. E. Lackey, *et al*, "Managing power and performance for system-on-chip designs using voltage islands," *ICCAD*, 2002, pp. 195-202.

[14] S. Mukhopadhyay, H. Mahmoodi, K. Roy, "Modeling of failure probability and satistical design of SRAM array for yield enhancement in nanoscale CMOS," *IEEE TCAD*, vol. 24, no. 12, pp. 1859-1880, 2005.

[15] A. Kawasumi, *et al*, "A single-power-supply 0.7V 1GHz 45nm SRAM with an asymmetrical unit-β-ratio memory cell," ISSCC, 2008, pp. 382-383.

[16] K. Zhang, *et al*, "A 3-GHz 70-Mb SRAM in 65-nm CMOS technology with integrated column-based dynamic power supply," *IEEE JSSC*, vol. 41, no. 1, pp. 146-151, 2006.

[17] Y. Lai, S. Huang and H. Hsu, "Resilient self-VDD-Tuning scheme with speed-margining for low-power SRAM," *IEEE JSSC*, vol. 44, no. 10, pp. 2817-2823, 2009.

[18] Predictive Technology Model, http://ptm.asu.edu/

[19] L. Hong, *et al*, "Modeling of PMOS NBTI effect considering temperature variation," *ISQED*, 2007, pp. 139-144.

[20] A. Maiti, *et al*, "A large scale characterization of RO-PUF," *HOST*, 2010, pp. 94-99.

[21] Y. Dodis, L. Reyzin and A. Smith, "Fuzzy extractors: how to generate strong keys from biometrics and other noisy data," *EUROCRYPT*, 2004, pp. 523-540.

[22] M. S. Gupta, *et al*, "Understanding voltage variations in chips multiprocessors using a distributed power-delivery network," *DATE*, 2007, pp. 624-629.

[23] H. Nambu, *et al*, "A 1.8-ns access, 550-MHz, 4.6-Mb CMOS SRAM," *IEEE JSSC*, vol. 33, no. 11, pp. 1650-1658, 1998.

APPENDIX

A. INFLUENCE OF VOLTAGE REDUCTION ON PERIPHERAL CIRCUITS

The VDD scaling increases T_{WR} of SRAM cells to generate write failure for signature, as well as the latency of paths in the peripheral circuits (e.g. row decoder). Fig. 11 shows a typical structure of row decoder in the synchronous SRAM with clock Φ to separate the pre-charge ($\Phi = 0$) and evaluation phase ($\Phi = 1$) [23]. When $\Phi = 0$, the output of WL is low to make the access transistors in the off state. In addition, the signal propagation on the two paths (red and blue) marked in Fig.11 is required to be completed when $\Phi = 0$ in each cycle. However, if it does not occur when Φ becomes high to start evaluation phase, the WL may not rise up in that cycle; or produce a glitch shorter than the normal duration as shown in Fig. 12. The short glitch may result in the write failure for all the activated SRAM cells. To address it, the address can stay unchanged for more than one cycle to provide sufficient time for decoding. Fig.12 shows the WL duration can recover the nominal value for write in the second cycle. Since in each cycle BL and BLB are re-charged to '1', the multiple-cycle operations would not influence the write failure probability on the evaluation phase.

B. SENSITIVITY OF FAILURES UNDER VDD SCALING

We assume the peripheral circuits (e.g. row decoder and bit line conditioning) share the supply VDD_{peri} and SRAM cells have the supply VDD_{cell}. Simply, if $VDD_{peri} = VDD_{cell}$, it is considered as a single supply SRAM; otherwise a dual supply SRAM. Next, the sensitivity of four types of failures to the VDD_{peri} and VDD_{cell} scaling is analyzed. Fig. 13(a) shows the simplified structure related to read and access failure when reading '1'. Capacitor C needs to be discharged through node QB. However, the charge of capacitor C makes the voltage of QB rise to certain level (V_{rd}) transiently. The read failure occurs, if V_{rd} is higher than the trip point of PL-NL. Fig. 14(a) shows that V_{rd} goes up with VDD_{peri} in the Hspice simulation. Hence, the supply scaling on VDD_{peri} cannot aggravate the read failure. Still based on Fig. 13(a), considering the larger resistance on AXR with the smaller VDD_{peri}, the interval of minimum distance (e.g. 0.1VDD) between BL and BLB is augmented, which is verified in Fig. 14(b). As a result, more access failures can be expected.

P_{WF} increases with T_{WR} in (1). In Fig. 13(b) that shows the simplified structure when writing '0', T_{WR} is related to both VDD_{peri} and VDD_{cell}. The Hspice simulation in Fig. 15 shows that T_{WR} grows with VDD_{peri} reduction or VDD_{cell} augment. Moreover, compared with that when VDD_{cell} is changed from 1.00 V to 0.80 V, T_{WR} is increased

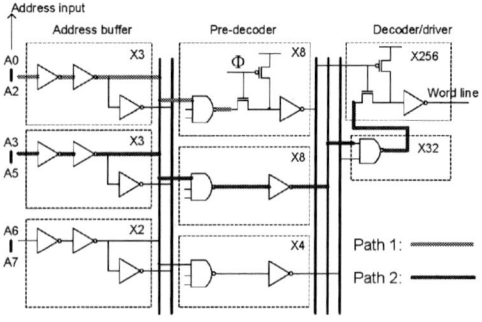

Figure 11: Typical architecture of row decoder in SRAM [23].

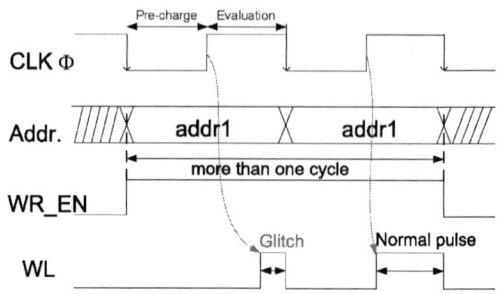

Figure 12: Waveform with more than one cycle decoding latency.

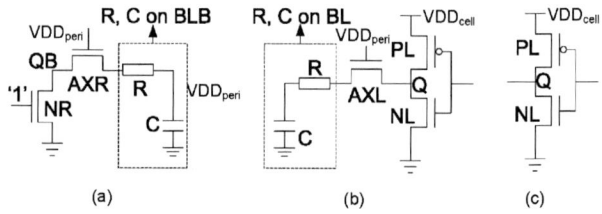

Figure 13: (a) Read and access failure; (b) write failure; and (c) hold failure.

much more with the same change scale on VDD_{peri}. Therefore, T_{WR} is more sensitive to the VDD_{peri} variation.

For the hold failure (Q='1' for example) as shown in Fig. 13(c), the voltage of Q gradually reduces to the standby VDD_{cell}, depending on the leakage current through NL below the trip point of inverter PR-NR determined by the device parameter (e.g. W, L and V_{th}). Hence, the exploitation of hold failure may severally rely on the length of standby duration. In addition, the residual charge on Q also impacts the randomness of stored value after VDD_{cell} is resumed to the nominal value. If the above two problems can be handled properly, exploiting the hold failure as PUF is also a potential choice. Based on the above analysis, the access failure and write failure are more proper to be employed to induce failure in the supply scaling for PUF.

C. DERIVATION OF SHANNON ENTROPY

The challenges of RESP include the cell locations, write contents ('0' or '1') and supply voltage VDD. In this section, we consider the maximum entropy achieved by tuning VDD for cells storing '1' when writing '0'.

First, we consider the ith set of supply voltage, denoted as vector $\underline{VDD}^{(i)} = (VDD_1^{(i)}, VDD_2^{(i)}, ..., VDD_m^{(i)})$, is provided for the overall m chips; specifically, $VDD_j^{(i)}$ is for chip j yielding $P_{WF} = p_j^{(i)}$. Denote $WF^{(i)}$ as the event of write under $\underline{VDD}^{(i)}$; $WF^{(i)} = 1$ means the write failure occurs, while $WF^{(i)} = 0$ corresponds to the successful write. Given the random intra-die process variation, the write failure happens on each cell of chip j with probability $p_j^{(i)}$. Assuming the uniform distribution for chip selection, the probability of choosing chip j, $j = 1, 2, ..., m$, is $1/m$. The write failure probability for chip j is $P(chip\,j, WF^{(i)} = 1) = (1/m) \cdot p_j^{(i)}$. We can obtain $P(WF^{(i)} = 1) = \sum_{j=1}^{m} P(chip\,j) \cdot P(WF^{(i)} = 1|chip\,j) = \sum_{j=1}^{m}(1/m) \cdot p_j^{(i)}$. Hence, the entropy of each cell under the ith supply vector is

$$H(WF^{(i)}) = -P(WF^{(i)} = 1) \cdot log_2(P(WF^{(i)} = 1)) - P(WF^{(i)} = 0) \cdot log_2(P(WF^{(i)} = 0)) \quad (7)$$

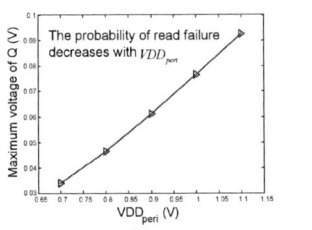
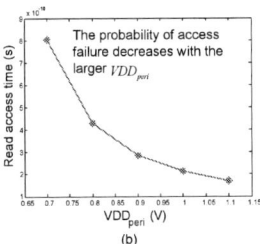

Figure 14: (a) Maximum voltage of Q during read; and (b) read access time.

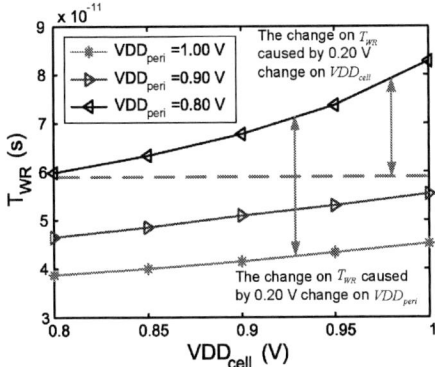

Figure 15: Change of T_{WR} for different VDD_{peri} and VDD_{cell}.

where $P(WF^{(i)} = 0) = 1 - P(WF^{(i)} = 1)$. $H(WF^{(i)})$ can be maximized as 1 bit when $P(WF^{(i)} = 0) = P(WF^{(i)} = 1) = 0.5$. Next, we supplement another supply vector $P(WF^{(i')})$, $i \neq i'$ and derive the joint entropy $H(WF^{(i)}, WF^{(i')})$. It can be obtained that $H(WF^{(i)}, WF^{(i')}) = H(WF^{(i)}) + H(WF^{(i')}|WF^{(i)})$. Considering both $H(WF^{(i')}|WF^{(i)})$ and $H(WF^{(i)})$ are no more than 1 bit, $H(WF^{(i)}, WF^{(i')}) \leq 2$ bit. Further,

$$H(WF^{(i')}|WF^{(i)}) =$$
$$- P(WF^{(i)} = 0) \cdot H(WF^{(i')}|WF^{(i)} = 0) \quad (8)$$
$$- P(WF^{(i)} = 1) \cdot H(WF^{(i')}|WF^{(i)} = 1)$$

Without losing generality, we first consider $VDD_j^{(i')} < VDD_j^{(i)}$, $j = 1, 2, ..., m$. Then the term $H(WF^{(i')}|WF^{(i)} = 1)$ in (8) become zero, due to $P(WF^{(i')} = 1|WF^{(i)} = 1) = 1$ under the reduced supply voltage. (8) can be simplified as $H(WF^{(i')}|WF^{(i)}) = P(WF^{(i)} = 0)H(WF^{(i')}|WF^{(i)} = 0)$. For $H(WF^{(i)}) = 1$ bit, $P(WF^{(i)} = 0)$ is set as 0.5. The maximum $H(WF^{(i')}|WF^{(i)} = 0)$ can be achieved by $P(WF^{(i')} = 0|WF^{(i)} = 0) = P(WF^{(i')} = 1|WF^{(i)} = 0) = 0.5$, which is shown in Fig. 16 as the condition $area1 = area2$ for chip j under $VDD_j^{(i')}$ and $VDD_j^{(i)}$. Hence, it can be derived that $H(WF^{(i')}, WF^{(i)})$ is maximized as 1.5 bits.

We further derive $H(WF^{(1)}, WF^{(2)}, ..., WF^{(d)})$ of d supply voltage sets, $\underline{VDD^{(1)}}, \underline{VDD^{(2)}}, \underline{VDD^{(3)}}, ...\underline{VDD^{(d)}}$, satisfying $VDD_j^{(1)} > VDD_j^{(2)} > VDD_j^{(3)} > ... > VDD_j^{(d)}$,

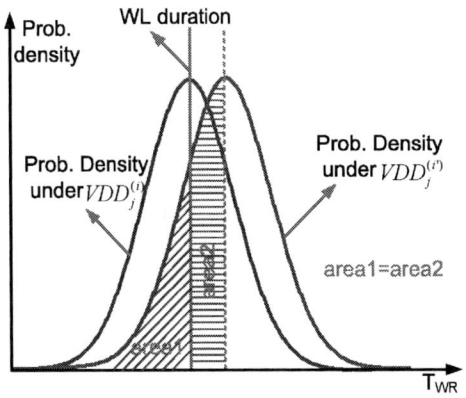

Figure 16: Illustration of maximizing $P(WF^{(i)} = 0)H(WF^{(i')}|WF^{(i)} = 0)$.

$j = 1, 2, ..., m$. As a result,

$$H(WF^{(1)}, WF^{(2)}, ..., WF^{(d)}) = H(WF^{(1)}) +$$
$$H(WF^{(2)}|WF^{(1)}) + H(WF^{(3)}|WF^{(1)}, WF^{(2)}) + \quad (9)$$
$$H(WF^{(4)}|WF^{(1)}, WF^{(2)}, WF^{(3)}) + ...$$

where $H(WF^{(3)}|WF^{(1)}, WF^{(2)}) = (0.5)^2 \times 1 = 0.25$ bit, $H(WF^{(4)}|WF^{(1)}, WF^{(2)}, WF^{(3)}) = (0.5)^3 \times 1 = 0.125$ bit. Hence, the entropy limit when $d \to \infty$ is

$$Lim_{d \to \infty} H(WF^{(1)}, WF^{(2)}, ..., WF^{(d)}) = \frac{1}{1 - 0.5} = 2 \; bits$$

By repeating the above procedure for $VDD_j^{(1)} < VDD_j^{(2)} < ... < VDD_j^{(d)}$, $j = 1, 2, ..., d$, extra 1-bit entropy is provided as $d \to \infty$. Therefore, the entropy upper bound is 3 bits per SRAM cell storing '1' when writing '0'.

For writing '1' into the cell storing '0', another 3 bits can be expected, since writing '0' and '1' deals with different transistors in the SRAM cell. Hence, each SRAM cell can ideally generate up to 6-bit entropy for write with tuning VDD.

D. DERIVATION OF P_{HIT} UNDER RESONATING CURRENT

The step current and resonating current are existent in real chips. The step current is caused by the sudden arouse of large logic gates from the standby or start-up state of chips. Since we can start to obtain the signature after such initialization process, the analysis is ignored in the paper. As periodic pulses, resonating currents have frequencies in the resonant band of the PDN [22]. Small current pulses may produce large peak-to-peak swings around the nominal voltage (voltage ripple) even in the steady state. According to Section 6.2, the supply fluctuation at time t_i, namely $\Delta V(t_i)$, $i = 0, 1, ...$ is assumed an ergodicity process. Statistically independent $\Delta V(t_i)$ follows the identical distribution with CDF $P_V(\cdot)$ in the range $[V_r - \Delta V_1, V_r + \Delta V_1]$. Assume the supply fluctuation range leading to small intradie HD is $[-\Delta V_e, \Delta V_e]$. For one measurement of signature at certain time, the probability to sample VDD within $[V_r - \Delta V_e, V_r + \Delta V_e]$ is $\Delta P_r = P_V(V_r + \Delta V_e) - P_V(V_r - \Delta V_e)$. Hence, for N times statistically independent measurements, such probability (P_{hit}) can be expressed as

$$P_{hit} = 1 - (1 - \Delta P_r)^N \quad (10)$$

From Section 6.2, the influence from resonating currents is

427

$\Delta V(t)$ modeled as

$$\Delta V(t) = \Delta V_1 sin(\omega t + \theta) \tag{11}$$

Where ΔV_1 and ω are the oscillation peak magnitude and frequency respectively; θ is the initial phase. From $-\Delta V_e \le \Delta V(t) \le \Delta V_e$, we can obtain the inequality as $-\Delta V_e/\Delta V_1 \le sin(\omega t + \theta) \le \Delta V_e/\Delta V_1$. The solutions of $sin(\omega t + \theta) = \Delta V_e/\Delta V_1$ and $sin(\omega t + \theta) = -\Delta V_e/\Delta V_1$ in the first and fourth quadrants are

$$t = \begin{cases} (arcsin(\Delta V_e/\Delta V_1) - \theta)/\omega & first\ quadrant \\ (arcsin(-\Delta V_e/\Delta V_1) - \theta)/\omega & fourth\ quadrant \end{cases} \tag{12}$$

Hence, the sample time t is in $[0, (arcsin(\Delta V_e/\Delta V_1) - \theta)/\omega]$, $[\pi + (arcsin(-\Delta V_e/\Delta V_1) - \theta)/\omega, \pi + (arcsin(\Delta V_e/\Delta V_1) - \theta)/\omega]$ and $[2\pi + (arcsin(-\Delta V_e/\Delta V_1) - \theta)/\omega, 2\pi]$. Further, we assume t uniformly locates in each cycle. As a result, $\Delta P_r = (2/\pi)arcsin(\Delta V_e/\Delta V_1)$. ΔP_r increases with $\Delta V_e/\Delta V_1$, which means the smaller supply fluctuation or better tolerance capability of SRAM cells (larger ΔV_e) can gain the larger P_{hit}. Further, (10) can be re-written as

$$P_{hit} = 1 - (1 - (2/\pi)arcsin(\Delta V_e/\Delta V_1))^N \tag{13}$$

E. POSSIBLE INFORMATION LEAKAGE OF REVEALING P_{WF}

In this section, we present a concise derivation to show the possible min-entropy reduction with the revealed p_{wf}, when the attackers know part of the signature. First, let's assume that for the m-bit signature, each bit is statistically independent. The min-entropy of each bit position excluding the first m_1 bits without revealing p_{wf} is

$$H_\infty(R_i) = -log_2(max(p_{wf}, 1 - p_{wf})) \tag{14}$$

$R_i = 1$ if write failure occurs on bit position i; otherwise $R_i = 0$, $i = m_1 + 1, m_1 + 2, ...m$. When the attacker has known p_{wf} and the write failure occurs on m'_1 bits (random variable M'_1) among the first m_1 bits, $mp_{wf} - m'_1$ bits should have write failure among the latter $m - m_1$ bits of the probability $p_c(m'_1) = (mp_{wf} - m'_1)/(m - m_1)$, $m_1 \ne 0$. In addition, $m'_1 \ge max(0, mp_{wf} - (m - m_1)) = m_l$ and $m'_1 \le min(m_1, mp_{wf}) = m_u$ are the lower and upper bound respectively. As a result, $H_\infty(R_i|m'_1) = -log_2(max(p_c, 1 - p_c))$. The conditional min-entropy comes from [21]:

$$\begin{aligned} H_\infty(R_i|M'_1) &= -log_2(\mathbb{E}_{m'_1 \leftarrow M'_1}[2^{-H_\infty(R_i|m'_1)}]) \\ &= -log_2(\mathbb{E}_{m'_1 \leftarrow M'_1}[max(p_c, 1 - p_c)]) \\ &= -log_2(\sum_{i=m_l}^{m_u} S_i \cdot max(p_c(i), 1 - p_c(i))) \end{aligned} \tag{15}$$

where $S_i = \frac{\binom{m_1}{i}(p_{wf})^i(1-p_{wf})^{m_1-i}}{\sum_{i=m_l}^{m_u} \binom{m_1}{i}(p_{wf})^i(1-p_{wf})^{m_1-i}}$. We consider $p_{wf} \ge 0.5$ (similar to $p_{wf} < 0.5$) for discussing the relationship between $H_\infty(R_i)$ and $H_\infty(R_i|M'_1)$. The key derivation is as follows:

$$\begin{aligned} &-log_2(\sum_{i=0}^{m_1} S_i \cdot max(p_c(i), 1 - p_c(i))) \\ &= H_\infty(R_i|M'_1) + \Delta \le -log_2(\sum_{i=0}^{m_1} S_i \cdot p_c(i)) \\ &= log_2 p_{wf} = H_\infty(R_i) \end{aligned} \tag{16}$$

Δ is a relaxed variable for (16), which is zero for $m_l = 0$ and $m_u = m_1$. Based on (16), if $\Delta = 0$ and $max(p_c(i), 1 - p_c(i)) = p_c(i)$, $H_\infty(R_i|M'_1) = H_\infty(R_i)$ always holds, which means no loss of min-entropy with revealing p_{wf}. Or $\Delta > 0$ results in $H_\infty(R_i|M'_1) < H_\infty(R_i)$. Specifically, when $p_{wf} =$

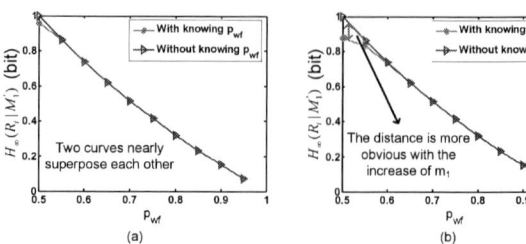

Figure 17: Min-entropy of $p_{wf} \in [0.5, 0.95]$ (a) $m_1 = 20$; and (b) $m_1 = 60$.

0.5, $max(p_c(i), 1 - p_c(i))$ should always be larger than 0.5. Hence, $H_\infty(R_i|M'_1) < H_\infty(R_i)$ leads to the information leakage. Fig.17 further shows that the information leakage is also related to m_1. When $m_1 = 20$ in Fig.17(a), the loss of min-entropy caused by revealing p_{wf} is negligible. The main difference locates in the range around $p_{wf} = 0.5$. However, it becomes more obvious as the augment of m_1 shown in Fig. 17(b) for $m_1 = 60$.

VeriTrust: Verification for Hardware Trust

Jie Zhang, Feng Yuan, Lingxiao Wei, Zelong Sun, and Qiang Xu

CUhk REliable Computing Laboratory (CURE)
Department of Computer Science & Engineering
The Chinese University of Hong Kong, Shatin, N.T., Hong Kong
Email: {jzhang, fyuan, lxwei, zlsun, qxu}@cse.cuhk.edu.hk

ABSTRACT

Hardware Trojans (HTs) implemented by adversaries serve as backdoors to subvert or augment the normal operation of infected devices, which may lead to functionality changes, sensitive information leakages, or Denial of Service attacks. To tackle such threats, this paper proposes a novel verification technique for hardware trust, namely VeriTrust, which facilitates to detect HTs inserted at design stage. Based on the observation that HTs are usually activated by dedicated trigger inputs that are not sensitized with verification test cases, VeriTrust automatically identifies such potential HT trigger inputs by examining verification corners. The key difference between VeriTrust and existing HT detection techniques is that VeriTrust is insensitive to the implementation style of HTs. Experimental results show that VeriTrust is able to detect all HTs evaluated in this paper (constructed based on various HT design methodologies shown in the literature) at the cost of moderate extra verification time, which is not possible with existing solutions.

1. INTRODUCTION

With globalization of the semiconductor industry, today's integrated circuit (IC) designs involve many third-parties during the design and manufacturing process, and hence they are vulnerable to a wide range of malicious alterations, namely hardware Trojans (HTs) [1]. For example, [2] reported a hardware backdoor found in a military grade FPGA, and King *et al.* [3] showed how easy it is to implement a HT in general-purpose processor, which grants privileged access to all memory elements of the system. Therefore, HTs are serious threats to military, financial, and other critical systems [4, 5].

HTs can be inserted in ICs in almost any stage, e.g., RTL design, logic synthesis, physical design, and manufacturing process. As it is not economically feasible to make the IC design and fabrication process completely trustworthy (even for military products), it is essential to develop verification techniques to tackle the challenging HT detection problem. Ideally, we would like to be able to detect a HT by activating it and observing its malicious behavior. In practice, however, since we are not knowledgeable about the location, the trigger condition and the malicious functionalities of the HT, it is very difficult, if not impossible, to directly activate it, especially considering that attackers would typically design a rare event to trigger the HT.

Most prior works on HT detection are based on *side-channel analysis (SCA)* (e.g., [6–10]). The idea behind is that a HT will affect some side-channel signatures, such as path delay and supply current,

even if it is not functionally activated. A common assumption of these works is that HTs are inserted into some random ICs post-fabrication (instead of all) and there exists a trustworthy golden IC that has been thoroughly tested used for signature comparison or characterization. Consequently, these methods are not applicable to detect HTs inserted at design time, which will appear in every fabricated IC product.

Generally speaking, however, the likelihood of HTs being inserted at design time is much higher than that being inserted at manufacturing stage, because adversaries do not need to access foundry facilities to implement HTs. To the best of our knowledge, Hicks *et al.* [11] made the only attempt to detect HTs inserted at design time in the literature. Based on the observation that tricky HTs are usually not activated by test cases at design time (otherwise their malicious behavior will be already manifested), the HT detection problem can be formulated as how to identify "unused circuits" in the system.

However, there is no rigorous definition for "unused circuits" and hence the general *unused circuit identification (UCI)* problem is an open problem without clear solutions. In [11], the authors defined one type of unused circuits as follows. If any pair of related signals are equal throughout all test cases, the intervening circuits between them are regarded as unused ones and hence potential HTs, which can be replaced by a wire. Clearly, such restricted definition of unused circuits can only cover a small set of possible HTs. Later, [12, 13] presented how to automatically construct HTs that can evade the HT detection algorithm shown in [11].

Generally speaking, a HT is composed of its activation mechanism (referred to as *trigger*) and its malicious function (referred to as *payload*). In order to pass functional test and trust validations, stealthy HTs usually employ certain trigger condition that is controlled by dedicated trigger inputs and difficult to be activated with verification test cases. Based on this observation, in this paper, we propose a novel verification technique for hardware trust, namely *VeriTrust*, to identify the malicious trigger inputs for HT detection, by exploiting the fact that trigger conditions for HTs are not satisfied with verification test cases. The main contributions of this paper include:

- We classify HTs into two categories, *bug-based HTs* and *parasite-based HTs*, based on their impacts on the normal functionalities of the circuit, and discuss their corresponding characteristics.

- We present the so-called *VeriTrust* technique to detect parasite-based HTs by identifying the dedicated trigger inputs used in HTs. Unlike existing HT detection algorithms, *VeriTrust* is insensitive to the implementation style of HTs and hence prevents attackers from defeating it by simple HT modifications.

- We propose several techniques to reduce the memory usage and runtime of *VeriTrust* to make it scalable to large circuits.

The remainder of this paper is organized as follows. Section 2 presents preliminaries of this work. In Section 3, we characterize HTs inserted at design time into two types, namely *bug-based HTs* and *parasite-based HTs*. Next, we detail *VeriTrust* technique for detecting parasite-based HTs in Section 4. Experimental results are then presented in Section 5. Finally, Section 6 concludes this paper.

Permission to make digital or hard copies of all or part of this work for personal or classroom use is granted without fee provided that copies are not made or distributed for profit or commercial advantage and that copies bear this notice and the full citation on the first page. To copy otherwise, to republish, to post on servers or to redistribute to lists, requires prior specific permission and/or a fee.
DAC'13, May 29-June 07 2013, Austin, TX, USA.

2. PRELIMINARIES

2.1 Hardware Trust Challenges

Traditionally, the hardware layer of computing systems is often implicitly regarded as trustworthy. This assumption turns out to be quite naive [2–5], and several governments have expressed serious concerns about IC security [14, 15]. Designing HTs that are able to evade traditional IC verification tests is in fact not a very challenging task. This is because, the objective of these techniques (e.g., simulation and emulation) is to ensure an IC performs its specified functionalities. They do not intend to detect extra functionalities introduced into the design. Given the huge state-space that HTs can hide within a reasonably sized circuit, attackers can easily employ a trigger condition that has extremely low probability to be activated with verification tests. Various HT designs have been shown in the literature (e.g., [16, 17]) in recent years, and the Trust-Hub website [18] has released a set of HT benchmark circuits with different triggers and payloads.

Ideally, we would like to prevent HTs from ever being inserted into ICs or ever being triggered at run time. There have been some recent research efforts to achieve the above objectives via design obfuscation and/or isolation [19–22]. These solutions facilitate to mitigate some HT threats, but the associated design cost is quite high and there is no guarantee that ICs would be HT-free with these design methodologies.

2.2 Verification for Hardware Trust

Most existing HT detection techniques consider HTs inserted during fabrication and use side-channel analysis for HT detection (e.g., power-based analysis [6], timing-based analysis [7], and current-based analysis [8]). To reduce the sensitivity of SCA-based HT detection methods on the process variation, several *gate-level characterization* (GLC) techniques were proposed for HT detection recently [9, 10].

For HTs inserted at the design stage, they usually keep dormant when applying verification tests because otherwise their malicious behaviors would have manifested themselves. From this perspective, if part of circuits in a design is not sensitized with verification test cases, it is likely that such unused circuitry contains a HT inserted by attackers. Consequently, the HT detection problem can be formulated as an unused circuit identification problem [11]. *It is important to note that we can define many kinds of "unused circuits" and develop the corresponding UCI algorithms, but whether a particular UCI algorithm is effective or not depends heavily on the definition itself.*

One way to define "unused circuit" is based on the code coverage metrics used in verification. During circuit verification, we have a number of widely-used code coverage metrics: line coverage, condition coverage, toggle coverage, finite state machine (FSM) coverage, branch coverage, and path coverage. We can simply define "unused circuits" as uncovered parts with respect to code coverage metrics and focus on them for HT detection. The above metrics facilitate to identify some HTs, but attackers can easily defeat such HT detection techniques by coding RTL in a different style as detailed in Appendix B.

Hicks *et al.* [11] defined another type of "unused circuit" as follows. Consider a signal pair (s, t), where t is dependent on s. If $t = s$ with all verification test cases, the intermediate circuit between s and t is regarded as "unused circuit". With the above definition, the UCI algorithm traces all signal pairs during the verification and reports those for which the property $s = t$ holds throughout all test cases as places where potential HTs may lie.

For a HT whose payload is implemented separately from the circuit's normal function, the UCI algorithm presented in [11] is guaranteed to find it. This is because, there would exist a dedicated signal representing the circuit's normal function and it is equal to the final output of the circuit throughout all the verification test cases when the HT is not triggered. Because of this, [11] is able to detect some HTs that evade the earlier-mentioned coverage-oriented HT detection method. However, the fact that the effectiveness of [11] relies on HT implementation style enables the simple attack. That is, it is fairly easy to modify the implementation of the HT so that no signal pairs

are equal to each other during verification (an example is given in Section 3.2), which has been shown in [12, 13].

Theoretically speaking, if we have a trustworthy high-level model of the design, we can resort to formal verification (FV) techniques [23] to check its equivalence with the questionable design for HT detection. In practice, however, FV techniques are usually not scalable to large circuits and the trusted high-level model may not be available.

2.3 Threat Model

As in [11], our threat model is that the hardware design can be covertly compromised by HTs inserted into the RTL code or the netlist, implemented by rogue designers in the design team or existing in third-party intellectual property cores to be integrated into the system. The motivation of HT insertion could be financial or general malice.

We assume the design verification procedure is not compromised and all verification test cases are trustworthy. We further assume a HT would be caught by verification test as long as it is triggered.

3. HT CLASSIFICATION

In this section, we classify HTs into two categories according to their impacts on the normal functionalities of the circuit. Attackers could either create certain HTs that directly modify the normal functionalities of the circuit, or insert certain HTs that introduce additional malicious behavior but keep all the normal functionalities of the circuit. They are named *bug-based HT* and *parasite-based HT* respectively in this paper.

For the ease of discussion, we have the following definitions:

DEFINITION 1. *A **functional input** is an input that is used by the circuit's specified normal functionality.*

DEFINITION 2. *A **trigger input** is an input that is used in the condition under which the HT is activated. Note that, functional inputs can serve as trigger inputs for HTs [12].*

3.1 Bug-Based HT

A bug-based HT changes the circuit in a manner that causes it to lose some of its normal functionalities. Consider an original design in Fig. 1(a) whose normal function is $f_n = d_1 d_2$. An attacker may change it to a malicious function, $f_m = \bar{d}_1 d_2$, by adding an additional inverter, as shown in Fig. 1(b). With this malicious change, the circuit has lost certain functionalities, i.e., the two circuits behave differently when $d_2 = 1$. Their corresponding K-Maps are shown in Fig. 2(a) and Fig. 2(b). By comparing the two K-Maps, we can observe that some entries of the normal function have been modified by the malicious function as highlighted in gray. For bug-based HT, some functional inputs serve as trigger inputs to the HT, e.g., for the circuit shown in Fig. 1(b), d_2 is both a functional input and a trigger input.

From a different perspective, the bug-based HT can be simply regarded as a design bug (with malicious intention though), as the design in fact does not realize all of its normal functionalities designated by the specification. As a result, the extensive simulation/emulation is likely to detect this type of HTs. From this perspective, bug-based HT is usually not a good choice for attackers in terms of the stealthy requirement, and almost all HT designs appeared in the literature (e.g., [3,11–13,16–18]) belong to the parasite-based type, as discussed in the following.

3.2 Parasite-Based HT

A parasite-based HT exists along with the original circuit, and does not cause the original design to lose *any* normal functionalities. Again, consider an original circuit whose normal function is $f_n = d_1 d_2$. Suppose an attacker wants to insert a HT whose malicious function is $f_m = \bar{d}_1 d_2$ into the design. To control when the design runs the normal function and when it runs malicious function, the attacker could employ some additional inputs as trigger inputs, t_1 and t_2. In order to escape from trust validation, trigger inputs are usually carefully selected and the trigger condition is designed to be an extremely rare event occurred with verification tests.

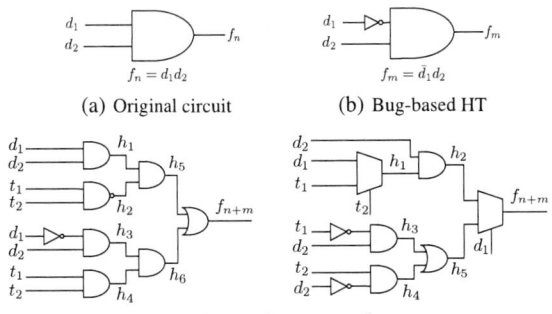

(a) Original circuit   (b) Bug-based HT

$f_n = d_1 d_2$ $f_m = \bar{d}_1 d_2$

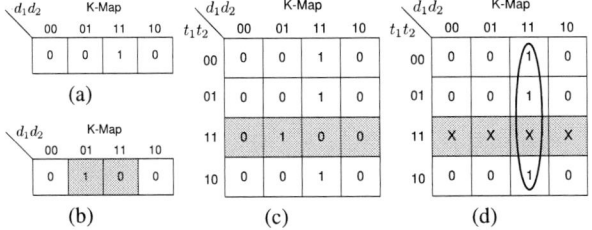

$f_{n+m} = \bar{t}_1 d_1 d_2 + \bar{t}_2 d_1 d_2 + t_1 t_2 \bar{d}_1 d_2$

(c) Parasite-based HT where $\{t_1, t_2\} = \{1, 1\}$ is the trigger condition

Figure 1: HT classification with a simple example.

Figure 2: (a) K-Map of original circuit in Fig. 1(a); (b) K-Map of bug-based HT in Fig. 1(b); (c) K-Map of parasite-based HT in Fig. 1(c); (d) K-Map of parasite-based HT in Fig. 1(c) by setting entities of the malicious function as don't cares.

Let us examine the K-Map of the parasite-based HT-inserted circuit, as shown in Fig. 2(c). The third row represents the malicious function while other rows show the normal function. By comparing it with the K-Map of the original circuit (see Fig. 2(a)), we can observe that the parasite-based HT enlarges the K-Map size with additional inputs so that it can keep the original function while embedding the malicious function. The circuit can then perform the normal function and the malicious function alternately, controlled by trigger inputs.

With the above, we obtain the following two lemmas which we rely on to detect parasite-based HTs.

LEMMA 1. *Consider a signal that attackers expect to modify its value with a parasite-based HT, at least one dedicated trigger input must be employed to activate this parasite-based HT.*

LEMMA 2. *Suppose the value of a signal, denoted by S, can be manipulated by a parasite-based HT. Any signals that are logically-driven by signal S have at least one dedicated trigger input.*

The proofs of *Lemma 1* and *Lemma 2* are shown in Appendix A.

Lemma 1 enables us to focus on the HT trigger signal identification for HT detection. Then, with *Lemma 2*, it is not necessary to consider every signal in the circuit as HT-affected signal. Instead, we only need to consider inputs of state elements (e.g., flip-flops) and primary outputs of the circuit, which dramatically reduces the search space of our solution (detailed in Section 4).

Note that, parasite-based HTs may or may not be detected by the UCI technique presented in [11], depending on how the HT is implemented. Fig. 1(c) presents two possible implementations for the earlier example. The one at the lefthand side can be detected by [11], since f_{n+m} is equal to h_5 (the normal function) under all non-trigger conditions. The implementation shown at the righthand side, however, will evade [11], because the HT is not implemented separately from the circuit normal function and none of the signal pairs in the circuit are equal under all non-trigger conditions.

4. THE VERITRUST SOLUTION

In this section, we detail the proposed *VeriTrust* technique for detecting parasite-based HTs inserted at design time. For the convenience of presentation, HTs mentioned in the rest of the paper means parasite-based HTs unless otherwise specified.

4.1 Overview

Consider a signal whose driving combinational logic cone contains a HT. According to *Lemma 1*, attackers must employ at least one dedicated trigger input to manipulate the value of this signal. *VeriTrust* detects the HT by identifying such dedicated trigger inputs according to the following lemma.

LEMMA 3. *For a signal affected by a parasite-based HT, if we set all entries of the malicious functionalities of the HT as don't-cares, the dedicated trigger inputs used to activate HTs become redundant[1].*

The proof of *Lemma 3* is presented in Appendix A. Let us take the K-Map shown in Fig. 2(d) to illustrate *Lemma 3*. This K-Map represents the circuit in Fig. 1(c) where all entries of the malicious function are considered as don't-cares. By logic simplification, we obtain the original normal function to be $d_1 d_2$ and hence the dedicated trigger inputs, t_1 and t_2, become redundant.

Lemma 3 enables us to identify HT trigger inputs for a particular signal by setting entries of the malicious function as don't-cares. Such HT detection method has the advantage of being insensitive to the HT implementation style (when compared to existing HT detection techniques), because the entries that represent the malicious functionalities of a HT do not change with the implementations.

Although we cannot know which entry belongs to the malicious function a priori, what we do know is that these malicious entries must have not been activated during verification tests (otherwise the HT would have been detected already). In other words, any activated entry is HT-free, which is the premise of our *VeriTrust* solution.

Suppose the verification for the normal function of a circuit is complete, then the un-activated entries are composed of those unreachable entries in functional mode and entries from malicious function. Unreachable entries have no effects on the circuit outputs and hence can be safely ignored by setting them as don't cares. As a result, by setting all the un-activated entries as don't-cares, we can determine those redundant inputs as trigger inputs for HTs. However, in practice, verification tests are usually incomplete, which means un-activated entries may also belong to normal function. Under such circumstance, if we set all un-activated entries as don't-cares to identify redundant inputs, the found ones may include both functional inputs and trigger inputs, and designers need to further examine them to identify true HT trigger inputs, if any. It is important to note that, we may include some functional inputs as potential trigger inputs due to incomplete verification tests, but we will *never* miss any dedicated trigger inputs when they do exist. Therefore, *VeriTrust* guarantees to detect parasite-based HTs, which is not possible with existing HT detection techniques.

The earlier discussion focuses on combinationally-triggered HTs, now let us examine whether *VeriTrust* is able to detect those sequentially-triggered HTs. Generally speaking, there are two kinds of sequential triggers widely used in the literature: the counter-based trigger and the pattern-based trigger. They are typically implemented in a similar manner that one or more dedicated signals are asserted to trigger the HT when the counter reaches the pre-defined counter value or the specific trigger pattern appears [18]. Because these dedicated signals driven by the actual trigger inputs are redundant for the normal function within the combinational logic cone driving the HT-affected signal (according to *Lemma 3*), *VeriTrust* is capable to detect such HTs.

To sum up, consider a RTL design or a synthesized gate netlist that may contain HTs, our *VeriTrust* technique detects HTs by looking for redundant inputs after setting all un-activated entries during verification tests to be don't-cares. From this perspective, *Veritrust* can be

---

[1] An input is redundant if its value change has no impact on any circuit output.

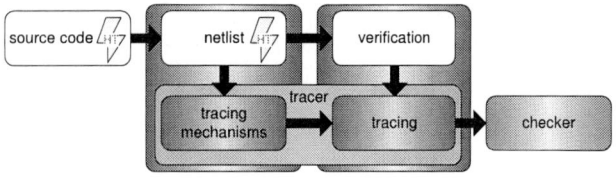

Figure 3: The overview of *VeriTrust*

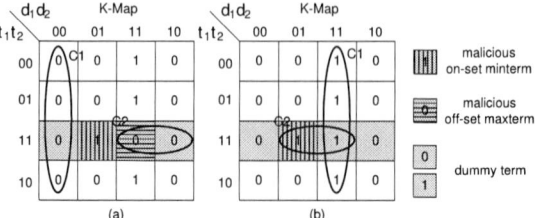

Figure 4: Two HT-affected circuits triggered by $\{t_1, t_2\} = \{1, 1\}$.

considered as an "unused input identification" technique. The overall framework of *VeriTrust* is shown in Fig. 3, which contains two parts: the *tracer* and the *checker*. The tracer traces verification tests to identify those signals that contain un-activated entries by tracing mechanisms that are constructed according to the design netlist. According to *Lemma 2*, it is not necessary to consider every signal in the circuit for HT detection and the signals of interest during tracing are the inputs to all flip-flops and primary outputs. Then, the checker analyzes these signals and determines whether any of them indeed contain redundant inputs and hence are potentially affected by HTs. The details of the tracer and the checker are shown in the following subsections.

4.2 Tracer

Consider a particular signal whose fan-in logic cone may contain a HT, the responsibility of the tracer is to find out whether it contains any un-activated entries after verification tests. A straightforward method is to record the activation history of each and every entry, but it would require unaffordable memory space for large circuits and incur high runtime overhead. To resolve this problem, instead of tracing the activation history of each and every logic entry, we propose to trace in a much more compressed form.

Before discussing the details, let us revisit some basics of Boolean functions. In general, any combinational circuit can be represented in the form of sum-of-products (SOP) and product-of-sums (POS). SOP uses OR operation to combine those on-set minterms, while POS uses AND operation to combine those off-set maxterms. Two minterms (maxterms) are *adjacent* if they have only one different literal.

Next, let us define three new terms, *malicious on-set minterm*, *malicious off-set maxterm* and *dummy term* that compose malicious function as follows.

DEFINITION 3. *The **malicious on-set minterm** is the on-set minterm in the malicious function whose adjacent minterms in the normal function are off-set.*

DEFINITION 4. *The **malicious off-set maxterm** is the off-set maxterm in the malicious function whose adjacent maxterms in the normal function are on-set.*

DEFINITION 5. *The **dummy term** is the on-set minterm or off-set maxterm in the malicious function whose adjacent minterms or maxterms in the normal function are also on-set or off-set.*

With the above definitions, only the malicious on-set minterms and malicious off-set maxterms have malicious behavior. *Consequently, it is not necessary to set dummy terms as don't cares to indentify dedicated trigger inputs by* Lemma 3. Fig. 4 shows the K-Maps of two example HT-infected circuits (t_1 and t_2 are trigger inputs) to illustrate the above terms. The entry filled with vertical lines is malicious on-set minterm, as its neighboring entries in the normal function are all logic '0's. The entry filled with horizontal lines is malicious off-set maxterm, as its neighboring entries in the normal function are all logic '1's. The remaining entries without vertical lines and horizontal lines in the malicious function are dummy terms, as they have the same values with their neighboring entries in the normal function. From the above definitions, when simplifying the circuit into the minimal form of SOP (POS), we have the following observations:

- Malicious on-set minterms and malicious off-set maxterms can only be combined with terms in the malicious function. This is because all the adjacent minterms (maxterms) of malicious on-set minterms (malicious off-set maxterms) in the normal function are off-set (on-set). For example, in Fig. 4 (a), one malicious off-set maxterm is combined with one dummy term (circled by C2), and in Fig. 4 (b), one malicious on-set maxterm is combined with one dummy term (circled by C2).

- Dummy terms can be combined with terms in the normal function as well as terms in the malicious function. For example, the circle C1 in Fig. 4 (a) and Fig. 4 (b) shows that the dummy term is combined with terms in the normal function; the circle C2 in Fig. 4 (a) and Fig. 4 (b) shows that the dummy term is combined with terms in the malicious function.

From the above, simplified products or sums containing malicious on-set minterms or malicious off-set maxterms cannot be activated during the verification. In other words, during the tracing process, we can record the activation history of products and sums instead of that of each logic entry, and hence the memory requirement and runtime overhead of our tracer can be dramatically reduced. The simplified products and sums can be obtained by simplifying the circuit to the minimum SOP and POS form, by leveraging the capability of logic synthesis tool. Note that, the proposed tracing methodology is applicable even if the circuit is not simplified to the minimum SOP and POS form. For the extreme case when we do not perform any simplification, the proposed tracer is simply degraded to tracing all entries.

The tracing procedure might still be time-consuming if the number of products and sums to be traced is large. To resolve this problem, we adopt the following two methods to reduce tracing overhead. First, we periodically remove those sums and products that have been activated. Second, we periodically remove those unactivated sums and products whose signal is determined to be HT-free with our checker.

4.3 Checker

The tracer outputs a number of signals that have un-activated products or sums after applying verification tests. The checker then checks whether a particular signal is driven by any redundant input by assigning the corresponding un-activated products and sums to be don't-cares. If it is, it would be a suspicious HT-affected signal; otherwise it is guaranteed to be HT-free.

To identify whether a signal is driven by redundant inputs could be time-consuming if its fan-in logic cone is large. To mitigate this problem, we adopt three methods for redundant input identification, which differ in the checking capability and time complexity.

Checker 1 simply checks whether there is any redundant input by removing un-activated products/sums from the signal's SOP/POS representation. Take the circuit in Fig. 4 (a) as an example. One SOP can be represented as: $f = d_1 d_2 + t_1 t_2 d_2$. If we remove the un-activated $t_1 t_2 d_2$ from the SOP, the logic function becomes $f = d_1 d_2$ with redundant inputs t_1 and t_2. This efficient checking mechanism is effective in many cases, but it cannot guarantee complete identification of redundant inputs. This is because, whether checker 1 can find out redundant inputs depends on the SOP/POS representation of the signal. For example, if the circuit in Fig. 4 (b) is represented as $f = \bar{t}_1 d_1 d_2 + \bar{t}_2 d_1 d_2 + t_1 t_2 d_2$, then removing the un-activated $t_1 t_2 d_2$ cannot leave t_1 and t_2 redundant.

Checker 2 leverages logic synthesis to re-simplify the function by considering un-activated products and sums as don't-cares. If an input does not appear in the synthesized circuit, it is a redundant input. This method is able to find most redundant inputs, but still cannot guarantee to find all since synthesis tool cannot guarantee optimality by employing heuristic algorithm for logic minimization.

Checker 3 verifies all inputs that are used in the un-activated products/sums one by one. If the change of an input would not cause the change of the function in all input patterns under the condition that un-activated products and sums are set as don't cares, it should be a redundant input. Checker 3 can guarantee to find out all redundant inputs, but it is more time-consuming than Checker 1 and Checker 2.

To ensure complete identification capability while keeping computational time low, we run checker 1, checker 2 and checker 3 in a consecutive manner. That is, if a more efficient checker (e.g., checker 1) finds out a redundant input for a particular signal with un-activated products/sums, we mark it as a suspicious HT-affected signal and move to process the next signal of interest. Otherwise, we use the next checker for redundant input identification. If all the three checkers cannot find redundant inputs for the signal of interest, it is guaranteed to be HT-free. Eventually, the checker returns a list of suspicious signals that are potentially affected by HTs.

5. EXPERIMENTAL RESULTS

5.1 Experimental Setup

The HTs used in our experiments are obtained from two sources: the Trust-Hub website [18] and some related papers [3, 11–13, 16], detailed in Appendix B. We do not directly conduct experiments on those circuits in which HTs are originally inserted, because verification tests required by both [11] and *VeriTrust* for HT detection are not available. Instead, we have selected a SoC design from Open-Cores [24], containing a 32-bit RISC microprocessor namely *OpenRisc* and many peripherals such as UART, USB and MAC, as the hardware platform for HT insertion and detection. We adopt the 17 test cases bundled with this design for verification.

For those RTL HTs, we carefully transplant them onto our experimental platform, keeping their triggers and payloads as discussed in Appendix B. For those HTs for circuit netlist, we use Synopsys Design Compiler to obtain the netlist of the design, and then we insert these HTs by copying gates used by the HT into the netlist and connect them to the targeted signals. To evaluate the effectiveness of *VeriTrust*, we compare it with [11] and code coverage metrics.

5.2 Results and Discussion

Generally speaking, a HT detection algorithm reports a list of candidate places where HTs may lie and requires designers to inspect further to identify whether these suspicious places indeed contain HTs. On one hand, if a true HT evades from the detection algorithm and does not appear in the candidate list, this would cause catastrophic effect; on the other hand, if the candidate list is very large, it will require much manpower to conduct further inspection. We therefore show results on the above two aspects first, and then present the runtime overhead of *VeriTrust*.

5.2.1 The Detection Capability

Fig. 5 presents the HTs identified by *VeriTrust* and UCI techniques after applying all the 17 test cases. Note that, a HT is considered to be detected if part of its trigger and/or payload is shown in the candidate list, reported by the corresponding HT detection algorithm.

First, let us look at the HT detection capability of UCI techniques for RTL HTs, as shown in Fig. 5(a). For HTs in G1, we observe that condition coverage detects T3 and T5 and FSM coverage detects T17 only, because most of the HTs in G1 are implemented without conditional expressions and FSMs. The line coverage, toggle coverage, branch coverage and path coverage detect all of them except T19, because T19 is implemented as a pure combinational logic block and all

Group	Index	Line	Cond	FSM	Toggle	Branch	Path	[11]	VeriTrust
G1	T1	√			√	√	√	√	√
	T3	√	√		√	√	√	√	√
	T5	√	√		√	√	√	√	√
	T7	√			√	√	√	√	√
	T9	√			√	√	√	√	√
	T11	√			√	√	√	√	√
	T13	√			√	√	√	√	√
	T15	√			√	√	√	√	√
	T17	√		√	√	√	√	√	√
	T19							√	√
	T21	√			√	√	√	√	√
	T23	√			√	√	√	√	√
G2	T2								√
	T4								√
	T8								√
	T10								√
	T12	√			√	√	√	√	√
	T14								√
G3	T6								√
	T16								√
	T18				√			√	√
	T20								√
	T22								√
	T24								√
Sum		12	2	1	13	12	12	14	24

√ means the HT is detected by this method.

(a)

	T25	T26	T27	T28	T29	T30	T31	T32	T33	Sum
[11]	√		√		√		√		√	5
VeriTrust	√	√	√	√	√	√	√	√	√	9

√ means the HT is detected by this method.

(b)

Figure 5: The summary of HTs identified by UCI techniques and *VeriTrust*

signals used by HTs have both 1-to-0 and 0-to-1 transitions. [11] detects all HTs in G1 because the final output is equal to the output of the normal function throughout the test cases. For HTs in G2, only T12 is detected by some code coverage metrics and [11]. This is because one signal partially indicating the trigger condition of T12 is quite difficult to be sensitized during verification and hence keeps constant. Theoretically speaking, however, HTs in G2 are likely to evade such UCI techniques, because all signals partially indicating the trigger condition could be set separately without activating the HT and there are no two signals that can hold equal under all non-trigger conditions. It is the fact that the lack of sufficient verification test cases exposes T12. The condition coverage and FSM coverage miss T12 because there are no conditions and FSMs in T12. Finally, for HTs in G3, T18 is detected by some code coverages and [11] due to the same reason as T12. Compared to T12, T18 evades line coverage, branch coverage and path coverage further. This is because T12 is implemented by the basic AND, OR and NOT operators.

From the above, we can observe that the different HT implementation style has a high impact on the detection capability of UCI techniques and none of them are able to detect all HTs. On the contrary, *VeriTrust* is insensitive to HT implementation styles and detects all the HTs based on the fact that they all contain dedicated trigger inputs.

Fig. 5(b) presents the HT detection result for HTs inserted into circuit netlist. As code coverage metrics are not meaningful for circuit netlist, we only present results for [11] and *VeriTrust*. It can be observed that *VeriTrust* can detect all the HTs while [11] misses T26, T28, T30 and T32 due to the fact that none of pairs of signals for HTs keep equal throughout the verification tests.

5.2.2 The Number of Suspicious Candidates

Fig. 6 presents the number of suspicious candidates reported by [11] and *VeriTrust*. By comparing the three curves, we observe that the number of suspicious candidates reported by *VeriTrust* is much smaller than those reported by [11] initially, especially when [11] is

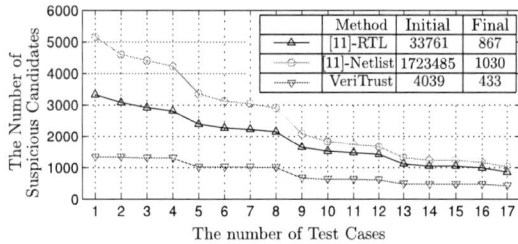

Figure 6: The number of suspicious candidates reported by [11] and *VeriTrust* with the increase number of test cases

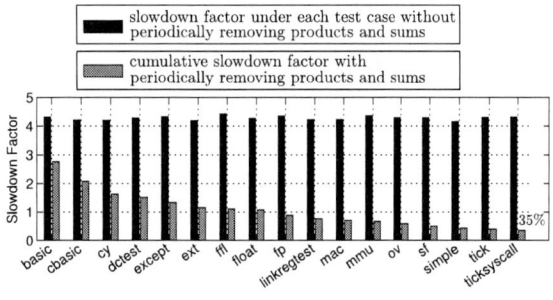

Figure 7: The slowdown factor of *VeriTrust* compared with the *base*

applied to the netlist. This is because the number of state elements traced by *VeriTrust* is much smaller than the number of signal pairs traced by [11] *VeriTrust*. On the other hand, the number of suspicious candidates reported by *VeriTrust* decreases slower. This is because, as soon as a signal pair is with different values, [11] can abandon it while *VeriTrust* requires to accumulate sufficient activated products and sums of the signal to deem it as HT-free.

Further trust validation is needed to determine whether the final list of suspicious candidates are indeed affected by HTs. As shown in Fig. 6, we can observe that the total number of suspicious candidates reported by *VeriTrust* is smaller than that reported by [11] but they are in the same order.

5.2.3 Runtime Overhead

Fig. 7 presents the slowdown factor of *VeriTrust* when compared against the case of functional verification without HT detection (referred to as *base*). It can be observed that the slowdown factor without periodically removing sums and products is quite high, about 4.25, as shown in the black bars. With periodically removing sums and products, however, the slowdown factors reduce with the application of more test cases (due to less traced signals). As shown in the gray bars, *VeriTrust* requires about 35% more runtime to finish 17 test cases.

Finally, Fig. 8 compares the runtime overhead of [11] and *VeriTrust* for netlist HTs. For the fair comparison, we remove signal pairs of [11] periodically as well. As can be seen, the runtime of [11] is larger than *VeriTrust* due to the high number of to-be-traced signal pairs.

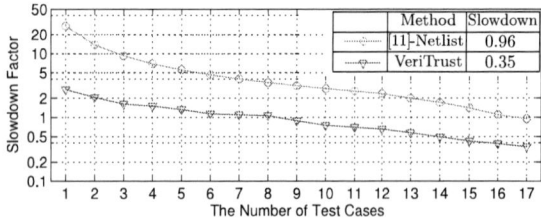

Figure 8: The slowdown factors of [11] and *VeriTrust* with the increase number of test cases compared with the *base*

6. CONCLUSION

In this paper, we propose a novel HT detection technique for HTs inserted at the design stage, namely *VeriTrust*, to automatically identify HTs trigger inputs by examining verification corners. The main advantage of *VeriTrust* compared to existing HT detection techniques is that *VeriTrust* is insensitive to HT implementation styles. Experimental results demonstrate that all the HTs that we have evaluated based on existing HT designs appeared in the literature are detectable with *VeriTrust*.

HT design and HT detection are like arms race, wherein attackers constantly update their tactics to intrude a system while defenders respond with more security measures to protect the system. We discuss the limitations of *VeriTrust* in Appendix C and we plan to investigate whether they can be used to evade *VeriTrust*.

7. ACKNOWLEDGEMENT

We thank Professor Srini Divadas for his insightful comments that greatly improved the paper. This work was supported in part by a CUHK Direct Grant No. 2050488.

8. REFERENCES

[1] M. Tehranipoor and F. Koushanfar. A survey of hardware Trojan taxonomy and detection. *IEEE Design & Test of Computers*, vol.27, no.1, 2010.
[2] S. Skorobogatov and C. Woods. Breakthrough silicon scanning discovers backdoor in military chip. In *Proc. International Conference on Cryptographic Hardware and Embedded Systems*, pp. 23–40, 2012.
[3] S. T. King, J. Tucek, A. Cozzie, C. Grier, W. Jiang, and Y. Zhou. Designing and implementing malicious hardware. In *Proc. USENIX Workshop on Large-Scale Exploits and Emergent Threats*, 2008.
[4] J. Markoff. Old trick threatens the newest weapons. In *The New York Times*, p. D1, Oct. 27, 2009.
[5] S. Adee. The hunt for the kill switch. *IEEE Spectrum*, pp. 34–39, 2008.
[6] D. Agrawal, et al. Trojan detection using IC fingerprinting. In *Proc. IEEE Symposium on Security and Privacy*, pp. 296–310, 2007.
[7] J. Li and J. Lach. At-speed delay characterization for IC authentication and trojan horse detection. In *Proc. IEEE International Workshop on Hardware-Oriented Security and Trust*, pp. 8–14, 2008.
[8] D. Du, S. Narasimhan, R. S. Chakraborty, and S. Bhunia. Self-referencing: a scalable side-channel approach for hardware trojan detection. In *Proc. International Conference on Cryptographic Hardware and Embedded Systems*, pp. 173–187, 2010.
[9] Y. Alkabani and F. Koushanfar. Consistency-based characterization for IC trojan detection. In *Proc. IEEE/ACM International Conference on Computer-Aided Design*, pp. 123–127, 2009.
[10] S. Wei, S. Meguerdichian, and M. Potkonjak. Gate-level characterization: foundations and hardware security applications. In *Proc. ACM/IEEE Design Automation Conference*, pp. 222–227, 2010.
[11] M. Hicks, et al. Overcoming an untrusted computing base: detecting and removing malicious hardware automatically. In *Proc. IEEE Symposium on Security and Privacy*, pp. 159–172, 2010.
[12] C. Sturton, M. Hicks, D. Wagner, and S. T. King. Defeating UCI: building stealthy and malicious hardware. In *Proc. IEEE Symposium on Security and Privacy*, pp. 64–77, 2011.
[13] J. Zhang and Q. Xu. On Hardware Trojan Design and Implementation at RTL. *Proc. IEEE International Symposium on Hardware-Oriented Security and Trust*, to appear, 2013.
[14] U. S. Dept. of Defense. Defense Science Board Task Force on High Performance Microchip Supply. http://www.acq.osd.mil/dsb/reports/ADA435563.pdf, 2005.
[15] M. Beaumont, B. Hopkins, and T. Newby. Hardware trojans-prevention, detection, countermeasures (a literature review), 2011.
[16] S. Wei, K. Li, F. Koushanfar, and M. Potkonjak. Hardware Trojan horse benchmark via optimal creation and placement of malicious circuitry. In *Proc. ACM/IEEE Design Automation Conference*, pp. 90–95, 2012.
[17] Y. Jin, N. Kupp, and Y. Makris. Experiences in hardware trojan design and implementation. In *Proc. IEEE International Workshop on Hardware-Oriented Security and Trust*, pp. 50–57, 2009.
[18] Trust-Hub Website. https://www.trust-hub.org/.
[19] R. Chakraborty and S. Bhunia. Security against hardware trojan through a novel application of design obfuscation. In *Proc. IEEE/ACM International Conference on Computer-Aided Design*, pp. 113–116, 2009.
[20] A. Waksman and S. Sethumadhavan. Silencing hardware backdoors. In *Proc. IEEE Symposium on Security and Privacy*, pp. 49–63, 2011.
[21] A. Waksman and S. Sethumadhavan. Tamper evident microprocessors. In *Proc. IEEE Symposium on Security and Privacy*, pp. 173–188, 2010.
[22] T. Huffmire, et al. Moats and drawbridges: An isolation primitive for reconfigurable hardware based systems. In *Proc. IEEE Symposium on Security and Privacy*, pp. 281–295, 2007.
[23] S. Vasudevan, J.A. Abraham, V. Viswanath, and J. Tu, Automatic decomposition for sequential equivalence checking of system level and RTL descriptions. In *Proc. ACM/IEEE International Conference on Formal Methods and Models for Co-Design*, pp. 71–80, 2006.
[24] OpenCores Website. http://opencores.org/.

APPENDIX

A. LEMMAS AND PROOFS

LEMMA 1. *Consider a signal that attackers expect to modify its value with a parasite-based HT, at least one dedicated trigger input must be employed to activate this parasite-based HT.*

PROOF. Suppose there exists a signal that is affected by a HT but does not have any dedicated trigger input. In this case, all inputs are functional inputs to this signal. Then, this HT must change the signal's normal function under some functional input patterns in order to be effective. Since any change would lead to the loss of original functionalities, it can only be a bug-based HT instead of a parasite-based HT. In other words, the parasite-based HT must be designed with at least one dedicated trigger input. ∎

LEMMA 2. *Suppose the value of a signal, denoted by S, can be manipulated by a parasite-based HT. Any signals that are logically-driven by signal S have at least one dedicated trigger input.*

PROOF. For any signal that is logically-driven by signal S, its value can be manipulated by the parasite-based HT as well. Then, according to *Lemma 1*, it should have at least one dedicated trigger input in order to activate the HT. ∎

LEMMA 3. *For a signal affected by a parasite-based HT, if we set all entries of the malicious functionalities of the HT as don't-cares, the dedicated trigger inputs used to activate HTs become redundant.*

PROOF. Consider a signal affected by a parasite-based HT whose function could be represented by $f = C_n f_n + C_m f_m$, wherein f_n and f_m denote the normal function of the circuit and the malicious function of the HT, which are controlled by the non-trigger condition C_n and the trigger condition C_m, respectively. If we set all entries of the malicious function $C_m f_m$ as don't-cares, from the perspective of the design, we can use $C_m f_n$ to replace $C_m f_m$. Then, the new function becomes $f' = (C_n + C_m) f_n$, which means it exhibits the normal function f_n under both trigger and non-trigger conditions. As a result, dedicated trigger inputs become redundant. ∎

B. HARDWARE TROJANS USED IN EXPERIMENTS

The HTs used in our experiments are obtained from two sources: the Trust-Hub website [18] and some related papers [3, 11–13, 16], as summarized in Table 1. To be specific, there are 33 HTs, wherein 24 HTs are inserted into the HDL source code at the register-transfer level (RTL) while the other 9 HTs are inserted into netlist. We do not include other HTs appeared in the literature, because we can find their corresponding types from what we used after examination. We believe these HTs are sufficient to evaluate UCI techniques and the proposed *VeriTrust* technique thoroughly.

We rename these HTs as shown in the first column of Table 1. Column 2 presents the name of designs where HTs are originally inserted. Column 4 and Column 5 demonstrate the trigger and the payload of HTs, respectively. Based on their triggers and payloads, we have the following observations:

- The trigger of these HTs can be roughly classified into two categories: ❶ counter-based trigger and ❷ pattern-based trigger. For counter-based trigger, the counter can be pulsed by any signals (e.g., clock, instruction, data signal). For pattern-based trigger, the trigger observes a specific pattern or a sequence of specific patterns defined by the attacker. Signal driving the counter or the pattern are usually independent from the HT payload to lower the probability of the trigger being activated.

- The HT payloads have quite diverse malicious functionalities, e.g., compromising address or data register, leaking secret information, reducing performance, depending on the objective of the attacker.

B.1 RTL HTs

T1–T20 are HT benchmark circuits in the form of RTL source code from Trust-Hub [18], originally inserted into three designs: MC8051, RISC, and RS232. As shown in [13], these HTs can be easily detected by both coverage metrics and [11], because all code lines controlled by the trigger condition of the HT that is indicated by a specific signal are never executed during the verification and the signal affected by the HT is always driven by one signal indicating the normal function.

While the above shows the effectiveness of UCI techniques for HT detection, it is still possible to make these HTs evade UCI techniques by simple HT design modifications. We therefore intentionally revise some HT implementations for comparison between UCI techniques and *VeriTrust* by using two code models proposed by [13]. Compared to the original HT implementation used by Trust-Hub, these two code models partition both the trigger condition and the normal function and adopt multiple signals indicating them. In this way, all code lines controlled by part of the trigger condition can be executed under non-trigger condition, and the signal affected by the HT can be driven by multiple signals alternately. The difference between two code models from [13] is that the trigger in the second code model is constructed with basic AND, OR and NOT operators.

Besides the 20 HTs from Trust-Hub, we use another 4 HTs from previous works. For T21 and T23 from [11], only high-level description on their triggers and payloads are available, and we choose to use the code model of HTs from Trust-Hub to construct them. T22 from [12] is a combinationally-triggered HT that can be mapped to the second code model in [13]. T24 is constructed by ourselves based on the malicious circuit provided by [12], and it is also a combinationally-triggered HT.

With the above, the 24 HTs in the form of RTL source code can be classified into three groups, G1, G2 and G3, based on the respective implementation methods.

G1: HTs constructed with their original implementations whose indexes are odd, including T1, T3, T5, T7, T9, T11, T13, T15, T17, T19, T21 and T23.

G2: HTs constructed with code model one in [13], including T2, T4, T8, T10, T12 and T14.

G3: HTs constructed with code model two in [13], including T6, T16, T18, T20, T22 and T24.

B.2 HTs for Netlist

Among the 9 HTs for netlist, 8 are from Trust-Hub [18]. While there are a number of HT benchmarks for netlist shown in [18], their trigger mechanisms are quite similar and we select 8 HTs with different kinds of payloads. [11] can detect all the original HTs, because the final function is always the same as the normal function under all non-trigger conditions. Similarly, we select 4 HTs, T26, T28, T30 and T32 and do some modifications on them based on the code model of HTs for Netlist that is obtained by transferring the code model in [13] into the netlist. Therefore, theoretically, HTs constructed based on this code model can evade [11] in all non-trigger conditions.

In addition, we generate one rare switching HT based on [16], namely T33. In [16], Wei *et al.* developed the one-gate HT trigger to power on HT payloads. The main consideration of [16] when creating the HT is its leakage and timing impact on the design as it targets on the HT inserted at the post-fabrication. As we consider HTs inserted at design stage, we generate the one-gate HT trigger whose switching probability is the lowest among all gates, and we use this trigger to control the payload to change the value of one flip-flop.

C. LIMITATIONS OF VERITRUST

VeriTrust has been shown to be able to detect all the HTs that we have evaluated in this paper. However, this does not mean no HTs can defeat it. In particular, attackers may exploit the assumptions used in *VeriTrust* to evade it. We therefore discuss its limitations in this section.

Table 1: The summary of HTs used in our experiments

Index	Circuit	Level	Trigger	Payload
T1	MC8051-T200	RTL	❷ idle mode state	activate internal timer
T2	MC8051-T300	RTL	❷ specific data through UART	block receiving any message
T3	MC8051-T400	RTL	❷ a specific sequence of commands	disable interrupt
T4	MC8051-T500	RTL	❷ a specific sequence of commands	compromise received data
T5	MC8051-T600	RTL	❷ interrupt on INT0 pin	modify PC to disable jump
T6	MC8051-T700	RTL	❷ a specific command	compromise data
T7	MC8051-T800	RTL	❷ specific data through UART	manipulate stack pointer
T8	RISC-T100	RTL	❶ the number of specific instructions	change memory address
T9	RISC-T200	RTL	❶ the number of specific instructions	replace instructions with sleep command
T10	RISC-T300	RTL	❶ the number of specific instructions	transmit data to external storage
T11	RISC-T400	RTL	❶ the number of specific instructions	manipulate the address
T12	RS232-T100	RTL	❷ specific data through UART	stick a signal
T13	RS232-T200	RTL	❷ specific data	compromise performance counter
T14	RS232-T300	RTL	❶ transmitting time	compromise transmitted data
T15	RS232-T400	RTL	❷ specific data	compromise received data
T16	RS232-T500	RTL	❶ execution time	stick a signal
T17	RS232-T600	RTL	❷ a specific sequence of data	stick a signal & compromise data
T18	RS232-T700	RTL	❷ a specific sequence of data	stick a signal
T19	RS232-T800	RTL	❷ specific data from UART	manipulate the output signal
T20	RS232-T900	RTL	❷ specific data from UART	block the transmission
T21	Leon3 [11]	RTL	❷ a specific sequence of bus data	access protected memories
T22	Leon3 [12]	RTL	❷ a specific sequence of instructions	compromise the supervisor mode
T23	Leon3 [11]	RTL	❷ a specific sequence of bus data	execute arbitrary code
T24	OpenRisc	RTL	❶ a specific counter value	compromise a register
T25	s15850-T100	Netlist	❷ specific values of flip-flops	leak internal signal
T26	s35932-T100	Netlist	❷ specific values of flip-flops	enable the scan chain
T27	s35932-T200	Netlist	❷ specific values of flip-flops	mask four gates
T28	s35932-T300	Netlist	❷ specific values of flip-flops	slow down the path
T29	s38417-T100	Netlist	❷ specific values of signals	control an internal signal
T30	s38417-T200	Netlist	❷ specific values of signals	propagate erroneous value
T31	s38417-T300	Netlist	❷ specific values of signals	leak value through side-channel
T32	s38584-T200	Netlist	❶ a specific counter value	leak value to primary output
T33	OpenRisc	Netlist	❷ a specific data pattern	change the value of the flip-flop

❶: counter-based trigger   ❷: pattern-based trigger

Firstly, *VeriTrust* is not able to detect bug-based HTs because it tries to find the trigger input that is redundant in terms of circuit normal functions. However, bug-based HTs are realized by using only functional inputs, thus bypassing *VeriTrust*. As discussed earlier, however, bug-based HTs can survive only in the hope of incomplete verification. Since attackers cannot control the design of verification test cases, the threat caused by bug-based HTs is usually small.

Secondly, *VeriTrust* would miss those HTs whose trigger is always on. This kind of HTs is usually used to compromise parameters of the design (e.g., timing, power or reliability) without introducing new functionalities to the design. For example, a HT may simply introduce some extra inverters on a circuit path to increase its delay. As *VeriTrust* focuses on detecting HTs that have function-level malicious behavior, these types of HTs can evade it. On the other hand, it is usually difficult to insert HTs to modify the circuit parameters at the design stage, because computer-aided design tools used in the later design stage (e.g., logic synthesis and physical design) may remove such impact with circuit optimization.

Finally, similar to [11], we assume that a HT is detected as long as it is functionally-activated. In practice, however, the verification test cases may miss identifying the malicious behavior of HTs. Therefore, it would be beneficial to review and/or redesign verification test cases for trust validation from this perspective. On the other hand, this problem is a less concern because attackers usually would not bet on careless verification to hide HT payloads.

RASTER: Runtime Adaptive Spatial/Temporal Error Resiliency for Embedded Processors

Tuo Li[†], Muhammad Shafique[‡], Jude Angelo Ambrose[†], Semeen Rehman[‡],
Jörg Henkel[‡], and Sri Parameswaran[†]

[†]School of Computer Science and Engineering, University of New South Wales, Australia
{tuol,ajangelo,sridevan}@cse.unsw.edu.au
[‡]Chair for Embedded Systems, Karlsruhe Institute of Technology, Germany
{muhammad.shafique,henkel}@kit.edu;semeen.rehman@student.kit.edu

ABSTRACT

Applying error recovery monotonously can either compromise the
real-time constraint, or worsen the power/energy envelope. Neither of these violations can be realistically accepted in embedded
system design, which expects ultra efficient realization of a given
application. In this paper, we propose a HW/SW methodology that
exploits both application specific characteristics and Spatial/Tem-
poral redundancy. Our methodology combines design-time and
runtime optimizations, to enable the resultant embedded processor
to perform runtime adaptive error recovery operations, precisely
targeting the reliability-wise critical instruction executions. The
proposed error recovery functionality can dynamically 1) evaluate
the reliability cost economy (in terms of execution-time and dynamic power), 2) determine the most profitable scheme, and 3)
adapt to the corresponding error recovery scheme, which is composed of spatial and temporal redundancy based error recovery operations. The experimental results have shown that our methodology at best can achieve fifty times greater reliability while maintaining the execution time and power deadlines, when compared to
the state of the art.

Categories and Subject Descriptors

C.3 [**Computer System Organization**]: Special-Purpose and
Application-Based Systems—*real-time and embedded systems*; C.4
[**Computer System Organization**]: Performance of Systems—
fault tolerance

General Terms

Design, Performance, Reliability

Keywords

ASIP, Soft Error, Checkpoint Recovery, Redundancy, Runtime Adaptation

1. INTRODUCTION AND MOTIVATION

Soft error has become one of the major reliability challenges for
CMOS based electronic systems [12]. In order to resolve the adverse effect of soft error, error recovery functionality must be accommodated in the underlying system. Error recovery functionality essentially has to induce a considerable amount of cost in either

Permission to make digital or hard copies of all or part of this work for
personal or classroom use is granted without fee provided that copies are
not made or distributed for profit or commercial advantage and that copies
bear this notice and the full citation on the first page. To copy otherwise, to
republish, to post on servers or to redistribute to lists, requires prior specific
permission and/or a fee.
DAC '13, May 29 - June 07 2013, Austin, TX, USA.

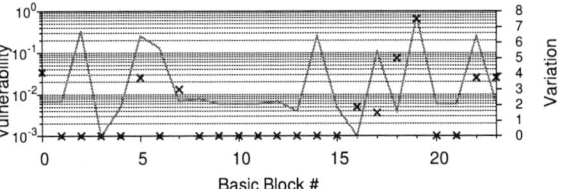

Figure 1: Error Vulnerability and Time Variation in ADPCM

logic gates (i.e., hardware or *spatial redundancy*) or time (i.e., operation or *temporal redundancy*). Therefore, engaging error recovery mechanism for embedded computing systems, where very tight
real-time and power envelope constraints must be satisfied, can be
a challenging design problem.

Traditional studies [3, 5, 16] have largely achieved soft-error resiliency by monotonously leveraging error recovery functionalities
without the consideration of the application software and hardware.
Recent advances [17, 21] have shown promise by the use of runtime
adaptive recovery techniques with one monotonous redundancy.
Those approaches have limitations. These limitations are: 1) un-
awareness of runtime workload-dependent variations; 2) invariant
and inflexible scheme at runtime; and, 3) do not on combine both
spatial and temporal redundancies. Consequently, adopting state-
of-the-art approaches can result in unrealistic and non-optimal cost
efficiency. Typically, this inefficiency can be seen in two example
scenarios:

- Wasting error recovery, which introduces the cost, for protecting the reliable executions that are inherently not vulnerable to soft errors; and,

- Forcing error recovery, at the cost of the violating the stringent real-time and power/energy constraints.

Fig. 1 illustrates the imbalance of relative soft-error vulnerability
amongst the basic blocks in an ADPCM function. The green points
represent the possible variation in execution-time amongst those
blocks. This observation suggests that, exploring an adaptive reliability solution, which takes advantage of the application-specific
error vulnerability features while jointly considering both spatial
and temporal redundancies, can use less resources or cost while
achieving higher reliability. In order to overcome the hurdles that
limit the applicability of error recovery techniques for embedded
systems, we propose a novel approach that provides runtime adaptive error recovery for embedded processors. Our approach focuses
on addressing the above discussed problem in two ways:

- *Reliability-Aware Static Analysis :* this analysis allows the
error recovery functionality to precisely focus on "Reliability Hot-Spot", where the soft errors have high probability to
manifest at the system level.

- *Dynamic Budgeting and Adapting Reliability:* this aspect aims to let the system strategically engage one realistically optimal scheme w.r.t. the specific runtime scenario (e.g., slack occurrence indicates increasing reliability; while missing deadline situations suggests the opposite).

Our Novel Contribution in a nutshell:

In this paper, we propose RASTER approach that is motivated by our observation stated above. RASTER addresses the soft error recovery by a "divide-and-conquer" manner: error recovery functionality is decisively leveraged on those executions, in terms of instructions, which are more prone to soft errors (through design time analysis). Moreover, the error recovery functionality, which utilizes spatial or temporal redundancies, is specialized to adapt to diverse configurations (in terms of spatial and temporal redundancies) based on the available runtime resources in terms of power and execution-time slack. Therefore, RASTER approach enables more optimally efficient and effective error recovery, particularly in embedded processors.

The rest of the paper is structured as follows. Section 2 provides a discussion of related work. Section 3 elaborates our system model and problem formulation. Section 4 and 5 depict the concept and implementation of RASTER approach respectively. Section 6 provides an experimental study that is followed by the conclusion in Section 7.

2. RELATED WORK

Existing error recovery techniques, by the usage of redundancies, can be categorized into two groups. Spatial redundancy based error recovery techniques [3, 5, 8, 19] use more space (mainly in terms of state logics) to save the correct state (in most of the runtime) and restore the system when an error is captured. In parallel, temporal redundancy based error recovery techniques [16] are known to replay the same instructions or operations for a number of iterations (usually 3) and use majority voting to improve reliability.

On the spatial redundancy side, Cache-aided rollback error recovery (CARER) [5] is implemented in the cache replacement policy. This approach exploits the spatial redundancy of the native cache system. The un-justified data is stored in the special cache lines and the checkpoint must be fixed at the time when cache is filled. This technique does not introduce separate state logic; but increases memory traffic. More recently, Sequoia [3] addresses HW error recovery in the context of multi-processor systems, in a similar way. Later, SWICH [19] further improved the cache based error recovery by increasing the rollback window size. For embedded processors, [8] customized the system's architecture to allow instructions to do checkpoint and recovery, while inducing the spatial redundancy (which was implemented separately).

Software-implemented fault tolerance (SWIFT) [16] explores the use of temporal redundancy. This approach modifies the compiler to generate additional lines of code, which creates temporal redundancy, for majority vote. To recover errors, it requires almost double the number of code lines. Lately, TRUMP [16] proposed the exploitation of AN-coding, which can replace the majority-vote for arithmetic instructions, and hence reduce overhead. The limitation is that AN-code cannot propagate through logical operations.

Recent research [17, 21] has focused on making adaptation individually for power and energy to achieve more efficient reliability systems. Existing studies have not considered exploiting: 1) the runtime variations or slack in power and execution-time that can be exploited as the source of spatial and temporal redundancies; 2) the inequality of instructions in terms of soft-error vulnerability, which urges different treatments for different instructions which promises cost efficiency; and, 3) jointly exploiting both these redundancies.

3. SYSTEM MODEL AND PROBLEM FORMULATION

In this section we present the model of the underlying system and the formulation of the target problem. The application A's software program S can be represented as a directed graph $G(N_A, E_A)$,

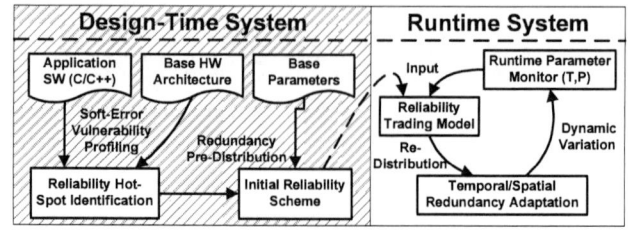

Figure 2: Brief Overview of RASTER Approach

where N_A is the set of basic blocks, and E_A the set of the dependencies between each pair of $n_A \in N_A$. Further, each n_A is a directed acyclic graph (DAG), where each sub-node i_n^A is an instruction, and the dependencies between sub-nodes are simple (the control flow only begins from the entry node i_{entry} and ends at the exit sub-node i_{exit}).

We further formulate that, given the specific processor architecture H, A has three main parameters: execution time T, power P, and reliability R. Correspondingly, the time deadline $T_{deadline}$ and power constraint $P_{deadline}$ are fixed as input parameters for design entry. Finally, the HW-SW system for A can be briefly described as a tuple $A = S \times H = (T, P, R)$ where T is determined by the number of executed instructions θ_{EI} and workload characteristic at runtime; while P is mainly characterized by its dynamic factor (which still dominates in current embedded system technologies [6]) $P_{dynamic} = C \cdot V^2 \cdot f \propto \theta_{LG}$ where θ_{LG} denotes the number of logic gates (i.e., space) that are involved into the execution at a time instant. Moreover, we also assume P is variable to the workload characteristic at runtime. We assume R is the ability to recover from the soft error effect upon its occurrences, which is often represented as single-bit flip [2] in architectural state (e.g., registerfile).

The relationship among parameters T, P, and R can be summarized w.r.t. the characteristic of a particular error recovery mechanism:

- Temporal Redundancy Based: $\Delta R \leftrightarrow \lambda_1 \cdot \Delta T$, $|\Delta R| \propto |\Delta T|$, which means that using the temporal redundancy based technique, such as instruction replay [16], to improve reliability, would primarily increase execution time.

- Spatial Redundancy Based: $\Delta R \leftrightarrow \lambda_2 \cdot \Delta P$, $|\Delta R| \propto |\Delta P|$, which means that using the spatial redundancy based technique, such as triple modular redundancy (TMR) [20] or checkpoint recovery [8], to improve reliability, would primarily cost dynamic power.

where Δ means the increased or decreased amount of the object parameter; while λ_1 and λ_2 are dependent on the base HW architecture and the characteristic of a particular temporal and spatial redundancy based error recovery technique.

Therefore, for a target application A_{base} with variations δ_T and δ_P induced by runtime workloads, the problem can be formulated as:

- **Given** H_{base}, S_{base}, with the base parameters T_{base}, P_{base}, and R_{base},

- **Maximize** ΔR for the optimized resultant system $A' = (T', P', R')$,

- **Let** $T' \leq T_{deadline}$ and $P' \leq P_{deadline}$, where $T' = T_{base} + \lambda_1 \cdot \Delta T + \delta_T$ and $P' = P_{base} + \lambda_2 \cdot \Delta P + \delta_P$.

4. RASTER APPROACH

4.1 Principal Concept and System Overview

The key idea of RASTER approach is strategically utilizing the design-time preprocess and runtime adaptation to achieve the efficient usage of power/execution-time slack, to increase reliability. In other words, RASTER's goal can be summarized as:

```
1  #Step1: collect the worst-case vulnerability
   for i in N:
3    for i in raw_input:
       i_wc = max(i_cur,i_nxt)
5  #Step2: annotate the graph nodes with average
       vulnerability
   for n in N:
7    n_sum = sum(i_wc)
     n_avg = avg(n_sum,n_size)
9    N_anno[n] = n_avg
   #Step3: categorize the graph nodes into two
       subsets: Hot and Cold
11 for n in N_anno:
     if n_avg > median:
13     S_hot.append(n)
     else:
15     S_cold.append(n)
```

Listing 1: Algorithm of Hot-Spot Identification

- Exploiting as much as $\delta\{P, T\}$ to increase the spatial or temporal redundancies (ΔR) in the system;

- Prioritizing the most vulnerable executions (i.e., $\Delta R \rightarrow$ *reliability hot-spots*) in terms of instructions at runtime.

Fig. 2 shows the overview of RASTER approach. **Design-time system** essentially manages reliability-aware static analysis, which provides necessary information for the runtime system; while **run-time system** adapts by reliability trading both temporal and spatial redundancies, to create a system with maximum reliability hot-spots with the most desirable error-recovery functionality (i.e., time or space).

We adopt the instruction vulnerability index (IVI) tool [14] to profile the program with the base HW architecture specification (instruction set and micro-architecture). The output from this tool is used to further generate the reliability hot-spot in the program. We also assume the functional infrastructure of P and T estimation exists. In the scope of this paper, we implement an instruction-count based technique based on [13] to estimate P and T.

4.2 Reliability Hot-Spot Identification

Definition of Reliability Hot-Spot: We define *reliability*, in the scope of this paper, as the ability to recover from soft error in architectural state. Being consistent with the findings from preceding research [1, 10, 15, 18], we apply two fundamental rules so as to define the reliability hot-spot:

- One instruction i is the primitive object of reliability analysis within an application A with S_A and H_A, depending on a number of factors such as pipeline residence period, relevant micro-architecture area, etc.

- Extending the soft-error vulnerability of each instruction i_n^A, the reliability of the basic block n_A, or even the SW program S_A, can be estimated statically.

Therefore, we define that a *reliability hot-spot* is a basic block η that has vulnerability to soft error for a specific application. Further, depending on the application and the base system, there might be a number of reliability hot-spots, which can be noted as a set $\Gamma \subset N_A$. Using the basic block as the hot-spot unit has the following advantages: 1) to beneficially engage either one of the two types of redundancies[1], and 2) to finely manage the adaptation of the redundancies for optimized cost efficiency.

Algorithm of Hot-Spot Identification: Listing 1 shows the brief algorithm in python-style pseudo-code that guides the identification process. Provided the vulnerability information, which is a raw trace, we first refine the input (Line 1-4), to obtain a annotated graph that has reliability information (Line 5-8), and identify the hot-spots $\eta \in \Gamma$, i.e., "S_hot" (Line 9-14). Conceptually, hot-spot blocks suggest a rigorous reliability effort; whereas the cold-spots $\xi \in \Lambda$, i.e., "S_cold" indicates less rigorous treatment.

[1]Spatial redundancy, e.g., checkpoint-recovery [8], requires a considerable time period in between the checkpoints, which is the gain.

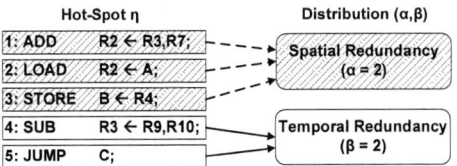

Figure 3: Example Initial Redundancy Distribution

4.3 Initial Redundancy Distribution

We apply two very contrasting initial schemes respectively for hot-spot set and cold-spot set:

- **Hot-Spot:** Distributed with complete reliability coverage (in terms of redundancy-based recovery functionality) initially. At runtime, this scheme can be finely scaled down or up, provided the runtime reliability trading happens to lose or gain. We note the distribution as

$$\mathcal{D} = (\alpha, \beta), \quad |\mathcal{D}| = \alpha + \beta$$

  where α is the quantity of spatial redundancy and β, the temporal redundancy. The total amount is represented as $|\mathcal{D}|$, which indicates the scale or level of the reliability for a particular hot-spot. If the distribution is complete (all instructions are protected), we note it differently as $\lceil \mathcal{D} \rceil = |\mathcal{D}|$.

- **Cold-Spot:** Distributed without any redundancy initially (no instruction is protected). With the run-time adaptation, the redundancy might be coarsely assigned or remains zero, provided the sufficiency of the slack in runtime parameters. We note zero-redundancy distribution as $\mathcal{D}_0 = (0, 0)$ and any positive-redundancy distribution as $\mathcal{D}_1 = (\alpha_1, \beta_1)$. Similarly, the total amount of distribution can be noted as $|\mathcal{D}_1|$; while the complete distribution, $\lceil \mathcal{D} \rceil$.

Specifically, for each reliability hot-spot, the optimized block-specific initial scheme is fixed. Each scheme specifies the exact distribution (in terms of instructions) of the temporal and spatial redundancies. Essentially, this process aims to achieve a locally optimal solution to a knapsack problem for every individual reliability hot-spot. Our solution uses two rules of thumb: Minimize the total amount of redundancies covering the hot-spot; Balance the temporal and spatial redundancies.

Fig. 3 illustrates an example of initial reliability distribution for one hot-spot η. The distribution configures the first three instructions to perform spatial redundancy based error recovery, which accounts for α spatial cost [2]; while the last two instructions, temporal redundancy based error recovery, accounts for β temporal cost.

4.4 Reliability Budget Trading

Reliability Budget Trading is one of the two key functionalities in runtime part of RASTER approach, which aims to educate the adaptation on temporal and spatial redundancies for error recovery (i.e., the other key runtime functionality) in each basic block. This functionality is built on the notion of reliability budget that records the capability of engaging error recovery for individual basic blocks at runtime. By using reliability budget, the other important part, the trading mechanism, determines the most desirable distribution (i.e., temporal/spatial redundancy) of error recovery in the corresponding basic block.

Definition of Reliability Budget: We define that reliability budget $\mathcal{R}$ is an quantification of the amount of reliability owned by the runtime system, in terms of temporal and spatial redundancies. On on hand, reliability budget can be obtained from the variation (or slack) in parameters P and T. On the other hand, it can be spent to increase reliability (i.e., performing error recovery) in the cost of execution time and/or power consumption at runtime. Following

[2]$\alpha = 2$ because Register 2 (R2) only needs to be checkpointed once. The detailed analysis on spatial cost can be seen in [8], which is adapted for spatial redundancy based recovery in this paper.

Algorithm 1: Functionality of Reliability Budget Trading

Input: Initial reliability budget $\mathcal{R}$, initial reliability distribution $\mathcal{D}$, complete reliability distribution $\lceil\mathcal{D}\rceil$, runtime time parameter T, runtime power parameter P, coefficient factors $\lambda_{1,2}$

Output: Reliability budget $\mathcal{R}'$, re-distribution $\mathcal{D}'$

```
 1  begin //At the entrance of a basic block n = {η, ξ}
        /* Step1:  Adjust the budget            */
 2      δ{P,T} ← monitor{P,T}
 3      ΔR_S ← δP/λ_1, ΔR_T ← δT/λ_2
 4      R ← (R_S ← R_S + ΔR_S, R_T ← R_T + ΔR_T)
 5      |R| ← R_S + R_T
        /* Step 2:  Trade the budget            */
 6      if |R| < ⌈D⌉ then //Insufficient budget: trade-in budget by
        sacrificing/redistributing reliability
 7          if n = η ∈ Γ then //Hot-spot: reduce distribution
 8              D' ← (α' ← R_S, β' ← R_T)
 9              R' ← (R_S ← 0, R_T ← 0)
10          else //Cold-spot: zero-redundancy distribution
11              D' ← D_0 = (0, 0)
12              R' ← (R_S, R_T)
13      else //|R| ≥ ⌈D⌉; Sufficient budget: trade-out budget for leveling
        up reliability distribution
14          if n = η ∈ Γ then //Hot-spot: finely redistributed with
            spatial/temporal exchange
15              switch Δ do //Difference between R and D
16                  case Δ = (+, +) //Use initial distribution
17                      D' ← (α' ← α, β' ← β)
18                      R' ← (R_S − α, R_T − β)
19                  case Δ = (+, −) //Redistribute: exchange spatial
                    to temporal redundancy
20                      D' ← (α' ← α + (β − R_T), β' ← R_T)
21                      R' ← (R_S − α', 0)
22                  case Δ = (−, +) //Redistribute: exchange
                    temporal to spatial redundancy
23                      D' ← (α' ← R_S, β' ← β + (α − R_S))
24                      R' ← (0, R_T − β')
25          else //Cold-spot: coarsely assigned without spatial/temporal
            exchange
26              D' ← (α' ← min(α_1, R_S), β' ← min(β_1, R_T))
27              R' ← (R_S − α', R_T − β')
28      return R', D'
```

the basic relationship of the parameters stated in the system model, $\mathcal{R}$ has two main factors: temporal redundancy based reliability $\mathcal{R}_S$ and spatial one $\mathcal{R}_T$. The *reliability unit grain* is one instruction that can be performed with a certain redundancy based error recovery technique. Formally, the reliability budget can be formulated as:

$$\mathcal{R} = (\mathcal{R}_S, \mathcal{R}_T), \quad |\mathcal{R}| = \mathcal{R}_S + \mathcal{R}_T$$

Algorithm of Budget Trading: The essential concept of reliability budget trading is to realistically organizing the configuration of the error recovery functionality for managing reliability in the undergoing basic block. Algorithm 1 describes the functionality of reliability budget trading, which is comprised of two primary components: **budget adjustment** (Line 1-5) and **budget redistribution** (Line 6-27).

The process of **adjusting reliability budget** aims to re-assess the budget $\mathcal{R}$ at runtime upon entering a basic block n. This step involves two factors: 1) accumulated runtime variation δ in P and T parameters during the last block, and 2) the remaining reliability budget from the previous blocks. Based on λ_1 and λ_2, which can be obtained in design-time analysis, the actual accumulated amount of resources in terms of spatial and temporal redundancies, can be interpreted. This amount is then added to the realistically remained reliability budget that is left from previous period, to yield the available reliability budget to the current basic block.

The process of **redistributing the budget**, namely budget trading, is composed based on two principles w.r.t. the identity (i.e., hot η or cold ξ) of the underlying basic block:

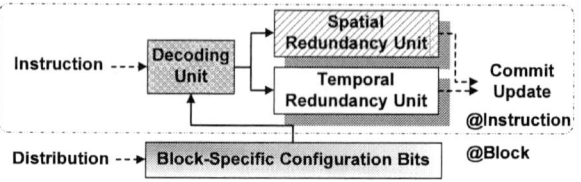

Figure 4: Block Diagram of Spatial Temporal Redundancy Adaptation

- For hot-spot $n = \eta$: First-order priority; Maximize reliability by always leveraging the budget trading at finegranularity; Allow exchange between spatial and temporal redundancy.

- For cold-spot $n = \xi$: Second-order priority; Throttle reliability whenever the budget is scarce; Prohibit exchange between spatial and temporal redundancy.

Specifically, *hot-spot budget trading functionality* is described through Line 7-9 and Line 14-24 in Algorithm 1. Firstly, in case that the budget is not enough to support the maximum distribution ($|\mathcal{R}| < \lceil\mathcal{D}\rceil$), then the initial distribution $\mathcal{D}$ must be scaled down. In this way, the redundancy is re-organized to fully exploit the entire available budget; until the budget becomes zero. Secondly, in case that the budget is sufficient ($|\mathcal{R}| \geq \lceil\mathcal{D}\rceil$), the necessity to exchange spatial and temporal redundancy of distribution is evaluated accordingly to the discrepancy Δ between $\mathcal{R}$ to $\mathcal{D}$. We denote three situations as $(+, +)$, $(+, -)$, and $(-, +)$, where, as same as the pattern of the notation $\mathcal{R}$ and $\mathcal{D}$, the first element in the discrepancy pair is P-induced reliability budget that can be used for spatial redundancy $\mathcal{R}_S$, the second element is T-induced reliability budget that corresponds to temporal redundancy $\mathcal{R}_T$, "+" means a positive discrepancy (sufficient budget), and "−" means a negative discrepancy (insufficient budget). Based on the situations in the difference, the budget trading would be conducted differently: 1) Complete sufficiency, i.e., $(+, +)$, leads to initial distribution; 2) If the budget to the distribution is bounded by one element, i.e., $(+, -)$ or $(-, +)$, make the exchange from the sufficient $(+)$ element to yield a new distribution that satisfies the budget.

In parallel, *cold-spot budget trading functionality* is described through Line 10-12 and Line 25-27 in Algorithm 1. Differing to the hot-spot counterpart, the cold-spot distribution is coarsely managed. At first, in case the budget is insufficient ($|\mathcal{R}| < \lceil\mathcal{D}\rceil$), the zero-distribution $\mathcal{D}_0$ would be active, which means all the error recovery actions would be throttled in the execution of the current block. In this way, when budget is scarce, the savings can be made in cold-spots for use in hot-spots. Lastly, in case the budget is abundant ($|\mathcal{R}| \geq \lceil\mathcal{D}\rceil$), the redistribution would be formed up, in which the maximally available amount is given to each element, without exchange. This means that very asymmetric amount of budget would bound one element in distribution while satisfying another element. This heuristic method aims to supply complementary increase in reliability to cold-spots when the budget becomes very abundant, and more importantly, to save the budget in either spatial or temporal element to hot-spots when the budget shrinks due to negative P and/or T variation for catching the application deadline. Note that the scenario of negative budget might exist, and in this case, the respective redundancy would be kept not available to be distributed, until it becomes positive, due to the coming slacks.

4.5 Spatial/Temporal Redundancy Adaptation

After budget trading, the behavior of the instructions must be able to adapt to the optimized distribution, which arguably better suits the runtime scenario which exhibits execution time variations. To address this problem, spatial temporal redundancy adaptation mechanism is designed to reconfigure the execution of corresponding instructions (i.e., those located in the current basic block). Fig. 4 illustrates the organization of this mechanism. We use configuration bits, which can be stored in a lookup table or other state logic, to configure the redundancy units. Configuration bits are

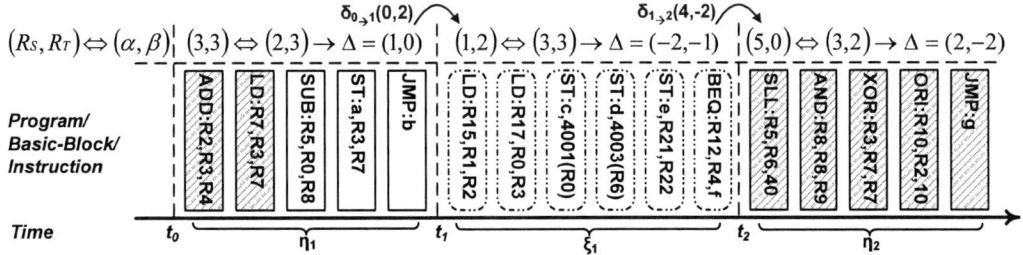

Figure 5: Example of RASTER Approach (Patterned Rectangular: Spatial; Solid: Temporal; Round & Dashed: None.)

specified at the entrance of the current block by taking the distribution as the input. To translate the configuration bits into the specific control signals, corresponding decoding mechanism is required, which is invoked upon the execution of each instruction, since the instruction is the reliability grain. The function of the decoding mechanism is essentially a functional mapping $f : Config \times Instr \rightarrow Op \in \{S, T, N\}$ and can be summarized as:

- **Spatial Redundancy Adaptation (S):** In case the current instruction is specified for spatial redundancy, the control signals are set to activate the spatial redundancy based error recovery operations, while deactivating the temporal redundancy one.

- **Temporal Redundancy Adaptation (T):** In case the current instruction is specified for temporal redundancy, the temporal redundancy based error recovery operation is set to perform, while the spatial counterpart is switched off.

- **Non-redundancy (N):** Neither redundancy based operations are activated, the instruction is executed with normal operations.

4.6 Example

Fig. 5 demonstrates the RASTER approach working in a certain series of three example runtime scenarios (i.e., three periods starting at t_0 and segmented by t_1 and t_2). In the first scenario (t_0-t_1), the basic block under execution is a reliability hot-spot η_1. The reliability after adjustment is equal to $(3, 3)$, which derives $|\mathcal{R}| = 6$ to be sufficient in both elements for the maximum initial distribution $(2, 3)$ for η_1. Therefore, the budget trading is straightforward, in which the distribution is complete without any exchange between spatial and temporal redundancies. In the second scenario (t_1-t_2), due to the remaining last period, and the result of the power and time variation $\delta_{0\rightarrow1}(0, 2)$, the budget $|\mathcal{R}|$ rises to $(1, 2)$. Provided that the underlying basic block is a cold-spot ξ_1, and the maximum distribution exceeds the available budget, the reliability is sacrificed to secure the budget. Hence, the processor adapts to totally switch off the reliability operations. In the third scenario (t_2 to end), by absorbing the unused budget and by the changing of P and T, initially $\mathcal{R}$ for hot-spot η_2 is raised to a highly asymmetric value $(5, 0)$. This situation leads to an exchange happening in budget trading, so as to form a new distribution $(5, 0)$, different to the initial $(2, 3)$. The redistribution is completely compatible with the available budget. Consequently, all the five instructions in η_2 are configured to S spatial redundancy based error recovery.

5. SYSTEM SYNTHESIS AND IMPLEMENTATION

Fig. 6 explains RASTER HW-SW system (i.e., the resultant processor) generation, using an approach based on architectural description language (ADL) modeling [9]. Our system synthesis methodology (Fig. 6(a)) is extended from previous ASIP studies [11], where the ADL-to-RTL synthesis is conducted via ASIPmeister[3]. Through

[3]http://www.asip-solutions.com/en/asip_meister.html

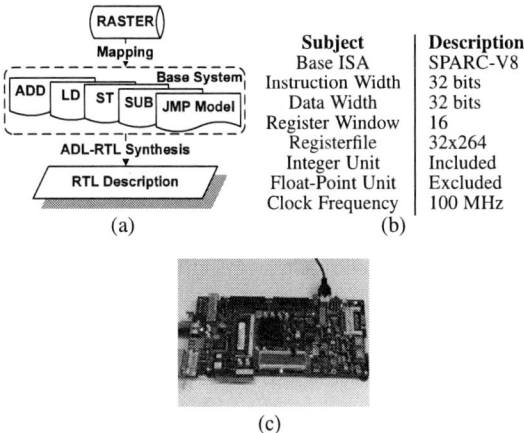

Figure 6: RASTER System Synthesis: (a) Flowchart; (b) Parameter; (c) FPGA Verification

Subject	Description
Base ISA	SPARC-V8
Instruction Width	32 bits
Data Width	32 bits
Register Window	16
Registerfile	32x264
Integer Unit	Included
Float-Point Unit	Excluded
Clock Frequency	100 MHz

customizing the input specification, i.e., the template ADL description of the baseline system (SPARC-v8), the description of runtime adaptation and error recovery mechanisms are mapped to the corresponding instruction models. Each instruction model is a graph representation, which is extended with the notion of a pipeline stage. The implementation parameters are summarized in Fig. 6(b). In this study, we select SPARC-v8 architecture as the baseline processor, with 32-bit instruction and data, and 16 register windows, excluding Float Point Unit. The configuration table and critical design time information is implemented as lookup table (in terms of multiple arrays) and registers (single data). The table size is proportional to the basic block number of the underlying application program. The design is verified using an Xilinx ml605 board with virtex 6 chip, with a utilization of 11880 slice registers and 48714 slice LUTs.

6. EXPERIMENT AND RESULT

For testing the reliability, we conduct HDL simulation with fault injection using ModelSim SE simulator[4]. We choose single-bit flip, which represents the most common soft error occurrence, also named as single event upset (SEU). The simulation is composed of a number (10000, determined by the method in [7]) of repetitions of random fault injections. In each repetition, one random fault is injected in to the system at an random instruction (similar to the method used in [8]). In order to accurately simulate the soft error's feature, the error probability is influenced by the corresponding area and residency time of the instruction [14]. We choose four mibench benchmark applications: adpcm, crc32, sha, and susan [4]. Each application is mainly exercised with its kernel function, so as to represent the application specific feature.

[4]http://model.com/

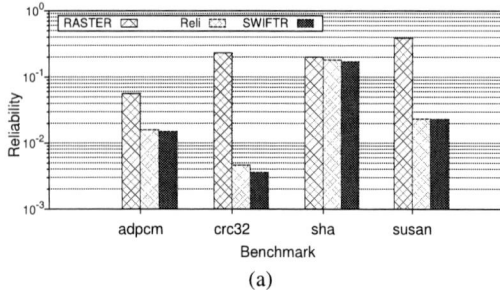

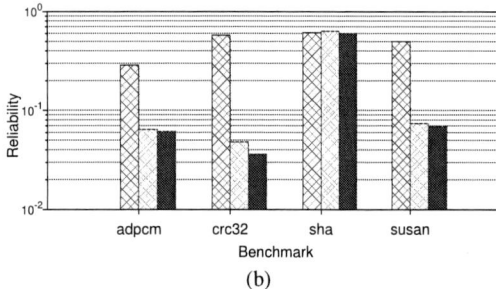

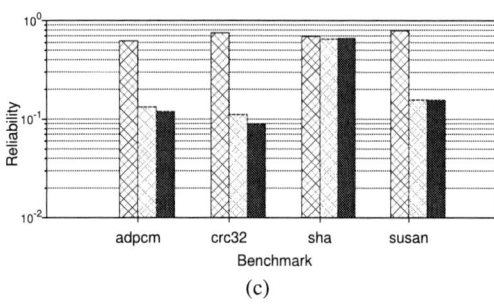

Figure 7: Reliability Test: (a) 25% slack;(b) 50% slack; (c) 75% slack

Fig. 7 shows the result of fault injection test. We compare the reliability of our approach with two single redundancy based error recovery technique: 1) a HW implementation of SWIFT-R [16], which uses temporal redundancy to replay the instruction and vote out the error; 2) A checkpoint recovery technique, namely Reli [8], which uses spatial redundancy periodically. We scale the slack in execution time and power to three different levels: 25%, 50%, and 75%, from the worst-case estimation (i.e. 1x time and power due to workload data). The slacks are proportionally assigned at static time to each basic block of the program. In each level, we allow SWIFT-R and Reli to perform error recovery until it is constrained by the given slack. The reliability is derived via the comparison between the number of corrected errors to the total number of errors.

In most of situations, RASTER leads to reliability gain, while one exception can be seen in the application sha, where RASTER and other two have similar reliability. The possible reason is that sha has very scarce distribution of error-prone instructions among blocks in the early part of the application. In other situations through 25% to 75%, RASTER can at most achieve 50x reliability gain (i.e., in crc32 at 50% level) over the other two techniques.

7. CONCLUSION

In this paper, we have proposed the RASTER approach which allows runtime adaptive error recovery functionality. Our methodology comprises of both spatial and temporal redundancy based error recovery. RASTER approach consists of design time and run-

time systems, which contributes to the realization of an enhanced embedded processor. The resultant processor is equipped with the functionality to budget and adapt the reliability by using spatial and temporal redundancies, based on the runtime situation. The experiment results have shown that RASTER can better optimize the reliability in comparison to existing approaches, given different levels of application slacks. At the best case in our experiment, RASTER can lead with almost 50x reliability gain (e.g., at crc32) in comparison.

Acknowledgment

This work is supported in part by the German Research Foundation (DFG) as part of the priority program "Dependable Embedded Systems" (SPP 1500 - spp1500.itec.kit.edu). The authors would like to thank Roshan Ragel and Swarnalatha Radhakrishnan for their contribution in implementation development.

8. REFERENCES

[1] H. Asadi, M. B. Tahoori, and C. Tirumurti. Estimating error propagation probabilities with bounded variances. In *DFT*, pages 41–49, 2007.
[2] R. Baumann. Soft errors in advanced computer systems. *IEEE Design & Test of Computers*, 22(3):258–266, 2005.
[3] P. Bernstein. Sequoia: a fault-tolerant tightly coupled multiprocessor for transaction processing. *Computer*, 21(2):37 –45, feb. 1988.
[4] M. R. Guthaus, J. Ringenberg, D. Ernst, T. Mudge, R. Brown, and T. Austin. MiBench: a free, commercially representative embedded benchmark suite. In *IEEE International Symposium on Workload Characterization*, 2001.
[5] D. Hunt and P. Marinos. A general purpose cache-aided rollback error recovery (CARER) technique. In *proceedings of the 17th international symposium on fault-tolerant coputing systems*, pages 170–175, 1987.
[6] IRC. *International Technology Roadmap for Semiconductor 2007 Edition Design*, 2007.
[7] R. Leveugle, A. Calvez, P. Maistri, and P. Vanhauwaert. Statistical fault injection: quantified error and confidence. In *Proceedings of the Conference on Design, Automation and Test in Europe*, DATE '09, pages 502–506, 3001 Leuven, Belgium, Belgium, 2009. European Design and Automation Association.
[8] T. Li, R. Ragel, and S. Parameswaran. Reli: Hardware/software checkpoint and recovery scheme for embedded processors. In *Design, Automation Test in Europe Conference Exhibition (DATE), 2012*, pages 875 –880, march 2012.
[9] P. Mishra and N. Dutt. *Processor Description Languages, Volume 1*. Morgan Kaufmann Publishers Inc., San Francisco, CA, USA, 2008.
[10] S. Mukherjee, C. Weaver, J. Emer, S. Reinhardt, and T. Austin. A systematic methodology to compute the architectural vulnerability factors for a high-performance microprocessor. In *Microarchitecture, 2003. MICRO-36. Proceedings. 36th Annual IEEE/ACM International Symposium on*, pages 29 – 40, dec. 2003.
[11] J. Peddersen, S. L. Shee, A. Janapsatya, and S. Parameswaran. Rapid embedded hardware/software system generation. *VLSI Design, International Conference on*, 0:111–116, 2005.
[12] J. M. Rabaey and S. Malik. Challenges and solutions for late- and post-silicon design. *IEEE Des. Test*, 25:296–302, July 2008.
[13] K. Rajamani, H. Hanson, J. Rubio, S. Ghiasi, and F. Rawson. Application-aware power management. In *Workload Characterization, 2006 IEEE International Symposium on*, pages 39 –48, oct. 2006.
[14] S. Rehman, M. Shafique, and J. Henkel. Instruction scheduling for reliability-aware compilation. In *Design Automation Conference (DAC), 2012 49th ACM/EDAC/IEEE*, pages 1288 –1296, june 2012.
[15] S. Rehman, M. Shafique, F. Kriebel, and J. Henkel. Reliable software for unreliable hardware: embedded code generation aiming at reliability. In *Proceedings of the seventh IEEE/ACM/IFIP international conference on Hardware/software codesign and system synthesis*, CODES+ISSS '11, pages 237–246, New York, NY, USA, 2011. ACM.
[16] G. A. Reis, J. Chang, and D. I. August. Automatic instruction-level software-only recovery. *IEEE Micro*, 27:36–47, January 2007.
[17] J. Sartori and R. Kumar. Architecting processors to allow voltage/reliability tradeoffs. In *CASES*, pages 115–124, 2011.
[18] V. Sridharan and D. R. Kaeli. Using hardware vulnerability factors to enhance avf analysis. In *Proceedings of the 37th annual international symposium on Computer architecture*, ISCA '10, pages 461–472, New York, NY, USA, 2010. ACM.
[19] R. Teodorescu, J. Nakano, and J. Torrellas. SWICH: A prototype for efficient cache-level checkpointing and rollback. *IEEE Micro*, 26:28–40, 2006.
[20] J. F. Wakerly. Transient failures in triple modular redundancy systems with sequential modules. *IEEE Trans. Computers*, 24(5):570–573, 1975.
[21] B. Zhao, H. Aydin, and D. Zhu. Enhanced reliability-aware power management through shared recovery technique. In *ICCAD*, pages 63–70, 2009.

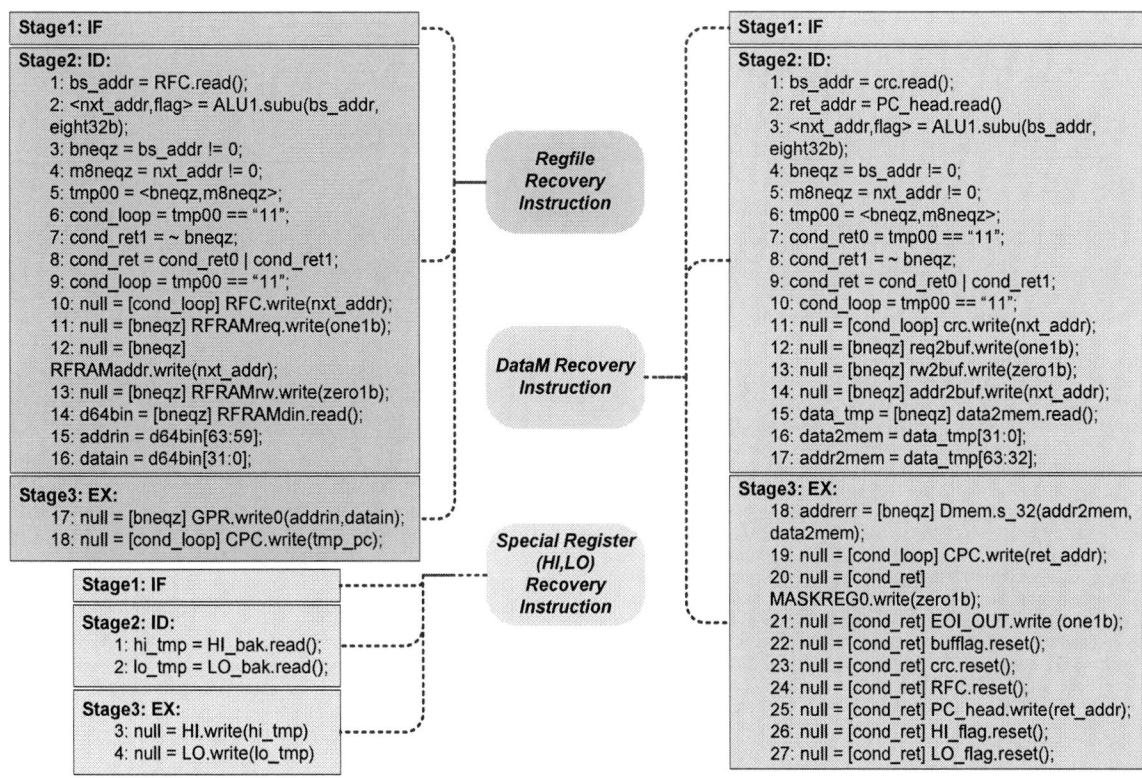

Figure 8: Example Description in ADL

APPENDIX

A. MAIN NOTATIONS USED IN THIS PAPER

$\mathcal{D}$: the reliability distribution of the spatial and temporal redundancies. The distribution can be obtained at runtime via reliability budget for error recovery.

α: the spatial element of reliability distribution, which can be used as spatial redundancy for checkpoint recovery or triple modular redundancy.

β: the temporal element of reliability distribution, which can be used as temporal redundancy for instruction replays.

$|\mathcal{D}|$: the quantity of the distribution, which means the total number of instructions that can be protected by spatial and temporal redundancy based error recovery.

$\mathcal{R}$: the reliability budget that is proportional to the amount of slack existing at runtime. The amount of budget indicates the number of instructions that can be potentially protected by error recovery. The budget can be distributed as $\mathcal{D}$ to a basic block, and thus be used for error recovery.

$\mathcal{R}_S$: the spatial element of the reliability budget. It can be distributed as α at runtime.

$\mathcal{R}_T$: the temporal element of the reliability budget. It can be distributed as β at runtime.

$|\mathcal{R}|$: the quantity of the total reliability budget as the sum of the two elements. Corresponding to that of reliability distribution.

B. INSTRUCTION VULNERABILITY INDEX

Instruction Vulnerability Index (IVI) is a reliability component attached to an instruction to indicate its vulnerability in the program execution. A higher IVI would suggest that the instruction is more susceptible for failure due to faults. An instruction is analysed based on its resource usage and dependencies between its previous instruction in the execution. For example, an instruction which uses a register with longer liveliness will be more vulnerable. Equation 1 depicts the formulation of IVI for an instruction i. c is a component in a processor, such as registerfile, divider, adder, etc., and A_c is its area in gates. The IVI related to each component used by the instruction is accumulated to compute the total IVI_i. P_{fault} is the probability of faults.

$$IVI_i = \frac{\sum_{c \in Proc} IVI_{ic} * A_c * P_{fault}(c)}{\sum_{c \in Proc} A_c} \quad (1)$$

The IVI of a component, IVI_{ic}, for a specific instruction is calculated using Equation 2, where the liveliness of the component is referred to as $VulnerabilityPeriod$ and the vulnerable bits in that component as $Bits_{ACE}$. This IVI for a component is normalized against all the bits in the component.

$$IVI_{ic} = \frac{VulnerabilityPeriod_{ic} * Bits_{ACE-c}}{\sum_{c \in Proc} TotalBits_c} \quad (2)$$

C. ADL DESCRIPTION EXAMPLE

We use ASIPmeister tool's ADL description language to model and customize the system, in terms of instructions. Fig.8 shows an example of instruction modeling with checkpoint recovery enhancement. The dark green (up left) blocks compose the recovery functionality, through pipeline stages, for registerfile. The grey blocks (up right) describe the functionality of memory content recovery. The blue blocks (down left) model the error recovery for special status registers. Each statement means a data transfer or a execution on a functional unit (e.g., ALU, MUX, etc.).

ABCD-L: Approximating Continuous Linear Systems Using Boolean Models

Aadithya V. Karthik[‡] and Jaijeet Roychowdhury

Department of Electrical Engineering and Computer Sciences, The University of California, Berkeley, CA, USA

[‡]Contact author. Email: aadithya@berkeley.edu

Abstract—We present ABCD-L, a scalable technique for Analog/Mixed Signal (AMS) modelling/verification that captures the continuous dynamics of Linear Time-Invariant (LTI) systems, using purely Boolean approximations, to any desired level of accuracy. ABCD-L's models can be used in conjunction with existing techniques for Boolean synthesis/verification/fast logic simulation, or with hybrid systems frameworks, to represent LTI dynamics without incurring the penalty of adding continuous variables. Unlike existing state-enumeration approaches like DAE2FSM [1], ABCD-L scales practically linearly with system size. We apply ABCD-L to I/O links composed of RC/RLGC units, capturing important analog effects like inter-symbol interference, overshoot/undershoot, ringing, *etc.* – all using purely Boolean models. We also present a continuous-time differential equalizer example, where ABCD-L accurately reproduces key design-relevant AMS metrics, including the eye diagram correction achieved by the circuit. Furthermore, for real-world LTI systems, we demonstrate that ABCD-L can be applied in conjunction with Model Order Reduction (MOR) techniques; we use this to produce accurate Boolean models of an industry-scale power grid network (with 25849 nodes) made available by IBM. We also demonstrate that Boolean simulation using ABCD-L's models offers considerable speed-up over standard circuit simulation using linear multi-step numerical methods.

I. INTRODUCTION

In today's advanced process technologies (32nm and below), Analog/Mixed-Signal (AMS) components (*e.g.*, interconnect, I/O and equalization circuitry, PLLs, DLLs, *etc.*) are becoming key bottlenecks that determine system-level performance [2], [3]. Moreover, an increasingly significant proportion of overall design bugs are now attributable to on-chip AMS components. For example, a recent internal study at Intel concluded that AMS modules account for over 20% of all design bugs in cutting edge microprocessors. Furthermore, such bugs tend to be difficult and costly to identify and correct, typically requiring extensive time-consuming SPICE-level simulations.

For early detection and timely correction of the above AMS-related design bugs, it is desirable to carry out functional validation and formal verification of AMS components *at or near SPICE-level accuracy*. However, in most existing approaches to AMS verification (see the accompanying supplement for an overview), the underlying model that is verified is usually a highly simplified abstraction that does not attempt to capture any of the SPICE-level subtleties (*e.g.*, layout-dependent parasitics, cross-talk, inter-symbol interference, ringing) that are responsible for bringing about design bugs/loss of performance. Thus, while the simplified models currently in use by AMS verification tools can be useful for gaining intuition about the circuit's operation as a whole (as intended by the designer), they are of limited use when it comes to *debugging* AMS designs or *issuing performance guarantees*.

Here, it is useful to draw a distinction between two kinds of formal verification techniques: *Boolean techniques* and *Hybrid systems techniques*. Boolean techniques (*e.g.*, [4]) represent each underlying circuit signal as a discrete (usually binary) quantity, whereas hybrid systems techniques (*e.g.*, [5]–[11]) offer the capability to represent signals as either discrete or continuous-valued quantities. However, the ability to represent and reason about continuous variables often comes at an enormous computational cost, which renders hybrid systems techniques typically orders of magnitude slower than their Boolean counterparts. Indeed, while Boolean techniques are routinely used in the industry to verify circuits with millions of logic gates, even state-of-the-art hybrid systems techniques are unable to verify systems with more than a few (*e.g.*, 5 to 10) continuous variables.

The *limited scalability* of hybrid systems techniques is the main reason why AMS circuits are typically not modelled/verified at SPICE-level accuracy; instead, existing hybrid systems methodologies are forced to adopt over-simplified "behavioural" AMS component models that often do not bear close resemblance to SPICE. This drastically limits their applicability in the context of AMS debugging/performance verification: without SPICE-accurate modelling, the predictions made by hybrid systems based AMS verification engines are not reliable enough for designers. As a result, the prevailing practice amongst AMS designers today is to carry out time-consuming SPICE simulations rather than place their trust in AMS verification tools. To overcome such "designer skepticism", we believe that it is necessary to significantly scale up existing hybrid systems techniques, so that they embrace SPICE-accurate models even for large AMS designs.

In this paper, we propose a technique (called ABCD-L[1]) to bridge the gap between SPICE-level detail and the models used by AMS verification engines, for an important subclass of AMS circuits, namely, Linear Time Invariant (LTI) systems[2]. The key idea behind ABCD-L is to approximate the continuous-time, continuous-valued dynamics of LTI systems using purely Boolean/discrete models. That is, given a set of differential equations for an LTI system (or alternatively, measured data, transfer function characteristics, scattering parameters, reduced order models, *etc.*), ABCD-L is a "push-button" style technique that produces as output a Boolean circuit abstraction (comprised entirely of Boolean logic elements such as registers, counters, *etc.*), that compactly encodes the analog behaviour of the given system, in a completely scalable fashion. Another important feature of ABCD-L is that it can approximate LTI systems to *any desired level of accuracy* (see §II).

Briefly, ABCD-L works by *discretizing* each underlying circuit signal, as well as the circuit's inputs, using as many bits as necessary to achieve the desired accuracy. These discretized values are stored in Boolean *registers*. Given the contents of each register at a particular time instant, ABCD-L uses Boolean logic to approximate the *next time instant* at which these contents must be updated (*e.g.*, in response to changing input). Thus, ABCD-L *transforms* the underlying LTI system into *an event-based discrete formulation*, realizable as a Boolean circuit (more details can be found in §II).

The above approach offers several compelling features. Firstly, because ABCD-L produces purely Boolean models, it is well-suited for use in conjunction with existing techniques for Boolean synthesis, verification, high speed logic simulation, *etc*. By reducing LTI dynamics to Boolean form, ABCD-L makes it possible to leverage powerful Boolean techniques for model checking/reachability analysis of LTI systems[3], as well as Boolean systems coupled with LTI dynamics (*e.g.*, high-speed digital logic with parasitic interconnect). Secondly, in the context of AMS modelling/verification, we know that existing hybrid systems approaches are unable to cope with more than a few continuous variables, whereas they can comfortably handle thousands of purely Boolean variables. Therefore, by enabling LTI dynamics to be accurately represented using "cheap" Boolean variables, ABCD-L frees up "precious" continuous variables for other purposes (*e.g.*, to accurately model non-linear dynamics). Therefore, with ABCD-L, it may be possible to expand the scope of hybrid systems approaches to much larger systems than they can handle at present. Although this notion has been theoretically studied before (*e.g.*, see [12]), we believe that ABCD-L constitutes the first practical approach for Booleanizing continuous LTI dynamics in an accurate, systematic, and scalable manner.

---

[1]**A**ccurate **B**ooleanization of **C**ontinuous **D**ynamics - **L**inear

[2]LTI systems constitute a fundamental class of AMS systems, including, for example, on-chip and off-chip interconnect, clock tree networks, filters, linearized small-signal circuits, channel models, *etc.*

[3]In this paper, we confine ourselves to the question of how to construct purely Boolean models that accurately capture analog LTI dynamics. *Formal verification* involving such models is an important next step, and one that is the subject of ongoing research. In this paper, however, we do not seek to address the verification problem.

Permission to make digital or hard copies of all or part of this work for personal or classroom use is granted without fee provided that copies are not made or distributed for profit or commercial advantage and that copies bear this notice and the full citation on the first page. To copy otherwise, to republish, to post on servers or to redistribute to lists, requires prior specific permission and/or a fee.

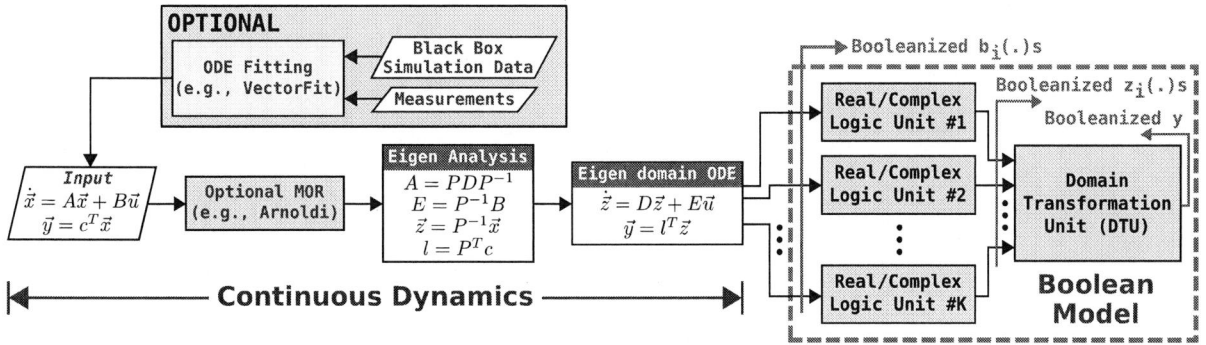

Fig. 1. The ABCD-L flow for producing a Boolean approximation that captures the analog dynamics of an LTI system. As shown in the figure, ABCD-L can also accept black-box simulation data as input, and it also integrates well with LTI MOR techniques.

Furthermore, ABCD-L also offers significant scalability advantages over explicit state-enumeration techniques like DAE2FSM [1], which produce Finite State Machines (FSMs) in State Transition Graph (STG) form, requiring exponential time and space complexity with respect to LTI system size. By contrast, ABCD-L's implicit, *circuit-based* models are exponentially more compact than STGs. As a result, ABCD-L scales practically linearly with respect to both system size and the desired fidelity of the Boolean approximation, without blowing up in either runtime or model size. In addition, the above event-based Boolean formulation lends itself to a new methodology for accurate, high-speed LTI system simulation (covered in §II). This methodology, as we show in §III, can offer considerable speed-up over traditional circuit simulation methods based on linear multi-step integration [13]. Moreover, ABCD-L has been designed to handle stiff systems efficiently, and it can take full advantage of LTI Model Order Reduction (MOR) techniques to Booleanize large LTI systems in a scalable manner.

We have applied ABCD-L to representative LTI circuits such as I/O links composed of RC and RLGC units. For these systems, we show that the Boolean models produced by ABCD-L are able to accurately reproduce the original system's continuous dynamics, including important performance-limiting analog effects such as inter-symbol interference, overshoot/undershoot, ringing, *etc*. In addition, we have applied ABCD-L to produce purely Boolean, small-signal models of a more complex, non-linear AMS system: an LTI channel followed by a differential equalizer (linearized using small-signal device models). In this case, too, ABCD-L is able to accurately capture and reproduce the circuit's dynamics in SPICE-level detail, using purely Boolean models all along. Further, ABCD-L is also able to capture higher-level design relevant AMS metrics, such as the eye diagram correction achieved by the equalizer. In addition, to accurately Booleanize industry-scale real-world LTI systems in a computationally viable manner, ABCD-L can be applied in conjunction with LTI MOR techniques; we demonstrate this for a 25849-node benchmark power grid network made available by IBM[4].

II. ABCD-L's CORE: A NEW TECHNIQUE FOR BOOLEANIZING LTI SYSTEMS

In this section, we describe the key ideas behind ABCD-L.

Fig. 1 depicts the ABCD-L flow, which takes as input an LTI system, and produces as output a Boolean approximation for it. For simplicity, we assume that the LTI system is specified as an ODE[5], of the form:

$$\dot{\vec{x}} = A\vec{x} + B\vec{u}, \quad \vec{y} = c^T\vec{x}, \tag{1}$$

where $\vec{x}$ is the system's (continuous) analog state (a vector of voltages and currents), A is a real square matrix, $\vec{u}$ is the system's (time-varying) input, and $\vec{y}$ its corresponding output. We note that, if the ODE is not directly available, but instead only measured data (*e.g.*, from AC excitation at several frequencies), or S-parameters, or black-box transfer function characteristics, are available, ABCD-L can still obtain the requisite ODE by applying standard fitting techniques (*e.g.*, VectorFit [14], table-based methods [15], *etc.*).

As indicated in Fig. 1, ABCD-L begins with an eigenanalysis [16] of the above ODE system, which produces a new ODE system of the form:

$$\dot{\vec{z}} = D\vec{z} + E\vec{u}, \quad \vec{y} = l^T\vec{z}, \tag{2}$$

[4]Please see the supplement at the end of this paper.

[5]ABCD-L is also capable of Booleanizing more general systems of the form $Q\dot{\vec{x}} = A\vec{x} + B\vec{u}$, $\vec{y} = c^T\vec{x}$, where Q may or may not be invertible. However, due to space constraints, we limit our discussion here to LTI systems in ODE form.

where D is a square diagonal matrix containing the eigenvalues of A. The matrices $[A, B, c]$ are related to the matrices $[D, E, l]$ through the eigenvector matrix P (the relevant equations can be found in Fig. 1).

The ith equation of the new ODE is a "de-coupled" scalar linear differential equation of the form:

$$\dot{z}_i = \lambda_i z_i + b_i(t), \tag{3}$$

where λ_i is the matrix entry $D_{i,i}$ and $b_i(.)$ is the ith entry of the vector $E\vec{u}(.)$. Given the initial condition $z_i(t_0)$, the solution to the above equation is known analytically, and is given by:

$$z_i(t) = z_i(t_0)e^{\lambda_i(t-t_0)} + \int_{t_0}^{t} b_i(\tau)e^{\lambda_i(t-\tau)}d\tau \tag{4}$$

At this point, we would like to make some observations:

○ On some (extremely rare, Lebesgue measure zero) occasions, the given LTI system may not be diagonalizable, *i.e.*, the matrix D above, instead of being diagonal, may take a Jordan form. It is possible (though tedious) to develop a general theory for Booleanizing such systems; however, because this almost never happens in practice, we do not consider it in this paper.

○ In some cases, the size of the given system makes eigenanalysis impractical. In such situations, we first use an LTI MOR technique (*e.g.*, Arnoldi iteration [17], [18]), to reduce the system size, and then subject the reduced system to eigenanalysis. The theory of LTI MOR is well-developed, and many large LTI systems encountered in practice can successfully be reduced using the MOR techniques available today. Also, as Fig. 1 shows, it is relatively straightforward to integrate virtually any MOR technique into the ABCD-L flow; we demonstrate this for a 25849-node benchmark power grid network made available by IBM, in the supplemental material at the end of the paper.

○ Many real-world LTI systems are stiff, *i.e.*, their underlying signals evolve at widely different timescales because the systems' eigenvalues span several orders of magnitude. ABCD-L can efficiently handle such systems by using different timescales to Booleanize the dynamics corresponding to different eigenvalues. However, due to space constraints, and for notational simplicity, we do not elaborate on this here.

Resuming the ABCD-L flow, the key idea behind ABCD-L is to *transform* the analytical solution of Eq. (4) into *Boolean operations* that can be expressed using digital logic constructs (registers, counters, *etc.*). This transformation is achieved by a set of Boolean *Logic Units (LUs)*, one for each component of $\vec{z}$. Each LU is either a *real LU (RLU)* or a *complex LU (CLU)*, depending on the corresponding eigenvalue. The output sequence of the ith such LU is a (multi-bit) Boolean approximation of the ith component of $\vec{z}(t)$. In addition, a combinational *Domain Transformation Unit (DTU)* combines the outputs of the LUs into a multi-bit Boolean approximation of $y(t)$ (which can be mapped back into a piecewise constant analog signal as a post-processing step). Below, we describe how the DTU and the individual LUs are structured.

The DTU's output is simply a Booleanized linear combination of its inputs. This is a combinational function whose Boolean specification/synthesis has been well-studied (*e.g.*, see [19]).

The LUs, on the other hand, are sequential systems, and their construction is more challenging. Each LU implements, using purely Boolean logic, a scalar linear differential equation $\dot{z} = \lambda z + b(t)$. The signals $z(t)$ and $b(t)$ are encoded as bit-vectors of length m (where m is a parameter passed to ABCD-L, called the *signal resolution*). The LU is designed so that its

445

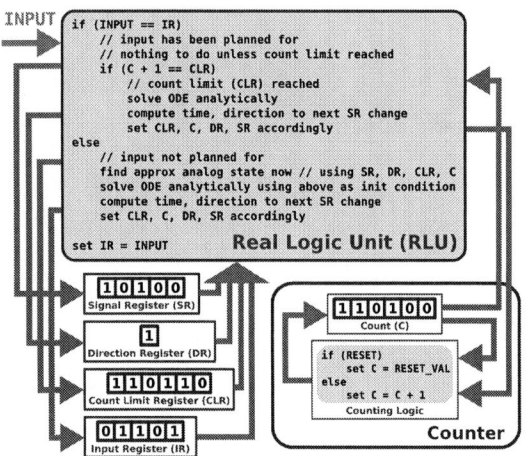

Fig. 2. Sequential logic schematic for discretizing a real scalar linear differential equation $\dot{z} = \lambda z + b(t)$ in the eigendomain.

bit-vector approximation to $z(t)$, when mapped back into the analog domain, closely matches the actual system response $z(t)$ for all input sequences $b(t)$. This is achieved by the logic structures depicted in Figs. 2 and 3, for real and complex eigenvalues respectively.

Let us first consider the real eigenvalue scenario, *i.e.*, the problem of Booleanizing $\dot{z} = \lambda z + b(t)$, where all quantities are real. Fig. 2 shows a Boolean schematic for this, which includes (a) a *Signal Register* SR that maintains an m-bit representation of z, (b) a 1-bit *Direction Register* DR, that denotes whether z is increasing/decreasing, (c) a *Count Limit Register* CLR that indicates the time at which SR must be incremented/decremented, (d) a set/reset counter with a count C, which measures time by counting up to the limit CLR, and (e) an m-bit *Input Register* IR that stores the input b. The whole unit is clocked at a pre-determined, fixed time-step Δ (in practice, it is usually straightforward to choose Δ to ensure that it is small enough to capture the dynamics of the scalar system above). For stiff systems, one can boost computational efficiency by using different Δs for the different LUs.

The above unit works as follows: as long as the input $b(t)$ remains constant, it is, in some sense, already "planned for". That is, the above structure stores enough information to know when, and in which direction, the register SR must be incremented/decremented. In this case, the count C keeps ticking up until it eventually becomes equal to the count limit CLR, at which time the register SR (and all the other registers as well) are updated accordingly. Analytical expressions, based on Eq. (4), are known for the "time to next change in SR mod Δ" operation, which can be either computed on the fly, or stored in a truth-table, *etc.*

For non-constant $b(t)$, as Fig. 2 indicates, the logic unit responds by considering the latest sampled input to be a new DC input, and updates all the registers using the analytical "time to next change mod Δ" function, making an intelligent estimate about the current value of z using the contents of registers SR, C, CLR, and DR (note that the count can be reset to any value by passing the RESET signal to the counter).

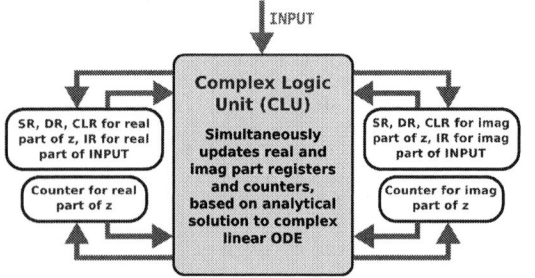

Fig. 3. Sequential logic implementation schematic for discretizing a complex scalar linear differential equation $\dot{z} = \lambda z + b(t)$ in the eigendomain.

For the complex eigenvalue case, a CLU (Fig. 3) essentially consists of two copies of each RLU register, storing the real and imaginary parts of each underlying signal. Whenever the registers need to be updated (for example, if the input has changed or if the limit CLR is reached by one of the counters), both the real and the imaginary sets of registers are updated simultaneously, based on the analytical solution given by Eq. (4).

From the above discussion, ABCD-L's accuracy clearly increases with the signal resolution m (which is a designer-specified parameter that determines how finely the underlying signals are discretized). In principle, this allows ABCD-L to abstract the given LTI system to any desired level of accuracy. In practice, for a designer, it is usually straightforward to determine an appropriate value for m through trial and error.

We also note that time-domain simulations involving ABCD-L's Boolean models can be very efficient, because they can be carried out entirely in the logical/Boolean domain (if necessary, using specialized logic simulation tools), without requiring any differential equation solving. To speed up simulation even further, we can devise an algorithm that jumps directly to the time instant specified by CLR, instead of incrementing the count C at every time-step. Indeed, we have implemented this algorithm, and in §III, we demonstrate that it can be significantly faster than conventional circuit simulation methods such as linear multi-step integration (even after accounting for the time taken up by eigenanalysis, model generation, *etc.*).

III. RESULTS

Having described the core techniques behind ABCD-L, we now apply them to Booleanize LTI systems that are of interest to AMS designers, including, (1) I/O links modelled using RC/RLGC chains, and (2) an "LTI channel followed by a differential equalizer" circuit, linearized using small-signal analysis. We show that the Boolean models produced by ABCD-L are able to accurately capture the continuous-time dynamics of such systems, including important analog effects such as inter-symbol interference (ISI), overshoot/undershoot, ringing, *etc.* Also, we show that ABCD-L is able to faithfully reproduce higher-level AMS-design metrics for such systems, including the *entire shape of the eye diagram opening* at key circuit nodes. Furthermore, from a computational efficiency perspective, we show that ABCD-L can offer considerable simulation speed-ups over traditional LTI ODE/DAE simulation methods like linear multi-step integration (especially for larger LTI systems). Note that the examples in this section do not use LTI MOR; for examples involving MOR, please see the supplemental material.

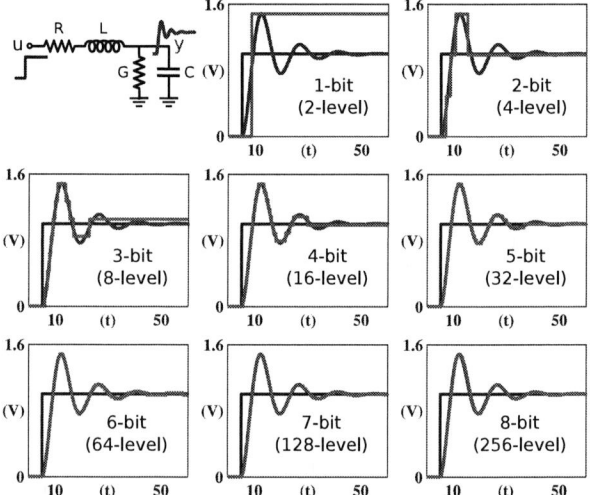

Fig. 4. ABCD-L applied to an RLGC filter, to produce Boolean models of arbitrarily high accuracy. The continuous system's response to a step input $u(t)$ (in black) is computed both analytically (blue), and using ABCD-L (green). In the plots above, the X-axis denotes time in RC units, and the Y-axis denotes voltages in Volts.

Before presenting the examples described above, we would first like to highlight an important feature offered by ABCD-L: in spite of using purely Boolean models, ABCD-L can reproduce the continuous-time dynamics of LTI systems with *arbitrarily high accuracy*, simply by increasing the number of bits used to represent the underlying circuit signals. Fig. 4 illustrates this using an RLGC filter (shown at the top left of the figure). This is a linear system of size 2, whose eigenvalues are both complex. As described in §II, ABCD-L discretizes the circuit's input $u(t)$ (in this example, a unit step function), its internal voltages/currents, and its output $y(t)$, into bit vectors of length m, where higher values of m correspond to finer quantization of the underlying analog signals (hence greater accuracy). The figure shows that, as we increase m from 1 to 8, the response predicted by ABCD-L's Boolean model (the green waveform) matches the actual system's response (the blue waveform) more and more accurately, duplicating important features such as

446

overshoot and ringing. This is true for LTI systems in general: ABCD-L's Boolean approximations can be as close to the original system as desired.

We now apply ABCD-L to chains of RC/RLGC units (Fig. 5); these are often used to model high-speed I/O links, interconnect networks, on-chip communication channels, *etc.* [20], [21].

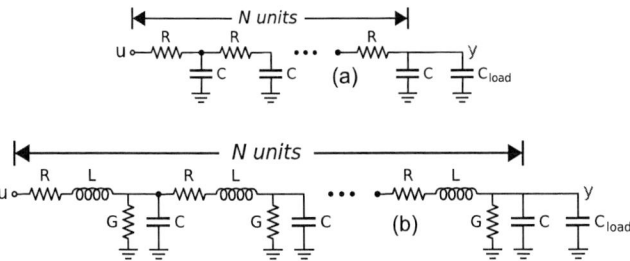

Fig. 5. RC and RLGC chains of length N, each driving a load capacitance C_{load}

Our experiments on RC/RLGC chains run as follows:

○ We build an RC/RLGC chain, and apply several long, randomly generated bit patterns $u(t)$ at its input. To model both small and large inter-symbol interference (ISI), we vary the unit-interval T (the time that elapses between successive bits at the input), which is the inverse of the bitrate. Note that ISI decreases with increasing T, and vice-versa.

○ We use ABCD-L to predict the system's responses $y(t)$ to the above inputs. Between experiments, we vary ABCD-L's signal resolution parameter m (the number of bits used by ABCD-L to quantize the underlying circuit's eigendomain signals).

○ We compare ABCD-L's time-domain predictions against those of an ODE/DAE solver, plotting them on top of one another.

○ Finally, since I/O link/interconnect engineering often makes extensive use of eye diagrams [22], we also convert ABCD-L's predictions into eye diagram form, and compare against eye diagrams generated by an ODE/DAE solver.

Fig. 6 depicts the results obtained by applying 4-bit ABCD-L to a 10-unit RC chain, and to a 10-unit RLGC chain, under conditions of both small ISI and large ISI. We note that, because all signals are quantized using 4 bits, the output of the system, as predicted by ABCD-L, consists of at most $2^4 = 16$ different levels. Moreover, as seen from the figure, in all cases, the 16-level predictions made by ABCD-L (drawn in green) closely match the system's actual continuous-time responses (drawn in blue). Therefore, the Booleanized models produced by ABCD-L appear to be good approximations of the underlying continuous LTI systems.

In parts (a) and (c) of Fig. 6, the bitrate of the applied input (the black waveform labelled $u(t)$) is low enough that the resulting ISI is small. This is readily seen from the system's responses: whenever the input bit is high (low), the output response has enough time to rise (fall) to a reasonably high (low) level before the next bit arrives. On the other hand, in parts (b) and (d) of the figure, we increased the input bitrates to a point where the induced ISI became significant. This can also be seen visually from the figure – even if an input bit is high (low), the output responses often do not have enough time to rise (fall) before the next bit arrives. As the figure shows, all these effects are captured quite accurately by ABCD-L.

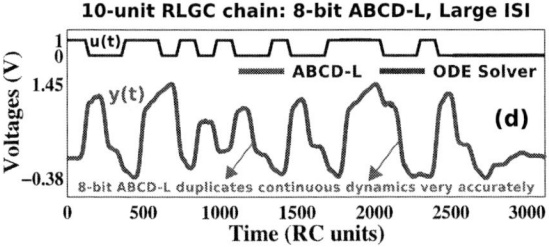

Fig. 7. Applying 8-bit ABCD-L to a 10-unit RLGC chain, for the same inputs as in Fig. 6(d). With the increased signal resolution, it is seen that the deviations between ABCD-L's prediction and the system's actual response are significantly reduced.

In Fig. 6(d), we have drawn attention (with a red circle) to a small time-interval where there is some deviation between ABCD-L's prediction and the system's response. Although such deviations tend to be "self-correcting" (as seen from the figure), it may be desirable to reduce the magnitude of these

deviations. As described in §II, this can be achieved simply by increasing the number of bits used by ABCD-L to discretize the underlying waveforms (this corresponds to changing a single parameter that is passed as an argument to ABCD-L). This is illustrated in Fig. 7, where we have applied 8-bit ABCD-L (instead of 4-bit ABCD-L) to the same RLGC chain, for the same inputs as in Fig. 6 (d). From the figure, it is readily seen that the deviations between ABCD-L and the original system have been significantly reduced. This supports our earlier assertion that ABCD-L can model any LTI system with arbitrarily high accuracy.

Having illustrated ABCD-L's ability to closely match the underlying system's responses for short input patterns, we now simulate the ABCD-L-generated Boolean models on much longer input patterns (thousands of bits), and convert the resulting predictions into *eye diagram form*; this representation is extensively used in I/O link/interconnect design, modelling, analysis, and simulation, because it provides the design architect with a quick snapshot of key system properties [22]. Due to space constraints, we present eye diagrams only for the RLGC chain (and not the RC chain).

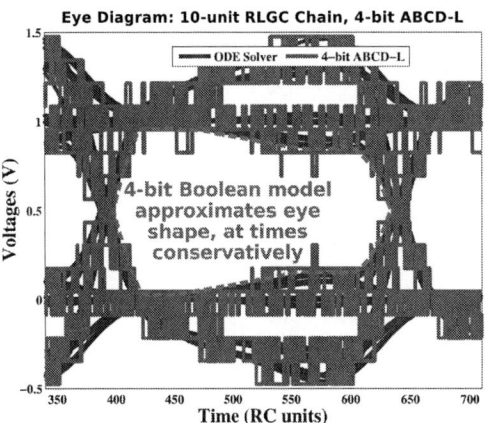

Fig. 8. Eye diagrams produced by 4-bit ABCD-L (green), and by an ODE solver (blue), for the 10-unit RLGC chain.

Fig. 8 depicts the eye diagram produced by applying 4-bit ABCD-L (in green) to the 10-unit RLGC chain of Fig. 6. This eye diagram is overlaid on top of the eye diagram produced by an ODE solver (which is in blue). The red dashed contour traces the shape of the eye opening, as predicted by the ODE solver. As seen from the figure, ABCD-L's 4-bit Boolean model is able to reproduce the entire shape of the eye opening with good accuracy.

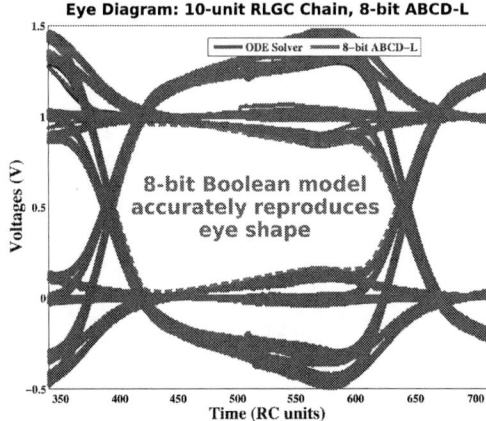

Fig. 9. Eye diagrams produced by 8-bit ABCD-L (green), and by an ODE solver (blue), for the 10-unit RLGC chain. It is seen that the shape of the eye opening is reproduced with increased accuracy compared to Fig. 8.

However, the eye opening produced by 4-bit ABCD-L is at times a bit conservative. While this may not be a problem for many applications, it may sometimes be necessary to obtain a more accurate representation of the eye opening. As we have indicated before, such increased accuracy can be achieved simply by asking ABCD-L to use more bits to discretize the underlying circuit's signals. This is illustrated in Fig. 9, which depicts the eye diagram produced by 8-bit ABCD-L for the same RLGC chain. Indeed, as seen from the figure, the 8-bit ABCD-L model almost perfectly reproduces

447

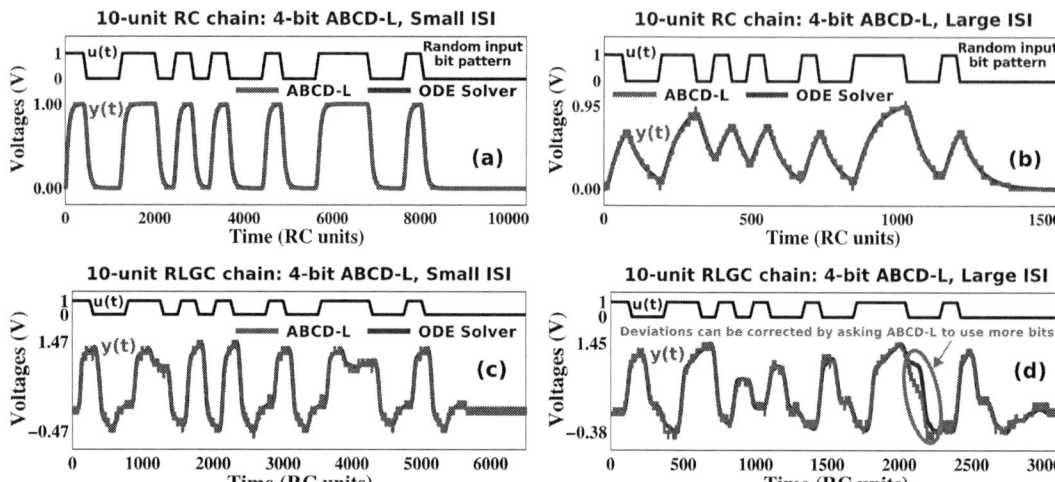

Fig. 6. Applying 4-bit and 8-bit ABCD-L to a 10-unit RC chain. The resulting Boolean models are simulated under conditions of both small ISI (parts (a), (c)) and large ISI (parts (b), (d)), and the Boolean models' predictions (the green waveforms) are compared against actual system responses (the blue waveforms), for a randomly generated input pattern (the black waveforms labelled $u(t)$).

the shape of the entire eye opening (as compared to the red dashed contour obtained by ODE simulation). This also confirms our assertion that ABCD-L can approximate LTI systems to any desired level of accuracy.

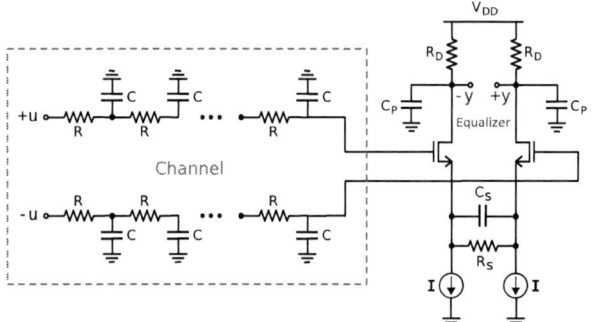

Fig. 10. LTI channel followed by a differential equalizer.

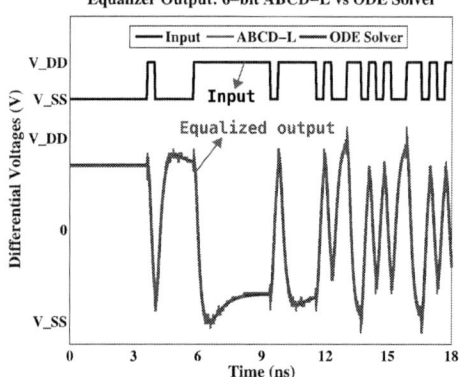

Fig. 11. ABCD-L accurately reproduces the time-domain continuous dynamics of the equalizer.

Having applied ABCD-L to RC/RLGC chains, we now consider a more complex example relevant to AMS-design: an LTI channel followed by an equalization circuit, as illustrated in Fig. 10. We note that, in this circuit, both the channel (modelled as an RC chain) and the equalizer use *differential signalling*, i.e., the circuit's inputs and outputs are represented by the *difference* between two voltages (instead of a single voltage)[6]. The equalizer plays a critical role in this circuit: it *partially reverses* the distortion (ISI) produced by the channel, so that one can transmit bits across the channel at much higher speeds than would be possible otherwise. For example, if the

---

[6]Differential signalling has important advantages over single-ended signalling, including better noise resilience, improved resistance to external interference, *etc.*

channel's cut-off frequency is 1 GHz, then reliable transmission can happen only at bitrates at or below 1Gbps. However, if the combined "channel plus equalizer" system has an effective cut-off frequency at 3 GHz, then one can triple the throughput without suffering distortion.

We also note that the circuit in Fig. 10 is non-linear. Therefore, we apply ABCD-L not to the original circuit, but to a small-signal linearization of the original circuit around its quiescent operating point, using small-signal device models obtained from, *e.g.*, the book by Sedra and Smith [23].

Fig. 11 illustrates the application of 6-bit ABCD-L to the small-signal linearized "channel plus equalizer" circuit (henceforth simply referred to as "the system" above (with the channel being a 5-unit RC chain). The blue waveform of Fig. 11 was obtained by using an ODE solver to simulate the system, for a random bit pattern applied at the input (the black waveform). As before, we see from the figure that the Boolean model produced by ABCD-L is able to accurately duplicate the time-domain behaviour of the system.

For equalizers, an important AMS-relevant design metric is the eye diagram correction produced by the circuit. In typical AMS applications, the eye diagram at the input to the equalizer (i.e., the channel output) has a very small or even non-existent eye opening (Fig. 12(a)). The equalizer offsets some of the ISI produced by the channel, which can considerably widen the eye opening; for example, Fig. 12(b) shows the eye diagram produced by a small-signal SPICE simulation of the above system, using SpiceOPUS [24]. Parts (c) and (d) of Fig. 12 depict the eye diagrams produced by applying 6-bit and 8-bit ABCD-L to the above "channel plus equalizer" system, overlaid on top of the (blue) SPICE eye diagram. As the figures show, the eye diagrams obtained from ABCD-L's Boolean models are able to accurately reproduce the eye diagram correction achieved by the equalizer. Thus, we have demonstrated that ABCD-L is a viable technique to Booleanize the continuous dynamics of AMS systems.

Having presented results pertaining to ABCD-L's accuracy, we now consider its computational efficiency. As outlined in §II, the Boolean models produced by ABCD-L lend themselves to efficient time-domain simulation carried out entirely in the discrete/logical domain, without having to solve differential equations. Even after taking into account the time taken to generate the ABCD-L models, ABCD-L can still be many times faster than conventional circuit simulation techniques like linear multi-step integration. This is illustrated in Fig. 13, which compares the total time taken by 4-bit, 5-bit, 6-bit, and 8-bit ABCD-L (total runtime includes pre-processing, model generation, simulation, and post-processing) against the time taken by Backward Euler integration, for simulating RC chains of various lengths on a long, randomly generated input bit pattern. As the figure shows, ABCD-L does offer a significant speedup advantage. Moreover, as the LTI system size increases, this advantage becomes even more pronounced[7]. In addition, as discussed in §II (and illustrated in the supplemental material), it is straightforward to integrate linear MOR techniques (*e.g.*, Arnoldi iteration [17], [18]) into

---

[7]All the ABCD-L simulations of Fig. 13 have been carried out in C++, on a system equipped with a 6-core 3.2 GHz AMD® Phenom™ II X6 1090T processor, and with a total of 16GB (shared) memory.

448

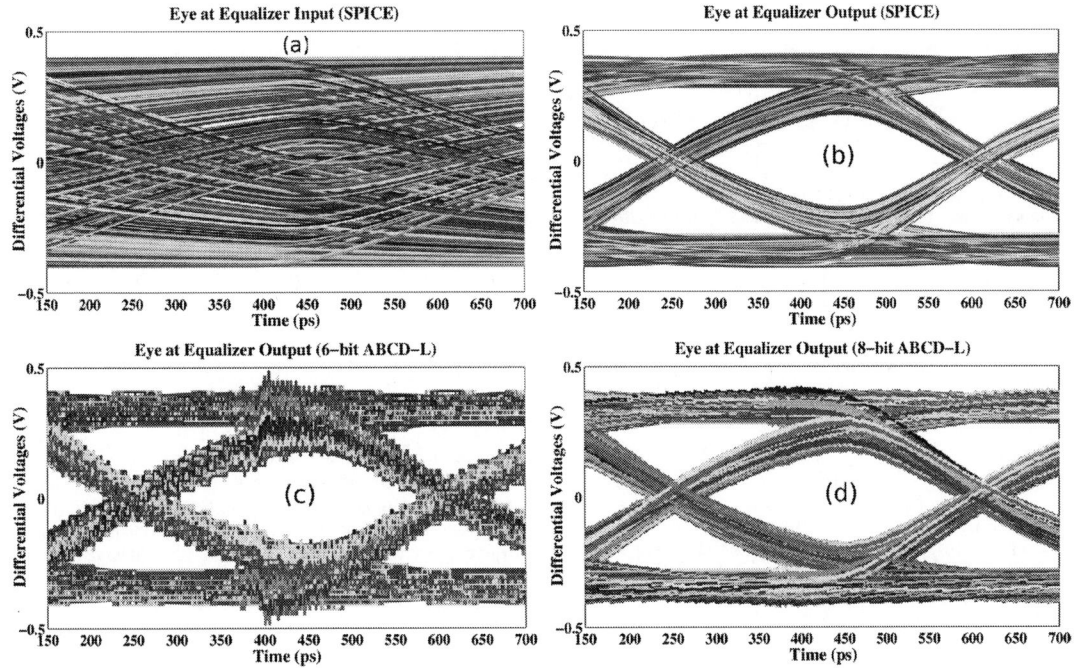

Fig. 12.   ABCD-L accurately reproduces the entire shape of the eye diagram at the equalizer's output.

ABCD-L, which can further improve its runtime.

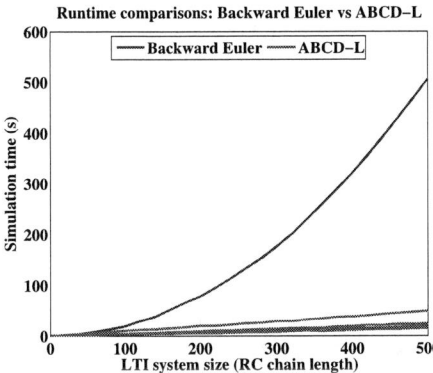

Fig. 13.   ABCD-L can offer considerable simulation speed-up over traditional circuit simulation techniques like Backward Euler integration. The figure illustrates this for RC chains of varying length, and for 4-bit, 5-bit, 6-bit, and 8-bit ABCD-L.

IV. SUMMARY, CONCLUSIONS, AND FUTURE WORK

In this paper, we have developed and demonstrated ABCD-L, a technique that automatically produces Boolean approximations of continuous LTI systems, to any desired level of accuracy, in a completely scalable fashion. We have applied ABCD-L to representative LTI systems such as RC/RLGC chains, where it captures important analog effects like inter-symbol interference, ringing, *etc.* We have also demonstrated ABCD-L on a small-signal linearized "channel plus equalizer" circuit, where it is able to reproduce key design-relevant AMS metrics, including the eye diagram correction achieved by the circuit. Also, we have shown that ABCD-L generated models can offer significant simulation speed-up over conventional circuit simulation techniques like linear multi-step integration.

Through ABCD-L, we have demonstrated that systems specified in purely Boolean form have the ability to capture continuous-time LTI dynamics with excellent accuracy and scalability. However, it is important to develop this idea further, and bring ABCD-L to closure with formal verification methods, Boolean SAT solvers, *etc.* The first step in this regard is to develop techniques for efficiently synthesizing the combinational logic present in ABCD-L models. To this end, we are actively exploring new ways to exploit the structure of the underlying floating point computations, using specialized tools like ABC [4], to come up with compact, gate-level descriptions of ABCD-L models.

Our long-term future plans are shaped by our belief that ABCD-L possibly espouses a new route to the longstanding problem of SPICE-accurate AMS verification. By Booleanizing continuous-time LTI dynamics, ABCD-L capitalizes on the ability of existing Boolean/hybrid systems modelling/verification techniques to handle large numbers of Boolean/discrete variables. At the same time, ABCD-L attempts to steer clear of the main weakness of existing formal methods, namely, their scalability limitations while analyzing systems with more than a few continuous variables. Therefore, in future, we would like to work on (1) extending ABCD-L to Booleanize strongly non-linear systems as well (*e.g.*, mixers, oscillators, *etc.*), and (2) integrating ABCD-L with state-of-the-art techniques for Boolean and hybrid systems verification, to develop a new AMS verification methodology founded on Booleanizing continuous dynamics.

REFERENCES

[1] K. V. Aadithya and J. Roychowdhury. DAE2FSM: Automatic generation of accurate discrete-time logical abstractions for continuous-time circuit dynamics. In *DAC '12: Proceedings of the 49th Design Automation Conference*, pages 311–316, 2012.
[2] G. Taylor. Future of analog design and upcoming challenges in nanometer CMOS. Keynote address at the 2010 International Conference on VLSI Design.
[3] R. Parker. Analog design challenges in the new era of process scaling. At the 2012 International Workshop on Design Automation for AMS Circuits (co-located with ICCAD).
[4] R. K. Brayton and A. Mishchenko. ABC: An academic industrial-strength verification tool. In *CAV '10: Proceedings of the 22nd International Conference on Computer Aided Verification*, pages 24–40, 2010.
[5] T. A. Henzinger, P. H. Ho, and H. Wong-Toi. HyTech: A model checker for hybrid systems. *International Journal on Software Tools for Technology Transfer*, 1(1):110–122, 1997.
[6] C. Tomlin, I. Mitchell, A. M. Bayen, and M. Oishi. Computational techniques for the verification of hybrid systems. *Proceedings of the IEEE*, 91(7):986–1001, 2003.
[7] A. Chutinan and B. H. Krogh. Computational techniques for hybrid system verification. *IEEE Transactions on Automatic Control*, 48(1):64–75, 2003.
[8] G. Al-Sammane, M. H. Zaki, and S. Tahar. A symbolic methodology for the verification of AMS designs. In *DATE '07: Proceedings of the ACM Conference on Design, Automation and Test in Europe*, pages 249–254, 2007.
[9] G. Frehse. PHAVer: Algorithmic verification of hybrid systems past HyTech. *International Journal on Software Tools for Technology Transfer*, 10(3):263–279, 2008.
[10] S. Little. *Efficient Modeling and Verification of Analog/Mixed-Signal Circuits Using Labeled Hybrid Petri Nets*. PhD thesis, University of Utah, 2008.
[11] M. Althoff, A. Rajhans, B. H. Krogh, S. Yaldiz, X. Li, and L. Pileggi. Formal verification of Phase Locked Loops using reachability analysis and continuization. In *ICCAD '10: Proceedings of the IEEE/ACM International Conference on Computer-Aided Design*, pages 659–666, 2010.
[12] R. Alur, T. A. Henzinger, G. Lafferriere, and G. J. Pappas. Discrete abstractions of hybrid systems. *Proceedings of the IEEE*, 88(7):971–984, 2000.
[13] C. W. Gear. *Numerical initial value problem in ordinary differential equations*. Prentice Hall, Inc., 1971.
[14] B. Gustavsen and A. Semlyen. Rational approximation of frequency domain responses by vector fitting. *IEEE Transactions on Power Delivery*, 14(3):1052–1061, 1999.
[15] C. P. Coelho, J. Phillips, and L. M. Silveira. A convex programming approach for generating guaranteed passive approximations to tabulated frequency data. *IEEE Transactions on Computer-Aided Design of Integrated Circuits and Systems*, 23(2):293–301, 2006.
[16] G. Strang. *Linear algebra and its applications*. Thomson, Brooks/Cole, 2006.
[17] W. E. Arnoldi. The principle of minimized iterations in the solution of the matrix eigenvalue problem. *The Quarterly of Applied Mathematics*, 9(1):17–29, 1951.
[18] W. H. A. Schilders, H. A. van der Vorst, and J. Rommes. *Model Order Reduction: Theory, research aspects and applications*, volume 13 of *Mathematics in Industry*. Springer Verlag, 2008.
[19] N. Brisebarre, F. De Dinechin, and J. M. Muller. Integer and floating-point constant multipliers for FPGAs. In *ASAP' 08: Proceedings of the 19th IEEE International Conference on Application Specific Systems, Architectures and Processors*, pages 239–244, 2008.
[20] P. K. Hanumolu, G. Y. Wei, and U. K. Moon. Equalizers for high-speed serial links. *International Journal of High Speed Electronics and Systems*, 15(2):429–458, 2005.
[21] B. Razavi. *Design of integrated circuits for optical communications*. Springer, Netherlands, 2003.
[22] G. Balamurugan, B. Casper, J. E. Jaussi, M. Mansuri, F. O'Mahony, and J. Kennedy. Modelling and analysis of high-speed I/O links. *IEEE Transactions on Advanced Packaging*, 32(2):237–247, 2009.
[23] A. S. Sedra and K. C. Smith. *Microelectronic circuits*. Oxford University Press, 2007.
[24] http://www.spiceopus.si/.

ABCD-L: Approximating Continuous Linear Systems Using Boolean Models (Supplement)

Abstract—**In this supplement, we provide additional context for ABCD-L and place our contributions in perspective, relative to the existing body of literature on topics like AMS modelling/verification, Boolean and hybrid systems frameworks, *etc.* Further, we demonstrate that ABCD-L can be applied in conjunction with Model Order Reduction (MOR) techniques, to Booleanize large LTI systems whose direct eigendecomposition may be computationally infeasible. For example, we combine ABCD-L with Arnoldi iteration based MOR to efficiently produce accurate Boolean models of a real-world power grid network (with 25849 nodes) obtained from a benchmark set made available by IBM. Due to space constraints, we were unable to include such material within our main manuscript.**

I. ABCD-L IN THE CONTEXT OF EXISTING FORMAL TECHNIQUES

Much of the existing body of work on the formal analysis and modelling of AMS systems has been carried out by the Boolean and hybrid systems verification communities. This literature is too vast to cover in full detail here; however, we will provide a brief overview highlighting the common features shared by existing approaches, and how ABCD-L complements them.

One trait that is shared by almost all existing formal verification systems is that they work with simplified behavioural models for AMS components – models that do not bear close resemblance to SPICE. Indeed, many general frameworks have been proposed for formal verification and reachability analysis of dynamical systems that involve both discrete and continuous variables; however, the verification involving continuous quantities often scales much more poorly than verification involving purely Boolean/discrete quantities. This limits the applicability of the proposed techniques/frameworks to behavioural models of AMS systems, rather than models that achieve SPICE-level accuracy.

For example, consider the work by Ghosh and Vemuri [1], who applied the PVS proof checking tool to simplified analog circuit models (*e.g.*, using idealized OpAmps, transistors with constant transconductance, *etc.*). While this was an important step towards AMS verification, its range of applications was limited by computational challenges arising from the need to formally verify arithmetic over real numbers.

Due to the inherent scalability limitations of verifying continuous systems, many other techniques were developed to abstract analog dynamics using highly simplified behavioural models, carefully tailored to specific application domains/circuit classes. For example, Hanna and others [2], [3] developed new approaches to formally verify digital circuits suffering from analog non-idealities.

There is also another class of AMS verification approaches; these methods try to partition the continuous analog state space of voltages and currents into discrete domains, and the idea is to encode transitions between these domains using Boolean data structures like Finite State Machines, Binary Decision Diagrams, *etc.* For example, the work by Kurshan and Macmillan [4], Hedrich et. al. [5], [6], *etc.* fall into this category. The formal verification and model checking of such abstractions can be carried out using either off-the-shelf techniques with only small modifications (*e.g.*, as done by Kurshan), or by specially augmenting existing CTL model-checking tools with extra AMS-relevant features (as espoused by Hedrich and others). However, in spite of researchers' best efforts along these lines, their techniques were scalable only to small designs (*e.g.*, a single gate [4], or a small tunnel diode [5]), and were also limited in their power to model real-world analog phenomena that were of interest to AMS designers.

With a view to scaling up verification techniques to real AMS designs, several new modelling frameworks, with associated verification methodologies, were introduced. Together, these fall under the umbrella of "hybrid systems verification", a topic that has been extensively studied in the literature, and mathematically formalised by researchers like Alur, Nerode, and Henzinger [7]–[10].

The above modelling/formalisation efforts were critical because many reachability questions on general hybrid systems are, in fact, undecidable. Therefore, for AMS verification, it is necessary to restrict one's domain to classes of hybrid systems that are known to be decidable; the above formalisation efforts helped develop a strong *theory of hybrid systems*, that

enabled researchers to ask more meaningful questions, which in turn enabled new advances in reachability analysis/model checking of AMS systems.

For instance, Al-Sammane et. al. used recurrence relations and difference equations to simplify analog blocks [11], which were later verified using interval arithmetic and Taylor approximations [12]. This methodology was successfully used to check the stability of a third order $\Delta\Sigma$ modulator (modelled behaviourally). Also of significance is the d/dt tool, developed exclusively for model checking continuous linear systems [13], although limited to rather small system sizes.

Other novel hybrid systems modelling frameworks (that restrict themselves to decidable hybrid systems classes) for AMS include: guarded state machines as applied to flash memories, verified using the ACL2 theorem prover [14], linear hybrid automata verified using tools like HyTech [15] and Phaver [16], Polyhedral invariant hybrid automata, verified using flowpipes and the theory of quotient transition systems, by tools like Checkmate [17], labelled hybrid petri-nets, verified using difference bound matrices as part of the LEMA toolkit developed by Myers et. al. [18]–[21], and many more that we do not mention here due to space constraints. In addition, several new algorithmic refinements have improved the accuracy and efficiency of hybrid systems' reachability analysis. These include zonotopes [22], hybrid restriction diagrams [23], and other over-approximation techniques (*e.g.*, using support functions [24], continuization methods [25], *etc.*).

Thus, much effort has been devoted to developing and fine-tuning Boolean and hybrid systems modelling frameworks and verification engines. However, it is our belief that the question of *SPICE-accurate modelling*, which ensures that the circuit models used by verification engines actually reflect underlying analog reality, has not received adequate scrutiny. Without SPICE-accurate modelling, the predictions made by verification engines are questionable and dangerous for designers to rely on. Indeed, this is an important reason why the prevailing practice amongst AMS designers today is to carry out time-consuming SPICE simulations rather than place their trust in AMS verification tools.

To overcome such "designer skepticism", we believe that it is necessary to significantly scale up existing hybrid systems techniques, so that they embrace SPICE-accurate models even for large AMS designs. However, this creates scalability problems. To circumvent such scalability issues, we suggest that continuous variables (which introduce major computational challenges that are orders of magnitude more severe than purely discrete/Boolean variables) be used as sparingly as possible in the modelling of AMS components.

Thus, in our view, there is a need to develop new techniques that accurately capture the behaviour of continuous systems using purely discrete/Boolean variables, even though, in theory, existing hybrid systems approaches can represent and reason about continuous quantities. Armed with such techniques, we believe that much of an AMS system's behaviour will be representable using "cheap" purely discrete/Boolean variables, which frees up the "precious" continuous variables for use only when absolutely necessary. This may help arrest the scalability issues inherent to existing hybrid systems approaches, and may one day enable the verification of large AMS systems at or near SPICE-level accuracy.

We view ABCD-L as a step in the above direction, which addresses the problem of Booleanizing analog dynamics, for LTI systems in particular. In future, we would like to extend ABCD-L to non-linear systems as well, and to integrate ABCD-L with existing Boolean and hybrid systems frameworks for AMS verification, to expand the scope of existing techniques to handle much larger systems than they can do so at present. This is our hope for the future of AMS verification, and is also the larger context behind ABCD-L.

II. ABCD-L COMBINED WITH MODEL ORDER REDUCTION

As described in §II of the main manuscript, ABCD-L requires eigendecomposition of LTI systems as part of the Booleanization process. However, as LTI system size increases, eigendecomposition can quickly become computationally infeasible. To address this problem, we suggest an approach involving Model Order Reduction (MOR). The idea is to first obtain a Reduced Order Model (ROM) of the original LTI system; many well-established techniques are available for this, including explicit moment matching methods such as Asymptotic Waveform Evaluation (AWE) [26], implicit Krylov subspace methods such as congruent transformation [27], Padé approximation via the Lanczos method [28], guaranteed-stable methods based on Arnoldi iteration [29], [30], *etc.* The next step is to apply ABCD-L to the ROM, which is typically much smaller than the original LTI system, and therefore not a barrier for eigendecomposition.

We now apply the above approach to a real-world power grid network, obtained from a benchmark set made available by IBM [31], [32]. This LTI network has 25849 nodes, making eigenanalysis slow and impractical. Therefore, we first carry out Arnoldi iteration [33] based MOR to reduce this system to a more manageable size. Indeed, as we show below, a ROM of size ∼20 suffices to capture the dynamics of this system for most frequencies of interest. We then Booleanize the ROM using ABCD-L, and show that the resulting Boolean model is able to accurately reproduce the behaviour of the original system, including such attributes as the power grid's voltage swings in the ground plane and in the V_{DD} plane.

A. Arnoldi ROMs for the power grid

Given an LTI system L_{orig} of size n, and a desired ROM size p (where $p \ll n$), the method of Arnoldi iteration produces an LTI system L_{ROM} of size p, that approximates the behaviour of the original system L_{orig}. The key idea behind Arnoldi MOR is to *match moments*, *i.e.*, the reduced order model L_{ROM} is constructed in such a way that the first p moments of its transfer function are identical to those of the original system L_{orig}. In addition to being fast, Arnoldi iteration has favorable numerical properties (in the context of fixed-precision computation), as opposed to other techniques like Padé approximation [33], [34].

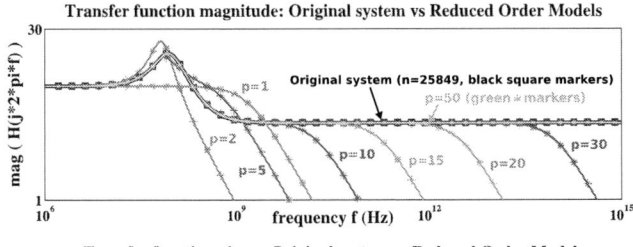

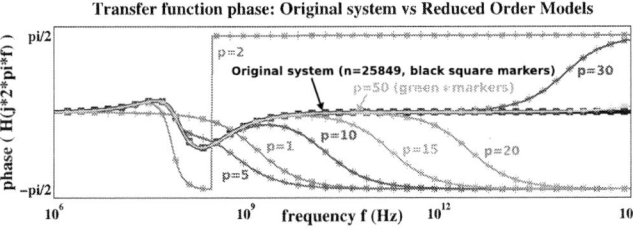

Fig. 1. Arnoldi ROM applied to the IBM power grid network. As ROM size p increases, the reduced order models produced by Arnoldi iteration become better and better approximations of the original system.

To determine a suitable ROM size p for the IBM power grid, Fig. 1 compares the LTI frequency-domain transfer function (both magnitude and phase) of the original power grid against those of several different Arnoldi ROMs, corresponding to different sizes p, over a wide frequency range (1 Megahertz to 1000 Terahertz). The black waveform (with black square markers) represents the original system's transfer function, while the '*' marked color waveforms correspond to the Arnoldi ROMs (with sizes as labelled in the figure). The figure clearly brings out the effectiveness of Arnoldi ROM for this network; for example, even though the original system is of size 25849, a ROM of size ∼20 suffices to accurately capture the system's behaviour for input excitations upto 100GHz. Therefore, an Arnoldi ROM of size ∼20 is more than adequate for most typical power grid applications[1].

---

[1]We note that Fig. 1 depicts the transfer function for only one output node of the power grid. In reality, the transfer functions corresponding to all output nodes (both in the power plane and in the ground plane) must be examined before concluding that ROM size $p = 20$ is adequate. For the given network, we have confirmed that this is indeed the case; however, due to space constraints, we do not include all the transfer function plots here.

B. ABCD-L + Arnoldi MOR applied to the power grid

We now apply ABCD-L to produce purely Boolean approximations of the ROMs generated above. As noted earlier, these ROMs are much smaller systems compared to the original power grid, and hence pose no difficulty for eigenanalysis.

Fig. 2 shows that the Boolean models produced by ABCD-L are able to accurately reproduce the transient dynamics of the Arnoldi ROMs generated for the IBM power grid. Parts (a), (b), (c), and (d) of the figure correspond to ROM sizes 5, 8, 10, and 20 respectively. In each case, the power grid was excited by the same input waveform $u(t)$ – a superposition of two damped sinusoidal excitations at 10GHz, one positive and the other negative (see Fig. 3).

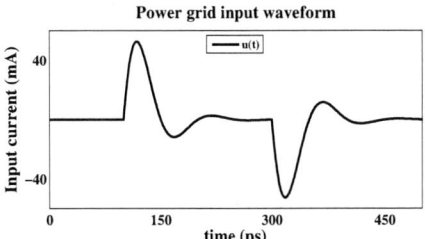

Fig. 3. Input current waveform applied to the IBM power grid: a superposition of two damped sinusoidal excitations at 10GHz, one positive and the other negative.

Each part of Fig. 2 depicts the following: (1) two output waveforms (one output from the power plane and one from the ground plane) produced by the original power grid (in dark gray), (2) the same outputs, as predicted by the respective reduced order model (in blue), and (3) the outputs as predicted by 5-bit ABCD-L applied to the corresponding ROM (in green).

From the figure, it is clear that the Boolean models generated by ABCD-L always closely reproduce the corresponding ROM's response. Moreover, as the ROM size increases, this response in turn becomes a very good approximation to the response of the original power grid system, both in the power plane and in the ground plane.

Furthermore, as expected, the combination of ABCD-L with Arnoldi MOR resulted in significant computational savings over direct eigendecomposition of the original system. For instance, even with $p = 100$, the Arnoldi step took only about 10 minutes, on a 64-bit Linux machine equipped with a 3.2GHz AMD® Phenom™ II X6 1090T processor and 16GB RAM. Moreover, once the Arnoldi iteration was completed, the time required for ABCD-L (including pre-processing, model generation, transient simulation, and post-processing) was well under a minute. This shows that ABCD-L, in conjunction with MOR, is indeed a viable technique that can be applied to accurately Booleanize even large LTI systems.

REFERENCES

[1] A. Ghosh and R. Vemuri. Formal verification of synthesized analog designs. In *ICCD '99: Proceedings of the IEEE International Conference on Computer Design*, pages 40–45, 1999.

[2] K. Hanna. Reasoning about real circuits. In *Proceedings of the 7th International Workshop on Higher Order Logic Theorem Proving and Its Applications*, pages 235–253, 1994.

[3] K. Hanna. Reasoning about analog-level implementations of digital systems. *Formal Methods in System Design*, 16(2):127–158, 2000.

[4] R. P. Kurshan and K. L. McMillan. Analysis of digital circuits through symbolic reduction. *IEEE Transactions on Computer-Aided Design of Integrated Circuits and Systems*, 10(11):1356–1371, 1991.

[5] W. Hartong, L. Hedrich, and E. Barke. Model checking algorithms for analog verification. In *DAC '02: Proceedings of the 39th annual ACM Design Automation Conference*, pages 542–547, 2002.

[6] S. Steinhorst and L. Hedrich. Model checking of analog systems using an analog specification language. In *DATE '08: Proceedings of the ACM Conference on Design, Automation and Test in Europe*, pages 324–329, 2008.

[7] R. Alur, C. Courcoubetis, T. A. Henzinger, and P. H. Ho. Hybrid automata: An algorithmic approach to the specification and verification of hybrid systems. *Hybrid Systems*, pages 209–229, 1993.

[8] A. Nerode and W. Kohn. Models for hybrid systems: Automata, topologies, controllability, observability. *Hybrid Systems*, pages 317–356, 1993.

[9] R. Alur, C. Courcoubetis, N. Halbwachs, T. A. Henzinger, P. H. Ho, X. Nicollin, A. Olivero, J. Sifakis, and S. Yovine. The algorithmic analysis of hybrid systems. *Theoretical Computer Science*, 138(1):3–34, 1995.

[10] T. A. Henzinger. The theory of hybrid automata. In *LICS '96: Proceedings of the 11th Annual IEEE Symposium on Logic in Computer Science*, pages 278–292, 1996.

[11] G. Al-Sammane, M. H. Zaki, and S. Tahar. A symbolic methodology for the verification of AMS designs. In *DATE '07: Proceedings of the ACM Conference on Design, Automation and Test in Europe*, pages 249–254, 2007.

[12] M. H. Zaki, G. Al-Sammane, S. Tahar, and G. Bois. Combining symbolic simulation and interval arithmetic for the verification of AMS designs. In *FMCAD '07: Formal Methods in Computer Aided Design*, pages 207–215, 2007.

[13] E. Asarin, T. Dang, and O. Maler. d/dt: A verification tool for hybrid systems. In *CDC '01: Proceedings of the 40th IEEE Conference on Decision and Control*, pages 2893–2898, 2001.

[14] S. Ray, J. Bhadra, T. Portlock, and R. Syzdek. Modeling and verification of industrial flash memories. In *ISQED '10: Proceedings of the 11th International Symposium on Quality Electronic Design*, pages 705–712, 2010.

[15] T. A. Henzinger, P. H. Ho, and H. Wong-Toi. HyTech: A model checker for hybrid systems. *International Journal on Software Tools for Technology Transfer*, 1(1):110–122, 1997.

[16] G. Frehse. PHAVer: Algorithmic verification of hybrid systems past HyTech. *International Journal on Software Tools for Technology Transfer*, 10(3):263–279, 2008.

[17] A. Chutinan and B. H. Krogh. Computational techniques for hybrid system verification. *IEEE Transactions on Automatic Control*, 48(1):64–75, 2003.

[18] S. Little, N. Seegmiller, D. Walter, C. Myers, and T. Yoneda. Verification of ams circuits using labeled hybrid petri nets. In *ICCAD '06: Proceedings of the IEEE/ACM International Conference on Computer-Aided Design*, pages 275–282, 2006.

[19] S. Little, D. Walter, K. Jones, and C. Myers. AMS circuit verification using models generated from simulation traces. *Automated Technology for Verification and Analysis*, pages 114–128, 2007.

[20] S. Little. *Efficient Modeling and Verification of Analog/Mixed-Signal Circuits Using Labeled Hybrid Petri Nets*. PhD thesis, University of Utah, 2008.

[21] S. Batchu. *Automatic Extraction of Behavioral Models from Simulations of Analog/Mixed-Signal (AMS) Circuits*. Master's thesis, University of Utah, 2010.

[22] M. Althoff, O. Stursberg, and M. Buss. Computing reachable sets of hybrid systems using a combination of zonotopes and polytopes. *Nonlinear Analysis: Hybrid Systems*, 4(2):233–249, 2010.

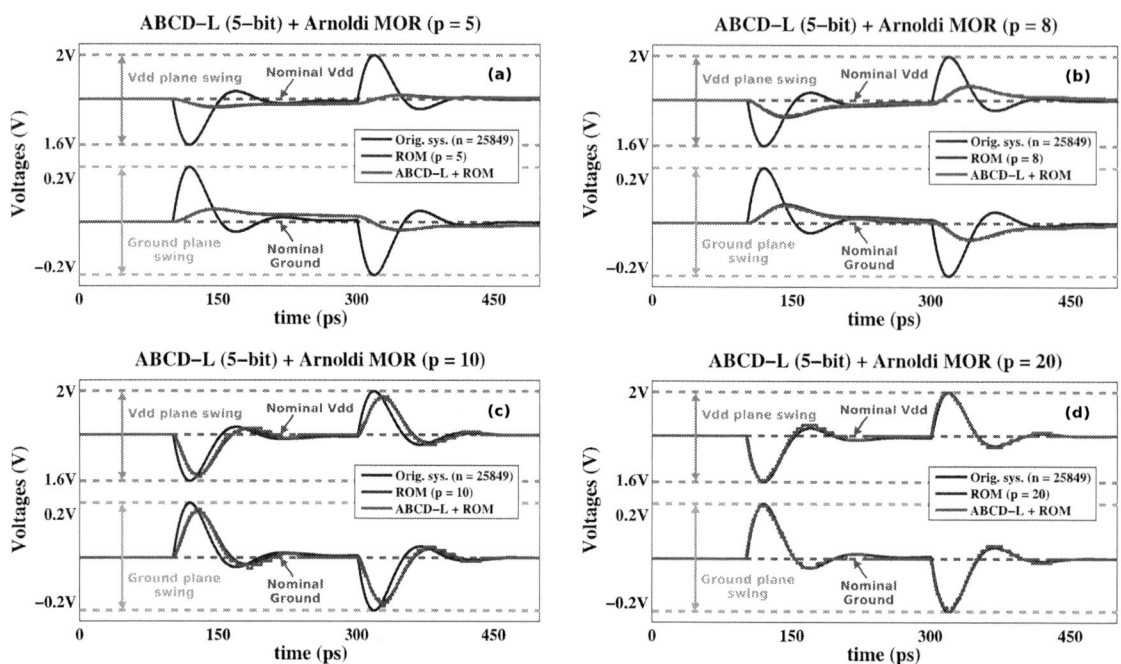

Fig. 2. ABCD-L plus Arnoldi MOR applied to the IBM power grid, for different ROM sizes p. The plots depict the network's response to the input $u(t)$ of Fig. 3. In each case, it is seen that the Boolean model generated by ABCD-L closely approximates the ROM's response. And as ROM size increases, this response becomes a very good approximation to the response of the original power grid system.

[23] F. Wang. Symbolic parametric safety analysis of linear hybrid systems with BDD-like data-structures. *IEEE Transactions on Software Engineering*, 31(1):38–51, 2005.

[24] C. Le Guernic and A. Girard. Reachability analysis of hybrid systems using support functions. In *CAV '09: Proceedings of the International Conference on Computer Aided Verification*, pages 540–554, 2009.

[25] M. Althoff, A. Rajhans, B. H. Krogh, S. Yaldiz. X. Li, and L. Pileggi. Formal verification of Phase Locked Loops using reachability analysis and continuization. In *ICCAD '10: Proceedings of the IEEE/ACM International Conference on Computer-Aided Design*, pages 659–666, 2010.

[26] L. T. Pillage and R. A. Rohrer. Asymptotic waveform evaluation for timing analysis. *IEEE Transactions on Computer-Aided Design of Integrated Circuits and Systems*, 9(4):352–366, 1990.

[27] K. J. Kerns, I. L. Wemple, and A. T. Yang. Stable and efficient reduction of substrate model networks using congruence transforms. In *ICCAD '95: Proceedings of the IEEE/ACM International Conference on Computer-Aided design*, pages 207–214, 1995.

[28] P. Feldmann and R. W. Freund. Efficient linear circuit analysis by Padé approximation via the Lanczos process. *IEEE Transactions on Computer-Aided Design of Integrated Circuits and Systems*, 14(5):639–649, 1995.

[29] L. Silveira, M. Kamon, I. Elfadel, and J. White. A coordinate-transformed Arnoldi algorithm for generating guaranteed stable reduced-order models of RLC circuits. *Computer Methods in Applied Mechanics and Engineering*, 169(3):377–389, 1999.

[30] A. Odabasioglu, M. Celik, and L. T. Pileggi. Prima: Passive reduced-order interconnect macromodeling algorithm. *IEEE Transactions on Computer-Aided Design of Integrated Circuits and Systems*, 17(8):645–654, 1998.

[31] S. R. Nassif. Power grid analysis benchmarks. In *ASPDAC '08: Proceedings of the 13th IEEE/ACM Asia and South Pacific Design Automation Conference*, pages 376–381, 2008.

[32] Download link for the IBM power grid benchmark set: http://dropzone.tamu.edu/~pli/PGBench/.

[33] W. E. Arnoldi. The principle of minimized iterations in the solution of the matrix eigenvalue problem. *The Quarterly of Applied Mathematics*, 9(1):17–29, 1951.

[34] W. H. A. Schilders, H. A. van der Vorst, and J. Rommes. *Model Order Reduction: Theory, research aspects and applications*, volume 13 of *Mathematics in Industry*. Springer Verlag, 2008.

Bayesian Model Fusion: Large-Scale Performance Modeling of Analog and Mixed-Signal Circuits by Reusing Early-Stage Data

Fa Wang[1], Wangyang Zhang[1], Shupeng Sun[1], Xin Li[1] and Chenjie Gu[2]

[1]ECE Department, Carnegie Mellon University, Pittsburgh, PA 15213
[2]Strategic CAD Labs, Intel Corporation, Hillsboro, OR 97124
{fwang1, wyzhang, shupengs, xinli}@ece.cmu.edu, chenjie.gu@intel.com

ABSTRACT

Efficient high-dimensional performance modeling of today's complex analog and mixed-signal (AMS) circuits with large-scale process variations is an important yet challenging task. In this paper, we propose a novel performance modeling algorithm that is referred to as Bayesian Model Fusion (BMF). Our key idea is to borrow the simulation data generated from an early stage (e.g., schematic level) to facilitate efficient high-dimensional performance modeling at a late stage (e.g., post layout) with low computational cost. Such a goal is achieved by statistically modeling the performance correlation between early and late stages through Bayesian inference. Several circuit examples designed in a commercial 32nm CMOS process demonstrate that BMF achieves up to 9× runtime speedup over the traditional modeling technique without surrendering any accuracy.

1. INTRODUCTION

The aggressive scaling of integrated circuits (ICs) leads to large-scale process variations that cannot be easily reduced by foundries. Process variations manifest themselves as the uncertainties associated with the geometrical and electrical parameters of semiconductor devices. These device-level variations significantly impact the parametric yield of analog and mixed-signal (AMS) circuits and, hence, must be appropriately modeled, analyzed and optimized at all levels of design hierarchy [1]-[2].

To address this variability issue, various techniques for performance modeling have been developed during the past two decades [3]-[8]. The objective is to approximate the circuit-level performance (e.g., gain of an analog amplifier) as an analytical (e.g., linear, quadratic, etc) function of device-level variations (e.g., ΔV_{TH}, ΔT_{OX}, etc). Once such a performance model is available, it can be applied to a number of important applications such as estimating parametric yield [9], extracting worst-case corner [10], optimizing circuit design [11]-[15], etc.

While performance modeling was extensively studied in the past, the evolution of today's AMS circuits has posed a number of new challenges in this area. In particular, the recent adoption of several emerging design methodologies (e.g., reconfigurable analog design, adaptive post-silicon tuning, etc) leads to highly complex AMS systems that integrate numerous nanoscale devices. The remarkable increase of AMS circuit size results in a two-fold consequence.

- *High-dimensional variation space*: A large number of device-level random variables must be used to model the process variations associated with a large-scale AMS system. For example, about 40 independent random variables are required to model the device mismatches of a single transistor for a commercial 32nm CMOS process. If an AMS system contains 10^4 transistors, there are about 4×10^5 random variables in total to capture the corresponding device-level variations, resulting in a high-dimensional variation space. In addition, it is extremely difficult, if not impossible, to pre-select a subset of these random variables for variation analysis, since the impact of device mismatches is circuit- and performance-dependent.

- *Expensive circuit simulation*: The computational cost of circuit simulation substantially increases, as the AMS circuit size becomes increasingly large. For instance, it may take a few days or even a few weeks to run the transistor-level simulation of a large AMS circuit such as phase-locked loop or high-speed link.

These recent trends of today's AMS circuits make performance modeling extremely difficult. On one hand, a large number of simulation samples must be generated in order to fit a high-dimensional model. On the other hand, creating a single sampling point by transistor-level simulation can take a large amount of computational time. The challenging issue here is how to make performance modeling computationally *affordable* for today's large-scale AMS circuits. This fundamental issue has not been appropriately addressed by the traditional performance modeling techniques, e.g., the recent sparse regression algorithm based on Orthogonal Matching Pursuit (OMP) [8].

In this paper, we propose a new *Bayesian Model Fusion* (BMF) technique to facilitate large-scale performance modeling of AMS circuits. The proposed BMF method is motivated by the fact that today's AMS circuits are often designed via a multi-stage flow. Namely, an AMS design often spans three core stages: (i) schematic design, (ii) layout design, and (iii) chip manufacturing and testing. At each stage, simulation or measurement data are collected to validate the circuit design, before moving to the next stage. The traditional performance modeling techniques rely on the data at a single stage only and they completely ignore the data that are generated at other stages. The key idea of BMF, however, is to *reuse* the early-stage data when fitting a late-stage performance model. As such, the performance modeling cost can be substantially reduced.

Mathematically, the proposed BMF method is derived from the theory of Bayesian inference. Starting from a set of early-stage (e.g., schematic-level) sampling points, BMF first approximates an early-stage performance model based on these samples. The early-stage model is used as a template to define our prior knowledge for late-stage (e.g., post-layout) performance modeling. Specifically, a prior distribution is statistically defined for the late-stage model coefficients. The prior knowledge is then combined with very few late-stage sampling points to solve the

---

Permission to make digital or hard copies of all or part of this work for personal or classroom use is granted without fee provided that copies are not made or distributed for profit or commercial advantage and that copies bear this notice and the full citation on the first page. To copy otherwise, to republish, to post on servers or to redistribute to lists, requires prior specific permission and/or a fee.
DAC'13, May 29 - June 07 2013, Austin, TX, USA.

late-stage model coefficients via Bayesian inference. From this point of view, by *fusing* the early-stage and late-stage performance models through Bayesian inference, we only need a small number of late-stage sampling points to fit a high-dimensional late-stage model, thereby significantly reducing the computational cost for performance modeling. As will be demonstrated by our numerical examples in Section 5, BMF achieves up to 9× runtime speedup over the traditional modeling technique without surrendering any accuracy.

BMF was previously proposed for parametric yield estimation of AMS circuits in [16] where Bayesian inference was used to estimate the probability distribution of AMS performance metrics. In this paper, we further extend the idea of BMF to performance modeling. It is important to emphasize that the formulation of our prior knowledge for performance modeling is completely different from that shown in [16], as will be discussed in Section 3.

The remainder of this paper is organized as follows. In Section 2, we review the important background on performance modeling, and then derive our proposed BMF method in Section 3. Several implementation issues are discussed in Section 4. The efficacy of BMF is demonstrated by several circuit examples in Section 5. Finally, we conclude in Section 6.

2. BACKGROUND

Given an AMS circuit (e.g., an analog amplifier), its performance (e.g., gain) may vary due to device-level variations (e.g., ΔV_{TH}, ΔT_{OX}, etc.). The objective of performance modeling is to approximate the circuit performance as an analytical function of the device-level variations:

$$f(\mathbf{x}) \approx \sum_{m=1}^{M} \alpha_m \cdot g_m(\mathbf{x}) \tag{1}$$

where f represents the performance of interest, $\mathbf{x}$ is a vector containing the random variables to model device-level variations, $\{\alpha_m; m = 1, 2, ..., M\}$ denote the model coefficients, $\{g_m(\mathbf{x}); m = 1, 2, ..., M\}$ are the basis functions (e.g., linear or quadratic polynomials), and M is the total number of basis functions.

In order to determine the performance model in (1), we need to find the model coefficients $\{\alpha_m; m = 1, 2, ..., M\}$. Towards this goal, the traditional least-squares fitting method first generates a set of sampling points and then solves the model coefficients from the following linear equation [18]:

$$\mathbf{G} \cdot \boldsymbol{\alpha} = \mathbf{f} \tag{2}$$

where

$$\mathbf{G} = \begin{bmatrix} g_1(\mathbf{x}^{(1)}) & g_2(\mathbf{x}^{(1)}) & \cdots & g_M(\mathbf{x}^{(1)}) \\ g_1(\mathbf{x}^{(2)}) & g_2(\mathbf{x}^{(2)}) & \cdots & g_M(\mathbf{x}^{(2)}) \\ \vdots & \vdots & \vdots & \vdots \\ g_1(\mathbf{x}^{(K)}) & g_2(\mathbf{x}^{(K)}) & & g_M(\mathbf{x}^{(K)}) \end{bmatrix} \tag{3}$$

$$\boldsymbol{\alpha} = \begin{bmatrix} \alpha_1 & \alpha_2 & \cdots & \alpha_M \end{bmatrix}^T \tag{4}$$

$$\mathbf{f} = \begin{bmatrix} f^{(1)} & f^{(2)} & \cdots & f^{(K)} \end{bmatrix}^T. \tag{5}$$

In (3)-(5), $\mathbf{x}^{(k)}$ and $f^{(k)}$ are the values of $\mathbf{x}$ and $f(\mathbf{x})$ at the kth sampling point respectively, and K represents the total number of sampling points. The number of sampling points (i.e., K) should be greater than the number of unknown coefficients (i.e., M). As such, the linear equation in (2) is overdetermined and the unknown model coefficients $\{\alpha_m; m = 1, 2, ..., M\}$ are found by solving its least-squares solution.

When the aforementioned least-squares fitting method is applied to fit a high-dimensional performance model with many

unknown model coefficients, it requires a large number of sampling points to form the overdetermined linear equation in (2). Note that each sampling point is generated by running an expensive transistor-level simulation. It, in turn, implies that the least-squares fitting approach can be extremely expensive for high-dimensional performance modeling.

Recently, sparse regression has been developed to address this complexity issue [8]. The key idea is not to solve an overdetermined linear equation. Instead, the unknown model coefficients are uniquely determined by solving an underdetermined linear equation. This goal is achieved by exploiting the fact that most model coefficients of a high-dimensional performance model are close to zero. In other words, the unknown model coefficients carry a unique *sparse* pattern. The sparse regression algorithms were particularly developed to solve these sparse coefficients from a small number of sampling points. As such, the simulation cost of generating the required sampling points is greatly reduced.

While sparse regression has been successfully applied to many practical applications, it still requires a large number of (e.g., 10^3) sampling points to fit a high-dimensional performance model [8]. Therefore, it remains ill-equipped for modeling large-scale AMS circuits where running a single transistor-level simulation to generate one sampling point may take a few days or even a few weeks. Motivated by this observation, we will propose a new BMF technique in this paper to further reduce the number of required simulation samples and, hence, the computational cost for large-scale performance modeling.

3. BAYESIAN MODEL FUSION

Similar to sparse regression, the proposed BMF method relies on the assumption that most model coefficients of a high-dimensional performance model are close to zero. However, unlike the traditional sparse regression approach that fits the sparse performance model based on the simulation data at a single stage only (e.g., post-layout simulation data), BMF attempts to identify the underlying sparse pattern by re-using the early-stage data (e.g., schematic-level simulation data) in order to efficiently fit a late-stage (e.g., post-layout) performance model. In particular, BMF consists of the following two major steps: (i) statistically defining the prior knowledge of the sparse pattern based on the early-stage simulation data, and (ii) optimally determining the late-stage performance model by combining the prior knowledge and very few late-stage simulation samples. In this section, we will discuss the mathematical formulation of these two steps and highlight the novelty.

3.1 Prior Knowledge Definition

We consider two different performance models: the early-stage model $f_E(\mathbf{x})$ and the late-stage model $f_L(\mathbf{x})$:

$$f_E(\mathbf{x}) \approx \sum_{m=1}^{M} \alpha_{E,m} \cdot g_m(\mathbf{x}) \tag{6}$$

$$f_L(\mathbf{x}) \approx \sum_{m=1}^{M} \alpha_{L,m} \cdot g_m(\mathbf{x}) \tag{7}$$

where $\{\alpha_{E,m}; m = 1, 2, ..., M\}$ and $\{\alpha_{L,m}; m = 1, 2, ..., M\}$ represent the early-stage and late-stage model coefficients, respectively. In (6)-(7), we assume that the early-stage model $f_E(\mathbf{x})$ and the late-stage model $f_L(\mathbf{x})$ share the same basis functions. More complicated cases where $f_E(\mathbf{x})$ and $f_L(\mathbf{x})$ are approximated by different basis functions will be further discussed in Section 4.1.

The early-stage model $f_E(\mathbf{x})$ is fitted from the early-stage simulation data. In practice, the early-stage simulation data are collected to validate the early-stage design, before we move to the next stage. For this reason, we should already know the early-stage model $f_E(\mathbf{x})$ before fitting the late-stage model $f_L(\mathbf{x})$. Namely, we assume that the early-stage model coefficients $\{\alpha_{E,m}; m = 1, 2, ..., M\}$ are provided as the input to our proposed BMF method for late-stage performance modeling.

Given the early-stage model $f_E(\mathbf{x})$, we first extract the prior knowledge that can be used to facilitate efficient late-stage modeling. To this end, we propose to learn the underlying sparse pattern for the late-stage model $f_L(\mathbf{x})$ based on the early-stage model coefficients $\{\alpha_{E,m}; m = 1, 2, ..., M\}$. Remember that both the early-stage and late-stage models are fitted for the same performance metric of the same circuit. Their model coefficients should be similar. Namely, if the early-stage model coefficient $\alpha_{E,m}$ has a large (or small) magnitude, it is likely that the late-stage model coefficient $\alpha_{L,m}$ also has a large (or small) magnitude. Such prior knowledge should be mathematically encoded into our proposed performance modeling flow.

In this paper, we statistically represent the prior knowledge as a probability density function (PDF) that is referred to as the *prior distribution* [19]. In particular, we model each late-stage model coefficient as a zero-mean Gaussian distribution:

$$pdf\left(\alpha_{L,m}\right) = \frac{1}{\sqrt{2\pi}\cdot\sigma_m}\cdot\exp\left(-\frac{\alpha_{L,m}^2}{2\cdot\sigma_m^2}\right) \sim N\left(0,\sigma_m^2\right) \quad (8)$$
$$\left(m = 1, 2, \cdots, M\right)$$

where the standard deviation σ_m is a parameter that encodes the magnitude information of the model coefficient $\alpha_{L,m}$. If the standard deviation σ_m is small, the prior distribution $pdf(\alpha_{L,m})$ is narrowly peaked around zero, implying that the coefficient $\alpha_{L,m}$ is possibly close to zero. Otherwise, if the standard deviation σ_m is large, the prior distribution $pdf(\alpha_{L,m})$ widely spreads over a large range and the coefficient $\alpha_{L,m}$ can possibly take a value that is far away from zero. Figure 1 shows a simple example of our proposed prior distribution for two model coefficients $\alpha_{L,1}$ and $\alpha_{L,2}$ where σ_1 is small and σ_2 is large.

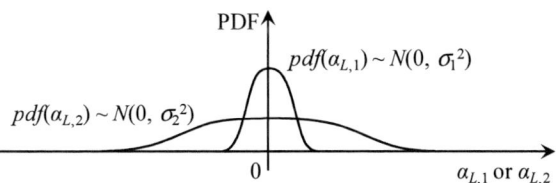

Figure 1. A simple example of our proposed prior distribution is shown for two model coefficients $\alpha_{L,1}$ and $\alpha_{L,2}$. The coefficient $\alpha_{L,1}$ is possibly close to zero, since its prior distribution is narrowly peaked around zero. The coefficient $\alpha_{L,2}$ can possibly be far away from zero, since its prior distribution widely spreads over a large range.

Given (8), we need to appropriately determine the standard deviation σ_m to fully specify the prior distribution $pdf(\alpha_{L,m})$. The value of σ_m should be optimized so that the probability distribution $pdf(\alpha_{L,m})$ correctly represents our prior knowledge. In other words, by appropriately choosing the value of σ_m, the prior distribution $pdf(\alpha_{L,m})$ should take a large value (i.e., a high probability) at the location where the actual late-stage model coefficient $\alpha_{L,m}$ occurs. However, we only know the early-stage model coefficient $\alpha_{E,m}$, instead of the late-stage model coefficient

$\alpha_{L,m}$, at this moment. Remember that $\alpha_{E,m}$ and $\alpha_{L,m}$ are expected to be similar. Hence, the prior distribution $pdf(\alpha_{L,m})$ should also take a large value at $\alpha_{L,m} = \alpha_{E,m}$. Based on this criterion, the optimal prior distribution $pdf(\alpha_{L,m})$ can be found by maximizing the probability for $\alpha_{E,m}$ to occur:

$$\max_{\sigma_m} \quad pdf\left(\alpha_{L,m} = \alpha_{E,m}\right) \quad (m = 1, 2, \cdots, M). \quad (9)$$

Namely, given the early-stage model coefficient $\alpha_{E,m}$, the optimal standard deviation σ_m is determined by the *maximum likelihood estimation* (MLE) in (9).

To solve σ_m from (9), we consider the following first-order optimality condition:

$$\frac{d}{d\sigma_m}pdf\left(\alpha_{L,m} = \alpha_{E,m}\right) = 0 \quad (m = 1, 2, \cdots, M). \quad (10)$$

Substituting (8) into (10) yields:

$$\frac{1}{\sqrt{2\pi}\cdot\sigma_m}\cdot\exp\left(-\frac{\alpha_{E,m}^2}{2\cdot\sigma_m^2}\right)\cdot\left(\frac{\alpha_{E,m}^2}{\sigma_m^3}-\frac{1}{\sigma_m}\right) = 0 . \quad (11)$$
$$\left(m = 1, 2, \cdots, M\right)$$

The optimal value of σ_m is equal to:

$$\sigma_m = \left|\alpha_{E,m}\right| \quad (m = 1, 2, \cdots, M). \quad (12)$$

Eq. (12) reveals an important fact that the optimal standard deviation σ_m is simply equal to the absolute value of the early-stage model coefficient $|\alpha_{E,m}|$. This observation is consistent with our intuition. Namely, if the early-stage model coefficient $\alpha_{E,m}$ has a large (or small) magnitude, the late-stage model coefficient $\alpha_{L,m}$ should also have a large (or small) magnitude and, hence, the standard deviation σ_m should be large (or small).

To complete the definition of the prior distribution for all late-stage model coefficients $\{\alpha_{L,m}; m = 1, 2, ..., M\}$, we further assume that these coefficients are statistically independent and their joint distribution is represented as:

$$pdf\left(\mathbf{\alpha}_L\right) = \frac{1}{\left(\sqrt{2\pi}\right)^M\cdot\prod_{m=1}^{M}\sigma_m}\cdot\exp\left(-\sum_{m=1}^{M}\frac{\alpha_{L,m}^2}{2\cdot\sigma_m^2}\right) \quad (13)$$

where

$$\mathbf{\alpha}_L = \begin{bmatrix}\alpha_{L,1} & \alpha_{L,2} & \cdots & \alpha_{L,M}\end{bmatrix}^T \quad (14)$$

contains all late-stage model coefficients. The independence assumption in (13) simply implies that we do not know the correlation information among these coefficients as our prior knowledge. The correlation information will be learned from the late-stage simulation data, when the posterior distribution is calculated by the Bayesian inference in Section 3.2.

Finally, it is important to mention that the prior knowledge can be possibly defined as a distribution that is different from (8). For example, the prior distribution is specified as a Gaussian distribution with non-zero mean in [16]. It, however, does not encode the sparse pattern of model coefficients, as is the case of this paper. The efficacy of different prior definitions is case-dependent. It remains an open question how to determine the optimal prior distribution for a specific performance modeling problem where the circuit and performance of interest are given. This problem will be further studied in our future research.

3.2 Maximum-A-Posteriori Estimation

Once the prior distribution $pdf(\mathbf{\alpha}_L)$ is derived in (13), we will combine $pdf(\mathbf{\alpha}_L)$ with K late-stage simulation samples $\{(\mathbf{x}^{(k)}, f_L^{(k)}); k = 1, 2, ..., K\}$, where $\mathbf{x}^{(k)}$ and $f_L^{(k)}$ are the values of $\mathbf{x}$ and $f_L(\mathbf{x})$ at the kth sampling point respectively, to solve the late-stage model coefficient $\boldsymbol{\alpha}_L$ by *maximum-a-posteriori* (MAP) estimation. The key idea of MAP is to find the *posterior distribution* [19], i.e., the conditional PDF *pdf*$(\boldsymbol{\alpha}_L \mid \mathbf{f}_L)$ where

$$\mathbf{f}_L = \begin{bmatrix} f_L^{(1)} & f_L^{(2)} & \cdots & f_L^{(K)} \end{bmatrix}^T \quad (15)$$

contains all late-stage simulation samples that are collected. Intuitively, the posterior distribution *pdf*$(\boldsymbol{\alpha}_L \mid \mathbf{f}_L)$ indicates the remaining uncertainty of $\boldsymbol{\alpha}_L$, after we observe K late-stage simulation samples. Here, since $\boldsymbol{\alpha}_L$ is a random variable, it is described by a probability distribution, instead of a deterministic value. MAP attempts to find the optimal value of $\boldsymbol{\alpha}_L$ to maximize the posterior distribution *pdf*$(\boldsymbol{\alpha}_L \mid \mathbf{f}_L)$. Namely, it aims to find the solution $\boldsymbol{\alpha}_L$ that is most likely to occur according to the posterior distribution.

Based on Bayes' theorem, the posterior distribution *pdf*$(\boldsymbol{\alpha}_L \mid \mathbf{f}_L)$ is proportional to the prior distribution *pdf*$(\boldsymbol{\alpha}_L)$ multiplied by the likelihood function *pdf*$(\mathbf{f}_L \mid \boldsymbol{\alpha}_L)$ [19]:

$$pdf\left(\boldsymbol{\alpha}_L \middle| \mathbf{f}_L\right) \propto pdf\left(\boldsymbol{\alpha}_L\right) \cdot pdf\left(\mathbf{f}_L \middle| \boldsymbol{\alpha}_L\right). \quad (16)$$

The prior distribution *pdf*$(\boldsymbol{\alpha}_L)$ is already defined in (13). To derive the likelihood function *pdf*$(\mathbf{f}_L \mid \boldsymbol{\alpha}_L)$, we further assume that the error for the late-stage performance model $f_L(\mathbf{x})$ follows a zero-mean Gaussian distribution and, hence, the approximate equality in (7) can be re-written as:

$$f_L(\mathbf{x}) = \sum_{m=1}^{M} \alpha_{L,m} \cdot g_m(\mathbf{x}) + \varepsilon_L \quad (17)$$

where ε_L denotes the modeling error with the distribution:

$$pdf\left(\varepsilon_L\right) = \frac{1}{\sqrt{2\pi} \cdot \sigma_0} \cdot \exp\left(-\frac{\varepsilon_L^2}{2 \cdot \sigma_0^2}\right) \sim N\left(0, \sigma_0^2\right). \quad (18)$$

In (18), the standard deviation σ_0 controls the magnitude of the modeling error. Its value can be optimally determined by using the cross-validation technique that will be discussed in Section 4.2.

Given (17)-(18), since the modeling error at the kth simulation sample $(\mathbf{x}^{(k)}, f_L^{(k)})$ is simply one sampling point of the random variable ε_L, it follows the Gaussian distribution:

$$f_L^{(k)} - \sum_{m=1}^{M} \alpha_{L,m} \cdot g_m\left(\mathbf{x}^{(k)}\right) \sim N\left(0, \sigma_0^2\right). \quad (19)$$

Therefore, the probability of observing the kth sampling point is:

$$pdf\left(f_L^{(k)} \middle| \boldsymbol{\alpha}_L\right) = \frac{\exp\left\{-\frac{1}{2 \cdot \sigma_0^2} \cdot \left[f_L^{(k)} - \sum_{m=1}^{M} \alpha_{L,m} \cdot g_m\left(\mathbf{x}^{(k)}\right)\right]^2\right\}}{\sqrt{2\pi} \cdot \sigma_0}. \quad (20)$$

Assume that all sampling points are independently generated, we can write the likelihood function *pdf*$(\mathbf{f}_L \mid \boldsymbol{\alpha}_L)$ as:

$$pdf\left(\mathbf{f}_L \middle| \boldsymbol{\alpha}_L\right) = \prod_{k=1}^{K} pdf\left(f_L^{(k)} \middle| \boldsymbol{\alpha}_L\right). \quad (21)$$

Combining (13), (16) and (20)-(21), it is straightforward to prove that the posterior distribution *pdf*$(\boldsymbol{\alpha}_L \mid \mathbf{f}_L)$ is Gaussian and its covariance matrix $\boldsymbol{\Sigma}_L$ and mean vector $\boldsymbol{\mu}_L$ are [17], [19]:

$$\boldsymbol{\Sigma}_L = \left[\sigma_0^{-2} \cdot \mathbf{G}^T \cdot \mathbf{G} + diag\left(\sigma_1^{-2}, \sigma_2^{-2}, \cdots, \sigma_M^{-2}\right)\right]^{-1} \quad (22)$$

$$\boldsymbol{\mu}_L = \sigma_0^{-2} \cdot \boldsymbol{\Sigma}_L \cdot \mathbf{G}^T \cdot \mathbf{f}_L \quad (23)$$

where $\mathbf{G}$ and $\mathbf{f}_L$ are defined by (3) and (15) respectively, and diag($\bullet$) represents an operator to construct a diagonal matrix. Since the Gaussian PDF *pdf*$(\boldsymbol{\alpha}_L \mid \mathbf{f}_L)$ reaches its maximum at the mean value, the MAP solution $\boldsymbol{\alpha}_L$ is equal to the mean vector $\boldsymbol{\mu}_L$:

$$\boldsymbol{\alpha}_L = \sigma_0^{-2} \cdot \boldsymbol{\Sigma}_L \cdot \mathbf{G}^T \cdot \mathbf{f}_L . \quad (24)$$

In other words, Eq. (24) shows the optimal coefficients solved by our proposed BMF method for the late-stage performance model $f_L(\mathbf{x})$.

While the basic idea of prior knowledge definition and maximum-a-posteriori estimation is illustrated in this section, several implementation issues must be carefully considered in order to make BMF of practical utility. These implementation details will be further discussed in the next section.

4. IMPLEMENTATION ISSUES

To make the proposed BMF method of practical utility, two implementation issues, (i) missing prior knowledge and (ii) cross-validation, must be carefully considered. In this section, we will discuss these implementation issues in detail.

4.1 Missing Prior Knowledge

The BMF method derived in Section 3 assumes that the early-stage model $f_E(\mathbf{x})$ and the late-stage model $f_L(\mathbf{x})$ share the same basis functions. In practice, this assumption may not always hold, because the early-stage model does not necessarily capture all the detailed behaviors of a circuit. For instance, it is well-known that layout parasitics will be added to the post-layout netlist (late stage) during layout extraction. The variations of these parasitics must be modeled by a number of new random variables that are completely ignored at the schematic level (early stage). The late-stage post-layout model $f_L(\mathbf{x})$ should contain additional basis functions corresponding to the new random variables that are not found from the early-stage schematic model $f_E(\mathbf{x})$. In this case, the early-stage model $f_E(\mathbf{x})$ does not carry any prior knowledge about the late-stage model coefficients associated with these additional basis functions. In other words, the prior knowledge for these late-stage model coefficients is missing.

To appropriately handle the cases with missing prior knowledge, we re-visit the prior distribution *pdf*$(\alpha_{L,m})$ defined in (8). As mentioned in Section 3.1, the standard deviation σ_m of the Gaussian distribution *pdf*$(\alpha_{L,m})$ encodes the magnitude information of the late-stage model coefficient $\alpha_{L,m}$. If there is no prior knowledge available for $\alpha_{L,m}$, it implies that the late-stage model coefficient $\alpha_{L,m}$ can possibly take any value with equal probability. Hence, the standard deviation σ_m should be set to $+\infty$:

$$\sigma_m = +\infty \quad (25)$$

so that the prior distribution is nearly constant over a wide range. Note that when calculating the posterior distribution in (22)-(23), only the value of σ_m^{-1} is needed. Hence, the infinite standard deviation in (25) would not cause any numerical problem for solving the late-stage model coefficients.

4.2 Cross-Validation

As mentioned in Section 3.2, the standard deviation σ_0 of the modeling error in (18) must be determined. Otherwise, without knowing σ_0, the late-stage model coefficients $\boldsymbol{\alpha}_L$ cannot be determined by the MAP solution in (24). The objective here is to find the optimal value of σ_0 so that the modeling error is minimized. Towards this goal, we must accurately estimate the modeling error for different σ_0 values and then select the optimal σ_0 with minimal error.

To quantitatively estimate the modeling error for a given σ_0 value, we adopt the idea of N-fold cross validation from the statistics community [19]. Namely, we partition the entire data set into N groups. Modeling error is estimated from N independent runs. In each run, one of the N groups is used to estimate the modeling error and all other groups are used to calculate the model coefficients. Note that the training data for coefficient estimation and the testing data for error estimation are not overlapped. Hence, over-fitting can be easily detected. In addition, different groups should be selected for error estimation in different runs. As such, each run results in an error value e_n ($n = 1$, 2, ..., N) that is measured from a unique group of data points. The final modeling error is computed as the average of $\{e_n; n = 1, 2, ..., N\}$, i.e., $e = (e_1 + e_2 + ... + e_N)/N$. More details on cross-validation can be found in [19].

4.3 Summary

Algorithm 1 summarizes the major steps of our proposed BMF method. It consists of two core components: (i) prior distribution definition, and (ii) MAP estimation. The efficacy of BMF will be further demonstrated by our numerical examples in the next section.

Algorithm 1: Bayesian Model Fusion (BMF)
1. Start from the early-stage performance model $f_E(\mathbf{x})$ in (6).
2. Define the prior distribution for the late-stage model coefficients $\{\alpha_{L,m}; m = 1, 2, ..., M\}$ by (12)-(13) and (25).
3. Collect K late-stage simulation samples $\{(\mathbf{x}^{(k)}, f_L^{(k)}); k = 1, 2, ..., K\}$.
4. Solve the late-stage model coefficients $\{\alpha_{L,m}; m = 1, 2, ..., M\}$ based on (24) where σ_0 is determined by cross-validation.

5. NUMERICAL EXAMPLES

In this section, several circuit examples designed in a commercial 32nm CMOS process are used to demonstrate the efficacy of the proposed BMF method. Our objective is to build post-layout performance models for these circuits. For testing and comparison purposes, two different performance modeling techniques are implemented: (i) the traditional sparse regression method based on OMP [8], and (ii) the proposed BMF method. Here, the OMP algorithm is chosen for comparison, since it is one of the state-of-the-art techniques in the literature. When implementing BMF, we use the schematic-level simulation data to define our prior knowledge for post-layout performance modeling.

In each example, two different data sets, referred to as the training set and the testing set respectively, are generated by random sampling based on post-layout transistor-level simulation. The training set is used for coefficient fitting, including cross-validation. The testing set contains 300 independent random samples that are used for model validation. All numerical experiments are run on a 2.9GHz Linux server with 4GB memory.

5.1 Ring Oscillator

Shown in Figure 2(a) is the simplified circuit schematic of a ring oscillator designed in a commercial 32nm CMOS process. In this example, there are 7177 independent random variables in total to model device-level process variations, including both inter-die variations and random mismatches. Our objective is to approximate three post-layout performance metrics, power, frequency and phase noise, as linear functions of these 7177 random variables.

Figure 2(b)-(d) show how the performance modeling error varies with the number of post-layout training samples. Note that for both OMP and BMF, the modeling error decays as the number of samples increases. However, given the same number of post-

layout training samples, BMF is able to achieve substantially higher accuracy than OMP, especially if only few samples are available.

Table 1 further compares the modeling error and cost for OMP and BMF. The total cost for performance modeling consists of two major portions: (i) simulation cost (i.e., the cost of running a transistor-level simulator to generate all post-layout samples in the training set), and (ii) fitting cost (i.e., the cost of solving all unknown model coefficients). As shown in Table 1, the total modeling cost is dominated by transistor-level simulation in this example. BMF achieves 9× runtime speed-up over OMP without surrendering any accuracy.

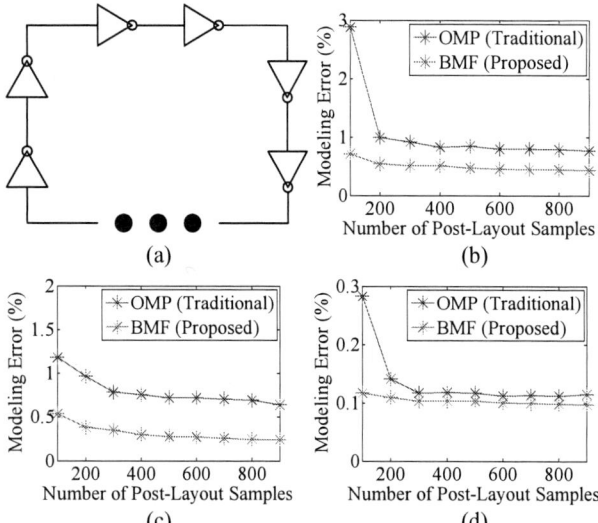

Figure 2. A ring oscillator designed in a commercial 32nm CMOS process is used as an example for performance modeling where BMF requires significantly less post-layout samples than OMP to achieve the same accuracy: (a) simplified circuit schematic of the ring oscillator, (b) modeling error for power, (c) modeling error for frequency, and (d) modeling error for phase noise.

Table 1. Performance modeling error and cost for ring oscillator

	OMP (Traditional)	BMF (Proposed)
# of post-layout training samples	900	100
Modeling error for power	0.77%	0.72%
Modeling error for frequency	0.65%	0.54%
Modeling error for phase noise	0.12%	0.12%
Simulation cost (Hour)	12.58	1.40
Fitting cost (Second)	5.75	1.69
Total modeling cost (Hour)	12.58	1.40

5.2 SRAM Read Path

Figure 3(a) shows the simplified circuit schematic of an SRAM read path designed in a commercial 32nm CMOS process. In this example, there are 66117 independent random variables to model device-level process variations. Read delay is our circuit performance of interest, and it is approximated as a linear function of all device-level random variables.

Figure 3(b) and Table 2 compare the modeling error and cost for OMP and BMF. Similar to the ring oscillator example, two important observations can be made here. First, BMF requires substantially less post-layout training samples than OMP, in order

457

to achieve the same modeling accuracy. In this example, BMF reduces the number of required samples from 400 to 100 without surrendering any accuracy. Figure 4 further plots the histograms of modeling error for both OMP and BMF with 400 and 100 post-layout training samples, respectively. Second, but more importantly, since the total modeling cost is dominated by transistor-level simulation, BMF successfully reduces the total modeling cost by reducing the number of required post-layout training samples. Compared to OMP, BMF achieves 4× runtime speed-up, as shown in Table 2.

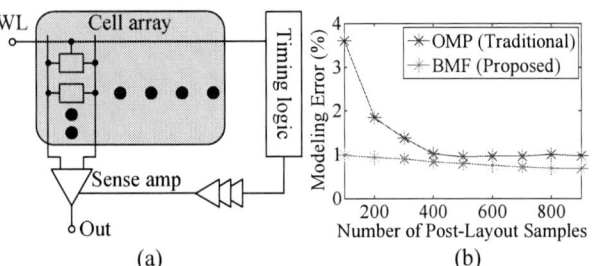

	(a)		(b)

Figure 3. A simplified SRAM read path designed in a commercial 32nm CMOS process is used as an example for performance modeling where BMF requires significantly less post-layout samples than OMP to achieve the same accuracy: (a) simplified circuit schematic of the SRAM read path, and (b) modeling error for read delay.

Table 2. Performance modeling error and cost for SRAM

	OMP (Traditional)	BMF (Proposed)
# of post-layout training samples	400	100
Modeling error for read delay	1.02%	0.99%
Simulation cost (Hour)	38.77	9.69
Fitting cost (Second)	3.56	2.11
Total modeling cost (Hour)	38.77	9.69

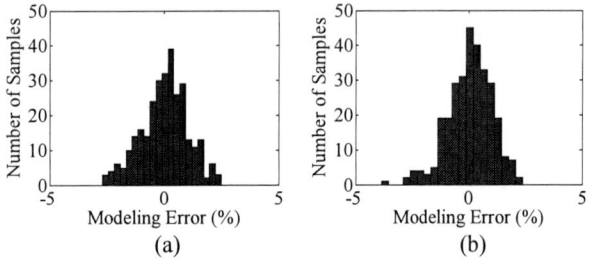

	(a)		(b)

Figure 4. Histograms of modeling error are estimated from the testing set for read delay: (a) modeling error of OMP with 400 post-layout training samples, and (b) modeling error of BMF with 100 post-layout training samples.

6. CONCLUSIONS

In this paper, a novel BMF algorithm is proposed for efficient high-dimensional performance modeling of complex AMS circuits with consideration of large-scale process variations. BMF borrows the early-stage (e.g., schematic-level) simulation data to learn a model template that is statistically represented as a prior distribution. Next, the model template encoding our prior knowledge is further calibrated by very few late-stage (e.g., post-layout) simulation samples to accurately create a late-stage performance model. As such, the computational cost of high-dimensional performance modeling can be substantially reduced, since only few transistor-level simulations are required at the late

stage. As is demonstrated by our circuit examples designed in a commercial 32nm CMOS process, the proposed BMF method achieves up to 9× runtime speedup compared to the traditional modeling technique. In our future work, we will further apply BMF to several practical applications such as statistical analysis of large-scale AMS systems.

7. ACKNOWLEDGEMENTS

This work has been supported in part by National Science Foundation and Intel Corporation.

8. REFERENCES

[1] Semiconductor Industry Associate, *International Technology Roadmap for Semiconductors*, 2011.

[2] X. Li, J. Le and L. Pileggi, *Statistical Performance Modeling and Optimization*, Now Publishers, 2007.

[3] Z. Feng and P. Li, "Performance-oriented statistical parameter reduction of parameterized systems via reduced rank regression," *IEEE ICCAD*, pp. 868-875, 2006.

[4] A. Singhee and R. Rutenbar, "Beyond low-order statistical response surfaces: latent variable regression for efficient, highly nonlinear fitting," *IEEE DAC*, pp. 256-261, 2007.

[5] A. Mitev, M. Marefat, D. Ma and J. Wang, "Principle Hessian direction based parameter reduction for interconnect networks with process variation," *IEEE ICCAD*, pp. 632-637, 2007.

[6] T. McConaghy and G. Gielen, "Template-free symbolic performance modeling of analog circuits via canonical-form functions and genetic programming," *IEEE Trans. on CAD*, vol. 28, no. 8, pp. 1162-1175, Aug. 2009.

[7] T. McConaghy, "High-dimensional statistical modeling and analysis of custom integrated circuits," *IEEE CICC*, 2011.

[8] X. Li, "Finding deterministic solution from underdetermined equation: large-scale performance modeling of analog/RF circuits," *IEEE Trans. on CAD*, vol. 29, no. 11, pp. 1661-1668, Nov. 2010.

[9] X. Li, J. Le, P. Gopalakrishnan and L. Pileggi, "Asymptotic probability extraction for nonnormal performance distributions," *IEEE Trans. on CAD*, vol. 26, no. 1, pp. 16-37, Jan. 2007.

[10] M. Sengupta, S. Saxena, L. Daldoss, G. Kramer, S. Minehane and J. Cheng, "Application-specific worst case corners using response surfaces and statistical models," *IEEE Trans. on CAD*, vol. 24, no. 9, pp. 1372-1380, 2005.

[11] Z. Wang and S. Director, "An efficient yield optimization method using a two step linear approximation of circuit performance," *IEEE EDAC*, pp. 567-571, 1994.

[12] A. Dharchoudhury and S. Kang, "Worse-case analysis and optimization of VLSI circuit performance," *IEEE Trans. on CAD*, vol. 14, no. 4, pp. 481-492, Apr. 1995.

[13] G. Debyser and G. Gielen, "Efficient analog circuit synthesis with simultaneous yield and robustness optimization," *IEEE ICCAD*, pp. 308-311, 1998.

[14] F. Schenkel, M. Pronath, S. Zizala, R. Schwencker, H. Graeb and K. Antreich, "Mismatch analysis and direct yield optimization by spec-wise linearization and feasibility-guided search," *IEEE DAC*, pp. 858-863, 2001.

[15] X. Li, P. Gopalakrishnan, Y. Xu and L. Pileggi, "Robust analog/RF circuit design with projection-based performance modeling," *IEEE Trans. on CAD*, vol. 26, no. 1, pp. 2-15, Jan. 2007.

[16] X. Li, W. Zhang, F. Wang, S. Sun and C. Gu, "Efficient parametric yield estimation of analog/mixed-signal circuits via Bayesian model fusion," *IEEE ICCAD*, 2012.

[17] S. Ji, Y. Xue, and L. Carin, "Bayesian compressive sensing," *IEEE Trans. on Signal Processing*, pp. 2346-2356, Jun. 2008.

[18] R. Myers and D. Montgomery, *Response Surface Methodology: Process and Product Optimization Using Designed Experiments*, Wiley-Interscience, 2002.

[19] C. Bishop, *Pattern Recognition and Machine Learning*, Springer, 2006.

Efficient Moment Estimation with Extremely Small Sample Size via Bayesian Inference for Analog/Mixed-Signal Validation

Chenjie Gu
Intel Strategic CAD Labs
chenjie.gu@intel.com

Eli Chiprout
Intel Strategic CAD Labs
eli.chiprout@intel.com

Xin Li
Carnegie Mellon University
xinli@cmu.edu

ABSTRACT

A critical problem in pre-Silicon and post-Silicon validation of analog/mixed-signal circuits is to estimate the distribution of circuit performances, from which the probability of failure and parametric yield can be estimated at all circuit configurations and corners. With extremely small sample size, traditional estimators are only capable of achieving a very low confidence level, leading to either over-validation or under-validation. In this paper, we propose a multi-population moment estimation method that significantly improves estimation accuracy under small sample size. In fact, the proposed estimator is theoretically guaranteed to outperform usual moment estimators. The key idea is to exploit the fact that simulation and measurement data collected under different circuit configurations and corners can be correlated, and are conditionally independent. We exploit such correlation among different populations by employing a Bayesian framework, i.e., by learning a prior distribution and applying maximum a posteriori estimation using the prior. We apply the proposed method to several datasets including post-silicon measurements of a commercial high-speed I/O link, and demonstrate an average error reduction of up to $2\times$, which can be equivalently translated to significant reduction of validation time and cost.

1. INTRODUCTION

In various product validation disciplines (e.g., pre-Silicon simulation-based validation, post-Silicon measurement-based validation), it is critical to make statistically valid predictions of circuit performances. This requirement boils down to the problem of estimating the probability distribution of circuit performance metrics of interest. From this distribution, we may also compute the probability of failure (PoF) or yield. The common practice is to estimate the moments of a distribution. In particular, if the distribution of performance metrics is Gaussian, the distribution is fully characterized by its first two moments, i.e., mean and variance. With abundant data, sample mean and sample variance converge to the actual mean and variance, as guaranteed by the law of large numbers and central limit theorem [1].

However, in practice, simulation and measurement are both time and cost consuming [2, 3, 4]. For example, post-layout simulation can be slow, especially for circuits such as SRAM/PLL where extremely small time steps are required for high accuracy. As another example, during post-Silicon validation, due to the time-line of product releases, only a limited amount of measurement may be performed within the post-Silicon time-frame. In addition, the measurement of performance metrics, such as Bit-Error-Ratio and Time/Voltage Margins of high-speed I/O links, takes a long time, and requires expensive equipment (such as BER testers) [5, 6, 7].

Furthermore, for the validation of products, there are many corners and configurations to be covered. As an example, in I/O link validation, in addition to common process, voltage and temperature (PVT) corners, we must also validate against different board/add-in card configurations, input patterns, different equalization settings, etc.. Therefore, with the time and cost constraints of simulation and measurement, an extremely small number (1 to 5) of data is available at each corner or configuration.

We call the above problem the small-sample-size problem. This problem makes most statistical analysis tools/algorithms not applicable because they are built upon the assumption that "enough" data is available for valid statistical estimation. When the assumption is broken, we obtain low confidence in the estimated quantities. In another word, this means that we may either under-validate or over-validate the circuit. Similar to over-design and under-design, over-validation and under-validation are as harmful, if not more, in terms of cost and time-to-market. Unfortunately, there is few existing satisfying solution to get around this problem. To the best of our knowledge, the usual practice is to increase the sample size as much as possible to reach a certain confidence level, or to set an empirical guard-band on top of the estimation. There is a recent work [8] that considers a similar problem, but for performance modeling. Another recently published technique [9] solves a similar problem for post-layout performance distribution estimation, but with mildly small number of samples (50 or more).

It is also important to point out that in many situations, it is necessary to estimate the distribution at each corner/configuration, for which we only have 1 to 5 samples. For example, to validate I/O interfaces such as PCIE[10] and DDR[11], it is critical to make sure that the interface works properly with different boards and add-in cards/DIMMs.

Permission to make digital or hard copies of all or part of this work for personal or classroom use is granted without fee provided that copies are not made or distributed for profit or commercial advantage and that copies bear this notice and the full citation on the first page. To copy otherwise, to republish, to post on servers or to redistribute to lists, requires prior specific permission and/or a fee.
DAC'13, May 29 – June 07 2013, Austin, TX, USA.

Therefore, for each board and add-in card, the distribution of the BER must be estimated separately. It is inappropriate to mix the measurements under different configurations, because even with a low overall PoF, we may obtain a very high PoF at a particular configuration. In this case, combining data from all configurations does not help us to increase the sample size. In fact, estimating the overall distribution can lead to misleading validation results.

In this paper, we propose a technique to efficiently estimate the mean and standard deviation of circuit performance distributions under the small-sample-size constraint. The key idea of the method is to exploit correlation in data collected at multiple populations to improve the accuracy of the proposed estimator. In particular, we emphasize that data collected at different design stages, different configurations and different corners are not independent, but are correlated. Taking advantage of this non-intuitive fact leads to a theoretically guaranteed better estimator. In comparison to sample mean/standard deviation estimators, our method achieves an average error reduction of up to $2\times$, for examples obtained from measurement of commercial designs.

Mathematically, we employ Bayesian inference[12] to fuse conditionally independent data. The method is composed of two steps. First, the Maximum Likelihood (ML) method is used to learn a prior distribution of mean/standard deviation from data collected at multiple populations. Second, the prior learned in the first step is used to obtain the Maximum A Posteriori (MAP) estimation of mean and standard deviation. The two steps are formulated as two optimization problems. Based on this formulation, we further propose a relaxed algorithm, to alleviate the computational burden.

While this paper focuses on derivations for the mean and standard deviation estimation, our formulation is general, and it incorporates estimation of moments of any order. Our formulation is also general to cover many application scenarios, depending on the availability of different data sets. In particular, the following two scenarios are commonly seen in practice:

1. Given early stage data, or empirical results, estimate the mean and standard deviation at a targeted configuration.
2. Given data at multiple configurations, estimate the mean and standard deviation at each configuration.

The rest of paper is organized as follows. Sec. 2 formulates the problem and describes the small-sample-size problem. Sec. 3 discusses and derives the multi-population estimation algorithm based on the Bayesian framework. Sec. 4 presents experimental results on several datasets to demonstrate the advantages of the proposed method.

2. BACKGROUND AND PROBLEM FORMULATION

For simplicity, in this paper, we consider the problem of estimating a single performance metric, denoted by x, which depends on many parameters such as process parameters, voltage, temperature, board, add-in card, etc.. The performance metric x also depends (indirectly) on time, because a subset of the parameters, such as process parameters, also change over time.

As an example application, we consider the problem of post-Silicon validation of I/O interfaces. In this application,

a configuration is defined by fixing the values of a subset of the parameters. By considering variability of all other parameters, x has a distribution at each configuration. For example, a configuration of an I/O link can be defined by the combination of a specific board and a specific add-in card. The variability of time/voltage margin (of the eye diagram) is caused by parameter variations such as PVT variations. Measurement of margins is repeated at each configuration for each Silicon stepping, and the goal of validation is to ensure that PoF meets the specification at each stepping and at each configuration.

2.1 Problem Formulation

To formalize the above description, we define a *population* to be a specific (corner, configuration, stepping) combination, and suppose that there are P populations. For each population, we define a random variable x_i, $(i = 1, \cdots, P)$ to model the variability of the performance metric at the corresponding (corner, configuration, stepping) combination, and x_i satisfies a Gaussian distribution $x_i \sim N(\mu_i, \sigma_i^2)$ where μ_i is the mean and σ_i^2 is the variance. For notational convenience, we define $\boldsymbol{\mu} = [\mu_1, \cdots, \mu_P]^T$ and $\boldsymbol{\sigma} = [\sigma_1 \cdots, \sigma_P]^T$.

For each population, we obtain a set of independent observations $\mathcal{X}_i = \{x_{i,1}, \cdots, x_{i,N_i}\}$, where N_i is the sample size of the i-th population. (For simplicity, we consider the case where $N_1 = \cdots = N_P = N$ throughout the paper. Extension to the more general case is straightforward. Each element in $\mathcal{X}_i$ corresponds to one independent measurement at the i-th population. The problem we aim to address is to estimate μ_i's and σ_i's given the observations $\{\mathcal{X}_1, \cdots, \mathcal{X}_P\}$, with the special constraint that N_i's are very small.

2.2 Low Confidence under Small Sample Size

For a specific population, the most widely used estimator for mean and variance is the sample mean $\bar{x}_i$ and sample variance S_i, respectively,

$$\bar{x}_i = \frac{1}{N_i} \sum_{j=1}^{N_i} x_{i,j}, \quad S_i = \frac{1}{N_i - 1} \sum_{j=1}^{N_i} (x_{i,j} - \bar{x}_i)^2. \quad (1)$$

Since $\bar{x}_i \sim N(\mu_i, \frac{\sigma_i^2}{N_i})$ and $S_i \sim \frac{\sigma_i^2}{N_i - 1} \chi^2_{N_i - 1}$, we obtain

$$\text{Std}(\bar{x}_i) = \frac{1}{\sqrt{N_i}} \sigma_i, \quad \text{Std}(S_i) = \frac{\sqrt{2}}{\sqrt{N_i - 1}} \sigma_i^2. \quad (2)$$

If the standard deviation of an unbiased estimator is used as a measure of accuracy and confidence level, Eqn. (2) shows that the accuracy of both sample mean and variance estimators depend on N_i. As N_i approaches infinity, the error converges to 0. However, when N_i is small, both estimators suffer from significant error.

3. MULTI-POPULATION MOMENT ESTIMATION

3.1 Overview

As is evident in Sec. 2.2, if each population is treated independently, there is little room for improvement. In contrast, our method views data at different populations as correlated, and it tries to exploit such correlation to improve the accuracy of the estimator.

Mathematically, out method employs a Bayesian framework, and consists of two steps, as shown in Fig. 1. First, it learns a prior distribution of $p(\boldsymbol{\mu}, \boldsymbol{\sigma})$ from data at all populations, using maximum likelihood estimation. Second, it applies Maximum A Posteriori estimation to each population using the prior distribution learned from the first step.

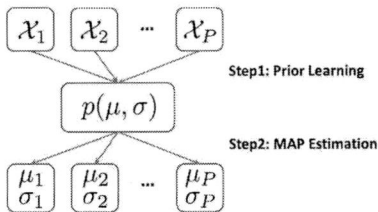

Figure 1: Proposed method consists of two steps.

To give an intuitive idea of why methods other than sample mean/variance can be much better, we consider two special examples. It can be shown that the estimators described in the two examples can be thought of as special cases of our proposed method.

EXAMPLE 3.1   (UNEQUAL MEAN, EQUAL VARIANCE). *Assume that μ_i's are different, and $\sigma_1 = \cdots = \sigma_P = \sigma$, and consider the problem of estimating σ^2. Since $S_i \sim \frac{\sigma^2}{N-1}\chi^2_{N-1}$, we obtain an unbiased estimator for σ^2*

$$\frac{1}{P}[S_1 + \cdots + S_P] \sim \frac{1}{P}\frac{\sigma^2}{N-1}\chi^2_{NP-P}, \qquad (3)$$

from which $Std(\frac{1}{P}[S_1 + \cdots + S_P]) = \sigma^2\sqrt{\frac{2}{P(N-1)}}$. Hence, the estimation error decreases as P increases, and is smaller than $Std(S_i)$.

EXAMPLE 3.2   (EQUAL MEAN, UNEQUAL VARIANCE). *Assume that $\mu_1 = \cdots = \mu_P = \mu$, and σ_i's are different, and consider the problem of estimating μ. Since $\bar{x}_i \sim N(\mu, \frac{\sigma_i^2}{N})$, we obtain an unbiased estimator for μ*

$$\frac{1}{P}[\bar{x}_1 + \cdots + \bar{x}_P] \sim N(\mu, \frac{1}{P^2}[\frac{\sigma_1^2}{N} + \cdots + \frac{\sigma_P^2}{N}]). \quad (4)$$

As P increases, the variance of $\frac{1}{P}[\bar{x}_1 + \cdots + \bar{x}_P]$ decreases. This shows that when there are many populations, we may achieve a very accurate estimate of μ.

Note, however, Eqn. (4) is not the "best" estimator. Intuitively, consider an estimator of μ which is a linear combination of $\bar{x}_i$'s. Then, more weight should be given to $\bar{x}_i$ if σ_i is smaller. However, we omit the derivation since the actual expression is rather involved.

3.2   Choice of Prior Distributions

Intuitively, prior distributions for μ_i's and σ_i's, denoted by $p(\mu_i)$ and $p(\sigma_i)$ respectively, describe our *belief* about the correlation among μ_i's and σ_i's. We stress that μ_i's and σ_i's are fixed quantities at each population, and we simply model the variation across populations by imposing a prior distribution. In Example 3.1, $\sigma_1 = \cdots = \sigma_P$ corresponds to a Dirac distribution $p(\sigma_i) = \delta(\sigma_i - \sigma)$. In Example 3.2, $\mu_1 = \cdots = \mu_P$ corresponds to a Dirac distribution $p(\mu_i) = \delta(\mu_i - \mu)$. In a real application, however, it is too strong to claim a priori that μ_i's and σ_i's at all populations are the same.

Rather, it is often seen that μ_i's and σ_i's at different populations are similar, but not equal. This observation makes

a lot of sense, especially for circuits designed to account for variability. For example, many circuits have compensation loops and self-reconfigurable features that cancel out the effects due to certain variability, which effectively pushes μ_i's towards each other. On the other hand, the variance in the circuit performance is usually caused by a small set of parameters (such as critical process parameters, temperature, voltage), and the dependency at different configurations tends to be similar, which effectively pushes σ_i towards each other.

Based on the above observation, we choose to use uniform priors for μ_i's and σ_i's, *i.e.*,

$$\mu_i \sim U(a,b), \quad \sigma_i \sim U(c,d), \qquad (5)$$

where $a, b, c, d \in \mathbf{R}, c, d \geq 0$. Note that the Dirac prior is an extreme case of the uniform prior when $|b - a|$ and $|d - c|$ become 0.

While the derivations in the rest of the paper will be based on the choice of uniform prior, there is no limitation to incorporate other types of prior distributions in our method in Fig. 1. For example, one may use a Gaussian prior, or even any arbitrary probability distribution. Furthermore, even μ_i and σ_i can be correlated, in which case we may define an arbitrary joint distribution $p(\boldsymbol{\mu}, \boldsymbol{\sigma})$ as the prior distribution. However, in order for the method to work well, the prior distribution must roughly reflect the relationships of μ_i's and σ_i's in reality.

We stress again that although we apply a prior distribution to $\boldsymbol{\mu}$ and $\boldsymbol{\sigma}$, μ_i's and σ_i's are fixed quantities for each population, rather than random variables. The prior distribution is simply a statistical tool to describe how μ_i's and σ_i's are correlated.

It will be shown later that using a uniform prior distribution effectively applies a bound on the estimated quantities. Therefore, the process of learning a uniform prior can be thought of obtaining a bound on the quantities to be estimated, and the probability distribution can be thought of as a mathematical tool to model correlation.

3.3   Learning a Prior Distribution

The first step in our method is to learn a prior distribution from data collected at all populations. We employ the maximum likelihood approach to learn the prior $p(\mu_i, \sigma_i|\boldsymbol{\theta})$, where $\boldsymbol{\theta}$ are *hyper-parameters* of the prior distribution. This problem can be formulated as an optimization problem

$$\underset{\boldsymbol{\theta}}{\text{maximize}} \quad p(\mathcal{X}_1, \cdots, \mathcal{X}_P|\boldsymbol{\theta}), \qquad (6)$$

where $p(\mathcal{X}_1, \cdots \mathcal{X}_P|\boldsymbol{\theta})$ is the likelihood function. We may either use a nonlinear optimizer to solve for the optimal $\boldsymbol{\theta}$, or we may derive closed-form solutions by solving

$$\frac{d}{d\boldsymbol{\theta}}p(\mathcal{X}_1, \cdots, \mathcal{X}_P|\boldsymbol{\theta}) = 0. \qquad (7)$$

In our formulation, the likelihood function can be computed by

$$p(\mathcal{X}_1, \cdots, \mathcal{X}_P | \boldsymbol{\theta})$$

$$= \int_{\boldsymbol{\mu}, \boldsymbol{\sigma}} p(\mathcal{X}_1, \cdots, \mathcal{X}_P | \boldsymbol{\mu}, \boldsymbol{\sigma}) p(\boldsymbol{\mu}, \boldsymbol{\sigma} | \boldsymbol{\theta}) d\boldsymbol{\mu} d\boldsymbol{\sigma}$$

$$= \int_{\boldsymbol{\mu}, \boldsymbol{\sigma}} \left(\prod_{i=1}^{P} p(\mathcal{X}_i | \mu_i, \sigma_i) \right) \left(\prod_{i=1}^{P} p(\mu_i, \sigma_i | \boldsymbol{\theta}) \right) d\boldsymbol{\mu} d\boldsymbol{\sigma} \quad (8)$$

$$= \prod_{i=1}^{P} \int_{\mu_i, \sigma_i} p(\mathcal{X}_i | \mu_i, \sigma_i) p(\mu_i, \sigma_i | \boldsymbol{\theta}) d\mu_i d\sigma_i,$$

where the second equality is due to two conditional independences, $(\mathcal{X}_1 \perp \cdots \perp \mathcal{X}_P | \boldsymbol{\mu}, \boldsymbol{\sigma})$[1] and $(\{\mu_1, \sigma_1\} \perp \cdots \perp \{\mu_P, \sigma_P\} | \boldsymbol{\theta})$. The integral Eqn. (8) can be computed by numerical integration, or we may derive its closed-form expression for special prior distributions.

In our formulation, we choose a uniform prior on μ_i's, as defined in Eqn. (5). To write it in the form of $p(\mu_i, \sigma_i | \boldsymbol{\theta})$, we define $\boldsymbol{\theta} = [a, b, c, d]^T$. We further assume that μ_i and σ_i are independent given $\boldsymbol{\theta}$.

Therefore, $p(\mu_i, \sigma_i | \boldsymbol{\theta}) = p(\mu_i | a, b) p(\sigma_i | c, d)$, i.e.,

$$p(\mu_i, \sigma_i | \boldsymbol{\theta}) = \begin{cases} \frac{1}{b-a} \frac{1}{d-c}, & \text{if } a \le \mu_i \le b, c \le \sigma_i \le d, \\ 0, & \text{otherwise.} \end{cases} \quad (9)$$

For each population, x_i satisfies the Gaussian distribution $N(\mu_i, \sigma_i^2)$, and therefore

$$p(\mathcal{X}_i | \mu_i, \sigma_i) = \prod_{j=1}^{N} \frac{1}{\sqrt{2\pi}\sigma_i} \exp \left\{ -\frac{1}{2} \frac{(x_{i,j} - \mu_i)^2}{\sigma_i^2} \right\}. \quad (10)$$

Inserting Eqn. (9) and Eqn. (10) into Eqn. (8), we need to compute for each i,

$$\int_c^d p(\sigma_i | c, d) d\sigma_i \int_a^b p(\mu_i | a, b) d\mu_i p(\mathcal{X}_i | \mu_i, \sigma_i), \quad (11)$$

where the integral with respect to μ_i is

$$\int_a^b p(\mu_i | a, b) d\mu_i p(\mathcal{X}_i | \mu_i, \sigma_i)$$
$$= \frac{1}{Z} \left\{ \Phi \left(\frac{b - \bar{x}_i}{\sigma / \sqrt{N_i}} \right) - \Phi \left(\frac{a - \bar{x}_i}{\sigma / \sqrt{N_i}} \right) \right\}, \quad (12)$$

where Z is a normalizing constant, and $\Phi(\cdot)$ is the CDF of the standard normal distribution. (Unfortunately, the integral in terms of σ_i is more involved, and we compute it by numerical methods.) Eqn. (11) can then be inserted into Eqn. (8) to compute the likelihood function.

3.4 Maximum A Posteriori Estimation of μ and σ

Once the prior $p(\mu_i, \sigma_i | \boldsymbol{\theta})$ is learned, MAP estimation can be applied to obtain a point estimate of μ_i's and σ_i's. MAP formulation searches for the values of μ_i's and σ_i's that maximize the posterior distribution, i.e., it solves

$$\underset{\mu_i, \sigma_i}{\text{maximize}} \quad p(\mu_i, \sigma_i | \mathcal{X}_i). \quad (13)$$

According to Bayes' rule,

$$p(\mu_i, \sigma_i | \mathcal{X}_i) \propto p(\mathcal{X}_i | \mu_i, \sigma_i) p(\mu_i, \sigma_i), \quad (14)$$

---

[1] The notation $(A \perp B | C)$ means that A and B are conditionally independent given C.

where $p(\mathcal{X}_i | \mu_i, \sigma_i)$ is derived in Eqn. (10), and $p(\mu_i, \sigma_i)$ is learned as described in Sec. 3.3.

For uniform priors of μ_i and σ_i, the right-hand side of Eqn. (14) is

$$\frac{1}{b-a} \frac{1}{d-c} p(\mathcal{X}_i | \mu_i, \sigma_i), \quad \text{if } \mu_i \in [a, b] \text{ and } \sigma_i \in [c, d]. \quad (15)$$

Therefore, MAP is equivalent to maximum likelihood estimation on the support $\mu_i \in [a, b]$ and $\sigma_i \in [c, d]$. The solution is simply

$$\mu_{i, MAP} = \begin{cases} a & \text{if } \mu_{i, MLE} < a \\ \mu_{i, MLE} & \text{if } a \le \mu_{i, MLE} \le b \\ b & \text{if } \mu_{i, MLE} > b \end{cases}, \quad (16)$$

$$\sigma_{i, MAP} = \begin{cases} c & \text{if } \sigma_{i, MLE} < c \\ \sigma_{i, MLE} & \text{if } c \le \sigma_{i, MLE} \le d \\ d & \text{if } \sigma_{i, MLE} > d \end{cases}, \quad (17)$$

where $\mu_{i, MLE}$ and $\sigma_{i, MLE}$[2] are equal to the sample mean and standard deviation, respectively[1].

3.5 Algorithm and Relaxation

Summarizing Sec. 3.3 and Sec. 3.4, our proposed algorithm consists of two steps, as shown in Algorithm 1.

Algorithm 1 Multi-Population Moment Estimation

Given: $\mathcal{X}_1, \cdots, \mathcal{X}_P$.
Outputs: (μ_i, σ_i), $i = 1, \cdots, P$.
1: Solve $\underset{\boldsymbol{\theta}}{\text{maximize}} \, p(\mathcal{X}_1, \cdots, \mathcal{X}_P | \boldsymbol{\theta})$ (Eqn. (6)) for $\boldsymbol{\theta}$
2: **for** $i = 1 \to P$ **do**
3:     Solve $\underset{\mu_i, \sigma_i}{\text{maximize}} \, p(\mu_i, \sigma_i | \mathcal{X}_i)$ (Eqn. (13)) for (μ_i, σ_i)
4: **end for**

As mentioned, the computation of the integral for σ_i in Eqn. (11) is quite involved. To migrate this problem, we may relax the optimization to an easier one

$$\underset{a, b}{\text{maximize}} \quad p(\mathcal{X}_1, \cdots, \mathcal{X}_P | a, b, \boldsymbol{\sigma}), \quad (18)$$

where $\boldsymbol{\sigma}$ is known, and only hyper-parameters (a, b) of the mean prior are searched for. However, in this formulation σ_i obtained in Eqn. (13) is dependent on μ_i (and hence (a, b)). The algorithm has to be modified to ensure that all the estimated quantities converge.

The modified algorithm with the above relaxation is shown in Algorithm 2. In step 1, we use sample mean/variance computed at each population as an initial guess for $\boldsymbol{\mu}$ and $\boldsymbol{\sigma}$, and the guess for (a, b) is computed by $a = \min(\mu_1, \cdots, \mu_P)$ and $b = \max(\mu_1, \cdots, \mu_P)$. Then we iteratively solve for (a, b), μ_i's and σ_i's until the convergence criteria in Algorithm 2 is satisfied.

3.6 Connections to Empirical Bayes Estimators

The ideas presented in this paper follow the philosophy of a class of Bayesian estimators, called *Empirical Bayes estimators* (EB)[13]. EB applies Bayes' rule to obtain either a point estimation or a posterior distribution of the parameters to be estimated. Unlike standard Bayesian methods that specify an arbitrary prior, EB learns the prior distribution from data. In particular, if a Gaussian prior is used for

---

[2] $\sigma_{i, MLE}$ is a biased estimator. To eliminate the bias, we may replace $\sigma_{i, MLE}$ in Eqn. (17) by its unbiased estimator.

Algorithm 2 Multi-Population Moment Estimation with Relaxation

Given: $\mathcal{X}_1, \cdots, \mathcal{X}_P$; ϵ (tolerance for convergence).
Outputs: (μ_i, σ_i), $i = 1, \cdots, P$.
1: Compute the initial guess for $a, b, \boldsymbol{\mu}, \boldsymbol{\sigma}$.
2: **repeat**
3:      $a_{\text{old}} = a, b_{\text{old}} = b, \boldsymbol{\mu}_{\text{old}} = \boldsymbol{\mu}, \boldsymbol{\sigma}_{\text{old}} = \boldsymbol{\sigma}$.
4:      Solve $\underset{a,b}{\text{maximize}} \, p(\mathcal{X}_1, \cdots, \mathcal{X}_P | a, b, \boldsymbol{\sigma})$ (Eqn. (18)) for
       (a, b)
5:      **for** $i = 1 \to P$ **do**
6:        Solve $\underset{\mu_i, \sigma_i}{\text{maximize}} \, p(\mu_i, \sigma_i | \mathcal{X}_i, a, b)$ (Eqn. (13)) for
         (μ_i, σ_i)
7:      **end for**
8: **until** $|a - a_{\text{old}}|^2 + |b - b_{\text{old}}|^2 + ||\boldsymbol{\mu} - \boldsymbol{\mu}_{\text{old}}||_2^2 + ||\boldsymbol{\sigma} - \boldsymbol{\sigma}_{\text{old}}||_2^2 < \epsilon$

the mean, EB gives the so-called *James-Stein estimator*[14] for the mean.

Particularly, a nice feature of the James-Stein estimator is that it is "superior" to the sample mean estimate, in the sense that the expected sum of mean square error of μ_i's at all populations is smaller than that of the sample mean estimator, *i.e.*

$$E\{\sum_{i=1}^{P}(\mu_i - \mu_i^{JS})^2\} < E\{\sum_{i=1}^{P}(\mu_i - \bar{x}_i)^2\}, \qquad (19)$$

where μ_i is the actual mean, μ_i^{JS} is the James-Stein estimator and $\bar{x}_i$ is the sample mean. One can show that if the Gaussian prior on μ_i's is used in our method, we obtain an estimator very similar to the James-Stein estimator, and Eqn. (19) still holds.

Unlike the James-Stein estimator, our method allows for more general prior distributions. Specifically, we have derived the case for uniform priors. We will show in Sec. 4 that our method can significantly out-perform sample mean/variance estimators.

3.7 Possible Limitations

Although our method may obtain a theoretically better overall estimate according to conclusions such as Eqn. (19), it can be the case, theoretically, that for a specific population, our method introduces a large bias.

As an extreme example, consider 100 populations, each with 1 observation, and $\mu_1 = \cdots = \mu_{99} = 0, \mu_{100} = 1$, $\sigma_1 = \cdots = \sigma_{100} = 1$. Effectively, our method will shrink the estimated mean towards 0. Therefore, for the 100-th population, the bias can be large.

However, due to the reasons mentioned in Sec. 3.2, such extremely pathological cases are unlikely to happen. Even if it happens, the outliers can be easily identified in a preprocessing step, and therefore accuracy will not be compromised by outliers.

3.8 Practical Implementation

It should be noted that the optimization problems in Algorithm 1 and Algorithm 2 may not be convex, and may have multiple local optimal points. There is no guarantee that our method will find the global optima. However, since initial guesses are estimated from the same data, the optimizer has a good guess start with, and is less affected by local optimal points.

To alleviate the computational cost associated with solving the optimization problems, we may impose an empirical prior distribution, instead of learning one from data. For example, experienced designers may have a good idea of the range of σ_i at each population – in this case, a uniform prior for σ_i's can be applied. However, empirical priors should be used with great caution, since it may incur unexpected bias. To be less biased, one may apply cross-validation [15] to check the validity of the empirical prior.

4. EXPERIMENTAL RESULTS

In this section, we apply the proposed method to two examples to show its accuracy and efficiency compared to sample mean/variance estimators. The first example is a set of artificial datasets to illustrate the advantage of our proposed method. The second example is a data set composed of time margin (eye width) measurements of a commercial high-speed I/O link. For notational convenience, we denote $\hat{\mu}_i$ and $\hat{\sigma}_i$ to be the estimation computed by our method, and $\mu_{i,\text{sample}}$ and $\sigma_{i,\text{sample}}$ to be the sample mean and sample standard deviation.

4.1 Illustrative Examples

In this example, we generate two sets of artificial data to illustrate the advantages of the proposed method in comparison with sample mean and variance estimators.

The data is generated as follows:

1. Choose a, b, c, d, P, N,
2. Randomly sample $\mu_i \sim U(a, b)$ and $\sigma_i \sim U(c, d)$ for $i = 1, \cdots, P$,
3. For each population, sample $x_{i,j} \sim N(\mu_i, \sigma_i)$ for $j = 1, \cdots, N$.

To compare $(\hat{\mu}_i, \hat{\sigma}_i)$ and $(\mu_{i,\text{sample}}, \sigma_{i,\text{sample}})$, we independently sample $\mathcal{X}_i$ (with μ_i's and σ_i's fixed) 500 times, and compute both estimators. We compare histograms of both mean/standard deviation estimators, as well as histograms of overall error for mean ϵ_μ and for standard deviation ϵ_σ, defined by

$$\epsilon_\mu = \sqrt{\frac{1}{P}\sum_{i=1}^{P}|\mu_i - \mu_{i,\text{est}}|^2}, \quad \epsilon_\sigma = \sqrt{\frac{1}{P}\sum_{i=1}^{P}|\sigma_i - \sigma_{i,\text{est}}|^2}.$$
$$(20)$$

4.1.1 Dataset 1

In the first dataset, we set $a = 0.9$, $b = 1.1$, $c = 0.9$, $d = 1.1$, $N = 5$, $P = 20$ which corresponds to the scenario where μ_i's are similar, σ_i's are similar, and the standard deviation of μ_i's is comparable to the value of σ_i's. Fig. 2 show the histograms of the mean and standard deviation estimations for 500 repeats at one population where $\mu_i = 0.9067$, $\sigma_i = 0.9786$. It is seen that the variance of $\hat{\mu}$ is much smaller than that of μ_{sample}. By using a prior learned from multiple populations, we successfully reduce the variance of the estimator. On the other hand, $\hat{\sigma}$ is only slightly better than σ_{sample}. This is partly due to the fact that in Algorithm 2, the prior for σ_i's is removed, and therefore σ_i's are estimated separately. However, due to the accuracy improvement of μ_i, the accuracy of σ_i is also improved.

It should also be mentioned that we choose to show the histogram at this configuration because its μ_i is the smallest among 20 populations, and therefore is likely to be biased by the prior distribution. However, the bias is almost negligible as can be seen from Fig. 2a.

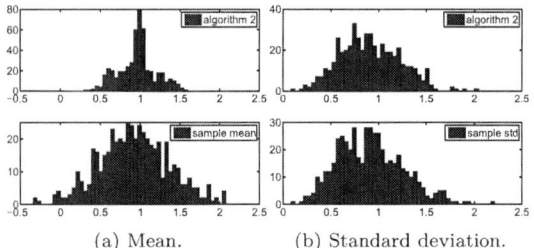

(a) Mean. (b) Standard deviation.

Figure 2: Histograms of both estimators for population ($\mu_i = 0.9067, \sigma_i = 0.9786$).

Fig. 3 shows histograms of ϵ_μ and ϵ_σ of both methods. From the histograms, we can compute $E(\epsilon_{\hat{\mu}}) = 0.2375$, $E(\epsilon_{\mu_{\text{sample}}}) = 0.4499$, $E(\epsilon_{\hat{\sigma}}) = 0.3191$ and $E(\epsilon_{\sigma_{\text{sample}}}) = 0.3431$. Hence, on average, our method achieves 2× accuracy improvement on μ. Moreover, the peak of ϵ_μ appears around 0.05, which implies that most likely, our method may achieve much more than 2× accuracy improvement.

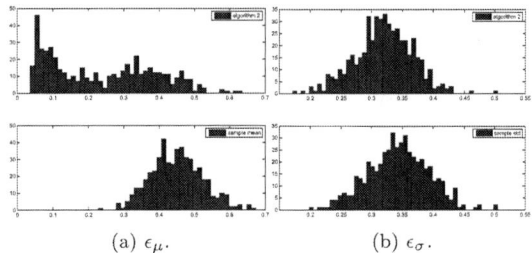

(a) ϵ_μ. (b) ϵ_σ.

Figure 3: Histograms of ϵ_μ and ϵ_σ, $P = 20$.

We also study the effect of the number of population on the estimation error. Fig. 4 shows histograms for $P = 100$ (a, b, c, d, N are the same as in dataset 1). Similar to the previous case, we can compute $E(\epsilon_{\hat{\mu}}) = 0.2130$, $E(\epsilon_{\mu_{\text{sample}}}) = 0.4478$, $E(\epsilon_{\hat{\sigma}}) = 0.3206$ and $E(\epsilon_{\sigma_{\text{sample}}}) = 0.3467$. The average accuracy for $P = 100$ is slightly better than that of $P = 20$. However, if we compare Fig. 4 with Fig. 3, we observe that with P larger, the peak at $\epsilon_\mu \simeq 0.05$ is higher, and there is a much clearer separation between the histograms of two estimators. In particular, for mean estimation, our method almost dominates the sample mean estimator, i.e., with high probability, our method obtains less error than the sample mean estimator.

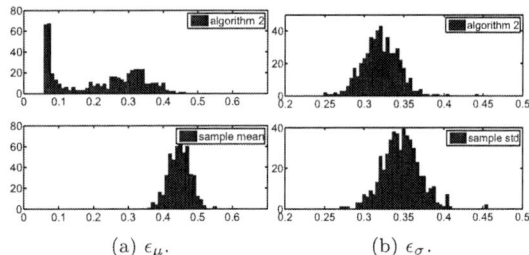

(a) ϵ_μ. (b) ϵ_σ.

Figure 4: Histograms of ϵ_μ and ϵ_σ, $P = 100$.

4.1.2 Dataset 2

In the second dataset, we set $a = 0$, $b = 2$, $c = 0.9$, $d = 1.1$, $N = 5$, $P = 20$. The major difference from the first

dataset is that the standard deviation of the mean at all populations is much larger, and is comparable to the standard deviation for the Gaussian distributions. Intuitively, if $(b-a)$ is large, the data at different population is less correlated. However, for this dataset, our method still out-performs sample mean/standard deviation estimators. Fig. 5 shows the histogram of ϵ_μ and ϵ_σ of both methods, from which we can compute $E(\epsilon_{\hat{\mu}}) = 0.3936$, $E(\epsilon_{\mu_{\text{sample}}}) = 0.4499$, $E(\epsilon_{\hat{\sigma}}) = 0.3329$ and $E(\epsilon_{\sigma_{\text{sample}}}) = 0.3431$. Although the improvement is not as evident as dataset1, we still achieve a relative error improvement of 5% and 1% for mean and standard deviation, respectively. For $P = 100$, we observe similar trends as in dataset1.

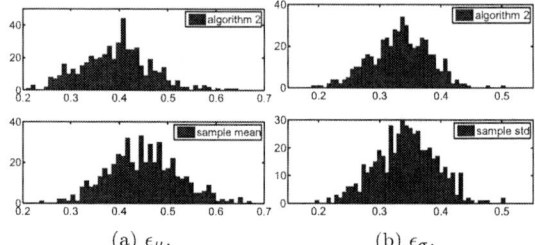

(a) ϵ_μ. (b) ϵ_σ.

Figure 5: Histograms of ϵ_μ and ϵ_σ, $P = 20$.

4.2 Validation of High-Speed I/O Links

In I/O link validation, one critical performance metric is Bit-Error-Ratio (BER). For the state-of-the-art high-speed links, the BER is extremely small. For example, in the latest PCIE specification [10], $\text{BER}_{\text{spec}} = 10^{-12}$ with 8Gb/sec data rate. This makes BER measurement a very time-consuming process. An alternative is to measure the eye width and eye height (a.k.a., time margin (TM) and voltage margin (VM), respectively) of the eye diagram at the receiver, which can be converted to BER under reasonable assumptions. Margin measurement, although much faster than direct BER measurement, is still expensive in terms of time and cost. For a limited time period, only a small number of data can be measured for each configuration.

In this example, we have measured the time margin of 50 dies (randomly sampled) for 8 different configurations. (Note that we measured 50 dies simply for the purpose of validating our algorithm.) The mean and standard deviation at different configurations are shown in Fig. 6. We have also observed from the histogram that the distribution of time margin can be well approximated by Gaussian distributions.

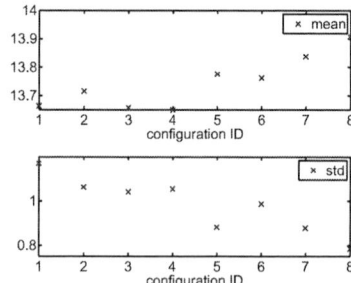

Figure 6: Mean and standard deviation at 8 configurations.

To compare the results of our method and sample mean/standard deviation estimators, we sample N_i data for each configuration from a distribution fitted from the 50 measurements,

464

and apply both methods. We repeat this experiment for 500 times, and compare the statistics of ϵ_μ and ϵ_σ.

The histogram of ϵ_μ for $N_i = 3$ is shown in Fig. 7a. Similar to dataset1, our method out-performs the sample mean estimator. We also study how ϵ_μ is affected by the sample size at each configuration. Fig. 7b shows the histogram of ϵ_μ for both methods for $N_i = 11$. With N_i larger, the accuracy of both $\hat{\mu}$ and μ_{sample} is improved, and ϵ_μ of $\hat{\mu}$ is peaked around 0.1.

The trend with respect to N_i can be better illustrated by Fig. 8 which shows ϵ_μ and ϵ_σ as a function of N_i. Particularly for ϵ_μ, we observe a consistently $1.5\times$ accuracy improvement over sample mean estimator. It is worth mentioning that as N_i becomes very large, ϵ_μ of both $\hat{\mu}$ and μ_{sample} converge towards 0, and there is little advantage of applying our method. However, if N_i is small, our method is much more accurate.

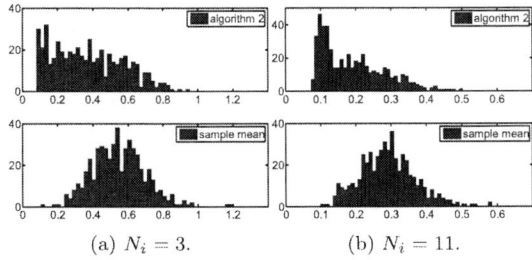

(a) $N_i = 3$.   (b) $N_i = 11$.

Figure 7: Histogram of ϵ_μ.

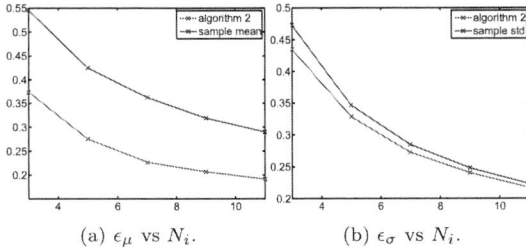

(a) ϵ_μ vs N_i.   (b) ϵ_σ vs N_i.

Figure 8: ϵ_μ and ϵ_σ decreases as N_i increases.

5. CONCLUSION

In this paper, we have proposed an efficient method for mean/standard deviation estimation under extremely small sample size. This problem is commonly seen in practice, and directly affects the time and cost associated with both pre-silicon and post-silicon validation, especially for complex analog/mixed-signal circuits. The validation of our method on several datasets, including measurement of commercial I/O links, shows that our method is consistently better than the sample mean/standard deviation estimators, and can achieve up to $2\times$ average accuracy improvement. Furthermore, the accuracy improvement can also be equivalently translated to a potentially large test/validation time reduction.

6. REFERENCES

[1] A. Papoulis and S. Pillai, *Probability, Random Variables and Stochastic Processes*. McGraw-Hill Science/Engineering/Math, 2001.

[2] G. Balamurugan, B. Casper, J. Jaussi, M. Mansuri, F. O'Mahony, and J. Kennedy, "Modeling and

Analysis of High-Speed I/O Links," *Advanced Packaging, IEEE Transactions on*, vol. 32, no. 2, pp. 237 –247, May 2009.

[3] J. Keshava, N. Hakim, and C. Prudvi, "Post-Silicon Validation Challenges: How EDA and Academia can Help," in *Design Automation Conference (DAC), 2010 47th ACM/IEEE*, June 2010, pp. 3 –7.

[4] C. Gu, "Challenges in Post-Silicon Validation of High-Speed I/O Links," in *Computer-Aided Design (ICCAD), 2012 IEEE/ACM International Conference on*. IEEE, 2012.

[5] Intel Corp., "Intel Platform and Component Validation." [Online]. Available: http://download.intel.com/design/chipsets/labtour/PVPT_WhitePaper.pdf

[6] E. E. Lior Shkolnitsky, "Electrical System-Validation Methodology for Embedded DisplayPort," June 2010. [Online]. Available: http://download.intel.com/design/intarch/PAPERS/323931.pdf

[7] K. Gambill, "System Margin Validation," December 2008. [Online]. Available: http://download.intel.com/design/intarch/papers/321078.pdf

[8] W. Zhang, T. Chen, M. Ting, and X. Li, "Toward Efficient Large-Scale Performance Modeling of Integrated Circuits via Multi-Mode/Multi-Corner Sparse Regression," in *Design Automation Conference (DAC), 2010 47th ACM/IEEE*. IEEE, 2010, pp. 897–902.

[9] X. Li, W. Zhang, F. Wang, S. Sun, and C. Gu, "Efficient Parametric Yield Estimation of Analog/Mixed-Signal Circuits via Bayesian Model Fusion," in *Computer-Aided Design (ICCAD), 2012 IEEE/ACM International Conference on*. IEEE, 2012.

[10] [Online]. Available: http://www.pcisig.com

[11] [Online]. Available: http://www.jedec.org

[12] C. Bishop, *Pattern Recognition and Machine Learning*. Springer, New York, 2006, vol. 4.

[13] G. Casella, "An Introduction to Empirical Bayes Data Analysis," *The American Statistician*, vol. 39, no. 2, pp. 83–87, 1985.

[14] B. Efron and C. Morris, "Data Analysis using Stein's Estimator and Its Generalizations," *Journal of the American Statistical Association*, vol. 70, no. 350, pp. 311–319, 1975.

[15] S. Arlot and A. Celisse, "A Survey of Cross-Validation Procedures for Model Selection," *Statistics Surveys*, vol. 4, pp. 40–79, 2010.

Verification of Digitally-Intensive Analog Circuits via Kernel Ridge Regression and Hybrid Reachability Analysis

Honghuang Lin
Texas A&M University
linhh@neo.tamu.edu

Peng Li
Texas A&M University
pli@tamu.edu

Chris J. Myers
University of Utah
myers@ece.utah.edu

ABSTRACT

The emergence of digitally-intensive analog circuits introduces new challenges to formal verification due to increased digital design content, and non-ideal digital effects such as finite resolution, round-off error and overflow. We propose a machine learning approach to convert digital blocks to conservative analog approximations via the use of *kernel ridge regression*. These learned models are then adopted in a hybrid formal reachability analysis framework where the support function based manipulations are developed to efficiently handle the large linear portion of the design and the more general *satisfiability modulo theories* technique is applied to the remaining nonlinear portion. The efficiency of the proposed method is demonstrated for the locked time verification of a *digitally intensive phase locked loop*.

1. INTRODUCTION

Designing *analog/mixed-signal* (AMS) circuits in highly scaled CMOS technologies is hampered by increasing *Process-Voltage-Temperature* (PVT) variations and worsening device characteristics. As a result, the so-called digitally-assisted or *digitally-intensive analog* (DIA) design methodology has emerged, which minimizes the pure analog content of the design while relying more on digital processing [1]. However, inclusion of increased digital content in such designs adds new complications to the existing challenges of AMS circuit design verification, which are the result of effects such as nonlinear dynamical characteristics and complex interactions between digital and analog signals. This work aims to develop verification techniques for DIA circuits under a formal reachability analysis framework [2] [3] [4].

The interaction between digital (Boolean) and continuous-valued analog signals is intensified in DIA circuits, presenting a key challenge in verification. In addition, to fully capture "digital" effects such as finite resolution (inherent in any analog-to-digital conversion), round-off error and overflow (inherent in additions/multiplications due to finite word length effects), additional state variables need to be introduced, blowing up the dimensionality of the state space and slowing down the reachability analysis.

The first main idea of this work is to "unify" the two

Permission to make digital or hard copies of all or part of this work for personal or classroom use is granted without fee provided that copies are not made or distributed for profit or commercial advantage and that copies bear this notice and the full citation on the first page. To copy otherwise, to republish, to post on servers or to redistribute to lists, requires prior specific permission and/or a fee.

types of signals by converting digital signals into approximate analog signals. More precisely, this paper leverages machine learning to find the potential error of the conversion on a given digital block and bound its output using a continuous analog signal and a tight error interval. It turns out kernel based learning methods such as support vector machines [5] [6] are useful tools for this purpose. In particular, this paper adopts kernel ridge regression [7] [8] and the related confidence interval computing algorithm [9] to construct such "analog" models for the digital block and estimate the error interval, the latter of which ensures the conservativeness of this conversion as required by formal verification. The accuracy of this "digital-to-analog" modeling, e.g. the length of the error interval, is tightly controlled in the learning process. Note that approximating (digital) quantization effects by including additive quantization noise has been used to provide empirical noise analysis for analog-to-digital converter designs for decades. However, to the best of our knowledge, this is the first time systematically learned "analog" models for digital blocks has been used in formal verification.

The second contribution of this work is to further speed-up reachability analysis by partitioning an AMS design into a linear and a nonlinear subsystem and adopting efficient support-function based state space manipulations. This paper over-approximates the reachable state space by using tight polyhedral convex sets based on the support function representation. The use of support functions alleviates the resolution problem of the box state space discretization adopted in the simulation-assisted *Satisfiability Modulo Theories* (SMT) based reachability analysis developed for general AMS circuits [4] [10]. Support functions have been used before for reachability analysis [11], where the entire system is assumed to be linear and efficient progression of support functions are used to track the linear transition of the state space. To verify more general nonlinear AMS circuits, this paper develops a modified support-function approach for the linear sub-system while adopting the more general SMT technique for the remaining nonlinear subsystem. This paper also proposes specific support function manipulation techniques to efficiently track the reachable state space of the linear sub-system.

While the second school of our ideas can be applied to generic AMS circuits [12], our experience has shown them to be particularly appealing to DIA circuits. DIA circuits are constructed to have high digital content with minimum use of pure analog-based processing. Digital blocks such as filters are designed to be "linear" and implemented in robust digital logic. However, this linearity disappears in the presence of round-off errors due to finite word length effects. The use of the machine learned conservative analog models re-establishes this lost linearity and allows application of the proposed support function approach to a more dominant portion of the design, which is modeled as the linear sub-system. This paper demonstrates the proposed techniques to the challenging task of locked time verification of a *digitally-intensive phase-locked loop* (DI-PLL).

2. VERIFICATION CHALLENGES

The DI-PLL studied in this paper is shown in Fig. 1. The circuit involves an analog feedback path from the *digital controlled oscillator* (DCO) to the *phase detector*. The phase detector uses an accumulator and a *time-to-digital converter* (TDC) to detect the only analog variable, which is the phase difference $\Delta\phi$ between the reference clock REF and the output signal CKV, and then outputs the phase difference to the loop filter to control the DCO. Neglecting the digital effects and assuming all functional blocks have an ideal continuous characteristics, the verification of the system can be performed on a model involving a few state variables such as ones to capture the input/output/internal signals of the loop filter. However, to fully capture those digital effects, more state variables are needed and hence the model has a much higher complexity.

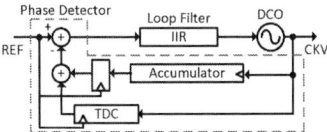

Figure 1: Block diagram for a DI-PLL.

For example, a TDC shown in Fig. 2(a) is used to measure the fractional phase difference between CKV and REF. Theoretically, the output of the TDC should be proportional to the fractional phase difference, making the transition of the phase detector linear. However, the finite delay of the inverters limits the achievable resolution of the TDC. As a result, the phase detector with a finite TDC resolution has a staircase transition curve instead of an ideal linear one.

Digital filters such as FIR or IIR filters are often designed from a linear z-domain transfer function. To model the ideal transfer function of a second order IIR loop filter shown in Fig. 2(b), four state variables are sufficient. But to fully model the finite word length effects such as overflow and round-off error of the filter, 8 more variables should be assigned to the output nodes of all the internal adders and multipliers. The dimension of the system is thus greatly increased.

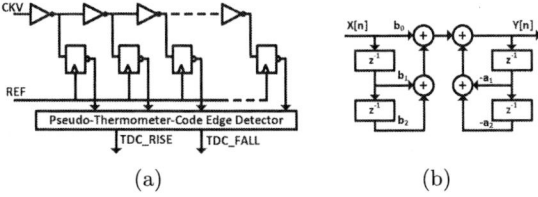

(a)   (b)

Figure 2: (a) Time-to-digital converter [13]; (b) Second order IIR filter.

To address the above verification challenges, this paper proposes to extract an abstract continuous model with a lower dimensionality for a given DIA design through the use of learning-based regression. The correctness of our approach is guaranteed by the use of error intervals that are part of the abstract model.

3. MODEL ABSTRACTION

3.1 Digital Abstraction with Analog Variables

An AMS system consists of analog and digital variables along with continuous and discrete transitions and state mapping functions. It can generally be defined as:

DEFINITION 1. *An AMS system is a tuple* $H_{AMS} = (X_a, X_d, R_{res}, F_a, F_b)$ *where:*

- $X_a \subseteq \mathbb{R}^n$ *is a set of continuous variables;*

- $X_d \subset \mathbb{R}^m$ *is a set of discrete variables;*

- $R_{res} \in \mathbb{R}^m$ *is the resolution of the discrete variables;*

- $F_a : (X_a, X_d) \to \mathbb{R}^n$ *is the mapping functions to the continuous state variables;*

- $F_d : (X_a, X_d, R_{res}) \to \mathbb{R}^m$ *is the mapping functions to the discrete state variables.*

As discussed in the previous section, the size of X_d and F_d in a DIA system may be large which may blow up the dimension of the state space for verification. From the design perspective, to focus on the key properties of the digital functional blocks, it is common to approximate mapping functions in F_d with continuous mapping functions while independent error variables are added into X_a to cover the approximation error. Fox example, a TDC is routinely modeled to have a linear conversion characteristics and additive quantization errors, the latter of which is used to analyze the TDC induced noise [14].

Motivated by the above design analysis technique, our method converts key variables in X_d into analog variables and approximates F_d with continuous mappings to simplify the system. However, the goal of our method is to find a conservative approximation for the entire reachable state space rather than analyzing the "average" noise behavior of the circuit. This prompts us to find conservative error intervals, ideally tight, for our abstract models.

Let us define the abstracted model produced by our proposed method as:

DEFINITION 2. *An abstraction of* H_{AMS} *is a tuple* $H_C = (X, F, E_u, E_l)$ *where:*

- $X \subseteq \mathbb{R}^N$ *consists of* X_a *and continuous approximation of* X_d;

- $F : X \to \mathbb{R}^N$ *is the set of mapping functions that consists of* F_a *and continuous approximation of* F_d;

- $E_u : X \to \mathbb{R}^N$ *is the upper bound of the error intervals of* $F(X)$;

- $E_l : X \to \mathbb{R}^N$ *is the lower bound of the error intervals of* $F(X)$.

To extract an abstract continuous model from the DI-PLL, all the digital variables in X_d such as the input/output of the phase detector and the loop filter are approximated by the corresponding continuous variables in X. Then, for all the digital blocks, the transition function will be approximated by the ideal characteristic. For example, the transition of the phase detector is modeled as a linear function that produces an output proportional to the detected phase difference. The transition of the filter is modeled by its ideal z-domain transfer function, which is also a linear mapping function that maps the input and internal storage of the filter to its output at the next sampling clock cycle.

Error accumulation is an inherent problem in reachability analysis. If the system abstraction comes with loose error intervals, i.e. large $E_u - E_l$, the reachability analysis may converge slower or, even worse, may no longer converge. If the error intervals are too narrow, i.e. small $E_u - E_l$, the abstraction may lose conservativeness and the result of the reachability analysis is not ensured to cover all the possible cases. So it is important and essential to find a tight interval that covers most of the errors with small over approximation. The next subsection leverages *kernel ridge regression* (KRR) to compute such tight intervals.

3.2 Error Interval Estimation via KRR

KRR [7], a.k.a. *least squares support vector regression* (LS-SVR) [8] is a very effective statistical learning method and has been applied to error estimation. It is formulated as:

$$\min_{w,b,e} w^T w + \gamma \sum_{i=1}^{n} e_i^2 \qquad (1)$$

subject to

$$Y_i = w^T \phi(X_i) + b + e_i, i = 1, ..., n. \qquad (2)$$

where (X_i, Y_i) are training data and ϕ is the mapping function that maps X_i into a higher dimensional space. By using Lagrange multipliers, one can solve the problem by defining a kernel function $K(X_i, X_j) = \phi(X_i) \cdot \phi(X_j)$ instead of defining an exact ϕ. This model produces the following decision function:

$$\hat{m}(x) = \sum_{i=1}^{n} \alpha_i K(x, X_i) + b_i \qquad (3)$$

Given enough training data and a suitable parameter γ, KRR can train an accurate model for error prediction. Considering the system abstraction for DI-PLL, errors may come from the continuation of digital variables and the mapping function approximation by an ideal continuous characteristic. For instance, let $x_i \in X$ denote the output of the phase detector in the abstract continuous model for the DI-PLL, our method assumes that the abstracted mapping function of the phase detector is $x_i = F_i(X)$ and its original mapping function is $x_i = F_{ad,i}(X)$. This method samples some random system states $X_1, X_2, ..., X_n$ and computes the error of the abstraction on x_i as $Y_j = F_i(X_j) - F_{ad,i}(X_j)$. Then, this method can get an error estimation of x_i by applying KRR on the training data set (Y_j, X_j) such that $j = 1, 2, ..., n$. By repeating the sampling and training process, error prediction functions are produced by KRR for every state variable.

To find a tight interval of the prediction, an efficient algorithm for computing confidence intervals of KRR prediction has been given in [9]. The algorithm treats KRR as a linear smoother to derive the formulation of the bias and variance of the prediction. A linear smoother is defined as an estimator $\hat{m}$ of m in which there exists a smoother vector $L(x) = (l_1(x), ..., l_n(x))$ such that:

$$\hat{m}(x) = \sum_{i=1}^{n} l_i(x) Y_i, \forall x \in \mathbb{R}^d \qquad (4)$$

KRR prediction is a kind of biased estimation and thus there is bias between the predicted value and the center of the

prediction interval. The bias is given as

$$\hat{B}(\hat{m}(x)) = L(x)^T \hat{m} - \hat{m}(x) \qquad (5)$$

where $\hat{m} = (\hat{m}(X_1), ..., \hat{m}(X_n))^T$. The variance of the prediction at x is given as

$$\hat{V}(\hat{m}(x)) = L(x)^T \hat{\Sigma}^2 L(x) \qquad (6)$$

where $\hat{\Sigma}^2 = diag(\hat{\sigma}^2(X_1), ..., \hat{\sigma}^2(X_n))$ and

$$\hat{\sigma}^2(x) = \frac{L(x)^T diag(\hat{\epsilon}\hat{\epsilon}^T)}{1 + L(x)^T (SS^T - S - S^T)} \qquad (7)$$

S denotes the smoother matrix of the initial smooth and $\hat{\epsilon}$ denotes the residuals of $\hat{m}(X_1), ..., \hat{m}(X_n)$.

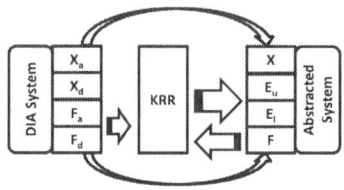

Figure 3: Error interval prediction via KRR.

The KRR based abstraction process is shown in Fig. 3. At first, all the digital variables are approximated by continuazation and ideal continuous characteristics like the linearized transition of the phase detector are used to approximate digital transitions. Then, for every extracted variables in X, KRR is trained to get the error prediction. After KRR training, the upper and lower bounds are obtained by $(4)\sim(7)$.

4. HYBRID REACHABILITY ANALYSIS

This section proposes our hybrid method for the verification of abstracted systems. The basic idea is to divide and conquer. Our method divides the abstracted system into a linear and a nonlinear subsystem. The reachable state space of the linear system can be obtained by applying the manipulation of support functions. Simulations and SMT techniques are employed in exploring the reachable state space of the nonlinear subsystem. After the two reachable subspaces are obtained, the method extends them to the same dimension with the whole system and gets the reachable space of the whole system by computing the intersection of the two subspaces.

4.1 System Partition

After the DIA system is converted to the abstracted system, this method picks up variables from X that are involved in linear transitions to form a subset X_L and variables that are involved in nonlinear transitions to form a subset X_{NL}. If variables in X_L and X_{NL} are independent with each other, then $X_L \cap X_{NL} \neq \phi$. However, there often exist variables in X_L correlated to variables in X_{NL}. Therefore, either subset should contain additional correlated variables to cover those correlations. Besides, F is divided into a linear mapping, F_L, and a nonlinear mapping, F_{NL}.

In the model of DI-PLL shown in Fig. 4, the phase detector and the loop filter are linearized with system abstraction while the DCO remains nonlinear. Assume that the loop filter is a second order IIR digital filter, then at time t, the state of the system is $X(t) = \{x_i(t)\}^T, i = 1, 2, 3, 4, 5$, where $x_1(t) = \Delta\phi(t)$, $x_2(t) = V_1(t), x_3(t) = V_1(t - \Delta t), x_4(t) =$

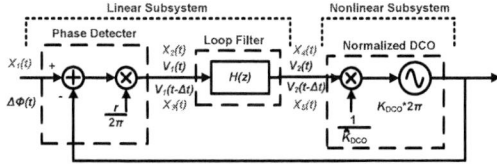

Figure 4: System partition of DI-PLL.

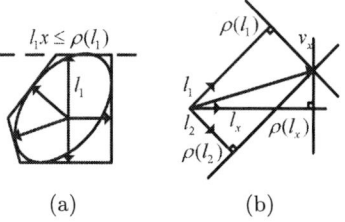

(a)   (b)

Figure 5: (a) Support function representation on 2D space; (b) Approximate support function value of arbitrary direction l_x.

$V_2(t)$, and $x_5(t) = V_2(t - \Delta t)$. From t to $t + \Delta t$, the transition of the system can be partitioned as a linear transition:

$$X(t) \to (x_2(t + \Delta t), x_3(t + \Delta t), x_4(t + \Delta t), x_5(t + \Delta t))^T \quad (8)$$

and a nonlinear transition:

$$(x_1(t), x_4(t))^T \to (x_1(t + \Delta t)) \quad (9)$$

After the partition, our method solves the reachable state space of the two subsystems with different methods. For the linear subsystem, once our method gets the constraint state space of t, it can directly compute the 4-dimensional reachable state space by support function space representation and corresponding properties. For the nonlinear subsystem, our method uses an SMT-based formal method to conservatively compute the reachable space. After exploring the reachable space of the two subsystems, our method compute the intersection of the two subspaces to obtain the reachable space of the entire system. Note that $x_1(t + \Delta t)$ is correlated with $x_2(t + \Delta t)$ and $x_5(t + \Delta t)$ since they are all determined by $x_1(t)$ and $x_4(t)$, the nonlinear subsystem reachable space should include x_1, x_2 and x_5, otherwise the intersection computation may cause great overapproximation.

4.2 Support Functions

Our method uses a support function [15] ρ_Ω to represent a convex reachable state space $\Omega \subseteq \mathbb{R}^d$. It is defined as:

$$\rho_\Omega(l) = max_{x \in \Omega} l \cdot x, \forall l \in \mathbb{R}^d \quad (10)$$

According to the definition, Ω is actually represented by an intersection of half spaces:

$$\Omega = \bigcap_{l \in \mathbb{R}^d} \{x \in \mathbb{R}^d | l \cdot x \le \rho_\Omega(l)\} \quad (11)$$

As shown in Fig. 5(a), $\rho(l_1)$ is actually defining a half space $l_1 \cdot x \le \rho(l_1)$. If the formula of $\rho(l)$ is known, or support function values $\rho(l_i)$ for infinite number of l_i are stored, then the convex space can be exactly represented without error. If just a finite number of l_i and their corresponding support function values are recorded, the space is over approximated by a polyhedron. For example, in Fig. 5(a), if five vectors l_i and their corresponding support function values are recorded, the ellipse is actually over approximated by a pentagon. As the number of l increases, over approximation is reduced and the polyhedron finally converges to the ellipse. Our approach leverages this observation to compute an over-approximation for the reachable state space with high efficiency and low storage.

4.3 Analysis of the Linear Subsystem

For the linear subsystem, let $X(t) \in \mathbb{R}^n$ denote the state of the whole system at t and $X_L(t) \in \mathbb{R}^m$ denote the state of the linear subsystem at $t + \Delta t$, then the transition can be formulated as:

$$X_L(t + \Delta t) = AX(t) \quad (12)$$

where the size of A is $m \times n$. To solve the reachable space of this system, support functions have a useful property to be exploited:

$$\rho_{AU}(l) = \rho_U(A^T l) \quad (13)$$

where $U \subseteq \mathbb{R}^d$ and A is an arbitrary matrix.

If the initial state space Ω_0 is represented by an accurate support function formula $\rho_{\Omega_0}(l)$, reachable spaces Ω_t at any moment can be solved by mapping any $l \in \Omega_t$ to $l' \in \Omega_0$ and hence an accurate support function representation is maintained. However, in the linear subsystem of DI-PLL, the formula form of the support function representation is difficult to use since the reachable state space computation contains the intersection operation between the linear and nonlinear reachable state space and vectors in Ω_t can only be mapped to $\Omega_{t-\Delta t}$. Hence, it is impractical to represent the reachable state spaces with symbolic formula of support functions.

Our approach, more practically, uses a finite number of normalized direction vectors l and their corresponding support function values to represent the reachable state space. As mentioned earlier, using a finite number of vectors, l, is actually producing a tight polyhedral over-approximation of Ω. For the convenience of the following computations, our method evenly samples normalized l from a d-dimensional spherical coordinate system.

To represent a reachable space Ω_t, our method stores a list of normalized vectors l_i and their corresponding support function values $\rho(l_i)$ and uses them to compute the support function values of $\Omega_{t+\Delta t}$. The problem is that for a vector $l \in \Omega_{t+\Delta t}$, $l_x = A^T l_x \in \Omega_t$ may not be stored in the representation of Ω_t. Therefore, our method uses the following algorithm to approximate $\rho_{\Omega_t}(l_x)$.

In a 2D space shown in Fig. 5(b), for an unknown vector l_x, our approach can first find two closest vectors from the evenly sampled known direction vector list. Denote the two vectors as l_1 and l_2, and the intersection point of planes $l_1 x = \rho(l_1)$ and $l_1 x = \rho(l_1)$ as v_x, according to the definition of support function, $\rho(l_x) \le l_x \cdot v_x$. Thus our method can first compute the intersection point v_x and then use $l_x \cdot v_x$ to approximate $\rho(l_x)$.

More generally, for arbitrary $l_x \in \mathbb{R}^d$, to approximate $\rho_\Omega(l_x)$ with a list of stored direction vectors and their corresponding support function values, our method can first select d closest vectors $l_1, ..., l_d$ from the direction vector list, and then compute the intersection point v_x of d corresponding hyperplanes. The selection of $l_1, ..., l_d$ can be easily achieved by mapping l_x into a spherical coordinate system since the stored vectors are sampled evenly from that system. Since v_x lies on the boundary of the corresponding half space of $l_1, ..., l_d$, it can be computed by solving:

469

$$(l_1^T, l_2^T, ..., l_d^T)^T v_x = (\rho_\Omega(l_1), \rho_\Omega(l_2), ..., \rho_\Omega(l_d))^T \quad (14)$$

After v_x is computed, $\rho_\Omega(l_x)$ can be approximated by $v_x \cdot l_x$. With this algorithm, the reachable state space of the linear subsystem can be solved with a support function representation of finite direction vectors.

To make the reachable state space conservative, error intervals obtained by the KRR-based method should be considered. Since the system transition is linear, the reachable points of the states on the boundary of the current space still appear on the boundary of the reachable space, our method can simply extend the boundary according to error intervals of on-boundary states.

4.4 Analysis of the Nonlinear Subsystem

For the nonlinear subsystem, $X_{NL}(t+\Delta t) = F_{NL}(X_{NL}(t))$, our method leverages the SMT-based framework proposed in [4] to explore the reachable state space. At first, simulations are used to quickly explore most of the reachable space, and then a SAT solver is invoked to find the remaining reachable space. This paper improves the method by using support function representations. Another property of support functions is useful here:

$$\rho_{CH(U,V)}(l) = max(\rho_U(l), \rho_V(l)) \quad (15)$$

where $CH(U,V)$ denotes the convex hull of U and V. In the first stage, each time our method gets a reachable state from the simulation, it enlarges that state into a small convex set according to its error interval estimation given by the previous KRR-based method. Then, our method computes the convex hull of the current reachable state space and this small convex set to update the reachable space. After a bunch of simulations, the result covers most of the reachable state space. Then, our method uses a SAT solver to check whether the reachable state space is totally covered or not. If not, the solver provides a counterexample that is reachable but still not covered. The counterexample is added into the convex hull and SAT checking is repeated until the solver cannot find any counterexample. The final convex hull conservatively covers the reachable state space.

The complexity of the original method in [4] exponentially increases if the resolution of its space discretization or the dimension of its state space increases. The resolution problem is solved by using support function representations. For the dimension problem, since the variables that are involved in the linear system can be directly computed according the previous state space, the dimension of the SMT framework is lower, which improves the runtime as well.

4.5 Hybrid Reachability Analysis

In the complete flow of the hybrid reachability analysis, after a simplified model is produced using abstraction, the system is partitioned to obtain a linear and a nonlinear subsystem. Then, the support function based method is employed to solve the linear subsystem and the SMT-based method is used to explore the reachable space of the nonlinear subsystem.

After the reachable state space exploration of the two subsystems are completed, the intersection of the two subspaces should be computed to obtain the reachable space of the whole system. At first, our method should extend the two subspaces to the same dimension, which can be performed in a similar way as approximating $\rho(l_x)$ for an unrecorded l_x. After that, for convex sets represented by support functions, the intersection can be achieved by the MIN operation:

$$\rho_{U \cap V} = min(\rho_U(l), \rho_V(l)) \quad (16)$$

As discussed in [16], the min operation can be used to compute the intersection of the two convex sets, but the result is not a strictly defined support function due to possible redundant halfspaces. In our representation, the problem does not affect our algorithm since our method is not using a strict support function representation but an approximation with a finite number of l instead.

5. EXPERIMENTS

This section demonstrates the proposed technique by verifying a DI-PLL.

5.1 Error Estimation of System Abstraction

The system extraction process abstracts continuous models of a DI-PLL that involves a phase detector, an IIR loop filter and a DCO. The abstracted phase detector is a simple linear functional block while the digital phase detector has a finite resolution. The abstracted IIR loop filter is represented by an ideal continuous-valued z-domain transfer function while the real digital filter may suffer from finite word length effect. The DCO is modeled by a continuous VCO model with a nonlinear transition function.

Our method uses a KRR solver provided in [17] to train regression models of different variables. This experiment observes the error prediction intervals of the phase difference $\Delta\phi(t)$ and the output of the IIR filter $V_2(t)$. The experiment randomly samples 100 possible states of the system, and runs simulations on both the original system and the abstracted system, respectively. The difference in $\Delta\phi(t+\Delta t)$ and $V_2(t+\Delta t)$ of the two systems are gathered as training data, and KRR models with the parameter $\gamma = 1e - 4$ are trained on these 100 training datasets. Fig. 6 shows the width of error prediction intervals for different word length.

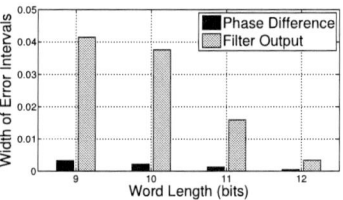

Figure 6: Width of error interval prediction with 95% confidence for different word length.

5.2 Enhancement with System Partition

The reachable state space exploration of the nonlinear subsystem first runs simulations to quickly cover most of the reachable space and then invokes a SAT solver provided in [18] to recursively find the reachable points outside the current approximation. In this flow, the SAT solver takes most of the runtime. Let us compare the complexity and the runtime of a single running of a SAT solver to illustrate the runtime enhancement of system partition.

Our method represents these two spaces with support functions, whose l vectors are evenly sampled every $\pi/12$ on every spherical coordinate dimension. For an abstracted model of the DI-PLL that has not been partitioned, both the input and output spaces of the abstracted model of the DI-PLL are 5-dimensional. For a partitioned system, the input space is a 2-dimensional space while the output space is a 3-dimensional space as discussed earlier. Our experiment performs a single running of the SAT solver on the partitioned and unpartitioned systems, respectively, and the results are shown in Table. 1.

470

Table 1: Comparison of single SAT solver running on partitioned and unpartitioned systems.

	Partitioned	Unpartitioned
SAT Variables	5	10
Support Function Constraints	532	64420
Runtime (sec)	4.1868	1942.9
Extra Runtime (sec)	21.163	-
Minimum Speedup	73	-

The number of SAT constraints for the unpartitioned system that are brought by support function representation is about $120\times$ more than those for the partitioned system. For a single SAT solver run, a partitioned system may gain $440\times$ speedup in runtime. Even if the additional runtime cost by abstraction and linear subsystem reachability analysis is taken into account, since for each time step the SAT solver is invoked at least once, the minimum speedup is still more than $73\times$.

5.3 Result of Hybrid Reachability Analysis

We apply the proposed method to verify the lock time specification of the DI-PLL of Fig. 4 described in the previous sections. We set the word length of the original system as 11 bits and then abstract a model with KRR based error interval estimation on $\Delta\phi(t)$, $V_1(t)$ and $V_2(t)$.

For most digital circuits, the initial state is often reset to be zero and thus we define the initial state space as $\Delta\phi \in [-\pi, \pi]$ and $V_1(0) = V_2(0) = V_1(-\Delta t) = V_2(-\Delta t) = 0$. The reachable space of $\Delta\phi$ obtained by our hybrid reachability analysis method is shown in Fig. 7. The time step $\Delta t = 1ns$ and it takes 18 hours for 250 steps in total. Compared to our hybrid method, direct SMT method is impractical since it cannot run to a completion within one week.

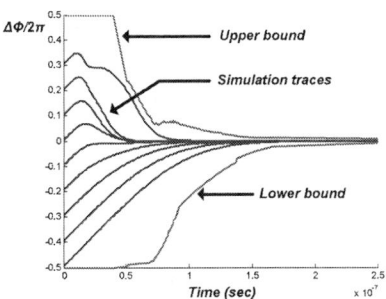

Figure 7: Results of the hybrid reachability analysis method.

The reachable space of $\Delta\phi$ conservatively covers all those evenly sampled simulation traces and finally converges to a narrow space. The result indicates that the phase of the output signal is locked and, if we define the lock condition as $|\Delta\phi| \leq 0.01\pi$, the lock time of the DI-PLL would be smaller than $0.25\mu s$.

6. CONCLUSION

This paper proposes a kernel ridge regression based system abstraction method to abstract a continuous and conservative system from a digitally-intensive system. To efficiently perform reachability analysis on the abstracted model, a hybrid method is presented to partition the system into linear and nonlinear subsystems and solve them with different strategies. The proposed method is demonstrated in the verification of a digitally-intensive phase-locked loop.

7. ACKNOWLEDGMENTS

This material is based upon work supported by the National Science Foundation under Grant No. 1117660 and 1117515, and the Semiconductor Research Corporation and Texas Analog Center of Excellence under Contract 2008-HC-1836.

8. REFERENCES

[1] B. Murmann. A/d converter trends: Power dissipation, scaling and digitally assisted architectures. In *Custom Integrated Circuits Conference, 2008. CICC 2008. IEEE*, pages 105–112. IEEE, 2008.

[2] M.H. Zaki, S. Tahar, and G. Bois. Formal verification of analog and mixed signal designs: A survey. *Microelectronics Journal*, 39(12):1395–1404, 2008.

[3] M. Althoff, A. Rajhans, B.H. Krogh, S. Yaldiz, X. Li, and L. Pileggi. Formal verification of phase-locked loops using reachability analysis and continuization. In *Proceedings of the International Conference on Computer-Aided Design*, pages 659–666. IEEE Press, 2010.

[4] Leyi Yin, Yue Deng, and Peng Li. Verifying dynamic properties of nonlinear mixed-signal circuits via efficient smt-based techniques. In *Proceedings of the International Conference on Computer-Aided Design*. IEEE Press, 2012.

[5] F. De Bernardinis, MI Jordan, and A. SangiovanniVincentelli. Support vector machines for analog circuit performance representation. In *Design Automation Conference, 2003. Proceedings*, pages 964–969. IEEE, 2003.

[6] Honghuang Lin and Peng Li. Classifying circuit performance using active-learning guided support vector machines. In *Proceedings of the International Conference on Computer-Aided Design*. IEEE Press, 2012.

[7] C. Saunders, A. Gammerman, and V. Vovk. Ridge regression learning algorithm in dual variables. In *(ICML-1998) Proceedings of the 15th International Conference on Machine Learning*, pages 515–521. Morgan Kaufmann, 1998.

[8] J.A.K. Suykens, J. De Brabanter, L. Lukas, and J. Vandewalle. Weighted least squares support vector machines: robustness and sparse approximation. *Neurocomputing*, 48(1):85–105, 2002.

[9] K. De Brabanter, J. De Brabanter, J.A.K. Suykens, and B. De Moor. Approximate confidence and prediction intervals for least squares support vector regression. *Neural Networks, IEEE Transactions on*, 22(1):110–120, 2011.

[10] David Walter, Scott Little, Chris Myers, Nicholas Seegmiller, and Tomohiro Yoneda. Verification of analog/mixed-signal circuits using symbolic methods. *Computer-Aided Design of Integrated Circuits and Systems, IEEE Transactions on*, 27(12):2223–2235, 2008.

[11] A. Girard, C. Le Guernic, et al. Efficient reachability analysis for linear systems using support functions. In *Proc. of the 17th IFAC World Congress*, pages 8966–8971, 2008.

[12] Honghuang Lin and Peng Li. Reachability analysis for ams verification using hybrid support function and smt-based method. In *Frontiers in Analog CAD (FAC) Workshop*, 2013.

[13] R.B. Staszewski, K. Muhammad, D. Leipold, C.M. Hung, Y.C. Ho, J.L. Wallberg, C. Fernando, K. Maggio, R. Staszewski, T. Jung, et al. All-digital tx frequency synthesizer and discrete-time receiver for bluetooth radio in 130-nm cmos. *Solid-State Circuits, IEEE Journal of*, 39(12):2278–2291, 2004.

[14] R.B. Staszewski and P.T. Balsara. *All-Digital Frequency Synthesizer in Deep-Submicron CMOS*. Wiley-Interscience, 2006.

[15] D.P. Bertsekas, A. Nedi, A.E. Ozdaglar, et al. Convex analysis and optimization. 2003.

[16] P.K. Ghosh and K.V. Kumar. Support function representation of convex bodies, its application in geometric computing, and some related representations. *Computer Vision and Image Understanding*, 72(3):379–403, 1998.

[17] Davis E. King. Dlib-ml: A machine learning toolkit. *Journal of Machine Learning Research*, 10:1755–1758, 2009.

[18] ISAT: Tight integration of satisfiability & constraint solving. http://isat.gforge.avacs.org/.

Machine-Learning-Based Hotspot Detection Using Topological Classification and Critical Feature Extraction *

Yen-Ting Yu[1], Geng-He Lin[1], Iris Hui-Ru Jiang[1], and Charles Chiang[2]

[1]Dept. of Electronics Engineering and Inst. of Electronics, National Chiao Tung University, Hsinchu, Taiwan
[2]Synopsys, Inc., Mountain View, CA, USA

ABSTRACT

Because of the widening sub-wavelength lithography gap in advanced fabrication technology, lithography hotspot detection has become an essential task in design for manufacturability. Current state-of-the-art works unite pattern matching and machine learning engines. Unlike them, we fully exploit the strengths of machine learning using novel techniques. By combing topological classification and critical feature extraction, our hotspot detection framework achieves very high accuracy. Furthermore, to speed up the evaluation, we verify only possible layout clips instead of full-layout scanning. After detection, we filter hotspots to reduce the false alarm. Experimental results show that the proposed framework is very accurate and demonstrates a rapid training convergence. Moreover, our framework outperforms the 2012 CAD Contest at ICCAD winner on accuracy and false alarm.

Categories and Subject Descriptors

B.7.2 [**Integrated Circuits**]: Design Aids

General Terms

Algorithms, Design.

Keywords

Design for manufacturability, lithography hotspot, hotspot detection, fuzzy pattern matching, machine learning, support vector machine.

1. INTRODUCTION

In advanced process technology, the ever-growing sub-wavelength lithography gap causes unwanted shape distortions of the printed layout patterns [1]. Although design rule checking (DRC) and reticle/resolution enhancement techniques (RET), such as optical proximity correction (OPC) and subresolution assist features (SRAF), can alleviate the printability problem, many regions on a layout may still be susceptible to lithography process. These regions, so-called lithography hotspots, have to be detected and corrected before mask synthesis.

Hotspot detection, therefore, is an essential task in physical verification. In this paper, we investigate the hotspot detection problem: Given a training data set of hotspot and non-hotspot

---

* This work was partially supported by Synopsys and NSC of Taiwan under Grant No's. NSC 101-2220-E-009-044 and NSC 101-2628-E-009-012-MY2.

Permission to make digital or hard copies of all or part of this work for personal or classroom use is granted without fee provided that copies are not made or distributed for profit or commercial advantage and that copies bear this notice and the full citation on the first page. To copy otherwise, to republish, to post on servers or to redistribute to lists, requires prior specific permission and/or a fee.
DAC'13, May 29 – June 07, 2013, Austin, TX, USA.

---

patterns and a testing layout, identify hotspots in the testing layout such that the accuracy is maximized and the false alarm is minimized.

Hotspot detection has attracted increasing attention in recent years, e.g., [2-12]. These works can be classified into four major categories, (1) lithography simulation, (2) pattern matching, (3) machine learning, and (4) hybrid. The full lithography simulation provides the most accurate detection result. However, the simulation suffers from an extremely high computational complexity and long runtime [2]. Pattern matching is the fastest hotspot detection approach in literature. This approach uses a specific representation to encode the topology of a hotspot pattern and/or layout, e.g., a dual graph in [3], strings in [4-5], modified transitive closure graph in [6], and improved tangent space in [7]. Then some searching algorithm is applied to identify hotspots. Pattern matching is good at detecting pre-characterized hotspot patterns but has a limited flexibility to recognize previously unseen ones. In contrast, machine learning is good at detecting unknown hotspots but needs special treatments to suppress the false alarm. Machine learning methods involve two phases: First, the training phase constructs a machine learning model, and second, the evaluating phase verifies a testing layout. Typically, supervised learning models are adopted, e.g., artificial neural network (ANN) and support vector machine (SVM). Recent works [8-9] further develop hierarchical learning and multi-level learning. The hybrid approach unites both pattern matching and machine learning engines (even with a lithography simulator) to enhance accuracy and reduce false alarm but may consume longer runtimes, e.g., [10-12].

In fact, prior endeavors have not fully exploited the strengths of machine learning because of the following issues: An adequate level of fuzziness/tolerance should be added to identify potential hotspots undefined in the training set without increasing false alarm. To achieve high accuracy and low false alarm, a balanced population between hotspot and non-hotspot samples is desired.

To properly address the above issues, in this paper, we propose a novel machine-learning-based hotspot detection framework with delicate techniques. In the training phase, our goal is to achieve a good training quality but maintain a high flexibility. First of all, we classify the known hotspot and non-hotspot patterns into clusters according to their topologies. Secondly, topological and non-topological critical features are extracted from each cluster. Thirdly, based on iterative multiple kernel learning, we construct a specific SVM kernel for each cluster. Because of topological classification, each kernel can concentrate on the critical features specific to its corresponding cluster, thus providing a flexibility to identify previously unseen patterns. Furthermore, topological classification also facilitates hotspot and non-hotspot population balancing. Compared with a single huge SVM kernel, our multiple kernel learning achieves high accuracy.

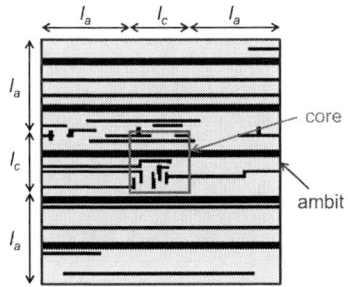

Figure 1. A hotspot or non-hotspot pattern [13].

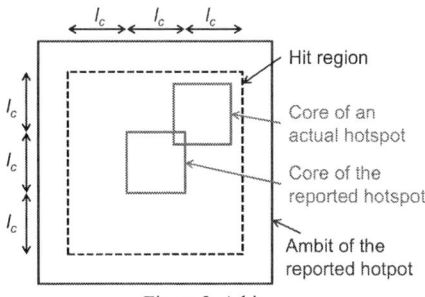

Figure 2. A hit.

On the other hand, a testing layout contains tremendous sites that need to be evaluated. Therefore, in the evaluation phase, to avoid time-consuming full-layout scanning, we extract only possible layout clips based on the polygon density. After evaluation, we further filter the detected hotspots to reduce the false alarm without sacrificing the accuracy.

Experiments are conducted on six 32nm and 28nm industrial designs [13]. Experimental results show that our framework is very accurate. We also demonstrate a rapid training convergence, i.e., achieving high accuracy using a small amount of training data. Moreover, our framework outperforms the 2012 CAD Contest at ICCAD winner on the overall performance, including accuracy and false alarm, with very competitive runtime.

The remainder of this paper is organized as follows. Section 2 describes the problem formulation. Section 3 details our hotspot detection framework. Section 4 lists experimental results. Finally, Section 5 concludes this work.

2. PROBLEM FORMULATION

As mentioned in Section 1, hotspot detection is an essential task in physical verification. We follow the problem formulation provided by 2012 CAD Contest at ICCAD in fuzzy pattern matching for physical verification [13]. For clarity, we have the following definition.

The Hotspot Detection Problem: Given a training data set of hotspot and non-hotspot patterns and a testing layout, identify hotspots in the testing layout. The primary objective is to maximize the accuracy, while the secondary objective is to minimize the false alarm.

Definition 1: A *hotspot* is a layout pattern that is susceptible to lithographic process and may induce the printability issue at the fabrication stage.

Definition 2: A *hit* is a correctly identified hotspot. *Accuracy* is the ratio of the number of total hits over the number of all actual hotspots.

Definition 3: An *extra* is a non-hotspot that is identified as a hotspot. The *false alarm* is the ratio of the number of total extras over the number of all actual hotspots.

Since one hotspot can kill a design, the primary objective is accuracy. The false alarm defined here represents the false positives (how many non-hotspots are reported as hotspots), while '1.0 – accuracy' reflects the false negatives (how many hotspots are missed).

As shown in Figure 1, a hotspot or non-hotspot pattern in the training data set is a layout clip defined by a core and its ambit, where the core is the central part of this clip providing its significant characteristics, while the ambit is the peripheral part of

this clip providing supplementary information. The training data set, provided by foundry (or lithography simulation), is highly imbalanced, i.e., the non-hotspot patterns greatly outnumber the hotspot patterns.

As shown in Figure 2, a reported hotspot is considered as a hit if the core of the reported hotspot overlaps with the core of an actual hotspot. For a hit, the clip of the reported hotspot fully covers the core of an actual hotspot, and two clips overlap at least a certain amount of area.

3. OUR APPROACH

In this section, we detail our hotspot detection framework.

3.1 Overview

Figure 3 shows the overview of our machine-learning-based hotspot detection framework. To fully exploit the strengths of machine learning, we introduce adequate fuzziness (Section 3.4.3), rebalance hotspot and non-hotspot population (Section 3.4.3), and develop an efficient evaluation scheme (Sections 3.5 and 3.6).

In the training phase, our goal is to achieve a good training quality but maintain a high flexibility. First of all, we classify the known hotspot and non-hotspot patterns into clusters according to their topologies. Secondly, topological and non-topological critical features are extracted from each cluster. Thirdly, based on iterative multiple kernel learning, we construct a specific SVM kernel for each cluster. With topological classification, each kernel can concentrate on the critical features specific to its corresponding cluster as well as provides a flexibility to identify previously unseen patterns. Topological classification also facilitates hotspot and non-hotspot population balancing. Compared with a single huge SVM kernel, our multiple kernel learning achieves high accuracy. On the other hand, in the

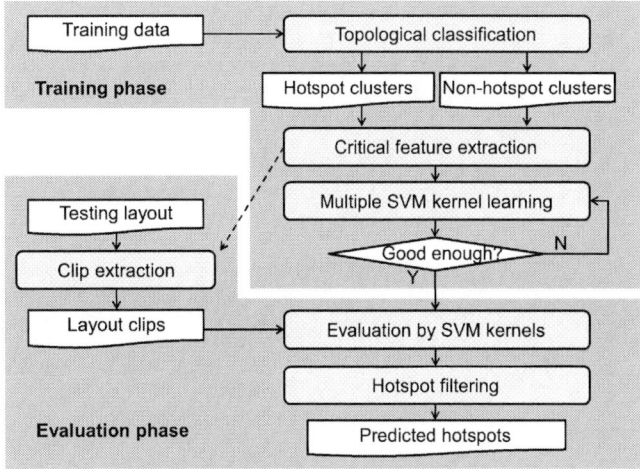

Figure 3. Our hotspot detection framework.

evaluation phase, to avoid time-consuming full-layout scanning, we extract only possible layout clips based on the polygon density. After evaluation, we further filter the detected hotspots to reduce redundant hotspot clips. Our hotspot filtering can greatly reduce the false alarm without sacrificing the accuracy.

3.2 Topological Classification

We observe that some training patterns have similar shapes and some are quite different. Hence, to facilitate the subsequent machine learning kernel training, hotspot/non-hotspot patterns in the training data set are classified into clusters based on topology. After topology classification, the patterns within one cluster have very similar geometrical characteristics (critical features).

We devise two-level topological classification: string-based and density-based classification. Consider four patterns A, B, C, D listed in Figure 4(a). String-based classification first splits these patterns into two clusters $\{A, D\}$ and $\{B, C\}$ based on topology (B and C are both crosses), and density-based classification further divides $\{B, C\}$ into clusters $\{B\}$ and $\{C\}$ based on polygon distribution. Topological classification makes each machine learning kernel concentrate on the critical features specific to its corresponding cluster as well as facilitates hotspot and non-hotspot population balancing (see Section 3.4.3).

3.2.1 String-Based Classification

By extending the string representation used by [4], we propose four directional strings to capture the topology of one pattern. To generate the string for the downward direction, the pattern is first vertically sliced along polygon edges, e.g., two slices are generated for Figure 5(a). For each slice, the boundary is labeled as "1", a polygon block is labeled as "1", and a space block is labeled as "0." Thus, each slice corresponds to a binary sequence, and then this sequence is converted to a decimal number. The downward slicing of Figure 5(a) generates a string, $<3, 10>$ ($=<11_2, 1010_2>$), recorded at the bottom side. Similarly, we have the other three strings recorded at right, top, left sides. (see Figure 5(b). Any two strings recorded at adjacent sides fully capture the topology of a pattern.

We verify whether two patterns have the same topology as follows. The four directional strings are generated for these two patterns. Any two strings of adjacent sides of one pattern are selected. Two composite strings are generated by concatenating the strings of the other pattern counterclockwise and clockwise. (The string of the beginning side should be added at the end.) We have the following theorem.

Theorem 1: Considering eight possible orientations[1], two patterns have the same topology if and only if any two strings at adjacent sides of one pattern exist in the counterclockwise or clockwise composite string of the other pattern.

For example, we select two strings at adjacent sides of the pattern in Figure 5(b), say $<5, 3, 5, 3, 10>$ for the left and bottom sides. We have the composite strings of the pattern in Figure 5(c) as follows:

Counterclockwise: $<6, 3, 6, 10, 3, 5, 3, 5, 3, 10, 6, 3, 6>$, and
Clockwise: $<10, 3, 5, 3, 5, 3, 10, 6, 3, 6, 10, 3>$.

$<5, 3, 5, 3, 10>$ exists in the counterclockwise composite string. Hence, two patterns given in Figure 5(b)(c) have the same

topology. Theorem 1 guarantees that the clusters generated by string-based classification are unique.

3.2.2 Density-Based Classification

After string-based classification, patterns within one cluster have the same topology. Even so, in some cases, two patterns with the same topology may still have very different geometrical characteristics. For example, one could be a hotspot, while the other is a non-hotspot under discrete process forbidden rules.

For a pattern p_i, its layout clip is first pixelated, and the polygon density of each pixel d_k is calculated [9]. (see Figure 5(d)) The distance $\rho(p_i, p_j)$ between two patterns p_i and p_j is defined by the summation of the pixel density difference over all pixels based on the same orientation:

$$\rho(p_i, p_j) = \min_{\tau \in D_8} \sum_k |d_k(p_i) - d_k(\tau(p_j))|, \quad (1)$$

where τ is the orientation, and D_8 represents the set of eight possible orientations. Based on the distance metric, the cluster radius used by density-based classification is defined as follows.

$$R = \max(R_0, \max_{i,j} \rho(p_i, p_j)/10), \quad (2)$$

where R_0 is the user-defined radius threshold. We adopt an efficient yet effective clustering method: For an investigated pattern, we check whether this pattern is covered by some existing

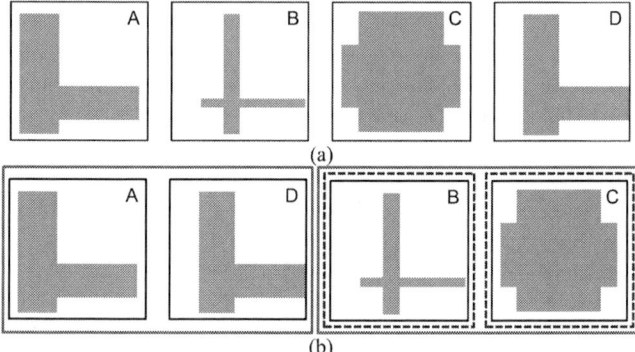

(a)

(b)

Figure 4. Topological classification. (a) Four patterns A, B, C, D. (b) Generated clusters: $\{A, D\}$, $\{B\}$, $\{C\}$.

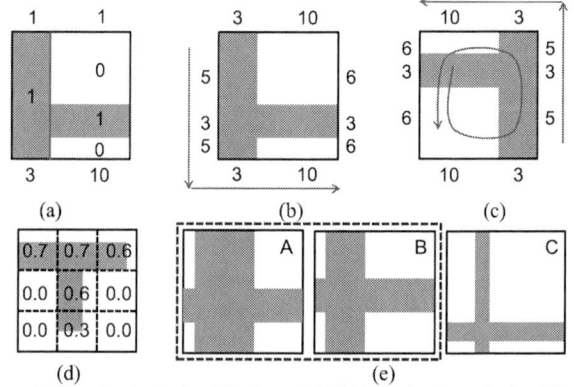

Figure 5. Topological classification. (a)(b)(c) String-based classification. (d)(e) Density-based classification.

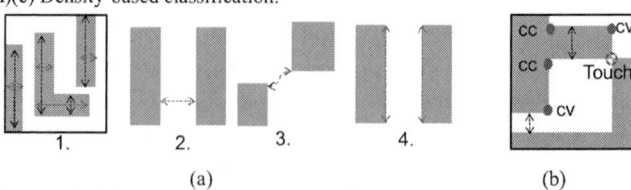

(a)

(b)

Figure 6. Critical features. (a) Topological features. (b) Non-topological features.

---

[1] Eight orientations include combinations of four rotations ($0°$, $90°$, $180°$, $270°$) and two mirrors (horizontal and vertical mirrors).

474

cluster. (A pattern is covered by a cluster if the distance between this pattern and the centroid (representative) of the cluster is less than or equal to the radius value.) If so, the pattern is added into the covering cluster. Otherwise, this pattern becomes the centroid of a new cluster. This flow is repeated for all patterns. In addition, the clustering process can be further improved by recalculating the centroid once a pattern is added to some cluster. As shown in Figure 5(e), two clusters, {A, B} and {C}, are generated by density-based classification.

3.3 Critical Feature Extraction

We extract topological (geometry-related) and non-topological (lithography-process-related) critical features of each pattern.

We adopt modified transitive closure graph proposed by [6] to extract the topological critical features. As shown in Figure 6(a), four types of topological features are extracted: (1) Horizontal and vertical distance between a pair of internally facing polygon edges, (2) horizontal and vertical distance between a pair of externally facing polygon edges, (3) diagonal distance of two convex corners, and (4) horizontal and vertical edge length of a polygon. Considering eight possible orientations, two sets of topological features are generated to preserve the vertical and horizontal relationships among extracted features [6].

On the other hand, we define five types of non-topological features as shown in Figure 6(b): (1) The number of corners (convex plus concave), (2) the number of touched points, (3) the minimum distance between a pair of internally facing polygon edges, (4) the minimum distance between a pair of externally facing polygon edges, and (5) the polygon density.

By topological classification, the number of critical features is identical for all patterns in a cluster. The equivalent feature number facilitates the subsequent SVM kernel training.

3.4 Iterative Multiple SVM-Kernel Learning

To provide the flexibility to identify unseen hotspots, we leverage on machine learning. We devise iterative multiple SVM kernel learning to fully exploit the strengths of machine learning.

3.4.1 C-SVM with Radial Basis Kernel Function

In machine learning, SVM is a popular supervised learning model. A two-class SVM transforms the training data to a high-dimensional space and calculates a hyperplane to separate the data into two classes with a maximum margin. If the SVM kernel function is a symmetric positive semidefinite function, then SVM guarantees a global optimum solution. SVM has showed superior performance in handling a small training data set, non-linear and high dimensional classification issues. Based on this evidence, in this paper, we use two-class soft-margin C-type SVM and adopt the radial basis function as our kernel to detect hotspots and non-hotspots [14-15]. Given training data x_n, $n = 1..N$, with label t_n (+1 or −1 for two-class SVM). The dual form of the quadratic programming formulation of C-type SVM is given as follows.

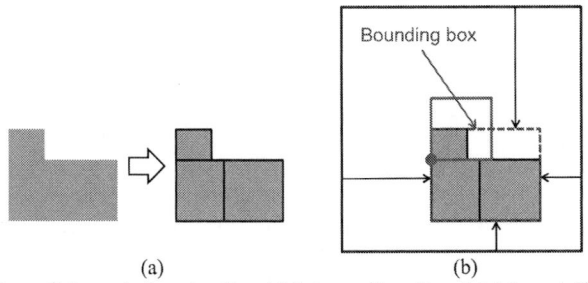

(a)   (b)

Figure 8. Layout clip extraction. (a) Polygon dissection. (b) A layout clip.

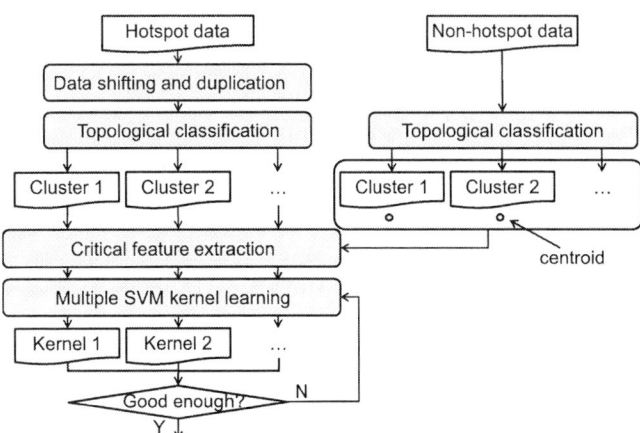

Figure 7. The training phase.

$$\max f(a) = \sum_{n=1}^{N} a_n - \frac{1}{2} \sum_{n=1}^{N} \sum_{m=1}^{N} a_n a_m t_n t_m k(x_n, x_m)$$

subject to

$$0 \le a_n \le C, \forall n = 1..N,$$
$$\sum_{n=1}^{N} a_n t_n = 0,$$
$$k(x_n, x_m) = \exp(-\gamma \|x_n - x_m\|^2),$$
$$a = (a_1, ..., a_N)^{\mathrm{T}}, \qquad\qquad (3)$$

where C controls the trade-off between the slack variable penalty and the margin, $k(x_n, x_m)$ is the Gaussian radial basis kernel function, and a_n is the Lagrange multiplier. The Gaussian radial basis kernel function is symmetric positive semidefinite thus leading to an optimal classification. Imbalanced population may destroy the soft margin and degrade the training quality.

3.4.2 Iterative Learning

Appropriate values of C and γ may result in a good training quality of an SVM kernel. Therefore, as shown in Figure 3, we introduce a self-training process to iteratively adapt C and γ parameters. The initial values of C and γ are 1000 and 0.01, respectively. C and γ are doubled if the stopping criterion is not satisfied. The stopping criterion of iterative learning is that the number of self-training iterations exceeds a user-defined bound or the hotspot/non-hotspot detection accuracy rate (with respect to the training data) exceeds a user-defined training accuracy, say 90%.

3.4.3 Population Balancing and Multiple Kernels

Figure 7 details our training phase, including topological classification, population balancing and multiple SVM kernel learning steps.

Imbalanced population between hotspot and non-hotspot samples may degrade the accuracy and increase the false alarm. Therefore, to balance population, we upsample hotspot training patterns and downsample non-hotspot training patterns. To upsample hotspot training patterns and compensate the layout clip extraction error (Section 3.5), we slightly shift each hotspot training pattern upwards, downwards, leftwards and rightwards to create several counterparts before topological classification. Multiple hotspot and non-hotspot clusters are generated after topological classification. To downsample non-hotspot training patterns, only the centroid of each non-hotspot cluster is selected. Our non-hotspot training data set is formed by all non-hotspot centroids. It is clear that resampling can effectively balance hotspot and non-hotspot population.

Since each hotspot cluster has its own characteristics, we construct one SVM kernel to learn the specific critical features of a cluster. Each SVM kernel is constructed based on one hotspot cluster and the newly formed non-hotspot training data set. We then have multiple SVM kernels. At the evaluation phase, a layout clip is flagged as a hotspot if one kernel classifies it as a hotspot. This mechanism makes each kernel focus on the critical features specific to its own hotspot cluster thus achieving a good training quality.

Moreover, data shifting generates several derivatives from a training pattern thus introducing adequate fuzziness into each cluster. Hence, our kernels have a flexibility to identify previously unseen patterns. Later, our results show that compared with a single huge SVM kernel, our multiple kernel learning achieves high accuracy.

3.5 Layout Clip Extraction

A testing layout contains tremendous sites that need to be evaluated. To avoid time-consuming full-layout scanning, we extract only possible layout clips based on the polygon distribution.

As shown in Figure 8(a), each layout polygon is first horizontally sliced into rectangles. These rectangles are then cut into smaller pieces if their widths or heights are greater than the hotspot core side length (l_c in Figure 2).

As shown in Figure 8(b), a core and an ambit are set with respect to the bottom left corner of each rectangle. The corresponding layout clip is extracted if the polygon distribution within this clip meets the user-specified requirements; otherwise, the clip is discarded. The polygon distribution here means the polygon density, the polygon count and the distances between the clip boundary and the bounding box that covers all polygon rectangles in the clip (indicated by four arrows in Figure 8(b)). The user-defined requirements are positively correlated to the critical features extracted for SVM kernels. It can be seen that if the polygon distribution requirements are met, each polygon must be included by at least one layout clip. Moreover, the possible misalignment between an extracted clip and an actual hotspot can be compensated by data shifting described in Section 3.4.3.

3.6 Hotspot Filtering

As shown in Figure 9(a), after SVM kernel evaluation, reported hotspot cores strongly overlap in an area with high polygon density. We apply hotspot filtering to reduce the redundancy. Moreover, our hotspot filtering can greatly reduce the false alarm without sacrificing the accuracy.

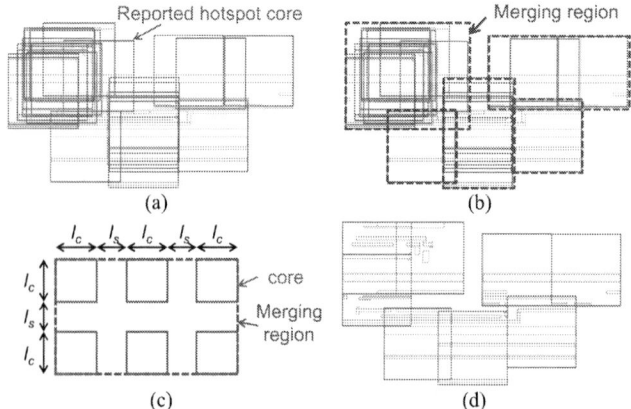

Figure 9. (a) Hotspots reported by our SVM kernel. (b) Clip merging of (a). (c) Clip reframing. (d) Clip reframing of (b).

TABLE 1. 2012 CAD CONTEST AT ICCAD BENCHMARK STATISTICS.

Training data			Testing layout			
Name	#hs	#nhs	Name	#hs	area (um²)	process
MX_benchmark1_clip	99	340	Array_benchmark1	226	12,516	32nm
MX_benchmark2_clip	176	5,285	Array_benchmark2	499	106,954	28nm
MX_benchmark3_clip	923	4,643	Array_benchmark3	1,847	122,565	28nm
MX_benchmark4_clip	98	4,452	Array_benchmark4	192	82,010	28nm
MX_benchmark5_clip	26	2,716	Array_benchmark5	42	49,583	28nm
			MX_blind_partial	55	224,975	32nm

#hs: number of hotspots; #nhs: number of non-hotspots.
The core size is $1.2 \times 1.2 um^2$, while the clip size is $4.8 \times 4.8 um^2$.

TABLE 2. COMPARISON WITH 2012 CAD CONTEST WINNERS.

Testing layout (Training data)	Methods	#hit	#extra	accuracy	hit/extra	Runtime
Array_benchmark1 (MX_benchmark1_clip)	1st place	212	1,826	93.81%	1.16E-01	0m05.1s
	2nd place	98	188	43.36%	5.21E-01	1m50.2s
	3rd place	157	728	69.47%	2.16E-01	0m06.7s
	ours	214	1,493	94.69%	1.43E-01	0m38.1s
Array_benchmark2 (MX_benchmark2_clip)	1st place	489	20,383	98.00%	2.40E-02	8m11.9s
	2nd place	108	548	21.64%	1.97E-01	23m40.8s
	3rd place	337	5,878	67.54%	5.73E-02	6m10.2s
	ours	490	11,834	98.20%	4.14E-02	3m54.4s
Array_benchmark3 (MX_benchmark3_clip)	1st place	1,696	20,764	91.82%	8.17E-02	18m44.0s
	2nd place	1,491	9,579	80.73%	1.56E-01	118m56.8s
	3rd place	1,840	71,328	99.62%	2.58E-02	7m58.1s
	ours	1,697	13,850	91.88%	1.23E-01	14m57.7s
Array_benchmark4 (MX_benchmark4_clip)	1st place	161	3,726	83.85%	4.32E-02	1m15.9s
	2nd place	124	956	64.58%	1.30E-01	21m57.9s
	3rd place	152	13,582	79.17%	1.14E-02	1m42.9s
	ours	165	3,664	85.94%	4.50E-02	5m56.3s
Array_benchmark5 (MX_benchmark5_clip)	1st place	39	2,014	92.86%	1.94E-02	0m26.6s
	2nd place	26	31	61.90%	8.39E-01	5m25.6s
	3rd place	20	245	47.62%	8.16E-02	0m40.0s
	ours	39	1,205	92.86%	3.24E-02	0m20.0s

1st place is executed on a platform with 2 Intel Xeon 2.3 GHz CPUs and with 64 GB memory, while 2nd and 3rd places are executed on 4 Intel Xeon 2.0 GHz CPUs and with 72 GB memory.

First of all, we merge reported hotspot cores into several regions; we merge a hotspot clip into an existing merging region if its core overlaps with some hotspot core of the region. A merging region is the minimum bounding box covering all hotspot cores in this region. (see Figure 9(b))

Secondly, a merging region containing more than four hotspot cores is reframed. The goal of reframing is to minimize the number of reported hotspots without missing any possible actual hotspots. Figure 9(c) shows the clip reframing, where the distance l_s between two reframed cores should be less than the core side length l_c. Our clip reframing guarantees that the core of an arbitrary actual hotspot is overlapped by at least one reframed core. Moreover, we further remove redundant clips located in the overlapping area of two merging regions. A hotspot core is discarded under two conditions: (1) All polygons within this core are covered by other hotspot cores, and (2) each corner of this core overlaps with other hotspot cores inside some merging region. Figure 9(d) shows the reframing result of Figure 9(a).

4. EXPERIMENTAL RESULTS

Our algorithm was implemented in the C++ programming language with a GDSII library Anuvad [16] and the SVM library LIBSVM [17]. We executed the program on a platform with two Intel Xeon 2.3 GHz CPUs and with 64 GB memory. Experiments are conducted on six 32nm and 28nm industrial designs released by [13] as listed in Table 1, with a highly imbalanced population between hotspot and non-hotspot training patterns. We do three sets of experiments to compare the overall performance with the

2012 CAD contest winners, demonstrate the effectiveness of our multiple SVM kernel training and hotspot filtering, and show our rapid training convergence.

In the first set of experiments, we compare our approach with 2012 CAD contest at ICCAD winners. Table 2 summarizes the experimental results. Overall, we outperform the first place winner on accuracy, false alarm, and the hit/extra rate. For Array_benchmark3, compared with the third place winner, we have lower accuracy but with a significantly lower false alarm.

In the second set of experiments, as listed in Table 3, we demonstrate the effectiveness of our approach. 'Single SVM' means the baseline SVM which uses one single huge SVM kernel (i.e., without topological classification and hotspot filtering); 'ours_wo_filtering' means our multiple SVM kernel training without hotspot filtering (i.e., with topological classification but without hotspot filtering); 'ours' means our whole framework. In our experiments, we use the following parameters to demonstrate our flow: The respective initial values of C and γ of our SVM kernel are 1000 and 0.01, the stopping criterion of self-training is 90% accuracy, data shifting is 120nm ($=l_c/10$), the maximum distance between the clip boundary and the bounding box of clip extraction is 1440nm, the minimum overlapping of clip merging is 20%, and the separating distance of core reframing is 1150nm. First of all, our critical features are effective. For example, single SVM achieves over 78% accuracy for Array_benchmark3 and Array_benchmark5. Secondly, our topological classification and population balancing indeed works well, and thus our multiple SVM kernel learning has adequate fuzziness and delivers very high accuracy, 85.9~98.2%. Thirdly, our hotspot filtering greatly reduces the false alarm for all cases without sacrificing accuracy.

In the third set of experiments, we show the impact of training data on accuracy as listed in Table 4. 'Data' means the ratio of the used training pattern count over the whole training pattern count. It can be seen that using different training data may achieve higher accuracy and lower false alarm, e.g., Array_benchmark2 and MX_blind_partial. Secondly, we have a rapid convergence on our training quality. We may use a small amount of training data to achieve high accuracy, especially for Array_benchmark3 and Array_benchmark5, thus shortening the runtime.

5. Conclusion

In this paper, we proposed a novel machine-learning-based hotspot detection framework. By combining topological classification and critical feature extraction, our machine learning kernel achieved high accuracy. On the other hand, our clip extraction and hotspot filtering effectively speeded up the evaluation phase and greatly reduced the false alarm. Experimental results showed that our framework not only is very accurate but also has a rapid accuracy convergence using a small amount of training data. Moreover, our framework outperformed the 2012 CAD Contest at ICCAD winner on accuracy and false alarm with very competitive runtime.

6. REFERENCES

[1] International Technology Roadmap for Semiconductors. 2011.
[2] P. Gupta et al. Lithography simulation-based full-chip design analyses. In Proc. SPIE vol. 6156, 2006.
[3] A. B. Kahng et al. Fast dual graph based hotspot detection. In Proc. SPIE, vol. 6349, 2006.
[4] H. Yao et al. Efficient process-hotspot detection using range pattern matching. In Proc. ICCAD, pp.625–632, 2006.
[5] J. Xu et al. Accurate detection for process-hotspots with vias and incomplete specification. In Proc. ICCAD, pp. 839–846, 2007.

TABLE 3. DETAILED COMPARISON ON OUR FEATURES.

Benchmark	Methods	#hit	#extra	accuracy	Runtime
Array_benchmark1 (MX_benchmark1_clip)	1st place	212	1,826	93.81%	0m05.1s
	single SVM	164	1,126	72.57%	0m02.7s
	ours_wo_filtering	214	2,729	94.69%	0m37.0s
	ours	214	1,493	94.69%	0m38.1s
Array_benchmark2 (MX_benchmark2_clip)	1st place	489	20,383	98.00%	8m11.9s
	single SVM	288	2,828	57.72%	3m42.8s
	ours_wo_filtering	490	22,775	98.20%	3m22.0s
	ours	490	11,834	98.20%	3m54.4s
Array_benchmark3 (MX_benchmark3_clip)	1st place	1,696	20,764	91.82%	18m44.0s
	single SVM	1,600	31,811	86.63%	7m42.8s
	ours_wo_filtering	1,697	51,067	91.88%	13m34.2s
	ours	1,697	13,850	91.88%	14m57.7s
Array_benchmark4 (MX_benchmark4_clip)	1st place	161	3,726	83.85%	1m15.9s
	single SVM	119	1,388	61.98%	0m26.7s
	ours_wo_filtering	165	5,936	85.94%	5m52.4s
	ours	165	3,664	85.94%	5m56.3s
Array_benchmark5 (MX_benchmark5_clip)	1st place	39	2,014	92.86%	0m26.6s
	single SVM	33	1,227	78.57%	0m13.0s
	ours_wo_filtering	39	2,136	92.86%	0m18.9s
	ours	39	1,205	92.86%	0m20.0s
MX_blind_partial (MX_benchmark1_clip)	1st place	51	66,818	92.73%	2m31.7s
	single SVM	38	31,148	69.09%	1m18.1s
	ours_wo_filtering	51	89,254	92.73%	2m59.7s
	ours	51	55,080	92.73%	5m04.6s

TABLE 4. ACCURACY AND TRAINING DATA.

Benchmark	Methods	Data	#hit	#extra	accuracy	Runtime
Array_benchmark1 (MX_benchmark1_clip)	1st place	100.0%	212	1,826	93.81%	0m05.1s
	ours	75.0%	214	1,476	94.69%	0m54.5s
Array_benchmark2 (Array_benchmark3,4)	1st place	100.0%	489	20,383	98.00%	8m11.9s
	ours	0.6%	494	18,256	99.00%	4m16.5s
Array_benchmark3 (MX_benchmark3_clip)	1st place	100.0%	1,696	20,764	91.82%	18m44.0s
	ours	1.0%	1,712	16,565	92.69%	6m09.0s
Array_benchmark4 (MX_benchmark4_clip)	1st place	100.0%	161	3,726	83.85%	1m15.9s
	ours	97.0%	164	2,946	85.42%	1m15.2s
Array_benchmark5 (MX_benchmark5_clip)	1st place	100.0%	39	2,014	92.86%	0m26.6s
	ours	95.0%	40	1,320	95.24%	0m19.3s
MX_blind_partial (Array_benchmark3)	1st place	100.0%	50	49,223	90.91%	15m04.9s
	ours	100.0%	52	43,810	94.55%	15m05.8s

[6] Y.-T. Yu et al. Accurate process-hotspot detection using critical design rule extraction. In Proc. DAC, pp. 1167–1172, 2012.
[7] J. Guo et al. Improved tangent space based distance metric for accurate lithographic hotspot classification. In Proc. DAC, pp. 1173–1178, 2012.
[8] D. Ding et al. High performance lithographic hotspot detection using hierarchically refined machine learning. In Proc. ASP-DAC, pp. 775–780, 2011.
[9] J.-Y. Wuu et al. Rapid layout pattern classification. In Proc. ASP-DAC, pp. 781–786, 2011.
[10] J.-Y. Wuu et al. Efficient approach to early detection of lithographic hotspots using machine learning systems and pattern matching. In Proc. SPIE vol. 7974, 2011.
[11] S. Mostafa et al. Multi-selection method for physical design verification applications. In Proc. SPIE vol. 7974, 2011.
[12] D. Ding et al. EPIC: Efficient prediction of IC manufacturing hotspots with a unified meta-classification formulation. In Proc. ASP-DAC, pp. 263–270, 2012.
[13] J. A. Torres. ICCAD-2012 CAD contest in fuzzy pattern matching for physical verification and benchmark suite. In Proc. ICCAD, pp. 349–350, 2012. http://cad_contest.cs.nctu.edu.tw/CAD-contest-at-ICCAD2012/problems/p3/p3.html.
[14] B. E. Boser et al. A training algorithm for optimal margin classifiers. In Proc. COLT, pp. 144–152, 1992.
[15] C. Cortes and V. Vapnik. Support-vector networks. Machine Learning, vol. 20, pp. 273–297, 1995.
[16] SoftJin Technologies Pvt. Ltd. Anuvad: A free suite of GDSII and OASIS libraries. http://www.softjin.com.
[17] C.-C. Chang and C.-J. Lin. LIBSVM: A library for support vector machines. http://www.csie.ntu.edu.tw/~cjlin/libsvm.

A Novel Fuzzy Matching Model for Lithography Hotspot Detection

Sheng-Yuan Lin, Jing-Yi Chen, Jin-Cheng Li, Wan-yu Wen, and Shih-Chieh Chang
Department of Computer Science, National Tsing Hua University, HsinChu 30010, Taiwan, R.O.C.
sylin.twn@gmail.com, scchang@cs.nthu.edu.tw

ABSTRACT

In advanced IC manufacturing, as the gap between lithography optical wavelength and feature size increases, it becomes challenging to detect problematic layout patterns called *lithography hotspot*. In this paper, we propose a novel fuzzy matching model which can dynamically tune appropriate fuzzy regions around known hotspots. Based on this model, we develop a fast algorithm for lithography hotspot detection with very low chances of false-alarm. Our results are very encouraging with under 0.56 CPU-hrs/mm^2 runtime.

Categories and Subject Descriptors

B.7.2 **[Integrated Circuits]**: Design Aids – *Layout, Verification*

General Terms

Algorithms, Design, Reliability

Keywords

Design for Manufacturability, Hotspot Detection, Lithography Hotspot, Machine Learning, Fuzzy Matching.

1. INTRODUCTION

As modern IC feature size shrinks, chip manufacturing is limited to printability of optical lithography [1]. The limitation is caused by manufacturing variability of the layout with sub-wavelength feature size. To detect and avoid problematic layout patterns (called *lithography hotspots*), many manufacturability-aware methods such as design rule check (DRC), optical proximity correction (OPC) and other resolution enhancement techniques (RETs) have been developed.

Traditional ways to detect lithography hotspots currently face new challenges. As feature size becomes smaller than the optical wavelength used in the lithography technique, more layout objects are influenced in the optical radius. In other words, the detection method should consider larger area to check whether a design layout is problematic. For example, despite the central pattern of the layout of Figure 1(a) is the same as that of Figure 1(b), only the layout of Figure 1(b) is identified as a hotspot.

Permission to make digital or hard copies of all or part of this work for personal or classroom use is granted without fee provided that copies are not made or distributed for profit or commercial advantage and that copies bear this notice and the full citation on the first page. To copy otherwise, to republish, to post on servers or to redistribute to lists, requires prior specific permission and/or a fee.
DAC '13, May 29 - June 07 2013, Austin, TX, USA.

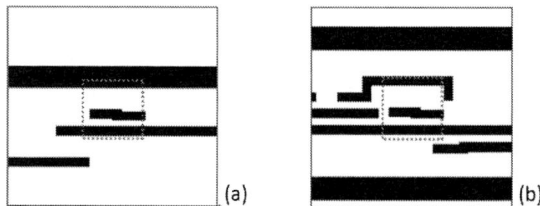

Figure 1. Layout pattern (a) Non-hotspot (b) Hotspot

The previous works to detect a "large-area" lithography hotspot can be characterized into two types. The first type uses full-layout lithography simulation [2-3] which is accurate but suffers from high computation complexity. For this reason, most state-of-the-art techniques use geometry methods for better efficiency. Figure 2 shows the flow of a geometry method which collects a set of known hotspot or non-hotspot patterns and then builds a classifying model. Then given an unknown layout pattern, the classifying model can recognize whether the unknown pattern is a hotspot or non-hotspot.

The related works using the geometry method fall into two categories. (1) Pattern matching approaches [4-8] have full detection capability on previously identified hotspots but are less efficient in classifying unseen patterns. (2) Machine learning techniques such as artificial neural network (ANN) [9-10, 14] and support vector machine (SVM) [11-14] may predict whether an unseen pattern is hotspot or not. Nevertheless, it is hard to accurately fine-tune a machine learning model. Therefore, these works suffer from the problem of false alarm, i.e., many non-hotspots are identified as hotspots.

Recently, to combine the advantages of the above two categories, hybrid hotspot detection methods are proposed. Ding, et al. [15] presented a meta-classifier which unites multiple machine

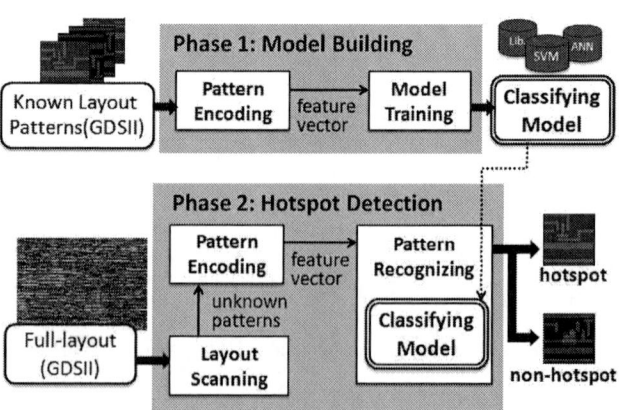

Figure 2. Geometry hotspot detection flow

learning classifiers and pattern matcher using a weighting function. Wuu, et al. [16] and Salma, et al. [17] proposed a detection flow that applies pattern matching first and then employs machine learning as the secondary checker. Although these methods obtain a better accuracy compared to previous works, they still suffer from high false-alarm rate. Furthermore, to achieve such accuracy, these hybrid models use multiple classifiers to check numerous potential problematic locations in the full-layout. As mentioned in [18], such hybrid models perform about 10 to 100 times slower than pattern matching approaches.

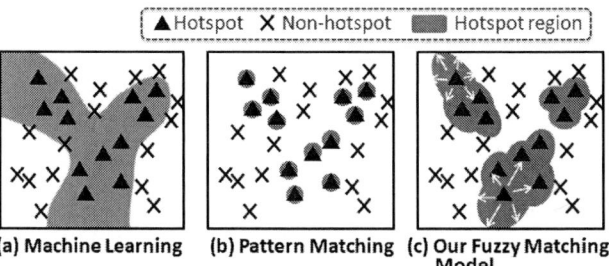

Figure 3. A 2D-space example of hotspot region decision

In this paper, we propose a novel fuzzy matching model which allows us to perform very fast lithography hotspot detection. Our model naturally integrates both pattern matching and machine leaning techniques together. Figure 3 demonstrates an example with 2-dimensional space contains known layout patterns of hotspots and non-hotspots. From these samples, Figure 3(a) shows the results of a machine learning technique which divides the space into two regions for hotspot and non-hotspot. Figure 3(b) shows the results of a pattern matching approach which records all known-hotspot samples. The results of our fuzzy matching model are shown in Figure 3(c) which dynamically tunes the fuzzy region around a known hotspot. Because the proposed fuzzy matching is based on pattern matching, our method not only can safely detect previously identified hotspots but also runs much faster than hybrid models with very low chances of false alarm. To the best knowledge of the authors, this is the first time to apply a fuzzy algorithm in the pattern recognizing stage of this problem. Experimental results show that our fuzzy matching model have 72.4% accuracy of hotspot detection with 1.9K false-alarm count/mm^2 which are both significantly better than those results in [19].

The remainder of the paper is organized as follows. In Section 2, we review hotspot detection flow and pattern encoding method for the following article. In Section 3, we discuss our fuzzy matching model in detail. In Section 4, our fuzzy matching detection flow and a fast full-layout detection algorithm are presented. Section 5 shows the experimental results and analysis. Section 6 concludes this paper.

2. PRELIMINARIES

This section discusses background information for hotspot detection.

2.1 Hotspot Detection Flow

Figure 2 shows a conventional flow for the geometry hotspot detection. The flow consists of two phases: the model building phase and the hotspot detection phase.

The objective of first phase is to build a classifying model. In this phase, a set of known hotspot or non-hotspot patterns is collected first. Then, a *pattern encoding* method is applied. The method extracts and encodes the essential features which characterize the geometry properties of these layout patterns. After the pattern encoding stage, feature vectors are obtained and are used as training data to train various models such as pattern matcher, artificial neural network, or support vector machine model.

The objective of the second phase is to detect hotspots in a full-layout. In this phase, a layout scanning algorithm is applied to the full-layout for tagging potential problematic locations as the unknown layout patterns which should be carefully analyzed. Then, the same layout feature encoding is applied for transferring these unknown patterns to feature vectors. After that, in the pattern recognizing stage, these vectors are fed into the classifying model trained in the first phase to determine whether the unknown layout pattern is a hotspot or non-hotspot.

2.2 Pattern Encoding

In this section, we review a method of layout pattern encoding called *density-based encoding* in [12] [13]. Figure 4 illustrates the basic concept. Given a layout pattern of a predefined grid, the method calculates the layout covering density of each grid. After that, the covering densities of grids of a layout pattern are encoded as an ordered feature vector. The ordered feature vector of the layout pattern is then mapped into a node in the multi-dimensional space.

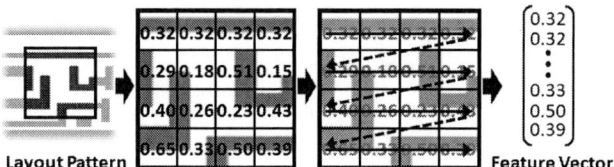

Figure 4. Density-based layout pattern encoding

For clarity, in this paper, a layout pattern is encoded through the following three steps:

1) Define grid on the pattern and calculate the covering densities.

2) Encode the densities as an ordered feature vector

3) Map the feature vector to be a node in the multi-dimensional space.

Note that pattern encoding methods such as [7, 12, 13] use fuzzy concept to extract geometry properties.

3. FUZZY MATCHING MODEL

In the section, we present the flow of our fuzzy matching model in Figure 5.

The training of our fuzzy matching model iterates the following four steps. Step 1 shows how to group the mapped hotspot patterns in the multi-dimensional space in Section 3.1. Step 2 extracts characteristics of a hotspot group in Section 3.2. In Step 3, we design a correction method to lower false-alarm rate in Section 3.3. Step 4 presents the fuzzy region growing algorithm to decide the fuzzy region around known hotspots in Section 3.4. Finally, section 3.5 describes the condition to end the iteration. Once the model is built, Section 3.6 explains the procedure to recognize whether an unknown pattern is a hotspot.

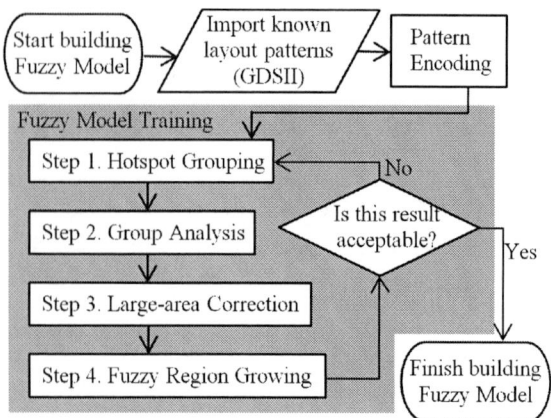

Figure 5. Building flow of our fuzzy matching model

3.1 Hotspot Grouping

The objective of hotspot grouping is to partition hotspots into groups. Hotspots with certain similarity are assigned in the same group. The similarity is measured by *Cityblock distance* defined below.

DEFINITION 1. We define the ***Cityblock distance, D_c*** to be the distance between two nodes in the multi-dimensional space.

$$D_c(a, b) = \sum_{i=1}^{\#dimension}|a_i - b_i|, \text{ for } a. b \in \text{nodes} \quad \text{EQ (1)}$$

If D_c between two hotspots is smaller than an important variable called *group distance*, these two hotspots are assigned into the same group. The group distance is iteratively updated and will be discussed in Section 3.5. Using the Cityblock distance, two hotspots which are conceptually close in the multi-dimensional space are assigned in the same group.

3.2 Group Analysis

In the previous section, we partition hotspots into a set of groups. In this section, for a group of hotspots, we extract a special characteristic for the group. Before describing how the characteristic is mathematically computed, we present our basic ideas. First, if all hotspots in the same group are closer in certain dimensions, these dimensions are conceptually more representative for the group. For example, in Figure 6(a), eight hotspots marked as triangles are in the same group. It can be found that the eight hotspots are closer in *i*-dimension than in *j*-dimension. In this paper, we consider *i*-dimension to be more representative than *j*-dimension for the group of hotspots. Later, as in Figure 6(b), when determining the fuzzy region for the eight hotspots, we allow "fuzzier" in *j*-dimension than in *i*-dimension.

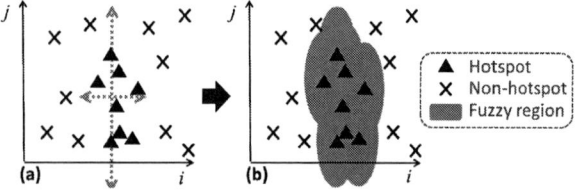

Figure 6. An example of representative dimension

To describe the characteristic of a group, we use the *dimensional weight* defined as follows. A higher weight implies the corresponding dimension is more representative. Let us consider one hotspot group.

DEFINITION 2. The ***Dimensional weight $W_d(i)$*** for dimension i is defined as follows.

$$disp(i) = \sum |a_i - b_i| \left[\begin{array}{l} \text{for all } a. b \in \text{hotspot in the same group} \\ \\ \text{for } i \in 1 \text{ to } S \end{array} \right. \quad \text{EQ (2)}$$

$$W_d(i) = \max. disp + 1 - disp(i) \quad \text{for } i \in 1 \text{ to } S \quad \text{EQ (3)}$$

where max.disp is the maximum of all *disp(i)*'s.

From Equation EQ(3), one can find that a dimensional weight is larger if hotspots in the group are closer in this dimension.

3.3 Large-area Correction

In this section, we discuss how to modify the dimensional weight to take into account the large-area effect within a single layout pattern.

As mentioned in introduction, current detection methods must consider a layout pattern in large area. Within a pattern, the inner area is more important than outer area. For example, patterns in Figure 7(a) and Figure (b) are hotspots which have a similar pattern in the core area. Patterns in Figure (a) and Figure (c) are also similar but pattern in Figure (c) is a non-hotspot since the core area is quite different. To take into account the large-area effect, we assign more weight for an inner area than an outer area of a pattern. We define the *range weight* as follows.

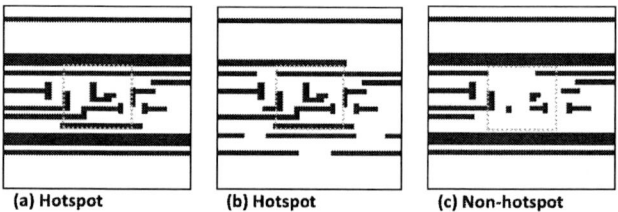

(a) Hotspot   (b) Hotspot   (c) Non-hotspot

Figure 7. An example of large-area effect

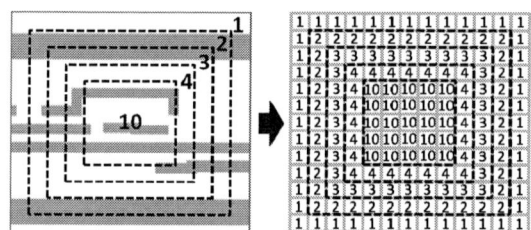

Figure 8. An example of large-area correction method

DEFINITION 3. The ***Range weight $W_r(i)$*** for dimension i is a pre-defined constant.

In our experiment, in Figure 8, we split a pattern into five regions and give different weights according to their distances to the center. The region with the longest distance is assigned the lowest weight while the shortest one is assigned the highest weight. Each grid in Figure 8 represents a dimension in the multi-dimensional space described in Section 2.2.

To simultaneously consider the dimensional weight and the range weight, we modify the Cityblock distance to be the *Cityblock weighted distance* defined below. Let us assume the number of dimensions is S.

DEFINITION 4. We define ***Cityblock weighted distance D_{cw}*** to be the distance between two nodes considering the dimensional weight and the range weight in the multi-dimensional space as follows.

$$D_{cw}(a,b) = \frac{\sum_{i=1}^{S}(|a_i - b_i| * W_d(i) * W_r(i))}{\sum_{i=1}^{S} W_d(i) * \sum_{i=1}^{S} W_r(i)}, \text{ for } a, b \in \text{nodes} \quad \text{EQ (4)}$$

In the following, we use the Cityblock weighted distance D_{cw} in Equation EQ(4) to represent the "effective" distance between every two nodes in the multi-dimensional space. In other words, when we say the distance of two nodes which can be hotspots or non-hotspots, we refer to the Cityblock weighted distance of these two nodes.

3.4 Fuzzy Region Growing

In this section, we describe how to obtain the *fuzzy region* of a hotspot group. An example of the fuzzy region is shown as the highlighted area in Figure 9. Any layout pattern whose feature vector located inside the fuzzy region will be considered as a hotspot. As a result, the larger area of the fuzzy region has, the fuzzier the hotspot group will be.

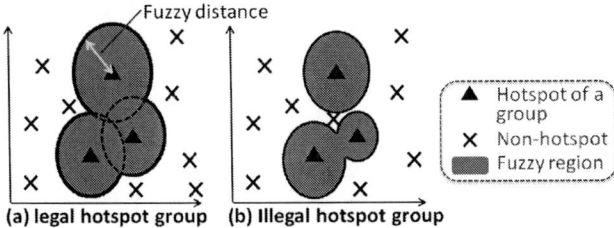

Figure 9. An example of the fuzzy region of a group

The fuzzy region of a hotspot group is the union of fuzzy regions of all hotspots. For example, in Figure 9(a), the fuzzy region of the three hotspots are the union of the three circles, each circle of which represents the fuzzy region of a hotspot.

The fuzzy region of a hotspot is determined by the "*fuzzy distance*" of the hotspot. We now describe how the fuzzy distance of a hotspot is defined. First, the fuzzy distance is NOT determined by a set of equations, instead the value is determined by an iterative algorithm.

DEFINITION 5. We define the *fuzzy distance, D_{fuzzy}* of a hotspot to be the largest distance satisfying the following two rules. Again, the distance is measured as the Cityblock weighted distance in EQ(4).

Rule 1) Any space within the fuzzy distance of a hotspot belongs to the fuzzy region of the hotspot.

Rule 2) The fuzzy region never covers a non-hotspot.

With these two rules, the fuzzy distance of a hotspot can grow gradually from zero until Rule 2 is violated. The largest value will be the fuzzy distance for the hotspot.

Once all the fuzzy distances are found, it is possible that the resulting fuzzy region form several disconnected regions as in Figure 9(b). When the resulting fuzzy region is not connected, we call the corresponding hotspot group illegal. If there is an illegal hotspot group, we need to reduce the group distance described in Section 3.5 to reduce hotspots in a group. Our fuzzy region growing algorithm is outlined in Algorithm 1 of Figure 10.

3.5 Determination of Group Distance

Note that we iterate step 1 to step 4 to train our fuzzy matching model in Figure 5. In this section, we discuss when to stop the iteration and how to determine the group distance mentioned in Section 3.1.

The group distance is an important index to partition hotspots into groups. The larger the group distance is, the more number of hotspots are assigned to the same group. However, when there are many hotspots in the same group, we may encounter the illegal problem described in Section 3.4. Initially, we set the value of the group distance to be one. We increase the value of the group distance in next iteration. We continue the process until the group distance causes the illegal problem.

The group distance, fuzzy distances and fuzzy regions all are trained and dynamically tuned in our fuzzy model. When more known hotspots and non-hotspots are provided as data into our traning model, our model have better accuracy and less false-alarm.

3.6 Pattern Recognizing

Given a fuzzy matching model and an unknown pattern *UN* in the feature vector format with size *S*, the pattern is recognized as a hotspot if any one of the following inequalities holds:

$$D_{cw}(UN, \{H\}_i) \leq \{ D_{fuzzy} \}_i \quad \text{for i} = 1 \text{ to } \#H$$

Otherwise, the unknown pattern is a non-hotspot. In other words, if the distance between the unknown pattern and any hotspot is less than or equal to the fuzzy distance of the corresponding hotspot, this unknown pattern is a hotspot; otherwise, the pattern is a non-hotspot.

Algorithm 1: Fuzzy region growing

Inputs: *Hotspots(H), Non-hotspots(NH), W_d , W_f*
Outputs: *D_{fuzzy}( ) for all Hotspots*

Initialization: *D_{fuzzy}( ) = 0, global_D = 0;*

for each *a ∈ H do*
    $D_{fuzzy}(a) = 0$;
    while all *b ∈ NH*, and $D_{cw}(a, b) > D_{fuzzy}$(a) *do*
        D_{fuzzy}(a) increases;
    end while
end for

for each *a ∈ H do*
    if D_{fuzzy}(a) cannot cover any *H* in the same group *then*
        Return illegal;
    end if
end for

Find *global_D* = max(D_{fuzzy}( ));
Base on Rule 2, let D_{fuzzy}( )of alone hotspot grow until *global_D*

Figure 10. Algorithm of fuzzy region growing

4. FAST FULL-LAYOUT DETECTION

In this section, we introduce our fuzzy matching detection flow for recognizing hotspots in the full-layout.

4.1 Fuzzy Matching Detection Flow

Figure 11 shows our hotspot detection flow consists of two phases. In the first phase, our fuzzy matching model is built as described in Section 3. In the second phase, a hotspot candidate search algorithm presented in Section 4.2 is applied to quickly tag potential problematic locations as the unknown patterns. And then our fuzzy matching model determines whether these patterns are hotspots in the full-layout.

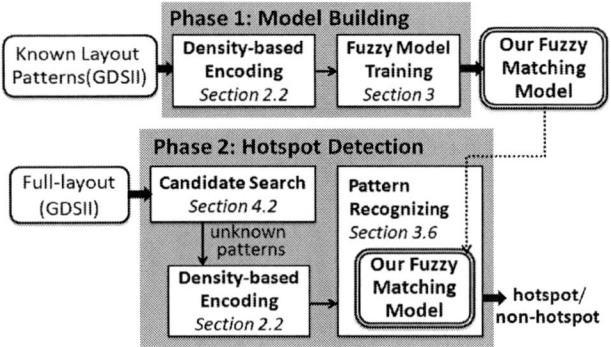

Figure 11. The overall flow of our fuzzy matching model

4.2 Hotspot Candidate Search

We discuss our hotspot candidate search algorithm. The objective of candidate search is to find those suspicious positions which should be analyzed carefully later. Our basic idea is to find the represented polygon of each known hotspot and then build a hash pattern library to speed up. The represented polygons are defined as the polygon with the most vertices which are close to the center of the hotspot. For example in Figure 12, there are two represented polygons of known hotspots. When scanning the full-layout, we look for these represented polygons. If a polygon in the full-layout is similar to a represented polygon, layout area around this polygon is tagged as an unknown pattern for further analysis.

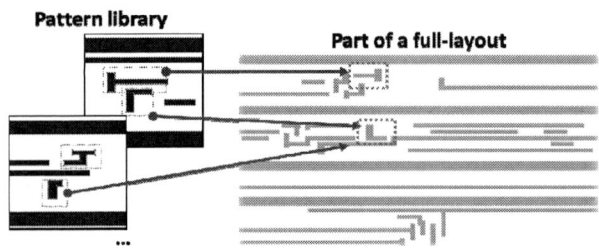

Figure 12. An example of hotspot candidate search

5. EXPERIMENTAL RESULTS

We implemented our fuzzy matching model and the hotspot candidate searching algorithm in C and performed experiments on several benchmark layouts. The benchmark layouts consist of one 28nm circuit and four 32nm circuits provided by 2012-ICCAD CAD contest [19, 20]. Table 1 shows detail information of these benchmark layouts. In the experiment, a set of training data and the corresponding clipped benchmark layouts were the input data of our program. We tested two versions of our model. The first one is named as **Orig** and the second one is named as **Limit**. To lower false-alarm rate, the version **Limit** limits the maximum

value of the group distance described in Section 3 while the version *Orig* does not constrain the group distance. The qualities evaluated for comparisons are accuracy, false-alarm and runtime as in [19]. All experiments were performed on the Linux workstation with 2.66GHz quad-core CPU and 8GB memory.

$$Accuracy = \frac{\#(Correctly\ recognized\ Hotspot)}{\#Real\ Hotspot}$$

$$False\text{-}alarm = \frac{\#(Non\text{-}hotspot\ but\ recognized\ as\ Hotspot)}{Layout\ Area(mm^2)}$$

Table 1. Benchmark layouts for experiments [20]

Benchmark	B1	B2	B3	B4	B5
Technology	28nm	32nm	32nm	32nm	32nm
Hotspot Count	223	508	1763	175	41
Layout Area (μm^2)	12516	106954	122565	82010	49583

Table 2 summarizes our experimental results. Row 1 shows the name of a benchmark layout. Row 2 and row 3 show the results of accuracy for version *Orig* and version *Limit*, respectively. Row 4 and row 5 show the amount of false-alarm per mm^2 for *Orig* and *Limit*, respectively. The last two rows show the runtime. On average, the *Orig* version achieves 72.41% accuracy with 1.9K false-alarm count/mm^2, and the *Limit* version achieves 50.51% accuracy with 0.5K false-alarm count/mm^2. The runtimes of all cases are within 0.56 hours/mm^2.

Table 2. Hotspot detection results of our method

Benchmark		B1	B2	B3	B4	B5	Avg.
Accuracy (%)	Orig	82.1	75.8	68.8	72.0	63.4	72.4
	Limit	62.3	43.3	42.5	52.6	53.7	50.9
False-alarm (per mm^2)	Orig	3356	1842	2407	1488	444	1907
	Limit	1358	701	57	500	80	539
Runtime (hrs/mm^2)	Orig	0.32	0.48	0.56	0.35	0.25	0.39
	Limit	0.32	0.47	0.55	0.33	0.25	0.38

In Table 3, we compare our method with top 3 teams of 2012-ICCAD CAD contest winners. All results are available from the contest website [19]. The results in the last row are the results of a commercial tool provided by the contest topic chair. Again, we compare all tools with accuracy, false-alarm, and runtime. According to these data, Figure 13 demonstrates the trade-off between accuracy and false-alarm of all teams. Results indicate our method achieves the best accuracy while maintaining a good balance between false-alarm and accuracy. As mentioned in [19], a method should achieve around 80% accuracy to be practical in industry. In addition, our detection flow performs faster than the top 3 teams and the commercial tool.

Table 3. Comparison with 2012-ICCAD CAD contest [19]

Methods	Accuracy	False-alarm (per mm^2)	Runtime (hrs/mm^2)
Ours (*Orig*)	**72.41%**	**1907.17**	**0.39**
Ours (*Limit*)	**50.88%**	**539.28**	**0.38**
1st place	29.05%	562.95	0.74
2nd place	21.39%	214.20	8.10
3rd place	19.80%	362.30	0.56
Commercial tool	46.93%	5039.49	2.25

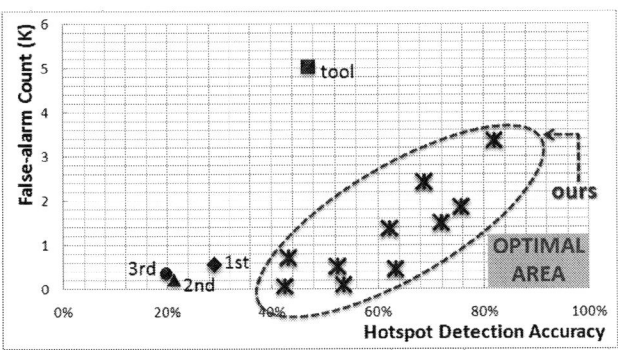

Figure 13. Comparison of accuracy and false-alarm

The experimental results show that our hotspot detection flow is better in both runtime and accuracy because of the following. First, we use a matching-based approach which is faster by its nature. Secondly, we apply the candidate search algorithm that efficiently filters out most locations in the full-layout. And lastly but most importantly, we design an efficient fuzzy matching model to achieve high accuracy.

6. CONCLUSIONS

In this paper, we present a fast lithography hotspot detection flow with high accuracy and low false-alarm rate. A fuzzy matching model is designed to dynamically determine the appropriate fuzzy region around known hotspots. On average, our model achieves 72.41% accuracy with 1.9K false-alarm count/mm^2 under 0.56 CPU-hrs/mm^2 which are better than 2012-ICCAD CAD contest results.

7. REFERENCES

[1] International Technology Roadmap for Semiconductors 2011.

[2] Juhwan Kim, Minghui Fan, 2003. Hotspot detection on post-OPC layout using full-chip simulation-based verification tool: a case study with aerial image simulation. In *Proc. of SPIE*. Vol. 5256.

[3] Michel Cote, Philippe Hurat, 2004. Layout printability optimization using a silicon simulation methodology. In *Proc. Inter. Symp. on Quality Electronic Design*. 159-164.

[4] Andrew B. Kahng, Chul-Hong Park, and Xu Xu, 2006. Fast dual graph-based hotspot detection. In *Proc. of SPIE*. Vol. 6349, 63490H.

[5] Jingyu Xu, Subarna Sinha, and Charles C. Chiang, 2007. Accurate detection for process-hotspots with vias and incomplete specification. In *Proc. of ICCAD*. 839-846.

[6] Vito Dai, Jie Yang, Norma Rodriguez, and Luigi Capodieci, 2007. DRC Plus: augmenting standard DRC with pattern matching on 2D geometries. In *Proc. of SPIE*. Vol. 6521, 65210A.

[7] H. Yao, S. Sinha, J. Xu, C. Chiang, Y. Cai, and X. Hong, 2008. Efficient range pattern matching algorithm for process-

hotspot detection. In *IET Circuits, Devices and Systems*. 2-15.

[8] Justin Ghan, Ning Ma, Sandipan Mishra, Costas Spanos, Kameshwar Poolla, 2009. Clustering and pattern matching for an automatic hotspot classification and detection system. In *Proc. of SPIE*. Vol. 7275, 727516.

[9] Norimasa Nagase, Kouichi Suzuki, Kazuhiko Takahashi, Masahiko Minemura, Satoshi Yamauchi, and Tomoyuki Okada, 2007. Study of hotspot detection using neural networks judgment. In *Proc. of SPIE*. Vol. 6607.

[10] Duo Ding, Xiang Wu, Joydeep Ghosh, and David Z. Pan, 2009. Machine learning based lithographic hotspot detection with critical-feature extraction and classification. In *IEEE Inter. Conf. on IC Design and Technology*. 219-222.

[11] Dragoljub Gagi Drmanac, Frank Liu, and Li-C Wang, 2009. Predicting variability in nanoscale lithography processes. In *DAC*. 545-550.

[12] Jen-Yi Wuu, Fedor G. Pikus, Andres Torres, and Malgorzata Marek-Sadowska, 2009. Detecting context sensitive hotspots in standard cell libraries. In *Proc. of SPIE*. Vol. 7275, 727515.

[13] Jen-Yi Wuu, Fedor G. Pikus, Andres Torres, and Malgorzata Marek-Sadowska, 2011. Rapid layout pattern classification. In *Proc. of ASP-DAC*. 781-786.

[14] Duo Ding, Andres J. Torres, Fedor G. Pikus, and David Z. Pan, 2011. High performance lithographic hotspot detection using hierarchically refined machine learning. In *Proc. of ASP-DAC*. 775-780.

[15] Duo Ding, Bei Yu, Joydeep Ghosh, and David Z. Pan, 2012. EPIC: Efficient prediction of IC manufacturing hotspots with a unified meta-classification formulation. In *Proc. of the ASP-DAC*. 263-270.

[16] Jen-Yi Wuu, Fedor G. Pikusb, and Malgorzata Marek-Sadowska, 2011. Efficient approach to early detection of lithographic hotspots using machine learning systems and pattern matching. In *Proc. of SPIE*. Vol. 7974, 79740U.

[17] Salma Mostafa, J. Andres Torres, Peter Rezk, Kareem Madkour, 2011. Multi-selection method for physical design verification applications. In *Proc. of SPIE*. Vol. 7974, 797407.

[18] J. Andres Torres, 2012. ICCAD-2012 CAD contest in fuzzy pattern matching for physical verification and benchmark suite. In *2012 ICCAD Special Session*.

[19] J. Andres Torres, 2012. ICCAD-2012 CAD contest in fuzzy pattern matching for physical verification and benchmark suite. Retrieved November 30, 2012, from http://cad_contest.cs.nctu.edu.tw/CAD-contest-at-ICCAD2012/ICCAD_P3_results_teamno.pdf

[20] J. Andres Torres, 2012. Fuzzy pattern match for physical verification. Retrieved November 30, 2012, from http://cad_contest.cs.nctu.edu.tw/CAD-contest-at-ICCAD2012/problems/p3/p3.html

An Efficient Layout Decomposition Approach for Triple Patterning Lithography

Jian Kuang
Dept. of Computer Science and Engineering
The Chinese University of Hong Kong
Shatin, NT, Hong Kong
jkuang@cse.cuhk.edu.hk

Evangeline F. Y. Young
Dept. of Computer Science and Engineering
The Chinese University of Hong Kong
Shatin, NT, Hong Kong
fyyoung@cse.cuhk.edu.hk

ABSTRACT

Triple Patterning Lithography (TPL) is widely recognized as a promising solution for 14/10nm technology node. In this paper, we propose an efficient layout decomposition approach for TPL, with the objective to minimize the number of conflicts and stitches. Based on our analysis of actual benchmarks, we found that the whole layout can be reduced into several types of small feature clusters, by some simplification methods, and the small clusters can be solved very efficiently. We also present a new stitch finding algorithm to find all possible legal stitch positions in TPL. Experimental results show that the proposed approach is very effective in practice, which can achieve significant reduction of manufacturing cost, compared to the previous work.

Categories and Subject Descriptors

B.7.2 [**Integrated Circuits**]: Design Aids

General Terms

Algorithms, Design

Keywords

Triple Patterning Lithography, Layout Decomposition, Manufacturability

1. INTRODUCTION

The next generation and ideal lithography methods, such as Extreme Ultra-Violet (EUV), still face some critical technological challenges, especially on the manufacturing equipment side, and their availabilities are further delayed. As a consequence, multiple patterning lithography, which decompose one single layer into multiple masks and require the decomposition before manufacturing, is regarded as a good alternative.

In multiple patterning lithography, a conflict occurs when the distance between two features is less than a threshold cs_{min}, namely the minimum coloring spacing, and two conflicting features should be assigned to different masks by

Permission to make digital or hard copies of all or part of this work for personal or classroom use is granted without fee provided that copies are not made or distributed for profit or commercial advantage and that copies bear this notice and the full citation on the first page. To copy otherwise, to republish, to post on servers or to redistribute to lists, requires prior specific permission and/or a fee.
DAC'13, May 29 - June 07 2013, Austin, TX, USA.

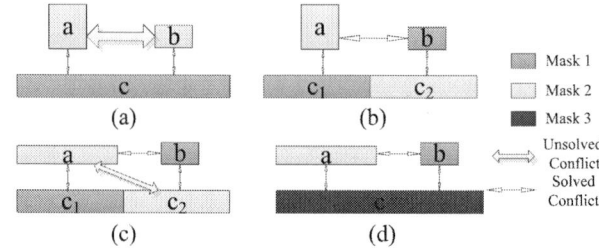

Figure 1: Multiple patterning lithograph

layout decomposition. Nonetheless we sometimes cannot do this for all features with a limited number of masks. For example, in Figure 1(a), since the three features are conflicting with each other, we cannot assign them to only two masks and one conflict still exits after decomposition. An obvious objective of decomposition is to minimize the number of remaining conflicts. For this purpose, stitch may be inserted to slice a feature into two pieces, so that they can be assigned to different masks. As shown in Figure 1(b), the conflict-free decomposition can be achieved with one stitch. However, stitch is manufacturing costly, so its number is preferred to be minimized.

Double Patterning Lithography (DPL), the simplest form of multiple patterning lithography, decomposes a layout into two masks and repeats the exposure/etching process twice. DPL with conflict and stitch minimization has attracted years of extensive research. The most up-to-date result was contributed by Tang et al. [8], who presented an optimal method in polynomial time. However, because native conflicts may exist in a layout, DPL may result in a significant number of unsolvable conflicts. As Figure 1(c) shows, the conflict cannot be resolved even with stitch. Although DPL has made the sub-22nm production a reality, with further decrease of the minimum feature size, it is thought to get to its limit, especially for 14nm and beyond technology nodes. TPL that decomposes a layout into three instead of two masks and thus can deal with very dense and complex layouts to attain fewer stitches and conflicts, is a considerable substitute and a natural extension of DPL. The layout in Figure 1(c) can be easily decomposed to three masks without any stitches or conflicts, as shown in Figure 1(d).

There are not many researches on the layout decomposition for TPL. The problem was formulated as an ILP by Yu et al. in [14], which is the first systematic study on this topic. Because of the low scalability of ILP, they introduced a semidefinite programming approximation. However, Fang et al. [3] pointed out that since this work started with all the candidate stitches obtained by the projection method that is only appropriate for DPL, there is a high chance for them to miss legal TPL stitches. Thus a heuristic was proposed in [3], based on the assumption that the conflict related to a feature with higher maximum overlap density of projections is harder to be resolved by inserting stitch. However, this assumption may not be true in some cases and we will give a counter-example in Section 4.3 to illustrate this. Besides, in terms of stitch finding, the color scanning method used in [3] failed to reflect the requirement of overlap margin, which is important as stitch is very sensitive to overlay error [4, 6]. Most recently, Tian *et al.* [11] proposed a triple patterning algorithm based on standard cell libraries and row structure layout, which may not be applicable to a general layout. Since it adopted the same method in [14] to identify stitch positions, it has the same problem of missing stitches.

In this paper, we study the decomposition problem for TPL, our main contributions can be summarized as follows:

- We present a new stitch position identification method, which can find all legal stitch positions in TPL.

- We use graph simplification techniques to reduce the problem to some simple subproblems. We find that most of the subproblems fall into only a few geometric structures that can be solved, thus we can match them with those structures and solve them very efficiently.

- We propose an effective heuristic to solve those unmatched subproblems.

- We present a simple approach to identify native conflicts in a layout.

- Our decomposition approach performs very well for practical benchmarks, with remarkable reduction in conflict numbers and stitch numbers comparing with the most up-to-date result on this TPL decomposition problem.

The rest of the paper is organized as follows. Section 2 introduces some preliminaries for the TPL problem. Section 3 states the new stitch finding method. Section 4 describes the whole decomposition approach. Section 5 reports experimental results and Section 6 concludes this paper.

2. PRELIMINARIES

Preliminaries of layout decomposition for TPL are introduced in this section, including formal problem formulation and the differences between decomposition for DPL and decomposition for TPL.

2.1 Problem Formulation

As DPL decomposition is formulated as a two-coloring problem, we formulate the decomposition for TPL as a three-coloring problem. We use three colors, namely blue, gray and orange, to represent three different masks. Every feature in the layout is of the shape of a rectilinear polygon. Given a layout specified by features, a conflict graph can be constructed. Conflict graph is an undirected graph with nodes representing features in the layout, and an edge between two nodes means that the two corresponding features are within the minimum coloring spacing cs_{min} from each other. As Figure 2(b) shows, a conflict graph with four nodes can be constructed according to the layout in Figure 2(a). A feature can be split by inserting stitch. However, stitch insertion has

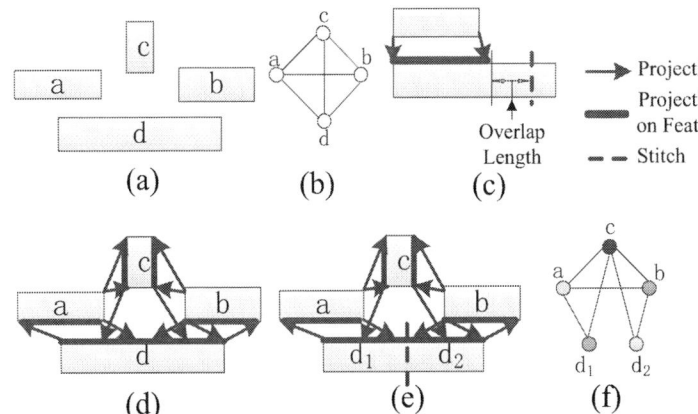

Figure 2: Conflict graph construction and stitch finding

two constraints: *overlap length* and *minimum feature size*. Overlap length (Figure 2(c)) is the length that a stitch position can move horizontally or vertically without causing any new conflicts between newly generated features and the others. The requirement is that the overlap length is not less than a threshold, called *overlap margin*. For the minimum feature size constraint, the size of the generated feature cannot be less than the minimum feature size. Now we can give the problem formulation of layout decomposition for TPL.

Problem 1. Given a layout, a minimum coloring spacing cs_{min}, an overlap margin m_o and a minimum feature size fs_{min}, our objective is to assign one mask out of three for each feature, while the numbers of conflicts and stitches are minimized and all the constraints are satisfied.

2.2 Comparisons Between DPL and TPL

Although with one extra mask, decomposition for TPL is much more complicated than that for DPL, mainly for the following reasons: 1. The minimum coloring spacing for TPL is typically larger than that for DPL, which leads to a conflict graph with more edges. 2. A graph without odd cycle is 2-colorable. However, to decide 3-colorability of a graph is NP-complete. 3. To 2-color a 2-colorable graph can be done efficiently, e.g., depth-first search, but coloring a 3-colorable graph with even four colors is NP-complete [5]. 4. All candidate stitches for DPL can be easily identified by projection, whereas the same method is not applicable for TPL, which has been proved in [3]. There is no general method for stitch finding for TPL in the literature to the best of our knowledge. 5. Native conflict, conflict that cannot be resolved even with stitch, can be identified in DPL as odd cycle in conflict graph with all candidate stitches, while this is not the case for TPL. It is still an open problem of how to identify native conflict in TPL.

3. STITCH FINDING

We present an example for which the DPL projection method fails to find legal TPL stitch. This kind of layout, forming a K_4 conflict graph, is among the most popular ones according to our observation. As Figure 2(d) shows, projections cover all features, thus no stitch is allowed if stitch is forbidden to overlap with any projection (as for DPL stitch). However, a stitch overlapping with the projections of c on

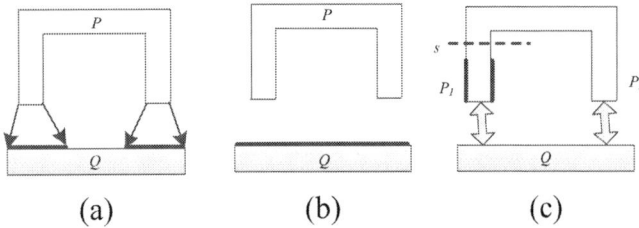

Figure 3: Projection merging and failed stitch position

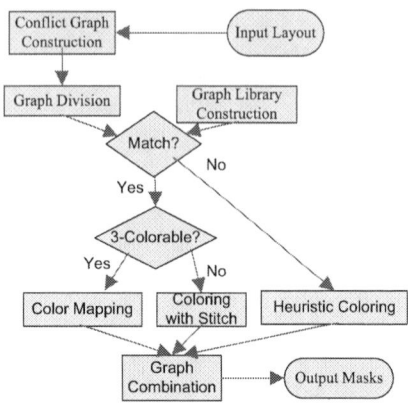

Figure 4: The flow of the decomposition approach

d can break the complete graph into a 3-colorable one, as Figure 2(e) and 2(f) show. If a stitch is overlapped with the projection of a polygon, both of the two split nodes of the stitch will still conflict with that polygon. In this example, because the stitch is overlapped with the projection of c, both d_1 and d_2 will conflict with c after inserting the stitch.

Algorithm 1 Stitch Finding in TPL

Input: Polygon P, conflict graph G
Output: All candidate stitches on P
for Every pairwise edges a and b with the lengths of a and b not less than $2 * fs_{min}$ **do**
  for Every polygon Q conflicting with P **do**
    Calculate projections from Q onto a and b
  end for
  Segment $op(a,b)$ according to the endpoints of all the projections
  for Every segmentation SE **do**
    Choose the position of stitch s within SE so that the distances between s and the projections on the two sides of SE, d_1 and d_2, are equal, split the polygon P and get two generated polygons
    if (1) no fs_{min} violations occur in the generated polygons **and** (2) both d_1 and d_2 are not less than m_o **and** (3) s is not near a corner **and** (4) neither of the generated polygons is such that it has no conflict edges **then**
      Store the stitch
    end if
  end for
end for

We first give some definitions for the terms used in our stitch finding method. The rectilinear polygon of a feature consists of one or more rectangles. Two parallel edges a and b of a polygon are called *pairwise edges* if they belong to the same rectangle. Note that projections on an edge from the same polygon may be separated (Figure 3(a)), but they need to be merged (Figure 3(b)), since the projection has actually covered the whole edge. The overlapping part of two pairwise horizontal (vertical) edges a and b in the x (y) direction is denoted as $op(a,b)$. Polygon cannot be split near a corner (distance within m_o) since corner stitch may cause significant side effect on printability [1, 6]. The pseudocode of our stitch finding method is listed in Algorithm 1. Here we prove the correctness of our method.

OBSERVATION 1. *Within a segment in Algorithm 1, every point has the same property of overlapping with projections from other polygons (the property refers to the projections it overlaps with), it will thus result in the same effect by*

splitting at any point in a segment, if no stitch constraints are considered.

THEOREM 1. *The stitches found by Algorithm 1 are complete.*

Proof: According to Observation 1, Algorithm 1 can exhaust all possible stitch positions by checking every segment delimited by the endpoints of the projections. Besides, by checking conditions (1)-(3), it can prune away stitches that violate constraints about m_o, fs_{min} or corner stitch. By checking condition (4), it can prune away useless stitches that will not help to resolve any conflict. (The reason is that adding isolated nodes into a conflict graph will not affect the 3-colorability of the graph). Therefore, we can conclude that Algorithm 1 can find all legal and effective candidate TPL stitches. Q.E.D.

Note that we have to test the conflict properties of the two generated polygons directly, instead of inferring them from the projections on the two sides of the stitch, because of the existence of U-shape polygon. As Figure 3(c) shows, both the generated polygon P_1 and P_2 conflict with Q if we split at s, although the projections of Q only appear on the lower side of s.

4. THE DECOMPOSITION APPROACH

In this section, we give details of our decomposition approach for TPL.

4.1 Overview

The flow of our decomposition approach for TPL is shown in Figure 4. First, conflict graph is constructed according to the input layout. Then, five graph simplification techniques are applied to divide the graph into a set of simple graphs. A graph library consisting of all possible graphs (with some constraints) of a limited number of nodes has been constructed beforehand. After that, we try to match all the simple graphs with the graphs in the library. For the matched ones, if they are 3-colorable, we simply map the colors from the graphs in the library to them; otherwise we will color them after stitch insertion. For the unmatched ones, we will color them by an effective heuristic as described below.

4.2 Graph Division

We construct the conflict graph, as shown in Section 2.1, and incorporate five graph decomposition techniques to divide the constructed graph. Three of which were proposed

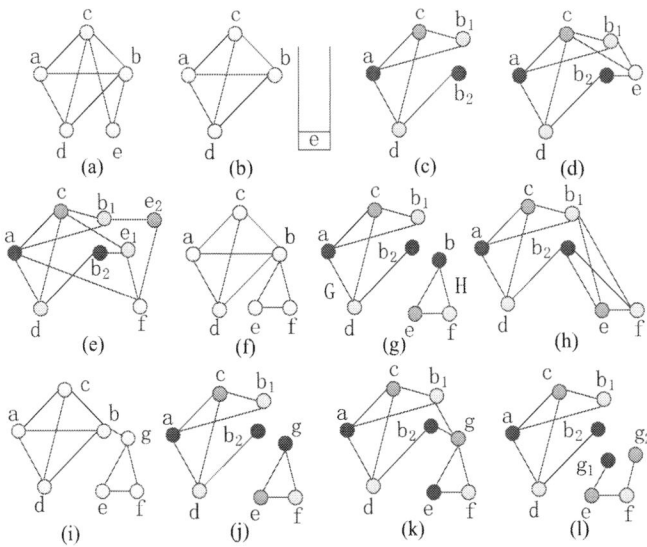

Figure 5: Examples for graph division

in [14]: *Independent Component Computation, Nodes with Degree Less than Three Removal* and *2-Edge-Connected Component Computation*. The other two were proposed in [3]: *Biconnected Component Computation* and *3-Edge-Connected Component Computation*. In the following, we will describe them briefly and point out their potential problems.

4.2.1 Independent Component Computation

Independent component of a graph is a maximal subgraph in which any two nodes are connected by at least one path. Independent component computation has been applied to many previous studies [4, 13, 15]. It is obvious that two independent components are coloring-independent.

4.2.2 Nodes with Degree Less than Three Removal

When we 3-color a graph, nodes with degree less than three can be removed temporarily and pushed to a stack. We can continue removing until every remaining node is with degree at least three. After coloring these remaining nodes, we can pop out all the nodes in the stack one by one, color them and add them back to the graph. Since every node has at most two neighbours when it is removed, there is at least one available color for it when it is added back. If all the nodes of a graph are removed, the graph can be colored easily. This method was adopted to reduce the graph in [14]. However, when some nodes of the graph cannot be removed, the removal process is not guaranteed to be safe. As Figure 5(a) and 5(b) show, node e is removed and pushed to the stack. The remaining nodes are colored with a stitch inserted into b that splits b into two nodes (Figure 5(c)). After that, the degree of node e is increased to 3 when it is being added back, and there is no available color for it (Figure 5(d)). A possible solution is to insert a stitch into e (Figure 5(e)) but this may create chain effect to other nodes like node f in Figure 5(e).

4.2.3 Biconnected Component Computation

A node is called an articulation node if its removal disconnects the graph. Biconnected component of a graph is a maximal connected subgraph without articulation nodes. Biconnected components can be 3-colored independently and combined together with color rotation. This method may

also be unsafe if stitch is inserted into the articulation node in at least one of the two components. As Figure 5(f) and 5(g) show, articulation node b is split in component G. After that, we will fail to combine the two components. As Figure 5(h) shows, there will be conflict if we just insert a stitch into the combined articulation node.

4.2.4 2-edge-connected and 3-edge-connected Components Computations

An edge is called a bridge if its removal disconnects the graph. A 2-edge-connected component of a graph is a maximal connected subgraph without bridges. A pair of edges is called a cut-pair if its removal disconnects the graph. A 3-edge-connected component of a 2-edge-connected graph is a maximal connected subgraph without cut-pairs. Both 2-edge-connected component and 3-edge-connected component can be 3-colored independently and combined together with color rotation to make the colors of the two nodes connected by the bridge or the cut-pair different. These two computations are safe if at most one of the two nodes connected by the bridge or an edge of the cut-pair is split. For example, two components of the graph in Figure 5(i) are computed and colored with a stitch inserted into node b, as shown in Figure 5(j). The two components can be combined with color rotation to make the color of node g different from the colors of nodes b_1 and b_2, as shown in Figure 5(k). However, if both nodes connected by the bridge or the cut-pair are split, the two components may not be able to be combined. As shown in Figure 5(l) (obtained from Figure 5(i)), both nodes b and g are split and combining them will cause a conflict.

4.2.5 Flow of Graph Division

The flow of our graph division is shown in Figure 6. First, we compute independent components, and remove nodes with degree less than three in every component. As this node removal can further disconnect some components, we will apply the independent component computation again. Then we compute 2-edge-connected components, biconnected components and 3-edge-connected components one after another. Note that endpoints of a bridge are also articulation nodes. For example, in Figure 5(i), nodes b and g, which are endpoints of bridge (b,g), are also articulation nodes. Recall that both nodes b and g cannot be split (by inserting stitch) if we divide the graph with these articulation nodes, whereas one of them can be split if we divide the graph with the bridge (b,g). Therefore we will compute 2-edge-connected components first. These component computations may generate nodes with degree one or two. For example, the degree of node g in Figure 5(j) becomes two after the 2-edge-connected component computation. Therefore we will always try to remove nodes with degree less than 3 again after each computation.

We compute independent components with depth-first search. Bridges, 2-edge connected components, articulation nodes and biconnected components are computed according to the algorithms in [9, 10]. Cut-pairs and 3-edge connected components are computed by the algorithm in [12]. All of them can be done in linear time $O(|E|+|V|)$, where $|E|$ and $|V|$ are the numbers of edges and the number of nodes in the conflict graph $G=(E,V)$ respectively.

4.3 Graph Library Construction

We construct a graph library that contains all biconnected graphs with four, five or six nodes according to the algo-

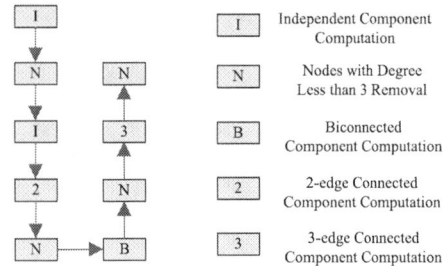

Figure 6: Flow of graph division

rithm in [7], and remove those with nodes of degree less than three. There are 23 such graphs in total. Obviously, all of them have no bridges nor cut-pairs. Nine of them can be 3-colored easily[1]. For example, the graphs in Figure 7(a) and 7(b) are 3-colorable ones in the library with five and six nodes respectively. For those non-3-colorable ones, it is also easy to decide which nodes need to split (by inserting stitch). For example, the graph in Figure 7(c) is not 3-colorable, but it can be 3-colored with a stitch properly inserted into a degree-4 node, such as node a, to split it into nodes a_1 and a_2 as shown in Figure 7(d). Whether a stitch can be properly inserted depends on which nodes a_1 and a_2 are connected with. For example, if node a in Figure 7(c) is split to give Figure 7(e), the graph is still not 3-colorable. Some graphs need two or more stitches to be 3-colorable. For example, the graph in Figure 7(f) is 3-colorable with two stitches inserted into nodes a and b respectively. With this library, after dividing a conflict graph, the subgraphs obtained will be matched with the graphs in the library. From the experiments, we found that almost all the subgraphs will be covered. For each matched subgraph, we will get the splitting and coloring information from the library directly. The library, once constructed, can be used for all benchmarks.

Here we present an example to show that the assumption in [3] is not always true. Consider Figure 7(g) that is an actual layout of the conflict graph in Figure 7(c). In Figure 7(g), different projections are shown with different colors. When we consider the conflict between e and a, the maximum overlap density of a is 3, since all the projections from b, c and d overlap with each other on a. Similarly, when we consider the conflict between e and b, the maximum overlap density of b is 2. According to the assumption, splitting b is better than splitting a. However, by splitting a, we can get a one-stitch solution (Figure 7(d)), but no matter how we split b, the graph remains non-3-colorable (Figure 7(h)).

4.4 Graph Matching and Coloring

We adopt a polynomial-time algorithm on graph isomorphism in [2] to match graphs. If a subgraph is matched with a 3-colorable one, we can color it simply by mapping the colors according to the nodes rearrangement. Otherwise, we will try inserting stitches. We will first try all possible 1-stitch solutions. If it fails, we will try to solve it with more stitches.

Because the graph division may not be safe, as explained in Section 4.2, we *disallow* stitches to be inserted at articulation nodes and at both endpoints of bridges or cut-pairs. Besides,

---

[1]We observed from experiments that these 3-colorable components cover most of the subgraphs obtained after dividing the conflict graphs.

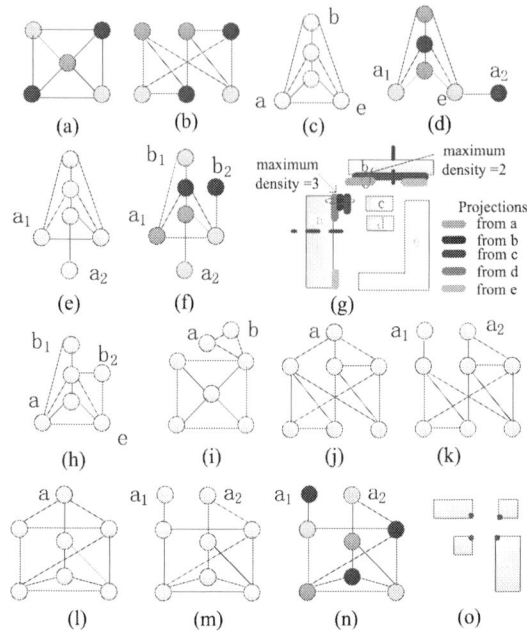

Figure 7: Examples for coloring of graphs in the library, heuristic coloring and native conflict

if the degree of a node is two when it is being removed in the "*nodes with degree less than three removal*" process, we will not insert a stitch that overlaps with its projection. For example, we will not insert a stitch to split node b into nodes b_1 and b_2, as shown in Figure 5(c), because they both conflict with node e (note that node e has degree two when it is being removed). To achieve this, when we find stitches by projection onto a node a with algorithm 1, we also consider the projections of the removed nodes that have degree two and have conflict with a when being removed.

Note that the order of removing nodes will have some impacts on the result. As shown in Figure 7(i), for the two degree-two nodes, a and b, if a is removed first, the degree of a when it is being removed is two, whereas if b is removed first, the degree is one. In our current implementation, we will just follow the order of the features given in the benchmark.

4.5 Heuristic Coloring

Subgraphs with seven or more nodes will not be matched and we use a heuristic to color them. We observed that there are not many such cases and the subgraphs are typically sparse, i.e., many nodes have low-degree. In our heuristics, we will try to split (by inserting stitches) those low-degree nodes in order to generate nodes with degree less than three. Since nodes with degree less than three can be removed temporarily, we may be able to convert the graph to one in the library, or simply all the nodes are removed step by step (because of degree less than three) and thus can be colored easily. An example is shown in Figure 7(j) and 7(k). We can split node a into two nodes a_1 and a_2. After the splitting and nodes removal, this graph is converted to the graph shown in Figure 7(b), which is 3-colorable. Another example is shown in Figure 7(l). Similarly, we can split node a into two nodes, as shown in Figure 7(m). After that, all nodes of the graph can be removed, thus we can 3-color it as shown in Figure 7(n). Similar to the coloring of matched subgraphs, we forbid unsafe stitches in our heuristics.

488

Table 1: Statistics and decomposition results

	Statistics						From [11]		From [3]				Ours			
Circuit	G#	M#	Ratio	3C#	Ratio	NC	C#	S#	C#	S#	Cost	cpu(s)	C#	S#	Cost	cpu(s)
C432	4	4	1.000	0	0.000	0	-	-	0	6	6	0.01	0	4	4	0.01
C499	3	3	1.000	3	1.000	0	0	0	0	0	0	0.01	0	0	0	0.01
C880	7	7	1.000	0	0.000	0	0	7	1	15	25	0.01	0	7	7	0.01
C1355	4	3	0.750	1	0.250	0	0	3	1	7	17	0.02	0	3	3	0.01
C1908	3	3	1.000	2	0.667	0	0	1	1	0	10	0.04	0	1	1	0.01
C2670	9	8	0.889	3	0.333	0	0	6	2	14	34	0.06	0	6	6	0.04
C3540	14	14	1.000	5	0.357	1	-	-	2	15	35	0.08	1	8	18	0.05
C5315	15	15	1.000	6	0.400	0	-	-	3	11	41	0.11	0	9	9	0.05
C6288	213	213	1.000	8	0.038	2	-	-	19	341	531	0.13	14	191	331	0.25
C7552	34	34	1.000	12	0.353	0	-	-	3	46	76	0.17	0	22	22	0.10
S1488	8	8	1.000	6	0.750	0	0	2	0	4	4	0.03	0	2	2	0.01
S38417	627	625	0.997	553	0.882	19	-	-	20	122	322	0.62	19	55	245	0.42
S35932	1831	1829	0.999	1746	0.954	44	-	-	46	103	563	2.13	44	41	481	0.82
S38584	1819	1818	0.999	1667	0.916	36	-	-	36	280	640	2.26	36	116	476	0.77
S15850	1576	1573	0.998	1443	0.916	34	-	-	36	201	561	2.14	34	100	440	0.76
Avg.	411.13	410.47	0.998	363.67	0.885	-	-	-	1.14	2.07	1.39	2.36	1.00	1.00	1.00	1.00

4.6 Native Conflicts

In order to understand the difficulty of each benchmark, we devised a simple method to compute a lower bound on the number of native conflicts. One common case of native conflict is the existence of a K_4 graph between four points in the layout (e.g., red corners in Figure 7(o)). It is obvious that this kind of conflict subgraph can never be resolved by inserting stitches.

5. EXPERIMENTAL RESULTS

We implemented the proposed approach in C++, on a 2.39 GHz Linux machine with 48 GB memory. To evaluate the approach, we tested the ISCAS-85 & 89 benchmarks provided by the authors of [14]. We used the same setting of cs_{min} as previous studies [3, 11]. Since the setting of m_o is not clearly stated in [3, 11], we set it to 10nm, which is the same as that in [14] and is believed to be the same m_o used by all these works since comparisons with each other have been shown in these papers.

The statistics and decomposition results of the benchmarks are shown in Table 1. G#, M#, 3C#, NC denote subgraph number, matched subgraph number, 3-colorable subgraph number and lower bound of native conflicts, respectively. It can be seen that most of the subgraphs can be matched and 3-colored and the total ratios of matched ones and 3-colorable ones are nearly 0.998 and 0.885 respectively. We compare our results with previous studies. C# denotes the number of conflicts, S# is for the number of stitches, cpu is the running time for the decomposition process and cost is computed as $10 \times C\# + S\#$ (same as in [3]) since conflicts typically cause much higher manufacturing cost. As shown in the table, we can always achieve as good as [11] for the benchmarks that they can solve and we have actually solved three more benchmarks (C432, C5315 and C7552). Note that the method in [11] will only give a solution when there is a non-conflict solution and that is why no solutions are reported for the other benchmarks. For the additionally solved benchmarks, since the work [11] considers candidate stitches using DPL projections, some of the possible TPL stitches are actually missed. Compared with [3], we can reduce the number of conflicts by 12%, number of stitches by 52%, and cost by 28%. Note that our numbers of conflicts are actually very close to the native conflict numbers. For running time, we can achieve a 2.36× speed-up, which clearly demonstrates our efficiency.

6. CONCLUSION

In this paper, we propose an approach of decomposition for TPL. Based on graph division and matching, our approach can be very efficient. To solve unmatched subgraphs, we propose an effective heuristic. Our approach can find all legal stitches in TPL to resolve more conflicts. Experimental results verify its effectiveness. We expect this result to benefit the industry on TPL. In the near future, we will study on a good ordering of removing nodes with degree less than three in order to give better results.

7. REFERENCES

[1] D. Abercrombie, P. Lacour, O. El-Sewefy, A. Volkov, K. Arb, C. Reid, Q. Li, and P. Ghosh. Double patterning from design enablement to verification. In *Pro. SPIE*, 2011.

[2] A. Dharwadker and J. Tevet. The graph isomorphism algorithm. In *Pro. The Structure Semiotics Research Group S.E.R.R.*, 2009.

[3] S. Fang, Y. Chang, and W. Chen. A novel layout decomposition algorithm for triple patterning lithography. In *Pro. DAC*, 2012.

[4] A. B. Kahng, C. H. Park, X. Xu, and H. Yao. Layout decomposition for double patterning lithography. In *Pro. ICCAD*, pages 465–472, 2008.

[5] S. Khanna, N. Linial, and S. Safra. On the hardness of approximating the chromatic number. In *Pro. Israel Symposium on the Theory and Computing Systems*, pages 250–260, 1993.

[6] Q. Ma, H. Zhang, and D. F. Wong. Triple patterning aware routing and its comparision with double patterning aware routing in 14nm technology. In *Pro. DAC*, 2012.

[7] D. Stolee. Isomorph-free generation of 2-connected graphs with applications. *CSE Technical Reports, University of Nebraska - Lincoln*, 2011.

[8] X. Tang and M. Cho. Optimal layout decomposition for double patterning technology. In *Pro. ICCAD*, 2011.

[9] R. Tarjan. Depth-first search and linear graph algorithms. In *Pro. SWAT*, pages 114–121, 1971.

[10] R. Tarjan. A note on finding the bridges of a graph. *Information Processing Letters*, 40:125–142, 2007.

[11] H. Tian, H. Zhang, Q. Ma, Z. Xiao, and D. F. Wong. A polynomial time triple patterning algorithm for cell based row-structure layout. In *Pro. ICCAD*, 2012.

[12] Y. H. Tsin. A simple 3-edge-connected component algorithm. *Theory of Computation Systems*, 2:160–161, 1974.

[13] J.-S. Yang, K. Lu, M. Cho, K. Yuan, and D. Pan. A new graph-theoretic, multi-objective layout decomposition framework for double patterning lithography. In *Pro. ASPDAC*, 2010.

[14] B. Yu, K. Yuan, B. Zhang, D. Ding, and D. Z. Pan. Layout decomposition for triple patterning lithography. In *Pro. ICCAD*, 2011.

[15] K. Yuan, J.-S. Yang, and D. Pan. Double patterning layout decomposition for simultaneous conflict and stitch minimization. In *Pro. ISPD*, 2009.

E-BLOW: E-Beam Lithography Overlapping aware Stencil Planning for MCC System

Bei Yu, Kun Yuan†, Jhih-Rong Gao, David Z. Pan
ECE Department, Univ. of Texas at Austin, Austin, TX, USA
†Cadence Design Systems, Inc., San Jose, CA, USA
{bei, jrgao, dpan}@cerc.utexas.edu

ABSTRACT

Electron beam lithography (EBL) is a promising maskless solution for the technology beyond 14nm logic node. To overcome its throughput limitation, recently the traditional EBL system is extended into MCC system. In this paper, we present E-BLOW, a tool to solve the overlapping aware stencil planning (OSP) problems in MCC system. E-BLOW is integrated with several novel speedup techniques, i.e., successive relaxation, dynamic programming and KD-Tree based clustering, to achieve a good performance in terms of runtime and solution quality. Experimental results show that, compared with previous works, E-BLOW demonstrates better performance for both conventional EBL system and MCC system.

Categories and Subject Descriptors

B.7.2 [**Hardware, Integrated Circuit**]: Design Aids

General Terms

Algorithms, Design, Performance

Keywords

Electron Beam Lithography (EBL), Overlapping aware Stencil Planning (OSP), Multi-Column Cell (MCC) System

1. INTRODUCTION

As the minimum feature size continues to scale to sub-22nm, the conventional 193nm optical photolithography technology is facing great challenge in manufacturing [1]. In the near future, double/multiple patterning lithography (DPL /MPL) has become one of viable lithography techniques for 22nm and 14nm logic node [2–4] . In the longer future, i.e., for the logic node beyond 14nm, extreme ultra violet (EUV) and electric beam lithography (EBL) are promising candidates for lithographic processes. However, EUV suffers from the delay due to the tremendous technical barriers such as lack of power sources, resists, and defect-free masks [5].

Permission to make digital or hard copies of all or part of this work for personal or classroom use is granted without fee provided that copies are not made or distributed for profit or commercial advantage and that copies bear this notice and the full citation on the first page. To copy otherwise, to republish, to post on servers or to redistribute to lists, requires prior specific permission and/or a fee.

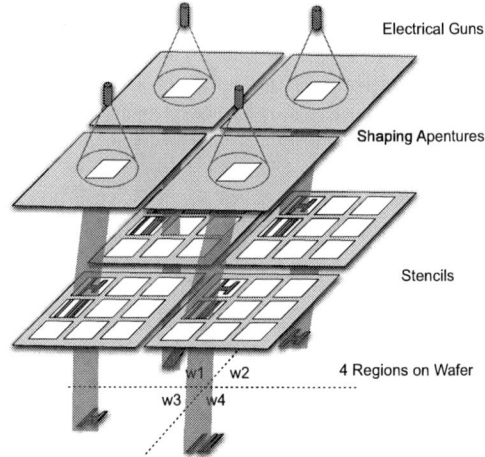

Figure 1: Printing process of MCC system.

EBL system, on the other hand, has been developed for several decades [6]. One of the conventional EBL systems is based on character projection (CP) mode. Some complex shapes, called *characters*, are prepared on the stencil. The key idea is that if a pattern is pre-designed on the stencil, it can be printed in one electronic shot, otherwise it needs to be fractured into a set of rectangles and printed one by one through variable shaped beam (VSB). Compared with purely VSB mode, in the CP mode the throughput can be improved significantly. Compared with the traditional lithographic methodologies, EBL has several advantages. (1) Electron beam can be easily focused into nanometer diameter with charged particle beam, which can avoid suffering from the diffraction limitation of light. (2) The price of a photomask set is getting unaffordable. As a maskless technology, EBL can reduce the manufacturing cost. (3) EBL allows a great flexibility for fast turnaround times and even late design modifications to correct or adapt a given chip layout. Because of all these advantages, EBL is being used in mask making, small volume LSI production, and R&D to develop the technological nodes ahead of mass production.

Even with decades of development, the key limitation of the EBL system has been and still is the low throughput. Recently, multi-column cell (MCC) system is proposed as an extension of conventional EBL system, where several independent character projections (CP) are used to further speed-up the writing process [7, 8]. Each CP is applied on one section of wafer, and all CPs can work parallelly to achieve better throughput. Due to the design complex-

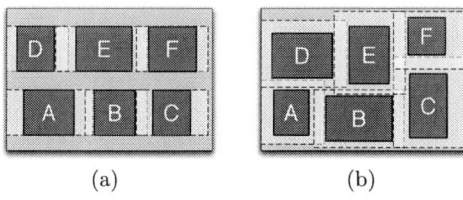

Figure 2: (a) 1D-OSP. (b) 2D-OSP.

ity and cost consideration, different CPs share one stencil design [9]. One example of MCC printing process is illustrated in Fig. 1, where four CPs are bundled to generate a MCC system. The whole wafer is divided into four regions, w_1, w_2, w_3 and w_4, and each region is printed through one CP. Note that the whole writing time of the MCC system is determined by the maximum writing time of the four regions. For modern design, because of the numerous distinct circuit patterns, only limited number of patterns can be employed on stencil. Therefore, the area constraint of stencil is the bottleneck. To improve the throughput, the stencil should be carefully designed/manufactured to contain the most repeated cells or patterns.

Much previous works focus on the design optimization for conventional EBL system [10–14]. Stencil planning, which is one of the most challenges, has earned much attentions. When blank overlapping is not considered, the stencil planning can be formulated as a character selection problem, where integer linear programming (ILP) was applied to select group of characters for throughput maximization [10]. Recently, Yuan et al. in [12] investigated on the overlapping aware stencil planning (OSP) problem.

However, no existing stencil planning work has done for the MCC system. Compared with conventional EBL system, MCC system introduces two main challenges. First, the objective is new: in MCC system the wafer is divided into several regions, and each region is written by one CP. Therefore the new OSP should minimize the maximal writing times of all regions. While in conventional EBL system, the objective is simply minimize the wafer writing time. Besides, the stencil for an MCC system can contain more than 4000 characters, previous methodologies for EBL system may suffer from runtime penalty.

The OSP problem can be divided into two sub-problems: *1D-OSP* and *2D-OSP* [12]. When standard cells with same height are selected into stencil, the problem is referred as 1D-OSP. As shown in Fig. 2(a), each character implements one standard cell, and the enclosed circuit patterns of all the characters have the same height. Note that here we only show the horizontal blanks, and the vertical blanks are not represented because they are identical. In 2D-OSP, the blanking spaces of characters are non-uniform along both horizontal and vertical directions. By this way, stencil can contain both complex via patterns and regular wires. Fig. 2(b) illustrates a stencil design example for 2D-OSP.

This paper presents E-BLOW, the first study for OSP problem in MCC system. E-BLOW integrates several novel techniques to achieve near-optimal solution in reasonable time. The main contributions of this paper are stated as follows: (1) First study for stencil planning problem in MCC system. (2) Shows that OSP problem for both EBL and MCC systems are NP-hard. (3) Proposes a simplified formulation for 1D-OSP, and proves its rounding lower bound theoretically. (4) A successive relaxation algorithm to find

a near optimal solution. (5) KD-Tree based clustering algorithm for speedup in 2D-OSP.

The remainder of this paper is organized as follows. Section 2 provides problem formulation. Section 3 presents algorithm details to resolve 1D-OSP problem in E-BLOW, while section 4 details the E-BLOW solutions to 2D-OSP problem. Section 5 reports experimental results, followed by the conclusion in Section 6.

2. PROBLEM FORMULATION

2.1 OSP Problem Formulation

In an MCC system with P CPs, the whole wafer is divided into P regions $\{w_1, w_2, \ldots, w_P\}$, and each region is written by one particular CP. We assume cell extraction [15] has been resolved first. In other words, a set of character candidates $C^C = \{c_1, \cdots, c_n\}$ has already been given to the MCC system. For each character candidate $c_i \in C^C$, its writing time through VSB mode is denoted as n_i, while its writing time through CP mode is 1.

The regions of wafer have different layout patterns, and the throughputs would be also different. Suppose character candidate c_i repeats t_{ic} times on region w_c. Let a_i indicate selection of character candidate c_i as follows.

$$a_i = \begin{cases} 1, & \text{candidate } c_i \text{ is selected on stencil} \\ 0, & \text{otherwise} \end{cases}$$

If c_i is prepared on stencil, the total writing time of patterns c_i on region w_c is $t_{ic} \cdot 1$. Otherwise, c_i should be printed through VSB. Since region w_c comprises t_{ic} candidate c_i, the writing time would be $t_{ic} \cdot n_i$. Therefore, for region w_c the total writing time T_c is as follows:

$$\begin{aligned} T_c &= \sum_{i=1}^{n} a_i \cdot (t_{ic} \cdot 1) + \sum_{i=1}^{n} (1 - a_i) \cdot (t_{ic} \cdot n_i) \\ &= \sum_{i=1}^{n} t_{ic} \cdot n_i - \sum_{i=1}^{n} t_{ic} \cdot (n_i - 1) \cdot a_i = T_c^{VSB} - \sum_{i=1}^{n} R_{ic} \cdot a_i \end{aligned}$$

where $T_c^{VSB} = \sum_{i=1}^{n} t_{ic} \cdot n_i$, and $R_{ic} = t_{ic} \cdot (n_i - 1)$. T_c^{VSB} represents the writing time on w_c when only VSB is applied, and R_{ic} can be viewed as the writing time reduction of candidate c_i on region w_c. In MCC system, both T_c^{VSB} and R_{ic} are constants. Therefore, the total writing time of the MCC system is formulated as follows:

$$\begin{aligned} T_{total} &= \max\{T_c\} \\ &= \max\{T_c^{VSB} - \sum_{i=1}^{n} R_{ic} \cdot a_i\}, \forall c \in P \quad (1) \end{aligned}$$

Problem 1. *Overlapping aware Stencil Planning (OSP) for MCC system: Given a set of character candidate C^C, select a subset C^{CP} out of C^C as characters, and place them on the stencil. The objective is to minimize the total writing time T_{total} expressed by (1), while the placement of C^{CP} is bounded by the outline of stencil. The width and height of stencil is W and H, respectively.*

For convenience, we use the term OSP to refer OSP for MCC system in the rest of this paper.

2.2 NP-Hardness

Lemma 1. *1D-OSP problem is NP-hard.*

491

Let us consider a special and simper case of 1D-OSP, where each candidate c_i has zero blank space, and CP number is 1. Then the problem can be reduced from a multiple knapsack problem, which is a well known NP-hard problem [16].

Lemma 2. *2D-OSP problem is NP-hard.*

Let us consider a special case of 2D-OSP, where each candidate c_i has zero blank space, and CP number is 1. The 2D-OSP problem includes two subproblems: candidate selection and candidate packing. After some candidates are selected on the stencil, the candidates packing problem can be reduced from a strip packing problem [17], which is NP-hard.

Combining Lemma 1 and Lemma 2, we can achieve the conclusion that OSP problem, even for conventional EBL system, is NP-hard.

3. E-BLOW FOR 1D-OSP

When each character implements one standard cell, the enclosed circuit patterns of all the characters have the same height. Corresponding OSP problem is called 1D-OSP, which can be viewed as a combination of character selection and single row ordering problems [12]. Different from two heuristic steps proposed in [12], we show that the two problems can be solved simultaneously through ILP formulation (2).

$$\min \quad T_{total} \tag{2}$$

$$\text{s.t} \quad T_{total} \geq T_c^{VSB} - \sum_{i=1}^{n} (\sum_{k=1}^{M} R_{ic} \cdot a_{ik}), \ \forall c \in P \tag{2a}$$

$$x_i + w_i \leq W, \qquad \forall i \in N \tag{2b}$$

$$\sum_{k}^{m} a_{ik} \leq 1, \qquad \forall k \in M \tag{2c}$$

$$x_i + w_{ij} - x_j \leq W(2 + p_{ij} - a_{ik} - a_{jk}) \tag{2d}$$

$$x_j + w_{ji} - x_i \leq W(3 - p_{ij} - a_{ik} - a_{jk}) \tag{2e}$$

$$a_{ik}, a_{jk}, p_{ij} : 0 - 1 \text{ variable} \tag{2f}$$

In (2) W is the width constraint of stencil, M is the number of rows, and w_i is width of character c_i. x_i is the x-position of c_i. If and only if c_i is assigned to kth row, $a_{ik} = 1$. In other words, a_{ik} determines the y-position of c_i. $w_{ij} = w_i - o_{ij}^h$ and $w_{ji} = w_i - o_{ji}^h$, where o_{ij}^h is the overlapping when candidates c_i and c_j are packed together. Constraints $(2d)$ $(2e)$ are used to check position relationship between c_i and c_j. For kth row, it is easy to see that only when $a_{ik} = a_{jk} = 1$, i.e. both character i and character j are assigned to row j, then only one of the two constraints $(2d)$ $(2e)$ will be active. If either of them are not assigned to the row, neither of the constraints are active. The number of variables for (2) is $O(N^2)$, where N is the number of character candidates.

Since ILP is a well known NP-hard problem, directly solving it may suffer from long runtime penalty. One straightforward speedup method is to relax ILP (2) into linear programming (LP) as following: replacing constraints $(2f)$ by $0 \leq a_{ik}, a_{jk}, p_{ij} \leq 1$. It is obvious that the LP solution provides a lower bound to the ILP solution. However, we observe that the solution of relaxed LP would be like this: for each i, $\sum_j a_{ij} = 1$ and all the p_{ij} are assigned 0.5. Although the objective function is minimized and all the constraints are satisfied, this LP relaxation provides no useful

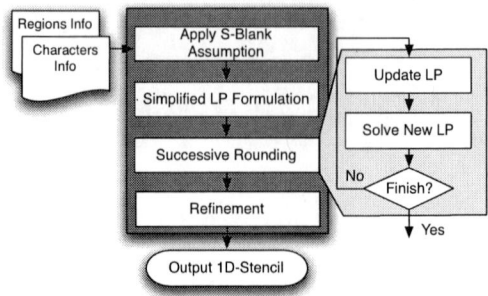

Figure 3: E-BLOW overall flow for 1D-OSP.

information to guide future rounding, i.e., all the character candidates are selected and no ordering relationship is determined.

To overcome the limitation of above rounding, E-BLOW proposes a novel iterative solving framework to search near-optimal solution in reasonable runtime. The main idea is to modify the ILP formulation, so that the corresponding LP relaxation can provide good lower bound theoretically. As shown in Fig. 3, the overall flow includes three parts: Simplified ILP formulation, Successive Rounding and Refinement. At section 3.2 the simplified formulation will be discussed, and its LP Rounding lower bound will be proved. Function $SuccRounding()$ is the successive rounding method, which will be introduced at section 3.3. At last, section 3.4 proposes a dynamic programming based refinement.

3.1 Symmetrical Blank (S-Blank) Assumption

Our simplified formulation is based on a symmetrical blank assumption: the blanks of each character is symmetry, left slack equals to right slack. Note that for different characters i and j, their slacks s_i and s_j can be different.

At first glance this assumption may lose optimality, however, it provides several practical and theoretical benefits. (1) Single row ordering [12] was transferred into Hamilton Cycle problem, which is a well known NP-hard problem and even particular solver is quite expensive. Under the assumption, this ordering problem can be optimally solved in $O(n)$. (2) The ILP formulation can be simplified to provide a reasonable rounding bound theoretically. Compared with previous heuristic framework [12], the proved rounding bound provides a better guideline for a global view search. (3) To compensate the inaccuracy in the asymmetrical blank cases, E-BLOW provides a refinement to further improve the throughput.

Given p character candidates, single row ordering problem adjusts the relative locations to minimize the total width. Under symmetrical blank assumption, this problem can be optimally solved by a two steps greedy approach. First, all characters are sorted decreasingly by blanking space s_i; second, they are inserted one by one. Each one can insert at either left end or right end.

Theorem 1: *Under S-Blank assumption, the greedy approach can get maximum overlapping space* $\sum_i s_i - \max\{s_i\}$.

In practical, we set $s_i = \lceil (sl_i + sr_i)/2 \rceil$, where sl_i and sr_i are c_i's left slack and right slack, respectively.

3.2 Simplified Formulation

To further simplify (2), we modify the objective function through assigning each character c_i with one profit value

$profit_i$. Then based on the Theorem 1, the formulation (2) can be simplified as follow:

$$\max \sum_i \sum_j a_{ij} \cdot profit_i \qquad (3)$$

$$s.t. \quad \sum_i (w_i - s_i) \cdot a_{ij} \leq W - B_j, \forall j \qquad (3a)$$

$$B_j \geq s_i \cdot a_{ij}, \forall i \qquad (3b)$$

$$\sum_j a_{ij} \leq 1, \quad \forall c_i \in C^C \qquad (3c)$$

$$a_{ij} = 0 \ \text{ or } \ 1 \qquad (3d)$$

(3a) and (3b) are based on Theorem 1 to calculate the row width, where (3b) is to linearize max operation. Here B_j can be viewed as the maximum blank space of all the characters on row r_j. (3c) means each character can be assigned into at most one row. It's easy to see that the number of variables is $O(nm)$. Generally speaking, single character number n is much larger than row number m, so compared with basic ILP formulation (2), the variable number of (3) can be reduced dramatically.

Furthermore, theoretically the simplified formulation (3) can achieve reasonable LP rounding lower bound. To explain this, let us first look at a similar program (3′) as follows:

$$\max \sum_i \sum_j (w_i - s_i) \cdot a_{ij} \cdot ratio_i \qquad (3')$$

$$s.t. \quad \sum_i (w_i - s_i) \cdot a_{ij} \leq W - max_s \qquad (3a')$$

$$(3c) - (3d)$$

where $ratio_i = profit_i / (w_i - s_i)$, and max_s is the maximum horizontal slack length of every character, i.e. $max_s = \max\{s_i | i = 1, 2, \ldots, n\}$. Program (3′) is a well known multiple knapsack problem [16].

Lemma 3. *If each $ratio_i$ is the same, the multiple knapsack problem (3′) can find a $1/2-$approximation algorithm using LP Rounding method.*

For brevity we omit the proof, detailed explanations can be found in [18]. It shall be noted that if all $ratio_i$ are the same, program (3′) can be approximated to a max-flow problem. Based on Lemma 3, if we denote α as $\max\{ratio_i\}$ /$\min\{ratio_i\}$, we can achieve the following theorem:

Theorem 2: *The LP Rounding solution of (3) can be a $0.5/\alpha-$ approximation to program (3′).*

Due to space limit, the detailed proof is omitted. The only difference between (3) and (3′) is that the right side values at (3a) and (3a′). Blank spacing is relatively small comparing with the row length, we can get that $W - max_s \approx W - B_j$. Then based on Theorem 2, we can conclude that program (3) has a reasonable rounding bound.

3.3 Successive Relaxation

Because of the reasonable LP rounding property shown in Theorem 2, we propose a successive relaxation algorithm to solve program (3) iteratively. The ILP formulation (3) becomes an LP if we relax the discrete constraint to a continuous constraint as: $0 \leq a_{ij} \leq 1$. The successive relaxation algorithm is shown in Algorithm 1. At first we set all a_{ij} to variables since any a_{ij} is not guided to rows. The LP is updated and solved iteratively. For each new LP solution, we

---

Algorithm 1 SuccRounding(th_{inv})

Input: ILP Formulation (3)
1: set all a_{ij} to variables;
2: **repeat**
3:     update $profit_i$ for all variables a_{ij};
4:     solve relaxed LP of (3);
5:     **repeat**
6:         find $a_{pq} = \max\{a_{ij}$, and c_i can insert into row $r_j\}$;
7:         **for all** $a_{ij} \geq a_{pq} \times th_{inv}$ **do**
8:             **if** c_i can be assigned to row r_j **then**
9:                 $a_{ij} = 1$ and set it to a non-variable;
10:                Update capacity of row r_j;
11:         **end if**
12:         **end for**
13:     **until** cannot find a_{pq}
14: **until**

---

search the maximal a_{pq} (line 6). Then for all a_{ij} that is close the the maximal value a_{pq}, we try to pack c_i into row r_j, and set it as non-variable. Note that since several a_{ij} are assigned permanent value, the number of variables in updated LP formulation would continue to decrease. This procedure repeats until no appropriate a_{ij} can be found. One key step of the Algorithm 1 is the $profit_i$ update (line 3). For each character c_i, we set its $profit_i$ as follows:

$$profit_i = \sum_c \frac{t_c}{t_{max}} \cdot (n_i - 1) \cdot t_{ic} \qquad (4)$$

where t_c is current writing time of region w_c, and $t_{max} = \max\{t_c, \forall c \in P\}$. Through applying the $profit_i$, the region w_c with longer writing time would be considered more during the LP formulation. During successive relaxation, if c_i hasn't been assigned to any row, $profit_i$ would continue to updated, so that the total writing time of the whole MCC system can be minimized.

3.4 Refinement

Simplified formulation and successive relaxation are under the symmetrical blank assumption. Although it can be effectively solved, for asymmetrical cases it would lose some optimality. To compensate the losing, we present a dynamic programming based refinement procedure. As discussed above, for k characters, single row ordering can have 2^{k-1} possible solutions. Under symmetrical blank space assumption, all these orderings get the same length. But for the asymmetrical cases, it does not hold anymore. Our dynamic programming based algorithm *Refine*(k) finds the best solution from these 2^{k-1} options. The detailed is shown in Algorithm 2. At first, if $k > 1$, then *Refine*(k) will recursively call *Refine*(k-1) to generate all old partial solutions. All these partial solutions will be updated by adding candidate c_k (lines 6-8). Note that maintaining all solutions is impractical and unnecessary, because many of them are inferior to others. In *SolutionPruning*(), all solutions are checked, if one solution S_A is inferior to another solution S_B, S_A would be pruned to save computation cost. For each solution a triplet (w, l, r) is constructed to store the information of width, left slack and right slack. We define the *inferior* relationship as follow. For two solutions $S_A = (w_a, l_a, r_a)$ and $S_B = (w_b, l_b, r_b)$, S_B is inferior to S_A if and only if $w_a \geq w_b$, $l_a \leq l_b$ and $r_a \leq r_b$.

After *Refine(k)* for each row, if more available spaces are generated, a greedy insertion approach similar to [12] would

Algorithm 2 Refine(k)

1: **if** k = 1 **then**
2: Generate partial solution (w_1, sl_1, sr_1);
3: **else**
4: Refine(k-1);
5: **for** each partial solution (w, l, r) **do**
6:   $(w_1, l_1, r_1) = (w + w_k - \min(sr_k, l), sl_k, r)$;
7:   $(w_2, l_2, r_2) = (w + w_k - \min(sl_k, r), l, sr_k)$;
8:   Replace (w, l, r) by (w_1, l_1, r_1) and (w_2, l_2, r_2);
9:   **if** solution set size $\geq$ threshold **then**
10:    SolutionPruning();
11:   **end if**
12: **end for**
13: **end if**

be proposed to further improve the throughput.

4. E-BLOW FOR 2D-OSP

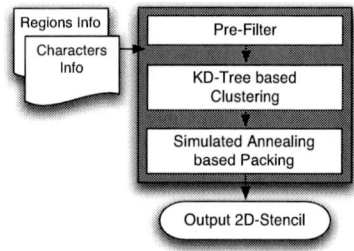

Figure 4: E-BLOW overall flow for 2D-OSP.

Now we consider a more general case: the blanking spaces of characters are non-uniform along both horizontal and vertical directions. This problem is referred as 2D-OSP problem. In [12] the 2D-OSP problem was transformed into a floorplanning problem. However, several key differences between traditional floorplanning and OSP were ignored. (1) In OSP there is no wirelength to be considered, while at floorplanning wirelength is a major optimization objective. (2) Compared with complex IP cores, lots of characters may have similar sizes. (3) Traditional floorplanner could not handle the problem size of modern MCC design. To deal with all these properties, a approximation packing framework is proposed (see Fig. 4). Given the input character candidates, the pre-filter process is first applied to remove characters with bad profit (defined in (4)). Then the second step is a KD-Tree based clustering algorithm to effectively speed-up the design process. Followed by the final floorplanner to pack all candidates.

4.1 KD-Tree based Clustering

Clustering is a well studied problem, and there are many of works and applications in VLSI [19]. However, previous methodologies cannot be directly applied here. (1) Traditional clustering is based on netlist, which provides the all clustering options. Generally speaking, netlist is sparse, but in OSP the connection relationships are so complex that any two characters can be clustered, and totally there are $O(n^2)$ clustering options. (2) Given two candidates c_i and c_j, there are several clustering options. For example, horizontal clustering and vertical clustering may have different overlapping space.

Algorithm 3 KD-Tree based Clustering

Input: set of candidates C^C.
1: **repeat**
2: Sort all candidates by $profit_i$;
3: Set each candidates c_i to unclustered;
4: **for all** unclustered candidate c_i **do**
5:   Find pair (c_i, c_j) with similar blank spaces and profits;
6:   Cluster (c_i, c_j), label them as clustered;
7: **end for**
8: Update candidate information;
9: **until** reach clustering threshold

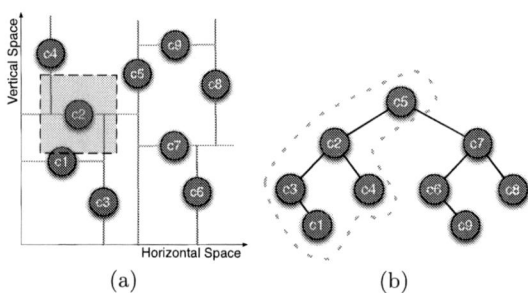

(a)   (b)

Figure 5: KD-Tree based region searching.

The details of our clustering procedure are shown in Algorithm 3. The clustering is repeated until the clustered candidate number reaches the clustering threshold. Initially all the candidates are sorted by $profit_i$, it means those candidates with more shot number reduction are tend to be clustered. Then clustering (lines 3-8) is carried out. For each candidate c_i, finding available c_j may need $O(n)$, and complexity of the horizontal clustering and vertical clustering are both $O(n^2)$. Then the complexity of the whole procedure is $O(n^2)$, where n is the number of candidates.

A KD-Tree [20] is used to speed-up the process of finding available pair (c_i, c_j). It provides fast $O(logn)$ region searching operations which keeping the time for insertion and deletion small: insertion, $O(logn)$; deletion of the root, $O(n(k-1)/k)$; deletion of a random node, $O(logn)$. Using KD-Tree, the complexity of the Algorithm 3 can be reduced to $O(nlogn)$. A simple example is shown in Fig. 5. For the sake of convenience, here for each candidate we only consider horizontal and vertical space. Given candidate c_2, to find another candidate with the similar space, it may need $O(n)$ to scan all other candidates. However, using the KD-Tree structure shown in Fig. 5(a), this finding procedure can be viewed as a region searching, which can be resolved in $O(logn)$. Particularly, as shown in Fig. 5(b), only candidates $c_1 - c_5$ are scanned.

4.2 Approximation Framework for 2D-OSP

In E-BLOW we adopt a simulated annealing based framework similar to that in [12]. To demonstrate the effectivity of our pre-filter and clustering methodologies, E-BLOW uses the same parameters with that in [12]. Sequence Pair [21] is used as a topology representation.

5. EXPERIMENTAL RESULTS

E-BLOW is implemented in the C++ programming language and executed on a Linux machine with two 3.0GHz

CPU and 32GB Memory. GUROBI [22] is used to solve linear programming. Eight benchmarks from [12] are tested. Besides, eight benchmarks (1M-x) are designed for 1D-OSP and the other eight (2M-x) are generated for the 2D-OSP problem. Character projection (CP) number are all set as 10. For small cases (1M-1, ..., 1M-4, 2M-1, ..., 2M-4) the character candidate number is 1000, and the stencil size is set as $1000\mu m \times 1000\mu m$. For larger cases (1M-5 , ..., 1M-8, 2M-5, ..., 2M-8) the character candidate number is 4000, and the stencil size is set as $2000\mu m \times 2000\mu m$. The size and blank width of each character is similar to that in [12].

5.1 Comparison for 1D-OSP

For 1D-OSP, Table 1 compares E-BLOW with greedy method and the heuristic framework in [12]. Note that the greedy method was also described in [12]. Column "char #" is number of character candidates, and column "CP#" is number of character projections. For each algorithm, we record "shot #", "char #" and "CPU(s)", where "shot #" is final number of shots and "char #" is number of characters on final stencil, "CPU(s)" reports the runtime. From table 1 we can see E-BLOW achieve better performance and runtime. Compared with E-BLOW, the greedy algorithm introduces 47% more shots number, and [12] would introduce 19% more shots number. Note that compared with heuristic method in [12], mathematical formulation can provide global view, even for traditional EBL system (1D-1, ..., 1D-4), E-BLOW achieves better shot number. Besides, E-BLOW can reduce 34.3% of runtime.

5.2 Comparison for 2D-OSP

For 2D-OSP, Table 2 gives the similar comparison. For each algorithm, we also record "shot #", "char #" and "CPU(s)", where the meanings are the same with that in Table 1. From the table we can see that for each test case, although the greedy algorithm is faster, its design results are not good that it would introduce 30% more shot number. Besides, compared with the work in [12], E-BLOW can achieve better performance that the shot number can be reduced by 14%. Meanwhile, because of the clustering method, E-BLOW can reach $2.8\times$ speed-up.

From both tables we can see that compared with [12], E-BLOW can achieve a better tradeoff between runtime and performance.

6. CONCLUSION

In this paper, we have proposed E-BLOW, a tool to solve OSP problem in MCC system. For 1D-OSP, a successive relaxation algorithm and a dynamic programming based refinement are proposed. For 2D-OSP, a KD-Tree based clustering method is integrated into simulated annealing framework. Experimental results show that compared with previous works, E-BLOW can achieve better performance in terms of shot number and runtime, for both MCC system and traditional EBL system. As EBL, including MCC system, are widely used for mask making and also gaining momentum for direct wafer writing, we believe a lot more research can be done for not only stencil planning, but also EBL aware design.

Acknowledgment

This work is supported in part by NSF and NSFC.

7. REFERENCES

[1] B. Yu, J.-R. Gao, D. Ding, Y. Ban, J.-s. Yang, K. Yuan, M. Cho, and D. Z. Pan, "Dealing with ic manufacturability in extreme scaling," in *IEEE/ACM International Conference on Computer-Aided Design (ICCAD)*, 2012, pp. 240–242.

[2] A. B. Kahng, C.-H. Park, X. Xu, and H. Yao, "Layout decomposition for double patterning lithography," in *IEEE/ACM International Conference on Computer-Aided Design (ICCAD)*, 2008, pp. 465–472.

[3] B. Yu, K. Yuan, B. Zhang, D. Ding, and D. Z. Pan, "Layout decomposition for triple patterning lithography," in *IEEE/ACM International Conference on Computer-Aided Design (ICCAD)*, 2011, pp. 1–8.

[4] K. Lucas, C. Cork, B. Yu, G. Luk-Pat, B. Painter, and D. Z. Pan, "Implications of triple patterning for 14 nm node design and patterning," in *Proc. of SPIE*, vol. 8327, 2012.

[5] Y. Arisawa, H. Aoyama, T. Uno, and T. Tanaka, "EUV flare correction for the half-pitch 22nm node," in *Proc. of SPIE*, vol. 7636, 2010.

[6] H. C. Pfeiffer, "New prospects for electron beams as tools for semiconductor lithography," in *Proc. of SPIE*, 2009.

[7] H. Yasuda, T. Haraguchi, and A. Yamada, "A proposal for an MCC (multi-column cell with lotus root lens) system to be used as a mask-making e-beam tool," in *Proc. of SPIE*, 2004.

[8] T. Maruyama, Y. Machida, S. Sugatani, H. Takita, H. Hoshino, T. Hino, M. Ito, A. Yamada, T. Iizuka, S. Komatsue, M. Ikeda, and K. Asada, "CP element based design for 14nm node EBDW high volume manufacturing," in *Proc. of SPIE*, 2012.

[9] M. Shoji, T. Inoue, and M. Yamabe, "Extraction and utilization of the repeating patterns for CP writing in mask making," in *Proc. of SPIE*, 2010.

[10] M. Sugihara, T. Takata, K. Nakamura, R. Inanami, H. Hayashi, K. Kishimoto, T. Hasebe, Y. Kawano, Y. Matsunaga, K. Murakami, and K. Okumura, "Cell library development methodology for throughput enhancement of character projection equipment," *IEICE Transactions on Electronics*, vol. E89-C, pp. 377–383, 2006.

[11] K. Yuan and D. Z. Pan, "E-Beam lithography throughput improvement with stencil planning and optimization," in *ACM International Symposium on Physical Design (ISPD)*, 2011.

[12] K. Yuan, B. Yu, and D. Z. Pan, "E-Beam lithography stencil planning and optimization with overlapped characters," *IEEE Transactions on Computer-Aided Design of Integrated Circuits and Systems (TCAD)*, vol. 31, no. 2, pp. 167–179, Feb. 2012.

[13] P. Du, W. Zhao, S.-H. Weng, C.-K. Cheng, and R. Graham, "Character design and stamp algorithms for character projection electron-beam lithography," in *IEEE/ACM Asia and South Pacific Design Automation Conference (ASPDAC)*, 2012.

[14] B. Yu, J.-R. Gao, and D. Z. Pan, "L-Shape based layout fracturing for e-beam lithography," in *IEEE/ACM Asia and South Pacific Design Automation Conference (ASPDAC)*, 2013.

[15] S. Manakli, H. Komami, M. Takizawa, T.Mitsuhashi, and L. Pain, "Cell projection use in mask-less lithography for 45nm & 32nm logic nodes," in *Proc. of SPIE*, 2009.

[16] S. Martello and P. Toth, *Knapsack problems: algorithms and computer implementations*. New York, NY, USA: John Wiley & Sons, Inc., 1990.

[17] C. Kenyon and E. Remila, "Approximate strip packing," in *Foundations of Computer Science, 1996. Proceedings., 37th Annual Symposium on*, oct 1996, pp. 31 –36.

[18] M. Dawande, J. Kalagnanam, P. Keskinocak, F. Salman, and R. Ravi, "Approximation algorithms for the multiple knapsack problem with assignment restrictions," *Journal of Combinatorial Optimization*, vol. 4, pp. 171–186, 2000.

[19] C. J. Alpert and A. B. Kahng, "Recent directions in netlist partitioning: a survey," *Integr. VLSI J.*, vol. 19, pp. 1–81, August 1995.

[20] J. L. Bentley, "Multidimensional binary search trees used for associative searching," *Commun. ACM*, vol. 18, pp. 509–517, September 1975.

[21] H. Murataand, K. Fujiyoshi, S. Nakatake, and Y. Kajitani, "VLSI module placement based on rectangle-packing by the sequence-pair," *IEEE Transactions on Computer-Aided Design of Integrated Circuits and Systems (TCAD)*, vol. 12, pp. 1518–1524, 1996.

[22] "GUROBI," http://www.gurobi.com/html/academic.html.

Table 1: Result Comparison for 1D-OSP

	char #	CP #	Greedy in [12]			[12]			E-BLOW		
			shot #	char #	CPU(s)	shot #	char #	CPU(s)	shot #	char #	CPU(s)
1D-1	1000	1	79193	876	0.2	50809	926	13.5	29536	934	2.2
1D-2	1000	1	122259	806	0.2	93465	854	11.8	44544	863	2
1D-3	1000	1	179822	708	0.2	152376	749	9.13	78704	758	2.7
1D-4	1000	1	223420	645	0.2	193494	687	7.7	107460	699	3.4
1M-1	1000	10	83786	876	0.2	53333	926	13.5	45243	938	4.3
1M-2	1000	10	123048	806	0.2	95963	854	11.8	81636	868	5.4
1M-3	1000	10	184950	708	0.2	156700	749	9.2	140079	769	10.8
1M-4	1000	10	225468	645	0.2	196686	687	7.7	179890	707	7.6
1M-5	4000	10	377864	3417	1.02	255208	3629	1477.3	227456	3650	59.2
1M-6	4000	10	542627	315	1.02	417456	3346	1182	373324	3388	65.1
1M-7	4000	10	760650	2809	1.02	644288	2986	876	570730	3044	58.68
1M-8	4000	10	930368	2565	1.01	809721	2734	730.7	734411	2799	65.3
Avg.	-	-	319454.6	1264.7	0.47	259958.3	1594.0	362.5	217751.1	1618.1	23.9
Ratio	-	-	**1.47**	0.78	**0.02**	**1.19**	0.99	**15.2**	**1.0**	1.0	**1.0**

Table 2: Result Comparison for 2D-OSP

	char #	CP #	Greedy in [12]			[12]			E-BLOW		
			shot #	char #	CPU(s)	shot #	char #	CPU(s)	shot #	char #	CPU(s)
2D-1	1000	1	159654	734	2.1	107876	826	329.6	105723	789	65.5
2D-2	1000	1	269940	576	2.4	166524	741	278.1	170934	657	52.5
2D-3	1000	1	290068	551	2.6	210496	686	296.7	178777	663	56.4
2D-4	1000	1	327890	499	2.7	240971	632	301.7	179981	605	54.7
2M-1	1000	1	168279	734	2.1	122017	811	313.7	91193	777	58.6
2M-2	1000	1	283702	576	2.4	187235	728	286.1	163327	661	48.7
2M-3	1000	1	298813	551	2.6	235788	653	289	162648	659	52.3
2M-4	1000	1	338610	499	2.7	270384	605	285.6	195469	590	53.3
2M-5	4000	10	824060	2704	19	700414	2913	3891	687287	2853	59
2M-6	4000	10	1044161	2388	20.2	898530	2624	4245	717236	2721	60.7
2M-7	4000	10	1264748	2101	21.9	1064789	2410	3925.5	921867	2409	57.1
2M-8	4000	10	824060	2704	18.9	1176700	2259	4550.0	1104724	2119	57.7
Avg.	-	-	507832.1	1218.1	8.3	448477	1324	1582.7	389930.5	1291.9	56.375
Ratio	-	-	**1.30**	0.94	**0.15**	**1.15**	1.02	**28.1**	**1.0**	1.0	**1.0**

Automatic Clustering of Wafer Spatial Signatures

Wangyang Zhang[1], Xin Li[1], Sharad Saxena[2], Andrzej Strojwas[1], Rob Rutenbar[3]

[1]ECE Department, Carnegie Mellon University, Pittsburgh, PA 15213
[2]PDF Solutions, 101 Renner Trail, Richardson, TX 75082
[3]CS Department, University of Illinois at Urbana-Champaign, Urbana IL 61801
{wyzhang, xinli, ajs}@ece.cmu.edu, sharad.saxena@pdf.com, rutenbar@illinois.edu

ABSTRACT

In this paper, we propose a methodology based on unsupervised learning for automatic clustering of wafer spatial signatures to aid yield improvement. Our proposed methodology is based on three steps. First, we apply sparse regression to automatically capture wafer spatial signatures by a small number of features. Next, we apply an unsupervised hierarchical clustering algorithm to divide wafers into a few clusters where all wafers within the same cluster are similar. Finally, we develop a modified L-method to determine the appropriate number of clusters from the hierarchical clustering result. The accuracy of the proposed methodology is demonstrated by several industrial data sets of silicon measurements.

1. INTRODUCTION

With the continued scaling of CMOS technology, process variation has become a critical issue for design and manufacture of integrated circuits [1]. Large-scale performance variability has been observed for integrated circuits at advanced technology nodes, resulting in significant yield loss. For this reason, reducing process variation to improve parametric yield is an extremely important task that is carried out throughout the lifecycle of any process and product.

In order to rapidly improve parametric yield, it is important to identify the key factors that significantly contribute to the yield loss [2]. To monitor the process characteristics, a number of test structures, such as ring oscillators [3] and transistor arrays [4], are placed within each chip or in the scribe line. An important observation is that different wafers may exhibit substantially different spatial signatures for the measurements of these test structures [5]. Different spatial signatures suggest that the underlying variation sources can be different for these wafers and, therefore, can be used to reveal a large number of yield-limiting factors, such as process shift/drift, mismatch between equipments, etc. If we can capture the spatial signature of each wafer with an accurate model, and further automatically partition all wafers into different groups based on such spatial signatures where each group carries a similar spatial signature, it would provide important insights to help process engineers for yield improvement. In particular, process engineers can rely on the information to prioritize different yield improvement strategies and focus on the variation sources associated with significant yield loss.

The problem of automatically grouping wafers with similar spatial signatures can be defined as a *clustering analysis* problem

Permission to make digital or hard copies of all or part of this work for personal or classroom use is granted without fee provided that copies are not made or distributed for profit or commercial advantage and that copies bear this notice and the full citation on the first page. To copy otherwise, to republish, to post on servers or to redistribute to lists, requires prior specific permission and/or a fee.
DAC'13, May 29 - June 07 2013, Austin, TX, USA.

in statistics. While clustering analysis has been extensively studied in the statistics community, a number of unique characteristics of our wafer clustering problem must be carefully considered in order to obtain accurate clustering results:

Large random variation: the performance measurements collected from test structures may be subject to large-scale random variation. As random variation becomes increasingly large with technology scaling [12], it obscures the spatial signature, thereby making the spatial signature non-trivial to identify.

Missing and outlier measurements: Defects in the manufacturing process, as well as measurement errors, may generate missing measurements. In this case, no data may be collected from a number of test structures, or outlier measurements collected from these test structures may significantly deviate from the regular variation range [7]. Meaningful clustering results cannot be obtained if the missing or outliner measurements are not properly handled.

Abnormal wafers: Because of equipment malfunction, there can be a small number of abnormal wafers whose spatial signatures are substantially different from the others [5]. We would like to automatically detect these abnormal wafers, rather than merging them into the main clusters. It, in turn, poses a unique challenge to the clustering algorithm, as will be explained in detail in Section 3.

Unknown number of clusters: Most clustering algorithms require knowing the number of clusters, or having user-defined parameters related to the number of clusters. In our wafer clustering application, the number of clusters cannot be known in advance. Therefore, additional efforts must be made to optimally determine the number of clusters from the measurement data.

Based on the aforementioned characteristics, we propose a new methodology for automatic clustering of wafer spatial signatures. Our proposed method consists of three steps. First, robust feature extraction based on sparse regression is performed on the measurement data, representing the spatial signature of each wafer by a small number of features. The impact of random variation is greatly reduced in our proposed feature space and, furthermore, the proposed feature extraction is extremely robust to missing and outlier measurements. Next, a clustering algorithm is performed on the extracted features. Since the number of clusters is not known in advance, the clustering algorithm does not directly generate the final clustering result. Instead, a set of possible clustering results are generated according to different settings of the clustering algorithm. Finally, a cluster selection algorithm is applied to automatically choose the optimal clustering result that best explains the data.

The remainder of the paper is organized as follows. In Section 2 we present our robust feature extraction algorithm, and then describe the clustering algorithm in Section 3. The algorithm for optimal cluster selection is presented in Section 4. The efficacy of our proposed method is demonstrated by several industrial examples in Section 5. Finally, we conclude in Section 6.

2. ROBUST FEATURE EXTRACTION

The goal of robust feature extraction is to represent the spatial signature of each wafer by a small number of features that minimize the impact of random variation, missing data and outlier measurements. We represent the parametric metric (e.g., ring oscillator frequency, leakage current, etc) measured from L wafers as a set of two-dimensional functions: $\{b_{(l)}(x, y); l = 1, 2, ..., L\}$, where l denotes the wafer label, and $x \in \{1, 2, ..., P\}$ and $y \in \{1, 2, ..., Q\}$ denote the spatial coordinates on the wafer. Each spatial variation function $b_{(l)}(x, y)$ contains two different components:

$$b_{(l)}(x, y) = s_{(l)}(x, y) + r_{(l)}(x, y) \quad (l = 1, 2, \cdots, L), \quad (1)$$

where $\{s_{(l)}(x, y); x = 1, 2, ..., P; y = 1, 2, ..., Q\}$ and $\{r_{(l)}(x, y); x = 1, 2, ..., P; y = 1, 2, ..., Q\}$ stand for the spatially correlated component and the uncorrelated random component, respectively. In order to reduce the impact of random variation, we would like to represent the spatial signature of each wafer by using its spatially correlated component only. Specifically, if the spatially correlated variation is modeled by the linear combination of λ basis functions:

$$b_{(l)}(x, y) = \sum_{j=1}^{\lambda} \eta_{(l),j} \cdot A_j(x, y) + r_{(l)}(x, y) \quad , \quad (2)$$

we define the features of the lth wafer as the following vector:

$$\eta_{(l)} = \begin{bmatrix} \eta_{(l),1} & \eta_{(l),2} & \cdots & \eta_{(l),\lambda} \end{bmatrix}^T. \quad (3)$$

By using the λ features in (3) to represent the spatial signature of the lth wafer, the uncorrelated random variation $r_{(l)}(x, y)$ would not impact the subsequent clustering process, and the clustering result would be made insensitive to random variation.

In practice, we do not know the wafer spatial signatures in advance. Therefore, we adopt the sparse regression idea in [8] to automatically select an appropriate set of basis functions from a dictionary that covers all possible spatial patterns. Such a dictionary can be customized by process engineers based on their prior knowledge. In the worst scenario, if no prior knowledge is available, a general dictionary containing Discrete Cosine Transform (DCT) [14] functions can be applied. The DCT functions are defined as:

$$A_{u,v}(x, y) = \alpha_u \cdot \beta_v \cdot \cos\frac{\pi(2x-1)(u-1)}{2 \cdot P} \cdot \cos\frac{\pi(2y-1)(v-1)}{2 \cdot Q} \quad , \quad (4)$$

$$(u = 1, 2, \cdots, P; v = 1, 2, \cdots, Q)$$

where

$$\alpha_u = \begin{cases} \sqrt{1/P} & (u = 1) \\ \sqrt{2/P} & (2 \leq u \leq P) \end{cases} \quad (5)$$

$$\beta_v = \begin{cases} \sqrt{1/Q} & (v = 1) \\ \sqrt{2/Q} & (2 \leq v \leq Q) \end{cases}. \quad (6)$$

The DCT coefficients, denoted as $\{\Box_{(l)}(u, v); u = 1, 2, ..., P; v = 1, 2, ..., Q\}$, represent the frequency-domain components of the spatial variation function $\{b_{(l)}(x, y); x = 1, 2, ..., P; y = 1, 2, ..., Q\}$. An important property of DCT is that if the spatial variation $b_{(l)}(x, y)$ exhibits a spatially correlated pattern, a vast majority of the DCT coefficients are close to zero. This unique property of sparseness has been observed in many image processing tasks and serves as the foundation of the compression algorithm for JPEG [14]. It has been recently explored by several works in the literature to model wafer-level spatial variation [6]-[8]. On the other hand, uncorrelated random variation can be characterized as white noise [8] and evenly distributed over all frequencies.

Therefore, the corresponding DCT coefficients are relatively small. It, in turn, implies that spatially correlated variation can be accurately represented by a small number of dominant DCT coefficients.

After selecting the dictionary of basis functions, the following sparse regression [8] is formulated to generate the features:

$$\begin{array}{ll} \underset{\eta_{(l)}}{\text{minimize}} & \left\| A_{(l)} \cdot \eta_{(l)} - B_{(l)} \right\|_2^2 \\ \text{subject to} & \left\| \eta_{(l)} \right\|_0 \leq \lambda \end{array}, \quad (7)$$

where $\|\bullet\|_2$ and $\|\bullet\|_0$ stand for the L2-norm (i.e., the square root of the summation of the squares of all elements) and the L0-norm (i.e., the number of non-zero elements) of a vector respectively,

$$B_{(l)} = \begin{bmatrix} b_{(l)}(x_{(l),1}, y_{(l),1}) & \cdots & b_{(l)}(x_{(l),N_{(l)}}, y_{(l),N_{(l)}}) \end{bmatrix}^T \quad (8)$$

represents the measurement data collected from $N_{(l)}$ different spatial locations $\{(x_{(l),i}, y_{(l),i}); i = 1, 2, ..., N_{(l)}\}$ of the lth wafer,

$$\eta_{(l)} = \begin{bmatrix} \eta_{(l),1} & \cdots & \eta_{(l),M} \end{bmatrix}^T \quad (9)$$

represents the unknown coefficients corresponding to the M basis functions in the dictionary. If the DCT dictionary is applied, the total number of basis functions (i.e., M) is equal to PQ. The matrix $A_{(l)}$ in (7) is $N_{(l)}$-by-M and is defined as:

$$A_{(l)} = \begin{bmatrix} A_{(l),1,1} & A_{(l),1,2} & \cdots & A_{(l),1,M} \\ A_{(l),2,1} & A_{(l),2,2} & \cdots & A_{(l),2,M} \\ \vdots & \vdots & \vdots & \vdots \\ A_{(l),N_{(l)},1} & A_{(l),N_{(l)},2} & \cdots & A_{(l),N_{(l)},M} \end{bmatrix}, \quad (10)$$

where $A_{(l),i,j}$ corresponds to the value of the jth basis function for the ith measurement on the lth wafer. The optimization in (7) attempts to use a small number of (i.e., λ) dominant basis functions to approximate the spatially correlated variation of the lth wafer. It can be efficiently solved by the numerical algorithm developed in [8] where the optimal value of λ can be automatically determined using cross-validation. Once the optimization in (7) is solved, the resulting λ dominant coefficients become the features of each wafer in (3). More details on sparse regression can be found in [6]-[8].

The sparse regression approach is extremely robust to missing measurements [6]-[8]. Moreover, it can be further made insensitive to measurement outliers by applying advanced outlier detection techniques [7], [15]. Therefore, by performing sparse regression to extract the important features, the subsequent clustering process can be appropriately shielded from the missing and outlier measurements, as will be discussed in detail in the following sections.

3. CLUSTERING ALGORITHM

After the features describing the wafer signatures are extracted, the next step is to apply a clustering algorithm to partition all wafers into multiple clusters. Many clustering algorithms have been proposed in the statistics community, such as k-means clustering [9], density-based clustering [10] and hierarchical clustering [13]. Each algorithm is based upon a specific assumption of the data and not all these algorithms are suitable for our wafer clustering application.

The traditional k-means method partitions the data into K clusters that minimize the following cost function:

$$\sum_{i=1}^{K} \sum_{l \in c_i} \left\| \eta_{(l)} - \mu_i \right\|_2^2, \quad (11)$$

where c_i denotes the index set of the wafers belonging to the ith cluster, $\eta_{(l)}$ is the feature vector of the lth wafer, and μ_i is the centroid of the ith cluster:

$$\mu_i = \frac{1}{|c_i|} \cdot \sum_{l \in c_i} \eta_{(l)} , \qquad (12)$$

where $|c_i|$ stands for the size of the set c_i. The number of clusters (i.e., K) is a parameter that must be specified by the user in advance. The clustering result can be efficiently found by the Expectation-Maximization (EM) algorithm [13] developed by the statistics community.

However, for our wafer clustering application, an important problem that prevents k-means from achieving accurate results is the existence of abnormal wafers. Abnormal wafers are a small number of wafers whose spatial signatures are substantially different from any of the main clusters, typically because of equipment malfunction. An ideal clustering algorithm should result in a number of separate clusters with very small sizes to reflect the abnormal wafers, rather than merging them into the main clusters. However, detecting such small clusters is often not possible with the k-means clustering algorithm [13], as will be demonstrated by our experimental results in Section 5.

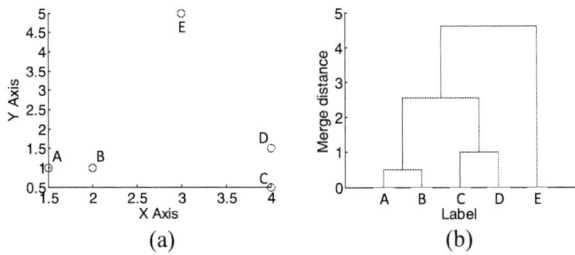

(a)   (b)

Figure 1. (a) A synthetic two-dimensional data set with 5 points $\{A, B, C, D, E\}$. (b) The dendrogram generated by hierarchical clustering.

An alternative algorithm that does not suffer from the aforementioned problem is hierarchical clustering [13]. Unlike the k-means method which explicitly minimizes a cost function, hierarchical clustering builds clusters in a greedy manner. Suppose that there are N data points in total. Hierarchical clustering first assigns an individual cluster for each point. Next, $N-1$ merging steps are performed iteratively, where two clusters that are closest in distance are merged in each step. Namely, the data points that are close will be merged first, and those that are far away will not be merged until the end of the iteration process.

To intuitively explain the idea of hierarchical clustering, we construct a synthetic data set with 5 points shown in Figure 1(a). When hierarchical clustering is applied, the first iteration step merges the data points A with B and the second iteration step merges the data points C with D. Next, the two clusters containing $\{A, B\}$ and $\{C, D\}$ respectively are merged in the third iteration step. Finally, the data point E is merged with $\{A, B, C, D\}$ in the last iteration step. The clustering result can be represented as a *dendrogram* shown in Figure 1(b).

Studying Figure 1(b) reveals an important fact that the height of each node reflects its *merge distance*, meaning the distance of two clusters that are merged to form this node. Note that close data points should be merged early, while distant data points will not be merged until the end. However, hierarchical clustering does not directly generate the final clustering result (i.e. the cluster labels for each data point). We will further discuss the algorithm to determine the cluster labels in Section 4.

An important component that must be determined for

hierarchical clustering is how to define the distance between clusters. While the distance between two individual data points can be simply defined by their Euclidean distance:

$$dist\big(\eta_{(l)}, \eta_{(k)}\big) = \big\|\eta_{(l)} - \eta_{(k)}\big\|_2 , \qquad (13)$$

the definition of distance between two clusters containing multiple data points is not unique. Different definitions of cluster distance have been proposed, and each of them corresponds to a different assumption about the cluster structure. We need to select the distance definition that best matches our goal for wafer clustering. In this work, the following definition is used, which calculates the distance between two clusters as the maximal distance between any two data points in the clusters:

$$dist\big(c_l, c_k\big) = \sup_{i \in c_l, j \in c_k} \big\|\eta_{(i)} - \eta_{(j)}\big\|_2 , \qquad (14)$$

where $\sup(\bullet)$ denotes the supremum (i.e., the least upper bound) of a set. The hierarchical clustering algorithm based on the distance metric in (14) is referred to as *complete-link hierarchical clustering* [13]. The physical meaning of (14) is that a cluster will be formed if and only if *all* members in the cluster are completely connected, i.e. within a small distance to each other. This definition matches our goal for wafer clustering: since all wafers in the same cluster should carry the same spatial signature, we want these wafers to be similar to each other.

To further explain why Eq. (14) is an appropriate choice for our application, we compare it with another commonly used definition based on minimal distance:

$$dist\big(c_l, c_k\big) = \inf_{i \in c_l, j \in c_k} \big\|\eta_{(i)} - \eta_{(j)}\big\|_2 , \qquad (15)$$

where $\inf(\bullet)$ denotes the infimum (i.e., the greatest lower bound) of a set. The hierarchical clustering algorithm based on the distance metric in (15) is referred to as *single-link hierarchical clustering* [13]. The assumption behind single-link hierarchical clustering is that two data points should belong to the same cluster, as long as there exists a path connecting these two data points such that any adjacent pair of points along this path is close in distance. As a result, single-link hierarchical clustering often generates elongated clusters, where distant data points are connected by a long path in between. While this type of cluster is suitable for many practical applications, it is undesirable for our wafer clustering. For instance, the change in process condition may not occur abruptly during the manufacturing process, but gradually drift from one state to another. Such a drift can happen because of, for example, equipment aging [16]. By employing single-link hierarchical clustering, we are unable to split the wafers into different clusters to reflect the drift of process condition. Note that several other clustering techniques, e.g., density-based clustering [10], are also based on the idea of forming a connecting path to define clusters. Therefore, they are not suitable for our wafer clustering application either.

In summary, complete-link hierarchical clustering can naturally break down a long string of data points into small clusters and, therefore, it is most suitable for our application in this paper. In Section 5, we will show several examples where the correct clusters detected by complete-link hierarchical clustering cannot be found by either single-link hierarchical clustering or k-means clustering.

4. CLUSTER SELECTION

In the previous section, we propose to apply complete-link hierarchical clustering for our application. However, as previously discussed, hierarchical clustering does not directly generate the cluster labels. To achieve automatic clustering and minimize human efforts, a cluster selection algorithm must be developed to automatically choose the appropriate cluster labels from the hierarchical clustering result.

The traditional approach to select the clusters from the hierarchical clustering result is based on the *inconsistency coefficient method* [17]. It visits each node in the dendrogram and compares its merge distance with the average merge distance of all nodes below it. The difference is quantitatively defined by the following inconsistency coefficient:

$$I_k = \frac{d_k - \mu_k}{\sigma_k}, \tag{16}$$

where I_k represents the inconsistency coefficient of the kth node, d_k is the merge distance of the kth node, μ_k is the average merge distance of the kth node and all nodes below it, and σ_k is the standard deviation of the merge distances of the kth node and all nodes below it. The nodes with inconsistency coefficient higher than a user-defined threshold are broken, yielding distinct clusters. This threshold value is often empirically assigned, and its optimal value can vary significantly over different applications or even different data sets. On the other hand, the clustering result is extremely sensitive to the aforementioned threshold value. Hence, it is non-trivial to develop a fully automatic clustering process based on the inconsistency coefficient method.

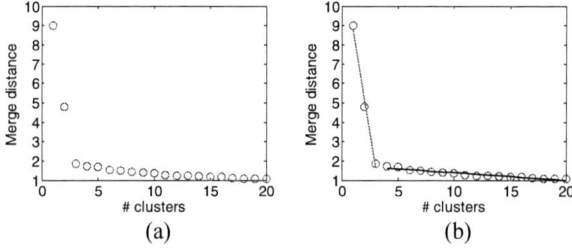

(a)   (b)

Figure 2. (a) The error curve of complete-link hierarchical clustering for a synthetic data set. (b) The optimal number of clusters can be found by fitting the curve with two lines.

An alternative approach to select the number of clusters that has gained popularity in recent years is based on the L-method [11]. The L-method is derived from the fact that for many clustering algorithms, it is possible to plot an error curve where the x-axis is the number of clusters and the y-axis is the evaluation metric internally used by the clustering algorithm. For hierarchical clustering, the evaluation metric of having i clusters is defined as the merge distance of the $(N-i)$-th merge [11]. While the error curve generally presents a decreasing trend, it typically has a sharp transition at the optimal clustering setup. For example, Figure 2(a) plots the error curve of complete-link hierarchical clustering for a synthetic data set with three clusters. It can be seen that the transition point is at $x = 3$. The L-method attempts to match human intuition by defining the criterion that if we find two consecutive lines that optimally fit the error curve, the intersection point of the two lines determines the transition point of the error curve. For example, Figure 2(b) accurately fits the error curve by two lines where one line fits the data within $x \in [1, 3]$ and the other line fits the data within $x \in [4, 20]$. It, in turn, determines that the data set should be partitioned into 3 clusters.

In what follows, we will first describe the mathematic formulation of the L-method and then discuss its limitation. Consider an error curve such as Figure 2(a) where the value of the

x-axis varies from $x = 1$ to $x = B$. We partition the data points into the left and right sequences at $x = c$. The left sequence has the data points with $x \in \{1, ..., c\}$ and the right sequence has the data points with $x \in \{c+1, ..., B\}$. Next, we find two optimal lines that minimize the mean-squares error to fit the left and right parts of the error curve respectively:

$$\underset{a_{lc}, b_{lc}}{\text{minimize}} \quad \left\| y_{lc} - a_{lc} - b_{lc} \cdot x_{lc} \right\|_2^2 \tag{17}$$

$$\underset{a_{rc}, b_{rc}}{\text{minimize}} \quad \left\| y_{rc} - a_{rc} - b_{rc} \cdot x_{rc} \right\|_2^2, \tag{18}$$

where

$$x_{lc} = \begin{bmatrix} 1 & 2 & \cdots & c \end{bmatrix}^T \tag{19}$$

$$x_{rc} = \begin{bmatrix} c+1 & c+2 & \cdots & B \end{bmatrix}^T, \tag{20}$$

y_{lc} and y_{rc} are the values of the evaluation metric at x_{lc} and x_{rc} respectively. Eq. (17) and (18) can be solved by least-squares fitting, yielding the following root-mean-squared error:

$$RMSE_{lc} = \frac{1}{\sqrt{c}} \cdot \left\| y_{lc} - a_{lc} - b_{lc} \cdot x_{lc} \right\|_2 \tag{21}$$

$$RMSE_{rc} = \frac{1}{\sqrt{B-c}} \cdot \left\| y_{rc} - a_{rc} - b_{rc} \cdot x_{rc} \right\|_2, \tag{22}$$

where a_{lc} and b_{lc} are the solution of (17), and a_{rc} and b_{rc} are the solution of (18). The total root-mean-squared error associated with the transition point $x = c$ is defined as the weighted sum of $RMSE_{lc}$ and $RMSE_{rc}$:

$$RMSE_c = \frac{c}{B} RMSE_{lc} + \frac{B-c}{B} RMSE_{rc}. \tag{23}$$

The optimal number of clusters is then defined by selecting the c value that minimizes the total error in (23):

$$\underset{c}{\text{minimize}} \quad RMSE_c. \tag{24}$$

In practice, the number of clusters is often much smaller than the number of data points. Therefore, when directly applying the criterion (24) to the entire data set, a large number of possible c values corresponding to extremely fine-grain clusters are irrelevant and may lead to an inaccurate result due to highly imbalanced left and right sequences. Therefore, the L-method is applied iteratively to the error curve. Starting from solving (24) for the entire error curve, each iteration step reduces the number of data points included in the next iteration to:

$$B_{next} = \max(2 \cdot c, 20), \tag{25}$$

where c is the optimal number of clusters determined in the current iteration step, and $2 \cdot c$ is the number to keep the left and right sequences balanced. The total number of data points is not permitted to drop below 20, which is an empirical number proposed in [11] in order to keep a reasonable number of data points to fit the lines. The L-method stops when the number of data points does not change over two successive iteration steps.

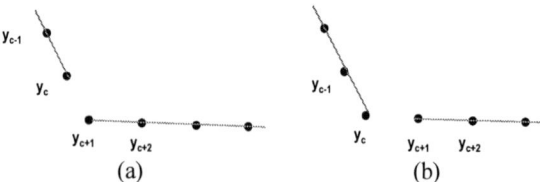

(a)   (b)

Figure 3. An synthetic example where the error can be minimized by either (a) $c = 2$ or (b) $c = 3$.

While the L-method attempts to match human intuition in finding the transition point of the error curve, we notice that its definition of the transition point is counter-intuitive. To explain its limitation, we construct a synthetic example in Figure 3, where the optimal solution with human inspection is $c = 3$. However, Figure 3(a) and (b) show the results fitted by setting $c = 2$ and $c = 3$ respectively. It can be seen that both solutions yield extremely small error. Therefore, the choice of $c = 2$ or $c = 3$ by the L-method is arbitrary in this example.

Based on the aforementioned observation, we propose to add a post-processing step to the traditional L-method to accurately determine the number of clusters. The key idea is to detect if a sharp transition occurs at the current point c or the next point $c+1$. The number of clusters is added by one, if the next point causes a sharp transition in the error curve. Specifically, we propose to use the following quantity to measure the transition rate:

$$s(c) = \left[\log(y_{c+1}) - \log(y_c)\right] - \left[\log(y_c) - \log(y_{c-1})\right], \quad (26)$$

where y_{c-1}, y_c and y_{c+1} are the values of the evaluation metric at $x = c-1$, $x = c$ and $x = c+1$, respectively. The number of clusters is increased by one, if $s(c+1)$ is greater than $s(c)$. Eq. (26) is essentially the second-order difference of the data series $\log(y)$. A large second-order difference means a significant change in the slope, thereby indicating an abrupt transition of the error curve. We take the logarithm for the evaluation metric y, because comparing the ratio between two consecutive data points is more intuitive than comparing their absolute difference. We summarize the major steps of the modified L-method for cluster selection in Algorithm 1.

Algorithm 1: Modified L-method for cluster selection

1. Start from a vector $y \in R^B$ representing the value of the evaluation metric associated with $x \in \{1, 2, ..., B\}$.
2. Find the optimal number of clusters (i.e., c) according to the criterion (24).
3. Compare $s(c)$ and $s(c+1)$ defined by (26). If $s_{c+1} > s_c$, then $c = c + 1$.
4. Calculate the value of B_{next} by (25) to determine the number of data points that should be included in the next iteration step.
5. If $B_{next} = B$, stop iteration. Otherwise, set $B = B_{next}$ and go to Step 1.

Note that Algorithm 1 is not restricted to hierarchical clustering only. Instead, it can be applied to select the optimal number of clusters for any clustering algorithm where the error curve (i.e., the evaluation metric vs. the number of clusters) can be generated. For instance, Algorithm 1 can be successfully applied to k-means clustering, where the evaluation metric is defined by the cost function in (11).

5. EXPERIMENTAL EXAMPLES

In the previous sections, we have proposed a wafer clustering methodology that mainly consists of three components: (i) robust feature extraction, (ii) complete-link hierarchical clustering, and (iii) cluster selection. In this section, we demonstrate the efficacy of the proposed methodology based on several industrial data sets for a commercial CMOS process below 90nm.

First, we consider the measurement data of NMOS drain saturation current (I_{dsat}) collected by the scribe-line test structures from 69 wafers. We apply the proposed methodology to cluster these wafers. In this example, four clusters, referred as Cluster "a", "b", "c" and "d", are identified where the numbers of wafers belonging to these four clusters are 34, 23, 9 and 3, respectively. Figure 4 shows the averaged wafer map for the four clusters. It can be seen that these clusters indeed contain distinct spatial

signatures. In particular, the wafers in Cluster "a" do not carry significant spatially correlated variation. The wafers in Cluster "b" present a strong edge effect and an increasing trend from the top-left corner to the bottom-right corner. The wafers in Cluster "c" have strong edge and center effects. Finally, the wafers in Cluster "d" have a large number of missing measurements at the bottom of the wafer.

In this example, the aforementioned spatial signatures cannot be accurately detected by k-means clustering or single-link hierarchical clustering. K-means clustering only detects three clusters where Cluster "b" and "c" in Figure 4 are merged into a single cluster. Therefore, it fails to detect the different spatial signatures presented by these two clusters. On the other hand, if single-link hierarchical clustering is applied, Cluster "a", "b" and "c" in Figure 4 are merged into one cluster. The fundamental reason is that there does not exist a clear boundary between these three clusters. To further validate this reason, we plot three different wafer maps in our data set in Figure 5. It can be seen that while there exist substantially different spatial signatures between Figure 5(a) and Figure 5(c), Figure 5(b) has a spatial signature that is similar to both Figure 5(a) and Figure 5(c). This observation may occur because of, for example, process drift. In this case, single-link hierarchical clustering will merge Figure 5(a) and Figure 5(c) into the same cluster because they are connected by Figure 5(b). Complete-link hierarchical clustering, however, requires all wafers in the same cluster to be similar and, therefore, does not suffer from this issue. Finally, we also verify that no satisfactory clustering result can be generated, if the inconsistency coefficient method is applied for cluster selection.

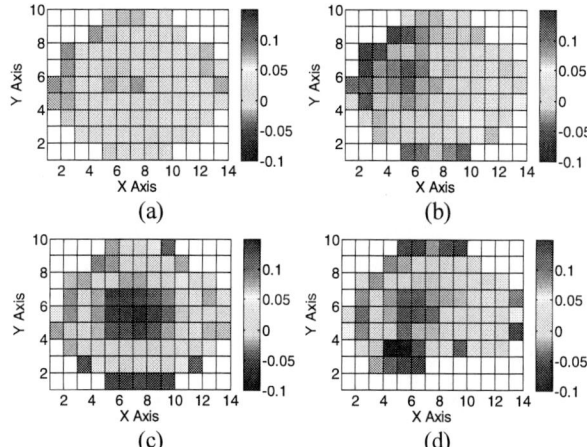

Figure 4. Averaged wafer maps (normalized) of four different clusters detected by the proposed methodology for the first measurement data set of drain saturation current (I_{dsat}).

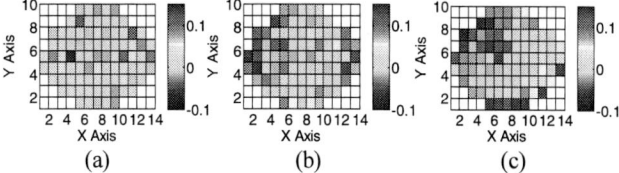

Figure 5. Three different wafers from the first measurement data set of drain saturation current (I_{dsat}).

Next, we further consider the I_{dsat} measurements from another data set with 82 wafers. In this data set, not all test structures are measured; instead, they are sampled in a "checkerboard" style to reduce the test cost. In this example, the proposed methodology again generates four clusters, referred to as Cluster "a", "b", "c" and "d". The numbers of wafers belonging to these four clusters are 43, 18, 20 and 1, respectively. Figure 6 shows the averaged wafer map for these four clusters. Inspecting Figure 6, it can be seen that although Cluster "b" presents larger spatially correlated variation than Cluster "a", the difference in spatial signature between these two clusters is not significant. Therefore, they can be simply merged into one cluster after manually inspecting Figure 6(a) and Figure 6(b). Note that even though the clustering result does not exactly match human intuition in this example, the aforementioned manual inspection requires little human effort. Cluster "c" and "d" detected by the proposed method carry completely different spatial signatures compared to Cluster "a" and "b". Namely, the wafers in Cluster "c" have a significant edge effect at the bottom-left corner of the wafer, and Cluster "d" contains an abnormal wafer with a completely different spatial signature compared to other wafers.

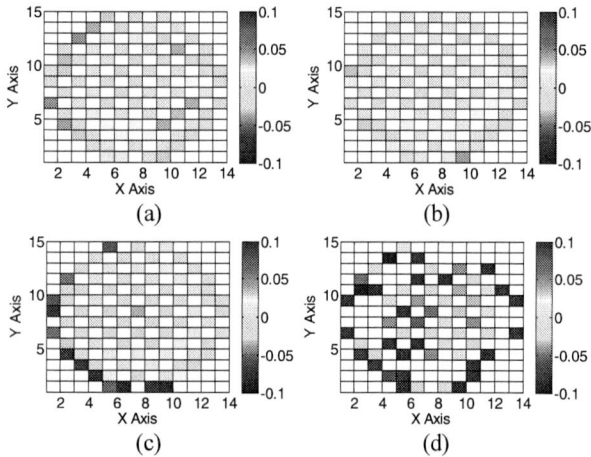

(a)   (b)

(c)   (d)

Figure 6. Averaged wafer maps (normalized) of four different clusters detected by the proposed methodology for the second measurement data set of drain saturation current (I_{dsat}).

In this example, applying k-means clustering or single-link hierarchical clustering fails to detect all the distinct signatures in Figure 6. In particular, k-means clustering generates two clusters where Cluster "a", "b" and "d" in Figure 6 are merged into one cluster and Cluster "c" in Figure 6 forms a separate cluster. While the k-means method merges Cluster "a" and "b", the abnormal wafer is also merged into this large cluster and cannot be detected by simple inspection. On the other hand, single-link hierarchical clustering generates two clusters where Cluster "a", "b" and "c" in Figure 6 are merged into one cluster and Cluster "d" in Figure 6 forms a separate cluster. Therefore, it fails to detect the wafers with edge effect. For these reasons, the proposed methodology with complete-link hierarchical clustering provides the best accuracy in this example.

6. CONCLUSIONS

In this paper, we develop an accurate three-step methodology for automatic clustering of wafer spatial signatures. First, the spatial signatures are automatically captured by a small number of features based on sparse regression. Second, complete-link hierarchical clustering is performed on the extracted features. Finally, a modified L-method is performed on the hierarchical clustering result for cluster selection. The efficacy of the proposed

methodology, as well as its superior accuracy over other alternative approaches, has been demonstrated by a number of industrial data sets of silicon measurements. In our future work, we will further apply the clustering results to identify the critical factors for yield improvement.

7. ACKNOWLEDGEMENTS

This work has been supported in part by the National Science Foundation.

8. REFERENCES

[1] Semiconductor Industry Associate, *International Technology Roadmap for Semiconductors*, 2011.

[2] N. Kupp, M. Slamani and Y. Makris, "Correlating inline data with final test outcomes in analog/RF devices," *IEEE DATE*, 2011.

[3] M. Bhushan, A. Gattiker, M. Ketchen and K. Das, "Ring oscillators for CMOS process tuning and variability control," *IEEE Trans. Semiconductor Manufacturing*, vol. 19, no. 1, pp. 10-18, Feb. 2006.

[4] S. Saxena, C. Hess, H. Karbasi, A. Rossoni, S. Tonello, P. McNamara, S. Lucherini, S. Minehane, C. Dolainsky and M. Quarantelli "Variation in transistor performance and leakage in nanometer-scale technologies," *IEEE Trans. Electron Devices*, vol. 55, pp. 131-144, Jan. 2008.

[5] A. Strojwas, "Conquering process variability: A key enabler for profitable manufacturing in advanced technology nodes," *IEEE International Symposium on Semiconductor Manufacturing*, pp. xxiii-xxxii, 2006.

[6] X. Li, R. Rutenbar and R. Blanton, "Virtual probe: a statistically optimal framework for minimum-cost silicon characterization of nanoscale integrated circuits," *IEEE ICCAD*, pp. 433-440, 2009.

[7] W. Zhang, X. Li, E. Acar, F. Liu and R. Rutenbar, "Multi-wafer virtual probe: minimum-cost variation characterization by exploring wafer-to-wafer correlation," *IEEE ICCAD*, pp. 47-54, 2010.

[8] W. Zhang, K. Balakrishnan, X. Li, D. Boning and R. Rutenbar, "Toward efficient spatial variation decomposition via sparse regression," *IEEE ICCAD*, pp. 162-169, 2011.

[9] J. MacQueen, "Some methods for classification and analysis of multivariate observations," *Berkeley Symposium on Mathematical Statistics and Probability*, pp. 281-297, 1967.

[10] M. Ester, H. Kriegel, J. Sander and X. Xu, "A density-based algorithm for discovering clusters in large spatial databases with noise," *International Conference on Knowledge Discovery and Data Mining*, pp. 226–231, 1996.

[11] S. Salvador and P. Chan, "Determining the number of clusters/segments in hierarchical clustering/segmentation algorithms," *International Conference on Tools with AI*, pp. 576-584, 2004.

[12] M. Orshansky, S. Nassif and D. Boning, *Design for Manufacturability and Statistical Design: A Constructive Approach*, Springer, 2007.

[13] P. Tan, M. Steinbach and V. Kumar, *Introduction to Data Mining*. Addison-Wesley, 2006.

[14] R. Gonzalez and R. Woods, *Digital Image Processing*, Prentice Hall, 2007.

[15] R. Maronna, R. Martin and V. Yohai, *Robust Statistics: Theory and Methods*, John Wiley and Sons, 2006.

[16] G. May and C. Spanos, *Fundamentals of Semiconductor Manufacturing and Process Control*, Wiley-IEEE Press, 2006.

[17] A. Jain and R. Dubes, *Algorithms for Clustering Data*, Prentice Hall, 1988.

Multidimensional Analog Test Metrics Estimation Using Extreme Value Theory and Statistical Blockade

Haralampos-G. Stratigopoulos
TIMA Laboratory (CNRS - Grenoble INP - UJF)
46 Av. Félix Viallet
38000 Grenoble, France
stratigo@imag.fr

Pierre Faubet
Infiniscale S.A.
186 chemin de l'Etoile
38330 Grenoble-Montbonnot, France
faubet@infiniscale.com

Yoann Courant
Infiniscale S.A.
186 chemin de l'Etoile
38330 Grenoble-Montbonnot, France
courant@infiniscale.com

Firas Mohamed
Infiniscale S.A.
186 chemin de l'Etoile
38330 Grenoble-Montbonnot, France
mohamed@infiniscale.com

ABSTRACT

The high cost of testing certain analog, mixed-signal, and RF circuits has driven in the recent years the development of alternative low-cost tests to replace the most costly or even all standard specification tests. However, there is a lack of solutions for evaluating the parametric test error, that is, the test error for circuits with process variations, resulting from this replacement. For this reason, test engineers are often reluctant to adopt alternative tests since it is not guaranteed that test cost reduction is not achieved at the expense of sacrificing test quality. In this paper, we present a technique to estimate the parametric test error fast and reliably with parts per million accuracy. The technique is based on extreme value theory and statistical blockade. Relying on a small number of targeted simulations, it is capable of providing accurate estimates of parametric test error in the general scenario where a set of alternative tests replaces all or a subset of standard specification tests.

Categories and Subject Descriptors

B.3.7 [**Integrated Circuits**]: Reliability and Testing—*test generation*

General Terms

Algorithms, Design, Measurement

Keywords

Analog/mixed-signal/RF IC testing, test metrics, extreme value theory, statistical blockade

Permission to make digital or hard copies of all or part of this work for personal or classroom use is granted without fee provided that copies are not made or distributed for profit or commercial advantage and that copies bear this notice and the full citation on the first page. To copy otherwise, to republish, to post on servers or to redistribute to lists, requires prior specific permission and/or a fee.
DAC'13, May 29 - June 07 2013, Austin, TX, USA.

1. INTRODUCTION

For certain classes of analog, mixed-signal, and RF circuits, manufacturing testing incurs a very high cost that amounts to a significant fraction of the overall manufacturing cost. To reduce the manufacturing test cost for such circuits, researchers and practitioners are continuously proposing alternative test approaches, including built-in test techniques [1, 2], where low-cost, information-rich, on-chip measurements are replacing the standard specification tests, and specification test compaction techniques [3, 4], where only a low-cost subset of specification tests are carried out and are subsequently used to infer the outcome of the rest of the specification tests. However, despite the large number and variety of alternative test approaches proposed to date, specification testing is still the prevalent approach in industry. The primary reason is that it is hard to evaluate alternative test approaches in terms of the test error that they may result in. Therefore, test engineers are reluctant to adopt such approaches since, although they may succeed to reduce test cost drastically, this may be at the expense of increased test error.

The test error may occur from accepting a circuit that fails one or more specifications, in which case we refer to *test escape*, or from failing a circuit that satisfies all the specifications, in which case we refer to *yield loss*. Furthermore, the test escape could relate to circuits that contain defects or to circuits with excessive process variations, in which case we refer to *parametric test escape*. While the capability of an alternative test to detect defects can be readily assessed given a list of probable defects, the estimation of the *parametric test error* which includes parametric test escape and yield loss is not straightforward [5].

Formally, let $P = [P_1, P_2, \cdots, P_{n_p}]$ denote the vector of performances of a circuit and let $[s_\ell^i, s_u^i]$ denote the acceptance specification limits for performance P_i, e.g. the performance acceptability region is $A_P = [s_\ell^1, s_u^1] \times \cdots \times [s_\ell^{n_p}, s_u^{n_p}]$. The standard specification tests consist of measuring directly P and accepting the circuit as functional if $P \in A_P$. Let also $T = [T_1, T_2, \cdots, T_{n_t}]$ denote a vector of alternative tests and let $[t_\ell^i, t_u^i]$ denote the test limits for test T_i, e.g. the test acceptability region is $A_T = [t_\ell^1, t_u^1] \times \cdots \times [t_\ell^{n_t}, t_u^{n_t}]$. The set T could be a subset of low-cost specification tests or could include other low-cost, non-specification tests. In this case, a circuit is deemed to be functional if $T \in A_T$.

Parametric test escape is the probability that a circuit is faulty due to process variations when it has actually been labeled as functional by the alternative tests

$$
\begin{aligned}
T_E &= \Pr\{P \notin A_P | T \in A_T\} \qquad (1) \\
&= \Pr\{\cup_{i=1}^{n_p} P_i \notin [s_\ell^i, s_u^i] \big| \cap_{i=1}^{n_t} T_i \in [t_\ell^i, t_u^i]\}.
\end{aligned}
$$

Yield loss is the probability of a circuit failing the alternative tests and being labeled as faulty when it is actually functional

$$
\begin{aligned}
Y_L &= \Pr\{T \notin A_T | P \in A_P\} \qquad (2) \\
&= \Pr\{\cup_{i=1}^{n_t} T_i \notin [t_\ell^i, t_u^i] \big| \cap_{i=1}^{n_p} P_i \in [s_\ell^i, s_u^i]\}.
\end{aligned}
$$

As is readily seen from (1) and (2), the two parametric *test metrics* T_E and Y_L are similar from a statistical point of view. This means that a technique to estimate T_E equally applies for the estimation of Y_L and vice versa. When the set T comprises only specification tests, e.g. when $T \subset P$, then $Y_L = 0$. Notice also that when in the set T there exist non-specification tests, then T_E and Y_L are contradictory objectives. By shrinking the test limits of the non-specification tests we will achieve a reduction of T_E, but simultaneously we may increase Y_L, and vice versa, by enlarging the test limits of the non-specification tests we reduce Y_L, but we may increase T_E.

In this paper, we discuss a method for estimating the test metrics in (1) and (2). The rest of the paper is structured as follows. Section 2 discusses the challenges and previous work. Section 3 shows the use of extreme value theory [6, 7] to derive an analytical model for a test metric and explains how to fit the parameters of the model. Section 4 shows the use of the statistical blockade technique [8, 9] to generate the required data for the fitting. Section 5 demonstrates the proposed method on an industrial case study. Section 6 concludes the paper.

2. CHALLENGE AND PREVIOUS WORK

The straightforward approach to estimate T_E and Y_L is to carry out both the alternative and the specification tests for a large number of fabricated circuits and then express T_E and Y_L with relative frequencies. This approach entails the risk of concluding too late that the alternative tests result in unacceptable T_E or Y_L. In this case, we will have wasted significant resources, increase test cost for a large number of circuits, etc., only to realize that a new low-cost test solution has to be devised from scratch. Thus, such an analysis needs to be carried out during the alternative test development phase based on a simulation campaign, in order to obtain a quick indication about the efficiency of the alternative tests.

The challenge in estimating T_E and Y_L using simulation is to keep the simulation time down at reasonable levels. The problem stems from the fact that T_E and Y_L have values that are typically in the order of a few hundreds parts per million (ppm). In other words, the probabilities in (1) and (2) are very small. Therefore, to obtain accurate estimates

of T_E and Y_L, it is required to run a very large number of simulations that is prohibitive for most practical circuits. Formally, let T_m be a test metric such as T_E or Y_L. If $\hat{T}_m$ is a Monte Carlo estimate of T_m obtained using N_{mc} simulations, then it is easy to show [10] that $\hat{T}_m$ is distributed as $\mathcal{N}(T_m, \sigma_{mc}^2)$ with variance

$$
\sigma_{mc}^2 = T_m(1 - T_m)/N_{mc}. \qquad (3)
$$

Therefore, $\hat{T}_m$ is contained with confidence $100 \cdot (1 - \alpha)\%$ within the interval

$$
[T_m - z_{\frac{\alpha}{2}} \sigma_{mc}, T_m + z_{\frac{\alpha}{2}} \sigma_{mc}], \qquad (4)
$$

where $z_{\frac{\alpha}{2}}$ is the $(1 - \frac{\alpha}{2})$ quantile of the standard normal distribution. If we would like to estimate T_m with an accuracy ϵT_m, i.e.

$$
z_{\frac{\alpha}{2}} \sigma_{mc} = \epsilon T_m, \qquad (5)
$$

then the required number of simulations can be found by substituting σ_{mc} from (3) in (5)

$$
N_{mc} = \left(\frac{z_{\frac{\alpha}{2}}}{\epsilon}\right)^2 \frac{1 - T_m}{T_m}. \qquad (6)
$$

For example, in a typical scenario, assuming a confidence level of 95%, $T_m = 100$ ppm, and $\epsilon = 0.1$, it turns out that $N_{mc} = 3.8 \cdot 10^6$ which is clearly a prohibitively large number of simulations. This demonstrates the need for fast statistical simulation methods that can replace efficiently the time-consuming Monte Carlo analysis.

To this end, several fast statistical simulation methods have been proposed to date for the purpose of estimating the parametric test error. In [11], the process parameters defined in the process design kit (PDK) (or *input parameters*) are mapped through regression functions to the performances and alternative tests (or *output parameters*). Thereafter, the regression functions replace circuit simulation. In particular, a point in the space of input parameters (e.g. a net-list) is sampled using the PDK and, instead of simulating this point, we map it to the output parameters using the regression functions. Since the regression functions might be inaccurate at the tails of the distribution of input parameters, in [12] it is proposed to identify the points that lie at the tails and simulate them to maintain the overall accuracy of the mapping.

In [13, 14, 15, 16], it is proposed to first estimate the distribution of output parameters and then sample it to rapidly generate a large volume of data that can be used to estimate test metrics with relative frequencies. In [13], the output parameters are assumed to follow a multivariate normal distribution which in many cases is a crude approximation. In [14], the Copula theory is employed to allow a broader use of the theory of multivariate normal distributions, however, still it is likely that the underlying assumptions do not hold. In [15, 16], a technique is proposed based on non-parametric kernel density estimation that is generally applicable regardless the form of the distribution of output parameters. The common disadvantage of the three techniques in [13, 14, 15, 16] is that the distribution of the output parameters is estimated by samples that lie in the main lobe and, thereby, the tails of the distribution where the T_E and Y_L events occur might not be well estimated.

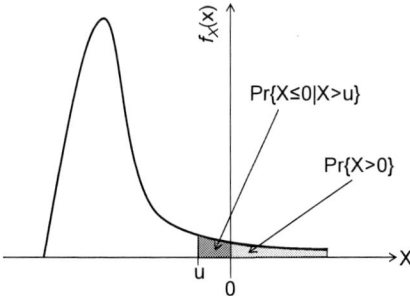

Figure 1: Probability density function of X.

In [17], an analytical mathematical formulation of a test metric is proposed focusing only at the tails of the distribution. The technique employs the extreme value theory [6, 7] and the statistical blockade technique [8, 9]. However, it is applicable only for the univariate problem, e.g. for estimating T_E and Y_L when a single alternative test is replacing a single specification test. In this paper, we extend the technique in [17] such that it is applicable for the general multivariate problem where a set of alternative tests is replacing a set of specification tests.

3. TEST METRICS MODEL

Following the notation in the introduction, we have shown that any parametric test metric can be expressed as

$$T_m = \Pr\{\cup_{i=1}^{n_z} Z_i \notin [z_\ell^i, z_u^i] \,\big|\, \cap_{i=1}^{n_w} W_i \in [w_\ell^i, w_u^i]\} \quad (7)$$

where $Z = [Z_1, Z_2, \cdots, Z_{n_z}]$ and $W = [W_1, W_2, \cdots, W_{n_w}]$ are sets of tests, $[z_\ell^i, z_u^i]$ are the acceptance limits for Z_i, and $[w_\ell^i, w_u^i]$ are the acceptance limits for W_i. If we define the random variable

$$V = \{Z \,\big|\, \cap_{i=1}^{n_w} W_i \in [w_\ell^i, w_u^i] \}, \quad (8)$$

then we can write (7) as

$$T_m = \Pr\{\cup_{i=1}^{n_z} V_i \notin [v_\ell^i, v_u^i]\}, \quad (9)$$

where $V = [V_1, V_2, \cdots, V_{n_z}]$ and $[v_\ell^i, v_u^i]$ are the acceptance limits for V_i. Furthermore, if we normalize V_i such that they have similar spread, then we can write

$$
\begin{aligned}
T_m &= \Pr\{V_1 \notin [v_\ell^1, v_u^1] \cup \cdots \cup V_{n_v} \notin [v_\ell^{n_z}, v_u^{n_z}]\} \\
&= \Pr\{\max(V_1 - v_u^1, v_\ell^1 - V_1) > 0 \cup \cdots \\
&\quad \cdots \cup \max(V_{n_z} - v_u^{n_z}, v_\ell^{n_z} - V_{n_z}) > 0\} \\
&= \Pr\{X > 0\}, \quad (10)
\end{aligned}
$$

where the random variable

$$
\begin{aligned}
X &= \max(V_1 - v_u^1, \cdots, V_{n_z} - v_u^{n_z}, \cdots \\
&\quad \cdots v_\ell^1 - V_1, \cdots, v_l^{n_z} - V_{n_z}) \quad (11)
\end{aligned}
$$

can be considered as a "dummy" performance. To obtain an observation of X, we first simulate the circuit and we obtain Z and W. If $\forall i\ W_i \in [w_\ell^i, w_u^i]$, then we define $V = Z$

and $[v_\ell^i, v_u^i] = [z_\ell^i, z_u^i]$ and we compute X, otherwise, if $\exists i$ such that $W_i \notin [w_\ell^i, w_u^i]$, then an observation of X cannot be obtained in this simulation.

We have shown so far that the problem of estimating a test metric T_m is equivalent to estimating the probability of a random variable X being larger than 0, as illustrated in Fig. 1. Since a test metric T_m can be as low as a few ppm, that is, 0 is an "extreme" value of X, a Monte Carlo analysis of a reasonable number of runs will result in untrustworthy estimates with large variance. Next, we use the extreme value theory to obtain an analytical mathematical expression for T_m.

Let $u < 0$. We can write

$$
\begin{aligned}
T_m &= \Pr\{X > 0 \cap X > u\} \\
&= \Pr\{X > 0 | X > u\}\Pr\{X > u\} \\
&= (1 - \Pr\{X \le 0 | X > u\})\Pr\{X > u\} \\
&= (1 - \Pr\{X - u \le -u | X > u\})\Pr\{X > u\} \quad (12)
\end{aligned}
$$

The main result of the extreme value theory states that for any distribution of X that has a smooth tail beyond u and for a large enough u, the tail distribution, that is, the distribution of the random variable

$$Y = \{X - u | X > u\} \quad (13)$$

denoted by

$$F_Y(y) = \Pr\{X - u \le y | X > u\}, \quad (14)$$

is a generalized Pareto [6, 7]

$$F_Y(y) = 1 - \left(1 + \frac{\xi y}{\sigma}\right)^{-1/\xi}, \quad (15)$$

where $-\infty < \xi < +\infty$ is the shape parameter and $\sigma > 0$ is the scale parameter. The support of the generalized Pareto distribution is $y \ge 0$ for $\xi \ge 0$ and $0 \le y \le -\sigma/\xi$ for $\xi < 0$. Therefore, if $\xi < 0$ and $-u > -\sigma/\xi$

$$F_Y(-u) = 1 \quad (16)$$

and equation (12) gives

$$T_m = 0. \quad (17)$$

In any other case

$$F_Y(-u) = 1 - \left(1 - \frac{\xi u}{\sigma}\right)^{-1/\xi} \quad (18)$$

and equation (12) gives

$$T_m = \left(1 - \frac{\xi u}{\sigma}\right)^{-1/\xi} \zeta_u, \quad (19)$$

where

$$\zeta_u = \Pr\{X > u\}. \quad (20)$$

We have shown so far that a test metric T_m can be expressed mathematically by the model in (19). To fit the model, we need to compute the three unknown parameters ξ, σ, and ζ_u.

For a Monte Carlo analysis with N runs, if k observations of X satisfy $X > u$, then the maximum likelihood (ML) estimate of ζ_u is $\hat{\zeta}_u = k/N$. If $y_1, y_2, ..., y_k$, $y_j = x_j - u > 0$, denote the corresponding k observations of Y, then the ML estimates $\hat{\xi}$ and $\hat{\sigma}$ of ξ and σ, respectively, are the values that maximize the log-likelihood function

$$\ell(\xi, \sigma) = \log \prod_{i=1}^{k} f_Y(y_i), \qquad (21)$$

where

$$f_Y(y) = \frac{1}{\sigma} \left(1 + \frac{\xi y}{\sigma} \right)^{-(1+1/\xi)} \qquad (22)$$

is the probability density function of Y. Once the ML estimates $\hat{\xi}$, $\hat{\sigma}$, and $\hat{\zeta}_u$ are obtained, a ML estimate $\hat{T}_m$ of T_m can be obtained

$$\hat{T}_m = \left(1 - \frac{\hat{\xi} u}{\hat{\sigma}} \right)^{-1/\hat{\xi}} \hat{\zeta}_u. \qquad (23)$$

For the derivation of confidence intervals the interested reader is referred to [7, 17].

4. STATISTICAL BLOCKADE

The parameter ζ_u can be estimated through a Monte Carlo analysis with a reasonable number of runs since u is a value that does not lie far at the tail of $f_X(X)$. However, the same is not true for ξ and σ since the resulting k observations of Y will probably be too few to allow accurate estimation with small variance. For the purpose of generating enough observations $X > u$ to estimate ξ and σ accurately and with small variance we use the statistical blockade technique described in [8, 9].

We recall that Monte Carlo consists of creating a circuit instance by sampling the input parameter distribution, simulating the circuit instance, and repeating this procedure until the required number of simulations is reached. Statistical blockade acknowledges the fact that a circuit instance can be rapidly generated and what is time consuming is in fact the simulation. Thus, the underlying idea behind statistical blockade is to insert a prediction block in the Monte Carlo procedure to assert each time after creating a circuit instance whether this instance is likely to be an extreme instance, that is, whether this instance is likely to result in $X > u$. If the answer is affirmative, we proceed with simulating the instance; otherwise, we block the simulation. Therefore, in this way we avoid spending simulation effort on instances that satisfy $X < u$ and are not useful for estimating ξ and σ.

The prediction block is implemented using classification boundaries in the space of input parameters. Let $p = [p_1, \cdots, p_m]$ denote the vector of input parameters. Next, we present the algorithm and we illustrate it with the help of a two-dimensional example in Fig. 2.

Step 1 Set $i = 1$.

Step 2 Obtain N Monte Carlo observations of p denoted by p^j, $j = 1, \cdots, N$, such that they result in N observations of X denoted by X^j, $j = 1, \cdots, N$. Fig. 2(a)

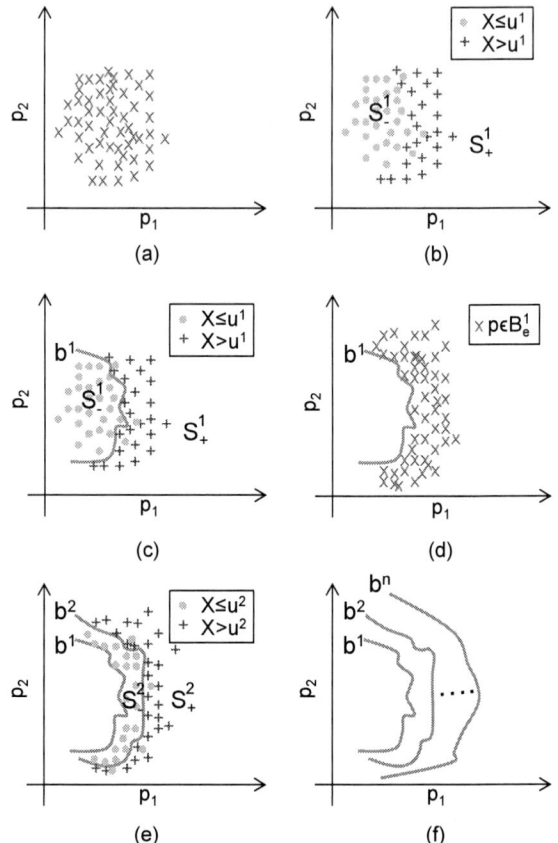

Figure 2: Illustration of the statistical blockade algorithm.

shows the projection of the N observations of p onto the $p_1 - p_2$ space.

Step 3 Consider the ordered sequence of the observations of X denoted by

$$X^{(1)} \leq \cdots \leq X^{(j)} \leq \cdots \leq X^{(N)} \qquad (24)$$

and calculate the $q\%$-quantile as

$$u^i = X^{(j)}, \quad j = \left\lceil \frac{N \cdot q}{100} \right\rceil, \qquad (25)$$

where $\lceil \cdot \rceil$ is the ceiling function.

Step 4 Define the training sets

$$S_-^i = \{p^j : X^j \leq u^i\} \qquad (26)$$
$$S_+^i = \{p^j : X^j > u^i\}. \qquad (27)$$

Fig. 2(b) shows the training sets S_-^1 and S_+^1 defined by the N observations of Fig. 2(a).

Step 5 Train a classifier to separate S_-^i from S_+^i. It is recommended to favor the class S_+^i, that is, penalize more misclassification of samples in the training set S_+^i than misclassification of samples in the training set S_-^i. Fig. 2(c) shows the allocation of boundary b^1 by the classifier.

Step 6a Obtain a Monte Carlo observation of p and present it to the classifier. If the classifier predicts that p belongs to the set S_+^i then simulate to obtain X and if $X > u^i$ then add p to the set B_e^i. Repeat this step until the set B_e^i has size N. Let N_{mc}^i be the total number of samples p presented to the classifier and N_s^i be the number of simulations in this step. Fig. 2(d) projects the samples of the set B_e^1 onto the $p_1 - p_2$ space.

Step 7 Set $i = i + 1$ and go to step 3. The samples p^j, $j = 1, \cdots, N$, are the samples in the set $B_e^{(i-1)}$ and X^j is the performance X of the circuit with input parameter vector p^j. Fig. 2(e) shows the training sets S_-^2 and S_+^2 defined by the samples of Fig. 2(d) and the allocated boundary b^2.

The algorithm proceeds and in each step the boundary is pushed more towards the tails of the distribution of p, as shown in Fig. 2(f). Thus, in each step we obtain $u^i > u^{i-1}$ and progressively we approximate a large enough u that is required for the model (19) to hold. In addition, we obtain a large number of observations $X > u$ that is required to estimate the parameters ξ and σ of the model accurately with low variance. In step 5, favoring the set S_+^i at the expense of the set S_-^i implies a larger simulation effort, but it is a safety net for ensuring that all extreme circuits are simulated, which is a requirement for estimating accurately the parameters of the model. In the end of the n-th iteration of the algorithm, an estimate of T_m can be obtained using the following steps.

Step 6b Use the observations $X > u$ to obtain the ML estimates $\hat{\xi}^n$ and $\hat{\sigma}^n$ of the parameters ξ and σ. The ML estimate of ζ_u is given by $\hat{\zeta}_u^n = N/N_{mc}^n$.

Step 6c If $\hat{\xi}^n < 0$ and $-u^n > -\hat{\sigma}^n/\hat{\xi}^n$, then $\hat{T}_m^n = 0$, otherwise $\hat{T}_m^n$ is given by (23) by substituting $\hat{\xi}^n$, $\hat{\sigma}^n$, and $\hat{\zeta}_u^n$.

In the end of the n-th iteration of the algorithm, the total number of simulations is

$$N_s(n) = N + \sum_{i=1}^{n} N_s^i. \tag{28}$$

The total number of Monte Carlo samples p used until the end of the n-th iteration of the algorithm is

$$N_{mc}(n) = N + \sum_{i=1}^{n} N_{mc}^i. \tag{29}$$

Thus, the number of simulations that are blocked until the completion of the n-th iteration is

$$N_{mc}(n) - N_s(n). \tag{30}$$

Table 1: Scenarios resulting in different T_E values. In each scenario, only the tests with "x" are carried out.

scenario	power	phase margin	PSRR @ 10MHz	f_0	PSRR @ 1MHz	gain	gain margin	T_E (in ppm)
1	x	x	x	x		x	x	38
2	x	x	x	x		x		145
3	x		x	x		x		259
4	x		x			x		488
5			x			x		657
6						x		1003

5. RESULTS

Our case study is a low-dropout regulator (LDO) designed using the 65nm CMOS065 technology by STMicroelectronics. The PDK has m=18 input parameters. The circuit comprises 94 MOSFET and more than a hundred passive elements (e.g. 32 diodes, 31 resistors, and 58 capacitors) and is characterized by $n_p = 7$ performances. The performances and their associated specification limits are

$$0.45 \text{ mW} \leq \text{power} \leq 0.532 \text{ mW} \tag{31}$$

$$53.3^o \leq \text{phase margin} \leq 65^o \tag{32}$$

$$10.9 \text{ dB} \leq \text{PSRR at 10MHz} \leq 13.5 \text{ dB} \tag{33}$$

$$458 \text{ KHz} \leq \text{unity gain frequency} \leq 750 \text{ KHz} \tag{34}$$

$$-6.31 \text{ dB} \leq \text{PSRR at 1MHz} \leq -5.34 \text{ dB} \tag{35}$$

$$99 \text{ dB} \leq \text{gain} \leq 104.25 \text{ dB} \tag{36}$$

$$50.6^o \leq \text{gain margin} \leq 90^o. \tag{37}$$

We assume 6 different scenarios where in each scenario a different subset of specification tests is eliminated and only the remaining specification tests are carried out. These scenarios are shown in Table 1 where with "x" we mark the performances that are tested, e.g. the set T. For example, in the first scenario, all the performances are tested except $PSRR$ at 1MHz. Since the actual specification tests are carried out, it is not possible that a functional circuit will be rejected, that is, there is no extra Y_L due to testing, e.g. $Y_L = 0$ in (2), and removing specification tests can only result in T_E.

We used a behavioral macromodel of the circuit obtained by the Lysis^{TM} tool of Infiniscale [18]. This tool combines nonlinear least square regression techniques with a heuristic algorithm to explore different classes of functions. Cross-validation techniques are used to select and validate the most pertinent behavioral macromodel. The resultant behavioral macromodel can be simulated very fast, in order to be able to compute T_E on the basis of millions of instances. The last column in Table 1 shows the resulting T_E for each scenario computed using 10 million simulations. As can be seen, the 6 scenarios result in different T_E values ranging from a few tens of ppm to hundreds of ppm. We consider the T_E values

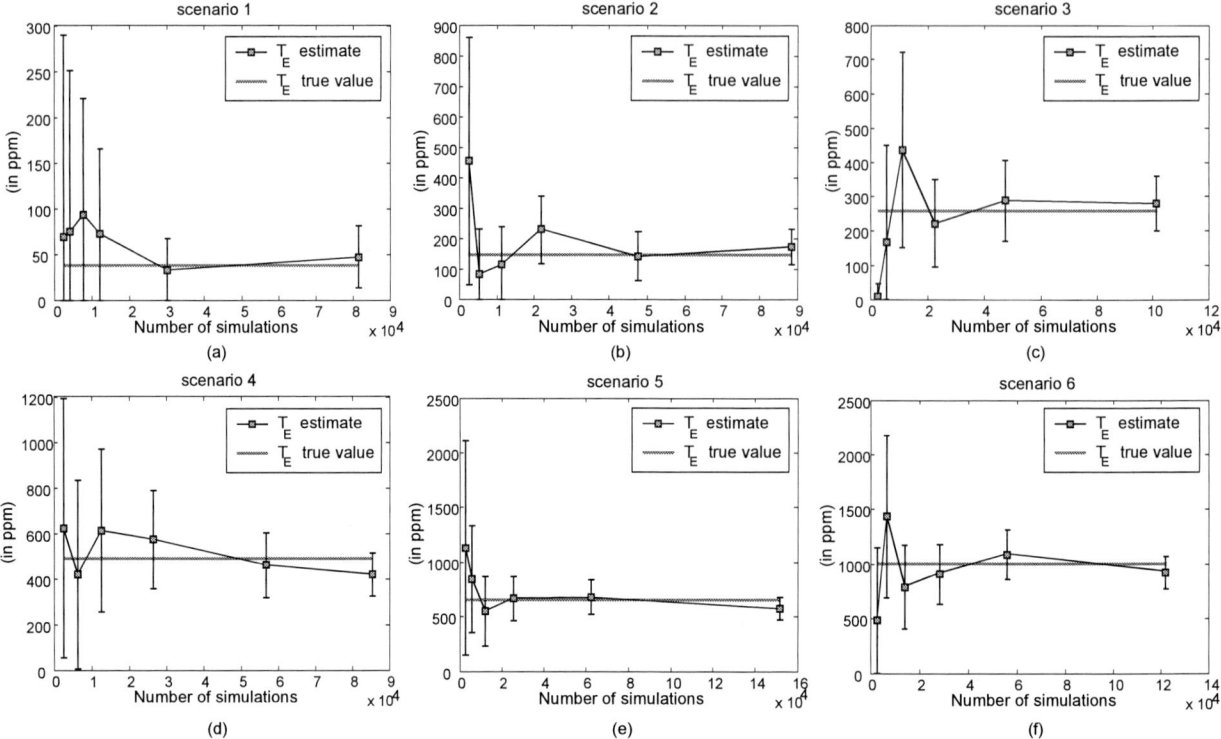

Figure 3: T_E estimates and 95% confidence intervals as a function of the number of simulations in the end of each iteration of the statistical blockade algorithm.

in Table 1 to be the true ones and we employ the technique proposed in this paper to provide estimates based on a small number of simulations.

The behavioral macromodel, the statistical blockade algorithm, and the mathematical calculations are implemented in MATLAB®. The settings for the statistical blockade algorithm are $N = 10^3$ and $q = 50$. The classifier that we employ is a decision classification tree implemented using the *classregtree* function in the MATLAB® statistics toolbox.

Fig. 3 shows the T_E estimates and 95% confidence intervals as a function of the number of simulations in the end of each iteration of the statistical blockade algorithm. The number of iterations is $n = 6$. The horizontal lines in Fig. 3 correspond to the true values of T_E in Table 1. The following observations can be made:

- For all scenarios, from the second iteration onwards, the true value of T_E always lies within the 95% confidence interval. The proposed technique is capable of estimating T_E values ranging from a few ppm to hundreds of ppm based on a number of simulations that is considerably smaller than the theoretical value in (6).

- By increasing the number of iterations, the 95% confidence intervals are shortened, that is, the region where the T_E lies is better confined. This is due to the fact that as the algorithm evolves we move closer to the distribution tails, thus the uncertainty of the estimates reduces.

- The number of simulations in the end of iteration i varies from one scenario to another. A trend can be observed that this number in general increases with the number of specifications tests that are eliminated. For example, in the end of the 6-*th* iteration, for scenario 1 where 1 specification test is left out $N_s(6) = 81365$ while for scenario 6 where 6 specification tests are left out $N_s(6) = 122195$. Intuitively, the reason is that it becomes harder to search for "extreme" circuits to fit the model in (23) when the dimensionality of the search space expands.

- One question that arises is when to stop the algorithm. With few iterations we risk to obtain inaccurate estimates (for example, for scenario 3, in the first iteration, T_E is underestimated) and wide confidence intervals that might not be useful to draw conclusions (for example, for scenario 4, in the first iteration, T_E is estimated to lie within the interval [53, 1192] with 95% confidence). The reason is that with few iterations we may not reach a large enough u for the model in (23) to hold or we may not simulate enough "extreme" circuits to fit the model with accuracy. With many iterations, we guarantee accurate estimates that are also confined in a short confidence interval, yet this is at the expense of a larger simulation effort. The choice depends on the number of simulations that we can afford to run in practice and the size of the estimated confidence interval that allows us to draw safe conclusions.

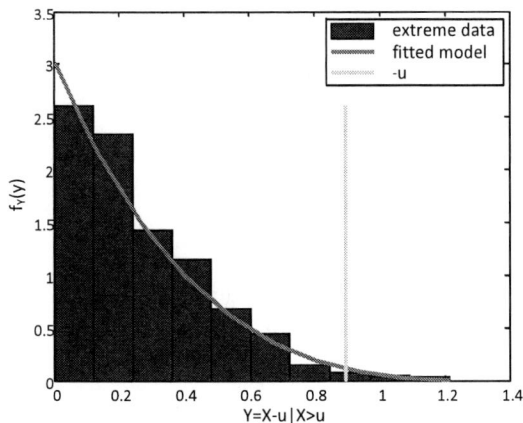

Figure 4: Histogram of tail distribution and model fitting.

Finally, as an example, Fig. 4 shows the histogram of the generated excesses $X - u$ and the fitted tail model at the end of the sixth iteration for scenario 1. As can be seen, the tail distribution of X is smooth, thus satisfying the assumption for the extreme value theory to hold, and the fitted model accurately approximates the envelope of the histogram.

6. CONCLUSIONS

We presented a technique for assisting making decisions on whether the standard specification tests of a circuit could be replaced by an alternative set of low-cost tests without sacrificing test accuracy. The technique is applied during the alternative test development phase and aims at providing estimates of parametric test metrics, that is, metrics associated with the test error for circuits with process variations. The estimates have tens to hundreds of ppm accuracy and are derived with relatively small computational effort. Estimates of parametric test metrics, together with area overhead and test cost reduction estimates, allow us to draw quick conclusions about the feasibility of replacing the standard specification tests with low-cost alternative tests and the benefits that such a replacement offers.

7. REFERENCES

[1] D. Mannath, D. Webster, V. Montano-Martinez, D. Cohen, S. Kush, T. Ganesan, and A. Sontakke, "Structural approach for built-in tests in RF devices," in *Proc. IEEE International Test Conference*, 2010, Paper 14.1.

[2] L. Abdallah, H.-G. Stratigopoulos, S. Mir, and C. Kelma, "Experiences with non-intrusive sensors for RF built-in test" in *Proc. IEEE International Test Conference*, 2012, Paper 17.1.

[3] S. Biswas, P. Li, R. D. (Shawn) Blanton, and L. Pileggi, "Specification test compaction for analog circuits and MEMS," in *Proc. Design, Automation & Test in Europe Conference*, 2005, pp. 164–169.

[4] H.-G. Stratigopoulos, P. Drineas, M. Slamani, and Y. Makris, "RF specification test compaction using learning machines," *IEEE Transactions on Very Large*

Scale Integration (VLSI) Systems, vol. 18, no. 6, pp. 998–1002, 2010.

[5] S. Sunter and N. Nagi, "Test metrics for analog parametric faults," in *Proc. IEEE VLSI Test Symposium*, 1999, pp. 226–34.

[6] P. Embrechts, C. Klüppelberg, and T. Mikosch, *Modelling Extremal Events for Insurance and Finance*, Stochastic Modeling and Applied Probability. Springer, 1997.

[7] S. Coles, *An Introduction to Statistical Modeling of Extreme Values*, Springer Series in Statistics. Springer, 2001.

[8] A. Singhee, J. Wang, B. H. Calhoun, and R. A. Rutenbar, "Recursive statistical blockade: An enhanced technique for rare event simulation with application to SRAM circuit design," in *Proc. IEEE International Conference on VLSI Design*, 2008, pp. 131–136.

[9] A. Singhee and R. A. Rutenbar, "Statistical blockade: Very fast statistical simulation and modeling of rare circuit events and its application to memory design," *IEEE Transactions on Computer-Aided Design of Integrated Circuits and Systems*, vol. 28, no. 8, pp. 1176–1189, 2009.

[10] D. E. Hocevar, M. R. Lightner, and T. N. Trick, "A study of variance reduction techniques for estimating circuit yields," *IEEE Transactions on Computer-Aided Design of Integrated Circuits and Systems*, vol. CAD-2, no. 3, pp. 180–192, 1983.

[11] L. Milor and A. L. Sangiovanni-Vincentelli, "Minimizing production test time to detect faults in analog circuits," *IEEE Transactions on Computer-Aided Design of Integrated Circuits and Systems*, vol. 13, no. 6, pp. 796–813, 1994.

[12] E. Yilmaz and S. Ozev, "Fast and accurate DPPM computation using model based filtering," in *Proc. IEEE European Test Symposium*, 2011, pp. 165–170.

[13] A. Bounceur, S. Mir, E. Simeu, and L. Rolindez, "Estimation of test metrics for the optimisation of analogue circuit testing," *Journal of Electronic Testing: Theory and Applications*, vol. 23, no. 6, pp. 471–484, 2007.

[14] A. Bounceur, S. Mir, and H.-G. Stratigopoulos, "Estimation of analog parametric test metrics using copulas," *IEEE Transactions on Computer-Aided Design of Integrated Circuits and Systems*, vol. 30, no. 9, pp. 1400–1410, 2011.

[15] H.-G. Stratigopoulos, S. Mir, and A. Bounceur, "Evaluation of analog/RF test measurements at the design stage," *IEEE Transactions on Computer-Aided Design of Integrated Circuits and Systems*, vol. 28, no. 4, pp. 582–590, 2009.

[16] N. Kupp, H.-G. Stratigopoulos, Y. Makris, and P. Drineas, "On proving the efficiency of alternative RF tests," in *Proc. IEEE/ACM International Conference on Computer-Aided Design*, 2011, pp. 762–767.

[17] H.-G. Stratigopoulos, "Test metrics model for analog test development," *IEEE Transactions on Computer-Aided Design of Integrated Circuits and Systems*, vol. 31, no. 7, pp. 1116–1128, 2012.

[18] Y. Courant, P. Hérédia, and F. Mohamed, "Nonlinear modelling for sub 65nm IC statistical analysis," in *2nd European Workshop on CMOS Variability*, 2011.

High-Throughput TSV Testing and Characterization for 3D Integration Using Thermal Mapping

Kapil Dev
School of Engineering
Brown University
Providence, RI 02912
kapil_dev@brown.edu

Gary Woods
ECE Department
Rice University
Houston, TX 77005
gary.woods@rice.edu

Sherief Reda
School of Engineering
Brown University
Providence, RI 02912
sherief_reda@brown.edu

ABSTRACT

We propose a new framework to detect structural defects and characterize the variability in the electrical resistance of through-silicon vias (TSVs) in 3D ICs. Our method offers a number of advantages that have been hard to achieve in the past. In particular, the proposed framework provides high throughput TSV testing at pre-bonding stage. A resistive liquid electrode is placed at the back side of the device to conduct electric current from TSVs. The current passing through TSVs leads to heat generation which can be captured by a remote, high-sensitivity thermal camera. The captured thermal signatures from the TSVs are then contrasted against reference thermal maps generated from known good die and/or electro-thermal simulations of models of good TSVs. A proposed automatic classification technique is capable of determining the status of TSVs based on their thermal signatures. We demonstrate the viability of the proposed technique using extensive simulation results on realistic TSV configurations.

ACM Categories & Subject Descriptors
B.8.1 [Performance and Reliability]: Reliability, Testing, and Fault-Tolerance.
General Terms: Reliability, Measurement, Performance.
Keywords: TSV testing, 3D ICs, thermal imaging.

1. INTRODUCTION

3D Integration with through-silicon vias is a promising technology that enables a number of improvements to electronic devices, including higher communication bandwidth, less communication power, higher integration density, and smaller form factor [4]. Despite these advantages, 3D integration faces a number of challenges including high spatial power densities and testing. Testing is a specially important issue, as it determines the eventual cost and acceptance of the technology [12, 5]. 3D integration offers a unique testing challenge, as full testing of TSVs is difficult to achieve before bonding [9, 6], and post-bonding testing implies the

possibility of packaging a good die with a faulty one, reducing the overall yield [10]. There is real need for pre-bonding test and characterization mechanisms that can enable high-throughput, reliable detection of TSV connectivity status.

In this paper, we propose a new paradigm for pre-bonding TSV testing and characterization using thermal imaging. Thermal imaging is an established technique for fault diagnosis in regular integrated circuits [2]. Our work extends the potential of this approach for the first time to TSV testing. The contributions of our paper are as follows.

1. We propose a new framework for pre-bonding TSV testing, where thermal imaging is used to test the electrical connectivity of TSVs. By using a resistive electrolyte solution on the back side of the device (e.g., in die or wafer form), electrical current can pass through the TSVs and solution. Heat generated from the current can be picked up by a thermal imaging equipment. The captured thermal signatures can be used to analyze whether a TSV is defective or operational, and can provide characterization information about defects.

2. To analyze the thermal signatures from a device, it is necessary to establish a *reference* thermal map, either through measurements on known good dies or through design modeling. We propose a finite element based electro-thermal modeling and simulation technique to establish the reference map.

3. To classify the status of TSVs, we propose a classification technique that takes as inputs the thermal signatures from the device under test and the thermal reference map, and automatically detects functional and defective TSVs using statistical clustering methods.

4. As a proof of concept, we simulate a test-design with 100 TSVs. We elucidate the tradeoff in TSV testing accuracy as a function of the number of faulty TSVs. We also demonstrate the ability to characterize some types of defects based on thermal signature.

The organization of this paper is as follows. Section 2 overviews previous related work in TSV and 3D IC testing. In Section 3, we describe our main proposed framework based on thermal imaging. In Subsection 3.1, we describe the main ideas for our electro-thermal modeling and simulation method, and we describe in Subsection 3.2 our proposed automatic classification technique. Our simulation results are provided in Section 4. Finally, Section 5 provides the main conclusions and future directions for this work.

Permission to make digital or hard copies of all or part of this work for personal or classroom use is granted without fee provided that copies are not made or distributed for profit or commercial advantage and that copies bear this notice and the full citation on the first page. To copy otherwise, to republish, to post on servers or to redistribute to lists, requires prior specific permission and/or a fee.
DAC '13, May 29 - June 07 2013, Austin, TX, USA.

2. RELATED WORK

Assembly of 2D ICs to build 3D ICs has become a reality and has potentially cleared the roadblock for keeping the momentum of Moore's law going. However, the 3D ICs would require modifications in all stages of the chip design. For example, the existing design for test methodologies might have to be supplemented with extra testing steps to address the new types of faults introduced from additional manufacturing steps required for fabricating 3D ICs. Therefore, testing 3D ICs is more challenging than testing 2D ICs due to increased complexity and higher testing cost of 3D ICs [5]. A typical 3D IC would have thousands of TSVs embedded in the substrate to provide electrical connection between stacked dies. The fabrication of these TSVs with high yield is still an active area of research in the fabrication community. It is typically cost effective to identify dies with defective TSVs early in the process to save the un-necessary cost incurred by later steps on defective dies [7].

To this end, a number of test and DFT solutions have been proposed to test the quality of TSVs at different stages of the manufacturing process; these include pre-bonding, post bonding and post-packaging stages [3, 9]. While Chi et al. proposed strategies to test ICs at post-bonding stage [3], Noia et al. implemented a new test-architecture in the logic to enable the testing and characterization of TSVs at pre-bonding stage [9]. It is generally assumed that all TSVs are controllable and observable through dedicated scan flip flops. The architecture proposed by Noia et al. would require probe tips to come in contact with TSVs. However, our proposed technique is based on imaging the die-surface and hence, the potential risk of TSV damage from probe tips is eliminated. Our technique requires to generate a simple test-pattern to put alternate TSVs at ground and V_{dd} potential using the scan flops available for each TSV.

While Smith et al. present a probing technology that could be used for probing TSVs with finer pitch [11], the possibility of implementing RF-based wireless techniques to test chips at both die and wafer levels has also been explored [6, 8]. In particular, laser direct testing techniques, currently used to detect defects in PCBs, could be used for contact-less testing of 3D ICs [6]. However, the laser based testing would be a slow process due to its inherent sequential nature. In this paper, we propose a high-throughput technique to test and characterize TSVs in a 3D IC at pre-bonding stage. The technique requires no expensive and potentially damaging mechanical contact to the TSVs. It does require a modest number of contacts to the front side of the die through existing test pads.

3. PROPOSED FRAMEWORK

We propose a new framework for testing and characterization of TSVs without requiring any form of bonding. In our framework, which is illustrated Figure 1, the back side of the wafer with exposed TSVs is contacted with a liquid that acts as a single backside electrode. The other side of the liquid is in contact with an infrared-transparent window. The infrared-transparent window has two functions: (i) it mechanically supports the wafer, and (ii) it allows transmitting of thermal emissions from the wafer side. The liquid has finite resistivity (to avoid electrical short among TSVs) so that electrical current flowing through the liquid will generate heat. A pattern of voltages is applied to the TSVs either directly through automated test equipment (ATE) contact-

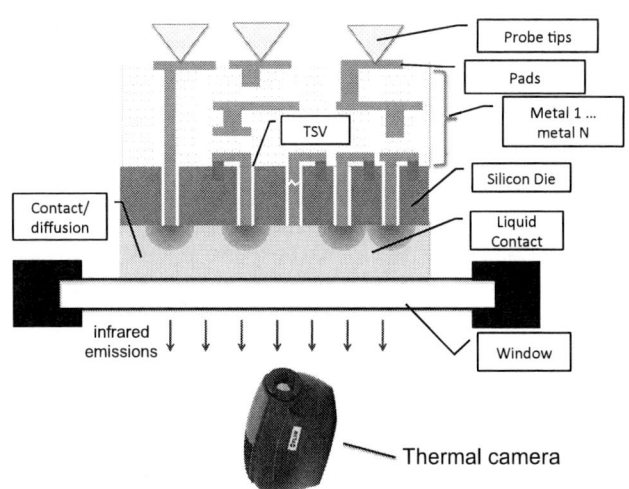

Figure 1: Proposed framework for high-throughput testing of TSVs at pre-bonding stage.

ing the wafer from the front side or through the use of scan chains or built-in self test. Current is conducted through the liquid from the TSVs that are at high voltage to the TSVs that are at ground. The induced current causes localized heating around the good TSVs due to spreading resistance. Open or highly resistive TSVs will not conduct current and no localized heating will be generated. The localized heating around good TSVs will cause temperature gradients, which can be measured by thermal imaging equipment viewing the liquid-window interface through the window. By comparing the measured steady-state thermal image from a device under test to a *reference steady-state thermal image* of a die with all good TSVs, we can pinpoint any open TSVs. Furthermore, modern thermal imaging cameras have thermal sensitivity of about 20 mK [1], which is less than the temperature difference caused by the induced current, as we will show below. Thus, by inspecting the strength of thermal emissions, we can characterize the resistance of the TSV.

The key element of our method is to provide a liquid electrode that has finite resistivity and that makes good electrical contact to the TSVs. Water-based electrolyte solutions can satisfy these requirements, as elaborated later in Section 3.1. We believe that electrochemically deposited compounds on the TSVs should be negligible or easily removable with a simple rinsing step. Furthermore, the liquid electrode and the infrared-transparent window must have sufficiently low thermal conductivity to maintain the thermal gradients generated from the flow of electrical currents. For these reasons, we utilize water-based solutions and a chalcogenide glass window in our setup. We assume that all TSVs can be set to a high or low voltage state. This is obviously true for TSVs connected to the power rails. Digital I/O lines can be set high or low via scan-in and/or BIST mechanism. If some TSVs are strictly inputs, continuity can still be tested by applying a bias voltage to the liquid electrode and inducing current through ESD diodes connected to the pins.

To illustrate the operation of the proposed technique, we construct a simple test case. We assume that there are only five TSVs, arranged in a diamond pattern with the center TSV connected to V_{dd} (1V) and the surrounding four TSVs are connected to ground as illustrated in Figure 2.a. Using modeling and simulation techniques (described in Section

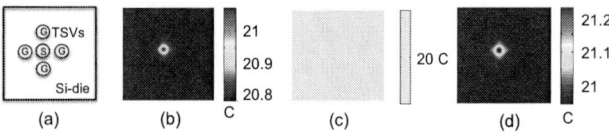

(a)   (b)   (c)   (d)

Figure 2: (a) A die with 5 TSVs arranged in the diamond pattern; (b-d) thermal-profiles at the top of ion solution: (b) when the center TSV is at 1V and other TSVs are at 0V, (c) when the center TSV is open and non-conductive, (d) same as (b), but the diameter of the center TSV is increased by 30%.

3.1), we provide the simulated thermal profile at the top of ion-solution in 2.b. (For simplicity, we refer to the window/solution interface as the "top" even though it appears at the bottom in Figure 1.) If the center TSV is defective, say completely open, then there is no current in any TSV and therefore the top surface of ion-solution will be at ambient temperature (20 C), as shown in Figure 2.c. Thus, we can obtain a thermal image by simulation of the good-TSV case and contrast this image to the thermal map captured from the actual die, and use the difference to test TSVs.

Next, we characterize the variability in the diameter of the center TSV. To that end we assume that all TSVs are conducting, but the center TSV has a smaller or larger diameter than nominal. Figure 2.d shows the thermal image of the top of the ion-solution when the diameter of the center TSV is 30% higher than its nominal value. We notice that the temperature at TSV location increases with increase in diameter of the TSV. This is because the increased diameter of the TSV at the liquid interface decreases the spreading resistance in the liquid, leading to more joule (V^2/R) heating in the vicinity of the TSV.

Our proposed method offers a number of advantages over current methods:

1. Scalability and Throughput. A thermal infrared camera can capture the thermal signatures of large fields of view spanning few mms with 3-5 μm resolution. Thus, our method can capture the status of several hundreds or even thousands of TSVs in parallel at the same time in a single image, enabling high-throughput testing. The size (few micron) and pitch (few 10s of microns) even in aggressive TSV roadmaps are resolvable by state-of-the-art short-wave and mid-wave infrared cameras.

2. Testing Cost: Our method requires standard thermal imaging equipment [5]. Further, only low-bandwidth ATE should be required to front-side contact of the device's power rails, ground, and some scan chain inputs. Thus, our setup eliminates the need for large number of pin contacts and pin electronics per TSV.

3. "Non-contact" nature: Although probe-technology has improved significantly in recent years [11], the problem of TSVs' ends becoming damaged from probe tips still exists. However, the proposed technique enables high-throughput testing of TSVs in a "contact-less" manner.

In the rest of the paper, we explain in details the model and simulation techniques required to generate the reference thermal map, together with techniques for automatic classification of TSVs.

3.1 Electro-thermal Modeling

Model Setup: Our proposed technique relies on capturing the thermal image of the die when electric current flowing in TSVs generates heat in the liquid contact solution. The thermal profile at the surface of the die depends on the location and number of TSVs that conduct electrical current. By applying a test pattern to the TSVs of the die under test, we can detect the defective TSVs immediately by comparing the measured thermal-profile with the thermal-profile of the die with good TSVs. In order to verify the feasibility of the proposed technique and also to perform comprehensive study of the proposed idea, we built a simulation model of the system using a well known Multiphysics simulation-software, COMSOL. The software has a finite-element based numerical solver as its core computational engine.

In order to solve the modeling problem using finite-element method (FEM), the complete geometry of the arrangement given in Figure 1 has to be divided into smaller elements in a process known as meshing. Creating a proper mesh is important for two reasons: (1) a properly-sized mesh enables accurate simulation of the required physical phenomena; and (2) it controls the convergence of the numerical solution. For these two reasons, we refined the mesh to appropriate sizes at different interfaces and corners by adding boundary-layers and by choosing the mesh-size individually for each domain. The mesh is refined iteratively until it has insignificant impact on the final solution.

In the proposed setup, we use a thin layer of electrically conductive ionic solution at the backside of the die, *i.e.* at the side where TSVs are exposed after substrate thinning. The ion solution is used to provide reliable path for the electric currents flowing among different TSVs when one side of the TSVs are applied with special test-patterns. Further, to keep the ion solution in place and for mechanical support, we put an infrared transparent window on top of the solution. The window material is chosen such that it is not only IR-transparent, but also has low electrical and thermal conductivities. The low thermal conductivity of the window ensures that the temperature gradients (distinguishable cold spots) generated at the top of ion-solution due to defective TSVs are not washed-out by the window. Similarly, the electrically non-conductive property of the window material helps in limiting the flow of electric current entirely inside the solution, which in turn generates detectable thermal hot spots inside the solution and exactly at the locations where TSVs are present. We propose using a state-of-the-art infrared camera to capture the thermal image at the surface of the ion-solution.

The properties of different materials used in our model are reported in Table 1. Mainly, the model has five different materials: silicon-die, copper-TSVs, ion-solution (e.g. NaCl solution with appropriate electrical conductivity), chalcogenide glass as an IR-transparent window, and air as ambient. Here, ρ denotes the density of the material in kg/m^3, k represents the thermal conductivity of the material in W/(m.K), C_p denotes the specific heat capacity of the material at constant pressure in J/(kg.K), ϵ_r represents the relative permittivity of the material, and σ denotes the electrical conductivity of the material in S/m. We assume that the ionic solution has the same heat capacity as pure water.

Model Simulation: In order to compute the thermal-profile of the system, essentially, we have to solve Joule heating from electrical currents and heat-transfer physics

Material properties	ρ (kg/m^3)	k (W/m.K)	C_p (J/kg.K)	ϵ_r	σ (S/m)
Silicon	2330	148	703	12.1	1e-12
Copper	8700	400	385	1	5.998e7
Ion-solution	1041.3	0.56	3930	81	5-10
Chalcogenide	4410	0.24	330	5.19	1e-9
Air	1.2041	0.024	1003.5	1	0

Table 1: Material properties, where the terms ρ, k, C_p, ϵ_r, and σ are defined in the text.

simultaneously. In particular, we solve the following set of steady-state equations using a finite-element solver to compute electric current density in the geometry:

$$\nabla . \mathbf{J} = Q_j \tag{1}$$
$$\mathbf{J} = \sigma \mathbf{E} + \mathbf{J}_e \tag{2}$$
$$\mathbf{E} = -\nabla V \tag{3}$$

where $\mathbf{J}$ denotes the current density in A/m^2, Q_j represents the current source in A/m^3, $\mathbf{E}$ denotes the electric field, $\mathbf{J}_e$ denotes the external current density in A/m^2, V represents the electric potential in Volts and ∇ is the differential operator in space domain. Here, Equation (1) denotes the current-continuity equation; Equation (2) is Ohm's law and Equation (3) relates electric potential and electric field under static conditions. We assume that the complete setup is placed in air and therefore, we apply electrical insulation boundary-conditions at all surfaces that are exposed to air. This is equivalent to setting $-\hat{n}.\mathbf{J} = 0$, where $\hat{n}$ is the unit normal-vector pointing outward to the boundary. Applying ground or non-zero potential at any TSV is equivalent to setting $V = 0$ or $V = V_{dd}$, respectively, at that surface, where V_{dd} is any non-zero voltage.

For simulating heat-transfer in the system, we solve the steady-state heat diffusion equation in COMSOL:

$$-\nabla . (k\nabla T) = Q, \tag{4}$$

where k is the thermal conductivity of the material, T is the temperature in Kelvin, and Q denotes the heat sources/sink in W/m^3. For Joule heating, as is the case in our system, Q comes from the resistive heating and is equal to $I^2 R/vol$, where I is the electric current, R is the electric resistance of the material, and vol is the volume of the meshed element. Further, we assume that the net thermal resistance of the reference die plus liquid and window assembly to ambient is equal to 10 K/W; this is consistent with values we have observed in the lab with similar setups, and is far larger than in commercial heat sinks for microprocessors which have less than 1K/W thermal resistance. For all simulations, we use 20 C as the ambient temperature in this paper.

In our setup, we capture thermal profile of the system in steady-state. Our simulations show that devices reach thermal steady-state within 1 second; since the data acquisition time can be significantly less than 1s per image, the acquisition time per capture is on the order of 1s.

3.2 Automatic TSV Classification

In order to test and characterize all TSVs simultaneously, we apply in simulation a test pattern to all TSVs such that TSVs are either at ground or V_{dd} potentials. The application of any desired test pattern is possible because all TSVs would have scan flops associated with them for design for testability purposes. Hence, each TSV is both controllable

and observable through boundary scan flops. We simulate the thermal profile at the surface of the ion-solution for a good die, and denote the corresponding thermal-profile as our reference thermal-image. We store this image and use it for testing all dies of the same configuration, i.e. same layout and dimensions for TSVs and substrate. To test any die with the same configuration, we apply the same pattern to all TSVs and capture thermal image of the top surface of the ion-solution using an infrared camera. Next, we subtract the reference thermal-image from the captured image. For example, if only one TSV is defective in the die, then that TSV would not carry any current and therefore there will not be any localized heating in the TSV and in the vicinity of this TSV in the ion-solution. Hence, it will be possible to detect the location of the defective TSV. It is worth mentioning that all TSVs, whether connected to ground or V_{dd} will carry electrical currents and generate heat, as long as they are not defective. When all TSVs are conducting, the temperature at all locations in the die would be higher than when one or more TSVs are non-conducting. Hence, the difference-image would have lower pixel values in the regions corresponding to defective TSVs. The more the number of defective TSVs in the die-under test, the higher would be the magnitude of temperatures in the difference-image. However, if the number of defective TSVs is large, it might become difficult to detect the cold-spots due to interaction between different TSVs through lateral heat diffusion.

We propose a fully automated process to identify, locate, test, and characterize the defective and good TSVs in the die. We use a standard k-means classification based method on the difference-image to classify the TSVs into two bins: good TSVs and defective TSVs. To this end, we divide the thermal image of the entire die/ion-solution-top-surface in to multiple areas, equal in number to the total number of TSVs in the die. Thus, each TSV will be represented by a multidimensional vector that represents its thermal signature; the dimension of the vector is equal to the number of pixels assigned to each TSV in a thermal map, where each pixel denotes a temperature value in the thermal map. We assume that all the TSVs have the same nominal diameter. Therefore, each TSV is represented by the same number of pixels in a thermal map. Also, the number of multidimensional vectors will be equal to the total number of TSVs in the die. These vectors are given as inputs to the k-means classifier to classify them into two bins. As confirmed by the simulation results (presented in Section 4) on a test-die, the k-means algorithm has very good classification performance.

We use two metrics for measuring the accuracy of the classified TSVs: detection accuracy and true-positive rate (TPR). The detection accuracy is computed by taking the ratio of true detection (sum of the number of good TSVs classified as good and the number of defective TSVs classified as defective) to the total number of TSVs in the die. The TPR is defined as the ratio of number of defective TSVs that are classified as defective to the total number of defective TSVs in the die. The k-means classification method used in this paper is a non-supervised method and therefore, there might be a scope of improving the detection-accuracy by building a new classifier with the help of training data.

Further, if TSVs have different diameters or different resistances than nominal, this would lead to differences in the associated thermal hot spots in the liquid, as explained in Section 4.

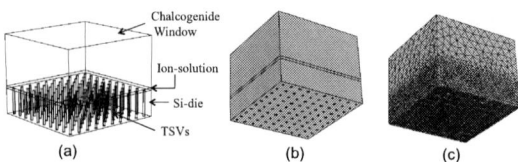

Figure 3: (a) Simulation model showing 100 TSVs, (b) un-meshed geometry, and (c) meshed geometry of the simulated model.

4. SIMULATION RESULTS

Setup Details: For our simulations, we assume the following dimensions for different domains of the system. The diameter, length, and pitch of good TSVs are assumed to be 5 μm, 50 μm, and 20 μm respectively. Thickness of the silicon die is kept the same as the length of TSVs because, before bonding two dies, the substrate is thinned until the TSVs are exposed. The thickness of the resistive ion solution is kept as 5 μm; a typical ion-solution that could be used in our setup is a solution of Sodium Chloride (NaCl) salt in water. Resistivity of the solution could be varied by changing the concentration of NaCl. We use 3% NaCl solution in our setup that has electrical conductivity of about 5 S/m. Fortunately, the NaCl solution has low thermal conductivity, which helps in maintaining the spatial gradients in the solution when good TSVs are carrying current and defective TSVs are non-conducting. Also, we use an IR-transparent chalcogenide glass window at the top of ion-solution for practical purposes as illustrated earlier in Figure 1. The temperature dependent infrared radiations generated from the top of ionic solution pass through this window before they are captured by the camera.

The geometry of the modeled system is shown in Figure 3.a; it shows 100 copper TSVs (10x10) fabricated in a 2-D grid pattern over the die-area. Figure 3.b and Figure 3.c show the un-meshed and meshed geometries of the system, simulated using a commercially available numerical solver. Moreover, appropriate boundary conditions and heat transfer coefficients are used at all surfaces of the system to approximate the real system.

TSV Testing: In order to test TSVs of the die, we use the technique described in section 3.2. We first simulate the reference thermal-profile profile at the top of the ion-solution when all 100 TSVs are non-defective. The simulated thermal-profile is shown in Figure 4.a; it represents the reference thermal image for the good die. Next, we simulated a few cases when multiple TSVs (selected randomly) are defective; in particular, we simulated the thermal-profiles when multiple TSVs are open, and hence, not conducting any current. For example, Figure 4.b shows the thermal profile when two TSVs, located at (3,7) and (7,10) coordinates in the x-y plane, are open. Similarly, Figure 4.c to Figure 4.f show the simulated thermal images at the top of ion-solution for cases when 4, 8, 16, and 32 TSVs are open respectively. It is worth mentioning again that in all cases we apply ground and 1V potential at all alternate TSVs, but the TSVs that are open do not carry any current and therefore, we have lower temperature at those locations.

Next, as described in Section 3.2, in order to automatically detect the location of faulty TSVs, we subtract the reference thermal-image (shown in Figure 4.a) from test thermal images. The images after taking the difference between reference thermal-image and test-images for different faulty-

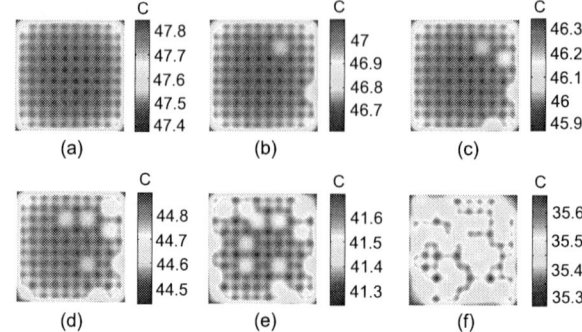

Figure 4: (a)-(f): Thermal maps when 0, 2, 4, 8, 16, and 32 TSVs are defective (open).

TSVs cases are shown in Figure 5. We simulated all cases, wherein any number of TSVs (0 to all) are defective. Figure 5.a to Figure 5.f show the difference-images for the cases when 2, 4, 8, 16, 32, and 64 TSVs (selected randomly) are assumed to be defective at a time. For a real system, the test images could be captured using a high-sensitivity thermal-imaging camera. The reference thermal-image could be obtained by either simulations or from the measurements on a known good die, whose all TSVs are non-defective. It is clear from Figure 5 that the difference image has lower temperature at locations corresponding to defective TSVs than at locations where good TSVs are located. For example, the difference in the mean temperature at locations above defective TSVs and the mean temperature at locations above good TSVs is more than 100mK when 32% of the TSVs are defective and it could be easily detected by the state-of-the-art mid-wave infra-red cameras, which have noise equivalent temperature difference (NETD) of about 20 mK. Based on the simulation results, we observe that the temperature gradients are higher than the NETD of IR cameras, even when 50% of the TSVs are defective.

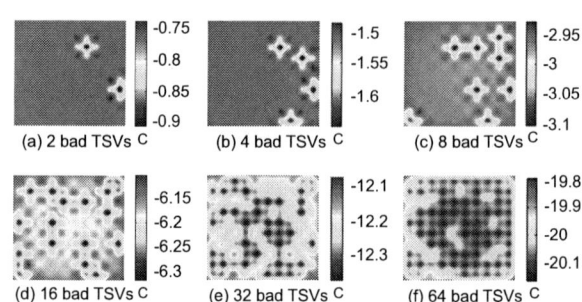

Figure 5: Difference between measured and true thermal maps when 2, 4, 8, 16, 32, and 64 TSVs are defective (open).

As described in Section 3.2, for automatic detection of faulty TSVs, we divide the difference thermal image into 100 different groups, leading to a thermal signature per TSV, and use k-means classification method to classify good and defective TSVs for each test-case. To verify the effectiveness of the proposed classification technique, we simulated the thermal profiles of the die by increasing the number of defective TSVs progressively from 1 to 100. We computed the true positive rate (TPR) and the detection accuracy for each case. Figure 6.a gives the TPR of our detection method as a function of the number of faulty TSVs. It is clear that

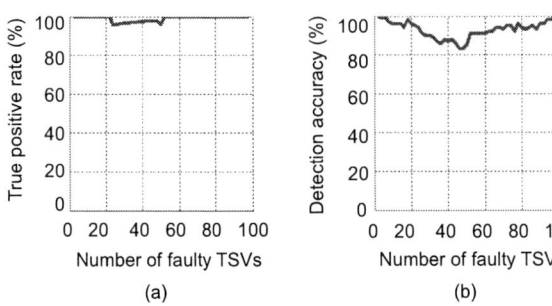

Figure 6: (a) True positive-rate (TPR), (b) Detection accuracy of the TSV classification using k-means clustering algorithm.

our proposed detection technique is able to detect the defective TSVs almost all the time. Further, as could be seen from Figure 6.b, the overall detection accuracy is quite high (>80%), even when 50% of the total number of TSVs are defective in the die. The results show that our proposed technique is not only fast and scalable, but also is very reliable in detecting the faulty TSVs in the die.

Characterizing Variations in TSV Diameters and Resistance: For simulating the characterization process, we performed two studies. First, we simulated the effect of a central TSV in the 100-element array having a diameter change of up to +/- 30% from nominal. Figure 7.a shows the resulting peak temperature vs. TSV diameter. As expected, larger diameters lead to lower spreading resistance and larger temperature peaks due to joule (V^2/R) heating. Every 10% change in diameter caused about a 33 mK change in peak temperature, which is above the NETD of a thermal camera. Second, we simulated increased resistance of a TSV compared to nominal in a 5-TSV case as described in Section 2. The result is shown in Figure 7.b, where difference between the maximum and the minimum temperature of a thermal map is denoted by ΔT. When the TSV resistance becomes comparable to the effective resistance of the fluid (on the order of 25 kOhm in our simulation) the thermal signal decreases due to a decreased voltage drop across the fluid. This effect allows characterization of the resistance of a TSV compared to a threshold value that can be tuned by varying some parameters of the experiment, such as the resistivity of the fluid.

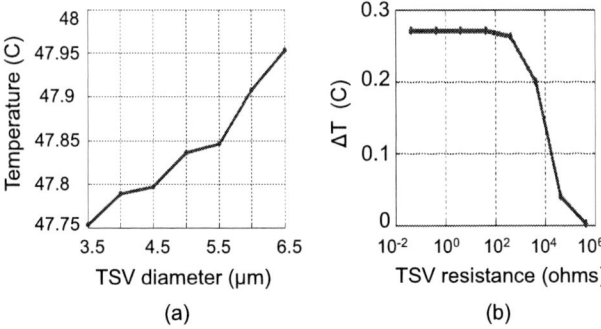

Figure 7: (a) Change in peak temperature due to variation in diameter of one of the TSVs for the die containing 100 TSVs. (b) Difference between the maximum and the minimum temperature of the thermal map (ΔT) when resistance of the center TSV changes due to partial defects for the die containing 5 TSVs.

5. CONCLUSIONS AND FUTURE WORK

In this work, we have proposed a novel and effective technique for high-throughput testing and characterization of TSVs in 3D ICs. In our method a high-sensitivity thermal camera is used to capture thermal signatures emitted due to heat generation from electric currents in TSVs. The thermal signatures are compared against a reference thermal image generated from simulation of a good device. We have proposed an electro-thermal simulation technique to generate the reference thermal map, as well as an automatic classification technique to analyze the differences between the thermal signatures and the reference map and to classify the status of TSVs. Simulation results show that our method is able to test TSVs and to characterize the resistance of TSVs.

Ongoing and Future Work. Our current method enables high-throughput testing of TSVs, but only one-by-one characterization. We plan to devise characterization techniques to analyze the resistances of multiple TSVs. This can be achieved through a combination of modeling and numerical inversion techniques, and/or through the use of multiple input patterns to the TSVs. In order to perform experimental validation of the proposed technique, we are also working on acquiring a representative TSV sample in our lab.

6. REFERENCES

[1] [Online]. Available: http://www.flir.com

[2] O. Breitenstein, W. Warta, and M. Langenkamp, *Lock-In Thermography: Basics and Use for Functional Diagnostics of Electronic Components*, 2nd ed. Springer Verlag, 2010.

[3] C. Chi, E. J. Marinissen, S. K. Goel, and C.-W. Wu, "Post-Bond Testing of 2.5D-SICs and 3D-SICs Containing a Passive Silicon Interposer Base," in *International Test Conference*, no. 17.3, 2011, pp. 1–10.

[4] W. R. Davis, J. Wilson, S. Mick, J. Xu, H. Hua, C. Mineo, A. Sule, M. Steer, and P.D.Franzon, "Demystifying 3D ICs: The Pros and Cons of Going Vertical," *IEEE Design & Test of Computers*, vol. 22(6), pp. 498–510, 2005.

[5] E. J. Marinissen, "Testing TSV-Based Three-Dimensional Stacked ICs," in *Design, Automation, and Test in Europe*, 2010, pp. 1689–1694.

[6] E. J. Marinissen *et al.*, "Contactless testing: Possibility or pipe-dream?" in *Design, Automation and Test in Europe*, 2009, pp. 676– 681.

[7] E. J. Marinissen and Y. Zorian, "Testing 3D Chips Containing Through-Silicon Vias," in *International Test Conference*, no. ET1.1, 2009, pp. 1–11.

[8] B. Moore *et al.*, "High Throughput Non-contact SiP Testing," in *International Test Conference*, no. 12.3, 2007, pp. 1–10.

[9] B. Noia and K. Chakrabarty, "Pre-Bond Probing of TSVs in 3D Stacked ICs," in *International Test Conference*, no. 17.1, 2011, pp. 1–10.

[10] S. Reda, G. Smith, and L. Smith, "Maximizing the Functional Yield of Wafer-to-Wafer 3D Integration," *IEEE Transactions on VLSI Systems*, vol. 17, no. 9, pp. 1357–1362, 2009.

[11] K. Smith, P. Hanaway, M. Jolley, R. Gleason, and E. Strid, "Evaluation of TSV and Micro-Bump Probing for Wide I/O Testing," in *International Test Conference*, no. 17.2, 2011, pp. 1–10.

[12] L. Smith, G. Smith, S. Hosali, and S. Arkalgud, "3-D Integration: It All Comes Down to Cost," in *3-D Architectures for Semiconductor Integration and Packaging*, 2007.

On Effective and Efficient In-Field TSV Repair for Stacked 3D ICs[*]

Li Jiang[†], Fangming Ye[‡], Qiang Xu[†], Krishnendu Chakrabarty[‡], and Bill Eklow[§]

[†]Department of CS&E, The Chinese University of Hong Kong, Shatin, N.T., Hong Kong
[‡]Deptartment of ECE, Duke University, Durham, NC
[§]Cisco Systems, San Jose, CA

ABSTRACT

Three-dimensional (3D) integration based on through-silicon-vias (TSVs) is rapidly gaining traction for industry adoption. However, manufacturing processes for TSVs have been shown to introduce new failure mechanisms. In particular, thermo-mechanical stress and electromigration introduce reliability threats for TSVs, e.g., voids and interfacial cracks, which can lead to hard-to-predict timing errors on critical paths with TSVs, thereby resulting in accelerated chip failure in the field. Burn-in for screening latent defects during manufacturing is expensive and its effectiveness for new TSV defect types has yet to be thoroughly characterized. We describe a reconfigurable in-field repair solution that is able to effectively tolerate latent TSV defects through the judicious use of spares. The proposed solution includes a reconfigurable repair architecture that enables spare TSV sharing between TSV grids, and the corresponding in-field repair algorithms. The effectiveness and efficiency of our proposed solution is evaluated using 3D benchmark designs.

1. INTRODUCTION

Three-dimensional integrated circuits (3D ICs) based on through-silicon vias (TSVs) have emerged as one of the most promising solutions to overcome interconnect bottleneck in CMOS scaling [1]. Comparing to planar ICs, 3D ICs offer many advantages, such as smaller footprint, heterogeneous integration capability, shorter interconnects, and higher memory bandwidth. However, TSV fabrication involves several disruptive manufacturing technologies, which leads to new types of defects [2]. These defects are often latent and difficult to screen during manufacturing test, but their impact can be significant during field operation, leading to reduced service life of 3D ICs [3, 4]. Burn-in for screening latent defects during manufacturing is expensive and its effectiveness for new TSV defect types has yet to be thoroughly characterized. Therefore, repair solutions are needed in order to exploit the potential of 3D ICs and facilitate commercialization.

During TSV fabrication, the temperature is first increased for copper electroplating and then brought down to the ambient temperature. Owing to the large difference in coefficients-of-thermal-expansion (CTE) of the copper TSVs and that of the silicon [5], however, tensile stress inevitably appears on the silicon [6]. Such thermal-mechanical stress is likely to cause TSV interfacial cracks (see Fig. 1) that is usually undetectable during manufacturing test [7]. The forces induced by residual stress in the 3D structure cause the crack to grow dur-

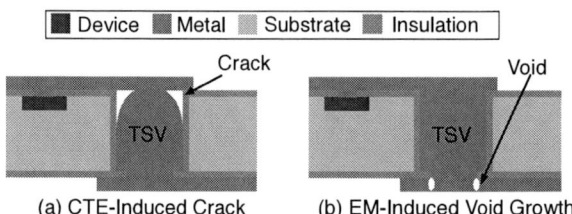

Figure 1: Illustration of some TSV latent defects

ing field operation, thereby increasing the delay of critical paths with TSVs (if any) and eventually forming an open defect [3]. Moreover, a number of recent works examined the classical electromigration (EM) failure mechanisms in 3D ICs, showing that TSVs are prone to EM-induced voiding effects [7–9] (see Fig. 1). Similar to TSV interfacial cracks caused by thermal-mechanical stress, EM-induced voids increase TSV resistance, causing path delay faults and eventually TSV open defects. Note that TSV-induced stress also reduces the reliability of nearby transistors and metal wires, and various analytical models and reliability-driven physical design techniques have been presented in the literature to mitigate this problem [2]. However, we limit the scope of this paper to the repair of TSV latent defects only.

One promising method to tolerate TSV failures is to add spare TSVs in the design for built-in self- repair (BISR). Various TSV redundancy allocation techniques and their corresponding repair algorithms have been proposed in the literature [10–20]. While effective for repairing TSV manufacturing defects occurred at $t = 0$, these solutions are not readily applicable for in-field repair of TSV latent defects that manifest themselves at $t > 0$. This is because the repair solution obtained with the *deterministic* repair algorithms used in these techniques may not satisfy the timing requirement of the circuit due to circuit aging, thereby rendering the repair solution less effective. To tackle the above problem, this paper presents a novel in-field TSV repair solution for stacked 3D ICs. The contributions of this work include the following:

- To the best of our knowledge, we present the first in-field TSV repair framework for 3D IC lifetime reliability enhancement.

- We propose an efficient TSV repair algorithm that is able to significantly improve the mean-time-to-failure (MTTF) of TSV grids through the judicious use of spares, as demonstrated by our experimental results.

- We enhance the TSV redundancy architecture in [18] by allowing redundancy sharing across neighboring TSV grids.

The remainder of this paper is organized as follows. Section 2 presents related works and further motivates this paper. In Section 3 and Section 4, we detail the proposed in-field TSV repair framework and the corresponding repair algorithm, respectively. Experimental results on 3D benchmark designs are next presented in Section 5. Finally, Section 6 concludes this paper.

---

[*]This work was supported in part by a research grant from Cisco Systems. The work of F. Ye and K. Chakrabarty was supported in part by the National Science Foundation (NSF) under grant no. CCF-1017391, and by the Semiconductor Research Corporation (SRC) under contract no. 2118.001.

Permission to make digital or hard copies of all or part of this work for personal or classroom use is granted without fee provided that copies are not made or distributed for profit or commercial advantage and that copies bear this notice and the full citation on the first page. To copy otherwise, to republish, to post on servers or to redistribute to lists, requires prior specific permission and/or a fee.
DAC '13, May 29 - June 07 2013, Austin, TX, USA

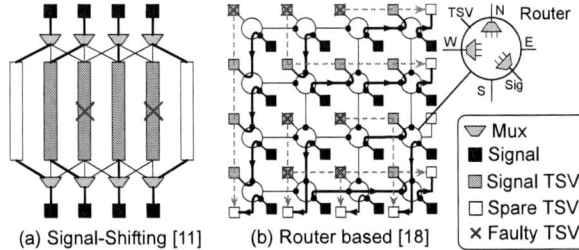

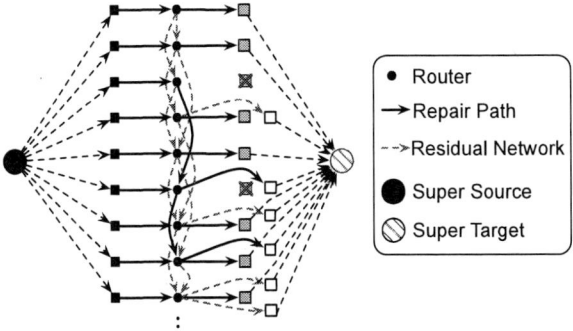

Figure 2: Existing TSV Repair Solutions

Figure 3: Maximum Flow Based Repair Algorithm.

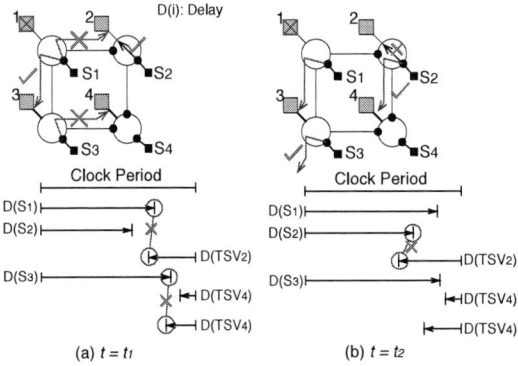

Figure 4: An example to motivate the need for careful in-field repair.

2. PRELIMINARIES AND MOTIVATION

In this section, we first review related prior work on TSV repair and defect tolerance to increase manufacturing yield. Following that, we discuss the need for explicitly targeting TSV latent defects in the development of a repair solution that can be used in the field.

2.1 Prior Work

Various TSV repair solutions have been proposed in the literature for manufacturing yield enhancement [10–18, 20]. The repair capability of these solutions vary according to their different redundancy architectures and the corresponding repair algorithms. In [10–13], one or more redundant TSVs are added for a group of TSVs and a defective TSV is swapped with a fault-free one via signal shifting (see Fig. 2(a)). In [14], spare TSV rows are added to a TSV array for repair to reduce the storage requirement of reconfiguration data. The above methods assume uniformly distributed TSV faults and use neighboring TSVs to replace faulty ones, if any. In practice, however, if one TSV is defective during fabrication, it is more likely that its neighboring TSVs are also defective due to clustering [21]. This issue has been considered in [15–17], wherein more practical TSV grouping strategies were proposed to tolerate clustered TSV faults.

Recently, a low-cost and flexible TSV redundancy architecture was proposed in [18, 20], which enables defective TSVs to be replaced with distant spares to tolerate clustered faults and it is shown to have higher repair capability than previous methods. Considering the fact that neighboring TSVs usually suffer from similar thermal and mechanical stress, we leverage this TSV redundancy architecture in our work to enable repair of clustered latent faults within a TSV grid. We next briefly discuss it in the following.

As shown in Fig. 2(b), the proposed architecture links TSVs with switches and wires, leading to a TSV grid. Redundant TSVs are put at two borders of the grid for repair. If one signal is disconnected due to a TSV fault, the switches linking two ends of the faulty TSV reroute the signal through a neighboring fault-free TSV. Since the fault-free TSV is "borrowed" by the previously-rerouted signal, the signal originally linked to it needs to be rerouted as well. This procedure continues until a spare TSV at the boundary is used.

Consequently, the TSV repair problem can be formulated as a problem of finding *edge-disjoint repair paths* for faulty TSVs. Con-

sider the TSV grid as a directed graph, wherein signals, TSVs and routers are represented as vertices, while the directed edges are used to link them. In order to route each signal to a fault-free TSV without any routing conflict, they first assign each edge in the directed graph with a unit capacity "1" to construct a directed flow network. Then, a super source vertex is added to the flow network, pointing to all the vertices that represent signals; while at the other side, all the vertices denoting fault free TSVs are pointing to a super target vertex merging all the fault-free TSVs into a target node (see Fig. 3). The original TSVs repair problem can then be solved using the maximum flow method and the TSV grid is repairable if and only if maximum flow value is equal to the number of signals.

2.2 Motivation for In-Field Repair

Unlike TSV repair at $t = 0$ for yield enhancement, the objective of in-field repair for TSV latent faults at $t > 0$ is to increase the MTTF of 3D ICs. This problem is especially difficult due to circuit aging. Pevious TSV repair solutions have focused on the replacement of defective TSVs with fault-free ones, i.e., the repair algorithms start from faulty TSVs and try to find repair paths to spares, without explicitly considering the impact of the repair solution on signal delays. Such repair methodology is generally applicable for detectable manufacturing defects such as opens and shorts when the distance between the failed TSV and its corresponding spare is not large.

However, both TSVs and other circuit elements wear out during field-operation. On one hand, it is likely that the "replacement-oriented" repair solution provided with existing methods violates signal timing requirements after shifting or rerouting, thereby leading to new "faulty TSVs". On the other hand, a faulty TSV linking to a particular signal might be a good one if it links to another signal instead. This is because, a TSV fault occurring online is not necessarily a catastrophic open/short defect, but often a delay fault that cannot meet the timing requirement of critical paths going through it due to circuit degradation. Consider the example TSV grid shown in Fig. 4(a). Signal S_1 needs to be rerouted due to the latent defect that is manifested on its corresponding TSV. However, it may fail again if it is rerouted to use TSV_2 originally linked to S_2, generating a "new" TSV fault even though this TSV is fault-free. Such fault propagation may eventually make the TSV grid irreparable, even though a more sustainable repair solution exists as shown in Fig. 4(b)).

Consequently, for in-field TSV repair, we should not focus only on faulty TSV replacement and simply find a repair path for each faulty TSV. Instead, we are to find the set of signal-TSV pairs that satisfy the timing requirement of every signal. Whether a signal and a particular TSV can be paired together is known only after we conduct online testing of those circuit paths going through the TSV, due to the difficulty to predict change of signal timing slacks with circuit aging. The above considerations have motivated the new in-field repair technique investigated in this paper.

3. IN-FIELD TSV REPAIR FRAMEWORK

In order to conduct in-field repair for TSV latent defects, we first need to be able to test and diagnose faulty TSVs in an online manner. To achieve these objectives, as in [22], we assume the existence of a processor core and non-volatile memory in the system for test and diagnosis purpose (see the conceptual architecture shown in Fig. 5). This assumption is reasonable because 3D logic-on-logic ICs or 3D logic-memory designs of the near future are likely to be large multiprocessor system-on-a-chip (MPSoC) designs. Such designs provide the most compelling motivation for high-density 3D integration. To be specific, the non-volatile memory stores the test and diagnosis patterns for TSV faults, our in-field repair algorithm and the repair signature for each TSV grid, while a processor core is called upon for online test and repair, triggered periodically or by events.

3.1 Online Test and Diagnosis

As discussed earlier, TSVs suffer from interfacial cracks and EM-induced voids and such latent defects usually manifest themselves as hard-to-predict timing errors on critical paths with TSVs. From this perspective, TSV BIST techniques (e.g., [23]) are insufficient for in-field test and diagnosis because they target on faults occurred in individual TSV structure (and often consider TSV open/short only) instead of delay faults of circuit paths with TSVs. For example, as discussed in Section 2.2, using a fault-free TSV to replace a faulty one does not necessarily lead to a valid repair solution because of the unknown signal timing slack changes with circuit aging.

Consequently, it is important to online test those critical paths that go through TSVs. To be specific, for each TSV, we need to pick one or more long paths that go through it and store the corresponding path delay test patterns in non-volatile memory (in a compressed form to reduce the storage requirement, whenever possible).

Note that, we try to overcome the delay fault on a particular path with TSVs by signal rerouting using other TSVs. Even though this strategy mainly targets TSV degradation/failure, it can also be used to target for path delay faults caused by the degradation of other on-path circuit elements. That is, as long as the identified repair solution is confirmed to be valid with online testing, it is not necessary to root-cause the path delay fault to a particular circuit element.

3.2 Spare TSV Sharing and Reconfiguration

Due to the clustering effects of latent faults, unless the redundancy ratio is quite high, we may still run into the situation that some faulty TSV grids lack spare TSVs while the others have many redundant TSVs. We therefore propose to enhance the TSV redundancy architecture presented in [18] by allowing spare TSV sharing between TSV grids, as shown in Fig. 5.

Given the above TSV redundancy architecture for a 3D IC, the design flow of the proposed in-field repair solution is as follows. With online testing triggered periodically or by events, if a particular path with TSVs is found to be faulty, our TSV repair algorithm (detailed in Section 4) is called upon to obtain a possible repair solution. Afterwards, we rerun online testing to check whether this solution is acceptable. The above procedure iterates until a valid repair solution is achieved. The 3D IC is regarded as being irreparable if the circuit is not free of path delay faults after tall the possible repair solutions have been considered.

4. PROPOSED REPAIR ALGORITHM

In this section, we first formulate the in-field repair problem and then present details of the proposed repair algorithm.

4.1 Problem Formulation

For a 3D IC with TSV redundancy architecture as shown in Fig. 5, the in-field TSV repair problem is formulated as follows:

Given the set of signals $\mathbf{S} = \{s_1, s_2, ..., s_n\}$ and the set of TSVs $\mathbf{T} = \{TSV_1, TSV_2, ..., TSV_m\}$ ($n < m$), our goal is to link every signal in $\mathbf{S}$ with a dedicated TSV in $\mathbf{T}$ under the following conditions: (i)

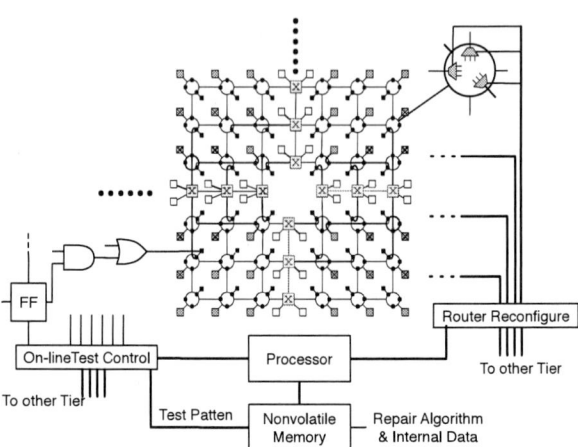

Figure 5: Illustration of the TSV redundancy architecture.

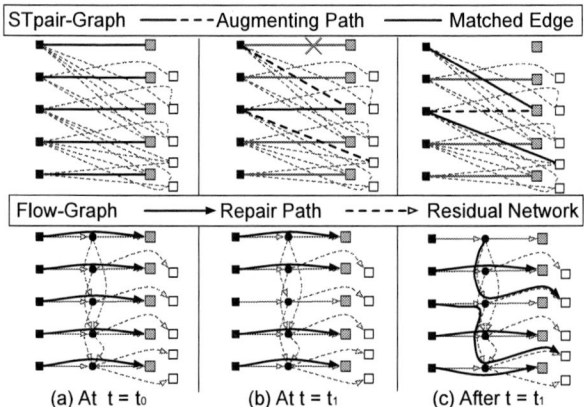

Figure 6: Illustration of the repair algorithm.

all signal-TSV pairs are routable with the given TSV redundancy architecture; (ii) it is confirmed with online testing with no timing violations.

As we need to invoke the online testing procedure whenever there is a possible repair solution, it is preferable to reduce the number of trials for valid repair.

4.2 In-field Repair Algorithm

In order to solve the above problem, we construct a bipartite graph to store all "possible" signal-TSV pairs, namely *STpair-graph* in this paper. Fig. 6 (a) presents an example *STpair-graph* at $t = 0$. In this graph, one side is the signal set $\mathbf{S}$ while the other side is the TSV set $\mathbf{T}$, and an edge exists for a possible signal-TSV pair that has the following two properties: (i). there is at least one routing path from the signal to the TSV in the flow graph; (ii). there is no *confirmed* timing violation for this signal-TSV pair. *STpair-graph* gets updated with online testing results, i.e., an edge is deleted if the corresponding signal-TSV pair fails path delay test, and a TSV and all its edges are removed when it has a catastrophic failure, e.g., a full open defect.

A valid repair solution is hence a *maximum matching* of the *STpair-graph* whose matching number is equal to n (i.e., every signal is paired with a dedicated TSV) and all of the signal-TSV pairs are both routable and confirmed to have no timing violations with online testing. With continuous circuit aging, one can imagine that the number of edges in *STpair-graph* keeps decreasing, and the 3D IC is irreparable when the matching number of a *STpair-graph* is less than n.

We use Fig. 6 to illustrate one possible repair algorithm. Fig. 6(a) presents the matching used in the 3D IC at $t = 0$. Suppose online

518

testing is performed at $t = t_1$, and we found one signal-TSV pair fails its test. We remove this edge from *STpair-graph* and thus the current matching is not maximum any more. In order to find another maximum matching, we resort to Berge's lemma [25], by iteratively finding the shortest *augmenting path*[1] from the unmatched signal to any free TSV. Such a method preserves the signal-TSV pairs in the earlier matching whenever possible and hence is more likely to be valid when compared to a solution based on a random maximum matching of the updated *STpair-graph*. In addition, routability checking is integrated into the above procedure for efficiency. That is, whenever we add an augmented path, we update the corresponding flow graph and check whether it can be routed in the residual network of the flow graph. To update the flow graph, we cancel the edges in the flow graph possessed by those signal-TSV pairs that are removed in the matching and move them to the residual graph (i.e., the sub-graph of the original one, composing edges with residual capacity) (Fig. 6(b)). Then, we can verify routability by finding edge-disjoint paths in the residual network for those signal-TSV pairs that are added into the new matching (Fig. 6(c)). If the matching solution is not routable, we find another augmenting path and iterate the above procedure. Otherwise, we invoke online testing to check whether this solution leads to any timing violation. If not, we have obtained a valid repair solution. Otherwise, we update *STpair-graph* by removing those edges whose corresponding signal-TSV pairs fail path delay tests, and repeat the above procedure on the updated *STpair-graph*.

While simple and effective, the above algorithm may invoke online testing many times due to the enumeration of matchings. Let us use M_i to denote the i_{th} maximum matching of the *STpair-graph* (containing the set of all signal-TSV pairs). The above repair algorithm iteratively finds a new maximum matching and performs online testing for it, until a matching (say, M_v) is shown to be valid. Hence, we need to perform v times of online testing. Generally speaking, however, there is usually a significant overlap of the signal-TSV pairs between M_i and M_{i+1} because we tend to preserve many existing valid signal-TSV pairs from the previous solution in each iteration. These preserved pairs are known to be fault-free with previous testing results, which do not need to be tested again.

Motivated by the above discussion, we propose a more efficient algorithm. Instead of checking one possible matching a time with online testing, we attempt to test "multiple matchings" simultaneously, whenever possible. For example, after testing M_0, for a new maximum matching M_1, we only need to perform online testing for those signal-TSV pairs in $M_1 \backslash M_0$, because the other signal-TSV pairs in $M_1 \bigcap M_0$ have been shown to be valid with previous testing results of M_0. Without loss of generality, let us consider another maximum matching M_2 (if any), there must be some signal-TSV pairs in $M_2 \backslash M_1$ (otherwise M_2 is not a new matching). If some of these signal-TSV pairs have not been tested (i.e., they do not belong to M_0) and they are routable together with those signal-TSV pairs in $M_1 \backslash M_0$, they can be tested simultaneously in one iteration. Such a method reduces the number of online testing because, if a signal-TSV pair in $M_2 \backslash M_1$ is shown to be invalid, not only we do not need to test M_2 any more, but also the corresponding edge is removed from the *STpair-graph* and reduces the possibility to find other invalid matchings.

4.3 Impact of TSV Redundancy Sharing

With TSV redundancy sharing between neighboring TSV grids, there might be conflict between the repair requirements between them. We use the example shown in Fig. 7 to explain how we resolve this issue. In this example, TSV Grid *A* and TSV Grid *B* are sharing spares TSVs in Fig. 7(a)). In this case, for both grids, their *STpair-graphs* contain the shared TSVs. Hence, the two STpair-graphs are connected as shown in Fig. 7(b). We still obtain maximum match-

---

[1]An augmenting path of a matching is defined as a path that starts and ends on free (unmatched) vertices, and alternates between edges in & not in the matching.

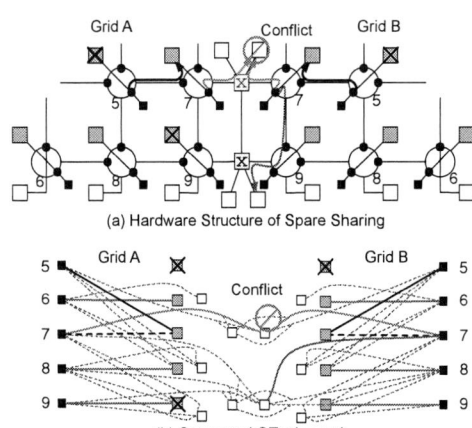

Figure 7: The impact of TSV redundancy sharing on repair algorithm.

ings for each grid by finding augmenting paths. When the two grids try to use the same spare TSV with their augmenting paths, a conflict arises (see red line in Fig. 7(b)). We then arbitrate which grid owns this spare TSV according to the fault maps of the two grids. In this example, Grid *A* has more faults than Grid *B* and hence this spare TSV will be assigned to Grid *A*. We remove this node from the STpair-graph of Grid *B* and it will look for a different matching for in-field repair.

5. EXPERIMENTAL RESULTS
5.1 Experimental Setup

To evaluate the effectiveness and efficiency of the proposed solution, we perform simulation studies and report results on MTTF and test times.

We use the maximum-flow based algorithm presented in [18] as the baseline solution for comparison. As [18] mainly deals with manufacturing defects and the original algorithm uses a static fault map as input, we make the following changes to generate two types of baseline in-field repair algorithms. The first type simply updates the fault map by marking the corresponding TSV to be "faulty" whenever online testing shows an invalid signal-TSV repair and utilizes the original algorithm to find a new repair solution (if possible), denoted as *MF*. For the second type, when a signal-TSV pair is shown to be invalid with online testing, it attempts to find another repair path for the faulty TSV instead of marking the TSV as "faulty", denoted as MF'. The proposed algorithm based on maximum matching with routability verification is denoted as *MV*, while the proposed repair algorithm with test time reduction is named as *MR*. The above results are obtained based on the TSV redundancy architecture in [18]. We further present the results based on the proposed TSV redundancy architecture with spare sharing capability, denoted as *MS*. We compare the MTTF of the above solutions, and a particular 3D IC is deemed to fail when no repair solution can be found for a path delay fault due to aging effects.

The circuits used in our experiments are the performance-optimized data encryption standard (DES) circuit and the fast-Fourier transform (FFT) circuit from the IWLS 2005 OpenCore benchmarks. The DES circuit contains 26,000 gates and 2,000 flops, while the FFT circuit contains 229,000 gates and 20,000 flops. The DES circuit was partitioned into two-, three-, and four-die stacks using the Nangate open cell library and a placement engine for timing optimization. Given the operational frequency of the benchmark 3D ICs, we extract the timing slacks for paths with TSVs. Due to the lack of reliability models for stress-induced TSV interfacial cracks in the public literature, we form our model based on an EM reliability model for TSVs and vary its parameter to reflect the impact of TSV interfacial cracks [9, 17]. We also consider initial TSV failures due to manufac-

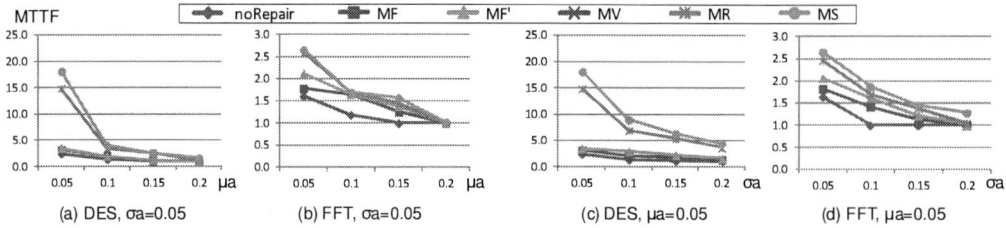

Figure 8: MTTF results in 4×4 TSV grid with varied aging coefficients and fixed Potential Crack or Void Defect Distribution (0.1 kΩm, 0.1 kΩm).

turing defects. Aging effects are characterized by additional latent delay in TSVs, reflected as resistance increase in terms of time t, calculated as

$$R(t) - R_0 = A\ln(\frac{t}{t_0}) \qquad (1)$$

where A is the slope of TSV degradation on a logarithmic scale, and t_0 is the time when the void becomes larger than their TSV section. Note that A and t_0 are affected by multiple parameters, such as the initial resistance R_0 of TSV, TSV barrier resistivity, TSV dimensions, and possibility of voids generated in TSVs. R_0 varies for different TSVs due to process variation and it is assumed to follow a Gaussian distribution. The parameter A indicates the aging rate, which is related to the workloads applied to the 3D IC which in turn determines the temperature and switching activities for TSVs. The dynamic changing of A due to workloads is aggregated in this paper and we use Gaussian distribution to obtain A.

5.2 Results and Analysis

Fig. 8(a)-(d) presents the normalized MTTF values, compared to the worst case without any redundancy for in-field repair. We have four configurations for aging coefficients with their mean values (μa) and variances (σa) varying between 0.05 kΩ/log(s) to 0.2 kΩ/log(s). This setting mimics the circuits under different stress. The distribution of the initial resistances R_0 of TSVs that represents the potential Crack or Void Defect Distribution are fixed with a mean value (μr) of 0.1 kΩ and variance (σr) of 0.1 kΩ.

First of all, it can be observed that the two proposed repair algorithms with the TSV redundancy architecture in [18] lead to much higher MTTF values when compared to the two baseline solutions. For example, for DES design, MTTFs are 14.8 for both MV and MR with aging coefficient of (0.05, 0.05), compared to 3.1 using MF and 3.5 using MF'. This is because we are able to search a much larger solution space by exploring all possible signal-TSV pairs while previous methods only target on repair path identification for faulty TSVs. The MTTF value of MF' is slightly better than that of MF because the latter solution regards the TSV from an invalid signal-TSV pair as "faulty", rendering an even smaller solution space. It should be noted that MV and MR have the same MTTF as the solution space are the same for these two algorithms. By adding spare sharing capability with the proposed architecture, the MTTF is increased to 18.2 under the same aging rate.

Secondly, we observe significant MTTF reduction as aging coefficients increase (see Fig. 8(a)-(d)), due to the higher TSV failure probability with increasing aging rates. The differences are even larger for the proposed two repair algorithms (MV and MS), wherein the

impetus of downtrends is reduced as the aging coefficient increases. This is expected as the solution space shrinks quickly as aging effect becomes more severe, rendering less repair efficiency for all repair algorithms. This indicates that, even with a better TSV redundancy architecture (with spare TSV redundancy sharing), we cannot achieve high MTTF values when the circuit is under severe aging effects.

Thirdly, we compare the results of DES design in Fig. 8(a)(c) and FFT design in Fig. 8(b)(d). While we can see similar trends for the results of FFT design, but the MTTF differences between the five algorithms are not as significant as that of DES design. This is mainly because the timing slacks of paths with TSVs in FFT design is much tighter, thus leading to less MTTF values.

Fig. 9(a)-(b) describe the corresponding test time (in terms of the number of performed online testing) of the circuit with the five repair methods under various aging coefficients, corresponding to Fig. 8(a)-(b). The proposed algorithms requires more test time compared to the baseline algorithms due to the fact that more on-line tests are conducted to achieve more successful repair. While the MTTF values for MV and MR are the same, the test times of MR are much smaller. This is because, we try to perform online testing for multiple possible matchings at the same time. With such test time reduction scheme, the test times of MS only increase slightly although it has a larger solution space to explore with spare redundancy sharing.

Fig. 10 shows the MTTF values of different repair methods when we vary the initial resistances of TSVs, which demonstrate the impact of undetectable cracks/voids during fabrication on the service life of 3D ICs. Due to space limit, we only report the results of DES circuit. We fix the aging coefficient as (0.05, 0.05), and vary the TSV initial resistances R_0 with its mean values from 0.1 kΩ to 0.4 kΩ and a fixed variance value in Fig. 10(a). While in Fig. 10(b), we also have four configurations for R_0 with the same variance value of 0.1 kΩ but different mean values ranging from 0.1 kΩ to 0.4 kΩ. From this figure, we can observe that the TSV initial resistance has minor impact on the MTTF values, when compared to the aging coefficients changes shown in Fig. 8. This is also expected because TSV voids/cracks that have passed burn-in test, have to grow large enough to affect circuit timing, which is determined more by the aging rates instead of their initial values.

Fig. 11 shows the MTTF values of the proposed repair methods with a 8×8 grid size for the repair architecture. The trends are similar to that in Fig. 8, however, the MTTF values of MR and MS are much larger than that in Fig. 8. The main reason is that, in this experiment the signal rerouting delay is not considered and hence we have a much larger repair solution space to explore with a larger TSV bundle.

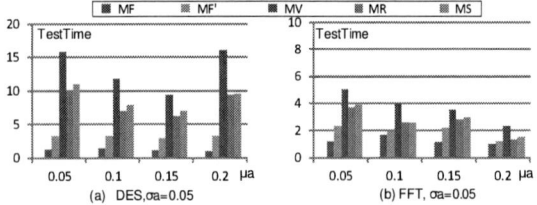

Figure 9: Test Time results in 4×4 TSV grid with varied aging coefficients and fixed Potential Crack or Void Defect Distribution (0.1 kΩm, 0.1 kΩm).

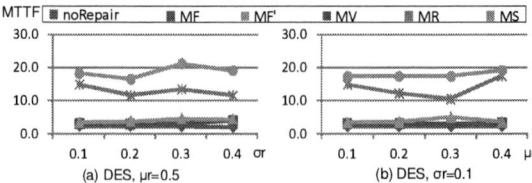

Figure 10: MTTF results for DES in 4×4 TSV grid with varied Potential Crack/void Defect Distribution and fixed aging coefficients (0.05 kΩ/log(s),0.05 kΩ/log(s)).

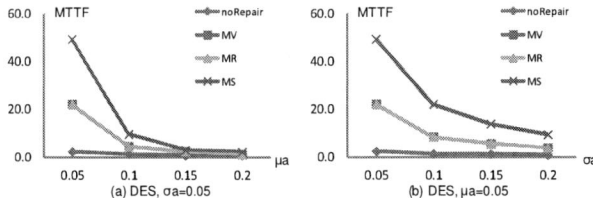

Figure 11: Experimental results in 8×8 TSV grid size repair architecture with varied aging coefficients and fixed Potential Crack/void Defect Distribution (0.1 kΩ, 0.1 kΩ).

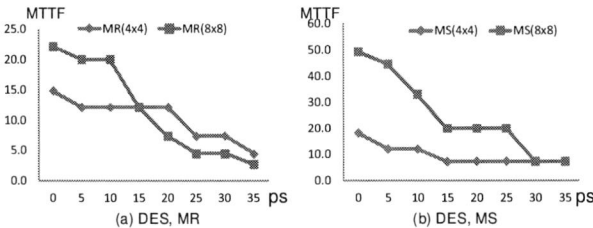

Figure 12: Experimental results with varied rerouting delay between two adjacent routers (*ps*) and fixed Aging Coefficents and Potential Crack/void Defect Distribution.

It should be noted that, the extra signal rerouting delay is already taken into consideration in the proposed repair architecture even though the previous simulation studies ignore it. As long as the rerouting delay of a signal-TSV pair exceeds the timing slack of the path containing this signal-TSV pair, the on-line test can detect a timing error and discard this pair from the solution space. Fig. 12 investigates the effect of this rerouting delays and shows the MTTF of the proposed repair methods for two different grid size in the repair architecture. For both methods, the architecture with 8×8 grid size performs better when the rerouting delay is small, because it has larger solution space for repair. As rerouting delay increases, the MTTF curves of the two architectures intersect at a point when it has become a bottleneck for signals to be able to reach many TSVs for repair. Beyond this point, the architecture with 4×4 grid size results in more successful repair with higher redundancy ratio. Compared to *MR*, the intersection point of the two architecture occurs later for *MS* because the shared redundant TSVs give each TSV grid more solution space to explore.

6. CONCLUSION

TSV-based 3D ICs have emerged as one of the most promising solutions to overcome interconnect bottleneck in CMOS scaling. The disruptive manufacturing process of TSVs, however, introduce new failure mechanisms such as stress-induced interfacial cracks and EM-induced voids. Such reliability threats reduce the service life of 3D ICs. In this paper, we have described a novel in-field repair solution that is able to effectively and efficiently tolerate latent TSV defects through the judicious use of spares. Experimental results on 3D benchmark circuits show that the proposed solution is able to significantly increase MTTF when compared to existing TSV repair techniques.

7. REFERENCES

[1] International Technology Roadmap for Semiconductors (ITRS'11), available at *http://www.itrs.net/*.

[2] D. Pan, et al. Design for manufacturability and reliability for TSV-based 3D ICs. In *IEEE/ACM Asia and South Pacific Design Automation Conference*, pages 750–755, 2012.

[3] A. Karmarkar, X. Xu, and V. Moroz. Performanace and reliability analysis of 3D-integration structures employing through silicon via (TSV). In *IEEE International Reliability Physics Symposium*, pages 682 –687, april 2009.

[4] K. N. Tu. Reliability Challenges in 3D IC Packaging Technology. In *Microelectronics Reliability*, pages 517-523, March 2011.

[5] T. Dao, D. Triyoso, M. Petras, and M. Canonico. Through silicon via stress characterization. In *Proc. IEEE International Conference on IC Design and Technology*, 2009.

[6] C. Selvanayagam, J. Lau, X. Zhang, S. Seah, K. Vaidyanathan, and T. C. Chai Nonlinear thermal stress/strain analysis of copper filled TSV and their flip-chip micro-bumps. In *Proc. IEEE Electronic Components and Technology Conference*, pages 1073–1081, 2008.

[7] S. Ryu, K. Lu, X. Zhang, J. Im, P. Ho, and R. Huang. Impact of near-surface thermal stresses on interfacial reliability of through-silicon vias for 3-D interconnects. *IEEE Transactions on Device and Materials Reliability*, 11(1):35–43, 2011.

[8] Y. Tan, C. Tan, X. Zhang, T. Chai, and D. Yu. Electromigration performance of through silicon via (TSV) – A modeling approach. *Microelectronics Reliability*, 50(9):1336–1340, 2010.

[9] T. Frank, C. Chappaz, P. Leduc, L. Arnaud, F. Lorut, S. Moreau, A. Thuaire, R. El Farhane, and L. Anghel. Resistance increase due to electromigration induced depletion under TSV. In *Proc. IEEE International Reliability Physics Symposium*, pages 3F.4.1–3F.4.6, 2011.

[10] A. Hsieh, T. Hwang, M. Chang, M. Tsai, C. Tseng, and H.-C. Li. TSV redundancy: Architecture and design issues in 3D IC. In *Proc. Design, Automation, and Test in Europe Conference Exhibition*, pages 166 –171, march 2010.

[11] U. Kang, H. Chung, S. Heo, D. Park, H. Lee, J. Kim, S. Ahn, S. Cha, J. Ahn, D. Kwon, et al. 8 GB 3-D DDR3 DRAM using through-silicon-via technology. *IEEE Journal of Solid-State Circuits*, 45(1):111–119, 2010.

[12] I. Loi, S. Mitra, T. Lee, S. Fujita, and L. Benini. A low-overhead fault tolerance scheme for TSV-based 3D network on chip links. In *Proc. International Conference on Computer-Aided Design*, pages 598–602, nov. 2008.

[13] M. Nicolaidis, V. Pasca and L. Anghel. Through-silicon-via built-in self-repair for aggressive 3D integration. In *IEEE International On-Line Testing Symposium (IOLTS)*, pages 91–96, 2012.

[14] Y.-J. Huang and J.-F. Li. Built-In Self-Repair Scheme for the TSVs in 3-D ICs. In *IEEE Transactions on Computer-Aided Design of Integrated Circuits and Systems*, 31(10): 1600–1613, 2012.

[15] Y. Zhao, S. Khursheed, and B. Al-Hashimi. Cost-Effective TSV Grouping for Yield Improvement of 3D-ICs. In *Proc. IEEE Asian Test Symposium*, 2011.

[16] J. Xie, Y. Wang, and Y. Xie. Yield-aware time-efficient testing and self-fixing design for TSV-based 3D ICs. In *IEEE/ACM Asia and South Pacific Design Automation Conference*, pages 738–743, 2012.

[17] F. Ye and K. Chakrabarty. TSV open defects in 3D integrated circuits: Characterization, test, and optimal spare allocation. In *Proc. IEEE/ACM Design Automation Conference*, pages 1024–1030, 2012.

[18] L. Jiang, Q. Xu, and B. Eklow. On effective TSV repair for 3D-stacked ICs. In *IEEE/ACM Proc. Design, Automation, and Test in Europe*, pages 6–11, 2012.

[19] L. Jiang, R. Ye, and Q. Xu. Yield enhancement for 3D-stacked memory by redundancy sharing across dies. In *Proc. International Conference on Computer-Aided Design*, pages 230–234, nov. 2010.

[20] L. Jiang, Q. Xu, and B. Eklow. On effective TSV repair for 3D-stacked ICs. *IEEE Transactions on Computer-Aided Design of Integrated Circuits and Systems*, to appear.

[21] G. Van der Plas, P. Limaye, I. Loi, et al. Design issues and considerations for low-cost 3-D TSV IC technology. *IEEE Journal of Solid-State Circuits*, 46(1):293–307, 2011.

[22] Y. Li, S. Makar, and S. Mitra. CASP: Concurrent autonomous chip self-test using stored test patterns. In *Proc. IEEE/ACM Design, Automation, and Test in Europe Conference and Exhibition*, pages 885–890, 2008.

[23] Huang, Yu-Jen, et al. A built-in self-test scheme for the post-bond test of TSVs in 3D ICs. In *Proc. IEEE VLSI Test Symposium (VTS)*, pages 20–25, 2011.

[24] S. Fortune, J. Hopcroft, and J. Wyllie. The directed subgraph homeomorphism problem. *Theoretical Computer Science*, 10(2):111–121, 1980.

[25] C. Berge. Two theorems in graph theory. In *National Academy of Sciences of the United States of America*, volume 43 of 9, pages 842–844, 1957.

Cloud Platforms and Embedded Computing – The Operating Systems of the Future

Jan S. Rellermeyer[1]   Seong-Won Lee[1,2]   Michael Kistler[1]

[1]IBM Austin Research Lab
Future Systems Group
{rellermeyer, mkistler}@us.ibm.com

[2]Seoul National University
School of Electrical Engineering and Computer Science,
swlee@altair.snu.ac.kr

ABSTRACT

The discussion on how to effectively program embedded systems has often in the past revolved around issues like the ideal instruction set architecture (ISA) or the best operating system. Much of this has been motivated by the inherently resource-constrained nature of embedded devices that mandates efficiency as the primary design principle.

In this paper, we advocate a change in the way we see and treat embedded systems. Not only have embedded systems become much more powerful and resources more affordable, we also see a trend towards making embedded devices more consumable, programmable, and customizable by end users. In fact, we see a strong similarity with recent developments in cloud computing.

We outline several challenges and opportunities in turning a language runtime system like the Java Virtual Machine into a cloud platform. We focus in particular on support for running multiple tenants concurrently within the platform. Multi-tenant support is essential for efficient resource utilization in cloud environments but can also improve application performance and overall user experience in embedded environments. We believe that today's modern language runtimes, with extensions to support multitenancy, can form the basis for a single continuous platform for emerging embedded applications backed by cloud-based service infrastructures.

Categories and Subject Descriptors

D.3.4 [**Programming Languages**]: Processors—*Run-time environments*

General Terms

Design, Performance

Permission to make digital or hard copies of all or part of this work for personal or classroom use is granted without fee provided that copies are not made or distributed for profit or commercial advantage and that copies bear this notice and the full citation on the first page. To copy otherwise, to republish, to post on servers or to redistribute to lists, requires prior specific permission and/or a fee.
DAC '13, May 29 - June 07 2013, Austin, TX, USA.

Keywords

Platform as a Service, Embedded Systems, Cloud Computing

1. INTRODUCTION

Traditional embedded systems are fixed-function devices that combine specialized hardware with special-purpose, low-level software and provide very limited or even no user interface elements. The software components of traditional embedded systems are often closely tied to the hardware implementation, in some cases written directly in the assembly language for the processor. Examples of traditional embedded systems are digital watches, MP3 players, and computer modems.

While many of today's embedded systems fit this traditional model, a new class of embedded systems is emerging that offers a much larger set of capabilities, increased flexibility, and a greatly enhanced user interface. This new class of embedded systems is made possible by the continued advance of hardware technologies, particularly in the areas of microprocessor performance, memory density, and reduced power consumption. Touch-screen and voice-recognition technologies have also helped drive many of the new user interface features in these new embedded systems. Common examples of this new type of embedded system are smart phones and tablets, in-car navigation and entertainment systems, and programmable wireless access points (e.g. running OpenWRT [29]).

Traditional cloud computing environments, on the other hand, offer a highly flexible and programmable environment. A typical cloud environment provides clients with *virtual machines* which can be programmed with a choice of operating system, middleware, and applications, all configured according to the client's specifications. The client typically also has a choice of virtual machines with different amounts of processing power, memory size, and storage configuration. This style of cloud environment has come to be known as Infrastructure-as-a-Service, or *IaaS*. Common examples of IaaS cloud environments are Amazon EC2 [12] and IBM's SmartCloud Entry [19].

Cloud environments are also evolving. While the traditional IaaS cloud offers very high flexibility and configurability, this comes with a large system administration burden.

Virtual machines need similar care in terms of maintenance and security patching as physical machines do. Furthermore, many new applications for the cloud are being developed in highly portable languages such as Java and Javascript. As a result, the flexibility and configurability of the IaaS virtual machine offers little benefit to these modern applications but still entails a high cost. These trends have driven a new paradigm in cloud computing referred to as Platform-as-a-Service, or *PaaS*. A PaaS cloud environment provides the client with one or more high-level application runtimes, such as the Java Runtime Environment (JRE), along with other services such as a database. a key-value store, or an authentication service. The PaaS cloud provider manages all of the underlying physical and virtual hardware, operating system images, file systems, and network configuration.

The trends in embedded systems and the trends in cloud computing platforms have striking similarities. In both cases, the platform is evolving towards higher functionality with a focus on application-level capabilities. In fact, many of the technologies between these two spaces are starting to converge. For example, Javascript was once almost exclusively a client-side language used primarily within web browsers, but through packages such as Node.js [27] it is now also being employed in web applications deployed in PaaS cloud environments (e.g., Microsoft's Windows Azure [25]). Similarly, Java, Ruby, Scala, and Python, are now quite commonly used in both embedded and cloud-based applications. Besides languages, other common infrastructures are also emerging, e.g., asynchronous and multi-channel network communication such as WebSockets [13]. This enables new opportunities for shifting parts of the application between the client and the cloud. In short, the embedded and PaaS cloud application runtime environments now have many significant areas of similarity.

What gives this trend special significance is that many emerging embedded applications utilize one or more web applications to provide their functionality, and these web applications are often deployed into PaaS cloud environments. The traditional view of this application design may be *client-server*, but the similarity of the application runtime environments between client and server allows us to view it as a more general *distributed* application. The difference is subtle but powerful, in that it allows a more flexible and dynamic distribution of functionality between embedded application and cloud-based service. Application data can be stored on the device or in the cloud, and the code to manage this can be nearly identical between the two platforms. Data analysis and transformation functions can be performed either on the device or in the cloud, with the decision made dynamically and possibly even involving transfer of the functions' implementation between the device and the cloud.

Differences certainly still remain between the embedded environment and PaaS cloud environments, e.g., the typical total amount of resources, the richness of user input and interaction, or the degree of multi-tenancy. While differences such as these are certainly significant, we believe there is often value in focusing on the similarities of the environments and exploiting these where possible. Having a common platform to program both the embedded device and the cloud helps to create a symbiotic relationship and a continuous user experience.

In the following section, we discuss mobile phones as an example of embedded systems which have already evolved into a more open and consumable platform that includes devices and the cloud. Mobile phones also provide a prime example of language runtime systems serving as a platform for both embedded devices and the cloud. In Section 3, we shed light on the systems issues involved in turning a language runtime system like the Java Virtual Machine [23] into a platform. While some of the architectural changes are motivated by the multi-tenant setup in the cloud, they can also help to reduce the footprint of the platform and make it better suited for embedded devices. Finally, in Section 4 we discuss the implications of using the principles of Platform-as-a-Service on the embedded system to create a continuous user experience between embedded device and the cloud.

2. LESSONS LEARNED FROM MOBILE PHONES

Mobile phones are a prominent example of a class of embedded systems that is already and aggressively moving towards a symbiotic relationship with the cloud. For instance, isolated on-device storage is becoming less important and many platforms already use it more as a cache for cloud-based storage. Furthermore, mobile phones have made the transition from closed devices programmed by experts and in low-level languages to vibrant platforms where a large community of developers creates an ecosystem of apps around the devices. The most popular platforms such as Android are based on high-level, interpreted or just-in-time compiled languages such as Java, C#, or JavaScript to bridge hardware heterogeneity and lower the burden of writing applications. The notable exception is Apple's iOS which uses Objective C, arguably still a low-level language. It has to be taken into account, though, that Apple is also one of the few examples of a mobile platform that is entirely controlled by a single company in that Apple develops both the software and the hardware. Since they do not make their software available to any other hardware platform heterogeneity is severely limited. However, even for the iOS platform there is a trend towards developing content in a platform-independent manner and using native capabilities only through plugins, e.g., as in PhoneGap [32].

Another notable aspect of smart phones is the interaction between the devices and the backend server infrastructure. Smart phone applications tend to be limited in the amount of processing performed on the device. Yet, they have emerged as a more seamless experience for accessing external content that is better integrated into the client platform as opposed to web pages running in a browser. Since mobile apps typically have strong dependencies to their backend infrastructure, developers are increasingly exploring ways of developing both client app and backend logic in a single process and using the same tools. For instance, it has become popular to use languages like JavaScript originally designed for client-side processing on the server side, e.g., with Node.js [27] or Vert.X [35]. This has lead to a new type of Platform as a Service (PaaS) specifically for mobile devices, often referred to as Mobile Backened as a Service (BaaS or MBaaS) which specifically targets the common services that mobile applications require.

While there are many potential advantages to a common platform that could support both embedded and PaaS cloud applications, there are also some significant challenges. One of these is support for *multi-tenancy*, which is the ability

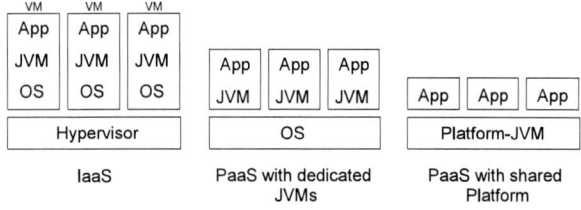

Figure 1: Options for Embedding Runtime Systems into Platforms

to support multiple independent active applications on the platform. Multi-tenancy is clearly a key feature for cloud environments, where the role of the cloud infrastructure is to achieve efficiency through high resource utilization by sharing physical and virtual resources to a potentially large set of active applications. Another dimension of multi-tenant support is the efficient handling of tenants and workloads that continuously come and go over time.

With embedded devices such as mobile phones increasing in processing capability and richness of user interface, support for multi-tenancy is poised to deliver significant value in this space as well, particularly if this support is provided in a consistent, seamless fashion across the embedded and cloud environments. Returning to our mobile phone example, many newer applications include support for push notifications, alerts, alarms, scheduled operations, and other features that generally require the app to remain running in the background. As a result, it is not uncommon today for a mobile phone to have tens of active applications sharing the resources of the device. All indications are that the number of active applications will grow rapidly in future years, so robust and efficient multi-tenancy is also a key requirement for future embedded platforms.

3. FROM RUNTIME SYSTEMS TO PLATFORMS

Language runtime systems such as the Java Virtual Machine [23] have reached a high level of optimization and by now run a significant share of enterprise workloads such as application servers (e.g., WebSphere [20]), analytics (e.g., Hadoop [1]), and search (e.g., Lucene [2]). As a result, many popular Platform as a Service offerings such as OpenShift [28], Google AppEngine [15], Heroku [17], and VMware CloudFoundry [10] include Java support. However, optimal resource utilization is important for cloud stacks in order to be cost effective, a similarity with embedded systems which are resource-constrained by design.

Figure 1 illustrates multiple ways to embed language runtime systems such as the JVM into platforms and make them available to multiple tenants at the same time. The first and most straightforward way is to build the platform atop IaaS and let every tenant get its own virtual machine image containing an operating system, the language runtime (in this case the JVM) and the application that the tenant wants to run (left part of Figure 1). This solution provides nearly perfect isolation between tenants and predictable performance governed by the hypervisor. However, it also has high resource overhead due to the dedicated OS instances

and the footprint of the virtual machines. As a result, typical servers can run only a small number of tenants on the same hardware, which limits the density and thereby the effective utilization of hardware resources. Many platforms therefore do not rely on hardware virtualization but instead resort to some form of OS-provided process-based isolation such as chroot jails [8] or Linux Containers (LXC) [24]. In this solution (middle of Figure 1) tenants share the operating system but still run dedicated instances of the language runtime, which imposes overhead. By intuition, it should be most resource-efficient to run all tenants on a single, shared runtime (right part of Figure 1). This certainly raises security concerns due to the lack of proper isolation but it represents a baseline for the performance of multi-tenancy.

We devised a series of experiments to explore the benefits and limitations of a shared JVM platform. All of our experiments use multiple concurrent instances of the DaCapo [6] Batik benchmark as the primary workload, where each benchmark instance represents a tenant. All experiments were performed on a 12 core Intel Xeon E5645 system with HyperThreading running at 2.40 GHz, 12 GiB of main memory, and running Oracle's HotSpot JVM build 1.6.0_33-b03 in server mode.

Figure 2 presents the average per-instance execution time of the DaCapo Batik benchmark when running on either a *dedicated JVM* platform, where each instance of the benchmark runs on a separate JVM, or a *shared JVM* platform, with all benchmark instances running in the same JVM. The results indicate that the dedicated JVM platform can support up to about 85 instances while still providing predictable performance. The candlestick error bars depict both the standard deviation between iterations of the same experiment (sticks), of which we conducted 5 per data point, as well as the average spread between the instances executed in each run (candle). After 85 instances, the system starts to thrash due to resource exhaustion and exhibits highly volatile behavior. Surprisingly, the shared JVM platform does not scale as well as expected. After about 20 concurrent benchmark instances the system exhibits volatile and inferior per-instance performance, significantly worse than running each instance on a dedicated JVM.

This experiment clearly illustrates that the JVM lacks support for efficient execution of multi-tenant workloads. The results suggest that increasing sharing between instances to utilize resources more effectively has actually created a

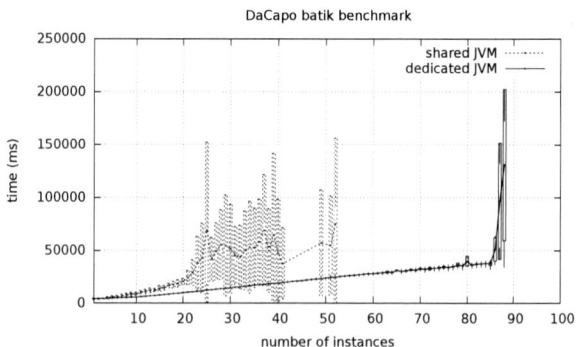

Figure 2: DaCapo Batik Benchmark

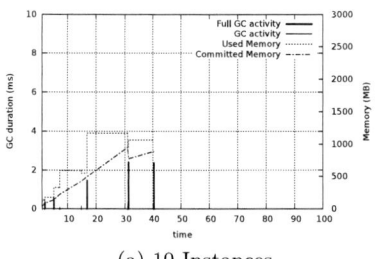 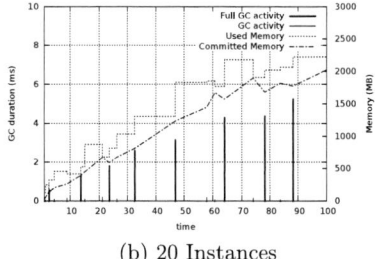 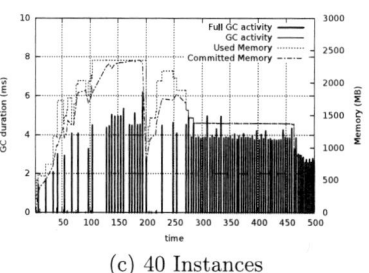

| (a) 10 Instances | (b) 20 Instances | (c) 40 Instances |

Figure 3: Garbage collection activity for the DaCapo Batik benchmark

high level of accidental sharing within the virtual machine that adversely affects performance. In the following sections we identify some of the key structural changes required for the JVM to support efficient multi-tenancy. By doing so, we can draw new lines in the design space for language runtime systems between tenant isolation where required and resource sharing where permitted. This will allow the JVM to use cloud resources more efficiently, and also to run more client applications within the scarce resources of an embedded system.

3.1 Tenant Isolation and Memory Management

Isolation in JVMs has been previously discussed, e.g., in the context of running multiple (desktop) applications on the same JVM. Sun's Multitasking Virtual Machine (MVM) [11] introduced the notion of *Isolates* which form containers of minimal isolation between tasks whereas the remaining JVM and its internal components are shared. The authors indicate that static fields, class initialization state, and instances of *java.lang.Class* are sufficient to replicate per task to achieve isolation whereas everything else (e.g., constant pool, interpreter, JIT compiler, etc.) can be shared. The isolates in the MVM, however, provide full isolation akin to OS processes which makes communication among isolates costly and sharing of common resources difficult. This is particularly an issue with fine-granular component models like OSGi [30] where inter-module communication is expected to happen frequently. Projects like i-JVM [14] have therefore looked into more lightweight forms of isolation by allowing threads to cross the boundaries of isolates and controlled sharing of objects for inter-isolate communication. The general challenge remains to share where desired, e.g., to avoid duplication of identical resources like the Java classpath, and to isolate where tenants show individual behavior. This is not only a security concern but has important implications on the resource consumption of applications and performance. One example where the performance aspect of isolation shows is communication, as the authors of the i-JVM have pointed out. Another one is memory management.

Garbage collection is a critical component in managed runtime systems and can have significant impact on the runtime performance of applications [5, 18]. In current JVMs, the garbage collector (GC) is unaware of independent tenants with no common object references, so it cannot operate on individual isolates but needs to traverse the entire heap each collection cycle. We performed an experiment to assess the impact of GC on a shared JVM platform. In this experiment, we run multiple instances of the benchmark

concurrently in a single shared JVM. Due to lack of isolation in the JVM, all tenants share a single heap which is managed by one instance of the garbage collector. This resembles the situation in i-JVM and the *old generation* on the heap of the MVM. Figure 3 presents time-series graphs that show the behavior of the GC during the execution of 10, 20, and 40 instances of the benchmark. GC activity is plotted as bars whose height indicates the duration of the activity in ms as shown on the left y-axis. Used and committed memory in MB are shown as lines plotted against the right y-axis. All runs use the default JVM heap size of 3GB, and GC statistics are obtained from the JVM through command line flags. When running 10 or 20 concurrent instances as in Figures 3(a) and 3(b), full collection cycles occur infrequently and complete in a short time. The trace for 40 instances (Figures 3(c)) shows a system that is overloaded with garbage collection towards the second half of the run, even though the memory pressure does not appear to be higher than for 20 instances.

This illustrates a scalability problem with larger heaps and underlines the need for better tenant isolation when using the Java Virtual Machine as a platform for multiple tenants. It also points out that the assumption that the majority of JVM components can be shared is sufficient for isolation but likely exhibits scalability and performance issues when running a high number of tenants. Instead, we propose a different JVM architecture where components like garbage collection are offered as a service (but not necessarily as a single instance) to multiple tenants.

3.2 Just-in-time Compilation

Just-in-time compilation is a proven technology for acceleration of high-level interpreted languages. For a single application, the rule of virtue is to concentrate on hot spots of the application with the compilation capacity as large as possible. Various approaches have been developed in research as well as in products for improving the performance of JIT compilation. Among these efforts are a selective compilation with bytecode interpreter [31, 33] and adaptive compilation with multi-level optimizing compilers [3, 34, 9]. Because the decision of when and what to compile has to be determined during runtime, the hot spot detection technique is one of the most important concerns in these optimization frameworks [4, 16, 7, 26].

Figure 4 shows the total JIT time (in seconds) and JIT throughput (in bytecodes processed per second) of the just-in-time compiler during the execution of multiple instances of the DaCapo Batik benchmark on a shared JVM platform. From our previous experiments we know that performance

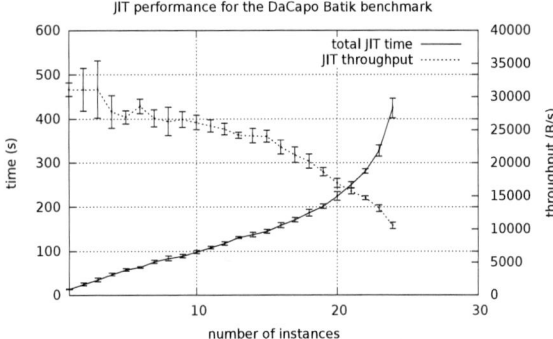

Figure 4: JIT Throughput for the Batik Benchmark

begins to degrade at around 20 instances, and in this graph we see that at this point the total JIT time is increasing exponentially while JIT throughput is significantly degraded. This is a clear indication that the shared JIT compiler becomes a bottleneck and contributes to the limited scalability of the JVM in a multi-tenant setup.

For our proposed cloud platform, we have to address a different strategy because each tenant has a resource and content sharing policy of its own and we are supposed to provide compilation resource in a more flexible manner. A promising design option is having isolated compiler instances, created on demand and assigned job allocation from task pools shared among tenants. Thus, as depicted in Figure 5, no duplicated compilation request for a shared item among tenants exists in the compilation task pool and tenants are served generated native code from the shared code cache or output of JIT compiler directly.

As our multi-tenancy model is structurally similar to a multi-threaded application running on a single JVM from a resource sharing point of view, we can anticipate our JIT compiler operation based on some previous work in such environments. Our cloud platform tries to achieve a goal to maximize resource sharing among tenants and it inevitably can occur situations such that multiple tenants together with a compiler instance are scheduled to be distributed CPU time. Kulkarni et. al. have demonstrated that application performance could be severely degraded as the number of threads per core is increased and a single compilation thread is dispossessed of its resource utilization by the JVM thread scheduler [21]. However, even under this circumstance, it makes the throughput stable sooner to guarantee some amount of CPU time for the compilation thread. Obviously, more than one JIT compiler instance can leverage the throughput of compilation requests especially when there exist multiple tenants starting up over abundant hardware resources. As a matter of fact, it is reported that multiple compilation threads can produce significant performance improvement on a many core environment [22].

3.3 Changes to the JVM

We have identified the need for structural changes in language runtime systems like the JVM in order to be used as multi-tenant platforms. Most importantly, proper support for tenant isolation needs to be added not only for security but also for performance reasons. At the same time, it is de-

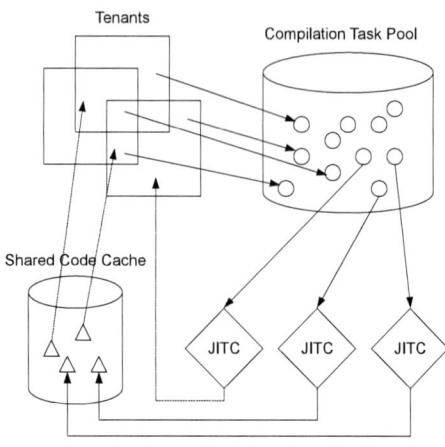

Figure 5: JIT Compiler Operation in Cloud Platform

sirable to share common infrastructure such as the classpath, class linking and loading, GC, JIT compiler, etc. However, especially the performance-critical components of the JVM do not scale well enough to be used as system singletons. Instead, what we envision is a dynamic pool of these components, each serving a group of tenants. We have illustrated this design approach in more detail for just-in-time compilation but the same principle can also be applied, e.g., to memory management/garbage collection or the classloading subsystem.

4. CONCLUSION: PAAS ON EMBEDDED DEVICES

In the previous sections, we have outlined some of our experience in using a traditional language runtime system like the JVM in a multi-tenant setup for cloud computing. The goal was to provide a higher density of tenants on a single server machine and we were able to show that the current JVM is not able to achieve this goal due to adverse effects when running multiple tenants on the same VM. We have discussed architectural changes to the JVM that are likely to improve this situation. In our ongoing work, we envision that language runtime systems like the JVM are not only becoming more scalable and amenable for running multiple tenants in PaaS settings but ultimately also become more lightweight in general. This is an important step towards running the same stack on both the backend server and the embedded device and provide a continuous platform experience between device and cloud. Here it will have even greater benefits because embedded resources are typically much more static and constrained in comparison to the cloud. If platforms will eventually be able to cover both the client devices and the cloud backend, thereby raising the level of abstraction significantly, they will become what the next generation of users and programmers will see as the operating system of the future for both embedded systems and the cloud.

5. REFERENCES

[1] Apache Hadoop. hadoop.apache.org.

[2] Apache Lucene. lucene.apache.org.

[3] M. Arnold, S. Fink, D. Grove, M. Hind, and P. F. Sweeney. Adaptive optimization in the Jalapeno JVM. In *Proceedings of the 15th ACM SIGPLAN conference on Object-oriented programming, systems, languages, and applications*, OOPSLA '00, pages 47–65, New York, NY, USA, 2000. ACM.

[4] M. Arnold and B. G. Ryder. A framework for reducing the cost of instrumented code. In *Proceedings of the ACM SIGPLAN 2001 conference on Programming language design and implementation*, PLDI '01, pages 168–179, New York, NY, USA, 2001. ACM.

[5] S. M. Blackburn, P. Cheng, and K. S. McKinley. Myths and realities: the performance impact of garbage collection. *SIGMETRICS Perform. Eval. Rev.*, 32(1):25–36, June 2004.

[6] S. M. Blackburn, R. Garner, C. Hoffmann, A. M. Khang, K. S. McKinley, R. Bentzur, A. Diwan, D. Feinberg, D. Frampton, S. Z. Guyer, M. Hirzel, A. Hosking, M. Jump, H. Lee, J. E. B. Moss, A. Phansalkar, D. Stefanović, T. VanDrunen, D. von Dincklage, and B. Wiedermann. The DaCapo Benchmarks: Java Benchmarking Development and Analysis. In *Proceedings of the ACM SIGPLAN Conference on Object-oriented Programming Systems, Languages, and Applications*, OOPSLA '06, pages 169–190, New York, NY, USA, 2006. ACM.

[7] D. Buytaert, A. Georges, M. Hind, M. Arnold, L. Eeckhout, and K. De Bosschere. Using hpm-sampling to drive dynamic compilation. In *Proceedings of the 22nd annual ACM SIGPLAN conference on Object-oriented programming systems and applications*, OOPSLA '07, pages 553–568, New York, NY, USA, 2007. ACM.

[8] chroot. http://www.freebsd.org/chroot/2.

[9] M. Cierniak, M. Eng, N. Glew, B. Lewis, and J. Stichnoth. The Open Runtime Platform: a flexible high-performance managed runtime environment: Research Articles. *Concurr. Comput. : Pract. Exper.*, 17(5-6):617–637, Apr. 2005.

[10] Cloud Foundry. http://www.cloudfoundry.com.

[11] G. Czajkowski and L. Daynés. Multitasking without comprimise: a virtual machine evolution. In *Proceedings of the 16th ACM SIGPLAN conference on Object-oriented programming, systems, languages, and applications*, OOPSLA '01, pages 125–138, New York, NY, USA, 2001. ACM.

[12] Amazon Elastic Compute Cloud (EC2). http://aws.amazon.com/ec2.

[13] I. Fette and A. Melnikov. The WebSocket Protocol. *RFC 6455*, December 2011.

[14] N. Geoffray, G. Thomas, G. Muller, P. Parrend, S. Frénot, and B. Folliot. I-JVM: a Java Virtual Machine for Component Isolation in OSGi. In *International Conference on Dependable Systems and Networks (DSN 2009)*, Estoril, Portugal, June 2009. IEEE Computer Society.

[15] Google AppEngine. https://developers.google.com/appengine.

[16] D. Gu and C. Verbrugge. Phase-based adaptive recompilation in a JVM. In *Proceedings of the 6th annual IEEE/ACM international symposium on Code generation and optimization*, CGO '08, pages 24–34, New York, NY, USA, 2008. ACM.

[17] Heroku Cloud Application Platform. http://www.heroku.com.

[18] M. Hertz and E. D. Berger. Quantifying the performance of garbage collection vs. explicit memory management. In *Proceedings of the 20th annual ACM SIGPLAN conference on Object-oriented programming, systems, languages, and applications*, OOPSLA '05, pages 313–326, New York, NY, USA, 2005. ACM.

[19] IBM Smart Cloud Entry. http://www.ibm.com/cloud.

[20] IBM WebSphere. www.ibm.com/software/websphere.

[21] P. Kulkarni, M. Arnold, and M. Hind. Dynamic compilation: the benefits of early investing. In *Proceedings of the 3rd international conference on Virtual execution environments*, VEE '07, pages 94–104, New York, NY, USA, 2007. ACM.

[22] P. A. Kulkarni. JIT compilation policy for modern machines. In *Proceedings of the 2011 ACM international conference on Object oriented programming systems languages and applications*, OOPSLA '11, pages 773–788, New York, NY, USA, 2011. ACM.

[23] T. Lindholm and F. Yellin. *Java Virtual Machine Specification*. Addison-Wesley Longman Publishing Co., Inc., Boston, MA, USA, 2nd edition, 1999.

[24] LXC - Linux Containers. http://lxc.sourceforge.net.

[25] Microsoft Windows Azure Node.js Developer Center. http://www.windowsazure.com/en-us/develop/nodejs.

[26] M. A. Namjoshi and P. A. Kulkarni. Novel online profiling for virtual machines. In *Proceedings of the 6th ACM SIGPLAN/SIGOPS international conference on Virtual execution environments*, VEE '10, pages 133–144, New York, NY, USA, 2010. ACM.

[27] Node.js. http://nodejs.org.

[28] OpenShift by RedHat. https://openshift.redhat.com.

[29] OpenWRT. https://openwrt.org.

[30] OSGi Alliance. *OSGi Core Release 5*, 2012.

[31] M. Paleczny, C. Vick, and C. Click. The Java HotSpot Server Compiler. In *Proceedings of the 2001 Symposium on JavaTM Virtual Machine Research and Technology Symposium - Volume 1*, JVM'01, Berkeley, CA, USA, 2001. USENIX Association.

[32] PhoneGap. http://phonegap.com.

[33] T. Suganuma, T. Yasue, M. Kawahito, H. Komatsu, and T. Nakatani. Design and Evaluation of Dynamic Optimizations for a Java Just-In-Time Compiler. *ACM Trans. Program. Lang. Syst.*, 27(4):732–785, July 2005.

[34] V. Sundaresan, D. Maier, P. Ramarao, and M. Stoodley. Experiences with Multi-threading and Dynamic Class Loading in a Java Just-In-Time Compiler. In *Proceedings of the International Symposium on Code Generation and Optimization*, CGO '06, pages 87–97, Washington, DC, USA, 2006. IEEE Computer Society.

[35] Vert.x. http://vertx.io.

Tessellation: Refactoring the OS around Explicit Resource Containers with Continuous Adaptation

Juan A. Colmenares[♦,♭], Gage Eads[♦], Steven Hofmeyr[†], Sarah Bird[♦], Miquel Moretó[♦],
David Chou[♦], Brian Gluzman[♦], Eric Roman[†], Davide B. Bartolini[♦], Nitesh Mor[♦],
Krste Asanović[♦], John D. Kubiatowicz[♦]
[♦]The Parallel Computing Laboratory, UC Berkeley, Berkeley, CA, USA
[†]Lawrence Berkeley National Laboratory, Berkeley, CA, USA
[♭]Samsung Research America - Silicon Valley, San Jose, CA, USA
juan.col@samsung.com, geads@eecs.berkeley.edu, shofmeyr@lbl.gov, sbird@eecs.berkeley.edu,
mmoreto@ac.upc.edu, {brian.gluzman,davidchou}@berkeley.edu, eroman@lbl.gov,
{dbb,mor,krste,kubitron}@eecs.berkeley.edu

ABSTRACT

Adaptive Resource-Centric Computing (ARCC) enables a simultaneous mix of high-throughput parallel, real-time, and interactive applications through automatic discovery of the correct mix of resource assignments necessary to achieve application requirements. This approach, embodied in the Tessellation manycore operating system, distributes resources to QoS domains called *cells*. Tessellation separates global decisions about the allocation of resources *to* cells from application-specific scheduling of resources *within* cells. We examine the implementation of ARCC in the Tessellation OS, highlight Tessellation's ability to provide predictable performance, and investigate the performance of Tessellation services within cells.

Categories and Subject Descriptors

D.4.1 [**Operating Systems**]: Process Management – Scheduling; D.4.7 [**Operating Systems**]: Organization and Design – Real-time and embedded systems; D.4.8 [**Operating Systems**]: Performance – Measurements, Monitors

General Terms

Multicore, parallel, quality of service, resource containers

Keywords

Adaptive resource management, performance isolation, quality of service

1. INTRODUCTION

Today's users demand lightning fast interaction with their portable devices while simultaneously displaying glitch-free multimedia and interacting with a variety of external information sources. Further, the growing number of mobile

Permission to make digital or hard copies of all or part of this work for personal or classroom use is granted without fee provided that copies are not made or distributed for profit or commercial advantage and that copies bear this notice and the full citation on the first page. To copy otherwise, to republish, to post on servers or to redistribute to lists, requires prior specific permission and/or a fee.

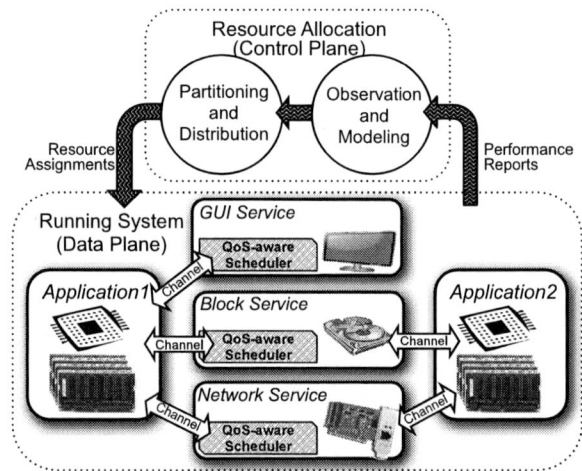

Figure 1: Adaptive Resource-Centric Computing: *Cells* provide performance isolation and guaranteed access to resources for applications and services. Cells are depicted as rounded boxes with solid lines. Resource allocations are automatically adjusted to maximize overall system utility. Resources wrapped with cells provide guaranteed services to other cells.

devices, sensors, and embedded systems makes efficiency a primary concern as systems demand an increasing amount of computation from the same battery capacity. Such requirements are very different from the high-throughput workloads that drove the design of operating systems in the past.

Modern applications consist of multiple interacting components, each with differing resource needs and quality-of-service (QoS) requirements. They often extend into the cloud and may include sensors and other embedded systems, requiring the operating system to provide responsiveness and performance predictability on a global scale. One example is that of a modern smartphone reproducing music retrieved on the fly from the cloud while rendering web pages with interactive multimedia contents. Other examples include Distributed Real-time Embedded (DRE) systems that control fly-by-wire airplanes and manufacturing plants; these systems need to ensure performance predictability across increasingly parallel components [13] and communication networks. Future DRE systems, such as autonomous vehicles and health-monitoring applications, will also have to manage large swaths of data from growing, interconnected "swarms" of sensors [22].

Thus, a new paradigm for the interaction between application components and systems software is clearly needed. Since resources are central to performance, predictability, and efficiency, we propose to utilize *Adaptive Resource-Centric Computing* (ARCC), as illustrated by Figure 1. In ARCC, resources are distributed to explicitly parallel, lightweight resource containers called *cells*, which provide stable execution environments for the software components running within them. Further, *composite* resources are constructed by wrapping cells around existing resources and exporting service interfaces. Applications can then be allocated QoS contracts from these services. This *service-oriented architecture* [34] approach enables performance predictability for a variety of more complex system services.

To reduce the burden on the programmer and to respond to changing environmental circumstances, our approach automatically adjusts resource allocations to meet application requirements. The stable environment of a cell makes it possible to experimentally observe user-defined progress metrics and predict how these metrics vary with resources – thus enabling accurate resource optimization. This approach addresses the *impedance-mismatch* between programmer and system that results from the fact that user-meaningful QoS metrics (*e.g.,* frame rate) are only indirectly related to resource allocations (*e.g.,* processor cycles).

In this paper, we illustrate the fundamental concepts of ARCC and evaluate it with Tessellation, a novel operating system for multi- and manycore systems. We base our evaluation on typical client applications, but the underlying methodology is equally applicable to embedded systems.

2. TESSELLATION ARCHITECTURE

In this section, we summarize some of the key components of Tessellation OS [11, 27], as illustrated by Figure 2. The Tessellation kernel is a thin, hypervisor-like layer that provides support for ARCC by implementing *cells* and providing interfaces for resource adaptation and cell composition. Tessellation uses *two-level scheduling* [24, 30, 11] to separate resource *allocation* from resource *usage*. This approach supports custom schedulers for efficient resource usage.

2.1 The Cell Model

Cells provide the basic unit of computation and protection in Tessellation. Cells are performance-isolated resource containers that export their resources to user level. The software running within each cell has full user-level control of the resources assigned to the cell, including CPU cores and memory pages. In the future, we plan to extend cells with multiple address spaces managed by the cell user-level runtime. We envision this facility as supporting more traditional UNIX-style processes or multi-component device drivers (*e.g.,* USB) within a cell.

Applications in Tessellation are created by composing cells via efficient and secure *channels*. Channels provide fast, user-level asynchronous message-passing between cells. Standard OS services (*e.g.,* network and file services) are hosted in cells and accessed via channels.

Tessellation OS virtualizes resources using *space-time partitioning* [36, 27, 28], a multiplexing technique that divides the hardware into a sequence of simultaneously-resident spatial partitions. With space-time partitioning, CPU cores and

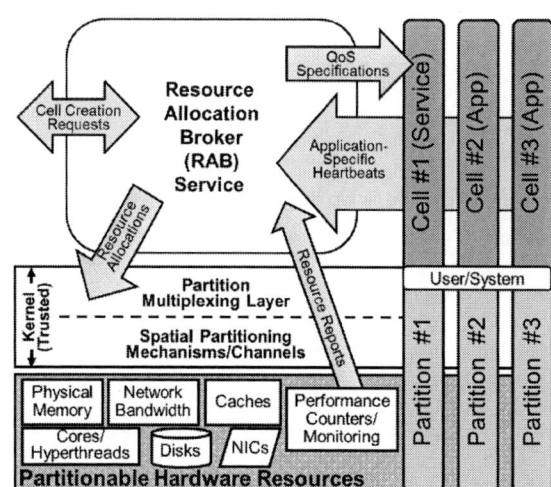

Figure 2: The Tessellation kernel implements *cells* through *spatial-partitioning*. The *Resource Allocation Broker* redistributes resources after consulting application-specific *heartbeats* and system-wide *resource reports*.

other resources are *gang-scheduled* [31, 14]. Cells thus provide to their hosted applications an environment that is very similar to a dedicated machine.

Partitionable resources include CPU cores, pages in memory, and guaranteed fractional services from other cells (*e.g.,* a throughput reservation of 150 Mbps from the network service). They may also include cache slices, portions of memory bandwidth, and fractions of the energy budget, when hardware support is available (*e.g.,* [4, 23, 38]).

Two-level scheduling [24, 30, 11] in Tessellation separates global decisions about resource allocation *to* cells (*first level*) from management and scheduling of resources *within* cells (*second level*). Resource redistribution occurs at a coarse time scale to amortize the decision-making cost and allow time for second-level scheduling decisions (made by each cell) to become effective.

The user-level runtime within each cell may utilize its resources (*e.g.,* hardware-thread contexts and memory pages) as it wishes – without interference from other cells. The cell's runtime can thus be customized for specific applications or application domains with, for instance, a particular scheduling algorithm and page replacement policy. Section 4 discusses second-level scheduling in detail.

2.2 Service-Oriented Architecture

Cells provide a convenient abstraction for building OS services with QoS guarantees. Such services can reside in dedicated cells, have exclusive control over devices, and encapsulate user-level device drivers (see Figure 1). Hence, each service can arbitrate access to its enclosed devices, and leverage its cell's performance isolation and customizable schedulers to offer service guarantees to other cells.[1] Services can shape data and event flows coming from external sources with unpredictable behavior and prevent other cells from being affected.

Two services in Tessellation that offer QoS guarantees are: the *Network Service*, which provides access to network adapters and guarantees that the data flows are processed

---

[1] In keeping with ARCC, we view the services offered by such *service cells* as additional resources to be managed by the adaptive resource allocation architecture.

with the agreed levels of throughput; and the *GUI Service*, which provides a windowing system with response-time guarantees for visual applications [19]. For details about the implementation and performance of these services, refer to Appendix B.

2.3 Adaptive Resource Allocation

Tessellation uses an adaptive resource-allocation approach to provide QoS guarantees to applications while maximizing efficiency in the system. The *Resource Allocation Broker* (RAB) is a *broker service* that distributes resources to cells while attempting to satisfy competing system-wide goals, such as deadlines met, energy efficiency, and throughput. Allocation decisions are communicated to the kernel and services for enforcement. The RAB Service uses system-wide goals, resource constraints, performance targets and current performance measurements as inputs to the optimization.

The RAB Service reallocates resources, for example, when a cell starts or finishes or when a cell significantly changes performance. The RAB Service can periodically adjust allocations; the reallocation frequency provides a tradeoff between adaptability (to changes in state) and stability (of user-level scheduling).

The RAB Service runs in its own cell and communicates with applications through channels. When a cell is started, it provides its QoS requirements to the RAB Service in the form of target performance goals, such as desired framerates. The RAB Service continuously monitors the cells' performance and compares it to target rates, adjusting resource allocations as required. To do this, the RAB Service utilizes two sources of information:

- Periodic performance reports, *heartbeats* [17], containing application-specific performance metrics from the cells (*e.g.*, the time to render a frame for a video app).
- System-wide performance counter values, such as cache-miss statistics and energy measurements.

The RAB Service provides a resource-allocation framework that supports rapid development and testing of new allocation policies. Section 5 demonstrates a few simple policies we developed as a proof of concept. The development time for each policy was under an hour. Using this framework we can explore the tradeoffs between enabling software components to meet their performance goals and optimizing resource distribution to achieve global objectives.

3. IMPLEMENTING THE CELL MODEL

As shown in Figure 2, the Tessellation kernel comprises two layers, the *Partition Multiplexing Layer* (or Mux Layer) and the *Spatial Partitioning Mechanisms Layer* (or Mechanism Layer). The Mechanism Layer performs spatial partitioning and provides resource guarantees by exploiting hardware partitioning mechanisms (when available) or through software emulation (*e.g.*, cache partitioning can be implemented using page coloring). Building on this support, the Mux Layer implements space-time partitioning and translates resource allocations from the RAB Service into an ordered time sequence of spatial partitions.

3.1 Types of Cells

The Mux Layer offers several time-multiplexing policies for cells to support applications (or parts thereof) with different timing requirements. Each multiplexing policy defines a cell type with a specific timing behavior. Tessellation pro-

vides: 1) non-multiplexed (non-muxed) cells with dedicated access to cores; 2) time-triggered (TT) cells, which are active during periodic time windows; 3) event-triggered (ET) cells, which are activated upon event arrivals, but never exceed their assigned fraction of processing time; and 4) best-effort (BE) cells with no time guarantees. The cell types are explained in greater detail in Appendix A.

These time-multiplexing policies allow users to easily specify the desired timing behavior for cells with a certain precision (currently 1 ms). The Mux Layer then ensures that, if feasible, a set of cells with different multiplexing policies harmoniously coexist and receive the specified time guarantees. In this way, Tessellation offers precise control over cells' timing behavior, one of the characteristics that differentiates Tessellation from traditional hypervisors and virtual machine monitors [5, 20].

3.2 Space-Time Partitioning

On each hardware thread there is a separate multiplexer (or *muxer*) that controls the multiplexing of cells on that thread. The muxers collectively implement gang scheduling [31] in a *decentralized* manner. They execute the *same* scheduling algorithm and rely on a high-precision global-time base [21] to simultaneously activate a cell on multiple hardware threads with minimum skew. In the common case, the muxers operate independently and do not communicate to coordinate the simultaneous activation of cells. The muxers thus implement an instance of *communication-avoiding* gang-scheduling.

For correct gang scheduling, the muxers need to maintain an identical view of the system's state whenever a scheduling decision is made. Hence, each muxer makes not only its own scheduling decisions but also reproduces the decisions made by other (related) muxers with overlapping schedules. In the worst case, each muxer must schedule the cell activations happening in every hardware thread in the system, but the RAB Service tries to avoid such unfavorable mappings.

The muxers implement gang-scheduling using a variant of *Earliest Deadline First* (EDF) [26], combined with the *Constant Bandwidth Server* (CBS) [3] reservation scheme in order to provide the variety of timing behaviors required by the cell types. EDF is used to implement TT cells while CBS is used for ET and BE cells. More details can be found in Appendix A.

3.3 Redistributing Resources among Cells

To request a redistribution of resources among cells (*e.g.*, resizing cells, changing timing parameters, or starting new cells), the RAB Service passes the new distribution to the Mux Layer via a system call (only accessible to this service). To implement resource-distribution changes, each muxer has two scheduler instances: one active and one inactive. The Mux Layer first validates the new resource distribution and, if successful, proceeds to serve the request. The Mux Layer next resets and prepares the inactive schedulers, and establishes the *global* time in the near future (*e.g.*, 1 ms later) at which the muxers will synchronously exchange their active and inactive schedulers. Then, the Mux Layer sends the muxers a message with the global time value and the system call returns. Finally, at the specified time, the muxers exchange their schedulers and perform other actions related to the relocation of cells (*e.g.*, re-routing device interrupts). This approach allows the Mux Layer to process resource-distribution requests almost entirely without disturbing the system's operation with only the overhead of switching schedulers and other cell-relocation actions. Note that if a subset of muxers is involved in a resource redistribution, only that subset performs the scheduler switch.

4. USER-LEVEL RUNTIME

One benefit of two-level scheduling is the ability to support different resource-management policies simultaneously. In Tessellation OS, cells provide their own, possibly highly-customized, user-level runtime system for processor (thread) scheduling and memory management. Further, each cell's runtime can control the delivery of events, such as timer and device interrupts, inter-cell message notifications, exceptions, and memory faults. This section describes the support Tessellation offers to implement user-level runtimes.

Our current Tessellation prototype includes two user-level thread scheduling frameworks: a *preemptive* one called PULSE (Preemptive User-Level SchEduling), and a *cooperative* one based on Lithe (LIquid THrEads) [33]. With either framework, a cell starts when a single entry point, `enter()`, is executed simultaneously on each core. After that, the kernel interferes with the cell's runtime only when: 1) the runtime receives events (*e.g.,* interrupts) it has registered for, 2) the cell is suspended and reactivated according to its time-multiplexing policy, and 3) the resources (*e.g.,* hardware threads) assigned to the cell change in response to RAB Service's requests. For instance, when an interrupt occurs during user-level code execution, the kernel saves the thread context and calls a registered interrupt handler, passing the saved context to the cell's user-level runtime. This way the cell's runtime can then choose whether to restore the previously running context or swap to a new one.

Since preemptive scheduling is commonly used in embedded and other types of computer systems, now we discuss PULSE, our framework for user-level preemptive scheduling, in more detail. PULSE is a simple framework, written in less than 800 lines of code (LOC). Creating a new user-level preemptive scheduler with PULSE is an easy task. The runtime must implement several callbacks, for instance: `enter()`, mentioned earlier; `tick(context)`, which is called whenever a timer tick occurs and receives the context of the interrupted thread; `yield()`, called when a thread yields; and `done()`, called when a thread terminates. The framework also provides functions for saving and restoring contexts, and other relevant operations.

PULSE's simplicity makes it easy to implement and customize schedulers – without having to patch the OS kernel, as is often the case with Linux. For example, we implemented a global round-robin scheduler with mutex and conditional-variable support, in ~850 LOC. We also wrote a global EDF scheduler with mutex support and priority-inversion control via dynamic deadline modification (DDM) [18], in less than 1000 LOC.

To support adaptive resource allocation, user-level runtimes must adjust to changes in the number of hardware threads assigned to their host cells. PULSE propagates changes in the number of hardware threads to the user-level scheduler. In the event that hardware threads are added, PULSE informs the runtime that additional resources are available. When a cell loses hardware threads (or harts), the framework delivers the extra application contexts (from the revoked hardware threads) to the user-level scheduler's

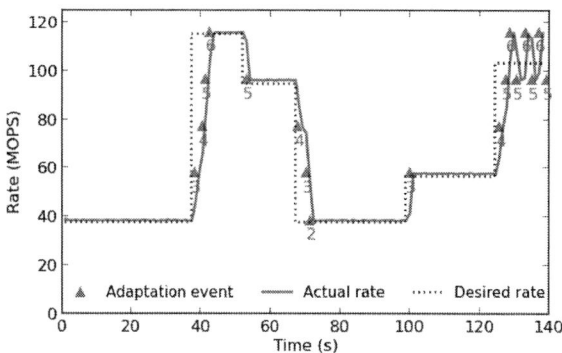

Figure 3: Adaptation of core counts while running the NAS EP benchmark. The number under each adaptation marker indicates the number of cores in the new allocation.

scheduling queue. If any of the saved contexts are scheduler contexts, PULSE uses a *non-preemptive auxiliary scheduler* that executes all scheduler contexts one by one until they enter application contexts. Then PULSE communicates these application contexts to the user-level scheduler with the callback `adapt(prev_num_harts, new_num_harts, context*)`. PULSE implements this procedure so that the user-level scheduler is completely unaware that adaptation is occurring until the `adapt` callback is executed. Consequently, we were able to add support for adaptation in the round-robin and global EDF schedulers simply by adding the `adapt` callback.

Although still a work in progress, Tessellation will soon support a user-level paging facility. The Resource Allocation Broker will allocate physical memory pages to cells after which the user-level runtime will manage these pages with customized page replacement and allocation policies, and use the block storage device(s) through a service interface[2] to implement paging, if desired. This approach, similar to self-paging in the Nemesis OS [16], can minimize uncontrolled interference between cells and enable better performance predictability and stronger QoS guarantees.

5. EXPERIMENTAL EVALUATION

In this section, we investigate the adaptive behavior of Tessellation OS through two simple experiments. The first shows the performance of the adaptation mechanisms in the kernel and user-level preemptive runtime when changing the number of cores assigned to a cell. The second, demonstrates the system's ability to adjust, via the RAB Service, the QoS guarantee that a service offers to a cell so that it can meet its performance goals.

5.1 Adjusting Core Allocation

This experiment is a simple feedback loop between a compute intensive application and the RAB Service. The application used was the NAS EP benchmark [2], which generates random numbers in an embarrassingly parallel manner. We chose EP because it scales perfectly and uses negligible memory; so we can easily confirm that it behaves as expected when varying number of cores.

EP was configured for 6 threads and run in a non-muxed cell with our global round-robin scheduler. The RAB Service

---

[2]We are currently developing a QoS-aware service that provides guaranteed client access to hard disks and other storage devices, called the Block Device Service.

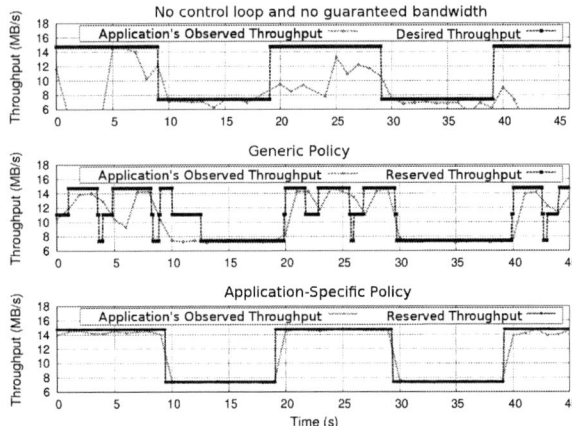

Figure 4: Adaptive network throughput no guarantees, with a general-purpose policy, and with an application-specific policy.

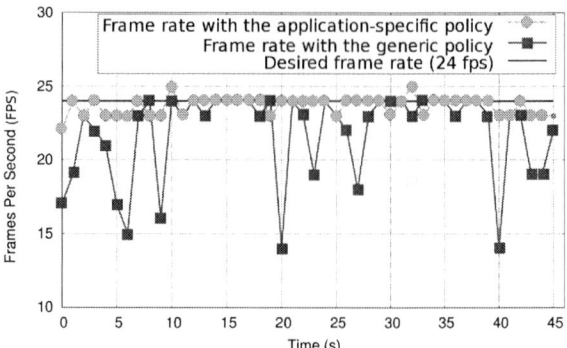

Figure 5: Achieved frames rates with the application-specific policy and the generic policy.

resided in a separate non-muxed cell running on a dedicated core. We modified EP to compute the rate R at which random numbers were generated, and set a goal of R' for the desired rate of random number generation. Each second, EP computed R and sent the value to the RAB Service.

We set 5 resource allocations, *operating points*, with core allocations of 2, 3, 4, 5, and 6. The RAB Service used a simple reactive policy to adjust the allocation of cores. When the performance is too low (*i.e.*, $R < R'$), it allocates the next larger operating point; when the performance is too high (*i.e.*, $R > R' + \epsilon$), it allocates the next smaller operating point. We used $\epsilon = 0.1$ to prevent oscillation between operating points.

We ran this experiment on an Intel system with two 2.66-GHz Xeon X5550 quad-core processors and hyper-threading disabled (*i.e.*, 8 hardware threads). The results can be seen in Figure 3, which shows the performance of EP as the desired rate R' changes periodically. There is a small lag between an adaptation event and the resulting performance change because EP only reports changes every second. We also see a noticeable lag in performance when a large change is required since the simple reactive policy shifts only one operating point at a time, and then waits for a new measurement before shifting again. Note, however, after the adaptation time it does settle on the correct rate illustrating the effectiveness of Tessellation's adaptation mechanisms. Although clearly, the reactive policy used here is too limited for general use as shown by the oscillations at $t \geq 125$ s.

5.2 Adjusting Service Guarantees

In this experiment, the RAB Service drove the Network Service to allocate network throughput to an application. We show how the RAB Service adjusted the throughput reservation for a video player against changes in the incoming video stream, so that the video player's performance goals were met, while enabling efficient use of the Network Service. We used two computers (1 Linux, 1 Tessellation), both with an Intel 3.4-GHz Core i7 quad-core processor with hyper-threading (*i.e.*, 8 hardware threads), and 4 GB of RAM, directly connected via 1-Gbps Ethernet adapters.

The Linux box was the video source and sent video frames to the Tessellation box over a TCP connection. The Linux box produced a stream of uncompressed frames at a constant rate of 24 frames per second (fps). The frame size was adjusted between 480x320 and 320x240 every 10 s to simulate changing video quality. Thus, large frames required 14.74 MB/s while small frames 7.37 MB/s.

On the Tessellation box, the video-player application received the video stream via the Network Service and passed the reconstructed frames to the GUI Service for display. The video player periodically sent performance reports to the RAB Service. Using this information, RAB Service chose the throughput reservation the Network Service provided to the video player. Additionally, two "bandwidth-hog" applications ran on the Tessellation box and contended with the video player for bandwidth by constantly sending UDP messages to the Linux box. The Network Service gave no bandwidth guarantees to these applications, which shared the excess of bandwidth. On the Tessellation box, applications and services resided in separate non-multiplexed cells.

Ideally, the RAB Service should allocate sufficient throughput for the video player to reach 24 fps without over provisioning. To this end, we experimented with two simple reactive policies: an *application-specific policy*, P_{App}, that exploits knowledge about the video stream's bandwidth demands; and a *general-purpose policy*, P_{Gen}, that operates only based on the video player's observed performance. P_{App} uses reports that explicitly specify the size of the frames being received. P_{Gen}, on the other hand, uses reports with the average frame inter-arrival time (in milliseconds per frame), and has three operating points: 15.0 MB/s, 11.0 MB/s, and 7.5 MB/s. When the RAB Service receives a performance report from the video player, the policy in use determines whether the application is falling short of its performance goal and needs higher throughput from the Network Service, or whether the applications is exceeding the performance goal by a pre-determined threshold could possibly run at a lower operating point. Both policies include hysteresis to prevent oscillations between operating points.

Figure 4 shows how the system reacts to changes in the incoming video stream for each policy type. The upper graph shows the video player's observed throughput with no guaranteed bandwidth reservation. In this case, the two "bandwidth-hog" applications consume too much bandwidth to allow the video player to receive its required amount. The lower graph shows the results of P_{App}, where the Network Service (via the RAB Service) rapidly adapts to the changing needs of the video player. For P_{Gen} (the center graph), the high-bandwidth intervals contain occasional low-reservation glitches when it probes the next lowest operating point. P_{Gen} is flexible enough for any scenario where a latency value (*e.g.*, frame inter-arrival time) is the key metric,

532

but is not optimized for the video-player application. However, adjusting the frequency of probing, number of operating points, and hysteresis in the RAB Service can improve the system's adaptation accuracy.

Figure 5 shows the video player's observed frame rate, which is the user-meaningful metric considered here. P_{App} achieves near-constant 24 fps with glitches near the high-bandwidth to low-bandwidth transition points. As expected, P_{Gen} falls short of 24 fps during the high-bandwidth intervals when the RAB Service probes lower operating points. This experiment demonstrates that the system, composed by the video player, the RAB Service, and the Network Service, can adapt to changes in the incoming video stream.

6. CONCLUSIONS

This paper introduced *Adaptive Resource-Centric Computing* (ARCC) and described and evaluated its implementation in the Tessellation OS. The key points for ARCC are:

- Resources distributed to QoS domains called *cells*, which are explicitly parallel, light-weight containers with guaranteed, user-level access to resources;
- User-level scheduling of resources within cells; and
- Adaptive allocation and distribution of resources to cells in order to meet QoS requirements efficiently.

We showed that this approach handles a simultaneous mix of high-throughput parallel, real-time, and interactive applications. Though we restricted our focus to client applications, the embedded systems domain presents a similar complex resource management problem that we believe Tessellation OS is capable of managing. We believe that ARCC is essential for designing systems that can meet the rigorous responsiveness and efficiency demands of modern applications.

7. ACKNOWLEDGMENTS

This research is supported by Microsoft (Award #024263), Intel (Award #024894), matching U.C. Discovery funding (Award #DIG07-102270), and DOE ASCR FastOS Grant #DE-FG02-08ER25849. Additional support comes from Par Lab affiliates National Instruments, Nokia, NVIDIA, Oracle, and Samsung. No part of this paper represents views and opinions of the sponsors mentioned above. We thank other Par Lab members for their collaboration and feedback. J. A. Colmenares participated in this work while he was a post-doctoral scholar at UC Berkeley. M. Moreto is supported by a MEC/Fulbright Fellowship.

8. REFERENCES

[1] Nano-X window system. http://www.microwindows.org/.
[2] NAS parallel benchmarks.
http://www.nas.nasa.gov/publications/npb.html.
[3] L. Abeni and G. Buttazzo. Resource reservations in dynamic real-time systems. *Real-Time Systems*, 27(2):123–165, 2004.
[4] B. Akesson et al. Predator: a predictable SDRAM memory controller. In *Proc. of CODES+ISSS*, 2007.
[5] P. Barham et al. Xen and the art of virtualization. In *Proc. of SOSP*, 2003.
[6] D. B. Bartolini et al. The Autonomic Operating System Research Project Achievements and Future Directions. In *Proc. of DAC*, 2013.
[7] S. Baruah et al. Implementing constant-bandwidth servers upon multiprocessors. In *Proc. of RTAS*, 2002.
[8] S. Baruah and G. Lipari. Executing aperiodic jobs in a multiprocessor constant-bandwidth server implementation. In *Proc. of ECRTS*, 2004.
[9] A. Baumann et al. The Multikernel: A new OS architecture for scalable multicore systems. In *Proc. of SOSP*, 2009.

[10] S. Boyd-Wickizer et al. Corey: an operating system for many cores. In *Proc. of OSDI*, 2008.
[11] J. A. Colmenares et al. Resource management in the Tessellation manycore OS. In *Proc. of HotPar*, 2010.
[12] D. R. Engler, M. F. Kaashoek, and J. O'Toole. Exokernel: An operating system architecture for application-level resource management. In *Proc. of SOSP*, 1995.
[13] P. Fischer. Multicore processors revolutionize real-time embedded systems. *Electronic Design*, December 2007.
[14] L. L. Fong et al. Gang scheduling for resource allocation in a cluster computing environment. Patent US 6345287, 1997.
[15] A. Gulati et al. mClock: handling throughput variability for hypervisor IO scheduling. In *Proc. of OSDI*, 2010.
[16] S. M. Hand. Self-paging in the nemesis operating system. In *In Proc. of OSDI*, 1999.
[17] H. Hoffmann et al. SEEC: a general and extensible framework for self-aware computing. Technical Report MIT-CSAIL-TR-2011-016, 2011.
[18] K. Jeffay. Scheduling sporadic tasks with shared resources in hard real-time systems. In *Proc. of RTSS*, 1992.
[19] A. Kim et al. A soft real-time parallel GUI service in Tessellation many-core OS. In *Proc. of CATA*, 2012.
[20] A. Kivity. kvm: the Linux virtual machine monitor. In *Proc. of OLS*, 2007.
[21] H. Kopetz. *Real-time systems: design principles for distributed embedded applications*. Springer, 1997.
[22] E. A. Lee et al. The TerraSwarm Research Center (TSRC) (A White Paper). Technical Report UCB/EECS-2012-207, EECS Department, University of California, Berkeley, Nov 2012.
[23] J. W. Lee et al. Globally-synchronized frames for guaranteed quality-of-service in on-chip networks. *SIGARCH Comput. Archit. News*, 36(3):89–100, June 2008.
[24] B. Leiner et al. A comparison of partitioning operating systems for integrated systems. In *Proc. of SAFECOMP*, 2007.
[25] J. Liedtke. On micro-kernel construction. *ACM SIGOPS Oper. Syst. Rev.*, 29:237–250, December 1995.
[26] C. L. Liu and J. W. Layland. Scheduling algorithms for multiprogramming in a hard-real-time environment. *Journal of the ACM*, 20(1):46–61, January 1973.
[27] R. Liu et al. Tessellation: Space-time partitioning in a manycore client OS. In *Proc. of HotPar*, 2009.
[28] L. Luo and M.-Y. Zhu. Partitioning based operating system: a formal model. *ACM SIGOPS Oper. Syst. Rev.*, 37(3), 2003.
[29] K. J. Nesbit et al. Multicore resource management. *IEEE Micro*, 28(3):6–16, 2008.
[30] R. Obermaisser and B. Leiner. Temporal and spatial partitioning of a time-triggered operating system based on real-time Linux. In *Proc. of ISORC*, 2008.
[31] J. Ousterhout. Scheduling techniques for concurrent systems. In *Proc. of ICDCS*, 1982.
[32] P. Padala et al. Automated control of multiple virtualized resources. In *Proc. of EuroSys*, 2009.
[33] H. Pan et al. Composing parallel software efficiently with Lithe. In *Proc. of PLDI*, 2010.
[34] M. P. Papazoglou and W.-J. Heuvel. Service oriented architectures: approaches, technologies and research issues. *The VLDB Journal*, 16(3):389–415, July 2007.
[35] B. Rhoden et al. Improving per-node efficiency in the datacenter with new os abstractions. In *Proc. of SOCC*, 2011.
[36] J. Rushby. Partitioning for avionics architectures: requirements, mechanisms, and assurance. Technical Report CR-1999-209347, NASA Langley Research Center, June 1999.
[37] B. Saha et al. Enabling scalability and performance in a large scale CMP environment. In *Proc. of EuroSys*, 2007.
[38] D. Sanchez and C. Kozyrakis. Vantage: scalable and efficient fine-grain cache partitioning. *SIGARCH Comput. Archit. News*, 39(3):57–68, June 2011.
[39] A. Sharifi et al. METE: meeting end-to-end qos in multicores through system-wide resource management. *SIGMETRICS Perform. Eval. Rev.*, 39(1):13–24, June 2011.
[40] D. D. Silva et al. K42: an infrastructure for operating system research. *SIGOPS Oper. Syst. Rev.*, 40(2):34–42, 2006.
[41] D. Wentzlaff and A. Agarwal. Factored operating systems (fos): the case for a scalable operating system for multicores. *ACM SIGOPS Oper. Syst. Rev.*, 43(2):76–85, 2009.

APPENDIX

A. CELL TYPES

Table 1 summarizes the cell types Tessellation provides. Non-multiplexed (non-muxed) cells are intended to host software components with stringent performance requirements that demand a high degree of performance isolation. Time-triggered (TT) cells are for hosting time-predictable components that can tolerate some latency. Event-triggered (ET) cells provide flexible event-handling as well as good responsiveness and resource utilization, which make them ideal for hosting OS services. Finally, best-effort (BE) cells are for components without strict timing constraints.

These time-multiplexing policies allow users to easily specify the desired timing behavior for cells with a certain precision (currently 1 ms). For example, TT and ET cells both take the parameters *period* and *active_time*, where *period* > *active_time*. In the case of an ET cell, its reserved fraction of processing time is given by (*active_time/period*).

The Partition Multiplexing Layer in the Tessellation kernel runs a separate multiplexer on each hardware thread of the system. The multiplexers, or *muxers*, control time-multiplexing of cells. The muxers collectively implement gang-scheduling using a variant of *Earliest Deadline First* (EDF) [26], combined with the *Constant Bandwidth Server* (CBS) [3] reservation scheme. The Partition Multiplexing Layer is then capable of providing all of the scheduling behaviors shown in Table 1. We chose EDF for TT cells because it enables the muxers to directly utilize the timing parameters specified for these cells. CBS, on the other hand, isolates each ET cell from other cell activations, and ensures each ET cell a fraction ($f = active\_time/period$) of processing capacity on each hardware thread assigned to the cell. Further, CBS offers ET cells responsiveness by allowing them to exploit the available slack without interfering with other cells. For short activation time (*e.g.,* for event processing in server cells), an ET cell is activated with an immediate deadline if it has not used up its time allocation.

Muxers could schedule BE cells using a hierarchical scheme, in which CBS reserves a small fraction of processing capacity for BE cells and a round-robin algorithm operates in the time slices given by CBS. For implementation simplicity, however, the muxers use only CBS to schedule BE cells. Unlike ET cells, which are activated by events, BE cells are always kept in the runnable queue. Each BE cell is given a fixed small reservation (*e.g.,* 2% with *active_time* = 5 ms and *period* = 100 ms) to ensure that it always makes progress.

B. OS SERVICES: TWO CASE STUDIES

Tessellation OS builds upon a service-oriented architecture (see Section 2.2), in which services encapsulate devices and, in general, resources. Each service can leverage its cell's performance isolation and customizable user-level schedulers to offer service guarantees to applications and services residing in other cells. Moreover, services may exploit parallelism to reduce service times or increase service throughput.

Each service in Tessellation comes with a library to facilitate the development of client applications. The client libraries offer friendly, high-level application programming interfaces (APIs) to manage connections and interact with the services (*i.e.,* they hide most of the details of inter-cell channel communication). Those libraries also allow applica-

tions to request the QoS guarantees they need from services.

Two services at a mature development state that provide QoS guarantees are the *Graphical User Interface (GUI) Service* and the *Network Service*. They are the focus of this appendix and we describe them in detail below. Additionally, Tessellation offers a *Console Service* that prints character strings from other cells to a serial console (via a serial port). This service has exclusive access to the console device, and for a client cell, printing a string is just sending a one-way message on a dedicated channel to the console service. Thus, cells do not need to contend and wait for accessing the console device, and the influence of printing console messages on each application's behavior can be controlled independently and minimized. The Console Service has been instrumental in collecting the experimental data presented in this paper.

Other services in active development include a Block Device Service and an Object Store Service. The Block Device Service provides block-level operations with additional QoS guarantees to applications or other services, and behaves as a swap pager for individual cells. The Object Store Service provides persistent storage through a simple interface that enables applications to put (and retrieve) variable sized objects in a flat namespace. Due to space constraints, we do not discuss these services further.

Next, we discuss the implementation of Tessellation's GUI Service and Network Service, and examine their ability to provide QoS guarantees to client cells.

B.1 GUI Service

Tessellation's GUI Service [19] provides a windowing system with response-time guarantees for visual applications. It resides in a dedicated cell, which encapsulates and has sole control of the framebuffer device. Therefore, applications can only draw to the screen by using the GUI Service.

The GUI Service is a rearchitected version of the Nano-X Window System [1]. It exploits a user-level *Earliest Deadline First* (EDF) scheduler to take advantage of multiple cores and ensure that rendering jobs with earlier deadlines are scheduled sooner. The GUI Service supplements the EDF scheduler with a *resource reservation scheme*, called Multiprocessor Constant Bandwidth Server (M-CBS) [7, 8], to provide different CPU reservations to different rendering tasks – a big distinction from traditional GUI systems.

We conduct an experiment to evaluate GUI Service's ability to provide QoS guarantees to visual applications. The experiment consists of eight video clients that send 8,000 computationally intensive rendering requests (frames) to the window system, half at a rate of 30 frames per second (fps), and half at 60 fps. To compare their performance, we use both the GUI Service on Tessellation and the original Nano-X system on Linux. The test platform is equipped with an Intel 3.4-GHz Core i7 quad-core processor, 4 GB of RAM, and hyper-threading enabled.

Figure 6 shows the result of our experiment. The traditional GUI system (represented by Nano-X running on Linux) runs on a single hardware thread and misses 65% of deadlines of the 60-fps requests. By contrast, even on one hardware thread, Tessellation misses only 0.1% of deadlines (GUIServ(1) in Figure 6), because it can reallocate some of the CPU reservation from the 30-fps streams. Reallocation is not necessary when the GUI Service uses more than one hardware thread, and the overall service times roughly halve when the hardware-thread count doubles. This suggests that

Cell Type	Description	Gang-Scheduling Algorithm
Non-Multiplexed (Non-Muxed)	The cell is given dedicated access to the hardware threads and the other managed resources.	Permanent activation.
Time-Triggered (TT)	The cell is active for some time during periodic time intervals.	Earliest Deadline First (EDF) [26].
Event-Triggered (ET)	The cell is activated upon the arrival of an event. Once activated, the cell remains "runnable" and is multiplexed with other cells until its user-level runtime requests the cell to yield all the resources via the `cell_yield()` system call. Once the cell yields it does not become runnable until another event arrives.	Constant Bandwidth Server (CBS) [3].
Best-Effort (BE)	These cells have no strong guarantees, but the kernel ensures that they have the chance to be activated (*i.e.*, make progress) and are multiplexed in a fair manner among themselves.	CBS with cells always available for activation and small reservations.

Table 1: Types of cells according to the time-multiplexing policies.

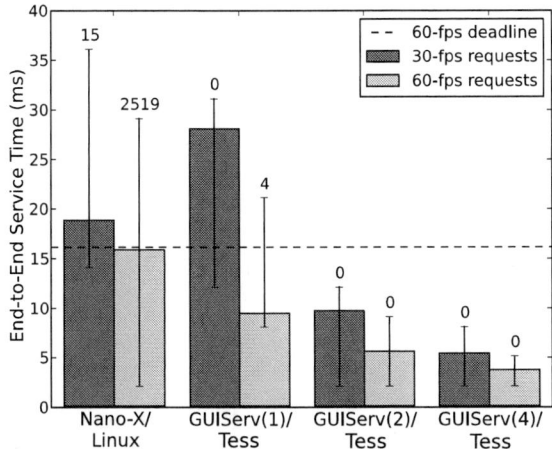

Figure 6: Service time for rendering requests. The numbers above the bars represent missed deadlines; the numbers in parentheses indicate the number of allocated hardware threads in Tessellation.

the GUI Service may scale well.

The GUI Service is a good example of the advantages of two-level scheduling in Tessellation. Implementing the GUI Service's customized scheduler on a monolithic kernel (*e.g.*, Linux) would require extensive kernel-side modifications, especially if we wanted to apply it to a single application. Modifying a general-purpose scheduler to meet the requirements of a specific application easily leads to performance issues for the other applications. However, on Tessellation, no kernel modification is required; the GUI Service's scheduler sits on top of the PULSE framework (Section 4), is only 445 lines of code, runs completely at user level, and applies only to the cell hosting the GUI Service.

So far, we have focused on a basic software-based rendering pipeline, which makes no use of acceleration capabilities in graphical processing units (GPUs). The reason is that using a CPU-based software rendering service is enough for exploring mechanisms for QoS guarantees; GPU acceleration would not add much on this side and its use is left to future work.

B.2 Network Service

Tessellation's Network Service provides access to network interface cards (NICs) through an API similar to the socket API. Its implementation is based on a modified multithreaded version of the lightweight TCP/IP protocol stack lwIP.[3] The Network Service allows the specification of *min-*

---

[3]http://savannah.nongnu.org/projects/lwip/

imum throughput reservations for data flows between NICs and client cells. The service guarantees that the data flows are processed with *at least* the specified levels of throughput, provided it is feasible to do so with the networking and computational resources available to the service (*e.g.*, the aggregate reservation should be less than or equal to the NIC's maximum capacity). Moreover, the Network Service distributes any excess throughput proportionally among the client cells via an adaptation of the mClock algorithm [15].

The Network Service enforces QoS guarantees by restricting channel communication. When an application opens a channel to send or receive data over the network through the Network Service API, it requests both a guaranteed and a proportional bandwidth. The Network Service's access control mechanism denies guaranteed-bandwidth requests that would cause the total guaranteed bandwidth to exceed the NIC's maximum capacity. The Network Service first enforces the request for guaranteed bandwidth and uses the proportional request to assign the slack of capacity. The bandwidth-guarantee enforcement takes place at channel message granularity, where each message contains an individual remote procedure call such as `recv()` or `send()`. While we stopped at the message granularity for implementation simplicity, bringing this enforcement at finer grain (*i.e.*, at channel data granularity) is a straightforward extension.

We evaluate the Network Service in a common client use case: a user wants an uninterrupted stream of video content (*e.g.*, Hulu) in a foreground application, while a background task (*e.g.*, Dropbox) generates bursty network traffic. The video stream requires 125 KB/s, which is the bandwidth of an H.264 480p Hulu stream, while the background application runs periodically and tries to use all the available capacity. To realize this experiment, we use two identical machines, each equipped with an Intel 3.4-GHz quad-core Core i7 processor and 4 GB of RAM. Both machines are directly connected over Ethernet via their Intel Pro/1000 1-Gbps NICs. One machine acts as the video source and runs Linux 3.1.9, while the other, which is the video player to be evaluated, runs Tessellation. The Linux box runs two separate socket applications that serve data respectively to the video-player application and the background application running in Tessellation. On the Tessellation box, each of the two test applications is assigned a dedicated hardware thread, while the network service uses three dedicated hardware threads.

Figure 7 shows the results of this experiment: the foreground application receives an average 125.2 KB/s throughput, while the background application periodically uses

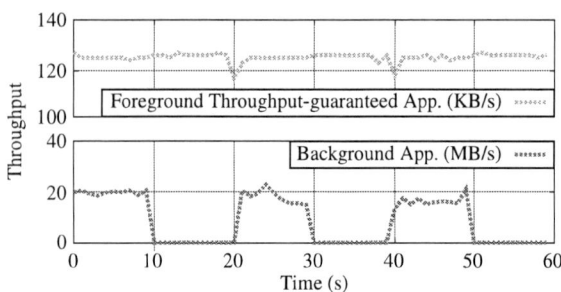

Figure 7: Guaranteeing network throughput in the presence of greedy applications. This graph shows that the greedy application is unable to deny service to the application with guaranteed service.

around 20 MB/s. The slight dip (around 5 KB/s) in the foreground application's throughput that occurs when the background application starts using the link is a result of a limitation in our modified mClock implementation. What happens is that the QoS state of previously dormant connections is out of sync with virtual time, resulting in boosted privilege for a brief instant. This problem has a known solution, described by Gulati et al. [15] but, since the effects of this issue are not disruptive, we did not put additional implementation effort in porting the solution. Despite the mClock artifact of our implementation, we achieve a standard deviation of only 1.65 KB/s for the throughput of the foreground application; this means that the user will not experience disruption in the video-streaming applications due to bandwidth-hungry background tasks.

The Network Service is another example of how the service-oriented architecture of Tessellation can help modern clients match users' needs. One of the key points is that, aside from one-time system calls to configure the NIC, the NIC driver[4] is completely contained in userspace. This structure allows the Network Service to avoid having to make relatively expensive system calls to access the transmit and receive buffers.

C. RELATED WORK

A number of research efforts have focused on the problem of adaptive resource allocation to meet QoS objectives in multi-application scenarios. Some previous work most relevant to this paper includes: AutoControl [32], SEEC [17], AcOS [6], and METE [39]. Next we present a comparison bewteen Tessellation and that previous work focusing on goals, software infrastructure, and underlying adaptation mechanisms.

AutoControl, SEEC, AcOS, METE, and other autonomic computing frameworks incorporate adaptation policies of various types. Policy-related aspects, such as formulation of the resource-allocation problem, resource-allocation decision engines using optimal control or machine learning techniques, and online performance-model estimators, are orthogonal to our discussion here. Those adaptation policies can be implemented in Tessellation's Resource Allocation Broker.

AutoControl [32] is a feedback-based resource-allocation system for shared virtualized infrastructure in data centers, where applications are hosted in virtual machines across

---

[4]Tessellation currently supports the Intel PRO/1000 PCI (E1000) and Realtek RL8168 adapters.

multiple nodes. It addresses the resource-management problem at the hypervisor level, and exploits Xen [5] to allocate CPU and disk I/O bandwidth to mitigate bottlenecks. AutoControl and Tessellation have similar characteristics, despite big differences in their target computing platforms – data centers for AutoControl and a single multicore node currently for Tessellation. Tessellation implements dynamic resource allocation at the cell level, and cells resemble some aspects of virtual machines [5, 20]. Cells, however, are intended to provide better performance isolation and more precise control over their timing behavior than traditional virtual machines; besides, cells do not host a complete OS – only a user-level runtime. In addition, AutoControl and Tessellation both consider I/O services as shared resources and dynamically allocate fractions of the services to applications.

SEEC [17] is a self-aware programming model designed to facilitate the development of adaptive computing systems on multicore platforms. It supports a decoupled approach in which application programmers specify the applications' goals and report the current progress toward those goals, while system programmers separately specify the set of actions system software (e.g., OS and runtime) and hardware can take to affect the applications. The SEEC framework implements an observe-decide-act (ODA) control loop to monitor applications and dynamically select actions to optimally meet their goals. SEEC currently supports three application goals: performance, accuracy, and power. As SEEC, Tessellation aims at providing a general and extensible framework for self-adapting computing, in particular through its RAB Service. Similar to SEEC's Application Heartbeat API, the RAB Service provides an API for applications hosted in cells to report their goals and performance over inter-cell channels. SEEC does not propose changes in the OS and currently uses mechanisms available in Linux to implement the actuator functions. On the contrary, our work on Tessellation focuses on rearchitecting the OS to provide better performance predictability and adaptive resource management.

AcOS [6] is a proposal for an autonomic resource-management layer to extend commodity OSs, such as Linux and FreeBSD. The authors investigate the application of autonomic-computing ideas at the OS level to automate resource allocation based on user-specified application performance goals and enforce system-level restrictions. They demonstrate different approaches to automate allocation of cores and processor time in order to meet those goals. Moreover, AcOS considers maximum processor temperature thresholds and implements a dynamic performance and thermal management (DPTM) control loop to cap temperature while still meeting the performance goals of a subset of the applications. With respect to SEEC, the scope of AcOS is closer to Tessellation's. The major difference is that AcOS only focuses on adaptation and extends commodity OSs, whereas Tessellation builds support for adaptive resource allocations into its novel cell model from the ground up. Also, neither AcOS nor SEEC in practice consider OS services as part of the resource-allocation problem.

METE [39] is a platform for end-to-end on-chip resource management for multicore processors. Its main goal is to dynamically provision hardware resources to applications to achieve performance targets. METE leverages feedback control to partition shared hardware resources among co-located applications and uses an autoregressive-moving-average (ARMA) model to capture applications' performance characteristics. METE considers cores, shared cache space, and off-chip memory bandwidth as partitionable resources. Since current processors do not support partitioning of all those resources, METE has been evaluated in a simulation environment. Tessellation shares with METE the interest in exploiting hardware partitioning mechanisms to guarantee end-to-end QoS. However, while METE assumes the availability of such mechanisms, Tessellation builds support for resource partitioning in a new OS model to enable experimentation on real hardware. We exploit hardware solutions when available and develop software solutions where hardware support is absent.

Tessellation has similarities to several recent manycore OSs. The use of message-passing communication via user-level channels is similar to Barrelfish [9]. However, Barrelfish is a multikernel OS that assumes no hardware assist for cache coherence, and does not focus on adaptive resource allocation. The way Tessellation constructs user-level services is similar to fos [41]. Services in Tessellation are QoS-aware and cells are partitioned based on applications rather than physical cores. Tessellation is similar to Corey [10] in that we also try to restrict sharing of kernel structures.

Tessellation adopts a microkernel philosophy [25], in which OS services are implemented in user-space and applications interact with them via message passing. Unlike in traditional microkernels, however, each service residing in a separate cell is explicitly parallel and performance-isolated, and includes an independent user-level runtime. The runtime customization in Tessellation is influenced by Exokernel [12]. However, Tessellation tries to mitigate some of the problems of exokernels by providing runtimes and services for the applications. Tessellation has some similarities to the K42 OS [40]. Both implement some OS services in user-space, but K42 uses protected procedure calls (PPCs) to access services, where Tessellation uses user-level channels.

Tessellation shares with the Nemesis OS [16] the emphasis on ensuring QoS for multimedia applications. Nemesis also uses an approach around OS services and message passing, but on uniprocessors.

Resource partitioning has also been presented in McRT [37] and Virtual Private Machines (VPM) [29]. The concepts of VPM and cells are similar, but VPM lacks inter-cell communication and has not been implemented yet. Gang-scheduling [31, 14] is a classic concept and has also been applied to other OSs – most similarly in Akaros [35]. However, unlike other systems, Tessellation supports cells with different timing behaviors.

The Autonomic Operating System Research Project – Achievements and Future Directions

Davide B. Bartolini, Riccardo Cattaneo, Gianluca C. Durelli, Martina Maggio*
Marco D. Santambrogio, Filippo Sironi
Politecnico di Milano, *Lund University
{bartolini, rcattaneo, durelli, santambrogio, sironi}@elet.polimi.it, martina.maggio@control.lth.se

ABSTRACT

Traditionally, hypervisors, operating systems, and runtime systems have been providing an abstraction layer over the bare-metal hardware. Traditional abstractions, however, do not consider for non-functional requirements such as system-level constraints or users' objectives. As these requirements are gaining increasing importance, researchers are looking into making user-specified and system-level objectives first-class citizens in the computer systems' realm.

This paper describes the *Autonomic Operating System (AcOS)* project; *AcOS* enhances commodity operating systems with an autonomic layer that enables self-* properties through adaptive resource allocation. With *AcOS*, we investigate intelligent resource allocation to achieve user-specified service-level objectives on application performance and to respect system-level thresholds on CPU temperature. We give a broad overview of *AcOS*, elaborate on its achievements, and discuss research perspectives.

Categories and Subject Descriptors

C.0 [**Computer Systems Organization**]: *Hardware/software interfaces; System architectures*; D.4.1 [**Operating Systems**]: Process Management—*Scheduling*; D.4.8 [**Operating Systems**]: Performance—*Measurements; Modeling and prediction; Monitors*

General Terms

Design, Management, Measurement, Performance

Keywords

Autonomic computing, Operating systems, Virtualization, Performance management, Dynamic thermal management

1. INTRODUCTION

In the last decade, the failure of Dennard's scaling law [11] determined the inability to leverage single-threaded performance improvements in order to keep doubling integrated circuits performance every two years, as stated by the established Joy's law. Meanwhile, transistors density has kept its exponential increase, as predicted by Moore's Law [22]. These two phenomena drove chip manufacturers towards embracing parallelism, leading to the preva-

lence of Chip-MultiProcessors (CMPs) and multi-processor system-on-chips (MPSoCs) throughout most computing systems segments.

On the one hand, mobile and embedded systems feature heterogeneous MPSoCs specialized for efficiency. For instance, the NVIDIA Tegra platform leverages the partner/companion core approach [20], while the ARM big.LITTLE chip implements the single-instruction set architecture (ISA) heterogeneous computing approach, coupling high-throughput and energy-efficient cores [17].

On the other hand, large-scale installations, like warehouse-scale computers, build on nodes equipped with homogeneous CMPs offering an increasing number of on-chip cores. For example, Tilera already offers 64-core solutions for general-purpose processing [5], while Intel and NVIDIA propose 100+thread solutions, like the Intel Xeon Phi and the NVIDIA Tesla, respectively, to accelerate embarrassingly parallel applications.

Thanks to Moore's law, chip manufacturers equip each new generation of CMPs and MPSoCs with an increased amount of on-chip resources (e.g., cores, caches, memory controllers). This unprecedented availability of on-chip resources encourages workload consolidation, realized by co-locating single and / or multi-threaded applications onto the same chip. Co-located applications share on-chip resources and require careful multiplexing in both space (i.e., placement on CPUs) and time (i.e., CPU bandwidth) to maintain performance predictability despite contention over on-chip shared resources [23]. The advent of virtualization and commodity hypervisors for widespread ISAs [3] brings additional complexity, as different users, with different service-level objectives (SLOs), can own the co-located applications. This scenario opens new research issues in the areas of resource allocation, which had settled on well-established techniques for time-shared single-core processors.

With the *Autonomic Operating System (AcOS)* project, we target these issues by looking for ways to automatize allocation of on-chip shared resources. We aim at enabling users to easily state SLOs and to automatically tune resource allocations in order to meet user-specified SLOs, while enforcing system-level constraints.

We focus on a specific system-level constraint: maximum processor temperature. This constraint is of primary concern for current and future CMPs and MPSoCs, since recent lithographic technologies cannot keep up the down-scaling of supply voltage with the up-scaling of clock frequency and transistor density, causing power density to increase and therefore reducing the capacity of packages of dissipating the resulting heat. Avoiding high processor temperature means avoiding to impair performance [10], energy efficiency [29], and reliability [30] of integrated circuits.

With *AcOS*, we research cheap and efficient software techniques to keep temperature within a system-specified threshold while minimizing the impact of this cap on performance.

The remainder of this paper first illustrates the high-level ap-

---

Permission to make digital or hard copies of all or part of this work for personal or classroom use is granted without fee provided that copies are not made or distributed for profit or commercial advantage and that copies bear this notice and the full citation on the first page. To copy otherwise, to republish, to post on servers or to redistribute to lists, requires prior specific permission and/or a fee.
DAC '13, May 29 - June 07 2013, Austin, TX, USA

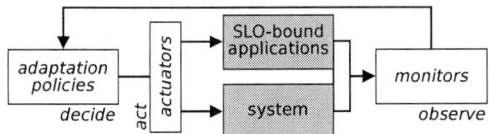

Figure 1: Interaction of the autonomic components with the computing system and the applications; these components realize the *observe*, *decide*, and *act* phases of the ODA control loop.

proach and methodology we adopt in *AcOS* (Section 2) and then focuses on performance (Section 3) and thermal (Section 4) management, going into details and validating *AcOS* through experimental results. We conclude by discussing related research (Section 5) and future directions and perspectives (Section 6). We provide further details on more specific aspects in Appendices A to C.

2. FOUNDATIONS AND METHODOLOGY

AcOS derives its overall methodology from *autonomic computing* [18]. In the last decade, autonomic computing has grown from a futuristic vision of computing systems autonomically taking care of themselves and of their own complexity to a multifaceted and pragmatic research field [9]. The aim of this research is automatically exploiting runtime information to ease user interaction with computing systems. From a theoretical standpoint, autonomic computing has the goal of enhancing computing systems with *self-** *properties*, as analyzed by Salehie and Tahvildari [25], who also propose a taxonomy for autonomic computing. According to this taxonomy, *AcOS* is a *closed*, *model-based* solution employing *continued monitoring*, *dynamic decision making*, and *external proactive adaptation*. *AcOS* extends commodity operating systems with self-* properties at the software level (both within the operating system and with a companion runtime system); more specifically, it enables *self-adaptive* and *self-managing* properties. *AcOS* enables these properties by employing feedback control to make active use of runtime information. Different representations exist for such control loops; in *AcOS*, we adopt the most compact of these abstractions: the *observe—decide—act* control loop (ODA).

We specialize the ODA control loop to highlight interactions of the autonomic layer provided by *AcOS* with the computing system and with the applications. Figure 1 shows the three *autonomic components* realizing the steps of the ODA control loop: (1) *monitors* realize the observe phase; (2) *adaptation policies* provide the decide phase; and *actuators* enact the act phase.

Monitors are components in charge of properly exposing runtime information; they can be either passive or active elements, depending on how they gather information [16]. For instance, we use a passive monitor to observe CPU temperature; this monitor simply retrieves data from model-specific registers. Instead, to observe the throughput of an application, we use an active monitor that implements an infrastructure to synthesize this metric, which is not directly available. Each monitor is also in charge of exposing an API to allow setting SLOs on the specific measurement it provides.

Adaptation policies elaborate the measurements and SLOs exposed by monitors to estimate the corrective action needed to drive the measurements towards meeting the objectives. Each adaptation policy uses a specific decision engine; we experimented with heuristics, analytic modeling, and control theory.

Actuators provide mechanisms that adaptation policies can use, through appropriate APIs, to enact corrective actions. For instance, we implemented an idle cycle injection actuator that adaptation policies can use to selectively preempt application threads in fa-

vor of the idle task with the goal of capping processor temperature without unnecessarily harming performance.

Figure 1 highlights interactions of monitors, adaptation policies, and actuators with the system as a whole and the applications. Monitors retrieve information on both system-wide parameters and application-specific measurements and SLOs. Adaptation policies elaborate this information and use actuators to affect system and application behavior. This paper illustrates how *AcOS* exploits this structure to enhance commodity operating systems with performance-aware resource allocation and temperature management.

Autonomic computing can be beneficial throughout the hardware/software stack. However, since the operating system, coupled with runtime systems, is traditionally in charge of managing system resources and can control both the hardware and the applications, we argue that this is the level where these techniques are most needed and can yield most benefits. Moreover, an operating system with autonomic capabilities can both serve as a convenient base to offer interfaces for autonomic applications and exploit additional self-* properties offered at the hardware level. For these reasons, we build *AcOS* as an extension to commodity operating systems.

3. PERFORMANCE MANAGEMENT

Modern computer architecture design follows the principle of optimizing for the common case. For instance, caches, coupled with prefetching, dramatically reduce memory latency for regular access patterns. While this strategy continuously improves performance for many applications, it also makes it unpredictable; this side effect can make advanced architectural features unsuitable for embedded systems, where what matters is often the worst-case execution time (WCET). Moreover, the advent of CMPs and MPSoCs leads to on-chip co-location of applications that must rely on shared hardware resources; this scenario further impairs performance predictability.

We seek answer to the following question: can we leverage autonomic computing to achieve predictable performance for applications co-located onto a multi-core processor? Well-established techniques exist to estimate the WCET of single-threaded applications on single-core processors [31] and recent research focuses on multi-core processors [12, 21]. This research relies on offline profiling to provide strong guarantees for time-critical systems. Instead, we want to dispense from offline profiling and provide users with an intuitive means of stating SLOs on execution time.

We tackle this challenge with *Metronome* [28] and *Metronome++*, respectively a heuristic and model-based feedback control policy; both these policies introduce performance-awareness in the Linux kernel and rely on the *Heart Rate Monitor* (HRM) to provide performance measurements and requirements.

3.1 Performance Metrics and Measurement

Our goal is to enforce a SLO defined on the execution time $\bar{t}$ to complete $\bar{n}$ of *units of work*; for instance, a video encoder may be required to process $\bar{n}$ frames in $\bar{t}$ seconds. In order to strike appropriate resource allocation without the need of offline profiling, we need a metric to estimate the execution time of an application at runtime. We leverage the known amount of units of work to define a proxy for the execution time: the *required throughput* to attain the SLO is $\bar{g} = \bar{n}/\bar{t}$. Equation (1) formalizes how we enforce the SLO: given that after k control steps the application completed m units of work, we keep the *global throughput* $g(k)$ up to step k close to $\bar{g}$.

$$\forall k, g(k) \equiv \bar{g} \quad \text{where} \quad g(k) = \frac{m}{k} \quad (1)$$

Global throughput is an application-specific high-level performance metric that easily allows users to state meaningful SLOs [14, 15,

23, 28]. We leverage this metric for automatic goal-oriented resource allocation, dispensing users and administrators from the laborious process of analyzing application properties (e.g., scalability) to manually determine resource allocations.

Application-specific high-level performance metrics require application support; in our case, applications need to provide progress information. For this reason, we developed *HRM* [28]: an active monitoring infrastructure to synthesize throughput measurements from *heartbeats*. Similarly to previous proposals [13], *HRM* exports a simple API for applications to emit a heartbeat whenever they complete a unit of work. Based on the time stamps of these signals, *HRM* efficiently provides adaptation policies with throughput measurements expressed in heartbeats/s; these measurements directly map to application-specific performance metrics such as frames/s for a video encoder or decoder. *HRM* also exports API calls for users to express SLOs, according to the methodology described in Section 2. The major novelties of *HRM* are support for both multi-threaded and multi-programmed applications, through the definition of monitoring *groups*, and the system-wide (i.e., both in user- and kernel-space) visibility of throughput measurements [28]. Appendix A, provides additional details regarding the usage of *HRM* to instrument various flavors of multi-threaded applications.

We implemented *HRM* both on top of Linux and FreeBSD.[1] Both implementations feature a split design where most of the infrastructure leaves in kernel-space and a small-footprint user-space library, namely *libhrm*, exports an API for applications and user-space adaptation policies. To pass information across address spaces, we exploit shared memory to carefully map shared memory pages and use cache-aligned data structures to avoid poor performance due to caching issues such as false sharing [28].

3.2 Metronome

With *Metronome*, we introduce performance-awareness by means of a non-invasive modifications to the *Completely Fair Scheduler* (*CFS*), which is the default scheduling class for the Linux kernel since version 2.6.23. CFS, as most schedulers in commodity operating systems, offers mechanisms (e.g., priorities and resource containers [2]) to allocate resources (e.g., CPUs and CPU bandwidth) to applications; however, the task of determining appropriate settings to respect SLOs is far from easy. To address this issue, *Metronome* automates CPU bandwidth allocation based, at each scheduling step k, on the performance error $e(k)$ between the desired ($\bar{g}$) and measured ($g(k)$) global throughput: $\forall k, e(k) = \bar{g} - g(k)$.

Metronome leverages *HRM* to retrieve throughput measurements and user-specified SLOs and uses this information to modify a single parameter of the CFS scheduler: the *virtual runtime* (*vruntime*). Since CFS picks tasks for execution in ascending order of *vruntime*, modifying this value implicitly defines CPU bandwidth allocation. *Metronome* uses a simple heuristic to tune the *vruntime* of tasks of SLO-bound applications with a scaling factor s that depends, for each SLO-bound application a, on the current performance error $e_a(k)$. If $e_a(k) > 0$, then a needs more CPU bandwidth and *Metronome* will weight the *vruntime* of its tasks with a factor $s = g(k)/\bar{g}$. Otherwise, if $e_a(k) <= 0$, a is delivering sufficient throughput to attain its SLO and *Metronome* will not affect the *vruntime* of its tasks.[2] Notice that, since in Linux and in most commodity operating systems the scheduler resides in kernel-space, *HRM*'s capability of exporting measurements system-wide is crucial for *Metronome* to be effective despite using a very simple heuristic.

---

[1]We support versions 2.6.35 and 3.2 of the Linux kernel and version 7.2 and 9.0 of the FreeBSD kernel.

[2]For additional details regarding design, implementation, and validation of *HRM* and *Metronome*, refer to [28].

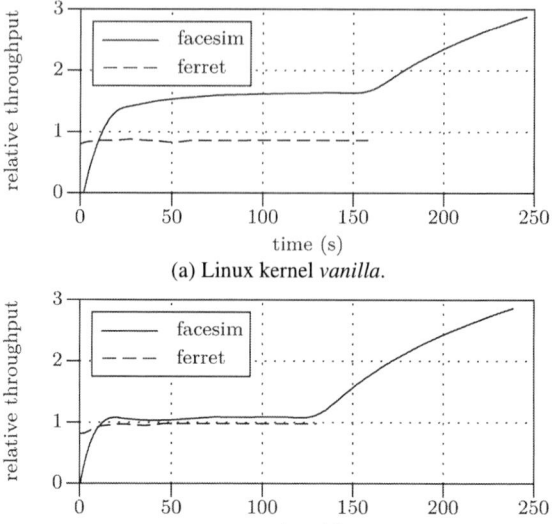

(a) Linux kernel *vanilla*.

(b) Linux kernel enhanced with *Metronome*.

Figure 2: Relative throughput of *facesim* and *ferret*; 1 represents the throughput required to attain per-application SLOs.

We validate *Metronome* on a workstation with an Intel Core i7-870 Processor (we disable Intel Hyper-Threading, Enhanced Speed-Step, and Turbo Boost Technologies), 4 GB of 1066 MHz Single Ranked DIMMs, and the Linux kernel 2.6.35 enhanced with *HRM* and *Metronome*.[2] We used two applications from the PARSEC 2.1 benchmark suite [6]: *facesim* and *ferret*. We measure the performance of *facesim* in frames/s, as it is instrumented with *libhrm* to emit a heartbeat per computed frame; *facesim* yields ≈ 0.67 frames/s when running with 4 threads. We instrumented *ferret* to emit a heartbeat per computed query; therefore, its performance reads in queries/s. When run with 4 threads, *ferret* yields ≈ 30 queries/s.

Figure 2 shows the dynamics of the global throughput of *facesim* and *ferret* relative to the respective SLOs; we arbitrarily choose a SLO of 0.22 frames/s for *facesim* and 19 queries/s for *ferret*. Figure 2a shows the two applications scheduled by the unmodified CFS. CFS partitions CPU time evenly among the applications till ≈ 150 s, when *ferret* terminates. Since CFS is a work-conserving scheduler, it never idles resources whenever there are runnable tasks; therefore, when *ferret* terminates, *facesim* can use the whole processor and its global throughput grows. Notice that, since CFS is not aware of the SLOs, it does nothing to enforce them: doing so would require manual intervention to increase the relative priority of *ferret*. *Metronome* performs this action automatically, as Figure 2b reports. The simple heuristic at the base of *Metronome* is able to adjust the relative priorities of the two applications to keep both close to a normalized performance of 1, which represents the respective SLOs. *Metronome* maintains all the desirable properties of CFS (e.g., non-starvation) and also the work-conserving behavior; therefore, *facesim* gets the full processor when *facesim* terminates, after approximately 130 s.

3.3 Metronome++

Metronome demonstrates how a simple heuristic can be enough to enable goal-oriented resource allocation by exploiting runtime performance feedback. With *Metronome++*, we devise and evaluate a more advanced adaptation policy to dynamically allocate CPUs to SLO-bound applications in a multi-core processor; again, our goal is achieving performance predictability to meet SLOs. Commodity operating systems provide various mechanisms (e.g.,

540

task pinning) to define the task to CPU mapping; however, just as for task priorities, these mechanisms only provide knobs that administrators are in charge of manually adjusting.

Similarly to other recent proposals [26], we estimate at runtime the scalability characteristics of applications; however, our goal is respecting SLOs and not maximizing performance. To estimate scalability characteristics, we use the least squares algorithm to fit a second order polynomial that correlates the number of allocated CPUs and the throughput measurements provided by *HRM*; Appendix B justifies this methodology through experimental results.

We build *Metronome++* based on a user/kernel-space split design and manage cross-address space communication by careful (i.e., cache-aware) sharing of mapped memory pages. The scalability characteristics estimation runs in user-mode in order to take advantage of linear algebra libraries, while the actuation (i.e., tasks migration among run queues to) runs in kernel-mode, so as to avoid the overhead of synchronous system calls and thus minimize runtime impacts. We devise the kernel-space side of *Metronome++* to care for task migration minimization and load balancing.

Since changing the mapping of tasks to CPUs can be expensive due to task migration, *Metronome++* takes into account the history of throughput measurements to avoid trashing a proper CPU allocation due to noise in the data. The drawback of this choice is reduced reactivity in case applications go through different execution phases. We address this issue by adding a prediction mechanism for execution phase transitions as a second adaptation level. To detect a transition, we use an exponential moving average of the ratio between the throughput and the resource allocation. We describe in more details and evaluate the prediction mechanism in Appendix C.

To evaluate *Metronome++*, we use the *x264* application from the PARSEC 2.1 benchmark suite [6] and co-locate two identical instances of the application on a workstation with an Intel Xeon Processor W3670 (we disable Intel Hyper-Threading, Enhanced SpeedStep, and Turbo Boost Technologies), 12 GB of 1333 MHz Single Ranked DIMMs, and the Linux kernel 3.2 enhanced with *HRM* and *Metronome++*. We instrument *x264* to emit a heartbeat per encoded frame: we measure its performance in frames/s.

Figure 3 shows the dynamics of the two instances of *x264* when run on *vanilla* Linux and managed by *Metronome++*. An important characteristic of *x264* is the presence of input-dependent execution phases: the native input of the PARSEC 2.1 benchmark suite presents a lighter-weight (i.e., higher performance) phase between the 70-th and the 300-th frames. The execution phases of *x264* emerge from Figure 3a: the global throughput is far from constant, even though the Linux kernel *vanilla* allocates resources evenly between the two instances (marked $x264_a$ and $x264_b$). Figure 3b shows the performance of the two instances when using *Metronome++* to dynamically allocate CPUs to match SLOs (we arbitrarily choose 8 and 12 frames/s). These results show that *Metronome++* is able to drive both instances to respect their SLO, effectively estimating scalability characteristics to strike proper CPU allocation allocations and responding to execution phase transitions; we validate our transition prediction mechanism in Appendix C.

Notice that *Metronome++* does not expose the same work-conserving behavior of *Metronome*, as it does not over-allocate resources. This feature potentially enables to activate power-saving techniques on idle resources. The use of a more complex analytic model with respect to *Metronome* makes *Metronome++* slower in converging to the desired global throughput due to the need to "warm up" the scalability characteristics and execution phase prediction; however *Metronome++* improves upon *Metronome* in terms of robustness to noise and response to execution phase transitions.

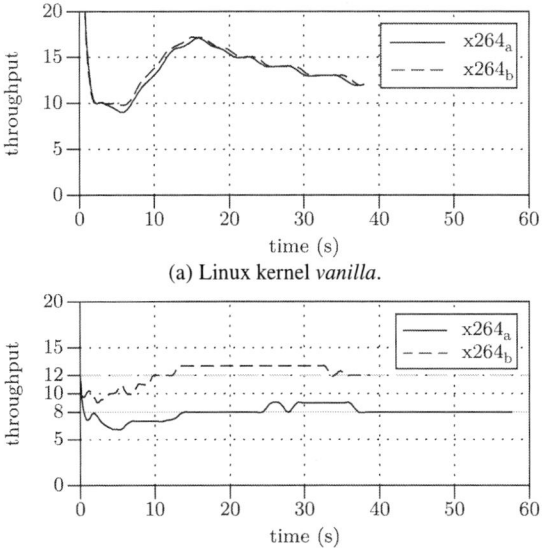

(a) Linux kernel *vanilla*.

(b) Linux kernel enhanced with *Metronome++*.

Figure 3: Throughput of two instances of *x264*; the green constant lines represent the throughput required to attain SLOs.

4. TEMPERATURE MANAGEMENT

We argue that providing autonomic performance-aware resource allocation is just one side of the coin with respect to the benefits embedded systems can draw from autonomic computing; on the other side is smart enforcement of system-level constraints. One of the major emerging system-level constraints in modern computing systems is capping CPU temperature. Maintaining CPUs cool is crucial for energy efficiency [29] and reliability [30], as high temperature increase of leakage power and leads to a reduction of the MTTF. Furthermore, high temperature in embedded systems may lead to usability issues (e.g, in hand-held devices).

Blindly enforcing temperature constraints, as done with classic dynamic thermal management (DTM) techniques, indiscriminately harms applications performance. With *AcOS*, we investigate how to intelligently enforce system-level temperature constraints while avoiding to break performance SLOs. For this purpose, we devise and evaluate *ADAptive Performance and Thermal ManagEment (ADAPTME)*: a feedback control framework for dynamic performance and thermal management. *ADAPTME* leverages control theory and idle cycle injection [1, 4] to co-manage CPU temperature and performance of multi-programmed workloads.

For *ADAPTME*, we need to observe both performance and temperature. We use *HRM* (see Section 3.1) for throughput measurements and per-CPU machine-specific registers available in modern multi-core processors for temperature measurements: per-CPU high-priority kernel-mode threads periodically sample and make available temperature measurements.

ADAPTME uses an adaptation policy leveraging discrete-time linear models for performance and temperature reported in Equations (2) and (3), respectively.

$$r_i(k+1) = r_i(k) + \eta_i \cdot p_i(k) \qquad (2)$$
$$T_j(k+1) = T_j(k) + \mu_j \cdot I_j(k) \qquad (3)$$

The model in Equation (2) assumes that performance $r_i(k+1)$ of application i at control step $k+1$ can be derived through a linear combination of its performance $r_i(k)$ and the priority $p_i(k)$ (e.g., *nice* value) of i's tasks, weighted with a parameter η_i, at the previous step. Equation (3) states the same relation for temperature

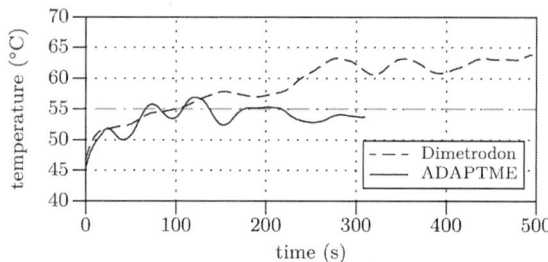

Figure 4: Average temperature for 5 consecutive runs of *swaptions* when capping with either *Dimetrodon* (50 % chance of injecting idle time) or *ADAPTME* (55 °C temperature constraint).

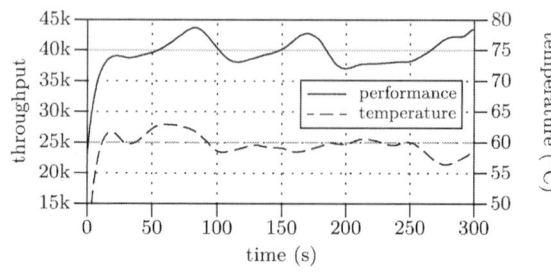

Figure 5: Throughput of one SLO-bound instance of *swaptions* among the 4 co-located and average temperature when using *ADAPTME* for performance (SLO: 40000 swaptions/s) and temperature (constraint: 60 °C).

$T_j(k+1)$ of CPU j, where $I_j(k)$ is the fraction of idle time injected during the previous control period.

Based on these two models, we synthesize two deadbeat adaptive controllers to respectively estimate priority and idle time required by each SLO-bound application and CPU. We couple each SLO-bound application with a priority controller and each CPU with an idle time controller. Since the control context is variable, finding static values for the parameters η_i and μ_j for each application and CPU is both impractical and ineffective. Therefore, we employ an adaptive filter, in this case an exponential moving average, to estimate parameters online.

Since idle cycle injection has the side effect of impairing applications performance, this action can conflict with the process of adjusting priorities to respect SLOs. For this reason, to avoid instability, we need to define a policy to coordinate controllers. We use a probabilistic solution: whenever a temperature controller requires the injection of idle cycles and a performance controller requires an application not to be preempted (which happens when the application is not respecting its SLO), *ADAPTME* actually injects idle time over that application with a tunable probability.[3]

We implemented *ADAPTME* in FreeBSD 7.2, which we also extend with a port of *HRM*. We use the *swaptions* applications from the PARSEC 2.1 benchmark suite [6] as our reference application to evaluate *ADAPTME*; we instrumented *swaptions* with *HRM* to measure performance in swaptions/s. We realize two different experiments to evaluate *ADAPTME*. First, we consider the temperature controller alone to test the ability of enforcing constraints; this experiment allows us to compare against *Dimetrodon* [1]: a state-of-the-art extension of FreeBSD 7.2 for preventive DTM. Second, we evaluate coupled temperature and performance controllers. Experimental results were collected on the workstations described in Sections 3.2 and 3.3, respectively.

Figure 4 shows the dynamics of average CPU temperature of *ADAPTME* and *Dimetrodon* in the first experiment. *Dimetrodon* employs probabilistic feedforward control and allows to specify an idle cycle injection probability; however, it does not provide any guarantee with respect to actual temperature capping. Instead, *ADAPTME* exploits feedback control and allows to specify a temperature constraint; moreover, *ADAPTME* is devised so as to minimize the impact of idle cycle injection on performance. These characteristics allow *ADAPTME* to attain better performance (i.e., faster execution time) and keep lower temperature than *Dimetrodon*.

Figure 5 shows the results of the second experiment: we co-locate four instances of *swaptions* running with 4 threads on an infinite input dataset and set a SLO (i.e., 40000 swaptions/s, which is twice as much as each instance achieves by default) for one them and a temperature constraint (i.e., 60 °C). Experimental re-

sults show that *ADAPTME* is able to both cap temperature and prioritize the SLO-bound application to attain its SLO.

5. RELATED WORK

With *AcOS*, we investigate how to leverage autonomic computing at the operating system and runtime level to ease the management of computing systems. In this paper, we focus on performance-aware resource allocation for multi-core processors (with *Metronome* and *Metronome++*) and intelligent thermal capping (with *ADAPTME*). The rest of this Section discusses related work, first focusing on each of these two topics and then on wider projects.

Application Heartbeats [13] was an early proposal for a framework allowing applications to easily export measurements and specify SLOs on their performance in high-level metrics. These ideas gave momentum to initial work on *HRM* and *Metronome* [28]. *HRM* improves *Application Heartbeats* in the observation phase, introducing the concept of *groups* and leveraging a design split across user and kernel-space to improve efficiency and enable system-wide visibility of throughput measurements. This last feature enables *Metronome* to achieve performance-aware CPU bandwidth allocation by means of a heuristic acting on applications *vruntime*. *Metronome++* shares the idea of performance-aware CPU allocation with *PDPA* [8] and the use of application scalability characteristics estimates like *SBMP* [26]. However, both these works pursue performance maximization, while *Metronome++* aims for SLOs satisfaction and performance predictability. Moreover, we leverage a high-level application-specific performance metric, which leads to the advantages discussed in Appendix B.

A relevant but orthogonal technique with respect to *ADAPTME* is thermal-aware scheduling; relevant works are *Heat-and-Run* [24] and *ThreshHot* [32]. *Heat-and-Run* [24] exploits simultaneous multithreading to place on the same core tasks requiring different functional units and it manages tasks migration to balance temperature across a CMP. *ThreshHot* [32] schedules tasks ordered from the "hottest" (most CPU-intensive) to the "coldest" (most I/O-intensive); this schedule guarantees to minimize temperature at the end of an epoch. The goal of thermal-aware schedulers is minimizing temperature without degrading performance. *ADAPTME*, which enforces temperature requirements through DTM, is orthogonal to thermal-aware scheduling: DTM tackles situations where minimizing temperature is not enough and capping is necessary.

HybDTM [19] exploits the *hot* and *cold* tasks classification for DTM: whenever temperature exceeds the threshold, it throttles "hot" tasks first by lowering their priority. *HybDTM* is meant for single-core processors and many of its considerations do not apply to multi-core processors. *Dimetrodon* Bailis et al. [1] is a framework on top of FreeBSD that leverages idle cycle injection to decrease temperature with a probabilistic feedforward approach. We com-

---

[3]For additional details regarding the design, implementation, and validation of *ADAPTME*, refer to [4].

pare *ADAPTME* against *Dimetrodon* in Section 4.

METE [27] is a control-theoretical framework for CMPs to meet QoS by managing CPU, cache ways, and memory bandwidth allocation for multi-threaded applications. *AcOS* focuses on only CPU and CPU bandwidth allocation, but also considers temperature capping; taking into consideration automatic allocation of additional resources is one of the future directions for *AcOS*. *SEEC* [14, 15] is a runtime system that performs resource allocation to respect SLOs by exploiting control theory and machine learning; *SEEC* focuses on balancing performance and power consumption requirements, while *AcOS* can balance performance and thermal requirements. *Tessellation* [7] is an operating system for multi and many-core processors and client computing systems based on the concept of adaptive resource-centric computing (ARCC). *Tessellation* restructures the operating system around QoS-guaranteed resource containers called *cells*. *AcOS* is orthogonal to *Tessellation*: we focus on feedback control, while *Tessellation* focuses on providing adaptation mechanism within the operating itself. *AcOS* could exploit an operating system like *Tessellation* as a base for further research.

6. PERSPECTIVES

We started the *AcOS* project to seek an answer to a question: *can we enhance commodity operating systems with an autonomic layer so as to respect user-specified SLOs and enforce system-level constraints?*

With the control loops (i.e., ODA loops) we propose and validate in this paper, we contribute to moving a step towards an affirmative answer. *Metronome* and *Metronome++* demonstrate the ability to respect user-specified SLOs on performance measurements by means of CPU bandwidth and CPU allocation, while *ADAPTME* is able to enforce a system-level temperature constraints while still accounting for performance. However, several open problems require further research before we can definitively answer this question.

An interesting direction is evaluating SLOs defined on different performance metrics. For instance, instead of requesting a bound on WCET, users may define the desired QoS on latency or real-time constraints. Feedback control techniques to automatically attain such requirements with on-chip shared resources may need adaptation policies based on different mechanisms and algorithms.

Possibly, different QoS definitions may require the management of a wider set of resources (e.g., cache ways, memory bandwidth, file system cache, disk bandwidth, network bandwidth, etc.). One of the challenges towards this direction is enabling mechanisms to effectively manage such resources at runtime. If coordinate management of multiple resources was demonstrated in a simulation environment [27], actual hardware and software mechanisms are needed to experiment with similar adaptation policies on commodity computing systems [7].

Having an increasing pool of resources to manage and an increasing number of control loops leads to a third compelling challenge: properly orchestrating a large number of possibly conflicting adaptation policies. With *ADAPTME*, we propose a probabilistic heuristics to define the interaction of two conflicting adaptation policies aimed at respecting performance SLOs and enforcing a temperature constraint. A solution of this kind, however, does not scale well with an increasing number of control loops: we need to research a more systematic methodology.

7. ACKNOWLEDGMENTS

The authors would like to thank Simone Campanoni, Fabio Cancarè, Henry Hoffmann, Giovanni F. Del Nero, and Donatella Sciuto for their contributions to the *Autonomic Operating System* project.

8. REFERENCES

[1] P. Bailis *et al.*, "Dimetrodon: Processor-Level Preventive Thermal Management via Idle Cycle Injection," in *Proc. of DAC*, 2011.

[2] G. Banga *et al.*, "Resource Containers: A New Facility for Resource Management in Server Systems," in *Proc. of OSDI*, 1999.

[3] P. Barham *et al.*, "Xen and the Art of Virtualization," in *Proc. of SOSP*, 2003.

[4] D. B. Bartolini *et al.*, "A Framework for Thermal and Performance Management," in *Proc. of MAD*, 2012.

[5] S. Bell *et al.*, "TILE64 - Processor: A 64-Core SoC with Mesh Interconnect," in *Proc. of ISSCC*, 2008.

[6] C. Bienia, "Benchmarking Modern Multiprocessors," Ph.D. dissertation, Princeton University, 2011.

[7] J. A. Colmenares *et al.*, "Tessellation: Refactoring the OS around Explicit Resource Containers with Continuous Adaptation," in *Proc. of DAC*, 2013.

[8] J. Corbalan *et al.*, "Performance-Driven Processor Allocation," in *Proc. of OSDI*, 2000.

[9] S. Dobson *et al.*, "Fulfilling the Vision of Autonomic Computing," *IEEE Computer*, vol. 43, no. 1, 2010.

[10] N. El-Sayed *et al.*, "Temperature Management in Data Centers: Why Some (Might) Like It Hot," in *Proc. of SIGMETRICS*, 2012.

[11] H. Esmaeilzadeh *et al.*, "Power Challenges May End the Multicore Era," *Commun. ACM*, vol. 56, no. 2, 2013.

[12] G. Giannopoulou *et al.*, "Timed Model Checking with Abstractions: Towards Worst-Case Response Time Analysis in Resource-Sharing Manycore Systems," in *Proc. of EMSOFT*, 2012.

[13] H. Hoffmann *et al.*, "Application Heartbeats: A Generic Interface for Specifying Program Performance and Goals in Autonomous Computing Environments," in *Proc. of ICAC*, 2010.

[14] ——, "SEEC: A General and Extensible Framework for Self-Aware Computing," Massachusetts Institute of Technology, Tech. Rep., 2011.

[15] ——, "Self-aware Computing in the Angstrom Processor," in *Proc. of DAC*, 2012.

[16] M. C. Huebscher and J. a. McCann, "A survey of Autonomic Computing – degrees, models and applications," *ACM Comp. Surv.*, vol. 40, no. 3, 2008.

[17] B. Jeff, "Big.little system architecture from arm: Saving power through heterogeneous multiprocessing and task context migration," in *Proc. of DAC*, 2012.

[18] J. O. Kephart and D. M. Chess, "The Vision of Autonomic Computing," *IEEE Computer*, vol. 36, no. 1, 2003.

[19] A. Kumar *et al.*, "HybDTM: A Coordinated Hardware-Software Approach for Dynamic Thermal Management," in *Proc. of DAC*, 2006.

[20] E. Lau *et al.*, "Multicore Performance Optimization Using Partner Cores," in *Proc. of HotPar*, 2011.

[21] M. Lv *et al.*, "Combining Abstract Interpretation with Model Checking for Timing Analysis of Multicore Software," in *Proc. of RTSS*, 2010.

[22] G. E. Moore, "Cramming More Components Onto Integrated Circuits," *Electronics*, 1965.

[23] R. Nathuji *et al.*, "Q-Clouds: Managing Performance Interference Effects for QoS-Aware Clouds," in *Proc. of EuroSys*, 2010.

[24] M. D. Powell *et al.*, "Heat-and-Run: Leveraging SMT and CMP to Manage Power Density Through the Operating System," in *Proc. of ASPLOS*, 2004.

[25] M. Salehie and L. Tahvildari, "Self-Adaptive Software: Landscape and Research Challenges," *ACM Trans. Auton. Adapt. Syst.*, vol. 4, no. 2, 2009.

[26] H. Sasaki *et al.*, "Scalability-Based Manycore Partitioning," in *Proc. of PACT*, 2012.

[27] A. Sharifi *et al.*, "METE: Meeting End-to-End QoS in Multicores through System-Wide Resource Management," in *Proc. of SIGMETRICS*, 2011.

[28] F. Sironi *et al.*, "Metronome: Operating System Level Performance Management via Self-Adaptive Computing," in *Proc. of DAC*, 2012.

[29] K. Skadron *et al.*, "Temperature-Aware Microarchitecture: Modeling and Implementation," *ACM Trans. Archit. Code Optim.*, vol. 1, no. 1, 2004.

[30] J. Srinivasan *et al.*, "The Case for Lifetime Reliability-Aware Microprocessors," in *Proc. of ISCA*, 2004.

[31] R. Wilhelm *et al.*, "The Worst-Case Execution-Time Problem – Overview of Methods and Survey of Tools," *ACM Trans. Embed. Comput. Syst.*, vol. 7, no. 3, 2008.

[32] X. Zhou *et al.*, "Performance-Aware Thermal Management via Task Scheduling," *ACM Trans. Archit. Code Optim.*, vol. 7, no. 1, 2010.

APPENDIX

A. HRM USAGE AND VALIDATION

HRM [28] is an active monitor providing: *libhrm*, which is a simple API to instrument parallel applications, and system-wide available application-specific throughput measurements and requirements (i.e., SLOs). This section provides several use cases for *libhrm*, which is employed to instrument parallel applications with diverse multi-threading models, and demonstrates the efficiency of *HRM* by evaluating its runtime impact.

A.1 Instrumenting Parallel Applications

HRM organizes instrumented applications in *groups*, where each group is a set of tasks cooperating on a certain activity (e.g., encoding a video) that is bound to a SLO. Therefore, to properly use *libhrm*, a multi-threaded application must: (1) *attach* its tasks (i.e., threads) to a group; (2) *set* the SLO;[4] (3) emit one *heartbeat* upon completion of a unit of work; and (4) *detach* its tasks from the group upon termination.

We analyze the instrumentation of 11 out of 13 multi-threaded applications from the PARSEC 2.1 benchmark suite [6] employing diverse multi-threading models, which allows to show the flexibility and ease of use of *libhrm*.

We focus on the following applications: *blackscholes, bodytrack, canneal, dedup, facesim, ferret, fluidanimate, raytrace, streamcluster, swaptions,* and *x264*. These 11 applications can be grouped in four categories according to their multi-threading model:

- *category 1—blackscholes, canneal, fluidanimate, streamcluster,* and *swaptions* use "fork & join" of workers;
- *category 2—bodytrack, facesim,* and *raytrace* leverage pools of workers running different jobs in parallel;
- *category 3—dedup* and *ferret* use a pipeline with pools of workers serving the stages;
- *category 4—x264* employs "spawn & kill" of workers to realize a virtual pipeline.

For each category, we give additional details regarding the structure and instrumentation of one application.

Applications in *category 1* are straightforward to instrument, as they use a simple multi-threading model. The sequence diagram in Figure 6 shows the structure and instrumentation of these applications. The main thread of the applications is responsible for forking (i.e., `pthread_create(3)`) the worker threads and joining them (i.e., `pthread_join(3)`) when they terminate. The first worker thread attaches to (and implicitly creates) the group, and it sets the SLO; the other worker threads attach to the group and, just like the first worker thread, start their computation. Figure 7 visualizes the typical structure of computation of a worker thread, which runs in a loop terminating a unit of work, and subsequently emitting a *heartbeat* for each iteration. When the application is terminating, before re-joining, the worker threads detach from the group.

Applications in *category 2* use a pool of worker threads running parallel kernels; therefore, none of the threads completes a unit of work alone. Figure 8 represents the common structure of these applications. The main thread, which acts as a dispatcher, is the first to attach to the group. Due to the structure of these applications, which is represented in Figure 9, the main thread emits heartbeats. However, we still attach all the worker threads to the group to make adaptation policies aware that they are actually relevant.

Applications of *category 3* employ many pools of worker threads organized in a pipeline. In these applications, the main thread is responsible for forking the pools of worker threads and waiting for

---

[4] We also allow users and administrators to change the SLO.

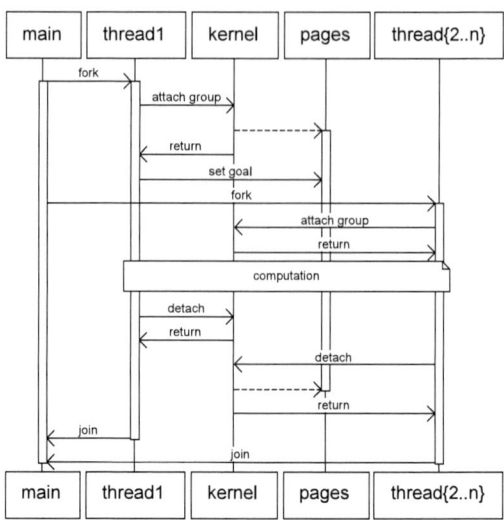

Figure 6: Structure of the applications in category 1.

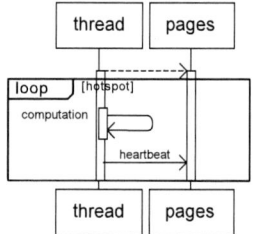

Figure 7: Computation the applications in category 1.

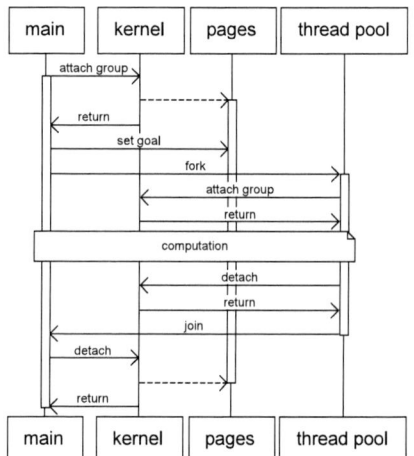

Figure 8: Structure of the applications in category 2.

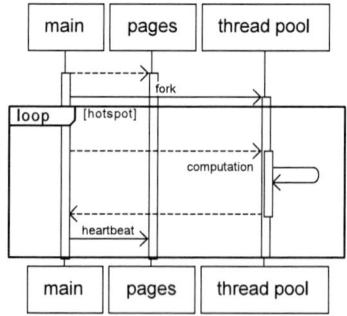

Figure 9: Computation of the applications in category 2.

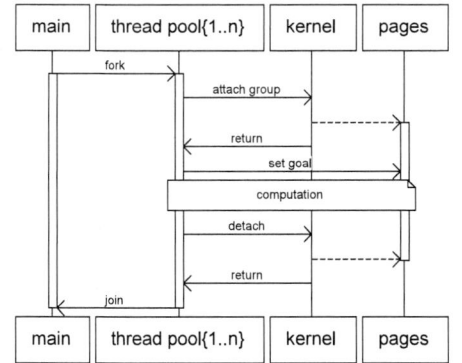

Figure 10: Structure of the applications in category 3.

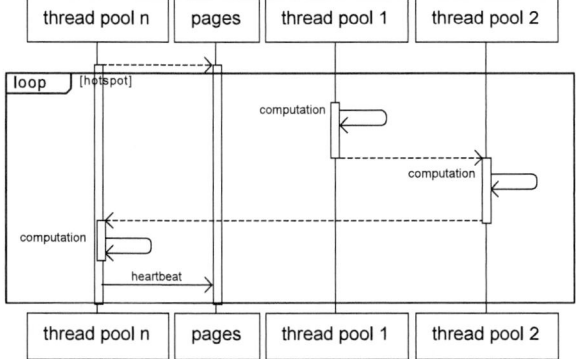

Figure 11: Computation of the applications in category 3.

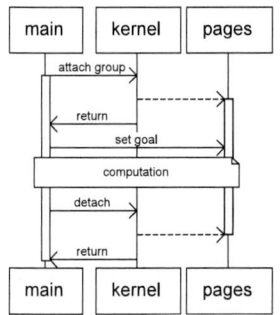

Figure 12: Structure of the applications in category 4.

them to join upon application completion. All the spawned threads attach to the group (the first thread automatically creates the group and sets the SLO); Figure 10 shows the general structure of these applications. The worker threads contained in the *n*-th pool (i.e., the last stage of the pipeline) are the ones committing each unit of work and are responsible for emitting heartbeats, as Figure 11 illustrates.

The last category, i.e., *category 4*, contains only one application, namely *x264*. *x264* creates a virtual pipeline based on a "spawn & kill" multi-threading model, which makes the instrumentation straightforward. Figure 12 illustrates the structure of the instrumentation of *x264*. The main thread is responsible for creating and attaching to the group. Figure 13 focuses on the computation phase: the main thread spawns many different worker threads that re-join when their computation ends. The main thread maintains the notion of advancement (i.e., encoding of frames in *x264*); hence, it is responsible for emitting heartbeats. Just as for applications in *category 2*, we still attach all the worker threads to the group to inform

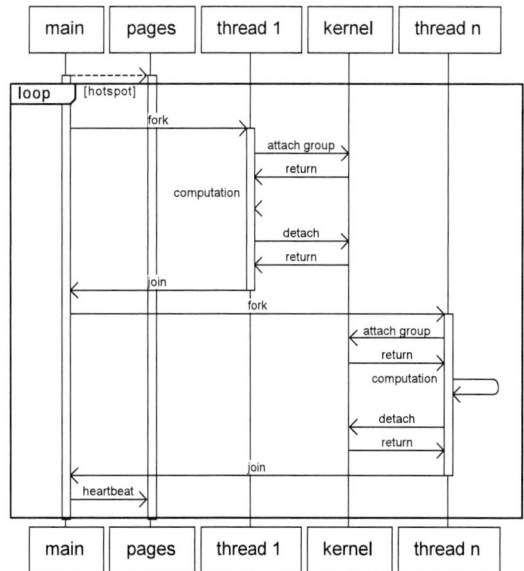

Figure 13: Computation of the applications in category 4.

adaptation policies about their relevance.

A.2 Evaluating the Runtime Impact

We evaluate the overhead of *HRM* on all the applications we instrumented from the PARSEC 2.1 benchmark suite.

Table 1 reports average execution time and its standard deviation of 100 consecutive runs of unmodified (i.e., *vanilla*) and instrumented applications with the native input and the computed runtime impact (i.e., overhead).

Experimental results where collected on a workstation with an Intel Xeon Processor W3570 (we disable Intel Hyper-Threading, Enhanced SpeedStep, and Turbo Boost Technologies), 12 GB of 1333 MHz Single Ranked DIMMs, and the Linux kernel 3.2 enhanced with *HRM*. We configured *HRM* to compute throughput measurements every 100 ms.

The highest runtime impact we measured is 2.80% for *dedup*; with the exception of *x264*, higher runtime impacts (e.g., *bodytrack* and *dedup*) coincide with short execution times and we argue this is due to "non-amortized" costs of creating the group and attaching worker threads, which are the most expensive operations. According to experimental results we can state that *HRM* is efficient and imposes negligible runtime impact.

Table 1: Comparison between *vanilla* and instrumented applications from the PARSEC 2.1 benchmark suite

category	application	vanilla		instrumented		overhead
		avg. (ms)	std. (ms)	avg. (ms)	std. (ms)	
1	*blackscholes*	68731.67	1998.33	68902.53	221.21	0.25%
	canneal	96405.94	1846.36	96913.76	488.74	0.53%
	fluidanimate	95785.44	627.38	96077.83	103.19	0.31%
	streamcluster	147536.15	2393.04	147460.57	333.19	-0.05%
	swaptions	75308.29	308.39	75508.16	249.35	0.27%
2	bodytrack	52849.39	412.03	53732.33	878.61	1.67%
	facesim	145175.15	2256.19	145408.80	787.26	0.16%
	raytrace	124036.75	901.47	124441.34	750.43	0.33%
3	dedup	33509.96	955.21	34448.43	1187.31	2.80%
	ferret	113626.21	527.36	114106.41	218.52	0.42%
4	*x264*	32657.13	252.06	32713.23	255.53	0.17%

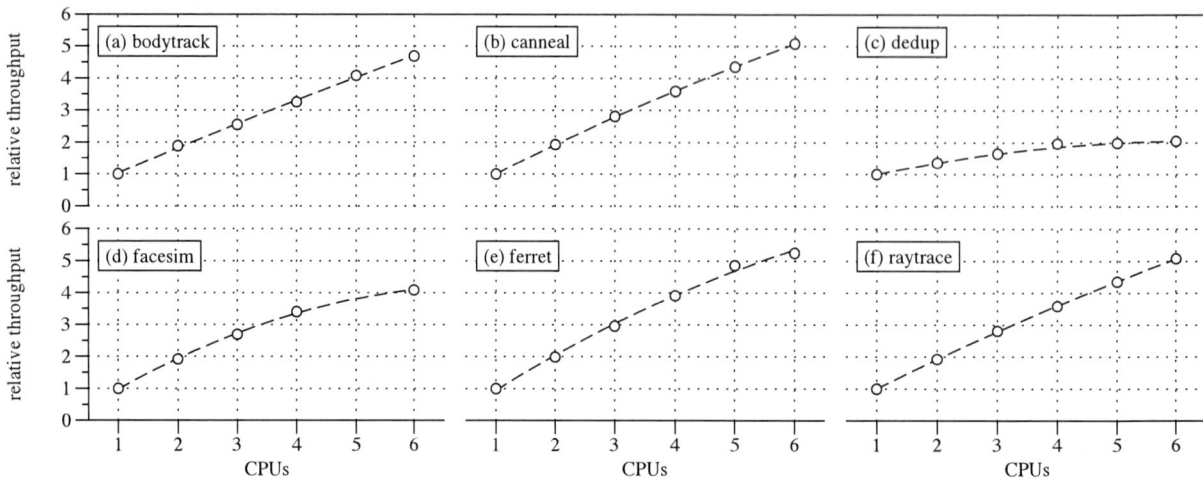

Figure 14: Scalability characteristics of 6 out of 13 applications from the PARSEC 2.1 benchmark suite showing sub-linear behavior.

B. SCALABILITY CHARACTERISTICS

This section elaborates on scalability characteristics and justifies our model exploiting a second order polynomial.

Figure 14 displays the scalability characteristics of 6 out of 13 applications from the PARSEC 2.1 benchmark suite [6]. The x-axes indicate the number n of allocated CPUs (in each experiment, we execute applications with n threads); y-axes indicate the relative (to the minimum) throughput measurements. Experimental results were collected on the workstation described in Section 3.3.

Most applications (i.e., *canneal*—Figure 14b, *ferret*—Figure 14e, and *raytrace*—Figure 14f) show quasi-linear scalability characteristics, with more than $5\times$ speedup with 6 CPUs. However, other applications (i.e., *dedup* and *facesim*—Figure 14c and d) present sub-linear scalability characteristics even with a relatively small number of CPUs. Previous research [26] analyzed an intersecting subset of applications reported that scalability characteristics bend drastically with 12 or more CPUs.

Previous work [8, 26] describes scalability characteristics through variations of Amdahl's law trying to account for the overheads introduced by synchronization primitives. Instead, we model scalability characteristics through a second order polynomial that puts in direct relationship the number of allocated CPUs with the throughput measurement. We justify our choice by fitting the data in Figure 14 with first and second order polynomials. Each data point is the average over experiments repeated until the width of the 95 % confidence interval was below 1 %. Fitting with a second order polynomial (i.e., our model) yields, for all applications, a coefficient of determination $R^2 \geq 0.99$; instead, the same metric with a first order polynomial, which is comparable to using Amdahl's law, varies more (e.g., down to $R^2 \approx 0.94$ for *dedup*). Therefore, our choice is justified for the applications we consider.

AcOS exploits this model with *Metronome++* (see Section 3.3), which estimates scalability characteristics at runtime by periodically collecting high-level throughput measurements through *HRM* and fitting them with the least squares algorithm. Conversely to previous research [26], we use high-level application-specific metrics instead of machine-specific metrics such as the number of retired instructions; this choice derives from two main reasons. First, high-level application-specific metrics are meaningful to users, who can easily state SLOs (see Section 3.1). Second, the number of retired instructions is not constant across different CPU allocations [26] and is also sensitive to applications employing non-sleeping syn-

chronization primitives (e.g., spinlocks); instead, the number of *heartbeats* an application emits is constant across different CPU allocations for a given dataset size.

To estimate scalability characteristics, *Metronome++* initially allocates 1 CPU to each instrumented application and collects the first data point (i.e., the number of allocated CPUs, throughput measurement pair). Then, it varies the allocations to collect two additional data points in order to have the three initial data points required to run the least squares algorithm and fit the second order polynomial reported in Equation (4).

$$r = c + b \cdot p + a \cdot p^2 \qquad \text{where} \qquad a < 0 \qquad (4)$$

We model the throughput measurement r of an application as a quadratic function of the number p of allocated CPUs; a, b, and c are parameters describing the scalability characteristic. The initial quasi-linearity of scalability characteristics is captured by b, while the final flattening is captured by a, whose influence becomes stronger as number of allocated CPUs grows.

Metronome++ adjusts the fitting with additional data points whenever the "environment" changes (i.e., the number of instrumented applications grows or shrinks); in this way, it can catch the effects of contention over on-chip shared resources. It is worth noticing that, even though *Metronome++* employs a second order polynomial for modeling purpose, there exists a single feasible solution $\bar{p}$ given a throughput requirement $\bar{r}$, which substitutes r when Equation (4) is used to predict the right CPU allocation $\bar{p}$.

C. EXECUTION PHASES

Two different events can trigger *Metronome++* to adjust CPU allocation. The first event (as already mentioned in Appendix B) is a change in the multi-programmed workload due to an instrumented application starting or finishing. This behavior is natural since changing the "environment" may alter the way co-located applications interact with on-chip shared resources. The second event triggering CPU allocation adjustment is a change in the execution phase of an application. Applications can go through execution phases with different scalability characteristics (e.g., CPU and I/O-bound execution phases have dramatically different scalability characteristics) or different performance for a given resource allocation. *Metronome++* attempts to address both of these issues.

To address the first issue, *Metronome++* re-evaluates the scalability characteristics of instrumented applications when their through-

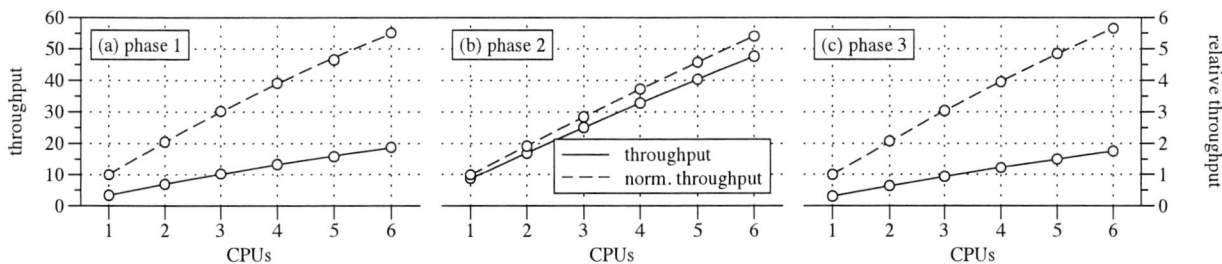

Figure 15: Scalability characteristics of the different execution phases of x264 from the PARSEC 2.1 benchmark suite.

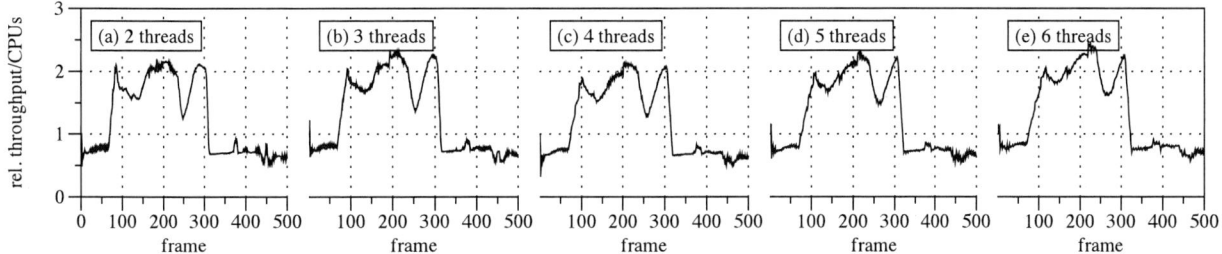

Figure 16: Relative ratio between the window throughput (100 ms) and the CPU allocation for different runs of x264 with 2 to 6 threads; the ratio remains roughly constant across CPU allocations, making it a robust metric to detect execution phases.

put measurements and requirements diverge, even though the "environment" did not change. It archives the previous the scalability characteristic and starts over as if the diverging application has just started. *Metronome++* addresses the second issue by weighing CPU allocations by means of a *workload predictor* to consider for changing execution phases.

The *x264* application in the PARSEC 2.1 benchmark suite [6] is a representative application presenting input-dependent execution phases. Figure 15a, b, and c show three execution phases, which *x264* goes through with the native input. The *x*-axes indicate the number n of allocated CPUs (in each experiment, we execute applications with n threads); *y*-axes indicate both the absolute (left) and relative (to the minimum, right) throughput measurements. Each data point is the average over experiments repeated until the width of the 95 % confidence interval was below 1 %. Experimental results were collected on the workstation described in Section 3.3.

The relative throughput measurements show that the scalability characteristics of *x264* do not change significantly across execution phases, thus leaving Equation (4) fitting stable. However, the absolute throughput measurements show a sharp performance increase in the second execution phase suggesting the presence of input-dependent execution phases.

To further analyze the execution phases of *x264*, we collect its frame-by-frame window (100 ms) throughput measurements; *HRM* provides such metric [28]. The window throughput metric is more sensitive (depending on the window length) to short-term trends than the global throughput metric and can highlight execution phase transitions. Figure 16 shows the frame-by-frame relative (to the initial) ratio between the window throughput measurements and the number of allocated CPUs for five different runs of *x264* from 2 to 6 threads. Experimental results were collected on the workstation described in Section 3.3.

The relative ratio increases sharply at the 70-th frame and decreases with similar intensity at the 300-th frame. The rising edge of the ratio indicates a transition from a high complexity subset to a low complexity subset of the dataset, while the falling edge

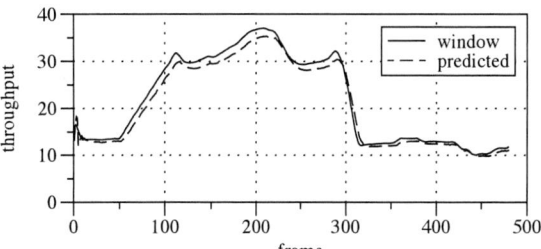

Figure 17: Window (1 s) predicted throughput for x264 running with 4 threads.

represents the opposite. Since the characteristics of the ratio (i.e., "shape" and "intensity") are roughly the same across the different CPU allocations, this metric proves to be a good proxy for execution phase detection, regardless of the current allocation.

For this reason, *Metronome++* uses an exponential moving average[5] of the values of this relative ratio as a workload predictor to detect execution phase transitions. *Metronome++* weighs throughput requirements with the workload predictor to realize the second adaption level (i.e., execution phase adaptation); the result goes through the first adaptation level that instead leverages the scalability characteristic as described in Appendix B.

We conclude with the evaluation of the accuracy of the workload predictor for a run of *x264* with 4 threads. Figure 17 shows the frames on the *x*-axis and both the window (1 s) and the predicted throughput on the *y*-axis. Experimental results were collected on the workstation described in Section 3.3. The predicted throughput, which is computed multiplying the workload predictor with the output of the scalability characteristic of the first execution phase, tracks almost perfectly the window throughput proving the accuracy of our approach.

---

[5]The use of an exponential moving average helps smoothing the occasional noise in the window throughput.

Role of Power Grid in Side Channel Attack and Power-Grid-Aware Secure Design

Xinmu Wang[1], Wen Yueh[2], Debapriya Basu Roy[3], Seetharam Narasimhan[1], Yu Zheng[1],
Saibal Mukhopadhyay[2], Debdeep Mukhopadhyay[3], and Swarup Bhunia[1]

[1]Case Western Reserve University, Cleveland, Ohio, USA
[2]Georgia Institute of Technology, Atlanta, Georgia, USA
[3]Indian Institute of Technology, Kharagpur, West Bengal, India
Email: xxw58@case.edu

ABSTRACT

Side-channel attack (SCA) is a method in which an attacker aims at extracting secret information from crypto chips by analyzing physical parameters (e.g. power). SCA has emerged as a serious threat to many mathematically unbreakable cryptography systems. From an attacker's point of view, the difficulty of mounting SCA largely depends on Signal-to-Noise Ratio (SNR) of the side-channel information. It has been shown that SNR primarily depends on algorithmic and circuit-level implementation, measurement noise, as well as device thermal noise. However, to the best of our knowledge, there has not been any study on the effect of power delivery network (PDN) on SCA resistance. We note that the PDN plays a significant role in SNR of measured supply current. Furthermore, SCA resistance strongly depends on the operating frequency due to RLC structure of a power grid. In this paper, we analyze the effect of power grid on SCA and provide quantitative results to demonstrate the frequency-dependent SCA resistance due to PDN-induced noise. This property can potentially be exploited by an attacker to facilitate the attack by operating a device at favorable frequency points. On the other hand, from a designer's perspective, one can explore countermeasures to secure the device at all operating frequencies while minimizing the design overhead. Based on this observation, we propose a frequency-dependent noise-injection based compensation technique to efficiently protect against SCA. Simulation results using realistic PDN model as well as experimental measurements using FPGA test board validate the observations on role of PDN in SCA and the efficacy of the proposed compensation approach.

Categories and Subject Descriptors

K.6.5 [**Security and Protection**]: Physical security

General Terms

Design, Security

Keywords

Side-channel attack (SCA), DPA, SCA resistance, power delivery network, noise injection

---

Permission to make digital or hard copies of all or part of this work for personal or classroom use is granted without fee provided that copies are not made or distributed for profit or commercial advantage and that copies bear this notice and the full citation on the first page. To copy otherwise, to republish, to post on servers or to redistribute to lists, requires prior specific permission and/or a fee.
DAC '13, May 29 - June 07 2013, Austin, TX USA

1. INTRODUCTION

The development of cryptography has greatly improved the security of encryption systems with respect to mathematical cryptanalysis. Traditional cryptanalysis considers only the input and output logic values to break the cipher. In this case, mathematical strength renders many modern ciphers rather invulnerable since the only possible brute-force attacks are not feasible in terms of computing time. However, these mathematically secure ciphers fail to claim the infeasibility of cryptanalysis due to *Side Channel Attack* (SCA). In SCA, secret information like the encryption key can be leaked through the physical parameters of the ciphers in various forms such as power profile [1], timing information [2], [3], and electromagnetic emissions [4]. Among different forms of SCAs, Power Analysis Attack (PAA) has emerged as one of the most significant attacks and has been widely studied. In PAA, the confidential information is extracted from the supply current signature of a crypto-chip. The relevant secret information is immersed amidst significant amount of noise; and the difficulty of successfully mounting an SCA attack, usually measured by the number of measurements required to disclose the secret information, largely depends on the Signal-to-Noise Ratio (SNR) of the side-channel information. Previous investigations [5], [6] have shown that the SNR is mainly determined by the algorithmic and circuit-level implementation of a chip. For example, pipelined implementation of Advanced Encryption Standard (AES) would be more difficult for SCA compared with non-pipelined AES [7]. Furthermore, device thermal noise [6] and measurement noise due to imperfect measurement instrument and operations can also considerably lower the SNR.

We note that the power delivery network (PDN) of a crypto-chip plays an important role in affecting the SNR of the measured supply current by causing a non-linear distortion in the supply current. It intrinsically limits propagation of useful information to supply current. However, to the best of our knowledge, there has been no study on analyzing the effect of PDN on the effectiveness of SCA. In fact, extracting information directly from the ideal power profile can be overly optimistic in modeling the true behavior of a chip. Accurate characterization of PDN's role in SCA can be significant in precise security analysis and design of secure crypto chips. In particular, it can provide the following benefits: 1) it provides a designer with more realistic measure of SCA resistance at design time; 2) it enables integration of the right level of protection; and 3) as we observe later, the PDN causes variation in SCA resistance with operating frequency, which helps a designer to choose right frequency of operation and/or accomplish balanced protection across frequency spectrum.

In particular, the paper makes the following key contributions.

1. It analyzes the effect of power grid on SCA and provide mathematical analysis as well as quantitative results to demonstrate the frequency dependent SCA resistance due to PDN-induced noise. Since PDN can be modeled as an RLC network, it re-shapes the supply current in a frequency dependent manner, effectively producing different SCA resistance levels with varying operating frequencies.

2. It performs validation of PDN's impact on SCA resistance through simulation verifications using realistic model of PDN. It shows a strong dependence of SCA resistance on operating frequency. Moreover, it demonstrates the effect through experimental measurements using SASEBO, a standard evaluation board for SCA [8].

3. Based on the observation on the role of PDN, it proposes a power-grid aware design methodology for crypto-chips using a noise injection approach that balances SCA resistance at different frequency points. We characterize the hardware overhead and analyze the effectiveness of the approach with both simulations and experiments in SASEBO.

The remainder of the paper is organized as follows. Section 2 provides background and motivational observations on the impact of PDN on SCA. Section 3 analyzes the frequency-dependent effect on SCA due to PDN. A corresponding noise injector circuit is proposed in Section 4. Simulation and experimental results are presented in Section 5 and Section 6 to demonstrate the frequency-dependent nature of PDN and prove the effectiveness of the noise injection mechanism. We conclude and provide future directions in Section 7.

2. BACKGROUND AND MOTIVATION

2.1 Previously Studied Noise Sources for SCA

The information beneficial for extracting confidential information from crypto chips in SCA attack comes from the dynamic switching current of circuit components (logic gates) when processing certain intermediate key-related results. Circuitry corresponding to the remaining part of the switching current can be viewed as noise sources. As has been studied in previous research [5] [6], there are several major noise contributors in SCA, including intrinsic noise sources due to the physical implementation, and external noise sources which are caused by environmental factors and imperfect measurement instruments/operations. The most significant intrinsic noise is contributed by dynamic switching of the circuit besides the attacked gates. For example, when performing SCA on 128-bit AES, the secret key is revealed byte by byte, namely the attack is performed on each Substitution-Box (S-Box) at a time, while considering as noise the switching of the remaining 15 S-Boxes as well as the irrelevant gates in the S-Box under consideration. Intuitively one can suggest that the spectral power of such systematic noise is much larger than the attack information itself, which is, however, conquered by statistical analysis in SCA methods. Another intrinsic noise is the jitter on the attacked gates [5]. This could cause the misalignment during power trace accumulation in statistical analysis of the power traces and thus lower the efficiency of the attack. Besides, intrinsic noise also includes S-Box inter-bit correlation and the thermal noise due to random movement of charge carriers within conductors [6]. External noise is caused by unstable environment factors e.g. varying temperature, imperfect measurement instrument e.g. A/D used to sample the power signals could introduce quantization noise [6], as well as imperfect operations by human users.

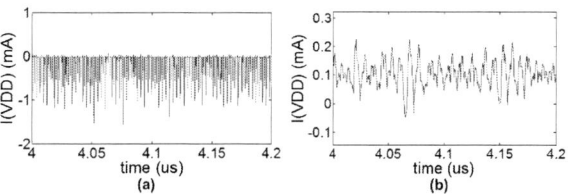

Figure 1: AES supply current (a) with ideal power supply; (b) with PDN.

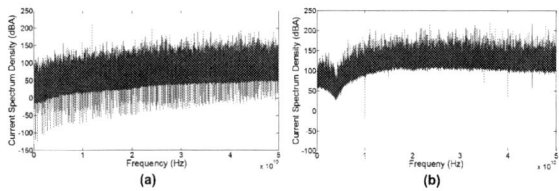

Figure 2: Frequency spectrum of AES supply current (a) with ideal power supply; (b) with PDN.

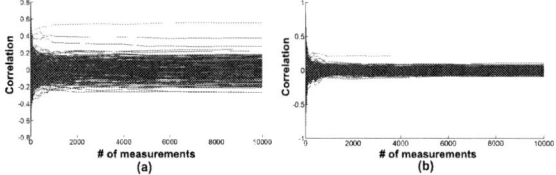

Figure 3: Correlation between predicted and measured power at 333.3 MHz clock frequency (a) with ideal power supply; (b) with PDN.

2.2 Motivation

The power grid impedance has a frequency dependent transfer function which affects the supply current. This changes the supply current profile when it passes through the PDN from inside the chip to the external pin for measurement. The effect causes distortion to the supply current waveforms. Fig. 1 and Fig. 2 show the time-domain and frequency spectral difference, respectively, between the supply current simulated with ideal power supply and the PDN model with 45nm Predictable Technology Model (PTM) [9]. Such distortion can create difficulty for SCA. Fig. 3 demonstrates the effect of PDN when Correlation Power Analysis (CPA) [10] is performed on a cipher to crack the key (blue curves correspond to wrong keys and red curves correspond to the correct key), showing decreased correlation coefficients and increased Measurement To Disclosure (MTD), i.e. the number of power traces required to extract the key, in the presence of PDN. More importantly, due to the RLC properties of PDN, the masking effect differs with different operating frequencies. Just like the circuit intrinsic noise can be viewed as a linear additive noise in SCA, PDN can be considered as a noise source which causes a non-linear distortion on power traces.

3. FREQUENCY-DEPENDENT EFFECT OF PDN ON SCA

To analyze the SCA resistance introduced by the power grid, Correlation Power Analysis (CPA) is used throughout the paper as the statistical tool to perform SCA. Previous study [11] categorizes the noise sources in SCA into two types: 1) Additive power consumed by circuit components other than the attacked gates, which effectively decreases the SNR of SCA; 2) Random disarrangement of the moment when the target switching activity of the attack happens, which prevents effective data alignment when performing correlation. The number of encryptions that need to be sampled to reveal the key (MTD) is determined mainly by the correlation ρ_{k_c,t_c}, where k_c is the correct key and t_c is the moment when the attacked gates switch [20]. Generally less samples of power traces are needed in the case of larger ρ_{k_c,t_c}. The effect of the two noise sources listed above, characterized as SNR and $\hat{p}$, respectively, on ρ_{k_c,t_c} are formulated as in [11]:

$$\rho_{max} = \frac{\rho(H,Q)}{\sqrt{1+\frac{1}{SNR}}} * \hat{p} * \sqrt{\frac{Var(P)}{Var(\hat{P})}} \qquad (1)$$

ρ_{max} is the maximum correlation coefficient between H (hypothetical power consumption) and the measured power consumption. Q denotes the power consumption of the attacked gates when processing the target intermediate result; and N refers to the additive noise. $\hat{t}_c$ refers to the moment with the largest probability $\hat{p}$ that locates the power consumption of the attacked gates. P is the sum of power at t_c and $\hat{P}$ is the sampled sum of power at moment $\hat{t}_c$. SNR is defined as $\frac{Var(Q)}{Var(N)}$. Note the SNR DC component has no impact on SCA hence the equation only captures the variance [11].

The lower bound of the power samples that is required to crack the key can be computed approximately as follows [11]:

$$S = 3 + 8\left(\frac{Z_\alpha}{ln(\frac{1+\rho_{max}}{1-\rho_{max}})}\right)^2 \qquad (2)$$

The quantile Z_a is the distance between the distributions $\rho = 0$ and $\rho = \rho_{max}$. From this we can again confirm the negative dependency of MTD on the maximum correlation coefficient.

The power grid imposes a frequency-domain distortion on the supply current due to its RLC property. The frequency dependent impedance responses also differ for the attacked gates (Q) and uncorrelated gates (N) due to the cell placements. Under these assumptions, Q and N are functions of the sampling frequency and location, which can be expressed as follows:

$$Q = \sum_{\mathbf{x}=i,j} \text{PDN}(f_s, \mathbf{x}) \cdot q_{\mathbf{x}}, \qquad (3)$$

$$N = \sum_{\mathbf{y}=i,j} \text{PDN}(f_s, \mathbf{y}) \cdot n_{\mathbf{y}} \qquad (4)$$

$\text{PDN}(f_s, \mathbf{x})$ models the PDN distortion factor at frequency f_s and grid location coordinate $\mathbf{x}$. The gate power $q_{\mathbf{x}}$ and $n_{\mathbf{x}}$ are the individual gates contributing to the collective instantious current Q and N, respectively. The frequency-dependent effect modulates the attacked gates' current amplitude and noise SNR depending on the design floorplan. In addition to the primary additive effect directly impacting the SNR, a secondary effect from supply swing may rearrange t_c. During a high switching activity cycle, the current demand from gates lowers the instantaneous supply VDD, and localized supply droop affects switching moment as the gate switching speed is directly propotional to VDD at high voltage. Such effect may be localized on a reasonable sized PDN and may increase the current spread $Var(\hat{P})$ and decrease $\hat{p}$. These parameter changes collaboratively decrease the actual ρ_{max} and thus lead to a requirement of a larger sample size S to recover the key, effectively increasing the difficulty of SCA.

Saint-Laurent et al. have shown that supply noise affects the circuit timing in both device-dominated or interconnect-dominated scenarios [12]. An accurate analytical model requires parameterizing the supply network topology and technology dependent models. In order to reduce the complexity, one may collect the current profile from the functional

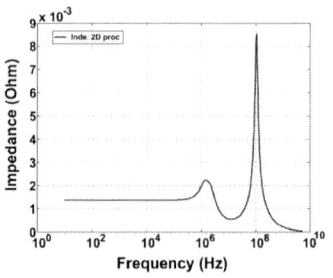

Figure 4: Impulse response of PDN.

core operating under ideal VDD and then apply the current profile to the power grid, utilizing advanced circuit simulators to determine the interactions. Such methdology may be extended to precomputed supply impedance at each grid node and apply to function $\text{PDN}(f_s, \mathbf{x})$; e.g. Fig. 4 provides a single impulse response sweep of the PDN model which may be used to estimate the frequency dependent factor for a particular PDN region. This allows the attacker to estimate SNR, but falls short to estimate the effect of $\sqrt{Var(P)/Var(\hat{P})}$ and $\hat{p}$ in Eq. (1). The model simplifies the supply-to-transistor interaction to a frequency dependent component, hence cannot capture this interaction introduced by the time-varying supply noise. Therefore, to quantify this frequency-dependent noise injected from the PDN, we built a framework to simulate the crypto system (AES) and PDN together at different operating frequencies. Correlation power analysis is performed on the external supply current corresponding to different operating frequencies.

4. NOISE INJECTOR CIRCUIT

Based on the observation of the frequency-dependent SCA resistance of cryptographic chips, we propose a frequency-dependent noise-injection based compensation technique to equalize the SCA resistance at all frequency points with the one exhibiting the maximum SCA resistance. As shown in Fig. 5, the *noise injector (NI)* consists of a frequency detector, a control unit, a digitally-controlled delay line, an array of noise injection units, and an optional initialization finite state machine (FSM). The frequency detector takes the system clock as the input and pass the detected frequency value to the control unit. The control unit plays two roles: 1) it determines the number of NI units to enable; 2) it determines the phase delay of the NI clock with respect to the system clock. An NI unit can be implemented with an FSM using deliberately assigned state encoding to cause varying switching current over clock cycles or a linear feedback shift register (LFSR). The number of enabled NI units determines the amplitude of the NI-induced additive noise. A digitally-controlled delay line is employed to generate a phase delay between NI clock and the system clock. A phase delay between the NI and system clock is necessary because within each clock cycle, only a limited duration of the switching

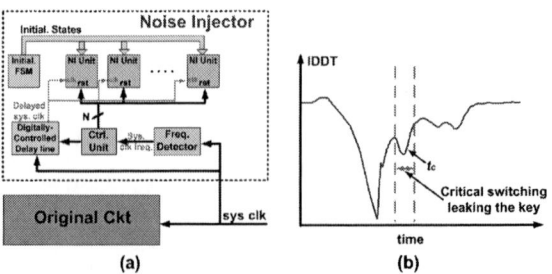

Figure 5: Block diagram of the proposed noise injector circuitry to compensate frequency-dependent SCA resistance.

550

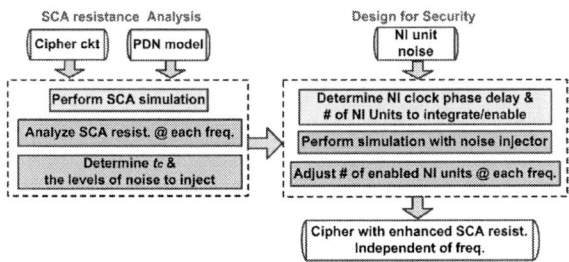

Figure 6: Noise injector design flow.

current (the part that corresponds to switching of the attacked gates) is of interest for SCA. Using a delayed version of the system clock for NI allows one to focus the additive noise to the critical time duration of a cycle thus mask the information-leaking switching. From another perspective, a lightweight NI circuit whose switching current duration is merely as long as that of the attacked components can be adequate, since the rest of the original circuit switching is irrelevant. More importantly, since the NI draws extra current from the power supply, it may cause increased supply voltage droop, which can make the functional circuits switch slower therefore leading to circuit performance degradation. Short duration of NI is desirable to minimize the negative influence on circuit performance. The function of the control unit is essentially a mapping between the system clock frequency and the number of enabled NI units/NI clock phase delay.

This information can be obtained from a characterization of a chip during the design stage, and stored in a look-up table in the control unit. The optimal NI phase delay is first captured by figuring out the moment that has the most key-related switching and the average switching duration of the NI unit. In CPA, since we do not know the time instant of the critical switching, correlation is performed between the hypothetical power consumption with the measured power at each time sampling point for each hypothetical key. This will lead to a correlation coefficient matrix R of size $T \times K$ (T is # of time samples, K is # of hypothetical keys). Because hypothetical power consumption predicted by the correct key is strongly correlated with the measured power at the moment of the critical switching, the two will lead to the maximum correlation coefficient in R (More information on CPA can be found in Section 9.4). Therefore by simply observing the position of the highest value in R, one can figure out the correct key as well as the index of the time moment of the critical switching t_c [13]. With known switching duration of the NI unit, a preferable NI clock phase delay can be derived to align the center (near the peak value) of the noise current with t_c. Fig. 5(b) illustrates the critical switching duration and t_c of one encryption. Actual alignment may experience some deviations due to intra-die process variations, which is expected to be minimal for reasonable logic depth (e.g. >6). With the optimal NI clock phase delay, an iterative test can be performed to characterize the number of NI units to be enabled. Some extent of conservativeness is necessary to tolerate post-silicon deviations due to manufacturing factors.

After obtaining the number of NI units for each frequency, we re-perform the simulation with PDN with the calculated number of NI units enabled. The difference in SCA resistance can be corrected according to the simulation results. The flow of designing NI for a generic cipher is provided in Fig. 6. Details on the NI design is provided in Section 9.3.

5. SIMULATION RESULTS

HSPICE simulation was performed using 45nm CMOS Predictable Technology Model (PTM) [9] on an AES crypto operator with PDN. The PDN model is described in Section

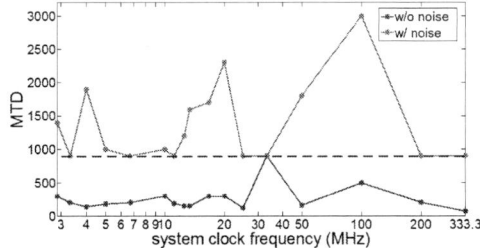

Figure 7: Frequency-dependent MTD of Side-Channel Attack (SCA).

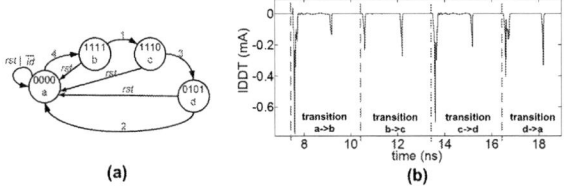

Figure 8: Noise injection FSM: (a) state transition diagram; (b) switching current.

9.2. In our simulation, we use a simplified cryptographic operator corresponding to the AES datapath for encrypting one byte plaintext. This is sufficient to demonstrate the effectiveness of the proposed NI structure on a generic cryptographic chip that resembles AES datapath. This is because it is equivalent to removing the irrelevant datapath element in SCA to reduce the noise level when recovering a particular key-byte. The crucial non-linear component S-Box and the downstream logic for key-related operations are all preserved, which is the target of most SCA attacks [14]. The methodology can also be readily applied to any other cipher. Moreover, we only simulate the last round in each encryption by inputting to the cipher the results of the second last round obtained from functional simulation. The description on the simulated datapath is provided in Section 9.4.

The blue curve in Fig. 7 demonstrates the frequency dependent Measurement-To-Disclosure (MTD) obtained from HSPICE simulation. MTD refers to the minimum number of power traces required to recover the key and is a commonly used measure for SCA resistance. The highest SCA resistance appears at 33.33 MHz clock frequency. We call it as *SCA resistance reference point*. The overall MTDs are fairly low (below 1000) compared to the MTD level of practical SCAs [15], which is due to the simplified crypto operator used. Multiple peaks and valleys appear at other frequencies and the entire curve exhibits a non-monotonic trend due to the complex RLC property of the power grid. Our goal is to bring the SCA resistance at all frequencies above the reference point in order to compensate the PDN-induced SCA resistance imbalance.

To achieve this, we implement a noise injection circuit that introduces frequency-dependent IDDT (transient power supply current) noise as discussed in Section 4. A simple 4-bit Finite State Machine (FSM) is implemented as the NI unit to match the smaller crypto operator size. More complex cipher circuits may choose larger FSMs to achieve a bigger range of noise current variation. Fig. 8(a) illustrates the state transition diagram of the FSM. The *rst* signal will reset the FSM to state a, and *ld* signal allows initializing the FSM to any valid state. When both signals are deasserted, the FSM behaves like a counter, i.e. repetitively goes through a loop of states. The state encoding is deliberately assigned in order to cause significantly varying switching current over cycles, assuming switching of state elements dominates the overall IDDT. Such variation is expected to be uncorrelated with the hypothetical power consumption yet repeats deterministically with the FSM state looping. When the number of enabled FSMs is large enough so that the total amplitude of the noise current is comparable with that of the original circuit current, the key-related switching may be masked to a large extent thus improving SCA resistance. The switching current of an FSM during each state transition is shown in Fig. 8(b). The corresponding numbers of flipping state elements are marked on the transition arrow in Fig. 8(a).

The MTDs after inclusion of the NI is shown as the red curve in Fig. 7. MTDs at all frequencies are pulled up to or above the reference point. They are not uniformly at the reference point because the MTD at particular frequencies is more sensitive to noise than at others. Furthermore, the injected noise current is not guaranteed to align with the switching of the key-related components in a perfectly uniform way. However, all frequencies achieved SCA resistance no worse than the reference point with minimal hardware overhead. The MTD values along with the number of NI units enabled at each frequency are provided in Table 1.

Table 2 provides the area and power overhead introduced by the NI. The NI consists of an array of 3 FSMs; while the control unit is composed of a SRAM based look-up table storing the frequency-dependent clock phase delay and NI units enabling information, as well as a decoder for generating the FSM array enable signals. The area percentage overhead is relatively large because of the tiny crypto operator we used in our simulation. In fact, the overhead exhibits good scalability with the increasing size of the circuit. For the FSM array, as a large system behaves equivalently as a noise injector itself for the key-byte being attacked, the extra noise level (therefore the size of the FSM array) required to compensate the PDN-induced SCA resistance imbalance could be at the same level as that of a simple crypto operator. For the control unit, the majority part is the look-up table (98.6% of the control unit area) storing two types of information: 1) Digits for controlling NI clock phase delay; 2) Values for setting the NI units enable signals. The required phase delay range might increase with the circuit size as the critical path delay (thus the key-related component switching) may increase. However, this increase should be tiny compared with the extent of circuit size scaling up, as modern IC designs always minimize the critical path delay to a satisfactory level with design techniques such as pipelining to meet the performance requirement. We may consider the phase delay range as almost independent of the circuit size. With a fixed resolution, the bit-width of the delay selecting signal should remain unchanged. Though the bit-width of the NI units enable signal might increase for a larger circuit in accordance with a larger NI unit array; with binary encoding the increase is on the order of $O(log_2(p))$, where p is NI array size growth rate and is far less than that of the circuit size. For example, a control unit in a complete AES would only incur less than 0.86% area overhead.

The power consumption overhead is dominated by the FSM array, which is, however, frequency dependent. The values shown in the table correspond to the worst case where all three FSMs are enabled. In reality, at most frequencies, one FSM is sufficient and the dynamic and leakage power overhead would be 11.2% and 5.7% (unused FSMs can be gated to minimized the leakage), respectively. In addition, the percentage overhead will reduce for the same reason as for area with circuit size scaling up. The control unit consumes minimal power because the look-up table is only written once with the pre-characterized value during system power-up. Consequently, the output decoder does not have switching activities either. This overhead could be further minimized by reusing the on-chip SRAMs for the look-up table.

Large instantaneous current drawn from the power supply may potentially create considerable voltage drop and lead to logic circuit delay failure or embedded memory cell

Table 2: Area and power overhead of the NI

Param/Ckt	Simple cipher	FSM array (overhead)	Control unit (overhead)
Area(μm^2)	942.8	133.8(+14.2%)	219.6(+23.3%)
AvgPwr(μW)	104.8	< 35.2(< 33.6%)	1.63(+1.6%)
AvgLkg(μW)	22.4	< 3.8(< 16.9%)	0.10(+0.4%)

hold-failure. In our experiment, the NI is designed to operate within the crypto core noise margin and the supply noise did not cause any functional failure. Also, since the noise current is superimposed to a short duration of the cipher switching current, the cipher is subject to extra supply noise for a limited duration within each cycle. This type of noise injection ensures the functional core does not suffer from significant performance degradation. In larger cryptographic circuit, the percentage overhead induced by NI will reduce significantly as discussed above, and the additional supply noise will also be smaller when a larger PDN is used.

6. EXPERIMENTAL RESULTS

Correlation Power Analysis (CPA) based SCA was performed on normal AES (with 128-bit key) and AES with NI at various frequencies on *Side-Channel Attack Standard Evaluation BOard (SASEBO)* [8]. More details about the experimental setup can be found in Section 9.5. Table 3 shows the CPA analysis for normal AES. Each cell of the table contains the average key ranking at a particular clock frequency with certain number of power traces. For each S-Box, we calculate the correlation coefficients for all hypothetical key bytes and sort them in descending order, where key ranking indicates the position of the correct key byte. Averaging the key rankings for all the S-Boxes leads us to the average key rankings that are shown in the table. Zero value of average key ranking indicates that all the 16 key bytes are recovered. Non-zero value means for certain S-Box(es) the key byte(s) is not recovered. The closer the average key ranking is to zero, the more key bytes are correctly recovered. Observation can be made from Table 3 that MTD varies significantly with the clock frequency, indicating a frequency-dependent SCA resistance. For example, MTD is 5000 at 10 MHz and 40 MHz yet 9000 at 30 MHz. 30 MHz is considered as the SCA reference point as it exhibits the highest SCA resistance. Because the property of SASEBO power grid is completely different from the PDN model used in our simulation, the way in which SCA resistance varies with frequency does not need to correlate with that of the simulation results.

Table 4 contains key rankings of AES with NI, where the NI unit is implemented with a 16-bit LFSR (see Section 9.3). With maximum 4 NI units enabled, SCA resistance at all frequencies are improved to above the reference point (9000 traces). In fact, because AES circuit has different placement and routing when being mapped alone to SASEBO for pre-characterization and with NI for measurement, the noise current could not align perfectly with the critical switching of AES. In Application Specific Integrated Circuits (ASICs), we would expect to achieve compensation with smaller NI.

7. CONCLUSION

We have presented a study on the role of PDN on side-channel attacks in crypto chips. Through mathematical analysis and quantitative results we have shown that PDN can significantly affect SCA resistance and the impact is frequency-dependent. Simulation verification is performed using a PDN modeled as RLC network with equivalent distributed power mesh derived from Pentium 4 processor. Experimental study on AES is performed using a standard SCA evaluation board. Both simulation and experimental results consistently show significant frequency-specific role of PDN in SCA. An attacker, who gains physical access to a crypto

Table 1: PDN-induced Frequency-dependent SCA resistance and MTD

Freq. (MHz)	2.86	3.33	4	5	6.67	10	11.11	12.5	13.33	16.67	20	25	33.33	50	100	200	333.33
MTD	300	200	140	180	200	300	190	150	150	300	300	120	900	160	500	200	70
# enabled FSMs	1	1	1	1	2	1	3	3	2	1	3	2	*ref*	3	1	1	1
MTD w/ noise	1400	900	1900	1000	900	1000	900	1200	1600	1700	2300	900	*ref*	1800	3000	900	900

Table 3: DPA on normal AES at various frequencies

Traces	10 MHz	15 MHz	20 MHz	25 MHz	30 MHz	40 MHz	45 MHz	50 MHz	55 MHz	60 MHz
1000	26.75	38.6875	42.6875	43.1875	39.5625	15.375	39.125	36.3125	55.0625	72.5625
2000	11.938	14.8125	18.6875	23.4375	21.0625	16.125	11.1875	18.0625	26	12.5
3000	2.125	7.1875	9.25	2.9375	3.875	1.25	1.3125	5	17.5625	1.75
4000	0.25	1.625	3.375	1.125	0.875	0.375	0.9375	0.4375	4.25	0.875
5000	0	0.25	0.5	0.25	0.375	0	0.4375	0.1875	2.5625	0.625
6000		0	0.3125	0	0.4375		0.3125	0.0625	0	0.4375
7000			0.125		0.5		0.0625	0		0.25
8000			0		0.125		0			0.1875
9000										0

Table 4: DPA on noisy AES at various frequencies

Traces	10 MHz	15 MHz	20 MHz	25 MHz	30 MHz	40 MHz	45 MHz	50 MHz	55 MHz	60 MHz
1000	43.5	37.3125	40.1875	24.875	54.3125	38.4375	66.75	53.3125	43.4375	58.5
2000	17.25	11.3125	18.5	17.5625	27.5625	17.4375	39.875	17.1875	26.3125	21.1875
3000	3.5	16.3125	19.875	7.6875	4.625	11	17.1875	12.875	6.1875	13
4000	1.9375	7.3125	19.125	9.375	4.3125	6.75	5.25	3.1875	2.9375	9.125
5000	1.375	1.75	6.9375	6.375	1.125	3.8125	0.375	3.0625	1.0625	4.1875
6000	1.375	0.5	7.3125	2.1875	0.4375	7.0625	0.1875	3.75	3.875	2.375
7000	1.625	0.125	5.75	1.3125	0.625	7.0625	0.125	0.1875	1.75	2.125
8000	0.5625	0.4375	1.5625	0.125	0.3125	10.1875	0.3125	0.875	1.75	1.25
9000	0.1875	0.125	1.6875	0.13	0.5625	7.75	0.1875	0.4375	0.5625	1.3125

system, can leverage on this trait to work under a favorable clock frequency for stealing the key. Accurate estimation of PDN's role in SCA would help prevent such a scenario and enable designer of a crypto chip to incorporate right level of countermeasures across the frequency spectrum. We have presented a secure design approach based on frequency-dependent noise injection to compensate for the PDN-induced SCA resistance imbalance by introducing commensurate masking noise. Through simulations and hardware measurements we show that such an approach can improve the overall SCA resistance at all frequencies with relatively low hardware overhead. Future work would focus on evaluating the effect of PDN through current measurements in custom crypto-chip; developing an integrated PDN-aware design flow for automatic synthesis of compensation circuitry; and extension of the technique to cryptographic chips with various SCA countermeasures.

8. REFERENCES

[1] P. C. Kocher *et al.*, "Differential Power Analysis," *CRYPTO*, 1999.

[2] P. C. Kocher *et al.*, "Timing Attacks on Implementations of Diffie-Hellman, RSA, DSS, and Other Systems," *CRYPTO*, 1996.

[3] J. F. Dhem *et al.*, "Practical Implementation of the Timing Attack," *CARDIS*, 1998.

[4] W. V. Eck *et al.*, "Electromagnetic Radiation from Video Display Units: An Eavesdropping Risk" *Computers and Security*, v. 4, 1985.

[5] S. Guilley *et al.*, "Differential Power Analysis Model and Some Results," *CARDIS*, 2004.

[6] T. S. Messerges *et al.*, "Investigations of Power Analysis Attacks on Smartcards," *USENIX Workshop on Smartcard Technology*, 1999.

[7] F.-X. Standaert *et al.*, "Power Analysis of an FPGA Implementation of Rijndael: Is Pipelining a DPA Countermeasure?" *CHES*, 2004.

[8] Available Online: http://staff.aist.go.jp/akashi.satoh/SASEBO/en/index.html

[9] Available Online: http://ptm.asu.edu/

[10] E. Brier *et al.*, "Correlation Power Analysis with a Leakage Model," *CHES*, 2004.

[11] S. Mangard *et al.*, "Hardware Countermeasures against DPA - A Statistical Analysis of Their Effectiveness," *CT-RSA*, 2004.

[12] M. Saint-Laurent *et al.*, "Impact of power-supply noise on timing in high-frequency microprocessors," *IEEE Transactions on Advanced Packaging*, Feb. 2004.

[13] S. Mangard *et al.*, "Power Analysis Attacks- Revealing the Secrets of Smart Cards," Springer 2007.

[14] K. Tanimura *et al.*, "HDRL: Homogeneous Dual-Rail Logic for DPA Attack Resistive Secure Circuit Design," IEEE Embedded System Letters, Vol. 4, No. 3, Sep 2012.

[15] Z. Chen *et al.*, "Using Virtual Secure Circuit to Protect Embedded Software from Side-Channel Attacks," *IEEE Transactions on Computers*, 2011.

[16] M. Joye *et al.*, "On Second-Order Differential Power Analysis," *CHES*, 2005.

[17] K. Tiri *et al.*, "Side-Channel Attack Pitfalls," *DAC*, 2007.

[18] M. S. Gupta *et al.*, "Understanding Voltage Variations in Chip Multiprocessors using a Distributed Power-Delivery Network," *DATE & Exhibition*, 2007.

[19] K. Tiri *et al.*, "Simulation Models for Side-Channel Leaks," *DAC*, 2005.

[20] T. S. Messerges *et al.*, "Examining Smart-Card Security under the Threat of Power Analysis Attacks," *IEEE transactions on Computers*, 51(5), 2002.

[21] A. Prabhakaran *et al.*, "Side-Channel Analysis of Block Ciphers Using CERG-GMU Interface on SASEBO-GII," Master's Thesis, ECE Department, George Mason University, Fairfax, Virginia, USA, May 2011.

[22] N.H.E.Weste *et al.*, "CMOS VLSI Design - a Circuits and Systems Perspective," 4th edition, Addison-Wesley.

[23] B. W. Garlepp *et al.*, "A Portable Digital DLL for High-Speed CMOS Interface Circuits," IEEE Journal of Solid-State Circuits, Vol. 34, No. 5, May 1999.

[24] Y. Moon *et al.*, "An All-Analog Multiphase Delay-Locked Loop Using a Replica Delay Line for Wide-Range Operation and Low-Jitter Performance," IEEE Journal of Solid-State Circuits, Vol. 35, No. 3, March 2000.

9. APPENDIX

9.1 Side-Channel Attack (SCA)

Side-Channel Attack (SCA) refers to the process of extracting the encryption/decryption key from a cryptographic device by monitoring certain physical parameters, e.g. power consumption or electromagnetic emission, of the device during its operation. In this paper, we only consider SCAs through power consumption. Generally, the secret key is figured out through statistical analysis of the recorded power traces of the device and certain hypothetical power consumption predicted based on certain model. Since SCA was introduced, many approaches of statistical analysis have been developed to achieve powerful SCA, e.g. in presence of SCA-resistant design techniques. These approaches include Single-Power Analysis (SPA) [1], Differential-Power Analysis (DPA) [1], higher-order DPA [16], and Correlation Power Analysis (CPA) [10]. We use CPA throughout the paper, which is a potent SCA analysis method because of the accurate alignment of the target switching moment when correlation coefficient is computed.

Fig. 9 illustrates the basic concept of SCA on AES. The power of the device is measured through the voltage drop across a resistor in series with the voltage source. Plaintexts (PTs) are applied at the input and Ciphertexts (CTs) are computed by the cryptographic engine with the secret key stored inside the device. The target moment of transient supply current is usually associated with processing of certain intermediate result. The key is extracted piece by piece [15]. For example, in the illustrative diagram in Fig. 9, one byte of the key $K[7:0]$ can be derived based on knowledge of CTs. In particular, a hypothetical intermediate result (IR) can be computed based on one hypothetical $K[7:0]$ and the known CT, with the publicly known AES algorithm. By enumerating 256 possible $K[7:0]$, all possible hypothetical IRs can be obtained. In this demonstrated architecture, the target switching activity happens in the last round (cycle) of AES, caused by overwriting of registers holding values of IRs with new values of CTs. Thus the Hamming Distance between IR and CT can serve as the hypothetical power consumption for one encryption with well empirical verification. Since the exact time when the target switching takes place is unknown, correlation can be performed between the hypothetical power and each sampling point during the encryption to find out the peak correlation coefficient. This would answer both the correct key byte and the moment of the target switching activity.

9.2 PDN Model

The power network used in this work references the 2D distributed design of [18]. The RLC network uses an equivalent distributed power mesh derived from the lumped impedance model of a Pentium 4 processor. The off-chip impedances

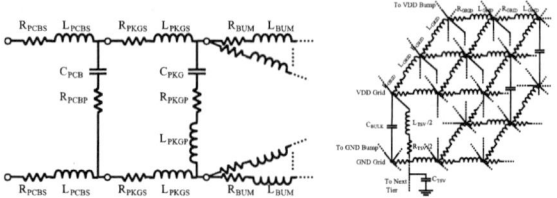

Figure 10: PND model: (a) Off-chip PDN; (b) on-chip 3D PDN.

Table 5: PDN model parameters [18]

Parameters	R(OHM)	L(H)	C(F)
PCB	$94\mu(s)/166.6\mu(p)$	21p	240μ
PKG	$1000\mu(s)/541.5\mu(p)$	120p	26μ
BUM	300m	0.5p	-
GRID	50m	5.6f	71.5p

are modeled with RLC ladders in Fig. 10(a). The first segment of the ladder models the board-level lump impedance and the second segment of the ladder models the package impedance. The package ladder is evenly distributed to the points on the on-die grid with partially-lumped solder bump impedance. The RLC distribution of the on chip power distribution network is shown in Fig. 10(b). On the 12x12 grid points, the gates are distributed on PDN according to the synthesis component number. Table 5 provides RLC parameters of the PDN model used in this work. The Hspice simulator performs transistor level simulation on an identical testbench for circuits with and without supply network.

9.3 Details of Noise Injector Design

To realize a tunable delay between the NI clock and the system clock, a *Digitally-Controlled Delay Line (DCDL)* is used [22]. DCDL is a chain of delay elements, which is, in our case, inverting buffers. A DCDL contains even number of inverting buffers, each of which has a tunable propagation delay. Fig. 11 demonstrates two implementations of DCDL inverting buffer. In Fig. 11(a) the controllable propagation delay is realized through tunable load capacitance of the inverting buffer; while Fig. 11(b) is a current-starved inverter with tunable resistance in series with the power supply. In reality, ICs usually have on-chip *Phase-Locked Loops (PLLs)* or *Delay-Locked Loops (DLLs)* for clock generation. In this case, the required DCDL can leverage the on-chip DLLs, which usually have superior performance, i.e. able to provide infinite phase range and very high resolution (on the order of 10ps) [23], and operate over a wide frequency range with low jitters [24]. More importantly, their performance does not suffer remarkably from process variations. This would make a high-quality controllable delay line without incurring area or power overhead.

The purpose of the NI unit array is to incorporate a certain level of switching current each cycle; and the switching current should vary significantly between cycles in order to mask the SCA relevant switching. The switching current amplitude should be comparable to that of the original circuit; however, the switching duration can be as short as that of the attacked gates, therefore the overall power consumption overhead can be maintained rather low. This means we need a circuit that can cause significant amount of switching within short durations and varying amplitude among cycles. This can be achieved by using FSMs with arbitrary state encoding. Since the functionality of the FSM is not of interest, the state transition can be designed to be counter-like with arbitrary state encoding. Therefore, the next state logic is reasonably simple because it is not a function of the pri-

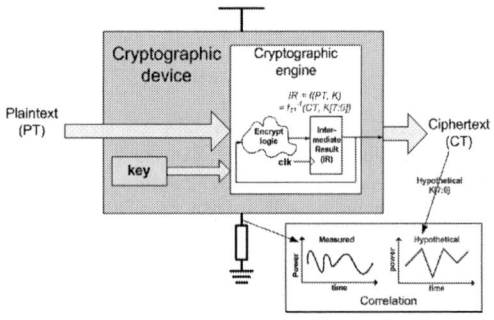

Figure 9: SCA attack on a cryptographic device.

mary inputs except for state initialization. This is enabled by the low requirement on the switching duration. Here "arbitrary" means the Hamming distances between each pair of adjacent states should vary as much as possible. Because if the next state logic is simple enough, the flip-flop switching will dominate the NI unit switching current. By carefully selecting the state encoding, an NI unit can exhibit desirable degree of random switching variations. In particular, we implement two types of NI unit: 1) FSM with state encoding to achieve adjacent state Hamming Distance variation; 2) Linear Feedback Shift Register (LFSR). We choose LFSR of a small size (16-bit in our case) to provide good granularity of the injected noise. An optional initialization FSM can be used to enable NI units with different states. Example LFSR structures are given in Fig. 12.

9.4 Model of SCA on AES

AES is widely used as a standard encryption system, which has three variations marked by different key lengths: 128, 192, and 256 bits. Larger key length provides higher strength against mathematical cryptanalysis, but also incurs more rounds in the computation. Fig. 13(a) illustrates the encryption datapath of an AES with 128-bit key, which specifies the four operations that the computation is composed of: SubBytes, ShiftRows, MixColumns and AddRoundKey (shown as the XOR circuit). Among these operations, SubBytes is the only non-linear component, which is the basis for preventing mathematical cryptanalysis. The entire encryption flow of AES with 128-bit key involves one initial AddRoundKey and 11 rounds of repeated four operations (the last round does not contain a MixColumns operation) before the final cipher text (CT) is generated. In each round, the round key that is used to XOR with the intermediate cipher texts, is derived from the input key through key scheduling. However, since there is a known linear relationship between the input key and the round key, the input key can be easily computed after obtaining the round key (e.g. K11 in our case). In AES, operations are performed with the unit of one byte, therefore SCA can break one byte of the key at one time. Although computations corresponding to the rest of the key, which correspond to considerably larger circuit part compared to that related to SCA, introduce significant amount of noise for SCA, the key byte under consideration can still be correctly extracted through practically achievable number of measurements and statistical analysis. This forms the primary reason that SCA exponentially reduces the computation time compared with brute-force attack, since it lowers the number of trials from 2^{128} to 16×2^8.

In this work, we use correlation power analysis (CPA) [10] as the SCA method to extract the key. The principle of correlation test is that the power profile predicted by the correct key guess will have the maximum correlation with the measured power profile among those predicted by all key guesses. The predicted power profile is defined by the Hamming Distance of R1 between round 11 (with value of

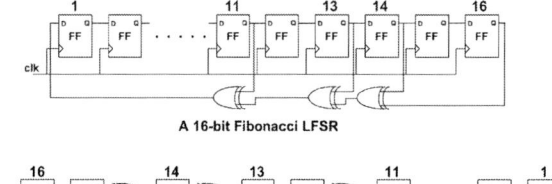

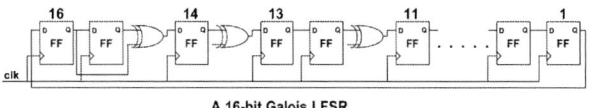

Figure 12: Example LFSR structures.

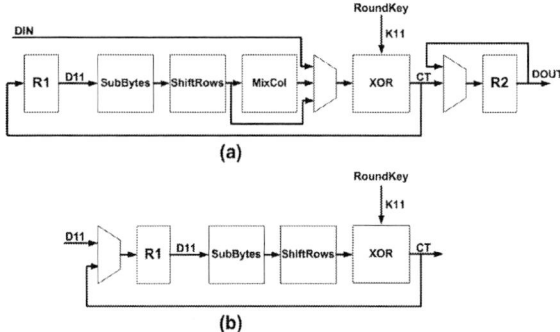

Figure 13: AES datapath: (a) Encryption datapath of AES [17]; (b) Simplified circuit used in simulation.

D11) and the next cycle (with value CT). This is because that on one hand, with known cipher text (CT), D11 can be back-traced given guessed K11. Hence, the Hamming distance is a non-linear function of the guessed K11. Despite the existence of intrinsic and external noise, the correct key can be revealed by correlation of large number of cycles of measurements. The computation of correlation test to reveal the correct key can be expressed in as $max_{K11} corr(P_{model}, P_{measurement})$ [19], where

$$P_{model} = HammingDistance(Sub^{-1}(ShiftRow^{-1}(K11 \oplus C11)), C11)$$

Since we do not know the time instance when the switching of the attacked registers occurs, correlation should be performed for each time sample (considering the switching of the attacked gate aligns to some extent in all encryptions). Thus the maximum correlation coefficient can reveal the moment of the key-related switching along with the correct key. The circuit model of the simulated datapath is given in Fig. 13(b).

9.5 Power Attack Setup

Successful power attack on a specific cryptographic algorithm requires proper understanding of the vulnerabilities of an implementation of the algorithm and developing attack strategies exploiting those vulnerabilities. However, successful attack also depends upon correct acquisition of the power traces. Hence developing a proper setup which will enable us to collect power traces accurately is of primary importance. Block diagram of the power attack setup developed by us is shown in Fig. 14. Fig. 15 and 16 illustrate the experimental setup and SASEBO under experimentation, respectively.

1. **SASEBO BOARD:** *Side-Channel Attack Standard Evaluation BOard (SASEBO)* [8] has two FPGAs on board. One is known as control FPGA (Spartan 3A

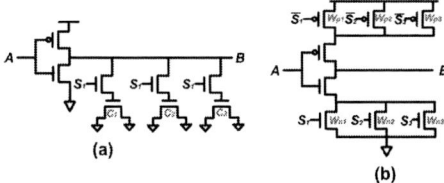

Figure 11: Element of digitally controlled delay line with: (a) tunable load capacitance; (b) tunable current-starved inverter.

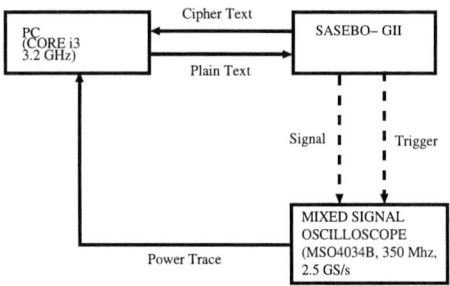

Figure 14: Block diagram of power attack setup.

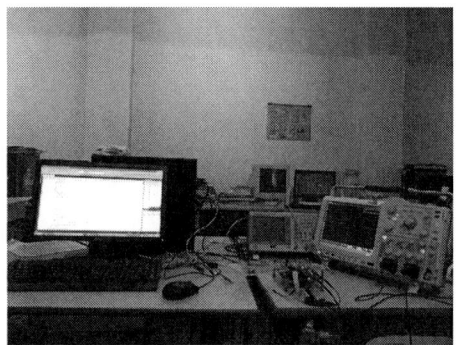

Figure 15: Experimental setup for CPA based power attack.

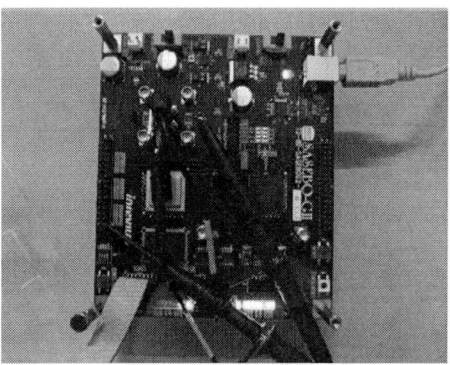

Figure 16: SASEBO under experimentation.

XC3S400A) and another as cryptographic FPGA (Virtex 5 xc5vlx50). Control FPGA contains the codes for communication with CPU [21]. It acts as a controller, which provides input and required control signals to cryptographic FPGA. Cryptographic FPGA contains the implementation of the cryptographic algorithm to be attacked. Communication with CPU is done through a USB cable. Plain texts are given to this board through CPU and cipher texts are sent to CPU for verification. Signals from this board is given to oscilloscope to plot power traces.

2. **Oscilloscope:** The oscilloscope is used for acquiring power traces. Upon receiving trigger signal, power traces is obtained and is sent to CPU. Triggering is very important for obtaining power traces, as it indicates start of the encryption and helps us to identify the desired power traces.

3. **CPU:** The heart of this setup is the CPU. The whole setup along with the operation of the other components are controlled by the CPU. CPU/SASEBO and CPU/oscilloscope interfaces are developed using $C\#$ codes. Plain texts are sent to FPGA and cipher texts are received for the verification. Power traces are received from the oscilloscope and are used to attack the crypto chip.

In the experiment, the control FPGA is fed by an external clock signal generated by a function generator. Power traces at different frequencies are obtained by varying the generated clock frequency.

NumChecker: Detecting Kernel Control-flow Modifying Rootkits by Using Hardware Performance Counters

Xueyang Wang
Polytechnic Institute of New York University
6 Metrotech Center, Brooklyn, NY, USA
xwang09@students.poly.edu

Ramesh Karri
Polytechnic Institute of New York University
6 Metrotech Center, Brooklyn, NY, USA
rkarri@poly.edu

ABSTRACT

This paper presents NumChecker, a new Virtual Machine Monitor (VMM) based framework to detect control-flow modifying kernel rootkits in a guest Virtual Machine (VM). NumChecker detects malicious modifications to a system call in the guest VM by checking the number of certain hardware events that occur during the system call's execution. To automatically count these events, NumChecker leverages the Hardware Performance Counters (HPCs), which exist in most modern processors. By using HPCs, the checking cost is significantly reduced and the tamper-resistance is enhanced. We implement a prototype of NumChecker on Linux with the Kernel-based Virtual Machine (KVM). Our evaluation demonstrates its practicality and effectiveness.

Categories and Subject Descriptors D.4.6 [**Operating Systems**]: Security and Protection-*invasive software*
General Terms Security
Keywords Kernel Rootkits, Virtualization, Hardware Performance Counters

1. INTRODUCTION

Kernel rootkits are formidable threats to computer systems. They are stealthy and can have unrestricted access to system resources. By subverting the operating system (OS) kernels directly, kernel rootkits are used by attackers to hide their presence, open backdoors, gain root privilege and disable defense mechanisms [8, 24].

Kernel rootkits perform their malicious activities in two ways: modifying the non-control data in the kernel data structures and hijacking the kernel control-flow to conceal resources from system monitoring utilities. In this work, we focus on the kernel rootkits that modify the kernel control-flow because the majority of kernel rootkits are of this type and they pose the most threat to system security. A recent analysis [16] indicates that more than 95% of Linux kernel rootkits persistently violate control-flow integrity. Control-flow modification makes the detection difficult because we do not know what the rootkits will modify. These rootkits may hijack the kernel static control transfers, such as changing the text of kernel functions or modifying the entries of the system call table. A

representative example is SucKIT rootkit [22], which replaces the system call table with its own copy, and then uses its own system call table to redirect to the malicious system calls. The control-flow modifying rootkits may also hijack dynamic control transfers such as dynamic function pointers. Adore-ng [2] is a rootkit of this type. It manipulates the function pointers at the virtual file system layer to redirect the execution flow to malicious handler routines, which can hide information by filtering the data.

1.1 Detection Techniques and Limitations

There has been a long line of research on defending against control-flow modifying rootkits. Host-based rootkit detection techniques run inside the target they are protecting, and hence are called "in-the-box" techniques. For example, Rkhunter [6] and Kstat [5] detect the malicious kernel control-flow modifications by comparing the kernel text or its hash and the contents of critical jump tables to a previously observed clean state. The main problem with the "in-the-box" techniques is that the detection tools themselves might be tampered with by advanced kernel rootkits, which have high privilege and can access the kernel memory.

With the development of virtualization, the Virtual Machine Monitor (VMM) based "out-of-the-box" detection techniques have been widely studied. These techniques move the detection facilities out of the target Virtual Machine (VM) and deploy them in the VMM. The isolation provided by the virtualization environment significantly improves the tamper-resistance of the detection facilities because they are not accessible to rootkits inside the guest VMs.

Most of the VMM-based rootkit detection techniques observe the static and dynamic kernel objects of a guest VM at the VMM level by directly acquiring the contents of the physical memory. However, there is a "semantic gap" between the external and internal observation. To extract meaningful information about the guest state from the low level view of the physical memory state, the detection tools require detailed knowledge of the guest OS implementation. For example, to retrieve the information of a guest VM's process list, the detection tools need to know where this particular data structure is laid out in the guest kernel memory. The location may vary from one implementation to another. Acquiring this detailed knowledge can be a tough task especially when the kernel source code is not available.

In regards to security, because the knowledge of the guest OS that the detection tools rely upon is not bound to the observed memory state, these techniques are subject to advanced attacks that directly modify the layout of the guest kernel data structures [9].

1.2 Introducing NumChecker

To overcome the challenges that the current "out-of-the-box" detection techniques face, we propose an "execution-oriented" VMM-based kernel rootkit detection framework called NumChecker. Num-

---

Permission to make digital or hard copies of all or part of this work for personal or classroom use is granted without fee provided that copies are not made or distributed for profit or commercial advantage and that copies bear this notice and the full citation on the first page. To copy otherwise, to republish, to post on servers or to redistribute to lists, requires prior specific permission and/or a fee.
DAC '13, May 29 - June 07 2013, Austin, TX, USA.

Checker performs integrity checking at a higher level. It validates the whole execution of a guest kernel function without checking any individual object on the execution path. **NumChecker models a kernel function with the number of certain hardware events that occur during the execution. Such hardware events include total instructions, branches, returns, floating point operations, etc.** If the control-flow of a kernel function is maliciously modified, the number of these hardware events that occur during the execution will be different.

To count the guest hardware events from the host side, NumChecker utilizes the Hardware Performance Counters (HPCs), which exist in most modern processors as a part of the processor's performance monitoring unit (PMU). **The HPCs were originally used for performance tuning. NumChecker leverages them for the system security purpose.** Because the events are automatically counted by the HPCs, the checking latency and the performance overhead are significantly reduced. Also, the security is enhanced because the HPCs count the events without a guest's awareness, and they are inaccessible to a guest VM.

We implement a prototype of NumChecker on the Linux platform with the Kernel-based Virtual Machine (KVM) [4]. We evaluate NumChecker on a number of real-world kernel rootkits. The results demonstrate that NumChecker can efficiently detect all the kernel rootkits with very low cost.

The rest of this paper is organized as follows: Section 2 describes the background and related work. Section 3 gives the overview of our design. Section 4 presents the implementation details. The evaluation results are shown in Section 5. Section 6 is the conclusion. Additional implementation details are given in the appendix.

2. BACKGROUND AND RELATED WORK

2.1 Hardware Performance Counters

HPCs are a set of special-purpose registers built into modern microprocessors' PMU to store the counts of hardware-related activities. HPCs were originally designed for performance debugging of complex software systems. They work along with event selectors which specify the certain hardware events, and the digital logic which increases a counter after a hardware event occurs. Relying on HPC-based profilers, the developers can easily understand the runtime behavior of a program and tune its performance. HPC-based profilers provide access to detailed performance information with much lower overhead than software profilers. Further, no source code modifications are needed.

HPC-based profilers are currently built into almost every popular operating system. Linux *Perf* [7] is a new implementation of performance counter support for Linux. It is based on the Linux kernel subsystem *Perf_event*, which has been built into 2.6+ systems. The user space *Perf* tool interacts with the kernel *Perf_event* by invoking a system call. It provides users a set of commands to analyze performance and trace data. When running in counting modes, *Perf* can collect specified hardware events on a per-process, per-CPU, and system-wide basis.

2.2 Related Work

Enhancing security with virtualization The use of virtualization technologies for enhancing system security has been studied for a long time. Garfinkel and Rosenblum [11] first introduced virtual machine introspection to detect intrusion. It leverages the virtual machine monitor to isolate the intrusion detection service from the monitored guest. XenAccess [17], VMwatcher [14], and VMWall [23] are virtual machine introspection techniques using memory acquisition. These techniques obtain the guest states from

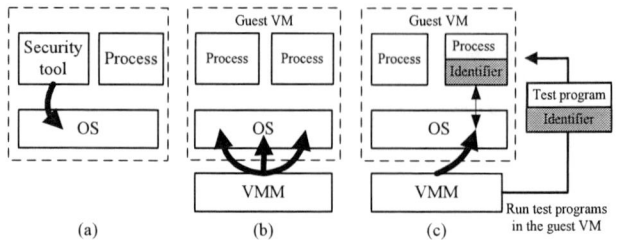

Figure 1: Comparison of in-the-box (a), out-of-the-box (b), and in-and-out-of-the-box (c) techniques.

the host side by accessing guest memory pages. As discussed in Section 1, to bridge the semantic gap, accurate kernel data structure layout or kernel symbols are required. Lares [18] monitors a guest VM by placing its hooking component inside the guest OS and protecting it from the VMM. These hooks would be triggered whenever certain monitored events were executed by the guest OS. This technique requires modification to the guest OSes, making it not applicable to close-source OSes like Windows.

Execution path analysis Patchfinder [21] is another closely related work that also uses execution path analysis for kernel rootkit detection. It counts the number of executed instructions by setting the processor to single step mode. In this mode, a debug exception (DB) will be generated by the processor after every execution of the instruction. The number counted during the execution of certain kernel functions will be analyzed to determine if the functions are maliciously modified. The vulnerability of this technique is that the counting and analysis facilities themselves might be manipulated by an advanced kernel rootkit which has the highest privilege and full access to kernel memory. From another perspective, running in the processor's single step mode leads to very high performance overhead.

HPC-based integrity checking HPCs are originally designed for the purposes of performance debugging. Performing system security analysis is a new use of HPCs and has not been studied much. A scheme proposed in [15] uses HPCs for integrity checking of programs. It targets malicious modifications to the user space programs, and assumes the OS kernel is trusted. Also, it cannot be directly used for virtualization systems. Our design is to detect rootkits in a guest VM's kernel space for virtualization systems.

3. NUMCHECKER OVERVIEW

3.1 Threat Model

We target a kernel rootkit which has the highest privilege inside the guest VM. The rootkit has full read and write access to guest VM's memory space, so it can perform arbitrary malicious activities inside the guest VM's kernel space. In order to hide its presence in the guest VM, the kernel rootkit modifies the kernel control-flow and executes its own malicious code. We assume that the VMM is trustworthy. And the rootkit cannot break out of the guest VM and compromise the underlying VMM [11, 14, 20].

3.2 System Call Analysis with HPCs

To detect control-flow modifying kernel rootkits, NumChecker focuses on validating the execution of system calls. System calls are the main interface that a user program uses to interact with the kernel. In order to achieve stealth, a common action that a kernel rootkit performs is to fool the user monitoring utilities (like *ps*, *ls*, *netstat* in Linux). These monitoring utilities retrieve the information about the system states by invoking some system calls. The rootkits usually manipulate the normal execution of these system calls to prevent the monitoring tools from obtaining the correct information. For example, the Linux *ps* command will return the status of all the running processes. The system calls invoked by the *ps* command include *sys_open, sys_close, sys_read, sys_lseek, sys_stat64, sys_fstat64, sys_getdents64, sys_old_mmap*, etc. To hide itself and other malicious processes, a rootkit modifies these system calls so that the information about the malicious processes will not appear in the list returned by *ps*. The modifications usually result in a different number of monitored hardware events from the uninfected execution. They are measured by NumChecker.

For a given system call, the number of hardware events that occur during the execution varies when different inputs are applied. To determine if an unusual number of events is caused by the malicious modification to the system call, the inputs to a monitored system call must be given. NumChecker invokes a monitored system call by executing a pre-generated test program in the guest VM. The counts of monitored events are then compared with those of the corresponding unmodified system call invoked by the same test program. By doing so, the noise from applying different inputs can be avoided.

Unlike the "in-the-box" techniques which perform the checking inside the monitored target (Figure 1(a)), or the "out-of-the-box" techniques which only depend on the observations from outside of the target (Figure 1(b)), NumChecker performs an "in-and-out-of-the-box" checking, shown in Figure 1(c): it first runs a test program inside the monitored guest VM. The test program will invoke monitored system calls (in-the-box). The host then accesses the HPCs to retrieve the guest state (counts of monitored events) from outside the guest VM (out-of-the-box). The combination takes advantage of both the meaningful information of the "in-the-box" checking and tamper-resistance of the "out-of-the-box" checking.

Because multiple programs are running concurrently in the guest VM, NumChecker has to identify the test programs. To overcome the semantic gap of the observation from the host side, NumChecker adds guest-transparent identifiers to the test programs to relate the guest VM's state observed from the outside of the guest VM and the execution inside the guest VM. These identifiers are updated randomly and dynamically by the host. For more details, please see Appendix A.

4. NUMCHECKER IMPLEMENTATION

In NumChecker, we use KVM to build our virtualization environment. KVM is a full virtualization solution for Linux on hardware containing virtualization extensions that can run unmodified guest images. The processor with hardware virtualization extensions has two different modes: host mode and guest mode. Execution of virtualization-sensitive instructions in guest mode will trap to the host, which is called VM-exit. In this way, the host can manage the guests' accesses to virtualized resources.

To profile the execution of system calls in a guest VM using HPCs, the profiler in the host should have the following capabilities: (1) it should be aware of the occurrence of system calls in a guest VM; (2) it should be able to trigger the HPCs. The existing HPC-based profiling tools cannot meet our design requirements because they are not able to capture the beginning and end of a system call in a guest VM. So the number of hardware events obtained by a profiling tool cannot be exactly pinned to the execution of a monitored system call.

To resolve this issue, NumChecker connects the profiling tool with the VMM, which is capable of intercepting system calls in the guest VM. NumChecker can be implemented with any HPC-based profiler. Our proof-of-concept design is based on the Linux *Perf*.

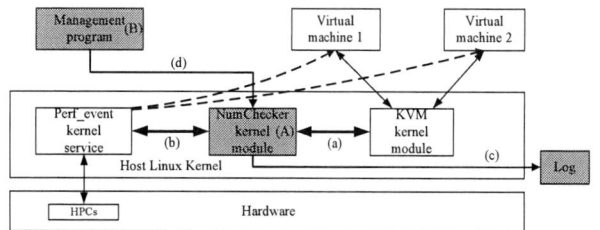

Figure 2: High-level structure of NumChecker design.

As shown in Figure 2, NumChecker has two main components: a lightweight module (A) in the host kernel between the KVM kernel module and the *Perf_event* kernel service, and a management program (B) running in the host user space. The kernel module of NumChecker performs two functions: first, it cooperates with the KVM to intercept monitored system calls in a guest VM (a). Second, it communicates with *Perf_event* kernel service to initialize, enable/disable, read, and close HPCs (b). The counted numbers are output to a log file (c). The management program is used to dynamically configure the module of NumChecker in the kernel by modifying the parameters through the *sysctl* system call (d). The configuration includes which system calls to intercept, which hardware events to count, etc.

A guest VM in KVM is seen as a single process from the host's point of view. NumChecker calls the *Perf_event* kernel service to launch a per-process profiling task and enables the HPCs only when a monitored system call is run in the guest VM. By doing so, the counted events are exactly contributed by the execution of the monitored system call in the specific guest VM.

4.1 Two-phase Rootkit Detection

NumChecker rootkit detection has two phases, shown in Figure 3: in the offline profiling phase, system calls of the trusted guest OSes are measured; in the online runtime checking phase, system calls of a running monitored guest OS are measured and compared with that of the corresponding trusted OS.

4.1.1 Offline Profiling

In this phase, the test programs are executed in guest VMs with trusted OSes installed. The host logs in to the guest VM through the network between the host and the guest (for example, using SSH), loads executables of the test programs, and launches NumChecker. The configuration parameters specific to the monitored system call, such as the system call number and the type of hardware events measured, are passed to NumChecker. Then the test programs are executed in the guest VM. To improve the accuracy of the measurement, the execution of a test program is repeated several times. On the host side, the hardware events corresponding to the monitored system calls are counted. (More technical details about HPC-based measurement of a system call are presented in Appendix B.) When the measurement is complete, the system call interception is disabled. The results are stored as the "clean copy" to be used at runtime.

Since a system call may have various implementations in the OSes with different kernel versions, for each particular OS we need to create a separate "clean copy." Generating a database containing clean copies of commonly used OSes is very fast because the number of kernel versions of commonly used OSes is limited. Unlike other comparison-based techniques, which take a long time to read large amounts of memory, for a given OS, NumChecker can profile one system call and create the "clean copy" in a few seconds.

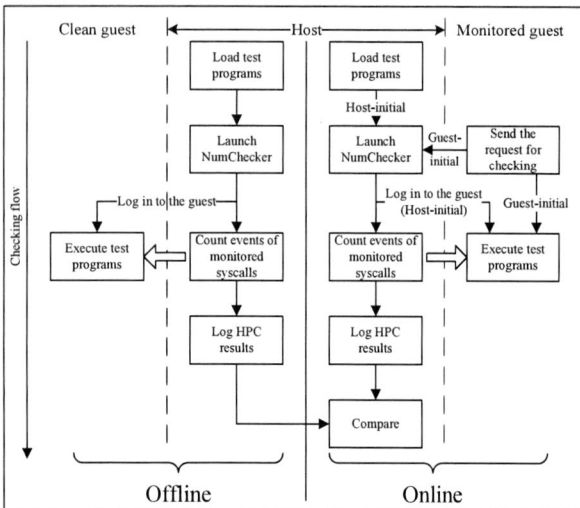

Figure 3: Offline profiling phase (left-hand side) and online checking phase (right-hand side) of NumChecker rootkit detection.

Moreover, the offline profiling only needs to be performed once for a particular OS.

4.1.2 Online Checking

The execution path of a monitored system call is dynamically measured at runtime. The steps in this phase are similar to those in the offline profiling. The test programs are loaded when a guest VM is created. As mentioned above, to maintain consistency, the test programs used in the runtime checking are identical to the ones used to generate the "clean copy." The system call profiling can then be dynamically invoked in either host-initial mode or guest-initial mode. For the host-initial mode, whenever the host administrator wants to launch a check, the host configures and enables NumChecker. Then the host logs in to the guest VM and executes the test programs in the guest VM. When the execution is done, the host turns off the system call interception.

Guest-initial mode is used for a guest user who wants to check if the OS is maliciously modified. In this mode, the guest user first sends a request to the host through the network. The management program in the host then checks the availability of HPCs and allocates unused ones to the guest. After the counters are allocated and ready to use, the host sends the acknowledgement to the guest. When the acknowledgement is received, the guest then runs the test programs to invoke monitored system calls.

4.2 Security Analysis

With the isolation provided by virtualization and the benefits of using HPCs, the "in-and-out-of-the-box" execution path analysis is very secure and tamper-resistant. Here, we discuss some possible attacks and show how they can be defended by our technique.

Scenario #1: The rootkit may try to tamper with the counting process. If the event counting is inside the guest VM, the kernel rootkit may disable the counters when its own code is executed and resume the counting when the control-flow returns to the normal execution. In this case, the malicious actions will not be detected since the counts remain the same as the unmodified execution. In our design, the hardware events are counted by the host. The HPCs are out of reach to the rootkits. Another way a rootkit could tamper with the counting process is that it may suspend the thread that runs the monitored system call and pass a pointer to another thread. The

malicious code is then executed in the unmonitored thread where the events are not counted by NumChecker. However, suspending a monitored thread will cause a VM-exit that can be captured by the host. Malicious activities are suggested when this type of VM-exit is repeatedly observed during a check.

Scenario #2: The rootkit may tamper with the analysis process. Even though the counters are working properly and count all the true numbers, a rootkit may directly manipulate the analysis. Consider Patchfinder, the "in-the-box" execution path analysis technique, as an example. Since the counts are stored in the memory, the kernel rootkits who have full access to the memory can simply modify the actual counted number. For our VMM-based design, the counted numbers are read from HPCs by the trusted host and all the analyses are performed by the host. The guest kernel rootkits cannot interfere with the analyses because they do not have access to the host memory.

Scenario #3: The rootkit may try to predict the "good" number. Specifically, if the rootkit can predict the exact number of hardware events that occur during the execution of a system call, it could carefully modify the system call to generate the same number as the original one. However, given a system call, the number of hardware events generated in the execution depends on the inputs to the system call. In our design, the inputs to the monitored system calls are applied in the pre-generated test programs. A test program is used as the "secret key" in a particular check and it is updated dynamically by the host. A rootkit is not able to predict and generate a valid number of a monitored system call in a particular check because the number varies when different test programs are applied.

Scenario #4: The rootkit may undo modifications. A rootkit is used by attackers to provide long-term stealth for malicious activities. If a clever rootkit is aware of the occurrence of a check, it can undo modifications when the check is performed and activate itself again when the check is over. In NumChecker, the detection processes are running in the host without a guest's awareness. The only thing the guest can see is the execution of a test program. However, from the guest's point of view, the execution of a test program is no different from the execution of other programs. So a guest is not able to know when it is being monitored. Additionally, we can randomize the intervals between checks to avoid attackers' prediction of the checking period.

5. EVALUATION

5.1 Detection Capability

To evaluate the effectiveness of NumChecker, we test our technique with eight real-world kernel rootkits on two different guest OSes. The host runs Ubuntu 11.10 with Linux kernel 3.0.16. The two guest OSes are Redhat 7.3 with Linux kernel 2.4.18 and Fedora core 4 with Linux kernel 2.6.11. Table 1 shows our experimental results. For each rootkit, we check the modifications it performs to five system calls, *sys_open*, *sys_close*, *sys_read*, *sys_getdents64*, and *sys_stat64*, with the corresponding test programs. Three hardware events, retired instructions (INST), retired returns (RN), and retired branches (BR), are monitored simultaneously for the execution of each system call. The percentages present the deviations of counts from uninfected executions.

To determine whether a system call is maliciously modified, one situation we must consider is **false positives**. Because of the complexity of an OS kernel, the noise is unavoidable. Even though identical test programs are applied, it cannot be guaranteed that the number of events is exactly the same for every single run. This noise can be reduced by increasing the number of times each test

560

Table 1: NumChecker detection capabilities. The numbers are deviations (%) from uninfected executions. Any deviation of more than 5% suggests a malicious modification. For each rootkit, the bold number indicates the largest deviation.

Guest OS	Rootkit	Events counted	System calls monitored					Detected?
			sys_open	sys_close	sys_read	sys_getde-nts64	sys_stat64	
Linux 2.4	SucKIT 1.3b	INST	836.1	8.6	59.5	242.9	284.3	Yes
		RN	676.5	50.0	150.0	483.3	383.3	
		BR	**1294.2**	72.0	33.3	1028.1	292.9	
	Adore 0.42	INST	99.4	10.3	0.0	427.7	91.9	Yes
		RN	123.5	25.0	0.0	650.0	161.1	
		BR	119.9	24.0	0.0	**1313.1**	162.9	
	Sk2rc2	INST	363.4	52.4	79.5	39.8	63.1	Yes
		RN	**488.2**	50.0	166.7	95.8	166.7	
		BR	359.2	128.0	76.9	66.9	98.6	
	Superkit	INST	827.8	10.8	59.5	244.4	283.1	Yes
		RN	535.3	50.0	233.3	483.3	383.3	
		BR	**1399.5**	28.0	61.5	1014.4	295.2	
Linux 2.6	Enyelkm 1.1	INST	0.8	2.3	41.0	62.0	-2.3	Yes
		RN	4.0	12.5	28.1	54.9	4.0	
		BR	1.7	2.6	55.7	**76.7**	1.1	
	Phalanx b6	INST	8.5	0.0	**201.5**	35.0	-2.5	Yes
		RN	14.0	0.0	56.3	17.6	0.0	
		BR	19.5	-1.7	165.1	69.2	-0.5	
	Sebek 3.2	INST	9.4	0.0	10.3	0.0	-0.7	Yes
		RN	8.0	0.0	**18.8**	0.0	0.0	
		BR	13.8	0.9	2.4	0.0	-0.5	
	Adore-ng	INST	0.0	0.0	0.0	289.0	-0.6	Yes
		RN	0.0	0.0	0.0	80.4	4.0	
		BR	0.0	2.6	2.4	**524.4**	-0.5	

program is run. In our experiment, each test program is repeated 500 times. We observe that the noise is less than 5% for the execution of a normal system call, no matter if the system load is light or heavy. So a deviation of more than 5% suggests a malicious modification.

From Table 1, we can see that in order to introduce their own functionality, the rootkits usually significantly modify the original system calls. The difference in the number of events between normal and infected executions is very notable. The Superkit rootkit modifies the system calls very heavily. The largest deviation is from the number of branches of the *sys_open* function, which is 1399.5%. The test on Sebek 3.2 gives smaller deviations. The largest deviation from the test on Sebek 3.2 is 18.8%, which is still much larger than the noise threshold of 5%.

5.2 Checking Latency

We perform the experiments of the checking latency on a PowerSpec platform with a 2.3GHz AMD Quad-Core Opteron 1356 CPU, which has 4 HPCs on each core. The host is running 32-bit Ubuntu 11.10 (kernel version 3.0.16) with 8GB RAM and 4-core configuration; The guest VMs are running 32-bit Redhat 7.3 (kernel version 2.4.18) and Fedora Core 4 (kernel version 2.6.11), with 512MB RAM and 1-core configuration.

Since NumChecker uses an "in-and-out-of-the-box" technique, the detection procedure has two separate parts, in the guest and host respectively. The total checking latency is the sum of the time for executing the test programs in the guest VM and the time for analyzing the counted numbers in the host. The checking latency depends on the number of test programs to be executed, the number of hardware events to be observed, and the specific guest OS version.

For each monitored system call, the analysis in the host takes 4.86 ms when three hardware events are observed simultaneously. Table 2 shows the execution times of each test program in two different guest VMs. To reduce false positives, each test program contains 500 iterations to repeatedly invoke the corresponding system call. The time is calculated from the total execution, and the numbers are averaged over 20 runs. Because test programs are very simple, the execution time is short. The average execution time of a test program on Redhat 7.3 and Fedora Core 4 are 45.6 ms and 59.5 ms, respectively.

A typical test of NumChecker checks all of the five system calls mentioned above with four corresponding test programs (*sys_open* and *sys_close* are checked in one test program). Table 3 presents the checking time of NumChecker for the Fedora Core 4 guest, and the comparison with other techniques. Because some results are from others' experiments on different platforms, for each implementation, we list the CPU frequency and guest memory size, which are related more to checking latency. The first three techniques, Rkhunter 1.2.8, Chkrootkit 0.48 [3], and Patchfinder, are host-based techniques running inside the target. MAVMM [12], VMwatcher, XenAccess, and OSck [13] are VMM-based techniques. For MAVMM, VMwatcher, and XenAccess, the checking time depends on the size of the memory to be examined because they require memory dumping. For VMwatcher, the checking time also depends on whether the kernel symbols are available. Examining a 512M raw window memory image takes 32 seconds while for Linux, the analysis can be finished within 500 ms.

NumChecker takes 24.3 ms in the host (5 system calls are analyzed) and 238 ms in the guest (4 test programs are executed). The total time to finish a typical test is 262.3 ms, regardless of the memory size of the guest VM.

Table 2: Execution times of different test programs on Redhat 7.3 and Fedora Core 4.

	Redhat 7.3	Fedora Core 4
test_open_and_close	44.9 ms	52.7 ms
test_read	50.5 ms	69.1 ms
test_getdents64	61.0 ms	75.7 ms
test_stat64	27.2 ms	40.5 ms
average	45.6 ms	59.5 ms

Table 3: Checking latency of NumChecker and other rootkit detection techniques

	CPU frequency/ Guest memory	Checking latency
Rkhunter 1.2.8	2.3GHz/512MB	40 sec
Chkrootkit 0.48	2.3GHz/512MB	3.8 sec
Patchfinder	N/A	15 sec
MAVMM	900MHz/256MB	271 ms/MB
VMwatcher	N/A /512MB	32 sec or 500 ms
XenAccess	2.33GHz/8GB	1 ms/MB
OSck	2.8GHz/2GB	525 ms
NumChecker	2.3GHz/512MB	262.3 ms

5.3 Guest Performance Overhead

Our next experiment is to test the performance overhead on the guest system operations when NumChecker is invoked periodically. The experiment is performed in a guest VM running Fedora Core 4 system, and the hardware configuration is the same as other experiments mentioned above.

We use 9 UnixBench [1] benchmarks to calculate the guest system performance, and the overall presents the average. NumChecker is invoked every 5 and 10 seconds, respectively, and the results are compared to the normal guest performance without NumChecker. Figure 4 shows the results of the experiment. All the numbers are the average of 10 runs.

The average overhead of the guest system performance is 2.8% when NumChecker is invoked every 5 seconds and 1.3% when NumChecker is invoked every 10 seconds. The benchmarks of file copy and system calls report more overhead than others. This is because NumChecker invokes and intercepts system calls. And writing the counted numbers to log files consumes system I/O resources.

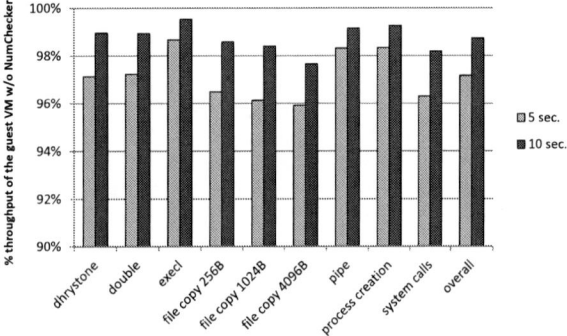

Figure 4: Throughput degradation of the guest VM when NumChecker is invoked every 5 and 10 seconds.

6. CONCLUSION

In this paper, we present NumChecker, a VMM-based framework to detect control-flow modifying kernel rootkits in guest VMs. NumChecker performs the checking by validating the execution of system calls in the guest VM. The validation is based on the number of specified hardware events that occur during the execution of a guest system call. These hardware events are automatically counted by HPCs. We implement a prototype of NumChecker on Linux with the KVM virtualization environment, and our evaluation demonstrates its practicality and effectiveness.

7. ACKNOWLEDGEMENT

This material is based upon work funded by AFRL under contract No. FA8750-10-2-0101. Any opinions, findings and conclusions or recommendations expressed in this material are those of the author(s) and do not necessarily reflect the views of AFRL.

8. REFERENCES

[1] www.tux.org/pub/benchmarks/System/unixbench/.
[2] The adore-ng rootkit. http://stealth.7350.org.
[3] Chkrootkit. http://packetstormsecurity.org/files/62258/chkrootkit-0.48.tar.gz.html.
[4] Kernel based virtual machine. http://www.linux-kvm.org/page/Main_Page.
[5] Kstat - kernel security therapy anti-trolls. http://www.s0ftpj.org/en/tools.html.
[6] Rkhunter. http://packetstormsecurity.org/files/44153/rkhunter-1.2.8.tar.gz.html.
[7] Performance counters for linux. http://lwn.net/Articles/310176, 2010.
[8] F. Azmandian, M. Moffie, M. Alshawabkeh, J. G. Dy, J. A. Aslam, and D. R. Kaeli. Virtual machine monitor-based lightweight intrusion detection. ACM SIGOPS Operating Systems Review, 45(2):38–53, July 2011.
[9] S. Bahram, X. Jiang, Z. Wang, M. Grace, J. Li, and D. Xu. Dksm: Subverting virtual machine introspection for fun and profit. In Proceedings of the 29th IEEE International Symposium on Reliable Distributed Systems, Oct. 2010.
[10] A. Dinaburg, P. Royal, M. Sharif, and W. Lee. Ether: Malware analysis via hardware virtualization extensions. In Proceedings of ACM conference on Computer and Communications Security, Oct. 2008.
[11] T. Garfinkel and M. Rosenblum. A virtual machine introspection based architecture for intrusion detection. In Proceedings of Network and Distributed Systems Security Symposium, pages 191–206, 2003.
[12] A. Godiyal, A. Nguyen, and N. Schear. A lightweight hypervisor for malware analysis. http://ivanlef0u.fr/repo/todo/HyperVisorMalware.pdf.
[13] O. S. Hofmann, A. M. Dunn, S. Kim, I. Roy, and E. Witchel. Ensuring operating system kernel integrity with osck. In Architectural Support for Programming Languages and Operating Systems, March 2011.
[14] X. Jiang, X. Wang, and D. Xu. Stealthy malware detection through vmm-based out-of-the-box semantic view reconstruction. In Proceedings of ACM conference on Computer and communications security, Nov. 2007.
[15] C. Malone, M. Zahran, and R.Karri. Are hardware performance counters a cost effective way for integrity checking of programs? In The Sixth ACM Workshop on Scalable Trusted Computing, Oct. 2011.
[16] J. N. L. Petroni and M. Hicks. Automated detection of persistent kernel control-flow attacks. In Proceedings of ACM conference on Computer and Communications Security, pages 103–115, 2007.
[17] B. Payne, M. de Carbone, and W. Lee. Secure and flexible monitoring of virtual machines. In Computer Security Applications Conference, Dec. 2007.
[18] B. Payne, M. S. M. Carbone, and W. Lee. Lares: An architecture for secure active monitoring using virtualization. In Proceedings of the IEEE Symposium on Security and Privacy, May 2008.
[19] J. Pfoh, C. Schneider, and C. Eckert. Nitro: Hardware-based system call tracing for virtual machines. Advances in Information and Computer Security, 7038:96–112, Nov 2011.
[20] R. Riley, X. Jiang, and D. Xu. Guest-transparent prevention of kernel rootkits with vmm-based memory shadowing. In Proceedings of International Symposium on Recent Advances in Intrusion Detection, 2008.
[21] J. Rutkowska. Execution path analysis: finding kernel based rootkits. Phrack Article, 11, 2002.
[22] Sd and Devik. Linux on-the-fly kernel patching without lkm. Phrack Magazine, 11, Jan. 2004.
[23] A. Srivastava and J. Giffin. Tamper-resistant, application-aware blocking of malicious network connections. In Proceedings of International Symposium on Recent Advances in Intrusion Detection, pages 39–58, 2008.
[24] Z. Wang, X. Jiang, W. Cui, and P. Ning. Countering kernel rootkits with lightweight hook protection. In Proceedings of the 16th ACM conference on Computer and Communications Security, pages 545–554, Nov. 2009.

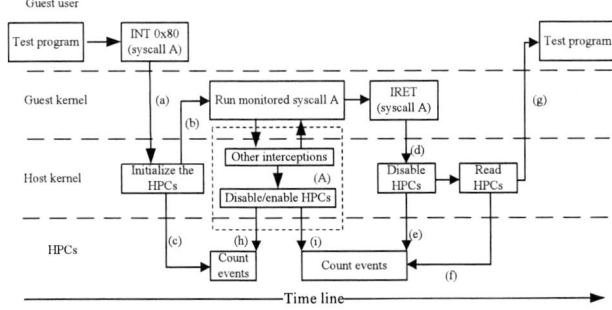

Figure 5: HPC-based measurement of a system call in the guest VM.

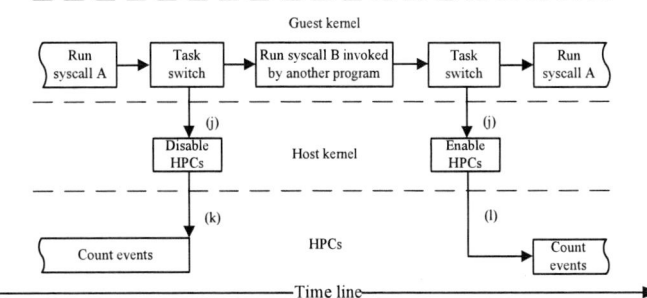

Figure 6: Kernel preemption handling in NumChecker.

APPENDIX

A. IDENTIFYING TEST PROGRAMS

To avoid the noise from the execution of other processes, Num-Checker needs to distinguish the system calls invoked by the test programs from ones invoked by other programs. This is done in two steps: identifying which system calls belong to the same process and identifying which process belongs to the test program.

To identify which system calls are invoked in the same process, the host can check either the CR3 register or the task descriptor pointer. To determine which process belongs to a test program, NumChecker generates an identifier for a test program with a special sequence of system calls. Specifically, a sequence of system calls (for example, one *sys_getuid* followed by two *sys_stat64* then three *sys_getuid* followed by four *sys_stat64*) is attached at the beginning of the test program. The process of the test program can be identified when NumChecker intercepts such a sequence in the same process. By carefully composing the sequence with some infrequently used system calls, it is almost impossible that another process can generate the same sequence by chance. The special system call sequence is only known by the host as a secret identifier, so an attacker in the guest VM is not able to use this to identify the test programs.

B. MEASUREMENT OF A SYSTEM CALL

System calls are implemented in two ways: interrupt-based system calls and sysenter-based system calls. The interception of guest system calls in both types is supported by processors with virtualization extensions, directly or with the help of other techniques [10, 19].

Figure 5 illustrates the procedure of measuring an interrupt-based system call in the guest VM. The test program invokes a monitored system call by executing an INT 0x80 instruction. This execution causes a VM-exit (arrow a in Figure 5) if the INT instruction interception is enabled in the host. When the system control is transferred to the host, NumChecker is launched to first determine whether the current system call needs to be monitored. If so, NumChecker passes configuration parameters to the *Perf_event* to initialize HPCs. These parameters include the counting mode, the process ID of the monitored guest VM, the type of hardware events, etc. After the HPCs are set up and right before entering the guest VM (b), NumChecker sends a signal to *Perf_event* to turn on the HPCs (c). Then the HPCs count the specified events when the system call is running in the guest kernel until reaching the end of the execution, which is an IRET instruction. System control is transfered to the host again (d). NumChecker turns off the HPCs (e) immediately after the CPU switches to host mode. The HPCs stop

counting when the host code is being executed. The counted numbers are kept in the HPCs. NumChecker then reads the numbers through *Perf_event* (f) and outputs them into a log file. After that, the control is transferred back to the guest (g), and the execution of the program is resumed.

B.1 Handling Other Interceptions

When a guest VM is in kernel mode executing a system call, other virtualization-sensitive activities (such as I/O operations and external interrupts) of the guest VM may also cause a VM-exit (see block A in Figure 5). These activities interrupt the guest VM's executions and will be intercepted by the host. If the HPCs keep counting when the VM-exit is being handled in the host, the events generated by the execution of the host code will be included in the final counts. To remove this noise, every time a VM-exit occurs during the execution of a guest system call, NumChecker suspends the HPCs (arrow h in Figure 5) by sending a disabling signal to *Perf_event* before the VMM handles the VM-exit. The HPCs are resumed (i) when the handling is finished.

Note that when an external interrupt returns, it also executes an IRET instruction. So if there are external interrupts taking place during the execution of a system call, more than one IRET instruction could be intercepted. It is necessary to find out the IRET corresponding to the monitored system call. This can be determined by checking the value of the stack pointer after returning from the interrupt. An external interrupt will return to the guest kernel space while a system call will return to the guest user space.

B.2 Handling Kernel Preemption

Traditional Linux kernels are not preemptible. When a task is running in kernel mode, it cannot be switched out until its completion, even though a higher-priority task is ready. Note that although the running task can still be interrupted by an external interrupt, a task switch cannot take place. In Linux 2.6 and later, a preemptible kernel option has been provided. This allows a higher-priority task to interrupt a running lower-priority task in the kernel. If kernel preemption is enabled, the HPC-based measurement becomes more complicated, shown in Figure 6. Whenever a monitored system call is suspended and switched out of the processor, the HPCs need to be suspended as well. Fortunately, a guest task switch can be intercepted (arrow j in Figure 6) by the VMM. When a task switch traps to the VMM, NumChecker disables the HPCs (k). It resumes the HPCs when the monitored system call is scheduled again (l). In this case, the execution of a system call is split into several "pieces." These "pieces" can be determined by checking the task descriptor pointer of the task currently running on the CPU from the value of the ESP register. The "pieces" with the same task descriptor pointer belong to the identical system call.

High-Performance Hardware Monitors to Protect Network Processors from Data Plane Attacks

Harikrishnan Chandrikakutty, Deepak Unnikrishnan, Russell Tessier and Tilman Wolf
Department of Electrical and Computer Engineering
University of Massachusetts, Amherst, MA, USA
{chandrikakut,unnikrishnan,tessier,wolf}@ecs.umass.edu

ABSTRACT

The Internet represents an essential communication infrastructure that needs to be protected from malicious attacks. Modern network routers are typically implemented using embedded multi-core network processors that are inherently vulnerable to attack. Hardware monitor subsystems, which can verify the behavior of a router's packet processing system at runtime, can be used to identify and respond to an ever-changing range of attacks. While hardware monitors have primarily been described in the context of general-purpose computing, our work focuses on two important aspects that are relevant to the embedded networking domain: We present the design and prototype implementation of a high-performance monitor that can track each processor instruction with low memory overhead. Additionally, our monitor is capable of defending against attacks on processors with a Harvard architecture, the dominant contemporary network processor organization. We demonstrate that our monitor architecture provides no network slowdown in the absence of an attack and provides the capability to drop attack packets without otherwise affecting regular network traffic when an attack occurs.

1. INTRODUCTION

The Internet is a critical infrastructure component in today's society. Many aspects of personal communication, business transactions, entertainment, digital government, etc. rely on the availability and correct operation of the Internet. In this work, we focus on the security of packet processing functions that are necessary to handle packet forwarding on a network router. The packet processing component of a modern router is typically implemented using a network processor (NP) system. A network processor has multiple simple embedded processor cores that can be programmed to handle network traffic. When changes are necessary, the software on this network processor can be changed to adapt the operation of the router. Unlike traditional routers that have been based on application-specific

---

Permission to make digital or hard copies of all or part of this work for personal or classroom use is granted without fee provided that copies are not made or distributed for profit or commercial advantage and that copies bear this notice and the full citation on the first page. To copy otherwise, to republish, to post on servers or to redistribute to lists, requires prior specific permission and/or a fee.
DAC'13, May 29 – June 07, 2013, Austin, Texas, USA.

---

integrated circuit (ASIC) technology, network-processor-based routers can be adapted in their functionality. However, this flexibility raises an interesting security problem: routers that use software-programmable packet processors are vulnerable to attacks [5]. Vulnerable packet processing code can be exploited using malformed data packets to launch an in-network denial-of-service attack. This type of attack on contemporary NPs is addressed by this paper.

To reduce the vulnerability of packet processors in routers (as well as embedded processors in general), digital circuits that monitor run-time processor operation have been proposed [3, 5, 13]. These hardware monitors use information about correct processor software execution to track the instructions executed by a processor core. If an attack on the processor occurs, a deviation from expected, programmed behavior is detected and a recovery process is initiated. Compared to software-based protection mechanisms (e.g., virus scanners), hardware-based monitors require less performance overhead and react faster. In the networking domain, low overhead and fast detection speed are particularly important. Therefore, there have been ongoing efforts to further improve NP protection mechanisms to better support the applications supported by NPs and the processor organization typically exhibited by NPs.

In this paper, we address two key problems that have not been previously addressed in protecting network processors from software attacks. An effective NP monitoring system must verify every instruction that is executed by the processor and thus needs to operate at very high speeds. This instruction-based monitoring operation can be viewed as a finite automaton with a fixed number of acceptable paths. Prior work in hardware monitoring [5, 13] has been based on non-deterministic finite automaton (NFA) implementations, which potentially require high memory bandwidth when tracking multiple parallel states, or coarse verification at the level of basic blocks [3], which may not detect attacks before they are executed. For our first contribution, we present an instruction-level monitoring solution for NPs based on deterministic finite automaton (DFA) implementation, which overcomes the shortcomings of both prior techniques.

Prior work in embedded NP security has assumed a von Neumann processor architecture with a combined instruction and data memory, where an attack can execute code from the processor stack [5]. Most network processors, however, are based on Harvard architectures that separate instruction and data memory making stack-based code execution impossible. Therefore, it is questionable whether data plane attacks are even possible in networks. For our second contribution, we present an attack example that demonstrates the existence of Harvard architecture attacks and we show that our monitoring system is effective in defending NPs against them.

The specific contributions of our paper are:

1. Design of a high-performance hardware monitoring system for NPs: Our pipelined design can perform instruction verification with a single memory read per instruction and thus can operate at speeds sufficient to maintain line rate networking data transfer.

2. Algorithm for construction of a deterministic monitoring graph: We present a method to convert the monitoring graph of NP instructions, which initially is non-deterministic due to control-flow changes (e.g. branches), into a deterministic automaton. The representation of the DFA is compacted to allow for a highly efficient implementation in the hardware monitor.

3. Demonstration of an attack on and defense of a Harvard architecture network processor: We demonstrate an in-network attack through the data plane of the network that exploits an integer overflow vulnerability to smash the processor stack and launch a return-to-library attack. This attack propagates the attack packet and crashes the processor system. We also show that our hardware monitor is effective in defending against this attack and allowing for continued NP-based router operation after attack identification and recovery.

The remainder of the paper presents these contributions in detail.

2. RELATED WORK

Programmability in the packet processing systems of routers has been used increasingly widely over the past decade. Most major router vendors employ network processors in their products (e.g., Cisco QuantumFlow [6], Cavium Octeon [4]). While the programmability of these devices is hidden from network users, it is used by vendors to extend system functionality. It can be expected that routers will continue to have programmable packet processing components, especially with network virtualization [2] emerging as promising technology for the future Internet.

While network security as a whole has received much recent attention (e.g., end-system vulnerabilities leading to botnets [9], worm propagation [14], etc.), there has been little focus on vulnerabilities in the networking infrastructure itself. Cui et al. [7] have surveyed vulnerabilities in the control plane of networks, where an attacker can potentially gain access to the router system. In the data plane, Chasaki et al. [5] have shown an example of how a simple integer overflow vulnerability can be exploited to launch a denial-of-service from within the network. In this case, a single malformed User Datagram Protocol (UDP) packet triggers the vulnerability, changes the network processor's operation, and causes a flood of attack packets to be sent by the system. We adapt this attack example to a processor system based on a Harvard architecture in our work to demonstrate the effectiveness of our monitoring system in detecting and stopping such data plane attacks in a practical networking environment.

Table 1: Comparison of Monitoring Approaches.

Monitor	granularity	implementation	underlying architecture
Chasaki et al. [5]	instruction	NFA	von Neumann
Arora et al. [3]	basic block	DFA	von Neumann
This paper	instruction	DFA	Harvard

Protection mechanisms for embedded processors have been proposed based on hardware monitors in general [1, 3, 5, 10, 13, 15, 16]. These monitors differ by the level of monitoring granularity (function calls, basic blocks, individual instructions) and if they require changes to source code or if they are based on program binaries. We only focus on approaches that do not require changes to the processor binaries. The main novelty of the monitoring system we present in this paper is highlighted in Table 1. In contrast to related work, we can monitor at the level of individual instructions and do so using a DFA, which can be implemented with high performance.

3. HARDWARE MONITOR SYSTEM ARCHITECTURE

The system architecture of the network processor system with security monitor is shown in Figure 1. The network processor shown on the left of the figure is based on a conventional Harvard architecture with separate data memory for network packets and processing state and instruction memory for packet processing code. For simplicity, only a single processor core is shown; the system can easily be extended for multiple processor cores. The processing monitor on the right side of the figure verifies the operation of the processor instruction-by-instruction. For every instruction that is executed on the processor core, a hash value of the executed operation is reported to the monitor. The monitor uses the comparison logic to compare the reported hash value to the information that is stored in the monitoring graph. The monitoring graph is derived by offline analysis of the packet processing code binary.

Any attack on the system necessarily needs to change the operation of the processor core (otherwise the attack is not effective). This deviation leads to the processor reporting hash values that do not match with the monitoring graph. The comparison logic can detect this deviation and reset the processor in response. In networking, such a reset and recovery operation is very simple: The current packet is dropped (i.e., the packet buffer is cleared), the processing state is reset (i.e., the stack is reset), and processing continues with the next packet. Since most packet processing operations are not stateful and there is no guarantee that packets are reliably delivered, no further recovery actions are necessary.

The monitoring graph used by the hardware monitor is a state machine, where each state represents a specific processor instruction. The state machine is derived from the packet processing code as illustrated in Figure 2. Each processor instruction corresponds to a state. The edges between states are labeled with information relating to next valid instruction that can be executed after the current instruction. In case of control flow operations, there may be multiple outgoing edges from each state (each being a valid transition). In our system, we use a 32-bit processor (i.e., open source embedded Plasma processor based on the MIPS instruction

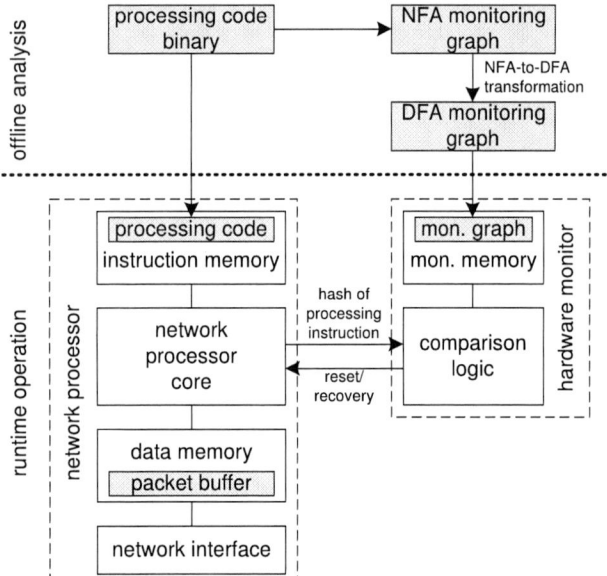

Figure 1: System architecture of network processor with security monitor.

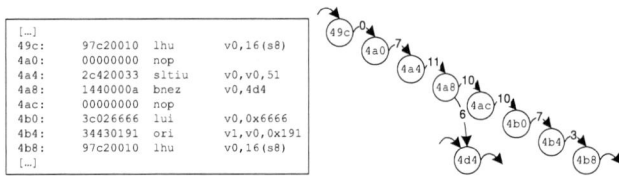

Figure 2: State machine generation from processing binary.

set). The monitoring system uses a 4-bit hash of the next instruction to label edges in the monitoring graph (as has been recommended in [13]). A hash (instead of the full 32-bit instruction) is used to reduce the size of the monitoring graph and thus to reduce the implementation overhead of the hardware monitor while still allowing instruction-by-instruction monitoring. The use of a hash (or any other method that uses a many-to-one mapping), however, leads to two fundamental problems:

- Attack detection ambiguity: The many-to-one mapping that occurs in a hash function of the monitor may make it possible for an attacker to remain undetected. This would require that the attack performs operations that lead to a sequence of hash values that matches the monitoring information of valid code. Mao et al. have shown that this probability decreases geometrically with the length of the attack code and thus is unlikely to lead to practical attacks [13] (in particular when the hash function is not known to the attacker). We do not consider this issue further in this paper.

- Nondeterminism during monitoring: The many-to-one mapping also leads to nondeterminism in the monitoring graph. There may be a control flow instruction where each of the next instructions has the same hash value. As a result, the corresponding node in the monitoring graph has two outgoing edges with the

same hash value (as illustrated in Figure 3). Since this nondeterminism can continue for multiple such control flow operations, it can lead to complex implementations [5], potentially slowing monitor performance.

In the following section, we show how we can address the latter problem by converting the nondeterministic monitoring graph into a deterministic monitoring graph, which is easier to use in high-performance implementations.

4. DETERMINISTIC PROCESSOR MONITORING

To realize a deterministic instruction-level monitor, we first convert the NFA monitoring graph described in the previous section to a DFA monitoring graph. We then describe how to implement a monitoring system that uses this DFA graph.

4.1 Construction of Deterministic Monitoring Graph

Tracking nondeterministic finite automata is difficult to implement in practice since the automaton can have multiple active states. This leads to high bandwidth requirements between the monitoring logic and the memory that maintains the NFA since next-state information for all active states has to be fetched in each iteration. When using a DFA, in contrast, only one state is active and implementation becomes much easier.

To convert an NFA to a DFA, a standard powerset construction algorithm can be used [11]. This algorithm computes all possible state sets in which the automaton can be situated (i.e., the powerset). Based on the powerset, a DFA is then constructed. Figure 4 shows the DFA that corresponds to the NFA shown in Figure 3. Note that state {3,5} represents the sets of states to where state 2 can branch when hash value c is observed.

One potential problem with NFA-to-DFA conversions is that the number of states in the DFA can grow exponentially over the number of states in the NFA. However, the monitoring NFAs constructed from binary code do not exhibit this pathological behavior. Our experiments indicate that this increase is small and does not lead to drastically larger state machines (see Section 7). Thus, this approach is effective for creating deterministic hardware monitors.

4.2 Implementation of Monitoring System

A key challenge in the implementation of our hardware monitoring system is how to represent the monitoring DFA in memory. The comparison logic needs to be able to retrieve the information about next state transitions for every instruction that it tracks. Thus, state transitions need to be implemented with no more than one memory access per instruction (to keep up with the network processor core) and be as compact as possible (to minimize the implementation overhead of the monitor).

The information that needs to be stored in the monitoring memory is illustrated on the left side of Figure 5. Each state represents an instruction and an outgoing transition edge from this state represents the hash value of the next expected instruction in the execution sequence. For example, state c has two next states, d and e, with hash values 11 and 3, respectively.

566

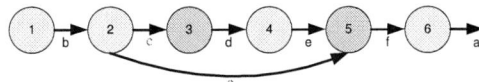

Figure 3: Nondeterministic monitoring graph.

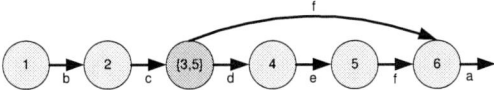

Figure 4: Deterministic monitoring graph after NFA-to-DFA conversion.

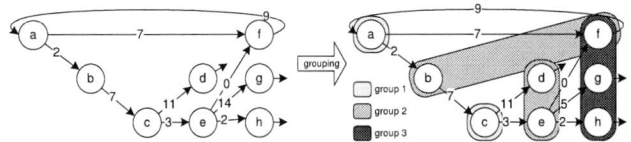

Figure 5: Grouping of DFA states.

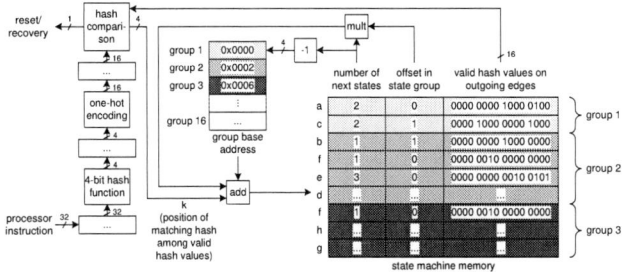

Figure 6: Memory representation of DFA monitoring graph.

A naïve way to store the state machine in RAM would be to store each state and all its possible edge transitions. This would require 2^h entries per state for an h-bit hash. Since most states have only one or two outgoing edges, a large number of edge transitions would never be used, leading to inefficient memory use. Assuming that only two outgoing transitions exist for each state is also not feasible due to the cases where powerset construction creates states with up to 2^h outgoing edges. Finally, for performance reasons we should only use one memory access per state transition, which precludes a design where states with more than two outgoing edges are handled as special cases.

Our main idea to compactly represent DFA states with varying numbers of outgoing edges is to encode all the necessary information in a single table entry and to group states by the number of outgoing edges. The main challenge in achieving compactness is to allocate exactly the amount of memory that is needed for each state to store next state information while still being able to index this memory without degrading to linear search. In our representation, we group states together if they have the same previous state. A state belongs to group g if the previous state has g outgoing edges. For a monitor with a 4-bit hash value, there are 16 possible groups. For example, in Figure 5 on the right side, groups are shown with different colors. Note that a state can belong to multiple groups (e.g., state f belongs to group 2 (because a has two outgoing edges, one to b and one to f) and to group 3 (because e has three outgoing edges)).

The memory layout and basic operation of our DFA monitor system is shown in Figure 6. The memory contains tuples of {number of next states, offset in state group, valid hash values on outgoing edges} and is logically divided into groups. The base addresses for each group are stored in a register file with 16 entries. Within a group, the sets of states that share the previous state are grouped together (e.g., b and f are together and d and e are together). Within a set, states are ordered by the hash value on their incoming edge (e.g., e before d because hash value 3 is smaller than hash value 11). The hash comparison block performs two functions: it determines if the one-hot coded hash bit is set in the 16-bit value read from memory and it determines k, which is the position of the matching hash value among the valid hash values read from memory.

To illustrate the operation of the monitor, we describe an example transition. Assume the monitor is in state a and the processor reports an instruction that leads to a hash value of 7. To perform the transition, the memory row labeled a is read. The tuple in this row indicates that there are two outgoing edges. The valid hash values of these two edges

are stored in the 16-bit vector. To verify that the transition is valid, the hash comparison unit checks if bit 7 is set in the bit vector (which it is). If this bit is not set, then an invalid transition takes place, indicating an attack, and the processor is reset. After the check, the next state (i.e., state f) in the DFA needs to be found in memory. To determine the address of that state, the base address of the group of the next state is looked up in the register file (i.e., 0x0002 since the next state belongs to group 2). To this base address, the product of the set size (i.e., group number) and the offset in the state group is added (to index the correct set within the group). Finally, k is added, which is the position of the matching hash in the bit vector (in our case 1 since 2 is the first matching hash (i.e., $k=0$) and 7 is the second matching hash (i.e., $k=1$)). Thus the memory location of state f is $0x0002 + 2 \times 0 + 1 = 0x003$.

Note that any state transition takes only one memory read from state machine memory and a lookup into a fixed-size register file. The DFA is represented compactly without wasting any memory slots (states shown with dots in Figure 6 point to other states not shown in our example). Thus, this representation lends itself to a high-performance implementation.

5. HARVARD ARCHITECTURE ATTACKS

Even though general memory error techniques (integer overflow, heap overflow etc.) cannot be used to generate code injection attacks, Francillion et al. [8] demonstrated that code injection attacks are still feasible on a Harvard architecture processor using a return-oriented programming technique. Here, an attacker takes control of return instructions in the stack to chain attack code from an existing library function. Since the code is already present in executable memory, the attack will not be prevented from running. In this section, we describe how such an attack can be constructed for the networking environment and how our monitor can detect it.

Figure 7 shows portions of congestion management protocol (CM) and a IPV4 packet forwarding application used to build an attack on the network processor system. The

567

```
int mybuf[60];
unsigned short sum;
Pack (in,out);
sum= len1 + len2;
if(sum > MAX_PKT_SIZE) {
return -1;}
else {
memset(mybuf, buf1, len1);
memset((mybuf+len1), buf1, len2);
return 0;
}
```
CM protocol

```
{
unsigned int d;
unsigned int port;
_u32 ip_dst;
ip_dst= ip_dst_hi + ip_dst_low;
port = (ip_dst & 0x000000ff);
port = (port << 16);
d = (d1 | port);
pkt_dbg(0x137,  d);
pkt_dbg(0x157,  port);
put_pkt(0,d);
```
IPV4 application

Figure 7: Vulnerable application code.

congestion management protocol inserts a custom protocol header in the packet header space between the IP header and the UDP header. During this operation, the code needs to make sure the new packet size does not exceed the maximum datagram length (the boxed instruction in the CM code). Exploiting an integer overflow vulnerability, the boundary check in the CM code can be circumvented and the stack can be smashed. To do so, an attacker sends a malformed UDP packet with a size 0xfffe (decimal value 65534), which will pass the maximum packet size check (since $65334 + 12 = 10$, due to integer overflow). As a result, the packet payload is copied over the stack. The packet payload of the attack packet is crafted in such a way that the return address is overwritten to direct the control flow to the IPv4 packet forwarding application (which is library code on the processor core) and the value of the *ip_dst_low* field is 0xff. The port information gets updated with this value (the boxed instruction in the IPv4 code), forwarding the attack packet to *all* the outgoing ports and then crashing the processor system. As a result, the attack packet gets forwarded to all outgoing interfaces before the system crashes, thus propagating the attack through the network.

Since our hardware monitor has no valid edge between the states in the middle of the CM application and the IPv4 application, this attack is detected. As soon as the control flow changes, the hash values reported by the processor no longer match the monitoring information and the system is reset, dropping the malicious packet.

6. PROTOTYPE SYSTEM IMPLEMENTATION

Although an end-system would likely be implemented in fixed logic, we have prototyped the described network processor and hardware monitoring system on a Stratix IV GX230 FPGA located on an Altera DE4 board. The router infrastructure surrounding the NP core is taken from the NetFPGA reference router, which has been migrated to the Stratix IV family. The DE4 board has four 1 Gbps Ethernet interfaces for packet input/output. In our prototype implementation, the single-core network processor is implemented as a soft core and the monitor is implemented in FPGA logic (using Quartus for synthesis, place and route). Only the memory initialization files need to be reconfigured on a per-application basis.

To run networking code on the processor plus monitor system, the code is first passed through a standard MIPS-GCC compiler flow to generate assembly-level instructions. The output of the compiler allows for the identification of branch instructions and their target addresses. In our current im-

Table 2: Evaluation of monitoring approaches for our new DFA approach and a previous NFA-only approach. The maximum number of memory accesses for our approach is 1 for all benchmarks.

| Netw. appli- cation | No. of instr. | Chasaki [5] | | Ours | | |
		NFA states	Max. mem. access	DFA states	Mem. entries	Mem. over- head
frag	573	573	3	592	627	9.4%
mtc	2427	2427	3	2460	2584	6.4%
red	802	802	2	808	857	6.8%
wfq	905	905	2	921	978	8.0%

plementation, all possible branch targets and return instructions are analyzed at compile time. The monitor can handle an arbitrary number of indirect branches to statically known targets (e.g., return addresses) since the NFA representation allows any number of outgoing branches. (Our monitor cannot handle indirect branches to statically unknown targets that are resolved at run time, but such programming constructs did not appear in any of 11 benchmark applications that we looked at.)

The NFA-to-DFA conversion starts with a nondeterministic NFA representation obtained from the compiler information. Through powerset construction, a DFA is constructed. This DFA is then converted into a memory initialization file using the process described in Section 4 and is loaded into the monitor when the processing binary is installed in the processor. To evaluate our system, four benchmarks from the NPbench suite [12] were processed with this flow.

7. EVALUATION RESULTS

7.1 Monitoring Graphs

The results of generating instruction-level monitoring graphs for both our approach and a previous approach [5] are illustrated in Table 2. The number of entries in the state machine memory (Figure 6) for each benchmark are shown in the *Mem. entries* column. A clear benefit of the new approach is speed. In all cases, only one access to the monitor memory is required for any benchmark (including the four shown here). The previous NFA-based approach requires up to three memory accesses for the benchmarks tested and potentially up to 16 for other benchmarks. The conversion from an NFA to a DFA does incur a memory overhead of 7.7% on average for the benchmarks.

Table 3: Resource Utilization

Resources	Secure monitor	Network proc.	DE4 interface	Available in FPGA
LUTs	140	3,792	37,803	182,400
FFs	26	2,120	38,444	182,400
Mem. bits	131,072	201,216	2,550,800	14,625,792

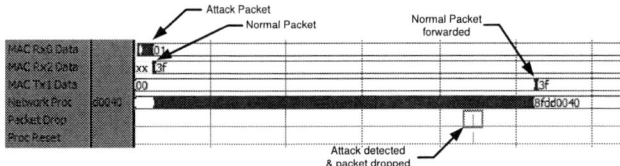

Figure 8: Simulation waveforms showing the identification of an attack packet and the successful forwarding of the subsequent packet. This behavior was confirmed using hardware.

7.2 Monitoring Speed and Effectiveness

Our network processor and monitoring system were successfully implemented on the DE4 platform. The lookup table (LUT), flip flop (FF), and memory resources required for the network processor core, monitor, and other interface circuitry for the router (e.g. buffers, input arbiter, queuing control, etc) are shown in Table 3. The NP memory includes space for up to 4096 monitor memory entries. All circuitry operated at 125 MHz, the same clock speed for the system without the monitor. Experiments in simulation and in the lab on FPGA hardware showed that the processor is able to forward packets ranging in size from 64 to 1500 bytes per packet at the same rate under monitoring as without monitoring (e.g. no slowdown for monitoring). For hardware experiments, packets were generated and transmitted to the DE4 with the NP and monitor at a 1 Gbps line rate by a separate DE4 card serving as a packet generator. This same card was used to receive the processed packets from the card with the NP.

In a final experiment, we tested the ability of the monitor-based system to detect and recover from an attack. The vulnerable application code shown in Figure 7 was implemented and used with the NP to send copies of a packet to all ports of the router and then crash the router. We confirmed this behavior for a system without a monitor both in simulation and in hardware. As shown in Figure 8, after the monitor was added to the system, the attack packet was successfully identified, the NP was reset, and subsequent regular packets were routed successfully. This behavior was verified using our DE4 hardware setup.

8. SUMMARY AND FUTURE WORK

The effective use of the Internet depends on reliable network routers that are impervious to attack. In this paper, we have described a high-performance monitor for a network processor that requires only a single memory lookup per network processor instruction. This single memory lookup is maintained regardless of the complexity of the NP program using an NFA-to-DFA translation of the monitoring graph. Our monitor, which tracks individual NP instructions, has been verified in hardware using an NP with a Harvard architecture. The presence of monitoring does not slow down NP operation since it is performed outside of the operational paths of the NP. In the future, we plan to evaluate our monitoring approach using a multi-core network processor.

Acknowledgements

This material is based upon work supported by the National Science Foundation under Grant No. 1115999. We grate-fully acknowledge Altera Corporation's donation of the DE4 boards and Quartus software.

9. REFERENCES

[1] ABADI, M., BUDIU, M., ERLINGSSON, Ú., AND LIGATTI, J. Control-flow integrity principles, implementations, and applications. In *ACM Conference on Computer and Communication Security (CCS)* (Alexandria, VA, Nov. 2005), pp. 340–353.

[2] ANDERSON, T., PETERSON, L., SHENKER, S., AND TURNER, J. Overcoming the Internet impasse through virtualization. *Computer 38*, 4 (Apr. 2005), 34–41.

[3] ARORA, D., RAVI, S., RAGHUNATHAN, A., AND JHA, N. K. Secure embedded processing through hardware-assisted run-time monitoring. In *Proc. of the Design, Automation and Test in Europe Conference and Exhibition (DATE'05)* (Munich, Germany, Mar. 2005), pp. 178–183.

[4] CAVIUM NETWORKS. *OCTEON Plus CN58XX 4 to 16-Core MIPS64-Based SoCs.* Mountain View, CA, 2008.

[5] CHASAKI, D., AND WOLF, T. Attacks and defenses in the data plane of networks. *IEEE Transactions on Dependable and Secure Computing 9*, 6 (Nov. 2012), 798–810.

[6] CISCO SYSTEMS, INC. *The Cisco QuantumFlow Processor: Cisco's Next Generation Network Processor.* San Jose, CA, Feb. 2008.

[7] CUI, A., SONG, Y., PRABHU, P. V., AND STOLFO, S. J. Brave new world: Pervasive insecurity of embedded network devices. In *Proc. of 12th International Symposium on Recent Advances in Intrusion Detection (RAID)* (Saint-Malo, France, Sept. 2009), vol. 5758 of *Lecture Notes in Computer Science*, pp. 378–380.

[8] FRANCILLON, A., AND CASTELLUCCIA, C. Code injection attacks on Harvard-architecture devices. In *Proc. of the 15th ACM Conference on Computer and Communications Security (CSS)* (Alexandria, VA, Oct. 2008), pp. 15–26.

[9] GEER, D. Malicious bots threaten network security. *Computer 38*, 1 (2005), 18–20.

[10] GOGNIAT, G., WOLF, T., BURLESON, W., DIGUET, J.-P., BOSSUET, L., AND VASLIN, R. Reconfigurable hardware for high-security/high-performance embedded systems: the SAFES perspective. *IEEE Transactions on Very Large Scale Integration (VLSI) Systems 16*, 2 (Feb. 2008), 144–155.

[11] HOPCROFT, J. E., AND ULLMAN, J. D. *Introduction to Automata Theory, Languages, and Computation.* Addison-Wesley, 1979.

[12] LEE, B. K., AND JOHN, L. K. NpBench: A benchmark suite for control plane and data plane applications for network processors. In *Proc. of IEEE International Conference on Computer Design (ICCD)* (San Jose, CA, Oct. 2003), pp. 226–233.

[13] MAO, S., AND WOLF, T. Hardware support for secure processing in embedded systems. *IEEE Transactions on Computers 59*, 6 (June 2010), 847–854.

[14] MOORE, D., SHANNON, C., AND BROWN, J. Code-Red: a case study on the spread and victims of an Internet worm. In *IMW '02: Proceedings of the 2nd ACM SIGCOMM Workshop on Internet measurement* (Marseille, France, Nov. 2002), pp. 273–284.

[15] RAGEL, R. G., AND PARAMESWARAN, S. IMPRES: integrated monitoring for processor reliability and security. In *Proc. of the 43rd Annual Conference on Design Automation (DAC)* (San Francisco, CA, USA, July 2006), pp. 502–505.

[16] ZAMBRENO, J., CHOUDHARY, A., SIMHA, R., NARAHARI, B., AND MEMON, N. SAFE-OPS: An approach to embedded software security. *Transactions on Embedded Computing Sys. 4*, 1 (Feb. 2005), 189–210.

Compiler-based Side Channel Vulnerability Analysis and Optimized Countermeasures Application

Giovanni Agosta, Alessandro Barenghi, Massimo Maggi, and Gerardo Pelosi
Dipartimento di Elettronica, Informazione e Bioingegneria – DEIB, Politecnico di Milano
Piazza Leonardo da Vinci, 32 – 20133 Milano, Italy
name.surname@polimi.it

ABSTRACT

Modern embedded systems manage sensitive data increasingly often through cryptographic primitives. In this context, side-channel attacks, such as power analysis, represent a concrete threat, regardless of the mathematical strength of a cipher. Evaluating the resistance against power analysis of cryptographic implementations and preventing it, are tasks usually ascribed to the expertise of the system designer. This paper introduces a new security-oriented data-flow analysis assessing the vulnerability level of a cipher with bit-level accuracy. A general and extensible compiler-based tool was implemented to assess the instruction resistance against power-based side-channels. The tool automatically instantiates the essential masking countermeasures, yielding a ×2.5 performance speedup w.r.t. protecting the entire code.

Categories and Subject Descriptors

C.3 [**Special-Purpose and Application Based Systems**]:
Microprocessor/microcomputer applications;
C.5.3[**Computer System Implementation**]:
Microcomputers[portable devices];

General Terms

Security

Keywords

Power Analysis Attacks, Software Countermeasures, Static Analysis

1. INTRODUCTION

Cryptographic primitives are becoming commonplace in embedded hardware security. At the same time, the growing threat of *Side-Channel Attacks* (SCAs) has lead to an increasing interest in designing SCA resistant implementations. SCAs take advantage of the fact that instantaneous power consumption, encryption time, results of artificially induced erroneous computations and/or electromagnetic emissions of an embedded device depend on the processed data and the performed operations. The effectiveness of these techniques in inferring the value of secret parameters, stored in a device in a non-accessible way, justifies the attention towards the implementation choices of HW/SW components as a critical dimension of embedded system design. For instance, the techniques proposed to perform power analysis attacks [8] and their countermeasures have been widely refined in terms of both precision and effectiveness [1]. The classic workflow for power-based SCAs aims at recovering the value of the secret parameter (i.e., the secret key) one portion at a time. This is possible since, during a cryptographic computation, the algorithm combines the secret key bits with the intermediate values involving a limited quantity of them at a time.

Permission to make digital or hard copies of all or part of this work for personal or classroom use is granted without fee provided that copies are not made or distributed for profit or commercial advantage and that copies bear this notice and the full citation on the first page. To copy otherwise, to republish, to post on servers or to redistribute to lists, requires prior specific permission and/or a fee.

To perform a power-based SCA, the first step is to measure the power consumption of the targeted device for a large number of computations with different input messages. Subsequently, an intermediate operation of the algorithm employing a small portion of the secret key is selected, and its results are guessed for all the possible values of the key portion. From these hypotheses on the results, a series of predictions (one for each possible value of the secret key portion) of the power consumption are made. Finally, the predicted consumption values are compared with the actual measured ones through the use of statistical means to find out which prediction fits best. Power-based SCAs affect both hardware and software implementations of cryptographic primitives. Many techniques have been designed to counter these attacks at logic style- or architectural- level, since energy consumption variations are strictly related to them [9]. However, as developing a dedicated hardware solution is expensive, software libraries are commonplace either alone, as a fall-back solution to hardware failures, or to provide complementary functionality with respect to those offered by the hardware platform. These secure software libraries are typically realized in low-level languages such as the target assembly or C language. Even Java-enabled smartcard processors actually expose a hardware support for the execution of Java bytecode, thus making the bytecode their own assembly language. In this work, we address the *protection of software implementations of symmetric ciphers against power analysis*. A very common countermeasure to protect implementations of these ciphers against SCA is to randomize the way sensitive variables are computed through *masking* techniques [6, 9]. The principle is to add one or more random values (*masks*) to every sensitive intermediate variable occurring during the computation. In a masked implementation, each sensitive intermediate value is represented as split in a number of *shares* (containing the randomized sensitive value and the masks employed), which are then separately processed. To this end, the target algorithm is modified to process each share and recombine them at the end of the computation. This technique effectively hinders the attacker from formulating a correct power consumption model, as the instantaneous power consumption is independent from the processed value. Typically, masking techniques are categorized by the number of masks d employed for each sensitive value, which is known as *order* of the masking. A d-th-order masking can always be theoretically broken by a $(d+1)$-th-order attack, i.e. an attack exploiting the combination of $d+1$ measurements of different instructions, during an execution, to build a mask-independent prediction [9, 13, 14]. In practice, the difficulty of carrying out a d-th-order attack increases exponentially with d, due to the difficulty of guessing which time instant is the one when the sensitive computations happen [3]. Even though a high order masking is crucial to ensure good security margins, only a few $d>1$ order masking schemes exist. They impose major performance overheads, and the masking is applied to each cipher with ad-hoc techniques [6, 9, 13, 14].

Contributions

As a major contribution, we provide the definition of a *security-oriented data flow analysis* (SDFA), with forward and backward variants, allowing a precise assessment of vulnerability on every instruction of a low-level intermediate language against power analysis attacks. As a second contribution, we employ the information provided by SDFA to automatically apply a masking protection to the vulnerable portions of the implementation of the target cryptographic primitive. We designed a set of compiler passes, implemented in the widely adopted LLVM compiler framework, parametric with respect to the number of masks to be applied, thus providing a trade-off between performance and protection against higher order attacks, and allowing the use of any language accepted by the LLVM frontend (among which, C). We exploit the fine grain of the information collected by the SDFA to significantly reduce the overhead imposed by the masking techniques, limiting their application only to the instruction of the algorithm which are computationally amenable to power analysis, for a desired security margin. The modular structure of the compiler passes allow the use of the SDFA analysis coupled with other countermeasure approaches. The new compiler passes work on the LLVM intermediate representation, a low-level (RISC-like) representation simple enough to be directly mapped to the actual assembly of most architectures, thus leveraging the compiler infrastructure without the loss of precision associated with high level representations.

Organization of the paper

Section 2 describes the related works. Section 3 summarizes the automated security analysis and countermeasures application as parts of a compiler workflow. Section 4 introduces our security-oriented data flow analysis and Section 5 describes the masking countermeasures and their deployment. Section 6 reports the experimental evaluation, while Section 7 draws our conclusions highlighting future directions.

2. RELATED WORK

The first work tackling the problem of automatically protect software implementations of cryptographic algorithm from power analysis attacks is presented by Bayrak *et al.* in [2]. The authors identify the instructions which are most vulnerable to power analysis running their target implementation and profiling the power consumption of the underlying platform. In this way, they identify the most vulnerable clock cycles of the program execution and associate to each of them the corresponding assembly instruction together with a "sensitivity value". Instructions, whose sensitivity is greater than a chosen threshold are replaced by an appropriate code snippet, which realizes a random pre-charging of either the registers or the memory cells. Note that even if the proposed workflow is general enough, the implemented code transformation step is specific for devices whose power consumption is proportional to the Hamming Distance between two consecutive execution cycles. In this paper, we put forward a novel and fully automated analysis to identify vulnerable instructions. Indeed, the presented technique does not need any profiling information about the power consumption of the target device while it is running the implementation of a cryptographic primitive. The main distinctive feature of our methodology is that it is not affected by completeness and/or accuracy problems bound to the collection of the power measurements, and can be applied even when a development board or cycle accurate simulator is not available, such as in the early stages of development of a processor. More recently, Moss *et al.* [10] proposed a first attempt at automating the process of inserting a 1st order masking scheme in the code of AES using an *ad-hoc* translator. Their scheme relies primarily on *type inference*, a kind of static analysis which is strongly dependent on the source language. To this end, the authors of [10] designed their own Domain Specific Language (DSL) with a specialized type system, which allows type inference. In practice the DSL source code must contain an explicit annotation for variables to be protected (depending on the programmer choices), as there is no automatic evaluation of the security margin bound to each instruction of the program to be executed. In addition, it is worth noting that most encryption primitives, especially for application in embedded systems, are not developed with a DSL, being instead available primarily in C.

3. SECURITY-ORIENTED COMPILER

A modern compiler is a pipeline of analysis and transformation passes processing the program code and transforming its source representation to the desired target representation [11]. The very first and last passes respectively take as input and produce as output the source and target languages. For the representation of all other inputs and outputs, one or more *intermediate representations* (IRs) are used. Analysis passes

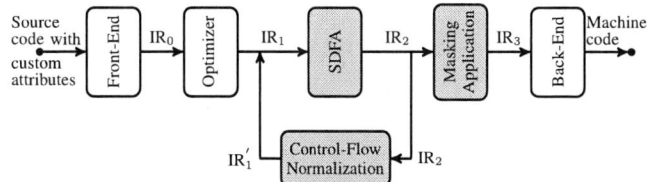

Figure 1: Security-oriented compiler pipeline

compute information from the input IR, and enrich the code through adding either metadata or annotations. Transformation passes, on the other hand, use the information contained in the annotations to modify and optimize the code. IRs have been widely studied in the field of compiler construction to balance the need to keep each analysis or transformation pass as easy as possible, as well as limit the total number of different IRs. A very popular class of IRs is the *Static Single Assignment* (SSA) form, defined as a representation where every variable is defined (i.e., assigned a value) only once [11]. It is important to note that IRs allow also to work at a very low level of abstraction, especially when RISC architectures are involved, as the Back-End passes do not generate vulnerable machine code instructions from the protected IR. This is because register allocation can, at most, spill values to memory, but all sensitive values are already protected, and the instruction selection does not use multiple machine instructions to map a single IR instruction. Figure 1 shows the modified pipeline of our security-aware compiler. The source-language code is extended with custom attributes to allow the developer to provide information to the compiler about the variables containing the "key material" and the "plaintext data" (f.i., the GNU extension mechanism for the C language). The decorated source program is parsed by the *Front-End* to produce the IR, which is optimized using standard optimization passes (f.i., the -O2 option of either clang or gcc compiler). The optimized IR is analyzed by a new *Security-oriented Data-Flow Analysis* (SDFA) pass, which adds metadata to each defined variable to identify its level of vulnerability. The SDFA can also identify control flow issues that would prevent a precise analysis and protection of the code. In such a case, the analysis pass invokes a *Control Flow Normalization* pass to transform both loop and conditional structures and to produce a "normalized" version of the input IR, suitable for the computation of vulnerability information in a subsequent execution of the SDFA pass. The vulnerability information is then used by the *Masking Application* pass, which modifies the normalized IR code through applying masking countermeasures where appropriate. The output IR is then translated to the target assembly by the standard *Back-End* pass. The next two sections detail the techniques implemented by the SDFA, Control-Flow Normalization and Masking Application passes.

4. VULNERABLE CODE DETECTION

After lowering the high-level code into an architecture agnostic intermediate representation (IR), it is possible to represent the program as a *control-flow graph* [11] defined as follows:

DEFINITION 4.1 (CONTROL-FLOW GRAPH). *A Control-Flow Graph (CFG) is a directed graph $\mathcal{G}(\mathcal{B}, \mathcal{E})$ with one node $i \in \mathcal{B}$ for each statement of the program ($stat_i$), augmented with two additional nodes i_{in}, i_{out}. An edge $(i, i') \in \mathcal{E}$ is added if the statement $stat_{i'}$ is executed immediately after the statement $stat_i$, and each node has at most two immediate successors. For the first statement ($stat_0$) there is an edge (i_{in}, i_0), while an edge (j, i_{out}) is added for each node j bound to a statement ($stat_j$) preceding an exit point of the program.*

In modern production-grade compilers (e.g., LLVM, GCC 4.x), the IR is emitted in "static single assignment" (SSA) form [11]. Indeed, for each non-jump instruction of the IR a fresh virtual register will be used to store the result, thus effectively yielding a one-to-one correspondence between the computed variable values and IR instructions.

To define a *security-oriented data-flow analysis*, aimed at detecting the amount of key material involved in the computation of an intermediate value, we will consider a CFG built from an IR in SSA form. The goal of this analysis is to identify a set of nodes of the CFG representing the portions of the program amenable to passive side-channel attacks. The choice of building a single-instruction-per-node CFG is justified by the fact that the application of countermeasures implies a significant performance penalty and should be done sparingly. An instruction is deemed to be vulnerable if computing a model of its behavior for each possible value of the key bits from which its output value depends is computationally feasible. Computing the aforementioned model is the ground on which passive side-channel attacks are built, as its predictions are matched against the measured behavior of the considered device. To trace which computations are affected by some key bits in the program, we need to introduce some terminology related to *data-flow analysis* (DFA). DFA aims at gathering information about the possible set of values calculated at each statement of a program, employing the CFG to determine the propagation paths of each computed value [7, 11]. In our case, the information to be traced is the data-dependence between any bit computed by a program instruction and every bit of the cryptographic key. Such a choice is mandated by the need to consider possible side-channel attack models predicting the behavior of the computation of a single bit within a w-wide value [8]. The aforementioned relation is modeled through an n-bit Boolean lattice $(\mathbb{BV}^n, \sqcup, \sqcap)$, where the elements of the support set

$$\mathbb{BV}^n = \{v_0, \dots, v_{2^n-1}\} = \{\langle 00 \dots 0 \rangle, \dots, \langle 11 \dots 1 \rangle\}$$

represent all the possible combinations of key bits from which a bit of an intermediate result depends on, thus n equals the key size of the cipher under exam. The bottom of the lattice $\perp$ is represented by the element $\langle 0 \dots 0 \rangle$, which indicates that no key bits are involved, while the top $\top$ element is $\langle 1 \dots 1 \rangle$, denoting that all the key bits are involved. The characteristic partial order relations $\succeq$ and $\preceq$ over the lattice elements are defined as follows:

$$v \succeq v' \Leftrightarrow \exists\, v'' \,|\, v' \sqcup v'' = v$$
$$v \preceq v' \Leftrightarrow \exists\, v'' \,|\, v' \sqcap v'' = v \qquad v, v', v'' \in \mathbb{BV}^n$$

with the $\sqcup$ operation being defined as the common bitwise inclusive-*or*, and the $\sqcap$ operator being the bitwise *and*.

Our SDFA computes how many key bits are involved in the computation of each intermediate value, i.e., due to the SSA nature of the IR from which the graph is obtained, how many bits are involved in the output of each bit composing the outcome of any instruction I of the program. To this end, we compute the key propagation for every bit of any size(I)-bit wide intermediate result through associating a *leakage vector* $V_I = (v_{\text{size}(I)-1}, \dots, v_t, \dots, v_0)$ of size(I) elements $v_t \in \mathbb{BV}^n$ to each node of the CFG, which represents a single SSA instruction. Each v_t represents the key bits involved in the computation of the t-th bit of the corresponding intermediate value output by I. We define the "meet" and "join" operations on the leakage vectors (denoted as $\vee$ and $\wedge$, respectively), as the extensions of the aforementioned $\sqcup$ and $\sqcap$. Given two leakage vectors $V_I = (v_{s-1}, \dots, v_0)$ and $V_J = (v'_{s-1}, \dots, v'_0)$ of equal size $s = \text{size}(I) = \text{size}(J)$, we define the "meet" composition law between $V_I, V_J \in (\mathbb{BV}^n)^s$ as $V_I \vee V_J = (v_{s-1} \sqcup v'_{s-1}, \dots, v_0 \sqcup v'_0)$. Dually, we define the "join" composition law as: $V_I \wedge V_J = (v_{s-1} \sqcap v'_{s-1}, \dots, v_0 \sqcap v'_0)$.

Given an instruction, its resistance to passive SCA is formally defined as follows:

DEFINITION 4.2 (INSTRUCTION RESISTANCE). *Consider an IR instruction* I *with a* size(I)-*bit output value, and the associated leakage vector* $V_I = (v_{\text{size}(I)-1}, \dots, v_t, \dots, v_0) \in (\mathbb{BV}^n)^{\text{size}(I)}$. *Denoting the Hamming weight of a bit-vector* $v_t \in V_I$ *as* $\text{HW}(v_t)$, *the instruction resistance is defined as:* $\min_{v_t \in V_I : v_t \neq \perp} \{\text{HW}(v_t), +\infty\}$, *that is, the minimum number of key bits influencing a bit of the output value of a sensitive* I. *An instruction that does not depend on any key bit is conventionally associated to a resistance value equal to* ∞.

To automatically evaluate the resistance of an instruction I, it is necessary to consider the leakage vector associated to each instruction preceding it and take into account which definitions are used by it. This information is captured by the notion of *In-Set* of the instruction. The propagation of resistance information through the specific transformation operated by I is captured by the notion of *Out-Set* of the instruction.

DEFINITION 4.3 (IN-SET). *Given an instruction* I, *the input set* $in(I)$ *is defined as the set of the leakage vectors associated to all the immediate predecessors of* I *on the CFG* $\mathcal{G}(\mathcal{B}, \mathcal{E})$:

$$in(I) \stackrel{\text{def}}{=} \left\{ V_J \mid J \in \mathcal{B}, J \in pred(I), V_J \in (\mathbb{BV}^n)^{\text{size}(J)} \right\}$$

DEFINITION 4.4 (OUT-SET). *Given an instruction* I, *the output set* $out(I)$ *is defined as the set of the leakage vectors associated to every immediate predecessor of* I *on the CFG* $\mathcal{G}(\mathcal{B}, \mathcal{E})$ *plus the one of* I, $V_I \in (\mathbb{BV}^n)^{\text{size}(I)}$:

$$out(I) \stackrel{\text{def}}{=} \{V_I\} \cup \left\{ V_J \mid J \in \mathcal{B}, J \in pred(I), V_J \in (\mathbb{BV}^n)^{\text{size}(J)} \right\}$$

The overall procedure to enact the SDFA is split into two phases: a local analysis and a global one [7]. A local DFA is limited to maximal instruction sequences without jumps (basic blocks) in the control flow. The global DFA works across different basic blocks on the whole CFG.

4.1 Local Security-oriented DFA

DEFINITION 4.5 (LOCAL SECURITY-ORIENTED DFA).
Each instruction I *within a basic block is characterized by an opcode,* $op(I)$, *an In-set:* $in(I)$, *and an Out-Set:* $out(I)$. *The effect of the execution of* I *is modelled through a transformation function* $\mathcal{F}_{op(I)}(\cdot)$ *taking as input its In-set. Therefore, for any instruction* I *the following equations can be stated:*

$$in(I) = \begin{cases} \emptyset, & \text{if } pred(I) = \emptyset \\ out(J), & \text{if } pred(I) = \{J\} \end{cases}$$
$$out(I) = \mathcal{F}_{op(I)}(in(I))$$

The SDFA solves the set of simultaneous equations derived from the instructions in the basic block through subsequent approximations until a fixed-point is reached. In the particular case of the local SDFA, the convergence is achieved in a single step. The behavior of the transformation function depends on the opcode of the instruction, as the propagation of the key dependencies depends on its nature. Note that, the semantics of each instruction determine also the bit-size (size(I)) of its output value. Thus, to compute the corresponding Out-Set it may be necessary to produce a properly sized leakage vector. We denote as $\text{RESIZE}_I(V_J)$ the adaptation of the leakage vector V_J to the same size of the instruction I as follows:

$$\text{RESIZE}_I(V_J) = \begin{cases} (v_{\text{size}(I)-1}, \dots, v_t, \dots, v_0), \\ \quad v_t \in V_J, \quad \text{if size}(I) \leq \text{size}(J); \\ (\perp_{\text{size}(I)-\text{size}(J)-1}, \dots, \perp_{\text{size}(J)}, \dots, v_t, \dots), \\ \quad v_t \in V_J, \quad \text{if size}(I) > \text{size}(J) \end{cases}$$

Let $\text{OPERANDS}(I)$ be the set variables used by I as operands. As the IR is in SSA form, each variable is defined only once, so, with a small notation abuse, we will use the form $J \in \text{OPERANDS}(I)$ to denote that instruction J defines one of the operands of I. For each instruction class of the IR, our analysis assumes the transformation function to be defined as:

$$out(I) = \mathcal{F}_{op(I)}(in(I)) \stackrel{\text{def}}{=} in(I) \cup \{V_I\}$$

where the leakage vector of the current instruction V_I is computed according the formulae presented hereafter.

Arithmetic, Bitwise-logic and `cmp` instructions
These instructions can be partitioned in two sets depending on the computation of their leakage vectors.

The first set includes all instructions with an opcode, $op(I)$, specifying a bitwise operation (f.i., `not`, `and`, `or`, `xor`), an `add` or `sub` operation, with the exception of the `and` and `or` with an immediate operand, as well as `shift`, `zero-` and `sign-extension`. The evaluation of the leakage vector bound either to an `add` or to a `sub` operation is done through considering them as a `xor` operation. This assumption neglects the influence of the carry/borrow propagation in the computation of the result. This is justified by the fact that the most favorable situation for an attacker is when there is no carry propagation (i.e., when the influence of the key bits on each bit of the final outcome is minimized). The computation of the leakage vector of any of the aforementioned instructions is the composition of the leakage vectors in their In-Sets, so that the output bit dependencies (from the key bits) are the ones of the corresponding bits of the input operands added together: $V_\text{I} = \bigvee_{\text{J} \in \text{OPERANDS}(\text{I})} \text{RESIZE}_\text{I}(V_\text{J})$. The second set of instructions includes `mul`, `div`, `mod`, and `cmp` operations. Multiplication, division and modulo operations diffuse the information contained in the operand bits, so that every bit of the output depends on every bit of the inputs. Let I be any of these instructions, and let $\text{J} \in \text{OPERANDS}(\text{I})$ be the instructions computing the operand values of I, with $V_\text{J} = (v_{\text{size}(\text{J})-1}, \ldots, v_t, \ldots, v_0)$ being the corresponding leakage vectors. The leakage vector of the instruction result is computed so that, for each of its bits, the dependencies (from the key bits) of all operands bits are added together: $V_\text{I} = \bigvee_{\text{J} \in \text{OPERANDS}(\text{I})} \text{RESIZE}_\text{I}(\widehat{V}_\text{J})$, where $\widehat{V}_\text{J} = (\widehat{v}_{\text{size}(\text{J})-1}, \ldots, \widehat{v}_t, \ldots, \widehat{v}_0), \forall t \mid \widehat{v}_t = \bigsqcup_{0 \leq t < \text{size}(\text{J})} v_t$. Note that, when considering a `cmp` instruction the outcome computed by the instruction is reduced to a single bit.

Bitwise `and` and `or` instructions with an immediate operand
Denoting as imm_i the i-th bit ($0 \leq i < \text{size}(imm)$) of the immediate operand, we need to define a support leakage vector V_{imm}:

$$V_{imm} = \begin{cases} (\ldots, \langle imm_i, \ldots, imm_i \rangle, \ldots), \\ \quad \text{with } 0 \leq i < \text{size}(imm), \text{ if } op(\text{I}) = \text{and} \\ (\ldots, \langle \neg imm_i, \ldots, \neg imm_i \rangle, \ldots), \\ \quad \text{with } 0 \leq i < \text{size}(imm), \text{ if } op(\text{I}) = \text{or} \end{cases}$$

to model the dependency-cancelling effect of the absorbing elements of bitwise `or` and `and` operations (1 and 0, respectively) on the input leakage vector V_J. The output leakage vector V_I is thus obtained removing the cancelled dependencies from the input ones as follows:

$$V_\text{I} = \text{RESIZE}_\text{I}(V_{imm}) \wedge \bigvee_{\text{J} \in \text{OPERANDS}(\text{I})} \text{RESIZE}_\text{I}(V_\text{J})$$

`shift` instructions with an immediate operand
Let $\text{J} \in \text{OPERANDS}(\text{I})$ be the instruction producing the non-immediate operand of I, and $V_\text{J} = (v_{\text{size}(\text{J})-1}, \ldots, v_0)$ the corresponding leakage vector. The leakage vector associated to I is $V_\text{I} = \text{RESIZE}_\text{I}(\widehat{V})$:

$$\widehat{V} = \begin{cases} (v_{\text{size}(\text{J})-1-imm}, \ldots, v_0, \bot_{imm}, \ldots, \bot_0), \\ \quad \text{if } op(\text{I}) = \text{shl}, \text{ashl} \\ (\bot_{\text{size}(\text{J})-1}, \ldots, \bot_{\text{size}(\text{J})-1-imm}, v_{\text{size}(\text{J})-1}, \ldots, v_{imm}), \\ \quad \text{if } op(\text{I}) = \text{shr}, \text{ashr} \end{cases}$$

The computation of $\widehat{V}$ takes into account the fact that the bits of the output are a permutation of the input ones, possibly discarding some.

Data-dependent `shift` instructions
In this case the non-immediate operands imply considering the outcome of the instruction as an unpredictable result. The corresponding leakage vector is conservatively estimated through removing every dependence from the key bits: $V_\text{I} = (\bot_{\text{size}(\text{I})-1}, \ldots, \bot_0)$.

`store` instruction
`store` operations do not produce any new value. Thus, the following equation applies: $out(\text{I}) = in(\text{I})$ as there is no leakage vector.

`load` instruction
The operands of the `load` instruction can compute an address value that possibly depends on the key bits. Thus, every output bit is considered as dependent on every bit of the address.

Given $\text{J} \in \text{OPERANDS}(\text{load})$, with $V_\text{J} = (v_{\text{size}(\text{J})-1}, \ldots, v_0)$, the information leakage is: $V_{\text{load}} = \bigvee_{\text{J} \in \text{OPERANDS}(\text{load})} \text{RESIZE}_\text{I}(\widehat{V}_\text{J})$, where

$\widehat{V}_\text{J} = (\widehat{v}_{\text{size}(\text{J})-1}, \ldots, \widehat{v}_t, \ldots, \widehat{v}_0), \forall t \mid \widehat{v}_t = \bigsqcup_{0 \leq t < \text{size}(\text{J})} v_t$. If the address depends on the key and the loaded value also contain some key material the above leakage vector is used as a conservative approximation of the actual one.

`zero-` and `sign-extension` instructions
These two instructions are usually employed in an IR when a change of data type occurs. In this respect, each of them can be managed as an instruction, I, with a non-immediate operand that must be extended up to known size. Let $\text{J} \in \text{OPERANDS}(\text{I})$ be the instruction producing the non-immediate operand of I, and $V_\text{J} = (v_{\text{size}(\text{J})-1}, \ldots, v_0)$ the corresponding leakage vector. The data associated to the instruction are simply computed as:

$$V_\text{I} = \begin{cases} (\bot_{\text{size}(\text{I})-1}, \ldots, \bot_{\text{size}(\text{J})}, v_{\text{size}(\text{J})-1}, \ldots, v_0), \\ \quad \text{if } op(\text{I}) = \text{zero-extension} \\ (v_{\text{size}(\text{I})-1}, \ldots, v_t, \ldots, v_{\text{size}(\text{J})}, v_{\text{size}(\text{J})-1}, \ldots, v_0), \\ \quad \text{with } v_t = v_{\text{size}(\text{J})-1}, \text{ where size}(\text{J}) \leq t < \text{size}(\text{I}), \\ \quad \text{if } op(\text{I}) = \text{sign-extension} \end{cases}$$

4.2 Global Security-oriented DFA and Control Flow Normalization

Given the local SDFA, it is possible to construct a global SDFA through extending the data-flow equations to cover the case where an instruction I has multiple immediate predecessors (i.e., $|pred(\text{I})| > 1$). To this end, we will define the relation between the In-set $in(\text{I})$ of each instruction with multiple predecessors and its Out-set, combining the contribution of the $pred(\text{I})$ through the so-called *confluence* operator. In data-flow analysis techniques, the *confluence* operator is employed to obtain a conservative information regarding the data-flow, as it is not possible to fully predict which value among the ones present in the out-sets of $pred(\text{I})$ will be employed by I. This is obtained through preserving only the data-flow information common to all the incoming execution paths, that is, applying a so-called "meet-over-all-paths" policy. In our context, we derive the information associated to the output of an instruction with multiple predecessors through combining them with the "meet" operation on the leakage vectors. More formally our global SDFA is defined as follows.

DEFINITION 4.6 (GLOBAL SECURITY-ORIENTED DFA).
Let $\mathcal{G}(\mathcal{B}, \mathcal{E})$ be a control-flow graph and let $\text{I} \in \mathcal{B}$ be an instruction with $|pred(\text{I})| \geq 1$, then the equations defining its In-set and Out-set are given as follows:

$$in(\text{I}) = \begin{cases} \emptyset, & \text{if } pred(\text{I}) = \emptyset \\ \bigcup_{\text{H} \in \mathcal{B}} \left\{ \bigwedge_{V_\text{H} \in out(\text{J}), \text{J} \in pred(\text{I})} V_\text{H} \right\}, & \text{otherwise} \end{cases}$$

$$out(\text{I}) = \mathcal{F}_{op(\text{I})}(in(\text{I}))$$

This method would lead to consider a bit of the output of I as protected by a key bit, if-and-only-if all the corresponding bits of the outputs of its predecessors are. Although this information is a safe conservative estimate of the instruction resistance, significant enhancements can be obtained through applying some transformations to the IR. We collect such transformations into a Control Flow Normalization pass. Cryptographic algorithms are designed so that the key material diffuses over the cipher state in an increasing fashion at each iteration of the loop (i.e. at each cipher round). Thus, if we detect that the application of the confluence operator ($\wedge$) at the beginning of a loop construct is causing information loss (f.i. when only one predecessor of I depends on the key bits), we apply a *loop peeling* transformation to the CFG and restart the analysis. The loop peeling extracts one iteration from the loop block, thus effectively procrastinating the confluence in the control flow and, possibly, allowing the computation of a lossless confluence. This procedure allows to recognize the minimal number of loop iterations to be unrolled. We note that if, at the end of the Global SDFA we detect that the key dependencies do not reach all the live output values (i.e., in this case, the ciphertext), we can automatically point out a cipher design flaw, as this implies that the key is not combined with the whole cipher state. Another control flow confluence results from conditional statements. In this case we apply a Control Flow Normalization pass, employing a variant of the standard *if-conversion* [11]. If-conversion, to produce efficient code, can only be applied to branches that have a small number of instructions, which is the case of cryptographic algorithms. The common if-conversion transforms a conditional statement either by means of predicated instructions or through computing both branches and employing a `select` instruction to pick the result. We want to transform the construct into a condition free sequence of bitwise logic instructions. For instance, consider the following conditional expression in C language: `res=(r<0) ? r^c : r;`. Denoting as `%r` and `%c` the virtual registers for the variable `r` and `c`, the previous C statement is translated in SSA IR form as shown in the left-pane of the following code snippets:

```
%1=icmp slt i8 %r,0        %1=ashr i8 %r,7
%2=xor i8 %r,i8 %c         %2=xor i8 %r,i8 %c
%res=select i1 %1,i8 %2,i8 %r    %res=xor i8 %res,%r
```

Note that the comparison with the zero value (checking if `%r` is negative) is substituted with the arithmetic right shift to recover the sign bit of the variable `r`.

4.3 Backward Security-oriented DFA

As it is typical to attack cryptographic algorithms through predicting an intermediate value of the computation preceding the output (the known ciphertext), we will now provide the description of a Backward SDFA to identify the resistance of instructions in this case. Thus, combining the results of the Forward- and Backward-SDFA we will be able to consider all the possible known values the attacker can exploit.

Backward DFAs are constructed similarly to the forward ones, through reversing the relation between In-set and Out-set. However, our SDFA is tailored to the approach used by the attacker to find out portions of the cryptographic key. The general form of the Backward SDFA equations is as follows:

$$
\begin{aligned}
out(\text{I}) &= \begin{cases} \emptyset, & \text{if } succ(\text{I})=\emptyset \\ \displaystyle\bigcup_{H\in\mathcal{B}} \left\{ \bigwedge_{V_H\in in(J),\, J\in succ(\text{I})} V_H \right\}, & \text{otherwise} \end{cases} \\
in(\text{I}) &= \mathcal{R}_{op(\text{I})}(out(\text{I}))
\end{aligned}
$$

For the arithmetic-logic instructions, the definition of the transformation function $\mathcal{R}_{op(\text{I})}$ is given by: $in(\text{I})=\mathcal{R}_{op(\text{I})}(out(\text{I})) \stackrel{\text{def}}{=} out(\text{I}) \cup \{V_\text{I}\}$, where $V_\text{I}=\bigvee_{J\in\text{USE}(\text{I})} \text{RESIZE}_\text{I}(V_J)$. The main difference from the forward SDFA is the case where one of the operands of I contains some key material. Accordingly, the transformation function is modified as follows:

$$
in(\text{I}) = \mathcal{R}_{op(\text{I})}(out(\text{I})) \stackrel{\text{def}}{=} out(\text{I}) \cup \{V_\text{I} \vee \text{RESIZE}_\text{I}(V_\text{K})\}
$$

where $V_\text{K}=(v_{\text{SIZE}(\text{K})-1},\ldots,v_t,\ldots,v_0)$ is the contribution of each bit of the key (v_t denotes a bit vector that has a single bit set at position t from the least to the most significant). This modification takes into account the fact that an attacker will need to make an hypothesis on the whole key material involved either directly (i.e. as an operand) or indirectly (i.e. in the computation of values depending on the result) with the instruction under exam. Thus, the Backward SDFA has one forward path (the one regarding the use of the key), an uncommon, but not unheard of case for DFA [7]. As the Backward SDFA aims at finding the dependence relations considering the point of view of an attacker in possess of the outputs of the algorithm, the relations among the key bits imposed by the key schedule should be considered in reverse order. To this end, the analysis considers the last values produced by the key schedule as the actual "initial" key and derives the relations with the rest of the key material backwards. The key schedule is identified as the set of instructions that employ, directly or indirectly, only key material and no plaintext. Then, a visit of the whole cipher CFG, starting from the instructions computing the output, is performed to identify instructions that use the key schedule values. The first key

Table 1: Complexity of bitwise masked operations as a function of the masking order d and lookup table size l

Operation	Complexity of masked operation	Ref.
`xor`	$3(d+1)$ `xor`	[6, 13]
`not`	1 `not`	[6]
`and`	$2d(d+1)$ `xor` $+(d+1)^2$ `and`	[6]
`or`	$2d(d+1)$ `xor` $+(d+1)^2$ `and` $+3$ `not`	$a \vee b =$ $\neg((\neg a) \wedge (\neg b))$
table lookup	$2ld$ `xor` $+ld$ `store` $+(ld+1)$ `load`	[14]

schedule values to be found are marked as the "initial" key. The amount of these values which should be considered as "initial", lest the whole key schedule be marked, is bound by their total size being at least the size of the user key, and by them being dependent on all the user key bits. This last condition can be easily checked as a full key schedule forward DFA has been performed at the beginning of this "initial" key identification step. After these "initial" key values are identified, the Backward SDFA computes the dependence from them as imposed by the data-flow equations.

5. MASKING COUNTERMEASURES

Masking aims at invalidating the link between the predicted power consumption, associated to an intermediate operation, and the corresponding power measurement. In a masked implementation, each sensitive intermediate value is concealed through splitting it in a number $d+1$ of shares, which are separately processed [6, 9]. The unprotected computation is substituted by three phases: an initial share-splitting, a transformation of the original computation into one processing all the $d+1$ shares and a final recombination, which must yield the same result as the unprotected computation. Table 1 shows the computational costs to mask bitwise operations as a function of the scheme order d. For multi-bit arithmetic operations it is possible to perform conversions between Boolean masked values and arithmetic masked ones and viceversa [5]. The state-of-the-art masking scheme for generic Boolean functions is proven to be secure up to a d-th order attack [6] for any choice of $d\geq 1$. In case the Boolean function is available in the form of a lookup table, the masking of the looked-up values is safe up to the 2nd order according to [4]. The key idea is that, whenever two share-split operands are combined together, fresh random values should be inserted in the computation of the resulting output shares. As the masking countermeasure is particularly computationally demanding (see Table 1) applying it as sparingly as possible, without lowering the security margin of the cipher, is paramount.

Automated Masking Application

We applied the previous masking schemes through our mask application compiler pass, employing the information provided by the SDFA to select which instructions should be protected. Concerning Forward SDFA, we deem an instruction at risk if both its resistance as per Definition 4.2 is lower than a user-selectable threshold, and the computation of its operands depend, directly or indirectly, from the definition of the plaintext. Analogously, in case a Backward SDFA is performed, the instruction is deemed at risk if both the computation of the ciphertext depends on the definition of its output and its resistance is below the aforementioned threshold. The masking application pass (as in Figure 1) visits the CFG in order, and, for each instruction, acts as follows. In case the instruction output needs to be protected, it checks whether the instruction operands are available in masked form or not. In the former case, it simply emits the masked operation corresponding to the unprotected one under exam. In the latter case, it first emits the code required to split the instruction operands into $d+1$ shares, followed by the masked operation code. In case the output of the instruction being visited does not need to be protected, the compiler pass checks whether its operands are masked or not. In case the operands are masked, the pass inserts the share recombination step to obtain an unmasked value. Subsequently, all the uses of the output of the instruction under exam are corrected so that they employ the result of the share recombination.

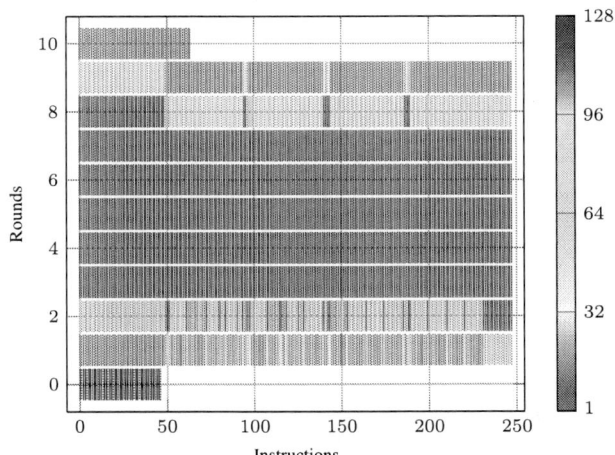

Figure 2: Resistance of sensitive instructions of the AES-128 encryption primitive (`ShiftRow`, `SubByte` as a tabulated S-box, computational `MixColumn`, `AddRoundKey`) in number of key bits protecting the weakest bit of the corresponding outcome. The Key-schedule part is not shown as its instructions are not sensitive

In case the operands of the instruction are already unprotected, the pass continues the visit of the CFG.

6. EXPERIMENTAL EVALUATION

To validate our approach we implemented the SDFA analysis and masking application compiler passes in the current trunk version of the LLVM codebase (Release Candidate in November 2012). We employed two C implementations of the AES cipher, one using the NIST standard 8-bit S-box and the other one with an equivalent optimized computational version. The C code is lowered into the LLVM IR and all the passes for the -O3 grade optimizations are performed before our SDFA and Masking Application passes are applied. Figure 2 reports the results of running the forward SDFA analysis on AES-128 with tabulated S-box: the initial ADDROUNDKEY (round 0) and 10 rounds of the algorithm are depicted starting from bottom up, with the x-axis representing number of instructions for each of them. The results of the same analysis on the computational S-box variant are analogous. The figure shows that, after the end of the second round, no intermediate values need to be protected regardless of the chosen security level as the whole 128 bits of the key influence all the outputs of the operations, thus rendering any SCA as hard as an exhaustive search of the keyspace. Analogous results are obtained from the Backward SDFA, consequentially we can avoid the computationally taxing application of masking techniques to a large part of the AES cipher. The instructions marked in black at the beginning of the cipher are the ones concerning the loading of the plaintext, which have infinite resistance, as the SDFA detects them as not depending on any key bits, but depending on the inputs. By contrast, the end of the cipher does not have any black-marked instruction as the instructions producing the ciphertext are the result of the last key addition: consequentially our backward SDFA correctly deems its result depending on the key bits and marks them with a low resistance, since they should be protected. We chose as a target platform for the benchmarks the "Pandaboard ES development platform". It is endowed with a TI OMAP4460 SoC, containing a dual core Cortex A9-MPCore clocked at 1.2 GHz, 1 MiB L2 cache and 1 GiB DDR2 RAM, running Linaro 12.09 Linux distribution (`armv7l` target). Table 2 reports the results in terms of performance and number of IR instructions of the automatically protected AES implementation obtained through our toolchain, comparing it with the naïve application of the same countermeasure techniques to the whole cipher, as reported [13] and [14]. Both solutions in [14] and in [13] are manually applied and tailored to an 8-bit AES implementation with tabulated and computational version of the S-box, respectively. The timings reported in Table 2 for our solution consider the protection of all the instructions for which the resistance value is lower than 80, a common choice

Table 2: Performance and code size of AES-128 algorithm with both computational and tabulated S-boxes as a function of the number of employed masks, d. All the instructions with resistance lower than 80 are masked. The number of IR instructions includes also the Key-Schedule operations

S-box mode	d	Perf. [μs]	Size [IR-Ins]	Perf. [μs]	Size [IR-Ins]	Speedup
Comp.		This work		[13]		
	Unprot.	60.8	985	60.8	985	–
	1	333.3	5105	870.7	3304	×2.61
	2	983.6	9042	2450.8	6763	×2.49
	3	1983.3	13938	5005.0	11170	×2.52
Tab.		This work		[14]		
	Unprot.	2.2	480	2.2	480	–
	1	94.9	2744	233.9	920	×2.46
	2	132.5	3960	313.7	1660	×2.37

of level of computational effort to deem a cipher secure. Applying the protection to all the instructions with resistance lower than 128 still yields comparable figures (a performance loss lower than 10% with respect to the results in Table 2), without sacrificing any security margin. Our results for the masking applied to the AES variant employing tabulated S-Boxes are 2.46×−2.61× faster than their counterpart for a 1-st order masking and 2.37×−2.49× for the second order masking with an acceptable instruction count increase due to the partial unrolling of the loops. The code size ratio among IR instruction sizes and the one corresponding to the actually emitted codes (i.e., with a specific ISA) is substantially the same.

7. CONCLUDING REMARKS

We presented a comprehensive security-oriented data-flow analysis and a compiler-based tool to automatically instantiate the essential set of masking countermeasures. Further countermeasure application passes can be easily integrated in the modular structure of the compiler pipeline. Future works will adapt the proposed bit-level security analysis also for HDLs, and extend the automatic application of countermeasures also to hardware circuit designs.

8. REFERENCES

[1] G. Agosta, A. Barenghi, and G. Pelosi. A Code Morphing Methodology to Automate Power Analysis Countermeasures. In *Proc. of the 49th Design Automation Conference*, DAC '12, pages 77–82. ACM, 2012.

[2] A. G. Bayrak, F. Regazzoni, P. Brisk, F.-X. Standaert, and P. Ienne. A First Step towards Automatic Application of Power Analysis Countermeasures. In *Proc. of the 48th Design Automation Conference*, DAC '11, pages 230–235. ACM, 2011.

[3] S. Chari, C. S. Jutla, J. R. Rao, and P. Rohatgi. Towards Sound Approaches to Counteract Power-Analysis Attacks. In M. J. Wiener, editor, *CRYPTO*, vol. 1666 of *LNCS*, pages 398–412. Springer, 1999.

[4] J.-S. Coron, E. Prouff, and M. Rivain. Side Channel Cryptanalysis of a Higher Order Masking Scheme. In P. Paillier and I. Verbauwhede, editors, *CHES*, vol. 4727 of *LNCS*, pages 28–44. Springer, 2007.

[5] B. Debraize. Efficient and Provably Secure Methods for Switching from Arithmetic to Boolean Masking. In Prouff and Schaumont [12], pages 107–121.

[6] Y. Ishai, A. Sahai, and D. Wagner. Private Circuits: Securing Hardware against Probing Attacks. In D. Boneh, editor, *CRYPTO*, vol. 2729 of *LNCS*, pages 463–481. Springer, 2003.

[7] U. Khedker, A. Sanyal, and B. Karkare. *Data Flow Analysis: Theory and Practice*. CRC Press, Inc., Boca Raton, FL, USA, 1st edition, 2009.

[8] P. C. Kocher, J. Jaffe, and B. Jun. Differential Power Analysis. In *Proc. of CRYPTO'99*, pages 388–397. Springer, 1999.

[9] S. Mangard, E. Oswald, and T. Popp. *Power Analysis Attacks: Revealing the Secrets of Smart Cards (Advances in Information Security)*. Springer, 2007.

[10] A. Moss, E. Oswald, D. Page, and M. Tunstall. Compiler Assisted Masking. In Prouff and Schaumont [12], pages 58–75.

[11] S. S. Muchnick. *Advanced Compiler Design and Implementation*. Morgan Kaufmann Publishers Inc., San Francisco, CA, USA, 1997.

[12] E. Prouff and P. Schaumont, editors. *Cryptographic Hardware and Embedded Systems - CHES 2012. Proc.*, vol. 7428 of *LNCS*. Springer, 2012.

[13] M. Rivain and E. Prouff. Provably Secure Higher-Order Masking of AES. In S. M. F.-X. Standaert, editor, *CHES*, vol. 6225 of *LNCS*, pages 413–427. Springer, 2010.

[14] K. Schramm and C. Paar. Higher Order Masking of the AES. In *Proc. of CT-RSA'06*, pages 208–225. Springer, 2006.

Lighting the Dark Silicon by Exploiting Heterogeneity on Future Processors

Ying Zhang   Lu Peng   Xin Fu[†]   Yue Hu

Division of Electrical & Computer Engineering
School of Electrical Engineering and Computer Science
Louisiana State University
{yzhan29, lpeng, yhu14}@lsu.edu

[†]Electrical Engineering and Computer Science
School of Engineering
University of Kansas
xinfu@ittc.ku.edu

ABSTRACT

As we embrace the deep submicron era, dark silicon caused by the failure of Dennard scaling impedes us from attaining commensurate performance benefit from the increased number of transistors. To alleviate the dark silicon and effectively leverage the advantage of decreased feature size, we consider a set of design paradigms by exploiting heterogeneity in the processor manufacturing. We conduct a thorough investigation on these design patterns from different evaluation perspectives including performance, energy-efficiency, and cost-efficiency. Our observations can provide insightful guidance to the design of future processors in the presence of dark silicon.

Categories and Subject Descriptors

C.1 [**PROCESSOR ARCHITECTURE**]: Heterogeneous systems; C.4 [**PERFORMANCE OF SYSTEMS**]: Design studies

General Terms

Design, Experimentation.

Keywords

Dark silicon, emerging device, heterogeneous

1. Introduction

Processor manufacturers have complied with Moore's Law to double the transistor count and performance on each new generation product in past decades. However, as we embrace the deep submicron era, Dennard scaling which describes the continuous decrease on the supply and threshold voltage of a transistor at each new technology node has stalled [8][17], leading to an ever increasing power density on modern processors. On the other hand, the maximum processor power consumption should be always enclosed within a reasonable envelope despite the manufacturing technology, due to physical constraints including heat dissipation and power delivery. Under this limitation, a large portion of integrated transistors on a future processor must be significantly underclocked or even completely turned off in order to satisfy the power constraint and maintain a safe working temperature. This phenomenon, which is termed the "dark silicon", is recognized to be one of the most critical constraints that prevent us from obtaining commensurate performance benefit from the increased number of transistors.

Dark silicon might be exacerbated as Moore's Law continues to dominate the processor development. Figure 1 illustrates the scal-

Permission to make digital or hard copies of all or part of this work for personal or classroom use is granted without fee provided that copies are not made or distributed for profit or commercial advantage and that copies bear this notice and the full citation on the first page. To copy otherwise, to republish, to post on servers or to redistribute to lists, requires prior specific permission and/or a fee.

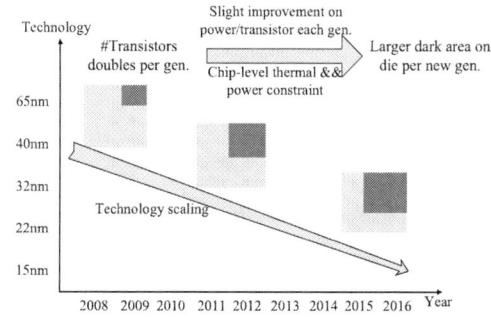

Figure 1. Increasing dark area with technology scaling

ing trend of the amount of "dark" transistors according to the ITRS roadmap [3]. As can be seen, the percentage of the dark area on a chip is exponentially expanding at each generation. This results in a chip with up to 93% of all transistors inactive in a few years from now [23]. Therefore, seeking new design dimensions to efficiently utilize the chip-level resource including power and area is important for us to obtain sustainable performance improvement in the future. Prior works have proposed a few solutions to address the dark silicon problem from certain aspects [8][9][17][24][25]. However, most of these works mainly concentrate on a specific solution, lacking general justifications of multiple design options. Considering that an initial guidance to the design of future processors in the presence of dark silicon is highly desired, we conduct a comprehensive assessment of new design dimensions with special concentration on heterogeneity in the early stage of processor manufacturing.

Our target processor is a chip multiprocessor (CMP) with fixed power and area budget. The first dimension that will be evaluated is *device heterogeneity*. Since dark silicon is essentially caused by the slow improvement in CMOS device's switch power, emerging low-power materials might be used to build processors in order to illuminate the dark area. However, many power-saving devices manufactured with nano-technology manifest a series of drawbacks such as long switch delay [11]. Due to this limitation, it is inappropriate to use such devices to completely replace the traditional CMOS in processor manufacturing. To effectively alleviate the power constraint without suffering from significant performance degradation, integrating cores made of different materials on the same die emerges as an attractive design option. A few works have justified the feasibility of hybrid-device CMP at circuit level [13][19][20][21] while some of them further demonstrate the advantage of the resultant processors in performance improvement [13]. Nevertheless, these works are mainly conducted on a fixed platform and thus the optimal design configuration which provides desirable balance among disparate evaluation metrics remains an open question. On the other hand, architectural heterogeneity (e.g., including both big and small cores on a processor) has been proved an effective solution to energy efficiency improvement [14][9]. Therefore, jointly applying the device heterogeneity and architectural heterogeneity becomes a promising option to further exploit their advantages over conventional designs, hence the second design dimension "*two-fold heterogeneity*". In general, by evaluating the described new design dimensions in detail, our study makes the following key observations:

- We demonstrate that using diverse materials in the chip fabrication is effective in relieving the dark silicon problem. By integrating more cores made of slower and power-saving device and relatively few cores built with faster yet power-consuming device, more processor cores can be booted up. Therefore, the advantages of both materials are leveraged, assisting us to produce processors that deliver impressive energy- and cost-efficiency.

- We observe that architectural heterogeneity is capable of offering higher cost-efficiency in addition to the well-known energy-efficiency over conventional designs, because including small low-power cores is able to reduce the peak chip temperature and thus decreasing the cooling expense. This further confirms the importance of building CMPs with different types of cores in the presence of dark silicon.

- We explore processor designs with two-fold heterogeneity with regards to both manufacturing devices and core architectures. We show that building complex out-of-order cores with power-saving device while manufacturing small in-order cores with relatively power-consuming material is able to deliver extra benefit on energy- and cost-efficiency, thus appearing as the optimal design option.

2. Methodology

2.1 Metric

In this section, we describe the metrics for the evaluation of different configurations. Note that we characterize multiple aspects including performance, energy efficiency, thermal features and cost-efficiency for each design configuration in order to make a comprehensive investigation.

We choose the *total execution time* for performance evaluation. For the energy-efficiency and thermal feature, we use *energy-delay product* (ED) and *peak temperature* for assessment. Besides these three extensively discussed metrics, we also include cost-efficiency as the fourth factor for investigation. In this work, we define the cost efficiency as *MIPS/dollar*. The considered cost is composed of the die cost and cooling expense, where the former part can be calculated with the following equations [16]:

$$Die\ cost = \frac{wafer\ cost}{Dies\ per\ wafer \times Die\ yield} \quad (1)$$

$$Dies\ per\ wafer = \frac{\pi \times \left(\frac{wafer\_diam}{2}\right)^2}{Die\ area} - \frac{\pi \times wafer\_diam}{\sqrt{2 \times Die\ area}} - Test\ dies \quad (2)$$

$$Die\ yield = wafer\ yield \times \left\{1 + \frac{Defects\ per\ unit\ area \times Die\ area}{\alpha}\right\}^{-\alpha} \quad (3)$$

Table 1. Parameter values for die cost calculation.

Parameter	Value
Wafer cost	$4900
Wafer diameter	300mm
Wafer yield	0.9
Defects per unit area	0.4/cm^2
Alpha	3

Table 1 lists the values of referred parameters derived from recently released data in industry [5][16]. The cooling cost is computed based on a model that is introduced in a prior work [28]:

$$C_{cooling} = K_c t + c \quad (4)$$

In general, this cost is determined by the peak temperature achieved during the execution. High temperature t corresponds to larger coefficient K_c and results in higher cooling cost as a consequence. Characterizing the cost-efficiency is necessary for comput-

Table 2. Architectural parameters for system components.

Component	Parameter	Value
Big core	Pipeline type	out-of-order
	Processor width	4
	ALU/FPU	4/4
	ROB/RF	160/160
	L1I cache size	32KB
	L1D cache size	32KB
	L1 associativity	4
Small core	Pipeline type	in-order
	Processor width	1
	ALU/FPU	1/1
	L1I cache size	8KB
	L1D cache size	8KB
	L1 associativity	2
Other parameters	L2 cache size	4MB
	L2 associativity	8
	Cache block size	32B
	Technology	22nm
	Frequency (High-K)	3G
	Chip area	100mm^2
	TDP	60W

Table 3. Estimated area and power for system components.

Component	Peak power	Area
Big core	5.6W (High-K)	7.6mm^2
	4.8W (NEMS-CMOS)	
Small core	1.1W (High-K)	1.97mm^2
	0.8W (NEMS-CMOS)	
L2 cache	0.8W/MB	3mm^2/ MB
Interconnect	5W	4mm^2
Other components	11W	23mm^2

Table 4. Selected applications for simulation.

Category	Benchmark Suite	Applications (Kernels)
Homogeneous	SPLASH-2	Barnes, FMM, Radix, Raytrace, Water-spatial, waterNS
	PARSEC	Blackscholes, Swaptions
	ALPBench	MPGDec, MPGEnc
Heterogeneous	Computation-intensive	h264, dealII, namd, spcrand, sjeng, omnetpp, gobmk, hmmer, bzip2
	Memory-intensive	mcf, libquantum, milc, leslie3d, perlbench, lbm, soplex, astar

er architects to identify the optimal design configurations, thus deserving careful consideration.

2.2 Simulation Environment and Workloads

We use a modified SESC [18], a widely used cycle-accurate simulator for architectural study, to conduct our investigation. We choose McPat 1.0 [15] for power and area estimation and Hotspot 5.0 [4] for temperature calculation. Note that we assume a 22nm technology in this work, thus we set the system budget based on an Intel Ivy Bridge processor [2]. In specific, the area of the target chip should not exceed 100mm^2 and the maximal power consumption is 60W.

Recall that our design space includes configurations which integrate both big and small cores on the same chip. For this purpose, we assume a complex out-of-order core and a simple in-order core whose parameters are listed in Table 2. Table 3 lists the estimated area and peak power for each component on the chip. Given these conditions, the number of cores that can be accommodated is determined by the following expressions:

$$Area\ constraint: N_b \times A_b + N_s \times A_s + A_{all\ other} \leq 100$$

$$Power\ constraint: N_b \times P_b + N_s \times P_s + P_{all\ other} \leq 60$$

where variables N_b and N_s denote the number of big cores and number of small cores respectively. Constants A_b and P_b indicate the area and peak power for a big core as listed in Table 3. Similar interpretations apply to other symbols such as A_s and P_s.

The workloads used for our exploration is based on the specific architecture in study. Multi-threaded programs are generally used for CMPs on which all cores have identical architecture (in the study of device heterogeneity); on the other hand, when both big and small cores are integrated, we consider that "heterogeneous"

Table 5. Features of materials considered in this work.

Material	Features
High-K	Reduce leakage power to 20% of the dynamic power
NEMS-CMOS	OR gate: 20% higher delay, reducing 60% switching power
	SRAM cell: 25% higher delay, saving 85% leakage energy

workloads are more appropriate for the investigation and thus use combinations of programs from SPEC CPU2006 as a substitute. For those parallel applications, the number of threads for execution always equals to the core count of the underlying CMP and all programs are executed till completion in order to guarantee that identical task is performed. We choose a total of 10 programs from SPLASH-2, PARSEC and ALPBench for the simulation. The reason for not including other workloads is that their intrinsic characteristics (e.g., requiring 2^n threads) prohibit the execution on many configurations. As for the SPEC mixes, each of them includes 30 individual programs (the maximum core count in all evaluated configurations). We simulate 100 million instructions after fast-forwarding the initial 1.5 billion for each individual program within a mix. This also ensures that identical tasks are performed across different configurations. Note that when the core count is less than 30, part of programs will be launched after some cores finish their tasks assigned earlier. Also, considering that program feature such as memory intensity determines the computation efficiency on heterogeneous CMPs, we briefly classify the programs from SPEC CPU 2006 into two categories, namely computation-intensive and memory-intensive, based on their L2 miss ratios. Table 4 lists all selected benchmarks used in this study.

3. Device Heterogeneity

3.1 New Device and Architectural Implication

The slight improvement in transistor power density is fundamentally caused by the physical characteristics of MOSFET [23]. Due to this limitation, it is intuitive to recognize that breakthroughs in semiconductor technology are the antidote to dark silicon in essence. In this work, we consider two representative emerging devices, namely High-K dielectrical [1] and Nano-electro-mechanical switch (NEMS) [6][11], to exploit the device heterogeneity and combat dark silicon.

High-K dielectrical refers to a device that replaces the silicon dioxide in semiconductor manufacture. The letter K stands for dielectrical constant, indicating how much charge the material can hold. High-K is capable of significantly decreasing the leakage current (i.e., < 1% of SiO_2) and has already been adopted by leading processor manufacturers [1]. In general, as an important substitute of conventional devices in current industry, it deserves a careful evaluation.

The NEMS material, on the other hand, is a candidate for future processor development because it is built on physical switch and is not limited by the drawbacks of MOSFET. NEMS is able to reduce the leakage current by orders of magnitude, however, it demonstrates a significantly longer switch delay compared to conventional devices, implying large performance degradation on the resultant processor. Taking this into consideration, researchers propose a hybrid device that combines NEMS and CMOS together. Dadgour et al. [6] elaborate the features of NEMS-CMOS circuits in detail and demonstrate the potential of this hybrid device in future processor manufacturing. Therefore, we consider NEMS-CMOS as an alternative material in this work. We carefully calibrate the parameters based on recent documents [1][6][11] for High-K and NEMS-CMOS and list the important features in Table 5.

Although the purpose of this section is not to make comparison among emerging devices, a glance at their characteristics can enlighten us on architectural innovation for the next generation CMP.

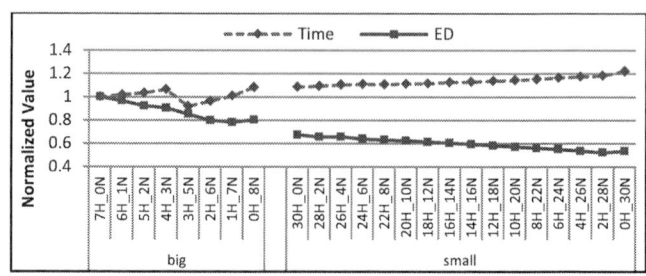

Figure 2. Average execution time and ED of multi-threaded applications running on mix-device CMPs.

Specific to High-K and NEMS-CMOS, the latter material switches at a lower rate than the former one but offering extra saving for both dynamic and leakage energy. Note that using other alternative materials such as Tunnel-FET (TFET) will introduce similar design trade-off. For instance, TFET cannot match the performance of CMOS under normal voltage, but it is beneficial for power saving [19]. Therefore, our conclusion made in this section can be generalized to scenarios where devices other than High-K and NEMS-CMOS are used for processor manufacturing. Nevertheless, this implies that integrating High-K cores and NEMS-CMOS cores on the same chip would deliver a processor that works more efficiently than a CMP manufactured with an exclusive device. Keeping this in mind, we evaluate a set of design configurations, with which a portion of integrated cores are built with High-K while the remaining ones with NEMS-CMOS. We compare such mix-device configurations with CMPs built with a single device alone (i.e., all High-K cores or NEMS-CMOS cores) and aim at identifying the better design choice.

3.2 Result Analysis

3.2.1 Average performance and ED

We consider two categories of CMPs to characterize the impact of device selection. The first group of chip-multiprocessors is composed of big out-of-order cores while the ratio of High-K cores over NEMS-CMOS cores is varying. Based on the power and area constraints depicted in section 2.2, the total number of big cores that can be accommodated on die is either 7 or 8. The reason of the varying core count is as follows. When all cores are manufactured with High-K, the power constraint restricts the maximal number of cores to be 7 although there is enough space for an extra core; as more NEMS-CMOS cores which consume relatively lower power are integrated to replace High-K cores, the area constraint becomes the determinative factor and confines the core count to be 8. On the other aspect, when all cores are small in-order ones, the core count is always limited by the area constraint and should not exceed 30.

We run multi-threaded applications with these configurations for evaluation. Figure 2 plots the average performance and energy-efficiency of these applications. All results are normalized to that corresponding to the *7H_0N* configuration in the "big" category, where the chip contains 7 out-of-order cores made of High-K. Note that in later sections of this paper, we also show results in this normalized fashion. The notation *xH_yN* means a total of *x* High-K cores and *y* NEMS-CMOS cores are installed. Also recall that the performance is measured in execution time, thus smaller values indicate better performance. As can be observed, in the "big" category, the execution time gradually increases at first and demonstrates a significant reduction from 4H_3N to 3H_5N, after which the curve rises again. The reason of the performance degradation (e.g., from 7H_0N to 4H_3N, and the segment between 3H_5N and 0H_8N) is that NEMS-CMOS cores execute at a lower rate than the High-K counterparts; therefore, increasing the number of NEMS-CMOS cores tends to prolong the overall execution time. The performance improvement at 3H_5N comes from the extra core in this configuration, with which the applications are executed

578

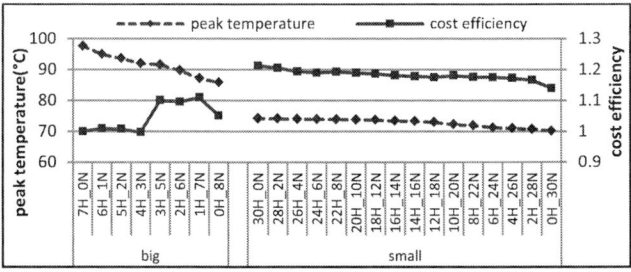

Figure 3. Average peak temperature and cost efficiency of multi-threaded benchmarks running on mix-device CMPs.

with one more thread. Note that in the extreme case where all cores are made of NEMS-CMOS (0H_8N), the processor takes even longer time to finish the execution compared to the 7-core configurations although it is equipped with an extra core. This is because that the slow execution on the master thread becomes the performance bottleneck and elongates the execution duration. As for the "small" category, the execution time gradually increases as more NEMS-CMOS cores are included since the core count is fixed to 30 irrespective of the manufacturing device.

The energy-efficiency demonstrates a different variation from the performance change. In general, the energy-delay product is decreasing as more NEMS-CMOS cores are equipped. This is because that the energy saving from NEMS-CMOS cores outweighs the corresponding performance degradation while running these parallel applications, thus using more such cores is beneficial to improving the energy-efficiency. The only exception is observed at the switch from 1H_7N to 0H_8N in the "big" category (or 2H_28N to 0H_30N in "small"), where the energy-delay demonstrates a slight increase. This is due to the fact that the performance degradation contributes more to the variation of ED for programs with long serial phase. With the 0H_8N configuration, the sequential stages are executed on the NEMS-CMOS cores, thus resulting in significant performance loss and higher ED.

In summary, for a CMP which only consists of big cores, including relatively more NEMS-CMOS cores and a few faster High-K cores is the preferable design paradigm than building a chip with processor cores made of a single device. Specifically, the 3H_5N configuration is able to shorten the execution time by an average of 8.9% while reducing the ED by 14.2% compared to the 7H_0N design. The ED-optimal configuration (i.e., 1H_7N) can save the ED by up to 20.8% with ignorable performance loss in comparison with 7H_0N. For the small-core-oriented architecture, the highest energy-efficiency is delivered by the configuration 2H_28N, meaning the optimal balance between performance and energy consumption is also achieved on a CMP with a large amount of NEMS-CMOS cores and a few High-K cores.

3.2.2 Thermal feature and cost-efficiency

Peak temperature and cost-efficiency are another two important metrics to evaluate a design configuration. We demonstrate the results of these two features for the proposed configurations in Figure 3. As shown in the figure, the temperature drops significantly as we employ more NEMS-CMOS big cores. The reason is that the power density on a NEMS-CMOS core is remarkably smaller than that of a High-K counterpart, thus a NEMS-CMOS core is relatively "cooler" compared to a High-K one. As more cool components are integrated on die, thermal coupling tends to be alleviated and the peak steady temperature is gradually decreased. Therefore, the coolest chip is the one where all cores are manufactured with NEMS-CMOS. On the other aspect, lower temperature results in lower cooling cost. This means that we are essentially trading off "performance" for "low cost" when we replace a NEMS-CMOS core for a High-K core. In this scenario, the cost-efficiency

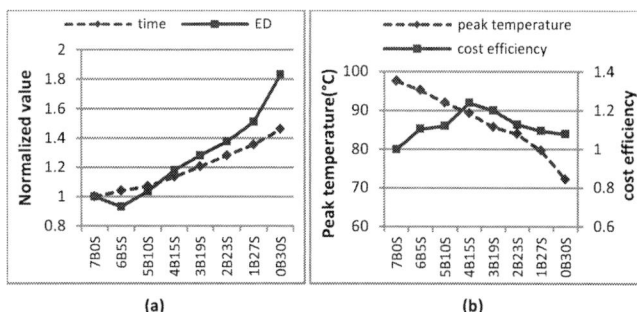

(a)   (b)

Figure 4. Execution information for computation-intensive workloads on high-K heterogeneous CMPs (a) normalized performance and ED (b) temperature and cost-efficiency.

reaches the peak value at 1H_7N where the performance and cost can be optimally balanced. Note that the increment of cost-efficiency from 4H_3N to 3H_5N is resulted from the performance boost. The curve corresponds to the "small" category is more smooth. The reason is that the in-order cores consume much smaller power than big cores and thus generate less heat. This results in relatively mild temperature variation across configurations. In this situation, the cost-efficiency does not largely vary when we change the manufacturing devices. Nevertheless, generally speaking, it is still reasonable to conclude that hybrid-device CMPs outperform chips built with a single device alone. Furthermore, to achieve the optimal balance among performance, energy consumption and total cost, a CMP should be equipped with more power-saving cores (NEMS-CMOS) and a small amount of faster yet power-consuming (High-K) cores.

4. Two-fold Heterogeneity

4.1 More Observations on Architectural Heterogeneity

Existing works have shown that executing a program on processors with different architecture may result in quite distinctive energy efficiency [14]. For example, a program with fairly low instruction-level parallelism might be more suitable to run on a simple in-order core instead of a big complex one for higher energy efficiency. This observation drives the development of architectural heterogeneous CMPs where integrated cores demonstrate different performance, area, and power features. In this subsection, we use the execution of computation-intensive workloads on a series of High-K heterogeneous CMPs as an example to illustrate that architectural heterogeneity also results in better cost-efficiency. Note that we run SPEC program mixes for the evaluation of architectural heterogeneity.

We first briefly analyze the performance and ED variations which are shown in Figure 4(a) to corroborate conclusions made in prior works. The notation *x*B*y*S indicates that *x* big cores and *y* small cores are integrated on the chip. Recall that the core counts are determined by both area and power constraint as described in section 2.2. From the figure we observe that the total execution time of the computation-intensive workloads keeps increasing as the number of big cores is reduced. This is due to the fact that the execution speed of such programs on big cores is remarkably faster than that on small in-order cores. For example, the relative performance (i.e., time on small core/time on big core) of *dealII* is around 6.02. This means that running a set of programs on a big core sequentially takes even shorter time than running them on a few small cores in parallel. However, the energy-delay product reaches the minimal value when 6 big and 5 small cores are installed on the chip. This is because the energy saving on small cores contributes more to the improvement in energy-efficiency at this point. Nevertheless, this scaling trend proves that architectural heterogeneity is effective in increasing the energy-efficiency.

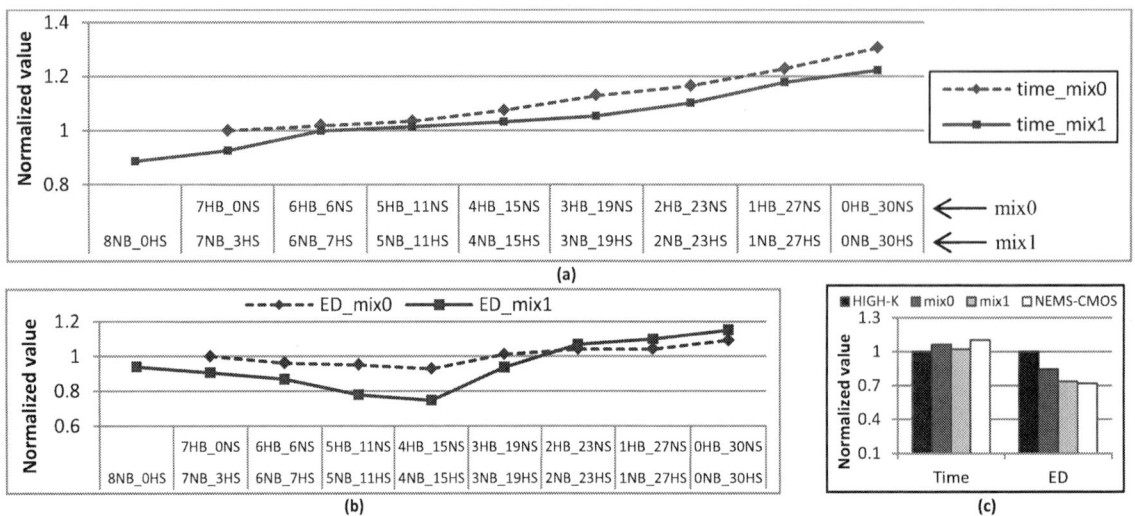

Figure 5. Execution information for computation-intensive workloads running on mix-device heterogeneous CMPs: (a) performance (b) energy-delay product (c) comparison among material-dependent optimal configurations.

Figure 4(b) plots the variations of temperature and cost-efficiency for computation-intensive workloads running on High-K heterogeneous CMPs. As can be observed, the temperature drastically drops as we gradually remove big cores to accommodate more small cores. This is straightforward to understand since small cores are much simpler and consume less power than big cores. The common hotspots in an out-of-order processor such as the instruction issue queue have been eliminated from small cores, thus replacing big cores with small cores is effective to decrease the chip temperature and save the cooling cost. However, computation-intensive workloads favor big cores for better performance, implying that the performance will be degraded as we reduce the number of big cores. In this situation, the interplay between performance and temperature results in a non-monotonic variation of the cost efficiency that it first increases to the peak value at 4B15S and then drops as the big core count is further decreased. In specific, the 4B15S configuration is able to cool the chip by 7.5°C while improving the cost-efficiency by 23.9% compared to the 7B0S organization. In one word, architectural heterogeneity delivers better cost-efficiency compared to homogeneous designs.

4.2 Performance and ED

After justifying the advantage of architectural heterogeneous CMPs with respect to energy-efficiency and cost-efficiency, it is natural for us to introduce the second design dimension, two-fold heterogeneity, with which both device-heterogeneity and architectural asymmetry are jointly adopted. More specifically, we consider a set of configurations where both the material and complexities are different among integrated cores. We assess two kinds of organizations: big High-K cores along with small NEMS-CMOS cores and the opposite.

Figure 5(a) plots the performance scaling of computation-intensive programs with these two design patterns. Note that all results are normalized to that in the 7HB_0NS case. The upper labels on the horizontal axis correspond to the first architecture where big cores are made of High-K and small cores are manufactured with NEMS-CMOS (mix0 or xHB_yNS); accordingly, the lower labels correspond to the opposite architecture which includes big NEMS-CMOS and small High-K processors (mix1 or xNB_yHS). As can be observed, configurations with the second pattern, namely xNB_yHS, always outperform the counterparts from the first category. This can be explained in two aspects. First, since NEMS-CMOS cores are relatively power-saving, the second design pattern accommodates more processors when the core count is power-limited. Due to this reason, the total number of cores is larger in the xNB_yHS designs, thus these configurations take shorter time to finish executing the program combination. This

corresponds to the scenarios where the number of big cores is no smaller than 6. Second, as the constraint factor shifts to chip area, the core counts in both design patterns become identical (from 5B_11S). In this situation, the global execution time basically depends on the performance of small cores because of their larger amounts. For instance, in the 2B_23S configuration, how fast the programs run on small cores determines the overall performance in essence, because the number of small cores is remarkably larger than that of big cores. Since those in-order processors are made of High-K, the chips designed with the second pattern still offer better performance.

Figure 5(b) demonstrates the variation of the energy-efficiency for the same program set running with considered configurations. Note that the interplay between the performance/energy of different cores makes the variation of ED non-monotonically. For both blending patterns, we note that the energy-delay product gradually decreases at first until the minimal value is reached at 4B_15S, after which the efficiency is getting worse. More specifically, the xNB_yHS delivers better energy-efficiency than the xHB_yNS when the configuration is varied from 8 big cores to 3 big cores. This is due to the shorter execution time and less energy consumption on big NEMS-CMOS cores. As small cores begin dominating the chip in 2B_23S and beyond, their relatively large energy consumptions mitigate the performance benefit and make the ED rise again.

To more clearly illustrate the benefit of such two-fold heterogeneity, we identify the most energy-efficient configurations from four different design patterns, namely High-K for all cores, xHB_yNS (mix0), xNB_yHS (mix1) and NEMS-CMOS for all cores, and make comparison among these material-dependent optima. For computation-intensive workloads, we choose 6B_5S according to Figure 4(a) and 6B_7S for High-K and NEMS-CMOS, respectively. Note that the evaluation results of architectural heterogeneity with NEMS-CMOS are not included in the paper due to space limitation. Nevertheless, 6B_5S and 6B_7S deliver the optimal energy-efficiency for High-K processors and NEMS-CMOS ones. We then select 4B_15S for HB_NS and NB_HS based on Figure 5(b). We normalize the execution time and ED to those corresponding to the optimal High-K processor and demonstrate the result in Figure 5(c). As can be observed, the CMP with 4 NEMS-CMOS big cores and 15 High-K small cores (4NB_15HS) is the global optimal configuration. It improves the energy-efficiency by 27% with only 4.3% performance degradation compared to the optimal High-K CMP. We conduct similar comparison for memory-intensive workloads and graph the result in the appendix.

580

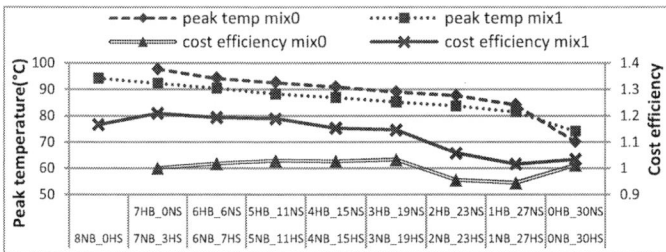

Figure 6. Peak temperature and cost-efficiency of computation-intensive workloads running on mix-device heterogeneous CMPs.

4.3 Thermal Effects and Cost-efficiency

Figure 6 plots the peak temperature and cost-efficiency of these two-fold heterogeneous CMPs while running computation-intensive workloads. As we have observed previously, NEMS-CMOS cores result in lower temperature than High-K cores and small cores are much cooler than big ones. Consequently, the second design pattern (i.e., xNB_yHS) tends to be cooler than its alternative (xHB_yNS), because the hotspot on die which is usually located in the out-of-order processor has lower temperature. Recall that the xNB_yHS also delivers better performance. Therefore, its cost-efficiency is significantly higher than that offered by xHB_yNS configurations. As can be seen, for computation-intensive workloads, the cost-efficiency reaches the peak value at 7NB_3HS configuration, which improves the efficiency by 20.9% compared to the 7HB_0NS case. For memory-intensive workloads, (graphs are in the appendix), the optimal configuration outperforms the baseline case by up to 66.7%. In conclusion, our observations made in this section demonstrate that the mix1 design paradigm (xNB_yHS, or big NEMS-CMOS cores along with small High-K cores) stands as the optimal among all evaluated configurations, since it can more efficiently balance the execution performance, energy consumption and total cost.

5. Related Work

Dark silicon emerges as an increasingly important issue that menaces the scaling of Moore's Law in the deep submicron era and beyond. Due to this reason, researchers recently start to investigate this problem and propose several solutions to alleviate the conundrum. A group from UCSD has made significant progress on using dark silicon for processor improvement. They develop conservation cores [24] and Quasi-specific cores [25] for increasing the computation energy-efficiency in different scenarios. In [9], Gupta et al. demonstrate the potential of heterogeneous CMP for energy-efficiency improvement. Systems built with near-threshold voltage processors (NTV) [7][26] are also effective approaches.

While most of these studies focus on a single solution individually, few works make attempt to address the dark silicon problem from a broader perspective. Esmaeilzadeh et al. [8] use an analytical model to predict the processor scaling for next few generations. They demonstrate that dark silicon will be heavily exacerbated as manufacture technology keeps shrinking. Taylor [23] reviews the current status of dark silicon and briefly describes four solutions from the high level. Hardavellas et al. [10] pay specific attention to the server processors and perform an exploration of throughput-oriented processors.

As for the hybrid device study, Saripalli et al. [19][20] discuss the feasibility of technology-heterogeneous cores and demonstrate the design of mix-device memory. Wu et al. [27] presents the advantage of hybrid-device cache. Kultursay [13] and Swaminathan [21] respectively introduce a few runtime schemes to improve performance and energy efficiency on CMOS-TFET hybrid CMPs. Our work deviates from the aforementioned in that we conduct a more comprehensive study to combat dark silicon in the early stage

of processor manufacturing. We propose to utilize device heterogeneity and architectural heterogeneity simultaneously to optimally utilize the chip resource and well balance the performance, energy consumption and total cost.

6. Conclusion

As dark silicon has begun to hazard the scaling of Moore's Law and prohibits us benefiting from the increasing number of transistors, new design technologies are in high demand to address this problem. This is especially important in the early stage of processor manufacturing where issues such as architectural organization and device selections need to be carefully considered. For this purpose, our work evaluates a series of design configurations by exploiting the device heterogeneity and architectural asymmetry in the processor manufacturing. Our evaluation results demonstrate that building heterogeneous chip multiprocessors with different materials is more preferable than conventional designs since it can efficiently utilize the chip level resource and deliver the optimal balance among performance, energy consumption and cost.

References

[1] Intel Corporation. High-K and Metal Gate Transistor Research. http://www.intel.com/pressroom/kits/advancedtech/doodle/ref_HiK-MG/high-k.htm

[2] Intel Corporation. Ivy Bridge Products. http://ark.intel.com/products/codename/29902/Ivy-Bridge

[3] International Technology Roadmap for Semiconductors. http://www.itrs.net/

[4] Hotspot 5.0 Temperature Modeling Tool. http://lava.cs.virginia.edu/HotSpot/.

[5] Global Semiconductor Alliance. http://www.gsaglobal.org

[6] H. F. Dadgour and K. Banerjee. Design and analysis of hybrid NEMS-CMOS circuits for ultra low-power applications. In DAC'07.

[7] R. G. Dreslinski, M. Wieckowski, D. Blaauw,D. Sylvester, and T.Mudge. Near-threshold computing: reclaiming Moore's law through energy efficient circuit. Proceedings of the IEEE, special issue on ultra-low power circuit technology, Feb. 2010.

[8] H. Esmaeilzadeh, E. Blem, R. St. Amant, K. Sankaralingam, D. Burger. Dark silicon and the end of multicore scaling. In ISCA'11.

[9] V. Gupta et al. Using heterogeneous cores to provide a high dynamic power range on over-provisioned processors. In Dark Silicon Workshop in conjunction with ISCA, Jun. 2012.

[10] N. Hardavellas, M. Ferdman, B. Falsafi, A. Ailamaki. Toward dark silicon in servers. In IEEE Computer Society, 2011.

[11] R. Jammy. Materials, process and integration options for emerging technologies. SEMATECH/ISMI symposium, 2009.

[12] P. L-Kamran et al. Scale-out processors. In ISCA'12.

[13] E. Kultursay et al. Performance enhancement under power constraints using heterogeneous CMOS-TFET multicores. In CODES+ISSS'12.

[14] R. Kumar, K. I. Farkas, N. P. Jouppi, P. Ranganathan, D.M. Tullsen. Single-ISA Heterogeneous Multi-Core Architectures: The Potential for Processor Power Reduction. In MICRO'03.

[15] S. Li et al. McPAT: an integrated power, area, and timing modeling framework for multicore and manycore architectures. In MICRO'09.

[16] J. M. Rabaey, A. Chandrakasan and B. Nikolic. Digital Integrated Circuits, 2nd edition.

[17] A. Raghavan et al. Computational Sprinting. In HPCA'12.

[18] J. Renau et al. SESC Simulator.

[19] V. Saripalli et al. Exploiting heterogeneity for energy efficiency in chip multiprocessors. In IEEE Transactions on Emerging and Selected topics in Circuits and Systems, Jun. 2011.

[20] V. Saripalli, A.K.Mishra, S. Datta and V.Narayanan. An energy-efficient heterogeneous CMP based on hybrid TFET-CMOS cores, in DAC'11.

[21] K. Swaminathan et al. Improving energy efficiency of multi-threaded applications using heterogeneous CMOS-TFET multicores. In ISLPED'11.

[22] S. Swanson et al. Area-performance trade-offs in tiled dataflow architectures. In ISCA'06.

[23] M.B.Taylor. Is dark silicon useful? In DAC'12.

[24] G. Venkatesh, J Sampson, N. Goulding, S. Garcia. Conservation cores: reducing the energy of mature computations. In ASPLOS'10.

[25] G. Venkatesh et al. QSCores: Trading dark silicon for scalable energy efficiency with quasi-specific cores. In MICRO'11.

[26] L. Wang, K. Skadron, and B. H. Calhoun. Dark vs. Dim silicon and near-threshold computing. In Dark Silicon Workshop in conjunction with ISCA, Jun. 2012.

[27] X. Wu et al. Hybrid cache architecture with disparate memory technologies. In ISCA'09.

[28] J. Zhao, X. Dong and Y. Xie. Cost-aware three-dimensional (3D) many-core multiprocessor design. In DAC'10.

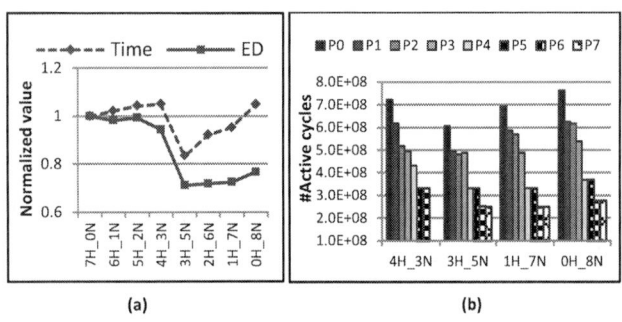

(a)

(b)

Figure 7. Execution information of MPGEnc: (a) time and ED (b) per-core active cycles while running with selected configurations.

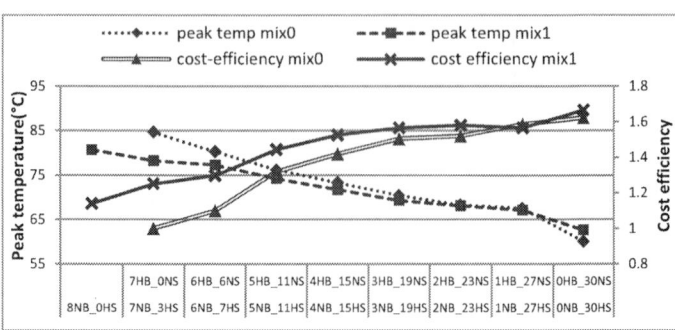

Figure 9. Peak temperature and cost-efficiency of memory-intensive workloads running on mix-device heterogeneous CMPs.

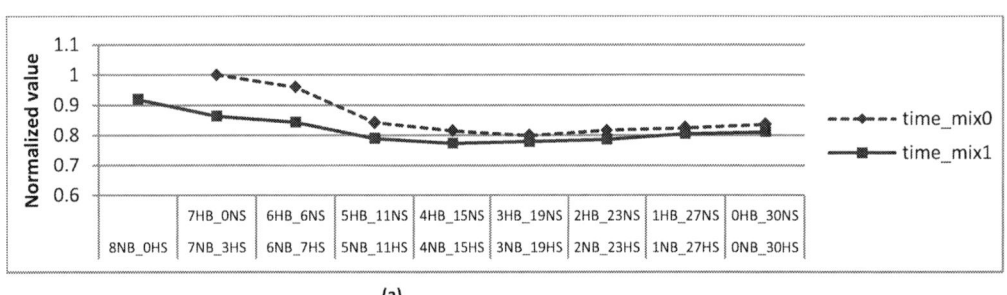

(a)

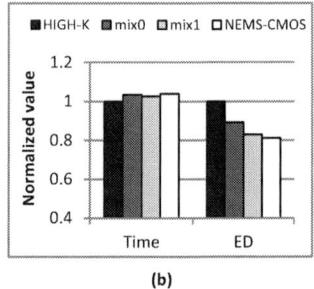

(b)

Figure 8. Execution information for memory-intensive workloads running on mix-device heterogeneous CMPs: (a) performance (b) comparison among material-dependent optimal configurations.

APPENDIX

Case Study for Device Heterogeneity

To further understand the performance scaling trend shown in Figure 2, we choose a representative application (*MPGEnc*) from the program set for analysis and demonstrate the results in Figure 7. Note that we only show the results on CMPs with big cores. The MPGEnc benchmark implements a parallel version of MPEG-2 encoder. In this application, the threads are respectively forked and joined at the beginning and end of the encoding for each frame. Each thread is responsible for encoding a set of macroblocks of a frame while thread 0 always operates on its dedicated buffer. The task assigned to each thread is not identical, thus the time spent by each thread also varies. Plot (a) demonstrates the performance and ED scaling while Plot (b) shows the active cycles of each core during the execution of this program with four configurations. The total execution time is determined by the main thread running on the first processor (P0), and the performance of the parallel stage can be generally estimated from the active cycles of P1. As can be observed, since the number of threads is increased from 7 to 8, the 3H_5N configuration takes much shorter time than 4H_3N to finish the encoding due to the acceleration in parallel stage, hence the remarkable performance improvement at 3H_5N. For the latter three configurations where the core counts are identical, the performance degradation is caused by the decreasing of faster cores (High-K). In specific, the 1H_7N organization includes only one High-K core (P0) while three such cores are equipped in 3H_5N; as a consequence, the parallel stage needs longer time to complete on the CMP configured as 1H_7N, thus lowering the overall performance. On the other hand, the performance degradation from 1H_7N to 0H_8N essentially stems from the slow execution of the sequential stage. This is especially critical for programs with long initialization and finalization.

More Results of Mix-device Heterogeneous CMP

We have shown that mix-device heterogeneous CMP is benefitial to improving the energy- and cost-efficiency for computation-intensive workloads. In this subsection, we will present the result of memory-intensive workloads in order to further justify the conclusion that the design paradigm mix1 is the globally optimal. Figure 8(a) demonstrates the performance comparison between mix0 and mix1 while Figure 8(b) illustrates the performance and energy-efficiency comparison among four material-dependent optimal configurations. Generally, we observe a similar trend that the mix1 design paradigm is more preferable than mix0 by delivering better performance. However, compared with the scaling behavior shown in Figure 5(a), Figure 8(a) demonstrates that memory-intensive workloads favor more small cores, hence more total number of cores, for shorter execution time. The reason is that running memory-bound programs on big cores will not significantly accelerate the execution as opposed to computation-intensive ones. Therefore, executing more programs concurrently can effectively reduce the time for completing all tasks compared to running them sequentially on few big cores. On the other hand, from Figure 8(b), we observe a trend similar to that shown in Figure 5(c). Specifically, the most energy-efficient configuration in the mix1 category outperforms the optimal High-K CMP by 17% in energy-efficiency with less than 4% performance loss. Figure 9 plots the thermal and cost-efficiency results for memory-intensive workloads running on mix-device heterogeneous CMPs. Not surprisingly, the mix1 design paradigm results in a cooler chip than mix0 in most cases, thus delivering up to 66.7% higher cost-efficiency compared to the baseline configuration. In one word, our conclusion that building big out-of-order cores with NEMS-CMOS and manufacturing small in-order cores with High-K is able to achieve the optimal balance among performance, energy consumption and total cost also holds for the memory-intensive applications.

582

Simultaneous Multithreading Support in Embedded Distributed Memory MPSoCs

Rafael Garibotti[1], Luciano Ost[1], Remi Busseuil[1], Mamady kourouma[1], Chris Adeniyi-Jones[2], Gilles Sassatelli[1], Michel Robert[1],

[1] LIRMM (CNRS-University of Montpellier II) – 161 rue Ada, Cedex 05 - 34095 Montpellier, France
{garibotti, ost, busseuil, kourouma, sassatelli, robert}@lirmm.fr

[2] ARM, Ltd. - Cambridge, Cambridgeshire, GB
Chris.Adeniyi-Jones@arm.com

ABSTRACT

Scalability and programmability are important issues in large homogeneous MPSoCs. Such architectures often rely on explicit message-passing among processors, each of which possessing a local private memory. This paper presents a low-overhead hardware/software distributed shared memory approach that makes such architectures multithreading-capable. The proposed solution is implemented into an open-source message-passing MPSoC through developing a POSIX-like thread API, which shows excellent scalability using application kernels used for benchmarking in shared-memory systems. This approach efficiently draws strengths from the on-chip distributed private memory that opens the way to exposing the multithreading programmability/capabilities of that component as a general-purpose accelerator.

Categories and Subject Descriptors

B.7.1 [**Integrated Circuits**]: Types and Design Styles – advanced technologies, VLSI (very large scale integration).

General Terms

Design, Experimentation, Performance, Verification.

Keywords

Programmability, Multithreading, Distributed memory organization, NoC-based MPSoCs.

1. INTRODUCTION

MPSoC have become the de-facto platform template in many application domains ranging from consumer market products such as smart phones to hard real-time systems. In most cases, high performance allied to low power consumption [1] is required. Due to the increasing complexity of both platform architecture and application software, programming such multiprocessor systems is a challenge and demands for better programming model facilities [2]. Parallel programming models for embedded systems have long been explored and some used in high-performance computing (HPC) have been ported and adapted to embedded systems, such

Permission to make digital or hard copies of all or part of this work for personal or classroom use is granted without fee provided that copies are not made or distributed for profit or commercial advantage and that copies bear this notice and the full citation on the first page. To copy otherwise, or republish, to post on servers or to redistribute to lists, requires prior specific permission and/or a fee.

as the Message Passing Interface (MPI). The use of such APIs allows software designers to model explicit parallelism and synchronization between tasks/processor nodes at a reasonable level of abstraction while allowing better utilization of hardware resources [3].

The *objective* of this paper is exploring the opportunity of developing a POSIX-like threads API (Pthreads) [4], very popular in general-purpose and high-performance computing, onto a distributed private memory NoC-based MPSoC. This API commonly used for simultaneous multithreading (SMT) on symmetric multiprocessing (SMP) systems assumes a coherent shared memory architecture that each processor may access at any time. As the target architecture is of distributed private memory type, the proposed approach relies on additional hardware that makes it possible to expose a logically shared memory architecture based on physically distributed memories. Different from ccNUMA architectures (cache-coherent Non-Uniform Memory Machines) that use complex cache-coherence protocols for enforcing strict memory consistency, our approach targets embedded devices and therefore aims at minimizing additional hardware through a software-oriented approach. Since the resulting architecture retains its purely distributed nature, clusters operating in this distributed shared memory mode (referred to as vSMP for Virtual Symmetric Multiprocessing) are of arbitrary size and geometry, and can be decided at run-time for better suiting application needs.

The *contributions* of this paper may be summarized as follows: (*i*) providing multithreading capability onto a RTL distributed private memory NoC-based MPSoC, (*ii*) implementation of a hardware module that enables vSMP clusters definition at run-time, along with a software-based memory consistency approach. These contributions allow us to evaluate the resulting system, demonstrating its efficiency in terms of scalability of platform and application, as well as better throughput for several application/benchmarks scenarios.

This paper is organized as follow: Section 2 presents related work in MPSoCs programmability. Section 3 describes the basic concepts inherent to the adopted NoC-based MPSoC along with proposed API and hardware. Section 4 describes the experimental setup. Section 5 gives results in term of performance scalability and analyzes the overhead created. Finally, Section concludes and points out future work directions.

2. RELATED WORK

Several studies have been aimed at improving MPSoCs programmability by including/extending different APIs primitives, as well as proposing tools that can help designers to program parallel embedded applications. For instance, Ceng et al. [3] propose the MAPS framework that can be employed to guide application designers during the parallelization process. A similar framework that allows to explore different parallelization alternatives at high-level is described in [5]. Both approaches generate a parallel version of a sequential C code, which is validated on the top of a shared-memory system. In turn, Marongiu et al. [2] extended the OpenMP API, allowing allocating array data over multiple distributed scratchpad memories (SPMs) based on profile information to optimize data allocation. Two MPSoC models were implemented in a SystemC full-system simulator in order to validate the proposed OpenMP API (e.g. primitives, compiler). The proposed approach differs from Marongiu's in four main aspects: (*i*) processors are interconnected by a cross-bar, while a mesh NoC is employed in the present work, which makes our system more scalable while consuming less energy and area when compared to a cross-bar; (*ii*) no OS/library support for thread creation and management is available in Marangiu's approach, (*iii*) only benchmarks that fit entirely in cache memories were used in [2] since no cache coherence protocol is supported in his work, while our system implements cache coherence protocol, allowing to execute applications that deal with larger memory footprint (e.g. MJPEG) and (*iv*) the proposed Pthreads approach was evaluated in a real NoC-based MPSoC that has already been prototyped in different FPGAs and also ported to ASIC to evaluate the area overhead, while Marangiu's approach uses a simulator to validate his idea. Ophelders [6] proposes a software cache coherence protocol that was validated on a dual-core ARM9 architecture using the SPLASH2 benchmark. Ophelders work differs from ours in two aspects. First it considers a system with MMU, and a 4-set associative cache, allowing high-level memory management; while our approach relies on compact scalar MMU-less processor architecture with direct mapped caches, which impacts on less memory utilization. Second, our approach to memory coherence is software-based and relies on flushes and invalidations, not allowing remote execution that implies in less memory access.

To the best of our knowledge this work presents the first purely distributed memory NoC-based MPSoC that supports multithreading through POSIX thread API while dealing with cache/memory coherence at OS level. The proposed approach improves the system programmability; while better scalability and performance can be achieved once high visibility permits improving cache coherence and management.

3. MULTITHREADING PLATFORM

3.1 Platform Description

OpenScale[1] is a homogeneous message-passing NoC-based MPSoC with distributed private memories. Each node comprises a 32 bit pipelined CPU with configurable instruction and data caches. This core implements the Microblaze ISA architecture and features a timer, an interrupt controller, a RAM and optionally an UART [7], as illustrated in Figure 1.

---

1 Available for download at: www.lirmm.fr/ADAC

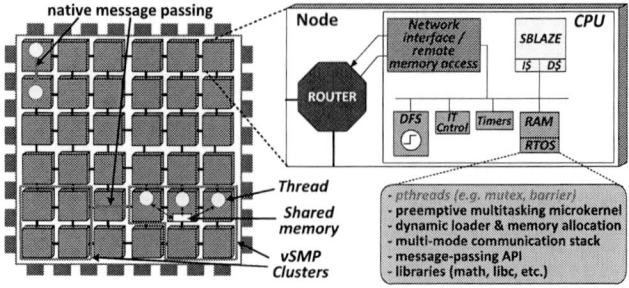

Figure 1 - Adopted NoC-based MPSoC architecture.

The platform RTOS is a pre-emptive priority-based micro-kernel. Each processing element (PE) runs its RTOS independently, and communications and synchronizations are made through a light implementation of the MPI message-passing library. Applications are represented through Khan Process Network, a standard representation in such a message-passing oriented system. Examples of services supported in the RTOS are: (*i*) run-time dynamic applications loading, (*ii*) preemptive round-robin scheduler based on thread credits, (*iii*) system-monitoring mechanisms (e.g. processor workload) and (*iv*) API with drivers support (e.g. UART). The platform supports a number of features such as task-migration and distributed decision-making capabilities [7].

3.2 Multithreading hardware support

In order to enable simultaneous multithreading (SMT) on any multiprocessor architecture, a logically shared memory space must be exposed to the software. Moreover, as the chosen architecture is organized into a distributed private memory, the hardware must also be modified to enable access to remote memories from any node participating to the vSMP cluster (Virtual Symmetric Multiprocessing). Figure 1 shows an example of vSMP clusters that may be defined at run-time using arbitrary size and geometry. They may communicate with each other using message passing, but be careful with the placement, because as shown in Figure 1, perhaps the PE between clusters may incur a communication performance penalty, depending if this PE is using the NoC to communicate with another PE or not.

The proposed module that serves this purpose, called remote memory access (RMA), is composed of an RMA-Send and an RMA-Reply, which together enable remote memory access through the NoC, more two asynchronous FIFOs are used for communication between both modules and the NoC. The RMA module contains a cache miss handling protocol that comprises three main steps: (*i*) whenever a cache miss occurs, the CPU issues a cache line request routed through the NoC, (*ii*) the host RMA reads the desired cache line (i.e. instruction/data), which is sent back to the remote CPU that resumes thread execution as soon as cache line is received (*iii*), as illustrated in Figure 2.

The remote memory access protocol has a latency of 182 clock cycles at zero NoC load, from the cache line request to the remote thread execution, which is depicted in details in Figure 3, remembering that this latency only occurs when trying to access the shared memory and not accessing the local memory during a cache miss. The protocol and its implementation have been optimized for latency and not throughput, as cache miss traffic is rather more latency-sensitive. Low to moderate latencies are ensured up to an end-to-end bandwidth of 90MB/s at 500MHz.

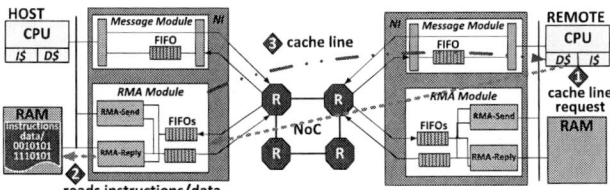

Figure 2 - Cache miss RMA protocol.

Note that the host RMA is kept busy during 64 cycles: since there is one input fifo and output fifo, a new request can be serviced before the cache line write to NoC (64 cycles), which corresponds to cache line request acquisition (15 cycles) and local memory read (14+35 cycles), as shown in Figure 3. This results in a maximum theoretical bandwidth of 250MB/s when several requests are aliased.

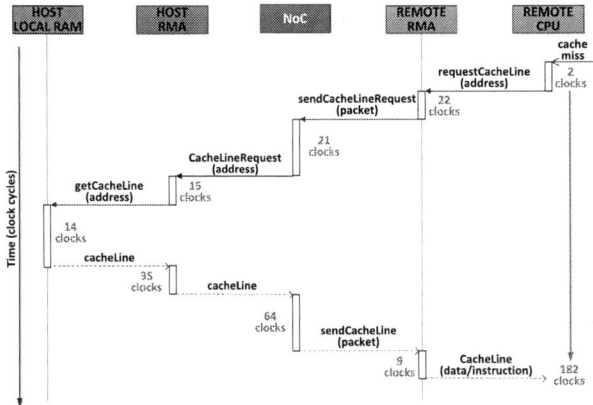

Figure 3 - Cache miss RMA protocol in a READ transaction.

3.3 Multithreading software support

The in-house RTOS described in 3.1 was modified to support a configurable processor memory mapping that permits specifying the ratio of local private data versus local shared data, made visible to all processors in the cluster as shown in the Figure 4.

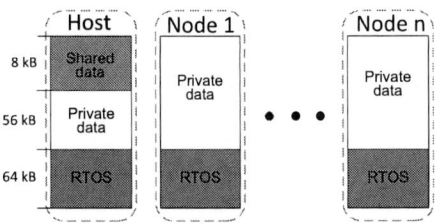

Figure 4 - DSM node memory organization.

Moreover, a subset of the widely used POSIX multithreading API has been ported into the microkernel, as shown in Table 1. The implementation of this API was focused on three categories: thread creation, mutexes and barriers for thread mutual exclusions and synchronizations. The memory consistency model relies only on the memory coherence guaranteed at each API function call: which corresponds to a relaxed memory consistency model, i.e. having a coherent memory whenever synchronization is made. The protocol operates as follows:

- During a thread creation, data caches are flushed on the caller side and invalidated on the executer side.

- During a mutex lock, cache lines that possibly contain shared data that can be accessed between the locking and unlocking are invalidated.

- During a mutex unlock, those same lines are flushed.

- During the barrier calls, the shared memory is also flushed and invalidated.

Table 1 - Implemented Pthread primitives.

Pthread primitives	Description
pthread_create()	Creates a new thread in the calling process
pthread_exit()	Terminates the calling thread
pthread_join()	Waits for the specified thread to terminate
pthread_mutex_init()	Initializes the specified mutex
pthread_mutex_destroy()	Destroys the specified mutex
pthread_mutex_lock()	Locks the mutex object
pthread_mutex_unlock()	Unlocks the mutex object
pthread_barrier_init()	Initialize a barrier object
pthread_barrier_wait()	Synchronizes participating threads at the barrier pointed to by the barrier argument

This is the smallest protocol that ensures cache coherency hardware and software support: the invalidation and flush operations of any cache line are performed only if the cache line tag corresponds to the address specified by the instruction. This condition avoids unnecessary cache flushes / invalidations of cache lines containing unrelated data.

4. EXPERIMENTAL SETUP

4.1 Application kernels

Three application kernels often employed in multithreaded architectures benchmarking were selected: Smith Waterman (DNA sequence-alignment matching application), MJPEG encoder and FFT (Fast Fourier Transform). A pthread implementation was developed for each, in which threads granularity is sound. MJPEG threads process entire images, whereas Smith-Waterman implementation relies on worker threads that run sequence alignment algorithm on different data sets. FFT is made of a number of threads, each of which performs a number of butterfly operations for minimizing data transfers. Table 2 gives benchmark-specific figures expressed in terms of single-thread processing time (using 16kB of cache), computation to data ratio and code size. Due to the different natures of those applications, execution time varies greatly and ranges from less than 2 million clock cycles (4ms at 500MHz) to more than 13 million clock cycles (26ms at 500MHz).

Table 2 - Performance information of adopted applications.

Application name	Single thread execution time	Computation to Data Ratio	Instruction size of a thread
MJPEG	5.3 Mega cycles	5.34	52kB
Smith Waterman	13.2 Mega cycles	7.08	3.8kB
FFT	1.9 Mega cycles	4.18	5kB

The second important factor provided by this table is the computation to data ratio. This number is the ratio between the number of actual compute instructions (e.g. add, mul, jump, etc.)

over the number of memory instruction (load / store) executed by a thread, which gives an overview of the importance of instruction loading in regards of data loading. Hence, an application having a high computation to data loading ratio would be instruction accesses dominated, with therefore proportionally less data accesses. Note that Smith Waterman has the highest computation data, when compared to the others. Fourth column in Table 2 shows the thread code size of each application. A thread with large code size would potentially lead to more instruction cache misses, because a smaller fraction of it would fit in the cache.

4.2 Reference Platforms

GEM5 Simulator [8] was used to produce the reference platforms. It was chosen because its good tradeoff between simulation speed and accuracy, besides modeling a real Realview Platform Baseboard Explore for Cortex-A9 (ARMv7 A-profile ISA) [9].

To create the desired ARM system, the CPU model and Linux bootloader have been modified to enable configurations with more than 4 cores and different interconnection network topologies were evaluated: (*a*) bus-based and (*b*) mesh. The reference platforms are configured as follows: (*i*) up to 8 ARM Cortex-A9, (*ii*) CPU running at 500MHz, (*iii*) Linux Kernel 2.6.38, (*iv*) 16kB private L1 data and instruction caches, (*v*) 32bits channel width, (*vi*) DDR physical memory running at 400MHz and (*vii*) 256MB unified L2 cache (only for mesh network) where the MOESI Hammer [10] is the cache coherence protocol used in these topologies. It is noteworthy that to avoid any traffic with DDR physical memory during benchmark execution was selected this huge L2 cache size.

4.3 Platform Setup

The platform is configured as follows: (*i*) 3x3 processor array, NoC with 4 position input buffers and 32 bits channel width, (*ii*) processor cache size was set to 4kB, 8kB and 16kB, 8 words per lines, direct mapped caches, (*iii*) 500MHz frequency for both processor nodes and NoC routers and (*iv*) CPU with hardware multiplier, divider and barrel shifter. Figure 5 shows the thread mapping used for the experiments. In those, shared data reside in the top-left processor node, and are accessed by the threads running in other nodes (one per node).

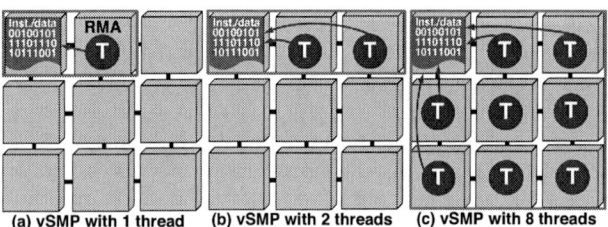

(a) vSMP with 1 thread   (b) vSMP with 2 threads   (c) vSMP with 8 threads

Figure 5 - Adopted scenario.

These mappings incur higher cache miss latencies as all traffic converges and flows through south and east ports of the top-left processor node. Results have however shown that mapping influence on performance remains below 5% in our setup, because of the NoC link bandwidth (1GB/s) corresponding to less than 20% bandwidth usage in all used configurations. All results were gathered on a synthesizable RTL VHDL description of the architecture. Note that all features inherent to prototype (e.g. run-time vSMP cluster definition) were also evaluated in FPGA (i.e. simplistic scenarios due to the memory limitation).

5. RESULTS

5.1 Speedup

The first experiment evaluates the platform scalability considering all applications as shown in Figure 7. Architecture scalability was evaluated and compared to different interconnection network topologies made from 1 to 8 ARMv7 CPU cores with cache sizes matching those of our vSMP MPSoC. In order to maximize speedup, one thread per node was used in all experiments, thereby avoiding performance penalties resulting from context switching. Figure 7a shows near-perfect linear speedup for the Smith-Waterman application, for all platforms. As being very compute oriented (as shown in the third column of Table 2), few data communication occur and thread code fit in the local caches for all tested configurations (as shown in the fourth column of Table 2), resulting in limited cache miss rate, but even under these conditions, our proposed MPSoC has a better performance compared to the mesh reference platform in which reached his maximum speed up of 6.3. In turn, b shows a similar speedup for cache sizes of 8kB and 16kB in the MJPEG, this can be easily explained by observing the behavior of the application shown in Figure 6, where it looks totally scalable, since the barriers would be found only at the end of each processing frame. But even though a scalable application, the shared-bus platform reaches a plateau from 5 cores, while the mesh platform reaches a plateau from 6 cores, they are due to the bus saturation. The same behavior is also observed with the proposed accelerator when using cache sizes of 4kB, but this will be explained in details in Section 5.2.

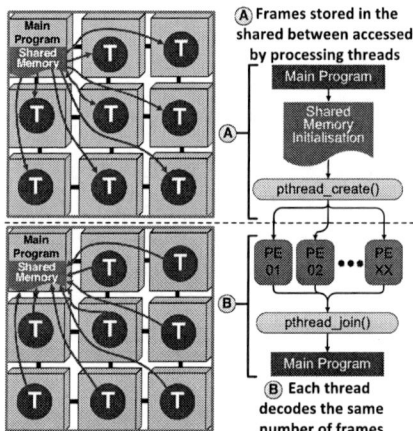

Figure 6 - Behavior of MJPEG Pthread implementation.

Figure 7c shows the performance scalability of FFT, regarding the adopted scenario (Figure 5). Due to the sequential synchronizations, the parallelization capability of FFT is low. Thus, with a 16kB cache configuration, the application can only achieve a speedup of 4.6. However, such results show a better performance when compared to other systems, which no one overcomes a speed up of 2.8. Different from the previous benchmarks, the FFT application was employed to explore synchronization primitives (e.g. Mutexes and Barrier) when multiple threads issue concurrent accesses to same data/variables (e.g. during a read/write operation). In this situation, Mutexes are used to allow exclusive Pthread data access and Barrier are used to coordinate the parallel execution of threads during their synchronization.

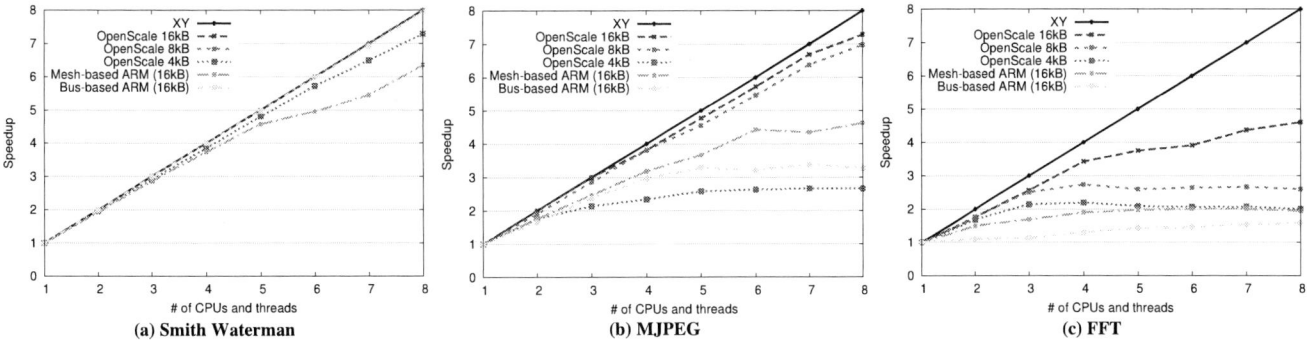

(a) Smith Waterman   **(b) MJPEG**   **(c) FFT**

Figure 7 - Architecture scalability, considering different applications and varying the cache memory size in 4kB, 8kB and 16 kB.

Another example to demonstrate this synchronization is shown in Figure 8, where the Mutex primitive is used to identify each thread (Thread_id) before starting their execution by using a single variable (shared_var) into the Host shared memory (i.e. instructions / data). This part of the application is inherently sequential due to the time of creation and identification of threads.

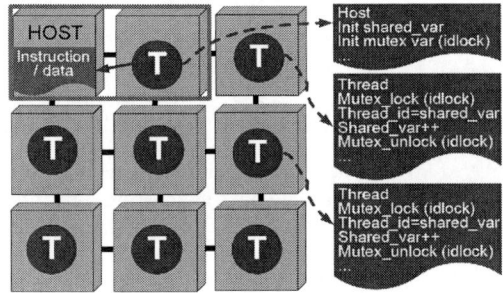

Figure 8 - Example of Mutex variable implementation.

5.2 RMA and NoC Throughput

Figure 9 shows the average bandwidth during MJPEG execution for the two used NoC links at the host level (south and east) as well as the RMA, which is the aggregated bandwidth of those. Thread mapping plays an important role in the NoC usage, as explained in Section 4.3.

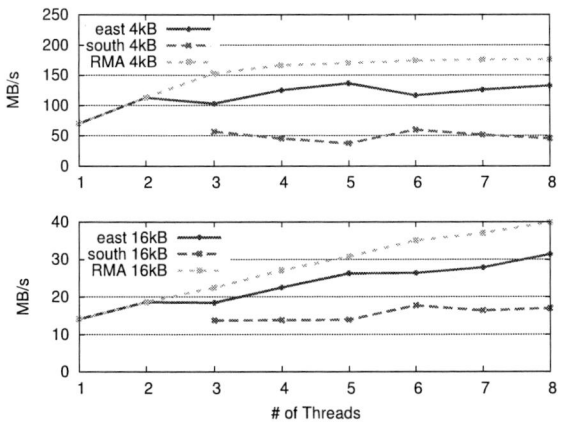

Figure 9 - Average bandwidth for the MJPEG application.

It can be observed that south link is unused for 1 and 2 threads, because of the decided mapping. But as the number of threads grows, the south link begins to be massively used (XY routing)

causing most of the host data flows through this south NoC link. Although overall the bandwidth grows linearly versus the number of threads for 16kB cache sizes, which is similar to 8kB. Using 4kB cache sizes results in a significant bump in bandwidth usage, mostly because of much increased instruction cache miss rate. A plateau is observed from 4 threads at around 200MB/s, which is about 80% of the maximum theoretical bandwidth of the RMA module (250MB/s). This explains the plateau observed in speedup in Figure 7b, because of the RMA saturation.

Figure 10 shows the average bandwidth for the FFT application. Similar to 4kB MJPEG scenario, saturation threshold around 200MB/s is achieved when more than 4 threads are executed considering 4kB cache configuration (similar to 8kB cache). This explains the plateau observed for speedup illustrated in Figure 7c. However, the average bandwidth of the 16kB scenario does not reach the saturation threshold, which infers that the maximum speed up of 4.6 (Figure 7c) is not related to the communication, but rather to the application behavior explained in Section 5.1 and shown in Figure 6.

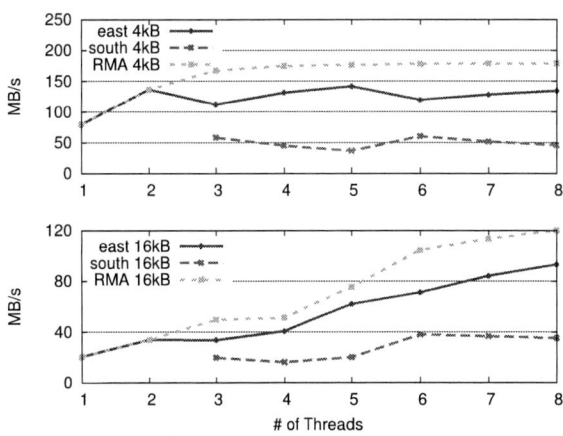

Figure 10 - Average bandwidth for the FFT application.

5.3 Miss Rate and Cache Miss Latency

Figure 11 shows both the average instruction and data cache miss latencies for MJPEG. Excellent scalability is observed for 2 cache configurations (8kB and 16kB), whereas 4kB configuration shows a steep latency increase due to the RMA saturation (Section 5.2). The average cache miss latencies for (a) instructions and (b) data for the FFT application for 4kB, 8kB and 16kB cache configuration is illustrated in Figure 12.

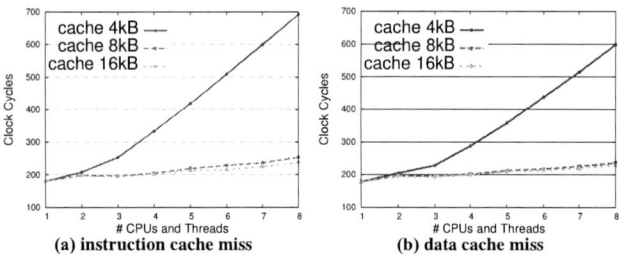

(a) instruction cache miss   **(b) data cache miss**

Figure 11 – Average instructions and data cache miss latencies for MJPEG.

Note that the saturation factor is related to the instruction cache miss latency. For instance, for the 4kB and 8kB scenarios the instruction cache miss latency increases almost linearly from 4 to 8 threads. In turn, due to the small number of data cache miss the latency does not have the same behavior.

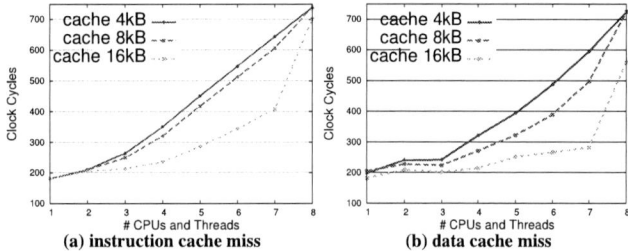

(a) instruction cache miss   **(b) data cache miss**

Figure 12 – Average instructions and data cache miss latencies for FFT.

5.4 Communication Overhead

As pointed out during the introduction, the placement of the desired cluster may interfere in the system performance. To demonstrate this situation the following scenarios are proposed: (*i*) 3 different clusters are defined at run-time by running the same FFT application without any communication between them, (*ii*) the application running in the first and third clusters has been modified to exchange results between them, while the second cluster will not run, and (*iii*) is a mix between the first two experiments, as shown in the Figure 13.

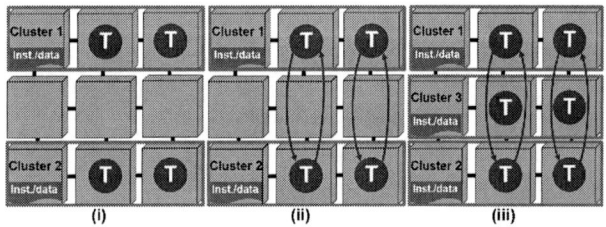

(i)   **(ii)**   **(iii)**

Figure 13 – Scenarios used to explore communication interference.

Table 3 shows that the communication overhead is 9.82%, and when the second cluster is running the first and third clusters are affected by the system and the execution time increases more 19.32%, showing qualitatively as interference can impact on the system.

Table 3 – Execution time considering scenarios with and without communication interference between clusters.

Scenarios	Execution Time (*Clock cycles*)	Interference
i	2356281	------
ii	2578409	9.42 %
iii	3076780	19.32 %

5.5 Placement Overhead

Another experiment which shows the placement overhead is demonstrated in Figure 14, where three scenarios are proposed varying the host position: (*i*) default scenario, where two threads will communicate through a port while all other threads will communicate using another, (*ii*) similar to previous, this scenario will decrease NPU-Host distance from 4 to 3 hops and (*iii*) the best scenario, where NPU-Host distance is balanced and in the worst case is just 2 hops away.

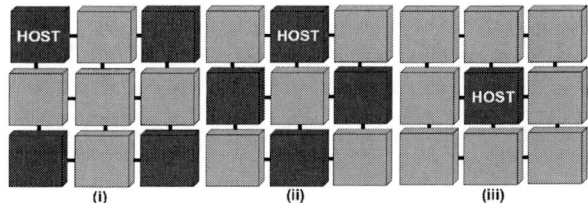

(i)   **(ii)**   **(iii)**

Figure 14 – Scenarios used to explore host position.

As displayed in Table 4, the best gain using cache size of 8kB or 16kB is about 3% by changing only the host position, this can be explained by the fact that the MJPEG application is scalable and the NoC was not saturated in these scenarios. But when the NoC is saturated (4kB cache size), the placement can be very weighty, showing a gain of almost 10% using the best configuration.

Table 4 – Larger differences exploring the host position.

Cache Size	Performance best case		
	Host_00	Host_10	Host_11
4 kB	100%	98.81%	90.51%
8 kB	100%	97.10%	96.84%
16 kB	100%	98.62%	96.90%

5.6 Area Overhead

To evaluate the area overhead of the proposed idea, two approaches were used: (*i*) ASIC and (*ii*) FPGA.

In these results, the processor architecture was shifted to a floating-point capable low power embedded processor [11]. Moreover, the synthesizable version has no optional IO (e.g. no UART). Table 5 shows that this contribution adds an area overhead between 5,72% and 15,59%, using respectively 256kB and 64kB of RAM in a 40nm CMOS Technology.

An additional analysis was performed using a Xilinx Spartan-3 FPGA (XC3S1000), where system with the RMA module incurs an area overhead ranging from 7% to 15% also depending on the memory size.

588

Table 5 – PE area evaluated at 40nm CMOS Technology.

Memory Size	Low-Power core with FP	Low-Power core with FP + RMA	Area Overhead
64 kB	$0.2803\ mm^2$	$0.3240\ mm^2$	15.59 %
128 kB	$0.4392\ mm^2$	$0.4829\ mm^2$	9.94 %
256 kB	$0.7650\ mm^2$	$0.8088\ mm^2$	5.72 %

6. CONCLUSION

This paper presents a multithreading approach for a distributed memory NoC-based MPSoC. The proposed solution was validated on FPGA and ASIC and incurs a limited area overhead (around 10%), because it relies on a relaxed memory consistency model that permits avoiding complex cache coherence protocols. This combined to the intrinsically low latencies of on chip NoC communications results in a promising solution that shows good performance scalability and enables facilitated programming through the use of POSIX thread API.

Future explorations will aim at investigating power efficiency and scalability with hundreds of processor nodes, besides also evaluate qualitatively our approach against NUMA systems. Some enhancements of this design are also envisioned, such as aliasing of both data and instruction spread over several processor nodes so as to better balance memory accesses, avoid code replication (such as microkernel) and achieve higher performance.

7. ACKNOWLEDGEMENT

The research leading to these results has received fund- ing from the European Communitys Seventh Frame- work Programme (FP7/2007-2013) under the Mont-Blanc Project: http://www.montblanc-project.eu, grant agree- ment no 288777.

8. REFERENCES

[1] Marculescu, R., Ogras, U.Y., Li-Shiuan Peh, Jerger, N.E. and Hoskote, Y. 2009. Outstanding Research Problems in NoC Design: System, Microarchitecture, and Circuit Perspectives." *IEEE Transactions on Computer-Aided Design of Integrated Circuits and Systems*, vol. 28(1), 3-21.

[2] Marongiu, A. and Benini, L. 2010. An OpenMP Compiler for Efficient Use of Distributed Scratchpad Memory in MPSoCs. *IEEE Transactions on Computers*, vol. 62(1), 222-236.

[3] Ceng, J. 2008. MAPS: An Integrated Framework for MPSoC Application Parallelization. In *Design Automation Conference* (DAC), 754-759.

[4] 9945-1I. 1996. The POSIX threads standard.

[5] Baert, R. 2009. Exploring Parallelizations of Applications for MPSoC platforms using MPA. In *Design, Automation and Test in Europe Conference* (DATE), 1148-1153.

[6] Ophelders, F. 2009. A Tuneable Software Cache Coherence Protocol for Heterogeneous MPSoCs. Master's Thesis, Eindhoven University of Technology.

[7] Busseuil, R., Barthe, L., Almeida, G. M., Ost, L., Bruguier, F., Sassatelli, G., Benoît, P., Robert, M. and Torres, L. 2011. Open-scale: A scalable, open-source noc-based mpsoc for design space exploration. *International Conference* on *Reconfigurable Computing and FPGAs*. 357–362.

[8] Binkert N. 2011. The gem5 simulator. *ACM SIGARCH Computer Architecture News*, vol.39 (2).

[9] ARM. 2011. RealView Platform Baseboard Explore for Cortex-A9 User Guide, DUI0440B.

[10] AMD, 2002. AMD Opteron Shared Memory MP Systems.

[11] ARM, 2009. Cortex-M4 Technical Reference Manual, DDI0439C.

APPLE: Adaptive Performance-Predictable Low-Energy Caches for Reliable Hybrid Voltage Operation

Bojan Maric[§,‡], Jaume Abella[§], Mateo Valero[§,‡]

[§]Barcelona Supercomputing Center (BSC-CNS)   [‡]Universitat Politecnica de Catalunya (UPC)

{bojan.maric, jaume.abella, mateo.valero}@bsc.es

ABSTRACT

Semiconductor technology evolution enables the design of resource-constrained battery-powered ultra-low-cost chips required for *new market segments* such as environment, urban life and body monitoring. Caches have been shown to be the main energy and area consumer in those chips.

This paper proposes simple, hybrid-operation (high Vcc, ultra-low Vcc), single-Vcc domain Adaptive Performance-Predictable Low-Energy (APPLE) L1 cache designs based on replacing energy-hungry SRAM cells by more energy-efficient and smaller cells enhanced with extra cache lines set up in an adapted victim cache to still enable strong performance guarantees. APPLE caches are proven to largely outperform existing solutions in terms of energy and area efficiency.

Categories and Subject Descriptors

B.3.2 [**Memory structures**]: Design styles—*Cache memories*; B.8.1 [**Performance and Reliability**]: Reliability, Testing, Fault-Tolerance

General Terms

Design, Performance, Reliability

Keywords

Cache, Faults, Predictable Performance, Low Energy

1. INTRODUCTION

Aggressive silicon geometry scaling opens the door to *new market segments*, encompassing a vast array of emerging applications such as environment sensors to monitor wind, sea level, temperature, tsunamis, biomedical and healthcare sensors to monitor the body, etc. In particular, technology evolution enables the design of battery-powered ultra-low-cost (e.g., below 1 USD) computing devices to achieve the main requirements for this new market segment: (i) ultra-low energy consumption in order to extend battery lifetime, (ii) very simple system design for increased yield and reduced cost and (iii) strong functional and timing guarantees for running critical applications on top.

Those new applications are *low duty cycle*, thus characterized by long time periods (which vary for different applications) of minimal data processing interleaved with short bursts of intense computation over larger data sets [20]. Typically, those computing systems have two operation modes with different needs and different optimal supply voltages (Vcc): (i) high-performance and low-power operation mode under high or moderate voltage (HP mode for short) during

relatively short periods of time to react to some infrequent particular events (e.g., 0.01% - 1% of time) and (ii) low performance, ultra-low energy and reliable operation mode under near-/sub-threshold (NST) voltage (ULE mode for short) during most of the time (e.g., 99% - 99.99% of time) until infrequent events arise [20]. Multiple Vcc domains may be used to implement HP and ULE modes, but they impose an unaffordable high design cost and complexity for our target market. Alternatively, cheap solutions based on a single-Vcc domain have been demonstrated recently [13].

Cache memories are used in those systems to increase performance by reducing the number of slow and energy-hungry memory accesses. However, caches become the main energy consumer on the chip [9]. Those caches use large memory cells to achieve high levels of reliability even at ULE mode as needed by critical applications run on top [13]. Decreasing the size of the memory cells for higher energy efficiency at the expense of higher failure rates is unacceptable in this environment. Faulty entries, basically randomly distributed, should be then disabled [23, 1] or replaced by error detection and correction schemes [23, 4]. However, albeit all those disable-based and correction methods are effective from an average performance perspective and provide functional guarantees, they suffer from a common problem, i.e., *they do not provide strong timing guarantees* required for worst-case execution time (WCET) estimation as needed for critical applications in our target market [22]. Therefore, our aim is devising new energy-efficient fault-tolerant caches without decreasing reliability levels to still provide strong performance guarantees.

In this paper, we propose efficient, but simple Adaptive Performance-Predictable Low-Energy (APPLE) single-Vcc domain L1 cache designs for reliable hybrid voltage operation, which meet *all* stringent needs of our target market. In particular, APPLE caches rely on replacing large memory cells by more energy-efficient and smaller cells enhanced with extra cache lines set up in a cache-assist structure, i.e., an adapted victim cache, to allow **extra associativity for some cache sets** that may need it due to disabled faulty cache lines. Experimental results show that APPLE caches achieve significant average energy and area reductions (42% and 12% for energy at HP and ULE mode respectively and 23% for the area) with a negligible *average* performance impact (less than 1.7% on average) with respect to existing solutions [13, 26, 8] without jeopardizing reliability levels to still provide strong performance guarantees.

The rest of the paper is organized as follows. Section 2 discusses related work. Section 3 presents APPLE cache designs. Section 4 describes our evaluation methodology and reports experimental results. Section 5 concludes the paper.

2. RELATED WORK

Differential 6T (6 transistors) SRAM cells have been commonly used for high voltage operation. However, many circuit-level techniques investigate the benefits of using alternative types of SRAM cells such as 8T [16], Schmitt-Trigger 10T (10T) [11], etc. to target different low voltage and robustness

---

Permission to make digital or hard copies of all or part of this work for personal or classroom use is granted without fee provided that copies are not made or distributed for profit or commercial advantage and that copies bear this notice and the full citation on the first page. To copy otherwise, to republish, to post on servers or to redistribute to lists, requires prior specific permission and/or a fee.

DAC '13, May 29 - June 07 2013, Austin, TX, USA

scenarios. Unfortunately, such memory cells substantially increase area and energy at high voltage w.r.t. 6T cells, which is unaffordable in embedded cache design if used extensively.

To the best of our knowledge, three state-of-the-art cache designs, which provide *functional and timing guarantees*, have been proposed in the past [26, 8, 13]. Zhou et al. [26] propose downsizing 6T cells of on-chip caches combined with error correction codes. Ghasemi et al. [8] propose mixing heterogeneous cell sizes of the *same* SRAM cell types. Maric et al. [13] propose hybrid voltage operation, single-Vcc domain cache designs, suitable for our target market and, therefore, are used as our baseline. The main drawback of these designs is that they use exclusively large SRAM cells in order to provide reliable high and NST voltage operation. This fact is particularly true for designs proposed in [26, 8] because they are devised for the high-performance market and high voltage operation. However, we put them in the context of hybrid-voltage operation implemented with a single-Vcc domain and use them for comparison purposes. As shown later, APPLE caches outperform those designs in all metrics.

The simplest way to achieve higher energy efficiency is decreasing the size of SRAM cells at the expense of higher failure rates which is particularly critical at NST voltage. Faulty cache entries should be then disabled or replaced. Techniques based on replacing faulty cache entries [23, 4] introduce significant overheads due to bypassing and signal re-routing. Some recent work [2] is somehow in spirit similar to ours by using extra hardware to provide time predictability. However, authors substantially complicate cache operation. Moreover, those designs are intended for single voltage operation. Techniques based on simply disabling faulty storage [23, 1] may provide noticeable performance variation for a given program depending on the faults location because the distribution of faulty bits is random. Such techniques are shown to be effective from an average performance perspective and provide functional correctness, *but fail to provide strong timing guarantees required for WCET estimation, as needed for critical applications in our target market (e.g., monitoring of the human body vital signs such as heart attacks, strokes, etc.)* [22].

In order to overcome those issues, we propose APPLE cache designs, which enable SRAM cells downsizing for higher energy efficiency without jeopardizing reliability levels to still provide strong performance guarantees.

3. APPLE CACHE

In this section, we describes the chosen baseline cache architecture and the proposed APPLE cache designs.

3.1 Baseline Cache Architecture

We have chosen a set-associative cache organization and LRU replacement policy as the target of our study, given that most L1 caches in existing embedded chips implement them, although significant parts of our study can be easily reused for other cache organizations and replacement policies.

We use a hybrid-operation, single-Vcc domain cache design particularly suited for our target market [13] as the starting point. The cache is designed in such a way that some of the cache ways are optimized to satisfy high performance requirements during high Vcc operation (HP ways) whereas the rest of the ways provide ultra-low energy consumption and reliability during NST Vcc operation (ULE ways). During ULE mode, data processing is expected to be minimal and workloads are much smaller than during HP mode [20]. Workload discrepancy across HP and ULE mode justifies reducing the hardware resources at ULE mode. Since HP ways would experience many faults at NST Vcc and thus would not provide reliable operation, they are turned off at ULE mode. However, all cache ways are enabled at HP mode

to fit larger workloads and provide high performance. ULE ways are reused at HP mode in spite of their inefficiency at high Vcc because they reduce the number of slow and energy-hungry memory accesses [14].

In particular, we use an 8KB 6T+10T hybrid cache as the baseline [13] with 8 ways and 32B/line, where 6 ways are implemented with differential 6T cells and 2 ways with 10T cells [11], although our proposal is not limited to this configuration.

3.2 APPLE: Adaptive Performance-Predictable Low-Energy Cache Design

APPLE cache design relies on replacing large and energy-hungry (*strong*) SRAM cells by energy-efficient and smaller (*weak*) cells in a selected cache subset (e.g., some cache ways in a set-associative cache) enhanced with extra cache lines set up in a cache-assist structure, i.e., an adapted victim cache [10] to keep the same reliability levels despite the potentially disabled faulty cache lines. We illustrate our cache designs with a scenario where *strong* 6T cells ($6T_S$) in **all** HP ways in the baseline are replaced by *weak* 6T cells ($6T_W$) (see Figure 1(a)). Analogously, *strong* 10T cells ($10T_S$) in **all** ULE ways can be replaced by *weak* 10T cells ($10T_W$).

Reliability of ULE ways at HP mode is not an issue because $10T_W$ cells are still largely more robust than $6T_W$ ones at high voltage. However, some faults may be expected in HP ways due to decreased robustness of the $6T_W$ cells. Then, faulty cache lines must be disabled which makes performance unpredictable.

To avoid this, we set up a new structure in the cache in order **to have extra associativity for some cache sets that may need it due to disabled faulty cache line(s)**, thus guaranteeing the same number of fault-free cache lines per set as in the baseline. We refer to this structure as High Performance Victim Cache (HPVC). The HPVC is an adapted victim cache with spare entries. Conventionally, victim caches are fully-associative and their entries are comprised of a *valid bit* and *tag and data* space to store a cache line. Given that the *set number* is already part of *tag* (see Figure 1(a)), the HPVC requires one extra field, i.e., *lock bits* - indicating whether a particular victim cache entry is being used for replacement of a faulty cache line. Similarly to [10], the HPVC is accessed in parallel with the L1 cache, thus without any impact on the L1 cache latency.

Similarly to existing fault-tolerant state-of-the-art approaches [23, 1], tags are extended with one (faultiness) bit, indicating whether the data block is faulty and there are mechanisms in place to detect and disable faulty storage as well as configure the extra fields of the HPVC properly at boot time. Fault detection, correction and diagnosis are out of the scope of this paper. Replacement bits, dirty line and valid bits are assumed to be hardened (e.g., with $10T_S$) as their relative impact on cache energy and area is negligible.

The number of acceptable faulty cache lines can be less or equal to the number of extra HPVC entries. Therefore, using a number of extra entries increases the fault tolerance of the cache. The HPVC is implemented with $6T_W$ cells, so the faults may also be expected in the HPVC itself. In case of a faulty entry in the HPVC, it will be disabled and there will be one less HPVC entry. Therefore, the HPVC replaces faulty entries in both L1 cache and HPVC itself. We consider this issue when calculating the number of HPVC entries required, as described later.

The cache works as follows:

1.) Cache hit. In this scenario, a hit cache line is served as usual either in the L1 cache or in the HPVC (accessed in parallel). In order to keep replacement (i.e., LRU) information consistent as in a fault-free system, there are two cases

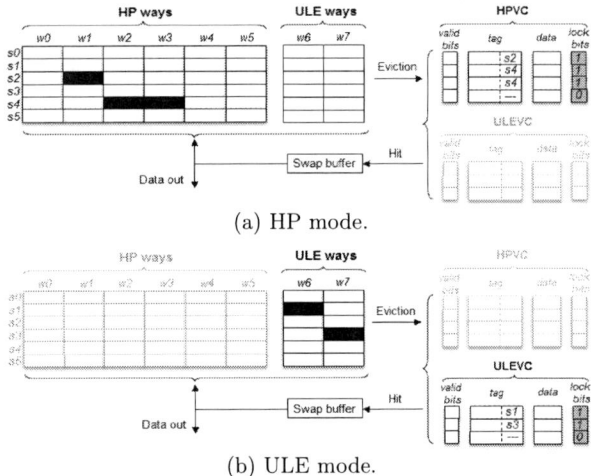

(a) HP mode.

(b) ULE mode.

Figure 1: Example of the APPLE cache. Black fields stand for faulty cache lines. Grey fields stand for the extra information required by the HPVC (ULEVC) for performance guarantees. Pale blocks are disabled ones.

depending on whether a hit occurs in the L1 cache or in the HPVC:

(a) If a hit occurs in the L1 cache, LRU information is updated as in a fault-free system.

(b) If a hit occurs in the HPVC (either locked or non-locked entry), the hit cache line is swapped with the LRU Fault-Free (LRU-FF[1]) cache line in the corresponding cache set. During the regular access, hit cache line is both gated onto the data bus and written to the swap buffer. Then, swapping is done in two phases: (i) the LRU-FF cache line is written to the HPVC into the hit cache line position and (ii) swap buffer content is written to the LRU-FF position in the corresponding cache set. In order to perform swapping successfully, we stall any access to the cache during swapping for the latency of 2 cache accesses. Preventing access to the cache during some cycles is as easy as keeping the cache ports busy to prevent the port arbiter from issuing new accesses. Swapping ensures that the Most Recently Used (MRU) line resides in the L1 cache while the LRU is in the HPVC simultaneously. Since the LRU lines typically experience a few hits, the performance impact is expected to be small.

2.) Cache miss. In case of a miss, a new line from memory is fetched and placed into the LRU-FF position in the corresponding cache set and becomes the MRU line. The LRU-FF line is moved to the HPVC. If there is any locked entry for that set, the LRU-FF locked line is replaced with the line coming from the cache. Otherwise, a non-locked LRU-FF line is replaced. The line replaced in the HPVC is simply evicted to the memory.

Note that such data movement can be afforded because it happens in parallel with retrieving data from memory and its delay (in the order of 2 cycles) is largely below the latency needed for data from the main memory to reach the L1 cache.

At ULE mode, HP ways and the HPVC are disabled (see Figure 1(b)). Following the same philosophy described for HP mode, ULE ways implemented with $10T_W$ cells are enhanced with the Ultra-Low Energy Victim Cache (ULEVC) structure, analogously to the HPVC at HP mode.

3.2.1 Performance Impact

Impact of the victim cache on *average* performance is small

---

[1]The LRU-FF cache line can be selected by slightly-modified LRU stack, similar to that used in the existing line-disabled approaches [23, 4] (see Figure 2). For example, if the LRU cache line is faulty, then the LRU-1 line becomes the LRU-FF. The overhead for the L1 cache is less than 0.1% in transistor count.

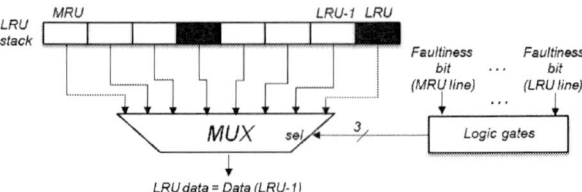

Figure 2: Modified LRU stack. Black fields stand for faulty cache lines.

because cache operation ensures that the LRU lines reside in the victim cache, which typically experience few hits. Regarding weak cells, HSPICE simulations show that the access time variation between strong and weak cells is less than 5% at high voltage, so the impact of those weak cells in the overall cache latency at HP mode is either null or can be easily accommodated shaving it from other cache components (at the expense of some extra power in those components). However, this impact is larger at ULE mode. There is a trade-off between increasing cache latency, which has limited impact in performance, or making weak cells not so weak, thus sacrificing part of the energy savings.

In terms of guaranteed (WCET) performance, our cache architecture provides exactly the same number of available fault-free cache lines per set as in the baseline, thus guaranteeing the same WCET performance (same hits and same misses). Therefore, WCET estimation is not more complex than for the baseline architecture.

Turning off HP ways and HPVC at ULE mode (and ULEVC at HP mode) is done by using the gated-Vdd technique [19]. The processor itself is responsible for gating or ungating the corresponding cache block and *writting back dirty lines* on a Vcc change.

Our proposed design is not limited to any particular Vcc level, SRAM cell type or technology node because SRAM cells exhibit the same tradeoff between cell size and failure probability.

3.2.2 Alternative Configurations

For the sake of clarity, we refer to the considered configuration as: $a+b+c+d$, where a and b represent the number of *strong* and *weak* HP ways respectively, and c and d the number of *strong* and *weak* ULE ways respectively. Therefore, our baseline is $6+0+2+0$ whereas the configuration which has been used for illustration of the APPLE design is $0+6+0+2$.

Alternatively, a smaller number of the cache ways may be weakened. For example, instead of weakening all HP and ULE ways ($0+6+0+2$ in Figure 1), only six ways (e.g., ways w2-w7) can be replaced ($2+4+0+2$ configuration).

3.2.3 Implementation Details

Remind that both HP and ULE ways are enabled at HP mode whereas only ULE ways are active at ULE mode. Each SRAM cell is sized by using the analysis based on importance sampling proposed by Chen et al. [7] assuming 6σ random variations in V_{TH} for both high and NST Vcc considering read, write and hold failures in 32nm technology node.

We first describe the cell sizing in the baseline. For the chosen NST Vcc (e.g., 0.35V) and reduced frequency at ULE mode, we size the $10T_S$ cells to provide a 99.9% cache yield. Then, for the chosen high Vcc (e.g., 1V) and increased frequency at HP mode, the $6T_S$ cells are sized to match the same bit failure rate (P_{f6T_S}) as $10T_S$ cells at ULE mode. The yield of *strong* HP ways chosen to be **weakened** is:

$$Y_{strong} = (1 - P_{f6T_S})^{N_{6T_S}} \qquad (1)$$

where N_{6T_S} is number of $6T_S$ cells in the HP ways chosen.

Next, we determine the size of $6T_W$ cells in order to replace $6T_S$ cells in HP ways chosen to be weakened. The algorithm

Table 1: Relative (to $6T_S$) cell area and bit failure probabilities of memory cells used in the baseline (grey-colored) and APPLE configurations targeting 99.9% yield for an 8KB, 8-way, 32B/line cache with a 6-entry HPVC and 3-entry ULEVC.

Configuration	Bit failure probability						Relative cell area					
	$6T_S$	$6T_W$	$10T_S$		$10T_W$		$6T_S$	$6T_W$	$10T_S$		$10T_W$	
			1V	0.35V	1V	0.35V			1V	0.35V	1V	0.35V
6+0+2+0	2.7×10^{-9}	—	1.2×10^{-12}	2.7×10^{-9}	—	—	1	—	1.92		—	
0+6+0+2	—	3.5×10^{-5}	—	—	2.3×10^{-11}	4.6×10^{-6}	—	0.68	—		1.78	
2+4+0+2	2.7×10^{-9}	9.1×10^{-5}	—	—	2.3×10^{-11}	4.6×10^{-6}	—	0.62	—		1.78	

is presented in Figure 3. We first set the minimal transistors sizes possible for $6T_W$ cells (3λ width for all transistors) and then obtain the bit failure probability (P_{f6T_W}) by using Chen's analysis [7]. The yield of *weak* HP ways is then:

$$Y_{weak} = \sum_{i=0}^{N_{hpvc}} \left[\binom{N_{cl} + N_{hpvc}}{i} P_l^{N_{cl}+N_{hpvc}-i} (1 - P_l)^i \right] \quad (2)$$

where N_{cl} is the number of cache lines in HP ways, N_{hpvc} is the number of entries in HPVC and P_l is the probability of a cache line to be fault-free. P_l can be calculated as:

$$P_l = (1 - P_{f6T_W})^{N_{bitsinline}} \quad (3)$$

where $N_{bitsinline}$ is the number of bits in the cache line. Note that we consider the faults in both L1 cache and HPVC itself. We use 6-entries HPVC although our design is not limited to this particular case. N_{hpvc} sensitivity is studied in the evaluation section.

If the obtained yield is lower than required, i.e. Y_{strong}, transistors sizes (widths) are increased by the step value equal to 0.5λ and yield is calculated again. Once yield is high enough, we have an optimal cell size.

The same procedure is used for replacing $10T_S$ cells by $10T_W$ cells in ULE ways when operate at ULE mode, assuming a 3-entry ULEVC. Table 1 presents the obtained bit failure probabilities for all memory cells used at HP mode (e.g., 1V) and ULE mode (e.g., 0.35V) as well as the relative cell sizes.

4. EVALUATION

This section presents the evaluation methodology and results in terms of performance, energy and area for proposed APPLE designs as well as for those used for comparison purposes.

4.1 Methodology

We have chosen a very simple processor architecture with one core and in-order execution, resembling a recently fabricated Intel® processor for hybrid Vcc operation although not suited for the ultra-low-cost market [9] (see Table 2). Both on-chip L1 data (DL1) and instruction (IL1) caches implement the proposed design.

In order to provide statistically meaningful results, we have generated 164 different cache samples for each APPLE cache configuration for a target cache yield. Using 164 different cache samples means that the confidence of our results is 90% with a confidence interval of 10% [15]. The faultiness

for each bit in each cache sample is computed randomly and independently of other bits, matching a given faulty bit rate. Therefore, we generate 164 different processors with a similar number of faults and different faulty bit distribution in DL1 and IL1. We have run each benchmark for each one of the 164 processor instances and present arithmetic mean (average) performance and energy results per benchmark.

4.1.1 Benchmarks

To the best of our knowledge, a set of benchmarks specific for the domain that we target does not exist. We have chosen MediaBench [12] because they fit very well the expected needs of the ultra-low-cost segment: an abundant data processing during HP mode and relatively small workloads at ULE mode [20]. We classify benchmarks into two categories, depending on the cache requirements: (i) *SmallBench* - workloads fit into very small cache sizes (e.g., 2KB) due to small data volume (adpcm_c, adpcm_d, epic_c and epic_d) and (ii) *BigBench* - larger cache space is required to fit the workload due to large data volume (g721_c, g721_d, gsm_c, gsm_d, mpeg2_c and mpeg2_d). *SmallBench* benchmarks are used during ULE operation whereas *BigBench* ones are used during HP operation.

4.1.2 Operating Modes

Our system has two distinct operating modes: HP and ULE. We have set Vcc to 1V and 350mV for HP and ULE modes respectively. Operating frequencies are set to 1GHz for HP mode, and 5MHz for ULE mode, which is in line with the Intel® processor for hybrid Vcc operations [9].

4.1.3 System Modeling

L1 cache memories have been modeled using CACTI 6.5, a flexible and accurate cache delay, energy, power and area simulator [17]. To support two different operating modes, we have extended CACTI tool in order to implement accurate energy models for 6T and 10T SRAM cells when operating at high and NST Vcc by adapting capacitances, resistances and geometry. Details about cache modeling are given in Appendix A.

Several hybrid cache microarchitectures have been implemented using heterogeneous SRAM cell types at a coarse granularity as explained in Section 3. In order to understand the impact of different cache designs on the whole chip, we

---

1. Calculate the yield of the HP ways to be replaced (Y_{strong})
2. Set minimal tranzistor sizes possible (i.e. 3λ width) for $6T_W$ cell for targeted technology node (our is 32nm)
3. Calculate $6T_W$ cell's bit failure probability P_{f6T_W} by using Chen's analysis [7] assuming 6σ random variations in V_{th} for target Vcc considering read, write and hold failures
4. Calculate the yield of the chosen section enhanced with HPVC (Y_{weak})
5. **If** ($Y_{weak} < Y_{strong}$)
5a.   Increase tranzistor sizes (widths) by 0.5λ
5b.   Go to step 3
6. **Else**
6a.   Optimal cell size is obtained

Figure 3: Replacement algorithm.

Table 2: Processor configuration.

Parameter	Description
Core	in-order
Fetch, Decode, Issue, Commit rate	2 instr/cycle
Window Size	8-entry fetch, issue and load/store queue
Functional Units	1 INT ALU (1 cycle), 1 INT Mult/Div (3 cycles mult, 15 cycles div); 1 FP ALU (3 cycles), 1 FP Mult/Div (4 cycles mult, 17 cycles div)
Register file	32 INT (32 bits) + 32 FP (64 bits)
L1 Instruction and Data Cache	8 KB, 8-way, 32 byte per line (2 cycles access)
Main memory	off-chip 8MB SRAM, Latency: 20/14 cycles at HP/ULE mode
ITLB, DTLB	16 entries fully-associative, Miss penalty: 20/14 cycles at HP/ULE mode
Branch Predictor	Hybrid 256B Gshare, BTB with 64 entries and 4-way, 16 entry RAS, 4-entry MSHR. Disabled at ULE mode.
Vcc, frequency	HP mode: 1V and 1GHz, ULE mode: 0.35V and 5MHz
Technology	32nm

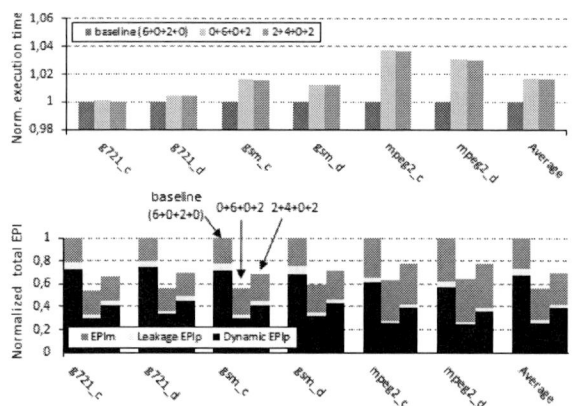

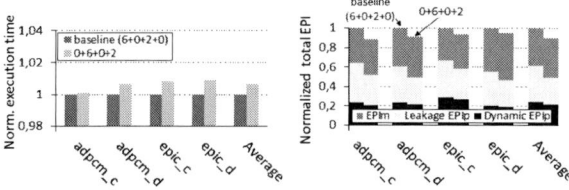

Figure 4: Normalized execution time and total EPI breakdown at HP mode.

Figure 5: Normalized execution time and total EPI breakdown at ULE mode.

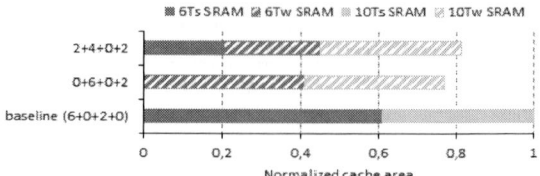

Figure 6: Normalized cache area breakdown.

ULE mode. HP ways and HPVC are turned off at this mode, whereas ULEVC is turned on to operate together with ULE ways (see Figure 1(b)). Results for 2+4+0+2 configurations are omitted because they are identical to those for 0+6+0+2 at ULE mode.

Figure 5 shows the normalized ET and total EPI breakdowns across all benchmarks at ULE mode. Our design exhibits small variation in ET (0.6% on average) basically due to swapping when hits occur in the ULEVC. In terms of EPI, results show a 12% reduction. Both $LeakageEPI_p$ and $DynamicEPI_p$ are reduced due to reduced cache area. Conversely to HP mode, $LeakageEPI_p$ savings are significant because leakage energy becomes a dominant energy component at NST Vcc. EPI_m remains constant because our design has the same guaranteed performance as the baseline.

EPI per hit in HPVC/ULEVC (extra read/write energy for swapping also accounted) is higher for benchmarks which exhibit homogenous hit distribution across the cache lines (i.e. gsm_c, gsm_d, mpeg2_c and mpeg2_d at HP mode and epic_c, epic_d at ULE mode). Nevertheless, such EPI is less than 5% of total EPI at both modes.

4.2.2 Area

Figure 6 presents normalized cache area breakdown for all configurations considered relative to the baseline. Proposed APPLE designs achieve area savings of 23% and 17% for 0+6+0+2 and 2+4+0+2 configurations respectively.

4.3 N_{hpvc} and N_{ulevc} Sensitivity Study

To better understand the effects of the number of HPVC and ULEVC entries, we vary N_{hpvc} (4, 6, 8 and 16) at HP mode. Similarly, we also vary N_{ulevc} (2, 3, 6 and 10) at ULE mode. Table 3 presents performance and energy results averaged across all applications at HP and ULE mode for the 0+6+0+2 configuration for all (N_{hpvc},N_{ulevc}) combinations. All results are normalized with respect to the default (6,3) case (highlighted in Table 3).

From the performance perspective, larger N_{hpvc} and N_{ulevc} values (e.g., 16 and 10 respectively) are not preferable because they increase the number of hits in the HPVC/ULEVC and thus, produce more swaps. From the total EPI perspective, larger N_{hpvc} and N_{ulevc} values increase the fault tolerance of the cache and enable more aggressive SRAM cell downsizing. However, for very high values ($N_{hpvc} = 16$ and $N_{ulevc} = 10$), swap overheads outweights the energy benefits achieved by cell downsizing. Conversely, lower values (i.e. $N_{hpvc} = 4$ and $N_{ulevc} = 2$) offer low fault tolerance and excessively conservative SRAM cell downsizing and thus, higher total EPI. Similar trends are observed for 2+4+0+2 configuration, but results are not reported due to lack of space.

4.4 Comparison with Existing Approaches

Next, we present a detailed comparison between our proposed APPLE caches and existing cache designs with deterministic behavior [26, 8, 13]. In order to perform an accurate comparison, we have implemented those designs for both DL1 and IL1 caches. Caches have been designed to have the same yield at ULE mode (i.e. 99.9%).

We have implemented designs proposed by Zhou et al. [26], where all 8 cache ways are always enabled at both modes

have incorporated our custom-modified CACTI tool into the MPSim [3] full-chip simulator (see Appendix B).

Additional energy and area consumption of the HPVC and ULEVC are measured with CACTI and these values are included in our results. We account in our simulations for the extra read/write energy introduced by swapping as well as additional latency of 2 cache accesses to perform swaps correctly (see Section 3).

4.2 Results

In this subsection, we present overall execution time (ET) and total energy per instruction (EPI) for the whole processor at HP and ULE modes for 0+6+0+2 and 2+4+0+2 APPLE cache configurations. Along with this, cache area results are reported. For the sake of clarity, results have been normalized with respect to the baseline cache, given that the execution time and energy vary noticeably across benchmarks. In order to provide more insights, total EPI is broken down into the following categories: EPI in main memory (EPI_m), leakage EPI in processor ($LeakageEPI_p$) and dynamic EPI in processor ($DynamicEPI_p$). L1 cache yield is 99.9%.

4.2.1 Performance and Energy

HP mode. Figure 4 shows the normalized ET and total EPI across all benchmarks at HP mode. Since caches are the main energy consumer in our extremely simple processors, cache behavior dominates full processor behavior. Results show that the variation in ET is 1.7% on average. This is because some benchmarks have quite homogenous hit distribution across the cache lines (gsm_c, gsm_d, mpeg2_c and mpeg2_d) thus, hit HPVC frequently thus swapping lines.

In terms of EPI, results show that the proposed designs achieve a reduction of 42% and 30% on average for the 0+6+0+2 and 2+4+0+2 configurations respectively. Such reductions come mainly due to smaller transistor sizes needed for $6T_W$ and $10T_W$ cells, thus keeping node capacitances lower and reducing $DynamicEPI_p$. $LeakageEPI_p$ is also reduced, but its relative impact on total EPI is small given that dynamic energy is the dominant energy factor at high voltage. EPI_m remains constant because guaranteed performance of considered configurations is the same as in the baseline.

594

Table 3: Normalized average ET and total EPI when varying the number of HPVC/ULEVC entries for 0+6+0+2 configuration with a 99.9% cache yield.

HP mode		(N_{hpvc}, N_{ulevc})			
		(4,3)	**(6,3)**	(8,3)	(16,3)
	Execution Time	0.988	**1**	1.033	1.074
	Total EPI	1.027	**1**	0.937	1.068
ULE mode		N_{ulevc}			
		2	**3**	6	10
	Execution Time	0.993	**1**	1.013	1.061
	Total EPI	1.018	**1**	0.951	1.049

Table 4: Normalized average ET, total EPI and cache area when comparing existing deterministic caches with APPLE designs targeting a 99.9% cache yield.

	Norm. ET		Norm. total EPI		Norm.
	HP mode	ULE mode	HP mode	ULE mode	cache area
APPLE (0+6+0+2)	1	1	1	1	1
Zhou et al. [26]	1.08	1.13	2.07	4.17	2.27
Ghasemi et al. [8]	0.983	1.49	1.88	2.91	1.96
Baseline [13]	0.984	0.993	1.72	1.14	1.29

and protected with Single Error Correction Double Error Detection (SECDED) codes [6] at cache line granularity. We have accounted energy and area overheads introduced by SECDED check bits in our simulations as well as additional latency of one clock cycle for SECDED encoding/decoding, but we have not accounted the energy consumed by encoding/decoding circuits. The size of 6T cells is calculated according to the methodology proposed in [26] to match the target cache yield at ULE mode.

Regarding the designs proposed by Ghasemi et al. [8], they are in spirit similar to our baseline. The only difference is that larger 6T cells are used in ULE ways instead of the 10T cells. Therefore, at HP mode all cache ways are enabled whereas at ULE mode only 2 cache ways implemented with large 6T cells keep operating. The size of large 6T cells is calculated analogously to the size of the 10T cells in the baseline cache, as explained in Section 3. According to our simulations, the access time variations at 0.35V for SRAM arrays implemented with those large 6T cells is around 40%, relative to SRAM arrays implemented with $10T_W$ cells, so we assume a 3-cycle cache latency instead of the regular 2-cycle latency at ULE mode. Access time variations at 1V are negligible, so a 2-cycle latency is assumed.

Table 4 presents ET, total EPI and cache area results. ET and total EPI results are averaged across all applications at HP and ULE mode. All results are normalized with respect to the proposed 0+6+0+2 APPLE configuration (grey-colored).

Results show that the design in [26] exhibits an ET degradation of 8% and 13% at HP mode and ULE mode respectively due to the extra cycle for SECDED encoding/decoding. On the other hand, APPLE cache has larger ET than the design in [8] at HP mode due to swapping overheads (around 1.7%). However, the design in [8] exhibits significantly larger ET (around 49%) at ULE mode due to the additional cycle for each cache access.

The main drawback of designs in [26] and [8] is their significant area overhead with respect to our proposed design (up to 127%). This directly translates into higher energy consumption at both modes. In other words, considered designs are overdesigned to operate reliably at ULE mode.

Comparison with designs proposed in [13] is already done in section 4.2.

5. CONCLUSIONS

We propose efficient, but simple Adaptive Performance-Predictable Low-Energy (APPLE) single-Vcc domain L1 cache designs for reliable hybrid voltage operation, which meet *all* specific and stringent needs of battery-powered ultra-low-cost (e.g., below 1 USD) systems. The APPLE cache design relies on replacing large memory cells by more energy-efficient and smaller cells enhanced with extra cache lines set up in cache-assist structure, i.e., an adapted victim cache, to allow extra associativity for some cache sets that may need it due to disabled faulty cache lines. Experimental results show that APPLE caches achieve significant average energy and area reductions (up to 42% for energy and 23% for area) with a negligible *average* performance impact (less than 1.7%

on average) with respect to existing solutions without jeopardizing reliability levels to still provide *strong performance guarantees*.

Acknowledgements

This work has been partially supported by the Spanish Ministry of Science and Innovation under grant TIN2012-34557, HiPEAC and the UPC under grant FPI-UPC.

6. REFERENCES

[1] J. Abella et al. Low vccmin fault-tolerant cache with highly predictable performance. In *MICRO*, 2009.

[2] J. Abella et al. Rvc-based time-predictable faulty caches for safety-critical systems. In *IOLTS*, 2011.

[3] C. Acosta, F. J. Cazorla, A. Ramirez, and M. Valero. The MPsim Simulation Tool. Technical Report UPC-DAC-RR-CAP-2009-15, in UPC, 2009.

[4] A. Ansari et al. Archipelago: A polymorphic cache design for enabling robust near-threshold operation. In *HPCA*, 2011.

[5] D. M. Brooks et al. Wattch: A framework for architectural-level power analysis and optimizations. In *ISCA*, 2000.

[6] C. L. Chen and M. Y. Hsiao. Error-correcting codes for semiconductor memory applications: A atate-of-the-art review. *IBM Journal of Research and Development*, 28(2), 1984.

[7] G. Chen et al. Yield-driven near-threshold SRAM design. In *ICCAD*, 2007.

[8] H. Ghasemi et al. Low-voltage on-chip cache architecture using heterogeneous cell sizes for high-performance processors. In *HPCA*, 2011.

[9] S. Jain et al. A 280mv-to-1.2v wide-operating-range ia-32 processor in 32nm cmos. In *ISSCC, dig. Tech. Papers*, 2012.

[10] N. P. Jouppi. Improving direct-mapped cache performance by the addition of a small fully-associative cache and prefetch buffers. In *ISCA*, 1990.

[11] J. Kulkarni et al. A 160 mV, fully differential, robust schmitt trigger based sub-threshold SRAM. In *ISLPED*, 2007.

[12] C. Lee et al. Mediabench: A tool for evaluating and synthesizing multimedia and communication systems. In *MICRO*, 1997.

[13] B. Maric et al. Hybrid high-performance low-power and ultra-low energy reliable caches. In *ACM CF*, 2011.

[14] B. Maric et al. Adam: An efficient data management mechanism for hybrid high and ultra-low voltage operation caches. In *GLSVLSI*, 2012.

[15] D. Moore and G. McCabe. Introduction to the practice of statistics. In *W. H. Freeman and Co.*, 1989.

[16] Y. Morita et al. An area-conscious low-voltage-oriented 8t-sram design under dvs environment. In *IEEE Symposium on VLSI Circuits*, 2007.

[17] N. Muralimanohar, R. Balasubramonian, and N. Jouppi. CACTI 6.0: A tool to understand large caches. *HP Tech Report HPL-2009-85*, 2009.

[18] S. Narendra et al. Full-chip sub-threshold leakage power prediction model for sub-0.18um cmos. In *ISLPED*, 2002.

[19] M. Powell et al. Gated-vdd: A circuit technique to reduce leakage in deep-submicron cache memories. In *ISLPED*, 2000.

[20] R. Szewczyk et al. Lessons from a sensor network expedition. In *European Workshop on Sensor Networks*, 2004.

[21] D. Tullsen et al. Simultaneous multithreading: Maximizing on-chip parallelism. In *ISCA*, 1995.

[22] R. Wilhelm et al. The worst-case execution time problem: overview of methods and survey of tools. *Trans. on Embedded Computing Systems*, 7(3):1–53, 2008.

[23] C. Wilkerson et al. Trading off cache capacity for reliability to enable low voltage operation. In *ISCA*, 2008.

[24] Y. Zhang et al. Hotleakage: A temperature aware model of subthreshold and gate leakage for architects. Number TR-CS-2003-05, 2002.

[25] W. Zhao and Y. Cao. New generation of predictive technology model for sub-45nm design exploration. In *ISQED*, 2006.

[26] S.-T. Zhou et al. Minimizing total area of low-voltage sram arrays through joint optimization of cell size, redundancy, and ecc. In *ICCD*, 2010.

APPENDIX

This section provides details about cache and processor modeling.

A. CACHE MODELING

We first describe how cache modeling has been performed with CACTI tool. Then, we describe how the model is validated using HSPICE.

A.1 Modeling with CACTI

The technology node considered is 32nm. 6T SRAM cells are already modeled in CACTI tool, so we have extended it with delay, power and area models for 10T SRAM cells by adapting capacitances, resistances and geometry. Here, we provide more details about dynamic and leakage power models for caches including 10T SRAM cells (the whole cache or some parts, e.g., some cache ways).

The dynamic power model for 10T cells is anologous to that already implemented for 6T cells. In fact, implementation of the 10T cells is quite similar to 6T ones due to its fully differential architecture. Following the same philosophy, our model tracks the physical capacitance of each stage of the cache model and calculates dynamic power consumed at each stage. Basically, cache dynamic power dissipation is comprised of wordline capacitance dissipation, bitline capacitance dissipation and short-circuit power consumption. Since capacitance plays an important role for dynamic power, we take into account the following capacitances: parasitic capacitances of transistors in SRAM cell, capacitances of the access transistors, capacitance of a pre-charge transistor and capacitances of a column select transistor and a wordline driver. Capacitances of the wordline/bitline wires and wires in decoders are modeled as a distributed RC network.

Given that the impact of process variations is high for the 32nm technology node, especially at NST Vcc (ULE mode), our cache leakage power model is updated to take into account process variations. We model random within-die variations in V_{TH} using the analytical model proposed in [18]. In particular, the cache is decomposed into smaller building blocks and total leakage power is the sum of leakage power in each block. Leakage current of one block, including within-die variations (I_{leak}), is then estimated as follows:

$$I_{leak}^p = \frac{I_{leak-p}^{mean} w_p}{k_p} \frac{1}{\sigma_p \sqrt{2\pi}} \int_{V_{TH_{min}}}^{V_{TH_{max}}} e^{-\frac{(V_{TH}-\mu)^2}{2\sigma_p^2}} e^{-\frac{(\mu-V_{TH})}{a}} dV_{TH} \tag{4}$$

$$I_{leak}^n = \frac{I_{leak-n}^{mean} w_n}{k_n} \frac{1}{\sigma_n \sqrt{2\pi}} \int_{V_{TH_{min}}}^{V_{TH_{max}}} e^{-\frac{(V_{TH}-\mu)^2}{2\sigma_n^2}} e^{-\frac{(\mu-V_{TH})}{a}} dV_{TH} \tag{5}$$

$$I_{leak} = I_{leak}^p + I_{leak}^n \tag{6}$$

where w_p and w_n are the total PMOS and NMOS devices widths in the block; k_p and k_n are factors that determine the fraction of PMOS and NMOS devices widths that are in off state; μ and σ are mean and standard deviation of V_{TH}. a is equal to $n\phi_t$, where ϕ_t is the thermal voltage and $n = 1 + (C_d/C_{ox})$. I_{leak}^{mean} is leakage current of a block with the mean V_{TH} and can be calculated by multiplying the device width and basic leakage per gate (I_{leak}^{gate}). I_{leak}^{gate} is defined as:

$$I_{leak}^{gate} = \beta e^{b(v_{dd}-V_{dd0})} V_t^2 (1 - e^{-\frac{V_{dd}}{V_t}}) e^{\frac{-|V_{TH}|-V_{off}}{nV_t}}. \tag{7}$$

We refer the reader to [24] for a description of the terms. According to [18], integrals in equations (4) and (5) can be simplified, so I_{leak} can be expressed as:

$$I_{leak} = \frac{I_{leak-p}^{mean} w_p}{k_p} e^{\frac{\sigma_p^2}{2\lambda_p^2}} + \frac{I_{leak-n}^{mean} w_n}{k_n} e^{\frac{\sigma_n^2}{2\lambda_n^2}} \tag{8}$$

where λ_p and λ_n are constants that relate channel lengths of PMOS and NMOS transistors to their corresponding subthreshold leakage current.

All SRAM cells have been sized as described in Section 3. The smallest rectangle where the cache fits is chosen in area calculation to keep layout regularity.

A.2 HSPICE Validation

The accuracy of the power model implemented in CACTI is validated using HSPICE. We have first created the 10T SRAM cell model using the low-power 32nm Predictive Technology Model [25] with 24mV and 30mV standard deviation[2] of V_{TH} for each NMOS and PMOS transistor respectively. Then we have modeled cache SRAM arrays (i.e. subarrays of the hybrid cache), comprised of one cell type/size (e.g., $6T_S$, $6T_W$, $10T_W$, etc.) including all peripheral circuits such as decoders, wordline and output buffers, pre-charge circuitry, column multiplexer, etc. The SRAM array is created by replicating a single SRAM cell as many times as needed (MxN times where M and N stand for the number of rows and columns respectively) instead of creating MxN different SRAM cells. Although some accuracy is lost, this is not an issue because power and delay are dominated by bitlines and wordlines rather than SRAM cells, and bitlines/wordlines are accurately modeled. On the other hand, SRAM cell simplification reduces drastically simulation time, which would be in the range of many hours (or even days) otherwise. We have compared dynamic and leakage power and read/write access time of the SRAM array with the corresponding SRAM array in CACTI.

We have created digital input vectors for several clock cycles (frequencies in Table 2 for HP and ULE mode) to perform SRAM array write and read operations. We exercise the array with a reasonable number (i.e. 200) of different write and read patterns in order to measure read, write and short-circuit[3] dynamic and leakage power. Finally, we have measured read and write access times for the created SRAM array. Read access time is measured as the difference between the time the address bit's voltage reaches Vdd/2 and the time the output (32 bit data value) of the read buffer reaches 90% of its final value. However, write access time is measured as the difference between the time the address bit's voltage reaches Vdd/2 and the voltage of bitnodes inside the cell reach 90% of their final values. Values obtained are averaged and those values are used as the results of HSPICE simulations.

Values obtained from CACTI for the corresponding SRAM array show to be accurate within 7% variation (on average) to the reference HSPICE models for all metrics considered. The maximum error observed is 18%. This is because the authors of the analytical models for leakage power [18] make some empirical assumptions when transforming the integrals in equations (4) and (5) to the expression defined in equation (8). Moreover, CACTI performs estimation within an inner-loop of a optimization flow designed to choose the best

---

[2] Such values are equivalent to 6σ deviations in 32nm technology node.

[3] The value written into the bitcell during write operation may be different from the previously stored value. In that case, there will be energy consumption to toggle the cell bit. During toggling, there is a small period when both the PMOS and NMOS of the cell inverter are conducting which causes short-circuit power.

performing cache configuration for area, power, and latency.

According to our simulations, the read/write access time variation between SRAM arrays comprised of the $6T_W$ and $10T_W$ cells respectively is less than 5% at 1V, so the impact of using different SRAM cell sizes at HP mode on total delay is negligible.

B. PROCESSOR MODELING

We have used MPSim [3] full-chip simulator, an enhanced version of SMTSim [21] extended with power models analogous to those of Wattch [5]. Wattch adds activity counters in the simulator, and estimates the energy consumption of the different structures using the CACTI tool. Therefore, we have used our enhanced updated version of CACTI tool. The main processor units that Wattch models are:

- Array Structures: Data and instruction caches, cache tag arrays, all register files, TLBs, BTB, register alias table, branch predictors and large portions of the issue queue and the load/store queue.

- Combinational Logic and Wires: Functional units, dependency check logic at decode stage, issue queue selection logic and result buses.

- Clocking: Clock buffers, clock wires, etc.

Note that the majority of resources in our simple single-core processor is devoted to SRAM array-like structures (e.g., only L1 caches occupy more than 50% of the total on-chip area). All those structures are modeled with our CACTI tool, whose accuracy is validated against HSPICE models, so it can be stated that our processor model is accurate enough. All SRAM arrays except L1 caches have been implemented using $10T_S$ cells, so they operate properly at any Vcc level considered.

An Optimized Page Translation for Mobile Virtualization

Yuan-Cheng Lee and Chih-Wen Hsueh
Institute of Networking and Multimedia
National Taiwan University
Taipei, Taiwan
{d00944007, cwhsueh}@csie.ntu.edu.tw

ABSTRACT

Recently, the significant progress in embedded microprocessors made it practical to support virtualization on mobile devices. Unfortunately, memory management has not been well studied in consideration of the characteristics of mobile virtualization due to the challenging aspects on both hardware and software. With a formal proof on a novel design of page translation at the second stage, we propose a novel design of page translation on both hardware and software with higher translation speed, lower variance of latency, and lower hardware complexity. The experimental results suggest it can reduce memory access by 51.37% to 56.01% as compared with the traditional approach.

Keywords

mobile virtualization, memory management, page translation, translation lookaside buffer

1. INTRODUCTION

Virtualization is the technique that allows multiple instances of operating systems being executed on the same physical machine simultaneously. In recent years, virtualization has been widely adopted on server side. Server consolidation is one of such applications, which replaces several existing physical machines with a single virtualization-enabled machine to save costs. There are many other situations where virtualization can be applied to server-side environment, such as load balancing [13]. However, there was not much dedicated research about mobile virtualization. Until now, most virtualization techniques are derived from the x86 platform. Nevertheless, the applications of mobile virtualization begin emerging. For example, it might be preferable to execute more than one OS on a mobile device simultaneously to enhance user experience and security [7]. Another possible use is seamless integration with cloud servers based on the same virtual interface [5].

The difference between server-side virtualization and mobile virtualization might not be so obvious under consid-

eration of the capability of microprocessors. For ARM-based microprocessors, which are the most common ones for mobile devices, it is quite usual to have multiple processor cores running up to 2.5GHz. Actually, the most significant difference is the design criteria and the usage model. On server side, the throughput is usually the main consideration for virtualization to utilize the hardware resources as much as possible. The fairness of scheduling among virtual machines (VMs) is also important because services executing in these VMs generally have to meet some latency requirements, such as the response time of a webpage request. In contrast, for mobile virtualization, users will focus on a foreground VM, i.e., the VM shown on the screen, at one time. Other background VMs might be executed at a lower priority or suspended totally. In addition, there are various peripheral devices, such as GPS receivers, audio devices, etc., in a mobile device. VMs might need to share these devices and thus it raises the problem that how to handle the I/O requests in a timely manner.

However, research about the characteristics of mobile virtualization is just in the early stages [6]. In this paper, we will focus on memory management, which is one of the most critical parts of the design of a virtualization environment. We propose a novel design that manipulates hardware and software to improve the performance, in terms of translation speed and variance of latency, with lower hardware complexity for mobile virtualization. Instead of using a second memory management unit (MMU) as in the traditional approach, our page translation (oPT) adopts a specialized TLB with a bounded size and guarantees no cache miss for that TLB. We also prove the correctness mathematically and analyze the performance with experimental data. The results suggest it can reduce memory access by 51.37% to 56.01% at the second translation stage as compared with the traditional approach.

The rest of this paper is organized as follows. In Section 2, we describe a common method of memory management in virtualization environments. Our method is then presented in Section 3 followed by a proof of correctness in Section 4. The evaluation is given in Section 5, and finally this paper is concluded in Section 6.

2. BACKGROUND

2.1 Terminology

A virtualization environment is a system of a specific hardware platform and the corresponding software supporting virtualization. To fulfill the functionality of virtualization,

---

Permission to make digital or hard copies of all or part of this work for personal or classroom use is granted without fee provided that copies are not made or distributed for profit or commercial advantage and that copies bear this notice and the full citation on the first page. To copy otherwise, to republish, to post on servers or to redistribute to lists, requires prior specific permission and/or a fee.
DAC'13, May 29 - June 07 2013, Austin, TX, USA.

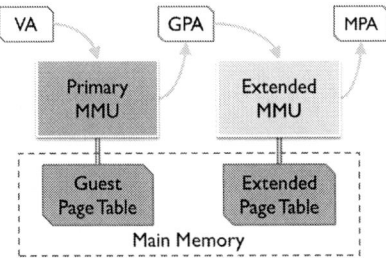

Figure 1: A Simplified Diagram of Nested Paging

we need an additional layer of software known as a virtual machine monitor (VMM). The role of a VMM is to manage virtual machines, which are instances of operating systems along with their own system resources. The OS running in a VM is known as a guest OS. Each guest OS will be executed in an independent environment consisting of virtual processors, virtual physical memory, virtual peripheral devices, and so on. The virtual physical memory is usually called guest physical memory to avoid ambiguity with virtual memory. Similarly, we call the real physical memory as machine physical memory.

2.2 Memory Management

There are several common techniques for memory management in virtualization environments [4]. Hardware-assisted paging (HAP) is probably the most promising one because, with the help of hardware support, it can deliver very high performance. Among the HAP methods, there is a strightforwad yet general design called nested paging. A simplified block diagram of nested paging is shown in Figure 1. The first MMU, called primary MMU in this paper, translates a virtual address (VA) to the corresponding guest physical address (GPA) based on the guest page table in the same way as an ordinary MMU. The newly added MMU, known as extended MMU, takes GPAs as the input and output the final machine physical addresses (MPAs) based on another page table called extended page table (EPT). An extended page table is similar to an ordinary page table except that the address mapping is from GPA space to MAP space. Conceptually, a page translation requires two lookups, i.e., from VA to GPA and then to MPA. In fact, it will require many memory references because guest page tables is specified in GPA space. To improve the performance of page translation, we might add more translation caches such as nested TLB (NTLB) and page walk cache (PWC).

The overall translation process is illustrated in Figure 2a. When a memory access is generated, the processor will first look up the TLB for a match. If the TLB is hit, the processor will get the MPA immediately. When the TLB is missed, the primary MMU begins traversing its page table. For each level of the traversal, it first checks whether the descriptor is in the PWC. If not, it needs to retrive the descriptor from the mina memory, and thus generates a memory reference in GPA space. After the VA has been translated to the corresponding GPA, the GPA is then translated to the final MPA. Figure 2b illustrates the process of translation from GPA to MPA. The address is first looked up in the NTLB. A hit in the NTLB will avoid the traversal of the EPT. If the NTLB is missed, the extended MMU will begin traversing

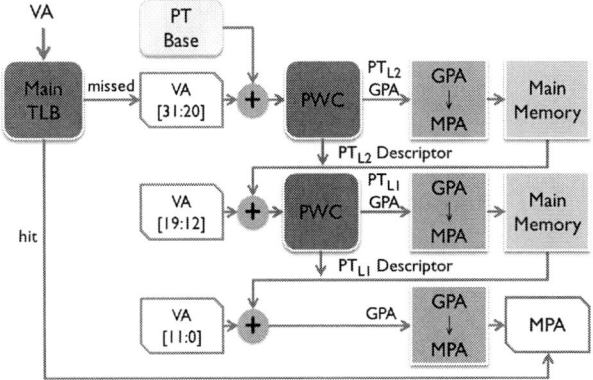

(a) The Overall Translation Process

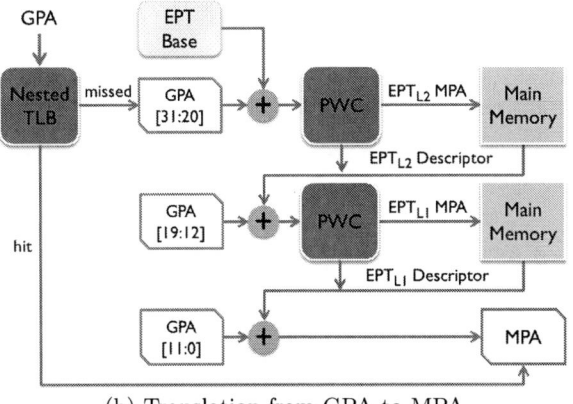

(b) Translation from GPA to MPA

Figure 2: Translation Process of Nested Paging

the EPT in a similar manner as in the primary MMU.

Obviously, this method has two noticeable drawbacks. First, the complexity increases as multiple levels of translation caches are introduced, especially for management. These caches will also causes additional energy consumption, which is also an important consideration for mobile virtualization. The other drawback is that the lookup latency will increase when there are two stages of page walks and multiple caches. The uncertainty of lookup cycles makes it more difficult to support real-time systems.

2.3 Related Work

For mobile virtualization, Hwang et al. proved the concept by presenting a prototype for the ARM architecture [9]. Recently, people also began studying how to utilize multiple processors [10], and how to deal with real-time constraints in virtualization environments [11]. Bhargava et al. extended the translation caches to cover the second MMU that manages machine physical memory [3]. Barr et al. discussed the efficiency of the translation caches [2]. Hoang et al. used a hashed page table to reduce the expected cost of page table lookup in the second MMU and their result shown that the performance of this method is similar to that of nested paging [8]. Unfortunately, these papers did not consider mobile virtualization. Even for the dedicated research such as Cells [1], it did not present detailed discussion of memory management considering the characteristics of mobile virtualization.

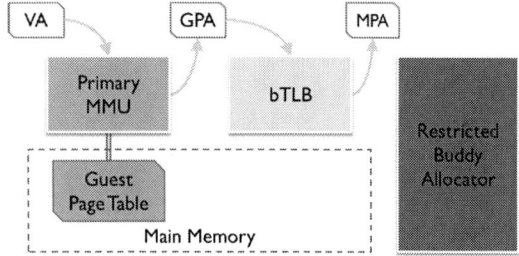

Figure 3: The Block Diagram of Our Page Translation

3. DESIGN AND IMPLEMENTATION

Without loss of generality, we assume the maximum number of virtual machines is a constant because a mobile device usually can support only a few virtual machines due to the limitation of system resources, especially the physical memory and the storage space. We also assume the platform adopts two-level page tables with 32-bit address space to simplify the discussion.

3.1 Our Page Translation

Our page translation consists of a hardware component called *buddy TLB* (bTLB) and a software component called *restricted buddy allocator*. The idea is to increase the performance of page translation with large, continuous, and specially-organized memory allocation. Consequently, there might be slight performance penalties of memory compaction for handling infrequent events such as restarting a VM. However, according to the usage model of mobile devices, it happens very rarely. Instead, the VMs run without restarting most of the time. In addition, with pseudo physical memory allocated in large and continuous regions, the restricted buddy allocator takes this advantage to further optimize the use of bTLB.

The block diagram of **oPT** is shown in Figure 3. Like the traditional HAP methods, it implements the two-stage page translation in hardware and the first stage remains the same. Instead of using a second hardware MMU,our design adopts a specialized TLB, i.e., bTLB, for the second stage of page translation. The design of bTLB is similar to the TLB of the MIPS architecture, which is managed by software and is capable of mapping variable-sized pages [12]. To make the size of bTLB as small as possible for lower power consumption, lower hardware cost, and zero cache miss ratio, we introduce the restricted buddy allocator. It imposes an extra allocation restriction as compared with a general buddy allocator so that we can determine the upper bound of the size of bTLB and guarantee there will not be any cache miss in bTLB. It allocates continuous regions of machine physical memory, known as chunks, to VMs as a general binary buddy allocator where the size of allocated memory regions is always in powers of two. The restricted rule is that any two chunks of the same size will not be allocated to the same VM, which can be met easily because it is very rarely to restart VMs in mobile virtualization.

The translation process is illustrated in Figure 4. The processor first looks up an address mapping in the main TLB for a given VA. As usual, the primary MMU will begin a traversal of the page table when the cache is missed. During the traversal, several GPAs will be generated to retrieve

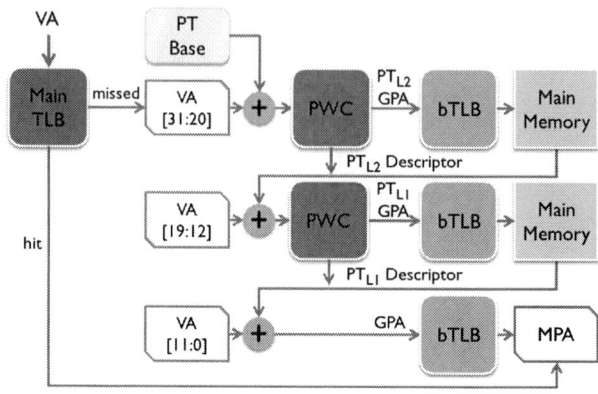

Figure 4: Our Page Translation

descriptors not in the PWC. These GPAs are then searched in the bTLB. If it cannot find a valid address mapping in bTLB, it will raise an exception and transfer the control to the VMM. In such cases, the VM might be terminated due to the access violation.

As compared with the nested paging method, **oPT** has several advantages. First, it has shorter and more predictable lookup latency because it does not have to access the main memory during the second stage of page translation. Hence, it is more suitable for supporting real-time systems. Another advantage is that it does not require extra space for EPTs in the main memory. The VMM can maintain only one list of address mappings in the main memory with tens of entries and thus it has higher efficiency of memory usage. In addition, the PWC only needs to cache descriptors of guest page tables and thus the effective size is increased as compared with the traditional method. Although the extended MMU might support variable-sized pages, such as superpages, to reduce the usage of the PWC in nested paging, it cannot guarantee zero miss ratio at the second stage of page translation. It still needs to look up EPTs and retrieve descriptors from the main memory when the mapping is not in the caches. It is obvious that our design will have a lower implementation cost because of the simplicity of the design, According to the comparison above, our design is better than nested paging for mobile virtualization.

3.2 Implementation Considerations

3.2.1 Buddy TLB

Here we present the general design of bTLB. The fields of a bTLB entry are described as follows.

- VM Identifier (VMID) – This field specifies the target GPA space. The role of VMID is similar to address space identifier (ASID) as in traditional TLB, except it is for GPA space instead of virtual address space.

- Guest Physical Address (GPA) – This field specifies the prefix of guest physical address to be compared. Combining with VMID forms the tag of this entry.

- Mask Size (SZ) – This field specifies the number of bits from the most significant bit (MSB) that should be compared for a match, i.e., the actual length of the prefix of the GPA field.

- Machine Physical Address (MPA) – This field specifies the start address in the MPA space of this address mapping.

- Present (P) – This field specifies whether this entry is presented or not. An entry with this field set to zero will not be compared during the matching process.

- Global Mapping (G) – This field specifies whether this address mapping is a global address mapping or not. When this field is set to one, the VMID field will not be compared during the matching process. This is useful for mapping common chunks to all VMs at once.

- Permission (PERM) – This field specifies the access permission of the chunk. The permission includes read/write for both VM and VMM.

3.2.2 Admission Control and Memory Management

A request to start a VM will be rejected if there is no sufficient machine physical memory. However, the free space might become too fragmented so that the restricted buddy allocator cannot fulfill the request. In such cases, it can perform memory compaction in the background, which usually takes a few seconds depending on the requested size of pseudo physical memory.

After a VM is started, it sometimes might want to adjust the amount of allocated memory dynamically. As long as the allocation rule is not violated, the requests always can be fulfilled. In the worst case, it might trig the memory compaction procedure. Fortunately, memory adjustment is usually done asynchronously without explicit timing constraints.

Memory sharing among VMs is another important concern in memory management. Each shared region of memory requires at least one extra bTLB entry. With a proper sharing scheme, each VM can establish only one shared region with another VM. Similarly, the copy-on-write feature also requires additional bTLB entries. However, it is mainly used by VMMs to optimize the instantiation of multiple VMs from a shared image. It is seldom used in mobile virtualization. As for memory-mapped I/O, VMs should not have the ability to access physical I/O region directly for secure isolation. Instead, it is more appropriate to delegate the I/O requests to the VMM through the existing shared memory region.

4. PROOF OF CORRECTNESS

To achieve zero miss at the second stage of page translation, we have to guarantee bTLB can hold all address mapping entries under the worst case regardless of temporal and spatial locality. If the maximum number of bTLB entries under the worst case can be found, we can determine the minimum size of bTLB for **oPT**. In this section, we describe how to prove the problem. The detailed proof and calculation are omitted due to the space constraints.

A bTLB entry represents an address mapping from a chunk of guest physical memory to the corresponding chunk of machine physical memory, and the set of bTLB entries represents an allocation instance of the buddy allocator. The characteristic of the restricted buddy allocator can be described by the number of chunk types classified by size, which can be determined from the allocation unit of the allocator and the width of GPA space. Without loss of

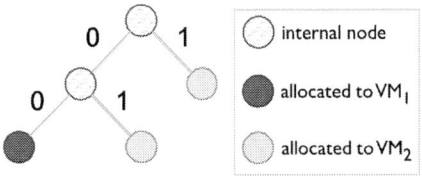

Figure 5: An Allocation Instance for Two VMs

Table 1: The Content of the bTLB

GPA	MPA	SZ	VMID	Size[*]
0x0000_0000	0x0000_0000	2	VM_1	1GB
0x0000_0000	0x8000_0000	1	VM_2	2GB
0x8000_0000	0x4000_0000	2	VM_2	1GB

[*] This field is for clarity, not part of a bTLB entry.

generality, the VMM can be viewed as another VM to the allocator. Given the number of VMs and the number of chunk types, we would like to find the maximum number of bTLB entries under the worst case so that we can determine the minimum size of bTLB for **oPT** page translation.

To make the discussion more straightforward, we map the original problem to another problem called *leaf l-k-coloring*. The leaf coloring problem is one kind of tree coloring problems, where we try to color as many leaf nodes as possible under the constraint that any two leaf nodes at the same level cannot have the same color. The formal definition of this problem is described as follows and the relation to the original problem is given in Lemma 1.

DEFINITION 1. **Leaf k-coloring:**
*Given a rooted tree T, **leaf k-coloring** of T is a mapping from leaf nodes to k colors such that any two colored leaf nodes at the same level do not have the same color. The **leaf l-k-coloring problem** is to determine the maximum number of colored leaf nodes for leaf k-coloring on rooted trees of l levels.*

LEMMA 1. *An allocation instance of a restricted buddy allocator can be represented by a leaf k-coloring of a binary tree, and vice versa.*

Figure 5 illustrates an example of the allocation instance for two VMs under the worst case with 4GB machine physical memory in 32-bit address space and allocated in 1GB units. The content of bTLB is listed in Table 1. In the worst case, one VM has one chunk of memory and other free memory is allocated to the other VM. In the corresponding leaf 2-coloring, it is obvious that the maximum number of colored leaf nodes is less than four because a binary tree of 3 levels will have four leaf nodes at most, while it can have only two colored leaf nodes in the such tree. In fact, coloring the leaf nodes from the bottom as many as possible will result in the maximum number of colored leaf nodes, and we call such colorings as the optimal leaf k-colorings.

For an optimal leaf k-coloring colored by the above method, colored nodes reside at consecutive levels in the bottom of that tree. Hence, the number of colored nodes will be less or equal to the number of colors multiplying the number of levels having colored leaf nodes. Specifically, given an optimal leaf k-coloring C of a binary tree of l levels,

601

the number of colored leaf nodes can be bounded by the following equation.

$$k \cdot \lceil (l - log_2 k) \rceil, \qquad (1)$$

where k is less or equal to the maximum number of leaf nodes at the lowest level of the tree, i.e., $k \leq 2^{l-1}$ and $\lceil (l - log_2 k) \rceil$ can be derived as the number of levels having colored nodes. Finally, we have the following theorem.

THEOREM 1. *A buddy TLB of $k \cdot \lceil (l - log_2 k) \rceil$ entries will not have any cache miss for k VMs with l types of chunks.*

The proof of Theorem 1 immediately follows from Lemma 1 and Equation (1). According to Theorem 1, **oPT** is indeed a feasible solution to page translation for mobile virtualization. In fact, it only requires 65 entries, $5 \cdot \lceil (13 - log_2 5) \rceil + (5 \cdot 4)/2 = 65$, when there are four VMs, and one VMM, with minimum memory sharing and the size of each chunk is large or equal to 1MB, which results in 13 types of chunks, i.e., from 1MB to 4GB.

5. EVALUATION

To understand the behavior of **oPT**, we conduct several experiments in an emulated environment and compare the results with nested paging. We choose the Android emulator because Android is one of the most popular software platforms for mobile devices. We execute the Android platform in a VMM-independent environment to avoid the implementation variability of VMMs [3]. The benchmark tool *V8 Benchmark Suite* is a tool for tuning JavaScript engines based on realistic test cases such as the frequently used regular expressions found in the most popular websites. The reason we choose this benchmark tool is that it should be able to generate reasonable workloads representing the daily usage of users because surfing the Internet is a common activity when using mobile devices.

We first add a module to the emulator for tracking TLB usage. It takes VAs as inputs and outputs missed addresses to a trace file. It also produces the statistics of TLB usage when the emulation is terminated. This module implements a unified TLB cache adopting the least-recently-used (LRU) replacement policy because LRU is expected to have the best performance under the consideration of locality. The size and the set associativity of the TLB are configurable parameters. Without loss of generality, we exclude micro TLB, which are optional components in the architecture, from the emulation to simplify the discussion.

We then implement a page translation simulator to analyze the performance of translation caches based on the trace files. The simulator is capable of simulating both nested paging and **oPT**. In this paper, we consider three different variants of nested paging. The first implementation, designated as **EPT^N**, adopts a nested TLB to reduce the times of page translation at the second stage. Another implementation, designated as **EPT^P**, stores frequently used page table descriptors and EPT descriptors in a page walk caches to reduce the number of memory accesses to the page tables. The last one, designated as **EPT^{NP}**, implements the most common design, which combines both NTLB and PWC. For **oPT**, it adopts a PWC to cache page table descriptors. The bTLB consists of 65 entries, which is sufficient for most applications as discussed in the previous

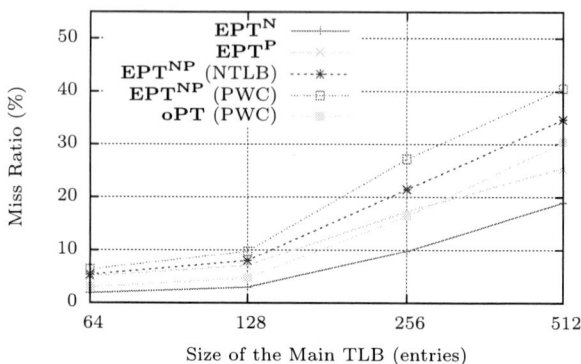

Figure 6: Miss Ratio of Translation Caches

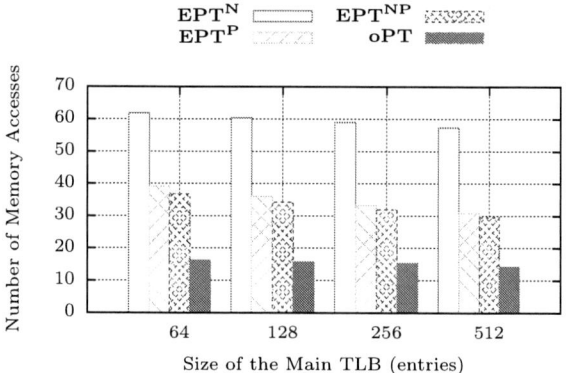

Figure 7: The Number of Memory Accesses

section. In the simulator, we assume no VM restarts and thus use predefined and static mappings to map GPAs to MPAs for both EPT and bTLB.

First, we would like to explore the relation between the size of the main TLB and the miss ratio of other translation caches. In the experiment, the size and the set associativity of NTLB and PWC are set to the popular configuration, i.e., 512 entries and 4-way set associativity, as in practical hardware. As we expected, the miss ratio of the main TLB decreases as the size or the set associativity increase. The result of other caches with the set associativity of the main TLB set to two is shown in Figure 6. The miss ratio of bTLB is not shown because it is always zero. However, the miss ratio of other caches goes in the opposite direction.

However, it is insufficient to compare the performance of page translation of these methods based on miss ratio only because the number of memory accesses induced by a cache miss is different for these methods. For example, a miss in NTLB will induce two memory accesses, while a miss in PWC will induce only one memory access for looking up EPTs. Hence, we then study the actual number of memory accesses of each method under a specific configuration of the main TLB, i.e., 512 entries and four-way set associativity.

Figure 7 shows the number of memory accesses incurred by the page translation process per million memory references including instruction fetch and load/store operations. We assume the configuration of NTLB and PWC is the same as in **EPT^{NP}** to simplify the discussion. Although

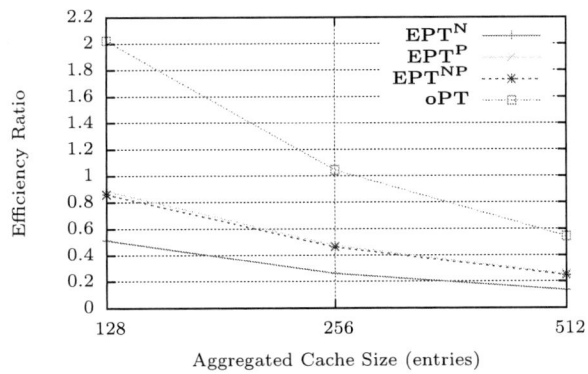

Figure 8: Efficiency Ratio

the miss ratio of the NTLB method is lower than that of others, it incurs the most number of memory accesses because each miss in the main TLB will result in at least two memory accesses to retrieve the descriptors of the page table. Misses in the NTLB will result in more memory accesses for looking up the EPT. In contrast, **oPT** always incurs the least number of memory accesses, which are caused by looking up guest page tables, and thus delivers the highest performance. According to the result, it reduces memory access by 51.37% to 56.01% as compared with EPT^{NP}.

It is interesting to find that EPT^P and EPT^{NP} have similar number of memory accesses, while the latter one has one extra cache. To study the efficiency under the consideration of cache size, we define the efficiency ratio between two methods as the ratio of the numbers of memory accesses multiplying the ratio of the aggregated cache sizes. Figure 8 shows the comparison between various cases and the base case, i.e., EPT^N with a size of 64 entries. It suggests the efficiency ratio of EPT^{NP} is less than EPT^P for the same configuration. This means a single large page walk cache is preferred over two small separate nested TLB and page walk cache when the aggregated size is the same. For a specific configuration, **oPT** always has the highest efficiency ratio, which means it requires the smallest number of cache entries to achieve the same number of memory accesses than other methods.

6. CONCLUSION

In this paper, we present a novel design of page translation for mobile virtualization under consideration of the usage model of mobile devices. The translation performance is improved with slight overhead of VM admission control. By proving the upper bound of bTLB size, the hardware cost is also reduced. Unlike nested paging, which usually requires tens to hundreds of TLB entries to lower cache miss ratio to an acceptable level, **oPT** requires only 65 entries of buddy TLB to achieve zero miss on itself for a VMM with four VMs. According to the experiment results, **oPT** can reduce memory access by 51.37% to 56.01% during the translation process at the second stage. Hence, **oPT** is more suitable for mobile virtualization than other existing methods of hardware-assisted paging because it can achieve higher performance with lower cost. In the future, we may apply **oPT** to virtualization environments where the maximum number of VMs is a variable. At that time, even

if there might be some cache misses in bTLB, we believe it will still achieve higher performance than the traditional method because the continuity of allocated guest physical memory has been taken into account.

Acknowledgment

This research was supported in part by grants from the ROC National Science Council, NSC 101-2628-E-002-009-and NTU Excellent Research Projects: 102R890822.

7. REFERENCES

[1] J. Andrus, C. Dall, A. V. Hof, O. Laadan, and J. Nieh. Cells: a virtual mobile smartphone architecture. In *Proceedings of the 23rd ACM Symposium on Operating Systems Principles*, SOSP '11, pages 173–187, New York, NY, USA, 2011. ACM.

[2] T. W. Barr, A. L. Cox, and S. Rixner. Translation caching: skip, don't walk (the page table). *SIGARCH Comput. Archit. News*, 38(3):48–59, June 2010.

[3] R. Bhargava, B. Serebrin, F. Spadini, and S. Manne. Accelerating two-dimensional page walks for virtualized systems. *SIGPLAN Not.*, 43(3):26–35, Mar. 2008.

[4] D. Chisnall. *The definitive guide to the xen hypervisor.* Prentice Hall Press, Upper Saddle River, NJ, USA, first edition, 2007.

[5] B.-G. Chun and P. Maniatis. Augmented smartphone applications through clone cloud execution. In *Proceedings of the 12th conference on Hot topics in operating systems*, HotOS'09, pages 8–8, Berkeley, CA, USA, 2009. USENIX Association.

[6] J. Fornaeus. Device hypervisors. In *Proceedings of the 47th Design Automation Conference*, DAC '10, pages 114–119, New York, NY, USA, 2010. ACM.

[7] G. Heiser. Virtualizing embedded systems: why bother? In *Proceedings of the 48th Design Automation Conference*, DAC '11, pages 901–905, New York, NY, USA, 2011. ACM.

[8] G. Hoang, C. Bae, J. Lange, L. Zhang, P. Dinda, and R. Joseph. A Case for Alternative Nested Paging Models for Virtualized Systems. *IEEE Comput. Archit. Lett.*, 9(1):17–20, Jan. 2010.

[9] J.-Y. Hwang, S.-B. Suh, S.-K. Heo, C.-J. Park, J.-M. Ryu, S.-Y. Park, and C.-R. Kim. Xen on ARM: System Virtualization Using Xen Hypervisor for ARM-Based Secure Mobile Phones. In *CCNC 2008*, pages 257 –261, Jan. 2008.

[10] H. Inoue, A. Ikeno, M. Kondo, J. Sakai, and M. Edahiro. Virtus: a new processor virtualization architecture for security-oriented next-generation mobile terminals. In *Proceedings of the 43rd annual Design Automation Conference*, DAC '06, pages 484–489, New York, NY, USA, 2006. ACM.

[11] M. Lee, A. S. Krishnakumar, P. Krishnan, N. Singh, and S. Yajnik. Supporting soft real-time tasks in the xen hypervisor. *SIGPLAN Not.*, 45(7):97–108, Mar. 2010.

[12] MIPS Technologies. *MIPSR Architecture For Programmers Volume III*, Apr. 2011.

[13] W. Vogels. Beyond Server Consolidation. *Queue*, 6(1):20–26, Jan. 2008.

Scalable Vectorless Power Grid Current Integrity Verification

Zhuo Feng
Department of ECE
Michigan Technological University
Houghton, MI, 49931
zhuofeng@mtu.edu

ABSTRACT

To deal with the growing phenomenon of electromigration (EM), power grid current integrity verification becomes indispensable to designing reliable power delivery networks (PDNs). Unlike previous works that focus on vectorless voltage integrity verification of power grids, in this work, for the first time we present a scalable vectorless power grid current integrity verification framework. By taking advantage of multilevel power grid verifications, large-scale power grid current integrity verification tasks can be achieved in a very efficient way. Additionally, a novel EM-aware geometric power grid reduction method is proposed to well preserve the similar geometric and electrical properties of the original grid on the coarse-level power grids, which allows to quickly identify the potential "hot wires" that may carry greater-than-desired currents in a given power grid design. The proposed multilevel power grid verification algorithm provides flexible tradeoffs between the current integrity verification cost and solution quality, while the desired upper/lower bounds for worst case currents flowing through a wire can also be computed efficiently. Extensive experimental results show that our current integrity verification approach can efficiently handle very large power grid designs with good solution quality.

Categories and Subject Descriptors

B.7.2 [**Design Aids**]: simulation—*Integrated Circuits*

General Terms

Performance, Algorithms, Verification

Keywords

Power grid, electromigration, multigrid

1. INTRODUCTION

Power grid electromigration (EM) verification tries to find the worst current densities under a given temperature, which can be achieved by finding the worst current flowing through a wire. Efficient current integrity verification methods are key to designing reliable power delivery networks (PDNs) for nowadays nanometer integrated circuits (ICs) designs.

Permission to make digital or hard copies of all or part of this work for personal or classroom use is granted without fee provided that copies are not made or distributed for profit or commercial advantage and that copies bear this notice and the full citation on the first page. To copy otherwise, to republish, to post on servers or to redistribute to lists, requires prior specific permission and/or a fee.
DAC'13 May 29- June 07, 2013, TX, Austin USA

In the past decade, to avoid prohibitively high cost of traditional vector-based power grid voltage integrity verification methods that rely on running numerous power grid simulations, a variety of vectorless verification techniques has been proposed [1–5].

In this work, we show that vectorless power grid current integrity verification problems share very similar properties as the original multilevel PDE-constranied optimization technique [6]. We also propose a multilevel vectorless power grid current integrity verification method that takes advantages of fast EM-aware power grid reduction method and circuit adjoint sensitivity analysis [7]. Our approach achieves desired scalability by attacking a set of power grid verification problems from the coarsest to finest grid levels, and more importantly, allowing to flexibly trade off the solution quality and computational efficiency by effectively controlling the number of current sources included in the linear program (LP). In this work, although we mainly focus on DC current verification problems, the proposed approach can potentially be extended to handle transient current verification problems. The major contributions of this work include:

1. Based on the key ideas of multilevel PDE-constrained optimization methods [6], we propose a novel multilevel vectorless power grid current integrity verification method that scales well with large power grid designs and provides insightful upper/lower bounds for the verification solutions.

2. We present a novel EM-aware power grid reduction technique to facilitate efficient coarse-level power grid verifications. To further improve the solution quality obtained on coarse-level grids, an iterative coarse grid correction scheme is introduced that enables the coarse grids to match important electrical behaviors with the original power grid.

3. We propose to quickly perform a series of coarse-level power grid current verifications and subsequently reuse part of the prior coarse-grid verification results in the fine-grid verification. Such a multilevel verification approach can quickly identify the "hot wires" on the coarsest-level grid, while incremental solution refinements can be successively performed on the finer to finest grids, which effectively trades off the overall verification cost and the final solution quality.

The rest of this paper is organized as follows. In Section 2, we briefly review the power grid modeling and analysis as well as prior vectorless power grid verification methods. In Section 3, we describe the proposed multilevel power grid current integrity verification approach in details. Section 4 demonstrates extensive experimental results of a variety of power grid verifications to validate the proposed approach, which is followed by the conclusion of this work in Section 5.

2. BACKGROUND

2.1 Power Grid Modeling and Analysis

DC analysis for a power grid with n nodes can be performed using nodal analysis (NA) as follows [8]:

$$Gx = b, \qquad (1)$$

where $G \in \mathbb{R}^{n \times n}$ is the conductance matrix that includes all interconnected resistors, $x \in \mathbb{R}^{n \times 1}$ is a vector including all node voltage unknowns, and $b \in \mathbb{R}^{n \times 1}$ denotes the right hand side (RHS) that includes information on excitation sources and boundary conditions.

2.2 Vectorless Power Grid Voltage Verification Approaches

The concept of vectorless power grid verification has been presented in [4]. It has been shown that power grid voltage integrity verification is equivalent to solving the following LPs [2]:

$$maximize: \quad v_i = e_i^T G^{-1} b, \quad for \quad i = 1, ..., n, \qquad (2)$$

subject to the local and global current constraints:

$$b^L \le b \le b^U, 0 \le Qb \le b_g, \qquad (3)$$

where e_i denotes a vector in which the i-th element is 1 and others are 0. G is an M-matrix whose inverse only includes non-negative elements. b_i^U and b_i^L denote the upper and lower bounds for current source b_i, respectively. Q is a matrix with only 1s and 0s describing the global current constraints. The local and global current constraints for different grid levels can be obtained from realistic circuit design information or early design specifications, such as the power consumption, voltage, and switching activities [2–4,9].

Recent popular works on power grid voltage integrity verification adopt a modified sparse approximate inverse (SPAI) technique to compute the approximate matrix inverse of G matrix using least square (LS) optimizations [2, 9]. Subsequently, they formulate the voltage integrity verification problems as LPs with the reduced sets of current sources according to the approximate matrix inverse. This technique has been shown to produce good results for power grid verification when grid sizes are not very large (during early-stage power grid designs). However, the SPAI technique may not scale comfortably to handle large grid verification problems (as shown in [2]).

3. SCALABLE POWER GRID CURRENT INTEGRITY VERIFICATION

3.1 Vectorless Power Grid Current Integrity Verification

To identify potential EM failures for a power grid design, we need to find the maximum current density of each wire. In this work, we assume that wire dimensions are known in advance, then the power grid current verification task is to *find the maximum current flowing through each power grid wire under local and global current constraints*. Using the same notations as (2), we propose the following LPs to find the maximum voltage difference between nodes i and j, which is equivalent to finding the maximum current that flows between node i and node j through the wire with a resistance $R_{i,j}$ for power grid current integrity verification:

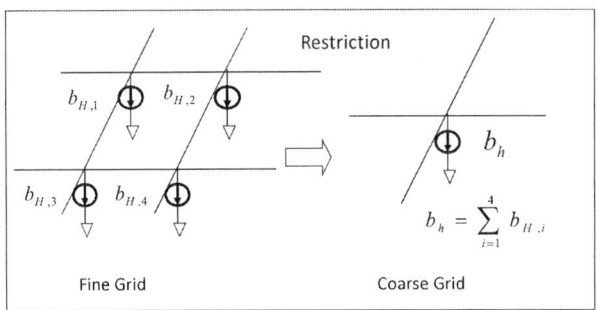

Figure 1: Restriction operator for 2D grid.

$$maximize: \quad (v_i - v_j) = (e_i - e_j)^T E^{-1} b, \quad for \ all \ grid \ wires, \qquad (4)$$

subject to the local and global current constraints:

$$b^L \le b \le b^U, 0 \le Qb \le b_g. \qquad (5)$$

It should be noted that since any current flowing through a wire may have two directions, so a same set of LPs can be applied to find maximum voltage difference between nodes j and i. Compared to prior power grid voltage integrity verification problem, instead of finding the maximum voltage drops across all nodes, current verification aims to find the maximum voltage differences between two nodes connected through a resistive wire. Consequently, extremely high computational cost is expected considering large-scale power grid analysis and the LP with a large number of variables. In this work, we will attack the DC current verification problem, whereas more general dynamic verification problems will be investigated in our future work.

3.2 Overview of Our Approach

Based on the ideas of traditional PDE-constrained multigrid optimization methods (see the supplementary materials for more details), we propose a novel multilevel power grid current integrity verification framework that includes the following steps:

1. Create replicas of coarsest to finest grid problems based on the original power grid design using the proposed EM-aware power grid reduction method [8, 10].

2. Construct local and global current constraints for coarse level grids using the restriction operations depicted in Fig. 1.

3. Perform global current integrity verifications: find the worst wire currents on the coarsest grid level using the proposed LP formulation (in Section 3.1) for all the wires. Identify the top wires (regions) with largest current densities for finer grid verifications, and store their verification results (excitation current distributions).

4. Perform refined verifications: verify the corresponding wires (regions) that have been identified on the coarser grids with partial solution obtained in the previous coarse grid verification. The partial solution can be obtained using the prolongation operation depicted in Fig. 2.

605

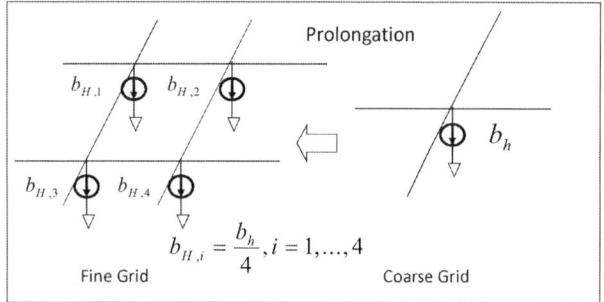

Figure 2: Prolongation operator for 2D grid.

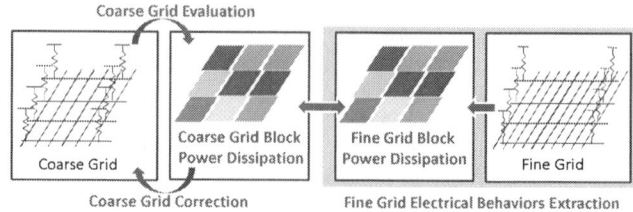

Figure 3: Coarse grid correction scheme.

3.3 EM-Aware Power Grid Reduction for Coarse Grid Generation

Power grid reduction methods can be adopted to create coarse-level grids for power grid verifications. For instance, during power grid voltage integrity verifications, since the nodes (regions) that exhibit the worst voltage drops can be similarly preserved on the reduced (coarse) grids, the coarse-grid verification result can effectively facilitate fine-grid verification tasks. However, for power grid current integrity verifications, traditional power grid reduction methods may cause troubles in revealing potential "hot wires" on the reduced grid if the geometric properties of the original power grid layout are not preserved after the reduction. To better preserve the geometric properties of the original power grids, one can apply geometric multigrid reduction methods to greatly reduce the original large power grids into much smaller ones through node and wire aggregations [8, 11]. Subsequently, the reduced grids can be used during coarse-level power grid current integrity verifications that are usually much cheaper since solving the linear systems of equations in (1) and the LP problems in (4) will involve much fewer variables than solving the original problem.

However, one major drawback of the existing geometric multigrid reduction methods [8, 11] is that the reduction error can not be easily controlled. As a result, if the coarse (reduced) grids can not well preserve the electrical properties of the original power grids, significant errors can be introduced during the coarse-grid verification procedures, which may further influence the fine-grid verification accuracy. In this work, based on our recent research [10] we propose an *EM-aware power grid reduction* technique that combines the prior geometric multigrid reduction procedure with a coarse grid correction scheme to significantly improve the power grid reduction accuracy. The basic idea of our approach is to scale up/down some of the power grid wires such that the new coarse grid can well capture the key electrical behaviors, such as the power dissipations (a.k.a Joule heat) and voltage drops, of the original power grid.

Considering nowadays flip-chip power grid designs, locality effects [12] are typically significant, which allows to partition and analyze the small power grid blocks separately. To more intuitively describe the coarse grid correction scheme, we will first neglect the complicated coupling effects between power grid blocks in the following discussion. Then the power dissipation of a power grid block can be computed by

$$P_{Joule} = x^T G x, \qquad (6)$$

where $x = G^{-1}b$, which is followed by

$$P_{Joule} = b^T G^{-T} G G^{-1} b = b^T G^{-1} b. \qquad (7)$$

So if we size up all the wires in the block by a *block scal-*

ing factor τ, the block voltage drops after the wire sizing become $x' = G^{-1}b/\tau$, while the block power dissipation becomes

$$P'_{Joule} = b^T G^{-1} b/\tau = P_{Joule}/\tau. \qquad (8)$$

The above derivations indicate that we can use the block power dissipation (voltage drop) ratios of the coarse grid and the original grid to determine the block scaling factors τ that can be further used to scale up/down the coarse grid wires.

Considering the couplings between power grid blocks, the coarse grid correction problem can be formulated into the following nonlinear optimization problem that tries to best match the original grid's block power dissipations as well as block average voltage drops on the coarse grid:

$$minimize : \phi(\tau_1, ..., \tau_m) =$$
$$\sum_{i=1}^{m} \left(x_{ri}^T G_{ri} x_{ri} \tau_i - x_i^T G_i x_i \right)^2 + \alpha \sum_{i=1}^{m} \left(\frac{|x_{ri}|_1}{n_{ri}} - \frac{|x_i|_1}{n_i} \right)^2 \qquad (9)$$

subject to the constraint:

$$G_r(\tau_1, ..., \tau_m) \begin{bmatrix} x_{r1} \\ \vdots \\ x_{rm} \end{bmatrix} = b_r, G \begin{bmatrix} x_1 \\ \vdots \\ x_m \end{bmatrix} = b, \qquad (10)$$

where G_{ri} (G_i) is the conductance matrix of the coarse (fine) grid block i, m is the number of blocks, τ_i for $i = 1, ..., m$ are the block scaling factors to be found, α is a weighting coefficient, and n_{ri} (n_i) is the number of nodes in the coarse (fine) grid block. It should be noted that the coarse grid conductance matrix $G_r(\tau_1, ..., \tau_m)$ is now parameterized, and the block grid solutions x_{ri} will become a function of $\tau_1, ..., \tau_m$.

However, there is no efficient method for solving the above nonlinear optimization problems. Fortunately, for flip-chip power grids, locality effects enable us to relax the above optimization problem, which allows to apply the following iterative coarse grid corrections (as illustrated in Fig. 3):

1. Partition the coarse and the original grid into multiple power grid blocks such that each block includes a few VDD/GND pads.

2. Perform DC analysis for the original grid and the reduced (coarse) grid, and compute each block's power dissipation and average voltage drop.

3. Size up/down the block wires in the coarse grid using the weighted block scaling factor $\tau_i = \left(\frac{x_{ri}^T G_{ri} x_{ri}}{x_i^T G_i x_i} \right) \omega + \left(\frac{|x_{ri}|_1 n_i}{|x_i|_1 n_{ri}} \right) (1 - \omega)$.

4. Repeat steps 3 and 4 until convergence. Output the updated coarse grid circuit.

In step 4, ω is a weighting coefficient between 0 and 1. We have implemented the above coarse grid correction scheme and observed that solution errors can be reduced by over $3X$ when compared with the original coarse grid obtained using geometric multigrid reduction scheme [8, 11] (see the supplementary materials for power grid reduction results).

It should be noted that unlike prior power grid reduction methods that can only preserve the electrical properties of the original grid, the proposed EM-aware power grid reduction scheme can also preserve the geometric properties of the original power grid in that the wire dimensions in the reduced power grid can be easily calculated after the reduction process. For example, since the coarse grid has the same geometric (lateral) dimensions (width and length) as the original grid but much fewer nodes/wires. Consider a resistor $R_{i,j}$ between nodes i and j in the reduced grid. $R_{i,j}$ can be considered as an equivalent wire resistor for a set of wires coming from possibly multiple metal layers of the original power grid. Since the distance between nodes i and j is preserved on the reduced grid, one can easily compute the effective wire cross-section area based on $R_{i,j}$, and therefore estimate the wire current density. This nice property allows to quickly estimate the wire current densities even for the coarse-level grids. As a result, the coarse-grid current integrity verification results can effectively facilitate identifying the worst current densities and "hot wires" in the original power grid.

3.4 Critical Regions for Linear Programs

To further reduce the current verification cost in the LP (4), the number of current variables should be well controlled. However, power grid current verification problems usually involve hundreds of thousands of current sources, which may take excessively long CPU time when using standard LP solvers. To effectively control the number of variables in LPs, we first define the *critical region* C_{crt} based on power grid electrical properties such as the wire current (voltage difference across the wire) sensitivities w.r.t. the underlying current sources. C_{crt} includes the most critical current sources that will significantly contribute to the wire current of our interest, such as the one shown in Fig. 4. C_{crt} can be obtained by estimating the amplitudes of the wire current sensitivities w.r.t. all current sources. For instance, if we are given a threshold sensitivity value s_{th} or normalized sensitivity value $\epsilon_{crt} = s_{th}/s_{max}$ where s_{max} is the maximum sensitivity value of all current sources, C_{crt} can be obtained in the following way: if a current source has a sensitivity value greater than the threshold value s_{th}, then it is included into C_{crt} and thus the LP for power grid current integrity verification, since its current value change will substantially influence the worst current density of the wire/region under verification. On the other hand, if a current source's sensitivity value is much smaller than the threshold value, it can be safely excluded from C_{crt} as it will not be an important variable for finding the worst current densities. In this work, we set ϵ_{crt} to be around $5e-3$, which allows to include much fewer current sources (10% of the total currents) into the LP without loss of much accuracy.

Since power grid current sensitivities can be efficiently computed using adjoint sensitivity analysis [7] by reusing the matrix factors (from one-time Cholesky or LU decomposition) of the conductance matrix G, finding C_{crt} is efficient. For instance, if we want to compute the wire current (voltage difference divided by the resistance) sensitivity used in current integrity verification (4), we simply set $b = e_i - e_j = [0, ..., 1, 0, ..., -1, ..., 0]^T$ to compute the sensitivity vector s. C_{crt} can be identified subsequently based on these sensitivity values on the original and coarse-level grids easily, as shown in Fig. 4.

Once the critical region C_{crt} is determined, an LP can be formulated based on previous adjoint sensitivity analy-

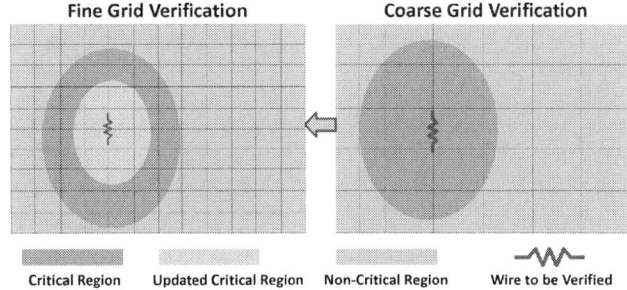

Figure 4: Critical and non-critical regions for power grid current integrity verification.

sis result, in which only the current sources that fall into the critical region C_{crt} are considered during the verification. If a circuit block sits across the boundary of C_{crt}, the current sources of that circuit block should also be included into C_{crt}. Assume that the adjoint sensitivity vector (s) is computed in advance, then the power grid current integrity verification (finding the worst voltage difference v_{wst}) for a specific wire within C_{crt} can be formulated into the following LP:

$$maximize: \quad \bar{v}_{wst} = \sum_{\forall b_i \in C_{crt}} s_i b_i, \qquad (11)$$

subject to the local and global current constraints:

$$b^L \le b \le b^U, 0 \le Qb \le b_g, \qquad (12)$$

where b^U and b^L denote the upper and lower bounds for current sources b, and Q is a matrix that includes the global current constraints. It should be noted that the proposed critical region verification approach based on power grid electrical properties such as adjoint sensitivity information will greatly reduce the number of current sources involved in the LP, resulting in much more efficient power grid verification. In our extensive experiments, we find that for the original power grid including only 10% (when using $\epsilon_{crt} = 5e-3$) of the current sources in LP will suffice for finding accurate worst case wire current densities.

3.5 Upper/Lower Bounds of Worst Currents

The proposed power grid current integrity verification also provides the lower and upper bounds of the worst wire currents. Since the current sources with small sensitivity values will not be included in the LP, the maximum extra voltage difference v_e they may contribute can be approximately computed based on the total current and the sensitivity threshold s_{th}:

$$v_e = \max_{\forall b_i \notin C_{crt}} \left(\sum_{i=1}^n s_i b_i \right) \le \max_{\forall b_i \notin C_{crt}} \left(\sum_{i=1}^n s_{th} b_i \right) \le s_{th}|b - b_p|_1,$$
$$(13)$$

where b_p is the vector that only includes the current sources in the critical region C_{crt}. We assume that s_{th} is much smaller than the peak sensitivity values observed in C_{crt}, then the upper bound of the worst case voltage difference can be approximately computed by:

$$v_{wst} \approx \bar{v}_{wst} + v_e \approx \bar{v}_{wst} + s_{th}|b - b_p|_1. \qquad (14)$$

From the above results, it can be concluded that the realistic worst case voltage difference v_{wst}^* will approximately satisfies:

$$\bar{v}_{wst} \le v_{wst}^* \le v_{wst}. \qquad (15)$$

It should be noted that a larger threshold sensitivity value s_{th} will result in fewer current sources to be included in the LP thus less computational cost, but the upper and lower bounds of worst case voltage difference v_{wst}^* will also become more conservative (looser upper/lower bounds). On the contrary, if a smaller s_{th} is chosen, more current sources will be involved in LP that gives much tighter bounds for the worst case voltage difference. Fortunately, using the proposed method, tradeoffs between the computational cost and the final solution quality can be effectively achieved by choosing different threshold sensitivity values s_{th}.

3.6 Power Grid Current Integrity Verification Algorithm

Once we have identified a few key "hot wires" or regions on the coarsest grid, the verification results can be interpolated (prolonged) to the finer grids to facilitate the next-level power grid verification (as shown in Fig. 2). Assuming that a "hot wire" has been identified after performing exhaustive current integrity verifications on the coarsest grid, and one would like to find the worst current densities of this wire on the finer to finest grids, the following issues should be addressed first: as we move from the coarsest grid to the finer grids, the number of decision variables (current sources) that fall into the critical region C_{crt} will grow dramatically, making the resultant LP increasingly expensive to solve. To take advantage of previous coarse-level grid verifications, we further define the *finer-grid solution refinement* step and *updated critical region* (denoted by C'_{crt}).

Algorithm 1 Multilevel power grid current integrity verification for a "hotspot" wire.

Input: the local and global current constraints b^U, b^L and Q in (11), the original power grid netlist, the sensitivity scaling factor $\beta > 1.0$, and the mapping operators between fine and coarse grids V_h^H and V_H^h, and the coarsest grid level K.

Output: the current vector that results in the worst case voltage difference (current density) across the wire.

1: Initialization phase:
   a) Create the multilevel coarsest to finest grid models (level 0 to level K grids) [8,11].
   b) Set up their local and global current constraints b^U, b^L and Q for all grid levels based on the inter-grid operator V_H^h.
   c) Perform coarse grid corrections described in Section 3.3 to improve the coarse grid solution quality.
2: Global verification phase for level K grid:
   a) Perform adjoint sensitivity analysis to compute sensitivity vector s^K w.r.t. current source b^K.
   b) Identify C_{crt}^K based on sensitivity threshold ϵ_K.
   c) Set up LP solver to obtain the solution b_{wst}^K using (11).
   d) Compute the upper bounds by:
   $v_{wst}^K = \bar{v}_{wst}^K + \epsilon_K * s_{max}^K |b^K - b_p^K|_1.$
3: Finer-grid solution refinement phase for finer to finest grids:
4: **for** $(k = K - 1; k \geq 0; k - -)$ **do**
5:    Interpolate coarser level $k + 1$ solution vector to level k grid by: $\bar{b}_{wst}^k = V_{k+1}^k b_{wst}^{k+1}$.
6:    Perform adjoint sensitivity analysis.
7:    Determine the updated critical region $C_{crt}^{\prime k}$ for finer-grid solution refinement using sensitivity threshold $\epsilon_k = \beta \epsilon_{k+1}$.
8:    Set up LP solver using (11) to obtain the solution vector $\tilde{b}_{wst}^k$ within the updated critical region $C_{crt}^{\prime k}$.
9:    For the current sources that fall outside $C_{crt}^{\prime k}$, reuse the interpolated coarse grid solution vector $\bar{b}_{wst}^k$.
10:   Compute the refined worst case current vector by: $b_{wst}^k = \bar{b}_{wst}^k + \tilde{b}_{wst}^k$.
11: **end for**
12: Return the final current verification results.

After finishing the global grid verification task on the coarsest grid for a specific wire, the current vector b_{wst}^h that results in the worst voltage difference across the wire can be interpolated to the finer grid for next-level power grid verification:

$$\bar{b}_{wst}^H = V_h^H b_{wst}^h, \qquad (16)$$

where V_h^H denotes the inter-grid operator shown in Fig. 2 interpolating the coarse grid vector to the finer grid vector. Since the prolonged solution vector $\bar{b}_{wst}^H$ can be very close to the true solution b_{wst}^H. This suggests that during the finer grid verification, instead of doing full-blown grid verification, we can relax the power grid verification problem by performing *finer-grid solution refinement* only within the updated critical region C'_{crt} that is computed in a similar way as C_{crt} (a smaller region as shown in Fig. 4) by setting a larger sensitivity threshold value.

The proposed multilevel power grid current integrity verification algorithm has been described in Algorithm 1. The "hot wires" will be first identified through global verifications on the coarsest grid, and finer-grid solution refinements can be subsequently performed to compute the worst wire current densities of the original grids.

4. EXPERIMENTAL RESULTS

Extensive experiments have been conducted to validate the proposed multilevel power grid current integrity verification approach that has been implemented using C++ and CUDA. The LP is solved by the solver proposed in [13], while the power grid adjoint sensitivities are computed using the GPU-based power grid simulator [8, 14]. The hardware platform is a Linux PC with Intel Core 2 Quad CPU running at 2.66 GHz clock frequency with an NVIDIA GTX 285 GPU. All runtime results are measured in seconds. A set of flip-chip power grids are generated using the typical wire resistances as well as the current source distributions of industrial designs [15].

To build the LPs for power grid current verification, adjoint sensitivity analysis is first performed to identify the critical region C_{crt} by examining the sensitivity values. Next, current sources can be determined for the LP subsequently. In Fig. 5, we show the relative sensitivity distributions (scaled with the largest sensitivity value) in logarithmic scale for the level 2 grid of a power grid design. As discussed in Algorithm 1, once the power grid verification proceeds to the finer grid level, the updated critical region C'_{crt} will shrink to a smaller one based on the scaling factor β. Subsequently, the finer-grid refinement using LP solver will be performed considering the current sources in the updated critical region C'_{crt} as shown in Fig. 4. It is also observed in our experiments that as the verification proceeds from the coarsest grid to the finest grid, more and more current sources have to be included into the LP when a fixed critical region is used, resulting in drastically increased computational cost. On the other hand, by introducing the gradually shrinking (updated) critical regions on the finer to finest grids, the number of variables involved in LPs can be effectively reduced, leading to much lower verification cost.

We show more comprehensive results of the proposed multilevel power grid current integrity verification method in Table 1. Different sets of sensitivity threshold values ϵ_{glb} (threshold for global grid verifications on the coarsest grid) and scaling factors β (for finer grids) are used for all test cases. It is obvious that when using smaller ϵ_{glb} and β, the number of current source variables in the LP will be greater, which makes the overall verification more accurate but less efficient. It is also observed that when using $\epsilon_{glb} = 5e - 3$ and $\beta = 2$, the upper bound of the worst case current can

608

Table 1: Results of the proposed multilevel power grid current integrity verification. N_{nodes}, N_{cur}, and N_{lev} are the numbers of power grid nodes, current sources of the finest grid, verification levels, respectively. ϵ_{glb} is the normalized threshold sensitivity for global verification, and β is the sensitivity scaling factor. T_{glb}, T_{ref} and T_{tot} are the global verification time checking $1,000$ wires on the coarsest grid, verification refinement time for verifying the top five candidate wires with worst current densities on the finer grids, and the total verification time. Err is the relative error of the computed upper bound (v_{wst}) of worst case results compared with the solution obtained using $\epsilon_{glb} = 5e-3$ and $\beta = 2$.

	Power grid design specifications			$\epsilon_{glb} = 5e-3, \beta = 2$			$\epsilon_{glb} = 10e-3, \beta = 2$				$\epsilon_{glb} = 20e-3, \beta = 1.5$			
CKT	N_{nodes}	N_{cur}	N_{lev}	T_{glb}	T_{ref}	T_{tot}	T_{glb}	T_{ref}	T_{tot}	Err	T_{glb}	T_{ref}	T_{tot}	Err
$CKT1$	44K	10K	2	164s	15s	179s	123s	11s	134s	10%	74s	7s	81s	15%
$CKT2$	67K	18K	2	150s	24s	174s	135s	13s	148s	15%	83s	8s	91s	20%
$CKT3$	131K	32K	2	181s	23s	203s	159s	12s	171s	12%	114s	12s	126s	17%
$CKT4$	168K	39K	2	452s	186s	638s	245s	52s	297s	18%	158s	34s	192s	22%
$CKT5$	256K	78K	2	890s	320s	1,210s	494s	148s	642s	11%	255s	94s	349s	16%
$CKT6$	511K	151K	3	320s	779s	1,099s	302s	493s	795s	19%	212s	253s	468s	24%
$CKT7$	1.02M	300K	2	2,023s	2,035s	4,058s	1,534s	1,225s	2,759s	14%	847s	606s	1,453s	19%

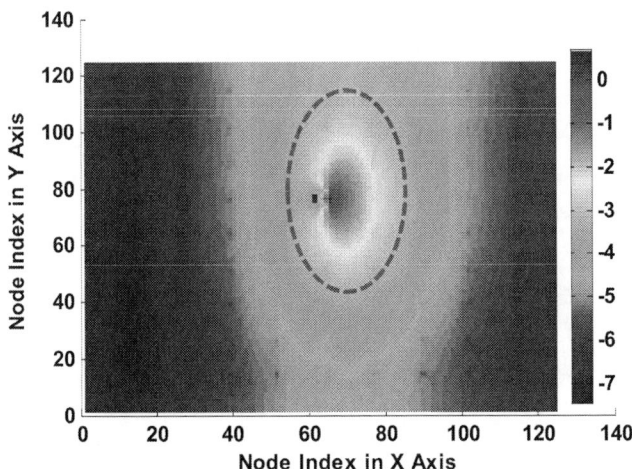

Figure 5: The logarithmic scale sensitivity map for a wire located at (65, 76). The critical region with $\epsilon_{crt} = 1e-3$ is illustrated with a dashed circle in red.

be very accurate. Therefore, the solution errors shown in the table are obtained by comparing the results with the ones obtained with $\epsilon_{glb} = 5e-3$ and $\beta = 2$. We want to emphasize that by replacing the LP solver with some state-of-the-art solvers, much less runtime can be expected.

5. CONCLUSIONS

For the first time, we propose a multilevel power grid current integrity verification method to efficiently compute the worst current across a specific wire. Through power grid current verifications on the coarsest grid and the subsequent solution refinement procedures on the finer to finest grids, power grid current integrity verification can be more efficiently performed, since coarse grid verifications can help identify the "hot wires" that may retain the worst current densities on the finer to finest grids. Additionally, an efficient EM-aware power grid reduction method is proposed to substantially improve the coarse grid solution quality. The proposed framework also provides favorable upper/lower bounds for verification solutions. Our results show that our approach can solve very large-scale power grid current integrity verification problems efficiently: finding the upper/lower bounds of the worst current density for a flip-chip power grid design with one million nodes takes less than two hours.

6. REFERENCES

[1] D. Kouroussis and I. Ferzli and F. Najm. Incremental partitioning-based vectorless power grid verification. In *Proc. IEEE/ACM ICCAD*, pages 358–364, 2005.

[2] N. Ghani and F. Najm. Fast Vectorless Power Grid Verification Using an Approximate Inverse Technique. In *Proc. IEEE/ACM DAC*, pages 184–189, 2009.

[3] X. Xiong and J. Wang. An Efficient Dual Algorithm for Vectorless Power Grid Verification under Linear Current Constraints. In *Proc. IEEE/ACM DAC*, pages 837–842, 2010.

[4] F. Najm. Overview of vectorless/early power grid verification. In *Proc. of IEEE/ACM ICCAD*, pages 670–677, 2012.

[5] Z. Feng. Scalable multilevel vectorless power grid voltage integrity verification. *IEEE Trans. on VLSI Systems, to appear*, 2013.

[6] R. Lewis and S. Nash. Model problems for the multigrid optimization of systems governed by differential equations. *SIAM J. Sci. Comput.*, 26(6):1811–1837, 2005.

[7] L. Pillage, R. Rohrer, and C. Visweswariah. *Electronic circuit & system simulation methods.* McGraw-Hill, 1995.

[8] Z. Feng. Parallel on-chip power distribution network analysis on multi-core-multi-gpu platforms. *IEEE Trans. on VLSI Systems*, 19(10):1823–1836, June 2011.

[9] M. Avci and F. Najm. Early p/g grid voltage integrity verification. In *Proc. of IEEE/ACM ICCAD*, pages 816–823, 2010.

[10] Z. Feng. Large-scale flip-chip power grid reduction with geometric templates. In *Proc. of DATE, to appear*, 2013.

[11] J. N. Kozhaya, S. R. Nassif, and F. N. Najm. A multigrid-like technique for power grid analysis. *IEEE Trans. on Computer-Aided Design*, 21(10):1148–1160, 2002.

[12] E. Chiprout. Fast flip-chip power grid analysis via locality and grid shells. In *Proc. IEEE/ACM ICCAD*, pages 485–488, 2004.

[13] M. Berkelaar. *Mixed integer linear programming solver lp solve.* [Online]. Available: http://lpsolve.sourceforge.net/5.5/, 2005.

[14] Z. Feng and Z. Zeng. Parallel multigrid preconditioning on graphics processing units (GPUs) for robust power grid analysis. In *Proc. IEEE/ACM DAC*, pages 661–666, 2010.

[15] S. R. Nassif. Power grid analysis benchmarks. In *Proc. IEEE/ACM ASPDAC*, pages 376–381, 2008.

Supplementary Material

6.1 PDE-Constrained Multilevel Optimization

PDE-constrained multilevel optimization methods have been proposed to solve general nonlinear optimization problems which are governed by system of partial differential equations (PDEs) [6], which is shown as follows:

$$maximize: \quad F(p) = f\left(v(p), p\right), \tag{17}$$

subject to the PDE constraint:

$$S(v, p) = 0, \tag{18}$$

where p can be a set of design variables or input variables, and $S(v, p)$ describes a system of partial differential equations that relate the PDE's solution v with the input parameter p. Since the linear system of equations for power grid analysis can be considered as a partial differential equation (PDE) [11], the the voltage vector v and RHS vector b in (1) can be considered as the model parameters v and p in (18), while the worst case voltage difference of two nodes sought in the power grid current verification formulation (4) can be considered as the objective function $F(p)$. Obviously, there are good similarities between the standard power grid verification problems and the aforementioned PDE-constrained optimization problems, though power grid verification usually solves an LP problem instead of a nonlinear problem. However, prior PDE-constrained multilevel optimization methods do not allow local and global inequality constraints, we can not directly apply the multilevel optimization method to power grid verification problems. Fortunately, ideas behind traditional PDE-constrained multilevel optimization methods will benefit our power grid current verification problems. A brief introduction to typical PDE-constrained multilevel optimization procedures is described as follows.

Assume that the coarser to finer (finer to coarser) problem mapping operators V_h^H (V_H^h) for different grid levels have been previously defined. Then the conventional PDE-constrained multilevel optimization methods include the following key steps [6]:

- If on the coarsest level problem,
  $maximize \ F_h(p_h) = f_h(v(p_h), p_h)$,
  with initial estimate $p_h^{(0)}$, to obtain $p_h^{(1)}$.

- Otherwise:

  1. Partially $maximize$ $F_H(p_H)$ with initial estimate $p_H^{(0)}$, to obtain $p_{H,1}$.
  2. Map the fine level problem solution to the coarse level problem by $p_{h,1} = V_H^h p_{H,1}$, and compute the gradient difference: $s_h = \nabla F_h(p_{h,1}) - V_H^h \nabla F_H(p_{H,1})$.
  3. Recursively apply multilevel optimization with initial solution guess $p_{h,1}$ to solve the following coarse problem: $maximize$ $F_h(p_h) - s_h^T p_h$, subject to the constraints: $p_h^L \le p_h \le p_h^U$ to get a refined solution $p_{h,2}$.
  4. Compute the new search direction $e_H = V_h^H(p_{h,2} - p_{h,1})$, apply line search $p_{H,2} = p_{H,1} + \alpha e_H$ and $maximize$ $F_H(p_H)$, with initial solution $p_{H,2}$ to obtain $p_H^{(1)}$.

"Partially maximize" in Step 1) is similar to the smoothing operation in traditional multigrid algorithm for numerically solving PDE problems, while the line search phase improves solution at every optimization iteration. The extra bound constraint $p_h^L \le p_h \le p_h^U$ has been introduced to improve the convergence. It is interesting to see that the above optimization framework adopts the multigrid V-cycle scheme

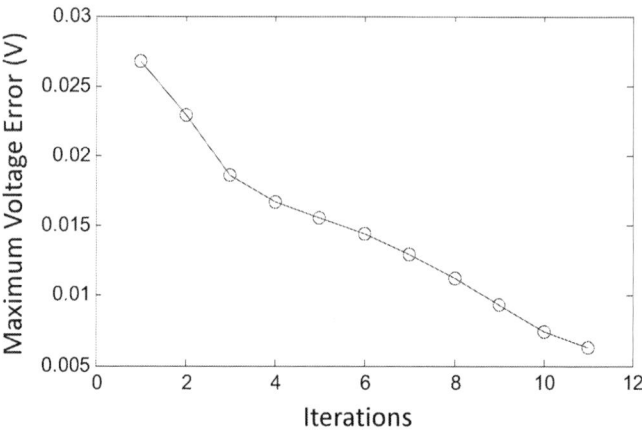

Figure 6: **Maximum solution errors after performing block iterative reduced grid corrections for an industrial power grid design.**

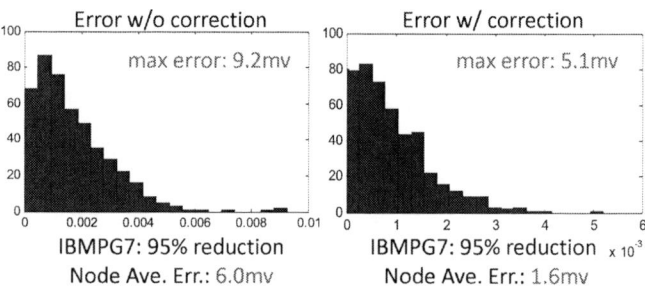

Figure 7: **DC analysis solution error distributions of pad nodes w/ and w/o reduced grid corrections for an industrial power grid design.**

(which is key for multigrid numerical solver), though other cycle formats such as the W-cycle or the full multigrid cycle can be applied in the similar manner.

6.2 Results of the EM-Aware Power Grid Reduction Method

In this section, we validate the EM-aware power grid reduction method proposed in Section 3.3 for a set of industrial power grid test cases [15]. The power grid analysis for the full grids and the reduced grids are performed on CPU-GPU platforms based on the iterative algorithms proposed in [8, 14]. The hardware platform is a Linux PC with Intel Core 2 Quad CPU running at 2.66 GHz clock frequency with an NVIDIA GTX 285 GPU.

In Fig. 6 and Fig. 7, we show the results of the proposed iterative grid correction scheme for industrial test cases. As shown, after only 10 iterations, the maximum voltage errors can be reduced to a much smaller one when compared with the initial reduced power grid model errors (without grid corrections). In the last, we show the block scaling factors for an industrial power grid design in Fig. 8. As observed, the wires on reduced power grid have been modified drastically from the original reduced grid obtained by using the naive geometric multigrid reduction (element/node aggregation) method. This also indicates the necessity of the proposed iterative grid correction procedure for achieving more accurate EM-aware power grid reductions.

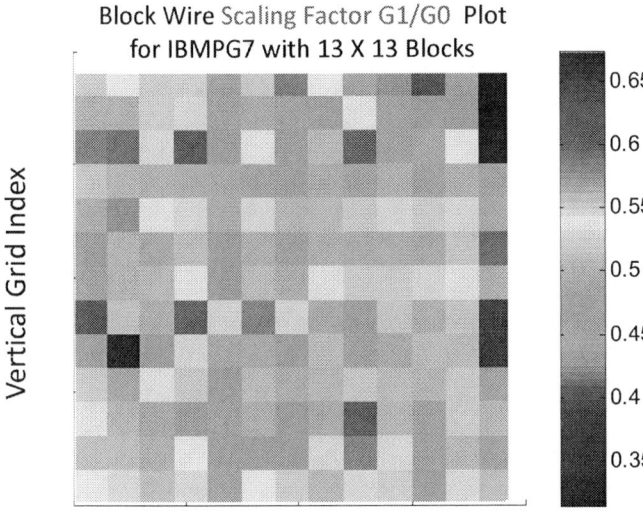

Figure 8: Block wire scaling factor distribution on the reduced grid. The factor G1/G0 is the ratio of block wire conductances before and after iterative block reduced grid corrections.

Constraint Abstraction for Vectorless Power Grid Verification

Xuanxing Xiong and Jia Wang
Department of Electrical and Computer Engineering
Illinois Institute of Technology, Chicago, IL 60616, USA
xxiong3@hawk.iit.edu, jwang@ece.iit.edu

ABSTRACT

Vectorless power grid verification is a formal approach to analyze power supply noises across the chip without detailed current waveforms. It is typically formulated and solved as linear programs, which demand intensive computational power, especially for large-scale power grids. In this paper, we propose a constraint abstraction technique to reduce the computation cost of vectorless verification. The boundary condition of a subgrid is modeled by boundary constraints, which enable efficient calculation of conservative bounds of power supply noises in a divide-and-conquer manner. Experimental results show that the proposed approach achieves significant speedup over prior art while maintaining good solution quality.

Categories and Subject Descriptors

B.7.2 [**Integrated Circuits**]: Design Aids

General Terms

Algorithms, Verification, Performance

Keywords

Power grid, voltage drop, vectorless verification

1. INTRODUCTION

As technology scaling continues, the performance and reliability of integrated circuits become increasingly susceptible to power supply noises, such as IR drops and Ldi/dt drops in the on-chip power grid. Reduced supply voltage levels in the grid can increase the gate delay, leading to timing violations and even logic failures. In order to ensure a reliable chip design, it is indispensable to verify that the power grid is robust, i.e., the power supply noises are acceptable for all possible runtime situations.

Nowadays, it is common practice to verify power grids by simulation. Typically, an equivalent RC/RLC circuit model of the grid is extracted from the layout, and designers perform simulations to evaluate the power supply noises based on the current waveforms drawn by the circuit. However, simulation-based verification has the following limitations. First, it is computationally prohibitive to simulate all possible current waveforms. In practice, some typical current waveforms are used for verification. Such current waveforms are either too pessimistic, or realistic but we are taking the risk of missing some important corner case. Second,

Permission to make digital or hard copies of all or part of this work for personal or classroom use is granted without fee provided that copies are not made or distributed for profit or commercial advantage and that copies bear this notice and the full citation on the first page. To copy otherwise, to republish, to post on servers or to redistribute to lists, requires prior specific permission and/or a fee.
DAC'13, May 29 - June 07, 2013, Austin, TX, USA.

simulation-based approaches require detailed circuit implementation to provide the current waveforms, thus do not allow power grid verification at an early design stage, when grid correction can be easily incorporated.

To overcome these limitations, *vectorless power grid verification* was proposed in [13], and further studied in many later works. Instead of enumerating all possible current waveforms, it evaluates power supply noises based on partial current specifications defined by *current constraints*. Grid verification is often formulated as linear programs of finding the worst-case power supply noises at each node. The initial study [13] considered the DC analysis model, and it was extended to verify RC power grids in [8]. Since solving the linear programs is very expensive, [3] proposed to generate a reduced-size linear program for each node by using an approximate inverse technique; [17] and [19] designed convex dual algorithms to solve the linear programs efficiently; [1] utilized the dominance relations among nodal voltage drops to reduce the number of linear programs to be solved; [12] and [5] used macromodeling technique to simplify the linear programs for efficient incremental verification. Besides, [6] proposed to verify VDD network and ground network together with additional *current conservation constraints* in order to consider the mutual effects in between.

Moreover, vectorless verification of RLC power grids was proposed in [2]. A fast approach to compute conservative bounds of power supply noises was proposed in [4], and *transient current constraints* were introduced in [20,21] for more realistic noise predictions. In [7], the authors proposed *hierarchical power constraints* to model current excitation, and the resultant linear programs for vectorless verification can be solved by a sorting-deletion algorithm efficiently. Model order reduction was used in subsequent work [16] to further reduce the cost of formulating the linear programs.

Despite these available approaches, the computation cost of vectorless verification remains much higher than that of power grid simulation. Typically, for full-chip power grid verification, both the number of linear programs to be solved and the size of each linear program are proportional to the size of the grid. As a result, the computation cost increases dramatically when the grid size becomes larger. Macromodeling is an effective technique to reduce the computation cost for verifying large-scale power grids, but building the macromodels can also be expensive, and relatively small performance gain can be achieved as studied in [5].

In this paper, we consider vectorless verification of RC power grids following previous studies [3, 17, 19], and propose to reduce the computation cost by *constraint abstraction*. Since a localized region of the power grid can be verified based on the *boundary condition* (i.e., the power supply noises at the neighboring nodes), we propose novel *boundary constraints* to define a partial specification of boundary condition, and perform full-chip grid verification in a divide-and-conquer manner to compute conservative bounds of power supply noises. The boundary constraints provide a high-level abstraction of the grid environment for the region of interest, and enables dramatic performance gain by significantly reducing the computation cost for verification. Experimental results with both synthetic power grids and IBM power grid benchmarks show that the proposed approach achieves up to about 17X speedup over the prior art [19], and the overestimation of power supply noises is reasonably small. In particular, a power grid with 562K nodes can be verified in about 1 hour.

The rest of the paper is organized as follows. The problem formulation and previous approaches are introduced in Section 2. The constraint abstraction approach is proposed in Section 3. After experimental results are shown in Section 4, we conclude the paper in Section 5.

2. PRELIMINARIES

2.1 Problem Formulation

Consider vectorless verification of power grids with an RC model, where each branch of the grid is represented by a resistor, and each node is connected to ground through a capacitor. External power supplies are modeled as ideal voltage sources connected to VDD pads, and some nodes have current sources (to ground), which represent the current drawn by the circuit. Let n be the number of nodes (except VDD pads) in the grid, $\mathbf{v}(t)$ be the $n \times 1$ vector of time-varying voltage drops at nodes, and $\mathbf{i}(t)$ be the $n \times 1$ vector of time-varying current sources connected to the grid (the elements corresponding to the nodes without current sources attached are constant 0s). Then the system equation of the grid can be written as

$$G\mathbf{v}(t) + C\dot{\mathbf{v}}(t) = \mathbf{i}(t), \qquad (1)$$

where G is the $n \times n$ conductance matrix, and C is an $n \times n$ diagonal matrix of nodal capacitances.

The aforementioned model represents a VDD network of the power gird. In fact, for the ground network, similar model can also be built, and equation (1) still holds if $\mathbf{v}(t)$ is redefined to be the vector of ground bounces. We refer to voltage drops in the VDD network and ground bounces in the ground network as *voltage noises*. Generally, the relationship between voltage noises $\mathbf{v}(t)$ and current excitations $\mathbf{i}(t)$ always satisfies the system equation (1), and voltage noises can be evaluated if some current specification is available.

We employ *current constraints* [13] to capture the feasible set of current excitations, so that power grid can be verified without detailed current waveforms, i.e., a vectorless approach. Specifically, two types of constraints are adopted: *local constraints* and *global constraints*. Local constraints define an upper bound on each current source,

$$0 \leq \mathbf{i}(t) \leq \mathbf{I}_L, \forall t,$$

where $\mathbf{I}_L \geq 0$ is an $n \times 1$ vector of peak current values. Global constraints define upper bounds on the sums of groups of current sources, and model the peak current drawn by circuit blocks. Let m be the number of global constraints, then global constraints can be expressed in matrix form as

$$U\mathbf{i}(t) \leq \mathbf{I}_G, \forall t,$$

where U is an $m \times n$ incidence matrix consisting of 0s and

1s, and $\mathbf{I}_G \geq 0$ is an $m \times 1$ vector.

Vectorless verification can be performed for either DC analysis model [13] or transient analysis model (with time step Δt) [8]. As identified in [19], the following maxVN-LCC (maximum Voltage Noise under Linear Current Constraints) problem is the key problem for vectorless verification of RC power grids.

PROBLEM 1 (MAXVN-LCC). *Consider an RC power grid of n nodes with conductance matrix G and capacitance matrix C. Let $A = G$ for DC analysis model or $A = G + \frac{C}{\Delta t}$ for transient analysis model with time step Δt. Given local and global current constraints with parameters $\mathbf{I}_L$, $\mathbf{I}_G$, and U, solve for each node $1 \leq j \leq n$,*

$$Maximize\ v_j \quad s.t. \qquad (2)$$
$$A\mathbf{v} = \mathbf{i}, 0 \leq \mathbf{i} \leq \mathbf{I}_L, U\mathbf{i} \leq \mathbf{I}_G.$$

Here $\mathbf{v}$ and $\mathbf{i}$ are the decision variables of voltage noises and current sources, respectively, and v_j is the jth element of $\mathbf{v}$.

As A is known to be an $n \times n$ symmetric positive definite **M**-matrix, so that A is invertible, A^{-1} is also symmetric and satisfies $A^{-1} \geq 0$. Then the optimization problem (2) can be decomposed into two sub-problems as follows:

$$\text{I: Compute } \mathbf{c}_j \text{ by solving} \quad A\mathbf{x} = \mathbf{e}_j, \qquad (3)$$
$$\text{II: Maximize } v_j = \mathbf{c}_j^T \mathbf{i} \quad s.t. \qquad (4)$$
$$0 \leq \mathbf{i} \leq \mathbf{I}_L, U\mathbf{i} \leq \mathbf{I}_G,$$

where $\mathbf{c}_j$ is an $n \times 1$ vector of coefficients, and $\mathbf{e}_j$ is an $n \times 1$ vector of 0s except for its jth element being 1. It is to be noted that $\mathbf{c}_j$ is the jth column of A^{-1}. The first sub-problem is equivalent to power grid DC analysis with current vector $\mathbf{e}_j$, and the second sub-problem is a linear program.

2.2 Previous Approaches

Obviously, a direct and exact approach involves solving sub-problems (3) and (4) with standard linear system solvers (or accurate power grid analysis algorithms) and linear programming (LP) solvers, respectively. However, such an approach consumes too much runtime to be practical for large-scale power grids. In order to expedite full-chip power grid verification, [3] proposed to compute an approximate $\mathbf{c}_j$ with small number of nonzero elements, so that the linear program (4) can be much simplified and solved efficiently. Later works [17, 19] proposed convex dual algorithms to solve the linear program (4) fast, and utilized a random walk based preconditioned conjugate gradient (PCG) power grid analyzer to calculate $\mathbf{c}_j$. The dual algorithms achieve significant speedup over the standard LP solver, and the resultant verification approach provides better runtime efficiency than [3] while ensuring better solution accuracy. However, as the problem sizes of (3) and (4) are proportional to the number of nodes in the power grid, the computation cost of these approaches is often too high to enable time-efficient verification of large-scale power grids. Hence, it is of great interest to reduce the computation cost by exploring more efficient techniques.

3. CONSTRAINT ABSTRACTION

3.1 Methodology

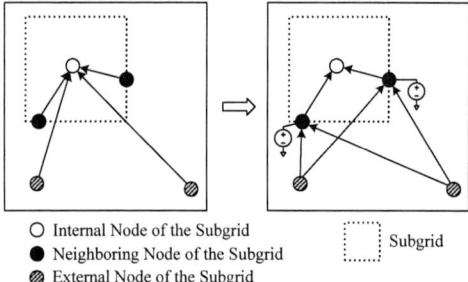

Figure 1: Illustration of conventional approaches (left) and the proposed constraint abstraction approach (right) for verifying an internal node inside a subgrid. The arrows represent the logical relation between nodes.

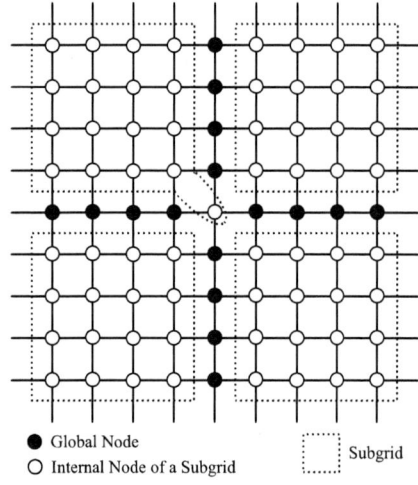

Figure 2: A simple partitioned power grid.

Consider a localized region of the power grid, referred to as a *subgrid*. For each subgrid, we refer to the nodes inside the subgrid as *internal nodes*, the nodes which are outside the subgrid but connected with some internal nodes as *neighboring nodes*, and other nodes in the power grid as *external nodes*.

As illustrated in Figure 1, conventional approaches verify an internal node of a subgrid by considering the whole grid structure, including all neighboring nodes, external nodes, and the corresponding voltage supplies and current sources attached. The corresponding problem sizes of (3) and (4) increase as the size of the power grid increases. In order to verify the subgrid efficiently, we propose to treat neighboring nodes as *uncertain voltage sources* (see Figure 1), which can be modeled by *boundary constraints* (detailed in the next subsection) for realistic scenarios. As a result, the subgrid can be verified based on proper boundary constraints, without involving the external grid structure explicitly. The resultant problem sizes are roughly equal to the size of the subgrid, enabling significant reduction in computation cost. This approach is called *constraint abstraction*, since boundary constraints provide a high-level abstraction of the boundary condition of the subgrid.

3.2 Grid Partitioning & Boundary Constraints

We apply constraint abstraction for full-chip power grid verification in a *divide-and-conquer* manner. As illustrated in Figure 2, the grid is partitioned into several disjoint *subgrids*, which are split by a small set of *global nodes* (i.e., the neighboring nodes of subgrids). Such a partition can be obtained by using the power grid partitioning technique proposed in [18]. Typically, a proper partition results in relatively small number of global nodes, and most nodes are internal nodes of subgrids.

For each subgrid, we use *boundary constraints* to model its boundary condition, i.e., the voltage noises at its neighboring global nodes, which are connected with the internal nodes inside the subgrid. Let $\widehat{m}$ be the number of neighboring global nodes of a subgrid, and $\mathbf{v}_{ex}$ be the $\widehat{m} \times 1$ vector of voltage noises at these nodes. Then the boundary constraints are represented as

$$0 \leq \mathbf{v}_{ex} \leq \mathbf{v}_\ell, \text{ and } \sum_{1 \leq j \leq \widehat{m}} v_{ex,j} \leq v_g, \tag{5}$$

where $v_{ex,j}$ is the jth element of $\mathbf{v}_{ex}$, $\mathbf{v}_\ell$ is an upper bound vector on the voltage noises $\mathbf{v}_{ex}$, and v_g is an upper bound on the sum of these voltage noises.

In practice, $\mathbf{v}_\ell$ can be either the exact worst-case voltage noises at neighboring global nodes, or some upper bounds, which are computed by verifying global nodes. v_g can be computed by solving the following linear program.

$$\text{Maximize} \sum_{1 \leq j \leq \widehat{m}} v_{ex,j} \quad \text{s.t.} \tag{6}$$
$$A\mathbf{v} = \mathbf{i}, 0 \leq \mathbf{i} \leq \mathbf{I}_L, U\mathbf{i} \leq \mathbf{I}_G.$$

Similar to solving the sub-problems (3) and (4) to compute the worst-case voltage noise, we calculate a coefficient vector to represent the objective function as an affine function of current sources. This can be achieved by solving a linear system $A\mathbf{x} = \mathbf{e}_g$, where $\mathbf{e}_g$ is a vector of 0s except that its elements corresponding to the neighboring global nodes are set to 1. With such a coefficient vector in hand, the optimal value of (6) can be obtained by solving a linear program like (4). Here, the computation cost is equivalent to verifying a node by solving (3) and (4).

Based on the above discussion, a partitioned power grid can be verified in the following steps.

1. For each global node, compute its worst-case voltage noise (or an upper bound of voltage noise).
2. For each subgrid, build boundary constraints, and then evaluate the worst-case voltage noises at internal nodes subject to boundary constraints.

In the first step, as there are only relatively small amount of global nodes, previous methods [3, 17, 19] can be applied. The verification of subgrids in the second step is detailed in the next subsection.

3.3 Verification of Subgrids

As summarized in Problem 1, we need to evaluate the worst-case voltage noises based on the generalized system equation $A\mathbf{v} = \mathbf{i}$ for both DC and transient models. Consider a subgrid with $\widehat{n}$ internal nodes and $\widehat{m}$ neighboring global nodes. The equation $A\mathbf{v} = \mathbf{i}$ is reduced to

$$[A_{in} \quad A_{ex}] \begin{bmatrix} \mathbf{v}_{in} \\ \mathbf{v}_{ex} \end{bmatrix} = \mathbf{i}_{in}, \tag{7}$$

where A_{in} is the $\widehat{n} \times \widehat{n}$ conductance matrix of the subgrid, A_{ex} is a non-positive $\widehat{n} \times \widehat{m}$ matrix representing the conductance links between internal nodes and neighboring global nodes, $\mathbf{v}_{in}$ and $\mathbf{i}_{in}$ are the vectors of voltage noises and current sources at internal nodes, respectively.

Since A_{in} is also a symmetric positive definite **M**-matrix, it is invertible, and A_{in}^{-1} is symmetric and non-negative. Then equation (7) can be rearranged as

$$\mathbf{v}_{in} = A_{in}^{-1}\mathbf{i}_{in} - A_{in}^{-1}A_{ex}\mathbf{v}_{ex}. \qquad (8)$$

Let $v_{in,j}$ be the jth element of $\mathbf{v}_{in}$, $\mathbf{c}_{in,j}$ be the jth column of A_{in}^{-1}, and $\mathbf{c}_{ex,j}$ be the transpose of the jth row of $-A_{in}^{-1}A_{ex}$. (Note that both $\mathbf{c}_{in,j}$ and $\mathbf{c}_{ex,j}$ are non-negative, and $\mathbf{c}_{ex,j}^T = -\mathbf{c}_{in,j}^T A_{ex}$.) We have

$$v_{in,j} = \mathbf{c}_{in,j}^T\mathbf{i}_{in} + \mathbf{c}_{ex,j}^T\mathbf{v}_{ex}, \qquad (9)$$

where the internal current vector $\mathbf{i}_{in}$ is defined by local and global current constraints, and the boundary condition $\mathbf{v}_{ex}$ is restricted by boundary constraints.

As a result, the subgrid can be verified by solving the following linear program for each internal node $1 \leq j \leq \widehat{m}$.

$$\text{Maximize } v_{in,j} = \mathbf{c}_{in,j}^T\mathbf{i}_{in} + \mathbf{c}_{ex,j}^T\mathbf{v}_{ex} \quad \text{s.t.} \quad (10)$$
$$0 \leq \mathbf{i}_{in} \leq \widehat{\mathbf{I}}_L, \widehat{U}\mathbf{i}_{in} \leq \widehat{\mathbf{I}}_G,$$
$$0 \leq \mathbf{v}_{ex} \leq \mathbf{v}_\ell, \sum_{1 \leq j \leq \widehat{m}} v_{ex,j} \leq v_g,$$

where $\widehat{\mathbf{I}}_L$, $\widehat{\mathbf{I}}_G$ and $\widehat{U}$ are matrices for local and global current constraints related to the subgrid. It is to be noted that the decision variables $\mathbf{i}_{in}$ and $\mathbf{v}_{ex}$ are independent, thus (10) can be further decomposed into two linear programs and solved separately.

$$\text{Maximize } \mathbf{c}_{in,j}^T\mathbf{i}_{in} \quad \text{s.t.} \quad 0 \leq \mathbf{i}_{in} \leq \widehat{\mathbf{I}}_L, \widehat{U}\mathbf{i}_{in} \leq \widehat{\mathbf{I}}_G, \qquad (11)$$
$$\text{Maximize } \mathbf{c}_{ex,j}^T\mathbf{v}_{ex} \quad \text{s.t.} \quad 0 \leq \mathbf{v}_{ex} \leq \mathbf{v}_\ell, \sum_{1 \leq j \leq \widehat{m}} v_{ex,j} \leq v_g. \qquad (12)$$

LEMMA 1. *Let $v_{in,j}^*$ be the exact worst-case voltage noise at internal node j (i.e., the optimal value of the linear program (2)), and $v_{in,j}^+$ be the worst-case voltage noise computed through constraint abstraction (i.e., the optimal value of the linear program (10)), then $v_{in,j}^* \leq v_{in,j}^+$.*

PROOF. Let $\mathbf{i}_{in}^*$ and $\mathbf{v}_{ex}^*$ be the local current vector inside the subgrid and the voltage noises at neighboring global nodes corresponding to $v_{in,j}^*$, respectively, so that

$$v_{in,j}^* = \mathbf{c}_{in,j}^T\mathbf{i}_{in}^* + \mathbf{c}_{ex,j}^T\mathbf{v}_{ex}^*.$$

Let $\mathbf{i}_{in}^+$ and $\mathbf{v}_{ex}^+$ be the optimal solution of (10), so that

$$v_{in,j}^+ = \mathbf{c}_{in,j}^T\mathbf{i}_{in}^+ + \mathbf{c}_{ex,j}^T\mathbf{v}_{ex}^+.$$

Clearly, we have $\mathbf{c}_{in,j}^T\mathbf{i}_{in}^* \leq \mathbf{c}_{in,j}^T\mathbf{i}_{in}^+$, and $\mathbf{c}_{ex,j}^T\mathbf{v}_{ex}^* \leq \mathbf{c}_{ex,j}^T\mathbf{v}_{ex}^+$. It follows that $v_{in,j}^* \leq v_{in,j}^+$. □

This verification approach derives an upper bound of voltage noise at each internal node inside the subgrid. The difference between the computed upper bound $v_{in,j}^+$ and the exact worst-case voltage noise $v_{in,j}^*$ is called *overestimation*. As we will show in the experimental results, the amount of overestimation is very small, i.e., the computed voltage noise bound is fairly tight and realistic.

Except for solving (6) to obtain v_g, verifying the subgrid mainly involves computing A_{in}^{-1} and $-A_{in}^{-1}A_{ex}$, and solving the linear programs (11) and (12) for each internal node. By keeping the size of the subgrid within some

bound, the Cholesky decomposition of A_{in} can be computed, and then A_{in}^{-1} can be calculated row by row (or column by column since A_{in}^{-1} is symmetric) by solving a linear system like (3) through forward and back substitutions. As A_{ex} is a sparse matrix representing the links between internal nodes and neighboring global nodes, the matrix multiplication $-A_{in}^{-1}A_{ex}$ can be performed efficiently. In practice, both A_{in}^{-1} and $-A_{in}^{-1}A_{ex}$ are not stored explicitly, their rows (i.e., $\mathbf{c}_{in,j}^T$ and $\mathbf{c}_{ex,j}^T$) are computed, used and discarded at runtime. The linear program (11) can be solved by standard LP solvers if the subgrid size is within some reasonable bound. As the linear program (12) only has one constraint defining an upper bound of the sum of all variables, it can be efficiently solved by sorting the coefficients in non-increasing order, and setting the corresponding variable to be the maximum feasible value sequentially.

3.4 Analysis of Computation Cost

In this subsection, we present the computation advantage of the proposed constraint abstraction approach over a direct one. Since the computation for $-A_{in}^{-1}A_{ex}$ and (12) can be performed fairly efficiently, the major computation cost for verifying a subgrid is to solve (6), to calculate A_{in}^{-1}, and to solve linear program (11) for each internal node. In contrast, a direct approach computes A^{-1} (by solving (3)) and solves linear program (4) for all the nodes.

Suppose the cost of computing the Cholesky decomposition is $f_1(N)$, the cost of one forward and one back substitution is $f_2(N)$, and the cost of solving a linear program (like (4)) is $f_3(N)$, where N is the size of the matrix, and the number of decision variables in the linear program. Typically, $f_3(N) \gg f_1(N) \gg f_2(N)$.

Consider a direct approach with Cholesky decomposition for computing A^{-1}. The total computation cost is given by

$$f_1(n) + nf_2(n) + nf_3(n). \qquad (13)$$

Let n_0 be the number of global nodes, $n_j, 1 \leq j \leq k$ be the number of internal nodes inside each subgrid, where $\sum_{0 \leq j \leq k} n_j = n$. The cost of verifying global nodes with the direct approach is $f_1(n) + n_0 f_2(n) + n_0 f_3(n)$. For each subgrid, the cost of solving (6) is $f_2(n) + f_3(n)$, the cost of computing A_{in}^{-1} and solving linear program (11) is $f_1(n_j) + n_j f_2(n_j) + n_j f_3(n_j)$. Hence, the computation cost of the proposed approach can be approximated as

$$f_1(n) + (n_0 + k)f_2(n) + (n_0 + k)f_3(n)$$
$$+ \sum_{1 \leq j \leq k} \left(f_1(n_j) + n_j f_2(n_j) + n_j f_3(n_j) \right). \qquad (14)$$

Expressions (13) and (14) provide a rough estimation of computation costs based on the size of the power grid and its subgrids. Generally, the time complexity of Cholesky decomposition is $O(N^3)$, the time complexity of forward and back substitutions is $O(N^2)$, and the complexity of linear programming by standard solvers can be in even higher order (e.g., $O(N^4)$). In practice, the sparsity of the conductance matrix A (as well as the constraint matrix U), combined with efficient reordering, enables actual computation costs to be less than such theoretical bounds, but the cost of the direct approach still remains greater than quadratic. Hence, the computation cost of the proposed approach will be much smaller than that of the direct approach if the power gird is partitioned properly.

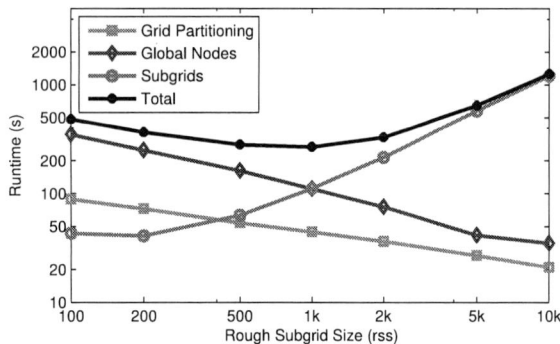

Figure 3: Runtime for verifying a synthetic power grid "pg4000" ($90,643$ nodes) with different rough subgrid sizes (rss).

4. EXPERIMENTS

4.1 Experimental Setup

The proposed constraint abstraction approach has been implemented in C++. hMETIS [10] is utilized to partition the power grid into subgrids with relatively small number of global nodes. We use a user-specified *rough subgrid size* called rss to control the size of subgrids, and the number of subgrids is calculated by $round(\frac{n}{rss})$, where n is the number of nodes (after merging nodes connected by shorts) in the grid. The DualVN algorithm proposed in [19] is employed to verify global nodes, and to solve the linear program (6) for building boundary constraints, where its error tolerance for solving linear programs is set to be 0.1mV. CHOLMOD [15] and GotoBLAS2 [9] are employed to solve the involved linear systems through Cholesky factorization, and MOSEK [14] is used to solve the linear program (11) for verifying each subgrid. We choose the simplex method of MOSEK, since it is slightly faster than the interior point method for our experiments.

For performance comparison, we implemented the DirectVN algorithm of [19] by solving linear system (3) and linear program (4) with CHOLMOD and MOSEK, respectively. The DualVN algorithm [19] has also been implemented for verifying the whole power grid, while linear system (3) is solved by CHOLMOD. Experiments are carried out on a 64-bit Linux server with 2.67GHz Intel Xeon X5650 processor and 64GB memory. Although the server has 12 cores, only one core is used for experiments.

We employ the synthetic power grids used in [19] for performance tests, and also experiment with IBM power grid benchmarks [11]. As IBM power grids have multiple networks, we choose to verify the ground network in our experiments, while their VDD networks can be verified separately. Local current constraints are extracted from the grid description, and global current constraints are generated by scaling down the total amount of current drawn by groups of current sources. For each power grid, we specify 4 global constraints in our tests.

4.2 Exploring Rough Subgrid Size

Since the computation cost of the proposed approach is highly dependent on the sizes of subgrids as discussed in Section 3.4, we experiment with different rough subgrid sizes rss (ranging from 100 to 10K) for performance analysis.

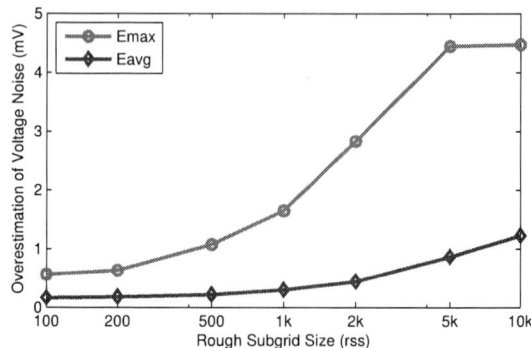

Figure 4: Solution accuracy of verifying a synthetic power grid "pg4000" ($90,643$ nodes) with different rough subgrid sizes (rss). E_{max}/E_{avg}: the maximum/average overestimation of voltage noises in mV.

Note that the actual sizes of most subgrids would be slightly smaller than rss, because global nodes do not belong to any subgrid. The total runtime of the proposed approach can be roughly divided into three parts: the runtime to partition the power grid, the runtime to solve global nodes, and the runtime to verify subgrids. Figure 3 plots these runtime components for verifying a synthetic power grid with different rss. As rss increases, both the number of subgrids and the number of resultant global nodes decrease, so it takes less runtime to partition the power grid, and to verify global nodes. On the contrary, both the subgrid sizes and the total number of internal nodes increase as rss, thus solving the subgrids consumes more runtime when rss becomes larger. Obviously, there is a tradeoff between the computation costs of global nodes and subgrids. If the runtime for grid partitioning is sufficiently small, then the minimum runtime is achieved at the best tradeoff point, where the runtime of global nodes and the runtime of subgrids are roughly equal, e.g., the optimal rss is 1K for the test case shown in Figure 3.

Recall that constraint abstraction results in overestimation of the worst-case voltage noises as stated in Lemma 1. We also evaluate the effects of different rss on the amount of overestimation. As shown in Figure 4, both the maximum and the average overestimation increase when a larger rss is specified. This phenomenon is due to the fact that the boundary constraints become less effective when the subgrid sizes become larger.

To determine a proper rss, both runtime and solution accuracy need to be considered. Fortunately, the overestimation for most rss settings are sufficiently small, so that the rss which provides the best runtime efficiency can be employed for production runs.

4.3 Performance Results

The performance data are presented in Table 1. The runtime of DirectVN and DualVN are reported under columns 4 and 5 for performance comparison. The golden solution is produced by DirectVN for analysis of overestimation. Since DiretVN takes too much runtime to verify large power grids, the reported data with $\approx$s are estimations from 1000 random nodes. The DualVN algorithm is fairly efficient, and can solve the largest grid "pg10000" with 562K nodes in 19 hours. The proposed constraint abstraction approach further improves the runtime efficiency of vectorless verifica-

Table 1: Performance Results of the Proposed Constraint Abstraction Approach. n: the number of nodes after merging the nodes connected by shorts; v_{max}: the maximum voltage noise across the grid in mV; rss: rough subgrid size; n_0: the number of global nodes; E_{max}/E_{avg}: the maximum/average overestimation of voltage noises in mV; the runtime units "s", "m", "h" and "d" represent seconds, minutes, hours, and days, respectively.

Power Grids			DirectVN	DualVN [19]	Constraint Abstraction					Speedup Relative to	
Name	n	v_{max}	Runtime	Runtime	rss	n_0	E_{max}	E_{avg}	Runtime	DirectVN	DualVN
pg1000	5875	46.31	2.66 m	15.40 s	200	546	2.32	0.42	6.57 s	24.31	2.35
pg2000	22939	39.91	53.47 m	2.52 m	500	1459	1.51	0.28	33.22 s	96.58	4.55
pg2500	35668	28.80	2.06 h	5.85 m	500	2287	0.94	0.16	1.00 m	123.61	5.85
pg3000	51195	43.63	4.83 h	12.27 m	500	3392	0.81	0.19	1.71 m	168.98	7.16
pg4000	90643	54.38	15.93 h	38.73 m	1000	4270	1.65	0.29	4.16 m	**229.72**	**9.31**
pg5000	141283	$\approx$45.91	$\approx$2.15 d	1.25 h	1000	6855	$\approx$1.02	$\approx$0.44	7.48 m	$\approx$**413.33**	**10.02**
pg10000	562363	$\approx$23.11	$\approx$44.07 d	18.58 h	2000	19427	$\approx$1.88	$\approx$1.44	1.10 h	$\approx$**963.08**	**16.92**
ibmpg1	10242	677.67	4.26 h	26.09 s	200	775	6.16	0.37	12.84 s	19.89	2.03
ibmpg2	65228	357.92	3.95 h	16.65 m	500	4513	5.00	0.44	2.45 m	96.90	6.81
ibmpg3	150687	$\approx$179.91	$\approx$3.47 d	1.29 h	1000	6853	$\approx$1.58	$\approx$0.42	8.93 m	$\approx$**559.49**	**8.70**
ibmpg4	478094	$\approx$3.42	$\approx$14.12 d	15.37 h	2000	20103	$\approx$0.19	$\approx$0.12	1.24 h	$\approx$**273.33**	**12.40**
ibmpg5	291382	$\approx$42.44	$\approx$19.51 d	6.13 h	1000	13674	$\approx$0.59	$\approx$0.12	29.29 m	$\approx$**959.33**	**12.56**
ibmpg6	430337	$\approx$109.96	$\approx$55.97 d	12.47 h	2000	11975	$\approx$1.20	$\approx$0.57	46.27 m	$\approx$**1741.73**	**16.17**

tion, achieving up to about 17X speedup over DualVN. As a result, "pg10000" can be verified in about 1 hour. Generally, grid partitioning takes a few seconds to a few minutes depending on the grid size and the number of desired subgrids. Most runtime of the proposed approach is spent on solving global nodes and subgrids. Moreover, for most test cases, the maximum overestimation is within 3mV, and the average overestimation is much smaller than 1mV. In addition, it is worth noting that the proposed approach is more effective when the grid size becomes larger, making it suitable for large-scale power grid verification.

The major merit of constraint abstraction is that it enables significant reduction in computation cost with small amount of overestimation. In practice, it can be used for fast estimation of power supply noises across the chip. The risky regions can be identified in small amount of runtime, and then more accurate methods can applied if it is necessary.

5. CONCLUSION

In this paper, we presented constraint abstraction for vectorless verification of power grids. Boundary constraints are built to model the boundary condition of subgrids, enabling efficient verification of power grids in a divide-and-conquer manner. Experimental results confirmed the efficiency of the proposed approach, and it may also be extended to transient verification of RLC power grids.

6. REFERENCES

[1] N. Abdul Ghani and F. N. Najm. Power grid verification using node and branch dominance. In *Proc. Design Automation Conf. (DAC)*, pages 682–687, Jun. 2011.

[2] N. H. Abdul Ghani and F. N. Najm. Handling inductance in early power grid verification. In *Proc. Int. Conf. Computer-Aided Design (ICCAD)*, pages 127–134, Nov. 2006.

[3] N. H. Abdul Ghani and F. N. Najm. Fast vectorless power grid verification using an approximate inverse technique. In *Proc. Design Automation Conf. (DAC)*, pages 184–189, Jul. 2009.

[4] N. H. Abdul Ghani and F. N. Najm. Fast vectorless power grid verification under an RLC model. *IEEE Trans. Computer-Aided Design*, 30(5):691–703, May 2011.

[5] Abhishek and F. N. Najm. Incremental power grid verification. In *Proc. Design Automation Conf. (DAC)*, pages 151–156, Jun. 2012.

[6] M. Avci and F. N. Najm. Early P/G grid voltage integrity verification. In *Proc. Int. Conf. Computer-Aided Design (ICCAD)*, pages 816–823, Nov. 2010.

[7] C.-K. Cheng, P. Du, A. B. Kahng, G. K. H. Pang, Y. Wang, and N. Wong. More realistic power grid verification based on hierarchical current and power constraints. In *Proc. Int. symp. Physical Design (ISPD)*, pages 159–166, Mar. 2011.

[8] I. A. Ferzli, F. N. Najm, and L. Kruse. A geometric approach for early power grid verification using current constraints. In *Proc. Int. Conf. Computer-Aided Design (ICCAD)*, pages 40–47, Nov. 2007.

[9] GotoBLAS2. http://www.tacc.utexas.edu/tacc-projects/gotoblas2.

[10] hMETIS. http://glaros.dtc.umn.edu/gkhome/metis/hmetis/overview.

[11] IBM Power Grid Benchmarks. http://dropzone.tamu.edu/ pli/pgbench/.

[12] D. Kouroussis, I. A. Ferzli, and F. N. Najm. Incremental partitioning-based vectorless power grid verification. In *Proc. Int. Conf. Computer-Aided Design (ICCAD)*, pages 358–364, Nov. 2005.

[13] D. Kouroussis and F. N. Najm. A static pattern-independent technique for power grid voltage integrity verification. In *Proc. Design Automation Conf. (DAC)*, pages 99–104, Jun. 2003.

[14] MOSEK. http://www.mosek.com/.

[15] SuiteSparse. http://www.cise.ufl.edu/research/sparse/suitesparse/.

[16] Y. Wang, X. Hu, C.-K. Cheng, G. K. H. Pang, and N. Wong. A realistic early-stage power grid verification algorithm based on hierarchical constraints. *IEEE Trans. Computer-Aided Design*, 31(1):109–120, Jan. 2012.

[17] X. Xiong and J. Wang. An efficient dual algorithm for vectorless power grid verification under linear current constraints. In *Proc. Design Automation Conf. (DAC)*, pages 837–842, Jun. 2010.

[18] X. Xiong and J. Wang. A hierarchical matrix inversion algorithm for vectorless power grid verification. In *Proc. Int. Conf. Computer-Aided Design (ICCAD)*, pages 543–550, 2010.

[19] X. Xiong and J. Wang. Dual algorithms for vectorless power grid verification under linear current constraints. *IEEE Trans. Computer-Aided Design*, 30(10):1469–1482, Oct. 2011.

[20] X. Xiong and J. Wang. Vectorless verification of RLC power grids with transient current constraints. In *Proc. Int. Conf. Computer-Aided Design (ICCAD)*, pages 548–554, 2011.

[21] X. Xiong and J. Wang. Verifying RLC power grids with transient current constraints. *IEEE Trans. Computer-Aided Design*, 2013.

The Impact of Electromigration in Copper Interconnects on Power Grid Integrity

Vivek Mishra and Sachin S. Sapatnekar

ECE Department, University of Minnesota, Minneapolis, MN

Abstract—**Electromigration (EM), a growing problem in on-chip interconnects, can cause wire resistances in a circuit to increase under stress, to the point of creating open circuits. Classical circuit-level EM models have two drawbacks: first, they do not accurately capture the physics of degradation in copper dual-damascene (CuDD) metallization, and second, they fail to model the inherent resilience in a circuit that keeps it functioning even after a wire fails. This work overcomes both limitations. For a single wire, our probabilistic analysis encapsulates known realities about CuDD wires, e.g., that some regions of these wires are more susceptible to EM than others, and that void formation/growth show statistical behavior. We apply these ideas to the analysis of on-chip power grids and demonstrate the inherent robustness of these grids that maintains supply integrity under some EM failures.**

Keywords: **Electromigration, process variation, robustness, power grid**

I. INTRODUCTION

Electromigration (EM) in interconnects occurs due to the movement of metal atoms, activated by momentum transfer from collisions with free electrons [1]. When bounded by a blocking boundary such as a barrier layer, this movement causes a depletion of atoms at the cathode end and a surplus at the anode; this depletion eventually leads to void nucleation and subsequent growth [2]. Since the critical stress for void nucleation is very small for copper dual damascene (CuDD) structures, voids can form early in the lifetime of a design [3].

There is a large gap between what is known about the physics of EM in CuDD wires and the knowledge used at the circuit level. Traditional EM analysis is based on failure criteria measured under accelerated aging. An interconnect whose resistance crosses a predetermined threshold under stress is deemed to have failed, and the time-to-failure parameters are extrapolated to normal operating conditions using Black's equation [4]. This analysis, supplemented with the Blech-length thresholding criterion [5] that defines wires that are immortal under EM, is used by circuit designers to derive maximum current density limit rules on individual wires.

There are several problems with such an approach. First, in a real circuit, the impact of such failures is context-dependent. In some cases, a large failure may be tolerated due to the inherent resilience in the circuit, e.g., due to redundancy in a power grid, where the failure of one wire may be compensated by current flow through other paths. Therefore, the use of a single threshold for the resistance change may either be excessively conservative, or not conservative enough, depending on how the threshold is chosen and how robust the circuit is. Second, for CuDD interconnects, previous work [3] has shown that the Blech-length approach, where wires with a sufficiently small jL product (j is the current density, L is the length) can be considered immortal, is invalid, and it has been observed that some lines fail *probabilistically* even if they satisfy the Blech criterion on their jL value. The root cause of this difference is that the critical stress for void nucleation in Cu is $10\times$ lower than that for Al, implying that it is

Permission to make digital or hard copies of all or part of this work for personal or classroom use is granted without fee provided that copies are not made or distributed for profit or commercial advantage and that copies bear this notice and the full citation on the first page. To copy otherwise, to republish, to post on servers or to redistribute to lists, requires prior specific permission and/or a fee.
DAC 13, May 29 - June 07 2013, Austin, TX, USA.

possible for voids to nucleate soon, before providing the opportunity for opposing back-stresses to build up to balance them. Third, there are known effects such as the current divergence effect (discussed in Section III-D) that are not widely considered at the circuit level.

These peculiarities for CuDD metallization indicate the need to develop models for EM to enable probabilistic circuit analysis in a context-sensitive way. The probabilistic viewpoint reflects both the fact that mechanisms for EM are stochastic, and that the number of interconnects on a chip is large enough that such statistical effects may show up in different parts of the chip. There have been few prior works in this direction: the work in [6] built up on [7] to consider some EM issues beyond the conventional Black's equation, but was based on the problematic Blech length criterion.

In this work, we first present an analytical model to predict the distribution of void growth and consequently, the resistance change in a wire. Next, we demonstrate how this affects the probabilistic distribution of voltage drops in standard power grid benchmarks.

II. DETERMINISTIC EM MODELS FOR CU INTERCONNECTS

The fundamental phenomenon of EM consists of forces that drive atoms from the cathode to the anode. This produces regions of uneven concentration, i.e., depletion and accumulation, which lead to diffusion through various possible mechanisms: grain boundary, volume, surface, and/or interface diffusion.

EM failure occurs in Cu interconnects in two phases:

- *Void nucleation*: After a wire has been stressed, the depletion of atoms at the cathode creates a tensile stress. Once a critical stress threshold value has been crossed, the void nucleates.
- *Void growth*: After nucleation, further movement of metal atoms from the void results in void growth. This results in increased wire resistance due to the effectively reduced cross-section. If the void grows large enough, it may result in a break in the wire, resulting in either an open circuit or a vastly increased resistance, in cases where the current through the wire can flow through the higher-resistivity barrier layer of the CuDD interconnect.

We begin by introducing deterministic models for the void nucleation and growth phases. The foundation for modern models for EM, incorporating the impact of stress, is based on the stress evolution model in [8], which details a set of differential equations describing the interplay between electron wind force and back stresses in the interconnect. To model void nucleation, we use this stress evolution model, extended to incorporate thermal stress effects using the formulation as presented in [9]. The time, t_n, at which a void nucleates is given by

$$t_n = \frac{K_{t_n}}{D_{\textit{eff}}} \tag{1}$$

$$\text{where} \quad K_{t_n} = \frac{\pi}{4}\left(\frac{(\sigma_c - \sigma_{th})^2 \Omega k_B T}{(eZ^\star_{\textit{eff}} \rho\, j)^2 B}\right)$$

The symbol K_{t_n} groups together a number of terms to reflect the dependency of t_n on the effective diffusivity, $D_{\textit{eff}}$. In the detailed expression for K_{t_n}, σ_c is the effective critical stress for void nucleation; σ_{th} is a term that accounts for the effects of thermal

618

stresses; Ω is atomic volume; k_B is Boltzmann's constant; T is the temperature; $q^\star = eZ^\star_{eff}$ is the effective charge, where e is the elementary charge on an electron and $Z^\star_{eff}$ is the apparent effective charge number; $E = \rho j$ is the electric field, where ρ is the resistivity of copper and j is the current density in the wire; B is the effective bulk modulus for the Cu–dielectric system.

In a CuDD process, a trench is first etched into the interlayer dielectric, and a Ta-based liner is deposited therein to prevent Cu from diffusing through. Next, the Cu used to construct the interconnect is deposited, and finally, the lines are capped above. The diffusivity, D_{eff}, for EM can be considered as a sum of contributions of atomic transport along various diffusion paths: the Cu capping interface I between Cu and the Ta liner, the surface S, the grain boundaries GB, and the bulk B. The product of effective diffusivity and the effective atomic number along a CuDD interconnect can be written as [10]:

$$Z^\star_{eff} D_{eff} = Z^\star_I D_I \left(\frac{\delta_I}{h}\right) + Z^\star_S D_S \delta_S \left(\frac{1}{h}\right) + \\ Z^\star_{GB} D_{GB} \delta_{GB} \left(\frac{1}{d} - \frac{1}{w}\right) + n_B D_B$$

where δ_I, δ_S, δ_{GB}, and δ_B denote the width of the capping interface, surface, grain boundary, and bulk, respectively, h is the line height, d is the grain size, and n_B is the fraction of atoms diffusing through the bulk. For thin copper interconnects at nanometer-scale geometries, which are the subject of this work, the primary diffusion paths for void nucleation is along the surface [10]–[13] and grain boundaries play a significant role only in much wider wires that show polycrystalline grain structures. However, once the void nucleates the surface of the void which is an open copper surface acts as a fast diffusion path [12]. Accordingly, we consider only the dominant surface diffusivity term for each stage and neglect the other terms.

After nucleation at time t_n, given by (1), the void starts to grow. Various void growth kinetics have been observed in CuDD structures, depending on the direction of the current. The scenario where the electron flow is downwards, shown in Fig. 1(a), corresponds to the via-above case, and results in potential void formations at the via or in the wire, as illustrated in the figure. For the case where the electron flow is upwards, illustrated in Fig. 1(b), an upstream void is potentially formed in the upper wire, typically at a corner of the wire or within the wire, as shown. The mechanics of void formation in each case – *span growth* (for both the via-above and via-below cases), when the void spans the entire interconnect, and *slit growth*, where it forms along the via (for the via-above case) – is different and necessitates a different model.

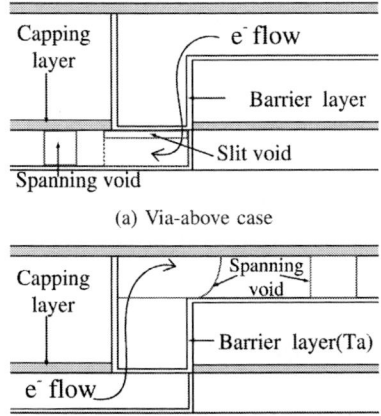

(a) Via-above case

(b) Via-below case

Fig. 1. Mechanisms of void evolution for the via-above and via-below cases.

In either case, once a void is formed, the primary mechanism of increase in the void size is due to drift with a constant drift velocity v_d [2], [14], which is related to the effective mobility and the driving force expressed as the Nernst-Einstein relation given by:

$$v_d = \left(\frac{D_{eff}}{k_B T}\right) e Z^\star_{eff} \rho\, j \qquad (2)$$

If the void nucleates at time t_n, then at an observation time t_o, the void has been growing for a length of time, $t_o - t_n$. The length of the void increases due to drift, as given by:

$$L_{void}(t_o) = v_d \cdot (t_o - t_n) = \left(\frac{D_{eff}}{k_B T}\right) e Z^\star_{eff} \rho\, j\, (t_o - t_n) \qquad (3)$$

III. PROBABILISTIC MODELING OF EM FAILURE

The conventional explanation of EM in Al interconnects was predicated on the interaction between the electron wind force and the back stress force. For some interconnects, with low current and/or small length, the two forces could be in equilibrium in the steady state, so that critical stress σ_c for void nucleation is never reached, implying that these wires are *immortal* to EM effects. For a wire of length L with current density j, it was shown that the criterion for immortality was $jL \le (jL)_{crit}$, where $(jL)_{crit}$ is a property of the material and the fabrication process. However, as noted in Section I, it has been observed that for CuDD interconnects, the immortality property does not hold, and lines are apt to show probabilistic behavior [3].

A. Probabilistic Models for Activation Energy

The diffusivity is related, through an Arrhenius relationship, to temperature T and the activation energy E_a as

$$D_{eff} = D_0 \exp\left(-\frac{E_a}{k_B T}\right) \qquad (4)$$

where D_0 is a constant. Recent work has observed EM failure is correlated to uncertainties in the microstructure and physical parameters of an interconnect, which relates to the statistical distribution of the normally-distributed activation energy.

Strictly speaking, since the activation energy is a property of the microstructure, it can vary within the wire depending on the grain boundary orientation. For instance EM activation energy can vary between grains depending on the orientation of the grain with respect to each other and with respect to the interfacial layer [25], [26]. However, at a macroscopic level, it is reasonable to assume that the effective activation energy is same for a wire and varies only between the wires [2], [12], [15]–[18]. Therefore, we work with the idea of the "effective activation energy" for each wire, which is an averaged activation energy value for that wire. As observed above, the activation energy is normally distributed, and so we model the effective activation energy, E_a, for each wire using an independent Gaussian random variable.

B. Statistical Models for Void Dimensions

Since effective activation energy, E_a, for a wire follows a Gaussian distribution, it is obvious from (4) that diffusivity, which contains E_a in an exponential term, follows a lognormal distribution.

In our discussion below, for a distribution $Z = N(\mu, \sigma)$ with mean μ and standard deviation σ, we denote a lognormal $X = e^Z$ as $\text{LogN}(\mu, \sigma)$. Therefore, if $E_a = N(\mu, \sigma)$, then $D_{eff} = \text{LogN}(\mu_{D_{eff}}, \sigma_{D_{eff}})$, where

$$\mu_{D_{eff}} = \log D_0 - \frac{\mu}{k_B T} \qquad (5)$$

$$\sigma_{D_{eff}} = \frac{\sigma}{k_B T}$$

As discussed in Section II, the mechanisms responsible for D_{eff} are different in the nucleation and growth phases: interface diffusivity for nucleation, and surface diffusivity for growth. We refer to the effective diffusivity for the nucleation and growth phases as $D_{eff,n}$ and $D_{eff,g}$, respectively.

Nucleation: The expression for nucleation time t_n was provided in Equation (1). From this, it is clear that $t_n = \text{LogN}(\mu_{t_n}, \sigma_{t_n})$,

$$\mu_{t_n} = \log(K_{t_n}) - \mu_{D_{eff,n}} \tag{6}$$
$$\sigma_{t_n} = \sigma_{D_{eff,n}}$$

The proof of this is straightforward, and relies on the observation that the distribution of a reciprocal of a lognormal $\log N(\mu, \sigma)$ is another lognormal characterized as $\log N(-\mu, \sigma)$.

Growth: During void growth, the length of a void evolves with time according to Equation (3). Grouping together all deterministic parameters in this equation, if a void nucleates, then its length at observation time t_o is given by:

$$L_{void}(t_o) = \left(\frac{D_{eff,g}}{k_B T}\right) e Z_{eff}^\star \rho_{cu}\, j\, (t_o - t_n) \tag{7}$$
$$= c_1 D_{eff,g} - c_2 t_n D_{eff,g} \tag{8}$$

Here, c_1 and c_2 are deterministic constants. The first term, $c_1 D_{eff,g}$, is clearly lognormal since $D_{eff,g}$ is lognormal; the second term, $c_2 t_n D_{eff,g}$ is a scaled product of lognormals, which is also a lognormal. Therefore, L_{void} is a difference of two lognormals, $c_1 D_{eff,g}$ and $c_2 t_n D_{eff,g}$, and it can be approximated by a lognormal using the widely-used Wilkinson approximation [19].

If μ_X, σ_X (μ_Y, σ_Y) are the mean (standard deviation) of the underlying normal distribution for $c_1 D$ ($c_2 t_n D$), then

$$\mu_X = \log c_1 + \mu_{D_{eff,n}}; \sigma_X = \sigma_{D_{eff,n}}$$
$$\mu_Y = \log c_2 + \mu_{D_{eff,g}} + \mu_{t_n}$$
$$\sigma_Y = \sqrt{\sigma^2_{D_{eff,g}} + \sigma^2_{t_n}}$$

where μ_{t_n} and σ_{t_n} are given by Equation 6. This provides us with an analytical expression for the distribution of the random variable, $L_{void}(t_o)$. For this lognormal distribution, $\log N(\mu_{L_{void}}, \sigma_{L_{void}})$, we can compute the parameters of the distribution using Wilkinson approximation as follows:

$$u_1 = e^{\left(\mu_X + \sigma_X^2/2\right)} - e^{\left(\mu_Y + \sigma_Y^2/2\right)}$$
$$u_2 = e^{\left(2\mu_X + 2\sigma_X^2\right)} + e^{\left(2\mu_Y + 2\sigma_Y^2\right)} - 2e^{\left(\mu_X + \mu_Y + (\sigma_X^2 + \sigma_Y^2)/2\right)}$$
$$\mu_{L_{void}} = 2\log(u_1) - \log(u_2)/2$$
$$\sigma^2_{L_{void}} = \log(u_2) - 2\log(u_1)$$

C. Probability Distribution of the Resistance Change due to EM

We will now use the void length distribution to evaluate the distribution of resistance change for different scenarios of void growth. Our resistance evolution model considers separately the cases of the *span growth* and *slit growth* mechanisms shown in Fig. 1. For the case of the span void, the change in resistance, ΔR, is [14]:

$$\frac{\Delta R}{R_o} = \left(\frac{\rho_{Ta}}{\rho_{Cu}}\frac{A_{Cu}}{A_{Ta}} - 1\right)\frac{L_{void}}{L_{wire}} \tag{9}$$

where ρ_{Cu} and ρ_{Ta} are, respectively, the resistivities of copper and Tantalum, $R_o = \rho_{Cu} L_{wire}/A_{Cu}$ is the resistance of the interconnect wire segment, which is assumed to have length L_{wire} and cross-sectional area A_{Cu}, and A_{Ta} is the cumulative cross-sectional area of the tantalum barrier. Recall that the void length, L_{void}, was shown in Section III-B to be lognormally distributed after nucleation.

Since all other terms are constants, it can be seen that for span growth voids, for both the via-above and via-below cases, ΔR is lognormal with the same σ as L_{void}, but with a shifted mean.

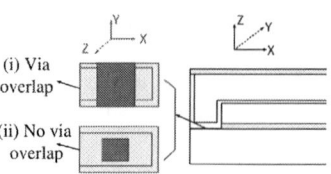

Fig. 2. CuDD via-liner alignment to limit resistance increase under EM [20].

For the slit growth scenario, as shown in Fig. 2, the via size may be chosen so that the via overlaps with the liner (case (i)) or not (case (ii)). As reported in [20], in the former case, the liner provides a conductive path that enables a continued connection, and this may be used to ensure electrical connectivity after a slit void is formed. A derivation similar to that in [14] can be used to show:

$$\frac{\Delta R}{R_o} = \left(\frac{\rho_{Ta}}{\rho_{Cu}}\frac{A_{Cu}}{A_{Ta}} - 1\right)\frac{H_{void}}{L_{wire}} \tag{10}$$

where H_{void} is the height of the slit void. We use this to ignore the impact of slit growth voids in our evaluation of ΔR: although slit voids tend to form earlier than span voids, it is easy to build redundancy into the power grid to guard against slit voids by inserting redundant vias; in fact, this is often done anyway. Therefore, we focus our attention on the impact of span voids.

To write the expressions for the mean and standard deviation of ΔR more simply, we introduce the notation

$$k_R = \left(\frac{\rho_{Ta}}{\rho_{Cu}}\frac{A_{Cu}}{A_{Ta}} - 1\right)\frac{1}{L_{wire}} \tag{11}$$

The resistance change distribution can then be expressed as ΔR $\text{LogN}(\mu_{\Delta R}, \sigma_{\Delta R})$, where

$$\mu_{\Delta R} = \mu_{L_{void}} + \log k_R + \log R_o \tag{12}$$
$$\sigma_{\Delta R} = \sigma_{L_{void}} \tag{13}$$

We now summarize the conditions that must be satisfied to achieve a resistance change of ΔR at an observation time t_o. First, a void must nucleate, and then this nucleated void must grow to the point where the wire resistance increases by ΔR. Using these notions, we can now determine the probability that a given wire will have a resistance change ΔR as:

$$\Pr(\Delta R) = \begin{cases} \Pr(\Delta R \mid \text{nucleation}) \cdot \Pr(\text{nucleation}) & \Delta R > 0 \\ 1 - \Pr(\text{nucleation}) & \Delta R = 0 \end{cases} \tag{14}$$

For the first case, the first term, for ΔR for a nucleated void is given by $\text{LogN}(\mu_{\Delta R}, \sigma_{\Delta R})$ as derived above, with the mean and standard deviation given by Equation (13). For the second, the nucleation probability is given by $\text{LogN}(\mu_{t_n}, \sigma_{t_n})$ in Equation (6), and this quantity corresponds to the probability that the nucleation time, t_n, is less than the observation time, t_o, i.e., the CDF of t_n at t_o. Therefore, the right-hand-side is a lognormal times a constant, i.e., the PDF of ΔR has the shape of a lognormal function. The second case corresponds to the scenario where the void does not nucleate, with a probability of $1 - CDF(t_n = t_o)$.

D. Incorporating the Effect of Current Divergence

The conventional approach to estimating EM failure is based on a current density-based model. Under this model, for two wires of equal length, the one with the larger current density should fail sooner. However, the work in [21] demonstrated experimentally that this is not always the case, by showing test circuit where one wire has twice the current density as another, but experiences consistently later failures. This is consistent with other reported work where the *current divergence* effect comes into play: for example, [22] shows fabricated test structures where the failure rates on a wire segment depends not only on the current density on the segment, but also on those on adjacent segments that share via(s) with this segment.

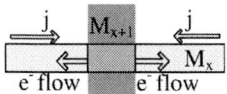

Fig. 3. A via-tree structure where the effective current density is more than the density of current flowing through the wire.

We compute the effective current density on a wire by considering the magnitude and directions of currents in neighboring wires. The effective current density for a wire is computed in terms of the flux-divergence criterion, consistent with [21]. This is illustrated in Figure 3, which shows two wires on metal layers M_x and $M_x + 1$ connected by a via. A via in a CuDD interconnect structure acts as a blocking layer so that metal atoms are not permitted to migrate through it. Therefore, any flux that would have gone to the via is transmitted to a neighboring wire.

In the example shown here, the current on both segments of the wire on layer M_x flows towards the via, i.e., the direction of electron flow is away from the via for both segments. Assuming equal current densities j on each of the two segments, this implies that there is an effective divergence, which can eventually lead to void nucleation and growth, equivalent to a current density of $2j$ on both wires. In other words, as compared to the case where the left-hand segment is missing and the right-hand segment has the same current density of j, the expected rate of atomic transfer is doubled at this node. Using the via node vector notion [21], an effective current density of $2j$ is used for this wire instead of the actual current density of j.

E. Monte Carlo Analysis of Power Grids Using Importance Sampling

We now use our probabilistic resistance model to perform Monte Carlo analysis of power grids in the presence of resistance variations. Our PDF for the resistance change, derived in Section III-C builds a simple circuit-level abstraction for complex physical phenomena, facilitating simplified analysis at the circuit level by considering ΔR as a random variable. However, given that EM is (and should be) a relatively unlikely event, it is essential for our Monte Carlo analysis to be biased appropriately: a truly random set of samples would probably see no resistance change in most (and possibly, for a small set of samples, no) wires. Most importantly, such an approach would see a large number of samples go to waste as they provide little meaningful information.

To overcome this, we use the notion of importance sampling, which biases the distribution, but "unbiases" it as it interprets the results of sampling. Importance sampling is a Monte Carlo method that computes the expected value of a function $f(x)$ of a random variable x, which is specified in terms of a distribution $p(x)$. This method is particularly useful when $p(x)$ is skewed or unevenly distributed, i.e., some values of x have a low probability of occurrence and are not sampled frequently enough, causing sampling errors. Importance sampling resolves this by sampling according to a function $q(x)$ that is uniformly distributed over the range of x, and then correcting the error due to sampling from this different distribution by adding appropriate weights to $f(x)$. For example, the expectation of $f(x)$ under the distribution $p(x)$, denoted $E_p[f(x)]$, is computed as:

$$
\begin{aligned}
E_p[f(x)] &= \int f(x)p(x) = \int f(x)p(x).q(x)/q(x) \\
&= \int w(x)q(x) = E_q[w(x)]
\end{aligned}
$$

where $w(x) = f(x)/q(x)$ and $E_p[f(x)]$ is the expectation of $f(x)$ under the new distribution $q(x)$.

In this work, we use a sampling distribution $q(x)$, which is a uniform distribution that stretches from 0 to the tail of the lognormal distribution of ΔR: the values of this lognormal go from $\Delta R = 0$ to the $(\mu + 3\sigma)$ point of the underlying Gaussian, $\log(\Delta R)$. If K is the span of this distribution, then every point has a uniform probability of $1/K$. The method samples points on this uniform distribution, feeds them into a power grid simulator based on DC modified nodal analysis, and determines the voltage distribution at each node. The voltages are then translated back to the original distribution by scaling them by to the original lognormal distribution using the $w(x)$ factor.

IV. RESULTS

A. Calibration of Correctness under Accelerated Aging

1) Failures in a Single Wire: To calibrate the correctness of our models, we first work under assumptions similar to [2], which computes t_g, the time at which L_{void} becomes equal to L_{via}, i.e., allowing the void to grow until it spans across the length of the via, so that we can compare our predicted values against their published Finite Element Analysis (FEA) simulations. Here, we use the statistical framework derived in Section III under accelerated aging under temperature and current stress.

Parameters for accelerated aging are set to ensure a fair comparison, drawing parameter values from [2] where available. We use $T = 295°C$ and $j = 1.33MA/cm^2$ (reflecting temperature and current stress), $\sigma_c = 41MPa$, $Z^\star_{eff} = 5$, $\rho_{Cu} = 2.5 \times 10^{-8}\Omega m$, $L_{via} = 0.07\mu m$, and $D_{eff} = 6.7 \times 10^{-9}cm^2/s$. Some parameters that were unavailable were extracted from the literature. Specifically, the mean of $E_a = 0.47eV$ [18], the standard deviation for the underlying Gaussian in the lognormal E_a was extracted from the Arrhenius plot of E_a vs. $1/T$ in [12] as $0.005eV$, and $B = 1GPa$ [1].

We run a Monte Carlo (MC) simulation to validate our analytical predictions of distribution of t_n and L_{void}. Our MC simulation uses 10^6 samples on a normal distribution of activation energy. Fig. 4 shows the values obtained from from the analytical model and the MC simulation. As in [2], the PDF for both the nucleation time and growth time is observed to follow a lognormal distribution. The close match that is seen between these curves is expected, since our analytical formulation makes no approximations.

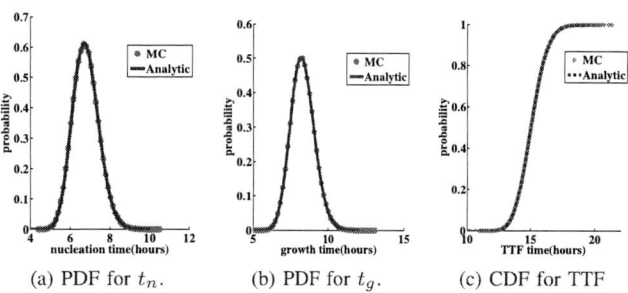

(a) PDF for t_n. (b) PDF for t_g. (c) CDF for TTF

Fig. 4. Comparing analytical vs. MC distributions under accelerated aging.

Table I lists the expected mean and standard deviation for the failure parameters t_n and t_g obtained by our model against the FEA-based values mentioned in [2]. The values show a reasonable but not perfect match. The discrepancies could be attributed to factors such as the unavailability of some parameters in [2], and differences in the simulation setup, e.g., explicit consideration of grain size variation on a microstructural level. Clearly, our method is much faster than FEA since it merely involves the evaluation of an analytical expression.

Phase	Nucleation		Growth	
μ, σ	μ_{t_n}	σ_{t_n}	μ_{t_g}	σ_{t_g}
From [2]	8.5h	0.38h	8h	0.7h
Analytical	7.27h	0.74h	8.44h	0.86h

TABLE I

COMPARISON OF OUR ANALYTICAL METHOD WITH [2].

2) Statistical vs. deterministic approach: Continuing under the assumptions in [2], where failure is defined as the time when $L_{void} = L_{via}$, the time to failure (TTF) is the sum of the nucleation time, t_n, and the growth time, t_g. We use the distribution of t_n and t_g to plot the distribution of TTF, and this distribution provides insights about the importance of incorporating the statistical behavior when modeling the effect of EM in circuits. From the CDF shown in Fig. 4(c), the time to failure, the 0.27%, 50% percentile and 99.73% points under accelerated aging are 12.73h, 15.68h, and 19.11h, respectively. This has two implications. First, it means that every wire has a nonzero probability of failure, which is not linked to its Blech length of jL. Indeed, wires that satisfy the Blech length criterion will fail: this has been observed experimentally in many of the references cited in this paper. Second, there is very low probability that the wire will fail in any manufactured part before 12.73h, and a probability that is so small implies that the wire is effectively immortal.

B. Applying the Single-Wire Model at Normal Operating Conditions

Next, we evaluate the resistance change ratio, $\Delta R/R$, of a wire according to our probabilistic formulation in Equation (14). We use a similar setup as described in Section IV-A.1, but we change the temperature to normal operating conditions at $25°C$ and the current density to $0.5MA/cm^2$, and we return to the assumption that slit voids are not significant since we assume that redundant vias are used. As expected, the use of normal operating conditions for aging analysis results in a reduction in the rate of EM degradation as compared to an accelerated aging case, where TTF is of the order of several hours, to a scenario where the TTF values are of the order of several years.

Using Equation (14), we obtain the probability distribution of the $\Delta R/R$ at an observation time of 12 years. Fig. 5 compares the CDF of the change in $\Delta R/R$ as predicted by our formulation against MC simulations. The small mismatch in the two CDFs can be attributed to the error generated by our approximate moment matching method for estimating the difference of two lognormals.

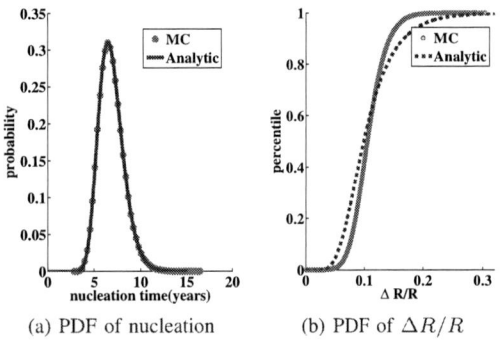

(a) PDF of nucleation   (b) PDF of $\Delta R/R$

Fig. 5. Distribution of t_n and $\Delta R/R$ under normal operation.

C. Power grid simulation

Having verified our probabilistic resistance change formulation for a single wire against published data due to slit voids, we use our model to analyze the effect of resistance change due to EM at the circuit level. The analysis is based on DC analysis of a set of power grid benchmarks from [23], enumerated in Table II. To perform a Monte Carlo simulation over the values of ΔR, we implement a statistical importance sampling MC approach, described in Section III-E, in C++ and MATLAB. We analyze the interconnects for EM risk in the power grid and simulate the distribution of voltage drops. We run our stochastic MC simulation for 1200 iterations for the benchmarks. Using the runtime per iteration from Table II we can estimate realistic runtimes if a specialized power grid simulator were used; in our implementation, for convenience, we have used the matrix solver from MATLAB. This table indicates that the MC simulation can be carried out in a computationally efficient manner with a better solver.

Name	Total # wires	Runtime [24]	Memory in MB [24]	Expected runtime 1200 iterations
PG1	30027	0.20s	4MB	4min
PG2	208325	1.42s	72MB	29min
PG3	1401572	8.29s	172MB	166min
PG4	1560645	19.35s	606MB	387min
PG5	1076848	9.36s	296MB	187min

TABLE II

LIST OF P/G BENCHMARKS EVALUATED IN THIS PAPER.

We set the observation time to 2.5 years for PG1–PG4, consistent with the results reported in [6]. For PG5, which shows a low nominal voltage drop after 2.5 years, the observation time is set to 5 years. All circuits are evaluated at a temperature of $105°C$.

To compare results of our statistical framework against previous work with deterministic mortality approach, we predict the set of mortal wires using the Blech-length [5] criterion. In the second and third columns of Table III, for each benchmark, we compare the results of the newly formulated statistical approach ("Stat.") with those from a deterministic Blech-length approach ("Blech"). As stated earlier, for our approach, "mortality" is defined in terms of a 3σ deviation from the mean in the underlying normal of the lognormal. From the table, it can be seen that our implementation shows that a larger number of wires must be considered mortal under the probabilistic formulation, as compared to the deterministic Blech-length based approach. This is entirely expected, and makes the case for not using the Blech length criterion for CuDD interconnects.

Next, we evaluate the variation in the IR drop due to the statistical distribution of EM. This distribution is observed to be non-Gaussian, and we compute the spread of the distribution by taking the difference (referred to as ΔV) between the 99.73 and 0.27 percentile points. We characterize the normalized spread by expressing ΔV as a fraction of the median (i.e., 50 percentile point) of V.

The next column of Table III presents the largest realistic normalized resistance change, estimated as the point that is three standard deviations from the mean of the underlying Gaussian of the lognormal ΔR. This data indicates the spread in resistance change due to EM. The subsequent columns show the variance in $\Delta V/V$, and the corresponding median, $V_{50\%}$, for the nodes in the network that have the largest variance and largest median, respectively. In each case, our approach provides a *precise* metric for the impact of EM on the variation in power grid voltage.

Most significantly, this table demonstrates that contrary to the assumptions in many other works on power grid analysis that assume that a failing wire causes a failed circuit, the power grid may continue to work even if a single wire fails. This is exemplified, for example, by PG2, where, in spite of a large spread in the resistance change in the wire, the spread for the IR drop is small.

Finally, we run MC simulations for PG5 at different observation times, t_o, to analyze how the worst case IR drop varies with time. For three values of t_o, Fig. 6 shows the CDF of IR drop for the node having largest median IR drop value. For $t_o = 5$ years, the CDF is to

Ckt	Number of mortal wires		Largest $\Delta R/R$ (in %)	At largest variance node		At largest median node	
	Stat.	Blech		$V_{50\%}$	$\Delta V/V$	$V_{50\%}$	$\Delta V/V$
PG1	16932	13272	218.4	0.73V	15.4%	0.89V	8.2%
PG2	81393	26723	113.6	0.39V	4.7%	0.50V	0.8%
PG3	63231	34998	65.9	0.19V	6.2%	0.24V	0.1%
PG4	140133	79737	36.4	0.006V	7.8%	0.01V	4.7%
PG5	131094	38746	120.4	0.035V	11.5%	0.07V	1.7%

TABLE III

RESULTS OF MONTE CARLO SIMULATION

the left, and it moves rightwards as the observation time increases. For a given threshold value (e.g., 70mV) on the x-axis, it is clear that the probability of seeing this value of IR drop increases with time, since a larger fraction crosses the threshold value.

Fig. 6 provides further insight about the circuit behaviour with respect to time. If we fix the threshold IR drop as 70mV, there can be many wire failures that result in an increase in the resistance, but almost all samples have IR drop below 70mV at $t_o = 5$ years. For a slightly higher threshold value of 80mV (off the scale), there is a high probability that the power grid will still be functional, for an even higher observation time of 7.5 years – even though the circuit does see wires that fail within 7.5 years. This indicates that for a given specification, the circuit lifetime, which is characterized by a threshold IR drop, can be longer than the lifetime of the EM-degraded wire, i.e., an individual wire whose resistance exceeds a specific threshold. In other words, the power grid is robust to some EM failures in individual wires.

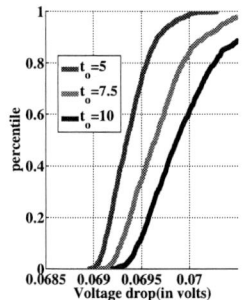

Fig. 6. CDF plots for worst node IR drop at various observation times.

V. CONCLUSION

We have developed a method for EM analysis of power grids taking into account two effects that were neglected in past work: first, that EM is a probabilistic phenomenon for CuDD interconnects, and second, that the power grid has inherent resilience to EM failures. Our results indicate that both effects are substantial.

REFERENCES

[1] C. M. Tan, *Electromigration in ULSI Interconnections*. World Scientific, Dec. 2010.
[2] R. L. de Orio, H. Ceric, and S. Selberherr, "A compact model for early electromigration failures of copper dual-damascene interconnects," *Microelectronics Reliability*, vol. 59, no. 9–11, pp. 1573–1577, 2011.
[3] S. P. Hau-Riege, "Probabilistic immortality of Cu damascene interconnects," *Journal of Applied Physics*, vol. 91, no. 4, pp. 2014–2022, 2002.
[4] J. R. Black, "Electromigration failure modes in aluminum metallization for semiconductor devices," *Proceedings of the IEEE*, vol. 57, pp. 1587–1594, Sept. 1969.
[5] I. A. Blech, "Electromigration in thin aluminum films on titanium nitride," *Journal of Applied Physics*, vol. 47, no. 4, pp. 1203–1208, 1976.
[6] D. Li and M. Marek-Sadowska, "Variation-aware electromigration analysis of power/ground networks," *Proceedings of the IEEE/ACM International Conference on Computer-Aided Design*, pp. 571–576, Nov. 2011.
[7] S. M. Alam, C. L. Gan, F. L. Wei, C. V. Thompson, and D. E. Troxel, "Circuit-level reliability requirements for Cu metallization," *IEEE Transactions on Devices and Materials Reliability*, vol. 5, no. 3, pp. 522–531, 2005.

[8] M. A. Korhonen, P. Borgesen, K. N. Tu, and C. Y. Li, "Stress evolution due to electromigration in confined metal lines," *Journal of Applied Physics*, vol. 73, no. 8, pp. 3790–3799, 1993.
[9] A. Heryanto, K. L. Pey, Y. K. Lim, W. Liu, N. Raghavan, J. Wei, C. L. Gan, M. K. Lim, and J. B. Tan, "The effect of stress migration on electromigration in dual damascene copper interconnects," *Journal of Applied Physics*, vol. 109, no. 1, pp. 013716–1–013716–9, 2011.
[10] C. K. Hu, R. Rosenberg, and K. Y. Lee, "Electromigration path in Cu thin-film lines," *Applied Physics Letters*, vol. 74, no. 20, pp. 2945–2947, 1999.
[11] E. Liniger, L. Gignac, C.-K. Hu, and S. Kaldor, "In situ study of void growth kinetics in electroplated Cu lines," *Journal of Applied Physics*, vol. 92, no. 4, pp. 1803–1810, 2002.
[12] Z. S. Choi, R. M. Monig, and C. V. Thompson, "Activation energy and prefactor for surface electromigration and void drift in Cu interconnects," *Journal of Applied Physics*, vol. 102, no. 8, pp. 083509–1–083509–9, 2007.
[13] C. Christiansen, B. Li, M. Angyal, T. Kane, V. M. Y. Y. Wang, and S. Yao, "Electromigration-resistance enhancement with CoWP or CuMn for advanced Cu interconnects," in *IEEE International Reliability Physics Symposium*, pp. 3E.3.1 –3E.3.5, 2011.
[14] M. Hauschildt, M. Gall, S. Thrasher, P. Justison, R. Hernandez, and H. Kawasaki, "Statistical analysis of electromigration lifetimes and void evolution," *Journal of Applied Physics*, vol. 101, no. 4, pp. 682–687, 2007.
[15] L. Doyen, X. Federspiel, L. Arnaud, F. Terrier, Y. Wouters, and V. Girault, "Electromigration multistress pattern technique for copper drift velocity and black's parameters extraction," in *IEEE International Reliability Workshop Final Report*, pp. 74–78, Oct. 2007.
[16] V. M. Dwyer, "Modeling the electromigration failure time distribution in short copper interconnects," *Journal of Applied Physics*, vol. 1004, no. 5, pp. 053708–1–053708–6, 2008.
[17] J. R. Lloyd and J. Kitchin, "The electromigration failure distribution," *Journal of Applied Physics*, vol. 69, no. 4, pp. 2117–2127, 1991.
[18] B. H. Jo and R. W. Vook, "In-situ ultra-high vacuum studies of electromigration in copper films," *Thin Solid Films*, vol. 262, no. 1–2, pp. 129–134, 1995.
[19] A. A. Abu-Dayya and N. C. Beaulieu, "Comparison of methods of computing correlated lognormal sum distributions and outages for digital wireless applications," in *IEEE Vehicular Technology Conference*, pp. 175–179, 1994.
[20] B. Li, T. D. Sullivan, and T. C. Lee, "Line depletion electromigration characterization of Cu interconnects," *IEEE Transactions on Device and Materials Reliability*, vol. 4, pp. 80–85, Mar. 2004.
[21] Y.-J. Park, P. Jain, and S. Krishnan, "New electromigration validation: Via node vector method dual-damascene Cu interconnect trees," in *IEEE International Reliability Physics Symposium*, pp. 6A.1.1–6A.1.7, 2003.
[22] C. L. Gan, C. V. Thompson, K. L. Pey, and W. K. Choi, "Experimental characterization and modeling of the reliability of three-terminal dual-damascene Cu interconnect trees," *Journal of Applied Physics*, vol. 94, no. 2, pp. 1222–1228, 2003.
[23] S. R. Nassif, "Power grid analysis benchmarks," in *Proceedings of the Asia-South Pacific Design Automation Conference*, 2008.
[24] Z. Zeng, T. Xu, Z. Feng, and P. Li, "Fast static analysis of power-grids: algorithms and implementation," in *Proceedings of the IEEE/ACM International Conference on Computer-Aided Design*, pp. 488–493, 2011.
[25] Z. S. Choi, R. M. Monig, and C. V. Thompson, "Dependence of the electromigration flux on the crystallographic orientations of different grains in polycrystalline copper interconnects," *Applied Physics Letters*, vol. 102, no. 8, pp. 241913-1–241913-3, 2007.
[26] R. L. de Orio, H. Ceric, and S. Selberherr, "The effect of copper grain size statistics on the electromigration lifetime distribution," *Simulation of Semiconductor Processes and Devices*, pp. 1–4, 2009.

TinySPICE: A Parallel SPICE Simulator on GPU for Massively Repeated Small Circuit Simulations

Lengfei Han
Department of ECE
Michigan Tech. University
Houghton, MI, 49931
lengfeih@mtu.edu

Xueqian Zhao
Department of ECE
Michigan Tech. University
Houghton, MI, 49931
xueqianz@mtu.edu

Zhuo Feng
Department of ECE
Michigan Tech. University
Houghton, MI, 49931
zhuofeng@mtu.edu

ABSTRACT

In nowadays variation-aware IC designs, cell characterizations and SRAM memory yield analysis require many thousands or even millions of repeated SPICE simulations for relatively small nonlinear circuits. In this work, we present a massively parallel SPICE simulator on GPU, TinySPICE, for efficiently analyzing small nonlinear circuits, such as standard cell designs, SRAMs, etc. In order to gain high accuracy and efficiency, we present GPU-based parametric three-dimensional (3D) LUTs for fast device evaluations. A series of GPU-friendly data structures and algorithm flows have been proposed in TinySPICE to fully utilize the GPU hardware resources, and minimize data communications between the GPU and CPU. Our GPU implementation allows for a large number of small circuit simulations in GPU's shared memory that involves novel circuit linearization and matrix solution techniques, and eliminates most of the GPU device memory accesses during the Newton-Raphson (NR) iterations, which enables extremely high-throughput SPICE simulations on GPU. Compared with CPU-based TinySPICE simulator, GPU-based TinySPICE achieves up to $138X$ speedups for parametric SRAM yield analysis without loss of accuracy.

Categories and Subject Descriptors

B.7.2 [**Design Aids**]: simulation—*Integrated Circuits*

General Terms

Performance, Algorithms, Verification

Keywords

Variation-aware analysis, SPICE simulation, GPU computing

1. INTRODUCTION

Reliability and yield analysis of embedded SRAM memory modules are critical to designs of modern microprocessors, 3D-ICs, and mixed-signal SOCs. However, nanoscale SRAM designs are greatly challenged by prohibitively high computational cost due to the extremely large number of repeated SPICE simulations considering parametric variations [1–4]. Additionally, present-day variation-aware design methodologies require extremely fast cell/driver characterization capability capturing important process, voltage supply, and temperature (PVT) variations [5,6], which also demands

Permission to make digital or hard copies of all or part of this work for personal or classroom use is granted without fee provided that copies are not made or distributed for profit or commercial advantage and that copies bear this notice and the full citation on the first page. To copy otherwise, to republish, to post on servers or to redistribute to lists, requires prior specific permission and/or a fee.
DAC'13, May 29 - June 07 2013, Austin, Tx, USA.

for much more powerful simulation methodologies. For instance, SRAM readability, writability and stability analysis considering threshold voltage (V_{th}), effective channel length (L_{eff}) and power supply variations requires tens of millions of repeated SPICE simulations for a given design, while variation-aware cell modeling and characterizations also involve constructing look-up tables (LUTs) for capturing all fast/slow corners that requires running many thousands of SPICE simulations [5–7].

Recent multiprocessors with heterogeneous architectures have emerged as mainstream computing platforms, which typically integrate a variety of processing elements of different computing performance, programming flexibility and energy efficiency characteristics. Heterogeneous computing platforms, such as IBM/Sony Cell architectures, nowadays personal computers (PCs) with multi-core CPUs and many-core GPUs, and the latest mobile heterogenous microprocessors (e.g. APU from AMD [8], Tegra from Nvidia [9], etc), can theoretically provide unprecedented high performance and high energy efficiency in the meantime. With such heterogeneous computing architectures, VLSI CAD developers will have good opportunities to fully utilize these unique computing resources, thereby targeting much greater performance and energy efficiency.

Although there have been works that target accelerating SPICE simulations by performing device evaluations on GPU's hundreds of streaming processors and sparse matrix solves on CPU [10], only a small fraction of the computations can be accelerated on GPU, while the overall simulation performance is still limited by the relatively low communication bandwidth and large latency between the CPU and GPU. As a result, only $2X$ speedups have be obtained when compared with the CPU-based SPICE simulator [10]. Although a GPU-based LU algorithm has been recently proposed to efficiently factorize sparse circuit matrices in [11], it is still not clear how to accelerate the entire computations involved in general-purpose SPICE simulations on GPU considering present-day GPU computing limitations.

In this work, we present a massively parallel SPICE simulator on GPU, TinySPICE, which accelerates the entire SPICE simulation computations on GPU without introducing excessive CPU-GPU data communications and device memory accesses. TinySPICE can analyze small nonlinear circuits in GPU's shared memory, as shown in Fig. 1 and thus gains unprecedentedly high computational throughput. We develop novel GPU-friendly data structures and efficient algorithm flow for every kernel function of the SPICE algorithm that includes device evaluations, matrix construction, linear system solving and Newton-Raphson (NR) iterations. A series of novel techniques also have been proposed in TinySPICE to more efficiently utilize the GPU hardware resources such as on-chip shared memory and registers, and optimize GPU memory accesses. TinySPICE is capable of solving thousands of small circuit simulation problems in GPU's shared memory concurrently, and achieves unprecedented high-performance massively parallel SPICE simulations on GPU. Compared with CPU-based TinySPICE

624

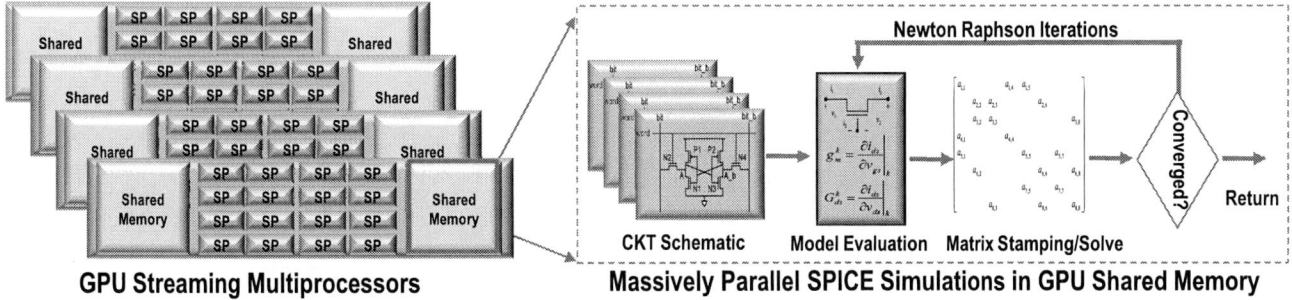

Figure 1: TinySPICE: massively parallel SPICE simulation program on GPUs

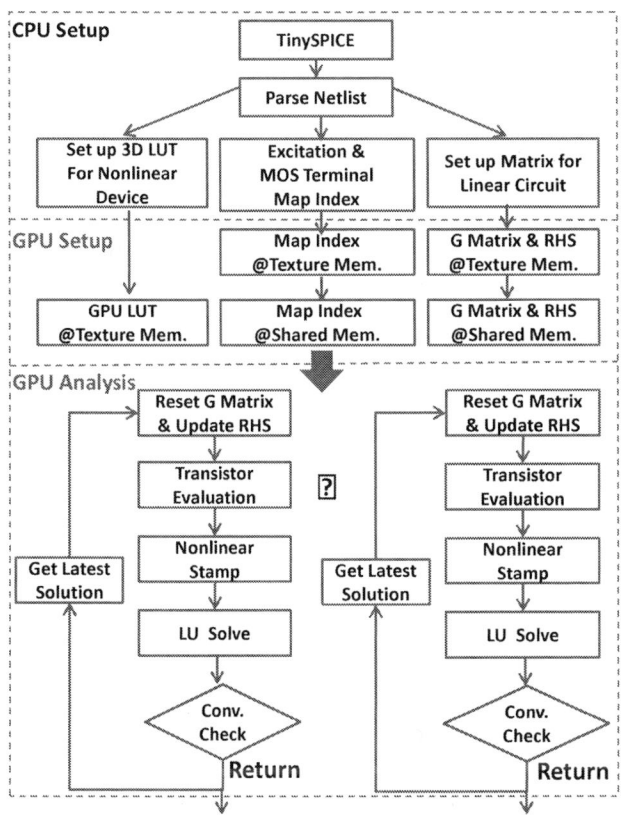

Figure 2: The algorithm flow of TinySPICE.

implementation, TinySPICE achieves up to $138X$ speedups for a variety of circuit analysis problems without loss of accuracy.

2. OVERVIEW OF TINYSPICE

TinySPICE is a SPICE-accurate nonlinear circuit simulator that leverages many-core GPU for fast repeated small circuit analysis (see the attached supplementary materials for more details of nonlinear circuit simulation algorithms and GPU computing). To leverage the powers of GPU's hundreds of streaming processors (SP) for data parallel computing, we propose a novel massively parallel algorithm with GPU-friendly data structures to accelerate the repeated small nonlinear circuit simulations. Since memory consumption for analyzing such small circuits is usually very low, dense modified nodal analysis (MNA) matrix structures [12] and direct solution methods, such as LU factorization technique, have

been applied in this work without sacrificing efficiency and accuracy.

It is also important to note that by running each circuit simulation with one GPU thread, many thousands of independent SPICE simulations can be executed on GPU concurrently. Moreover, since only a small memory space is required for each circuit simulation, the data required by each circuit simulation can be fully stored in GPU's shared memory, as illustrated in Fig. 1. Since data accesses from GPU's device memory may take much longer time than loading data directly from GPU's shared memory, shared memory has been carefully utilized in this work to achieve much faster simulations and extremely high computing throughput. As a result, our TinySPICE engine can run in a much faster way on GPU than conventional SPICE simulators running on CPUs, especially for massively repeated Monte Carlo small circuit simulations.

3. TINYSPICE ON GPU

In this section, the proposed TinySPICE simulator on GPU for massively parallel small circuit simulations will be described in details. The TinySPICE algorithm flow includes three key steps: the CPU-setup phase, the GPU-setup phase and the GPU-analysis phase, as illustrated in Fig. 2, where G denotes the system MNA matrix and RHS stands for right hand side vector.

3.1 CPU-Setup Phase

The main task of the CPU-setup phase is to set up GPU-friendly data structures for SPICE simulations on GPU, such as the look-up tables (LUTs) for nonlinear devices, the indices mapping vectors, the linear-element matrices, the right hand side (RHS) vectors and the solution vectors.

3.1.1 Parametric 3D LUTs for Device Evaluations

For small circuits, due to the very limited number of nodes, device evaluations typically dominate SPICE simulation cost. In order to more effectively parallelize device evaluations on GPU during circuit simulations, 3D LUTs for evaluating transistors have been adopted in this work (see the supplementary sections for more details). Compared with standard BSIM4 model evaluations which involve more complicated computationss, device evaluations using 3D LUTs require much less computational time. Such 3D LUTs can be efficiently built on CPU and then transferred to GPU before SPICE simulations start. To capture the impact of process variations, such as effective channel length L_{eff} and threshold voltage V_{th} variations, parameterized 3D LUTs are constructed to facilitate fast variation-aware SPICE simulations, as described in the supplementary materials.

3.1.2 Storage of Circuit Elements

It should be noted that, in order to obtain the stamping locations of nonlinear elements in the system MNA matrix, it is necessary to store the terminal indices of each nonlinear device. In this work, we propose to store terminal indices of all transistors into a long index-mapping vector, as shown in Fig. 3. In the Mos_map vector, Idx stands for the corresponding LUT storage index for a transistor. d,g,s, and b represent the indices of each transistor's terminals in the MNA matrix, respectively. With such mapping information, device evaluation results from the LUTs can be directly written into the system MNA matrix as well as the RHS vector.

Since the elements of linear devices such as resistors, capacitors and inductors have fixed values, their corresponding stamped elements in the MNA matrix will not change throughout the entire SPICE simulation. Consequently, they can be pre-evaluated and stored into a linear-element matrix. During each NR iteration, once the linear-element matrix is combined with the nonlinear-element matrix obtained from nonlinear device evaluations, direct solution methods, such as LU factorization, can be applied to compute the solutions.

Since GPU's data parallel computing scheme is not suitable for processing sparse matrices, our TinySPICE simulator adopts dense matrix structure in this work. We emphasize that for small circuit simulations, the memory consumption and computational cost for storing and processing the dense MNA matrices is still acceptable.

3.1.3 Summary of CPU-Setup Phase

We conclude the CPU-setup phase for TinySPICE as follows:

- TinySPICE first builds parametric 3D LUTs for all transistors according to a user-defined accuracy level. A suitable discretization step size can be selected based on the circuit design information and specific simulation requirements: more accurate LUTs typically require greater memory space and transistor characterization time. The parametric 3D LUTs are stored in a long 1D vector (as shown in Fig. 3) that will be transferred to GPU's device memory once before the massively parallel SPICE simulations start.

- TinySPICE creates the 1D terminal index-mapping vectors Mos_map to store the node indices for all nonlinear devices, as shown in Fig. 3. Mos_map is used to help stamp nonlinear devices into the system MNA matrices, which has to be constructed and transferred to GPU for one time.

- The linear-element matrices will also be stored in a 1D vector and sent to GPU memory. Once GPU kernel functions are launched, linear-element matrices will be loaded into GPU's shared memory at the initial step and will be combined with the nonlinear-element matrices to form the final MNA matrices for subsequent NR iterations.

- TinySPICE creates the VS_map vectors including information on all excitation sources such as voltage and current sources. Node a and node b denote regular terminal nodes. In modified nodal analysis (MNA), each voltage source requires including a Pseudo node into the index-mapping vector for representing the current flowing through the device, as shown in Fig. 3.

- Similarly, TinySPICE creates VS_step vectors including all the values of time-varying voltage and current sources at each time step of transient simulations. VS_step will be then combined with constant excitation vector to form the final RHS vectors.

3.2 GPU-Setup and GPU-Analysis Phases

The main task of the GPU-setup phase is to prepare proper simulation environment for the subsequent circuit analysis on GPU, which includes device memory allocations, and data transmission

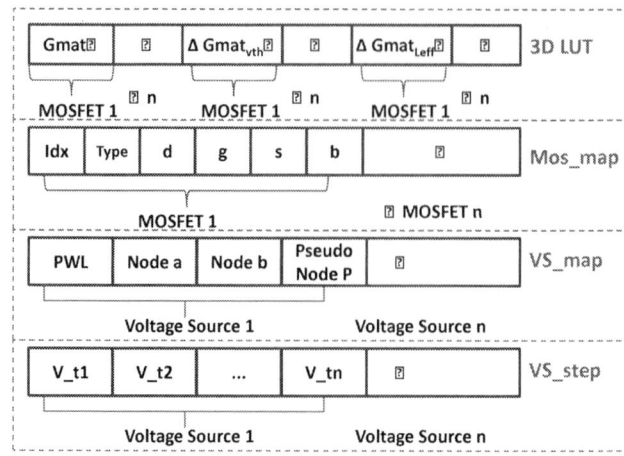

Figure 3: Vectors for storing LUTs, Mosfets, and excitation sources on GPU.

from host (CPU) to device (GPU). CUDA devices have several types of memories that exhibit different data access latencies and bandwidths which may greatly influence the GPU kernel execution performance (see the supplementary materials for more details).

3.2.1 GPU Memory Usages

TinySPICE has been designed to carefully utilize GPU's on-chip memory resources as follows.

- *Read-only GPU memory:* The parametric 3D LUTs, linear-element matrices, PWL voltage source values and RHS vectors are read-only for all GPU threads. Therefore, it is preferred to store them in GPU's read-only texture memory, such that the memory data access latency can be effectively reduced.

- *Read-write GPU memory:* The index-mapping vectors for nonlinear devices and voltage sources are shared among all the circuits, thus they need to be stored in GPU's shared memory. If GPU's shared memory is not sufficient, such mapping vectors can also be stored in GPU's registers. Additionally, nonlinear-element matrices and RHS vectors are stored in GPU's registers since their values are changing frequently.

3.2.2 GPU Data Organization for Coalesced Device Memory Access

In order to obtain the best parallel GPU computing performance, coalesced device memory (global memory) accesses should be satisfied for all GPU threads. Since the solution vectors will not be very frequently accessed during NR iterations, but only be used for storing and sending results back to the host. Consequently, we store them in GPU's global memory. Since TinySPICE works on many repeated circuit simulation tasks with different input excitations and circuit parameters, each solution vector stored on GPU's global memory can be reused for many consecutive simulations on GPU's shared memory.

To enable coalesced device memory accesses, we organize the memory storage of solution vectors in such a way, that for all n circuits, the memory space of all the n solution vectors are continuous, as shown in Fig. 4, where T_k denotes the k-th GPU thread, and $xi.m$ denotes the m-th element of the solution vector of circuit i. This GPU-friendly data storage obviously allows for efficient

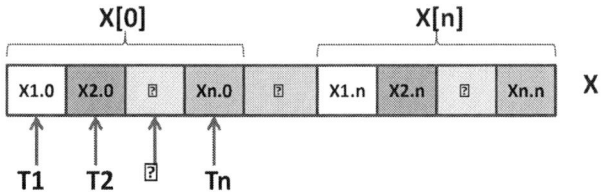

Figure 4: The solution vector data access pattern on GPU.

coalesced global memory accesses, which can significantly reduce the GPU device memory access overhead.

3.2.3 GPU Thread Organization for TinySPICE

Since each GPU's streaming multiprocessor (SM) has very limited memory resources, the number of circuits to be analyzed at the same time should be carefully determined based on the circuit sizes and on-chip memory usage (e.g. Nvidia GeForce GTX480 GPU has 15 SMs, each of which includes $48k$ shared memory and $32k$ registers). The limited memory can impact the number of GPU threads (SPICE simulations) running on each SM. To achieve the best simulation performance, TinySPICE first finds out the optimal thread block sizes and grid sizes by evaluating simple memory-cost functions (for computing the maximum number of circuits that can be analyzed in one SM). Subsequently, the proper thread organization and assignment are determined, and the final simulation code can be compiled for a given circuit design. It is worth noting that different circuit analysis problems may result in different GPU thread settings, and therefore different speedups compared to CPU-based SPICE simulations.

3.3 Algorithm Flow for TinySPICE

The algorithm flow of TinySPICE is summarized in Algorithm 1. At the beginning of each NR iteration, TinySPICE evaluates all nonlinear devices (linearizes the system) using LUT-based trilinear interpolations according to the latest solution results. After the device evaluations, the computed elements for nonlinear devices are stamped into nonlinear-element matrices based on the terminal indices stored in the index-mapping vectors. RHS vectors also need to be updated based on the latest solution results. The final MNA matrices can be created by combining the nonlinear-element matrices with the linear-element matrices that has been previously built and stored in GPU's texture memory. Subsequently, GPU-based LU decomposition algorithm is applied to factorize the MNA matrices.

Algorithm 1 Newton-Raphson (NR) Iteration Algorithm Flow on GPU

Allocate system MNA matrix and RHS in registers for each GPU thread.
Load linear-element matrix, RHS vectors, index-mapping vectors from GPU's texture memory to shared memory.
for $i = 1 \rightarrow n$ NR iterations **do**
    1. Reset system MNA matrix and RHS vector by loading initial data from shared memory.
    2. Evaluate nonlinear devices.
    3. Stamp system MNA matrix and compute the RHS vector.
    4. Factorize system MNA matrix of each circuit and solve for the solution vector.
    5. Apply a damping factor for the solution if needed.
end for
if NR does not converge **then**
    Perform another n iterations of steps 1-5.
end if
Return solution if NR converged. Otherwise return an error flag.

It should be noted that, in order to reduce GPU thread divergence considering GPU's single-instruction-multiple-thread (SIMT) scheme,

Table 1: Experimental setup of test cases. "NL_Num" denotes the number of nonlinear devices, "Node_Num" represents the number of nodes in the circuit, "Vs_Num" represents the number of independent voltage sources, and "Unk_Num" denotes the number of unknowns of the nonlinear system.

Circuit	NL_Num	Node_Num	Vs_Num	Unk_Num
6T-SRAM	6	8	5	12
D-Latch	8	9	5	13
D-Flip-Flop	16	12	5	16
Invertor-Chain	32	20	3	22
4:1 Mux	24	27	9	35

the convergence condition is not checked during every NR step. Instead, we check the convergence after several NR iterations. Although this method will result in some overhead, it may efficiently reduce the divergence of GPU threads.

4. EXPERIMENT RESULT

4.1 Experimental Setup

In this work, several widely used digital circuits have been tested using TinySPICE on GPU. To demonstrate the benefit of our GPU-based TinySPICE simulator, traditional CPU-based SPICE simulation methods and TinySPICE on CPU are implemented and evaluated. Detailed characteristics of test cases are summarized in Table 1. We set up both the first-order and second-order parametric 3D LUTs in our experiments. These LUTs have been tested using different resolutions. Throughout the following experiments, we use a high LUT resolution to guarantee that the final solution of TinySPICE is matching the SPICE solution. Under the high resolution, the first-order LUTs totally cost 27MB memory for a single transistor, while using the second-order LUTs will double the memory cost[1]. It should be also noted that, in the experimental results, the 3D LUTs setup time is not included. The average time for generating the first-order LUTs for a single transistor is around 0.435s, while including the second-order LUTs will double the setup time. Compared with the whole SPICE simulation runtime, the 3D LUTs setup time is typically much smaller. Furthermore, the LUTs generation process can be easily parallelized using multi-core CPUs to reduce the LUTs setup time. Since the accuracy level with first-order parametric LUTs is very satisfactory in our experiments, all the following experiments results are obtained based on the first-order LUTs to reduce the memory and runtime cost. All experiments have been performed on Ubuntu8.04 64-bit with 2.66GHz quad-core CPU, 6GB DRAM memory, and one Nvidia GeForce GTX480 GPU with $1.5GB$ device memory.

4.2 Experimental Results

4.2.1 Accuracy of Parametric 3D LUT

A static random access memory (SRAM) cell is simulated to show the accuracy of the parametric 3D LUTs. For each test, we sweep the input from 0 to VDD. At each sweeping point, 1000 ΔV_{th} and ΔL_{eff} variation parameters are generated randomly and independently for each transistor following a normal distribution. For each normal-distribution parameter, 10% of the nominal value is set to be the standard deviation σ. 1000 circuit DC simulations are performed. CPU-based SPICE simulator using the original BSIM4 model evaluations generates the reference results, and are

---

[1]Our second-order LUTs for transistors have neglected cross-term impacts to reduce the LUT characterization time and memory, as described in the supplementary materials.

Table 2: Runtime results of TinySPICE for 1.5M Monte Carlo DC simulations. "CPU-LUT" denotes the runtime for LUT-based SPICE simulation on CPU, "CPU BSIM4" denotes the runtime for SPICE simulation with BSIM4 models on CPU, "GPU-LUT" denotes the runtime for proposed TinySPICE on GPU. Speedups are calculated by comparing to the "CPU BSIM4"

Circuit	CPU BSIM4(s)	CPU LUT(s)	GPU LUT(s)
6T-SRAM	768.153	403.200(1.9X)	2.902(264X)
D-Latch	1212.979	527.155(2.3X)	5.727(211X)
D-Flip-Flop	2027.827	982.579(2.1X)	10.677(189X)
Invertor Chain	4377.600	1981.440(2.2X)	41.863(104X)
4:1 Mux	3686.400	1812.480(2.0X)	81.366(45X)

Table 3: Runtime results of TinySPICE for 1.5M Monte Carlo TR simulations. "CPU-LUT" denotes the runtime for LUT-based SPICE simulation on CPU, "CPU BSIM4" denotes the runtime for SPICE simulation with BSIM4 models on CPU, "GPU-LUT" denotes the runtime for proposed TinySPICE on GPU. Speedups are calculated by comparing to the "CPU BSIM4"

Circuit	CPU BSIM4(s)	CPU LUT(s)	GPU LUT(s)
6T-SRAM	30720	7679.82(4.0X)	163.59(187X)
D-Latch	41472	10751.95(3.9X)	186.42(222X)
D-Flip-Flop	69120	18432.15(3.8X)	341.3(202X)
Invertor Chain	121344	41472(2.9X)	755.76(160X)
4:1 Mux	256512	33792.15(7.6X)	2658.3(96X)

compared with TinySPICE simulators implemented for CPU and GPU computing platforms.

Fig. 5 shows the I-V characteristics of an nMOS transistor. In the figure, asterisks represent the I-V characteristics obtained using BSIM4 model evaluations and the circles represent the results obtained using parametric 3D LUTs. As observed, the results obtained from parametric LUTs are very close to the results generated using BSIM4 models. In our experiment, several different V_{gs} values are chosen, such as $0.3, 0.7, 1.0$, to validate the model accuracy.

Fig. 6 demonstrates the DC simulation results (for an internal node voltage) of the parametric SRAM analysis. The solid line in red is the base line. The results show that our TinySPICE simulator matches well with the original SPICE simulator, and can capture the parametric variations accurately. The average relative error is measured as 0.29%. The second-order LUTs have also been tested for DC simulation. The average relative error has dropped to 0.289%.

4.2.2 Runtime Results

First, we show the DC and transient simulation runtime results of our TinySPICE tool by comparing them with the results obtained by CPU-based simulators. The runtime results of all simulators are obtained by running $1,536,000$ simulations of different circuits with different excitations and circuit design parameters.

As observed in Table 2 and Table 3, CPU-based SPICE simulator using LUTs can achieve up to $2X$ speedups for DC simulations and $7X$ speedups for transient simulations when compared with traditional SPICE simulator "CPU BSIM4". The reason is that the device evaluation cost for parametric 3D LUTs interpolation is much cheaper than the evaluation of BSIM4 models. Moreover, compared to CPU-based SPICE simulator using LUTs, when performing DC simulations using TinySPICE on GPU, we can achieve up to $138X$ speedups. TinySPICE on GPU runs up to $264X$ faster than traditional SPICE simulator "CPU BSIM4". For transient simulations, TinySPICE on GPU runs up to $222X$ faster than the tra-

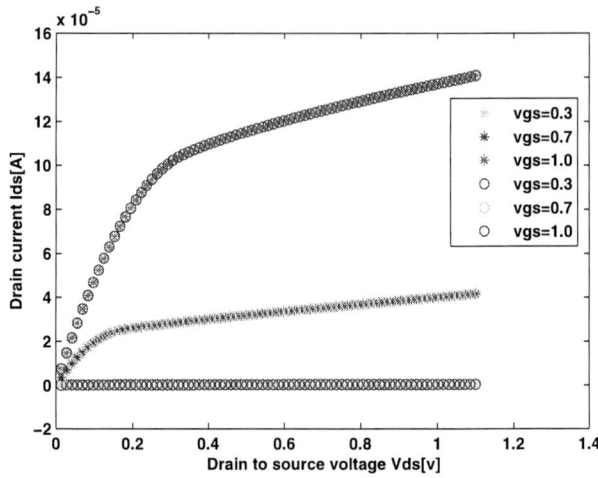

Figure 5: The I-V characteristics obtained by parametric 3D LUT and Bsim4 model evaluations. Circles denote the LUT evaluation results.

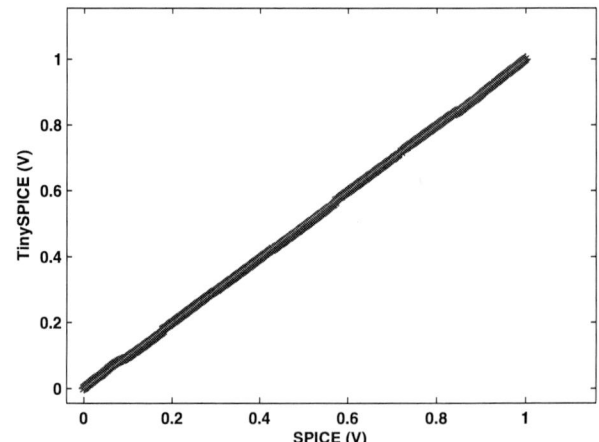

Figure 6: Scatter plot of the DC simulation results for SRAM circuits obtained by TinySPICE and the original Bsim4 SPICE simulator.

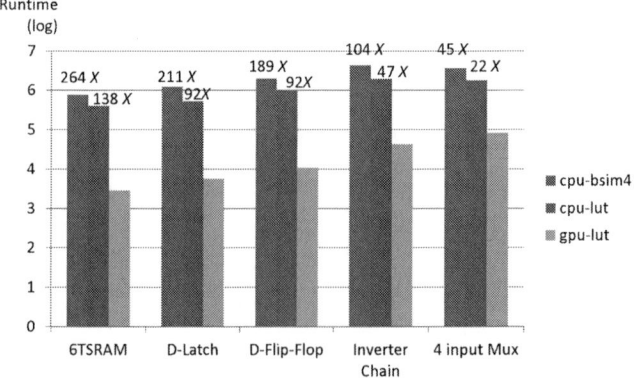

Figure 7: Comparison of DC Simulation Runtime

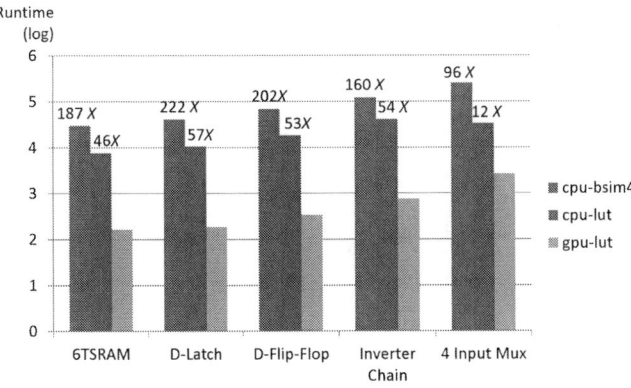

Figure 8: Comparison of Transient Simulation Runtime

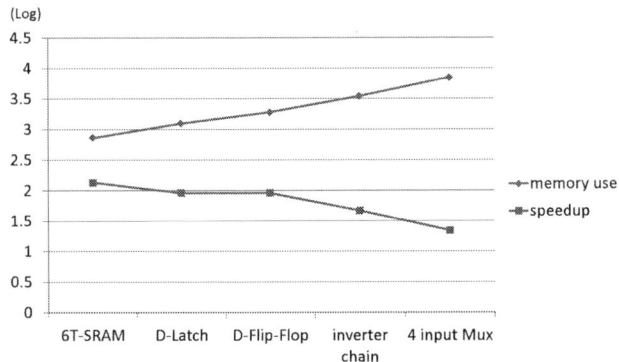

Figure 9: Memory usage (shared memory + registers) v.s. speedups

ditional SPICE simulator (as shown in Fig. 8).

It should be noted that, once the circuit problem size increases, the memory consumption of each GPU thread will also increase. As a result, the total number of GPU threads will decrease due to the limited GPU on-chip memory resources, such as registers and shared memory. For instance, the "4:1 Mux" test case has 35 unknowns, and the speedups obtained by GPU is only $22X$ in DC simulations, as illustrated in Fig. 7. This corresponds to a much lower simulation performance on GPU than the result obtained from the "6T-SRAM" circuit that has only 12 unknowns.

In the following, the relationship between the runtime speedups and GPU on-chip memory consumption (shared memory and registers) will be analyzed. As illustrated in Fig. 9, the blue curve denotes the memory usage for different circuits, and the red curve denotes the speedups of GPU-based SPICE simulator using LUTs obtained by comparing it with CPU-based SPICE simulator using LUTs. We observe that, when the number of unknowns of a circuit increases linearly, the memory consumption will dramatically increase, which is due to the storage requirement of the dense MNA matrices. Obviously, for each GPU thread, the dominant on-chip GPU memory is consumed for storing the system MNA matrices. Since the total registers available for each SM is very limited, once more on-chip memory is consumed by a single GPU thread, much fewer GPU threads can be assigned onto a GPU's SM. As a result, the GPU computing resources may not be fully utilized or there may not exist enough active GPU threads, which in turn dramatically reduces the runtime speedups.

5. CONCLUSIONS

In this work, we present a massively parallel SPICE-accurate nonlinear circuit simulation engine, TinySPICE, for variation-aware embedded memory and standard cell analysis by leveraging the emerging parallel GPU computing platforms. By accelerating the entire flow of SPICE simulation algorithm on GPU's on-chip memory, such as shared memory and registers, and employing parametric 3D LUTs, SRAM yield analysis and standard cell variation-aware characterizations can be performed in a much faster way than ever before. Compared with standard CPU-based SPICE simulation engines, our extensive experimental results show that TinySPICE simulation engine achieves up to $264X$ speedups for parametric SRAM simulations without sacrificing solution accuracy. Although TinySPICE is especially designed for small circuit simulations (with less than 20 unknowns), and its performance can be limited by GPU's on-chip memory resources for larger circuit analysis problems, TinySPICE still achieves up to $22X$ speedups in simulating a circuit with 35 unknowns.

6. REFERENCES

[1] R. Kanj, R. V. Joshi, and S. R. Nassif. Mixture importance sampling and its application to the analysis of SRAM designs in the presence of rare failure events. In *Proc. IEEE/ACM DAC*, pages 69–72, 2006.

[2] A. Bansal, R. N. Singh, R. Kanj, S. Mukhopadhyay, J. Lee, E. Acar, A. Singhee, K. Kim, C. Chuang, S. R. Nassif, F. Heng, and K. K. Das. Yield estimation of SRAM circuits using "Virtual SRAM Fab". In *Proc. IEEE/ACM ICCAD*, pages 631–636, 2009.

[3] J. Wang, S. Yaldiz, X. Li, and L. T. Pileggi. SRAM parametric failure analysis. In *Proc. IEEE/ACM DAC*, pages 496–501, 2009.

[4] J. Wang, A. Singhee, R. A. Rutenbar, and B. H. Calhoun. Two Fast Methods for Estimating the Minimum Standby Supply Voltage for Large SRAMs. *IEEE Trans. on Computer-Aided Design*, 29(12):1908–1920, 2010.

[5] C. Amin, C. Kashyap, N. Menezes, K. Killpack, and E. Chiprout. A multi-port current source model for multiple-input switching effects in CMOS library cells. In *Proc. IEEE/ACM DAC*, pages 247–252, 2006.

[6] P. Li, Z. Feng, and E. Acar. Characterizing Multistage Nonlinear Drivers and Variability for Accurate Timing and Noise Analysis. *IEEE Trans. on Very Large Scale Integration (VLSI) Systems*, 15(11):1205–1214, 2007.

[7] N. Menezes and C. V. Kashyap and C. S. Amin. A "true" electrical cell model for timing, noise, and power grid verification. In *Proc. IEEE/ACM DAC*, pages 462–467, 2008.

[8] AMD Corporation. AMD FusionŹ Family of APUs: Enabling a Superior, Immersive PC Experience. *AMD whitepaper*, [Online]. Available: http://sites.amd.com/us/fusion/apu/Pages/fusion.aspx, 2011.

[9] Nvidia Corporation. Bringing High-End Graphics to Handheld Devices. *Nvidia whitepaper*, 2011.

[10] K. Gulati, J. F. Croix, S. P. Khatri, and R. Shastry. Fast circuit simulation on graphics processing units. In *Proc. IEEE/ACM ASPDAC*, pages 403–408, 2009.

[11] L. Ren, X. Chen, Y. Wang, C. Zhang, and H. Yang. Sparse LU factorization for parallel circuit simulation on GPU. In *Proc. IEEE/ACM DAC*, pages 1125–1130, 2012.

[12] L. Pillage, R. Rohrer, and C. Visweswariah. *Electronic circuit & system simulation methods.* McGraw-Hill, 1995.

[13] *Nvidia CUDA programming guide.* [Online]. Available: http://www.nvidia.com/object/cuda.html, 2007.

[14] Nvidia Corporation. *Fermi compute architecture white paper.* [Online]. Available: http://www.nvidia.com/object/fermi_architecture.html, 2010.

Supplementary Material

S.1 Nonlinear Circuit Simulation Approaches

General nonlinear electronic circuit simulation techniques rely on NR method to solve the following nonlinear differential equations [12]:

$$f\left(x\left(t\right)\right) + \frac{d}{dt}q\left(x\left(t\right)\right) + u\left(t\right) = 0, \qquad (1)$$

where $f(\cdot)$ and $q(\cdot)$ denote the static and dynamic nonlinearities, $x(t)$ is a vector including nodal voltages as well as branch currents, and $u(t)$ is the input excitation vector. Sophisticated numerical methods can be used to solve the above nonlinear differential equations by first linearizing the nonlinear circuit system at a given solution point, and subsequently solving the corresponding linear matrix problems. For instance, after linearizing the system, conductance matrix $G\left(x^k\right) = \left.\frac{\delta f}{\delta x}\right|_{x^k}$ and capacitance matrix $C\left(x^k\right) = \left.\frac{\delta q}{\delta x}\right|_{x^k}$ can be easily obtained which are typically asymmetric matrices. The dominant computational cost for solving small circuit problems is mainly due to the nonlinear device evaluations, while for much larger circuits solving the asymmetric Jacobian matrices using direct solution method can be much more expensive due to the exponentially increased runtime and memory cost.

S.2 Massively Parallel GPU Computing

S.2.1 Recent GPUs

The recent Nvidia Geforce GTX 285 GPU includes 30 streaming multiprocessors (SMs) and each SM has eight streaming processors (SPs) that share the same instruction unit as well as the 32 KB on-chip shared memory, as shown in Fig. 10. According to CUDA programming model [13], 32 threads are formed into a warp, and will execute the same instruction every four clock cycles, resulting in a very light overhead (one instruction issuing is followed by 32 thread executions). When a kernel function is launched on GPU, the task (data) is further divided into many thread blocks (1D, 2D or 3D) based on the problem size and available on-chip hardware resources. Each thread block may include multiple warps of threads. Subsequently, each SM will work on a few thread (data) blocks with its eight SPs. Recent Fermi GPU from Nvidia has increased the number of streaming processors (SPs) in each streaming multiprocessor (SM) from 8 to 32, boosting the total number of streaming processors to 512 [14]. The new GPU model also supports high performance double-precision computing and concurrent kernel executions. Up to 16 kernels can be launched concurrently on the 16 SMs for Fermi GPUs, while in previous GPU architectures only one kernel can be launched at the same time on GPU, which allows for more flexible and efficient GPU computing.

S.2.2 Key Issues in Efficient GPU Computing

GPU's on-chip memory (shared memory and registers) is very fast, but the available on-chip memory resource can be quite limited, whereas the off-chip device memory (global memory) is sufficiently large but can be much slower than on-chip memories. Additionally, coalesced GPU global memory accesses are important since random memory accesses are typically much slower. The device memory bandwidth can be up to $100Gb/s$ if accessed in a coalesced pattern but may also reduce to $10X$ lower if accessed in a random manner [13]. If random memory access is needed, texture memory on GPU (like the L1 and L2 caches for CPU) should be used, though a good memory access pattern is still desired such that threads of a warp can access the neighboring memory locations.

GPU's hardware and software properties impose the following challenges when developing streaming data parallel computing algorithms: (1) the dependencies among different tasks (data) should

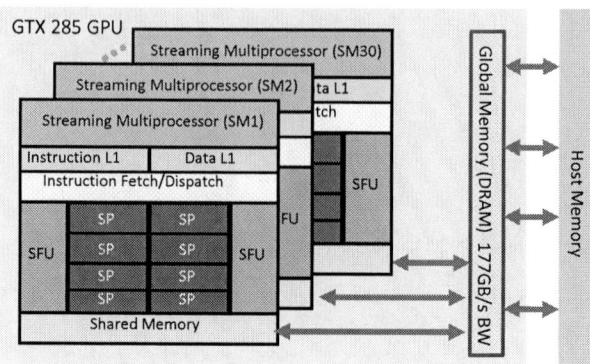

Figure 10: The GTX 285 GPU architecture.

be minimized, (2) excessive global data sharing and shared memory (register) bank conflicts should be avoided, (3) the arithmetic intensity that is defined as the number of floating point operations per data reading/writing should be maximized, and (4) the algorithm control flow should be simplified.

S.3 Parametric 3D Look-up Tables

In order to meet requirements in both accuracy and runtime efficiency, parametric 3D LUT models will be constructed for evaluating transistors during circuit simulations. LUT-based evaluation of a smooth function derived from the truncated Taylor expansion can be formulated as follows:

$$T_d\left(x\right) = f\left(c\right) + \sum_{k=1}^{d} \frac{f^{(k)}\left(c\right)}{k!}\left(x-c\right)^k \qquad (2)$$

where $f\left(c\right)$ denotes the evaluation function and $f^{(k)}\left(c\right)$ denotes the k-th order derivatives at reference point c, x is the evaluation point, and d is the degree of the Taylor polynomial. The approximated evaluation can be carried out by looking up a precalculated LUT for coefficients associated with $\left(x-c\right)^k$. For the second-order Taylor polynomial expansion, which means $d = 2$, we can get the second-order parametric 3D LUTs evaluation function:

$$\begin{aligned}
LUT = LUT_{base} &+ LUT_{V_{th}} \cdot \Delta V_{th} + LUT_{L_{eff}} \cdot \Delta L_{eff} \\
&+ LUT_{V_{th2}} \cdot \Delta V_{th}^2 + LUT_{L_{eff2}} \cdot \Delta L_{eff}^2 \quad (3) \\
&+ \Delta V_{th} \cdot \Delta L_{eff} \cdot LUT_{V_{th}L_{eff}}
\end{aligned}$$

where LUT_{base} represents the base LUT generated based on the transistor nominal parameters. $LUT_{V_{th}}$ and $LUT_{L_{eff}}$ are the first-order coefficient LUTs for transistor threshold voltage and effective channel length respectively. Similarly $LUT_{V_{th2}}$ and $LUT_{L_{eff2}}$ are the second-order coefficient LUTs. $LUT_{V_{th}L_{eff}}$ is the coefficient LUT derived from the partial derivatives of V_{th} and L_{eff}. In order to reduce the complexity, this cross-term is ignored in our implementation. ΔV_{th} and ΔL_{eff} denote the variations of the threshold voltage and effective channel length. So the base LUT and two coefficient LUTs compose the whole parametric 3D LUTs of a transistor. The number and order of coefficient LUTs can be adjusted according to the number of input parameters and accuracy requirement. However it is not always necessary to introduce the higher order LUTs for each parameter. Benefited from these parametric LUTs that can capture the variations of transistor parameters, we do not need to update the LUTs for every parametric SPICE simulation. In other words, only one-time data transferring of the parametric LUTs from CPU to GPU is required, which can greatly reduce the overhead of CPU-GPU communications.

The proposed TinySPICE first parses standard SPICE-like circuit netlist, and evaluates the BSIM4 transistor models to build parametric 3D LUTs for all nonlinear transistors. When building the parametric 3D LUTs, we use the $\Delta V_{th}, \Delta L_{eff}$, V_{ds}, V_{gs} and V_{bs} as the input variables, where V_{ds}, V_{gs} and V_{bs} denote the terminal voltages of MOSFET devices. To get coefficient LUTs, $LUT_{V_{th}}$, $LUT_{V_{L_{eff}}}$, $LUT_{V_{th2}}$ and $LUT_{V_{L_{eff2}}}$ are also calculated after generating the LUT_{base}. The parametric 3D LUTs outputs include all the required elements for stamping the conductance and capacitance matrices obtained from linearizing (1) during SPICE simulations, such as conductance, capacitance, currents and charges.

After extracting all the data required by these parametric 3D LUTs using thousands of BSIM4 model evaluations, we store all the data into a long vector to allow GPU's coalesced device memory accesses, as shown in Fig. 3. Considering the huge amount of data (more than forty elements) computed in one transistor evaluation, we store the data in such a way that good data locality can be well preserved to ensure GPU's efficient texture memory accesses during LUTs' trilinear data interpolations using neighboring eight points.

Since device evaluations using 3D LUTs are based on eight-point trilinear data interpolations, device evaluated by LUTs requires much less computational time than the BSIM4 model evaluations that involve very complex formulas. We observe that for most digital circuit modeling and analysis applications, the accuracy level obtained using LUT-based SPICE simulator (with first-order parametric LUTs) is very satisfactory, though for analog circuits the convergence may become more difficult.

An Optimal Algorithm of Adjustable Delay Buffer Insertion for Solving Clock Skew Variation Problem

Juyeon Kim[1]
juyeon@ssl.snu.ac.kr

Deokjin Joo[1]
jdj@ssl.snu.ac.kr

Taewhan Kim[1,2]
tkim@ssl.snu.ac.kr

[1]School of Electrical Engineering and Computer Science, Seoul National University, Seoul, Korea
[2]Nano Systems Institute (NSI), Seoul National University, Seoul, Korea

ABSTRACT

Meeting clock skew constraint is one of the most important tasks in the synthesis of clock trees. Moreover, the problem becomes much hard to tackle as the delay of clock signals varies dynamically during execution. Recently, it is shown that adjustable delay buffer (ADB) whose delay can be adjusted dynamically can solve the clock skew variation problem effectively. However, inserting ADBs requires non-negligible area and control overhead. Thus, all previous works have invariably aimed at minimizing the number of ADBs to be inserted, particularly under the environment of multiple power modes in which the operating voltage applied to some modules varies as the power mode changes. In this work, unlike the previous works which have solved the ADB minimization problem heuristically or locally optimally, we propose an elegant and easily adoptable solution to overcome the limitation of the previous works. Precisely, we propose *an* $O(n \log n)$ *time* (bottom-up traversal) *algorithm that* (1) *optimally solves the problem of minimizing the number of ADBs to be inserted with continuous delay of ADBs* and (2) *enables solving the ADB insertion problem with discrete delay of ADBs to be greatly simple and predictable.* In addition, we propose (3) *a systematic solution to* an important extension to *the problem of buffer sizing combined with the ADB insertion* to further reduce the ADBs to be used.

1. INTRODUCTION

Clock is one of the most important signals on a chip, as all the synchronous components on the chip such as flip-flops (FFs) rely on it. Clock tree is a commonly used structure of circuits that distributes the clock signal from the clock source to all the *clock sinks* (e.g., FFs), where the clock signal is required. It is imperative that the maximum of the arrival time difference between the clock sinks, which is known as *clock skew*, should be maintained under a certain bounded value typically within 10% of the clock period, as a large clock skew may cause timing violation on the circuits.

Many research works on the clock tree optimization such

as clock routing, clock buffer insertion/sizing, and wire sizing have been performed to control or minimize the clock skew [1–7]. While these approaches were effective, advanced low power design techniques introduced new challenges to the clock skew control problem. Specifically, for multiple power mode designs, where the supply voltage to the circuit components varies dynamically depending on modes, the clock arrival time also varies dynamically.

Even though the previous works can consider the clock skew constraint on every power mode, it would be highly likely that the resulting clock tree uses a substantially long wirelength or there exists no clock tree that satisfies the clock skew constraint on every power mode. On the other hand, post-silicon tuning (e.g., [8–11]) such as inserting Adjustable Delay Buffers (ADBs) is a widely used method to deal with the timing problem caused by process and environment variations. Because the delay of an ADB can be controlled by its delay control inputs [12], the clock skew variation caused by process variation can be tuned by properly inserting ADBs after the manufacturing stage has been completed. The idea of using ADBs in multiple power modes is to replace some of normal clock buffers with ADBs so that the clock skew constraint on each power mode can be met; when the power mode changes during execution, for example from power mode *mode-1* to power mode *mode-2*, the delays of ADBs in clock tree that have been adjusted under *mode-1* are readjusted to meet the clock skew constraint under *mode-2*. Since ADB logic component is much bigger than normal buffer and it requires control line as well as switching logic, the set of related problems to be solved for the ADB-based clock skew optimization in multiple power modes are allocating a minimum number of ADBs, finding the normal buffers (or locations) in the clock tree that are to be replaced by ADBs, and determining the delay value of ADBs to be assigned on each power mode. We call the these problems collectively *ADB insertion problem*.

Su *et al.* [13,14] proposed a linear-time optimal algorithm for the delay assignment problem and exploits the algorithm to solve the rest of two subproblems of the ADB insertion problem heuristically in a greedy manner. Lin, Lin, and Ho [15] proposed an efficient algorithm of two-stage approach which performs a top-down ADB insertion followed by a bottom-up ADB elimination. Even though the approach reduces the run time over that in [13, 14], it still does not guarantee an optimality. Lim and Kim [16] proposed a linear-time algorithm for the ADB insertion problem where they solved the problem optimally for *each* power mode. However, merely collecting the optimal results on in-

Permission to make digital or hard copies of all or part of this work for personal or classroom use is granted without fee provided that copies are not made or distributed for profit or commercial advantage and that copies bear this notice and the full citation on the first page. To copy otherwise, to republish, to post on servers or to redistribute to lists, requires prior specific permission and/or a fee.
DAC'13, May 29 - June 07 2013, Austin, TX, USA.

dividual power modes does not mean globally optimal for all power modes. In this work, we revisit the ADB insertion problem and propose a set of solutions to overcome the limitation of the previous works. More precisely, we propose (1) *an* O($n \log n$) *time algorithm that optimally solves the problem of minimizing the number of ADBs to be inserted for all power modes with continuous delay of ADBs* and (2) *enables solving the ADB insertion problem with discrete delay of ADBs to be greatly simple and predictable.* In addition, we propose an effective solution to an important extended problem: (3) *the ADB insertion problem combined with buffer sizing.*

2. ADB STRUCTURE AND INSERTION OF ADB

Fig. 1 shows the structure of a capacitor bank based implementation of ADB [17]. This implementation of a well known capacitor bank based ADB consists of two inverters at the input and output ports, and in the middle there is an array of capacitors with switch transistors attached. The switches are controlled by the capacitor bank controller, which controls the number of active capacitors according to the control bits.

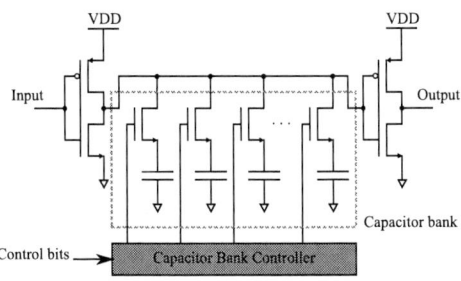

Figure 1: The structure of a capacitor bank based ADB.

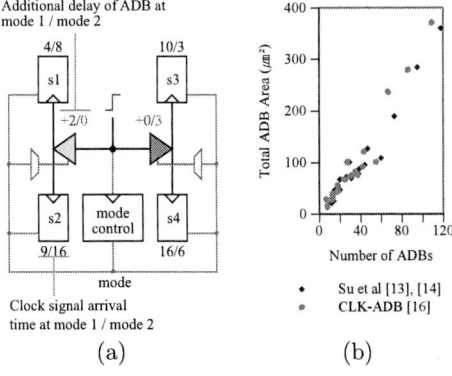

Figure 2: (a) An example of clock tree with the replacement of two clock buffers with ADBs. (b) The relationship between the number of ADBs and the total ADB area (including logic overhead) used by [13, 14, 16].

Fig. 2(a) shows an example of clock tree that has four sinks $s1$, $s2$, $s3$, and $s4$, two ADBs replacing two clock buffers, and ADB control logic. Suppose there are two power modes *mode-1* and *mode-2* in this design. Then, the two numbers

separated by a slash next to each sink indicate the clock signal arrival times in *mode-1* and *mode-2*. When the clock skew bound is given to 10, the clock tree causes clock skew violations in both modes if ADBs were not used. With the replacement of two clock buffers by ADBs, the two numbers next to each ADB indicate the delay increments (simply called *delay values*) in *mode-1* and *mode-2*: The ADB on the left adds delay of 2 in *mode-1*, thus the clock signal arrival time at $s1$ in *mode-1* becomes 6. Likewise, the ADB on the right adds delay of 3 in *mode-2*, increasing the arrival time at $s3$ in *mode-2* to 6. To control the ADBs' delay, a mode signal is required. In addition, depending on the implementation of the ADBs, control logic that converts the mode signal to ADB's bank controller input is needed. This additional overhead incurred by the insertion of ADBs is also shown in Fig. 2(a).

Fig. 2(b) shows a scatter plot of the number of ADBs inserted versus the total ADB area including the overhead, obtained by implementing the algorithms in the previous work [16]. The plot indicates that the number of ADBs has a strong correlation with the total sum of the area of ADBs, justifying that the primary objective of the ADB insertion problem is to minimize the number of ADBs to be inserted.

3. PROBLEM DEFINITION AND MOTIVATION

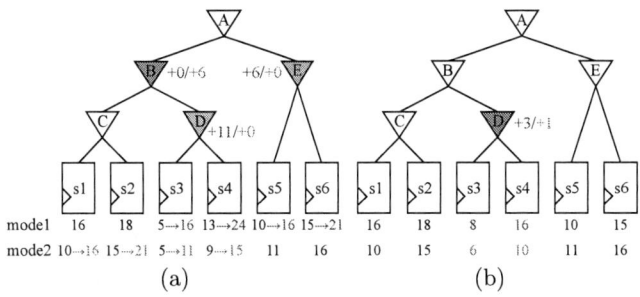

Figure 3: A motivational example for ADB allocation and delay assignment. (a) A clock tree with clock skew violation when skew bound is given to 10 units of delay. The initial clock signal arrival times are shown in black numbers. With optimization methods that rely on earliest arrival time synchronization (i.e., [16]), three ADBs are allocated in place of buffers B, D and E. The adjusted arrival times are shown in red or blue numbers. (b) An optimal allocation that uses only one ADB.

PROBLEM 1. *ADB insertion problem: Given a synthesized clock tree, arrival times of clock sinks in each power mode, and clock skew bound* κ, *replace the least number of clock buffers with ADBs and assign delays to the ADBs to satisfy* κ *in all power modes.*

Two common features of the previous ADB insertion algorithm [13, 14, 16] are that they resolve the clock skew violation by synchronizing *the earliest arrival times* of (two) subtrees of interest where they set the delay value of ADB on a root of one of the subtrees to the difference of the earliest arrival times of the subtrees, and the delay value adjustment is performed *mode by mode*. While this method of delay value assignment does minimize the clock skew, their method of applying mode by mode yields sub-optimal results that use more ADBs than necessary. For example, Fig. 3 shows a comparison of two results of ADB allocation where Fig. 3(a) corresponds to the result by the sub-optimal ADB allocation algorithm in [16] while Fig. 3(b) corresponds to an optimal result for the same input clock tree as that in Fig. 3(a).

4. ADB INSERTION ALGORITHM

Table 1: Notations

Symbol	Description
$n_{(\cdot)}$	A node in a clock tree, which is either a buffer or a sink;
T_{n_i}	The subtree rooted at node n_i;
$arr_{n_i,m}$	Arrival time at sink node n_i at $mode\text{-}m$;
$lst_{n_i,m}$	The latest arrival time of the subtree rooted at node n_i in $mode\text{-}m$;
κ	The given clock skew bound to meet;
$\alpha_{n_i,m}$	Delay value (i.e., increment) of ADB located at node n_i in $mode\text{-}m$;
H_{n_i}	Set of child nodes of n_i not to be replaced by ADBs.

4.1 The Proposed Optimal Algorithm

First, we demonstrate the procedure of our algorithm for the continuous delay of ADBs, called ADB-PULLUP, step-by-step using an example to see how the algorithm works. (The definitions of the notations used in the presentation are in Table 1.)

Let us consider the clock signal arrival times shown in the clock tree in Fig. 4(a). Let $\kappa = 10$. Then, ADB-PULLUP initially assumes that each sink has a distinct fictitious ADB at the front of it. The blue numbers at the bottom of each sink s_i indicate the delay values i.e. $\alpha_{s_i,1}$ in $mode\text{-}1$ and $\alpha_{s_i,2}$ in $mode\text{-}2$ of the ADB in s_i, $i = 1, \cdots, 10$. We compute the delay value by

$$\alpha_{s_i,m} = max\{0, \; lst_{root,m} - \kappa - arr_{s_i,m}\} \quad (1)$$

where $root$ represents the clock source (root) node of the clock tree. For example, $\alpha_{s1,1} = max\{0, 20\text{-}10\text{-}7\} = 3$ and $\alpha_{s1,2} = max\{0, 20\text{-}10\text{-}8\} = 2$. Note that the value by $Eq.(1)$ for each sink s_i corresponds to the least increase of delay required on the fictitious ADB in s_i to meet the clock skew constraint. Then, ADB-PULLUP performs a bottom-up traversal on the clock tree to move up (i.e., pull up) the ADBs towards the root of clock tree.

The decision of inserting an ADB to n_k which is a non-sink and whose α value has been assigned is made according to the evaluation result of the inequality:

$$\alpha_{n_k,m} > lst_{root,m} - lst_{n_i,m} \quad (2)$$

where n_i is the parent node of n_k.

If the inequality is true for at least one power mode, an ADB is inserted. For example, since $\alpha_{b4,2} \; (= 2) > lst_{root,2} - lst_{b2,2} \; (= 20\text{-}20 = 0)$, an ADB is inserted to $b4$. However, since $\alpha_{b5,1} \; (= 0) \le lst_{root,1} - lst_{b2,1} \; (= 20\text{-}16 = 4)$ and $\alpha_{b5,2} \; (= 0) \le lst_{root,2} - lst_{b2,2} \; (= 20\text{-}20 = 0)$, no ADB is inserted to $b5$. Once the decision of inserting ADBs to all children of n_i is made, the α value of n_i is updated by

$$\alpha_{n_i,m} = max\{\alpha_{n_k,m} : n_k \in H_{n_i}\} \quad (3)$$

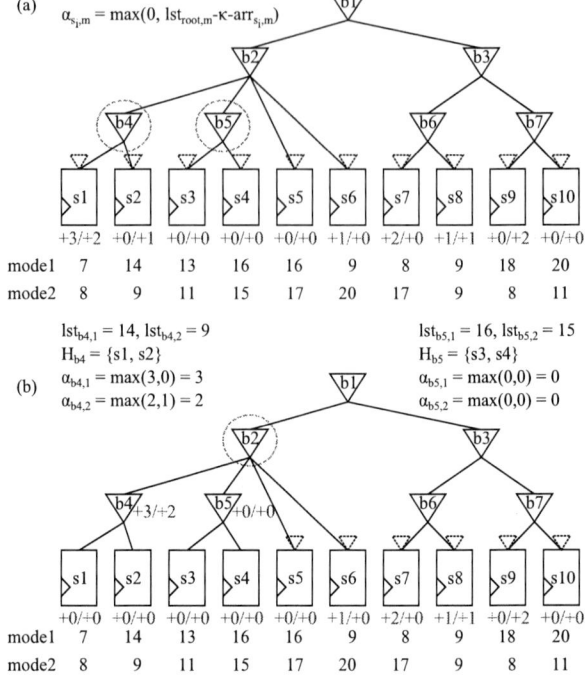

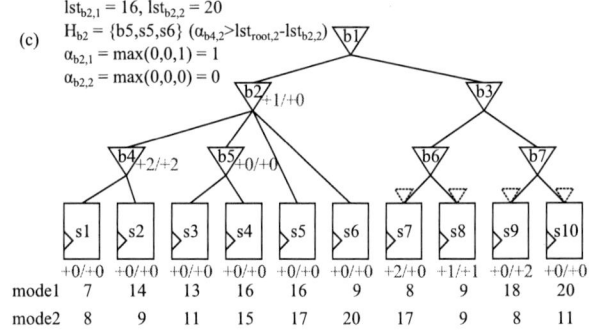

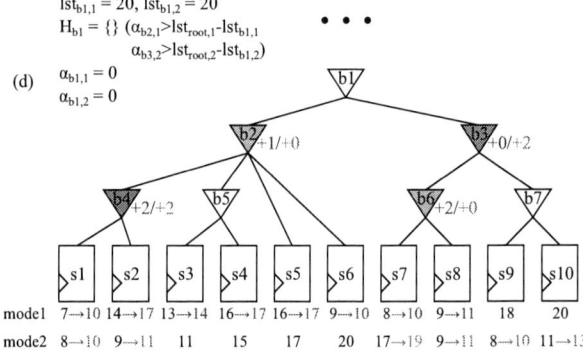

Figure 4: Example showing step-by-step procedure of ADB-PULLUP: (a) A clock tree T before the ADB insertion by ADB-PULLUP with $\kappa = 10$; allocating $\alpha_{n_i,m}$ for each sink n_i and mode m. (b) After the process of clock subtrees rooted at $b4$ and $b5$. (All children n_k of each subtree rooted at n_i satisfy $\alpha_{n_k,m} \le lst_{root,m} - lst_{n_i,m}$ for all modes. Thus, no ADB is inserted.) (c) After the process of clock subtree rooted at $b2$. ($\alpha_{b4,2} > lst_{root,2} - lst_{b2,2}$, thus, an ADB is inserted at $b4$.) (d) The complete subtree T after the ADB insertion by ADB-PULLUP.

634

where H_{n_i}[1] represents the set of n_i's children that are either sinks or non-sinks, but not the nodes with ADB. For example, since $H_{b2} = \{b5, s5, s6\}$, $\alpha_{b2,1} = \max\{\alpha_{b5,1}, \alpha_{s5,1}, \alpha_{s6,1}\}$ $= max\{0, 0, 1\} = 1$ and $\alpha_{b2,2} = \max\{\alpha_{b5,2}, \alpha_{s1,2}, \alpha_{s2,2}\} = max\{0, 0, 0\} = 0$. (See $b2$ in Fig. 4(c).)

At this stage, from node n_i where its α values are set, we perform delay-resetting on every child, n_k, of n_i by calling function READJUST described in Fig. 5. READJUST subtracts $\alpha_{n_i,m}$ from the sum of delays on each path from a child of n_i to its descendent sink, or set to 0 if $\alpha_{n_i,m}$ is bigger than the previous sum of delays. For example, $\alpha_{b4,1} = 3 - min\{1, 3\} = 2$ and $\alpha_{b4,2} = 2 - min\{0, 2\} = 2$. Fig. 4(c) shows the results of delay readjustment when the delay value of $b2$ is computed by $Eq.(3)$. Subtree T_{b3} is processed likewise. After all the nodes are processed, ADB-PULLUP reports the result of ADB insertion with the updated arrival times as shown in Fig. 4(d).

The flow of ADB-PULLUP is depicted in Fig. 5. In the initialization phase, the $\alpha_{n_i,m}$ value of each sink n_i is assigned to the minimum value by which $arr_{n_i,m} + \alpha_{n_i,m}$ is not shorter than $lst_{root,m} - \kappa$. This fixes the skew violations by assuming the allocation of a fictitious ADB to each sink. The next phase is "pulling up" these ADBs to non-sink locations of the clock tree, by performing PULLUP operation in a topological order. Consider a non-sink node n_i to be processed in the flow. Each child, n_k, of n_i, is checked to see if an ADB is needed according to the evaluation of $\alpha_{n_k,m} > lst_{root,m} - lst_{n_i,m}$. If the evaluation is true, an ADB is inserted to n_k, otherwise, the maximum α value (initially 0) to be assigned to n_i is updated if needed. Once the process PULLUP at the bottom loop in Fig. 5 is done, the α values at the descendants of n_i are recursively re-set according to function READJUST. The time complexity of ADB-PULLUP is bounded by $O(KN \log N)$ where K is the number of power modes and N is the number of nodes of the input clock tree. Since K is usually very small, the complexity is reduced to $O(N \log N)$. The following summarizes the properties and theorems of ADB-PULLUP. (All the proofs are left out due to the space limitation.)

PROPERTY 1. *The arrival times at sinks produced by* ADB-PULLUP *never exceed* $lst_{root,m}$ *for every mode m.*

THEOREM 1. *The result produced by* ADB-PULLUP *has a positive value of α in a sink if and only if it is impossible for the given clock tree to meet the clock skew bound with ADB allocation.*

Note that Property 1, which is a feature that enables to keep the total size of capacitor banks in ADBs within a certain limit, does not hold for the previous ADB insertion algorithms. In addition, Theorem 1 indicates that if there is at least one solution, ADB-PULLUP will always find an ADB insertion solution such that the α values of all sinks are 0.

THEOREM 2. *After the execution of* ADB-PULLUP *on n_i, subtree T_{n_i} of clock tree T rooted at n_i has been inserted with a minimum number of ADBs while meeting the clock skew constraint for T_{n_i}.*

By theorem 2, for T_{root} ADB-PULLUP minimally inserts ADBs while meeting the clock skew constraint.

---

[1]If $H_{n_i} = \phi$, then $\alpha_{n_i,m}$ is set to 0 for every mode m.

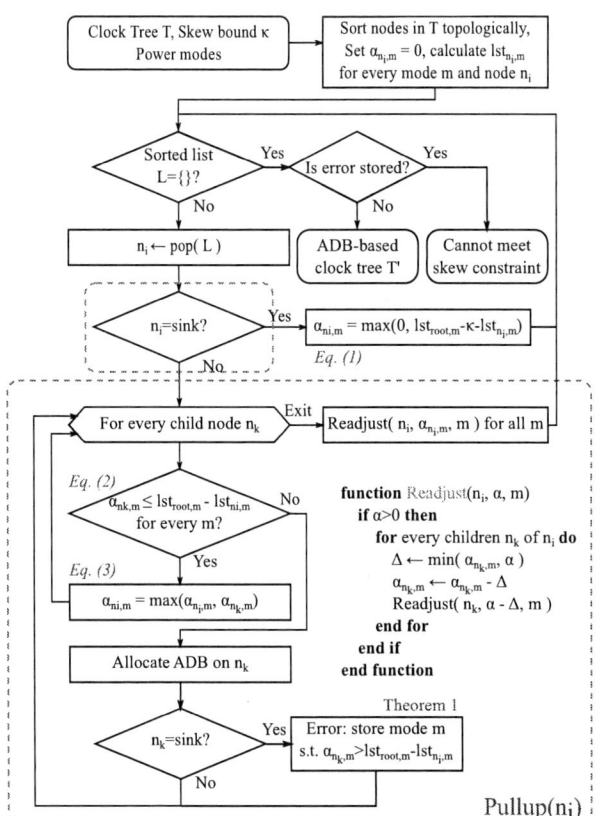

Figure 5: The flow of ADB-PULLUP.

4.2 Supporting Discrete ADB Delay

ADB-PULLUP can be easily fitted to support the discrete delay of ADBs, which we call ADB-PULLUP-Q. If we want to quantize ADB delays with a unit of Q, we simply assign $\alpha_{s_i,m}$ of sink s_i to $\alpha_{s_i,m} = \lceil (lst_{root,m} - \kappa - lst_{i,m})/Q \rceil \times Q$ rather than using $lst_{root,m} - \kappa - lst_{n_i,m}$ in $Eq.(1)$.

THEOREM 3. *If the result produced by* ADB-PULLUP-Q *meets the clock constraint, all the ADBs inserted always use the discrete delays.*

Note that Theorem 3 does not mean that like to ADB-PULLUP, ADB-PULLUP-Q always finds a valid solution if there exists under the discrete delay of ADBs. We use the following strategy: (1) apply ADB-PULLUP to the input clock tree; (2) if ADB-PULLUP signals "$\alpha > 0$ *for some sink*", report "*the problem is unsolvable*" (according to Theorem 1) and stop; (3) apply ADB-PULLUP-Q to the input clock tree; (4) if ADB-PULLUP-Q returns "*no $\alpha > 0$ for any sink*", the valid ADB insertion (according to Theorem 3) is found and stop; (5) identify a sink with $\alpha > 0$ and increase its arrival time by Q (based on Property 1) by conducting wire detouring or wire resizing at the sink; (6) if the resulting time at the sink exceeds the longest latency of the initial clock tree, report "*fail to find a solution or the problem is unsolvable*" and stop; (7) go to (3).

The idea behind this strategy is that since ADB-PULLUP-Q can detect the location where the ADB insertion fails and knows the reason why it fails, by locally tuning the wires at the detected location, the next iteration can be performed with a higher chance of finding a valid solution of ADB insertion.

5. EXTENSION: INTEGRATION OF BUFFER SIZING

We can think of buffer sizing as an ADB insertion imposed by the restriction that the α values in power modes are pre-defined. For example, when a buffer b_i in the input clock tree is going to be replaced by a buffer buf_j in the buffer library $\mathcal{L}$ (rather than an ADB), the delay number in each power mode may be increased or decreased, but the number is fixed, which means un-controllable, unlike ADB. Let $\beta^j_{n_i,m}$ be the delay increase or delay decrease in power mode m caused by the replacement of buffer b_i in the input clock tree by $buf_j \in \mathcal{L}$. We can compute all β values from the input clock tree and $\mathcal{L}$. Now, we want to substitute the minimal ADBs determined by ADB-PULLUP (or ADB-PULLUP-Q) with as many buffers in $\mathcal{L}$ as possible to further reduce the number of ADBs to be inserted in the clock tree while still meeting the clock skew constraint for every power mode. Since we have all the β and α, values in every node of the clock tree in all power modes, a naive solution is to generate all the combinations of buffer sizing as well as ADB insertion for all nodes, and choose the one that uses the least number of ADBs while meeting the clock skew constraint. However, its computation time grows exponentially as the problem size increases. To be practically feasible, we propose a simple but effective iterative method:

1. For each node n_i in the clock tree, in which ADB-PULLUP (or ADB-PULLUP-Q) has decided that an ADB should be inserted in the node, for each buffer $buf_j \in \mathcal{L}$, we compute

$$\delta^j_{n_i} = \sum_{m=1}^{K} (\alpha_{n_i,m} - \beta^j_{n_i,m})^2 \qquad (4)$$

where K is the number of modes. For example, if $\alpha_{n_1,1} = +3$, $\alpha_{n_1,2} = +1$, $\beta^1_{n_1,1} = +3$, $\beta^1_{n_1,2} = +2$, $\beta^2_{n_1,1} = +1$, and $\beta^2_{n_1,2} = -1$, then, $\delta^1_{n_1} = (3-3)^2 + (1-2)^2 = 1$ and $\delta^2_{n_1} = (3-1)^2 + (1-(-1))^2 = 8$.

2. Select the pair of node and buffer sizing such that the corresponding δ value is minimal and it satisfies the clock skew and latency constraints. The buffer in the selected node is then resized accordingly. For the previous example, selecting buf_1 is preferred to that of buf_2 for resizing in node n_1 since $\delta^1_{n_1} < \delta^2_{n_1}$. The iteration stops when there is no pair that satisfies the skew and latency constraints or the resizing causes the number of ADBs to increase.

3. Update the arrival times at clock sinks according to the buffer resizing performed in step 2.

Note that the rationale behind the use of δ is that as the smaller the value of δ in a node is, the more the corresponding buffer sizing is likely to close to the ADB that has been inserted to the node, thus, the buffer sizing taking over the role of the ADB with a minimal impact on the overall timing of the clock tree. We call the ADB insertion algorithm combined with buffer sizing ADB-PULLUP-BS for the continuous delay of ADB.

6. EXPERIMENTAL RESULTS

The proposed algorithm ADB-PULLUP (continuous delay), ADB-PULLUP-Q (discrete delay), and ADB-PULLUP-BS (combining buffer sizing) have been implemented in Python 3 language on a Linux machine with 16 cores of 2.67Ghz Intel Xeon CPU and 51GB memory. ISCAS'95 and ITC'99 benchmarks were synthesized with *Synopsys IC Compiler* with 45nm Nangate Open Cell Library. ISPD'09 bench-

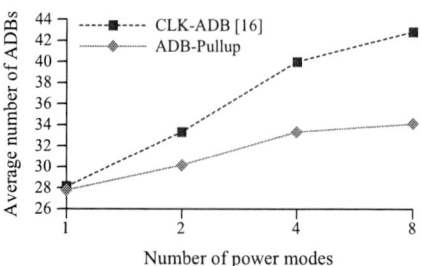

Figure 6: The changes of the average number of ADBs used by CLK-ADB [16] and ADB-PULLUP by varying the number of power modes used.

marks were synthesized using the algorithm in [18]. Each benchmark was partitioned into 6 to 10 power domains which are able to operate in two different supply voltage levels, 0.95V and 1.1V.

Table 2 summarizes the results produced by applying CLK-ADB [16] (continuous delay), CLK-ADB-RD (discrete delay) [16], ADB-PULLUP, ADB-PULLUP-Q and ADB-PULLUP-BS to the benchmark clock trees using four power modes. The columns in the left part of Table 2 represent the number of flip-flop, the number of clock buffers, the worst clock skew, the worst clock latency in the four power modes of the input clock trees, and the clock skew constraint. The columns in the middle part show the results by CLK-ADB [16], ADB-PULLUP, ADB-PULLUP-BS. It is observed that ADB-PULLUP uses consistently less number of ADBs compared to CLK-ADB. In addition, ADB-PULLUP-BS further reduces the number of ADBs with a slight area saving over that by ADB-PULLUP. The results shown in the right part indicates that ADB-PULLUP-Q uses considerably less ADBs than CLK-ADB-RD. This is because CLK-ADB-RD relies on re-iteration with tighter skew bound when clock skew violation occurs after delay quantization while ADB-PULLUP-Q can use quantized delay directly during its bottom-up phase.

Fig. 6 shows the average numbers of ADBs inserted by CLK-ADB [16] and our ADB-PULLUP when the number of modes varies. Clearly, ADB-PULLUP always uses less ADBs in all situations. The gap between the results increases as we increase the number of modes used since it is less likely that the ADB allocation in one mode coincides with the allocation in another mode. However, another factor to be consider is that as the number of modes increases, more buffers would be replaced with ADBs, which increases the chance of the coincidence. The actual gap is a complex function of these two factors.

7. CONCLUSIONS

In this paper, we proposed a polynomial-time optimal algorithm to the problem of ADB insertion on clock trees for the continuous ADB delay. Then, based on the algorithm, we proposed a much simple and predictable solution to the ADB insertion problem for the discrete ADB delay. In addition, we proposed an effective solution to the combined problem of ADB insertion and buffer sizing. From the experimental results on benchmarks, it was shown that compared to the results by the best known ADB insertion algorithm, our proposed algorithms reduced the number of ADBs by 13.5%, 15.4%, and 31.6% (15.0% total area reduced) on average when continuous ADB delay, discrete ADB delay, and the integration of buffer sizing were used, respectively.

Table 2: Comparison of results produced by CLK-ADB [16], CLK-ADB-RD [16], ADB-Pullup, ADB-Pullup-Q and ADB-Pullup-BS.

Bench-mark Circuit	#FFs/#Bufs	Original Skew/Lat. (ps)	Skew bound (ps)	Continuous delay						Discrete delay			
				CLK-ADB [16]		ADB-Pullup		ADB-Pullup-BS		CLK-ADB-RD [16]		ADB-Pullup-Q	
				#ADBs	Area	#ADBs	Area	#ADBs	Area	#ADBs	Area	#ADBs	Area
s35932	1728/97	264.1/545.1	30	27	156.1	25	151.5	20	135.4	42	228.1	25	151.7
			40	25	147.0	23	140.0	19	126.9	26	151.7	23	140.2
			50	25	144.5	23	137.8	19	124.7	25	145.2	23	137.9
s38417	1564/89	387.1/612.1	30	31	212.1	27	196.1	22	180.6	36	235.5	28	200.9
			40	28	197.9	25	184.6	20	169.0	31	211.8	26	189.4
			50	26	186.5	23	173.1	18	164.4	29	200.8	23	173.3
s38584	1168/66	299.8/552.8	30	22	138.3	20	127.3	14	123.2	22	138.2	20	127.4
			40	18	118.9	16	107.3	11	107.4	21	133.4	17	112.0
			50	18	118.8	16	105.7	11	104.6	18	118.9	16	105.8
B17	1312/89	287.7/654.7	30	29	174.8	25	160.0	19	157.7	35	203.5	26	164.7
			40	26	159.5	22	143.8	15	139.0	30	179.7	22	143.9
			50	26	158.4	22	141.7	15	135.4	26	158.4	22	141.8
B18	2752/173	405.1/825.1	30	150	1010.1	120	896.4	104	849.1	155	1033.8	120	897.2
			40	147	988.9	118	872.3	99	817.0	153	1024.6	118	873.1
			50	144	974.1	118	856.6	90	772.5	149	1003.4	118	857.4
B22	583/42	354.2/690.2	30	32	202.8	24	171.5	21	191.3	33	207.6	24	171.7
			40	32	202.7	24	169.1	21	186.9	32	202.8	24	169.3
			50	31	197.6	24	165.3	21	179.9	32	202.7	24	165.5
F31	273/345	268.8/1268.5	30	13	80.5	13	77.1	11	72.1	13	80.5	13	77.2
			40	13	80.5	13	75.8	7	57.2	13	80.5	13	75.9
			50	7	50.9	7	47.4	7	47.4	7	50.9	7	47.4
F34	157/218	211.2/1137.5	30	30	171.8	24	136.9	21	128.5	30	171.8	24	137.0
			40	30	171.5	24	135.1	21	126.3	30	171.8	24	135.3
			50	30	171.5	24	133.3	18	112.9	30	171.5	24	133.5
Average Relative Values				1	1	0.86	0.89	0.68	0.85	1.07	1.05	0.87	0.90

* The columns indicated by "Area" represent the sum of the areas of ADBs, ADB control logic and resized buffers in μm^2.

8. ACKNOWLEDGMENTS

This work was supported by Basic Science Research Program through National Research Foundation (NRF) grant (No.2011-0029805), the Center for Integrated Smart Sensors funded by the Ministry of Education, Science and Technology as Global Frontier Project (CISS 2011-0031863) and supported by the MKE (Ministry of Knowledge Economy), Korea, under ITRC (Information Technology Research Center) support program supervised by NIPA (National IT Industry Promotion Agency) (NIPA-2012-H0301-12-1011).

9. REFERENCES

[1] C. J. Alpert, A. Devgan, and S. T. Quay, "Buffer insertion with accurate gate and interconnect delay computation," in *DAC*, 1999.

[2] J. Cong, C. Koh, and K. Leung, "Simultaneous buffer and wire sizing for performance and power optimization," in *ISLPED*, 1996.

[3] C. C. N. Chu and M. D. F. Wong, "An efficient and optimal algorithm for simultaneous buffer and wire sizing," *IEEE TCAD*, 1999.

[4] I.-M. Liu, T.-L. Chou, A. Aziz, and M. D. F. Wong, "Zero-skew clock tree construction by simultaneous routing, wire sizing and buffer insertion," in *ISPD*, 2000.

[5] T. Okamoto and J. Cong, "Buffered steiner tree construction with wire sizing for interconnect layout optimization," in *ICCAD*, 1996.

[6] J.-L. Tsai, T.-H. Chen, and C.-P. Chen, "Zero skew clock-tree optimization with buffer insertion/sizing and wire sizing," *IEEE TCAD*, 2004.

[7] K. Wang, Y. Ran, H. Jiang, and M. Marek-Sadowska, "General skew constrained clock network sizing based on sequential linear programming," *IEEE TCAD*, 2005.

[8] S. Hu and J. Hu, "Unified adaptivity optimization of clock and logic signals," in *ICCAD*, 2007.

[9] V. Khandelwal and A. Srivastava, "Variability-driven formulation for simultaneous gate sizing and post-silicon tunability allocation," in *ISPD*, 2007.

[10] J.-L. Tsai and L. Zhang, "Statistical timing analysis driven post-silicon-tunable clock-tree synthesis," in *ICCAD*, 2005.

[11] E. Takahashi, Y. Kasai, M. Murakawa, and T. Higuchi, "A post-silicon clock timing adjustment using genetic algorithms," in *Symposium on VLSI Circuits*, 2003.

[12] S. Tam, S. Rusu, U. Nagarji Desai, R. Kim, J. Zhang, and I. Young, "Clock generation and distribution for the first IA-64 microprocessor," *IEEE JSSC*, 2000.

[13] Y.-S. Su, W.-K. Hon, C.-C. Yang, S.-C. Chang, and Y.-J. Chang, "Value assignment of adjustable delay buffers for clock skew minimization in multi-voltage mode designs," in *ICCAD*, 2009.

[14] ——, "Clock skew minimization in multi-voltage mode designs using adjustable delay buffers," *IEEE TCAD*, 2010.

[15] K.-Y. Lin, H.-T. Lin, and T.-Y. Ho, "An efficient algorithm of adjustable delay buffer insertion for clock skew minimization in multiple dynamic supply voltage designs," in *ASPDAC*, 2011.

[16] K.-H. Lim and T. Kim, "An optimal algorithm for allocation, placement, and delay assignment of adjustable delay buffers for clock skew minimization in multi-voltage mode designs," in *ASPDAC*, 2011.

[17] N. J. A. Kapoor and S. P. Khatri, "A novel clock distribution and dynamic de-skewing methodology," in *ICCAD*, 2004.

[18] T.-Y. Kim and T. Kim, "Clock tree synthesis for TSV-based 3D IC designs," *ACM ToDAES*, 2011.

Smart Non-Default Routing for Clock Power Reduction

Andrew B. Kahng[†‡], Seokhyeong Kang[†] and Hyein Lee[†]

[†]ECE and [‡]CSE Departments, University of California at San Diego

abk@ucsd.edu, shkang@vlsicad.ucsd.edu, hyeinlee@ucsd.edu

ABSTRACT

At advanced process nodes, *non-default routing rules* (NDRs) are integral to clock network synthesis methodologies. NDRs apply wider wire widths and spacings to address electromigration constraints, and to reduce parasitic and delay variations. However, wider wires result in larger driven capacitance and dynamic power. In this work, we quantify the potential for capacitance and power reduction through the application of *"smart" NDR* (SNDR) that substitute narrower-width NDRs on selected clock network segments, while maintaining skew, slew, delay and EM reliability criteria. We propose a practical methodology to apply smart NDRs in standard clock tree synthesis flows. Our studies with a 32/28nm library and open-source benchmarks confirm substantial (average of 9.2%) clock wire capacitance reduction and an average of 4.9% clock switching power savings over the current fixed-NDR methodology, without loss of QoR in the clock distribution.

Categories and Subject Descriptors

B.7.2 [**Hardware**]: INTEGRATED CIRCUITS—*Design Aids*; J.6 [**Computer Applications**]: COMPUTER-AIDED ENGINEERING

General Terms

Algorithms, Design, Performance

Keywords

Clock Network Synthesis, Clock Network Optimization, Power Minimization

1. INTRODUCTION

Clock distribution is well-known to have a large impact on integrated-circuit performance, area and power consumption. From the 40nm node onward, *non-default routing rules* (NDRs) have become an integral element of clock tree synthesis (CTS) methodology, as a means of reducing electromigration (EM) violations and delay variations. NDRs specify per-net, per-layer requirements for the router to use wiring geometries that differ from the default single-width, single-spacing (1W1S) configuration. Example NDRs might include (2W2S) (double-width, double-spacing), (1W3S), (4W2S), (3W3S), etc., where W denotes width and S denotes spacing. Figure 1 illustrates sample NDRs, along with their respective routing track costs when wire segments are centered on track gridlines, as in the outputs of modern detailed routers.

At today's leading-edge process nodes, NDRs are employed in CTS for several basic reasons.

Permission to make digital or hard copies of all or part of this work for personal or classroom use is granted without fee provided that copies are not made or distributed for profit or commercial advantage and that copies bear this notice and the full citation on the first page. To copy otherwise, to republish, to post on servers or to redistribute to lists, requires prior specific permission and/or a fee.
DAC'13, May 29 - June 07 2013, Austin, TX, USA.

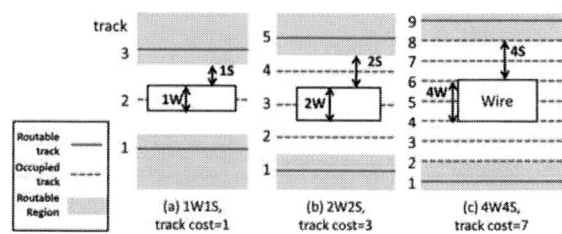

Figure 1: Illustration of three example NDRs, showing the track cost (number of tracks unavailable to (1W1S) wires) given the required spacing to neighbor wires. We assume that the detailed router centers each wire segment on a track gridline, and that track pitch = 2 × minimum width = 2 × minimum spacing; (a) (1W1S) NDR, with track cost 1; (b) (2W2S) NDR, with track cost 3; and (c) (4W4S) NDR, with track cost 7.

- Signal electromigration (EM) limits are violated by minimum-width wires when large buffers (e.g., 32X) are used to drive large fanouts (e.g., anywhere from 16 to 40 loads for each clock buffer instance in a typical buffered clock tree solution). To satisfy EM limits, wider wiring must be used.[1]

- Smaller geometries and more resistive interconnects, in conjunction with multi-patterning in lithography, result in higher parasitic and delay variability. Wider wires are less sensitive to these variations.

- Coupling capacitance, and therefore coupling-induced delay uncertainty, will be reduced with larger spacing [11].

- Though wider wires have higher capacitances, overall delays tend to be less due to reduced resistances.

For all practical purposes, modern use of NDRs means that IC designs intentionally spend extra wire capacitance nearly everywhere in the clock distribution network. Within a typical clock tree solution, the clock *subnet* driven by a given clock buffer[2] will be routed entirely with an NDR, except for the few microns of (1W1S) wiring needed to connect to input pins of the buffer's fanouts (loads). This increases capacitance and clock dynamic power. Given the need to reduce overall IC power consumption without sacrificing performance – particularly in mobile applications – we revisit the classic idea of optimizing wire width to reduce wire capacitance and dynamic power.

Motivating Studies

Consider a clock subnet driven by a large clock buffer. It is reasonable for the wiring of the subnet incident to the source (driver) to be wide: a large amount of downstream capacitance is being switched,

---

[1]In light of random variation models and on-chip variation-aware (OCV) signoff, clock distribution methodologies usually seek to minimize source-to-sink insertion delays. This implies a trend toward fewer levels and larger (stronger) drivers in the clock topology.

[2]We say that a buffered *clock tree* consists of a number of *clock subnets*. Each subnet is driven by a buffer and has fanouts that are either other clock buffers or clock sinks (flip-flops). A routed subnet consists of a number of *edges*.

and this takes more time (with more current flow) with more downstream load. However, as we follow the subnet's wiring topology away from the source, the number of downstream loads continues to decrease with every branching of the topology, until each leaf segment in the subnet's routing tree has only one downstream load. Figure 2 summarizes statistics for 64-sink clock subnets routed by *Cadence Encounter DIS V10.1* [24]. The number of downstream loads (reflecting the average current) is highest near the source (driver) buffer, but rapidly decreases for the overwhelming majority of the clock subnet wirelength. Thus, there is no (electromigration-reliability) reason for a 2W or 3W NDR to persist for the entire routing topology of a given clock subnet.

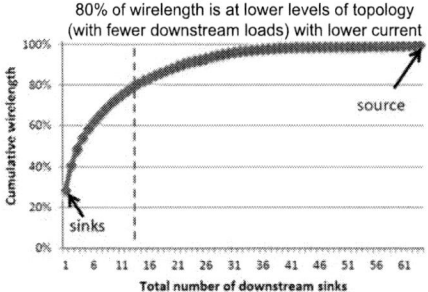

Figure 2: Study of 64-sink subnets routed by *Cadence Encounter DIS V10.1* [24]. Approximately 80% of clock subnet wirelength is in the lower levels of the tree, with few downstream loads and lower currents. The wire segment incident to the source has highest current but on average is only a small fraction of a subnet's total wirelength.

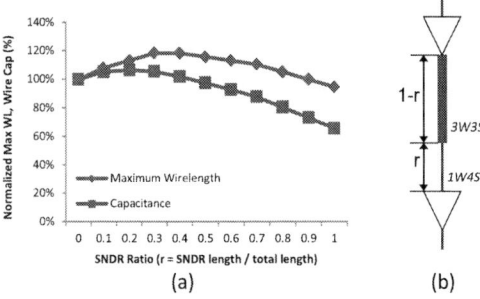

Figure 3: (a) Electrical performance when an equivalent-pitch SNDR (1W4S) is applied to a fraction r of a wire with a given original NDR (3W3S). Shown are maximum wirelength possible while satisfying a prescribed slew constraint, and total wire capacitance (both values normalized, and plotted against the y-axis). The ratio $r = 0.9$ achieves 27% capacitance reduction without incurring any reduction of maximum wirelength. (b) Illustration of tapering of wire from the original NDR (blue color) to the SNDR (red color). SPICE simulation is used to determine delay and slew with different SNDR ratios r.

Further motivation is obtained from SPICE experiments that evaluate the potential for SNDR-based capacitance reduction *without loss of electrical performance*. We study a symmetric H-tree [1] with a large buffer as driver, and 16 identical buffers as sinks. We sweep an *SNDR ratio*, r (i.e., wirelength with a given SNDR, divided by total wirelength – see Figure 3 (b)), to observe the impacts of SNDR wire tapering from different fixed original NDR widths and spacings. We assume that the signal transition at the input pin of the driver cell has 50ps slew time, and we find the maximum wirelength which maintains the same slew time at the end of the wire (i.e., at sink input pins), as well as the corresponding total wire capacitance.[3] For example, Figure 3 shows maximum wirelength under the 50ps maximum slew time constraint, and wire capacitance (both values normalized and plotted against the y-axis), when a (1W4S) SNDR is applied to save capacitance from an equivalent-pitch (3W3S) original NDR. The library and interconnect technol-

ogy used to generate the figure are from a public-domain 32/28nm PDK [26]; 32X buffers are used for driver and sinks. In this case, application of the SNDR can reduce wire capacitance by 27% "for free" - that is, with zero decrease in maximum driven wirelength, at $r = 0.9$. Alternatively, wire capacitance can be reduced by 34% at the cost of 5% decrease in maximum driven wirelength, at $r = 1.0$.

With the above studies as our starting point, we explore the dichotomy between today's *fixed NDR* methodology and a possible *smart NDR* methodology. The cartoon of Figure 4 shows that *fixed NDRs* on clock subnets result in larger driven capacitance, potentially leading to larger currents, larger drivers, and increased dynamic power. (Indeed, if the larger drivers violate signal EM limits, even the larger (wider) NDRs may be required.) By contrast, *smart NDRs* taper wire widths as the number of downstream loads decreases, reducing driven capacitance, dynamic power, and the number and size of buffers.

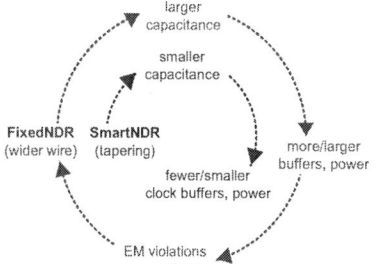

Figure 4: Intuitive dichotomy between today's "fixed NDR" methodology and a potential "smart NDR" methodology.

This Work

In this work, we assess the potential for capacitance and power reductions through use of *"smart" NDRs* (SNDRs) that substitute narrower-width NDRs for selected clock net segments while maintaining skew, slew, delay and EM reliability criteria. We formulate the optimal application of SNDRs for a given clock subnet as a quadratically constrained program. We also demonstrate the practicality of a flow that applies SNDRs at the end of the CTS phase in a commercial place-and-route tool.

The main contributions of our work are summarized as follows.

- We perform studies of practical clock routing instances and NDRs to establish the potential for substantial dynamic power reductions using SNDRs.

- We formulate optimal application of SNDRs as a quadratically constrained program to minimize wire capacitance under skew, slew, delay and EM constraints, and we efficiently solve this problem for each subnet of a given clock tree.

- We extend our SNDR solution approach to the entire clock tree by propagating skew constraints from downstream subnets to upstream subnets.

- We also propose a practical flow to apply SNDRs transparently post-CTS within a standard commercial place-and-route tool.

- We empirically confirm an average of 16% clock wire capacitance reduction, and 5% total clock power reduction, achieved by our proposed technique, as compared to traditional CTS approaches with fixed NDRs.

The remainder of this paper is organized as follows. In Section 2, we give an overview of related literature. Section 3 formulates the SNDR problem, and Section 4 describes our wire-tapering approach. Section 5 provides experimental results and analysis. We give conclusions and ongoing research directions in Section 6.

---

[3]Because the H-tree is symmetric, greater wirelength means that the sinks of the H-tree are spaced farther apart.

2. RELATED WORK

Wire width optimization (tapering) for clock and signal distribution has been extensively studied since the early 1990s. In this section, we survey related literature according to three main categories: (1) works on wire width optimization in clock trees; (2) electromigration-constrained wire sizing methods; and (3) coupling noise-driven wire sizing methods.

(1) Wire sizing in clock trees. Many works ([19], [9], [23], [15], [14], [17]) have applied wire sizing to minimize skew in clock trees. Tsai et al. [19] propose a dynamic programming method for simultaneous buffer insertion and wire sizing to optimize delay and power of a given zero-skew or useful-skew clock tree. They calculate a feasible delay-capacitance region for all nodes in a bottom-up phase, then determine buffer locations/widths and wire widths. Guthaus et al. [9] propose sequential linear programming as well as quadratic programming based clock buffer/wire sizing to minimize skew. The former technique uses first-order sensitivities for a small region of buffer/wire size solutions to represent a nonlinear objective using a set of linear functions. Zhu et al. [23] perform wire sizing to minimize skew using Gauss-Marquardt least-squares minimization. Pullela et al. [15] use wire width widening to reduce clock skew, delay and process variability impact. Their method starts with a minimum-delay tree and then optimizes skew by widening wires based on a delay sensitivity to wire width derived from Elmore delay [8]. In subsequent work [17], they suggest an analytical moment-sensitivity based methodology, using the moments of the transfer function of an RC tree circuit, to simultaneously reduce skew and slew. Liu et al. [14] also propose simultaneous clock routing, wire sizing and buffer insertion based on the Deferred-Merge Embedding algorithm [2] [3].

To our understanding, previous literature on clock tree wire sizing emphasizes timing optimization; exceptions are [19] and [9], the latter of which incorporates a power constraint.[4] In contrast to previous works, use of NDRs in clock routing is now largely driven by *signal EM reliability* limits. Moreover, previous works have various limitations such as continuous sizing that limit application within conventional physical design methodologies.

(2) Electromigration-constrained wire sizing. The literature on EM-constrained wire sizing has centered on power/ground networks. Tan et al. [20] suggest sequential linear programming-based wire sizing that considers IR drop, EM and other design constraints while minimizing area. A nonlinear function of constraints is relaxed to a sequence of linear programs which always enables convergence to an optimum solution. Wu et al. [22] propose a power/ground wire sizing algorithm with IR drop, EM and minimum-width constraints. A penalty method is used to solve the nonlinear problem so as to achieve area minimization of the power/ground network. For clock trees, Pullela et al. [16] consider an EM constraint in their low-power clock tree design methodology. They optimize a clock tree with buffer insertion by decreasing wire width while satisfying bounds on process variation-dependent skew and current density. Among all previous works of which we are aware, the work of [16] is the closest to our present target; however, buffer insertion is not our objective as it potentially draws more power and causes more EM violations in a vicious cycle (recall Figure 4).

(3) Noise-driven wire sizing. Last, wire sizing has been analyzed in conjunction with spacing to address coupling noise sensitivity. Cong et al. [5] propose symmetric and asymmetric wire sizing and spacing to minimize the weighted sum of coupling capacitance-induced delay at all sinks. Lagrangian relaxation is used to solve

the same problem as in [4], but it is shown that proper wire sizing can further reduce delay when coupling capacitance is considered. In a subsequent work, Cong et al. [6] study a simultaneous wire spacing problem for multiple nets, to deal with crosstalk noise. In our methodology, although we do not address the noise problem directly, by increasing wire spacing, we reduce the coupling capacitance (and coupling-induced delay variation) as well.

3. PROBLEM FORMULATION

We seek to apply different NDRs (i.e., *smart NDRs*) to each wire segment of a given routed clock net, to reduce total wire capacitance and dynamic power subject to reliability and electrical requirements. More specifically, we want to minimize tree capacitance without violating slew, skew, clock source-to-sink insertion delay, and EM reliability constraints. Our focus is on an "iso-area" use model that is *transparent* to existing CTS flows: we start with a routed CTS solution, then identify a set of wire segment width reductions that save power without affecting any (skew, slew, etc.) solution metric.[5] Below, we describe such a transparent flow, which imports implemented SNDRs back into the P&R tool as DRC-clean modified DEF (Design Exchange Format, [30]) for extraction and performance analysis, with no ECO routing needed for any other nets.

Table 1 presents the notations that we use to describe our problem formulation and algorithm. Upper bounds on slew (transition) time, skew, and insertion delay are respectively denoted by U_S, U_K, and U_L. We use e to indicate an edge (between clock buffers, clock sinks, or Steiner points) in the clock tree. Given a clock tree T with edges $e \in T$, a set N of allowed SNDRs (indexed as n) with maximum current limit $E_{e,n}$ for edge e with SNDR $n \in N$, and electrical performance bounds U_S, U_K and U_L, our SNDR optimization seeks to map each edge width w_e to an SNDR $n \in N$ while minimizing total wire capacitance, subject to the constraints $E_{e,n}$, U_S, U_K and U_L.

$$\textbf{Minimize:} \quad \sum_{e \in T} C_e$$

Subject to:

$$S_v \le U_S, \quad I_e \le E_{e,w_e}, \quad L_v \le U_L, \quad K_{u,v} \le U_K,$$
$$w_e \in N, \quad w_e > w_{desc(e)}, \qquad (\forall v, e \in T) \qquad (1)$$

Table 1: Notation

Notation	Meaning
C_e	capacitance of edge e
S_v	slew at the node v
I_e	average current of edge e
w_e	NDR of edge e, $w_e \in N$
x_e	the alternative NDR variable, $x_e = ln(w_e)$
$D_{u,v}$	delay from node u to node v
$K_{u,v}$	skew between node u and node v
L_v	clock latency at sink v
T	given clock tree
N	set of NDRs
E_{e,w_e}	maximum current limit for edge e with NDR $w_e \in N$
U_S	maximum slew constraint
U_K	maximum skew constraint
U_L	maximum clock latency constraint
$desc(e)$	set of all downstream sinks of e

We conclude this section with two comments on the extensibility of our SNDR problem formulation. First, the benefits of SNDR optimization can encompass area, routing congestion, and/or coupling between routes, since the use of SNDRs decreases the track consumption of the clock routing. Realizing these reductions requires ECO routing, and possibly ECO placement as well; we do not implement such a flow in our present work. However, we do report results in Section 5.2 suggesting that a "reduced-area" SNDR

---

[4]Of separate interest is the literature on wire sizing in general signal routing trees. These works are exemplified by Chen et al. [4], which targets signal routing and does not consider skew. The authors of [4] apply Lagrangian relaxation for gate and wire sizing to minimize area subject to a maximum delay bound.

---

[5]Section 5.2 also considers a use model where track usage is allowed to decrease as a result of SNDR application.

optimization, where wire widths are reduced and spacings can take on any value that does not result in wire capacitance exceeding that of the original fixed NDRs, can also significantly reduce track consumption. Second, analysis and enforcement of electrical constraints readily extend to include coupling noise, noise-induced delay variation, process variation, and a number of other concerns. However, the basic optimization will remain the same as what we study, and we leave such extensions to future work.

4. SNDR WIRE SIZING

4.1 Wire RC Delay Model for SNDR

RC modeling of wire is given by Equation (2), where l_e, w_e and s_e are the length, width and spacing of edge e, respectively.

$$R_e = \rho \cdot \frac{l_e}{w_e}, \quad C_e = \varepsilon \cdot \frac{l_e w_e}{s_e} \tag{2}$$

We assume $w_e + s_e$ is a constant Y equal to a fixed track pitch. Then, $s_e = Y - w_e$. To obtain linear formulations in w_e, we approximate R per μm and C per μm as functions of $x_e = ln(w_e)$[6]. Then, R_e and C_e can be approximated as

$$R_e = (\alpha_R \cdot x_e + \beta_R) \cdot l_e, \quad C_e = (\alpha_C \cdot x_e + \beta_C) \cdot l_e \tag{3}$$

where α_R, β_R, α_C and β_C are fitting coefficients which we obtain by linear regression. Figure 5 shows that our suggested model is fairly accurate with measurements[7]. If the range of wire width increases, our model may not be applicable. However, these models are reasonable for the general wire width range limited by recent process technologies.

We use the Elmore delay model [8] to calculate the delay of clock tree. The delay between node u and v is

$$D_{u,v} = \sum_{e \in P_{u \to v}} \left(R_e \cdot \sum_{i \in desc(e)} C_i \right) \tag{4}$$

where $P_{u \to v}$ is the path from node u to v.

By substituting clock source s to u in Equations (3) and (4), we get the clock latency at node v as

$$L_v = \sum_{i \in P_{s \to v}} ((\alpha_R \cdot x_i + \beta_R) \cdot l_i \sum_{j \in desc(i)} (\alpha_C \cdot x_j + \beta_C) \cdot l_j) \tag{5}$$

For wire slew calculation, we apply the PERI model [10]. The slew at node v, where s is the clock source, is

$$S_v = \sqrt{S_s{}^2 + \ln 9 \cdot D_{s,v}{}^2} \tag{6}$$

where S_s is the output slew of clock buffer at the clock source. The output slew of clock buffers depends on the input slew and the output capacitances. Since the input slew of clock buffer is the output slew of the upstream net which is constrained by U_S, we can assume U_S to be the worst case value. With a constant input slew U_S, we use a piecewise linear model to approximate the output slew as a function of the total output capacitance.

$$S_s = \alpha_S \cdot \sum_{e \in desc(s)} C_e + \beta_S \tag{7}$$

where α_S and β_S are fitting coefficients. We assume the total output capacitance before optimization as the initial capacitance. This assumption is pessimistic since the total output capacitance will be minimized after applying SNDR. However, for large sizes of buffers, which are typically used in clock trees, the output slew is not very sensitive to the output capacitance. Thus, with reasonable

pessimism, we obtain a constant S_s with given U_S and the initial total capacitance.[8]

$$K_{u,v} = |D_{s,u} - D_{s,v}| \tag{8}$$

The skew constraint should be checked for all pairs of source-to-sink timing paths with the upper bound U_K.

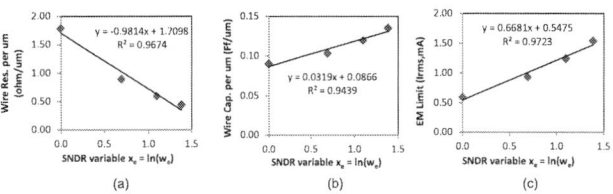

Figure 5: (a) Wire resistance per unit length (b) wire capacitance per unit length, and (c) EM limit (I_{RMS}); each is fitted as a linear function of $x_e = ln(w_e)$, where w_e is the wire width in the limited range that we are using.

4.2 EM Current Model and EM Rule

For EM constraints, we use a simplified I_{RMS} model derived from Black's Equation [12]. If V_{dd} is the operating voltage, F is the operating frequency and sw is the switching activity on the net,

$$I_{RMS}(e) = \sum_{i \in desc(e)} C_i \cdot V_{dd} \cdot F \cdot \sqrt{sw} \tag{9}$$

where $I_{RMS}(e)$ is the root mean square current of the edge e. We use Synopsys 32/28nm PDK [26], where the I_{RMS} limit is defined as a polynomial function of wire width and the metal layer,

$$EMLimit(w) = \alpha_E w^2 + \beta_E w + \gamma_E \tag{10}$$

where α_E, β_E and γ_E are fitting coefficients. We obtain a linear function of x_e as shown in Equation (11), where α'_E and β'_E are fitting coefficients. For the ranges of SNDRs considered, this model accurately captures characterized values as shown in Figure 5.

$$E_{e,w_e} = \alpha'_E x_e + \beta'_E \tag{11}$$

4.3 Iterative Linear Programming (LP)

To avoid quadratic constraints arising from the Elmore delay model, we separate the sizing problem into two (alternating) linear programs by fixing the R values and the C values in alternation. Our method recalls the rescaled simple iteration of [21], and has the following steps.

Delay constraints are formulated based on Equation (5). First, we fix x_i with a constant $x_{e,0}$ (= initial NDR value) and formulate a linear function of x_j ($Constr_C$). In a similar way, we fix x_j with a constant x_0 and formulate a linear function of x_i ($Constr_R$). And then, we solve the problem with both the constraints ($Constr_R$ and $Constr_C$) simultaneously using the objective function (Equation (1)). Second, we formulate the constraints in the same way as the first step, but with different initial values $x_{e,1}$, which are the solutions derived from the previous step. We iteratively solve the problem until all x_e for each edge e are determined. In our experiments, our approach obtained results that are essentially identical to those of the QCP-based method, but with runtime reductions of $6 \times$ to $30 \times$[9].

4.4 Applying SNDR to an Entire Clock Tree

In subsections (Section 4.1 $\sim$ 4.3), we have formulated the SNDR problem, and proposed an iterative method to optimize a subtree of clock (a single net). To optimize the entire clock tree, we perform our SNDR method from the downstream to upstream subnets of clock tree. Algorithm 1 presents pseudocode of our SNDR flow for

---

[6]We note that R and C are respectively proportional to, and inversely proportional to, w_e. Taking logs transforms multiplier and divider into first-order terms.

[7]The maximum errors that we have seen are -22% for R and -4% for C, for the SNDR sets that we study.

[8]We do not perform buffer sizing since it can lead to larger delay variations and degrade EM when buffers are upsized.

[9]We have compared the runtime with testcase wb_dma_top; the overall runtimes of the QCP-based method and LP-based method are 170 minutes and 29 minutes respectively with a similar quality of solution.

the entire clock tree. We apply SNDR to each subtree with EM, slew constraints, skew constraints and delay margin to generate a tapered clock tree. $D_{init(u,v)}$ is the initial delay from node u to v before applying SNDR. We collect the set of delay constraints (Line 6) with the wire delays. To calculate clock skew, we define $D_{min(u)}$ and $D_{max(u)}$, which are the minimum and maximum path delay from node u to its leaf nodes (flip-flops). $path\_delay_{min}(u,v)$ and $path\_delay_{max}(u,v)$ are the minimum and maximum path delay from u to leaf nodes through node v, and they are used for the skew constraints (Line 17). After collecting the delay and skew constraints for each subtree, we apply SNDR ($SNDR_{sub}$) to the subtree t (Line 22). $SNDR_{sub}$ finds an optimal NDR for each wire segment in the subtree by solving the problem of Equation (1) using the Iterative LP described in Section 4.3. Then, we update the $D_{min(u)}$ and $D_{max(u)}$, which are calculated recursively from downstream subnets to upstream subnets (Line 23-26). For $cell\_delay(v)$, we get the worst buffer delay which is calculated with the initial capacitances before optimization. EM and slew constraints are checked net by net independently in $SNDR_{sub}$.

Algorithm 1 SNDR Wire Sizing for Clock Tree

Procedure $SNDR(T, \{E\}, U_K, U_S, M)$
Input : clock tree T, a set of EM constraints $\{E\}$, skew constraints U_K, slew constraints U_S and delay margin M;
Output : tapered clock tree with SNDR
1: $T_n \leftarrow$ all subtrees (= routed subnets) in T;
2: **while** $T_n \neq \emptyset$ **do**
3:     Pick a subtree t with maximum level;
4:     $u \leftarrow$ source node of t;
5:     **for all** sink nodes v of t **do**
6:         $DelayConst_n \leftarrow DelayConst_n \cup \{D_{u,v} < D_{init(u,v)} + M\}$;
7:         $SlewConst_n \leftarrow SlewConst_n \cup \{S_v < U_S\}$;
8:         **if** v is a clock port of flip-flop **then**
9:             $path\_delay_{max}(u,v) \leftarrow D_{u,v}$;
10:           $path\_delay_{min}(u,v) \leftarrow D_{u,v}$;
11:         **else**
12:           $path\_delay_{max}(u,v) \leftarrow D_{max(v)} + cell\_delay(v) + D_{u,v}$;
13:           $path\_delay_{min}(u,v) \leftarrow D_{min(v)} + cell\_delay(v) + D_{u,v}$;
14:         **end if**
15:     **end for**
16:     **for all** two sink nodes v and w pair of t **do**
17:         $SkewConst_n \leftarrow SkewConst_n \cup \{path\_delay_{max}(u,v) - path\_delay_{min}(u,w) < S\}$;
18:     **end for**
19:     **for all** edges e of t **do**
20:         $EMConst_n \leftarrow EMConst_n \cup \{I_e < E_{e,w_e}\}$;
21:     **end for**
22:     $SNDR_{sub}(t, DelayConst_n, SkewConst_n, SlewConst_n, EMConst_n)$;
23:     **for all** sink nodes v of t **do**
24:         $D_{max(u)} \leftarrow max(D_{max(u)}, path\_delay_{max}(u,v))$;
25:         $D_{min(u)} \leftarrow min(D_{min(u)}, path\_delay_{min}(u,v))$;
26:     **end for**
27:     $T_n \leftarrow T_n - t$;
28: **end while**

Algorithm 1 can be used to minimize skew by setting the skew constraints to near zero and/or adding weights for aggressive optimization in the objective function. *SNDR* then finds solutions which minimize skew under given EM, slew and delay constraints.

4.5 Further Optimization with More SNDRs

Recall that our proposed methodology focuses on transparency to existing CTS flows, i.e., we perform the wire sizing after routing, potentially just before design closure and signoff. Thus, we limit our selection of SNDRs to constant track cost, so as not to harm the original design with noise coupling or with routing ECOs. This being said, if we allow additional options for NDRs, for example, smaller spacing rules, then other optimizations such as reduction of wire congestion with ECO routing become possible. Figure 6 shows the capacitance per μm with various SNDRs (the values are extracted from the capacitance tables used in *Cadence Encounter DIS V10.1* [24].). Though coupling capacitance increases as spac-

ing decreases, the total capacitance of each SNDR is below that of the original NDR (4W5S) while maintaining the original spacing without (3W3S) case. Clearly, there is an available tradeoff between area and coupling noise along with various spacing options. Given this, it is possible to tune the objective function to maximize the benefits from SNDR across a range of power-, area- and noise-constrained designs.

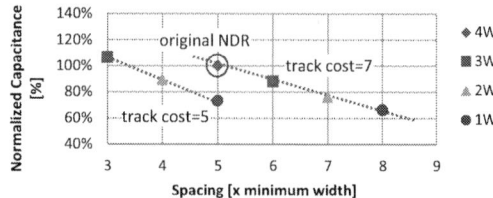

Figure 6: % reduction of capacitance per unit length with various SNDRs.

4.6 Implementation Flow

We propose a practical flow that implements SNDRs using a commercial place-and-route tool, as illustrated in Figure 7. For an input design, we perform clock tree synthesis with a given fixed NDR. After routing, we extract the locations of buffers and flip-flops, which are the sources and sinks of clock nets, along with all clock wire segments, from the routed DEF [30] of the implemented design. From this information, we construct the optimization instances (Equation (1)) with appropriate timing (delay, skew, slew) and EM constraints. Solving each optimization instance entails (i) calculating a matrix of coefficients for the equations governing the solution of width variables (one variable for every edge in each given subnet) with timing, EM reliability and tapering constraints, and (ii) applying Iterative LP to solve these problem instances and obtain wire sizing solutions for every edge of each given subnet. We then update the original DEF file with the obtained wire sizing solutions for each subnet. The modified DEF is read back into the P&R tool, and RC values are recalculated. With the updated RC values, we check for any electrical/timing or EM reliability violations; if there are any violations caused by an SNDR on a wire segment, we revert that segment to its original NDR.

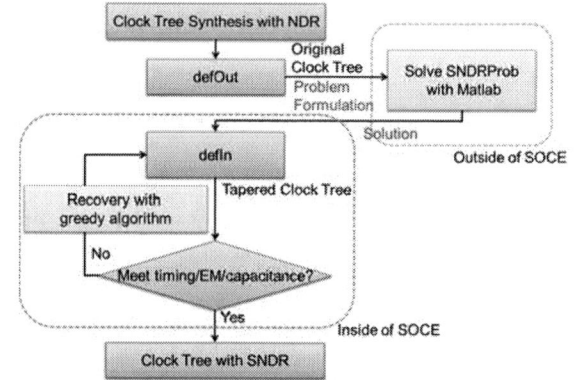

Figure 7: Overall implementation flow.

5. EXPERIMENTAL SETUP AND RESULTS

5.1 Experimental Setup

We extract wire resistance and capacitance for different wire NDRs, starting from capacitance table, technology LEF and ITF standard inputs from the relevant PDK. For our experiments, we use nine open-source designs from the *OpenCores* website [29]. We use the *Synopsys 32/28nm PDK* [26] cell library for the design implementation. We synthesize the designs using *Synopsys DesignCompiler vF-2011.09* [27] and perform place-and-route with

642

Cadence Encounter DIS v10.1 [24]. We solve the wire sizing problem formulated above using *Mathworks MATLAB R2012b* [25]. Insertion delay constraints are set by adding a small (5%) margin to the original insertion delay of each sink; this improves flexibility of the SNDR assignment with negligible impact on the final insertion delay, as shown in Table 4. We use the maximum skew of the original design as the skew constraint, and the maximum transition time of the original design with a margin (less than 5% of clock period) as the maximum transition time (slew) constraint. EM constraints are obtained from the Synopsys 32/28nm PDK.[10] For simplicity, we assume that the track constraint is a constant, i.e., we perform an "iso-area" optimization; the corresponding set of NDRs is (4W5S), (3W6S), (2W7S) and (1W8S). We allow the optimization to use only smaller-width NDRs to replace the original NDR of the clock tree, i.e., we only taper the wire. The clock tree is initially routed with the maximum wire width NDR (4W5S). We note that in this particular technology, use of a smaller-width NDR in the initial implementation can result in EM violations after CTS.

5.2 Experimental Results

We have performed the SNDR optimization on nine benchmark designs to assess capacitance and power reductions afforded by use of SNDRs. Table 2 shows the number of instances, sinks and clock buffers for each testcase.

Table 2: Testcases: number of instances, sinks and clock buffers.

testcase	#instances	#sinks	#clock buffers
aes_cipher_top	21766	530	11
eth_top	11342	1743	37
jpeg_encoder	67185	4512	100
mc_top	6525	934	27
mpeg2_top	13026	2931	59
tv80s	6478	359	25
usbf_top	11446	1515	31
wb_conmax_top	32559	770	25
wb_dma_top	3035	523	15

Figure 8 shows the proportions of total clock wirelength routed with each NDR after applying our wire sizing optimization. From the results, we see that more than 80% of the wiring is replaced by smaller NDRs.

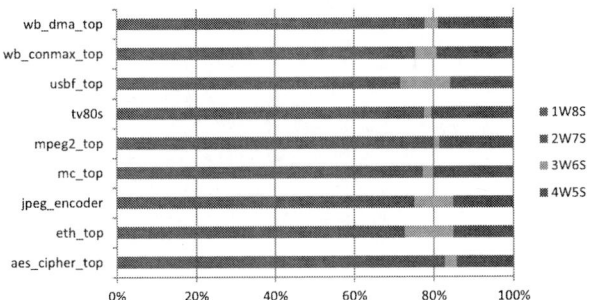

Figure 8: Proportions of total clock tree wirelength routed with different NDRs after optimization.

Table 4 shows the post-optimization implemented results and clock network power reduction versus the original (conventional) designs which use fixed NDRs of (4W5S) in Table 4(a) and (2W4S) in Table 4(b). Runtime for the *MATLAB* implementation of our algorithm is between 10 seconds and ~100 minutes per subnet on a 2.5GHz Intel Xeon processor; this runtime varies widely depending on the number of sinks, the number of wire segments, and the number of constraints. Each testcase requires sequential optimization of subnets, but it is simple to parallelize the solution of subnets that

---

[10]Maximum skew constraints of 3-18ps are used except for jpeg_encoder (110ps). Slew constraints of 47-100ps are used except for eth_top, jpeg_encoder, mpeg2_top (> 200ps). EM constraints are calculated using Equation (11).

---

are independent of each other. Importing modified DEF back into the P&R tool never requires more than 10 seconds for any testcase. The "delta" values in the rightmost three columns of the table are reported in *picoseconds*, i.e., the insertion delay, slew and skew changes are negligible (moreover, negative delta values represent improvements in these parameters). We see that our SNDR flow can reduce clock wire capacitance over the conventional designs by up to 10.3% (average 9.2%) when (4W5S) is used as the fixed NDR. With the reduced wire capacitance, we achieve up to 5.7% (average 4.9%) and 4.2% (average 3.4%) clock switching power and total power savings[11] with essentially zero EM violations or timing degradations. We have also applied SNDRs for the case that (2W4S) is used as the fixed NDR. In this case, SNDRs have just two options, i.e., (2W4S) or (1W5S). Wire capacitances reduce by up to 5.9% (4.7% on average), which yields up to 2.8% (average 2.2%) and 2.1% (average 1.5%) clock switching power and total power savings. We note that our results correspond to a viable flow implemented in a widely-used commercial EDA tool: all of the post-optimization SNDR-based clock routing is DRC-clean, and while the optimization is guided by our abstractions of delay and slew, all of the reported extraction and timing results, EM checks, etc. are according to the same commercial tool.

Table 3: Results using SNDRs with an upper bound on track cost of 7.

testcase	track cost reduction (%)	capacitance reduction (%)
aes_cipher_top	53.8	12.15
eth_top	46.8	13.14
jpeg_encoder	48.1	12.37
mc_top	49.1	12.45
mpeg2_top	52.5	12.14
tv80s	48.0	11.49
usbf_top	48.1	11.11
wb_conmax_top	49.7	13.16
wb_dma_top	52.7	14.28

Last, to assess the potential for area recovery using various SNDRs, we perform an additional study using a set of SNDRs which maintains track cost less than the original fixed NDR. That is, we explore "reduced-area" SNDRs of (1W2S), (2W4S) and (3W6S). While the original set of SNDRs all require 7 tracks, the new set of SNDRs requires ≤ 7 tracks. For example, a (1W2S) SNDR might be used where a (1W8S) SNDR was used in the previous experiment. We perform the same optimization using characterized RC models for the new set of SNDRs, with results summarized in Table 3. We see in the table that the new set of SNDRs actually achieves slightly improved capacitance reduction, while also saving up to 53.8% of routing tracks (49.9% on average), calculated as a weighted average along the entire routed wirelength of the clock tree.

6. CONCLUSIONS

The notion of wire tapering for clock tree (area, power, skew) optimization has been studied in the literature for nearly 20 years. However, tapering flows have not achieved traction in production IC implementation methodologies, possibly due to the lack of perceived benefit in combination with the complexity of flow development. For several reasons, we believe that wire width selection in CTS has become worth revisiting: clock power reduction is increasingly critical in a power-limited world; variability has driven CTS optimizations toward larger drivers and high-fanout clock nets; and EM reliability limits in particular (along with the desire to avoid time-consuming EM-fix steps in physical design) have motivated the widespread use of NDRs in modern CTS methodologies.

---

[11]Clock buffers and FFs are not changed. The power savings are achieved only with wire capacitance reduction.

Table 4: SNDR results: Capacitance, clock power, insertion delay, skew, and maximum slew. Δ values < 0 are improvements.

(a) With the default NDR (4W5S)

test case	wire cap. (pF)	switching power (mW)	total power (mW)	wire cap. reduction	switching power reduction	total power reduction	Δ clock latency (ps)	Δ clock skew (ps)	Δ max slew (ps)	EM vio.
aes_cipher_top	0.62	0.742	0.986	8.69%	5.41%	4.22%	-6.00	-3.00	0.40	1
eth_top	1.26	1.883	2.589	8.88%	4.46%	3.24%	-1.70	0.00	-0.30	0
jpeg_encoder	4.07	5.068	6.926	9.72%	5.43%	4.16%	-9.70	-19.00	-12.80	1
mc_top	0.68	0.933	1.445	9.40%	4.61%	3.11%	-4.80	-2.00	0.10	0
mpeg2_top	2.47	4.116	5.633	7.76%	4.20%	3.18%	-3.90	-2.00	0.70	0
tv80s	0.36	0.403	0.821	9.31%	5.35%	2.75%	-4.10	-2.00	0.80	0
usbf_top	1.16	2.016	2.780	8.44%	4.37%	3.24%	-1.70	-1.00	0.50	0
wb_conmax_top	0.71	1.031	1.587	10.11%	5.65%	3.72%	-10.10	-4.00	0.40	0
wb_dma_top	0.34	0.918	1.447	10.32%	4.94%	3.25%	-3.50	-2.00	0.50	0

(b) With the default NDR (2W4S)

test case	wire cap. (pF)	switching power (mW)	total power (mW)	wire cap. reduction	switching power reduction	total power reduction	Δ clock latency (ps)	Δ clock skew (ps)	Δ max slew (ps)	EM vio.
aes_cipher_top	0.49	0.651	0.892	4.85%	2.76%	2.10%	-2.30	-2.00	3.10	0
eth_top	1.24	1.866	2.657	4.13%	2.04%	1.54%	-1.80	-11.00	-0.80	0
jpeg_encoder	3.23	4.478	6.121	5.15%	2.57%	1.83%	-6.20	-15.00	-3.30	0
mc_top	0.53	0.831	1.338	4.28%	1.78%	1.12%	-1.80	-1.00	0.00	0
mpeg2_top	1.88	3.586	5.045	3.88%	1.84%	1.35%	-2.40	-4.00	-2.10	0
tv80s	0.27	0.346	0.760	3.71%	1.82%	0.88%	-1.20	0.00	0.50	0
usbf_top	0.89	1.773	2.537	5.88%	2.65%	1.85%	-5.00	-3.00	0.30	0
wb_conmax_top	0.57	0.919	1.471	4.86%	2.37%	1.56%	-3.60	-2.00	-0.20	0
wb_dma_top	0.27	0.829	1.354	5.10%	2.13%	1.40%	-1.50	0.00	0.30	1

In this work, we have assessed the potential for capacitance and power reduction from "smart NDRs" that substitute narrower-width NDRs for selected clock segments while maintaining all skew, slew, insertion delay and EM reliability criteria. We formulate a wire sizing problem of choosing per-segment NDRs to minimize the capacitance of the clock tree subject to electrical and reliability constraints, and we propose an effective solution to this problem for each subnet. We then extend the SNDR approach to the entire clock tree by propagating skew constraints from downstream to upstream subnets. We also propose a practical methodology to apply SNDRs without disruption of standard clock network synthesis flows.

With a 32/28nm library and open-source benchmark netlists, an "iso-area" execution of our SNDR methodology (scripted in a widely-used commercial EDA tool, with all extraction and analysis performed in the same tool) achieves up to 5.7% clock switching power savings (4.9% savings on average) and up to 10.3% (9.2% on average) clock network capacitance reduction versus the fixed-NDR (4W5S) methodology. A "reduced-area" execution of the SNDR methodology shows that substantial wiring resource (i.e., track usage) reductions are possible with lesser capacitance savings. Our ongoing work adds more detailed signal integrity analyses to both the optimization and validation aspects of our SNDR methodology. We are also extending our implementation to work with other commercial P&R (CTS) tools, and extending validations to additional technology libraries and design testcases. Additional research directions include the use of SNDRs not only to reduce power and routing costs, but also to control local skews at all levels of the buffered clock tree hierarchy for improved chip timing and robustness.

7. REFERENCES

[1] H. B. Bakoglu, *Circuits, Interconnections, and Packaging for VLSI*, Addison-Wesley, 1990.

[2] K. D. Boese and A. B. Kahng, "Zero-Skew Clock Routing Trees With Minimum Wirelength", *Proc. ASIC Conf.*, 1992, pp. 17-21.

[3] T.-H. Chao, Y.-C. Hsu and J.-M. Ho, "Zero Skew Clock Net Routing", *Proc. DAC*, 1992, pp. 518-523.

[4] C.-P. Chen, C. C. N. Chu and D. F. Wong, "Fast and Exact Simultaneous Gate and Wire Sizing by Lagrangian Relaxation", *IEEE TCAD* 18(7) (1999), pp. 1014-1025.

[5] J. Cong, L. He, C.-K. Koh and D. Z. Pan, "Interconnect Sizing and Spacing with Consideration of Coupling Capacitance", *IEEE TCAD* 20(9) (2001), pp. 1164-1169.

[6] J. Cong, D. Z. Pan and P. V. Srinivas, "Improved Crosstalk Modeling for Noise Constrained Interconnect Optimization", *Proc. ASP-DAC*, 2001, pp. 373-378.

[7] M. A. El-Mousy and E. G. Friedman, "Exponentially Tapered H-tree Clock Distribution Networks", *IEEE TVLSI* 13(8) (2005), pp. 971-975.

[8] W. C. Elmore, "The Transient Response of Damped Linear Networks with Particular Regard to Wideband Amplifiers", *J. Applied Physics* 19(1) (1948), pp. 55-63.

[9] M. R. Guthaus, D. Sylvester and R. B. Brown, "Clock Buffer and Wire Sizing Using Sequential Programming", *Proc. DAC*, 2006, pp. 1041-1046.

[10] S. Hu, C. J. Alpert, J. Hu, S. Karandikar, Z. Li, W. Shi and C. N. Sze, "Fast Algorithms For Slew Constrained Minimum Cost Buffering", *IEEE TCAD* 26(11) (2007), pp. 2009-2022.

[11] A. B. Kahng, S. Muddu, E. Sarto and R. Sharma, "Interconnect Tuning Strategies for High-performance ICs", *Proc. DATE*, 1998, pp. 471-478.

[12] A. B. Kahng, S. Nath and T. S. Rosing, "On Potential Design Impacts of Electromigration Awareness", *Proc. ASP-DAC*, 2013, to appear.

[13] R. Kay and L. T. Pileggi, "EWA: Efficient Wiring-sizing Algorithm for Signal Nets and Clock Nets", *IEEE TCAD* 17(1) (1998), pp. 40-49.

[14] I.-M. Liu, T.-L. Chou, A. Aziz and D. F. Wong, "Zero-Skew Clock Tree Construction by Simultaneous Routing, Wire Sizing and Buffer Insertion", *Proc. ISPD*, 2000, pp. 33-38.

[15] S. Pullela, N. Menezes and L. T. Pillage, "Reliable Non-Zero Skew Clock Tree Using Wire Width Optimization", *Proc. DAC*, 1993, pp. 165-170.

[16] S. Pullela, N. Menezes and L. T. Pillage, "Low Power IC Clock Tree Design", *Proc. CICC*, 1995, pp. 263-266.

[17] S. Pullela, N. Menezes and L. T. Pillage, "Moment-Sensitivity-Based Wire Sizing for Skew Reduction in On-Chip Clock Net", *IEEE TCAD* 16(2) (1997), pp. 210-215.

[18] T. Sakurai, "Approximation of Wiring Delay in MOSFET LSI", *IEEE JSSC* 18(4) (1983), pp. 418-426.

[19] J.-L. Tsai, T.-H. Chen and C.-P. Chen, "Zero Skew Clock-Tree Optimization With Buffer Insertion/Sizing and Wire Sizing", *IEEE TCAD* 23(4) (2004), pp. 565-572.

[20] S. X. D. Tan, C. J. R. Shi and J.-C. Lee, "Reliability-Constrained Area Optimization of VLSI Power/Ground Networks via Sequence of Linear Programmings", *IEEE TCAD* 22(12) (2003), pp. 1678-1684.

[21] R. A. Thisted, *Elements of Statistical Computing: Numerical Computation*, Chapman & Hall / CRC, 1988.

[22] X. Wu, X. Hong, Y. Cai, C. K. Cheng, J. Gu and W. Dai, "Area Minimization of Power Distribution Network Using Efficient Nonlinear Programming Techniques", *Proc. ICCAD*, 2001, pp. 153-157.

[23] Q. Zhu and W. W.-M. Dai, "High-Speed Clock Network Sizing Optimization Based on Distributed RC and Lossy RLC Interconnect Models", *IEEE TCAD* 15(9) (1996), pp. 1106-1118.

[24] *Cadence Encounter DSI User's Manual*, http://www.cadence.com .

[25] *Mathworks MATLAB User's Manual*, http://www.mathworks.com .

[26] *Synopsys 32/28nm Open PDK*, http://www.synopsys.com .

[27] *Synopsys DesignCompiler User's Manual*, http://www.synopsys.com .

[28] *Synopsys HSPICE User's Manual*, http://www.synopsys.com .

[29] *OpenCores: Open Source IP-Cores*, http://www.opencores.org .

[30] *LEF DEF Guide*, http://www.si2.org.openeda.si2.org/projects/lefdef .

AUTHOR INDEX

Abadir, M. 847
Abella, J. 590, 1276
Abousamra, A. 237
Abraham, J. 711
Aceituno, P. 110
Acharyya, D. 410
Adeniyi-Jones, C. 583
Agosta, G. 570
Agrawal, P. 913
Akesson, B. 161
Alaghi, A. 945
Alexander, C. 219
Alle, M. 355
Alpert, C. 645
Amaru, L. 328, 868
Ambrose, J. 437, 931
Anagnostopoulos, I. 1154
Ancajas, D. 275, 721, 1084
Andalam, S. 665, 1013
Angione, C. 298
Asadi, H. 705
Asanovic, K. 528
Ascheid, G. 404
Asenov, A. 219
Atienza, D. 365
Auras, D. 404
Axer, P. 1184
Ayoub, R. 808, 1220
Badaroglu, M. 169
Baert, R. 169
Bai, K. 1023
Ballal, B. 169
Bamakhrama, M. 1170
Banerjee, D. 389
Banerjee, P. 62
Bank, J. 859
Barenghi, A. 570
Bario, P. 169
Bartolini, A. 11
Bartolini, D. 528, 538
Bartzas, A. 1154
Basu, A. 137
Batterywala, S. 181
Bauer, L. 695
Becker, M. 988
Benazouz, M. 17
Benini, L. 11, 104
Beretta, I. 365
Bernstein, G. 756
Bhadra, J. 847
Bhattacharya, S. 181
Bhunia, S. 420
Bird, S. 528
Bobba, S. 868
Bodenmiller, B. 986

Bodin, B. 17
Bombieri, N. 1075
Borkar, S. 1098
Brisk, P. 319
Bruschi, F. 365
Burleson, W. 86
Busseuil, R. 583
Cai, W. 1044
Calhoun, B. 1101
Calimera, A. 781
Carapezza, G. 298
Carlo, S. 775, 1234
Carloni, L. 348, 1075, 1160
Cattaneo, R. 538
Catthoor, F. 913
Cazorla, F. 1276
Chakrabarty, K. 307, 516
Chakraborty, K. 275, 721, 1084
Chakraborty, R. 410
Chakraborty, S. 665, 671, 686, 988
Chakradhar, S. 799
Chandrachoodan, N. 1257
Chandrasekar, K. 161
Chandrikakutty, H. 564
Chang, N. 680
Chang, S. 478
Chang, W. 665
Chang, Y. 24, 30, 36, 62, 175, 187, 1038, 1117, 1132
Charbon, E. 882
Chatterjee, A. 389
Chattopadhyay, A. 921
Chattopadhyay, S. 213
Chava, B. 169
Chen, D. 68
Chen, F. 878
Chen, G. 1138
Chen, H. 872
Chen, I. 878
Chen, J. 478, 899
Chen, T. 24, 1038
Chen, W. 847
Chen, X. 404, 808
Chen, Y. 42, 1038
Cher, C. 711
Chiang, C. 472
Chien, H. 24, 30
Chippa, V. 799
Chiprout, E. 459
Cho, H. 711
Choi, J. 889
Chou, D. 528
Choudhury, M. 334
Chow, W. 1044
Colmenares, J. 528
Cong, J. 54, 68, 78

AUTHOR INDEX

Cong, K.	201
Corbalan, M.	1107
Costanza, J.	298
Courant, Y.	503
Cozzens, B.	721
Craig, K.	1101
Cristal, A.	377
Croes, K.	169
Csaba, G.	756
Dally, W.	659
Daneshtalab, M.	269
Das, A.	815
Das, C.	247
Debole, M.	939
Derrien, S.	355
Dev, K.	510
Devarakond, S.	389
Dieny, B.	886
Dietrich, B.	988
Ding, H.	1003
Dousti, M.	291
Drechsler, R.	822
Du, Y.	653
Durelli, G.	538
Dutt, N.	695, 1226
Eads, G.	528
Ebrahimi, M.	705
Eklow, B.	516
Enz, C.	137
Ernst, R.	1184
Faeder, J.	48
Fahmy, S.	665
Fan, D.	763
Fan, J.	98
Fang, G.	958
Fang, S.	175
Fariborzi, H.	878
Fattah, M.	269
Faubet, P.	503
Feng, Z.	604, 624
Flynn, D.	1093
Franke, B.	145
Fu, B.	263
Fu, X.	576
Fummi, F.	1075
Gaillardon, P.	328, 868
Gao, J.	490
Garg, S.	1191
Garibotti, R.	583
Ge, Y.	396
Geier, M.	988
Georakos, G.	686
Ghosh, S.	992
Gielen, G.	872
Girault, A.	1013

Girbal, S.	1276
Givargis, T.	1059
Gluzman, B.	528
Goncalves, O.	886
Goossens, K.	161
Goswami, D.	671, 988
Gould, M.	145
Graeb, H.	225
Grasset, A.	1276
Gratz, P.	808
Grissom, D.	319
Gross, W.	383
Große, D.	822
Grossman, J.	859
Grupp, L.	1111
Gu, C.	453, 459
Gupta, P.	695
Gupta, R.	104
Ha, H.	889
Ha, S.	889
Hamzeh, M.	119
Han, L.	624
Han, S.	968
Han, Y.	263
Hao, K.	828
Haralampos-G., S.	503
Hartman, M.	913
Hashemian, M.	420
Hasholzner, R.	1178
Hassoun, S.	314
Hayes, J.	945
He, X.	1044
Henkel, J.	1, 110, 437, 695, 899
Herdt, V.	822
Hills, G.	746, 872
Ho, K.	36
Ho, T.	307
Ho, Y.	187
Hofmeyr, S.	528
Hornayoun, H.	1226
Hsu, M.	1038
Hsu, S.	193
Hsueh, C.	598
Hu, J.	808, 1031
Hu, M.	42
Hu, X.	756
Hu, Y.	576
Huang, C.	834, 840, 1038
Huang, P.	1117
Huang, T.	42, 1044
Hujsa, T.	17
Hutin, L.	878
Hwu, W.	68
Irwin, M.	939
Jahn, J.	899

AUTHOR INDEX

Jang, J. 951
Javaid, H. 151
Jha, N. 92
Jiang, I. 472
Jiang, J. 342
Jiang, L. 516
Jiao, D. 974
Jone, W. 207, 793
Jones, A. 237
Joo, D. 632
Jouppi, N. 769
Juels, A. 86
Jung, Y. 1160
Kahng, A. 231, 638
Kang, S. 638
Kapur, R. 213
Karakostas, V. 377
Karnik, T. 1098
Karri, R. 557
Karthik, A. 444
Kauer, M. 665, 671, 1065
Keckler, S. 659
Kestur, S. 939
Keval, A. 1107
Khalid, A. 921
Kim, H. 808
Kim, J. 632, 889, 951
Kim, M. 1031
Kim, T. 632
Kim, Y. 680
Kinsman, A. 853
Kishinevsky, M. 808
Kistler, M. 522
Kleeberger, V. 225
Ko, H. 853
Kobbe, S. 899
Koh, C. 1050
Kourouma, M. 583
Koushanfar, F. 86
Krishnawarny, S. 986
Kuan, Y. 24
Kuang, J. 484, 1044
Kubiatowicz, J. 528
Kumar, A. 1, 815
Kundu, S. 213
Kuruvilla, V. 1257
Lai, C. 834
Lam, K. 1044
Larnech, C. 410
Le, H. 822
Leblebici, Y. 868
Lee, H. 638
Lee, K. 314
Lee, R. 878
Lee, S. 522

Lee, Y. 598, 736
Lei, L. 201
Leupers, R. 404
Li, C. 945, 1138
Li, H. 42, 371
Li, J. 193, 478
Li, P. 78, 466
Li, T. 437
Li, X. 263, 453, 459, 497
Li, Y. 645, 968, 1050, 1249
Li, Z. 645
Liang, Y. 68, 1003
Liao, K. 193
Liljeberg, P. 269
Lim, S. 736, 1242
Limbrick, D. 736
Lin, G. 472
Lin, H. 466
Lin, S. 478
Lin, T. 62
Lingamneni, A. 137
Lio, P. 298
Lisk, D. 1107
Liu, B. 42
Liu, C. 1242
Liu, D. 129
Liu, H. 348, 974, 1075
Liu, I. 175
Liu, L. 129
Liu, R. 1138
Liu, S. 756
Liu, T. 169, 878
Liu, W. 645, 1050
Liu, Y. 207
Lu, H. 263
Lu, J. 1023
Lukasiewycz, M. 665, 671, 1065
Luk-Pat, G. 653
Luo, Y. 307
Ma, Q. 653
Macii, E. 781
Mackin, C. 746
Maggi, M. 570
Maggio, M. 538
Malachowsky, C. 659
Mallik, A. 169
Manoj, S. 1207
Mao, Z. 42
Marchi, M. 868
Marculescu, D. 48, 1191
Maric, B. 590
Markov, I. 1031
Masrur, A. 671
Melham, R. 237
Mercati, P. 11

AUTHOR INDEX

Mercha, A. 169
Micheli, G. 328, 868
Miller, B. 1059
Miloslavsky, A. 653
Min, S. 151
Mirkhani, S. 711
Mishra, A. 247
Mishra, V. 618
Miskov-Zivanov, N. 48
Mitra, S. 711, 746, 872
Mitra, T. 1003, 1198
Mohamed, F. 503
Mor, N. 528
Moreto, M. 528, 1276
Morvan, A. 355
Mukhopadhyay, S. 775, 1234
Mundhenk, P. 665
Munier-Kordon, A. 17
Muralimanohar, N. 769
Muthukaruppan, T. 1198
Mutlu, O. 247
Myers, C. 466
Nacci, A. 365
Najar, W. 980
Naranayaswami, S. 1065
Narasimhan, S. 548
Narayanan, V. 257, 939
Nassif, S. 695
Nath, R. 1220
Nathanael, R. 878
Nemirovsky, M. 377
Nickerson, J. 275
Nicolici, N. 853
Nicosia, G. 298
Niemier, M. 756
Niu, D. 769
Nowak, M. 1107
Ogras, U. 808
Oh, H. 889
Oliveira, R. 994
Onizawa, N. 383
Orshansky, M. 314
Ost, L. 583
Ou, H. 24, 30, 36
Ouyang, J. 257
Pagani, S. 899
Palem, K. 137
Pan, D. 334, 490, 1249
Pant, M. 1098
Papakonstantinou, A. 68
Parameswaran, S. 151, 437, 931
Parandhaman, A. 787
Park, J. 1160
Park, M. 939, 951
Park, S. 680

Paterna, F. 11
Paul, G. 921
Pedram, M. 285, 291
Pe'Er, D. 986
Pelosi, G. 570
Pendina, G. 886
Peng, L. 576
Peng, Y. 1242
Perre, L. 913
Petracca, M. 1160
Piaget, J. 1257
Piguet, C. 137
Pimentel, A. 907
Plosila, J. 269
Plusquellic, J. 410
Poncino, M. 781
Porod, W. 756
Potkonjak, M. 990
Prenat, G. 886
Pricopi, M. 1198
Puri, R. 334
Qiu, Q. 396
Qu, G. 1270
Quan, W. 907
Quinones, E. 1276
Radojcic, R. 1107
Raghavan, P. 913
Raghunathan, A. 92, 787, 799
Raghunathan, B. 1191
Rahimi, A. 104
Rai, D. 1144
Rajagopalan, S. 181
Rakossy, Z. 921
Ramasubramanian, S. 787
Ramesh, S. 671
Rana, V. 365
Ray, S. 828
Reda, S. 510, 1213
Regazzoni, F. 882
Rehman, S. 110, 437
Reineke, J. 1013
Rellermeyer, J. 522
Reparaz, O. 98
Rethy, J. 872
Riddet, C. 219
Robert, M. 583
Roman, E. 528
Roop, P. 1013
Rosales, R. 1178
Rosing, T. 11, 1220
Rostami, M. 86
Roy, D. 548
Roy, K. 763, 799
Roy, S. 275, 334, 721, 1084
Roychowdhury, J. 444

AUTHOR INDEX

Rozic, V.	98
Rutenbar, R.	497
Ryckaert, J.	169
Sacchetto, D.	868
Saeedi, M.	285
Sagstetter, F.	665
Sakurai, T.	875
Salodkar, N.	181
Santambrogio, M.	538
Santos, L.	994
Sapatnekar, S.	618
Sassatelli, G.	583
Saxena, S.	497
Schlichtmann, U.	225, 686
Schneider, R.	686, 988
Schor, L.	1144
Schurmans, S.	404
Sciuto, D.	365
Sekitani, T.	875
Sen, S.	389
Sengupta, I.	213
Shafaei, A.	285
Shafique, M.	1, 110, 437, 695
Shahzad, K.	921
Shanker, S.	665
Sharad, M.	763
Sharma, N.	913
Shaw, D.	859
Shiely, J.	653
Shrivastava, A.	119, 1023
Shulaker, M.	746, 872
Singh, A.	1, 815
Sinha, D.	1257
Sinha, R.	1013
Sironi, F.	538
Someya, T.	875
Song, H.	653
Song, T.	1242
Soudris, D.	1154
Stefanov, T.	1170
Steinhorst, S.	665, 671, 1065
Stoimenov, N.	257, 1144
Stojanovic, V.	878
Strojwas, A.	497
Sun, S.	453
Sun, Z.	371, 429, 793
Swanson, S.	1111
Sze, C.	645
Tahoori, M.	695, 705
Tajik, H.	1226
Takamiya, M.	875
Teich, J.	1178
Tessier, R.	564
Thiele, L.	257, 1144
Tiang, L.	931
Tomic, S.	377
Toms, T.	1107
Topham, N.	145
Towie, E.	219
Towles, B.	859
Trivedi, A.	775
Tsao, C.	1117, 1132
Tsao, H.	36
Tseng, H.	1111
Tsoutouras, V.	1154
Tu, K.	342
Turakhia, Y.	1191
Unnikrishnan, D.	564
Unsal, O.	377
Vahid, F.	1059
Valero, M.	590
Varga, E.	756
Venkataramani, S.	787
Venkataramani, V.	1198
Verbauwhede, I.	98
Verkest, D.	169
Villarreal, J.	980
Vishin, S.	1198
Viswanathan, N.	645
Visweswariah, C.	1257
Vrudhula, S.	119
Wagstaff, H.	145
Wang, B.	1178
Wang, C.	1123
Wang, F.	453
Wang, J.	612
Wang, K.	1207
Wang, L.	404, 847
Wang, X.	548, 557
Wang, Y.	78
Wang, Z.	729
Waszecki, P.	665
Wehn, N.	161, 695
Wei, H.	746
Wei, L.	429
Wei, S.	129, 990
Wei, Y.	645
Weis, C.	161
Wen, W.	478
Wolf, T.	564
Wong, H.	746, 872
Wong, M.	653
Wong, W.	1123
Woods, G.	510
Wu, B.	840, 968
Wu, C.	834
Wu, W.	371
Wuerges, E.	994
Xiao, B.	54
Xie, F.	201, 828

AUTHOR INDEX

Xie, Y. 257, 769
Xiong, X. 612
Xu, C. 769
Xu, Q. 207, 429, 516, 793, 1264
Xu, Y. 1178
Xu, Z. 808
Yan, G. 263
Yang, C. 1138
Yang, M. 1117, 1132
Yang, Z. 828
Ye, F. 516
Ye, R. 793
Ye, Z. 968
Yehia, S. 1276
Yin, C. 1270
Yin, S. 129
Yokota, T. 875
Young, E. 484, 1044
Yousofshahi, M. 314
Yu, B. 490
Yu, H. 1207
Yu, Y. 472
Yuan, F. 207, 429, 793, 1264
Yuan, K. 490
Yueh, W. 548, 1234
Yunge, D. 988
Zhai, J. 1170
Zhan, J. 257
Zhan, X. 1213
Zhang, C. 78, 939
Zhang, D. 404
Zhang, J. 429, 746
Zhang, M. 92
Zhang, P. 78
Zhang, W. 42, 453, 497
Zhang, Y. 396, 576
Zhao, X. 624
Zheng, Y. 420, 548
Zhou, B. 974
Zuber, P. 169

2013 50th ACM/EDAC/IEEE Design Automation Conference

(DAC 2013)

Austin, Texas, USA
29 May – 7 June 2013

Pages 645-1285

IEEE Catalog Number:	CFP13DAC-PRT
ISBN:	978-1-4503-2071-9

Copyright © 2013, Association for Computing Machinery
All Rights Reserved

***This publication is a representation of what appears in the IEEE Digital Libraries. Some format issues inherent in the e-media version may also appear in this print version.

IEEE Catalog Number: CFP13DAC-PRT
ISBN 13:   978-1-4503-2071-9
ISSN: 0738-100X

Additional Copies of This Publication Are Available From:

Curran Associates, Inc
57 Morehouse Lane
Red Hook, NY 12571 USA
Phone: (845) 758-0400
Fax: (845) 758-2633
E-mail: curran@proceedings.com
Web: www.proceedings.com

TABLE OF CONTENTS

MAPPING ON MULTI/MANY-CORE SYSTEMS: SURVEY OF CURRENT AND EMERGING TRENDS 1
A. Singh, M. Shafique, A. Kumar, J. Henkel

WORKLOAD AND USER EXPERIENCE-AWARE DYNAMIC RELIABILITY MANAGEMENT IN MULTICORE PROCESSORS 11
P. Mercati, A. Bartolini, T. Rosing, L. Benini, F. Paterna

LIVENESS EVALUATION OF A CYCLO-STATIC DATAFLOW GRAPH 17
M. Benazouz, A. Munier-Kordon, T. Hujsa, B. Bodin

DOUBLE PATTERNING LITHOGRAPHY-AWARE ANALOG PLACEMENT 24
H. Chien, H. Ou, T. Chen, Y. Kuan, Y. Chang

SIMULTANEOUS ANALOG PLACEMENT AND ROUTING WITH CURRENT FLOW AND CURRENT DENSITY CONSIDERATIONS 30
H. Ou, H. Chien, Y. Chang

COUPLING-AWARE LENGTH-RATIO-MATCHING ROUTING FOR CAPACITOR ARRAYS IN ANALOG INTEGRATED CIRCUITS 36
K. Ho, H. Ou, Y. Chang, H. Tsao

DIGITAL-ASSISTED NOISE-ELIMINATING TRAINING FOR MEMRISTOR CROSSBAR-BASED ANALOG NEUROMORPHIC COMPUTING ENGINE 42
B. Liu, M. Hu, H. Li, Z. Mao, Y. Chen, T. Huang, W. Zhang

DYNAMIC BEHAVIOR OF CELL SIGNALING NETWORKS – MODEL DESIGN AND ANALYSIS AUTOMATION 48
N. Miskov-Zivanov, D. Marculescu, J. Faeder

DEFECT TOLERANCE IN NANODEVICE-BASED PROGRAMMABLE INTERCONNECTS: UTILIZATION BEYOND AVOIDANCE 54
J. Cong, B. Xiao

AN EFFICIENT AND EFFECTIVE ANALYTICAL PLACER FOR FPGAS 62
T. Lin, P. Banerjee, Y. Chang

THROUGHPUT-ORIENTED KERNEL PORTING ONTO FPGAS 68
A. Papakonstantinou, J. Cong, D. Chen, Y. Liang, W. Hwu

MEMORY PARTITIONING FOR MULTIDIMENSIONAL ARRAYS IN HIGH-LEVEL SYNTHESIS 78
Y. Wang, P. Li, P. Zhang, C. Zhang, J. Cong

BALANCING SECURITY AND UTILITY IN MEDICAL DEVICES? 86
M. Rostami, W. Burleson, F. Koushanfar, A. Juels

TOWARDS TRUSTWORTHY MEDICAL DEVICES AND BODY AREA NETWORKS 92
M. Zhang, A. Raghunathan, N. Jha

LOW-ENERGY ENCRYPTION FOR MEDICAL DEVICES: SECURITY ADDS AN EXTRA DESIGN DIMENSION 98
J. Fan, O. Reparaz, V. Rozic, I. Verbauwhede

AGING-AWARE COMPILER-DIRECTED VLIW ASSIGNMENT FOR GPGPU ARCHITECTURES 104
A. Rahimi, L. Benini, R. Gupta

EXPLOITING PROGRAM-LEVEL MASKING AND ERROR PROPAGATION FOR CONSTRAINED RELIABILITY OPTIMIZATION 110
M. Shafique, S. Rehman, P. Aceituno, J. Henkel

REGIMAP: REGISTER-AWARE APPLICATION MAPPING ON COARSE-GRAINED RECONFIGURABLE ARCHITECTURES (CGRAS) 119
M. Hamzeh, A. Shrivastava, S. Vrudhula

POLYHEDRAL MODEL BASED MAPPING OPTIMIZATION OF LOOP NESTS FOR CGRAS 129
D. Liu, S. Yin, L. Liu, S. Wei

IMPROVING ENERGY GAINS OF INEXACT DSP HARDWARE THROUGH RECIPROCATIVE ERROR COMPENSATION 137
A. Lingamneni, A. Basu, K. Palem, C. Piguet, C. Enz

EARLY PARTIAL EVALUATION IN A JIT-COMPILED, RETARGETABLE INSTRUCTION SET SIMULATOR GENERATED FROM A HIGH-LEVEL ARCHITECTURE DESCRIPTION 145
H. Wagstaff, M. Gould, B. Franke, N. Topham

XDRA: EXPLORATION AND OPTIMIZATION OF LAST LEVEL CACHE FOR ENERGY REDUCTION IN DDR DRAMS 151
S. Min, H. Javaid, S. Parameswaran

TOWARDS VARIATION-AWARE SYSTEM-LEVEL POWER ESTIMATION OF DRAMS: AN EMPIRICAL APPROACH 161
K. Chandrasekar, C. Weis, B. Akesson, N. Wehn, K. Goossens

TEASE: A SYSTEMATIC ANALYSIS FRAMEWORK FOR EARLY EVALUATION OF FINFET-BASED ADVANCED TECHNOLOGY NODES 169
A. Mallik, P. Zuber, T. Liu, B. Chava, B. Ballal, P. Bario, R. Baert, K. Croes, J. Ryckaert, M. Badaroglu, A. Mercha, D. Verkest

STITCH-AWARE ROUTING FOR MULTIPLE E-BEAM LITHOGRAPHY 175
S. Fang, I. Liu, Y. Chang

AUTOMATIC DESIGN RULE CORRECTION IN PRESENCE OF MULTIPLE GRIDS AND TRACK PATTERNS 181
N. Salodkar, S. Rajagopalan, S. Batterywala, S. Bhattacharya

MULTIPLE CHIP PLANNING FOR CHIP-INTERPOSER CODESIGN 187
Y. Ho, Y. Chang

GPU-BASED N-DETECT TRANSITION FAULT ATPG 193
K. Liao, S. Hsu, J. Li

POST-SILICON CONFORMANCE CHECKING WITH VIRTUAL PROTOTYPES 201
L. Lei, F. Xie, K. Cong

ON TESTING TIMING-SPECULATIVE CIRCUITS 207
F. Yuan, Y. Liu, W. Jone, Q. Xu

AN ATE ASSISTED DFD TECHNIQUE FOR VOLUME DIAGNOSIS OF SCAN CHAINS 213
S. Kundu, S. Chattopadhyay, I. Sengupta, R. Kapur

PREDICTING FUTURE TECHNOLOGY PERFORMANCE 219
A. Asenov, C. Alexander, C. Riddet, E. Towie

PREDICTING FUTURE PRODUCT PERFORMANCE: MODELING AND EVALUATION OF STANDARD CELLS IN FINFET TECHNOLOGIES 225
V. Kleeberger, H. Graeb, U. Schlichtmann

THE ITRS DESIGN TECHNOLOGY AND SYSTEM DRIVERS ROADMAP: PROCESS AND STATUS 231
A. Kahng

PROACTIVE CIRCUIT ALLOCATION IN MULTIPLANE NOCS 237
A. Abousamra, A. Jones, R. Melham

A HETEROGENEOUS MULTIPLE NETWORK-ON-CHIP DESIGN: AN APPLICATION-AWARE APPROACH 247
A. Mishra, O. Mutlu, C. Das

DESIGNING ENERGY-EFFICIENT NOC FOR REAL-TIME EMBEDDED SYSTEMS THROUGH SLACK OPTIMIZATION 257
J. Zhan, N. Stoimenov, J. Ouyang, L. Thiele, V. Narayanan, Y. Xie

RISO: RELAXED NETWORK-ON-CHIP ISOLATION FOR CLOUD PROCESSORS 263
H. Lu, G. Yan, Y. Han, B. Fu, X. Li

SMART HILL CLIMBING FOR AGILE DYNAMIC MAPPING IN MANY-CORE SYSTEMS 269
M. Fattah, M. Daneshtalab, P. Liljeberg, J. Plosila

HCI-TOLERANT NOC ROUTER MICROARCHITECTURE 275
D. Ancajas, J. Nickerson, K. Chakraborty, S. Roy

OPTIMIZATION OF QUANTUM CIRCUITS FOR INTERACTION DISTANCE IN LINEAR NEAREST NEIGHBOR ARCHITECTURES 285
A. Shafaei, M. Saeedi, M. Pedram

LEQA: LATENCY ESTIMATION FOR A QUANTUM ALGORITHM MAPPED TO A QUANTUM CIRCUIT FABRIC 291
M. Dousti, M. Pedram

PARETO EPSILON-DOMINANCE AND IDENTIFIABLE SOLUTIONS FOR BIOCAD MODELING 298
C. Angione, J. Costanza, P. Lio, G. Nicosia, G. Carapezza

DESIGN OF CYBERPHYSICAL DIGITAL MICROFLUIDIC BIOCHIPS UNDER COMPLETION-TIME UNCERTAINTIES IN FLUIDIC OPERATIONS 307
Y. Luo, K. Chakrabarty, T. Ho

GENE MODIFICATION IDENTIFICATION UNDER FLUX CAPACITY UNCERTAINTY 314
M. Yousofshahi, M. Orshansky, K. Lee, S. Hassoun

A FIELD-PROGRAMMABLE PIN-CONSTRAINED DIGITAL MICROFLUIDIC BIOCHIP 319
D. Grissom, P. Brisk

BDS-MAJ: A BDD-BASED LOGIC SYNTHESIS TOOL EXPLOITING MAJORITY LOGIC DECOMPOSITION 328
L. Amaru, P. Gaillardson, G. Micheli

TOWARDS OPTIMAL PERFORMANCE-AREA TRADE-OFF IN ADDERS BY SYNTHESIS OF PARALLEL PREFIX STRUCTURES 334
S. Roy, M. Choudhury, R. Puri, D. Pan

SYNTHESIS OF FEEDBACK DECODERS FOR INITIALIZED ENCODERS 342
K. Tu, J. Jiang

ON LEARNING-BASED METHODS FOR DESIGN-SPACE EXPLORATION WITH HIGH-LEVEL SYNTHESIS 348
H. Liu, L. Carloni

RUNTIME DEPENDENCY ANALYSIS FOR LOOP PIPELINING IN HIGH-LEVEL SYNTHESIS 355
M. Alle, A. Morvan, S. Derrien

A HIGH-LEVEL SYNTHESIS FLOW FOR THE IMPLEMENTATION OF ITERATIVE STENCIL LOOP ALGORITHMS ON FPGA DEVICES 365
A. Nacci, V. Rana, F. Bruschi, D. Sciuto, I. Beretta, D. Atienza

CROSS-LAYER RACETRACK MEMORY DESIGN FOR ULTRA HIGH DENSITY AND LOW POWER CONSUMPTION 371
Z. Sun, W. Wu, H. Li

IMPROVING THE ENERGY EFFICIENCY OF HARDWARE-ASSISTED WATCHPOINT SYSTEMS 377
V. Karakostas, S. Tomic, O. Unsal, M. Nemirovsky, A. Cristal

LOW-POWER AREA-EFFICIENT LARGE-SCALE IP LOOKUP ENGINE BASED ON BINARY-WEIGHTED CLUSTERED NETWORKS 383
N. Onizawa, W. Gross

REAL-TIME USE-AWARE ADAPTIVE MIMO RF RECEIVER SYSTEMS FOR ENERGY EFFICIENCY UNDER BER CONSTRAINTS 389
D. Banerjee, S. Devarakond, S. Sen, A. Chatterjee

IMPROVING CHARGING EFFICIENCY WITH WORKLOAD SCHEDULING IN ENERGY HARVESTING EMBEDDED SYSTEMS 396
Y. Zhang, Y. Ge, Q. Qiu

CREATION OF ESL POWER MODELS FOR COMMUNICATION ARCHITECTURES USING AUTOMATIC CALIBRATION 404
S. Schurmans, D. Zhang, D. Auras, R. Leupers, G. Ascheid, X. Chen, L. Wang

A TRANSMISSION GATE PHYSICAL UNCLONABLE FUNCTION AND ON-CHIP VOLTAGETO-DIGITAL CONVERSION TECHNIQUE 410
R. Chakraborty, C. Larnech, D. Acharyya, J. Plusquellic

RESP: A ROBUST PHYSICAL UNCLONABLE FUNCTION RETROFITTED INTO EMBEDDED SRAM ARRAY 420
Y. Zheng, M. Hashemian, S. Bhunia

VERITRUST: VERIFICATION FOR HARDWARE TRUST 429
J. Zhang, F. Yuan, L. Wei, Z. Sun, Q. Xu

RASTER: RUNTIME ADAPTIVE SPATIAL/TEMPORAL ERROR RESILIENCY FOR EMBEDDED PROCESSORS 437
T. Li, M. Shafique, J. Ambrose, S. Rehman, J. Henkel, S. Parameswaran

ABCD-L: APPROXIMATING CONTINUOUS LINEAR SYSTEMS USING BOOLEAN MODELS 444
A. Karthik, J. Roychowdhury

BAYESIAN MODEL FUSION: LARGE-SCALE PERFORMANCE MODELING OF ANALOG AND MIXED-SIGNAL CIRCUITS BY REUSING EARLY-STAGE DATA 453
F. Wang, W. Zhang, S. Sun, X. Li, C. Gu

EFFICIENT MOMENT ESTIMATION WITH EXTREMELY SMALL SAMPLE SIZE VIA BAYESIAN INFERENCE FOR ANALOG/MIXED-SIGNAL VALIDATION 459
C. Gu, E. Chiprout, X. Li

VERIFICATION OF DIGITALLY-INTENSIVE ANALOG CIRCUITS VIA KERNEL RIDGE REGRESSION AND HYBRID REACHABILITY ANALYSIS 466
H. Lin, P. Li, C. Myers

MACHINE-LEARNING-BASED HOTSPOT DETECTION USING TOPOLOGICAL CLASSIFICATION AND CRITICAL FEATURE EXTRACTION 472
Y. Yu, G. Lin, I. Jiang, C. Chiang

A NOVEL FUZZY MATCHING MODEL FOR LITHOGRAPHY HOTSPOT DETECTION 478
S. Lin, J. Chen, J. Li, W. Wen, S. Chang

AN EFFICIENT LAYOUT DECOMPOSITION APPROACH FOR TRIPLE PATTERNING LITHOGRAPHY 484
J. Kuang, E. Young

E-BLOW: E-BEAM LITHOGRAPHY OVERLAPPING AWARE STENCIL PLANNING FOR MCC SYSTEM 490
B. Yu, K. Yuan, J. Gao, D. Pan

AUTOMATIC CLUSTERING OF WAFER SPATIAL SIGNATURES 497
W. Zhang, X. Li, S. Saxena, A. Strojwas, R. Rutenbar

MULTIDIMENSIONAL ANALOG TEST METRICS ESTIMATION USING EXTREME VALUE THEORY AND STATISTICAL BLOCKADE 503
S. Haralampos-G., P. Faubet, F. Mohamed, Y. Courant

HIGH-THROUGHPUT TSV TESTING AND CHARACTERIZATION FOR 3D INTEGRATION USING THERMAL MAPPING 510
K. Dev, G. Woods, S. Reda

ON EFFECTIVE AND EFFICIENT IN-FIELD TSV REPAIR FOR STACKED 3D ICS 516
L. Jiang, F. Ye, Q. Xu, K. Chakrabarty, B. Eklow

CLOUD PLATFORMS AND EMBEDDED COMPUTING – THE OPERATING SYSTEMS OF THE FUTURE 522
J. Rellermeyer, S. Lee, M. Kistler

TESSELLATION: REFACTORING THE OS AROUND EXPLICIT RESOURCE CONTAINERS WITH CONTINUOUS ADAPTATION 528
J. Colmenares, G. Eads, S. Hofmeyr, S. Bird, M. Moreto, D. Chou, B. Gluzman, E. Roman, D. Bartolini, N. Mor, K. Asanovic, J. Kubiatowicz

THE AUTONOMIC OPERATING SYSTEM RESEARCH PROJECT – ACHIEVEMENTS AND FUTURE DIRECTIONS 538
D. Bartolini, R. Cattaneo, G. Durelli, M. Maggio, M. Santambrogio, F. Sironi

ROLE OF POWER GRID IN SIDE CHANNEL ATTACK AND POWER-GRID-AWARE SECURE DESIGN 548
X. Wang, W. Yueh, D. Roy, S. Narasimhan, Y. Zheng

NUMCHECKER: DETECTING KERNEL CONTROL-FLOW MODIFYING ROOTKITS BY USING HARDWARE PERFORMANCE COUNTERS 557
X. Wang, R. Karri

HIGH-PERFORMANCE HARDWARE MONITORS TO PROTECT NETWORK PROCESSORS FROM DATA PLANE ATTACKS 564
H. Chandrikakutty, D. Unnikrishnan, R. Tessier, T. Wolf

COMPILER-BASED SIDE CHANNEL VULNERABILITY ANALYSIS AND OPTIMIZED COUNTERMEASURES APPLICATION 570
G. Agosta, A. Barenghi, M. Maggi, G. Pelosi

LIGHTING THE DARK SILICON BY EXPLOITING HETEROGENEITY ON FUTURE PROCESSORS 576
Y. Zhang, L. Peng, X. Fu, Y. Hu

SIMULTANEOUS MULTITHREADING SUPPORT IN EMBEDDED DISTRIBUTED MEMORY MPSOCS 583
R. Garibotti, L. Ost, R. Busseuil, M. Kourouma, C. Adeniyi-Jones, G. Sassatelli, M. Robert

APPLE: ADAPTIVE PERFORMANCE-PREDICTABLE LOW-ENERGY CACHES FOR RELIABLE HYBRID VOLTAGE OPERATION 590
B. Maric, J. Abella, M. Valero

AN OPTIMIZED PAGE TRANSLATION FOR MOBILE VIRTUALIZATION 598
Y. Lee, C. Hsueh

SCALABLE VECTORLESS POWER GRID CURRENT INTEGRITY VERIFICATION 604
Z. Feng

CONSTRAINT ABSTRACTION FOR VECTORLESS POWER GRID VERIFICATION 612
X. Xiong, J. Wang

THE IMPACT OF ELECTROMIGRATION IN COPPER INTERCONNECTS ON POWER GRID INTEGRITY 618
V. Mishra, S. Sapatnekar

TINYSPICE: A PARALLEL SPICE SIMULATOR ON GPU FOR MASSIVELY REPEATED SMALL CIRCUIT SIMULATIONS 624
L. Han, X. Zhao, Z. Feng

AN OPTIMAL ALGORITHM OF ADJUSTABLE DELAY BUFFER INSERTION FOR SOLVING CLOCK SKEW VARIATION PROBLEM 632
    J. Kim, D. Joo, T. Kim

SMART NON-DEFAULT ROUTING FOR CLOCK POWER REDUCTION 638
    A. Kahng, S. Kang, H. Lee

ROUTING CONGESTION ESTIMATION WITH REAL DESIGN CONSTRAINTS 645
    W. Liu, Y. Wei, C. Sze, C. Alpert, Z. Li, Y. Li, N. Viswanathan

SPACER-IS-DIELECTRIC-COMPLIANT DETAILED ROUTING FOR SELF-ALIGNED DOUBLE PATTERNING LITHOGRAPHY 653
    Y. Du, Q. Ma, H. Song, J. Shiely, G. Luk-Pat, A. Miloslavsky, M. Wong

21ST CENTURY DIGITAL DESIGN TOOLS 659
    W. Dally, C. Malachowsky, S. Keckler

SYSTEM ARCHITECTURE AND SOFTWARE DESIGN FOR ELECTRIC VEHICLES 665
    M. Lukasiewycz, S. Steinhorst, S. Andalam, F. Sagstetter, P. Waszecki, W. Chang, M. Kauer, P. Mundhenk, S. Shanker, S. Fahmy, S. Chakraborty

MODEL-BASED DEVELOPMENT AND VERIFICATION OF CONTROL SOFTWARE FOR ELECTRIC VEHICLES 671
    D. Goswami, M. Lukasiewycz, M. Kauer, S. Steinhorst, A. Masrur, S. Chakraborty, S. Ramesh

HYBRID ENERGY STORAGE SYSTEMS AND BATTERY MANAGEMENT FOR ELECTRIC VEHICLES 680
    S. Park, Y. Kim, N. Chang

RELIABILITY CHALLENGES FOR ELECTRIC VEHICLES: FROM DEVICES TO ARCHITECTURE AND SYSTEMS SOFTWARE 686
    G. Georakos, U. Schlichtmann, R. Schneider, S. Chakraborty

RELIABLE ON-CHIP SYSTEMS IN THE NANO-ERA: LESSONS LEARNT AND FUTURE TRENDS 695
    J. Henkel, L. Bauer, N. Dutt, P. Gupta, S. Nassif, M. Shafique, M. Tahoori, N. Wehn

A LAYOUT-BASED APPROACH FOR MULTIPLE EVENT TRANSIENT ANALYSIS 705
    M. Ebrahimi, H. Asadi, M. Tahoori

QUANTITATIVE EVALUATION OF SOFT ERROR INJECTION TECHNIQUES FOR ROBUST SYSTEM DESIGN 711
    H. Cho, S. Mirkhani, C. Cher, J. Abraham, S. Mitra

EFFICIENTLY TOLERATING TIMING VIOLATIONS IN PIPELINED MICROPROCESSORS 721
    K. Chakraborty, B. Cozzens, S. Roy, D. Ancajas

HIERARCHICAL DECODING OF DOUBLE ERROR CORRECTING CODES FOR HIGH SPEED RELIABLE MEMORIES 729
    Z. Wang

POWER BENEFIT STUDY FOR ULTRA-HIGH DENSITY TRANSISTOR-LEVEL MONOLITHIC 3D ICS 736
    Y. Lee, D. Limbrick, S. Lim

RAPID EXPLORATION OF PROCESSING AND DESIGN GUIDELINES TO OVERCOME CARBON NANOTUBE VARIATIONS 746
    G. Hills, J. Zhang, C. Mackin, M. Shulaker, H. Wei, H. Wong, S. Mitra

MINIMUM-ENERGY STATE GUIDED PHYSICAL DESIGN FOR NANOMAGNET LOGIC 756
    S. Liu, G. Csaba, X. Hu, E. Varga, M. Niemier, G. Bernstein, W. Porod

ULTRA LOW POWER ASSOCIATIVE COMPUTING WITH SPIN NEURONS AND RESISTIVE CROSSBAR MEMORY 763
    M. Sharad, D. Fan, K. Roy

UNDERSTANDING THE TRADE-OFFS IN MULTI-LEVEL CELL RERAM MEMORY DESIGN 769
    C. Xu, D. Niu, N. Muralimanohar, N. Jouppi, Y. Xie

EXPLORING TUNNEL-FET FOR ULTRA LOW POWER ANALOG APPLICATIONS: A CASE STUDY ON OPERATIONAL TRANSCONDUCTANCE AMPLIFIER 775
    A. Trivedi, S. Carlo, S. Mukhopadhyay

ENERGY-OPTIMAL SRAM SUPPLY VOLTAGE SCHEDULING UNDER LIFETIME AND ERROR CONSTRAINTS 781
    A. Calimera, E. Macii, M. Poncino

RELAX-AND-RETIME: A METHODOLOGY FOR ENERGY-EFFICIENT RECOVERY BASED DESIGN 787
    S. Ramasubramanian, S. Venkataramani, A. Parandhaman, A. Raghunathan

POST-PLACEMENT VOLTAGE ISLAND GENERATION FOR TIMING-SPECULATIVE CIRCUITS 793
    R. Ye, F. Yuan, Z. Sun, W. Jone, Q. Xu

ANALYSIS AND CHARACTERIZATION OF INHERENT APPLICATION RESILIENCE FOR APPROXIMATE COMPUTING 799
V. Chippa, S. Chakradhar, K. Roy, A. Raghunathan

DYNAMIC VOLTAGE AND FREQUENCY SCALING FOR SHARED RESOURCES IN MULTICORE PROCESSOR DESIGNS 808
X. Chen, Z. Xu, H. Kim, P. Gratz, J. Hu, M. Kishinevsky, U. Ogras, R. Ayoub

ENERGY OPTIMIZATION BY EXPLOITING EXECUTION SLACKS IN STREAMING APPLICATIONS ON MULTIPROCESSOR SYSTEMS 815
A. Singh, A. Das, A. Kumar

VERIFYING SYSTEMC USING AN INTERMEDIATE VERIFICATION LANGUAGE AND SYMBOLIC SIMULATION 822
H. Le, D. Große, V. Herdt, R. Drechsler

HANDLING DESIGN AND IMPLEMENTATION OPTIMIZATIONS IN EQUIVALENCE CHECKING FOR BEHAVIORAL SYNTHESIS 828
Z. Yang, K. Hao, S. Ray, F. Xie

A COUNTEREXAMPLE-GUIDED INTERPOLANT GENERATION ALGORITHM FOR SAT-BASED MODEL CHECKING 834
C. Wu, C. Lai, C. Huang

A ROBUST CONSTRAINT SOLVING FRAMEWORK FOR MULTIPLE CONSTRAINT SETS IN CONSTRAINED RANDOM VERIFICATION 840
B. Wu, C. Huang

SIMULATION KNOWLEDGE EXTRACTION AND REUSE IN CONSTRAINED RANDOM PROCESSOR VERIFICATION 847
W. Chen, L. Wang, J. Bhadra, M. Abadir

HARDWARE-EFFICIENT ON-CHIP GENERATION OF TIME-EXTENSIVE CONSTRAINED-RANDOM SEQUENCES FOR IN-SYSTEM VALIDATION 853
A. Kinsman, H. Ko, N. Nicolici

THE ROLE OF CASCADE, A CYCLE-BASED SIMULATION INFRASTRUCTURE, IN DESIGNING THE ANTON SPECIAL-PURPOSE SUPERCOMPUTERS 859
J. Grossman, B. Towles, J. Bank, D. Shaw

TOWARDS STRUCTURED ASICS USING POLARITY-TUNABLE SI NANOWIRE TRANSISTORS 868
P. Gaillardon, M. Marchi, L. Amaru, S. Bobba, D. Sacchetto, Y. Leblebici, G. Micheli

SACHA: THE STANFORD CARBON NANOTUBE CONTROLLED HANDSHAKING ROBOT 872
M. Shulaker, J. Rethy, G. Hills, H. Chen, G. Gielen, H. Wong, S. Mitra

ELECTRICAL ARTIFICIAL SKIN USING ULTRAFLEXIBLE ORGANIC TRANSISTOR 875
T. Sekitani, T. Sakurai, T. Yokota, T. Someya, M. Takamiya

RELAYS DO NOT LEAK – CMOS DOES 878
H. Fariborzi, F. Chen, R. Nathanael, I. Chen, L. Hutin, R. Lee, T. Liu, V. Stojanovic

SINGLE-PHOTON IMAGE SENSORS 882
E. Charbon, F. Regazzoni

NON-VOLATILE FPGAS BASED ON SPINTRONIC DEVICES 886
O. Goncalves, G. Prenat, G. Pendina, B. Dieny

A NOVEL ANALYTICAL METHOD FOR WORST CASE RESPONSE TIME ESTIMATION OF DISTRIBUTED EMBEDDED SYSTEMS 889
J. Kim, H. Oh, J. Choi, H. Ha, S. Ha

OPTIMIZATIONS FOR CONFIGURING AND MAPPING SOFTWARE PIPELINES IN MANY CORE SYSTEMS 899
J. Jahn, S. Pagani, S. Kobbe, J. Chen, J. Henkel

A SCENARIO-BASED RUN-TIME TASK MAPPING ALGORITHM FOR MPSOCS 907
W. Quan, A. Pimentel

EARLY EXPLORATION FOR PLATFORM ARCHITECTURE INSTANTIATION WITH MULTI-MODE APPLICATION PARTITIONING 913
P. Agrawal, P. Raghavan, M. Hartman, N. Sharma, L. Perre, F. Catthoor

COARX: A COPROCESSOR FOR ARX-BASED CRYPTOGRAPHIC ALGORITHMS 921
K. Shahzad, A. Khalid, Z. Rakossy, G. Paul, A. Chattopadhyay

RECONFIGURABLE PIPELINED COPROCESSOR FOR MULTI-MODE COMMUNICATION TRANSMISSION 931
L. Tiang, J. Ambrose, S. Parameswaran

ACCELERATORS FOR BIOLOGICALLY-INSPIRED ATTENTION AND RECOGNITION 939
M. Park, C. Zhang, M. Debole, S. Kestur, V. Narayanan, M. Irwin

STOCHASTIC CIRCUITS FOR REAL-TIME IMAGE-PROCESSING APPLICATIONS 945
A. Alaghi, C. Li, J. Hayes

AN EVENT-DRIVEN SIMULATION METHODOLOGY FOR INTEGRATED SWITCHING POWER SUPPLIES IN SYSTEMVERILOG 951
J. Jang, M. Park, J. Kim

A NEW TIME-STEPPING METHOD FOR CIRCUIT SIMULATION 958
G. Fang

TIME-DOMAIN SEGMENTATION BASED MASSIVELY PARALLEL SIMULATION FOR ADCS 968
Z. Ye, B. Wu, S. Han, Y. Li

A DIRECT FINITE ELEMENT SOLVER OF LINEAR COMPLEXITY FOR LARGE-SCALE 3-D CIRCUIT EXTRACTION IN MULTIPLE DIELECTRICS 974
B. Zhou, H. Liu, D. Jiao

FPGA CODE ACCELERATORS - THE COMPILER PERSPECTIVE 980
W. Najar, J. Villarreal

CAN CAD CURE CANCER? 986
S. Krishnawarny, B. Bodenmiller, D. Pe'er

LET'S PUT THE CAR IN YOUR PHONE! 988
M. Geier, M. Becker, D. Yunge, B. Dietrich, R. Schneider, D. Goswami, S. Chakraborty

THE UNDETECTABLE AND UNPROVABLE HARDWARE TROJAN HORSE 990
S. Wei, M. Potkonjak

PATH TO A TERABYTE OF ON-CHIP MEMORY FOR PETABIT PER SECOND BANDWIDTH WITH < 5WATTS OF POWER 992
S. Ghosh

RECONCILING REAL-TIME GUARANTEES AND ENERGY EFFICIENCY THROUGH UNLOCKED-CACHE PREFETCHING 994
E. Wuerges, R. Oliveira, L. Santos

INTEGRATED INSTRUCTION CACHE ANALYSIS AND LOCKING IN MULTITASKING REAL-TIME SYSTEMS 1003
H. Ding, Y. Liang, T. Mitra

PRECISE TIMING ANALYSIS FOR DIRECT-MAPPED CACHES 1013
S. Andalam, A. Girault, R. Sinha, P. Roop, J. Reineke

SSDM: SMART STACK DATA MANAGEMENT FOR SOFTWARE MANAGED MULTICORES (SMMS) 1023
J. Lu, K. Bai, A. Shrivastava

TAMING THE COMPLEXITY OF COORDINATED PLACE AND ROUTE 1031
J. Hu, M. Kim, I. Markov

ROUTABILITY-DRIVEN PLACEMENT FOR HIERARCHICAL MIXED-SIZE CIRCUIT DESIGNS 1038
M. Hsu, Y. Chen, C. Huang, T. Chen, Y. Chang

RIPPLE 2.0: HIGH QUALITY ROUTABILITY-DRIVEN PLACEMENT VIA GLOBAL ROUTER INTEGRATION 1044
X. He, T. Huang, W. Chow, J. Kuang, K. Lam, W. Cai, E. Young

OPTIMIZATION OF PLACEMENT SOLUTIONS FOR ROUTABILITY 1050
W. Liu, C. Koh, Y. Li

EXPLORATION WITH UPGRADEABLE MODELS USING STATISTICAL METHODS FOR PHYSICAL MODEL EMULATION 1059
B. Miller, F. Vahid, T. Givargis

MODULAR SYSTEM-LEVEL ARCHITECTURE FOR CONCURRENT CELL BALANCING 1065
M. Kauer, S. Naranayaswami, S. Steinhorst, M. Lukasiewycz

A METHOD TO ABSTRACT RTL IP BLOCKS INTO C++ CODE AND ENABLE HIGH-LEVEL SYNTHESIS 1075
N. Bombieri, H. Liu, F. Fummi, L. Carloni

DMR3D: DYNAMIC MEMORY RELOCATION IN 3D MULTICORE SYSTEMS 1084
D. Ancajas, K. Chakraborty, S. Roy

POWER GATING APPLIED TO MP-SOCS FOR STANDBY-MODE POWER MANAGEMENT 1093
D. Flynn

POWER MANAGEMENT AND DELIVERY FOR HIGH-PERFORMANCE MICROPROCESSORS 1098
T. Karnik, M. Pant, S. Borkar

FLEXIBLE ON-CHIP POWER DELIVERY FOR ENERGY EFFICIENT HETEROGENEOUS SYSTEMS 1101
B. Calhoun, K. Craig

POWER AND SIGNAL INTEGRITY CHALLENGES IN 3D SYSTEMS................1107
M. Corbalan, A. Keval, T. Toms, D. Lisk, R. Radojcic, M. Nowak

UNDERPOWERING NAND FLASH: PROFITS AND PERILS................1111
H. Tseng, L. Grupp, S. Swanson

NEW ERA: NEW EFFICIENT RELIABILITY-AWARE WEAR LEVELING FOR ENDURANCE ENHANCEMENT OF FLASH STORAGE DEVICES................1117
M. Yang, Y. Chang, C. Tsao, P. Huang

SAW: SYSTEM-ASSISTED WEAR LEVELING ON THE WRITE ENDURANCE OF NAND FLASH DEVICES................1123
C. Wang, W. Wong

PERFORMANCE ENHANCEMENT OF GARBAGE COLLECTION FOR FLASH STORAGE DEVICES: AN EFFICIENT VICTIM BLOCK SELECTION DESIGN................1132
C. Tsao, Y. Chang, M. Yang

DURACACHE: A DURABLE SSD CACHE USING MLC NAND FLASH................1138
R. Liu, C. Yang, C. Li, G. Chen

DISTRIBUTED STABLE STATES FOR PROCESS NETWORKS – ALGORITHM, ANALYSIS, AND EXPERIMENTS ON INTEL SCC................1144
D. Rai, L. Schor, N. Stoimenov, L. Thiele

DISTRIBUTED RUN-TIME RESOURCE MANAGEMENT FOR MALLEABLE APPLICATIONS ON MANY-CORE PLATFORMS................1154
I. Anagnostopoulos, V. Tsoutouras, A. Bartzas, D. Soudris

NETSHIP: A NETWORKED VIRTUAL PLATFORM FOR LARGE-SCALE HETEROGENEOUS DISTRIBUTED EMBEDDED SYSTEMS................1160
Y. Jung, J. Park, M. Petracca, L. Carloni

EXPLOITING JUST-ENOUGH PARALLELISM WHEN MAPPING STREAMING APPLICATIONS IN HARD REAL-TIME SYSTEMS................1170
J. Zhai, M. Bamakhrama, T. Stefanov

ON ROBUST TASK-ACCURATE PERFORMANCE ESTIMATION................1178
Y. Xu, B. Wang, R. Hasholzner, R. Rosales, J. Teich

STOCHASTIC RESPONSE-TIME GUARANTEE FOR NON-PREEMPTIVE, FIXED-PRIORITY SCHEDULING UNDER ERRORS................1184
P. Axer, R. Ernst

HADES: ARCHITECTURAL SYNTHESIS FOR HETEROGENEOUS DARK SILICON CHIP MULTI-PROCESSORS................1191
Y. Turakhia, B. Raghunathan, S. Garg, D. Marculescu

HIERARCHICAL POWER MANAGEMENT FOR ASYMMETRIC MULTI-CORE IN DARK SILICON ERA................1198
T. Muthukaruppan, M. Pricopi, V. Venkataramani, T. Mitra, S. Vishin

PEAK POWER REDUCTION AND WORKLOAD BALANCING BY SPACE-TIME MULTIPLEXING BASED DEMAND-SUPPLY MATCHING FOR 3D THOUSAND-CORE MICROPROCESSOR................1207
S. Manoj, K. Wang, H. Yu

TECHNIQUES FOR ENERGY-EFFICIENT POWER BUDGETING IN DATA CENTERS................1213
X. Zhan, S. Reda

TEMPERATURE AWARE THREAD BLOCK SCHEDULING IN GPGPUS................1220
R. Nath, R. Ayoub, T. Rosing

VAWOM: TEMPERATURE AND PROCESS VARIATION AWARE WEAROUT MANAGEMENT IN 3D MULTICORE ARCHITECTURE................1226
H. Tajik, H. Hornayoun, N. Dutt

ON THE POTENTIAL OF 3D INTEGRATION OF INDUCTIVE DC-DC CONVERTER FOR HIGH-PERFORMANCE POWER DELIVERY................1234
S. Carlo, W. Yueh, S. Mukhopadhyay

FULL-CHIP MULTIPLE TSV-TO-TSV COUPLING EXTRACTION AND OPTIMIZATION IN 3D ICS................1242
T. Song, C. Liu, Y. Peng, S. Lim

AN ACCURATE SEMI-ANALYTICAL FRAMEWORK FOR FULL-CHIP TSV-INDUCED STRESS MODELING................1249
Y. Li, D. Pan

SPEEDING UP COMPUTATION OF THE MAX/MIN OF A SET OF GAUSSIANS FOR STATISTICAL TIMING ANALYSIS AND OPTIMIZATION................1257
V. Kuruvilla, D. Sinha, J. Piaget, C. Visweswariah, N. Chandrachoodan

INTIMEFIX: A LOW-COST AND SCALABLE TECHNIQUE FOR IN-SITU TIMING ERROR MASKING IN LOGIC CIRCUITS..1264
F. Yuan, Q. Xu
IMPROVING PUF SECURITY WITH REGRESSION-BASED DISTILLER..1270
C. Yin, G. Qu
ON THE CONVERGENCE OF MAINSTREAM AND MISSION-CRITICAL MARKETS...................1276
S. Girbal, M. Moreto, A Grasset, J. Abella, E. Quinones, F. Cazorla, S. Yehia
Author Index

Routing Congestion Estimation with Real Design Constraints

Wen-Hao Liu[1,4], Yaoguang Wei[2], Cliff Sze[3], Charles J. Alpert[3],
Zhuo Li[3], Yih-Lang Li[1], Natarajan Viswanathan[3]

[1]Department of Computer Science, National Chiao-Tung University, Hsin-Chu, Taiwan
[2]Department of Electrical and Computer Engineering, University of Minnesota, Minneapolis, MN, USA
[3]IBM Austin Research Lab, Austin, TX, USA
[4]Department of Computer Science, National Tsing-Hua University, Hsin-Chu, Taiwan
dnoldnol@gmail.com; weiyg@umn.edu; {csze, alpert, lizhuo}@us.ibm.com; ylli@cs.nctu.edu.tw; nviswan@us.ibm.com

Abstract *– To address the routability issue, routing congestion estimators (RCE) become essential in industrial design flow. Recently, several RCEs [1-4] based on global routing engines are developed, but they typically ignore the effects of routing on timing so that the identified routing paths may be overlong and thus impractical. To be aware of the timing issues, our proposed global-routing-based RCE obeys the layer directive and scenic constraints to respectively limit the routing layers and the maximum routing wirelength of the potentially timing-critical nets. To handle the scenic constrains, we propose a novel method based on a relaxation-legalization scheme. Also, because the work in [5] reveals that congestion ratio is a better indicator than overflow to evaluate routability, this work focuses on minimizing the congestion ratio rather than overflows. As will be shown, the problem of minimizing congestion ratio is more complicated than minimizing overflows, so we develop a new rip-up and rerouting scheme to reduce congestion and further to approach a target congestion ratio. Moreover, to fit the demands of practical uses, this work presents a control utility to trade off runtime and quality, which is an essential function to an industrial RCE tool. Experiments reveal that the proposed RCE is faster and more accurate than another industrial global-routing-based RCE.*

Categories and Subject Descriptors
B.7.2 [**Integrated Circuits**]: Design Aids - Routing.

General Terms
Algorithms, Design.

Keywords
Congestion estimation, global routing, routability.

1. INTRODUCTION

In advanced technology nodes, routability has become a critical issue because considerable blockages, interconnects and complex designs rules worsen the routing congestion. To avoid wasting time on routing unroutable designs, a routing congestion estimator (RCE) can help designers to judge a design whether is routable in the early stages to speed up the design closure. Moreover, a RCE can cooperate with placers or optimizers to do optimization with consideration of routability. Therefore, the demand of a fast and accurate RCE is stringent in industry.

Two methods are generally adopted to estimate the routing congestion. The probabilistic-based RCEs [6,7] are fast, but typically fail to capture actual routing behavior and thus have low estimation accuracy. The global-routing-based RCEs [1-4] can identify more precisely routing congestion information than probabilistic-based RCEs, but they are still not accurate enough since timing and local

Permission to make digital or hard copies of part or all of this work for personal or classroom use is granted without fee provided that copies are not made or distributed for profit or commercial advantage and that copies bear this notice and the full citation on the first page. Copyrights for components of this work owned by others than ACM must be honored. Abstracting with credit is permitted. To copy otherwise, to republish, to post on servers or to redistribute to lists, requires prior specific permission and/or a fee.

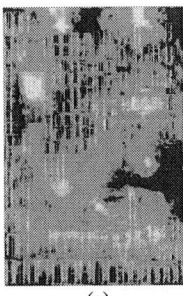  
     (a)   (b)   (c)

Fig. 1. Congestion maps obtained by different RCEs: (a) Proposed RCE ignoring the scenic constraint; (b) Proposed RCE considering the scenic constraint; (c) a RCE from a full-blown industrial router.

congestion issues are ignored. The goal of this work is to develop a fast and accurate global-routing-based RCE that takes timing and local congestion issues into account.

To be aware of timing issue, the author of [8] suggests that designers formulate scenic and layer directive constraints for the timing-critical nets before feeding these nets into a router or RCE. For a net, the scenic constraint restricts its maximum routing wirelength, and the layer directive constraint specifies a range of metal layers that the net can legally route on. A net obeying the scenic constraint can avoid detouring too much and thus causing the timing violations. The layer directive constraint forces the timing-critical nets route on the higher metal layers since the wires on the higher layers have smaller delay in advanced technology nodes. Some papers [9-11] address the global routing problem with the layer directive constraint. However, how to handle the scenic constraint is seldom discussed. Figures 1(a) and 1(b) show the congestion map of the proposed RCE without and with consideration of the scenic constraint, and Fig. 1(c) shows the congestion map obtained by a congestion estimator from a full-blown industrial router which is very accurate but is much slower than the proposed RCE. Figure 1 reveals that the proposed RCE ignoring the scenic constraint cannot correctly estimate the congestion hot spots in the top-right corner, because the nets in Fig. 1(a) detour too much to dissolve the congestion. However, when the timing issue is considered, the real router cannot detour so much to dissolve the congestion in the top-right corner.

In traditional global routing problem, overflow happens when a region's routing demand exceeds its routing capacity, and global routers focus on minimizing overflows. Notably, a region in global routing model usually means a global-cell (g-cell) or a grid edge. However, the duty of a RCE is more than that. For example, even given two overflow-free designs, a RCE should evaluate which design is easier to route. Accordingly, a better objective for RCE is to reduce congestion to approach a **target congestion ratio** (TCR). The **congestion ratio** of a region is the ratio of the region's routing demand to the region's routing capacity. To an industrial router, a region is easy to route if the region's maximal congestion ratio is lower than TCR, while a region's routing difficulty increases beyond linearly as the region's congestion ratio increases over TCR. Compared to minimizing overflows, we find that the routing problem becomes more complicated with the objective to minimize congestion ratio. For example, given two overflowed nets in a region with limited routing resource; we can get an overflow-free routing result when our objective is to minimize overflows. However, when the objective is to reduce congestion ratio to approach TCR (typically smaller than 1), the first routed net may use too many routing resource in order to meet TCR, resulting in the second routed net has no routing resource to solve its overflow, not to mention approaching TCR.

The authors of [5, 17] indicate that local congestion causes a big mismatch between global routing and detailed routing but most global-routing-based RCEs [1-4] ignore the effect of local congestion. Namely, a good routing result obtained by [1-4] may not imply that a feasible detailed routing solution exists. Therefore, to obtain more accurate routability estimation in terms of detailed routing, we formulate the local congestion estimation metric presented by [5] into the routing model of the proposed RCE.

A control utility to trade off runtime and quality is important to industrial RCE tools, also users require a control utility to limit the runtime of RCEs not being crazy for over-congested designs, i.e., users hope the runtime for the same scale designs even with different congestion conditions to be similar and stable. However, the existing academic RCE tools hardly satisfy this requirement. CGRIP [4] uses a timeout constraint to be the termination condition. Since different machines have different performance, CGRIP gets nondeterministic results when it runs on different machines. BFG-R [1] and NCTU-GR [2] set a maximum routing iteration to be the termination condition. However, the routing efforts in each iteration for designs with different congestion levels are quite different, so it is difficult to find a unique number of maximum routing iteration that works for all the designs, in order to control the runtime well.

This work presents a global-routing-based RCE that includes a unified routing flow to consider all the following factors: local congestion, TCR, scenic and layer directive constraints, and runtime control. This work has following contributions:

(a) A relaxation-legalization method is presented to handle the scenic constraint; it can escape from the local optimum to obtain better solution quality.

(b) This work presents a TCR-driven rip-up and rerouting (R&R) scheme, the proposed RCE with this scheme would solve peak congestion first in a limited runtime budget. If the runtime budget is sufficient, the proposed RCE would gradually reduce congestion to approach TCR.

(c) By the proposed throughput controlling method, users can trade off runtime and quality to get deterministic routing results. The runtime would linearly increase and the routing quality is monotonically improved as users increase the value of a tunable parameter.

The rest of paper is organized as follows. Section 2 describes the problem of global-routing-based RCE and introduces previous works. Section 3 presents the proposed global-routing-based RCE to tackle industrial requirements and constraints. Section 4 summarizes the experimental results. Finally, Section 5 draws conclusions.

2. PRELIMINARIES

2.1. Problem Description

The proposed global-routing-based RCE can be treated as a very fast global router. In global routing problem, the placement is typically partitioned into an array of uniform g-cells to form a 3-dimension (3D) grid graph $G(V, E)$, where V denotes the set of g-cells, and E refers to the set of grid edges. Each grid edge is termed by the proximity of the related g-cells to its two end nodes. The capacity $c(e)$ indicates the number of routing tracks on grid edge e. The blocked capacity $b(e)$ refers to the number of routing tracks

blocked by blockages on e. The routing demand $d(e)$ indicates the number of global routing paths passing through e. The overflow of e is defined as $max(0, b(e)+d(e)-c(e))$, and most global routers focus on minimizing overflows. In this work, we redefine overflow of e as $max(0, b(e)+d(e)+w(e)-c(e))$ where $w(e)$ is the estimated number of routing tracks on e that may be used for local routes, and $w(e)$ is obtained by the method in [5]. Note that, in modern designs, different layers may have varying wire width and wire spacing, so the capacity of the grid edges on different layers can be different.

Rather than minimizing overflows, the objective of this work is to efficiently minimize congestion to approach TCR as close as possible. This work sets TCR to 80% because, in the industrial environment where we test our RCE, 20% of the routing capacity is reserved to route non-signal nets such as clock nets later in the design flow. Moreover, each net in the routing result has to obey the scenic and layer directive constraints. The scenic constraint restricts that the wirelength of each net n cannot exceed an upper bound $B(n)$, and the layer directive constraint restricts that n only can route on the layers between $tl(n)$ and $bl(n)$ where $tl(n)$ and $bl(n)$ can be any layer among all layers, but $tl(n)>bl(n)$. In addition, to satisfy the industrial demands, a control utility to trade off RCE's runtime and routing quality is required.

2.2. Congestion Evaluation Metrics

This work adopts a net-based metric and a grid-edge-based metric to evaluate the routing congestion. Given a routing result, we first calculate the congestion ratio $g(e)$ for each grid edge e by the following equation,

$$g(e) = [d(e) + b(e) + w(e)]/c(e). \qquad (1)$$

After that, net-based congestion metric $WCI(x)$ is defined as the number of nets whose congestion is greater or equal to x%, the congestion of a net is the maximum congestion ratio among all grid edges traversed by the net. Moreover, grid-edge-based congestion metric $ACE(y\%)$ is computed by averaging the congestion ratio of the top y% congested grid edges. In our experiments, we set $y\in\{0.5, 1, 2, 5, 10, 20\}$. $ACE(0.5\%)$ provides a local view for the peak congestion, while $ACE(20\%)$ provides a global view for the average congestion.

2.3. Previous Works

The global router in [3] runs fast enough to serve as a fast RCE tool, which develops two efficient routing algorithms called unilateral monotonic routing and hybrid unilateral monotonic (HUM) routing, and presents a bounding box expansion scheme to avoid over-expansion. The routing kernel of the proposed RCE is based on the innovations in [3] to achieve high performance. However, [3] cannot handle scenic and layer directive constraints, and does not consider the issues of local congestion and TCR.

The global router in [3] adopts a 2D routing with layer assignment framework to solve 3D global routing problem. However, this framework may struggle for the layer directive constraint because it has no precise layer information during 2D routing. For example, a design has 9 layers and a region has congestion between layers 5 and 9. A net with layer directive constraint {1, 9} can legally pass through this region, but a net with layer directive constraint {7, 8} would suffer congestion when it passes through this region. However, the 2D routing stage in [3] cannot distinguish these two cases.

The global router GLADE [9,10] maintains a virtual demand data structure during 2D routing to query the estimated 3D congestion between two specified layers. In contrast, the work in [11] adopts a grouping method to handle the layer directive constraint, which classifies the nets with the same layer directive constraint into a group, and then sorts each group in increasing order according to the range of each group's layer directive. Next, each group is sequentially processed during 2D routing. For each group, only one

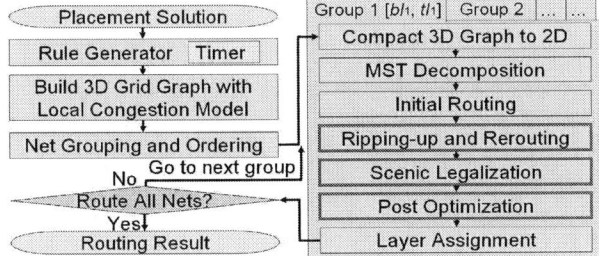

Fig. 2. Design flow of the proposed RCE

layer directive constraint needs to be addressed, that simplifies the problem. Although GLADE [9,10] can explore larger solution space to get better results, [11] is faster. Due to the runtime consideration, the grouping method [11] is adopted in this work.

3. PROPOSED RCE

Figure 2 shows the design flow of the proposed RCE, in which the red boxes highlight the stages including our innovations. At first, given a placement solution and a netlist, a rule generator is invoked to generate the layer directive and scenic constraints for each net. The rule generator uses an industrial timer to analyze the timing slack for each net, and then the method in [12] is used to assign layer directive constraints for the timing-critical nets. Moreover, for each net n, the scenic constraint $B(n)$ is identified by each net's minimum spanning tree (MST) length multiplying a variable p_n that is between 1.05 and 1.4. Typically, p_n of most nets (>90%) is around 1.15 in the industrial environment used in this work and some timing-critical nets have smaller p_n. Because the rule generator is not the focus in this work, the details of the rule generator are skipped. After that, a 3D grid graph is built and the local-congestion-aware factor $w(e)$ in Eq. (1) is calculated by the pin-density method in [5].

Before performing routing, the grouping method [11] is used to classify nets into several groups and then sorts groups. Subsequently, the proposed RCE routes nets group by group. At the beginning of processing a group with layer directive $\{bl_i, tl_i\}$, the partial 3D grid graph from layer bl_i to tl_i is projected on the 2D grid graph and each net in this group is decomposed into two-pin subnets based on the topology of the MST. Note that, we call a two-pin subnet as a *segment* in this paper. If a segment s belongs to a net n, n is the *parent net* of s. In this work, we use MST instead of Steiner tree to decompose the nets, because the works in [1-3] indicate that MST offers better flexibility than Steiner tree to avoid blockages or congestion. Also, we found that the runtime and wirelength of FLUTE [13] (a well-known Steiner tree generator) are worse than MST for some modern industrial designs, because these designs contain considerable high fan-out nets and MST runs faster and produce shorter wirelength for very high fan-out nets than FLUTE. Accordingly, the proposed RCE adopts the algorithm in [14] to efficiently build MST. After that, the initial routing stage generates an initial routing result by pattern routing and monotonic routing. Because pattern routing and monotonic routing do not make detours, every net in the initial routing result must obey the scenic constraint. Next, the ripping-up and rerouting (R&R) stage iteratively reroutes the congested segments until either the congestion of any net is not greater than TCR or the termination condition is satisfied. The definition of congested segments and the termination condition will be introduced later.

In the R&R stage, the scenic constraint is relaxed as a soft constraint, namely the routing solutions in this stage can violate the scenic constraint. Then, the scenic legalization stage reroutes the nets with the scenic violation to force them to satisfy the scenic constraint, but this stage may increase congestion. In the post optimization stage, the congested segments are rerouted under the hard scenic constraint to reduce the congestion worsened by the scenic legalization stage.

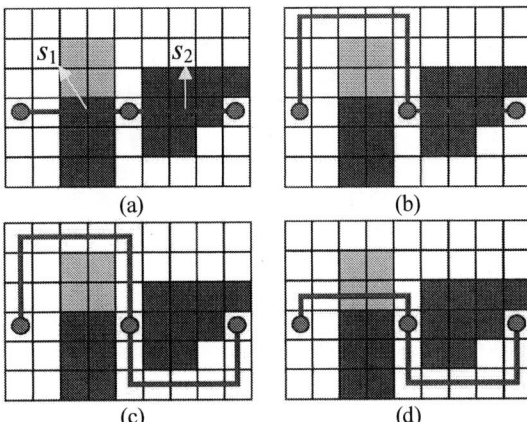

Fig. 3. (a) the initial routing result of a net; (b) a segment exhausts all detour quotas, resulting another segment has no detour quotas to bypass congestion regions; (c) the routing result of the proposed R&R stage; (d) the routing result after the scenic legalization stage.

Using this relaxation-legalization method to handle the scenic constraint can get better results than always treating the scenic constraint as a hard constraint. Finally, we adopt a fast heuristic layer assignment algorithm presented in [15] to map the 2D routing result to the 3D grid graph. The proposed RCE sequentially processes each group until the routing solutions of all nets are obtained.

3.1. Relaxation-Legalization Method to Handle Scenic Constraint

Most global routers decompose multi-pin nets into several segments and then route each segment individually, since it can reduce the multi-terminal routing problem to a two-point routing problem. However, it is not easy to handle the scenic constraint when using this method. For instance, Fig. 3(a) shows an initial routing result of a three-pin net n with two segments s_1 and s_2, in which the light and dark gray rectangles respectively denote the light congestion and heavy congestion regions. The wirelength of n is 8 and the scenic constraint restricts that the wirelength of n cannot exceed 14, namely n has 6 detour quotas. If s_1 is routed earlier than s_2 and is allowed to use all detour quotas, we will get a routing result shown in Fig. 3(b), in which s_2 cannot bypass the congestion regions because the detour quotas have run out. Suppose we averagely allocate the detour quotas to s_1 and s_2 to avoid s_1 exhausting all detour quotas, both s_1 and s_2 cannot bypass the congestion regions because they both have no enough detour quotas.

This work presents a relaxation-legalization method to handle the scenic constraint. At first, we allow the routing solutions with scenic violations for reducing congestion. Then, the routing cost of each grid edge is dynamically adjusted to encourage the routing solutions gradually fitting the scenic constraint. In this work, $r_c(e,s)$ denotes the routing cost of grid edge e for segment s, which consists of $c_c(e)$ and $w_c(s)$, where $c_c(e)$ is the congestion cost on grid edge e and $w_c(s)$ is the wirelength cost of segment s passing through any grid edge.

To avoid that the routing wirelength of segment s exceeds its scenic constraint, we can adjust the ratio between $c_c(e)$ and $w_c(s)$ to indirectly control the routing wirelength of s. For example, assuming the initial detour-free routing path of s passes through a congested grid edge e and a least-cost routing algorithm is adopted to reroute s. If $c_c(e)$ dominates $r_c(e,s)$, the algorithm may identify a detoured path to bypass the congestion on e. If $w_c(s)$ dominates $r_c(e,s)$, the algorithm may identify a path passing through e to save wirelength. Notably, the least-cost routing algorithm here can be any shortest path finding algorithm such like A* search or Dijkstra's algorithm, and this work uses unilateral monotonic and HUM routing.

647

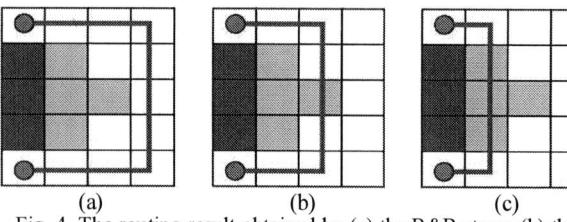

(a)　　　　　　(b)　　　　　　(c)

Fig. 4. The routing result obtained by (a) the R&R stage; (b) the first iteration of the soft legalization phase; (c) the second iteration of the soft legalization phase.

For the case in Fig. 3(a), the proposed RCE first generates a routing solution with the scenic violation for reducing congestion (Fig. 3(c)). Next, the proposed RCE iteratively reroutes s_1 and s_2, and gradually increases $w_c(s_1)$ and $w_c(s_2)$ to drive them to route shorter until net n satisfies its scenic constraint. In our example, because s_1 faces a lighter congestion region than s_2, driving s_1 to take the shorter path is easier than driving s_2. Thus, we can get the routing result in Fig. 3(d).

3.1.1 Ripping-up and Rerouting Stage

This stage treats the scenic constraint as a soft constraint; each net can violate its scenic constraint to solve congestion. However, too many nets with scenic violations would make the subsequent scenic legalization stage sacrifice congestion a lot to satisfy the scenic constraint. Accordingly, if the parent net n of segment s has scenic violations, this stage increases $w_c(s)$ before rerouting s to prevent the new routing solution violating the scenic constraint again. This stage formulates $w_c(s)$ as follows:

$$w_c(s) = \alpha + \beta \times (v_s)^2 + \gamma \times (v_n)^2, \qquad (2)$$

where α, β and γ are user defined constants, we set them to be 500, 20 and 50 in our implementation, respectively. Notations v_n and v_s initially are zero. Once the wirelength of n exceeds $B(n)$ after rerouting s, v_n increases by one. Once the wirelength of s exceeds $b(s)$ after rerouting s, v_s increases by one, where $b(s)$ is the suggested wirelength for s, that is linearly proportional to the ratio of initial wirelength of s and n, where the initial wirelength means the wirelength of a net or a segment at the beginning of the R&R stage. For example, $B(n)$ is 30 and n consists of three segments s_1, s_2 and s_3 whose initial wirelength are 6, 10 and 4, respectively. The $b(s_1)$, $b(s_2)$ and $b(s_3)$ are 9 (=6*(30)/(6+10+4)), 15 (=10*3/2) and 6 (=4*3/2), respectively.

In this stage, $c_c(e)$ also increases if e frequently becomes congested, which makes the least-cost routing algorithms more actively bypass e to reduce congestion. So, we cannot guarantee that the routing wirelength of s always becomes shorter when $w_c(s)$ and $c_c(e)$ both are increased. The formulation of $c_c(e)$ will be introduced in section 3.2.

The routing order of segments would influence the quality of routing results and the runtime of the proposed RCE. In this work, each routing stage routes long segments earlier than short segments, because short segments usually need to make additional detours to bypass congestions that may result in scenic violations.

3.1.2 Scenic Legalization Stage

This stage consists of the soft and hard legalization phases. At the beginning of the soft legalization, an array is built to contain all segments whose parent net voilates the scenic constraint. Then, the array is iteratively scanned to check each segment s in the array. If the parent net of s has the scenic violation and the wirelength of s is longer than the Manhattan distance between its two terminals, we increase $w_c(s)$ and then reroute s. In this stage, $c_c(e)$ would not continually increase even e frequently becomes congested, so we can guarantee that the routing wirelength of s monotonically decreases as s is rerouted and $w_c(s)$ is increased.

The goal of the soft legalization is to reduce the number of nets with scenic violations in a few iterations but does not sacrifice congestion too much. How to properly increase $w_c(s)$ is challenging. Increasing $w_c(s)$ too much would worsen congestion a lot. If the increment of $w_c(s)$ compared to $c_c(e)$ is slight, the reduction of the scenic violations is also slight in each iteration.

The soft legalization uses a cost accumulation method to update $w_c(s)$ properly. Before rerouting s, we trace the path of s and compute the path cost of s, $pathCost(s)$, which is the sum of the routing cost of each grid edge passed by the path of s, and then $w_c(s)$ is updated by the following equation.

$$w_c(s) = \lceil pathCost(s)/manh(s) \rceil, \qquad (3)$$

where $manh(s)$ denotes the Manhattan distance between the two terminals of s. Figure 4 shows an example to illustrate how Eq. (3) works. Figure 4(a) shows a routing solution of s with the scenic violation at the beginning of the soft legalization, in which $w_c(s)$ is 1. Assume $c_c(e)$ is 4 for edge e which is light congested (light grey edges in the figure); if e is very congested (dark grey edges), $c_c(e)$ is 10; otherwise (white edges), $c_c(e)$ is 0. Before rerouting s, we trace the path of s in Fig. 4(a) to obtain $pathCost(s)$ that is 10. Based on Eq. (3), $w_c(s)$ is updated to 2.5 (=10/4) and then rerouting s with new $w_c(s)$ obtains the routing solution in Fig. 4(b). If the routing solution in Fig. 4(b) still violates the scenic constraint, $w_c(s)$ is updated to 6 (=(8*2.5+4)/4) and then s is rerouted again in the next iteration. After that, the routing solution in Fig. 4(c) is obtained. The example in Fig. 4 reveals that the wirelength of s monotonically decreases as the routing iterations increases.

In our implementation, the soft legalization performs three iterations, which can eliminate most scenic violations. However, there are few remained scenic violations that need to be solved by the hard legalization. In the hard legalization phase, we adopt the monotonic routing to reroute the segments whose parent net has the scenic violation. Because monotonic routing makes no detour, a net must conform to its scenic constraint when all segments of the net are routed by monotonic routing. After the hard legalization phase, all nets must obey the scenic constraints.

3.1.3 Post Refinement Stage

The goal of the post refinement stage is to reroute the congested segments to reduce the congestion which may be worsened by the scenic legalization stage. For example, after the R&R stage, two neighboring segments s_1 and s_2 both have congestion-free routing paths but the parent net of s_1 violates the scenic constraint. After the scenic legalization stage, s_1 is rerouted to be shorter, in which causes s_1 now to compete the routing resource with s_2 and thus causes the congestion. In this case, rerouting s_2 may dissolve the congestion.

In the post refinement stage, the routing path of each segment is traced once. A segment is rerouted if the segment passes through a congestion region and has not been rerouted in the scenic legalization stage. To ensure that the routing solutions in this stage always obey the scenic constraint, if the new routing solution obtained by rerouting violates the scenic constraint, we discard the new routing solution and restore the segment to the old routing solution. Because the segments that have been rerouted in the scenic legalization stage commonly have no detour quota to legally bypass congestion, the post refinement stage does not reroute these segments to save time.

3.2. TCR-driven R&R Scheme

The problem of minimizing congestion to approach TCR (80% in this work) is more complicated than minimizing overflows. For example, given a limited runtime budget, a RCE may spend most time on rerouting the nets whose congestion is between 80% and 100% to meet TCR, finally has no time to reroute the nets whose congestion is over 100%. In another case, given a region with limited routing resources, the early routed nets may consume too many routing resources when they target to meet TCR, causing the later routed nets have no routing resources to reduce congestion. Based on

648

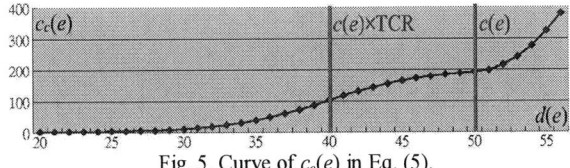

Fig. 5. Curve of $c_c(e)$ in Eq. (5).

TABLE 1 DESIGN INFORMATION

	Grid size	#nets(10^5)		Grid size	#nets(10^5)
Ind1	357×535×9	13.3	Ind4	227×740×11	16.7
Ind2	1068×710×10	40.5	Ind5	444×518×10	16.6
Ind3	1068×710×10	41.7	Ind6	534×407×10	9.2

above cases, we can summarize two requirements for TCR-driven RCE: (1) route the nets with peak congestion first in the limited runtime budget; (2) reduce congestion averagely. For example, two nets with 90% congestion are better than one with 80% and another with 100%.

The R&R stage adopts a TCR-driven R&R scheme to satisfy abovementioned two requirements. In the R&R stage, if a segment passes through a grid edge e satisfying the following inequation, the segment is treated as a **congested segment**. The congested segment will be rip-upped and rerouted.

$$g(e) > \max(1 - p \times \lambda, TCR) , \qquad (4)$$

where p is an iteration count and initially is zero, p increases by one as iteration increases; λ is a user defined constant, we set λ to 0.02. According to Eq. (4), the first iteration of the R&R stage would reroute the nets whose congestion is greater than 100%, the second iteration would reroute the nets whose congestion is greater than 98%, and so on. The R&R stage iteratively reroutes the congested segments until either no net has congestion greater than TCR or termination condition meets. This R&R scheme can solve the peak congestion first to meet requirement (1).

To meet requirement (2), we formulate the congestion cost $c_c(e)$ as follows:

$$
\begin{aligned}
c_c(e) &= [1 + \frac{C_1}{1 + C_2{}^{C_3 \times to(e)}} + C_4 \times (o(e))^2] \times [1 + (h_e)^2] \\
o(e) &= \max(0, d(e) + w(e) + b(e) - c(e)) \\
to(e) &= d(e) + w(e) - \max(0, c(e) \times TCR - b(e))
\end{aligned}
\qquad , \qquad (5)
$$

where C_1, C_2, C_3 and C_4 are user defined constants, we set them to 200, 2.72, -0.3 and 5, respectively. History cost h_e is initialized to zero for each grid edge e at the beginning of the R&R stage. If the routing solution of s passes through e after rerouting s and $to(e)>0$, h_e increases by one. Notably, $to(e)>0$ means that the congestion ratio of e is greater than TCR. Figure 5 shows that the curve of $c_c(e)$ when $c(e)$, $w(e)$, $b(e)$, TCR and h_e are 50, 0, 0, 80% and 0, in which x-axis and y-axis respectively denotes $d(e)$ and $c_c(e)$. Using Eq. (5) to formulate $c_c(e)$ encourages that routing algorithms avoid overflows but do not use too much routing resource for pushing congestion down to TCR. However, if the congestion of e is continuously over TCR in the R&R stage, $c_c(e)$ gradually increases since h_e increases. This makes the routing algorithms become more active to push congestion down to TCR when the routing iteration increases. Notably, h_e would not increase in the scenic legalization and post refinement stages.

3.3. Throughput Controlling

A control utility to trade off runtime and quality is essential to an industrial RCE tool. Based on the feedback from users, we summarize three features that an industrial RCE tool requires: (1) when users increase the value of a tunable parameter t_u, the congestion in the routing result should monotonically improve and the runtime of the RCE tool linearly increases; (2) with the same value of t_u, the RCE tool can automatically give a design more runtime budget to route if the design is larger (with more nets or larger routing grid size); (3) with the same value of t_u, the runtime of a RCE tool can be stable for any congestion circumstance. Because, a RCE tool may cooperate with other tools to optimize a design, but other tools may worsen congestion when they address other issues. Feature (3) ensures that the runtime of the optimization flow can be stable and predictable.

The R&R stage mainly influences the runtime and the routing quality in the proposed RCE; hence this work presents a throughput controlling method to set a termination condition for the R&R stage. In the proposed RCE, the R&R stage for group R terminates if the following condition holds.

$$tvg > t_u \times \mu \times \sum_{s \in S_R} numGcell(s), \qquad (6)$$

where t_u is the tunable parameter to control the tradeoff between runtime and quality, S_R denotes the set of all segments in group R, and μ is a user scaling factor. In our implementation, u is set to 0.5. Moreover, $numGcell(s)$ denotes the number of g-cells within the initial bounding box of s which is the minimum rectangle enclosing the terminals of s. The right term in Eq. (6) can be initialized in the MST decomposition stage. Notation tvg denotes the number of visited g-cells in the R&R stage. Initially, the value of tvg is zero. During the R&R stage, when the path of a segment is traced, tvg increases by the length of the traced path. If the segment is a congested segment, the bounding box of the segment will be expanded and then the segment will be rerouted by the unilateral monotonic and HUM routing algorithms [3] within the bounding box. Because the time complexity of unilateral monotonic and HUM routing is linear to the number of g-cells in the bounding box, tvg increases by the number of g-cells in the bounding box after a segment is rerouted. We can treat tvg as an indicator for the computation throughput, and we can adjust t_u to control the upper bound of the computation throughput and easily satisfy the abovementioned three features.

4. EXPERIMENTAL RESULTS

The proposed algorithms are implemented in C/C++ on a linux server with four 2.27 GHz Xeon E7-8860 CPUs. In this work, there are many tunable parameters that are set based on empirical results. However, although tuning these parameters can obtain better results, we only adopt a single set of parameters for every design used in the following experiments. Table 1 shows the grid size and the net number of the adopted industrial designs.

This work compares the proposed RCE (TPR$_{all}$) with a fast industrial congestion estimator (CA$_{ind}$) and a full-blown congestion estimator (GR$_{ind}$) from an industrial router in Table 2. TPR$_{all}$ includes all innovations presented in this paper; we set parameter t_u in Eq. (6) to 5 to strike a good balance between the routing quality and runtime. Due to the paper's space limitation, the effectiveness of individual innovations will be detailed in the supplement part. CA$_{ind}$, similar to that in [5], is a fast global-routing-based RCE based on the edge-shifting algorithm [16] to avoid invoking time-consuming maze routing algorithms. The edge-shifting method is fast but may struggle for the very congested situations because the solution space of the edge-shifting method is restricted. GR$_{ind}$, with a complex local congestion model to consider the local routability, runs most accurate and will be used to judge the quality of all results but is much slower than TPR$_{all}$ and CA$_{ind}$. Notably, CA$_{ind}$ and GR$_{ind}$ both can handle layer directive and scenic constraints, and take the local congestion and TCR issues into account. In terms of the scenic constraint, CA$_{ind}$ always treats the scenic constraint as a hard constraint, and the situation where a segment may exhaust all detour quotas like the example in Fig. 3(b) may happen. In contrast, GR$_{ind}$ gradually relaxes the scenic constraint for the rerouted nets, so the routing results of GR$_{ind}$ may violate the scenic constraint.

Table 2 adopts a net-based metric WCI and a grid-edge-based metric ACE to evaluate the routing results of TPR$_{all}$, CA$_{ind}$ and GR$_{ind}$, in which ACE metrics are shown as percentage (%). Table 2 shows

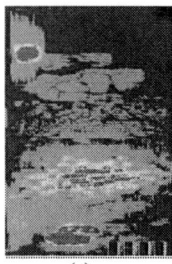 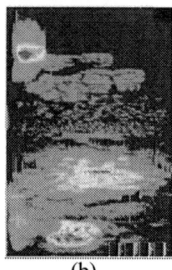 

(a)   (b)   (c)

Fig. 6. Congestions maps of Ind4 obtained by (a) CA_{ind}, (b) TPR_{all} and (c) GR_{ind}.

that TPR_{all} respectively runs 2.6× and 108.5× faster than CA_{ind} and GR_{ind} on average. (TPR_{all} and CA_{ind} use one thread, while GR_{ind} uses four threads). In addition, the routing results obtained by TPR_{all} have better congestion than that obtained by CA_{ind}, and the congestion analysis of the routing results obtained by TPR_{all} is similar to that of GR_{ind}. Table 2 reveals that TPR_{all} can achieve fast and accurate congestion estimation. Figure 6 shows the congestion maps obtained by TPR_{all}, CA_{ind} and GR_{ind} for design Ind4.

5. CONCLUSIONS

This work presents a unified routing flow to develop a fast and accurate global-routing-based RCE that takes TCR, local congestion, scenic and layer directive constraints into account. The goal of this work is to minimize congestion ratio rather than overflows, the proposed TCR-driven R&R scheme can offer a satisfactory routing quality in a limited runtime budget. Further, a relaxation-legalization method is presented to handle the scenic constraint, which can escape from the local optimum to get better solution quality. Finally, by the proposed throughput controlling method, users can trade off the runtime and quality of the proposed RCE to get deterministic routing results. Experiments reveal that the proposed RCE is faster and more accurate than another industrial global-routing-based RCE.

REFERENCES

[1] J. Hu *et al*, "Completing high-quality global routes," in *Proc. ISPD*, pp. 35-41, 2010.

[2] W.-H. Liu *et al.*, "Multi-threaded collision-aware global routing with bounded-length maze routing," in *Proc. DAC*, pp. 200-205, 2010.

[3] W.-H. Liu *et al.*, "A fast maze-free routing congestion estimator with hybrid unilateral monotonic routing," in *Proc. ICCAD*, pp. 713-719, 2012.

[4] H. Shojaei *et al.*, "Congestion analysis for global routing via integer programming," in *Proc. ICCAD*, pp. 256-262, 2011.

[5] Y. Wei *et al.*, "GLARE: global and local wiring aware routability evaluation," in *Proc. DAC*, pp. 768-773, 2012.

[6] J. Lou *et al.*, "Estimating routing congestion using probabilistic analysis," *IEEE TCAD*, 21(1), pp. 32-41, 2002.

[7] J. Westra *et al.*, "Probabilistic congestion prediction", in *Proc. ISPD*, pp. 204-209, 2004.

[8] M. D. Moffitt, "Global routing revisited," in *Proc. ICCAD*, pp. 805-808, 2009.

[9] Y.-J. Chang *et al.*, "GLADE: A modern global router considering layer directives," in *Proc. ICCAD*, pp.319-323, 2010.

[10] T.-H. Lee *et al.*, "An enhanced global router with consideration of general layer directives," in *Proc. ISPD*, pp.53-60, 2011.

[11] M. D. Moffitt and C. N. Sze, "Wire synthesizable global routing for timing closure," in *Proc. ASP-DAC*, pp. 545-550, 2011.

[12] Y. Wei *et al.*, "CATALYST: planning layer directives for effective design closure," in *Proc. DATE*, pp. 1873-1878, 2013.

[13] C. Chu and Y.-C. Wong, "FLUTE: fast lookup table based rectilinear steiner minimal tree algorithm for VLSI design," *IEEE TCAD*, 27(1), pp. 70-83, 2008.

[14] Hai Zhou *et al.*, "Efficient minimum spanning tree construction without Delaunay triangulation," *Information Processing Letter*, pp. 271-276, 2002.

[15] K.-R. Dai *et al.*, "NCTU-GR: efficient simulated evolution-based rerouting and congestion-relaxed layer assignment on 3-D global routing", *IEEE TVLSI*, 20(3), pp. 459-472, 2012.

[16] M. D. Moffitt, "MaizeRouter: engineering an effective global router," *IEEE TCAD*, 27(11), pp. 2017-2026, 2008.

[17] W.-H. Liu *et al.*, "Case study for placement solutions in ISPD11 and DAC12 routability-driven placement contests," in *Proc. ISPD*, to appear, 2013.

TABLE 2   ROUTING RESULTS COMPARISON BETWEEN THE PROPOSED RCE, INDUSTRIAL CONGESTION ANALYZER AND REAL ROUTER

		Wall time(s)	WCI(100)	WCI(90)	ACE(0.5%)	ACE(1%)	ACE(2%)	ACE(5%)	ACE(10%)	ACE(20%)	WL (10^7)
Ind1	TPR_{all}	188	2021	18781	92.93	88.88	85.59	82.18	80.84	79.47	27.03
	CA_{ind}	368	13736	46925	98.30	93.49	89.63	85.05	82.41	80.17	26.94
	GR_{ind}	21222	1047	14335	88.66	86.27	84.29	82.85	81.64	80.32	27.05
Ind2	TPR_{all}	436	767	52226	92.70	90.90	88.19	83.72	80.83	76.92	16.01
	CA_{ind}	1987	6845	42893	94.24	91.54	88.50	83.99	80.86	78.47	16.30
	GR_{ind}	14649	41	10514	87.66	86.34	85.22	82.98	80.83	76.08	16.08
Ind3	TPR_{all}	422	4257	166705	95.98	94.34	92.38	88.62	84.69	80.90	18.27
	CA_{ind}	1697	16055	157971	97.34	95.37	93.03	89.09	85.38	81.40	17.90
	GR_{ind}	10376	895	29303	89.96	87.92	86.42	84.20	82.48	79.21	17.93
Ind4	TPR_{all}	101	27	44100	92.68	90.95	88.58	84.27	81.46	74.48	8.14
	CA_{ind}	162	22747	78090	101.57	98.62	95.00	88.20	83.70	76.12	7.99
	GR_{ind}	11064	104	9543	87.93	86.05	84.14	81.24	78.01	69.22	8.42
Ind5	TPR_{all}	140	149	130556	93.80	92.58	91.30	89.07	86.85	83.83	11.82
	CA_{ind}	216	4123	277376	95.03	93.77	92.37	90.59	87.91	84.32	12.05
	GR_{ind}	47261	1141	89904	94.62	93.48	92.18	90.91	88.82	85.66	11.39
Ind6	TPR_{all}	51	1	1598	88.47	85.32	82.92	80.63	78.08	71.46	4.69
	CA_{ind}	108	1478	3309	92.38	87.44	83.08	80.24	78.11	72.11	4.54
	GR_{ind}	1672	19	2788	86.33	84.77	83.22	80.63	76.61	67.44	4.56
Ratio	TPR_{all}	1	1	1	1	1	1	1	1	1	1
	CA_{ind}	2.633	394.608	1.706	1.040	1.032	1.024	1.017	1.011	1.012	0.994
	GR_{ind}	108.476	5.215	0.632	0.962	0.967	0.975	0.989	0.991	0.979	0.993

TABLE 3 DESIGN INFORMATION

	Grid size	#nets(10^5)		Grid size	#nets(10^5)
Ind7	800×415×10	9.0	**Ind10**	357×535× 9	13.3
Ind8	774×570×10	8.0	**Ind11**	444×518×10	16.6
Ind9	704×413×10	10.6	**Ind12**	425×516×10	13.9

TABLE 4 DIFFERENT VERSIONS OF THE PROPOSED RCE

TPR_1	Without consideration of the scenic constraint
TPR_2	TPR_1 + Scenic legalization
TPR_3	TPR_2 + Default Eq. (2)
TPR_{all}	TPR_3 + Post Refinement

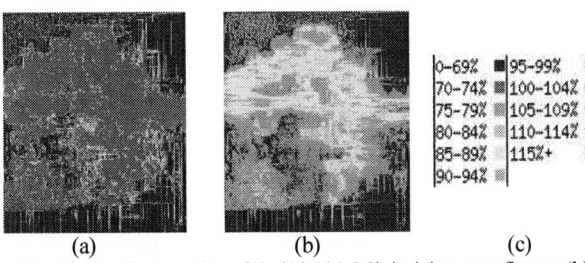

(a)   (b)   (c)

Fig. 7. Routing results of Ind11. (a) Minimizing overflows; (b) minimizing congestion ratio to approach 80%; (c) color scheme

6. SUPPLEMENT

To evaluate individual innovations in this work, we run more detailed experiments in this section. However, due to the licensing issue, the following experiments cannot adopt the designs shown in Table 1. Therefore, the following experiments adopt another suit of industrial designs shown in Table 3 and perform on a 2.4 GHz Xeon-based linux server with E5620 CPU.

6.1. Effectiveness of using Relaxation-Legalization Method to Handle the Scenic Constraint

To evaluate the effectiveness of each stage in our relaxation-legalization method, we built several versions of the proposed RCE to handle the scenic constraint. Table 4 lists the differences between each version, in which TPR_1 does not have scenic legalization and post refinement stages, and set β and γ in Eq. (2) to zero to totally ignore the scenic constraint in its entire flow. In contrast, TPR_2, TPR_3 and TPR_{all} use the different levels of the relaxation-legalization method to handle the scenic constraint. Notably, t_u in Eq. (6) is set to 5 for each version.

Table 5 shows the routing results obtained by each version, in which SV denotes the number of nets with the scenic violation. The scenic legalization stage in TPR_2 can eliminate all scenic violations but worsens the congestion. Averagely, TPR_2 gets better ACE values but worse WCI(100) than TPR_1. Although TPR_2 for design Ind11 obtains the results with better WCI(100) than TPR_1, TPR_2 has worse ACE(0.5%) than that of TPR_1. This phenomenon implies that only relying on either net-based metric or grid-edge-based metric to evaluate the congestion may get a biased view, while using both metrics provides a more comprehensive view.

TPR_3 has the consideration of the scenic constraint in the R&R stage, which reduces the number of nets that need to be legalized in the scenic legalization stage. Hence, the runtime and overhead of congestion degradation in the scenic legalization diminish. Compared to TPR_2, TPR_3 spends shorter runtime to obtain better congestion results. Finally, TPR_{all} includes the post optimization stage to further improve routing results, so TPR_{all} obtains better results than TPR_3.

TABLE 6 ROUTE IND11 WITH DIFFERENT OBJECTIVES

	WCI (10^3)		ACE					
	100%	90%	0.5%	1%	2%	5%	10%	20%
OV	0.5	148.5	99.9	99.6	99.2	98.2	96.3	90.5
CO	4.5	50.2	99.7	97.8	95.3	90.8	87.0	83.2

TABLE 8 ROUTE IND11 WITH EXTRA BLOCKAGES

	WCI (10^3)		ACE					
	100%	90%	0.5%	1%	2%	5%	10%	20%
Ex0	0.08	77.0	93.8	92.4	91.3	89.2	86.6	83.3
Ex2	0.13	87.1	95.8	94.2	93.2	91.1	88.3	85.2
Ex5	2.50	137.7	98.4	97.2	96.1	94.0	91.3	87.6

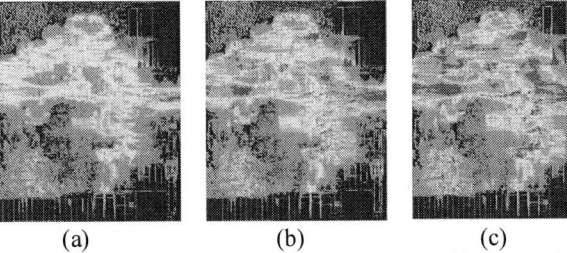

(a)   (b)   (c)

Fig. 8. Routing results of Ind11 (a) without extra blockages; (b) with 2% extra blockages; (c) with 5% extra blockages.

6.2. Minimizing Overflow v.s. Minimizing Congestion

Compared to minimizing overflows, minimizing congestion ratio to approach TCR, say 80%, offers more useful congestion information to help designers to identify the locations of hotspots. Figures 7(a) and 7(b) show the congestion maps of design Ind11 obtained by TPR_{all} addressing on overflow minimization and congestion ratio minimization, respectively. Obviously, Fig. 7(b) is more helpful to distinguish the hotspots' locations than Fig. 7(a). In Table 6, rows **OV** and **CO** list the congestion analysis for Figs. 7(a) and 7(b), respectively. Notably, TPR_{all} both executes around 45 seconds to get the results in rows OV and CO, but OV has lower WCI(100) than CO because OV focuses on eliminating overflows.

6.3. Effectiveness of Throughput Controlling

Table 7 shows TPR_{all} with different values of t_u to trade off routing quality and runtime. Table 7 shows that the ACE metric monotonically improves and the runtime of TPR_{all} almost linearly increases when t_u increases. Also, WCI metric of the routing results except for design Ind9 monotonically improves when t_u increases. Table 7 reveals that the throughput controlling method can practically trade off the runtime and quality.

To demonstrate that TPR_{all} with the same value of t_u offers the stable runtime for any congestion circumstance, we add extra blockages on each grid edge in Ind11 and then use TPR_{all} with t_u=5 to route Ind11 with extra blockages. In Table 8, rows **Ex0**, **Ex2** and **Ex5** respectively show the congestion analyses for the routing results of Ind11 with extra blockages that block 0%, 2% and 5% capacity to each grid edge uniformly, in which the runtimes of Ex0. Ex2 and Ex5 are respectively 45.6, 44.2 and 47.4 seconds. Table 8 indicates that the runtime of TPR_{all} is stable for different congestion circumstances, which can avoid the runtime of TPR_{all} being crazy when other tools that cooperate with TPR_{all} worsen the congestion largely. Figure 8 shows the congestion maps of Ind11 with 0%, 2% and 5% extra blockages.

TABLE 5   EFFECTIVENESS OF USING THE PROPOSED RELAXATION-LEGALIZATION METHOD TO HANDLE THE SCENIC CONSTRAINT

	Versions	CPU(s)	WCI(100)	WCI(90)	ACE(0.5%)	ACE(1%)	ACE(2%)	ACE(5%)	ACE(10%)	ACE(20%)	WL (10^7)	SV
Ind7	TPR_1	40.42	1182	26950	95.50	92.92	90.36	85.51	82.60	79.71	11.18	4102
	TPR_2	46.29	4307	25206	97.63	94.45	91.03	85.46	82.51	79.40	11.06	0
	TPR_3	42.46	3554	25367	97.19	94.09	90.82	85.40	82.49	79.41	11.06	0
	TPR_{all}	45.63	3257	24216	96.61	93.52	90.25	85.03	82.30	79.29	11.06	0
Ind8	TPR_1	43.19	2660	22103	98.25	94.58	91.14	85.27	82.02	76.85	10.94	4368
	TPR_2	44.12	5889	21407	99.08	95.88	91.63	85.13	81.72	76.34	10.83	0
	TPR_3	43.86	5556	21580	98.77	95.65	91.55	85.14	81.74	76.38	10.84	0
	TPR_{all}	44.54	5000	18985	97.93	94.75	90.69	84.61	81.42	76.14	10.83	0
Ind9	TPR_1	30.61	752	9401	103.36	97.01	91.23	84.76	81.74	77.31	10.73	1909
	TPR_2	36.16	2305	10494	98.85	94.36	89.37	83.89	81.21	76.79	10.66	0
	TPR_3	31.13	2106	10488	98.89	94.34	89.35	83.89	81.20	76.78	10.66	0
	TPR_{all}	32.29	1098	9383	97.31	93.76	89.24	83.87	81.22	76.87	10.67	0
Ind10	TPR_1	84.31	1343	13384	92.22	89.31	86.70	83.31	81.47	79.58	24.15	3883
	TPR_2	86.50	1788	16781	92.75	89.63	86.80	83.30	81.45	79.52	24.09	0
	TPR_3	85.89	1491	16683	92.61	89.53	86.75	83.29	81.44	79.52	24.10	0
	TPR_{all}	85.04	1681	13589	91.88	88.60	85.80	82.66	81.10	79.34	24.13	0
Ind11	TPR_1	43.24	157	86159	94.23	92.65	91.49	89.63	87.15	83.81	15.14	10955
	TPR_2	45.33	123	77926	94.60	92.84	91.58	89.42	86.67	83.35	14.78	0
	TPR_3	44.08	123	77921	94.59	92.84	91.58	89.42	86.67	83.35	14.78	0
	TPR_{all}	45.55	77	77080	93.82	92.41	91.32	89.23	86.56	83.29	14.78	0
Ind12	TPR_1	27.07	4359	52158	103.23	99.35	95.89	91.18	87.48	83.58	12.75	5286
	TPR_2	28.81	5353	51490	99.82	97.93	95.40	91.03	87.29	83.42	12.58	0
	TPR_3	28.36	5162	51690	99.83	97.90	95.33	90.99	87.26	83.41	12.58	0
	TPR_{all}	30.49	4480	50166	99.74	97.84	95.26	90.83	86.99	83.24	12.59	0
Ratio	TPR_1	1	1	1	1	1	1	1	1	1	1	h
	TPR_2	1.081	2.044	1.028	0.994	0.999	0.998	0.997	0.997	0.996	0.989	
	TPR_3	1.028	1.829	1.029	0.992	0.998	0.997	0.997	0.997	0.996	0.989	
	TPR_{all}	1.067	1.478	0.938	0.985	0.992	0.992	0.993	0.994	0.994	0.989	

TABLE 7   EFFECTIVENESS OF USING THROUGHPUT CONTROLLING METHOD TO TRADE OFF RUNTIME AND ROUTING QUALITY

	TPR_{all}	CPU(s)	WCI(100)	WCI(90)	ACE(0.5%)	ACE(1%)	ACE(2%)	ACE(5%)	ACE(10%)	ACE(20%)	WL (10^7)
Ind7	$t_u = 3$	39.56	4209	29202	97.54	94.84	91.76	86.25	82.93	79.64	11.00
	$t_u = 5$	45.63	3257	24216	96.61	93.52	90.25	85.03	82.30	79.29	11.06
	$t_u = 7$	52.88	2941	18437	96.02	92.68	89.25	84.18	81.84	79.00	11.11
	$t_u = 9$	60.42	2751	14633	95.58	92.08	88.44	83.69	81.58	78.84	11.15
Ind8	$t_u = 3$	39.31	5807	23629	98.59	95.94	92.04	85.42	81.85	76.32	10.79
	$t_u = 5$	44.54	5000	18985	97.93	94.75	90.69	84.61	81.42	76.14	10.83
	$t_u = 7$	51.39	4834	16185	97.71	94.32	89.89	84.13	81.14	75.99	10.87
	$t_u = 9$	58.68	4705	14280	97.63	94.09	89.32	83.83	80.94	75.88	10.90
Ind9	$t_u = 3$	31.43	1142	12121	97.48	94.46	90.60	84.60	81.61	77.07	10.65
	$t_u = 5$	32.29	1098	9383	97.31	93.76	89.24	83.87	81.22	76.87	10.67
	$t_u = 7$	33.08	1004	6774	97.01	92.89	88.12	83.35	80.92	76.71	10.69
	$t_u = 9$	33.65	1021	4227	96.63	91.96	87.04	82.86	80.62	76.55	10.71
Ind10	$t_u = 3$	70.38	2608	36888	95.31	92.30	88.84	84.30	81.95	79.72	24.01
	$t_u = 5$	85.04	1681	13589	91.88	88.60	85.80	82.66	81.10	79.34	24.13
	$t_u = 7$	100.08	1411	10996	90.21	86.54	83.98	81.60	80.51	79.09	24.26
	$t_u = 9$	107.93	1395	10130	89.70	86.06	83.40	81.33	80.36	79.03	24.30
Ind11	$t_u = 3$	41.30	84	80837	94.68	93.31	91.92	89.90	86.95	83.48	14.72
	$t_u = 5$	45.55	77	77080	93.82	92.41	91.32	89.23	86.56	83.29	14.78
	$t_u = 7$	50.36	72	67575	93.32	92.07	91.06	88.69	86.18	83.09	14.82
	$t_u = 9$	55.95	72	59844	92.64	91.54	90.71	88.19	85.85	82.93	14.89
Ind12	$t_u = 3$	29.83	5260	52432	99.81	98.21	95.76	91.29	87.47	83.52	12.56
	$t_u = 5$	30.49	4480	50166	99.74	97.84	95.26	90.83	86.99	83.24	12.59
	$t_u = 7$	33.63	2716	47227	99.45	97.32	94.59	90.28	86.61	83.04	12.62
	$t_u = 9$	36.48	2388	43859	99.38	97.11	94.16	89.85	86.27	82.86	12.64
Ratio	$t_u = 3$	1	1	1	1	1	1	1	1	1	1
	$t_u = 5$	1.108	0.835	0.781	0.989	0.985	0.985	0.989	0.994	0.997	1.004
	$t_u = 7$	1.244	0.721	0.652	0.983	0.976	0.974	0.982	0.989	0.994	1.007
	$t_u = 9$	1.367	0.701	0.551	0.979	0.971	0.967	0.977	0.986	0.992	1.010

Spacer-Is-Dielectric-Compliant Detailed Routing for Self-Aligned Double Patterning Lithography

Yuelin Du[†], Qiang Ma[‡], Hua Song[‡], James Shiely[‡], Gerard Luk-Pat[‡],
Alexander Miloslavsky[‡] and Martin D. F. Wong[†]
[†]Dept. of ECE, University of Illinois at Urbana-Champaign
[‡]Synopsys Inc.
Email: du6@illinois.edu

ABSTRACT

Self-aligned double patterning (SADP) lithography is a leading technology for $10nm$ node Metal layer fabrication. In order to achieve successful decomposition, SADP-compliant design becomes a necessity. Spacer-Is-Dielectric (SID) is the most popular flavor of SADP with higher flexibility in design. This paper makes a careful study on the challenges for SID-compliant detailed routing and proposes a graph model to capture the decomposition violations and SID intrinsic residue issues. Then a negotiated congestion based scheme is adopted to solve the overall routing problem. The proposed SID-compliant detailed routing algorithm simultaneously assigns colors to the routed wires, which provides valuable information guiding SID decomposition. In addition, if one pin has multiple candidate locations, the optimal one will be automatically determined during detailed routing. The decomposability of the conflict-free routing layers produced by our detailed router is verified by a commercial SADP decomposition tool.

Keywords

SADP, SID-Compliant Detailed Routing

1. INTRODUCTION

At $10nm$ technology node, SADP lithography is one of the most promising candidates for Metal layer fabrication. Compared to Litho-Etch-Litho-Etch (LELE) double patterning lithography (DPL), SADP has a great advantage in overlay tolerance and much lower line-width roughness (LWR). SID is one popular flavor of SADP with higher flexibility in design. Many works have been done to study the decomposition for SID process [1, 2, 3, 4]. Similar to other DPL, in SID the target layout is also decomposed into two masks – mandrel mask and trim mask. Each mandrel pattern is surrounded by spacer, and the area which is covered by the trim mask but not covered by spacer will become the final patterns [2]. Fig. 1 shows a valid SID decomposition of a

toy layout.

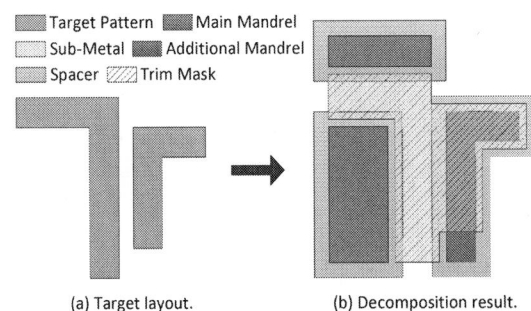

(a) Target layout. (b) Decomposition result.

Figure 1: An example to show SID decomposition [5].

From Fig. 1 we can observe that the patterns in the mandrel mask or trim mask are not always directly from the target layout. The mandrels directly from the target patterns are defined as **main mandrels**. The remaining patterns on the mandrel mask are defined as **additional mandrels**. The target patterns which do not exist on the mandrel mask are defined as **sub-metals**.

Zhang [2] reports that only a limited number of cells from a large cell library can be directly decomposed for SID, suggesting that SID-compliant design is a pre-requisite for successful decomposition. [6] examines the challenges for SID-compliant design, such as forbidden spacing, anti-parallel line-ends and residue issues arising from contour simulation. [8] is the first work adopting SADP-based guidelines in detailed routing, but the decomposability of the routing layers cannot be guaranteed. [9] claims that the routing and decomposition are solved simultaneously by the proposed detailed routing algorithm. However, it fails to consider some major challenges faced by the SID process, such as anti-parallel line-ends conflicts and residue problems, and decomposition violations also exist on conflict-free layouts produced by the router.

This paper studies the SID-compliant detailed routing problem. Since the SID process is more adoptable to the largely unidirectional routing layers, we assume that SID-compliant detailed routing is applied to Metal 2 or higher layers where each layer has a preferred direction. Depending on the design of Metal 1, one input/output pin may have multiple available locations, providing more flexibility for detailed routing on higher layers.

Permission to make digital or hard copies of all or part of this work for personal or classroom use is granted without fee provided that copies are not made or distributed for profit or commercial advantage and that copies bear this notice and the full citation on the first page. To copy otherwise, to republish, to post on servers or to redistribute to lists, requires prior specific permission and/or a fee.
DAC '13, May 29 - June 07 2013, Austin, TX, USA.

In this paper, we examine the challenges for SID-compliant detailed routing, and propose a graph model that correctly captures the decomposition violations and SID intrinsic residue issues. We then develop a negotiated congestion based routing scheme to resolve all conflicts. Our SID-compliant detailed routing algorithm has the following features.

- Single net routing can be optimally computed by performing the shortest path algorithm on the proposed graph model.

- Optimal pin locations are simultaneously determined while routing a net.

- Conflicts are effectively resolved over iterations of rip-up and reroute in the proposed negotiated congestion based routing scheme.

- The proposed SID-compliant detailed routing algorithm simultaneously assigns colors to the routed wires, providing valuable information guiding SID decomposition.

- Conflict-free routing layers produced by our detailed router are verified as 100% decomposable by Synopsys Proteus.

The rest of this paper is organized as follows. Section 2 examines the challenges for SID-compliant detailed routing. The SID-compliant detailed routing problem is formulated in Section 3. Section 4 presents the graph model and our negotiated congestion based routing scheme to solve the problem. Section 5 shows the experimental results, and finally Section 6 concludes the paper.

2. PRELIMINARY: MAIN CHALLENGES IN SID-COMPLIANT DETAILED ROUTING

Due to the special step of spacer deposition, SID process faces several intrinsic challenges in the design aspect. In this section, we will discuss the main challenges in SID-compliant detailed routing.

2.1 Avoid Forbidden Spacing

In SID process, the allowed spacing values between two adjacent wires should be either equal to the spacer width or large enough to satisfy the minimum spacing requirement of the trim mask [6]. Any values in between are strictly forbidden. When the wire space is exactly the spacer width, the two adjacent wires can be defined within the same pattern on the trim mask and separated from each other by spacer. Otherwise, if two wires are far away enough from each other, they can be defined by two separate trim patterns.

2.2 Avoid Odd Cycles

Similarly to LELE DPL, the SID decomposition can be formulated as a two-coloring problem [10]; hence odd cycles will introduce problems in decomposition. In LELE, odd cycles can be resolved by introducing stitches. However, in SID, since all mandrels are surrounded by spacer, splitting a single wire into two will introduce a gap in the final pattern, which breaks the wire and causes logic failures. Therefore, no stitches are allowed in SID decomposition, and in consequence, odd cycles must be avoided in SID decomposable layouts.

2.3 Prohibited Anti-Parallel Line-Ends

When the space between two wires is equal to the spacer width, one of them has to be defined by main mandrel, the other by sub-metal. The trim mask will cover both wires as well as the spacer between them. If the main mandrel and the sub-metal on adjacent tracks form a pair of line-ends in opposite directions, as illustrated in Fig. 2(a) [6], the end-by-end overlapping must be larger than a certain length in order to fulfill the minimum trim width requirement. Therefore, for two wires on adjacent tracks with anti-parallel line-ends, they should either have enough end-by-end overlapping (i.e. ≈ minimum trim width) or enough end-to-end distance (i.e. ≈ minimum trim space) such that they can be defined by two separate patterns on the trim mask, as shown by Fig. 2(b). Either line-end falling into the prohibited region will result in a failure in SID decomposition.

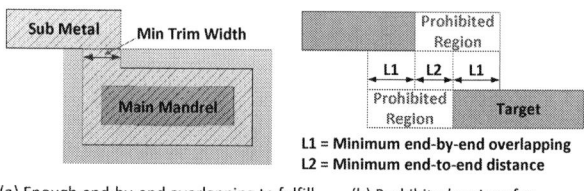

(a) Enough end-by-end overlapping to fulfill the minimum trim width requirement.

(b) Prohibited regions for anti-parallel line ends.

L1 = Minimum end-by-end overlapping
L2 = Minimum end-to-end distance

Figure 2: Design rule for anti-parallel line-ends.

2.4 Sub-Metal Residue Artifacts

With high spacer width uniformity, the shape of the spacer gets rounded at convex mandrel corners even with perfect OPC for the mandrel mask [7], as shown in Fig. 3(a). As a result, when a 2D pattern is defined by sub-metal, large residue will be left at its concave corner due to the spacer rounding at the adjacent convex mandrel corner, as illustrated in Fig. 3(b). Instead, Fig. 3(c) shows that the spacer shape stays sharp at concave mandrel corners, and hence clean final patterns can be obtained if the 2D pattern is defined by main mandrel, as illustrated in Fig. 3(d).

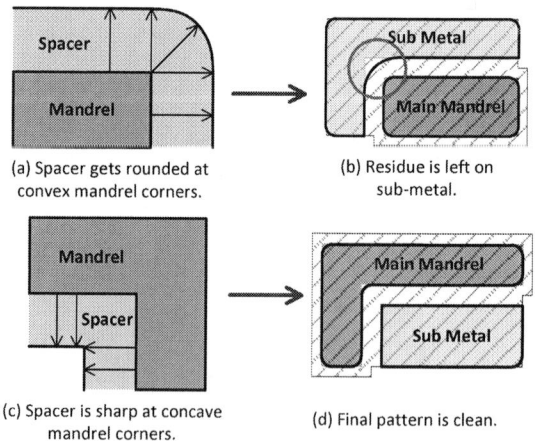

(a) Spacer gets rounded at convex mandrel corners.

(b) Residue is left on sub-metal.

(c) Spacer is sharp at concave mandrel corners.

(d) Final pattern is clean.

Figure 3: Define the 2D pattern by main mandrel to avoid sub-metal residue.

654

To completely avoid the residue artifacts, all 2D patterns in the design must be defined by main mandrels during SID decomposition, which is extremely difficult to be achieved. Therefore, it has to be considered in the design phase. We define the jogs on sub-metals as **sm-jogs**. Then the number of sm-jogs should be minimized during detailed routing.

3. PROBLEM FORMULATION

Since the SID process is more adoptable to the largely uni-directional routing layers, we assume the SID-compliant detailed routing algorithm is applied to Metal 2 or higher layers where each layer has a preferred direction, and the other direction perpendicular to the preferred direction is defined as **non-preferred direction** of the layer. Depending on the design of Metal 1, an input/output pin may have more than one candidate locations, providing more flexibility for detailed routing on higher layers. With such assumptions, we formulate the SID-compliant detailed routing problem in this section considering the challenges mentioned in the previous section.

In order to efficiently avoid odd cycles, in the non-preferred direction of each layer, the routing tracks are assigned as main mandrel tracks and sub-metal tracks alternatively with track space equal to the spacer width, and odd-track jogs on the same layer are strictly forbidden. In this way, all wires on main mandrel tracks become main mandrels during the step of decomposition, the rest being sub-metals. The scheme of simultaneous color assignment provides valuable information to guide the subsequent procedure of SID decomposition. Furthermore, the alternative track assignment automatically avoids spacing conflicts in the non-preferred direction. We only need to guarantee that the minimum spacing rules are not violated along each routing track and no prohibited anti-parallel line-ends occur, in order to produce decomposable routing layers.

In each layer, there are 2 options for main mandrel/sub-metal track assignment (either odd/even track is assigned as main mandrel/sub-metal track or vice versa). Since the total number of routing layers fabricated by the SID process is very limited, we can always enumerate the possible assignment combinations for all layers in order to achieve the optimal routing result. Therefore in our problem formulation, we assume the track assignments for each layer have been fixed and define the SID-compliant detailed routing problem as follows.

Definition 1. **SID-Compliant Detailed Routing**
Given a netlist with candidate source/target pin locations for every net, a routing grid, a main mandrel/sub-metal track assignment strategy and the minimum spacing requirement, detailed routing with simultaneous pin location determination is performed such that no odd-track jog, spacing violation or prohibited anti-parallel line-ends occur, and the number of sm-jogs is minimized.

4. PROBLEM SOLUTION

A negotiated congestion based scheme [11] is adopted in our SID-compliant detailed routing algorithm. In order to reduce the adverse effect of improper net ordering, wire crossing and wire spacing conflicts are initially allowed, and then resolved over iterations of rip-up and reroute. The key subproblem of the negotiated congestion based routing scheme is how to perform maze routing for a single net on the routing graph in the presence of a set of previously routed nets. In this section, we first define the subproblem of the SID-compliant detailed routing problem and propose a graph model where the subproblem can be optimally solved; then we present the overall negotiated congestion based routing scheme.

4.1 Subproblem Definition

When routing a net, it is desired to compute a path p which produces the minimum number of crossing conflicts, spacing conflicts and sm-jogs. Of course, the wire length, as a conventional metric, also needs to be minimized. Note that the unit length wire segment in the non-preferred direction of a layer should cost more than that in the preferred direction in order to preserve the unidirectional property. Therefore, the weighted sum $l_w^p + \alpha * v_c^p + \beta * v_s^p + \gamma j_s^p$ is a good cost metric to minimize when routing a single net, where l_w^p, v_c^p, v_s^p and j_s^p denote the weighted wire length, the number of vertices with crossing conflicts, the number of vertices with spacing conflicts and the number of sm-jogs produced by the path p computed, respectively, and α, β and γ are user defined parameters that specify the relative importance between them. We define the SID-compliant maze routing problem below.

Definition 2. **SID-Compliant Maze Routing**
Given a set of previously routed nets as well as the candidate source/target pin locations of a net, the objective is to determine the optimal source/target pin locations and compute a path between the two pins such that the weighted sum $l_w^p + \alpha * v_c^p + \beta * v_s^p + \gamma j_s^p$ is minimized.

4.2 Subproblem Solution

In this subsection, we propose a graph model that correctly captures the cost of crossing, spacing conflicts and sm-jogs, and show that the SID-compliant maze routing problem can be optimally solved by performing the shortest path algorithm on the proposed graph model. We then demonstrate that our algorithm is able to automatically extend line-ends and remove anti-parallel line-ends conflicts through simple post-processing.

4.2.1 Expanded Routing Graph Model

Suppose we are given a routing grid G with preferred direction and main mandrel/sub-metal track assignment for each layer, which can be viewed as a routing graph if we regard every segment intersection as a vertex and segments between vertices as edges. In order to capture the cost of sm-jogs, we split each vertex v of G into 4 vertices and construct an expanded routing graph model on them, as illustrated in Fig. 4(a). Fig. 4(b) shows four types of edges to capture the cost of sm-jogs and wrong-way wires. The detailed construction is described as follows.

- Each split vertex works as a switch box with four vertices connecting each other.

- All edges are categorized into four types, namely, e_s, e_0, e_1 and e_w. Inside a switch box located at a sub-metal track, two vertices in the diagonal direction are connected by a type e_s edge with cost γ (the cost of a sm-jog). The rest of the edges inside switch boxes, as shown in yellow in Fig. 4(b), are classified as type e_0 edges with 0 cost. This means that inside a switch box,

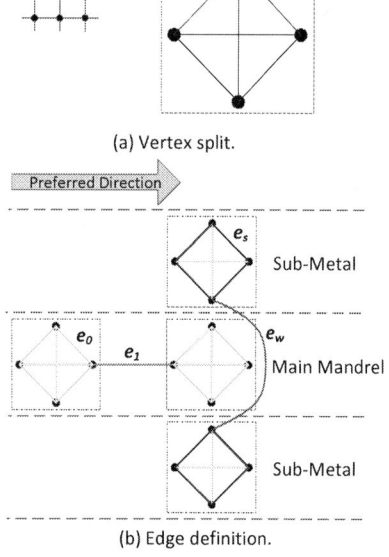

(a) Vertex split.

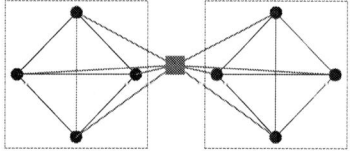

(b) Edge definition.

Figure 4: The definitions of vertices and edges in the expanded routing graph model.

a wire can travel freely on a main mandrel track; however, on a sub-metal track, it has to pay the sm-jog cost in order to travel diagonally, because such travel introduces an sm-jog which is undesired. Outside the switch boxes, along the preferred direction, a pair of vertices at neighboring switch boxes are connected by a type e_1 edge with cost 1, as shown in blue in Fig. 4(b). In order to forbid odd-track jogs, in the non-preferred direction, only the vertices located at every other track are connected by type e_w edges with twice the wrong-way wire length cost $2 \times c_w$, as shown in red in Fig. 4(b).

- On top of the switch box model, an extra vertex is added to represent a pin. Fig. 5 illustrates the graph model for a pin with two candidate locations. For each candidate pin location, four extra edges are added connecting the pin to the vertices inside the switch box.

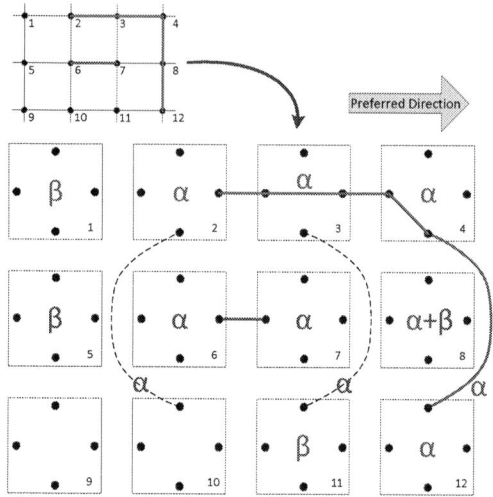

Figure 5: The graph model for a pin with two candidate locations.

- The cost of crossing and spacing conflicts is captured by assigning congestion cost to the verities and edges. Vertices from the same switch box always share the same congestion cost. Before the routing starts, all vertices are initialized with 0 cost. Then after a vertex v is occupied by a routed net, the cost of all vertices in the same switch box as well as the wrong-way edge passing that switch box is increased by α, which is the

cost of a crossing conflict. And along the same track of v, the vertices within its spacing conflict region but not occupied by the current net have cost increased by β – the cost of a spacing conflict. Fig. 6 shows the congestion cost assignment after two wires of different colors have been routed. Note that the wire length cost is not displayed.

Figure 6: The wire crossing and spacing conflict cost assigned to an expanded routing graph with two pre-routed wires.

So far we have constructed an expanded routing graph G' from the original routing grid G. When routing one net, we simply apply Dijkstra's shortest path algorithm on G' to find the optimal path p' between two pins on G', which corresponds to a path p on G. In addition, it is obvious that the shortest path on G' passes exactly one candidate switch box for the source pin, and one for the target pin. So the optimal pin locations can be determined at the same time. Therefore, we conclude that the SID-compliant maze routing problem can be solved optimally by performing the shortest path algorithm.

4.2.2 Forbidding Prohibited Anti-Parallel Line-Ends

As mentioned previously, prohibited anti-parallel line-ends violate SADP design rules and hence are forbidden in a decomposable layout. Fig. 7 illustrates three scenarios of anti-parallel line-ends conflicts on a routing grid G. In order to avoid these conflicts during single net routing, we modify our expanded routing graph to disallow the routing scenarios shown in Fig. 7. Fig. 8 gives an example of the detailed modification.

In Fig. 8(a), suppose the blue wire is a pre-routed wire in G and the rectangular regions show the prohibited regions for anti-parallel line-ends. Then for any switch box located within the prohibited region, two types of edges are blocked, as illustrated in Fig. 8(b). The first type of edge (in blue) connects a vertex inside the switch box to a pin or a via, generating the first or the second routing scenario in Fig. 7. The second type of edge (in green) connects two vertices inside the switch box, generating the third scenario in Fig. 7. However, only blocking these problematic edges may not

656

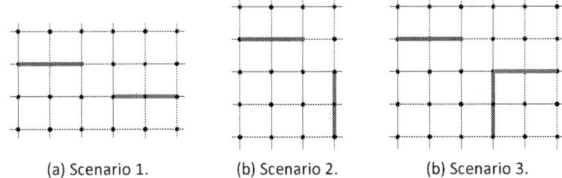

(a) Scenario 1.　　(b) Scenario 2.　　(b) Scenario 3.

Figure 7: Three scenarios of anti-parallel line-ends conflicts on a routing grid.

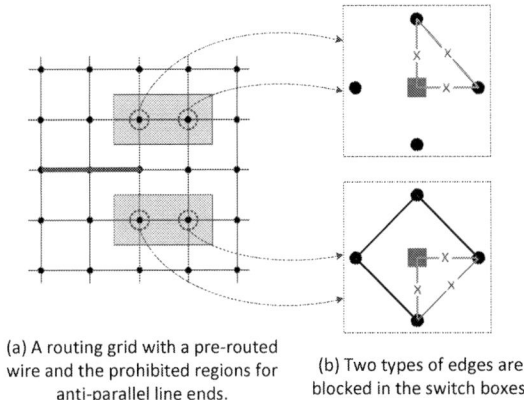

(a) A routing grid with a pre-routed wire and the prohibited regions for anti-parallel line ends.

(b) Two types of edges are blocked in the switch boxes.

Figure 8: Graph model modification to avoid prohibited anti-parallel line-ends.

work correctly. For example, the right vertex in the second switch box of Fig. 8(b) may still reach the bottom vertex through a detour inside the box (shown in black) even if the green edge is blocked. To avoid such inside-box detours, we further split a vertex v in the graph model into two vertices v_{in} and v_{out}, and make the edges directed, as illustrated in Fig. 9. The input vertex on one boundary is connected to the three output vertices located at different boundaries. Note that only the connections from/to the left boundary are displayed in Fig. 9.

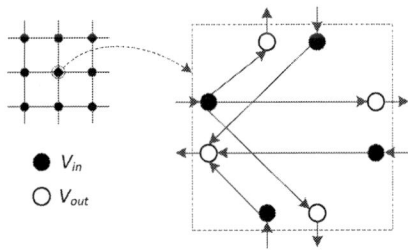

● V_{in}
○ V_{out}

Figure 9: The graph model that disallows inside-box detour.

By blocking certain edges in the new graph model, prohibited anti-parallel line-ends will not show up while routing a net; instead, a looped walk in G may be obtained by the shortest path algorithm as a routing path, as illustrated in Fig. 10(a), where net 1, 2 and 3 are pre-routed on G. When routing net 4, since the candidate location of its target pin

lies within the prohibited region of wire 3. in the corresponding switch box, the edge connecting the input vertex on the right boundary to the target pin has been blocked. Therefore, the shortest path will go through a loop and connect to the target pin from the input vertex on the left boundary. When tracing the path on G, we simply remove all wire segments inside the loop and obtain a routed wire with extended line-end, as shown in Fig. 10(b), which automatically avoids the anti-parallel line-ends conflict.

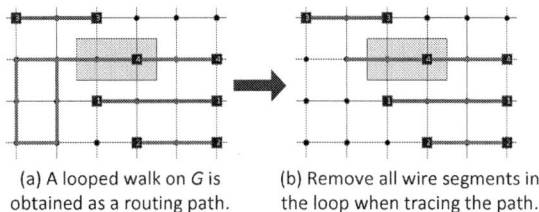

(a) A looped walk on G is obtained as a routing path.

(b) Remove all wire segments in the loop when tracing the path.

Figure 10: The strategy of automatic line-end extension to avoid anti-parallel line-ends conflicts.

4.3 Overall Routing Scheme

In this subsection, we present the overall negotiated congestion based routing scheme for the SID-compliant detailed routing problem, where crossing and spacing conflicts are resolved over iterations of rip-up and reroute. We let the nets negotiate for routing resources by adding history costs to vertices and wrong-way edges in G'. The cost of a vertex v is computed by the following formula:

$$cost(v) = \alpha \times h_c \times n_c + \beta \times h_s \times n_s, \quad (1)$$

where n_c/n_s denotes the number of pre-routed nets having crossing/spacing conflict with the current vertex, and h_c/h_s denotes the history cost for crossing/spacing conflicts. All history cost is initialized as 1. Similarly, the cost of a wrong-way edge e is computed by the following formula:

$$cost(e) = 2 \times c_w + \alpha \times h_c \times n_c, \quad (2)$$

where the first term describes its weighted wire length cost, and the second term describes the crossing conflict cost of the vertex it passes.

The scheme works as follows. We first route all nets sequentially in a random order on G'. Then as long as conflicts exist, iterations of rip-up and reroute will be performed. When a net i is ripped-up and rerouted, we first remove its current route, unblock the edges causing anti-parallel line-ends conflict with the current route, and update the cost values on the vertices and wrong-way edges it has impact on. Then the shortest path algorithm is performed to compute a new path for net i, and newly impacted vertices and edges are updated accordingly. In addition, if the new path causes any conflicts with previously routed nets, the history cost (h_c or h_s) on the corresponding vertices and edges is incremented by 1. In this way, the conflicting vertices and edges grow more expensive over iterations, and those nets with more options will tend to choose alternative routes in subsequent iterations, so that the conflicts can potentially be resolved. In our implementation, this procedure will terminate when either no conflict exists or enough iterations have been performed.

657

Table 1: Comparison of automatic/random pin location determination in SID-compliant detailed routing

Test Cases	♯ Net	Size (μm^2)	♯ Conflict		♯ SM-Jogs		♯ Via		Wire Length (μm)		Runtime(s)	
			Auto.	Rand.	Auto.	Rand.	Auto.	Rand.	Auto.	Rand.	Auto.	Rand.
test1	1k	66.6	0	68	199	227	854	1682	381.6	513.4	78	298
test2	2k	132.7	0	189	374	482	2104	3388	735.6	1021.6	465	866
test3	4k	368.6	0	17	683	894	3254	5894	1392.5	1894.8	1984	3037
test4	8k	829.4	0	33	1329	1728	6706	11686	2766.4	3777.3	7081	12506
test5	16k	1866.2	0	60	2593	3467	13594	23008	5535.2	7475.9	38958	54136

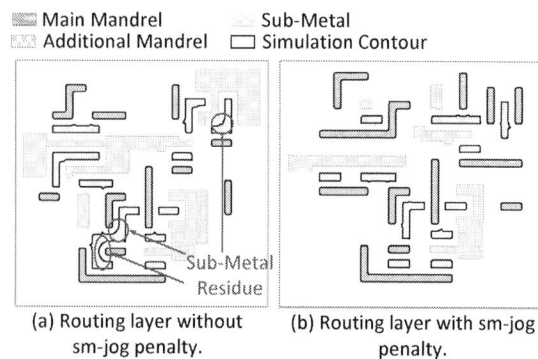

Main Mandrel   Sub-Metal
Additional Mandrel   Simulation Contour

Sub-Metal Residue

(a) Routing layer without sm-jog penalty.

(b) Routing layer with sm-jog penalty.

Figure 11: Comparison of the simulation results with and without sm-jog penalty.

5. EXPERIMENTAL RESULTS

We implement our algorithm in C++ on a Linux machine with 3.0GHz CPU and 16GB RAM. Experiments are performed with $10nm$ node benchmarks where both wire width and spacer width are $24nm$. All the conflict-free routing layers produced by our detailed router are verified by Synopsys Proteus as 100% decomposable.

Fig. 11 compares the simulation results on the routing layers with and without sm-jog penalty. From Fig. 11(a) we can easily see that without sm-jog minimization, a lot of residue is left at concave sub-metal corners. However, the sm-jog penalty introduced in our SID-compliant routing scheme helps to clean up such residue effectively, as shown in Fig. 11(b). The remaining 'spur' shaped residue can be simply removed by post-processing such as mandrel extension and mandrel merging [7].

We then perform the experiments on a set of benchmarks with different scales and show the advantage of our automatic pin location determination strategy in routability, sm-jog number, via number, wire length and runtime. We first randomly choose one of the candidate locations for each pin and run the router with the fixed pin locations. Then we set free all candidate locations and let the router decide where to place the pins. In the experiments, each pin has 3 candidate locations on average, and the maximum iterations is 50. Table 1 shows the comparison results, from which we can conclude that simultaneously determining pin locations during detailed routing has great advantages over random selection in all aspects.

6. CONCLUSIONS

This paper proposes an expanded graph model to solve the SID-compliant detailed routing problem. The challenges faced by the SID process such as forbidden spacing, odd cycles, anti-parallel line-ends conflicts and sub-metal residue issues have been considered in the proposed graph model. In addition, color assignment and pin locations can be simultaneously determined during the detailed routing. An overall negotiated congestion based routing scheme is developed to resolve wire crossing and design rule conflicts over iterations of rip-up and reroute, and all conflict-free routing layers produced by our detailed router have been verified as 100% SID decomposable.

7. REFERENCES

[1] Y. Ban, A. Miloslavsky, K. Lucas, S.-H. Choi, C.-H. Park, and D. Z. Pan, Layout decomposition of self-aligned double patterning for 2D random logic patterning. *Proc. SPIE*, Vol. 7974, p. 79740L, 2011.

[2] H. Zhang, Y. Du, M. D. F. Wong, and R. O. Topaloglu, Self-aligned double patterning decomposition for overlay minimization and hot spot detection. *Proc. DAC*, pp. 71 - 76, 2011.

[3] H. Zhang, Y. Du, M. D. F. Wong, R. O. Topaloglu, and W. Conley, Effective decomposition algorithm for self-aligned double patterning lithography. *Proc. SPIE*, Vol. 7973, p. 79730J, 2011.

[4] Z. Xiao, H. Zhang, Y. Du, and M. D. F. Wong, A polynomial time exact algorithm for self-aligned double patterning layout decomposition. *Proc. ISPD*, pp. 17 - 24, 2012.

[5] Y. Ma, J. Sweis, C. Bencher, H. Dai, Y. Chen, et al., Decomposition strategies for self-aligned double patterning. *Proc. SPIE*, Vol. 7641, p. 76410T, 2010.

[6] G. Luk-Pat, A. Miloslavsky, B. Painter, L. Lin, P. D. Bisschop, and K. Lucas, Design compliance for spacer is dielectric (SID) patterning. *Proc. SPIE*, Vol. 7641, p. 83260D, 2012.

[7] Y. Du, H. Song, J. Shiely and M. D. F. Wong, Improved Spacer-Is-Dielectric (SID) Decomposition with Model Based Verification. *Proc. SPIE*, Vol. 8684, p. 8684-13, 2013.

[8] M. Mirsaeedi, J. A. Torres, and M. Anis, Self-aligned double patterning (SADP) friendly detailed routing. *Proc. SPIE*, Vol. 7974, p. 79740O-1, 2011.

[9] J.-R. Gao and D. Z. Pan, Flexible self-aligned double patterning aware detailed routing with prescribed layout planning. *Proc. ISPD*, pp. 25 - 32, 2012.

[10] A. B. Kahng, C.-H. Park, X. Xu, and H. Yao, Layout decomposition for double patterning lithography. *Proc. ICCAD*, pp. 465 - 472, 2008.

[11] L. McMurchie, and C. E. Pathfinder, A negotiation-based performance-driven router for FPGAs. *Proc. FPGA*, pp. 111 - 117, 1995.

21st Century Digital Design Tools

William J. Dally
NVIDIA & Stanford University
2701 San Tomas Expressway
Santa Clara, CA 95050
408-486-2000
bdally@nvidia.com

Chris Malachowsky
NVIDIA
2701 San Tomas Expressway
Santa Clara, CA 95050
408-486-2000
chris@nvidia.com

Stephen W. Keckler
NVIDIA & UT-Austin
2701 San Tomas Expressway
Santa Clara, CA 95050
408-486-2000
skeckler@nvidia.com

ABSTRACT

Most chips today are designed with 20th century CAD tools. These tools, and the abstractions they are based on, were originally intended to handle designs of millions of gates or less. They are not up to the task of handling today's billion-gate designs. The result is months of delay and considerable labor from final RTL to tapeout. Surprises in timing closure, global congestion, and power consumption are common. Even taking an existing design to a new process node is a time-consuming and laborious process.

Twenty-first century CAD tools should be based on higher-level abstractions to enable billion-gate chips to go from final RTL to tapeout in days, not months. Key to attaining this increase in productivity is raising the level of design and using simple, standard interfaces. Designs should be composed from high-level modules – processors, MODEMs, CODECs, memory subsystems, and I/O subsystems – rather than gates and flip-flops. Each module, which we expect to contain 100 thousand to 10 million gates, is easily laid out by today's tools, is placed as a unit, and communicates over a NoC via a standard interface. Restricting modules to standard sizes and aspect ratios further simplifies physical design. We expect even a large chip to contain at most a few thousand such modules and expect the physical design and chip-assembly to take a few days with minimal labor after completion of the module-level design.

Categories and Subject Descriptors

B.7.2 [**Integrated Circuit**]: Design aids – *layout, placement and routing, verification.*

General Terms

Design, Standardization

Keywords

Design automation, NoC, Chiplet, Modularity, Digital design

1. INTRODUCTION

Many designers today are able to realize complex systems with billions of transistors in a few weeks by composing previously designed pieces of intellectual property (IP). The physical design

Permission to make digital or hard copies of all or part of this work for personal or classroom use is granted without fee provided that copies are not made or distributed for profit or commercial advantage and that copies bear this notice and the full citation on the first page. To copy otherwise, or republish, to post on servers or to redistribute to lists, requires prior specific permission and/or a fee.
DAC '13, May 29-June 7, 2013, Austin, TX, USA

takes only a few hours and changes can be accommodated with a minimum of rework. These designers are working at the board or system level. In contrast, designers working at the chip level take months to complete a design of similar complexity.

The system designer is able to achieve a high level of productivity because the packaging of the components they are composing enforces modularity. Modularity implies information hiding and a fixed, often standard, interface. The system designer typically sees only the specification of a chip they are using. They cannot see or alter the implementation of the chip. Each chip can only interact with the rest of the system over its package pins, a set of fixed and often standard interfaces. The modularity enforced by packaging enables a system designer to use a complex component without incurring a design cost associated with its internal complexity.

In this paper, we suggest that chip designers can achieve productivity comparable to system designers if they adopt a comparable level of modularity. A chip design has no packaging constraints to enforce this modularity found at the system level. The designer must adopt a discipline and a methodology that enforces modularity – and resist the temptation to violate it.

The ideas we describe here are at an early stage. We have not yet implemented any commercial chips using this approach. We provide a number of specific examples to show what might be possible and to give some substance to the proposal. We share these ideas here to encourage others in the design community to help flesh out and then adopt this approach in a quest to build an ecosystem for more productive SoC design. While others have advocated raising the level of abstraction for chip design [1], we propose specific methods for a modular design discipline. The advantages of modular design and information hiding have long been understood in the software engineering community [2]. They apply equally well to hardware design.

We propose two artificial constraints to impose modularity on a design and simplify composition. First, all modules are one of a few standard heights (for example 0.5mm, 1mm, and 2mm) placed in rows, with no global signals routed through a module. Second, modules are restricted to communicate over a network-on-chip (NoC) [3, 4] using standard interfaces. A few flavors of the standard interface would be provided to accommodate a range of bandwidth requirements. Restricting modules to standard heights greatly simplifies the bin-packing problem of module placement. Restricting global signals to dedicated wiring channels enables separate design and verification of modules. Restricting communication to a standard packetized interface enforces information hiding, avoiding the *rat's nest* that can develop when modules are able to see and exploit the internals of other modules.

The cost of self-imposed modularity is likely to be a chip that is less optimal along some axis than one that violates modularity. It may be slightly larger, slightly slower, or consume slightly more power than a chip that is carefully handcrafted without imposing this discipline. However, this slightly sub-optimal chip will get to market months before its handcrafted cousin. The revenue generated and market position gained by earlier market delivery could easily justify, for instance, consuming a bit more die area. If reclaiming the additional area is important, a handcrafted design can follow the modular design as a cost reduction step.

Semiconductor manufacturers have already shown a willingness to adopt modularity at the expense of optimality in the use of standard cells. Making all cells a standard height and enforcing fully-restored CMOS electrical interface constraints imposes an area penalty. We have seen cases where redesigning a standard-cell unit using full-custom design can save upwards of 50% in area [5]. Yet this is rarely done in practice because full-custom design at the transistor or gate level is simply too costly. Interestingly, recent work indicates that the gap between automated standard cell and custom design methodologies may be closing [6].

The design discipline we are proposing is simply up-leveling the concept of standard cells. Like standard cells, we propose cells with a common height to simplify composition. The difference is that our proposed row height is 0.5mm while a standard cell's height is 6 to 12 wire pitches, about 1μm in a typical 28nm process. A unit of modularity that is 6-12 wire pitches in height was the right abstraction when 10^3 to 10^5 such units fit on an edge of a chip. Now that we are producing 10^8 to 10^9-gate chips, we need to scale up unit modularity to a unit that is at least 0.5mm on a side, resulting in designs that have perhaps a thousand of these new units on a chip.

As we up-level standard cells, the need for a standard interface changes. With standard cells, the standard interface is a fully-restored static CMOS logic signal. Pre-charged signals, dynamic nodes, pass-gate signals, and flip-flop storage nodes are not allowed to be exposed outside the cell. Flip-flop and latch cells typically use several inverters solely to provide this isolation. Enforcing this standard signaling convention ensures that any standard cell can talk to any other standard cell without fear of dynamic charge sharing or other subtle circuit effects. As we up-level to modules in our proposal, the standard interface becomes a packetized transport protocol, namely a NoC interface. This interface ensures that any module can talk to any other module without worrying about subtle protocol effects or timing surprises when integrated at the chip-level.

The remainder of this paper explores the concept of high-level modular design in more detail, focusing on the design elements of chiplets, standard interfaces, NoC, and I/O. We conclude with a case study of an experimental chip that employed some of these principles and a call to arms to the chip design community.

2. CHIPLETS

Figure 1 shows how we envision modules or *chiplets* being placed on a portion of a chip. The figure shows one corner of a chip with I/O chiplets along the periphery and rows of single, double, and quad-height modules. Dedicated wiring tracks are provided between the module rows. To preserve modularity, routing through modules is prohibited.

All modules are a fixed height but have an arbitrary width. As with standard cells, this approach facilitates simple placement in rows while allowing modules that span a wide range of areas. We

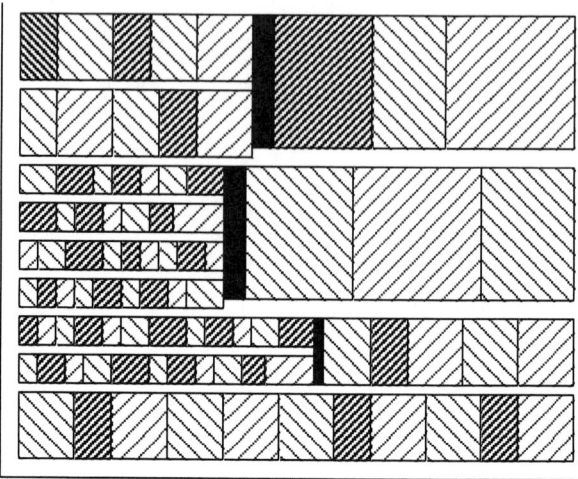

Figure 1: Layout of chiplets on a modular SoC.

envision perhaps three to four module heights to allow a wide range of module sizes without requiring extreme aspect ratios that tend to result in inefficient layout. Chiplet heights can be arranged as a multiple of the power grid so that vias from the power grid to the chiplet are in the same position in each row.

Assuming modules of 0.5, 1, and 2mm in height (single, double, and quad height), the design can accommodate areas from 0.06mm^2 (0.5mm x 0.125mm) to 16mm^2 (2mm x 8mm) with aspect ratios no worse than 4:1. In a typical 28nm process, this range of areas would enable chiplets to include 190k to 48M gates. Modules smaller than 0.06mm^2 (190k gates) can be combined with other small modules into a single module sharing a NoC interface, using conventional design techniques. Modules larger than 16mm^2 (48M gates) can be placed as a large macro or be decomposed into multiple, smaller modules.

The top 5% of a chiplet's height (25, 50, or 100μm for single, double, and quad-height chiplets) can be reserved for a wiring channel. The NoC logic and its channels are implemented in this area. A 25μm wiring channel for single-height modules provides over 250 wiring tracks per layer, sufficient bandwidth for a row of single-height modules. Should additional bandwidth be required, a row of modules may be omitted and its space used entirely for wiring. NoC connections between wiring channels are accomplished by inserting NoC wiring modules into each row. The connection from each module to the NoC takes place entirely along the upper and/or lower edge of the module using one or more of the fixed interfaces described below.

Providing dedicated wiring area simplifies layout and timing closure, allowing each module's layout to be finalized and timing verified independent of global wiring. It also makes the global wiring more predictable. Timing of the NoC is not dependent on the design of any module.

The center of a module only has connections to power, ground, and external I/O. Chiplets can be powered via global power and ground grids that are distributed on upper layers of metal and supplied via an array of supply balls.

I/O chiplets are modules that include I/O drivers, pads, and signal balls that connect to the package. The presence of the signal balls, and any special supply balls needed to support them may interrupt the power grid and the array of power supply balls. Otherwise an I/O chiplet is no different from any other chiplet. We anticipate that I/O chiplets will include not just the I/O circuit elements, but also the logic associated with those pads. For example, a memory controller would be included in the chiplet containing SDDR I/O pads, and a PCIe root complex would be contained in the chiplet containing PCIe PHYs. While I/O chiplets will typically be placed along the periphery of the chip, area I/O is also possible. The implications of I/O chiplets are described in more detail in Section 5 below.

As shown in Figure 1, we expect the tallest rows of chiplets to be placed toward the center of the chip so that the larger wiring channels associated with these chiplets are available in the center of the die, where the highest wiring density is expected. At the point where row heights change, a NoC chiplet is inserted (solid black in Figure 1) to route packets between the different channels.

3. STANDARD INTERFACES

One of the motivations of advocating for standard interfaces in our modularized design proposal is to ensure that relatively straightforward automation can fully construct that top-level design ensuring functional correctness, while avoiding timing and routing surprises.

3.1 The Rat's Nest

Perhaps the hardest step of converting an existing design to the modular approach we advocate is eliminating the rat's nest of connections between modules on a modern SoC and replacing this wiring with disciplined communication over standard interfaces. Without the discipline of modularity, it is not unusual to have thousands of connections between modules. A designer working on one module needing to know the status of an internal component of another module is tempted to just reach inside that module and grab the relevant signal. To the designer, this option appears inexpensive as it only requires typing a signal name in Verilog or adding a wire to a module specification. In practice, it is quite costly to both the logical and physical design of the chip.

Violating strict modularity increases the logical design complexity in several ways. Accessing internal signals of modules violates the principle of *information hiding* and in doing so makes module designs fragile and brittle. Changing or verifying a design becomes challenging. Substituting a new module for an old module may be very difficult when several other modules on a SoC depend on internal signals and potentially subtle or non-obvious behavioral properties. Except in rare circumstances, accessing internal signals of a module is simply bad design.

Violating strict modularity also incurs physical design costs, including area and design time. Random wiring between modules is costly in terms of area because it has a low duty factor. A separate wiring track (or tracks for multi bit signals) is allocated for each signal accessed in this manner, even though the typical signal is relevant (both changing and observed) a tiny fraction of the time. Except for signals with very high duty factor, a far more area-efficient approach multiplexes many signals over a shared communication structure, such as a NoC, rather than allocating dedicated wiring to each signal. Random wiring also makes full-chip timing closure more difficult and time consuming, as internal module signals may have little local timing margin and translate into top-level timing violations.

3.2 A Modular Interface

Instead of unstructured module interfaces and access to internal module signals, we propose that modules communicate with other modules by sending packets over a NoC. To illustrate how this would work, we will propose one such packet-based NoC design.

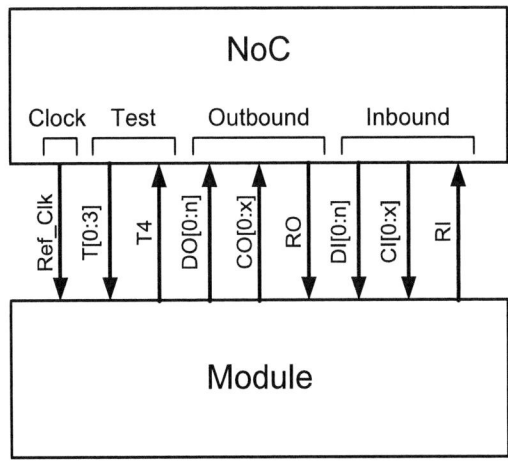

Figure 2: Standard module interface.

In this design, a packet consists of one or more 32-bit *flits*. The destination of the packet is encoded in the first flit. The remainder of the packet is user-defined. Figure 2 shows the standard module interface. The interface consists of separate outbound (CO, RO, DO), inbound (CI, RI, DI), test, and clock signal groups. The inbound and outbound signal groups are further divided into data (DO and DI), control (CO and CI), and flow control (RO and RI). To allow the bandwidth to be configured for modules with different needs, the data fields (DO and DI) can be configured to be 32, 64, 128, or 256 bits wide, enabling modules to send 1, 2, 4, or 8 flits each cycle. The least significant 16 bits of the first flit of a message encodes the message destination. The remainder of the bits is defined by the two endpoints and is not interpreted by the network. The control (CO and CI) signal groups encode how many flits are valid, whether the last valid flit terminates the packet, and on which virtual channel the packet is sent [7]. The flow control lines (RO and RI) flow in the direction opposite to the corresponding data and control lines to signal when the data and control are accepted by the interface.

To send a packet, a module places the first flit(s) on the DO lines and sets the CO lines appropriately. The data and control are held until the RO line is sensed high at the end of a cycle, indicating that the data has been accepted. A packet that is longer than the width of the DO lines is sent over multiple cycles. Flits from different virtual channels can be interleaved on a cycle-by-cycle basis by setting the CO lines appropriately. Modules receive arriving packets using an identical interface but in the opposite direction.

The test interface is comparable to JTAG and allows a tester to connect to and independently test each module via the NoC. The NoC must also include a capability to test the NoC itself and the module to NoC interfaces.

The clock field provides a standard reference clock to the module; we expect that a 1GHz reference will be suitable for most SoCs. The NoC operates globally at this 1GHz clock rate and the module to NoC interface is synchronous with this clock. Each module may operate using an arbitrary clock and arbitrary timing convention but is responsible for generating its local clock (or clocks). It is also responsible for synchronizing its local signals with the NoC interface, for example by using a FIFO synchronizer. A standard synchronizer would be available for inclusion in each module. This approach to timing allows

661

Table 1: Example NoC communication.

	Source		Destination	
Cycle	CO	DO	CI	DI
1	2, No, 3	A, B		
2	1, Yes, 3	C		
...				
i			1, No, 3	A
i+1			1, No, 3	B
i+2			1, Yes, 3	C

complete flexibility in module design (including asynchronous) while providing a global timing reference that can be used to generate a local clock.

Table 1 illustrates how an encoder chiplet with a 2-flit (64-bit) interface sends a 3-flit (96-bit) message to a memory controller with a 1-flit (32-bit interface). On the first cycle, the encoder puts two flits (A and B) on the DO signals and sets CO to indicate 2 valid flits, no termination, and virtual channel 3. The low 16-bits of flit A encode the network address of the memory controller. On the second cycle, the remaining flit (C) is placed on the low DO lines and the CO signals are set to encode 1 flit, packet termination, and virtual channel 3.

A few cycles later, the packet arrives on the CI and DI pins of the memory controller. Three cycles are required for the packet to be delivered at the far end over the 32-bit controller interface. The network converts the packet from 64-bit wide to 32-bit wide at the destination NoC interface.

Providing a variable width interface enables the NoC to efficiently support a wide variety of interface bandwidths ranging from 4GB/s to 32GB/s, with a 1GHz NoC clock. Providing automatic conversion between widths in the network allows any module to talk to any other module regardless of their width. To simplify width conversion, the physical layout of the multi-flit data buses are bit interleaved. If interface bandwidth higher than 32GB/s is required, multiple parallel 32GB/s NoC interfaces can be instantiated on a module.

4. NETWORK-ON-CHIP (NoC)

We envision that the NoC for a particular SoC will be customized based on the anticipated traffic matrix for that NoC and the quality of service (QoS) required for each traffic flow. The traffic matrix may be explicitly specified or may be extracted from simulations of the SoC. Starting with the traffic matrix and the placement of modules, a NoC synthesis tool generates a NoC that provides the required throughput for each flow with a minimum of latency and energy. The tool configures a NoC from a library of channels, routers, NICs, and controllers. The NoC synthesis tool also assigns virtual channels to particular flows to meet QoS requirements of the application.

We expect the NoC channels to be implemented with optimized circuits that employ equalization and low-swing signaling to achieve substantially lower energy and latency than a wire driven by a standard full-swing CMOS gate. The use of low-swing signaling is enabled by dedicated wiring channels, which facilitates control of wire geometry and crosstalk. We expect the performance of these optimized wires to more than offset the increased energy and latency due to the insertion of routers in the communication path.

Some SoCs include isochronous flows that have fixed bandwidth requirements with tight latency constraints. We expect the NoC to handle such flows by provisioning channels with sufficient bandwidth to meet the worst-case requirements of all simultaneous flows and then assigning isochronous flows to a high-priority virtual channel.

Some interactions between modules on a SoC involve simple events. For example, one module may need to know when a queue in another module reaches a threshold. In a traditional rat's nest SoC design, a dedicated wire would be run between the two modules with this indication. On a NoC with standard interfaces, such events are sent as simple single-flit packets on the highest-priority virtual channel. To minimize latency, an event packet can be launched in the same cycle that the event is detected.

5. I/O CHIPLETS

One of the most visibly and structurally obvious holdovers from the early days of IC design is the I/O ring. This ring is the traditional repository of all the I/O buffer and power/ground pads for the device. The I/O padring requires up-leveling in a world that has full subsystem-sized modularity.

One of the obvious problems with the I/O ring structure is that it separates the functional block with the I/O need from the actual I/O pads. This independence provided desirable isolation of all the challenges of high energy and often electrically-hostile external interfaces from the more standardized and less tolerant small-signal domain of the internal logic. However the costs include long wires, non-standard module interfaces, inter-module timing challenges, global wiring congestion, unnaturally constrained floor plans and aspect ratios, and a considerable amount of exposed detailed module specific knowledge and behavior. None of these properties match our goal of design modularity with encapsulated functionality communicating via standard interfaces.

Instead we propose to abandon the global I/O ring structure altogether, in favor of an I/O chiplet architecture which embeds the I/O pads with its interface logic. At the chip level, I/O pads are spread throughout the whole of the device using area I/O, rather than being constrained to the periphery. Within a chiplet, the I/O can still be distributed peripherally. However, the main goal is to make the physical implementation of the module completely self-contained. This approach will require, for instance, that each module design solve its own di/dt and noise issues, provide for its own connections to power and ground resources, and accommodate standardized test infrastructures. While in some respect, we have just moved I/O design challenges from the chip to the chiplet, making chiplets and their I/O truly composable will facilitate much easier chip assembly and full-chip electrical verification.

6. A CASE STUDY

The TRIPS processor was a research prototype chip that employed much of the design style described above [8]. The TRIPS chip was designed to demonstrate distributed processor and memory architectures and included two processors and a distributed non-uniform NUCA cache. As shown in chip floorplan of Figure 3, the chip was implemented using 106 instances of 11 different tiles (chiplets). The diagram shows the tile boundaries, annotated with tile names, along with the outlines of the major RAM structures within each tile. Each processor was composed of 30 instances of five types of tiles, connected via a NoC and a small number of control networks. The five tiles (GT, RT, ET, DT, and IT) represented the major functions of the distributed processor, including global control, register file, execution units,

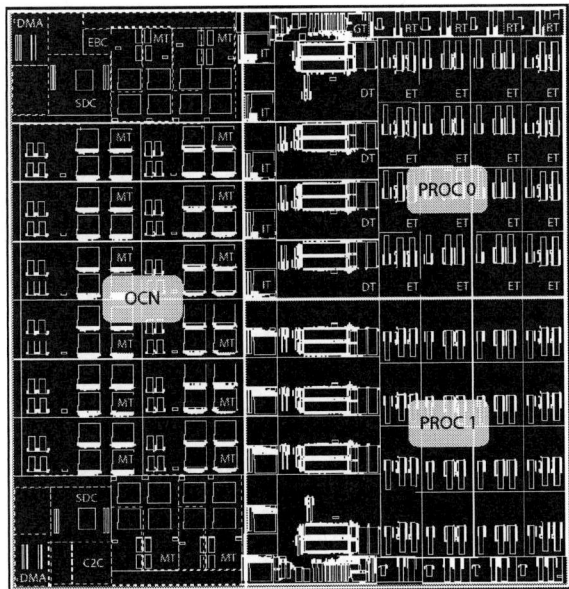

**Figure 3: Floorplan of a tiled chiplet architecture.
© 2006 IEEE. Reprinted, with permission, from [8].**

data caches, and instruction caches. The tiles ranged in size from 1mm2 to 9mm2 (in a 130nm process) and the design intended to align the tiles in both the X and Y dimensions. The on-chip NUCA memory system on the left side of the chip was composed of 16 level-2 cache tiles, memory controller tiles, DMA engines, and external communication interfaces.

The logical connections between the tiles in each processor included a general purpose NoC for data operand transmission, as well as several control networks to implement the distributed uniprocessor control protocols, such as instruction distribution. Each tile contained a NoC router as well as pipelined channels for the control protocol networks. All tiles connected via abutment with no inter-tile global wiring. The memory system only included a NoC (separate from the operand NoCs in the processors) and the tiles there were intended to connect via abutment.

In light of the design methodology described above, the TRIPS design achieved some of the physical design productivity goals. The tiled design in fact did eliminate all global wiring, making chip assembly trivial. Likewise, global timing closure was easy because of the clean timing interfaces at each tile boundary. In most areas of the chip, the tiles were sufficiently aligned to enable connection via abutment.

However, a number of challenges prevented the design from being as modular as intended. First, the aspect ratios of the tiles were not always well matched in both dimensions. For example, the DT needed to match the width of the GT and the height of the ET, but the fit was not perfect. Second, the tiling approach broke down in the corners of the memory system because space and aspect ratio constraints required the MTs on the top and bottom to be arranged vertically (instead of horizontally) so that the different controllers could fit in a square space, rather than a thin rectangular space. Four different tiles (the two memory controllers, the chip-to-chip router, and the I/O controller) each required a substantial number of chip pins, which far exceeded the size of each of controller. Instead, the pins were spread throughout the chip with top-level wiring connecting the controllers to their I/Os.

Finally, the design intended to perform a single physical design for each type of tile, with the layout and intra-tile wiring replicated during chip assembly. However, because the I/Os were spread throughout the chip, the I/O pattern caused non-uniform obstructions in different physical instances of the same tile. As a result, each tile was actually placed and routed individually, subject to the different I/O blockages within each instance. However, because all of the tile instances were independent, the placement and routing of each block could be performed in parallel. The entire chip physical design including placement, routing, and chip assembly was automated and could be performed in about a day.

While the TRIPS chip demonstrates a number of advantages of raising the level of abstraction for full-chip logical and physical design, it also highlights a number of the challenges, including imperfect chip component tiling and a need for both data and control networks. Arguably, the TRIPS chip is a hard design to tile because each tile contains only a small portion of the design, such as a slice of the level-1 data cache or a slice of the register file. Contemporary SoCs composed of commodity building blocks likely require fewer if any communication channels that could not be transmitted via a general purpose NoC. However, composing a two-dimensional tiled design in which tiles align in both dimensions may still be difficult.

Many of the differences between the TRIPS design and the approach we propose here are due to differences in requirements. Aligning modules in two dimensions and connecting by abutment is an ideal way to achieve modularity for a single design. Aligning uniform height modules in one dimension and routing the NoC in wiring channels external to the modules is better suited to building an ecosystem capable of quickly generating a wide range of SoCs.

7. COMPROMISES

While we expect the modular approach described above to serve the vast majority of SoCs, two areas may require compromises to the proposed methodology.

Out-of-band signals: The vast majority of signaling between modules can be accomplished over the NoC. Even most low-latency event signals can be efficiently packetized, with appropriate use of virtual channels for providing priority and QoS. However, in rare cases modules may need to exchange signals directly, bypassing the NoC. Such signals may arise in things like interrupt signals when partitioning a complex block into multiple tightly coupled chiplets. This practice of sending *out-of-band* signals should be strongly discouraged lest designers resort to it to avoid the effort required to packetize their interfaces. However, when needed, the methodology we describe here can be extended to allow direct connection of signals between modules. By requiring that such signals be sampled by the NoC clock at both the outbound and inbound interfaces and be routed entirely in the wiring channels, module verification/timing closure and global verification/timing closure can still be decoupled.

Captive vs. non-captive signal balls: Making all of the signal balls used by an I/O module *captive* (contained within the boundaries of the module) keeps the effect of these balls local to the module. Modules can be composed in any way without the ball pattern of one module affecting the functionality of another module. However, this restriction becomes costly in area when a module has a large number of balls but only a small area for circuits and logic. The alternative is to allow the module to route signals via a redistribution layer to non-captive balls, risking disruption to the power and ground grids of any modules placed under the balls.

8. A CALL TO ARMS

We articulate our vision of the future of SoC design with the goal of engaging the broader design community to adopt a similar vision and work with us to build the ecosystem needed to realize it. Four major components are required to make this vision a reality: module placement software, a NoC generator, IP modules, and chiplet verification tools.

Module placement: The module placement software is perhaps the simplest element as it is just a standard-cell placement system scaled up to larger modules and modified to deal with multiple row heights. The inter-module traffic matrix can be used instead of a netlist to derive module affinity. The standard techniques of graph partitioning for coarse placement followed by iterative refinement should yield a good placement. Given the small number of modules being placed (10^3) we expect a good solution to be reached within only a few hours of CPU time.

NoC generator: The NoC generator is perhaps the most complex element of the ecosystem. The NoC generator uses the module placement and the traffic matrix to derive the communication loading of key *cuts* of the SoC. It then determines a topology to provision sufficient communication across each cut at minimum cost. We expect this step to employ a standard topology such as a flattened butterfly as the underlying substrate, and adapt it to provide more bandwidth where needed and less bandwidth where there is little demand. The final step of the NoC generator instantiates library modules for channels, routers, and NICs to realize the selected topology.

IP modules: The most important part of the ecosystem is the IP because a critical mass of it is required to make the ecosystem viable. Chiplets for processors, on-chip memory, off-chip memory controllers, common peripheral interfaces (PCIe, SATA, USB, Ethernet, etc.), and common CODECs and MODEMs are needed to enable SoC designers to quickly assemble chips from a library of modules. IP modules may be soft (synthesizable Verilog) or hard (placed and routed cells). A proper IP module will include the module functionality packaged behind a standard NoC interface of an appropriate width. Hard IP is simply soft IP synthesized for a particular process and with placement, routing, timing closure, and test generation completed. The hard IP is ready to be placed in a modular SoC.

We expect that most of the required IP already exists and simply needs to be tied to a standard interface. For a given user, some IP will be generated in house, some IP will be provided by tool vendors, and some will be licensed by third parties. Our vision is that simpler composition of modules will lower the barrier to creating a vibrant market of standard high-level digital IP chiplets.

Chiplet verification tools: The challenges to widespread adoption of this approach lies in inter-module interconnects and establishing the necessary conventions and standards to allow automation to be responsible for all the heavy lifting. Besides functional connection requirements within the constraints of standardized module heights, formal specifications for ensuring appropriate power/ground resources and test interfaces to each module will be needed and require verification.

New analysis tools will be required to focus on the chiplets to ensure that they are built correctly and follow established rules so that they can be safely re-instantiated in any circumstance.

Compositions of chiplets should be *correct by construction*. Chiplet composition tools and appropriate checkers will be needed to ensure that all rules are followed and proper inter-chiplet connections are made. Given the independent nature of these chiplets, new automated test generation and test application methodologies and tools will be required. Such tools would allow the testing requirements of a module to be encapsulated so that those integrating a chiplet can again be isolated from the details of the test, yet be assured a quality result each and every time the module is used.

We have heard both engineers and venture capitalists lament the high cost of designing a SoC today. Estimates are that $50M is required to get a complex digital SoC to first prototype [9]. Some attribute the dearth of fabless semiconductor startups and the slowdown of innovation in the field to this high cost. While some of this cost can be attributed to mask sets and wafer starts, the bulk comes from design and verification. Our vision of a modular SoC built from standard chiplets offers a path toward greatly reducing the non-recurring cost of a SoC. In doing, so we hope that it will spur a new generation of fabless semiconductor startups and encourage more innovation in chip architecture and design.

9. ACKNOWLEDGMENTS

We thank our colleagues at NVIDIA whose work on numerous SoCs has shaped our thoughts on this topic.

10. REFERENCES

[1] Borkar, S. 2009. Design perspectives on 22nm CMOS and beyond. In *Proceedings of the Design Automation Conference*, July 2009, pp. 93-94.

[2] Parnas, D. L. 1972. On the criteria to be used in decomposing systems into modules. *Communications of the ACM, 15*(12) 1972, pp. 1053-1058.

[3] Dally, W. J. and Towles, B. 2001. Route packets, not wires: on-chip interconnection networks. In *Proceedings of the Design Automation Conference*, June 2001, pp. 684-689.

[4] Dally, W. J. and Towles, B. P. 2003. *Principles and Practices of Interconnection Networks*. Morgan Kaufmann.

[5] Dally, W.J. and Chang, A. 2000. The role of custom design in ASIC chips. In *Proceedings of the Design Automation Conference*, June 2000, pp. 643-647.

[6] Ueno, K. et al. 2007. A design methodology realizing an over GHz synthesizable streaming processing unit. In *Proceedings of the IEEE Symposium on VLSI Circuits*, June 2007, pp. 48-49.

[7] Dally, W. J. 1992. Virtual-channel flow control. *IEEE Transactions on Parallel and Distributed Systems, 3*(2), 1992, pp. 194-205.

[8] Sankaralingam, K., et al. 2006. Distributed microarchitectural protocols in the TRIPS prototype processor. In *Proceedings of the International Symposium on Microarchitecture*, December 2006, pp. 480-491.

[9] Goering, R. (2009) *Are SoC Development Costs Significantly Underestimated?*, http://www.cadence.com/Community/blogs/ii/archive/2009/09/24/are-soc-development-costs-significantly-underestimated.aspx

System Architecture and Software Design for Electric Vehicles

Martin Lukasiewycz, Sebastian Steinhorst, Sidharta Andalam, Florian Sagstetter,
Peter Waszecki, Wanli Chang, Matthias Kauer, Philipp Mundhenk
TUM CREATE, Singapore
martin.lukasiewycz@tum-create.edu.sg

Shreejith Shanker, Suhaib A. Fahmy
Nanyang Technological University, Singapore
sfahmy@ntu.edu.sg

Samarjit Chakraborty
TU Munich, Germany
samarjit@tum.de

ABSTRACT

This paper gives an overview of the system architecture and software design challenges for Electric Vehicles (EVs). First, we introduce the EV-specific components and their control, considering the battery, electric motor, and electric powertrain. Moreover, technologies that will help to advance safety and energy efficiency of EVs such as drive-by-wire and information systems are discussed. Regarding the system architecture, we present challenges in the domain of communication and computation platforms. A paradigm shift towards time-triggered in-vehicle communication systems becomes inevitable for the sake of determinism, making the introduction of new bus systems and protocols necessary. At the same time, novel computational devices promise high processing power at low cost which will make a reduction in the number of Electronic Control Units (ECUs) possible. As a result, the software design has to be performed in a holistic manner, considering the controlled component while transparently abstracting the underlying hardware architecture. For this purpose, we show how middleware and verification techniques can help to reduce the design and test complexity. At the same time, with the growing connectivity of EVs, security has to become a major design objective, considering possible threats and a security-aware design as discussed in this paper.

Categories and Subject Descriptors

C.0 [**Computer Systems Organization**]: General—*System architectures*

General Terms

Design

Keywords

Electric Vehicle, Software Design, System Architecture

Permission to make digital or hard copies of all or part of this work for personal or classroom use is granted without fee provided that copies are not made or distributed for profit or commercial advantage and that copies bear this notice and the full citation on the first page. To copy otherwise, to republish, to post on servers or to redistribute to lists, requires prior specific permission and/or a fee.
DAC '13, May 29–June 7 2013, Austin, TX, USA.

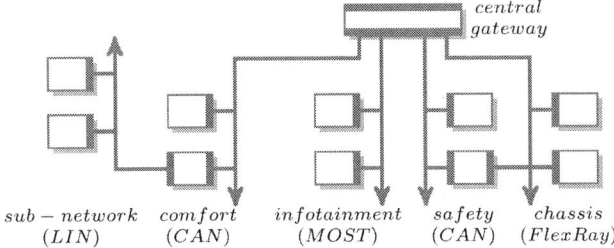

Figure 1: Illustration of a typical in-vehicle network architecture of a modern automobile. Various ECUs (▭) are interconnected via different buses. Gateways (▭) are used to interconnect the buses.

1. INTRODUCTION

EVs are widely accepted as emerging and sustainable solution to environmental and transportation challenges in growing mega-cities. In contrast to Internal Combustion Engine (ICE) cars, EVs comprise several components like the battery, electric motor, or powertrain that bring along new implementation, integration, and control challenges. Moreover, the implementation of drive-by-wire control and novel information systems may increase the safety and energy efficiency of EVs significantly. While these new components and systems bring along several fundamental challenges, EVs at the same time may serve as a platform to drive a paradigm change in the system architecture and software design for road vehicles.

Current top-of-the-range vehicles use embedded system architectures that consist of up to 100 ECUs with several heterogeneous buses that are interconnected by one or more gateways as illustrated in Figure 1. This complex network is a result of an incremental design over the last decades where new functionality is often introduced by adding separate hardware devices. The reasons for this *federated* approach can be found in the structure of the automotive industry where the car manufacturers obtain new functions from several different and competing suppliers.

This publication is made possible by the Singapore National Research Foundation under its Campus for Research Excellence And Technological Enterprise (CREATE) programme.

While the federated approach has been feasible in the recent years, it is reaching its limits with the growing complexity of in-vehicle networks and the lack of installation space, particularly in EVs. With a growing computational power of novel processing units, the major trend goes towards a consolidation of ECUs and a unification of the in-vehicle network. This *integrated* approach will require an entirely different design methodology where functionality has to be developed for a shared underlying architecture.

In this paper, we give an overview of the challenges and present first results in the system architecture and software design for EVs. Section 2 gives an introduction to EV components, including the battery, electric motor, and powertrain. In Section 3, the trends in communication and computation devices are presented. The software design challenges, comprising the control design as well as the upcoming security in vehicles, are presented in Section 4. Finally, Section 5 makes concluding remarks.

2. ELECTRIC VEHICLE COMPONENTS

In the following, the essential components of EVs such as the battery, electric motor, and electric powertrain are introduced. Moreover, drive-by-wire and information systems are discussed. While it is to some extent possible to convert an ICE vehicle to an EV by replacing the powertrain, a tailored design is considered as more sustainable. Due to the shorter driving ranges and their emission-free operation, it is projected that EVs will play a major role in growing megacities. As a result, a design of the vehicle and its components has to take this scenario into account.

Battery Pack. In EVs, the battery pack is currently the most essential and expensive component while it is also a major bottleneck restricting the driving range of the vehicle. These Electrical Energy Storages (EESs) require both high energy and power density to optimize the utilization of allocated weight and volume of the battery. In this context, Lithium-Ion (Li-Ion) batteries are widely considered to dominate other battery chemistries [1]. However, Li-Ion batteries are sensitive to their operating parameters and exceeding specified bounds such as overcharging or undercharging cells negatively impacts the reliability by causing damage to the battery. In the worst case, this damage leads to a thermal runaway resulting in fire or explosion that poses a serious safety issue.

Therefore, sophisticated Battery Management Systems (BMSs) are applied to maintain the battery in a safe and healthy operating state. Figure 2 illustrates a state-of-the-art BMS architecture in a hierarchical approach. This incorporates monitoring the State of Charge (SoC) of battery cells by measuring voltage levels and the current drawn from the cells, as well as the temperature of cells.

As series-connected battery cells charge and discharge unevenly, Cell Balancing (CB) is required to maintain an equalized overall SoC for the battery. This equalization is achieved by either passive or active CB. Passive approaches that are state-of-the-art discharge cells with a higher SoC over a resistor to the charge level of the cell with the lowest SoC. In contrast, active approaches transfer charges between cells to avoid the waste of energy, increasing the driving range as well as the lifetime of the battery. A recent advancement in the area of active cell balancing system design is presented in [2].

Electric Motor. Synchronous motors are commonly used in EVs due to their high efficiency and light weight [4]. As shown in Figure 3, the motor is driven by sinusoidal waveforms that are controlled by the inverter. The inverter is using six

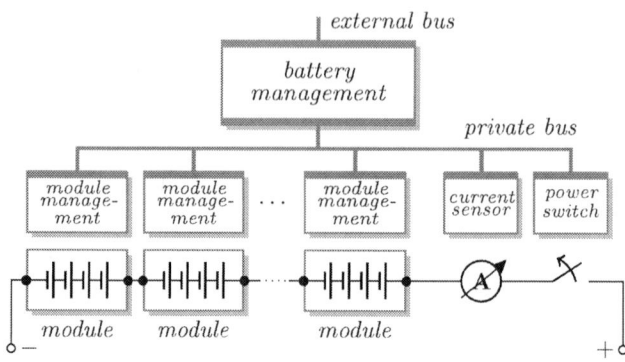

Figure 2: Illustration of the Battery Management System (BMS) consisting of a hierarchical architecture with battery cell modules controlled by module management devices, see [3].

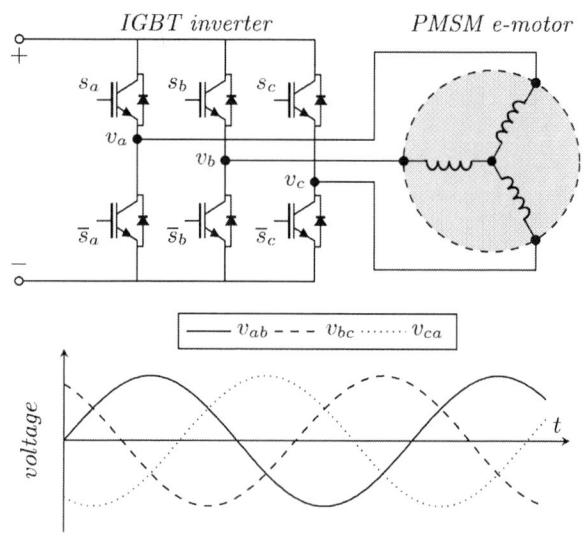

Figure 3: Illustration of the control of a Permanent Magnet Synchronous Motor (PMSM) using an inverter consisting of six Insulated Gate Bipolar Transistors (IGBTs) $s_a, \overline{s}_a, s_b, \overline{s}_b, s_c, \overline{s}_c$. The IGBTs are controlled such that the three voltages are sinusoidal waveforms that are phase-shifted by $\frac{2}{3}\pi$.

Insulated Gate Bipolar Transistors (IGBTs) to convert the DC source from the battery pack to AC current with the desired frequency and magnitude to drive the motor. For this purpose, the IGBTs work as electronic switches that are controlled by Pulse Width Modulation (PWM) signals using the space-vector modulation technique [5]. Here, proper switching sequences for all six IGBTs guarantee that the output is the desired three-phase alternating current with specific phase shifts.

The efficient and reliable control of electric motors in EVs can become a challenging task. For instance, due to harsh operating environments like high voltages, switching frequency, and changing temperatures, IGBTs might fail. One or more faults in the IGBT package make the output current no longer sinusoidal which drives the motor into unpredicted

666

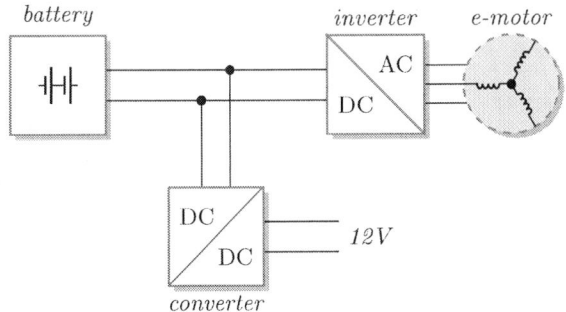

Figure 4: Illustration of a basic electric powertrain of a full electric vehicle.

operating modes and thus may jeopardize passenger lives. Therefore, a fault-tolerant control strategy is desirable to ensure normal operation of the motor under one fault or a combination of different faults. The solution should be optimized to require minimal redundant hardware while covering a high amount of faulty situations. The controller needs to calculate post-fault PWM sequences and react in real-time, requiring efficient algorithms to ensure that the motor returns to the best possible operation quickly enough such that safety requirements of the vehicle are not compromised.

Electric Powertrain. The electric powertrain is a distinctive feature of EVs. It is understood that the transition from ICE cars to EVs cannot be accomplished by simply replacing the ICE with an electric motor and the fuel tank with a battery. As illustrated in Figure 4, an architecture for the powertrain of an EV has to be developed for the motor, battery, converters, and inverters which needs a sophisticated control to operate in an effective and efficient way. Energy flow control is of particular importance in this context [6]. This incorporates the management of the energy storage as well as the energy consumers and providers.

As both the electric motor and the battery can operate as energy sources and consumers, depending on the driving state such as acceleration or regenerative braking, the control and the timing of energy flows are crucial design tasks that significantly affect the energy efficiency of the overall system. To increase the efficiency, a distributed real-time control becomes inevitable. Moreover, a conversion between different voltage levels may be required in the powertrain. Therefore, an optimization of the different conversion levels using DC-DC converters is equally important for the energy efficiency of the architecture.

Information Systems. One major challenge in EVs is the limited driving range due to the relatively small battery capacity. To cope with this drawback, an information system becomes necessary that ensures that driving ranges are never exceeded. At the same time, it is projected that information systems will be tightly coupled with next-generation entertainment systems.

Information systems in EVs need to be more sustainable than systems using fixed mounted screens that require a high amount of maintenance and are outdated within a short period of time in comparison to the fast progress in consumer electronics [7]. One way to achieve this goal is by establishing a native connection between the user devices such as smartphones or tablets and the vehicle. These user devices might interface the vehicle to access information, entertainment, and control related functionality. Implementations of this

approach might consist of one central server component in the vehicle, offering wireless open access for user devices. Via the central vehicle server, the devices can be integrated with Vehicle-to-Vehicle (V2V) and Vehicle-to-Grid (V2G) infrastructures that are particularly important for EVs. Providing information on available charging stations to drivers can be further qualified by taking into account the locations, energy-consumption and destinations of all vehicles, as well as the number and location of charging stations. On the other hand, such an open system brings along several challenges in the domain of security that have to be addressed appropriately to guarantee that safety-critical applications are under no circumstances compromised.

Drive-by-wire. While drive-by-wire is a technology emerging also for ICE cars, it is of particular importance for EVs. In EVs, the acceleration is already not performed mechanically and as current battery technology only allows a limited range, energy recuperation during braking is essential to extend the driving range [8]. The energy savings using this mechanism as well as using prudent acceleration could be optimized via sophisticated and supportive control, enhancing the desired driver inputs. Therefore, a mechanical decoupling between the braking pedal and the brakes becomes necessary as provided by a brake-by-wire implementation [9]. Besides that, EVs are more straightforward to actuate by electronics than ICE cars.

However, since drive-by-wire is highly safety-critical, it needs to be designed in a fault-tolerant fashion, introducing a certain amount of redundancy in the control system. With most software errors being of systematic nature, straightforward component duplication may not be sufficient to reduce the associated risk to a sufficiently small level [10]. Functions may have to be implemented by different programmers or at least run on non-identical hardware. Additional care needs to be taken to fail-proof actuating and sensing devices. As a result, the introduction of this technology in the automotive industry where cost sensitivity prevails is facing certain obstacles. In recent years, the advent of more sensing devices and cheaper as well as more efficient ECUs have sparked the interest again in the context of EVs.

3. SYSTEM ARCHITECTURE

In the following, challenges and trends in the system architecture in EVs are presented. We give an overview of the in-vehicle communication as well as next-generation computational devices in EVs. A major trend goes towards the consolidation of ECUs, relying on a homogeneous network. This will require significantly more powerful computational devices as well as deterministic bus systems with a high bandwidth. In the automotive domain, this would require a paradigm change where EVs might serve as a platform for an entire redesign of the system architecture. This transition from *federated* systems with one function per single-core ECU to *integrated* systems, comprising several applications on one powerful ECU can impede the growing complexity of automotive architectures by reducing the overall number of controllers and thus increase the flexibility and reliability [11].

3.1 Communication

Protocols. Automotive architectures today make use of many different heterogeneous bus systems such as Controller Area Network (CAN), Local Interconnect Network (LIN), FlexRay, and Media Oriented Systems Transport (MOST), see Figure 1. As a result of the complexity of this in-vehicle network, integration and configuration became time and cost intensive tasks that may even prohibit distributed functions that rely on predictable communication with high bandwidths. Therefore, one of the key challenges for next-generation architectures is the design of a unified in-vehicle network that allows to integrate and configure components and subnetworks transparently and independently of each other. The basis for such architectures are deterministic communication protocols. One suitable solution introduced in recent years is FlexRay which is based on a hybrid protocol supporting both time-triggered and event-triggered communication with a bandwidth of 10MBit/s [12]. For next-generation vehicles and particularly EVs, Ethernet with a bandwidth of 100MBit/s and more may become the first choice as a cost-efficient and light-weight implementation, using a physical layer implementation that allows twisted-unshielded pair cables [13]. As standard Ethernet is non-deterministic and therefore unsuitable for time-critical applications, extensions to support stringent real-time requirements become necessary. In this context, the Audio/Video Bridging (AVB) protocol is currently used in the automotive industry [14] for the implementation of entertainment systems, relying on a synchronization with Precision Time Protocol (PTP) [15] which might also serve as basis for other time-triggered Ethernet protocols.

Besides the communication between the components in the in-vehicle network, EVs also benefit from V2G and to some extend V2V communication. For this purpose, protocols like Bluetooth or IEEE 802.11 (WiFi) may become a common feature in the communication infrastructure. In this context, Service Oriented Architectures (SOAs) have to be employed to enable a provision of information and control services [16].

Time-triggered Scheduling. EVs implement many functions that require a deterministic communication such as battery management, electric motor control, and upcoming drive-by-wire implementations. Additional safety-critical functions will even increase the need for real-time communication with very strict end-to-end timing, such that time-triggered protocols will further gain importance. As schedules are defined at design time, time-triggered protocols ensure predictability and provide a deterministic communication. In particular, the time-synchronization between ECUs allows to obtain a global schedule, considering all tasks and messages.

While synchronous time-triggered scheduling allows to significantly reduce the end-to-end timing delays of applications, obtaining schedules for all tasks and messages concurrently is highly complex and existing approaches only provide limited scalability [17]. A remedy might be an integration approach where the configuration for each component or subsystem might be defined independently and integrated into a global schedule in the integration phase [18]. This methodology is also in accordance with the design approach in the automotive industry where individual components are designed and tested with a valid configuration before being integrated in a later stage.

3.2 Computation

Multi-core. The ever growing demand for innovation and new functionality in modern cars necessitates a paradigm shift in automotive system architectures in terms of hardware and software. In particular, the introduction of drive-by-wire applications and high-performance BMSs for EVs as well as new information systems clearly redefines the requirements of current hardware platforms in terms of computation, communication, and overall topology. This change, in turn, calls for the implementation of multi-core ECUs for which three main reasons can be pointed out. First of all, gaining higher

computational performance without increasing the power consumption and heat dissipation can only be achieved by adding new cores rather than increasing clock frequencies of single-core processors [19]. Moreover, the enhanced performance per watt ratio supports the strict energy requirements in EVs and the higher level of parallelism provided by multi-core systems allows for the compliance with automotive safety standards like ISO 26262 [20]. Finally, a major advantage is that software from single-core ECUs can be ported to multi-core ECUs more easily than to architectural different computational platforms, like FPGAs.

Besides a reliable high-performance hardware, safety-critical applications as used for next-generation driver assistance systems would strongly benefit from real-time capable multi-core Operating Systems (OSs). By providing comprehensive software partitioning and resource sharing mechanisms an OS can guarantee the desired functionality. The challenge is to design a specialized multi-core OS which provides not only hard real-time guarantees for safety-critical tasks but also mechanisms for segregating trusted and non-trusted code and strategies for intelligent cache utilization. Furthermore, to ensure a fully predictable real-time behavior for both the multi-core ECU and the corresponding communication within the system architecture, only time-triggered execution models come into consideration. Although current multi-core OSs, like *PharOS* [21] or *Barrelfish* [22], are not able to fulfill the aforementioned requirements, they can highlight general design trends for the development of a reliable and deterministic multi-core OSs for system architectures of EVs.

GPU. Modern vehicles comprise a significant amount of image and sensor processing to increase the safety of all road users. In this context, there are specific challenges for EVs which for instance have almost soundless engines, making them extremely dangerous in situations where pedestrians come into close proximity with the vehicle. A remedy might be a radar or camera-based pedestrian recognition that is coupled with a warning system that generates audible alerts to pedestrians.

One possible implementation of the warning system can be achieved using cameras and image processing techniques. Cameras capture a snapshot of the environment and then the image is processed in real-time to detect any pedestrians. This image processing demands a lot of computational resources which can be provided by Graphical Processing Units (GPUs) [23]. When compared to a CPU, due to more hardware-level parallelism, a GPU is significantly faster at processing an image [23]. Thus, GPUs would make a good choice for implementing safety-critical functions that require a high amount of parallel processing.

GPU programming frameworks such as OpenCL and CUDA are generally used for programming GPUs [24]. However, due to the inherent parallelism, programming a GPU is much more complicated than programming a CPU. The programmer must be aware of the underlying hardware architecture and structure the program carefully to avoid multiple threads accessing the same memory locations at the same time, resulting in either inconsistent data or deadlock during memory access.

FPGA. Field Programmable Gate Arrays (FPGAs) have found favor in a number of application domains as a platform for accelerating complex algorithms. They offer the benefits of a custom hardware implementation at a fraction of the fixed cost of designing a custom Integrated Circuit (IC). Hence, they find use in medium-volume markets or those where evolving standards may require regular changes to the hardware.

668

While FPGAs are currently not often used in the automotive domain, they bring along several advantages for EVs. For computationally intensive tasks, within an embedded systems power budget, FPGAs are often the only sensible choice. Furthermore, since computation is implemented spatially, it is possible to completely isolate distinct tasks while maintaining their individual determinism. This could help to reduce the number of ECUs and at the same time guarantee the isolation between various safety-critical applications. Note that Application-Specific Integrated Circuits (ASICs) are cheaper and even more energy-efficient, but they might not provide the necessary flexibility that is required in case many functions are implemented on one ECUs.

An additional capability that is unique to FPGAs is that, as volatile devices, they can be reconfigured. This has so far primarily been used to allow for design iteration, or incorporation of improvements in subsequent revisions. However, this capability can also be leveraged at runtime, in the form of dynamic reconfiguration, allowing different applications to be implemented at different times. All these capabilities make the incorporation of FPGAs in EVs a promising prospect [25].

Advanced techniques like Partial Reconfiguration (PR) extend the capabilities of FPGAs for use in safety-critical applications. PR allows us to define fault-tolerant embedded computing units that can recover from faults by selectively reconfiguring the faulty module alone, while a redundant mode with lower specifications takes over control during the recovery process [26].

4. SOFTWARE DESIGN

A consolidated system architecture will require an entirely different function and software design approach in EVs. A central component of this software design approach might be a middleware that enables a more flexible development and operation. In the following, we discuss challenges and approaches in the area of software design for EVs. First, we present techniques to enable an efficient implementation of control functions using several different techniques. Finally, the issue of security in an increasingly connected vehicle is discussed.

4.1 Control

Middleware Approach. Current premium-cars already require millions lines of code [27] and it is projected that EVs will further increase the amount of software in vehicles. A major reason is the requirement to create additional sources of revenue for car manufacturers to cope with the high costs of batteries as well as the loss of the profitable service of the ICE. Such a source of revenue for EVs could be made possible by allowing additional purchasing of functionalities while the vehicle is already in operation. In turn, this will require a significantly more flexible system and software design to cope with the growing complexity of distributed control.

For the sake of higher flexibility and shorter time-to-market, a lean middleware approach is a potential solution. This middleware may abstract the underlying hardware and operating systems and enable a unified platform for the design of software while providing support for virtualization such that various tasks can be executed in parallel on the same hardware in an isolated fashion. As a result, software tasks may be distributed in a more flexible way, supporting the desired reduction of ECUs.

While flexibility is a major goal in software design for vehicles, the determinism of functions and the entire system has to be ensured. This might be achieved by incorporating the mentioned time-triggered mechanisms into the middleware as well as verification techniques.

Verification of distributed control systems. Many components in EVs require a precise and responsive control. Battery packs and upcoming drive-by-wire applications have a spatially distributed control where sensors, controllers, and actuators cannot be implemented on a single device or ECU, respectively. This leads to significant communication delays in the control loops that further complicate the correct design of the control functions. As a remedy, verification approaches might be used to formally guarantee the correct functionality of safety-critical control functions. This approach represents a significant improvement in current design flows where, although controller models are formally verified, their implementation on a distributed architecture is validated in an ad-hoc fashion with extensive testing and integration efforts.

The classic control-theoretic approach relies on idealized assumptions. Typically, computation and even communication are assumed to function perfectly and without delay. This is to a certain extent reasonable for single ECU systems where no communication takes place. Recently, verification techniques have been applied to tackle this problem. In [28], it was shown that an ω-regular language can be used as interface between the performance requirements for the control system and the transmission timings on a communication network. It is more expressive than an interface that only relies on combinations of periods and deadlines. In [29], this model is further extended such that an automaton framework for streaming systems is adapted to describe the transmission of control messages. It is then verified that the communication remains within the allowed patterns using model checking. In turn, this guarantees the initial performance requirement. A major challenge of these verification approaches that are very versatile remains the scalability.

Precise Timing Analysis. Safety-critical systems require guarantees on the functionality as well as the timing characteristics of programs. This requires modelling timing behaviour of micro-architectural features such as memory hierarchies, pipelines and buses to compute the Worst Case Execution Time (WCET) of a program. The ability to compute precise timing analysis is important and is dependent on the architectural features. For example, caches with replacement policies like LRU provide the best predictability, while PLRU and FIFO are much harder to analyse [30].

Static analysis of caches with various allocation and replacement algorithms have been studied in great detail [30, 31]. In general, there is a trade-off between the precision (tightness of the estimate) and the scalability (analysis time). For example, [31] captures the precise behavior of caches. However, this technique does not scale for large programs. In contrast, the approach in [30] avoids the state-explosion using an abstraction that scales for very large programs at the expense of reduced precision.

An alternative to caches are ScratchPad Memories (SPMs) that are fully software controlled caches. In SPMs, the allocation and replacement decisions are made in software, guided by compile time decisions. Recent work on SPMs focuses on developing software allocation algorithms and/or designing tailored architectures with SPMs [32]. SPMs are allocated statically and are easier to analyze, but they provide comparatively lower average and sometimes worst case performance when compared to caches. However, for successful implementation of safety-critical systems, it is important to emphasize on *predictability* rather than *performance*. Thus, SPMs are an ideal design choice for implementations of safety-critical systems.

4.2 Security

Threats. A recent analysis of automotive architectures has shown that current series vehicles are often insufficiently protected against attacks aiming to temper the system. For instance, researchers were able to gain access to the in-vehicle network via Bluetooth and implemented a virus [33, 34]. While EVs have many security vulnerabilities in common with ICE vehicles, additional security threats are arising: Latest generation charging plugs implement a communication protocol to allow information exchange between the BMS and the charging station, e.g., for billing or for future V2G applications. This might allow man-in-the-middle attacks where the attacker attaches a connector between the charging plug of the car and the charging station. For a more detailed discussion of the security threats for electric vehicles see [35].

Security-aware Design. Due to a significantly higher connectivity of EVs, security has to become a major design objective for automotive architectures. Besides protecting potential access points, the charging plug or V2G communication interfaces with authentication approaches such as challenge response [36], additional security measures for the in-vehicle network become necessary. For instance, a secure in-vehicle communication is required that encrypts messages transmitted between ECUs, including an authentication of the sender. While existing buses like CAN are unsuitable for a secure communication due to the limited message size, upcoming protocols in the automotive domains such as Ethernet with IPSec [37] offer an established security solution. Additionally, as security leaks are commonly introduced by careless integration of secure components, a holistic design approach is essential to obtain a secure automotive architecture. The basis for such an architecture might be the discussed middleware which allows isolating the components and software functions from each other such that one compromised component does not affect the whole system.

5. CONCLUDING REMARKS

Already today, a vast majority of innovations in vehicles is driven by electronics and software. However, the current architectures grew incrementally over the past decades and the integration approach to implement new functions by adding new hardware devices is reaching its limits. Here, EVs may serve as a platform to implement a consolidated hardware architecture using middleware approaches to drastically simplify the integration, providing a solution to cope with the growing pressure to innovate in the automotive industry. At the same time, there exist several challenges such as the limited driving range of EVs that need to be addressed and solved properly in order not to become obstacles in this potential paradigm change in the automotive industry.

6. REFERENCES

[1] M. Brandl et al. Batteries and battery management systems for electric vehicles. In *Proc. of DATE*, pages 971 –976, 2012.

[2] M. Kauer, S. Narayanaswami, S. Steinhorst, M. Lukasiewycz, S. Chakraborty, and L. Hedrich. Modular system-level architecture for concurrent cell balancing. In *Proc. of DAC 2013*, 2013.

[3] M. Brandl et al. Batteries and battery management systems for electric vehicles. In *Proc. of DATE*, pages 971 –976, 2012.

[4] I. Boldea. Control issues in adjustable speed drives. *IEEE Industrial Electronics Magazine*, 2(3):32–50, Sep 2008.

[5] K. Zhou and D. Wang. Relationship between space-vector modulation and three-phase carrier-based PWM: A comprehensive analysis. *IEEE Transactions on Industrial Electronics*, 49(1):186–196, Feb 2002.

[6] H. Yoo, S. Sul, Y. Park, and J. Jeong. System integration and power-flow management for a series hybrid electric vehicle using supercapacitors and batteries. *IEEE Transactions on Industry Applications*, 44(1):108 –114, 2008.

[7] A. Schmidt, A.K. Dey, A. L. Kun, and W. Spiessl. Automotive user interfaces: human computer interaction in the car. In *Ext. Abstracts CHI*, pages 3177–3180, 2010.

[8] C. C. Chan. The state of the art of electric, hybrid, and fuel cell vehicles. *Proceedings of the IEEE*, 95(4):704–718, 2007.

[9] E.a. Bretz. By-wire cars turn the corner. *IEEE Spectrum*, 38(4):68–73, 2001.

[10] R. Isermann, R. Schwarz, and S. Stolzl. Fault-tolerant drive-by-wire systems. *IEEE Control Systems*, 22(5):64 – 81, October 2002.

[11] P. Peti, R. Obermaisser, F. Tagliabo, a. Marino, and S. Cerchio. An integrated architecture for future car generations. In *Proc. of ISORC*, pages 2–13, 2005.

[12] FlexRay Consortium. FlexRay communications systems - protocol specification. http://www.flexray.com.

[13] Broadcom. BroadR-Reach Ethernet hardware, 2011.

[14] H.T. Lim, L. Volker, and D. Herrscher. Challenges in a future IP/Ethernet-based in-car network for real-time applications. In *Proc. of DAC*, pages 7–12, 2011.

[15] IEEE standard for a precision clock synchronization protocol for networked measurement and control systems, 2008.

[16] T.J. Giuli, D. Watson, and K.V. Prasad. The last inch at 70 miles per hour. *IEEE Pervasive Computing*, 5(4):20–27, October 2006.

[17] M. Lukasiewycz, R. Schneider, D. Goswami, and S. Chakraborty. Modular scheduling of distributed heterogeneous time-triggered automotive systems. In *Proc. of ASP-DAC*, pages 665–670, 2012.

[18] F. Sagstetter, M. Lukasiewycz, and S. Chakraborty. Schedule integration for time-triggered systems. In *Proc. of ASP-DAC*, 2013.

[19] J. Wolf, M. Gerdes, F. Kluge, S. Uhrig, J. Mische, S. Metzlaff, C. Rochange, H. Casset', P. Sainrat, and T. Ungerer. RTOS support for parallel execution of hard real-time applications on the MERASA multi-core processor. In *Proc. of ISORC*, pages 193–201, 2010.

[20] N. Navet, A. Monot, B. Bavoux, and F. Simonot-Lion. Multi-source and multicore automotive ecus-os protection mechanisms and scheduling. In *Proc. of ISIE*, pages 3734–3741, 2010.

[21] C. Aussaguès, D. Chabrol, V. David, D. Roux, N. Willey, A. Tournadre, and M. Graniou. PharOS, a multicore os ready for safety-related automotive systems: Results and future prospects. In *Proc. of ERTS*, 2010.

[22] A. Baumann, P. Barham, P.E. Dagand, T. Harris, R. Isaacs, S. Peter, T. Roscoe, A. Schüpbach, and A. Singhania. The multikernel: a new OS architecture for scalable multicore systems. In *Proc. of SOSP*, pages 29–44, 2009.

[23] B. Bilgic, B.K.P. Horn, and I. Masaki. Fast human detection with cascaded ensembles on the GPU. In *Proc. of IV*, pages 325 –332, June 2010.

[24] F. Jianbin Fang, A.L. Varbanescu, and H. Sips. A comprehensive performance comparison of cuda and opencl. In *Proc. of ICPP*, pages 216 –225, 2011.

[25] S. Shreejith, S. A. Fahmy, and M. Lukasiewycz. Reconfigurable computing in next-generation automotive networks. *IEEE Embedded Systems Letters*, 5(1):12–15, 2013.

[26] S. Shreejith, K. Vipin, S. A. Fahmy, and M. Lukasiewycz. An approach for redundancy in FlexRay networks using FPGA partial reconfiguration. In *Proc. of DATE*, 2013.

[27] R. Charette. This car runs on code. *IEEE Spectrum*, 46(3):3, 2009.

[28] R. Alur and G. Weiss. Regular specifications of resource requirements for embedded control software. In *Proc. of RTAS*, 2008.

[29] M. Kauer, S. Steinhorst, D. Goswami, R. Schneider, M. Lukasiewycz, and S. Chakraborty. Formal verification of distributed controllers using time-stamped event count automata. In *Proc. of ASP-DAC*, 2013.

[30] H. Theiling, C. Ferdinand, and R. Wilhelm. Fast and precise WCET prediction by separated cache and path analyses. *Journal of Real-Time Systems*, 18:157–179, 1999.

[31] N. Singh, T. Mitra, and A. Roychoudhury. Accurate estimation of cache-related preemption delay. In *Proc. of CODES+ISSS*, pages 201–206, 2003.

[32] I. Liu, J. Reineke, D. Broman, M. Zimmer, and E. Lee. A PRET microarchitecture implementation with repeatable timing and competitive performance. In *Proc. of ICCD*, 2012.

[33] S. Checkoway, D. McCoy, B. Kantor, D. Anderson, H. Shacham, S. Savage, K. Koscher, A. Czeskis, F. Roesner, and T. Kohno. Comprehensive experimental analyses of automotive attack surfaces. In *Proc. of Usenix Security*, 2011.

[34] K. Koscher, A. Czeskis, F. Roesner, S. Patel, T. Kohno, S. Checkoway, D. Mccoy, B. Kantor, D. Anderson, H. Shacham, and S. Savage. Experimental security analysis of a modern automobile. In *IEEE Symposium on Security and Privacy*, pages 447–462, 2010.

[35] F. Sagstetter et al. Security challenges in automotive hardware/software architecture design. In *Proc. of DATE*, 2013.

[36] R. Falk and S. Fries. Electric vehicle charging infrastructure - security considerations and approaches. In *Proc. of INTERNET*, pages 58–64, 2012.

[37] Internet Engineering Task Force. RFC 4301 security architecture for the internet protocol, 2005.

Model-Based Development and Verification of Control Software for Electric Vehicles

Dip Goswami[1], Martin Lukasiewycz[2], Matthias Kauer[2], Sebastian Steinhorst[2],
Alejandro Masrur[1], Samarjit Chakraborty[1] and S. Ramesh[3]

[1]Institute for Real-Time Computer Systems, TU Munich, Germany
[2]TUM CREATE, Singapore
[3]General Motors Corp., USA

ABSTRACT

Most innovations in the automotive domain are realized by electronics and software. Modern cars have up to 100 Electronic Control Units (ECUs) that implement a variety of control applications in a distributed fashion. The tasks are mapped onto different ECUs, communicating via a heterogeneous network, comprising communication buses like CAN, FlexRay, and Ethernet. For electric vehicles, software functions play an essential role, replacing hydraulic and mechanic control systems. While model-based software development and verification are already used extensively in the automotive domain, their importance significantly increases in electric vehicles as safety-critical functions might no longer rely on mechanical (fall-back) solutions. The need for reducing costs, size, and weight in electric vehicles has also resulted in a considerable interest in topics such as the consolidation of ECUs as well as *efficient* implementation of control software. In this paper we discuss two broad issues related to model-based software development and verification in electric vehicles. The first is concerned with how to ensure that model-level semantics are preserved in an implementation, which has important implications on the verification and certification of control software. The second issue is related to techniques for reducing the computational and communication demands of distributed automotive control algorithms. For both these topics we provide a broad introduction to the problem followed by a discussion on state-of-the-art techniques.

Categories and Subject Descriptors

D.2.10 [**Software Engineering**]: Design—*Methodologies*

General Terms

Theory, Design, Performance, Verification

Permission to make digital or hard copies of all or part of this work for personal or classroom use is granted without fee provided that copies are not made or distributed for profit or commercial advantage and that copies bear this notice and the full citation on the first page. To copy otherwise, to republish, to post on servers or to redistribute to lists, requires prior specific permission and/or a fee.
DAC'13, May 29 - June 07 2013, Austin, TX, USA

Keywords

Electric vehicles, control systems, model-based design, control/architecture co-design

1. INTRODUCTION

The volume of functionality implemented in software has been steadily increasing in the automotive domain over the past decades. In electric vehicles the importance of software functions will even further increase as mechanic and hydraulic systems will be replaced by software-based control. This will reduce the overall weight of the car and in turn increase the driving range which is of utmost importance in electric vehicles, compensating the limitations of state-of-the-art batteries. At the same time, software-based control may increase efficiency, for instance, for regenerative breaking, etc.

The shift towards software-based control increases the computational demand significantly. Current design methods in the automotive domain follow a federated architecture approach, where each function is implemented on a separate ECU. While this enables the Original Equipment Manufacturer (OEM) to outsource the development to various suppliers – with the role of the OEM being that of a function integrator – it increases the number of ECUs in the car as well as the cabling harness. In the case of electric vehicles this becomes a major problem, not only because of the complexity and the distributed nature of the resulting architecture, but also due to the additional weight of the ECUs and the cables. As a result, there is an increasing push to move towards integrated architectures, where multiple software functions are integrated onto a single ECU. An application in this architecture is distributed into different tasks, each of which runs on a different ECU, communicating via shared buses.

While integrated architectures combine many advantages, they also bring along many challenges due to the higher complexity in the software implementation and integration. As a result, the need for appropriate verification and certification techniques becomes more prominent. Whereas *model-based* software development and verification are already practiced in the automotive domain, their importance is significantly higher in electric vehicles, primarily due to the lack of mechanical fall-back solutions as, for example, in drive-by-wire implementations. In this paper we discuss two important issues in this context. The first is concerned with how to ensure that model-level semantics are preserved in an implementation, particularly for complex distributed architectures with multiple ECUs communicating via a heterogeneous network using different bus protocols. This has important implications on the verification and certification when using model-based design techniques. The second issue is concerned with *efficient* implementations of control software, such that their computational and communication demands are reduced.

The paper is organized as follows. We first present an overview of how to model feedback control software and the implementation platform in Section 2 followed by an introduction of embedded control in Section 3. This is in turn followed by a discussion of how an implementation may be synthesized from high-level control models (in Section 4). In particular, we show how schedules in automotive-specific bus protocols like CAN or FlexRay might be synthesized in order to meet control performance requirements. The second part of the paper shows how control performance goals may be realized even when certain feedback signals are dropped (in Section 5). This reduces both computational and communication demands, and therefore results in cost-effective implementations. Towards this, we first show how to quantify the bounds on the number of feedback signal drops for given control performance requirements. We then show how such bounds may be verified for a given architecture, i.e., whether the number of signals dropped by an architecture is within the limits of what may be tolerated by the application.

2. SYSTEM MODEL

2.1 Feedback control model

Often the system model is considered to be linear time-invariant (LTI) in the following form

$$
\begin{aligned}
\dot{x}(t) &= A_c x(t) + B_c u(t), \\
y(t) &= C_c x(t),
\end{aligned}
\tag{1}
$$

where $x(t)$ is the $n \times 1$ vector for *state variables*, $u(t)$ is the *control input* to the system and $y(t)$ is the *output*. A_c is a $n \times n$ system matrix, B_c and C_c are input and output matrices of appropriate dimension. Subsequently, the continuous-time system is sampled at a constant sampling interval h with zero-order-hold (ZOH). The resultant system is a discrete-time system of the form

$$
\begin{aligned}
x[k+1] &= Ax[k] + Bu[k], \\
y[k] &= Cx[k],
\end{aligned}
\tag{2}
$$

where

$$
A = e^{A_c h}, B = \int_0^h (e^{A_c t} dt) \cdot B_c, C = C_c.
\tag{3}
$$

Such feedback control systems have two physical components: *actuators* (to apply the input $u[k]$ to the system) and *sensors* (to read states $x[k]$ from the system). In a distributed architecture such as those in the automotive domain, the actuators and the sensors are often spatially distributed and connected to different ECUs which communicate via a bus system. A feedback *controller* is an algorithm to compute $u[k]$ as a function of the states $x[k]$ or output $y[k]$ (feedback signals) such that $x[k]$ or $y[k]$ behave according to the specified control goals – we assume all states $x[k]$ to be *measurable*.

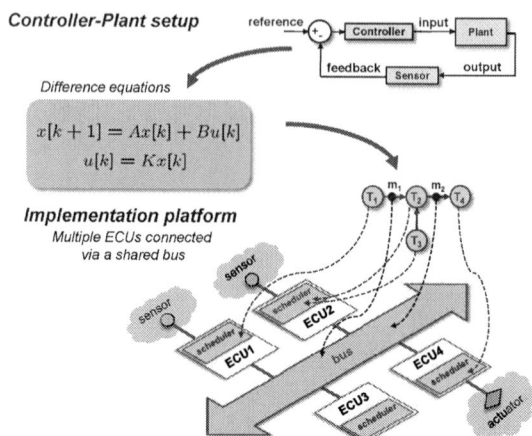

Figure 1: Distributed implementation platform.

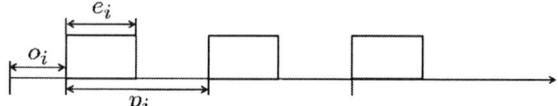

Figure 2: Periodic task model.

In this work we use full-state feedback controllers of the form shown in the following equation:

$$
u[k] = Kx[k],
\tag{4}
$$

where K are $1 \times n$ *state-feedback gains*. The controller design essentially boils down to choosing K such that the system (i) is *stable*, i.e., the closed-loop system matrix $(A+BK)$ has all the *eigenvalues* within the *unit circle*, and (ii) meets performance requirements in steady-state and transient phases, e.g., minimization of the tracking error and the settling time of the control system.

2.2 Implementation platform

Let us consider an automotive in-vehicle network where several ECUs are connected via a bus as depicted in Fig. 1. The software implementation of distributed control applications running on such architectures is typically realized by model-based software development. This involves automatic code generation from high-level control models such as MATLAB/Simulink, along with the bus driver stack and the operating system (OS). For this, the high-level control models are partitioned into several software tasks that need to be mapped onto different ECUs. Subsequently, the generated code, e.g., C-code, is cross-compiled for the target hardware and flashed to the dedicated ECUs.

Automotive ECU. In general, an ECU consists of (i) a host microcontroller running the OS which schedules and executes application tasks T_i and system tasks (e.g., for communication purposes, etc.), (ii) a communication controller that implements the communication protocol (e.g., FlexRay, CAN) and (iii) bus drivers that realize the conversion of the logical bit stream into physical signals propagated on the communication bus.

The task model depends on the scheduling policy implemented by the OS. The scheduling policy can either be preemptive (e.g., OSEK) or non-preemptive (e.g., eCos). Moreover, the ECUs can either be synchronous (e.g., FlexRay) or asynchronous (e.g., CAN) to the bus schedules.

The control tasks are often periodic and time-triggered. In such cases, the OS provides a task dispatcher which allows cyclic task execution. Let a dispatch event for a task T_i (see Fig. 2) be defined by the tuple $T_i = \{o_i, p_i, e_i\}$:

- The task offset o_i specifies the duration from start time to the first task invocation of T_i.

- The task period p_i specifies the time between two consecutive activations of T_i.

- The worst-case execution time e_i specifies T_i's maximum execution time.

Hence, the k-th instance of task T_i is triggered at

$$t_i^k = o_i + kp_i, \ k \in \mathbb{N}_0. \tag{5}$$

The scheduler of the OS processes all tasks according to a dispatch table in a cyclic manner where the length of the dispatch table is determined by the *hyperperiod* $H = \underset{\forall i}{lcm}(p_i)$. Consequently, all task invocation times can be computed offline and stored in the dispatch table. Further, the k-th instance of T_i finishes at latest at

$$\tilde{t}_i^k = o_i + kp_i + e_i, \ k \in \mathbb{N}_0. \tag{6}$$

Automotive Bus. Today's cars usually consist of a wide range of different bus systems connected via gateways. Local Interconnected Network (LIN), Controller Area Network (CAN), FlexRay, Media Oriented Systems Transport (MOST) and Ethernet are some of the bus protocols commonly used. Moreover, different bus systems are used in different functional domains depending on the requirements on bandwidth and temporal behavior. For example, the power-train and the chassis domain often use FlexRay due to their stringent timing constraints. On the other hand, MOST is suitable for the infotainment domain. Moreover, bus protocols can be purely time-driven, event-driven and hybrid in nature. In the following we illustrate the nature of a typical automotive bus system using FlexRay as an example.

FlexRay is a hybrid communication protocol which is organized as a sequence of 64 bus cycles that are periodically repeated [1]. Each cycle is of fixed duration T_{bus} and consists of a static and a dynamic segment. The communication is time-driven in nature on the static segment whereas communication over the dynamic segment is event-driven. Messages m_i are transmitted on the FlexRay bus either on the static or the dynamic segment according to their predefined bus schedules. We denote a bus schedule by the tuple $\Theta_i = \{S_i, B_i, R_i\}$ (Fig. 3):

- The communication slot S_i determines the time window in which a message may be transmitted. $S_1 = 3$ for message m_1 in Fig. 3.

- The base cycle $B_i \in \{0, 1, ..., 63\}$ specifies the first cycle during which S_i is available. For instance, in Fig. 3, m_2 is assigned $B_2 = 1$ since the first cycle in which $S_2 = 6$ is available is *cycle 1*.

- The repetition rate $R_i \in \{1, 2, 4, 8, 16, 32, 64\}$ determines the number of cycles between two consecutive transmissions. For example, m_1 is assigned repetition

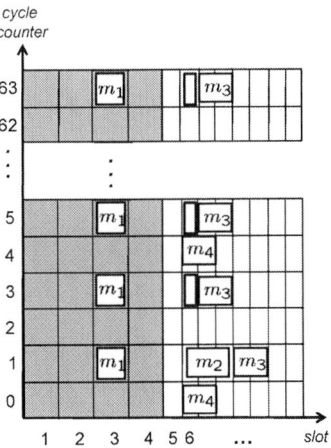

Figure 3: FlexRay Schedule: Gray boxes indicate static segment slots.

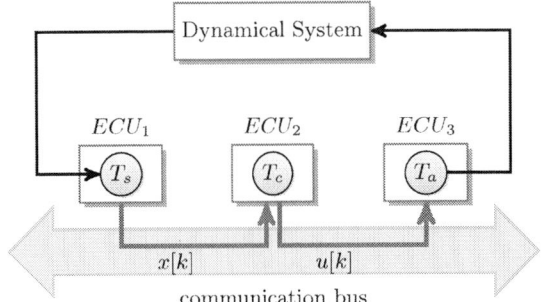

Figure 4: Distributed control application.

rate $R_1 = 2$ and hence S_1 is available in every second cycle. Similarly, m_4 is assigned $R_4 = 4$, therefore S_4 is available in every fourth cycle. Clearly, $B_i < R_i$ must hold.

3. EMBEDDED CONTROL

We analyze in this section how to implement control software based on the models discussed above. In general, a control application C_l is *partitioned* into three tasks (see Fig. 4): (i) the sensor task T_s – reading $x[k]$ from sensors, (ii) the controller task T_c – computing $u[k]$ using (4) and, (iii) the actuator task T_a – applying $u[k]$ to the dynamic system. In an embedded control implementation T_c and T_a are typically time-triggered periodic tasks. Often, the period of T_c and T_a is equal to the sampling period h of the corresponding control application. The triggering paradigm of the sensor task T_s depends on the sensing mechanism. Generally, the sensor tasks have to constantly monitor some states of the control plant and are therefore *interrupt-driven*. T_s can be triggered many times within a sampling period with a negligible execution time. Further, the application tasks are *mapped* onto different ECUs and communicated over a bus. Fig. 4 shows an example of task mapping where T_s, T_c and T_a are mapped onto separate ECUs. Here, the system states

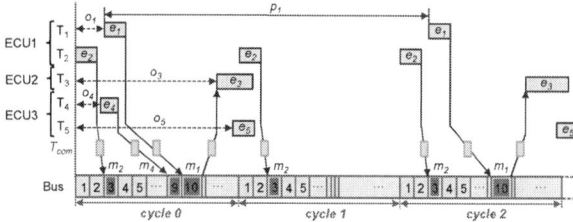

Figure 5: An example of control implementation.

$x[k]$, measured by T_s, are sent over the bus to T_c, which then computes $u[k]$ and sends it to T_a.

3.1 Implementation semantics

Generalizing the above description, each control application can be partitioned into a set of n control-related tasks T_i, $i \in \{1, ..., n\}$. Furthermore, depending on the task mapping, schedules Θ_i are assigned to messages m_j on the communication bus where $j \in \{1, ..., \kappa\}$. Thus, a control application $\mathcal{C}_l$ has a $(n + \kappa)$-tuple of platform parameters $\Phi_l = \{T_i, \Theta_i\}$. For example, let a control application $\mathcal{C}_l$ be partitioned into $n = 5$ tasks that trigger $\kappa = 3$ messages as shown in Fig. 5 – note that m_1, m_2 and m_4 are the only messages sent by T_1, T_2 and T_4 respectively. Then the corresponding platform parameters are denoted by the 8-tuple:

$$\Phi_l = \{T_1, T_2, T_3, T_4, T_5, \Theta_1, \Theta_2, \Theta_4\}.$$

Recall that $T_i = \{o_i, p_i, e_i\}$, and $\Theta_i = \{S_i, B_i, R_i\}$ are also tuples themselves.

3.2 Design questions

In this context, the pertinent design questions that have gained significant attention from both academia and industry are the following:

- **How to design Φ_l and $\mathcal{C}_l$ such that all the high-level control goals are met?** Traditionally, the design of Φ_l and $\mathcal{C}_l$ are performed in two isolated phases. That is, $\mathcal{C}_l$ is designed assuming a possible worst-case design of Φ_l and similarly, Φ_l is designed by translating the performance requirements of $\mathcal{C}_l$ into overly conservative timing constraints. Such isolated design phases introduce pessimism at various layers making the overall design very conservative. Hence, such design flows can potentially become a major bottleneck for electric cars in particular, due to design parameters like weight of cabling, number of ECUs, battery lifetime, etc. that might be worsened in this case. In contrast, the design of Φ_l and $\mathcal{C}_l$ can mutually be directed by each other. Hence, a part of Φ_l is first designed. Next, a part of $\mathcal{C}_l$ is designed exploiting/utilizing the information on the already designed part of Φ_l and so on. In recent literature [18, 16, 19, 17], it is shown that such an interactive design approach opens up scope for significant design optimization. We illustrate this joint design approach with an example in Section 4.

- **How to improve the overall design efficiency in terms of resources and performance?** In such platforms, a control performance is achieved at the cost of certain computational and communication resources. An efficient implementation consumes lesser

resource towards achieving the same control performance. Hence, an interesting approach is to consider the trade-off between control performance and *quality of resource*. For example, consider a distributed implementation such as the one shown in Fig. 4. With a purely time-driven communication system, it is possible to assure a very short delay in the feedback loop and achieve very good performance. Thus, a high resource quality (because of the time-driven nature in this case) provides better performance. On the other hand, with an event-driven communication at a lower priority level, the control loop occasionally suffers a long delay and jitter which degrades the control performance.

In a flurry of recent works [2, 20, 11, 12, 10], the control-related messages are assigned a lower priority communication schedule (lower quality resource). In such scenarios, the feedback delay occasionally becomes very large while *mostly* not exceeding a threshold d_{th}. When the delay exceeds the threshold d_{th}, the feedback signal is either *dropped* or overwritten. These cases are often considered to be "deadline misses" by the control message. Allowing such deadline misses in the design process makes it less stringent compared to the case where deadlines need to be met in all occasions. Thus, it is possible to reduce the amount of pessimism otherwise introduced by a worst-case based design. This approach raises further two research questions: How frequently should such dropping be allowed by the control application and how to formally validate that the implementation platform never violates that constraint? We illustrate these aspects in Section 5.

4. CONTROL/PLATFORM CO-DESIGN

As explained above, the design phase in automotive control systems is traditionally separated from the development phase. This makes it difficult to provide guarantees for high-level control requirements. To overcome this problem and, at the same time, achieve a resource-efficient system, both control and platform-related parameters (such as the schedules on ECUs and communication bus) have to be co-designed.

To illustrate this technique, we consider in this section a deterministic time-triggered system where the sensor-to-actuator delay does not suffer from any jitter. We first present a scheduling approach that determines feasible schedules satisfying all constraints or deadlines. Second, we show how the control parameters can be optimized iteratively, using the scheduling approach for each iteration.

4.1 Scheduling Approach

Based on control requirements, a deadline on the delay τ can be deduced for each application. Here, each path π of data-dependent tasks has to satisfy this deadline.

In time-triggered systems, a schedule is defined by the offset o_i of each task T_i for a globally synchronized time. Determining a feasible schedule that satisfies all deadlines might become a complex problem that makes a manual design impractical. For this reason, it can be converted to a mathematical programming problem such as Integer Linear Programming (ILP). In the following, one such solution is outlined. Here, the variables to be optimized are the offsets which are bounded by the period of each task such that $\forall T_i$:

$$0 \leq o_i \leq p_i. \tag{7}$$

For each pair of tasks that is running on the same resource, only a single task can be executed at a time.

$\forall T_i, T_j$(on the same resource):

$$(o_i + e_i + k \cdot p_i)\%H \quad \leq \quad o_j + \tilde{k} \cdot p_j \vee \tag{8}$$

$$(o_j + e_j + \tilde{k} \cdot p_j)\%H \quad \leq \quad o_i + k \cdot p_i, \tag{9}$$

where $k = \{0, .., \frac{H}{p_i} - 1\}$ and $\tilde{k} = \{0, .., \frac{H}{p_j} - 1\}$. Here, we assume a non-preemptive scheduler which might be extended to preemptive schedulers as proposed in [14].

Eq. (8) and (9) ensure that no instances/jobs from tasks that are running on the same resource preempt each other. In this context, all jobs k and $\tilde{k}$ in the hyperperiod H are considered.

To determine the offsets o_i of communication tasks sending messages via a bus, the specific protocol has to be considered. An efficient formulation for the FlexRay bus is presented in [14]. Given the offsets of all tasks and messages, the waiting time w_{T_i, T_j} between each pair of data-dependent tasks T_i and T_j can be determined.

$\forall T_i, T_j$(with data-dependencies):

$$0 \leq w_{T_i, T_j} \quad \leq \quad p_i \tag{10}$$

$$w_{T_i, T_j} \quad = \quad (o_j - (o_i + e_i) + H)\%H \tag{11}$$

Here, the waiting time is always less than the period of the current application. Finally, the end-to-end delay of each application (and path within the application) has to be less than the specified constraint or deadline.

$\forall \tau_X$(deadline), π(paths of tasks):

$$\sum_{T_i \in \pi} + \sum_{T_i, T_j \in \pi} w_{T_i, T_j} \leq \tau_X \tag{12}$$

Eq. (7) to (12) have to be linearized appropriately to enable an optimization with an ILP. This linearization requires the introduction of additional constraints and variables [14].

4.2 Optimization

The optimization of control parameters is performed iteratively as described in the following. For each iteration, the period p_i for each T_i is chosen from a predefined set of feasible periods. In the model presented above, the periods within one application are considered to be equal. However, this assumption might also be relaxed to allow different task periods within a single control application. Given all periods, the controller gains K can be determined under the assumption that the sensor-to-actuator delay constraint equals the period.

For this given configuration, a scheduling approach as discussed above has to be performed to determine a feasible schedule. If a feasible schedule exists, the performance of the current implementation can be determined using, for instance, the quadratic performance function for each control application:

$$J(h, \tau) = \sum_{k=0}^{N} \int_{kh}^{(k+1)h} [u(t)^2 + x(t)'x(t)]dt, \tag{13}$$

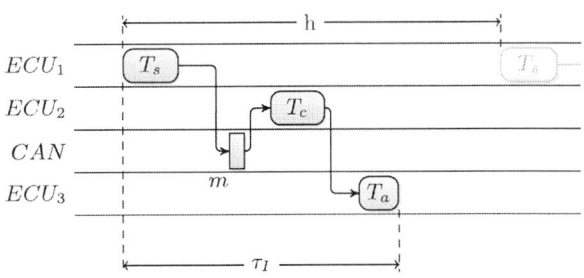

Figure 6: Timing requirements for the control messages: Case I, $\tau_I < h$.

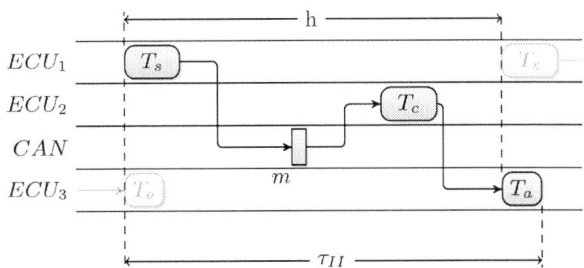

Figure 7: Timing requirements for the control messages: Case II, $\tau_{II} \approx h$.

where N is the total number of samples under consideration. Additional design objectives might be the utilization of ECUs or buses which in turn might find implementations with more economical hardware still satisfying the defined control quality constraints.

The optimization is performed iteratively such that the system is gradually improved. In case the design space (the set of available periods for all applications) is small, an exhaustive search can be performed as presented in [9]. If the design space becomes larger, an exhaustive search becomes impractical and randomized search algorithms like simulated annealing might be applied.

5. FEEDBACK DELAY AND SIGNAL DROP

In this section we are concerned with the second research question stated above. Using information from the control models, we relax some of the typical conservative assumptions that are usually made in the design and development process. In particular, we show that is possible to meet performance requirements even if some control-related messages are lost in the communication bus.

In general, the feedback signal has to go through a series of executions, i.e., $T_s \rightarrow T_c \rightarrow Bus \rightarrow T_a$ in a feedback loop. The delay τ is the time required to complete the entire execution sequence which highly depends on Φ_l. Thus, the delay depends on the triggering patterns of tasks T_s, T_c and T_a as well as the message schedules on the bus. There are two obvious possibilities (Fig. 6 and Fig. 7).

Case I: In this case, the triggering pattern of tasks T_s, T_c and T_a is such that the offset between T_s and T_a is less than one sampling interval. Tasks T_s, T_c and T_a are triggered periodically with period h. The entire series of executions

675

$T_s \rightarrow T_c \rightarrow Bus \rightarrow T_a$ is mostly expected to be finished by $\tau_I < h$ (see Fig. 6). Therefore, the *deadline* τ_I for the feedback signal is lesser than one sampling interval with such Φ_l. When $\tau_I << h$, the delay is assumed to be *zero* and the control law (4) is therefore *realizable* when the feedback loop meets deadline τ_I. This case is more difficult to analyze but it greatly exceeds the basic case's performance.

Case II: In this case, the triggering of T_s and T_a is synchronized with *zero* offset. Tasks T_s, T_c and T_a are triggered periodically with period h. The entire series of executions $T_s \rightarrow T_c \rightarrow Bus \rightarrow T_a$ is mostly expected to be finished by $\tau_{II} \approx h$ (see Fig. 7). Therefore, the *deadline* τ_{II} for the feedback signal is approximately equal to the sampling period h with such Φ_l. Since the feedback signal is delayed by one sampling interval, the control law (4) is not realizable and therefore, it is modified as follows:

$$u[k] = K \cdot x[k-1]. \tag{14}$$

In this case, the control law (14) is realizable when the feedback loop meets deadline τ_{II}.

From Cases I and II, we have seen that the execution of a feedback loop, i.e., $T_s \rightarrow T_c \rightarrow Bus \rightarrow T_a$, is associated with a deadline (τ_I and τ_{II}). Since the control tasks are usually time-triggered, the schedule of message m on the bus plays a key role in this context . In the case of priority-based arbitration on the communication bus, the transmission of message m might get delayed and reach ECU_2 after T_c has already been triggered – see Fig. 6 and Fig. 7. Such a timing scenario leads to a deadline miss, i.e., $\tau > \tau_I$ or $\tau > \tau_{II}$.

In a traditional approach, the communication schedule for message m is designed such that the deadline is always met. Towards this, the worst-case end-to-end delay of the feedback loop needs to be computed and this should essentially be within the deadline. As already mentioned, there has been a renewed interest in allowing certain deadline violations in the above context since the worst-case delay estimation is often overly conservative and only occurs very rarely.

5.1 Closed-loop system

Based on the above cases, the control law needs to be appropriately adapted. For example, for the Case I in Fig. 6, the following control law is adapted in [2, 20]:

$$u[k] = \begin{cases} Kx[k] & , \text{if } \tau \leq \tau_I \\ 0 & , \text{if } \tau > \tau_I \end{cases}$$

Here, it is typically assumed that the gain matrix K is designed for the lossless transmission case, e.g., via the LQG framework.

Similarly, for the Case II in Fig. 7, the following control law is adapted in [10, 11]:

$$u[k] = \begin{cases} Kx[k-1] & , \text{if } \tau \leq \tau_{II} \\ 0 & , \text{if } \tau > \tau_{II} \end{cases}$$

Of course, a more general case would be to consider deadlines of any length including multiple sampling periods [12].

In the above strategy, the control input $u[k]$ is applied only when the control message meets its deadline and $u[k]$ is set to *zero* otherwise. Based on this control strategy we have two systems: (i) when the deadline is met, $u[k]$ is applied and the resulting closed-loop system becomes A_c and, (ii) when

the deadline is violated, $u[k] = 0$ and the resulting open-loop system is A. Note that the resulting system matrices A_c and A might have higher dimension due to the presence of feedback delay as in (14). Both for cases I and II, the closed-loop system becomes,

$$x[k+1] = A_\sigma x[k], \tag{15}$$

where $A_\sigma = A_c$ when the deadline is met and $A_\sigma = A$ in the case of deadline violation. Depending on the nature of the *delay pattern*, the closed-loop system keeps switching between A_c and A. Over a duration of l sampling periods, the closed-loop system is given by,

$$x[k+l] = A_{\sigma_{k+l}} \cdots A_{\sigma_{k+1}} A_{\sigma_k} x[k]. \tag{16}$$

The requirements from the control side mainly deal with stability and performance of the closed loop system (16). Each element of the overall system matrix sequence $A_{\sigma_{k+l}} \cdots A_{\sigma_k}$ results from either meeting or violating the deadlines.

5.2 Stability-based requirements

Clearly, the closed-loop system (16) is a *switched* system where the switching happens between $A_\sigma = A_c$ and $A_\sigma = A$ depending on the delay pattern. The question is: *What is the maximum frequency allowed for deadline violations with guaranteed stability of system (16)?*

A general answer to the above question follows from [21] with a slightly modified formulation of the control strategy,

$$u[k] = \begin{cases} Kx[k] & , \text{if } \tau \leq \tau_I \\ Kx[k-1] & , \text{if } \tau > \tau_I \end{cases}, \tag{17}$$

with $A_\sigma = A_1$ when $\tau \leq \tau_I$ and $A_\sigma = A_2$ when $\tau > \tau_I$.

THEOREM 1. (Stability condition) *Consider the closed-loop system (16) with control input (17) and assume that A_1 is Schur stable.*

- *If the open-loop system A is marginally stable, the system (16) is exponentially stable for $0 < r \leq 1$.*

- *If the open-loop system A is unstable, the system (16) is exponentially stable for*

$$\frac{1}{1 - \gamma_1/\gamma_2} < r \leq 1 \tag{18}$$

where r is the rate of met deadlines or the ratio of successful transmissions over the infinite horizon and

$$\gamma_1 = \ln \lambda_{max}^2(A_1), \gamma_2 = \ln \lambda_{max}^2(A).$$

The above theory holds true for any value of l in (16). For example, suppose we obtain $r = 0.1$ for a given choice of parameters in (16). With $l = 1000$, the above theorem implies that *any* 100 samples are allowed to violate their deadlines with guaranteed exponential stability. On the one hand, the above condition is very relaxed and generic enough to be applied to any system. On the other hand, such a condition is not suitable for analyzing performance-critical applications since it does not guarantee any performance (e.g., a desired settling time, etc.). Furthermore, the control law (17) is realizable if the worst-case delay of a feedback loop is bounded by one sampling interval. As already mentioned, it might be complex and overly pessimistic to design an architecture with only the worst-case delay estimation in mind.

A more structured requirement on stability was introduced in [12] where the systems are required to be stable with a desired *stability margin*. As discussed, not all feedback messages suffer the worst-case delay. With this observation, a deadline or threshold delay d_{th} (lesser than the worst-case delay) is chosen such that it is met by the *most* of the feedback messages. For example, one can choose $d_{th} = h$ in the case shown in Fig. 7. The control requirement is then represented as *delay frequency metric*.

DEFINITION 1 (DELAY FREQUENCY METRIC (d_{th}, n)). *If every feedback message with delay larger than d_{th} is followed by at least n feedback messages with delay no more than d_{th},* the *delay frequency metric is said to be* (d_{th}, n).

For the delay frequency metric with $n = 3$, every sample violating the deadline is followed by at least three samples where the deadline is met. The overall system can then be represented as follows,

$$x[k + 2 + n_i + n_j] = AA_c^{n_i} \times AA_c^{n_j} x[k], \quad (19)$$

where $n_{i,j} \geq 3$. The stability is assured with a given stability margin by showing the existence of a Common Quadratic Lyapunov Function (CQLF) [15] between systems $AA_c^{n_i}$ and $AA_c^{n_j}$ for all combinations of n_i and n_j.

5.3 Performance-based requirements

In an assortment of recent works [2, 11, 20], the control requirements are specified using a notion of *exponential stability*,

$$\frac{\|x[k + l]\|}{\|x[k]\|} < \epsilon, \quad (20)$$

where $\|.\|$ denotes 2-norm. Hence, to ensure that the plant remains exponentially stable, any error must be reduced at least by a factor of ϵ in l sampling periods, i.e., $l \times h$ time. For example, $l = 5, \epsilon = 0.75$ means that any error signal must be reduced by at least 25% in five samples to maintain exponential stability of the system. It should be noted that the above notion of exponential stability is *stronger* compared to its definition found in control theory literature [8].

Coming back to the control requirement on exponential stability (20) and considering the closed-loop system (16), we obtain the following relation,

$$x_{k+l} = A_{\sigma_{k+l}} \cdots A_{\sigma_k} x_k,$$
$$\Rightarrow \frac{\|x_{k+l}\|}{\|x_k\|} \leq \|A_{\sigma_{k+l}} \cdots A_{\sigma_k}\|$$
$$\Rightarrow \|A_{\sigma_{k+l}} \cdots A_{\sigma_k}\| < \epsilon. \quad (21)$$

It follows that the exponential stability requirement can be re-written to (see [20, 2]):

$$ES(l, \epsilon) = \{\sigma_i \in \{o, c\}^\omega : \|A_{\sigma_{k+l}} \cdots A_{\sigma_{k+1}}\| < \epsilon \quad \forall k \in \mathbb{N}\}$$

$ES(l, \epsilon)$ is the language of strings σ over the alphabet $\{o, c\}$ corresponding to switching patterns of A and A_c that ensure that a possible error in the system is reduced by at least factor ϵ in l sampling periods. It was shown in [20, 2] that this language is ω-regular and how a Büchi automaton can be constructed to represent it.

Hence, ensuring the control performance requirement of exponential stability boils down to verifying that the system's switching between A and A_c remains within the acceptable patterns as specified by condition (21).

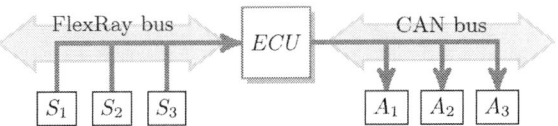

Figure 8: Example architecture.

The computation of this language of acceptable patterns or equivalently its Büchi automaton can be done by a brute-force search as described in [20, 2]. Model-checking this language pattern-by-pattern can become tedious however. To avoid this, a deadline constraint was used in [11]:

DEFINITION 2 $((f, H)$-FIRM DEADLINE). *A stream of control messages is said to fulfill the (f, H)-firm deadline with respect to period h if at least f out of any H consecutive samples meet their deadline.*

Note that this definition takes a sliding window perspective. The idea is that among all possible patterns, one can rule out all the unacceptable ones by requiring a combination of (f, H)-firm deadlines. Translating the resulting set of (f, H)-firm deadlines to LTL is then straightforward. Once the allowable feedback signal drop patterns have been specified, the next step is to capture the timing characteristics or behaviors of the implementation platform and subsequently checking if they constitute a subset of the behaviors that the controller can tolerate. This guarantees that the *implementation* of the control system on this particular platform will meet the performance requirement.

5.4 Timing properties of architectures

Towards characterizing timing behaviors of the architectures, various techniques have been proposed in the area of real-time and embedded systems, as well as in the formal methods literature. Analyzing embedded platforms in the particular context of implementing distributed controllers has been studied in [11, 12]. In [12], the Real-Time Calculus [5] modeling formalism has been combined with the tool Uppaal [13] for verification of timed automata. Real-Time Calculus relies on specifying upper and lower bounds on the number of messages that might arrive at a communication resource over different time interval lengths. These arrival curves are typically denoted by $\alpha^u(\Delta)$ and $\alpha^l(\Delta)$ respectively. Similarly, the communication resource is modeled using upper and lower bounds on the number of messages it can transmit; let these service curves be denoted by $\beta^u(\Delta)$ and $\beta^l(\Delta)$ respectively. These bounds may then be used to compute the worst-case delays suffered by messages, in addition to properties like buffer requirements. The exact same techniques apply to both computation and communication resources alike.

When multiple message streams are to be scheduled on a communication resource, scheduling or resource arbitration policies like TDMA, fixed priority or EDF may also be modeled using Real-Time Calculus, where the *service bounds* $\beta^u(\Delta)$ and $\beta^l(\Delta)$ for the entire resource are transformed into similar bounds for each message stream (see [5] for details). Fig. 8 shows an architecture where sensor readings are transmitted over a FlexRay bus to a controller implemented on the ECU. Control messages are then transmitted over the CAN bus to actuators. The sensor-to-actuator delays experienced by the individual messages certainly influence the

677

stability and QoC of the controller. In order to compute this end-to-end delay, the service bounds of the individual architectural components (FlexRay and CAN buses and the ECU) need to be composed in order to obtain the service bound of the overall architecture. This is given by:

$$\beta^{\text{end-to-end}} = \beta^{\text{FR}} \otimes \beta^{\text{ECU}} \otimes \beta^{\text{CAN}}, \qquad (22)$$

where $\otimes$ is the convolution operation as defined in Real-Time Calculus [5]. In order to estimate not only the maximum delay suffered by control messages, but also the *frequency* with which messages violate their deadline (derived from control-theoretic analysis, as described in the previous section) constraints, in [12] the service bounds or timing behaviors of the resource were transformed to a timed automata model that was then verified using UPPAAL.

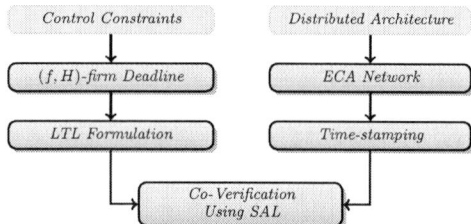

Figure 9: ECA-based co-verification workflow.

A similar technique was adopted in [11]. Its verification workflow is outlined in Fig. 9. First, the control performance specification is translated to LTL as described in the previous section. The communication hardware is then modeled as a network of Event Count Automata (ECA). ECA are a more flexible model for the streams that RTC is also modeling. Formally, ECA are tuples (see [6] for details)

$$\mathcal{A} = (S, s_{in}, X, V_{in}, Inv, \rho, \rightarrow) \qquad (23)$$

where S is a set of states and s_{in} is the initial state, X is a set of count variables with V_{in} being their initial valuation. $Inv : S \rightarrow \Phi(X)$, the Invariant Constraint Function, describes the possible valuations for every state whereas $\rho : S \rightarrow \mathbb{N} \times \mathbb{N}$, the rate function, indicates how many events can occur in a state and how the count variables evolve. Finally, $\rightarrow \subset S \times \Phi(X) \times 2^X \times S$, the transition relation, describes how the system can evolve from state to state.

The language of an ECA consists of strings that record the event counts that occur along the path. This can be more readily seen from an example like the periodic with jitter arrival automaton in Fig. 10. There, the automaton starts in the initial state $s_{in} = A$ with initial valuation $V_{in} = \{x = 0\}$. It then moves to B with either $x = 0$ or $x = 1$. In the third step, it progresses to C, recording again $x = 0$ or $x = 1$ depending on what the valuation of x was in the second step before resetting and starting over. This produces strings of 100 or 010 that are repeated infinitely often. These strings indicate how many events arrive or how much service is available in a particular time step. This is an extension of the RTC concept of arrival and service curves. For any pair of upper and lower arrival curves α^u and α^l, there is an ECA for which it holds:

$$L_{ECA} = \left\{ R(t) : \alpha^l(t) \leq R(t) \leq \alpha^u(t) \right\} \qquad (24)$$

The reverse is not true – ECA are in fact more expressive since they can accommodate state-based rate changes.

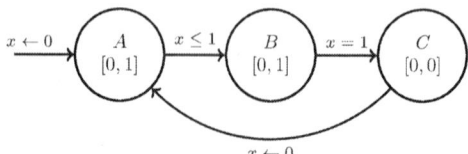

Figure 10: Periodic with jitter arrival ECA ($p = 3, j = 2$).

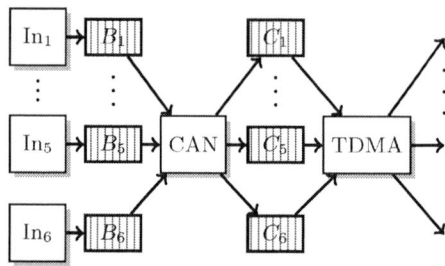

Figure 11: Running example modeled as ECA network.

On the next level, multiple ECA can be connected in a network of ECA by introducing buffers in between them. The network starts with arrival ECA that define the system inputs that are deposited into the first buffers. From there, service ECA process data in their input buffers to their output buffers – see Fig. 11 for an example.

To keep track of the delay for complicated patterns across the system, time-stamps for the messages are introduced (see Fig. 12). This allows running an evaluation automaton at the end that produces outputs that are compatible with the $\{o, c\}$ patterns from $ES(l, \epsilon)$.

5.5 Interface-based approach

Both the timed automaton approach from [12] and the timestamped ECA method from [11] are steps towards a common goal. The aim is to provide a richer interface for the specification of hardware/software requirements.

The next step in line with the intention of [2] is to formally specify the miss patterns of the architecture as a language $L_{architecture}$. It could then be combined with approaches that rely on *interface theory* [4, 7, 3]. More precisely, it would be possible to compare a language $L_{controller}$ that models the requirements of the control system, e.g., $ES(l, \epsilon)$. These two languages constitute the *interfaces* of the controller and the architecture. Checking the *compatibility* of these two interfaces now boil down to the problem of checking for language inclusion, i.e., whether

$$L_{controller} \subseteq L_{architecture}$$

Satisfaction of this inclusion implies that the controller may be implemented on the given architecture. This could then more readily be propagated through the system to verify multiple control applications. The work reported in [2, 20] leads into this direction, but it was written from a scheduling point of view. In other words, the requirements of multiple control systems were established and it was evaluated if all of them can be executed on a common processor.

678

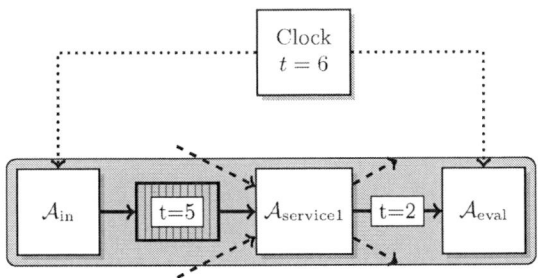

Figure 12: A stream of time-stamped messages in a network of ECA.

6. CONCLUDING REMARKS

Whereas, to a great extent, combustion cars still rely on mechanical fall-back systems, all safety-critical control applications in electric vehicles are exclusively implemented based on embedded electronics and software. As a result, it is necessary to carefully design such systems, in particularly, in the context of electric cars.

Nowadays, since model-based design methods dominate the design and development of automotive control systems, there is an increasing need for techniques that allow guaranteeing that the properties specified by high-level models translate into correct implementations. Towards this, it is necessary to utilize information about the underlying architecture at the model level and vice versa, i.e., to design the architecture using information provided by models.

In this paper we presented different techniques for modeling control performance requirements. Moreover, we discussed how to use those performance parameters to optimize an implementation and, in particular, the schedules on the ECUs and on typical automotive buses such as FlexRay. We showed that an iterative procedure increasingly involving information provided by models and that the implementation leads to more resource-efficient designs. Finally, we discussed different model-based techniques to formally verify properties of control systems such as timing, stability, and performance on a given architecture.

7. REFERENCES

[1] FlexRay protocol specification version 2.1. http://www.flexray.com.

[2] R. Alur and G. Weiss. Regular specifications of resource requirements for embedded control software. In *IEEE Real-Time and Embedded Technology and Applications Symposium (RTAS)*, 2008.

[3] P. Bhaduri and S. Ramesh. Interface synthesis and protocol conversion. *Formal Asp. Comput.*, 20(2):205–224, 2008.

[4] A. Chakrabarti, L. de Alfaro, T. A. Henzinger, and M. Stoelinga. Resource interfaces. In *EMSOFT*, pages 117–133, 2003.

[5] S. Chakraborty, S. Künzli, and L. Thiele. A general framework for analysing system properties in platform-based embedded system designs. In *DATE*, 2003.

[6] S. Chakraborty, L. T. X. Phan, and P. S. Thiagarajan. Event count automata: A state-based model for

stream processing systems. In *26th IEEE Real-Time Systems Symposium (RTSS)*, pages 87–98, 2005.

[7] L. de Alfaro and T. A. Henzinger. Interface theories for component-based design. In *EMSOFT*, pages 148–165, 2001.

[8] R. C. Dorf and R. H. Bishop. *Modern Control Systems*. Addison Wesley, 1995.

[9] D. Goswami, M. Lukasiewycz, R. Schneider, and S. Chakraborty. Time-triggered implementations of mixed-criticality automotive software. In *Design, Automation and Test in Europe (DATE)*, 2012.

[10] D. Goswami, R. Schneider, and S. Chakraborty. Co-design of Cyber-Physical Systems via controllers with flexible delay constraints. In *Asia and South Pacific Design Automation Conference (ASP-DAC)*, 2011.

[11] M. Kauer, S. Steinhorst, D. Goswami, R. Schneider, M. Lukasiewycz, and S. Chakraborty. Formal verification of distributed controllers using time-stamped Event Count Automata. In *Asia and South Pacific Design Automation Conference (ASP-DAC)*, 2013.

[12] P. Kumar, D. Goswami, S. Chakraborty, A. Annaswamy, K. Lampka, and L. Thiele. A hybrid approach to cyber-physical systems verification. In *Design Automation Conference (DAC)*. ACM, 2012.

[13] K. G. Larsen, P. Pettersson, and W. Yi. Uppaal in a nutshell. *STTT*, 1(1-2):134–152, 1997.

[14] M. Lukasiewycz, R. Schneider, D. Goswami, and S. Chakraborty. Modular scheduling of distributed heterogeneous time-triggered automotive systems. In *Asia and South Pacific Design Automation Conference (ASP-DAC)*, pages 665–670, 2012.

[15] O. Mason and R. Shorten. On common quadratic Lyapunov functions for stable discrete-time LTI systems. *IMA Journal of Applied Mathematics*, 69(3):271–283, 2002.

[16] P. Naghshtabrizi and J. Hespanha. Analysis of distributed control systems with shared communication and computation resource. In *ACC*, 2009.

[17] S. Samii, P. Eles, Z. Peng, and A. Cervin. Design optimization and synthesis of FlexRay parameters for embedded control applications. In *DELTA*, 2011.

[18] R. Schneider, D. Goswami, S. Zafar, M. Lukasiewycz, and S. Chakraborty. Constraint-driven synthesis and tool-support for FlexRay-Based automotive control systems. In *International Conference on Hardware/Software Codesign and System Synthesis (CODES+ISSS)*, 2011.

[19] H. Voit, R. Schneider, D. Goswami, A. Annaswamy, and S. Chakraborty. Optimizing hierarchical schedules for improved control performance. In *International Symposium on Industrial Embedded Systems (SIES)*, 2010.

[20] G. Weiss and R. Alur. Automata based interfaces for control and scheduling. *Hybrid Systems: Computation and Control (HSCC)*, 2007.

[21] W. Zhang, M. S. Branicky, and S. M. Phillips. Stability of networked control systems. *IEEE Control Systems*, 21:84–99, 2001.

Hybrid Energy Storage Systems and Battery Management for Electric Vehicles*

Sangyoung Park, Younghyun Kim and Naehyuck Chang[†]
Seoul National University, Korea
{sypark, yhkim, naehyuck}@elpl.snu.ac.kr

ABSTRACT

Electric vehicles (EV) are considered as a strong alternative of internal combustion engine vehicles expecting lower carbon emission. However, their actual benefits are not yet clearly verified while the energy efficiency can be improved in many ways. The carbon emission benefits from EV is largely diminished if we charge EV with electricity from petroleum power plants due to power loss during generation, transmission, conversion and charging. On the other hand, regenerative braking is direct power conversion from the wheel to battery and one of the most important processes that can enhance energy efficiency of EV. Power loss during regenerative braking can be reduced by hybrid energy storage system (HESS) such that supercapacitors accept high power as batteries have small rate capability.

Conventional charge management does not systematically exchange charge between the supercapacitor and battery. However, asymmetry in acceleration and deceleration as well as battery charging and discharging capability make the supercapacitor state of charge (SoC) management override the efficiency optimization. Unlike previous works, we show how charge migration during idle and cruise/stopping time can be beneficial in terms of energy efficiency and cruise range. Systematic charge migration decouples SoC management and charging efficiency optimization giving a higher degree of freedom to charging efficiency optimization. We demonstrate the proposed charge migration between the supercapacitor and battery improves energy efficiency by 19.4%.

Categories and Subject Descriptors

C.3 [**Special-Purpose and Application-Based Systems**]: *Real-time and embedded systems*

---

*This work is supported in part by the Mid-Career Researcher Program through NRF grant funded by the MEST (No. 20120005640). The ICT and ISRC at Seoul National University provides research facilities for this study.
[†]Corresponding author.

Permission to make digital or hard copies of all or part of this work for personal or classroom use is granted without fee provided that copies are not made or distributed for profit or commercial advantage and that copies bear this notice and the full citation on the first page. To copy otherwise, to republish, to post on servers or to redistribute to lists, requires prior specific permission and/or a fee.
DAC '13, May 29 - June 07, 2013, Austin, TX, USA.

General Terms

Algorithm, Design, Performance

Keywords

Electric vehicle, Battery-supercapacitor hybrid, Regenerative braking, Charging/discharging asymmetry

1. INTRODUCTION

Electric vehicles (EV) are rapidly gaining popularity as demand for cleaner means of transportation increases. Most countries actively promote deployment of EV. Governments offer subsidies and tax credits to EV manufacturers and customers to give a boost to EV market. For example, the US Government provides federal tax credits to EV consumers according to battery capacity of the vehicle such that Chevrolet Volt and Tesla vehicles are eligible for one-time $7,500 tax credit.

However, the actual benefit from EV is not yet clearly confirmed in terms of the entire life cycle of EV and electricity generation for EV charging. A recent analysis points out that the maximum annual profit considering the real power grid electricity price and battery degradation is only $10 to $120 per EV, which is definitely not sufficient to attract the customers due to the higher EV price comparing with internal combustion engine vehicle [1]. Average power plant boiler and turbine efficiency is at around 33%, power transmission efficiency is at around 93%, battery charger shows at around 70% efficiency, and the battery charging efficiency is at around 90% [2, 3]. The overall efficiency is at around 23%, which is not meaningfully higher than the efficiency of internal combustion engines, which is known as 20% or higher.

Braking energy is often above 30% of traction energy and goes up to 80% in heavy city traffic [4]. Regenerative braking is effective battery recharging with the energy directly coming from the wheels to the battery unlike the long lossy energy trip from the power plants for plugin recharging. Thus, efficient harvesting of regenerative braking energy is the key to maximize the annual profit of EV ownership. A brief calculation shows that increasing the regenerative braking efficiency by 10% is equivalent to 3% to 8% of improvement in the gas mileage.

This paper focuses on three important factors in regenerative braking. The first one is limited rate capability of the battery [5]. Batteries in EV are subject to peak current, which is often 10 times higher than the average current. FreedomCAR and Vehicle Technologies (FCVT) have defined requirements of an energy storage system (ESS) for EV and HEV (hybrid EV) such as total energy, power capacity, total life time, and so on [6]. The second factor is the maximum power transfer during the regenerative braking. The traction motor is a non-ideal power source with appreciable internal impedance that requires the maximum power transfer control.

This paper aims at a higher energy efficiency during a sequence of driving cycles consisting of braking, stopping/cruising, and acceleration. We enhance the regenerative braking energy harvesting efficiency applying the maximum power transfer tracking (MPTT) [7, 8]. The previous regenerative braking architecture commonly connect the battery bank directly to the DC bus. This makes the traction motor voltage output clamped to the battery terminal voltage, which does not allow the maximum power point tracking (MPPT) or MPTT[1]. We introduce a regenerative braking architecture that allows arbitrary voltage control on the DC bus for MPTT.

We perform the near optimal charge replacement [9] for acceleration and compensate the limited rate capability of the battery. The arbitrary voltage control capability on the DC bus is also beneficial for the optimal acceleration. Most of all, we decouple the acceleration and deceleration optimization, which was coupled by the state of charge (SoC) management of the supercapacitor in the previous works. We apply charge migration [10] from the supercapacitor to battery while stopping or cruising, and the charge migration manages the supercapacitor SoC without sacrificing the energy loss during acceleration and regenerative braking. As for the third factor, we specially focus on the battery asymmetry in charging and discharging [11] and tailer the charge migration.

The proposed method more efficiently enhances effective gas mileage of EV with accurate prediction of the vehicle route and traffic conditions. Manual driving nowadays is largely relied on a GPS navigator, a semi-autonomous driving such as adaptive cruise control utilizes the traffic condition via onboard radars, and autonomous driving even more utilizes computerized traffic information. Therefore, it is not surprising to have traffic and driving information and predict the near future driving patterns available.

The proposed method allows for inherent asymmetry in ESS charging/discharging behaviors during acceleration and deceleration of the EV. We make use of the idle and cruise periods of the EV to perform proactive charge migration. Our results show that the proposed technique increases energy efficiency by 19.4%.

2. BACKGROUND

2.1 Hybrid Energy Storage System for EV

Hybrid ESS (HESS) is an energy storage composed of multiple heterogeneous energy storage devices. The key components are ESS banks, a charge transfer interconnect (CTI), converters for the power sources and load devices, and a microprocessor-based charge management policy controller. An ESS bank typically consists of homogenous ESS elements organized into a two-dimensional array to meet the power/energy capacity and the voltage rating. There is also a bidirectional charger (or two unidirectional chargers connected in opposite directions) that controls the charge and discharge current of the ESS array (or equivalently, the current flowing into and out of the ESS bank.) Because the SoC, terminal output voltage, and power rating of different ESS arrays may not be compatible with each other, direct connection among ESS arrays is generally not feasible. The primary function of CTI is to provide charge transfer paths among storage banks, power sources and load devices. There are various ways to implement CTI including a single DC bus wire, a segmented DC bus, multiple DC buses, or a more complicated interconnect network such as a mesh network. Converters for the power sources and load devices can be any of chargers, DC-DC converters, AC-DC rectifiers, and DC-AC inverters as appropriate. These components are not different from the components in a homogeneous ESS system. Charge management policy controller is a microprocessor-based controller in charge of the CTI current flowing from or to each ESS bank and the CTI voltage according to the elaborated charge management policies.

A combination of a battery and an energy storage element with a higher power capacity can be a good complementary setup both for efficiency and cost in ESS for regenerative braking. Among them, battery-supercapacitor HESS is considered a promising solution to mitigate rate capability problem of batteries while meeting other ESS constraints. Adding appropriate amount of supercapacitor could increase the overall energy efficiency and thus the cruise range. Despite its benefits, supercapacitor is still expensive and cause severe volumetric overhead in EV. Therefore, the key issue in HESS is determination of the supercapacitor capacitance and SoC management.

Previous works mostly focused on supercapacitor SoC so that supercapacitors are not fully charged during regenerative barking and not fully depleted during acceleration. Unfortunately, optimization of the HESS charging efficiency is not accordance with the ideal supercapacitor SoC management [8]. In other words, previous works perform SoC management of the supercapacitor without systematic migration charge between the battery and the supercapacitor. Therefore, SoC management of the supercapacitor generally override the efficiency optimization.

A heuristic approach can maintain the supercapacitor SoC inversely proportional to the vehicle speed [12], and use of machine learning to avoid modeling complexity of the entire regenerative braking process [13]. Such complicated problems are often solved by intuitive methods based on expert rules. This type of heuristic approach enhance load balancing among the fuel cell, battery, and supercapacitor [14]. Cost of power converter is additional overhead for the SoC management of supercapacitor, and sometimes, size of the power converter has a higher priority over energy efficiency in the supercapacitor SoC management [15]. Such SoC maintenance significantly restricts the efficient charging and discharging during acceleration and braking [8]. Charge migration gives a greater freedom in the SoC management of the supercapacitor. Charge migration in HESS have been intensively studied recently: definition and optimization of HESS from the view point of computer-aided design and design automation [16], various optimization objectives such as energy efficiency, battery lifetime, peak power minimization for charge allocation (charging HESS) [17], replacement (discharging HESS) [9], and migration (internal transfer) [10], which are named after computer memory management.

2.2 MPTT for Regenerative Braking

Regenerative braking is the key feature in EV to enhance energy efficiency as it allows reuse of kinetic energy during braking. It is widely adopted in commercial EV and HEV including Toyota Prius, Honda Insight, Tesla Roadster, and Chevrolet Volt. Figure 1(a) shows a typical regenerative braking system. Regenerative brakes in EV involve using an electric motor as an electric generator and stores energy in energy storage for later use. Not all of the kinetic energy is recovered using regenerative braking due to the following reasons. Regenerative brakes alone cannot make the vehicle to a complete stop, and it only works on wheels with an electric motor, whereas braking force is often required from other wheels. Figure 1(b) shows a typical braking pattern. Regenerative braking force plus hydraulic braking force should match the driver

---

[1]MPPT maximizes power output from the power source. MPTT considers power converter efficiency and maximizes the power going into the battery.

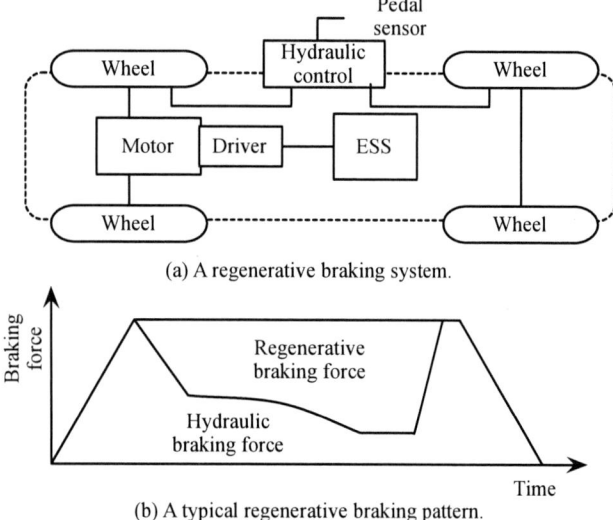

(a) A regenerative braking system.

(b) A typical regenerative braking pattern.

Figure 1: Regenerative braking for EV.

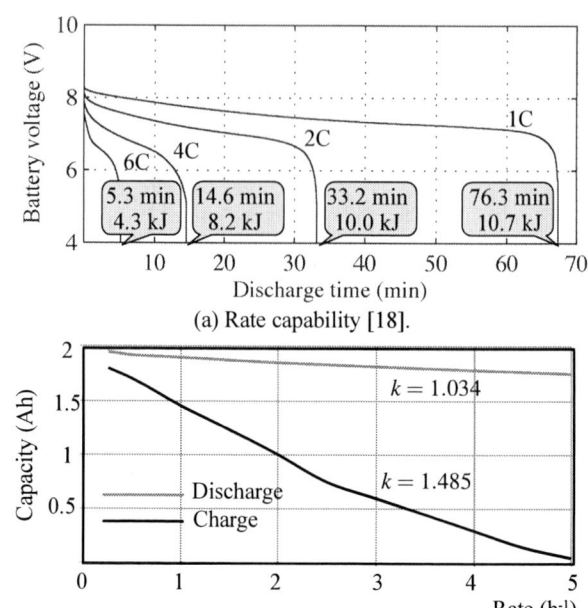

(a) Rate capability [18].

(b) Peukert plot of a 1.9 Ah 18650 Li-ion cell [5].

Figure 2: Battery characteristics.

demand from pedal sensor. In this paper, we do not control the portion of regenerative braking force and hydraulic braking force, and thus we assume there is a well-defined control algorithm that maximizes regenerative braking energy while meeting the driver's demands. Thus, hydraulic brakes should always be used together with regenerative brakes, so some portion of kinetic energy will still be dissipated as heat. However, the portion of regenerative braking force to the hydraulic braking force is not very controllable because it should satisfy the driver's demands and match pressure on the pedal sensor.

2.3 Battery Rate Capability

Although numerous types of devices such as supercapacitors and flywheels are considered for ESS, batteries are the primary energy storage in EV. ESS management algorithms should carefully count on the battery characteristics to achieve high energy efficiency. A dominant phenomenon that explicitly dictates battery charging/discharging efficiency is the rate capability. The rate capability is modeled in the Peukert's formula, which is an empirical equation to evaluate the relationship between the charging/discharging efficiency by the battery current.

$$C_p = I^k t, \tag{1}$$

where C_p is the battery capacity at a nominal discharge current in Ampere-hour, I is the battery current relative to the nominal battery current, and t is time in hours. Figure 2(a) illustrates the rate capability of a Li-ion battery by the discharging current ranging from 1C to 6C. A higher C-rating current significantly decreases the amount usable energy extracted from the battery [18]. We observe the same phenomena for the charging operation.

Rate capability of the same battery is not the same for the charging and discharging. We see distinct asymmetry in charging and discharging as shown in Figure 2(b). Peukert's constant is 1.485 while for discharging it is 1.034 in the case of 1.9 Ah 18650 lithium ion cells [11]. However, such asymmetry is not well investigated and considered in the previous EV ESS management papers.

2.4 EV Energy Efficiency

The energy efficiency of the EV power train depends on efficiency of its consisting components starting from the traction force on the wheels to axle, motor, inverter, charger, and finally, to ESS

banks. Speed and traction force of the vehicle is converted to motor torque and rpm by the following steps. Traction force is equal to total running resistance which is described as,

$$R_{tot} = R_R + R_A + R_G + R_I + R_B, \tag{2}$$

where R_R, R_A, R_G, R_I, and R_B are rolling resistance, aerodynamic resistance, gradient resistance, inertia resistance, and brake force provided by hydraulic brakes, respectively. Simple models exists for calculating the resistance values using vehicle mass, drag coefficients, drag area, vehicle speed, and so on [19].

$$R_R = C_{rr}W, \tag{3}$$

$$R_A = \frac{1}{2}\rho C_d A v^2, \tag{4}$$

$$R_G = W\sin\theta, \tag{5}$$

$$R_I = ma, \tag{6}$$

where C_{rr}, W, C_d, A, v, θ, m, and a is the rolling resistance coefficient, vehicle weight, air density, drag coefficient, car frontal area, vehicle speed, gradient angle, vehicle mass, and vehicle acceleration, respectively. We calculate the motor torque and angular velocity from the traction force, wheel size and axle ratio information. Many EV have single speed constant ratio transmission, which makes the calculation the easier.

$$R_{tot} - R_B = \tau_w d_w/2, \tag{7}$$

$$\tau_m = \tau_w/G, \tag{8}$$

$$\omega_m = \frac{v}{\pi d_w} \cdot 2\pi = \frac{2v}{d_w}G, \tag{9}$$

where $F_{traction}$, τ_w, τ_m, G, ω_m, d_w are traction force, wheel torque, motor torque, axle ratio, motor angular velocity, and wheel diameter.

The scope of our optimization is focused on electrical components. A motor driver is typically insulated-gate bipolar transistor (IGBT)-based inverters, and battery and supercapacitor chargers are transistor based converters, which have their own efficiency map. Moreover, ESS is subject to the device's cycle efficiency as will be shown in the next section. We should consider all the losses in the regenerative braking process and systematically minimize it to achieve the MPTT [8]. The energy efficiency of the regenerative braking process depends on the motor generation efficiency, motor driver efficiency, ESS charger efficiency, and the cycle efficiency of the ESS components. Three-phase brushless DC motors (BLDC) are generally used for EV. The electrical model of the BLDC motor we used is given as,

$$V_k = Ri_k + (L-M)\frac{di_k}{dt} + E_k,$$ (10)

$$E_k = K_k \omega_m F(\theta_e + \theta_k),$$ (11)

$$T_k = K_t i_k F(\theta_e),$$ (12)

where V_k, E_k, T_k, i_k are the voltage, back-EMF voltage, torque, and current of k-th phase, R, L are the resistance, inductance of each phase, M is the mutual inductance, K_t and K_e are the torque constant and back-EMF constant, ω_m and θ_e are the angular speed and angle of the rotor, $F(\theta_e)$ is the back-EMF reference as function of rotor, θ_k is the phase difference between phases [20]. We calculate the relationship between torque, angular speed, and motor input voltage and current.

Conversion efficiency of inverters and chargers in EV power train is not constant. We describe them as a function of V_{in}, V_{out}, I_{out}, which are input voltage, output voltage and output current with reasonable accuracy [21]:

$$\eta = f(V_{in}, V_{out}, I_{out}).$$ (13)

3. CHARGE MANAGEMENT IN EV

Charge management such as charge allocation, replacement, and migration requires careful determination of sources, destinations, amount of current, CTI voltage, and so on, in order to maximize the energy efficiency, which is defined as

$$\eta_{transfer} = \frac{\text{Total energy transferred to the destination}}{\text{Total energy extracted from the sources}},$$ (14)

where the sources are power sources or discharging ESS banks, and the destinations or load devices or charging ESS banks. According to the observation from recent related works [9, 10, 17, 22], charging efficiency is strongly dependent on the type of the bank, the magnitudes of the charging currents, SoCs of the EES banks, voltage and current characteristics of the external power source, and so on. Excessive mismatch between the input voltage level and the EES bank terminal voltage results in unnecessarily large power loss in the chargers. Severe mismatch between the input current and the destination EES bank charging current results in a high IR loss and rate capacity effect. The destination EES banks must be compatible with the input power source in terms of the energy capacity as well.

The voltage on the CTI significantly change the efficiency of the charges, which should be carefully determined by the input source voltage and the destination bank voltage. The optimal charging/discharging current and the CTI voltage changes over time as charge allocation progresses. We continue to monitor, calculate the optimal setup and control the charge allocation process accordingly [9, 10, 17, 22]. Elaborated charge management policies improves efficiency up to 30% comparing with conventional homogeneous ESS systems [9].

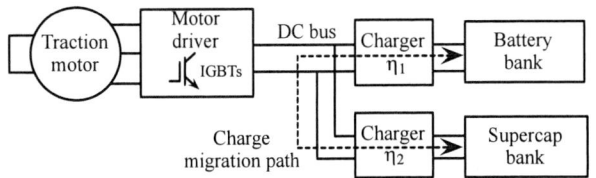

Figure 3: EV HESS topology and charge migration.

A typical battery-supercapacitor HESS topology for an EV is shown in Figure 3. The battery and supercapacitor banks are connected to the shared DC bus via chargers. Some previous works consider direct parallel connection of battery and supercapacitor. However, the topology poses a lot of stress on the battery bank, and offers no freedom of control for systematic optimization, so exclude it from discussion.

Charge migration is also necessary in the HESS for an EV. Charge migration can be beneficial under certain circumstances of EV. For example, in cold start up situation, supercapacitor would be empty and sudden acceleration results in drawing large current from the battery bank. After driving is over, electrical energy remains in the supercapacitor, which is susceptible to high self-discharge rate of supercapacitors. For both cases, it is better to migrate charge from one to another for energy efficiency.

However, cold start and after-drive conditions are not the only scenarios that migration can take effect. A recent study found out that there exists optimal distribution of charge current during the regenerative braking phase [8]. Unlike common belief, utilizing supercapacitor as much as possible is not the optimal because large supercapacitor SoC fluctuation leads to degradation in charger efficiency. We made an observation from Section 2.3 that there is significant asymmetry in battery charging and discharging. The optimal discharge current of both banks during acceleration differs from the optimal charging current during regenerative braking, which implies imbalance in ratio of net power between the battery bank and supercapacitor bank. Migration during vehicle idle and cruise times can resolve this imbalance and enhance the overall driving energy efficiency.

4. CHARGE MANAGEMENT EFFICIENCY ENHANCEMENT

4.1 Charge Management Efficiency Enhancement Problem

We formulate EV energy efficiency enhancement problem as optimization for given driven profile. Enhancement of the energy efficiency is equivalent to increasing the SoC remaining in the HESS at the end of given driving profile is executed. We make a discrete time approach and divide the driving profile into N equal time slots. The define E_{HESS} as the optimization objective defined as

$$E_{HESS}[N] = E_{bat}[N] + E_{cap}[N],$$ (15)

where $E_{bat}[N]$, $E_{cap}[N]$, N are energy remaining in battery bank at Time Slot N, energy remaining in the supercapacitor at Time Slot N.

We define efficiency of the two chargers and motor driver as a function of battery voltage (V_{bat}), battery current (I_{bat}), supercapacitor voltage (V_{cap}), supercapacitor current (I_{cap}), DC bus voltage (V_{bus}), motor RMS voltage per phase (V_k), and motor RMS current per phase (i_k). We assume that the hydraulic braking force profile during braking is given so that the torque and angular velocity of the motor and thus the voltage and current of the motor can

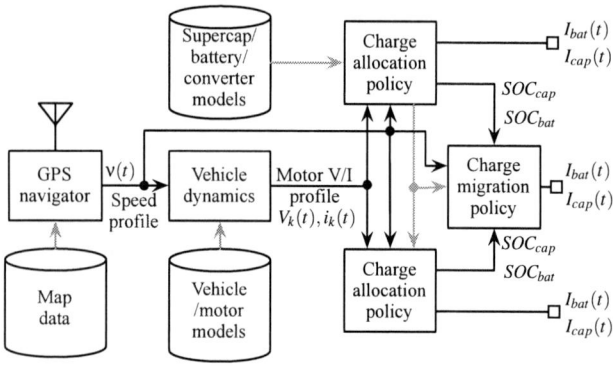

Figure 4: EV HESS management framework.

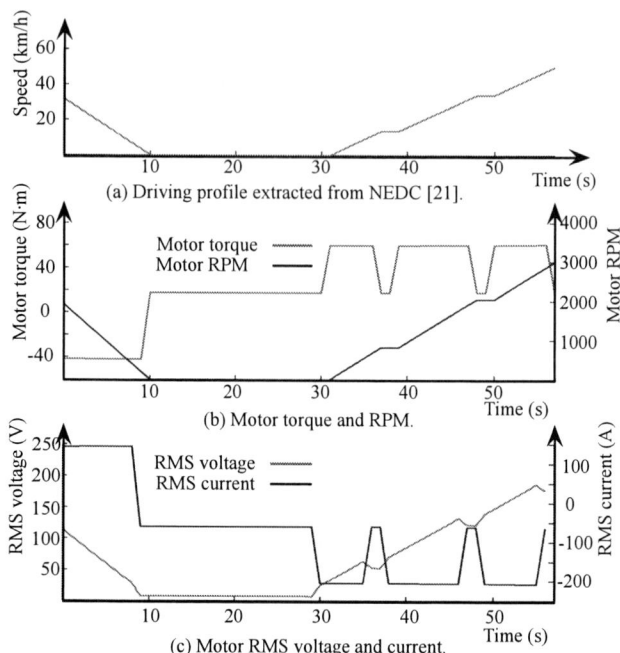

(a) Driving profile extracted from NEDC [21].

(b) Motor torque and RPM.

(c) Motor RMS voltage and current.

Figure 5: Driving cycle and motor trace.

be obtained from the vehicle speed profile using vehicle dynamics and the torque-current model of the motor in Section 2.4.

4.2 EV HESS Management Policy

We deliberately take advantage of charge migration in EV HESS charge management. Figure 4 demonstrates the framework of the proposed EV HESS management. First, we independently optimize the charge allocation and charge replacement during regenerative braking and vehicle acceleration. Such independent optimization is feasible thanks to charge migration. We perform charge allocation considering the motor the power source and find the optimal profile of the DC bus voltage, supercapacitor bank current, and battery bank current. The ESS in an EV is subject to strict weight and volume constraints. Therefore, it is not recommended to install an excessively large supercapacitor in an EV. We assume that the capacity of the supercapacitor is designed to accommodate average energy capacity for only one time acceleration or deceleration. We thus adjust the SoC of the supercapacitor to its minimum level before the regenerative braking and apply charge allocation algorithm presented in [17]. The algorithm considers most non-ideal characteristics of the power converter and battery and derives the optimal supercapacitor and battery charge currents. Likewise, we perform charge replacement considering the motor as the load device and find the optimal profile of the DC bus voltage, supercapacitor bank current, and battery bank current using the algorithm in [9]. We maintain the SoC of the supercapacitor to be the minimum limit at the end of acceleration to maximize its capacity utilization[2]. The optimal charge allocation and replacement of EV are not symmetrical even though we apply a symmetric deceleration and acceleration profiles. The final supercapacitor and battery SoC at the end of charge allocation is generally different from the initial supercapacitor and battery SoC at the beginning of charge replacement. The major reasons that cause such asymmetry include battery asymmetry discussed in Section 2.3, existence of hydraulic brake force, and so on.

However, the proposed method is able to provide superior energy efficiency because charge migration fills the gap between charge replacement during acceleration and charge allocation during deceleration supercapacitor SoC requirement. The charge migration algorithm tries to evenly distribute the migration current as much as possible over time while considering the efficiency of the converters. We set the initial conditions of the charge migration problem as the final status of the charge allocation problem, and the terminal conditions as the initial status of the charge replacement problem, and apply charge migration algorithm in [10].

---

[2]We cannot fully empty the supercapacitor because the power converter requires a certain minimum input voltage.

The framework solves charge management problem ahead of time to find input conditions for the charge migration policy. The underlying assumption here is that we know the vehicle behavior in the near future using GPS navigator and semi-autonomous driving features discussed in Section 1.

5. EXPERIMENTS

We validate the proposed approach by simulation through a commercial EV on standard driving cycles. The target vehicle is a 5-door hatchback full-EV Nissan Leaf. Nissan Leaf is equipped with an 80 kW, 280 N·m electric motor, a 24 kWh Li-ion battery, weighs 1521 kg, has an axle ratio of 7.94:1, and drag coefficient of 0.28. Other parameters required to calculate the electrical outputs of the motor from the driving profile are extracted from other cars of similar size. The size of the supercapacitor is 15 F, which is similar to the capacity used in previous works on battery-supercapacitor HESS for EV [13]. It consists of 200 series connection of 3000 F supercapacitors to exhibit maximum voltage of 500 V. The driving profile we used for simulation is shown in Figure 5(a). It is a part of the driving cycle, ECE-15 UDC (Urban Driving Cycles), in standard driving profile NEDC (New European Driving Cycle) [23]. Figures 5(b) and 5(c) show the motor torque, motor RPM, RMS voltage, and the RMS current of the electric motor calculated from the equations in Section 2.4. The baseline policy for comparison is balanced charging/discharging during deceleration and acceleration without active migration. Figure 6 shows the experimental results for the proposed and baseline policies. The ESS receives the regenerative braking energy from 0 s to 10 s. The charging profile of the ESS controlled by MPTT technique during the period is the same both for the proposed and baseline. However, the proposed policy performs gradual migration from the battery to supercapacitor during 10 s to 30 s to prepare for acceleration in 30 s to 50 s as opposed to the baseline policy. The acceleration of the EV is much supported by supercapacitor discharge power, which beneficial for the overall energy efficiency. The proposed policy consumes 19.4% less energy (1643 kJ) than the baseline (2013 kJ) for the same profile.

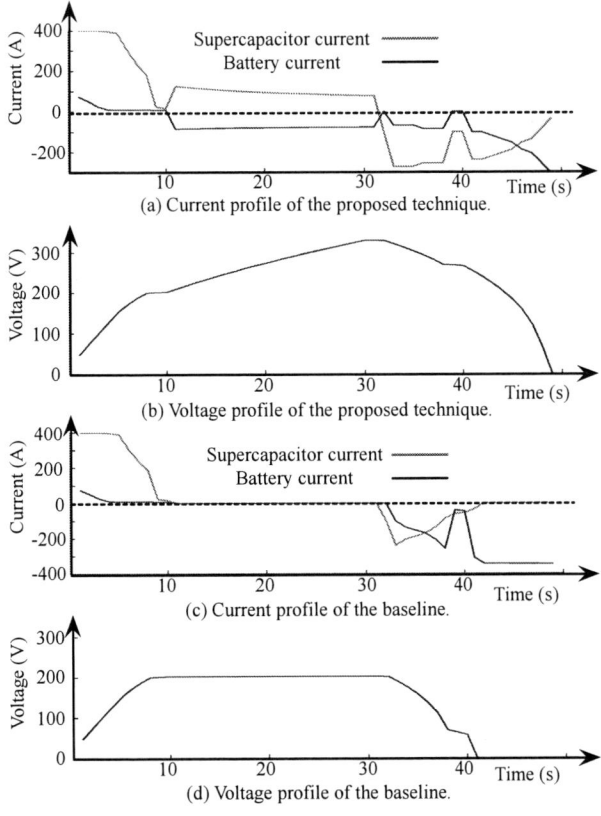

Figure 6: Experimental results ($C = 15$ F).

7. REFERENCES

[1] S. B. Peterson, J. Whitacre, and J. Apt, "The economics of using plug-in hybrid electric vehicle battery packs for grid storage," *J. of Power Sources*, 2010.

[2] D. K. Bellman, "Power plant efficiency outlook," in *Working Document of the NPC Global Oil & Gas Study*, 2007.

[3] US Energy Information Adminstration, "State electricity profiles," in *http://www.eia.gov/electricity/state/*, 2013.

[4] H. Yeo, D. Kim, S. Hwang, and H. Kim, "Regenerative braking algorithm for a HEV with CVT ratio control during deceleration," in *Proc. of SAE CVT Congress*, 2004.

[5] D. Doerffel and S. A. Sharkh, "A critical review of using the peukert equation for determining the remaining capacity of lead-acid and lithium-ion batteries," *J. of Power Sources*, 2006.

[6] S. Lukic, J. Cao, R. Bansal, F. Rodriguez, and A. Emadi, "Energy storage systems for automotive applications," *IEEE T. on Industrial Electronics*, 2008.

[7] Y. Kim, Y. Wang, N. Chang, and M. Pedram, "Maximum power transfer tracking for a photovoltaic-supercapacitor energy system," in *Proc. of ISLPED*, 2010.

[8] S. Chakraborty, M. Lukasiewycz, C. Buckl, S. Fahmy, N. Chang, S. Park, Y. Kim, P. Leteinturier, and H. Adlkofer, "Embedded systems and software challenges in electric vehicles," in *Proc. of DATE*, 2012.

[9] Q. Xie, Y. Wang, Y. Kim, D. Shin, N. Chang, and M. Pedram, "Charge replacement in hybrid electrical energy storage systems," in *Proc. of ASP-DAC*, 2012.

[10] Y. Wang, Y. Kim, Q. Xie, N. Chang, and M. Pedram, "Charge migration efficiency optimization in hybrid electrical energy storage (HEES) systems," in *Proc. of ISLPED*, 2011.

[11] M. Dubarry, C. Truchot, M. Cugnet, B. Y. Liaw, K. Gering, S. Sazhin, D. Jamison, and C. Michelbacher, "Evaluation of commercial lithium-ion cells based on composite positive electrode for plug-in hybrid electric vehicle applications. part I: Initial characterizations," *J. of Power Sources*, 2011.

[12] R. Carter and A. Cruden, "Strategies for control of a battery/supercapacitor system in an electric vehicle," in *Proc. of SPEEDAM*, 2008.

[13] M. Ortuzar, J. Moreno, and J. Dixon, "Ultracapacitor-based auxiliary energy system for an electric vehicle: Implementation and evaluation," *IEEE T. on Industrial Electronics*, 2007.

[14] E. Schaltz, A. Khaligh, and P. Rasmussen, "Influence of battery/ultracapacitor energy-storage sizing on battery lifetime in a fuel cell hybrid electric vehicle," *IEEE T. on Vehicular Technology*, 2009.

[15] J. Cao and A. Emadi, "A new battery/ultracapacitor hybrid energy storage system for electric, hybrid, and plug-in hybrid electric vehicles," *IEEE T. on Power Electronics*, 2012.

[16] M. Pedram, N. Chang, Y. Kim, and Y. Wang, "Hybrid electrical energy storage systems," in *Proc. of ISLPED*, 2010.

[17] Q. Xie, Y. Wang, Y. Kim, N. Chang, and M. Pedram, "Charge allocation for hybrid electrical energy storage systems," in *Proc. of CODES+ISSS*, 2011.

[18] D. Shin, Y. Wang, Y. Kim, J. Seo, N. Chang, and M. Pedram, "Battery-supercapacitor hybrid system for high-rate pulsed load applications," in *Proc. of DATE*, 2011.

[19] J. Y. Wong, *Theory of ground vehicles*. Wiley-Interscience, 2001.

[20] A. Tashakori, M. Ektesabi, and N. Hosseinzadeh, "Modeling of bldc motor with ideal back-emf for automotive applications," in *in Proc. WCE*, 2011.

[21] Y. Choi, N. Chang, and T. Kim, "DC-DC converter-aware power management for low-power embedded systems," *IEEE T. on CAD*, 2007.

[22] Y. Wang, Q. Xie, M. Pedram, Y. Kim, N. Chang, and M. Poncino, "Multiple-source and multiple-destination charge migration in hybrid electrical energy storage systems," in *Proc. of DATE*, 2012.

[23] *NEDC, European Council Directive, 70/220/EEC with amendments.*

6. CONCLUSIONS

Electric Vehicles (EV) still should be invested more to make EV more commercially competitive. Energy efficiency enhancement is one of the most demanding requirements to make EV commercially competitive. There are many ways to enhance EV energy efficiency, especially for electricity generation, transmission and conversion, which are major power loss. However, they are not only related to EV development but nation-wide infrastructure renovation.

This paper introduces systematic enhancement of regenerative braking efficiency for hybrid energy storage systems (HESS) in EV. Energy efficiency enhancement of the regenerative braking gives significant impact on carbon emission because it is energy harvesting directly from the wheels to EV HESS unlike plugin charging from the grid electricity coming from fuel through power plant, transformers, transmission lines, and distribution lines. The proposed method decouples ESS charging efficiency optimization from the supercapacitor state of charge (SoC) management, which generally sacrifices the charging efficiency in previous works due to the limited supercapacitor capacitance. We characterize the asymmetric factors during acceleration and regenerative braking including the battery rate capacity asymmetry in charging and discharging. We show that systematic charge migration between the supercapacitor and battery among idle and cruise times enhances the energy efficiency significantly by balancing the asymmetry. Experimental results show that the proposed approach achieves 19.4% energy efficiency improvement. Such efficiency gain comes from more active and efficient use of a precious energy storage element, the supercapacitor, through the proposed method.

685

Reliability Challenges for Electric Vehicles: From Devices to Architecture and Systems Software

Georg Georgakos
Infineon Technologies
Neubiberg, Germany

Ulf Schlichtmann
Institute for Electronic Design Automation
Tech. Universität München, Germany

Reinhard Schneider
Samarjit Chakraborty
Institute for Real-Time Computer Systems
Tech. Universität München, Germany

ABSTRACT

Today, modern high-end cars have close to 100 electronic control units (ECUs) that are used to implement a variety of applications ranging from safety-critical control to driver assistance and comfort-related functionalities. The total sum of these applications is several million lines of software code. The ECUs are connected to different sensors and actuators and communicate via a variety of communication buses like CAN, FlexRay and now also Ethernet. In the case of electric vehicles, both the amount and the importance of such electronics and software are even higher. Here, a number of hydraulic or pneumatic controls are replaced by corresponding software-implemented controllers in order to reduce the overall weight of the car and hence to improve its driving range. Until recently, most of the software and system design in the automotive domain – as in many other domains – relied on an always correctly functioning or a *zero-defect* hardware implementation platform. However, as the device geometries of integrated circuits continue to shrink, this assumption is increasingly not true. Incorporating large safety margins in the design process results in very pessimistic design and expensive processors. Further, the processors in cars – in contrast to those in many consumer electronics devices like mobile phones – are exposed to harsh environments, extreme temperature variations, and often, strong electromagnetic fields. Hence, their reliability is even more questionable and must be explicitly accounted for in all layers of design abstraction – starting from circuit design to architecture design, to software design and runtime management and monitoring. In this paper we outline some of these issues, currently followed practices, and the challenges that lie ahead of us in the automotive and electric vehicles domain.

Categories and Subject Descriptors

B.4.5 [**Hardware**]: Reliability, Testing, and Fault-Tolerance – *Diagnostics, Error-checking, Hardware reliability, Redundant design.*

General Terms

Performance, Low Power, Design, Reliability, Verification.

Permission to make digital or hard copies of all or part of this work for personal or classroom use is granted without fee provided that copies are not made or distributed for profit or commercial advantage and that copies bear this notice and the full citation on the first page. To copy otherwise, or republish, to post on servers or to redistribute to lists, requires prior specific permission and/or a fee.
DAC '13, May 29 – June 07 2013, Austin, Texas, USA.

Keywords

Electric vehicles, automotive electronics, process variations, aging, embedded systems, software, cross-layer.

1. INTRODUCTION

The volume of electronics and software in modern cars is increasing at a tremendous rate. Today, most of the innovation in the automotive domain is in the area of electronics and software rather than in mechanical engineering, which used to be the case until a few years ago. High-end cars now have reached the mark of close to 100 ECUs, containing almost 250 processors and GPUs [1]. These ECUs are connected to various communication buses like CAN, LIN, MOST, FlexRay and Ethernet and are used to run various applications related to safety-critical control, driver assistance and comfort-related functionalities. These applications sum up to several million lines of software code and are expected to grow at an exponential rate as more driver assistance functions are being introduced everyday. In fact, *autonomous driving*, that is technologically feasible since the last couple of years, is expected to become a reality very soon. Such applications rely on a number of cameras and radars that produce large volumes of streaming data, which needs to be processed in real-time thereby further increasing the need for computation power inside the car.

Electric vehicles: In the case of electric vehicles, electronics and software – both in volume, as well as in importance – are even larger. Since battery life and hence driving range is of crucial importance in electric cars, a number of pneumatic and hydraulic controllers in such cars are replaced by software-based control in order to reduce the overall weight of the car. This further increases the need for computation power, but also introduces more ECUs and cabling. Current design methods in the automotive domain follow a *federated architecture*, where each function is implemented on a separate ECU. This enables the OEM (original equipment manufacturer) to outsource the different functions to different Tier 1 suppliers – with the role of the OEM being that of function integrator. But it also increases the number of ECUs in the car and the volume of cables needed to connect these ECUs. For electric cars this is especially a problem, not only because of the complexity and the distributed nature of the resulting architecture, but also because of the overall weight of the ECUs and the cables, which are now non-negligible. As a result, there is an increasing push to move towards more *integrated* architectures, where multiple software functions are integrated onto a single ECU. An application therefore is now distributed into different tasks, running on different ECUs and communicating via shared communication buses. Such integrated architectures and the move towards fewer ECUs and shorter cables introduce more powerful multicore ECUs that use the latest processor design and fabrication technologies. Further, very soon it will also no longer be cost-effective to use simple microcontrollers in the ECUs but powerful commodity processors from the consumer electronics domain will be used.

The reliability problem: However, such processors from the consumer electronics domain are currently faced with a number of reliability-related problems stemming from process variations, aging, and radiation-induced soft errors, which could be safely neglected in the low-end microcontrollers in older ECUs that had much larger device geometries and used old fabrication technologies. Hence, design methods and software development techniques in the automotive domain always assumed a zero-defect and fully reliable hardware layer, which is increasingly no longer true.

To make matters worse, the electronics in a car are exposed to harsh conditions, extreme temperature variations, and often, strong electromagnetic fields, which further aggravates the reliability problem. Further, the electronics in subsystems like *battery monitoring and management* in electric vehicles are always "on" for the entire lifetime of the car, which is in the range of 10-15 years and sometimes even more. This makes issues like aging an important concern. Automotive being a highly cost sensitive domain, large safety margins to overcome these problems are not feasible and more integrated and intelligent solutions that tackle the reliability issue at multiple layers of design abstraction are necessary. While current processor design for automotive ICs is in the 65nm domain, the industry is rapidly moving towards processors with smaller geometries (40nm and below in the near future). For devices with these geometries power consumption is a major issue. Hence, power management techniques like dynamic voltage and frequency scaling – that are routinely used in consumer electronics devices like mobile phones – will have to be introduced in automotive ECUs (currently the power consumed by the electronics in a car is largely neglected and hence no runtime power management techniques are deployed). Once this happens, the issue of reliability will become a bigger challenge since the processor will have to operate at both high and low supply voltages, which is explained in more detail in Section 2.

Furthermore, in electric vehicles, conventional control systems with mechanical backup systems are likely to be replaced by solely electronic components, so called *X-by-wire systems* such as brake-by-wire or steer-by-wire. This leads to significantly reduced weight and space and enables a tight integration of several applications. On the other hand, such systems, e.g., brake-by-wire, are highly safety-critical and impose stringent reliability requirements on the used hardware and software. Hence, fault-tolerant and reliable software and platform design will become inevitable key issues to realize X-by-wire systems.

Today, processors in automotive ECUs are increasingly adopting multicore architectures to keep pace with growing performance and reliability demands. Similarly, cost and quality requirements necessitate efficient model-based development techniques to realize seamless model-based software development involving automatic code generation. Generally, the system design process is typically distributed among several layers, starting at a high-level application layer implementing the core functionality of the system at a high level of abstraction, e.g., in form of Simulink models or as textual representations, down to the microarchitecture layer on which the embedded software is finally implemented. In this context, the specification, design, test and verification phases across the different layers are driven by various automotive standards such as OSEK/VDX, AUTOSAR, ISO26262, CMMI (to name but a few), each imposing individual constraints and requirements.

Further, the growing complexity due to increased ECU consolidation and associated integration of more and more SW-components with different levels of criticality on complex multicore architectures requires integrated design approaches in hardware and software to guarantee *freedom from interference* between SW-components and to meet specified performance requirements. In this context, interference between SW-components may be due to concurrent memory accesses to shared memory regions or because of preemption of tasks scheduled on shared processing resources resulting in increased execution times.

In order to realize safety-related architectures, dedicated multicore platforms operating in dual core *lock-step mode* have become popular. That is, two cores (master and checker) execute the same code while being synchronized, to detect potential errors produced by a faulty core. This is especially important to satisfy high functional safety requirements, e.g., in a braking ECU. On the other hand, multicore architectures of course provide parallel processing on multiple cores giving rise to increased computational performance. Similarly, dedicated hardware units such as DMA controllers, DSPs, and GPUs enable additional execution speed-up. Further, virtualization technologies are being studied with the goal to implement an effective layer of isolation between single user-level applications with different criticality levels.

The communication network between the ECUs is built up of bus systems such as LIN, CAN, MOST, FlexRay, Ethernet and one or more gateways to interconnect the different network domains. Since gateways often represent single-point-of-failures in the communication network special attention needs to be paid to designing fault-tolerant and redundant communication architectures in order to improve the reliability in the entire network.

Organization: In this paper we discuss the various reliability issues and why they arise (Sect. 2) as well as current practices to mitigate them (Sect. 3). Sect. 4 discusses reliability especially in the automotive context. The challenges that lie ahead of us are addressed in this paper as well in Sections 3 and 4.

2. RELIABILITY: BACKGROUND AND CIRCUIT IMPACT

This section will provide some background on the causes and effects for the reliability challenges that we address in this paper. The following section describes mitigation techniques which represent the current state of the art.

Manufacturing variations have been with us since the early days of ICs. They are, together with the need to consider the range of operating conditions (supply voltage V_{DD} and operating temperature T), the reason why the corner-based design methodology has been introduced. They happen due to imperfections in the manufacturing process, which either have fundamental physical reasons, or are a result of tradeoffs between manufacturing cost and quality. Typically they are variations of physical parameters, which in turn result in variations of electrical parameters. These affect circuit properties that are of interest to the designer (e.g., delay, static/dynamic power). Some examples of manufacturing variations are e.g. fluctuations in doping profiles of the transistor channel (resulting in variations of threshold voltage V_{th}), variations in gate oxide thickness Tox (V_{th} change), width/thickness of metal lines (changes in resistance and therefore wire delay as well as electromigration risk) [2][3].

Various classifications of these variations exist [4]. An often used classification refers to their spatial correlation: wafer-to-wafer (all parameters on one wafer are identical, but differences exist between wafers), die-to-die (parameters within one die are (essentially) identical, but differ between dies), and within-die. The first two classes are often referred to as global variations, whereas the within-die variations are also known as local variations. The poster-child for within-die variations is random dopant fluctuations (RDF) of the transistor gate channel doping, which affect V_{th}. From 32nm onwards, the number of dopant atoms is below 100 [5]. As the dopant atoms are deposited by processes, their exact number can differ significantly from one transistor to the next, even between adjacent transistors. Minimum-sized transistors are affected the most.

The trend is that manufacturing variations increase from one technology generation to the next [6][7][8]. This is especially true for their relative values, since the nominal values of parameters typically decrease with successive process technology generations [2]. However, this is not a monotonous trend. Sometimes innovations in process technology, new materials or improved transistor structures result in a significant reduction in variations. But the overall trend of increasing relative variations remains. Especially worrying is the fact that within-die variations are increasing as a percentage of overall variations [9]. This is a problem since the standard corner-based design methodology was developed for wafer-to-wafer and die-to-die variations and cannot properly handle within-die variations.

Defects during the manufacturing process are usually caused by impurities of the process materials or particle contamination during the process itself. These defects might directly cause a functional fail, e.g., an open of an interconnect or contact. But even worse they might create a weak spot in a device, such as a local gate oxide thinning of a MOS transistor or a partly open via hole. These latent faults may not be detected during initial product test, and transform to a functional fail or a parameter drift quite early during the lifetime of the device in the field. Most of the reliability mechanisms described in the following can be influenced or accelerated by these defects which are also described as extrinsic effects in contrast to the intrinsic behavior which describes the reliability mechanism in an ideal case without any type of an overlaying defect.

Aging effects are changes in transistor or wire parameters over the lifetime of the device. Today, the following are the most relevant effects:

Negative / Positive Bias Temperature Instability (NBTI / PBTI). NBTI affects PMOS transistors, PBTI NMOS transistors. In the past, PBTI was not an issue, but with the introduction of high-k metal gates below 40nm it is evolving into a problem as well. The aging happens when the transistor is in inversion (the stress condition), e.g. for the PMOS when the gate terminal is negatively biased regarding source or drain. The resulting changes in the transistor can be modeled as a degradation of V_{th}. The severity of the degradation depends on many factors, most importantly V_{DD}, T, and percentage of the time the transistor is in inversion. Transistor degradation due to NBTI or PBTI can be reversed when the stress condition is removed. This is known as recovery.

Recovery is taking place slower than degradation, and only partially reverses the previous degradation.

Figure 1 and Figure 2 show some influences on NBTI [10]. The circuit designer is interested in the transistor delay degradation caused by an aging effect such as NBTI. This delay degradation has two components: NBTI causes a shift of transistor V_{th} (Figure 1), which then in turn results in a degradation of the delay of the logic gate containing the respective transistor (Figure 2). As the figures show, V_{DD} and T influence both components. However, V_{DD} has different impact on the two components: the V_{th} degradation increases with increasing V_{DD}, the sensitivity of delay to V_{th} increases with decreasing V_{DD}. This can be especially critical in a scenario where different supply voltages V_{DD} are used for purposes of power reduction. Assume that an IC is operated mostly using a high V_{DD} value for high performance modes. This will result in strong transistor degradation. When the IC then operates with lower V_{DD} in a low-power mode, the impact of the V_{th} shift is amplified by the increased sensitivity of delay to V_{th} changes (see Figure 2). This demonstrates that techniques for power reduction cannot be considered in isolation. Their effects on reliability due to aging need to be taken into account.

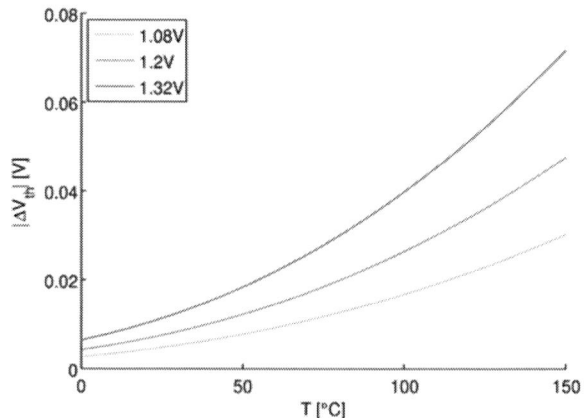

Figure 1: Exemplary V_{th} shift of a PMOS transistor in a 90nm node caused by NBTI stress representing a realistic end of live scenario and a 100% duty cycle in dependence on T, V_{DD}

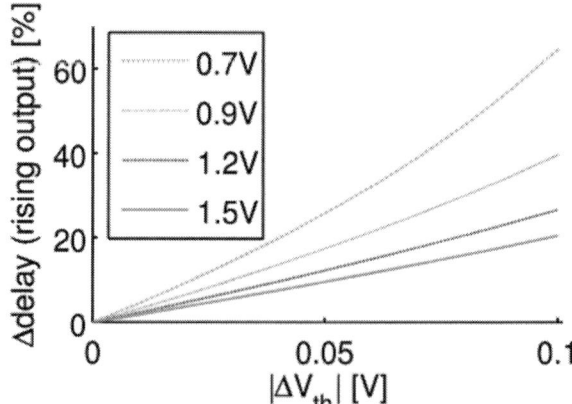

Figure 2: Delay degradation of a 90nm inverter in dependence on V_{th} shift and V_{DD} at an operating temperature of 85°

Hot Carrier Injection (HCI) affects both PMOS and NMOS transistors. Essentially, carriers (electrons or holes) are injected from the transistor channel into the gate oxide, resulting in a permanent change of the transistor characteristics, typically modeled as a degradation of the on-current I_{on}. This happens during the transition from the off- to the on-state of the transistor, so the switching frequency of the transistor strongly affects the degradation. The impact of HCI on digital circuits is similar to NBTI and PBTI as described above.

Time Dependent Dielectric Breakdown (TDDB) affects both PMOS and NMOS transistors. However, the failure rate of NMOS transistors due to TDDB is much higher. TDDB is defined as the localized loss of isolating properties of the gate oxide, which leads to an increased gate leakage current. Typically the impact of this aging mechanism is modeled as a degradation of the gate current I_g. Unlike NBTI, there is no recovery from TDDB. There are two oxide breakdown (OBD) modes to distinguish: Soft breakdown (SBD) and Hard Breakdown (HBD). The underlying physical origins for both modes are the same, and they are distinguished by the resistance of the conducting path, or, in other words, the severity of breakdown. SBD has relatively higher resistance and lower breakdown current. In the SBD stage the transistor drifts in energy and delay, but is still functional. The HBD mode is always preceded by SBD. In the HBD case, the resistance is much lower and the current is exponentially higher. This fault is called catastrophic and results in a permanent malfunction of the respective circuit block.

As the gate oxide thickness is shrinking with successive process technologies, transistors are becoming more prone to SBD, which has become one of the most important reliability issues threatening the further reduction of transistor feature sizes. The principles of scaling of gate area, oxide thickness, and voltage for both soft and hard breakdowns have been investigated [17]. The study of relation between breakdown mode and its location suggests that the spot, located in the gate and drain/source overlapping region, will have the most severe impact on circuit functionality [18]. On a circuit level, the impact of TDDB is being studied mainly on SRAM cells, as they are most sensitive to TDDB [19][20].

Electromigration (EM) refers to mass transport in metal interconnects like wires and via holes [11]. With sufficiently high current densities, electrons may eventually cause atoms to move. Over time this can result in an increase in resistance as the interconnect cross-section becomes narrower. Eventually it can lead to an open, likely causing a catastrophic circuit failure. For a given current, the severity of EM increases with decreasing interconnect width. Therefore, EM problems can be avoided by sufficiently sizing wires or number of via holes. But of course this approach is at odds with the desire to keep wires as thin as possible to reduce IC area.

Stressmigration (SM) is similar to electromigration but the mass transport is not due to an electron flow but due to diffusion of vacancies in interconnects through mechanical stress or temperature gradients. An agglomeration of these vacancies may eventually create voids under or in via holes. This also leads to an increase in resistance first and might eventually cause a permanent open of an interconnect [24].

Soft Errors or **Single Event Upsets (SEUs)** cause intermittent, non-permanent failures in a circuit. An SEU happens when a charged particle strikes a semiconductor, resulting in a buildup of ionized material, which in turn causes a current. The charged particle might result from radioactive decay of packaging materials or from cosmic particles in the atmosphere. As the incidence of cosmic particles increases with height above the earth's surface, in the past electronics for air and space applications were most at risk. However, with decreasing feature sizes, also electronics used in terrestrial applications are increasingly affected. While the literature is not uniform as to whether smaller feature sizes by themselves increase SEU risk, certainly the exponential increase in number of components on a chip due to shrinking feature sizes increases the risk for an IC of a given die size. Today, no safety-critical IC is designed without some sort of SEU protection anymore.

In the past, most concern has been about particle strikes directly hitting a storage cell, resulting in the flip of a single bit in memory. Meanwhile more than one bit may flip in a memory due to a single event. And increasingly, single event transients (SETs, pulses resulting from a particle strike which then propagate through the circuit) are also a source of concern, since they can also lead to an incorrect value in a storage element (e.g. a flipflop) if the transient reaches such a storage element at the time it latches.

Both SEUs and SETs may not necessarily cause a malfunction of a circuit. There are many types of masking that might happen. E.g. a flipped memory bit might be overwritten by a new value before it is accessed by a read operation. It is challenging to determine how much at risk a certain part of a circuit is. Research is conducted e.g. to efficiently determine the Architectural Vulnerability Factor (AVF) – which describes the masking of errors - of a circuit [33].

Electromagnetic Interference (EMI) is not a classical reliability issue but also requires a certain robustness of the design regarding high electromagnetic and pulse disturbances. They are typically injected or generated and distributed in complex systems consisting of one or several printed circuit boards, connectors and cables. The noise is either received from a harsh environment which contains radio frequency or pulse noise or is generated by the systems themselves mostly due to switching noise. Coupling of noise is due to jointly used interconnects of an emitting and receiving part of the system in a conducting way or through traces of interconnects, packages, connectors or cables acting as an antenna. Usually no physical destruction of the components occurs. Only a temporarily or permanent malfunction of the system is observed.

A special case of interference is Electrostatic Discharge (ESD). The according noise pulse is in this case typically provided through the chip pins and caused by the handling of the chip during production or system assembly or by many kinds of overstress conditions during chip operation. An ESD exposure can also happen during end-user operation in case of a plug-in of cables and change of batteries or maintenance in an unprotected area. A special kind of electrostatic discharge during the chip manufacturing process might occur due to resist charging during implantation or due to plasma processes which create non-uniformities of charge distribution across a wafer. The charge is coupled into the chip by antennas, simple traces of metal or diffusion. A corresponding discharge might lead to various parameter drifts of active devices, increased TDDB risk, increased gate leakage or immediate physical destruction of diodes, transistor channels or dielectrics.

Above, the circuit impact of the described reliability effects was considered from the perspective of a digital circuit design. Complex SoCs may also contain analog and mixed signal circuits. Reliability effects that lead to a permanent physical destruction, do not distinguish between digital or analog circuits. However, the operating conditions of the devices are different, leading to different acceleration of the related reliability effects. In digital circuits MOS transistors are usually operated at full supply voltage levels and current flow is limited to the switching transients whereas in analog circuits the supply voltage of a single MOS transistor is usually smaller due to voltage drop at serially connected devices, but the current flow spans a wide range of different pulse shapes, up to a constant current over time. Reliability effects, which lead to a parameter degradation of a device, are more difficult to consider for analog circuits. As above the operating conditions are different compared to digital circuits and therefore also the acceleration factors. Additionally not only V_{th} and I_{on} drifts impact the analog circuit behavior, also parameters like g_m and g_{ds} might suffer from aging and change the circuit response over time. And finally the wide variety of analog circuit classes shows a big difference in sensitivity to certain parameter changes. This is mainly due to design techniques utilizing relative parameters instead of absolute ones relying on a good matching of these parameters to cope with variations.

3. RELIABILITY-AWARE DESIGN TECHNIQUES

This section describes reliability-aware design techniques that are considered to be "state of the art" for advanced IC designs, and also gives an outlook. For the purposes of this paper, we define „state of the art" as techniques that are used for designing automotive ICs in 65nm technology. In the rather conservative automotive field, major vendors are currently qualifying this process technology. Design activities have started in 40nm, but are still in early stages. The focus of the descriptions in this section is on broadly applicable techniques, not on approaches specific to automotive ICs.

3.1 Device and Circuit Level Techniques

Manufacturing variations have traditionally been addressed by corner-based (best case / worst case) design techniques [12]. Increasingly, such techniques are running out of steam, however, due to a number of reasons:

- Variations are increasing. This results in ever more performance being lost to generous corner guard banding.
- More parameters are being considered, resulting in an exponential increase in the number of corners. E.g. some manufacturers consider variations in wire thickness on different metal levels in addition to transistor speed variation.
- Corner-based design methodologies can not cope with purely random within-die variations. These variations can only be addressed by safety-guardbands – and by the hope that they will typically average out if a path is long enough. Unfortunately, hoping for the best is not a very reliable design technique. To date, the industry has managed to get by with tweaking existing design techniques. E.g. Static Timing Analysis is being enhanced with various forms of "On-Chip-Variation (OCV)" techniques to address purely random within-die variations [13]. These work for the time being, but are really just band-aids to postpone the demise of established design techniques a bit further into the future.

Statistical design techniques for digital circuits have seen a lot of attention in the past 10 years [4]. But most of them have not yet gained any significant traction in typical industrial design flows, with the exception of in-house flows at a few of the world's largest IDMs. As the effort required to introduce such techniques into design flows (new tools, significantly more detailed library characterization required, very high computational effort required for some approaches) is very substantial, it remains to be seen to which degree statistical design techniques will see industrial adoption in the future.

In analog circuit design Monte-Carlo simulations result in more realistic behavior than corner-based methods, with the disadvantage of highly increased design time. Worst-Case-Distance methods [14] are much faster but also very complex and are therefore not yet established in the analog design community. With increasing complexity and sensitivity of the circuits towards variations this situation has to change in future to enable an efficient and accurate design process.

Extrinsic faults can't be addressed directly during the design process. A critical area analysis and layout post-processing like wire spread and fattening techniques can reduce the number of faults. Current strategy is to activate these latent faults with a burn-in stress and screen them out with a respective test program before final shipment. This procedure is going to be more and more challenging in the future because burn-in is time consuming and therefore quite expensive and on the other hand due to increased system complexity it is very challenging to find realistic burn-in stress scenarios which activate all possible latent faults.

Aging can provide an exemplary perspective on the evolution of design techniques to address reliability challenges (DfR, Design for Reliability). In the past, many of these reliability challenges were not addressed specifically during design at all. Overall guard bands were being used in chip design, and it was assumed that these guard bands would cover aging effects, if any consideration was given to them at all.

As a next step up in sophistication, the size of these guard bands was verified by measurements on individual components (usually transistors) under stress conditions. These measurements typically result in conservative worst-case values which are used to determine the size of guard bands. However, they usually are quite conservative, resulting in unnecessary design effort and/or increased IC area and delay. On the other hand, they might not be sufficient to cover extreme cases – and a single extremely slow path can be sufficient to render an entire IC useless.

A major enhancement in DfR is then to perform a detailed, circuit-specific timing analysis incorporating aging. As transistor level analysis can only handle rather small circuit sizes due to the required computational effort, a prerequisite for such analysis is a modeling of aging effects at gate level and higher abstraction levels. Efficient and versatile techniques for such modeling have only recently been proposed [16]. Such specific analysis techniques are also a prerequisite to optimize the circuit against aging effects. The literature reports primarily on variations of optimization techniques that have been used e.g. in timing and power optimization before, such as pin reordering and gate restructuring [15].

Such analyses and optimizations performed during design will in the future need to be enhanced by run-time monitors, which observe an IC during operation and enable an immediate reaction to monitoring results. Such reactions can take various forms:

- Issue a warning that an IC is reaching a specification limit, such that the corresponding ECU can be replaced. While this ensures safety, replacing an ECU is very costly.
- Increase the supply voltage so that the original frequency can be maintained despite the aging-related degradation. This results in increased power consumption (and increased future aging as well).
- Perform a graceful degradation of the performance of the IC (e.g. by reducing the workload so that the IC can perform the remaining functionality with a lower frequency). E.g. in a multi-core system, reallocate some tasks among cores so that an aged core can still be used despite offering lower performance. This requires that some spare capacity be built into the system. Also, reallocating task among cores, processors or ECUs might pose special challenges in the automotive domain due to certification issues.

An even more advanced technique of addressing aging is to take recovery effects into consideration as well. For NBTI and PBTI, recovery takes place if the stress condition is removed. This can be utilized in conjunction with online aging monitors. For example, in a multi-core system, if strong aging is sensed in a specific core, tasks can preferentially be moved to other cores to give the strongly aged core a chance to recover. However, such techniques require that specific causes of aging can be identified, rather than just the effects of the aging be noticed.

Regarding **TDDB**, the modeling of this aging effect has become a topic of much interest recently, e.g. because in ultrathin oxide transistors the time between SBD and HBD is very long [22], and the transistors can undergo even multiple SBDs [23]. Most often the breakdown is modeled as a voltage-dependent resistance between gate and drain/source to consider the worst-case scenario. Based on this, an analytical model has been developed to simulate the TDDB-based timing degradation in combinational cells [28]. However the model has potential for improvement, as it does not model the resistance change over time.

Via manufacturing risks are addressed by inserting redundant vias into the circuit. In situations where a circuit contains sufficient space for additional vias, this has no drawbacks. When wires need to be enlarged to allow the insertion of additional vias, the tradeoff between increased manufacturing cost and improved reliability needs to be considered. Therefore intelligent rule based methods to add redundant vias only where they are necessary and to verify a design by identifying only really critical single vias are needed in the future.

Soft Errors are addressed by techniques specific to different circuit components.

Memories are typically protected by parity codes or error-correcting codes (ECC), a special type of redundancy. Alternatively, transistors can be upsized, or an 8T architecture instead of the standard 6T SRAM cell can be introduced.

For flipflops in the circuit logic, ECC is not easily possible. The standard solution for such flipflops is to harden them, e.g. by upsizing the transistors. Also, double or even triple modular redundancy (DMR, TMR) or even more sophisticated techniques

are possible. They can be applied to individual flipflops, or to entire larger circuit modules.

All of these techniques for protecting flipflops in logic against SEUs result in (usually significant) area and delay penalties. An important current research area is therefore to determine which flipflops are most critical to circuit operation and definitely need to be protected, and which other flipflops could be left without protection without catastrophic consequences for a circuit. The concept of the AVF is relevant here. Research has been reported using both simulation and formal techniques to determine the importance of a flipflop. The formal techniques are much less mature than simulation-based techniques today, however [29][30].

General techniques that are used to address many reliability issues are:

- Redundancy (e.g. DMR, TMR) – typically very costly
- Parity and error-correcting codes – cost-efficient implementations are typically limited to regular structures such as memories
- Circuit architectures such as RAZOR [31] or Pre-Error Adaptive Voltage Scaling [32]

3.2 System-Level Techniques

Software and system-level design techniques in the automotive domain usually follow a cross-layer design approach. Here, high-level models are used to specify (often various control) applications. Such a model-based design approach, in contrast to using, for example, handwritten code, allows formal verification and certification of the safety-critical functionality. These models are used to generate software code, which is then partitioned and mapped onto a distributed architecture consisting of various ECUs connected by communication buses.

One of the major challenges in this design flow stems from the fact that the high-level controller models more often than not ignore many platform architecture details (i.e., details of the platform on which the synthesized code is to be implemented). In other words, the models make a number of idealistic assumptions – like control functions are computed in zero/negligible time, there is no delay between sensing and actuation, etc. – which are increasingly not true in modern distributed automotive architectures. Hence, control performance properties that are proven at the model level do not hold true in the actual implementation, thereby requiring considerable integration and debugging efforts and raising questions on the safety/reliability of the resulting system.

In order to address this issue, the abovementioned design flow has to be suitably modified to take into account relevant platform architecture level details during the design of the high-level controller models. Similarly, the architecture design also should be aware of control performance and delay constraints. In other words, techniques from different layers of design abstraction should be combined together rather than designing these layers independently of each other. Such *cross-layer* design approaches are also discussed later in the following section.

So far, our discussion was based on the assumption that all the components in the platform architecture function *correctly*. From our previous reliability-related discussions, we know that this assumption is increasingly not true. To cope with an unreliable implementation platform, the cross-layer design approach outlined above has to be extended to make the reliability information from the architecture level visible at the software and system levels. This way, critical parts of the computation/communication may be appropriately replicated or protected, and certain software or system-level functions may be designed to be more robust to architecture or device level failures/errors.

3.3 Interaction Between Design Layers

Protecting an IC against reliability challenges purely on individual layers of the design hierarchy is increasingly considered to be very costly and not the most efficient solution. E.g. hardening an IC against SEUs could be done by hardening each memory cell (larger transistors, 8T cell), or by providing redundancy and/or ECC or by computing checksums in SW. However, a growing consensus is evolving that the most cost-efficient techniques for analyzing and optimizing against reliability challenges involve cooperation between multiple design layers [21]. Major research projects are under way in Asia, Europe and the US to evaluate cross-layer techniques in addressing reliability (e.g. [25], [26] for an effort in Germany).

"Cross-layer" refers to the idea that efforts on different levels (layers) of the design hierarchy are combined to achieve an overall optimal tradeoff between required resources and resulting improvement in design quality. Cross-layer approaches can be employed both during IC design and during IC operation.

An example for cross-layer optimization during the design phase of an IC is protection against SEUs in logic. This can be achieved by hardening all flipflops. Possibly a more efficient cost-benefit tradeoff can be achieved by analyzing the design, identifying the most mission-critical flipflops (however this term might be defined) and then hardening only those. RAZOR or similar techniques might be implemented. Alternatively, entire critical modules could be replicated (DMR, TMR).

During IC operation, the operating system could employ various techniques for error detection (e.g. computation of checksum) and error recovery (e.g. checkpointing and roll-back). There are obvious tradeoffs here for the resources and the time required for both error detection and recovery. To which degree real-time requirements need to be fulfilled plays an important role in deciding which techniques to apply.

A specific example for cross-layer considerations is to consider how the execution of different instructions of a processor influences the results of circuit timing analysis [27]. This knowledge, gained during the design phase, can be used in multiple ways:

- change the design such that the most critical instructions become less critical (likely at the expense of IC area)

- supply the compiler with information about criticality of instructions such that this information can be considered when deciding which processor instructions to use in compiling a program

- finally, consider making changes to a program during its execution if e.g. aging of an IC is sensed and a risk of a certain instruction failing soon appears.

In general, the more information about an application is available, the more specific (and therefore cost-efficient) approaches can be chosen to improve reliability. Typically, in an embedded system such as most automotive ECUs are, the environment is more constrained than in a general purpose CPU. If higher layers of the design hierarchy have information on (i) usage frequencies, and (ii) reliability requirements, then this information may be used for design and analysis at the lower layers of design.

On the other hand, e.g. reliability requirements at the device layer will imply thermal constraints (e.g., changes in thermal profile). These thermal constraints will in turn imply constraints on higher layers, e.g. on task mapping, task migration, and in the future dynamic frequency scaling.

4. RELIABILITY IN THE AUTOMOTIVE CONTEXT

Aging is becoming a major reliability concern especially for automotive electronics as the automotive environment poses specific challenges. In many consumer markets, electronics-based products are used only a few years before they are discarded – often before aging degradation becomes relevant. In automotive, IC manufacturers need to guarantee specified functionality for 2-5 years operating time depending on application and temperature range and up to 15 years in standby mode, and desire them for even longer time to avoid reputational risks. At the same time, their ICs are sometimes used in very harsh conditions (e.g. temperatures up to 150°C and for special purposes also up to 175°C at reduced life times), and almost continuous (e.g. taxis being used in multiple shifts; battery management electronics in electrical vehicles) which amplifies the aging.

Another major concern are power saving methodologies. Up to now power was not a very big issue for automotive electronics due to the existence of a strong battery in the car, which was mainly used for the ignition process of the combustion engine. Early car electronics therefore did not utilize any low power design techniques. As the number of ECUs in a modern vehicle is now approaching 100, with a further increasing tendency, power meanwhile has become an important issue. Low power techniques as dynamic voltage scaling, several kinds of sleep and power down modes or even module switch-off techniques will be used more frequently with the drawback of increased sensitivity to reliability effects as described in Section 2 (see also Figure 2).

On top of that we still have to deal with different voltage domains in a car. This increases the design effort for the communication between these domains due to low power requirements on the one hand and reliability requirements like EMI and aging on the other hand. The biggest challenge in future will be to handle the complexity of this problem. EMI and ESD up to now have been optimized for each chip separately. This methodology will not be sufficient for above described systems due to the strong interaction of the components. A modeling of each component on an abstract level and according high level system simulations are mandatory to be successful.

As discussed previously, there exist different communication buses to connect the ECUs. Among these, in particular, the FlexRay protocol has gained wide acceptance for safety-critical domains as it provides the infrastructure to design reliable communication networks. We give some examples [34] of such reliability-related features in the following.

Each FlexRay controller provides two communication channels for redundant data transmissions and a hardware bus guardian for schedule monitoring. In terms of network topologies the protocol offers flexible solutions such as bus, star, and hybrid structures to design fault-tolerant and redundant backbone architectures. The FlexRay frame contains two CRC fields; an 11-bit header CRC, and a separate 24-bit CRC which is calculated for the entire frame and able to detect up to five arbitrary bit errors during any frame transmissions. Further, eight samples per bit are available and a majority voting mechanism enables a filter for suppressing any glitches.

In addition to the above described protocol features, switched FlexRay networks and frame packing techniques have been recently studied in [37] and [40] with the goal to isolate *babbling idiots* and short circuits to single branches and to adopt frame retransmissions techniques for faulty frames, respectively.

One broad category of design approaches to cope with reliability issues could explore tradeoffs between the *accuracy* of computations and the associated *computation time*. Since most of the applications are various controllers (implementing, e.g., safety-critical, driver assistance or comfort related functionality), control performance depends on (i) the sampling period (which in turn depends on the computation time or the time taken to compute the control law for each sample), (ii) and the accuracy of the computed control signal. Depending on the chosen tradeoff between these two at the control design stage, the architecture could be suitably designed. For example, during the compilation phase, instructions could be chosen whose accuracy/reliability is higher but result in a less efficient code that runs slower. Alternatively, certain other choices of instructions may result in more efficient code (i.e., smaller running time) but the outcome of certain instruction execution is unreliable because of the possibly one or more reasons that were discussed previously. Such reliability-aware compilation techniques have been studied recently [35], but they have not been combined with higher levels of design abstraction such as the levels at which controller models are designed and analyzed.

Similarly, the above mentioned tradeoff analysis between sampling periods of control algorithms and the accuracy of the feedback control signal may be used to harden certain instructions whose reliable execution is essential for meeting control performance objectives. Again, such selective hardening of instructions along with custom instruction set design has been explored in the past from a reliability perspective [36] but has not been combined with model-based (control) algorithm design.

Finally, such tradeoffs between reliability and computation time may also be explored in the context of cache memories. Caches occupy more than half of the chip area in today's microprocessors and their reliability is therefore a critical design issue. Since the charge stored in a memory cell (such as an SRAM cell) decreases with each process generation, the accuracy of the information stored in the memory cells is becoming increasingly vulnerable to soft errors. Execution correctness characterization (such as those based on the AVF) have recently been refined using a number of models such as PARMA [38] and MACAU [39]. These models capture the effects of soft-errors such as single-bit upsets and temporal multi-bit upsets on a cache memory and the impact of protection/correction codes on the error incurred by a given application.

It will be meaningful to utilize these models in conjunction with code reordering and cache locking techniques. By reordering the code of a control application, the lifetime of various variables in the cache memory can be modified. On one hand this influences the execution time of the code and on the other hand it will influence the accuracy of the computed control signals (because of the change in the vulnerability of the application to soft-errors).

Similar tradeoffs may also be explored by *locking* parts of the cache. Cache locking has been explored in the past for improving the *predictability* of real-time applications. While caches result in improving the execution time of programs in the average case, they also introduce significant variability in the execution time, thereby introducing jitter in some cases and making the computation of worst-case execution times (WCET) more difficult. Cache locking makes the computation of WCET more straightforward, thereby increasing the predictability of the application at the cost of deteriorating its average case performance. However, cache locking has not been commonly used for increasing *reliability*. By suitably analyzing the impact of sampling delay and accuracy on the performance of a control application, appropriate cache locking mechanisms will be useful in the context of designing safety-critical automotive control software.

5. SUMMARY

Advances in automotive technology are increasingly driven by electronics and software. As the amount of electronics in cars increases, and automotive electronics moves to advanced process nodes of 40nm and below, reliability concerns become a major issue. These are amplified in the automotive domain, as it is both very cost-sensitive and safety-conscious at the same time. In this paper, we describe major reliability challenges and discuss both established and emerging techniques to handle them. We especially point out the need for cross-layer optimization from transistor level all the way to software to conquer reliability challenges.

Acknowledgements

This work was supported partially by the German Research Foundation (DFG) as part of the priority program "Dependable Embedded Systems" (SPP 1500 – http://spp1500.itec.kit.edu).

6. REFERENCES

[1] E. Frickenstein. Mikroelektronik fährt BMW ConnectedDrive. In *3. Symposium Mikroelektronik*, Berlin, Germany, September 25-25, 2012.

[2] S. R. Nassif. Modeling and forecasting of manufacturing variation (embedded tutorial). In *ASP-DAC*, 2001.

[3] D. Boning and S. R. Nassif. Models of process variations in device and interconnect. In *Design of High-Performance Microprocessor Circuits*, A. Chandrakasan, Ed. Piscataway, NJ: IEEE Press, 2000.

[4] D. Blaauw, et al. Statistical timing analysis: from basic principles to state of the art. IEEE Transactions on CAD, 27 (4): 589-607, Apr. 2008.

[5] K. Kuhn et al. Managing process variation in Intel's 45nm CMOS technology. In *Intel Technology Journal*, 12(2): 131-144, June 2008.

[6] S. R. Nassif, N. Mehta, and Y. Cao. A resilience roadmap. In *DATE*, 2010.

[7] S. R. Nassif, V. B. Kleeberger, and U. Schlichtmann. Goldilocks failures: not too soft, not too hard. In In *Proc. of Reliability Physics Symposium*, 2011.

[8] K. Bernstein et al. High-performance CMOS variability in the 65-nm regime and beyond. In IBM Journal of Research and Development, 50(4.5): 433-449, 2006.

[9] S. R. Nassif. Modeling and analysis of manufacturing variations. In *CICC*, 2001.

[10] D. Lorenz, M. Barke, and U. Schlichtmann. Aging analysis at gate and macro cell level. In *ICCAD*, 2010.

[11] K. N. Tu. Recent advances on electromigration in very-large-scale-integration of Interconnects. In. J. Appl. Phys, 94(9), September 2003.

[12] N. Weste and D. Harris, CMOS VLSI Design: A Circuits and Systems Perspective, Addison-Wesley, 2009.

[13] Synopsys PrimeTime Advanced OCV Technology

[14] H. Graeb. *Analog Design Centering and Sizing.* Springer, 2007.

[15] K.-C. Wu and D. Marculescu: Aging-aware timing analysis and optimization considering path sensitization. In *DATE*, 2011.

[16] D. Lorenz, M. Barke, and U. Schlichtmann. Efficiently analyzing the impact of aging effects on large integrated circuits. In *Microelectronics Reliability* 52(8), 1546-1552, August 2012.

[17] M. Alam, B. Weir, and P. Silverman. A study of soft and hard breakdown - Part II: Principles of area, thickness, and voltage scaling. In *IEEE Transactions on Electron Devices*,. 49(2): 239 –246, February 2002.

[18] R. Degraeve et al. Relation between breakdown mode and location in short-channel nMOSFETs and its impact on reliability specifications. In *Proc. of Reliability Physics Symposium,* 2001.

[19] R. Rodriguez et al. The impact of gate oxide breakdown on SRAM stability. In *IEEE Electron Device Letters*, 23(2), September 2002.

[20] B. Kaczer et al., Gate oxide breakdown in FET devices and circuits: from nanoscale physics to system-level reliability. In *Microelectronics Reliability*, 47(4-5), April-May 2007.

[21] N. P. Carter, H. Naeimi, and D. S. Gardner. Design techniques for cross-layer resilience. In *DATE*, 2010.

[22] Y.-H. Lee et al. Prediction of logic product failure due to thin-gate oxide breakdown. In *Proc. of Reliability Physics Symposium*, 2006.

[23] M. A. Alam et al. Statistically independent soft breakdowns redefine oxide reliability specifications. In *Proc. of Int. Elec. Dev. Meeting (IEDM)*, 2002.

[24] H. Matsuyama et al.: Investigation of stress-induced voiding inside and under vias in copper interconnects with "wing" pattern. In *Proc. of Reliability Physics Symposium*, 2008.

[25] J. Henkel et al. Design and architectures for dependable embedded systems. *International Conference on Hardware/Software Co-design and System Synthesis (CODES+ISSS)*, 2011.

[26] A. Herkersdorf et al. Cross-layer dependability modeling and abstraction in systems on chip. In *Workshop on Silicon Errors in Logic - System Effects (SELSE)*, 2013.

[27] V. B. Kleeberger et al. Program-aware circuit level timing analysis. In *International Symposium on Integrated Circuits (ISIC)*, 2011.

[28] M. Choudhury et al. Analytical model for TDDB-based performance degradation in combinational logic. In *IEEE Design, Automation, and Test in Europe (DATE)*, 2010.

[29] R. Hartl et al. Architectural vulnerability factor estimation with backwards analysis. In *13th Euromicro Conference on Digital System Design*, 2010.

[30] S. A. Seshia, W. Li, and S. Mitra. Verification-guided soft error resilience. In *DATE* 2007.

[31] D. Ernst et al. RAZOR: A low-power pipeline based on circuit-level timing speculation. In *Micro-36*, 2003.

[32] M. Wirnshofer et al. On-line supply voltage scaling based on in-situ delay monitoring to adapt for PVTA variations. In *Journal of Circuits, Systems, and Computers*, 21(8), 2012.

[33] S. Mukerjee et al. Measuring architectural vulnerability factors. In *IEEE Micro*, 23(6): 70-75, 2003.

[34] Rausch, M.. FlexRay: Grundlagen, Funktionsweise, Anwendung. In Carl Hanser Verlag GmbH & CO. KG.

[35] S. Rehman et al. Reliable software for unreliable hardware: Embedded code generation aiming at reliability. In *CODES + ISSS*, 2011.

[36] U. D. Bordoloi et al. Reliability-aware instruction set customization for ASIPs with hardened logic. In *RTCSA*, 2012.

[37] B. Tanasa et al. Reliability-aware frame packing for the static segment of FlexRay. In *International Conference on Embedded Software (EMSOFT)*, 2011.

[38] J. Suh, et al. Soft error benchmarking of L2 caches with PARMA. In *SIGMETRICS*, 39(1), 2011.

[39] J. Suh, M. Annavaram, and M. Dubois. MACAU: A Markov model for reliability evaluations of caches under single-bit and multi-bit upsets. In *HPCA*, 2012.

[40] P. Milbredt et al. Switched FlexRay increasing the effective bandwidth and safety of FlexRay networks. In *15th International Conference on Emerging Technology and Factory Automation (EFTA), 2010.*

Reliable On-Chip Systems in the Nano-Era: Lessons Learnt and Future Trends

Jörg Henkel*, Lars Bauer*, Nikil Dutt†, Puneet Gupta‡, Sani Nassif°,
Muhammad Shafique*, Mehdi Tahoori*, Norbert Wehn⊤

*KIT Karlsruhe, †UCI Irvine, ‡UCLA Los Angeles, °IBM Austin, ⊤Uni Kaiserslautern

ABSTRACT

Reliability concerns due to technology scaling have been a major focus of researchers and designers for several technology nodes. Therefore, many new techniques for enhancing and optimizing reliability have emerged particularly within the last five to ten years. This perspective paper introduces the most prominent reliability concerns from today's points of view and roughly recapitulates the progress in the community so far. The focus of this paper is on perspective trends from the industrial as well as academic points of view that suggest a way for coping with reliability challenges in upcoming technology nodes.

1. INTRODUCTION

Around a decade ago research increased focus on reliability for on-chip design with the move to nanoscale At that time, technology roadmaps provided evidence that upcoming technology nodes would heavily suffer from reliability problems since feature sizes would reach limits at which certain effects would seriously jeopardize the correct functionality of circuits. An example of such an effect is Random Dopant Fluctuations, whereby the number of dopant atoms in the channel of a MOSFET becomes so small (e.g., 40) that the addition or subtraction of a single atom can make a significant difference to the behavior of the device. Even though many of these atomic-level effects had been known and measurable for a long time, their impact had not reached the point where they would impact circuit functionality. An inflection point occurred around a decade ago and started then to change the mindset of researchers and designers to plan for dependability as a major design challenge in upcoming technology nodes (also referred to as the nano-CMOS era). We have seen heavy research and development efforts towards dependability since then. The contribution of this *perspective paper* is as follows:

- Introducing the most prominent effects (from today's view) that seriously jeopardize reliability. This helps novices to understand the basic problems. In particular, we will introduce variability, aging and temperature effects, and soft errors (Section 1).
- Summarizing and categorizing state-of-the-art techniques at the hardware-level, software-level and application-level (without claiming comprehensiveness since that is not the main focus of this perspective paper) (Section 2).
- Providing perspectives of future trends and approaches of how to combat the inherent reliability problems. This is the main contribution of this paper as we introduce new ways to cope with these problems that go far beyond the current state-of-the-art. Some catchwords are *cross-layer approaches, self-organization, run-time adap-*

tion and more. These perspectives are a result of research in that field conducted within approximately the last ten years and they reflect views from industry as well as from academia. These perspectives may help to guide the way to cope with reliability challenges in upcoming technology nodes (Section 3).

The discussions within this paper are focused towards higher levels of abstraction to increase and optimize for reliability. Physical and device level techniques have been the primary focus a decade ago and it has been recognized that higher abstraction levels like, for example, the software stack might contribute its share to increase reliability even if the source of unreliability is the device physics due to ever shrinking feature sizes. That said, we believe that *all* abstraction levels should contribute its share to achieve a high degree of reliability in future technology nodes.

In summary, this paper recapitulates a decade of research in reliability and it focuses on drawing a way to future research and development in dependability from lessons learnt in the past.

1.1 Manufacturing Variability: Sources and Magnitude

Dimensional scaling has exceeded the tolerances of equipment used to manufacture semiconductor circuits, resulting in ever increasing variaiblity in power and performance.

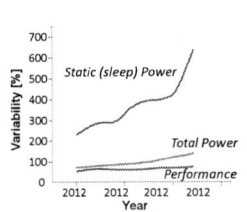
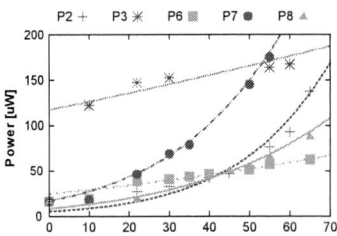

Figure 1: Circuit variability as predicted by ITRS [1]

Figure 2: Sleep power variability across temperature for ARM Cortex M3 processor [2]

There are two sources of hardware variation:

- *Semiconductor Manufacturing.* Random as well as systematic variation at all length scales (die-to-die and within-die) has become a scaling bottleneck. In addition to increasing magnitudes (see Figure 1), the nature of variability is also changing (e.g., bimodality due to double-patterning [3]). For instance, measurements on 10 off-the-shelf embedded processors (results for 5 of them are shown in Figure 2) show over 9× variation in leakage power at room temperature.
- *Vendor.* Parts with almost identical specifications can have substantially different power, performance or reliability characteristics, as shown in Figure 3. This variability is a concern as single vendor sourcing is difficult for large-volume systems.

Summary: Variation among parts with identical specifications is large and is expected to grow, which is a significant

---

Permission to make digital or hard copies of all or part of this work for personal or classroom use is granted without fee provided that copies are not made or distributed for profit or commercial advantage and that copies bear this notice and the full citation on the first page. To copy otherwise, to republish, to post on servers or to redistribute to lists, requires prior specific permission and/or a fee.
DAC '13, May 29 - June 07 2013, Austin, TX, USA.

concern with respect to reliability as variability directly impacts major design constraints like power and performance.

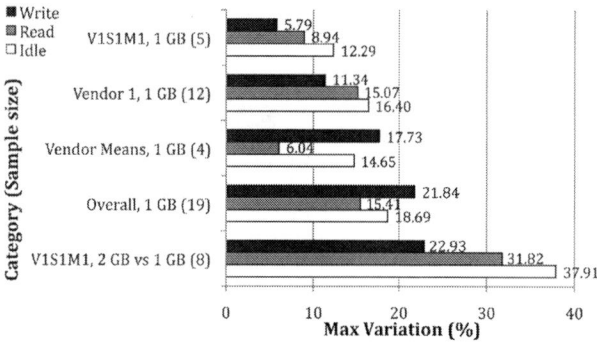

Figure 3: Maximum variations in Write, Read, and Idle Power by DIMM Category, 30°C [4]

1.2 Aging

It is well known that modern devices are susceptible to temporal degradation. This is primarily due to the slow scaling of power supply voltages, which when coupled with shrinking device dimensions causes ever increasing vertical and horizontal electric fields. These high fields lead to phenomena like *Hot Carriers* and *Negative Bias Temperature Instability* both of which cause permanent degradation of devices. Of these phenomena, NBTI has become most prominent and has received widespread attention (see for example [5–7]). While the exact mechanism involved is still a topic of active research, it is believed that high fields in the gate region cause the activation of *traps* in the gate material which when filled create a fixed charge that changes the surface potential and in turn causes the threshold voltage to *shift*.

NBTI exhibits itself on two time scales. The first is a short time constant (ns regime) phenomena whereby a device under high gate voltage stress will exhibit a threshold voltage higher than normal for a short period of time, with subsequent return to normal after the stress is removed. The second is a slow and steady change in the threshold voltage over time as traps get permanently filled. The phenomena is more pronounced for P-Channel devices, and Figure 4 shows a plot of threshold voltage increase after seven years of operation for a range of technologies.

Summary: Modern devices degrade, referred to as *aging*. NBTI is the most prominent aging effect.

1.3 Impact of Temperature

All major aging effects exhibit a temperature dependency i.e., they stimulate and/or accelerate aging. The dependency is expressed through the Arrhenius' Law [8]. An aging effect λ_{EFF} has the property:

$$\lambda_{EFF} \propto e^{\frac{-E_a}{kT}} \qquad (1)$$

where T is the temperature and k is Boltzmann's constant. The activation energy E_a is specific for a certain aging process. Currently (i.e. 2012 where 22 nm has been commercialized) the most critical aging effect is NBTI (Figure 5 shows the influence of temperature on NBTI) as stated before in Section 1.2. Others are: *electromigration (EM)*, *Time-Dependent Dielectric Breakdown (TDDP)*, and *Hot Carrier Injection (HCI)*. EM [9] is caused by the erosion of metal interconnects through ion movement, TDDB [10] results in conductive paths due to the breakdown of the dielectric through the formation of traps caused by high electric fields, HCI [10] is caused when hot carriers in a source-drain current attain sufficient energy to be injected into to gate oxide to form traps. Apart from the property expressed in

Equation (1), λ_{EM} is also affected by the *thermal gradients*[1] on a chip. In this case the time dependency of electromigration becomes:

$$\lambda_{EM} \propto e^{\frac{-E_a}{k(T+\Delta T_{joule})}} \qquad (2)$$

where ΔT_{joule} is the difference in heat energy resulting from local power consumption and not from heat conducted from elsewhere in the chip.

An additional effect which is responsible for shortening chip lifetime is *thermal cycling* which induces stress through periodic heating and cooling, modeled through the Coffin-Manson equation [11]:

$$N = C\left(\frac{1}{\Delta T}\right)^q \qquad (3)$$

where ΔT is the change in temperature, C is a material constant, and N represents the expected number of cycles until a failure occurs. The exponent q is the experimentally determined Coffin-Mason exponent with $q \in [1,3]$.

Summary: Temperature stimulates and accelerates aging. Thermal gradients (spatial and temporal) play a key role.

1.4 Soft Errors

It is a common belief that with continuous downscaling of CMOS technologies the susceptibility of memories and logic to radiation coming from atmospheric neutrons or by onchip radioactive impurities increases. This is due to the fact that the soft error rate has an exponential relationship with the critical charge Q_{crit} which is is the minimum amount of charge that can flip a data value in a memory cell or in logic [12]. But the soft error also depends on the sensitive depletion area which decreases with scaling. In the past, the reduction in the critical charge caused by lower supply voltage has been more than offset by the reduction of the sensitive area and by technology improvement. Recent investigations have shown that the trend does no longer appear to be true, at least for SRAMs below 40 nm [13]. In general, SRAMs are more vulnerable to soft errors than logic, since memory cells lack transient masking mechanisms and they are much more dense. This density makes SRAM cells much more susceptible to process induced transistor variability which strongly impacts Q_{crit}. Moreover, recent measurements have shown the importance of Multiple bit/multiple Cell Upsets (MCU) which results in an increased MCU occurrence to the total number of upsets [14].

But despite the fact that the soft error rate for a single memory cell or latch decreased, the capacity on a chip increased faster than the soft error rate change. For example, the transition from the 130 nm to the 65 nm technology node reduced the soft error rate by about a factor of 2 in SRAMs [14]. But at the same time memory capacity increased faster resulting in an increase of the system soft error rate. This is depicted in Table 1 taken from [13] which shows the single event upset (SEU) rate per microprocessor in various technologies. Beyond that, power efficiency forces designers to reduce the voltage via sophisticated techniques like dynamic voltage scaling or near subthreshold voltage which decreases Q_{crit} and consequently increases the soft error rate.

Summary: Soft errors reverse a long term trend and show an increase. Furthermore, multi-cell upsets become much more frequent.

2. RESEARCH IN RELIABILITY

Since reliability has been recognized to become a major on-chip design challenge around a decade ago, heavy research efforts started at various levels of abstraction. In this section we are providing a glimpse of reliability-enhancing

---

[1]The *spatial thermal gradient* expresses the variation of temperate with respect to distance measured in $[°C/mm]$ whereas the *temporal thermal gradient* expresses the variation of temperature with respect to time measured in $[°C/s]$.

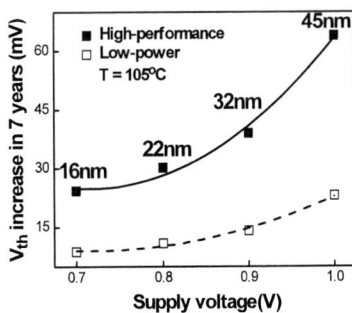

Figure 4: The prediction of V_{th} increase in 7 years due to NBTI

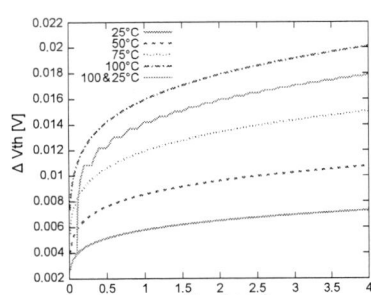
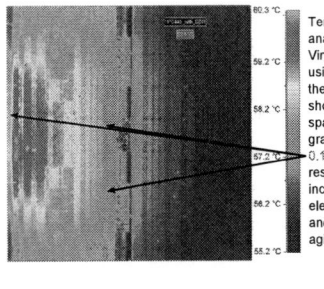

Temperature analysis of Virtex-5 FPGA using infrared thermal camera showing peak spatial thermal gradients of 0.12°C/µm resulting in an increase of electromigration and accelerated aging

Figure 5: Impact of temperature: shown here is the NBTI-induced threshold voltage shift at various temperatures (left) and and a thermal image of an FPGA chip (right)

Tech-nology (nm)	Relative SEU rate in $FITs/kbit$	Approx. Mbit per micro-processor	Relative uncorrected SEU rate per micro-processor (kFIT)
180	3.0	1.52	4.3
130	2.4	3.28	7.9
90	1.0	33.6	33.6
65	0.7	44.3	30.5
40	0.94	71.0	67.0

Table 1: Soft Error Rates in Microprocessors

and reliability-optimizing techniques without the claim of a comprehensive overview of state-of-the-art. The purpose is to provide an introduction to the large variety of techniques and approaches that have been proposed so far with focus on higher abstraction levels since research has steadily moved towards higher levels to leverage the efficiency of the traditional lower-lever techniques at physical and device level.

Representative topics in this section are therefore dedicated to the hardware-level, software-level, OS (thermal management), and application-level with focus on prominent research projects.

2.1 Hardware-Level Mitigation

The existing hardware-level techniques range from hardware redundancy (like TMR/DMR [15–18]) to pipeline protection with shadow latches [19] and designs with reduced architectural vulnerability factor [20]. Logic duplication and TMR have been used in commercial systems such as Tandem's NonStop system [21]. Various sequential elements with circuit-level hardening, e.g., DICE [22], RCC [23], BISER [24], BCDMR [25], and selective protection [26] provide different degrees of error resilience with varying costs.

Some architecture level techniques such as self-checking designs [27] are cost-effective solutions, in which the protected circuit generates outputs encoded in an error detecting code (parity, Berger, or arithmetic code) and the circuit outputs are controlled by a checker circuit to detect and maybe correct possible errors [28]. Residue codes have been used for arithmetic circuits such as multipliers [29].

Another group of techniques used to protect memories, buses, or other microprocessor array structures (e.g., register file) are *parity* and *error correction codes* [30]. Several techniques to reduce their performance penalty are presented in [31]. Built-in current sensors are an alternative approach to detect soft errors [32] that can be applied to all memories. Examples of microarchitecture and architecture level techniques include DIVA [33] and Argus [34].

Various canaries and monitoring circuits [35, 36], circuit-level guardbanding techniques [37, 38], and on-line self-test and diagnostics [39, 40] are examples of other classes of resiliency techniques targeting specific failure mechanisms, such as early life failures, voltage droop and aging.

Summary: Existing hardware solutions focus on error sensing, detection, and masking. Error recovery is typically performed at higher (e.g., software) layers. However, the impacts of reliability failure mechanisms at hardware must be

better evaluated at multiple levels for more cost-effective solutions. Therefore, there is a need for *cross-layer* resiliency to protect only what actually matters.

2.2 OS-Level – Thermal Management

As discussed in Section 1.3, many reliability concerns like aging are stimulated and/or accelerated by temperature, making thermal management a key technique. While Dynamic Voltage and Frequency Scaling (DVFS) and power gating have been widely used to reduce the power consumption, they are only capable to effectively reduce the peak temperature but they unfortunately are not suitable to combat the important spatial/temporal thermal gradients [41]. For example, in [42] the hottest core is switched off and the coldest is activated to control the chip temperature. However, this method leads to thermal cycling and thus shortens the chip lifetime. A more effective method with respect to aging is proposed in [43]. It employs a closed-loop PI controller to increase the performance and throughput of the many-core systems under thermal constraints. A further controller-based approach proposed in [44] implements extremum-seeking control to optimize temperature distribution. The obtained reduction in peak temperature is $9°C$ and the reduction in thermal spatial variation is from $6°C$ to $1°C$ and therefore it effectively decelerates aging.

Distributing the computational workload (i.e., tasks) through different cores can largely mitigate the consumed power density resulting in balancing both the peak temperature and the aging-related spatial/temporal thermal gradients. The approach presented in [41] uses offline Integer Linear Programming (ILP) in order to optimally scheduling tasks to the available cores in a Multi-Processor Systems. While this approach achieves optimal thermal results, it requires the task workload to be known *a priori*.

Summary: To prevent and/or decelerate various aging effects, thermal management techniques tailored to reducing spatial/temporal thermal gradients are necessary. Traditional DVFS techniques are not suited for that purpose.

2.3 Software-Level Mitigation

Prominent software-level reliability techniques primarily rely redundant code execution such as aimed in SWIFT/-CRAFT [45–47] by embedding redundant instructions, comparison instructions, and control flow checking instructions using EDDI [47–49], check pointing and replication [16]. SWIFT-R employs majority voting or AN-code for recovery. These software-based schemes duplicate all the instructions and incur a performance and/or memory overhead (more than 2×), etc. Besides these, other prominent compile time approaches are: control flow checking [47, 50, 51], register vulnerability reduction [52], wear-out and aging-aware schemes to improve the reliability [53, 54], etc.

Several compiler-level techniques have been proposed as well. State-of-the-art reliability-aware instruction scheduling approaches [52, 55–57] reorganize the instruction profile of a program at the cost of some performance degradation and some memory overhead compared to instruction redun-

697

dancy techniques. The technique ISSE [55] reschedules a program's assembly code to minimize the operands' vulnerable periods via exploiting the slack time. Since the available slack after the performance-optimized scheduling is limited, other instruction scheduling techniques [52, 55] provide limited reliability improvements. In order to protect the control flow, basic block signatures have been introduced [58].

Summary: Software level techniques alleviate or complement hardware level techniques or are deployed when hardware redundancy is prohibitive. However, software-level reliability techniques do not ensure 100% correct execution and require hardware-level support, thus motivating the need for cross-layer reliability (see also Section 3.1) to provide a better tradeoff between reliability and overhead in performance/power/area.

2.4 Application-specific Reliability Enhancement

Application-level dependability is a very important mechanism to minimize the implementation overhead for reliability. A large number of important and relevant applications dealing with uncertain information show a certain inherent error resilience, which does not require the underlying implementation platform to be completely error-free. This tolerance toward various kinds of errors can be attributed to one of the following two categories:

- **Algorithmic resilience** is given when a certain amount of errors can be tolerated by the algorithm itself without rendering the outcome of the calculations invalid [59, 60]. This is typical of probabilistic and iterative algorithms, which can be found in a class of applications often subsumed under the term Recognition, Mining and Synthesis (RMS) [61] and of wireless communication systems [62, 63], one of the fundamental technologies of today's information society. Also all algorithms that can tolerate statistical behavior, such as fixed-point DSP algorithms and numerical calculations exhibit algorithmic resilience.
- **Cognitive resilience** stems from an application's interaction with a human observer. E.g., video and audio processing. Errors are tolerable, as long as the user cannot discern quality difference, or accepts them as a trade-off for, say increased battery life time [63].

Taking into account the application-inherent resilience at the architectural and software level clears the way for a further reduction of the overhead for resilience by providing only a sufficient hardware reliability for a given application, while still achieving the desired application reliability. Compared to lower level resilience techniques like error-detection sequential by, e.g., Razor [64, 65] or tunable replica circuits, techniques on higher levels are very application dependent and require a real cross-layer approach. A typical example for a static application dependent protection mechanism is selective protection. It is well known that in many signal processing algorithms the higher-order bits have a much larger influence on the overall system than the low-order bits. So, it is very often sufficient to protect only the higher-order bits [62]. Many applications exhibit a large dynamic at run-time. This dynamic is well explored to optimize power and energy by, e.g., dynamic voltage and frequency scaling. Here, application specific quality-of-service (QoS) is traded-off against power efficiency at run-time. E.g., in video processing, the compression rate can be traded-off against power saving. But dynamic application behavior can be exploited for application-specific reliability enhancements, as well. This yields adaptive reliability tuned by the application at runtime according to the required application performance, monitored disturbances and operating conditions of the underlying hardware. For this purpose, so called resilience actuators can be defined (on hardware and/or software level) which are dynamic protection mechanisms to increase the error resilience on system level [66, 67]. They exist on various abstraction levels:

- *Do nothing:* this is the case for very low error rates where no impact of the errors on the system level is observed.
- *Change hardware operating point:* e.g., changing the voltage and/or frequency operating point which makes the system more robust at the cost of increased power/energy.
- *Change algorithmic parameters:* many algorithms have parameters which can be changed at run-time. E.g., most of the algorithms in wireless communication are iterative like Turbo-Code decoding in LTE. If the communication channel has a high signal-to-noise-ratio (SNR), less iterations are required compared to a low SNR channel. So, if the channel shows a high SNR, the decoding iterations can be reduced without changing the system performance. The timing margin obtained by the reduction of the iterations can be used to increase the reliability.
- *Change algorithms:* Besides changing only algorithmic parameters we can also change the complete algorithm itself. Often, there is a choice of different algorithms, starting from optimal with high complexity down to suboptimal algorithms with very low complexity. This offers a trade-off between QoS and resilience robustness.

In general, different actuators or combinations of actuators are used to increase the resilience robustness. It is preferable to use actuators which do not require changes in the system components themselves and can be performed on software level. So changes in the algorithm or algorithmic parameters are excellent candidates for software, resulting in variability-resistant software concepts [67]. Obviously each actuator offers a different trade-off between hardware reliability, QoS and implementation performance, e.g., energy and throughput. These trade-offs have then to be characterized.

Summary of Research in Reliability: We have focused on some representative abstraction layers to introduce to state-of-the-art in reliability. In a more comprehensive overview, the whole stack of layers from *Device* through *Hardware Architecture, System Software, Compiler* and *Application* should be covered since all layers can potentially contribute to increase the reliability (for example through cross-layer approaches like in Section 3.1.)

3. RELIABILITY PERSPECTIVES

In this section we present our vision of the road to reliability within the next decade formulated as a set of *perspectives*. These perspectives are not intended to be orthogonal meaning that, for example, a method introduced in one perspective might be a subset of a more general concept presented within another perspective. In that sense, the following perspectives also do not represent any kind of categorization. The perspectives, however, do represent statements that contain promising methods and concepts from different angles: from hardware and software, and from industry and academia. The perspectives have their origin in, in parts, larger-scale ongoing projects on reliability where at least some initial lessons have been learned that are believed to guide the way to the future reliability research. It should also be mentioned that these perspectives primarily focus on methods and concepts at higher abstraction levels since this is a recent trend as opposed to a decade ago when physical and device level techniques were the exclusive focus[2].

Our Perspectives:

P1: Cross-Layer Leverage:
"Cross-layer approaches can leverage reliability through techniques that are pro-actively designed with respect to techniques at other layers. This opens a so far neglected but efficient means to increase reliability".

P2: Run-Time Sense:
"To address ever-increasing spatial and temporal variations, future systems will inherently sense and adapt".

---

[2]This by no means implies that lower techniques, like device, in future will have lower relevance.

P3: Run-Time Adaptation and Self-Organization:
"In order to achieve error resiliency in complex future many core systems, run-time adaptation and concepts/means of self-organization are needed".

P4: Exploiting Application Resilience:
"Applications' inherent resilience paves the path for cost-efficient error resilience; however, it requires fine-grained control on reliability methods".

P5: CAD for Reliability:
"Reliability needs to be considered as the major constraint in an integrated CAD flow".

P6: Scaling and Cost for Resilience:
"The cost for resiliency might become non-affordable in future technology nodes".

P7: Reliability for CPSoCs:
"Future of complex chips will evolve into Cyber-physical Systems-on-Chip (CPSoCs) - this requires seamless integration of optimizations for reliability, safety, and security related issues".

In the following we elaborate on these perspectives in more depth and provide –in parts– some early evidence supporting our perspectives.

3.1 P1 – Cross-Layer Leverage

A robust and dependable on-chip design needs to consider reliability at all levels of abstraction in order to leverage on the opportunity that techniques designed for one level have knowledge of other levels' techniques. As opposed to early work a decade ago, it is not sufficient anymore to confine all reliability concerns at the hardware layers, as faults ultimately propagate to the software layer, and in between there are several sources of fault masking that may contribute towards cost-efficient cross-layer reliability solutions. Since each system layer comes with its own reliability tradeoffs, it is crucial to leverage multiple system layers in an integrated and collaborative fashion for joint optimization.

Key prerequisites in enabling cross-layer resiliency are cross-layer modeling of various reliability effects and cross-layer optimizations in order to be able to consider failure generation and failure propagation through multiple layers from *Device* all the way up to the *Application* layer.

3.1.1 Cross-Layer Reliability Modeling & Analysis

In the following, we discuss cross-layer software error analysis which is used to develop accurate software program-level reliability and resilience models.

CLASS: Cross-layer Soft Error Rate Analysis: Existing soft error rate (SER) analysis techniques mostly target one abstraction level [20, 68–74]. Lower-level techniques consider logic, electrical and timing window masking, and therefore suitable for irregular (random logic such as functional units and control logic) structures. Architecture and microarchitecture level methods consider only *Architectural Vulnerability Factor* (AVF) for regular (e.g. register files, caches, memories) structures. Hence existing methods fail to provide an accurate system-level analysis considering circuit to architecture level masking at various layers, and lead to an unacceptable inaccuracy.

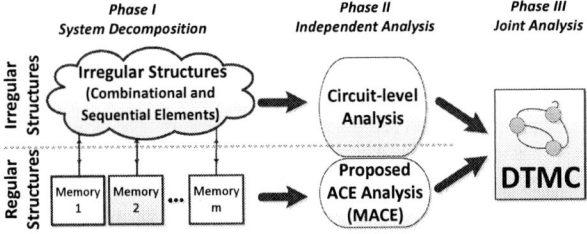

Figure 6: Conceptional Block Diagram of CLASS

Combined Logic and Architectural Soft error Sensitivity

analysis (CLASS) is a cross-layer approach for computing the soft error vulnerability of the entire (microprocessor) system. This hybrid approach uses a circuit-level SER analysis to model the propagation of errors in combinational and sequential logic blocks, and a Memory Architecturally Correct Execution (MACE) analysis to compute the vulnerability of memories, including the scenarios in which an erroneous value is written in the memory hierarchy and several cycles later either masked or used by the program, i.e. *Memory Vulnerability Factor* (MVF). The main idea is perform profiling of application at the (micro-)architecture levels and extracting all memory accesses and architectural masking (MACE analysis), and then performing circuit- and logic-level analysis with those annotations. The conceptional block diagram of this approach is shown in Figure 6.

Evaluation of the OR1200 processor using the CLASS approach is five orders of magnitude faster than *statistical fault injection* while its inaccuracy is less than 7%. This is more than 5 times more accurate than single-layer analysis either at logic or architecture levels. This cross-layer soft-error modeling framework can be combined with the cross-layer soft error mitigation framework, presented next.

To employ efficient software-level reliability methods, the challenge is to bridge the gap between the hardware and software by quantifying the effects of hardware-level faults at the instruction level, while considering the knowledge of low-level fault models (as discussed above) and processor layout. Moreover, it is important to understand which instructions –when fail– lead to which type of errors in the software program. Researchers like in [75] have developed various software program reliability models to quantify three reliability-related properties namely vulnerability, masking, and resilience of application software at different granularities (like instruction, basic block, and function-level):

- *Instruction Vulnerability Index (IVI)* quantifies the effects of faults in different processor components (spatial) during the execution of different instructions over time (temporal), types of errors, critical and non-critical instructions, and vulnerable bit analysis.
- *Instruction-Level Error Masking Index* quantifies the error masking probabilities at instruction level depending upon the data- and control flow properties of a program.
- *Function Resilience* that provides the probabilistic measure of functional correctness (output quality) of a program in the presence of failures.

3.1.2 Cross-Layer Reliability: Code Generation and Execution

Code generation by means of reliability-driven compilation should entail:

- Reliability-Aware Software Transformations like: Various software transformations i.e. reliability-guided loop unrolling, data type optimization, on-fly table computations, for (a) reducing the IVI; and (b) lowering the probabilities of program failures.
- Reliability-Aware Instruction Scheduling should target at improving the reliability of a software program given a user-provided tolerable performance overhead.

To increase reliability through code generation, the flexibility of compilers to generate multiple versions of functions should be provided such that the one with the lowest error probability, given a certain fault model, can finally be selected for execution at run-time.

Results in Figure 7 illustrate that a reliability-driven compiler is capable to provide up to 60% lower software program failures running on an unreliable hardware.

3.1.3 Underdesigned and Opportunistic Computing

*U*nderdesigned and *O*pportunistic (UnO) computing machines (see Figure 8) provide a unified way of addressing variations due to manufacturing, operating conditions, and even reliability failure mechanisms in the field: difficult-to-predict spatiotemporal variations in the underlying hard-

699

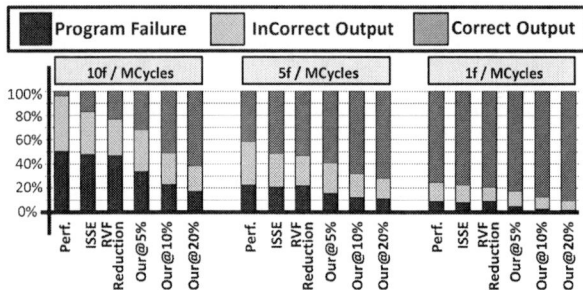

Figure 7: Advantage of cross-layer leverage: Comparing a Reliability-Driven Compilation [76] to other state-of-the-art (that does not leverage on cross-layer) ISSE [55] and Register Filer Vulnerability Reduction Scheme [77] for the adpcm application

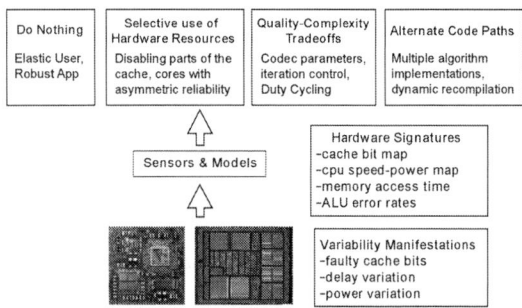

Figure 8: Underdesigned and opportunistic computing using cross-layer concepts [78]

ware, instead of being hidden behind conservative specifications, are fully exposed to and mitigated by multiple layers of software [78].

Underdesigned machines could use parametrically under-provisioned circuits (e.g., voltage over-scaling as in [79]) or be implemented with explicitly altered functional description (e.g., [80]). They may allow erroneous operation and rely upon applications level of tolerance to limited errors (as in [81]); correct all errors (e.g., [65]); or operate hardware within distinct but correct operation limits (e.g., [2, 82]).

Consider an UnO machine example *ViPZone* [83] a system-level solution (see Figure 9) that opportunistically exploits DRAM power variation through physical address zoning in the operating system (OS). ViPZonE is composed of a variability-aware software stack that allows developers to indicate to the OS the expected dominant usage patterns (write or read) as well as level of utilization (high, medium, or low) through high-level APIs. ViPZonE's variability-aware page allocator, implemented in the Linux kernel, is responsible for interpreting these high-level requests for memory and transparently mapping them to physical address zones with different power consumptions. Experimental results across various configurations running PARSEC workloads show an average of 13.1% memory power savings.

There are several challenges in realizing the UnO vision: (1) efficient ways to generate hardware signatures and exposing them to software; (2) identifying effective, non-intrusive methods across the software stack to leverage monitored variations, and (3) taking advantage of software adaptability in hardware design flows.

Summary: Cross-layer approaches to leverage on reliability techniques applied throughout various layers from *Device* all the way up to the *Application* layer are key as they are indispensable for efficiently handling reliability in future on-chip systems.

3.2 P2 – Run-Time Sense

Adaptive mechanisms in hardware and/or software can optimize the trade off between errors, energy and performance based on the feedback from fabrication-time or run-

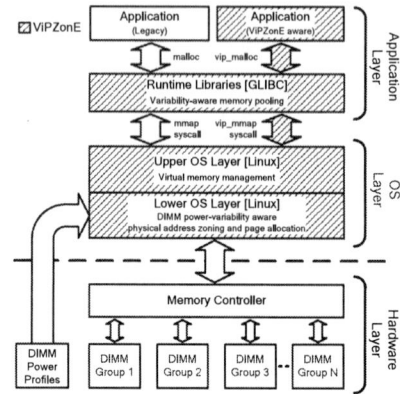

Figure 9: ViPZonE Software Architecture

time circuit performance/power/error monitors. Hardware monitoring can be done in one of the following ways.

- *Production Test.* Signatures can be measured and stored (to make them software accessible) using traditional test methods.
- *Replica Monitors.* Structures that are generally "decoupled" from the functional design (but can be aware of it) are inserted to capture hardware characteristics of interest (e.g., achievable frequency [84], leakage power [85] or aging [86]).
- *In-situ Monitors.* Replica monitors though cheap and non-intrusive cannot predict local variations. Structures that are embedded in the circuit can get better accuracy (e.g., delay fault detection flip-flop designs [64] or delay detection probes [87]).
- *Online Self-Test.* On-line self-test and diagnostics techniques allow a system to test itself concurrently during normal operation with little downtime visible to the end user (e.g., [88]).
- *Software-Based Inference.* A compiler can use the functional specification to automatically insert (in each component) software-implemented test operations to dynamically probe and determine which parts are working (e.g., [89]).

As an example, consider two alternative approaches to performance monitoring. [84] proposes a methodology to automatically synthesize *multiple* design-dependent ring oscillators as smart canary structures which can reliably predict achievable chip frequency but with margins for local variations. The key idea is to identify clusters of critical paths which behave similarly with variation and construct a canary structure per cluster. Simulation results for 45 nm benchmarks show required margins decrease as number of monitors (e.g., number of clusters) increase from 1 to 12. Early silicon measurements indicate that multiple design-dependent ring oscillators can reduce needed delay margin beyond the predicted performance significantly ($\sim 25\%$) compared to generic inverter based ROs. To further improve the prediction (albeit at a higher overhead), [87] proposes a novel in situ timing slack monitoring methodology, namely SlackProbe, which monitors in situ timing slacks of carefully selected circuit nets (i.e. including path endpoints as well as internal nodes). By shifting the monitoring points to the internal nodes, the slack monitors can cover more critical paths with the cost of some additional timing margin, thereby reducing monitoring overheads by over $10\times$ compared to monitoring all timing endpoints.

Summary: In future we will see an increased number of on-chip sensors. Designing and managing a hierarchy of heterogeneous sensors to get accuracy with minimum overheads is important. Systems will integrate a variety of fabrication-time test and runtime monitors to enable adaptation. Challenges lie in designing low-overhead monitoring schemes and exploring their tradeoffs with design margining approaches.

700

3.3 P3 – Run-Time Adaptation and Self-Organization

Run-time adaption will be another key component to increase the reliability of future on-chip systems. If only design-time techniques would be employed the designs would have to be more and more conservative as they would have to reserve more and more resources in order to provide some infrastructure to make up for any kind of failure of components that exhibit signs of unreliability only after the system has started operating (due to, for example, aging and temperature etc.; see also Section 1). But this, in turn, would render future on-chip systems infeasible from a cost point of view (see also Section 3.6) since the inherent unreliability will further increase with future technology nodes.

A good example for why run-time adaption is necessary is the thermal problem that we will use for the rest of this section. Temperature increase when computational activity is increased. This is due to the distribution of computational workloads that can hardly be predicted at design-time: as a consequence, power densities (power per area) go up and in turn increase temperature and temperature gradients. Runtime adaption techniques can then be employed to balance the workload throughout the many core system and thus decreasing peak temperatures and gradients.

Since future on-chip system not only exhibit a higher and unpredictable behavior in terms of reliability but also become more complex – many cores system with hundreds or even thousands of cores – the question arises of how to facilitate efficient and low-overhead run-time adaptation in complex many-core systems. Our perspective is that concepts of **self-organization** will need to be employed for that purpose. A method for providing a scalable light-weight infrastructure for making distributed decisions is to employ *agent-based* systems. These allow a self-adaptive and decentralized approach that can handle global complexity by pairing local decisions with the cooperation among fellow agents thereby achieving scalability.

A fully-distributed agent-based approach is presented in [90] where each core has exactly one *local agent*. Each agent negotiates with it's neighboring agent, i.e. those of adjacent cores, in order to distribute a global power budget based on performance constraints and measured temperatures. Through multiple iterations the budget can be propagated over larger regions of the many-core system with the goal of balancing temperature and thereby reducing peak temperatures, thermal gradients, and the related aging effects. While scalability is key to cope with complexity, a centralized approach with unlimited resources will still result in a higher quality thermal management solution due to its global view of the system. This is illustrated in Figure 10 where both the central reactive approach – i.e., one where thermal management is performed when a specified temperature threshold is reached – and the central proactive approach – i.e., one where future temperatures are predicted and thermal management is performed before a threshold is reached – obtain lower peak temperatures than the fully-distributed approach.

Summary: Run-time adaption is indispensable for managing reliability and self-organization is a promising means to get there since the complexity of future systems grows rapidly with hundreds or even thousand cores on a chip.

3.4 P4 – Exploiting Application Resilience

The importance of application-specific reliability was already identified in Section 2.4. There are many applications, i.e., all applications dealing with imprecise or incomplete information have an inherent error resilience, which can pave the path for cost-efficient error resilience. However fine-grained reliability methods are inevitable. The inherent error-resilience of these applications is usually limited to the data processing. Therefore, the control flow is not error resilient. Furthermore, often the application resilience is

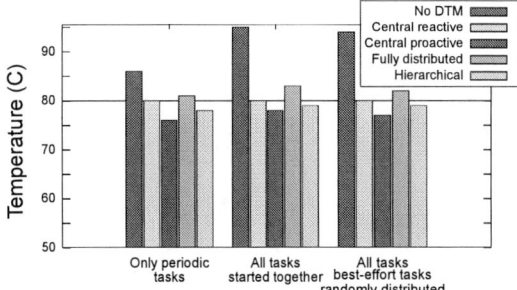

Figure 10: Run-Time Adaption: Peak temperature reduction through means of self-organization [91]

dynamic and depends, e.g., on statistical properties of data content or environmental conditions. So, fine-grained reliability mechanisms are mandatory that can be activated and deactivated and application information has to be forwarded to different layers to guide the reliability.

Wireless application is a good example to illustrate applications' inherent resilience. Communication systems are designed to recover the originally transmitted data sequence in spite of errors that are induced during transmission over a noisy wireless channel. These channel errors are corrected by advanced forward error corrections techniques in today's communication systems. Errors induced by hardware can now be considered as yet another error source in communication systems and can be treated similarly to channel errors, i.e., the algorithms used for correcting channel errors can partially be re-used to recover hardware errors. Under bad channel conditions, i.e. large channel noise, hardware errors with low error rates are not at all visible at the system since the overall system behavior is dominated by the noisy channel errors. Figure 11 taken from [62] shows the system behavior of a WiMAX decoder: Frame error rate (FER) is traced over channel noise (E_b/N_0). The dashed blue graph shows the correct behavior of the decoder without hardware errors. Such behavior is specified in the WiMAX standard. The red graph shows the decoder's behavior when hardware errors are injected in the memories (RAMs), the communication network (NW) and the functional units (CFUs). Obviously the decoder does not work at all. Fine-grained reliability mechanisms were used for the various building blocks exploiting the probabilistic behavior of the channel decoding algorithms to make the decoder error-resilient. Techniques like triplication with majority voting for the control parts, sign bit protection only for calculation etc. were applied. The green graph shows the behavior of the error-resilient decoder. The small deviation from the correct behavior is within quantization noise and the estimation error of the channel condition. The area overhead for the error resilient decoder is only about 20%. If the MTBF is quite high (purple graph), protection of the controller only is sufficient yielding an overhead of about 5%.

Summary: Application-specific reliability should be exploited whenever possible. However it requires profound application knowledge, fine-grained reliability methods and resilience propagation throughout different layers.

3.5 P5 – CAD for Reliability

Reliability needs to be considered as a major constraints in the CAD flow. As an example, here we present a timing analysis flow for consideration of various runtime variation effects. In the nanometer era, short-term and long-term runtime variations due to workload-dependent voltage and temperature variations as well as transistor aging introduce remarkable uncertainty and unpredictability to nanoscale VLSI designs.

BTI gradually shifts the threshold voltage of the transistors [92], thereby increases the circuit delay over the time (*long-term effects*). BTI-induced timing degradation highly depends on operating context parameters including supply

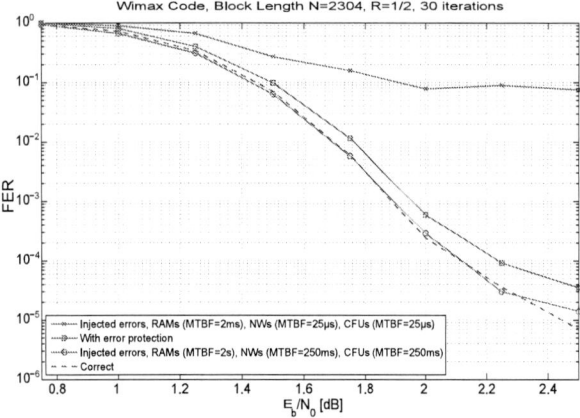

Figure 11: Reliable WiMAX Decoder as an example for exploiting application-inherent resilience

voltage, temperature, and input patterns [93] which are non-uniform and significantly vary from gate to gate and time to time [94]. Runtime variations of the temperature and the supply voltage due to workload signature have emerged as another dominant factor in circuit delay uncertainty (*short-term effects*) [95].

A timing analysis framework should accurately capture the combined effects of various workload-dependent runtime variations happening at different time scales, making the link between system-level runtime effects and circuit-level design. Such a framework should be fully integrated with existing commercial EDA toolset, making it scalable for very large designs. The key idea is to provide a link between workload-dependent runtime variations and timing analysis during design. This could be achieved by performing system-level workload profiling and accurately projecting the effects at the circuit-level delay estimation through circuit-level simulation of system-level profiles by considering the combined effects of voltage and temperature together with BTI effect during the timing analysis. This method significantly reduces the inaccuracy of the safety margin when each source of timing variation is considered independently.

Table 2: Relative circuit delay increase (w.r.t. **-V-T-BTI**) due to runtime variations ($Error = (Holistic - Margin)/Holistic$)

Circuit	#cells	+V-T -BTI	-V+T -BTI	-V-T +BTI	Margin	Holistic	Error	Time [s]
b17	27k	6%	6%	6%	17%	22%	25%	654
b18	88k	9%	6%	6%	21%	25%	16%	978
b19	165k	8%	7%	8%	22%	32%	29%	1071
b22	40k	9%	7%	6%	17%	23%	24%	658
dsp	42k	2%	6%	17%	25%	28%	13%	444
leon2	995k	3%	9%	11%	23%	29%	20%	3245
leon3mp	721k	3%	7%	15%	25%	30%	18%	2458
vga_lcd	114k	5%	16%	21%	41%	48%	14%	1059
risc	61k	10%	10%	13%	33%	39%	16%	754
des_perf	84k	2%	19%	19%	40%	44%	10%	1060
average							17%	

Table 2 shows the circuit relative delay increase (w.r.t. *-V-T-BTI*) due to runtime variations with different schemes in which different sources of runtime variations are considered in isolation. Early results show that independent analysis of temperature, voltage, and BTI leads to, one average, 17% inaccuracy in circuit delay estimation.

Summary: Reliability constraints need to be supported in an integrated CAD flow for design optimization and analysis. In this flow, the interdependence of various reliability effects must accurately be considered to avoid pessimistic overdesign.

3.6 P6 – Scaling and Cost for Resilience

We have observed that scaling to future technologies introduces a variety of reliability and resilience issues that have

to be dealt with (see Section 1); and we have proposed that these issues *must* be dealt with in a cross-layer fashion (see Section 3.1) where other levels of design, e.g. the circuit, operating system and application levels, become involved in correcting resilience faults. It is clear that moving to more advanced nodes brings both opportunity as well as cost, and one must ensure that the net result is positive in order to make such a move viable. In this section, we propose a simple qualitative model[3] for the impact and cost of resilience to put this issue into perspective.

In creating this simple cost model, we will make the following assumptions:

- The size and cost per wafer are constant. This is not a very accurate assumption since advanced technologies generally require additional manufacturing steps (e.g. double patterning), but it is optimistic in the sense that it makes the best possible case for scaling. We will denote the cost per wafer by C_W and the size (area) of the wafer by A_W.
- The price of our product chip is constant. Meaning we are only making the choice as to which technology to make the chip in. We will denote the price per chip by P_C, and the area of the chip by A_C.
- The cost and area of design are proportional to the number of transistor. More transistors require more area and more design resources, and we assume that –for a given basic architecture– the two are linearly related.

The number of chips per wafer is $\frac{A_W}{A_C}$. The number of *working* chips per wafer incorporates the yield Y which is assumed to be an increasing function of *volume*. The cost of manufacturing a working chip can thus be expressed as:

$$C_C = \frac{C_W A_C}{A_W Y}. \qquad (4)$$

We denote the cost of designing the chip by C_D. If we make N chips, the profit per chip at a given technology node will be:

$$P_C - C_C - \frac{C_D}{N}. \qquad (5)$$

Let us consider how the profit per chip will change if we were to use a more advanced technology node. A more advanced node would scale the area by a factor of S. But a more advanced node may require the addition of on-chip resources to handle resilience (an example might be the addition of parity protection or on-chip monitors). Assume the more advanced node requires an increase in device count by a factor R.

The area of the chip in this new technology node will be $A_C RS$, thus the new profit per chip would be:

$$profit = P_C - C_C RS - R\frac{C_D}{N}. \qquad (6)$$

Summary: From this result we can observe that:

1. Large volume chips benefit from scaling only if the cost of resilience is smaller than the *shrink factor*.
2. Small volume chips cannot afford as much resilience as high volume chips.
3. Reducing the cost of designing-in resilience (e.g. by creating automatic flows that can do this with little designer interaction) can make a significant difference for low volume chips.

3.7 P7 – Reliability for CPSoCs

The future of complex chips will evolve into so-called Cyber-physical Systems-on-Chip (CPSoCs). These CPSoCs will be sensor/actuator-rich, with multiple layers of hardware and software self-adaptive loops, comprising sense-de-

---

[3]Though this model is very simple and not accounting for various cost factors like masking, it indeed is sufficient to make some general statements with respect to the cost of resilience as scaling continues.

cide-actuate intelligence embedded into the fabric to manage a wide range of application-level, environmental, and platform (manufacturing) variability. The role of hardware/software virtualization and sensor fusion becomes very important in reasoning about, managing, and effectively achieving specific design goals, be they individual or combinations of metrics such as energy, reliability, security, thermal, etc.

Summary: In future we will see entire Cyber-Physical Systems integrated on a chip. This will increase complexity and pose additional constraints to reliability.

4. CONCLUSIONS

Reliability turns into *the* major design constraint for on-chip systems as scaling moves on. After almost a decade of research, the problems are far from being solved. Whereas early reliability-ensuring/increasing techniques almost exclusively focused on physical and device level, we have observed that higher abstraction levels (e.g., all the way up to the application software) have and increasingly will contribute its share to enhance reliability.

The first part of the paper gave an overview of the currently most prominent reliability concerns like variability, aging, temperature effects and soft errors. The second part provided (a certainly not comprehensive) state-of-the-art of techniques at hardware-level, software-level and application-level. The third part provided our perspectives obtained through larger-scale ongoing projects on reliability where the authors from industry and academia are involved in. Our perspectives reach from already practiced techniques like exploiting cross-layer techniques to visionary ones like future on-chip systems will evolve into complex Cyber-physical Systems-on-Chip.

Though our seven perspectives are neither orthogonal nor comprehensive, we believe that they will cover a large part of the challenges that we are going to face within the next decade of technology scaling.

5. ACKNOWLEDGMENTS

This work is supported in parts by the German Research Foundation (DFG) as part of the priority program "Dependable Embedded Systems" (SPP 1500 – http://spp1500.itec.kit.edu) and the NSF Variability Expedition (Grants CCF-1029030, CCF-1029783).

References

[1] "Int'l technology roadmap for semiconductors", 2009.
[2] L. Wanner *et al.*, "Hardware variability-aware duty cycling for embedded sensors", *IEEE Transactions on Very Large Scale Integration Systems*, vol. PP, no. 99, 2012.
[3] T.-B. Chan, R. Ghaida, and P. Gupta, "Electrical modeling of lithographic imperfections", in *International Conference on VLSI Design*, 2010, pp. 423–428.
[4] M. Gottscho, A. Kagalwalla, and P. Gupta, "Power variability in contemporary drams", *IEEE Embedded Systems Letters*, vol. 4, no. 2, pp. 37–40, 2012.
[5] W. Wang *et al.*, "Compact modeling and simulation of circuit reliability for 65nm cmos technology", *IEEE Trans. on Device and Materials Reliability*, vol. 7, no. 4, 2007.
[6] R. Zheng *et al.*, "Circuit aging prediction for low-power operation", in *Custom Integr. Circ. Conf.*, 2009, pp. 427–430.
[7] J. B. Velamala *et al.*, "Aging statistics based on trapping/detrapping: Silicon evidence, modeling and long-term prediction", in *Int'l Reliability Physics Symposium*, 2012.
[8] M. White, "Microelectronics reliability : physics-of-failure based modeling and lifetime evaluation", *JPL Publ.*, 2008.
[9] H. Nguyen *et al.*, "Effect of thermal gradients on the electro-migration life-time in power electronics", *IEEE Int'l Symposium on Reliability Physics*, pp. 619–620, 2004.
[10] J. B. Bernstein *et al.*, "Electronic circuit reliability modeling", *Microelect. Reliab.*, vol. 46, no. 12, pp. 1957–1979, 2006.
[11] J. Srinivasan *et al.*, "The case for lifetime reliability-aware microprocessors", *SIGARCH Computer Archrchitecture News*, pp. 276–287, 2004.
[12] S. Jahinuzzaman, M. Sharifkhani, and M. Sachdev, "An analytical model for soft error critical charge of nanometric SRAMs", *IEEE Transactions on Very Large Scale Integration Systems*, vol. 17, no. 9, pp. 1187–1195, 2009.

[13] A. Dixit and A. Wood, "The impact of new technology on soft error rates", in *IEEE Int. Reliab. Physics Symp.*, 2011.
[14] J. L. Autran *et al.*, "Altitude and underground real-time SER characterization of CMOS 65nm SRAM", in *European Conf. Radiation and its Effects Components and Systems (RADECS)*, 2008, pp. 519–524.
[15] S. Mitra and E. J. McCluskey, "Word voter: A new voter design for triple modular redundant systems", in *VTS*, 2000, pp. 465–470.
[16] V. Izosimov *et al.*, "Synthesis of fault-tolerant embedded systems with checkpointing and replication", in *Intl. Work. on Electronic Design, Test and Appl.*, 2006, pp. 440–447.
[17] R. Lyions and W. Vanderkulk, "The use of triple modular redundancy to improve computer reliability", *IBM Journal of Research*, vol. 6, no. 2, pp. 200–209, 1962.
[18] R. Vadlamani *et al.*, "Multicore soft error rate stabilization using adaptive dual modular redundancy", in *Conference on Design, Automation and Test in Europe*, 2010, pp. 27–32.
[19] D. Ernst *et al.*, "Razor: circuit-level correction of timing errors for low-power operation", *IEEE Micro*, vol. 24, no. 6, pp. 10–20, 2004.
[20] S. S. Mukherjee *et al.*, "A systematic methodology to compute the architectural vulnerability factors for a high-performance microprocessor", in *IEEE/ACM Int'l Symp. on Microarchitecture*, 2003, pp. 29–40.
[21] C. Kong, "A hardware overview of the NonStop Himalaya K10000 server", *Tandem Syst. Review*, vol. 10, no. 1, 1994.
[22] T. Calin, M. Nicolaidis, and R. Velazco, "Upset hardened memory design for submicron CMOS technology", *IEEE Trans. Nuclear Science*, vol. 43, no. 6, pp. 2874–2878, 1996.
[23] N. Seifert *et al.*, "On the radiation-induced soft error performance of hardened sequential elements in advanced bulk CMOS technologies", in *IEEE International Reliability Physics Symposium (IRPS)*, 2010, pp. 188–197.
[24] S. Mitra *et al.*, "Robust system design with built-in soft-error resilience", *IEEE Computer*, vol. 38, no. 2, pp. 43–52, 2005.
[25] J. Furuta *et al.*, "A 65nm bistable cross-coupled dual modular redundancy Flip-Flop capable of protecting soft errors on the C-element", in *IEEE Sym. on VLSIC*, 2010, pp. 123–124.
[26] O. Ruano, J. A. Maestro, and P. Reviriego, "A methodology for automatic insertion of selective TMR in digital circuits affected by SEUs", *IEEE Transactions on Nuclear Science*, pp. 2091–2102, 2009.
[27] T. Balen *et al.*, "A self-checking scheme to mitigate single event upset effects in SRAM-based FPAAs", *IEEE Trans. on Nuclear Science*, vol. 56, no. 4, pp. 1950–1957, 2009.
[28] S. Ghosh, P. Ndai, and K. Roy, "A novel low overhead fault tolerant Kogge-Stone adder using adaptive clocking", in *DATE*, 2008, pp. 366–371.
[29] H. Ando *et al.*, "A 1.3-GHz fifth-generation SPARC64 microprocessor", *IEEE Journal of Solid-State Circuits*, vol. 38, no. 11, pp. 1896–1905, 2003.
[30] C. Chen and M. Hsiao, "Error-correcting codes for semiconductor memory applications: A state-of-the-art review", *IBM Journal of Research and Development*, vol. 28, pp. 124–134, 1984.
[31] M. Nicolaidis, "Desig for soft error mitigation", *IEEE Trans. Device and Materials Reliability (TDMR)*, 2005.
[32] E. H. Neto *et al.*, "Using bulk built-in current sensors to detect soft errors", *IEEE Micro*, vol. 26, no. 5, pp. 10–18, 2006.
[33] T. Austin, "DIVA: A reliable substrate for deep submicron microarchitecture design", in *International Symposium on Microarchitecture (MICRO)*, 1999, pp. 196–207.
[34] A. Meixner, M. Bauer, and D. Sorin, "Argus: Low-cost, comprehensive error detection in simple cores", in *Int'l Symposium on Microarchitecture (MICRO)*, 2007.
[35] A. Drake *et al.*, "A distributed critical-path timing monitor for a 65nm high-performance microprocessor", in *Solid-State Circuits Conference, 2007. ISSCC 2007. Digest of Technical Papers. IEEE International*. IEEE, 2007, pp. 398–399.
[36] J. Tschanz *et al.*, "A 45nm resilient and adaptive microprocessor core for dynamic variation tolerance", in *IEEE Int'l Solid-State Circuits Conference*, 2010, pp. 282–283.
[37] S. Kumar, C. Kim, and S. Sapatnekar, "Adaptive techniques for overcoming performance degradation due to aging in digital circuits", in *Asia and South Pacific Design Automation Conference*, 2009, pp. 284–289.
[38] E. Mintarno *et al.*, "Self-tuning for maximized lifetime energy-efficiency in the presence of circuit aging", *IEEE Transactions on Computer-Aided Design of Integrated Circuits and Systems*, vol. 30, no. 5, pp. 760–773, 2011.
[39] K. Constantinides *et al.*, "Software-based online detection of hardware defects mechanisms, architectural support, and evaluation", in *IEEE/ACM International Symposium on Mi-

703

croarchitecture, 2007, pp. 97–108.

[40] Y. Li *et al.*, "Concurrent autonomous self-test for uncore components in system-on-chips", in *VLSI Test Symposium (VTS)*, 2010, pp. 232–237.

[41] A. K. Coskun *et al.*, "Temperature-aware MPSoC scheduling for reducing hot spots and gradients", *Asia and South Pacific Design Automation Conf.*, pp. 49–54, 2008.

[42] K. Skadron *et al.*, "Temperature-aware microarchitecture", in *Int'l Symp. on Computer Architecture*, 2003, pp. 2–13.

[43] J. Donald and M. Martonosi, "Techniques for multicore thermal management: Classification and new exploration", in *Int'l Symp. on Computer Architecture*, 2006, pp. 78–88.

[44] T. Ebi, H. Amrouch, and J. Henkel, "COOL: control-based optimization of load-balancing for thermal behavior", *Int'l Conf. CODES+ISSS*, pp. 255–264, 2012.

[45] G. Reis, J. Chang, and D. August, "Automatic instruction-level software-only recovery", *IEEE Micro*, vol. 27, no. 1, pp. 36–47, 2007.

[46] J. Hu *et al.*, "Compiler-directed instruction duplication for soft error detection", in *DATE*, 2005, pp. 1056–1057.

[47] G. A. Reis *et al.*, "Software-controlled fault tolerance", *ACM Trans. Archit. Code Optim.*, vol. 2, no. 4, pp. 366–396, 2005.

[48] N. Oh, P. Shirvani, and E. McCluskey, "Error detection by duplicated instructions in super-scalar processors", *IEEE Transactions on Reliability*, vol. 51, no. 1, pp. 63–75, 2002.

[49] J. Hu, S. Wang, and S. Ziavras, "In-register duplication: Exploiting narrow-width value for improving register file reliability", in *International Conference on Dependable Systems and Networks*, 2006, pp. 281–290.

[50] R. Venkatasubramanian, J. Hayes, and B. Murray, "Low-cost on-line fault detection using control flow assertions", in *IEEE On-Line Testing Symposium*, 2003, pp. 137–143.

[51] P. Shirvani, N. Saxena, and E. McCluskey, "Software-implemented EDAC protection against SEUs", *IEEE Trans. on Reliability*, vol. 49, no. 3, pp. 273–284, 2000.

[52] J. Yan and W. Zhang, "Compiler-guided register reliability improvement against soft errors", in *ACM international conference on Embedded software*, 2005, pp. 203–209.

[53] A. Masrur *et al.*, "Schedulability analysis for processors with aging-aware autonomic frequency scaling", in *IEEE International Conference on Embedded and Real-Time Computing Systems and Applications*, Seoul, Korea, 2012.

[54] F. Ahmed *et al.*, "Wearout-aware compiler-directed register assignment for embedded systems", in *International Symposium on Quality Electronic Design*, 2012, pp. 33–40.

[55] J. Xu, Q. Tan, and R. Shen, "The instruction scheduling for soft errors based on data flow analysis", in *IEEE Pacific Rim Int'l Symp. on Dependable Comp.*, 2009, pp. 372–378.

[56] A. Benso *et al.*, "A C/C++ source-to-source compiler for dependable applications", in *International Conference on Dependable Systems and Networks*, 2000, pp. 71–78.

[57] X. Fu *et al.*, "Optimizing issue queue reliability to soft errors on simultaneous multithreaded architectures", in *Int'l Conference on Parallel Processing*, 2008, pp. 190–197.

[58] E. Borin *et al.*, "Software-based transparent and comprehensive control-flow error detection", in *Int'l Symposium on Code Generation and Optimization*, 2006, pp. 333–345.

[59] J. George *et al.*, "Probabilistic arithmetic and energy efficient embedded signal processing", in *Int'l Conf. on Compilers, Architect. and Synthesis for Emb. Syst.*, 2006, pp. 158–168.

[60] N. R. Shanbhag *et al.*, "Stochastic computation", in *ACM/IEEE Design Automation Conf.*, 2010, pp. 859–864.

[61] L. Leem *et al.*, "ERSA: Error resilient system architecture for probabilistic applications", in *Int'l Conf. on Design, Automation and Test in Europe*, 2010, pp. 1560–1565.

[62] M. May, M. Alles, and N. Wehn, "A case study in reliability-aware design: A resilient LDPC code decoder", in *Design, Automation and Test in Europe*, 2008, pp. 456–461.

[63] A. Khajeh *et al.*, "Cross-layer co-exploration of exploiting error resilience for video over wireless applications", in *Workshop ESTIMedia*, 2008, pp. 13–18.

[64] D. Ernst *et al.*, "Razor: A low-power pipeline based on circuit-level timing speculation", in *International Symposium on Microarchitecture*, 2003, pp. 7–18.

[65] T. Austin *et al.*, "Making typical silicon matter with razor", *IEEE Computer*, vol. 37, no. 3, pp. 57–65, 2004.

[66] C. Brehm *et al.*, "A case study on error resilient architectures for wireless communication", in *Architecture of Computing Systems*, 2012, pp. 13–24.

[67] "NSF Variability Expedition", http://variability.org.

[68] A. Biswas *et al.*, "Computing architectural vulnerability factors for address-based structures", in *International Symposium on Computer Architecture*, 2005, pp. 532–543.

[69] B. Zhang, W. Wang, and M. Orshansky, "FASER: Fast analysis of soft error susceptibility for cell-based designs", in *Int'l Symp. on Quality Elec. Design*, 2006, pp. 755–760.

[70] S. Krishnaswamy *et al.*, "Accurate reliability evaluation and enhancement via probabilistic transfer matrices", in *Design, Autom. and Test in Europe Conference*, 2005, pp. 282–287.

[71] G. Norman *et al.*, "Evaluating the reliability of NAND multiplexing with PRISM", *IEEE Trans. on CAD of Integr. Circuits and Sys.*, vol. 24, no. 10, pp. 1629–1637, 2005.

[72] D. Bhaduri *et al.*, "Scalable techniques and tools for reliability analysis of large circuits", in *IEEE International Conference on VLSI Design*, 2007, pp. 705–710.

[73] G. Asadi and M. B. Tahoori, "An analytical approach for soft error rate estimation in digital circuits", in *Int'l Symp. on Circuits and Systems*, 2005, pp. 2991–2994.

[74] L. Chen and M. B. Tahoori, "An efficient probability framework for error propagation and correlation estimation", in *IEEE Int'l On-Line Testing Symposium*, 2012, pp. 170–175.

[75] S. Rehman *et al.*, "Reliable software for unreliable hardware: Embedded code generation aiming at reliability", in *IEEE International Conference on Hardware/Software Codesign and System Synthesis*, 2011, pp. 237–246.

[76] S. Rehman, M. Shafique, and J. Henkel, "Instruction scheduling for reliability-aware compilation", in *IEEE Design Automation Conference*, 2012, pp. 1288–1296.

[77] J. Yan and W. Zhang, "Compiler-guided register reliability improvement against soft errors", in *International Conference on Embedded Software*, 2005, pp. 203–209.

[78] P. Gupta et al., "Underdesigned and opportunistic computing in presence of hardware variability", *IEEE Transactions on CAD*, 2013.

[79] V. K. Chippa *et al.*, "Scalable effort hardware design: exploiting algorithmic resilience for energy efficiency", in *Design Automation Conference*, 2010, pp. 555–560.

[80] P. Kulkarni, P. Gupta, and M. D. Ercegovac, "Trading accuracy for power in a multiplier architecture", *Journal of Low Power Electronics*, vol. 7, no. 4, pp. 490–501, 2011.

[81] A. Kahng *et al.*, "Designing a processor from the ground up to allow voltage/reliability tradeoffs", in *IEEE Int'l Symp. on High Perf. Computer Architecture*, 2010, pp. 1–11.

[82] A. Pant, P. Gupta, and M. van der Schaar, "AppAdapt: Opportunistic application adaptation in presence of hardware variation", *IEEE Transactions on Very Large Scale Integration Systems*, vol. 20, no. 11, pp. 1986–1996, 2012.

[83] L. A. D. Bathen *et al.*, "ViPZonE: OS-level memory variability-driven physical address zoning for energy savings", in *CODES+ISSS*, 2012, pp. 33–42.

[84] T.-B. Chan *et al.*, "DDRO: A novel performance monitoring methodology based on design-dependent ring oscillators", in *Int'l Symp. on Quality Electronic Design*, 2012, pp. 633–640.

[85] C. Kim *et al.*, "An on-die CMOS leakage current sensor for measuring process variation in sub-90nm generations", in *International Conference on Integrated Circuit Design and Technology*, 2005, pp. 221–222.

[86] P. Singh *et al.*, "Dynamic NBTI management using a 45 nm multi-degradation sensor", *IEEE Transactions on Circuits and Systems I*, vol. 58, no. 9, pp. 2026–2037, 2011.

[87] L. Lai *et al.*, "SlackProbe: A low overhead in situ on-line timing slack monitoring methodology", in *DATE*, 2013.

[88] H. Inoue, Y. Li, and S. Mitra, "VAST: Virtualization-assisted concurrent autonomous self-test", in *IEEE International Test Conference*, 2008, pp. 1–10.

[89] S. Sahoo *et al.*, "Using likely program invariants to detect hardware errors", in *IEEE International Conference on Dependable Systems and Networks*, 2008, pp. 70–79.

[90] T. Ebi, M. A. Al Faruque, and J. Henkel, "TAPE: thermal-aware agent-based power economy for multi/many-core architectures", in *International Conference on Computer-Aided Design*, 2009, pp. 302–309.

[91] T. Ebi *et al.*, "Economic learning for thermal-aware power budgeting in many-core architectures", in *Int'l Conf. on HW/SW Codes. and Sys. Synth.*, 2011, pp. 189–196.

[92] S. Bhardwaj *et al.*, "Predictive modeling of the NBTI effect for reliable design", in *IEEE Custom Integrated Circuits Conference*, 2006, pp. 189–192.

[93] S. Krishnappa, H. Singh, and H. Mahmoodi, "Incorporating effects of process, voltage, and temperature variation in BTI model for circuit design", in *IEEE Latin American Symp. on Circuits and Systems*, 2010, pp. 236–239.

[94] W. Wang *et al.*, "The impact of NBTI on the performance of combinational and sequential circuits", in *Design Automation Conference*, 2007, pp. 364–369.

[95] P. Li, "Critical path analysis considering temperature, power supply variations and temperature induced leakage", in *Int'l Symposium on Quality Electronic Design*, 2006, pp. 254–259.

A Layout-based Approach for Multiple Event Transient Analysis

Mojtaba Ebrahimi
Karlsruhe Institute of Technology
Karlsruhe, Germany
mojtaba.ebrahimi@kit.edu

Hossein Asadi
Sharif University of Technology
Tehran, Iran
asadi@sharif.edu

Mehdi B. Tahoori
Karlsruhe Institute of Technology
Karlsruhe, Germany
mehdi.tahoori@kit.edu

ABSTRACT

With the emerging nanoscale CMOS technology, *Multiple Event Transients* (METs) originated from radiation strikes are expected to become more frequent than *Single Event Transients* (SETs). In this paper, a fast and accurate layout-based *Soft Error Rate* (SER) estimation technique with consideration of both SET and MET fault models is proposed. Unlike previous techniques in which the adjacent MET sites are obtained from logic-level netlist, we perform a comprehensive layout analysis to extract MET adjacent cells. It is shown that layout-based technique is the only effective solution for identification of adjacent cells as netlist-based techniques significantly underestimate the overall SER.

Categories and Subject Descriptors

B.8.1 [**Reliability, Testing, and Fault-Tolerance**]

General Terms

Reliability

Keywords

Transient errors, Soft errors, Error propagation

1. INTRODUCTION

By downscaling of transistor feature size and operating voltage together with increased device count per chip, the susceptibility of circuits to soft errors has significantly increased in the past years [1, 2]. In the absence of protection mechanisms, the system *Soft Error Rate* (SER) will grow in direct proportion to the number of cells in the design [2, 3].

Transient errors caused by a single particle strike in combinational gates and sequential elements (i.e., memory cells, latches and flip-flops) are called *Single Event Transient* (SET) and *Single Event Upset* (SEU), respectively. SET and SEU fault models have been widely studied over the recent years [4, 5, 6, 7, 8, 9]. With smaller device geometries in nanoscale technologies, it is very likely that a high energy particle strike affects several adjacent cells in a circuit resulting in *Multiple Event Transients* (MET) in combinational gates or *Multiple Bit Upsets* (MBU) in sequential elements [10, 11, 12, 13].

In previous technology nodes, soft errors had a considerable effect only in sequential elements which were significantly mitigated by means of *Error Correcting Codes* (ECCs) [14, 15, 16] and built-in soft error tolerance [17, 18, 19, 20]. Recent experiments reveal that the contribution of combinational gates is considerable in nanoscale technologies [21, 22]. Furthermore, it is claimed that a remarkable fraction of particle strikes results in MET [13, 23]. In order to cost-effectively mitigate soft errors in the presence of both SETs and METs, their impacts at the layout and logic levels must be accurately modeled.

The analytical techniques presented in [24] and [25] address the MET fault model in logic circuits. The *Error Probability Propagation* (EPP)-based technique presented in [24] propagates the error probabilities from the error sites towards primary outputs and sequential elements. The technique presented in [25] is based on *Boolean Decision Diagrams* (BDDs) and provides more accuracy at the expense of runtime, compared to the earlier method [7]. The major shortcoming of these techniques in estimating SER due to METs is that they use logic-level netlist for identification of MET error sites, neglecting layout-level adjacency of error sites. Such assumption can significantly underestimate the circuit SER as we will experimentally demonstrate later in this paper. Additionally, the distribution of affected error sites and the number of affected cells completely depends on the layout-level details and cannot be extracted from the netlist. For example, our experiments show that a considerable fraction of particles simultaneously affects both combinational gates and flip-flops, which is completely ignored in previous techniques. Ignoring such cases can further increase the SER inaccuracy. In this paper, occurrence of multiple transient errors at sequential elements, combinational gates, or combination of them is called *Multiple Transients* (MT).

In this paper, a fast and accurate technique is proposed for SER estimation considering the effect of MTs at the circuit layout. In the proposed technique, the surface affected by a particle in combinational and sequential logic is estimated with oval shapes obtained from available MBU patterns in memory arrays. Considering MBU patterns occurrence probability, ovals are randomly placed at different locations inside the circuit and the list of affected cells are extracted. Then, logic level analysis with multiple error propagation is performed to obtain the overall SER. Analysis of ISCAS'89 and ITC'99 benchmark circuits reveal that less than 10% of the netlist adjacent cells are physically adjacent in the layout. Also, more than 60% of physically adjacent cells are not adjacent in the netlist. Experimental results show that neglecting layout adjacency can cause inaccuracy up to 36.04% in the circuit overall SER.

The rest of this paper is organized as follows: Section 2 motivates for layout-based MT modeling and investigates the validity of the netlist-based adjacency assumption used in previous techniques. In Section 3, the proposed layout-based approach is explained. Section 4 demonstrates the experimental results, and finally, the conclusion is given in Section 5.

2. MOTIVATION FOR LAYOUT-BASED MT ANALYSIS

An important step in MT analysis is identifying physically adjacent cells as error sites. Although, the layout of the circuit is necessary for accurate identification of adjacent cells, the previous MT analysis techniques presented in [24] and [25] employ some heuristic approaches to extract the list of adjacent cells from the netlist. In these techniques, four categories including a gate and its fan-in (GFI), a gate and its fan-out (GFO), common fan-ins of a gate (CFI), and common fan-out of a gate (CFO) are considered as adjacent nodes for MT error sites. Fig. 1 shows several examples of netlist-based adjacencies. In these techniques, a gate is first selected as primary error site and then its MT pair is randomly selected among its netlist adjacent cells.

In order to check the accuracy and layout-relevance of this model, i.e., extraction of adjacent cells from the logic netlist, the layout of several circuits selected from ISCAS'89 and ITC'99 benchmarks have been comprehensively analyzed (the details of this analysis framework is provided in Section 4). In this regard, all possible adjacency pairs for different netlist adjacency categories are first extracted from the netlist and then the physical adjacency of each pair in the circuit layout is investigated. Since the order of adjacency pairs is not important in this investigation, both GFI and GFO categories are equal and are assumed as a single category called GFI/GFO. In the example given in Figure 1, E is the fanout of gate B and gate B is a fanin of gate E. As result, pair (B,E) belongs to both GFI and GFO categories. Figure 2 shows the results of this experiment. As it can be seen, on average, only less than 10% of netlist adjacent pairs are also adjacent in the circuit layout. Also, the probability of physical adjacency of GFI/GFO category is much higher than that of CFI and CFO categories.

There is another experiment conducted in which all layout adjacencies on the circuit layout are first extracted and then for each physical adjacency, its netlist adjacency category is investigated (Figure 3). It can be inferred from Figure 3 that more than 60% of the physical adjacencies do not belong to the previously defined netlist adjacency categories.

From these two experiments, it becomes clear that there is no statistically meaningful correlation between netlist and

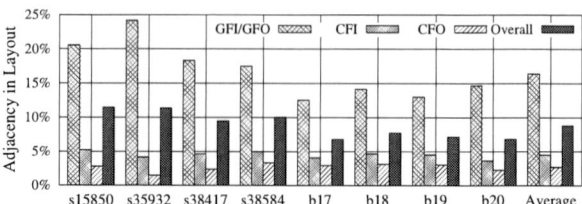

Figure 2: Relevance of different adjacency scenarios considered in the netlist to layout adjacency

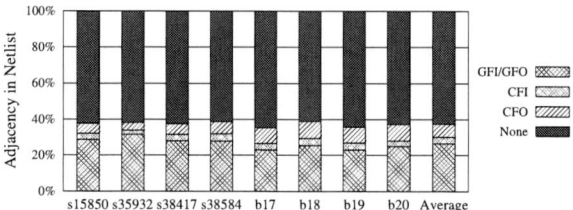

Figure 3: Relevance of extracted adjacencies from layout to netlist-level MT error models

layout adjacencies. This means that for accurate MT analysis, netlist-level analysis and MT error site abstraction is not sufficient and layout-level analysis must be performed. To address this issue, we propose a fast and accurate layout-based MT modeling technique and then the computed layout-based SERs are compared with the SERs obtained by the previously proposed netlist-based techniques.

3. PROPOSED LAYOUT-BASED MT MODELING

In this section, the proposed layout-based SER estimation approach with consideration of MT fault model is explained. This approach has two main steps, layout-based MT error site extraction and multiple error propagation at logic-level.

3.1 Layout-based MT Error Site Extraction

3.1.1 MT Error Site Extraction Using MBU Analysis

The first step for accurate MT modeling is extracting physically adjacent error sites from the circuit layout. This requires to have MT patterns projected in the circuit layout and their occurrence probability. However, due to the observation limits of logic, especially combinational blocks, and their irregularity as compared to memory arrays (such as SRAMs), to the best of our knowledge, no field results about MT patterns on combinational and sequential logic has been reported in the literature. Since memory arrays are much more regular and dense than logic structures and also have a full observability, the affected area can be accurately estimated. In this work, we try to use available MBU patterns in memory arrays for identification of MT error sites in logic circuits. In this method, the surface affected by a strike in a memory array is first extracted and then all logic cells covered by a similar surface are assumed to be MT error sites.

For this purpose, the affected area for each MBU pattern is first extracted. Predominant MBU patterns in memory arrays have been comprehensively studied using neutron beam-based accelerated SER estimation [10, 11]. The num-

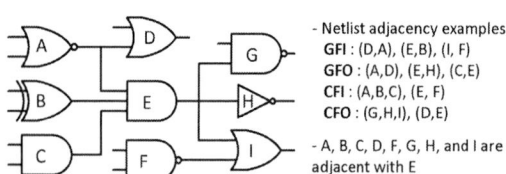

Figure 1: Examples of different adjacency scenarios considered in the netlist

- Netlist adjacency examples
GFI : (D,A), (E,B), (I, F)
GFO : (A,D), (E,H), (C,E)
CFI : (A,B,C), (E, F)
CFO : (G,H,I), (D,E)

- A, B, C, D, F, G, H, and I are adjacent with E

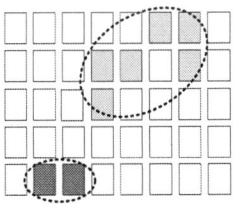
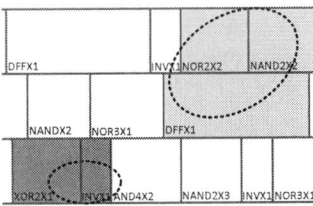

a) MBU patterns in SRAM   b) Equivalent MT error sites in circuit layout

Figure 4: Extraction of MT error sites from existing MBU patterns

ber of affected bits by a single strike depends on several parameters including particle type and its energy, strike angle, cell type, cell size, and output load [12]. Given memory cell dimension as well as vertical and horizontal distance between adjacent cells, for each MBU pattern the surface affected by the strike can be calculated. Most of the MBU patterns, especially MBUs with more than eight bits, can be effectively covered by an oval surface. As a result, in our approach as shown in Figure 4, all cells affected by an MBU are first surrounded by an oval. Then, during SER estimation the same oval is transferred to random locations inside the logic layout and all cells affected by the MT are listed as error sites.

3.1.2 Library Characterization

In the circuit layout, only a subset of cells which have an overlap with the oval surface are considered as error sites. A cell has overlap with an oval surface if at least one of the *sensitive zones* to soft errors (N-diffusion and P-diffusion) falls within the surface. In fact, when a particle strikes a cell, it causes the additional charge to be collected in the diffusion parts of the transistor which in turn disturbs the normal operation of the transistor. The charge collected to the diffusion parts which are connected to VDD and GND pins is evacuated and does not affect the circuit behaviour. The other parts of the diffusion can be disturbed by collection of additional charge. Figure 5 shows how sensitive zones for a NAND gate layout are extracted based on this explanation. Using this approach, all cells inside the technology library are characterized and sensitive zones coordinations are extracted.

3.1.3 Overall Flow

The main steps of the proposed layout-based SER estimation are summarized in Algorithm 1. Initially, the list

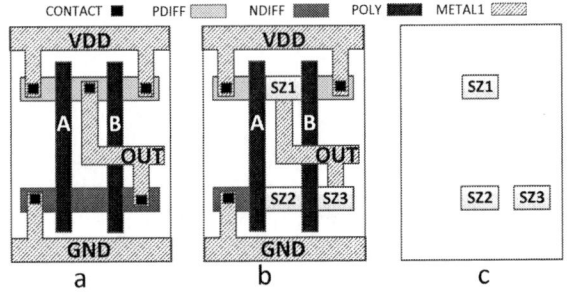

Figure 5: Sensitive Zone (SZ) extraction for a NAND gate: a) cell layout b) identification of diffusion parts which are not connected to supply voltage c) sensitive zones (adapted from [26])

Algorithm 1: Layout-based SER estimation

1 Extract a surrounding oval for each MBU pattern
2 Extract sensitive zones of each cell by library characterization
3 Divide layout into smaller grids and extract the list of cells in each grid
4 **while** *sampling error < predefined value AND number of cells covered by at least one MT < 99.9%* **do**
5   Randomly select an oval based on their occurrence probability
6   Place the oval in a random location on the layout
7   Find the list of grids which have overlap with the oval
8   Search overlapped grids cell lists and construct overlapped cells list
9   Remove cells without overlapped sensitive zones from cell list
10   Mark all cells in overlapped cell list as covered by an MT
11   Propagate MT at logic-level and calculate failure probability
12 **end**
13 Report average failure probability

---

of ovals and their occurrence probability are extracted from existing MBU patterns (line 1) and then technology library is characterized for identification of sensitive zones (line 2). These two steps are performed once in advance and their results are used for all circuits to be analyzed in the same technology and library settings.

Due to large number of cells and sensitive zones inside industrial-size circuits, a hierarchical approach is employed to minimize the time needed for identification of error sites affected by an MT (line 3). In this approach, the entire layout area is divided into smaller grids and the list of cells inside each grid is extracted. During SER estimation, instead of searching among large number of cells, in the first step, for each grid it is checked whether it has an overlap with the oval surface and the list of layout grids overlapped by the oval surface is extracted. Then, the list of overlapped cells is extracted by investigation of cells inside overlapped layout grids. At the end, those cells which have no overlapping sensitive zone with the oval surface are eliminated from the target cell list. The remaining cells will be used as candidate MT fault sites (line 7-9).

3.2 Multiple Error Propagation

In the layout-based MT error sites extraction, it is quite possible that flips-flops and combinational gates are simultaneously affected by an MT. This issue is completely ignored in the previous work. In such scenarios, a transient pulse is produced at the output of affected cells while the value stored in flip-flops are logically inverted. To handle such cases, a fast and accurate propagation mechanism is required.

During multiple error propagation, unified treatment of three timing masking factors, i.e. logical, electrical, and latching-window, is essential for accurate SER estimation [7, 4]. The four-value logic $(0, 1, 0^e, 1^e)$ [24] which offers an effective trade-off between runtime and accuracy, is employed to compute the logical masking factor. This technique can efficiently handle the effect of single error propagation in re-convergent paths as well as the effect of multiple errors propagation in convergent paths. For electrical masking factor, the equation-based transfer function presented in [27] is adopted. This techniques models a transient pulse using a trapezoidal model and can accurately compute the electrical attenuation. Latching-window masking model is based on the well-known and widely used equation presented in [28].

It is quite possible that an MT does not propagate to the primary outputs in the first cycle, rather it may latched in some flip-flops and propagates to the primary outputs in the subsequent cycles. Experimental results in [29] reveal that failure probability saturates in few cycles (normally less than 10 cycles) after error occurrence. Multi-cycle error propagation is also taken into account in our framework.

While propagating errors along combinational gates, all three masking factors should be considered in the first cycle. At the end of the first cycle, the error is captured in the flip-flops or eliminated (masked) from the system. In the subsequent cycles, only logical masking factor can prevent the error from propagation and as a result, the other masking factors are ignored. In contrast, when a strike affects a flip-flop, in all cycles including the first cycle, only logical masking is taken into account. In case of simultaneous error occurrence at both logic gates and flip-flops, all three masking factors have been considered in the first cycle. However, the width of the output transient pulse of erroneous flip-flops is set to be equal to the clock period to overcome the latching-window masking factor for such errors.

3.3 Combined Layout and Logic SER Analysis

Since there are lots of oval shapes and each oval can be placed in different locations of the circuit layout, there are infinite MT scenarios even for very small circuits. Therefore, we use a Monte-Carlo simulation-based approach to extract the overall SER of the circuit with respect to MT. In this approach, in each iteration, based on the MBU patterns occurrence probability, one of them is randomly selected and its corresponding oval will be placed in a random location on the layout. After extracting the list of affected cells using the hierarchical approach, the errors are propagated from the error site and the failure probability for this MT is calculated. This continues until reaching a predefined accuracy level. An equation to compute the sampling error of Monte-Carlo simulations with respect to the number of iterations and current failure probability is provided in [30]. The MT analysis terminates when the sampling error is less than the predefined value and the number of cells contributed by at least one MT exceeds 99.9%. The second condition is used to become sure that most of the cells in the layout has been considered during SER estimation.

4. EXPERIMENTAL RESULTS

Using the proposed layout-based MT error site extraction and combined combinational and sequential multiple error propagation at logic-level, we have performed an extensive analysis on the impact of particle energy on the MT error sites. Also, the impact of netlist adjacency assumption on the overall SER of the circuit is investigated.

4.1 Work Flow

In order to show the scalability of the proposed approach, we have evaluated largest available benchmark circuits in IS-CAS'89 and ITC'99 benchmarks suites. For each benchmark, the HDL description of the circuit is first synthesized using a Synopsis Design Compiler [31] with respect to Nangate 45 nm library. Then the layout of the netlist is extracted using SoC Encounter [32]. In the experiments, the layout is divided into $30 \times 30 \ \mu m^2$ grids. Each grid includes around 800 cells.

The MBU patterns for particles with 22, 37, 95, and 144 Mev provided by [10] are used during the layout-based MT extraction. This information is given to our layout-based SER estimation to calculate the overall SER of the circuit

according to Algorithm 1. In our framework, SER estimation analysis terminates when the maximum inaccuracy of the Monte-Carlo is less than 0.5%.

The failure in this paper defined according to [29] as the probability of propagation from error sites to primary outputs during first few cycles after error occurrence. The error is propagated for 10 cycles and during error propagation, all three masking factors have been considered.

4.2 MBU Patterns and MT Error Sites

As mentioned earlier, in order to extract MT error sites, the area affected by MBU patterns are first extracted and then a surrender oval for each MBU pattern is constructed. These ovals are used for identification of MT error sites. For this purpose, detailed information about different MBU patterns in a memory array is necessary for identification of MT error sites.

Radaelli et. al. [10] have reported a detailed information about predominant MBU patterns in a 150 nm technology SRAM device and their occurrence probability for particles with 22, 47, 95, and 144 Mev energy. Considering the SRAM cell dimensions, the area affected by each MBU surrounding oval can be accurately estimated. For these cases, the oval shapes and their occurrence probability (same as the occurrence probability of the corresponding MBU pattern) are computed. Table 1 shows the average area affected by each particle energy obtained by wighted averaging of oval surfaces based on their occurrence probability.

Particle Energy (Mev)	Average Affected Area (μm^2)
22	1.178
47	1.902
95	2.903
144	4.613

Table 1: Average area affected with different particle energies

The area affected by a particle strike is mostly a function of particle energy, while the strength of the transient pulse mostly depends on other parameters such as diffusion volume (width, length, depth) and load capacitance [2]. As a result, the affected area information acquired for a 150 nm SRAM technology can also be used for the logic area affected by a particle strike with the same energy in the 45 nm technology. Please note that although the affected area remains constant, however, due to the technology downscaling, the number of affected cells increases in smaller technologies.

By randomly locating these ovals on the circuit layout according to their occurrence probability, different combinations of affected combinational gates and flip-flops are extracted and identified as MT error sites. Figure 6 shows the occurrence probability of different gate/flip-flop combinations for particle strikes with 22, 47, 95, and 144 Mev energy. As it is expected, by increasing the particle strike energy, the occurrence probabilities of SET and SEU decreases significantly and MT becomes predominant. In previous netlist-based techniques, it is assumed that a particle strike leads to either MET on combinational logic or MBU on sequential cells. Also, the number and type of affected cells was a function of the particle energy not layout. However, the results shown in Figure 6 clearly indicate that 1) both combinational and sequential cells can be affected by a single particle strike. 2) the number, type, and combination of affected cells depend on the layout structure as well.

708

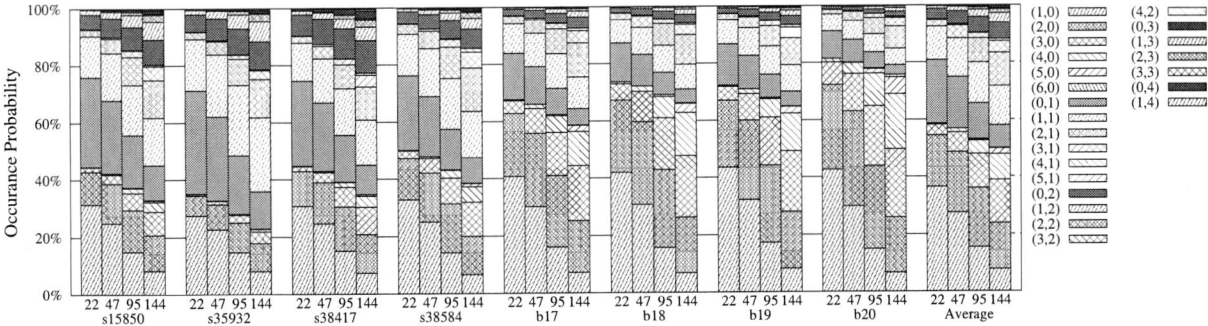

Figure 6: Different gate and flip-flops combinations affected by particle strikes with 22, 47, 95, and 144 Mev energy. For example, (3,2) means three gates and two flip-flops are affected by the particle

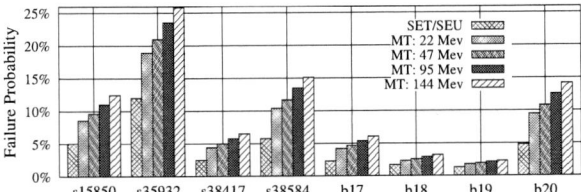

Figure 7: Failure probabilities for different particle energies

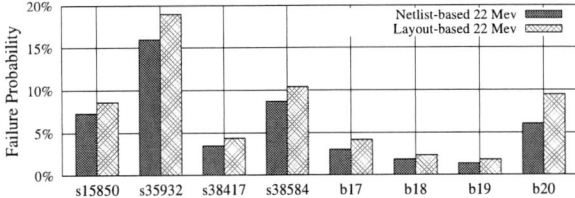

Figure 8: Comparison of overall SER obtained by netlist-based and layout-based approaches for particle with 22 Mev energy

4.3 Impact of SET/SEU Versus MT Model on Overall SER

In order to show the importance of MTs, overall SERs extracted for particles with 22, 47, 95, and 144 Mev energy are compared with the case that the simple SET/SEU model is considered (Figure 7). In case of SET/SEU, a single error is injected in each gate/flip-flop and the average of failure probabilities of all cells are reported as the circuit failure probability. All error sites are extracted from layout and propagated using the propagation method explained in Section 3.2. The results shown in Figure 7 also reveal that the circuit SER does not linearly increase with the particle energy. As an example, on average, the SER in the presence of 47 Mev particles is only 1.15X greater than when considering 22 Mev particles. This can be also explained by the number of affected cells for different particle energies reported in Figure 6.

4.4 Impact of netlist-adjacency assumption on SER

In order to investigate the effect of netlist adjacency on the overall SER, we have implemented a netlist-based approach. The error propagation method of the netlist-based approach is similar to the one explained in Section 3.2. Although different combinations of affected cells and their occurrence probability are unknown at the netlist-level, in order to have a fair comparison, the same occurrence probabilities is also used in the netlist-based approach. Figure 8 reports the failure probability obtained by both netlist- and layout-based approaches. As it can be seen, the netlist-based approach always underestimates the overall failure probability. Our analysis reveals that there are two main reasons for this underestimation. First, when there are simultaneous errors at the outputs of the CFI pairs, these transient pulses reach at the same time to the inputs of the fanout gate. In this

case, the propagated transients are either completely masked, attenuated, or at least converged to one transient pulse. However, as shown in Figure 1, most of CFI pairs are not physically adjacent in the layout. Second, the forward cones of netlist adjacency pairs are highly overlapped and share similar paths from error sites to the circuit outputs. This can increase the chance that several errors are masked due to one kind of masking (e.g., logical masking in a common gate in the forward cone of both error sites). When MT occurs in error sites which have non-overlapping forward cones, they are independent and the probability of masking is much lower.

On average, netlist-based MT analysis has inaccuracy of 22.34% which is as high as 36.04% for b20 benchmark. Please note that since the information regarding the occurrence probability of different affected gate/flip-flops does not exist at the netlist level, the inaccuracy of those techniques could be even higher.

4.5 Runtime

In order to evaluate the scalability of the proposed layout-based approach to estimate the SER of large circuits, the runtime of layout-based and netlist-based approaches are reported in Table 2. All experiments are done on a workstation with Intel Xeon E5540 2.53GHz and 16GB RAM. As it can be seen, the runtime of layout-based technique is comparable to that of netlist-based technique, i.e., only 15.7% increase in runtime is imposed for layout analysis and extracting MT error sites. The low runtime of the proposed technique is due to the hierarchical layout analysis employed in the proposed approach as detailed in Section 3.1.

5. CONCLUSIONS

In this paper, a fast and accurate layout-based SER estimation technique was presented. Unlike previous techniques

Benchmark	Elements		Runtime [Seconds]		Overhead
	Gates	Flip-flops	Netlist-based	Proposed	
s15850	2,418	513	3.45	3.84	11.3%
s35932	5,328	1,728	11.38	13.28	16.7%
s38417	6,935	1,564	19.23	21.02	9.3%
s38584	7,958	1,275	38.11	47.62	25.0%
b17	17,971	1,317	117.84	132.97	12.8%
b18	54,151	3,020	676.04	782.34	15.7%
b19	99,907	6,042	1836.44	2157.75	17.6%
b20	12,631	430	478.59	562.68	17.6%
Average					**15.7%**

Table 2: Comparison of runtime between netlist- and layout-based approaches

in which the adjacent MT sites are obtained from logic-level netlist, we perform a comprehensive layout analysis to extract MT error sites. It is shown that the layout-based approach is the only viable solution for identification of adjacent cells as netlist-based techniques underestimate the overall SER of the circuit by up to 36.04%. Experimental results show that the layout-based approach has modest runtime and it is scalable for industrial-size circuits.

6. ACKNOWLEDGMENTS

This work was partly supported by the German Research Foundation (DFG) as part of the national focal program "Dependable Embedded Systems" (SPP-1500, http://spp1500.ira.uka.de).

7. REFERENCES

[1] R.C. Baumann. Radiation-induced soft errors in advanced semiconductor technologies. *IEEE Transactions on Device and Materials Reliability*, 5(3):305–316, 2005.

[2] E. Ibe, H. Taniguchi, Y. Yahagi, K.-i. Shimbo, and T. Toba. Impact of scaling on neutron-induced soft error in srams from a 250 nm to a 22 nm design rule. *IEEE Transactions on Electron Devices*, 57(7):1527–1538, 2010.

[3] A. Dixit and A. Wood. The impact of new technology on soft error rates. In *Proceedings of the International Reliability Physics Symposium*, pages 5B.4.1–5B.4.7, 2011.

[4] H. Asadi, M.B. Tahoori, M. Fazeli, and S.G. Miremadi. Efficient algorithms to accurately compute derating factors of digital circuits. *Microelectronics Reliability*, pages 1215–1226, 2012.

[5] M. Ebrahimi, L. Chen, H. Asadi, and M. B. Tahoori. Class: Combined logic and architectural soft error sensitivity analysis. In *Asia and South Pacific Design Automation Conference*, pages 601–607, 2013.

[6] H. K. Lee, K. Lilja, M. Bounasser, P. Relangi, I. R. Linscott, U. S. Inan, and S. Mitra. Leap: Layout design through error-aware transistor positioning for soft-error resilient sequential cell design. In *IEEE International Reliability Physics Symposium*, pages 203–212, 2010.

[7] N. Miskov-Zivanov and D. Marculescu. Modeling and optimization for soft-error reliability of sequential circuits. *IEEE Transaction on CAD of Integrated Circuits and Systems*, 27(5):803–816, 2008.

[8] S. Kiamehr, M. Ebrahimi, F. Firouzi, and M. B Tahoori. Chip-level modeling and analysis of electrical masking of soft errors. In *VLSI Test Symposium*, pages –, 2013.

[9] A. Mohammadi, M. Ebrahimi, A. Ejlali, and S. G. Miremadi. SCFIT: A FPGA-based Fault Injection Technique for SEU Fault Model. In *Proceedings of Design, Automation and Test in Europe Conference*, pages 586–589, 2012.

[10] D. Radaelli, H. Puchner, S. Wong, and S. Daniel. Investigation of multi-bit upsets in a 150 nm technology SRAM device. *IEEE Transactions on Nuclear Science*, 52(6):2433–2437, 2005.

[11] D. Giot, P. Roche, G. Gasiot, J.-L. Autran, and R. Harboe-Sorensen. Heavy ion testing and 3-d simulations of multiple cell upset in 65 nm standard srams. *IEEE Transactions on Nuclear Science*, 55(4):2048–2054, 2008.

[12] J.A. Maestro and P. Reviriego. Study of the effects of mbus on the reliability of a 150 nm sram device. In *IEEE Design Automation Conference*, pages 930–935, 2008.

[13] R. Harada, Y. Mitsuyama, M. Hashimoto, and T. Onoye. Neutron induced single event multiple transients with voltage scaling and body biasing. In *IEEE International Reliability Physics Symposium*, pages 3C.4.1–3C.4.5, 2011.

[14] Z. Ming, X.L. Yi, L. Chang, and Z.J. Wei. Reliability of memories protected by multi-bit error correction codes against mbus. *IEEE Transactions on Nuclear Science*, 58(1):289–295, 2011.

[15] P. Reviriego, J.A. Maestro, and C. Cervantes. Reliability analysis of memories suffering multiple bit upsets. *IEEE Transactions on Device and Materials Reliability*, 7(4):592–601, 2007.

[16] D. Rossi, N. Timoncini, M. Spica, and C. Metra. Error correcting code analysis for cache memory high reliability and performance. In *Design, Automation & Test in Europe Conference*, pages 1–6, 2011.

[17] M. Zhang, S. Mitra, T. M. Mak, N. Seifert, N. J. Wang, Q. Shi, K. S. Kim, N. R. Shanbhag, and S. J. Patel. Sequential element design with built-in soft error resilience. *IEEE Transactions on Very Large Scale Integration Systems*, 14(12):1368–1378, 2006.

[18] R. Yamamoto, C. Hamanaka, J. Furuta, K. Kobayashi, and H. Onodera. An area-efficient 65 nm radiation-hard dual-modular flip-flop to avoid multiple cell upsets. *IEEE Transactions on Nuclear Science*, 58(6):3053–3059, 2011.

[19] S. Lin, H. Yang, and R. Luo. A new family of sequential elements with built-in soft error tolerance for dual-vdd systems. *IEEE Transactions on Very Large Scale Integration Systems*, 16(10):1372–1384, 2008.

[20] M. Omana, D. Rossi, and C. Metra. High-performance robust latches. *IEEE Transactions on Computers*, 59(11):1455–1465, 2010.

[21] B. Gill, N. Seifert, and V. Zia. Comparison of alpha-particle and neutron-induced combinational and sequential logic error rates at the 32nm technology node. In *IEEE International Reliability Physics Symposium*, pages 199–205, 2009.

[22] N. N. Mahatme, S. Jagannathan, T. D. Loveless, L. W. Massengill, B. L. Bhuva, S.-J. Wen, and R. Wong. Comparison of combinational and sequential error rates for a deep submicron process. *IEEE Transactions on Nuclear Science*, 58(6):2719–2725, 2011.

[23] D. Rossi, M. Omana, F. Toma, and C. Metra. Multiple transient faults in logic: an issue for next generation ics? In *Proceedings of International Symposium on Defect and Fault Tolerance in VLSI Systems*, pages 352–360, 2005.

[24] M. Fazeli, S.N. Ahmadian, S.G. Miremadi, H. Asadi, and M.B. Tahoori. Soft error rate estimation of digital circuits in the presence of multiple event transients. In *Design, Automation Test in Europe Conference Exhibition*, pages 1–6, 2011.

[25] N. Miskov-Zivanov and D. Marculescu. Multiple transient faults in combinational and sequential circuits: A systematic approach. *IEEE Transactions on Computer-Aided Design of Integrated Circuits and Systems*, 29(10):1614–1627, 2010.

[26] C. Rusu, A. Bougerol, L. Anghel, C. Weulerse, N. Buard, S. Benhammadi, N. Renaud, G. Hubert, F. Wrobel, T. Carriere, and R. Gaillard. Multiple event transient induced by nuclear reactions in cmos logic cells. In *IEEE International On-Line Testing Symposium*, pages 137–145, 2007.

[27] R. Rajaraman, J. S. Kim, N. Vijaykrishnan, Y. Xie, and M. J. Irwin. Seat-la: A soft error analysis tool for combinational logic. In *Proceedings of the 19th International Conference on VLSI Design*, pages –, 2006.

[28] P. Shivakumar, M. Kistler, S. W. Keckler, D. Burger, and L. Alvisi. Modeling the effect of technology trends on the soft error rate of combinational logic. In *Proceedings of the International Conference on Dependable Systems and Networks*, pages 23–26, 2002.

[29] M. Fazeli, S. G. Miremadi, H. Asadi, and S. N. Ahmadian. A fast and accurate multi-cycle soft error rate estimation approach to resilient embedded systems design. In *Proceedings of the International Conference on Dependable Systems and Networks*, pages 131–140, 2010.

[30] R. Leveugle, A. Calvez, P. Maistri, and P. Vanhauwaert. Statistical fault injection: Quantified error and confidence. In *Proceedings of Design, Automation and Test in Europe Conference*, pages 502–506, 2009.

[31] Synopsys Design Compiler. http://www.synopsys.com, 2012.

[32] SOC Encounter. http://www.cadence.com, 2012.

Quantitative Evaluation of Soft Error Injection Techniques for Robust System Design

Hyungmin Cho[1], Shahrzad Mirkhani[3], Chen-Yong Cher[4], Jacob A. Abraham[3], Subhasish Mitra[1,2]

[1]Department of EE and [2]Department of CS   [3] Computer Engineering Research Center   [4]IBM T. J. Watson Research Center,
Stanford University, Stanford, CA, USA   The University of Texas at Austin, Austin, TX, USA   Yorktown Heights, NY, USA

Abstract

Choosing the correct error injection technique is of primary importance in simulation-based design and evaluation of robust systems that are resilient to soft errors. Many low-level (e.g., flip-flop-level) error injection techniques are generally used for small systems due to long execution times and significant memory requirements. High-level error injections at the architecture or memory levels are generally fast but can be inaccurate. Unfortunately, there exists very little research literature on quantitative analysis of the inaccuracies associated with high-level error injection techniques. In this paper, we use simulation and emulation results to understand the accuracy trade-offs associated with a variety of high-level error injection techniques. A detailed analysis of error propagation explains the causes of high degrees of inaccuracies associated with error injection techniques at higher levels of abstraction.

1. Introduction

Radiation-induced transient errors (*soft errors*) are important for digital systems in advanced technologies [Sanda 08, Seifert 12]. While radiation testing is generally successful in evaluating the soft error resilience of digital systems [Michalak 12, Sanda 08], simulation-based error injection techniques are also important at various stages of robust system design:

1. To analyze the application-level effects of errors.
2. To quantify the effectiveness of various error resilience techniques.
3. To make decisions about the set of error resilience techniques that must be used to protect a given design from soft errors.

In this paper, we focus on soft errors in flip-flops (*flip-flop soft errors*) because design techniques to protect them are generally expensive. Coding techniques are routinely used for protecting on-chip memories. Combinational logic circuits are significantly less susceptible to soft errors [Seifert 12].

Error injections at the flip-flop level can accurately capture the effects of flip-flop soft errors [Ramachandran 08, Sanda 08]. Such injections generally rely on slow RTL simulations, sometimes with hardware acceleration or emulation. In contrast, error injections at higher abstraction layers are much less precise but can be very fast. Error injection techniques at high-level abstraction layers are important for understanding the application-level erroneous behaviors. The following abstraction layers are widely used (Table 1):

1. Software-level: Errors are often represented as single bit-flips in software variables, e.g., [Chen 08, Yim 10].
2. Architecture-level: Errors are injected into states defined by the Instruction Set Architecture (ISA). Single bit-flips in the architectural registers are often used, e.g., [Feng 10, Pattabiraman 11, Racunas 07, Zhang 10].
3. Micro-architecture-level: Error injection is performed using a detailed micro-architectural simulator. The internal states of the simulator are targets of error injection. Depending on the simulator implementation, such error injection targets may not always correspond to actual hardware components.

Permission to make digital or hard copies of all or part of this work for personal or classroom use is granted without fee provided that copies are not made or distributed for profit or commercial advantage and that copies bear this notice and the full citation on the first page. To copy otherwise, to republish, to post on servers or to redistribute to lists, requires prior specific permission and/or a fee.

Table 1. Error injection techniques at various layers of abstraction.

Abstraction layer	Example platform	Performance (cycles / sec.)
Software	x86 processor [Yim 10]	3×10^9
Architecture	TSIM SPARC simulator [Leon]	6×10^7
Micro-architecture	gem5 simulator [Gem5]	3×10^6 (Simple CPU) 2×10^5 (Detailed CPU)
Flip-flop	IVM Alpha-like processor RTL simulation [Maniatakos 11]	6×10^2
Flip-flop (Emulation)	OpenSPARC T1 FPGA emulation [Pellegrini 12]	10^7

To select a suitable error injection technique that meets the target accuracy and execution time requirements, one must address the following two essential aspects:

1. **Quantify** the inaccuracies of results obtained from error injection at various layers of abstraction.
2. **Analyze** the sources of these inaccuracies.

There exist very few publications that quantitatively address these questions. [Rimen 94] discusses inaccuracies of pin-level error injections that model only a small fraction (9-12%) of flip-flop errors. The authors also conclude that results from flip-flop error injections can match those from a special register-level error injection technique which injects register-level effects profiled from flip-flop error injections. However, the comparison is limited to a simple processor for which 80% of all flip-flops belong to user-visible registers. [Rebaudengo 02] reports that register-file error injections can result in up to 400% inaccuracies for a version of the LEON processor [Leon]. However, the authors do not quantitatively analyze the sources of such inaccuracies.

[Arlat 03, Sanda 08] compare results obtained from actual error injection experiments with those obtained from error injection simulations / emulation. [Arlat 03] compares radiation, pin-level stuck-at faults, and electromagnetic interference experiments versus error injection simulations into program code and data. Although the authors report that error injections into program code and data can generate (erroneous) outcomes similar to actual error experiments, the observed rates of these outcomes can differ from actual error experiments. [Sanda 08] compares radiation tests versus flip-flop soft error injection, and concludes that results obtained from flip-flop soft error injections closely match those obtained from radiation experiments.

Some publications, e.g., [Kalbarczyk 99, Kanawati 93, Maniatakos 11], profile high-level effects resulting from low-level errors, and use these high-level effects for quick error injection simulations. It has been pointed out in [Kanawati 93, Miskov-Zivanov 10, and numerous other papers on testing and high-level fault models] that a single flip-flop error can propagate through the system resulting in multiple error effects at the architecture- or software-level. However, there exists little work on systematic methodologies for deriving such high-level effects and for quantifying their effectiveness.

The lack of quantitative understanding of the accuracy trade-offs and their causes hinders progress in the evaluation and design of robust systems. This problem becomes especially pronounced in the context of cross-layer resilience, where multiple error resilience techniques from different layers of the system stack cooperate to achieve cost-effective error resilience [DeHon 10, Mitra 10]. To achieve effective cross-layer resilience, error injection techniques must be able to capture low-level details accurately, simulate real-world applications in a scalable manner, and enable correct design decisions quickly.

In this paper, we make the following contributions:

1. We quantify the inaccuracies of various error injection techniques through detailed FPGA-based emulation of the LEON3 in-order SPARC processor. Our results show high levels of inaccuracies (up to an order of magnitude) associated with high-level soft error injection techniques compared to flip-flop soft error injection.

2. In order to explain the sources of inaccuracies associated with high-level error injection techniques, we introduce a methodology which enables us to track, observe, and analyze how errors propagate through the system: from flip-flops all the way to the application outputs.

3. We explain why high-level error injection techniques directly model only a very small subset of system-level behaviors that can arise from flip-flop-level errors.

4. We demonstrate the generality of our results through RTL simulations of a super-scalar and out-of-order processor [Wang 04].

Section 2 describes our error injection methodology. Section 3 quantifies the inaccuracies associated with various error injection techniques. Section 4 analyzes the sources of inaccuracies. Section 5 concludes this paper. Appendices A-G present additional details.

2. Error Injection Methodology

We created an FPGA-based error injection system by mapping a LEON3 processor (in-order and SPARC-based [Leon]) on the BEE3 emulation system [Davis 09] using Xilinx Virtex-5 FPGAs (Appendix A). The LEON3 processor is a good choice for experimentation because the entire system, including L1 and L2 caches and the DRAM controller, can be mapped on the emulation platform. As a result, a large number of error injections can be performed. Moreover, in-order processor cores are often used in multi- and many-core SoCs [Borkar 11, Howard 10, OpenSPARC]. We designed the system with appropriate hardware support to track the propagation of injected errors through various layers of the system stack (details in Sec. 4.1). Such a setup enables us to compare various error injection techniques and analyze their inaccuracies using the same consistent environment for the same set of applications. In order to minimize the sensitivity of our results to the LEON3 architecture, we also present results using another set of error injections (fewer than LEON3) through RTL simulations of the IVM processor, which is a super-scalar and out-of-order processor [Wang 04].

2.1. Benchmark Applications

We used 11 out of 12 applications from the widely-used SPECINT 2000 benchmark suite[1]. We used the MinneSPEC workload for the input dataset [KleinOsowski 02]. The execution times of the benchmark applications range from 1.4×10^8 cycles to 3.5×10^9 cycles, and the cycle-per-instruction (CPI) values range from 1.54 to 3.36 (details in Appendix B).

2.2. Error Injection Samples

It is impossible to inject all possible error scenarios (injection target × execution cycle). Hence, for each error injection technique and each application, we collected results from 40,000 or more error injection runs. (For error propagation and tracking analysis in Sec. 4, we report results from 320,000 flip-flop error injection runs for the LEON3 processor emulation, and from 160,000 flip-flop error injection runs for the IVM processor simulation.) Our error injection system initializes all the system states to their default reset values before each error injection run. A state is set to zero if its reset value is undefined, e.g., cache arrays and DRAM.

We implicitly assume that each flip-flop has the same (raw) soft error rate, similar to [Ramachandran 08, Seifert 10, Wang 04, 07], for the ease of reporting results. For situations where this assumption may not apply, the observed results from error injections may change. However, the error injection methodology used in this paper still can be applied for those situations. To determine the confidence intervals of the error injection outcomes, we use a derivation similar to [Choi 90]

and many other publications. With a sample size of 40,000 error injections, the 95% confidence interval is smaller than ±0.1% when the observed outcome rate is 1%. This derivation implicitly assumes that the system behavior with respect to soft errors is statistically similar during the execution, i.e., the probability of a certain outcome does not change according to a particular phase of the execution.

2.3. Error Injection Techniques

We compare five error injection techniques across three abstraction layers. To ensure proper initialization, no error is injected during the first 10,000 clock cycles (warm-up period).

1. **Flip-flop:** For each error injection experiment, the content of a randomly-chosen flip-flop is flipped during a randomly-chosen clock cycle. SRAM structures, e.g., register-file and cache, are not included because they are generally protected using ECC and parity[2]. Each flip-flop in the processor is an error injection target (1,250 flip-flops for LEON3, 13,877 flip-flops for IVM). The results obtained from flip-flop error injections are treated as "ground truth" because they closely mimic actual soft error effects [Sanda 08].

2. Register-file: Error injection into software-visible registers is widely used. Depending on the target system architecture and simulator capabilities, various error injection studies use slightly different register-file error injection techniques.

2a. **Register Uniform (RegU):** For each error injection experiment, a single-bit error is injected into a randomly-chosen bit location of a randomly-chosen register during a randomly-chosen clock cycle. The target register set includes all general-purpose registers, stack pointer, and branch pointer. This type of error injection is used in [Feng 10, Zhang 10]

2b. **Register Write (RegW):** For each error injection experiment, during a randomly-chosen clock cycle, a single-bit error is injected into a randomly-chosen bit location of a register being written into during that clock cycle. If no register is being written into during that clock cycle, the error injector waits for the next instruction that writes into a register, and injects error into a randomly-chosen bit location of that register. The target register set is the same as that for RegU. This type of error injection is used in [Pattabiraman 11, Racunas 07].

3. Program Variable: Errors are injected into application software (program) variables: global data, heap, and stack.

3a. **Program Variable Uniform (VarU):** For each error injection experiment, a single-bit error is injected into a randomly-chosen bit location of a randomly-chosen program variable during a randomly-chosen clock cycle. The target program variables include all variables in memory (stack, global data, and heap) at the chosen error injection cycle; i.e., the target set includes memory locations actually used by the program. No error injection is performed into freed heap objects or intermediate variables eliminated during compilation. This type of error injection is used in [Chen 08, Yim 10].

3b. **Program Variable Write (VarW):** For each error injection experiment, during a randomly-chosen clock cycle, a single-bit error is injected into a randomly-chosen bit location of a program variable being written into during that clock cycle. If no program variable is being written into during that clock cycle, the error injector waits for the next instruction that writes into a program variable, and injects an error into a randomly-chosen bit location of that variable. The target variable set is the same as that for VarU. A similar error injection technique is used in [Chen 06, Gu 04].

3. Results: Error Injection Inaccuracies

The *outcome* of an error injection run can be categorized into one of the following categories [Sanda 08, Wang 04, 07]:

---

[1] We excluded the perlbmk application because it requires extensive file system support that was not modeled in our emulation system.

---

[2] Error injection results that include register-file errors can be derived by combining flip-flop error injection results and RegU error injection results (defined in 2.3.2a).

1. **Vanished**: The application terminates normally, and at the end of the execution, the output files and all architectural states match with those obtained from the error-free run.
2. **Application Output Not Affected (ONA):** The application terminates normally without any error indication, and, at the end of the execution, the output files from the erroneous run match those obtained from the error-free run. However, one or more remaining bits of the architectural state differ from those obtained from the error-free run.
3. **Application Output Mismatch (OMM):** The application terminates normally without any error indication. However, at the end of the execution, the output files of the application are different from those obtained from the error-free run. The remaining architectural state bits may or may not match with those of the error-free run. This category is often referred to as silent data corruption (SDC) as well [Sanda 08, Michalak 12].
4. **Unexpected Termination (UT):** The application terminates abnormally with error indication. These include error reporting interrupts, e.g., divide-by-zero, invalid instruction, or memory access violation, and application-detected errors, e.g., exit() function calls with error codes.
5. **Hang**: The application does not produce any result or does not terminate within a specified timeout limit set to 2× the nominal execution time[3].

Figure 1 compares the observed rate of each outcome category (*outcome rate*) obtained from 40,000 error injections for each of the 11 applications using each of the 5 error injection techniques. The 95% confidence intervals from the error injection samples are shown as the error bars. (We verified the integrity of our error injection results by comparing results obtained from 10 subsamples, 4,000 error injection runs each, for each error injection technique. The differences in the outcome rates across these subsamples are less than 2%.)

To compare the results for various injection techniques, consider the OMM rate of the crafty application in Fig. 1c as an example. 1.3% of flip-flop error injections result in OMM. However, high-level error injections result in different rates for the same outcome: 2.7% for RegU, 6.7% for RegW, 0.53% for VarU, and 16.1% for VarW. These differences are well beyond the confidence intervals (e.g., the 95% confidence interval for the OMM rate of the crafty application is ±0.11% for flip-flop error injection and ±0.24% for RegW error injection).

Since outcome rates vary across applications even for the same error injection technique (e.g., the OMM rate obtained from flip-flop error injection varies from 0.02% for the *parser* application to 1.78% for the *twolf* application), the inaccuracy levels are compared using *normalized outcome rates* with respect to the corresponding flip-flop error injection results. For example, for the *parser* application, the 0.04% OMM rate obtained from RegU error injection is 2× that of the corresponding flip-flop error injection result. Figure 1 also compares these normalized outcome rates using geometric means according to the following expression [Fleming 86]:

$$G.\,Mean(x, t) = \sqrt[n]{\prod_{i=1}^{n} \frac{Rate(x,t,i)}{Rate(x,flip\text{-}flop,i)}} \quad (3.1)$$

where x is an outcome, t is an error injection technique, and n is the number of benchmark applications. $Rate(x,t,i)$ is the observed rate of outcome x when error injection technique t is applied for application i.

While geometric means show overall inaccuracy levels, they may not capture how inaccuracy levels vary across various applications. For example, consider the UT outcome type in Fig. 1d. The geometric mean of normalized UT outcome rates for RegW error injection is 1.15× that of flip-flop error injection. However, the normalized UT outcome rates for RegW error injection (with respect to flip-flop error injection) vary from 0.5× to 3× across applications.

---
[3] The nominal execution time is measured on warmed-up caches (without error injection).

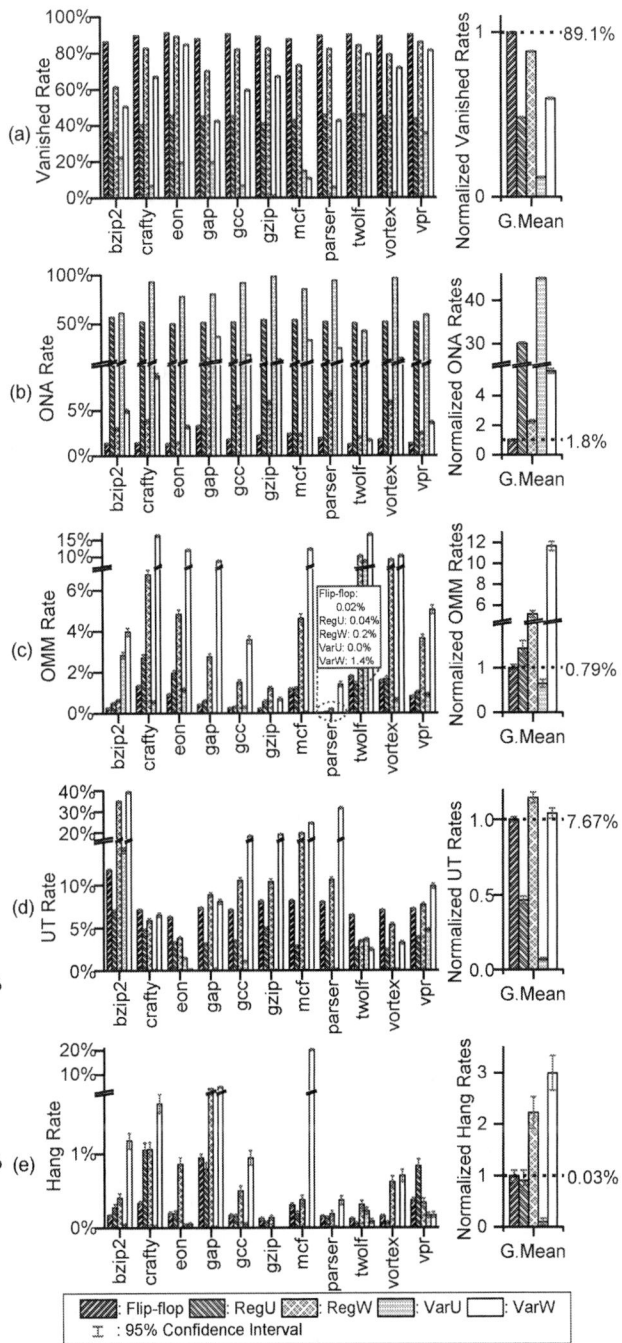

Figure 1. Comparison of the observed rate of each outcome type obtained from various error injection techniques for the LEON3 processor. (a) Vanished. (b) ONA. (c) OMM. (d) UT. (e) Hang.

Key observations from the above results are:

1. Existing high-level error injection techniques, **that inject single errors into registers or program variables**, can result in high degrees of inaccuracies by more than an order of magnitude. For example, the geometric means of normalized outcome rates obtained from VarU error injection can range from 0.07× (for the UT outcome type) to 45× (for the ONA outcome type) when compared to the corresponding flip-flop error injection results.
2. There is no single trend (e.g., always optimistic or always pessimistic) that can explain the inaccuracies associated with existing high-level injection techniques. For example, the studied high-level error injection techniques generally tend to overestimate

713

OMM outcome rates, but the RegU and VarU error injection techniques tend to underestimate UT outcome rates.

3. RegU and VarU error injection techniques result in very high degrees of ONA outcome rates. These techniques select error injection targets uniformly over the entire target space, and, hence, may inject errors into locations that may not be accessed during the rest of the application execution (resulting in ONA outcomes at the end of the execution). These situations can potentially contribute to the very high ONA outcome rates for RegU and VarU error injection techniques.

Error injection results obtained from RTL simulations of the IVM processor confirm similar trends as shown in Appendix C.

4. Results: Inaccuracy Analysis

To understand the causes of inaccuracies associated with high-level error injection techniques (Sec. 3), we track how flip-flop errors propagate and affect high-level system states. This is enabled by special hardware support that we implemented in our FPGA-based emulation system for the LEON3 processor (details in Sec 4.1). The inaccuracies generally result from the following limitations of existing high-level error models:

1. They generally inject a single error into a single location, e.g., a single register or a single program variable, during a randomly-chosen clock cycle.

2. The target of error injection does not cover all software-visible architectural states. For example, error injection targets are often limited to general-purpose registers while flip-flop errors can propagate to other software-visible states, e.g., integer condition codes and interrupt modes (details in Table 2).

4.1. Flip-flop Error Propagation Tracking

In our emulation system, we implemented two copies of the LEON3 processor core, the *erroneous core* and the *golden core* (Fig. 2). Flip-flop errors are injected only into the erroneous core. The golden core shares the same register-file, memory (L1 and L2 caches, off-FPGA DRAM), and all the other modules on the emulated system, e.g., the interrupt controller and the Ethernet controller, with the erroneous core.

We implemented a set of *propagation checkers* to detect if a flip-flop error in the erroneous core results in a mismatch with the golden core (i.e., *error propagation* occurs). The *R-checker (Register-checker)* and the *M-checker* (*Memory-checker*) detect error propagations to the register-file and memory, respectively. The address inputs, the data inputs, and the read and write enable signals to the register-file from both cores are compared by the R-checker at every cycle to detect if there is any mismatch. Since load and store operations access program variables through the memory interface, the M-checker compares the address inputs, the data inputs, and the read and write enable signals to the memory from both cores to track error propagations to program variables in the memory.

There exist software-visible architectural states other than the register-file and memory that can also affect application execution, e.g., program counter and integer condition codes (details in Table 2). The *F-checker (Flip-flop-checker)* detects any mismatch between the flip-flops storing these states in the erroneous core and the corresponding flip-flops in the golden core.

Flip-flop errors may propagate to various signals belonging to the L1 cache controller (in addition to the signals checked by the M-checker) and the interrupt controller, and affect application execution. The *S-checker (Signal-checker)* detects any mismatch in the signals to these two modules from both cores, e.g., cache flush signal to the L1 cache controller and interrupt clear signal to the interrupt controller. (Signals checked by the M-checker are not checked by the S-checker.) The rest of the modules in the emulated system, e.g., the Ethernet controller and the general-purpose timer, have memory-mapped interfaces and are accessed through load and store operations only. The address space for these modules is not accessible by user applications, and any access to these modules triggers memory access violations resulting in UT outcome. Details of various propagation types detected by the F- and S-checkers are shown in Table 2.

Suppose that, upon injection of a flip-flop error during clock cycle i, a mismatch is detected by the R-checker during clock cycle j (and no mismatch is detected by any other checker). This detection indicates the propagation of the injected flip-flop error to the register-file. If no more mismatches are detected until the end of program execution, the effect of the injected flip-flop error during clock cycle i is equivalent to the injection of register error(s) during clock cycle j. Since the golden core and the erroneous core share the same (erroneous) register-file and memory, subsequent register-file or program variable corruptions caused by <u>this error</u> are no longer tracked by any of the propagation checkers.

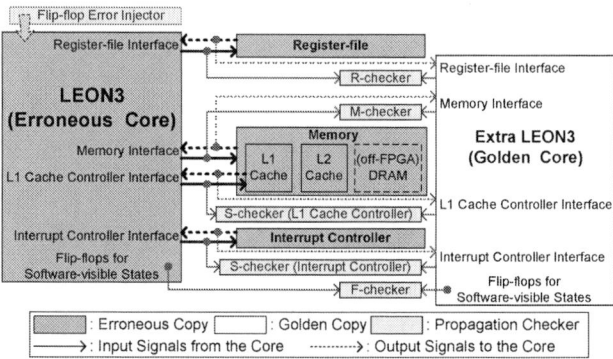

Figure 2. Error propagation detection logic on FPGA.

4.2. Flip-flop Error Propagation Types

The effect of an injected flip-flop error can be categorized into one of the following types (Fig. 3) depending on its propagation **during** (not necessarily at the end of) program execution.

1. **Masked**: The flip-flop error **never** propagates to any software-visible state at any point during the execution, i.e., no propagation checker in Fig. 2 detects a mismatch. A masked flip-flop error results in the Vanished outcome only. The converse is not necessarily true.

2. **Register-only (R-only)**: The flip-flop error propagates only to the register-file and corrupts register-file contents, i.e., only the R-checker is triggered (once or multiple times) during program execution. For example, if a flip-flop error injected during cycle i directly propagates to a register during cycle j, the R-checker is triggered during cycle j. If the erroneous flip-flop also corrupts additional flip-flops, and those additional flip-flop errors propagate to the register-file during clock cycles $k, l,...$, the R-checker can be triggered multiple times during the execution. A mismatch in the read or write enable signal or the address input is categorized as an R-only propagation type because it can be modeled as single or multiple register-file error injections.

3. **Memory-only (M-only)**: The flip-flop error propagates only to the memory and corrupts program variable contents, i.e., only the M-checker is triggered (once or multiple times) during program execution.

4. **Other-only (O-only)**: The flip-flop error results in error propagations that result in the firing of F- and S-Checkers (once or multiple times) only; i.e., R- or M-checkers are never triggered. These propagation types are summarized in Table 2.

5. **Combination**: The flip-flop error results in error propagations that cause two or more of the R-, M-, or F- and S-checkers to be triggered.

Figure 3 shows a Venn diagram with the percentages of various propagation types resulting from 3,520,000 flip-flop error injections (320,000 flip-flop error injections for each of 11 benchmark applications). High-level error injection techniques directly model only a small fraction of all flip-flop error propagation types. Register-file error injection techniques directly model ② (2.18% of all flip-flop error injections or 6.99% of all non-masked flip-flop error injections), and program variable error injection techniques directly model ③ (1.53% of all flip-flop error injections or 4.9% of all non-masked flip-flop error injections). The rest of the non-masked errors result in either the

714

other-only type ④ or the combination type ⑤-⑧ (27.4% of all flip-flop error injections or 88.1% of all non-masked flip-flop error injections).

Suppose that a flip-flop error during clock cycle i results in a propagation type ④-⑧. Existing high-level error injection techniques do not directly model this flip-flop error. However, it is possible that a single error in the register-file or program variable during **some** clock cycle c can produce the same behavior (especially an outcome at the end of program execution) as this flip-flop error. Such a high-level error may be referred to as an *equivalent error candidate*. This concept is related to fault equivalence in digital testing [McCluskey 71]. For a high-level error injection model to be accurate, there must exist many equivalent error candidates for each flip-flop error. Otherwise, it will be highly unlikely that an injected high-level error at a randomly-chosen location during a randomly-chosen clock cycle will match the effect of a given flip-flop error. A detailed analysis of equivalent error candidates is beyond the scope of this paper.

Appendix D presents an analysis of how various propagation types affect outcome rates.

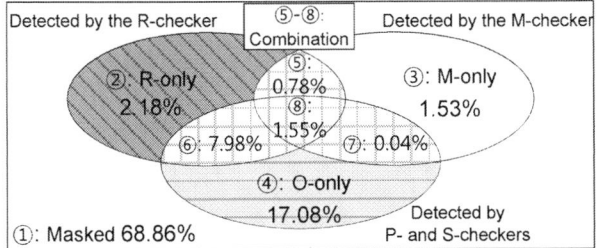

Figure 3. Flip-flop error propagation types and their observed rates.

Table 2. List of flip-flop error propagation types detected by P- and S-checkers and their observed rates[4].

Detected by	Propagation types	Rates	Processor state corruption types	Rates
S-checker	Incorrect external interrupt handling	25.2%	Integer condition codes	20.0%
S-checker	Incorrect cache flush	0.5%	Peripheral states	4.4%
F-checker	Program counter corruption	31.2%	Supervisor mode	4.6%
F-checker	Processor state corruption	59.7%	Processor internal interrupt mode	22.9%
			Register window	33.9%
			Multiplier result (Y-reg.)	25.7%

4.3. Flip-flop Error Propagation Patterns

Even if a flip-flop error results in R- or M-only propagations, it may not be directly modeled by a single-bit error in the register-file or program variable. This is because the resulting differences in the register-file or memory (*propagation pattern*) may not be the same as a single-bit error. This may happen due to the following reasons:

1. R- (or M-) checker can be triggered multiple times (**Multi-instance** propagation).
2. Even if the R- (or M-) checker is triggered only once (**Single-instance** propagation), the following cases can happen:
   2a. Multiple bits of the data input to the register-file (or memory) can mismatch (**Multi-bit** propagation).
   2b. Write address input mismatches.
   2c. Write enable signal mismatches.
   2d. Read address input mismatches.
   2e. Read enable signal mismatches.
   2f. Combinations of 2 or more of cases 2a-2e. Situations resulting in a single-bit mismatch in the data input to the register-file (or memory) together with any one or more of cases 2b-2e are also included in this category.

Table E1 in Appendix E presents the observed rates of error propagation patterns for the R-only and M-only propagation types. For most cases of the R-only propagation type, the R-checker is triggered

---
[4] The rates do not add up to exactly 100% because F- and S-checkers can be triggered multiple times for some flip-flop error injections.

only once (87.2%). Multi-instance propagation is fairly common for M-only propagation (53.2%). Single-instance propagations that may not be directly modeled by a single-bit error injection in the register-file or program variable (cases 2a-2f) account for 57.2% and 38.6% of R- and M-only propagations, respectively. The rest of the R-only (M-only) error propagations, which result in single-instance propagation and a single-bit mismatch in the data input to the register-file (memory), can be directly modeled by a single-bit error injection to the register-file (program variable). Such single-bit data-only (*SBD*) propagation pattern accounts for 30% (8.2%) of R-only (M-only) propagations. Considering the percentages of R-only and M-only propagations themselves (Fig. 3), R-only SBD accounts for only 0.65% of all flip-flop error injections (or 2.1% of all non-masked flip-flop error injections), and M-only SBD accounts for only 0.12% of all flip-flop error injections (or 0.4% of all non-masked flip-flop error injections).

4.4. Results Cross-Check

In this section, we analyze the outcomes obtained from flip-flop error injections resulting in R- and M-only SBD (Sec. 4.2 and 4.3) versus outcomes obtained from RegW and VarW error injections. We expect the inaccuracy of outcome rates resulting from RegW (VarW) error injections to be lower when compared to flip-flop error injections resulting in R-only (M-only) SBD, rather than the entire set of flip-flop error injections (Sec. 3). While this analysis can be somewhat conservative, given the discussion on equivalent error candidates in Sec. 4.2, it provides insights into inaccuracies of high-level error injections resulting from the fact that they model a small fraction of all flip-flop errors. As discussed in Sec. 3, RegU and VarU error injections may inject errors into locations not accessed during the rest of the execution. Hence, we focus on RegW and VarW error injections.

Table 3 reports the estimated inaccuracy of RegW (VarW) error injections for a given outcome x using the following expression.

$$\frac{|G.Mean(x,H) - G.Mean(x,F)|}{G.Mean(x,F)} \quad (4.1)$$

where H is the RegW or VarW error injection technique, F corresponds to flip-flop error injections being compared to (either the entire set of flip-flop error injections or flip-flop error injections resulting in R- or M-only SBD), and $G.Mean(x, t)$ is the geometric mean of normalized outcome rates obtained using (3.1). The accuracy improvements with SBD are also estimated in Table 3.

As expected, accuracies improve when high-level error injection results are compared with respect to flip-flop errors resulting in SBD: by 7.3× on average (geometric mean) for RegW and 2.28× on average for VarW. The accuracy improvements are also verified using the error injection results for the IVM processor (Appendix F). Details of cross-check results for each application and corresponding geometric means are presented in Appendix G.

Table 3. Inaccuracies of outcome rates resulting from high-level error injections for the LEON3 processor.

Outcome	RegW inaccuracies			VarW inaccuracies		
	vs. flip-flop error injections (i)	vs. flip-flop error injections resulting in R-only SBD (ii)	Accuracy improvement (i / ii)	vs. flip-flop error injections (iii)	vs. flip-flop error injections resulting in M-only SBD (iv)	Accuracy improvement (iii / iv)
Vanished	11.6%	5.87%	1.98×	40.1%	28.7%	1.40×
ONA	126%	9.38%	13.4×	468%	81.6%	5.74×
OMM	416%	12.8%	32.5×	1065%	83.7%	12.7×
UT	14.9%	22.9%	0.65×	4.18%	42.5%	0.10×
Hang	122%	3.28%	37.2×	199%	32.4%	6.14×
	Geometric mean		7.31×	Geometric mean		2.29×

5. Conclusion

Existing high-level error injection techniques, that inject single-bit errors at randomly-chosen register and memory locations during randomly-chosen clock cycles, can be highly inaccurate when compared to flip-flop error injection techniques. This paper

715

demonstrates this point for the LEON3 in-order processor core as well as for a complex out-of-order Alpha-like IVM processor core. This paper also quantifies the causes of these inaccuracies through a detailed analysis of error propagation through various layers of the system stack. The presented results provide insights that can potentially help create new classes of high-level error models with significantly higher accuracies. While the feasibility of high-level error models that are accurate for any arbitrary digital system is unclear, one can possibly use our results to derive accurate high-level error models that are tailored for certain families of digital systems.

This paper focuses on the accuracy aspects of existing high-level error injection techniques. However, depending on the application, accuracy is not necessarily a requirement. For example, an inaccurate error injection technique can be very useful as long as it is effective in driving the correct design decisions for building robust systems. One such example exists in digital system testing literature. Stuck-at faults are highly inaccurate in modeling actual manufacturing defects, but are highly effective as test metrics that drive automatic test pattern generation and design for testability techniques [McCluskey 00]. Future research must explore and quantify this aspect of high-level error models.

While we focused on soft errors in this paper, future work must address other sources of errors as well, e.g., errors induced by manufacturing and environmental variations, early-life failures, circuit aging, and voltage over-scaling.

6. Acknowledgment

Stanford researchers were supported in part by the Focus Center Research Program Gigascale Systems Research Center, the Semiconductor Technology Advanced Research Network SONIC, the National Science Foundation, the Defense Threat Reduction Agency, and the Semiconductor Research Corporation. Stanford and IBM researchers were supported in part by the Defense Advanced Research Projects Agency (Contract No. HR0011-13-C-0022). The views expressed are those of the authors and do not reflect the official policy or position of the Department of Defense or the U.S. Government. This document is Approved for Public Release, Distribution Unlimited. The authors thank Dr. Nicholas J. Wang, Dr. Farzan Fallah, and Eric Cheng for their advice, and Synopsys for supporting the simulation environment at the University of Texas at Austin.

7. References

[Arlat 03] J. Arlat et al., "Comparison of Physical and Software-Implemented Fault Injection Techniques," *IEEE Trans. Computers*, vol. 52, no. 9, pp. 1115–1133, Sept. 2003.

[Borkar 11] S. Borkar and A. A. Chien, "The Future of Microprocessors," *Commun. ACM*, vol. 54, no. 5, pp. 67–77, May 2011.

[Chen 06] G. Chen, G. Chen, M. Kandemir, N. Vijaykrishnan, and M. J. Irwin, "Object Duplication for Improving Reliability," *Proc. Asia and South Pacific Design Automation Conf.*, pp. 140–145, 2006.

[Chen 08] D. Chen, G. Jacques-Silva, Z. Kalbarczyk, R. K. Iyer, and B. Mealey, "Error Behavior Comparison of Multiple Computing Systems: A Case Study Using Linux on Pentium, Solaris on SPARC, and AIX on POWER," *Proc. IEEE Pac. Rim Intl. Symp. Dependable Computing*, pp. 339–346, 2008.

[Choi 90] G. S. Choi, R. K. Iyer, and V. A. Carreno, "Simulated Fault Injection: A Methodology to Evaluate Fault Tolerant Microprocessor Architectures," *IEEE Trans. Reliability*, vol. 39, no. 4, pp. 486–491, Oct. 1990.

[Davis 09] J. D. Davis, C. P. Thacker, and C. Chang, "BEE3: Revitalizing Computer Architecture Research," *Microsoft Tech. Rep.* MSR-TR-2009-45, 2009.

[DeHon 10] A. DeHon, H. M. Quinn, and N. P. Carter, "Vision for Cross-Layer Optimization to Address the Dual Challenges of Energy and Reliability," *Proc. Design, Automation and Test in Europe*, pp.1017–1022, 2010.

[Feng 10] S. Feng, S. Gupta, A. Ansari, and S. Mahlke, "Shoestring: Probabilistic Soft Error Reliability on the Cheap," *Proc. Intl. Conf. Architectural Support for Programming Languages and Operating Systems*, pp. 385–396, 2010.

[Fleming 86] P. J. Fleming and J. J. Wallace, "How not to lie with statistics: the correct way to summarize benchmark results," *Commun. ACM*, vol. 29, no. 3, pp. 218-221, March 1986.

[Gem5] "The gem5 Simulator System," http://www.m5sim.org

[Gu 04] W. Gu, Z. Kalbarczyk, R. K. Iyer, "Error Sensitivity of the Linux Kernel Executing on PowerPC G4 and Pentium 4 Processors," *Proc. Intl. Conf. on Dependable Systems and Networks*, pp. 887–896, 2004.

[Howard 10] J. Howard et al., "A 48-Core IA-32 Message-Passing Processor with DVFS in 45nm CMOS," *Proc. IEEE Intl. Solid-State Circuits Conf.*, pp. 108–109, 2010.

[Kalbarczyk 99] Z. Kalbarczyk et al., "Hierarchical Simulation Approach to Accurate Fault Modeling for System Dependability Evaluation," *IEEE Trans. Software Engineering*, vol. 25, no. 5, pp. 619–632, Sept. –Oct. 1999.

[Kanawati 93] G. A. Kanawati, N. A. Kanawati, and J. A. Abraham, "EMAX: An Automatic Extractor of High-Level Error Models," *Proc. AIAA Computing Aerospace Conf.*, pp. 1297–1306, 1993.

[KleinOsowski 02] AJ KleinOsowski, D. J. Lilja, "MinneSPEC: A New SPEC Benchmark Workload for Simulation-Based Computer Architecture Research," *IEEE Computer Architecture Letters*, vol. 1, no. 1, p. 7, Jan. –Dec. 2002.

[Leon] Aeroflex Gaisler, "Leon3 Processor," http://www.gaisler.com.

[McCluskey 71] E. J. McCluskey and F. W. Clegg, "Fault Equivalence in Combinational Logic Networks," *IEEE Trans. Computers*, vol. 20, no. 11, pp. 1286–1293, Nov. 1971.

[McCluskey 00] E. J. McCluskey and C.-W. Tseng, "Stuck-Fault Tests vs. Actual Defects," *IEEE Intl. Test Conf.*, pp. 336-343, 2000.

[Maniatakos 11] M. Maniatakos, N. Karimi, C. Tirumurti, A. Jas, and Y. Makris, "Instruction-Level Impact Analysis of Low-Level Faults in a Modern Microprocessor Controller," *IEEE Trans. Computers*, vol. 60, no. 9, pp. 1260–1273, Sept. 2011.

[Michalak 12] S. E. Michalak et al., "Assessment of the Impact of Cosmic-Ray-Induced Neutrons on Hardware in the Roadrunner Supercomputer," *IEEE Trans. Device and Materials Reliability*, vol. 12, no. 2, pp. 445–454, June 2012.

[Miskov-Zivanov 10] N. Miskov-Zivanov, D. Marculescu, "Multiple Transient Faults in Combinational and Sequential Circuits: A Systematic Approach," *IEEE Trans. Comput.-Aided Des. Integr. Circuits and Syst.*, vol. 29, no. 10, pp. 1614–1627, Oct. 2010.

[Mitra 10] S. Mitra, K. Brelsford, and P. N. Sanda, "Cross-Layer Resilience Challenges: Metrics and Optimization," *Proc. Design, Automation and Test in Europe*, pp. 1029–1034, 2010.

[OpenSPARC] "OpenSPARC: World's First Free 64-bit Microprocessor," http://www.opensparc.net.

[Pellegrini 12] A. Pellegrini et al., "CrashTest'ing SWAT: Accurate, Gate-Level Evaluation of Symptom-Based Resiliency Solutions," *Proc. Design, Automation and Test in Europe*, pp. 1106–1109, 2012.

[Pattabiraman 11] K. Pattabiraman, G. P. Saggese, D. Chen, Z. T. Kalbarczyk, and R. K. Iyer "Automated Derivation of Application-Specific Error Detectors Using Dynamic Analysis," *IEEE Trans. Dependable and Secure Computing*, vol. 8, no. 5, pp. 640–655, Sept.–Oct. 2011.

[Ramachandran 08] P. Ramachandran, P. Kudva, J. Kellington, J. Schumann, and P. Sanda, "Statistical Fault Injection," *Proc. IEEE Intl. Conf. Dependable Systems and Networks*, pp. 122–127, 2008.

[Racunas 07] P. Racunas, K. Constantinides, S. Manne, and S. S. Mukherjee, "Perturbation-based Fault Screening," *Proc. IEEE Intl. Symp. High Performance Computer Architecture*, pp. 169–180, 2007.

[Rebaudengo 02] M. Rebaudengo, M. S. Reorda, and M. Violante, "Analysis of SEU effects in a pipelined processor," *Proc. IEEE Intl. On-Line Testing Workshop*, pp. 112–116, 2002.

[Rimen 94] M. Rimen, J. Ohlsson, and J. Torin, "On microprocessor error behavior modeling," *Proc. IEEE Intl. Symp. Fault-Tolerant Computing*, pp. 76–85, 1994.

[Sanda 08] P. N. Sanda et al., "Soft-error resilience of the IBM POWER6 processor," *IBM Journal of Research and Development*, vol. 52, no. 3, pp. 275–284, May 2008.

[Seifert 10] N. Seifert, "Radiation-induced soft errors: A chip-level modeling per- spective," *Foundat. Trends® in Electron. Design Autom.*, vol. 4, no. 2–3, pp. 99–221, Feb. 2010.

[Seifert 12] N. Seifert et al., "Soft Error Susceptibilities of 22 nm Tri-Gate Devices," *IEEE Trans. Nucl. Sci.*, vol. 59, no. 6, pp. 2666–2673, Dec. 2012.

[Wang 04] N. J. Wang, J. Quek, T. M. Rafacz, and S. J. Patel, "Characterizing the Effects of Transient Faults on a High-Performance Processor Pipeline," *Proc. Intl. Conf. on Dependable Systems and Networks*, pp. 61–70, 2004.

[Wang 07] N. J. Wang, A. Mahesri, and S. J. Patel, "Examining ACE Analysis Reliability Estimates Using Fault-Injection," *Proc. Intl. Symp. Computer Architecture*, pp. 460–469, 2007.

[Yim 10] K. S. Yim, Z. Kalbarczyk, and R. K. Iyer, "Measurement-based Analysis of Fault and Error Sensitivities of Dynamic Memory," *Proc. IEEE/IFIP Intl. Conf. on Dependable Systems and Networks*, pp. 431–436, 2010.

[Zhang 10] Y. Zhang, J. W. Lee, N. P. Johnson, and D. I. August, "DAFT: Decoupled Acyclic Fault Tolerance," *Proc. Intl. Conf. Parallel Architectures and Compilation Techniques*, pp. 87–98, 2010.

Appendix A. BEE3 Emulation System

Figure A1. BEE3 emulation system using Virtex-5 FPGAs.

Appendix B. SPECINT 2000 Benchmark Applications on LEON3

Table B1. SPECINT 2000 benchmark applications.

Name	Execution time (cycles)	CPI
bzip2	2,429M	1.97
crafty	294M	2.43
eon	3,479M	1.54
gap	145M	2.29
gcc	216M	2.71
gzip	2,753M	2.25
mcf	627M	3.37
parser	683M	1.84
twolf	751M	1.66
vortex	222M	2.60
vpr	424M	1.66

Appendix C. Results: Error Injection Inaccuracies for IVM

Similar to Sec. 3, we present a comparison of inaccuracies associated with various error injection techniques for the IVM processor. We use the same benchmark applications [5] and error injection techniques as for the LEON3 error injections. Figure C1 compares the observed rate of each outcome obtained from 40,000 error injections for each of the 9 applications using each of the 5 error injection techniques. The results show trends similar to the LEON3 error injections.

1. Existing high-level error injection techniques, **that inject single errors into registers or program variables**, can result in high degrees of inaccuracies by more than an order of magnitude. For example, the geometric means of normalized outcome rates obtained from VarU error injection can range from $0.07\times$ (for the Hang outcome type) to $22\times$ (for the ONA outcome type) when compared to the corresponding flip-flop error injection results.

2. There is no single trend (e.g., always optimistic or always pessimistic) that can explain the inaccuracies associated with existing high-level error injection techniques. For example, the RegU error injection technique generally tends to underestimate Hang outcome rates, but it tends to overestimate OMM outcome rates. Also, the RegU error injection technique generally tends to underestimate UT outcome rates while RegW error injection technique tends to result in overestimated UT outcome rates.

3. RegU and VarU error injection techniques result in very high ONA outcome rates.

Appendix D. Outcomes versus Flip-flop Error Propagation Types

Flip-flop errors resulting in various error propagation types (Sec. 4.2) ultimately produce various outcome rates at the end of program execution. In Fig. D1, for each propagation type, we report its percentage that results in a given outcome. For example, 100% of flip-flop error injections that result in masked propagation type (①) produce the Vanished outcome type only because those flip-flop errors does not have any application-level effects.

---

[5] We excluded *eon* and *twolf* since these applications very frequently use floating point instructions which are not supported by the existing IVM processor model.

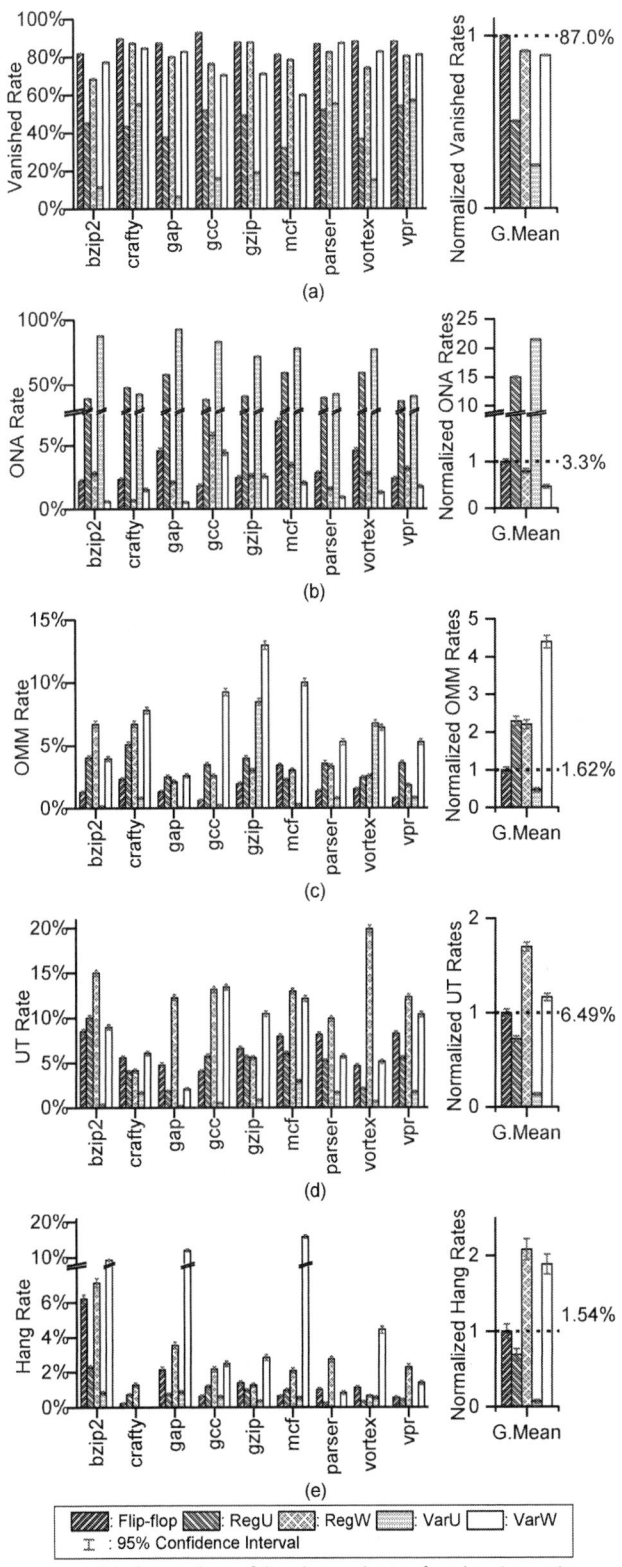

Figure C1. Comparison of the observed rate of each outcome type obtained from various error injection techniques for the IVM processor. (a) Vanished. (b) ONA. (c) OMM. (d) UT. (e) Hang.

In Table D1, for each outcome type, we report the percentage of that outcome contributed by each propagation type. For the Vanished outcome, the masked propagation type (①) is the dominant cause, but other propagation types also contribute as well. For the remaining outcomes, propagation type ⑥ is the dominant cause. For the ONA outcome, for example, correct modeling of error propagation types ②-⑥ is very important. For the UT outcome, however, error propagation type ⑧ plays an important role.

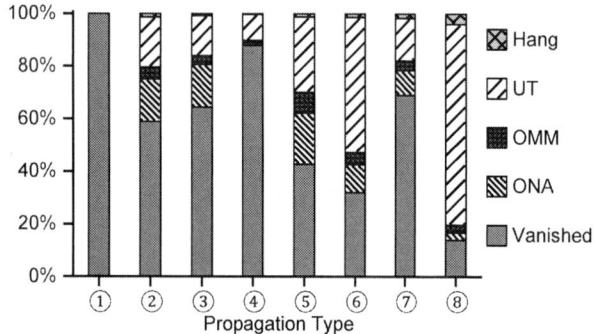

Figure D1. Percentage of a propagation type resulting in an outcome.

Table D1. Percentage of an outcome contributed by a propagation type.

Propagation Type \ Outcome	Vanished	ONA	OMM	UT	Hang
Masked (①)	77.17%	0.00%	0.00%	0.00%	0.00%
R-only (②)	1.46%	18.92%	12.73%	5.27%	11.07%
M-only (③)	1.11%	14.49%	5.74%	3.08%	4.00%
O-only (④)	16.81%	8.66%	21.91%	21.73%	19.00%
Combination (⑤)	0.40%	8.12%	7.27%	2.83%	4.76%
Combination (⑥)	2.78%	47.17%	46.16%	51.78%	37.54%
Combination (⑦)	0.03%	0.24%	0.18%	0.08%	0.27%
Combination (⑧)	0.23%	2.40%	6.01%	15.24%	23.35%
Total (①-⑧)	100%	100%	100%	100%	100%

Appendix E. Observed Rates of Flip-flop Error Propagation Patterns

Table E1 shows the observed rates of error propagation patterns (Sec. 4.3) for the R-only and M-only propagation types.

Table E1. Observed rates of flip-flop error propagation patterns for R- and M-only propagation types.

Propagation pattern \ Propagation type		R-only	M-only
1. Multi-instance propagation		12.8%	53.2%
2. Single-instance propagation	2a	3.7%	2.1%
	2b	35.5%	19.3%
	2c	11.2%	5.6%
	2d	6.0%	7.5%
	2e	0.8%	2.3%
	2f	0.0%	1.8%
	SBD	30.0%	8.2%

Appendix F. Results Cross-Check for the IVM Processor

In this section, we present a cross-check of the results for the IVM processor (similar to Sec. 4.4). Flip-flop error injection results presented in this section are obtained from 160,000 flip-flop error injection runs for the IVM processor. We analyze the outcomes obtained from flip-flop error injections resulting in R- and M-only SBD (Sec. 4.2 and 4.3) vs. outcomes obtained from RegW and VarW error injections. For the IVM processor, R-only SBD accounts for only 0.62% of all flip-flop error injections (or 4.1% of all non-masked flip-flop error injections), and M-only SBD accounts for only 0.73% of all

flip-flop error injections (or 4.9% of all non-masked flip-flop error injections).

We expect the inaccuracy levels of outcome rates resulting from RegW (VarW) error injections to be lower when compared to flip-flop error injections resulting in R-only (M-only) SBD, rather than the entire set of flip-flop error injections (Appendix C). Table F1 reports the estimated inaccuracy of RegW (VarW) error injections using (4.1). The accuracy improvement with SBD is also estimated in Table F1.

As expected, accuracy levels improve when high-level error injection results are compared with respect to flip-flop errors resulting in SBD: by 4.29× on average (geometric mean) for RegW error injection results and 2.89× on average for VarW error injection results.

Table F1. Inaccuracies of outcome rates resulting from high-level error injections for the IVM processor.

Outcome	RegW inaccuracies			VarW inaccuracies		
	vs. flip-flop error injections (i)	vs. flip-flop error injections resulting in R-only SBD (ii)	Accuracy improve-ment (i / ii)	vs. flip-flop error injections (iii)	vs. flip-flop error injections resulting in M-only SBD (iv)	Accuracy improve-ment (iii / iv)
Vanished	8.86%	3.72%	2.38×	11.4%	3.93%	2.90×
ONA	20.4%	43.7%	0.47×	53.5%	53.4%	1.00×
OMM	121%	5.14%	23.5×	340%	11.8%	28.8×
UT	69.6%	9.73%	7.15×	16.4%	10.8%	1.52×
Hang	108%	13.9%	7.77×	88.6%	55.9%	1.58×
	Geometric mean		4.29×	Geometric mean		2.89×

Appendix G. Results Cross-Check Details

In this section, for each application, we compare the observed rate of each outcome type obtained from the entire set of flip-flop error injections, flip-flop error injections resulting in R-only (M-only) SBD, and RegW (VarW) error injections. This comparison shows accuracy improvements for each application that may not be captured by the geometric means in Sec. 4.4 and Appendix F.

Figure G1 shows the observed outcome rates for the entire set of flip-flop error injections, flip-flop error injections resulting in R-only SBD, and RegW error injections for the LEON3 processor. The geometric means of normalized outcome rates (with respect to corresponding flip-flop error injection results) are also shown. Figure G2 shows the corresponding comparison for flip-flop error injections resulting in M-only SBD and VarW error injections for the LEON3 processor. Figures G3-G4 present corresponding results for the IVM processor.

For example, the geometric mean of normalized OMM outcome rates in Fig. G1c is 1 for the entire set of flip-flop error injections, 5.9 for flip-flop error injections resulting in R-only SBD, and 5.1 for RegW error injection. The confidence intervals for flip-flop error injections resulting in R-only (M-only) SBD are larger than those for the entire set of flip-flop error injections. This is because the number of flip-flop error injections resulting in R-only (M-only) SBD is very small even with the 320,000 flip-flop error injection samples per each application (less than 3,000 samples per application). Although the IVM results are obtained from 160,000 flip-flop error injection samples per each application, the IVM results may show smaller confidence intervals compared to the LEON3 results. This is because:

1. Confidence intervals tend to be larger if the outcome rate is small, e.g., the Hang outcome rate is 0.03% for LEON3 flip-flop error injection and 1.54% for IVM flip-flop error injection.
2. M-only SBDs are more common in IVM flip-flop error injections (0.12% for LEON3 and 0.73% for IVM), i.e., more M-only SBD samples for the IVM results.

718

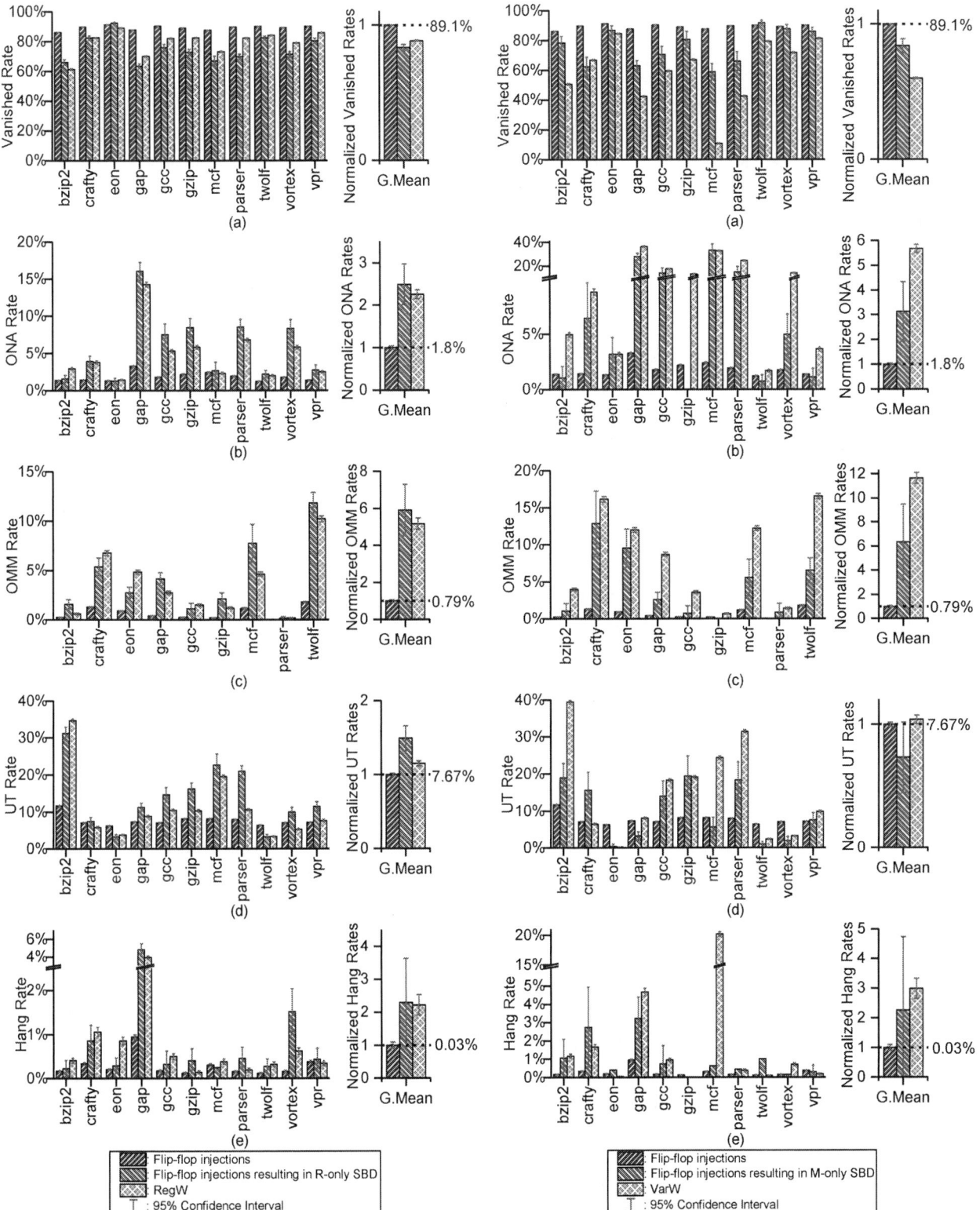

Figure G1. Comparison of observed rate of each outcome type obtained from the entire set of flip-flop error injections, flip-flop error injections resulting in R-only SBD, and RegW error injections for the LEON3 processor. (a) Vanished. (b) ONA. (c) OMM. (d) UT. (e) Hang.

Figure G2. Comparison of observed rate of each outcome type obtained from the entire set of flip-flop error injections, flip-flop error injections resulting in M-only SBD, and VarW error injections for the LEON3 processor. (a) Vanished. (b) ONA. (c) OMM. (d) UT. (e) Hang.

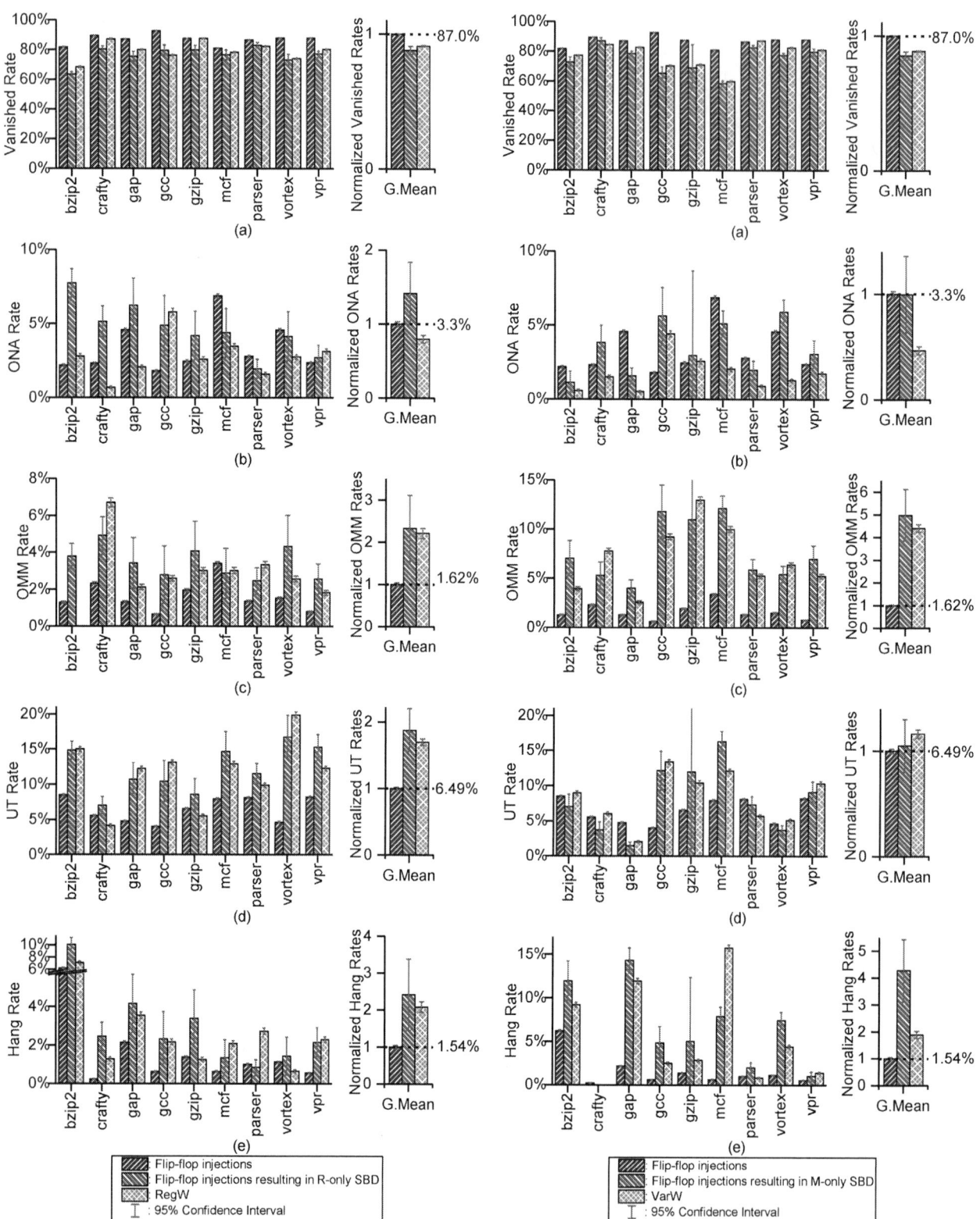

Figure G3. Comparison of observed rate of each outcome type obtained from the entire set of flip-flop error injections, flip-flop error injections resulting in R-only SBD, and RegW error injections for the IVM processor. (a) Vanished. (b) ONA. (c) OMM. (d) UT. (e) Hang.

Figure G4. Comparison of observed rate of each outcome type obtained from the entire set of flip-flop error injections, flip-flop error injections resulting in M-only SBD, and VarW error injections for the IVM processor. (a) Vanished. (b) ONA. (c) OMM. (d) UT. (e) Hang.

720

Efficiently Tolerating Timing Violations in Pipelined Microprocessors

Koushik Chakraborty   Brennan Cozzens   Sanghamitra Roy   Dean M. Ancajas

USU BRIDGE LAB, Electrical and Computer Engineering, Utah State University
{koushik.chakraborty, sanghamitra.roy}@usu.edu

ABSTRACT

Early prediction of an upcoming timing violation presents a tremendous opportunity to mask the performance overhead of tolerating these faults. In this paper, we explore several techniques for optimizing instruction scheduling in an Out-of-Order pipeline, exploiting this new perspective in robust system design. Compared to recently proposed stall based techniques for tolerating predictable timing violations, we demonstrate a massive reduction in performance overhead, while supporting correct execution in faulty environments (64–97% across different benchmarks).

Categories and Subject Descriptors

B.8.1 [**Hardware**]: Reliability, Testing and Fault Tolerance

General Terms

Reliability

Keywords

Timing Faults, Path Sensitization, Instruction Scheduling.

1. INTRODUCTION

Growing unreliability in electronic systems is reshaping the design approaches of the computing world. In this domain, timing violations—an artifact of rapid technology scaling—embody a central reliability challenge [1, 2]. Guided by a combined effect of static (process variation and wearout) and temporal (thermal, voltage or utilization) variation, timing violations can occur sporadically [1, 2]. Consequently, runtime error detection and correction techniques have been a topic of major research in recent years [3–7]. Existing works in this area either provide a very high fault coverage at the expense of a large performance overhead, or provide a poor fault coverage with a low performance penalty [3, 6–11].

In this paper, we demonstrate that it is possible to approach the performance of fault-free execution, while tolerating timing errors in an Out-of-Order (OoO) microprocessor pipeline. We establish a foundation for low-overhead timing-error tolerance using a violation aware instruction scheduling framework. This unique framework is based on the recently observed predictability of timing errors from specific instructions causing them [12, 13]. We can use the instruction Program Counter (PC), a unique instruction identifier, for predicting an upcoming timing violation, several clock cycles in advance. While tolerating an unexpected timing error entails a large performance overhead in pipelined architectures [3], we observe that scheduling an instruction with a predictable timing error becomes logically equivalent to a variable latency operation.

To exploit this property, our proposed violation aware instruction scheduling framework ensures two fundamental principles: (a) the faulty instruction occupies one additional cycle in the same pipe stage, and no new input is fed in the resources occupied by that instruction during this time; and (b) the dependent instructions behind the faulty instruction are held back by one extra cycle. Our proposed techniques can thus mask the penalty of timing violations from the system performance by confining the error overhead to the faulty instruction and its dependents only. Our proposed technique avoids stalling the whole pipeline as was done in recent works [12, 13], thereby marginalizing the system level overhead of tolerating timing violations.

Enabled by our violation aware scheduling techniques, microprocessors can operate at a tighter frequency, where predictable errors frequently occur and are tolerated with minimal performance loss. The main contributions of our paper are outlined next.

- We propose a series of low overhead micro-architectural enhancements coupled with instruction scheduling techniques to drastically limit the performance overhead of predictable timing faults (Sections 2, 3).
- Using a rigorous circuit-architectural simulation (Sections 4 and 5), combining synthesized hardware with full system simulation, we demonstrate dramatic reduction in the performance overhead from faulty execution (64–97%) compared to stall based schemes [12, 13]. Our proposed schemes have negligible power/area overhead giving an energy-efficient alternative for robust pipelines (Section S3).
- We study the locality of sensitized paths from multiple dynamic instances of a given instruction in several microprocessor structural blocks. Our analysis, employing the Fabscalar infrastructure [14], demonstrates the predictability of timing faults in these components (Section S1).

2. ROBUST PIPELINE DESIGN OVERVIEW

In this section, we present an overview of our proposed timing error tolerant Out-of-Order (OoO) pipelined microprocessor. The goal of our proposed techniques is to approach the performance of fault-free execution while tolerating timing errors. We outline an overview of our design in Section 2.1 and summarize how timing errors are handled in various stages of the pipeline (Section 2.2).

Permission to make digital or hard copies of all or part of this work for personal or classroom use is granted without fee provided that copies are not made or distributed for profit or commercial advantage and that copies bear this notice and the full citation on the first page. To copy otherwise, to republish, to post on servers or to redistribute to lists, requires prior specific permission and/or a fee.
DAC 2013, May 29–June 07, Austin, TX, USA

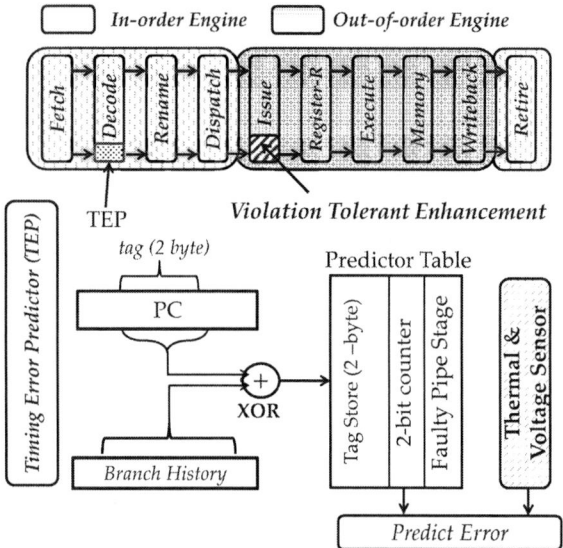

Figure 1: An Overview of our proposed techniques showing key enhancements. The decode stage is augmented with a Timing Error Predictor (TEP). This predictor information is propagated in subsequent pipe stages. The Issue stage is enhanced for violation tolerant scheduling.

2.1 Pipeline Overview

Figure 1 shows the overview of our pipelined micro-architecture. An OoO processor typically consists of a front end spanning from *Fetch* to *Dispatch*, where instructions proceed in-order, the OoO core engine spanning from *Issue* to *Writeback*, and an in-order *Retire* stage. An individual stage in this pipeline may also span across multiple clock cycles [14].

We augment this pipeline with several micro-architectural enhancements. The decode stage is enhanced with a Timing Error Predictor (TEP), which dynamically learns and predicts timing errors in the pipeline from a decoded instruction PC. The TEP is accessed in parallel to decode avoiding any impact on the critical path. The TEP prediction is subsequently propagated with the instruction meta-data as it traverses through various pipeline stages. Subsequently, the issue stage recognizes an instruction with a predicted timing error and activates violation aware instruction scheduling. For this purpose, the issue stage micro-architecture is augmented with a Violation Tolerant Enhancement (VTE) detailed in Section 3. If an instruction meta-data does not indicate a timing violation, instructions proceed normally. We next discuss our TEP design, and then briefly outline how timing error mis-predictions are handled.

2.1.1 Timing Error Predictor

Our TEP combines features from the Most Recent Entry (MRE) predictor proposed by Xin et al. with the Timing Violation Predictor (TVP) proposed by Roy et al. [12, 13]. Figure 1 shows our TEP design. Each entry in the predictor table contains a 2 byte tag obtained from the PC. The entries in the table are indexed using a combination of bits in the PC and the recent branch outcomes. A 2-bit saturating counter in each entry keeps track of the potential of a timing error in the system. A non-zero value in the saturating counter indicates a possible timing violation. We also keep track of the faulty pipe stage associated with an error causing instruction. The prediction also considers favorable conditions for timing errors through the use of thermal and voltage sensors.

2.1.2 Handling Mis-prediction

If an instruction incurs a timing violation without early prediction, an error recovery is triggered using instruction replay, similar to *Razor* [15]. This recovery corrects a fault that is not handled by our violation aware scheduling framework. Instruction replays are rare, but incur a large performance overhead.

2.2 Tolerating Timing Violations in Specific Pipe Stages

Timing errors may happen in the in-order engine or the OoO engine of the processor. Our experiments, as well as, recent works have indicated that the likelihood of timing errors is significantly more in the OoO engine [16]. Therefore, our proposed techniques are primarily focused on efficiently tolerating timing errors in the OoO core. However, for the sake of completeness, we outline how we handle errors in the in-order engine next.

In-order Engine: The violation aware scheduling framework is not applicable to the in-order part of the pipeline. For the rename, dispatch and retire stages, we use the predicted violation from the TEP to enable a stall signal at the appropriate stage. This stall signal allows the faulty pipe stage to complete in two clock cycles, while the input to all other stages are recirculated to avoid forward flow of instructions during that cycle. The stall signal can be enabled using existing circuitry in modern microprocessors with minimal modification. The TEP cannot be used to mitigate timing violations in the fetch and decode stages of the pipeline. Any violations in these two stages are mitigated using instruction replay with our error recovery circuitry discussed in Section 2.1.2. Such replays are however rare, as the fetch and decode stages have substantially lower fluctuations of temperature and voltage making timing violations rare [17].

OoO Engine: We use our proposed timing violation aware instruction scheduling framework for tolerating timing errors in the OoO engine, discussed next.

.

3. VIOLATION AWARE SCHEDULING

In this section, we describe our violation tolerant enhancement (VTE) and scheduling algorithms in a pipelined OoO microprocessor. Our goal is to efficiently tolerate timing violations in the OoO engine shown in Figure 1, radically improving upon stall based techniques [12, 13].

3.1 Violation Aware Scheduler Overview

From an instruction scheduling perspective, a pipe stage execution with a timing violation becomes equivalent to a variable latency operation *when that violation can be predicted early*. Consequently, we can suitably alter the instruction scheduling in a manner such that: (a) the faulty instruction occupies one additional cycle in the same pipe stage, and no new input is fed in the resources occupied by that instruction during this time; and (b) the dependent instructions behind the faulty instruction are held back by one extra cycle. At its core, these *scheduling features* require modification to the pipe resource management and the communication of dependency between two instructions (detailed in Section 3.2). Eventual impact of these corrective measures on the processor performance is determined by the existing *architectural slack* of the faulty instruction [18] (e.g., increased latency on some instructions may have negligible impact on the system performance).

3.2 Violation Tolerant Enhancements

To ensure correct execution with timing violations, we need to make micro-architectural modifications in the scheduler logic in the issue stage. Other stages within the out-of-order engine require supporting modifications. First, we describe the VTE in the Issue stage. Then we describe how predictable timing violations are tolerated in all stages within the OoO engine (Section 3.3). Three major aspects of the VTE are: (a) Issue Queue Entry; (b) Tag Broadcast Logic and (c) Issue Slot Management.

3.2.1 Issue Queue Alteration

The issue queue entries are augmented to include a single-bit that indicates the *fault prediction* of an instruction. Furthermore, another field indicates the faulty pipe stage, so that the pipe stage logic is modified when that instruction enters the faulty stage. Combined together, a 4-bit field is sufficient to encode the error prediction information for each instruction.

3.2.2 Tag Broadcast Logic

When an instruction is scheduled, engaging the Register Read, Execute and Memory stages, the scheduler logic keeps track of its expected completion time. Subsequently, in the cycle the instruction completes, the instruction tag is broadcast to the issue queue. Waiting instructions then perform a tag match with this result tag, to evaluate if their operands are ready. Based on whether the completion is triggered from the Memory stage (load instructions) or Execute stage (ALU instructions), we use a countdown to keep track of its completion time. In case of a faulty instruction, we increment the completion counter, based on its expected delay. Thus, the tag broadcast is delayed by one cycle.

3.2.3 Issue Slot Management

In each cycle, the issue stage in the OoO engine prepares a packet consisting of several instructions (equal to the pipeline width W), which is then propagated through the later stages of the pipeline. We denote the position of a particular instruction in this packet as an issue slot in our description.

When a faulty instruction is issued, the issue slot occupied by that instruction must be managed carefully. This is necessary to avoid sending a new instruction in the same slot to the faulty stage before the faulty instruction has sufficient time to complete its computation. To accomplish this task, we keep track of the issue slot occupied by a faulty instruction. In the subsequent cycle, we freeze the slot to disallow issuing another instruction behind the faulty instruction. We next discuss, how each pipe stage in the OoO engine tolerates timing violations.

3.3 Tolerating Timing Errors in the OoO Pipe Stages

3.3.1 Issue

Issue can have multiple pipe stages. However, the wakeup/select stage inside the issue is particularly prone to timing errors due to the use of content addressable memory (CAM) logic in the wakeup. In our experiments, also corroborated by others [16], we find that almost all timing errors happen in the wakeup/select stage. The issue stage is responsible for handling timing errors in the other OoO pipe stages. However, a timing error in the issue itself can cause a pipeline deadlock when back-end errors are relying on correct operation of the issue.

To avoid such a pipeline deadlock, we adopt a low-complexity technique that trades off marginal performance loss for complexity reduction. After an instruction with a predictable timing error in the issue is inserted in the *issue queue* from the dispatch stage, we track the functional unit or memory port where the faulty instruction will be scheduled. Once this faulty instruction is scheduled, we freeze

the corresponding issue slot for the functional unit or memory port in the subsequent cycle. Consequently, when this faulty instruction broadcasts its tag, the input to the wakeup select lane will remain steady for two cycles, thereby providing sufficient time to complete the logic computation.

3.3.2 Register Read

When an instruction has a predictable timing error in the register read stage, the issue queue blocks the respective register read port, where the faulty instruction is assigned, for one additional cycle. This blocking allows the instruction to complete register read in two cycles, and avoids using the read port in the next cycle after the faulty instruction enters the register read stage. The scheduling cycle for the execute/memory stage for this faulty instruction is adjusted to handle this additional delay from the register read stage.

3.3.3 Execute

The execute stage is composed of various functional units. The key to functional unit management is to ensure that the *instructions are issued to functional units when they are ready to process new instructions.* To keep track of this information, we use a *Functional Unit State Register (FUSR)*. Each bit in the FUSR keeps track of one functional unit, and indicates if a new instruction can be issued to that unit in the next cycle. We now discuss two major classes of functional units based on their completion delay: single cycle and multi-cycle.

Single-Cycle Latency: Functional units with a single cycle latency can process new instructions in every cycle. However, when a faulty instruction is scheduled, its FUSR bit is turned off for one cycle to disallow issuing a new instruction in the next cycle.

Multi-cycle Latency: A multi-cycle functional unit may or may not be pipelined. In the case it is not pipelined, the FUSR is adjusted to indicate the busy state for one extra cycle beyond its expected completion time. A fully pipelined functional unit can process new instructions every cycle in the absence of timing errors. However, a timing error can happen in any of the internal pipeline stages. To effectively handle these multi-cycle pipelined units, we temporarily avoid issuing new instructions behind a faulty one. We resume issuing new instructions to that unit only after the faulty instruction completes. This approach is agnostic of the exact error pipe stage within the multi-cycle execution unit, thereby saving design complexity at the cost of a marginal performance loss.

3.3.4 Memory

Similar to the issue, the memory stage is also susceptible to errors due to the presence of CAM logic in the load-store queue. In particular, when the CAM search results in several tag matches, we observe additional delay in this stage, potentially causing timing errors. When the issue queue schedules a predictable faulty instruction to the memory, it estimates the cycle when the CAM match will be performed. Based on this estimation, the issue queue avoids issuing a load/store instruction behind the faulty one to prevent another CAM match in the cycle right after the faulty instruction. Consequently, the faulty instruction can continue to do the CAM match for two cycles. The writeback stage of this faulty instruction is delayed in the memory stage by one cycle to preserve correct execution.

3.3.5 Writeback

The writeback stage is relatively less susceptible to errors, compared to other stages discussed above. However, to achieve fault coverage in the entire OoO engine, we propose some enhancements to the Writeback stage. Every cycle, this stage receives W pack-

Instructions	Scheduling							
I₁: ADD R3, R1, R2	clock cycle	1	2	3	4	5	6	7
I₂: SUB R5, R1, R4 *	Select	I₁	I₂	Stall	I₃	I₄	-	-
I₃: ADD R6, R5, R2	Register Read	-	I₁	I₂	-	I₃	I₄	-
I₄: SUB R9, R8, R7	Execution	-	-	I₁	I₂	-	I₃	I₄
	FUSR	1	0	1	1	1	1	1
*Faulty instruction	Tag Broadcast	-	-	I₁	-	I₂	I₃	I₄

Figure 2: An example showing the scheduler logic modification.

ets, each of which consist of information necessary to complete the writeback of an in-flight instruction (W is the issue width). If one of the instructions has a potential timing error in the writeback, then its corresponding input slot is frozen by the issue queue in the next cycle, allowing the input to recirculate.

3.4 An Illustrative Example

Figure 2 shows an operational example of scheduling instructions around a faulty instruction. We assume a functional unit with one-cycle completion time, so that new instructions can be scheduled in every cycle, while scheduled instructions complete in the same cycle. The instruction I_2 is predicted to have a fault in the execution unit. This instruction is selected in cycle 2, to be executed on the functional unit on cycle 4. However, as it is faulty, it takes one additional cycle. Moreover, as no new instruction can be scheduled on cycle 5 on that functional unit, the FUSR is marked 0 at the end of cycle 2 to avoid selecting a new instruction for that functional unit in cycle 3 (representing issue slot freezing discussed in Section 3.2.3). The tag broadcast logic is delayed by one cycle, so that dependent instruction I_3 is held back for one cycle.

3.5 Violation Aware Scheduling Algorithms

Beyond ensuring correct execution in the presence of predictable timing violations, the next design issue pertains the selection priority of instructions with operands ready. We explore three different algorithms for instruction scheduling, all of which confine the penalty to a faulty instruction and its dependents, and aim to minimize the system level performance overhead of a timing fault:

- Age based selection (ABS)
- Faulty First Selection (FFS)
- Criticality Driven Selection (CDS)

The age based policy, ABS, uses a timestamp—implemented using a 6-bit module-64 counter—to select instructions to schedule, among those that are operand ready. The faulty first policy, FFS, attempts to schedule instructions with faults early, so as to release their dependent instructions sooner. The criticality driven policy dynamically estimates the criticality of a predictable faulty instruction (detailed in Section 3.5.2), and attempts to eagerly select those instructions that have a higher criticality. We next outline our proposed issue queue modifications to implement these policies, and subsequently discuss our approach for dynamic criticality estimation.

3.5.1 Selection Logic Enhancement (SLE)

Figure 3 shows the implementation of the SLE for the proposed priority schemes discussed above. Each issue queue entry keeps track of three major aspects necessary for selection: (i) operand ready; (ii) timestamp; and (iii) a 4-bit field indicating the fault prediction (Section 3.2.1) and criticality of an instruction. The faulty bit is also used to manage functional units and the tag broadcast logic, as discussed in Section 3.2. All instructions with operand ready, bids for selection (Figure 3). The ABS policy sets the grant

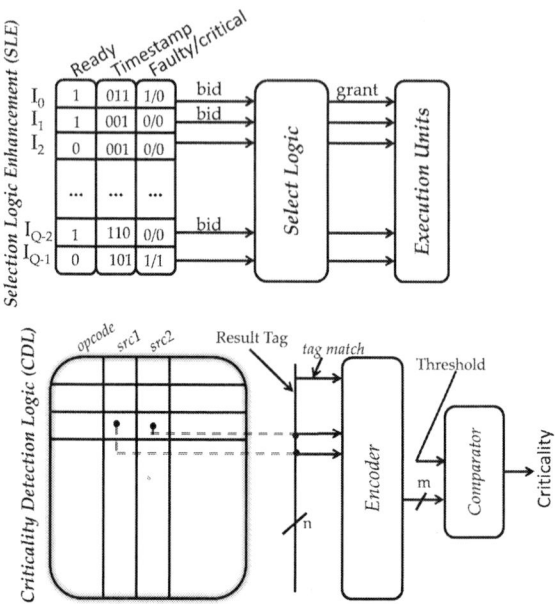

Figure 3: SLE and CDL Implementation.

line for the instructions with the lowest timestamp (oldest instructions). The FFS policy sets the grant line for instructions with faulty bit set. When none of the instructions are faulty, then it uses the timestamp to select instructions (similar to ABS). The CDS policy eagerly selects faulty instructions that are expected to be critical. Again, similar to FFS, if no such instructions (faulty and critical) exist, then it uses the timestamp.

3.5.2 Criticality Detection Logic (CDL)

Precisely estimating the criticality of an instruction in hardware is challenging, as the hardware has limited information about the dynamic data flow of a program [18]. Instead, we use a low-complexity technique to estimate the instruction criticality by tracking the number of dependent instructions behind a given instruction in the issue queue. Figure 3 shows the implementation of our proposed scheme in a reservation station. When an instruction broadcasts its result tag, we track the number of tag matches in the reservation station. These tag matches are fed to an encoder, and then compared with a predefined *Criticality Threshold (CT)*. This CT dictates the minimum number of dependent instructions that must be present in the issue queue to consider a given instruction to be critical. Once we determine this instruction criticality, we store this information with the timing error predictor (Section 2.1.1). In our experiments, we find that a CT of *8* gives the best outcome.

4. METHODOLOGY

In this section, we describe our extensive circuit-architectural methodology for performance tradeoff analysis of our proposed techniques.

4.1 Circuit Implementation

We implement our proposed scheduling techniques within the Fabscalar infrastructure [14]. For the purpose of this paper, we use the *Core-1* configuration, which represents an out-of-order pipeline capable of fetching, issuing and committing 4 instructions each cycle. The pipeline has single-cycle (e.g. simple ALU) as well as multi-cycle (e.g. complex ALU) functional units. The misprediction loop for this pipeline is 10 stages, spanning fetch to execute. We synthesize our implementation with the Synopsys Design Compiler using a 45nm FreePDK library. Energy results are gathered by combining architectural usage information with power characteristics from the synthesized hardware.

4.2 Architectural Simulation

We use full-system simulation built on top of WindRiver SIMICS [19]. We use our own detailed timing model to enforce the timing characteristics of a 4-wide out-of-order microprocessor, identical to the Core-1 configuration mentioned above. The core uses a two-level cache hierarchy where L1 (32KB 4-way split Instruction and Data) has a single cycle latency, while the 16-way 8MB L2 and the main memory are accessed in 25 and 240 cycles, respectively. We use the TEP as our predictor design (Section 2.1.1). For both fault-free execution and Error Padding scheme (discussed in Section 5) [13], we use the age based instruction selection policy. We use several SPEC CPU2006 benchmarks, and focus our architectural simulation on representative phases extracted using the SimPoint toolset [20]. Each phase corresponds to 1 million committed instructions.

4.3 Fault Simulation

To simulate timing faults, we embed gate delay information in the architectural simulation. The effect of process variation and aging on the circuit timing is obtained by our in-house statistical timing tool that uses SPICE characterized gate delay distributions [21]. To model process variation, we assume that the transistor length, width and oxide thickness behave as Gaussian distributions with ±20% deviation across the nominal values [1, 22].

For the purpose of this paper, we focus on timing violations in the OoO engine of the processor, spanning from *Issue* to *Writeback*. Together, these stages comprise the heart of the control and datapath in a pipelined microprocessor, and also contain the timing critical stages in the microprocessor [16].

Depending on the program input, different instructions incur different delays based on the delays in individual gates in the sensitized paths. We alter the supply voltage to create two different faulty environments: high fault rate (0.97V), and low fault rate (1.04V). Faults are assumed to occur when the 95% confidence interval of the stage delay exceeds the cycle time ($\mu + 2\sigma$). The baseline machines have zero fault rate when executing at 1.1V supply voltage. Most of the timing violations are accurately predicted and tolerated with one of the comparative schemes. However, when timing faults occur without early prediction, we initiate error recovery using instruction replay, similar to Razor [3].

5. EXPERIMENTAL RESULTS

In this section, we present experimental results of our proposed schemes for optimizing scheduling around predictably faulty instructions. Our goal is to study the power-performance overhead incurred *during* faulty execution.

Comparative Schemes: We study the following schemes:

- **Razor:** This scheme fires an instruction replay for all errors in the system [3].
- **Error Padding (EP):** This is our baseline scheme that introduces stall cycles for predicted errors, similar to [12, 13].
- **ABS, FFS, CDS:** These are our proposed schemes for violation aware scheduling described in Section 3.5.

5.1 Fault Rates

Depending on specific paths sensitized during program execution, different benchmark programs exhibit different fault rates while

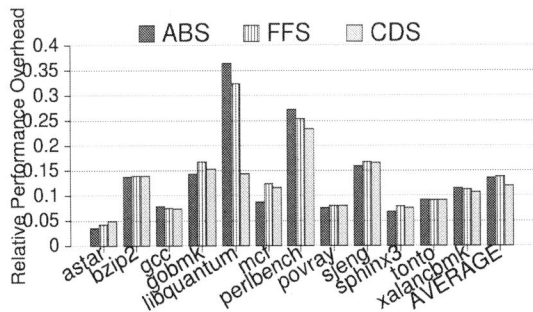

Figure 4: Performance Overhead Comparison During Faulty Execution at low fault rate (1.04V) normalized to EP [12, 13]. (Lower is better.)

operating at the same supply voltage. In Table 1, we report the average fault rates seen in the OoO engine, collectively, when operating under 1.04V and 0.97V, respectively. We also report the performance overhead over fault-free execution seen with Razor [3] and EP schemes. Overheads are shown as a tuple, representing the percentage of performance and energy efficiency degradation, respectively. Performance is estimated using *Instruction per cycle (IPC)*, while energy efficiency is estimated using energy-delay product. Certain benchmarks like *sjeng*, with higher inherent instruction level parallelism, show greater susceptibility to timing violations. On the other hand, benchmarks like *libquantum*, with greater data stalls, show substantially lower performance impact from occasional timing violations. As Razor has substantially higher overhead than EP for performance and energy efficiency, we provide all our subsequent results normalized to the EP scheme.

5.2 Performance Overhead Comparison

Figures 4 and 5 present the relative performance and ED (Energy-Delay) overhead comparison of our proposed schemes, normalized to the baseline *Error Padding* (EP) scheme during lower fault rate. All of our schemes are remarkably effective in marginalizing the performance overhead during timing violations when compared with the baseline. During a lower fault rate ($V_{DD} = 1.04V$), on an average our schemes reduce the performance overheads by 87% compared to *Error Padding* (Figure 4). Likewise, on an average, our schemes reduce the ED (Energy Delay) overhead by 82% compared to EP (Figure 5). For example, in *astar*, ABS is able to dramatically erase the performance under faulty environment by 97%, delivering performance similar to fault-free execution. ABS also shows similar improvements in ED overhead. On the other hand, in *libquantum*, CDS is particularly effective as it can eliminate 86% of the timing violation penalty, compared to 64% in ABS. Our low complexity criticality assessment is highly effective in the particular data flow pattern in *libquantum*, thereby resulting in a substantial performance advantage. Our schemes are also highly effective during high fault rates (results in the higher fault rate (V_{DD} = 0.97V) are presented in Section S2).

6. RELATED WORK

Recent works in tackling timing faults in pipelined microprocessors can be broadly classified into three groups: reactive, proactive, and predictive. Reactive techniques primarily focus on precise fault detection. Once detected, they ensure correct execution through costly instruction replay [3, 15]. Despite this large performance overhead for error correction, these techniques are often necessary to achieve full fault coverage. Proactive techniques aim to reduce the correction overhead by taking corrective action just before the

Benchmark	Fault-Free IPC	V_{DD}=0.97V			V_{DD}=1.04V		
		FR (%)	Razor Overhead	EP Overhead	FR (%)	Razor Overhead	EP Overhead
astar	0.69	6.74	(31.2, 45.6)	(5.17,6.45)	2.01	(10.2,14.6)	(1.29,1.7)
bzip2	1.48	8.92	(43.2, 56.8)	(12.35,16.5)	2.24	(17.4,25.6)	(3.1,3.7)
gcc	1.34	8.43	(47.2, 61.3)	(8.57,10.3)	1.5	(19.4,29.6)	(2.14,2.6)
gobmk	1.68	8.64	(47.3, 53.3)	(12.65,16.3)	2.16	(18.2,24.5)	(3.16,3.95)
libquantum	0.51	10.54	(25.3, 32.5)	(4.5,5.7)	2.1	(6.8, 10.2)	(1.12, 1.5)
mcf	0.34	6.45	(30.1, 42.3)	(1.96,2.8)	1.73	(9.5,12.6)	(0.49, 0.85)
perlbench	1.31	7.21	(45.7, 54.7)	(6.52,7.1)	1.8	(15.6,21.2)	(1.63, 2.1)
povray	1.941	6.31	(51.2, 75.4)	(7.58,9.1)	1.57	(24.5, 32.5)	(1.89, 2.25)
sjeng	1.93	9.19	(58.6, 72.5)	(15.19,17.8)	2.29	(23.5,29.8)	(3.79, 4.83)
sphinx3	1.30	6.95	(52.5, 67.4)	(5.45,5.9)	1.73	(17.2, 22.5)	(1.36, 1.78)
tonto	1.41	5.59	(45.6, 65.7)	(5.04,6.5)	1.39	(16.5, 21.4)	(1.25, 2.6)
xalancbmk	0.51	7.95	(34.5, 45.2)	(3.093.8)	1.99	(12.5, 15.6)	(0.77, 1.02)

Table 1: Benchmark Fault Rates (FR), and fault tolerance overhead employing Razor [3] and Error Padding [12,13].

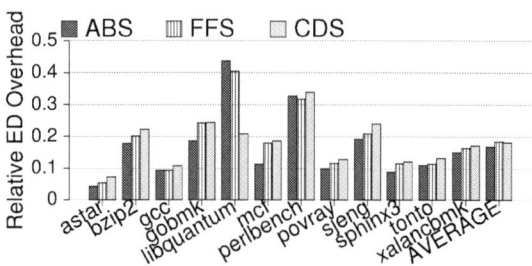

Figure 5: ED Overhead Comparison During Faulty Execution at low fault rate (1.04V) normalized to EP. (Lower is better.)

fault occurrence: in the same clock cycle where timing faults are about to happen [11]. Upcoming faults are anticipated using timing sensors. However, a lack of sufficient time limits the scope of corrective techniques, and hurts their fault coverage [12]. Recent work on predictive techniques aim to predict an upcoming timing fault, several clock cycles in advance [12, 13]. However, neither of these works exploit the immense potential of predicting timing faults. In this work, we explore micro-architectural techniques to radically marginalize the overhead from timing faults, and demonstrate massive improvements over the existing techniques based on early prediction.

7. CONCLUSION

Predicting timing violations offers a tremendous leverage in marginalizing the performance overhead of tolerating recurring timing violations. In this paper, we propose three techniques to schedule instructions with predicted timing violations. Our goal is to confine the performance impact of timing faults on the faulty instructions, while eliminating its impact on other independent instructions. Compared to recently proposed techniques to tolerate predictable timing violations, our proposed schemes dramatically reduce the performance overhead by 64–97% across different faulty environments, while reducing the ED overhead by 58-96%.

Acknowledgments

This work was supported in part by National Science Foundation grants CNS-1117425 and CAREER-1253024, and donation from the Micron Foundation.

8. REFERENCES

[1] S. Sarangi, B. Greskamp and others, "Varius: A model of process variation and resulting timing errors for microarchitects," *IEEE Transactions on Semiconductor Manufacturing*, vol. 21, no. 1, pp. 3 –13, 2008.

[2] S. Pan, Y. Hu, and X. Li, "Ivf: Characterizing the vulnerability of microprocessor structures to intermittent faults," in *Proc. of DATE*, pp. 238–243, 2010.

[3] S. Das, C. Tokunaga and others, "Razorii: In situ error detection and correction for pvt and ser tolerance," *J. of Solid-State Circ.*, vol. 44, pp. 32–48, jan. 2009.

[4] S. Das, D. Roberts and others, "A self-tuning dvs processor using delay-error detection and correction," *Solid-State Circuits, IEEE Journal of*, vol. 41, pp. 792 – 804, april 2006.

[5] K. Bowman, J. Tschanz and others, "Circuit techniques for dynamic variation tolerance," in *Proc. of DAC*, pp. 4–7, 2009.

[6] A. B. Kahng, S. Kang and others, "Designing a processor from the ground up to allow voltage/reliability tradeoffs," in *HPCA*, pp. 1–11, 2010.

[7] A. B. Kahng, S. Kang and others, "Recovery-driven design: a power minimization methodology for error-tolerant processor modules," in *Proc. of DAC*, pp. 825–830, 2010.

[8] B. Greskamp, L. Wan and others, "Blueshift: Designing processors for timing speculation from the ground up," in *HPCA*, pp. 213–224, 2009.

[9] A. Tiwari, S. R. Sarangi, and J. Torrellas, "Recycle: pipeline adaptation to tolerate process variation," in *Proc. of ISCA*, pp. 323–334, 2007.

[10] J. Long and S. O. Memik, "Automated design of self-adjusting pipelines," in *Proc. of DAC*, pp. 211–216, 2008.

[11] M. Ghasemazar and M. Pedram, "Minimizing the energy cost of throughput in a linear pipeline by opportunistic time borrowing," in *Proc. of ICCAD*, pp. 155–160, 2008.

[12] S. Roy and K. Chakraborty, "Predicting timing violations through instruction level path sensitization analysis," in *Proc. of DAC*, pp. 1074–1081, 2012.

[13] J. Xin and R. Joseph, "Identifying and predicting timing-critical instructions to boost timing speculation," in *Proc. of MICRO*, pp. 128–139, 2011.

[14] N. K. Choudhary, S. V. Wadhavkar and others, "Fabscalar: composing synthesizable rtl designs of arbitrary cores within a canonical superscalar template," in *Proc. of ISCA*, pp. 11–22, 2011.

[15] D. Ernst, N. S. Kim and others, "Razor: A low-power pipeline based on circuit-level timing speculation," in *Proc. of MICRO*, pp. 7–18, 2003.

[16] J. Sartori and R. Kumar, "Compiling for energy efficiency on timing speculative processors," in *Proc. of DAC*, pp. 1301–1308, 2012.

[17] F. J. Mesa-Martinez, J. Nayfach-Battilana, and J. Renau, "Power model validation through thermal measurements," in *Proc. of ISCA*, pp. 302–311, 2007.

[18] B. A. Fields, R. Bodík, and M. D. Hill, "Slack: Maximizing performance under technological constraints," in *Proc. of ISCA*, pp. 48–58, 2002.

[19] P. S. Magnusson, M. Christensson and others, "Simics: A full system simulation platform," *IEEE Computer*, vol. 35, pp. 50–58, Feb 2002.

[20] T. Sherwood, E. Perelman, and B. Calder, "Basic block distribution analysis to find periodic behavior and simulation points in applications," in *PACT*, pp. 3–14, 2001.

[21] S. Kothawade, K. Chakraborty and others, "Analysis of intermittent timing fault vulnerability," *Microelectronics Reliability*, vol. 52, pp. 1515–1522, July 2012.

[22] W. Zhao, F. Liu and others, "Rigorous extraction of process variations for 65-nm cmos design," *IEEE Transactions on Semiconductor Manufacturing*, vol. 22, no. 1, pp. 196 –203, 2009.

Supplemental Materials

S1. INSTRUCTION LEVEL PREDICTABILITY OF TIMING VIOLATIONS

Timing violation predictability, has been studied by two recent works [12, 13]. In this section, we verify this intriguing property using a more elaborate and detailed methodology. We next present our cross-layer analysis combining information from the application, architecture and circuit layers. Our goal is to show the underlying causes of specific instructions causing repeated timing violations—the primary reason behind timing violation predictability. Subsequently, we identify the tremendous opportunity of tolerating a timing violation using violation aware instruction scheduling techniques that hasn't been exploited by prior works.

S1.1 Underlying Cause Behind Timing Violation Predictability

When a certain static instruction executes repeatedly, its many dynamic instances sensitize strikingly similar logic paths in a given circuit block. Consequently, there is a high commonality in the critical paths sensitized by many dynamic instances of a static PC. Hence, if a certain PC causes a timing violation in a circuit block, future occurrences of that PC is highly likely to cause timing violations under identical temperature and voltage conditions. This phenomenon helps us to use the instruction PC to predict an upcoming timing violation in a pipe stage, several clock cycles early.

S1.2 Cross-Layer Methodology

Figure 6 presents an overview of our extensive cross-layer methodology. We use the Fabscalar system design environment that allows us to create and verify superscalar microprocessor cores [14]. Using Fabscalar, we extract synthesizable RTL designs of a few critical components of a microprocessor core (details in Section S1.2.2). The Fabscalar environment helps us to combine architectural simulation of real programs with gate level logic simulation using the Cadence NC-Verilog functional verification tool. We simulate the execution of six SPEC2000 integer benchmarks on each core component to obtain the inputs corresponding to specific instructions.

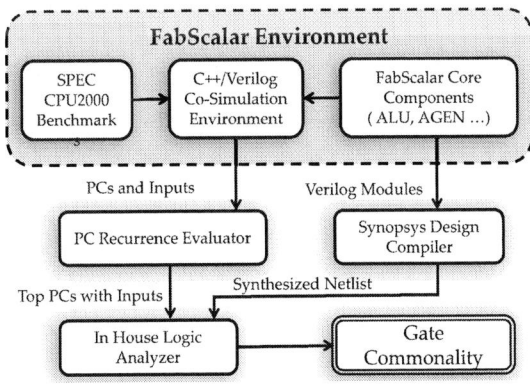

Figure 6: Cross-layer Methodology.

We study the timing violation predictability of instructions, each identified by a unique Program Counter (PC). Along with each PC, we collect its respective inputs for each benchmark. For each PC, we also identify the preceding instruction PC that sets the internal logic state of a microprocessor component. Finally, our in house

Module	Issue Queue Select	ALU	AGEN	Forward Check
# Gates	189	4728	491	428
Logic Depth	33	46	43	15

Table 3: Details of Synthesized Processor Components.

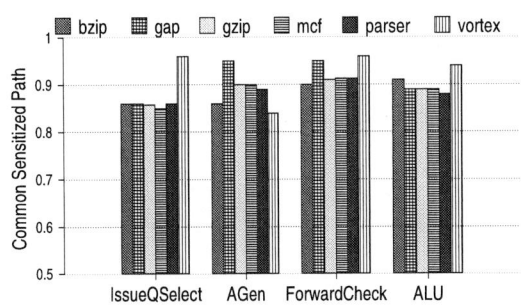

Figure 7: Commonality in sensitized paths in four components of a microprocessor core.

logic analyzer combines inputs from several repeated instances of a PC with a synthesized microprocessor component.

Commonality Estimation: We estimate the gate level commonality in the sensitized paths by many dynamic instances of a given static PC using the following expression. If ϕ is the set of gates in a circuit that change state in every dynamic instance of a static PC, and ψ is the set of gates that change state in at least one dynamic instance of the same static PC, then the commonality in sensitized gates for that PC is calculated using the ratio $\frac{\phi}{\psi}$.

S1.2.1 Core Microarchitecture

Using the Fabscalar Core-1, we generate synthesizable RTL design of a 4-wide out-of-order microprocessor core with 32 entry instruction queue, 96 entry physical register file and fetch-to-execute pipeline depth of 10 (Figure 1).

S1.2.2 Core Components

The following four microprocessor components are selected for this study. Together, they cover a wide spectrum of micro-architectural events for studying instruction level commonality in sensitized paths.

- 32-bit *Simple ALU*: We select this component as it contains a high logic depth compared to most other structures in a microprocessor core. Consequently, the ALU provides an interesting structure to study commonality in sensitized paths. It also offers a way to compare our results with the existing work.

- *Issue Queue Select*: This unit implements the instruction selection logic in the processor. Given a request vector from the existing instructions in the issue queue, it picks up to four instructions to be scheduled. The selection logic sets the request grant line for the selected instructions. Due to frequently repeated patterns in instruction selection, we expect a high degree of commonality in this structural component.

- *Address Generation Unit* (AGEN): This module represents the effective address computation necessary during typical load and store instructions. A given static instruction can compute different addresses in its various instances. However, often these effective addresses differ by a single bit (e.g., while looping through an array structure), resulting in excellent similarity in the sensitized logic paths.

- *Forward Check Logic*: This unit controls the latches in the by-

Scheme	Overhead (Scheduler only)			Overhead (core-level)		
	Area	Dynamic Power	Leakage Power	Area	Dynamic Power	Leakage Power
ABS	0.77%	0.57%	0.87%	0.03%	0.05%	0.01%
FFS	0.77%	0.57%	0.87%	0.03%	0.05%	0.01%
CDS	6.35%	1.56%	6.80%	0.24%	0.13%	0.08%

Table 2: Area and Power overhead of proposed VTE.

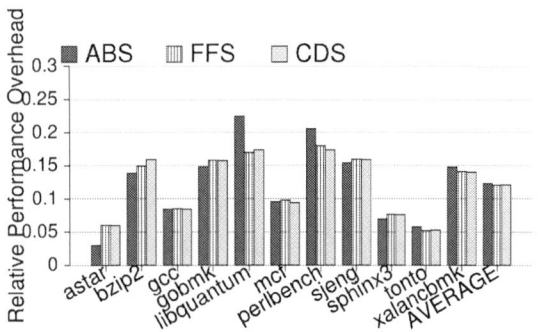

Figure 8: Performance Overhead Comparison During Faulty Execution at high fault rate (0.97V), normalized to EP [12,13]. (Lower is better.)

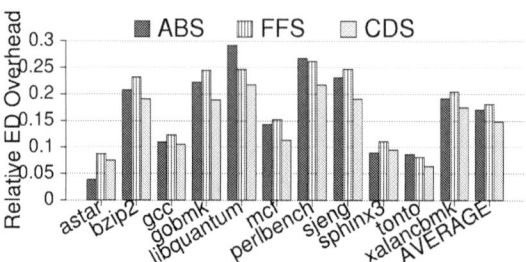

Figure 9: ED Overhead Comparison During Faulty Execution at high fault rate (0.97V), normalized to EP [12,13]. (Lower is better.)

pass network to ensure correct execution of back-to-back dependent instructions. If dependency conditions are met, then the output from a functional unit is latched directly to one of the inputs in the same or another functional unit. From a given instruction perspective, these logic steps will compute identical latching of outputs as long as scheduling decisions remain identical. Since the code path followed in a program often recur, instructions behind a given instruction tend to recur frequently, leading to identical scheduling decisions. Thus, we expect a high degree of similarity in the sensitized paths.

These components are synthesized using the Synopsys Design Compiler tool. Table 3 presents the characteristics of the synthesized processor components. The size and complexity of these structures vary substantially. For example, *Simple ALU* has the largest size (4728 gates) and greatest logic depth. In comparison, the *Forward Check* module is substantially smaller. Together, these structures represent a range of sizes and complexities of sub-modules expected in modern pipelined microprocessors.

S1.3 Results

Figure 7 presents the commonality in sensitized paths for several benchmark programs in the four selected components. The results show the weighted average, based on frequencies of each instruction, of all dynamic instances from the static instructions exercising the units. We see a substantially high commonality in the paths sensitized across a wide range of real programs. On an average, we observe a 87.4%, 89%, 92.4% and 90% commonality in the issue queue select, address generator, forward check logic and ALU, respectively. Certain benchmarks like *vortex* show an extremely high commonality (e.g., 96% in the issue queue) as it operates on a smaller range of input values.

S1.4 Opportunity For Predictive Scheduling

The collective high commonality in the sub-modules indicates a high timing violation predictability from instruction PCs. This predictability opens up a whole new class of violation mitigation

techniques, radically marginalizing the overhead of tolerating timing violations. Using precise information about a timing violation, several clock cycles ahead, it is possible to design violation aware scheduling techniques as presented in this work, to tolerate the violation with a minimum performance overhead.

S2. HIGH FAULT RATE ENVIRONMENT

During a higher fault rate (V_{DD} = 0.97V), our scheme reduces the performance overhead by 88% on an average across various benchmarks (Figure 8). In certain benchmarks like *libquantum*, both FFS and CDS are more effective compared to ABS, achieving 83% overhead reduction compared to 78% reduction. On the other hand, in *astar*, ABS outperforms both FFS and CDS. On an average, our schemes reduce the ED overhead by 83% (Figure 9).

S3. AREA AND POWER OVERHEAD

Table 2 presents the overhead of our proposed schemes compared to the scheduler in the baseline machine that tolerates timing violations through error padding. Our proposed schemes ABS and FFS utilize the same fundamental logic in scheduling and tracking functional unit status, while CDS needs additional logic to dynamically assess the criticality of instructions. At the entire core level the scheduler consumes 3.9% area, 8.9% dynamic power, and 1.2% leakage power. Thus, the overheads at the entire core level are minimal in our schemes (e.g., 0.24% area overhead, 0.13% dynamic power overhead, and 0.08% leakage power overhead in CDS).

728

Hierarchical Decoding of Double Error Correcting Codes for High Speed Reliable Memories

Zhen Wang

Mediatek Wireless Inc. , Woburn , USA

wang.zhen.mtk@gmail.com

Abstract—As the technology moves into the nano-realm, traditional single-error-correcting, double-error-detecting (SEC-DED) codes are no longer sufficient for protecting memories against transient errors due to the increased multi-bit error rate. The well known double-error-correcting BCH codes and the classical decoding method for BCH codes based on Berlekamp-Massey algorithm and Chien search cannot be directly adopted to replace SEC-DED codes because of their much larger decoding latency. In this paper, we propose the hierarchical double-error-correcting (HDEC) code. The construction methods and the decoder architecture for the codes are described. The presented error correcting algorithm takes only 1 clock cycle to finish if no error or a single-bit error occurs. When there are multi-bit errors, the decoding latency is $O(log_2m)$ clock cycles for codes defined over $GF(2^m)$. This is much smaller than the latency for decoding BCH codes using Berlekamp Massey algorithm and Chien search, which is $O(k)$ clock cycles – k is the number of information bits for the code and $m \sim O(log_2k)$. Synthesis results show that the proposed $(79, 64)$ HDEC code requires only 80% of the area and consumes $< 70\%$ of the power compared to the classical $(78, 64)$ BCH code. For a large bit distortion rate ($10^{-3} \sim 10^{-2}$), the average decoding latency for the $(79, 64)$ HDEC code is only $36\% \sim 60\%$ of the latency for decoding the $(78, 64)$ BCH code.

I. INTRODUCTION

Memories are crucial parts in modern digital embedded systems, which can occupy over 50% of the whole chip area [1]. This large area of the chip is especially vulnerable to soft errors caused by single energetic particles like high-energy neutrons and alpha particles. While most of the soft errors only cause single-event upsets (SEUs), the probability of multi-bit upsets (MBUs) – especially double-bit errors – is drastically increased as the technology moves into the nano-realm [2].

Error correcting codes are widely adopted to protect memories against soft errors. The most commonly used codes are single-error-correcting, double-error-detecting (SEC-DED) codes, e.g. Hsiao codes [3]. In many situations correcting only single-bit errors is no longer sufficient due to the increased MBU rate. Double-bit errors occurring to cache memories, for instance, may drastically deteriorate the performance of the processor due to the extra clock cycles required to re-fetch the data from external memories under a high MBU rate.

The decoder for the classical double-error-correcting BCH codes [4] has higher area complexity and consumes more power than the decoder for SEC-DED codes. The conventional decoding method for BCH codes based on Berlekamp-Massey algorithm and Chien search requires $O(k)$ clock cycles to complete while the decoding algorithm for SEC-DED codes takes only 1 clock cycle. For applications like cache memories, the average latency for decoding BCH codes can still be larger than SEC-DED codes even though the latter needs to re-fetch data from external memories more often. Thereby, BCH codes cannot be directly used to replace SEC-DED codes to protect memories against MBUs in speed-critical applications.

There are many works discussing how to protect memories against multi-bit errors. In [5], the authors adopted a pure combinational logic approach to implement the decoder for BCH codes. All syndromes corresponding to different error patterns were pre-computed and stored in a ROM-based lookup table (LUT). The decoding can be finished in one clock cycle. However, the decoder shown in [5] still has $50\% \sim 70\%$ latency penalty compared to the decoder for the Hsiao codes. Moreover, the area overhead for the decoder increases exponentially and can be more than 10 times larger than that of the Hsiao codes. In [6], the authors proposed to use a 2-D coding technique to correct multi-bit errors in memories. The proposed technique requires a few hundred or thousand cycles for the error correction depending on the number of rows in memories and is not suitable for speed-critical applications. In [7], the authors re-designed the parity check matrix of SEC-DED codes so that the codes can correct adjacent double-bit errors. These codes, however, cannot correct non-adjacent double-bit errors and are not fully DEC codes. Similar method was used by the authors in [8] to construct codes with the same parameters as double-error-correcting BCH codes, which can also correct some burst errors whose multiplicities are larger than 2. But the authors in [8] did not simplify or analyze the decoding complexity of the presented codes.

Although MBU rate for memories increases in nano-realm technologies, the probability of single-bit errors is still much higher than multi-bit errors. A number of papers proposed to improve the decoding efficiency for multi-bit error correcting codes by distinguishing between single-bit errors and multi-bit errors. In [9], the authors proposed to partially compute the syndromes for double-error-correcting BCH codes to detect errors before conducting the error correcting algorithm. In [10], the authors proposed to correct single-bit errors using BCH codes by comparing syndromes to columns of the parity check matrix and correct multi-bit errors using classical decoding methodology for BCH codes. The single-error-correcting stage presented in [10] can be further simplified for double-error-correcting codes by methods presented in Section II. In [11], the authors showed a hierarchical decoding architecture for large memories based on concatenating SEC-DED and multi-bit error correcting BCH codes. We show in Section III that the architecture described in [11] can be further improved by studying the code structure of BCH codes.

In this paper we modify the BCH codes to construct the hierarchical double-error-correcting codes (HDEC). Compared to works presented in [9], [10], [11], the primary contributions of the paper are two-fold. First, we improved the single-error-correcting stage of the code. When there is no error or a single-bit error in the information bits of the code, the HDEC code has exactly the same decoding procedure, decoding latency and similar dynamic power

Permission to make digital or hard copies of all or part of this work for personal or classroom use is granted without fee provided that copies are not made or distributed for profit or commercial advantage and that copies bear this notice and the full citation on the first page. To copy otherwise, to republish, to post on servers or to redistribute to lists, requires prior specific permission and/or a fee.
DAC '13, May 29 - June 07 2013, Austin, TX, USA.

consumption as the SEC-DED code. Second, when there are double-bit errors, we reduce the number of clock cycles required for decoding by applying alternative decoding methodology based on solving quadratic equations in Galois field. Suppose the code is defined over $GF(2^m)$, the proposed decoding algorithm requires only $O(log_2 m)$ clock cycles to correct double-bit errors. This is much better than the $O(k)$ clock cycles (k is the number of information bits of the code) required by the more general error correcting method based on Berlekamp-Massey algorithm and Chien search for BCH codes that can correct not only double-bit errors but also errors with higher multiplicities.

The rest part of the paper is organized as follows. In Section II, we show the basic constructions of the HDEC codes and the hierarchical decoding algorithm. In Section III, a double-error-correcting methodology for memories with long data-width is presented. The decoder architecture and the decoding complexity for the proposed codes are studied in Section IV. The decoding latency and the synthesis results for the area and the average power consumption for the $(72, 64)$ Hsiao code, the $(78, 64)$ BCH code and the $(79, 64)$ HDEC code are shown and compared in Section V.

II. HIERARCHICAL DECODING FOR MODIFIED DOUBLE-ERROR-CORRECTING BCH CODES

The property of a (n, k, d) linear code is determined by its $r \times n$ parity check matrix H, where $r = n - k$ is the number of redundant bits. Let h_i be the i_{th} column of H. For any (n, k, d) linear code, h_i is a r-bit vector that corresponds to an element in the Galois field $GF(2^r)$. For a possibly distorted codeword $\tilde{c}$, the syndrome of a linear code can be computed as $S = H\tilde{c}$. No error is detected if $S = \mathbf{0} \in GF(2^r)$.

Let α be a primitive element in $GF(2^m)$. For any length $n \leq 2^m - 1$, the parity check matrix for a Hamming code can be taken as

$$H = [\alpha^0 \alpha \cdots \alpha^{n-1}]. \qquad (1)$$

The number of redundant bits and information bits for the Hamming code is $r = m$ and $k = n - r$ respectively.

For the same number of information bits k, the parity check matrix of a double-error-correcting BCH code over $GF(2^m)$ can be obtained from H as follows, assuming $n + m \leq 2^m - 1$.

$$B^\Delta = \left[\begin{array}{c} H^\Delta \\ L \end{array} \right] = \left[\begin{array}{cccc} \alpha^0 & \alpha & \cdots & \alpha^{n-1+m} \\ \alpha^0 & \alpha^3 & \cdots & \alpha^{3(n-1+m)} \end{array} \right], \qquad (2)$$

where

$$H^\Delta = [H \mid \alpha^n \cdots \alpha^{n-1+m}]. \qquad (3)$$

The number of redundant bits for the BCH code is $2m$.

For any possibly distorted codeword $\tilde{c}$, the syndrome of the BCH code can be computed as

$$S = B^\Delta \tilde{c} = \left[\begin{array}{c} H^\Delta \tilde{c} \\ L\tilde{c} \end{array} \right] = \left[\begin{array}{c} S_1 \\ S_2 \end{array} \right]. \qquad (4)$$

Intuitively, due to the similarity between the parity check matrices of BCH codes and Hamming codes, the decoding of a double-error-correcting BCH code can be potentially split into two steps as follows.

1) Compute S_1 and apply the single-error-correcting algorithm for the Hamming code;
2) If an error is detected but cannot be corrected by the Hamming code, compute S_2 and apply the double-error-correcting algorithm for the BCH code.

The above straightforward hierarchical decoding method for BCH codes, however, has the following problems.

First, since the Hamming code only has distance 3, it is possible that a double-bit error will be miscorrected as a single-bit error by

the error correcting algorithm for the Hamming code. As a result, the second step of the error correcting algorithm will not be triggered and the double-bit error will he miscorrected by the BCH code.

Second, when there is a single-bit error occurring to the code, the above decoding method has slightly higher decoding complexity than the Hamming codes protecting the same number of information bits due to the longer code length.

The first problem can be solved by starting with an extended Hamming code with distance 4. To solve the second problem, we note that in practice the detection/correction of errors in the redundant bits are not important since the redundant bits are not used for the following computations. Thereby, the codes can be modified to simplify the hierarchical decoding complexity as shown in the next Theorem.

Theorem 2.1: Let V be a $(n_1, k_1, 4)$ extended Hamming code with the parity check matrix

$$H = \left[\begin{array}{cccc} 0 & \alpha^0 & \cdots & \alpha^{n_1-2} \\ 1 & 1 & \cdots & 1 \end{array} \right], \qquad (5)$$

where α is a primitive element in $GF(2^m)$, $r_1 = n_1 - k_1 = m + 1$. Let

$$r_2 = \left\{ \begin{array}{ll} m & , n_1 + m \leq 2^m \\ m + 1 & , n_1 + m > 2^m \end{array} \right. . \qquad (6)$$

Let h_i be the i_{th} column of H and $\delta = (h_0)^3$ when $r_2 = m + 1$ and $\delta = \mathbf{0}$ otherwise, $\delta \in GF(2^{r_2})$. Construct a $(r_1 + r_2) \times (n_1 + r_2)$ matrix B^* as shown below.

$$B^* = \left[\begin{array}{c} H^* \\ L^* \end{array} \right] = \left[\begin{array}{cccc|ccc} \mathbf{0} & \alpha^0 & \cdots & \alpha^{n_1-2} & \mathbf{0} & \cdots & \mathbf{0} \\ 1 & 1 & \cdots & 1 & 0 & \cdots & 0 \\ \delta & \bar{\beta}_0 & \cdots & \bar{\beta}_{n_1-2} & & I & \end{array} \right], \qquad (7)$$

where I is a $r_2 \times r_2$ identity matrix and $\beta_i \in GF(2^{r_2}), 0 \leq i \leq n_1 - 2$ is defined by

$$\beta i = \left\{ \begin{array}{ll} \alpha^{3i} & , r_2 = m, \\ \left[\begin{array}{c} \alpha^i \\ 1 \end{array} \right]^3 & , r_2 = m + 1. \end{array} \right. \qquad (8)$$

The code C with B^* as the parity check matrix has length $n_1 + r_2$, dimension k_1 and can correct any single-bit and double-bit errors if at least one information bit is distorted.

Let $\tilde{c}$ be a possibly distorted codeword. The error correcting algorithm for a code C constructed by Theorem 2.1 can be divided into the following steps.

1) Compute the syndrome for the extended Hamming code $S_1 = H^* \tilde{c} \in GF(2^{m+1})$.
2) Conduct the error correcting algorithm for the extended Hamming code using the syndrome S_1. If there is no error or there is a single-bit error, the error will be successfully corrected by the extended Hamming code and the decoding procedure for C is completed. Otherwise go to step 3.
3) If an uncorrectable error is detected by the extended Hamming code, compute the syndrome $S_2 = L^* \tilde{c}$. Conduct the error correcting algorithm for a double-error-correcting BCH code defined over $GF(2^{r_2})$ using S_1 and S_2. (When $r_2 = m$, only the last m bits of S_1 is used.)

Following the decoding steps described above, the code C constructed by Theorem 2.1 can correct any single-bit or double-bit errors assuming at least one information bit is distorted. When $n_1 + m > 2^m$, codes constructed by Theorem 2.1 have the same number of redundant bits as a double-error-correcting BCH codes with the same k.

Different from BCH codes based on Berlekamp-Massey algorithm and Chien search, the components used for double-error-correcting for codes constructed by Theorem 2.1 are only active when there

730

are multi-errors. When there is no error or a single-bit error, the decoder for codes constructed by Theorem 2.1 has the same decoding latency and similar dynamic power consumption to the decoder of SEC-DED codes. The extra latency and dynamic power consumption for correcting double-bit errors are only needed when uncorrectable multi-bit errors are detected by the extended Hamming code. Thereby, the average decoding latency and power consumption of codes constructed by Theorem 2.1 are expected to be smaller than that of the BCH codes based on Berlekamp-Massey algorithm and Chien search (see Section V).

Traditional double-error-correcting using BCH codes is usually based on Berlekamp Massey algorithm and Chien search which requires $O(k)$ clock cycles to finish. An alternative method of correcting double-bit errors is shown in Section IV, which can take only $O(log_2 m)$ clock cycles.

Since errors among bits n_1 to $n_1 + r_2 - 1$ are ignored during the SEC-DED step, the decoding algorithm can also correct the information part of some errors with a multiplicity larger than 2 as shown in the following example.

Example 2.1: Let α be a primitive element in $GF(2^3)$ constructed by $x^3 + x + 1 = 0$. The parity check matrix H of a $(n_1 = 8, k_1 = 4, 4)$ extended Hamming code can be taken as

$$H = \begin{bmatrix} \mathbf{0} & \alpha^0 & \alpha^1 & \alpha^2 & \alpha^3 & \alpha^4 & \alpha^5 & \alpha^6 \\ 1 & 1 & 1 & 1 & 1 & 1 & 1 & 1 \end{bmatrix} \quad (9)$$

$$= \begin{bmatrix} 0 & 0 & 0 & 1 & 0 & 1 & 1 & 1 \\ 0 & 0 & 1 & 0 & 1 & 1 & 1 & 0 \\ 0 & 1 & 0 & 0 & 1 & 0 & 1 & 1 \\ 1 & 1 & 1 & 1 & 1 & 1 & 1 & 1 \end{bmatrix}. \quad (10)$$

Since $m = 3$ and $n_1 + m > 2^m$, we have $r_2 = m + 1 = 4 = r_1$. Suppose the double-error-correcting code is defined over $GF(2^4)$ constructed by $x^4 + x + 1 = 0$ with a normal base $\{\lambda^3, \lambda^6, \lambda^{12}, \lambda^9\}$ [12], where λ is the primitive element in $GF(2^4)$. We have

$$B^* = \begin{bmatrix} 00 & 0 & 1 & 0 & 1 & 1 & 1 & 0000 \\ 00 & 1 & 0 & 1 & 1 & 1 & 0 & 0000 \\ 01 & 0 & 0 & 1 & 0 & 1 & 1 & 0000 \\ 11 & 1 & 1 & 1 & 1 & 1 & 1 & 0000 \\ \delta & \beta_0 & \beta_1 & \beta_2 & \beta_3 & \beta_4 & \beta_5 & \beta_6 & I \end{bmatrix}$$

$$= \begin{bmatrix} 00010111 & 0000 \\ 00101110 & 0000 \\ 01001011 & 0000 \\ 11111111 & 0000 \\ 10100110 & 1000 \\ 00100010 & 0100 \\ 01101010 & 0010 \\ 00110011 & 0001 \end{bmatrix}. \quad (11)$$

$$(12)$$

The code has the same number of redundant bits as a $(12, 4, 5)$ double-error-correcting BCH code. Let $e = (e_0, e_1, \cdots, e_{11})$ be the error vector. Suppose e is a triple-bit error with $e_5 = e_{10} = e_{11} = 1$. Then e_{10} and e_{11} will be ignored by the error correcting algorithm for the extended Hamming code. The syndrome is $S_1 = (1101)^T$. The information part of the triple-bit error can be successfully corrected by the extended Hamming code.

If e is a double-bit error with $e_1 = e_2 = 1$, the error can be detected but cannot be corrected by the extended Hamming code. In this case S_2 is computed. We have $S_1 = (0011)^T + (0101)^T = (0110)^T$ and $S_2 = (0010)^T + (1111)^T = (1101)^T$. The error can be successfully corrected by any error correcting algorithm for BCH codes. (See Section IV-C.)

Remark 2.1: In theory, H does not have to be defined by (5). Most of the parity check matrices for SEC-DED codes that optimize the decoder area and/or power consumption can also be used, e.g. the parity check matrix for Hasio codes.

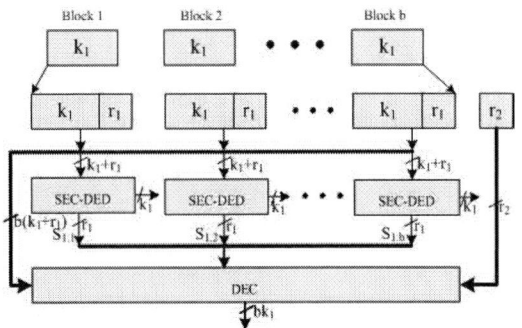

Fig. 1. The Architecture for Decoding Codes Generated by Corollary 3.1

To end this section, we summarize the correctable errors and the corresponding decoding complexity for codes constructed by Theorem 2.1 in Table I.

III. DOUBLE-ERROR-CORRECTING METHODOLOGIES FOR LONG DATA WIDTH

Generally speaking, the encoding and decoding latency for a code will grow as the length of the code increases. As a result, for memories with a long data width, e.g. a 512-bit cache line, a direct implementation of the double-error-correcting mechanism based on Theorem 2.1 may be unsuitable.

In this case, the memory data can be split into small blocks. Each block is protected by a separate SEC-DED code. A hierarchical double-error-correcting (HDEC) code can be constructed by adding extra redundant bits computed using all the memory data as shown in the next Corollary.

Corollary 3.1: Suppose b identical $(n_1, k_1, 4)$ SEC-DED codes with the parity check matrix H (see (5)) are used to protect bk_1 information bits. Define $\delta, \beta_i, 0 \le i \le n_1 - 2$ and r_2 as described in Theorem 2.1. Let

$$M = \begin{bmatrix} H & & & & \\ & H & & & \\ & & \ddots & & \\ & & & H & H^* \\ \underbrace{L & L & \cdots & L}_{b-1} & L^* \end{bmatrix}, \quad (13)$$

where H^* and L^* is defined by (7). The code C with M as the parity check matrix has length $bn_1 + r_2$, bk_1 information bits and can correct any single-bit or double-bit errors if at least one information bit is distorted.

The architecture of protecting a bk_1-bit data line using the code constructed by Corollary 3.1 is shown in Figure 1. Denote the syndromes for the b SEC-DED codes by $S_{1.1}, S_{1.2}, \cdots, S_{1.b}$. Let

$$S_2 = [\underbrace{LL \cdots L}_{b-1} L^*]\tilde{c} \quad (14)$$

be the syndrome corresponding to the r_2 extra redundant bits, where $\tilde{c}$ is the possibly distorted codeword and L^* is given in (7). Since $S_{1.i}, 1 \le i \le b$ can be computed in parallel, the HDEC codes constructed by Corollary 3.1 have the same decoding latency as the $(n_1, k_1, 4)$ SEC-DED code when there is at most 1 distorted bit. The computation of S_2 is only required when uncorrectable multi-bit errors occurring to a single SEC-DED code are detected.

The architecture shown in Figure 1 can correct some errors with multiplicities larger than 2. If a multi-bit error occurs to several SEC-DED codes and at most one bit is distorted in each of the code,

731

TABLE I

CORRECTABLE ERRORS $e = (e_0, e_1, \cdots, e_{n_1-1}, e_{n_1}, \cdots e_{n_1+r_2-1})$ FOR CODES CONSTRUCTED BY THEOREM 2.1

Multiplicity	Distorted Bit(s)	Correctable	Decoding Complexity
1	$e_i, 0 \leq i \leq n_1 - 1$	Yes	$\approx$ SEC-DED Code
1	$e_i, n_1 \leq i \leq n_1 + r_2 - 1$	No (Safe to ignore)	-
2	$e_i, e_j, 0 \leq i, j \leq n_1 - 1$	Yes	$\leq$ Double-Error-Correcting BCH Code
2	$e_i, 0 \leq i \leq n_1 - 1$ $e_j, n_1 \leq i \leq n_1 + r_2 - 1$	Yes	$\approx$ SEC-DED Code
2	$e_i, e_j, n_1 \leq i, j \leq n_1 + r_2 - 1$	No (Safe to ignore)	-
≥ 3	One bit among $e_0 \sim e_{n_1-1}$ Other bits among $e_{n_1} \sim e_{n_1+r_2-1}$	Yes	$\approx$ SEC-DED Code

then the multi-bit error can be successfully corrected during the single-error-correcting stage. The double-error-correcting procedure is invoked only if a multi-bit error occurs to the same SEC-DED code. Increasing b can shorten the length of the critical path for the decoder for SEC-DED codes and can reduce the probability of invoking the double-error-correcting procedure thus reduce the average decoding latency of the code. However, larger b also results in larger area overhead since the total number of redundant bits for the SEC-DED codes will increase.

Compared to the method described in [11], codes constructed by Corollary 3.1 require less redundant bits to protect memories with the same data width as shown in the next example.

Example 3.1: Suppose 512 information bits are split into 8 blocks. Each of the block contains a 64-bit double word and can be protected by a $(72, 64, 4)$ extended Hamming code. To correct double-bit errors, it is sufficient to add another 7 redundant bits to form a code with a length of $72 \times 8 + 7 = 583$ bits. The total number of redundant bits is $8 \times 8 + 7 = 71$. The code is suitable for the protection of a speed-critical cache memory in applications where the double-bit errors cannot be ignored, e.g. nano-scale microprocessors. As it was shown in [11], to correct double-bit errors in memories with the same data width, the method described in [11] requires 13 more redundant bits than codes constructed by Corollary 3.1.

IV. HARDWARE IMPLEMENTATION OF THE DECODER FOR HIERARCHICAL DOUBLE-ERROR-CORRECTING CODES

In this section, we will discuss the hardware implementation of the decoder for the HDEC codes constructed by Theorem 2.1 and Corollary 3.1.

As described in Section II, the decoding of the HDEC codes contains two steps – the SEC-DED step using extended Hamming codes and the DEC step using BCH codes. The decoder for the extended Hamming code mainly requires XOR operations and can be directly used for the SEC-DED step. The decoder for BCH codes usually consists of three blocks: the syndrome computation block, the error locator polynomial generation block and the Chien search block. We next show how to modify these blocks for the DEC step to reduce the area overhead and decoding latency for the decoder.

A. Syndrome Computation for the DEC step

Let $S = (S_1, S_2)$ be the syndrome for codes constructed by Theorem 2.1, where $S_1 = H^* \tilde{c} \in GF(2^{r_1}), S_2 = L^* \tilde{c} \in GF(2^{r_2})$ and $\tilde{c}$ is the possibly distorted codeword. S_1 is computed during the SEC-DED step and can be directly used for double-error-correcting if required. The syndrome S_2 is computed during the DEC step. In the rest part of the Section, we assume $r_2 = r_1 = m + 1$. When $r_1 = m+1, r_2 = m$, only the last m bits of S_1 are used in the DEC step (see Section II). In that case the hardware design described in this Section can be applied with minor modifications.

Traditional syndrome computation for long BCH codes is usually based on linear feedback shift registers (LFSR), which takes multiple clock cycles to finish [13]. For short double-error-correcting BCH codes, a direct implementation using XOR gates can provide a better decoding latency. For codes constructed by Corollary 3.1, although the possible error locations can be limited to n_1 bits by checking which $S_{1.i}, 1 \leq i \leq b$ is nonzero, the computation of S_2 needs all the $bn_1 + r_2$ bits of the codeword. Parallelism can be introduced into the computation of S_2 to reduce the length of the critical path if necessary.

B. Error Locator Polynomial Generation

After the syndromes are computed, the error locator polynomial is usually generated using the Berlekamp-Massey (BM) algorithm. The hardware implementation of the BM algorithm is well studied in the academic community [14], [15]. The inversionless Berlekamp-Massey algorithm [14] takes multiple clock cycles to finish. We note that for the HDEC code, the generation of the error locator polynomial can be simplified to take only one clock cycle as described below.

Assume a double-bit error occurs. Let h_i^* be the i_{th} column of H^*. When the i_{th} and the j_{th} bit of the codeword are distorted, we have $S_1 = h_i^* + h_j^*$ and $S_2 = (h_i^*)^3 + (h_j^*)^3$, where all the operations are in $GF(2^{r_2})$. Thereby, h_i^* and h_j^* are solutions of the following quadratic equation defined over $GF(2^{r_2})$.

$$\lambda^2 S_1 + \lambda S_1^2 + S_1^3 + S_2 = 0. \tag{15}$$

The square and cubic operation in $GF(2^{r_2})$ have low computational complexities. Obviously, all coefficients in the error locator polynomial defined in (15) can be generated in one clock cycle.

C. Solving the Error Locator Polynomial

The traditional decoding procedure for BCH codes uses Chien search [16] to exhaustively test whether an element in $GF(2^{r_2})$ is a root of the error locator polynomial. The Chien search takes $O(k)$ clock cycles to finish, where k is the number of information bits of the code.

For the HDEC codes, the error locator polynomial is always in the quadratic form. As a result, the problem of finding the error locations is equivalent to the problem of solving a quadratic equation over $GF(2^{r_2})$.

Let $\Lambda = \lambda S_1^{-1}$. Then (15) can be re-written as

$$\Lambda^2 + \Lambda + z = 0, \tag{16}$$

where $z = 1 + S_2 S_1^{-3}$. The computation of $z = 1 + S_2 S_1^{-3}$ mainly requires one multiplication and one inversion in $GF(2^{r_2})$. A simplified iterative inversion algorithm based on multiplication and square operation for elements in $GF(2^{r_2})$ was shown in [17]. For

732

any $z \in GF(2^{r_2})$, let $A = z$ and $B = z^2$. Then $z^{-1} = B$ after computing $B = (AB)^2$ for $r_2 - 2$ times.

It is well known that for any Galois field $GF(2^{r_2})$, there exists a normal basis $\{\theta, \theta^2, \theta^4, \cdots, \theta^{2^{r_2-1}}\}$. Any element in $z \in GF(2^{r_2})$ can be represented as

$$z = b_0\theta + b_1\theta^2 + \cdots + b_{m-1}\theta^{2^{r_2-1}}, b_i \in GF(2). \quad (17)$$

In a normal base Galois field, the square operation can be achieved by cyclic shifting. Thereby, the inversion algorithm shown in [17] only requires 1 multiplier to implement and takes $r_2 - 2$ clock cycles to finish.

Remark 4.1: The number of clock cycles for the inversion in a Normal base Galois field $GF(2^{r_2})$ can be further reduced to $O(log_2 r_2)$ by using the inversion algorithm proposed in [17], [18], [19]. These algorithms, however, usually have more complicated control logic and require more multipliers. A similar technique is used in Section V to reduce the number of clock cycles required for the inversion when the decoder is modeled in Verilog.

After computing $z = 1 + S_2 S_1^{-3}$, the trace $Tr(z) = \sum_{i=0}^{m-1} b_i$ is computed. Obviously, $Tr(z)$ is either 1 or 0. The quadratic equation (16) has two solutions when $Tr(z) = 0$ and no solutions when $Tr(z) = 1$ [4].

In a normal base Galois field, the calculation of $Tr(z)$ can be simplified to counting the number of $1's$ in the binary representation of z. If $Tr(z) = 1$, an uncorrectable multi-bit error occurs. If $Tr(z) = 0$, the roots for the quadratic equations can be derived using formulas presented in [20]. Especially, when r_2 is odd, the two roots for (16) are

$$\theta_1 = \sum_{i \in I} z^{2^i}, I = 0, 2, 4, \cdots, r_2 - 1, \quad (18)$$

$$\theta_2 = \theta_1 + 1. \quad (19)$$

The computation of (18) only requires adders and can be completed in one clock cycle in a normal base Galois field. After finding θ_1, we have $h_i^* = \theta_1 S_1$ and $h_j^* = h_i^* + S_1$. The two column vectors h_i^* and h_j^* can then be used as inputs (syndromes) to the decoder of the extended Hamming code to locate errors.

Example 2.1 (Continued) In Example 2.1, when $e_1 = e_2 = 1$, we have $S_1 = (0110)^T = \lambda^4$ and $S_1 = (1101)^T = \lambda^{13}$, where λ is the primitive element in $GF(2^4)$. Then

$$z = 1 + S_2 S_1^{-3} = (0110)^T = \lambda^4. \quad (20)$$

For $r_2 = 4$ and the above z, it was shown in [20] that the two solutions for (16) is $z^8 + z^{12} = (0010)^T = \lambda^6$ and $z^8 + z^{12} + 1 = (1101)^T = \lambda^{13}$. Thereby we have $h_i^* = \lambda^6 S_1 = \lambda^{10} = (0101)^T$ and $h_j^* = \lambda^{13} S_1 = \lambda^2 = (0011)^T$. The errors can be successfully corrected by the decoder of the extended Hamming code using h_i^* and h_j^*. Solving the quadratic equation in this case requires one multiplier to compute z^{12} and one adder to sum z^8 and z^{12}.

V. SYNTHESIS RESULTS OF THE AREA AND THE AVERAGE POWER CONSUMPTION FOR THE DECODER

The decoder for the $(72, 64)$ Hsiao code, the $(78, 64)$ BCH code and the $(79, 64)$ HDEC code constructed by Theorem 2.1 were modeled in Verilog and synthesized in Cadence RTL Compiler using the Nangate 45nm open cell library [21]. The clock speed was fixed to be 1GHz for all three designs. The area and the average power consumption for different decoders are shown in Table II.

The SEC-DED Hsiao code has the smallest area overhead and the smallest average power consumption among the three alternatives. The decoder for the $(78, 64)$ BCH code has a parallelism level of 8. A simplified Berlekamp-Massey algorithm was implemented for the

TABLE II

THE AREA AND THE AVERAGE POWER CONSUMPTION FOR THE $(72, 64)$ HSIAO CODE, THE $(79, 64)$ EXTENDED BCH CODE AND THE $(79, 64)$ HDEC CODE CONSTRUCTED BY THEOREM 2.1 (CLOCK = 1GHZ)

	Area (μm^2)	Power (nW)		
		Leakage	Dynamic	Total
Hsiao	861.84	10411.78	436933.85	447345.64
BCH	4444.06	66295.34	1531439.06	1597734.41
HDEC	3567.59	49489.14	1059119.09	1108608.23

TABLE III

THE ERROR CORRECTING LATENCY (NUMBER OF CLOCK CYCLES) FOR THE $(72, 64)$ HSIAO CODE, THE $(78, 64)$ BCH CODE AND THE $(79, 64)$ HDEC CODE CONSTRUCTED BY THEOREM 2.1

	No Error	Single	Double	Multiplicity > 2
Hsiao	1	1	-	-
BCH	1	10	10	-
HDEC	1	1	7	7*

* A small portion of errors whose multiplicity is larger than 2 can be corrected.

BCH code to reduce the area overhead and the power consumption. However, the area overhead and the average power consumption of the decoder for the double-error-correcting BCH code are still 5.1 times and 3.5 times larger than that of the Hsiao code.

As expected, the decoder for the $(79, 64)$ HDEC code constructed by Theorem 2.1 consumes less power and requires less area than the double-error-correcting BCH code. Compared to the BCH code, the decoder for the proposed HDEC code requires 80% of the area and consumes 70% of the total average power. Moreover, since the double-error-correcting circuits in the decoder for the proposed HDEC code is only active when a multi-bit error is detected, in practice the average dynamic power consumption of the decoder is much smaller than the data shown in Table II. As the probability of multi-bit errors decreases, the dynamic power consumption of the HDEC code becomes more and more similar to the dynamic power consumption of the Hsiao code since the active parts in the decoders for both alternatives for correcting single-bit errors are the same.

Table III shows the number of clock cycles required to detect or correct different types of errors using the three error-correcting codes. The $(78, 64)$ BCH code with a parallelism level of 8 using a simplified Berlekmap-Massey algorithm and Chien search always requires 10 clock cycles to correct errors, no matter whether it is a single-bit error or a double-bit error (1 for syndrome computation, 1 for Berlekamp-Massey algorithm and 8 for Chien search).

Different from BCH codes, the proposed HDEC code takes only 1 clock cycle to correct single-bit errors. If multi-bit errors are detected by the extended Hamming code, the DEC stage will be triggered. For the $(79, 64)$ HDEC code, $r_2 = 7$ and $S_1^{-3} = S_1^{2^2+2^3+2^4+2^5+2^6}$ in $GF(2^7)$. Since

$$S_1^{2^4+2^5} = S_1^{(2^2+2^3)^{2^2}}, \quad (21)$$

the constant term $z = 1 + S_2 S^{-3}$ in the error locator polynomial can be computed in 3 clock cycles as follows.

$$1: \quad A = S_1^{2^2} \times S_1^{2^3}, B = S_1^{2^6} \times S_2;$$

$$2: \quad A = A \times A^{2^2};$$

$$3: \quad z = A \times B + 1.$$

The computation requires 2 normal base multipliers given the fact that square operation in the normal base can be achieved by cyclic shifting. The computational complexity and the number of clock cycles required to solve the error locator polynomial $\Lambda^2 + \Lambda + z = 0$ depend on r_2 [20]. When r_2 is odd and all operations are in a normal base Galois field $GF(2^{r_2})$, the root of the error locator polynomial $\Lambda^2 + \Lambda + z = 0$ can be calculated in one clock cycle (see (18)). Another two clock cycles are required for locating the erroneous bits using the decoder for the extended Hamming code. Thereby, when multi-bit errors occur, the total number of clock cycles required for decoding the $(79, 64)$ HDEC code is 7 (1 for the single-error-correcting stage, 1 for the computation of S_2, 3 for the computation of the constant term and 2 for locating errors using the decoder for extended Hamming code).

Figure 2 shows the average decoding latency in the number of clock cycles for the three codes under different bit error rate p. When p is very small, all three codes take about 1 clock cycle to finish the decoding on average. As p increases, the number of clock cycles required by BCH code for decoding converges to 10. The decoding latency for Hsiao code is still one clock cycle. However, for memories such as caches, uncorrectable errors by Hsiao code may trigger a re-fetch of data from higher level caches or external memories, which can result in a worse system response time and increase the average decoding latency. The HDEC code requires less clock cycles for decoding than BCH code. When $p = 10^{-3}$, the HDEC code has an average decoding latency of 1.01 clock cycles, which is very close to that of Hisiao code. Under the same p, BCH code has an average decoding latency of 1.67 clock cycles, which is 67% larger than that of the Hsiao code and the HDEC code. Similarly, when $p = 10^{-2}$, the average decoding latency for BCH code is 5.89 clock cycles and is 2.77 times larger than that of the HDEC code.

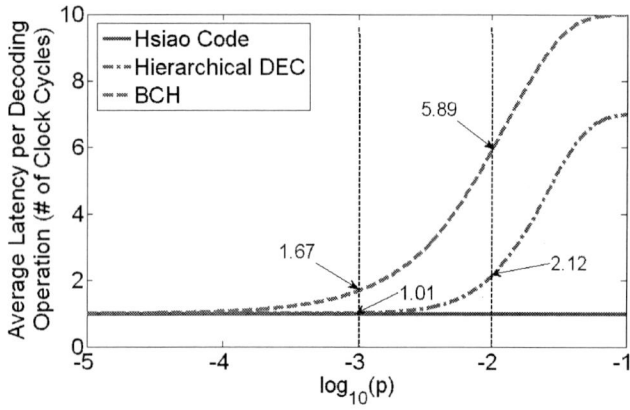

Fig. 2. The average decoding latency for the (72, 64) Hsiao Code, the (78,64) BCH code and the (79,64) HDEC code. In practice the average decoding latency for the (72,64) Hsiao Code can be drastically worse for large bit distortion rate due to the data re-fetch penalty.

VI. CONCLUSIONS

In this paper a hierarchical double-error-correcting algorithm and the decoder architecture for the hierarchical double-error-correcting (HDEC) codes are proposed. The algorithm takes only 1 clock cycle when there is no error or a single-bit error. When multi-bit errors occur, the algorithm requires $O(log_2 m)$ clock cycles to finish for codes defined over $GF(2^m)$. This is much smaller than the decoding latency for BCH codes using Berlekamp-Massey algorithm and Chien search, which is $O(k)$ clock cycles – k is the number of information bits and $m \sim O(log_2 k)$. The proposed algorithm can also correct the information part of some errors with multiplicities larger than 2 with no extra overhead. The presented decoder for the HDEC code

has similar dynamic power consumption to SEC-DED codes when at most 1-bit of the codeword is distorted. The average decoding latency for the presented codes using the proposed double-error-correcting algorithm is higher than SEC-DED codes but lower than the BCH codes using Berlekamp-Massey algorithm and Chian search. Synthesis results show that the $(79, 64)$ HDEC code requires 80% of the area and consumes $< 70\%$ of the total average power compared to the $(78, 64)$ BCH code. For a large bit distortion rate ($10^{-3} \sim 10^{-2}$), the average decoding latency for the $(79, 64)$ HDEC code is only $36\% \sim 60\%$ of the average latency for the $(78, 64)$ BCH code.

REFERENCES

[1] S.K.Moore, "Winner: Masters of memory," *IEEE Spectrum*, vol. 44, no. 1, pp. 45–49, Jan 2007.

[2] E. Ibe, H. Taniguchi, Y. Yahagi, K.-i. Shimbo, and T. Toba, "Impact of scaling on neutron-induced soft error in srams from a 250 nm to a 22 nm design rule," *Electron Devices, IEEE Transactions on*, vol. 57, no. 7, pp. 1527 –1538, july 2010.

[3] M. Y. Hsiao, "A class of optimal minimum odd-weight-column SEC-DED codes," *IBM Journal of Research and Development*, vol. 14, pp. 395–401, 1970.

[4] F. MacWilliams and N. Sloane, *The Theory of Error-Correcting Codes*. North-Holland, 1998.

[5] R. Naseer and J. Draper, "DEC ECC design to improve memory reliability in sub-100nm technologies," in *Electronics, Circuits and Systems, 2008. ICECS 2008. 15th IEEE International Conference on*, 31 2008-sept. 3 2008, pp. 586 –589.

[6] J. Kim, N. Hardavellas, K. Mai, B. Falsafi, and J. Hoe, "Multi-bit error tolerant caches using two-dimensional error coding," in *Microarchitecture, 2007. MICRO 2007. 40th Annual IEEE/ACM International Symposium on*, dec. 2007, pp. 197 –209.

[7] A. Dutta and N. Touba, "Multiple bit upset tolerant memory using a selective cycle avoidance based SEC-DED-DAEC code," in *VLSI Test Symposium, 2007. 25th IEEE*, pp. 349 –354.

[8] S. Shamshiri and K.-T. Cheng, "Error-locality-aware linear coding to correct multi-bit upsets in srams," in *Test Conference (ITC), 2010 IEEE International*, nov. 2010, pp. 1 –10.

[9] P. Reviriego, C. Argyrides, and J. A. Maestro, "Efficient error detection in double error correction BCH codes for memory applications," *Microelectronics Reliability*, vol. 52, no. 7, pp. 1528 – 1530, 2012.

[10] C. Wilkerson, A. R. Alameldeen, Z. Chishti, W. Wu, D. Somasekhar, and S.-l. Lu, "Reducing cache power with low-cost, multi-bit error-correcting codes," *SIGARCH Comput. Archit. News*, vol. 38, no. 3, pp. 83–93, Jun. 2010.

[11] E. J. Gieske, D. F. Greenberg, and R. Ramaraju, "Hierarchical error correction for large memories," 2012, Patent US 20120233498 A1.

[12] S. Gao, "Normal bases over finite fields," Ph.D. dissertation, University of Waterloo, 1993.

[13] T.-B. Pei and C. Zukowski, "High-speed parallel CRC circuits in VLSI," *Communications, IEEE Transactions on*, vol. 40, no. 4, pp. 653 –657, apr 1992.

[14] H. Burton, "Inversionless decoding of binary BCH codes," *Information Theory, IEEE Transactions on*, vol. 17, no. 4, pp. 464 – 466, jul 1971.

[15] S. Rizwan, "Retimed decomposed serial Berlekamp-Massey (BM) architecture for high-speed Reed-Solomon decoding," in *VLSI Design, 2008. VLSID 2008. 21st International Conference on*, jan. 2008, pp. 53 –58.

[16] J. Cho and W. Sung, "Strength-reduced parallel Chien search architecture for strong BCH codes," *Circuits and Systems II: Express Briefs, IEEE Transactions on*, vol. 55, no. 5, pp. 427–431, May 2008.

[17] M. Olofsson, "VLSI aspects on inversion in finite field," Ph.D. dissertation, Linkopings University, 2002.

[18] T. Itoh and S. Tsujii, "A fast algorithm for computing multiplicative inverses in gf(2m) using normal bases," *Information and Computation*, vol. 78, no. 3, pp. 171 – 177, 1988.

[19] N. Takagi, J. Yoshiki, and K. Takagi, "A fast algorithm for multiplicative inversion in GF(2m) using normal basis," *Computers, IEEE Transactions on*, vol. 50, no. 5, pp. 394 –398, may 2001.

[20] C.-L. CHEN, "Formulas for the solutions of quadratic equations over $GF(2^m)$," *Information Theory, IEEE Transactions on*, vol. 28, no. 5, pp. 792–794, Sep 1982.

[21] "Nangate 45nm open cell library," http://www.nangate.com.

APPENDIX

Proof: (Theorem 2.1) Any single-bit error in the information bits can be corrected by the extended Hamming code. If a double-bit error occurs, there are three different situations as described below.

- If the two distorted bits are both among bits n_1 to $n_1 + r_2 - 1$ (Assume that the index of the bits starts from 0.), the error will not be detected by the extended Hamming code. Double-error-correcting (DEC) procedure will not be triggered. However, since all the distorted bits are redundant bits, the error can be safely ignored.
- If one of the distorted bit is among bits 0 to $n_1 - 1$ and the other is among bits n_1 to $n_1 + r_2 - 1$, the former will be corrected by the extended Hamming code and the latter will be safely ignored.
- If the two distorted bits are both among bits 0 to $n_1 - 1$, the error will be detected by the extended Hamming code and the double-error-correcting procedure will be triggered.
  - If $r_2 = m$ and bit 0 (parity bit) is distorted, then $S_2 = S_1[1 : m]^3$, where $S_1[1 : m]$ represents the last m bits of $S_1 \in GF(2^{m+1})$. In this case $S_1[1 : m]$ can be used by the standard Hamming decoder to locate the single-bit error in the information bits.
  - If $r_2 = m + 1$ or bit 0 (parity bit) is not distorted, the double-bit errors can be successfully corrected by any decoder for the double-error-correcting BCH code.

∎

Proof: (Corollary 3.1) Any single-bit error in the information bits can be corrected by the SEC-DED code. Double-bit errors with at least one information bit distorted can be divided into the following classes.

- One bit of the error is in the extra redundant bits added for double-error-correcting. Then the single-bit error occurring to the information bits can be successfully corrected.
- The double-bit error occurs to two SEC-DED codes. In this case the two bits can be successfully corrected by the two codes separately.
- The double-bit error occurs to a single SEC-DED code. The error will be detected but cannot be corrected by the SEC-DED code. The double-error-correcting procedure will be triggered. Since the double-bit error only occurs to one SEC-DED code, it is equivalent to conduct the double-error-correcting procedure for a code with B^* as the parity check matrix. According to the proof of Theorem 2.1, the error can be successfully corrected.

∎

Power Benefit Study for Ultra-High Density Transistor-Level Monolithic 3D ICs

Young-Joon Lee, Daniel Limbrick, and Sung Kyu Lim
School of ECE, Georgia Institute of Technology, Atlanta, GA
yjlee@gatech.edu, daniel.limbrick@gatech.edu, limsk@ece.gatech.edu

ABSTRACT

The nano-scale 3D interconnects available in monolithic 3D IC technology enable ultra-high density device integration at the individual transistor-level. In this paper we demonstrate the power benefits of transistor-level monolithic 3D designs. We first build a cell library that consists of 3D gates and model their timing/power characteristics. Next, we build timing-closed, full-chip GDSII layouts and perform sign-off iso-performance power comparisons with 2D IC designs. We also study the characteristics of benchmark circuits that maximize the power benefits in monolithic 3D designs. Lastly, our study is extended to predict the power benefits of monolithic 3D designs built with future devices.

Categories and Subject Descriptors

B.8.2 [**Performance and Reliability**]: Performance Analysis and Design Aids

General Terms

Design

Keywords

3D IC, monolithic 3D, transistor-level, power analysis

1. INTRODUCTION

To better exploit the benefits from 3D die stacking, monolithic 3D technology is currently being investigated as a next generation technology. In a monolithic 3D IC, the device layers are fabricated sequentially. When the top layer is attached to the bottom layer, the top layer is a blank silicon. Alignment precision is determined by lithography stepper accuracy, which is around $10nm$ today. Also, the top layer can be made very thin, around $30nm$ [1]. Thus, monolithic inter-tier vias (**MIV**s) for vertical connections are very small—about two orders of magnitude smaller than through-silicon-via (TSV)—with almost negligible parasitic RC. With these small MIVs, designers can truly exploit the benefit of vertical dimension.

The early works for monolithic 3D ICs were technology-driven [6, 4, 9]. Recently, logic design methodologies for monolithic 3D ICs were demonstrated [2, 8, 7]. In these works, the authors presented various comparisons among monolithic 3D ICs and TSV-based 3D ICs and conventional 2D ICs in terms of footprint, timing, and power. However, timing was not closed in these works, which make the studies not practical. In addition, all these works assume that the timing and power characteristics of 3D monolithic gates are the same as 2D gates and did not demonstrate why that is a reasonable assumption. The authors also did not provide in-depth analyses and discussions on why monolithic 3D technology reduces power consumption and what factors affect the power reduction margin. This knowledge is crucial to maximize the benefit and justify on-going and future research on fabrication and design technologies for monolithic 3D ICs.

As discussed in [2, 8], monolithic 3D technology enables a very fine-grained 3D circuit partitioning. We can divide standard cells into PMOS and NMOS parts, place them in different layers, and connect them using MIVs, which we call transistor-level monolithic 3D integration (**T-MI**) in this paper. Or, as in TSV-based 3D ICs, we may place planar cells in different layers and connect them using MIVs, which is named gate-level monolithic 3D integration (**G-MI**). In this paper we focus on transistor-level integration that allows the highest integration density possible. The T-MI designs are different from G-MI: (1) Most of the 3D interconnects are embedded in the cells. (2) PMOS and NMOS transistors are on different layers, thus manufacturing processes can be optimized separately. (3) Physical layout (placement, routing, optimization, etc.) can be performed using existing 2D electronic design automation (EDA) tools, with modifications.

In this paper, we study the power benefit of T-MI based on timing-closed, detailed routing completed GDSII-level layouts and sign-off analysis on timing and power. *Our comprehensive work encompasses device and interconnect-level study, gate-level modeling and optimization, and full-chip layout constructions, optimization, and timing/power analysis for the current and future technology nodes.* With our layout-based simulations and in-depth analyses, we demonstrate how to maximize the power benefit of T-MI technology. For fair comparisons between 3D and 2D designs, timing is closed on all designs (iso-performance), and power consumption is compared. We also investigate the circuit characteristics that affect the power benefit of monolithic 3D ICs.

Our major contributions are as follows: (1) To the best of our knowledge, this is the first work to characterize the timing and power of the individual transistor-level monolithic 3D cells. We extract the internal parasitic RC of our T-MI cells and characterize their timing and power. We then compare T-MI cells with 2D counterparts. (2) We study the design aspects that significantly affect the power benefit of monolithic 3D ICs. We discuss what kind of logic circuits are suitable for power reduction in monolithic 3D ICs. In addition, we demonstrate that the power reduction rate also depends on the target clock period. (3) We build the libraries and full-chip layouts for monolithic 3D ICs implemented using 7nm devices. The goal is to predict the future trend of power saving with

---

Permission to make digital or hard copies of all or part of this work for personal or classroom use is granted without fee provided that copies are not made or distributed for profit or commercial advantage and that copies bear this notice and the full citation on the first page. To copy otherwise, to republish, to post on servers or to redistribute to lists, requires prior specific permission and/or a fee.

DAC'13, May 29 - June 07 2013, Austin, TX, USA.

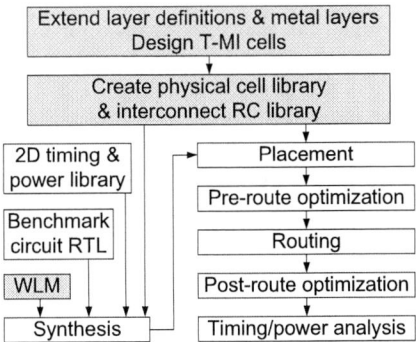

Figure 1: Overall design and analysis flow. Shaded boxes highlight differences in T-MI. The WLM means wire load model.

monolithic 3D technology and study how the smaller dimensions and varying parasitic RC affect the power benefit.

2. DESIGN AND ANALYSIS FLOW

One of the major benefits of T-MI is that existing 2D EDA tools can be used, with simple modifications if needed. We extensively use commercial EDA tools in this study. Our design and analysis flow, summarized in Fig. 1, consists of four parts: (1) library preparations, (2) synthesis, (3) layout, and (4) analysis. In the library preparation part, we prepare T-MI-specific library files. We synthesize the RTL codes of benchmark circuits using Synopsys Design Compiler.[1] In the layout part, we perform placement, routing, and optimizations using Cadence Encounter (v10.12). Finally, we perform static timing analysis and statistical power analysis.

Our major efforts for T-MI design flow are spent on T-MI cell library construction and characterization, T-MI interconnect structure modeling, and T-MI wire load modeling. We modify the technology files and design rules to account for additional layers on the bottom tier as well as additional metal layers on the top tier (see Section 3.3). Using Cadence Virtuoso, we create our T-MI cells by modifying existing 2D cells. The cells are then abstracted to create the T-MI physical cell library. We also build interconnect RC libraries using Cadence capTable generator and QRC Techgen. For synthesis, we create the T-MI wire load models (see Section 3.4) that guide synthesis optimizations.

During layout construction, we first run Encounter placer. The tool recognizes T-MI cells as the cells with pins on multiple layers. For routing, we set up Encounter to utilize the additional metal layers on bottom and top tiers. Since our T-MI cells contain routing blockages on the MIV layer, the router avoids 3D routing through the top tier part of the cells using MIVs. Using our T-MI interconnect library that reflects the T-MI metal layer structures and materials, we perform RC extraction on all the nets in the layout. Our full-chip timing/power optimizations and analyses for T-MI and 2D are the same, because the entire T-MI design (top/bottom tiers) is captured in a single Encounter session. We perform statistical power analysis with the switching activity of the primary inputs and sequential cell outputs at 0.2 and 0.1, respectively.[2]

3. 45NM TECHNOLOGY SETUP

3.1 Monolithic 3D Cell Design

---

[1]Our benchmark circuits and the synthesis results are shown in Section S4.
[2]The impact of switching activity is shown in Section S10.

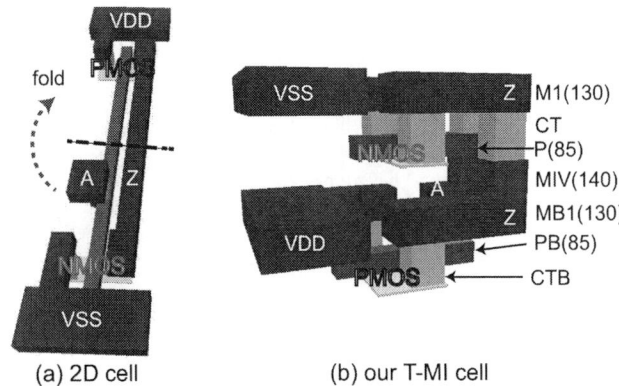

Figure 2: The layout of an inverter from (a) Nangate 45nm library, and (b) our T-MI library. P, M, and CT represent poly, metal, and contact. The suffix 'B' means the bottom tier. Top/bottom tier silicon substrate and p/nwells are not shown for simplicity. Numbers in parentheses mean thickness in nm.

We design our T-MI 3D cells using the (2D) standard cells in Nangate 45nm library [10] as our baseline. As shown in Fig. 2, we fold the 2D standard cells into 3D and create T-MI 3D cells. The thicknesses of top/bottom tier silicon substrates and inter-layer dielectric (ILD) are $30nm$ and $110nm$, respectively. The diameter of MIV is $70nm$. Note that by folding, each input/output pin is on both tiers. We prefer to place the PMOS transistors on the bottom tier and the NMOS on the top tier. In Nangate 45nm library, P/NMOS transistors show hole/electron mobility skew. To compensate the difference, in Nangate 45nm library, a PMOS is larger than the corresponding NMOS. Since extra silicon space on the top tier is required for MIVs (not on the bottom tier – see Fig. 2(b)), placing PMOS transistors on the bottom tier balances top/bottom silicon area usage. However, we should also consider manufacturing aspects in deciding the P/NMOS layer assignment.[3]

After folding the cell, VDD and VSS strips are overlapping, as shown in Fig. 2. The power to VDD on the bottom tier can be delivered down through arrays of MIVs, placed apart from the VSS strip. We may need extra space for these VDD MIVs. Yet, power delivery network design and IR-drop analysis are outside our scope. Also, since VDD and VSS strips are overlapping, it may act as a small decoupling capacitor. However, in the extracted cell internal RC data for our inverter cell, the coupling capacitance (or *cap*) between VDD and VSS strips is around $0.01fF$, which is small compared with other cell internal parasitic capacitances.

The transistor model in Nangate 45nm library is ASU PTM 45nm with bulk silicon technology. In monolithic 3D technology, because of the structure, top tier transistors are similar to silicon-on-insulator (SOI) devices [1]. However, in this study we assume the same transistor model for T-MI and 2D cells, because (1) the original Nangate 45nm library is based on bulk silicon technology, and (2) if we assume both devices and interconnect structures in T-MI are different from 2D, it becomes harder to understand which factor contributes to power reduction, by how much.

3.2 Comparison with 2D Cells

Our T-MI cells preserve the same transistor sizes as in the original 2D cells.[4] The T-MI cell height is $0.84\mu m$, which is 40% smaller than the original 2D cell height ($1.4\mu m$). Thus, cell foot-

---

[3]In sub-32nm nodes, thanks to advanced channel engineering techniques, the hole/electron mobility is about the same.
[4]Our T-MI cell layouts are presented in Fig. 5 in the supplement.

Table 1: Cell internal parasitic RC values. The 3D-c means 3D with top tier silicon modeled as a conductor.

cell	R ($k\Omega$)			C (fF)		
	2D	3D	3D-c	2D	3D	3D-c
INV	0.186	0.107	0.107	0.363	0.368	0.349
NAND2	0.372	0.237	0.237	0.561	0.586	0.547
MUX2	1.133	0.975	0.975	1.823	1.938	1.796
DFF	2.876	3.045	3.045	4.108	5.101	4.740

Table 2: Delay and internal power consumption of cells with various input slew and load capacitance conditions. The library uses different input slew settings for DFF. The values in the parentheses mean the percentage ratio of 3D to 2D.

cell	delay (ps)		power (fJ)	
	2D	3D	2D	3D
fast case: input slew=7.5ps (5ps for DFF), load cap.=0.8fF				
INV	17.2	16.9 (98.3%)	0.383	0.351 (91.6%)
NAND2	21.2	20.9 (98.6%)	0.616	0.583 (94.6%)
MUX2	59.8	58.2 (97.3%)	2.113	2.060 (97.5%)
DFF	108.8	113.4 (104.2%)	6.341	6.735 (106.2%)
medium case: input slew=37.5ps (28.1ps for DFF), load cap.=3.2fF				
INV	51.1	50.8 (99.4%)	0.362	0.343 (94.8%)
NAND2	56.2	55.9 (99.5%)	0.604	0.581 (96.2%)
MUX2	97.0	95.3 (98.2%)	2.239	2.168 (96.8%)
DFF	142.6	147.0 (103.1%)	6.358	6.756 (106.3%)
slow case: input slew=150ps (112.5ps for DFF), load cap.=12.8fF				
INV	188.3	188.0 (99.8%)	0.449	0.431 (96.0%)
NAND2	195.9	195.5 (99.8%)	0.698	0.675 (96.7%)
MUX2	215.1	212.5 (98.8%)	2.555	2.487 (97.3%)
DFF	237.4	243.3 (102.5%)	7.303	7.659 (104.9%)

Table 3: Summary of metal layers. Unit is nm.

level	metal layers	width	spacing	thickness
global	2D:M7-8, 3D:M10-11	400	400	800
intermediate	2D:M4-6, 3D:M7-9	140	140	280
local	2D:M2-3, 3D:M2-6	70	70	140
M1	2D:M1, 3D:MB1,M1	70	65	130

print reduces by 40%. The reasons why it is not 50% are (1) P/NMOS size mismatch incurs extra space on NMOS side, and (2) MIVs require extra space on the top tier.

When designing T-MI cells, care should be taken to reduce cell internal parasitic RC. As shown in Fig. 2(b), the connection from the PMOS on the bottom tier to the NMOS on the top tier needs to go through CTB, MB1, MIV, CT, M1, then CT to diffusion. This 3D path may become larger than the original 2D path and may increase cell internal parasitic RC. Similarly, the path from the PB on the bottom tier to the P on the top tier goes through multiple layers. To reduce cell internal parasitic RC, it is important to minimize the lengths of 3D paths. To achieve shorter 3D paths, we should place MIVs close to the connecting transistors. We also need to utilize direct source/drain (S/D) contacts (see Fig. 5(c) in the supplement). The direct S/D contacts reduce the detour in the 3D paths and unnecessary parasitic RC.

We examine the cell internal parasitic RC of 3D and 2D cells and the impact on timing/power. In previous works [2, 8, 7], the authors assumed that the delay and power of 3D cells are the same as 2D cells and used 2D timing/power library. In [1], the authors fabricated a transistor-level monolithic 3D IC and measured the top/bottom transistor performances. They reported that the differences between 3D transistors and baseline 2D transistors were negligible. Yet, the delay and power of cells are also affected by cell internal parasitic RC. From Fig. 2(b), we can conjecture that there are coupling capacitances among PB, CTB, MB1, MIV, CT, and M1. Using Mentor Graphics Calibre XRC with EM-simulation-based extraction rules, we extract these capacitance values as well as resistances and transistors from our T-MI cell layout. Then, we generate a SPICE netlist of the cell that consists of transistors and parasitic RC components.

Since Calibre XRC is designed for 2D ICs, it can only model one diffusion layer. Due to this tool limitation, top tier diffusion layer can be modeled as either dielectric or conductor. Even though the top tier silicon is doped (low resistivity) and the bodies of top tier trasistors are tied to the ground, we expect that some amount of electric field may penetrate the top tier silicon and coupling among top and bottom tier objects (M1, MB1, P, PB, etc.) may exist. When we assume that the top tier silicon is dielectric, the coupling between top and bottom tier objects would be overestimated; when it is conductor, the coupling would be underestimated. The real case would be between these two extreme cases.

The total cell internal RC values, extracted from the original 2D cells and our 3D (T-MI) cells, are shown in Table 1. For 3D case, the results with top tier silicon as both dielectric (3D) and conductor (3D-c) are shown. From the results, we observe the followings: (1) For INV, NAND2, and MUX2, the R values of 3D are noticeably smaller than 2D counterparts, because we reduce the length of poly and metal lines inside the cells, using 3D interconnects. (2) The C values of 3D are comparable with those of 2D – the 2D value is between 3D and 3D-c. (3) For DFF, both R and C of 3D are larger than 2D counterparts. Due to the complex internal connections, we could not create a 3D cell layout that match parasitic RC of 2D. In summary, depending on the cell layout complexity, the internal RC

ratio between 3D and 2D may vary.

Yet, the delay and power of the cells are more important metrics. We perform cell timing/power characterizations using commercial softwares. The SPICE netlists obtained from the previous RC extractions are fed into Cadence Encounter Library Characterizer, which runs SPICE simulations to characterize delay and power of cells under various input slew and load capacitance conditions. The delay/power of 3D and 2D cells are shown in Table 2. The values are obtained from the data tables in the characterized Liberty library. The delay is the cell internal delay including load effect, and the power is the dynamic power consumed within cell boundary (including short circuit power and power for gate/parasitic capacitances). We observe that for INV, NAND2, and MUX2, the delay and power of 3D are slightly better than 2D, whereas for DFF, they are a little worse. In addition, as the input slew and load capacitance condition changes from fast to slow case, the difference between T-MI and 2D becomes smaller. Note that depending on cell design quality and manufacturing technology, the results may change. We believe that with proper cell designs, the delay and power of 3D cells could be similar to 2D counterparts.

3.3 Monolithic Interconnect Setup

Our T-MI interconnect structure is an extension of the Nangate (2D) 45nm library. As shown in Table 3, we use 8 out of 10 metal layers in the Nangate 45nm. For T-MI, we make two modifications: We add (1) a new metal layer on the bottom tier (MB1), and (2) three local metal layers on the top tier (M4-6).[5]

With T-MI cell folding, the cells become 40% smaller than 2D (see Section 3.2). This results in about 40-50% smaller core footprint area. As a result, the cell pin density in T-MI becomes about 1.7-2X larger than in 2D, leading to a higher routing demand per unit area (or routing tile). To satisfy the high routing demand, we need to increase the routing capacity (#routing tracks per routing tile). The most area-efficient way is to add local metal layers, be-

---

[5] Our 2D and T-MI metal layers are shown in Fig. 9 in the supplement.

Table 4: Summary of layout results for 45nm node. The values represent the percentage difference of T-MI over 2D.

circuit name	footprint	total wirelen.	power			
			total	cell	net	leakage
FPU	-41.7%	-26.3%	-14.5%	-9.4%	-19.5%	-11.1%
AES	-42.4%	-23.6%	-10.9%	-7.6%	-13.9%	-9.5%
LDPC	-43.2%	-33.6%	-32.1%	-12.8%	-39.2%	-21.7%
DES	-40.9%	-21.5%	-4.1%	-1.6%	-7.7%	-1.4%
M256	-43.4%	-28.4%	-17.5%	-10.7%	-22.2%	-12.9%

cause of the small pitch. We found that adding 3 local metal layers increases routing capacity sufficiently.

Due to manufacturing issues (low thermal budget), in [2] the authors suggest tungsten is suitable for bottom tier metal. However, in this work we assume copper, because a copper-based manufacturing process may be developed. Besides, MB1 is mostly used for short interconnects such as within cells or short nets.[6] In our benchmark circuit M256 (see Table 12), the wirelength of MB1 (for net routing) is only 0.3% of the total wirelength. Thus, the impact of MB1 material on the timing and power of a whole circuit is minimal. When tungsten is used, IR-drop on the VDD strips could be an issue, which is outside our scope.

3.4 Monolithic 3D Wire Load Model

In T-MI designs, the wires are about 20-30% shorter than in 2D designs (see Table 4). We provide this information to the synthesis step by modifying wire load models (WLM). A WLM defines the statistical average of unit length resistance, capacitance, area of wires, as well as the fanout vs. wirelength tables. For each net, according to the fanout, the synthesis engine finds the corresponding wirelength and the capacitance/resistance/area from the WLM. We reflect the reduced wirelength of T-MI designs in the fanout vs. wirelength tables. With these WLMs, the synthesized netlists for 2D and T-MI are different.[7]

4. 45NM RESULTS

4.1 Design Analysis Results

The layout simulation results for 45nm node are summarized in Table 4.[8] With T-MI, the footprint reduces by 40.9-43.4%, which is larger than the cell footprint reduction rate, 40%. With T-MI, timing is better because of shorter wirelengths, and the optimizer may downsize cells and use less number of buffers while still meeting the target clock period. Thus, the footprint of the whole T-MI design could be further reduced than the individual cell footprint reduction rate. With T-MI, total wirelength reduces by 21.5-33.6%. Depending on the circuit characteristics, the wirelength reduction rate varies. We observe that the circuit with a larger wirelength reduction rate tends to show a larger power reduction rate. All designs met the timing. The power reduction was the largest in LDPC, 32.1%, whereas in DES, only 4.1%. In LDPC, the net power is much larger than the cell power, thus a large net power reduction with T-MI leads to a large total power reduction. We also observe that with T-MI, not only net power but also cell power reduces; with a better timing, cells are downsized and less number of buffers are used, to reduce cell power.

---

[6]The impact of MB1 on optimization quality is discussed in Section S5.

[7]Our WLM is further presented in Section S2. The impact of T-MI WLM on design quality is presented in Section S7.

[8]Our detailed layout results for 45nm node are presented in Section S6. GDSII layouts of our AES design are shown in Fig. 8 in the supplement.

Table 5: Summary of design results in our work and previous works. The [2]-3D means their INTRACEL method with timing driven + IPO, which corresponds to transistor-level monolithic 3D design. The [7]-3D means their 3TM setup.

circuit name	design type	total wire-length (m)	longest path delay (ns)	total power (mW)
AES	ours-2D	0.260	0.770	13.69
	ours-3D	0.199 (-23.5%)	0.775	12.20 (-10.9%)
	[7]-2D	0.271	1.310	13.7
	[7]-3D	0.214 (-21.0%)	1.165	12.8 (-6.6%)
LDPC	ours-2D	3.806	2.400	54.79
	ours-3D	2.528 (-33.6%)	2.388	37.22 (-32.1%)
	[2]-2D	1.83	2.461	1,554
	[2]-3D	1.60 (-12.6%)	2.421	1,461 (-6.0%)
DES	ours-2D	0.611	0.976	63.88
	ours-3D	0.479 (-21.6%)	0.968	61.24 (-4.1%)
	[2]-2D	0.671	1.132	620.2
	[2]-3D	0.581 (-13.4%)	0.971	608.2 (-1.9%)
	[7]-2D	0.849	1.086	134.9
	[7]-3D	0.682 (-19.7%)	0.923	130.7 (-3.1%)

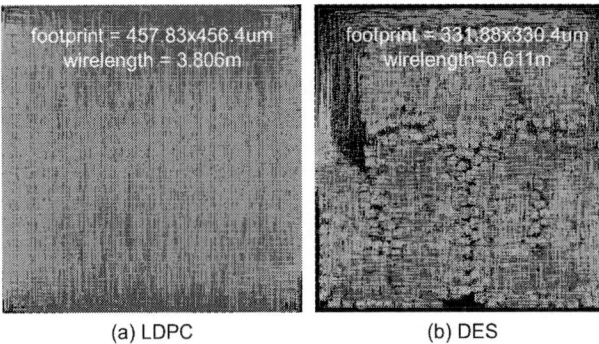

(a) LDPC   (b) DES

Figure 3: Snapshots of routing results for LDPC and DES.

4.2 Comparison with Existing Works

Our results and the results from previous works ([2][7]) are summarized in Table 5.[9] All three works use Nangate 45nm library as baseline 2D. The footprint reduction rate of 3D over 2D in this work, [2], and [7] are about 42.3%, 30%, and 40%, respectively. This footprint reduction rate mostly affects overall design quality of 3D designs, because the timing and power reduction in the monolithic 3D designs is from reduced footprint and wirelength. Our results show larger wirelength reduction than these previous works. In [2, 7], they intentionally chose small target clock periods, thus timing was not closed. Note that power values in different works vary by much. For AES and LDPC, our results show larger power reduction rate than previous works. Interestingly, in all three works, the power reduction rates for DES circuit are low (only 2-4%).

4.3 Circuit Characteristics Study

As shown in Table 4, LDPC and DES showed much different power reduction rate with T-MI. By contrasting these two designs, we explain for what kind of circuits T-MI provides large power benefit. With T-MI, the buffer count reduces by 48.6% (in LDPC) vs. 3.2% (in DES), total wirelength reduces by 33.6% vs. 21.5%, total power reduces by 32.1% vs. 4.1%, cell power reduces by 12.8% vs. 1.6%, and net power reduces by 39.2% vs. 7.7%. Compared with LDPC, the buffer count reduction for DES is very small, which leads to very small cell power reduction. Although the wirelength

---

[9]Note that the purpose of this study is not to directly compare the design quality of ours to the previous works; due to different setup, design, and analysis flow, it is not possible to provide fair comparisons.

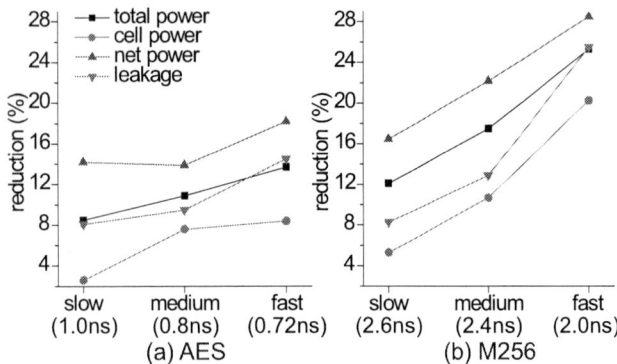

Figure 4: Power reduction rate (T-MI over 2D) under various target clock periods.

Table 6: Comparison of our 45nm and 7nm node setup.

	45nm	7nm
transistor	planar	multi-gate
VDD (V)	1.1	0.7
transistor length (drawn, nm)	50	11
transistor width	varies	fixed
back-end-of-line ILD k	2.5	2.2
M2 width (nm)	70	10.8
MIV diameter (nm)	70	10.8
ILD thickness (nm)	110	50
standard cell height (um)	1.4	0.218

Table 7: Summary of layout results for 7nm node.

circuit name	footprint	total wirelen.	power			
			total	cell	net	leakage
FPU	-47.0%	-34.2%	-37.3%	-32.4%	-44.4%	-21.0%
AES	-62.0%	-47.8%	-19.8%	-10.3%	-28.4%	-28.5%
LDPC	-42.9%	-27.7%	-19.1%	-3.7%	-26.6%	-3.5%
DES	-40.8%	-21.9%	-3.4%	-1.3%	-7.3%	-3.0%
M256	-44.6%	-23.0%	-17.8%	-14.1%	-23.0%	-2.4%

reduction in DES is not so small, the net power reduction rate is significantly smaller than LDPC. The net capacitance/power consists of wire and (cell input) pin parts.[10] For most nets in DES, wires are very short. This difference is also observed in Fig. 3. In DES layout, there are many small regions where cells are tightly connected inside but not so much to outside. For these short nets, pin capacitances dominate wire capacitances, thus reducing wirelength does not reduce net power as much. Although these two circuits are similar in size (#cells, nets) and average fanout, because of the inherent difference in circuit characteristics, the power benefit of T-MI differs by much.

4.4 Impact of Target Clock Period

The power benefit of T-MI also depends on the target clock period. For AES and M256, we vary the target clock period and perform full designs, from synthesis to layout optimizations. The power reduction rate is shown in Fig. 4. The trend is clear; when the target clock is faster, the power benefit of T-MI becomes larger. This is because at faster clock speeds, the timing of the 2D design becomes harder to meet than T-MI, because of longer wires. The optimization engine uses more buffers and larger cells, leading to steep increase in cell power. Thus, the cell power reduction rate increases noticeably as clock becomes faster. With faster clock speeds, core footprint and wirelengths also become larger, leading to larger net power reduction rate with T-MI.

5. 7NM TECHNOLOGY SETUP

Another major aspect that affects the power benefit of T-MI is the technology node. As the technology advances, devices and wires shrink at different speed, affecting timing/power of the circuit and changing power benefit of T-MI. According to the latest ITRS 2011 roadmap [5], 7nm node is near the end of the roadmap.[11] In ITRS projection for 7nm node, devices become dramatically efficient, however wires do not. The copper effective resistivity in 7nm is 3.7X larger than in 45nm, due to size effects (edge scattering, etc.).

We now predict how the power benefit of T-MI changes in the future 7nm node. The comparison between our 45nm and 7nm setup is shown in Table 6. Since there is no real 7nm node data available today, we scale down our 45nm library data as well as use data from ITRS projection. As a transistor model, we use ASU PTM-MG HP 7nm model [11]. The interconnect dimensions are scaled down to $(7/45)X = 0.156X$, and the interconnect RC libraries are

---

[10] We provide wire vs. pin power breakdown in Section S8.

[11] A summary of 45nm and 7nm node device and interconnect characteristics from ITRS projections are shown in Table 10 in the supplement.

rebuilt, with a lower dielectric k (=2.2). We scale down the physical shapes of cells to 0.156X. Based on preliminary SPICE simulations[12], we also scale down cell input capacitance to 0.179X, cell delay to 0.471X, output slew to 0.420X, cell power to 0.084X, and cell leakage power to 0.678X. We apply these scaling factors to the 45nm Liberty library and create our 7nm Liberty library. Since the transistors in 7nm node are not planar but multi-gate (e.g. FinFET), the coupling between top/bottom tier transistors would be much smaller. Thus, we can reduce ILD thickness to keep the aspect ratio of MIV reasonable.

The interconnect RC characteristics for 45nm and 7nm are obtained from the capTable built with Cadence Encounter, which runs EM simulations. The unit length resistances ($\Omega/\mu m$) of 45nm and 7nm nodes for a local metal layer (M2) are 3.57 and 638, respectively, whereas for a global metal layer (M8), 0.188 and 2.650, respectively. The unit length capacitances ($fF/\mu m$) of 45nm and 7nm nodes for M2 are 0.106 and 0.153, respectively, whereas for M8, 0.100 and 0.095, respectively. We observe that in 7nm node, the local metal layers become very resistive, due to the larger copper effective resistivity and the smaller metal width/thickness. Yet, in 7nm node, the wirelengths of the nets on local metal layers become shorter, thus the resistances of the net wires do not increase as dramatically. The capacitance per unit length increases for local metal layers, even though the dielectric k becomes smaller.

6. 7NM RESULTS

The layout simulation results for 7nm node are summarized in Table 7.[13] Compared with the results in Table 4, we see that the footprint reduction rate is larger, especially for AES where 62% footprint reduction was achieved. In the AES case, the target clock period is very small, $0.27ns$. For the 2D design, Encounter performed high-effort optimization techniques to meet the timing, while for T-MI design it did not. As a result, the buffer count of the T-MI design is 84.5% smaller. We also observed similar optimization differences for FPU. Wirelength reduction is 21.9-47.8%. In the FPU case, total power reduction is the largest, 37.3%. For DES, the power reduction is the smallest, 3.4%.

For LDPC, the power reduction rate in 7nm node is smaller than in 45nm. In LDPC, there are lots of long wires across the core area.

---

[12] Our 7nm cell characterizations are presented in Section S3.

[13] Our detailed layout results for 7nm node are presented in Section S6.

Table 8: Impact of lower cell pin cap in 7nm node. The '-p20/40/60' mean 20/40/60% reduced pin cap cases.

design	total WL (mm)	total power (mW)	cell (mW)	net (mW)	leak (mW)
DES-2D	81.2	15.11	9.49	5.03	0.60
DES-3D	63.5 (-21.9%)	14.60 (-3.4%)	9.36	4.67	0.58
DES-2D-p20	81.3	14.38	9.48	4.30	0.60
DES-3D-p20	63.5 (-21.9%)	14.12 (-1.8%)	9.42	4.09	0.60
DES-2D-p40	81.2	13.54	9.39	3.56	0.59
DES-3D-p40	63.2 (-21.8%)	13.17 (-2.7%)	9.31	3.27	0.59
DES-2D-p60	81.3	12.74	9.35	2.81	0.59
DES-3D-p60	63.5 (-21.9%)	12.45 (-2.3%)	9.32	2.55	0.59

Table 9: Impact of the lower metal resistivity in 7nm node for M256. The '-m' suffix means reduced metal resistivity.

design	total WL (mm)	total power (mW)	cell (mW)	net (mW)	leak (mW)
M256-2D	795	30.55	13.26	15.21	2.07
M256-3D	612 (-23.0%)	25.12 (-17.8%)	11.39	11.71	2.02
M256-2D-m	795	27.57	12.10	13.67	1.80
M256-3D-m	613 (-22.9%)	22.67 (-17.8%)	10.42	10.69	1.57

Considering the unit length metal resistance, the router prefers intermediate/global layers than local metal layers for long nets. However, in T-MI we added 3 metal layers to only local layers; on intermediate/global layers, T-MI suffers more routing congestion than 2D.[14] Thus, in 7nm node, the extremely high resistance on local layers (see Section 5) reduces the power reduction rate, because of worse timing (the local metal resistance was not so high in 45nm node.). In summary, depending on circuit characteristics, in 7nm node, the power benefit may become larger or smaller.

6.1 Impact of Pin Cap Reduction Rate

As mentioned in Section 5, when we compare 7nm node with 45nm node, the cell pin cap reduces by 82.1%, which is smaller than the wirelength reduction rate, about 85% (compare total wirelength of designs in Table 13 and 14). Thus, in 7nm node, the (pin cap)/(wire cap) ratio may become larger than in 45nm node. Then, the wire cap reduction with T-MI reduces the total net cap by a smaller percentage in 7nm node. However, depending on the materials and manufacturing technology, the pin cap of cells may reduce further than our projection. Thus, we explore how the power benefit of T-MI changes when pin cap reduces more.

For this study, we choose DES as the test circuit, because it showed the largest (pin cap)/(wire cap) ratio among our circuits. Thus, we expect to see larger impact with various pin cap settings. Our simulation results are summarized in Table 8. Surprisingly, the power benefit of T-MI does not increase with larger pin cap reduction rate. As pin cap reduces, the net power reduces. Then, the cell power becomes more dominating factor, because cell power does not decrease so much with smaller pin caps. Thus, the power reduction rate with T-MI becomes smaller.

6.2 Impact of Lower Metal Resistivity

As discussed in Section 5, in 7nm node, the effective resistivity of copper becomes very high. However, in the future, thanks to better interconnect materials and manufacturing process, the resistivity of interconnect may be lower than expected. In this scenario, we may expect that the timing benefit of 3D may become smaller, because the nets are longer in 2D designs and the lower resistivity would reduce delay of nets in 2D more than in 3D.

As a case study, we reduce the resistivity of local and intermediate layers by 50%.[15] We choose M256 as the test circuit, because it is the largest circuit among our benchmark circuits and more affected by net delay change.

The impact of the reduced metal resistivity is shown in Table 9. All designs met the timing. With lower resistivity, the power consumption reduces, because with better timing smaller cells are used. However, there is not much difference in wirelength and total power reduction *percentage*. The cell and net power reduction rate went down a little, however the leakage power reduction rate went up. Thus, we conclude that the lower metal resistivity does not necessarily lead to smaller power reductions in monolithic 3D ICs.

7. CONCLUSIONS

In transistor-level monolithic 3D ICs, reduced footprints lead to shorter wirelengths, better performances, and lower power consumptions. With carefully designed T-MI 3D cells, we performed layout simulations for the benchmark circuits and demonstrated up to 32.1% and 37.3% total power reductions in 45nm and 7nm nodes. In addition, we discussed other factors that affect the power benefit of T-MI, such as circuit characteristics and target clock periods. We expect to see larger power benefits with T-MI in future technology nodes, where wires become serious problems.

8. ACKNOWLEDGMENTS

This material is based upon the work supported by Intel, Qualcomm, and the CISS funded by the MEST Global Frontier Project of the South Korean Government (CISS-2-3).

9. REFERENCES

[1] P. Batude et al. Advances in 3D CMOS Sequential Integration. In *Proc. IEEE Int. Electron Devices Meeting*, pages 1–4, 2009.

[2] S. Bobba et al. CELONCEL: Effective Design Technique for 3-D Monolithic Integration targeting High Performance Integrated Circuits. In *Proc. Asia and South Pacific Design Automation Conf.*, pages 336–343, 2011.

[3] K. D. Boese, A. B. Kahng, and S. Mantik. On the Relevance of Wire Load Models. In *Proc. Int. Workshop on System-Level Interconnect Prediction*, pages 91–98, 2001.

[4] N. Golshani et al. Monolithic 3D Integration of SRAM and Image Sensor Using Two Layers of Single Grain Silicon. In *Proc. IEEE Int. Conf. on 3D System Integration*, pages 1–4, 2010.

[5] International Technology Roadmap for Semiconductors. ITRS 2011 Edition.

[6] S.-M. Jung et al. The Revolutionary and Truly 3-Dimensional 2 F^2 SRAM Technology with the smallest S^3 (Stacked Single-crystal Si) Cell, .1 um^2, and SSTFT (Stacked Single-crystal Thin Film Transistor) for Ultra High Density SRAM. In *Proc. Symposium on VLSI Technology*, pages 228–229, 2004.

[7] Y.-J. Lee, P. Morrow, and S. K. Lim. Ultra High Density Logic Designs Using Transistor-Level Monolithic 3D Integration. In *Proc. IEEE Int. Conf. on Computer-Aided Design*, pages 539–546, 2012.

[8] C. Liu and S. K. Lim. A Design Tradeoff Study with Monolithic 3D Integration. In *Proc. Int. Symp. on Quality Electronic Design*, pages 531–538, 2012.

[9] T. Naito et al. World's first monolithic 3D-FPGA with TFT SRAM over 90nm 9 layer Cu CMOS. In *Proc. Symposium on VLSI Technology*, pages 219–220, 2010.

[10] Nangate. Nangate 45nm Open Cell Library.

[11] S. Sinha et al. Exploring Sub-20nm FinFET Design with Predictive Technology Models. In *Proc. ACM Design Automation Conf.*, pages 283–288, 2012.

---

[14] The impact of a different metal layer setup is discussed in Section S9.

[15] The resistivity of global metal layers is not changed, because the wires on the global layers are large and the resistivity is not too high.

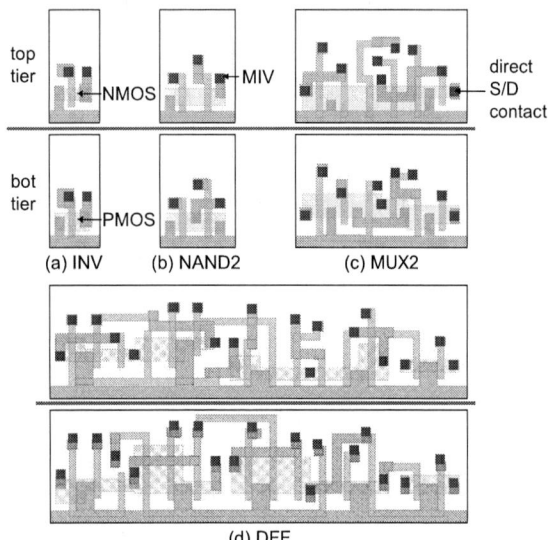

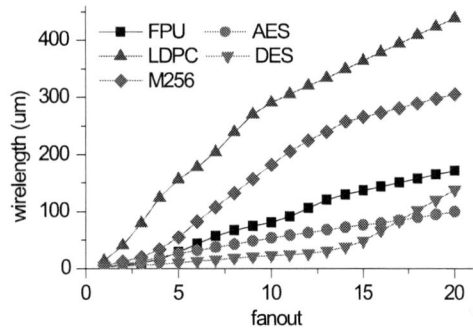

Figure 5: GDSII layouts of our T-MI cells. The S/D means source/drain. The p/nwell and implants are not shown for simplicity.

Table 10: Summary of the ITRS projection on high performance logic devices and interconnects. The 45nm and the 7nm projection data are from ITRS 2008 and 2011, respectively. The copper effective resistivity and unit length capacitance are for local/intermediate metal layers.

node	45nm	7nm
year	2010	2025
device type	bulk Si	multi-gate
NMOS drive current ($\mu A/\mu m$)	1,210	2,228
Cu effective resistivity ($\mu\Omega \cdot cm$)	4.08	15.02
Cu unit length capacitance ($fF/\mu m$)	0.19	0.15

Table 11: The 7nm cell characterization results. The cell delay, output slew, and cell power are obtained by averaging the rise/fall transition cases, when input slew is $19ps$ and load capacitance is $3.2 fF$.

	INV		NAND2		DFF	
	45nm	7nm	45nm	7nm	45nm	7nm
input cap (fF)	0.463	0.125	0.523	0.082	0.877	0.097
cell delay (ps)	44.27	25.56	49.24	30.50	124.70	27.07
output slew (ps)	31.35	15.13	35.89	19.29	34.55	8.25
cell power (fJ)	0.446	0.020	0.680	0.020	3.425	0.604
leakage (pW)	2,844	2,583	4,962	2,906	42,965	23,241

Figure 6: Fanout vs. wirelength in 2D wire load models.

SUPPLEMENT

S1 T-MI Cell Layouts

We created total 66 T-MI cells. Some of our T-MI cells are shown in Fig. 5. The internal connections of the DFF cell are rather complex. We found that direct S/D contact is helpful for reducing the cell internal parasitic RC of some cells. Note that we preserve the transistor locations of the baseline 2D cells; further reductions in cell internal parasitic RC may be possible if transistors are allowed to be relocated within a cell or the cells are completely redesigned.

S2 Wire Load Model for Monolithic 3D

The fanout vs. wirelength trends for our benchmark circuits are shown in Fig. 6. From preliminary layout simulations, per each circuit we extract a WLM for T-MI as well as 2D. Note that the curves of circuits are distinct, which is related to the circuit characteristics discussed in Section 4.3.

S3 Scaling Factors of 7nm Standard Cells

To obtain the scaling trends of 7nm cell characteristics, we first create SPICE netlists of 7nm cells. From the SPICE netlists of Nangate 45nm cells, the transistor models are replaced by ASU PTM-MG HP 7nm model [11]. The transistor fin height, width,

and length of the ASU model are 18, 7, and $11nm$, respectively. We assume the number of fins per MOS transistor is 1, because the original cells are of X1 strength; the results may change if we use multiple fins. We also scale the cell internal parasitic R and C components in the original SPICE netlists by 7.7X and 0.156X, respectively, because: (1) The resistance of metal interconnect is $R = \rho \cdot L/(Wt) = \rho_s \cdot L/W$. The sheet resistance ($\rho_s = \rho/t$) becomes 7.7X, because M1 thickness (t) is 0.156X and we increase effective resistivity (ρ) by 20% to account for size effects and barrier thickness. Both the length (L) and width (W) of cell internal interconnects become 0.156X. Thus, the R components become 7.7X of the original. (2) The unit length capacitance does not change much. And the length of cell internal interconnects becomes 0.156X. Thus, the C components become 0.156X of the original.

With the SPICE netlists of our 7nm cells, we run Cadence Encounter Library Characterizer (ELC) to obtain Liberty timing and power library. The ELC runs SPICE simulations for various input slew and load capacitance conditions and builds a library with timing and power data. The characterization results are shown in Table 11. Per each cell, we calculate the scaling ratio, then average them for all cells to obtain the final scaling trend.

S4 Benchmark Circuits and Synthesis Results

Our benchmark circuits and synthesis results for 45nm and 7nm nodes are summarized in Table 12. The FPU is a double precision floating point unit. The AES and the DES are encryption engines. The LDPC is a low-density parity-check engine for the IEEE 802.3an standard. And the M256 is a simple partial-sum-add-based 256bit integer multiplier. The circuits are in different sizes. Note that target clock periods for 7nm node are smaller than those for 45nm node. We use Synopsys Design Compiler (ver. F-2011.09) for synthesis. The synthesis results are from 2D results. All synthesized designs (2D, T-MI, in 45nm, 7nm) met target clock periods.

S5 Concerns in Layout Optimizations

In the post-route optimization step, the Encounter optimization engine tries to preserve routed wires. In T-MI designs, the MB1 wires and the routing MIVs block the cell placement, thus the op-

742

Table 12: Benchmark circuits and synthesis results.

	FPU	AES	LDPC	DES	M256
45nm node					
target clock period (ns)	1.8	0.8	2.4	1.0	2.4
#cells	9,694	13,891	38,289	51,162	202,877
cell area (μm^2)	19,123	16,756	60,590	85,526	293,636
#nets	11,345	14,218	44,153	54,724	222,569
average fanout	2.35	2.40	2.38	2.33	2.23
7nm node					
target clock period (ns)	0.72	0.27	0.9	0.3	1.0
#cells	11,378	12,541	37,322	50,833	191,543
cell area (μm^2)	447.1	362.3	1456.4	2061.3	6788.8
#nets	12,484	12,811	43,183	54,426	209,545
average fanout	2.44	2.57	2.41	2.33	2.30

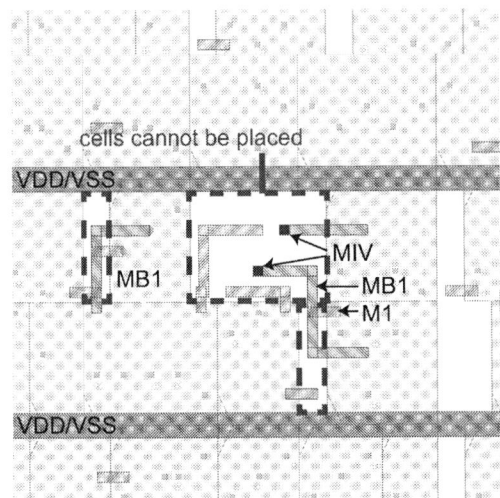

Figure 7: A zoom-in shot of T-MI design for AES. Skyblue rectangles are standard cells. For clarity, only MB1, M1, and MIV layers are shown.

timizer cannot place cells at (nor move cells to) such places. For example, in Fig. 7, the white spaces (dotted boxes) cannot be used for optimization such as buffering or gate sizing.

To see whether these MIV/MB1 blockages cause design quality degradation, we perform a layout simulation. For this case study, we use AES as the target circuit, because it showed a high placement utilization with lots of densely packed placement regions. From layout simulations, we observe that there are negligible differences in design quality, in terms of wirelength (+0.1%), timing (WNS = +25ps in original vs. +21ps without MB1 and MIV), and total power (-0.1%). Thus, we conclude that under our settings (placement, routing, optimization options, final utilization, etc.), the routings on MB1 and MIV do not degrade design quality noticeably. Note that the utilization of the above AES design is around 80%; we may see problems caused by the MIV/MB1 blockages when utilization is very high. However, in general, it is customary not to exceed the 80% utilization, due to various reasons (placement and routing quality, optimization quality, decap area, etc.)

S6 Detailed Layout Results

The detailed layout simulation results for 45nm node are shown in Table 13. We set the target utilization to around 80%, which is common in industry designs. Since we observed severe wire congestions in LDPC (see Fig. 3(a)), the target utilization was lowered to about 33%; the 2D design was barely routable with this setting. We also observed significant wire congestions in M256, thus the

Table 15: Layout results with/without our T-MI WLMs. The '-n' suffix means without our T-MI WLM.

design	total WL (mm)	WNS (ps)	total power (mW)
FPU-3D	149.1	+4	7.22
FPU-3D-n	152.0 (+1.9%)	+11	7.20 (-0.3%)
AES-3D	198.8	+25	12.20
AES-3D-n	199.0 (+0.1%)	+21	12.19 (-0.1%)
LDPC-3D	2527.8	+12	37.22
LDPC-3D-n	2782.2 (+10.1%)	+16	40.99 (+10.1%)
DES-3D	479.1	+32	61.24
DES-3D-n	481.7 (+0.5%)	+29	61.79 (+0.9%)
M256-3D	4760.2	0	160.5
M256-3D-n	5020.6 (+5.5%)	+3	166.8 (+3.9%)

Table 16: Wire vs. pin capacitance breakdown of LDPC and DES in 45nm node. The values are for the entire circuit.

design	total cap. (pF)		power (mW)	
	wire	pin	wire	pin
LDPC-2D	558.0	134.4	30.73	9.04
LDPC-3D	310.3	123.6	15.88	8.32
DES-2D	64.4	127.4	8.88	17.80
DES-3D	50.1	126.6	6.87	17.76

target utilization was lowered to 68%. All designs met the timing (WNS$\geq$0).

The detailed layout simulation results for 7nm node are shown in Table 14. We set similar target utilizations as for 45nm node. All designs met timing.

S7 Impact of T-MI Wire Load Model

As mentioned in Section 3.4, we create custom WLMs for T-MI designs. There have been debates on whether WLM is helpful or not to the final layout results [3]. Since our target circuits are small to medium sized, we may expect that WLM is helpful to some extent. To see the impact of the custom WLMs on design quality, we perform the synthesis for T-MI designs with not our T-MI WLMs but the 2D WLMs. As a result, the synthesized netlists for T-MI and 2D become similar. The layout results with/without custom WLM for T-MI designs are shown in Table 15. For FPU, AES, and DES, the design quality difference is negligible. However, for LDPC and M256, we observe significant increase in wirelength and total power without T-MI WLM. Thus, we conclude that for some designs, T-MI WLM models are helpful for obtaining larger power benefits with T-MI.

S8 Breakdown of Net Power

We break net power into wire and pin power components (net = wire + pin). Wire means metal wires and vias used for connecting cell pins, and pin means input pins of cells. As shown in Table 16, in LDPC, wire cap is much larger than pin cap, and so is wire power. Most of the net power reduction is from reduced wirelengths, as seen by the wire power reduction. In contrast, in DES, pin cap is much larger than wire cap. Thus, reduced wirelengths and wire power only reduces a small portion of the net power. In fact, most of the nets in DES are short, whereas most are long in LDPC; the average wirelength of LDPC-2D and DES-2D are 72.0μm and 10.5μm, respectively.

S9 Impact of the Metal Layer Setup

To see the impact of the metal layer setup on power benefit of T-MI, we modify the metal layer stack of T-MI. Instead of adding 3 local metal layers on the top tier, we add 2 to local and 2 to intermediate metal layers. The original and modified metal stacks

Table 13: Layout results of 2D and monolithic 3D designs for 45nm node. The #cells mean total number of cells, and #buffers mean the number of inverting/non-inverting buffers. The #cells include #buffers. The utilization means final cell placement density, after all optimizations. The WL and WNS mean wirelength and worst negative slack, respectively. Positive WNS value means timing is met with a positive slack. The values in parentheses show the percentage ratio to the 2D designs.

circuit name	design type	footprint (μm^2)	#cells	#buffers	utili-zation (%)	total WL (m)	WNS (ps)	total power (mW)	cell power (mW)	net power (mW)	leakage (mW)
FPU	2D	24,839 (100)	10,959	1,644 (100)	80.4	0.202 (100)	+6	8.44 (100)	3.98 (100)	4.21 (100)	0.25 (100)
	3D	14,476 (58.3)	9,922	1,240 (75.4)	79.5	0.149 (73.7)	+4	7.22 (85.5)	3.61 (90.6)	3.39 (80.5)	0.23 (88.9)
AES	2D	25,375 (100)	19,577	4,952 (100)	79.9	0.260 (100)	+30	13.69 (100)	6.36 (100)	6.94 (100)	0.40 (100)
	3D	14,613 (57.6)	18,996	5,157 (104.1)	79.7	0.199 (76.4)	+25	12.20 (89.1)	5.87 (92.4)	5.97 (86.1)	0.36 (90.5)
LDPC	2D	208,954 (100)	47,017	13,374 (100)	32.6	3.806 (100)	0	54.79 (100)	14.17 (100)	39.78 (100)	0.85 (100)
	3D	118,758 (56.8)	42,831	6,868 (51.4)	32.4	2.528 (66.4)	+12	37.22 (67.9)	12.36 (87.2)	24.20 (60.8)	0.66 (78.3)
DES	2D	109,652 (100)	54,402	8,436 (100)	79.9	0.611 (100)	+24	63.88 (100)	36.17 (100)	26.68 (100)	1.03 (100)
	3D	64,830 (59.1)	53,534	8,170 (96.8)	80.5	0.479 (78.5)	+32	61.24 (95.9)	35.60 (98.4)	24.62 (92.3)	1.02 (98.6)
M256	2D	478,077 (100)	245,935	62,970 (100)	68.2	6.647 (100)	0	194.6 (100)	74.73 (100)	115.2 (100)	4.70 (100)
	3D	270,748 (56.6)	216,956	48,125 (76.4)	67.3	4.760 (71.6)	0	160.5 (82.5)	66.70 (89.3)	89.66 (77.8)	4.10 (87.1)

Table 14: Layout results of 2D and monolithic 3D designs for 7nm node.

circuit name	design type	footprint (μm^2)	#cells	#buffers	utili-zation (%)	total WL (mm)	WNS (ps)	total power (mW)	cell power (mW)	net power (mW)	leakage (mW)
FPU	2D	639 (100)	17,306	3,931 (100)	80.9	33.1 (100)	+2	2.87 (100)	1.37 (100)	1.34 (100)	0.17 (100)
	3D	339 (53.0)	11,371	1,368 (34.8)	78.9	21.8 (65.8)	+1	1.80 (62.7)	0.92 (67.6)	0.74 (55.6)	0.13 (79.0)
AES	2D	724 (100)	29,153	11,496 (100)	79.2	45.5 (100)	+9	2.85 (100)	1.35 (100)	1.27 (100)	0.23 (100)
	3D	275 (38.0)	12,687	1,778 (15.5)	79.6	23.8 (52.2)	+6	2.29 (80.2)	1.21 (89.7)	0.91 (71.6)	0.16 (71.5)
LDPC	2D	5,208 (100)	47,503	11,689 (100)	30.9	608 (100)	+2	8.68 (100)	2.43 (100)	5.83 (100)	0.41 (100)
	3D	2,972 (57.1)	43,453	7,936 (67.9)	31.4	439 (72.3)	+4	7.02 (80.9)	2.34 (96.3)	4.28 (73.4)	0.40 (96.5)
DES	2D	2,612 (100)	50,878	6,851 (100)	79.1	81.2 (100)	0	15.11 (100)	9.49 (100)	5.03 (100)	0.60 (100)
	3D	1,546 (59.2)	50,758	6,693 (97.7)	80.1	63.5 (78.1)	0	14.60 (96.6)	9.36 (98.7)	4.67 (92.7)	0.58 (97.0)
M256	2D	11,411 (100)	255,364	59,153 (100)	68.6	795 (100)	+23	30.55 (100)	13.26 (100)	15.21 (100)	2.07 (100)
	3D	6,172 (55.4)	213,272	40,997 (69.3)	67.9	612 (77.0)	+14	25.12 (82.2)	11.39 (85.9)	11.71 (77.0)	2.02 (97.6)

Table 17: Impact of the different metal layer setup for T-MI. The '+M' suffix means the modified metal layer stack.

design	total WL (mm)	total power (mW)	cell (mW)	net (mW)	leak (mW)
LDPC-3D	439	7.02	2.34	4.28	0.40
LDPC-3D+M	432 (-1.6%)	6.85 (-2.4%)	2.27	4.23	0.36
M256-3D	612	25.12	11.39	11.71	2.02
M256-3D+M	618 (+1.0%)	24.42 (-2.8%)	11.11	11.47	1.83

are shown in Fig. 9. We use LDPC and M256 for this case study. The results are summarized in Table 17. With the modified metal layer structure, compared with our T-MI results, total wirelength of the design with modified metal layers decreases by 1.6% for LDPC and increases by 1.0% for M256. The cell power, net power, and leakage power reduces, and the total power of LDPC and M256 reduces by 2.4% and 2.8%, respectively. Thus, we conclude that the metal layer structure of T-MI affects power benefit and should be chosen carefully.

The local, intermediate, and global metal layer usage for LDPC and M256 designs are shown in Fig. 10. We observe that both local and intermediate layers are heavily used. On global layers, we see a lot of long wires. LDPC used more global metal than M256. Note that a net uses combinations of these layers; the line segments in the snapshot do not represent the whole net.

S10 Impact of Switching Activity Factor

Another major factor that affects the power consumption is the switching activity factor. The switching activity factor is defined as the number of signal transitions (0-1 or 1-0) per a given clock period. The power values of cells and nets are linearly proportional to the related switching activities. Depending on various factors (architecture, usage scenario, etc.), the actual switching activity values may vary. For statistical power analyses, we provide switching ac-

tivity factors to the primary input ports and the outputs of sequential cells (e.g. flipflop). Our default settings for primary inputs and sequential cell outputs are 0.2 and 0.1, respectively. Then, the given switching activity values are propagated to the rest of the circuit, based on the netlist connectivity and the functionality of cells.

Since the switching activities of primary inputs affects until the first sequential cells and these paths are usually short, changing the switching activity factor of primary inputs affects the power by a small amount. In this case study, we vary the switching activity factors of the sequential cell outputs only. The total power of 2D and 3D designs for M256 under various switching activity factors are shown in Fig. 11(a). Although the total power increases with a larger switching activity factor, the power reduction rate does not change much, as shown in Fig. 11(b). The other circuits also show negligible differences in power reduction rate under various switching activity factors. Thus, we conclude that the power benefit of T-MI is not largely affected by the switching activity level.

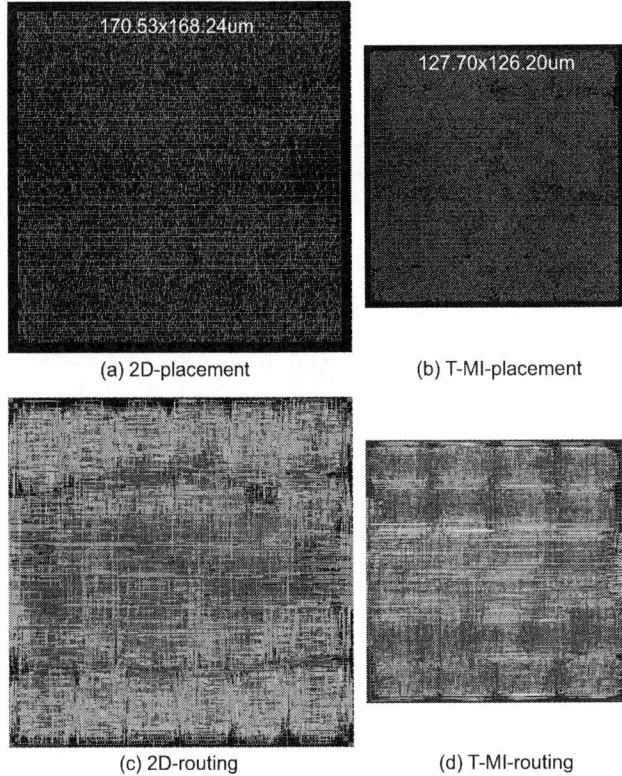

(a) 2D-placement   (b) T-MI-placement

(c) 2D-routing   (d) T-MI-routing

Figure 8: The placement and routing snapshots of AES designs. The figures reflect the relative sizes of 2D vs. T-MI designs.

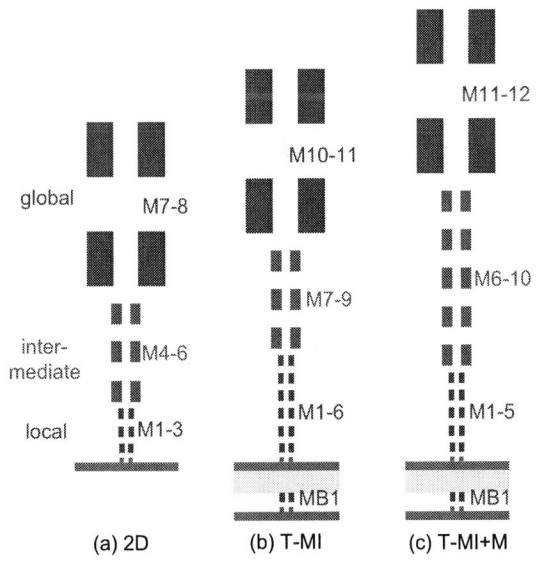

Figure 9: Metal layer stack diagrams for (a) 2D, (b) T-MI, and (c) T-MI+M. The '+M' means modified metal layer stack.

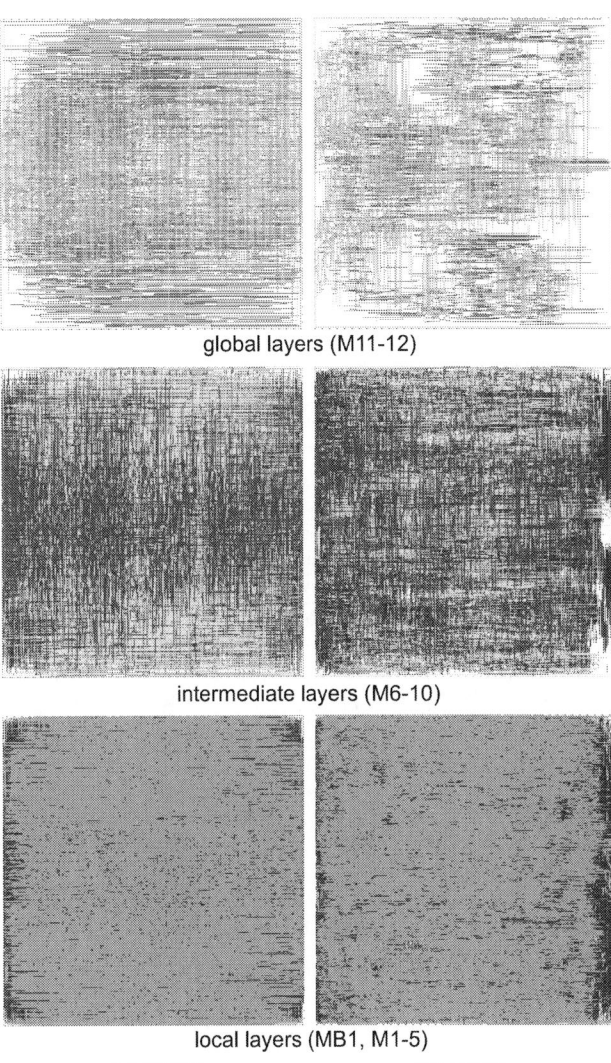

global layers (M11-12)

intermediate layers (M6-10)

local layers (MB1, M1-5)

(a) LDPC   (b) M256

Figure 10: GDSII snapshots of local, intermediate, and global metal layers for (a) LDPC and (b) M256.

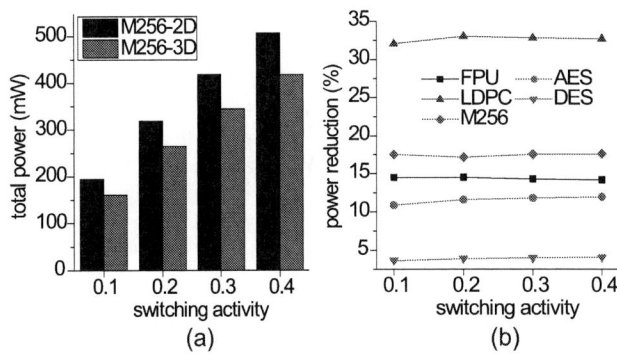

Figure 11: Power dependency on switching activity factor. (a) Total power of M256 with various switching activity factors, and (b) power reduction rate under various switching activity factor. All results are from 45nm node.

Rapid Exploration of Processing and Design Guidelines to Overcome Carbon Nanotube Variations

Gage Hills[1], Jie Zhang[2], Charles Mackin[3], Max Shulaker[1], Hai Wei[1], H.-S. Philip Wong[1], Subhasish Mitra[1]

1. Stanford University, CA   2. Google, Inc.   3. Massachusetts Institute of Technology, MA

ABSTRACT

Carbon nanotube field-effect transistors (*CNFETs*) are promising candidates for building energy-efficient digital systems at highly-scaled technology nodes. However, carbon nanotubes (*CNTs*) are inherently subject to variations that reduce circuit yield, increase susceptibility to noise, and severely degrade their anticipated energy and speed benefits. Joint exploration and optimization of CNT processing options and CNFET circuit design are required to overcome this outstanding challenge. Unfortunately, existing approaches for such exploration and optimization are computationally expensive, and mostly rely on trial-and-error-based ad-hoc techniques. In this paper, we present a systematic framework which quickly evaluates the impact of CNT variations on circuit delay and noise margin, and automatically explores the large space of CNT processing options to derive optimized CNT processing and CNFET circuit design guidelines. We demonstrate that: 1. Our new framework runs over 100X faster than existing approaches. 2. It accurately identifies the most important CNT processing parameters, together with CNFET circuit sizing, to minimize the impact of CNT variations while meeting circuit-level noise margin constraints.

Categories and Subject Descriptors

B.7 [Hardware]: Integrated Circuits

General Terms

Design, Performance, Reliability

Keywords

Carbon Nanotube, Carbon Nanotube Variations, Delay Optimization, Noise Margin Optimization

1. INTRODUCTION

Energy-efficiency, often expressed in terms of performance per watt [Laudon 05, Rivoire 07], is a key driver for a vast majority of digital systems. To achieve improved energy-efficiency, alternatives to Si-CMOS technologies are currently being explored. Carbon nanotubes (*CNTs*) are a promising emerging technology, and are staged as a potential supplement to Si-CMOS due to their expected energy-efficiency benefits. Experimentally, it has been shown that carbon nanotube field-effect transistors (*CNFETs*) with sub-10nm channel length can provide the highest drive current density compared to other technologies [Franklin 12a]. CNFETs have also been shown to operate at a low supply voltage of 0.4V [Ding 12]. Due to their superior electrostatics and quasi-ballistic transport properties [Appenzeller 08], CNFET-based digital systems can potentially achieve an order of magnitude improvement in energy-delay-product (*EDP*) over Si-CMOS at highly-scaled technology nodes [Wei 09, Chang 12].

A typical CNFET is shown in Figure 1. Multiple CNTs comprise the transistor channel, whose conductance is modulated by the gate. The gate, source, and drain are defined using traditional photolithography, and the source-drain separation is limited by the lithographic pitch. The inter-CNT spacing, however, is determined by CNT growth [Patil 09a] and can therefore be smaller than the minimum lithographic pitch. For high current drive, the target inter-CNT spacing is 4nm [Wei 09].

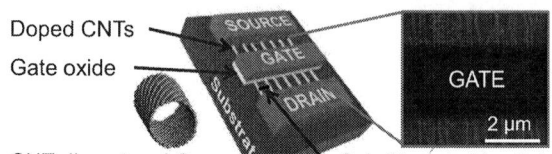

Figure 1. (Left) CNT; (center) CNFET structure; (right) scanning electron microscopy (*SEM*) image of the CNFET channel.

Despite several demonstrations of high-performance CNFETs [Franklin 12a, Park 12], realization of VLSI CNFET circuits has been prohibited by substantial imperfections inherent to CNTs: mis-positioned CNTs and metallic CNTs. Mis-positioned CNTs cause stray conducting paths, and metallic CNTs (resulting from imprecise control over CNT chirality) cause source-drain shorts, resulting in incorrect logic functionality. A unique combination of CNT processing solutions and CNFET circuit design, *the imperfection-immune design paradigm* [Zhang 12], overcomes these challenges to enable the realization of the first CNFET circuits and sub-systems [Patil 09b, Patil 11, Shulaker 13]. Two key enablers of these demonstrations are: (1) mis-positioned-immune layout design [Patil 08], and (2) VMR: VLSI-compatible Metallic CNT removal, which efficiently removes over 99.99% of metallic CNTs [Patil 09b, Wei 10].

Unfortunately, process variations specific to CNTs, such as the imprecise control over CNT chirality and non-uniform density of grown CNTs (details in Section 2), can significantly degrade ideal performance projections. They lead to large variations in CNFET circuit delays, increased susceptibility to noise, and significantly reduced circuit yield [Zhang 12]. One method to counteract these effects is to upsize all CNFETs. However, naïve upsizing incurs large energy costs that diminish CNFET technology benefits.

As an alternative, various CNT process improvement options, when combined with circuit design, provide an energy-efficient method of overcoming CNT variations. Without such strategies, CNT variations can degrade the potential speed benefits of CNFET circuits by 60% at the 16nm node (Section A8). By leveraging CNT process improvements, together with CNFET circuit design, the overall speed degradation can be limited to less than 5% with only 5% energy increase [Zhang 12].

However, co-optimization of CNT technology options and CNFET circuit design parameters is prohibitively time consuming, and has previously relied on trial-and-error-based search. We demonstrate, for the first time, a systematic and VLSI-scalable methodology to overcome the effects of CNT variations through joint exploration and optimization of CNT processing and CNFET circuit design. Our key contributions are:

(1) Techniques for quick evaluation of the impact of CNT variations on circuit delay and noise margin. These techniques run 100X faster than previous approaches. Experimental data on actual CNFET circuits validate the correctness of these techniques.

(2) A systematic methodology that automatically explores the large space of CNT processing options together with CNFET

Permission to make digital or hard copies of all or part of this work for personal or classroom use is granted without fee provided that copies are not made or distributed for profit or commercial advantage and that copies bear this notice and the full citation on the first page. To copy otherwise, to republish, to post on servers or to redistribute to lists, requires prior specific permission and/or a fee.
DAC '13, May 29 - June 07 2013, Austin, TX, USA.

circuit sizing. This is in sharp contrast to previous trial-and-error-based approaches.

(3) Demonstration of the effectiveness of our approach in deriving optimized CNT processing and circuit sizing guidelines that minimize the impact of CNT variations while meeting circuit-level noise margin constraints.

In this paper, we briefly introduce the circuit-level impact of CNT variations (Section 2), describe our methodology to overcome these variations (Section 3), and then present results to demonstrate the effectiveness of our approach (Section 4).

2. CNT VARIATIONS

In addition to process variations that exist for Si-CMOS transistors [Nassif 07], CNFETs are also subject to CNT-specific variations (details in Section A8). The variations that most significantly impact CNFET circuit performance are those that cause variations in *CNT count*: the number of CNTs in a CNFET after metallic CNT (*m-CNT*) removal (e.g., using VMR [Patil 09b]). CNT count variations (Figure 2, Section A8) result from both grown CNT density variations (non-uniform inter-CNT spacing) and m-CNT-induced variations (variations in the remaining CNT count after m-CNT removal).

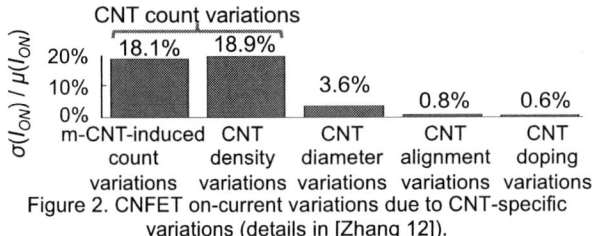

Figure 2. CNFET on-current variations due to CNT-specific variations (details in [Zhang 12]).

Section A8 describes the effects of CNT count variations on the following performance metrics:

(1) *Delay penalty*: the increase in the *95-percentile-delay* (T_{95}, the clock period that the circuit has a 95% probability of meeting) relative to the *nominal delay* (no variations).

(2) *Static Noise Margin (SNM)*: a measure of the noise margin of a logic gate pair.

(3) *Probability of Noise Margin Violation* (*PNMV*): the probability that **any** logic gate pair in a circuit will fail to meet SNM_R, a required SNM level.

Table 1 summarizes the key CNT processing parameters used in this paper to quantify the impact of CNT count variations.

Table 1. CNT processing parameters for CNT count variations.

Proc. Param	Definition	Ideal value	Experimental value
IDC	Index of Dispersion for CNT count [Zhang 09a] $IDC = \sigma_s^2 / \mu_s^2$, μ_s and σ_s^2 are the mean and variance of the grown-CNT spacing	0	0.5 [Zhang 09a]
p_m	Probability that a given CNT is an m-CNT	0%	<1% [Park 12]
p_{Rs}	Conditional probability that a CNT is removed, given that it is a semiconducting CNT (s-CNT)	0%	<5% [Patil 09b]
p_{Rm}	Conditional probability that a CNT is removed, given that it is an m-CNT	100%	>99.99% [Patil 09b]

We use our methodology to answer the following key questions: (1) Which processing parameters to improve? (2) By how much? Without such a systematic methodology, one might blindly pursue expensive CNT processing paths to improve parameters with

diminishing returns (e.g., the percentage of m-CNTs), while overlooking other processing parameters (e.g., CNT density variations) that might enable far bigger performance gains.

3. INTELLIGENT CO-OPTIMIZATION OF PROCESS IMPROVEMENTS AND CIRCUIT DESIGN

An existing approach to overcome CNT variations is based on trial-and-error [Zhang 11a]. An arbitrary combination of CNT process improvements and logic gate upsizing, hereby referred to as a single *design point*, is chosen, and then its delay penalty is calculated. This process repeats until it reaches a design point that satisfies a desired delay penalty with a small energy cost. Additionally, this approach utilizes highly accurate, yet computationally expensive, delay models to calculate delay penalties. It suffers from the following bottlenecks:

(1) The time required to calculate delay penalties limits the total number of design points that can be explored.

(2) The number of required simulations can be exponential in the number of CNT processing parameters.

Our new methodology overcomes these bottlenecks:

(1) We estimate delay penalties 100X faster than the previous approach, enabling exploration of many more design points. At the same time, we maintain enough accuracy to make correct design decisions.

(2) We use a gradient descent search algorithm, based on delay sensitivity information with respect to processing parameters, to systematically guide the exploration of design points.

3.1 Rapid Analysis of Circuit Delay Penalty

For rapid analysis of CNFET circuit delay variations, we use a simplified delay model, as is common in methods to estimate Si-CMOS circuit delays [Markovic 04]. We leverage the probabilistic framework in [Zhang 11a], which is based on a Monte Carlo statistical static timing analysis (*MC SSTA*) approach with two key enhancements:

(1) *Highly-efficient sampling method*: It is not trivial to analytically model the effects of CNT correlation (Section A8.3.1) at the circuit level. We instead partition the circuit area into *sampling regions*, each of which has its own **independent** CNT count. The CNT count of each CNFET is then the sum of the CNT counts of each sampling region that it overlaps (illustration in Section A9, Figure 14).

(2) *Variation-aware CNFET model*: The current and parasitic capacitances of CNFETs are modeled as affine functions of CNT counts in each sampling region [Zhang 11b].

We incorporate two additional enhancements for significantly faster computation time and for sensitivity analysis of CNFET circuit delay with respect to each processing parameter:

(1) *Gaussian approximation of CNT count distributions*: this allows us to factorize our delay model into two components, one of which does not depend on the values of the processing parameters and can therefore be pre-computed (details below). As a result, we can quickly estimate the sensitivity of delay to each of the processing parameters. The CNT count variables are thus elements of the set of real numbers ($\mathbb{R}$) instead of the set of non-negative integers ($\mathbb{Z}^+$) [Zhang 09a]. The validity of this approximation is verified in Section A9.

(2) *First-order approximation of delay variations*: we leverage the same delay model as in [Zhang 11b] (Section A10, [16]) to compute the maximum path delay of a circuit when no variations are present (*nominal delay*). We use a first-order approximation to estimate the impact of CNT count variations on delay. Unlike [Zhang 11b], we fix the input slew rate of each logic gate to its <u>nominal</u> value. This allows us to efficiently compute all of the gate delays in a circuit

747

simultaneously. These approximations have minimal impact on our design choices (Sections 4.1, A10). We refer to the model in [Zhang 11b] as the *non-linear delay model*, and to our model as the *linearized delay model*.

3.1.1 Full Circuit Delay Model

Let μ_R and σ_R be the mean and standard deviation of the sampling region CNT count distribution. Note that, these are functions of the processing parameters. The first step to estimate the delay penalty of a design point is to sample the CNT count for each sampling region and for each MC trial. Each sample is one entry in a matrix $N \in \mathbb{R}^{r \times n}$, where r is the total number of sampling regions and n is the total number of MC trials. We then compute the total capacitive load and drive current for each of the m gates (for each trial), via an affine mapping of the region CNT counts (based on the model in [Zhang 11b]). We express this transformation in matrix form, where $C_{Tot}, I_{Drive} \in \mathbb{R}^{m \times n}$:

$$C_{Tot} = A_{CLoad}N + b_{CLoad}\mathbf{1}^T, \qquad [1]$$
$$I_{Drive} = A_{IDrive}N + b_{IDrive}\mathbf{1}^T. \qquad [2]$$

Our delay models are *fully specified* by $A_{CLoad}, A_{IDrive} \in \mathbb{R}^{m \times r}$ and by $b_{CLoad}, b_{IDrive} \in \mathbb{R}^m$ ($\mathbf{1}$ is a vector with every entry equal to one), which contain the coefficients of the affine transformation from the sampling region CNT counts to the CNFET currents and parasitic capacitances [Zhang 11b].

Next, we factor out the processing parameters, a crucial step in achieving computational efficiency. We rewrite $N = \mu_R + \sigma_R X$, where each entry of $X \in \mathbb{R}^{r \times n}$ is distributed according to a *unit* Gaussian distribution, allowing [1] and [2] to be rewritten as:

$$C_{Tot} = \sigma_R A_{CLoad}X + (\mu_R A_{CLoad}\mathbf{1} + b_{CLoad})\mathbf{1}^T, \qquad [3]$$
$$I_{Drive} = \sigma_R A_{IDrive}X + (\mu_R A_{IDrive}\mathbf{1} + b_{IDrive})\mathbf{1}^T. \qquad [4]$$

Any product that does not involve μ_R or σ_R is independent of the processing parameters, and can be pre-computed. The dominant computational tasks are the matrix multiplications AX ($O(mn)$ since A is sparse [Eigen]). After calculating such terms (and after factoring in the multiplication of C_{Tot} and V_{DD}), we arrive at equivalent expressions for total charge and drive currents that require scalar operations:

$$Q_{Tot} = \sigma_R Q_{MC} + (\mu_R q_{Exp} + q_{Fix})\mathbf{1}^T, \qquad [5]$$
$$I_{Drive} = \sigma_R I_{MC} + (\mu_R i_{Exp} + i_{Fix})\mathbf{1}^T. \qquad [6]$$

We pre-compute $Q_{MC}, q_{Exp}, q_{Fix}, I_{MC}, i_{Exp},$ and i_{Fix} (Table 2), allowing each gate delay to be efficiently computed with only two multiplications, one division, and three additions per trial (only counting operations that must be individually computed for each trial, [7]). This includes the addition of $d_{Fix} \in \mathbb{R}^m$, a vector of fixed delays (e.g., input delays from external circuits):

$$D = \frac{\sigma_R Q_{MC} + (\mu_R q_{Exp} + q_{Fix})\mathbf{1}^T}{\sigma_R I_{MC} + (\mu_R i_{Exp} + i_{Fix})\mathbf{1}^T} + d_{Fix}\mathbf{1}^T. \qquad [7]$$

Table 2. Input terms to the full circuit delay model.

Par.	$\in$	Equation	Description
Q_{MC}	$\mathbb{R}^{m \times n}$	$V_{DD}A_{CLoad}X$	Variable gate charge per CNT
q_{Exp}	$\mathbb{R}^m$	$V_{DD}A_{CLoad}\mathbf{1}$	Expected gate charge per CNT
q_{Fix}	$\mathbb{R}^m$	$V_{DD}b_{CLoad}$	Fixed gate charge
I_{MC}	$\mathbb{R}^{m \times n}$	$A_{IDrive}X$	Variable gate current per CNT
i_{Exp}	$\mathbb{R}^m$	$A_{IDrive}\mathbf{1}$	Expected gate current per CNT
i_{Fix}	$\mathbb{R}^m$	b_{IDrive}	Fixed gate current

We then perform static timing analysis (*STA*) for each MC trial (and for the nominal case), and use the results to estimate T_{95} and the delay penalty. The total circuit energy is estimated with model of the form $E = \frac{1}{2}CV^2$ [Weste 05]:

$$E_{Tot} = \frac{1}{2}V_{DD}\mathbf{1}^T\left(\frac{1}{n}\sigma_R Q_{MC}\mathbf{1} + \mu_R q_{Exp} + q_{Fix}\right). \qquad [8]$$

3.1.2 Experimental Validation of the Linearized Delay Model

We compare our delay model output with experimentally measured current and delay characteristics of a set of CNFET inverters. Figure 3 shows SEMs of three different CNT densities ((a) 2 CNTs/μm, (b) 4 CNTs/μm, and (c) 8 CNTs/μm, achieved via multiple transfers of grown CNTs [Shulaker 11]), and of an array of CNFETs and CNFET circuits with vertically aligned-active layouts (Section A8.3.1 for aligned-active layouts).

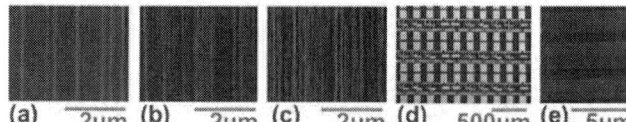

Figure 3. (a, b, c) Three CNT densities used to vary currents of (d) CNFET circuits. (e) Aligned-active CNFETs in an inverter.

Figure 4 shows the outputs of our current and delay models overlaid on experimentally measured data from a set of 30 CNFET inverters. For processing parameters, we use $IDC = 0.5$, $p_m = 0.33$, $p_{Rs} = 0.05$, and $p_{Rm} = 0.9999$, which have been characterized for our CNT growth and m-CNT removal process [Patil 09a, 09b]. The CNFET inverters all have identical channel widths, and thus with linearly increasing CNT density, we linearly increase the average CNT count. We find that our linearized delay model correctly predicts the experimentally observed reduction in both current variations and delay variations with increasing average CNT count, a key aspect of our probabilistic framework. We underestimate the current standard deviation by 2.4X, and hypothesize the reason to be due to additional sources of variations (e.g., contact resistance [Cao 12] and threshold voltage [Cao 12, Franklin 12c, Park 12]).

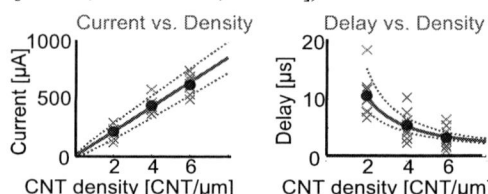

Figure 4. Current model (left) and delay model (right) overlaid on experimentally measured data. Measured data in red (averages in black). Models in blue (solid line: expected value, dotted line: 3σ).

3.2 Rapid Analysis of Circuit PNMV

Our method of analyzing circuit *PNMV* is similar to that of circuit delay penalty. The SNM is computed for each logic gate pair, and then *PNMV* is extracted from the SNM cumulative distribution function (*CDF*) at $SNM_R = V_{DD}/4$. The same highly-efficient MC sampling method is utilized (Section 3.1), and the final form of the SNM model for each logic gate pair is similar to that of the delay model for each logic gate (but is based on a variation-aware SNM model, Section A11). Calculating the SNM of each logic gate pair requires scalar operations. The required time to compute *PNMV* is on the same order as that for delay penalty (Sections 4.1, A11, A14.1).

3.3 Circuit Performance Metrics & Their Sensitivity to Processing Parameters

To overcome CNT variations, our target is to achieve small delay penalties without incurring high energy costs. This objective is expressed using the energy-delay-product (*EDP*) metric. We similarly define the energy-*PNMV*-product (*ENP*) metric to quantify the trade-off between noise resiliency and energy.

$$EDP = E_{Tot}T_{95}, \qquad [9]$$
$$ENP = E_{Tot}PNMV. \qquad [10]$$

748

Rapid computation of circuit delay penalty and *PNMV* overcomes the speed bottleneck of analyzing a single design point, but a method for intelligently exploring the entire space of CNT processing options is still required. A common measure of the sensitivity of an objective function (in this case, EDP or ENP) with respect to each of its input variables (the processing parameters) is its *gradient*. The EDP and ENP gradients are defined in [11]-[13] (details in Section A12.1), and are used to guide the optimization of EDP and ENP (Section 3.4):

$$\nabla E_{Tot} = \begin{bmatrix} \frac{\partial}{\partial IDC} \\ \frac{\partial}{\partial p_m} \\ \frac{\partial}{\partial p_{Rs}} \end{bmatrix} E_{Tot}, \quad \nabla T_{95} = \begin{bmatrix} \frac{\partial}{\partial IDC} \\ \frac{\partial}{\partial p_m} \\ \frac{\partial}{\partial p_{Rs}} \end{bmatrix} T_{95}, \quad \nabla PNMV = \begin{bmatrix} \frac{\partial}{\partial IDC} \\ \frac{\partial}{\partial p_m} \\ \frac{\partial}{\partial p_{Rs}} \end{bmatrix} PNMV, \quad \textbf{[11]}$$

$$\nabla EDP = \nabla E_{Tot} T_{95} + E_{Tot} \nabla T_{95}, \quad \textbf{[12]}$$

$$\nabla ENP = \nabla E_{Tot} PNMV + E_{Tot} \nabla PNMV. \quad \textbf{[13]}$$

Before advancing to the optimization algorithm, we summarize the steps taken so far. Figure 5 illustrates our methodology to calculate delay penalty, *EDP*, ∇*EDP*, *ENP*, and ∇*ENP* of a single design point. We use the non-linear model to accurately calculate the nominal circuit delay, and use the linearized model to estimate the delay variations (using the CNT count samples). We perform STA for each trial to estimate T_{95} and ∇T_{95}. *PNMV* and ∇*PNMV* are similarly computed via estimation of the SNM distribution.

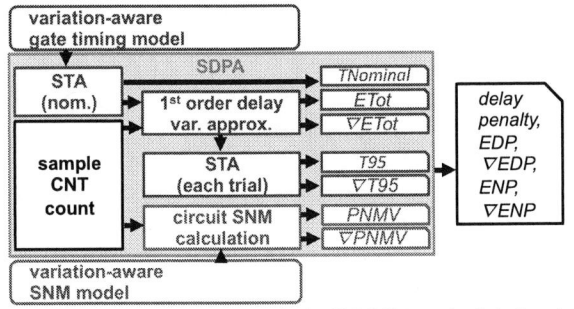

Figure 5. Single design point analysis (*SDPA*) to calculate the delay penalty, *EDP*, ∇*EDP*, *ENP*, and ∇*ENP* of a single design point.

3.4 Guided Exploration to Minimize CNT Variations

To overcome the bottlenecks of trial-and-error-based EDP optimization, we use a gradient descent-based strategy to systematically guide the reduction of EDP and ENP in the presence of CNT variations. After analyzing a single design point, we incrementally update the processing parameters by an amount proportional to their relative magnitudes in either the EDP or ENP gradient (Section A12.2). Thus, the EDP (or ENP) of the next design point is expected to be less than that of the current design point. This procedure continues until a design point with a target delay penalty (or *PNMV*) is reached.

Selective logic gate upsizing, when combined with CNT process improvements, is effective in overcoming variations [Zhang 11a]. Thus, in addition to performing gradient descent from our initial design point (with no upsizing and no processing parameter improvement), we also employ a gate upsizing algorithm: a parameterized fraction (*k*%) of the logic gates with the largest loads (in the nominal case) are upsized. By incrementally increasing *k* from the initial design point, we generate a set of design points, each with a unique delay penalty and energy. We refer to this set of points as the initial energy-delay tradeoff curve. Figure 6 illustrates the initial energy-delay tradeoff curve for the 16nm exu module [OpenSparc] (Section A13). T_{95} and $T_{Nominal}$ (the nominal delay) of each design point are calculated via the single design point analysis (*SDPA*) shown in Figure 5.

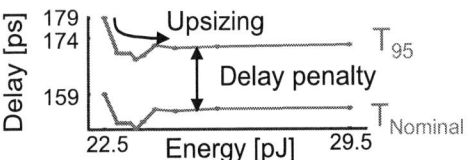

Figure 6. Initial energy-delay tradeoff curves, 16nm exu module.

Our full methodology, illustrated in Figure 7, combines upsizing and gradient descent to overcome the impact of CNT count variations on both delay penalty and *PNMV*. Starting from the initial design point (with $k = 0$), we target a design point that meets both a delay penalty constraint and a *PNMV* constraint (e.g., 1%, chosen by the designer). This is consequently a feasibility problem, in which we are searching for a design point that meets two constraints, and we utilize a variation of an alternating projections (*AP*) algorithm [Bauschke 94] to solve it. A standard AP algorithm continuously projects a point onto multiple constraints until all are satisfied. In our methodology, we use *gradient descent* instead of *projection*. As shown in Figure 7, we alternate between: (1) performing gradient descent until the target *PNMV* is satisfied (using ∇*ENP*), and (2) performing gradient descent until the target delay penalty is satisfied (using ∇*EDP*). From the initial design point (with initial processing parameters set by the designer), processing parameters (*PP*) are updated until both the delay penalty and *PNMV* constraints are satisfied. CNFET upsizing is then performed to generate a *new* design point, and the processing parameters are reset to their initial values. Now we are on a new point on the initial energy-delay tradeoff curve. Next, we find another design point that meets both constraints, then upsizing is performed again and the processing parameters are reset. This process continues until the energy increase from upsizing is too large (another design choice). At this point, our algorithm returns the design point with the lowest EDP (which satisfies both constraints). Finally, the delay penalty can be validated using the non-linear model [Zhang 11b].

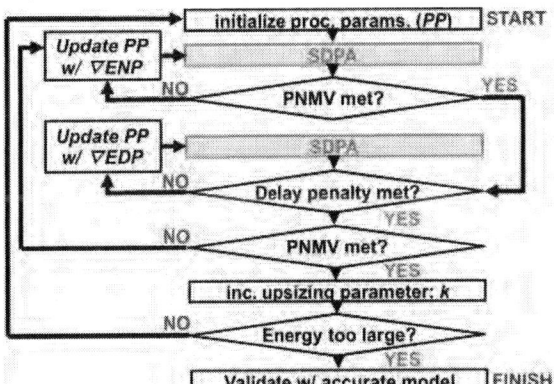

Figure 7. Methodology to satisfy delay penalty and *PNMV* requirements. SDPA details in Figure 5.

Additional details of our gradient descent implementation, including gradient calculation, processing parameter updates, and strategies to avoid local optima, are provided in Section A12.

4. RESULTS

We present two sets of results to demonstrate that we have overcome the bottlenecks of trial-and-error based approaches (Section 3). The first set demonstrates that we can analyze a design point over 100X faster than before, while still identifying design points with less than 2.5% EDP sub-optimality (Section 4.1). The second set demonstrates the ability of our gradient descent algorithm to simultaneously achieve a delay penalty ≤ 5% and *PNMV* ≤ 1% without exhaustive search (Section 4.2).

4.1 Model Validation

To validate our single design point analysis (Figure 5), we perform a trial-and-error-based optimization to find the design point with the lowest EDP. We analyzed 104 unique design points (chosen to resemble the design points in [Zhang 11a]), which include thirteen different values of the upsizing parameter, k, and eight unique sets of processing parameters. A detailed explanation of these design points is provided in Section A13.

Each design point is analyzed separately with our linearized delay model and with the non-linear delay model. Each method selects an optimal design, which are then compared using the EDP computed by the non-linear model. We define *EDP sub-optimality* to quantify the error in the linearized model's design choice:

$$EDP_{sub-opt} = \frac{EDP_{MeasuredByNonLin}^{(ChosenByLin)}}{EDP_{MeasuredByNonLin}^{(ChosenByNonLin)}} - 1. \quad [14]$$

Ideally, the two models should select the same design point (0% EDP sub-optimality). This is the case for six of our eight test modules (from the OpenSparc T2 core [OpenSparc], Section A13); the other two have $EDP_{sub-opt}$ less than 2.5%. All test cases achieve a speed-up of over 100X (Table 3). Figure 8 illustrates a case in which both models have chosen the same design point, and the EDP sub-optimality is 0%. Similar results for *PNMV* are presented in Section A14.1.

Table 3. EDP sub-optimality and simulation time (measured on a single 2.93GHz processor with no parallelization).

Mod	Gate count	Time, non-lin model	Time, lin model	Speed-up	EDP sub-opt
dec	6K	1.6 days	20 min	112X	0%
pmu	12K	3.4 days	40 min	121X	0%
gkt	11K	3.6 days	41 min	126X	0%
pku	15K	4.1 days	50 min	119X	0%
exu	25K	1 week	1.3 hr	124X	0%
lsu	53K	3 weeks	3.2 hr	160X	0%
tlu	83K	1 month	4.7 hr	162X	0.24%
fgu	130K	1.4 months	6.0 hr	172X	2.31%

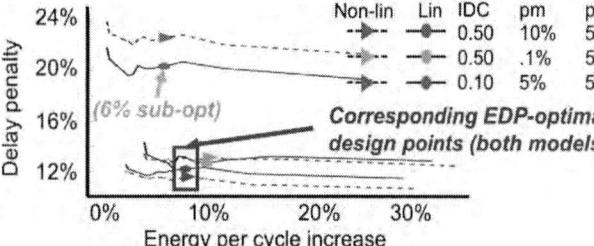

Figure 8. Delay penalty vs. energy per cycle increase for the 16nm lsu module. Solid lines represent delay penalties using the linearized model, and dotted lines represent the non-linear model.

4.2 Delay Gradient Descent

We now demonstrate the effectiveness of our gradient descent algorithm to simultaneously achieve a delay penalty $\leq 5\%$ and a $PNMV \leq 1\%$. The initial design point assumes no upsizing, and assumes current state-of-the-art values of the processing parameters (Table 1, $IDC = 0.50$, $p_m = 1\%$, $p_{Rs} = 5\%$). We first perform gradient descent using ∇ENP until the $PNMV$ constraint is satisfied (Figure 7). Intermediate values of the processing parameters after this step are provided in Section A14.2. The next step is to reduce the delay penalty to $\leq 5\%$. This is accomplished via gradient descent with ∇EDP (Figure 7), and the final results are shown in Table 4 (satisfied constraints in green). Figure 9 shows the relative improvement (definition in Section A14.2) of each processing parameter, which is a measure of its effectiveness in terms of achieving the delay penalty and $PNMV$ constraints. The relative improvement is highest for IDC in all cases,

indicating that IDC is a highly effective parameter to improve in order to reduce delay penalties and $PNMV$ in an energy-efficient manner. Figure 10 illustrates the gradient descent as ENP and EDP improve. Multiple gradient descent instances are initialized from the initial energy-delay tradeoff curve, and each descends to $\leq 5\%$ delay penalty with less than 10% energy increase (ΔE). The optimal design point determined from trial-and-error search (Section 4.1) has 8% delay penalty, and is shown as a reference.

Table 4. Gradient descent results after attaining both delay penalty $\leq 5\%$ and $PNMV \leq 1\%$. ΔE refers to the increase in energy over the initial design point. The upper bound of 0.5% on $PNMV$ is related to the number of MC trials (details in Section A14.1).

Mod.	ΔE	IDC	p_m	p_{Rs}	PNMV	Delay pen
dec	0.3%	0.05	1.0%	4.5%	<0.5%	4.4%
pmu	7.4%	0.09	0.9%	2.8%	<0.5%	5.0%
gkt	2.4%	0.07	0.9%	3.6%	<0.5%	4.7%
pku	0.7%	0.08	1.0%	4.3%	<0.5%	5.0%
exu	2.8%	0.13	0.9%	3.2%	<0.5%	4.9%
lsu	7.4%	0.04	0.7%	1.6%	<0.5%	4.7%
tlu	3.1%	0.18	1.0%	4.2%	0.9%	4.5%

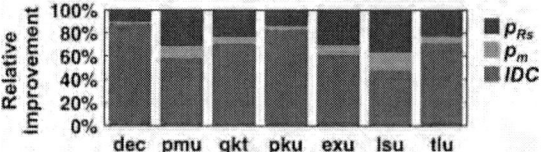

Figure 9. Relative improvement of each processing parameter to achieve both delay penalty $\leq 5\%$ and $PNMV \leq 1\%$ (Table 4).

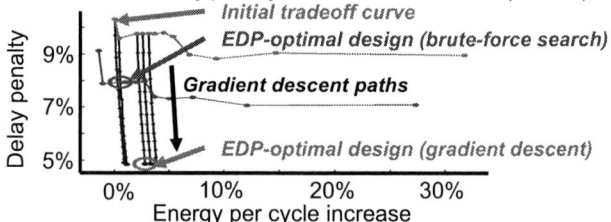

Figure 10. EDP and ENP gradient descent (16nm exu module) to simultaneously achieve delay penalty $\leq 5\%$ and $PNMV \leq 1\%$. Gradient descent paths (black) are initialized from the original energy-delay tradeoff curve (red) and descend until 5% delay penalty is achieved. The new EDP-optimal design is shown in pink.

From our results, we conclude:

(1) Our computationally efficient delay model runs over 100X faster than the non-linear model, and still has enough accuracy to identify designs with less than 2.5% EDP sub-optimality for all test cases.

(2) *PNMV* computation achieves similar efficiency.

(3) Gradient descent is a systematic and scalable method to meet both delay penalty and *PNMV* constraints.

(4) In contrast to traditional thinking (which focuses on reducing p_m to ultra-low values), gradient descent identifies reduction in IDC as a highly-effective method of meeting delay penalty and *PNMV* constraints. Reducing p_m past 1% suffers from diminishing returns at the 16nm node. Unlike trial-and-error approaches (e.g., in [Zhang 11a]), gradient descent establishes these facts in a highly rigorous manner.

Note that, the IDC value is reduced to less than 0.05 in some cases. If this is not technologically feasible, then the gradient descent algorithm can identify alternate sets of design points. For example, the update direction can be weighted towards p_m or p_{Rs} if IDC is difficult to improve, or the gradient descent algorithm can be forced to never update IDC past a hard-limit. These constraints can be provided as inputs, and are features of our flexible framework.

5. CONCLUSION

We have demonstrated, for the first time, a **systematic** methodology for joint exploration and optimization of CNT processing and CNFET circuit design to overcome the significant challenge of CNT variations. Our approach enables quick evaluation of delay variations and noise margin of CNFET VLSI circuits with over 100X speed-up compared to existing approaches. Experimental validation of our simplified delay model, that is key to achieving this large speedup, demonstrates the effectiveness of our methodology. Our new gradient descent-based framework accurately identifies the most important processing parameters, in conjunction with CNFET circuit sizing, to achieve high energy-efficiency while satisfying circuit-level noise margin constraints. Unlike existing trial-and-error techniques, our framework can automatically explore the large space of CNT processing options, and generate a variety of CNT processing routes depending on CNT processing technology constraints. Such automatic exploration is essential for a successful CNFET technology to avoid potential obstacles.

Future research directions include:

(1) Adaptation of our framework for highly-advanced technology nodes (e.g., sub-10nm) with proper device models. (We have not analyzed technology nodes beyond 16nm due to limitations of the CNFET SPICE model [SPICE].) For highly-scaled nodes, incorporation of CNT contact variations and threshold voltage variations, as well as other CNT processing techniques (e.g., CNT sorting [Arnold 06]), will be important.

(2) Validation of model parameters for high-density CNT growth techniques and for channel lengths significantly closer to the ballistic regime.

(3) Examination of the applicability of our framework for other emerging nanotechnologies. It is commonly believed that many emerging nanotechnologies are expected to exhibit substantial variations. Our methodology can potentially be adapted to overcome the challenges in those technologies.

6. ACKNOWLEDGMENT

We acknowledge NSF, FCRP C2S2, FCRP FENA, STARNet SONIC, the Stanford Graduate Fellowship and the Hertz Foundation Fellowship (for Max Shulaker). We thank Sean Keller of the California Institute of Technology for valuable discussions.

7. REFERENCES

[Appenzeller 08] Appenzeller, J., "Carbon nanotubes for high-performance electronics—Progress and prospect," *Proc. IEEE*, Vol. 96.2, pp. 201-211, 2008.

[Arnold 06] Arnold, M. S., *et al.*, "Sorting carbon nanotubes by electronic structure using density differentiation," *Nature Nanotech.*, Vol. 1.1, pp. 60-65, 2006.

[Bauschke 94] Bauschke H. H., and Borwein, J. M., "Dykstra's alternating projection algorithm for two sets," *Journal of Approximation Theory*, Vol. 79.3, pp. 418-443, 1994.

[Cao 12] Cao, Q., *et al.*, "Evaluation of Field-Effect Mobility and Contact Resistance of Transistors That Use Solution-Processed Single-Walled Carbon Nanotubes," *ACS nano*, Vol. 6.7, pp. 6471-6477, 2012.

[Chang 12] Chang, L., *et al.*, "IEDM Short Course," *IEDM*, 2012.

[Chu 04] Chu, C., "FLUTE: fast lookup table based wirelength estimation technique," *ICCAD*, Vol. 27.1, pp. 70-83, 2004.

[Ding 12] Ding, L., *et al.*, "Carbon Nanotube Based Ultra-Low Voltage Integrated Circuits: Scaling Down to 0.4 V," *Applied Physics Letters*, Vol. 100, pp. 263116-1 - 263116-5, 2012.

[Eigen] http://eigen.tuxfamily.org.

[Franklin 12a] Franklin, A. D., *et al.*, "Sub-10 nm Carbon nanotube transistor," *Nano letters*, Vol. 12.2, pp. 758-762, 2012.

[Franklin 12b] Franklin, A., *et al.*, "Scalable and fully self-aligned n-type carbon nanotube transistors with gate-all-around," *IEDM*, pp. 4-5, 2012.

[Franklin 12c] Franklin, A. D., *et al.*, "Variability in Carbon Nanotube Transistors: Improving Device-to-Device Consistency," *ACS nano*, Vol. 6.2, pp. 1109-1115, 2012.

[ITRS] *International Technology Roadmap for Semiconductors*.

[Kang 07] Kang, S. J., *et al.*, "High-performance electronics using dense, perfectly aligned arrays of single-walled carbon nanotubes," *Nature Nanotech.*, Vol. 2.4, pp. 230-236, 2007.

[Keller 11] Keller, S., *et al.*, "Reliable Minimum Energy CMOS Circuit Design," *European Workshop on CMOS Variability*, 2011.

[Laudon 05] Laudon, J., "Performance/watt: the new server focus," *ACM SIGARCH Computer Architecture News*, Vol. 33.4, pp. 5-13, 2005.

[Lin 09] Lin, A., *et al.*, "ACCNT—A metallic-CNT-tolerant design methodology for carbon-nanotube VLSI: Concepts and experimental demonstration," *IEEE Trans. Electron Devices*, Vol. 56.12, pp. 2969-2978, 2009.

[Lohstroh 83] Lohstroh, J., *et al.*, "Worst-case static noise margin criteria for logic circuits and their mathematical equivalence," *IEEE Journal of Solid-State Circuits*, Vol. 18.6, pp. 803-807, 1983.

[Markovic 04] Markovic, D., *et al.*, "Methods for true energy-performance optimization," *IEEE Journal of Solid-State Circuits*, Vol. 39.8, pp. 1282-1293, 2004.

[Nangate] http://www.nangate.com.

[Nassif 07] Nassif, S., *et al.* "High performance CMOS variability in the 65nm regime and beyond," *IEDM*, pp. 569-571, 2007.

[OpenSparc] http://www.opensparc.net/opensparc-t2.

[Park 12] Park, H., *et al.*, "High-density integration of carbon nanotubes via chemical self-assembly," *Nature Nanotech.*, Vol. 7.12, pp. 787-791, 2012.

[Patil 08] Patil, N., *et al.*, "Design methods for misaligned and mispositioned carbon-nanotube immune circuits," *IEEE Trans. CAD*, Vol. 27.10, pp. 1725-1736, 2008.

[Patil 09a] Patil, N., *et al.*, "Wafer-scale growth and transfer of aligned single-walled carbon nanotubes," *IEEE Trans. Nanotech.*, Vol. 8.4, pp. 498-504, 2009.

[Patil 09b] Patil, N., *et al.*, "VMR: VLSI-compatible metallic carbon nanotube removal for imperfection-immune cascaded multi-stage digital logic circuits using carbon nanotube FETs," *IEDM*, pp. 1-4, 2009.

[Patil 11] Patil, N., *et al.* "Scalable carbon nanotube computational and storage circuits immune to metallic and mispositioned carbon nanotubes," *IEEE Trans. Nanotech.*, Vol. 10.4, pp. 744-750, 2011.

[Paul 07] Paul, B. C., *et al.* "Impact of a process variation on nanowire and nanotube device performance," *IEEE Trans. Electron Devices*, Vol. 54.9, pp. 2369-2376, 2007.

[Raychowdhury 09] Raychowdhury, A., *et al.*, "Variation tolerance in a multichannel carbon-nanotube transistor for high-speed digital circuits," *IEEE Trans. Electron Devices*, Vol. 56.3, pp. 383-392, 2009.

[Rivoire 07] Rivoire, S., *et al.*, "JouleSort: a balanced energy-efficiency benchmark," *Proc. 2007 ACM SIGMOD international conference on Management of data*, pp. 365-367, 2007.

[Roy 08] Roy, J. A., and Markov, I. L., "High-performance routing at the nanometer scale," *IEEE Trans. CAD*, Vol. 27.6, pp. 1066-1077, 2008.

[Shulaker 11] Shulaker, M. M., *et al.*, "Linear increases in carbon nanotube density through multiple transfer technique," *Nano letters*, Vol. 11.5, pp. 1881-1886, 2011.

[Shulaker 13] Shulaker, M. M., *et al.*, "Experimental Demonstration of a Fully Digital Capacitive Sensor Interface Built Entirely Using Carbon Nanotube FETs," *ISSCC*, pp. 112-113, 2013.

[SPICE] http://nano.stanford.edu/models.php.

[Wei 09] Wei, L., *et al.*, "A non-iterative compact model for carbon nanotube FETs incorporating source exhaustion effects," *IEDM*, pp. 1-4, 2009.

[Wei 10] Wei, H., *et al.*, "Efficient Metallic Carbon Nanotube Removal Readily Scalable to Wafer-Level VLSI CNFET Circuits," *Proc. Symp. VLSI Tech.*, pp. 237-238, 2010.

[Weste 05] Weste, N. H., and Harris, D. M., "CMOS VLSI Design," Pearson/Addison Wesley, 2005.

[Shor 85] Shor, N. Z., "Minimization Methods for Non-differentiable Functions," Springer-Verlag, 1985.

[Zhang 09a] Zhang, J., *et al.*, "Carbon Nanotube Circuits in the Presence of Carbon Nanotube Density Variations," *DAC*, pp. 71-76, 2009.

[Zhang 09b] Zhang, J., *et al.*, "Probabilistic Analysis and Design of Metallic-Carbon-Nanotube-Tolerant Digital Logic Circuits," *IEEE Trans. CAD*, pp. 1307-1320, 2009.

[Zhang 10] Zhang, J., *et al.*, "Carbon nanotube correlation: promising opportunity for CNFET circuit yield enhancement," *DAC*, pp. 889-892, 2010.

[Zhang 11a] Zhang, J., *et al.*, "Overcoming carbon nanotube variations through co-optimized technology and circuit design," *IEDM*, pp. 4-6, 2011.

[Zhang 11b] Zhang, J., "Variation-aware design of carbon nanotube digital VLSI circuits," *Ph.D. Diss.*, Stanford University, 2011.

[Zhang 12] Zhang, J., *et al.*, "Carbon Nanotube Robust Digital VLSI," *IEEE Trans. CAD*, Vol. 31.4, pp. 453-471, 2012.

Appendix

A8. OVERVIEW OF CNT VARIATIONS

The CNT-specific variations that we consider are:

(1) *CNT-type variations*: CNTs can be either metallic (*m-CNT*) or semiconducting (*s-CNT*). The presence of m-CNTs leads to various circuit problems, including increased leakage current, susceptibility to noise, and incorrect functionality.

(2) *CNT density variations*: explained below.

(3) *CNT diameter variations*: the diameter of a CNT is a function of its chirality, and can lead to changes in its threshold voltage and maximum current.

(4) *CNT alignment variations*: mis-positioned CNTs cause random alignment angles with respect to the CNT growth direction.

(5) *CNT doping variations*: CNFETs require heavily doped source and drain extension regions to achieve small parasitic series resistance. Variations in the doping concentration lead to variations in series resistance.

CNT count variations are caused by:

(1) *CNT density variations*: precise positioning of CNTs is difficult to control. This causes variations in the inter-CNT spacing, which leads to variations in the CNT count of a CNFET.

(2) *m-CNT-induced variations*: due to CNT type variations, a given CNFET contains a random fraction of m-CNTs and s-CNTs. Consequently, there would be variations in the CNT count even with a perfect m-CNT removal technique. Additional CNT count variations are caused by imperfect m-CNT removal techniques, which may also inadvertently remove a small fraction of s-CNTs.

A8.1 Impact on Circuit Delay Variations

The maximum current of a CNFET with a single CNT as its channel is highly sensitive to CNT diameter variations [Paul 07]. However, VLSI digital circuits require CNFETs consisting of multiple CNTs to provide sufficient drive current. For these situations, the impact of CNT diameter, doping, and alignment variations is significantly reduced due to statistical averaging [Raychowdhury 09]. Instead, variations in CNFET on-current (I_{ON}) are dominated by variations in the CNT count [Zhang 11a].

To derive CNFET circuit delay distributions resulting from CNT count variations, Zhang *et al.* adapted a Monte Carlo statistical static timing analysis (MC SSTA) approach with two key changes: (1) a variation-aware timing model [Zhang 11b] for CNFET logic gates, based on a CNFET device model [SPICE], and (2) highly-efficient sampling, based on the unique asymmetric CNT correlation property (Section A8.3.1). Analysis was performed on the OpenSparc T2 processor core design [OpenSparc]. Figure 11 quantifies the delay penalty due to CNT count variations. The delay penalty is shown to reach 60% at the 16nm node [Zhang 11a], and is expected to increase due to decreased statistical averaging as technology continues to scale.

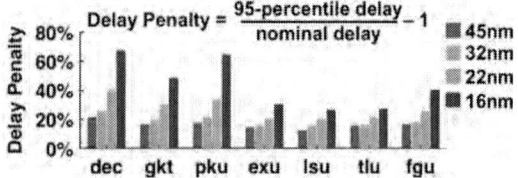

Figure 11. SSTA results over technology generations.

A8.2 Impact on Circuit Noise

Consider a logic gate pair, as shown in Figure 12 (for the case of two inverters). A common metric to quantify the noise resiliency of a logic gate pair is the *static noise margin* (SNM),

which is defined as the size of the maximum nested square between the voltage transfer curve (*VTC*) of the driving logic gate, and the mirrored VTC of its load [Lohstroh 83] (Figure 12).

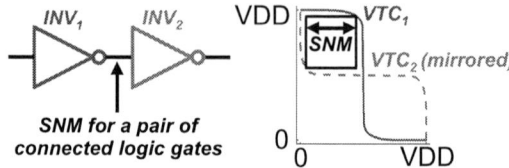

Figure 12. SNM illustration. (Left) pair of inverters; (right) VTC of the first inverter and mirrored VTC of the second inverter.

SNM is sensitive to I_{ON} variations. Let (G_i, G_j) be a logic gate pair, and let $SNM(G_i, G_j)$ be its SNM. Additionally, let SNM_R be the required SNM that (G_i, G_j) must satisfy (a commonly used value of SNM_R is one quarter of the supply voltage, $V_{DD}/4$). Then, given the CNT count distribution, and given a method to calculate $SNM(G_i, G_j)$ from the CNT count distribution (Section A11), we can compute the probability that a logic gate pair will fail to satisfy the SNM requirement, i.e. $P\{SNM(G_i, G_j) < SNM_R\}$. At the circuit-level, we define *PNMV* as the probability that **any** logic gate pair does not meet the SNM requirement, where C is the set of all connected logic gate pairs (G_i, G_j):

$$PNMV = 1 - \prod_{(G_i, G_j) \in C} P\{SNM(G_i, G_j) \geq SNM_R\}. \quad [15]$$

A8.3 Overcoming CNFET Delay Variations

Upsizing of CNFETs reduces delay variations. However, naïve upsizing can be expensive [Zhang 12]. To efficiently overcome the challenge of circuit delay penalties resulting from CNT count variations, a combination of CNT processing and CNFET circuit design is required.

A8.3.1 Layout Design Optimization

Layout design is a powerful method to mitigate CNT count variations, as it can utilize correlations in the CNT count among various CNFETs. Because CNTs are one-dimensional nanostructures, and because their lengths are typically much longer than a single CNFET [Kang 07, Patil 09a], the CNT counts of CNFETs can either be uncorrelated or highly correlated depending on their relative physical placement in a layout [Lin 09, Zhang 10] (Figure 13). This correlation can be "engineered" to overcome CNT variations through special aligned-active layout design, in which the active regions of CNFET standard cells in a library are aligned to one another to maximize correlation (Figure 13). Aligned-active layouts reduce delay penalties by up to 2.6X with only 4% increase in energy [Zhang 11a]. Aligned-active layouts incur minimal area increase, and the locations of I/O pins are mostly retained, resulting in negligible impact on inter-cell routing [Zhang 11b].

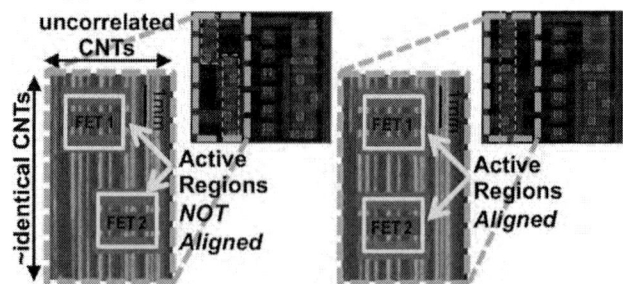

Figure 13. Aligned-active layout. AOI222_X1 standard cell [Nangate]. Before (left) and after (right).

A8.3.2 Processing Parameter Improvement

Processing parameters must be improved so that the delay penalty can be reduced to 5% or less for practical CNFET VLSI, as shown in [Zhang 11a] using a trial-and-error-based approach. This is the motivation behind our gradient descent algorithm (Section 3.4).

A9. CNT COUNT SAMPLING

Figure 14 illustrates the mapping from the region CNT count variables to the CNFET CNT count variables [Zhang 11a].

Figure 14. Relationship between the sampling region CNT counts and the CNFET CNT counts. P-type CNFETs are shown in blue and N-type CNFETs are shown in red. The sampling region variables $(N_1, N_2, ..., N_r)$ define the CNT count for each CNFET.

To validate the Gaussian approximation to the CNT count distribution, we run the single design point analysis (SDPA, Figure 5) for each of two cases: with discrete CNT count variables [Zhang 09a], and with the Gaussian approximation (Section 3.1). Figure 15 shows the cumulative distribution function (CDF) in each case. The Gaussian approximation overestimates the median delay (where the CDF is equal to 0.5) by only 0.2%, and underestimates the delay spread (measured as the width between the points where the CDF is equal to 0.95 and 0.05) by only 1.6%. We conclude that the Gaussian approximation is sufficient for our exploration purposes.

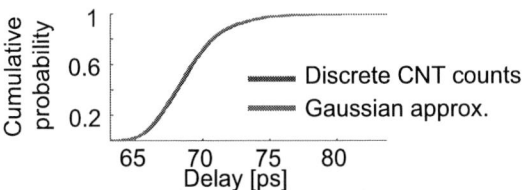

Figure 15. Cumulative distribution function (CDF) of maximum path delay for the 16nm dec module (2,000 MC trials).

A10. DELAY MODEL LINEARIZATION

Solving for d in the non-linear delay model used in [Zhang 11b] ([16]) is not trivial, and may involve a numerical method that requires significantly more computational effort than a model of the form $d = CV/i$ [Weste 05]:

$$d = \frac{(c_{Par} + c_{Load})V_{DD}}{i_1 \min\left(\frac{2d}{t_{slew}}, 1\right) + i_2}. \qquad [16]$$

Additionally, t_{slew} must be determined for each gate and must propagate through the circuit (as it affects the delay of subsequent gates), which further increases the simulation time. To obtain our linearized model ([17]), we linearize [16] with the following three-step approach:

$$d = \frac{(c_{Par} + c_{Load})V_{DD}}{i_{Drive}}. \qquad [17]$$

(1) Perform STA with the non-linear delay model to calculate $T_{Nominal}$. This also yields the nominal gate delays and slew rates for each gate ($d_{Nominal}$ and $t_{SlewNominal}$).

(2) Define a new parameter, i_{Drive}, which is an affine function of i_1 and i_2, and is therefore also an affine function of the region CNT counts (affine transforms preserve affinity):

$$i_{Drive} = i_1 \min\left(\frac{2d_{Nominal}}{t_{SlewNominal}}, 1\right) + i_2. \qquad [18]$$

(3) Replace the denominator in [16] with the value of i_{Drive}, yielding a first-order delay approximation of the form $d = CV/i$ ([17]), but which outputs the same value of d as the non-linear model in the nominal case.

We choose to linearize the delay model around the nominal case so that it is independent of a chosen set of values of the processing parameters. This enables delay factorization (Section 3.1) to further improve computational efficiency. To illustrate the effect of the first-order approximation to variations, we show the gate delay of a minimum-sized inverter as a function of its load capacitance (the capacitance of the next stage, including wire capacitance) in Figure 16. The nominal capacitive load is 284fF, and the nominal delay is 4.1ps. The delay error in this case is less than +/-1% error as the capacitance varies from its nominal value by +/-50%.

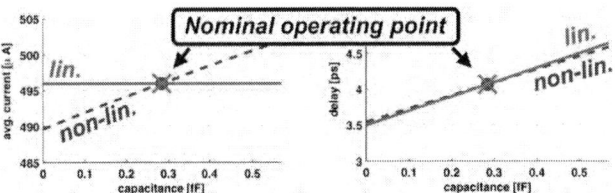

Figure 16. Current model (left) and delay model (right).

A11. STATIC NOISE MARGIN MODEL

We base our circuit $PNMV$ calculation on a methodology that has been shown to efficiently model the $PNMV$ of circuits consisting of a million logic gates [Keller 11]. In this approximation, SNM ([21]) is modeled as a function of the high/low output levels of a driving gate (V_{OH} and V_{OL}) and the high/low input voltage levels of its load (V_{IH} and V_{IL}). These four parameters indicate the points where the slope of the voltage transfer curve (VTC) is equal to -1 [Weste 05] (Figure 17(a)).

We refer to V_{OH}, V_{IH}, V_{IL}, and V_{OL} as the VTC parameters, and we model them separately for each input pin of each logic gate. The input variables to each model are the CNT counts of the CNFETs gated by the specified input pin. These CNT counts are denoted by n_{PU} and n_{PD}, for the pull-up and pull-down CNFETs, respectively. The VTC parameters are modeled according to [19], where T is a real-valued matrix ($T \in \mathbb{R}^{4 \times 2}$) that maps the CNT counts into the VTC parameters. Note that, the entire SNM model consists of many instances of T: one for each input pin of each gate. In each case, we extract the VTC parameters from the VTCs of 2,000 SPICE simulations with different sampled values of n_{PU} and n_{PD} (example VTC in Figure 17(a)). We then use linear regression to fit T to the VTC parameters. In [20], we assign specific variable names to the entries of T for ease of reference.

$$\begin{bmatrix} V_{OH} \\ V_{IH} \\ V_{IL} \\ V_{OL} \end{bmatrix} = T \begin{bmatrix} 1 \\ \log10(n_{PU}/n_{PD}) \end{bmatrix}, \qquad [19]$$

$$T = \begin{bmatrix} T_{VOH0} & T_{VOH1} \\ T_{VIH0} & T_{VIH1} \\ T_{VIL0} & T_{VIL1} \\ T_{VOL0} & T_{VOL1} \end{bmatrix}, \qquad [20]$$

$$SNM = \min\left(V_{OH} - V_{IH}, V_{IL} - V_{OL}\right). \qquad [21]$$

We observed that in all cases (for all input pins of all gates), T_{VOH1} and T_{VOL1} were approximately equal to zero. This is indicative of the fact that the output levels of a logic gate are not strong functions of the CNT count. Thus, to simplify our model, we set T_{VOH1} and T_{VOL1} equal to zero, and still observe a strong fit to the SPICE simulation data. To quantify the error in this fit, we use the root-mean-square error (*RMSE*, [29]), which is less than 9mV in all cases (Section A11.2). An example of our model, in the case of a minimum-sized inverter, is shown in Figure 17(b) (the model, T, is inset).

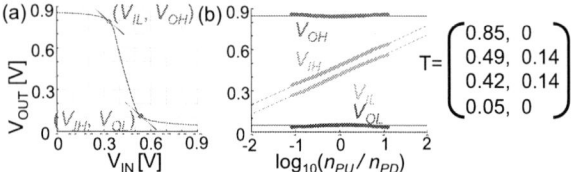

Figure 17. (a) Inverter VTC, indicating the VTC parameters at the points where the slope is -1. (b) SPICE simulation data for a minimum-sized inverter and overlaid model.

A11.1 Full Circuit PNMV Model

Let c be the number of connected logic gate pairs, and let $N_{PU}, N_{PD} \in \mathbb{R}^{c \times n}$ be matrices that contain the values of n_{PU} and n_{PD} (defined above) for each logic gate connection, and for each MC trial. Also, let $B_{PU} \in \{0,1\}^{c \times r}, B_{PD} \in \{0,1\}^{c \times r}$ be the matrices that map the region CNT counts into N_{PU} and N_{PD} (N is defined in Section 3.1):

$$N_{PU} = B_{PU}N, \qquad [22]$$
$$N_{PD} = B_{PD}N. \qquad [23]$$

Using all instances of T (each input pin, each logic gate), we formulate [24]-[27] to relate N to the SNM of each connected logic gate. $t_{VOH0}^{drive} \in \mathbb{R}^c, t_{VOL0}^{drive} \in \mathbb{R}^c$ contain the output voltage level for each connection (driving gate), $t_{VIH0}^{load} \in \mathbb{R}^c, t_{VIL0}^{load} \in \mathbb{R}^c$ contain the input voltage level for each connection (loading gate), and $t_{VIH1}^{load} \in \mathbb{R}^c, t_{VIL1}^{load} \in \mathbb{R}^c$ contain the dependence of the input voltage level on N. The values in each of these vectors correspond to the entries (denoted by the subscript) in the individual models (T, for each input pin of each logic gate [20]):

$$H = \frac{B_{PU}N}{B_{PD}N}, \qquad [24]$$
$$M_H = (t_{VOH0}^{drive} - t_{VIH0}^{load})\mathbf{1}^T - (t_{VIH1}^{load}\mathbf{1}^T).*\log_{10}(H), \qquad [25]$$
$$M_L = (t_{VIL0}^{load} - t_{VOL0}^{drive})\mathbf{1}^T + (t_{VIL1}^{load}\mathbf{1}^T).*\log_{10}(H), \qquad [26]$$
$$M_{HL} = \min(M_L, M_H). \qquad [27]$$

Here, the division, the logarithm, the min function, and the multiplication denoted by ".*" are all element-wise, yielding matrices $M_H, M_L, M_{HL} \in \mathbb{R}^{c \times n}$ whose columns give the noise margin samples for each connection.

As in the delay case, we can rewrite $N = \mu_R + \sigma_R X$, which allows us to factorize the SNM model, and to calculate H with scalar operations, using the parameters defined in Table 5:

$$H = \frac{\sigma_R s_{PU} + \mu_R s_{PU}\mathbf{1}^T}{\sigma_R s_{PD} + \mu_R s_{PD}\mathbf{1}^T}. \qquad [28]$$

Table 5. Input terms in the SNM model.

Par.	$\in$	Equation	Description
s_{PU}	$\mathbb{R}^{c \times n}$	$B_{PU}X$	Variable pull-up CNT count
s_{PU}	$\mathbb{R}^{c}$	$B_{PU}\mathbf{1}$	Expected pull-up CNT count
s_{PD}	$\mathbb{R}^{c \times n}$	$B_{PD}X$	Variable pull-down CNT count
s_{PD}	$\mathbb{R}^{c}$	$B_{PD}\mathbf{1}$	Expected pull-down CNT count

Once s_{PU} and s_{PD} have been pre-computed, calculating the SNM of each logic gate pair requires scalar operations: four multiplications, four additions, one division, one logarithm, and one min function per trial (only counting operations that must be individually computed for each trial, [25]-[28]). The CDF of the minimum SNM for each MC trial is then computed, from which we extract *PNMV*.

A11.2 Static Noise Margin Modeling Error

To quantify the accuracy of our SNM model, we use the RMS modeling error of each VTC parameter ([29], shown for the case

of V_{OH}). $V_{OH,i}^{(VTC)}$ is the value of V_{OH} extracted from the VTC for the ith SPICE simulation (of p), and $V_{OH,i}^{(model)}$ is its modeled value [19]:

$$RMSE = \sqrt{\frac{1}{p}\sum_{i=1}^{p}\left(V_{OH,i}^{(VTC)} - V_{OH,i}^{(model)}\right)^2}. \qquad [29]$$

We demonstrate an RMS modeling error of less than 9mV (1% of the 0.9V supply voltage [ITRS]) for all VTC parameters, and for all input pins of all gates, as shown in Figure 18.

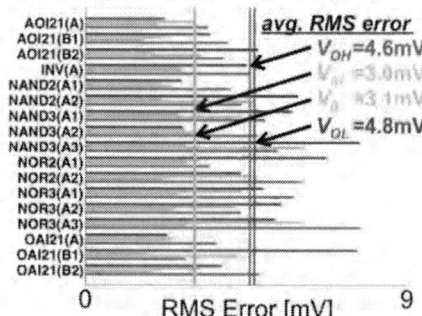

Figure 18. SNM model fit across a subset of the gates in our standard cell library [Zhang 11b].

A12. GRADIENT DESCENT IMPLEMENTATION

Here, we provide the details of our gradient descent algorithm.

A12.1 Gradient Calculation

A *critical path* is any path between an input and an output with delay equal to the maximum path delay. There can be multiple critical paths for a single MC trial. Immediately after STA of each MC trial (Figure 5), we numerically estimate ∇T_{95} ([11]) via the following process:

(1) For each MC trial, we record an arbitrary critical path. These paths are used to estimate ∇T_{95} using a *sub-gradient*, borrowing from the sub-gradient method for minimization of non-differentiable functions [Shor 85] (e.g., "max()" in STA).

(2) We estimate $\partial T_{95}/\partial \mu_R$ and $\partial T/\partial \sigma_R$ ([30]) as follows: we increase μ_R by an incremental amount ($\delta\mu_R = 10^{-6}$), and then recompute the path delay only for the arbitrarily chosen critical path of each MC trial (using [7]). We build the CDF of these path delays, and extract the delay where the CDF is equal to 0.95. This delay differs from T_{95} (Section 2) by an amount δT_{95}: we approximate $\partial T_{95}/\partial \mu_R$ as $\delta T_{95}/\delta\mu_R$. This step assumes that for each MC trial, the same paths remain critical paths after changing μ_R, which is only true in the limit as $\delta\mu_R \to 0$, and is an approximation for $\delta\mu_R = 10^{-6}$. We do the same for σ_R to estimate $\partial T_{95}/\partial \sigma_R$.

(3) We numerically estimate the partial derivatives of μ_R and σ_R with respect to each processing parameter ([30]) as follows. In the case of $\partial\mu_R/\partial p_{Rs}$, we increase p_{Rs} by an incremental amount ($\delta p_{Rs} = 10^{-6}$) and then recalculate μ_R. We estimate $\partial\mu_R/\partial p_{Rs}$ as the change in μ_R divided by δp_{Rs}. We do the same for σ_R and for the other processing parameters.

(4) We estimate the entries of ∇T_{95} according to (in the case of p_{Rs}, similar expressions are used for IDC and p_m):
$$\frac{\partial T_{95}}{\partial p_{Rs}} = \frac{\partial T_{95}}{\partial\mu_R}\frac{\partial\mu_R}{\partial p_{Rs}} + \frac{\partial T_{95}}{\partial\sigma_R}\frac{\partial\sigma_R}{\partial p_{Rs}}. \qquad [30]$$

We adopt a similar methodology to estimate $\nabla PNMV$ ([11]): a logic gate pair with SNM equal to the minimum SNM across all gate pairs is chosen (arbitrarily chosen if there is more than one) for each MC trial. Then the SNM CDF is recomputed with these logic gate pairs (similar to steps 2-3) to estimate $\nabla PNMV$ (similar to step 4). ∇E_{Tot} ([11]) is computed analytically according to [8].

754

A12.2 Gradient Descent Update

To update the processing parameters during gradient descent, we first normalize the gradient vector (∇EDP [12], or ∇ENP [13]) by its $\ell 1$-norm (the sum of the absolute values of its elements). This will force the sum of the updates of each processing parameter to be constant. We then take a step so that the improvement in each processing parameter is proportional to its relative magnitude in the normalized gradient, and the total improvement in processing parameters sums to 10%. Note that, small step-sizes require more simulations, and large step-sizes yield coarse granularity in exploration. This strategy (though others may be adopted) assumes that it is equally difficult to update each processing parameter by a fixed percentage. For example, if the elements of the normalized gradient vector are 0.7, 0.1, and 0.2, then we reduce IDC, p_m, and p_{Rs} by 7%, 1%, and 2% (of their current values), respectively.

A12.3 Mitigating the Effect of Local Optima

Given that our optimization is based on gradient-descent, we employ two strategies to avoid convergence to local optima (since our objective is not necessarily a convex function):

(1) *Initialize gradient descent from multiple design points on the initial energy-delay tradeoff curve*: each instance of gradient descent leads to a certain design point, and we choose the one with the lowest EDP. Even if all instances of gradient descent have converged to the same local optimum, we will never choose a worse design point by starting another instance of gradient descent.

(2) *Never increment a processing parameter away from its ideal value* (Table 1): The only case in which all variations are zero, by definition, is the nominal case (in which all processing parameters have their ideal values). This is consequently the global optimum in terms of minimizing the effect of variations. Any case in which the gradient points towards incrementing the value of a processing parameter away from its ideal value is indicative of local optima. We choose simply to not update that parameter in these cases.

A13. OPENSPARC CIRCUIT SIMULATION

Our eight test circuits are synthesized from the OpenSparc T2 processor core [OpenSparc] using the Synopsys Design Compiler. Our CNFET standard cell library [Zhang 11b] is derived from the Nangate 45nm Open Cell Library [Nangate] (dimensions are scaled for the 16nm case), and placement is determined via Capo [Roy 08]. Wire lengths are determined via FLUTE [Chu 04], and wire parasitics are estimated using parameters from the ITRS [ITRS]. All simulations are performed at the 16nm node assuming 250 CNT/µm and a supply voltage of 0.9V [ITRS]. Delay and SNM distributions are estimated from 2,000 MC trials.

Test circuits are evaluated in Sections 4 and A14 with 104 unique design points, including all combinations of:

(1) *Thirteen values of the upsizing parameter*: spaced between values that correspond to no upsizing (0% energy increase) and to upsizing with a 30% energy increase. We neglect designs with over a 30% energy increase due to upsizing.

(2) *Eight sets of processing parameters*: these are the eight sets of processing parameters analyzed in [Zhang 11a] (Table 6).

Table 6. Processing parameters used to validate the circuit delay and *PNMV* models (Sections 4, A14).

Test	0	1	2	3	4	5	6	7
IDC	.50	.50	.50	.50	.35	.10	.25	.25
p_m	.10	.05	.01	.001	.01	.05	.05	.05
p_{Rs}	.05	.05	.05	.05	.05	.05	.05	.025

A14. ADDITIONAL RESULTS

Additional results are provided in this section.

A14.1 PNMV Model Validation

We analyze the same 104 designs as in Section 4.1 to show that co-optimization of CNT technology and CNFET circuit design can significantly reduce *PNMV* in an energy-efficient manner. Without upsizing, and without guidelines for CNT process improvements, *PNMV* can reach 80% for the OpenSparc T2 core modules, assuming current state-of-the-art processing parameter values: $IDC = 0.50$, $p_m = 1\%$, $p_{Rs} = 5\%$. However, *PNMV* can be reduced to less than 1% with processing parameter improvement and minimal upsizing ($\Delta E \leq 5\%$), as shown via trial-and-error (Table 7). Table 7 also shows the *PNMV* model simulation time required to analyze the 104 design points, which is on the same order as that of the linearized delay model (Section 4.1). We use 0.5% as a conservative upper bound on *PNMV* since it is estimated via MC, and so the number of MC trials limits its resolution. A simulation with 2,000 MC trials yields a resolution of 0.05%, 10X smaller than our upper bound.

Table 7. *PNMV* model results.

Mod.	Total conn.	Orig. *PNMV*	Lowest *PNMV* ($\Delta E \leq 5\%$)	Time
dec	9K	25%	<0.5%	21 min
pmu	19K	32%	<0.5%	50 min
gkt	17K	30%	<0.5%	43 min
pku	23K	42%	<0.5%	58 min
exu	41K	44%	<0.5%	1.8 hr
lsu	86K	68%	<0.5%	3.6 hr
tlu	139K	77%	<0.5%	5.7 hr
fgu	223K	83%	<0.5%	8.6 hr

A14.2 Gradient Descent to Target PNMV

The first step in Section 4.2 to perform gradient descent using ∇ENP until *PNMV* $\leq 1\%$ is attained; Table 8 shows the intermediate values of the processing parameters and the delay penalty immediately after this constraint is met. Table 8 shows *PNMV* $\leq 1\%$ in all cases (green), but shows delay penalty $> 5\%$ (yellow) for all but one case. Figure 19 illustrates the relative improvement ([31]-[33]) of each processing parameter, which indicates the sensitivity of *PNMV* to each of them. *PNMV* is the most sensitive to *IDC*, as its average relative improvement (over all modules) is 80%, compared to 3% for p_m and 17% for p_{Rs}. The *relative improvement* (R) of each parameter is calculated in terms of its percentage improvement (I) and the total improvement (I_{Tot}):

$$I_{IDC} = 1 - \frac{IDC^{final}}{IDC^{init}}, I_{pm} = 1 - \frac{p_m^{final}}{p_m^{init}}, I_{pRs} = 1 - \frac{p_{Rs}^{final}}{p_{Rs}^{init}}, \quad [31]$$

$$I_{Tot} = I_{IDC} + I_{pm} + I_{pRs}, \quad [32]$$

$$R_{IDC} = \frac{I_{IDC}}{I_{Tot}}, R_{pm} = \frac{I_{pm}}{I_{Tot}}, R_{pRs} = \frac{I_{pRs}}{I_{Tot}}. \quad [33]$$

Table 8. Gradient descent results after attaining *PNMV* $\leq 1\%$.

Mod.	ΔE	IDC	p_m	p_{Rs}	*PNMV*	Delay pen
dec	0.3%	0.24	1.0%	4.5%	0.8%	6.5%
pmu	0.2%	0.21	1.0%	4.4%	0.7%	7.6%
gkt	0.3%	0.23	1.0%	4.5%	1.0%	7.0%
pku	0.4%	0.19	1.0%	4.3%	0.8%	6.5%
exu	0.2%	0.22	1.0%	4.4%	1.0%	6.2%
lsu	0.3%	0.19	1.0%	4.3%	0.9%	9.4%
tlu	0.2%	0.18	1.0%	4.2%	0.2%	4.5%

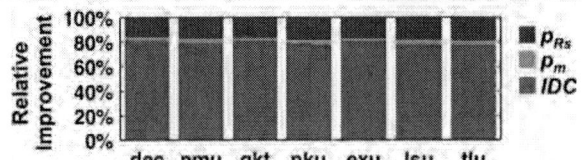

Figure 19. Relative improvement of each processing parameters to achieve *PNMV* $\leq 1\%$ (Table 8).

Minimum-Energy State Guided Physical Design for Nanomagnet Logic

Shiliang Liu[1], Gyorgy Csaba[2], Xiaobo Sharon Hu[1], Edit Varga[2], Michael T. Niemier[1],
Gary H. Bernstein[2], Wolfgang Porod[2]
Departments of Computer Science and Engineering[1], and Electrical Engineering[2]
University of Notre Dame {sliu5, gcsaba, shu, evarga1, mniemier, bernstein.1, porod}@nd.edu

ABSTRACT

Nanomagnet Logic (NML) accomplishes computation through magnetic dipole-dipole interactions. It has the potential for low-power dissipation, radiation hardness and non-volatility. NML circuits have been designed to process and move information via nearest neighbor, device-to-device coupling. However, the resultant layouts often fail to function correctly. This paper reveals an important cause of such failures showing that a robust NML layout must take into account not only nearest neighbor, but also the next nearest neighbor couplings. A new design method is then introduced to address this issue that leverages the minimum-energy states of an NML circuit to guide the layout process. Case studies show that the new method is efficient and effective in arriving at correct NML layouts.

1. INTRODUCTION

Nanomagnet Logic (NML) is a beyond-CMOS technology that realizes the Quantum-dot Cellular Automata (QCA) computing paradigm with nanoscale magnets. In NML, lithographically defined nanomagnets are used to accomplish logic operations. Data moves from one device (and one circuit component) to another via nearest neighbor dipole-dipole interactions. Wires, gates, and inverters have all been experimentally demonstrated at room temperature [1]. NML circuits can be extremely energy efficient [2]. Even with drive circuitry overhead, the energy-delay product of an NML circuit can be 10-1000 times lower than that of CMOS equivalents [3]. Furthermore, magnets with feature sizes above the superparamagnetic limit are non-volatile and can retain states without power. Devices should also be radiation hard.

To implement a desired logic function in NML, a typical design process starts by creating an arrangement of nanomagnets that can realize the logic function based on the *nearest one neighbor* (N1) coupling. The design process then proceeds to determine the actual physical layout of the NML circuit, i.e., finding a combination of geometric device parameters as well as spacings that leads to logically correct output. If detailed micromagnetic simulations are used in the second design step (which is often the case), it can be extremely time consuming. Some recent work, e.g., [4], attempts to improve the design process by leveraging the concept of phase diagram (akin to a "lookup table" for a particular size/material). However, generating phase diagrams itself can still be costly. Furthermore, it is not clear if the methodology can be extended to larger NML circuits.

The key challenge in the NML physical design process is guaranteeing state equivalence, i.e., guaranteeing that the fi-

Permission to make digital or hard copies of all or part of this work for personal or classroom use is granted without fee provided that copies are not made or distributed for profit or commercial advantage and that copies bear this notice and the full citation on the first page. To copy otherwise, to republish, to post on servers or to redistribute to lists, requires prior specific permission and/or a fee.
DAC '13, May 29 - June 07 2013, Austin, TX, USA.

nal magnetization state of each device in an NML circuit is the same as the corresponding device's magnetization state of the desired logic function. (From the system's energy state point of view, the final magnetization state should be the lowest energy state, which is also referred to as *the ground state of the system*.) If state equivalence guarantee cannot be achieved, the NML circuit would have a high probability of producing logically *incorrect* output. The requirement might seem to be trivial as the N1 coupling used in creating the NML ensemble should automatically lead to such guarantee. However, it has been observed repeatedly in experiments that the ground state (LE) of an NML ensemble is not always the same as the magnetization state of the desired logic function (e.g., [5, 6]). No consistent explanation exists in the literature.

One main contribution of this paper is the identification of the root cause of why an N1 coupling based design approach is not sufficient in guaranteeing the state equivalence discussed above. Specifically, we note that as the inter-device spacing increases, the magnetic dipole-dipole interactions between NML devices decays slower than the quadrupole-quadrupole interactions between traditional QCA devices [7]. This difference results in non-nearest neighbor couplings having a non-negligible effect on the functionality of NML circuits. However, to date, nearly all NML design efforts are intuitively based on N1 coupling. This is acceptable for simple circuits like an antiferromagnetically (AF) or ferromagnetically (FM) coupled lines. For more complex circuits (e.g., circuits with closely positioned majority gates together with AF/FM lines), coupling between devices beyond N1 can be significant, and thus can cause the ground state of an NML ensemble to be different from the magnetization state of the desired logic function.

The other main contribution of the paper is the introduction of an energy analysis based design flow. Given that an NML circuit derived solely from N1 coupling may not guarantee state equivalence, our design flow finds the energy difference between the ground state and the state corresponding to the desired logic function. Based on the energy difference, our method determines the NML physical layout (i.e., device sizes and spacing) that achieve the state equivalence guarantee. The significance of the energy landscape of an NML circuit is especially clear in the light of recent studies about the stability [8] and the temperature-activated switching [9] of NML circuits. The energy analysis used in this paper is static and much simpler than that in [8, 9].

Other existing work has also studied the NML physical layout problem. For example, the authors in [10] propose an efficient way to produce a signal and its complement at desired locations. To our best knowledge, no existing work has addressed the state equivalence problem.

2. BACKGROUND

In this section, we start with NML background, then review the energy concept in an NML circuit, and finally discuss the single domain model as a means to calculate NML ensemble energy.

2.1 Nanomagnet Logic and energy concept

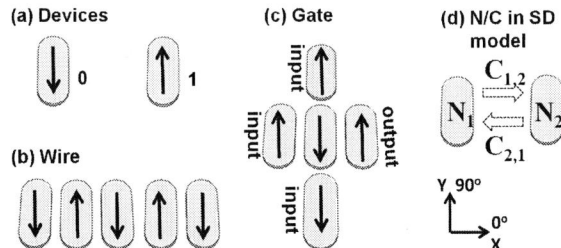

Figure 1: (a) Single-domain magnets with different magnetizations. (b) An AF ordered line. (c) A majority voting gate. (d) A schematic of N/C in the SD model.

Single-domain magnets can represent and store binary data (Fig. 1a). Ensembles of such magnets can also move and process information. A "wire" (Fig. 1b) can be formed from a line of magnets that are AF coupled with each other. FM coupled interconnect is also possible. Logic can be accomplished by majority voting gates (Fig. 1c). By setting one input to a logic 0 or 1, the majority gate can execute a NAND/NOR (or AND/OR) function. All the basic NML elements have been experimentally demonstrated [1]. With the functionally complete logic set, NML can implement any given Boolean logic function.

To facilitate the logical re-evaluation of an NML ensemble, the inputs are first set to the desired values. The rest of the devices then need to be put into a metastable state. This can be achieved through a hard-axis (i.e., the short axis of magnets) directed magnetic clock field. After the removal of this field, the ensemble settles into a new state based on the new input. The hard-axis directed clock field can be generated by current-driven wires. Recently, CMOS compatible clocks with the aforementioned functionability have been experimentally demonstrated [11]. The work presented here is also relevant to various voltage controlled clocking approaches [12, 13] that are generally being studied.

Energy plays an important role in understanding how an NML circuit works. There are a number of magnetic energy components in an NML circuit that should be considered. They include (i) exchange energy ($E_{exchange}$) [14] between magnetic moments inside a magnet, (ii) Zeeman energy (E_{Zeeman}) [14] due to the application of a magnetic field (e.g., a clock field), (iii) demagnetization energy (E_{demag}) [14] due to the shape anisotropy of a magnet, and (iv) coupling energy ($E_{coupling}$) [15, 16] between pairs of magnets. There are also other energy components, e.g., magnetocrystalline anisotropy [14]. These are aggregated together and are represented with the term E_{other}. The total energy of an NML circuit is the sum of the above, i.e.,

$$E_{total} = E_{exchange} + E_{Zeeman} + E_{demag} + E_{coupling} + E_{other} \quad (1)$$

All the energy components can be computed accurately through micromagnetic simulations (e.g., using micromagnetic tools like OOMMF [17]). However, these simulations can be quite expensive with respect to computing resources/time if they are used repeatedly.

In this paper, we are interested in the final magnetization states of NML circuits. The final magnetization state corresponds to the state that the devices finally settle into after the hard-axis directed magnetic clock field has been removed. In this case, E_{Zeeman} is zero and E_{demag} remains constant. Furthermore, magnetic devices in NML are in the SD regime, which means that $E_{exchange}$ is constant. We assume that E_{other} is constant. Therefore, $E_{coupling}$ is the only energy component that can change due to changes in the magnetization states of the devices. We discuss how to compute $E_{coupling}$ efficiently in the next subsection.

2.2 Coupling energy calculation based on single domain model

The total coupling energy, $E_{coupling}$, is the sum of all the dipole-dipole coupling energy in an NML circuit, i.e.,

$$E_{coupling} = \sum_{i=1}^{n} \sum_{j} E_{i,j}, \quad (2)$$

where n is the number of the magnets in the NML circuit, i is one magnet and j is another magnet. $E_{i,j}$ is the magnetic dipole-dipole coupling energy between magnets i and j.

The calculation of the magnetic dipole-dipole coupling energy can be achieved differently. For example, one can get this energy via micromagnetic simulations (e.g., OOMMF). However, it is much more efficient to use the single domain (SD) model [15] to do so.

The micromagnetic model deals with interactions between magnetic moments, which are usually on the scale of the material exchange length. The SD model, on the other hand, treats each magnetic device as a single magnetic moment when the device size resides in the SD regime. There are two key parameters in the SD model. One is the demagnetization tensor field (**N**), which is a function of the shape of a magnet [18]. The other is the coupling factor (**C**) between two magnets, and it represents the coupling strength between two magnets. The calculation of **C** is given in [15]. For two magnets with given shapes, **C** is a function of the relative spacing between the two magnets [16]. The schematic in Fig. 1d shows the key relationship within the SD model. Both **N** and **C** are 3×3 matrices as shown in Eqn. 3 and 4, and both can be calculated using the SD model.

$$\mathbf{N} = \begin{vmatrix} N_{xx} & N_{xy} & N_{xz} \\ N_{yx} & N_{yy} & N_{yz} \\ N_{zx} & N_{zy} & N_{zz} \end{vmatrix} \quad (3)$$

$$\mathbf{C} = \begin{vmatrix} C_{xx} & C_{xy} & C_{xz} \\ C_{yx} & C_{yy} & C_{yz} \\ C_{zx} & C_{zy} & C_{zz} \end{vmatrix} \quad (4)$$

(**N** can be used to compute E_{demag}. Since our focus is on $E_{coupling}$, we will skip the details on E_{demag}.)

C can be used to compute $E_{i,j}$ as follows

$$E_{i,j} = \mu_0(\mathbf{M}_i \bullet \mathbf{C}_{i \to j} \bullet \mathbf{M}_j V_j + \mathbf{M}_j \bullet \mathbf{C}_{j \to i} \bullet \mathbf{M}_i V_i)/(-2) \quad (5)$$

Note that $C_{i \to j}$ is the coupling factor from magnet i to magnet j while $C_{j \to i}$ is the coupling factor from magnet j to magnet i. If magnets i and j have the same shape and size, $C_{i \to j}$ is the same as $C_{j \to i}$. μ_0 is the magnetic permeability of free space. V_i and V_j are the volumes of magnets i and j, respectively. $\mathbf{M}_i$ and $\mathbf{M}_j$ are the magnetization states of magnets i and j, respectively. $\mathbf{M}_i$ and $\mathbf{M}_j$ are vectors generally represented by $[M_x, M_y, M_z]$. M_x and M_z are zero for a magnet at the ground demagnetization energy state with the easy axis along the y-axis. Therefore, if we use M_y to replace $\mathbf{M}$ in Eqn. 5, we can simplify **C** to C_{yy}. Thus, we have

$$E_{i,j} = \mu_0(M_{yi}C_{yy,i \to j}M_{yj}V_j + M_{yj}C_{yy,j \to i}M_{yi}V_i)/(-2). \quad (6)$$

As the number of magnets in an NML circuit increases, the number of coupling pairs increases quadratically ($O(n^2)$). However, because the dipole-dipole interactions between NML devices decay as the inter-device spacing increases, it is not necessary to calculate the coupling energy between all pairs of NML devices. It turns out that N1 and nearest two neighbor (N2) couplings are enough to estimate the total coupling energy, per supplement S.1.

3. THE IMPORTANCE OF NON-NEAREST NEIGHBOR COUPLING

To implement a desired logic function in NML, a typical

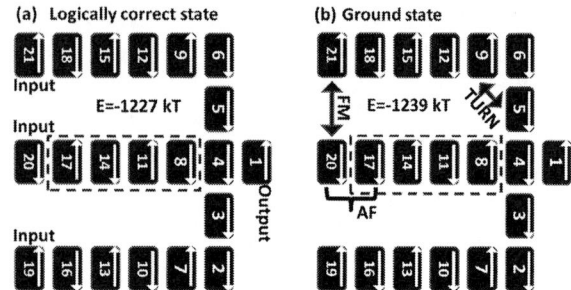

Figure 2: A majority gate whose LC state (a) has higher energy than its ground state (b). The magnetization difference between (a) and (b) is shown in the dotted rectangles.

design process starts by creating an arrangement of nanomagnets that can realize the logic function based on N1 coupling. The corresponding state is referred to as the *logically correct state*. Specifically, we have the following definition.

DEFINITION 1. *LC state: The logically correct (LC) state of an NML circuit is defined as the magnetization states of all the magnets in the NML circuit, which correspond to the correct logic result for a given input combination.*

To ensure that an NML ensemble settles into the LC state, after the clock field is removed, it is essential that this LC state has the smallest E_{total} (i.e., $E_{coupling}$). That is, a necessary condition for an NML circuit to function correctly, the LC state must be equivalent to the ground state.

For simple circuits (e.g., an AF/FM-coupled line), the state equivalence can be readily satisfied. Unfortunately, for complicated circuits, N1 coupling based designs cannot guarantee that the LC state is the ground state. Consider the majority gate example shown in Fig. 2. Fig. 2a shows the LC state of the NML ensemble for a specific input combination. Fig. 2b is a different state for the same input. If we calculate the energy using only N1 coupling, we have

$$E_{coupling(N1)} = \underbrace{E_{2,3} + E_{3,4} + ...}_{FM} + \underbrace{E_{21,18} + E_{18,15} + ...}_{AF} \quad (7)$$

(Based on the coupling types (AF and FM), we have categorized the terms in Eqn. 7 into AF and FM group, respectively.) Both states shown in Fig. 2 gives the same $E_{coupling(N1)}$ value. In fact, it can be easily verified that the LC state for the given input combination indeed has the lowest $E_{coupling(N1)}$. However, from Sec. 2, we know that N2 coupling should be included when computing energy values. In this case, we have

$$E_{coupling(N2)} = \underbrace{E_{2,4} + E_{3,5} + ...}_{FM} + \underbrace{E_{21,15} + E_{18,12} + ...}_{AF} \\ + \underbrace{E_{9,5} + E_{8,5} + ...}_{TURN} \quad (8)$$

(Besides AF and FM groups, we also have the TURN group in Eqn. 8. TURN only exists for N2 coupling.) With both the N1 and N2 couplings, the total coupling energy for the LC state and the ground state is -1227 kT and -1239 kT, respectively. (kT is an energy unit, i.e., 1 kT is 4.11 × 10^{-21} J at 300 K.) Clearly, the LC state is not the same as the ground state.

To find the cause of this "surprising" discrepancy, we introduce some concepts that will help explain the energy difference between the LC state and the ground state.

DEFINITION 2. *Conflict: A conflict is defined as two magnets not being in the ground state with respect to each other.*

According to the definition of conflict, we categorize conflicts into three types:

- AF conflict: a conflict from an AF ordered pair of magnets with FM magnetization states.
- FM conflict: a conflict from an FM ordered pair of magnets with AF magnetization states.
- TURN conflict: a conflict from a 90° diagonal pair of magnets with AF magnetization states, which has higher energy than FM magnetization states.

For example, devices 17 and 20 in Fig. 2b form an AF conflict for they are AF ordered magnets but with FM magnetization states. Devices 20 and 21 in Fig. 2b form an FM conflict for they are FM ordered magnets but with AF magnetization states. Devices 5 and 9 in Fig. 2b form a TURN conflict for they are a 90° diagonal pair of magnets but with AF magnetization states, which have higher energy than FM magnetization states. We use λ_{AF}, λ_{FM}, and λ_{TURN} to represent the number of AF, FM and TURN conflicts in an NML circuit.

The energy difference between the LC and ground states can be uniquely attributed to the numbers of conflicts defined above. To see why this is the case, let us now examine the coupling energy calculation in more detail. Since we know that $E_{coupling(N1)}(LC) = E_{coupling(N1)}(LE)$, we focus on $E_{coupling(N2)}$ only. We first compute the coupling energy contributed by the TURN group of magnets for the LC and LE states, and have the following

$$E_{coupling(N2,TURN)}(LC) = \lambda_{TURN}(LC)E_{TURN,AF2} \\ + \lambda'_{TURN}(LC)E_{TURN,FM2} \quad (9)$$

$$E_{coupling(N2,TURN)}(LE) = \lambda_{TURN}(LE)E_{TURN,AF2} \\ + \lambda'_{TURN}(LE)E_{TURN,FM2} \quad (10)$$

where $\lambda_{TURN}(LC) = 4$ (resp., $\lambda_{TURN}(LE) = 6$) is the number of AF coupled TURN magnet pairs and $\lambda'_{TURN}(LC) = 2$ (resp., $\lambda'_{TURN}(LE) = 0$) is the number of FM coupled TURN magnet pairs in Fig. 2a (resp., Fig. 2b). Let $\Delta E_{TURN} = E_{TURN,AF2} - E_{TURN,FM2}$, which is always positive. We then have

$$E_{coupling(N2,TURN)}(LC) = \lambda_{TURN}(LC)\Delta E_{TURN} + \\ (\lambda_{TURN}(LC) + \lambda'_{TURN}(LC))E_{TURN,FM2} \quad (11)$$

$$E_{coupling(N2,TURN)}(LE) = \lambda_{TURN}(LE)\Delta E_{TURN} + \\ (\lambda_{TURN}(LE) + \lambda'_{TURN}(LE))E_{TURN,FM2} \quad (12)$$

It can be readily verified that

$$\lambda_{TURN}(LC) + \lambda'_{TURN}(LC) = \lambda_{TURN}(LE) + \lambda'_{TURN}(LE)$$

Therefore, the energy difference between the LC state and the ground state due to TURNs comes from the difference in number of TURN conflicts, λ_{TURN}, between the two states. Similarly, we can derive the dependence of the energy difference on the numbers of AF and FM conflicts. The details are omitted.

Besides the number of conflicts, the values of ΔE_{TURN}, ΔE_{AF} (the energy difference due to an AF conflict) and ΔE_{FM} (the energy difference due to an FM conflict) also impact the total energy difference between the LC state and the ground state. These energy values are proportional to the $|C_{yy}|$ values of the corresonding conflicting magnets.

To ensure the LC state is the ground state, the undesirable higher order coupling should be reduced, In the next section, we present such an effort.

4. DESIGN FLOW

There are many ways to reduce higher order couplings and thus make the LC state have the lowest energy. For example, one can (i) add or remove magnets, and (ii) adjust the physical design parameters (i.e., shapes and sizes of magnets and spacings between magnets). The process is essentially a

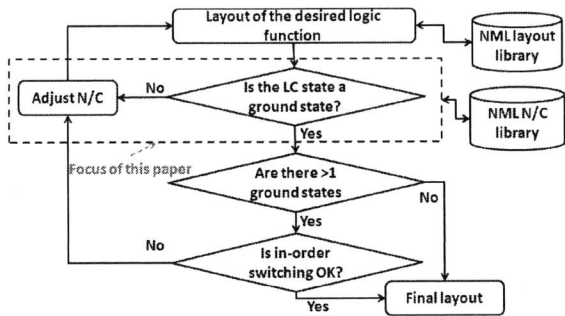

Figure 3: Proposed design flow. The focus of this paper is on the determination of the LC state being the ground state and the adjustment of N and C.

design space exploration. In this section, we introduce our energy analysis guided design exploration approach, which focuses on (ii). We start by describing the design flow and then discuss the key steps in more detail.

4.1 Design flow

The central theme of our design flow is to make the LC state of an NML circuit be the ground state. The entire flow is depicted in Fig. 3.

The design flow starts with an initial NML circuit layout corresponding to the desired logic function. For the NML layout, we check if the LC state is the ground state for each input combination. If the LC state is not the ground state, the N/C parameters are adjusted according to the energy difference between the LC state and the ground state to generate a new NML layout for further testing. The determination of the LC state being the ground state and the adjustment of N/C are two key steps in the flow and also the focus of this paper, which will be discussed in more detail in the next subsection.

In rare cases, multiple equivalent ground states may exist. In such a case, the design flow further tests whether magnets in the NML circuit switch in the desired order, referred to as in-order switching. In-order switching can be guaranteed by (i) clocking the NML circuit properly (to make the ground state unique), and/or (ii) adjusting N/C (as shown in Fig. 3). Since non-unique ground states are rare, we leave the details of this part of the design flow for future work.

The design flow relies on two libraries: (i) an NML layout library, and (ii) an N/C library. The NML layout library stores the set of elementary NML circuit layouts that have been verified to satisfy the state equivalence condition. These elementary layout modules can be used to construct the initial circuit layout of a desired logic function. The N/C library contains N values for magnets with various shapes and C values for pairs of magnets with varying spacings. The library is used to compute the energy of each state and to help guide the adjustment of N/C values.

The construction process of the N/C library is done incrementally by varying the shapes, sizes, and spacings of nanomagnets. One might feel that the N/C library can become extremely large. However, the shapes, sizes and spacings are confined by the fabrication process. Thus, the size of the N/C library would not be too large. For example, the magnets adopted in experiments today are usually around $(60\text{-}75)\text{x}(90\text{-}120)$ nm^2 rectangular prisms with round corners, while the spacing is around 10-30 nm. Furthermore, interpolation can be used to obtain certain N/C values not in the N/C library, although details of this are beyond the scope of this paper.

4.2 Key Steps

We now delve into the details of the two key steps in Fig. 3: (i) check if the LC state is the ground state, and (ii) adjust

the N/C parameters in the layout. The development of the algorithms in the two steps hinges on two observations. First, a large NML circuit is generally partitioned into clock zones to ensure circuit stability and improve throughput performance [8, 19]. The energy evaluation can consider each zone separately since no neighboring clock zones have their clocks removed simultaneously. Secondly, in each clock zone the number of magnets that have dependencies on one another are usually quite small (<100)[1].

4.2.1 Check state equivalence

To check whether the LC state is the ground state, we need to find the ground state first. For those magnets that are independent of one another, their coupling factors equal zero. Hence, they can be treated separately. Our effort is then on finding the ground state of each set of dependent magnets. Given that the number of dependent magnets in each set is relatively small, we simply use a brute-force approach to find the ground state for each input combination of the set of given dependent magnets. The approach is described in Alg. 1, where m is the number of input magnets, n is the number of non-input magnets, and c is the number of C's up to the N2 level.

Algorithm 1 Check if the LC state is the ground state

1: **Input** : m, n, c, **N/C** *library*
2: **Output** :$EQ_{flag} = 1$, $\Delta E = 0$
3: //For each input combination
4: **for** $p = 1 \to 2^m$ **do**
5:     //$\mathbf{S}_{LC}(p)$ holds the magnetization of each device
6:     //in the LC state, M_y's only
7:     $\mathbf{S}_{LC}(p) = LC\ State$
8:     //For each combination of device magnetization
9:     **for** $q = 1 \to 2^n$ **do**
10:         //For bit value 0/1, use $-M_s/M_s$
11:         $\mathbf{S} = BINARY(q)$
12:         $E_{total} = 0$
13:         //$\mathbf{S}_{LE}$: magnetization state of the ground state
14:         $\mathbf{S}_{LE} = \{-1, -1, ..., -1\}M_s$
15:         //$\mathbf{E}_{LE}$: energy of the ground state
16:         $E_{LE} = +\infty$
17:         //For each coupling factor (C_{yy} only)
18:         **for** $r = 1 \to c$ **do**
19:             Compute $E_{i,j}$ using Eqn. 6
20:             $E_{total} = E_{total} + E_{i,j}$
21:         **end for**
22:         **if** $\mathbf{S} == \mathbf{S}_{LC}$ **then**
23:             $E_{LC} = E_{total}$
24:         **end if**
25:         **if** $E_{LE} > E_{total}$ **then**
26:             //Save lower energy
27:             $E_{LE} = E_{total}$
28:             $\mathbf{S}_{LE} = \mathbf{S}$
29:         **end if**
30:     **end for**
31:     **if** $E_{LE} != E_{LC}$ **then**
32:         $EQ_{flag} = 0$
33:         $\Delta E = E_{LC} - E_{LE}$
34:         *break*
35:     **end if**
36: **end for**

Alg. 1 takes the number of magnets and the N/C library

---

[1]We say two magnets are dependent on each other if they are both along the signal paths for evaluating a parallel output. For example, for an n-bit parallel AND circuit, the magnets used for the bit 0 and those for the bit 1 output are not considered to be dependent.

as the input, and outputs the energy difference (ΔE) between the LC state and the ground states as well as a flag indicating whether the LC state is the ground state. The outermost *for* loop (Line 4) checks all input combinations. The second *for* loop (Line 9) assigns the magnetization state of each magnet according to the binary bit information from q. That is, the magnetization state of magnet i, M_{yi}, is set to M_s (resp., $-M_s$) if bit i of q is 1 (resp., 0), where M_s is the saturation magnetization of the magnet. (Note that we only consider that the magnets are at the ground demagnetization energy state, so only M_y is needed for each magnet.) The third *for* loop (Line 18) calculates the total coupling energy. It is possible to reduce some operations in Alg. 1 (e.g., the second *for* loop (Line 9)), but we leave this part for future work.

4.2.2 Adjust N/C parameters

The goal of adjusting the **N/C** parameters is to make the LC state be the ground state. This is equivalent to finding **N/C** values of certain magnets such that the energy corresponding to the magnetization configuration of the original "ground state" is higher than the energy corresponding to the magnetization configuration of the LC state. Note that to reduce the search space, it is better to adjust the **N/C** values of as few magnets as possible. Intuitively, if a magnet has a different magnetization state in the ground state from the magnetization state in the LC state and causes conflicts with its neighbors, it tends to contribute to the energy difference between the ground state and the LC state. We identify such a magnet as a hotspot. Specifically, we have the following definition.

DEFINITION 3. *Hotspot: A hotspot is a magnet in the ground state configuration that (i) has a different magnetization state from that of the corresponding magnet at the LC state, and (ii) forms a conflict with its nearest one neighbor.*

Our general idea to adjust **N/C** parameters is to focus on adjusting the sizes/shapes/spacings of the hotspots. We summarize the main steps for one input combination below.

(1) Find the hotspots (Def. 3) in the NML ensemble.

(2) Adjust the **N/C** parameters around the hotspots based on ΔE. This step is critical and more details follow later.

(3) Determine the new shape/size/spacing values based on the adjusted **N/C** to generate a new NML layout.

For a complete solution (i.e., for all input combinations), an intersection set of the adjusted **N/C** values is needed. If the intersection set is not empty, we find a solution. Otherwise, there is no solution.

Step (2), the intuition is to increase the energy difference due to the conflicts related to the hotspots in the ground state (i.e., ΔE_{AF}, ΔE_{FM} or ΔE_{TURN}), so that the energy of the ground state is increased. In other words, the energy of the LC state is decreased relatively. The final result is that the LC state becomes the ground state, if possible. According to Eqn. 6, the coupling energy (of the ground state), $E_{i,j}$, is an increasing function of $|C_{yy,i \to j}|$ (and $|C_{yy,j \to i}|$) of the conflicts related to the hotspots. Thus, increasing $|C_{yy,i \to j}|$ (and $|C_{yy,j \to i}|$) of the conflicts related to the hotspots increases the coupling energy of the conflicts related to the hotspots. Regarding the amount of increase in the $|C_{i \to j}|$ (and $|C_{j \to i}|$), we use ΔE as a guideline. In the presence of multiple conflicts related to hotspots, ΔE is divided evenly among them, so that each one is responsible to make up a portion of the difference. An example of this step will be given in Sec. 5.

Besides adjusting **C**, it is also possible to adjust **N** in order to increase the coupling energy of the conflicts related to the hotspots. Aside from increasing the energy of the

ground state, we can also decrease the energy of the LC state. We omit the details due to the page limit.

Step (3) relies on the **N/C** library. As stated earlier (in Sec. 2), **N** is a function of shape, and **C** is a function of spacing between two magnets with given shapes. The lookup step simply searches the **N/C** library to find the corresponding shape, size, and spacing based on the adjusted **N** and **C**, respectively.

5. CASE STUDY

This section presents a case study to demonstrate how our method can be used to transform the LC state of a relatively complex NML circuit into the ground state. The case study is based on a 25-device adder from the experiment in [5]. Following Alg. 1, we determined that the LC state is not the ground state. The energy values of the two states are shown in Figs. 4a, 4b. The errors in the ground state are circled in Fig. 4b. Fig. 4c shows a magnetic force microscope (MFM) image of the ground state of the 25-device adder with the errors circled. The LC state is shown in Fig. 4a. The minor difference between Fig. 4b and Fig. 4c is likely due to fabrication variations.

Next we show how to adjust the design so that the LC state becomes the ground state according to the steps in Sec. 4.2. Specifically, we have the following results.

(1) Magnets 11, 13 and 14 are the hotspots (as numbered in Fig. 4d). Magnets 11 through 24 have different magnetization states at the LC state and at the ground state, but only magnets 11, 13 and 14 have conflicts at the ground state. The conflicts are magnets (5,11), (6,13) and (8,14). (To see why these are conflicts, refer to Fig. 4b.)

(2) The energy of the conflicts related to the hotspots is

$$E_{hotspots} = E_{8,14} + E_{6,13} + E_{5,11},$$

where

$$
\begin{aligned}
E_{8,14} &= \mu_0 (M_{y,8} C_{yy,8 \to 14} M_{y,14} V \\
&\quad + M_{y,14} C_{yy,14 \to 8} M_{y,8} V)/(-2) \\
&= \mu_0 M_s^2 V (|C_{yy,8 \to 14}| + |C_{yy,14 \to 8}|)/2 \\
&= \mu_0 M_s^2 V |C_{yy,8 \to 14}|,
\end{aligned}
$$

$$E_{6,13} = \mu_0 M_s^2 V |C_{yy,6 \to 13}|, \quad E_{5,11} = \mu_0 M_s^2 V |C_{yy,5 \to 11}|.$$

Note that magnets 5, 6, 8, 11, 13, and 14 are identical in material, size and shape, so M_s and V are constant, and $|C_{yy,8 \to 14}|$ is the same as $|C_{yy,14 \to 8}|$. $E_{hotspots}$ can be further expressed as

$$
E_{hotspots} = \\
\mu_0 M_s^2 V (|C_{yy,8 \to 14}| + |C_{yy,6 \to 13}| + |C_{yy,5 \to 11}|). \tag{13}
$$

$E_{hotspots}$ is proportional to $|C_{yy,8 \to 14}| + |C_{yy,6 \to 13}| + |C_{yy,5 \to 11}|$. Here we adjust $|C_{yy}|$ of the three conflicts. Let

$$\Delta |C_{yy}| = \Delta |C_{yy,8 \to 14}| = \Delta |C_{yy,6 \to 13}| = \Delta |C_{yy,5 \to 11}|, \tag{14}$$

then $\Delta E_{hotspots}$ is deduced as

$$\Delta E_{hotspots} = 3\mu_0 M_s^2 V \Delta |C_{yy}| = \Delta E = 54kT, \tag{15}$$

which corresponds to dividing ΔE evenly by the three hotspots. Solving Eqn. 15, we get $\Delta |C_{yy}| = 0.0014$. The new $|C_{yy}|$ is

$$
\begin{aligned}
|C'_{yy,8 \to 14}| &= |C_{yy,8 \to 14}| + \Delta |C_{yy}| \\
&= 0.0065 + 0.0014 = 0.0079,
\end{aligned} \tag{16}
$$

which corresponds to the increase in $|C_{yy,8 \to 14}|$ from 0.0065 to 0.0079. $|C'_{yy,5 \to 11}|$ and $|C'_{yy,6 \to 13}|$ are the same as $|C'_{yy,8 \to 14}|$.

760

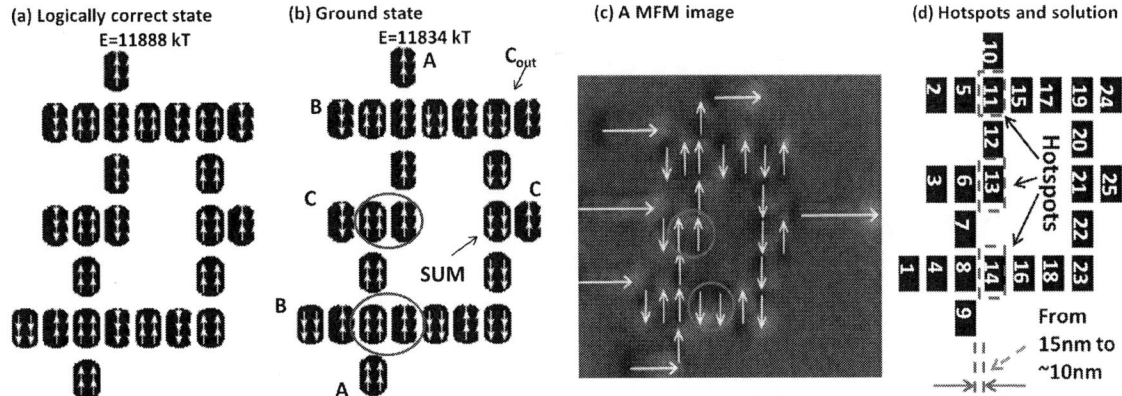

Figure 4: (a,b) The LC state and the ground state of a 25-device adder. (c) An MFM image of the ground state of a 25-device adder from the experiment. The magnets with errors are circled. (d) The LC state of the 25-device is the ground state after decreasing the critical spacing from 15 nm to ~10 nm. Note that, to calculate the energy terms, equivalent input fields were used to mimic the input magnets (with horizontal magnetization states in (c)) from the experiment.

(3) This step generates a new NML layout by finding new shapes/sizes/spacing from the **N/C** library that can lead to the desired **N/C** values. Specifically, the increase in $|C_{yy}|$ from 0.0065 to 0.0079 corresponds to decreasing the spacing from 15 nm to ~10 nm between the hotspots (magnets 11, 13, and 14) and their counterparts within the conflicts (magnets 5, 6, and 8, respectively), as shown in Fig. 4d.

The new NML layout satisfies the the necessary condition that the LC state is the ground state, with 47 kT lower energy than the next lowest energy state.

6. SUMMARY

We have identified an important problem with N1 coupling based designs. That is, the LC state might not be the ground state. Intuitive explanation and solution are given for the problem. Based on the energy analysis and identification of conflicts, we introduce a systematic method to achieve the state equivalence. Based on our design method, error rate in NML circuits can be reduced.

7. REFERENCES

[1] A. Imre and et al. Majority logic gate for Magnetic Quantum-Dot Cellular Automata. *Science*, 311(5758):205–208, January 2006.

[2] G. Csaba and et. al. Simulation of power gain and dissipation in field-coupled nanomagnet. *J. of Comp. Elec.*, 4(1/2):105–110, 2005.

[3] A. Dingler and et al. Performance and energy impact on locally controlled nml circuits. *ACM Journal on Emerging Technologies in Computing*, 7(1):1–24, 2011.

[4] I. Palit and et al. Systematic design of nanomagnet logic circuits. In *Design, Automation and Test in Europe*, DATE 13, page 1795, 2013.

[5] E. Varga and et al. Implementation of a nanomagnetic full adder circuit. In *Nanotechnology (IEEE-NANO), 2011 11th IEEE Conference on*, pages 1244 –1247, aug. 2011.

[6] D. Carlton and et al. Investigation of errors and defects in nanomagnetic logic circuits. *Nanotechnology, IEEE Transactions on*, PP(99):1–1, 2012.

[7] J.D. Jackson. *Classical Electrodynamics*. Wiley, 1998.

[8] A. Dingler and et al. Making non-volatile nanomagnet logic non-volatile. In *Proceedings of the 49th Annual Design Automation Conference*, DAC '12, pages 476–485, New York, NY, USA, 2012. ACM.

[9] D.B. Carlton and et al. Computing in thermal equilibrium with dipole-coupled nanomagnets. *Nanotechnology, IEEE Transactions on*, 10(6):1401 –1404, nov. 2011.

[10] J. Das and et al. Addressing the layout constraint problem when cascading logic gates in nanomagnetic logic. In *Nanotechnology (IEEE-NANO), 2012 12th IEEE Conference on*, pages 1 –4, aug. 2012.

[11] M. T. Alam and et al. On-chip clocking of nanomagnet logic lines and gates. *IEEE Transactions on Nanotechnology*, 11(2):273–286, 2012.

[12] Fashami Mohammad Salehi and et al. Magnetization dynamics, bennett clocking and associated energy dissipation in multiferroic logic. *Nanotechnology*, 22(15):155201, 2011.

[13] Y. H. Chu and et al. Electric-field control of local ferromagnetism using a magnetoelectric multiferroic. *Nature Materials*, 7:482, 2008.

[14] Mathias Getzlaff. Applications. In *Fundamentals of Magnetism*, pages 293–335. Springer Berlin Heidelberg, 2008.

[15] G. Csaba and et al. Development of cad tools for nanomagnetic logic devices. *Int. J. Circ. Theor. Appl.*, 2012.

[16] M. Beleggia and et al. General magnetostatic shape-shape interactions. *Journal of Magnetism and Magnetic Materials*, 285:L1–L10, January 2005.

[17] M. J. Donahue and D. G. Porter. OOMMF User's Guide, Version 1.0, Interagency Report NISTIR 6367. http://math.nist.gov/oommf.

[18] M. Beleggia and et al. On the computation of the demagnetization tensor field for an arbitrary particle shape using a Fourier space approach. *Journal of Magnetism and Magnetic Materials*, 263, July 2003.

[19] M. T. Niemier and et al. Clocking structures and power analysis for nanomagnet-based logic devices. *Int. Symp. on Low Power Elec. and Design (ISLPED)*, pages 26–31, 2007.

Paper Supplement

S.1. EXAMPLE OF COUPLING ENERGY CALCULATION BASED ON DIFFERENT LEVELS OF COUPLINGS

We use an AF coupled 5-magnet line to illustrate that N1 and N2 couplings are accurate enough for NML coupling energy calculations. The 5-magnet line is shown in the inset of Fig. 1 (in this page). The total coupling energy is calculated for each magnetization configuration (a total of $2^5=32$ states) based on three different coupling considerations. If only N1 coupling is included, we name

$$E_{coupling(N1)} = E_{1,2} + E_{2,3} + E_{3,4} + E_{4,5}. \quad (S1)$$

If both N1 and N2 couplings are included, we have

$$E_{coupling(N1+N2)} = E_{N1} + E_{N2} = E_{N1} + E_{1,3} + E_{2,5} + E_{3,5}. \quad (S2)$$

If all couplings are included, we have

$$E_{coupling(N*)} = E_{N1+N2} + E_{1,4} + E_{2,5} + E_{1,5}. \quad (S3)$$

Fig. 1 (in this page) depicts the total energy of an AF coupled line as a function of magnetization states. The figure shows that the coupling energy based on N1 and N2 couplings (N1+N2) is the same as or very close to that based on all neighbor coupling (N*). However, the coupling energy based on N1 coupling shows a nontrivial difference from others (N* and N1+N2). Therefore, we use the N1 and N2 couplings for the energy calculation for the rest of this paper.

S.2. PARAMETERS

The parameter values used in the simulations and calculations throughout this paper are given below, unless specified otherwise. The magnets are permalloy, the magnet size is 60x90x5 nm^3, and the x(y)-spacing is 15(30) nm, and the magnetization are either $[0, M_s, 0]$ or $[0, -M_s, 0]$ (corresponding to the ground demagnetization energy states). M_s of permalloy is 860 KA/m.

S.3. ACRONYMS USED IN THE PAPER

AF   Antiferromagnetic(ally)

C   Coupling factor

FM   Ferromagnetic(ally)

LC   Logically correct

M$_s$ Saturation magnetization

MFM   Magnetic force microscope

N   Demagnetization tensor field

N1   Nearest one neighbor

N2   Nearest two neighbor

N*   All neighbors

NML   Nanomagnet Logic

QCA   Quantum-dot Cellular Automata

SD   Single domain

ΔE   The energy difference between the LC state and the ground state

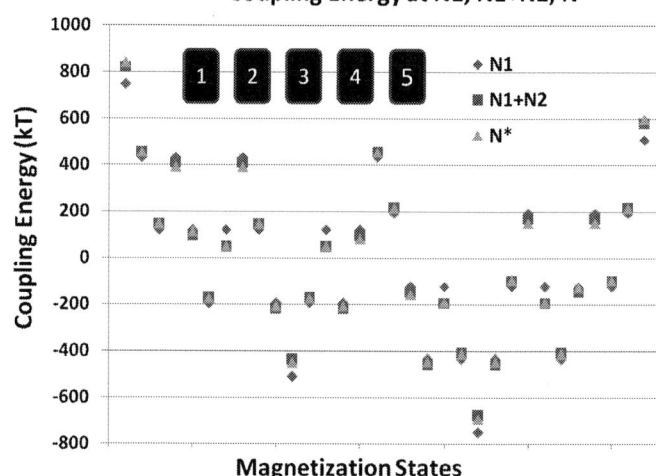

Figure 1: The coupling energy of a 5-device AF ordered line as a function of the magnetization states. There are three sets of data shown, coupling energy calculated based on (i) N1 coupling, (ii) N1+N2 coupling, and (iii) N* coupling. N1+N2 agrees well with N*, therefore, N1+N2 is used for the rest of this paper. There is a local y magnetic field (10 mT) applied on device 1 to mimic an input, leading to the asymmetry (or shift) of the left and right halves of the plot.

Ultra Low Power Associative Computing with Spin Neurons and Resistive Crossbar Memory

Mrigank Sharad, Deliang Fan, and Kaushik Roy

Department of Electrical and Computer Engineering, Purdue University, West Lafayette, IN, USA

(msharad, dfan, kaushik)@purdue.edu

ABSTRACT

Emerging resistive-crossbar memory (RCM) technology can be promising for computationally-expensive analog pattern-matching tasks. However, the use of CMOS analog-circuits with RCM would result in large power-consumption and poor scalability, thereby eschewing the benefits of RCM-based computation. We propose the use of low-voltage, fast-switching, magneto-metallic 'spin-neurons' for ultra low-power non-Boolean computing with RCM. We present the design of analog associative memory for face recognition using RCM, where, substituting conventional analog circuits with spin-neurons can achieve ~100x lower power. This makes the proposed design ~1000x more energy-efficient than a 45nm-CMOS digital ASIC, thereby significantly enhancing the prospects of RCM based computational hardware.

Categories and Subject Descriptors

B.7.1 [Integrated Circuits] Types and Design Styles – Advanced Technologies

General Terms

Design

Keywords

Magnets, Memory, Spin-Transfer Torque, Emerging Circuits and Devices, Spintronics

1. INTRODUCTION

In recent years several device solutions have been proposed for fabricating nano-scale programmable resistive elements, generally categorized under the term 'memristor' [1-9]. Of special interest are those which are amenable to integration with state of the art CMOS technology, like memristors based on Ag-Si filaments [6-8]. Such devices can be integrated into metallic crossbars to obtain high density resistive crossbar memory (RCM) [1-8]. Continuous range of resistance values obtainable in these devices can facilitate the design of multi-level, non-volatile memory [1-3].   The Resistive-Crossbar Memory (RCM) technology has led to interesting possibilities of combining memory with computation [1-5]. RCM can be highly suitable for a class of non-Boolean computing applications that involve pattern-matching [5, 11]. Such applications employ highly memory-intensive computing that may require correlation of a multidimensional input data with a large number of stored patterns or templates, in order to find the best match [11]. Use of conventional digital processing techniques for such tasks incurs high energy and real-estate cost, due to the sheer number of computations involved. Structurally, RCM can be a much closer fit for this class of associative computation. Owing to the direct use of nano-scale memory array for associative computing, it can provide very high degree of parallelism, apart from eliminating the overhead due to memory read.

Associative computing of practical complexity with RCM is essentially analog in nature, as it involves evaluating the degree of correlation between inputs and the stored data. As a result, most of the designs for associative hardware using RCM's proposed in recent

years, involved analog CMOS circuits for the processing task [9, 11]. Recent experiments on analog-computing with of multi-level Ag-Si memristors also employed analog operational amplifiers for current-mode processing [8]. However, application of multiple analog blocks for large scale RCM may lead to power hungry designs, due to large static power consumption of such circuits. This can eclipse the potential energy benefits of RCM for non-Boolean computing. Moreover, with technology scaling, the impact of process variations upon analog circuits becomes increasingly prominent, resulting in lower resolution for signal amplification and processing [16]. Hence, conventional analog circuits may fail to exploit the RCM technology for energy-efficient, non-Boolean computing.

The solution to this bottleneck may lie with alternate device technologies that can provide a better fit for the required non-Boolean, analog functionality, as compared to CMOS switches. Recent experiments on spin-torque devices have demonstrated high-speed switching of scaled nano-magnets with small currents [12-14]. Such magneto-metallic devices can operate at ultra low terminal voltages and can implement current-mode summation and comparison operations, at ultra low energy cost. Such current-mode spin switches or 'neurons' can be exploited in energy-efficient analog-mode computing [20-24]. In this work we present a design of RCM based associative memory using such "spin-neurons". In the proposed scheme, the spin-neurons form the core of hybrid processing elements (PE) that are employed in RCM based associative modules and achieve more than two orders of magnitude lower computation energy as compared to conventional mixed-signal (MS) CMOS circuits.

The rest of the paper is organized as follows. Section-II describes the application of RCM in non-Boolean computing and the design challenges associated with a mixed-signal implementation. The device model for spin neuron is introduced in section III. Design of analog associative memory module (AMM) using spin neurons applied to RCM modules is described in section IV. Section V discusses the performance and prospects of the proposed design scheme. Conclusions are given in section-VI.

2. COMPUTING WITH RESISTIVE CROSSBAR MEMORY (RCM)

Fig. 1 depicts a resistive crossbar memory. It constitutes of memristors (Ag-Si) with conductivity g_{ij}, interconnecting two sets of metal bars (i^{th} horizontal bar and j^{th} in-plane bar). Multi-level write techniques for memristors in crossbar arrays have been proposed and demonstrated in literature that can achieve precision up to 0.3% (equivalent to 8-bits) [1-2]. However, the energy-cost of the write operations may increase significantly for higher precision requirements [1-2]. In this work we have used 3% write accuracy (equivalent to 5-bits) for the memristors [8]. For a given write-precision, larger number of bits can be obtained by using parallel combination of multiple memristors to store a single analog value [4]. Note that, the class of non-Boolean pattern-matching computations, a prospective application of RCM technology, are inherently approximate and have relaxed precision constraints [11].

Memory-based pattern-matching applications generally apply some form of feature reduction technique to extract and store only the essential 'patterns' or 'features' corresponding to different data samples. The extracted patterns can be represented in the form of analog vectors that can be stored along individual columns of the RCM shown in fig.1 (note that the data is stored in the cross-point memristive element). In order to compute the correlation between an

Permission to make digital or hard copies of all or part of this work for personal or classroom use is granted without fee provided that copies are not made or distributed for profit or commercial advantage and that copies bear this notice and the full citation on the first page. To copy otherwise, to republish, to post on servers or to redistribute to lists, requires prior specific permission and/or a fee.

DAC '13, May 29 - June 07 2013, Austin, TX, USA.

input and the stored patterns, input voltages V_i (or currents I_i) corresponding to the input feature can be applied to the horizontal bars. Assuming the outward ends of the in-plane bars grounded, the current coming out of the j_{th} in-plane bar can be visualized as the dot product of the inputs V_i and the cross-bar conductance values g_{ij} (fig. 1). Hence, an RCM can directly evaluate correlation between an analog input vector and a number of stored patterns. This technique can be exploited in evaluating the degree of match (DOM) between an input and the stored patterns, the best match being the pattern corresponding to the highest correlation magnitude ($\sum_i V_i \, g_{ij}$).

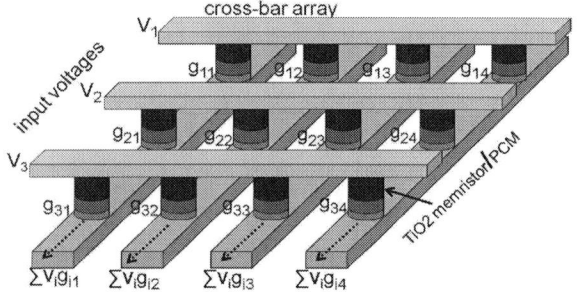

Fig. 1 A Resistive crossbar network used for evaluating correlation between inputs and stored data.

Fig. 2 depicts the feature extraction step for human face-images. In this work, we have used 10 different face-images for 40 individuals, for generating 40 stored data patterns. For an individual, each of the 10 face-images were normalized and down sized from 128x96, 8-bit pixels to 16x8, 5-bit pixels. Pixel wise average of the 10 reduced images was taken to generate 128-element (16x8), 32 level analog patterns corresponding to the 40 individual faces.

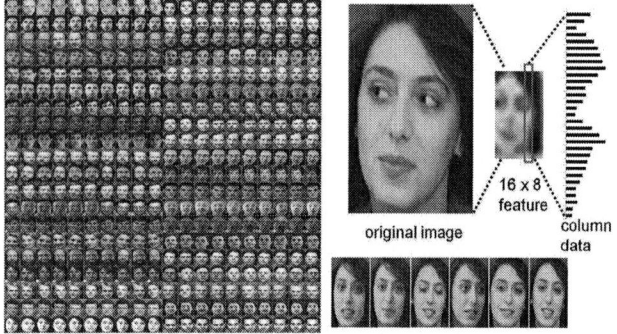

Fig. 2. 400 test images of 40 individuals (from ATT Cambridge Image Database [22]) and the feature reduction method used in this work.

The limit of image down-sizing was identified as the scaling factor below which matching accuracy for the 400 test images dropped significantly below the value achieved using the full size image (fig. 3a). For each set of downsizing factor and bit-size, current-mode correlation outputs were obtained using SPICE model of RCM.

Variations in input source as well as memristor values were incorporated to obtain realistic values for the current-outputs. For a given set of stored images, classification accuracy also depends upon the resolution of the detection unit used to determine the DOM figures for all the stored patterns. A resolution of 4% (5-bit) was chosen based on the observation that down to this value, the classification accuracy remained close to that achievable using ideal comparison (fig. 3b).

Resolving ~4% difference among the current-mode dot product requires a precision of 5-bits for the detection unit, responsible for identifying the winning pattern. Fig. 4 shows a conventional mixed-signal-CMOS solution for the detection unit. It constitutes of regulated current mirrors as the input stage, that offer low input-impedance and a near constant DC bias to the RCM. Following this, a winner-take-all (WTA) circuit receives the current inputs and determines the 'winner'. Several versions of WTA circuits have been proposed in literature, that

can be classified into two broad catagories, current-conveyer WTA (CC-WTA) [18], and binary tree WTA (BT-WTA) [18]; the later being more suitable for large number of inputs [17, 18]. BT-WTA employs a binary tree of 2-input comparison stages which involve copying and propagating the larger of the two current inputs to the output (fig 4) [17].

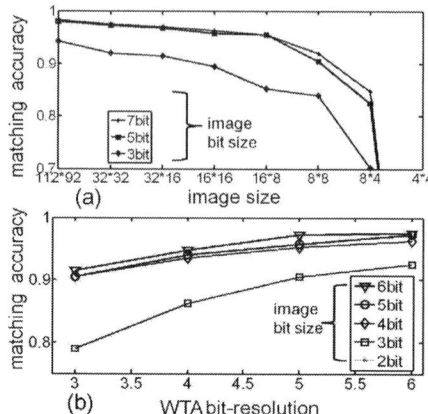

Fig. 3 (a) Training accuracy reduces with image down-sizing, (b) similar trend is obtained for reducing WTA resolution

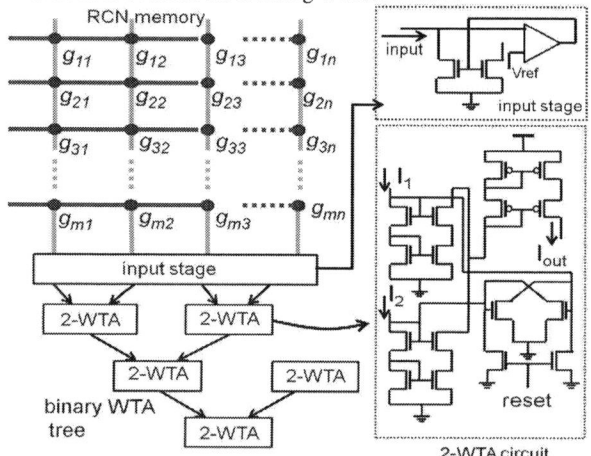

Fig. 4 A standard CMOS solution for associative memory module using binary tree winner-take-all circuit.

In general, the use of such analog WTA circuits leads to large static power consumption. In fact, the power consumption of an analog WTA unit can be several times larger than the RCM itself. Moreover, performance of such current-mirror based circuits is limited by random mismatches in the constituent transistors and other non-idealites like, channel length modulation, that introduce mismatch in different current paths [16]. In order to maintain a sufficiently high resolution, larger transistor dimensions (both length as well as width) and hence, larger cell area is needed. This is evident from some recent designs [18], that employed significantly large channel lengths for such circuits despite using relatively scaled CMOS technology. This leads to increased parasitic capacitance and hence, lower operating frequency for a given static power. Higher frequency and resolution can be achieved at the cost of increased input currents, ie., at the cost of larger power consumption [16]. Special techniques to enhance the precision of current mirrors have been proposed in literature [18], but they introduce significant overhead in terms of power consumption and area complexity. Voltage-mode processing can also be employed in RCM, however, it incurs additional overhead due to current to voltage conversion and subsequent amplifications. This results in larger mismatch, non-linearity and power consumption.

The above discussion suggests that the conventional mixed-signal CMOS design techniques may not be able to leverage the emerging nano-scale resistive memory technology for memory based computing. This motivates us to look towards alternate device technologies that can be more suitable for this purpose. In the next section we present the spin-based neuron model that can lead to efficient computing hardware based on RCM.

3. SPIN NEURON FOR RCM BASED COMPUTING

In this section, we describe the device operation of the spin based neuron model that is based on domain wall magnet (DWM) [12-15]. The circuit technique employed to interface the domain wall neuron (DWN) with CMOS units is also discussed.

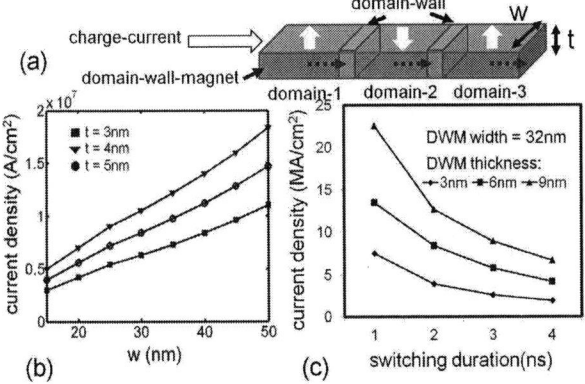

Fig. 5 (a) Domain wall magnet with three domains. (b) Scaling of DWM achieves reduction in critical switching current. (c) Smaller device dimensions achieve faster switching for a given write-current.

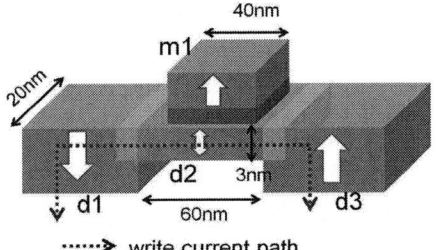

Fig.6 Device structure for domain wall neuron (DWN). The input current enters the device through d_1 and exits through d_2. The magnet m_1 associated with the MTJ is used to read the state of the free layer (d_2).

A domain wall magnet (DWM) constitutes of multiple *nano-magnet* domains separated by non-magnetic regions called domain-wall (DW) as shown in fig. 5a. DW can be moved along a magnetic nano-strip using current-injection. Hence, the spin polarity of the DWM strip at a given location can be switched, depending upon the polarity of its adjacent domains and the direction of current flow. Recent experiments have achieved switching current density of ~10^6A/cm^2 for nano-scale DWM strips for, and, a switching time of less than 1ns [12-14]. Thus, the polarity of a scaled *nano-magnet* strip of dimension 3x20x60nm^{3}can be switched using a small current of ~1µA. Moreover, the current threshold as well as the switching time of DWM scales down with device dimensions (fig. 5b) [15].

The device structure for a domain-wall neuron (DWN) is shown in fig. 6. It constitutes of a thin and short (3x20x60 nm^3) *nano-magnet* domain, d_2 (domain-2, the 'free-domain') connecting two anti-parallel *nano-magnet* domains of fixed polarity, d_1 (domain-1) and d_3 (domain-3). Domain-1 forms the input port, whereas, domain-3 is grounded. Spin-polarity of the free-domain (d_2) can be written parallel to d_1 or d_2 by injecting a small current along it from d_1 to d_2 and vice-versa [2]. Thus, the DWN can detect the polarity of the current flow at its input-node. Hence it acts as an ultra-low-voltage and compact current-

comparator that can be employed in energy efficient current-mode data processing [20-24]. A non-zero current threshold for DW motion however, results in a small hysteresis in the DWN switching characteristics (fig. 7a). It is desirable to reduce the threshold to get closer to the step transfer function of an ideal comparator. Apart from device-scaling, the use lower anisotropy barrier for the magnetic-material can be effective in lowering the switching threshold for computing applications [10].

A magnetic tunnel junction (MTJ) [15], formed between a fixed polarity magnet m_1 and d_2 is used to read the state of d_2. The effective resistance of the MTJ is smaller when m_1 and d_2 have the same spin-polarity and vice-versa ($R_{parallel}$=~5kΩ and $R_{anti-parallel}$~15kΩ). We employ a dynamic CMOS latch shown in fig. 7b to detect the MTJ state. One of its load branches it connected to the DWN MTJ whereas the other is connected to a reference MTJ whose resistance is midway between the two resistances of the DWN MTJ. The latch effectively compares the resistance between its two load branches through transient discharge currents. Since the transient read-current flows only for a short duration, it does not disturb the state of d_2.

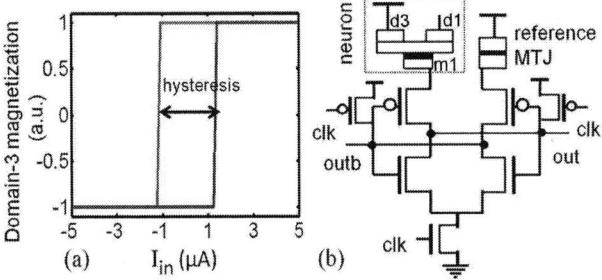

Fig. 7 (a) DWN transfer characteristics for anisotropy-energy-barrier, E_b = 20KT, (b) dynamic CMOS latch used to detect DWN's state.

Note that integration of similar spin-device structure with CMOS has already been demonstrated for memory applications [15]. Notably, the energy-barrier, the threshold current as well as dimensions for a memory device needs to be larger, in order to ensure long-term stability. However, for computing applications, aggressive scaling is desirable [10]. Application of similar devices in digital logic design has also been proposed earlier [10]. In this work however, we focused on an entirely different and under-explored potential of such a spin-torque device, and showed its benefit in analog-mode non-Boolean computing.

4. ASSOCIATIVE MEMORY MODULE USING SPIN NEURONS AND RCM

In the following subsections, we first describe the design of RCM-based correlation unit and its interfacing with DWNs. This is followed by circuit level description of spin-CMOS hybrid-PE based on DWN that achieves the WTA functionality at ultra low energy cost.
A. Network Design

Fig. 8a depicts the DWNs with their input (d_1 terminals) connected to RCM outputs. A DC voltage, V, is applied to the d_3 terminals of all the DWNs. Owing to the small resistance of the DWN devices; this effectively biases output ends of the RCM (connected to d_1 terminals) to the same voltage. As described in section-2, in order to perform associative matching of an input face-image with the data stored in the RCM, the input-image is down-sized to 16x8, 5-bit pixels. Each of the 128 digital values needs to be converted into analog voltages/current levels, to be applied to the RCM-input. The low-voltage operation of DWN can be exploited to implement, compact and energy efficient current-mode DAC using binary weighted deep-triode current-source (DTCS) PMOS transistors, as shown in fig. 8a. A DC supply of $V+\Delta V$ is applied to the source terminals of the DTCS, where ΔV is ~30mV. Ignoring the parasitic resistance of the metal crossbar, the drain to source voltage of the DTCS-DAC can be approximated to ΔV. The current $I_{in}(i)$, supplied by the i_{th} DAC can thus be written as $\Delta V.G_T(i)G_{TS}/(G_T(i) + G_{TS}))$, where $G_T(i)$ is the data dependent conductance of the i_{th} DAC and G_{TS} is the total conductance (of all the Ag-Si memristors) connected to a horizontal bar (dummy memristors are added for each horizontal input bar such that G_{ST} is equal for all horizontal bars). As a result, the current input through a memristor connecting the i_{th} input bar to the j_{th} output bar (in-plane) can be written as $I(i,j)=\Delta V.G_T(i)G_{ST}/(G_T(i)+G_{ST})(G(i,j)/G_{ST})$, where, $G(i,j)$ is the programmed conductance of the memristor.

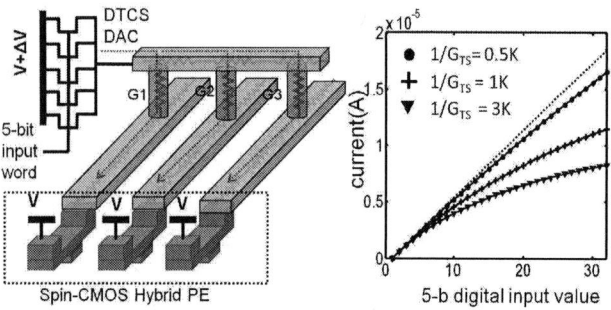

Fig. 8(a) RCM with a single DTCS input and three receiving DWN (b) non-linear characteristics of DTCN resulting due to series combination with Gs.

For accurate dot-product evaluation, the current $I(i,j)$ should be proportional to the product of $G_T(i)$ (ie, the DTCS conductance, proportional to the input data) and $G(i,j)$. Hence, a low value of G_{TS} (i.e. higher resistance values of the memristors) introduces non-linearity in the DTCS-DAC characteristics (fig. 8b). This leads to reduction in the detection margins (difference between the best and the second best match) for the current-mode dot product outputs for different input images (fig. 9a). As a result, the overall matching accuracy of the network reduces for a given WTA resolution. Ideally, choosing the lowest possible range of values for the memristor resistances (say 200Ω-6.4K Ω) would largely overcome the non-linearity (fig. 9b). However, for higher $G(i,j)$ (low resistance value for the memristors), voltage drop in the metal lines due to parasitic resistances result in corruption of the current signals, once again, leading to degradation in the detection-marging. Hence, the optimal range for the conductance values was found based on the maximum achievable read margin, as shown in fig. 9a. The design parameters like the image compression factor, data bit-width etc, discussed earlier, were therefore determined based on the simulation of RCM model, in order to ensure resolvable detection margin.

The range of current output from the DTCS-DAC needed is mainly determined by the choice of WTA resolution. If the DWNs are designed to have a threshold of ~1µA, the maximum value of the dot-product output must be greater than 32µA for a 5 bit resolution for the WTA (described later). This in turn, translates to the required range of DAC output current. For 128 element input vectors and 5-bit resolution for the WTA, the maximum value for DAC output required was found to be ~10 µA. This range of current can be obtained using different combination of DTCS sizing and the terminal voltage, ΔV. For a required amount of DAC current, it is desirable to push ΔV to the minimum possible value, in order to reduce the static power consumption in the RCM. This would imply exploiting the low-voltage operation of the DWNs to the maximum possible extent. The minimum value of ΔV is limited mainly by the parasitic voltage drops that degrade the detection margin and hence the matching accuracy (fig. 9b). For this design (RCM of size 128x40) ΔV of 30mV was found to be enough to preserve the matching accuracy close to the ideal case (with no-parasitic). The proposed scheme effectively biases the RCM across a small terminal voltage (ΔV), thereby ensures that the static

current flow in RCM takes place across a small terminal voltage of ~30mV (between two DC supplies V and $V+\Delta V$).

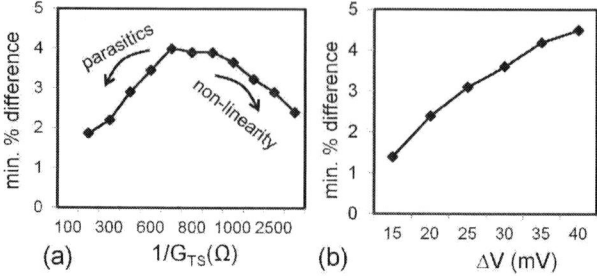

Fig. 9 (a) degradation in detection margin for a given input due to non-linearity (for low G_{TS}) and parasitic voltage drops (for high G_{TS}). (b) degradation in detection margin for the same input, for reducing ΔV, due to parasitic voltage drops.

Above, we noted that the application of DWN in the RCM offers the benefit of ultra-low-voltage operation that reduces the static power-consumption resulting from current-mode, analog computing. Next we describe the design of spin-CMOS hybrid WTA that performs the winner selection task with negligible static power consumption.

B. WTA Design

The DWN device essentially acts as a low voltage, high speed, high resolution current-mode comparator and hence can be exploited in digitizing analog current levels at ultra low energy cost [20]. The proposed WTA scheme, algorithmically depicted in fig. 10, exploits this fact and clubs a digitization step with a parallel 'winner-tracking' operation.

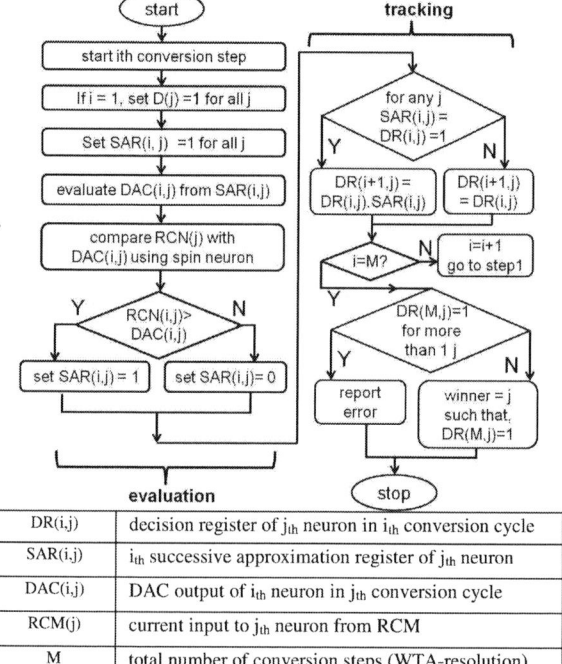

Fig. 10 WTA algorithm used in this work

DR(i,j)	decision register of j_{th} neuron in i_{th} conversion cycle
SAR(i,j)	i_{th} successive approximation register of j_{th} neuron
DAC(i,j)	DAC output of i_{th} neuron in j_{th} conversion cycle
RCM(j)	current input to j_{th} neuron from RCM
M	total number of conversion steps (WTA-resolution)

The first half of the flowchart can be identified as the standard algorithm for successive approximation register (SAR) ADC [20]. The data conversion algorithm employed in an SAR-ADC can be explained as follows. To begin the conversion, the approximation register (that stores the digitization result) is initialized to the mid-scale (i.e., all but the most significant bit is set to 0). At every cycle a digital to analog converter (DAC) produces an analog level corresponding to the digital value stored in the SAR and a comparator compares it with the analog

766

input using an analog comparator. If the comparator output is high, the current bit remains high, else it is turned low and the next lower bit is turned high. The process is repeated for all the bits. At the end of conversion, the SAR stores the digitized value corresponding to the analog input.

The circuit realization of this operation using DWN's is shown in fig. 11. Output currents of the RCM columns (in this case 40 columns storing the pattern vectors of 40 face-images) are received by individual DWN input nodes that are effectively clamped at a DC supply V, as described earlier. Each DWN has an associated DTCS-DAC, which is driven by the corresponding successive approximation register. The drain terminals of the DTCS transistors are a DC voltage V-ΔV. In each conversion cycle, the DWN device essentially compares the RCM output and the DAC output (and hence, acts as the comparator of the SAR block). The comparison result is detected by the latch described in fig. 7, and the result is used to modify the SAR logic using the scheme described above (though pass-gate based multiplexers P, driven by a global controller). Note that, in the overall scheme, the component of RCM output current sunk by the DTCS in the ADC's flow through across a DC level of 2ΔV.

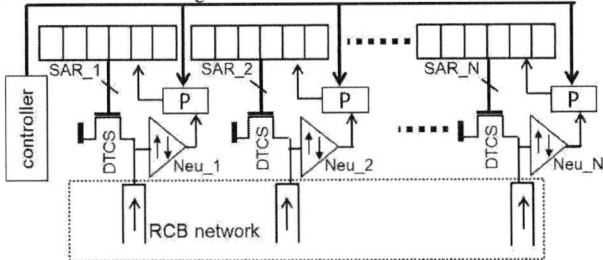

Fig. 11 Block diagram for SAR operation of the WTA circuit

The second half of the WTA algorithm operates in parallel with the first (i.e., the ADC operation). It can be explained with the help of the corresponding circuit diagram shown in fig. 12. Results of the first ADC conversion step obtained from the SAR are directly transferred to the tracking registers (TR) shown in the figure through the pass-gate multiplexing switch (PGS). Thus, at this stage, all the TR's with a high output correspond to the ADC results with MSB = '1'. Let us now, consider the second cycle operation. The detection line (DL) is first pre-charged to Vdd and the set of discharge registers (DR) driving it are cleared to low output. Next, if for at least one of the SAR's with high MSB, the second MSB also evaluates to '1', the corresponding DR is driven high by the associated AND gate. Thus, DL is discharged to ground and the write of all the TR's is enabled. All the TR's for which both, first and second MSB's evaluated to '1', stay high, but the rest are set to low. In simple terms, if at least one of the SAR's (5-bit) evaluated to '11000' in the second conversion cycle, the DL is discharged and all the TR's with SAR value '11000' stay high, while those with SAR value '10000' are set to low. In case all SAR's evaluated to '10000' in the second cycle, no change is made to the TR values. Thus, at the end of conversion cycle, if only one of the TR's remains high, it is identified as the winner and the corresponding SAR value is effectively the degree of match (DOM). In case a random image is input to the hardware, the proposed scheme will still identify the 'winning' pattern. But if the DOM is lower than a predetermined threshold, the winner is discarded, implying that the input image does not belong to the stored data set.

The winner-tracking circuitry described above is fully digital and does not consume any static power. Moreover, owing to the global digital control, it is easily scalable with number of input as well as required bit precision. The overall power consumption in the proposed design is drastically reduced as compared to a MS-CMOS realization (described in section-2), due to two main reasons; firstly, the power consumption in the RCM itself is significantly lowered due to low-voltage operation, and secondly, the fully digital WTA scheme avoids any additional static-power consumption. Note that the proposed WTA

scheme implemented in MS-CMOS would result in large power consumption, resulting from conventional ADC's. The low-voltage current-mode switching characteristics of DWN however, provides a compact and ultra-low-power digitization technique [20, 22]. The performance and prospects of the proposed design are discussed in some more detail in the next section.

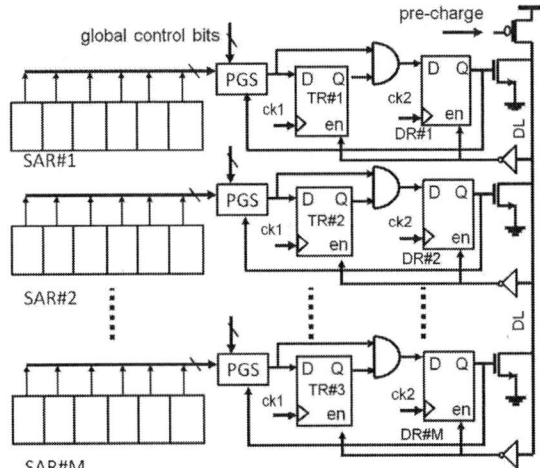

Fig.12 Circuit operation for the tracking part of the WTA algorithm

5. PERFORMANCE AND PROPECTS

In order to compare the performance of the proposed design with state of the art mixed signal (MS) CMOS design, we simulated two different CMOS BT-WTA topologies proposed in [17] and [18] respectively, using 45nm CMOS technology models. The first design is the standard BT-WTA, whereas, the second is a recently proposed modification. We also simulated a 45nm digital CMOS design that employed multiply and accumulate operations for evaluating the correlation between the 5-bit 128 element digital templates and input features of the same size.

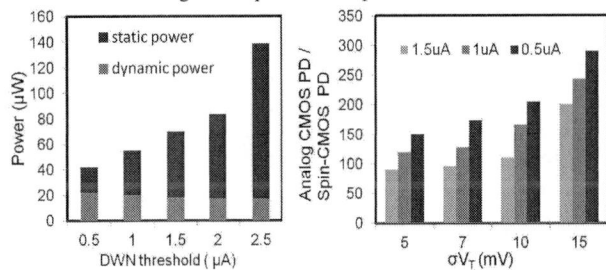

Fig. 13 (a) Power consumption of the proposed design with its static and dynamic components, for different values of DWN threshold, (b) ratio of power-delay (PD) product of MS-CMOS and the proposed design for increasing transistor variations.

Simulations for MS-CMOS designs show that the power consumption for the WTA unit dominates the total power. On the other hand, for the proposed scheme, there is negligible static power consumption in the WTA operation. However, since, the static power consumption in RCM is also significantly lowered, it becomes comparable to the dynamic switching power in the WTA. This is evident from the trend shown in fig. 13a. It also shows that the static power consumption in the DWN-based design can be significantly reduced by lowering the DWN switching threshold further. However, the dynamic power remains almost constant and starts to dominate for reduced DWN thresholds.

Plot in fig. 13b shows the impact of transistor process variations upon MS-CMOS designs. The power-delay products are plotted for a WTA resolution of 4%. Note that in the proposed WTA, the impact of transistor-variations in the DTCS-DAC is limited to just a single step, whereas, the MS-CMOS circuits suffer more due to the cumulative effect of multiple transistors in the signal path. As discussed in section-2, with larger variations, the accuracy and resolution of MS-CMOS circuits like, current-mirrors decreases steeply, necessitating the use of larger devices, which impairs the circuit performance.

Table-1 compares the proposed spin-CMOS design with MS-CMOS designs in [17] and [18], and with the 45 nm digital CMOS design. The results shown are for σV_T =5mV for minimum sized transistors, which is a near ideal case for MS-CMOS circuits. Results for three different WTA resolutions are given which show similar energy benefits of the proposed scheme, even for smaller WTA resolution. For analog designs, lower resolution constraints allows smaller transistors and hence better performance. Power consumption for the DWN based design, also reduces with resolution. Lower WTA resolution allows smaller DAC currents, resulting in reduced static power and lower switched capacitance for the smaller WTA blocks, leading to reduced dynamic power.

Table-1 Performance

		spin-CMOS PE	[18]	[17]	45nm Digital CMOS
Power	5-bit	65µW	5.5mW	8mW	4mW
	4-bit	45µW	2.9mW	5.0mW	2.8mW
	3-bit	32µW	2.3mW	3.2mW	1.2mW
Frequency		100 MHz	50MHz	50MHz	2.5MHz
Energy	5-bit	1	160	215	2460
	4-bit	1	140	221	2300
	3-bit	1	155	210	1100

Most interestingly, results for comparison with 45nm digital hardware shows ~1000x lower computing energy for the proposed design. Note that, this comparison does not include the overhead due to memory read in the digital design. As discussed earlier, digital hardware in general prove inefficient for the class of computation considered in this work. Another important point to be noted is that, the use of MS-CMOS circuits in RCM barely perform ~10x better than the digital implementation and hence achieve far less energy efficiency as compared to the proposed design. Thus, ultra-low energy analog-computing using spin-neurons can significantly enhance the prospect of RCM technology for computational hardware. The basic associative-module discussed in this work can be extended to a more generic architecture. For instance, very large number of images can be grouped into smaller clusters [25], that can be hierarchically stored in the multiple RCM modules. Individual patterns of larger dimensions can also be parititioned and stored in modular RCM-blocks. The proposed design scheme can be applicable to a wide class of non-Boolean computing architectures that also include different categories of neural networks [20]. For instance, the spin-RCMbased correlation modules presented in this work can provide energy efficient hardware solution to convolutional neural networks that are attractive for cognitive computing tasks, but involve very high computational cost.

A self explanatory pictorial depiction of the simulation-framework used in this work is given in fig.14. We used micromagnetic simulation model for DWN that was calibrated with experimental data on DWMs. Behavioral model based on statistical characteristics of the device were used in SPICE simulation to assess the system level functionality. Some important design parameters used are listed in table-2.

6. CONCLUSION

Emerging RCM technology holds great potentials for non-Boolean computing hardware. However, conventional mixed signal CMOS circuits may fail to leverage its benefits due to their large power consumption and poor scalability. We showed that the critical analog functionality needed in RCM based computing tasks can be provided by magneto-metallic spin-neurons at ultra low energy cost. The resulting design can achieve more than three orders of magnitude lower

energy cost as compared to a dedicated digital hardware. The use of spin-torque neurons can therefore boost the prospects of RCM as a computation tool.

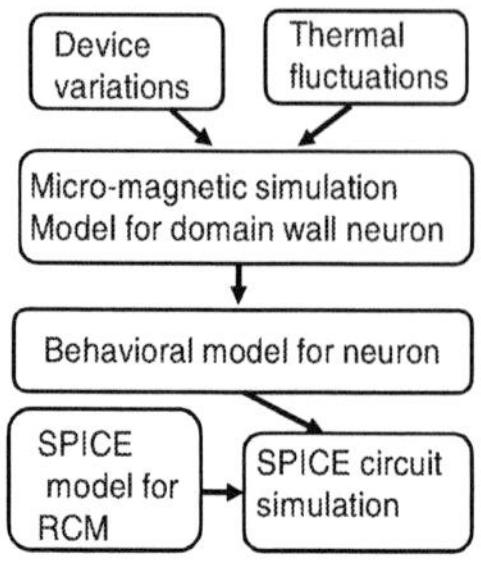

Table-2 Design parameters

Template size	16x8, 5-bit	Magnet material	NiFe
# template	40	free-layer size	3x22x60nm³
comparator resolution	5-bit		
Input data rate	100MHz	Ms	800 emu/cm³
		Ku2V	20KT
		Ic	1µA
crossbar parasitics	1Ω/µm, 0.4fF/µm	Tswitch	1.5ns
Crossbar material	Cu	Cross-bar material	Cu
memristor material	Ag-aSi	Resistance range	1kΩ to 32kΩ

Fig. 14 Simulation framework

ACKNOWLEDGEMENT

This work was supported by STARnet, a Semiconductor Research Corporation program sponsored by MARCO and DARPA.

REFERECNCES

[1]Sangho Shin et. al, "Memristor-Based Fine Resolution Programmable Resistance and Its Applications", ISCAS, 2009.

[2]R. Berdan et.al., "High precision analogue memristor state tuning" Electronics Letters, 2012

[3]Feng Miao et al., "Continuous Electrical Tuning of the Chemical Composition of TaOx- Based Memristors" ACS Nano, 2012

[4]K. Likharev et. al.,"Biologically Inspired Computing in CMOL CrossNets" , 2009

[5]J. Turelet. al. , " Neuromorphic architectures for nanoelectronic circuits", Int. J. Circ. Theor. Appl. 2004

[6]S. H. Jo et.al., "High-Density Crossbar Arrays Based on a Si Memristive System", Nano Letters, 2009

[7]S. H. Jo et.al, "CMOS Compatible Nanoscale Nonvolatile Resistance Switching Memory", ASC, 2008

[8]L. Gao et. al., "Analog-Input Analog-Weight Dot-Product Operation with Ag/a-Si/Pt Memristive Devices", VLSISOC, 2012

[9]B. Mouttet, " Proposal for Memristors in Signal Processing", NanoNet ,2009

[10]D. Morris et. al., "mLogic: Ultra-Low Voltage Non-Volatile Logic Circuits Using STT- MTJ Devices" DAC, 2012

[11]M. Hu et. al., "Hardware Realization of BSB Recall Function Using Memristor Crossbar Arrays" , DAC 2012

[12]C. K. Lim, "Domain wall displacement induced by subnanosecond pulsed current", App. Phy. Lett., 2004

[13]J. Vogel et. al., "Direct Observation of Massless Domain Wall Dynamics in Nanostripes with Perpendicular Magnetic Anisotropy", arXiv:1206.4967v1, 2012

[14]Ngo et al., " Direct Observation of Domain Wall Motion Induced by Low-Current Density in TbFeCo Wires", Applied Physics Express,2011

[15] S. Fukami et. al., "Low-current perpendicular domain wall motion cell for scalable high-speed MRAM," VLSI Tech. Symp, 2009

[16] P. R. Kinget, "Device Mismatch and Tradeoffs in the Design of Analog Circuits" JSSC, 2005

[17] Andreas et. al., " A CMOS Analog Winner-Take-All Network for Large-Scale Applications" , IEEE TCAS, 1998

[18] DŁugosz et. al., " Low power current-mode binary-tree asynchronous Min/Max circuit", Microelectronics Journal, 2009.

[19] Sani R. Nassif, "Process Variability at the 65nm node and Beyond" CICC, 2008

[20] M. Sharad, et. al, "Spin Neurons for ultra low power computational hardware", DRC , 2012

[21] M. Sharad, et. al, "Boolean and Non-Boolean Computing with SpinDevices", IEDM , 2012

[22] M. Sharad, et. al, "Ultra Low Energy Analog Image Processsing Using Spi Neurons", International. Syposium on Nanoscale Architecture, 2012.

[23] M. Sharad et al, " Cogitive Computing Using Spin-Based Neurons", DAC, 2012.

[24] M. Sharad et al, " Proposal for Neuromorphic Hardware Using Spin-Based Devices", arXiv:1206.3227v4 [cond-mat.dis-nn], 2012.

[25] Duda, R. O.; Hart, P. E.; and Stork, D. G. Pattern Classification; John Wiley & Sons: 2000.

[26] http://www.cl.cam.ac.uk/research/dtg/attarchive/facedatabase.html

Understanding The Trade-Offs In Multi-Level Cell ReRAM Memory Design

Cong Xu[†], Dimin Niu[†], Naveen Muralimanohar[‡], Norman P. Jouppi[‡], Yuan Xie[†],[§]

†Pennsylvania State University, {czx102,dun118,yuanxie}@cse.psu.edu
‡Hewlett-Packard Laboratory, {naveen.muralimanohar,Norm.Jouppi}@hp.com
§AMD Research, yuanxie@amd.com

ABSTRACT

Resistive Random Access Memory (ReRAM) is one of the most promising emerging memory technologies as a potential replacement for DRAM memory and/or NAND Flash. Multi-level cell (MLC) ReRAM, which can store multiple bits in a single ReRAM cell, can further improve density and reduce cost-per-bit, and therefore has recently been investigated extensively. However, the majority of the prior studies on MLC ReRAM are at the device level. The design implications for MLC ReRAM at the circuit and system levels remain to be explored. This paper aim to provide the first comprehensive investigation of the design trade-offs involved in MLC ReRAM. Our study indicates that different resistance allocation schemes, programming strategies, peripheral designs, and material selections profoundly affect the area, latency, power, and reliability of MLC ReRAM. Based on this analysis, we conduct two case studies: first we compare MLC ReRAM design against MLC phase-change memory (PCM) and multi-layer cross-point ReRAM design, and point out why multi-level ReRAM is appealing; second we further explore the design space for MLC ReRAM.

1. INTRODUCTION

Emerging non-volatile memory (NVM) technologies such as Spin-transfer torque RAM (STT-RAM), Phase-change Memory (PCM), and Resistive RAM (ReRAM) are considered as promising replacements for traditional DRAM-based main memory and NAND Flash storage, due to the advantages of high density, non-volatility, and zero standby power from memory cells. Among these memory technologies, ReRAM has been demonstrated with excellent scalability beyond the 10nm technology node [17] and with the capability of high density due to multi-level cells (MLC) and cross-point array structures. In particular, since the MLC ReRAM memory structure has reliability advantages over the cross-point ReRAM structure (see Section 2.A), it has recently been studied extensively.

While many experimental results of MLC ReRAM cells have been reported, there is still a gap between the device-level study and system-level design implications. The majority of the prior studies on MLC ReRAM are at the device level. The trade-offs in MLC ReRAM memory array design

Cong Xu and Dimin Niu are supported in part by SRC grants, NSF 1147388, 0903432, and 0643902. This material is based upon work supported by the Department of Energy under Award Number DE - SC0005026. The disclaimer can be found at http:www.hpl.hp.com/DoE-Disclaimer.html

Permission to make digital or hard copies of all or part of this work for personal or classroom use is granted without fee provided that copies are not made or distributed for profit or commercial advantage and that copies bear this notice and the full citation on the first page. To copy otherwise, to republish, to post on servers or to redistribute to lists, requires prior specific permission and/or a fee.
DAC '13 May 29 - June 07 2013, Austin, TX, USA

and programming strategies still need to be fully understood to design a better memory subsystem.

In this paper, we first give a brief introduction to ReRAM technology and the design challenges for MLC ReRAM. We then demonstrate that different resistance allocation schemes, programming strategies, peripheral circuitry designs and material selections can profoundly affect the area, latency, power and, reliability of MLC ReRAM. Two programming schemes, one optimized for write latency and another for low error rate, are proposed to improve the performance and reliability of MLC ReRAM design. Based on this analysis, we conduct two case studies. The first one compares MLC ReRAM design against MLC phase-change RAM (PCRAM) and cross-point ReRAM design, and the second explores the design space for MLC ReRAM. To the best of our knowledge, this is the first comprehensive investigation of the design trade-offs involved in MLC ReRAM memory array design.

2. PRELIMINARIES

In this section, we will describe the background of ReRAM technology including the multi-level switching mechanism of the cell and the characteristics of retention failure observed in some ReRAM devices.

2.1 ReRAM Basics

The basic structure of a ReRAM cell is a metal oxide layer sandwiched between two metal electrodes. The switching from high resistance state (HRS) to low resistance state (LRS) is defined as a SET operation while the opposite process is a RESET operation.

Projected as a low cost-per-bit memory technology, high density ReRAM can be improved in many ways: (a) Some ReRAM technologies exhibit a non-linear relationship between voltage and resistance. We can leverage this property to build a cross-point structure with a cell size of $4F^2$ and avoid having a dedicated access device in each cell. On the flip side, the lack of an access device results in sneak current and voltage drop problems. Even with a diode access device implemented in the cross-point structure, the array size can still be limited by the voltage drop [10]. (b) Another way to achieve high density is by building multiple layers stacked on top of each other. Note that this stacking refers to layers within a single die; 3D stacking of multiple dies is an orthogonal technique. Two-layer cross-point ReRAM prototypes have been demonstrated with reliable read and write operations [6]. While growing arrays in the vertical dimension increases density, the effective size of an array is still limited by the sneak current and the voltage drop problems. (c) Multi-level cells offer another opportunity to improve ReRAM density. Compared to the multi-layer approach, MLC relaxes the magnitude of the sneak current and voltage drop problems. But it requires more accurate tuning of the analog resistance value of each cell. Thus the resistance distribution and programming strategies in MLC

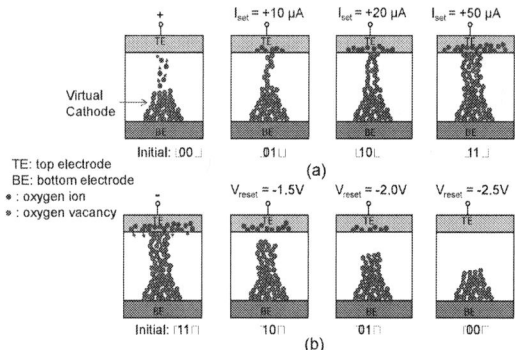

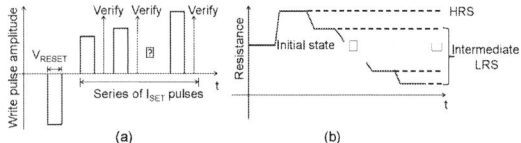

Figure 1: Multi-level switching in ReRAM: (a) H2L and (b) L2H programming.

Figure 2: Write-and-verify in H2L programming

Table 1: 3 strategies in allocating 8 resistance levels (unit: kΩ) ranging from $1k\Omega$ to from $10M\Omega$

State	111	110	101	100	011	010	001	000
$ISO\text{-}\Delta R$	1	1430	2860	4290	5710	7140	8570	10^4
$ISO\text{-}\Delta I$	1	1.17	1.40	1.75	2.33	3.50	7.00	10^4
$ISO\text{-}\Delta \log(R)$	1	3.73	13.9	51.8	193	719	2680	10^4

Table 2: Metrics of resistance allocation schemes

Metric	Minimum ΔI	t_{sense}	Average P_{write}/b
$ISO\text{-}\Delta R$	$16nA$	$513ns$	$0.05mW$
$ISO\text{-}\Delta I$	$143\mu A$	$13ns$	$0.31mW$
$ISO\text{-}\Delta \log(R)$	$270nA$	$29ns$	$0.13mW$

can profoundly affect area, latency, and power. Without a comprehensive model, the cost benefit of MLC over other approaches is not obvious.

2.2 Multi-level Switching in ReRAM Cell

When a voltage is applied across a ReRAM cell, depending upon the voltage polarity, one or more conductive filaments (CF) made out of oxygen vacancies are either formed or ruptured. Once the CFs are formed inside the metal oxide to bridge the top and bottom electrodes, current can flow through the CFs, and the cell is in a low resistance state (LRS). The larger the size of the CFs, the lower the resistance. Figure 1(a) illustrates the formation of the CFs in a bipolar ReRAM cell. Conversely, the rupture of the CFs disconnects the top electrode from the bottom electrode, resulting in a high resistance state (HRS) of the cell. When a positive current passes through the cell, the oxygen atoms are knocked out of the lattice and become negatively-charged oxygen ions. Under a positive electric field, the oxygen ions will drift towards the anode, leaving corresponding oxygen vacancies in the metal oxide layer.

Figure 1(a) also shows that the size of the CFs is directly related to the value of the current, and by changing the strength of the CFs, we can control the cell resistance. Thus we can program ReRAM to intermediate levels between the highest resistance state and the lowest resistance state by adjusting the programming current.

Programming to intermediate states are often done in steps, starting from either the highest resistance state (H2L programming) or the lowest resistance state (L2H programming). This multi-step process, illustrated in Figure 2, helps tolerate both temporal and spatial process variations.

2.3 Retention Failure in MLC ReRAM

Most emerging NVM technologies such as ReRAM, PCM, and STTRAM are immune to particle strikes, but they are still susceptible to transient errors due to other reasons. For instance, STT-RAM suffers from thermal fluctuations and PCRAM has both short-term and long-term resistance drift problems. In ReRAM, unlike the gradual drift seen in PCRAM, sudden transitions occur in which the resistance changes abruptly to the highest resistance state (called LRS retention failure [14, 19]) or lowest value (called HRS reten-

tion failure [3]). However, once we reprogram the cell, it resumes normal operation.

This retention failure is essentially explained by the same principle as the normal switching behavior of ReRAM. The key difference between the HRS retention failure and H2L programming is that the thermal activation that causes the retention failure is a random process, which occurs very rarely and requires much longer time than a typical write operation. The conclusion in [3] indicates that for ReRAM with a large SET current ($> 500\mu A$), strong CFs exist in the cell and thus only HRS retention failure is observed in these cells because the ruptured CFs are more like to be reconstructed. In contrast, for ReRAM with a small SET current ($< 100\mu A$) [19], only LRS retention failure is observed, because weakly formed CFs are more likely to be ruptured due to the random degeneration of oxygen vacancies in the CFs. The experimental results in [14] confirm that there exists a reverse linear dependence between the resistance value of the LRS and the average retention failure time: $t_{\text{failure}} \propto 1/R_{\text{LRS}}$. In this work, we considers a key reliability issue in MLC ReRAM design: after H2L programming higher LRS levels are associated with weak CFs and are vulnerable to LRS retention failure; while after L2H programming lower HRS levels are associated with strong CFs and are vulnerable to HRS retention failure.

3. TRADE-OFFS IN MLC RERAM DESIGN

3.1 Trade-offs in Resistance Allocation Schemes

In this section we study the impact of resistance allocation schemes on noise margin, sensing latency and programming power. Given the minimum resistance, maximum resistance and number of levels required, there are typically three schemes in determining the resistance values: $ISO\text{-}\Delta R$ where the resistance is linearly spaced, $ISO\text{-}\Delta I$ where the read current (inverse of resistance) is linearly spaced and $ISO\text{-}\Delta \log(R)$ where the resistance is geometrically spaced. Table 1 shows the distribution of resistances in a 8-level cell for various schemes. The eight resistance levels range from $R_{\min} = 1k\Omega$ to from $R_{\max} = 10M\Omega$. In the $ISO\text{-}\Delta R$ scheme, all the states except "111" have resistance values above $1M\Omega$, limiting the programming current below $1\mu A$ for most of the time. Therefore this scheme is preferred from the energy standpoint. However, the difference between the reading current of state "000" and "001" is only 16nA, making it challenging to develop sensing circuitry to identify the state reliably and quickly. In the $ISO\text{-}\Delta I$ scheme, all the states except "000" have resistance values below $10k\Omega$, maintaining a constant read current difference ($> 100\mu A$) between adjacent states. This scheme enables fast read speed for all cases but requires a large average write power during programming. For a fixed power budget, the maximum write throughput is limited by the number of cells that be programmed at the same. Therefore, the $ISO\text{-}\Delta I$ scheme is preferred from the sensing point of

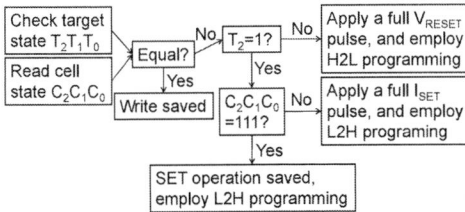

Figure 3: Flowchart of FPS

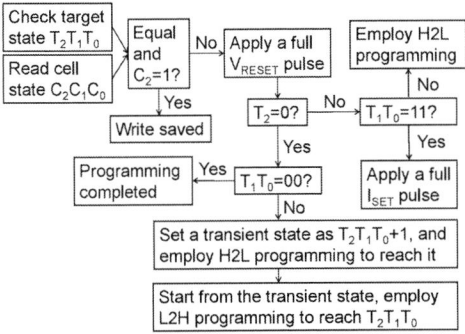

Figure 4: Flowchart of RPS

view but incurs high energy overhead. One straightforward trade-off between sensing latency and programming power is to choose resistance values with equal interval in log space, denoted as the $ISO\text{-}\Delta\log(R)$ scheme. The resistance values are then shifted upwards compared to the $ISO\text{-}\Delta I$ scheme, lowering the average programming power; the read current difference between adjacent levels increases compared to the $ISO\text{-}\Delta R$ scheme, relaxing the sensing requirement. Moreover, this scheme also benefits from the observation that the resistance distribution in MLC spreads is wider as the resistance value increases [17].

We calculate the write power per bit for the three schemes. We also simulate the sensing latency in an advanced current-mode MLC sensing circuitry proposed in [15] using the 22nm PTM model [20] in HSPICE. As part of the sensing circuitry, we assume each bitline is connected with 2048 cells. The results are presented in Table 2. We can see that the sensing latency t_{sense} of the $ISO\text{-}\Delta R$ scheme is much longer than that of $ISO\text{-}\Delta I$ and $ISO\text{-}\Delta\log(R)$ due to its much smaller read current difference - 16nA. Among the three schemes, $ISO\text{-}\Delta I$ has the highest average programming power (0.31mW/b) because of the aforementioned reason. The $ISO\text{-}\Delta\log(R)$ scheme is clearly a reasonable trade-off between sensing latency and average programming power.

3.2 Trade-offs in Programming Strategies

In this section we propose two programming strategies targeted either for performance and energy or reliability. The fast programming strategy (FPS) is optimized for write latency and energy with reduced reliability. It is devised for a memory subsystem whose performance is very critical and can tolerate occasional soft errors. These scenarios include, but not limited to, the server DRAM main memory where ECC is typically implemented to correct particle-induced soft errors. The second option called reliable programming strategy (RPS) is optimized for extremely low error rate but it trades-off write latency. It is more suitable for storage systems where data integrity is of most importance. For example, the data in a solid-state disk (SSD) or USB drive may need to be stored for years. Moreover, even microsecond-level write latency can be hidden by a large block size in NAND flash as long as the bandwidth requirements are met. To explain these policies, consider a 3-bit MLC cell with bits

$D_2D_1D_0$: D_2 is the most significant bit (MSB) and D_0 is the least significant bit (LSB). We use "111" to represent the state with lowest resistance and "000" to represent the state with the highest resistance.

3.2.1 Fast programming strategy

The idea behind FPS is to reduce the average write latency by choosing either H2L or L2H programming to reach the target states in as fewer iterations as possible. Figure 3 shows the steps involved in FPS. Before the actual programming starts, the current state of a cell $C_2C_1C_0$ is first read and compared with the target state $T_2T_1T_0$ to be programmed. If they are equal, the programming phase is skipped. Otherwise it is required to identify whether the target state is faster to program from the highest resistance state or the lowest resistance state, which can be simply achieved by checking the MSB of the target state: $T_2 = 0$ means that the target state is greater than the median value and we can reach the final state in fewer iterations through H2L programming. In that case, we first apply a full RESET voltage across the cell and program it to the highest resistance state. After that, we simply employ H2L programming to reach $T_2T_1T_0$. If $T_2 = 1$ then we check if the current state is already in the lowest resistance state. If $C_2C_1C_0 = 111$, we can avoid a dedicated SET operation that set the cell to the lowest resistance state. Then, we proceed with L2H programming to complete the write operation. Note that this optimization is not adopted before H2L programming. We do not skip the RESET operation even if $C_2C_1C_0 = 000$, since the highest resistance has very wide distribution. It is critical to RESET it to the maximum resistance value. The drawback of FPS is that some states may be vulnerable to retention failure. For example, if $T_2T_1T_0 = 110$, the final state is achieved by slightly rupturing the strongly formed CFs through L2H programming. As discussed in Section 2, this process can lead to LRS retention failure. In contrast, if $T_2T_1T_0 = 001$, weak CFs are reconstructed from the highest resistance state through H2L programming, and this can lead to the HRS retention failure.

3.2.2 Reliable programming strategy

RPS is devised to meet strict retention requirement (i.e. >10 years). This is done by associating LRS with strong CFs and HRS with weak CFs (ruptured). Figure 4. In RPS, a write operation is skipped only if the target state has low resistance and is also equal to the original cell state; otherwise a complete RESET operation is performed first. After that, T_2 is checked: if $T_2 = 1$ then H2L programming is employed to reach the target state (expect for the case that $T_2T_1T_0 = 111$ is directly SET by applying a full SET current pulse). If $T_2 = 0$, in theory, we can either do H2L programming or we can first performace a SET operation followed by L2H programming. But both of the approaches can impact retention. For example, H2L programming will form weak CFs and can lead to HRS retention failure. On the other hand, doing a SET followed by L2H programming results in ruptured strong CFs, which can lead to LRS retention failure. To overcome this, we first define a transient state which has one-level lower resistance than the target state, and use H2L programming to first reach the transient state. This state will have week CFs formed since $T_2 = 0$. Then we do L2H programming to reach by $T_2T_1T_0$ rupturing the weak CFs. By controlling the strength of the CFs in the programmed state, a stable state is created using RPS.

3.2.3 Monte Carlo Simulations

Table 3: Average iteration count required to switch from $C_2C_1C_0$ (left column) to $T_2T_1T_0$ (top row) in FPS (upper row) and RPS (lower row)

T (ns)	000	001	010	011	100	101	110	111
000	0	3.25	7.93	12.1	14.4	9.52	4.74	10.9
	1.01	26.9	22.1	16.3	15.6	20.9	25.8	2.11
001	1.07	0	9.01	13.0	14.6	9.77	4.98	1.13
	1.09	2.71	22.2	16.4	15.8	21.1	26.0	2.14
010	1.25	4.33	0	13.5	14.7	9.82	5.04	1.15
	1.27	27.4	22.6	16.6	16.0	2.11	2.64	2.16
011	1.31	4.48	9.54	0	15.0	10.1	5.27	1.21
	1.34	27.7	22.7	16.7	16.1	21.3	26.5	22.1
100	1.48	4.49	9.69	13.9	0	10.0	5.22	1.17
	14.9	27.5	22.1	16.5	0	21.4	26.7	2.18
101	1.49	4.53	9.72	14.0	14.9	0	5.17	1.11
	1.49	27.5	22.0	16.3	16.1	21.5	26.8	2.15
110	1.51	4.59	9.78	14.0	14.7	9.78	0	1.05
	1.52	27.3	22.0	16.2	16.2	21.5	0	2.09
111	1.52	4.68	9.83	14.3	14.6	9.71	5.05	0
	1.52	25.3	20.1	14.3	16.3	21.5	26.9	0

In order to the evaluate the trade-offs in the switching latency, energy and retention by the two different programming strategies, we conduct exhaustive Monte Carlo experiments using the SPICE compact model of ReRAM proposed in [4]. The duration of each I_{SET} or V_{RESET} pulse is set to 5ns. We assume ISO-$\Delta \log(R)$ for all experiments. For any give pair of states - switching from $C_2C_1C_0$ to $T_2T_1T_0$, we perform 1000 Monte Carlo simulations and record the programming iteration count. Based on the results in Table 3, we can calculate the average programming latency and energy. We can see that most switching in FPS completes in a few programming iterations. Actually the switching in the last two digits of a 3b/cell contributes to the major portion of the programming iterations in FPS. Our simulation results indicate that the iteration count of FPS has an upper bound of 16, which is consistent with the reported value in [15]. In RPS, if $C_2C_1C_0 \neq T_2T_1T_0$, writes are faster when $T_2T_1T_0 =000$ or 111. In all the other cases, the iteration count will be larger than 15. The worst-case programming latency in RPS is about twice that latency in FPS. Assuming equal switching probability between all the states, the average programming energy of FPS is 14pJ while that of RPS is 46pJ. Thus, FPS clearly has superior latency and energy, but it pays the penalty of reduced retention time in some programmed states. To estimate the retention time of the resistance state after FPS and RPS, we largely simplified the flow presented in [19] and did a first-order approximation of the retention under 150° Celsius. In our study, we found that some intermediates states after FPS show limited retention time ($10^5 - 10^7$ s) while no retention failure is observed in RPS under our simulation setup.

3.3 Trade-offs in Circuit Design

In order to better understand the circuit design trade-offs, we develop a circuit-level model to estimate the area, timing, dynamic energy and leakage power of MLC ReRAM. The generic area, timing, and power modeling is derived from NVSim [2], which is a framework to model different emerging NVM technologies including PCM, ReRAM, STT-RAM and Floating-body DRAM (FBDRAM). However, we made several enhancements to the original models,

- Some circuitry modules in MLC NVMs have different requirements from those originally designed for SLC. For example, the existing sense amplifier model in NVSim is based on single-step single-reference sensing circuitry, while MLC sensing needs either a multi-step (sequential) single-reference sensing scheme or single-step multi-reference (parallel) sensing circuitry. We

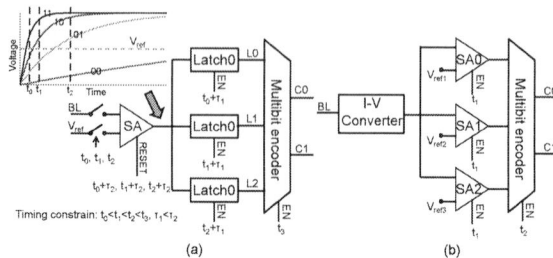

Figure 5: Schematic view of (a) Sequential and (b) Parallel sensing scheme for 2b/cell read operation

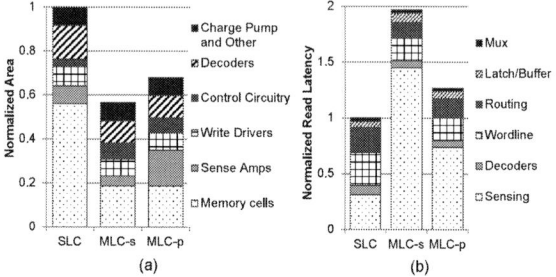

Figure 6: ReRAM area and latency breakdowns for SLC, MLC-s (3b/cell) with sequential sensing, MLC-p (3b/cell) with parallel sensing scheme

model these modules in an approach compatible with the previous model.

- MLC NVMs need specialized control circuits to properly handle their write iterations. We extend the sensing circuit to model this extra overhead, which models the flow chart in Figure 3 and 4.

- Some ReRAM technologies require a high RESET voltage, which can be larger than the core Vdd, and thus need dedicated charge pump circuitry. For cross-point ReRAM, the diode in series with the ReRAM cell may add to operating voltage by 1.8V [6], further increasing the complexity of the charge pump circuits. We use the model presented in [13] to calculate the area of the charge pump circuits.

Among circuit design options, the area and latency trade-offs between sequential and parallel sensing circuitry seem to be an interesting option to explore. In the sequential scheme, a single sense amplifier together with multiple latches are used, as shown in Figure 5(a). Using a 2b/cell read operation as an example, the enable signals of the three latches are activated at different time steps t_0, t_1 and t_2 to sample the output of the sense amplifier at the corresponding time point. After that, the output of the latches is passed to a 3bit-to-2bit encoder to produce the actual stored information of the selected cell. Figure 5(b) shows an alternative design - the parallel scheme, in which multiple sense amplifiers associated with different reference voltage levels are implemented. In the first stage, the read current of the selected cell is first converted to a voltage signal by passing through an I-V converter. In the second stage, the voltage signal is connected to the input of three voltage sense amplifiers and compared with the three different voltage levels simultaneously. The parallel sensing scheme reduces the sensing latency overhead but triples the number of sense amplifiers in for 2-bit cells.

We demonstrate the capability of our extended model by estimating the area and latency of a single-level ReRAM (SLC), a 3b/cell ReRAM with the sequential sensing scheme (MLC-s) and a 3b/cell ReRAM with the parallel sensing

Table 4: Technology survey of MLC ReRAM

Material	TiO_x	TaO_x	HfO_x	WO_x	CuO_x	ZrO_x
MLC levels	5	4	5	8	5	3
R_{off}/R_{on}	1000	1000	> 1000	> 20	> 100	5000
Endurance	> 10^6	> 10^8	> 10^8	> 10^4	-	> 10^2

scheme (MLC-p) for the same capacity. The results are normalized to the SLC ReRAM in shown Figure 6. As we can see, even though the area of memory cells in MLC ReRAM is one-third that of SLC ReRAM, the overall area reduction is only about 40%. This is because the peripheral circuits do not scale well with the cell size and result in poor area efficiency [15]. For example, the write drivers and control circuitry are more complex in MLC than in SLC, and their area overhead is higher in MLC. Moreover, the area of the charge pumps stays almost constant from SLC to MLC. In MLC-s, although the number of sense amplifiers is one-third that of the SLC cells, the area overhead in additional latches and timing control circuity prevent the sensing circuitry from scaling linearly with the number of cells. In MLC-p, each bitline is associated with 7 sets of sense amplifiers, making the area of sensing circuitry comparable to that of memory cells. Figure 6(b) illustrates that the wordline delay and routing delay in MLC is smaller than SLC due to the reduced array size and routing distance, but the read latency of MLC-s can be almost twice that of SLC. However, MLC-p has 30% smaller read latency than MLC-s because the parallel sensing reduces the sensing latency overhead.

3.4 Trade-offs in Material Selection

Many ReRAM material systems, such as TiO_x [18], HfO_x [8], WO_x [1], TaO_x [9], CuO [16], and ZrO_x [11] were reported to be capable of MLC operation. Table 5 summarizes the state-of-the-art MLC ReRAM metrics in different ReRAM technologies. There are many trade-offs in speed, power, endurance, and uniformity of different ReRAM material systems. For example, compared to the TiO_x-based ReRAM, the TaO_x-based ReRAM has two orders of magnitude higher endurance but needs a large RESET voltage ($> 5V$), and thus requires a more complex and higher overhead charge pump. The WO_x-based ReRAM has a tight resistance distribution but small R_{off}/R_{on} ratio, favoring the *ISO-ΔR* scheme. the ZrO_x-based ReRAM has a very large resistance ratio but poor endurance, preventing it from being adopted in frequently accessed random access memory. Based on this technology survey, we built a device library in our model and provided the flexibility of selecting different materials when exploring the MLC ReRAM design space.

4. CASE STUDIES

In this section we conduct two case studies. The first case study is to compare the cost per bit of three high-density memory technologies: 3b/cell ReRAM, 3b/cell PCM and 2-layer cross-point ReRAM. The second case study is to explore all the trade-offs in designing a 22nm 128Mb MLC ReRAM macro.

Cost-per-bit is the single most important factor when adopting a new memory technology for the main memory. We developed an area-centric cost model based on Nakatsuka's work [12]. We update the required input parameters from IC Knowledge LLC [5], which incorporates industrial DRAM data of past technology generations and projected data of advanced 20nm-class DRAM. We model both manufacturing and packaging costs. The manufacturing cost includes the substrate wafer cost and the fabrication cost. In order to model the extra fabrication steps in forming NVM cells, we

Table 5: Cell configurations

	3b/cell ReRAM	3b/cell PCM	2-layer Xpoint ReRAM
Access device	NMOS	NMOS	Diode
Cell Size (F^2)	6	8	4
Bit Density (F^2)	2	2.67	2
Max write current (μA)	25	90	25
Single write pulse (ns)	5	80	5

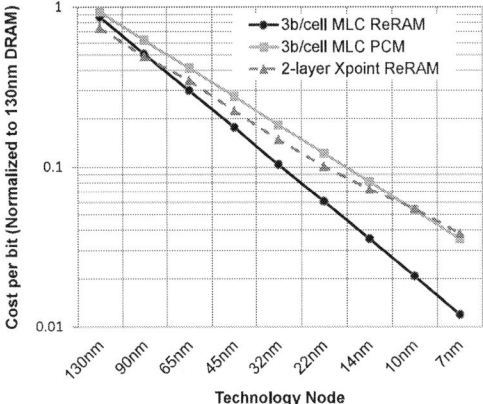

Figure 7: Cost-per-bit comparisons

decomposes the additional process steps into stack deposition, isolation, damascene, and more importantly, lithography steps. IC Knowledge LLC provides the user interface to add these extra fabrication process and the tool calculates the cost overhead automatically.

4.1 Cost-pet-bit Analysis

PCM is another widely explored emerging NVM technology. Similar to ReRAM, PCM can either be built in a cross-point array or 1T1R structure, and also has MLC capability [7]. However, PCM requires a larger programming current than ReRAM and may use a wider transistor to provide enough driving current. The cell configurations are listed in table 5. We assume a DRAM process and design rules when modeling ReRAM and PCM. The cell-level parameters such as write current amplitude and pulse width are extracted from the published literature: for ReRAM we directly used the values reported in [15]; for PCM we used an aggressive set of parameters surveyed in [7]. Another critical cell parameter in cross-point ReRAM is the current density in a half selected cell and we use $10^3 A/cm^2$ as reported in [6].

Figure 7 compares the cost-per-bit of the three memory technologies and normalizes the results to the cost-per-bit of the 130nm DRAM. As we can see, the cost-per-bit of the 3b/cell ReRAM is higher than that of the 2-layer cross-point ReRAM in 130nm and 90nm technology nodes, because the control and sensing circuitry in MLC ReRAM increases the area compared to the 2-layer cross-point ReRAM. However, the cost-per-bit of the 3b/cell ReRAM goes below the 2-layer cross-point ReRAM beyond the 90nm technology node due to several reasons. First, the wire resistance scales superlinearly with technology node, resulting in inefficient subarray size for cross-point ReRAM [10]. This lowers the area efficiency of the cross-point ReRAM and hurts the cost benefit of multi-layer stacking. Second, the operating voltage added by the access diode does not scale well with the technology node, increasing the complexity and area of the charge pumps. Third, multi-layer stacking adds extra fabrication cost and the portion of this part of thecost increases as the die size shrinks. We can see that in sub-10nm technology node even the 3b/cell PCM has a similar cost-per-bit as the

773

Table 6: MLC ReRAM macro design optimizations with different goals

Optimization goal	Area	Read latency	Write latency	Read energy	Write energy	Leakage power
Area (mm^2)	**0.257**	0.336	0.325	0.301	0.356	0.298
Read latency (ns)	55.6	**16.1**	22.5	31.1	25.7	49.3
Write latency (ns)	320	640	**160**	640	160	640
Read energy (nJ)	2.53	1.79	1.98	**1.41**	2.19	2.86
Write energy (nJ)	32.8	85.1	15.6	83.3	**4.1**	61.7
Leakage power (mW)	563	1120	980	1005	1033	**413**
Material System	HfO$_x$	TiO$_x$	HfO$_x$	TiO$_x$	HfO$_x$	TaO$_x$
Resistance allocation	$ISO\text{-}\Delta \log(R)$	$ISO\text{-}\Delta I$	$ISO\text{-}\Delta \log(R)$	$ISO\text{-}\Delta I$	$ISO\text{-}\Delta R$	$ISO\text{-}\Delta \log(R)$
Sense amp type	Sequential	Parallel	Parallel	Sequential	Parallel	Sequential
Sense amp placement	External	Internal	Internal	Internal	Internal	External
Programming Strategy	RPS	RPS	FPS	RPS	FPS	RPS
Wordline Voltage Boost	No	Yes	Yes	No	No	No

2-layer cross-point RRAM. In MLC PCM, the charge pump circuits occupy a significant portion of the die area when the technology scales. Beyond the 22nm technology node, the charge pump may occupy 10%~20% of the PCM chip area. Thus the difference in cost-per-bit of the MLC ReRAM and MLC PCM becomes larger as technology scales.

4.2 Design Space Exploration

MLC ReRAM research is still in an early stage and most studies are focused on the device level. We integrated all the design options mentioned in Section 3 in our model and enabled the tool to predict the full design spectrum of a projected a ReRAM macro given the design parameters, i.e. number of bits per cell, technology node, capacity, I/O width etc. Our model helps explore different implementations of MLC ReRAM and helps make early-stage design decisions by providing an estimation of cost, performance, and energy of different MLC designs. Table 6 tabulates the design space of a 22nm 128Mb 3b/cell ReRAM macro by listing the details of each design point. As the table shows, different trade-offs exist between material selections, resistance allocation schemes, programming strategies, sensing and other peripheral designs for different optimization targets. Based on this, architectural enhancements can be made by making the right trade offs for a target application.

5. CONCLUSION

ReRAM is one of the most promising emerging NVM technologies for next-generation memory/storage subsystems. Compared to multi-layer stacking, MLC offers the opportunity to further reduce the cost per bit of ReRAM. In this work, we developed an area, timing, power, and cost model of MLC ReRAM to study the trade-offs that exist between different resistance allocation schemes, programming strategies, peripheral designs, and material selections in MLC ReRAM design. The model also exposes these trade-offs to facilitate system-level memory design using MLC ReRAM technology.

6. REFERENCES

[1] W.-C. Chien et al. Multi-level 40nm WOx resistive memory with excellent reliability. In *Proceedings of the International Electron Devices Meeting*, Dec. 2011.

[2] X. Dong et al. NVSim: A circuit-level performance, energy, and area model for emerging nonvolatile memory. *IEEE Transactions on Computer-Aided Design of Integrated Circuits and Systems*, 31(7):994 –1007, 2012.

[3] B. Gao et al. Modeling of retention failure behavior in bipolar oxide-based resistive switching memory. *Electron Device Letters*, 32(3):276 –278, 2011.

[4] X. Guan et al. A spice compact model of metal oxide resistive switching memory with variations. *Electron Device Letters*, 33(10):1405 –1407, 2012.

[5] IC Knowledge LLC. IC cost model revision 1105.

[6] A. Kawahara et al. An 8Mb multi-layered cross-point ReRAM macro with 443MB/s write throughput. In *Proceedings of International Solid-State Circuits Conference*, Feb. 2012.

[7] B. Lee et al. Architecting phase change memory as a scalable DRAM alternative. In *Proceedings of the International Symposium on Computer Architecture*, 2009.

[8] H. Lee et al. Low-power and nanosecond switching in robust hafnium oxide resistive memory with a thin ti cap. *Electron Device Letters*, 31(1):44 –46, 2010.

[9] S. R. Lee et al. Multi-level switching of triple-layered TaOx RRAM with excellent reliability for storage class memory. In *Proceedings of Symposium on VLSI Technology*, June 2012.

[10] J. Liang et al. Scaling challenges for the cross-point resistive memory array to sub-10nm node - an interconnect perspective. In *Proceedings of International Memory Workshop*, May 2012.

[11] M. Liu et al. Multilevel resistive switching with ionic and metallic filaments. *Applied Physics Letters*, 94(23):233106, 2009.

[12] H. Nakatsuka. Derivation and implication of a novel DRAM bit cost model. *IEEE Transactions on Semiconductor Manufacturing*, 15(2):279–284, 2002.

[13] G. Palumbo and D. Pappalardo. Charge pump circuits: An overview on design strategies and topologies. *IEEE Circuits and Systems Magazine*, 10(1):31 –45, 2010.

[14] J. Park et al. Investigation of state stability of low-resistance state in resistive memory. *Electron Device Letters*, 31(5):485 –487, 2010.

[15] S.-S. Sheu et al. A 4Mb embedded SLC resistive-RAM macro with 7.2ns read-write random-access time and 160ns MLC-access capability. In *Proceedings of the International Solid-State Circuits Conference*, 2011.

[16] S.-Y. Wang et al. Multilevel resistive switching in Ti/CuxO/Pt memory devices. *Journal of Applied Physics*, 108(11):114110, 2010.

[17] H.-S. Wong et al. Metal oxide RRAM. *Proceedings of the IEEE*, 100(6):1951 –1970, 2012.

[18] C. Yoshida et al. High speed resistive switching in Pt/TiO2/TiN film for nonvolatile memory application. *Applied Physics Letters*, 91(22):223510, 2007.

[19] S. Yu et al. A monte carlo study of the low resistance state retention of HfOx based resistive switching memory. *Applied Physics Letters*, 100(4):043507, 2012.

[20] W. Zhao and Y. Cao. New generation of predictive technology model for sub-45nm design exploration. In *Proceedings of the International Symposium on Quality Electronic Design*, Mar. 2006.

Exploring Tunnel-FET for Ultra Low Power Analog Applications: A Case Study on Operational Transconductance Amplifier

Amit Ranjan Trivedi, Sergio Carlo, and Saibal Mukhopadhyay

School of ECE, Georgia Institute of Technology, Atlanta, GA-30322

<amitrt>,<sergio.carlo>,<saibal>@gatech.edu

ABSTRACT

This work studies the potentials and challenges of designing ultra-low-power analog circuits exploiting unique characteristics of Tunnel-FET (TFET). TFET can achieve ultra-low quiescent current (~pA). In the subthreshold operation, TFET exhibit subthreshold swing lower than 60mV/decade, and hence higher transconductance per bias current than the MOSFET. TFET also exhibit very weak temperature dependence, and higher output resistance. Among several challenges, TFET demonstrate higher Shot noise at low biasing current. Through design of TFET based Operational Transconductance Amplifier (OTA) these challenges and opportunities are discussed. For implantable bio-medical applications, TFET OTA based neural amplifier design is studied.

Categories and Subject Descriptors

B.7.2 [**Hardware, Integrated-circuits, Design Aids**]: Simulation

General Terms

Performance, Design, Theory

Keywords

Tunnel-FET, ultra-low power designs, operation transconductance amplifier, bio-medical designs

1. INTRODUCTION

The ultra-low-power but low-throughput sensors find critical applications in various areas such as bio-medical electronics, environmental monitoring as illustrated in Table 1. Although, the speed/bandwidth requirements for such applications are relaxed, energy/power constraints are stringent. The low-throughput operation provides opportunities for energy/power reduction. Analog designs are integral part of the most sensor systems, and can consume appreciable portion of system power [16]. Power dissipation of analog components can be contained by subthreshold operation. Transconductance per bias current (g_m/I_{DS}) of N-MOSFET, in 90nm CMOS technology, is demonstrated across bias conditions in Fig. 1. In the subthreshold region, the g_m/I_{DS} significantly enhances due to the exponential dependence of drain-current to gate voltage. However, due to lower drain current, the performance (such as g_m itself) reduces.

Permission to make digital or hard copies of all or part of this work for personal or classroom use is granted without fee provided that copies are not made or distributed for profit or commercial advantage and that copies bear this notice and the full citation on the first page. To copy otherwise, or republish, to post on servers or to redistribute to lists, requires prior specific permission and/or a fee.
DAC'13, May 29–June 07, 2013, Austin, TX, USA.

Table 1: Low speed/bandwidth sensor applications.

Application	Frequency Range	Ref
EEG Sensor	0.5 ~ 40 Hz	[19]
Hearing aids	6kHz	[20]
ENAP Sensor	0.1Hz ~ 10kHz	[19]
Seismometer	1kHz	[5]

Hence, when performance requirements are not stringent (as in bio-medical applications [19-20]) the subthreshold operation of MOSFET will be the energy optimal.

In the subthreshold operation, the subthreshold slope (SS) relates to the g_m/I_{DS} as

$$\frac{1}{SS} = \frac{\Delta \log_{10} I_{DS}}{\Delta V_{GS}} = \frac{1}{\log(10)}\frac{\partial \log(I_{DS})}{\partial V_{GS}} = \frac{1}{\log(10)}\frac{1}{I_{DS}}\frac{\partial I_{DS}}{\partial V_{GS}} \quad (1)$$

$$\frac{g_m}{I_{DS}} = \frac{\log(10)}{SS} \quad (2)$$

Thus, even higher g_m/I_{DS} in MOSFET is restrained by its SS. The SS in MOSFET is limited to 60mV/decade at room temperature. Novel device architectures are being explored with steeper SS (less than 60mV/decade) TFET [12], Ferroelectric FET [9], and Carbon Nano Tube transistors [10] to name a few. Ultra low quiescent current (~pA) of these devices is also attractive for low power analog applications. Hence, there is a pressing need for early design exploration of these technologies, and to comprehend their benefits and challenges in low power analog designs.

In this work, application of TFET for ultra low power analog designs is investigated using Technology CAD (TCAD) and circuit simulations. The electrical (current, capacitance, transconductance, and noise) characteristics of TFETs are studied using TCAD simulations. Another benefit of TFET is its reduced temperature sensitivity. This feature assists to the robustness of designs especially when compared to the subthreshold MOSFET

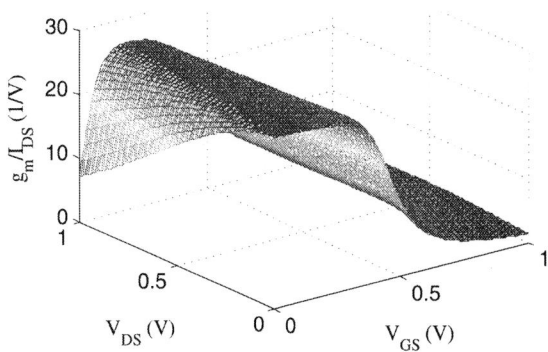

Figure 1: Bias dependences of g_m/I_{DS} for NMOS in 90nm technology.

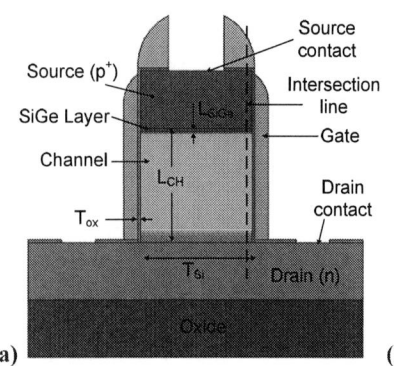
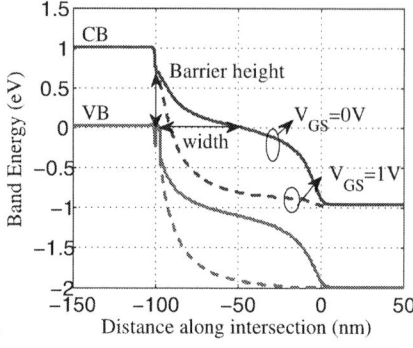

Table-2: TFET device specifications

Specification	N-TFET	P-TFET
L_{CH}	90nm	90nm
T_{si}	100nm	100nm
T_{ox}	2.5nm	2.5nm
L_{SiGe}	3nm	12nm
Ge mole fraction	0.5	0.3
Source-doping	10^{20}	10^{20}
Drain-doping	5×10^{18}	5×10^{18}
Channel-doping	10^{16}	10^{16}
Gate workfunction	4.1eV	5.1eV

Figure 2: **(a)** N-TFET schematic, and **(b)** band-diagram along channel intersection.

(where current depends exponentially to temperature). TFET based Operational Transconductance Amplifier (OTA) circuits are studied. OTA is a key building block in analog filters [8], amplifier [7], neural network [18], and non-linear synthesis [13]. Our analysis shows that power-bandwidth trade-off in TFET based OTA reaches to sub-nW regime, which is not exposed in MOSFET based designs due to MOSFET's excessive leakage. The noise analysis suggests that eventually the Shot noise imposes the limit on the power reduction of the analog operation of TFET.

2. TUNNEL-FET (TFET)

In this work, CMOS compatible Vertical channel TFET (V-TFET) structures are investigated. V-TFET has benefits of smaller footprint, self-decoupled source/drain implants and paves way for gate-all-around structure [6]. The N-type vertical TFET device is shown in Fig. 2a. Unlike MOSFET, in TFET the drain and source junctions are doped with different dopant types. The channel is undoped. The P-type counterpart is similar except the source is doped with the N+-type dopant, and drain with the P+-type dopant. To enhance the limited on current of the TFET device, a Germanium doped thin Silicon layer is deposited at the source-channel interface. Various structural specifications are listed in Table-2.

2.1 Current Conduction Mechanism

The current conduction mechanism in TFET is band-to-band tunneling (BTBT). In N-TFET, electrons tunnel from the source valence band to the drain conduction band (in P-TFET, holes from the source conduction band to the drain valence band). The BTBT probability (and thus the current) is controlled by changing the gate potential. In Fig. 2b, the conduction and valence band-energies of N-TFET along a vertical cross-section [shown in Fig. 2a] is shown for V_{GS}=0V, and V_{GS}=1V. At V_{GS}=0V, the tunneling barrier width is ~45nm, and thus due to very low tunneling probability, TFET achieves extremely low off current. At higher V_{GS} band energies in the channel are lowered, and the tunneling width reduces (~8nm for V_{GS}=1V). This increases the tunneling probability, and significant current conduction occurs. The temperature dependence of the BTBT current comes from the temperature dependence of the band-gap. However, since the temperature dependence of band-gap itself is weak, the TFET current demonstrates weak temperature dependence.

2.2 Electrical Characteristics

The electrical characteristics of N-TFET are shown in Fig. 3. The characteristics of P-TFET are similar. Key specifications of N&P-TFET are listed in the Table 3. Characteristics were simulated using TCAD simulator Synopsys Sentaurus. Non-local tunneling model is employed. Tunneling path is dynamically altered as per the bias conditions [14].

N-TFET device achieves ultra low off current (~fA), and the drain current increases exponentially with higher V_{GS} [Fig. 3a]. Exponential drain current dependence results into the exponential transconductance increase with V_{GS}. The point SS across V_{GS} is shown in Fig. 3b. The SS is steeper at low V_{GS}, and rapidly degrades at higher V_{GS}. Thus, better g_m/I_{DS} benefits of TFET over MOSFET are limited to low gate biases [Eq. 2]. Characteristic against drain bias (V_{DS}) is shown in Fig. 3c. A near perfect saturation is observed. The correlation of the saturation voltage ($V_{DS,SAT}$) against the gate voltage is shown in Fig. 3d. At low V_{GS}, the $V_{DS,SAT}$ is more than the V_{GS}, and the saturation in the TFET is delayed than in the MOSFET [12]. Capacitive characteristics of the N-TFET are shown in Fig. 3e-f. The gate to drain capacitance (C_{GD}) is dominant over the gate to source capacitance (C_{GS}). With increasing V_{DS}, C_{GD} enhances sharply when V_{DS} is large enough to set pinch-off in the channel. Due to the dominant C_{GD} in this region TFET will show enhanced Miller effect. A second rise of C_{GD} occurs when the source region overlapped with the gate [Fig. 2a] is inverted.

3. CIRCUIT SIMULATION

To explore TFET's capability for circuits, compact model is desired. Various compact modeling works for TFET's electrical characteristics are underway [3,15,1]. Using Kane's model for BTBT, a simplified description of the TFET's drain current was given [3]. However, this model doesn't account for the drain voltage dependence of the current. A more accurate analytical model with description of gate on-set voltage was given [15]. However, the inaccuracies of the model increase at higher gate voltage. Pseudo-two-dimensional model for double-gated Silicon channel TFET was given [1].

In this work we follow an alternative TCAD based approach. Electrical characteristics of TFET were extracted using two dimensional TCAD simulations with bias conditions finely varying over the operating range. Next, Verilog-A based table models were constructed by interpolating these values with quadratic spline. Based on these Verilog-A models, the spice simulations were performed to study analog characteristics of TFET based designs. Accurate correlation of I_{DS}-V_{GS} [Fig. 4a] and C_{GS}-V_{GS} [Fig. 4b] characteristics for N-TFET between Spice (by the above methodology) and TCAD simulations is shown in Fig. 4b. However, unlike a compact model, the above approach is not a

776

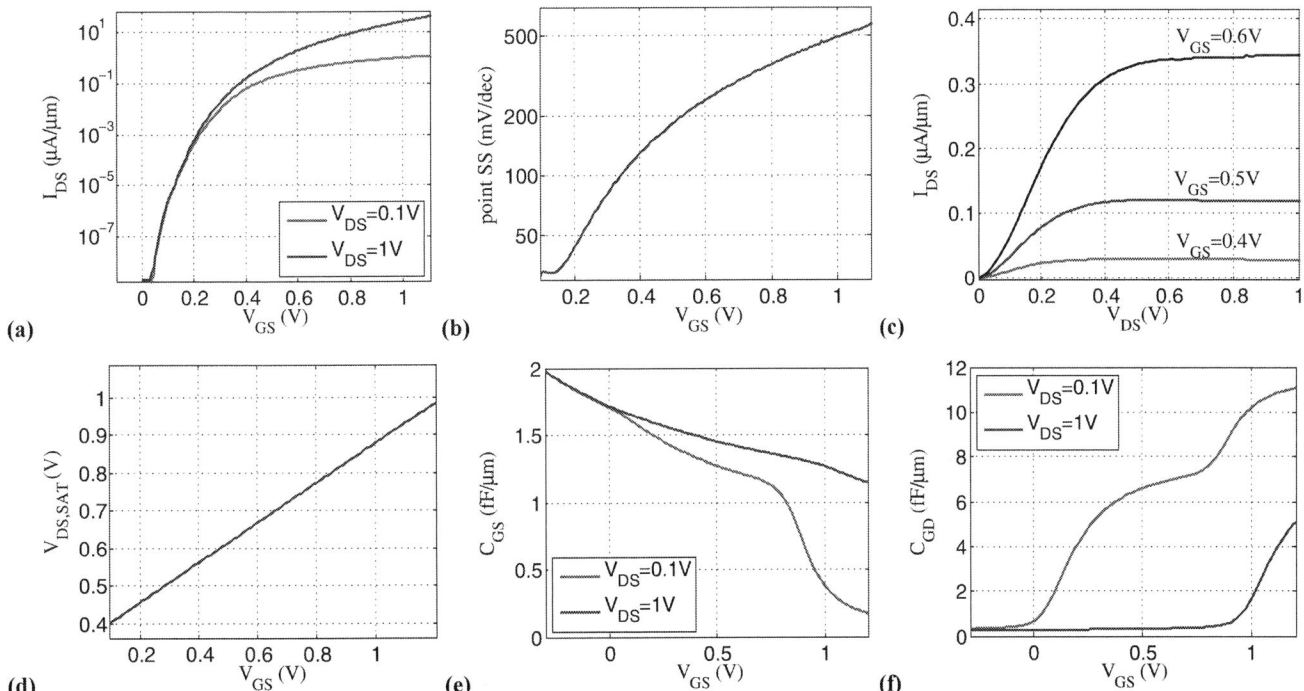

Figure 3: N-TFET electrical characteristics (a) I_{DS}-V_{GS} (b) SS-V_{GS} (c) I_{DS}-V_{DS}, (d) $V_{DS,SAT}$-V_{DS}, (e) C_{GS}-V_{GS}, and (f) C_{GD}-V_{GS}.

scalable solution, and its utility is limited to designs consisting of fewer channel lengths/widths.

Table-3: TFET Device Electrical Specifications [V_{DD}=1V]

Specification	N	P
$I_{DS,ON}$ (μA/ μm)	38	41
$I_{DS,OFF}$ (fA/ μm)	2	5.1
$I_{DS,LIN}$ (μA/ μm)	0.97	3.2

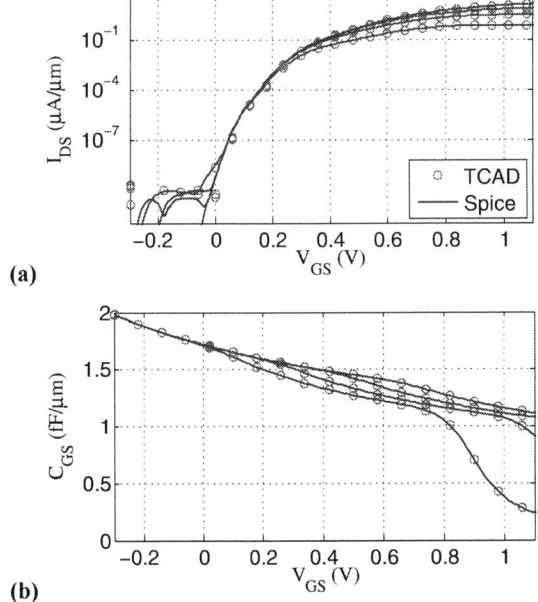

Figure 4: Correlation between TCAD and Spice characteristics (a) I_{DS}-V_{GS} and (b) C_{GS}-V_{GS}.

4. TUNNEL-FET BASED OTA

OTA is an integral component of several low power analog designs. In many designs, OTA is the only active component, such as in Gm-C filter synthesis [8], Neural amplifier [7], cellular neural network (CNN) [18], and non-linear functional synthesis [13], and claims to active biasing/operating power. Due to its widespread utilization, we use OTA to demonstrate capability and challenges of TFET based design.

Circuit schematic of TFET-OTA is demonstrated in Fig 5a. Transistors M_1 and M_2 generate differential transconductance current ($i_{d,1}$ and $i_{d,2}$) proportional to the differential input gate voltage (V_{AC}), and the current $i_{d,1}$ and $i_{d,2}$ is mirrored by the current mirrors M_3-M_5 and M_4-M_6. This mirrored current is added at the branch M_6-M_8 and supplied to the load at OUT. Due to its improved output resistance [Fig. 3c], TFET based design obviates the cascaded transistors at the output stage [7], and the associated biasing power. Various characteristics of the TFET-OTA are discussed below for varying power conditions, and compared to equivalent MOSFET-OTA designs. Power of the OTA design can be varied by changing V_{DD} or its total bias current (I_{OTA}). Since, in a complete system other analog components such as floating switches, biasing elements can limit voltage scaling [16], we restrain our exercise to varying I_{OTA} (and fixed V_{DD}=1V). To compensate its degraded output resistance, the channel length for MOSFET based designs was chosen to be $2\times L_{min}$, where L_{min} is the minimum channel length of the technology.

4.1 Transconductance (G_m)

The net trans-conductance gain (G_m) of OTA is given by

$$G_m = g_{m,1} \times M_R \tag{3}$$

Here, $g_{m,1}$ is the trans-conductance of the transistors M_1(=M_2), M_R is the current mirror ratio for the mirrors M_3-M_4 (=M_5-M_6). Hence, with respect to I_{OTA}

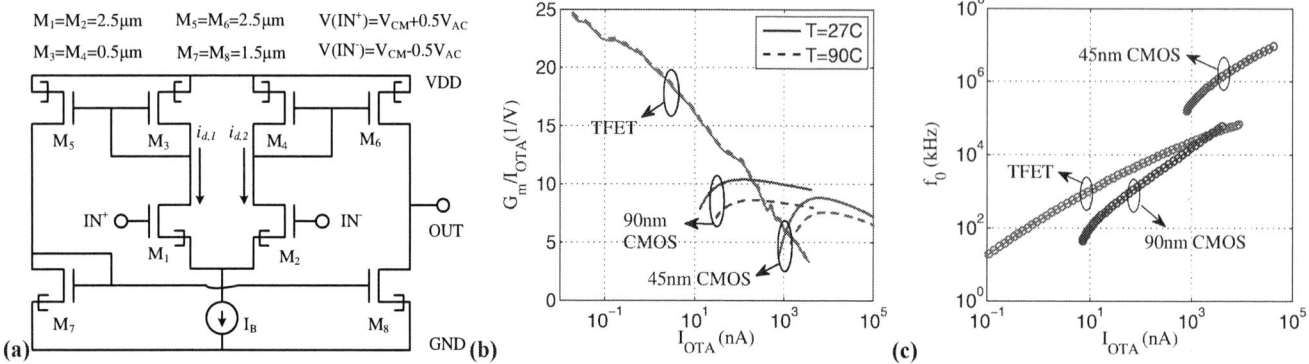

Figure 5: OTA: **(a)** schematic **(b)** G_m/I_{OTA} across OTA biasing current (I_{OTA}) **(c)** unity gain frequency (f_0) across I_{OTA}.

$$\frac{G_m}{I_{OTA}} = \frac{G_m}{I_B(M_R+1)} = \frac{g_{m,1}}{I_B} \times \frac{M_R}{M_R+1} \qquad (4)$$

For the TFET OTA design, G_m/I_{OTA} is analyzed by sweeping the I_B. With reduced I_B the DC gate-source bias for M_1 and M_2 moves to lower values, and it enhances its point SS and $g_{m,1}/I_{DS}$ [Fig. 3b, Eq. 2]. Hence, it is observed that the G_m/I_{OTA} for the TFET based designs increases at lower I_B (or I_{OTA}) as in Fig. 5b.

On the other hand, for MOSFET in the sub-threshold region

$$\frac{g_{m,1}}{I_{DS}} = \frac{1}{nV_T} = \frac{1}{n}\frac{q}{kT} \qquad (5)$$

Here, k is the Boltzmann constant, T is the temperature, and n is the ideality parameter related to its SS. Hence, with scaling I_B, for MOSFET based designs G_m/I_{OTA} stays relatively invariant as observed in Fig. 5b. At higher temperature, this ratio is degraded [Eq. 5]. At extremely low bias currents, due to drain induced barrier lowering (DIBL), the current of the branch M_5-M_7 and M_6-M_8 doesn't follow the mirror ratio enforced by the design. MOSFET based OTA designs succumb to excessive leakage, and G_m/I_{OTA} ratio drops. Hence, operating region with the TFET based designs is extended to the ultra-low energy regime and beyond achievable by the MOSFET. Nevertheless, at high power (and performance) conditions, when TFET SS degrades worse than the MOSFET [point SS in Fig. 3b], 90nm CMOS device outperforms TFET for power > 200nW, and 45nm for power > 1µW.

4.2 Unity gain frequency (f_0)

OTA is single pole dominant design, and gain bandwidth product is determined by the unity gain frequency (f_0). The f_0 of the TFET-OTA was analyzed for a capacitive load of 100fF and for varying I_{OTA} and compared against MOSFET-OTA. Higher G_m of TFET-OTA to MOSFET-OTA at lower power facilitates TFET-OTA to achieve better f_0. However, TFET based design suffers from

TFET's higher overlap capacitance (double gated structure with the gate overlapping the source, Fig. 2a) as well. At higher power scenarios, when TFET-OTA's G_m reduces with respect to MOSFET-OTA, MOSFET based design overtakes. At lower bias current, the f_0 of TFET-OTA reduces. Nevertheless, near kHz f_0 at sub-nW power expense in TFET based designs will be useful for several applications [5,16,19,20].

4.3 Linearity

Although, the subthreshold analog designs enjoy benefits of lower power and greater efficiency, a fundamental deterrent is reduced linearity. For the MOSFET based designs, in the subthreshold, the differential current $i_{d,1}$ ($=-i_{d,2}$) is given as

$$i_{d,1}(=-i_{d,2}) = \frac{I_B}{2}\tanh\left(\frac{V_{AC}}{2nV_T}\right) = \frac{I_B}{2}\left(\frac{V_{AC}}{2nV_T} - \frac{1}{3}\left(\frac{V_{AC}}{2nV_T}\right)^3 + \cdots\right) \quad [6]$$

Hence, transconductance generated by the transistors M_1 (M_2) is expressed as

$$g_{m,1}\left(=g_{m,2}\right) = \frac{I_B}{2}\frac{\partial}{\partial(V_{AC})/2}\tanh\left(\frac{V_{AC}}{2nV_T}\right) = I_B\left(\frac{1}{2nV_T} - \frac{3V_{AC}^2}{(2nV_T)^3} + \cdots\right) \quad [7]$$

Here, the linearity range is limited to the ac signal magnitude V_{AC} satisfying

$$\frac{3V_{AC}^2}{(2nV_T)^3} \ll \frac{1}{2nV_T} \qquad [8]$$

Depending on the n and V_T, linearity range is a few mV. For processes with higher n (and hence higher SS), the linearity range is higher. However, the fundamental challenge is that the linearity range is process limited. For TFET based designs, when operating region is modulated by the bias current, it exposes a different point SS for transistors M_1 (M_2). And, it provides unique

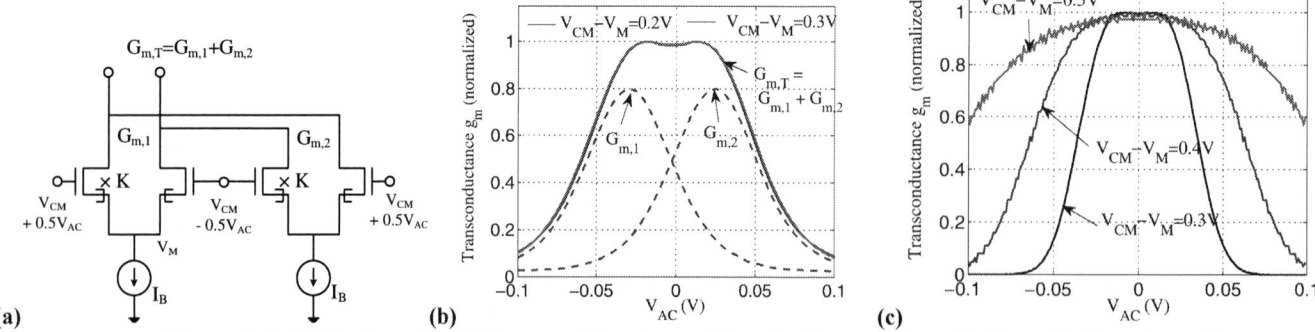

Figure 6: (a) OTA schematic with improved linearity, and linearity demonstration: **(b)** 90nm CMOS and **(c)** TFET.

778

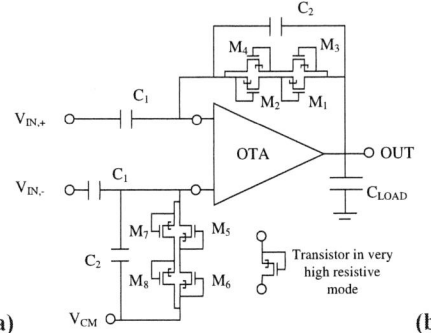
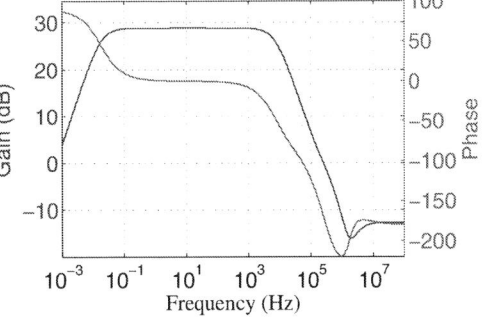

Table-4: Neural-amp design Specifications

Specification	Value
VDD	1V
Power	3.6nW
Gain	27.7dB
CMRR	64dB
PSRR	55dB
Low-f cutoff	36mHz
High-f cutoff	3.2kHz
Input ref. noise	3.1µVrms (Γ=1)

Figure 7: TFET based neural amplifier (a) schematic (b) AC response.

opportunity to modulate the linearity range by modulating bias condition across M_1 (M_2).

One of the popular low-power techniques to enhance the linearity of OTA is to use a cross-coupled structure as shown in Fig. 6a. Transconductance of each pair (i.e. $G_{m,1}$ and $G_{m,2}$) is shown in Fig. 8b. The net transconductance (i.e. $G_{m,T}$) achieves a higher linear range, where $G_{m,1}$ ($G_{m,2}$) compensates for $G_{m,2}$ ($G_{m,1}$) roll-off. Linearity range of operation is extended to ~40mV. However, a greater improvement is limited by intrinsic non-linearity of MOSFET in subthreshold region. Similar structure was experimented with TFET. By changing I_B several gate-source bias conditions (i.e., $V_{CM}-V_M$) were enforced. At higher $V_{CM}-V_M$ linearity range is significantly increased [Fig. 6c].

5. TUNNEL-FET BASED NEURAL AMPLIFIER

Implantable neural amplifiers have found widespread utilization in health-care and monitoring. Non-invasive implementation of neural amplifier poses stringent energy limitations. Under these power constraints, OTA based neural amplifier design was proposed by Harrison et al. [7], where the OTA functions in the subthreshold region. Greater details of this design can be found in the work within. And, we limit the scope to emphasize benefits of TFET based implementation of this design.

The circuit schematic of OTA based neural amplifier is demonstrated in Fig. 7a. Feedback configuration of OTA is realized by the capacitance C_1 and C_2, and the Mid-band gain is set by C_1/C_2. Gate and source of p-transistors M_{1-8} are connected together to realize highly resistive pseudo-resistance. The low frequency cut-off ($f_{c,low}$) is designed to filter off Flicker noise, and it is set by $1/(R_M C_2)$. Here, R_M is the incremental resistance generated by the pseudo-resistance M_{1-4}. Design requirements for $f_{c,low}$ to be in ~mHz requires $R_M C_2$ time constant to be very large. However, limited chip area for implantable applications can impose restriction on the physical area of C_2, and hence use of transistors in high resistive configuration was proposed [7].

We realize the similar design with TFET. The mid-band gain was designed to be 50, with C_1=500fF, and C_2=10fF. Various other characteristics of this design are listed in Table 4. TFET based OTA design due to its ultra low energy operability pushes the power requirement to ~3nW. Due to its lower off-current TFET based pseudo-resistance are more resistive, and even with C_2=10fF high pass cut-off frequency 36mHz is attained [Fig. 7b]. However, uni-directionality of TFET, requires two anti-parallel paths as shown. Lower size requirement on C_2 also relaxes power

requirement for the OTA which is now required to drive smaller load (C_1 and C_2).

6. NOISE CHARACTERISTICS

For the intended low frequency designs with TFET, the Thermal noise, Flicker noise, and Shot noise characteristics are important. Flicker noise characteristics of TFET were studied experimentally [4,17]. Similar to MOSFET, in V-TFET Flicker noise falls off exponentially with 1/f trend [4]. However, for horizontal TFET, $1/f^2$ roll off in frequency spectrum was observed [17]. The effect of Flicker noise can be minimized by incorporating a high pass filter to filter off excessive noise at lower frequencies [as in Neural amplifier design]. Also, large gate area can be chosen to minimize the impact of Flicker noise [7].

Thermal noise and Shot noise modeling and experiments of TFET have not received much attention. In the absence of relevant literature, we utilize tunnel diode noise studies to understand Thermal noise and Shot noise in TFET. Tunnel diodes have similar structure to TFET [except the gate terminal], where current conduction through degenerately doped junction occurs through BTBT. Shot noise in these structures occurs at the tunneling junction, where due to discreteness of carriers the number of electrons (holes) tunneling across the barrier fluctuates. Thermal noise occurs at the channel and drain region, where the carriers cross over the barrier with their thermal energy, and create fluctuation in the current. Equivalent noise models for the Tunnel diode was proposed [2] as seen in Fig. 8a. Here, i_{Th}^2 is the noise current density due to the Thermal noise, and i_S^2 is the noise current density due to the Shot noise. r_{Th} is the resistance associated with the Thermal noise, and r_S is the tunnel resistance associated with the Shot noise. The equivalent voltage noise at the drain terminal with load r_L can be given by

$$v_{D,noise}^2 = \left(i_S^2 \left(\frac{r_S r_L}{r_S + r_{Th} + r_L} \right)^2 + i_{Th}^2 \left(\frac{r_{Th} r_L}{r_S + r_{Th} + r_L} \right)^2 \right) \Delta f \quad (9)$$

Due to significant barrier height and width $r_S \gg r_{Th}$. Hence,

$$v_{D,noise}^2 \approx i_S^2 r_L^2 \quad (10)$$

Thus, the effect of Shot noise in these structures becomes more prominent. By similarity of current conduction mechanism in TFET to Tunnel-diode, we argue that equivalently the TFET Shot noise can be modeled as [2]

$$i_S^2 = 2q I_{DC} \Gamma(V_{GS}, V_{DS}) \quad (11)$$

Here, I_{DC} is the bias current through device. $\Gamma(V_{GS}, V_{DS})$ is termed as Fano factor [11], and it depends on the biasing conditions. Through modeling and experiments in Tunnel-diode, it was shown that Γ significantly increases at low bias current [11]. In

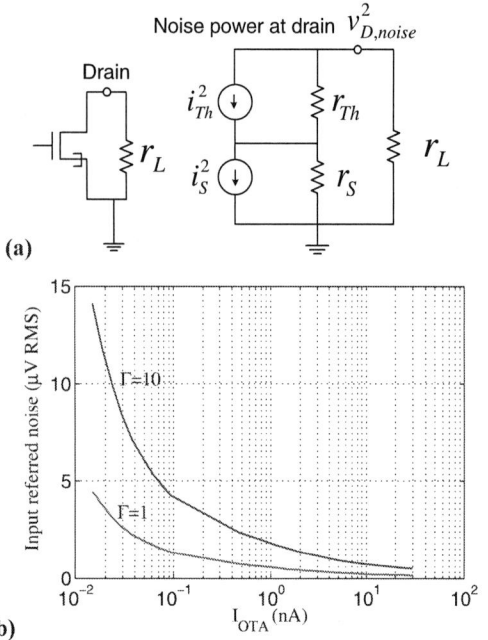

(a)

(b)

Figure 8: (a) TFET noise model based on Tunnel-diode and **(b)** input refered noise characteristics of TFET-OTA.

the absence of experimental/modeled understanding of Γ for TFET, we analyze the impact of Shot noise on OTA design analytically. Utilizing Shot noise model [Eq. 11], the input referred noise of the OTA [as in Fig. 7a] can be obtained as

$$v_{in,OTA}^2 = \frac{2qI_B}{g_{m,1}^2}\left[(\Gamma_1 + \Gamma_3) + \frac{(\Gamma_5 + \Gamma_7)}{M_R}\right] \quad (12)$$

Here, Γ_{1-7} represent Fano factor for the transistors M_{1-7}. Under, the simplistic assumptions (Γ_{1-7}=1 & 10), the input referred noise voltage $\overline{v_{in,OTA}^2}$ is shown in Fig. 8b. With reducing bias current I_B, the input referred noise increases. Thus, Shot noise imposes a limit on the minimum operating power when input signal amplitude is ~μVs. The impact of Shot noise can be reduced by designing high transconductance $g_{m,1}$ and mirror ratio M_R.

7. CONCLUSIONS

The bandwidth-power trade-off in ubiquitous sensors with an ultra-low energy demand can be limited by excessive leakage and limited SS in MOSFET. Steep-subthreshold devices with ultra low off-current and higher g_m/I_{ds} are promising for such applications. An early design exploration of TFET based OTA shows sub-nW operation is possible at kHz of bandwidth. Additional benefits of subthreshold-TFET based design are reduced temperature sensitivity and higher output resistance. TFET based pre-amplifier design with ~nW power raises intriguing opportunities for ultra-low-power analog circuits and sensors for low-throughput applications. The noise study shows that the Shot noise can become an obstacle to greater power reduction in TFET based designs.

8. ACKNOWLEDGMENTS

This work is based on materials supported in part by National Science Foundation (CNS-1054429 and ECCS-1002090).

9. REFERENCES

[1] Bardon, M., Neves, H., Puers, R., and Van, C. 2010. Pseudo-Two-Dimensional Model for Double-Gate Tunnel FETs Considering the Junction Depletion Regions. In *IEEE TED*.

[2] Barry Turner. 1962. Noise in the Tunnel Diode. *MS Thesis*.

[3] Bhuwalka, K., Schule, J., and Eisele, I. 2005. A Simulation Approach to Optimize the Electrical Parameters of a Vertical Tunnel FET. In *IEEE TED*.

[4] Bijesh, R., Mohata, D., Liu, H., and Dutta, S. 2012. Flicker Noise Characterization and Analytical Modeling of Homo and Hetero-Junction III-V Tunnel FETs. In *Device Research Conference*.

[5] Colibrys corp. "http://www.colibrys.com/e/page/183/".

[6] Gandhi, R., Chen, Z., Singh, N., and Banerjee, K. 2011. CMOS-Compatible Verticle-Silicon-Nanowire Gate-All-around p-Type Tunneling FETs with $\leq$ 50mV/decade Subthreshold Swing. In *IEEE EDL*.

[7] Harrison, R., and Charles, T.. 2003. A Low-Power Low-Noise CMOS Amplifier for Neural Recording Applications. In *IEEE JSSC*.

[8] Hori, S., Maeda, T., Matsuno, N. and Hida, H. 2004. Low-power widely tunable Gm-C filter with an adaptive DC-blocking, triode-biased MOSFET transconductor. In *IEEE JSSC*.

[9] Ionescu et al. 2011. Ultra low power: emerging devices and their benefits for Integrated Circuits. In *IEEE IEDM*.

[10] Javey, A., Guo, J., Wang, Q., Lundstrom, M., and Dai, H. 2003. Ballistic carbon nanotube field-effect transistor. In *Nature*.

[11] Kim et al. 2008. Noise Properties of Coherent Tunneling Processes in Resonant Interband Tunneling Diode. In *Journal of Korean Physical Society*.

[12] Mookerjea et al. 2009. Experimental Demonstration of 100nm Channel Length In0.53Ga0.47As-based Vertical Inter-band Tunnel Field Effect Transistors (TFETs) for Ultra Low-Power Logic and SRAM Applications. In *IEEE IEDM*.

[13] Sánchez-Sinencio, E., Ramirez-Angulo, J., Linares, B., and Rodriguez, A. 1989. Operational transconductance amplifier-based nonlinear function syntheses. In *IEEE JSSC*.

[14] Synopsys TCAD http://www.synopsys.com.

[15] Vandenberghe et al. 2008. Analytical Model for a Tunnel Field-Effect Transistor. In *IEEE TED*.

[16] Verma, N., and Chandrakasan, A. 2007. An Ultra Low Energy 12-bit Rate-resolution Scalable SAR ADC for Wireless Sensor Nodes. In *IEEE JSSC*.

[17] Wan, J., Le Royer, C., Zaslavsky, A., Crístoloveanu, S. 2010. Low-frequency noise behavior of tunneling field effect transistors. In *Applied Physics Letters*.

[18] Wang, L., De Gyvez, J., and Sánchez-Sinencio, E. 1998. Time multiplexed color image processing based on a CNN with cell-state outputs. In *IEEE TVLSI*.

[19] Wise, K. 2002. Wireless implantable Microsystems: coming breakthroughs in health care. In *VLSI Symposium*.

[20] Zierhofer et al. 1995. Electronic Design of a Cochlear Implant for Multichannel High-Rate Pulsatile Stimulation Strategies. In *IEEE Transaction on Rehabilitation Engineering*.

Energy-Optimal SRAM Supply Voltage Scheduling under Lifetime and Error Constraints

Andrea Calimera   Enrico Macii   Massimo Poncino

Politecnico di Torino
10129, Torino, Italy

{andrea.calimera, enrico.macii, massimo.poncino}@polito.it

ABSTRACT

This work addresses the energy efficiency of the memory architecture in safety-critical systems that have to guarantee a given level of service and a minimum lifetime. We specifically target SRAM structures in which decreased reliability manifests itself in terms of the aging induced by NBTI (Negative Bias Temperature Instability), and in which the level of service is represented by the bit-error rate (BER).

Our approach is based on the idea of determining an energy-optimal scheduling of supply voltages for the SRAM that satisfy the specified lifetime and BER constraints. The construction of the scheduling leverages semi-empirical models for the quantity of interest (aging, energy, memory performance, error rate) in terms of the supply voltage, and is determined through a search-based algorithm in the corresponding solution space.

The optimization framework is embedded into a design space exploration tool that allows browsing the energy/performance/reliability space for the various desired lifetime/error rate and by varying architectural parameters such operating frequency and memory size.

Categories and Subject Descriptors

B.6.3 [**Design Aids**]: Optimization

Keywords

SRAM, Aging, NBTI, Energy optimization, Reliability

1. INTRODUCTION

The growing issues of decreased yield and reliability resulting from nanoscale silicon processes require smart approaches for measuring and mitigating device degradation, which have significant implications for safety-critical applications (such as space, automotive, or medical ones) that typically have the twofold requirement of long intended lifetimes *and* a guaranteed level of service.

This work targets such safety-critical systems and proposes a strategy to meet the lifetime and error constraints with minimum energy cost. More specifically, we focus on a specific type of reliability threat, namely the *aging of CMOS devices*, and in particular the one due to NBTI (Negative Bias Temperature Instability), which has been regarded as

Permission to make digital or hard copies of all or part of this work for personal or classroom use is granted without fee provided that copies are not made or distributed for profit or commercial advantage and that copies bear this notice and the full citation on the first page. To copy otherwise, to republish, to post on servers or to redistribute to lists, requires prior specific permission and/or a fee.

DAC'13, May 29 June 07, 2013, Austin, TX, USA.

the most significant source of device aging in nano-scale technologies [1, 2]. We specifically target the aging of SRAMs in the memory sub-system, which is the component that mostly affects the overall system performance in terms of reliable and timely operations.

The well-known value dependence of NBTI (a pMOS ages - its threshold voltage increases - when a logic "0" is applied to its gate terminal, and it partially recovers aging with a logic "1") is marginally relevant in SRAMs: a SRAM cell ages regardless of the stored value because of its symmetric structure [3]. Management of *values* as done in circuits cannot therefore provide significant mitigation of SRAM aging. One opportunity to mitigate aging comes from the observation that the implementations of power managed states (i.e., via voltage scaling and power/ground gating) can be leveraged to reduce NBTI-induced aging [4, 5]. These solutions pursue the extension of lifetime of an SRAM structure by enhancing *the recovery* of idle power-managed units, and therefore refers to the *standby intervals*.

Other works have addressed the issue of mitigating aging *during active intervals* by appropriate use of voltage scaling as a sort of guard-banding mechanism [4, 6, 7]: upscaling voltage allows to have a faster implementation that masks the delay increased caused by aging. These works all sacrifice energy (due to higher V_{dd}) to increase lifetime. Furthermore, they refer to generic logic circuits (where aging directly affects performance) and are not applicable as is to SRAMs (in which it is rather the reliability of a SRAM cell that is affected by aging rather than its performance).

This work presents a similar DVS-based guardbanding applied to SRAM memories to mitigate their aging during *active* periods, with minimum energy cost. We term our strategy "voltage scheduling" because it consists of an assignment of SRAM supply voltages to time intervals so as to maintain a given error rate throughout the required system lifetime. The schedule is designed in such a way that the overall energy efficiency of the system is maximized. Since scaling V_{dd} affects memory access time and energy in opposite way, we use a specific energy/delay product metric, specifically, the *average energy per memory access*.

Construction of the optimal voltage schedule requires (1) appropriate models of SRAM energy, aging, and error rate as a function of V_{dd}, and (2) a tool that allows to explore the solution space and to determine the optimal voltage schedule. These are the two key contributions of this work.

Simulation results show that, by choosing the appropriate voltage assignment, is possible to meet the desired lifetime and error constraints while significantly improving the energy efficiency of the SRAM with respect to a conservative solution consisting of operating the memory at the nominal supply voltage.

2. BACKGROUND AND RELATED WORK
2.1 NBTI Physics and Models

The most widely adopted physical model to explain the NBTI phenomenon is the Reaction Diffusion (R-D) mechanism ([1]), which explains the temporal shift of V_{th} in terms of the breaking of $Si - H$ bonds at the at the Si/SiO$_2$ interface and the subsequent diffusion of hydrogen, which induces the formation of interface traps that contribute to the increase of threshold voltage of the device. This trap generation phase is called *stress* phase, and occurs when a pMOS is negatively biased ($V_{gs} < 0$, or a logic "0" applied to its gate terminal). When electrical stress is removed ($V_{gs} \approx 0$, or a logic "1" on the pMOS gate terminal), part of the free hydrogen atoms diffuse back and anneal the broken $Si - H$ bonds. In this *recovery* phase, the number of interface traps is reduced and the V_{th} partially recovered.

Two important properties of the R-D model allow simplifying the calculation of the actual V_{th} drift stress over time without requiring explicit timed simulation, otherwise unfeasible for the typical temporal horizons of NBTI (years).

  a) NBTI is roughly *frequency independent*: it has been shown experimentally that the final ΔV_{th} is independent of the switching frequency of the device. Hence only the duty cycle (i.e., fraction of stress/recovery time) will affect the aging [1].

  b) NBTI is mostly determined by the *cumulative* amount of stress and recovery time rather than the actual temporal profile. Therefore, a generic stress/recovery waveform can be modeled as a periodic one with a fixed frequency but same amount of stress time [3].

These properties allow treating NBTI effects in probabilistic terms as a function of the stress probability β, that is, the fraction of time the gate voltage is at the logic "0":

$$\Delta V_{th} = K \cdot (\beta \cdot t)^{1/4} \qquad (1)$$

where K lumps all the technological constants and t denotes time. The term βt can be seen as the *effective stress time*.

In an SRAM cell, the threshold voltage drift induced by NBTI does not truly affects the delay of an SRAM cell but rather it impacts its stability [8] . A conventionally accepted metric for SRAM cell stability is the static noise margin (SNM), which is traditionally represented graphically by the "butterfly curve" that plots the voltage transfer characteristic (VTC) of one of cell inverters and the inverse VTC of the other one. The SNM is visually represented by the side of the largest square that can be inscribed within the two VTC curves (Figure 1).

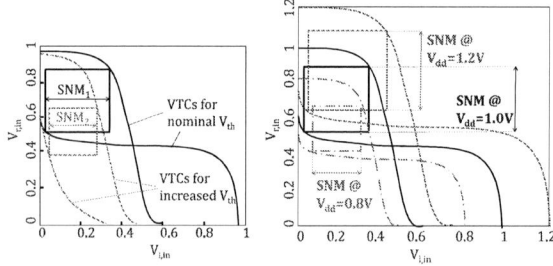

Figure 1: SNM vs. V_{th} Drift (Left) and V_{dd} (Right).

As a result of aging, the threshold voltage drift of PMOS transistors lowers the VTC curves; hence, the SNM de-

creases until it becomes so small that it does not allow safe operations anymore. (left plot in Figure 1).

Since our method is based on the scaling of V_{dd}, it is also important to understand the relation between SNM aging and supply voltage. On one hand, it is well-known that SNM decreases approximately linearly with decreasing V_{dd} [9]. The right plot of Figure 1 shows this effect for $V_{dd} = 0.8, 1.0$ and 1.2V.

On the other hand, SNM drift is weaker at smaller V_{dd}'s because of reduced stress voltage. This contrasting effects can be used to implement a "smart" guardbanding scheme. A static guardbanding in which an upscaled V_{dd} contrasts the SNM drift by starting from a larger SNM, would in fact be very inefficient energy-wise due to using a fixed higher V_{dd}. In alternative, we can devise a dynamic scheme in which V_{dd} is changed over time in such a way that the guardbanding is achieved at minimum energy cost.

2.2 Related Work

The effects of aging in SRAM memories has received special attention by the research community because of their criticality in determining the overall system performance.

One class of solution uses the fact that the best case aging for an SRAM cell occurs when both inverters age of the same quantity; this can be guaranteed by equalizing the probability of storing a "0" or a "1". In [3] the authors present hardware and software schemes to periodically invert the entire content of a memory so as to guarantee a perfectly balanced probability.

A second class of solutions aims at designing ad-hoc *NBTI-resilient cells*. An example is the "*recovery boosting*" solution of [10], in which both pMOSs in the cell are allowed to be put into the recovery mode by raising the ground voltage and bitlines to the nominal voltage.

The third class of solutions exploits the aging benefits provided by *low-energy states* ([11, 12]) and target the mitigation of aging *during standby intervals*. The work of [11] proposes power management solutions at the architectural level (based on both DVS and power gating) acting on entire memory blocks. The strategy proposed in [12] proposes an "averaging" technique called *dynamic indexing* to achieve a uniform distribution of idleness over cache lines by changing the cache indexing function over time. This causes all cache lines to age of the same quantity so that all of the idleness can be used for aging (and leakage) reduction.

The approach described in this work is complementary to these three classes of solutions, and is similar in scope to the *dynamic guardbanding* strategies used for generic circuits or cores [4, 6, 7]. The common idea behind these works is to dynamically upscale the supply voltage (to speed up the circuit) in order to achieve the desired performance over the desired lifetime while keeping the energy overhead is kept to a minimum. In practice, these methods revisit traditional DVFS using aging as a performance constraint.

Although the principle is similar, our work differs in three main aspects. First, in SRAMs it is the reliability that is affected by aging rather than performance. In particular, we have to consider that different supply voltages imply different starting SNM values, as described in Section 2.1. Therefore the aging profiles are intrinsically different from the delay degradation profile in circuits.

Second, as a consequence of the impact of aging on reliability in SRAMs, it is possible to express the "performance" constraints in terms of "quality of service" (e.g., bit-error rate) in the formulation of the problem.

Last, in SRAMs we do not need to consider possible interactions between the (low-frequency) voltage assignments required by guardbanding and the ones (high-frequency) used during normal operations: SRAMs are not usually operated with variable voltages. This point is especially critical in cores, and makes the application of voltage scheduling to SRAMs easier to accept.

3. VOLTAGE SCHEDULING

3.1 Problem Formulation

Consider a digital system that has to guarantee operations until some point in time T_{LT}, and with some bounded error level. The latter is typically expressed in terms of bit error rate (BER). Let us then assume that the memory is the critical component from the reliability perspective, and that it determines the lifetime of the system.

Although errors in a memory can occur due to many reasons (e.g, radiation, over-heating, faults), here we focus solely on errors caused by NBTI-induced aging, and therefore we assume that errors manifest themselves as a result of progressive decrease of the SNM. Since our strategy is based on voltage scaling, we can put in relation the SNM with the BER by combining their respective relations with supply voltage [13, 14, 15]. This allows to translate the error constraint (BER) into a given SNM value.

Figure 2 pictorially shows the possible scenarios. The diagrams represent the lifetime/reliability space, with time on the x-axis, and SNM values on the y-axis.

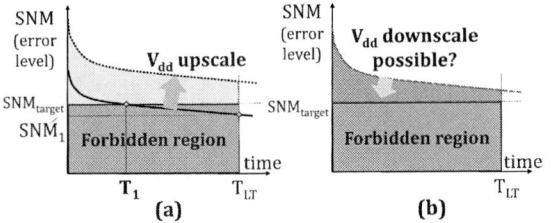

Figure 2: Voltage Assignment Scenario.

Memory operations in this space are represented by the SNM aging curves over time, as a function of V_{dd}. There is one curve for each V_{dd} value, and all the curves start from the time-zero SNM for that V_{dd} value and decrease over time according to the usual $t^{\frac{1}{4}}$ trend.

The gray rectangle labeled "forbidden region" represent the portion of the space delimited by the two constraints (desired lifetime T_{LT} and desired error level SNM_{target}). The SNM curves should never enter the "forbidden region".

Two are typical scenarios faced by a system designer that has to guarantee such a two-fold constraint. In the first one (Figure 2-(a)), operating the memory at the nominal V_{dd} value does not allow satisfying the constraints (solid curve): either the target SNM is guaranteed only up to time $T_1 < T_{LT}$, or we reach T_{LT} with some error level $SNM_1 < SNM_{target}$. In this case, without any insight on the impact of V_{dd} upscaling, the designer will likely upscale V_{dd} to some conservatively large value that allows meeting the constraints (dotted curve). Because of a larger V_{dd}, total energy will increase: the area between the aging curve and the forbidden region is a measure of the energy overhead.

In the second scenario (Figure 2-(b)), the nominal V_{dd} allows to meet the constraints. In this case, however, it could be possible that a smaller V_{dd} would still meet the constraints, thus saving some energy.

The problem we address in this work is to *determine an assignment over time of supply voltages to be applied to a memory so as to guarantee the lifetime/error constraints with optimal energy efficiency.*

3.2 Models

The key enabling technology for our strategy is the derivation of appropriate models for the quantities involved in the optimization algorithm as a function of supply voltage.

3.2.1 SNM Degradation Model

We model the SNM drift for an SRAM cell as follows:

$$SNM(t, V_{dd}) = SNM(0, V_{dd}) - \eta(V_{dd}) \cdot K \cdot t^{\frac{1}{4}} \quad (2)$$

where $SNM(0, V_{dd})$ is the SNM at time 0 for voltage V_{dd} and it is a linear function of V_{dd}, determined empirically from the analysis of our cell as $SNM(0, V_{dd}) = 0.368 \cdot V_{dd} + 0.02$. $\eta(V)$ is a correction factor that takes into account the dependence of NBTI stress on supply voltage. It is also derived empirically by extrapolation of the SNM drift curves, resulting into the following equation:

$$\eta(V_{dd}) = -0.194 \cdot V_{dd}^3 - 0.1119 \cdot V_{dd}^2 + 1.071 \cdot V_{dd} + 0.1314 \quad (3)$$

It is a monotonically increasing function, whose non-linearity emphasizes the fact that relief from aging provided is more evident for lower V_{dd} values, whereas the impact in the range 1.0–1.2 V is almost negligible.

3.2.2 Energy Model

Energy values have been obtained by characterizing various SRAM memory arrays on a 40nm technology from STM, with nominal supply voltage $V_{dd,nom} = 1.2V$. Dynamic power values are provided by the memory generator in $\mu W/MHz$, which allows considering frequency a parameter of the design exploration. The power vs. V_{dd} relation has been obtained by by re-scaling values with respect to $V_{dd,nom}$, by a factor $(\frac{V_{dd}}{V_{dd,nom}})^2$. Conversely, static power values are directly expressed in terms of leakage current, so scaling with respect to V_{dd} is immediate.

Table 1 reports the extrapolated model of total energy for the four memory sizes considered in our experiments.

Size	Power Model [μW]
4K	$19.78 \cdot V_{dd}^2 + 0.06393 \cdot V_{dd} + 0.00007143$
8K	$20.66 \cdot V_{dd}^2 + 0.08269 \cdot V_{dd} - 0.0000543$
16K	$21.73 \cdot V_{dd}^2 + 0.1276 \cdot V_{dd} - 0.001386$
32K	$24.62 \cdot V_{dd}^2 + 0.1986 \cdot V_{dd} + 0.0035$

Table 1: Power Models.

3.2.3 Cycle Time Model

Memory performance is measured in terms of its cycle time, which quantifies the time between two consecutive R/W operations. Since the technology data do not provide timing figures as function of supply voltage, we have derived such data on the SRAM cell used for characterizing SNM data and measured the read time for different supply voltages. Cycle times have then been obtained by summing read time to the delay of the other SRAM elements, a by post-calibration values against published data [14, 13]. Table 2 reports the extrapolated model for the four memory sizes.

783

Size	Cycle Time [ns]
4K	$0.9672 \cdot V_{dd}^{-6.795} + 0.7841$
8K	$1.022 \cdot V_{dd}^{-6.799} + 0.813$
16K	$1.038 \cdot V_{dd}^{-6.799} + 0.8445$
32K	$1.085 \cdot V_{dd}^{-6.798} + 0.8823$

Table 2: Cycle Time Models.

Since we need the *total number of accesses* over a given time interval, these equations are then translated into a discrete function $C()$ that maps cycle times into a number of cycles. This number is clearly frequency dependent, i.e., $C(V, f)$.

3.3 Optimization Metric

Since energy and cycle time have opposite dependencies on supply voltage, we use an optimization metric that expresses the energy/performance tradeoff, namely the **average energy per access**, computed as the ratio between total energy and total number of accesses over a time T.

When executing for a time T (in seconds) at voltage V_i, the total energy will be $E_{tot} = T \cdot P(V_i)$, where $P()$ is the model of Table 1. The total number of accesses over time T at voltage V_i will be $A_{tot} = (f \cdot T)/C(V_i, f)$, where $f \cdot T$ is the total number of cycles in T at frequency f, and $C()$ is the function described in the previous section.

The average energy per access over time T, at frequency f, and at voltage V_i will thus be $E_{tot}/A_{tot} = \frac{P(V_i) \cdot C(V_i)}{f}$.

3.4 Voltage Schedule Calculation

Let N be the number of available supply voltages, and $\mathbf{V} = (V_1, \ldots, V_N)$, with $V_1 < V_2 < \ldots V_N$ the corresponding N voltage levels. N is determined by technological constraints such as the maximum/minimum operating voltages and the resolution of the voltage generator. Each voltage identifies a corresponding $SNM(t, V)$ curve. The various curves belong to one of the three categories below, depending on their geometrical relation with the "forbidden" region (Figure 3):

a) *Curves that do not intersect the forbidden region*, i.e., intersecting the SNM_{target} value at times $t \geq T_{LT}$ We call this set $\mathbf{V}_{high}$, and m its cardinality. Operating the memory at any of these voltages guarantees that both error and lifetime constraints are met.

b) *Curves that intersect the forbidden region*, i.e., curves that intersect the SNM_{target} value at times $0 < t < T_{LT}$. We call this set $\mathbf{V}_{low}$, and n its cardinality.

c) *Curves that are entirely below the forbidden region*, i.e., that never intersect the SNM_{target} value. We call this set $\mathbf{V}_u$, and p its cardinality. These curves denotes voltage values that cannot be used at all.

In Figure 3, dash-dot black lines denote the $\mathbf{V}_u$ set, solid blue curves the $\mathbf{V}_{low}$ set, and the dashed red ones the $\mathbf{V}_{high}$ set. The cardinalities of the three sets depend on the (T_{LT}, SNM_{target}) pair, and determine the existence of a solution. In particular, if $m = 0$, the desired lifetime and error constraints are impossible to meet. On the opposite extreme, if $n = 0$ (and thus also $p = 0$) any of the available voltages can be used to meet the constraints. In between these two extremes, $n > 0$ implies that voltages in $\mathbf{V}_{low}$ can be used only for some part of the lifetime: these voltages are those that allow to calculate a non-trivial voltage schedule.

Let $\mathbf{T} = (T_1, \ldots, T_n)$ be the set of time points (**schedule points**) at which the curves in $\mathbf{V}_{low}$ intersect the value

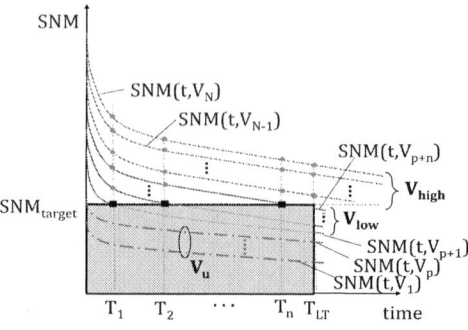

Figure 3: SNM Curves and Schedule Points.

SNM_{target} line (Figure 3). The T_i's can be easily calculated by solving the equation $SNM(V_i, T_i) = SNM_{target}$ for T_i, for each $V_i \in \mathbf{V}_{low}$.

The construction of the optimal supply voltage schedule corresponds to a path in the conceptual grid determined by the intersection of the the schedule points and the various SNM curves. At each schedule point, we should decide which SNM curve to follow (i.e., which voltage) in order to optimize the chosen cost function. Since at each schedule point we have different alternatives, the possible voltage schedules from 0 to T_{LT} can be represented by a tree: Nodes represent a grid point (a curve) and are weighted by the value of the cost function (energy per access), and edges represent the choice of a given voltage. The number of voltage option decreases at each schedule point as we move from 0 to T_{LT}; for this reason, it makes sense to build the decision tree backwards from T_{LT}. Therefore, the root of this tree represents the choice of voltage to be used from T_n to T_{LT}, the first level represents the choice of voltage T_{n-1} to T_n, and so on, until the last level in which the voltage from 0 to T_1 is selected. The height of such a tree is $n+1$, i.e., the number of schedule points n plus one, since the first voltage schedule decision occurs at T_{LT}. The degree of the tree is variable, and increases at each schedule point; at the generic schedule point T_{n-i} the degree is $m + i + 1$. Notice that the p voltages in $\mathbf{V}_u$ do not contribute in any way to the solution.

Since a path identifies a voltage scheduling solution, the number of leaves determines the size of the solution space. Such number is $\Pi_{i=0,\ldots,n}(m+i)$; a loose upper bound for this quantity in terms of N is $O(N^n)$. Given the relative small value of N (10 to 20), this space can be easily be enumerated exhaustively.

EXAMPLE 1. *Figure 4 shows a simple example with $N = 4$ voltages, ($p = 0, n = 2, m = 2$). Notice that there are $m \cdot (m+1) \cdot (m+2) = 3 \cdot 4 = 24$ leaves. The rightmost path in the tree shown with a dashed line corresponds to lowering the supply voltage at each schedule point (thick solid line in Figure 4-(c)), i.e., with minimum total energy. The leftmost path in the tree shown with a solid thick line corresponds to a schedule in which the maximum voltage V_4 is applied throughout the lifetime of the memory (dotted curve in in Figure 4-(c)), i.e., committing the maximum number of memory accesses in $[0, T_{LT}]$.*

3.5 Application of the Voltage Schedule

One important aspect concerns how the voltage changes are controlled. Since the schedule points T_i can be pre-calculated, the mechanism can be easily implemented in the

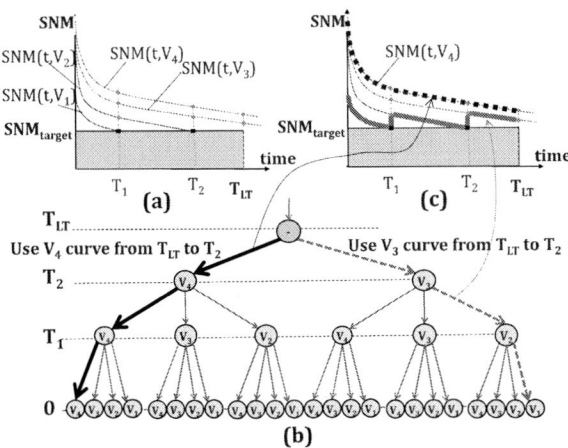

Figure 4: Example of Grid for $N = 4$ (a), Corresponding Solution Tree (b), and Two Examples Solutions (c).

operating system by maintaining a counter that keeps track of the elapsed cycles from time 0; when the first schedule point is reached, the next voltage values (stored in some table in the OS) is applied until the next schedule point, and so on until we reach the last schedule point T_n, in which the last voltage of the schedule is applied.

4. RESULTS

4.1 Exploration Framework

We have implemented the above strategy into a tool that calculates the voltage scheduling for the desired constraints (BER and lifetime) and a given set of parameters (available voltage range, memory size and frequency); the tool has then been embedded into a design exploration framework that allows to evaluate the dependency on the solution on the constraints and the parameters. We evaluated the proposed framework on a set of SRAM arrays whose timing, aging and power models have been derived from characterization on a 40nm CMOS technology by STM. The set of available voltages ranges from 0.7V up to 1.2V (the nominal voltage) with a step of 50mV. The range of memory sizes includes 4, 8, 16, and 32 KB (typical size of a memory in the first level of a memory hierarchy), and the range of frequencies spans from 100MHz to 1GHz (although there is no limitation in the frequency values). Concerning the constraints, i.e., the BER values considered are in the range [1e-7...1e-3], and lifetime values from 1 to 10 years.

4.2 Simulation Data

Tables 3 show results for four points in the design space, labeled Case A, B, C, and D, whose specific values are described in the four sub-tables of Tables 3. These points have been chosen as four samples of the design space, with the rational of using two values for each parameters and constraint (memory size, frequency, lifetime, BER). In each sub-table, the rows E/A (energy-per-access) and E (total energy) report (i) the savings with respect to a memory supplied at V_{dd}=1.2V (the nominal voltage), and (ii) the the optimal voltage schedule as a list of V_{dd} values, one for each schedule point, from 0 to T_{LT}.

We can notice how the voltage assignment allows substantial savings in both energy (from 14.53% to 32.04%) and energy-

per-access (from 14.47% to 85.83%), for the four cases of Tables 3. One notable exception is Case D, for which the tight constraints (BER= 1e-7 for 10 years) cannot be satisfied; this is reported as "No solution".

Concerning the shape of the resulting voltage schedules, we notice that they are monotonically increasing. When minimizing energy E alone, the schedule start from the smallest feasible voltage and is progressively increased by the minimum possible amount (i.e., the voltage conversion resolution) at each schedule point. Conversely, when optimizing energy-per-access E/A, the schedule is typically flat at beginning and increases only towards the end of the lifetime; this is because a too small voltage would excessively penalize the number of accesses. There might be situations however (e.g., Case A) in which the optimal schedules for E and E/A coincide. This happens when the minimum feasible voltage is also the one that allows maximum number of access. Concerning performance, the minimum energy-per-access solution improves the total number of accesses by 35.1%; in the best case the improvement reaches 236.3%, whereas in the worst case it drops to 0% (when min energy and min energy/access coincide, e.g., Case A).

Given the large design space, the best way to infer some general considerations on the properties of the optimal schedules is to explore the design space defined by the parameters and by the constraints. Figures 5, 6, and 7 show some typical plots that can be obtained with our exploration framework; the figures present energy and energy/access savings with respect to BER, lifetime, and frequency, respectively, all for a 4KB memory.

Concerning dependence on the BER, Figure 5 supports the intuition that a higher BER allows a more aggressive schedule, which, in turn, allows larger savings. Similarly, shorter lifetimes allow more aggressive schedules (Figure 6).

Figure 7 shows E and E/A savings as a function of frequency, for a lifetime of 3 years and a BER of 1e-6. One first observation is that E/A savings progressively decrease (energy is obviously independent of frequency). and tend to nullify at 900 MHz. This is due to the fact that while E stays constant, a smaller number of accesses is committed as frequency increases. The plot does not show values beyond 900MHz because for higher frequencies the access takes more than 1 cycle (the cycle time of the 4KB SRAM in this technology is about $1.06ns$), and the tool by default considers solutions with single-cycle access.

Another type of output provided by our tool is the Pareto analysis of the solution space. Figure 8 shows an example for a reference point, namely, a 4KB memory with $f = 100MHz$. The red curve represents the BER values (y-axis) that can be satisfied at different lifetimes (x-axis), and it divides the space into two regions: the red hatched area contains the solutions that are unfeasible, i.e., not achievable with the current set of supply voltages. The region above the curve is the feasibility region. For instance, Case D of Table 3 (BER=1e-7, LT=10yrs) falls in the unfeasible region. Plots like this one are very useful for system designers to immediately rule out unfeasible requirements.

5. CONCLUSIONS

We have proposed a technique to calculate an energy-optimal scheduling of the supply voltage of SRAM memories that operate under lifetime and reliability constraints. The methodology leverages semi-empirical models for the quantities of interest in terms of supply voltage, and a search-based op-

Case A: 4KB - 100MHz - 3 years		
	BER = 1e-7	BER = 1e-5
E/A	[0.90 0.95 1.00 1.05 1.10 1.15] 14.53%	[0.80 0.85 0.90 0.95 1.00 1.05] 32.04%
E	[0.90 0.95 1.00 1.05 1.10 1.15] 14.53%	[0.80 0.85 0.90 0.95 1.00 1.05] 32.04%

Case C: 32KB - 500MHz - 3 years		
	BER = 1e-7	BER = 1e-5
E/A	[1.00 1.00 1.00 1.05 1.10 1.15] 14.47%	[1.00 1.00 1.00 1.00 1.00 1.05] 29.12%
E	[0.90 0.95 1.00 1.05 1.10 1.15] 14.51%	[0.80 0.85 0.90 0.95 1.00 1.05] 31.98%

Case B: 4KB - 1000MHz - 3 years		
	BER = 1e-7	BER = 1e-5
E/A	[1.00 1.00 1.00 1.05 1.10] 82.90%	[1.00 1.00 1.00 1.00 1.00 1.05] 85.83%
E	[0.90 0.95 1.00 1.05 1.10 1.15] 14.53%	[0.80 0.85 0.90 0.95 1.00 1.05] 32.04%

Case D: 32KB - 500MHz - 10 years		
	BER = 1e-7	BER = 1e-5
E/A	[No Solution] -	[1.00 1.00 1.00 1.00 1.00 1.05 1.10 1.15] 21.11%
E	[No Solution] -	[0.80 0.85 0.90 0.95 1.00 1.05 1.10 1.15] 21.97%

Table 3: Voltage Scheduling and Energy Savings for 4 Different Testcases.

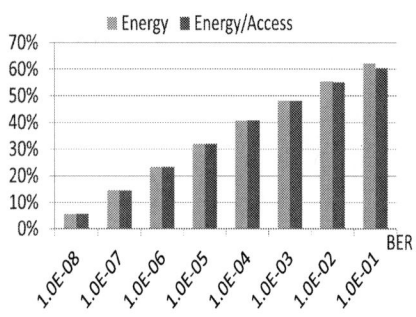

Figure 5: Energy savings vs. BER (4KB - 100MHz - LT 3yrs)

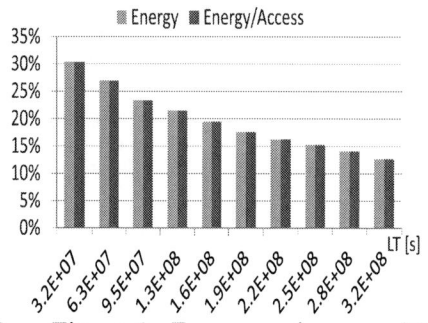

Figure 6: Energy savings vs. LT (4KB - 100MHz - BER 1e-6)

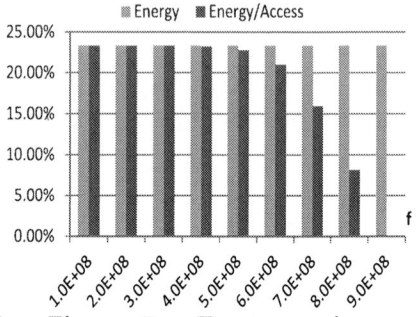

Figure 7: Energy savings vs. Freq. (4KB - LT 3yrs - BER 1e-6)

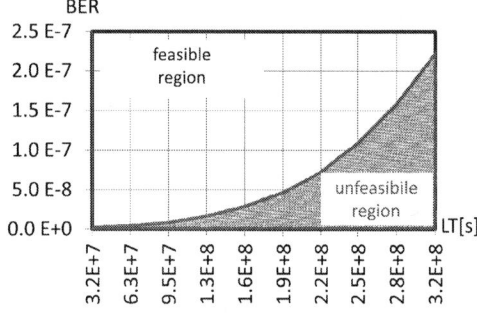

Figure 8: Pareto Curve (Size=4KB, f =100MHz).

timization algorithm based on an abstract model of the solutions space. The tool allows Pareto analysis of the design space in terms of parameters (available voltage range, memory size, and frequency) and and constraint values (BER and lifetime).

Results show how the resulting voltage schedule allows substantial energy savings with respect to the default case in which the memory is operated with the nominal supply voltage for the entire system lifetime.

6. REFERENCES

[1] M.A.Alam, "Reliability- and process-variation aware design of integrated circuits," *Microelectronics Reliability*, Vol. 48, No. 8, Aug. 2008, pp. 1114–1122.

[2] V. Huard, M. Denais, C. Parthasarathy, "NBTI degradation: From physical mechanisms to modelling," *Microelectronics Reliability*, Vol. 46, No. 1 , Jan. 2006, pp. 1–23.

[3] S.V. Kumar, K.H. Kim, S.S Sapatnekar, "Impact of NBTI on SRAM read stability and design for reliability," *ISQED'06*, Mar. 2006, pp. 213–218.

[4] L. Zhang, R. P. Dick, "Scheduled Voltage Scaling for Increasing Lifetime in the Presence of NBTI," *ASPDAC'09*, Jan. 2009, pp. 492–497.

[5] A. Calimera, E. Macii, M. Poncino, "NBTI-Aware Power Gating for Concurrent Leakage and Aging Optimization," *ISLPED'09*, pp. 127-132, Aug. 2009.

[6] M. Basoglu, M. Orshansky, M. Erez "NBTI-Aware DVFS: A New Approach to Saving Energy and Increasing Processor Lifetime," *ISLPED'10*, pp. 253-258, Aug. 2010.

[7] E. Mintarno, et al., "Self-Tuning for Maximized Lifetime Energy-Efficiency in the Presence of Circuit Aging," *IEEE Transactions on CAD*, Vol. 30, No. 5, May 2011, pp. 760–773.

[8] K.Kang, H. Kufluoglu, K. Roy, M.A. Alam, "Impact of Negative-Bias Temperature Instability in Nanoscale SRAM Array: Modeling and Analysis," *IEEE Transactions on CAD*, Vol. 26, No. 10, pp. 1770–1781, Oct. 2008.

[9] E. Seevinck, F.J. List, J. Lohstroh, "Static-noise margin analysis of MOS SRAM cells," *IEEE Journal of Solid-State Circuits*, Vol. 22, No. 5, pp. 748–754, May 1987.

[10] T. Siddiqua, S. Gurumurthi, "Recovery Boosting: A Technique to Enhance NBTI Recovery in SRAM Arrays," *ISVLSI'10*, Jul. 2010, pp. 616–629.

[11] A. Ricketts, et al., "Investigating the Impact of NBTI on Different Power Saving Cache Strategies," *DATE'10*, pp. 592–597, Mar. 2010.

[12] A. Calimera, M. Loghi, E. Macii, M. Poncino, " Dynamic indexing: Concurrent leakage and aging optimization for caches," *ISLPED'10*, pp.343–348, Aug. 2010.

[13] M. Qazi, K. Stawiasz, L. Chang, A.P. Chandrakasan, "6T SRAM Macro Operating Downto 0.57V with an AC-Coupled Sense Amplifier and Embedded Data-Retention Voltage Sensor in 45nm SOI CMOS," *IEEE Journal of Solid-State Circuits*, Vol.46, No.1, Jan. 2011.

[14] G. Chen, D. Sylvester, D. Blaauw, T. Mudge, "Yield-Driven Near-Threshold SRAM Design," *IEEE Transactions on VLSI*, Vol. 18, No. 1, Nov. 2010, pp. 1590–1598.

[15] Y. Kagiyama, et al. "Bit Error Rate Estimation in SRAM Considering Temperature Fluctuation," *ISQED'12*, Mar. 2012, pp. 516–519.

[16] M. R. Guthaus et al., "MiBench: A free, commercially representative embedded benchmark suite," *IEEE 4th Annual Workshop on Workload Characterization*, pp. 3–14, Dec. 2001.

Relax-and-Retime: A Methodology for Energy-Efficient Recovery Based Design[*]

Shankar Ganesh Ramasubramanian, Swagath Venkataramani, Adithya Parandhaman and Anand Raghunathan
School of Electrical and Computer Engineering, Purdue University
{sramasub,venkata0,aparandh,raghunathan}@purdue.edu

ABSTRACT

Recovery based design (RBD) is a promising approach for the design of energy-efficient circuits under variations. RBD instruments circuits with mechanisms to identify and correct timing violations, thereby allowing reduced guard bands or design margins. In addition, RBD enables aggressive voltage overscaling to a point where timing errors occur even under nominal conditions. A major barrier to the widespread adoption of RBD is that traditional design practices and synthesis tools result in circuits with so-called "path walls", leading to an explosion in the number of timing errors beyond a certain critical operating voltage. To alleviate this effect, previous techniques focused on combinational circuit optimizations such as sizing, use of dual V_{th} cells, re-structuring, *etc.* to favorably re-shape the path delay distribution. However, these techniques are limited by the inherent sequential structure of the circuit, which defines the boundaries of the combinational logic.

In this work, we explore a completely different approach to synthesize circuits for RBD. We propose the use of retiming, a well-known and powerful sequential optimization technique to redefine the boundaries of combinational logic, thereby creating new opportunities for RBD that cannot be explored by previous techniques. We make the key observation that, in retiming circuits with RBD (unlike classical retiming), it is acceptable for a few paths in the circuit to exceed the clock period. Using this insight, we propose a synthesis methodology, *Relax-and-Retime*, wherein the original circuit is relaxed by ignoring timing constraints on selected paths that are bottlenecks to retiming. When classical minimum period retiming is employed on this relaxed circuit, the path wall is shifted to a lower delay, thus allowing additional voltage overscaling. The *Relax-and-Retime* methodology judiciously selects bottleneck paths by trading off recovery overheads caused by timing errors due to these paths with the opportunities for retiming. We utilize the proposed methodology to synthesize a wide range of benchmarks including arithmetic circuits, ISCAS89 benchmarks and modules from the UltraSPARC T1 processor. Our results demonstrate 9-25% (average of 15.3%) improvement in overall energy compared to a well-optimized baseline with RBD.

Categories and Subject Descriptors

B.7.1 [**INTEGRATED CIRCUITS**]: VLSI (Very large scale integration)

General Terms

Algorithms, Design, Synthesis

Keywords

Recovery Based Design, Low Power Design, Retiming

---

[*]This work was supported in part by the National Science Foundation under grant no. 1018621 and 0916117

Permission to make digital or hard copies of all or part of this work for personal or classroom use is granted without fee provided that copies are not made or distributed for profit or commercial advantage and that copies bear this notice and the full citation on the first page. To copy otherwise, to republish, to post on servers or to redistribute to lists, requires prior specific permission and/or a fee.
DAC '13, May 29 - June 07 2013, Austin, TX, USA.

1. INTRODUCTION

Recovery based design (RBD) is a promising approach to designing energy-efficient circuits under variations in nanoscale technologies [1,2]. Unlike traditional designs, whose operating points (supply voltage/frequency) are determined from worst-case corner analysis, RBD allows more aggressive operating points by instrumenting circuits with the capability to detect and recover from the ensuing timing errors. Thus, RBD improves energy-efficiency by eliminating guard bands that are traditionally used to deal with variations and enabling circuits to operate at lower voltages. Additionally, RBD facilitates further energy savings by aggressively overscaling the supply voltage beyond the "zero margin" point *i.e.*, until timing errors occur even under nominal conditions. Net savings result when the energy reduction due to lower voltage operation outweighs the overheads of error recovery. Although RBD has shown great promise [1,2], several challenges need to be addressed to make it feasible in practice.

The major challenge to RBD is that conventional design techniques and synthesis tools produce circuits that are inherently not amenable to voltage overscaling. Designers create circuits in which different logic paths are inherently balanced (e.g., in pipelines or high performance arithmetic circuits). Furthermore, synthesis tools speed up longer paths while slowing down shorter paths through circuit re-structuring, sizing *etc.*, in order to reduce area/power while meeting performance constraints. This leads to the so-called "path wall", which is especially pronounced in large designs with huge numbers of paths. Such circuits can still benefit from elimination of guard bands, but cannot be overscaled beyond a "critical" operating point [3] due to an explosion in errors that leads to very high overheads for error recovery.

Due to the above challenge, circuits need to be designed to be amenable for RBD, rather than just applying RBD as an afterthought. Towards this end, several synthesis techniques [4-8] have been proposed that reshape the path delay distribution of a given combinational logic circuit to mitigate the path wall. They employ circuit optimizations such as cell sizing [6], using cells of different threshold voltages V_{th} [4], adaptive body biasing [5], re-ordering of inputs to gates [7], and restructuring the circuit to reduce delay for frequently occurring inputs [8] to achieve this objective.

All the above techniques are limited by the inherent structure of the circuit, as defined by the boundaries of the combinational logic. In this work, we explore a completely different approach to synthesizing circuits for RBD - rather than being limited by the given sequential circuit structure, we redefine the boundaries of combinational logic to create new opportunities for RBD, which cannot be explored by previous techniques. To achieve this, we propose to use retiming [9], a well-known and powerful sequential optimization technique that repositions flip-flops or latches in a circuit. Retiming has been used for clock period minimization [9], reducing the number of registers [9], reducing power [10], and lowering the cost of DFT [11]. We propose the use of retiming to improve the ability to operate circuits with RBD at lower voltages, without incurring significant overheads due to error recovery. We present a systematic methodology for the synthesis of circuits that are suitable for RBD.

Other issues with RBD that have been addressed include hardware overheads associated with error detection and recovery (typically, these consist of enhanced FFs, additional control logic for error detection and correction, and routing) [12], the overheads of buffer insertion to eliminate short paths (which cause errors to be incorrectly detected) [12], and the extension of RBD to handle within-die variations [13, 14].

In summary, the contributions of this work are as follows:

- We propose to use retiming, a powerful sequential optimization technique, to favorably alter the timing behavior of circuits under voltage overscaling, thus facilitating energy-efficient recovery based designs.
- Our methodology, *Relax-and-Retime*, maps the problem of retiming for RBD to a classical minimum period retiming problem, thereby allowing existing synthesis tools to be reused for this purpose.
- We implement and evaluate an automatic synthesis framework that embodies the proposed approach. Our experiments demonstrate significant savings in energy for a wide range of arithmetic blocks, circuits from the IS-CAS 89 benchmark suite, and modules from the Ultra-SPARC T1 processor.

The rest of the paper is organized as follows. Section 2 provides an overview of prior efforts that investigate recovery based design. Section 3 provides the necessary background on retiming. Section 4 describes the proposed approach and details the *Relax-and-Retime* methodology and algorithms used therein. Section 5 elaborates the experimental methodology used in our evaluation, and the results are presented in Section 6. Section 7 provides a brief summary and concludes the paper.

2. RELATED WORK

Previous efforts on recovery based designs have spanned various levels of design abstraction from circuits to systems. In this section, we provide a detailed account of techniques at the logic level - the focus of this work - and mention representative examples at other levels.

At the circuit level, in-situ error detection and recovery mechanisms such as Razor [1], EDS [2] and Razor II [15] have enabled the elimination of guard bands in designs by allowing voltage selection in response to timing errors. These mechanisms typically employ enhanced flip-flops that sample selected outputs at a delayed time in addition to the usual clock period, in order to detect the presence of timing errors. They also use error recovery circuitry based on techniques such as clock gating, counter-flow pipelining [1], *etc.* to correct these timing errors.

RBD techniques incur hardware overheads due to the use of more complex flip-flops and the additional circuitry for error detection and correction. More recently, [16] proposed a confidence driven computing model that extends the error detection window over multiple cycles. In addition, they incur peripheral hardware overheads, such as routing and buffer insertion, to eliminate short paths that cause spurious errors. Techniques to mitigate these overheads were investigated in [12].

While RBD helps eliminate design margins that are used to account for variations in process, voltage and temperature (PVT), further overscaling of supply voltage to achieve substantial energy reduction requires careful design at the logic level. This is due to the well-known phenomenon of "path walls", which is especially pronounced in large, well-optimized designs. Due to the presence of large numbers of near-critical paths, even marginal overscaling leads to excessive timing errors, greatly limiting the energy benefits of RBD [3].

To alleviate this effect, several techniques propose to favorably reshape the path delay distribution by using well-known combinational optimization methods. DynaTune [4] uses low-V_{th} cells on near critical paths that are frequently exercised. BlueShift [5] instead employs on-demand selective body biasing on those paths. The use of cell sizing to favorably redistribute slack was explored in [6]. Common Case Promotion [8] identifies inputs that occur more frequently and expedites their evaluation by re-structuring the circuit. Finally, [7] affects the path delay distribution by re-ordering fan-ins of gates based on their arrival times.

One common limitation to all the above techniques is that they are forced operate within the confines of combinational logic boundaries as defined by the circuit's sequential structure (locations of flip-flops). We overcome this limitation and expand the scope for optimization through retiming, thereby creating a different combinational structure and timing behavior that is better suited to recovery based design.

At higher levels of abstraction, micro-architectural techniques for processors instrumented with recovery mechanisms are presented in [14, 17]. A methodology to extend RBD to system-on-chips(SoC) by partitioning them into recovery islands was presented in [13].

3. BACKGROUND

Retiming [9] is a well-known technique to optimize sequential circuits by re-positioning their sequential elements (flip-flops/registers) while preserving the functional behavior at the outputs. Over the years, retiming has been targeted towards various design objectives such as minimizing register count and power, improving testability, *etc.* However, the most common application of retiming is to minimize the clock period of a circuit. Most commercial synthesis tools, implement minimum clock period retiming. This section briefly describes the minimum period retiming [9] algorithm, which is used in the proposed methodology.

3.1 Minimum Period Retiming

For retiming, the circuit is represented as a directed graph with the combinational elements (gates) constituting vertices, while the wires between gates form the edges. The weight of each edge is the number of registers located on the wire. Each vertex is associated with a value corresponding to the delay of the gate. The timing feasibility of a circuit for a clock period (T) is judged as follows: (i) compute, for all pairs of vertices (V_i, V_j), the least number of registers $W(V_i, V_j)$ present between all paths from V_i to V_j and the maximum delay $D(V_i, V_j)$ over all such paths having a register count $W(V_i, V_j)$, and (ii) check to ensure the ratio of maximum delay $D(V_i, V_j)$ to minimum register count $W(V_i, V_j)$ is $\leq T$ for all vertex pairs. The retiming is then formulated as an integer linear programming (ILP) problem with constrains ensuring the legality and timing feasibility of the design. The special structure of the retiming constraints allows the ILP to be solved optimally in polynomial time [9]. The minimum clock period T_{min} possible through retiming is identified by performing a binary search between the current clock period and the smallest element of D.

We note that retiming for minimum clock period is constrained by the longest/critical path in a circuit and rejects circuit configurations where even one path causes the design to be timing infeasible. We make the key observation that under RBD, it is acceptable for a few paths to exceed the clock period. Accordingly we formulate retiming for RBD by relaxing the constraints imposed in classical retiming. The methodology and heuristics proposed to this end are explained in the next section.

4. RETIMING FOR RBD

In this section we discuss the impact of retiming on RBD and outline our approach. We then describe the *Relax-and-Retime* methodology, and the algorithms used therein.

4.1 Impact of Retiming on RBD

Relax-and-Retime is based upon using retiming to alter the timing behavior of a given sequential circuit under voltage overscaling. Figures 1, 2, and 3 illustrate the impact of retiming on RBD using a simple example. Consider the sequential circuit shown in Figure 1. The circuit is optimized for performance and operates at a clock period of $4d$. It has 3 logic cones X, Y, Z, which contain Px, Py, and Pz paths ($Px \ll Py, Pz$) and have delays of $4d, 2d$, and $4d$ respectively. The path delay distribution, shown in Figure 1, contains a large number of paths that are critical ($Px + Pz$) and hence even marginal voltage overscaling leads to excessive errors.

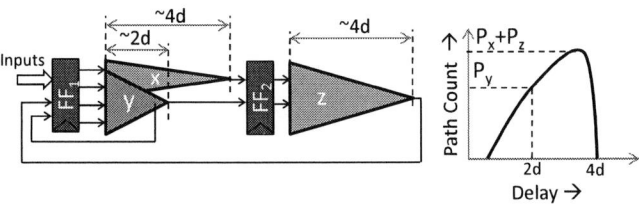

Figure 1: Performance optimized sequential circuit

Relax-and-Retime tries to create a more favorable path delay profile by re-positioning flip-flops (FFs), thus changing the combinational logic boundaries. For instance, the circuit shown in Figure 2 is a result of moving FF_2 forward by a delay d. The corresponding path delay distribution, shown in 2, is sub-optimal in terms of its longest path and incurs timing errors if operated at $4d$. However, since logic cone X has relatively few paths, the path wall is pushed in to a delay of $3d$ (due to logic cones Y and Z). This circuit is much more suitable for RBD and we can overscale the circuit without incurring huge penalties.

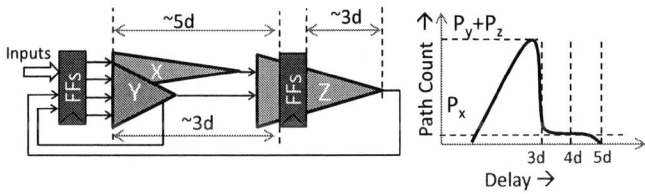

Figure 2: Retimed circuit 1: FF_2 moved forward

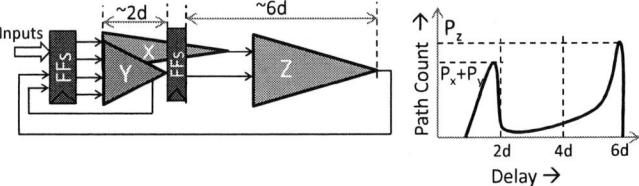

Figure 3: Retimed circuit 2: FF_2 moved backward

Although retiming has the potential to facilitate RBD a systematic methodology is required to guide it to select circuit structures with a graceful timing behavior. As shown in Figure 3, when $FF2$ is moved backwards instead of forwards, the path wall shifts to a higher delay, making the circuit worse for RBD. Also, in the above discussion, for the ease of explanation, we assumed that the likelihood of the timing errors is proportional to the number of paths that exceed the clock period. However, the probability of path activation should be accounted for as part of the methodology.

4.2 Relax-and-Retime: Approach

This section describes the approach for using retiming to improve the ability to overscale voltage under RBD. As mentioned in Section 3.1, the algorithm for minimum period retiming is constrained by the longest path(s) in the circuit and hence does not directly target the less than critical paths. We relax this constraint by eliminating paths that are bottlenecks to retiming from consideration. This presents the retiming tool with a relaxed version of the original problem, which often leads to new opportunities for retiming.

After retiming, the relaxed circuit operates at a lower clock period, implying that the path wall has been pushed to a lower delay compared to the original circuit, thus making it more amenable for RBD. This process can be iterated by identifying additional bottleneck paths to further favorably shift the location of path wall. However, the bottleneck paths that are ignored in the procedure may not meet the retimed clock period and hence could result in timing errors at overscaled voltages. Thus, the reduction in supply voltage as a result of relaxing and retiming should be weighed against the number of timing errors that the bottleneck paths introduce.

Applying this methodology to the previous example, the paths through logic cone X and Z are the longest in the circuit and hence are bottlenecks to minimum period retiming. However, since X has a small number of paths, very few timing errors are introduced when the paths in X are relaxed. Therefore, paths through logic cone X are ignored and minimum period retiming will automatically move $FF2$ forward to balance the path lengths of Y and Z, shifting the path wall favorably as shown in Figure 2. On the other hand, minimum period retiming performed after ignoring paths in Z will result in an unfavorable distribution as depicted in Figure 3.

Enumerating all paths in a design in order to identify bottleneck paths is exponential in the number of gates and thus infeasible for large designs. We address this issue by considering bottleneck gates instead of paths. We identify specific gates in the design that are bottlenecks to retiming and set all paths passing through the gates as false to relax the retiming problem. Although this leads to a coarser selection of groups of paths, our results indicate that relaxing at the granularity of bottleneck gates is effective and yields significant energy benefits.

4.3 Methodology

This subsection describes the proposed *Relax-and-Retime* methodology in detail and outlines the algorithm used to identify bottleneck gates for relaxation.

Algorithm 1 describes the overall *Relax-and-Retime* methodology, which takes a sequential circuit and a maximum error rate[1] as its inputs. It generates a functionally equivalent version of the circuit that is more energy-efficient under RBD, while meeting the performance constraint. The algorithm is iterative and in each iteration (lines 5-20), a set of top candidate gates RG_{Top} is first identified for relaxation. To obtain the best candidate among the gates in RG_{Top}, lines 8-14 evaluate the energy savings for each gate as follows: (i) all paths through the gate are marked as false paths, (ii) the relaxed circuit is retimed for minimum period while ignoring false paths, and (iii) an estimate of the minimum energy point under voltage overscaling (subject to the target error constraint) is obtained. The gate that yields the best savings is chosen and added to the list of previously relaxed gates. The iterations terminate when no additional energy benefits result while operating within the target error rate. The last retimed circuit is given as the output. Note that this algorithm can also be used to retime for the global minimum energy operating point,

---

[1]Since the clock period of the circuit is fixed, performance is proportional to the maximum error rate allowed at overscaled voltage.

Algorithm 1 Relax-and-Retime: Retiming for RBD

Input: Sequential Circuit: Ckt, Max. Error Rate: MER
Output: Retimed RBD Circuit: Ckt_{RBD}
1: **Begin**
2:    $E_{Ckt} = \min (E_{Ckt} \ni Error \leq MER)$
3:    List of Relaxed Gates: $RelGates = \{\ \}$
4:    **while** $\Delta E_{sav} \geq \delta$ **do**
5:      $RG_{Top} = identify\_relaxation\_candidates(Ckt, RelGates)$
6:      $RG_{Best} = \emptyset;\ E_{Best}=0;$
7:      **for** each RG: Gate $\in RG_{Top}$ **do**
8:        $RelGates_{temp} = RelGates \cup RG$
9:        $Ckt' = set\_false\_paths(Ckt, RelGates_{temp})$
10:       $retime\_for\_minimum\_period(Ckt')$
11:       $E_{Ckt'} = \min (E_{Ckt'} \ni Error \leq MER)$
12:       **if** $E_{Ckt'} > E_{Best}$ **then**
13:         $E_{Best} \leftarrow E_{Ckt'};\ RG_{Best} \leftarrow RG$
14:         $Ckt_{Best} \leftarrow Ckt'$
15:       **end if**
16:      **end for**
17:      $RelGates = RelGates \cup RG_{Best}$
18:      $E_{sav} = E_{Ckt} - E_{Best}$
19:      $\Delta E_{sav} = E_{sav}(current) - E_{sav}(previous)$
20:      $Ckt_{RBD} = Ckt_{Best}$
21:    **end while**
22:    **return** Ckt_{RBD}
23: **End**

Algorithm 2 Identify Relaxation Candidates

Input: Sequential Circuit: Ckt,
              Previously Relaxed Gates: $RelGates$
Output: Candidate gates for relaxation : RG_{Top}
1: **Begin**
2:    Read Ckt and Sort in topological order
3:    D: Delay of Ckt
4:    $CktCritPathFrac = \frac{\#\text{Paths in } Ckt \text{ with delay} \geq (D-\epsilon)}{\#\text{Paths in } Ckt}$
5:    FF_{list}: List of FFs in Ckt
6:    G_{list}: List of gates in Ckt
7:    $Ckt = set\_false\_paths(Ckt, RelGates)$
8:    Compute $G.Slack\ \forall\ G \in (G_{list} - RelGates)$
9:    Compute $FF.Slack_D, FF.Slack_Q\ \forall\ FF \in FF_{list}$
10:   $FF_{crit} = FF \in FF_{list} | FF.Slack_D, FF.Slack_Q \leq \epsilon$
11:   $RG_{candidates} = G \in (G_{list} - RelGates)\ | G.Slack \leq \epsilon$
12:   $FF_{crit}.TFIO$ = Trans. Fan-in/Fanout Gates $\forall FF_{crit}$
13:   $RG_{candidates} = RG_{candidates} \cap FF_{crit}.TFIO$
14:   **for** each G: Gates $\in RG_{candidates}$ **do**
15:     $G.FF_{crit}$ = # FFs $\in FF_{crit}$ impacted by relaxing G
16:     $G.SlackGen = \sum_{\forall FF_{crit}} (FF.Slack$
                                 $- FF.slack_{\text{relaxing } G})$
17:     $G.CritPathFrac = \frac{\#\text{Paths in } Ckt \text{ with delay} \geq (D-\epsilon)}{\#\text{Paths in } Ckt}$
18:     $G.Switch$ = Switching probability of gate G
19:     $G.Score = \alpha \left(\frac{G.FF_{crit} * G.SlackGen}{|FF_{crit}|} \right) +$
                        $(1 - \alpha) \left(\frac{G.CritPathFrac}{CktCritPathFrac} \frac{1}{G.Switch} \right)$
20:   **end for**
21:   RG_{Top} = Top 'K' $RG_{Candidates}$ ranked by $RG.Score$
22:   **return** RG_{Top}
23: **End**

---

disregarding performance, by setting the maximum error rate to infinity.

A key step in Algorithm 1 is the identification of potential candidates for relaxation, which is outlined in Algorithm 2. The heuristics in Algorithm 2 (also illustrated in Figure 4) evaluate the potential for retiming that a gate offers when relaxed, and the amount of timing errors that may be introduced as a result. First, given a circuit, gates that have little impact on its timing are pruned by constructing the ϵ-critical network of the circuit (line 11), which is comprised of gates whose timing slack is less than some small value ϵ (we choose ϵ to be half of the average delay of cells in the library). Next, flip-flops with both input (D) and output (Q) slacks less than ϵ are identified and labeled as critical flip-flops FF_{crit}. The flip-flops in FF_{crit} are bottlenecks to retiming because they cannot be moved forward or backward without causing a timing violation. So, the chosen candidate gate should relax the timing constraints at least on of some of the flip-flops in FF_{crit}. Hence, the network is further pruned by eliminating gates that are not in the transitive fan-in/fan-out cones of at least one of them (line 13). Note that a few FFs in the circuit may have structural restrictions that prevent their movement. Such FFs are identified and removed from FF_{crit}. Also, the timing constraints of all previously relaxed gates (and hence their exclusive fan-ins/fan-outs) are ignored in the above slack calculations.

After eliminating gates based on the above criteria, the remaining relaxation candidates $RG_{candidates}$ are ranked in lines 14-20, based on the following metrics. For each gate G in the candidate list, the number of critical flip-flops ($G.FF_{crit}$) that are impacted by relaxing the gate is calculated. This can be easily computed by comparing the slack at the flip-flops in FF_{Crit} with and without ignoring the timing constraints imposed by G. Next, the slack ($G.SlackGen$) generated by relaxing G is computed by summing the additional slacks introduced at the flip-flops in FF_{Crit}. $G.FF_{crit}$ and $G.SlackGen$ together indicate the potential for retiming that is created by relaxing G. The more elements of FF_{Crit} it impacts and more the slack it generates to facilitate their movement, better is the scope for retiming the relaxed circuit.

For example, in Figure 4, when all paths through G are set

as false, timing slack is introduced at two critical flip-flops - $FF2$ and $FF3$. The increase in slack is more pronounced in $FF3$, as the delay of logic cone E is very small compared to A and D. $FF3$ thus has a better chance to be moved forward. No slack is generated at $FF1$, as it is fed by another logic cone C that is critical. $FF4$ is a non-critical flip-flop and hence is not considered in the heuristics.

The metrics - $G.FF_{crit}$ and $G.SlackGen$ - provide the potential for retiming obtained by relaxing G and they together constitute the first part of the gate score ($G.Score$ - line 19) that is used to rank the $RG_{candidates}$.

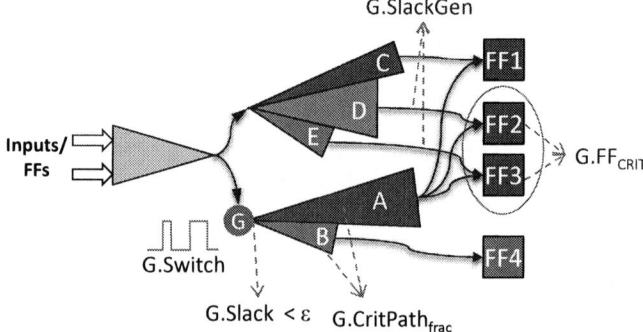

Figure 4: Heuristics used in selecting candidates gates for relaxation

Next, the potential error introduced by the relaxation is estimated. First, for the entire circuit, the critical path fraction $CktCritPathFrac$ i.e., the ratio of number of paths that are critical (paths with delay $\geq D - \epsilon$) to the total number of paths in the circuit is computed. Note that computing the number of paths, unlike path enumeration, is not exponential. Then, for each candidate gate G, its critical path fraction ($G.CritPathFrac$) is calculated as the ratio of critical paths

790

that pass through G to total paths that pass through *G*. If this fraction is larger, then relaxing this gate would: (i) foster retiming by relaxing more critical paths, but at the same time (ii) introduce a larger number of timing errors in the retimed circuit. To make an informed decision, we consider the activation probabilities of these critical paths. We use the switching activity $G.Switch$ of G as a simple proxy for the activation probabilities of the paths that pass through it. While this simple metric worked well in practice, more involved metrics that consider the activity of the transitive fan-ins/fan-outs of G, *etc.* may be used. The larger this activation probability, the more likely the paths through G are to be exercised, leading to a larger number of timing errors. For the circuit considered in Figure 4, the relaxation candidate gate (G) has a high $G.CritPathFrac$ and hence is a good candidate for relaxation only if it has a low switching probability. The second part of $G.Score$ uses $G.CritPathFrac$ and $G.Switch$ to estimate the potential errors introduced by relaxation. Note that $G.Score$ for all gates can be computed in a single topological and reverse topological traversal of the circuit. The $RG_{candidates}$ are ranked by their scores, and the top K gates (we chose K to be 5 in our experiments) are returned.

In summary, *Relax-and-Retime* efficiently identifies bottleneck gates in the circuit, and by relaxing, them enables retiming to create energy-efficient recovery based designs.

5. EXPERIMENTAL METHODOLOGY

To demonstrate the wide applicability of the proposed methodology, we performed experiments on a wide range of designs including arithmetic blocks, the larger circuits from the ISCAS89 benchmark suite, and modules from the UltraSPARC T1 processor, as shown in Table 2. The baseline circuits are aggressively optimized for performance, including retiming for minimum clock period and high-effort combinational synthesis.

Table 1: Circuits used in experiments

Name	Function	# Gate	I/O	FFs
Arithmetic Blocks				
MAC	Multiply and Accumulate	315	9/9	27
SAD	Sum of Absolute Differences	368	17/16	42
FIR	4-tap FIR Filter	968	9/13	125
EU_DIST	Euclidean Distance	1116	17/24	86
ISCAS89 Benchmarks				
s953	–	448	17/23	107
s1423	–	929	18/5	129
s9234	–	936	38/39	258
s13207	–	1116	65/152	233
s15850	–	2964	77/150	524
s35932	–	9301	36/320	1733
s38417	–	9691	36/320	1701
UltraSPARC T1 processor modules				
lsu_dctl	MEM L1 Dcache Control	778	457/508	871
lsu_excpctl	MEM L1 Exception Control	778	152/80	117
ffu_ctl	EX FFU Control	2398	223/262	687
ifu_dcl	FD Decode Control Logic	497	63/19	117
tlu_intctl	Interrupt Controller	261	50/45	44

The CAD flow used to implement the proposed methodology is shown in Figure 5. Synopsys Design Compiler Ultra [18] was used to synthesize and map the circuits to the NangateOpenCell library based on the FreePDK 45 nm technology node. To identify the bottleneck gates for relaxation, a custom tool was developed to implement Algorithm 2 (Section 4). Gate switching probability and slack were computed using Synopsys Power Compiler and Design Compiler, respectively. The retiming engine in Design Compiler was used to perform minimum period retiming on the relaxed circuits. We used functional simulation to obtain the golden output traces for randomly generated input vectors (these traces are used to compute the error rates). The V2LVS utility in CALIBRE [19] was used to obtain the SPICE netlist of the retimed circuits. Finally, transistor level simulations are performed on the traces using Nanosim [20] to

obtain the energy and error rate characteristics of the circuits at overscaled voltages.

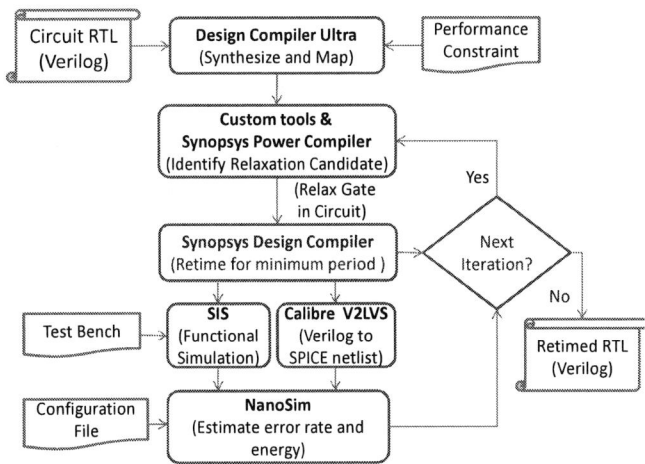

Figure 5: CAD flow used to implement the proposed Relax-and-Retime methodology

6. EXPERIMENTAL RESULTS

In this section, we present the results of various experiments that were conducted to evaluate the proposed methodology.

6.1 Energy Comparison

Table 2 compares the baseline circuit with the retimed circuit at their respective minimum energy points, subject to performance constraints as specified by the maximum error rate. For the experiments, the maximum error rate tolerable under overscaling is set to 2%. For each benchmark circuit, Columns 3-6 list the best operating voltage and the corresponding power for the baseline and retimed circuits. To account for the overheads associated with error detection and recovery circuitry, we increase the power for the original and retimed circuits by 3.1% as indicated in [1]. The percentage reductions in supply voltage and energy are listed in Columns 7 and 8. The *Relax-and-Retime* approach offers energy savings ranging from 9 to 25 % (average: 15.3%), demonstrating its ability to generate circuits that are more energy-efficient under RBD.

Table 2: Energy Comparison of Baseline and Retimed Circuits

Circuit	Delay (ns)	Baseline		Retimed		ΔV (%)	Energy (%)
		Voltage (V)	Power (mW)	Voltage (V)	Power (mW)		
MAC	0.27	1.07	1.75	0.96	1.48	10.28	15.15
SAD	0.34	1.06	2.45	0.98	2.15	7.54	12.25
FIR	0.30	1.07	5.09	0.94	4.08	12.15	19.66
EU_DIST	0.57	1.08	5.97	0.99	5.14	8.33	13.94
s953	0.33	1.08	1.91	1	1.68	7.4	12.11
s1423	0.48	1.07	3.61	0.96	2.84	10.28	21.19
s9234	0.35	1.08	5.06	1.01	4.43	6.48	12.37
s15850	0.44	1.07	12.40	0.99	10.74	7.47	13.35
s13207	0.28	1.1	5.93	1.02	4.84	7.27	18.22
s35932	0.27	1.09	52.04	0.99	44.55	9.17	14.4
s38417	0.50	1.09	45.64	0.97	38.49	11.01	15.67
lsu_dctl	0.36	1.09	20.81	1.03	18.93	5.5	9.02
lsu_excpctl	0.31	1.09	2.8	0.96	2.09	11.9	25.41
ffu_ctl	0.32	1.06	13.86	0.95	11.41	10.37	17.68
ifu_dcl	0.29	1.09	2.71	1.03	2.41	5.5	11.03
tlu_intctl	0.29	1.1	1.04	0.99	0.9	10	13.54

Next, the constraints on maximum error rate are completely relaxed to obtain the global minimum energy operating points for the baseline and retimed circuits. Figure 6 plots the energy *vs.* voltage characteristics for a few representative circuits. As seen from the graphs, when the circuits are voltage overscaled, a reduction in energy is observed until a certain point. When voltage is scaled further, the recovery overheads (due to increased timing errors) begin to dominate the savings in power from supply voltage reduction, and as a result, the energy begins to increase. This phenomenon is commonly observed in all recovery based designs.

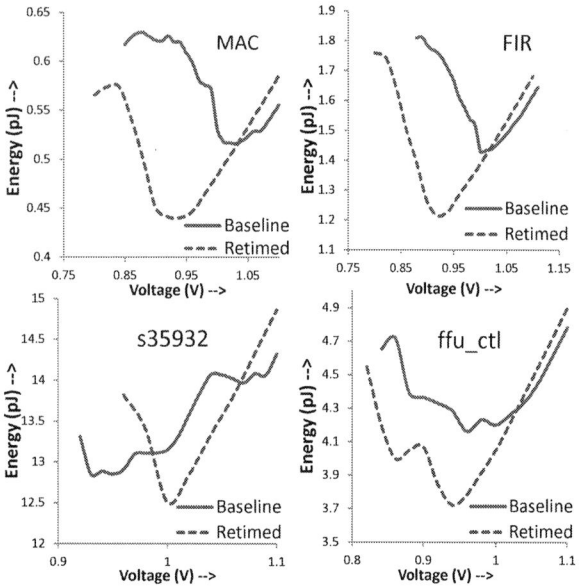

Figure 6: Energy at overscaled voltages for benchmark circuits

Comparing the behavior of the baseline and retimed circuits under voltage overscaling, it is seen that, at nominal operating voltage (V_{DD}), the baseline circuit consumes lower energy than the retimed circuit. This can be attributed to the following: (i) retiming may extend critical paths that were relaxed leading to timing errors even at nominal V_{DD}, and (ii) due to increase in FF count, retiming may cause the actual power of circuit at nominal V_{DD} to be higher. However, when the supply voltage is scaled, as shown in Figure 6, the retimed circuit quickly outperforms the baseline and has a substantially lower global minimum energy operating point. This is because, the retimed circuit has a more graceful timing behavior and hence allows additional voltage overscaling.

6.2 Error rate characteristics of retimed circuits

As shown in section 6.1, the additional overheads due to error recovery limit the energy benefits obtained from RBD. Hence, a circuit optimized for RBD should have lower error rates at lower voltages. Equivalently, the error rate *vs.* voltage curves should remain flat as long as possible before rising sharply. Figure 7 shows the error rate *vs.* voltage curves for 6 representative circuits. We see that the retimed circuit has better error rate characteristics across all benchmarks. This results from a judicious relaxation of bottleneck gates, which ensures that paths that may contribute to timing errors are infrequently exercised, thus resulting in negligible error rate. Ignoring the bottleneck paths in turn leads to increased voltage scaling potential, as the remaining paths may be retimed for a lower delay. This significantly reduces the voltage at which errors begin to rise sharply compared to the baseline circuit, enabling the retimed circuits to be operated at lower voltages without incurring significant penalties due to high error rate.

7. CONCLUSION

Recovery based design promises significant energy-efficiency by eliminating guard bands and allowing designs to operate at overscaled supply voltages. However, circuits generated using traditional design approaches are inherently not amenable to RBD, thus limiting its energy-efficiency. We propose to

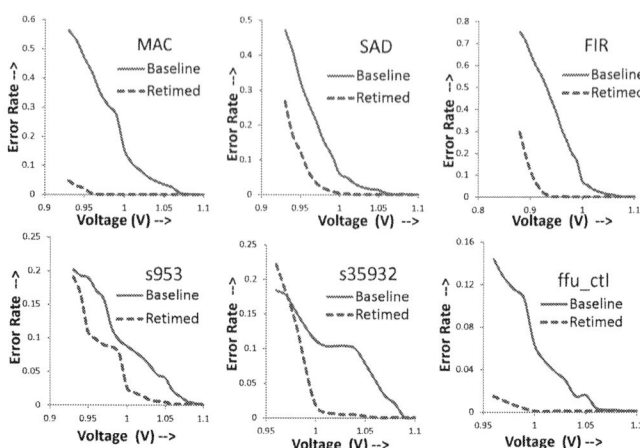

Figure 7: Error Rate *vs.* voltage characteristics for benchmark circuits

overcome this barrier by using retiming, a powerful sequential optimization technique, to design circuits that have graceful timing characteristics and hence can be aggressively voltage overscaled to tap the complete energy benefits offered by RBD. We reformulate our problem as a minimum period retiming problem with relaxed constraints, which allows us to use conventional synthesis tools to produce RBD-friendly circuits. We implemented an automatic synthesis methodology, *Relax-and-Retime*, and demonstrated its efficiency on a wide range of circuits with significant benefits in energy consumption.

8. REFERENCES

[1] D. Ernst et. al. Razor: A low-power pipeline based on circuit-level timing speculation. In *Proc. MICRO*, pages 7–18, Dec. 2003.

[2] K. Bowman et. al. Circuit techniques for dynamic variation tolerance. In *Proc. DAC*, pages 4–7, June 2009.

[3] J. Patel. CMOS Process Variations: A Critical Operation Point Hypothesis. *Online Presentation*. June 2008.

[4] L. Wan and D. Chen. Dynatune: Circuit-level optimization for timing speculation considering dynamic path behavior. In *Proc. ICCAD*, pages 172 –179, Nov. 2009.

[5] B.Greskamp et.al. Blueshift: Designing processors for timing speculation from the ground up. In *Proc. HPCA*, pages 213 –224, Feb. 2009.

[6] A.B. Kahng, S. Kang, R. Kumar, and J. Sartori. Slack redistribution for graceful degradation under voltage overscaling. In *Proc. ASP-DAC*, pages 825 –831, Jan. 2010.

[7] Yuxi Liu et. al. On the logic Synthesis for Timing Speculation. In *Proc. ICCAD*, Nov. 2012.

[8] Lu Wan and Deming Chen. CCP: Common case promotion for improved timing error resilience with energy efficiency. In *Proc. ISLPED*, pages 135–140, July 2012.

[9] Charles E. Leiserson and James B. Saxe. Retiming synchronous circuitry. *Algorithmica*, 6(1):5–35, 1991.

[10] J. Monteiro et. al. Retiming sequential circuits for low power. In *Proc. ICCAD*, pages 398–402, Nov. 1993.

[11] S.T. Chakradhar and S. Dey. Resynthesis and retiming for optimum partial scan. In *Proc. DAC*, pages 87–93, June 1994.

[12] Yuxi Liu, Feng Yuan, and Qiang Xu. Re-synthesis for cost-efficient circuit-level timing speculation. In *Proc. DAC*, pages 158–163, June 2011.

[13] V. Kozhikkottu et. al. Recovery-based design for variation-tolerant SoCs. In *Proc. DAC*, pages 826 –833, June 2012.

[14] M.S. Gupta, J.A. Rivers, P. Bose, G. Wei, and D. Brooks. Tribeca: Design for PVT variations with local recovery and fine-grained adaptation. In *Proc. MICRO*, pages 435 –446, Dec. 2009.

[15] D. Blaauw. et. al. Razor II: In Situ Error Detection and Correction for PVT and SER Tolerance. In *Proc. ISSCC*, pages 400 –622, Feb. 2008.

[16] Chia-Hsiang Chen et. al. A confidence-driven model for error-resilient computing. In *Proc. DATE*, March 2011.

[17] S. Sarangi et. al. EVAL: Utilizing processors with variation-induced timing errors. In *Proc. MICRO*, pages 423–434, Dec. 2008.

[18] Design Compiler Ultra. Synopsys Inc.

[19] Calibre. Mentor Graphics Inc.

[20] Nanosim. Synopsys Inc.

Post-Placement Voltage Island Generation for Timing-Speculative Circuits

Rong Ye[†], Feng Yuan[†], Zelong Sun[†], Wen-Ben Jone[§] and Qiang Xu[†‡]

[†]CUhk REliable Computing Laboratory (CURE)
Department of Computer Science & Engineering
The Chinese University of Hong Kong, Shatin, N.T., Hong Kong
[‡]Shenzhen Institutes of Advanced Technology, Chinese Academy of Sciences
Email: {rye, fyuan, zlsun, qxu}@cse.cuhk.edu.hk

[§]School of Electronic and Computing Systems
University of Cincinnati, USA
Email: {jonewb}@ucmail.uc.edu

ABSTRACT

Region-based multi-supply voltage (MSV) design, by which circuits are partitioned into multiple "voltage islands" and each island operates at a supply voltage that meets its own performance requirement, is an effective technique to tradeoff power and performance. Different from conventional voltage island generation techniques that work in a conservative manner to guarantee "always correct" computation, in this work, we investigate the MSV design problem for timing-speculative circuits, which achieves high energy-efficiency by allowing the occurrence of infrequent timing errors and correcting them online. A novel algorithm based on dynamic programming is developed to tackle this problem. Experimental results on various benchmark circuits demonstrate the effectiveness of the proposed methodology.

1. INTRODUCTION

Motivated by the fact that individual blocks of a circuit can have timing/power characteristics unique from the rest of the design, the concept of multi-supply voltage (MSV) design was introduced to trade off power consumption and performance, and has attracted lots of interests from both academia and industry [1–9]. In MSV designs, circuits are partitioned into multiple "voltage islands" and each island operates at a specified supply voltage that satisfies its performance requirement.

In conventional MSV designs, to meet the timing requirement of each voltage island, the corresponding supply voltage has to be high enough to drive the most timing-critical cell, even though the rest of cells may have much more relaxed timing requirements. Moreover, with the ever-increasing variation effects (e.g., process variation effects due to manufacturing imperfection and dynamic variation effects caused by voltage and temperature fluctuations) in nanometer technology, a large design guard band needs to be reserved to tolerate timing uncertainty. Due to the above, we have to be rather conser-

Permission to make digital or hard copies of all or part of this work for personal or classroom use is granted without fee provided that copies are not made or distributed for profit or commercial advantage and that copies bear this notice and the full citation on the first page. To copy otherwise, to republish, to post on servers or to redistribute to lists, requires prior specific permission and/or a fee.
DAC'13, May 29 - June 07 2013, Austin, TX, USA.

vative when assigning voltages for each island, reducing the possible power savings that can be achieved with MSV designs.

Recently, *timing speculation* (TS) techniques that allow the occurrence of infrequent timing errors and employ error detection and correction techniques to recover from them have emerged as a promising solution to achieve *error-resilient computing* [10–14]. Such "better than worst-case" designs allow the tradeoff between reliability and performance/power, thereby being much more energy-efficient when compared with conventional "worst-case-oriented" designs. Intel [15] has recently demonstrated in their test chip that a timing-speculative microprocessor is able to achieve more than 30% throughput gain when compared to a conventional microprocessor design.

Introducing timing speculation capability into circuits can naturally extend the flexibility of MSV designs to a new horizon, since we do not need to guarantee "always correct" operations any longer and the voltage assignment of islands can avoid being dominated by certain sparse timing-critical cells. How to conduct MSV design for timing-speculative circuits is hence an interesting problem, which, to the best of our knowledge, has not been explored in the literature yet.

Motivated by the above, in this work, we formulate the MSV problem for timing-speculative circuits and develop a novel algorithm based on dynamic programming to solve it. The proposed technique naturally supports "recovery island" design methodology described in [20], wherein each island can recover independent of the rest of the circuit. Experimental results on various benchmark circuits demonstrate that the proposed technique is able to achieve significant power reduction when compared to exiting MSV design techniques.

The remainder of this paper is organized as follows. In Section 2, we present the preliminaries and motivation of this work. The problem formulation and the corresponding algorithms are then detailed in Section 3 and Section 4, respectively. Next, Section 5 presents our experimental results based on various benchmark circuits. Finally, Section 6 concludes this paper.

2. PRELIMINARIES AND RELATED WORK

2.1 MSV Design

A large amount of work has been devoted to MSV designs in the literature and they are applied in various design stages, e.g., floorplanning stage [1, 2], post-floorplanning stage [3], placement stage [4, 5], and post-placement stage [6–9].

As pointed out in [7], conducting region-based MSV design before placement based on their logic boundaries, while "natural", is usually far from optimal. Instead, by using placement proximity (instead of logical) information for MSV design, the acquired solution can achieve much better power savings. Motivated by this observation, the authors proposed to utilize dynamic programming (DP) to generate voltage islands considering placement proximity. While DP provides optimal results, the computational complexity and memory requirement to conduct it at fine-grained granularity is not acceptable for a reasonable-sized circuit. Consequently, a heuristic algorithm is used to partition the circuit into $p \times q$ coarse grids first and DP is conducted at the coarse-grained level. While being more efficient, the effectiveness of this technique is inevitably constrained by the heuristic partitioning algorithm. In [8], the authors investigated how to generate an initial voltage assignment considering the physical proximity of high voltage cells as the input of [7]. After that, to tackle the problem that the freedom of voltage assignment is limited by the amount of available slacks on timing-critical paths, [9] performed incremental placement to improve timing on these paths. All the above works try to generate voltage islands with the guarantee that the timing requirements of all cells are satisfied.

2.2 Timing Speculation

Circuit-level timing speculation technique, being able to detect timing errors at online stage, react to the error quickly and recover from it by rolling back to a known-good pre-error state, has become one of the most promising solutions for variation-aware designs. Without loss of generality, let us discuss one of the most representative timing speculation techniques, Razor [10], to illustrate how resilient computation can be achieved with timing speculation. To detect timing errors on a critical path, the receiving end of the critical path, referred to as suspicious flip-flop, is replaced with Razor flip-flop (Razor-FF), which includes a main flip-flop (FF), an additional shadow latch and some control logic. The main flip-flop latches the output signal of the critical path at the clock edge with a possible timing error, while the shadow latch (controlled by a delayed clock signal) latches the signal a fraction of a cycle later, which guarantees to receive the correct value. Consequently, when the shadow latch and the main FF values do not agree, indicated by the comparator, timing error is detected. Then, by replaying instructions at lower frequency, the processor is able to recover from the timing error with a small re-execution cost.

Recently, Intel has demonstrated a timing-speculative microprocessor test chip in [15]. Their measurement results show that the resilient design enables 25% throughput gain over a conventional design by eliminating the guardband from circuit dynamic variations and an additional 7% throughput increase from exploiting the path-activation probabilities for timing error rate reduction. The above benefits have motivated a large amount of recent research efforts on design and optimization techniques for timing-speculative circuits (e.g., [16–19]).

3. MSV DESIGN FOR TIMING-SPECULATIVE CIRCUITS

The MSV design problem for timing-speculative circuits investigated in this work can be formulated as follows:

Problem: Given

- A timing-speculative circuit C, equipped with timing speculators, such as Razor [10];

- A circuit placement $\mathcal{P}$ with $m \times n$ grids, where each grid g_{ij} is placed at position (i, j);

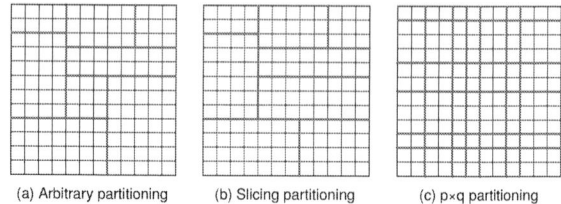

(a) Arbitrary partitioning   (b) Slicing partitioning   (c) p×q partitioning

Figure 1: Three types of rectangular partitioning.

- The probability function $F_{ij}(V_{dd})^1$ for timing errors to occur in grid g_{ij} with respect to V_{dd}, where V_{dd} is the supply voltage;

- The number of voltage islands N_{VI};

- The performance degradation constraint caused by re-execution, represented by throughput degradation ratio $\eta\%$;

to determine a circuit partitioning $\mathbf{P}$ and a voltage assignment $\mathbf{V}$ for voltage island generation, such that the power consumption P_{total} of targeted circuit C is minimized under the performance constraint.

As it is essential to conduct re-computation when timing errors occur, the power consumption of timing-speculative circuits is:

$$P_{total}(\mathbf{P}, \mathbf{V}) = P(\mathbf{P}, \mathbf{V}) \cdot (1 + error(\mathbf{P}, \mathbf{V}) \cdot penalty) \quad , \quad (1)$$

where $P(\cdot)$ is the power function (including dynamic power P_d and static power P_s) of circuit C in one clock cycle after circuit partitioning $\mathbf{P}$ and voltage assignment $\mathbf{V}$ are given, $error(\cdot)$ is the error probability function, $penalty$ is the cost including both the cycles of wasted execution that must be discarded and the time spent on checkpointing and re-execution. Meanwhile, we need to ensure the performance constraint:

$$Th(\mathbf{P}, \mathbf{V}) = \frac{1}{(1 + error(\mathbf{P}, \mathbf{V}) \cdot penalty)} > 1 - \eta\% \quad , \quad (2)$$

where $Th(\cdot)$ is the equivalent circuit throughput considering performance penalty for timing error correction.

Similar to prior works (e.g., [7–9]), we assume that only rectangular voltage islands are allowed, because voltage islands with arbitrary shapes generally lead to difficulty in power-supply network design. Note that, the hardware cost of MSV design (e.g., the overhead of voltage level shifters [1, 25]) is strongly related to the number of voltage islands, which is also considered in this work.

4. VOLTAGE ISLAND GENERATION

4.1 Partitioning Model

How to partition a circuit into rectangular voltage islands has been well studied in [7, 22]. As described in Fig. 1, arbitrary partitioning allows any partitioning with rectangular tiles, slicing partitioning performs slicing through recursive cuts, and $p \times q$ partitioning cuts the circuit into $p \times q$ coarse grids. [7] proved that the optimal slicing partitioning result is a 2-approximation for the optimal arbitrary partitioning. As a special type of slicing partitioning, $p \times q$ partitioning is used in [7] to provide the initial grids that are merged later to form voltage islands, which is also used in our work.

4.2 DP-Based Voltage Island Generation

To solve the proposed voltage island generation problem for timing-speculative circuits, we resort to a DP-based algorithm that enumerates all combinations of the horizontal and vertical cuts.

---

[1] The error probability function $F_{ij}(V_{dd})$ of each grid g_{ij} can be acquired by timing simulation of the targeted circuit with representative workloads.

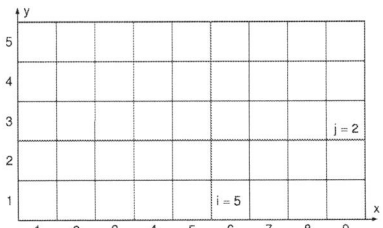

Figure 2: An example to show the enumeration process.

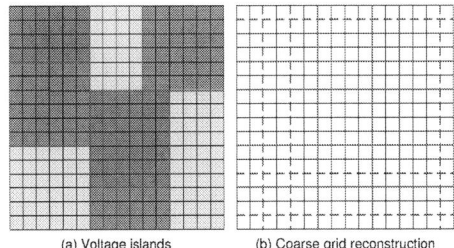

(a) Voltage islands   (b) Coarse grid reconstruction

Figure 3: An example to show the coarse grid reconstruction process.

Given the error probability function $F_{ij}(V_{dd})$, we can have the power consumption of each grid g_{ij} considering power penalties,

$$P_{ij}(V_{dd}) = (P_d(V_{dd}) + P_s(V_{dd})) \cdot (1 + F_{ij}(V_{dd}) \cdot penalty) \quad , \quad (3)$$

where $P_d(\cdot)$ is dynamic power function and $P_s(\cdot)$ is static power function. By solving Eq. 3, we can easily obtain the optimal supply voltage V_{dd}^* for which the power consumption of g_{ij} has the optimal value P_{ij}^*.

Let an $m \times n$ array $\mathbf{A}$ with $A_{ij} = P_{ij}^*$ represent the optimal power consumptions of all the grids, and $R(x_1, y_1; x_2, y_2)$ represent a rectangular region covering the grids $\{g_{ij} | x_1 \le i \le x_2, y_1 \le j \le y_2\}$. For a region $R(x_1, y_1; x_2, y_2)$, we can just replace the power and error probability functions in Eq. 3 with the corresponding terms of this region, and then use such an equation to describe the relationship between power and supply voltage. Similar to the case of a grid g_{ij}, we can also find out the optimal supply voltage V_{opt} for such a region. By denoting the optimal total power consumption of this region with all the grids in it driven by V_{opt} is P_R^*, we define the power wastage of a region $R(x_1, y_1; x_2, y_2)$ as,

$$W(R) = P_R^* - \sum_{g_{ij} \in R} P_{ij}^* \quad . \quad (5)$$

Therefore, we can have the power wastage of a partitioning $\mathbf{P} = \{R_i\}$ as follows,

$$W(\mathbf{P}) = \sum_{1 \le i \le N_{VI}} W(R_i) \quad , \quad (6)$$

where N_{VI} is the specified voltage island number.

With the above definitions, we can have the recursion under slicing partitioning as shown in Eq. 4. A simple example is described in Fig. 2 to show the enumeration procedure. In the 9×5 grids with s islands allowed, we can choose an either vertical (e.g., $i = 5$) or horizontal (e.g., $j = 2$) cut to partition it, and allow t and $(s-t)$ voltage islands in the newly-cut rectangular regions, respectively. This enumeration ensures DP to find the optimal partitioning.

Note that, since the error probability functions of grids $\{g_{ij}\}$ and regions $\{R_i\}$ are fed into the DP solver as inputs to calculate the optimal power consumptions, we assume the error occurrences in different grids are independen[2]. This allows us to calculate the error probability function $F_R(V_{dd})$ of a region R, given the error probabilities of the grids $\{g_{ij} | g_{ij} \in R\}$. For example, we can calculate the error probability of a region R consisting of two regions R_1 and R_2 according to Eq. 7 as follows,

$$F_R = F_{R_1} + F_{R_2} - F_{R_1} \cdot F_{R_2} \quad . \quad (7)$$

4.3 Coarse Grid Reconstruction

The circuit partitioning problem under slicing partitioning can be solved by DP optimally [7]. However, the placement size $m \times n$ at the cell-level is usually too large to employ DP directly in practical

applications. To avoid the huge time and memory costs, one intuitive and viable method is to partition the $m \times n$ grids into $p \times q$ coarse grids as shown in Fig. 1(c), and then apply DP to the coarse grids. Clearly, the effectiveness of the MSV design is limited by the heuristic coarse grid construction algorithm due to search space reduction. In [7], a heuristic-based partitioning algorithm according to [22] is used to construct the $p \times q$ coarse grids before voltage island generation. With such fixed coarse grids, only a constrained MSV design solution space can be explored. Different from their solution, we propose a novel coarse grid reconstruction algorithm to explore more solution space by reconstructing coarse grids and applying DP iteratively.

As discussed in Section 4.2, given an array $\mathbf{A}$ consisting of many grids, DP can achieve an optimal solution for this array $\mathbf{A}$ if enough runtime is allowed. With this property, if we ensure the optimal voltage island design of the last $p \times q$ partitioning is still kept as a solution point in a newly-constructed coarse grids, it is guaranteed to achieve a solution not worse than the last one. Let us explain it using the following example.

Suppose we would like to generate 8 voltage islands based on a 16×16 placement and we decide to use 7×8 coarse grids to save runtime, we can perform any partitioning to divide this 16×16 placement into coarse grids and then use the DP algorithm in Section 4.2 to generate voltage islands. By doing so, we can achieve an optimal voltage island design with the current 7×8 coarse grids. Without loss of generality, we assume the generated voltage island design[3] is the one depicted in Fig. 3(a). To construct a new 7×8 coarse girds for further exploration, it is obvious that we need to determine how to partition the 16×16 placement using 6 vertical lines and 7 horizontal. It is worth noting that, if we keep all the grid lines going through the boundaries of voltage islands as the new coarse grid lines (see the solid lines in Fig. 3(b)), we can make sure the current generated voltage islands (see Fig. 3(a)) is still achievable with newly-constructed coarse grids. In other words, given the 3 vertical lines and 4 horizontal lines that go through the boundaries of voltage islands, no matter how we assign the other 3 vertical lines and 3 horizontal lines (see the dashed lines in Fig. 3(b)) to partition the 16×16 placement, the voltage island design in Fig. 3(a) is one possible solution with the reconstructed coarse grids. As DP can always find out an optimal solution with given coarse grids, we should at least find a solution as good as the previous one and hence it is guaranteed to get a solution not worse than the design in Fig. 3(a) under the new 7×8 partitioning.

The above optimization process can be clarified using Fig. 4. The rectangle represents the entire solution space for DP to explore based on the original $m \times n$ fine-grained grids, and the ellipses represent the sub-spaces after partitioning into $p \times q$ coarse grids. Once the $p \times q$ coarse grids are obtained, we can use DP to achieve the optimal solution in the corresponding sub-space. Therefore, by reconstructing the sub-space and applying DP iteratively, we can get the optimal solution in each sub-space one by one: Point A, Point B, Point C, etc.

---

[2]This is a simple approximation to reduce computational complexity, and its impact is reflected in our experimental results.

[3]The voltage islands are represented by rectangular blocks and plotted out using solid lines.

$$W_s^*(R(x_1,y_1;x_2,y_2)) = \min_{1 \le t < s} \left\{ \min_{x_1 \le i < x_2, y_1 \le j < y_2} \left\{ \begin{array}{l} W_t^*(R(x_1,y_1;i,y_2)) + W_{s-t}^*(R(i+1,y_1;x_2,y_2)), \\ W_t^*(R(x_1,y_1;x_2,j)) + W_{s-t}^*(R(x_1,j+1;x_2,y_2)) \end{array} \right\} \right\} . \tag{4}$$

$$metric(L) = \sum_i (sd(R_i) - sd(R_{i1}) \cdot \frac{A(R_{i1})}{A(R_i)} - sd(R_{i2}) \cdot \frac{A(R_{i2})}{A(R_i)}) \cdot \frac{A(R_i)}{\sum_i A(R_i)} . \tag{8}$$

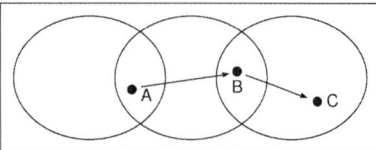

Figure 4: Solution space changes with iterative coarse grid reconstruction.

Bench.	TG #	TFF #	T_{cp} (ns)	(m,n,p,q)	Island #	Cost(%)
38584	21021	1426	6.96	$(20,20,10,10)$	5	5.14
s38417	23949	1636	6.12	$(20,20,10,10)$	5	6.76
des_perf	155746	9105	13.7	$(30,30,15,15)$	10	6.63
ethernet	164912	10752	11.28	$(30,30,15,15)$	10	7.46
AVERAGE						6.50

TG #, total gate count; TFF #, total FF count; T_{cp}, the operating clock cycle period; Island #, the specified voltage island number.

Table 1: Experimental setup.

4.4 Reconstruction Algorithm

To keep the previous partitioning inside the reconstructed solution space, we would like to use those lines going through the boundaries of voltage islands as coarse grid lines. However, in most cases, there are still some vertical and horizontal lines (see the dashed lines in Fig. 3(b)) left to obtain a different $p \times q$ partitioning, which can be used to explore new solution space. We propose a heuristic-based algorithm to obtain new $p \times q$ partitionings, which selects $(p - 1 - p_0)$ vertical coarse grid lines out of $(m - 1 - p_0)$ candidate lines and $(q - 1 - q_0)$ horizontal coarse grid lines out of $(n - 1 - q_0)$ candidate lines. Here, p_0 and q_0 are the number of vertical and horizontal lines determined by the boundaries of voltage islands.

The proposed heuristic algorithm is based on the intuition that, for MSV design, it tends to group those grids with similar voltage requirement together, in order to achieve more power savings. In previous works (e.g., [6, 7]), voltages that guarantee no timing violations are chosen. However, for timing-speculative circuits, it is preferable to use the "optimal" voltage values obtained by trading off reliability with power (see Section 4.2). In this work, to support the proposed heuristic algorithm, we use an evaluation metric to reflect the similarity of the grids that are partitioned into the same islands and we tend to select those grid lines with larger metric values during the coarse grid line selection process.

Given a circuit partitioning **P**, if the grid line L intersects n original islands $\{R_i | 1 \le i \le n\}$ to cut them into $2n$ new islands $\{R_{ij} | 1 \le i \le n, 1 \le j \le 2\}$, $metric(L_k)$ is defined as in Eq. 8, wherein $R_i = R_{i1} \cup R_{i2}$, $sd(R_i)$ is the standard deviation of all the optimal voltage values of the grids in region R_i, and $A(R_i)$ is the number of grids in it.

Note that, to avoid being trapped in local optimal points, we use ε-greedy to select the coarse grid lines for $p \times q$ partitioning. That is, we set up a probability parameter ε (e.g., $\varepsilon = 10\%$), and hence we have the probability of ε to select a grid line randomly, instead of the one with largest metric defined in Eq. 8.

5. EXPERIMENTAL RESULTS

5.1 Experimental Setup

To evaluate the effectiveness of the proposed voltage island generation methodology, we conduct experiments on several large IS-CAS'89 and IWLS'05 benchmark circuits. We synthesize these circuits on a 90nm technology, conduct physical design, and obtain timing information using commercial EDA tools. To take process variation effects into consideration, we perform Monte Carlo simulations to inject gate-level delay variations following Gaussian distribution.We conduct simulations with random inputs and each simulation is per-

formed with 100,000 cycles. By performing simulation for representative workloads and recording error rates occurring in the grids under various operational clock periods, we achieve error probability function $F_{ij}(V_{dd})$ for each grid. We employ the power and delay models used in [2, ?, 21] in our experiments. All the experiments are conducted on a *2.8GHz* PC with *4GB* RAM.

We perform offline timing analysis with false paths excluded according to [23] and use the reported maximum path delay as the operating clock cycle period during timing simulation. For reasonable comparison, a widely-accepted voltage island generation algorithm proposed in [7] is used as the baseline solution and denoted as $MSV_{baseline}$. Because our proposed reconstruction-based $p \times q$ partitioning algorithm is also applicable for the non-TS voltage island generation problem in [7], we replace the corresponding $p \times q$ partitioning algorithm in [7] with ours and keep the rest of algorithm unchanged. This MSV design scheme is denoted as $MSV_{reconstruciton}$. We apply timing speculation directly to the MSV design of $MSV_{baseline}$, and denote this solution as MSV_{ts}. That means, in MSV_{ts} we keep the MSV design of $MSV_{baseline}$ and then perform timing simulation with different voltage assignments to obtain the error probability functions, so that we can achieve the "optimal" voltage assignment and power consumption considering timing speculation. Our proposed solution is denoted as $MSV_{proposed}$. The range of supply voltages allowed for voltage islands to operate is 0.7V to 1.0V in our experiments.

In timing-speculative circuits, we need to add timing error detectors to the receiving end of critical paths. A simple scheme is to transform all the FFs, whose maximum path delays are larger than β of the clock period (e.g., $\beta = 80\%$), as Razor-FFs. Then, to avoid hold time violation on the shadow latch of Razor-FFs, we need to conduct short path padding and this is achieved by constraining paths that drive Razor-FFs with at least γ of the clock period (e.g., $\gamma = 50\%$) during synthesis. In this work, once a voltage island design is generated, we perform timing analysis using timing information with voltage scaling considered and then set up Razor-FFs and conduct short path padding using the obtained path delays. Both of these hardware costs are accounted for in our experiments and β and γ are set to be 80% and 50%, respectively. The hardware cost for equipping each Razor-FF is assumed to be 10 gates. The *penalty* in Eq. 1 is assumed to be 10 clock cycles similar to prior works (e.g., [14]).

5.2 Results and Discussion

In Table 1, we report the operating clock period obtained by excluding false paths according to [23], the used parameters (m, n, p, q), the specified voltage island number and the hardware cost to enable timing speculation for each benchmark circuit. To be specific, we set up the values of (m, n, p, q) as $(20, 20, 10, 10)$ for small-scale circuits

Bench.	$MSV_{baseline}$		$MSV_{reconstruction}$			MSV_{ts}				$MSV_{proposed}$					Runtime (s)
	power	σ	power	σ	Δ_1(%)	power	σ	ΔTh(%)	Δ_2(%)	power	σ	ΔTh(%)	Δ_3(%)	Δ_4(%)	
s38584	0.852	0.014	0.813	0.015	-4.58	0.793	0.014	-4.28	-6.92	0.689	0.016	-3.55	-13.11	-19.13	2.75
s38417	0.857	0.013	0.835	0.016	-2.57	0.825	0.018	-3.56	-3.73	0.781	0.020	-4.09	-5.33	-8.87	1.92
des_perf	0.862	0.019	0.806	0.017	-6.50	0.674	0.014	-5.52	-21.81	0.598	0.015	-7.03	-11.28	-30.63	17.35
ethernet	0.778	0.018	0.723	0.019	-7.07	0.631	0.015	-6.32	-18.89	0.581	0.012	-5.63	-7.92	-25.32	15.04
AVERAGE					-5.18			-4.92	-12.84			-5.08	-9.41	-20.99	

σ: standard deviation of *power*; Δ_1: *power* difference ratio between $MSV_{reconstruction}$ and $MSV_{baseline}$; Δ_2: *power* difference ratio between MSV_{ts} and $MSV_{baseline}$; Δ_3: *power* difference ratio between $MSV_{proposed}$ and MSV_{ts}; Δ_4: *power* difference ratio between $MSV_{proposed}$ and $MSV_{baseline}$; ΔTh, performance degradation ratio.

Table 2: Results on the proposed reconstruction-based $p \times q$ partitioning.

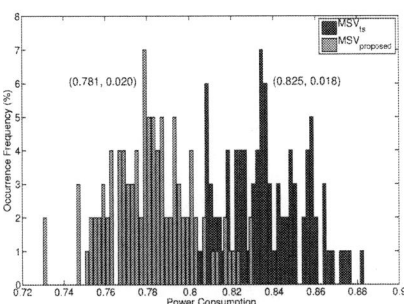

Figure 5: Monte Carlo simulation results.

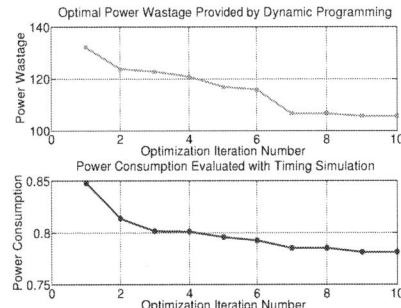

Figure 6: Power wastage and power consumption wrt. optimization iteration number.

(e.g., $s38584$ and $s38417$) and as $(30, 30, 15, 15)$ for large-scale circuits (e.g., *des_perf* and *ethernet*). The average hardware cost to equip the circuits with TS capability (including timing speculator and short path padding cost) is about 6.5%.

To verify the effectiveness of the proposed voltage island generation methodology, we, first of all, perform Monte Carlo simulation to produce 100 sample chips with different variation patterns for each benchmark circuit. In Table 2, we report the average power consumption[4] and its standard deviation σ for $MSV_{baseline}$, $MSV_{reconstruction}$, MSV_{ts} and $MSV_{proposed}$, respectively. It is important to note that, the reported results includes the power overhead of MSV design (e.g., level shifters) and power penalties to correct timing errors.

As can be seen from Table 2, compared to $MSV_{baseline}$, the proposed $MSV_{reconstruction}$ can achieve 5.18% power saving on average. This improvement comes from using our proposed $p \times q$ partitioning algorithm to replace the corresponding one in $MSV_{baseline}$ only, which demonstrates the effectiveness of our reconstruction algorithm. In other words, even for non-TS conventional circuits, our proposed solution lead to much more power-efficient MSV designs.

Besides, MSV_{ts} can achieve 12.84% power reduction on average when compared with $MSV_{baseline}$. This improvement reflects the efficacy of timing speculation itself, since in MSV_{ts} we just apply timing speculation directly to the MSV design of $MSV_{baseline}$. Compared with MSV_{ts}, our proposed methodology $MSV_{proposed}$ can further achieve 9.41% power reduction on average, which reflects the efficacy of explicitly considering timing speculation during the MSV design process. The runtime of the proposed algorithm (see Fig. **??**) is quite small.

$MSV_{proposed}$ achieve better results because (i) the proposed partitioning model and DP-based voltage island generation method facilitate to identify voltage islands with optimal supply voltages based on circuit slack distribution, which gives the first-level of power saving; (ii) once a voltage island has been formed, another level of power saving can be achieved by minimizing the timing error rates. Any voltage island with only a small number of critical paths (i.e., most circuit paths have relatively large slacks) can fully take advantage of this

power saving while maintaining the performance. At the same time, we can observe that the power reduction ratios of these four benchmark circuits are quite different, and we attribute this phenomenon to the unique timing characteristic of each circuit. Generally speaking, if a circuit has gradually-decreasing path delay distribution, the benefit brought by timing speculation can be larger than that of those circuits with a sharply-declining delay distribution. This is because, in the latter case, a large number of paths may fail at the same time in the design when voltage overscaling exceeds a critical point, which causes a steep increase of timing error rate [24].

MSV_{ts} and $MSV_{proposed}$ would suffer from performance degradation caused by infrequent timing errors. We report this performance degradation in Table 1 and denote it as ΔTh, compared to the case that the circuit uses the maximum path delay as its operational clock period. On average, MSV_{ts} and $MSV_{proposed}$ have 4.92% and 5.08% throughput degradation, respectively. However, it is important to note that, for $MSV_{baseline}$ without timing error correction capability, designers usually have to reserve a large timing guard band (e.g., 15% of maximum path delay) to tolerate variation-induced timing uncertainty and hence system throughput is degraded due to lower operational frequency [15]. From this perspective, if we consider the timing guardband existing in the non-TS solution $MSV_{baseline}$, the performance of MSV_{ts} and $MSV_{proposed}$ would be actually better than that of $MSV_{baseline}$.

To get more details of the proposed methodology, we take $s38417$ as an example in the following experiments. In Fig. 5, we show the results of MSV_{ts} and $MSV_{proposed}$ with process variation effects after performing Monte Carlo simulation. The corresponding mean value of power consumption and standard deviation for each case are depicted in the figure in the form of (μ, σ), which, again, demonstrates the benefits of $MSV_{proposed}$. In Fig. 6, we plot the curves to reflect the changes of both the power wastage provided by DP and the power consumption evaluated by timing simulation with error penalties taken into account. As can be seen, the power wastage is decreased all the time, which proves the effectiveness of the reconstruction algorithm to explore new solution space and guarantee the power wastage to be optimized step by step, as discussed in Section 4.3. Note that, this can be used to trade off the algorithm runtime with optimization quality during design process. Moreover, with respect to the

---

[4] Each power value has been normalized by using the power consumption of the case without MSV design as unit value.

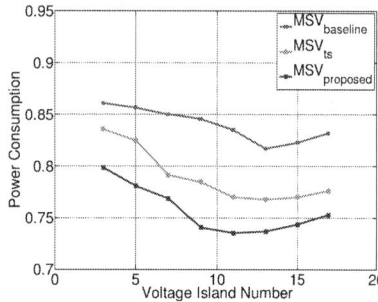

Figure 7: Power consumption wrt. voltage island number.

optimization iteration number, the two curves descends in the same manner. The similar trends of these two curves can prove the effectiveness of our proposed optimization process. To investigate the effects with different specified voltage island number, we vary the number of islands and get the power consumption curves of $MSV_{baseline}$, MSV_{ts} and $MSV_{proposed}$ as described in Fig. 7. Clearly, with different number of voltage islands, $MSV_{proposed}$ always outperforms the other solutions. It can be also observed that, with increasing number of allowed voltage islands in the MSV design, the power savings of all these solutions increase in the beginning, but decrease in the end. This is because, more voltage islands allow fine-grained voltage assignments that satisfy the performance constraint of each individual island, leading to better power savings. However, more voltage islands also incur higher cost for the supporting circuitries (e.g., level shifters). Consequently, when the number is too large, the benefit provided with fine-grained voltage assignment cannot compensate the associated power cost.

6. CONCLUSION

Region-based MSV design has been used as an effective technique to reduce power consumption and attracted lots of research interests. However, all of the previous MSV works try to guarantee "always correct" operations, which greatly limits the design flexibility. In this work, we formulate the MSV design problem for timing-speculative circuits, and propose a novel DP-based algorithm to generate voltage islands. Experimental results based on various benchmark circuits demonstrate that the proposed methodology is able to significantly reduce power consumption of timing-speculative circuits with acceptable performance degradation.

7. ACKNOWLEDGEMENT

This work was supported in part by the Hong Kong SAR Research Grants Council under General Research Fund No. CUHK418111 and No. CUHK418812.

8. REFERENCES

[1] W. K. Mak, J. W. Chen, Voltage Island Generation under Performance Requirement for SoC Designs. In *Proc. IEEE/ACM Asia South Pacific Design Automation Conference (ASP-DAC)*, pp. 798-803, 2007.

[2] Q. Ma, E. Young, Multivoltage Floorplan Design. In *IEEE Transactions on Computer-Aided Design of Integrated Circuits and Systems*, vol. 29, no. 4, pp. 607-617, April 2010.

[3] W. Lee, H. Liu, Y. Chang, An ILP Algorithm for Post-Floorplanning Voltage-Island Generation Considering Power-Network Planning. In *Proc. ACM/IEEE International Conference on Computer-Aided Design (ICCAD)*, pp. 650-655, 2007

[4] B. Liu, Y. Cai, Q. Zhou, X. Hong, Power Driven Placement with Layout Aware Supply Voltage Assignment for Voltage Island Generation in Dual-Vdd Designs. In *Proc. IEEE/ACM Asia South Pacific Design Automation Conference (ASP-DAC)*, pp. 582-587, 2006.

[5] L. Guo, Y. Cai, Q. Zhou, X. Hong, Logic and Layout Aware Voltage Island Generation for Low Power Design. In *Proc. IEEE/ACM Asia South Pacific Design Automation Conference (ASP-DAC)*, pp. 666-671, 2007.

[6] R. Ching, E. Young, K. Leung, C. Chu, Post-Placement Voltage Island Generation. In *Proc. ACM/IEEE International Conference on Computer-Aided Design (ICCAD)*, pp. 641-646, Nov. 2006.

[7] H. Wu, M, Wong, I. Liu, Y. Wang, Placement-Proximity-Based Voltage Island Grouping Under Performance Requirement. In *IEEE Transactions on Computer-Aided Design of Integrated Circuits and Systems*, vol.26, no.7, pp.1256-1269, July 2007.

[8] H. Wu, M. Wong, I. Liu, Timing-Constrained and Voltage-Island-Aware voltage assignment. In *Proc. ACM/IEEE Design Automation Conference (DAC)*, pp. 429-432, 2006.

[9] H. Wu, M. Wong, Incremental Improvement of Voltage Assignment. In *IEEE Transactions on Computer-Aided Design of Integrated Circuits and Systems*, vol.28, no.2, pp.217-230, Feb. 2009.

[10] D. Ernst, et al., Razor: a Low-Power Pipeline based on Circuit-Level Timing Speculation. In *Proc. IEEE/ACM International Symposium on Microarchitecture (MICRO)*, pp. 7-18, 2003.

[11] S. R. Sarangi, et al., VARIUS: a Model of Process Variation and Resulting Timing Errors for Microarchitectus. In *IEEE Transactions on Semiconductor Manufacturing*, vol. 21, pp. 3-13, Feb. 2008.

[12] S. Sarangi, B. Greskamp, A. Tiwari, and J. Torrellas, EVAL: Utilizing Processors with Variation-Induced Timing Errors. In *Proc. IEEE/ACM International Symposium on Microarchitecture (MICRO)*, pp. 423-434, 2008.

[13] T. Austin, V. Bertacco, D. Blaauw and T. Mudge, Opportunities and Challenges for Better than Worst-Case Design. In *Proc. IEEE/ACM Asia South Pacific Design Automation Conference (ASP-DAC)*, pp. 2-7, 2005.

[14] M. de Kruijf, S. Nomura, and K. Sankaralingam, A Unified Model for Timing Speculation: Evaluating the Impact of Technology Scaling, CMOS Design Style, and Fault Recovery Mechanism. In *Proc. IEEE/IFIP International Conference on Dependable Systems and Networks (DSN)*, pp. 487-496, 2010.

[15] K. Bowman, J. Tschanz, C. Wilkerson, S. Lu, T. Karnik, V. De, S. Borkar, Energy-Efficient and Metastability-Immune Resilient Circuits for Dynamic Variation Tolerance. In *IEEE Journal of Solid-State Circuits*, 44(1): 49–62, 2009.

[16] B. Greskamp, et al. Blueshift: Designing processors for timing speculation from the ground up. *Proc. IEEE International Symposium on High-Performance Computer Architecture (HPCA)*, pp. 213-224, 2009.

[17] A. B. Kahng, et al. Slack redistribution for graceful degradation under voltage overscaling. *Proc. IEEE/ACM Asia South Pacific Design Automation Conference (ASP-DAC)*, pp. 825-831, 2010.

[18] R. Ye, F. Yuan and Q. Xu. Online clock skew tuning for timing speculation. *Proc. ACM/IEEE International Conference on Computer-Aided Design (ICCAD)*, pp. 442–447, 2011.

[19] Y. Liu, et al. On logic synthesis for timing speculation. *Proc. ACM/IEEE International Conference on Computer-Aided Design (ICCAD)*, pp. 591–596, 2012.

[20] V. Kozhikkottu, S. Dey, A. Raghunathan, Recovery-based design for variation-tolerant SoCs. In *Proc. ACM/IEEE Design Automation Conference (DAC)*, pp. 826-833, 2012.

[21] K. Roy, S. Mukhopadhyay, H. Mahmoodi-Meimand, Leakage Current Mechanisms and Leakage Reduction Techniques in Deep-Submicrometer CMOS Circuits. In *Proceedings of the IEEE*, vol.91, no.2, pp. 305- 327, Feb. 2003.

[22] S. Muthukrishnan, T. Suel, Approximation Algorithms for Array Partitioning Problems. In *Journal of Algorithms*, Volume 54, Issue 1, pp. 85-104, Jan. 2005.

[23] F. Yuan and Q. Xu, On Timing-Independent False Path Identification. In *Proc. ACM/IEEE International Conference on Computer-Aided Design (ICCAD)*, pp. 532-535, 2010.

[24] J. Sartori and R. Kumar, Architecting Processors to Allow Voltage/Reliability Tradeoffs. In *Proc. ACM International Conference on Compilers Architecture and Synthesis for Embedded Systems (CASES)*, pp. 115-124, 2011.

[25] J. Lin, W. Cheng, C. Lee and R. C.J. Hsu, Voltage island-driven floorplanning considering level shifter placement. In *Proc. IEEE/ACM Asia South Pacific Design Automation Conference (ASP-DAC)*, pp. 443-448, 2012.

Analysis and Characterization of Inherent Application Resilience for Approximate Computing*

Vinay K. Chippa[†], Srimat T. Chakradhar[‡], Kaushik Roy[†] and Anand Raghunathan[†]

[†] School of Electrical and Computer Engineering, Purdue University
[‡] Systems Architecture Department, NEC Laboratories America
[†]{vchipp,kaushik,raghunathan}@purdue.edu, [‡]chak@nec-labs.com

ABSTRACT

Approximate computing is an emerging design paradigm that enables highly efficient hardware and software implementations by exploiting the inherent resilience of applications to in-exactness in their computations. Previous work in this area has demonstrated the potential for significant energy and performance improvements, but largely consists of ad hoc techniques that have been applied to a small number of applications. Taking approximate computing closer to mainstream adoption requires (i) a deeper understanding of inherent application resilience across a broader range of applications (ii) tools that can quantitatively establish the inherent resilience of an application, and (iii) methods to quickly assess the potential of various approximate computing techniques for a given application. We make two key contributions in this direction. Our primary contribution is the analysis and characterization of inherent application resilience present in a suite of 12 widely used applications from the domains of recognition, data mining, and search. Based on this analysis, we present several new insights into the nature of resilience and its relationship to various key application characteristics. To facilitate our analysis, we propose a systematic framework for Application Resilience Characterization (ARC) that (a) partitions an application into resilient and sensitive parts and (b) characterizes the resilient parts using approximation models that abstract a wide range of approximate computing techniques. We believe that the key insights that we present can help shape further research in the area of approximate computing, while automatic resilience characterization frameworks such as ARC can greatly aid designers in the adoption approximate computing.

Categories and Subject Descriptors

B.7.0 [**INTEGRATED CIRCUITS**]: General

General Terms

Algorithms, Design

Keywords

Inherent Application Resilience, Approximate Computing, Resilience Characterization

1. INTRODUCTION

Inherent application resilience is the property of an application to produce acceptable outputs despite some of its underlying computations being incorrect or approximate. It is prevalent in a broad spectrum of applications such as digital

---

*This material is based upon work supported in part by the National Science Foundation under Grant No. 1018621

Permission to make digital or hard copies of all or part of this work for personal or classroom use is granted without fee provided that copies are not made or distributed for profit or commercial advantage and that copies bear this notice and the full citation on the first page. To copy otherwise, to republish, to post on servers or to redistribute to lists, requires prior specific permission and/or a fee.
DAC'13, May 29 - June 07, 2013, Austin, TX, USA.

---

signal processing, image, audio, and video processing, graphics, wireless communications, web search, and data analytics. Emerging application domains such as Recognition, Mining and Synthesis (RMS) [1], which are expected to drive future computing platforms, also exhibit this property in abundance. The inherent resilience of these applications can be attributed to several factors: (i) significant redundancy is present in large, real-world data sets that they process, (ii) they employ computation patterns (such as statistical aggregation and iterative refinement) that intrinsically attenuate or correct errors due to approximations, and (iii) a range of outputs are equivalent (*i.e.*, no unique golden output exists), or small deviations in the output cannot be perceived by users.

For inherently resilient applications, functionality is defined on a continuous scale of *output quality*. Therefore, applications should produce outputs of acceptable quality rather than a unique "correct" output. Approximate Computing is a new design approach that leverages inherent resilience through optimizations that trade off output quality for improved performance, energy efficiency or other metrics. Effectively, approximate computing techniques relax the traditional requirement of exact (numerical or Boolean) equivalence between the specification and implementation.

Several previous efforts have explored approximate computing in software [8,9,15] and hardware [2-8,18,19], with promising results. Software techniques typically improve performance by skipping computations or reducing the use of costly operations such as inter-thread synchronization, whereas hardware techniques modify the design at various levels of abstraction to introduce tradeoffs between output quality and efficiency. These efforts have established the significant potential of approximate computing, and there is increasing interest in its use with the growth in inherently resilient applications. However, several challenges still need to be addressed before approximate computing can move from its initial stages of exploration to broader adoption.

First, the property of inherent application resilience and the various application characteristics that contribute to it need to be understood comprehensively across a broader spectrum of applications. Second, designers require tools that quantitatively evaluate the resilience of a given application, and identify the parts of the application that are amenable to approximate computing. Finally, there is a need for a systematic methodology that can help designers to quickly evaluate various approximate computing techniques for a given application, or a given technique across a wide range of applications.

In this work, we make two key contributions to broaden the scope and applicability of approximate computing. The first contribution of our work is the analysis and characterization of inherent application resilience present in a suite of 12 widely used recognition, mining and search applications. We demonstrate the high degree of resilience existing in these applications that emphasizes the scope and potential of approximate computing. We present several new insights into the nature of application resilience and its relationship to various application characteristics. These insights serve as guidelines for future research in the area of approximate computing.

Our second contribution is an Application Resilience Characterization (ARC) framework that (i) partitions a given appli-

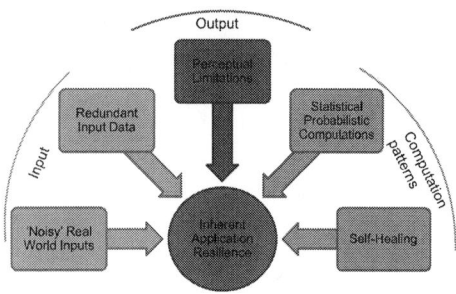

Figure 1: Various sources of inherent application resilience

cation into potentially resilient parts, which may be subject to approximate computing, and sensitive parts, which may not, and (ii) characterizes the potentially resilient parts in greater detail to evaluate the applicability of various approximate computing techniques. For this purpose, we propose the use of "approximation models" that efficiently abstract a wide range of approximate computing techniques to enable their quick evaluation. The proposed ARC methodology is realized using the valgrind dynamic binary instrumentation framework. We believe that tools such as the ARC framework will greatly aid designers in adopting approximate computing as an additional avenue for design optimization.

The rest of the paper is organized as follows: We present a qualitative analysis of application resilience in Section 2. We describe related efforts and place our contribution in their context in Section 3. The ARC framework is described in Section 4 and its implementation is detailed in Section 5. Section 6 describes various insights derived from the application of the ARC framework to our suite of benchmarks.

2. INHERENT APPLICATION RESILIENCE

Inherent Application Resilience is defined as the property of an application to produce acceptable outputs in spite of some of its underlying computations being incorrect (or approximate). The various sources that contribute to inherent application resilience are shown in Figure 1, and can be classified into three categories.

Inputs: These applications process input data that is noisy and redundant. The robustness to noise in the input data and the fact that similar data is processed several times (redundancy), often manifests as resilience to approximations.

Outputs: There does not exist a unique golden output, *i.e.*, a range of outputs are considered equally acceptable. Moreover, these applications generate output for consumption by humans, whose perceptual limitations imply that minor variations that cannot be discerned are acceptable.

Computation Patterns: These applications employ statistical computations that result in attenuation or cancellation of errors. Moreover, due to the iterative nature of computations in these applications, errors due to approximations in one iteration can potentially get healed/recovered in subsequent iterations.

3. RELATED WORK

Approximate computing has been applied to the design of hardware building blocks such as arithmetic units and entire datapaths, either through voltage overscaling [2, 3, 4] or through logic simplification [5, 6]. A systematic approach to apply approximate computing to hardware design at various levels of design abstraction was presented in [7, 8]. The application of approximate computing to programmable processors has been recently explored [9]. Approximate computing techniques in software [10, 11, 12] have been proposed to improve the performance and parallel scalability of applications on general-purpose computing platforms. These efforts evaluated specific approximate computing techniques for a limited set of applications, leaving open the question of whether approximate computing is applicable in a broader context.

Approximate computing is related to, but distinct from error-resilient computing, which seeks to lower the overheads

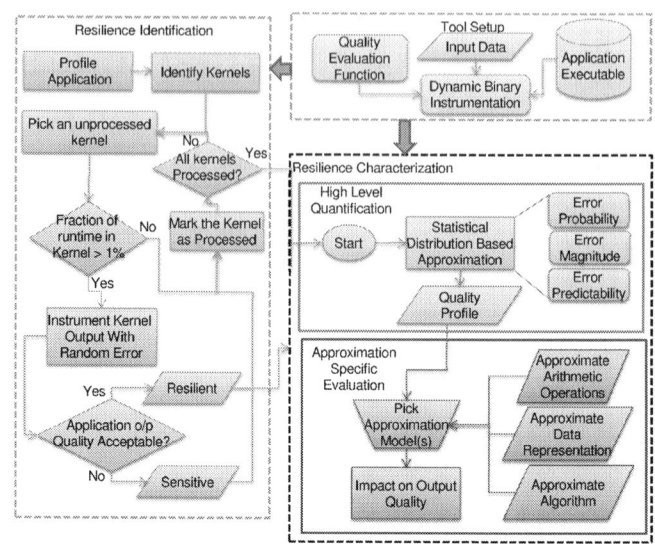

Figure 2: Overview of the ARC framework

of fault-tolerance from hardware defects or transient faults in hardware, by exploiting error masking at the application level, or by using low-cost algorithmic fault tolerance [13]. In approximate computing, "errors" are created intentionally, and are typically predictable (and can therefore be modeled). Automatic error injection frameworks that study the robustness of applications under faults have been extensively studied in the fault tolerance community [14]. A recent effort [15] employed an automatic error injection framework to study the resilience of signal processing applications using random bit flips. These techniques are inadequate for evaluating inherent application resilience, since they either (i) inject arbitrary faults (*e.g.*, random bit-flips), which often do not reflect the nature of approximate computing techniques, or (ii) consider any deviation from the golden numerical output of the program as unacceptable, oblivious to the continuum of output quality that we are concerned with.

The ARC framework proposed in this paper provides designers with a tool set to easily evaluate approximate computing for a new application. Additionally, the insights derived from our characterization of a broad range of applications provide guidelines for further efforts in the area of approximate computing.

4. APPLICATION RESILIENCE CHARACTERIZATION FRAMEWORK

In this section, we describe the proposed Application Resilience Characterization (ARC) framework that can be used to quantitatively evaluate the inherent resilience of applications. The framework, shown in Figure 2, consists of two major steps: (i) identification of potentially resilient computations and (ii) characterization of these computations through approximation models.

The inputs to the framework are the application program, a representative input data set and a user-defined quality evaluation function. The quality evaluation function processes the output of the application and evaluates the output quality as a numerical value. The quality evaluation function is application-specific and must be provided by the user; however the ARC framework itself is general. The outputs of the ARC framework include a list of the resilient computations in the application and the results of evaluating various approximation models on the resilient computations.

The overall approach taken in both steps of the ARC framework is to utilize dynamic binary instrumentation to introduce random errors or controlled approximations into specific computations as the application executes, and observe the resulting application behavior. We next describe the two steps of the ARC framework in detail.

4.1 Resilience Identification

Even the most resilient applications contain both resilient and sensitive computations. Approximate computing techniques should be targeted towards resilient computations while avoiding the sensitive ones. The first step of the ARC framework identifies potentially resilient computations by using Dynamic Binary Instrumentation (DBI) to inject errors into the results of computations as the application executes.

Although ARC uses software implementations of applications on a general-purpose platform (due to their widespread availability), our intent is really to evaluate the resilience of the algorithmic computations rather than all instructions in the software implementation. Therefore, we first partition the instructions in the program into computation kernels as follows. We consider innermost loops that account for over a specified fraction of the program's execution time (we used 1% in our experiments) as atomic kernels. As the program executes over the provided input data set, we add random errors to the program variables that are modified in a kernel and used outside it, *i.e.*, the kernel's outputs. If the application crashes or hangs, or produces an output that does not meet a relaxed output quality criterion, we mark the kernel as sensitive; otherwise, it is marked as potentially resilient. We use unconstrained random errors and a relaxed output quality criterion since the objective of this step is only to identify *potentially* resilient kernels; in the second step of the ARC framework, we use approximation models and the user-provided quality evaluation function to further evaluate and characterize these kernels.

For completeness, each instruction that lies outside the processed loops is considered as a separate kernel. However, in our experiments, virtually all resilient computations were found to be kernels generated from loops.

4.2 Resilience Characterization

Once potentially resilient kernels are identified in the first step of the ARC framework, the next step characterizes their resilience to provide insights into the applicability of various approximate computing techniques. The resilience characterization step uses the same strategy as the identification step *i.e.*, execute the application under DBI on the provided input data, inject errors in the kernels, and evaluate application behavior. However, there are two key differences. First, the errors introduced in the kernels are derived from approximation models that model the effects of various approximate computing techniques. The key objective of the approximation model is to (i) quantify the resilience using generic attributes of the approximations such as error probability, magnitude and predictability of errors introduced, and (ii) evaluate the impact of a single or a class of approximate computing techniques on the application output quality. Second, we use the quality evaluation function provided by the user. For a given approximation model, a quality profile is generated that characterizes the application output quality as a function of the model's parameters.

The approximation models used for resilience characterization are described in more detail in the following subsections.

4.2.1 High-level Approximation Models

Any approximate computing technique can be thought of as introducing errors in the computations that are being approximated. However, these errors are usually constrained by design such that the application output quality is not drastically impacted. To quantify an application's resilience at a high level (*i.e.*, independent of any specific approximate computing technique), we propose a *statistical approximation*

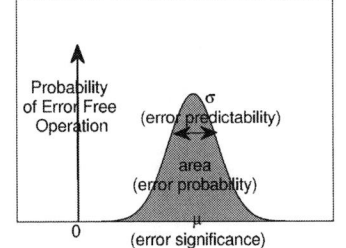

Figure 3: Statistical approximation model

model, in which the errors produced due to approximate computing techniques are modeled using a statistical distribution (Figure 3). The statistical approximation model is parameterized by three high level parameters: error probability, error magnitude and error predictability. Error probability determines the rate at which errors are produced by the approximation, and is denoted by the area under the statistical distribution (not including the error-free case). The error magnitude and error predictability constrain the numerical value of the error and correspond to the mean and variance of the error distribution. The ARC framework employs this model to generate a quality profile of the application as a function of these parameters. This model is very useful in the early stages of the design cycle, as it gives insights into the resilience of the application and helps narrow down design choices without the significant effort needed to implement various approximate computing techniques.

4.2.2 Technique-specific Approximation Models

The statistical approximation model proposed to quantify resilience may not adequately reflect a specific approximate computing scheme. Therefore, we propose three approximation models that abstract important classes of approximate computing techniques.

Approximation of Arithmetic Operations

Many approximate computing techniques are applied to arithmetic units such as adders [17], multipliers *etc.* We propose the bit error profile model to represent the effect of these approximations. A bit error profile specifies the probability that each bit in the output of an arithmetic operation has an error. During resilience characterization with this model, the outputs of arithmetic operations in resilient kernels are modified in accordance with the chosen bit error profiles. For example, the bit error profile of an approximate adder proposed in [16] is presented in Figure 4. A bit error profile model may optionally specify conditional error probabilities based on the values of the operation's inputs.

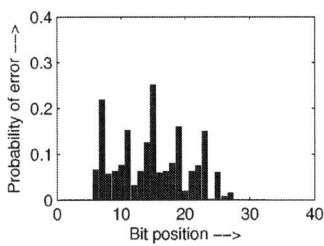

Figure 4: Bit error profile for an approximate adder [16]

Approximate data representation

Some approximate computing techniques exploit application resilience by employing approximate data representations. Bit truncation is one such commonly employed technique, where approximations are introduced by reducing the data width used to represent variables in hardware or software. Approximate data representations may also utilize different number systems such as logarithmic and residue number systems. To evaluate such techniques, we transform the kernel's inputs into the chosen approximate data representation, perform all kernel computations in the chosen representation, and transform the output back to the original representation.

Algorithm level approximations

In this model, we consider approximate computing techniques that modify the algorithm being implemented at a coarser granularity *e.g.*, an iteration of a loop might get skipped. We model the impact of these techniques as computation skipping - the kernel's execution is skipped with a specified probability.

5. EXPERIMENTAL METHODOLOGY

In this section, we describe the experimental setup used to implement and evaluate the proposed ARC framework. The basic functionality of ARC framework is implemented using valgrind, a popular dynamic binary instrumentation framework. valgrind enables easy and efficient instrumentation of a program by first translating it into an "Intermediate Representation"(IR) that is processor independent. Any tool im-

Table 1: The applications and the data-sets used to evaluate the proposed resilience characterization methodology

Application	Algorithm	% Runtime in resilient kernels	dominant kernel (Contribution to runtime)
Document Search	Semantic Search Index	90	Dot Product Computation (86)
Image Search	Feature Extraction	78	Dot Product Computation (71)
Hand Written Digit Classification	Support Vector Machines (SVM): Testing	94	Dot Product Computation (89)
Hand Written Digit Model Generation	Support Vector Machines (SVM): Training	97	Dot Product Computation (93)
Eye Detection	Generalized Learning Vector Quantization (GLVQ): Testing	89	Distance Computation (83)
Eye Model Generation	Generalized Learning Vector Quantization (GLVQ): Training	96	Distance Computation (92)
Image Segmentation	K-means Clustering	74	Distance Computation(66)
Census Data Modeling	Neural Networks: Multi Layer Back Propagation	62	Matrix Vector Multiplication (42)
Census Data Classification	Neural Networks: Forward Propagation	79	Matrix Vector Multiplication (64)
Nutrition and Health Information Analysis	Logistic Regression	65	Dot Product Computation (48)
Digit Recognition	K-Nearest Neighbors	96	Distance Computation (92)
Online Data Clustering	Stream Cluster	77	Distance Computation (68)

plemented using valgrind is allowed to instrument this IR and the instrumented IR is then executed on the host machine. By implementing the ARC framework using valgrind, we are able to easily apply the framework to any application without the need to modify the application source code.

We apply the ARC framework on a benchmark suite consisting of 12 widely used recognition, data mining and search applications, along with representative input data. All the applications are annotated with appropriate quality evaluation functions that translate application outputs into a numerical measure of output quality. The applications and the underlying algorithms of the benchmark suite are presented in Table 1. A detailed explanation of the applications and datasets, and the results of resilience characterization are presented in the supplemental section.

6. RESULTS AND INSIGHTS

We characterized the resilience of the applications in the benchmark suite using the ARC framework and present the results in this section. In the first step of the framework, we identify the resilient kernels of the application. The results for this step are presented in Table 1, column 3. It can be seen that, across all the applications in the benchmark suite, the fraction of the application's run-time that is spent in resilient kernels ranges from 67% to 96%. This demonstrates the high degree of inherent resilience present in these applications and underscores the potential for approximate computing. In most of these applications, there exists a single compute kernel that dominates the execution time. We present the dominant kernels for the benchmark suites in column 4 of the Table 1, and their contribution to program execution time.

On average, these applications spend 83% of their run-time in resilient kernels, out of which 74% belongs to the dominant kernel. Therefore, the bulk of the resilience can be exploited by focusing approximate computing design efforts on the dominant kernel. In order to apply suitable approximate computing techniques, it is important to understand application resilience in greater detail. While the raw data (results of resilience characterization) are presented in the supplemental section, we devote the remainder of this section to presenting several insights that we derived from our experiments.

6.1 Granularity of Approximation Matters

The efficiency of approximate computing techniques greatly depends on how the application is realized (dedicated hardware *vs.* software on a programmable processor). For example, consider an application with vector dot product as the dominant

kernel. The dot product kernel may be resilient in the sense that controlled approximations to its result lead to acceptable program outputs. A hardware module that implements dot products in an approximate manner [4] may therefore be utilized. However, realizing the dot product kernel on a general purpose processor introduces instructions for loop control and pointer arithmetic. Introducing approximations at the granularity of instructions in this software implementation may lead to a very different conclusion about application resilience.

To study this effect, we expanded all loop kernels that were identified by ARC into machine instructions and performed error injection at the instruction level. On an average, we observed that the scope of approximate computing reduces by a factor of 57% when approximations are applied to individual instructions within kernels. This observation is consistent with prior efforts that have characterized application resilience at the granularity of processor instructions [18, 19]. Therefore, it is important to evaluate inherent application resilience in the appropriate implementation context.

6.2 Fail Small or Fail Rare

Consider the quality profile shown in Figure 5 for the image search application. This quality profile was generated using the statistical approximation model, with varying values of error probability (error rate), error magnitude (error mean), and error predictability (error variance). The quality profile is presented in the form of a 3-D slice plot, where each slice depicts the output quality for a given error rate (x-axis) and each block in a slice represents the output quality for a specific error mean (y-axis) and variance (z-axis). The output quality is color coded such that white regions represent no degradation in the output quality and darker regions represent higher degradation. In the case of image search, the output quality of

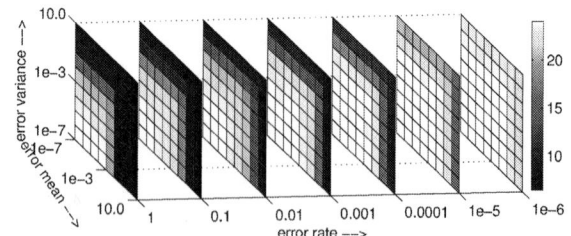

Figure 5: Resilience characterization of image search

is measured as the number of correct search results (out of the top 25) that match the reference golden output. It can be seen

802

from the white regions in Figure 5 that the output quality of the application remains acceptable in the following two cases:

- Fail small: If the magnitude of the errors introduced by the approximations remains small, the application can tolerate even very frequent errors (left bottom regions in the left most slice).

- Fail rare: An approximation can introduce errors of arbitrarily high magnitude and still result in acceptable output quality as long as the errors are introduced very rarely (right most slice in the slice plot).

This insight of "fail small or fail rare" was observed across all applications in the benchmark suite. Therefore, designers should aim to develop approximation schemes that are constrained in either the rate or the magnitude of the approximations introduced.

6.3 Computation Patterns that Enhance Resilience

As described in Section 2, computation patterns are an important factor that contribute to application resilience. To quantitatively study this phenomenon, we consider the eye detection application (using the GLVQ algorithm) and compare

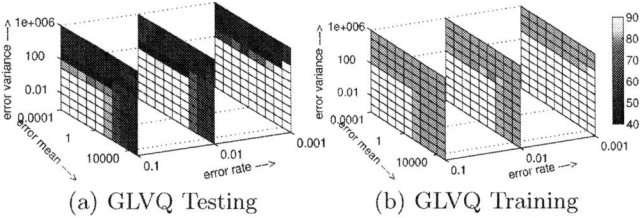

(a) GLVQ Testing   (b) GLVQ Training

Figure 6: Impact of iterative computations on application resilience for eye detection application

the resilience of the training phase that builds the eye detection model with the testing phase that employs the trained model to detect eyes. Although the computation hot spot is the same in both parts, *i.e*, distance computation, the training phase employs an iterative convergence algorithm that keeps refining the model in each iteration, while the testing phase does not. The comparison of the quality profiles is presented in Figure 6. It can be seen that for the same point in the approximation space, the training phase that employs iterative computations is more resilient than the testing phase, demonstrating the contribution of iterative computations to resilience. Similar characteristics were observed in other recognition applications such as handwriting digit recognition (Support Vector Machines), census data analysis (Neural Networks) that contain training and testing phases. The impact of computation

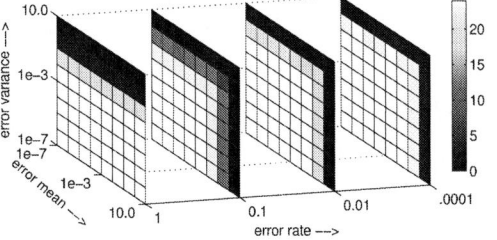

Figure 7: Impact of relative computations on resilience (Document search application)

patterns on resilience is observed in a different form in the document search application (implemented using Semantic Search Indexing). In this application, the input search query and the text documents in the database are represented in terms of vectors and the score of each document is calculated by computing the dot product of its vector with the input query. In the resilience characterization of this application, we observed that

if the variance of the errors introduced due to approximations is constrained to a small value and the errors are introduced in all the computations (rate = 1), then the magnitude of errors can be arbitrarily large without the application quality being impacted. This is because, if errors of similar magnitude are introduced in all the computations, the numerical order of the document scores remains unchanged. This can be seen in the bottom region of the first slice shown in Figure 7. The resilience in such applications can be exploited effectively by constraining the variability of errors introduced due to the approximations.

6.4 Scale of Data Matters

Resilience is a function of scale, since redundancy in the input data is one of the major contributors to resilience. In this section, we analyze the impact of the scale of input data on application resilience. We consider the training phase of hand-written digit recognition, and perform experiments to evaluate its resilience by varying the size of the input data. The results of this evaluation are presented in Figure 8. It can be seen that the resilience of the application improves, as denoted by the lower degradation in the output quality, as we increase the number of samples in the training data set. This result quantitatively demonstrates the impact of redundancy in input data on application resilience. It is essential that the design of approximate computing techniques and their evaluation be performed in the context of representative input data scales.

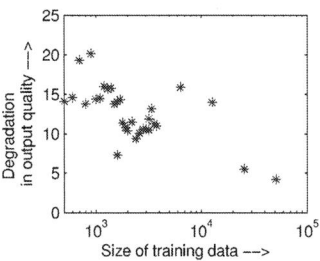

Figure 8: Impact of size of input data

6.5 Impact of Quality Metric

Inherent application resilience stems from the fact that application functionality is defined on a continuous scale of output quality. For many applications, there exists a choice between multiple metrics to measure output quality. We performed ex-

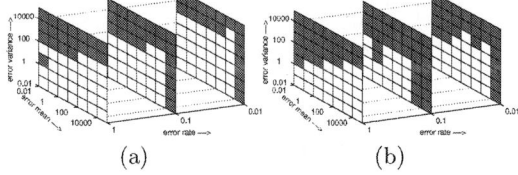

(a)   (b)

Figure 9: Impact of quality metric: (a)mean centroid distance *vs.* (b)percentage of mis-clustered points

periments to study how the choice of quality metric impacts the degree of resilience. We consider the image segmentation application implemented using k-means clustering, and compare two commonly used quality metrics, namely (1) percent mis-clustered points and (2) mean centroid distance. The results are presented in Figure 9 in the form of a slice plot with white regions representing the acceptable quality levels. We normalized the ranges of these quality metrics such that degradation in output quality of less than 1% could be deemed acceptable. It can be seen that the quality profile corresponding to the mean centroid distance metric has a significantly larger white region (acceptable output quality), suggesting that the application is able to tolerate more aggressive approximation if mean centroid distance is the quality metric rather than percentage of mis-clustered points. These results demonstrate that, in addition to the computation patterns and input data, the context in which the application is used (encoded in the quality metric) significantly impacts resilience.

803

6.6 Application-aware Approximation

Approximate computing techniques can be implemented and applied in an application-aware or application-agnostic manner. In order to compare the effectiveness of these techniques, we consider the image segmentation application implemented using k-means clustering, and consider 2 types of approximations, one that skips computations randomly and the other that

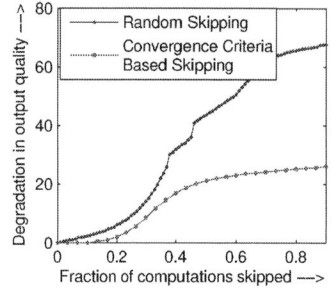

Figure 10: Impact of application semantics

skips computations in an application-aware manner using a technique called early termination [20]. The comparison, presented in Figure 10, shows that the early termination technique results in much better output quality compared to random computation skipping for the same number of computations being affected. Therefore, it is important to understand the semantics of the application and apply approximation techniques accordingly to optimally exploit inherent resilience.

6.7 Synergy between Approximation Techniques

Once the inherent resilience of an application is established, it can be exploited using several different approximate computing techniques. We performed experiments to analyze interactions between approximation techniques that are simultaneously applied to an application. We consider hand-written digit recognition using the SVM algorithm and simultaneously apply application aware computation skipping at the algorithm level

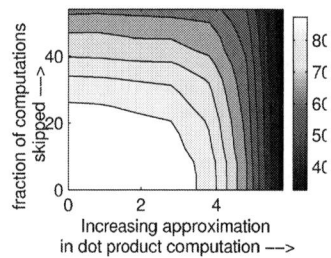

Figure 11: Synergy between algorithm level and arithmetic operation approximations

and a voltage scaled implementation [4] of the dot product computation. It can be observed that these two approximation techniques can be employed in a synergistic manner, i.e. the approximations at the algorithm level can be applied on top of approximations in arithmetic operations without further degrading the output quality represented by vertical and horizontal parts of the contours. However, in some regions (curved portions of the contours), employing both techniques together results in worse output quality compared to a single technique. Therefore, designers should consider combining multiple approximate computing techniques to maximally exploit application resilience.

In summary, we have characterized the resilience of a broad set of applications using the proposed ARC framework and quantitatively established the high degree of resilience present in a broad range of applications. We presented key insights that quantitatively evaluate the relationship between resilience and various application characteristics such as computation patterns, input data, quality metric etc.

7. CONCLUSIONS

In this paper, we proposed an Application Resilience Characterization (ARC) framework for the analysis and characterization of resilience. We implemented the ARC framework using the dynamic binary instrumentation tool valgrind and applied it to a suite of 12 recognition, mining and search applications. We quantitatively established the high degree of re-

silience present in these applications, demonstrating the scope for approximate computing techniques. We performed experiments to explore the relationship between inherent resilience and various application characteristics such as computation patterns, scale of input data, quality metric etc. We believe that these key insights, along with the ARC framework, can greatly aid designers in the adoption of approximate computing.

8. REFERENCES

[1] P. Dubey. A Platform 2015 Workload Model Recognition, Mining and Synthesis Moves Computers to the Era of Tera. White paper, Intel Corp., 2005.

[2] Rajamohana Hegde and Naresh R. Shanbhag. Energy-efficient signal processing via algorithmic noise-tolerance. In Proc. Int. Symp. on Low Power Electronics and Design, pages 30–35, 1999.

[3] Krishna V. Palem, Lakshmi N. Chakrapani, Zvi M. Kedem, Lingamneni Avinash, and Kirthi Krishna Muntimadugu. Sustaining moore's law in embedded computing through probabilistic and approximate design: retrospects and prospects. In CASES, pages 1–10, 2009.

[4] Debabrata Mohapatra et.al. Design of Voltage Scalable Metafunctions for Multimedia, Recoginition and Mining Applications. In Proc. DATE, pages 950–955, 2011.

[5] Vaibhav Gupta et.al. IMPACT: Imprecise Adders for Low-Power Approximate Computing. In Proc. ISLPED, pages 409–414, 2011.

[6] P. Kulkarni, P. Gupta, and M. Ercegovac. Trading accuracy for power with an underdesigned multiplier architecture. In VLSI Design, pages 346–351, 2011.

[7] V.K. Chippa, D. Mohapatra, A.Raghunathan, K. Roy, and S.T. Chakradhar. Scalable Effort Hardware Design: Exploiting Algorithmic Resilience for Energy Efficiency. In DAC'10.

[8] Vinay K. Chippa, Anand Raghunathan, Kaushik Roy, and Srimat T. Chakradhar. Dynamic Effort Scaling : Managing the Quality Efficiency Tradeoff. In Proceedings of the 48th Design Automation Conference (DAC'11), pages 603–608, San Diego, California, USA, 2011. ACM.

[9] Hadi Esmaeilzadeh et.al. Architecture Support for Disciplined Approximate Programming. SIGARCH Comput. Archit. News, 40(1):301–312, March 2012.

[10] S. T. Chakradhar and A. Raghunathan. Best-effort Computing: Re-thinking Parallel Software and Hardware. In Proc. DAC, pages 865–870, 2010.

[11] W. Baek and Trishul M. Chilimbi. Green: A Framework for Supporting Energy-Conscious Programming using Controlled Approximation. In Proc. PLDI, pages 198–209, 2010.

[12] Sidiroglou-Douskos et.al. Managing Performance vs. Accuracy Trade-offs with Loop Perforation. ESEC/FSE, pages 124–134, 2011.

[13] Larkhoon Leem et.al. ERSA: Error Resilient System Architecture for Probabilistic Applications. In DATE, pages 1560–1565, 2010.

[14] Mei-Chen Hsueh et.al. Fault Injection Techniques and Tools. Computer, 30(4):75–82, April 1997.

[15] Jason Cong and Karthik Gururaj. Assuring Application-Level Correctness Against Soft Errors. ICCAD, pages 150–157, 2011.

[16] S. Venkataramani et.al. SALSA: Systematic Logic Synthesis of Approximate Circuits. Proc. DAC, pages 796–801, 2012.

[17] R. Amirtharajah et.al. A Micro-Power Programmable DSP Using Approximate Signal Processing Based on Distributed Arithmetic. In JSSC, pages 337–347, 2004.

[18] S. Rehman, M. Shafique, F. Kriebel, and J. Henkel. Raise: Reliability-aware instruction scheduling for unreliable hardware. In Proc. ASP-DAC, pages 671–676, 2012.

[19] H. Duwe. Exploiting application level error resilience via deferred execution. Master's thesis, University of Illinois at Urbana Champaign, 2013.

[20] J. Meng et.al. Best-Effort Parallel Execution Framework for Recognition and Mining Applications. IPDPS, pages 1–12, 2009.

[21] Department of machine learning. www.nec-labs.com/research.

[22] Thorsten Joachims. Advances in Kernel Methods. chapter Making large-scale support vector machine learning practical, pages 169–184. MIT Press, 1999.

[23] Yann Lecun and Corinna Cortes. The MNIST Database of Handwritten Digits.

[24] D. Martin et.al. A Database of Human Segmented Natural Images and its Application to Evaluating Segmentation Algorithms and Measuring Ecological Statistics. In Proc. 8th Int'l Conf. Computer Vision, volume 2, pages 416–423, 2001.

[25] Richard M. Yoo, Anthony Romano, and Christos Kozyrakis. Phoenix Rebirth: Scalable MapReduce on a Large-Scale Shared-Memory System.

[26] Cheng-Tao Chu et.al. Map-Reduce for Machine Learning on Multicore. In Advances in Neural Information Processing Systems 19, pages 281–288. 2007.

[27] A. Frank and A. Asuncion. UCI Machine Learning Repository, 2010.

[28] Christian Bienia. Benchmarking Modern Multiprocessors. PhD thesis, Princeton University, January 2011.

SUPPLEMENTAL SECTION

A. BENCHMARK SUITE: OVERVIEW AND RESULTS

A.1 Document Search

- Algorithm: Semantic Search Index
- Source code: An industry scale implementation developed at NEC Labs, America [21]
- Data set: A subset of Wikipedia pages, America [21]
- Dimensionality: 100
- No of pages to search: 1863573
- Quality function: No of correct results in top 25 results
- Description: This application takes a text query as input and gives out the top 25 documents based on their similarity to the input query. The input query is first transformed into a vector of dimensionality 100. The application contains the vector representations of all the documents in the wiki sample. The dot product of the query vector with a document vector computes the similarity score between them. The documents are then sorted based on their similarity score and the top 25 documents are considered for our quality evaluation. The quality profile of this application is shown in Figure 7

A.2 Image Search

- Algorithm: Feature Extraction
- Source code: An industry scale implementation developed at NEC Labs, America [21]
- Data set: A database for 7700 pre-classified images taken from NEC Labs, America [21]
- Dimensionality: 128
- No of image categories: 765
- Quality function: No of correct results in top 25 results
- Description: This application takes a query image as input, analyzes its content and outputs image categories sorted based on their similarity to the query image. In this application, the image is first converted into a feature map that is then classified using an SVM classifier to determine the category the image is closest to. We consider the feature extraction step for resilience analysis. The quality profile of this application is presented in Figure 5

A.3 Handwritten Digit Recognition and Model Generation

- Algorithm: Support Vector Machines (Training and Testing)
- Source code: SVM light [22]
- Data set: MNIST database [23]
- Dimensionality: 784
- No of output classes: 10
- Training data size: 10,000
- Testing data size: 60,000
- Quality function: classification accuracy calculated in terms of percentage of correctly classified points
- Description: This application consists of two phases. The training phase generates the model of classification using a labeled training data. This phase of the application is formulated as a quadratic optimization problem and can be solved using off-the-shelf solvers [22]. The model

generated from the training phase is used in testing to classify a new unlabeled data point. The quality profile of training and testing are shown in Figures 12 and 13 respectively.

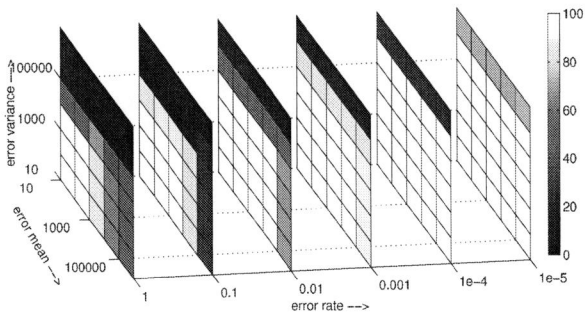

Figure 12: Hand written digit recognition: SVM Training

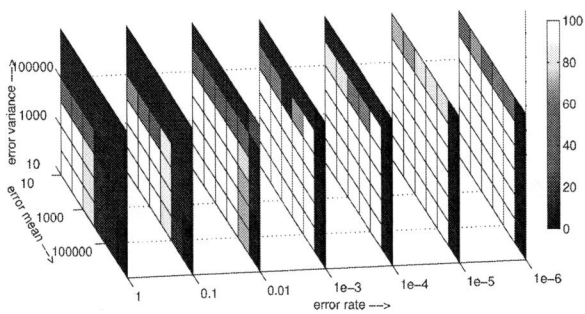

Figure 13: Hand written digit modeling: SVM Testing

A.4 Image Segmentation

- Algorithm: K-means Clustering
- Data set: Berkeley image segmentation dataset [24]
- Source code: An open source implementation taken from PHOENIX [25]
- Dimensionality: 3 (red, green and blue pixel values)
- No of clusters: 4
- Data size: 154401
- Quality function: Percentage of mis-clustered points and mean centroid distance
- Description: This application takes an image as input and segments it based on the Red, Green and Blue components of the pixels. The algorithm start off with random cluster centroids and assigns the pixels to centroids they are closest to based on Euclidean distance. The centroids are then recalculated as the mean of the points assigned to them. These two steps are performed iteratively until a convergence criterion is satisfied. The quality profile of this application for two quality metrics, mean centroid distance and % mis-clustered points is presented in 14 and 15.

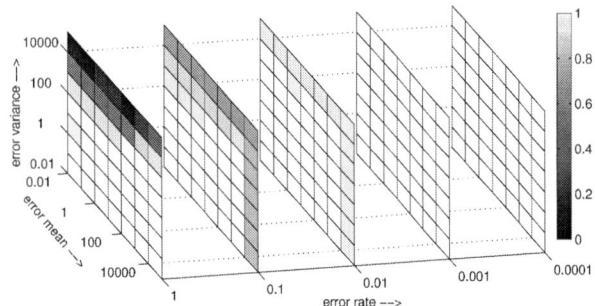

Figure 14: Image segmentation with mean centroid as quality metric: K-means Clustering

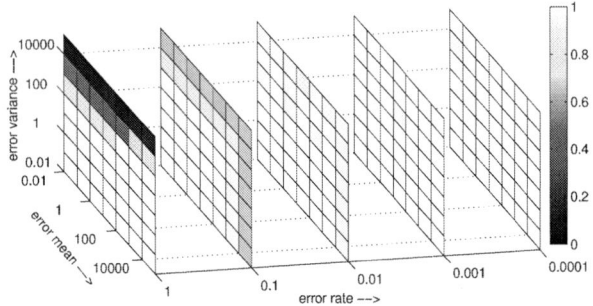

Figure 15: Image segmentation with percentage of misclustered points as quality metric: K-means Clustering

A.5 Eye Detection (Model Generation and Classification)

- Algorithm: Generalized learning vector quantization (Training and Testing)
- Source code: An industry scale implementation developed at NEC Labs, America [21]
- Data set: Set of eye and non-eye images from NEC Labs, America [21]
- Dimensionality: 500
- No of output classes: 2
- Training data size: 6000
- Testing data size: 1467
- Quality function: Classification accuracy represented in terms of percentage of correctly classified points.
- Description: This application consists of training and testing phases. In the training phase, the classification model is generated using a set of reference vectors which are updated using a labeled training data set. For each training vector, the closest reference vector from the same category as the training vector is modified such that it moves closer to the training vector. The closest reference vectors from the other categories are updated such that they move away from the training vector. The testing phase takes an unlabeled data as input and computes its distance from all the reference vectors. The input data is then labeled with the category of the reference vector it is closest to. The quality profiles of the training and testing phases of this application are depicted in Figures 6b and 6a

A.6 Census Data Classification (Model Generation and Classification)

- Algorithm: Neural Networks (Training and Testing)

- Source code: Single thread version from Mapreduce benchmark suite [26]
- Data set: UCI census database [27]
- Dimensionality: 14
- No of output classes: 2
- training data size: 32,560
- testing data size: 16,282
- Quality function: Classification accuracy represented in terms of percentage of correctly classified points.
- Description: This application employs Neural Networks to estimate the salary of a person based on the information in the census database. A 3-layer neural network is trained using back propagation algorithm to determine the weights of connections in neural network. In the testing phase, the feed forward algorithm is employed on the trained network to determine a person's salary. The quality profiles for the back propagation (training) and forward propagation (testing) parts of this application are presented in Figures 16 and 17

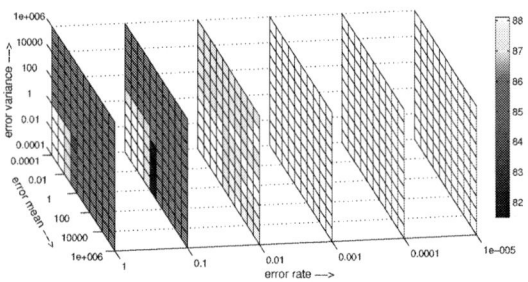

Figure 16: Census data modeling: Neural Networks Back Propagation

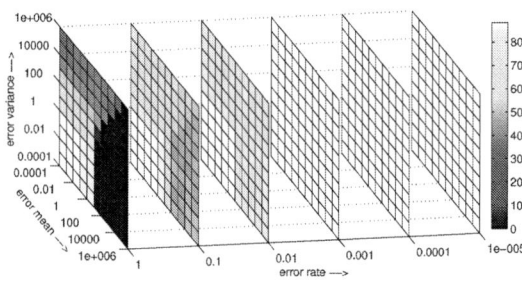

Figure 17: Census data analysis: Neural Networks Forward Propagation

A.7 Nutrition and Health Information Analysis

- Algorithm: Logistic Regression
- Source code: Single thread version from Mapreduce benchmark suite [26]
- Data set: National Health and Nutrition Examination Survey
- Data size: 17000
- Dimensionality: 15
- No of output classes: 2
- Quality function: Percentage classification accuracy
- Description: In this application, logistic regression is used to assess the likelihood of a disease or health condition as a function of risk factors. We use a quadratic optimization based implementation that generates a linear

806

model to minimize the error on the labeled training data. The dot product operation between the training data and the model is the dominant kernel computation in the quadratic optimization. The quality profile of this application for approximations in the dot product kernel is presented in Figure 18

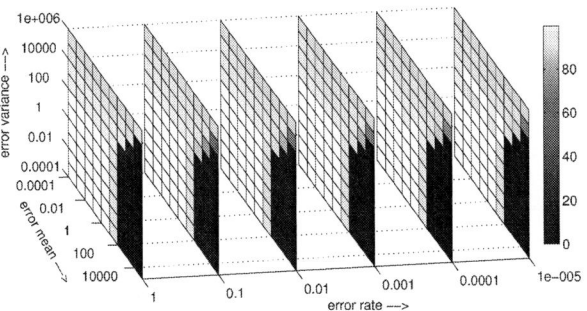

Figure 18: Health and nutritional information analysis: Logistic Regression

A.8   Digit Recognition

- Algorithm: K-Nearest Neighbors
- Source code: Self implementation
- Data set: UCI digit recognition database [27]
- Dimensionality: 64
- Training data size: 3823
- Testing data size: 1797
- No of output classes: 10
- Quality function: Percentage classification accuracy
- Description: This application uses the K-nearest neighbors algorithm to recognize digits from images. The algorithm computes the Euclidean distance between the test image and all training images. The training images are ranked based on the proximity to the test image, and the top K among them are identified. A majority vote among these nearest training images determines the classification. The quality profile, shown in Figure 19, illustrates the significant amount of resilience present in the distance computations.

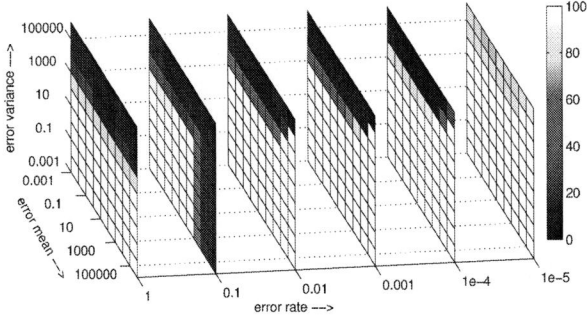

Figure 19: Digit recognition: K-Nearest Neighbors

A.9   Online Data Clustering

- Algorithm: Streamcluster
- Source code: Parsec benchmark suite [28]
- Data set: simmedium dataset provided with Parsec [28]
- Dimensionality: 64

- Data size: 8192
- No of output classes: Adaptive according to the input data
- Quality function: Mean centroid distance
- Description: Streamcluster finds the optimal number of clusters from a stream of input points where each input point is assigned to the closest cluster in terms of Euclidean distance from the cluster centroid. For each incoming set of points, the application employs heuristics [28] to determine the optimal number of clusters. The distance computation kernel is considered for approximation and the corresponding quality profile is presented in Figure 20.

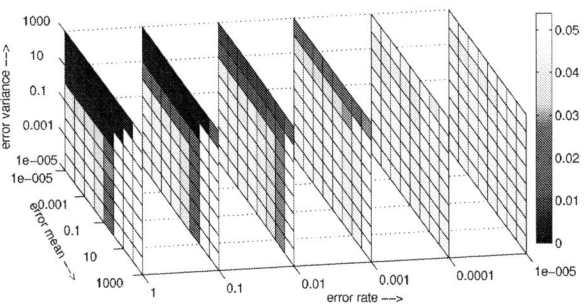

Figure 20: Online data clustering: Streamcluster

Dynamic Voltage and Frequency Scaling for Shared Resources in Multicore Processor Designs

Xi Chen*, Zheng Xu*, Hyungjun Kim*, Paul V. Gratz*, Jiang Hu*, Michael Kishinevsky**, Umit Ogras** and Raid Ayoub**
*Department of Electrical and Computer Engineering, Texas A&M University
**Strategic CAD Labs, Intel Corporation

ABSTRACT

As the core count in processor chips grows, so do the on-die, shared resources such as on-chip communication fabric and shared cache, which are of paramount importance for chip performance and power. This paper presents a method for dynamic voltage/frequency scaling of networks-on-chip and last level caches in multicore processor designs, where the shared resources form a single voltage/frequency domain. Several new techniques for monitoring and control are developed, and validated through full system simulations on the PARSEC benchmarks. These techniques reduce energy-delay product by 56% compared to a state-of-the-art prior work.

1. INTRODUCTION

Modern multicore processor designs face challenges across multiple fronts, including: the communication bottleneck, off-chip and on-chip, and the approaching fundamental limits of chip power density. Together these constraints conspire to reinforce one another. Recent designs have resorted to increasing cache size to circumvent the off-chip memory bottleneck. A large cache in turn demands an increase of on-chip communication bandwidth. Indeed, on-chip communication fabrics and shared, last-level caches (LLCs) have grown to occupy a large portion of the overall die area, as much as 30% of chip area in recent Intel chip multiprocessors [10]. Such growth inevitably exacerbates the power challenge.

Networks-on-Chip (NOC) are recognized as a scalable approach to addressing the increasing demand for on-chip communication bandwidth. One study shows that NOCs can achieve 82% energy savings compared to conventional bus design in a 16-core system [9]. Nonetheless, the NOC still accounts for a considerable portion of total chip power, e.g., 36% in MIT RAW architecture [19]. When the workload is small, some cores can be shut down to save leakage power. However, NOC and LLC need to stay active for serving the small workload and therefore their power proportion is even greater. The power-efficiency of NOC and LLC can be improved by Dynamic Voltage and Frequency Scaling (DVFS), with the rationale that power should be provided based on dynamic need instead of a constant level. DVFS has been in-

Permission to make digital or hard copies of all or part of this work for personal or classroom use is granted without fee provided that copies are not made or distributed for profit or commercial advantage and that copies bear this notice and the full citation on the first page. To copy otherwise, to republish, to post on servers or to redistribute to lists, requires prior specific permission and/or a fee.
DAC'13, May 29 - June 07 2013, Austin, TX, USA

tensively studied for individual microprocessor cores as well as the NOC [17, 12, 18, 15, 6, 13, 16, 3, 4] [1]. Much of this prior work assumes a core-centric voltage/frequency (V/F) domain partitioning. The shared resources (NOC and/or LLC) are then divided and allocated to the core-based partitions according to physical proximity. While such configuration allows a large freedom of V/F tunings, the inter-domain interfacing overhead can be quite large. Furthermore, as these shared resources are utilized as a whole, with cache line interleaving homogenizing traffic and cache slice occupancy, per-slice V/F tunings makes little sense.

This work focuses on a realistic scenario where the entire NOC and LLC belong to a single V/F domain. As such, the interfacing overhead can be largely prevented and there is a coherent policy covering the whole of these shared resources. To the best of our knowledge, only two works [12, 4] have addressed DVFS for such scenario. Liang and Jantsch propose a rule-based control scheme, using network load as the measured system performance metric [12]. Chen et al. examine the motivation and advantages of the single shared V/F domain in detail [4]. They use a PI controler based on AMAT (Average Memory Access Time) and a low-overhead AMAT monitoring technique is proposed. Although both works demonstrate the benefit of DVFS, there are two critical hurdles that have not been well solved. First, the impact of the NOC/LLC V/F level on the chip energy-performance tradeoff is not straightforward. These prior works shy away from this problem by evaluating only parts of the chip system. Second, the chip energy-performance tradeoff is dynamic at runtime while the controls of these prior approaches are based on fixed reference points.

In this paper, we present remarkable progress on overcoming these hurdles. Three new methods are proposed and investigated. First, a throughput driven controller with dynamic reference point is examined. Second is a model assisted PI controller based on a new metric that bridges the gap between the NOC/LLC V/F level and the chip energy-performance tradeoff. The last one is a PI controller with a dynamic reference point based on the new metric. These methods are evaluated in full system simulation on the PARSEC benchmarks [1]. The experimental results show that our techniques can reduce NOC/LLC energy by $\sim$ 80% and $\sim$ 50% compared to a baseline fixed V/F level (no power management) and the state-of-the-art prior work [4], respectively. Simultaneously, we achieve our target of $\leq$ 5% performance loss. Compared to the competing design [4], the

---

[1] Although supply voltage change may affect SRAM read/write stability, Kirolos and Massoud show that DVFS for SRAM is feasible [8].

energy-delay product is reduced by 56%.

2. RELATED WORK

Shang et al. present a pioneering work on dynamic voltage scaling in NOCs [17]. They tune voltage levels for individual links separately according to their utilization history. Mishra et al. propose DVFS techniques for NOC routers [13]. They monitor input queue occupancy of a router, based on which the upstream router changes its V/F level. Son et al. perform DVFS on both CPUs and network links [18]. They target to parallel linear system solving and the V/F levels are decided according to task criticality. Guang et al. propose a voltage island based approach [6], where router queue occupancies are monitored and island V/F levels are tuned accordingly. Rahimi et al. take a similar rule-based approach according to link utilization and queue occupancy [16]. Ogras et al. propose a formal state-space control approach also for voltage island based designs [15]. Bogdan et al. introduce an optimal control method using a fractional state model [3]. In drowsy caches, dynamic voltage scaling is applied at certain cache lines at a time for reducing leakage power [5]. To the best of our knowledge, there is no published work on DVFS which focuses on shared caches in multicore chips.

Liang and Jantsch present a DVFS controller that attempts to maintain the network load near its saturation point, where the load is the number of flits in the network [12]. At each control interval, its policy is to increase (decrease) network V/F level by one step if the network load is significantly greater (less) than the saturation point. This method neglects the fact that chip performance also depends on the distribution besides the amount of network load. A non-uniform distribution may imply certain congestion hotspots, which may significantly degrade chip performance. Even with consideration of the distribution, network load does not always matter. For example, many *store* operations induce large network load but they are not critical to the overall chip performance. Liang and Jantsch's method may respond slowly for bursty traffic as only one step V/F change is allowed in each control interval.

Chen et al. introduce a PI controller based on AMAT [4]. AMAT, as they formulate it, including the effects of the private caches, NOC, LLC and off-chip memory, reflects network load and contention inherently, providing an approximation of the latency seen by typical core memory references. Therefore, it captures a more global system effects than a purely network-based metric such as Liang and Jantsch's approach [12]. Their AMAT metric, however, does not truly separate out the LLC and NOC utility to the core from the effects of the off-chip memory. Thus, applications which frequently miss in the LLC, causing off-chip memory accesses, will lead to high AMAT values, and thus high LLC and NOC frequencies, despite the LLC utility being low in this case. Chen et al.'s DVFS policy uses a PI controller, subsuming that of Liang and Jantsch. This work also describes implementation techniques on how to monitor AMAT of multicore with low overhead.

Both Liang and Jantsch's [12] and Chen et al.'s [4] approaches suffer from another weakness, they have no systematic approach to decide the reference point for their controllers. In Chen et al.'s work, the reference point is decided empirically according to offline simulations. However, it is very difficult, if not impossible, for a fixed reference point to be appropriate for different kinds of applications. Moreover, neither work provides performance results from full-system simulation to validate their approach.

3. PRELIMINARIES

3.1 Problem Description

We consider a common case in multicore processor design where the entire chip is composed of an array of tiles. Each tile contains a processor core and private caches. The communication fabric is a 2D mesh NOC with one router residing in each tile. There is a shared LLC partitioned into slices and distributed uniformly among tiles. The NOC and the LLC together are referred to as the *uncore* system in this paper. The system is illustrated in Figure 1 where NI denotes Network Interface with cores.

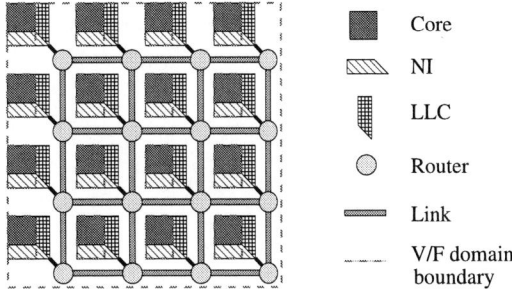

Figure 1: A multicore processor design where the uncore (NOC+LLC) forms a single V/F domain.

The problem we attempt to solve is formulated as follows. **Uncore Dynamic Voltage and Frequency Scaling:** *within a set time window, find the voltage/frequency level for the uncore such that the uncore energy dissipation is minimized while the chip performance, in terms of total application runtime, has negligible or user-specified degradation.*

Please note that our problem formulation has a key difference from previous works on NOC DVFS, which try to optimize the performance of NOC itself, not considering directly its utility to the system. In contrast, we have the more challenging goal of optimizing uncore energy under the constraint of entire system performance. We present the first work we are aware of in this area with such formulation. The uncore energy we try to minimize includes both dynamic and leakage energy and their models are well-known.

3.2 Options for DVFS Policy

Broadly speaking, there are two categories of approaches for DVFS: open-loop control and closed-loop control. Open-loop control decides control variables based on the current system state and a system model obtained either theoretically or through machine learning. The behavior of a multicore system is typically very complex. Even if a decent model is available, its behavior depends on environmental parameters that are highly dynamic and thus are very difficult to reliably predict.

Closed-loop control adjusts control variables with consideration of observed output. It includes several options: rule-based, PID (Proportional-Integral-Differential) control, state-space model-based control and optimal control. In a rule-based approach, a set of *ad hoc* rules are determined, such as simply increase (decrease) uncore V/F level if the network performance is poor (good) according to given metric. Based on the observed output error, PID control adjusts the system to track a target output. State-space model-based control formally synthesizes a control policy that guarantees system stability [15]. Optimal control [3] decides system operations by solving an optimization problem and sometimes can be applied with a state-space model.

We adopt PID control due to its simplicity, flexibility, low implementation overhead and guaranteed stability. In a discrete-time system, where the output is a time-varying function y_i and the control variable is u_i for control interval i, the error function is defined by $e_i = y_{ref} - y_i$, where y_{ref} is the reference point or target output. Since differential control is sensitive to data noise, we drop the differential term. A PI controller can be described by:

$$u_i = u_{i-1} + K_I \cdot e_i + K_P \cdot (e_i - e_{i-1}) \qquad (1)$$

where K_P and K_I are constant parameters. Please note that the problem described in Section 3.1 is an optimization problem, which implies a dynamic goal instead of a steady target in typical PI controls. To bridge this gap, it is very important to use a dynamic reference point, as shown in subsequent sections, instead of fixed reference as in [4].

4. DESIGN DESCRIPTION

4.1 Throughput-Driven DVFS

One technique we propose is to use a throughput metric such that a naturally dynamic reference is enabled for the DVFS PI control.

Throughput is the amount of data processed by the uncore per unit time. It can be measured by $R_{U,Out}$, which is the rate of data flowing out of the uncore to cores. Injection rate $R_{U,In}$, on the other hand, is defined as the cores' requested data rate. Figure 2 depicts the throughputs for various injection rates with respect to varying uncore frequency. As the uncore frequency increases, we observe that the throughput increases asymptotically to the given injection rate. In a transient period, e.g., control interval i, if the uncore frequency is not high enough, the throughput $R_{U,Out,i}$ can be different from data injection rate $R_{U,In,i}$ of the same interval. We have then a PI controller with state (output) variable $R_{U,Out}$ and reference point $R_{U,In}$, which is dynamically decided during operations. Intuitively, such controller ensures that $R_{U,Out}$ becomes $R_{U,In}$, and it possibly saves energy by setting the uncore V/F to a lower level than the maximum level.

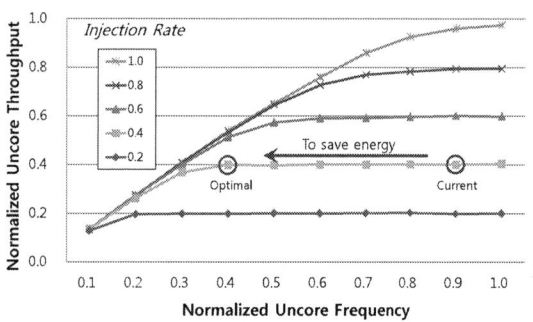

Figure 2: Throughput vs. uncore frequency.

Finding the optimal uncore V/F level here, however, is not trivial when the uncore V/F level is higher than necessary. For example, in Figure 2, where the given injection rate is 0.4 and the current normalized uncore frequency is 0.9, the controller must reduce the frequency to 0.4 to save energy. In such a case, unfortunately, the error $(R_{U,In} - R_{U,Out})$ becomes zero and the PI controller will not change the uncore V/F level. To avoid this, we magnify $R_{U,Out}$ by a very small percentage δ, e.g., 1%, resulting in a slightly negative value

of the error. By doing so, the controller moves the V/F level down to the saturation point for the given injection rate without significantly hurting the throughput.

The proposed controller requires care to evaluate the injection rate $R_{U,In}$, especially when the program phase changes to increase $R_{U,In}$. Typically, a core stalls after issuing a number of requests yet to be served. The maximum value of outstanding requests is determined by the capacity of its MSHR (Miss Status Handling Register) for out-of-order processors, or it is 1 in case of in-order processors. At this point, the core has to wait till a request is served by the uncore, and then it will resume injecting further requests. If there are a sizable number of such stalled cores, the overall injection rate does not grow enough to increase the uncore frequency, although each core actually starts generating more requests than before. To address this issue, the number of stalled cores and the duration of waiting are considered. If the number of cycles when there are more than N cores waiting their uncore requests is M, and the size of control interval is P, the effective injection rate, which is also the reference request rate for the controller, is estimated by $R_{ref,i} = R_{U,In,i} \cdot (1 + \frac{M}{P})$ where i is the index of control interval. Then, the overall error function for the throughput-driven controller is:

$$e_i = R_{U,In,i} \cdot \left(1 + \frac{M}{P}\right) - R_{U,Out,i} \cdot (1 + \delta) \qquad (2)$$

where δ is a small number, e.g., 1%. One can use a greater value of N to achieve more energy savings but at the higher performance cost. The value of N is set to 3 in our experiment for the best energy-performance trade off.

4.2 Latency-Based DVFS

4.2.1 Composite Uncore Metric – Critical Latency

The throughput-driven DVFS described in Section 4.1 largely solves the dynamic reference problem. Although the metric used by the technique captures the performance of the uncore well, it is still oblivious of the overall chip performance. Ideally, the metric should reflect both the uncore performance and its criticality to the overall chip performance. We therefore define a new metric - *critical latency*, expressed by

$$\Gamma = \eta \cdot \lambda_U \qquad (3)$$

where η is the criticality factor and λ_U is the uncore latency.

The uncore latency should account for the latency in the network and LLC. Subtracting the reply return from request inject time, one can obtain the packet latency; however, this latency may contain off-chip memory latency in the event of an LLC miss. This memory latency should be excluded from consideration since it is not affected by the uncore DVFS. In certain cases, the LLC miss can be very high and the overall data latency is dominated by the memory latency. In this case, increasing uncore V/F does not help to improve chip performance while causes more power dissipation [2]. The uncore latency can be described by

$$\lambda_U = \frac{\left(\sum_{j=1}^{N_{packets}} \lambda_{packet,j}\right) - \lambda_{Mem} \cdot N_{LLC\_Misses}}{N_{packets}} \qquad (4)$$

---

[2]We intentionally do not differentiate between packets which lead to coherence traffic and those which do not. Coherence related traffic can increase LLC latency and is sensitive to uncore V/F state.

where $\lambda_{packet,j}$ is the total round-trip latency for packet j, λ_{Mem} is the memory access latency, N_{LLC_Misses} is the number of LLC misses in a control interval and $N_{packets}$ is the number of packets in the same interval.

In microprocessors, both *store* and *load* data induce network traffic and potentially LLC access. These two types of packet requests have different impact on the overall chip performance. Often, a long latency *load* can block the execution of instructions that need the data, and therefore is performance critical. In contrast, a *store* operation can often run in parallel with subsequent instructions and is rarely critical. Thus, the criticality factor of uncore performance includes *Loads_Fraction*, which is the number of *load* instructions per cycle. We scale *Loads_Fraction* by the L1 miss rate, assuming L1 is the only level of private cache, because loads which hit in the private caches never enter the uncore and are not affected by uncore performance. Therefore, we have

$$\eta = L1\_Miss \times Loads\_Fraction \qquad (5)$$

4.2.2 Model Assisted PI Controller

The error function in the critical latency-based PI controller is

$$e_i = \Gamma_{ref} - \Gamma_i \qquad (6)$$

where Γ_i is the monitored critical latency for control interval i and Γ_{ref} is the reference point for the PI controller. Although Γ should be correlated with the overall chip performance, its target value is not obvious. Of course, one may select one empirically from offline simulations. However, the offline testcases might not behave the same as online cases.

To address this problem, we propose a model assisted PI control method. From Equation (3-5), we can see that Γ is approximately a linear function of λ_{packet}, which is in turn proportional to the uncore clock period T_U. Hence, we have

$$\Gamma = \alpha \cdot T_U + \beta \qquad (7)$$

where α and β are coefficients independent of T_U. Within each program phase, the program execution behaviors are generally consistent so that the variations of α and β are often limited. Based on $(T_{U,i}, \Gamma_i)$ of current interval and $(T_{U,i-1}, \Gamma_{i-1})$ of the previous interval, we can estimate the values for α and β. Then, we can predict the $\Gamma(T_U)$ function of the next control interval.

This prediction can guide reference V/F toward a more aggressive or more conservative level. We first find three different reference points empirically, one is normal, another one is aggressive and the other one is conservative. Then, we dynamically choose among them at runtime according to the uncore frequency $\frac{1}{T_{U,i+1}}$ computed from the predicted $\Gamma(T_U)$ function. By default, the normal reference point is employed. If the model based frequency is significantly higher (lower), the reference point is changed to the aggressive (conservative) one. If there is no frequency change in two consecutive control intervals, we cannot obtain an update on α and β values. In this case, we continue to use the reference of the previous control interval.

4.2.3 Latency-Based PI Control with Dynamic Reference

The model-assisted controller is conceptually superior to that proposed by Chen et al [4], where one fixed reference point is used. The improvement, however, can be limited as the model-based prediction may be inaccurate. To fix this problem, we examine an alternate PI controller that allows a truly dynamic reference point. For the uncore latency

λ_U, defined in Equation (4), we can determine our desired target. The λ_U mainly consists of the propagation latency in the network, the serialization latency, the queuing latency in the network and the LLC access latency. When there is no congestion, the queuing latency should approach zero. Ideally, we want to keep the network lightly loaded. Thus, we can define a reference uncore latency as

$$\lambda_{ref} = (1 + \rho) \cdot (2 \cdot (\lambda_{hop} \cdot N_{hops} + L_{packet}) + \lambda_{LLC}) \qquad (8)$$

where ρ is a constant selected to be 0.1, λ_{hop} is the propagation (without queuing) latency per hop, N_{hops} is the average number of hops for packets in a uniformly random traffic, L_{packet} is the average packet length and λ_{LLC} is the LLC access latency. The packet length L_{packet} is in terms of flits and to account for the serialization latency. The coefficient 2 in Equation (8) is to cover the round-trip. The underlying reason for including the ρ is the same as having the δ in Equation (2). The δ allows the throughput to be slightly lower than the injection rate, i.e., very limited network congestion. Likewise, the ρ allows the queuing latency to be slightly above zero, which is also equivalent to a very small network congestion. Please note that λ_{ref} is a constant and can be pre-characterized offline. Overall, the reference for the critical latency becomes:

$$\Gamma_{ref,i} = \eta_i \cdot \lambda_{ref} \qquad (9)$$

As the criticality factor η_i varies from interval to interval, this reference is dynamic with respect to packets at runtime.

4.3 Stability of PI Control with Dynamic Reference

Whether the reference is static or dynamic does not affect the stability analysis for a PI control. Thus, the stability analysis proposed by Chen et al. [4] can be directly applied here to guarantee the stability of our PI controllers.

4.4 Design Implementation

The implementation of the proposed DVFS methods mainly include: (1) information collection at each tile; (2) a central controller that aggregates the collected information and performs control policy computation; (3) information transportation from tiles to the central controller.

For the critical latency-based DVFS controls, several registers are required to save relevant information at each tile. The bit-width of each register is decided by the range of the data to be saved. Here we show a design for our experiment setting. We use three 16-bit registers to save the numbers of *load* instructions, private cache hits and private cache misses, respectively. Additionally, there is a 20-bit register for accumulating the total request latency. Another 12-bit register is used to count the number of LLC misses. A 16-bit register is needed to save the uncore request count. Last, there is a 16-bit counter to keep track of control interval. Overall, 112 bits of registers are required for each tile.

In modern multicore processor designs, e.g., Intel's Nehalem architecture [11], there exists a Power Control Unit (PCU), which is a small processor dedicated to chip power management. Thus, the control of our DVFS can be implemented using PCU without additional hardware. The PCU retains a lookup table with each entry containing the data for a given tile. The PCU also needs to keep the reference points and parameters for the PI control. Typically the PCU has plenty of storage space for these needs. At the end of each control interval, the PCU computes the critical

811

Parameter	Values
Core Frequency	1GHz
#processing cores	16
L1 data cache	2-way 256Kb, 2 core cycle latency
L2 cache (LLC)	16-way, 2MB/bank, 32MB/total, 10 uncore cycle latency
Directory cache	MESI, 4 uncore cycle latency
NoC	4 × 4 2D mesh, X-Y DOR, 4 flits depth/VC
Voltage/Frequency	10 levels, voltage: 1V–2V, frequency: 250MHz–1GHz
V/F transition	100 core cycles per step

Table 1: Simulation setup.

latency and uncore V/F level. The computations are several arithmetic operations and thus can be carried out quickly.

For the transmission of status information from the cores to the PCU, we use a method similar to that proposed by Chen et al. [4]. When a packet is sent out from a tile, the 96 bits information (by excluding the control interval counter) is scaled to 64 bits and is embedded in the header flit. If the packet reaches or passes by the tile where the PCU resides, the data is scaled back to 96 bits and downloaded to the lookup table. Within each control interval, later data from a tile overwrites the old data from the same interval. Since we do not use additional network or send dedicated packets, the data transportation overhead is fairly low. Chen et al. [4] showed that a single monitor tile can obtain sufficient sample data in a control interval of $50K$ clock cycles.

For the throughput-driven DVFS, the registers are required only at the PCU to count the number of requests issued (14 bits) and the number of requests served (14 bits). These counters are incremented whenever its corresponding event occurs. We have a register for the number of cores currently stalled (4 bits) which counts the number of signals from each tile enabled when the core is stalled. We need another register (16 bits) to evaluate M of Equation (2) which is incremented when the number of currently stalled cores exceeds a certain number. Because this technique requires the PCU have more timely measurement data, the PCU updates these registers via 3 directly connected lines from each tile. For each tile, each line signals when a packet request is issued, a request is served or the core is stalled, respectively. Overall, the implementation overhead of the throughput-driven DVFS in smaller CMPs is much lower than that of the latency-based controls, though its scalability in very large CMPs may be limited by the requirement of directly connected status wires.

5. EVALUATION

5.1 Experiment Setup

The baseline architecture in our experiments is a 16-tile chip multiprocessor with a 2-level cache hierarchy similar to the Sun SPARC T3 processor. Here, each tile is composed of a processing core with 1-level of private cache, a network interface, an NOC router and a partition of the shared L2 cache (LLC) and directory. Each core is an in-order Alpha ISA processor. Table 1 summarizes our experimental configurations and parameters. We use Gem5 [2] full system simulator with PARSEC shared-memory multiprocessor benchmarks [1]. For each benchmark, the entire application is run in full-system mode; the results obtained are based upon statistics from the region of interest (ROI), which is usually hundreds of millions of cycles long. Gem5

"Ruby" memory hierarchy (L1-LLC+directory) and "Garnet" network simulator are used for all results. Frequency scaling for uncore is emulated by changing the latency of each uncore component. For instance, the latencies of each router pipeline stage, link traversal time and the LLC access time are doubled when the uncore frequency is reduced to 50%. As off-chip memory is not affected by uncore DVFS, its access latency is assumed to be a constant. Since the focus of this work is uncore DVFS, we assume constant core frequency. We expect the proposed techniques should work well with core DVFS as this would be expressed through decreased L1 demand, decreasing the utility of the uncore. The control interval for the uncore DVFS is 50K core clock cycles. According to our experience and existing literature, such interval size allows sufficient time for the uncore V/F change to settle, and is sufficiently small to capture fine-grain program phase behavior. Both dynamic and leakage power are considered in the experiment. We use ORION 2.0 [7] and CACTI 6.0 [14] based on $65nm$ technology for the power models of NOC and LLC, respectively. The overall performance is evaluated as the execution time for the ROI of each application.

In this work, we compare the following methods:
Baseline constantly high uncore V/F level.
AMAT+PI Chen, et al.'s method [4].
CL+PI PI control based on the Critical Latency described in Section 4.2.1.
CL+ModelAssist the Critical Latency-driven, model assisted PI control described in Section 4.2.2.
Throughput the throughput driven PI control with dynamic reference method described in Section 4.1.
OurBest the Critical Latency-driven, PI control with dynamic reference described in Section 4.2.3.

5.2 Energy and Performance Comparisons

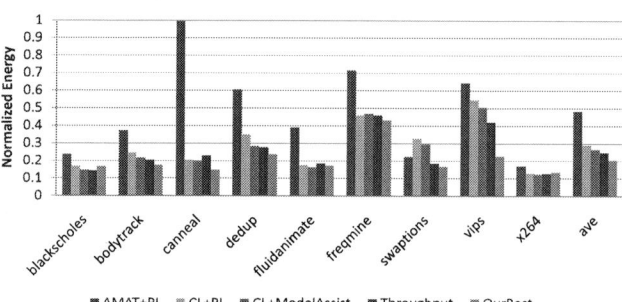

Figure 3: Normalized energy (with the baseline result as 1) for PARSEC benchmarks.

Figure 3 shows the energy comparison for the different methods. The figure shows, the critical latency defined by Equation (3) leads to significantly more power savings than that from AMAT [4]. This is especially obvious for the case *canneal*. The model-assist technique can further reduce power dissipation. Our best method can provide additional 50% power reduction over prior techniques [4].

Figure 4 provides a comparison of the performance impact of the different methods. Except *dedup*, the performance degradation from all of our techniques is quite limited. In the worst case of *dedup*, model assist and our best control progressively improve the performance. Overall, the performance loss from all our methods is around only 5%. In Table 2, the normalized energy-delay product among all cases are listed. The progressive improvement from the new metric (critical latency), model assist and dynamic reference can

812

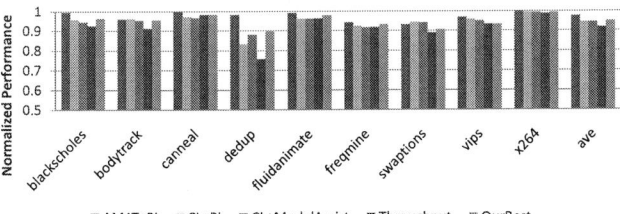

Figure 4: Normalized system performance (with the baseline result as 1) for PARSEC benchmarks.

be observed. Compared to AMAT+PI [4], our best method reduces the energy-delay product by 56%.

Method	$Energy \times Delay$
Baseline	1.0
AMAT+PI	0.5
CL+PI	0.31
CL+ModelAssist	0.28
Throughput	0.27
OurBest	0.22

Table 2: Normalized energy-delay product on average for all PARSEC cases.

To analyze controller sensitivity to parameters, we varied the parameters to obtain different solutions for the throughput-based and our best methods. The results are plotted in Figure 5 together with those from AMAT+PI and CL+ModelAssist. Points in Figure 5 indicate the average results among all benchmarks using the same set of parameters. Our target performance loss is 5%, thus we choose the point from the OurBest curve which is closest to 1.05. The solutions at lower-left envelope for each technique are Pareto optimal as they imply either high performance or low energy consumption. One can see that OurBest, the latency-based PI control with dynamic reference, achieves the best performance-energy tradeoff regardless of parameter set point. On the other hand, please note that the throughput-driven DVFS has much lower implementation overhead as shown in Section 4.4.

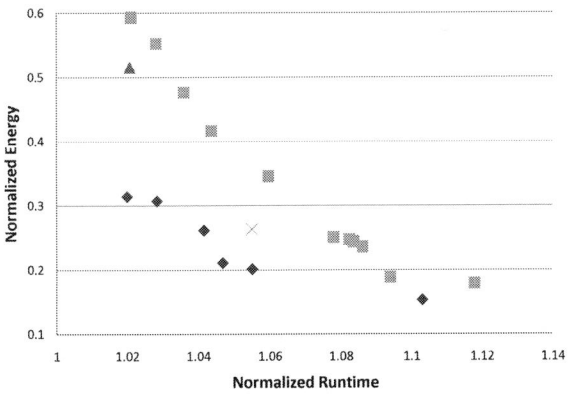

Figure 5: Enery-runtime tradeoff comparison.

6. CONCLUSIONS

In this work, we investigated DVFS for shared resources (NOC/LLC) in multicore processor designs. Several metrics and policies are developed. The proposed techniques are

evaluated on public domain architecture benchmarks with full-system simulation. The results show quite large energy savings and improvement over recent previous work.

7. ACKNOWLEDGMENTS

This work is supported by Intel.

References

[1] Bienia, C., Kumar, S., Singh, J.P., and Li, K. The PARSEC Benchmark Suite: Characterization and Architectural Implications. In *PACT*. 2008.

[2] Binkert, N., Beckmann, B., Black, G., Reinhardt, S.K., et al. The gem5 simulator. *ACM Computer Architecture News*, 39(2):1–7, May 2011.

[3] Bogdan, P., Marculescu, R., Jain, S., and Gavila, R.T. An optimal control approach to power management for multi-voltage and frequency islands multiprocessor platforms under highly variable workloads. In *NOCS*, pages 35–42. 2012.

[4] Chen, X., Xu, Z., Kim, H., Gratz, P., et al. In-network monitoring and control policy for dvfs of cmp networks-on-chip and last level caches. In *NOCS*, pages 43–50. 2012.

[5] Flautner, K., Kim, N.S., Martin, S., Blaauw, D., et al. Drowsy caches: simple techniques for reducing leakage power. In *ISCA*, pages 148–157. 2002.

[6] Guang, L., Nigussie, E., Koskinen, L., and Tenhunen, H. Autonomous DVFS on supply islands for energy-constrained NoC communication. *Lecture Notes in Computer Science: Architecture of Computing Systems*, 5455/2009:183–194, 2009.

[7] Kahng, A.B., Li, B., Peh, L.S., and Samadi, K. ORION 2.0: a power-area simulator for interconnection networks. *TVLSI*, 20(1):191–196, January 2012.

[8] Kirolos, S. and Massoud, Y. Adaptive SRAM design for dynamic voltage scaling VLSI systems. In *MWSCAS*, pages 1297–1300. 2007.

[9] Konstantakopoulos, T., Eastep, J., Psota, J., and Agarwal, A. Energy scalability of on-chip interconnection networks in multicore architectures. Technical report, MIT Computer Science and Artificial Intelligence Laboratory, November 2007.

[10] Kowaliski, C. Gelsinger reveals details of Nehalem, Larrabee, Dunnington, 2008.

[11] Kumar, R. and Hinton, G. A family of 45nm IA processors. In *ISSCC*, pages 58–59. 2009.

[12] Liang, G. and Jantsch, A. Adaptive power management for the on-chip communication network. In *Proeedings of the Euromicro Conference on Digital System Design*. 2006.

[13] Mishra, A.K., Das, R., Eachempati, S., Iyer, R., et al. A case for dynamic frequency tuning in on-chip networks. In *MICRO*, pages 292–303. 2009.

[14] Muralimanohar, N., Balasubramonian, R., and Jouppi, N.P. CACTI 6.0: a tool to model large caches. Technical report, HP Laboratories, 2009.

[15] Ogras, U.Y., Marculescu, R., and Marculescu, D. Variation-adaptive feedback control for networks-on-chip with multiple clock domains. In *DAC*, pages 614–619. 2008.

[16] Rahimi, A., Salehi, M.E., Mohammadi, S., and Fakhraie, S.M. Low-energy GALS NoC with FIFO-monitoring dynamic voltage scaling. *Microelectronics Journal*, 42(6):889–896, June 2011.

[17] Shang, L., Peh, L., and Jha, N.K. Power-efficient interconnection networks: dynamic voltage scaling with links. *IEEE Computer Architecture Letters*, 1(1), 2002.

[18] Son, S.W., Malkowski, K., Chen, G., Kandemir, M., et al. Integrated link/CPU voltage scaling for reducing energy consumption of parallel sparse maxtrix applications. In *IPDPS*. 2006.

[19] Wang, H., Peh, L.S., and Malik, S. Power-driven design of router microarchitectures in on-chip networks. In *MICRO*, pages 105–116. 2003.

SUPPLEMENTAL MATERIAL

In this section we provide some additional analysis, exploring how the control algorithm works and its sensitivity to control interval size.

In the DVFS control, one practical issue is how to decide the control interval size for the control. We use $50K$ core clock cycles, which is long enough for V/F change and short enough for fine-grained power control. We performed simulations on two PARSEC benchmarks with different control interval sizes and the results are summarized in Table 3. One can see that the results are not sensitive to moderate interval size changes.

| Interval size | Energy $\times$ Delay | |
(# cycles)	freqmine	x264
12K	0.468	0.151
25K	0.470	0.145
50K	0.462	0.136
75K	0.479	0.147
100K	0.500	0.145

Table 3: Impact of different control interval sizes. The Energy $\times$ Delay values are normalized with the baseline result as 1.

The model assisted PI control described in Section 4.2.2 is based on Equation (7), which is loosely dervied from Equation (3-5). We tried to obtain some supporting evidence through experiments. For the PARSEC benchmark *fluidanimate*, we simulate with different constant uncore V/F levels throughout the entire ROI. The average critical latency versus uncore clock period results are plotted in Figure 6. The results confirm that critical latency has approximately linear dependence on the uncore period.

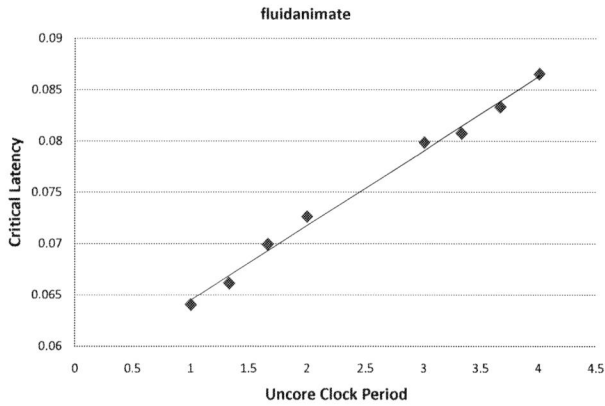

Figure 6: Critical latency vs. uncore clock period.

In Section 4.2.2 and 4.2.3, we argue that it is very hard to accurately predict the system behaviors. Figure 7, 8 and 9 show some simulation results for a segment of *x264*, an application from the PARSEC benchmarks. Its horizontal axis is in terms of control interval of $50K$ clock cycles, i.e., each data point is an average over $50K$ cycles. Despite such smoothing, the data still show drastic changes from interval to interval. These results confirm that it is very difficult to accurately predict the behavior of a multicore system.

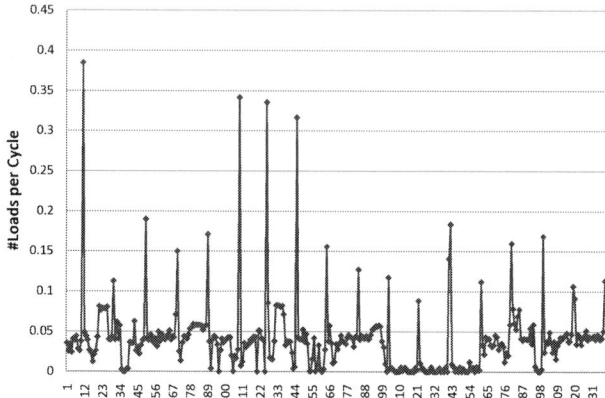

Figure 7: Loads fraction over control intervals.

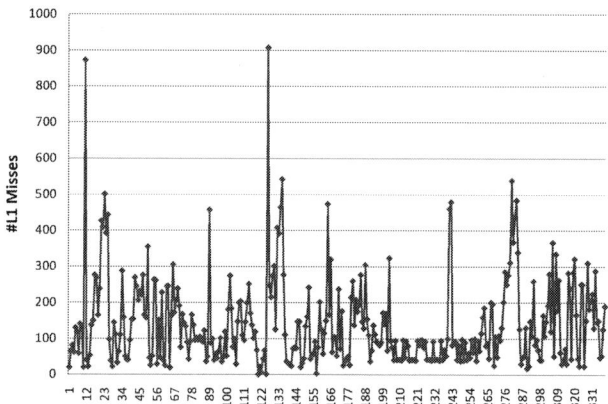

Figure 8: The number of L1 misses over control intervals.

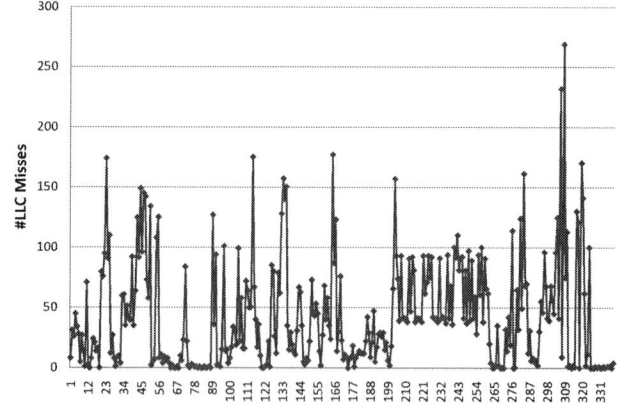

Figure 9: The number of LLC misses over control intervals.

Energy Optimization by Exploiting Execution Slacks in Streaming Applications on Multiprocessor Systems

Amit Kumar Singh, Anup Das, Akash Kumar
Department of Electrical and Computer Engineering, National University of Singapore, Singapore
{eleaks,akdas,akash}@nus.edu.sg

ABSTRACT

Dynamic voltage and frequency scaling (DVFS) offers great potential for optimizing the energy efficiency of Multiprocessor Systems-on-Chip (MPSoCs). The conventional approaches for processor voltage and frequency adjustment are not suitable for streaming multimedia applications due to the cyclic nature of dependencies in the executing tasks which can potentially violate the throughput constraints. In this paper, we propose a methodology that applies DVFS for such cyclic dependent tasks. The methodology involves an off-line analysis that assumes worst-case execution times of tasks to identify the executions that can be slowed down and an on-line analysis to utilize the slacks arising from tasks that finish their execution before the worst-case execution times. Thus, the methodology minimizes energy consumption during both off-line and on-line analysis while satisfying the throughput constraints. Experiments based on models of real-life streaming multimedia applications show that the proposed methodology reduces the overall energy consumption by 43% when compared to existing approaches.

Categories and Subject Descriptors

C.3 [Special-purpose and application-based systems]: Real-time systems and embedded systems

General Terms

Algorithms, Design, Management, Performance

Keywords

Multiprocessor Systems-on-Chip, streaming applications, energy consumption, throughput constraint

1. INTRODUCTION

Modern embedded systems (e.g., mobile phones, tablets) need to support a number of streaming multimedia applications. In order to satisfy the ever increasing performance (throughput) constraints of the applications, MPSoC based systems are being designed. Since such systems are usually operated by stand-alone power supply like battery, minimizing energy consumption during their design and operation is important in order to increase the operational time.

Several efforts have been made to minimize energy consumption of battery-operated MPSoCs. These efforts use system-level energy optimization techniques such as efficient scheduling and DVFS. Many advanced processors support DVFS with multiple voltage levels, which are used in both general-purpose and embedded computing domains [1–4,17]. For applying DVFS in MPSoCs, the voltage and frequency of one or more processors is adjusted depending upon the workload of processors while satisfying the throughput constraint [5]. It has been observed that lowering the voltage

by half might lead to eight times reduction in power consumptions with the linear reduction in maximum operating frequency of a CMOS circuit [7]. The DVFS approaches have also been validated [29]. Executing at lower frequency results in stretched (slowed down) execution, which should not violate the timing constraints such as throughput.

Existing DVFS techniques apply off-line [18] [14] or on-line [8] [23] voltage and frequency scaling. Off-line DVFS techniques estimate the voltage scaling (VS) levels at compile time (design-time) with the knowledge of specific task timings, e.g. worst-case task execution times. On-line DVFS techniques make the voltage and frequency scaling decisions at run-time based on the slack arising from tasks finishing before the worst-case execution-times (WCETs). The off-line DVFS techniques cannot exploit the slacks arising at run-time and the on-line DVFS strategies utilize only limited information to keep a minimum run-time computation overhead. Further, most of the existing DVFS strategies are applicable only to applications described as independent tasks and task graphs represented as directed acyclic graphs (DAGs). Additionally, they don't take the actual VS overhead into account. Such strategies are not applicable for streaming applications that normally exhibit *cyclic* dependencies amongst the tasks and need to be executed in modern embedded systems such as smart phones and tablets. In [24] and [20], DVFS techniques that are applicable on the applications having cyclic dependencies amongst tasks are presented but the techniques have several limitations. For example, in [24], applications need to be described as *homogeneous* synchronous dataflow graphs (HSDFs) [19] that impose high computation complexity, tasks are bound by WCET and only static-slack (off-line slack) created by the difference between application's desired and obtained throughput is used for energy reduction. In [20], only on-line DVFS is applied while considering a uni-processor system.

Contribution: This paper addresses shortcomings of existing DVFS techniques and proposes a methodology that applies both off-line and on-line DVFS to reduce the energy consumption for applications containing cyclic dependent tasks. In the off-line analysis, the tasks are assumed to have WCETs and DVFS is applied in two phases. In the first phase, the executions of tasks are stretched (slowed down) by applying suitable voltage scaling (VS) without violating the throughput constraint. The second phase analyzes execution traces of application tasks & their dependencies to identify the parallel executions that can be slowed down without violating the throughput constraint and applies appropriate VS. The updated execution trace by applying the off-line analysis is used to apply on-line DVFS where VS is further applied based on the dynamically created slacks due to tasks finishing earlier than their WCETs. VS assumes linear frequency scaling and the methodology takes VS overhead captured from manufacturer's datasheet into account, which is explained in section 5. We evaluate the proposed methodology for streaming multimedia applications represented as SDFs. The methodology applies VS directly on SDFs without converting them into HSDFs in order to reduce the computation cost and it is general enough to be applied to applications described as independent tasks and DAGs as well.

The remainder of the paper is organized as follows. Section 2 reviews the literature in the direction of off-line and

---

Permission to make digital or hard copies of all or part of this work for personal or classroom use is granted without fee provided that copies are not made or distributed for profit or commercial advantage and that copies bear this notice and the full citation on the first page. To copy otherwise, to republish, to post on servers or to redistribute to lists, requires prior specific permission and/or a fee.
DAC'13 May 29 - June 07 2013, Austin, TX, USA.

on-line DVFS. Section 3 introduces the preliminaries necessary to understand the work. Section 4 presents the proposed DVFS methodology. The experimental results to evaluate our methodology are presented in Section 5. Section 6 concludes the paper.

2. RELATED WORK

Amongst the earliest works to apply DVFS, Yao et. al. [32] proposed an algorithm to compute optimal static slow down schedule for a set of tasks. Several extensions were undertaken for periodic and aperiodic tasks [6] [26]. Inter-task DVFS [34] [33] and intra-task DVFS [27] [20] have been applied to execute a task at a single and multiple voltage levels, respectively. Recent work also incorporates leakage power aware DVFS [16] [9]. Most of the aforementioned DVFS strategies target a uni-processor system and apply off-line or on-line DVFS to reduce the energy consumption [31]. Simply extending them for MPSoCs leads to increased complexity and inefficiency.

A large body of research exists for applying off-line DVFS while targeting MPSoC [14, 18, 21, 24, 25]. These strategies assume fixed execution time (e.g. WCET) for each task and thus cannot be applied to reduce energy consumption at run-time where tasks have varying execution times.

Dynamic slack reclamation techniques to apply on-line DVFS while targeting MPSoC have also been proposed [8, 23, 35]. These techniques utilize dynamically created slack (due to earlier finish of the tasks) to reduce overall energy consumption. In order to maximize utilization of the slack created on a processor, it is shared amongst other processors or forwarded to later executing tasks. These on-line DVFS strategies don't use any off-line analysis and thus impose high run-time scheduling overhead and reduce energy consumption only due to dynamically created slacks.

On-line DVFS strategies using off-line analysis are presented for MPSoC [10, 11, 22]. The off-line analysis aims at minimizing expected energy consumption. In [10], an off-line analysis phase calculates expected future slack and computation of future tasks by taking average and worst case execution timings. In [22] and [11], the off-line analysis constructs static schedules to be used at run-time. The schedule is constructed by considering critical path and task execution orders. Since these strategies use off-line analysis, run-time scheduling overhead gets reduced and energy consumption is optimized during both design-time and run-time. However, they cannot be efficiently applied to applications containing cyclic dependent tasks as they are developed while targeting independent tasks or DAGs.

In contrast to above strategies, our approach applies both off-line and on-line DVFS for MPSoCs while targeting applications containing cyclic dependent tasks, which are modeled as Synchronous Dataflow Graphs (SDFGs) [19]. In [20] and [24], the applications are modeled as SDF but energy is reduced by applying either off-line [24] or on-line [20] DVFS. Further, the strategy in [20] considers uni-processor system and in [24] is applicable to homogeneous SDFs (HSDFs) that need to be derived from SDFs. Our approach is applicable to applications represented as independent tasks and DAGs as well.

3. PRELIMINARIES

This section provides a brief overview of the MPSoC platform & application model and challenges involved in applying DVFS.

3.1 MPSoC Platform & Application Model

The MPSoC platform is modeled as tile-based architecture [12]. Fig. 1 shows an example platform containing three tiles having DVFS capabilities. Each tile consists of a processor (P), a local memory (M, size in bits) and a network interface (NI). The DVFS controller adjusts voltage and frequency of processors to reduce the overall energy consumption similar to the one supported in Intel's XScale processor [2]. In order to facilitate communication amongst tiles, they are connected to an interconnection network through the NI. The interconnection network provides

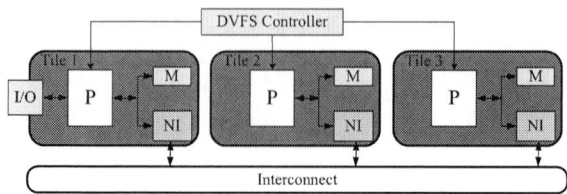

Figure 1: Example MPSoC platform.

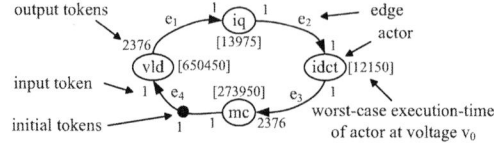

Figure 2: SDFG model of an H.263 decoder.

end-to-end connections between the tiles. However, the latencies of connections can be modeled according to different network-on-chips (NoCs).

The streaming multimedia applications with timing constraints are modeled as SDFGs [19]. Fig. 2 shows SDFG model of H.263 decoder that contains cyclic dependent tasks. The nodes (*vld, iq, idct* & *mc*) and edges (e_1, e_2, e_3 & e_4) model tasks and dependencies respectively. The nodes are referred to as *actors* that communicate with *tokens* sent from one actor to another through the edges. Each actor has its attributes WCET and memory requirement when mapped on a tile operating at a particular voltage level. Each edge has following attributes: size of a token, memory needed on the tile when connected actors are allocated to the same tile, memory needed in source and destination tiles when connected actors are allocated to different tiles and respective bandwidth requirements between the tiles. An actor *fires* (executes) when there are sufficient input tokens on all of its input edges and sufficient buffer space on all of its output connections, and in turn the actor consumes a fixed amount of tokens from the input edges and produces a fixed amount of tokens on the output edges. These token amounts are referred to as *rates*. An edge may contain *initial tokens*.

Throughput of an application is determined as the inverse of the long term period that is calculated as the average time needed for one iteration of the application. An iteration is defined as the minimum non-zero execution such that the original state of the SDFG is obtained. Period for the example H.263 decoder is equal to the summation of Exec-Time(*vld*), $2376 \times$ExecTime(*iq*), $2376 \times$ExecTime(*idct*) and ExecTime(*mc*), where ExecTime is the WCET of respective actors. This period does not include network and memory access delays. It should be noted that actors *iq* and *idct* have to execute 2376 times in one iteration and the number of executions is referred to as *repetition vector* of the actor. An SDFG with a throughput of 1000 Hz has a period of 1 millisecond (ms), i.e. takes 1 ms to complete one iteration.

3.2 DVFS for Applications modeled as Cyclic SDF Graphs

The DVFS process is applied for a given application to MPSoC platform mapping. In a mapping, actors are bound to tiles and edges to memory inside tiles or to connections in the platform. Applying voltage scaling (VS) on one or more tiles results in stretched (slowed down) execution of the bound actors. Modern processors support multiple voltage levels and thus several VS options exist for the same given mapping. Evaluating the throughput obtained with different VS options on each processor is time consuming. This imposes a challenge to rapidly identify the VS options that will lead to throughput satisfying scaling. Further, for a given application and MPSoC platform, as there are several possible mappings (actors to tiles allocations), the complete evaluation with different VS options on each processor might not be feasible within an acceptable time. In order to reduce

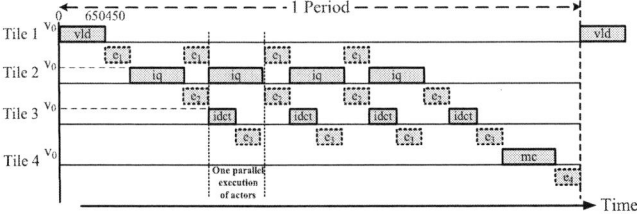

Figure 3: Execution trace of actors/edges of H.263 decoder.

the evaluation time, efficient mappings (providing optimal throughput) need to be identified and only appropriate VS options need to be applied.

In case of applications containing cyclic dependent actors, applying VS on a tile is very challenging as one needs to capture the VS effect on the execution of actors mapped on other tiles by taking the cyclic dependencies into account. Most of the existing DVFS strategies do not take such cyclic dependencies into account and thus cannot be applied to streaming multimedia applications such as H.263 decoder (Fig. 2). Fig. 3 shows the execution trace of the H.263 decoder mapped on a 4-tile MPSoC such that each actor is mapped on a separate processor tile operating at voltage v_0. The connections containing edges also operate at the same voltage. Each rectangle represents an execution such that its height corresponds to the operating voltage (pertaining to a frequency) and its length corresponds to the execution time. First, actor *vld* fires (executes) as it has sufficient input tokens on its incoming edge e_4. Thereafter, it generates 2376 tokens to be transferred through e_1 to process them one by one by *iq*. The transfer of tokens through edges and their processing by different actors follows the shown trace. For easier understanding, the shown trace considers rates as 4 in places of 2376 and thus actors *vld*, *iq*, *idct* & *mc* fire 1, 4, 4 & 1 times respectively during one period. Actor *vld* fires again after finishing the execution of actor *mc* and similar execution patterns are followed in the upcoming periods due to cyclic dependencies. The existing DVFS strategies cannot be applied on such executions.

4. PROPOSED DVFS METHODOLOGY

This section describes our DVFS methodology. In contrast to conventional existing DVFS methodologies, our methodology differs in following aspects: *1)* applies DVFS for applications containing cyclic dependent actors, *2)* applies DVFS by analyzing the execution traces of actors/edges, *3)* for a given application and MPSoC platform, applies off-line DVFS only for the optimal mappings in order to perform faster evaluation, *4)* off-line DVFS results are used to apply on-line DVFS for the dynamically created slacks, and *5)* considers actual VS overhead.

An overview of our DVFS flow is presented in Fig. 4. The DVFS flow optimizes energy consumption for each application to be supported on a MPSoC platform. The applications are evaluated one after another by using the same DVFS flow. The flow takes an application (*Application Model*) & a platform (*Platform Model*) as input and performs design-time and run-time energy optimization for different mappings. At design-time, the flow first evaluates mappings for their throughput and energy consumption. The mappings are generated by using the strategy proposed in [28] as it discards evaluation of inefficient mappings (providing less throughput) and performs faster evaluation without missing the efficient mappings. Then, off-line analysis is applied for the Pareto-optimal mappings that represent resource-throughput trade-offs. At run-time, first, the best mapping (having maximum throughput) is selected depending upon the throughput constraint and available platform resources to configure the platform. Then, on-line DVFS is applied to utilize the dynamically created slacks. Throughput and energy consumption computation for each mapping are done as follows.

Throughput Computation: The throughput for a map-

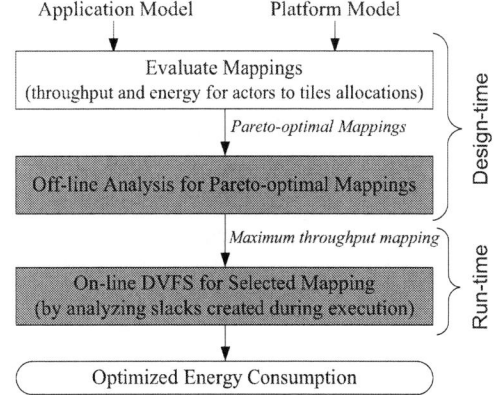

Figure 4: Proposed DVFS flow.

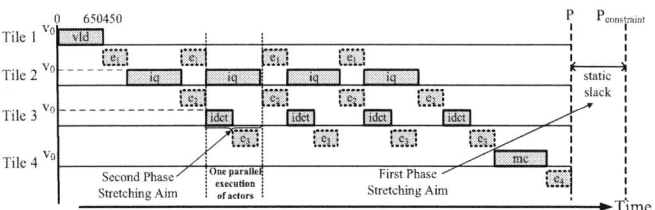

Figure 5: Aim of first and second phase off-line analysis.

ping is computed by taking the resource allocations of actors/edges on the platform into account. For each platform tile, first, static-order schedule that orders the execution of bound actors is constructed. Then, all the binding and scheduling decisions are modeled in a graph called binding-aware SDFG. Finally, throughput is computed by self-timed state-space exploration of the binding-aware SDFG [13]. During the self-timed execution, states visited are examined and stored until a recurrent state is found. The throughput is computed from the periodic part of the state-space.

Energy Consumption Computation: The total energy consumption for a mapping is computed as the sum of communication and computation energy for one iteration (periodic execution) of the application. *Communication energy* is required to transfer data (tokens) from source tile to destination tile and the energy required to process the transferred token on the destination tile is referred to as the *computation energy*. The detailed approach to compute total energy consumption is provided in Appendix A.

4.1 Off-line Analysis

The off-line analysis strategy applies voltage scaling (VS) for each Pareto-optimal mapping in two different phases. The first phase aims to utilize the static slack while maximizing the energy savings as shown in Fig. 5. The static slack is defined as the difference between period ($P_{constraint}$) corresponding to throughput constraint and period (P) corresponding to obtained throughput, where period is equal to 1/throughput. The timing constraint $P_{constraint}$ has also been referred to as deadline constraint. The second phase aims to stretch the slower parallel executions.

4.1.1 First Phase Analysis

In the first phase, the lowest possible VS levels for different tiles are identified while satisfying the throughput constraint and maximizing the energy savings. Towards this, we have proposed a Greedy and an Integer Linear Programming (ILP) formulation based approach.

Greedy Approach

The greedy approach is presented in Algorithm 1. The algorithm takes throughput constraint (τ), VS levels (V) and WCETs of actors at different VS levels (ET) as input and identifies the VS levels to be applied. For the given Pareto-optimal mapping, first, the tiles containing actor(s) are selected. Then, for each selected tile, different available VS

ALGORITHM 1: Greedy Analysis

Input: throughput constraint τ, $V = \{v_i | \forall i \in [1, \cdots, n]\}$,
    $ET = \{WCET[a] \to t_{v_i | \forall i} | \forall actors\}$.
Output: VS levels of tiles.
Select tiles containing actor(s);
repeat
  | **for** *each selected tile t whose VS level is not fixed* **do**
  | | **for** *each VS level v_i* **do**
  | | | Apply VS v_i on t and compute throughput *thrMap*;
  | | | **if** *thrMap $> \tau$* **then**
  | | | | Calculate energy savings $ES[t][vi]$ from equation 1;
  | | | **end**
  | | **end**
  | **end**
  | Find tile t_f & VS level v_f corresponding to maximum ES[][];
  | Fix voltage of t_f to v_f;
until *VS levels of all selected tiles are not fixed;*

---

levels are applied and throughput of the mapping (*thrMap*) is computed. If an applied VS on a tile satisfies the throughput constraint then energy savings for all the actors mapped on the tile $(a_0,...,a_m)$ during one periodic execution is calculated from equation 1, where v_0 and v_i represents the initial and applied voltages respectively. Thereafter, the tile and its VS level corresponding to maximum energy savings is found. The same process is repeated to find VS levels of other tiles.

$$ES[t][v_i] = \sum_{a=a_0}^{a_m} repV[a] \times [(ET[a] \to t_{v_0}) \times (pow \to t_{v_0}) -$$
$$(ET[a] \to t_{v_i}) \times (pow \to t_{v_i})] \tag{1}$$

The **complexity** of the Greedy analysis in terms of number of used (selected) tiles n in the mapping and available VS levels l has been evaluated. For a given value of n and l, the worst-case complexity (C) is calculated as the *maximum number of throughput computations* by equation 2. The complexity of the analysis is $O(ln^2)$.

$$C = n \times l + (n-1) \times l + ... + 2 \times l + 1 \times l$$
$$= l \times \sum_{p=1}^{n} p = l \left(\frac{n^2}{2} + \frac{n}{2} \right) \tag{2}$$

ILP Formulation
To facilitate the ILP formulation, two tables – *slack use table* and *energy savings table* are defined (see Appendix B). There are n rows and l columns for both the table, where n and l are the number of used tiles and supported voltage levels respectively. Each entry (i, j) of the slack used table (SU) is computed as follows.

$$SU(i,j) = computePeriod(t_i \to v_j | t_k \to v_0) - P$$
$$\forall k \in [1 \cdots n], k \neq i \tag{3}$$
$$P = computePeriod(t_i \to v_0, \forall i \in [1 \cdots n])$$

where *computePeriod* function computes period as 1/throughput with tile t_i assigned voltage level v_j and all other tiles set to voltage v_0 (the highest supported voltage level in the platform). The slack used is obtained by subtracting this period with the period obtained by setting all tile voltages to v_0. The energy savings table (ES) is computed in a similar way as shown below.

$$ES(i,j) = E - computeEnergy(t_i \to v_j | t_k \to v_0)$$
$$\forall k \in [1 \cdots n], k \neq i \tag{4}$$
$$E = computeEnergy(t_i \to v_0, \forall i \in [1 \cdots n])$$

Binary Variables: X_{ij}, $i \in [1, n]$, $j \in [1, l]$
Objective: Maximize z $= \sum_{ij} X_{ij} \times ES(i, j)$
Constraints:
One element from each row and column of the table are to be selected.

$$\sum_{i=1}^{n} X_{ij} = 1; \quad \sum_{j=1}^{l} X_{ij} = 1 \tag{5}$$

Total slack distributed on the tiles must be less than or equal to the available slack.

$$\sum_{ij} X_{ij} \times SU(i,j) \leq D - P \tag{6}$$

where D is deadline ($P_{constraint}$) of the given application.

The *number of throughput (period) computations* in the ILP approach is $n \times l$, whereas for the Greedy approach it is $O(ln^2)$. The ILP approach uses $n \times l$ throughput computations to fill the slack used table.

4.1.2 Second Phase Analysis

In the second phase, the executions that can be further slowed down without violating the deadline ($P_{constraint}$) are identified to apply the appropriate VS while maximizing the utilization of the remained static slack. The remaining slack is defined as difference between $P_{constraint}$ and period obtained after applying first phase analysis. The second phase of the off-line DVFS technique is presented in Algorithm 2. The algorithm takes captured execution traces of actors/edges operating at different voltage levels after applying first phase analysis, WCETs of actors at available VS levels & static slack as input and provides updated execution traces after applying additional VS. The execution traces for each actor and edge is captured as the start time, end time and operating voltage of the active executions (firings) during one periodic execution. The algorithm identifies the executions that can be slowed down and applies VS on them. The smaller parallel executions are the ones that can be stretched without significantly stretching the overall execution trace. In Fig. 3, parallel executions of *idct* (with *iq*) can be stretched till the executions of *iq* by applying a suitable voltage scaling if the deadline does not get violated.

The algorithm first finds actors executing in parallel by analyzing the execution traces. Then, the actors are sorted in ascending order of their WCETs in order to apply voltage scaling from the actor having lowest WCET to the actors having higher WCET until all the parallel executing actors are covered. The execution trace of the actor having lowest WCET is stretched by applying an appropriate voltage scaling provided the dependency of the actor is neither longer than connected outgoing edge execution (i.e., just after the outgoing edge firing, there should not be any dependent firing) in some of the actor firings, nor on other actors mapped on the same tile. The stretching (slow down) of all the firings is done based on the execution of the parallel executing actor having next higher WCET and the static slack. Stretching slower parallel executions does not affect the overall throughput and thus the same has been considered. Then, the algorithm selects the actor with next higher WCET and applies voltage scaling on all the earlier selected actors in the same manner. The same process is repeated till the parallel executing actor having highest WCET is not selected as the actor having next higher WCET. Thus, the algorithm applies recurrent voltage scaling on most of the parallel executing actors, resulting in maximum energy consumption reduction.

The demonstration of the off-line analysis for the example H.263 decoder (Fig. 2) is presented in Appendix C. The analysis applies same voltage scaling for all firings of an actor on a tile in order to avoid the overhead of keeping track of the actor firings with different WCETs and VS levels to be used at run-time. This consideration reduces the voltage switching overhead as well.

4.2 On-line DVFS

The off-line processed traces with voltage level information are used to apply on-line DVFS based on dynamically created slacks. The on-line technique applies further voltage scaling to utilize dynamically created slacks due to finishing of the actors earlier to their WCETs at run-time. The technique is presented in Algorithm 3. For a given application

ALGORITHM 2: Trace-based Analysis

Input: Execution traces operating at different voltages,
$ET = \{WCET[a] \rightarrow t_{v_i}|_{\forall i}|\forall actors\}$ and static slack.
Output: Updated execution traces with new operating voltages.
Find actors executing in parallel and their parallel executions;
Sort the actors based on their WCETs at the operating voltages;
Select actor a having lowest WCET to stretch its execution;
repeat
    Select actor with next higher WCET as current actor a_c;
    for *each earlier selected actor a_i except a_c* **do**
        if *firings of other actors are not dependent on firing of*
        outgoing edge of a_i AND firing of a_i is not dependent on
        other actor on the same tile **then**
            Determine stretching of a_i firings by considering its
            parallel firings with a_c and static slack;
            Apply appropriate VS for a_i firings based on the
            determined stretching;
        end
    end
until *all parallel executing actors not covered*;

ALGORITHM 3: On-line DVFS

for *each firing on tile t* **do**
    Calculate $AET = finish\_time - start\_time$;
    Compute $slack += (WCET - AET)$;
    if $slack > WCVST$ **then**
        Compute reduced speed f' for the next firing by Equation 7;
        Apply VS level v_i for next firing on t such that *(speed at*
        $v_i) \geq f'$;
        $slack = 0$;
    end
end

---

to be supported onto a platform at run-time, the technique applies voltage scaling by using off-line processed execution traces for the best (maximum) throughput mapping that is selected from the Pareto-optimal mappings based on the available platform tiles and throughput constraint. The execution traces contain information of actors' & edges' firings by considering WCETs and their operating voltages on the tiles. At run-time, these operating voltages are used for the first firing on the different tiles. Each tile invokes the algorithm whenever the tile finishes a firing of the mapped actor(s).

The implementation of on-line DVFS considers a Ready Queue (RQ) that contains all ready firings of actors, an array to keep initial speed of tile for the firings, and an array to store start time of the firings, for each tile. For each tile, all the firings are put into RQ in the order of their execution priorities. When a firing is finished on a tile, a slack (worst-case execution time (WCET) - actual execution time (ACT)) gets created and the tile starts with the next firing from the RQ by following Algorithm 3. If the created slack is greater than the worst-case voltage switching time (WCVST), then the tile calculates its speed (f') to execute the next firing by Equation 7, where f' and f represent the reduced and earlier speed respectively. Otherwise, the created slack is accumulated to be used for further firings (Algorithm 3). Such consideration avoids non-beneficial DVFS. Calculation of f' considers WCVST in order to take voltage switching overhead into account. An appropriate voltage scaling is applied for the next firing based on f' so as to avoid violation of the deadline. As the voltage and frequency levels are discrete, a scaling level having frequency higher and closer to f' is applied.

$$f' = f \times \frac{WCET}{WCET + slack + WCVST} \qquad (7)$$

The runtime overhead of the algorithm depends upon the number of used tiles (m) by the application and the number of firings (n) on the different tiles. DVFS is not applied on the first firing on each tile as it starts with the off-line suggested voltage/speed. Therefore, the algorithm has worst-case runtime overhead for a maximum of (n-m) firings, where

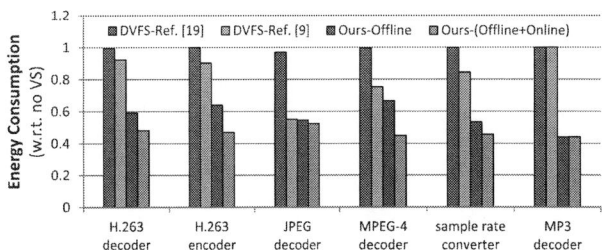

Figure 6: Energy consumption for different applications.

total time for each firing consists of time taken to calculate AET, slack, reduced speed f', and WCVST.

The algorithm does not use slack on one tile for another tile to avoid the overhead of managing its effect on the dependent execution on different tiles. Thus, the algorithm applies efficient voltage scaling leading to reduced energy consumption. The on-line DVFS for different firings of actor *idct* of H.263 decoder is demonstrated in Appendix D.

5. PERFORMANCE EVALUATION

The proposed DVFS methodology has been implemented as an extension of the publicly available SDF[3] tool set [30]. As a benchmark to evaluate the quality of the methodology, models of real-life streaming multimedia applications H.263 decoder (4 actors), H.263 encoder (5 actors), MPEG-4 decoder (5 actors), JPEG decoder (6 actors), sample rate converter (6 actors) and MP3 decoder (14 actors) have been considered. For each application, its throughput requirement and WCETs of actors are known a priory and specified in the application model. The methodology considers DVFS-capable processors in the platform. Each platform tile contains StrongARM processor [1], which supports four voltage-frequency levels: 1.5V–206MHz, 1.4V–192MHz, 1.2V–162MHz and 1.1V–133MHz. The worst-case voltage switching time (WCVST) for each processor is considered as the voltage switching time of StrongARM processor [1]. The energy consumed during the voltage transition is considered negligible in comparison to the overall energy consumption. The AET of an actor firing is expressed as fraction (α) of it's WCET. For different actor firings, we vary α randomly from 0.6 to 1.0.

We present results obtained from our DVFS methodology and compare them with existing methodologies proposed in [20] and [10] in terms of total energy consumption. In order to realize energy savings by applying different DVFS methodologies, energy consumption has been estimated without applying voltage scaling (VS) as well. The strategy in [20] considers applications modeled as SDFGs but is applicable to a uni-processor system. It has been extended to multiprocessor system for a fair comparison. In [10], applications modeled as DAGs are considered. Based on the execution behavior of DAGs, the application models are translated into corresponding SDFGs in order to provide a fair comparison.

Fig. 6 shows energy consumption by different DVFS methodologies with respect to (w.r.t) energy consumption without applying VS. Our approach utilizes the Greedy algorithm for the first phase offline optimization. For each application, energy consumption is estimated for the mapping using the same number of tiles as the number of actors in the application. It can be observed that our methodology shows higher reduction in energy consumption for all the applications. Reduction in energy consumption by the methodologies in [20] and [10] is not significant as they have high penalty to consider the voltage switching time. On an average, our methodology reduces energy consumption by 52% and 43% when compared to methodologies in [20] and [10], respectively.

Next, we analyzed the effect of number of used tiles by the applications on the reduction in energy consumption. Fig. 7 shows energy consumption for the best (maximum throughput) mappings using different number of tiles for

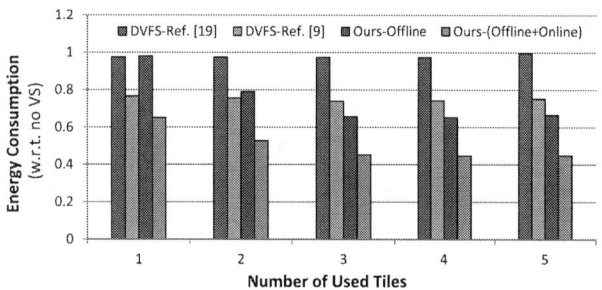

Figure 7: Energy consumption for different number of used titles.

MPEG-4 decoder when different DVFS methodologies are employed. The energy consumptions are shown with respect to no VS energy consumption. The shown results by our approach utilize ILP formulation for the first phase offline optimization. It can be observed that total energy savings (reduction) by our approach increases with the number of used tiles as the combinations of VS options on different tiles get increased, which facilitates for higher energy savings.

Effect of α has been analyzed on the energy savings. The evaluated energy consumption by different methodologies with varying α is presented in Appendix E. The runtime overhead of the on-line DVFS algorithm has been computed. For the H.263 decoder mapped on four tiles, the runtime overhead is 0.8 milliseconds when VS gets applied on 50 firings out of 190 total firings. Experiments have also been performed to analyze the effect of number of voltage levels on the energy savings by the Greedy and ILP approach proposed for the first phase off-line analysis. The number of voltage levels of strongARM [1] are restricted by taking 5 to 15 equally spaced voltage levels between its minimum (1.1V) and maximum (1.5V) operating voltage. Table 1 shows the effect of number of voltage levels on the energy savings for different applications. A couple of observations can be made from Table 1. First, at lower number of supported voltage levels, Greedy approach provides more energy savings as compared to the ILP approach for few applications (e.g. MPEG-4 decoder at 5 voltage levels). Second, ILP approach provides more energy savings than Greedy at higher number of supported voltage levels (e.g. 10 & 15) for all the applications. With large number of supported voltage levels, the Greedy approach finds a local maxima by fixing the voltage levels of a few tiles at the minimum and higher for other tiles, whereas ILP approach performs better (more uniform) distribution of voltage levels on the tiles providing higher energy savings. It has been observed that Greedy provides more energy savings than ILP for most of the applications when number of supported voltage levels is less than 5. For such case (VS levels < 5), trace-based analysis has been applied after the ILP approach and it has been observed that energy savings become closer to that of the Greedy approach.

6. CONCLUSION

We present a novel DVFS methodology for streaming applications that contain actors having cyclic dependencies. Moreover, the methodology is applicable to acyclic task graphs as well. We show that the methodology applies voltage scaling both at design-time and run-time while satisfying the temporal deadlines, and thus provides significant energy savings when compared to existing DVFS methodologies. In future, we plan to apply voltage scaling on execution of application edges while considering an interconnect supporting multiple voltage levels. We also plan to consider heterogeneous platform to accelerate some executions towards creating longer slacks, which can be used to apply further voltage scalings. These considerations will facilitate for further energy savings.

7. ACKNOWLEDGMENTS

This work is supported by Singapore Ministry of Education Academic Research Fund Tier 1 under grant number

Table 1: Energy savings (nJ) at different number of supported voltage levels

	5 voltage levels		10 voltage levels		15 voltage levels	
	ILP	Greedy	ILP	Greedy	ILP	Greedy
H.263 decoder	640	518	53135	474	52628	486
H.263 encoder	644	644	14153	889	5369	880
MPEG-4 decoder	1431	1723	87471	1561	86626	1546
JPEG decoder	36	36	1022	681	1448	337

R-263-000-655-133.

8. REFERENCES

[1] Marvell, StrongARM 1100 processor, 1997. http://www.marvell.com/.

[2] Intel XScale 80200 Processor, 2001. http://www.intel.com/.

[3] Transmeta, Transmeta Crusoe Processor, 2001. http://www.transmeta.com/.

[4] ARM1176 Processor, 2004. http://www.arm.com/.

[5] A. Alimonda et al. A feedback-based approach to dvfs in data-flow applications. *IEEE TCAD*, pages 1691–1704, 2009.

[6] H. Aydin et al. Power-aware scheduling for periodic real-time tasks. *IEEE Trans. on Comput.*, 53(5):584 – 600, may 2004.

[7] T. D. Burd and R. W. Brodersen. Energy efficient cmos microprocessor design. In *IEEE HICSS*, pages 288–297, 1995.

[8] J.-J. Chen et al. Slack reclamation for real-time task scheduling over dynamic voltage scaling multiprocessors. In *IEEE SUTC*, pages 358–367, 2006.

[9] J.-J. Chen and T.-W. Kuo. Procrastination determination for periodic real-time tasks in leakage-aware dynamic voltage scaling systems. In *IEEE ICCAD*, pages 289–294, 2007.

[10] P. Choudhury et al. Online dynamic voltage scaling using task graph mapping analysis for multiprocessors. In *IEEE VLSID*, pages 89–94, 2007.

[11] J. Cong and K. Gururaj. Energy efficient multiprocessor task scheduling under input-dependent variation. In *DATE*, pages 411 –416, 2009.

[12] D. Culler et al. *Parallel computer architecture: a hardware/software approach*. Morgan Kaufmann Pub, 1999.

[13] A. H. Ghamarian et al. Throughput Analysis of Synchronous Data Flow Graphs. In *IEEE ACSD*, pages 25–36, 2006.

[14] F. Gruian. System-level design methods for low-energy architectures containing variable voltage processors. In *Springer PACS*, pages 1–12, 2001.

[15] J. Hu and R. Marculescu. Energy-aware communication and task scheduling for network-on-chip architectures under real-time constraints. In *DATE*, pages 234–239, 2004.

[16] R. Jejurikar and R. Gupta. Dynamic slack reclamation with procrastination scheduling in real-time embedded systems. In *DAC*, pages 111–116, 2005.

[17] J. A. Kahle et al. Introduction to the cell multiprocessor. *IBM J. Res. Dev.*, 49:589–604, 2005.

[18] P. Langen and B. Juurlink. Leakage-aware multiprocessor scheduling. *J. Signal Process. Syst.*, 57:73–88, 2009.

[19] E. A. Lee and D. G. Messerschmitt. Static scheduling of synchronous data flow programs for digital signal processing. *IEEE Trans. Comput.*, 36:24–35, 1987.

[20] S. Lee et al. An intra-task dynamic voltage scaling method for soc design with hierarchical fsm and synchronous dataflow model. In *ISLPED*, pages 84 – 87, 2002.

[21] K. Li. Performance analysis of power-aware task scheduling algorithms on multiprocessor computers with dynamic voltage and speed. *IEEE TPDS*, 19(11):1484 –1497, nov. 2008.

[22] J. Luo and N. K. Jha. Static and dynamic variable voltage scheduling algorithms for real-time heterogeneous distributed embedded systems. In *IEEE ASP-DAC*, pages 719–726, 2002.

[23] P. Malani et al. Adaptive scheduling and voltage scaling for multiprocessor real-time applications with non-deterministic workload. In *DATE*, pages 652–657, 2008.

[24] A. Nelson et al. Power minimisation for real-time dataflow applications. In *IEEE DSD*, pages 117–124, 2011.

[25] D. Shin and J. Kim. Power-aware scheduling of conditional task graphs in real-time multiprocessor systems. In *ISLPED*, pages 408–413, 2003.

[26] D. Shin and J. Kim. Dynamic voltage scaling of periodic and aperiodic tasks in priority-driven systems. In *IEEE ASP-DAC*, pages 653–658, 2004.

[27] D. Shin and J. Kim. Optimizing intratask voltage scheduling using profile and data-flow information. *IEEE TCAD*, 26(2):369 –385, feb. 2007.

[28] A. K. Singh et al. A Hybrid Strategy for Mapping Multiple Throughput-constrained Applications on MPSoCs. In *ACM CASES*, pages 175–184, 2011.

[29] D. C. Snowdon et al. Power management and dynamic voltage scaling: Myths and facts. In *PARC*, New Jersey, USA, 2005.

[30] S. Stuijk et al. SDF³: SDF For Free. In *IEEE ACSD*, pages 276–278, 2006.

[31] W. Wang et al. Energy-aware dynamic slack allocation for real-time multitasking systems. *Sust. Comp.: Infor. and Sys.*, 2:128 – 137, 2012.

[32] F. Yao et al. A scheduling model for reduced cpu energy. In *IEEE FOCS*, pages 374–, 1995.

[33] S. Zhang et al. Approximation algorithms for power minimization of earliest deadline first and rate monotonic schedules. In *ISLPED*, pages 225–230, 2007.

[34] X. Zhong and C.-Z. Xu. System-wide energy minimization for real-time tasks: Lower bound and approximation. *ACM Trans. Embed. Comput. Syst.*, 7:28:1–28:24, 2008.

[35] D. Zhu et al. Scheduling with dynamic voltage/speed adjustment using slack reclamation in multiprocessor real-time systems. *IEEE TPDS*, 14:686–700, 2003.

APPENDIX

A. TOTAL ENERGY CONSUMPTION

Communication energy is required to transfer tokens from one tile to another through the connections. In between two tiles, the communication has to take place when actors mapped on them need to communicate with each other. The communication energy for each edge e mapped to a connection c is estimated as product of the number of tokens (in bits) to be transferred through c, delay (D) and power consumption (P_{bit}) of c for transferring one bit while operating at voltage v. Total communication energy for all the edges is estimated from equation 8. The number of transferred tokens for an edge is computed as the product of repetition vector ($repV$) of source (or destination) actor and source (or destination) port rate (equation 9). The power required to transfer one bit is estimated from [15].

$$E_{comm} = \sum [\{nrTokens[e] \times tokenSize[e]\} \times (D \to c_v) \times (P_{bit} \to c_v)] \quad (8)$$

$$nrTokens[e] = repV[e \to srcActor] \times (e \to srcPortRate) \quad (9)$$

Computation energy is required to process the transferred token on the destination tile after it is received and able to fire (execute) the mapped actor. Computation energy for each actor a mapped to tile t is estimated as product of the number of executions of a ($repV[a]$), execution time ($ET[a]$) and power consumption (pow) on tile t operating at voltage v. Total computation energy for all actors is estimated from equation 10. Power consumption on tile is estimated by equation 11, where C, v and f denote average load capacitance, supply voltage and operating frequency, respectively.

$$E_{comp} = \sum [repV[a] \times (ET[a] \to t_v) \times (pow \to t_v)] \quad (10)$$

$$pow \to t_v = C \times v^2 \times f \quad (11)$$

Total energy consumption is measured as sum of communication and computation energy. The total energy consumption does not include static energy. In our approach, we focus on mapping of applications on the architecture after it is designed. So we cannot optimize static energy consumption and focus on optimizing only dynamic energy consumption.

B. ILP FORMULATION USED TABLES

The two tables **slack use table** and **energy savings table** as defined in Table 2 and Table 3 respectively are used to formulate the ILP.

Table 2: Slack used table

Tiles	Slack used in ms			
	v_0	v_1	$\cdots$	v_{l-1}
t_0	0	140	$\ldots$	200
t_1	0	35	$\ldots$	150
$\vdots$	$\vdots$	$\vdots$	$\ddots$	$\vdots$
t_n	0	100	$\ldots$	140

Table 3: Energy savings table

Tiles	Energy savings in nJ			
	v_0	v_1	$\cdots$	v_{l-1}
t_0	0	11,000	$\ldots$	40,0000
t_1	0	135,000	$\ldots$	150,000
$\vdots$	$\vdots$	$\vdots$	$\ddots$	$\vdots$
t_n	0	15,0000	$\ldots$	50,000

C. OFF-LINE ANALYSIS DEMONSTRATION

The demonstration considers one actor to one tile mapping (Fig. 3). Fig. 8 shows the execution traces after applying the off-line analysis (first & second phase). The first phase identifies the VS levels to be applied on different tiles while satisfying the deadline. The identified VS levels for each tile are assumed as v_1. The second phase analyzes the execution traces operating at voltage v_1 and provides the updated traces satisfying the deadline. The analysis (Algorithm 2) finds parallel executing actors as iq & $idct$, and applies voltage scaling v_2 ($< v_1$) for actor $idct$ as it has lower WCET. Executions of $idct$ are stretched till the execution of iq since the outgoing edge of $idct$ does not affect most of other executions.

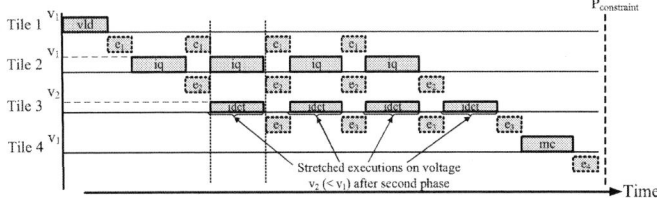

Figure 8: Execution trace of actors/edges of H.263 decoder after applying off-line DVFS.

D. ON-LINE ANALYSIS DEMONSTRATION

Fig. 9 shows the demonstration of on-line DVFS for different firings of actor $idct$ of H.263 decoder mapped on a tile in order to apply further voltage scaling. It can be seen that if a firing finishes earlier than its WCET then created slack is used to apply DVFS on the next firing. The voltage scaling for the next firing is decided based on the reduced speed for the firing calculated by Equation 7. The voltage switching overhead has also been considered for the next firing. It can also be observed that overall time for all firings get reduced, i.e. execution trace gets shrunk. Thus, better throughput is obtained in addition to reduced energy consumption after applying on-line DVFS.

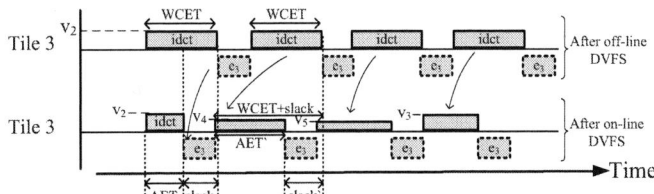

Figure 9: Applying on-line DVFS for different firings of an actor.

E. EVALUATION WITH DIFFERENT α

Effect of α has been analyzed on the energy savings. Experiments have been performed for different fixed values of α ranging from 0.1 to 1.0 at an interval of 0.1. At a fixed α, all actor firings assume the same value of α. Fig. 10 shows energy consumption by different methodologies for the maximum throughput mapping of MPEG-4 decoder that uses 3 tiles. A couple of observations can be made from Fig. 10. First, energy savings (reduction) is higher at lower values of α and decreases with higher values of α. Second, at α equal to 1, the methodologies in [20] and [10] do not provide energy savings as no slack is created at run-time, which contributes to the energy savings. However, our methodology provides energy savings by the off-line analysis.

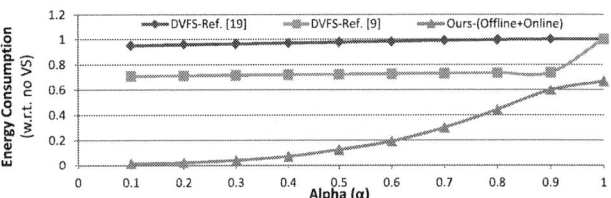

Figure 10: Energy consumption at different α.

Verifying SystemC using an Intermediate Verification Language and Symbolic Simulation*

Hoang M. Le[1]   Daniel Große[2]   Vladimir Herdt[1]   Rolf Drechsler[1,3]

[1]Institute of Computer Science, University of Bremen, 28359 Bremen, Germany
[2]solvertec GmbH, 28359 Bremen, Germany
[3]Cyber-Physical Systems, DFKI GmbH, 28359 Bremen, Germany
{hle, vherdt, drechsle}@informatik.uni-bremen.de   grosse@solvertec.de

ABSTRACT

Formal verification of SystemC is challenging. Before dealing with symbolic inputs and the concurrency semantics, a front-end is required to translate the design to a formal model. The lack of such front-ends has hampered the development of efficient back-ends so far.

In this paper, we propose an isolated approach by using an *Intermediate Verification Language* (IVL). This enables a SystemC-to-IVL translator (frond-end) and an IVL verifier (back-end) to be developed independently. We present a compact but general IVL that together with an extensive benchmark set will facilitate future research.

Furthermore, we propose an efficient symbolic simulator integrating Partial Order Reduction. Experimental comparison with existing approaches has shown its potential.

1. INTRODUCTION

The system modeling language SystemC [16, 12] is being widely adopted to create golden models in the *Electronic System Level* (ESL) design and verification flow [2]. The golden models are developed using a behavioral/algorithmic style in combination with abstract communication based on *Transaction Level Modeling* (TLM) [16]. Ensuring the correctness of these models is of major importance since undetected errors become very expensive in later design steps. To verify the abstract SystemC models, the straight-forward approach is simulation offered already by the free event-driven simulation kernel shipped with the SystemC class library [1]. Substantial improvements have been proposed by supporting the validation of (TLM) assertions, see e.g. [4, 8, 9, 23]. To further enhance simulation coverage, methods based on *Partial Order Reduction* (POR) have been proposed [19, 3]. They allow to explore all possible scheduling sequences of SystemC processes for a given data input. However, representative inputs are still needed. Therefore, formal ap-

proaches for SystemC TLM have been devised. But due to the object-oriented nature of SystemC *and* its sophisticated synchronous and asynchronous simulation semantics, formal verification is very challenging [25].

We review existing formal approaches in detail in Section 2. With some exceptions[1], they use freely available SystemC parsers and thus are hampered by their limitations as detailed in [20]. At the same time most of the existing approaches translate the SystemC design into a formal representation. These representations are similar in their expressiveness which motivates the first major contribution of this paper: the *Intermediate Verification Language* (IVL). The properties of IVL and the resulting advantages include:

- Compact, intuitive and readable language: IVL has been designed in such a way that both manual and automatic transformations from SystemC are possible.

- Independent development of front-end and back-end: IVL enables to focus on the problem that the user wants to address.

- Open language and support: IVL is open and a free parser is provided. Moreover, all freely available benchmarks used by existing formal verification approaches for SystemC have been transformed into an extensive IVL benchmark set. This accelerates research in particular with respect to new formal approaches.

Based on IVL the second contribution of this paper is an *efficient symbolic simulator*. The novelty of our simulator is to combine and adapt two efficient verification techniques – POR [11, 10] and symbolic execution [5] – under the simulation semantics of SystemC. While POR prunes redundant process scheduling sequences, symbolic execution efficiently explores all conditional execution paths of each individual process in conjunction with symbolic inputs. Subsequently, the simulator covers all possible inputs and scheduling sequences of the design exhaustively. It supports both static and dynamic POR and is configurable with respect to the search algorithm for state traversal. For the first time we provide a full experimental comparison of all available state-of-the-art formal approaches. This comparison also demonstrates clearly the potential of our proposed approach.

The rest of this paper is organized as follows. In Section 2 we review related work mainly on formal verification of SystemC. Section 3 gives a brief introduction to SystemC. Then, in Section 4 the IVL is introduced. The symbolic simulator is presented in Section 5. Section 6 gives the experimental results. Finally, the paper is concluded in Section 7.

---

*This work was supported in part the German Federal Ministry of Education and Research (BMBF) within the project SANITAS under contract no. 16M3088 and by the German Research Foundation (DFG) within the Reinhart Koselleck project DR 287/23-1.

Permission to make digital or hard copies of all or part of this work for personal or classroom use is granted without fee provided that copies are not made or distributed for profit or commercial advantage and that copies bear this notice and the full citation on the first page. To copy otherwise, to republish, to post on servers or to redistribute to lists, requires prior specific permission and/or a fee.
DAC '13, May 29 - June 07 2013, Austin, TX, USA.

---

[1]The parser is either undocumented or a proprietary tool.

2. RELATED WORK

A handful of formal verification approaches for SystemC TLM have been proposed. Early efforts, for example [21, 17, 24], have very limited scalability or do not model the SystemC simulation semantics thoroughly [18]. Among the more recent approaches, the following four are the most promising and currently represent the state-of-the-art.

STATE, first proposed in [15], translates SystemC designs to timed automata. With STATE it is not possible to verify properties on SystemC designs directly. Instead, they have to be formulated on the automata and can then be checked using the UPPAAL model checker.

SCIVER [13] translates SystemC designs into sequential C models first. Temporal properties using an extension of PSL [22] can be formulated and integrated into the C model during generation. Then, C model checkers can be applied to check for assertion violations. High-level induction on the generated C model has been proposed to achieve completeness and efficiency. However, no dedicated techniques to prune redundant scheduling sequences are provided.

KRATOS [7] translates SystemC designs into threaded C models. Then, the ESST algorithm is employed, which combines an explicit scheduler and symbolic lazy abstraction. POR techniques are also integrated into the explicit scheduler. For property specification, simple C assertions are supported. The main performance bottleneck of KRATOS is the potentially slow abstraction refinements.

SDSS [6] formalizes the semantics of SystemC designs in terms of Kripke structures. Then, BMC and induction can be applied in a similar manner as SCIVER. The main difference is that the scheduler is not involved in the encoding of SDSS. It is rather explicitly executed to generate an SMT formula that covers the whole state space. Still, no dedicated techniques to handle equivalent scheduling sequences are supported. SDSS allows simple properties reasoning about variable values at the beginning of each evaluation phase.

With respect to our proposed IVL, the threaded C programs used by KRATOS are the most close representation. However, KRATOS just employs this representation as a means to enable explicit process scheduling in its model checking algorithm. It is not designed as an IVL and thus for example it is not documented whether the full language or which subset of C can be used. KRATOS itself appears to support only a very small fraction of C. The parser of KRATOS for this representation is also not available. Furthermore, the SystemC-related constructs cannot be cleanly separated, e.g. processes and channel updates are identified by function name prefixes, events are declared as enum values, etc. The new SystemC constructs for process control such as *suspend* and *resume* (cf. SystemC 2.3) are also not supported.

3. BACKGROUND ON SYSTEMC

In the following only the essential aspects of SystemC are described. SystemC has been implemented as a C++ class library, which includes an event-driven simulation kernel. The structure of the system is described with ports and modules, whereas the behavior is described in processes which are triggered by events and communicate through channels. A process gains the *runnable* status when one or more events of its sensitivity list have been notified. The simulation kernel selects one of the runnable processes and gives this process the control. The execution of a process is non-preemptive, i.e. the kernel receives the control back if the process has finished its execution or suspends itself by call-

ing *wait()*. SystemC provides three types of processes with *SC_THREAD* being the most general type, i.e. the other two can be modeled by using SC_THREAD. For event-based synchronization, SystemC offers many variants of *wait()* and *notify()* such as *wait(time)*, *wait(event)*, *event.notify(delay)*, *event.notify()*, etc.

The simulation semantics of SystemC can be summarized as follows [16]: First, the system is elaborated, i.e. instantiation of modules and binding of channels and ports is carried out. Then, there are the following steps to process:

1. *Initialization*: Processes are made runnable.

2. *Evaluation*: A runnable process is executed or resumes its execution. In case of immediate notification, a waiting process becomes runnable immediately. This step is repeated until no more processes are runnable.

3. *Update*: Updates of channels are performed. These updates have been requested during the evaluation phase.

4. *Delta notification*: If there are delta notifications, the waiting processes are made runnable, and then the simulation is continued with the Evaluation step.

5. *Timed notification*: If there are timed notifications, the simulation time is advanced to the earliest one, the waiting processes are made runnable, and the simulation is continued with the Evaluation step. Otherwise the simulation is stopped.

In the next section, we define the IVL based on this simulation semantics.

4. INTERMEDIATE VERIFICATION LANGUAGE

The IVL is the stepping stone between a front-end and a back-end. Ideally, it should be compact and easily manageable but at the same time powerful enough to allow the translation of SystemC designs. Our view is that a back-end should focus purely on the behavior of the considered SystemC design. This behavior is fully captured by the SystemC processes under the simulation semantics of the SystemC kernel. Therefore, a front-end should first perform the elaboration phase, i.e. determine the binding of ports and channels. Then it should extract and map the design behavior to the IVL, whose elements are detailed in the following.

Based on the simulation semantics described above, we identify the three basic components of the SystemC kernel: *SC_THREAD*, *sc_event* and *channel update*. These are adopted to be *kernel primitives* of the IVL: *thread*, *event* and *update*, respectively. Associated to them are the following primitive functions:

- *suspend* and *resume* to suspend and resume a thread, respectively;

- *wait* and *notify* to wait for and notify an event (the notification can be either immediate or delayed depending on the function arguments);

- *request_update* to request an update to be performed during the update phase.

These primitives form the backbone of the kernel. Other SystemC constructs such as *sc_signal*, *sc_mutex*, static sensitivity, etc. can be modeled using this backbone.

```
1  SC_MODULE(Module) {          19
2    sc_core::sc_event e;        20    void B() {
3    uint x, a, b;               21      e.wait();
4                                22      b = x / 2;
5    SC_CTOR(Module)             23    }
6    : x(rand()), a(0)           24
7    , b(0) {                    25    void C() {
8    SC_THREAD(A);               26      e.notify();
9    SC_THREAD(B);               27    }
10   SC_THREAD(C);               28  };
11   }                           29
12                               30  int sc_main() {
13   void A() {                  31    Module m("top");
14     if (x % 2)                32    sc_start();
15       a = 1;                  33    assert(2 * m.b + m.a
16     else                                == m.x);
17       a = 0;                  34    return 0;
18   }                           35  }
```

Figure 1: A SystemC example

```
1  event e                      15  thread B begin
2    uint x = ?<uint>           16    wait e
3  uint a = 0                   17    b = x / 2
4  uint b = 0                   18  end
5                               19
6  thread A begin               20  thread C begin
7    if x % 2 goto elseif       21    notify e
8      a = 0                    22  end
9      goto endif               23
10   elseif:                    24  main begin
11     a = 1                    25    start
12   endif:                     26    assert 2 * b + a == x
13 end                          27  end
14
```

Figure 2: The example in IVL

The behavior of a *thread* or an *update* is defined by a function. Functions which are neither *threads* nor *updates* can also be declared. Every function possesses a body which is a list of statements. We allow only assignments, (conditional) goto statements and function calls. Every structural control statement (*if-then-else*, *while-do*, *switch-case*, etc.) can be mapped to conditional *goto* statements (this task should also be performed by the front-end). Therefore, the representation of a function body as a list of statements is general and at the same time much more manageable for a back-end.

As *data primitives* the IVL supports Boolean and integer data types of C++ together with all arithmetic and logic operators. Furthermore, arrays and pointers of primitive types are also supported. Additionally, bit-vectors of finite width can be declared. This enables the modeling of SystemC data types such as *sc_int* or *sc_uint* in the IVL.

For verification purpose, the IVL provides *assert* and *assume*. More expressive temporal properties can be translated to FSMs and embedded into an IVL description by a front-end. Symbolic values of primitive types are also supported.

SystemC Example.

Figure 1 shows a simple SystemC example. The main purpose of the example is to demonstrate some elements of the IVL. The design has one module and three SC_THREADs A, B and C. Thread A sets variable a to 0, if x is divisible by 2, and to 1 otherwise (Line 14-17). Variable x is initialized with a random integer value on Line 6 (i.e. it models an input). Thread B waits for the notification of event e and sets $b = x / 2$ subsequently (Line 21-22). Thread C performs an immediate notification of event e (Line 26). If thread B is not already waiting for it, the notification is lost. After the simulation the value of variable a and b should be $x \% 2$ and $x / 2$, respectively. Thus the assertion $(2*b+a == x)$ is expected to hold (Line 33). Nevertheless, there exist counterexamples, for example the scheduling sequence CAB leads to a violation of the assertion. The reason is that b has not been set correctly due to the lost notification.

IVL Example.

Figure 2 depicts the same example in IVL. As can be seen the SystemC module is "unpacked", i.e. variables, functions, and threads of the module are now global declarations. The calls to *wait* and *notify* are directly mapped to statements of the same name. The *if-then-else* block of thread A is converted to a combination of conditional and unconditional goto statements (Line 7-12). Variable x is initialized with a symbolic integer value (Line 2) and can have any value in

the range of *unsigned int*. The statement *start* on Line 25 starts the simulation.

In short, the IVL is kept minimal but expressive enough for the purpose of formal verification. It covers all benchmarks used by existing formal verification approaches for SystemC. It would only take little effort to adapt these approaches to support this IVL as their input language. That would lead to the availability of a checker suite for SystemC once a capable front-end is fully developed. A grammar and a parser for the IVL are provided at our website[2].

In the next section, we present a new efficient verifier combining symbolic execution and POR. We also refer to an IVL description as a SystemC design since both define the same behavior.

5. SYMBOLIC SIMULATION

In this section we present a symbolic simulator for the SystemC IVL. The simulator can discover assertion violations and other types of errors such as division by zero or memory access violation. Our approach enables exhaustive exploration of the state space by providing support for symbolic values and taking all possible scheduling sequences into consideration. As a pre-processing step, all non-primitive function calls are inlined.

Figure 3 shows the complete search tree (i.e. state space) for the example from last section. As can be seen, even this very simple design has a total of 14 possible execution paths. Only six of them violate the assertion (paths that end with a filled box). The circles and boxes correspond to *execution states* of the design. An execution state is split at a ◇ node in case of a conditional *goto* statement (explained later in Section 5.1)

Execution State.

An execution state contains values of all variables, states of all threads, a pending event notification list, a pending update list, a *path condition* and a Boolean flag indicating whether this is a split state. The state of a thread consists of a status (either *runnable*, *blocked* or *terminated*) and a *statement pointer* (SP) that determines the next statement to be executed. The path condition describes the constraint on the variables, which must be satisfied to reach this execution state from the initial one. Also note that variables have in general no concrete values, their values are rather expressions of symbolic and concrete values. Take the initial execution state in Figure 3 (the uppermost node) as an example: The variable x is initialized as a symbolic value, while a and b have the initial value of zero. Each thread A, B or C is *runnable* and has a SP pointing to the first statement of the thread body (Line 7, 16 and 21 in Figure 2,

---
[2]www.systemc-verification.org/sissi

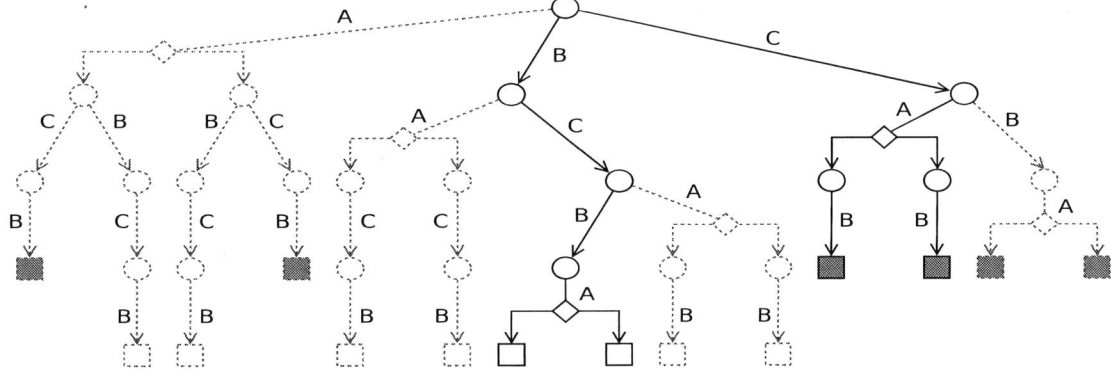

Figure 3: Search tree for the example

respectively). There are no pending event notifications, no pending updates and yet no path condition. Now, the execution of thread B will change its state to *blocked*, add *e* to the pending notification list, and update its SP to Line 17. All other values remains unchanged.

Every edge depicts a transition between two states and its label shows which thread has been selected and executed in this transition. The execution of B mentioned above thus corresponds to the outgoing edge with label B from the uppermost node. The differences between dashed and solid edges are explained later in Section 5.3. In the following we describe the main components of the simulator: scheduler, interpreter and employed POR techniques.

5.1 Scheduler

The scheduler manages the set of execution states. It selects an unvisited execution state and explores all interleavings in this state. That means for each *runnable* thread, the control is given to the interpreter to execute it. The interpreter implements symbolic execution and is responsible for the handling of symbolic values and the detection of errors during thread execution (see Section 5.2).

The scheduler receives the control back from the interpreter in one of the following two cases:

1. The execution reaches the end of the thread or a *wait* statement. This case results from the simulation semantics of SystemC. The current execution state is marked as visited and a new execution state with the updated status of the thread is added.

2. The thread execution reaches a conditional *goto* statement, whose condition is a symbolic expression. Because the condition can be *true* or *false* depending on the involved symbolic values, the execution state is marked as visited and then split into two new ones for both cases. In the *true* (*false*) case, the condition (its negation) is added to the path condition, the *split* flag is set such that the thread execution can be continued later by the interpreter directly after the conditional goto. Note that, because of this direct continuation (no other runnable thread is executed), the execution of the thread is actually not preempted under the SystemC simulation semantics.

If the selected execution state has no *runnable* threads, that means the evaluation phase is completed. The scheduler performs the other phases (i.e. update, delta and timed notification). If new *runnable* threads arise after that, the

exploration is continued, otherwise the end of an execution path is reached.

Now consider the first solid path from the left in Figure 3. From the initial execution state, thread B is executed and then blocked by the *wait* statement. Afterwards, a new execution state is produced and thread C is executed in this new state. This execution notifies event *e* and terminates. This notification makes B *runnable* again and it is executed in the next execution state. After the termination of B, A is executed. When the execution of A reaches the conditional *goto*, two new execution states are produced because the condition involves the symbolic value of *x*. There are no *runnable* threads and no pending notifications in both states, therefore both mark the end of an execution path.

For the state space exploration, the scheduler supports depth-first search, breath-first search, and iterative deepening. An interactive mode is also provided to enable the user to select only one *runnable* thread to execute from an execution state. This is useful for example to replay an erroneous execution path.

5.2 Interpreter

The interpreter implements symbolic execution. It executes a thread in an execution state by interpreting the statements of the thread. The execution can either be started or resumed at the statement determined by the thread SP. As mentioned in Section 4, there are three types of statements: assignments, *goto*, and function calls. For an assignment, the interpreter simply replaces the value of the left-hand side in the execution state with the right-hand side expression. In case of an unconditional *goto*, the execution continues at the specified label. Conditional *goto* is handled as follows: if the condition involves a symbolic value, the scheduler takes over as described above, otherwise the condition is evaluated and the execution continues at either the specified label or the next statement.

Since we have inlined non-primitive function calls, only calls to kernel and verification primitive functions are left to process. They are interpreted according to their semantics in SystemC as follows.

- *wait*: The thread execution is blocked and the control is given back to the scheduler as described above.

- *notify*: In case of an immediate notification, all waiting threads in the execution state are made *runnable*. Otherwise, the pending notification list is updated accordingly.

- *suspend*: The status of the to-be-suspended thread is changed to *blocked*.

- *resume*: The status of the to-be-resumed thread is changed to *runnable*.

- *request_update*: The requested update is added to the pending update list.

- *assert*: The conjunction of the path condition and the negation of the asserted expression is given to an SMT solver. If a solution can be found, the assertion is violated, and a counter-example is created from this solution and reported.

- *assume*: The given expression is added to the path condition.

In addition to assertions, the interpreter also checks for other types of errors such as division by zero or memory access violation.

5.3 Partial Order Reduction

The above described scheduler explores all possible thread scheduling sequences explicitly. Many of them can be reduced without affecting the verification result by applying POR techniques.

The basic idea is to divide a thread into several *transitions*. A transition is a list of statements that can be executed without interruption by the interpreter before the control is given back to the scheduler. A transition begins either at the start of the thread or after a *wait* statement. It ends either at the end of the thread or before another *wait* statement. Thus, every time a thread is executed, actually one of its transitions is being executed. A thread is *runnable* if one of its transitions is *runnable*.

After all transitions are identified, we establish a *dependency relation* between every pair of them. Intuitively, two transitions X and Y are dependent if two execution orders XY and YX lead to different results. We have the following cases of dependency:

- Both transitions access the same memory location identified by a variable, a pointer or an array, with at least one write access;

- A transition notifies an event immediately which the other transition waits for (e.g. thread B and C from Figure 2);

- A transition suspends the other transition by calling *suspend*.

Based on this dependency relation, the persistent set and sleep set techniques [11] are employed to identify equivalent thread scheduling sequences. These two techniques are orthogonal and can be combined to achieve better results. Basically, for each visited execution state, both techniques derive a subset of *runnable* transitions. Future exploration from this state is restricted to this transition subset.

The dependency relation can be either statically or dynamically determined. Dynamic dependency is calculated during the simulation and thus more precise than static dependency which is often an over-approximation. However, it produces a much bigger overhead in comparison to static dependency calculation. For a more detailed formal treatment of static and dynamic POR we refer to [11, 10, 19].

The dashed execution paths in Figure 3 are pruned by using static POR and must not be traversed. This reduction

can be intuitively explained as follows. Because A is independent of B and C, it is unimportant when A is executed, e.g. CAB, CBA and ACB are equivalent. In contrast, the order between B and C is important due to their dependency.

5.4 Limitations

Currently, loop detection is not implemented in our symbolic simulator. For models without symbolic inputs, the symbolic simulation becomes explicit model checking and thus well-known loop detection algorithms for example in SPIN can be used. But the general case with symbolic inputs is much more interesting and challenging. Therefore, loop detection is left for future work and consequently our simulator can only be applied to models that either terminate or contain bugs.

Symbolic execution can run into the path explosion problem in some cases. For software verification, advanced techniques for path merging and redundant path elimination have been proposed and implemented in modern tools such as KLEE [5]. Our simulator does not integrate such techniques yet. Nevertheless, its potential is demonstrated by the experiments in the next section.

6. EXPERIMENTAL RESULTS

We have implemented the proposed approach in a prototype called SISSI (SystemC IVL Symbolic Simulator) using Python (version 2.7.3rc2). Our implementation also uses an intermediate SMT layer [14] that allows to switch between different solvers or also run several solvers in parallel. However, for the experiments here we just employ Boolector. Furthermore, we use two variants SISSI-S and SISSI-D which perform static and dynamic POR, respectively.

All experiments have been conducted on an AMD Phenom 3.4 GHz machine with 8 GB RAM running Linux. Time limit for each run is set to 1200 seconds. Among the four state-of-the-art approaches mentioned in Section 2, SDSS and its benchmarks are to the best of our knowledge not available. We use benchmarks taken from the websites of KRATOS[3], SCIVER[4] and STATE[5], and develop some new benchmarks as well. For each benchmark, three equivalent models are needed (in IVL for SISSI, in threaded C for KRATOS, in SystemC for SCIVER and STATE). The checked properties are source code assertions which are supported by all approaches but STATE. Hence, we need to change the models slightly before giving them to STATE.

Table 1 shows a representative excerpt of the results. The first column gives the name of the benchmark. The next columns present for each benchmark the lines of code in IVL, the verification result (Safe or Unsafe), and the verification time needed by SISSI-S, SISSI-D, KRATOS, SCIVER and STATE, respectively. Furthermore, Table 1 is divided by the dashed line into two halves. The upper half shows results for benchmarks without symbolic inputs. We believe symbolic approaches are not the right tool for these models. But since they have been used in the past for the evaluation of such approaches, we still include them in the comparison. As can be seen, SCIVER and KRATOS do not perform well on these benchmarks. SISSI and STATE resort to explicit model checking and therefore are much faster in general.

The more important results for benchmarks with symbolic inputs are presented in the lower half of Table 1. Note that for these models, STATE explores a much smaller state space

---

[3]es.fbk.eu/tools/kratos
[4]www.systemc-verification.org/sciver
[5]www.pes.tu-berlin.de/state_project

Table 1: Comparison with state-of-the-art approaches (runtime in seconds)

Benchmark	LoC	Result	SISSI-S	SISSI-D	KRATOS [7]	SCIVER [13]	STATE [15]
kundu	54	S	9.25	12.15	1.07	9.70	0.04
transmitter.10	81	U	0.05	0.34	0.07	18.63	0.03
transmitter.50	361	U	0.24	4.70	304.54	time-out	0.22
transmitter.200	1411	U	1.38	106.30	mem-out	mem-out	12.80
mem-slave-tlm.4	207	S	0.18	0.20	140.24	13.38	0.03
mem-slave-tlm.5	218	S	0.20	0.23	223.78	20.18	0.04
token-ring-bug.10	94	U	0.05	0.11	0.74	6.31	0.53
token-ring-bug.50	414	U	0.25	2.58	mem-out	time-out	mem-out
token-ring-bug.200	1614	U	1.88	57.01	mem-out	mem-out	mem-out
mem-slave-tlm-sym.4	208	S	0.18	0.23	time-out	28.33	50.42
mem-slave-tlm-sym.5	219	S	0.29	0.29	time-out	56.29	62.77
simple-fifo-1c-2p	73	U	0.13	0.10	65.22	1.65	error
simple-fifo-2c-1p	72	U	0.08	0.12	39.26	1.26	error
jpeg	230	U	0.58	0.49	time-out	22.84	error
buffer-ws-p3	51	S	0.25	0.05	2.01	0.48	mem-out
buffer-ws-p4	55	S	2.62	0.06	14.33	3.13	mem-out
buffer-ws-p5	60	S	70.80	0.07	210.82	1.95	mem-out

in comparison to the other methods since its back-end does not support the full range of C++ *int*. Still, STATE does not perform/scale well in the presence of symbolic inputs and also it does not accept some of the models (indicated as *error*). The *token-ring-bug* and *mem-slave-tlm-sym* benchmarks are basically *transmitter* and *mem-slave-tlm*, respectively, with symbolic inputs. With the exception of SISSI, these are notably harder for the model checkers. Overall, SISSI delivers clearly the best performance by far on benchmarks with symbolic inputs.

The trade-off between static and dynamic POR discussed in Section 5.3 can also be observed in the *transmitter*, *token-ring-bug* and *buffer-ws* benchmarks.

7. CONCLUSIONS

This paper makes two contributions to the formal verification of SystemC TLM. First, we present a compact, intuitive and readable *Intermediate Verification Language* (IVL) for SystemC that enables the independent development of front-ends and back-ends. With the availability of the IVL, a free parser and an extensive benchmark set, research in particular with respect to new back-ends can be accelerated. Second, we propose a new efficient symbolic simulator integrating Partial Order Reduction and symbolic execution. This combination enables the effective exploration of all possible inputs and process scheduling sequences. The experimental comparison confirms the potential of our symbolic simulator. This is also to the best of our knowledge the most comprehensive comparison of available state-of-the-art approaches.

8. REFERENCES

[1] Accellera Systems Initiative. SystemC, 2012. Available at http://www.systemc.org.

[2] B. Bailey, G. Martin, and A. Piziali. *ESL Design and Verification: A Prescription for Electronic System Level Methodology*. Morgan Kaufmann/Elsevier, 2007.

[3] N. Blanc and D. Kroening. Race analysis for SystemC using model checking. *ACM Trans. on Design Automation of Electronic Systems*, 15:21:1–21:32, 2010.

[4] N. Bombieri, F. Fummi, and G. Pravadelli. Incremental ABV for functional validation of TL-to-RTL design refinement. In *DATE*, pages 882–887, 2007.

[5] C. Cadar, D. Dunbar, and D. R. Engler. KLEE: Unassisted and automatic generation of high-coverage tests for complex systems programs. In *OSDI*, pages 209–224, 2008.

[6] C.-N. Chou, Y.-S. Ho, C. Hsieh, and C.-Y. R. Huang. Symbolic model checking on SystemC designs. In *DAC*, pages 327–333, 2012.

[7] A. Cimatti, A. Griggio, A. Micheli, I. Narasamdya, and M. Roveri. Kratos - a software model checker for SystemC. In *CAV*, pages 310–316, 2011.

[8] W. Ecker, V. Esen, T. Steininger, M. Velten, and M. Hull. Implementation of a transaction level assertion framework in SystemC. In *DATE*, pages 894–899, 2007.

[9] L. Ferro and L. Pierre. ISIS: Runtime verification of TLM platforms. In *FDL*, pages 1–6, 2009.

[10] C. Flanagan and P. Godefroid. Dynamic partial-order reduction for model checking software. In *POPL*, pages 110–121, 2005.

[11] P. Godefroid. *Partial-Order Methods for the Verification of Concurrent Systems: An Approach to the State-Explosion Problem*. Springer, 1996.

[12] D. Große and R. Drechsler. *Quality-Driven SystemC Design*. Springer, 2010.

[13] D. Große, H. M. Le, and R. Drechsler. Proving transaction and system-level properties of untimed SystemC TLM designs. In *MEMOCODE*, pages 113–122, 2010.

[14] F. Haedicke, S. Frehse, G. Fey, D. Große, and R. Drechsler. metaSMT: Focus on your application not on solver integration. In *DIFTS*, pages 22–29, 2011.

[15] P. Herber, J. Fellmuth, and S. Glesner. Model checking SystemC designs using timed automata. In *CODES+ISSS*, pages 131–136, 2008.

[16] IEEE Std. 1666. *IEEE Standard SystemC Language Reference Manual*, 2011.

[17] D. Karlsson, P. Eles, and Z. Peng. Formal verification of SystemC designs using a petri-net based representation. In *DATE*, pages 1228–1233, 2006.

[18] D. Kroening and N. Sharygina. Formal verification of SystemC by automatic hardware/software partitioning. In *MEMOCODE*, pages 101–110, 2005.

[19] S. Kundu, M. Ganai, and R. Gupta. Partial order reduction for scalable testing of SystemC TLM designs. In *DAC*, pages 936–941, 2008.

[20] K. Marquet, B. Karkare, and M. Moy. A theoretical and experimental review of SystemC front-ends. In *FDL*, pages 124–129, 2010.

[21] M. Moy, F. Maraninchi, and L. Maillet-Contoz. LusSy: an open tool for the analysis of systems-on-a-chip at the transaction level. *Design Automation for Embedded Systems*, pages 73–104, 2006.

[22] D. Tabakov, M. Vardi, G. Kamhi, and E. Singerman. A temporal language for SystemC. In *FMCAD*, pages 1–9, 2008.

[23] D. Tabakov and M. Y. Vardi. Monitoring temporal SystemC properties. In *MEMOCODE*, pages 123–132, 2010.

[24] C. Traulsen, J. Cornet, M. Moy, and F. Maraninchi. A SystemC/TLM semantics in promela and its possible applications. In *SPIN*, pages 204–222, 2007.

[25] M. Y. Vardi. Formal techniques for SystemC verification. In *DAC*, pages 188–192, 2007.

Handling Design and Implementation Optimizations in Equivalence Checking for Behavioral Synthesis

Zhenkun Yang
Portland State University
zhenkun@cs.pdx.edu

Kecheng Hao
Portland State University
kecheng@cs.pdx.edu

Sandip Ray
University of Texas at Austin
sandip@cs.utexas.edu

Fei Xie
Portland State University
xie@cs.pdx.edu

ABSTRACT

Behavioral synthesis involves generating hardware design via compilation of its Electronic System Level (ESL) description to an RTL implementation. Equivalence checking is critical to ensure that the synthesized RTL conforms to its ESL specification. Such equivalence checking must effectively handle design and implementation optimizations. We identify two key optimizations that complicate equivalence checking for behavioral synthesis: (1) operation gating, and (2) global variables. We develop a sequential equivalence checking (SEC) framework to compare ESL designs with RTL in the presence of these optimizations. Our approach can handle designs with more than 32K LoC RTL synthesized from practical ESL designs. Furthermore, our evaluation found a bug in a commercial tool, underlining both the importance of SEC and the effectiveness of our approach.

Categories and Subject Descriptors

B.6.3 [**Design Aids**]: Design Aids—*Automatic synthesis, Optimization, Verification*

General Terms

Algorithms, Performance, Verification

Keywords

Equivalence checking, behavioral synthesis, optimization

1. INTRODUCTION

Electronic System Level (ESL) specifications provide a promising approach to deal with the high complexity of modern VLSI systems: design functionality is specified at a high level of abstraction (*e.g.*, with SystemC, C/C++, or domain-specific language), and compiled by a behavioral synthesis tool to RTL. Several behavioral synthesis tools are commercially available [8, 3, 2, 4]. However, their adoption

Permission to make digital or hard copies of all or part of this work for personal or classroom use is granted without fee provided that copies are not made or distributed for profit or commercial advantage and that copies bear this notice and the full citation on the first page. To copy otherwise, to republish, to post on servers or to redistribute to lists, requires prior specific permission and/or a fee.
DAC 2013 May 29 - June 07, 2013, Austin, Texas, USA

critically depends on our ability to *certify* the result of synthesis, *i.e.*, ensure that the synthesized RTL conforms to the ESL specification. This task is challenging because of the large difference in abstraction between the two.

In previous work [19, 11], we have developed a sequential equivalence checking (SEC) framework for behavioral synthesis. The key ingredients of the framework were (1) the use of a formal structure, *Clocked Control/Data Flow Graph* (CCDFG) as a uniform design abstraction, (2) a certified sequence of high-level transformations to reduce the abstraction gap, (3) an SEC algorithm based on dual-rail symbolic simulation between CCDFG and RTL, and (4) optimizations that enable compositional application of SEC exploiting internal cutpoints and modular structures. Experimental results reported successful certification of synthesized designs with tens of thousands of lines of RTL for ESL specifications of a number of cryptographic algorithms.

Unfortunately, the above approach cannot directly handle certification of designs from other domains that involve considerably less structure. In particular, one key requirement to achieve compositionality in SEC is the availability of equivalent internal operations or modules between the abstract CCDFG and the corresponding RTL, which are then used as *cutpoints*. However, we found that for many synthesized ESL designs, there are very few internal operations that preserve such equivalence in the presence of design and implementation optimizations, thus undermining compositionality and hence scalability.

In this paper, we present techniques for SEC between ESL designs and synthesized RTL, in the presence of optimizations that violate local equivalences of internal signals. Our framework can handle large synthesized designs from diverse domains, *e.g.*, we could successfully certify all designs in the CHStone benchmark [12] synthesized by a commercial synthesis tool. CHStone is a publicly available C-based ESL benchmark suite containing designs from four different categories; some designs have over 1600 lines of C, and generate over 32K lines of RTL when synthesized. We are not aware of any other SEC framework for behavioral synthesis that can handle synthesized designs of such diversity and scale. As a point of comparison, the framework described in previous work [11] (including all the SEC optimizations but without the techniques presented here) can certify only one of the twelve designs in the benchmark suite.

Our key observation is that there are two key sources of local inequivalence between CCDFG and RTL:

Operation Gating: Behavioral synthesis tools often optimize the RTL by introducing control structures or "guards" to ensure that certain operations are executed only when their results are relevant to downstream computation, and turned off otherwise. Such gated operations are functionally equivalent to the behavioral specification only under these guards. This makes such an operation difficult to identify; more problematically, it precludes the naive approach of using it as a cutpoint by verifying it in isolation and replacing it with an uninterpreted function in the CCDFG and RTL.

Global Variables: Global variables are used commonly in ESL as a *design optimization*: the user can then define some design functionalities as implicit side effects of other design modules, reducing the lines-of-code in ESL description and thus improving compactness. Unfortunately, global variables break the compositional approach of verifying modules compositionally, since the side effects on these variables must be accounted for during SEC.

Our key contributions are algorithms that enable compositional SEC for behavioral synthesis in the presence of the above optimizations.

1. We develop an algorithm for *relaxed* SEC that includes identification and compositional use of gated variables. The approach tolerates local, "irrelevant" inequivalences between gated variables and their RTL counterparts, as long as the inequivalences are resolved during symbolic simulation of downstream computation.

2. We develop an approach to modeling the side effects of global variables explicitly and show how the approach can then be used with modular analysis.

The algorithms, albeit not individually complex, have been carefully developed to (1) exploit the constraints and invariants available from the behavioral synthesis process, and (2) reinforce the available SEC optimizations, facilitating smooth integration. Finally, we found a subtle bug in an optimization of the behavioral synthesis tool itself, demonstrating both the need for certification of behaviorally synthesized designs and the importance of SEC in general and our framework in particular to achieve such certification.

2. BACKGROUND

2.1 Behavioral Synthesis

A behavioral synthesis tool takes a high-level behavioral description of a design and a library of hardware resources, and generates an RTL implementation. Similar to a generic compiler, it first performs lexical, syntax and semantic analysis, and builds an intermediate representation (IR) of the high-level description. A series of transformations is then applied to the IR, which can be categorized in three phases:

Compiler transformations form the first level. This includes transformations such as dead code elimination, constant propagation, loop unrolling, etc.

Scheduling transformations entail computing for each operation the clock cycle for its execution. The clock cycle must account for constraints in hardware resources as well as control and data flow. These transformations include pipelining loop iterations, grouping independent operations for concurrent execution, etc.

Resource binding and control synthesis maps a hardware resource to each operation, allocates registers for variables used across clock cycles, and generates a finite state machine (FSM) to implement the schedule.

After these transformations, the design can be expressed in RTL. Often manual tweaks are added to optimize for different parameters (*e.g.*, performance, power, etc.).

2.2 A Certification Framework

In previous work [19, 11], we proposed an SEC framework for certifying behaviorally synthesized RTL. A key idea was to apply SEC to compare the RTL with the design representation after high-level transformations have been applied to the ESL description. The framework introduced a formalization called CCDFG as a design abstraction. The CCDFG semantics entail (1) state-based semantics for individual operations, and (2) interpretation of control and data flows and scheduling constructs. The operations supported are those in the LLVM assembly language [17]. High-level transformations are certified by theorem proving. The transformed CCDFG is compared with RTL through SEC via dual-rail symbolic simulation. Three key optimizations were used to improve SEC performance by exploiting compositionality. *Cutpoints* reduce lengths of symbolic expressions by replacing verified sub-circuits with new symbolic values. *Cut-loop* partitions SEC for a loop into three checks to avoid expensive fix-point computation. *Modular analysis* optimizes SEC by replacing verified sub-modules by uninterpreted functions.

Unfortunately, as we described above, the compositional approaches require equivalence of internal operations or modules between CCDFG and RTL, which is broken in practice by design and implementation optimizations such as global variables and operation gating. Indeed, our attempts to apply the framework on diverse examples showed this was a blocking problem; the SEC optimizations were unusable, and hence the framework did not scale. This motivates developing an SEC approach to robustly handle design and implementation optimizations.

3. CHALLENGES FROM OPTIMIZATION

3.1 Operation Gating

The idea of operation gating is to add controlling predicates so that an operation is not executed when the value computed is irrelevant to downstream computation. Behavioral synthesis tools generate optimized RTL with operation gating to facilitate power-friendly hardware systems [7]. The transformation itself is complex, and its details are not germane to this paper. The characteristic of operation gating that is relevant to equivalence checking is that some operations have explicitly generated gating predicates in the synthesized RTL, when no such predicate appears in the CCDFG. The effects of the operation on the CCDFG and the RTL are then equivalent only when the gating predicate holds.

Consider synthesizing the code fragment shown in Fig. 1(a). According to the semantics of C, the multiplication operation in Line 3 (and the assignment of the result to c) must be executed regardless of the value of b. However, the result of multiplication is only relevant to the eventual return value f when the value of b is 1. In the RTL shown in Fig. 1(c),

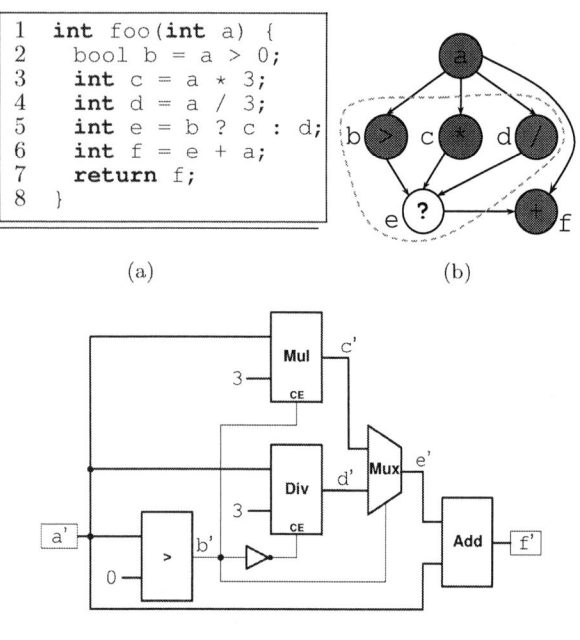

```
1  int foo(int a) {
2    bool b = a > 0;
3    int c = a * 3;
4    int d = a / 3;
5    int e = b ? c : d;
6    int f = e + a;
7    return f;
8  }
```

(a)   (b)

(c)

Figure 1: Operation gating example. (a) C code. (b) Data flow graph. (c) Schematic of generated RTL

the multiplication operation is therefore gated by condition b' so it is only executed when b' has the value 1.

Unfortunately, operation gating breaks compositionality. Recall from Section 2.2 that a key optimization involved in scaling up SEC for behavioral synthesis is the utilization of *cutpoints*. Cutpoints entail pre-verification of equivalence between corresponding internal variables in the CCDFG and the RTL, which are then replaced by (equivalent) symbolic variables. However, since the output of a gated operation is only equivalent when the gating condition is satisfied, its use as a cutpoint will cause the pre-verification to report inequivalence, breaking the compositional SEC flow.

To address this issue, we develop a *relaxed checking* algorithm for compositional SEC between a CCDFG G and a circuit M that tolerates local, "irrelevant" inequivalences for individual variables. The key idea is to continue dual-rail symbolic simulation even when a local inequivalence is encountered, but keep track of these inequivalences so that we can check if they are irrelevant during subsequent symbolic simulation. Algorithm 1 provides a high-level presentation of our approach. Here t_k denotes the scheduling step in clock cycle k, $EMap$ maps an operation op in CCDFG to combinational node in M, and x_k, s_k, i_k denote CCDFG state, circuit state, and inputs in clock cycle k respectively. At any point, the algorithm maintains a set, called $InEqSet$, of currently encountered variable inequivalences. For our example, in Fig. 1, $InEqSet$ will record the inequivalent pairs $\langle c, c' \rangle$ and $\langle d, d' \rangle$ between the CCDFG and the RTL when simulating Lines 3 and 4 respectively. During subsequent symbolic simulation, whenever an equivalence is discovered between variables in G and M, we check if that makes any of the inequivalences currently in $InEqSet$ irrelevant. For instance, when simulating Line 5 we find that e and e' are equivalent irrespective of the inequivalences between $\langle c, c' \rangle$ and $\langle d, d' \rangle$, making these two inequivalences irrelevant. When symbolic

simulation terminates, one of two outcomes is possible.

- $InEqSet$ is empty, meaning all inequivalences encountered have been resolved (*i.e.*, found irrelevant). The algorithm then reports G and M to be equivalent.

- $InEqSet$ still contains some inequivalences. This means that some operations found inequivalent during symbolic simulation remain relevant even after fix-point is reached. Thus the algorithm returns G and M to be inequivalent (and outputs the unresolved inequivalences).

Algorithm 1: RELAXED-CHECKING(G, M)

1 $k \leftarrow 0$ ⊳ Set clock cycle to 0
2 $InEqSet \leftarrow \emptyset$ ⊳ Empty inequivalence set
3 $G_{Info} \leftarrow$ FIND-GATING-INFO(G)
4 **while not** *(checking bound or fix-point reached)* **do**
5 $x_{k+1} \leftarrow$ SIM-CCDFG(G, t_k, x_k, i_k)
6 $s_{k+1} \leftarrow$ SIM-RTL($M, s_k, EMap(i_k)$)
7   **foreach** $op_g \in t_k$ **do**
8 $op_m \leftarrow EMap(op_g)$ ⊳ find the op in circuit M
9   **if not** IS-EQUAL(op_g, op_m) **then** ⊳ SMT query
10 $InEqSet \leftarrow InEqSet \cup \{\langle op_g, op_m \rangle\}$
11   **else**
12   RESOLVE-INEQ($InEqSet, G_{Info}, op_g, op_m$)
13 $k \leftarrow k + 1$
14 **if** $|InEqSet| = 0$ **then** ⊳ All inequivalences resolved
15   **return** true
16 **else**
17   **print** $InEqSet$ ⊳ Report all inequivalences
18   **return** false

Algorithm 1 makes use of two key subroutines, FIND-GATING-INFO and RESOLVE-INEQ to do the analysis of irrelevance of local inequivalences. To describe these subroutines we first need a key definition below. For this definition, recall that a Data Flow Graph (DFG) is a directed graph $G_D = (V, E)$, where each $v \in V$ is a variable in the program, each edge $(x, y) \in E$ represents a data dependency, meaning the value of variable y depends on the value of variable x. Furthermore, we will assume that each node in G_D is labeled with an operation (*e.g.*, add, mul, etc.).[1]

DEFINITION 1 (POST DOMINANCE). *Let G_D be a Data Flow Graph for a design, and u and v be two variables. We say that u is post-dominated by v in G_D iff $u \neq v$ and any path that starts from u goes through v.*

REMARK 1. *Post-dominance is a common concept in compiler literature [9], although it is typically defined with respect to the Control Flow Graph instead of the DFG as above. The definition extends to a CCDFG G by taking G_D to be the DFG component of G. Given a variable mapping $EMap$, we can also extend the notion to the circuit M: a variable u' in the circuit is post-dominated by v' if and only if (1) there are variables u and v in G that are mapped to u' and v' respectively, and (2) u is post-dominated by v. Thus we will often call $\langle u, u' \rangle$ to be post-dominated by $\langle v, v' \rangle$.*

---
[1] This assumption is valid in our case since the instructions in a CCDFG are in static single assignment (SSA) form; thus each variable can be uniquely associated with one operation.

The definition of post dominance guarantees that every path from u in G_D must go through v, e.g., in the example in Fig. 1(b), the variables c and d are post-dominated by e. Let $\langle u, u' \rangle$ be post-dominated by variables $\langle v, v' \rangle$ in G and M respectively. Then if v and v' are equivalent, it follows that from the perspective of any pair of corresponding variables $\langle x, x' \rangle$ that are descendants of $\langle v, v' \rangle$, the equivalence or inequivalence of $\langle u, u' \rangle$ does not matter. For instance, in Fig. 1, if e and e' are equivalent, then the inequivalence of c and c' is irrelevant. This observation leads to the theorem below that is an easy consequence of data flow.

THEOREM 1. *Suppose G is a CCDFG and M is a circuit such that the following hold: (1) variables $\langle v, v' \rangle$ are equivalent in G and M, and (2) $\langle u, u' \rangle$ are post-dominated by $\langle v, v' \rangle$ respectively. Let $\langle x, x' \rangle$ be arbitrary corresponding descendants of $\langle v, v' \rangle$. Then the equivalence between u and u' is irrelevant to the equivalence of x and x'.*

We now discuss the two subroutines.

FIND-GATING-INFO. This subroutine finds the potential gating information for a CCDFG G. A *potential gating information* is a list of pairs $\langle v, U \rangle$ where v is a variable and U is a set of variables such that each variable $u \in U$ is post-dominated by v. Theorem 1 guarantees that if v is equivalent to v' in G and M then the inequivalences of variables in U are irrelevant. Our implementation exploits the underlying LLVM constructs and information from the synthesis to efficiently determine relevant post dominance information. In particular, LLVM has a special `select` instruction of the form `y = select cond x1 x2`; the synthesis tool typically targets the condition variable of `select` instructions for operation gating.[2] Function FIND-GATING-INFO crawls over the data flow graph of CCDFG G, first identifying each `select` instruction; for each `y` it then finds all variables that are post-dominated by `y` recursively.

RESOLVE-INEQ. This function tries to resolve inequivalences in *InEqSet* using the gating information found by FIND-GATING-INFO. Let $\langle v, v' \rangle$ be determined to be equivalent during symbolic simulation. Then we find the set U such that $\langle v, U \rangle$ is a pair computed by FIND-GATING-INFO. From the above discussion, inequivalences involving variables in U are irrelevant, therefore dropped from *InEqSet*.

3.2 Global Variables

Modular design provides several advantages by breaking the design into modules. One key optimization presented in previous work [11] is modular analysis. The basic idea is to check each module individually in a bottom up manner.

- For each module M, check the equivalence of CCDFG and RTL.

- When checking module M' that calls M, replace the invocation of M in both CCDFG and RTL by equivalent uninterpreted functions.

However, global variable usages break this modular view, and one must account for side effects on these variables

---

[2]U need not be the *complete* set of variables post-dominated by v. This permits us to merely consider conditions in the LLVM `select` instruction as potential gating information. This runs the risk of possible spurious SEC failures. However, in our experience, this check has been sufficient.

while performing modular analysis. Note that while the side effects are implicit for high-level design descriptions (and hence CCDFGs), they are explicit on the synthesized RTL since the synthesis tool usually places the global variable on the interface when generating RTL.

Our solution is to compute an *extended signature* for a module that accounts for globals explicitly. Algorithm 2 shows how to compute the extended signature of a module. The key idea is to analyze the module to determine the globals used in the module. The parameters of the module are then extended to include *read-only* and *read-write* globals among the inputs and *write-only* and *read-write* globals among the outputs.[3]

Algorithm 2: GET-EXTENDED-SIGNATURE(f)

```
1  I ← PARAMETERS(f)
2  O ← OUTPUTS(f)
3  V_G ← FIND-ALL-GLOBALS(f)
4  foreach v ∈ V_G do
5      switch USAGE-TYPE(v) do
6          case R :  I ← I ∪ {v} ;              ▷ read-only
7          case W :  O ← O ∪ {v} ;              ▷ write-only
8          case RW :                            ▷ read-and-write
9              I ← I ∪ {v}
10             O ← O ∪ {v};
11 return ⟨I, O⟩
```

```
1   char A;      // global variable
2   int B;       // global variable
3   int C[2];    // global variable
4   void top() {
5     int i = 10;
6     bar(i);
7   }
8   void bar(int x) {
9     B = A + x;         // side effect
10    C[1] = C[0] + x;   // side effect
11  }
```

Figure 2: Global variable usage example

Fig. 2 shows an example of the computation. Module `top` includes `bar` as a sub-module which updates the global array `C`. Based on the parameters, the signature of `bar` is: `bar :: int -> void`. However, accounting for globals, the extended signature of `bar` by using Algorithm 2 is the following,

```
bar :: int->char->int[2] -> (int, int[2])
```

meaning that `bar` is represented as a function of three inputs (of the specified types), generating a pair of outputs.

Extended signatures account for global variables during modular analysis. Suppose that `bar` has been certified; when certifying `top` we replace `bar` with an uninterpreted function (say `BAR`) of three arguments, and the effect of the invocation of `bar` on the globals is given by `(B, C) = BAR(i, A, C)`.

---

[3]The algorithm assumes that local variables within a module have been standardized apart via renaming to avoid name conflicts with the globals and consequent variable capture. In our case capture avoidance is trivial since LLVM adopts different naming conventions for local and global variables.

Table 1: Summary of Evaluation on CHStone Benchmark

App. Domain	Design	Lines of code C	Lines of code RTL	C Functions	RTL Modules	Operation Gating	Global Variables[a] R	Global Variables[a] W	Global Variables[a] RW	Time (s)	Mem. (MB)
Arithmetic	DFADD	526	3722	17	5	Yes	4	0	1	174.9	169.34
	DFDIV	436	5192	19	4	Yes	4	0	1	6946.1	594.87
	DFMUL	376	3115	16	2	Yes	4	0	1	63.5	75.31
	DFSIN	755	11224	31	8	Yes	6	0	1	7151.3	603.50
Microprocessor	MIPS	232	2944	1	1	No	1	0	0	250.4	125.21
Media Processing	ADPCM	541	14935	15	5	No	15	19	63	68.2	105.45
	GSM	393	5598	12	4	Yes	4	0	0	49.6	83.07
	JPEG	1692	32846	30	17	Yes	30	14	17	2187.3	375.90
	MOTION	583	6168	13	5	Yes	9	0	4	1515.1	408.77
Security	AES	716	11869	11	7	Yes	4	0	5	170.7	106.59
	BLOWFISH	1406	17420	6	4	No	3	0	4	44.9	91.89
	SHA	1284	18819	8	4	No	3	0	4	6.0	89.04

[a]R means read-only, W means write-only, and RW means read-and-write.

4. EXPERIMENTAL RESULTS

4.1 Performance Evaluation

We have applied our framework to certify synthesized RTL for all the ESL designs in the CHStone benchmark. CHStone is a publicly available benchmark suite for behavioral synthesis, that includes twelve designs selected from different application domains. We used a commercial behavioral synthesis tool to synthesize the RTL. The most complex design in the benchmark is JPEG which has more than 32K lines of RTL code. For our experiments we have used the benchmark designs as is with one modification: two designs, JPEG and MOTION, used double pointers to represent two-dimensional arrays; these were modified to eliminate the double-pointer and represent the arrays explicitly. The reason has to do with the quirks of the synthesis tool used in this experiment. The synthesis tool inlines functions that have double pointers, thus flattening the module structure in the synthesized RTL. In addition to generating significantly larger RTL, this also destroys the module structure in the synthesized design. Since scalability of *modular analysis* (in the presence of design optimizations) is the key target of the experiments, we found the original designs unsuitable as targets for evaluation. The experiments were conducted on a workstation with 3GHz Intel Xeon processor and 8GB memory. For each design, we checked the equivalence between its CCDFG and RTL via dual-rail symbolic simulation, which symbolically simulates the CCDFG and RTL clock cycle by clock cycle. After each clock cycle, we checked the equality of mapped variables in the CCDFG and RTL by the MathSAT SMT solver [6]. We also applied cutpoints, cut-loop, and modular analysis optimizations when checking each design.

Table 1 shows the results of the experiments. The JPEG design takes about 36 minutes with 375.9 MB memory usage. The maximum certification time is required for DFSIN, which takes around 119 minutes with 603.5 MB memory usage. The experiment results demonstrate that independent of application domain our framework scales up to designs of practical complexity. No other SEC framework to our knowledge can handle behaviorally synthesized designs at this scale. Furthermore, only MIPS can be certified without handling operation gating and global variable optimizations.

4.2 A Behavioral Synthesis Bug

Our experiments found a bug in the synthesis tool during the certification of the MOTION design, which is a C implementation of a motion vector decoding algorithm for MPEG-2. Fig. 3(a) shows the source code fragment that triggers the bug. Here ld_Bfr is a global variable. In function Get_Bits, the return value Val is computed by right-shifting ld_Bfr. After Val is computed, ld_Bfr is updated in the subroutine Flush_Buffer. The update performed by

```
1   unsigned int ld_Bfr; // global ld_Bfr
2   void Flush_Buffer(int N) {
3       // modify the global variable
4       ld_Bfr = update(N, ld_Bfr);
5   }
6   unsigned int Get_Bits(int N){
7       unsigned int Val;
8       Val = ld_Bfr >> (32 - N);
9       Flush_Buffer(N);
10      return Val;
11  }
```

(a)

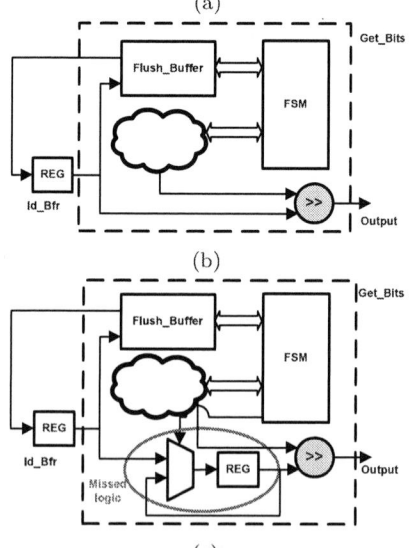

(b)

(c)

Figure 3: Bug found in MOTION example. (a) C source code (b) Wrong RTL (c) Correct RTL

`Flush_Buffer` does not affect the return value. Fig. 3(b) shows the RTL implementation synthesized by the behavioral synthesis tool. The global variable `ld_Bfr` is synthesized to a register outside of module `Get_Bits`. The output of `Get_Bits` is thus a combinational circuit with `ld_Bfr` as input. Therefore, when sub-module `Flush_Buffer` produces a new data for `ld_Bfr`, the new data is propagated to the output in the same clock cycle, leading to a wrong output.

The bug is caused because behavioral synthesis applies aggressive transformations to minimize resource usage. As can be seen by comparing Figs. 3(b) and 3(c), the synthesis tool in this case eliminates a register without correctly taking into account the side effect on the global variable. Such subtleties reinforce the need for SEC for certification of synthesized RTL designs. The bug has been confirmed by developers of the synthesis tool and fixed in a new release.

5. RELATED WORK

Koebl *et al.* [15] provides a good overview of research in SEC between high-level and RTL designs. Recently increasing sophistication of behavioral synthesis has resulted in several SEC optimizations to scale up certification of synthesized RTL [13, 16, 14, 11, 10]. For instance, Vasudevan *et al.* [20] introduce *sequential compare points* as a set of observable signals to be compared between high-level designs and RTL. There are commercial tools [1, 14] that can apply SEC between RTL and high-level (C/C++/SystemC) models. However, we have found no published results on approaches to handle design and implementation optimizations in any certification framework for behavioral synthesis.

There has however been research on handling such optimizations in SEC comparing RTL and netlist designs. Baumgartner *et al.* [5] discuss an approach for invariant generation to address the conditional equivalence checking problem for optimizations including clock gating and power gating. Moon *et al.* [18] propose equivalence checking techniques that exploit well-partitioned circuit structures.

6. CONCLUSION AND FUTURE WORK

We have presented a SEC framework to compare an ESL design with its behaviorally synthesized RTL implementation in the presence of design and implementation optimizations, *e.g.*, operation gating and global design variables. The framework scales to practical designs: it can handle all designs of the CHStone benchmark, some of which have more than 32K LoC synthesized RTL. We do not know of any other tool that can handle diverse designs at this scale, and the algorithms presented here are crucial to this scalability. In addition, certification found a bug in a commercial synthesis tool, underlining the importance of SEC for behavioral synthesis and the effectiveness of our framework. In future work, we plan to develop techniques for handling more sophisticated designs, including concurrent ESL specifications and complex, synthesized module interfaces.

7. ACKNOWLEDGMENTS

This research was partially supported by National Science Foundation Grants #CCF-0916772 and #CCF-0917188 and by a research grant from Intel Corporation. We thank Disha Puri, Naren Narasimhan, and Jin Yang for advice and help.

8. REFERENCES

[1] *Sequential Equivalence Checking: A new approach to functional verification of datapath and control logic changes*, 2007.

[2] *C-to-Silicon Compiler User Guide, 11.10*, 2011.

[3] *Catapult C Reference Manual*, 2011.

[4] *Cynthesizer Reference Guide, 4.1*, 2011.

[5] J. Baumgartner, H. Mony, M. L. Case, J. Sawada, and K. Yorav. Scalable conditional equivalence checking: An automated invariant-generation based approach. In *FMCAD*, 2009.

[6] R. Bruttomesso, A. Cimatti, A. Franzén, A. Griggio, and R. Sebastiani. The MathSAT 4 SMT solver. In *CAV*, 2008.

[7] J. Cong, B. Liu, R. Majumdar, and Z. Zhang. Behavior-level observability analysis for operation gating in low-power behavioral synthesis. *ACM Trans. Des. Autom. Electron. Syst.*, 16(1), 2010.

[8] J. Cong, B. Liu, S. Neuendorffer, J. Noguera, K. Vissers, and Z. Zhang. High-level synthesis for FPGAs: from prototyping to deployment. *Computer-Aided Design of Integrated Circuits and Systems, IEEE Transactions on*, 30(4):473–491, April 2011.

[9] J. Ferrante, K. J. Ottenstein, and J. D. Warren. The program dependence graph and its use in optimization. *ACM Trans. Program. Lang. Syst.*, 9(3), July 1987.

[10] K. Hao, S. Ray, and F. Xie. Equivalence checking for behaviorally synthesized pipelines. In *DAC*, 2012.

[11] K. Hao, F. Xie, S. Ray, and J. Yang. Optimizing equivalence checking for behavioral synthesis. In *DATE*, 2010.

[12] Y. Hara, H. Tomiyama, S. Honda, and H. Takada. Proposal and quantitative analysis of the CHStone benchmark program suite for practical C-based high-level synthesis. *Journal of Information Processing*, 17, 2009.

[13] A. J. Hu. High-level vs. RTL combinational equivalence: An introduction. In *International Conference on Computer Design*. IEEE, 2006.

[14] A. Koelbl, R. Jacoby, H. Jain, and C. Pixley. Solver technology for system-level to RTL equivalence checking. In *DATE*, 2009.

[15] A. Koelbl, Y. Lu, and A. Mathur. Embedded tutorial: formal equivalence checking between system-level models and RTL. In *ICCAD*, 2005.

[16] S. Kundu, S. Lerner, and R. Gupta. Validating high-level synthesis. In *CAV*, 2008.

[17] The LLVM Compiler Infrastructure. *LLVM Language Reference Manual (Version 2.7)*, 2010.

[18] I.-H. Moon, P. Bjesse, and C. Pixley. A compositional approach to the combination of combinational and sequential equivalence checking of circuits without known reset states. In *DATE*, 2007.

[19] S. Ray, K. Hao, F. Xie, and J. Yang. Formal verification for high-assurance behavioral synthesis. In *ATVA*, 2009.

[20] S. Vasudevan, J. Abraham, V. Viswanath, and J. Tu. Automatic decomposition for sequential equivalence checking of system level and RTL descriptions. In *MEMOCODE*, 2006.

A Counterexample-Guided Interpolant Generation Algorithm for SAT-based Model Checking[*]

Cheng-Yin Wu[†], Chi-An Wu[†], Chien-Yu Lai[†], and Chung-Yang (Ric) Huang[†‡]

[†]Graduate Institute of Electronics Engineering, National Taiwan University, Taipei 10617, Taiwan
[‡]Department of Electrical Engineering, National Taiwan University, Taipei 10617, Taiwan

ABSTRACT

Interpolation is an important and distinguished method popularly applied to recent synthesis and verification research topics. Existing approaches generate interpolants by analysing unsatisfiability proofs from SAT solvers. Unfortunately, the interpolant is predestinedly determined by how the unsatisfiability proof is logged. This particularly weakens the abstraction of interpolation-based model checking procedure. In this paper, a new approach to generate a variety of functionally different interpolants using simulation and SAT solving is proposed. We further seamlessly integrated the novel interpolant generation algorithm into the reinterpreted interpolation-based model checking procedure. Moreover, spurious counterexamples from the model checker further guide the generation of interpolants to refute excessive refinements. As an extra benefit, proof logging is not required for SAT solvers. Experiments show promising results of our interpolation-based model checker *NewITP* on solving a large set of HWMCC benchmarks.

Categories and Subject Descriptors

B.6.3 [**Logic Design**]: Design Aids—*Formal Verification*

General Terms

Model Checking, Verification

Keywords

Verification, Interpolation, Satisfiability, Generalization

1. INTRODUCTION

Given two inconsistent Boolean formulae A and B, an *interpolant* [5] is a Boolean formula I in terms of only common variables of A, B that over-approximates A and remains inconsistent with B. Interpolation receives both theoretic and practical attentions for generating an abstraction in a linear time complexity to the proof of unsatisfiability [17, 16].

From a practical perspective, interpolation has been broadly applied to a variety of research areas, especially for logic synthesis and verification, for instance, the computation of functional decomposition [12, 13], the reduction of dynamic power by enlarging clock gating functions [14], and engineering change order [21, 23].

The application of interpolation to formal verification is pioneered by McMillan's interpolation-based model checking algorithm [16]. It was acknowledged as the best single engine before the advent of property directed reachability (PDR, a.k.a IC3) [3, 9]. However, the interpolation-based algorithm remains the best technique to complete bounded model checking (BMC) [2] and complements PDR as well as other model checkers on verifying hard instances. Various researchers have enhanced the algorithm in different aspects, including [4, 15, 7].

In previous works, however, an interpolant is predestinedly determined by how the unsatisfiability proof is logged. Particularly, interpolants are regarded as blindly constructed. Therefore, for an interpolation-based model checker, interpolation techniques cannot assure a good abstraction for the verification process and thereby might result in excessive refinements frequently. As a consequence, existing interpolant generation algorithms may limit the effectiveness of the interpolation-based model checker.

We present a model checker, named *NewITP*, that naturally solves the problem. The features of *NewITP* are summarized as follows: (1) In contrast to previous works [17, 16, 11], the interpolants are generated using simulation and SAT solving. (2) The novel interpolant generation algorithm natively constructs functionally different interpolants. (3) Interpolant generation and reachability analysis in interpolation-based model checking algorithm are collaborated. (4) *NewITP* analyses spurious counterexamples from refinements and guides the generation of interpolants to avoid excessive refinements. Therefore the effectiveness of overall model checking procedure is improved.

This paper is organized as follows. Section 2 depicts the preliminaries. Our new interpolant generation algorithm is presented in Section 3, and the implementation of our interpolation-based model checker, *NewITP*, is described in Section 4. Experimental results are illustrated in Section 5. Finally, Section 6 concludes the paper.

2. PRELIMINARIES

As conventional notations, sets and set elements are represented as upper- and lower-case letters, respectively. Superscripts and subscripts of a set element indicate the time information and the numbering of the element, respectively.

We focus on applying the new interpolant generation algorithm to the model checking of safety properties on *finite state transition systems*. A finite state transition system M is a tuple $\langle F, S, I, \delta \rangle$ described by two propositional logic formulae: an initial condition $I(S)$ and a transition relation $\delta(F, S, S')$, where F is a set of free input variables while S and S' are sets of current and next state variables, respectively.

A *run* of M is a sequence of (F, S) pairs: $(f^0, s^0), (f^1, s^1), (f^2, s^2), \cdots$, such that $I(s^0)$ as well as $\delta(f^i, s^i, s^{i+1})$ for every adjacent state pair is satisfied. A state that appears in some run of the system is *reachable*. For any propositional logic

---

[*]This work was partially supported by the National Science Council of Taiwan ROC under Grant No. NSC-99-2221-E-002-211-MY3.

Permission to make digital or hard copies of all or part of this work for personal or classroom use is granted without fee provided that copies are not made or distributed for profit or commercial advantage and that copies bear this notice and the full citation on the first page. To copy otherwise, to republish, to post on servers or to redistribute to lists, requires prior specific permission and/or a fee.
DAC '13, May 29 - June 07 2013, Austin, TX, USA

formula P over F and S, a state s either satisfies P (denoted as $s \models P$) or satisfies $\neg P$ (denoted as $s \not\models P$). Model checking of a safety property $\mathbf{G}P$ on the system M confirms whether all reachable states of M satisfy P. If the *bad* state $\neg P$ is not **safe** (i.e. $\neg P$ is **reachable**), then there exists a state $s \not\models P$ in some *counterexample*.

2.1 Propositional Satisfiability

A *literal* l_i is either a Boolean variable v_i or its negation $\neg v_i$. A *cube* is a conjunction of literals while a *clause* is a disjunction of literals. A SAT instance is a conjunction of clauses. A SAT instance is *satisfiable* if and only if there exists an assignment to variables such that every clause is satisfied; otherwise, the instance is *unsatisfiable*. There exists a resolution proof (a sequence of resolution steps) for the unsatisfiability [18], and modern SAT solvers, including [10], are capable of generating a proof under proof logging.

Finally, a propositional logic formula can be converted into a SAT instance in linear time whilst the satisfiability is preserved [22].

2.2 Interpolation and Model Checking

Theorem 1. *[5] For any two propositional formulae ϕ_A, ϕ_B with $\phi_A \wedge \phi_B$ unsatisfiable, there exists a propositional formula ϕ'_A such that: (1) $\phi_A \Rightarrow \phi'_A$, (2) $\phi'_A \wedge \phi_B$ is unsatisfiable, and (3) ϕ'_A refers to only common variables of ϕ_A, ϕ_B.*

The new propositional formula ϕ'_A is an *interpolant* of ϕ_A, ϕ_B. Since the SAT instance of $\phi_A \wedge \phi_B$ is unsatisfiable, an interpolant ϕ'_A can thus be constructed from the resolution proof under proof logging.

In the previous work [16], McMillan proposed a pioneered interpolation-based model checking algorithm which is a breakthrough in SAT-based formal verification. The intuition to the algorithm is to over-approximate the reachability using interpolation, since the computation of the exact reachability is usually intractable.

The algorithm can be implemented in either *forward* or *backward* manner: Forward implementation constructs interpolants that over-approximate forward reachable states of initial states while the backward counterpart over-approximates backward reachable states (pre-images) of bad states (states that satisfy $\neg P$). The backward implementation of the algorithm involves three major steps:

1. BMC Step: Let $\neg P$ be the bad state formula and Img be the formula that characterizes a set of states. Initially, $Img = \neg P$. A bounded model checking [2] procedure is applied on the system M by checking the k-step reachability from initial states to states that satisfy Img. The $(k-1)$-step reachability from initial states is represented as

$$R^{k-1} = I(s^0) \wedge \bigwedge_{i=1}^{k-1} \delta(f^{i-1}, s^{i-1}, s^i) \qquad (1)$$

and the pre-image of Img is characterized as

$$Pre^{Img} = \delta(f^{k-1}, s^{k-1}, s^k) \wedge Img. \qquad (2)$$

Therefore,

$$BMC^k = R^{k-1} \wedge Pre^{Img} \qquad (3)$$

checks the k-step reachability from initial states to Img. If Formula (3) is satisfiable[1], $\neg P$ is **reachable** and a k-step *counterexample* is returned. Otherwise, R^{k-1} and Pre^{Img} are disjoint and the property P holds at the k-th time frame.

2. ITP Step: Given that Formula (3) is unsatisfiable, an interpolant Img' that over-approximates Pre^{Img} is constructed. From the definition of interpolant, $R^{k-1} \wedge Img'$ remains unsatisfiable.

---

[1]A special case for $k = 0$ is omitted here.

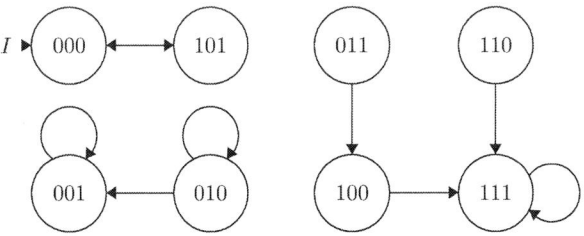

Figure 1: A transition system with three state variables and the only initial state 000.

3. Containment and Refinement Step: Consider the formula $Img' \Rightarrow Img$. If it is false, replace the present Img by $Img \vee Img'$ (disjunction of abstractions) and repeat Formula (3). The unsatisfiability of Formula (3) allows **ITP Step** to further abstract the pre-image of the updated Img. Otherwise, a potential spurious counterexample is found and thereby the algorithm aborts and restarts **BMC Step** in the next iteration by increasing the bounded depth k. On the contrary, if the formula $Img' \Rightarrow Img$ is true, an *inductive invariant* is found because a fixed point Img is reached, and $\neg P$ is proved **safe**.

The algorithm terminates either when a real counterexample is returned from **BMC Step** or when an inductive invariant is found. In practice, the above-mentioned procedure successfully over-approximates pre-images of $\neg P$ in any steps by means of interpolation. However, existing interpolant generation algorithms limit the effectiveness of model checking because they usually produce coarse over-approximations that result in excessive refinements.

3. COUNTEREXAMPLE-GUIDED INTERPOLANT GENERATION ALGORITHM

We propose a novel algorithm to generate interpolants using *ternary simulation*[2] and SAT solving. As an extra benefit, proof logging is not required by the SAT solver.

3.1 Generation of Interpolants from Counterexamples

Recall the **ITP Step** in Section 2.2: As Formula (3) is unsatisfiable, an interpolant Img' that over-approximates the pre-image of Img is constructed such that $R^{k-1} \wedge Img'$ remains unsatisfiable. Our algorithm generates an interpolant Img' by iterating three phases:

1. SAT Solving: Consider Formula (2), the pre-image of Img. Since Pre^{Img} is not empty, the formula is satisfiable and thus a state s^{k-1} is extracted from Pre^{Img}. Let $img \in Img$ be a state reachable from s^{k-1}.

2. SAT Generalization: In this phase, the algorithm generalizes the state s^{k-1} to a larger set of states using ternary simulation. The procedure is described as follows: Initially, set the cube C as the state s^{k-1} and simulate the satisfying assignment for one frame to reach img. Next, for each variable v_C of C, replace its value by the unknown symbol X and simulate again. If X does not appear in any variables of img, v_C is removed from C. Otherwise, undo the replacement for v_C, re-simulate and move on to the next state variable of C. Finally, the cube C represents a set of Pre^{Img} states[3]. A similar procedure is described in [8, 9].

3. UNSAT Generalization: Consider another formula $R^{k-1} \wedge C$. The formula is unsatisfiable because states of the

---

[2]Three-valued simulation: The augmentation of the unknown symbol X extends binary semantics by $X \wedge 0 = 0$ and $(X \wedge 1) = (X \wedge X) = \neg X = X$.

[3]For instance, the cube $0XX1$ represents four states: $0001, 0011, 0101, 0111$.

Table 1: Comparison between Characteristics of Interpolant Generation Algorithms.

	Ours	Previous [17, 16]
Representation	Disjunction of cubes	A Boolean function
Engines	Ternary simulation and SAT solving	A SAT solver with proof logging
Generate Iters	Many iterations	One iteration
Varied Interps	Easy	Limited

cube C are situated in Pre^{Img} and Formula (3) is unsatisfiable. By assuming literals of C as unit assumptions to SAT solvers, modern solvers (e.g. [10]) are capable of returning a subset of assumption literals used in the proof of unsatisfiability[4]. Consequently the cube C is further generalized by retaining only literals of C that are used in the proof of unsatisfiability.

At the end of an iteration, the final clause $\neg C$ is pushed into the solver to block states of C in future iterations. We call $\neg C$ a blocking clause and C a blocked cube. This enables **SAT Solving** phase to find fresh Pre^{Img} states. Finally, the resulting cube C is collected before the next iteration starts.

The procedure eventually terminates in the **SAT Solving** phase when Formula (2) becomes unsatisfiable. In the meanwhile, Pre^{Img} is fully precluded by blocking clauses. Since the number of pre-image states is finite, the termination of the above-mentioned procedure is guaranteed.

Let $Img' = C_0 \vee C_1 \vee \cdots$ be the disjunction of blocked cubes. We show that Img' is an interpolant of pre^{Img} and R^{k-1}. First, the algorithm terminates when $pre^{Img} \Rightarrow Img'$ is true. Second, because $R^{k-1} \wedge C_i$ is unsatisfiable for each cube C_i, $R^{k-1} \wedge Img'$ is unsatisfiable. Finally, every cube C_i in terms of only pre-image state variables, which are the common variables of pre^{Img} and R^{k-1}. Therefore, Img' is an interpolant of pre^{Img} and R^{k-1} according to Theorem 1.

Example 1. Figure 1 illustrates a transition system with three state variables. As customary, 0 and 1 stand for the negative and positive literal of the corresponding state variable, respectively. Assume that 111 is the only state that satisfies $\neg P$. Consider the interpolant of Pre^{Img} and R^0 with $R^0 = \{000\}$, $Img = \{111\}$, and $Pre^{Img} = \{100, 110, 111\}$. Assume that a fresh state 110 is returned from **SAT Solving** phase and it is generalized further to the cube $11X$ in **SAT Generalization** phase. Finally, $11X$ is generalized to $X1X$ after **UNSAT Generalization**. Then the cube $X1X$ is blocked and the algorithm enters the next iteration to find a fresh state. Clearly, the state 100 remains and assume that it is generalized to $1X0$ then $1XX$ in the remaining two phases. Since $\{1XX, X1X\}$ covers Pre^{Img}, the procedure terminates and $Img' = 1XX \vee X1X$ thus characterizes an interpolant of Pre^{Img} and R^0.

3.2 Comparisons Between Interpolant Generation Algorithms

We compare our interpolant generation algorithm with previous approaches in Table 1. One representative feature of our algorithm is the inherited capability of generating functionally different interpolants. It is achieved by adding some literals of the initial cube to the final cube that are removed after generalization in the last two phases. (Recall that in the last two phases, a cube C is generalized after some literals are removed. However, for a removable literal, we can choose whether to remove it or retain it in C.) Notice that the legitimacy of the resulting Img' is guaranteed because Img' still contains every Pre^{Img} state and remains

---

[4]It is believed much simpler in contrast to add proof logging to modern SAT solvers [8].

Algorithm 1 Interpolation-based Model Checking Algorithm of *NewITP*

Input: System M, bad state formula $\neg P$
Output: Result = $\{SAFE, REACHABLE\}$
Initialize: $k \leftarrow 0$
1.   **if** $(BMC(k, \neg P) == \text{SAT})$ // special case for $k = 0$
2.   **return** *REACHABLE*
3.   **Repeat**
4. $k \leftarrow k + 1, i \leftarrow 0$
5. $Img^0 \leftarrow \neg P$
6.   **Repeat**
7. $Img^{i+1} \leftarrow ITP(k, Img^i)$
8.   **if** $(Img^{i+1} == \text{INVALID})$ // reachable
9.   **if** $(i == 0)$ // $Img^i == \neg P$, real counterexample
10.   **return** *REACHABLE*
11.   **else** // $i \neq 0$, might be spurious
12.   *break*
13.   **if** $(Img^{i+1} == \text{NULL})$ // inductive invariant
14.   **return** *SAFE*
15.   **else** // find new interpolant
16. $i \leftarrow i + 1$

---

inconsistent with R^{k-1}. By intuition, it is possible to create functionally different interpolants in general.

To our knowledge, some prior works have been proposed to compute interpolants of different strengths [6, 19] by manipulating unsatisfiability proofs. However, they do not offer an explicit approach on choosing interpolants that can improve the model checking procedure. In the next section, we explain how to guide the above-mentioned interpolant generation procedure to benefit the interpolation-based model checking algorithm.

4. IMPLEMENTATION OF INTERPOLATION-BASED MODEL CHECKER

Our interpolation-based model checker *NewITP* reinterprets the original interpolation-based model checking algorithm using the above-mentioned interpolant generation techniques. Furthermore, we describe a more efficient implementation of the algorithm by integrating reachability analysis with interpolant generation. The algorithm of *NewITP* is shown in Algorithm 1.

4.1 The Reinterpreted Model Checking Algorithm

Algorithm 1 reinterprets the ordinary interpolation-based model checking flow as mentioned in Section 2.2. We highlight our implementation details that reinforce the algorithm as the new interpolant generation techniques are applied.

In Algorithm 1, *two* SAT solvers are exploited by *NewITP*[5]: (1) BMC-solver checks whether there exists a state of a cube C that is reachable from initial states at a bounded depth, i.e. the query $R^{k-1} \wedge C$, and (2) ITP-solver focuses on finding pre-image states in the **SAT Solving** phase of our interpolant generation flow.

Two main differences between reinterpreted and original interpolation-based model checking algorithms are explained as follows.

1. Present interpolant is not replaced by the disjunction of all computed interpolants in the **Containment and Refinement Step**:

Let IMG be $Img \vee Img'$, where Img and Img' are the present interpolant and the disjunction of former interpolants,

---

[5]Leverage tasks between two solvers in more intuitive and implementation-wise while using a single solver increases the difficulty to implementation.

respectively. After a refinement, Img is replaced by a IMG and the abstraction of Pre^{IMG} is constructed in the original flow. Because Pre^{IMG} can be represented as the disjunction of Pre^{Img} and $Pre^{Img'}$, the disjunction of their over-approximations forms an over-approximation of Pre^{IMG}. Observe that IMG is a legitimate over-approximation of $Pre^{Img'}$ and thus it is more practical to only abstract Pre^{Img}.

Since the Boolean space of IMG is larger than Img, this particularly lessens the burden of ITP-solver and thereby accelerates the generation of interpolants. Moreover, to refute finding repeated states, cubes that represent $Pre^{Img'}$ are blocked when computing an over-approximation of Pre^{Img}. Consequently, fresh states returned from **SAT Solving** phase are those new to all blocked cubes in the present and former interpolant generation iterations. As an extra benefit, the containment check of interpolants can be simplified to check the emptiness of the present interpolant.

2. Reachability analysis in **BMC Step** and the construction of interpolants in **ITP Step** are integrated:

Recall that our interpolants are represented as the disjunction of cubes that are generated in consecutive iterations. As **SAT Generalization** phase of the interpolant generation procedure completes, $NewITP$ immediately checks whether there is a $(k-1)$-step run to the cube C by the BMC-solver. If yes, a k-step *counterexample* to Img is returned, which is either real (as $Img = \neg P$) or potentially spurious. Otherwise, the interpolant generation procedure continues. Until all cubes are confirmed unreachable at the $(k-1)$-th step, the abstraction of Pre^{Img} and R^{k-1} returns.

Integrating reachability analysis and interpolant generation usually lessens the burden of BMC-solver because the large SAT instance of $R^{k-1} \wedge \delta(f^{k-1}, s^{k-1}, s^k) \wedge Img$ is partitioned into several smaller instances $R^{k-1} \wedge C$. In practice, there are even fewer SAT queries to BMC-solver because the **SAT Solving** phase merely finds states that do not appear in any existing interpolants.

Back to Algorithm 1, if Img^i is unreachable from the initial state at the k-th step (line 7), function $ITP(k, Img^i)$ returns an interpolant that over-approximates $Pre^{Img^i} \setminus Pre^{Img'}$ (states of Pre^{Img^i} that are new to $Pre^{Img'}$), where $Img' = Img^0 \vee Img^1 \vee \cdots \vee Img^{i-1}$. If the result is NULL, Img' represents an *inductive invariant* (lines 13-14). If Img^i is reachable otherwise, $ITP(k, Img^i)$ returns an INVALID result immediately as a cube C from the **SAT Generalization** phase is confirmed reachable (line 8). The counterexample is real if $i = 0$ (line 9) and otherwise it might be spurious (line 11). In the latter case, the inner loop breaks and k increases for the next iteration after all interpolants and blocked cubes are freed.

Example 2. Consider the model checking of the system in Figure 1 with $\neg P = 111$. Initially, there is no counterexample for $k = 0$, and the first interpolant $Img^1 = 1XX \vee X1X$ is generated as described in *Example1* when $k = 1$. Now consider the second interpolant $Img^2 = ITP(1, Img^1)$ with $R^0 = \{000\}$ and $Pre^{Img^1} = \{000, 010, 011, 100, 110, 111\}$. Because states in Img^1 are blocked in the last iteration, the ITP-solver returns the only fresh state 000. Then a spurious counterexample is returned because the reachability check $R^0 \wedge 000$ is satisfiable. All pre-images and blocked cubes are disposed after increasing k by one.

Assume that the first interpolant $Img^1 = 100 \vee X1X$ is computed under $R^1 = \{101\}$ in the next iteration ($k = 2$). Then states of $Pre^{Img^1} = \{010, 011, 100, 110, 111\}$ are already involved in Img^1. Therefore, there is no fresh state in Img^2 and the model checker confirms that $\neg P$ is **safe**.

4.2 Guiding Interpolant Generation by Spurious Counterexample Analysis

The toughest issue to interpolation-based model checkers is how to avoid excessive refinements, which are usually resulted from computing too coarse abstractions. The following example describes the problem.

Example 3. Assume that when $k = 2$ in *Example2*, Img^1 becomes $XX0 \vee X1X$. Because $R^1 = \{101\}$ and 101 is a pre-image state of Img^1, the reachability check is satisfiable and the procedure refines. Notice that the system has a fixed reachability $R^k = \{101\}$ for every odd k and $R^k = \{000\}$ for every even k. Unfortunately, the model checking algorithm can never terminate if the interpolant generation procedure returns $Img^1 = 1XX \vee X1X$ and $Img^1 = XX0 \vee X1X$ successively. It is a potential problem resulted from what functions of generated interpolants are, regardless of how they are generated.

A substantial solution to the problem is to extend R^k from the *k-step reachability* to *reachability within k steps* [16]. Empirically, this alteration improves the effectiveness of the model checker to solve more **safe** cases; however, it also sacrifices the efficiency on solving most instances, especially the **reachable** counterparts.

We proposed an alternative to leverage effectiveness and efficiency of the model checking algorithm. By analysing spurious counterexamples that lead to refinements, $NewITP$ *learns* the "cause" from refinements and *guides* the procedure of interpolant generation to avoid potential and excessive refinements: Evidently, a refinement occurs only if the reachability R^{k-1} and Pre^{Img^i} intersects. In the meanwhile, there is a reachable state s^i of Pre^{Img^i}. Moreover, there may exist a state s^{i-1} of Img^i reachable from s^i in one step. $NewITP$ thus traces the counterexample further until either Img^0 or some Img^j is confirmed unreachable.

These states $s^i, s^{i-1}, \cdots$ on the spurious counterexample are defined as the *states to refinement*, i.e. the "cause" to refinements. Mostly, *states to refinement* are not reachable to $\neg P$ and should be refuted from interpolants to avoid unnecessary refinements in the future. After each refinement, $NewITP$ accumulates all *states to refinement* to a set S_{ref} for *guiding* the generation of future interpolants against intersecting any of them. However, if Pre^{Img^i} does intersect with S_{ref}, an INVALID interpolation is returned immediately by the function $ITP(k, Img^i)$ in Algorithm 1.

Example 4. $NewITP$ (with guidance) solves the problem mentioned in *Example3*. At the first refinement, $R^0 = \{000\}$ is reachable to a pre-image state of $Img^1 = 1XX \vee X1X$ when $k = 1$ in *Example2*. Observe that the reachable state 101 is not a pre-image state of Img^0 (which is $\neg P$), and thus it is unreachable from Img^0. Therefore, $NewITP$ initializes S_{ref} to $\{000, 101\}$. As k increases to two, the new interpolant Img^1 under $R^1 = \{101\}$ is computed *with guidance*. Similarly, the state 100 is found and generalized to a cube $1X0$ using ternary simulation. Different from *Example1*, the cube cannot be generalized to $1XX$ because 101 is a *state to refinement*. Therefore, a fine-grained interpolant $Img^1 = 1X0 \vee X1X$ is returned. In the next interpolant generation iteration, there is no fresh state in Pre^{Img^1} excluded from Img^1 and thereby $NewITP$ terminates as $\neg P$ is confirmed **safe** successfully.

A simple but efficient approach to guide our interpolant generation engine from intersecting S_{ref} is described as follows. Since states of a cube returned from **SAT Generalization** phase are indeed pre-image states, INVALID is returned immediately if the cube intersects with S_{ref} (line 8 of Algorithm 1). Therefore, only **UNSAT Generalization** phase is extended as follows: Assume that a cube C is

837

returned by **SAT Generalization** phase and L is the set of literals of C. Let $cost(l_i)$ be the number of intersecting states between S_{ref} and $C \setminus l_i$, where $C \setminus l_i$ is the cube C with literal l_i removed. Let l_{min} be the literal with the minimum cost popped from L. If the literal l_{min} is involved in the proof of unsatisfiability, keep it in C as usual. Otherwise, l_{min} is removed from C if $cost(l_{min})$ is zero. Then, update costs for literals that remain in L and continue on the next literal with the minimum cost. The procedure ends when L becomes empty or the minimum cost is non-zero. As a result, C is generalized from the unsatisfiability while refuting *states to refinement*. Clearly, the initial cube C is always the finest generalization inconsistent with S_{ref}.

After considering *states to refinement* into *NewITP*, some interpolants are forcedly returned as INVALID. We claim that the *NewITP* algorithm remains sound and complete because (1) *Img* remains an over-approximation of pre-image states while cubes returned by **UNSAT Generalization** are inconsistent with R^{k-1}. (2) Early or forced refinements preserve the soundness and completeness of the algorithm.

5. EXPERIMENTAL RESULTS

We implemented both *NewITP* and the original interpolation-based model checking algorithm [16] (McMillan's ITP) in C++ with the underlying SAT solver MiniSAT [10]. The instances chosen were drawn from the single property track of HWMCC benchmark suite [1] (465 from HWMCC11 plus 175 new from HWMCC12). All experiments were conducted on a Linux machine with Xeon 2.5 GHz CPU. Each instance was solved under 900 seconds time-out and 8GB memory-out limits.

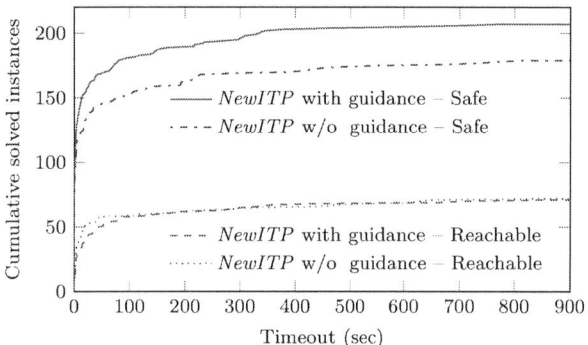

Figure 2: Result of *NewITP* with and without guidance.

The first experiment shows the effect of guidance to *NewITP* on solving all instances. Figure 2 illustrates the results on verifying both **safe** and **reachable** instances. The x- and y-axes represent the runtime and cumulative solved instances, respectively. On solving **safe** instances, *NewITP* with guidance successfully proves 28 more instances. On the **reachable** counterparts, since guidance introduces an inevitable but insignificant overhead to *NewITP*, only one instance is affected and becomes undecided. As a result, 27 more instances are solved by *NewITP* with guidance.

Figure 3 compares *NewITP* with McMillan's ITP on solving all instances. Totally *NewITP* (with guidance) solves 278 while McMillan's ITP solves 231 instances. Specifically there are 15 **reachable** and 32 **safe** more instances solved by *NewITP*, which shows that *NewITP* is superior in constructing adequate interpolants efficiently. The verification results of both checkers with R^k extended to the *reachability within k steps* are also included. They reveal that extending R^k does not help McMillan's ITP solve more (in fact, both **safe** and **reachable**) instances. For *NewITP*, since extending R^k eliminates the effect of guidance, we experimented

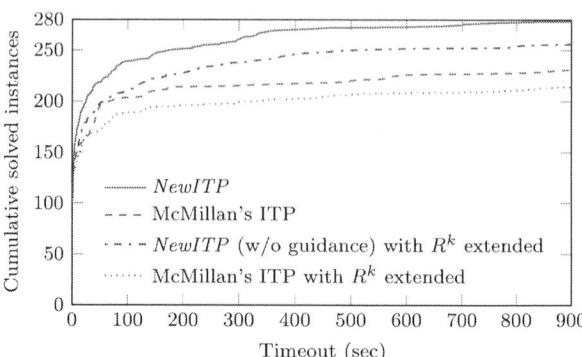

Figure 3: Result of interpolation-based model checking algorithms.

on R^k-extended *NewITP* without guidance. Compared to the results of McMillan's ITP, *NewITP* effectively solves the same number of **safe** and 22 more **reachable** instances than the R^k-extended version does.

The overall effect of extending R^k is not apparent to *NewITP* without guidance. Although we do not include the data here, extending R^k do help *NewITP* (w/o guidance) on solving 28 more **safe** instances; however, 23 **reachable** instances become undecided. This states that guidance strongly improves the effectiveness of our model checker while imposing an insignificant overhead to the efficiency.

Although considering reachability within k steps is a more complete solution to the excessive but avoidable refinement problem, it can be an overkill in practice. On the contrary, our approach on introducing guidance is more efficient and effective in general.

Another experiment on solving comparably hard instances was conducted. Instances that are unsolvable by other SAT-based model checkers, including property directed reachability (PDR , a.k.a IC3) [3, 9] and bounded model checking with induction (UMC) [2, 20], were retained. Figure 4 shows the result. The x- and y- axes correspond to the runtime results of *NewITP* and McMillan's ITP (both without extending R^k), respectively. A spot in the figure corresponds to the model checking result of an instance. Experiments show that *NewITP* complements other model checkers better than McMillan's ITP does: In the right figure, *NewITP* proves two more instances and performs better in runtime than McMillan's ITP. Moreover, in the left figure, *NewITP* discovered 3 counterexample runs that any other checkers can't. Experiments explain that UMC takes too much effort in the induction steps while PDR fails to investigate real counterexample runs efficiently on solving these **reachable** instances.

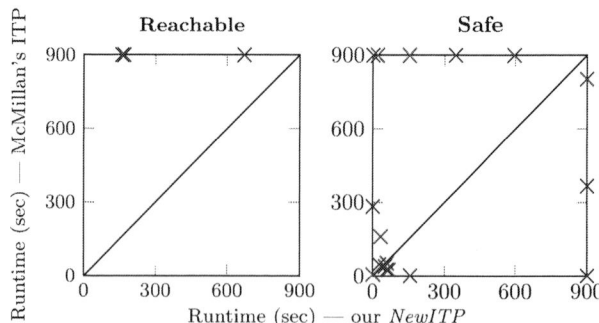

Figure 4: Runtime on solving hard instances.

We have researched on some potential improvements to *NewITP*. To accelerate *NewITP* on **reachable** instances, the bounded depth k can be safely increased by i instead of one after refinement in Algorithm 1. It is because the refinement at the i-th iteration implies that R^{k-1} is not reachable to $\neg P$ in i steps. (Refinement at $i=0$ returns a real counterexample.) Therefore, it is safe to increase k by i since the procedure has confirmed that there is no counterexample with depth smaller than $k+i$. Experimental results show that *NewITP* solves the same number of instances if i is increased dynamically; however, it indeed improves *NewITP* to solve seven more **reachable** instances.

In fact, we investigated some **reachable** instances that can only be disproved by *NewITP* under dynamic increased i. Because *NewITP* merely solves the one-step counterexample from undiscovered pre-image states to states discovered in the last iteration of interpolant generation, the SAT instance is usually small and easy to solve by modern solvers. We observed that on verifying these instances, i usually grows rapidly before refinement and then k is increased by a large i once. This not only blocks *NewITP* from checking worthless counterexamples smaller than $k+i$ steps but also increases the effectiveness of *NewITP* to investigate deep counterexamples that are unreachable by other model checking algorithms.

Theoretically, *NewITP* can be implemented in the reverse manner[6] by considering the over-approximation of forward images instead of backward pre-images. Two major alterations in our implementation for reverse *NewITP* are: (1) no **SAT Generalization** phase in interpolant generation procedure because ternary simulation can no longer be applied, and (2) ternary simulation can thus be applied to generalize more *states to refinement* for guidance. Because **SAT Generalization** accelerates the collection of pre-image states, the runtime of interpolant generation in the reversed implementation is detained and thereby the efficiency drops. Nevertheless, making *states to refinement* a larger set renders reverse *NewITP* refine more frequently and construct fine-grained interpolants. This enables reverse *NewITP* to prove some hard instances that the original implementation requires more refinement steps.

6. CONCLUSIONS AND FUTURE WORK

We have proposed a counterexample-guided interpolant generation algorithm that allows constructing functionally different interpolants using ternary simulation and SAT solving. Our interpolation-based model model checker, *NewITP*, not only reinterprets McMillan's model checking algorithm but also reinforces it by analysing spurious counterexamples for interpolant generation guidance. Experiments showed that *NewITP* outperforms McMillan's algorithm on solving both **safe** and **reachable** instances from a large benchmark suite. However, the performance of *NewITP* is still inferior to PDR, which indicates a possible research direction in the future to leverage different algorithms. As various applications of interpolation has been researched, we anticipate techniques in *NewITP* may further inspire and support researchers on solving new problems.

7. REFERENCES

[1] Hardware Model Checking Competition. *http://fmv.jku.at/hwmcc/*.

[2] A. Biere, A. Cimatti, E. Clarke, and Y. Zhu. Symbolic model checking without bdds. *Tools and Algorithms for the Construction and Analysis of Systems*, pages 193–207, 1999.

[3] A. Bradley. Sat-based model checking without unrolling. In *Verification, Model Checking, and Abstract Interpretation*, pages 70–87. Springer, 2011.

[4] G. Cabodi, M. Murciano, S. Nocco, and S. Quer. Stepping forward with interpolants in unbounded model checking. In *Computer-Aided Design, 2006. ICCAD'06. IEEE/ACM International Conference on*, pages 772–778. IEEE, 2006.

[5] W. Craig. Linear reasoning. a new form of the herbrand-gentzen theorem. *Journal of Symbolic Logic*, pages 250–268, 1957.

[6] V. D' Silva, D. Kroening, M. Purandare, and G. Weissenbacher. Interpolant strength. In *Verification, Model Checking, and Abstract Interpretation*, pages 129–145. Springer, 2010.

[7] V. D' Silva, M. Purandare, and D. Kroening. Approximation refinement for interpolation-based model checking. In *Verification, Model Checking, and Abstract Interpretation*, pages 68–82. Springer, 2008.

[8] N. Een, A. Mishchenko, and N. Amla. A single-instance incremental sat formulation of proof-and counterexample-based abstraction. In *Formal Methods in Computer-Aided Design (FMCAD), 2010*, pages 181–188. IEEE, 2010.

[9] N. Een, A. Mishchenko, and R. Brayton. Efficient implementation of property directed reachability. In *Formal Methods in Computer-Aided Design (FMCAD), 2011*, pages 125–134. IEEE, 2011.

[10] N. Eén and N. Sörensson. An extensible sat-solver. In *Theory and Applications of Satisfiability Testing*, pages 333–336. Springer, 2004.

[11] C.-J. Hsu, S.-L. Huang, C.-A. Wu, and C.-Y. Huang. Interpolant generation without constructing resolution graph. In *Computer-Aided Design-Digest of Technical Papers, 2009. ICCAD 2009. IEEE/ACM International Conference on*, pages 9–12. IEEE, 2009.

[12] R.-R. Lee, J.-H.-R. Jiang, and W.-L. Hung. Bi-decomposing large boolean functions via interpolation and satisfiability solving. In *Design Automation Conference, 2008. DAC 2008. 45th ACM/IEEE*, pages 636–641. IEEE, 2008.

[13] H.-P. Lin, J.-H.-R. Jiang, and R.-R. Lee. To sat or not to sat: Ashenhurst decomposition in a large scale. In *Computer-Aided Design, 2008. ICCAD 2008. IEEE/ACM International Conference on*, pages 32–37. IEEE, 2008.

[14] T.-H. Lin and C.-Y. Huang. Using sat-based craig interpolation to enlarge clock gating functions. In *Proceedings of the 48th Design Automation Conference*, pages 621–626. ACM, 2011.

[15] J. Marques-Silva. Interpolant learning and reuse in sat-based model checking. *Electronic Notes in Theoretical Computer Science*, 174(3):31–43, 2007.

[16] K. McMillan. Interpolation and sat-based model checking. In *Computer Aided Verification*, pages 1–13. Springer, 2003.

[17] P. Pudlák. Lower bounds for resolution and cutting plane proofs and monotone computations. *Journal of Symbolic Logic*, pages 981–998, 1997.

[18] J. Robinson. A machine-oriented logic based on the resolution principle. *Journal of the ACM (JACM)*, 12(1):23–41, 1965.

[19] S. Rollini, O. Sery, and N. Sharygina. Leveraging interpolant strength in model checking. In *Computer Aided Verification*, pages 193–209. Springer, 2012.

[20] M. Sheeran, S. Singh, and G. Stålmarck. Checking safety properties using induction and a sat-solver. In *Formal Methods in Computer-Aided Design*, pages 127–144. Springer, 2000.

[21] K.-F. Tang, C.-A. Wu, P.-K. Huang, and C.-Y. Huang. Interpolation-based incremental eco synthesis for multi-error logic rectification. In *Design Automation Conference (DAC), 2011 48th ACM/EDAC/IEEE*, pages 146–151. IEEE, 2011.

[22] G. S. Tseitin. On the complexity of derivation in propositional calculus. In J. Siekmann and G. Wrightson, editors, *Automation of Reasoning 2: Classical Papers on Computational Logic 1967-1970*, pages 466–483. Springer, Berlin, Heidelberg, 1983.

[23] B.-H. Wu, C.-J. Yang, C.-Y. Huang, and J.-H.-R. Jiang. A robust functional eco engine by sat proof minimization and interpolation techniques. In *Computer-Aided Design (ICCAD), 2010 IEEE/ACM International Conference on*, pages 729–734. IEEE, 2010.

---

[6]In this paper, the definition of the original and reverse implementations of *NewITP* is opposite to that of McMillan's ITP.

A Robust Constraint Solving Framework for Multiple Constraint Sets in Constrained Random Verification

Bo-Han Wu and Chung-Yang (Ric) Huang

Graduate Institute of Electronics Engineering/Department of Electrical Engineering, National Taiwan University

ABSTRACT

To verify system-wide properties on SoC designs in Constrained Random Verification (CRV), the default set of constraints to generate patterns could be overridden frequently through the complex testbench. It usually results in the degradation of pattern generation speed because of low hit-rate problems. In this paper, we propose a technique to preprocess the solution space under each constraint set. Regarding the similarity between constraint sets, the infeasible subspaces under a constraint set help identify the infeasible subspaces under another constraint set. The profiled results under each constraint set are then stored in a distinct range-splitting tree (RS-Tree). These trees accelerate pattern generation under multiple constraint sets and, simultaneously, ensure the produced patterns are evenly-distributed. In our experiments, our framework achieved 10X faster pattern generation speed than a state-of-art tool in average.

Keywords

Functional Verification, Constrained Random Verification (CRV)

1. INTRODUCTION

In modern VLSI verification plans, constrained random verification (CRV) plays an important role to ensure that the design-under-verification (DUV) conforms to the specification. In CRV, engineers focus on designing a set of constrained random scenarios to fulfill coverage specification. Specifically, those constrained scenarios or environmental restrictions are modeled as a set of constraints in a verification description language (VDL, e.g., SystemVerilog [1] and *e* [2]). The constraints are then automatically solved by a constraint solver and the produced valid stimuli should be evenly distributed to increase the probability to hit corner cases [1]. Therefore, the performance of the constraint solver becomes the key influential factor on the verification quality.

However, to verify system-wide specifications for SoC designs, the default set of constraints may be overridden frequently to reshape the distribution in the complex verification plans. For example in Figure 1(a), engineers can turn on/off specific constraints by the methods such as "*random_mode*" and "*constraint_mode*." Further, extra constraints can be instantly introduced in the randomization process by "*randomize with.*" These constraint controlling methods frequently appear in the testbench of modern verification methodologies such as *verification methodology manual* (VMM) [3] and *universal verification methodology* (UVM) [4].

To the best of our knowledge, this paper is the first work trying to conquer the challenges resulting from multiple sets of constraints. The

---

*This work was supported in part by National Science Council of ROC under grant NSC 99-2221-E-002-211-MY3.

Permission to make digital or hard copies of all or part of this work for personal or classroom use is granted without fee provided that copies are not made or distributed for profit or commercial advantage and that copies bear this notice and the full citation on the first page. To copy otherwise, to republish, to post on servers or to redistribute to lists, requires prior specific permission and/or a fee.
DAC '13, May 29 - June 07 2013, Austin, TX, USA.

```
1  class RndClass;
2    rand integer x, y;
3    constraint c1 { x+y≤100;}
4    constraint c2 { x≤60;}
5    constraint c3 { y≤60;}
6    constraint c4 { x≥10;}
7    constraint c5 { y≥10;}
8  endclass
9  program mainProg;
10   RndClass obj = new();
11   repeat (7000) obj.randomize();
12   obj.c4.constraint_mode(0);
13   repeat (5000) obj.randomize();
14   obj.c5.constraint_mode(0);
15   repeat (8000) obj.randomize();
16 endprogram
        (a) Example 1.
```

```
1. class RndClass;
2.   rand integer x, y;
3.   constraint c1 { x+y≤100;}
4.   constraint c2 { x≤60;}
5.   constraint c3 { y≤60;}
6. endclass
7. program mainProg;
8.   RndClass obj = new();
9.   repeat (1000) obj.randomize()
       with {(x≥10) && (y≥10);};
10. repeat (1000) obj.randomize();
11.endprogram
        (b) Example 2.
```

Figure 1. Two examples in SystemVerilog syntax to demonstrate constraint overriding by constraint controlling methods. (a) The method "*constraint_model*(0)" turns off constraints $c4$ and $c5$ in Lines 12 and 14 respectively. (b) The keyword "*with*" after the method "*randomize*" specifies an extra constraint {$(x≥10)$ && $(y≥10)$} in Line 9.

previous studies [5-16] are restricted to a single set of constraints and may cause either distribution violation or serious speed degradation if treating various sets of constraints independently.

For example, in [5], the *Interval Propagation* technique iteratively reduces sample space to increase hit-rates by acceptance and rejection (A&R) technique [6]. However, the generated patterns are biased, which violates the distribution requirement of constraint solvers. Thus, to meet the requirements of both distribution and speed, *Monte Carlo Markov Chain* (*MCMC*) based methods [7-10] guarantee convergence of evenness distribution after necessarily large amount of state transitions. However, the transition matrix needs to be reformulated and recomputed when handling different sets of constraints. Further the required pattern number for various sets of constraints could be insufficient for MCMC to achieve evenness distribution.

On the other hand, the formal pattern generators [11-14] such as SAT engines can perform incremental SAT solving to share the learned clauses between various sets of constraints. By inserting an extra Boolean control variable in each clause, we can switch between constraint sets. However, the resultant distribution is biased and unpredictable due to the complicated decision procedures.

Furthermore, the A&R with the RSSDE technique [15] is proposed to break the tradeoff between pattern generation speed and distribution under a single set of constraints. It attempts to identify as many infeasible subspaces as possible to enhance the hit-rates for random sampling. The profiles of solution space are entirely recorded in a range-splitting tree so that the top-down random walks on the range-splitting tree ensures even distribution of generated patterns. However, to deal with multiple sets of constraints, it is required to construct one range-splitting tree under each set of constraints. The construction time could prolong seriously to degrade the overall performance.

In this paper, we adopt the aforementioned idea of range-splitting trees [15-16] to facilitate pattern generation. However, instead of treating each set of constraints independently, our proposed technique reuses the identified infeasible subspaces under a certain set of constraints to infer the infeasible subspaces under another set of constraints. Specifically, a range-splitting tree is constructed incrementally from existing trees. Therefore, the runtime to construct all the required range-splitting trees significantly decreases. The pattern generation speed increases a lot since sufficient infeasible regions are pruned. Furthermore, the resultant generated patterns are still ensured evenly-distributed.

The remainder of this paper is organized as follows. Section 2 defines the problem formulation and presents an overview of our constraint solving framework. The detailed algorithms and theorems for incremental construction are explained in Section 3. Finally, Sections 4 and 5 demonstrate experimental results and conclusions respectively.

2. PROBLEM FORMULATION AND OVERVIEWS

2.1 Problem Formulation

Consider a set of variables $\boldsymbol{\mathcal{V}}$ such that each variable $v_i \in \boldsymbol{\mathcal{V}}$ can be in either bit-level or word-level. As mentioned, the constraint-controlling methods such as "*random_mode*," "*constraint_mode*" and "*randomize with*" result in various sets of constraints. We give the formal definition for those sets of constraints as follows.

Definition I. A *constraint set* $\mathcal{C}_i$ is defined as the set of constraints where $1 \le i \le n$ and at least one feasible pattern satisfying all the constraints in $\mathcal{C}_i$ will be produced. The variable n represents the number of the distinct constraint sets in the testbench.

Accordingly, in the testbench, we let $\boldsymbol{\mathcal{C}} = \{\mathcal{C}_1, \mathcal{C}_2 ..., \mathcal{C}_n\}$ denote the set of constraint sets in simulation order and $\boldsymbol{\mathcal{N}} = \{\mathcal{N}_1, \mathcal{N}_2 ..., \mathcal{N}_n\}$ denote the corresponding number of desired patterns. The constraints can contain general operators such as arithmetic, relational, logical, bit-wise, conditional operators to facilitate testbench implementation.

In Example 1, the constraint set in the randomization process in Line 11 is $\mathcal{C}_1 = \{c1, c2, c3, c4, c5\}$. After the constraint $c4$ is turned off in line 12, the constraint set in Line 13 becomes $\mathcal{C}_2 = \{c1, c2, c3, c5\}$. Similarly, the constraint set in Line 15 becomes $\mathcal{C}_3 = \{c1, c2, c3\}$. The required amount of feasible patterns for each constraint set is listed in Table I. The number of required patterns for $\mathcal{C}_1$ is $\mathcal{N}_1 = 7000$. Similarly, for $\mathcal{C}_2$ and $\mathcal{C}_3$, we have $\mathcal{N}_2 = 5000$ and $\mathcal{N}_3 = 8000$ respectively.

Therefore, the objective of the constraint solver is to efficiently generate $\mathcal{N}_i$ valid patterns for each constraint set $\mathcal{C}_i \in \boldsymbol{\mathcal{C}}$ and, simultaneously, the produced patterns should be evenly distributed.

TABLE I THE CORRESPONDING CONSTRAINT SETS AND THE NUMBER OF REQUIRED PATTERNS IN EXAMPLE 1

Constraint Set	Position	#patterns
$\mathcal{C}_1 = \{c1, c2, c3, c4, c5\}$	Line 11	$\mathcal{N}_1 = 7000$
$\mathcal{C}_2 = \{c1, c2, c3, c5\}$	Line 13	$\mathcal{N}_2 = 5000$
$\mathcal{C}_3 = \{c1, c2, c3\}$	Line 15	$\mathcal{N}_3 = 8000$

2.2 Flow of Our Constraint Solving Framework

In this paper, we propose a constraint solving framework to handle the testbench with multiple constraint sets. The proposed framework contains two stages: 1) preprocessing stage and 2) pattern generation stage as shown in Figure 2.

In the preprocessing stage, the input constraints are first simplified. Then we adopt the idea of range-splitting trees (RS-Tree) in the RSSDE technique [15] to profile solution space for solution density enhancement. However, it is required to profile the solution space of each constraint set for the testbench with multiple constraint sets. Instead of repeating this profiling procedure on every constraint set, we extract the *core set* of $\boldsymbol{\mathcal{C}}$, denoted as $\mathcal{C}_{core}$, in the constraint compilation step. A core set is the set of constraints which frequently appear in most constraint sets in $\boldsymbol{\mathcal{C}}$. The corresponding RS-tree $\mathcal{RST}_{core}$ is constructed to be the reference to accelerate the tree constructions of other constraint sets in $\boldsymbol{\mathcal{C}}$. If $\mathcal{C}_{core}$ is similar to a

constraint set $\mathcal{C}_i$, the information in $\mathcal{RST}_{core}$ can be effectively reused to construct the RS-Tree under $\mathcal{C}_i$. In Example 1, $\mathcal{C}_{core1} = \{c1, c2, c3, c4\}$ is better than $\mathcal{C}_{core2} = \{c1, c4\}$ to be the core set because $\mathcal{C}_{core1}$ is more similar to each constraint set in $\boldsymbol{\mathcal{C}}$.

After $\mathcal{RST}_{core}$ is constructed, we propose an algorithm to reuse the identified infeasible subspaces and estimated solution densities to incrementally construct the RS-tree $\mathcal{RST}_i$ under each $\mathcal{C}_i \in \boldsymbol{\mathcal{C}}$. The details will be explained in Section 3.

In the pattern generation stage, we utilize the range-splitting tree $\mathcal{RST}_i$ to efficiently produce $\mathcal{N}_i$ feasible patterns by top-down random walking on $\mathcal{RST}_i$ where $1 \le i \le n$. Due to the enhancement of solution densities by pruning infeasible regions, the runtime to obtain a feasible pattern is significantly reduced. In the meanwhile, we ensure the generated patterns are evenly distributed.

Please note that the number of core sets can be more than one in practice. That is, we can extract multiple core sets to enhance the similarity between the core set and other constraint sets. Nevertheless, to simplify the explanation in this paper, we assume there is only one single core constraint set. Besides, any constructed range-splitting tree can be a reference that provides information for reuse to build other range-splitting trees which are not yet constructed.

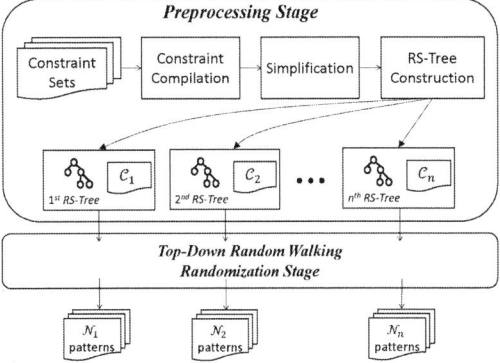

Figure 2. The flow chart of our constraint solving framework. The input constraints are first compiled and simplified in the preprocessing stage. The solution space is then profiled and the corresponding results are stored in one range-splitting tree per constraint set. In the pattern generation stage, the A&R technique is performed with top-down random walks on the RS-Trees to generate all the required patterns.

3. CONSTRAINT SOLVING ALGORITHM WITH MULTIPLE CONSTRAINT SETS

In this section, we illustrate the main algorithm to fast infer the solution distribution under each constraint sets by reusing existing profiles. Further, the proposed theorems ensure that the inference guarantees the precision of solution distribution in statistics so that the patterns can be evenly generated. We apply the proposed algorithm to Example 1 to construct the range-splitting trees $\mathcal{RST}_1$, $\mathcal{RST}_2$ and $\mathcal{RST}_3$ under the constraint sets $\mathcal{C}_1$, $\mathcal{C}_2$ and $\mathcal{C}_3$ respectively.

3.1 Range-Splitting Tree (RS-Tree) Construction

To model the solution distribution under a constraint set, we adopt the idea of RS-Trees in the RSSDE technique. The objective to construct RS-Trees is to identify as many infeasible subspaces as possible. If more infeasible subspaces are pruned (avoid sampling these subspaces), the solution density of the remaining feasible subspaces is higher. However, one range-splitting tree only corresponds to a certain constraint set.

Accordingly, in this paper, to reuse the identified infeasible subspaces and obtained solution densities, we modify and reformulate the definition of range-splitting tree ($\mathcal{RST}$) as follows:

a. The root represents the original space s.

b. For every node n_i in $\mathcal{RST}$, it represents a subspace s_i of s.

c. If n_i is not a leaf node and there are m child nodes of n_i, denoted as n_{ik} where $1 \leq k \leq m$ and $2 \leq m$, the corresponding subspaces is denoted as s_{ik} and the following properties hold.

  i) $\bigcup_{1 \leq k \leq m} s_{ik} = s_i$

  ii) $s_{ia} \cap s_{ib} = \varnothing$ where $1 \leq a, b \leq m$ and $a \neq b$

d. The set of random samples (an assignment of all variables $\mathcal{V}$) to estimate the solution density of space s_i is denoted as rs_i.

e. If n_i is not a leaf, the edge of n_i to n_{ik} is annotated by a *branching probability*, denoted as br_{ik}. It is formulated as:

$$br_{n_i \to n_{ik}} = \frac{\#\text{sols of } s_{ik}}{\#\text{sols of } s_i} = \frac{Sd(s_{ik}) \times SpaceSize(s_{ik})}{Sd(s_i) \times SpaceSize(s_i)} \quad (1)$$

where $Sd(s_i)$ defines the solution density of s_i from random samples rs_i.

The *Definition c* enables multiple cutting planes to partition a subspace simultaneously for effective identification of infeasible subspaces. The stored random samples in *Definition d* can be reused to estimate solution densities in another range-splitting tree. The *Definition e* ensures that the top-down random walks on $\mathcal{RST}$ in the pattern generation process matches evenness distribution.

Figure 3 shows an example of the relation between solution distribution and RS-Trees. In this case, there are two word-level variables, x and y with three constraints (cst_1, cst_2 and cst_3).

$$cst_1: (x - 3)^2 - (y - 3)^2 \geq 4$$
$$cst_2: 0 \leq y \leq 8$$
$$cst_3: 0 \leq x \leq 8$$

The corresponding solution distribution is plotted in Figure 3(a) and the resultant RS-Tree $\mathcal{RST}$ is shown in Figure 3(b). The subspace s_k corresponds to the node n_k in $\mathcal{RST}$. The grey area indicates that there is no feasible solution appearing in such region. The tagged information on the edge from n_i to n_j includes both the split range and the branching probability. The detailed construction procedure is explained in S1. From $\mathcal{RST}$, we observe that the solution densities of leaf nodes (n_5 and n_7) are higher than the root node (n_1). Therefore, the runtime to obtain a feasible pattern by A&R technique in s_5 and s_7 is reduced when compared to s_1.

Further, to ensure the solution density estimation is precise enough, the solution density and the required number of random samples in the subspace s_i should satisfy (2) defined in [15]. For example in Figure 3(b), the solution density for the corresponding subspace of the node n_1 is 33.3%. The confidence level is set to 95% so the corresponding z-score is 1.96. The minimum confidence interval is 3%. Then the required number of random samples ($|r_{s_1}|$) performed in this subspace s_1 should be greater than 4268 to guarantee the confidence level. Please note that, given an RS-Tree $\mathcal{RST}$, the solution density and the number of applied random samples for whole nodes in $\mathcal{RST}$ should all satisfy (2).

$$|r_{s_i}| \geq \left(\frac{Z}{w}\right)^2 \quad (2)$$

where $|r_{s_i}|$ represents the number of random samples. The Z indicates the corresponding z-score with respect to the given confidence level and w is the minimum confidence interval.

In summary, based on the above definition, a range-splitting tree models its corresponding solution space under a constraint set. The structure of an RS-Tree describes the corresponding partition status on the solution space. The solution density on each node reflects the distribution of feasible solutions in the entire space. Therefore, given an RS-Tree under a constraint set $\mathcal{C}_i$, the shape of the solution space for $\mathcal{C}_i$ is precisely approximated in statistics.

In the following subsection, from an existing RS-Tree, we describe the theorems to precisely infer the desired RS-Tree whose solution space is not yet profiled. Please note that, in our framework, we build the RS-Tree $\mathcal{RST}_{core}$ under the core set from scratch as the first existing RS-Tree.

3.2 Incremental RS-Tree Construction Technique under Multiple Constraint Sets

Given one existing RS-Tree under $\mathcal{C}_{exist}$, denoted as $\mathcal{RST}_{exist}$, we propose an algorithm to incrementally build the desired RS-Tree $\mathcal{RST}_{desired}$ under $\mathcal{C}_{desired}$ by reusing the tree structure and solution densities in $\mathcal{RST}_{exist}$. If the cardinality of the intersection set of $\mathcal{C}_{exist}$ and $\mathcal{C}_{desired}$ is larger, these two constraint sets are more similar. More information in $\mathcal{RST}_{exist}$ can be effectively reused to build $\mathcal{RST}_{desired}$. Therefore, in our framework, we attempt to select the profiled constraint set which is most similar to $\mathcal{C}_{desired}$ to incrementally construct $\mathcal{RST}_{desired}$.

For instance, we assume the core set is $\mathcal{C}_{core} = \{c1, c2, c3, c4\}$ in Example 1. Its corresponding solution distribution and $\mathcal{RST}_{core}$ are shown in Figure 4(a) and (b) respectively. Note that this assumption holds for all the following examples. Due to the high similarity between $\mathcal{C}_{core}$ and $\mathcal{C}_1$, the range-splitting tree $\mathcal{RST}_{core}$ is chosen to be the reference to construct $\mathcal{RST}_1$. The infeasible regions could be highly overlapped between $\mathcal{C}_{core}$ and $\mathcal{C}_1$. Therefore, in our framework, $\mathcal{RST}_1$ and $\mathcal{RST}_{core}$ are structurally identical in the beginning. Note that, in this step, the solution densities and branching probabilities in $\mathcal{RST}_1$ are not yet determined.

In the following subsections, the proposed algorithm targets to estimate solution densities of all subspaces in $\mathcal{RST}_{desired}$ in an incremental manner. A two-fold classification of this estimation is provided with respect to the difference between constraint sets ($\mathcal{C}_{desired}$ and $\mathcal{C}_{exist}$): 1) $\mathcal{C}_{desired}$ is the relaxation of $\mathcal{C}_{exist}$ and 2) $\mathcal{C}_{desired}$ is the restriction of $\mathcal{C}_{exist}$.

Please note that, to achieve the accuracy of solution density estimation, the number of applied random samples and resultant densities are required to satisfy (2) for every node in $\mathcal{RST}_{desired}$.

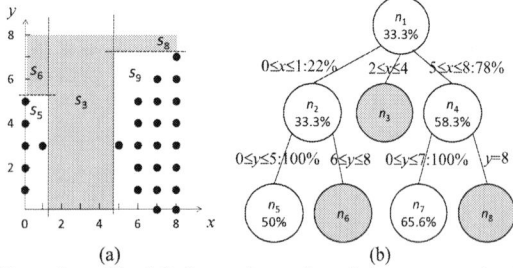

(a)   (b)

Figure 3. The left figure shows the solution space under the constraints cst_1, cst_2 and cst_3. The black dots represent feasible solutions and the grey area indicates infeasible spaces. The right figure demonstrates the corresponding range-splitting tree. The node n_i in the tree corresponds to the subspace s_i in the left figure. The dotted lines mean the cutting lines to partition the space.

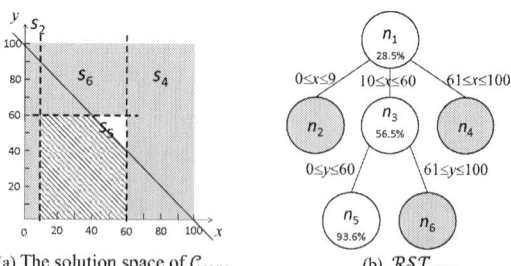

(a) The solution space of $\mathcal{C}_{core}$ (b) $\mathcal{RST}_{core}$

Figure 4. The hatched region in (a) indicates a feasible solution space under the constraint set $\mathcal{C}_{core}$. The corresponding range-splitting $\mathcal{RST}_{core}$ is plotted in (b). Three infeasible subspaces (s_2, s_4 and s_6) are identified. The solution density increases from 28.5% in s_1 to 93.6% in s_5.

3.2.1 Incremental RS-Tree Construction for Relaxation

For the first classification, the RS-Tree RST_{exist} is used to construct $RST_{desired}$ where the constraint set $C_{desired}$ is a subset of C_{exist}. From the point of view of solution space, the feasible space under $C_{desired}$ covers the feasible space under C_{exist}.

Theorem I. *Consider a constraint set C_{exist} and a constraint set $C_{desired}$ such that $C_{desired}$ is a subset of C_{exist}. The solution density for every node in $RST_{desired}$ is higher than the corresponding one in RST_{exist} where RST_{exist} and $RST_{desired}$ are in the same partition status and structurally identical.*

Proof:

Regarding that RST_{exist} and $RST_{desired}$ are structurally identical, we consider an arbitrary node n_{exist} in RST_{exist} and the corresponding node $n_{desired}$ in $RST_{desired}$.

1) If a random sample rs is feasible in n_{exist}
   → The random sample rs satisfies C_{exist}
   → The random sample rs also satisfies $C_{desired}$
   ($\because C_{desired} \subset C_{exist}$)
   → The random sample rs is also feasible in $n_{desired}$

2) If a random sample rs is infeasible in n_{exist}
   → The feasibility of rs is either infeasible or feasible in $n_{desired}$.

According to 1) and 2), the number of feasible random samples in $n_{desired}$ is ensured to be no less than the number in n_{exist}. Therefore, the solution density of $n_{desired}$ is higher than n_{exist}. ∎

Accordingly, by Theorem I, we can reuse the random samples stored in RST_{exist} to estimate the solution densities in $RST_{desired}$. That is, under the constraint set $C_{desired}$, we only need to re-examine the feasibility of the random samples which are invalid to C_{exist}. The efforts of solution density estimation are significantly reduced. Also, the desired confidence level defined by (2) still holds because of the property of higher solution densities by Theorem I.

However, an infeasible subspace in RST_{exist} may become feasible in $RST_{desired}$. Therefore, it is required to profile such subspace again with finer granularity and construct a sub-tree from the corresponding node in $RST_{desired}$.

For Example 1, assume that $C_{desired} = C_3 = \{c1, c2, c3\}$ and $C_{exist} = C_{core} = \{c1, c2, c3, c4\}$. The solution distribution under C_3 is shown in Figure 5(a). By Theorem I, we only re-examine the feasibility of the invalid random samples stored in RST_{core} to determine the solution densities in RST_3 instead of performing full solution density estimation. In this case, the result is shown as Figure 5(b). We observe that all the solution densities recorded on nodes (n_1 to n_6) in RST_3 are higher than the corresponding nodes in RST_{core}. However, the node n_2 which is infeasible in RST_{core} becomes feasible in RST_3. It is required to profile s_2 in detail to prune more infeasible subspaces (e.g., s_8). Therefore, the progress to construct the range-splitting tree RST_3 is significantly accelerated by reusing the random samples in RST_{core}.

3.2.2 Incremental RS-Tree Construction for Restriction

For the second classification, the RS-Tree RST_{exist} is used to construct $RST_{desired}$ where the constraint set $C_{desired}$ is a superset of C_{exist}. The feasible space under C_{exist} covers the feasible space under $C_{desired}$.

Theorem II. *Consider a constraint set C_{exist} and a constraint set $C_{desired}$ such that $C_{desired}$ is a superset of C_{exist}. The solution density for every node in $RST_{desired}$ is lower than or equal to the corresponding one in RST_{exist} where RST_{exist} and $RST_{desired}$ are in the same partition status and structurally identical.*

Proof:

From the proof of Theorem I, we realize the solution density of the RS-Tree with fewer constraints is higher than or equal to the RS-Tree with more constraints. Therefore, the solution density for every node in $RST_{desired}$ is lower than or equal to the corresponding one in RST_{exist}. ∎

Theorem III. *The infeasible nodes in RST_{exist} are also infeasible in $RST_{desired}$.*

Proof:

By theorem II, consider an arbitrary node n_{exist} in RST_{exist} and the corresponding node $n_{desired}$ in $RST_{desired}$. We have known that the solution density in $n_{desired}$ is lower than or equal to n_{exist}. If the solution density of n_{exist} is zero (infeasible), the corresponding subspace of $n_{desired}$ is also zero (infeasible). ∎

Similarly, by Theorems II and III, the infeasible nodes in RST_{exist}, imply that the corresponding nodes in $RST_{desired}$ are also infeasible. That is, the infeasible regions under C_{exist} are also infeasible under $C_{desired}$. However the confidence level property defined by (2) may be violated for some feasible nodes in $RST_{desired}$. In other words, we need to apply more random samples to estimate the solution density in the corresponding subspace and further profile such subspace to build a sub-tree if necessary.

For example in Figure 6, assume that $C_{desired} = C_1 = \{c1, c2, c3, c4, c5\}$ and $C_{exist} = C_{core} = \{c1, c2, c3, c4\}$. By Theorems II and III, the infeasible nodes (n_2, n_4, n_6) in RST_{core} retain infeasible in RST_1 as shown in Figure 6(b). This inference requires no formal procedures to prove the feasibility of n_2, n_4 and n_6. On the other hand, for the remaining feasible nodes (n_1, n_3 and n_5), we observe that their solution densities decrease. The precision defined by (2) is violated. More random samples and further profiles are required to compute solution densities and prune more infeasible subspaces (e.g., s_7), which results in the circled sub-tree in Figure 6(b). Nevertheless the extra profiles occupy only a small portion of the runtime to construct RST_{core} from scratch so that RST_1 can be fast constructed.

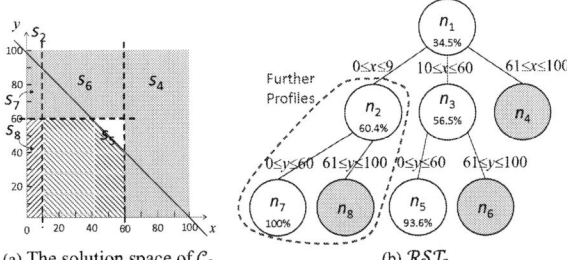

(a) The solution space of C_3 (b) RST_3

Figure 5. (a) shows the solution space under the constraint set C_3 and (b) plots its corresponding range-splitting tree RST_3. The tree structure (partition status) is reused from RST_{core}. However, the subspace s_2 is no longer infeasible. The further profiles in s_2 corresponds to the circled sub-tree (n_2, n_7 and n_8) in RST_3.

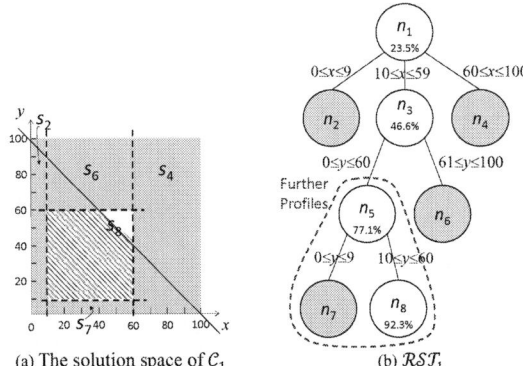

(a) The solution space of C_1 (b) RST_1

Figure 6. (a) shows the solution space under the constraint set C_1 and (b) plots its corresponding range-splitting tree RST_1. By Theorem III, the identified infeasible subspaces can be directly reused from RST_{core}. However, the solution density in s_5 decreases and the further profiles in s_5 corresponds to the circled sub-tree (n_5, n_7 and n_8) in RST_1.

843

Further the constructed RS-Trees such as RST_1 and RST_3 can be utilized to help build other RS-Trees as well. For example, to construct RST_2, the RS-Tree RST_1 is better than RST_{core} to be the reference for incremental construction because $C_2 = \{c1, c2, c3, c5\}$ is more similar to $C_1 = \{c1, c2, c3, c4, c5\}$ than $C_{core} = \{c1, c2, c3, c4\}$. The resultant RST_2 is shown in Figure 10 in S2.

In summary, given two distinct constraint sets (C_{exist} and $C_{desired}$), we apply a sequence of operations by adding and removing constraints from C_{exist} to make it become $C_{desired}$. The solution densities and the identified infeasible spaces can be reused by Theorems I to III. This mechanism reduces the construction time of RS-Trees so that, for complex testbench, it is efficient and scalable to completely profile all the possible distribution under multiple constraint sets.

After the solution densities in $RST_{desired}$ are entirely determined, the branching probabilities are computed by (1) to completely represent the distribution relation between subspaces. Further, the incremental construction of range-splitting trees ensures the profiled results precisely model the actual solution space in statistics. Therefore, in the pattern generation stage, we guarantee that the produced patterns are evenly distributed by top-down random walking on $RST_{desired}$. And, by pruning the identified infeasible subspaces for all the constraint sets, the pattern generation speed enhances a lot due to the increase of solution densities.

3.3 Pseudocode of our Incremental Construction Algorithm

Figure 7 shows the pseudocode of our framework for the proposed RS-Tree construction algorithm. In the beginning, we compile all the input constraint sets to obtain a core set C_{core} which contains the constraints frequently appearing in most constraint sets. Then we construct its corresponding RS-Tree RST_{core}. The variable $\boldsymbol{RST_{out}}$ stores all the required RS-Trees for C and it starts from an empty set. The variable $\boldsymbol{RST_{exist}}$ stores all the existing RS-Trees which can be further reused to perform incremental construction. In Line 4, it includes only RST_{core}.

In Line 5, we attempt to iteratively construct each RS-Tree RST_i under the constraint set C_i in C. In our algorithm, the function "$chooseRefRST$" determines the constraint set which is most similar to C_i. The corresponding RS-Tree $RST_{ref} \in \boldsymbol{RST_{exist}}$ is also determined in Line 6. By Theorems I to III, "$constructFrom$" reuses the structure and the information in RST_{ref} to fast incrementally build RST_i in Line 7. Then the RS-Tree RST_i is added to both sets $\boldsymbol{RST_{out}}$ and $\boldsymbol{RST_{exist}}$ in Lines 8 and 9 respectively. After all the required RS-Trees under the constraint sets C are constructed ($|\boldsymbol{RST_{out}}| = n$), this incremental RS-Tree construction process terminates and returns $\boldsymbol{RST_{out}}$.

Please note that, for real cases, $\boldsymbol{RST_{exist}}$ may contain multiple

---

Input: a set of constraint sets C
Output: a set of RS-Trees $\boldsymbol{RST_{out}}$

Function IncrementalRSTConstruction **begin**
1. $C_{core} \leftarrow$ compile(C);
2. $RST_{core} \leftarrow$ RstConstruct(C_{core});
3. $\boldsymbol{RST_{out}} \leftarrow \{\}$;
4. $\boldsymbol{RST_{exist}} \leftarrow \{RST_{core}\}$;
5. **foreach** C_i in C **begin**
6. $RST_{ref} \leftarrow$ chooseRefRST(C_i, C_{core}, C, $\boldsymbol{RST_{exist}}$);
7. $RST_i \leftarrow$ constructFrom(RST_{ref}, C_i);
8. $\boldsymbol{RST_{out}} \leftarrow \boldsymbol{RST_{out}} \cup RST_i$
9. $\boldsymbol{RST_{exist}} \leftarrow \boldsymbol{RST_{exist}} \cup RST_i$
10. **end**
11. **return** $\boldsymbol{RST_{out}}$
end

Figure 7. The pseudocode of the preprocessing stage in our constraint solving framework. The for-loop iteratively reuses the identified infeasible subspaces and solution density information to speed up the RS-Tree construction.

---

core sets in the beginning to effectively determine the most similar constraint sets for the function "$chooseRefRST$."

3.4 Constrained Pattern Generation Procedure

After all the required RS-Trees are incrementally constructed, the pattern generation process can be always performed with a RS-Tree to achieve high throughput of output patterns due to solution density enhancement.

For Example 1, the randomization process to generate 7000 patterns in Line 11 can be performed on RST_1. After the constraint set C_1 is overridden to C_2 in Line 12, we then apply RST_2 to produce another 5000 patterns in Line 13. In the end, after the constraint c_5 is turned off, the RS-Tree RST_3 is applied to generate 8000 patterns in Line 15.

Please note that, in practice, the number of constraint sets may be extremely large. In other words, we need to construct all the corresponding RS-Trees if we want to perform full profiles on solution space. To avoid such exhaustive construction, we only extract the critical constraint sets which are applied to generate a significantly large number of patterns since a large portion of pattern generation time is spent on those constraint sets.

In summary, the incremental construction of RS-Trees fast profiles the solution spaces under various constraint sets. We reuse the information from the existing RS-Trees to reduce the construction time of other RS-Trees. By pruning the identified infeasible subspaces recorded on the RS-Trees, the hit-rate for A&R technique in the pattern generation stage can enhance significantly. Moreover, the resultant distribution of generated pattern is ensured to be even.

4. EXPERIMENTAL RESULTS

All experiments are conducted on an Intel® Xeon® 2.00GHz workstation with 32GB memory. Our engine is set with confidence level ci=95%, max confidence interval w=0.01. We utilize MiniSAT-2.20 [17] as the SAT solver. We compared our framework with a state-of-the-art commercial tool, Synopsys VCS® C-2010.06. For fair comparison, test cases were randomly generated including both arithmetic and logical constraints.

4.1 Analyses of Pattern Generation Speed

The random test cases contain up to 40 word-level variables ($|V| \leq 40$) with 2 to 20 constraint sets. The constraints include general operators such as arithmetic, logic, relation operators and etc. For each case, 10^6 patterns were generated by both VCS® and our constrained random pattern generator. The corresponding results are plotted in Figure 8. The y-axis represents the runtime of VCS® in log-scale. On the other side, the x-axis is the runtime of ours in log-scale. For the data point above the dotted line, it means the pattern generation speed of ours is faster than VCS®.

As shown in Figure 8, we observe that the runtime to generate 10^6 patterns by our proposed framework mostly range from 100 to 1000 seconds. The runtime of VCS® ranges from 100 to 10000 seconds. The pattern generation speed of our framework outperforms VCS® for most cases and achieves about 10X faster than VCS® in average.

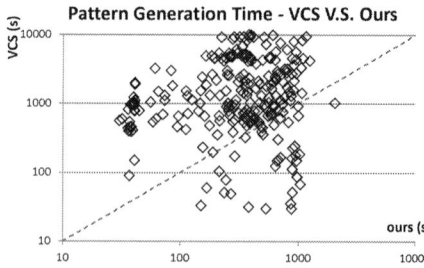

Figure 8. The comparison of pattern generation speed in log-scale between VCS® and our framework. The data point above the dotted line means that our framework is faster than VCS® and vice versa.

TABLE II. THE DETAILED INFORMATION FOR RSSDE TECHNIQUE, VCS® AND OUR FRAMEWORK.

Case	#vars	#cons	Arithmetic			Comp.		Logic			$|\mathcal{C}|$	RSSDE			Ours			VCS®
			*	+	-	<	=	&	|	^		Prep. (s)	Gen. (s)	Total (s)	Prep. (s)	Gen. (s)	Total (s)	Gen. (s)
1	45	12	5	4	6	4	8	0	0	0	9	4007.9	1405.4	5413.3	629.87	1388.8	2018.6	24485.6
2	40	8	4	4	4	3	5	0	0	0	12	3383.9	1166.8	4550.7	493	1158.1	1651.1	9442.3
3	49	12	5	4	6	4	8	0	0	0	9	4754.8	730.0	5484.8	527.24	738.5	1265.8	12449.5
4	55	12	5	4	6	4	8	0	0	0	9	5375.1	1776.3	7151.4	576.45	1619.4	2195.8	28769.6
5	34	5	0	0	1	2	3	2	3	0	7	771.2	355.0	1126.2	221.23	356.6	577.9	1762.5
6	51	6	1	0	1	2	4	2	3	1	6	1399.7	315.1	1714.8	313.24	314.3	627.6	728.3
7	47	11	3	2	1	3	8	2	4	2	16	8653.1	430.2	9083.3	926.04	437.6	1363.7	3791.0
8	36	5	0	0	1	2	3	2	3	0	7	1676.7	147.5	1824.2	180.14	150.1	330.2	1527.7
Avg.												3752.8	790.8	4543.6	483.4	770.4	1253.8	10369.6

4.2 Comparison between Different Techniques

To show the results in detail, we list the constraint information and runtime for ten cases from the previous experiments in Table II. Columns 2 and 3 indicate the number of variables and constraints respectively. Columns 4, 5 and 6 show the number of arithmetic, comparison and logic operators respectively. The number of constraint sets is shown in Column 7. For further comparison, we also implemented the RSSDE technique [15] to observe its preprocessing time on these cases. The pattern generation time of RSSDE technique, ours and VCS® are listed in Columns 8 to 10.

We observe that, for most cases, our framework outperforms both the RSSDE technique and VCS® (the average aggregate runtime: Ours:1253.8s < RSSDE:4543.6s < VCS®:10369.6s). By constructing the range-splitting trees for all constraint sets, our pattern generation speed can effectively increase. In addition, compared to the RSSDE technique, the resultant pattern generation time (Gen.) is approximately identical (ours:770.4s and RSSDE:790.8s) but we spent significantly less preprocessing time (Prep.) to construct multiple range-splitting trees (ours: 483.4s to RSSDE:3752.8s). For some cases (cases 6 to 8), the preprocessing time of the RSSDE technique without incremental construction is even larger than the overall runtime of VCS®. Therefore, the reusing strategy does effectively avoid the explosion of runtime in terms of the number of constraint sets.

4.3 Analyses of RS-Tree Construction Time

To discuss the relation between the constraint set number and the construction time of range-splitting trees, we generate several cases that the controlling factor is the number of constraint sets and the remaining (variables and constraints) is fixed the same. In Figure 9, the dark bar represents the RS-Tree construction time of ours and the light one means the construction time for the RSSDE technique.

We observe that, both construction time is positively related to the number of constraint sets. However, our construction time is largely less than the RSSDE technique. In addition, the construction time may not be strictly increasing to the number of constraint sets. It may result from the difference of solution space between constraint sets. The runtime to prove the feasibility of a subspace for a complex constraint set by SAT engines may sometimes largely increase the construction time. For instance, the construction time of the RSSDE technique for ten constraint sets is larger than 12 constraint sets.

In summary, by constructing multiple constraint sets, our framework succeeds to achieve about 10 times faster pattern generation speed than VCS® in average. Further, by reusing the partition status and solution densities, the construction time is largely reduced and is much less than the construction time of the RSSDE technique. In practice, our framework is scalable to be applied on real complex testbench with multiple constraint sets to speed up pattern generation.

5. CONCLUSIONS

A constraint solver to handle multiple constraints is a must to enhance the simulation quality on system-wide verification. To the best of our knowledge, this is the first work focusing on multiple constraint sets for constraint solvers. We propose a technique to fast

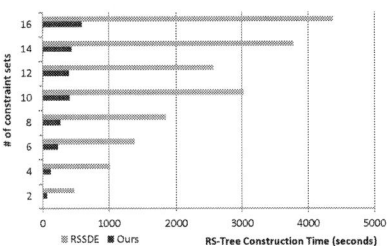

Figure 9. The relation between the number of constraint sets and the construction time of range-splitting trees regarding the RSSDE technique and our framework.

approximate the solution distribution under each constraint set. It reuses the profiles such as the solution densities and the identified infeasible subspaces in RST_{exist} to infer the desired range-splitting tree $RST_{desired}$ by Theorems I to III. Further, we ensure the estimated solution space is precise enough in statistics to guarantee the generated patterns are evenly distributed. Also, although the preprocessing time occupies only a small portion of the pattern generation time, the pattern generation process is significantly accelerated. In our experimental results, our framework outperformed a state-of-art tool with 10X faster pattern generation speed in average.

ACKNOWLEDGEMENT

The authors would like to thank all anonymous reviewers for their comments.

6. REFERENCES

[1] 1800-2009-IEEE Standard for System Verilog-Unified Hardware Design, Specification, and Verification Language, IEEE Computer Society, IEEE, New York, NY, IEEE Std 1800, 2005.

[2] IEEE Standard for the Functional Verification language 'e', IEEE Computer Society, IEEE, New York, NY, IEEE Std 1647, 2006.

[3] Verification Methodology Manual (VMM): http://www.vmm-sv.org/

[4] Universal Verification Methodology (UVM): http://www.uvmworld.org/

[5] M. A. Iyer, "RACE: A word-level ATPG-based constraints solver system for smart random simulation," ITC, pp. 299–308, 2003.

[6] L. Devroye, "Random variate generation for unimodal and monotone densities", Computing, v.32 n.1, pp. 43-68, 1984.

[7] W. Wei, J. Erenrich, and B. Selman, "Towards efficient sampling: exploiting random walk strategies," AAAI, pp. 670–676, 2004.

[8] B. Selman, H. A. Kautz, and B. Cohen, "Local search strategies for satisfiability testing," DMTCS, 1993.

[9] N. Kitchen and A. Kuehlmann, "Stimulus generation for constrained random simulation," ICCAD, pp. 258-265, 2007.

[10] N. Kitchen and A. Kuehlmann, "A Markov chain Monte Carlo sampler for mixed Boolean/integer constraints," CAV, pp. 446-461, 2009.

[11] J. Yuan, A. Aziz, C. Pixley, and K. Albin, "Simplifying Boolean constraint solving for random simulation-vector generation," TCAD, pp. 412–420, 2004.

[12] Y. Zhao, J. Bian, S. Deng and Z. Kong, "Random stimulus generation with self-tuning," CSCWD, pp. 62-65, 2009.

[13] S.M. Plaza, I. L. Markov and V. Bertacco, "Random stimulus generation using entropy and XOR constraints," DATE, pp. 664-669, 2008.

[14] S. Deng, Z. Kong, J. Bian and Y. Zhao, "Self-adjusting constrained random stimulus generation using splitting evenness evaluation and XOR constraints," ASPDAC, pp. 769-774, 2009.

[15] B.-H. Wu, C.-J. Yang, C.-C. Tso and C.-Y. (Ric) Huang, "Toward an Extremely-High-Throughput and Even-Distribution Pattern Generator for the Constrained Random Simulation Techniques," ICCAD, 2011.

[16] B.-H. Wu and C.-Y. (Ric) Huang, "A Robust General Constrained Random Pattern Generator for Constraints with Variable Ordering," ICCAD, 2012.

[17] MiniSAT 2.20 website: http://minisat.se/MiniSat.html

S1. DETAILS OF RANGE-SPLITTING TREE CONSTRUCTION

Consider the example shown in Figure 3. In order to identify as more infeasible subspaces as possible, we apply random samples to estimate the shape of solution distribution. In this example, the estimated solution density of the entire space is about 33.3%. The range-splitting tree starts from a root node (n_1), which represents the entire space. The estimated solution density and performed random samples are stored on this root node as well.

Moreover, the distribution of random samples is utilized to determine the cutting planes to partition the solution space. We restrict the cutting planes to being orthogonal to a certain axis. The computation of space size is simplified to the product of length in each dimension. In this case, the resultant cutting planes are $x=2$ and $x=5$, which are both orthogonal to x-axis. The original range of x is partitioned into three sub-ranges ($0 \leq x \leq 1$, $2 \leq x \leq 4$ and $5 \leq x \leq 8$) corresponding to the resultant three subspaces (s_2, s_3 and s_4). Please note that, unlike [15], we determine multiple cutting planes accordingly to the distribution of random samples to facilitate effective infeasible subspace identification.

After the entire space is partitioned, similarly, the solution densities of the resultant three subspaces are estimated by random sampling as well (s_2:33.3%, s_3:0% and s_4:58.3%). Simultaneously, three nodes (n_2, n_3 and n_4) are created to be the child nodes of n_1 and the solution densities as well as random samples are stored on the corresponding nodes. We observe that there is no feasible solution in the subspace s_3. A formal engine such as a SAT solver is used to prove its feasibility. Once s_3 is examined to be infeasible, it is unnecessary to profile s_3 in detailed in the following construction stage.

On the contrary, it is required to repeat the profiling procedure on the two feasible subspaces (s_2 and s_4) until the resultant solution densities of feasible leaf nodes are sufficiently high or the tree size exceeds a threshold value. After the solution density of each subspace is obtained, we apply (1) to compute branching probabilities on all edges. For example, for the edge from n_1 to n_2, the branching probability is calculated as

$$br_{n_1 \to n_2} = \frac{Sd(s_2) \times SpaceSize(s_2)}{Sd(s_1) \times SpaceSize(s_1)} = \frac{33.3\% \times 18}{33.3\% \times 81} \approx 22\%$$

where the space size can be simply computed by the product of length in each dimension since the cutting plane is orthogonal to x-axis. Accordingly, $SpaceSize(s_2) = 2 \times 9 = 18$ and $SpaceSize(s_1) = 9 \times 9 = 81$.

Further, the branching probability records the relation between subspaces. For example, the ratio of the solution number in s_2 to s_4 is $6/21 = 0.28$, which is equal to the ratio of branching probabilities ($br_{n_1 \to n_2}/br_{n_1 \to n_4} = 22/78 \approx 0.28$).

In the pattern generation stage, the range-splitting tree is utilized to generate feasible patterns by top-down random walking with respect to branching probabilities. For instance, the probability to generate a feasible pattern in s_7 is the product of the branching probabilities on the path from n_1 to n_7:

$$br_{n_1 \to n_4} \times br_{n_4 \to n_7} = 0.78 \times 1 = 0.78$$

Similarly, the probability to obtain a feasible pattern in n_5 is

$$br_{n_1 \to n_2} \times br_{n_2 \to n_5} = 0.22 \times 1 = 0.22$$

Assume the leaf node n_7 is reached. One feasible pattern in the subspace s_7 will be generated by A&R technique in our framework. The top-down random walking procedure is repeated until all the required patterns are produced. Since the branching probabilities precisely reflect the solution distribution, the generated patterns are ensured to be evenly distributed. Most importantly, the branching probabilities avoid sampling on the identified infeasible regions so that the pattern generation speed can significantly enhance.

S2. CONSTRUCTION OF $\mathcal{RST}_2$ FROM $\mathcal{RST}_1$

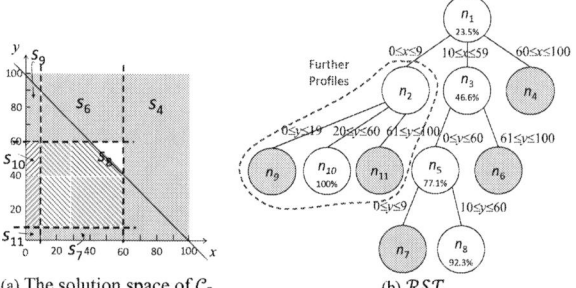

(a) The solution space of $\mathcal{C}_2$ (b) $\mathcal{RST}_2$

Figure 10. (a) shows The solution space under the constraint set $\mathcal{C}_2$ and (b) plots its corresponding range-splitting tree $\mathcal{RST}_2$. The tree structure (partition status) refers to $\mathcal{RST}_1$. However, the subspace s_2 is no longer infeasible. The further profiles in s_2 corresponds to the circled sub-tree (n_2, n_9 n_{10} and n_{11}) in $\mathcal{RST}_2$.

Simulation Knowledge Extraction and Reuse in Constrained Random Processor Verification

Wen Chen[1], Li-Chung Wang[1], Jay Bhadra[2], Magdy Abadir[2]

[1]University of California - Santa Barbara

[2]Freescale Semiconduction Inc.

ABSTRACT

This work proposes a methodology of knowledge extraction from constrained-random simulation data. Feature-based analysis is employed to extract rules describing the unique properties of novel assembly programs hitting special conditions. The knowledge learned can be reused to guide constrained-random test generation towards uncovered corners. The experiments are conducted based on the verification environment of a commercial processor design, in parallel with the on-going verification efforts. The experimental results show that by leveraging the knowledge extracted from constrained-random simulation, we can improve the test templates to activate the assertions that otherwise are difficult to activate by extensive simulation.

Categories and Subject Descriptors

B6.3 [**Logic Design**]: Design Aids—*Verification*

General Terms

Verification, Data Mining

Keywords

Functional Verification, Assertion, Coverage, Rule Learning

1. INTRODUCTION

The mainstream practice for functional verification of microprocessors today still heavily relies on extensive simulation. Functional verification starts with a verification plan, specifying the aspects of the design to verify [9]. In constrained-random verification, test cases are generated by constrained-random test generation, in which the test generator instantiates tests using templates written by verification engineers. The tests are simulated with checkers to check the correctness of the design. The completeness of simulation-based verification is measured by various coverage metrics. During the verification process, coverage results are analyzed to guide test generation. A satisfactory level of coverage must be met before tapeout.

This work is supported by Semiconductor Research Corporation, project 2012-TJ-2268.

Permission to make digital or hard copies of all or part of this work for personal or classroom use is granted without fee provided that copies are not made or distributed for profit or commercial advantage and that copies bear this notice and the full citation on the first page. To copy otherwise, to republish, to post on servers or to redistribute to lists, requires prior specific permission and/or a fee.

DAC '13 May 29 - June 07 2013, Austin, TX, USA

In a design cycle, the design evolves over time. Consequently, functional verification is an iterative process in which extensive simulation is run on a few relatively stable versions of the design. When a new version is released with accumulated changes over a period, the verification process restarts with the new version. From one iteration to another, two assets are kept. The first are the test templates refined and accumulated up to the previous iteration. The second are the important tests identified so far. For example, an important test can be the one activating a particular assertion of interest or capturing a bug in the previous design versions. These two assets embed the knowledge accumulated through the iterative verification process.

During the iterative process, it may not be effective to maintain the detailed structural coverage results from one iteration to the next due to major changes in the implementation. Therefore, functional coverage based on assertions (how many times an assertion is activated) is often used as the metric to evaluate the importance of tests and to guide test template refinement. Assertions are relatively stable and do not change as often as the design implementation.

In this work, we propose a novel learning methodology for extracting knowledge from important tests. The extracted knowledge then is reused for two purposes: (1) for producing more tests similar to those important ones and (2) for producing new important tests that, for example, can activate assertions not covered before. To develop such a learning methodology, we need to address three aspects: (1) what knowledge to extract, (2) how to extract and represent knowledge, and (3) how to reuse the extracted knowledge.

In this work, we applied the proposed methodology to verifying a dual-threaded low-power 64-bit Power Architecture-based processor core to be manufactured with a 28nm technology. Our experiments were conducted in parallel with the verification process where the design was not yet stable. The experimental results demonstrate the effectiveness of the methodology for the two intended purposes. More specifically, we show that after applying the extracted knowledge, a refined test template can effectively generate additional tests for activating an assertion that received low coverage before. Moreover, a refined test template can effectively generate tests for activating an assertion that was not covered before. Test template refinement using knowledge learned from simulation is naturally used in real-world verification processes, however, the knowledge is largely acquired by manual effort. In this work, automating the process of knowledge extraction is addressed by feature-based rule learning.

The rest of the paper is outlined as follows: Section 2

briefly reviews related works. Section 3 addresses the first aspect, i.e. what knowledge to extract. A feature-based rule learning methodology is presented in Section 4 to address the knowledge representation aspect. Section 5 discusses the knowledge extraction aspect using subgroup discovery rule learning. Section 6 illustrates how the knowledge can be reused. Experiment results are presented in Section 7. Section 8 concludes the paper.

2. RELATED WORKS

Coverage-directed test generation (CDTG) is an approach to dynamically analyze coverage results and automatically adapt the test generation process to improve coverage. Recent works proposed various techniques to learn from the simulation results. These approaches employ a variety of learning techniques such as Bayesian Networks [8], Markov Models [13], Genetic Algorithms [11] and Inductive Logic Programming [6]. However, automatically modifying the input to the test generator, based on the feedback from simulation, can be very difficult for complex designs. In a recent work [9] the authors proposed to learn test knowledge from micro-architectural behavior and embed the knowledge into test generator to produce more effective tests.

Early identification of the important tests to reduce simulation cost was proposed in [2][4]. Feature-based rule learning has been applied in the context of understanding design-silicon mismatch [1][3]. In contrast, this work studies the feasibility and effectiveness of applying feature-based analysis for extracting knowledge from important tests to improve functional (assertion) coverage.

3. WHAT KNOWLEDGE TO EXTRACT

This work considers knowledge extraction in the context of constrained-random verification for assertion coverage. Figure 1 illustrates a scenario of simulation with tests instantiated from a given test template that had been refined by the verification team up to the time of the experiment. The figure summarizes the statistics of covered assertions for the Load Store Unit (LSU) of the processor in a simulation of 3000 tests. The LSU is among the most complex and difficult-to-verify units in the design. Over 90% of the covered assertions were already activated by 50 or more tests. However, there existed other assertions activated only by 10 tests or fewer. Furthermore, there were assertions with zero coverage (not shown in the figure).

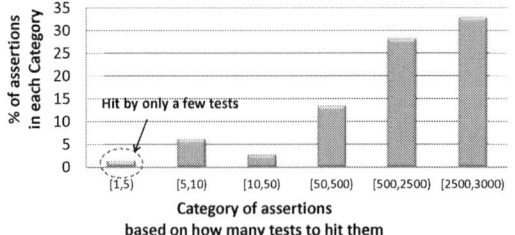

Figure 1: Histogram of covered assertions in LSU based on frequency of being activated

Our interest is in knowledge extraction for activating those assertions with low or zero coverage. The property stated by a complex assertion comprises multiple conditions. Learning the knowledge about the entire assertion directly could be difficult. Hence a divide-and-conquer strategy is employed.

The idea is to learn knowledge with respect to each condition and then, the knowledge can be combined for activating the assertion.

Knowledge extraction for a given condition is based on tests activating the condition. We call those tests the *novel* tests. In processor verification, a test is an assembly program. Figure 2 illustrates the learning goal. Suppose a novel assembly program is identified to trigger a special condition in the simulation, for example, a "coreflush" condition concerning the instructions already fetched but not yet committed. Then what we want to learn are descriptive *rules* explaining the properties in the novel tests that trigger the condition, for example, the rule being the existence of a mispredicted branch in the test. Such rules are then used as constraints to refine test templates for hitting the condition.

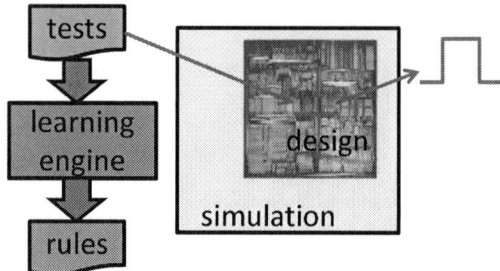

Figure 2: Illustration of the learning goal

To summarize, in our methodology we begin by extracting a list of conditions from the assertions for monitoring. Novel tests with respect to these conditions are identified in the simulation. The extracted knowledge is rules describing the special properties of the novel tests.

4. FEATURE GENERATION
4.1 Snippet-based Vector Representation

To extract knowledge from an assembly program, we first need an approach to automatically convert an assembly program into a representation suitable for applying the feature-based rule learning. A given assembly program may consist of hundreds of instructions. Our representation approach comprises two steps. The first step converts an assembly program into multiple *snippets* of instruction sequence of equal length k where k is a user-supplied input.

Figure 3 illustrates how this step, with a slide window size of 3, works on an example test with 6 instructions. Six snippets are extracted, where the ith snippet ends with the ith instruction in the test. Beginnings of the first two snippets are filled with dummy instructions.

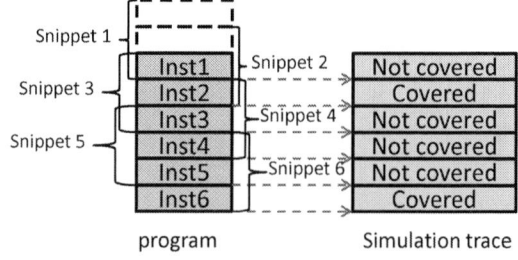

Figure 3: Illustration of the slide window approach

Each snippet is paired with the episode of simulation trace starting from the the commitment of the second-to-last instruction and ending at the commitment of the last instruction. In this way, each snippet is paired with a unique simulation episode. For a given condition to be monitored, the episode is used to decide if the condition is covered by the snippet. Figure 3 shows, in the example, the condition is covered by snippets 2 and 6.

The second step of the representation approach is to convert each snippet into a feature vector. A feature vector encodes a sequence of instructions based on a set of features. A feature can be an *occurrence* feature or a *descriptive* feature. An occurrence feature has a value of 0 or 1. A descriptive feature has a numerical value.

occurrence				descriptive				
o_1	o_2	...	o_n	d_1	d_2	...	d_m	Class
1	0	...	1	231	6	...	54	−1
1	1	...	0	78900	654	...	37	+1
0	1	...	0	256	800	...	24	−1
0	0	...	0	3	60	...	4096	−1
0	0	...	1	701	9754	...	7	−1
1	1	...	1	1	570	...	0	+1

Figure 4: Illustration of the transformed dataset

Figure 4 illustrates the look of a feature-encoded dataset based on the six snippets for a condition. The illustration shows n occurrence features and m descriptive features. The set of snippets are divided into two classes, the positive class with the condition being activated and the negative class without the condition being activated.

It is important to note that in the analysis, the negative class will also contain snippets obtained from non-novel tests. Hence, the size of the negative class is usually much larger than the size of the positive class. It is also important to note that such a feature-encoded dataset is constructed for each condition to be monitored.

4.2 Defining a Set of Features

Features are ISA dependent. A feature set defined for PowerPC ISA can be different from that for x86 architecture. Defining the feature set is treated as a one-time cost, although the feature set may be manually expanded during the verification process. The development of a good feature set is a process of incorporating domain knowledge into learning, thus it requires delicate design efforts. However, because the feature set only depends on the ISA, it can be reused for generations of design compatible to the ISA.

In this work, the feature set is defined based on the Power ISA. In the experiment presented later, we consider three categories of features:

- **State-based features:**
  - The contents of a set of special registers such as machine state register (MSR), exception syndrome register (XER), L1 cache control and status register (L1CSR), and etc.
- **Instruction-based features:**
  - Instruction types and data patterns of associated operands, the result of execution, and etc.
  - Information associated with load/store addresses, such as the virtual addresses and physical addresses, the attributes of the page which the addresses lie on, and etc.
- **Sequence-based features:**
  - Data dependency in a sequence of instructions, the distance between the dependent instructions, and etc.

- Address collision in a sequence of instructions, the distance between the collided instructions, and etc.

```
             stdx 4,28,15
(EA=0x00000000ee308888,RA=0x0000000020edb888)
             ldx 22,22,22
(EA=0x00000000fff1d908,RA=0x0000000020edb908)
             ldx 21,22,3
(EA=0x00000000fff1d888,RA=0x0000000020edb888)
```

Figure 5: Illustration of a test program snippet

Figure 5 illustrates an example showing a simplified view of a snippet from a novel test. The feature vector extracted from the third instruction is illustrated in Table 1. The subscript 3 denotes the features of the third instruction. EA_3 specifies the effective address, while RA_3 is the real address. op_type_3 refers to the instruction type. $collided_3$ is an occurrence feature indicating whether the instruction has address collision with any of previous instructions. $collision\_dist_3 = 1$ means there is one instruction between the third instruction and the closest previous collided instruction.

Feature	...	EA_3		RA_3	
Value	...	0x00000000fff1d888		0x0000000020edb888	
...	op_type_3	$collided_3$	$collision\_distance_3$	...	
...	ldx	1	1	...	

Table 1: Illustration of portion of a feature vector

4.3 Feature Discretizatition

In rule learning, a descriptive feature with numerical values is first partitioned into multiple bins to facilitate the rule search. For example, RA is a descriptive feature whose value can be partitioned into bins based on the cache line size or page size. In general, an entropy minimization heuristic developed by [7], can be employed for the partitioning such that a small range of feature values with rare occurrence is considered important and identified as a separate bin. A large range of feature values commonly-appearing in many samples are considered less important and grouped into the same bin. We use a discretization scheme based on the entropy minimization heuristic with additional constraints based on the known design features such as cache line and page sizes, etc.

5. KNOWLEDGE EXTRACTION BY RULE LEARNING

Given two classes of snippets, $S_{covered}$ (*positive* samples) and $S_{not-occurred}$ (*negative* samples), we are interested in finding the rules to describe the properties of positive samples $S_{covered}$. A rule is in the form of $Ante \Rightarrow S_{covered}$, where the class $S_{covered}$ appears in the rule consequent, and the rule antecedent $Ante$ is a conjunction of clauses $c_1 \wedge c_2 \ldots \wedge c_n$. Each clause involves a single feature. For an occurrence feature f, a clause can be either $f = 0$ or $f = 1$. For a descriptive feature f', a clause can be $f' = bin$ where bin is a bin number after the discretization described above. The $Ante$ is essentially a combination of important features selected to describe the properties on the positive samples. In principle, the $Ante$ should appear in zero or only very few negative samples. Moreover, an $Ante$ with a smaller number of clauses is preferred because such an $Ante$ is more general. An example rule based on features discussed in Table 1 is shown as follows:

$$op\_type_1 = stdx \quad \wedge \quad op\_type_3 = ldx \quad \wedge$$
$$collided_3 = 1 \quad \wedge \quad collision\_dist_3 = [1, 2) \quad \Rightarrow \quad S_{covered}$$
$$(1)$$

There are two classes of rule learning algorithms: classification rule learning and association rule learning. Classification rule learning is an approach for *predictive induction* (supervised learning), aimed at constructing a set of rules to be used for classification. Association rule learning is a form of *descriptive induction* (unsupervised learning), aimed at the discovery of rules which define interesting patterns in data. Subgroup discovery aims to address a task at the intersection of predictive and descriptive induction. For descriptive induction, it identifies groups of similar samples that should be analyzed collectively. Then, for a group of multiple similar samples, predictive induction is applied to extract rules. The search iterates between descriptive induction and predictive induction to find the optimal group boundaries and rules to describe each group.

Compared to classification rule learning, subgroup discovery is more suitable for the application. A class of positive samples hitting a particular condition can be due to multiple reasons. In classification rule learning, the positive samples are analyzed collectively. But because one subset of samples may be due to one reason and another subset may be to due a different reason, it becomes difficult to find a single rule to explain most of the samples, i.e. a single rule with high accuracy. This problem is resolved in subgroup discovery by grouping similar samples and searching rules to describe each group individually.

In this work, we implement a rule search engine similar to the CN2-SD algorithm proposed by [10], which adapted the classification rule learning CN2 algorithm [5] to subgroup discovery learning in order to achieve both predictive and descriptive induction.

The rule search engine performs a breadth-first search where the depth is characterized by the number of clauses. The evaluation metric of a rule is based on a weighted relative accuracy [12] as described below.

For a rule $Ante \Rightarrow S_{covered}$, the weighted relative accuracy $WRAcc$ is defined as follows:

$$WRAcc(Ante \Rightarrow S_{covered})$$
$$= p(Ante) \cdot (p(S_{covered}|Ante) - p(S_{covered})) \quad (2)$$

$p(Ante)$ is the frequency of the total samples satisfying the $Ante$. $p(S_{covered}|Ante)$ is the frequency of the positive samples satisfying the $Ante$. $p(S_{covered})$ is the frequency of the positive samples. The weighted relative accuracy consists of two components. The *relative accuracy* component $(p(S_{covered}|Ante) - p(S_{covered}))$ and the *generality* component $(p(Ante))$. Therefore, the weighted accuracy provides a tradeoff between the generality of the rule (rule coverage) and the relative accuracy.

In classification rule learning, covered samples are dropped to avoid finding the same rule again. However, a single sample may attribute to two reasons for hitting the condition. If such a sample is dropped after uncovering one reason, its information is lost for uncovering the other reason. To address this problem, the rule search engine uses a weighed covering heuristic. Instead of dropping a covered sample, it stores the covered sample with a weight indicating how many times the sample has been covered, i.e. how many rules have been

produced based on the sample. Then, in Equation (2) the frequencies are adjusted based on these weights. The output of the search is a ranked list of rules where the ranking can be based on several metrics [10].

6. KNOWLEDGE REUSE

6.1 Rule Validation and Refinement

From a ranked list of rules, a rule can be selected and validated by creating a test template macro satisfying the rule. A macro is a parameterized building block of a template, which specifies how instruction sequences are instantiated. For example, the rule in Equation (1) can be encoded into a macro illustrated in Figure 6, which will generate a pair of *stdx-ldx* collision with a random instruction between them.

```
sequence:
    var a = random()
    gen_inst(optype=stdx, addr=a)
    gen_inst()
    gen_inst(optype=ldx, addr=a)
```

Figure 6: Illustration of a test template macro

A rule is evaluated based on the frequency of the produced tests hitting the desired condition. A rule is considered to be meaningful if the frequency is higher than the ratio of the number of positive samples over the total number of samples in the original dataset. The larger the difference is, the more meaningful the rule is. In the learning process, a rule can be further refined based on additional positive samples produced in the rule validation process.

6.2 Rule Reuse

Rules and macros are reused to improve the coverage of complex assertions. A database is built to store the rules and macros for each condition to be monitored. When we want to produce tests to activate an assertion comprising multiple conditions, the corresponding macros for the conditions are retrieved from the database. These macros are combined to create more complex macros for activating the assertion.

In our methodology, combining macros follows a predefined set of built-in procedures that can be selected by the user. For example, one procedure combines macros by enumerating all the orderings without interleaving instructions from two macros. Another procedure combines macros based on a given fixed ordering by interleaving the instructions from two consecutive macros in the ordering. There are variants of interleaving schemes in the procedure to decide how instructions from two macros can be interleaved.

When creating compound macros, the constraints specified by individual macros should be preserved. For example, if we combine the macro in Figure 6 with another macro, the *stdx* instruction should still proceed the *ldx* instruction. While we can interleave instructions from another macro between the *stdx* and *ldx*, the number of intercepted instructions should not exceed one.

7. EXPERIMENT RESULTS

7.1 Experiment Environment

In this work, the experiments were conducted based on a dual-threaded low-power 64-bit Power Architecture-based processor core. It is targeted to be manufactured in a 28 nm technology. The processor core supports dual-thread capability that enables each core to act as two virtual cores. Each thread is two-way superscalar and maintains up to 16 out-of-order instructions in-flight through 10 parallel execution pipelines. The core is designed with a memory subsystem supporting up to a twelve-core SoC implementation.

The in-house simulation-based verification environment conforms to a state-of-the-art constrained-random verification flow. An in-house test generator is used to generate constrained-random test programs based on user-supplied test templates. During the test generation, architectural simulation is also performed and the simulation results are embedded in test programs. During the RTL simulation, the RTL simulation results are compared with the architectural simulation results for checking correctness.

The experimental results shown below focus on the assertion coverage of the Load Store Unit (LSU). LSU is one of the most complex and difficult-to-verify units in the design. Since the experiment was conducted in parallel with the ongoing verification efforts, the experiment started with the best test templates that had been refined by the verification team up to the time of the experiment. In the following, we describe three results in detail to demonstrate the effectiveness of the proposed learning methodology.

7.2 The First Result

The first result demonstrates the following: The learning began with novel tests activating an assertion A comprising a single condition c_1. Learning was to extract rules for hitting c_1. After the learning, two things were accomplished by the tests instantiated from the refined test template. First, the frequency of activating assertion A was substantially improved. Moreover, two additional assertions B, C (with zero coverage before) were covered.

The assertion B comprises a single condition c_2 that is highly correlated to the condition c_1. The assertion C comprises both conditions c_1 and c_2. The result demonstrates that learning from tests activating one assertion can lead to fortuitous coverage of other correlated assertions.

In the simulation run, 1000 tests were instantiated from a test template based on 114 types of memory instructions. Each test consists of 50 instructions. The simulation time for each test on a single machine took several minutes. When simulating using a server farm, 20-30 tests could be simulated simultaneously. The assertion A was covered by merely three tests. This assertion refers to the special condition c_1 concerning how certain queues in LSU are filled up.

We applied the learning methodology to extract rules from the three novel tests. One interesting rule we found can be interpreted as follows:

- There is a *lmw* (load multiple word) instruction.
- The page on which the real address of *lmw* lies, is not cache inhibited.
- The destination register of the *lmw* is before $G20$.

The rule is converted into a test template macro and used to generate another 200 tests, each comprising 50 instructions. Table 2 shows the comparison of coverage between the original 1000 tests and 200 new tests. As shown by the 4th row, the number of tests activating the assertion A increases from 3 to 33 in the new test set, thus boosting the frequency from 0.3% to 15.5%. Moreover, assertions B and C which were not activated by the original 1000 tests could be activated by 9 tests and 5 tests from the 200 new tests (the 5th and 6th rows).

test set	# of tests		% of tests	
	original	new	original	new
size	1000	200	1000	200
assertion A	3	33	0.3%	15.5%
assertion B	0	9	0	4.5%
assertion C	0	5	0	2.5%

Table 2: Comparison of assertion coverage between original 1000 tests and 200 new tests

7.3 The Second Result

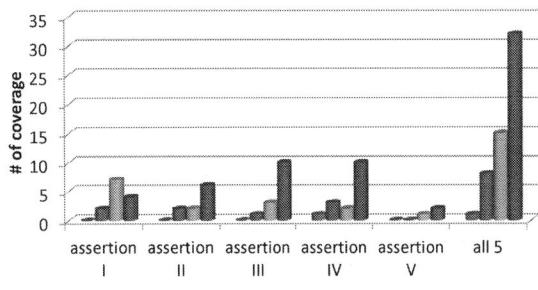

Figure 7: Assertion coverage improvement

Figure 7 summarizes the second result. In the original simulation run, 2000 tests were instantiated from a template based on 44 types of memory instructions and over 200 types of non-memory instructions. Each test consists of 100 instructions. In the simulation, only assertion IV was activated by one test. Assertion I, II, III, and V had zero coverage.

Assertion IV comprises two conditions c_3 and c_4. Other assertions comprise the same two conditions. However, the temporal constraints between the two conditions are different across the five assertions. c_3 and c_4 last only one or two clock cycles when asserted. The temporal constraints require them to occur within a small time window, which makes the assertions difficult to activate.

Learning was carried out based on the novel tests hitting conditions c_3 and c_4. Note that there were multiple tests hitting c_3 and c_4 individually. After the learning, rules were extracted for hitting c_3 and c_4, resulting in multiple macros for each condition.

Two macros m_1 and m_2 (for c_3 and c_4, respectively) were identified to be consistent with two respective segments of instructions in the one test activating assertion IV. The corresponding rules for these two macros are illustrated in Table 3. Hence, macros m_1 and m_2 were combined to produce a new template macro. Because in the test, instructions from m_1 was followed by instructions from m_2 without interleaving, in the combined macro, m_1 was followed by m_2 without instruction interleaving.

Rule for m_1	There is a *mulld* instruction and the two multiplicands are larger than 2^{32}
Rule for m_2	There is a *lfd* instruction and the instructions prior to the *lfd* are not memory instructions whose addresses collide with the *lfd*

Table 3: Rules for macros m_1 and m_2

The combined macro was used to produce 100 new tests. These new 100 tests led to higher coverage for assertions I to IV as shown with the legend "combined macro" in Figure 7. But assertion V remained at zero coverage.

The learning was re-applied with the additional 100 tests and new rules/macros were obtained for hitting c_3 and c_4.

Again, 100 new tests were produced. The result was denoted as "iteration 1" in Figure 7. Observe in "iteration one" that coverage for assertion I to IV was improved further. More importantly, assertion V could be covered. The process repeated in the "iteration 2" and we can observe further coverage improvement for assertions II to V.

7.4 The Third Result

The original simulation was based on 2000 tests instantiated from a given test template. For the 10 assertions of interest, none of them was covered. The simulation was expanded with tests instantiated from multiple other test templates. In total, over 30k tests were simulated without activating any of the 10 assertions. Then, the learning was carried out based on the 2000 tests in the first simulation run.

The 10 assertions comprise the same two conditions, c_5 and c_6, and the temporal constraint between c_5 and c_6 varies across the 10 assertions. Also, c_5 and c_6 are evanescent and are required to happen with a small time window. While none of the tests activate an assertion, multiple tests could hit each condition individually. Table 4 illustrates some of the rules learned for c_5 and c_6:

Rules for condition c_5	
Rule R_1	There is an *isync* instruction.
Rule R_2	There is a *mcrxr* instruction followed by a *mullwo* instruction. The *mullwo* instruction causes an overflow in XER.
...	...
Rules for condition c_6	
Rule Q_1	A *stmw* instruction, then a *stdx* instruction followed by a *ldx* with address collision.
...	...

Table 4: Illustration of rules for c_5 and c_6

A macro from a rule for c_5 was combined with instruction interleaving with a macro from a rule for c_6 to produce a combined macro. In total, 12 combined macros were produced based on 4 rules for c_5 and 3 rules for c_6. Then, 1200 tests were generated, 100 tests based on each combined macro. The third column of Table 5 shows that 6 out of the 10 assertions were covered by these 1200 tests.

test set	original	combined macros	refined macro
# of tests	>30k	1200	100
cycles($\times 10^4$)	>1000	40.8	4.23
assertion 1	0	1	2
assertion 2	0	0	1
assertion 3	0	0	1
assertion 4	0	16	56
assertion 5	0	0	1
assertion 6	0	1	2
assertion 7	0	0	1
assertion 8	0	16	56
assertion 9	0	15	61
assertion 10	0	26	77

Table 5: Comparison of assertion coverage

Analyzing the tests activating the assertion led to the conclusion that a combination of particular two rules (R_2 and Q_1) with a particular instruction interleaving scheme is especially effective. A new macro (illustrated in Figure 8) was created based on the two rules by fixing the particular interleaving scheme. Additional 100 tests were produced using the new macro and the result was shown in the 4th column of Table 5 where all 10 assertions were covered.

```
sequence:
    var a = random()
    gen_inst(optype=mcrxr)
    gen_inst(optype=stmw)
    gen_inst(optype=stdx, addr=a)
    gen_inst(optype=mullwo, result=overflow)
    gen_inst(optype=ldx, addr=a)
```

Figure 8: Illustration of the effective macro for activating the assertions

8. CONCLUSION

This work proposes a learning methodology to extract knowledge from simulation in constrained-random processor verification. A feature-based rule learning approach is developed for the knowledge extraction. The extracted knowledge is reused for test template refinement to improve assertion coverage. Experimental results demonstrate the effectiveness of the proposed learning methodology in various scenarios where assertion coverage could be further improved after a substantial verification effort had been spent.

9. REFERENCES

[1] P. Bastani and et al. Diagnosis of design-silicon timing mismatch with feature encoding and importance ranking - the methodology explained. In *IEEE International Test Conference*, pages 1 –10, oct. 2008.

[2] P.-H. Chang and et al. Online selection of effective functional test programs based on novelty detection. In *Computer-Aided Design (ICCAD), 2010 IEEE/ACM International Conference on*, pages 762 –769, nov. 2010.

[3] J. Chen and et al. Mining ac delay measurements for understanding speed-limiting paths. In *Test Conference (ITC), 2010 IEEE International*, pages 1 –10, nov. 2010.

[4] W. Chen and et al. Novel test detection to improve simulation efficiency–a commercial experiment. In *Computer-Aided Design (ICCAD), 2012 IEEE/ACM International Conference on*, nov. 2012.

[5] P. Clark and T. Niblett. The cn2 induction algorithm. *Mach. Learn.*, 3(4):261–283, Mar. 1989.

[6] K. Eder, P. Flach, and H.-W. Hsueh. Inductive logic programming. chapter Towards Automating Simulation-Based Design Verification Using ILP, pages 154–168. Springer-Verlag, Berlin, Heidelberg, 2007.

[7] U. M. Fayyad and K. B. Irani. Multi-interval discretization of continuous-valued attributes for classification learning. In *IJCAI*, pages 1022–1029, 1993.

[8] S. Fine and A. Ziv. Coverage directed test generation for functional verification using bayesian networks. In *Design Automation Conference, 2003. Proceedings*, pages 286 –291, june 2003.

[9] Y. Katz, M. Rimon, A. Ziv, and G. Shaked. Learning microarchitectural behaviors to improve stimuli generation quality. In *Design Automation Conference (DAC), 48th ACM/EDAC/IEEE*, pages 848 –853, 2011.

[10] N. Lavrač, B. Kavšek, P. Flach, and L. Todorovski. Subgroup discovery with cn2-sd. *J. Mach. Learn. Res.*, 5:153–188, Dec. 2004.

[11] G. Squillero. Microgp-an evolutionary assembly program generator. *Genetic Programming and Evolvable Machines*, 6(3):247–263, Sept. 2005.

[12] L. Todorovski, P. A. Flach, and N. Lavrac. Predictive performance of weghted relative accuracy. In *Proceedings of the 4th European Conference on Principles of Data Mining and Knowledge Discovery*, PKDD '00, pages 255–264, London, UK, UK, 2000. Springer-Verlag.

[13] I. Wagner, V. Bertacco, and T. Austin. Microprocessor verification via feedback-adjusted markov models. *Computer-Aided Design of Integrated Circuits and Systems, IEEE Transactions on*, 26(6):1126 –1138, june 2007.

Hardware-Efficient On-Chip Generation of Time-Extensive Constrained-Random Sequences for In-System Validation

Adam B. Kinsman, Ho Fai Ko and Nicola Nicolici
Department of Electrical and Computer Engineering
McMaster University, Hamilton, ON, L8S 4K1, Canada
Email: kinsmaab@mcmaster.ca, henryko@grads.mcmaster.ca, nicola@ece.mcmaster.ca

ABSTRACT

Linear Feedback Shift Registers (LFSRs) have been extensively used for compressed manufacturing test. They have been recently employed as a foundation for porting constrained-random stimuli from a pre-silicon verification environment to in-system validation. This work advances this concept by improving both the hardware efficiency and the duration of in-system validation experiments.

General Terms

Validation, Verification

Keywords

Constrained-Random Sequences, In-System Validation

1. INTRODUCTION

The objective of pre-silicon verification is to validate if the design implementation complies to its specification. Considering the complexity of the state-of-the-art designs, it is not tractable to formally prove the design correctness. Therefore, to increase the confidence of the validation process, pre-silicon verification tools commonly measure which design functionality has been exercised. These collected measurements are subsequently analyzed in order to guide the generation of new stimuli needed to reach the insufficiently exercised corner cases. A key point is that during pre-silicon verification what can be measured is limited by the simulation time and designs are often released to first silicon prototypes when the confidence level is deemed sufficient. Before committing to high-volume production, the validation process continues on the first silicon prototypes. Unlike simulation, these silicon prototypes are fast in operation, nonetheless the limited controllability and observability are hindering to the in-system validation process.

To deal with controllability/observability aspects of in-system validation, the basic infrastructure employed for manufacturing test can be reused for in-system validation. For example, scan chains have been successfully deployed for debugging silicon prototypes [15]. Nevertheless, before applying (and observing) patterns (and responses) through scan it is important to first learn what causes the bugs to manifest themselves and how to excite them. Therefore, recording the traces that caused the failing behaviour without

Permission to make digital or hard copies of all or part of this work for personal or classroom use is granted without fee provided that copies are not made or distributed for profit or commercial advantage and that copies bear this notice and the full citation on the first page. To copy otherwise, to republish, to post on servers or to redistribute to lists, requires prior specific permission and/or a fee.
DAC'13, May 29-June 7 2013, Austin, TX, USA.

stopping the real-time execution is essential to improve real-time observability and hence understand how to control the failure. Due to the inherent limitations of scan chains to capture data without interruption in real-time environments, embedded logic analysis has been adopted to improve the observability aspects of in-system validation. For this reason, different variants of on-chip trace memories [1, 4, 11, 12] are commonly employed together with event monitors [6] to detect events of interest and acquire traces in-system.

From one perspective it can be considered that the basic concepts for scan chain-based and logic analysis-based debugging have been inspired by the known techniques ordinarily used for manufacturing test and embedded system bring-up respectively. A different direction of work relies on building on the body of knowledge available in the pre-silicon verification domain. During the early phases of pre-silicon verification, when the number of bugs is relatively large, directed tests are iteratively refined to answer "does-it-work"-type of questions. Once the bug detection rate decreases, "are-we-done"-type of questions dominate over the "does-it-work"-type of questions, in which case one has to generate a large volume of functionally-compliant stimuli [8, 10] that attempt to exercise the insufficiently measured corner cases. Constrained-random stimuli combined with assertion checking are the state-of-the-art tools employed in practice during the pre-silicon verification phase. Porting such concepts from pre-silicon [2] to in-system validation has been done successfully for high-performance microprocessors [3]. The key observation is that in a hardware environment the quality of the stimuli is improved by the large number of patterns that can be applied in real-time. While this has been successfully deployed for processor-centric designs, for system-on-chip (SOC) devices, which are known to have application-specific blocks, such as high-speed peripheral controllers and custom hardware accelerators, there are indeed known techniques to implement assertions into hardware [7]; nevertheless, it is not yet adequately understood if and how one can port constrained-random stimuli to in-system validation of random logic blocks. Despite the fact that the fundamental knowledge on how to achieve this task in a systematic manner is not established, it has been demonstrated recently that for complex SOC designs there is indeed value in generating functional sequences from compact constrained-random generators placed in hardware during in-system validation [17].

Recently [9] has investigated how to adapt pseudo-random sequence generation, commonly used in built-in self-test (BIST) for very large scale integrated (VLSI) circuits, to generate pseudo-random, yet functionally-compliant, sequences that are consistent with the constraints described in a pre-silicon environment. The basic idea from [9] is to map the functionally-compliant sequences onto linear-feedback shift registers (LFSRs). The access to such embedded sequence generators can be facilitated by the emerg-

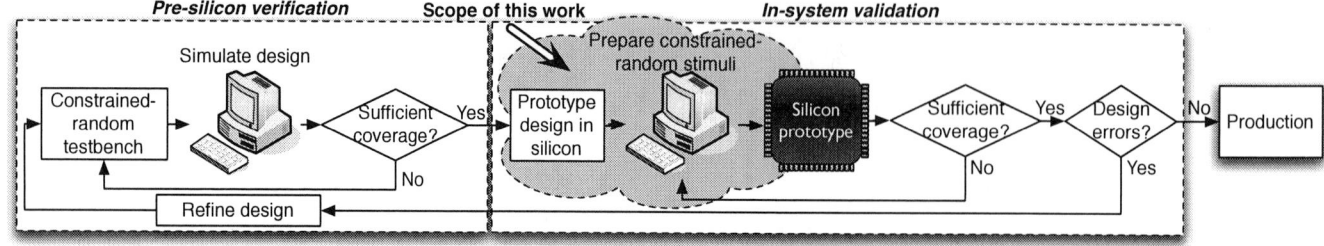

Figure 1: The scope of this work as related to bridging pre-silicon verification to in-system validation

ing standards for internal instrumentation [13]. Despite presenting the first systematic approach for porting constrained-random stimuli aspects from pre-silicon to in-system validation, the main limitations of the approach from [9] are the excessive area investment when employing large LFSRs, as required for time-extensive in-system validation, as well as intractable runtimes needed when preparing the seeds for large LFSRs during successive in-system experiments. It is the main focus of this paper to address the above shortcomings and enable fast preparation of validation data for in-system experiments, while also significantly reducing the size of the on-chip signal generators.

The scope of this work is illustrated in Figure 1. When the measured coverage is considered sufficient during pre-silicon verification, the first silicon prototypes are built. Regardless whether this prototype is a fully or a partially implemented design, or whether the validation is done in a hardware emulation environment or the target application boards, our objective is to facilitate the rapid development and effective in-system generation of the tests that rely on the constrained-random simulation infrastructure already developed for pre-silicon verification. It is important to note that in this paper we present a methodology for porting only the controllability aspects used in constrained-random verification. Observability hooks for coverage measurement or error detection are considered to be an orthogonal concern to this work and hence any known techniques (e.g., [1, 6, 7]) can be employed. While the design of hardware-efficient on-chip generators is based on the premise that sequences should run for very long times (note, the stimuli count applied in-system is several orders of magnitude larger than during pre-silicon), a key advantage over relying solely on stimuli from the native environment is the ability to control experiments. For example, by accounting for the seeds employed to update the LFSR states, different failing sequences can be reproduced by zooming into particular windows of interest. Such knowledge of the failing sequences can be subsequently used in a pre-silicon environment to further assist with root-causing the underlying problem.

The main contributions of this paper are on designing on-chip programmable signal generators and can be summarized as follows:

- Because generating long functional sequences is a key requirement for in-system validation, one has to determine a large number of LFSR seeds from which these constrained-random sequences can run for long periods of time. Our new approach relies on encoding the pre-silicon verification constraints into compact seed vector bases that can be stored on-chip, as well as automatically determined in-field through rapidly solving linear systems of equations in Galois Fields;

- We present a new hardware-efficient architecture from which k basis vectors stored on-chip can be expanded into 2^k LFSR seeds, while guaranteeing that the complexity of the hardware circuitry depends only linearly on the size of the LFSR.

2. IMPLEMENTATION FLOW

In order to generate time-extensive constrained-random stimuli for validating a silicon prototype using the efficient architecture that will be detailed in the following section, the set of stimuli constraints that are used during pre-silicon verification has to be transformed using the proposed implementation flow shown in Figure 2. The flow starts by generating stimuli cubes using the constraints from the testbench and information about the on-chip LFSR-based generator. The generated cubes will then be fed to a Galois field (GF2) solver to obtain the set of basis vectors for the chosen generators. These basis vectors will then be loaded onto the on-chip generators in the silicon prototype to generate time-extensive sequence of constrained-random stimuli in real-time. Using on-chip monitors such as hardware assertions, validation coverage of the design can be measured. This process repeats until sufficient coverage of the design is achieved, at which point, the design can be moved to production if no error has been found.

2.1 Stimuli cube generation

To illustrate how to transform the stimuli constraints from the testbench used during pre-silicon verification into stimuli cubes that can be used with a library of on-chip generators, let's consider a very simple example constraint of $a \geq b$, where a and b are 4-bit inputs to the design to be validated. Under this constraint, the possible number of 8-bit input patterns that can be applied to the design is $\sum_{i=1}^{16} i = 136$. This set of input patterns can then be transformed

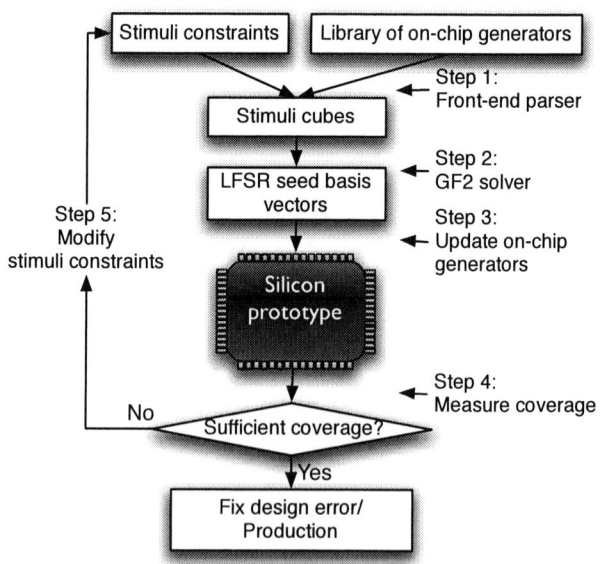

Figure 2: The proposed implementation flow

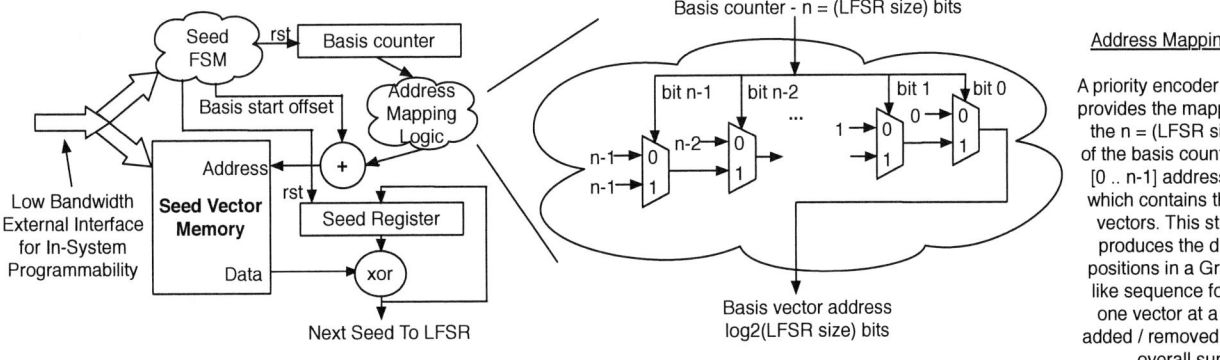

Figure 3: Hardware architecture for expanding seed basis vectors

into stimuli cubes using off-the-shelf logic minimization tools such as [14]. An example of an 8-bit cube from the above constraint would be $\{a, b\} = \{1XXX, 0XXX\}$, where X represents don't care bit. This single cube covers 64 of the 136 possible input patterns specified by the example constraint. For this example, there will be 23 stimuli cubes remaining after this transformation.

2.2 Seeding LFSR-based generators

With the stimuli cubes in hand, the next part of the flow is to encode each stimuli cube as seeds for LFSRs, in the form of a *seed vector basis* of vectors in GF2 for generating all the seeds which will produce an LFSR output that matches the stimuli cube in the specified positions. The theory of representing LFSR operation through matrix multiplication in GF2 has been well developed in the area of manufacturing test [9] and can be summarized in the statement that for an n-bit LFSR reseeded (i.e. initialized with a parallel load) with a seed vector s, the state q of the LFSR t clock cycles after reseeding is given by $q_t = L^t s$, where q, s are n-bit vectors in GF2, and L is an $n \times n$ matrix in GF2 which captures the state transitions of the LFSR (which by nature are linear). Further to this, it is common to map the n LFSR outputs to the m circuit inputs C through an XOR network defined by the $m \times n$ matrix in GF2 P, yielding $c_t = Pq_t = PL^t s$.

In light of the well established methods for deriving L and P for a given number of circuit inputs m and a desired LFSR size n, we can treat L and P as givens, which reduces the encoding process to merely solving a system of equations in GF2. As an example, consider the cube $\{1XXX, 0XXX\}$ discussed above where $m = 8$, and take $n = 8$ and P to be the identity matrix I. If we want to find a seed which satisfies this constraint at the time the LFSR is reseeded, we use $c_0 = [1XXX0XXX]^T$ in $c_0 = L^0 s$ (note that L^0 equals I). This system embodies 8 equations (rows) over 8 variables (columns - the bits of the seed $s_0..s_7$), and the 6 rows for this $c_0 = X$ (rows 1,2,3,5,6,7) can be disregarded because in plain language they say that it doesn't matter what the value of this bit of the LFSR output is. Thus this constraint is encapsulated by the two equations from row 0 and row 4 of the matrix equation, and any seed (combination of variables s_0 to s_7) which is linearly consistent with these two equations is guaranteed to produce an LFSR output which satisfies the stimuli cube at the time of reseeding.

If in addition to the above, we want the output of the LFSR to be consistent with the same stimuli cube *one clock cycle after reseeding*, we can write the equation $c_1 = L^1 S$, using the same value for c_1 as was used for c_0. The rows of the matrix for which $c_1 = X$ can be disregarded in the same way as for c_0, yielding two new equations, distinct from the ones for c_0, since L is now raised to the power

1 instead of 0. Any seed consistent only with these two equations will guarantee nothing about the clock cycle in which the LFSR is reseeded, but will guarantee that in the first clock cycle after reseeding that the LFSR output will be consistent with $[1XXX0XXX]^T$. If we now combine the two sets of two equations each into a single system over the same seed variables and solve again, we will obtain a set of linear dependencies over the seed variables which, if satisfied, will guarantee that the output of the LFSR will be consistent with $[1XXX0XXX]^T$ in *both the clock cycle of reseeding and the subsequent one*.

2.3 Expansion of all seeds

The method above links each stimuli cube (or sequence of stimuli cubes) to a system of GF2 equations over the LFSR seed variables; and the LFSR output sequence is guaranteed to satisfy that stimuli cube (or sequence of stimuli cubes) when the LFSR is seeded with any seed that is linearly consistent (i.e. valid) with the system of equations. Based on this system of equations, we would like to be able to generate *all* the valid seeds, as this is the key to generating time-extensive sequences from a compact representation. If we write the aggregated system of equations over the seed s discussed in the previous section as $As = b$, the encoding of all the valid seeds is accomplished by making use of the *nullspace* of $A : N(A)$, i.e. the space spanned by all solutions to $As = 0$. If we have a vector $s_0 | As_0 = b$, then we know that for any vector $s' \in N(A)$, $s_0 + s'$ is also a solution since $A(s_0 + s') = As_0 + As' = b + 0 = b$. Gaussian elimination produces the base solution (i.e. origin s_0) and generation of all valid seeds unfolds by combining s_0 with each s' in turn; this sequence is generated by iterating through all linear combinations of the basis vectors of $N(A)$. For a GF2 nullspace of rank = k, the basis will be k vectors yielding a sequence of 2^k seeds. The $k + 1$ vectors (basis + origin) on n bits plus the value of k itself (the rank) constitute the entire encoding for a given stimuli, and comprise the information which is passed to the on-chip generators in Part 3 of Figure 2. The in-system programmable architecture which expands this encoding in a hardware efficient way is discussed in Section 3, and experiments regarding expanded data vs. encoded data for two case studies are given in Section 4.

3. ARCHITECTURAL DETAILS

In Section 2, the method of obtaining seed basis vectors from stimuli cubes derived from high-level verification constraints was elaborated. In this section we detail the on-chip architecture which expands a time-extensive sequence from the compactly encoded basis vectors in a hardware efficient way.

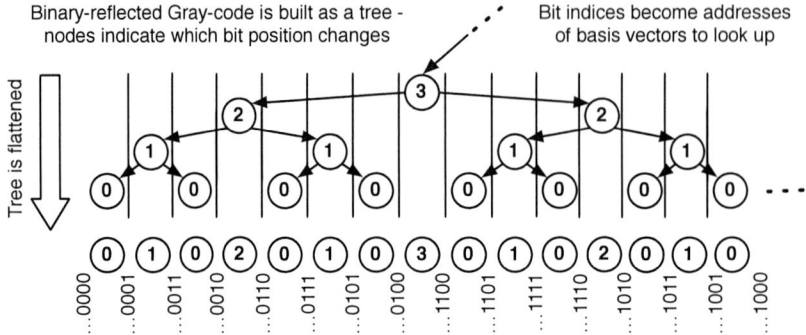

Figure 4: Tree of differing positions for Gray-code generation

3.1 Basis storage / expansion

Figure 3 shows the connection of the *Seed Vector Memory* where the encoded basis is stored to the *Seed Register* which contains at any point the currently generated seed in the sequence of all seeds encoded by the basis. The addressing is driven by the *Basis counter*, which traverses the entire sequence of $[0..2^k - 1]$ for k basis vectors, through the *Addressing Mapping Logic* which derives the next basis vector to load based on the current position in the entire sequence. The *Seed FSM* controls moving from one stored system to the next, with the current system residing at the *Basis start offset*. Finally, a low bandwidth interface is the means by which in-system programmability is provided, i.e. generator specific information (such as choice of LFSR polynomial) and the encoded bases themselves are uploaded via this interface.

The encoded bases are stored sequentially in the *Seed Vector Memory*, with each system starting with the origin of the expansion (Section 2.3). Next come the k basis vectors comprising the basis for a system of rank k. This way, the *Seed FSM* takes care of setting the *Basis start offset* to the location of the origin while loading a new system (to look up the origin from the memory) and then sets it to the location of the origin + 1 during expansion.

During operation, the origin is first loaded from the *Seed Vector Memory*. Then, as the *Basis counter* iterates through the space of 2^k combinations of the basis vectors, at any given time the *Seed Register* holds the origin plus the sum of basis vectors for which the corresponding bit of the value of the *Basis counter* is a 1. For example, in a system with 8-bit seeds, and 5 basis vectors (i.e. rank of 5), there are 32 combinations of the basis vectors represented by the values 00000 (the origin vector only) to 11111 (the origin vector plus the sum of all 5 basis vectors). The same end is achieved by the architecture proposed in [9], however, that approach loads all the basis vectors into registers which are selectively masked before summation by the bits which have value 0 of their equivalent to our *Basis counter*. In this way their method requires the entire matrix multiplication to be performed for the generation of each seed.

Two key observations are necessary to enable our architecture. The first is that any vector when XOR'd with itself will produce the zero vector. Thus if a basis vector is already present in a summation, a further addition of that basis vector will remove it from the summation. In light of this, if we can navigate the entire sequence by changing the current summation by only one basis vector at a time, that vector can be looked up during the generation of that seed. This is precisely the operation of a Gray-code sequence, thus by navigating the memory in a Gray-code sequence instead of a linear sequence, only one basis vector (indicated by whichever bit of the counter changes) is added/removed (the same operation under GF2 addition, i.e. XOR) at a time.

Thus, instead of requiring a large register array that scales quadratically with the LFSR size (as in [9]), only a few support registers

are required, with the complexity being pushed into the addressing logic and the far greater number of accesses required to the memory. The extra complexity in addressing will be covered in the next sub-section, but the memory accesses come for free since the memory is dedicated and thus may be accessed in any clock cycle, and cycles without an access are in effect wasted.

3.2 Addressing mapping logic

As mentioned above, a key insight for *greater efficiency* of the architecture for expansion is a Gray-code like traversal of the set of all combinations of the basis vectors as opposed to a simple linear one, where more than one basis vector must be added / removed to obtain an updated seed, as done by [9]. However, it is not simply Gray-code addressing which is required, but rather identification of which bit changes at any given point in the sequence, and the use of that bit to look up a given vector. For instance for a Gray-code sequence on 4 bits where 1111 is followed by 1011, the seed would be transitioning from being the origin plus the sum of all 4 basis vectors, to the origin plus the sum of basis vectors 3, 1 and 0. The addressing logic would need to produce the value 2, as it is bit 2 which is changing value, thereby causing basis vector $bv[2]$ to be looked up. The resultant sum would be $(origin + bv[3] + bv[2] + bv[1] + bv[0]) + (bv[2])$, and since under GF2, $x + x = 0$, the $bv[2]$ occurrences would cancel yielding the desired seed.

The problem reduces therefore to producing a numerical sequence which indicates the index of the changing bit during a Gray-code counting sequence. Figure 4 shows the tree based construction of the well known *Binary-reflected* Gray-code, indicating at each node which bit should change its value. The symmetric nature of the tree allows the code to be built recursively from n bits to n+1 bits by joining the roots of two copies of the n-bit tree by a node of value n+1. The process has been expanded to 4 bits, giving a Gray-code sequence of length 16 across the bottom of the figure.

When the tree is flattened, as in the bottom of the figure, the sequence of bit changes can be seen. Inspection of the pattern reveals that of the entire sequence, every other element indicates a change to bit zero. Removal of these zero nodes, leaves a sequence where every other element is a node of value 1. Thus the recursive construction procedure gives rise to a recursive procedure for extracting the location of the changing bit. The hardware structure which implements this function is a priority encoder, and it is detailed in the middle section of Figure 3. Starting from the least significant bit, when the value is zero (which happens every other element in the counter sequence) a zero is passed. In the remaining cases (i.e. the other input of the bit-zero multiplexer), every other element has value 1, this corresponds to bit 1 of the counter being 0. This pattern continues up to bit n-1 of the counter, thus creating the mapping from the n bits of the counter to the $log_2(n)$ bits needed to encode a position within the counter.

856

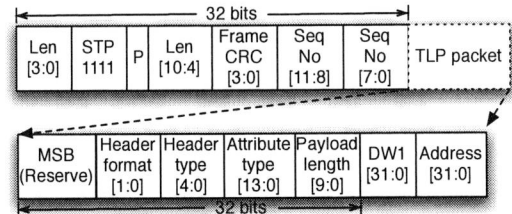

Figure 5: The format of a TLP packet from the 3rd generation PCI-express protocol[5]

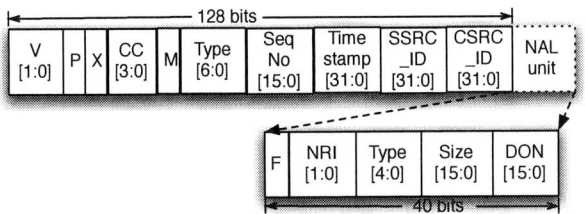

Figure 6: The RTP payload format for H.264 video[16]

4. EXPERIMENTAL RESULTS

The proposed methodology (explained in Section 2), together with the efficient architecture (detailed in Section 3), enable the generation of time-extensive constrained-random stimuli on-chip to aid in-system validation of digital circuits. To verify the effectiveness of the proposed technique in terms of area (discussed in Section 4.1) and length of the generated stimuli (analyzed in Section 4.2), it has been applied to generate patterns that conform to the constraints extracted from the third generation of PCI-express (PCIe) transaction layer protocol (TLP) [5], as well as the real-time transport protocol (RTP) payload format for H.264 video codec [16]. It should be noted that in the same way as reported in [9], the goal of these experiments is to show how long runs of constrained-random stimuli can be generated on-chip efficiently in real-time. Thus, the stimuli used for the experiments only contain the header part of the PCIe packets and the header part of the RTP packet for H.264. The payload part of the PCIe packets, as well as the RTP packets for H.264 are not considered in these experiments. This is because for PCIe, the payload can be entirely randomized as long as the header information conforms to the standard. On the other hand, since we are not targeting to generate conformance bitstream for H.264, we focus our experiments to generate the header part of the RTP packets with their corresponding network abstraction layer (NAL) units for providing packetized data for H.264 decoder.

For PCIe, the total number of 128-bit stimuli cubes obtained after the transformation described in Section 2.1 is 5119. Of the 128 bits, 27 bits are fully specified in all the cubes (i.e. care-bit density of 1), while another five bits have care-bit densities between 0.2 and 0.5. One example of fully specified bits is the 4-bit start-of-packet fixed to 0xF, and the 5-bit header type partially specified for PCIe. For H.264, 335 stimuli cubes of 168-bits are obtained, with 3 of the bits (the 2-bit V and 1-bit F) fully specified, 2 out of 3-bit of *Type* having care-bit densities over 0.9, and 37 other bits having care-bit density below 0.2. The detailed formats of the patterns for PCIe and H.264 are given in Figures 5 and 6.

4.1 Architecture area

Figure 7 shows the relationship between area and LFSR size both for the reference [9] and the proposed method. The quadratic relationship between LFSR size and area identified within [9] is evident in the figure. Note that results only up to an LFSR size of 28 (log_2(LFSR)=4.8) are reported in [9], thus the area values for

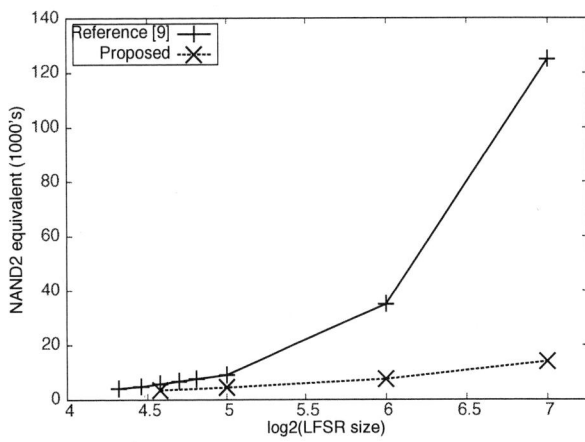

Figure 7: Area vs log_2(LFSR size) for proposed and [9]. Note that values ≥ 5.0 for [9] are estimated as discussed in the text.

LFSRs of 32, 64 and 128 bits given here for that method have been estimated. In scaling from 20 to 28 LFSR bits, [9] reports area increase from 4035 to 7586, or a factor of 1.88 increase in area for a factor 1.40 increase in LFSR size. We thus use $4035 \times (LFSRsize/20)^{1.8}$ as our estimation, matching [9] for a 20-bit LFSR, and underestimates the quadratic dependence discussed therein.

Regardless of precise values, the key difference between the proposed architecture and [9] is the shift from quadratic to near-linear dependence; indeed Figure 7 shows how LFSR size scaling is much closer to linear. This results from removal of the register array in the column logic of [9], which [9] points out as the dominating factor in the area required. This removal is made possible by the observation from Section 3 that the overall seed can be updated one basis vector at a time. In this way, at only the cost of increased sophistication of the control logic, this method leverages memory access opportunities which are wasted by [9] where the memory sits idle after the load phase at the start of any basis expansion.

4.2 Sequence generation

Table 1 shows the results for encoding stimuli cubes for both PCIe and H.264 onto linear basis vectors for varying sizes of LFSR. For each design, each stimuli cube has been encoded onto its own linear basis; the *total basis vectors* column indicates the total basis vectors across all stimuli cubes.

A key point to notice is that for both designs, as the LFSR size is increased, the number of basis vectors increases by the *number of patterns × additional LFSR bits*. For example, when moving from the 24-bit to the 32-bit LFSR for the H.264 design, the total basis vectors increases from 7040 to 9720, or by the amount $2680 = 335 \times (32 - 24)$. This occurs because once the basic linear dependencies which capture the specified bits in each cube are encoded, each additional bit appended to the LFSR brings a complete degree of freedom, encoded through an additional basis vector.

Table 1: Seed Results

Design	LFSR size	Total basis vectors	Seed data	Total Seeds	Avg bits / seed
PCIe	24	93165	2.46Mb	1.61e9	1.53e-3
	32	134117	4.55Mb	4.12e11	1.11e-5
	64	297925	19.5Mb	1.77e21	1.11e-14
	128	625541	80.9Mb	3.26e40	2.48e-33
H.264	24	7040	182kb	1.23e9	1.48e-4
	32	9720	327kb	3.15e11	1.04e-6
	64	20440	1.34Mb	1.35e21	9.89e-16
	128	41880	5.41Mb	2.50e40	2.17e-34

The implication of the above is that the nature of the specified bits in the cube (i.e. the number of linear dependencies which they impart through the selected polynomial / XOR network) imposes a lower bound on the required LFSR size. Beyond this lower bound, further LFSR bits double the number of seeds which will expand into a sequence consistent with the stimuli cube. From the converse perspective, given an LFSR of size n, each stimuli cube can be seen as "consuming" degrees of freedom in seed generation in accordance with the number of linear dependencies between its specified bits. The remaining k degrees of freedom can be encoded as a basis of k vectors of n bits, expanded by the architecture into 2^k seeds.

The above relationship produces a quadratic dependence of the basis size (in bits) on the LFSR size; the number of basis vectors (i.e. degrees of freedom) increases linearly with the LFSR, as does the size of each basis vector. For example, in both designs increasing the LFSR by factors of 2 (i.e. 32 to 64 to 128) increases the total seed data by factors of 4. Alongside the quadratic LFSR size vs. seed data dependence is the exponential dependence of number of seeds (and by proxy sequence generation length) on the LFSR size, as evidenced by the approximate squaring in the number of seeds as the LFSR size goes as before 32-64-128. *This exponential scaling is what enables the proposed method to generate time-extensive constrained-random stimuli on-chip.* Note that in some cases (i.e. 24-bit LFSR for both designs) the total seeds can even exceed the number of distinct seeds ($1e9 > 2^{24}$). This is analogous to an instance where thousands of 8-bit numbers are generated. The order of values in the sequence changes as values are repeated; there are many more *sequences* of patterns than *distinct* patterns.

A final important note regarding seed data is that while 80Mb (i.e. PCIe with 128-bit LFSR) of memory is by current standards considerably large, perhaps even impractical to be placed in-system for the sole purpose of validation, the basis vector memory need not in-fact to be so large. Given the in-system programmability of the method, it is sufficient to store one complete basis only, which is continuously updated during a validation session. For an n-bit LFSR, the maximum basis is on the order of n^2 memory bits. Specifically, an on-chip memory of $128 \times 129 = 16.5kb \approx 2KB$ could store an entire basis (n basis vectors plus the origin all on n bits). The requirement for this approach to work is that the basis vectors can be streamed into the system at a rate higher than they are consumed through generation of seeds. The final column of the table indicates the average number of bits per generated seed, thus providing a reference point for the rate at which new seed data must be streamed into the system during validation. Since we have organized the patterns the same way as in [9], each pattern is expanded over 5 clock cycles and so for a circuit operating at 1GHz the seeds must be generated at a rate of 200MHz. Considering even the most demanding rate of 1.53e-3 bits / seed (i.e. PCIe for a 24-bit LFSR), 1.53e-3×200e6 bits of seed data are consumed per second and the same amount must be streamed into the system per second to replace the consumed seeds. This modest rate of $\approx$ 300kb/s is well within reach of system configuration interfaces such as JTAG.

5. CONCLUSION

The work presented in this paper has shown how long in-system validation sequences can be generated from compact signal generators placed on-chip that need only to be re-initialized infrequently through low-bandwidth interfaces. By using the programmability feature for the on-chip signal generators the validation engineers can run successive in-system experiments. The methodology presented in this paper further advances the body of knowledge for porting of pre-silicon stimuli constraints to in-system validation environments in a cost-effective and systematic manner.

Acknowledgement.

The authors acknowledge the financial support of the University Research Office (URO) from Intel Corporation.

6. REFERENCES

[1] M. Abramovici. In-System Silicon Validation and Debug. *IEEE D & T of Computers*, 25(3):216–223, May-Jun 2008.

[2] A. Adir, E. Almog, L. Fournier, E. Marcus, M. Rimon, M. Vinov, and A. Ziv. Genesys-Pro: Innovations in Test Program Generation for Functional Processor Verification. *IEEE D & T of Computers*, 21(2):84–93, Mar-Apr 2004.

[3] A. Adir, S. Copty, S. Landa, A. Nahir, G. Shurek, A. Ziv, C. Meissner, and J. Schumann. A Unified Methodology for Pre-silicon Verification and Post-silicon Validation. In *Proc. IEEE/ACM DATE*, pages 1 –6, march 2011.

[4] E. Anis and N. Nicolici. On Using Lossless Compression of Debug Data in Embedded Logic Analysis. In *Proceedings of the IEEE ITC*, 2007. Paper 18.3.

[5] J. Ajanovic. PCI Express 3.0 Overview. In *Proceedings of Hot Chip: A Symposium on High Performance Chips*, 2009.

[6] T. Bojan, M. Arreola, E. Shlomo, and T. Shachar. Functional Coverage Measurements and Results in Post-Silicon Validation of Core™2 Duo Family. In *Proceedings of IEEE International High Level Design Validation and Test Workshop*, pages 145–150, 2007.

[7] M. Boule, J.-S. Chenard, and Z. Zilic. Debug Enhancements in Assertion-Checker Generation. *IET Computers and Digital Techniques*, 1(6):669–677, Nov 2007.

[8] M. Iyer. RACE: A Word-Level ATPG-based Constraints Solver System for Smart Random Simulation. In *Proceedings of the IEEE ITC*, pages 299–308, Sep 2003.

[9] A. B. Kinsman, H. F. Ko, and N. Nicolici. In-System Constrained-Random Stimuli Generation for Post-Silicon Validation. In *Proc. IEEE ITC*, 2012. Paper 3.3.

[10] N. Kitchen and A. Kuehlmann. Stimulus Generation for Constrained Random Simulation. In *Proceedings of the IEEE/ACM ICCAD*, pages 258–265, Nov 2007.

[11] H. F. Ko and N. Nicolici. Automated Trace Signals Identification and State Restoration for Improving Observability in Post-Silicon Validation. In *Proceedings of the IEEE/ACM DATE*, pages 1298-1303, Mar 2008.

[12] H. F. Ko, A. B. Kinsman, and N. Nicolici. Distributed Embedded Logic Analysis for Post-Silicon Validation of SOCs. In *Proceedings of the IEEE ITC*, 2008. Paper 16.3.

[13] J. Rearick, B. Eklow, K. Posse, A. Crouch, and B. Bennetts. IJTAG (Internal JTAG): A Step Toward a DFT Standard. In *Proc. IEEE ITC*, 2005. Paper 32.4.

[14] R. Rudell and A. Sangiovanni-Vincentelli. Multiple-Valued Minimization for PLA Optimization. *IEEE Transactions on CAD*, 6(5):727–750, Sep 1987.

[15] B. Vermeulen, T. Waayers, and S. Goel. Core-Based Scan Architecture for Silicon Debug. In *Proceedings of the IEEE ITC*, pages 638–647, Oct 2002.

[16] Y. Wang, R. Even, T. Kristensen, and R. Jesup. RFC 6184: RTP Payload Format for H.264 Video. Proposed standard, Internet Engineering Task Force (IETF), May 2011.

[17] Y. Wu, S. Thomson, D. Mutcher, and E. Hall. Built-In Functional Tests for Silicon Validation and System Integration of Telecom SoC Designs. *IEEE Transactions on VLSI Systems*, 19(4):629–637, April 2011.

The role of Cascade, a cycle-based simulation infrastructure, in designing the Anton special-purpose supercomputers

J.P. Grossman[1,†], Brian Towles[1], Joseph A. Bank[1], David E. Shaw[1,2,†]

[1] D. E. Shaw Research, New York, NY 10036, USA.

[2] Center for Computational Biology and Bioinformatics, Columbia University, New York, NY 10032, USA.

[†] Correspondence: JP.Grossman@DEShawResearch.com and David.Shaw@DEShawResearch.com

ABSTRACT

Cascade is a cycle-based C++ simulation infrastructure used in the design and verification of two successive versions of Anton, a specialized machine designed for high-speed molecular dynamics computation. Cascade was engineered to address the size and speed challenges inherent in simulating massively parallel special-purpose machines. It provides a lightweight programming interface, rich debugging support, tight Verilog integration, fast multithreaded execution, and low memory overhead. Here, we describe the core features of Cascade that proved most valuable for our simulation efforts.

Categories and Subject Descriptors

I.6.7 [**Simulation and Modeling**]: Simulation Support Systems—*Environments*; B.8.2 [**Performance and Reliability**]: Performance Analysis and Design Aids

General Terms

Algorithms, Performance, Design, Verification

Keywords

Cascade, Anton, cycle-based simulation, reflection

1. INTRODUCTION

Software simulation is an invaluable tool in the design and verification of complex hardware architectures: simulations written in a high-level language (typically C or C++) offer performance and flexibility far beyond that which can be achieved by Register-Transfer Level (RTL) simulation alone. Simulation infrastructures are generally classified as either event-driven or cycle-based. Event-driven simulators are well suited to coarse-grained hardware models, whereas cycle-based infrastructures are often preferred when greater accuracy is desired, and have been widely used in both industry and academia (e.g., [9,12,17,21,23]).

We relied extensively on simulation for the design and verification of Anton 1 [18] and Anton 2 (in development), two massively parallel special-purpose supercomputers for molecular dynamics computation. To address the challenges presented by simulating machines of this scale, we created Cascade—a cycle-based hardware simulation infrastructure. Much like SystemC

[20], Cascade is a C++ library in which hardware modules are modeled as classes (referred to herein as "components" to distinguish them from RTL modules) with explicitly declared input and output ports that can be connected to one another at construction time. Cascade, however, was designed with emphasis on performance, programmability, and scalability—key considerations for simulations comprising millions of components and ports.

This paper describes the core features of Cascade that allowed it to meet these design goals and serve as the basis for the Anton simulators. Some of these features—such as reflection, static scheduling, and Verilog co-simulation—exist in other simulation infrastructures, but have been engineered in novel ways within Cascade to minimize programmer effort. Other features are themselves novel, including run-time port validation, component deactivation, and specialized optimizations for synchronous connections. Together, these features reduce the amount of code required to implement hardware models, provide rich debugging support, and mitigate the performance overheads typically associated with high-fidelity cycle-based simulations.

2. DEVELOPMENT OF CASCADE

One of the earliest decisions we faced in the Anton project was whether to use an existing simulation infrastructure or develop our own. Our conclusion was that the benefits of developing an in-house infrastructure tailored to our requirements justified the time and effort required to do so. The result was Cascade, which has proven to be a valuable tool in our hardware design process. A full description of Cascade is provided elsewhere [3]; here we focus on the features that were most important for our simulation efforts.

2.1 Simulation-based methodology

In designing the Anton machines, we made use of simulation in a variety of ways. Initially, simulation was used for architectural exploration, allowing rapid prototyping and performance evaluation for many different design options. Hardware/software codesign was a critical part of this process as we established the boundaries between programmable embedded cores and special-purpose hardware accelerators. The use of a C++ simulation infrastructure permitted full integration of embedded software within the hardware simulator, often using "magic" interfaces that would later be refined into hardware APIs. Software written with these APIs could then either be directly linked into the simulator and executed natively, or cross-compiled and run on instruction set simulators [8].

As the design solidified, we began the work of RTL implementation, while simultaneously updating simulator components to match the interface and behavior of the corresponding Verilog modules. Bit-accurate interfaces allowed simulator components,

---

Permission to make digital or hard copies of all or part of this work for personal or classroom use is granted without fee provided that copies are not made or distributed for profit or commercial advantage and that copies bear this notice and the full citation on the first page. To copy otherwise, to republish, to post on servers or to redistribute to lists, requires prior specific permission and/or a fee.

DAC '13, May 29–June 07 2013, Austin, TX, USA.

Table 1. Number of distinct component types, component instances and port instances for the four simulation targets described in this paper.

	A1-64	A1-512	A2-62	A2-512
# component types	70	70	56	56
# components	23,559	188,423	276,163	2,209,283
# ports	116,992	935,936	1,976,768	15,814,144

already tested within full-machine simulations running embedded software, to be used as golden models in design-verification testbenches [7]. This provided substantial verification that the RTL modules were functioning correctly and would interact properly with one another and with the embedded software.

2.2 Simulation requirements

Throughout this paper we report on four common simulation targets: 64- and 512-node configurations of Anton 1 ("A1-64", "A1-512"), as well as 64- and 512-node configurations of Anton 2 ("A2-64", "A2-512"). Table 1 lists the numbers of components and ports in these simulations. The scale of the machines being simulated, together with our hardware design methodology, placed a number of requirements on the simulation infrastructure:

Language. Our need to integrate embedded software with the machine simulator led us to focus on C++ infrastructures; special-purpose languages such as Esterel [1] were not a viable option.

Ease of use. The scale of the simulator effort demanded an infrastructure with minimal programmer overhead and significant debugging support.

Verilog co-simulation. We desired the ability to seamlessly integrate simulator models into Verilog testbenches without the need to write specialized data-marshalling functions.

Performance. With simulation times measured in hours, good simulator performance was naturally important.

Checkpointing. One specific concern was the time required to debug errors in long simulations. We viewed the ability to checkpoint and restore simulations as essential in order to be able to replay failures (either in a debugger or with additional logging enabled) without having to restart the entire simulation.

Memory efficiency. The memory overhead of a simulation infrastructure is often inconsequential given the low cost of DRAM. In our case, however, simulations could contain millions of components and tens of millions of ports, so memory efficiency was an important consideration.

In addition to these requirements, we felt that our needs would best be met by a cycle-based (rather than event-driven) infrastructure. First, we wanted certain simulator models to be cycle accurate where timing was critical for either performance estimation or design verification. Second, cycle-based models with bit-accurate interfaces are more amenable to co-simulation with RTL. Finally, it has been observed that cycle-based simulations may offer better performance than event-driven simulations in practice [19].

2.3 Overview of Cascade

Each component in Cascade is a C++ class that inherits from a base `Component` class, and communication between components is effected via input and output ports. Ports can be directly

```
class Producer :
  public Component
{
  DECLARE_COMPONENT(Producer);
public:
  Producer (COMPONENT_CTOR) {}

  Output(char, o_ch);

  void reset ()
  {
    m_ch = "Hello DAC\n";
  }
  void update ()
  {
    o_ch = *m_ch;
    if (*m_ch)
      m_ch++;
  }
private:
  const char *m_ch;
};
```

```
class Consumer :
  public Component
{
  DECLARE_COMPONENT(Consumer);
public:
  Consumer (COMPONENT_CTOR) {}

  Input(char, i_ch);

  void update ()
  {
    if (i_ch)
      putchar(i_ch);
  }
};

int main ()
{
  Producer producer;
  Consumer consumer;
  consumer.i_ch << producer.o_ch;
  Sim::run(20 * CLOCK_PERIOD);
}
```

Figure 1. Simple Cascade example showing a consumer component connected to a producer component.

connected to one another, thus obviating the need for explicitly declared signals or wires. A Cascade simulation consists of three phases. In the "construction" phase, components are constructed and connections are established between input and output ports. In the "initialization" phase, Cascade performs significant preprocessing to set up and optimize the simulation. Finally, the actual hardware simulation takes place in the "simulation" phase, which alternates between rising clock edges and combinational evaluation as simulated time progresses.

The behavior of a component is modeled by one or more "update" functions. An update function reads some subset of the component input ports, updates the internal component state, and writes some subset of the component output ports. A simulation contains one or more clock domains and is driven by a series of rising clock edges: on each rising clock edge all register values within the clock domain are copied from register input ports to register output ports, then all component update functions within the clock domain are invoked to simulate combinational evaluation and to update component state.

Figure 1 shows a very simple example in which a producer component sends text to a consumer component one character at a time. The first three lines of `main()` instantiate the components and connect their ports (using the overloaded << operator). The last line runs the simulation for 20 clock cycles; Cascade automatically initializes the simulation and calls the producer's `reset()` function before the first rising clock edge. The output of the simulation is "Hello DAC\n".

2.4 Comparison with SystemC

SystemC was the main alternative that we considered, and it is instructive to compare it to Cascade on a representative benchmark to understand why we did not feel it would meet our needs. We modeled a simple 8×8 on-chip mesh network in both Cascade and SystemC, with one client per router capable of sending and receiving messages. We then simulated 512 instances of these "chips" for 10,000 clock cycles with randomized network traffic. The simulation comprised a total of 67,072 components and 1,703,936 ports. We used version 2.2.0 of SystemC, and both implementations were compiled using version 4.5.2 of gcc with -O3. Table 2 summarizes the comparison: the Cascade implementation required 39% fewer lines of code, consumed 14.7 times less memory, and ran 14.7 times faster with multithreading.

Table 2. Comparison between SystemC and Cascade for an on-chip network benchmark with 67,072 components and 1,703,936 ports, simulated for 10,000 clock cycles.

	SystemC	Cascade
Lines of code	361	220
Memory (MB)	1249	85
Seconds (single-threaded)	677	76
Seconds (multithreaded)	N/A	46

Cascade reduces the number of lines of code by automatically constructing component and port names, eliminating the need for explicit signals, and providing connection operations for matching interfaces (see supplemental section S1 for details). The large difference in memory usage is due to component and port overheads. On a 64-bit machine, the per-component overhead in SystemC is over 4500 bytes; in Cascade it is 36 bytes, comprising a virtual function table pointer (8 bytes), "parent", "first child" and "next sibling" pointers (8 bytes each), and a 4-byte integer containing additional flags and data. Similarly, the per-port overhead in SystemC for a single-byte port is over 380 bytes, compared to 9 bytes in Cascade (1 byte for a separately allocated value shared by connected ports, and 8 bytes for a pointer to this value). This alone makes SystemC infeasible for our simulations; the 512-node Anton 2 simulation would require over 15 GB of memory for the component and port overhead alone.

3. CORE CASCADE FEATURES

This section describes the core features of Cascade that allowed it to meet our simulation requirements.

3.1 Reflection

Reflection refers to a program's ability to inspect and operate on its own types at run time (e.g., to iterate over the members of a class or structure). In the context of a simulation infrastructure, reflection allows the infrastructure to automatically garner structural information regarding the programmer-supplied hardware models. While some languages including C#, Python and Java provide native support for reflection, C++ does not. In order to benefit from reflection, a simulation infrastructure written in C++ must therefore supply its own implementation.

Cascade uses a set of macros (including, in Figure 1, the macros DECLARE_COMPONENT, COMPONENT_CTOR, Input and Output) to automatically capture the component hierarchy, the set of ports within each component, and the names of all components and ports. These macros are minimally intrusive, and the

programmer does not need to explicitly specify port names separately from the port declarations: the macros define functions that are called automatically to register port names with the reflection infrastructure (see supplemental section S2 for implementation details). This technique is quite general, and it could be used to add lightweight reflection to any C++ library.

The component hierarchy is dynamically represented as a tree using the component "parent", "first child" and "next sibling" pointers. The remaining information is captured in a true type-based reflection system with no per-instance storage overhead, leading to very lightweight component and port implementations.

3.1.1 Benefits of reflection

Cascade's reflection system greatly enhances debugging support by allowing error and warning messages to specify the full hierarchical name of a component or port. This often exposes the source of a bug without having to step through the simulation in a debugger. Reflection also allows Cascade to automatically generate Value Change Dump (VCD) files tracing all port values within a subtree of the component hierarchy.

3.1.2 Interface binding

One application of Cascade's reflection system that was particularly valuable to our design verification effort is *interface binding*, a mechanism that allows Cascade components to be instantiated within Verilog simulations in the same manner as ordinary Verilog modules. If a Verilog wrapper module and a Cascade component have the same interfaces, Cascade can automatically establish a binding between corresponding ports so that data is transparently marshaled between Verilog and C++. We relied on interface binding in order to reuse simulator components as golden models within Verilog testbenches.

Figure 2 illustrates the use of interface binding for a simple adder component. An empty Verilog module is created with the same interface as the C++ component. The body of the Verilog module consists of a single call to $create_cmodule; this Cascade system task creates the component (as specified by the DECLARE_CMODULE macro) then uses reflection information to iterate over the component ports and bind them to the Verilog ports, verifying that the sizes, directions and names all match.

3.2 Static update schedule

As the simulation components are constructed, Cascade keeps track of combinational dependencies between update functions. This defines a directed graph which is topologically sorted to produce a static update schedule. If combinational cycles are

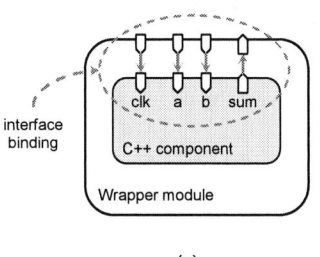

(a)

```
class Adder : public Component
{
    DECLARE_COMPONENT(Adder);
public:
    Adder (COMPONENT_CTOR) {}

    Clock (clk);
    Input (u16, i_a);
    Input (u16, i_b);
    Output(u17, o_sum);

    void update () { o_sum = i_a + i_b; }
};
DECLARE_CMODULE(adder, new Adder);
```

(b)

```
module CAdder
(
    input  logic      clk,
    input  logic [15:0] i_a,
    input  logic [15:0] i_b,
    output logic [16:0] o_sum
);

    initial $create_cmodule("adder");

endmodule
```

(c)

Figure 2. Interface binding. (a) C++ ports are automatically bound to corresponding Verilog wrapper module ports. (b) Definition of the Adder component. (c) Verilog wrapper module, which instantiates the Adder component. On a rising `clk` edge Cascade copies `i_a` and `i_b` to C++, then calls `Adder::update()`, then copies `o_sum` back to Verilog.

Table 3. Performance improvements due to programmer-driven component deactivation optimizations.

	A1-64	A1-512	A2-64	A2-512
% updates eliminated	33	40	75	80
% speedup	15	15	23	28

Table 4. Performance improvements due to automatic register elimination and fast register copies.

	A1-64	A1-512	A2-64	A2-512
% register bytes eliminated	52	52	52	52
% speedup	17	17	11	13

present, then the topological sort fails with a detailed error message, and the developer must modify the implementation to eliminate the cycle. This can be accomplished either by splitting one or more update functions into separate update functions, or by replacing combinational connections with synchronous connections (which will be described in Section 3.2).

3.2.1 Specifying dependencies

In an event-driven simulation infrastructure such as SystemC, input sensitivities must be specified for each process. Because this information alone is insufficient to determine the combinational dependencies between update functions, Cascade requires the programmer to declare both the ports that are read *and* the ports that are written by an update function. Cascade's default assumption, which handles the vast majority of cases, is that a single update function named "update" reads all of a component's inputs and writes all of its outputs; explicit read/write declarations are required only when this assumption does not hold. Reflection is used to obtain the list of component ports, so in the common case no programmer effort is required to specify dependencies. Of the 56 distinct component types modeled for Anton 2, only 8 required explicit read/write declarations.

Referring again to the code in Figure 1, Cascade will automatically detect the update functions `Producer::update()` and `Consumer::update()`; furthermore it will assume that the producer update function writes `Producer::o_ch` and that the consumer update function reads `Consumer::i_ch`. The connection between these ports therefore creates a combinational dependency between the update functions, so a static update schedule will be generated in which `producer.update()` is called before `consumer.update()`.

3.2.2 Component deactivation

The use of a static schedule is known to improve simulator performance and avoid "unnecessary wakeups" [2,13,15,17]. Still, calling every update function on every cycle is a significant source of overhead in cycle-based simulations, particularly for components that are quiescent. Cascade addresses this issue by providing every component with an "active" flag; update functions are only called for active components. The programmer can add code to an update function to determine if the component has become quiescent and, if so, deactivate it.

With this optimization in place, it remains to automatically reactivate components based on the values of certain programmer-specified input ports (typically "valid" bits indicating the arrival of new data). A natural approach is to associate activation with port writes, that is, when a producer component sets its "valid" output port, the act of writing the port causes any consumer components to become activated. This approach, however, would increase the execution overhead for all port writes (due to the need to check for components that must be activated) as well as the per-port storage overhead (in order to specify this set of components).

Instead, Cascade leverages both the static update schedule and the knowledge of the set of ports written by each update function: if a component is conditionally activated based on a certain input port, then Cascade checks the value of the port immediately after invoking the update function that writes it. This allows components to be automatically reactivated without increasing the performance or memory overheads associated with ports. It also avoids testing ports when the producer component is itself inactive, thus maintaining the property that the overall simulation overhead is proportional to the number of active components.

Most of the components in our simulations were instrumented to deactivate themselves when quiescent. We found that this optimization eliminated up to 80% of the update function invocations and improved the overall simulation performance by as much as 28% (Table 3).

3.3 Synchronous connections

Another significant source of overhead in cycle-based simulations is the need to copy register values from their input ports to their output ports on every rising clock edge. This has both a performance impact (the time required to perform the copy) and a storage impact (the need to store two copies of each register value). In a statically scheduled cycle-based simulation, where each update function is guaranteed to be called at most once per cycle, it is generally unnecessary to model registers within components: internal state can simply be stored in member variables and modified appropriately by the update functions. Only registers that exist between components need to be explicitly modeled.

Cascade has native support for *synchronous connections* between ports: if a port A is connected to a port B using the overloaded less-than-or-equal-to operator ("A <= B"), a register is placed between the ports so that writes to A are only visible from B on the next rising clock edge. Internally, storage is allocated for two values, with port A pointing to one value, port B pointing to the other, and a copy occurring on each rising clock edge.

Giving the simulation infrastructure full ownership of registers enables two optimizations to automatically reduce register overhead. First, Cascade leverages the static update schedule to eliminate many explicit registers altogether. If the first write of a register input port always occurs after the last read of the register output port, then the register is unnecessary and can be replaced by a single uncopied value without affecting simulation behavior. Cascade attempts to generate an update schedule that eliminates as many bytes of register storage as possible. Second, Cascade optimizes register value copies by allocating two large identical blocks of memory, one for register input ports and one for register output ports, then using a single `memcpy` to copy values on the rising clock edge. In addition to dramatically reducing the performance overhead of register copies, this also improves memory efficiency by eliminating the need for a global data structure used to iterate over the registers. These optimizations consistently eliminated 52% of the register storage, and improved overall simulation performance by up to 17% (Table 4).

3.4 Multithreaded simulation engine

Cascade uses a simple yet effective multithreading strategy: when multiple clock domains have a simultaneous rising clock edge, the clock domains are partitioned among the available threads so that their register copies and combinational evaluations can be performed in parallel on different cores. The simulation itself is deterministic and functionally independent of the number of threads; multithreading is strictly a performance improvement.

This approach requires some cooperation from the programmer in the form of artificially dividing clock domains into two or more domains with identical clocks. So long as there are no combinational connections between components in the subdivided domains (Cascade requires connections between clock domains to be synchronous), this transformation is straightforward and has no impact on the simulation. In our simulations, there was a natural partition: each node was placed in a separate clock domain.

Multithreading was not implemented in the initial version of Cascade, and is only supported within our Anton 2 simulator. On a dual-socket quad-core machine, we observed speedups of up to 6.7 fold with 8 threads (Figure 3). These speedups are much larger than the speedup observed for the network benchmark in Table 2, which contains fairly simple update functions and is memory-bandwidth-limited. The Anton 2 simulator is compute-limited and therefore benefits more from multithreading.

3.5 Debugging support

Strong debugging support was an essential requirement for our development process. Cascade implements three debugging mechanisms that are tailored to hardware simulation: run-time port validation, conditional tracing, and simulation checkpointing.

3.5.1 Port Validation

In debug builds, every Cascade port has an associated "valid" flag, immediately preceding the port value in memory, that is set when the port is written, checked when the port is read, and cleared on each rising clock edge. This mechanism automatically detects a variety of common errors:

Invalid outputs. A component's update function(s) may neglect to set one or more outputs, resulting in undefined values.

Uninitialized outputs. An output register that is not initialized during reset will have an undefined value on the first cycle.

Bad evaluation order. If components with combinational dependencies are not evaluated in the correct order, then the consumer component will read a stale value from the previous cycle.

Missing connections. One or more required connections between ports may be omitted at construction time.

Incorrect handshaking. Certain ports may be valid only on certain cycles depending on an interface's protocol. A producer may neglect to set these ports when it is supposed to, or a consumer may attempt to read these ports when it shouldn't.

When any one of these errors occurs, Cascade produces a warning specifying the full hierarchical name of the invalid port, from which the source of the problem can be quickly deduced. Without port validation, these errors tend to produce silent failures that are difficult to diagnose and correct. A cycle-based infrastructure supports port validation by providing a well-defined point in time (the rising clock edge) at which the port valid flags

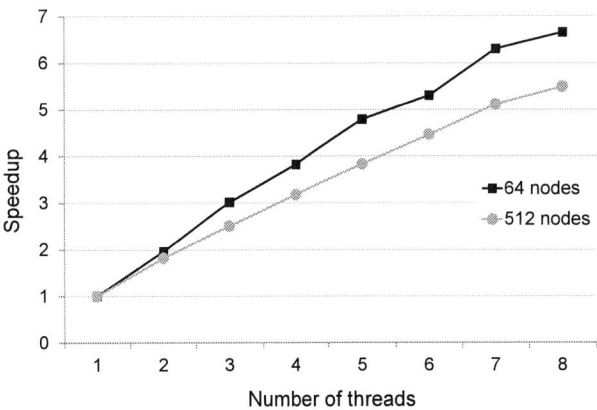

Figure 3. Performance improvements due to multithreading for the Anton 2 simulator. Simulations were run with 1–8 threads on a dual-socket quad-core machine.

should be cleared. In an event-driven simulation, by contrast, there is no way to distinguish a port whose value is stale from a port whose value is simply unchanged.

3.5.2 Tracing

One of the most basic debugging techniques is using printf statements to generate output that can be inspected to determine the source of a problem. Cascade extends this technique for hardware simulations with *tracing*, which is conditional output that can be enabled on a per-component basis. For example, if an error occurs within a specific component (identified by the error message), then the simulation can be re-run with tracing enabled for that component. All trace output generated by a component is prefixed with both the current simulation time and the component's name.

3.5.3 Simulation checkpointing

Cascade supports simulation checkpointing wherein the entire state of a simulation can be stored to or retrieved from a compressed archive. Furthermore, simulation checkpoints are compatible across debug and release builds. A long simulation can be run in release mode generating periodic checkpoints; one of these checkpoints can then be restored within a debugger in order to inspect simulation state or step through an event of interest.

All port and register state is automatically archived; the programmer is only responsible for archiving internal state by implementing an `archive()` function for each component. These functions are called automatically by Cascade whenever a simulation is written to or read from a checkpoint.

4. RELATED WORK

SystemC is a widely used simulation infrastructure, but it is known to have performance issues due to its focus on generality [1,4,6,15,22]. In [15] the performance of SystemC was improved using "acyclic scheduling", which attempts to statically schedule process wakeups based on the graph of combinational dependencies, relying on the event-driven SystemC engine to handle cycles in this graph. A similar technique was proposed in [10] and used in the Liberty Simulation Environment [14]. In [13], combinational cycles were aggressively eliminated using a tool that parses and automatically rewrites SystemC modules, splitting larger processes into multiple smaller ones.

Requiring the programmer to eliminate all combinational cycles from the dependency graph allows the use of a fully static schedule, further improving performance by simplifying the simulation infrastructure and guaranteeing that each process is woken up at most once per clock cycle [2,16,17]. Cascade builds on this approach with support for component deactivation, and by leveraging the static schedule to automatically eliminate registers.

The use of reflection within a simulation infrastructure was proposed in [5] and implemented by providing component descriptions in an interface-description language. SystemC provides a heavyweight instance-based (rather than type-based) naming mechanism that requires the programmer to explicitly supply string names to port constructors. By contrast, Cascade's reflection system is automatic and minimally invasive, requiring only a slight modification to the port declaration syntax.

5. DISCUSSION

Cascade played a central role in the development of both Anton 1 and Anton 2. For each machine, we used our Cascade-based simulator for architectural exploration and validation, performance estimation, design verification, and embedded software development. Cascade's support for Verilog co-simulation was particularly valuable. In addition to reusing simulator components as golden models within Verilog testbenches, we were able to run full 512-node machine simulations with one C++ node replaced by an RTL node. Running our molecular dynamics software in this mixed simulation environment provided a more stringent test of the hardware than block-level testbenches, and uncovered a number of hardware and software defects.

Not all of the design requirements listed in Section 2.2 were evident at the outset of the project, and a number of design tradeoffs were made as Cascade underwent several major revisions. For example, earlier versions of Cascade supported *threaded components* (similar to SystemC's SC_THREAD) based on the QuickThreads library [11]. The use of QuickThreads, however, is incompatible with checkpointing. As our need to debug long simulations grew, we abandoned threaded components in favor of the ability to save and restore simulations.

The Cascade features discussed in Section 3 are those that were found to be most useful for our design efforts; all of them could be incorporated into various other simulation infrastructures. Component deactivation, synchronous connections and port validation are more specific to cycle-based simulation, while reflection, multithreading, tracing and checkpointing apply equally well to event-driven infrastructures. Cascade itself is available for download at [3].

6. REFERENCES

[1] Gérard Berry and Georges Gonthier, "The Esterel synchronous programming language: Design, semantics, implementation", Science of Computer Programming, Vol. 11, No. 2, Nov. 1992, pp. 87–152.

[2] Richard Buchmann and Alain Greiner, "A fully static scheduling approach for fast cycle accurate SystemC simulation of MPSoCs", *2007 International Conference on Microelectronics*, Cairo, Egypt, Dec. 29–31, 2007, pp. 101–104.

[3] Cascade, http://www.deshawresearch.com/resources_cascade.html

[4] L. Charest, E.M. Aboulhamid, C. Pilkington and P. Paulin, "SystemC performance evaluation using a pipelined DLX multiprocessor", *2002 Design, Automation and Test in Europe*, Paris, France, Mar. 4–8, 2002, pp. 8–12.

[5] Frederic Doucet, Sandeep Shukla and Rajesh Gupta, "Introspection in system-level language frameworks: meta-level vs. integrated",

2003 Design, Automation and Test in Europe, Munich, Germany, Mar. 3–7, 2003, pp. 382–387.

[6] Wolfgang Ecker, Volkan Esen, Lars Schönberg et al., "Impact of description language, abstraction layer, and value representation on simulation performance", *2007 Design, Automation and Test in Europe*, Nice, France, Apr. 16–20, 2007, pp. 767–772.

[7] J.P. Grossman, John K. Salmon, C. Richard Ho et al., "Hierarchical simulation-based verification of Anton, a special-purpose parallel machine", *26th International Conference on Computer Design*, Lake Tahoe, CA, Oct. 12–15, 2008, pp. 340–347.

[8] J.P. Grossman, Cliff Young, Joseph A. Bank et al., "Simulation and embedded software development for Anton, a parallel machine with heterogeneous multicore ASICs", *6th International Conference on Hardware/Software Codesign and System Synthesis*, Atlanta, GA, Oct. 19–24, 2008, pp. 125–130.

[9] Dale E. Hocevar, Ching-Yu Hung, Dan Pickens and Sundararajan Sriram, "Top-down design using cycle based simulation: an MPEG A/V decoder example", *8th Great Lakes Symposium on VLSI*, Lafayette, LA, Feb. 19–21, 1998, pp. 400–405.

[10] Denis Hommais and Frédéric Pétrot, "Efficient combinational loops handling for cycle precise simulation of system on a chip", *24th EUROMICRO*, Västerås, Sweden, Aug. 25–27, 1998, pp. 51–54.

[11] David Keppel, "Tools and techniques for building fast portable threads packages", Technical Report UWCSE 93-05-06, University of Washington Dept. of Computer Science and Engineering, 1993.

[12] Brucek Khailany, William J. Dally, Andrew Chang et al., "VLSI design and verification of the Imagine processor", *2002 IEEE International Conference on Computer Design*, Freiburg, Germany, Sept. 16–18, 2002, pp. 289–294.

[13] Youssef N. Naguib and Rafik S. Guindi, "Speeding up SystemC simulation through process splitting", *2007 Design, Automation and Test in Europe*, Nice, France, Apr. 16–20, 2007, pp. 111–116.

[14] David A. Penry and David I. August, "Optimizations for a simulator construction system supporting reusable components", *40th Design Automation Conference*, Anaheim, CA, Jun. 2–6, 2003, pp. 926–931.

[15] Daniel Gracia Pérez, Gilles Mouchard and Olivier Temam, "A new optimized implementation of the SystemC engine using acyclic scheduling", *2004 Design, Automation and Test in Europe*, Paris, France, Feb. 16–20, 2004, pp. 552–557.

[16] Frédéric Pétrot, Denis Hommais and Alain Greiner, "Cycle precise core based hardware/software system simulation with predictable event propagation", *23rd EUROMICRO Conference*, Budapest, Hungary, Sept. 1–4, 1997, pp. 182–187.

[17] Luc Séméria, Andrew Seawright, Renu Mehra et al., "RTL C-based methodology for designing and verifying a multi-threaded processor", *39th Design Automation Conference*, New Orleans, LA, Jun. 10–14, 2002, pp. 123–128.

[18] David E. Shaw, Martin M. Deneroff, Ron O. Dror et al., "Anton, a special-purpose machine for molecular dynamics simulation", *34th Annual International Symposium on Computer Architecture*, San Diego, CA, Jun. 9–13, 2007, pp. 1–12.

[19] Chulho Shin, Peter Grun, Nizar Romdhane et al., "Enabling heterogeneous cycle-based and event-driven simulation in a design flow integrated using the SPIRIT consortium specifications", Design Automation for Embedded Systems, Vol. 11, Nos. 2–3, Sept. 2007, pp. 119–140.

[20] SystemC, http://www.systemc.org

[21] Scott Taylor, Michael Quinn, Darren Brown et al., "Functional verification of a multiple-issue, out-of-order, superscalar Alpha processor—the DEC Alpha 21264 microprocessor", *35th Design Automation Conference*, San Francisco, CA, Jun. 15–19, 1998, pp. 638–643.

[22] Dr. Greg Tumbush and Mark Hupp, "Dramatically increase the performance of SystemC simulations", *2007 Design and Verification Conference and Exhibition*, San Jose, CA, Feb. 21–23, 2007.

[23] Joon Seo Yim, Yoon Ho Hwang, Chang Jae Park et al., "A C-based RTL design verification methodology for complex microprocessor", *34th Design Automation Conference*, Anaheim, CA, Jun. 9–13, 1997, pp. 83–8.

S1. Comparison with SystemC

We modeled an on-chip mesh network benchmark in both Cascade and SystemC, and found that the Cascade implementation required fewer lines of code (220) than the SystemC implementation (361). Here we compare portions the source code for the two implementations to explain this difference.

1. *Packet header port type.* In Cascade, all port types must be "bags of bits": they cannot have constructors, destructors, or virtual functions. One of the ways in which SystemC is more general is that any C++ type may be used as a port type. To support this generality, all port types must define a comparison operator, a streaming operator, and an `sc_trace` function. Our benchmark includes a `PacketHeader` port type, defined in the Cascade implementation as

```
struct PacketHeader
{
    byte dest_x;
    byte dest_y;
};
```

The corresponding SystemC definition, including the required operators, is as follows:

```
struct PacketHeader
{
    byte dest_x;
    byte dest_y;

    bool operator== (const PacketHeader &rhs) const
    {
        return dest_x == rhs.dest_x &&
               dest_y == rhs.dest_y;
    }
};

inline void sc_trace (sc_trace_file *,
                      const PacketHeader &,
                      const string &)
{
}

std::ostream &operator<< (std::ostream &os,
                          const PacketHeader &header)
{
    return os;
}
```

`sc_trace` and the streaming operator are unused, but the (null) implementations are required for the code to compile.

2. *Tracing and assertions.* Cascade is built on a library that adds context names to assertion failure messages and supports conditional logging (*tracing*) based on context names. We incorporated these features in the SystemC implementation by defining an `sc_component` base class that inherits from both `sc_module` and the library's `TraceContext` helper class:

```
class sc_component : public sc_module,
                     public TraceContext
{
public:
    sc_component (sc_module_name name) :
        sc_module(name),
        TraceContext((const char *) name)
    {
    }
};
```

As an example of the utility of these features, our initial implementation of the `PacketHeader` comparison operator contained a typo that caused the following failure:

```
$ sc_router
Error: Client(0,3)
Assertion failed: m_inSeq[idx] == int(i_data)
sc_router.cpp:92: void Client::update()
```

We re-ran the simulation with tracing enabled for the client identified in the error message:

```
$ sc_router -trace "Client(0,3)"
Client(0,3): [7 ns] Sending message to 23 (7,2)
Client(0,3): [9 ns] Received message from 16 for 24
Client(0,3): [10 ns] Sending message to 58 (2,7)
Client(0,3): [13 ns] Received message from 2 for 24
Client(0,3): [14 ns] Received message from 39 for 24
Client(0,3): [15 ns] Received message from 2 for 0
Error: Client(0,3)
Assertion failed: m_inSeq[idx] == int(i_data)
sc_router.cpp:92: void Client::update()
```

This revealed that the client, whose numerical ID is 24, was receiving a message that should have been delivered to Client(0,0). From this starting point we were able to track down the source of the error—a bug in the comparison operator that was causing a stale packet header to be reused in the network.

3. *Network port interface.* In the Cascade implementation, the credit-based network port interface is defined as follows:

```
struct NetPort : public Interface
{
    DECLARE_INTERFACE_WITH_CTOR(NetPort);

    NetPort (INTERFACE_CTOR) : credits(4)
    {
        o_valid.setType(PORT_PULSE);
        o_credit.setType(PORT_PULSE);
    }

    void disable ()
    {
        i_valid.wireToConst(0);
        i_credit.wireToConst(0);
    }

    Input(bit,          i_valid);
    Input(PacketHeader, i_header);
    Input(uint64,       i_data);

    Output(bit,          o_valid);
    Output(PacketHeader, o_header);
    Output(uint64,       o_data);

    Input(bit,  i_credit);
    Output(bit, o_credit);

    int credits;
};
```

In the constructor, the `o_valid` and `o_credit` outputs are given the type `PORT_PULSE`, which causes them to be reset to zero on every rising clock edge. The `disable()` function is used to tie off unconnected network ports by forcing the `i_valid` and `i_credit` inputs to zero.

The `Interface` base class, along with the macros `DECLARE_INTERFACE_WITH_CTOR` and `INTERFACE_CTOR`, define a Cascade interface. Cascade provides connection operations for matching interfaces, so for example the synchronous on-chip network connections in the *x* dimension are established using the following code:

```
syncConnect(m_routers(x-1,y).port[DIR_XPOS],
            m_routers(x,y).port[DIR_XNEG]);
```

The System C network port interface definition, including helper functions to connect and disable network ports, is as follows:

```
struct NetSignals
{
    sc_signal<bit>          valid;
    sc_signal<PacketHeader> header;
    sc_signal<uint64>       data;
    sc_signal<bit>          credit;
};

struct NetPort
{
    NetPort (const string &name) :
        i_valid(*(name + "::i_valid")),
        i_header(*(name + "::i_header")),
        i_data(*(name + "::i_data")),
        o_valid(*(name + "::o_valid")),
        o_header(*(name + "::o_header")),
        o_data(*(name + "::o_data")),
        i_credit(*(name + "::i_credit")),
        o_credit(*(name + "::o_credit")),
        credits(4)
    {
    }

    void bind_in (NetSignals &signals)
    {
        i_valid(signals.valid);
        i_header(signals.header);
        i_data(signals.data);
        o_credit(signals.credit);
    }

    void bind_out (NetSignals &signals)
    {
        o_valid(signals.valid);
        o_header(signals.header);
        o_data(signals.data);
        i_credit(signals.credit);
    }

    sc_in<bit>          i_valid;
    sc_in<PacketHeader> i_header;
    sc_in<uint64>       i_data;

    sc_out<bit>          o_valid;
    sc_out<PacketHeader> o_header;
    sc_out<uint64>       o_data;

    sc_in<bit>  i_credit;
    sc_out<bit> o_credit;

    int credits;
};

void connect (NetPort *p1, NetPort *p2,
              NetSignals &s1to2, NetSignals &s2to1)
{
    p1->bind_in(s2to1);
    p1->bind_out(s1to2);
    p2->bind_out(s2to1);
    p2->bind_in(s1to2);
}

void disable (NetPort *p, NetSignals &to,
              NetSignals &from)
{
    p->bind_in(to);
    p->bind_out(from);
    to.valid = 0;
    from.credit = 0;
}
```

There are three sources of overhead in this code, all of which Cascade eliminates. First, the port names must be explicitly specified in the NetPort constructor. Second, signals are required to connect network ports or to tie off disabled network ports. Third, helper functions must be provided to bind signals to ports and to connect two ports.

4. Router model. The two router class definitions are very similar, as can be seen in a side-by-side comparison:

```
// Cascade                          // SystemC
class Router                        class Router
 : public Component                  : public sc_component
{                                    {
    DECLARE_COMPONENT(Router);       public:
public:                                sc_in<byte> i_x;
    Input(byte, i_x);                  sc_in<byte> i_y;
    Input(byte, i_y);                  sc_in<bool> clk;
    RouterPort port[5];                RouterPort *port[5];

    Router (COMPONENT_CTOR);           Router (sc_module_name name);
                                       ~Router ();
    void update ();
                                       void update ();
private:
    int m_rrPriority;                private:
};                                     int m_rrPriority;
                                     };
```

There are two differences worth noting. First, the Cascade model contains a static array of router ports (the RouterPort class inherits from NetPort), whereas the SystemC model contains an array of pointers: each router port must be individually constructed so that the interface names can be provided. This is also the reason that the SystemC model requires a destructor (to delete the router ports). Second, an explicit clock port is unnecessary within Cascade components; a component with no clock port is automatically placed in the same clock domain as its parent.

The update() functions in the two implementations are nearly identical; the only difference is that the SystemC implementation explicitly clears the o_valid and o_credit outputs on every cycle. The destructor in the SystemC implementation simply deletes the router ports. The constructors are as follows:

```
// Cascade
Router::Router (IMPL_CTOR) : m_rrPriority(0)
{
}

// SystemC
Router::Router (sc_module_name name) :
    Component(name),
    i_x("i_x"),
    i_y("i_y"),
    clk("clk"),
    m_rrPriority(0)
{
    for (int i = 0 ; i < 5 ; i++)
        port[i] = new RouterPort(str("Port%d", i));

    SC_HAS_PROCESS(Router);
    SC_METHOD(update);
    sensitive << clk.pos();
}
```

Most of the extra code in the SystemC implementation is related to port and interface names. In addition, the update function must be explicitly registered and made sensitive to the rising clock edge; in Cascade this is automatic.

Summary. Table S1 gives a detailed breakdown for the extra 141 lines of code in the SystemC implementation.

Table S1. Breakdown of additional lines of code required in the SystemC implementation of the network benchmark.

Description	Lines of Code
Component/port names	56
Signals	27
Interface connections	16
User-defined port type	14
Tracing/assertions	11
Method declarations	8
Other	9

S2. Lightweight reflection for C++ classes

Reflection mechanisms for C++ typically require the programmer to explicitly specify class-member names separately from the class-member declarations. In Cascade, by contrast, the port macros simultaneously declare the ports and register them with the reflection infrastructure. This is accomplished by defining functions that are automatically called before a component is constructed. Defining these functions is easy, whereas calling them automatically requires some engineering.

Consider a simple adder with two inputs and one output:

```
class Adder : public Component
{
    DECLARE_COMPONENT(Adder);

    Input(int,  i_a);
    Input(int,  i_b);
    Output(int, o_sum);
    ...
};
```

Our strategy is to have the preprocessor transform the above component definition into something resembling the following:

```
class Adder : public Component
{
    static void preConstruct ()
    {
        _preConstruct0();
    }
    static void _preConstruct3 (...) {}

    Input<int> i_a;
    static void _preConstruct0 ()
    {
        register reflection information for i_a
        _preConstruct1();
    }
    Input<int> i_b;
    static void _preConstruct1 ()
    {
        register reflection information for i_b
        _preConstruct2();
    }
    Output<int> o_sum;
    static void _preConstruct2 ()
    {
        register reflection information for i_c
        _preConstruct3();
    }
    ...
};
```

The key elements are (1) a sequence of _preConstruct functions, one per port, such that each one calls the next; (2) an empty _preConstruct function called by the last port's function to terminate the chain of function calls; and (3) a top-level preConstruct function invoked by the infrastructure that calls _preConstruct0() to register all ports.

The actual macro definitions used to convert this strategy into a working implementation are shown in Figure S1. DECLARE_REFLECTION_FUNCTIONS is a helper macro that appears in the expansion of DECLARE_COMPONENT, and DECLARE_NAMED_PORT is a helper macro that appears in the expansions of the Input and Output port macros.

The templated _Counter structure defines a compile-time function that maps a port's line number to its zero-based index within the component. In particular, the index of a port is given by _Counter<__LINE__,0>::count.

The templated _Index<n> structure is used to define a different function for each port. Instead of defining functions with different names, the macros overload the same function name (_preConstruct) with different argument types (pointers to _Index<n>). Each function calls the next simply by casting the descriptor argument to a pointer to _Index<n+1>.

The last call to _preConstruct() does not match the signature of any port function and is therefore captured by the empty version defined by DECLARE_REFLECTION_FUNCTIONS.

Within each _preConstruct function, the getsize struct is used to support statically-sized single-dimensional arrays, e.g.,

```
Input(int, i_data[4]);
```

In particular, the expression offsetof(getsize, name) statically evaluates to the size of the array, or to zero if the port is not an array.

The descriptor object is responsible for assembling reflection information for the component type. Each port registers itself with this object within its _preConstruct function, supplying the byte offset of the port within the component, the name of the port, the port array size (or zero for a scalar port), and the size of the port in bytes.

These macros place minimal burden on the programmer and could easily be adapted to any C++ library that would benefit from reflection.

```
#define DECLARE_REFLECTION_FUNCTIONS()                                                        \
    template <int line, int x> struct _Counter { enum { count = _Counter<line-1,x>::count }; }; \
    template <int x> struct _Counter<__LINE__,x> { enum { count = -1 }; };                     \
    template <int n> struct _Index : public Cascade::InterfaceDescriptor {};                   \
    static void _preConstruct (...) {}                                                          \
    static void preConstruct (Cascade::InterfaceDescriptor *descriptor)                        \
    {                                                                                          \
        _preConstruct((_Index<0>*) descriptor);                                                \
    }

#define DECLARE_NAMED_PORT(type, name)                                                                    \
    type name;                                                                                            \
    template <int x> struct _Counter<__LINE__,x> { enum { count = _Counter<__LINE__-1,x>::count + 1 }; }; \
    static void _preConstruct (_Index<_Counter<__LINE__,0>::count> *descriptor)                          \
    {                                                                                                    \
        struct getsize { byte name; };                                                                   \
        descriptor->addPortName(offsetof(_thistype, name), #name, offsetof(getsize, name), sizeof(type)); \
        _preConstruct((_Index<_Counter<__LINE__,0>::count + 1> *) descriptor);                           \
    }
```

Figure S1. Reflection macros. DECLARE_REFLECTION_FUNCTIONS defines the helper structures _Counter and _Index, as well as the main preConstruct function. DECLARE_NAMED_PORT defines the per-port _preConstruct function that registers the port with the reflection infrastructure, then calls the next port's function.

Towards Structured ASICs using Polarity-Tunable Si Nanowire Transistors

Pierre-Emmanuel Gaillardon[1], Michele De Marchi[1], Luca Amarù[1], Shashikanth Bobba[1], Davide Sacchetto[1,2], Yusuf Leblebici[2], Giovanni De Micheli[1]

[1] Integrated Systems Laboratory (LSI), EPFL [2] Microelectronic Systems Laboratory (LSM), EPFL

CH-1015 Lausanne, Switzerland

pierre-emmanuel.gaillardon@epfl.ch

Abstract

In addition to scaling semiconductor devices down to their physical limit, novel devices show enhanced functionality compared to conventional CMOS. At advanced technology nodes, many devices exhibit ambipolar behavior, i.e., they show n- and p-type characteristics simultaneously. This phenomenon can be tamed using double-gate structures. In this paper, we present a complete framework relying on *Double-Gate-all-around Vertically stacked NanoWire FETs* (DG-NWFETs). Such device enables a compact realization of arithmetic logic functions and presents unprecedented interest for structured ASIC applications.

Categories and Subject Descriptors

[Hardware] Emerging technologies – Circuit substrates

General Terms

Design, Performance

Keywords

Nanowire transistors; controllable polarity; regular fabrics; XOR logic synthesis

1. Introduction

During the last four decades, the increase of computing power was achieved by reducing the dimensions of semiconductor devices down to the nanoscale. Today, the semiconductor industry is approaching the ultimate limits of conventional silicon-based *Integrated Circuits* (IC) and researchers are focusing their effort on identifying possible approaches that will enable the continuation of Moore's scaling law. Keeping the trend towards "More Moore", FinFET transistors are successfully replacing planar CMOS transistors at the 22-nm technology node [1]. Further, they are expected to evolve in the next few years to *vertically-stacked Silicon NanoWires Field Effect Transistors* (SiNWFETs) [2]. Indeed, by splitting the 2-D thin film channel in a collection of 1-D *Gate-All-Around* (GAA) structures, the device exhibits higher $Ion/Ioff$ ratio and reduced leakage current [3].

In addition to this traditional path, the "More than Moore" approach emerged in the last decade and consists of adding functionality to keep increasing its computation power. Widely used at the system level, this approach is also valid at the device level, where devices can present an enhanced-functionality. In particular, several novel devices demonstrate controllable polarity [4-7]. At advanced technology nodes, an increasingly larger number of devices are affected by Schottky contacts at the source and drain interfaces. Hence, such devices face an ambipolar behavior, i.e., that the devices exhibit n- and p-type characteristics simultaneously. While technologists aim at suppressing the ambipolar behavior of the devices through additional process steps, the construction of independent double-gate structures can tame it. As a result, the device polarity can be electrostatically programmed to be either n- or p-type. The functionality of such a device is biconditional on both gate values and enables a compressed realization of XOR-based logic functions, which are not implementable in CMOS in a compact form [8], as well as the realization of ultra-fine grain configurable logic cells [9].

In this paper, we present a complete framework associated with the technology of *Double-Gate SiNWFET (DG-SiNWFET)* from technology to CAD tools. The remainder of the paper is organized as follows. In Section 2, we report on our DG-SiNWFET technology. In Section 3, we present its opportunities for realizing arithmetic logic gates. Section 4 introduces the interests of regular arrangements to mitigate the impact of the additional gate, while Section 5 discusses the opportunities from a CAD perspective. The paper is concluded in Section 6.

1. Vertically-Stacked Double Gate NWFETs

Transistors with controllable polarity consist of double-independent gate FETs having one gate regulating the Schottky barriers on source/drain junctions. In this work, we highlight a top-down fabricated, vertically-stacked SiNW FET, featuring two GAA electrodes (Fig. 1).

Vertically-stacked GAA SiNWs represent a natural evolution of FinFET structures, providing the best electrostatic control over the channel and consequently superior scalability properties [10]. In our device, one gate electrode, the *Control Gate* (CG) acts conventionally by turning *on* and *off* the device. The other electrode, the *Polarity Gate* (PG), acts on the side regions of the device, in proximity of the *Source/Drain* (S/D) Schottky junctions, switching the device polarity dynamically between n- and p-type (Fig. **Error! Reference source not found.**). The voltage ranges applied to the two gates are comparable, and the input and output voltage levels are compatible, resulting in directly cascadable logic gates.

Permission to make digital or hard copies of all or part of this work for personal or classroom use is granted without fee provided that copies are not made or distributed for profit or commercial advantage and that copies bear this notice and the full citation on the first page. To copy otherwise, or republish, to post on servers or to redistribute to lists, requires prior specific permission and/or a fee.
DAC'13, May 29 – June 07 2013, Austin, TX, USA.

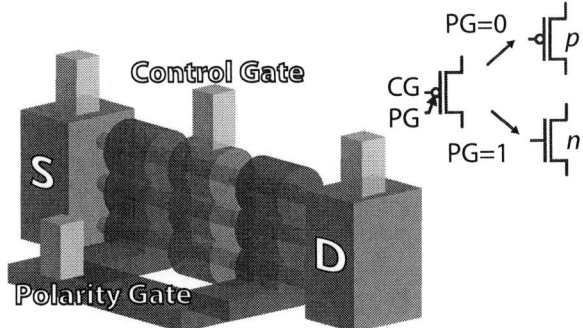

Figure 1. 3D sketch of the SiNWFETs featuring 2 independent gates and its associated symbol.

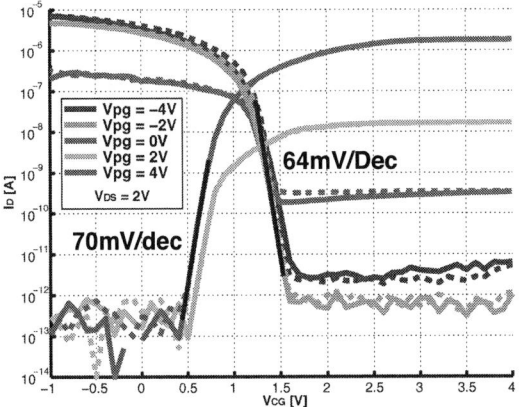

Figure 2. I_{DS}-V_{CG} logarithmic plot of a measured device for serveral V_{PG} voltages. Curves extracted at V_{DS} = 2V [10].

2. Compact Arithmetic Logic Gates

Compared to unipolar transistors, that intrinsically embed the INV function, DG-SiNWFETs can implement natively the XOR operation. Therefore, the realization of arithmetic logic operators can reach a high level of compactness. In particular, with this technology, a full-swing 2-input XOR, reported in Fig. 3a, requires 4 transistors while the traditional full-swing static CMOS implementation uses 8 transistors [11]. Extending the logic design style from static to pass-transistor, the 3-input XOR realization introduced in [12] is obtained, and depicted by Fig. 3b. The same implementation scheme can be used to implement the majority voting operation. Hence, a 4-transistor 3-input majority logic gate is proposed in [13] and reported in Fig. 3c. Note that in static CMOS, the same gate has 10 devices in place of 4 [11]. The compact implementation of both 3-input XOR and 3-input majority function offers an advantageous realization for both the *Sum* and *Carry* functions of *Full-Adder* (FA) gates. Therefore, the FA is competitively realized with 8 devices, input inverters apart, as depicted in Fig. 3d. For comparison, the static (transmission-gate) CMOS counterpart has 28 (14) transistors [11].

3. Towards Stuctured ASICs

Regularity is one of the key features to increase the yield of integrated circuits at advanced technology nodes [14], while keeping the routing complexity under control. The introduced technology presents a unique opportunity to rejuvenate the field. With the electrostatic device configuration, the frontier between

n- and *p*-branches does not exist anymore. This simplifies the technology realization drastically and enables even more regularity. However, the polarity selection comes at the cost of extra routing. To mitigate this extra-cost, we describe a regular array of elementary logic blocks, called *Sea-of-Tiles* (SoT). This structure, depicted in Fig. 4, was presented as an optimal layout fabric for ambipolar SiNWFET [15].

As introduced in [15], both unate and binate logic functions fit on the regular tile structure reported in Fig. 6a. This structure consists of 4 transistors grouped into pairs. Each pair shares a common source/drain terminal, as well as its polarity gate connection. The connection is done at the polysilicon level to minimize the wiring complexity. The two pairs are aligned in parallel and share their control gates.

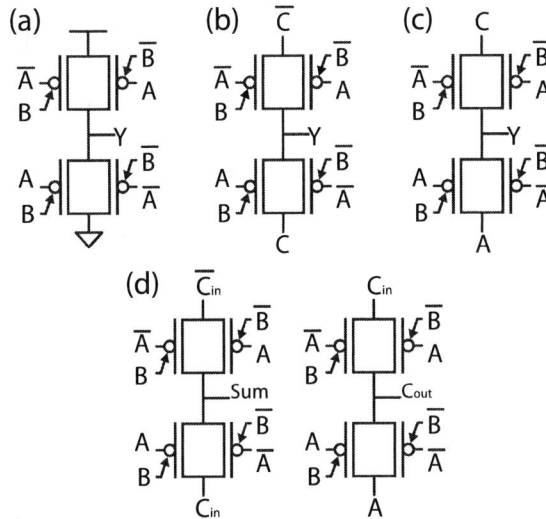

Figure 3. 2-input XOR (a), 3-input XOR (b), 3-input MAJ, 1-bit Full-Adder (d) implementations with DG-SiNWFETs.

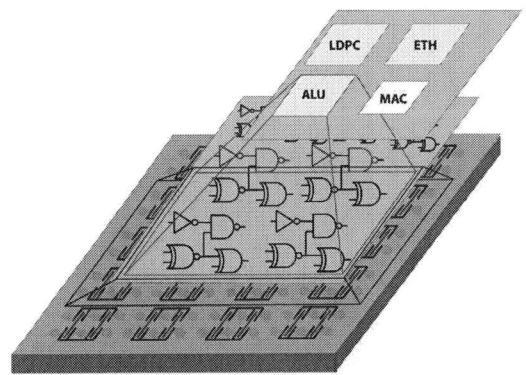

Figure 4. Conceptual representation of a regular *sea-of-tiles*. Tiles are configured to realize logic functions that are part of a complex system [16].

Unate logic functions are obtained by biasing the PGs of the *Pull-Up-Network* (PUN) and *Pull-Down-Network* (PDN) to Gnd and V_{DD} respectively. Hence, all the transistors in the PUN (and PDN) can be grouped together, as shown in Fig. 6b where a 2-input NAND gate example is provided. In the case of binate functions, the polarity gates in the PUN (and PDN) cannot be grouped, leading to extra routing effort. However, the transistor sharing the same polarity gates can be grouped. This is possible, as the

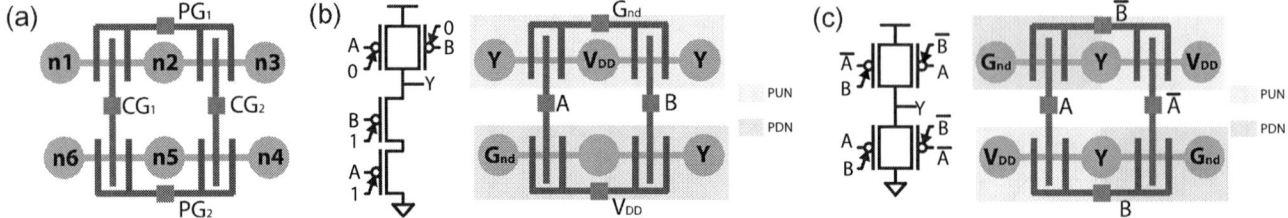

Figure 5. Regular Tile structure sketch from top view (a); Static 2-input NAND gate mapping (b – left: schematic – right: mapped view); Static 2-input XOR gate (c – left: schematic – right: mapped view).

concept of PUN/PDN with segregated *n*- or *p*-type transistors does not hold with the presented technology. An efficient implementation is shown in Fig. 6c with a 2-input XOR gate illustration.

4. CAD Opportunities and Challenges

In this section, we present the novel opportunities and challenges in *Electronic Design Automation* (EDA) targeting logic circuits based on controllable polarity DG-SiNWFETs.

4.1 Logic Synthesis

DG-SiNWFETs intrinsically embed the biconditional (XNOR) logical connective thanks to the on-line re-configuration of the device polarity. Consequently, XOR/XNOR-based logic is remarkably compact and became as efficient as NAND/NOR operations. In order to harness this novel paradigm during automated circuit synthesis, both XOR/XNOR and AND/OR functions must be evidenced and exploited at the same time. A first attempt to achieve effective logic synthesis targeting controllable polarity transistors is MIXSyn [17]. The key feature of MIXSyn is a hybrid logic optimization. A step further in the synthesis of DG-SiNWFETs based circuits is presented in [13] where *Biconditional Binary Decision Diagrams* (BBDDs) are introduced. BBDD is a canonical logic representation form based on the biconditional (XNOR) expansion, which is defined as:

$$f(x,y,..,z) = (x \oplus y)f(y',y,..,z) + (x\overline{\oplus}y)f(y,y,..,z)$$

The one-to-one correspondence between the functionality of DG-SiNWFETs and the core expansion of BBDDs enables an efficient mapping of the devices onto BBDD structures. In this way, the representation compactness of BBDDs is preserved in the final logic circuit.

Recently, it was shown that also majority logic has an efficient implementation with controllable polarity devices [13]. This highlights the interest to extend the capabilities of previous synthesis tools to deal with majority functions, as introduced in [18].

4.2 Physical Synthesis

Despite MIXSyn and BBDDs being efficient means to synthesize logic circuits based on controllable polarity DG-SiNWFETs, they imply challenges at the physical synthesis level. Indeed, both these synthesis flows require on the fly generation of logic cells while currently standard cells are characterized and designed off-line. Regular layout fabrics represent a promising solution to soften this problem. In the context of SoT, direct mapping of logic cells onto a regular array is enabled and therefore supports

natively library-free synthesis flows at a limited utilization-efficiency overhead cost.

5. Conclusion

In this paper, we presented a complete design framework of DG-SiNWFET technology involving process, design and automated tools. In particular, we showed how transistors with controllable polarity are interesting from an arithmetic logic perspective and we proposed a credible route for its very large scale integration, through the Sea-of-Tiles organization. Finally, we provided insights on specific CAD research that is motivated by this technology.

6. Acknowledgments

This work has been partly supported by the grand ERC-2009-AdG-246810.

7. References

[1] C. Auth *et al.*, "A 22nm high performance and low-power CMOS technology featuring fully-depleted tri-gate transistors, self-aligned contacts and high density MIM capacitors," VLSI Tech. Symp., 2012.

[2] S. D. Suk *et al.*, "High performance 5nm radius twin silicon nanowire mosfet (tsnwfet): fabrication on bulk si wafer, characteristics, and reliability," IEDM Tech. Dig., 2005.

[3] S. Bangsaruntip *et al.*, "High performance and highly uniform gate-all-around silicon nanowire MOSFETs with wire size dependent scaling," IEDM Tech. Dig., 2009.

[4] J. Appenzeller, J. Knoch, E. Tutuc, M. Reuter and S. Guha, "Dual-gate silicon nanowire transistors with nickel silicide contacts," IEDM Tech. Dig., 2006.

[5] A. Heinzig, S. Slesazeck, F. Kreupl, T. Mikolajick and W. M. Weber, "Reconfigurable Silicon Nanowire Transistors," Nano Letters, vol. 12, pp. 119-124, 2011.

[6] Y.-M. Lin, J. Appenzeller, J. Knoch and P. Avouris "High-Performance Carbon Nanotube Field-Effect Transistor With Tunable Polarities," IEEE Trans. Nanotechnology, vol. 4, pp. 481-489, 2005.

[7] N. Harada, K. Yagi, S. Sato and N. Yokoyama, "A polarity-controllable graphene inverter," Applied Physics Letters, vol. 96, pp. 12102, 2010.

[8] M.H. Ben Jamaa, K. Mohanram and G. De Micheli, "Novel library of logic gates with ambipolar CNTFETs: Opportunities for multi-level logic synthesis," DATE Tech. Dig., 2009.

[9] I. O'Connor *et al.*, "CNTFET modeling and reconfigurable logic-circuit design," IEEE Trans. on CAS, vol. 54, pp. 2365-2379, 2007.

[10] M. De Marchi *et al.*, "Polarity control in Double-Gate, Gate-All-Around Vertically Stacked Silicon Nanowire FETs," IEDM Tech. Dig., 2012.

[11] J.M. Rabaey, A.P. Chandrakasan and B. Nikolic, "Digital Integrated Circuits: A Design Perspective," Prentice Hall, 2003

[12] A. Zukovski, Y. Xuebei and K. Mohanram, "Universal logic modules based on double-gate carbon nanotube transistors," DAC. Tech. Dig., 2011.

[13] L. Amarù, P.-E. Gaillardon and G. De Micheli, "Biconditional BDD: A Novel Canonical BDD targeting the Synthesis of XOR-rich Circuits," DATE Tech. Dig., 2013.

[14] T. Jhaveri *et al.*, "Maximization of layout printability/manufacturability by extreme layout regularity," J. of Micro/Nanolith. MEMS, vol.6, 2007.

[15] S. Bobba, M. De Marchi, Y. Leblebici an G. De Micheli, "Physical Synthesis onto a Sea-of-Tiles with Double-Gate Silicon Nanowire Transistors," DAC Tech. Dig., 2012.

[16] P.-E. Gaillardon *et al.*, "Vertically Stacked Double Gate Nanowires FETs with Controllable Polarity: From Devices to Regular ASICs," DATE Tech. Dig., 2013.

[17] L. Amarù, P.-E. Gaillardon and G. De Micheli, "MIXSyn: An Area-Efficient Logic Synthesis Methodology for Mixed XOR-AND/OR Dominated Circuits," ASP-DAC Tech. Dig., 2013.

[18] L. Amarù, P.-E. Gaillardon and G. De Micheli, "BDS-MAJ: A BDD Logic Synthesis Tool Exploiting Majority Logic Decomposition," DAC Tech. Dig., 2013.

Sacha: the Stanford Carbon Nanotube Controlled Handshaking Robot

Max Shulaker[1], Jelle Van Rethy[2], Gage Hills[1], Hong-Yu Chen[1], Georges Gielen[2], H.-S. Philip Wong[1], Subhasish Mitra[1]

[1] Stanford University, Stanford, CA, [2] KU Leuven, Heverlee, Belgium

{maxms, ghills, hongyuc, hspwong, subh}@stanford.edu

{Jelle.VanRethy,Georges.Gielen}@esat.kuleuven.be

ABSTRACT

Low-power applications, such as sensing, are becoming increasingly important and demanding in terms of minimizing energy consumption, driving the search for new and innovative interface architectures and technologies. Carbon Nanotube FETs (CNFETs) are excellent candidates for further energy reduction, as CNFET-based digital circuits are projected to potentially achieve an order of magnitude improvement in energy-delay product at highly scaled technology nodes. This paper presents an overview of the first demonstration of a complete sub-system, a sensor interface circuit, implemented entirely using CNFETs. The demonstrated sub-system is an all-digital capacitive sensor to digital converter. The CNFET sensor interface is demonstrated by using the CNFET circuitry to interface with a sensor used to control a handshaking robot.

Categories and Subject Descriptors

B.10.3 [Emerging technologies] : Carbon based electronics

General Terms

Experimentation, design

Keywords

CNT, CNFET, digital systems

1. INTRODUCTION

While advances in silicon-CMOS continue to be made, alternative technologies are currently being explored. Carbon nanotube field effect transistors (*CNFETs*, shown in Figure 1) are a promising emerging technology, due to their expected benefits in performance and energy efficiency [Franklin 12, Wei 09]. For instance, it has been projected that ideal CNFETs can potentially provide an order of magnitude improvement in energy-delay-product compared to competing technologies such as Si-CMOS [Franklin 12]. Experimental demonstrations at the single device level have shown increasingly high-performance CNFETs [Chai 12, Cao 13]. Small-scale stand-alone logic elements have also been demonstrated [Cao 08, Patil 11, Ding 12]. However, due to inherent imperfections associated with carbon nanotubes (*CNTs*), larger and more complex circuits and sub-systems have not been possible until now. To show that such increased level of integration is achievable using CNFETs, this paper presents an overview of the first experimental demonstration of a complete sub-system consisting of a capacitive sensor to digital interface circuit, built entirely out of CNFETs [Shulaker 13a].

Permission to make digital or hard copies of all or part of this work for personal or classroom use is granted without fee provided that copies are not made or distributed for profit or commercial advantage and that copies bear this notice and the full citation on the first page. To copy otherwise, to republish, to post on servers or to redistribute to lists, requires prior specific permission and/or a fee.

DAC '13, May 29 - June 07 2013, Austin, TX, USA.

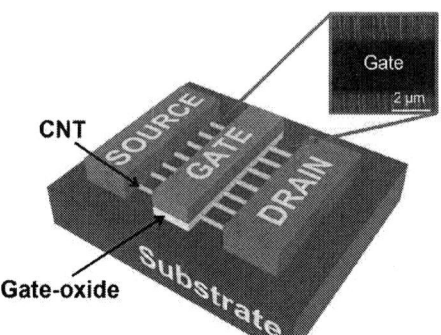

Figure 1: A single CNFET. Scanning electron microscopy (SEM) image shows gate and channel region of a CNFET, with CNTs extending into channel region.

2. OVERCOMING OBSTACLES

While CNFETs hold significant promise, major technological obstacles have prohibited complex circuit demonstrations in the past. Two major obstacles facing CNFETs have been mis-positioned CNTs and metallic CNTs (Figure 2). These inherent imperfections can cause incorrect logic functionality and excess leakage power, rendering large-scale and complex circuits infeasible.

Mis-positioned CNTs   Metallic CNTs

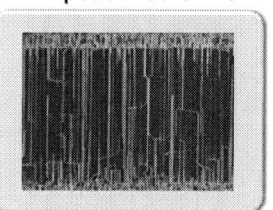 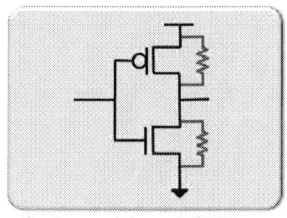

Figure 2. left: SEM image of mis-positioned CNTs. right: schematic of how the presence of metallic CNTs create source-drain shorts within CNFETs.

To overcome these obstacles, we employ the imperfection-immune design paradigm, which addresses both of these fundamental challenges [Zhang 12]. To overcome mis-positioned CNTs, we perform both aligned CNT growth and use mis-positioned CNT immune design [Patil 08a, Patil 08b]. The aligned CNT growth achieves 99.5% alignment of CNTs, while the mis-positioned CNT immune design ensures the circuit layout is immune to the remaining mis-positioned CNTs. To remove any grown metallic CNTs, we use the VLSI Metallic CNT Removal (VMR) technique [Patil 09]. VMR is a chip-scale electrical breakdown technique performed during CNFET circuit fabrication (in conjunction with CNFET-based library cell design). These imperfection-immune design techniques are encapsulated inside standard library cells of digital circuits. Hence, the imperfection-immune design paradigm is VLSI compatible, and follows standard VLSI manufacturing and library-based digital design methodologies. The imperfection-immune design paradigm enables the successful realization of CNFET-based integrated sub-systems, such as the presented sensor interface circuit.

3. SENSOR INTERFACE CIRCUIT
3.1 Architecture
The implemented sub-system is a complete, fully digital capacitive sensor interface, and is inspired by [Danneels 11]. The block diagram of the sensor-to-digital converter is shown in Figure 3. It consists of two main basic blocks: a frequency modulating block and a digital phase-lock-loop (*PLL*). The frequency-modulating block is composed of a sensor-controlled oscillator (*SCO*), which converts a change in the capacitance value of the sensor to a change in the frequency of oscillation. The digital PLL is composed of a single-bit phase detector and a single-bit digitally-controlled oscillator (*DCO*) in the feedback loop, resulting in a first-order PLL. The operation is explained below. The structure consists of digital logic gates only, and has inherent first-order integration of quantization noise.

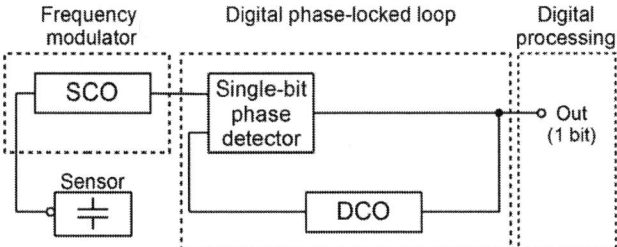

Figure 3: Block diagram of the sensor interface circuit.

3.2 Working Principle
The working principle of the sensor-to-digital circuit is as follows. The SCO first converts the sensor capacitance into the frequency domain, by converting a change in capacitance value into a corresponding change in oscillation frequency. This change in oscillation frequency, which incorporates the sensor information, is then converted to the digital domain by demodulating it with the digital PLL. This is done by comparing the input SCO frequency to the frequency of the DCO using the single-bit phase detector. The resulting single-bit phase difference information is then fed back to the DCO, which responds by either connecting or disconnecting a fixed capacitive load to the DCO. This fixed capacitive load is chosen to be the upper-bound of the sensor value. The feedback signal from the single-bit phase detector is thus used to slow down or to speed up the frequency of the DCO. If the PLL is locked to the SCO, the average frequency of the DCO matches that of the SCO, and thus the two controls of the oscillators are correlated. Since both oscillators are implemented identically, the control of the SCO, the sensor value, is measured by the single-bit control signal of the DCO. Hence the value of the sensor is digitized in this single-bit control signal. In order to increase the resolution of the single-bit output, the digital output can be oversampled, similar to a $\Delta\Sigma$-modulator [Danneels 11].

3.3 Circuit Implementation
The circuit implementation is depicted in Figure 4. It is implemented in a fully digital manner, which leads to robust circuitry compared to an analog counterpart. Both the DCO and SCO are implemented as two (nominally) identical 9-stage inverter-based ring oscillators. The frequency of the SCO is controlled by the sensor C_{sensor}, while the frequency of the DCO is controlled by a fixed capacitance C_{fixed}, whose value is equal to the maximum possible value of the sensor. The capacitances function as extra capacitive load on one stage of the ring oscillator –and thus influence the frequency of the oscillators. The single-bit phase

detector is implemented as a D-latch and its output serves as both the input of the DCO and the output of the entire system.

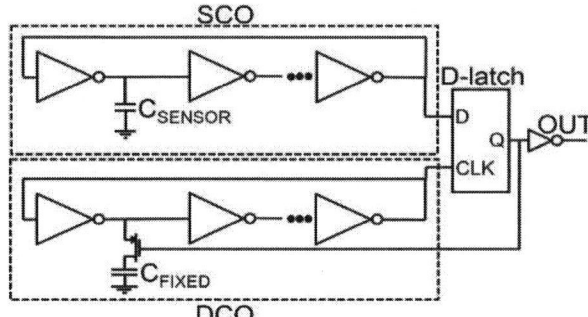

Figure 4: Circuit implementation of the capacitive sensor interface circuit.

4. RESULTS
The CNFET-based sensor interface was fabricated at the Stanford Nanofabrication Facility, following the process described in [Shulaker 11]. The fabricated CNFET circuit is shown in Figure 5. The sensors used during the measurements are external capacitive sensors.

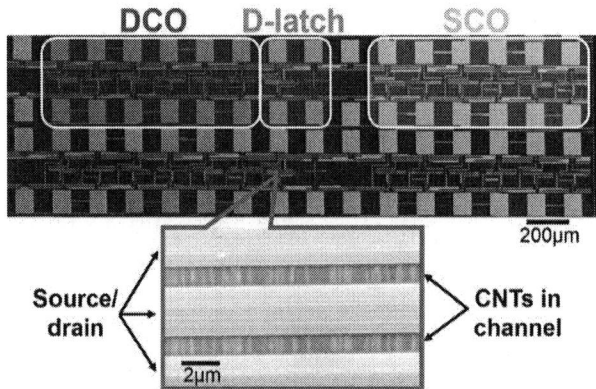

Figure 5: SEM images of the CNFET-based sensor interface circuit. top: two full circuits. bottom: magnified view of stacked CNFETs with CNTs bridging the channel.

4.1 Experimental Measurements
The 9-stage ring oscillators function as the core of the interface circuit, and correct capacitance-to-frequency conversion is essential for the interface circuit to work. Figure 6 shows a repeatable monotonically decreasing frequency versus increasing capacitive sensor values for multiple measured SCO samples.

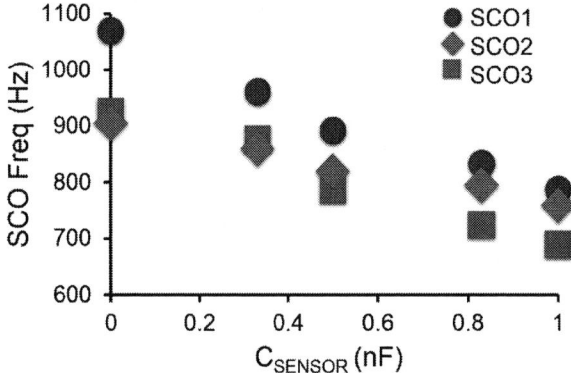

Figure 6: Monotonic relationship between frequency and sensor value for three SCOs.

The entire sensor interface circuit is validated by measuring the digital output for different sensor values. We take our output to be the average duty cycle of the single-bit control signal of the DCO, which is the output of the single-bit phase detector. As shown in Figure 7, with increasing sensor values, the average duty cycle of the output increases. Compared to the expected output based on the Stanford CNFET Spice Model [Deng 07], we find the mismatch between the measured and expected outputs to be ~5.5% (calculated as the root-mean-square error). Thus the implemented circuit functions correctly.

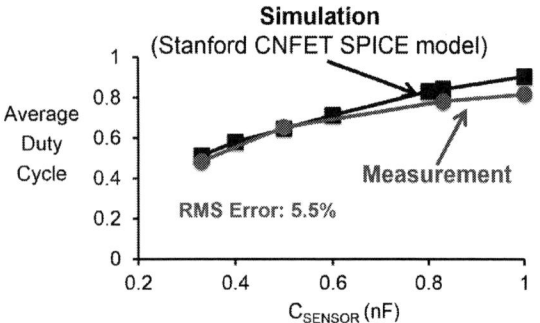

Figure 7: Measured output of the capacitive sensor interface circuit vs. with simulated (expected) output.

4.2 Demonstration

To demonstrate correct operation of the implemented CNFET sub-system, the sensor interface is integrated with a human interface for live demonstrations. Sacha, the Stanford CNT Controlled Handshaker, is an actuated hand, controlled by the CNFET circuit (Figure 8). When a hand holds Sacha, a capacitor is switched into the CNFET circuit at the SCO, acting as a change in sensor value. The CNFET circuit converts this change to a digital equivalent. Post-processing of the single-bit digital signal (over time) produces the average duty cycle. Based on this average duty cycle value, the motor to actuate Sacha's hand is either turned on or off, making Sacha shake hands ([Shulaker 13b] for a demonstration video).

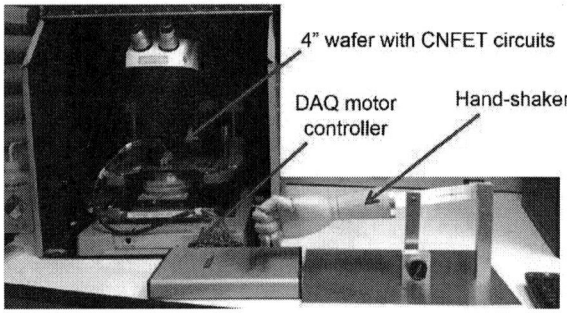

Figure 8: Sacha: the Stanford CNT Controlled Handshaking Robot.

5. ACKNOWLEDGMENT

The authors acknowledge the support of NSF, FCRP C2S2, FCRP FENA, STARNet SONIC, FWO Flanders (for Jelle Van Rethy), and the Stanford Graduate Fellowship and the Hertz Foundation Fellowship (for Max Shulaker).

6. REFERENCES

[Cao 08] Q. Cao and J. A. Rogers, "Random networks and aligned arrays of single-walled carbon nanotubes for electronic device applications," Nano Res., vol. 1, no. 4, pp. 259–272, 2008.

[Cao 13] Q. Cao, S.J. Han, G.S. Tulevski, Y. Zhu, D.D. Lu, W. Haensch, "Arrays of single-walled carbon nanotubes with full surface coverage for high-performance electronics," Nature Nanotechnology, vol. 8, no. 3, pp. 180-186, 2013.

[Chai 12] Y. Chai, A. Hazeghi, K. Takei, H.Y. Chen, P.C. Chan, A. Javey, H.S. Wong, "Low-resistance electrical contact to carbon nanotubes with graphitic interfacial layer," in IEEE Trans. Electron Devices, vol. 59, no. 1, pp. 12-19, 2012.

[Danneels 11] H. Danneels, K. Coddens, G. Gielen, "A fully-digital, 0.3V, 270nW capacitive sensor interface without external references," in IEEE Proc. ESSCIRC, pp. 287-290, Sep. 2011.

[Deng 07] J. Deng, H.S. Wong, "A compact SPICE model for carbon-nanotube field-effect transistors including nonidealities and its applications – part II: full device model and circuit performance benchmarking," in IEEE Trans. Electron Devices, vol. 54, no. 12, pp. 3195-3205, 2007.

[Ding 12] L. Ding, S. Liang, T. Pei, Z. Zhang, S. Wang, W. Zhou, L.M. Peng, "Carbon nanotube based ultra-low voltage integrated circuits: Scaling down to .4V," Applied Physics Letters, vol. 100, no. 26, pp. 263116-263116, 2012.

[Franklin 12] A.D. Franklin, M. Luisier, S.J. Han, G. Tulevski, C.M. Breslin, L. Gignac, M.S. Lundstrom, W. Haensch, "Sub-10nm Carbon nanotube transistor," Nano letters, vol. 12, no. 2, pp. 758-762, 2012.

[Patil 08a] N. Patil, A. Lin, E. R. Myers, H.-S. P. Wong, and S. Mitra, "Integrated wafer-scale growth and transfer of directional carbon nanotubes and misaligned-carbon-nanotube-immune logic structures," in Proc. Symp. VLSI Tech., pp. 205–206, Jun. 2008.

[Patil 08b] N. Patil, J. Deng, A. Lin, H.S. Wong, S. Mitra, "Design methods for misaligned and mis-positioned carbon-nanotube-immune circuits," IEEE Trans. Comput. Aided Des. Integr. Circuits Syst., vol. 27, no. 10, pp. 1725–1736, Oct. 2008.

[Patil 09] N. Patil, A. Lin, J. Zhang, H. Wei, K. Anderson, H.-S. P. Wong, and S. Mitra, "VMR: VLSI-Compatible metallic carbon nanotube removal for imperfection-immune cascaded multi-stage digital logic circuits using carbon nanotube FETs," in Proc. Int. Electron. Devices Meet., pp. 573–576, Dec. 2009.

[Patil 11] N. Patil, A. Lin, J. Zhang, H. Wei, K. Anderson, H.-S. P. Wong, and S. Mitra, "Scalable carbon nanotube computational and storage circuits immune to metallic and mis-positioned carbon nanotubes," IEEE Trans. Nanotechnol., vol. 10, no. 4, pp. 744–750, Jul. 2011.

[Shulaker 11] M. Shulaker, H. Wei, N. Patil, J. Provine, H. Y. Chen, H. S. Wong, and S. Mitra, "Linear increases in carbon nanotube density through multiple transfer technique," Nano letter, vol. 11, no. 5, pp. 1881–1886, May 2011.

[Shulaker 13a] M. Shulaker, J. Van Rethy, G. Hills, H.Y.Chen, G. Gielen, H.S. Wong, S.Mitra, "Experimental Demonstration of a Fully Digital Capacitive Sensor Interface Build Entirely Using Carbon Nanotube FETs," in Proc. Intl. Solid State Circuits Conf., pp. 112-113, 2013.

[Shulaker 13b] M. Shulaker, "Sacha: Stanford Carbon Nanotube Controlled Handshaking Robot," 19 March, 2013. YouTube.

[Wei 09] L. Wei, D. J. Frank, L. Chang, and H.-S. P. Wong, "A non-iterative compact model for carbon nanotube FETs incorporating source exhaustion effects," in Proc. Int. Electron Devices Meet., pp. 917–920, Dec. 2009.

[Zhang 12] J. Zhang, L. Wei, N. Patil, A. Lin, H. Wei, H.-S.P. Wong and S. Mitra, "Robust Digital VLSI using Carbon Nanotubes," Keynote paper, IEEE Trans. CAD, vol. 31, no. 4, pp. 453-471, 2012.

Electrical artificial skin using ultraflexible organic transistor

Tsuyoshi Sekitani
University of Tokyo
JST/ERATO
7-3-1, Hongo, Bunkyo-ku,
Tokyo, JAPAN
+81-3-5841-0413
sekitani@ee.t.u-tokyo.ac.jp

Tomoyuki Yokota
University of Tokyo
JST/ERATO
7-3-1, Hongo, Bunkyo-ku,
Tokyo, JAPAN
+81-3-5841-0413
yokota@ntech.t.u-tokyo.ac.jp

Makoto Takamiya
University of Tokyo
JST/ERATO
4-6-1, Komaba, Meguro-ku,
Tokyo, JAPAN
+81-3-5452-6253
mtaka@iis.u-tokyo.ac.jp

Takayasu Sakurai
University of Tokyo
JST/ERATO
4-6-1, Komaba, Meguro-ku,
Tokyo, JAPAN
+81-3-5452-6253
tsakurai@iis.u-tokyo.ac.jp

Takao Someya
University of Tokyo
JST/ERATO
7-3-1, Hongo, Bunkyo-ku,
Tokyo, JAPAN
+81-3-5841-0413
someya@ee.t.u-tokyo.ac.jp

ABSTRACT

We demonstrate ultrathin, ultraflexible, large-area pressure sensors based on an organic transistor integrated circuit. A 10-μm-thick plastic film with an organic transistor active matrix is developed that can be bent to a bending radius of less than 1 mm to create an electrical artificial skin (E-skin). The thin-film, flexible pressure-sensor matrix is implemented on a curved surface, and the spatial distribution of pressure is successfully obtained in real time.

Categories and Subject Descriptors

B.7 INTEGRATED CIRCUITS

General Terms

Design, Performance

Keywords

Organic transistor, Flexible sensor, Active matrix.

Permission to make digital or hard copies of all or part of this work for personal or classroom use is granted without fee provided that copies are not made or distributed for profit or commercial advantage and that copies bear this notice and the full citation on the first page. To copy otherwise, to republish, to post on servers or to redistribute to lists, requires prior specific permission and/or a fee.

DAC '13, May 29 - June 07 2013, Austin, TX, USA.

1. INTRODUCTION

Electronics for actively controlling the electrons in solids greatly advanced in the 20th century by the use of semiconductors and has become an essential fundamental technology in human lives and society. In the 21st century, further advances in electronics are required from various viewpoints to achieve high compatibility with the global environment and human lives. For example, improving the usability of machines and electronic devices, particularly for the elderly and children, is being focused on in addition to improving the performance of devices in terms of the operation speed and memory capacity. Moreover, it is necessary to improve the performance and lower manufacturing costs while reducing CO_2 emissions.

With this background, it is becoming more important to explore inorganic semiconductors such as silicon as well as new electronic functional materials with characteristics complementary to those of silicon. In particular, we have great expectations for organic molecular materials, which are mainly composed of carbon, and carbon nanomaterials such as graphene and carbon nanotubes for use as functional materials because abundant elements can be used for their synthesis. Furthermore, these materials can be processed by printing technologies with little environmental impact. Supported by intensive research and development in nanotechnology-related fields, organic electronics have markedly advanced in recent years. For example, organic semiconductors have been practically applied as organic photoconductors (OPCs) and even organic electroluminescent light-emitting diodes (ELLEDs). In research, organic solar cells have achieved an energy conversion efficiency of 8%, and several companies have already started distributing their solar cell samples. With the aim of commercializing this technology, research and development on organic solar cells, from their materials to their systems, is being carried out to increase their reliability, stability, and energy conversion efficiency. However, these applications are merely a small part of the vast potential of organic semiconductors. Thin-film transistors (TFTs) with organic semiconductor channel layers have been developed, and their performance has improved, increasing the expectations of new applications that exploit the flexibility of organic materials.

As an application of organic materials to transistors, research on molecular electronics with the aim of realizing ultrahigh density memories using single molecules had been intensively carried out. Since the successful fabrication of TFTs using various molecular materials, the research trend has shifted from improving the performance of silicon-based transistors, such as integration density and operating speed, to the utilization of features complementary to those of silicon-based transistors.

Now, what are the characteristics complementary to those of silicon-based transistors? In the case of organic transistors, the complementary characteristics can be summarized as follows (these characteristics are common to most molecular electronic materials except organic semiconductors): (1) lightweight, high bendability, high shock resistance, and the ability to be grown on large-area substrates (i.e., ease of fabrication on large substrates) owing to low-temperature fabrication processes on plastic substrates, and (2) low environmental impact and high-throughput production owing to the use of printing processes. New technologies based on characteristics (1) and (2) are referred to as flexible and printed electronics, respectively, and have led to increased expectations for realizing new applied fields of electronics.

What applications can be realized using the above characteristics complementary to those of silicon? How can we create more user-friendly machines by exploiting the flexibility of organic materials? How can electronics compatible with the global environment be realized using molecular electronic materials? Currently, competitive worldwide research activities with the aim of answering these questions are ongoing.

In this paper, we describe the development of new electronics focusing on the successive advances in TFTs based on molecular electronic materials. A large-area, flexible pressure-sensor matrix is fabricated on a plastic film that consists of integrated rubbery pressure sensors and organic field-effect transistors that form an active matrix for obtaining pressure images [1]. The effective sensor area is 80×80 mm^2, and the periodicity is 5.08 mm (Fig. 1). An organic semiconductor and a polymer are used as a channel layer and a gate dielectric layer for the organic transistors, respectively, and the mobility reaches as high as 2.0 cm^2/Vs. The mechanical flexibility of the transistors is shown with a bending radius of less than 1 mm (Fig. 2). The thin-film, flexible pressure sensor matrix is implemented on arbitrary surfaces, and the spatial distribution of pressure is successfully obtained in real time. The present large-area pressure sensor demonstrates the feasibility of applying the organic transistor technology to flexible area sensors. This provides new applications for organic transistors including artificial skins.

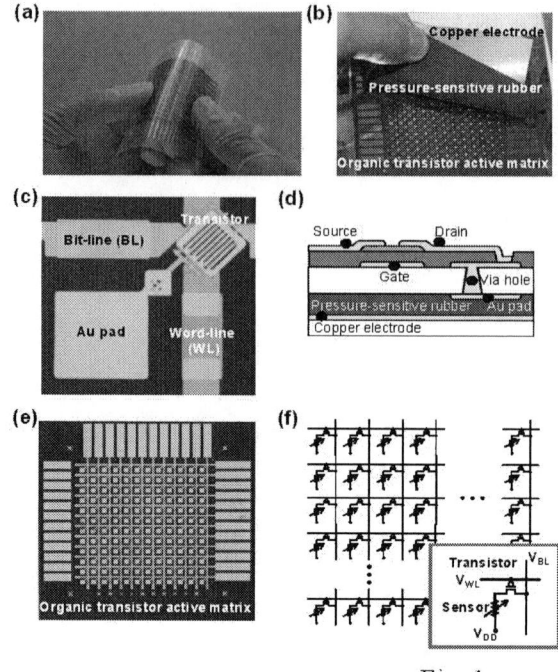

Fig. 1

Figure 1: (a) Flexible pressure sensor using an organic transistor active matrix. (b) Picture of the pressure sensor comprising an organic transistor active matrix, a pressure-sensitive rubber, and a film with a copper electrode. (c) Magnified picture of one organic transistor. The gate and source electrodes are interconnected using word lines (WL) and bit lines (BL), respectively. (d) Cross-sectional illustration of one sensor cell of the pressure sensor sheet. (e) Picture of the entire organic transistor active matrix. (f) Schematic circuit diagram of an organic transistor active matrix. The inset shows one sensor cell of the pressure sensor.

2. DEVICE FABRICATION & CHARACTERIZATION

The base film (substrate) is a flexible polyimide or polyethylene naphthalate (PEN) film with a thickness of 10 μm. First, a 50-nm-thick Au film is evaporated through a metal shadow mask to form the gate electrodes and word lines for the interconnected gate electrodes. The surface is coated with a 400-nm-thick parylene gate dielectric layer. Next, the 50-nm-thick organic semiconductor, dinaphtho[2,3-b:2',3'-f]thieno[3,2-b]thiophene (DNTT), is evaporated as a p-type channel [2] and a 50-nm-thick Au film is evaporated as the source and drain electrodes of the organic transistors. The channel length L and width W are 50 μm and 2.0 mm, respectively. The entire surface of the organic-transistor active matrix is coated with 10-μm-thick parylene for encapsulation. A CO_2 process laser is used to create via holes through the gate dielectric layer. A 200-nm-thick Au film is again evaporated to form the top electrodes that can access each electrode of the organic transistors. In the last step, a pressure-sensitive rubbery sheet and a copper electrode suspended by a PEN film are laminated on the top side of the organic transistor active matrix, resulting in the integration of pressure sensors with transistors. The resistance of the rubbery sheet changes when pressure is applied to the bottom side of the device. Because the rubbery sheet and copper electrodes do not require patterning, the last lamination process is alignment-free and could be potentially very low in cost.

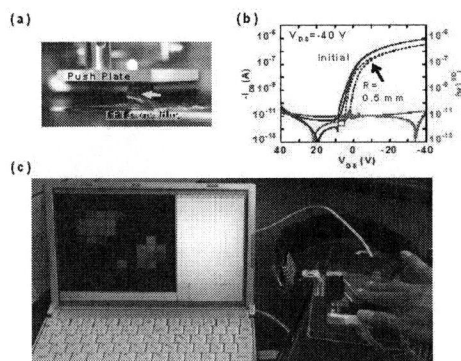

Figure 2: Electrical characteristics of an organic transistor-based pressure sensor. (a) Picture of bending apparatus for flexible transistors. (b) Transistor characteristics of an organic transistor before and during bent to 0.5 mm in bending radius. (c) Spatial distribution of the pressure measured using the flexible pressure-sensor active matrix.

The DC characteristics of the organic transistors are measured in the ambient environment prior to the application of the rubber pressure sensor film using a semiconductor parameter analyzer. The manufactured DNTT transistors exhibit p-type conduction, and the typical DC characteristics are shown in Figs. 2. The field-effect mobility reaches a value as high as 2 cm^2/Vs in the saturation regime, and the on/off ratio is approximately 10^6 at 40 V. The transistors function even bent to 0.5 mm in bending radius, where mobility was decreased to 1 cm^2/Vs (Fig. 2b). The mobility of the present device is far superior to that of amorphous silicon (typically 1.0 cm^2/Vs) and is the highest ever reported for flexible organic transistors prepared on plastic films with polymeric gate insulators, to the best of our knowledge. However, an operating bias of 40 V is not realistic for artificial skin applications; however, the device is still functional with a relatively large

mobility (0.1 cm^2/Vs) when an 8-V operating bias is applied. The initial transistor yields exceed 99%, and device failure mainly occurs owing to the leakage current through the gate dielectric layers. The measured drain current variation over the entire processed area is less than 10%, which is very important for obtaining a highly efficient sensor matrix.

The complete devices including the rubber pressure sensors are characterized by applying a uniform pressure using several objects. The circuit diagram of a sensor cell is shown in the inset of Fig. 1f. The configuration is similar to a memory cell and pixel of a charge-coupled device (CCD), where the gate electrodes of each line are connected to a word line (V_{WL}), and the drain electrodes of each line are connected to a bit line (V_{BL}). As the pressure is applied to the bottom side of the device and varies from 0 to 10 kPa ($\sim$100 gf/cm^2), the resistance of the rubbery sheet varies from 10 MΩ to 1 kΩ. In addition, the transconductance and the measured current increase over this range of applied pressures. A spatial map for pressure is obtained by individually applying V_{DS} = -40 for each bit line and V_{GS} = -40 for each word line to each sensor cell using the pressure from two fingers. The spatial distribution of the pressure is clearly observed in real time.

3. ACKNOWLEDGMENTS

This study was partially supported by a Grant-in-Aid for Scientific Research (KAKENHI WAKATE S & WAKATE A) and the Special Coordination Funds for Promoting Science and Technology. We also thank Daisankasei Co., Ltd. for providing us with high-purity parylene (diX-SR).

REFERENCES

[1] Someya, T., Sekitani, T., Iba, S., Kato, Y., Kawaguchi, H., and Sakurai, T. "A large-area, flexible pressure sensor matrix with organic field-effect transistors for artificial skin applications", *Proceedings of the National Academy of Sciences of the United States of America*, 101 (2004) 9966-9970.

[2] Someya, T., Kato, Y., Sekitani, T., Iba, S., Noguchi, Y., Murase, Y., Kawaguchi,H., and Sakurai, T. "Conformable, flexible, large-area networks of pressure and thermal sensors with organic transistor active matrixes", *Proceedings of the National Academy of Sciences of the United States of America*, 102 (2005) 12321-12325.

[3] Yamamo, T. and Takimiya, K. "Facile Synthesis of Highly π-Extended Heteroarenes, Dinaphtho[2,3-b:2',3'-f]chalcogenopheno[3,2-b]chalcogenophenes, and Their Application to Field-Effect Transistors", *J.Am.Chem.Soc.*129 (2007) 2224-2225.

Relays Do Not Leak – CMOS Does

Hossein Fariborzi[1], Fred Chen[2], Rhesa Nathanael[3], I-Ru Chen[3], Louis Hutin[3], Rinus Lee[4], Tsu-Jae King Liu[3], Vladimir Stojanovic[1]

[1]Massachusetts Institute of Technology, [2]On-Chip Power Corp., [3] University of California, Berkeley, [4]SEMATECH

ABSTRACT

This paper describes the micro-architectural and circuit design techniques for building complex VLSI circuits with micro-electromechanical (MEM) relays and presents experimental results to demonstrate the viability of this technology. By tailoring the circuits and micro-architecture to the relay device characteristics, the performance of the relay-based multiplier is improved by an order of magnitude over any known static CMOS style implementation, and by ~4x over CMOS pass-gate equivalent implementations. A 16-bit relay multiplier is shown to offer ~10x lower energy per operation at sub-10 MOPS throughputs when compared to an optimized CMOS multiplier at an equivalent 90 nm technology node. The functionality of the primary multiplier building block, a full (7:3) compressor built with 46 scaled MEM-relays, which is the largest working MEM-relay circuit reported to date, is also demonstrated.

Keywords

Ideal Switches, MEM-relays, Energy-Aware VLSI Design, Multipliers

1. INTRODUCTION

In the past four decades, CMOS technology scaling has resulted in drastic improvements in energy efficiency, cost per function and performance of integrated circuits. In recent years, the increasing proportion of sub-threshold leakage current, attributed to its dependence on the non-scalable thermal voltage (k_BT/q), has slowed the scaling of the threshold voltage (V_T) and consequently the scaling of the supply voltage (V_{DD}). The resulting increase of power density has overshadowed the performance benefits of transistor scaling, forcing a move toward parallelism. However, the significance of leakage energy in CMOS functional units will eventually make this approach ineffective, as each CMOS block has a well-defined minimum energy point [1] at which an incremental decrease in leakage current is exactly offset by a corresponding loss of performance. Parallelism is unable to reduce energy per operation beyond this point.

Recently, MEM-relays with negligible leakage and abrupt switching behavior have been proposed as a solution to overcome the minimum energy limitation of CMOS circuits [2]. We have previously demonstrated the circuit design principles and the feasibility of the MEM-relay technology for several key circuit components of integrated VLSI systems, such as basic logic, flops, memory and I/O circuits [3]-[5].

Although the mechanical movement makes relays slower than CMOS transistors, we have developed a MEM-relay catered circuit design methodology to narrow the performance gap at the system level by implementing the functional units as large complex logic gates instead of staged logic, and hence minimizing the mechanical delay on the critical path, as demonstrated on relay-based Manchester-carry adder chains [6]. In this work, we extend and adapt these principles to multiplier structures which are commonly known as the most complex arithmetic blocks. We develop the multiplier micro-architecture and circuit techniques tailored to the relay device properties. These new micro-architectures are optimized around larger compressor circuits to minimize the mechanical delay. The larger pass-gate style compressor circuits are also optimized to provide single mechanical delay operation and minimize the number of devices.

In order to show the potential energy efficiency gains over CMOS implementations, an optimized multiplier simulated using a projected 90-nm MEM-relay model is benchmarked and shown to have an order of magnitude improved energy-delay tradeoff over a wide range of frequencies, as compared with an optimized CMOS multiplier in the same technology node. In the last section, some experimental roadblocks and methods to overcome them are described, and the functionality of the optimized (7:3) compressor built using a new generation of scaled MEM-relays is demonstrated.

2. MEM-RELAY STRUCTURE AND OPERATION

Figure 1 shows isometric and cross-sectional views of the original 4-terminal "crab leg" relay design [2]. The general composition and function of the relay is similar to a MOSFET in that each has four terminals. The relay consists of a poly-SiGe gate structure suspended by spring-like folded flexures above the tungsten body, drain, and source electrodes. The channel is attached beneath the gate's center plate via an insulating aluminum oxide (Al_2O_3) layer which acts as a gate dielectric. The basic operational states of the MEM-relay are also illustrated in Figure 1. The device is actuated by applying a voltage (V_{gb}) between the movable gate electrode

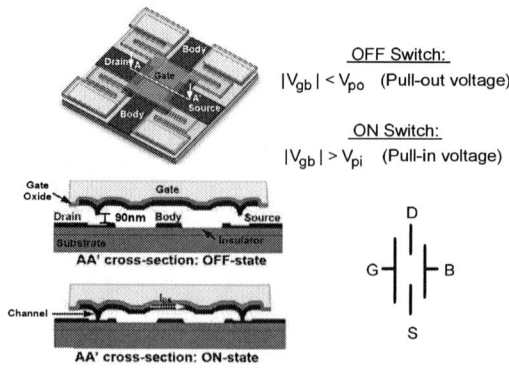

Figure 1. Diagram, circuit symbol and operating states of a 4-terminal MEM-relay

---

Permission to make digital or hard copies of all or part of this work for personal or classroom use is granted without fee provided that copies are not made or distributed for profit or commercial advantage and that copies bear this notice and the full citation on the first page. To copy otherwise, or republish, to post on servers or to redistribute to lists, requires prior specific permission and/or a fee.
DAC '13, May 29 - June 07 2013, Austin, TX, USA.

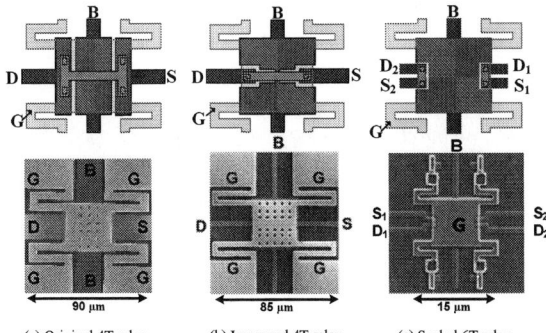

Figure 2. MEM-relay evolution: layout and SEM images [4], [6], [11]

(a) Original 4T relay   (b) Improved 4T relay   (c) Scaled 6T relay

and the body electrode beneath it. If the applied voltage difference is sufficiently large (greater than the "pull-in voltage" V_{PI}), the electrostatic force overcomes the mechanical spring restoring force of the folded flexures so that the gate stack is "pulled-in" such that the channel comes into contact with the source and drain to allow the flow of current between these electrodes. If the gate-to-body voltage falls below a certain release voltage, the spring restoring force overcomes the electrostatic force plus contact adhesive forces to pull the channel out of contact with the drain and source, severing the current path between these electrodes. Based on these basic operation principles, the MEM-relay is expected to have very steep subthreshold slope and virtually zero leakage in the off-state, with minimum energy theoretically set by the contact adhesive forces.

6Since its introduction, the MEM-relay has gone through various revisions to improve functionality, performance and reliability. Figure 2 summarizes the device design evolution in recent years. The main improvements made to the original 4T relay (Figure 2(a)) was the reduction of the gate-drain, gate-source and channel-body overlap areas to eliminate the corresponding parasitic actuation and undesired electrostatic attractive forces. A more recent generation of MEM-relays, shown in Figure 2(c), is 30 times smaller than the second generation 4T, and has 6 terminals: two independent source-drain pairs and a common gate-body pair. This enhancement approximately halves the number of relays required to implement most logic functions. The 6T MEM-relay enables considerable reduction in area cost and gate switching energy, compared to the 4T MEM-relay.

3. MEM-RELAY CIRCUIT DESIGN

In relay-based digital logic circuits all mechanical movement should happen simultaneously, if possible, to minimize the impact of the relatively slow mechanical delay [3], thus favoring a tailored pass-gate design. Figures 3(a),(b) illustrate the difference between CMOS and MEM-relay logic design styles. A simple substitution of CMOS transistors with relays in a standard CMOS 4-input AND logic circuit would result in 4 mechanical delays as each signal hitting a gate triggers an additional mechanical delay. The optimized relay design shown in Figure 3(b) incurs only one mechanical delay since all mechanical movements happen at the same time. Thus, given a logic function, the design strategy is to stack as many MEM-relays in series as possible. The upper bound on the number of MEM-relay devices in a stack is reached when the electrical and mechanical delays are equal [7]. Figure 3(c) compares the electrical delay of a device stack with the mechanical delay, for our current 4T, scaled 6T and predictive 90nm MEM-relays. The mechanical delay is obtained for a reasonable range of V_{GB} overdrives and shows that this design

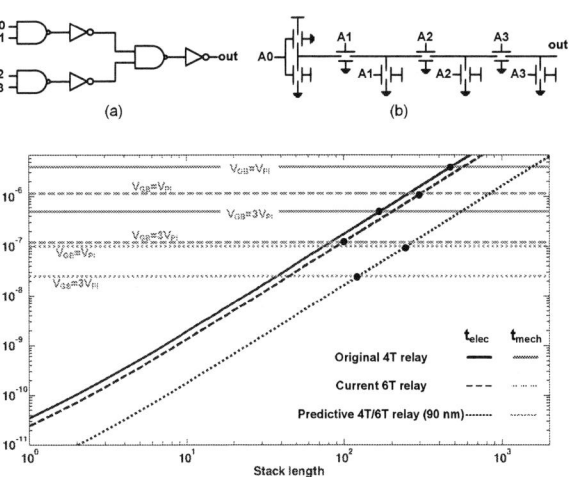

Figure 3. A 4-input AND built with CMOS (a) and MEM-relays (b). Mechanical vs. electrical delay for 4T, scaled 6T and predictive 90nm relays (c)

approach is extendable to hundreds of series-connected devices and consequently encompasses most practical logic functions.

4. MEM-RELAY MULTIPLIER DESIGN

The first stage of multiplication is to generate the partial products matrix. We have implemented this with two different approaches: a simple AND network, and radix-4 Booth encoding [7]. The main focus of the design and optimization is on the second stage which is partial product compression. The most straightforward method would be the "logarithmic reduction," which puts the partial products in groups of two and adds them up until the final result is found. This method is not ideal, as it creates $1+logN$ mechanical delays. Although we would like the partial products reduction function to incur one mechanical delay, the most advanced commercial tools or even the relay-based synthesis tool we have developed cannot achieve this goal without using hundreds of thousands of MEM-relays. In order to avoid such unrealistic designs, the logic needs to be deliberately partitioned, meaning that extra mechanical delay should be inserted into the system to enable simplification of the logic. Figure 4 shows the microarchitectures of an example 6-bit multiplier implemented with two optimized partial products reduction techniques explored

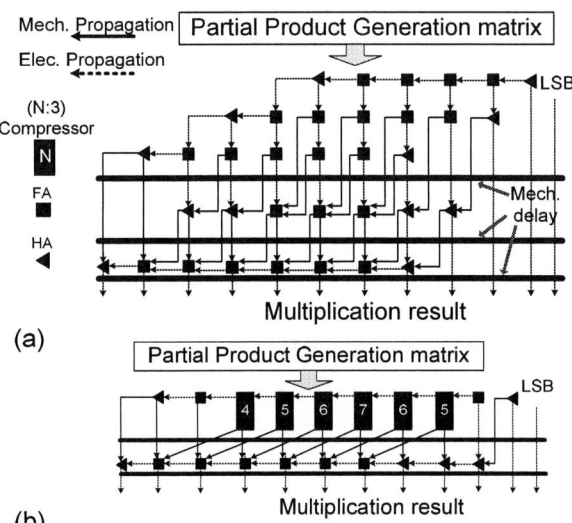

Figure 4. The sample 6-bit multiplier built with (a) half- and full-adders, and (b) larger compressors

in [7]. While the first approach, illustrated in Figure 4(a), uses small blocks such as half- and full-adders to achieve simplicity and lower relay counts, the second approach (Figure 4(b)) utilizes larger compressor blocks to decrease the total number of reduction steps and hence reduces the total mechanical delay by 33% by using 30% more relays.

In [7], the implementation of MEM-relay multiplier components using the 4T relay is discussed. The updated schematics for the components built with scaled 6T relays are shown in Figures 5 and 6(a-c). Since the 6T device has 2 source/drain pairs, implementation of multipliers with 6T MEM-relays enables ~40% reduction in the total relay count. The key design consideration for all components (adders, compressors, *etc.*) is that an electrical path from at least one input and the output needs to be existent, solely through source/drain terminals of the stacked relays. This enables stacking of components with no mechanical delay penalty. A direct MEM-relay translation of a standard CMOS (7:3) compressor (Figure 6(d)) would result in accumulation of 19 mechanical delays in the critical path of a 32-bit multiplier, while the optimized compressor design (Figure 6(a-c) incurs only 5 mechanical delays.

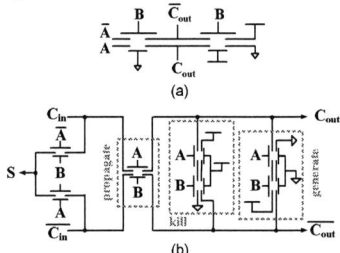

(a)

(b)

Figure 5. (a) half-adder and (b) full-adder built with 6T relay

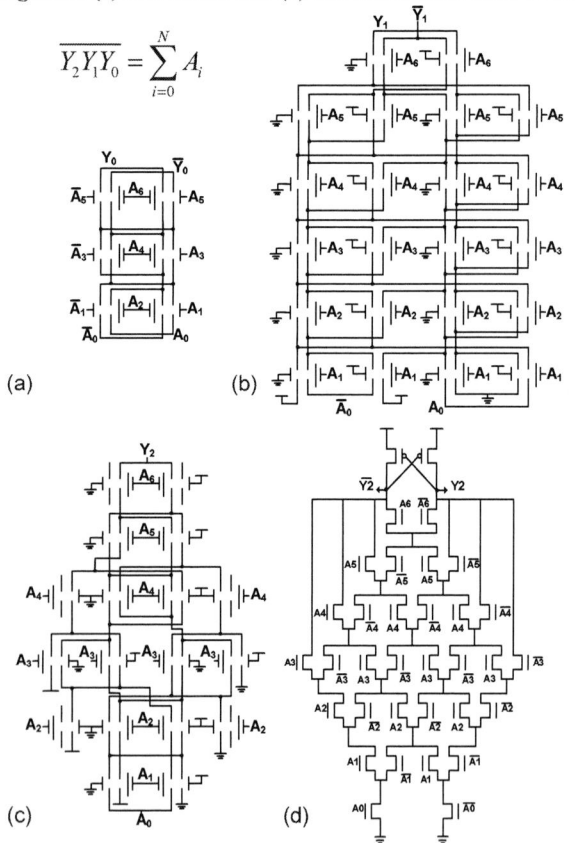

Figure 6. (a-c) the (7:3) compressor built with the 6T MEM-relay, (d) the MSB of the CMOS compressor [7]

5. THE (7:3) COMPRESSOR TEST RESULTS AND EXPERIMENTAL ISSUES

A practical reliability challenge, since the first generation of MEM-relays, has been the growth of native oxide on the tungsten contacting electrode surfaces of the relays which leads to increased R_{ON} and, eventually, relay stuck-open failure after ~200 million cycles of operation [8]. Although ongoing efforts toward contact material engineering and hermetic packaging of relay chips will eventually remedy this problem, our experimental demonstrations performed in open ambient were challenged by this contact oxidation issue. Before any current can flow through the relay, the oxide layer needs to be "broken". This process requires a sufficiently high source-drain voltage while the relay is actuated and since breaking the oxide for more than 3 relays in series requires high voltages, it becomes complicated for larger circuits like the (7:3) compressor. We addressed this problem by developing an autonomous oxide-breaking procedure which searches for shorter stacks in the circuit, activates them (*i.e.* breaks the oxide on all relays on the path), and then uses the shorter active paths to apply the stimuli to the inactive relays/paths and eventually activates the entire circuit.

The die photo and the operation of the multiplier structures of Figure 6 are shown in Figure 7. The small size of the 6T MEM-relay, availability of two metal routing layers, and the fact that the 6T compressor needs only 46 MEM-relays, compared to the 98 relays for the 4T compressor [7], lead to 40x smaller circuit area. Following the oxide breaking procedure, a full-set of random *A0-6* input vectors, ranging from *0* to *127*, is applied to the compressor, actuating the relay gates and activating different paths in the sub-circuits. The required V_{gb} is 9V. The measured output code perfectly matches the expected values, demonstrating the correct functionality. This circuit is the largest functional scaled MEM-relay circuit reported to date.

6. ENERGY/DELAY ESTIMATES OF MEM-RELAY VS. CMOS MULTIPLIERS

In order to show the potential energy benefits of MEM-relays for the future microcomputers, a 16-bit multiplier built with 90-nm

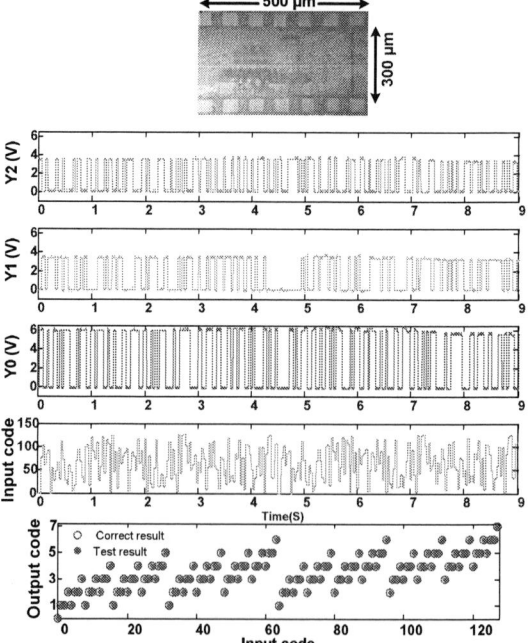

Figure 7. Die photo and experimental results of the 6T (7:3) compressor

880

equivalent MEM-relays is simulated and compared to an optimized 16-bit CMOS multiplier in the same process node. The CMOS multiplier employs optimally tiled compressor tree architecture (OTCT) with radix-4 Booth encoding and an arrival-profile aware completion adder [9]. The total area of this multiplier is 0.03mm^2. The energy/delay curves have been plotted for various reported operation voltages and frequencies.

The first MEM-relay-based multiplier is designed with 5610, 4T relays in a predictive 90-nm process where each relay occupies 11 μm^2. The total area of this multiplier is 0.087 mm^2. The second multiplier is designed with 3211, 6T relays in a predictive 90-nm process, resulting in a total area of 0.046 mm^2. The energy/delay curves of both implementations are shown for the operating voltage in the range of 2 V_{PI} to 5 V_{PI}. The simulations are based on the parameters of the MEM-relay model described in [6], and mechanical delay derivation in [10].

The energy-delay trade-offs of both CMOS and MEM-relay 16-bit multipliers are shown in Figure 8. The CMOS multiplier reaches its minimum energy point for delays greater than 50 ns. As a result, the scaled MEM-relay multipliers on average offer ~10x better energy-efficiency over CMOS multipliers for sub-10 MOPS operation. As predicted, the 6T implementation is able to achieve even better energy/op figures due to lower switching energy and smaller overall area. As illustrated in Figure 8, a further trade-off of increasing area for 16-parallel multiplications enables operation in the GOPS region, while preserving the energy efficiency. Another interesting observation is that even with the current 6T relay, the sample 16-bit multiplier is able to achieve ~1.5-3x lower energy than CMOS in 0.1-1MOPS range, showing the potential of current relay as CMOS alternative in applications where energy savings has priority over area and performance.

7. CONCLUSION

This paper has described how micro-architecture and circuit design can be tailored to MEM-relay device properties for implementation of complex arithmetic units. Design analysis of MEM-relay multipliers shows the performance benefits of higher ratio compressors, while suggesting the use of simple half- and full-adders when area is constrained. Further enhancement of relay-based synthesis methods and CAD tools will enable optimized partitioning of more complex MEM-relay circuits and systems, such as microprocessors.

The operation of the main building block of the multiplier, the (7:3) compressor which is the largest scaled MEM-relay circuit reported to date, is experimentally demonstrated. Simulation results of 16-bit relay multipliers built in predictive 90-nm relay processes predicts 5-20x improvement in energy-efficiency over

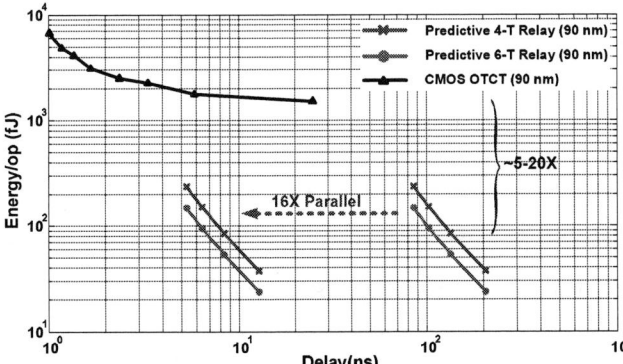

Figure 8. Energy/throughput comparison of CMOS and predictive 90-nm (4T and 6T) MEM-relay 16-bit multipliers

CMOS designs in the sub-10 MOPS performance range and suggests that parallelism can be employed to extend these benefits to GOPS operation region. The relative performance of the multiplier enhancements confirms that the energy-efficiency gains can be extended to larger arithmetic blocks, suggesting that complete VLSI systems such as microprocessors can expect to see similar energy/performance improvements from adopting MEM-relay technology.

8. ACKNOWLEDGMENTS

We gratefully acknowledge the contributions of the whole DARPA NEMS team at UC Berkeley, UC Los Angeles and MIT, especially Matthew Spencer, Elad Alon, Hei Kam, Vincent Pott, Jaeseok Jeon, Cheng Wang, Kevin Dwan and Dejan Marković. This work was supported by the DARPA NEMS program and in part by the C2S2 and MSD FCRP, MIT CICS, and NSF Infrastructure Grant No. 0403427.

9. REFERENCES

[1] Calhoun, B.H., Wang, A., and Chandrakasan, A. "Modeling and sizing for minimum energy operation in subthreshold circuits," *IEEE J. Solid-State Circuits*, vol. 40, no. 9, pp. 1778–1786, Sep. 2005.

[2] Nathanael, R. Pott, V., Kam, H., Jeon, J., and Liu, T. J-. K. "4-terminal relay technology for complementary logic," in *Proc. IEEE Int. Electron Device Meeting Tech. Dig.*, Dec. 2009, pp. 223–226.

[3] Chen, F., Kam, H., Markovic, D., Liu, T.-J. K., Stojanovic, V. and Alon, E. "Integrated circuit design with NEM relays," in *Proc. IEEE/ACM Int. Conf. Computer-Aided Design*, Nov. 2008, pp. 750–757.

[4] Chen, F. *et al.* "Demonstration of integrated micro-electro-mechanical switch circuits for VLSI applications," in *Proc. IEEE Int. Solid-State Circuits Conf. Tech. Dig.*, Feb. 2010, pp. 150–151.

[5] Fariborzi, H. *et al.* "Analysis and Demonstration of MEM-relay Power Gating," *IEEE Custom Integrated Circuits Conf.*, 2010, San Jose, CA.

[6] Spencer, M. *et al.* "Demonstration of Integrated Micro-electro-mechanical Relay Circuits for VLSI Applications," *IEEE J. Solid-State Circuits*, vol. 46, no. 1, pp. 308-320, Jan. 2011.

[7] Fariborzi, H., Chen, F., Nathanael, R., Jeon, J., Liu, T.K. and Stojanović, V. "Design and Demonstration of Micro-electro-mechanical Relay Multipliers," in *Proc. IEEE Asian Solid State Circuit Conf.*, Jeju, S. Korea, 2011, pp. 117-120.

[8] Chen, Y., Nathanael, R., Jeon, J., Yaung, J., Hutin, L., and Liu T.-J.K. "Characterization of Contact Resistance Stability in MEM-relays With Tungsten Electrodes", *IEEE J. MEMS.* vol. 21, no. 3, pp. 511-513, June 2012.

[9] Hsu, S.K. *et al.* "A 110 GOPS/W 16-bit multiplier and reconfigurable PLA loop in 90-nm CMOS," *IEEE J. Solid-State Circuits*, vol. 41, no. 1, pp. 256–264, Jan. 2006.

[10] Kam, H., Liu, T.K., Stojanovic, V., Markovic, D. and Alon, E. "Design, optimization, and scaling of MEM-relays for ultra-low-power digital logic," *IEEE Trans. Electron Devices*, vol. 58, no. 1, pp. 236-250, Jan. 2010.

[11] Chen, I.R. *et al.* "Scaled Micro-Relay Structure with Low Strain Gradient for Reduced Operating Voltage," *ECS Trans.*, vol 45, no. 6, pp 101-106, 2012.

Single-Photon Image Sensors

Edoardo Charbon
TU Delft
2628 CD Delft
Netherlands
+31 15 278 3667
e.charbon@tudelft.nl

Francesco Regazzoni
TU Delft
2628 CD Delft
Netherlands
+31 15 278 3663
f.regazzoni@tudelft.nl

ABSTRACT[1]

The main goal of this paper is to expose the EDA community to the emerging class of circuits operating with single quanta of energy (e.g. photons or electrical carriers). We describe recent developments in the field of single-photon detection and single-photon imaging based on the avalanche effect. Single-photon detection is useful in a number of applications, from time-of-flight based 3D vision systems to fluorescence lifetime imaging microscopy, from low-light cameras to quantum random number generators, from positron emission tomography to time-resolved Raman spectroscopy. These applications have speed and accuracy requirements that conventional systems cannot provide if not at a very high cost. EDA has not yet adapted to the revolution introduced by avalanching devices and, though tools capable of simulating these devices exist, there is little or no capability to do so in a coherent flow, let alone at system level. We challenge CAD designers to fill this gap and prepare them to the circuits of the future, quantum in nature but built in standard CMOS technology.

Keywords

Single-photon detection, photon counting, single-photon avalanche diode, SPAD, silicon photomultiplier, SiPM, time-resolved imaging, EDA

1. INTRODUCTION

Photon counting is a useful tool in many scientific and biomedical imaging sensors; it usually requires single-photon detection capability, as well as functionality such as counting and time-of-arrival evaluations. When photon counting is available in large arrays of independently operating pixels, it may enable emerging imaging modalities. Examples include fluorescence lifetime imaging microscopy (FLIM), fluorescence correlation spectroscopy (FCS), time-resolved Raman spectroscopy, time-of-flight cameras, etc. Photon-counting imagers can be useful in nuclear medicine, in particular in positron emission tomography (PET), single-photon emission computed tomography (SPECT), and in Gamma Cameras for 3D visualization of radionucleotides in living tissue. Photon-counting detection can also be used in embedded security techniques, where the quantum nature of light is used in the generation of true random numbers for encryption and other information hiding purposes.

Devices for photon counting have been in existence for some time; they have been introduced in non solid-state form, i.e. photomultiplier tubes (PMTs) and microchannel or multichannel plates (MCPs) already in the 1930s. These devices share a few

properties: a large active area (cm²), a high fill factor, and a single channel that generally requires amplification; they are also bulky and require vacuum for normal operation. Recently, compact photon counting devices have emerged, known as silicon photomultipliers (SiPMs), that operate in normal atmosphere and at room temperature. SiPMs have a single output, however their multi-channel counterparts are based on arrays of single-photon avalanche diodes (SPADs) that provide single-photon detection in tens or hundreds of thousands of locations independently. In addition to a (x,y)-position information of a photon hit, SPADs can also provide the time-of-arrival of a photon at picosecond resolutions millions of times-per-second. SPADs and SPAD arrays may be fabricated in dedicated silicon processes or in standard CMOS, thus enabling the implementation of megapixel cameras operating in single-photon regime.

Cova and McIntyre were among the first to advocate SPADs as an effective technology for fast timing applications in the 1980s [1,2]. High-resolution, time-resolved photon detection, as time-correlated single-photon counting (TCSPC) are natural applications for SPADs. Since the demonstration of CMOS SPADs at the beginning of the millennium, it has become possible to create true imagers [3].

This paper reports on innovations that have followed since 2005 in the domain of CMOS SPAD imagers with the intention of exposing the EDA community to the challenges of the design of this class of circuits, while emphasizing the limits of the current CAD tools to design such circuits.

2. SINGLE-PHOTON AVALANCHE DIODE (SPAD)

A SPAD is essentially a p-n junction that relies on impact ionization to create a large number of photon-generated electrons and holes from a single electron-hole pair. Fig. 1 shows the *steady-state* I-V characteristics of a typical p-n diode.

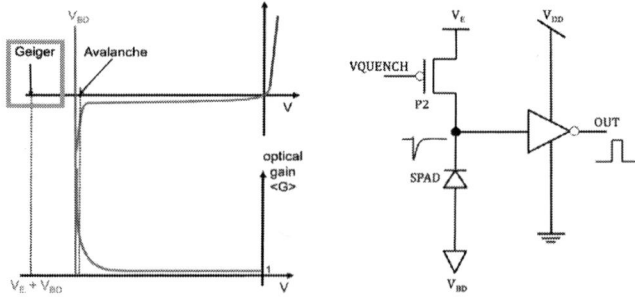

Fig. 1. Steady-state I-V characteristics for a p-n junction with Geiger and avalanche mode of operation (left). Passively quenched SPAD (right). V_E is known as the excess bias voltage at which a diode must be biased in Geiger mode. A comparator or inverter is used to shape the output pulse of the detector.

---

[1] Permission to make digital or hard copies of all or part of this work for personal or classroom use is granted without fee provided that copies are not made or distributed for profit or commercial advantage and that copies bear this notice and the full citation on the first page. To copy otherwise, or republish, to post on servers or to redistribute to lists, requires prior specific permission and/or a fee.

What this curve does not show is the pseudo-steady-state behavior in the breakdown operating condition. A voltage above the breakdown voltage can be applied so long as no carriers exist in the diode's depletion region. As soon as a carrier is injected into the depletion region, impact ionization may cause an avalanche, and the diode will shift operating points to the steady state [4].

In general, electrical simulators, such as SPICE, SPECTRE™, NanoSim™, and others do not take into account the pseudo-steady-state behavior of the junction at and above breakdown, thus making it impossible to simulate the actual behavior of a SPAD in the transient. Although macroscopic models have been built in Verilog-A and other languages, an accurate modeling of SPADs is still elusive.

When biased above breakdown, the SPAD is said to operate in Geiger mode; in this mode of operation, it is capable of detecting single photons, since the avalanche current generated upon photon detection can be easily converted to a digital signal, as shown in Fig. 1. The figure also shows the circuitry used for quenching the avalanche and to recharge the diode to its initial idle state. There exists a variety of avalanche quenching techniques, partitioned in active and passive methods. The literature on these variants is extensive [5]. In active methods, the avalanche is detected and stopped by acting on the bias. In passive methods, the p-n junction bias is self-adjusted using, for example, a ballast resistor. These mechanisms are clearly analog in nature and require accurate modeling capabilities to estimate speed and current requirements at a microscopic level. This characterization is especially important when hundreds or thousands of SPADs operate simultaneously on chip.

When regenerated in a comparator or an inverter, the signal becomes digital and thus acceleration techniques can be used to simulate the signal afterwards. However, conventional mixed-signal simulators are inefficient, as the actual analog segment of the circuit is actually very small. Thus, a new generation of modeling and simulation tools is sought.

One of the main requirements for a SPAD is that its junction does not cause premature edge breakdown (PEB). This phenomenon has the effect of limiting the zone where an avalanche can occur to the edge of the sensitive area, thereby strongly reducing the sensitivity of the SPAD by effectively slashing the fill factor. Several measures can be taken to prevent PEB; these measures generally involve the use of guard rings to reduce the electric field at the edge of the device or to increase the breakdown voltage locally. Fig. 2(a)-(d) show some of the structures used to achieve the goal. Shallow trench isolation (STI) was also demonstrated [6]. STI is used to delimit the junction, provided that it is surrounded by a multi-layer of doped silicon so as to force recombination of those charges generated in the defect-rich STI as shown in structure (d) [7]. The other structures may or may not be compatible with a CMOS process; this is an important requirement to construct images sensors, as understood early on by Rochas [8] and proposed by us and by others [3,7,9,10].

With proper knowledge of the doping profiles of the various CMOS layers and a device simulator, it is possible to predict whether a given structure is likely to achieve the goal. This is done implicitly by looking at a simulated electric field profile to see where it exceeds the critical value for a sustained avalanche ($\approx 3 \times 10^5$V/cm in silicon).

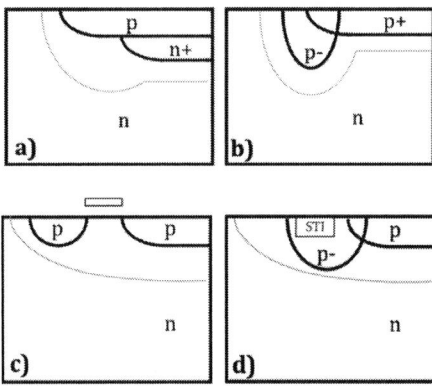

Fig. 2. Example of possible junctions with various types of guard rings: a) enhancement mode; b) explicit; c) implicit; d) STI based.

Fig. 3 shows a study of a guard ring performed before fabrication that shows the effectiveness of the guard ring (shown on the half-cross-section of the device).

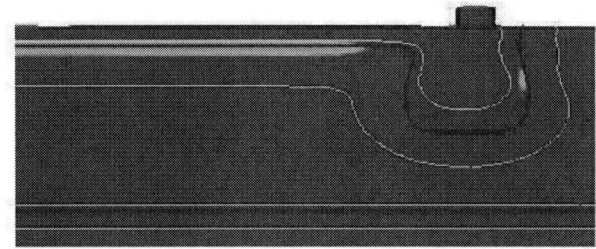

Fig. 3. Device simulation of the electric field distribution in a guard ring of a SPAD [7].

Though effective, this approach does not yield a guarantee that the SPAD will be free of PEB. Thus device simulation should be coupled with the conditions for a sustained avalanche in form of the well-known equation

$$1 < \int_{z_0}^{z_1} \alpha \, dz,$$

where α is the mean ionization per free carrier, and z_0 and z_1 are the limits of the depletion region.

Several more parameters in a SPAD are hard to predict. For example, the probability that a single photon's generated carriers are detected, called the photon detection probability (PDP), requires a deep knowledge of the optical stack and of the junction's quantum efficiency. Noise's sources include tunneling and fabrication defects, which ease valence-to-conduction band transitions, such as thermally generated or tunneling carriers. Dark counts are characterized by the dark count rate (DCR) and are difficult to predict with high degrees of certainty due to the non-deterministic nature of the noise sources and their localization.

The dead time is referred to as the time required in a detection cycle, generally in the ns~µs range. The dead time determines the maximum count rate a SPAD can support. In active quenching, such maximum count rate is the inverse of the dead time; when passive quenching is used the maximum count rate is divided by e. ($e = 2.718281...$) The ratio between maximum count rate and DCR gives an indication of the dynamic range that in SPAD imagers is usually over 80dB. Also in this case, accurate modeling of the quenching and recharge mechanisms are critical and often overlooked by designers or simply unavailable due to limitations of the simulation tools.

3. SPAD IMAGE SENSORS

Creating large arrays of essentially independent digital pulse generators (the SPADs) implies the design of efficient data readout mechanisms that are not different from conventional imagers in terms of functionality and complexity, though of purely digital nature. Thus, they are more similar to clock trees than readout circuits. The simplest readout architecture implementing photon-counting on-chip in combination with random-access single-photon detection, was demonstrated for the first time in [3]. In this readout scheme, all time-sensitive operations had to be performed sequentially. The micrograph of the chip is shown in Fig. 4(a).

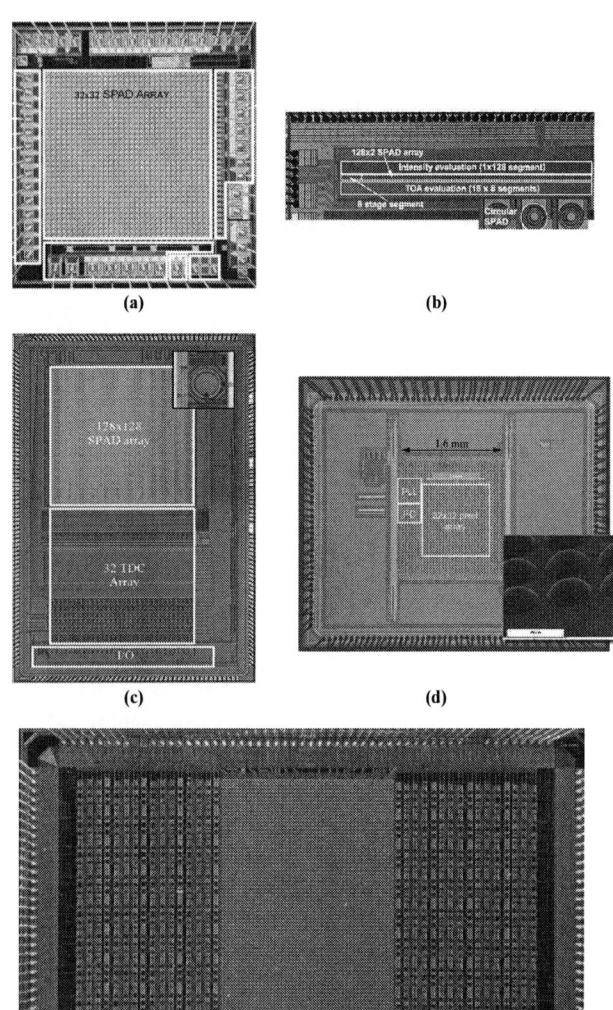

(a)

(b)

(c)

(d)

(e)

Fig. 4. SPAD arrays and readout architectures. (a) Random access 32x32 SPAD array [3]; (b) latchless access 128x2 SPAD array [11]; (c) event-driven access with column-parallel TDCs [12]; (d) pixel-parallel TDCs with microlenses in the inset [14]; (e) event-driven access array with auto-generated digital circuits, whereby the SPADs were instantiated in Verilog™ [19].

The readout bottleneck was partially addressed by means of a latchless pipeline, a technique proposed in [11] and shown in Fig. 4(b), where a time-to-digital converter (TDC) was used at the column level to determine where in the column and when the photon was received. The challenge of this design was the characterization of an unusual technique to transmit digital data through a channel with constant and uniform delay. The first fully integrated SPAD array was reported in LASP [12,13], shown in Fig. 4(c), where column-parallel TDCs were used to process photon arrivals in an event-driven fashion. Again, the clock tree-like readout required particular care during the design phase due to timing requirements. Finally, in the project MEGAFRAME [14,15,16], a pixel-parallel array of 32x32 TDC-SPAD pixels was implemented. The chip reported in [14] is shown in Fig. 4(d). A larger version of the chip (160x128 TDC-SPAD pixels) was later reported in [17], while other column-parallel arrays have recently been reported in [18].

In MEGAFRAME, each TDC had resolutions varying from 52ps to 119ps, with a depth of 10b and a cycle time of 1μs. In these chips the differential non-linearity (DNL) and integral non-linearity (INL) could typically range from 1 to 4LSBs and in LASP, they were recently improved to ±0.1LSB and ±0.25LSB, respectively [13]. The design of the readout and of the pixel required care as they were performed separately and independently. Timing closure techniques were essential, while advanced analog simulation tools had to be used at the pixel level.

The design reported in [19] is the first attempt to treat the SPAD as a device that can be instantiated like any other digital component within a library and placed/routed as such. The fabricated design is shown in Fig. 4(e); it required only 2 weeks for design and it was successfully fabricated and tested.

4. LESSONS LEARNED AND FUTURE CHALLENGES

Since the first CMOS SPAD, the growth of SPAD image sensors in resolution, format, and functionality has been tremendous, often matching Moore's Law. Optical detection and processing can now be performed in massively parallel systems, thanks to very large scale integration and miniaturization. However, limitations still exist, especially in EDA tools, from device modeling (micro and macroscopically) to pixel modeling, from system-level simulation to formal verification, while optical stacks and optical concentrators are generally ignored [20]. Digital SiPM have emerged with arrays of mini SiPMs instead of single SPADs. An example of this trend has been reported by [21], where four 10x10 mini SiPMs has been implemented in 0.18μm HV CMOS technology in combination with a mirror to scan large areas in TCSPC mode.

Modeling and simulation, as well as design support, has become even more challenging with the migration of SPAD structures to nanoscale CMOS [22,23,24], exhibiting improved DCR and spectral efficiency, as well as compatibility with through silicon vias (TSVs) and backside processing. These advanced processes also bring improvements in time resolution, fill factor, pixel pitch as well as the capacity to integrate on-chip time-of-flight computation to ease I/O data rate demands.

In-pixel analog approaches to time-resolved image sensing offer smaller pitch, provided uniformity issues are addressed [25]; on the other hand, the emergence of III-V materials in fully CMOS compatible solutions may bring these materials to the mainstream. Examples of this trend are two independent works reporting the first Ge-on-Si SPADs fabricated in a way that is fully compatible with a conventional CMOS technology [26,27]. Clearly, new modeling and design tools will be required in this domain as well.

5. ACKNOWLEDGMENTS

We acknowledge all our doctoral students and post-doctoral fellows that over the years made the advances in single-photon detection and single-photon imaging possible. The Swiss National Science Foundation, NCCR MICS, the European Commission (FP6 and FP7 programs), Xilinx, and the European Space Agency are also acknowledged.

REFERENCES

1. S. Cova, A. Longoni, and A. Andreoni, "Towards Picosecond Resolution with Single-Photon Avalanche Diodes", *Rev. Sci. Instr.*, **52**(3), 408-412 (1981).

2. R.J. McIntyre, "Recent Developments in Silicon Avalanche Photodiodes", *Measurement*, **3**(4), 146-152 (1985).

3. C. Niclass, A. Rochas, P.A. Besse, and E. Charbon, "Design and Characterization of a CMOS 3-D Image Sensor based on Single Photon Avalanche Diodes", *IEEE J. of Solid-State Circuits*, **40**(9), 1847-1854 (2005).

4. A. Spinelli and A. Lacaita, "Physics and Numerical Simulation of Single Photon Avalanche Diodes", *IEEE Trans. on Electron Devices*, **44**, 1931-1943 (1997).

5. S. Cova, M. Ghioni, A. Lacaita, C. Samori, F. Zappa, "Avalanche Photodiodes and Quenching circuits for Single-Photon Detection", *Appl. Opt.*, **35**(12), 1956-1976 (1996).

6. H. Finkelstein, M. J. Hsu and S. C. Esener "STI-Bounded Single-Photon Avalanche Diode in a Deep-Submicrometer CMOS Technology," *IEEE Electron Device Lett.*, **27**(11), 887-889 (2006).

7. M. Gersbach *et al.*, "A Low-Noise Single-Photon Detector Implemented in a 130nm CMOS Imaging Process", *Solid-State Electronics*, **53**(7), 803-808 (2009).

8. A. Rochas *et al.*, "Single Photon Detector Fabricated in a Complementary Metal–oxide–semiconductor High-voltage Technology", *Rev. Sci. Instr.*, **74**(7), 3263-3270 (2003).

9. L. Pancheri and D. Stoppa, "Low-noise CMOS Single-photon Avalanche Diodes with 32ns Dead Time", *IEEE ESSCIRC*, (2007).

10. N. Faramarzpour, M.J. Deen, S. Shirani, and Q. Fang, "Fully Integrated Single Photon Avalanche Diode Detector in Standard CMOS 0.18-um Technology", *IEEE Trans. on Electron Devices*, **55**(3), 760-767 (2008).

11. M. Sergio, C. Niclass, E. Charbon, "A 128x2 CMOS Single Photon Streak Camera with Timing-Preserving Latchless Pipeline Readout", *IEEE Intl. Solid-State Circuits Conference*, 120-121 (2007).

12. C. Niclass, C. Favi, T. Kluter, M. Gersbach, and E. Charbon, "A 128x128 Single-Photon Image Sensor with Column-Level 10-bit Time-to-Digital Converter Array", *IEEE J. of Solid-State Circuits*, **43**(12), 2977-2989 (2008).

13. J.M. Pavia, C. Niclass, C. Favi, M. Wolf, E. Charbon, "3D Near-infrared Imaging based on a SPAD Image Sensor", *International Image Sensor Workshop*, (2011).

14. M. Gersbach *et al.*, "A Time-resolved, Low-Noise Single-Photon Image Sensor Fabricated in Deep-Submicron CMOS Technology Parallel 32x32 Time-to-Digital Converter Array Fabricated in a 130nm Imaging CMOS Technology", *IEEE J. of Solid-State Circuits*, **47**(6), (2012).

15. J.R. Richardson *et al.*, "A 32x32 50ps Resolution 10 bit Time to Digital Converter Array in 130nm CMOS for time Correlated Imaging", *IEEE Custom Integrated Circuits Conference*, 77-80 (2009).

16. D. Stoppa *et al.*, "A 32x32-Pixel Array with In-Pixel Photon Counting and Arrival Time Measurement in the Analog Domain", *IEEE European Solid-State Device Conference*, 204-207 (2009).

17. C. Veerappan *et al.*, "A 160x128 Single-Photon Image Sensor with On-Pixel, 55ps 10b Time-to-Digital Converter", *IEEE Intl. Solid-State Circuits Conference*, 312-314 (2011).

18. R.J. Walker, J.R. Richardson and R.K. Henderson; "A 128×96 Pixel Event-Driven Phase-Domain $\Delta\Sigma$-Based Fully Digital 3D Camera in 0.13μm CMOS Imaging Technology", *IEEE Intl. Solid-State Circuits Conference,* 410-412 (2011).

19. C. Niclass, M. Sergio, and E. Charbon, "A CMOS 64x48 Single Photon Avalanche Diode Array with Event-Driven Readout", *IEEE European Solid-State Circuit Conference* (ESSCIRC), (2006).

20. S. Donati, G. Martini, M. Norgia, "Microconcentrators to recover fill-factor in image photodetectors with pixel on-board processing circuits", *Opt. Express* **15**(26), 18066-18075 (2007).

21. C. Niclass, M. Soga, S. Kato, "A 0.18μm CMOS Single-Photon Sensor for Coaxial Laser Rangefinders", *ASSC*, (2010).

22. M.A. Karami, M. Gersbach, E. Charbon, "A New Single-photon Avalanche Diode in 90nm Standard CMOS Technology", *SPIE Optics+Photonics, NanoScience Engineering, Single-Photon Imaging*, (2010).

23. R.K. Henderson, E. Webster, R. Walker, J.A. Richardson, L.A. Grant, "A 3x3, 5um Pitch, 3-Transistor Single Photon Avalanche Diode Array with Integrated 11V Bias Generation in 90nm CMOS Technology", *IEEE International Electron Device Meeting*, 1421-1424 (2010).

24. E.A.G. Webster, J. A. Richardson, L.A. Grant, D. Renshaw, and R.K. Henderson, "An infra-red sensitive, low noise, single-photon avalanche diode in 90 nm CMOS," *International Image Sensor Workshop (IISW)*, Hokkaido, Japan, 8–11 June 2011.

25. L. Pancheri, N. Massari, F. Borghetti, D. Stoppa, "A 32x32 SPAD Pixel Array with Nanosecond Gating and Analog Readout", *International Image Sensor Workshop (IISW)*, Hokkaido, Japan, 8–11 June 2011.

26. A. Sammak, M. Aminian, L. Qi, W.D. de Boer, E. Charbon, L. Nanver, "A CMOS Compatible Ge-on-Si APD Operating in Proportional and Geiger Modes at Infrared Wavelengths", *International Electron Device Meeting*, (2011).

27. Z. Lu, Y. Kang, C. Hu, Q. Zhou, H.-D. Liu, J.C. Campbell, "Geiger-Mode Operation of Ge-on-Si Avalanche Photodiodes", *IEEE Journal of Quantum Electronics*, **47**(5), 731-735 (2011).

Non-Volatile FPGAs based on Spintronic Devices *

Olivier Goncalves, Guillaume Prenat, Gregory Di Pendina and Bernard Dieny
Spintec, CEA-INAC/CNRS/UJF/INPG, Grenoble, France
guillaume.prenat@cea.fr

ABSTRACT

This paper presents an innovative architecture for radiation-hardened FPGA (Field Programmable Gate Array). This architecture is based on the use of MTJs (Magnetic Tunnel Junctions), magnetic nanostructures used as basic elements of MRAM (Magnetic Random Access Memory). These devices are totally immune to radiations and can be used as a reference memory to perform "scrubbing" techniques, which consist in regularly reloading the configuration of the FPGA to fix the radiation induced errors that may have occured. This approach allows hardening the circuits at low cost in terms of area, while reducing the standby power consumption and offering new fonctionalities, like dynamic reconfiguration. A silicon demonstrator was implemented, including a 2-inputs LUT (Look Up Table) and tested using a digital tester, giving encouraging results.

Keywords

FPGA, LUT, MRAM, MTJ, DRAM, radiation hardning, soft error, scrubbing

1. INTRODUCTION

FPGAs (Field Programmable Gate Arrays) are circuits whose functionality can be changed without hardware modification. Since they do not require a specific mask set, they are very cheap for a small number of pieces and are widely used for low-volume production or prototyping. They can also be used for specific applications, for example for dynamic reconfiguration, which consists in changing the function during

---

*Permission to make digital or hard copies of all or part of this work for personal or classroom use is granted without fee provided that copies are not made or distributed for profit or commercial advantage and that copies bear this notice and the full citation on the first page. To copy otherwise, to republish, to post on servers or to redistribute to lists, requires prior specific permission and/or a fee.
DAC'13, May 29-June 07 2013, Austin, TX, USA.

operation to optimize the performance. A FPGA is essentially composed of memory, which determines the functionality of the circuit. For circuits used in space applications, the sensitivity to radiations is a major issue. Indeed, all the technologies based on the charge of the electron are very sensitive since a particle can switch the state of a transistor, or a capacitor, resulting in a local fault and possibly an error in the circuit operation. To protect the circuit, several hardening techniques are used, with a large impact on area and performance. The use of emerging technologies, which are intrinsically immune to radiations can ease the hardening of the circuits, while offering new functionalities and reduce power consumption [3].

2. STRUCTURE OF A FPGA

A FPGA is typically composed of elementary functions, called LUTs (Look Up Tables, Fig.1) whose functionality is determined by an operating code stored in a configuration memory. A LUT is generally associated with a Flip-Flop within a CLB (Configurable Logic Block) allowing sequential operation. These LUTs are connected together to build a complex function (Fig.2). The connections are made by a programmable interconnection network, in which transistors are used as switches for the routing. Each connection is activated or not, using a memory cell connected to the gate of the transistor. A FPGA is then essentially composed of memory and routing circuits. Two main technologies are used today: SRAM-based FPGAs are very fast to reconfigure, but they are volatile. An additional non-volatile memory (typically flash) is then required to permanently store the information when the power supply is off, resulting in a low density. Flash-based FPGAs are intrinsically non-volatile, dense, but they are slow and power consumming during reconfiguration. Moreover, these two technologies are sensitive to radiations and the effect of a particule can change the content of a memory cell and so permanently change the configuration of the circuit. SRAM memories are very sensitive by SEU (Single Event Upset) or soft errors, while Flash memories are more sensitive to TID (Total Ionizing Dose).

3. STANDARD HARDENING TECHNIQUE

Several techniques can be used to harden a FPGA, at technology level (use of SOI substrate for example), at layout level or at circuit/system level. At circuit level, it is possible to add redundancy to the circuit. In particular, the TMR (Triple Modular Redundancy) technique consists in duplicating a part of the circuit three times and use a majority

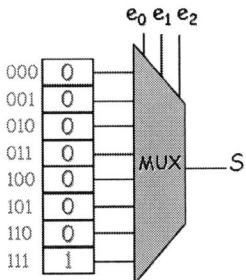

Figure 1: 3-inputs LUT. It is composed of a 3-inputs multiplexor the selection inputs of which are the inputs of the LUT. The multiplexor is connected to a 8-bits configuration memory in which the functionality is stored. In this case, we see that only the combination of inputs "111" will route a "1" logic value to the output of the LUT, corresponding to a "AND" logic function

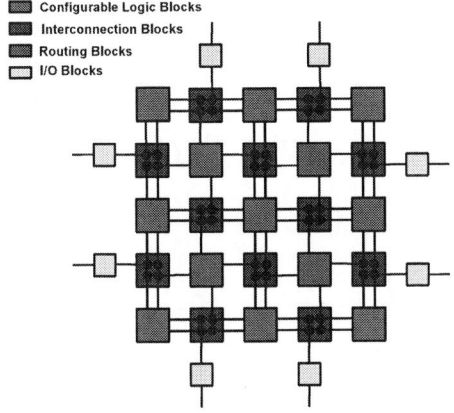

Figure 2: A full FPGA array made of CLBs interconnected by means of a programmable interconnection network. The FPGA can be divided in tiles, representing basic blocks that can be repeated to build the full circuit

voter to identify which output is correct. This assumes that only one circuit has been affected. To avoid errors accumulation, this technique is often associated to the "scrubbing" technique consisting in regularly reloading the configuration of the FPGA to correct errors that may have occured.

4. MAGNETIC TECHNOLOGY

MRAM (Magnetic Random Access Memory) [5] is an emerging non-volatile resistive memory technology, which combines non-volatility, hardness to radiations, speed, endurance and density. Its basic element is the MTJ (Magnetic Tunnel Junction) whose electric resistance depends on the magnetic state. To write the magnetic configuration, several techniques can be used, using an external magnetic field generated by current lines close to the junction or passing a current directly through it. The TAS (Thermally Assisted Switching) is a writing scheme based on magnetic field, that has been proposed by Spintec and is now used by the MRAM

compagny Crocus-technology [4]. This technique relies on the strong dependance of the transport and magnetic parameters upon the temperature. The writing sequence consists in passing a current through the junction to heat it by joule effect to free the magnetization of its SL (Storage Layer) which can then be written at low magnetic field (Fig.3). This approach offers a good thermal stability in advanced technology nodes and a good immunity to external magnetic fields. Whatever the writing scheme, the magnetic element itself is totally immune to radiations [2].

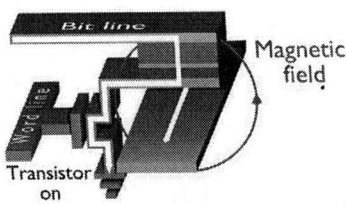

Figure 3: Schematic of a TAS MRAM. The transistor, when activated, generates a current through the MTJ. The MTJ heats up to the write temperature of about $230°C$ within 5ns. Then, a pulse of current is sent in the line underneath the MTJ to generate a magnetic field on the Storage Layer. The direction of the current determines the data to be written. After magnetization switching, the transistor is turned off. The MTJ cools down in about 15ns. The magnetic field is switched off after the cooling phase.

5. INNOVATIVE ARCHITECTURE

The idea proposed here (Fig.4) is to use a DRAM (Dynamic Random Access Memory) for the configuration and a MRAM memory as a reference memory. The DRAM is periodically refreshed using the content of the MRAM memory: the scrubbing and refreshing operations are thus combined. The reference part is composed of differential pairs MTJs associated with selection transistors to select which cell is read (sel0 to sel3). The selected twin cell is read using the latch operating as a sense amplifier. Two additional transistors are used to heat the MTJs, while the writing magnetic field is generated by an external driver (not represented here). The configuration memory is composed of capacitances connected to the gates of the transistors and selections transistors are used to select which cell is refreshed (s0 to s3). The same principle is used for the LUTs and for the interconnections. In this architecture, it is not necessary to add redundancy for the MRAM part, which is intrinsically immune to radiations. If redundancy is required for the configuration part, it will only concern DRAM cells, meaning a very small area overhead. In terms of power consumption, the non-volatility allows cutting off the power supply of unused blocks of the circuit, drastically reducing the power consumption (power gating). Since writing the magnetic devices does not affect the operation of the circuit until refreshing, it is possible to perform dynamic reconfiguration easily. A full tile of the FPGA was designed using a compact model of the MTJ developed at Spintec [1], so that a full FPGA can be easily built from these elementary repeatable building blocks.

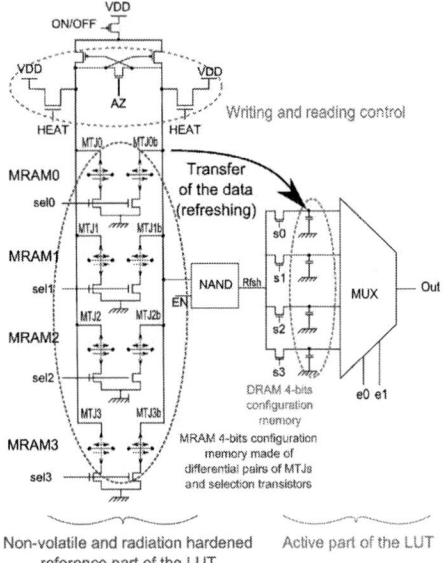

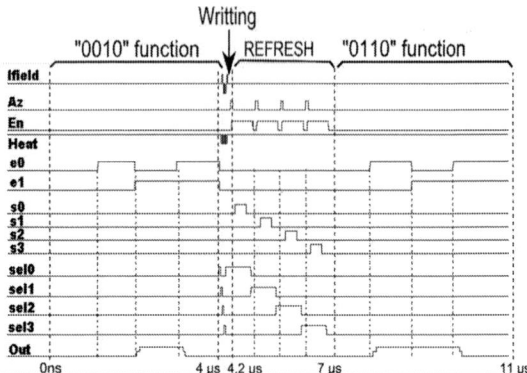

Figure 6: Functional test of the dynamic reconfiguration of the LUT. The configuration ("0010") is changed into the XOR ("0110") function.

Figure 4: Schematic of the 2-input LUT. The latch reads the data of the selected MTJs when the AutoZero (AZ) signal is high. When read, the data is formatted by the NAND gate. EN is activated when the data is read. An ON/OFF transistor switches off the MRAM block when unused between two refresh or programming phases.

6. SILICON DEMONSTRATOR

A 2-inputs LUT based on the described architecture has been implemented in a silicon demonstrator (Fig.5) using the hybrid magnetic/CMOS 130 nm technology from Crocus-technology and TowerJazz. The magnetic technology is based on the TAS writing scheme. The circuit has been tested and gives the expected results: all the 8 possible 2-inputs logic functions has been sequentially written in the MRAM. Each time, the configuration has been reloaded in the DRAM configuration memory and operation of the LUT checked for all the 4 combinations of inputs (Fig.6).

Figure 5: Picture of the silicon demonstrator

7. CONCLUSIONS

The LUT architecture presented here takes advantage of the magnetic technology to allow hardening a FPGA against radiation at low area overhead. It also allows dynamic reconfiguration while the non-volatility can be used to reduce the power consumption. In particular, MRAM is less sensitive than Flash to TID [6], what makes the architecture particularly suitable for space applications for example. The concept has been validated in silicon, proving the viability of the technology for logic applications. The next steps are now to implement a full FPGA and perform radiation hardness tests. Its performance should also be compared to equivalent FPGAs (in terms of applications and complexity) in standard CMOS technologies.

8. ACKNOWLEDGMENTS

TowerJazz and Crocus-Technology are deeply acknowledged for having given access to their technology in order to build this demonstrator and for their assistance in the tests. European funding is also acknowledged through ERC Adv grant HYMAGINE 246942.

9. REFERENCES

[1] M. Baraji, V.Javerliac, W.Guo, G.Prenat, and B.Dieny. Dynamic compact model of thermally assisted switching magnetic tunnel junctions. *Journal of Applied Physics*, 106(12), Dec. 2009.

[2] Y. Conraux, J. P. Nozières, V. D. Costa, M. Toulemonde, and K. Ounadjela. Effects of swift heavy ion bombardment on magnetic tunnel junction functional properties. *Journal of Applied Physics*, 93(10):7301–7303, May. 2003.

[3] Y. Guillemenet, L. Torres, G. Sassatelli, and I. Hassoune. A nonvolatile run-time fpga using thermally assisted switching mrams. In *International Conference on Field Programmable Logic and applications*, Sept. 2008.

[4] I.L.Prejbeanu, W. Kula, K. Ounadjela, R. Sousa, O. Redon, B. Dieny, and J. Nozières. Thermally assisted switching in exchange-biased storage layer magnetic tunnel junctions. *IEEE Transactions on Magnetics*, 40(4):2625–2627, July. 2004.

[5] S.Tehrani. Magnetoresistive random access memory using magnetic tunnel junctions. *proceedings of IEEE*, 91(5):3703–3714, May. 2003.

[6] Y. Troxel. In *Inc. Seminar Day, Military and Aerospace Programmable Logic Devices (MAPLD)*, NASA Goddard Space Flight Center, Aug. 2009.

A Novel Analytical Method for Worst Case Response Time Estimation of Distributed Embedded Systems

Jinwoo Kim
School of Electrical
Engineering and Computer
Science
Seoul National University
Seoul, Korea
jwkim@iris.snu.ac.kr

Hyunok Oh
Department of information
Systems
Hanyang University
Seoul, Korea
hoh@hanyang.ac.kr

Junchul Choi
School of Electrical
Engineering and Computer
Science
Seoul National University
Seoul, Korea
hinomk2@iris.snu.ac.kr

Hyojin Ha
School of Electrical
Engineering and Computer
Science
Seoul National University
Seoul, Korea
polaris@iris.snu.ac.kr

Soonhoi Ha
School of Electrical
Engineering and Computer
Science
Seoul National University
Seoul, Korea
sha@snu.ac.kr

ABSTRACT
In this paper, we propose a novel analytical method, called scheduling time bound analysis, to find a tight upper bound of the worst-case response time in a distributed real-time embedded system, considering execution time variations of tasks, jitter of input arrivals, and scheduling anomaly behavior in a multi-tasking system all together. By analyzing the graph topology and worst-case scheduling scenarios, we measure the conservative scheduling time bound of each task. The proposed method supports an arbitrary mixture of preemptive and non-preemptive processing elements. Its speed is comparable to compositional approaches while it gives a much tighter bound. The advantages of the proposed approach compared with related work were verified by experimental results with randomly generated task graphs and a real-life automotive application.

Keywords
Worst case response time, performance analysis, distributed embedded system.

1. INTRODUCTION
In the design of distributed embedded systems that execute real-time applications, it is critical to guarantee that the real time constraints are satisfied at all times. Since it is not feasible to examine all possible behaviors as the system complexity grows, measurement-based and simulation-based timing analysis methods cannot be used to estimate the worst-case performance. Thus analytical techniques have been extensively sought for to estimate the worst-case response time (WCRT) of an application at the system level.

Finding a tight upper bound of the WCRT is very desirable for design space exploration to compare the candidate architectures accurately and to avoid over-design. A model checking approach [1] and an ILP-based approach [2] have been proposed to accurately estimate the WCRT. But they both suffer from poor scalability since the time complexity is inherently exponential. On the other hand, compositional approaches have been developed to overcome this difficulty by performing the analysis in a modular manner; SymTA/S [3] and MPA [4] are two well-known approaches in this category. While they are fast and scalable, however, they suffer from over-estimation. In short, it remains a

Permission to make digital or hard copies of part or all of this work for personal or classroom use is granted without fee provided that copies are not made or distributed for profit or commercial advantage and that copies bear this notice and the full citation on the first page. Copyrights for components of this work owned by others than ACM must be honored. Abstracting with credit is permitted. To copy otherwise, to republish, to post on servers or to redistribute to lists, requires prior specific permission and/or a fee.
DAC '13, May 29 - June 07 2013, Austin, TX, USA

very challenging problem to determine a tight upper bound of the worst-case response time in a distributed real-time embedded system when execution time variations of tasks, jitter of input arrivals, and scheduling anomaly behavior [5] in a multi-tasking system , and heterogeneity of processing elements, are considered all together.

As a novel solution to this problem, we propose a holistic analytical method, called *scheduling time bound analysis* (STBA) that has a polynomial time complexity while producing a much tighter bound than compositional approaches. We assume that applications are specified by a set of task graphs where an arc between tasks represents execution dependency between two tasks. A task can start its execution only after all predecessor tasks are completed. A source task without any predecessor is activated periodically with arbitrary start offset. The WCRT of a task graph is measured from the arrival time of the source task to the latest finish time of sink tasks. We assume that the priorities of tasks and task mapping onto processing elements are given a priori. Note that finding an optimal mapping or priority assignment is another challenging problem, which is left as a future work.

As the name implies, the proposed technique performs scheduling of task graphs during analysis. We compute the scheduling time bound of each task which is characterized by two pairs of timing information: (earliest start time, latest start time) and (earliest finish time, latest finish time) by considering all possible scheduling scenarios of tasks on the mapped processing elements. The proposed technique has some similarities with a schedulability analysis method [6], denoted as Y&W method, in that both methods compute the latest finish time of a task considering the scheduling effect. Unlike Y&W method that supports only preemptive processing elements, the proposed technique supports an arbitrary mixture of preemptive and non-preemptive processing elements.

Experimental results confirm that the proposed technique gives near-optimal results for small-size benchmarks for which optimal results could be obtained. For large-size benchmarks, it provides tighter bounds than the previous compositional approaches while the analysis time is about the same order.

2. RELATED WORK
Analytical techniques for system-level performance estimation can be classified into two groups: holistic approach and compositional approach. A holistic approach integrates task scheduling and communication scheduling of a given system into a single analysis framework.

A group of researchers extended the schedulability analysis technique of a single processor to distributed systems. The schedulability analysis technique estimates the maximum delay of a certain task in a pessimistic scenario to compute the WCRT.

Pioneered by K. Tindell et al. [7], a series of techniques have been proposed to overcome the limitation of earlier ones. For example, precedence relation is considered in [14]. The most relevant work to the proposed approach is Yen and Wolf's [6]. They consider data dependency between tasks and variable execution times of tasks. They compute the timing bounds for all tasks on a distributed system. Therefore we select their approach as a baseline technique to compare with the proposed technique, but only for preemptive systems since they do not support non-preemptive processing elements. Non-preemptive processing elements are supported in the MAST suite [16] that includes several schedulability analysis techniques. But they support only chain-structure graphs where a task has a single input and/or a single output port.

For accurate WCRT estimation, a model checking technique is used in UPPAL [8] where tasks, processors, and their interactions are modeled by timed-automata [15]. Recently, an analysis technique based on integer linear programming (ILP) was proposed to estimate the worst-case performance of a non-preemptive multitasking MPSoC [2]. The technique proposed by Madl [9] enumerates all possible event execution paths and verifies them using discrete event simulation to analyze the schedulability of a distributed embedded system assuming non-preemptive scheduling. The main weakness of these approaches is poor scalability since the time complexity is inherently exponential.

Compositional approaches achieve scalability by performing the analysis in a modular manner. The behavior of a processing component is encapsulated inside and abstracted by a function of event streams at the component boundary. SymTA/S [3] uses a tuple (p, j, d) to characterize an event stream with the period (p), and jitter of arrival (j). If the jitter is larger than the period, the events may arrive in a bursty fashion and delay d indicates the minimum time gap between arrivals. MPA [4] defines the notion of arrival curves and service curves, and applies the real-time calculus to model the event stream processing of components. While compositional approaches are scalable and fast, they usually produce a loose bound on the worst-case performance since they do not fully utilize the dependency relation between task executions. To improve the solution quality, a hybrid approach between the MPA and UPPAAL has been proposed in [10] where different analysis techniques are used for intra-component analysis. The critical processing components for accurate estimation are analyzed by model checking and the others are analyzed by real-time calculus to make a good compromise between the accuracy and the computation speed.

Compared with the related work, the proposed STBA technique has the following characteristics.

- It supports both preemptive and non-preemptive processing elements and has no limitation on the task graph topology.
- It is as scalable as compositional approaches since the analysis time complexity is polynomial.
- It produces much tighter bound than compositional approaches.

3. Problem Definition

In this section, we describe the input task model and the system architecture assumed in this paper. The input application is represented as a set of acyclic task graphs, $\{\mathcal{T}_i\}$, as illustrated in Figure 1(a). In a task graph, $\mathcal{T} = \{\mathcal{V}, \mathcal{E}\}$, $\mathcal{V}$ represents a set of tasks and $\mathcal{E} = \{(\tau_1, \tau_2) | \tau_1, \tau_2 \in \mathcal{V}\}$ a set of edges to represent execution dependency between tasks; i.e. τ_2 cannot start its

execution before τ_1 finishes. A source task that has no predecessor is time-driven and assigned a tuple (p, j, d) as shown in Figure 1(a) next to task graph $\mathcal{T}$, adopting the notation of SymTA/S regarding the event arrival as the task arrival (see supplementary section 1 for illustration). A task graph may have multiple source tasks that have the same (p, j, d) parameters. We assume that the first instances of source tasks have arbitrary start offset, $\mathcal{T}^{\text{offset}}$. A task graph $\mathcal{T}_i$ is given a deadline $\mathcal{D}_i$ to meet once activated.

The system architecture is given as a processing element (PE) graph as depicted in Figure 1(c) where a node represents a processing element and an edge represents a connection between two processing elements. Note that a communication network is modeled as a separate PE. For instance, the PE graph of Figure 1(c) represents a system that consists of two processors (PE0 and PE2) connected to a bus (PE1).

A task is a basic mapping unit onto a processing element as depicted in Figure 1(a). We assume that task mapping is given and fixed. Each task τ is assigned a tuple [BCET, WCET] to represent the best and worst execution times of the task on the mapped PE. Note that tasks mapped to a communication network (PE1 in this example) represent the communication tasks that correspond to the message delivery between two computation tasks; for example τ_1 indicates message communication between two computation tasks, τ_0 and τ_2 that are mapped to different processing elements.

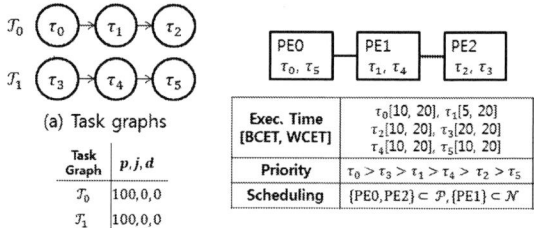

(a) Task graphs

(b) Activation parameters

(c) Mapping and task information

Figure 1. (a) Example of task graphs, (b) task graph information, (c) and mapping and task information

Each PE is given a scheduling policy. In this work, we assume two scheduling policies are available for each PE: fixed-priority preemptive scheduling (denoted by $\mathcal{P}$ in the figure) and a fixed-priority non-preemptive scheduling (denoted by $\mathcal{N}$). In Figure 1(a), PE0 executes the assigned tasks with a preemptive scheduling policy while PE1 serves the communication tasks in a non-preemptive fashion; a higher-priority message cannot preempt the current message delivery. We assume that all tasks mapped to a PE have distinct priorities to make the scheduling order deterministic on each processing element.

The response time of a task graph is defined as the time difference between the latest finish time of tasks in the task graph and the earliest arrival time of source tasks. To compute the response time of a task graph, we have to consider the effect of the other task graphs. Sometimes we need to consider the next period of the same task graph in case the response time becomes greater than the period. In the proposed approach, we explicitly instantiate a task graph as many times as necessary by creating a new task graph for each execution of the graph. How many times a task graph should be expanded will be discussed in section 5.

Let $\mathcal{G}(\mathcal{T})$ be the set of cloned task graphs associated with a task graph $\mathcal{T}$. We assume that the priority of a task is inherited to the cloned task graphs. Among the same tasks, the task of an earlier instance has a higher priority than that of a later instance. Then the WCRT for $\mathcal{T}$, denoted $\mathcal{R}_\mathcal{T}$, is defined as the maximum value among all response times of the task graph instance in $\mathcal{G}(\mathcal{T})$.

Now, we summarize the problem as follows:

INPUT: We are given a set of task graphs $\mathcal{A}$ and the PE graph with the mapping information of tasks to processing components. Execution time of each task is specified with a range [BCET, WCET] on the mapped PE. Each task graph is given a (p, j, d) tuple for activation and a deadline. Each PE is given a fixed-priority based scheduling policy. It can be preemptive or non-preemptive. The priorities of assigned tasks are uniquely defined in each PE.

PROBLEM: For a task graph $\mathcal{T}$, determine the WCRT, $\mathcal{R}_\mathcal{T}$.

4. Proposed Scheduling Time Bound Analysis

In the proposed STBA technique, we compute the scheduling time interval for each task τ_i, characterized by two pairs of timing information: $(\tau_i^{\min S}, \tau_i^{\max S})$ and $(\tau_i^{\min F}, \tau_i^{\max F})$ denoting the earliest and the latest start times, and the earliest and the latest finish times, respectively. When we compute the scheduling time bound of a task, we have to consider all possible scheduling scenarios that depend on the execution times of tasks, jitter values, and starting offsets of task graphs.

4.1 Base Time Bound Computation

The time interval, $(\tau_i^{\min S}, \tau_i^{\max S})$ and $(\tau_i^{\min F}, \tau_i^{\max F})$, should be decided conservatively, meaning that the finish (start) time should not be later than $\tau_i^{\max F}(\tau_i^{\max S})$ and the start (finish) time should not be earlier than $\tau_i^{\min S}(\tau_i^{\min F})$.

For a source task τ, the earliest and the latest arrival times, $\tau^{\min A}$ and $\tau^{\max A}$, are formulated as follows:

$$\tau^{\min A} = \min(\mathcal{T}^{\text{offset}}) \qquad \text{eq 1}$$
$$\tau^{\max A} = \max(\mathcal{T}^{\text{offset}}) + \mathcal{T}^{J} \qquad \text{eq 2}$$

where $\mathcal{T}^{\text{offset}}$ indicates the starting offset and $\mathcal{T}^J$ is the maximum jitter of the task graph.

When multiple task graphs are involved, we compute the WCRT of each task graph separately. The task graph under WCRT computation is called the *target* task graph. The starting offset of the target task graph is fixed to $P^{\max}$ that means the maximum period among all task graphs. The starting offsets of the other task graphs may vary up to their own period from $P^{\max}$ to consider all possible interruption that they can make to the target task graph execution. In short, the range of starting offset becomes as follows:

$$\mathcal{T}^{\text{offset}} = P^{\max} \qquad \text{if } \mathcal{T} \text{ is the target task graph.}$$
$$(P^{\max} - \mathcal{T}^p) \leq \mathcal{T}^{\text{offset}} \leq P^{\max} \quad \textbf{otherwise} \qquad \text{eq 3}$$

where $\mathcal{T}^p$ indicates the period of task graph $\mathcal{T}$. (Refer to supplementary section 2 for more explanation). If the starting offset of a task graph is given and fixed, $\mathcal{T}^p$ in eq 3 is replaced by the given offset.

We can determine the earliest and the latest arrival times by examining the topology of a task graph for a non-source task. A task is defined as arriving (becomes executable) when all its predecessors are executed. Let $pred[\tau_i]$ be the predecessor set of task τ_i. We can determine the earliest and the latest arrival times by examining the topology of a task graph.

For a task with $|pred[\tau]| \geq 1$, $\tau^{\min A}$ and $\tau^{\max A}$ become

$$\tau^{\min A} = \max_{\tau_p \in \text{pred}[\tau]}(\tau_p^{\min F}) \qquad \text{eq 4}$$
$$\tau^{\max A} = \max_{\tau_p \in \text{pred}[\tau]}(\tau_p^{\max F}) \qquad \text{eq 5}$$

Then, the earliest and the latest start times of a task can be formulated as follows:

$$\tau^{\min S} = \max(\tau^{\min A}, \alpha_\tau(\tau^{\min S})) \qquad \text{eq 6}$$
$$\tau^{\max S} = \max(\tau^{\max A}, \beta_\tau(\tau^{\max S})) \qquad \text{eq 7}$$

where $\alpha_\tau(t)$ and $\beta_\tau(t)$ represent the earliest and the latest times that the PE becomes available after time t. If the PE is executing another task when task τ arrives, it may have to wait until the current task is completed. Under a preemptive scheduling policy, it has to wait in case the priority of the current task is higher. Under a non-preemptive scheduling policy, it has to wait regardless of the priority of the current task. Thus we need different formulas for $\alpha_\tau(\tau^{\min S})$ and $\beta_\tau(\tau^{\max S})$ depending on the scheduling policy.

First, we can formulate $\alpha_\tau(\tau^{\min S})$ as follows:

$$\alpha_\tau(\tau^{\min S}) = \max_{\tau_s \in \mathcal{L}_\tau}(\tau_s^{\min F}), \qquad \text{eq 8}$$

where $\mathcal{L}_\tau = \left\{ \tau_s \middle| \tau_s^{\text{proc}} = \tau^{\text{proc}}, \tau_s^{\text{pri}} > \tau^{\text{pri}}, \tau_s^{\max S} \leq \tau^{\min S} \right\}$ for the preemptive scheduling policy

$$\mathcal{L}_\tau = \left\{ \tau_s \middle| \begin{array}{l} \tau_s^{\text{proc}} = \tau^{\text{proc}}, (\tau_s^{\text{pri}} > \tau^{\text{pri}}, \tau_s^{\max S} \leq \tau^{\min S}) \text{ or} \\ (\tau_s^{\text{pri}} < \tau^{\text{pri}}, \tau_s^{\max S} \leq \tau^{\min A} < \tau_s^{\min F}) \end{array} \right\}$$ for the non-preemptive scheduling policy where τ^{pri} and τ^{proc} denote the priority of task τ and the processing element task τ is mapped to, respectively.

Set $\mathcal{L}_\tau$ includes the higher priority tasks that will be completed before task τ starts by collecting the tasks whose latest start times are no later than the earliest start time of task τ. Note that we include a lower priority task that is being executed at the arrival time of task τ into set $\mathcal{L}_\tau$ for the non-preemptive scheduling case.

To compute $\beta_\tau(\tau^{\max S})$, on the other hand, we have to consider all tasks that can possibly start earlier than task τ. Hence $\beta_\tau(\tau^{\max S})$ is formulated as follows;

$$\beta_\tau(\tau^{\max S}) = \max_{\tau_s \in \Delta_\tau}(\tau_s^{\max F}) \qquad \text{eq 9}$$

where $\Delta_\tau = \left\{ \tau_s \middle| \tau_s^{\text{proc}} = \tau^{\text{proc}}, \tau_s^{\text{pri}} > \tau^{\text{pri}}, \tau_s^{\min S} \leq \tau^{\max S} \right\}$ for the preemptive scheduling policy

$$\Delta_\tau = \left\{ \tau_s \middle| \begin{array}{l} \tau_s^{\text{proc}} = \tau^{\text{proc}}, (\tau_s^{\text{pri}} > \tau^{\text{pri}}, \tau_s^{\min S} \leq \tau^{\max S}) \text{ or} \\ (\tau_s^{\text{pri}} < \tau^{\text{pri}}, \tau_s^{\max S} \leq \tau^{\max A} \leq \tau_s^{\max F}) \end{array} \right\}$$ for the non-preemptive scheduling policy.

Since we compare the earliest start time of a task in set Δ_τ with the latest start time of task τ, the set includes all tasks that can be completed before task τ. And we include a lower priority task that is being executed at the arrival time into set Δ_τ for the non-preemptive scheduling case. It is guaranteed that the PE becomes available after all tasks in set Δ_τ are completed.

Lemma 1. The start time of a task τ cannot be earlier than $\tau^{\min S}$ nor later than $\tau^{\max S}$. It means that the start time bounds formulated in eq 6 and eq 7 are defined conservatively.

(Proof is in supplementary section 3)

The earliest and the latest finish times of a task can be formulated as follows:

$$\tau^{\min F} = \tau^{\min S} + \tau^{\text{BCET}} + \gamma_\tau(\tau^{\min S}) \qquad \text{eq 10}$$

$$\tau^{maxF} = \tau^{maxS} + \tau^{WCET} + \delta_\tau(\tau^{maxS}) \qquad \text{eq 11}$$

where τ^{BCET} and τ^{WCET} represent the best-case and the worst-case execution times of task τ, and $\gamma_\tau(t)$ and $\delta_\tau(t)$ denote the minimum and the maximum time duration by which the task is preempted by higher priority tasks after time t. Under a non-preemptive policy, they are all zero. For a preemptive scheduling policy, $\gamma_\tau(\tau^{minS})$ and $\delta_\tau(\tau^{maxS})$ are formulated as follows:

$$\gamma_\tau(\tau^{minS}) = \sum_{\tau_s \in X_\tau} \tau_s^{BCET} \qquad \text{eq 12}$$

where $X_\tau = \left\{ \tau_s \middle| \begin{array}{c} \tau_s^{proc} = \tau^{proc},\ \tau_s^{pri} > \tau^{pri}, \\ \tau^{minS} \le \tau_s^{minS} \le \tau_s^{maxS} < \tau^{minF} \end{array} \right\}$, a set that includes the higher priority tasks that will be scheduled during execution of task τ.

$$\delta_\tau(\tau^{maxS}) = \sum_{\tau_s \in Y_\tau} \tau_s^{WCET} \qquad \text{eq 13}$$

where $Y_\tau = \left\{ \tau_s \middle| \begin{array}{c} \tau_s^{proc} = \tau^{proc},\ \tau_s^{pri} > \tau^{pri}, \\ \tau^{maxS} \le \tau_s^{maxF},\ \tau_s^{minS} \le \tau^{maxF} \end{array} \right\}$, a set that includes all higher priority tasks that can be scheduled during execution of task τ.

The finish time bounds formulated in eq 10 through eq 13 are defined conservatively. Thus we state the following lemma.

Lemma 2. The finish time of a task τ cannot be smaller than τ^{minF} nor larger than τ^{maxF}.

(Proof is in supplementary section 3)

Note that there is a cyclic dependency between the equations (eq 1 to eq 13); we need to know the latest start times of tasks to determine the earliest start times of tasks and vice versa. Therefore they should be solved as a whole. Since the equations are not linear, we solve them in an iterative fashion until the solution is converged to a fixed set of values.

After we determine all time bounds of tasks, we compute the WCRT of the target task graph $\mathcal{T}$ as follows:

$$WCRT(\mathcal{T}) = \max_{\tau \in \mathcal{T}} \tau^{maxF} - \min_{\tau \in \mathcal{T}} \tau^{minA} \qquad \text{eq 14}$$

Since the WCRT of a task graph is determined by the maximum finish times of tasks that are conservatively computed, the estimated WCRT obtained from the proposed analysis is not smaller than the real WCRT of the system by lemmas 1 and 2, which is the correctness condition of any WCRT estimation scheme. We summarize this fact as the following theorem:

Theorem 1. The WCRT estimated from the scheduling time bound analysis is no smaller than the real WCRT.

4.2 Making the Time Bounds Tighter

While the equations of section 4.1 define the conservative time bounds, there is room for optimization to make the time bounds tighter. First, we need to avoid multiple counts of a single task preemption. To this end, we examine the preemption history when computing the latest start time and the latest finish time. We define a preemptor set of task τ, which is denoted by τ^{pmtor}, a set of tasks that affect τ^{maxF} or τ^{maxS}. Initially τ^{pmtor} is an empty set. When we compute the latest arrival time according to eq 5, we inherit the preemptor set of a predecessor task to its preemptor set. That is,

$$\tau^{pmtor} = \tau_p^{pmtor} \cup \tau^{pmtor} \quad \text{if} \quad \forall_{\tau_p}(\tau_p^{proc} = \tau^{proc}) \qquad \text{eq 15}$$

where τ_p is the predecessor task of task τ. When we examine the scheduling delay caused by the tasks that belong to set Δ_τ in eq 9

and set Y_τ in eq 13, we consider only the tasks that do not belong to τ^{pmtor}.

Another possible source of overestimation is infeasible preemption that can be made from a higher priority descendant task. To remove infeasible preemption, we manage another set of tasks, called the exclusion set τ^{excl} for each task. The exclusion set contains the higher-priority tasks that depend on the task by the topology order. The exclusion set of each task can be constructed before performing the scheduling time bound analysis.

Taking into account both preemptor set and exclusion set, set Δ_τ in eq 9 and set Y_τ in eq 13 are modified again to the final forms as follows:

$$\Delta_\tau = \left\{ \tau_s \middle| \begin{array}{c} \tau_s^{proc} = \tau^{proc},\ \tau_s^{pri} > \tau^{pri}, \tau_s^{minS} \le \tau^{maxS}, \\ \tau_s \notin \tau^{pmtor},\ \tau_s \notin \tau^{excl} \end{array} \right\},$$

$$Y_\tau = \left\{ \tau_s \middle| \begin{array}{c} \tau_s^{proc} = \tau^{proc},\ \tau_s^{pri} > \tau^{pri}, \tau^{maxS} \le \tau_s^{maxF}, \\ \tau_s^{minS} \le \tau^{maxF}, \tau_s \notin \tau^{pmtor}, \tau_s \notin \tau^{excl} \end{array} \right\} \qquad \text{eq 16}$$

We can also tighten the lower bounds, the earliest start time and the earliest finish time, of tasks. Tightening the lower bounds can tighten the upper bounds further since we can reduce the number of tasks that belong to set Δ_τ in eq 9 and set Y_τ in eq 13. To tighten the lower bound of the start time, we construct a set of tasks τ^{sched} that includes the tasks that will be scheduled ahead of task τ in all cases. τ^{sched} is obtained by local scheduling inside each processing element. Then, the earliest start time cannot be smaller than the earliest finish time of them. Then, eq 8 is modified as follows,

$$\alpha_\tau(\tau^{minS}) = \max_{\tau_s \in \mathcal{L}_\tau \cup \tau^{sched}} (\tau_s^{minF}) \qquad \text{eq 17}$$

Some examples that show how use of τ^{pmtor}, τ^{excl}, and τ^{sched} can tighten the bounds are included in supplementary section 5.

Note that aforementioned optimization techniques do not change the proofs of Lemmas 1 and 2 and theorem 1.

Thus, the following theorem holds.

Theorem 2. The conservativeness of timing bound is preserved after modification of set Δ_τ in eq 9 and set Y_τ in eq 13 to eq 16 and eq 8 to eq 17.

(Proof is in supplementary section 3)

4.3 Overall Algorithm

As explained earlier, we compute the time bounds of all tasks iteratively until they are converged. It is observed that we need to know the time bounds of all higher priority tasks and predecessor tasks to determine the time bound of a task as the equations imply. Thus we sort all tasks by the topological order first and their priorities next in descending order; let Q be the resulting sorted list of tasks.

The iterative procedure is composed of nested loop. The innermost loop of the proposed STBA algorithm traverses list Q to compute the time bounds of tasks as many as possible. If there is any task whose bound cannot be computed in the loop, we iterate the traversal until the time bounds of all tasks are computed, which defines the second loop. At the top-level, we iterate the second loop until all time bounds do not change.

For each traversal of Q, we can determine the time bound of at least one new task. The worst case arises when the priorities of the tasks are assigned in the reverse order of the topological dependency in a linear graph. The number of the middle loop iterations, therefore, becomes $O(N^2)$ where N is the number of tasks after task graph expansion. Since the outermost loop resolves the scheduling dependency among tasks with no topological dependency, we need at most N iterations until converged. In fact, we need only a few iterations in most cases as confirmed by experiments. In summary, the overall time complexity of the scheduling time bound analysis algorithm is $O(N^3)$. It is noteworthy that the time complexity does not depend on the number of PEs and the scheduling policy of the PEs.

For the detailed description of the overall algorithm, refer to the supplementary section 4.

5. Task Graph Expansion

To compute the WCRT of a task graph, we have to consider the worst case scenario of intervention from the other task graphs that overlap their executions in time with the target task graph. When the periods of task graphs are different, during one iteration of the target task graph, multiple iterations of other task graphs can be overlapped in time. In addition, multiple iterations of the same task graph can be overlapped in time to cause *self-blocking*. To account for the worst-case scenarios, we expand each task graph by instantiating a separate task graph for each iteration before applying our analysis in section 4.

The key question is how many times a task graph is expanded. At first, we compute the number for the target task graph, χ_{target}. It is given as $\chi_{target} = \left\lceil \frac{\mathcal{R}_{\mathcal{T}} + jitter}{\mathcal{T}^p_{target}} \right\rceil$. Next we calculate the expansion factors for the other task graphs; for task graph $\mathcal{T}_g$ the expansion factor becomes $\chi_g = \left\lceil \frac{\chi_{target} \times \mathcal{T}^p_{target}}{\mathcal{T}^p_g} \right\rceil + 1$. Note that there is another cyclic dependency. After expansion, the scheduling time bound analysis computes the WCRT value. For graph expansion, however, WCRT, $\mathcal{R}_{\mathcal{T}}$, value is required to compute χ_{target}. To resolve this cyclic dependency, we initialize the WCRT of the target task graph with a critical path length and repeat the scheduling time bound analysis until the expansion factors are converged, i.e. $\mathcal{R}_{\mathcal{T}} \le \chi_{target} \times \mathcal{T}^p_{target}$.

The cloned task graph is considered as an independent task graph. The priority of a cloned task inherits a priority from an original task and a task invoked earlier has higher priority than one invoked later by the same original task. The cloned task τ_i which is i-th instance of a task has the arrival time as follows.

$$\tau_i^{minA} = \min(\mathcal{T}^{offset}) + \mathcal{T}^p \times i$$
$$\tau_i^{maxA} = \max(\mathcal{T}^{offset}) + \tau^J + \mathcal{T}^p \times i$$

6. Experiments

Our experiments were carried out on an Intel 3GHz I7 core machine with 8GB RAM. We use the Java language to implement the STBA algorithm. For comparison purpose, we implemented Y&W algorithm following the pseudo code in [6].

6.1 Simple Examples

In the first set of experiments, we compare STBA with Y&W, MPA, and optimal solution, using three simple examples of Figure 2. Table 1 shows the comparison results. Since Y&W supports preemptive processors only, all processors are assumed to use preemptive scheduling policies.

The Y&W algorithm does not handle efficiently the case where there are multiple predecessors as shown in Figure 2(a). If there is no common predecessor, the Y&W algorithm assumes that the

tasks can intervene each other; τ_1 can preempt τ_2 and τ_0 can preempt τ_3 at the same time, which is not possible. For the example of Figure 2(b), the Y&W algorithm over-estimates the maximum finish time of τ_2 by considering that τ_0 delays τ_4, τ_1, and τ_2, which cannot occur in real scenarios. Since Y&W does not allow the jitter of task arrival time, the result is not available for Figure 2(c) in Table 1

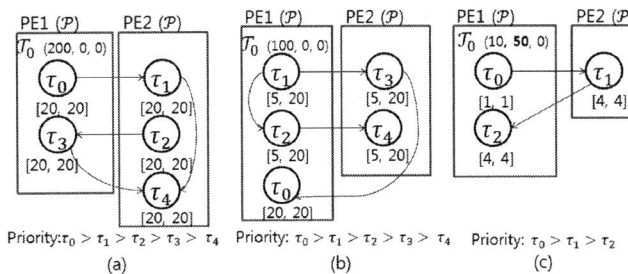

Figure 2. Simple examples that show different estimation results among difference approaches

Table 1 WCRT estimation results for simple graphs in Figure 2

	Y&W [6]	MPA [11]	Proposed	Optimal
Fig.2(a)	100	140	60	60
Fig.2(b)	100	100	80	80
Fig.2(c)	-	38	33	33

The MPA algorithm which is a well-known compositional approach provides the most over-estimated results for all example graphs. Since the event stream abstraction is overly pessimistic, infeasible and feasible preemptions cannot be distinguished. Since all task graphs in Table 1 have data-dependency arcs, it shows the poorest results. Note that the proposed analysis gives the optimal results for all examples used in this experiment.

6.2 Random Examples

For extensive performance evaluation, we generate 100 instances randomly for 4 different sets of experiments whose results are summarized in Table 2. The first set of experiments is performed for comparison with the Y&W method, assuming only preemptive processors. The task graphs are generated randomly in two groups, simple and large. For large(simple) configuration, the number of cores varies from 5 to 10 (1 to 5), the number of task graphs from 5 to 15 (1 to 3) and the total number of tasks lies from 50 to 100 (10). The execution time of each task is set randomly from 500 to 1500 time unit (1000 to 1200). The period of task graphs is also randomly chosen.

Table 2. Comparison for 100 random task graphs

Comparison		Win	Tie	Avg.	Max
Proposed vs Y&W	Large	100	0	89%	213%
	Simple	35	65	5%	44%
Optimal vs Proposed	Preemptive	27	73	2%	79%
	Non-preemptive	71	29	22%	109%
	Mixed scheduling	45	55	8%	48%

The next three sets are performed to compare the estimation results obtained from the proposed analysis with the optimal ones obtained from [2]. Since the time complexity of the optimal algorithm is exponential, only small configurations are used as the simple group above.

We compute the largest WCRT among the task graphs involved in each experiment. For all test sets, the proposed analysis gives no worse result than the Y&W method while it surely produces no better result than the optimal solution. The third column shows how many cases the proposed analysis produces a tighter bound than the Y&W method and a looser bound than the optimal solution. The forth column denotes the number of cases for which both analyses show an equivalent bound. The fifth and the sixth columns indicate the average and the maximum WCRT estimation gap between two approaches compared in each set of experiments. Note that the advantage of the proposed analysis over the Y&W method increases as the graph size increases. The performance gap between the proposed analysis and an optimal technique is relatively small compared with the gap between the proposed analysis and the Y&W method both for the average value and the maximum value. For each experiment with a large graph example, Y&W takes 0.5 second and proposed analysis takes 5.8 seconds on average, which confirms the scalability of the proposed technique. We observe that the performance gap between the proposed technique and the optimal one is larger for the non-preemptive case in general.

Table 3. Comparison with MAST [16] for 100 random task graphs

Comparison	Win	Tie	Lose	Avg. (%)	Max (%)
Preemptive	99	0	1	34	88
Non-Preemptive	87	3	10	9	49
Mixed	90	4	6	15	50

Table 3 shows the comparison results between STBA and MAST [16]. In MAST suite, we chose the *optimized offset based analysis* which is based on the technique introduced in [14]. Since the MAST supports only chain-structured graphs, all randomly generated task graphs are chain-structured. A configuration consists of 5 processing elements and 20 task graphs. And the total number of tasks is 100 in each configuration. The execution time range and the priority of each task, task mapping, and the period of each task graph are all randomly generated. As shown in the table, STBA outperforms MAST in most cases.

6.3 Case Study: An Automotive Application

We apply the proposed analysis algorithm to a real-life automotive application. Refer to the supplementary section 6 for the details of the system architecture. A target architecture has a central ECU (Electronic Control Unit) and 16 sensor and actuator ECUs. The ECUs are connected via one CAN bus. Communication task models a CAN frame that is a data packet with a fixed size and a fixed priority. The CAN bus is modeled as a fixed priority non-preemptive processing element. Task execution times and priorities are determined according to its size and frame id. In total, 50 frames are used in this system. The (p, j, d) information is obtained from the CAN database of the real system. The task graphs running on the ECUs and the task profiling information were also obtained from actual target platform. The ECUs use OSEK [12], a preemptive scheduler, to schedule the assigned tasks. The real data cannot be disclosed here due to confidentiality obligation. The task graphs perform the SCC/LKAS (Smart Cruise Control/Lane Keeping Assist System) algorithm, and three different periods are used: 10 ms, 20ms, 50ms. The senders and receivers have higher priority than the functional tasks for periodic frame transmission.

SymTA/S tool [13] is used to estimate the WCRT for comparison. The experimental results show that the proposed analysis produces 75% tighter results for ECU tasks than the SymTA/S approach on average. It takes about 3.7 seconds for the proposed analysis and 5 seconds in SymTA/S.

7. Conclusion

We propose a new analytical method, called scheduling time bound analysis (STBA), for worst case response time (WCRT) estimation of distributed embedded systems. Experimental results show it gives tighter bounds than well-known compositional approaches, SymTA/S and MPA while the analysis speed is comparable to those techniques. The proposed technique supports an arbitrary mixture of preemptive and non-preemptive processing elements.

While we assumed fixed-priority scheduling in the current implementation, other scheduling policies could be easily supported, since the proposed method performs scheduling for analysis. And we will find a way to tighten the WCRT bound more, especially for the non-preemptive systems in the future.

Acknowledgement

This work is supported jointly by MKE(Ministry of Knowledge Economy), Korea, under the ITRC support program supervised by the NIPA(National IT Industry Promotion Agency) (NIPA-2013-H0301-13-1011), Hyundai automotive Co., IT R&D program MKE/KEIT (No.10041608, Embedded system Software for New-memory based Smart Device) and Seoul Creative Human Development Program (HM120006).

References

[1] A. Brekling, M. R. Hanse., and J. Madsen, "Models and formal verification of multiprocessor system-on-chips," J. Logic Algebraic Program., 77(1-2), pp. 1–19, Sep.–Oct. 2008.

[2] J. Kim, H. Oh, H. Ha, S. Kang, J. Choi, and S. Ha, "An ILP-Based Worst-Case Performance Analysis Technique for Distributed Real-Time Embedded Systems," IEEE Real-Time Systems Symposium 2012, pp. 363-372, Dec. 2012.

[3] K. Richter, M. Jersak., and R. Ernst, "A formal approach to MpSoC performance verification," IEEE Computer., 36(4), pp. 60–67, Apr. 2003.

[4] L. Thiele, E. Wandeler, and S. Chakraborty, "Performance analysis of multiprocessor DSPs: A stream-oriented component model," IEEE Signal Process. Mag., 22(3), pp. 38–46, May 2005.

[5] R. L. Graham, "Bounds on multiprocessing timing anomalies," SIAM J. Appl. Math., 17(2), pp. 416–429, Mar. 1969.

[6] T. Y. Yen and W. Wolf, "Performance estimation for real-time distributed embedded systems," IEEE TPDS., 9(11), pp. 1125–1136, Nov. 1998.

[7] K. Tindell and J. Clark, "Holistic schedulability analysis of distributed hard real-time systems," Microprocessing and microprogramming, Vol. 40, pp. 117-134, April, 1994

[8] G. Behrmann, A. David, K. G. Larsen, P. Hakansson, P. Petterson, W. Yi, and M. Hendriks. "UPPAAL 4.0," QEST, 2006.

[9] G. Madl, N. Dutt, and S. Abdelwahed, "Performance estimation of distributed real-time embedded systems by discrete-event simulatioins," EMSOFT, pp. 183-102, Oct. 2007.

[10] K. Lampka, S. Perathoner, and L.Thiele, "Analytic real-time analysis and timed automata: a hybrid methodology for the performance analysis of embedded real-time systems,". Design Automation for Embedded Systems, 14(3), pp. 193-227, 2010.

[11] Ernesto Wandeler and Lothar Thiele, "Real-time calculus (RTC) toolbox," http://www.mpa.ethz.ch/Rtctoolbox, 2006

[12] OSEK VDX, http://www.osek-vdx.org

[13] R. Henia, A. Hamann, M. Jersak, R. Racu, K. Richter and R. Ernst, "System level performance analysis - the SymTA/S approach", IIEE Proceedings Computers and Digital Techniques, 2005.

[14] J C Palencia, Gonzalez Harbour M. "Exploiting precedence relations in the schedulability analysis of distributed real-time systems." In: Proceedings of the 20th IEEE Real Time Systems Symposium,1999.

[15] Martijn Hendriks, Marcel Verhoef, "Timed automata based analysis of embedded system architectures," IPDPS, 2006.

[16] MAST suite, http://mast.unican.es/

Supplementary Materials

1. Task Activation Example

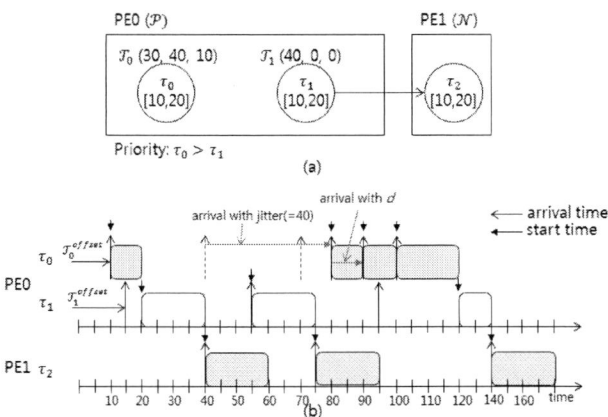

Figure 3. Example of task graph (a), and activation example (b)

Figure 3 (a) shows that two task graphs are mapped on two PEs. The activation condition of task graphs $\mathcal{T}_0$ and $\mathcal{T}_1$ are specified by (p,j,d) parameters that are (30, 40, 10) and (40,0,0) respectively. Task τ_2 has dependency to τ_1 which is mapped on PE1. An example scenario is depicted in Figure 3 (b) where the starting offsets of each task graph are 10 and 15 respectively. Task τ_1 arrives periodically every 40 time unit. Since task graph $\mathcal{T}_0$ has a jitter 40, the second arrival of instance τ_0 could arrive at 80 as illustrated in the graph. Then the third arrival cannot be earlier than 90 since the minimum distance between arrivals is given to 10.

2. Dynamic Offset (explanation of eq 3)

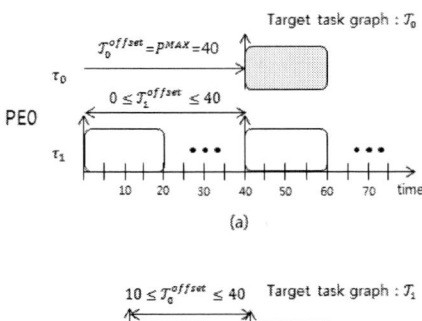

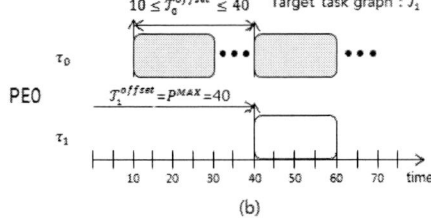

Figure 4. Starting offset range of task graphs in Figure 1 (a) when the target task graph is $\mathcal{T}_0$ and (b) when it is $\mathcal{T}_1$

To compute the WCRT of the target task graph, we should examine all possible scenarios that another task graph affects the schedule of the target task graph. So, it is necessary to start the other task earlier than the target task. In the current implementation of the STBA algorithm, we make all starting offsets positive. So we set the starting offset of the target task graph as the maximum of the periods of task graphs involved in the analysis. Then, we can consider all relative starting offset between the target task graph and another task graph with positive starting offset of the task graph.

Figure 4 shows the example how the starting offset range is determined for the task graphs of Figure 1 for two cases: one is when the target task graph is $\mathcal{T}_0$ and the other is when the target task graph is $\mathcal{T}_1$. The starting offset of the target task graph is set to 40, which is the maximum value of 30 and 40, in both cases. In the former case, the starting offset of τ_1 ranges from 0 to 40 while the starting offset of τ_0 ranges from 10 to 40.

3. Proofs of Lemmas and Theorem 2

Lemma 1. The start time of a task τ cannot be earlier than τ^{minS} nor later than τ^{maxS}. It means that the start time bounds formulated in eq 6 and eq 7 are defined conservatively.

(Proof) Let τ^{start} be the actual start time of τ. First, we prove that $\tau^{minS} \leq \tau^{start}$ by contradiction. Suppose, for some $\tau_s \in \mathcal{L}_\tau$, $\tau^{start} < \tau_s^{minF}$ which means τ can start earlier than eq. 6.

(Preemptive) Then $\tau_s^{minS} \leq \tau_s^{maxS} \leq \tau^{minS} < \tau_s^{minF} \leq \tau_s^{maxF}$, which implies that τ preempts τ_s. Hence $\tau_s^{pri} < \tau^{pri}$. It is contradictory to the assumption that $\tau_s \in \mathcal{L}_\tau$.

(Non-preemptive) (i) If $\tau_s^{pri} > \tau^{pri}, \tau_s^{maxS} \leq \tau^{minS} < \tau_s^{minF}$ implies that τ preempts, which is impossible. (ii) If $\tau_s^{pri} < \tau^{pri}, \tau_s^{maxS} \leq \tau^{minA}$ then $\tau_s^{maxS} \leq \tau^{minA} \leq \tau^{minS} < \tau_s^{minF}$. It implies that τ preempts τ_s, which is not possible under a non-preemptive scheduling policy.

Second, we prove that $\tau^{start} \leq \tau^{maxS}$. Suppose that there exists τ_t such that $\tau_t \notin \Delta_\tau$ and $\tau_t^{minS} < \tau^{maxS} < \tau^{start} = \tau_t^{maxF}$, which means τ starts immediately after τ_t completes. Then τ can start later than eq. 7.

(Preemptive) Since $\tau_t \notin \Delta_\tau$, the following inequality should hold; $\tau_t^{pri} < \tau^{pri}$ or $\tau_t^{minS} > \tau^{maxS}$. Since $\tau_t^{minS} \leq \tau_t^{maxF} = \tau^{start}$, $\tau_t^{pri} < \tau^{pri}$ should be satisfied, which means τ_t preempts τ; it is not possible. Therefore, no such τ_t exists.

(Non-preemptive) We consider two cases as follows;

(i) If $\tau_t^{pri} < \tau^{pri}$, the following inequality should hold; $\tau_t^{maxS} > \tau^{maxA}$ or $\tau^{maxA} \geq \tau_t^{maxF}$. If $\tau_t^{maxS} > \tau^{maxA}$, then τ should be scheduled before τ_t^{maxS}. It implies that $\tau^{start} < \tau_t^{maxS} < \tau_t^{maxF} = \tau^{start}$, which cannot be hold. On the other hand, if $\tau^{maxA} > \tau_t^{maxF}$ then, $\tau_t^{maxF} < \tau^{maxA} \leq \tau^{maxS} < \tau_t^{maxF}$, which is also impossible. Therefore, there is no τ_t such that $\tau_t \notin \Delta_\tau$ and $\tau_t^{pri} < \tau^{pri}$.

(ii) If $\tau_t^{pri} > \tau^{pri}$, $\tau_t^{minS} > \tau^{maxS}$ since $\tau_t \notin \Delta_\tau$. Then $\tau_t^{minS} < \tau^{maxS} < \tau_t^{minS}$, which is a contradiction.

By (i) and (ii), no such τ_t exists. Q.E.D.

Lemma 2. The finish time of a task τ cannot be smaller than τ^{minF} nor larger than τ^{maxF}.

(Proof) (Non-preemptive) If a task is non-preemptive then no task preempts it. Hence the finish time is the sum of the start time and the execution time.

(Preemptive) Let τ^{finish} be the actual finish time of τ.

First, we prove that $\tau^{minF} \leq \tau^{finish}$. Suppose that $\tau^{finish} < \tau^{minS} + \tau^{BCET} + \gamma_\tau(\tau^{minS})$ which means τ can finish earlier than eq. 10. Then there exists $\tau_s \in X_\tau$ which does not preempt τ. In order not to preempt τ, task τ_s finishes before task τ starts or τ_s starts after τ finishes, shortly $\tau_s^{start} > \tau^{minF}$ or $\tau_s^{finish} < \tau^{minS}$. Since $\tau_s \in X_\tau$, $\tau^{minS} \leq \tau_s^{minS} \leq \tau_s^{maxS} < \tau^{minF}$. Since $\tau_s^{start} \leq \tau_s^{maxS} < \tau^{minF}$, it does not hold that $\tau_s^{start} > \tau^{minF}$. Also since $\tau^{minS} \leq \tau_s^{minS} \leq \tau_s^{finish}$, it does not hold that $\tau_s^{finish} < \tau^{minS}$. There is no $\tau_s \in X_\tau$, which does not preempt τ.

Second, we prove that $\tau^{finish} \leq \tau^{maxF}$. Suppose that $\tau^{finish} > \tau^{maxF}$. It means that there exists τ_t such that $\tau_t \notin Y_\tau$ and τ_t preempts τ, being scheduled between τ^{maxS} and τ^{maxF}. Since $\tau_t \notin Y_\tau$, $\tau^{maxS} > \tau_t^{maxF}$ or $\tau_t^{minS} > \tau^{maxF}$.

(i) If $\tau^{maxS} > \tau_t^{maxF}$ then τ_t is scheduled before τ^{maxS}. Hence τ_t cannot preempt τ.

(ii) If $\tau_t^{minS} > \tau^{maxF}$ then τ_t is scheduled after τ^{maxF}. Hence τ_t cannot preempt τ.

By (i) and (ii), no such τ_t exists such that $\tau_t \notin Y_\tau$ and τ_t preempts τ. Q.E.D.

Theorem 2. The conservativeness of timing bound is preserved after modification of set Δ_τ in eq 9 and set Y_τ in eq 13 to eq 16 and eq 8 to eq 17.

(Proof) First, we will show that the start time of a task τ computed by eq 17 cannot be earlier than τ^{minS}. The only difference between eq 8 and eq 17 is that τ^{sched} is considered additionally in eq 17. Suppose that there exists $\tau_t \in \tau^{sched}$ such that $\tau^{start} < \tau_t^{minF}$. Task τ_t always finishes earlier than τ by definition of τ^{sched}, i.e. $\tau_t^{minF} \leq \tau^{start}$ since τ_t arrives earlier and has higher priority than τ, which is a contradiction to $\tau^{start} < \tau_t^{minF}$.

Second, we will show that the start time of task τ cannot be larger than τ^{maxS}. Suppose that there exists τ_t such that $\tau_t \notin \Delta_\tau$ in eq 16 and τ_t delays τ^{start} by preempting τ, which implies τ^{start} can be larger than τ^{maxS} computed by eq. 7 based on eq.16. It means that $\tau_t \in \tau^{pmtor} \cup \tau^{excl}$ to satisfy $\tau_t \notin \Delta_\tau$ by lemma 1. But τ_t cannot preempt τ by definition of τ^{pmtor} and τ^{excl}. It is a contradiction.

Last, we will show that the finish time of a task τ cannot be larger than τ^{maxF}. Suppose that there exists τ_t such that $\tau_t \notin Y_\tau$ in eq 16 and τ_t delays τ^{finish} by preempting τ, which implies τ^{finish} can be larger than τ^{maxF} computed by eq. 11 based on eq 16. It means that $\tau_t \in \tau^{pmtor} \cup \tau^{excl}$ to satisfy $\tau_t \notin Y_\tau$ by lemma 2. But τ_t cannot preempt τ by definition of τ^{pmtor} and τ^{excl}. It is a contradiction. Q.E.D.

4. Overall Algorithm of the Proposed STBA

Below is the overall algorithm of the proposed STBA.

Step0: $Q \leftarrow \mathcal{V}$. Q is a list of tasks sorted by the topological order first and their priorities next in descending order.
Step1: $S \leftarrow \{\}$. S is a list of tasks whose time bounds are all computed.
Step2: $W \leftarrow \{\}$. W is a waiting queue which contains tasks whose time bounds are not computable when it is visited.
Step3: $\tau \leftarrow Q_0$. Pop first element Q_0 from Q.
Step4: compute time bound of τ. If the time bound is computable then put τ into S and go to Step5; otherwise put τ into W and go to Step6.
Step5: compute time bound of tasks in list W again. Put tasks whose time bound is determined into list S and remove them from W.
Step6: if there is any unvisited task in Q, go to Step3
Step7: if any task was registered to the waiting queue W during Step3 to Step6, go to Step0. Otherwise go to Step8.
Step8: if time bound of any task is changed during Step1-Step7 then go to Step0. When the deadline of the target task graph is violated, finish the algorithm. Otherwise, return the WCRT of the target task graph and finish the algorithm.

It consists of three levels of nested loops. The outer-most loop (steps 0-8) checks whether the time bounds are all converged. The inner-most loop (steps 3-6) visits all tasks in the order of priority and computes the time bound of tasks according to the equations in 4.1 and 4.2. If we cannot determine the time bound of a task, we put the task into a waiting queue, W and move to the next task in Q. In case there is any waiting task in W, we examine whether the current task execution can resolve the waiting condition of any task in W. In the middle loop (steps 1-7) repeats the inner-most loop until no task is registered to the waiting queue. If the time bound of any task is changed during the middle loop, it sets the change flag to inform that the time bounds are not converged yet and repeats the outer-most loop.

The algorithm always terminates since the inner-most loop (steps 3-6) visits at least one per task. The timing bound of visited task does not decrease but always increases based on scheduling logic in section 4.1 and 4.2. If an application is not schedulable, the algorithm should finish without finding a fixed converged WCRT value. In that case, therefore, the algorithm terminates at step 8 with a finite value of deadline.

5. Examples for Tightening Time bounds

In this section, some illustrative examples show how to tighten the time bounds with the techniques explained in section 4.2. In all examples, we assume that the start offsets of all task graphs are zero for simple illustration.

For an example of Figure 5, there are two task graphs in which the execution time of task τ_0 varies. The base scheduling time bound analysis in section 4.1 computes the time bounds of the tasks as shown in Figure 5 (b) where the $WCRT(\mathcal{T}_1)$ is estimated to 45, the latest finish time of task τ_4. But the actual WCRT is 35. The reason of this over-estimation is that task τ_1 affects the latest finish time of τ_3 and the latest start time of τ_4; in other words, preemption delay caused by τ_1 is multiply counted in the WCRT computation of task graph $\mathcal{T}_1$.

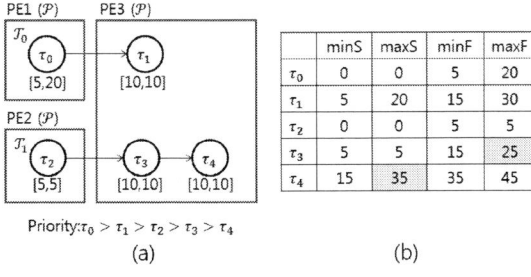

Priority: $\tau_0 > \tau_1 > \tau_2 > \tau_3 > \tau_4$

	minS	maxS	minF	maxF
τ_0	0	0	5	20
τ_1	5	20	15	30
τ_2	0	0	5	5
τ_3	5	5	15	25
τ_4	15	35	35	45

(a)　　　　　　　　(b)

Figure 5. An example to show why τ^{pmtor} is necessary

To prevent multiple counts of a single task preemption, we deploy eq 15 and eq 16. We update τ^{pmtor} to include the tasks that preempt task τ during computation of eq 9 and eq 13.

When the variation of task execution time is large, a loose bound can be made by ignoring infeasible preemptions even after we manage the preemptor task sets for all tasks. Consider an example of Figure 6 where the priorities of three tasks do not follow the topological order. At the first iteration of the inner-most loop (steps 2-5) in the base scheduling time bound analysis, the time bounds of the tasks are determined as shown in Figure 6 (b). But at the second iteration, task τ_0 affects the latest finish time of task τ_1 since the earliest start time of τ_0 is smaller than the latest finish time of τ_1; τ_0 belongs to set Y_{τ_1} in eq 13. Task τ_0 is inserted into the preemptor set of task τ_1 to avoid multiple preemptions. The preemptor set of task τ_2 also includes task τ_0, inherited from the preemptor set of its predecessor task τ_2. And the latest start time and the latest finish time of τ_3 are changed accordingly. At the 3rd iteration, the latest start time and the latest finish time of τ_0 are changed. The WCRT of the task graph is estimated to 100 finally because of task τ_4 which is connected to τ_2.

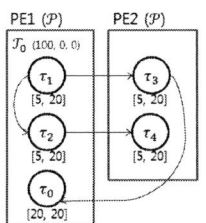

Priority: $\tau_0 > \tau_1 > \tau_2 > \tau_3 > \tau_4$

(a)

	minS	maxS	minF	maxF
τ_0	10	40	30	60
τ_1	0	0	5	20
τ_2	5	20	10	40
τ_3	5	20	10	40
τ_4	10	40	15	60

(b) 1st iteration

	minS	maxS	minF	maxF
τ_0	10	40	30	60
τ_1	0	0	5	40
τ_2	5	40	10	60
τ_3	5	40	10	60
τ_4	10	80	15	100

(c) 2nd iteration

	minS	maxS	minF	maxF
τ_0	10	60	30	80
τ_1	0	0	5	40
τ_2	5	40	10	60
τ_3	5	40	10	60
τ_4	10	80	15	100

(d) 3rd iteration

Figure 6. An example to show why τ^{excl} is necessary

In this example, the source of over-estimation is infeasible preemption by task τ_0 over its predecessor task τ_1. To exclude infeasible preemption possibility, we manage a set of tasks, called the exclusion set τ^{excl} for each task. The exclusion set of a task contains the higher-priority tasks that depend on the task by the topology order. Task τ_0 is included in the exclusion set of task τ_1 and that of task τ_2. The exclusion set of each task can be constructed before performing the scheduling time bound analysis.

Figure 7 shows the example that can tighten the lower bounds, the earliest start time and the earliest finish time, of tasks. Five tasks

are assigned to the same PE under a preemptive scheduling policy. The time bounds are computed as depicted in Figure 7 (b). Since the latest start times of task τ_3 and τ_4 are larger than the earliest release time of task τ_5, tasks τ_3 and τ_4 do not belong to set $\mathcal{L}_\tau$ of eq 8. As a result, the earliest start time of τ_5 is loosely computed ignoring the scheduling delay caused by tasks τ_3 and τ_4. It increases the latest finish time of task τ_2 in turn.

To avoid this problem, we construct a set of tasks τ^{sched} including the tasks that are scheduled ahead of task τ in all cases. In the example of Figure 7, tasks τ_3 and τ_4 belong to τ_5^{sched} since they are scheduled ahead of task τ_5 in all cases. Then, the earliest start time cannot be smaller than the earliest finish time of them.

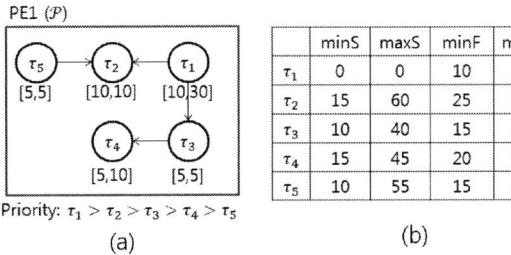

Priority: $\tau_1 > \tau_2 > \tau_3 > \tau_4 > \tau_5$

(a)

	minS	maxS	minF	maxF
τ_1	0	0	10	30
τ_2	15	60	25	70
τ_3	10	40	15	45
τ_4	15	45	20	55
τ_5	10	55	15	60

(b)

Figure 7. An example to show why τ^{sched} is necessary

897

6. Case Study

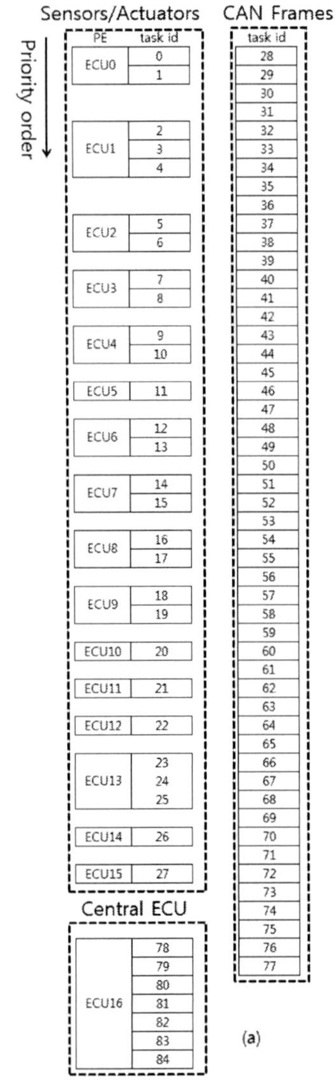

Sensors/Actuators   CAN Frames

Priority order

period	task id
10ms	0, 2, 9, 12, 14, 16, 23, 27, 78, 81, 82
20ms	1, 3, 5, 7, 11, 15, 17, 24, 26, 79, 83
50ms	6, 8, 22, 80, 84
100ms	4, 13, 20, 21, 25
200ms	18
1000ms	10, 19

(b)

dependency	
from(task id)	to (task id)
0	33, 34, 38, 43, 44, 45, 48, 49, 69
1	28, 52
2	37, 42, 46
3	54
4	74
5	53
6	15, 16, 66
8	73, 67
9	47
10	70, 71
11	55
12	51, 59, 60
14	35, 36, 40
15	39, 57, 61, 62, 63, 75
18	29
19	30, 31, 32
21	76
24	64, 65
25	77
26	58
27	50
78	41
79	56
80	68, 72

(c)

(a)

Figure 8. Architecture of SCC/LKAS system; (a) task mapping and priority assignment, (b) task periods, and (c) task dependency

Figure 8(a) shows the architecture diagram of SCC/LKAS. The core algorithm consists of two parts; smart cruise control (SCC) and lane keeping assistant system (LKAS). Both parts are running as periodic tasks on the central ECU which is marked as ECU16 in Figure 8 (a). Sensor and actuator ECUs are notated ECU0~ECU15 in Figure 8 (a) and they generate CAN frames periodically. The periods of tasks are shown in Figure 8 (b). The CAN frames are modeled as non-preemptive tasks that are activated by data dependency from computation tasks in the other processing elements.

Optimizations for Configuring and Mapping Software Pipelines in Many Core Systems

Janmartin Jahn, Santiago Pagani, Sebastian Kobbe, Jian-Jia Chen, Jörg Henkel

Karlsruhe Institute of Technology (KIT), Germany

{ jahn, santiago.pagani, sebastian.kobbe, j.chen, henkel }@kit.edu

Abstract—**Efficiently utilizing the computational resources of many core systems is one of the most prominent challenges. The problem worsens when resource requirements vary unpredictably and applications may be started/stopped at any time. To address this challenge, we propose two schemes that calculate and adapt task mappings at runtime: a centralized, optimal mapping scheme and a distributed, hierarchical mapping scheme that trades optimality for a high degree of scalability. Experiments on Intel's 48-core Single-Chip Cloud Computer and in a many core simulator show that a significant improvement in system performance can be achieved over current state-of-the-art.**

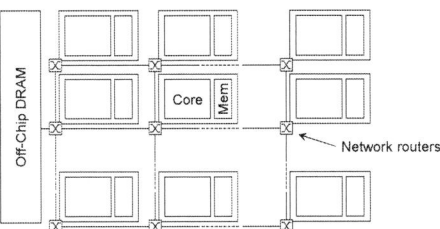

Fig. 1: Target architecture

I. INTRODUCTION AND NOVEL CONTRIBUTION

While many core systems offer the potential of vastly increasing computational performance as Moore's Law continues, in practice there are significant hurdles to actually utilize these resources and to profit from scalability. One key issue is the mapping of applications to the available cores. When considering *dynamic runtime scenarios*, i.e. when applications may be started, stopped, or when their *resource demands* (i.e. the computational demands of their tasks and the communication requirements between them) may vary unpredictably at any time (e.g. due to user interactions or changing input data), the problem of mapping tasks is extended from *finding* mappings to also *adapting* them at runtime. In such scnearios, it is crucial to adapt mappings to account for such changes in order to maintain high system performance: Section II shows an example how significantly changing resource demands of an application may lead, for an established task mapping, to a degradation of the system throughput of more than 50% compared to an adapted mapping. The same problem may arise when applications are started or stopped unpredictably (e.g. by the user). Such scenarios are increasingly common and thus need to be addressed [16]. To tackle this challenge, one solution is to employ so-called *malleable* applications that provide the flexibility of using more or less cores dynamically (e.g. [5], [20]). They can change their degree of parallelism at runtime so that a system may re-distribute the cores among applications to increase its performance [10]. Our approach focuses on software pipelines because they are a well-established means to parallelize a large class of complex applications. Especially stream-processing applications, among which are very common image/video and networking applications, are well suitable for software pipelining. Multiple approaches to extract software pipelines (semi-)automatically from sequential C code by parallelizing compilers [3], [13], [19] have been presented.

For the rest of this paper, we use the following definitions:

Software pipelines consist of multiple *stages*, each processing subsequent *iterations* on a stream of input data. Each stage is an individual *task* consisting of a working set, program code, and task state. The output data of one stage forms the input data of its direct successor. There is no further communication.

A **malleable software pipeline** can reduce the number of its stages (thus the number of cores used) at runtime by *fusing* consecutive stages so they are mapped to the same core (similar to fusing filters

in StreamIt [6]). Consequently, no on-chip communication is necessary between them. Fused stages can be split through *fissions* until the initial degree of parallelism is restored.

We use **throughput** as a performance metric because software-pipelined applications often run continuously until they are stopped (thus metrics like makespan are not applicable). We define the *throughput of an application* as the number of iterations it completes per second, and the *throughput of a system* as the averaged throughputs of all running applications. A formal definition follows in Section IV.

In this paper, **task migration** is used to denote the transfer of the execution of a task from one core to another. **Task remapping**, in contrast, refers to the abstraction of deciding about task migrations based on a system model (i.e. the underlying algorithm or heuristic). We target systems with many cores, private, distributed memories and Network-on-Chips. Figure 1 shows this exemplarily.

In this paper, our **novel contributions** are:

1) We present a centralized, optimal mapping scheme for malleable software pipelines.
2) As an extension, we present a distributed, hierarchical mapping scheme that trades the optimum for a high degree of scalability.

To illustrate the effectiveness of our schemes, we have implemented them on Intel's Single-Chip Cloud Computer (SCC) [9] and in a high-level many core system simulator. Our centralized scheme requires approx. 60ms for calculating optimal mappings for 48 cores, while our distributed scheme calculates near-optimal mappings for 1024 cores in less than 1ms[1].

The rest of this paper is organized as follows: Section II presents a motivational example, Section III discusses the state of the art, and Section IV presents the system model. We define the mapping problem in Section V and afterwards we present our centralized and our distributed solution (Sections VII and VIII). Section IX details our implementation, and Section X describes our experiments and comparison to the state of the art.

II. MOTIVATION

This section discusses the importance of adapting task mappings in dynamic runtime scenarios. Let us consider a simple example of a software-pipelined computer vision application (object tracking) with 8 stages mapped to a system with 4 cores[2]. Figure 2 (a) shows how the average runtime of each stage changes when adding multiple tracked objects to the input scene. Figure 2 (b) shows that an established (optimal) task mapping (Core 1: Stages 1-3, Core 2: Stages 4-5, Core

Permission to make digital or hard copies of all or part of this work for personal or classroom use is granted without fee provided that copies are not made or distributed for profit or commercial advantage and that copies bear this notice and the full citation on the first page. To copy otherwise, to republish, to post on servers or to redistribute to lists, requires prior specific permission and/or a fee.

DAC '13, May 29 - June 07 2013, Austin, TX, USA.

---

[1]Experiment conducted on a P45C core running at 800 MHz.

[2]For this example, we use 4 cores of Intel's SCC [9].

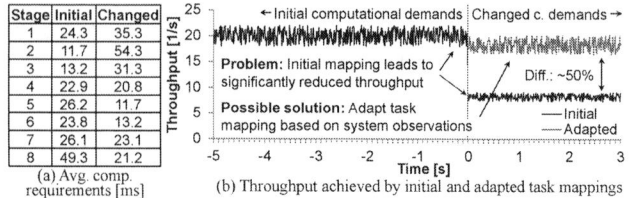

Stage	Initial	Changed
1	24.3	35.3
2	11.7	54.3
3	13.2	31.3
4	22.9	20.8
5	26.2	11.7
6	23.8	13.2
7	26.1	23.1
8	49.3	21.2

(a) Avg. comp. requirements [ms]

(b) Throughput achieved by initial and adapted task mappings

Fig. 2: Changing computational requirements and resulting throughputs

3: Stages 6-7, Core 4: Stage 8) achieves a throughput of approx. 20.13 iterations/second. Due to these changes, the average throughput drops to 8.35 iterations/second. A possible solution to this problem is to adapt the task mapping (Core 1: Stage 1, Core 2: Stage 2, Core 3: Stages 3-4, Core 4: Stages 5-8), which achieves a throughput of 18.40 iterations/second. Consequently, adapting the established task mapping based on observations about the (possibly unpredictable) resource demands can significantly improve the throughput of a system. A more complex example of a system with 128 cores running 35 instances of real-world applications concurrently is discussed in Section X-C.

However, adapting such mappings at runtime is a challenging problem because task mapping is NP-complete. Thus, calculating mappings for larger systems at runtime may require an infeasibly high overhead or may require heuristics that lead to suboptimal solutions.

III. RELATED WORK

The related work can be grouped into mapping schemes for software pipelines and for parallel applications in general, assuming either distributed or shared memories.

Mapping schemes specific to software-pipelined systems have been recently proposed: [16] suggests calculating a set of optimal mappings at design-time. This works well for a specific set of scenarios, but it does not aim at capturing cases where application resource demands are unknown at design time. This is, however, the case when they depend on user interactions or on (unpredictable) properties of the input data. Dynamic scheduling of stream-processing applications, which are a superset of software pipelines, to embedded multi-cores with scratchpad memories is proposed by [11]. The property deemed to be unpredictable, and hence targeted by this approach, is merely the availability of cores, while application resource demands are assumed to be static. With similar assumptions, [4] incorporates user behavior in runtime task mappings, but aims at minimizing communication energy and requires that the number of tasks is less or equal to the number of cores.

Task mapping of general parallel applications assuming distributed memories: [2] presents a heuristic runtime load balancing scheme for asynchronous, iterative algorithms (AIAC) in grid computing systems. Due to its focus, it does not take inter-task communication into account. It therefore may achieve inferior performance when tasks communicate heavily, as it is the case for many complex, real-world applications. In [10], a distributed heuristic for (re-)mapping of malleable applications using multiple agents is proposed. It relies on runtime observations and on offline profiles. However, their approach for achieving a scalable solution limits their decisions to local regions, which results in a lower throughput of the system. As [2] and [10] are most similar to our contribution, Section X compares them to our proposed schemes.

A statistical approach based on extreme value theory is presented in [14]. It generates a large random set of task assignments and has a runtime of 25 minutes to 2 hours, which we consider infeasible for dynamic scenarios that require to update mappings at runtime. In contrast to this, our restriction to (malleable) software pipelines allows to calculate optimal mappings in polynomial time, and near-optimal mappings in nearly constant time.

Task mapping for general parallel applications assuming shared memories: [12] and [15] propose runtime load balancing for symmetric multiprocessing systems. The authors of [17] propose to derive co-schedules based on offline profiles, with an extension to support different priority levels [18]. The focus of these schemes is on architectures with

few cores and they require a shared address space. They are thus not directly applicable to many core systems with distributed memories.

To summarize, state-of-the-art task mapping schemes either achieve inferior performance due to their broad scope, are not applicable to systems with distributed memories, or do not target dynamic runtime scenarios. However, it is important to address these scenarios for systems with many cores and distributed memories.

IV. SYSTEM MODEL

In the following, we discuss the system model we use for malleable software-pipelined applications. Each application k forms a pipeline P_k with N_k stages. Every stage S_j is characterized by c_j, e_j and o_j that denote the time consumed (in each iteration) for computation, for receiving the input data from its direct predecessor, and for transferring the output data to its direct successor. Figure 3 illustrates this model. For notational brevity, $e_i = o_{i-1} \forall i > 1$.

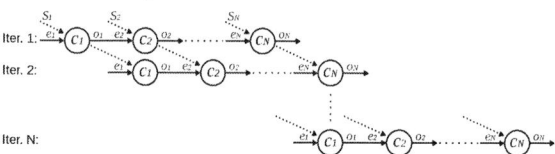

Fig. 3: Software pipeline model

In order to decide about the mapping of applications, it is important to model their throughput for a given mapping. To achieve this, we require that each core belongs to at most one application (i.e. cores may not be shared among applications). Furthermore, we need to determine their maximum throughput, which is limited by their slowest stage. We consequently denote the *maximal response time* R_k for pipeline P_k as:

$$R_k = \max_{1 \leq j \leq N_k} \{e_j + c_j + o_j\}. \tag{1}$$

Therefore, the maximum *throughput* of pipeline P_k is defined as $\frac{1s}{R_k}$.

We introduce the *malleability* property to software pipelines by defining the basic operation *fusion* (and the inverse operation *fission*), in which multiple consecutive pipeline stages are combined, similar to fusing filters in StreamIt [6].

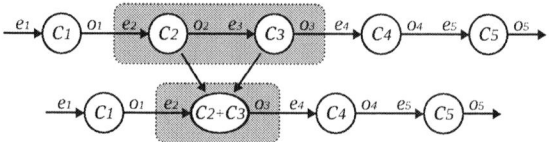

Fig. 4: Fusion of pipeline stages

A fusion of stages creates a new stage which combines the computational requirements of the original stages but does not require communication between them, as shown in Figure 4. This way, fusing stages may reduce the maximal response time R_k of a pipeline. Additionally, fusing stages changes the degree of parallelism of the application, which then runs on a smaller number of cores.

V. PROBLEM DEFINITION

We divide the problem of mapping malleable software pipelines into:

1) How to distribute the cores of a system among the applications (Section V-A) so that the overall system throughput is maximized.
2) How to assign the stages of an application to a given number of cores (Section V-B), thus providing their fusions.

A. Global Problem: Optimizing System Throughput

Given a set of K *weighted* (weights express priority levels) applications $P = \{P_1, P_2, \ldots, P_K\}$ with weights $W = \{w_1, w_2, \ldots, w_K\}$, each application P_k uses up to M_k cores and has a maximal response time R_k. The objective is to **maximize the overall weighted system throughput** by finding an optimal distribution of (up to) M available cores to the individual applications.

$$\text{Maximize} \left\{ \sum_{k=1}^{K} \frac{w_k}{R_k} \right\} \quad | \text{ such that } \sum_{k=1}^{K} M_k \leq M \qquad (2)$$

We present a centralized, optimal scheme in Section VII and a highly scalable, distributed scheme in Section VIII.

To solve this problem, however, we need to solve the sub-problem of fusing pipeline stages first:

B. Sub-Problem: Fusion of Pipeline Stages

The throughput of an application is affected by *how* the stages are fused. Thus, we define a sub-problem that minimizes the maximal response time of each pipeline P_k (with N_k stages) by fusing stages for an optimal throughput when utilizing at most M_k cores.

We present an algorithm to solve this problem in Section VI.

VI. FUSION OF PIPELINE STAGES

In order to find an optimal solution to the problem of Section V-B, all possible combinations of fusions need to be taken into consideration. An exhaustive search would result in an exponential time complexity, which may be unacceptable, especially for adapting mappings at runtime. We therefore propose an algorithm based on dynamic programming that derives optimal solutions for minimizing the maximal response time by using m cores to execute pipeline P_k.

Let $P_{k,j}$ be a *sub-pipeline* by considering only the pipeline stages from stage S_1 to stage S_j of pipeline P_k. The dynamic programming defines a recursive function $R_k(j, m)$ to store the optimal configurations for the maximal response time minimization for $P_{k,j}$ with (at most) m cores. That is, let $R_k(j, m)$ be the minimum maximal response time for executing $P_{k,j}$ on m cores. Moreover, we build table $F_k(\ell, j)$ for all ℓ, j such that $1 < \ell \leq j \leq N_k$ in which

$$F_k(\ell, j) = e_\ell + o_j + \sum_{h=\ell}^{j} c_h. \qquad (3)$$

Then, the initial boundary conditions for $R_k(j, 0)$ and $R_k(j, 1)$ are:

$$\begin{aligned} R_k(j, 0) &= \infty \qquad \forall j = 1 \dots N_k \\ R_k(j, 1) &= F_k(1, j) \quad \forall j = 1 \dots N_k \end{aligned} \qquad (4)$$

Furthermore, we define function $\text{minmax}RF_k(j, m)$ as:

$$\text{minmax}RF_k(j, m) = \min_{m-1 \leq \ell < j} \{\max\{R_k(\ell, m-1), F_k(\ell+1, j)\}\}. \qquad (5)$$

The recursive function for $R_k(j, m)$ with $m \geq 2$ is defined as:

$$R_k(j, m) = \begin{cases} R_k(j, m-1) & j < m \\ \min\{R_k(j, m-1), \text{minmax}RF_k(j, m)\} & j \geq m \end{cases} \qquad (6)$$

The dynamic programming starts by computing the resulting maximal response times utilizing only one core for the first $j = 1 \dots N_k$ stages. Then, the programming computes the maximal response times for the first $j = 1 \dots N_k$ stages, on up to two cores. Since the programming already stored the resulting maximal response times of using only one core for the first j stages, it can easily choose whether to use one or two cores (in one of the possible fusion combinations) for the same j stages. The process is repeated once again for three cores, knowing in advance if is optimal to use one or two cores for the first j stages, so it only needs to compare the previous result with any new possible fusion for the same j stages but now utilizing up to three cores. Thus, iteratively, an optimal solution is achieved because all combinations of stages and cores are considered, but the complexity is reduced since optimal solutions are stored in tables and do not need to be recomputed.

The space/time complexity is $O(N_k^2)$ for building the table F_k. The time complexity for building an entry $R_k(j, m)$ is $O(j) = O(N_k)$. The size of the table $R_k(j, m)$ is $O(M_k N_k)$. Therefore, the total time complexity is $O(M_k N_k^2)$. The maximal response time by using at most M_k cores for pipeline P_k is stored in $R_k(N_k, M_k)$. Algorithm 1 shows the pseudo-code for this dynamic programming.

Algorithm 1 Maximal Response Time Minimization

Input: The times e, c, and o for the N_k stages of pipeline P_k, and the maximum M_k cores available;
Output: The minimal maximal response time using at most M_k cores;
1: Initialize $F_k(\ell, j)$ according to Eq. (3), $\forall (\ell, j)$ such that $1 \leq \ell \leq j \leq N_k$;
2: **for** $m = 0$ to M_k **do**
3:     **for** $j = 1$ to N_k **do**
4:        **if** $m \leq 1$ **then**
5:           Build $R_k(j, m)$ according to Eq. (4);
6:        **else**
7:           Build $R_k(j, m)$ according to Eq. (6);
8:        **end if**
9:     **end for**
10: **end for**
11: return $R_k(N_k, M_k)$;

The actual fusions that lead to the optimal result can be derived by backtracking the dynamic programming table or by using an additional **tracking table** $TR_k(N_k, M_k)$ of size $O(M_k N_k)$. When building the $TR_k(j, m)$ table, each cell holds the j^* value of the sub-solution that makes the programming optimal. For the initial condition $m = 1$, $TR_k(j, m)$ is set to zero. When $j < m$, or when $j \geq m$ and $R_k(j, m-1)$ turned out to be minimal, then $TR_k(j, m) = j$. In the case where an additional core provides improvement, $TR_k(j, m)$ will be set to the index ℓ from Equation 5 that made this improvement possible and therefore $TR_k(j, m) \neq j$.

The fusions that give an optimal maximal response time can be derived from table $TR_k(N_k, M_k)$ as follows: starting from cell $(j, m) = (N_k, M_k)$, the table is traversed in the direction $(TR_k(j, m), m-1)$.

If $TR_k(j, m) = j$, this means that it is not possible to assign more cores to the pipeline since no finer granularity can be achieved or that no additional core may improve the throughput and the sub-solution that uses one less core was already optimal.

If $TR_k(j, m) \neq j$, an additional core provides improvement, so if $TR_k(j, m) + 1 = j$ then stage S_j is mapped to one core and if $TR_k(j, m) + 1 < j$ all stages between $TR_k(j, m) + 1$ and j (both inclusive) should be fused. Section S2 discusses a detailed example.

VII. CENTRALIZED SCHEME

With the dynamic programming of Section VI, we can decide how to maximize the overall weighted system throughput in a centralized manner. Suppose that $R_k(N_k, m)$ for $m = 1, 2, \dots, \min\{N_k, M\}$ has been built. For notational brevity, if $N_k < M$, we define $R_k(N_k, m) = R_k(N_k, N_k)$ for any $m \geq N_k$. Let $G(k, m)$ be the maximum centralized weighted system throughput for the first k pipelines based on any arbitrary order of pipelines on at most m cores. Moreover, when there is no feasible solution, i.e. $k > m$, the function $G(k, m)$ is defined to $-\infty$. Then, we know that the initial (boundary) condition for $G(1, m)$ is:

$$G(1, m) = \frac{w_1}{R_1(N_1, m)} \qquad \forall m = 1, 2, \dots, M \qquad (7)$$

The recursive function for $G(k, m)$ with $k \geq 2$ is expressed in Equation (8). The time complexity, provided that $R_k(N_k, m)$ is known, is $O(KM^2)$. Note that the last column of R_k, i.e. $R_k(N_k, m)$ $\forall m = 1, 2, \dots, M$, contains the application's weighted throughput and thus serves as its *speed-up vector*. Algorithm 2 shows a pseudo-code for this dynamic programming.

$$G(k, m) = \begin{cases} -\infty & k > m \\ \max_{k-1 \leq m' < m} \left\{ G(k-1, m') + \frac{w_k}{R_k(N_k, m-m')} \right\} & k \leq m \end{cases} \qquad (8)$$

An additional tracking table $TG(K, M)$ of size $O(KM)$ allows for easily deriving how many cores should be assigned to each pipeline. When building the $TG(k, m)$ table, each cell holds the m^* value of the sub-solution that makes the programming optimal. For the initial condition $k = 1$, $TG(k, m)$ is set to zero. When $k > m$, then

901

Algorithm 2 Maximizing Overall Weighted System Throughput

Input: The maximum number of available M cores. For every pipeline P_k, the weights w_k and tables $R_k(N_k, m)$ for $m = 1, 2, \ldots, M$;
Output: Maximum overall weighted system throughput for K pipelines, using at most M cores;
1: **for** $k = 1$ to K **do**
2:    **for** $m = 1$ to M **do**
3:       **if** $k = 1$ **then**
4:          Build $G(k, m)$ according to Equation (7);
5:       **else**
6:          Build $G(k, m)$ according to Equation (8);
7:       **end if**
8:    **end for**
9: **end for**
10: return $G(K, M)$;

$TG(k, m) = -1$. In the case were $k \leq m$, $TG(k, m)$ will be set to the value of m' from Equation 8 that made this sub-solution optimal.

Once table $TG(K, M)$ has been built, the number of cores for each pipeline can be derived from it: Starting from the final cell $(k, m) = (K, M)$, the table is traversed in the direction $(k - 1, TG(k, m))$. If $TG(k, m) = -1$, then there is no feasible solution for this set of values. In any other case, cores between $TG(k, m) + 1$ and m (both inclusive) should be assigned to application k. Section S3 shows a detailed example. It should be noted that our proposed scheme is not tied to the objective of maximizing the overall weighted system throughput: Equation 8 could be modified, e.g. to balance the throughput among applications, or to guarantee a minimum throughput.

VIII. DISTRIBUTED, HIERARCHICAL SCHEME

The scheme proposed in Section VII is designed in a centralized manner. This requires global system knowledge and computes the mapping for the entire system. This leads to a quadratic time complexity (with the number of cores), which may be infeasible for a large number of cores. To achieve a highly scalable solution (i.e. its overhead should not grow significantly with a growing number of cores or applications), this section (in combination with Section VI) proposes a distributed, hierarchical scheme for which we group the pipelines into several independent clusters. Clusters are grouped hierarchically into larger clusters and so on, therefore constructing a tree, as shown on Figure 5.

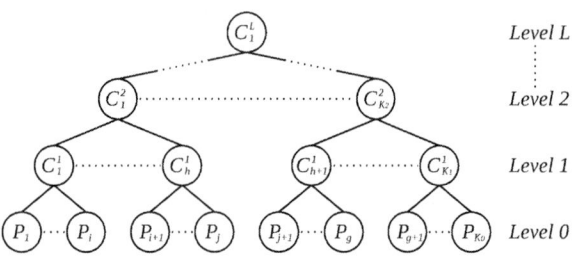

Fig. 5: Distributed solution

There are K_0 pipelines $P_1, P_2, \ldots, P_{K_0}$ on level 0, and these nodes are the leaves of the tree. The rest of the nodes are clusters and are expressed as C_i^ℓ, where indexes ℓ and i represent the level of the cluster in the tree and the index of the cluster in the level, respectively. All clusters in level 1 ($\ell = 1$) are the adjacent parents of the pipelines. There are L levels in the tree, where level L is the root of the tree, and each level ℓ holds K_ℓ nodes.

With this distributed model, the solution from Section VI is utilized to build the tables $R_k(N_k, M)$ for every pipeline P_k, where M continues to be the total amount of cores available in the system.

Each cluster C_i^1 (level 1) contains the information of the weights w_k and tables $R_k(N_k, M)$ of its children (pipelines) and utilizes the solutions of Sections VII to build the corresponding tables $G(K^*, M)$, where K^* is the number of child nodes of the cluster. According to this table, we record the best configuration for cluster C_i^1 by allocating

$m = 1, 2, \ldots, M$ cores to its children pipelines, independently upon the other clusters of the same level.

Similarly, the clusters C_i^2 (level 2) contain the information of table $G(K^*, M)$ of its child clusters C_i^1 (level 1). This applies likewise to all upper levels. In this way, each level distributes cores among its children based solely on this (limited) information. Consequently, the computational requirement is distributed hierarchically among the system. By limiting the frequency of propagating the aforementioned tables, our distributed scheme largely reduces the computational overhead, but is unable to achieve optimal mappings if the tables change more frequently than they are propagated.

We denote $G_i^\ell(k, m)$ as the table for the modified version of the dynamic programming in Section VII. Considering that w_1^*, R_1^*, N_1^* are the parameters of the first child (pipeline) of node C_i^1, node $C_{1*}^{\ell-1}$ is the first child of node C_i^ℓ, value $K_{\ell-1}^*$ is the number of children of node C_i^ℓ, and value $K_{\ell-2}^*$ is the number of children of node $C_{1*}^{\ell-1}$ and node $C_k^{\ell-1}$, then the initial conditions of $G_i^\ell(1, m)$ are:

$$G_i^\ell(1, m) = \frac{w_1^*}{R_1^*(N_1^*, m)} \quad \forall m = 1, 2, \ldots, M \quad \text{when } \ell = 1$$
$$G_i^\ell(1, m) = G_{1*}^{\ell-1}(K_{\ell-2}^*, m) \quad \forall m = 1, 2, \ldots, M \quad \text{when } \ell \geq 2, \quad (9)$$

the value of $G_i^\ell(k, m)$ is set to $-\infty$ whenever $k > m$, the recursive function when $\ell = 1$ and $k \leq m$ is:

$$G_i^\ell(k, m) = \max_{k-1 \leq m' < m} \left\{ G_i^\ell(k - 1, m') + \frac{w_k}{R_k(N_k, m-m')} \right\}, \quad (10)$$

the recursive function when $\ell \geq 2$ and $k \leq m$ is:

$$G_i^\ell(k, m) = \max_{k-1 \leq m' < m} \left\{ G_i^\ell(k - 1, m') + G_k^{\ell-1}(K_{\ell-2}^*, m - m') \right\}, \quad (11)$$

and finally, the result is found in cell $G_i^\ell(K_{\ell-1}^*, M)$.

It is important to notice that even though the root node makes decisions that affect every pipeline, this is still a distributed and scalable scheme, since every node only contains the partial information of its children.

IX. SYSTEM DETAILS

In the following, we discuss the components of our schemes and their implementation details. We have implemented both schemes on Intel's Single-Chip Cloud Computer (SCC) and in a high-level system simulator detailed in Section X-A. Both schemes employ several components written in C++ that communicate by exchanging network messages:

A. Components

The **centralized scheme** employs *application heads* and a *centralized controller*. Each application denotes one of its cores to form its *application head* (this core may execute a stage). The application head registers and signs-off the application with the centralized controller on starting and stopping of the application. To register an application, the application head sends a message including a unique identifier of the application (4 bytes), its number of stages (4 bytes), and an initial R_k table of Section VI (4 bytes $\times$ N_k stages $\times$ M_k cores). During runtime, application heads re-compute their R_k table when the values of e_i, c_i or o_i for one of their stages change and send this to the centralized controller. The centralized controller (re-)computes the optimal distribution of cores and sends a list of cores to each application. Based on this list, the application heads fuse and migrate their stages. The values e_i, c_i, and o_i are obtained by comparing the CPU tick counter before and after the corresponding operations are performed, thus they are updated once per iteration.

The **distributed scheme** employs *application heads* (see above) and *cluster heads*: for each cluster, a cluster head receives the R_k tables (4 bytes for each entry) from each of its children, which may be either cluster heads themselves, or application heads. However, instead of calculating global mappings, cluster heads only calculate core distributions for their children and propagate the combined R_k tables to their parent (see Section VIII).

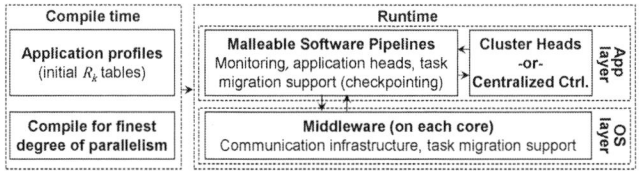

Fig. 6: Schematic overview of our implementation

B. Implementation of our Schemes

Figure 6 gives an overview of our implementation, which is divided into a compile-time and a run-time part. At compile-time, initial R_k tables are derived from profiling. Partitioning the application into a software pipeline defines its finest degree of parallelism. At runtime, our implementation is split into the application- and the OS-layer.

The application layer contains the components of Section IX-A and a checkpointing-based implementation of task migration (see below). Each core that is assigned to one application executes the same executable file, while its parameters control which of its stages are executed. A pseudo-code of this main procedure is shown in Section S1.

The OS layer provides the communication infrastructure (to allow stages to communicate via an MPI-like interface, orthogonal to their physical location and fusion) and supports task migrations.

On the SCC, our components are implemented as daemons for Intel's 3.1.4 ubuntu-based linux. The components of both schemes communicate via sockets and separate program- and control communication by two logical channels. To measure the communication overhead of our schemes, we log the communication volume in the control channel. For obtaining the computational overhead of the distributed scheme, we average core distribution/fusion calculations in each cluster-/application head over 1,000,000 times by comparing CPU ticks before and after.

C. Implementation of Task Migration

Task migration is carried out on application level through checkpointing after each iteration, with a lightweight support by the middleware (to start executables as needed). When the controlling scheme (both our centralized and distributed schemes) chooses to change the fusions or re-distributes cores among applications, the respective stages are notified by the middleware: When the corresponding stage reaches a checkpoint, it saves its state and requests the middleware at the destination core to start its executable file if the destination core formerly belonged to a different application. It then sends its state to the newly started executable, which then continues the execution of the (fused) stage procedure. The corresponding overheads of these operations are evaluated and discussed in Section X-E.

X. Experimental Results

A. System Setup

Our experiments have been conducted on Intel's Single-Chip Cloud Computer (SCC) [9] and using a high-level many core simulator. The SCC is a platform that integrates 48 x86 cores in 24 tiles (two cores each) on a single chip. The individual P54C cores (45nm process) run at 800 MHz, are connected via a 2 GHz network-on-chip with a bisection bandwidth of 2TB/s. Each core has 16 KB of instruction- and 16 KB of data cache, and 256 KB of unified instruction/data L2 cache. It runs a single-core Ubuntu Linux (kernel 3.1.4) on each core.

Our high-level many core simulator is written in C++, executes task traces collected on the SCC, and simulates the network-on-chip interconnect. The simulator delivers accurate information on the application

Name	Stages	Source
automotive	21	*see Section X-B*
h264ref	4	SPEC CPU 2006 [8]
lame	4	MiBench [7]
PGP	5	MiBench [7]
sphinx3	22	SPEC CPU 2006 [8]

TABLE I: Benchmark applications

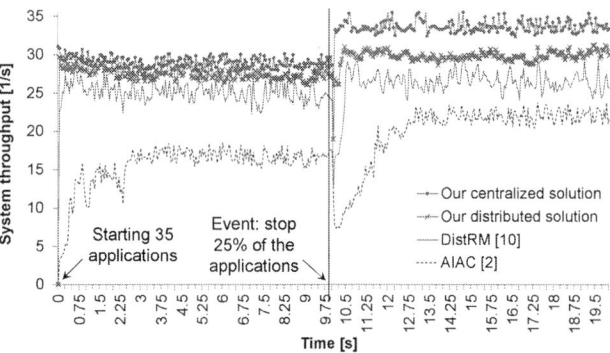

Fig. 7: Comparison of the achieved system throughputs

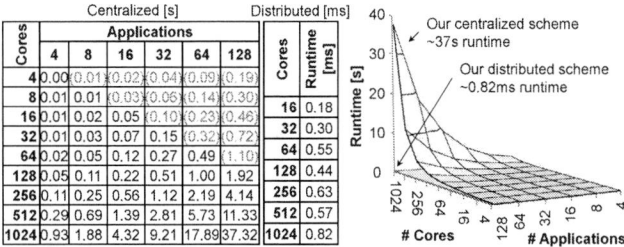

Fig. 8: Computational overhead of our schemes. Infeasible app./core combinations in brackets. Only one column for our distributed scheme as the runtime does not change significantly with the # of applications.

/ system throughputs and on the communication volumes / overheads (algorithm runtimes have been collected on the SCC). It runs on a six-core AMD Opteron^{TM} 8431 CPU (2.4 GHz) with 64 GB DDR3 RAM. The SCC allows measuring the computational overhead accurately, but as it integrates 48 cores, we cannot analyze the system throughputs and the communication overhead for larger systems. However, we measured the computational overhead on the SCC even for (virtually) large systems because these computations do not demand to dispose of the cores physically.

Measurements conducted on the SCC:

- Computational overhead for up to 1024 cores.
- Throughput of the centralized scheme for up to 48 cores.
- Fusion/fission overheads.

Experiments conducted using our simulator:

- Communication overhead.
- Throughput of our centralized and of our distributed schemes.

For the experiments, we spawn the benchmark applications listed in Table I multiple times so that the total number of stages in the system exceeds the number of cores by at least a factor of 3 (we chose this number arbitrarily to establish a considerable system load).

B. Benchmark Scenario

Table I shows an overview of the benchmark applications and their number of stages. The applications have been manually parallelized to form malleable software pipelines. We chose this set of applications because they are most suitable to form software pipelines. The implementation details of how we adapted the state-of-the-art schemes of [2], [10] to compare them against our schemes can be found in Section S4.

The automotive application is a vision-based application that takes its algorithms from the IVT library [1]. It performs stereo vision, image enhancement, object recognition (based on scale-invariant feature transform (SIFT) and Harris corner detection), morphological operations, and pattern matching algorithms to identify and track objects in a continuous stream of color stereo video data (648x480 pixels at 30fps). The other applications have been taken from the respective benchmark suites.

Application	Carried State [KB]				Overhead [ms]	
	Min	Max	Avg	σ	Old Core	New Core
automotive	1	32	19	15.21	0.63	22
h264ref [8]	13	53	27	22.73	1.07	76
lame [7]	9	10	9	1.32	0.18	19
PGP [7]	1	27	12	9.11	0.30	66
SPHINX3 [8]	12	22	17	4.21	0.51	44

TABLE II: Overheads of fusion/fission operations

C. Achieved System Throughput

In the following, we compare the throughput achieved by the distributed scheme with the centralized scheme (thus against optimal mapping) and with two state-of-the-art runtime remapping schemes, DistRM [10] and AIAC [2]. Figure 7 shows the average system throughput over 50 runs when running 7 instances of each benchmark application (35 applications, or 392 stages in total) on 128 cores, connected by a NoC mesh as featured by the SCC. To show how each scheme gradually improves the mapping of stages to cores, we initially start all stages on a single core and let the corresponding schemes improve the system throughput incrementally. After this is achieved, we randomly stop 25% of the applications at $t = 10$ seconds. While the centralized (thus optimal) scheme achieves an increased throughput of roughly 13.7%, the average system throughput drops for the other schemes from roughly ca. 17-29 iterations/second to ca. 9-17 iterations/second because without adapting the mapping, the cores that formerly executed the stopped applications are now idle. The schemes then improve the throughput by adapting the mapping of stages to cores. Our distributed scheme achieves a system throughput of 94.78% compared to the optimal (centralized) scheme. On average, the distributed scheme increases the throughput by 11.3% and 60.6% over [10] and [2], respectively.

D. Computational and Communication Overheads

Figure 8 shows how the computational overhead of our centralized scheme grows with a growing number of cores and applications. Up to a considerable problem size (e.g. 64 cores, 64 applications), optimal mappings can be calculated in less than 0.5s, which may be sufficiently fast for certain systems. However, this overhead is significantly larger for larger systems as the runtime grows quickly beyond 35 seconds.

The distributed scheme has a constant time complexity as each cluster head on level 1 C_h^1 only calculates the optimal distribution of the cores to its children (which does not grow with the problem sizes). Thus, its computational overhead is small (less than 0.1ms for 1024 cores).

Figure 9 compares the total communication overhead of our centralized and of our distributed schemes. This overhead includes status updates and notifications, the updates of the e_i, c_i, and o_i values, as well as the propagation of all tables and speed-up vectors. As this overhead merely reaches around 365.3 KiB/s (0.025% of the total communication for a system with 1024 cores, 275 applications) for our centralized scheme and roughly 138 KiB/s (0.009%) for our distributed scheme, we consider it as negligible.

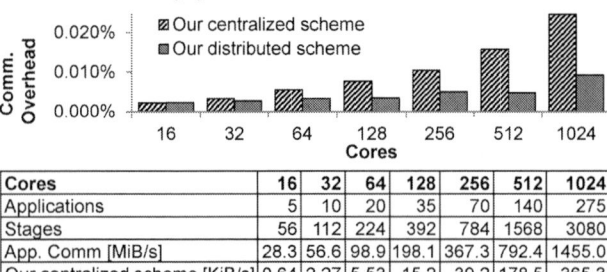

Cores	16	32	64	128	256	512	1024
Applications	5	10	20	35	70	140	275
Stages	56	112	224	392	784	1568	3080
App. Comm [MiB/s]	28.3	56.6	98.9	198.1	367.3	792.4	1455.0
Our centralized scheme [KiB/s]	0.64	2.27	5.53	15.2	39.2	178.5	365.3
Our distributed scheme [KiB/s]	0.69	1.58	3.34	6.79	19.0	54.2	138.0

Fig. 9: Comparison of communication overheads

E. Fusion/Fission (i.e. Task Migration) Overhead

Table II summarizes the overhead for fusions/fissions of two stages of each application (collected on Intel's Single-Chip Cloud Computer). When the fusion/fission operation incurs an old core (i.e. the application is already running on this core before this operation), the overhead is limited to transferring the carried state of the stage and is thus very small. Otherwise, the executable file of the application needs to be started by the middleware, which takes considerably more time. However, our experiments show that this is only the case in less than 5% of conducted fusion/fission operations. Hence, the overhead of our proposed schemes is small and thus, we find that our centralized scheme is well suitable for managing smaller many core systems, while our distributed schemes is well suitable for systems with hundreds of cores.

XI. Concluding Remarks

In this paper, we show how a high system throughput can be achieved and maintained even in large many core systems despite unpredictable, significant variances in the demand for both computational as well as for communication resources. This is achieved by optimizing the configurations (fusion of stages) and the distribution of cores among the applications at runtime. Additionally to proposing an optimal scheme, we show how optimality can be sacrificed to maintain near-optimal throughputs even for large systems with hundreds of cores.

XII. Acknowledgements

This work was partly supported by the German Research Foundation (DFG) as part of the Transregional Collaborative Research Centre "Invasive Computing" (SFB/TR 89).

References

[1] P. Azad, T. Gockel, and R. Dillmann. *Computer Vision - Principles and Practice*. Elektor Electronics, 2008.

[2] J. M. Bahi et al. Dynamic load balancing and efficient load estimators for asynchronous iterative algorithms. *IEEE Trans. Parallel Distrib. Syst.*, 16:289–299, April 2005.

[3] J. Cheng et al. MAPS: An Integrated Framework for MPSoC Application Parallelization. In *IEEE/ACM Des. Aut. Conf. (DAC)*, 2008.

[4] C.-L. Chou and R. Marculescu. User-Aware Dynamic Task Allocation in Networks-on-Chips. In *IEEE/ACM Des., Aut., and Test in Europe (DATE)*, 2008.

[5] D. G. Feitelson et al. Theory and Practice in Parallel Job Scheduling. In *Workshop on Job Sched. Strat. for Parallel Proc. (JSSPP)*, 1994.

[6] M. I. Gordon, W. Thies, and S. Amarasinghe. Exploiting coarse-grained task, data, and pipeline parallelism in stream programs. In *ACM Int. Conf. on Arch. Support for Prog. Lang. and Oper. Syst. (ASPLOS)*, 2006.

[7] M. R. Guthaus, J. S. Ringenberg, D. Ernst, T. M. Austin, T. Mudge, and R. B. Brown. MiBench: A Free, Commercially Representative Embedded Benchmark Suite. In *IEEE Workshop on Workload Charact. (WWC)*, 2001.

[8] J. L. Henning. SPEC CPU2006 Benchmark Descriptions. *SIGARCH Comput. Archit. News*, 34(4), Sept. 2006.

[9] J. Howard et al. A 48-Core IA-32 Message-Passing Processor with DVFS in 45nm CMOS. In *IEEE Int. Solid-State Circ. Conf. (ISSCC)*, 2010.

[10] S. Kobbe et al. DistRM: Distributed Resource Management for On-Chip Many-Core Systems. In *Int. Symp. on Hardw./Softw. Codesign and Syst. Synth. (CODES+ISSS)*, 2011.

[11] H. Lee, W. Che, and K. Chatha. Dynamic scheduling of stream programs on embedded multi-core processors. In *Int. Symp. on Hardw./Softw. Codesign and Syst. Synth. (CODES+ISSS)*, 2012.

[12] T. Li, D. Baumberger, D. A. Koufaty, and S. Hahn. Efficient Operating System Scheduling for Performance-Asymmetric Multi-Core Architectures. In *IEEE Int. Comp. Symp. (ICS)*, 2007.

[13] G. Ottoni et al. Automatic Thread Extraction with Decoupled Software Pipelining. In *IEEE/ACM Int. Symp. on Microarch. (MICRO)*, 2005.

[14] P. Radojković et al. Optimal Task Assignment in Multithreaded Processors: A Statistical Approach. In *ACM Int. Conf. on Arch. Support for Prog. Lang. and Oper. Syst. (ASPLOS)*, 2012.

[15] M. Rajagopalan et al. Thread scheduling for Multi-Core Platforms. In *HotOS*, 2007.

[16] L. Schor et al. Scenario-based design flow for mapping streaming applications onto on-chip many-core systems. In *IEEE Int. Conf. on Compilers, Arch., and Synth., for Embedded Syst. (CASES)*, 2012.

[17] A. Snavely and D. Tullsen. Symbiotic Jobscheduling for a Simultaneous Multithreading Processor. In *IEEE/ACM Int. Symp. on Microarch. (MICRO)*, pages 234–244, 2000.

[18] A. Snavely, D. M. Tullsen, and G. Voelker. Symbiotic jobscheduling with priorities for a simultaneous multithreading processor. In *SIGMETRICS*, 2002.

[19] W. Thies, V. Chandrasekhar, and S. Amarasinghe. A Practical Approach to Exploiting Coarse-Grained Pipeline Parallelism in C Programs. In *IEEE/ACM Int. Symp. on Microarch. (MICRO)*, 2007.

[20] J. Turek, J. L. Wolf, and P. S. Yu. Approximate Algorithms Scheduling Parallelizable Tasks. In *ACM Symp. on Par. Alg. and Arch. (SPAA)*, 1992.

SUPPLEMENTAL MATERIAL

S1. PIPELINE PROCEDURES

Each application corresponds to a single executable file. Two parameters, *FirstStage* and *Fusions*, controls which of its stages are executed. Both parameters are supplied from the command line when the application is started on a core, but may be changed by an application head during runtime to change the fusions/fissions of stages.

Example 1 Pseudo-code of a Pipeline Procedure

This is the main (entry point) procedure for an application which is executed on each core that is assigned to the application. *StageFunction* is the individual function that executes the functionality of stage i.

Input:

    FirstStage   First stage to be executed on this core
    Fusions   Number of stages to be executed on this core

```
1: while true do
2:   if FirstStage > 1 then
3:     data = ReceiveData( FirstStage − 1 );
4:   end if
5:   for i = FirstStage to Fusions do
6:     if Task Migration Triggered (send state) then
7:       SendState( Destination )
8:       continue
9:     end if
10:    if Task Migration Received (receive state) then
11:      ReceiveState( )
12:      continue
13:    end if
14:    data = StageFunction( data )
15:  end for
16:  if FirstStage + Fusions < Nk then
17:    SendData( FirstStage + Fusions + 1, data );
18:  end if
19: end while
```

The main pipeline procedure forms the entry point which is started on each core that is assigned to an application. Example 1 illustrates this: For each iteration, an iteration starts with receiving the required input data (if it is not the first stage) (lines 2-4). Then, all stages that are fused on the current core are executed sequentially (lines 5-15). Finally, the output data is sent to the succeeding stage, if applicable (lines 16-18).

We carry out our application-layer task migration in this loop: When one or more stages are migrated to a different core, they save and transmit their state. The middleware supports this when our schemes decide to assign a new core to an application (i.e. this core was not already assigned to this application) by starting the

S2. EXAMPLE OF FUSING PIPELINE STAGES

Given the pipeline k shown in Figure S-1 with $N_k = 4$ stages and having available up to $M_k = 4$ cores to assign to it, we first proceed to build table $F_k(l, j)$ according to Equation (3), as stated in Algorithm 1:

$$
\begin{array}{lll}
F_k(1,1) = 60 & F_k(2,2) = 110 & F_k(3,3) = 110 \\
F_k(1,2) = 150 & F_k(2,3) = 60 & F_k(3,4) = 140 \\
F_k(1,3) = 100 & F_k(2,4) = 90 & \\
F_k(1,4) = 130 & & F_k(4,4) = 70
\end{array}
$$

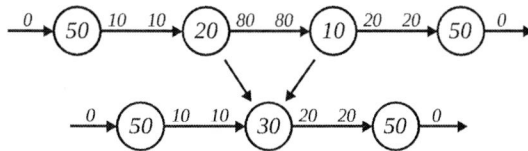

Fig. S-1: Pipeline example

The next step is to compute the initial conditions for the table $R_k(j, m)$ according to Equation (4), which in other words means to compute the maximal computational requirements for a sub-pipeline with j stages using up to m cores. Since this are initial conditions, the tracking table $TR_k(j, m)$ for $m = 0, 1$ has no previous value for j^*, and is therefore filled with zeros.

From now on, since $m \geq 2$, the table $R_k(j, m)$ is build according to Equation (6). In this particular example, when $m = 2$ the solution for every sub-pipeline chooses to use the result from $R_k(1,1)$ and to fuse the rest of the stages together in one core, thus, the tracking table $TR_k(j, 2)$ will be filled with $j^* = 1$ for any j.

The resultsare shown in Table S-I. The optimal solution can be derived from table $TR_k(j, m)$: Starting from cell $(j, m) = (4, 4)$ and traverse the table in the direction $(j^*, m − 1)$, one can derive that the optimal solution fuses stages S_2 and S_3, and leave stages S_1 and S_4 as they are.

$R_k(4, 4)$: Maximal response

m \ j	1	2	3	4
1	60	150	100	130
2	60	110	60	90
3	60	110	60	70
4	60	110	60	(70)

$TR_k(4, 4)$: Tracking

m \ j	1	2	3	4
1	(0)	0	0	0
2	1	1	(1)	1
3	1	2	3	(3)
4	1	2	3	(4)

TABLE S-I: Example tables from Algorithm 1

$R_1(3, 6)$

m	$R_1(3, m)$
1	130
2	90
3	70
4	70
5	70
6	70

$R_2(5, 6)$

m	$R_2(5, m)$
1	120
2	110
3	100
4	90
5	80
6	80

$R_3(4, 6)$

m	$R_3(4, m)$
1	300
2	300
3	80
4	40
5	40
6	40

TABLE S-II: Example tables of different pipelines

S3. EXAMPLE OF CONSTRUCTING TABLES

Example: Given the pipelines R_1, R_2 and R_3 shown in Table S-II, with weights $w_1 = w_2 = w_3 = 10000$ and having up to $M = 6$ available in the system, in order to find the maximal overall system throughput we first proceed to compute the initial conditions for the table $G(k, m)$ according to Equation (7). Since this are initial conditions, the tracking table $TG(k, m)$ for $k = 1$ has no previous value for m^*, and is therefore filled with zeros.

From now on, since $k \geq 2$, the table $G(k, m)$ is build according to Equation (8).

The fully completed results are shown in Table S-III. Looking at table $TG(k, m)$, starting from cell $(k, m) = (3, 6)$ and traversing the table in the direction $(k − 1, m^*)$, one can derive that the optimal solution will assign one core to pipeline R_1, one core to pipeline R_2 and four cores to pipeline R_3.

S4. ADAPTION OF THE STATE-OF-THE-ART SCHEMES OF [2], [10]

This section details how we adapted the state-of-the-art schemes of [2] and [10] in order to achieve a fair comparison to our proposed schemes.

$G(3, 6)$: Overall Performance

m \ k	1	2	3
1	76.92	−∞	−∞
2	111.11	160.26	−∞
3	142.86	194.44	193.59
4	142.86	226.19	227.78
5	142.86	233.76	285.26
6	142.86	242.86	(410.25)

$TG(3, 6)$: Tracking

m \ k	1	2	3
1	(0)	-1	-1
2	0	(1)	-1
3	0	2	2
4	0	3	3
5	0	3	2
6	0	3	(2)

TABLE S-III: Example tables from algorithm 2

905

A. AIAC [2]

AIAC exchanges workload between physically neighboring cores to balance the computational load evenly. To adapt this scheme for software-pipelined applications in many core systems, we exchange workload by migrating pipeline stages when the computational load is not balanced. This is achieved by comparing the load of adjacent cores and migrating a pipeline stage i when the difference of the summed computational demands among all stages on each core exceeds c_i. To achieve a fair comparison, we relax the assumption that only consecutive stages may be mapped to the same core. For our implementation of AIAC, a core may execute any stage from any application.

B. DistRM [10]

DistRM [10] distributes cores among applications, but relies on the applications to themselves decide how to distribute their tasks accordingly. Therefore, we use our optimal fusion algorithm from Section VI to achieve a fair comparison. Consequently, only the number of cores assigned to each application differs between DistRM and our schemes. Fusions of pipeline stages are carried out identically. We also adapt DistRM by using the speed-up vectors according to Section VI. As DistRM remains in local optima if the speed-up of an application does not increase with another core (even if this was the case for a larger number of additional cores), we report marginal improvements (we choose an $\epsilon = 5 * 10^{-4}$) as long as the number of cores does not exceed the number of stages of the corresponding application. Using the described adaptions, we can achieve a fair comparison with DistRM.

A Scenario-based Run-time Task Mapping Algorithm for MPSoCs

Wei Quan[†,‡]

†Informatics Institute
University of Amsterdam
The Netherlands
{w.quan,a.d.pimentel}@uva.nl

Andy D. Pimentel[†]

‡School of Computer Science
National University of Defense Technology
Hunan, China
quanwei02@gmail.com

ABSTRACT

The application workloads in modern MPSoC-based embedded systems are becoming increasingly dynamic. Different applications concurrently execute and contend for resources in such systems which could cause serious changes in the intensity and nature of the workload demands over time. To cope with the dynamism of application workloads at run time and improve the efficiency of the underlying system architecture, this paper presents a novel scenario-based run-time task mapping algorithm. This algorithm combines a static mapping strategy based on workload scenarios and a dynamic mapping strategy to achieve an overall improvement of system efficiency. We evaluated our algorithm using a homogeneous MPSoC system with three real applications. From the results, we found that our algorithm achieves an 11.3% performance improvement and a 13.9% energy saving compared to running the applications without using any run-time mapping algorithm. When comparing our algorithm to three other, well-known run-time mapping algorithms, it is superior to these algorithms in terms of quality of the mappings found while also reducing the overheads compared to most of these algorithms.

Categories and Subject Descriptors

C.4 [**Performance of Systems**]: Performance Attributes

General Terms

Algorithm, Design, Performance

Keywords

Embedded systems, KPN, MPSoC, task mapping, simulation

1. INTRODUCTION

Modern embedded systems, which are more and more based on MultiProcessor System-on-Chip (MPSoC) architectures, often require supporting an increasing number of applications and standards, where multiple applications can run simultaneously. For

Permission to make digital or hard copies of all or part of this work for personal or classroom use is granted without fee provided that copies are not made or distributed for profit or commercial advantage and that copies bear this notice and the full citation on the first page. To copy otherwise, to republish, to post on servers or to redistribute to lists, requires prior specific permission and/or a fee.
DAC'13 May 29 - June 07 2013, Austin, TX, USA.

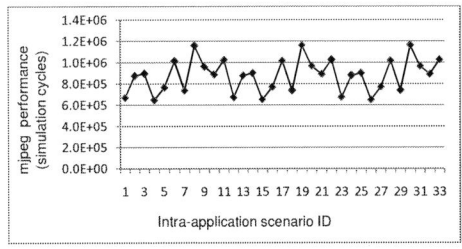

Figure 1: Intra-application scenario performance of MJPEG.

each single application, there are typically also different execution modes (or program phases) with different requirements. For example, a video application could dynamically lower its resolution to decrease its computational demands in order to save battery. As a consequence, the behavior of application workloads executing on the embedded system can change dramatically over time. Here, one can distinguish two forms of dynamic application behavior: inter-application dynamism and intra-application dynamism. These forms of dynamism are often captured using *scenarios* [13, 8]. This means that there are two different kinds of scenarios: inter-application scenarios to describe the simultaneously running applications in the system, and intra-application scenarios that define the different execution modes for each application. The combination of these inter- and intra-application scenarios are called *workload scenarios*, and specify the application workload in terms of the different applications that are concurrently executing and the mode of each application.

At design time of an embedded system, a designer could aim at finding the optimal mapping of application tasks to MPSoC processing resources for each inter- and intra-application scenario. However, when the number of applications and application modes increase, the total number of workload scenarios will explode exponentially. Considering, e.g., 10 applications with 5 execution modes for each application, there will be 60 million workload scenarios. If each scenario takes one second to find the optimal mapping at design time, then one would need nearly two years to obtain all the optimal mappings. Moreover, storing all these optimal mappings such that they can be used at run time by the system to remap tasks when a new scenario is detected would also be unrealistic as this would take up too much memory storage.

An approach to solve this problem is by clustering workload scenarios and only storing a single mapping per cluster of workload scenarios to facilitate run-time mapping [8]. Such clustering implies a significant space reduction needed to store the mappings. Moreover, so-called scenario-based design space exploration [17] can be deployed to efficiently find these mappings by only evalu-

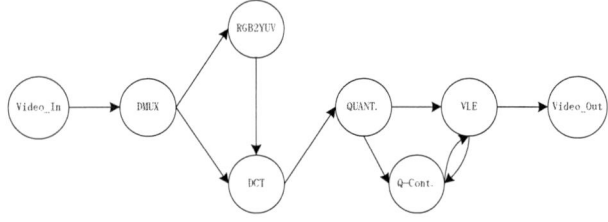

Figure 2: KPN for MJPEG application.

ating a representative subset of scenarios for each cluster. In this paper, we consider a clustering method[1] in which we find and store a single mapping for each inter-application scenario that yields, on average, best performance for all possible intra-application scenarios within the inter-application scenario. However, as we can see from the behavior of a Motion-JPEG (MJPEG) encoder application in Figure 1, using such a single mapping to represent an entire inter-application scenario shows considerable performance variations for the different intra-application scenarios that exist in this inter-application scenario. In this particular example, the inter-application scenario contains three simultaneously running multimedia applications: MJPEG, a MP3 decoder, and a Sobel filter for edge detection in images. The use of cluster-level mappings (i.e., mappings found to be good for an entire cluster of workload scenarios) can provide a run-time mapping system with enough information to quickly find an adequate mapping for a detected workload scenario but it will not immediately lead to finding the optimal system mapping for any identified workload scenario. Therefore, we propose a novel run-time Scenario-based Task Mapping algorithm (STM) that uses the cluster-level mapping information derived from design-time design space exploration (DSE) but, additionally, performs run-time mapping optimization by continuously monitoring the system and trying to perform (relatively small) mapping customizations to gradually further improve the system performance.

The remainder of this paper is organized as follows. Section 2 gives some prerequisites and the problem definition for this paper. Section 3 provides a detailed description of the scenario-based run-time mapping algorithm. Section 4 introduces the experimental environment and presents the results of our experiments. Section 5 discusses related work, after which Section 6 concludes the paper.

2. PREREQUISITES AND PROBLEM DEFINITION

In this section, we explain the necessary prerequisites for this work and provide a detailed problem definition.

2.1 Application Model

In this paper, we target the multimedia application domain, as was already illustrated in Figure 1. For this reason, we use the Kahn Process Network (KPN) model of computation [11] to specify application behaviour since this model of computation fits well to the streaming behaviour of multimedia applications. In a KPN, an application is described as a network of concurrent processes that are interconnected via FIFO channels. This means that an application can be represented as a directed graph $KPN = (P,F)$ where P is set of processes (tasks) p_i in the application and $f_{ij} \in F$ represents the FIFO channel between two processes p_i and p_j. Figure 2 shows the KPN of the MJPEG application.

---

[1]We note, however, that other clustering methods would also be possible and that our run-time mapping algorithm is independent on the clustering method used.

2.2 Architecture Model

In this work, we restrict ourselves to homogeneous MPSoC target architectures[2]. An architecture can be modeled as a graph $MPSoC = (PE,C)$, where PE is the set of processing elements used in the architecture and C is a multiset of pairs $c_{ij} = (pe_i, pe_j) \in PE \times PE$ representing a communication channel (like Bus, NOC, etc.) between processors pe_i and pe_j. Combining the definition of application and architecture models, the computation cost of task (process) p_i on processing element pe_j is expressed as T_i^j and the communication cost between tasks p_i and p_j on channel c_{xy} is $C_{ij}^{c_{xy}}$.

2.3 Task Mapping

The task mapping defines the corresponding relationship between the tasks in a KPN application and the underlying architecture resources. For a single application, given the KPN of this application and a target $MPSoC$, a correct mapping is a pair of unique assignments ($\mu : P \rightarrow PE$, $\eta : F \rightarrow C$) such that it satisfies $\forall f \in F, src(\eta(f)) = \mu(src(f)) \wedge dst(\eta(f)) = \mu(dst(f))$. In the case of a multi-application workload, the state of simultaneously running applications that are distinguished as inter- and intra-application scenarios should be considered in the task mapping. Let $A = \{app_0, app_1, ..., app_m\}$ be the set of all applications that can run on the system, and $M^i = \{md_0^i, md_1^i, ..., md_n^i\}$ be the set of possible execution modes for $app_i \in A$. Then, $SE = \{se_0, se_1, ..., se_{n_{inter}}\}$, with $se_i = \{app_0 = 0/1, ..., app_m = 0/1\}$ and $app_i \in A$, is the set of all inter-application scenarios. And $sa_j^i = \{app_0 = md_{j_0}^0, ..., app_m = md_{j_m}^m\}$, with $app_i \in A \wedge app_i = 1 \in se_i$ and $md_{j_x}^i \in M^i$, represents the j-th intra-application scenario in inter-application scenario $se_i \in SE$. The set of all workload scenarios can then be defined as the disjoint union $S = \sqcup_{i \in SE} SA^i$, with $SA^i = \{sa_1^i, sa_2^i, ..., sa_{n_{intra}}^i\}$.

As already explained in the previous section, we propose to perform the run-time mapping of applications in two stages. In the first stage, which is performed at design time, we cluster workload scenarios (similar to [8]) and perform DSE for each of these scenario clusters to find a mapping that shows the best average performance for that particular cluster. More specifically, in this paper, we consider each $se_i \in SE$ as a different cluster of scenarios (i.e., we cluster all intra-application scenarios of an inter-application scenario). The mappings derived from design-time DSE are stored so they can be used by the run-time mapping algorithm to re-map applications when a workload scenario is detected that belongs to a different scenario cluster. Since these statically determined mappings may not be optimal for the current active intra-application scenario, the second stage of the run-time mapping algorithm tries to perform (relatively small) mapping customizations to gradually further improve the system performance. In our goal to optimize mappings, we recognize two kinds of objectives: system-level objectives and application-dependent objectives. System-level objectives, denoted as $O_\alpha = \{O_{\alpha 0}, O_{\alpha 1}, ...\}$, define the system-wide metrics such as system energy consumption, total system execution time, etc. Application-dependent objectives, denoted as $O_\beta = \{O_{\beta 0}, O_{\beta 1}, ...\}$, are mainly used to define the performance requirements of each separate application like throughput, latency, etc. As will be explained in the next section, the first stage of our run-time mapping approach uses system-level objectives to find mappings per scenario cluster. Here, we use system energy consumption and total workload scenario execution time as metrics: $E_{s_i}, s_i \in S$ represents the system energy consumption of workload scenario s_i

---

[2]In subsequent work, we will show how scenario-based run-time mapping can also be applied to heterogeneous MPSoCs.

and $X_{s_i}, s_i \in S$ is the execution time of scenario s_i. For the second stage, during which the mapping is gradually optimized, we apply application-specific objectives – in our case throughput requirements for each application – for the optimization process. However, to measure the results of the run-time optimization process, we also use the system-level metrics E_{s_i} and X_{s_i}.

Under these definitions and given the $KPN = (P, F)$ for each application and an $MPSoC = (PE, C)$, our goal is to continuously customize the mapping at run time such that the system-level and/or application-specific objectives under every workload scenario $s_i \in S$ are satisfied.

3. SCENARIO-BASED TASK MAPPING

The STM algorithm, which is outlined in Algorithm 1, can be divided into a static part and a dynamic part. The static part is used to capture application dynamism at the granularity of inter-application scenarios. For each inter-application scenario $se_i \in SE$, we have determined – using design-time DSE – a mapping that on average performs best for all intra-application scenarios SA^i of se_i. That is, for each se_i we search for a mapping by solving the following multi-objective optimisation problem:

$$min[\sum_{sa_j^i \in SA^i} E_{sa_j}, \sum_{sa_j^i \in SA^i} X_{sa_j}]. \qquad (1)$$

To this end, we have deployed the scenario-based DSE approach presented in [17], which is based on the well-known NSGA-II genetic algorithm and allows for effectively pruning the design space by only evaluating a representative subset of intra-application scenarios of SA^i for each $se_i \in SE$. As this design-time DSE stage is not the main focus of this paper, we refer the interested reader to [17] for further details. The mappings derived from this design-time DSE are used by the STM algorithm as shown in lines 1-3 of Algorithm 1. When the system detects the execution of a different inter-application scenario, the static part of the STM algorithm will choose the corresponding mapping as derived from the design-time DSE stage and which has been stored in a so-called *scenario database*. Because this database only stores mappings for entire scenario clusters, its size can be controlled by choosing a proper granularity of scenario clusters (e.g., inter-application scenarios).

The dynamic part of our STM algorithm is active during the entire duration of an inter-application scenario. As explained in the previous section, it uses application-specific objectives, specified for each separate application, to continuously optimize the mapping. When the algorithm detects that an objective is unsatisfied, it will try to find a new task mapping for that particular application that missed the performance goal. If multiple applications miss their performance goal, then the STM algorithm will start optimizing the most problematic application first. The main steps of the dynamic part of the STM algorithm are described below.

3.1 Finding the Critical Task

The first step of the dynamic part of the STM algorithm is to find the so-called *critical task* for the application that missed its objective, as shown in lines 10-13 of Algorithm 1. The rationale behind this is that by remapping this critical task and possibly its neighbouring tasks (forming a bottleneck in the application), the resulting effect will be optimal. To find the critical task, the STM algorithm maintains three lists. The first list stores the task costs (TC). For every application, it contains the cost of the application's tasks, where the cost is determined by the sum of the execution and communication times of a task. These task costs are arranged in descending order in the list. The two other lists concern the storing of two other metrics for each task: the proportion of task cost in

---

Algorithm 1 STM algorithm

Input: $KPN_{app_0,...,app_m}$, $MPSoC$, O_α, O_β, μ, η
Output: $New(\mu, \eta)$
list: TC, CIC, CIB, PU
pCIC = δ_c, pCIB = δ_b

1: if detectScenario() == true : //new inter-application scenario
2:   $New(\mu, \eta)$ = getMapping();
3:   return $New(\mu, \eta)$;
4: else :
5:   results[] = getStatistics();
6:   if (i = objectiveUnsatisfied(results, O_α, O_β)) != -1:
7:     taskCost(KPN_{app_i}, results, TC, CIC, CIB);
8:     peUsage(results, PU);
9:     while(1) :
10:      if (apptype = getType(KPN_{app_i})) == DATA_PARALLEL :
11:       critical = findDPCritical(KPN_{app_i}, CIC, CIB, pCIB, pCIC);
12:      else :
13:       critical = findCritical(KPN_{app_i}, CIC, CIB, pCIB, pCIC);
14:      reason = findReason(critical, CIC, CIB, pCIB, pCIC);
15:      if reason == POOR_LOCALITY :
16:       MCC[] = minCircle(KPN_{app_i}, results, critical);
17:       if GetSubstitute(PU, μ, η, MCC, apptype) == true :
18:        return $New(\mu, \eta)$;
19:       else failed;
20:      else if reason == LOAD_IMBALANCE :
21:       if GetSubstitute(PU, μ, η, apptype) == true :
22:        return $New(\mu, \eta)$;
23:       else failed;
24:      else :
25:       pCIB += ε;
26:       pCIC -= ε;

---

the total busy time of the PE (i.e., processor) onto which the task is currently mapped (CIB), and the proportion of task communication time (read and write transactions) in the task cost (CIC).

Using the TC list, the algorithm checks the task at the top of the list to find the critical task, taking the following two conditions into account: 1) whether or not the task's CIB proportion is lower than a specific threshold, defined by *pCIB*. Here, the rationale is that a high-cost task receiving only a small fraction of processor time may imply that the processor is overloaded. If the task satisfies this condition, then this task is considered as the critical task and the process of finding the critical task ends. Otherwise, the algorithm continues to check the other tasks in the TC list with lower costs until it finds the critical task. If there is no task in the application that satisfies the first condition, then the second condition will be used: 2) Whether or not the CIC proportion is higher than the threshold *pCIC*. The algorithm checks all the tasks using this second condition just like it did for the first condition. If all the tasks do not satisfy these two conditions, then the algorithm will, respectively, increase and decrease the pCIB and pCIC thresholds by ε, after which the above process is restarted again.

For data parallel applications, the process of finding the critical task has one additional test as compared to regular applications. This extra test (performed in the function *findDPCritical*) involves the check whether or not all data-parallel tasks are mapped onto different PEs. If there are data-parallel tasks that are mapped onto the same processor, then those tasks with higher task costs will be treated as critical tasks. Otherwise, the process of finding the critical task will be the same as for regular applications.

3.2 Remapping the Critical Task

After the critical task has been found, the STM algorithm tries to analyze the reason for missing the application's performance goal. In this respect, we recognize two different reasons: *poor locality* and *load imbalance*. Here, we use the process of determining the critical task to also determine the reason for not meeting the performance goal: If the CIC proportion of the critical task is higher than the value of the current pCIC threshold, then the algorithm assumes that poor locality is the reason. Otherwise it takes load imbalance as the reason for not meeting the application demands. This means that poor locality has a higher priority than load imbalance as a reason for not meeting the application demands, which is helpful to reduce the energy consumption due to communications.

Subsequently, the function *GetSubstitute* in the STM algorithm can follow different strategies to find a target PE to which the critical task will be remapped. The selection of remapping strategy depends on the reason for not meeting the application's performance demands as well as on the type of application (data parallel or not). The strategies that are used to find the substitute PE for data-parallel applications are similar to the ones for regular applications except that one additional condition is taken into account for finding the substitute PE: the substitute PE should not be a PE onto which its parallel tasks are mapped.

3.2.1 Poor locality

In the case of poor locality, the STM algorithm will try to find a better mapping for the application in question based on a *minimal cost circle (MCC)* approach. A situation that has been identified as "poor locality" is mainly due to the communication overhead between tasks. Evidently, if the communicating frequency between two tasks is very high or the communicating data size is very large, then these two tasks should preferably be mapped onto the same PE or onto two different PEs that contain a more efficient interconnect between each other. The MCC strategy aims at redistributing the critical task and its neighbouring tasks over PEs such that communication overhead is reduced while trying to avoid creating new computational bottlenecks. To this end, it first finds the minimal cost circle based on equation (2) for the critical task p_i:

$$min(Circle\_Cost(p_i)_{mn}), \ with \ 0 \le n, m \le |P| \qquad (2)$$

where:

$$Circle\_Cost(p_i)_{mn} = \sum_{m \le i \le n} T_i^z + \sum_{0 \le i < |P|} \sum_{m \le j \le n} C_{ij}^{c_{xy}} \qquad (3)$$

where T_i^z denotes the execution time of task i for PE z onto which task i is currently mapped, and $C_{ij}^{c_{xy}}$ denotes the communication overhead between tasks i and j (see Section 2.2). This strategy is applicable for heterogeneous MPSoC architectures. However, in this paper, our focus is on homogeneous architectures using a shared bus interconnect. This means that each task will have a constant computational cost irrespective of the PE it is mapped on, and that communication overhead only involves internal communication within a single PE (i.e., when the communicating tasks are mapped to the same PE) or external communication between PEs via shared memory. Clearly, internal communication costs are much lower than external communication costs. Figure 3.a shows an example of an MCC (indicated by the red oval) that contains two tasks, including the critical task (red task), whereas Figure 3.b illustrates an MCC that only contains the critical task itself.

After the MCC of the critical task has been determined, the function *GetSubstitute* will choose a substitute PE for all the tasks included in the identified MCC to achieve a new mapping. For this purpose, the PU list is used, containing the processor utilisations

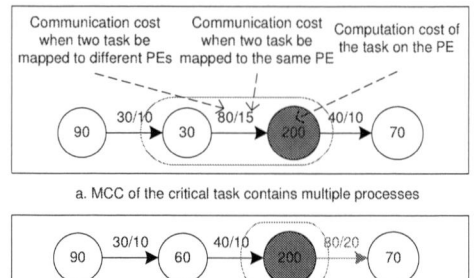

a. MCC of the critical task contains multiple processes

b. MCC of the critical task contains single process

Figure 3: Examples of an MCC for a critical task (red task).

for each PE. The substitute PE is the PE with the lowest utilization in the PU list that is different from the PE onto which the critical task is currently mapped. If the MCC solely consists of the critical task itself, then the critical task will be mapped onto the PE of a neighboring task that has the heaviest communication with the critical task. This is, e.g., shown in Figure 3.b, where the critical task will be mapped onto the same PE as the task with cost 70. Moreover, the substitute PE should be different than the PE the critical task is currently mapped on. Otherwise, the algorithm fails to find a new mapping. After the substitute PE has been found, the FIFO channels between the tasks that need to be remapped are either mapped as internal communication onto the new PE (if communicating tasks are mapped onto this PE) or onto the system bus.

3.2.2 Load imbalance

In the case a load imbalance has been identified as the reason for not meeting the application demands, a load balancing strategy is used to remap the critical task. The substitute PE should satisfy the condition that it is different from the current PE of the critical task and should have the lowest processor utilization in the PU list. If such a substitute does not exist, then the algorithm cannot find a better mapping.

4. EXPERIMENTS

4.1 Experimental Framework

To evaluate the efficiency of our STM algorithm and the mappings found at run time by this algorithm, we deploy the (open-source) Sesame system-level MPSoC simulator [14]. To this end, we have extended this simulator with our run-time resource scheduling framework, as illustrated in Figure 4. Our extension includes the Scenario DataBase (SDB), a System Monitor (SM) and a run-time Resource Scheduler (RS). The SDB is used to store the mappings for each inter-application scenario as derived from design-time DSE. The SM is in charge of recording the running statistics for each active application as well as monitoring system-wide statistics. The RS uses the run-time task mapping algorithm and the statistics provided by the SM to dynamically remap application tasks when needed, as explained in the previous section.

4.2 Experimental Results

In this subsection, we present several experimental results in which we investigate various aspects of our STM algorithm and compare it to three well-known mapping algorithms: First-Fit Bin-Packing (FFBP) [7] which has been frequently adapted to do task mapping by means of modelling it as a bin-packing problem, Output-Rate Balancing (ORB) [5] and Recursive BiPartition and Refining (RBPR) [18]. We modified these algorithms to fit our mapping

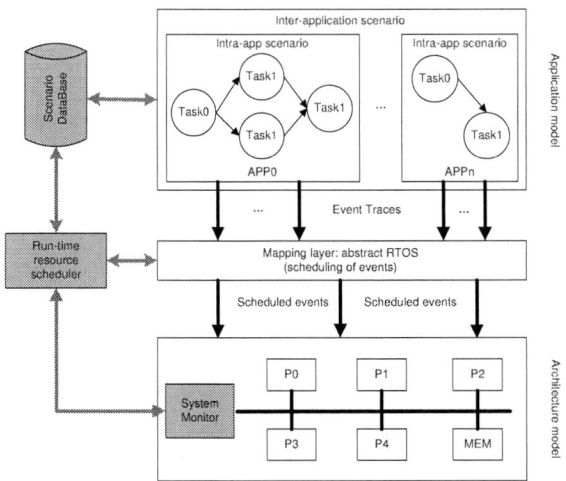

Figure 4: Extended Sesame framework.

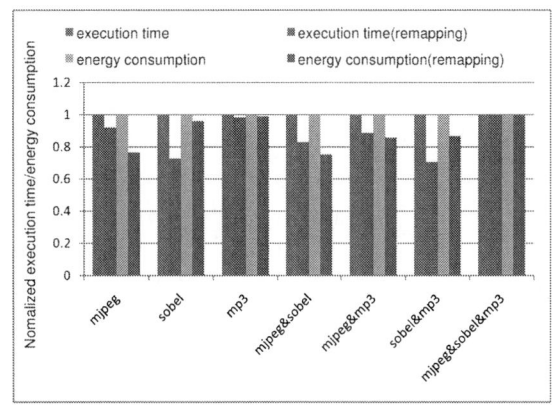

Figure 5: Performance and energy consumption of each inter-application scenario.

problem and extended them to also allow for mapping data-parallel applications by constraining the data-parallel tasks so that they have to be mapped onto different processing elements. For the FFBP algorithm, the PE with the lowest utilization is taken as the first-fit bin and the computational cost of each task in the target application is considered as the object that needs to be packed into the bins.

For our experiments, we use the three typical multi-media applications that were already introduced in Section 1: MJPEG, Sobel and MP3. The KPN of the MJPEG application contains 8 processes and 18 FIFO channels, Sobel contains 6 processes and 6 FIFO channels, and MP3 contains 27 processes and 52 FIFO channels. In the Sobel and MP3 applications, data parallelism is exploited. Moreover, MJPEG has 11 intra-application scenarios, MP3 has 3 intra-application scenarios, whereas Sobel only has 1 intra-application scenario. This results in a total of 95 different workload scenarios. At design time, we have determined the on-average best mapping for each possible inter-application scenario as explained in Section 3. With respect to the target architecture, we modeled a homogeneous MPSoC containing 5 processors, connected to a shared bus and memory. The model also includes the required components for our run-time scheduling framework.

As there are just three applications and each application contains a limited number of intra-application scenarios, we are able to exhaustively evaluate all workload scenarios. For each workload scenario, we have simulated the system using two methods: one is deploying *only the static part* of our STM algorithm to deal with the dynamism at the level of different inter-application scenarios, whereas the other one is running all the workload scenarios under a single, fixed mapping: the on-average best mapping found for the inter-application scenario in which all three applications are concurrently executing. The results of this experiment are shown in Figure 5. From this figure, we can see that the static part of our STM algorithm already yields both performance improvements and energy savings by dynamically adjusting the mapping based on the variation in inter-application scenarios. For this specific test case, the performance improvements for the different inter-application scenarios range from 1.69% to 29.49% and the energy savings range from 1.09% to 24.51%. Overall, for the execution of all 95 workload scenarios, the improvements in terms of performance and energy saving are 7.4% and 9.4%, respectively.

Figures 6.a and 6.b show the intra-application scenario execution times and energy consumption for the FFBP, ORB, RBPR and STM run-time mapping algorithms for a single inter-application scenario in which all three applications are concurrently executing. More-

over, these two graphs also contain the results when using optimal mappings (OPT) for each intra-application scenario (we derived these mappings in a design-time DSE experiment). The results in these two graphs have been ordered in a monotonically increasing fashion based on the results from the OPT mappings. Figures 6.c-e show the overall (for the entire inter-application scenario) performance, energy consumption and overhead. Here, the overhead includes the run-time calculation of new mappings as well as the migration of tasks. From Figure 6, we can see that our STM clearly performs better than the other algorithms in terms of the execution time of scenarios. For several intra-application scenarios, the STM algorithm even approaches the OPT results. With respect to energy consumption and overhead, the STM algorithm also performs well: it ranks second closely behind the ORB algorithm. The reason for a low overhead of ORB is that it only needs to migrate a few tasks in our experiment which means a very low task migration cost.

In our last experiment, we used the full STM algorithm, including the static and dynamic parts and thus combining the dynamism of inter-application as well as intra-application scenarios, to test all the 95 workload scenarios of our three applications. Our algorithm could achieve a 11.3% performance improvement and an energy saving of 13.9% compared to an approach in which we run the applications using the (static) on-average best mapping for the inter-application scenario in which all three applications are active. Comparing these results to those when only using the static part of our STM algorithm (improvements of 7.4% and 9.4%, respectively; see above), this means that the dynamic part of the STM algorithm is capable of significantly further improving the mappings.

5. RELATED RESEARCH

In recent years, much research has been performed in the area of run-time mapping for embedded systems. The authors of [6] propose a run-time mapping strategy that incorporates user behavior information in the resource allocation process. An agent based distributed application mapping approach for large MPSoCs is presented in [1]. The work of [9] proposes a run-time spatial mapping technique to map streaming applications on MPSoCs. In [3], dynamic task allocation strategies based on bin-packing algorithms for soft real-time applications are presented. A Dynamic Spiral Mapping (DSM) algorithm for mapping an application on an MP-SoC arranged in a 2-D mesh topology is proposed in [2]. The authors from [4] present network congestion-aware heuristics for mapping tasks on NoC-based MPSoCs at run-time. The work of [16] uses a Smart Nearest Neighbour approach to perform run-time task mapping. In [10], a run-time task allocator is presented that

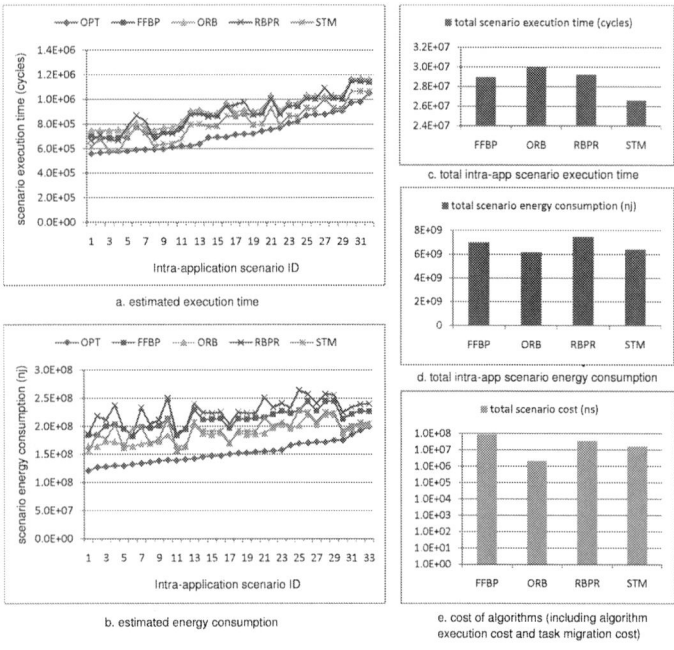

Figure 6: Comparing different run-time mapping algorithms.

uses an adaptive task allocation algorithm and adaptive clustering approach for efficient reduction of the communication load. Mariani et al. [12] proposed a run-time management framework in which Pareto-fronts with system configuration points for different applications are determined during design-time DSE, after which heuristics are used to dynamically select a proper system configuration at run time. Compared with these algorithms, our STM algorithm takes a scenario-based approach, and takes computational and communication behavior into account to make (re-)mapping decisions. Recently, Schor et al. [15] also proposed a scenario-based run-time mapping approach in which mappings derived from design-time DSE are stored for run-time mapping decisions, but they do not cluster mappings to reduce mapping storage nor do they dynamically optimize the mappings at run time.

6. CONCLUSION

We have proposed a run-time mapping algorithm for MPSoC-based embedded systems to improve their performance and energy consumption by capturing the dynamism of the application workloads executing on the system. This algorithm is based on the idea of application scenarios and consists of a design-time and run-time phase. The design-time phase produces mappings for clusters of application scenarios after which the run-time phase aims to optimize these mappings by continuously monitoring the system and trying to perform (relatively small) mapping customizations to gradually further improve the system performance. In various experiments, we have evaluated our algorithm and compared it with three other algorithms. The results show that our algorithm can yield considerable improvements as compared to just using a static mapping strategy. Comparing our algorithm with three other, well-known run-time mapping algorithms, it shows a better trade-off between the quality and the cost of the mappings found at run time.

7. REFERENCES

[1] M. A. Al Faruque, R. Krist, and J. Henkel. Adam: run-time agent-based distributed application mapping for on-chip communication. In *Proc. of DAC'08*, pages 760–765, 2008.

[2] M. Armin, K. Ahmad, and S. Samira. Dsm: A heuristic dynamic spiral mapping algorithm for network on chip. *IEICE Electronics Express*, 5(13):464–471, 2008.

[3] E. W. Brião, D. Barcelos, and F. R. Wagner. Dynamic task allocation strategies in mpsoc for soft real-time applications. In *Proc. of DATE'08*, pages 1386–1389, 2008.

[4] E. Carvalho and F. Moraes. Congestion-aware task mapping in heterogeneous MPSoCs. In *Int. Symposium on System-on-Chip*, pages 1–4, Nov. 2008.

[5] J. Castrillon, R. Leupers, and G. Ascheid. Maps: Mapping concurrent dataflow applications to heterogeneous mpsocs. *IEEE Trans.on Industrial Informatics*, PP(99):1, 2011.

[6] C.-L. Chou and R. Marculescu. User-aware dynamic task allocation in networks-on-chip. In *Proc. of DATE'08*, pages 1232–1237, 2008.

[7] E. G. Coffman, Jr., M. R. Garey, and D. S. Johnson. Approximation algorithms for bin packing: a survey. In *Approximation algorithms for NP-hard problems*, pages 46–93. PWS Publishing Co., 1997.

[8] S. V. Gheorghita, M. Palkovic, J. Hamers, A. Vandecappelle, S. Mamagkakis, T. Basten, L. Eeckhout, H. Corporaal, F. Catthoor, F. Vandeputte, and K. D. Bosschere. System-scenario-based design of dynamic embedded systems. *ACM Trans. Design Autom. Electr. Syst.*, 14(1), 2009.

[9] P. K. Hölzenspies, J. L. Hurink, J. Kuper, and G. J. Smit. Run-time spatial mapping of streaming applications to a heterogeneous multi-processor system-on-chip (mpsoc). In *Proc. of DATE'08*, pages 212–217, March 2008.

[10] J. Huang, A. Raabe, C. Buckl, and A. Knoll. A workflow for runtime adaptive task allocation on heterogeneous mpsocs. In *Proc. of DATE'11*, pages 1119–1134, 2011.

[11] G. Kahn. The semantics of a simple language for parallel programming. In *Information processing*, pages 471–475. North Holland, Amsterdam, Aug 1974.

[12] G. Mariani, P. Avasare, G. Vanmeerbeeck, C. Ykman-Couvreur, G. Palermo, C. Silvano, and V. Zaccaria. An industrial design space exploration framework for supporting run-time resource management on multi-core systems. In *Proc. of DATE'10*, pages 196–201, march 2010.

[13] J. M. Paul, D. E. Thomas, and A. Bobrek. Scenario-oriented design for single-chip heterogeneous multiprocessors. *IEEE Trans. VLSI Syst.*, 14(8):868–880, 2006.

[14] A. D. Pimentel, C. Erbas, and S. Polstra. A systematic approach to exploring embedded system architectures at multiple abstraction levels. *IEEE Trans. Computers*, 55(2):99–112, 2006.

[15] L. Schor, I. Bacivarov, D. Rai, H. Yang, S.-H. Kang, and L. Thiele. Scenario-based design flow for mapping streaming applications onto on-chip many-core systems. In *Proc. of CASES'12*, pages 71–80, 2012.

[16] A. K. Singh, W. Jigang, A. Kumar, and T. Srikanthan. Run-time mapping of multiple communicating tasks on mpsoc platforms. *Procedia CS*, 1(1):1019–1026, 2010.

[17] P. van Stralen and A. D. Pimentel. Scenario-based design space exploration of mpsocs. In *Proc. of IEEE ICCD'10*, October 2010.

[18] J. Yu, J. Yao, L. Bhuyan, and J. Yang. Program mapping onto network processors by recursive bipartitioning and refining. In *Proc. of DAC'07*, pages 805–810, june 2007.

Early Exploration for Platform Architecture Instantiation with Multi-mode Application Partitioning

Prashant Agrawal*†, Praveen Raghavan*, Matthias Hartman*†, Namita Sharma‡,
Liesbet Van der Perre*†, Francky Catthoor*†

*IMEC vzw., Leuven, Belgium, †KU Leuven, Leuven, Belgium, ‡IIT Delhi, Delhi, India
prashant.agrawal@imec.be

ABSTRACT

We present a systematic methodology for exploring application partitioning and assignment together with platform architecture instantiation. Streaming applications with multiple runtime modes are considered. The platform architecture is based on a domain specific MPSoC architecture template. We show results using complete inner modem physical layer processing of wireless applications, WLAN and LTE. We show that the proposed methodology obtains up to 30% energy improvement in energy with negligible area overheads as compared to straight-forward mapping to one processor, while meeting performance constraints, for a multi-mode WLAN 11n system and single-mode LTE system.

Categories and Subject Descriptors

C.3 [Computer Systems Organization]: Special-Purpose and Application-based Systems—*Real-time and embedded systems*

Keywords

Streaming applications, Heterogeneous MPSoC, Partitioning, Mapping, Co-Design

1. INTRODUCTION

It is imperative in today's growing constraint on low power design for embedded systems to go for heterogeneous MPSoC (Multiprocessor System-on-Chip) designs. Although heterogeneous MPSoCs have energy, performance and cost benefits, such platforms are increasingly becoming complex and often have irregular memory hierarchy organizations. With the advent of ASIP (Application Specific Instruction Set Processor) based MPSoC architectures, the complexity increases further as they have multiple heterogeneous configurable processors.

One of the key challenges in a heterogeneous MPSoC design is partitioning and assignment (P&A) of the application and the architecture instantiation of the underlying platform. Partitioning is the design decision for splitting the ap-

plication in to several parts. Assignment is the step of binding the application elements to the system resources. Architecture instantiation involves exploring architecture configuration parameters, both at the platform and the individual processors level.

The problem of P&A is relatively simple when dealing with a given fixed platform and has been researched [1][2]. The architecture instantiation for an already partitioned application is also straightforward, where mostly application assignment and scheduling is explored [3][4]. However, the design of a MPSoC architecture and P&A are not orthogonal to each other and are coupled to each other. In the co-design space, [5][6] traditionally partitioning decisions were limited to hardware-software partitioning only, where the design space is limited. Most of the works in ASIP based MPSoC designs with P&A exploration [7][8][9] either do not consider complete application and focus on specific parts only or carry out limited exploration. Further, the problem of P&A and architecture exploration gets amplified exponentially with applications with multiple modes (runtime profiles) [10] or even more in case of multiple applications [11]. Thus, P&A exploration "together" with platform architecture instantiation for mapping multi-mode streaming applications, is still an unsolved problem in the state of the art.

The key contributions of this paper are: (a) Proposes systematic methodology for early exploration of the P&A "together" with platform architecture exploration, (b) Proposes systematic approach for ASIP based MPSoC platform instantiation (c) Targets complete real-life multi-mode streaming applications (complete inner modem processing of the physical layer of wireless applications, WLAN 11n [12] and LTE-3GPP [13]) (d) Achieves better than worst-case energy efficiency for platform instantiation to support multi-mode applications

The proposed methodology has been shown to achieve energy gains with negligible area overheads by carrying out fine grained P&A exploration by considering the static and runtime dynamism across and within the modes. The methodology generates multiple heterogeneous partitions such that the tasks assigned to a partition are well matched in complexity, parallelism, duty cycle and hardware requirements, and do not have conflicting requirements. This ensures energy efficiency while minimizing the area overheads. We achieve energy gains of up to 30% for WLAN and LTE, with negligible area overheads and while meeting performance constraints, as compared to straight forward mapping to single processor. Although, this paper uses wireless ap-

---

Permission to make digital or hard copies of all or part of this work for personal or classroom use is granted without fee provided that copies are not made or distributed for profit or commercial advantage and that copies bear this notice and the full citation on the first page. To copy otherwise, to republish, to post on servers or to redistribute to lists, requires prior specific permission and/or a fee.
DAC '13 May 29-June 07 2013, Austin, TX, USA

plication drivers, the proposed methodology is generic and can be applied on other streaming multi-mode applications such as MPEG-2/MPEG-4, etc.

The paper is structured as follows: Section 2 gives an overview of the related work followed by description of the platform template in Section 3. Section 4 discusses details of the proposed methodology, followed by results and conclusions in Sections 5 and Section 6, respectively.

2. RELATED WORK

System scenarios [14][15] and use-case scenarios [16] based approaches have been proposed for mapping multi-mode applications, i.e. with runtime dynamism, on heterogeneous MPSoC architecture. However, these [3][4][17] approaches have the limitation in a way that they either do not explore partitioning of the application tasks on a heterogeneous architecture template or [2][16][18] assume a fixed underlying architecture and have limited scope for platform architecture exploration. In contrast, in our work we explore P&A of multi-mode applications along with exploring the instantiation of the platform from an application domain specific template (refer Table 1 in Appendix F). Moreover, our work is complementary to the scenario based approaches. We do not focus on creating the scenarios but consider representative scenarios for the application(s) and carry out the P&A and MPSoC architecture exploration for mapping these scenarios on the same platform. Our approach can handle multiple scenarios and is completely compatible with scenario based approaches.

Hardware/software (HW/SW) partitioning has been researched well [5][6]. Here, the P&A decisions are relatively straightforward as the application has to be partitioned across generic processors (SW) or custom hardware blocks (HW). The design space here is much limited as compared to P&A exploration on ASIP based MPSoC architectures, which is the focus of our work.

Most of the industrial mappings of wireless standards on SDR (Software Defined Radio) platforms, carried out for complete receivers [9], typically neither explore P&A nor analyze the trade-offs in details. Besides this, PHY layer implementations mostly focus on on efficient implementations for the computationally dominant blocks, such as MIMO detectors and FFT [7][8].

In this paper, we extend our previous work [19] which focused on P&A exploration for single mode and static behaviour of the application. Here, we present a methodology to handle dynamic runtime behavior of different types of multi-mode streaming applications. The dynamism of the input applications is assumed to be well-defined at the design time.

To the best of our knowledge, our work is the first work that systematically explores, early in the design flow, P&A and architecture together for mapping of multi-mode application(s) on a MPSoC platform. We are also not aware of any work which can achieve energy efficient mapping with negligible area overheads for such multi-mode applications, besides minimizing energy overheads for the lower complexity modes when mapped with worst case complexity modes on the same platform.

3. ARCHITECTURE TEMPLATE

We consider application domain specific architecture templates, where the template has been designed and optimized

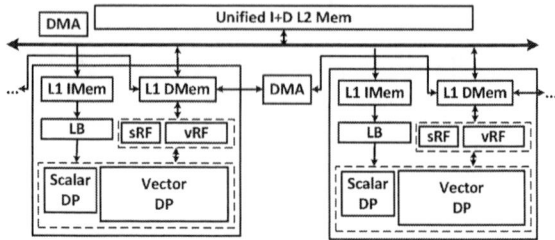

Figure 1: Architecture template

for applications of a particular domain. Figure 1 shows the high level architecture template considered in this work. This template is assumed to be designed for streaming applications like wireless standards and multimedia. The figure shows both the platform level and the processor level template. At the platform level there are multiple processors, all of which share a unified instruction and data level-2 (I+D L2) memory. The communication between the processors is point-to-point using DMA (Direct Memory Access) transfers. The number of processors, size of L2 memory and the bus width are configurable.

At the processor level, each processor consists of scalar and vector datapaths (DP), and local instruction and data memories. We assume single issue for scalar as well as vector datapaths. For the targeted multi-mode streaming applications, which are inherently data-parallel, we can get reasonably accurate estimates of area and energy by considering data-level parallelism (DLP) only. Although we do not consider instruction-level parallelism (ILP), the results of the methodology are valid even if ILP is considered. The pipeline depths for the scalar and the vector datapaths are 3 and 7, respectively. These have been obtained based on targeted application requirements at the targeted 40nm process technology. The local instruction (IMEM) and data (DMEM) memories are software-controlled, separated (scalar and vector) and hierarchical (i.e. each consisting of two levels, L1 and L0). The L0 instruction memory is essentially a loop buffer whereas L0 data memory is a Register File (RF). The datapath of the processor consists of computational operators specific to application domain requirements. For each processor, the memory sizes, the scalar and the vector word size, and the type and number of compute operators can be configured. We mostly focus on the vector datapath as it is the more dominant part of the design in terms of area, power and other metrics, especially for the application domains considered.

Although this template is basic and high level, it captures the essential elements and provides configuration parameters also provided in other academic templates such as SODA [20] and commercial ones such as Tensilia [21], CEVA [22].

4. PROPOSED METHODOLOGY

In this section, we discuss the proposed methodology for carrying out the P&A exploration, a high level view of which is shown in Fig. 2. The methodology explores the complete design space and generates a set of P&A schemes under given constraints (including performance). It estimates area and energy for each P&A scheme.

4.1 Application Characterization

Streaming applications such as wireless communication (WLAN, LTE), multimedia (MPEG-2, MPEG-4), etc. sup-

914

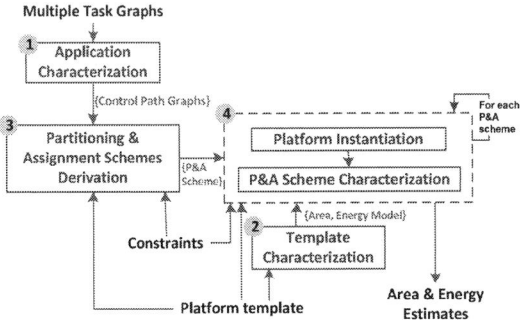

Figure 2: Proposed methodology for P&A exploration

port multiple modes (runtime profiles) and exhibit runtime dynamism. We handle this dynamism in the input applications in two steps. Firstly, the multiple modes of the application are clustered into a number of system scenarios [14]. Each scenario is modeled as a separate control and dataflow graph (CDFG) G_s. Secondly, for each scenario, if different types of the input data (such as preamble and payload symbols in case of WLAN or I, P, B type frames in case of MPEG-2) results in a different control and data flow of the execution, each such execution flow is considered as a separate control path graph (CPG) in G_s. Refer to Appendix D for an illustration. The execution of these CPGs is considered to be mutually exclusive in time. Splitting G_s into multiple CPGs abstracts away the heterogeneity in the control flow of G_s while preserving the data flow dependency, besides making the proposed methodology generalized enough to handle multiple modes of one or more applications.

The number of CPGs created will depend on the granularity of the nodes in the CDFG and the granularity of the input data. In the most simple case, the entire CDFG can be considered as a single CPG which will result in worst-case over-dimensioned system. To keep the number of CPGs manageable, we consider each task (node) in G_s to be a functional step in the application. A task is also considered as the lowest granularity of the partitioned application during the P&A exploration. The application is partitioned into these tasks based on the communication and computation complexities, and to minimize communication overhead. It is also assumed that the inter-task data transfers are blocked-transfer and at a symbol or frame level.

Each G_s is characterized separately by internally profiling each of its tasks to estimate its compute and data-memory related complexities for the processing of one unit of input data, which is a symbol or a frame in case of WLAN/LTE or MPEG-2/MPEG-4, respectively. The profiling is carried out on a simple template architecture with no parallelism. It is an one time effort for each input G_s.

Each task-level operation is mapped to one or more operators supported in the platform template. The set of different platform operators required and the number of times each of them is executed is calculated for each task to estimate its compute complexity. The memory sizes and the data-memory access complexity for each task are estimated using the sizes of the input, output and the intermediate data, required for the processing of one symbol/frame by the task. The inter-task data transfers are considered as the communication complexity of the tasks. The data transfers between the datapath and the memories (L1-D and RF) are consid-

ered as the data-memory access complexity of the tasks.

Using these estimates, each task in the CPG is annotated with a 5-tuple $< \Omega_n, \epsilon_n, \Psi_n, \phi_n, \rho_n >$, referred to as the *characterization tuple*, where Ω_n, ϵ_n and Ψ_n denote task complexity per iteration, number of iterations and the set of required hardware resources, respectively. ϕ_n and ρ_n denote the available degree of data-level parallelism at the application and the task level, respectively. Thus, ϕ_n is an upper bound on ρ_n. All the tasks in the same CPG will have the same values of ϵ_n and ϕ_n.

It may be noted that in this case, Ω_n is calculated assuming blocked transfer of data between the tasks. The methodology can be easily extended for the streaming transfer of data between the tasks by recalculating Ω_n based on the size of the data chunk that is received before the task starts processing.

4.2 Platform Template Characterization

In this step, the datapath operators, RF and L1-D memory components of the platform are characterized. Physical synthesis and back-annotated gate level simulations are carried out to build a component level area and energy library. This library is used to derive an empirical cost model, using regression analysis, to estimate area and energy for different platform components at different configurations. This step is also one time effort. It has to be repeated only if a new template is considered, or if new components are added to the existing template.

4.3 P&A Schemes Derivation

Here, we discuss the algorithm for systematically deriving all the P&A schemes (Algorithm 1). This step receives as input the set of all CPGs derived for all the modes of the input application. It may be noted that the proposed P&A exploration is complete under the assumptions that the tasks in the CPGs are not further partitioned and that only blocked data transfer is considered between the tasks.

These streaming applications offer parallelism at task-(TLP), data- (DLP) and instruction-level (ILP) and at different granularity levels. The proposed P&A algorithm explores TLP by pipelining the execution of the tasks in the CPGs. The extent of pipelining explored depends on the latency constraints imposed by the application specifications (refer Appendix E). Pipelining decisions have major impact on the platform configuration, cycle budget allocation across different processors in the platform, etc.

Coarse-grained DLP is explored at the task level by partitioning input data streams and processing them in parallel. Fine-grained DLP is explored at the processor level using vectorization/ SIMD (Single Instruction Multiple Data). Although ILP is not explored, it does not restrict the effectiveness of the proposed methodology, as discussed in Section 3.

The P&A algorithm iterates over different cases of pipelining (referred to as pipeline schemes) and derives P&A schemes for each of these cases, as described below. The algorithm has been implemented as a tool and it's time complexity is $O(n \cdot p \cdot s)$, where n is the total number of nodes in the input CPGs, p is the number of pipeline stages in a scheme and s is the number of pipeline schemes being explored.

4.3.1 Taskgroup Creation (Lines 2-4)

For each CPG, *create_taskgroups()* divides its tasks into disjoint sets called task groups based on the tasks' characterization information. The objective is to create task

ALGORITHM 1: P&A schemes derivation

input : Set of control path graphs (CPG) $\forall$ modes of the input application
output: Set of P&A schemes

1 **for** *each pipeline scheme* **do**
2   **for** *each CPG* **do**
3 | $\{tg\} \leftarrow$ create_taskgroups(CPG) ;
4   **end**
5    $\{tc\} \leftarrow$ create_taskclusters$(tg \forall CPG)$;
6   **for** *each tc* **do**
7 | $\{tsc\} \leftarrow$ create_tasksubclusters(tc) ;
8 | $\{p\} \leftarrow$ create_partitions$(tc, tsc \in tc)$;
9   **end**
10    $\{$P&A Schemes$\} \leftarrow$ create_pa-schemes$(p \forall tc)$;
11 **end**

ALGORITHM 2: *create_partitions()*

input : Set of task cluster (tc) and task sub-clusters (tsc)
output: Set of partitions

1 $tc \rightarrow$ single partition   // Level 1
2 **if** *number of tsc in tc* > 1 **then**
3   **for** *each tsc in tc* **do**
4       $tsc \rightarrow$ single partition ;   // Level 2
5   **if** $\rho_{tsc_{min}} > 1$ **then** $tsc \rightarrow 2$ to k partition // Level3a
6   **end**
7 **else**
8   **if** $\rho_{tc_{min}} > 1$ **then** $tc \rightarrow 2$ to k partitions   // Level 3b
9 **end**

---

groups with balanced complexity while preserving the data flow dependency among the tasks, and also to ensure that tasks with similar degree of parallelism are grouped together. This impacts the number of processors instantiated in the platform and configuration of each processor.

Each task group (tg) is annotated with a 6-tuple $< \Omega_{tg}, \epsilon_{tg}, \Psi_{tg}, \phi_{tg}, \rho_{tg}, \kappa_{tg} >$ obtained based on the characterization tuple of the tasks (n) grouped in tg. Ω_{tg} denotes the total complexity of tg and is also used as a measure of latency as the hardware is not yet defined. This is a reasonable assumption since we assume blocked data transfers between the tasks. Ψ_{tg} is the union of hardware resources requirement across all the tasks in tg. ρ_{tg} is the maximum degree of parallelization of tg, which is limited by the task with the minimum degree of parallelization (ρ_n) in tg. κ_{tg} denotes the degree of hardware parallelism required for the task to complete its processing in the allocated cycle budget. $\epsilon_{tg} = \epsilon_n$ and $\phi_{tg} = \phi_n$ as these are same for all the tasks in a CPG.

4.3.2 Task Cluster Creation (Line 5)

create_taskclusters() merges the task groups (tg) across all the CPGs and creates a set of task clusters (= number of pipeline stages). Each task cluster (tc) is assigned to a separate stage. Task groups of each CPG are assigned a cycle budget based on the latency of the pipeline stage and κ_{tg} for each task group is calculated.

Typically, each tc is a candidate for being mapped on a single processor as well as being partitioned further for exploring coarse-grain DLP. The objective of this step is to cluster task groups, across the CPGs, based on homogeneity in terms of complexity, hardware resource requirements, scope of parallelism, etc., while maintaining the data flow dependencies within CPGs. This prevents over-dimensioning of processor configuration by minimizing the gap between the worst- and the average-case. It also prevents restricting the parallelization opportunities of tc. Thus, this is a very crucial step in ensuring handling of inter- and intra-mode dynamism of the application in an energy and area efficient manner.

Each task cluster (tc) is also annotated with a 5-tuple $< \Psi_{tc}, \rho_{tc_{min}}, \rho_{tc_{max}}, Z_{tc}, TG_{tc} >$ based on the task groups assigned to it. Ψ_{tc} is the union of hardware resources required across all the tg in the tc. $\rho_{tc_{min}}$ and $\rho_{tc_{max}}$ are the minimum and the maximum degree of parallelism, respectively, for the tc. Z_{tc} is the set of unique 2-tuple $< \epsilon_{tg}, \kappa_{tg} >$ across all the tg and TG_{tc} is the set indicating all the tg clustered in the task cluster.

4.3.3 Task Sub-Cluster Creation (Line 7)

Although, *create_taskclusters()* attempts to create task clusters with homogeneous task groups, due to the data flow dependency constraints, it is possible that a task cluster may contain task groups having significant variance in the available DLP (ϕ_{tg}), frequency of execution (ϵ_{tg}) and the required hardware parallelism (κ_{tg}). The objective of *create_tasksubclusters()* is to analyze each task cluster, and if it contains task groups with large variation in ϕ_{tg}, ϵ_{tg}, κ_{tg} values, the task cluster is split into one or more task sub-clusters based on the clustering of ϕ_{tg}, ϵ_{tg}, κ_{tg} values.

Like, task clusters, each task sub-cluster is also a candidate for being mapped on a single processor as well as being partitioned further for exploring coarse-grain DLP. Each sub-cluster (tsc) is also annotated with a 5-tuple $< \Psi_{tsc}, \rho_{tsc_{min}}, \rho_{tsc_{max}}, Z_{tsc}, TG_{tsc} >$.

4.3.4 Partition Creation (Line 8)

create_partitions() creates one or more partitions for each task cluster or task sub-cluster (in case the cluster has been split into sub-clusters), as shown in Algorithm 2. It creates partitions at three granularity levels (denoted by Level 1, 2, 3a/3b in Algorithm 2) to cover the complete P&A space. Each partition is assigned to a separate processor. Thus, the number of partitions in a P&A scheme decides the platform configuration. If the platform template imposes constraints on the maximum number of processors, this can be considered in Algorithm 2 to limit the partitioning choices.

Level 3 partitions are created to explore coarse-grained DLP. The maximum number of level 3 partitions (k) created depends on the maximum value of the degree of parallelism ($\rho_{tc_{max}}$) of the task cluster, the application characteristics and platform template constraints.

Each partition (P) is also annotated with a 3-tuple $< \Upsilon_P, \kappa_P, \Psi_P >$. Υ_P is a set of 2-tuple $< tg, \varrho >$ where tg is the task group assigned to the partition and ϱ denotes the number of input data streams processed by this tg. κ_P is the required degree of hardware parallelism and Ψ_P is the set of the hardware resources required for the partition.

4.3.5 P&A Schemes Creation (Line 10)

Partitioning at each level, in the previous step, can be considered as separate intra-taskcluster P&A scheme. These schemes are combined across the task clusters to obtain the complete set of P&A schemes for each case of pipelining. Schemes across all the pipelining cases form the complete set of P&A schemes for the input application.

4.4 Platform Instantiation

Platform is instantiated for each P&A scheme separately. The methodology configures the number of processors at the platform level (=number of partitions in the P&A scheme, as mentioned in Section 4.3.4). At the processor level, it configures parameters for the type and number of compute operators, vector word size and data memory sizes (RF and L1). The methodology currently assumes a fixed communication structure and ignores level-2 memory hierarchy as it is not crucial for streaming applications being considered.

The entire platform for a P&A scheme is instantiated by instantiating each of its individual processors. A processor is instantiated based on the *characterization tuple* of the partition assigned to it. The degree of parallelization (PF) required to meet the latency requirements is considered equal to the partition's κ_P value. Since we consider DLP only, PF implies the vector wordsize for the datapath and the width of the RF and the L1-D memory. The set of compute operators specified by Ψ_P are instantiated at a SIMD width equal to PF. The required size of L1-D memory for the processor is considered equal to the maximum of the size of L1-D estimated for each task mapped on the processor. The depth of the RF has to be configured based on the application profiling and currently has to be provided by the designer.

4.5 P&A Scheme Characterization

The total area and energy of a P&A scheme are estimated by individually characterizing each processor in the scheme. This is done using processor's configuration parameters, tasks mapped to the processor and template's area and energy models.

The total area of a processor is estimated by calculating areas of datapath and the memories (RF and L1-D) separately. The datapath area is calculated based on the compute operators instantiated in the processor and their SIMD factor, whereas areas of the RF and L1-D memory are calculated using their sizes. For energy estimation, the compute, data-memory access and communication energies are calculated separately. These are calculated based on the set of tasks mapped to the processor, the amount of input data each task processes and the processor configuration. We do not discuss characterization in details due to space constraints in the paper.

5. RESULTS

5.1 Experimental Setup

We have considered a uniform quantization of 16 bits for all data variables. The word size and the frequency considered for the platform template are 16-bits and 700 MHz, respectively. The template has been characterized at 40nm G process technology. The datapath operators are implemented using commercial standard cells and L1-D memory using commercial memory generators. The register file is D flip-flop based. Physical synthesis back-annotated gate level simulations have been performed to obtain area and energy numbers for these components at different bit widths.

5.2 Application Drivers

We have used complete inner modem of the physical layer of wireless applications, WLAN 11n [12] and LTE-3GPP [13], as the application drivers (refer Appendices A and B). These have in-house developed industry quality code base. These are fairly large applications with several thousand lines of code. We consider two modes of WLAN 11n, i.e. 2×2 40MHz and 4×4 40MHz both with QAM-64 mod-

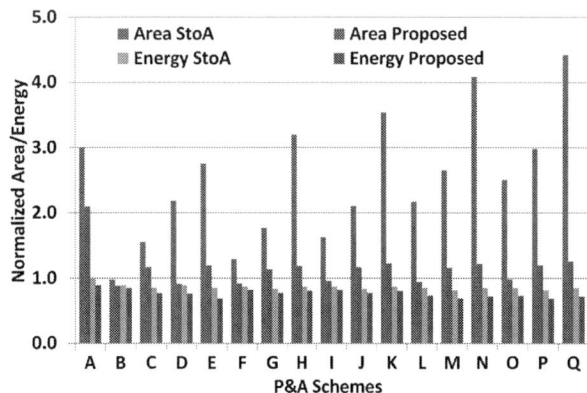

Figure 3: WLAN 4×4 mode (combined platform)

ulation, and single mode of LTE, i.e. 2×2 20MHz with QAM-256 modulation.

We carry out single-mode and multi-mode P&A exploration for LTE and WLAN, respectively. In each case, we explore area-energy trade-offs across different P&A schemes while meeting the required performance constraints. The 2×2 and the 4×4 modes of WLAN are considered as two different scenarios, where the 2×2 mode is an average complexity scenario whereas the 4×4 mode is a worst-case complexity scenario. It is also assumed that 2×2 mode is the more frequently occurring scenario for WLAN.

We chose WLAN 11n and LTE because they have very different characteristics and provide very different constraints and opportunities for P&A exploration (refer Appendix C). As mentioned in Section 1, although this paper uses wireless application drivers, the proposed methodology is generic and can be applied on other streaming multi-mode applications such as MPEG-2/MPEG-4, etc.

5.3 P&A Exploration Results

Design Space Coverage : In case of WLAN 11n, the methodology derives 18 P&A schemes with number of processors ranging from 1 to 11 by considering both the 4×4 and the 2×2 modes together. Whereas in case of the single mode of LTE, 29 P&A schemes are derived with number of processors ranging from 1 to 18. The design space obtained in each case is complete under the given constraints and the assumptions that the tasks (nodes) of the input CDFG are not partitioned further and that the transfer of data between the nodes is blocked and at a symbol level. If this exploration is carried out considering either fixed architecture or a given partitioning ,as is the case in the existing approaches (refer Table 1 in Appendix F), only a limited design space will be explored and the obtained P&A schemes are likely to be sub-optimal. In contrast, the proposed methodology obtains all the pareto optimal P&A schemes in a more predictable manner and with lower effort.

Energy Reduction with Negligible Area Overhead : Each P&A scheme derived is characterized for processing one input packet in each mode. Figure 3 shows the change in area and energy for the more complex 4×4 mode of WLAN 11n for different P&A schemes obtained using the proposed methodology and [17] (referred as the StoA approach here) which instantiates generic MPSoC platforms composed of homogeneous processors.

Here, both 2×2 and 4×4 modes have been mapped on the

917

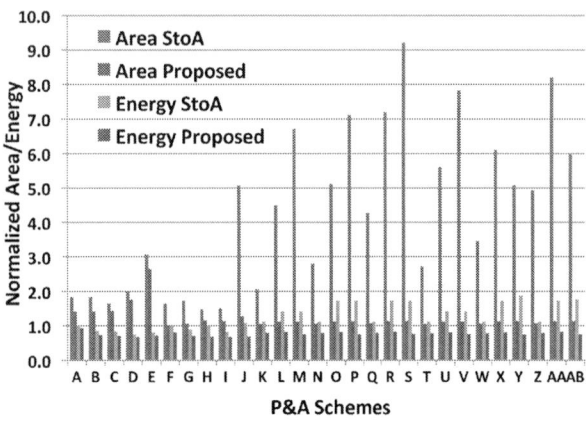

Figure 4: LTE 2×2 mode (independent platform)

same platform. The results have been normalized to single processor based implementation. The proposed methodology results in P&A schemes which are more energy efficient and at significantly lower area, as compared to the StoA approach. The proposed methodology achieves up to 30% energy gains as compared to single processor implementation and up to 20% energy gains as compared to the StoA approach, without incurring any area overheads (P&A schemes labelled as B, D, F, I, L and O). The area overheads are minimized in the proposed methodology because the tasks assigned to a partition are very well matched in complexity, parallelism, duty cycle and hardware requirements, and do not have conflicting requirements. P&A schemes E, M, P and Q also achieve energy reductions similar to the above mentioned P&A schemes but with at least 10-20% area overhead because in these schemes the lower complexity tasks were also parallelized. The methodology explores up to 3 pipeline stages beyond which the latency requirements of WLAN 11n cannot be met.

Figure 4 shows similar results for LTE where the proposed methodology also achieves up to 30% energy gains as compared to single processor implementation with area overhead of only 5-10%. LTE is a memory intensive application, unlike WLAN, and the memory dominates the total area. Our results are better up to $2.5\times$ and $8\times$ in energy and area, respectively as compared to the StoA approach [17] because the memory sizes and accesses are also taken into account while carrying out the P&A exploration.

Better than Worst Case Energy Efficiency : The area-energy trade-offs for the 2×2 mode, when mapped with 4×4 mode on the same platform, shows similar trends as the 4×4 mode. Hence, we do not show results for it separately. However, the trend changes when 2×2 mode is mapped independently on a platform. This is because the clustering of task groups and the partitioning of task clusters changes depending on other modes being mapped together. Besides, 2×2 mode is a lower complexity mode with lesser parallelization possibilities.

We compare the most efficient P&A scheme for 2×2 mode on independent platform with the pareto-optimal schemes obtained for 2×2 mode in the combined exploration with 4×4 mode. Although, the area overhead in each case is at least $2\times$, **the energy overheads lies between 0-7%.** Thus, the proposed methodology guarantees mapping of low complexity modes without or negligible energy overheads on

platform which also supports higher complexity modes.

6. CONCLUSIONS

We have proposed a systematic methodology for exploring application partitioning and assignment together with platform architecture instantiation. It has been shown for real-life streaming multi-mode wireless applications, WLAN 11n and LTE, that the proposed methodology obtains energy gains up to 30%, with negligible area overheads and while meeting performance constraints, as compared to single processor mapping solution. The methodology also enables mapping of lower complexity modes together with higher complexity modes on the same platform without energy overheads.

7. REFERENCES

[1] A.K. Singh, T. Srikanthan, A. Kumar, and W. Jigang. Communication-aware heuristics for run-time task mapping on noc-based mpsoc platforms. *J. Syst. Arch.*, 56:242–255, 2010.

[2] A.M.A. Hussien, A.M. Eltawil, R. Amin, and J. Martin. Energy aware task mapping algorithm for heterogeneous mpsoc based architectures. In *Proc. of ICCD*, pages 449 –450, 2011.

[3] P. van Stralen and A. Pimentel. Scenario-based design space exploration of mpsocs. In *Proc. of ICCD*, pages 305–312, 2010.

[4] P. van Stralen and Andy Pimentel. Fast scenario-based design space exploration using feature selection. In *Proc. of ARCS*, pages 1–7, 2012.

[5] J. Wu, Q. Sun, and T. Srikanthan. Algorithmic aspects for multiple-choice hardware/software partitioning. *Computers & Operations Research*, 39(12):3281–3292, 2012.

[6] M. Abdelhalim and S. Habib. An integrated high-level hardware/software partitioning methodology. *ACM TODAES*, 15:19–50, 2011.

[7] B. Zhang, H. Liu, H. Zhao, F. Mo, and T. Chen. Domain specific architecture for next generation wireless communication. In *Proc of DATE*, pages 1414–1419, 2010.

[8] M. Wani, Z. Miljanic, P. Spasojevic, and J. Redington. Asip data plane processor for multi-standard interleaving and de-interleaving. In *Proc. of ASILOMAR*, pages 1259 –1263, 2010.

[9] T. Suzuki and et. al. High-throughput, low-power software-defined radio using reconfigurable processors. *IEEE Micro*, 31(6):19–28, 2011.

[10] J. Cong and et. al. Mc-sim: An efficient simulation tool for mpsoc designs. In *Proc. of ICCAD*, pages 364 –371, 2008.

[11] H. Nikolov and et. al. Daedalus: Toward composable multimedia mp-soc design. In *Proc. of DAC*, pages 574–579, 2008.

[12] *IEEE 802.11n-2009 Standard Document,Amendment 5,29 Oct 2009.*

[13] *3GPP TSG-RAN, Physical Channels and Modulation, v8.1.0 (2007-11).*

[14] S. V. Gheorghita and et. al. System-scenario-based design of dynamic embedded systems. *ACM TODAES*, 14:3:1–3:45, 2009.

[15] J.M. Paul, D.E. Thomas, and A. Bobrek. Scenario-oriented design for single-chip heterogeneous multiprocessors. *IEEE TVLSI*, 14(8):868 –880, 2006.

[16] A. Schranzhofer, Jian-Jian Chen, and L. Thiele. Dynamic power-aware mapping of applications onto heterogeneous mpsoc platforms. *IEEE TII*, 6(4):692 –707, 2010.

[17] A.K. Singh, A. Kumar, and T. Srikanthan. A hybrid strategy for mapping multiple throughput-constrained applications on mpsocs. In *Proc. of CASES*, pages 175–184, 2011.

[18] S. Stuijk, M. Geilen, and T. Basten. A predictable multiprocessor design flow for streaming applications with dynamic behaviour. In *Proc. of DSD*, pages 548–555, 2010.

[19] P. Agrawal and et. al. Partitioning and assignment exploration for multiple modes of ieee 802.11n modem on heterogeneous mpsoc platforms. In *Proc. of DSD*, pages 608–615, 2012.

[20] I. Verbauwhede and C. Nicol. Low power dsp's for wireless communications. In *Proc. of ISLPED*, pages 303 – 310, 2000.

[21] Tensilica Inc., http://www.tensilica.com/products/dsps/connx-baseband-engine.htm. *Tensilica ConnX Baseband Engine.*

[22] CEVA Inc., http://www.ceva-dsp.com/CEVA-MM3000.html. *CEVA-MM3000.*

APPENDIX

A. WLAN 11N APPLICATION DESCRIPTION

An 802.11n packet, as shown in Fig. 5, consists of two parts, i.e. (a) preamble with predetermined data, and (b) payload consisting of the actual data symbols. Both the preamble and the payload consists of multiple $4\mu s$ symbols. The preamble contains the header information for the packet, such as number of transmit and receive antennas, channel bandwidth, coding and modulation schemes, number of payload symbols, etc. The payload contains the transmitted data.

802.11n has several modes based on the number of transmitting and receiving antennas, modulation schemes, data-coding rate, etc. We have considered two modes of 802.11n in this work, i.e. 4×4 and 2×2 MIMO 40MHz with 64-QAM modulation. The number of preamble symbols is 7 and 5 for the 4×4 and 2×2 modes, respectively, whereas the number of payload symbols can vary from 1 to 2048 [12].

In a $N \times N$ MIMO mode, each symbol is a set of $N \times Q$ complex values where N is the number of transmit and receive antennas and Q is the number of sub-carriers per symbol which is dependent on the channel bandwidth. In our case, N is 2 and 4 for the 2×2 and 4×4 modes, respectively and Q is 128 for the channel bandwidth of 40 MHz.

The block diagram for the inner-modem processing is shown in Fig. 6. All the shaded blocks carry out preamble processing whereas the rest carry out payload processing. The complexity of a 802.11n packet processing depends on the selected mode and the number of payload symbols in the packet.

B. LTE APPLICATION DESCRIPTION

The input data of LTE has a very different format as compared to WLAN. In case of LTE, the transmission is based on frames of 10ms duration. Each frame consists of 10 sub-frames, each of which further consists of 2 slots, as shown in Fig. 7. A sub-frame is the smallest unit of time which is allocated to a user. Each slot consists of 7 symbols. Thus, a LTE frame consists of multiple $71\mu s$ symbols.

LTE also has several modes similar to WLAN 11n. We have considered only a single mode, i.e. 2×2 MIMO 20 MHz with 256-QAM modulation.

A LTE symbol can also be considered as a set of $N \times Q$ complex values, where N is the number of antennas and Q is the number of sub-carriers. In our case, N is 2 for the 2×2 mode and Q is 1200 for the channel bandwidth of 20 MHz. Unlike WLAN, LTE does not have dedicated symbols carrying only the header information. All the 7 symbols in a slot carry the transmitted data and the symbols 0 and 4 also carry the header information.

The block diagram for the LTE inner-modem processing is shown in Fig. 8. The figure shows the processing for the 7 symbols of each slot, the complexity of which mostly depends on the selected mode.

C. COMPARISON OF WLAN 11N AND LTE

WLAN has very tight latency constraints whereas LTE is very data dominated. As a result of this, WLAN requires higher computational resources whereas LTE requires higher memory resources. Both WLAN and LTE have packet based input data but WLAN packets have large number of smaller symbols whereas LTE packets have lower number of larger symbols. Thus, processing on the input data in case of WLAN is largely uniform resulting in a very regular control and data flow (CDF) whereas in case of LTE it is non-uniform resulting in an irregular CDF. WLAN can thus be more effectively pipelined than LTE, which results in higher utilization of the hardware resources. Besides these, parallel input data streams can be processed in parallel for large part in the WLAN application chain. In case of LTE, the parallel input data streams are merged after conversion from time to frequency domain and the granularity of data-level parallelism changes, thus impacting the possibilities of partitioning and parallelization of the LTE tasks. Thus, WLAN and LTE provide very different constraints and opportunities for P&A during the mapping.

D. INPUT APPLICATION MODELING

The 2×2 and 4×4 modes of WLAN are considered as two different scenarios, where the 2×2 mode is an average complexity scenario whereas the 4×4 mode is an worst-case complexity scenario. It is also assumed that 2×2 mode is the more frequently occurring scenario for WLAN. Fig. 9 shows the CDFG ($G_{4 \times 4}$) for the 4×4 mode scenario. In this graph, the control and data flow are denoted by the dotted green and the solid blue edges, respectively. Fig. 10 shows the CPGs created for $G_{4 \times 4}$ and the input data (symbols) which each control path graph processes.

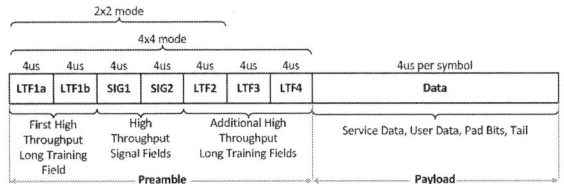

Figure 5: 802.11n packet structure

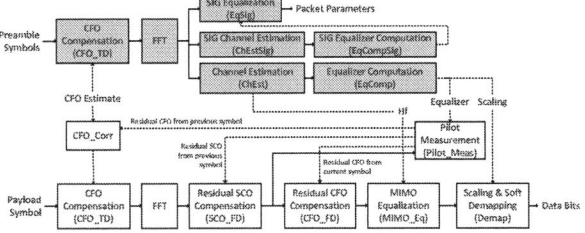

Figure 6: 802.11n inner-modem processing

E. EXTEND OF PIPELINE CALCULATION

For streaming applications, the processing of input data can be pipelined where the tasks in the processing chain are partitioned across different pipeline stages. However, pipelining increases the overall latency of the entire processing. The total latency for processing N input data on a pipelined architecture with S stages can be given by Eqn. 1.

$$L_{pipe} = N + (S - 1) \cdot L_{stage} \qquad (1)$$

where, L_{pipe} and L_{stage} are the total and per stage latency, respectively. The application constraint on the allowable latency will impose an upper bound on the extent of pipelining, i.e. the number of pipeline stages, as shown in Eqn. 3.

$$L_{pipe} \leq (N \cdot L_{data} + L_{app}) \qquad (2)$$

$$\implies S \leq \frac{N \cdot L_{data} + L_{app}}{L_{stage}} - N + 1 \qquad (3)$$

where, L_{app} is the total allowable latency and L_{data} is the allowable latency for processing each input data, at the application level.

For example, in case of WLAN 11n, a new symbol arrives at every $4\mu s$. We consider that each symbol must be processed within $4\mu s$ ($=L_{data}$) and $L_{stage} = L_{data}$. The WLAN 11n standard specifies a latency constraint, referred to as SIFS timing budget, which has been considered to be $8\mu s$ ($L_{app} = 2 \cdot L_{data}$) for the inner-modem processing, in this work. Thus from Eqn. 3, the maximum number of pipeline stages is constraint to 3. Thus, for the P&A exploration of WLAN, we consider three cases of pipelining, i.e. with one-, two- and three-stages. We derive P&A schemes for each of these cases.

F. COMPARISON WITH EXISTING APPROACHES

In Table 1, we compare the features of the proposed approaches to other existing approaches. These approaches do not explore partitioning. Each of them considers one or more application as a set of given tasks and explore allocation and binding of these tasks on the underlying architecture. In the proposed approach, instead of starting with a given set of tasks, we systematically split the application into a set of tasks and then systematically derive different partitions, considering multiple parameters such as available parallelism, hardware resource requirements, complexity, etc. These partitions are then allocated on the underlying architecture. The existing approaches consider architectures which are either fixed or have limited configurability. In contrast, in the proposed methodology we consider template based domain specific architecture where we explore the instantiation of the platform architecture along with partitioning and assignment. The methodology instantiates a heterogeneous MPSoC platform for each P&A scheme explored where the number of processors and configuration of each processor is also explored.

Properties	[16]	[18]	[17]	[3]	[4]	[2]	Proposed
Application	Multi-mode	Multi-mode	Single-mode	Multi-mode	Multi-mode	Multi-mode	Multi-mode
Partitioning	No	No	No	No	No	No	Yes
Platform	Heterogeneous	Homogeneous	Homogeneous	Heterogeneous	Heterogeneous	Heterogeneous	Heterogeneous
Flexibility	Limited	Fixed	Limited	Limited	Limited	Limited	High

Table 1: Comparison of proposed methodology with existing approaches

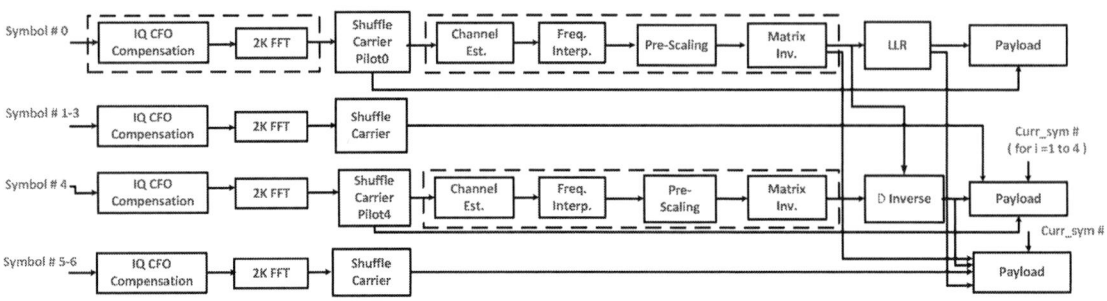

Figure 8: LTE inner modem processing

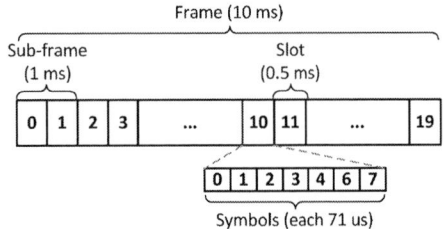

Figure 7: LTE frame structure

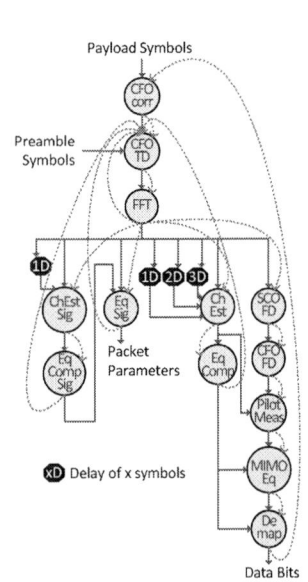

Figure 9: Task graph for 4×4 mode ($G_{4 \times 4}$)

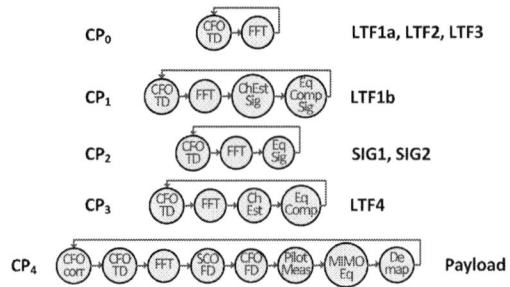

Figure 10: Control path graphs for $G_{4 \times 4}$

CoARX: A Coprocessor for ARX-based Cryptographic Algorithms

Khawar Shahzad[1], Ayesha Khalid[1], Zoltán Endre Rákossy[1], Goutam Paul[2,*], Anupam Chattopadhyay[1]

[1]MPSoC Architectures, RWTH Aachen University, Germany
khawar.shahzad@rwth-aachen.de
{khalid,rakossy,anupam}@umic.rwth-aachen.de
[2]Department of CSE, Javavpur University, Kolkata, India
goutam.paul@ieee.org

Abstract

Cryptographic coprocessors are inherent part of modern System-on-Chips. It serves dual purpose - efficient execution of cryptographic kernels and supporting protocols for preventing IP-piracy. Flexibility in such coprocessors is required to provide protection against emerging cryptanalytic schemes and to support different cryptographic functions like encryption and authentication. In this context, a novel crypto-coprocessor, named CoARX, supporting multiple cryptographic algorithms based on Addition (A), Rotation (R) and eXclusive-or (X) operations is proposed. CoARX supports diverse ARX-based cryptographic primitives. We show that compared to dedicated hardware implementations and general-purpose microprocessors, it offers excellent performance-flexibility trade-off including adaptability to resist generic cryptanalysis.

Categories and Subject Descriptors

B.7[**Integrated Circuits**]: [General,Types and Design Styles];
C.3[**Special-purpose and Application-based Systems**]: [Microprocessor/ microcomputer applications, Real-time and embedded systems]

General Terms

Design, Performance

Keywords

ARX, Cryptography, Coprocessor, CGRA

1. Introduction

Security is an essential part of today's information systems. In order to guarantee the privacy of a user, cryptographic algorithms are intrinsic for a wide range of application domains, e.g., wireless transmission, multimedia and image processing. Additionally, cryptographic protocols are employed in modern SoCs to prevent IP piracy threats [28]. To ensure efficient execution of cryptographic algorithms ranging from private and public key encryptions to authentication schemes, cryptographic accelerators are increasingly being used. These accelerators are implemented as dedicated hardware with little flexibility [17] or as a flexible processor with algorithm-specific customizations [32, 12]. Flexibility in the accelerator is an

---

*Corresponding Author. This work was done in part while he was visiting RWTH Aachen, Germany as an Alexander von Humboldt Fellow.

Permission to make digital or hard copies of all or part of this work for personal or classroom use is granted without fee provided that copies are not made or distributed for profit or commercial advantage and that copies bear this notice and the full citation on the first page. To copy otherwise, to republish, to post on servers or to redistribute to lists, requires prior specific permission and/or a fee.
DAC '13, May 29 - June 07 2013, Austin, TX, USA.

---

important design dimension for several reasons. *Firstly*, it provides a common implementation supporting different cryptographic operations required by a standard. For example, the communication standards GSM, 3GPP, Bluetooth, IEEE 802.11 and ISO/IEC 29192 recommends usage of cryptographic algorithms belonging to classes of stream ciphers, block ciphers, message authentication codes (MAC) as well as public key cryptography. *Secondly*, it can continuously protect against evolving cryptanalytic methods by adopting appropriate design changes. For example, to increase the resistance against cryptanalysis, a proposed countermeasure against attack on Threefish-512 requires only some additional operations [21]. Moreover, throughput requirements often require designers to switch between different versions of the same algorithm.

Despite a number of a design proposals over flexible cryptographic accelerators, flexibility during architectural design is restricted to determination of common operators and it is rarely explored from the algorithm designers' perspective. The idea of combining computational workloads under a class that captures a pattern of computation and communication is presented at [4] and also promoted by Intel Recognition-Mining-Synthesis (RMS) view [14].

However, these constructions are not always sufficient to include the algorithm designers' knowledge. For example, in symmetric-key cryptography, certain classes can be identified, e.g., ARX functions [35], Fiestel networks [24], Substitution-Permutation networks [19] and sponge constructions [9]. Such classes provide a template or a construction method, which can be employed to design a cryptographic algorithm. Generic architecture templates or flexible coprocessors for these classes have not been proposed earlier. This paper makes a first attempt into such design by proposing a coprocessor for ARX-based cryptographic algorithms.

The term AXR (later renamed to ARX) was coined by Ralf-Philipp Weinmann [35]. The algorithmic simplicity, efficient implementations in software, absence of timing attacks of ARX cipher contribute to their popularity [26]. Further, it is shown in [21, Section 5] that A, R and X operations are *functionally complete* in the sense that any function can be implemented with these three operations. A suitable sequence of ARX operations may lead to a secure cryptographic primitive. Therefore, design of a generic ARX architecture is of prime importance which, to the best of our knowledge, has not been attempted before the current work. Cryptanalysis of ARX based ciphers has been extensively reported in literature [21, 25, 23, 26] though, no major cryptanalytic breakthrough to threaten the security is till date known. As a result, diverse ARX-based algorithms made to the final rounds of recent design competitions for symmetric-key cryptosystems [3] and hash functions [2].

To begin with, we select five prominent ARX-based algorithms. Three of those, namely, HC-128 [36], Salsa20 [7] and ChaCha [8] belong to stream cipher category and two, namely, BLAKE [6] and Skein [15] are hash functions. A commercial high-level processor design environment [1] is chosen for our design flow. We performed a detailed design space exploration before making each design decision. The final coprocessor implementation, termed CoARX, is synthesized with a standard cell library for obtaining performance results.

The performance is benchmarked against published implementation results for each of the algorithms on dedicated hardware and flexible processors. The advantage of algorithmic view in the architecture is demonstrated with intuitive design alterations for improved resiliency. In summary, our contributions are as following.

- A detailed design space exploration for a novel programmable ARX coprocessor.
- Mapping of 5 common ARX algorithms on the coprocessor and enhancements of the design with algorithm-specific protections and optimizations.
- Detailed performance results and benchmarking against dedicated as well as programmable implementations.

2. Related Work

Acceleration efforts of various ARX cryptographic algorithms are reported on various platforms such as ASICs, FPGAs, GPUs and GPPs. Inclusion of custom instruction set extensions on 16-bit microcontrollers was also proposed for SHA-3 finalists (including Skein and BLAKE) [12] based on their implementation studies [20]. Section 5 compares the acceleration efforts of the algorithms under consideration when mapped on GPPs, customized microcontrollers and multi-core systems (GPUs and IBM cell architecture). Due to the absence of a single VLSI implementation flexible enough to map any ARX algorithm, we consider implementation of individual algorithms. We refer below the reported ASIC implementations that standout in throughput. The reader is referred to Appendix A for a summary of FPGA implementations of the algorithms.

BLAKE: Out of the various dedicated hardware implementations of hash function BLAKE [6, 16, 22, 17], the most noteworthy in terms of throughput comes from the authors of the original algorithm [17]. A throughput of 20 Gbps for 8G-BLAKE-512 on 90nm CMOS technology while consuming 128 kGE area is reported.

Skein: Tillich *et al.* [33] reports area and throughput results of the implementation of Skein hash function using a 180nm standard cell library. Walker *et al.* used a 32nm standard cell technology and claimed to achieve a 5X throughput improvement over Tillich's work [34], after an appropriate scaling. More recent skein-512-256 implementation results claim a throughput of 3 Gbps with 66 KGates on 130nm CMOS technology [16] and 6.7 Gbps with 43 KGates on 90nm CMOS technology [22].

HC-128: The only ASIC based implementation for HC-128 proposed at [11] reports a throughput of 22.88 Gbps while consuming an area of 12.65 kGE with 21 KBytes of dual ported RAM [11] using 65nm standard cell library.

Salsa20 and ChaCha: Various implementations of the two stream ciphers are presented at [18], with a reported peak throughput of 6.5 Gbps using 40 KGates of area for Salsa20 and ChaCha.

From the algorithm point of view, ARX-based cryptography received close scrutiny of the cryptographers and cryptanalysts. General [25, 21] and individual cryptanalysis [5] for ARX-based algorithms is performed. Software toolkit for performing differential cryptanalysis on ARX-based constructions is proposed at [27].

3. Design Space Exploration

The microarchitecture of CoARX is decided by a simple observation that, all ARX-based algorithms perform a mix of addition (A), rotation (R) and xor (X) operations in different order in each round of the algorithm. Each of the stream ciphers HC-128, Salsa20 and ChaCha has an internal state that is randomized by a secret key and at every round of keystream generation the state is updated using ARX operations. The hash functions BLAKE and Skein compress the input message into fixed length output string and the operation again happens through several iterations each involving the basic ARX operations. For detailed description of the algorithms, one may refer to [36, 7, 8, 6, 15] (we give a summary of the algorithms in the Appendix B). One should note that the order of ARX operations, the number of rounds, the word-lengths of variables that are processed

vary from one algorithm to another and hence poses a difficult challenge for generic architecture design.

To have the maximum flexibility and performance, the design need to support 3^3 different functional units for the triple operations. The same flexibility can be obtained from a reconfigurable architecture with these three operators. The algorithms also exhibit data-oriented computing pattern thereby, justifying the selection of Coarse-Grained Reconfigurable Architecture (CGRA) [13]. CGRAs can be implemented with wide range of design choices. In the following the rationale of specific CGRA design choices are elaborated, considering all the structural elements in bottom up fashion.

Functional Units Arrangement: Any number of functional units (A, R and X) may be arranged in any order inside a Processing Element (PE). We consider three architectures, showing an increasing trend of complexity and efficiency, allowing up to one, two or three operations per cycle as shown in Fig. 1(a), Fig. 1(b) and Fig. 1(c) respectively. They were modeled using a high-level synthesis environment [1] and synthesized with Synopsys Design Compiler using 90nm technology with maximum achievable clock frequency as reported in the Table 1.

PE_1 allows one operation per cycle and consequently, requires a single write port to the register file. The second architecture, PE_2, is designed to exploit the fact that a 64-bit A and R blocks in hardware contribute almost equally to the critical path while the contribution of X is significantly less. Hence the decrease in the clock frequency, compared to PE_1 is not significant. All the three units may be used separately, like PE_1, or in a non-overlapping simultaneous combination, e.g., AX and R or A and RX. PE_3 is more flexible, allowing all possible non overlapping combinations of one, two and three operations in a single cycle, e.g., AR and X or AXR or XR and A. The flexibility is achieved at the expense complex interconnects and a layer of multiplexers before each FU. Consequently, PE_3 has the least operating frequency compared to other architectures.

The round operations of each of the algorithms are mapped to the three architectures considered and cycle count and functional units used per cycle are calculated. For facilitating the critical choice amongst these architectures the following metrics are evaluated.

- Computational time is calculated from the clock frequency and the number of cycles for each algorithm.

- Resource Utilization (RU) gives the average utilization of functional units in the cluster grid. It is computed as,

$$RU_{algo} = \sum_{i=0}^{N} \sum_{j=0}^{C} \sum_{k=0}^{M} \frac{U_{ijk}}{(N * C * M)} \quad (1)$$

Where C is the number of clusters, M is the number of FUs inside a cluster (3 in PE_2) and N is the total number of cycles. $U_{ijk} = 1$ if the kth FU in j^{th} cluster is participating in the ith cycle of the algorithm, otherwise it is 0.

The RU of PE_1 is worst since it can not exceed 33.3% and only one operation can be performed in a single cycle. Consequently the mapping of ARX algorithms requires a large number of instructions. On the other extreme, PE_3 shows the best RU figures. However, its multiplexers and read/write ports affect the critical path, resulting in relatively worse execution time compared to PE_2 in few cases as given in Table 1. Furthermore, the large configurability options of PE_3 requires a large configuration memory word. PE_2 is a compromise of complexity and flexibility. Due to its reasonable RU and computational time comparable to that of PE_3, we pick PE_2 to be our design choice. Note that, the choice of number of registers is also influenced by the choice of FU arrangement, i.e., more intermediate storage is needed for PE_1 than for PE_3 or for PE_2.

Register File and Read/Write ports: The FUs as well as the local registers of clusters are 64-bit wide to support double-word ARX algorithms (Skein and BLAKE-64), however, mapping of 32-bit algorithms (HC-128, Salsa20, ChaCha) remains trivial. The number of

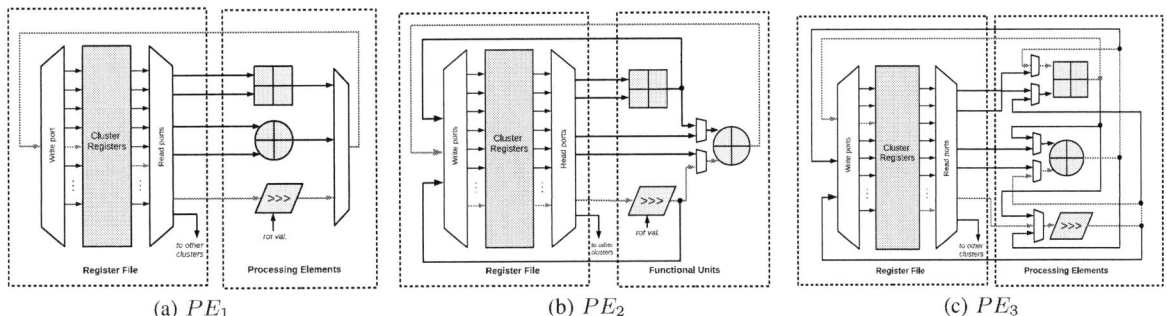

(a) PE_1 (b) PE_2 (c) PE_3

Figure 1. Processing Elements with Different Functional Unit Arrangement

Table 1. Estimated Performance and Resource Utilization Results

	Freq. (MHz)	Possible Operations	BLAKE-512 (128 Bytes)		Skein-512-512 (64 Bytes)		ChaCha (64 Bytes)		Salsa20 (64 Bytes)		HC-128 (264 Bytes)		Average	
			time (ns)	RU %	time (ns)	RU %	time (ns)	RU %	time (ns)	RU %	time (ns)	RU %	time (ns)	RU %
PE_1	1075	A,R,X	431.53	33.33	245.52	33.33	234.36	33.33	234.36	33.33	265.98	15.38	282.35	29.74
PE_2	913	AX,RX	323.96	53.32	131.33	86.07	188.25	50	188.25	50	216.70	27.77	209.70	53.43
PE_3	610	AR,RA,XA,XR ARX,AXRR,XA RAX,XAR,XRA	301.86	77.77	196.87	86.07	150.93	100	150.93	100	324.83	27.77	225.09	78.31

local registers is dictated by BLAKE, which requires 6 registers for the computation of its core function.

Number of Clusters: The choice of number of clusters inside a CGRA influences the throughput and resource utilization, resulting in resource under-utilization when number of clusters is too large and low parallelism and consequently low throughput when number of clusters is too small. Among the target ARX algorithms, BLAKE, Salsa20 and ChaCha consist of up to 4 parallel executions of their core operation. Flavors of Skein can have 2-8 parallel ARX operations, whereas for HC-128, 3 parallel ARX operations can take place. Hence a 4 cluster grid (2x2) is chosen as a trade-off between throughput and resource utilization.

Memory Hierarchy: Inside each cluster, a 2-level memory hierarchy is maintained by a small register file and a large SRAM. The memory holds the plaintext, ciphertext, message hash, Initialization Vector and Key. This dual-ported SRAM has 64 bit word length and has 1024 elements. Its size is dictated by HC-128, that requires two secret S-Boxes, each with 512 elements. Out of the 6 registers in register file, two namely *m0, m1* hold the value to be written or read from port 0 and port 1 of the memory, respectively. Movement of memory values to any other register incurs one cycle latency.

Inter Cluster Interconnects: The number and nature of interconnects between clusters is influenced by the permutation of ARX operations in the subsequent rounds of ARX algorithms. For BLAKE, Salsa and ChaCha each cluster calculates one core function (referred as *G* function in the algorithms). At the completion of *G* functions on the columns, these algorithms proceed diagonally for which each cluster requires exactly one value from every other cluster. Conse-

outgoing register values from each cluster. For algorithms requiring more permutation and data exchange between clusters, data availability is ensured by one cycle overhead.

The top-level architecture of the final CGRA-based ARX coprocessor is presented in Fig. 2. The design is distributed over two pipeline stages. *State Automaton* stage fetches the instruction from the *configuration memory*. The instruction is organized as a 4-tuple. A *Qualifier*, a *True* address and a *False* address are the three fields, which handle looping and conditional jumps during the execution. The rest of the instruction are the configuration bits for the CGRA that control the functionality of the clusters in the *Execute* stage.

4. Mapping of the ARX Algorithms

On the proposed CGRA coprocessor, any ARX-based cryptographic algorithm can be mapped. The reconfigurablility is controlled by the instructions in the configuration memory. The configuration of the entire CGRA has 269-bit word length and constitutes 46 words. Configuration words are manually written and are updated to perform a different combination of ARX operation every cycle. For switching the current algorithm running on the CGRA, the configuration memory is updated. Fig. 3 shows a breakdown of a configuration word and a discussion of various fields follows.

Qualifier Control: The iterative nature of the cipher algorithms is handled by a 4-bit *Qualifier*. The selection of next address to be *True* address or *False* address is determined by the qualifier condition. For conditional jumps (e.g., evaluating number of rounds, message/ block size), if the condition is evaluated true then a jump is made to the 8-bit true address or else to the false address. An unconditional jump is evaluated always as the qualifier being true.

Cluster Configuration: The bit fields in the instruction word of configuration memory are used to configure the input multiplexers and the processing elements of the entire CGRA. Each cluster requires 61 bits of configuration word. The dual ported local memory in each cluster requires a 2-bit *cmd* and a 10 bit *address* for 1024 word memory to be written and read from. Register *m0* and *m1* are tied to communicate with port0 and port 1 of the memory, respectively. *FU config.* includes the specification of source and destination registers of the functional units namely A, R and X.

- *misc.* bits: specify if the operation is 32-bit or 64-bit wide. Future extension may include specifying Rotator FU being used as a shifter or not.
- *src. operands*: The multiplexers for each of the two operands of A and X is 8×1, R requires only one operand and a 6-bit rotation value. The rotation value and direction of the 64 bit

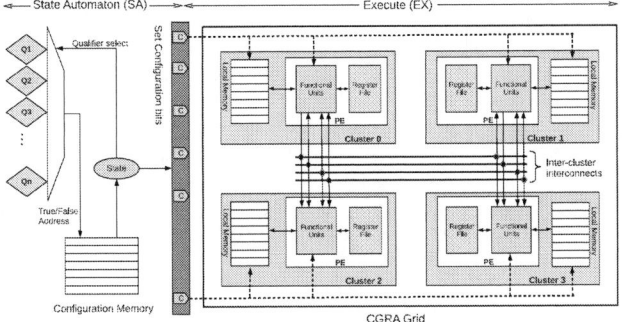

Figure 2. Block Diagram of CoARX

quently, a MESH style interconnect is chosen with 3 incoming and 3

923

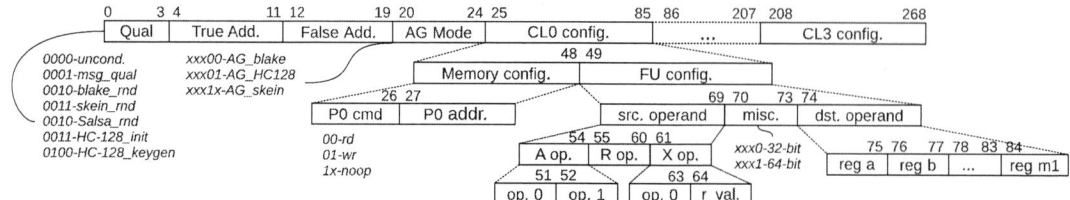

Figure 3. Configuration Word Details

operand is specified by a 6-bit *r_val* field. The operands may include registers from other clusters as well.

- *dst. operand*: For each of the 6 registers in the register file, selection lines for write demultiplexers requires 2-bits. The registers may be updated by the Functional Units output or simply bypassed values.

4.1 Algorithm Specific Modes

Some ARX algorithms require specific modes to be indicated in configuration word to carry out non-ARX operations.

BLAKE Address Generation: For BLAKE, the addresses for loading *msg* and *constants* from memory, are not static and they change in each round. One solution for implementation of this scenario is direct loading of data memory in a cluster by *msg* and *constants* during initialization phase. But this approach renders the system vulnerable to side channel attacks. Instead a specific address generation mode is defined to carry out BLAKE address generation.

HC-128 Address Generation: Some operations in HC-128 keystream generation are not strictly ARX. They are modulo-512 subtraction, byte processing in h_1 and h_2 functions and counter increment/addition and shift operations in the expansion step of initialization. To cater them, specific hardware blocks are being used.

Skein Key Generation: Skein requires injecting a subkey into the states by adding it with the outputs of MIX functions every fourth round. Skein key generation mode enables hardware block *skein subkey generator* that is implemented in hardware as shift registers, along with adders and is capable of generating one subkey in a single cycle.

Table 2. CoARX Synthesis Results

Area (kGE)			Memory (kBytes)		Freq. (MHz)
Combinational	Sequential	Total	Configuration	Data	(Core)
82.6	12.4	95	1.5	32	700

5. Implementation and Benchmarking

The proposed architecture CGRA is developed using Language for Instruction Set Architectures (LISA) description [1]. LISA can be used to generate RTL description of an architecture along with software tools including Compiler, Assembler, Linker, Profiler and Debugger. At present, the CGRA is programmed with hand written assembly language but the configuration word generation process can be easily automated due to the regular arrangement of FUs in a PE. LISA processor description (5K lines of code) was processed by Synopsys Processor Designer to generate Verilog RTL (38K lines of code). We used Synopsys Design Compiler, topographical mode, Faraday 90nm CMOS technology library for the synthesis of HDL and results are presented in Table 2. The area estimates of sequential and combinational logic is given in equivalent NAND gates. Each of the four clusters have 8 KBytes of dual ported SRAM as local memory. Performance of different ARX algorithms on CoARX is presented in Table 3.

5.1 Comparison with ASICs

Different ASIC implementations considered for comparison are tabulated with their reported throughput in Table 4. For BLAKE, Henzen *et al.* report the ASIC with highest throughput [17] so far. For Skein, results reported for Skein-512-256 with highest throughput are considered [22]. Implementation results for HC-128, taking keystream generation of 0.75 cycles/Byte (same as our implementation) is considered for comparison [11]. For Salsa and ChaCha comparison, we consider the implementation from [18].

For a fair comparison, the estimated throughput of the ASICs of ARX algorithms under consideration has been added in Table 4 after direct technology scaling to 90nm. CoARX delivers throughput in the same order of magnitude. On a closer investigation, some architectural aspects of CoARX, which affects the critical path of the design are as following.

- **Barrel Shifter**: For CoARX, a 64-bit Barrel rotator is used to support flexible rotation amounts. Considered ASICs for comparison (Table 4) use wire routing for fixed rotations instead.

- **Datapath Width**: For catering algorithms with 64-bit datapath and scalability, CoARX datapath is 64-bit wide. Consequently, ASIC implementations of ARX-based algorithms with 32-bit datapath have a smaller critical path and outperform CoARX in throughput. This can be made configurable to reduce the throughput gap.

- **Flexible Interconnects**: To support various possibilities of source operands from local and neighboring clusters registers, the input multiplexers of FUs in each of cluster are 8x1. Similarly, the write ports of the registers may take outputs of various FUs and each has a 4x1 multiplexer for possible inputs. These multiplexers increase the critical path of the design.

- **Unfolding Transformation**: According to the unfolding transformation [20], a design achieves better throughput if it is unfolded, which is the case in most of the ASIC implementations. ASICs for BLAKE implement an entire G function in one cycle [17]. For Skein, eight rounds are unfolded and implemented in one cycle [33]. Salsa20 and ChaCha also use similar kinds of transformations [18]. To incorporate enough flexibility, CoARX utilizes modular functional units and hence does not take advantage of unfolding transformation.

The area of the individual ASIC implementations cannot be compared vis-a-vis the area of CoARX due to its added flexibility. Sum of area of the best-performing ASICs results in 181 kGE, the area of CoARX is 47% less in comparison. A more suitable metric for comparison is the area-efficiency (throughput per area) reported in Table 4. The individual ASICs have significantly higher throughput per area, when compared against CoARX. This is the expected *flexibility gap*. The fact that CoARX implementation is efficient can be shown by considering the same metric for a hypothetical ASIC combining the best implementations. For this measure, the collective area (181 kGE) of individual ASICs is chosen. CoARX performs comparably for all the designs, except BLAKE. The BLAKE implementation reported in [17] performed an efficient round rescheduling technique to reduce the critical path after unrolling multiple subsequent ARX operations. For CoARX, such optimization is not possible since, the critical path is constrained by the operations within a PE.

Fig. 4 compares CoARX throughput of ARX algorithms mapped on it, with Intel M 1600 MHz processor and ARM926EJ-S 1200 MHz processor and easily outperforms these platforms by a factor of 1.6 to 22 times.

In the absence of a CoARX like flexible crypto core, a fair comparison in terms of power is not possible. Power-figures (in mW), in similar process-technology, are available only for some individual ASICs, e.g. BLAKE: 10.84 [22], 15.65 [38], Skein: 17.17 [22], 39.71 [38] and Salsa20: 8.42 [37]. These values are better than (though the in same order of magnitude as) CoARX (Table 4). That is because these ASICs have algorithm-specific optimizations in their design and operate at a clock frequency lower than that of CoARX. Redundancy to

924

Table 3. Performance of different ARX algorithms on CoARX

	BLAKE-512	Skein-512	HC-128	Salsa20/20	ChaCha/20
Input Block size (Bytes)	128	64	-	-	-
Output Block size (Bytes)	64	64	-	64	64
No of rounds (r)	16	72	-	20	20
Initialization	5 cycles	7 cycles	46241 setups/sec	4 cycles	4 cycles
Round Calculation	$21 \times r$ cycles	$r+((r/4)+1)\times 2$ cycles	0.75 cycles/Byte	$17\times(r/2)$ cycles	$17\times(r/2)$ cycles
Finalization	9 cycles	6 cycles	-	12 cycles	13 cycles
Total (cycles)	350	123		186	187
Throughput (Gbps)	2.05	2.91	7.47	1.93	1.92
Power (mW)	83	89	53	71	61
Energy (mJ/Gbit)	41	30	7	33	29

Table 4. Comparison with ASICs (Throughput scaled to 90nm)

Implementation Reference	Area (kGE)	Frequency (MHz)	Throughput (Gbps) ASIC	Throughput (Gbps) CoARX	Throughput/Area (kbps/GE) Individual	Throughput/Area (kbps/GE) Combined	Throughput/Area (kbps/GE) CoARX
4G-BLAKE-512(16 rounds) [17]	79	532	16.5	2.05	208.86	91.16	21.57
Skein-512 [22]	43.13	251	6.73	2.91	156.13	37.18	30.63
HC-128 Parallel keystream [11]	13.66	1670	12.86	7.47	941.43	71.04	78.63
Salsa20 4xS-QR [18]	22.81	365	4.67	1.93	204.64	25.79	20.31
ChaCha 4xS-QR [18]	22.44	366	4.67	1.92	208.19	25.81	20.21

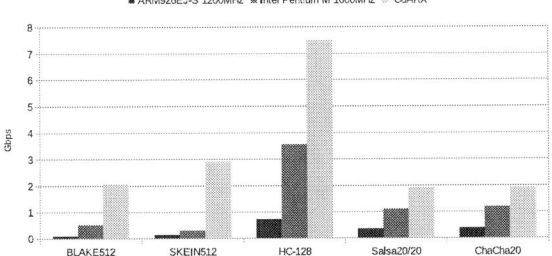

Figure 4. Comparison with GPP and Embedded Processor

achieve flexibility in CoARX design also contributes to higher power figures.

5.2 Comparison with Customized Microcontroller

Constantin *et al.* proposed custom instruction extensions for efficient implementation of all SHA-3 hash function competition finalists including BLAKE and Skein [12]. A 16-bit PIC24 microcontroller is used as a starting point. For BLAKE, the throughput improvement is mainly attributed to custom instructions for rotation and address generation. For Skein, a single custom instruction is added for performing 64-bit left rotate in two cycles. In order to make sure that the device remains usable in resource constrained environments, 2×16-bit barrel shifters have been used to perform 64-bit rotations. The reference microcontroller core occupies an area of 23 kGE. The customizations led to an area overhead of 10%. BLAKE and Skein can be mapped on the above mentioned extended architecture with 155 Cycles/Byte (10.32Mbps throughput) and 158 Cycles/Byte (10.12Mbps throughput) respectively. In terms of throughput and area-efficiency (throughput/area), CoARX easily outperforms this. Furthermore, CoARX offers more flexibility (e.g. different rotation amounts) by focusing on ARX class of algorithms.

5.3 Comparison with Multicore Implementations

Several ARX based cryptographic functions have been undertaken for throughput benchmarking on different multicore architectures, e.g., IBM Cell architecture and GPGPUs [30, 31]. A fair comparison between CoARX and a multicore system is hard, since the throughput scaling of multicore architectures is not linear due to the distribution of storage, varying latency for different storage and limitation of active number of threads per core. Considering these restrictions, multiple parallel data-streams for encryption/authentication are invoked in [30] in order to maximize the usage of the available cores. For Cell architecture, the 4-way SIMD instructions for each Synergistic Processing Elements (SPEs) are exploited for parallelism. With increasing size of the data block, more number of parallel threads are

deployed resulting in higher encryption/authentication speed. To provide a fair comparison, the highest achievable throughput (denoted *TP*), the number of active Scalar Processor (SP) cores for GPU and active streams for an SPE are reported in the following Table 5.

Table 5. Performance on Multicore Architectures

	SPE [30] TP (Gbps)	SPE [30] Streams	GPGPU [30, 31] TP (Gbps)	GPGPU [30, 31] Active SP	CoARX TP (Gbps)
BLAKE-32	5.1	4	36.80	240	2.05
Skein-512	1.9	2	22.10	240	2.91
HC-128	-	-	2.26	240	7.47
ChaCha	-	-	42.40	240	1.93
Salsa20	-	-	10.86	48	1.93

Though it is hard to compare multicore architectures' performance with that of CoARX without offering similar scalability, several points can be made. First, CoARX offers 4 parallel ARX operations per cycle matching a 4-way SIMD instruction of SPE and 4 active SPs in GPGPU. Second, the individual SP cores of [30, 31] are synthesized at a frequency of 1242 MHz and 1350 MHz respectively, which are much higher than that of CoARX. Finally, for both BLAKE-32 and Skein-512, the implementation of [30] considers 4 different messages for parallel hashing, fixing the message size suiting the GPGPU requirements. In practice, the message size can vary. We experimented with a single large message for CoARX, allowing any message size as input. Since both BLAKE-32 and Skein-512 involve chaining with the hash value of last message chunk, the internal dependency between the concurrent threads are ignored in [30], resulting in more throughput. Considering the above facts, CoARX outperforms both GPGPU and SPE on a throughput per core basis.

5.4 Flexibility Study: Threefish-512 and Skein-512 Design Variants

Cryptanalysis of existing algorithms and design of new algorithms that resist the known attacks are always at arms race. CoARX is flexible enough to adapt variants and offer resistance against generic attacks. We demonstrate this through a case study with Threefish-512 and Skein-512. Threefish-512, an ARX-based block cipher, has been mapped on CoARX as the building block of Skein-512. A general attack to ARX-based cryptosystems and in particular to Threefish-512 has been reported in [21]. This attack makes a reduced round variant of Skein-512 to be vulnerable. We show that with CoARX, addition of few instructions is sufficient to thwart such attacks. At first, the attack model presented in [21] is studied.

According to Lemma 1 and the attack model in Section 4.2 of [21], the probability of getting a rotational pair for each addition is $2^{-1.415}$ for each addition in MIX and $2^{-0.28}$ for each addition in *subkey* addition part. Since there are a total of $72 \times 4 = 288$ additions in MIX and $18 \times 8 = 144$ additions in *subkey* addition, the probability of the complete attack is given by $2^{(-1.415)\times288+(-0.28)\times144} = 2^{-448}$, giving a complexity of 2^{448}. The 512-bit key has 8 words of 64-bits and the three leftmost bits in each key-word are assumed to be known in the attack model, the effective key-length is $8 \times 61 = 488$ bits. Thus, according to the attack of [21], instead of 2^{488} attack complexity of random guessing, one has 2^{448} attack complexity. To counteract the above attack, we need to increase the number of additions. Suppose, we increase the total number of additions in the MIX by a_1 and the total number of additions in the *subkey* addition by a_2. Then we must have $1.415 \times (288 + a_1) + 0.28 \times (144 + a_2) \geq 488$. Simplifying the above inequality, we obtain

$$1.415 \times a_1 + 0.28 \times a_2 \geq 40. \qquad (2)$$

Now, we propose two design variants to achieve the inequality (2). In the first variant, the total number of rounds is increased by a multiple of 4, say, by $4x$. Since we get 8 extra additions from *subkey* addition after every 4 rounds, we have $a_1 = 16x$ and $a_2 = 8x$. Substituting in 2 and simplifying, we get $x >= 1.61$. Since x has to be an integer, we take $x = 2$, meaning that 8 extra rounds are needed. We propose to continue the same *subkey* generation algorithm for those 8 extra rounds. In the second variant, only the number of additions in MIX is increased without increasing the *subkey* additions. This means, $a_2 = 0$. Thus, from inequality (2), we get $a_1 \geq 29$ (after rounding). For symmetry of computation, $a_1 = 36$ is taken and those 36 additions are distributed as follows. Instead of 4 rounds followed by a *subkey* addition, we propose to use 5 and 4 rounds alternately followed by a *subkey* addition. In this way, there will be a total of $5 \times 9 + 4 \times 9 = 81$ rounds interleaved by the usual *subkey* additions. The above two design variants of Skein-512 (Threefish-512) are mapped on the CoARX without much degradation in throughput. The first design variant requires 12 extra cycles $(8 + 2 \times 2)$ and the second design variant requires 9 extra cycles causing a throughput degradation of 8.8% and 6.8% respectively. For designing the second design variant additional permutation and rotation constants are also required to perform the 5th alternate round, which are added to the configuration memory. The entire design modification is performed within few hours.

Very recently, a differential power attack has been reported [10] on Skein. The same paper also proposes a countermeasure which does not alter the basic ARX structure and therefore our architecture can be easily adapted to include this as well.

6. Conclusion and Outlook

In the context of flexible and efficient cryptographic accelerators, this paper studies the design of CoARX, a flexible coprocessor for ARX-based algorithms. CoARX shows comparable area-efficiency against dedicated hardware accelerators and significantly higher throughput compared to off-the-shelf processor-based implementations. The algorithmic perspective of flexibility provides unique advantage against general and specific cryptanalysis.

The detailed physical design and studies for general prevention of side channel attacks are on our future roadmap. The idea of class-specific cryptographic accelerator will be further probed for other design primitives like Fiestel networks and sponge-based constructions.

Acknowledgment

We would like to sincerely thank Rishiraj Bhattacharyya for drawing our attention to the class of ARX-based cryptosystems as a common design kernel.

7. References

[1] LISA 2.0. Available at www.synopsys.com/Systems/BlockDesign/ProcessorDev.

[2] SHA-3 Cryptographic Hash Algorithm Competition. Available at http://csrc.nist.gov/groups/ST/hash/sha-3/index.html.

[3] eSTREAM: the ECRYPT Stream Cipher Project. Available at http://www.ecrypt.eu.org/stream/index.html.

[4] K. Asanovic et al. The landscape of parallel computing research: A view from Berkeley. Technical report, UCB/EECS-2006-183, University of California, Berkeley, 2006.

[5] J. Aumasson et al. Tuple cryptanalysis of ARX with application to BLAKE and Skein. ECRYPT 2 Hash Workshop, 2011.

[6] J. Aumasson, L. Henzen, W. Meier and R. Phan. SHA-3 proposal BLAKE ver 1.3, 2010. Available at https://www.131002.net/blake.

[7] D. J. Bernstein. The salsa20 family of stream ciphers. Available at http://cr.yp.to/papers.html#salsafamily, December 2007.

[8] D. J. Bernstein. ChaCha, a variant of Salsa20. Available at http://cr.yp.to/papers.html#chacha, January 2008.

[9] G. Bertoni, J. Daemen, M. Peeters and G. Van Assche. Sponge functions. In Ecrypt Hash Workshop 2007. Available at http://csrc.nist.gov/pki/HashWorkshop/Public_Comments/2007_May.html.

[10] C. Boura, S. Leveque and D. Vigilant. Side-Channel Analysis of Grostl and Skein. In IEEE Symposium on Security and Privacy Workshops 2012, pages 16–26.

[11] A. Chattopadhyay, A. Khalid, S. Maitra and S. Raizada. Designing high-throughput hardware accelerator for stream cipher HC-128. In IEEE International Symposium on Circuits and Systems 2012, pages 1448–1451.

[12] J. Constantin, A. Burg and F. Gürkaynak. Investigating the potential of custom instruction set extensions for SHA-3 candidates on a 16-bit microcontroller architecture. Cryptology ePrint Archive, Report 2012/050. 2012. Available at http://eprint.iacr.org/2012/050.

[13] A. DeHon. The density advantage of configurable computing. In *Computer*, vol. 33 (4), pages 41–49, 2000.

[14] P. Dubey and S. Engineer. Teraflops for the masses: Killer apps of tomorrow. In Workshop on Edge Computing Using New Commodity Architectures, 2006. Available at http://gamma.cs.unc.edu/EDGE/SLIDES/dubey.pdf.

[15] N. Ferguson et al. The Skein Hash Function Family, Version 1.3. http://www.skein-hash.info/sites/default/files/skein1.3.pdf, October 2010.

[16] X. Guo et al. ASIC implementations of five SHA-3 finalists. In IEEE DATE 2012, pages 1006–1011.

[17] L. Henzen, J.-P. Aumasson, W. Meier and R.-W. Phan. VLSI Characterization of the Cryptographic Hash Function BLAKE. In *IEEE Transactions on VLSI Systems*, vol. 19 (10), pages 1746–1754, 2011.

[18] L. Henzen, F. Carbognani, N. Felber and W. Fichtner. VLSI hardware evaluation of the stream ciphers Salsa20 and ChaCha, and the compression function Rumba. In 2nd International Conference on Signals, Circuits and Systems 2008, pages 1–5.

[19] J. Katz and Y. Lindell. *Introduction to Modern Cryptography*. CRC Press, 2007.

[20] J. K. Kobayashi, Ikegami, S. Matsuo, K. Sakiyama and K. Ohta. Evaluation of Hardware Performance for the SHA-3 Candidates using SASEBO-GII. Available at http://eprint.iacr.org/2010/010.

[21] D. Khovratovich and I. Nikolić. Rotational cryptanalysis of ARX. In Fast Software Encryption 2010, LNCS vol. 6147, Springer, pages 333–346.

[22] M. Knezevic et al. Fair and consistent hardware evaluation of fourteen round two SHA-3 candidates. In *IEEE Transactions on VLSI Systems*, vol. 20 (5), pages 827–840, 2012.

[23] G. Leurent. ARXtools: A toolkit for ARX analysis. In: The Third SHA-3 Candidate Conference. Available at http://www.di.ens.fr/~leurent/arxtools.html.

[24] M. Luby and C. Rackoff. How to Construct Pseudorandom Permutations and Pseudorandom Functions. In *SIAM Journal on Computing*, vol. 17 (2), pages 373–386, 1988.

[25] K. Mckay and P. Adviser-Vora. Analysis of ARX round functions in secure hash functions. PhD thesis, George Washington University, 2011.

[26] N. Mouha. ARX-based cryptography. Available at https://www.cosic.esat.kuleuven.be/ecrypt/courses/albena11/slides/nicky_mouha_arx-slides.pdf.

[27] N. Mouha, V. Velichkov, C. De Canniere and B. Preneel. Toolkit for the Differential Cryptanalysis of ARX-based Cryptographic Constructions. In Workshop on Tools for Cryptanalysis 2010, pages 125–126.

[28] J. Roy, F. Koushanfar and I. Markov. Epic: Ending piracy of integrated circuits. In IEEE DATE 2008, pages 1069–1074.

[29] A. Shimizu and S. Miyaguchi. Fast data encipherment algorithm FEAL. In EUROCRYPT 1987, LNCS vol. 304, Springer, pages 267–278.

[30] D. Stefan. Analysis and Implementation of eSTREAM and SHA-3 Cryptographic Algorithms. Master's thesis, Cooper Union College, 2011. Available at http://www.scs.stanford.edu/~deian/pubs//stefan:2011:analysis.pdf.

[31] S. Neves. Cryptography in GPUs. Master's thesis, University of Coimbra, 2009. Available at http://eden.dei.uc.pt/~sneves/pubs/2009-sn-msc.pdf.

[32] S. Tillich. Instruction Set Extensions for Support of Cryptography on Embedded Systems. PhD thesis, Graz University of Technology, Austria, 2008. Available at https://online.tugraz.at/tug_online/voe_main2.getvolltext?pCurrPk=39243.

[33] S. Tillich. Hardware Implementation of the SHA-3 candidate Skein. In Cryptology ePrint Archive Report 2009/159. Available at http://eprint.iacr.org/2009/159.

[34] J. Walker, F. Sheikh, S. Mathew and R. Krishnamurthy. A Skein-512 hardware implementation. 2010. Available at http://csrc.nist.gov/groups/ST/hash/sha-3/Round2/Aug2010/documents/papers/WALKER_skein-intel-hwd.pdf.

[35] R.-P. Weinmann. AXR - Crypto Made from Modular Additions, XORs. In Dagstuhl Seminar 09031, January 2009. Available at http://www.dagstuhl.de/Materials/Files/09/09031/09031.WeinmannRalfPhilipp.Slides.pdf.

[36] H. Wu. The stream cipher HC-128. Available at http://www.ecrypt.eu.org/stream/p3ciphers/hc/hc128_p3.pdf.

[37] Good, T and Benaissa, M Hardware results for selected stream cipher candidates. In State of the Art of Stream Ciphers, 2007, pages 191–204.

[38] M. Srivastav, X. Guo, S. Huang, D. Ganta, M. B. Henry, L. Nazhandali and P. Schaumont. Design and Benchmarking of an ASIC with Five SHA-3 Finalist

Table 6. FPGA based Implementations of ARX Algorithms

Reference	FPGA Device	Area (CLB / LE)	Block RAMs	Operating Frequency (MHz)	Throughput (Mbps)
Salsa20-sr [F1]	Xilinx Virtex-II 2V250fg256	194 CLB	4	250	38
Salsa20-qr [F2]	Altera Cyclone EP1C20F400C6	2356 LE	-	55	343
Salsa20-sr [F2]	Altera Cyclone EP1C20F400C7	3400 LE	-	40	931
Salsa20-dr [F2]	Altera Cyclone EP1C20F400C8	3510 LE	-	30	1280
Salsa20[F3]	Xilinx Spartan 3 xc3s50pq208-5	1615 CLB	-	23.5	213
ChaCha_config1 [F4]	Xilinx Virtex-6 XC6VLX75T-2	49 CLB	2	362	595, 422, 266
ChaCha_config2 [F4]	Xilinx Virtex-6 XC6VLX75T-2	77CLB	2	316	520, 368, 232
ChaCha_config2 [F4]	Xilinx Virtex-6 XC6VLX75T-2	77CLB	2	345	569, 403, 254
BLAKE-32 [F6]	Xilinx Virtex-3 xc3s50-5	124 CLB	2	190	115
BLAKE-32 [F5]	Xilinx Virtex-3 xc3s50-5	360 CLB	2	135	315
BLAKE-64 [F7]	Xilinx Virtex-6 xc6vlx75t-1	117 CLB	-	274	105
BLAKE-64 [F5]	Xilinx Virtex-6 xc6vlx75t-1	146 CLB	1	189	277
BLAKE-64 [F6]	Xilinx Virtex-5 xc5vlx50-2	108 CLB	3	358	314
Skein [F7]	Xilinx Virtex-6 xc6vlx75t-1	240 CLB	-	160	179
Skein [F5]	Xilinx Virtex-6 xc6vlx75t-1	162 CLB	1	166	34.9
Skein [F8]	Xilinx Virtex-5 xc5v	555 CLB	-	271	237
Skein [F9]	Xilinx Virtex-5 xc5v1x110-3	821 CLB	not specified	119	1610

Candidate. In ECRYPT 2 Hash Workshop, 2011.

APPENDIX

A. FPGA Implementations of ARX Algorithms

Most of the FPGA implementations target high throughput or low area as their design goals. The reuse of functional units in various ARX based cryptographic algorithms has not be exploited for a flexible implementation. The only work that is noteworthy in terms of flexibility for ARX based Skein and BLAKE is reported by Nuray et al. [F4]. Their reported FPGA implementations of these hash functions also support the primitive cipher functions on which they are based on; i.e., a unified core for BLAKE and ChaCha (a stream cipher) and one for Skein and Threefish (a tweakable block cipher). For a unified BLAKE and ChaCha coprocessor when mapped on XC6VLX75T-2 using 144 CLBs and 3 Block RAMs the throughput reported for BLAKE-32 and BLAKE-64 was 288 and 255 Mbps respectively [F4]. The same coprocessor could produce keystream for 8, 12 and 20 rounds of ChaCha at 1102, 780 and 492 Mbps respectively [F4]. Skein and Threefish coprocessor consumes relatively more slices on the same device and support Skein-512-512, Skein-256-256 along with various flavors of Threefish i.e., Threefish-256, Threefish-512 and Threefish-1024 [F4].

Due to the absence of any other flexible implementation, this Section summarizes the reported individual FPGA implementations of the 5 ARX based cryptographic algorithms undertaken in CoARX. A comparison of these implementations in terms of throughput and resource utilization of the target device is done in Table 6.

HC-128

HC-128 is an synchronous stream cipher in eSTREAM finalists. Since it belongs to software profile, no FPGA implementations are being reported for it so far.

Salsa20

For Salsa20, the earliest FPGA implementation results were reported by Junjie et al.[F1]. They proposed a compact hardware implementation of Salsa20 comprising of a single *QuarterRound* function block. The global clock was 250 MHz, and for the *QuarterRound* block it was 125 MHz. Gaj et al. also implemented salsa20 using a single quarterround architecture in combinational logic and using eight clock cycles to implement the entire *DoubleRound* [3]. Evaluation of various possible architectures for sasla20 namely Salsa20-dr (unrolled *DoubleRound* iterative architecture), Salsa20-sr (single round iterative architecture), Salsa20-qr (*QuarterRound* resource shared iterative architecture) were undertaken for implementation on an Altera device [F2]. These three possibilities provide various design points in area-performance trade-off as specified in Table 6.

ChaCha

ChaCha has not been separately undertaken for an FPGA implementation. Since ChaCha makes the building block of BLAKE hash function, a combined lightweight coprocessor is designed for BLAKE and ChaCha using deep pipelining for high clock frequency [F4]. Various piepline configurations of ChaCha are reported for its 8, 12 and 20 round variants as specified in Table 6.

BLAKE

Kaps et al. presented a lightweight implementation of SHA-3 finalists, including Skein and BLAKE on various FPGAs [F5]. The initial state is stored in a four way distributed RAM for ease in access by the 1/2 G-function implemented. For BLAKE-32, the achieved area efficiency

is comparable to the work by Beuchat et al. [F6]. For BLAKE-64, the area efficiency of the design proposed by Kaps et al. [F5] is more than twice than the one reported by Kerckhof et al. [F7] on the same FPGA device, due to the use of a 32 bit data width instead of 64 in the former.

Skein

For a lightweight implementation of Skein, Kaps et al. employed resource reuse by folding 4 Mix functions into 1 and within the Mix function reused a 32-bit adder to perform 64-bit additions. The adder is also used for key injections [F5]. Consequently the area is reduced but the number of clock cycles increase significantly when compared to other lightweight 64 bit [F7] and 32 bit implementations [F8].

A.1 FPGA Implementations References

[F1] J. Yan, H. M. Heys. Hardware Implementation of the Salsa20 and Phelix Stream Ciphers. In *Canadian Conference on Electrical and Computer Engineering (CCECE)* 2007, pages 1125-1128.

[F2] M. Rogawski. Hardware evaluation of eSTREAM Candidates: Grain, Lex, Mickey128, Salsa20 and Trivium. In *State of the Art of Stream Ciphers Workshop (SASC)* 2007, eSTREAM, ECRYPT Stream Cipher Project Report, volume 25. Available at http://www.ecrypt.eu.org/stream.

[F3] K. Gaj, G. Southern and R. Bachimanchi. Comparison of hardware performance of selected Phase II eSTREAM candidates. In *State of the Art of Stream Ciphers Workshop (SASC)* 2007, eSTREAM, ECRYPT Stream Cipher Project Report, volume 26. Available at http://www.ecrypt.eu.org/stream.

[F4] N. At, J. L. Beuchat, E. Okamoto, I. San and T. Yamazaki. Compact Hardware Implementations of ChaCha, BLAKE, Threefish, and Skein on FPGA. In Cryptology ePrint Archive Report 2013/113. Available at http://eprint.iacr.org/2013/113.

[F5] J. P. Kaps, P. Yalla, K. Surapathi, B. Habib, S. Vadlamudi, S. Gurung and J. Pham. Lightweight Implementations of SHA-3 Candidates on FPGAs. In *Progress in Cryptology-INDOCRYPT* 2011, pages 270-289.

[F6] J. L. Beuchat, E. Okamoto, T. Yamazaki. Compact Implementations of BLAKE-32 and BLAKE-64 on FPGA. In *International Conference on Field-Programmable Technology (FPT)* 2010, pages 170-177.

[F7] S. Kerckhof, F. Durvaux, N. Veyrat-Charvillon, F. Regazzoni and G. de Dormale and F.X. Standaert. Compact FPGA implementations of the five SHA-3 finalists. In *Smart Card Research and Advanced Applications* 2011, pages 217-233.

[F8] B. Jungk. Compact implementations of Grostl, JH and Skein for FPGAs. In *ECRYPT II Hash Workshop* 2011. Available at www.ecrypt.eu.org/hash2011/proceedings/hash2011_09.pdf.

[F9] K. Latif, M. Tariq, A. Aziz, A. Mahboob. Efficient hardware implementation of secure hash algorithm (SHA-3) finalist-Skein. In *Frontiers in Computer Education* 2012, pages 933-940.

B. Overview of ARX-based Algorithms

ARX based cryptographic algorithms are limited not just to one class of cryptographic functions. Several hash functions, stream ciphers and block ciphers are based on ARX. A Few noticeable examples include FEAL [29] and Threefish (block ciphers), Salsa20, ChaCha, HC-128 (stream ciphers) BLAKE and Skein (hash functions). Including bitwise boolean functions with ARX includes more hash functions like MD4, MD5, SHA-1 to ARX family. 6 out of the 14 second-round candidates of NIST SHA-3 hash function competition are ARX based: Blue Midnight Wish, CubeHash, Shabal, SIMD, BLAKE and Skein, out of which two reached final round [2]. Also 2 out of 7 finalists of the eSTREAM project (Salsa20, HC-128) are ARX based. Algorithmic descriptions of the ARX algorithms chosen for mapping on our coprocessor are given below.

B.1 BLAKE Hash Functions Family

BLAKE is a cryptographic hash function chosen as one of the finalists of the SHA-3 hash function competition. It performs well in software as well as in hardware [17]. As per the initial specifications, its has four major variants as given in Table 7.

Table 7. BLAKE HASH functions

Algorithm	Word	Message	Block	Digest	Salt
BLAKE-224	32	$< 2^{64}$	512	224	128
BLAKE-256	32	$< 2^{64}$	512	256	128
BLAKE-384	64	$< 2^{128}$	1024	384	256
BLAKE-512	64	$< 2^{128}$	1024	512	256

BLAKE follows a HAIFA iteration mode which is an improved version of Merkle-Damgård paradigm. The BLAKE compression function takes the following input values:

- an 8-word chaining value, $h = h_0, h_1..., h_7$
- a 16-word message block, $m = m_0, m_1..., m_{15}$
- a 4-word salt, $s = s_0, s_1..., s_3$
- a 2-word counter, $t = t_0, t_1$

Apart from the above mentioned input values, BLAKE uses 16 constants $c = c_0, c_1..., c_{15}$. The BLAKE's compression function has three steps:

- Initialization
- Rounds calculation
- Finalization

In the initialization round, a 4x4 matrix is initialized such that results of initialization vary with varying inputs. Chaining hash value or initialization value, salt, counter, and constants are used in the initialization stage. The input to the initialization process is as follows:

$$\begin{pmatrix} v_0 & v_1 & v_2 & v_3 \\ v_4 & v_5 & v_6 & v_7 \\ v_8 & v_9 & v_{10} & v_{11} \\ v_{12} & v_{13} & v_{14} & v_{15} \end{pmatrix}$$

After the initialization the initial state is given by:

$$\begin{pmatrix} h_0 & h_1 & h_2 & h_3 \\ h_4 & h_5 & h_6 & h_7 \\ s_0 \oplus c_0 & s_1 \oplus c_1 & s_2 \oplus c_2 & s_3 \oplus c_3 \\ t_0 \oplus c_4 & t_0 \oplus c_5 & t_1 \oplus c_6 & t_1 \oplus c_7 \end{pmatrix}$$

Where h_i are the initialization or the chaining values, s_i are the salt, c_i are the constants and t_i are the counter values. After the initialization stage, the round stage begins and depending upon the BLAKE variant, the round function iterates for a number of rounds. A round function performs the transformation in the following manner:

$$G_0(v_0, v_4, v_8, v_{12}) \quad G_1(v_1, v_5, v_9, v_{13})$$
$$G_2(v_2, v_6, v_{10}, v_{14}) \quad G_3(v_3, v_7, v_{11}, v_{15})$$

Fig. 5 shows a column and diagonal step on which the following transformation is applied:

$$G_4(v_0, v_5, v_{10}, v_{15}) \quad G_5(v_1, v_6, v_{11}, v_{12})$$
$$G_6(v_2, v_7, v_8, v_{13}) \quad G_7(v_3, v_4, v_9, v_{14})$$

Where the function $G_i(a, b, c, d)$ is as follows:

$$a = a + b + (m_{\sigma_r(2*i)} \oplus c_{\sigma_r(2*i+1)})$$
$$d = (d \oplus a) \ggg R_1$$
$$c = c + d$$
$$b = (b \oplus c) \ggg R_2$$
$$a = a + b + (m_{\sigma_r(2*i+1)} \oplus c_{\sigma_r(2*i)})$$
$$d = (d \oplus a) \ggg R_3$$
$$c = c + d$$
$$b = (b \oplus c) \ggg R_4$$

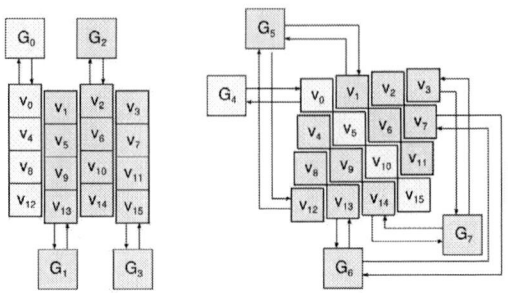

Figure 5. Diagonal and column steps of BLAKE [6]

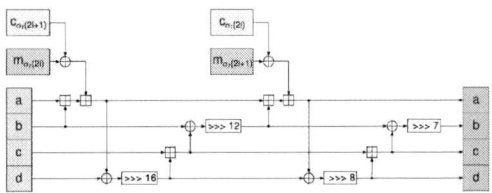

Figure 6. G function of BLAKE hash [6]

Fig. 6 shows a G_i function. Where $\sigma_0, ..., \sigma_{10}$ are ten permutations of the numbers $0, ..., 15$, r denotes the round number, R_i denotes the rotation constants which are dependent on the variants of BLAKE, c_i are the constants and m_i are the messages. The first four invocations of the G function (G_0, G_1, G_2, G_3) can operate in parallel and the next invocations (G_4, G_5, G_6, G_7) can operate in parallel after that. After each round a chaining value h is calculated and it is fed to the new round. After a specified number of rounds (depending on BLAKE variants) the round function is complete and the finalization step starts. The input to the finalization stage is the initial state $v_0, ..., v_{15}$, salt $s_0, ..., s_3$ and the chaining value $h_0, ..., h_7$. The finalization stage performs the following steps:

$$h'_0 = h_0 \oplus s_0 \oplus v_0 \oplus v_8$$
$$h'_1 = h_1 \oplus s_1 \oplus v_1 \oplus v_9$$
$$h'_2 = h_2 \oplus s_2 \oplus v_2 \oplus v_{10}$$
$$h'_3 = h_3 \oplus s_3 \oplus v_3 \oplus v_{11}$$
$$h'_4 = h_4 \oplus s_0 \oplus v_4 \oplus v_{12}$$
$$h'_5 = h_5 \oplus s_1 \oplus v_5 \oplus v_{13}$$
$$h'_6 = h_6 \oplus s_2 \oplus v_6 \oplus v_{14}$$
$$h'_7 = h_7 \oplus s_3 \oplus v_7 \oplus v_{15}$$

Where h'_i denotes the hashed value that can be the final output or the chaining value. To hash a message that has more than 16 words, the message is broken into blocks, each of 16 words. After that the compression function is applied to it, the resultant h'_i is chained to the next iteration of the compression function. The process continues till we reach the end of message. This process is given as:

$$h^0 = \mathbf{IV}$$
$$\text{for } i = 0, ..., N - 1$$
$$h^{i+1} = \mathbf{compress}(h^i, m^i, s, l^i) \tag{3}$$
$$\text{return } h^N$$

where h is the chaining value, N is the number of message chunks each of 16 words, s denotes the salt, and h denotes the output of the compression function.

B.2 Skein Hash Functions Family

Skein is also one of the five finalists of the SHA-3 hash competition [15]. The datapath of Skein is 64 bits, which makes it an excellent candidate to be implemented on 64 bit microprocessors. Threefish, a tweakable block cipher is at the core of the Skein hash function. Threefish block cipher has three inputs, key, tweak and plain text. The tweak is 128 bits for all block sizes. Based on the internal state size of the cipher, there are three variants of Threefish, Threefish-256, Threefish-512 and Threefish-1024. Threefish can have multiple state sizes and multiple output sizes.

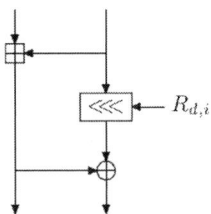

Figure 7. A MIX function of Skein [15]

The compression function of Skein consists of Threefish block cipher in Matayas-Meyer-Oseas (MMO) construction form. Threefish in MMO mode along with tweak specification defines the Unique Block Iteration mode. The UBI mode converts variable length inputs to fixed length outputs. Threefish uses a large number of simple rounds instead of small number of complex rounds. The core of Threefish is a MIX function that is made up of an Addition, Rotation and Xor. Fig. 7 shows a MIX function.

In case of Skein-512, the primary candidate of Skein family, four MIX functions are required to constitute one round of Skein-512. Fig. 8 shows the construction of Skein-512 using MIX functions.

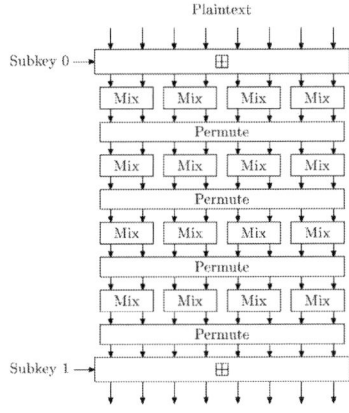

Figure 8. 4 Rounds of Skein-512 Hash [15]

After every four rounds, a subkey has to be injected. The subkey is injected by adding it with the outputs of MIX functions after every fourth round. A subkey is generated using keywords, tweakwords and subkey number in a subkey generator. Skein-256 has two MIX operations in every round and Skein-1024 has 8 MIX operations in every round. Skein-512 has 72 rounds. Fig. 9 shows the key generator for Skein-512.

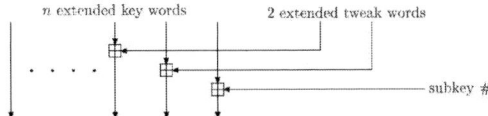

Figure 9. Keygenerator of Skein [15]

In normal hashing mode, Skein has three invocations of the UBI. One invocation is for the configuration, one for the message and the last invocation is for the output transform. Fig. 10 shows Skein in a normal hashing mode. The rotation constants of Skein repeat after every eight rounds and thus VLSI hardware are made, consisting of eight rounds along with two subkey injections.

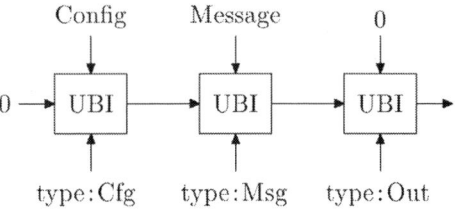

Figure 10. Skein in normal hashing mode [15]

B.3   HC-128 Stream Cipher

HC-128 is one of the finalists of eStream project [36]. The cipher comprises of two secret tables. A non-linear feedback function is used to update the elements inside the tables. During the update, three consecutive steps can be computed in parallel. HC-128 makes use of the following mathematical functions:

$+$: addition modulo 2^{32}.

$\boxminus$: subtraction modulo 512.

$\oplus$: bit-wise exclusive OR.

$\|$: bit-string concatenation.

$\gg$: right shift operator (defined on 32-bit numbers).

$\ll$: left shift operator (defined on 32-bit numbers).

$\ggg$: right rotation operator (defined on 32-bit numbers).

$\lll$: left rotation operator (defined on 32-bit numbers).

Two internal state arrays P and Q are used in HC-128, each with 512 many 32-bit elements. A 128-bit key array $K[0, \ldots, 3]$ and a 128-bit initialization vector $IV[0, \ldots, 3]$ are used, each entry being a 32-bit element. Let s_t denote the keystream word generated at the t-th step, $t = 0, 1, 2, \ldots$.

The following six functions are used in HC-128.

$$f_1(x) = (x \ggg 7) \oplus (x \ggg 18) \oplus (x \gg 3),$$

$$f_2(x) = (x \ggg 17) \oplus (x \ggg 19) \oplus (x \gg 10),$$

$$g_1(x, y, z) = ((x \ggg 10) \oplus (z \ggg 23)) + (y \ggg 8),$$

$$g_2(x, y, z) = ((x \lll 10) \oplus (z \lll 23)) + (y \lll 8),$$

$$h_1(x) = Q[x^{(0)}] + Q[256 + x^{(2)}],$$

$$h_2(x) = P[x^{(0)}] + P[256 + x^{(2)}],$$

where $x = x^{(3)}\|x^{(2)}\|x^{(1)}\|x^{(0)}$ is a 32-bit word, with $x^{(0)}, x^{(1)}, x^{(2)}$ and $x^{(3)}$ being the four bytes from right to left.

The key and IV setup of HC-128 recursively loads the P and Q array from expanded key and IV and run the cipher for 1024 steps to use the outputs to replace the table elements. It happens in four steps as follows.

Let $K[i + 4] = K[i]$ and $IV[i + 4] = IV[i]$ for $0 \leq i \leq 3$.
The key and IV are expanded into an array $W[0, \ldots, 1279]$ as follows:
$$
\begin{aligned}
W[i] \quad = \quad & K[i], && \text{for } 0 \leq i \leq 7; \\
= \quad & IV[i - 8], && \text{for } 8 \leq i \leq 15; \\
= \quad & f_2(W[i - 2]) + W[i - 7] \\
& + f_1(W[i - 15]) + W[i - 16] + i, && \text{for } 16 \leq i \leq 1279.
\end{aligned}
$$
Update the tables P and Q with the array W as follows:
$P[i] = W[i + 256]$, for $0 \leq i \leq 511$.
$Q[i] = W[i + 768]$, for $0 \leq i \leq 511$.
Run the cipher 1024 steps and use the outputs to replace the table elements as follows:
For $i = 0$ to 511, do
$P[i] = (P[i] + g_1(P[i \boxminus 3], P[i \boxminus 10], P[i \boxminus 511])) \oplus h_1(P[i \boxminus 12]);$
For $i = 0$ to 511, do
$Q[i] = (Q[i] + g_2(Q[i \boxminus 3], Q[i \boxminus 10], Q[i \boxminus 511])) \oplus h_2(Q[i \boxminus 12]);$

The keystream is generated using the following algorithm.

```
i = 0;
repeat until enough keystream bits are generated
{
    j = i mod 512;
    if (i mod 1024) < 512
    {
        P[j] = P[j] + g₁(P[j ⊟ 3], P[j ⊟ 10], P[j ⊟ 511]);
        sᵢ = h₁(P[j ⊟ 12]) ⊕ P[j];
    }
    else
    {
        Q[j] = Q[j] + g₂(Q[j ⊟ 3], Q[j ⊟ 10], Q[j ⊟ 511]);
        sᵢ = h₂(Q[j ⊟ 12]) ⊕ Q[j];
    }
    end-if
    i = i + 1;
}
end-repeat
```

B.4 Salsa20/r Stream Cipher

Salsa20 is a stream cipher that makes use of hash function in counter mode to generate keystream bits (S) that are mixed with the plaintext (P) by xoring to generate ciphertext (C) [7]. Block size of its inputs and outputs is 64 bytes. The input counter is incremented for every preceding block of plaintext. Salsa20 accepts four types of inputs, each consisting of words of 32 bit granularity. Firstly, an input key, that is either 256-bit $(k_0, k_1, ..., k_7)$ or 128-bit $(k_4 = k_0, ..., k_7 = k_3)$ in size; further a 64-bit nonce (n_0, n_1) and a 64-bit counter (t_0, t_1); finally four words of pre-defined constants ϕ_i, whose values are dependent upon the key size. These inputs are arranged in a predefined order in a 4x4 vector, called the initialization vector (X) in the rest of the discussion as shown below.

$$X = \begin{pmatrix} x_0 & x_1 & x_2 & x_3 \\ x_4 & x_5 & x_6 & x_7 \\ x_8 & x_9 & x_{10} & x_{11} \\ x_{12} & x_{13} & x_{14} & x_{15} \end{pmatrix} = \begin{pmatrix} \phi_0 & k_0 & k_1 & k_2 \\ k_3 & \phi_1 & n_0 & n_1 \\ t_0 & t_1 & \phi_2 & k_4 \\ k_5 & k_6 & k_7 & \phi_3 \end{pmatrix}$$

The initialization vector is subjected to a series of rounds composed of additions, cyclic rotations and xors, to achieve a random permutation. Originally the number of rounds was set to 20 (Salsa20/r, r = 20); however, the version of cipher included in the eSTREAM portfolio was reduced to 12 rounds, for performance reasons.

Let X, S and P be 4x4 arrays of 16-words each enumerated as $(x_0, x_1, ..., x_{15})$, representing initialized vector, Salsa20 keystream and plaintext respectively. Then Salsa20/r function for keystream generation can be represented mathematically as:

$$S = Salsa20_k(X)$$
$$Salsa20_k(X) = DoubleRound^r(X) + X$$
$$DoubleRound(X) = RowRound(ColumnRound(X))$$

Each *DoubleRound* comprises of four *QuarterRounds* performed on the columns of the initialization vector X, followed by four *QuarterRounds* performed on the rows of the output.

$$ColumnRound(X) = (y_0, y_1, ..., y_{15}) = Y$$
$$(y_0, y_4, y_8, y_{12}) = QuarterRound(x_0, x_4, x_8, x_{12}),$$
$$(y_5, y_9, y_{13}, y_1) = QuarterRound(x_5, x_9, x_{13}, x_1),$$
$$(y_{10}, y_{14}, y_2, y_6) = QuarterRound(x_{10}, x_{14}, x_2, x_6),$$
$$(y_{15}, y_3, y_7, y_{11}) = QuarterRound(x_{15}, x_3, x_7, x_{11}).$$

$$RowRound(Y) = (z_0, z_1, ..., z_{15}) = Z$$
$$(z_0, z_1, z_2, z_3) = QuarterRound(y_0, y_1, y_2, y_3),$$
$$(z_5, z_6, z_7, z_4) = QuarterRound(y_5, y_6, y_7, y_4),$$
$$(z_{10}, z_{11}, z_8, z_9) = QuarterRound(y_{10}, y_{11}, y_8, y_9),$$
$$(z_{15}, z_{12}, z_{13}, z_{14}) = QuarterRound(y_{15}, y_{12}, y_{13}, y_{14}).$$

Where each $QuarterRound(a, b, c, d)$ consists of four $ARX rounds$, named so since they comprise of additions (A), cyclic rotations (R) and xoring (X) operations only as given:

$$b = b \oplus ((a + d) \lll 7),$$
$$c = c \oplus ((b + a) \lll 9),$$
$$d = d \oplus ((c + b) \lll 13),$$
$$a = a \oplus ((d + c) \lll 18).$$

A 4x4 matrix C, representing a 16-word ciphertext block is calculated simply by xoring it with the keystream ($C = S \oplus P$).

B.5 ChaCha20/r Stream Cipher

ChaCha, a variant of Salsa20, was proposed by the author of Salsa20. The main aim of its creation was to improve the diffusion per round in Salsa20 which does not come at the expense of extra operations [7]. Being similar to Salsa20, ChaCha also has reduced round variants. The major differences between Salsa20 and ChaCha are following:

- Initialization matrix
- Inputs to the *DoubleRound* function
- *QuarterRound* function

The initialization matrix in ChaCha differs from Salsa20 in the following manner:

$$X = \begin{pmatrix} x_0 & x_1 & x_2 & x_3 \\ x_4 & x_5 & x_6 & x_7 \\ x_8 & x_9 & x_{10} & x_{11} \\ x_{12} & x_{13} & x_{14} & x_{15} \end{pmatrix} = \begin{pmatrix} \phi_0 & \phi_1 & \phi_2 & \phi_3 \\ k_0 & k_1 & k_2 & k_3 \\ k_4 & k_5 & k_6 & k_7 \\ t_0 & t_1 & n_0 & n_1 \end{pmatrix}$$

Where k_i, ϕ_i, n_i and t_i are key, constants, nonce and counter words each of 32 bit granularity, respectively. The *DoubleRound* calculation in ChaCha operates in the following order:

$$ColumnRound(X) = (y_0, y_1, ..., y_{15}) = Y$$
$$(y_0, y_4, y_8, y_{12}) = QuarterRound(x_0, x_4, x_8, x_{12}),$$
$$(y_5, y_9, y_{13}, y_1) = QuarterRound(x_1, x_5, x_9, x_{13}),$$
$$(y_{10}, y_{14}, y_2, y_6) = QuarterRound(x_2, x_6, x_{10}, x_{14}),$$
$$(y_{15}, y_3, y_7, y_{11}) = QuarterRound(x_3, x_7, x_{11}, x_{15}).$$

$$RowRound(Y) = (z_0, z_1, ..., z_{15}) = Z$$
$$(z_0, z_1, z_2, z_3) = QuarterRound(y_0, y_5, y_{10}, y_{15}),$$
$$(z_5, z_6, z_7, z_4) = QuarterRound(y_1, y_6, y_{11}, y_{12}),$$
$$(z_{10}, z_{11}, z_8, z_9) = QuarterRound(y_2, y_7, y_8, y_{13}),$$
$$(z_{15}, z_{12}, z_{13}, z_{14}) = QuarterRound(y_3, y_4, y_9, y_{14}).$$

Where X is the 4x4 array of 16-words enumerated as $(x_0, x_1, ..., x_{15})$ representing initialized vector.

The third difference between Salsa20 and ChaCha is the structure of the *QuarterRound*.

$$a = a + b$$
$$d = (d \oplus a) \lll 16$$
$$c = c + d$$
$$b = (b \oplus c) \lll 12$$
$$a = a + b$$
$$d = (d \oplus a) \lll 8$$
$$c = c + d$$
$$b = (b \oplus c) \lll 7$$

Reconfigurable Pipelined Coprocessor for Multi-mode Communication Transmission

Liang Tang,   Jude Angelo Ambrose,   Sri Parameswaran

School of Computer Science and Engineering, University of New South Wales, Sydney, Australia
{liangt, ajangelo, sridevan}@cse.unsw.edu.au

ABSTRACT

The need to integrate multiple wireless communication protocols into a single low-cost, low-power hardware platform is prompted by the increasing number of emerging communication protocols and applications. This paper presents a novel application specific platform for integrating multiple wireless communication transmission baseband protocols in a pipelined coprocessor, which can be programmed to support various baseband protocols. This coprocessor can dynamically select the suitable pipeline stages for each baseband protocol. Moreover, each carefully designed stage is able to perform a certain signal processing function in a reconfigurable fashion. The proposed platform is flexible (compared to ASICs) and is suitable for mobile applications (compared to FPGAs and processors). The area footprint of the coprocessor is smaller than an ASIC or FPGA implementation of multiple individual protocols, while the overhead of throughput is 34% worse than ASICs and 32% better than FPGAs. The power consumption is 2.7X worse than ASICs but 40X better than FPGAs on average. The proposed platform outperforms processor implementation in all area, throughput and power consumption. Moreover, fast protocol switching is supported. Wireless LAN (WLAN) 802.11a, WLAN 802.11b and Ultra Wide Band (UWB) transmission circuits are developed and mapped to the pipelined coprocessor to prove the efficacy of our proposal.

1. INTRODUCTION

Numerous wireless communication protocols have been proposed recently, each targeting a different application domain, such as WCDMA and GSM for wide area communication, WLAN for high speed, medium range communication and UWB for high speed, short distance communication. To meet stringent market demands, modern mobile devices have to combine a number of these communication protocols. For example, a modern phone will usually support at least Bluetooth, 802.11 and GSM protocols. It is even predicted that multi-mode communication will be the norm in [1]. Small area and low power consumption are critical for the success of a mobile product, thus ASIC chips are preferred when integrating multiple communication protocols for mass production of mo-

Permission to make digital or hard copies of all or part of this work for personal or classroom use is granted without fee provided that copies are not made or distributed for profit or commercial advantage and that copies bear this notice and the full citation on the first page. To copy otherwise, to republish, to post on servers or to redistribute to lists, requires prior specific permission and/or a fee.
WOODSTOCK DAC'13, May 29-June 07 2013, Austin, TX, USA.

bile devices. Current state-of-the-art communication solutions for mobile devices are able to integrate several protocols on a single ASIC chip. Broadcom 4330 chip [2] is used in the 3rd generation Apple iPad to provide multi-mode short range communication, integrating protocols such as 802.11a/b/g/n and Bluetooth. Although all these protocols are included in a single chip, each protocol is still implemented as an individual hardware block, and most components sit idle as only one of the protocols will be active at any one time. Thus, sharing amongst units can greatly reduce area consumption, and any possible component sharing across multiple communication protocols is highly beneficial. We studied numerous communication baseband protocols, including 802.11a, 802.11b, 802.11n, UWB, Bluetooth, WiMAX, DVB, ISDB and DTMB, and found that all of these protocols can be represented by a single pipelined structure (allowing component sharing). The transmission portion of these protocols is illustrated in Figure 1, including all major functional stages.

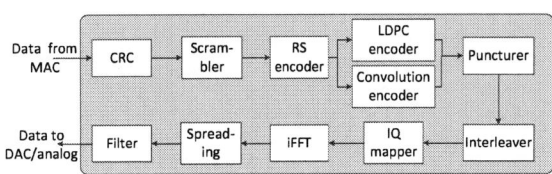

Figure 1: General Pipeline Structure of Wireless Communication Baseband for Transmission

Future wireless terminals are expected to be programmed to support hitherto unknown, upcoming protocols allowing for longer lifetime of a single design. This flexible programmability will enable a single terminal to support various communication protocols (time multiplexed), provided that the computation meets the required throughput. This unified single platform can be mass manufactured to reduce Non-Recurring Engineering (NRE) cost, compared to ASICs which contain only a limited number of specification-ready protocols.

Software Defined Radios (SDRs) have been proposed by researchers to meet this flexibility, and various approaches have been used for SDR implementations. Some of these platforms, such as FPGA, DSP or processor array [3, 4, 5, 6, 7], are flexible, but consume a significant amount of power, have large footprints, and are infeasible for mobile terminals. Typically such solutions are useful for base stations. Cost effective solutions with such flexible circuits for mobile terminals usually result in inadequate performance. For example, it is difficult to achieve the 480 Mbps data rate [8] required for UWB using FPGAs, DSPs or processor arrays in low power mobile terminals. Some of the newer protocols reach 1Gbps data rate, and certainly require the use of ASICs. One example of such a system is the WLAN 802.11ac [9]. Some other approaches, such as datapath merging [10], are area effective, but only the specification-ready protocols can be supported and future protocols cannot be implemented.

In order to conceive a reconfigurable power/area efficient and high throughput solution, for the first time, we propose a novel, two-level reconfigurable, pipelined coprocessor architecture in this paper. This reconfigurable coprocessor is expected to be implemented as an ASIC, rather than an FPGA. Thus high performance is achieved making this platform suitable for mobile applications. Each stage in the pipeline, shown in Figure 1, fulfills a certain signal processing function in the communication protocol, such as convolutional encoding, scrambling, etc. The first level configuration is performed at inter-stage, dynamically enabling and disabling pipeline stages, based on the functionalities required by the wireless protocol. The second level performs the intra-stage reconfigurability, allowing changes to each stage to adapt it to differing protocols. For example, various convolutional encoders are supported in the convolutional encoder stage, such as the ones in 802.11a, UWB, WiMAX.

Outline

The rest of this paper is organized as follows. In section 2, the current research on SDR is reviewed. In section 3, the reconfigurable pipelined coprocessor architecture is proposed. Section 4 and Section 5 show the experimental setup and results respectively. The conclusion is given in Section 6.

2. RELATED WORK

FPGA implementations have been studied for SDR enactments [11, 12, 13] Since FPGAs' cost, power consumption, and timing delays are considerably high when compared to ASIC designs [14], they cannot be used in mobile devices. General purpose DSP solutions have been proposed for SDR by [15]. Communication signal processing oriented DSP solutions, such as SODA from Lin et al. [5], have been proposed to improve the computation ability of the general purpose DSPs. SODA can meet the processing requirements of WCDMA and 802.11a. However, modern high speed standards, such as 480Mbps Ultra-Wideband (UWB) or 1Gbps 802.11ac, are not supported, as the timing constraints are violated. Due to the complexity of the modern wireless communication signal processing requirements, a single (or a limited number of) CPU/DSP may not be enough to complete the signal coding/decoding necessary within the required throughput. Thus, processor arrays were proposed as an implementation medium for SDR baseband circuits [6]. However, this is a costly and power hungry solution and is only suitable for base stations instead of hand held devices [16].

Datapath merging solutions search the shareable components between individual communication circuits, and insert MUXes into the merged circuit [17, 10]. Although this method has small area and timing overhead, the merged circuit is fixed to the predefined individual circuits and future protocols cannot be supported.

There are proposals targeting reconfigurable individual communication function blocks, such as reconfigurable convolutional encoder [18], reconfigurable interleaver [19, 20], and reconfigurable IQ mapper [21], etc. These manually designed reconfigurable functional blocks achieve high performance when implemented in ASIC while having similar programmability to FPGAs and processors. However, the design of a complete baseband by reconfigurable functional blocks has not been proposed as yet. In this paper we overcome these shortcomings to create a complete reconfigurable baseband platform.

We propose a pipelined coprocessor architecture for the entire

baseband by the use of reconfigurable communication functional blocks. The coprocessor is controlled by a general purpose processor (GPP) which sets parameters for each communication protcol. Moreover, a special mechanism is in place for the GPP to handle hitherto unknown protocols which are not supported by the coprocessor. Each stage in the coprocessor is reconfigurable. The reconfigurable convolutional encoder in [22] eliminated the common sub-expression and reduced the calculation complexity of DSP. However, according to Table 3 in [22], 382 clock cycles are needed to transmit 216 bits in 802.11a 54Mbps data rate. This large clock cycle/data bits ratio requires high clock frequency to maintain 54Mbps data rate, which results in high power consumption. We designed a reconfigurable Shifter-XOR based circuit to process arbitrary polynomial convolutional encoder in FEC stage of the proposed coprocessor. For the same 216 transmission bits, only 108 cycles are needed in our proposal. The reconfigurable interleavers in [19, 20] are only suitable for WLAN. We utilize the extended version of the reconfigurable interleaver [23] so that both WLAN and UWB can be supported. The reconfigurable IQ mapper in [21] can only support BPSK, QPSK, and QAM modulation mapping. A reconfigurable CPU from [24] is embedded into our IQ mapper to process any derivation based on BPSK, QPSK, QAM based modulation mapping, including $\pi/4$QPSK, DQPSK, Complementary Code Keying (CCK) [25], and Dual-Carrier Modulation (DCM) [8]. We also designed reconfigurable blocks which are yet to appear in any research literature, such as reconfigurable scrambler, reconfigurable CRC and reconfigurable puncturer.

Our Contributions

- For the first time, a novel pipelined coprocessor is developed to construct reconfigurable communication baseband circuit for transmission. The coprocessor is programmed using a two-level reconfiguration approach by a GPP, providing appropriate parameters.

- Novel interface between GPP and coprocessor is proposed to use GPP's computing power when hitherto unknown baseband computation is not supported in coprocessor.

- Multiple reconfigurable communication function blocks are utilized and developed to build the coprocessor, including scrambler, convolutional encoder, puncturer, etc.

3. THE COPROCESSOR PLATFORM

Our proposed reconfigurable hardware platform, as shown in Figure 2, provides a reconfigurable pipelined coprocessor for various wireless communication baseband protocols, interfacing with a General Purpose Processor (GPP). Communication protocols differ in noise tolerance levels and data rate reqirements depending on the nature of applications. The stages/function blocks illustrated in the pipelined coprocessor architecture (such as CRC, Scrambler, FEC-encoder, Puncturer, Interleaver, IQ Mapper, FFT, Spreading and Filter) are necessary components for transmission (based on the wireless communication theory [26]). Hence we keep these components but augmented the functional blocks to adapt to different protocols.

There are 10 pipelined stages in the coprocessor. Each baseband function block in Figure 1 can be mapped to a corresponding stage in the coprocessor except the Reed-Solomon encoder (RS encoder) and LDPC encoder which are not supported in our current coprocessor, due to their infrequent need. The GPP executes these unsupported protocols via the coprocessor/GPP interface, which is handled by the Control stage. The coprocessor's output, which is

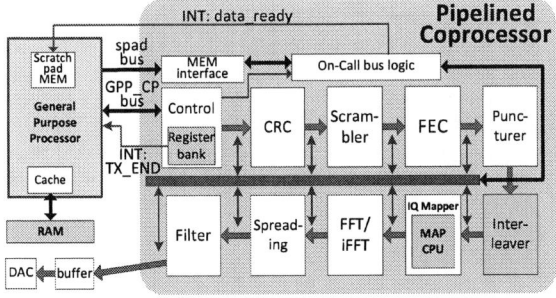

Figure 2: Our Pipelined Coprocessor Platform

from the Filter stage, is fed into a digital/analog converter through a buffer. Thus, the whole baseband transmission chain is covered by our platform (note that the receiving can be similarly designed but not the focus of this paper). The *On-Call bus logic* block assists the unsupported functions to be executed in GPP. The *GPP_CP bus* is used to transfer commands from GPP to coprocessor, and status from coprocessor to GPP. There are three important GPP commands: *Load_Param* instructs the coprocessor to load parameters; *TX_Start* initiates loading of payload data to the coprocessor and activates the coprocessor pipeline for transmission; and *SPD_Ready* command is used for *On-Call bus logic* operations. The payload data between GPP and coprocessor are transferred by *spad bus* which connects the scratchpad memory in GPP and the memory interface module in coprocessor. Two interrupts are provided by the coprocessor for efficient GPP response. First interrupt *TX_END*, indicates whether the current packet frame transmission is finished, another interrupt *data_ready* which is for *On-Call bus logic* operation again. We utilized previous designs for the MAP CPU [24] and Interleaver [23] components.

Reconfigurations to the pipelined coprocessor are provided by the GPP. The two-level reconfigurations are: 1), inter-stage reconfiguration, such that stages can be disabled and enabled; and 2), intra-stage reconfiguration, where functional blocks are carefully designed to support various protocols by setting parameters within it.

3.1 Inter-Stage Reconfiguration Level

All functional blocks are implemented as stages in the pipelined coprocessor to accelerate the communication throughput. However, only a subset of functional blocks are needed for each protocol (e.g., convolutional encoder is essential in 802.11a and UWB, but not used in 802.11b; spreading is a key component in 802.11b, though not used in 802.11a and UWB). Each stage in the pipeline can be disabled and bypassed by setting parameters.

Besides the main datapath for payload packet data (signal *spad_bus* in Figure 2), an extra data processing interface, named *On-Call bus logic*, is provided to control the communication between coprocessor and GPP to allow more flexible signal processing. If a future protocol is created with a specific baseband function block, not supported by our coprocessor, this specific function block can be implemented in the GPP using software. For example, an RS encoder block is used for the UWB packet header and this block should be located between the scrambler and convolutional encoder. However, the RS encoder has not been implemented in the coprocessor, hence executed by the GPP. The *On-Call bus* exchanges data between GPP and coprocessor through the scratchpad memory, and its operation is as follows: 1), because scratchpad memory is used to exchange data between coprocessor and GPP, the parameters for

transferring data into the scratchpad memory, such as size and starting address of the transferring data block, need to be set by the GPP to the coprocessor before baseband frame transmission; 2), GPP needs to set the access point which is the location for missing signal processing capability in pipelined coprocessor to fetch data for processing on the GPP (for example, the access point is located on the output of the scrambler stage in the scenario where the RS encoder is missing). The coprocessor needs this information to select the data to/from specific stage via MUXes; 3), the output of the stage before the access point will be selected by the coprocessor and stored in the scratchpad memory and an interrupt signal (*data_ready*) will be asserted to inform the GPP when the space in the scratchpad memory is full (or the coprocessor dumps all processed data at the completion of the last data block of frame); 4), the GPP will perform the corresponding signal processing and a special coprocessor command, called *SPD_Ready*, will be issued to inform the coprocessor via a coprocessor instruction when the GPP has finished signal processing; and 5), once the coprocessor receives an *SPD_Ready* command, the GPP processed data are fed into the pipeline stage, just after the access point, to ensure the completeness of the pipeline. In the scenario where the RS encoder is missing, the coprocessor feeds data from the scratchpad memory to the input of the convolutional encoder. By this *On-Call bus* implementation, only one extra physical signal, *data_ready* interrupt, is added for GPP and coprocessor communication, to minimize the impact on the GPP modification.

3.2 Intra-Stage Reconfiguration Level

For effective multi-mode baseband methodology, reconfigurable baseband functional blocks are created and implemented in each stage of the pipelined coprocessor. The whole SDR baseband transmission path has been established in our architecture including reconfigurable convolutional encoder, reconfigurable interleaver, reconfigurable IQ mapper, reconfigurable scrambler, etc. We utilize previously proposed designs for interleaver [23] and IQ mapper [24] (i.e., a processor/CPU), implemented the designs for reconfigurability (see Support Material). Due to restricted space, we only detail the reconfigurable convolutional encoder. Other components, such as the scrambler, CRC and puncturer are designed in a similar fashion.

3.2.1 Reconfigurable convolutional encoder

A convolutional encoder is used for error correction and is commonly used in wireless systems. The convolutional encoder can be implemented by a series of concatenated shift registers, and certain registers' outputs are wired to XOR cells to generate encoded data. All convolutional encoders share a similar structure, but the differences are: the number of shift registers, the source of XOR operations, and the number of output ports (branches). Figure 3 (a) and (b) depict the block diagram of the serial implementation of these two convolutional encoders. To improve throughput, it is a common practice to implement convolutional encoder in parallel, usually 8-bits width wide. The parallel circuit of one output branch of 802.11a is illustrated in figure 3 (c). The 8-bit input data is combined with the register output, which is the buffered input data with one clock cycle delay, to a 16-bit bus which connects to the shifters. The final result can be derived by the shifters' output and XORs. The parallel circuits for UWB branches have a similar structure, the only difference being the configuration of shifter logic.

We studied the structure of parallel processing implementations of convolutional encoders and found that the XORs and shifters form a unique *XOR matrix* for each protocol. However, the inputs for each tier of XOR operations in the *XOR matrix* are merely a

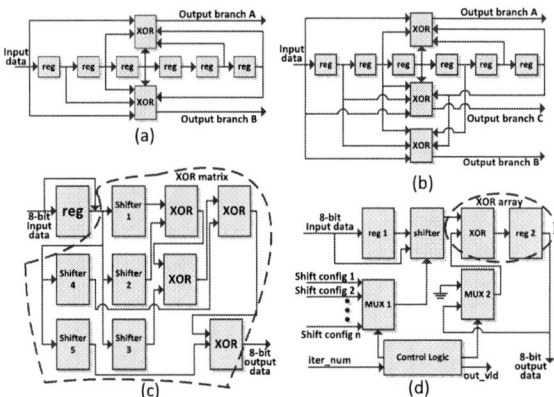

Figure 3: 802.11a and UWB convolutional encoder implementation (a) 802.11a serial encoder (b) UWB serial encoder (c) one output branch of 802.11a parallel encoder (d) one output branch of reconfigurable encoder

shifted version of the buffered input bytes. Thus, this *XOR matrix* can be split into multiple stages such that the reconfigurability can be efficient. An eight-bit register (*reg 2* in Figure 3 (d)) is added after the eight-bit XOR operator, to hold the XOR result. The combination of this register and XOR is called *XOR array* which has two inputs: one is *MUX 2* controlled feedback from the *reg 2*; another is from the barrel shifter circuit which can generate any shifted version of the input bytes according to the predefined parameters selected by *MUX 1*. Note that the operation of *XOR array* will be transparent during the first iteration controlled by *MUX 2*, as one of the *XOR array*'s inputs will be zero due to the selection of *MUX 2* and the output of the XOR will be identical to the shifter circuit output. The *Control Logic* selects the MUXes and manages the number of iterations according to the input *iter_num* signal. The detailed reconfigurable *XOR matrix* implementation by shifter and *XOR array* is shown in the Figure 3 (d). The designs of the reconfigurable scrambler and CRC are similar to the method used for reconfigurable convolutional encoder.

3.3 Programming of the coprocessor & mode switching

To program the coprocessor, all parameters for functional blocks need be transferred from the GPP to the coprocessor before the packet data transmission in each protocol. The coprocessor stores all parameters in a register bank which is located in the control stage as shown in Figure 2. During mode switching, the parameters for new protocols need to be set from the GPP again. Based on our current coprocessor design, there are about 600 bits for parameters. It can take about 20 clock cycles in our system for these parameters to be transferred between GPP scratchpad memory and coprocessor. Mode switching period can be further shrunk if the register bank in the control sub module of the coprocessor can be doubled and then two sets of parameters from both protocols can be stored simultaneously. Thus the turn-around time of mode switching is only determined by the stage flushing time.

A communication flowchart between the GPP and the coprocessor is shown in Figure 4. At first, the GPP selects the active protocol and stores the parameters in the memory location 1 and sends *Load_Param* command to the coprocessor via *GPP_CP bus* if these parameters have not been loaded to the coprocessor (switching to new protocol). Once the coprocessor receives *Load_Param* command, it will load the parameters from the memory location 1 to

the register bank and set the pipeline accordingly. Secondly, the GPP stores the transmission packet data in the memory location 2 and issues *TX_Start* command. When the coprocessor receives this command, it will exit from its low power idle state and load a block of packet data from the memory location 2 to the pipeline stage for processing. If there are unprocessed data in memory location 2, the coprocessor will repeatedly load and process until all packet data are processed. During the coprocessor pipeline processing, if the GPP's computing power is needed for certain signal processing, *On-Call bus* activities will be initiated between the GPP and the coprocessor (details can be found in section 3.1). The memory location 3 is reserved for *On-Call bus*. Finally, the *TX_END* interrupt will be asserted by the coprocessor if all packet data are processed. Once the GPP detects this interrupt, the GPP finishes the current packet transmission and can send a new packet or switch to a new protocol by sending new parameters to the coprocessor. Please note that all memories here are in the scratchpad memory of the GPP.

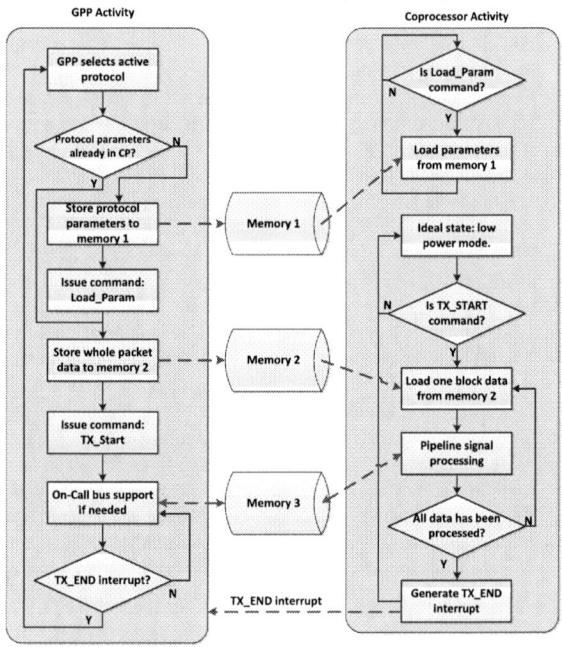

Figure 4: GPP and coprocessor communication flowchart

4. EXPERIMENTAL SETUP

To evaluate our coprocessor design, the GPP and coprocessor transmission circuits were implemented. By simulating different instructions in the GPP, the whole baseband packet payload transmission chain from WLAN [25] (including 802.11a and 802.11b) and UWB [8] were demonstrated. Wireless LAN (802.11 protocol), one of the most popular wireless protocols, and UWB, a high speed wireless communication protocol, which can support 480Mbps data rates, were selected for evaluation. Modulations in the 802.11 family use Orthogonal Frequency Division Multiplexing (OFDM) in 802.11a and Direct Sequence Spread Spectrum (DSSS) in 802.11b. Thus both 802.11a and 802.11b protocols were chosen, as OFDM and DSSS are fundamental technologies in 802.11 protocol family.

The performance of the coprocessor is compared to ASIC implementation of individual protocols, FPGA, software implementation with the Xtensa procesor from Tensilica, and the work in [10] which is the only whole transmission chain for multi-mode protocols. Since the CRC, iFFT and Filter blocks are excluded in [10], these blocks are also removed in our testbench for fair comparison.

We developed the baseband packet payload transmission circuit of these three protocols in RTL. The GPP was developed using the ASIPMeister tool suite [27]. After that the reconfigurable pipelined coprocessor was designed to support 802.11a, 802.11b and UWB. Note that all data rates in these protocols are supported, including data rates of 6, 9, 12, 18, 24, 36, 48 and 54Mbps (indicated as 11a R0 upto 11a R7) for 802.11a, 1, 2, 5.5 and 11Mbps (depicted as 11b R0 to 11b R3) for 802.11b, 53.3, 80, 106.7, 160, 200, 320, 400 and 480Mbps (UWB R0 to UWB R7) for UWB.

The test vectors for the system were generated using Matlab [28]. These test vectors are then fed to individual circuits in RTL simulation (ModelSim in our system), to verify functional correctness. The GPP instructions were manually implemented to control the coprocessor. Finally the test vectors generated by Matlab are inserted into the coprocessor via the GPP. The output results from the simulation of the individual circuits were verified with the outputs from simulating the coprocessor. This proves the correctness of our coprocessor design and implementation. Area and timing results were generated by Synopsys Design Compiler with TSMC 65nm technology, and power consumption was measured by Synopsys Prime Time and Power Compiler. Xilinx Virtex5 XC5VLX50 is selected for FPGA implementation as it is also fabricated using 65nm process, thus providing a fair comparison. The area, timing and power results of the FPGA are from Xilinx ISE and XPower tools. Tensilica tool set is utilized to evaluate the software version of the pipeline implementation on the Xtensa LX processor.

5. RESULTS AND ANALYSIS

5.1 Analysis of Hardware Area

Transmission circuits are separately implemented for all the 802.11a, 802.11b and UWB protocols in the coprocessor, as individual ASICs, in FPGA, datapath merging from [10] and in Xtensa processor. The area results of these implementations are summarized in Figure 5. The y-axis indicates the equivalent number of ASIC gates for each design implementation. The x-axis shows all the experimented designs. Due to lack of a proper RAM model, we estimated the equivalent gate count of RAM by the RAM size. Suppose 6-transistor SRAM is used, then each RAM bit is equivalent to 1.5 gate [29]. The FPGA gate count is derived from the number of used LUTs. The largest individual design, UWB, uses 508 6-input LUTs on Xilinx Virtex-5 FPGA. Each 6-input LUT is equivalent to 194 gates [24], thus the 508 LUTs are equivalent to 98K gates (=508 × 194).

The coprocessor occupies the smallest area, i.e., 11.9K gates without RAM, or 17.3K gates with estimated RAM. Datapath merging and ASIC achieve similar result, 18K and 21.6K gates respectively. The gate number of the Xtensa processor is 64.8K. The FPGA implementation consumes the most area, i.e. 98K gates.

5.2 Analysis of Throughput

The throughput of coprocessor is determined by the maximum working frequency and number of processing bits per cycle. The maximum working frequency is constrained by the longest datapath delay. Figure 6 depicts the maximum throughput of coprocessor, ASIC, FPGA, datapath merging and Xtensa implementations. The required throughput from each protocol/data rate is also plotted (the line with label "protocol").

The maximum working frequencies of coprocessor is 294 MHz, as the longest datapath delay is 3.4ns. The throughput for coproces-

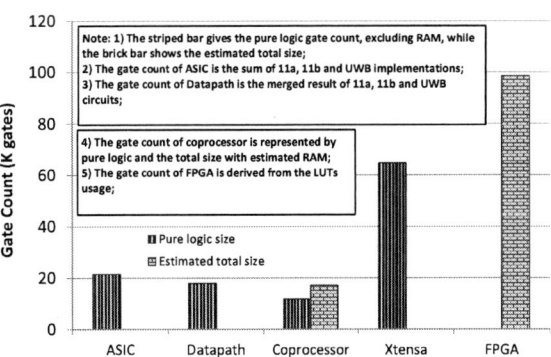

Figure 5: Area Evaluation of Various Transmission Designs

sor 802.11a R0 and R1 rate are 1176Mbps (1176=294 × 8 × 1/2) and 1764Mbps (1764=294 × 8 × 3/4) as the coding rates are 1/2 and 3/4 and 8 input bits are processed in parallel. The throughput calculations for the ASIC, datapath merging and FPGA are similar. The highest working frequency for Xtensa is 1300MHz [30]. To get the throughput of Xtensa, the number of cycles to process each bit is needed. For example, 538 cycles are used to process one input bit in 802.11a R0 data rate, thus the Xtensa throughput is only 2.4Mbps (2.4=1300/538).

As shown in Figure 6, all the implementations can support the required throughput of the each interleaving scheme, except Xtensa. Only these schemes (11b R0-R2) of Xtensa meet the required throughput. ASIC has the best throughput as it can work at a higher frequency and eight bits can be processed in parallel. Datapath merging has very similar performance as ASIC. Coprocessor outperforms FPGA and Xtensa.

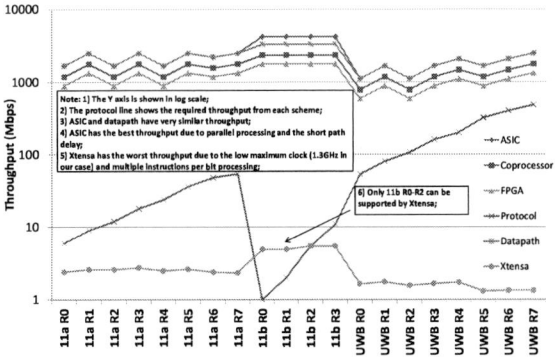

Figure 6: Throughput Evaluation of various Transmission Protocols

5.3 Analysis of power consumption

Low power consumption is a critical factor in mobile devices, the power consumption of the coprocessor has been reduced due to two reasons. First, the logic size of the coprocessor is small as indicated in Figure 5, thus the static power consumption can be reduced compared to other reconfigurable designs. Second, because parallel implementation is supported by the coprocessor, the clock frequency can be reduced while maintaining the required throughput from different protocols. As clock frequency is one of the main factors in the consumption of dynamic power, low clock frequency results in low dynamic power consumption.

Figure 7 shows power consumption comparisons between ASIC, coprocessor, FPGA and Xtensa implementations. Datapath merging is excluded here as the power consumption is not reported in [10]. Clock gating is applied to ASIC implementation. The lines depicting ASIC, coprocessor and Xtensa's power consumption include both static and dynamic power. The static power of FPGA is excluded due to the difficulty in extracting the information for a small part of the FPGA, thus only the dynamic power is reported. Flip-flops are used to model RAM in our experiment. We believe the power consumption of coprocessor will be further reduced when the real RAM is applied.

The ASIC has the lowest power consumption in Figure 7, on average 424uW. Although the average power consumption of the coprocessor is 1.1mW, coprocessor outperforms all FPGA and Xtensa solutions. Xtensa consumes the highest power, and only three schemes can meet throughput constraints.

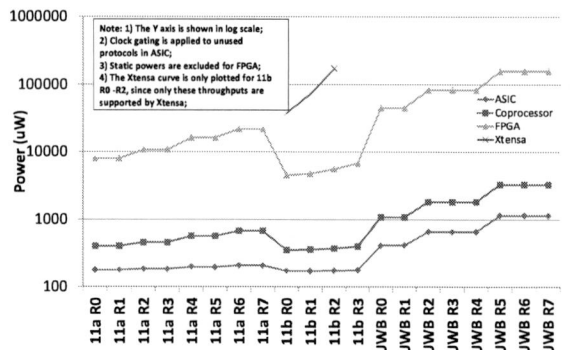

Figure 7: Power Evaluation of various Transmission Protocols

5.4 Discussion

The wireless communication baseband tailored reconfigurable method is provided in this paper. The proposed methodology has greater flexibility than ASICs, with near ASIC performance. It is less flexible than DSP and FPGA, which can support any protocol as long as the computation resources are sufficient. However, since most wireless protocols are similar, the architecture provided in this paper will form an ideal platform for flexible baseband communication chips.

There are limitations in our proposal. Only the baseband transmission path is studied. The receiving path usually contributes more to area and power consumption in the baseband. Our next step is to extend our methodology to the receiving path. The power consumption of the pipelined coprocessor is high when compared to single protocol ASIC implementation. By improving the parallelism of data processing in the pipelined coprocessor, the working frequency can be reduced, thus the power consumption of coprocessor can be reduced. This high parallelism coprocessor is under development.

6. CONCLUSIONS

In this paper, a two-level reconfigurable pipelined coprocessor is presented to build a reconfigurable baseband transmission platform in the context of the SDR. We state that this is a highly efficient and flexible solution for multi-mode baseband integration. This methodology can save area when integrating multiple protocols without worsening the circuit timing and power performance, while allowing fast mode switching.

7. REFERENCES

[1] D. A. Reed, J. R. Larus, and D. Gannon, "Imagining the future: Thoughts on computing," *IEEE Computer*, vol. 45, no. 1, pp. 25–30, 2012.

[2] "Bcm4330 product brief." Available at: www.broadcom.com.

[3] L. S. Nagurney, "Software defined radio in the electrical and computer engineering curriculum," in *Proceedings of the 39th IEEE international conference on Frontiers in education conference*, FIE'09, 2009.

[4] N. Himanshu Shekhar, C.B.Mahto, "FPGA Implementation of Tunable FFT For SDR Receiver," in *International Journal of Computer Science and Network Security*, pp. 186–190, 2009.

[5] Y. Lin, H. Lee, M. Woh, Y. Harel, S. Mahlke, T. Mudge, C. Chakrabarti, and K. Flautner, "Soda: A high-performance dsp architecture for software-defined radio," *IEEE Micro*, vol. 27, pp. 114–123, 2007.

[6] A. Duller, D. Towner, G. Panesar, A. Gray, and W. Robbins, "picoArray Technology: The Tool's Story," in *DATE*, 2005.

[7] M.I.Taj, O.Hammami, and M.Akil, "SDR waveform components implementation on single FPGA multiprocessor platform," in *ICECS*, pp. 790 – 793, Dec. 2010.

[8] "Multiband ofdm physical layer specification," 2005. Available at: http://www.wimedia.org/.

[9] "Official ieee 802.11 working group project timelines," 2011. Available at: http://www.ieee802.org/11/Reports/802.11_Timelines.htm.

[10] L. Tang, J. Peddersen, and S. Parameswaran, "A rapid methodology for multi-mode communication circuit generation," in *Proceedings of the 2012 25th International Conference on VLSI Design*, 2012.

[11] P. Coulton and D. Carline, "An SDR inspired design for the FPGA implementation of 802.11a baseband system," in *IEEE Symposium on Consumer Electronics*, 2004.

[12] M. Cummings and S. Haruyama, "FPGA in the Software Radio," *IEEE Communications Magazine*, vol. 37, pp. 108–112, 1999.

[13] C. Dick and F. Harris, "Configurable logic for digital communications it's about time," in *Signals, Systems, and Computers, Conference*, 1999.

[14] I. Kuon and J. Rose, "Measuring the gap between FPGAs and ASICs," in *FPGA '06: Proceedings of the 2006 ACM/SIGDA 14th international symposium on Field programmable gate arrays*, (New York, NY, USA), pp. 21–30, ACM, 2006.

[15] H. Karimi, N. Anderson, and P. McAndrew, "Digital signal processing aspects of software definable radios," in *IEE Colloquium on Adaptable and Multistandard Mobile Radio Terminals*, 1998.

[16] D. Pulley and R. Baines, "Software defined baseband processing for 3G base stations," in *4th International Conference on 3G Mobile Communication Technologie*, 2003.

[17] N. Moreano, E. Borin, C. D. Souza, and G. Araujo, "Efficient datapath merging for partially reconfigurable architectures," in *IEEE Transactions on Computer Aided Design of Integrated Circuits and Systems*, pp. 969–980, 2005.

[18] B. Krill and A. Amira, "Efficient reconfigurable architectures of generic cyclic convolution," in *B. Krill, A. Amira*, 2011.

[19] C. R. Sanchez-Ortiz, R. Parra-Michel, and M. Guzman-Renteria, "Design and implementation of a multi-standard interleaver for 802.11a, 802.11n, 802.16e & dvb standards," *Reconfigurable Computing and FPGAs, International Conference on*, vol. 0, pp. 379–384, 2008.

[20] B. K. Upadhyaya and S. K. Sanyal, "Design of a novel fsm based reconfigurable multimode interleaver for wlan application," in *International Conference on Devices and Communications*, 2011.

[21] K. Smitha, A. P. Vinod, and R. Mahesh, "Reconfigurable area and power efficient i-q mapper for adaptive modulation," in *International Midwest Symposium on Circuits and Systems*, 2011.

[22] J.-C. Lin, C. Yu, M.-H. Yen, P.-A. Hsiung, S.-J. Chen, and Y. H. Hu, "Parallel implementation of convolution encoder for software defined radio on dsp architecture," in *ICSAMOS*, pp. 180–186, 2009.

[23] L. Tang, J. A. Ambrose, and S. Parameswaran, "Variable increment step based reconfigurable interleaver for multimode communication application," in *ISCAS'13: The IEEE International Symposium on Circuits and Systems*, 2013.

[24] L. Tang, J. A. Ambrose, and S. Parameswaran, "MAPro: A Tiny Processor for Reconfigurable Baseband Modulation Mapping," in *Proceedings of the 2013 26th International Conference on VLSI Design*, 2013.

[25] "Wireless LAN Medium Access Control (MAC) and Physical Layer (PHY) specifications," 2011. Available at: http://standards.ieee.org/.

[26] J. Proakis, *Digital communications*. McGraw-Hill series in electrical and computer engineering, McGraw-Hill, 2001.

[27] "ASIP Solutions ASIPMeister." Available at: http://www.asip-solutions.com/en/asip_meister.html/.

[28] "Mathworks matlab." Available at: http://www.mathworks.com/.

[29] "Transistor count." Available at: http://en.wikipedia.org/wiki/Transistor_count.

[30] "Xtensa Processor." Tensilica Inc. (http://www.tensilica.com).

Support Material
Reconfigurable interleaver

An interleaver reorders the transmission bits to distribute the burst errors on multiple symbols, so that the error rate per symbol is reduced and thus it is more likely that all symbols can be decoded successfully by an error correction circuit. In different protocols, the interleaver reordering patterns are different. For example, the interleaver in 802.11a can be expressed as two permutations in Equation 1 and Equation 2. The index of the bit before the first permutation shall be denoted by k; i shall be the index after the first permutation; j shall be the index after the second permutation. Note that N_{CBPS} is the number of coded bits per OFDM symbol and N_{BPSC} is the number of bits in each OFDM subcarrier [25]. Both these two parameters vary between different data rate in 802.11a and different interleaving patterns are generated.

$$i = \frac{N_{CBPS}}{16} \times (k \bmod 16) + \lfloor \frac{k}{16} \rfloor \qquad (1)$$
$$k = 0, 1, ..., N_{CBPS} - 1$$

$$j = s \times \lfloor \frac{i}{s} \rfloor + (i + N_{CBPS} - \lfloor (\frac{16 * i}{N_{CBPS}}) \rfloor) \bmod s \qquad (2)$$
$$j = 0, 1, ..., N_{CBPS} - 1, s = \max(\frac{N_{BPSC}}{2}, 1)$$

The execution of these equations without optimization is computation intensive as there are modular and division operations associated with each data bit. The interleaver in UWB has the similar equations to Equation 1 and Equation 2 with different parameters. However, there is a third permutation in UWB for cyclic shifting. A variable increment step (VIS) based interleaver is proposed in the reconfigurable pipelined coprocessor and shown in Figure 8 (a). Two memory blocks are used (RAM 0 and RAM 1) to eliminate data halting due to RAM R/W conflict, one is for data writing and another one is for reading, and their roles will be alternated when they are full/empty. Two sets of address generation circuits are available, one set is used for RAM 0, and another set is used for RAM 1. One set of address generation is the proposed variable increment step circuit (combination of *B0 ADDR Gen* and *ADDR Accumulate* blocks in Figure 8 (a)), another set is merely a counter for sequential address generation (*seq addr Gen* component in Figure 8 (a)). The role of these two sets of address generation circuit is controlled by two MUXes (*MUX addr0* and *MUX addr1*).

Two identical Step Adjust (SA) components, to perform the address step adjustment by addition or subtraction operations, are used in *B0 ADDR Gen* block. One of them, *SA b0*, is used to generate the new address according to the Equations 1 and 2 for block interleaving, or the RAM reading address pattern for convolutional interleaving. One of *SA b0*'s input is from *MUX b0*, which determines the increment step according to the preset parameters, with the control of *modular counter b0_1*. Another input of *SA b0* is from *ADDR Accumulate* block. *SA shift* is after *SA b0* and it is used to add the cyclic shift value, which is controlled by MUX *MUX shift* and counter *modular counter b0_2*. The output of *SA shift* is fed to flip-flop *FF*, through *MUX Reset*, which is under the control of *control logic*. The output of FF is the current generated address, this address is used for RAM accessing, and fed back to *SA b0* for the next address generation.

The required speed for interleaving processing varies greatly in different protocols, such as 12 Mbps in WLAN, and 640 Mbps in UWB. A parallel implementation is desired to support high speed protocols. Furthermore, the clock frequency can be reduced for parallel implementation, lowering the dynamic power consumption (although the static power will be increased due to larger area). Thus, a parallel implementation of VIS interleaver is proposed in

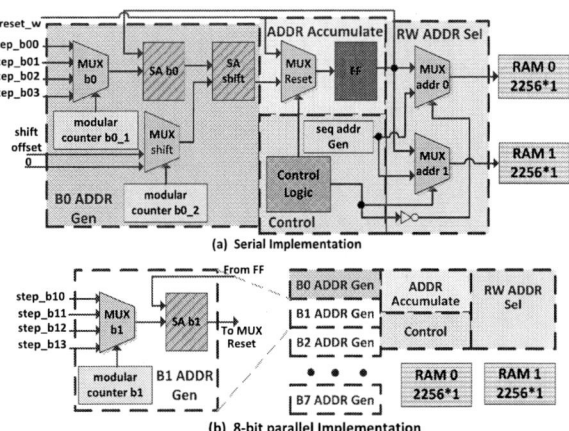

Figure 8: Reconfigurable interleaver block diagram

Table 1: MAPro Stage Description

Stage	Description
IF	Instruction fetch.
ID	Instruction decoding.
EX	Calculation execution.
PG	Phase generation for BPSK and QPSK.
IQ	Final IQ signal generation.

this paper.

The block diagram of the 8 bits parallel implementation of VIS interleaver is illustrated in Figure 8 (b). The parallel implementation is based on the serial implementation, all the *B0 ADDR Gen*, *ADDR Accumulate*, *Control* and *RW ADDR Sel* blocks are same in Figure 8 (a) and (b). The *B1 ADDR Gen - B7 ADDR Gen* blocks are added to generate the address for bit1 - bit7. All the implementations of these seven new blocks are identical, the only difference is the MUX input parameters, which define the different steps for each bit. One of these new blocks, *B1 ADDR Gen* is given in detail. Comparing to *B0 ADDR Gen*, the *SA shift* and its control logics are removed in *B1 ADDR Gen*. To support eight independent data RAM accesses, RAM 0 and RAM1 have to be implemented using eight-port RAMs.

Reconfigurable IQ mapper

IQ mapper is used to map data bits to certain amplitude and phase for analog transmissions, and it is an indispensable function block in all wireless baseband circuits. The IQ mapping schemes are mostly BPSK, QPSK, 16-QAM, 64-QAM and their derivatives.

To effectively support various calculations, a tiny two-bit data bus pipelined processor, called MAP CPU, is created for the reconfigurable pipelined coprocessor. Various IQ mapping schemes, including BPSK, DBPSK, QPSK, DQPSK, Offset QPSK, π/4-QPSK, CCK, 8PSK, 16-QAM, 64-QAM and DCM, are supported in our methodology due to the highly programmable MAP CPU. The MAP CPU has five pipelined stages listed in Table 1. And there are only six types of instructions (each instruction: 12-bit) for this CPU (given in Table 2), thus the instruction decoding circuit can be simplified. The block diagram of MAP CPU is illustrated in Figure 9.

The first stage, *IF*, fetches instructions from the instruction memory (whose size is only 16*12 bits as only maximal 16 instructions can be stored and each instruction is 12 bits wide). For each modulation mapping scheme, the instructions to produce one symbol (if each symbol is decided by different input data) or a set of symbols

937

Table 2: MAPro Instruction Description

Instruction	Description
MOV	Move data from source register or immediate value to destination register.
LSADD	Combined with left shifting and addition operations.
MADD	Perform multiple addition by at most four source registers.
PSK	Perform BPSK/DBPSK/QPSK/DQPSK phase generation.
BSEL	Select specified input bits and store them in register file.
NOP	Nop instruction.

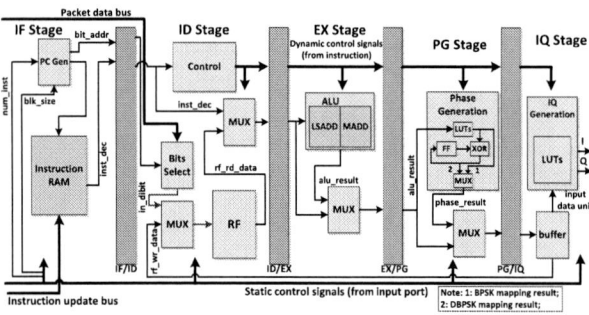

Figure 9: MAP CPU block diagram

(if these symbols are computed by the same input data), are identical in MAPro. Since the same sequence of instructions are applied for different input data in iterations, we name one such iteration as *unit processing*. For example, one symbol is calculated by three instructions in a QPSK *unit processing*, while eight symbols are computed by 14 instructions in a CCK *unit processing*. The maximum number of instructions for each unit processing is 16 and the all studied modulation schemes in this paper can be supported. Furthermore, the instructions for each *unit processing* are identical if the modulation scheme is not changed. Thus, the Program Counter (PC) generation circuit in the IF stage is different from a PC in a traditional processor: when the PC reaches a predefined threshold (*num_inst*), referring to the number of instructions for *unit processing*, the PC will be reset to zero to fetch instructions from the first address again. Such reset in PC indicates the execution of the next *unit processing*. When the number of times the PC reset reaches another predefined value (*blk_size*, meaning the number of *unit processing* for each transmission frame) set by *static control signals*, the PC is stopped and a NOP instruction (or stall) is issued from the IF stage. This stops MAPro and indicates that the payload data for the current frame has been modulated. By this implementation, the traditional jump and condition checking instructions are not used, one of the features used to reduce the complexity of MAPro. The modulation mapping input bit address, *bit_addr*, is also generated in the IF stage. When the PC reaches the *blk_size*, the *bit_addr* will be increased by a defined step size (such as 1 for BPSK, and 2 for QPSK) to denote that the current *unit processing* has completed execution. The new input data for the next *unit processing* is then loaded.

The second stage, *ID*, decodes the instruction via its *Control* submodule and prepares the packet bits for modulation mapping by *Bits Select* submodule. Various control signals will be generated in the *Control* submodule by analyzing the input instruction. These control signals will be transferred to the following stages via the pipeline registers. The *Bits Select* submodule selects the correct processing packet bits from the 200 bits width *packet data bus* which is external to MAPro, according to the address (*bit_addr*) provided by the *IF* stage. A tiny *register file* (RF) is constructed to store the intermediate results during processing, such as $\varphi 1$ - $\varphi 4$ in

CCK modulation. Each register in *RF* is only two bits wide, and there are eight registers, thus the size of *RF* is 8*2 bits. The input of *RF* can be either from the *in_dibit* signal which is the selected modulation mapping input bits from the *Bits Select* submodule, or from the *rf_wr_data* which is the result of the PG stage. The *RF*'s output, 8-bit wide *rf_rd_data* signal, including four registers' read out results (each of them is 2-bit wide), is fed to the EX stage via a MUX.

The third stage, *EX*, performs calculations by the *ALU*. This very simple *ALU* has four inputs from *RF* and can perform only three operations: *LSADD*, *MADD* and *NOP*. The definition of these three operations can be found in Table 2. The output of *ALU* will be fed to the *PG* stage via a MUX which selects the ALU data or the data from the *ID* stage, depending on if the *ALU* result or the *ID* stage result is needed for the processing which follows.

The *PG* stage generates the IQ components phase by the *Phase Generation* submodule. The reconfigurable BPSK and QPSK modulation mappings are achieved by a LUT which gives the output phase based on the input signal. The LUT content can be modified for a different modulation scheme, by setting *static control signals*, thus allowing BPSK, QPSK modulations to be supported. Additional flip-flops and XOR operators are added to support the differential BPSK (DBPSK) and QPSK (DQPSK) modulations. For certain modulation schemes, such as 16QAM in WiMAX, the coding for the same modulation scheme will be interchanged between even and odd IQ symbols. To support this feature, two LUTs are implemented to provide two IQ symbols simultaneously. A selection signal, associated with symbol counter, chooses the correct IQ symbol for the output. Moreover, if the *ALU*'s output is for the intermediate result, the *Phase Generation* submodule will be skipped and the *ALU*'s output will be connected to the next stage directly via a MUX.

The final stage is the *IQ* stage which maps the generated phase to the output IQ symbols. Note that the data path is only 2-bit wide, and the width is not sufficient to generate a symbol which requires more than two bits for mapping at a single time instance (e.g., when four input bits are required for 16QAM). To overcome this issue, a buffer is inserted in the *IQ* stage to store the input bits until all bits are ready for IQ symbol generation. All the bits for each IQ symbol mapping, referred to as *input data unit*, are fed into *IQ Generation* submodule. Another LUT in *IQ Generation*, to define the amplitude and the final phase for each *input data unit*, is utilized to generate the final IQ symbol. Similar to the interleaved mapping in *PG*, the LUT mapping in *IQ Generation* can be modified for even/odd symbols, such as the DCM modulation in UWB. Thus, two LUTs are implemented. Moreover, if the input data of the *IQ* stage is for the intermediate result instead of IQ symbol mapping, this result will be transmitted to the *ID* stage, to be stored in *RF*.

In contrast to a traditional processor, MAPro is not only controlled by instructions, but also the input *static control signals* from outside of MAPro providing the control parameters for modulation mapping. Another difference between a traditional processor and MAPro is the data memory accessing method. As the input data of MAPro are the output of the stage before the Modulation Mapping in Figure 2, a buffer (which is outside of MAPro) is created to hold a block of processing data and it feeds these data into MAPro via *packet data bus*. Then the *Bits Select* submodule in *ID* can correctly select the processing bits with the activation of the *BSEL* instruction. The output of MAPro will be fed the stage after the Modulation Mapping directly. Thus, data memory is not used in MAPro.

After all these optimizations on the MAP CPU, it can support all IQ mapping methods discussed in this section maintaining the throughputs defined by the individual protocol specifications. The synthesized size of MAP CPU is only 2K Gates (excluding the instruction RAM) and very suitable for mobile applications.

938

Accelerators for Biologically-Inspired Attention and Recognition

Mi Sun Park[*], Chuanjun Zhang[†], Michael DeBole[‡], Srinidhi Kestur[*],
Vijaykrishnan Narayanan[*], Mary Jane Irwin[*]

[*]The Pennsylvania State University	[†]Intel Corporation	[‡]IBM Corporation
University Park, PA 16802, USA	Pittsburgh, PA 15213, USA	Poughkeepsie, NY 12601, USA
{mup183, kesturvy, vijay, mji}@cse.psu.edu	chuanjun.zhang@intel.com	mvdebole@us.ibm.com

ABSTRACT

Video and image content has begun to play a growing role in many applications, ranging from video games to autonomous self-driving vehicles. In this paper, we present accelerators for gist-based scene recognition, saliency-based attention, and HMAX-based object recognition that have multiple uses and are based on the current understanding of the vision systems found in the visual cortex of the mammalian brain. By integrating them into a two-level hierarchical system, we improve recognition accuracy and reduce computational time.

Results of our accelerator prototype on a multi-FPGA system show real-time performance and high recognition accuracy with large speedups over existing CPU, GPU and FPGA implementations.

Categories and Subject Descriptors

C.3 [**Special Purpose And Application-Based Systems**]: Signal processing systems

General Terms

Design, Experimentation, Performance

Keywords

FPGA prototyping, Hardware Acceleration, Object and Scene Recognition, Visual Attention

1. INTRODUCTION

The human brain has a remarkable ability to obtain high-level visual information and subsequently use that information to extract complex visual relationships between objects and their surroundings. While the exact biological processes that allow the brain to possess such function are the subject of numerous studies, neuroscientists have made progress towards understanding the brain's cognitive abilities and have successfully been able to produce models of cognitive function which are highly in agreement with experimental data [17] [9] [23] [6]. In particular, this paper focuses on three categories of models pertaining to scene understanding (gist)[12], visual-attention (saliency)[21], and object identification (hierarchical models and X, HMAX) [20].

All three categories play an important role in modelling the brain's ability to understand a scene and piece together a complete picture. As a result, it is also important to understand how information collected from one area may influence that of another, rather than treating them in isolation. For instance, the gist model is a biologically-inspired scene recognition that extracts summary statistics over an entire image [21]. Alternatively, objects themselves can be found through the saliency visual attention model [12], which attempts to automatically identify regions within scenes which are visually perceived to be foreground objects. These detected objects may then be identified via HMAX, a feed-forward hierarchical object recognition model. During object recognition, scene understanding can be used to tune a priori probabilities to help improve recognition accuracy for objects.

One key area of interest has been the ability to execute these biologically-inspired algorithms efficiently in hardware, specifically for use in environments that have energy and power constraints. For example, these architectures are pertinent to analyzing video streams in real-time for artificial vision applications including mobile augmented reality, autonomous vehicle navigation, remote elder care monitoring, mobile robotics, and security. Heterogeneous System on Chips (SoC) are an attractive platform for these types of applications as they are comprised of both general purpose cores and domain-specific accelerators. For example, the recent introduction of the FPGA-Atom based platform integrates Atom cores with an on-chip FPGA providing the capability to run control oriented tasks on a low-power CPU and computationally dependent tasks on application specific hardware.

This paper presents two real-time hardware accelerators. The first configurable gist-saliency accelerator can be dynamically configured to support two key video analytics algorithms which extract the gist of the scene and identify visual saliency. The second reconfigurable HMAX accelerator is used to classify objects. We integrate these accelerators into a two-level hierarchical accelerator system to improve recognition accuracy and reduce computations by focusing on only the interesting areas instead of using an exhaustive search in a scene with scene awareness (understanding).

This paper provides the following two major contributions:

- A run-time configurable architecture is proposed for supporting both gist-based scene recognition and saliency-based visual attention - this includes the first hardware implementation of the gist model.

- An integrated two-level hierarchical accelerator system is proposed for fine-grained recognition with improvement in accuracy.

Permission to make digital or hard copies of all or part of this work for personal or classroom use is granted without fee provided that copies are not made or distributed for profit or commercial advantage and that copies bear this notice and the full citation on the first page. To copy otherwise, to republish, to post on servers or to redistribute to lists, requires prior specific permission and/or a fee.
DAC '13 May 29 - June 07 2013, Austin, TX, USA

We have developed the accelerator prototype on a Dinigroup multi-FPGA platform and experimental results demonstrate large speedups over other contemporary solutions on different platforms.

2. RELATED WORK

A variety of hardware accelerators for biologically-inspired algorithms have been developed. Visual attention algorithms such as Attention based on Information Maximization (AIM) [8] and a bottom-up saliency model [7] have also been accelerated using FPGAs. The AIM algorithm defines the saliency of visual content as the measure of local information within a scene. The intent is to maximize information of sampled visual-attentions by identifying regions which are visually perceived to be unique from the background.

There have been several HMAX accelerators presented in [10] [19] [18]. The main focus of these papers is to design a high performance run-time reconfigurable accelerator for the computationally intensive S2 stage in the HMAX model. Recent efforts on accelerating a HMAX-based neuromorphic system for universal recognition was presented in [14]. Acceleration frameworks for biologically plausible spiking neural networks [22] and convolution neural networks for synthetic vision systems [11] have been developed.

Recently, closely related work was presented in [13], which presents a FPGA framework for an end-to-end attention and recognition system using saliency and HMAX accelerators. However, there is no prior work on hardware acceleration of the gist-based scene recognition algorithm. Our work presents the first hardware implementation of gist and further use of scene understanding from the gist accelerator as a cue to improve recognition accuracy.

3. MODELS OF NEUROMORPHIC VISION

In this section, we describe three widely accepted biologically-inspired attention and recognition models.

3.1 Gist-Saliency Models for Scene Recognition and Attention

The gist recognition model and saliency attention model share the significant portion of computational stages as shown in Figure 1. These models begin by extracting feature maps from orientation, color and intensity channels at multiple spatial scales ($1 : 2^0$ to $1 : 2^8$) from an input color image. The initial image pyramid is generated by interpolating each subsequent scale using a 5x5 Gaussian filter. Within the orientation and color channels, there are sub-channels which are computed to extract features specific to orientation ($\theta = 0$, 45, 90, and 135°) and color combinations (Red-Green and Blue-Yellow), respectively. The color and intensity feature maps (and orientation feature maps for saliency), $M(c, s)$, are obtained by computing a center-surround difference (CSD) with each sub-channel. As shown in Equation 1,

$$M_i(c,s) = |O_i(c) - Interp_{s-c}(O_i(s))| \qquad (1)$$

where $O(c)$ is a Gaussian filter output at scale c and $i = 6$, the CSD is found by computing the difference between a center (fine) scale c and a surround (coarse) scale s, where $c = 2, 3, 4$ and $s = c + d$, with $d = 3, 4$. This is done by interpolating the surround to the finer scale and then performing a point by point subtraction.

The orientation feature maps for gist, $M(c)$, are computed by a set of Gabor filters (at each θ) applied to the grayscale image representation at four scales ($c = 0, 1, 2, 3$) with $i = 4$ as shown in equation 2.

$$M_i(c) = Gabor(\theta_i, c) \qquad (2)$$

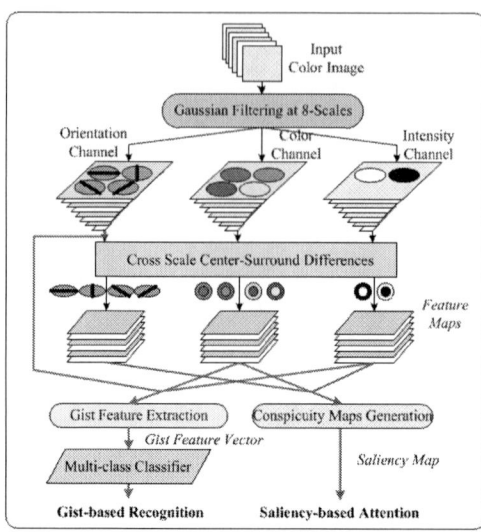

Figure 1: Gist shared with the Saliency-based Attention [21] [12]

After the feature maps are computed, the gist model extracts a partial feature vector by finding the mean within subregions defined by a 4x4 grid over each of 34 feature maps (4x4 for orientation, 2x6 for color and 6 for intensity). Each of the partial feature vectors contains 16 values and are then combined to form a 544 gist-feature vector, which is passed to a classifier for scene classification. The saliency model, on the other hand, generates three conspicuity maps for the three channels by normalizing the each of 42 feature maps (4x6 for orientation, 2x6 for color and 6 for intensity) and computing a weighted sum across scales at scale $c = 4$. The three conspicuity maps are normalized and summed to form a saliency map. For an in-depth treatment of Gist and Saliency, see [21] [12].

3.2 HMAX Model for Object Recognition

The HMAX recognition model consists of four stages of S1 (Gabor filter), C1 (Local pooling), S2 (Template matching) and C2 (Global pooling). These stages extract a feature vector and send it to a classifier for object classification.

The S1 stage performs a convolution on an image pyramid of 12 scales from an input grayscale image with 2D Gabor filters at multiple orientations at different scales. The C1 stage is computed by convolving the S1 pyramids with a 3D Max filter at every 10x10 units in position and 2 units deep in scale to obtain scale and position invariance [16][20].

The S2 stage extracts intermediate features by performing template matching at every position and scale of the C1 pyramids with 4075 *sparse* prototype patches. The patches are of size n x n, where $n \in \{4, 8, 12, 16\}$, and each coefficient in the patch has its preferred orientation. The response of the C1 pyramids X to these prototype patches P is computed by the Gaussian Radial Basis Function (GRBF),

$$R(X, P) = exp(-\frac{\| X - P \|^2}{2\sigma^2 \alpha}) \qquad (3)$$

where $\sigma = 1$ and $\alpha = (n/4)^2$. The S2 stage consumes more than 90% of execution time of HMAX software implementation running on CPU [10].

The C2 stage computes a global max across scales for each of the 4075 S2 pyramids to obtain a 4075 C2-feature vector, each element of which is the maximum response to a given patch. For an in-depth treatment of HMAX, see [16].

4. HARDWARE DESIGN

In this section, we describe the hardware design of our two accelerators. The run-time configurable gist-saliency accelerator is designed for scene recognition and attention, while the reconfigurable HMAX accelerator is for object recognition.

4.1 Configurable Architecture of Gist-Saliency Accelerator

The proposed configurable architecture of a gist-saliency accelerator pipeline depicted in Figure 2 has three parts: (i) The configurable common blocks for gist and saliency, (ii) Gist-dedicated blocks and (iii) Saliency-dedicated blocks. The common blocks include image decomposition into I, RG and BY components, Gaussian and Laplacian pyramids generation using successive 2D convolutions and down-sampling, separable steerable filters and center-surround difference modules. Each of these modules can be dynamically configured using domain-specific instructions which have feature-specific control information for 7 different channel computations (4 for orientation, 2 for color and 1 for intensity). Each channel is computed one at a time by executing the pipeline, therefore it requires 7 iterations to complete the gist or saliency computation if using one pipeline.

Although the same pipeline is used for both gist and saliency computation, the data flow and working set of data (due to different algorithmic parameters) are different. The different data flow sequences for the various channels of gist and saliency and algorithmic parameters used in the common blocks are shown in Figure 3.

4.1.1 Gist-dedicated Blocks

The gist-dedicated blocks include subregion mean operation and feature concatenation modules. The subregion mean operator is used to calculate the mean of a 2D n x n grid over each feature map. Up to 6 different scales of parallel input streams from the previous CSD module requires multiple, parallel subregion mean computations. Therefore, the mean operator instruction format has fields to indicate the desired size of the grid, markers for indicating the start or end of a stream from a different scale or different pipeline, and the size of the incoming image. For example, in our gist configuration, we compute the mean values for 16 subregions in each of the 2D feature maps. Each subregion is a 2D grid whose size varies for each channel and each scale within a channel. Finally, the feature concatenator appends thirty four extracted 16-partial feature vectors to generate a final 544 gist feature vector.

In general, 96 mean operators (adders, shifters and buffers) are required for the mean of 16 subregions of each of 6 parallel feature maps. However, since the feature maps are streaming into the mean operation modules in a raster-scan pattern, we only use 24 mean operators by extracting the mean of each 4 row subregions of 4x4 subregions sequentially. This optimization improves the resource utilization of the gist-dedicated blocks by a factor of 4.

4.1.2 Saliency-dedicated Blocks

There are maximum normalization, across-scale addition and accumulation modules in the saliency-dedicated blocks. The maximum normalizer and across-scale adder are used to normalize each of 42 feature maps and compute across-scale addition to generate three conspicuity maps at scale 4. Then the accumulator sums the conspicuity maps across the three channels to form a saliency map, where the maximum value of the saliency map indicates the most salient image region.

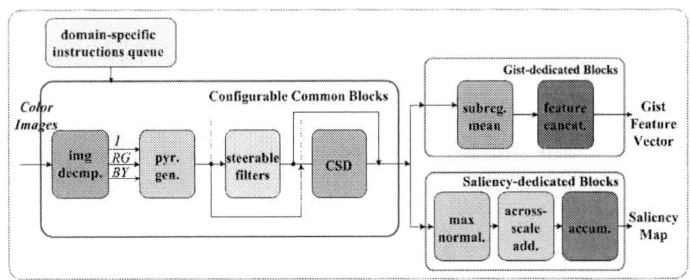

Figure 2: Configurable Gist-Saliency Pipeline

	Pyr. Gen.		Steerable filter	CSD	GIST	SALIENCY
	Gauss.	Lapl.				
O-Gist					img. decmp.	$I = \frac{r+g+b}{3}$ $R = r - \frac{g+b}{2}$ $G = g - \frac{r+b}{2}$ $B = b - \frac{r+g}{2}$ $Y = r + g - 2(\|r-g\| + b)$	$I = \frac{r+g+b}{3}$ $R = r - \frac{g+b}{2}$ $G = g - \frac{r+b}{2}$ $B = b - \frac{r+g}{2}$ $Y = \frac{r+g}{2} - \frac{\|r-g\|}{2} - b$
C-Gist							
I-Gist					pyr. gen.	Gauss. Pyr. for CI: I(σ), RG(σ), BY(σ) Laplacian Pyr. for O: O(σ, θ) $\sigma = [0...8], \theta = \{0°, 45°, 90°, 135°\}$ 5x5 Gaussian filter	Gauss. Pyr. for CI: I(σ), RG(σ), BY(σ) Laplacian Pyr. for O: O(σ, θ) $\sigma = [0...8], \theta = \{0°, 45°, 90°, 135°\}$ 5x5 Guassian filter
O-Sal.							
C-Sal.					CSD	$s = c + d$ $c \in \{2, 3, 4\}, d \in \{3, 4\}$ **6 csd-scales for CI:** $\{(2,5), (2,6), (3,6), (3,7), (4,7), (4,8)\}$ **4 c-scales for O:** $\{0,1,2,3\}$	$s = c + d$ $c \in \{2, 3, 4\}, d \in \{3, 4\}$ **6 csd scales for CIO:** $\{(2,5), (2,6), (3,6), (3,7), (4,7), (4,8)\}$
I-Sal.							

Figure 3: Flows and Parameters for Gist and Saliency in the Common Blocks

4.2 Reconfigurable Architecture of HMAX Accelerator

Since the S2 stage of the HMAX model is found to be the most time consuming [10], we focus on accelerating the critical S2 stage with succeeding C2 stage using hardware. We implemented our HMAX S2C2 accelerator by leveraging the state-of-the-art HMAX S2 architectures [19] [18]. Our accelerator design is based on a 2D systolic primitive and can be dynamically reconfigured to compute convolutions of different patch sizes of n x n, where $n \in \{4, 8, 12, 16\}$.

We use sixteen 4x4 primitives to form a reconfigurable S2C2 accelerator, along with C2 module for across-scale max operation and two memory subsystems for buffering C1 image pyramids (4 pyramids of 11 scales) and 4075 patches. The number of primitives used in the reconfigurable accelerator is determined by the biggest patch size, while the primitive size is determined by the smallest patch size used in the GRBF computation. Therefore, our accelerator can be run-time reconfigured to perform one 16x16 convolution, one 12x12 convolution, four 8x8 convolution or sixteen 4x4 convolution per cycle.

4.3 System Integration

Recognition systems without front-end attention require an exhaustive search over the entire image, especially for a high resolution broad-area image, where a target object is too small, or is not placed at the center of the image. This is computationally expensive and inefficient. Furthermore, since classification accuracy is inversely correlated to the number of classes, it is hard to obtain optimum accuracy only using one recognition system. Therefore, we present a two-level hierarchical accelerator system by integrating our bio-inspired attention and recognition accelerators, using the configurable gist-saliency accelerator as first-level processing and then using the HMAX accelerator as second-level processing for fine-grained recognition.

As shown in Figure 4, we compute both the saliency and gist features in parallel to use the gist for scene understanding ("place") and the saliency for finding a region of interest (ROI). When the gist feature vector and saliency map are generated from the hardware, the gist feature vector is used for scene classification. The HMAX accelerator performs recognition on the pre-identified ROI from the saliency map to generate a C2 feature vector. The extracted C2 feature vector is then passed for object classification using a Regularized Least-Squares (RLS) classifier.

After we obtain raw outputs from the object classifier, we rank them in descending order of probability and select top 3 candidate object classes. Then we choose the highest ranking object class which belongs to the gist predicted scene category, if found in the candidate classes. Otherwise, it classifies as the first class from the candidates which has the highest probability. Our experimental results show from 9% to 14% improvement in recognition accuracy.

In summary, our hierarchical system is (a) computation-efficient by only computing on the interesting area of a visual scene instead of exhaustively searching, and (b) accuracy-enhancing by utilizing visual scene understanding as a cue to drive better recognition.

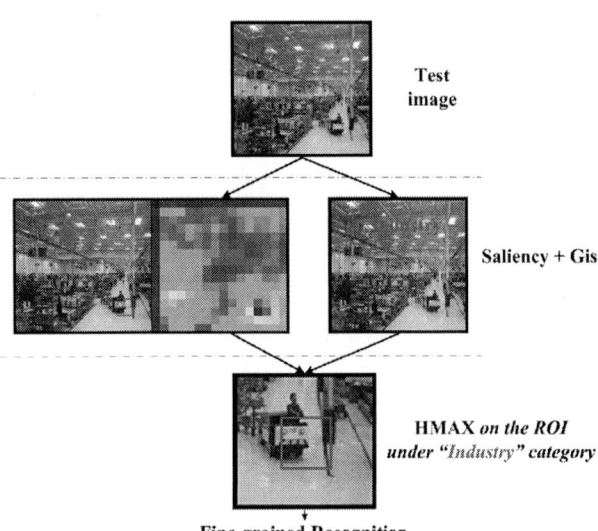

Figure 4: Two-level Hierarchical Accelerator System for Fine-grained Recognition - Saliency and gist results on a test image, from the Stanford Scenes dataset [4], are shown. The result from the saliency shows the most salient location on the image labeled with a red-box along with the corresponding saliency map. The gist result shows the predicted scene category labeled on the test image.

5. EXPERIMENTAL RESULTS

The Dinigroup DNV6F6-PCIe multi-FPGA platform [3] was selected for prototyping our accelerators for biologically-inspired attention and recognition algorithms. The board is comprised of six compute FPGAs, each one of which are Xilinx Virtex6 SX475T devices and operates at 100MHz. Host communication is provided via a PCIe x8 link between one of the compute FPGAs and a software system built via the DNSEAM-PCIe [2].

Input images are provided directly to the accelerators either through video files or live video data from a camera with

minimal pre-processing. The accelerators use a 24-bit fixed-point numerical representation which provides real-time performance with an error rate of $1 \sim 3 \times 10^{-5}$, compared to floating-point MATLAB software references. For power consumption comparison, we measure the board power with a power meter, while we use Thermal Design Power (TDP) numbers for the reference platforms.

In order to support a multi-FPGA prototype of our accelerators, we use two FPGAs each for gist-based scene recognition and HMAX-based object recognition, one for saliency-based visual attention and the other FPGA as network interface. By fully utilizing the multi-FPGA platform, we support a real-time neuromorphic system to process visual information similarly to the visual cortex.

5.1 Gist-saliency Accelerator

Our gist-saliency accelerator can be dynamically configured at run-time to support both gist-based scene recognition and saliency-based attention by sharing the configurable logic blocks, therefore, we provide the resource utilization for the common, gist-dedicated, saliency-dedicated blocks and overall gist-saliency pipeline in Table 1. The common blocks are RAM intensive, since input images are buffered on-chip for the image decomposition. Gist-dedicated blocks consume little resources since the subregion mean operation module is optimized by a factor of 4, while saliency-dedicated blocks need large size of logic and RAM for normalizing each feature map, which is required to store incoming 6 feature maps in parallel.

Table 1: Resource Utilization of (one pipeline) Gist-Saliency Accelerator on Virtex6 SX475T

	Slice Regs	Slice LUTs	BRAMs	DSP48s
Common Blocks	72,760 (12%)	85,998 (29%)	458 (43%)	376 (19%)
Gist-dedicated	7,018 (1%)	8,554 (3%)	12 (1%)	0 (0%)
Saliency-dedicated	67,186 (11%)	62,440 (21%)	159 (15 %)	316 (16%)
Overall Pipeline	146,973 (25%)	156,992 (53%)	629 (59 %)	692 (34%)

5.1.1 Gist-based Scene Recognition

Figure 5 shows the performance results with degree of parallelism (number of FPGAs) for different frame sizes. The performance scales linearly with both frame size and the number of utilized FPGAs. The accelerator architecture on each FPGA is identical and only different custom instructions are loaded into pipelines on multiple FPGAs. This graph shows that our design is scalable and using more resources implies a higher degree of parallelism from the accelerator.

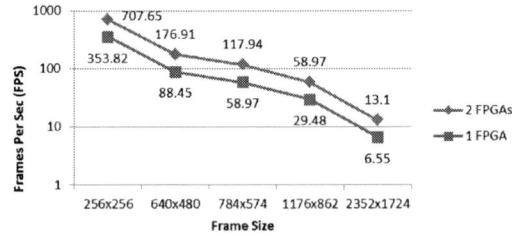

Figure 5: Performance Scaling for Different Frame Size

The result from the gist system for scene classification is shown in Figure 6. The gist feature vector from the gist accelerator is used by a RLS classifier to classify images from Stanford scenes dataset [4]. The gist system is tested with 6 different classes (beaches, buildings, forests, highways, industry and mountains). 30 random images from the scene dataset are used for training and 30 non-overlapping images are used for testing from each class. As shown in Table 2, the classification accuracy is higher with fewer classes, since it becomes easier for the classifier to distinguish a single class from the other classes using one-vs-all strategy.

Figure 6: Scene Recognition Result on Stanford Scenes dataset [4].

Table 2: Scene Classification Accuracy

# of classes	# of test images	# of correct images	Accuracy
3	90	75	83.3 %
4	120	99	82.5 %
5	150	121	80.7 %
6	180	145	80.6 %

Table 3 shows trade-off between performance and classification accuracy for two different versions of the gist computation. The version for color images (CIO-gist) computes all three channels, while the version for grayscale images (IO-gist) computes only intensity and orientation channels to extract the gist feature vector. The IO-gist provides 1.3X speedup over the CIO-gist for an image size of 800x600 but underperforms in terms of classification accuracy. The reason for the IO-gist speedup is that two iterations of color channel computations are avoided. If dedicated color channel pipelines exist, it can be reconfigured for other channel computations. Due to the configurability of the accelerator, the version of gist computed is based solely on preference (FPS or Accuracy) without changing the hardware implementation.

Table 3: GIST Computation of (800x600) Color and Grayscale Images

	FramesPerSec (FPS)	Accuracy			
		3 classes	4 classes	5 classes	6 classes
CIO-gist	117.9	83.3 %	82.5 %	80.7 %	80.6 %
IO-gist	156.5	81.1 %	76.7 %	75.3 %	67.2 %

In Table 4, we compare the performance of our accelerator with existing CPU implementations for gist computation for 800x600 color images. Our gist accelerator is prototyped on two Virtex6 FPGAs and shows 38X and 393X speedups and 188X and 30X higher performance-per-Watt (FPS-per-Watt) over Intel high-performance Core i7-3960x and ultra low-power Atom platforms respectively.

5.1.2 *Saliency-based Visual Attention*

Table 5 shows the attention performance comparison between our accelerator and existing GPU and FPGA implementations [24][13] for 640x480 color images. Our accelerator executes on one Virtex6 FPGA and allows speedups of 2X and 1.1X over the GPU and FPGA implementations.

The saliency accelerator provides better throughput compared to the gist accelerator's throughput, because the gist model computes on bigger spatial scales from the same input image for orientation channel computation.

5.2 HMAX Accelerator

Table 4: Comparison of Gist Implementations (800x600)

	CPU [5]	CPU [5]	Our FPGA
	Intel Core i7-3960x	Intel Atom	2 x Virtex6 SX475T
Frequency	3.30 GHz	1.30 GHz	100 MHz
Precision	floating-point	floating-point	fixed-point (24bit)
Power	130 W	2 W	26 W
Frame Rate	3.07 fps	0.30 fps	117.94 fps
FPS-per-Watt	0.024	0.15	4.53

Table 5: Comparison of Saliency Implementations (640x480)

	GPU [24]	FPGA [13]	Our FPGA
	Geforce 8800GTX	Virtex6 SX475T	Virtex6 SX475T
Frequency	1.35 GHz	100 MHz	100 MHz
Precision	floating-point	fixed-point	fixed-point (24bit)
Power	155 W	12.6 W	13 W
Frame Rate	94.3 fps	169.5 fps	193.7 fps
FPS-per-Watt	0.61	13.45	14.9

In Table 6, we compare the performance of our hardware accelerator with existing GPU and FPGA implementations [15][19] for HMAX S2C2 computation for 256x256 grayscale images. Our accelerator implemented based on [18] runs on two Virtex6 FPGAs and obtains speedups of 31X and 2X over the GPU and FPGA platforms respectively.

Table 6: Comparison of Hmax Implementations (256x256)

	GPU [15]	FPGA [19]	Our FPGA
	Telsla C1060	2 x Virtex6 SX475T	2 x Virtex6 SX475T
Frequency	1.3 GHz	100 MHz	100 MHz
Precision	floating-point	fixed-point	fixed-point (24bit)
Power	187.8 W	65 W	21.4 W
Frame Rate	2.88 fps	45.84 fps	90.50 fps
FPS-per-Watt	0.015	0.705	4.228

5.2.1 *Recognition Accuracy*

The results from our system on images from Caltech 101 object dataset [1] for multi-class object classification is shown in Figure 7. The C2 feature vector from the HMAX accelerator is used by a RLS classifier to classify images into 12 classes.

Table 7 shows improvement in object recognition accuracy from scene understanding, which is learned from the first-level gist-based scene recognition processing in our hierarchical system. Classification probability based ranking mechanism is applied to select the highest ranking object belonging to the gist predicted scene category, if found in the top 3 candidate classes. For example, if we have ferry, car and ketch as the candidate classes in an order with the predicted scene category as highways, then our system classifies it as a car, instead of a ferry. Our hierarchical system has been validated with 20 images from each object class from up to 12 different classes. The results show 9% to 14% improvement in accuracy over HMAX-only recognition system and the throughput of our hierarchical system is determined by the HMAX system.

6. CONCLUSION

As video and image content becomes increasingly prevalent, there will be a profound increase in the number of applications which attempt to leverage this information in a variety of ways. A common trend of these applications is to increase the amount of automation where scene understanding plays a growing role. We presented and integrated a two-level hierarchical accelerator system for fine-grained recognition, which improves accuracy from 9% to 14%. Results show that our accelerators provide

Figure 7: Object Recognition Result on Caltech 101 Object dataset[1].

Table 7: Object Classification Accuracy Improvement from scene understanding

# of scene categories	# of object classes per category	HMAX-only Accuracy	Gist + HMAX New Accuracy
4	3	69.17 %	77.92 %
	2	77.50 %	84.38 %
3	3	72.22 %	82.22 %
	2	83.33 %	92.50 %

Scene Category	Object Class
highways	car
	stop sign
	motorbikes
beaches	ketch
	ferry
	crab
forests	panda
	leopards
	beaver
buildings	cup
	laptop
	chair

real-time performance, energy efficiency and from 1.1X to 393X speedups over existing CPU, GPU and FPGA implementations, which are significant factors for embedded systems.

7. ACKNOWLEDGMENT

This work was supported in part by NSF Awards 0916887, 1205618, 1213052, 1147388 and the Intel Science and Technology Center on Embedded Computing.

8. REFERENCES

[1] Caltech 101 Database for Object Classification. http://www.vision.caltech.edu/Image_Datasets/Caltech101/.

[2] Dinigroup DNSEAM-PCIE. http://www.dinigroup.com/new/DNSEAM_PCIE.php.

[3] Dinigroup DNV6F6-PCIE Documentation. http://www.dinigroup.com/product/data/DNV6F6PCIe/files/DNV6F6PCIe_v14_lo.pdf.

[4] Stanford Dataset for Scene Classification. http://vision.stanford.edu/fmriscenes/resources.html.

[5] USC iLab for GIST C++ Implementation. http://ilab.usc.edu/toolkit/documentation.shtml.

[6] C. Ackerman and L. Itti. Robot steering with spectral image information. *Robotics, IEEE Transactions on*, 21(2):247 – 251, april 2005.

[7] P. Akselrod, F. Zhao, I. Derekli, C. Farabet, B. Martini, Y. LeCun, and E. Culurciello. Hardware accelerated visual attention algorithm. In *Information Sciences and Systems (CISS), 2011 45th Annual Conference on*, pages 1 –6, march 2011.

[8] S. Bae, Y. Cho, S. Park, K. M. Irick, Y. Jin, and V. Narayanan. An FPGA implementation of information theoretic visual-saliency system and its optimization. In *Intl. Symp. on Field Programmable Custom Computing Machines*, FCCM, pages 41–48, 2011.

[9] I. Biederman. Do background depth gradients facilitate object identification. *Perception*, 10:573–578, 1982.

[10] M. DeBole, A. Maashri, M. Cotter, C.-L. Yu, C. Chakrabarti, and V. Narayanan. A Framework for Accelerating Neuromorphic-Vision Algorithms on FPGAs. In *Computer-Aided Design (ICCAD), 2011 IEEE/ACM International Conference on*, nov. 2011.

[11] C. Farabet, B. Martini, P. Akselrod, S. Talay, Y. LeCun, and E. Culurciello. Hardware accelerated convolutional neural networks for synthetic vision systems. pages 257 –260, may. 2010.

[12] L. Itti, C. Koch, and E. Niebur. A model of saliency-based visual attention for rapid scene analysis. *Pattern Analysis and Machine Intelligence, IEEE Transactions on*, 20(11):1254 –1259, nov. 1998.

[13] S. Kestur, M. Park, J. Sabarad, D. Dantara, V. Narayanan, Y. Chen, and D. Khosla. Emulating Mammalian Vision on Reconfigurable Hardware. In *Intl. Symp. on Field Programmable Custom Computing Machines FCCM'12*, May 2012.

[14] A. Maashri, M. DeBole, M. Cotter, N. Chandramoorthy, Y. Xiao, V. Narayanan, and C. Chakrabarti. Accelerating neuromorphic vision algorithms for recognition. In *Design Automation Conference (DAC), 2012 49th ACM/EDAC/IEEE*, pages 579 –584, june 2012.

[15] J. Mutch, U. Knoblich, and T. Poggio. CNS: a GPU-based framework for simulating cortically-organized networks. Technical Report MIT-CSAIL-TR-2010-013 / CBCL-286, Massachusetts Institute of Technology, Cambridge, MA, February 2010.

[16] J. Mutch and D. G. Lowe. Object class recognition and localization using sparse features with limited receptive fields. *Intl. J. Comput. Vision*, 80:45–57, October 2008.

[17] A. Oliva and P. Schyns. Coarse blobs or fine edges? evidence that information diagnosticity changes the perception of complex visual stimuli. *Cognit Psychol*, 34(1):72–107, 1997.

[18] M. Park, S. Kestur, J. Sabarad, V. Narayanan, and M. Irwin. An FPGA-based Accelerator for Cortical Object Classification. In *Proc. of Design Automation and Test Conference and Exhibition DATE'12*, Mar 2012.

[19] J. Sabarad, S. Kestur, M. Park, D. Dantara, V. Narayanan, Y. Chen, and D. Khosla. A Reconfigurable Accelerator for Neuromorphic Object Recognition. In *Proc. of Asia South Pacific Design Automation Conference ASPDAC'12*, Jan 2012.

[20] T. Serre, L. Wolf, S. Bileschi, M. Riesenhuber, and T. Poggio. Robust object recognition with cortex-like mechanisms. *Pattern Analysis and Machine Intelligence, IEEE Tran on*, 29(3):411 –426, march 2007.

[21] C. Siagian and L. Itti. Rapid biologically-inspired scene classification using features shared with visual attention. *Pattern Analysis and Machine Intelligence, IEEE Transactions on*, 29(2):300 –312, feb. 2007.

[22] D. Thomas and W. Luk. Fpga accelerated simulation of biologically plausible spiking neural networks. In *Field Programmable Custom Computing Machines, 2009. FCCM '09. 17th IEEE Symposium on*, pages 45 –52, april 2009.

[23] A. Torralba. Modeling global scene factors in attention. *JOSA - A*, 20:1407–1418, 2003.

[24] T. Xu, T. Pototschnig, K. Kühnlenz, and M. Buss. A high-speed multi-gpu implementation of bottom-up attention using cuda. In *ICRA'09: Proceedings of the 2009 IEEE international conference on Robotics and Automation*, pages 1120–1126, Piscataway, NJ, USA, 2009. IEEE Press.

Stochastic Circuits for Real-Time Image-Processing Applications

Armin Alaghi, Cheng Li and John P. Hayes
Advanced Computer Architecture Laboratory
Department of Electrical Engineering and Computer Science
University of Michigan, Ann Arbor, MI, 48109, USA
{alaghi, elfchris, jhayes}@ umich.edu

ABSTRACT

Real-time image-processing applications impose severe design constraints in terms of area and power. Examples of interest include retinal implants for vision restoration and on-the-fly feature extraction. This work addresses the design of image-processing circuits using stochastic computing techniques. We show how stochastic circuits can be integrated at the pixel level with image sensors, thus supporting efficient real-time (pre)processing of images. We present the design of several representative circuits, which demonstrate that stochastic designs can be significantly smaller, faster, more power-efficient, and more noise-tolerant than conventional ones. Furthermore, the stochastic designs naturally produce images with progressive quality improvement.

Categories and Subject Descriptors

B.2 Arithmetic and Logic Structures, B.6 Logic Design, C.3 Special-Purpose and Application-Based Systems.

General Terms

Design.

Keywords

Emerging Technologies, Image Processing, Real-Time Computing, Stochastic Computing, Vision Chips.

1. INTRODUCTION

Advances in semiconductor technology have enabled many exciting new applications of embedded computers. They have also exposed problems and opportunities that cannot be easily addressed using conventional design approaches. An example that motivates our work is the provision of retinal implants for the visually impaired [15]. This involves designing an integrated circuit (IC) chip that can be surgically placed on a dysfunctional retina to sense images (or process images sent wirelessly from an external camera) and convert an array of pixel streams to streams of neural-style electrical signals that stimulate useful visual sensations. The implanted IC is linked to an external power supply and must not dissipate more than a few mW/mm^2 to avoid heat damage to the eye [17].

Permission to make digital or hard copies of all or part of this work for personal or classroom use is granted without fee provided that copies are not made or distributed for profit or commercial advantage and that copies bear this notice and the full citation on the first page. To copy otherwise, or republish, to post on servers or to redistribute to lists, requires prior specific permission and/or a fee.
DAC'13, May 29–June 07, 2013, Austin, TX, USA.

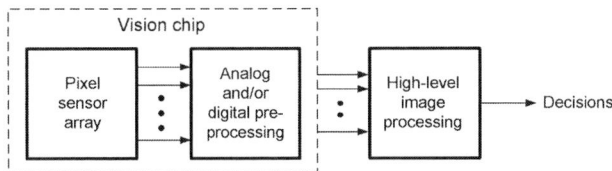

Figure 1. Image-processing system employing a vision chip.

Due to the huge amounts of data in pixel streams, real-time image processing usually requires extensive hardware and/or software resources [8]. If the hardware support is sufficiently small, some of it can be integrated with the imaging-sensing circuits to form a so-called "vision chip," as indicated in Figure 1. Such chips serve as the preprocessing front end of an image-processing system [4] [14].

Vision chips are loosely classified as analog or digital (pulse domain), depending on the type of circuitry used in the preprocessing stage to convert the sensed analog input signals to digital form for final processing. Typical preprocessing circuits are analog-to-digital converters (ADCs), noise filters, and edge detectors [3] [8] [11]. These steps may require many operations per pixel, and consume most of the power of the system [21]. The design of vision chips is very challenging since it involves complex trade-offs among chip area, power, speed and accuracy. It also requires some degree of parallel processing, which can be at the level of individual pixels, groups of pixels, or the overall system [21].

We propose to use stochastic computing (SC) for real-time image preprocessing. This is a method of computing with bit-streams at very low hardware cost [2][7]. A stream of N bits containing N_1 1s and $N - N_1$ 0s denotes the stochastic number (SN) $x = N_1/N$, which is treated as a probability. For example, 0111, 1101, and 10110111 all denote $x = 3/4 = 0.75$. Stochastic circuits are small and so have very low power requirements. For example, multiplication of two N-bit SNs x_1 and x_2 to form the arithmetic product $x_1 \times x_2$ can be done in N clock cycles by means of a single AND gate. Besides low area and power, SC has the advantages of high error tolerance (bit flips have little impact on signal probability) and support for massive, low-level parallelism. Its main disadvantage is the need for very long bit-streams to achieve high precision. However, as we will show, this problem can be greatly mitigated by exploiting a special property of SC we call *progressive precision*, where result quality gets better as the computation proceeds.

SC and real-time image processing share some key properties. They both handle streaming analog data (image intensities or probabilities), process the data digitally, and have good noise tolerance. Several proposed vision chips encode the sensed light signals using pulse-frequency modulation (PFM) [9] [11] [20].

This means that pixel information is conveyed by the frequency of a pulse train, as in biological neural networks and SC circuits. SC thus has the potential to meet most of the challenging requirements of the retinal implant application mentioned earlier: streaming neural-style data, very small circuit size, extremely low power, and insensitivity to noise.

Although known for years [7], SC has only recently gained attention with the emergence of applications that can take full advantage of its unique features, such as support for massive parallelism. A notable example is decoding the low-density parity check (LDPC) codes employed in the IEEE WiFi and other communication standards [10]; this is an application that requires massive amounts of fast, but relatively simple, parallel processing. Naderi et al. [16] have used SC to implement an LDPC decoder chip that has performance comparable to conventional (weighted-binary) designs, but is significantly smaller. Li and Lilja [12] and Qian et al. [18] have shown that SC can outperform binary computing in some processing tasks involving stored images. Ma et al. [13] show that SC is useful in fault-tolerant image processing. To apply SC to stored images, data conversion between the weighted-binary and stochastic domains is necessary. This conversion is costly; in some cases, it can consume up to 80 percent of SC circuit area [18]. As we will show, real-time image processing avoids much of this cost. The use of SC in real-time vision chips was briefly discussed by Hammadou et al. in 2003 [9], but otherwise has received very little attention. SC application has also been hindered by lack of a general design methodology, and by limited understanding of its underlying theory.

This paper presents designs for various image-processing tasks like real-time edge detection and gamma correction. The new designs are evaluated by emulation on FPGAs under normal and noisy conditions. The results show that SC can provide high performance and huge savings in area and power. They also illustrate SC's progressive precision and noise-tolerance properties. The main contributions of the paper are:

1. Demonstration of the suitability of SC for real-time image processing with demanding design constraints.

2. Novel stochastic circuits that are far smaller and more efficient than both their conventional counterparts and (where they exist) previous SC designs.

3. Two orthogonal methods of producing images with progressive quality at minimal cost.

2. STOCHASTIC COMPUTING
We begin by reviewing stochastic computing and its relevant properties. As noted already, SC circuits tend to be very small. Figure 2a shows a stochastic adder [7] that implements the scaled addition $z = (x_1 + x_2)/2$. Scaling is needed to ensure that the sum lies in the interval [0,1]. This SC adder is just a 2-way multiplexer with a random bit-stream r of probability value 1/2 applied to its control (select) input.

A typical SN x has many different representations, and the ones chosen for the various inputs of an SC circuit usually need to be uncorrelated and derived from independent stochastic number generators (SNGs). Correlated inputs are generally believed to produce inaccuracies in the computation. A simple example is seen in the use of an AND gate to multiply SN x by itself to obtain x^2. If identical bit-streams are used for the two copies of x (implying maximum correlation), the AND gate produces x instead of x^2, a large error. So circuits with correlated inputs are rarely used in the

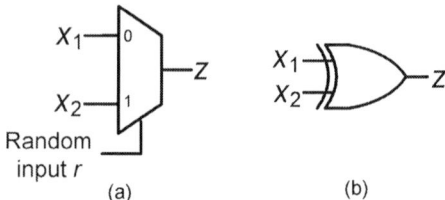

Figure 2. (a) Stochastic adder and (b) subtracter.

SC literature. However, we observe that correlation does *not* necessarily lead to inaccuracy, if properly managed. In particular, in circuits like the SC adder, inputs can be correlated without affecting the accuracy of the result. This can greatly mitigate the complications of random number generation.

In some cases, correlated inputs can change the functionality of a circuit to another, and perhaps more desirable, operation. Figure 2b shows an XOR gate that, assuming independent inputs, performs the function

$$z = x_1 \times (1 - x_2) + x_2 \times (1 - x_1)$$

However, when fed with correlated inputs where x_1 and x_2 have maximum overlap of 1s, we can show that the circuit computes $z = |x_1 - x_2|$, i.e., it acts as a type of subtracter. As we will see, this operation is very useful in SC image processing.

The major drawback of SC is that very long bit-streams are required for high-precision calculations. Precision is defined as the number of bits needed to represent a given number x in conventional weighted-binary format. An n-bit binary number maps to an SN of length of $N = 2^n$. Hence, the SN has precision of roughly $\log_2 N$, and takes $N = 2^n$ clock cycles to generate or process. This overhead makes SC impractical for most high-precision digital computations. About 8 bits of precision suffice for most image-processing tasks, implying a maximum SN length of $2^8 = 256$ bits—a reasonable size.

The speed of stochastic circuits can be increased by exploiting the *progressive precision* properties of SC: if properly chosen, the first few bits of an SN can yield a rough approximation to the final number. For instance, consider the following bit-stream representing the number 9/16:

$$0\ 1\ 1\ 0\ 1\ 1\ 0\ 0\ 1\ 0\ 1\ 1\ 0\ 1\ 0\ 1$$

The first two, four and eight bits, i.e., 01, 0110 and 01101100, all represent the number 1/2, which is a good approximation of the final number 9/16. This progressive precision property of SC can be exploited if a decision can be made quickly from a particular image. For example, in the edge-detection application discussed later, many sharp edges are detected as early as four clock cycles into the computation.

Stochastic image-processing circuits have been proposed [12] [18] that are smaller than conventional designs, but are not particularly efficient. As noted above, the use of many conversion units degrades the performance of those designs. Our approach integrates format conversion into stochastic image processors in a way that minimizes the overhead of conversion circuits. Even if conversion costs are ignored, our designs are smaller and more efficient than those of [12] and [18].

3. SYSTEM OVERVIEW
Vision chips vary widely in how their sensors and processing circuits are laid out. In the simplest form, one processor handles all the pixels in series and no parallel processing occurs. At the other extreme, each pixel has a processing element (PE) of its own,

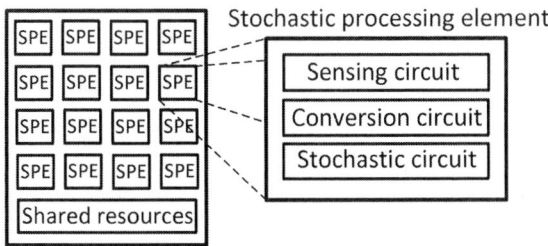

Figure 3. Top-level view of an SC-based vision chip and its stochastic processing elements (SPEs).

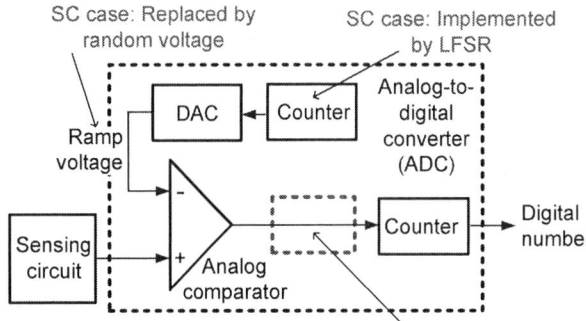

Stochastic circuit for edge-detection, etc. inserted here

Figure 4. Conventional ADC circuit for a vision chip with the changes needed for analog-to-stochastic conversion.

providing maximum parallelism. Since conventional digital PEs can be large, this approach does not scale well [21]. For real-time applications, one processor per pixel is desirable and, as we show, is achievable using stochastic computing techniques.

We propose a vision chip with maximum parallelism using stochastic processing elements (SPEs) that are very small and scale well. Our designs are also applicable to cases where processing circuits are shared among pixels. Figure 3 shows a high-level view of the proposed chip and its SPEs. For clarity, a 4×4 pixel array is shown, but it is possible to have many more pixels on chip. In addition to the SPEs, the chip has shared resources that manage random number generation and include a few counters based on LFSRs (linear feedback shift-registers). The area cost of these resources is minor since they are small and their cost does not change with the pixel count.

As Figure 3 shows, vision chips have image sensors that convert the perceived light intensity to an analog electrical voltage. To enable digital processing, this analog signal must be converted to digital form using a conventional ADC or, in the SC case, an analog-to-stochastic converter. As noted earlier, the cost of an analog-to-stochastic converter is very similar to that of a conventional ADC, which is depicted in Figure 4. In the conventional case, the analog voltage from the sensor is converted to a digital number using a ramp-compare technique. This requires an analog comparator fed by the sensor voltage, and a ramp voltage generated by a counter and a digital-to-analog converter (DAC). The comparator directly triggers a second counter which produces the desired digital output. In the SC case, the sensor voltage is converted to a stochastic number by comparing it to a random voltage generated by an LFSR-based counter and a DAC. A stochastic number appears at the output of the comparator and can then be processed by an application-specific stochastic circuit, such as an edge detector. The second counter is used to convert the final result to weighted binary form. It should be clear from Figure 4 that analog-to-stochastic conversion imposes little overhead as it employs essentially the same ADC circuits found in any digital vision chip.

Although analog comparators are well understood, they still present some circuit design challenges; for instance, low-area comparators are susceptible to noise. It is feasible to place comparators of suitable quality and size at every pixel [6]. In conventional digital image processors, a noise reduction step such as median filtering is needed [12]. In the proposed SC vision chip, however, the impact of noise is minimal thanks to the error tolerance of stochastic numbers, and a separate noise suppression step is unnecessary.

4. IMAGE PROCESSING OPERATIONS

This section discusses two basic image preprocessing categories, namely pixel-wise operations and windowing operations, examples of which are implemented later. We then present two ways to produce images with progressive quality improvement, which greatly speed up stochastic processing.

4.1 Basic Operations

Pixel-wise operations modify a pixel's intensity value x independent of the values at other pixels. They typically implement a real-valued function $f(x)$ that is used to adjust intensity values. A well-known example is gamma correction, which is used to compensate for non-linearities in recording or display devices, or to increase pixel contrast [8]. One of the simpler gamma-correction functions is $f(x) = x^{0.45}$. To synthesize stochastic circuits that implement functions like $f(x)$, we use the synthesis method of [1]. This approach produces efficient circuits for a broad class of arithmetic functions.

A second category of image-processing operations of concern are windowing operations, where a weighted moving-average operation is performed on a small window of pixels, either to extract features of the image or to enhance its quality. Examples of such operations are edge detection, sharpening and blurring [8]. The pixel windows are typically of size 2×2, 3×3, or 5×5. In order to design operations of this type, we mainly use the components of Figure 2. An m-to-1 multiplexer with a random select input performs averaging operations of the form $z = \frac{1}{m}(x_1 + x_2 + \cdots + x_m)$; if negative weights are present, subtraction can be implemented by XOR gates.

4.2 Spatial Progressive Quality Improvement

Generating images that progressively improve is an important task in image processing because it enables a trade-off between accuracy and computation that can be exploited in several ways. Image standards such as JPEG2000 [5] encode images of various qualities simultaneously. A conventional method of reducing the quality of an image is to reduce its number of pixels, i.e., its resolution. Figure 5 shows an image with several resolution levels; clearly, the quality diminishes as the resolution decreases.

Processing images with multiple resolutions imposes some computational overhead. However, we show by an example, that in the SC case, this overhead is minimal. Assume that a given image is to be processed at its original resolution, and at a lower-quality version with 16 times less resolution. In the latter case, intensity signals from 16 neighboring pixels of the original image are averaged to produce a super-pixel.

Figure 6 shows this averaging process implemented by a 16-to-1 multiplexer. This circuit processes each input individually, and records its results in the corresponding counter. Meanwhile, it

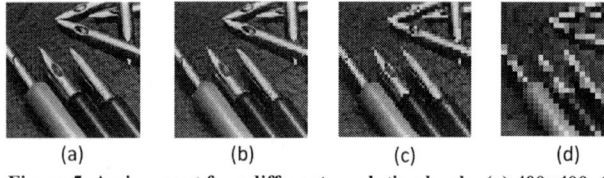

(a)   (b)   (c)   (d)

Figure 5. An image at four different resolution levels: (a) 400×400, (b) 100×100, (c) 50×50, and (d) 25×25 pixels.

(a)   (b)   (c)   (d)

Figure 7. Progressive precision results for edge detection: (a) input image; output image after (b) 4, (c) 32, and (d) 256 clock cycles.

performs the same computation on the low-resolution super-pixel and records that result in a separate counter. As seen in the figure, the overhead of a super-pixel computation is the additional counter, implying a very low cost.

4.3 Temporal Progressive Quality Improvement

As noted earlier, stochastic numbers have the progressive precision property, meaning that short sub-sequences of an SN can provide low-precision estimates of its value. This property can also be used to obtain images of different qualities because we can have SN-encoded pixels with different precisions. This approach is orthogonal to the previous spatial-resolution method, and, since it is an inherent property of SC, it comes at essentially no cost. One simply uses the values appearing at the output counters of Figure 6 at successive points of time.

Figure 7 shows how this property can be exploited in image processing. An edge-detection operation is being performed on an image. The input image has 8 bits of precision (the precision of an image corresponds to its gray-scale resolution [8]), and hence requires SN bit-streams of length 256. However, if the output image is checked at different points of time, it can be seen that as early as 4 clock cycles into the computation, many edges of the input image are detected, and after 32 clock cycles, almost all the edges are detected.

5. STOCHASTIC EDGE DETECTION

Edge detection is useful in image processing and computer vision because it allows objects to be extracted from an image by highlighting their edges. In the retinal implant application, a real-time edge-detecting circuit generates high-contrast images of the environment that greatly help a vision-impaired person to navigate correctly and avoid obstacles. We now consider the design of high-efficiency SC circuits for this task.

5.1 Circuit Design

Many edge-detection algorithms are known [8], and a few have been implemented with (non-real-time) SC [12]. Here we use the Roberts cross algorithm [8]. It computes a moving average on a window of size 2×2 for each pixel $x_{i,j}$ at row i and j of the image, and generates an output value $z_{i,j}$ according to the following formula.

$$z_{i,j} = 0.5 \times (|x_{i,j} - x_{i+1,j+1}| + |x_{i,j+1} - x_{i+1,j}|) \quad (1)$$

A stochastic implementation of this operation has been proposed by Li and Lilja [12], but it uses relatively large sequential circuits to approximate the absolute value function. Instead, we use the simple combinational SC components of Figure 2, which lead to a design that is more than 20 times smaller than that of [12], but has similar (or even better) performance. Figure 8a shows the proposed stochastic circuit for edge detection. It uses just two XOR gates to implement the subtractions in (1) and a multiplexer to perform the addition. As discussed, the inputs must be correlated, which is assured by assigning them a common random number source. The multiplexer's select input is fed with a random input r, which is produced at minimal cost since it is shared among the SPEs. In contrast, the corresponding binary design (Figure 8b), assuming 8-bit precision, contains several big arithmetic units such as adders and subtracters.

As proof-of-concept, we implemented and validated the SC edge-detection circuit (and the other examples in this paper) on a Xilinx Virtex-5 FPGA chip. Figure 9 illustrates the FPGA board (XUPV5-LX110T) used for our experiments, along with a representative image-processing task. It is important to note that the FPGA implementation was only used to verify the functionality of the circuits, and not for performance comparison. The output generated by the circuit is validated by comparing it to an expected output generated via conventional approaches.

In order to compare the proposed edge-detection design with previous work, we used the SIS synthesis CAD tools [19]. Table 1 summarizes the results. All the numbers are estimated values based on a generic library of cells using $0.35\mu m$ CMOS technology. The delay of each circuit is obtained by multiplying the clock period by the number of cycles required to perform the operation. The conventional binary design shown in Figure 8b implements (1) in a single clock cycle. The SC designs, on the other hand, require 256 cycles for the full 8 bits of precision. The dynamic power consumption of the circuits are also estimated using the SIS tools [19].

The results reveal that the proposed edge-detection circuit is strictly better than the previous SC design [12]. Our SC design is about 30 times smaller than the conventional designs, and only 3 times slower. From the area-delay numbers, we see that the SC design has a significant cost advantage. The dynamic power consumption of each circuit is also reported in Table 1, and show that for a given clock frequency (in this case 20MHz), SC has much lower power consumption. However, since a stochastic circuit with fixed precision runs for a longer time, its energy

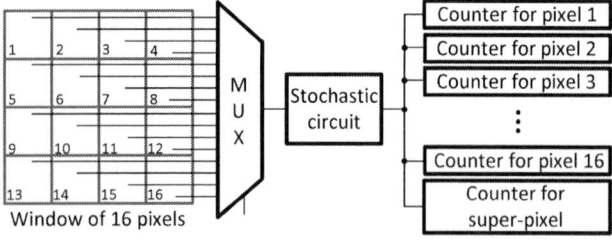

Figure 6. Stochastic processing of 16 pixels individually and as a super-pixel.

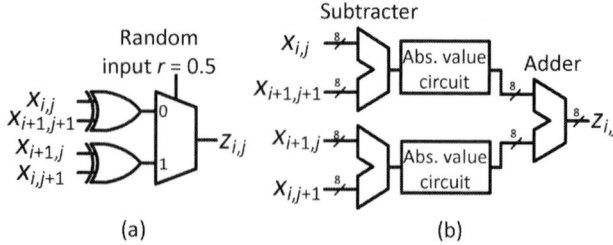

(a)   (b)

Figure 8. Edge detectors: (a) stochastic and (b) conventional designs.

Figure 9. FPGA setup for emulating image-processing tasks, in this case, gamma correction.

consumption eventually becomes higher than that of a conventional design. We do not report the leakage power/ energy of each circuit, but since leakage power is directly proportional to area, we can conclude that the leakage power of SC circuits is lower than the conventional case.

Table 1. Synthesis results for the edge detection circuits.

Implementation	Area (μm^2)	Delay (ns)	Power @20MHz (μW)	Energy (nJ)	Area × Delay (μm^2×ps)
Conventional weighted-binary	6928	19.49	7767.9	0.39	135.03
Previous SC design [12]	4312	1300	2213.7	28.34	5607.67
SC design proposed here	200	58.88	88.7	1.14	11.78

5.2 Error-Tolerant Behavior

Stochastic circuits are inherently noise tolerant, and their performance is not significantly degraded if the inputs contain noise, or even if the circuit components are noisy and unreliable [12] [18]. This is potentially a huge advantage, since pixel-level circuits need to be extremely small, with electrical characteristics that are difficult to control precisely. In this section, we demonstrate this benefit by a qualitative comparison between our SC-based edge detector and a conventional binary implementation when a noisy image is perceived by the sensors.

We model input noise as a Gaussian random variable with mean value 0 and standard deviation σ, which is added to the voltage signal generated by the sensor, i.e., right before the analog-to-digital or analog-to-stochastic conversion stage. We define the noise level as the ratio of σ to the full voltage swing of the sensor, so a 10% noise level means that the noise is 1/10 of the voltage range of the sensor.

Figure 10 illustrates the simulated performance of the three design methods at various noise levels. The performance of the conventional weighted-binary approach significantly degrades as the noise level reaches 5%; at noise levels of 10% and 20%, the output images become useless. The second implementation again employs a conventional binary approach, but with noise reduction implemented by a median filter [8]. This costly noise-reduction step improves the results, but significant quality degradation is still seen at higher noise levels. On the other hand, the SC implementation is almost unaffected by noise and is able to detect the edges even in the 20% noise case. The impact of noise on the SC circuit appears as a gray background, a result also seen in [18].

5.3 Progressive Precision

Also of interest is the performance of the edge-detection circuit when producing images with progressive quality improvement. The examples in Figure 7 suggest that the runtime of the edge-detection circuit can be further reduced (by a factor of 8), without compromising accuracy. This implies that edge detection requires

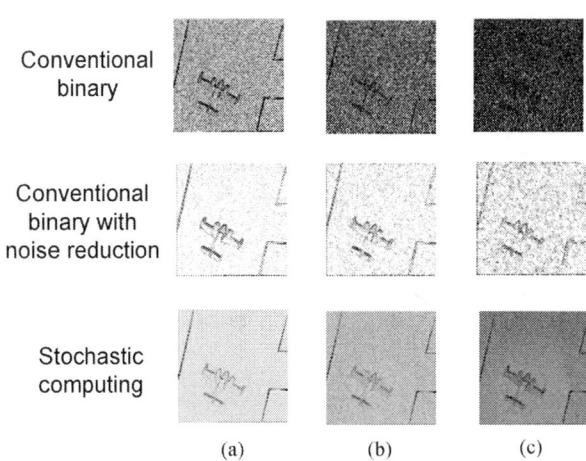

Figure 10. Edge-detection performance for three implementation methods with noise levels of (a) 5%, (b) 10% and (c) 20%.

less precision (than the original 8 bits), so for a fair comparison, we also implemented a low-precision version of the conventional edge detection circuit. As can be seen from Table 2, the stochastic design is strictly better than the conventional one. Also, the stochastic edge-detection is so efficient that can operate in real-time (15 frames per second) at 1nW power consumption. This number might be further reduced by switching to sub-threshold technologies.

Table 2. Synthesis results for low-precision edge-detection circuits.

Implementation	Area (μm^2)	Delay (ns)	Power @20MHz (μW)	Energy (nJ)	Area × Delay (μm^2×ps)
Conventional weighted-binary	4344	7.66	5156.8	0.26	33.28
SC design proposed here	200	7.36	88.7	0.14	1.47

6. OTHER IMAGE-PROCESSING OPERATIONS

Besides edge detectors, we designed several other stochastic image-processing circuits and evaluated their performance compared to alternative designs. We used the synthesis method of [1] to obtain the gamma-correction circuit in Figure 11a. A flip-flop is used in this design in order to produce a second uncorrelated copy of the input bit-stream. The function implemented approximates the target function $z_{i,j} = x_{i,j}^{0.45}$ but produces acceptable results, as seen in Figure 9. Figure 11b shows a conventional binary implementation of the same function. Like our edge-detection circuit, the stochastic gamma-correction circuit has a random input of probability 0.25 produced by a vision chip's shared resources.

Table 3 shows the synthesis results of our gamma-correction circuit, along with a conventional design of the same precision, and a previous SC design [18]. The design of [18] has better accuracy,

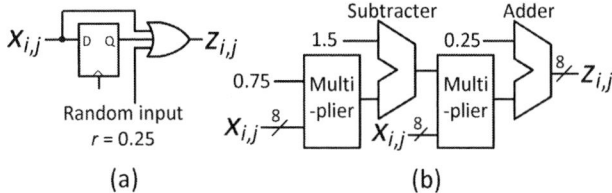

Figure 11. Gamma correctors: (a) stochastic and (b) conventional.

949

Table 3. Synthesis results for gamma-correction, blurring, and gradient calculation circuits.

Task	Design method	Area (μm²)	Delay (ns)	Power 20MHz (μW)	Energy (nJ)	Area × Delay (μm²×ps)
Gamma correction	Conventional binary	10576	27.2	21486	1.07	287.77
Gamma correction	Previous SC method [18]	1416	5365	970.3	49.68	7597.9
Gamma correction	Our SC method	168	15.4	55.8	0.71	2.58
Blurring	Conventional binary	19464	32.6	13196	0.66	634.92
Blurring	Our SC method	664	589	433.4	5.55	390.96
Gradient calculation	Conventional binary	1520	3.4	716.6	0.04	5.20
Gradient calculation	Our SC method	72	51.2	26.9	0.34	3.69

but is much costlier. The table also includes synthesis results for two other image-processing tasks, namely, blurring and gradient calculation. These results are consistent with those we obtained for edge detection.

7. CONCLUSIONS

We have shown that stochastic computing (SC) is practical for high-performance vision chips. Complex image-processing tasks can be implemented with only a few gates, thus enabling massively parallel processing at the pixel level and real-time operation. We presented designs for stochastic image-processing circuits that outperform existing designs (both conventional and SC) in most aspects. In particular, we have designed an edge-detection circuit that is strictly better than equivalent conventional designs. This highly efficient circuit seems ideal for vision chips and retinal implants. We also designed other representative image-processing circuits and made detailed comparisons with alternative implementations. We conclude that stochastic circuits are, in general, much smaller than conventional designs, and are much more efficient in terms of power consumption and area-delay product.

Furthermore, we demonstrated two different ways to process images with progressive quality improvement at minimal cost overhead. These properties compensate for SC's longer computation times. We successfully implemented and validated the discussed image-processing tasks on an FPGA development system. Finally, we showed that unlike conventional designs, the SC circuits can process very noisy images with almost no performance degradation.

8. ACKNOWLEDGEMENTS

This work was supported by Grant CCF-1017142 from the U.S. National Science Foundation.

9. REFERENCES

[1] A. Alaghi & J.P. Hayes, "A spectral transform approach to stochastic circuits," *Proc. Intl. Conf. Computer Design*, pp.315-321, 2012.

[2] A. Alaghi & J.P. Hayes, "Survey of stochastic computing," to appear in *ACM Trans. Embedded Computing Systems*, 2012.

[3] F. Andoh *et al.*, "A digital pixel image sensor for real-time readout," *IEEE Trans. Electron. Dev*, **47**, pp. 2123-2127, 2000.

[4] Centeye Inc. "Introduction to current Centeye vision chips" http://centeye.com/technology/vision-chips/, Feb. 2011.

[5] C. Christopoulos *et al.*, "The JPEG2000 still image coding system: an overview," *IEEE Tran. Consumer Electronics, IEEE Transactions on* , **46**, 4, pp. 1103-1127, 2000.

[6] P. Dudek and P.J. Hicks, "A general-purpose processor-per-pixel analog SIMD vision chip," *IEEE Trans. Ccts. & Sys. I*, **52**, pp.13-20, 2005.

[7] B.R. Gaines, "Stochastic computing systems," *Advances in Information Systems Science*, **2**, pp. 37-172, 1969.

[8] R.C. Gonzalez and R.E. Woods, *Digital Image Processing*, 2nd ed., Prentice Hall, 2002.

[9] T. Hammadou *et al.*, "A 96 × 64 intelligent digital pixel array with extended binary stochastic arithmetic," *Proc. Intl. Symp. Ccts. & Sys.* (ISCAS), pp. IV-772–IV-775, 2003.

[10] IEEE, *Standard 802.11n for Info. Technology Telecommunications & Info. Exchange between Systems Local & Metropolitan Area Networks.* http://standards.ieee.org, 2009.

[11] K. Kagawa *et al.*, "Pulse-domain digital image processing for vision chips employing low-voltage operation in deep-submicrometer technologies," *IEEE Jour. Sel. Topics in Quantum Electronics*, **10**, pp. 816-828, 2004.

[12] P. Li and D.J. Lilja, "Using stochastic computing to implement digital image processing algorithms," *Proc. Intl. Conf. Computer Design*, pp. 154-161, 2011.

[13] C. Ma *et al.*, "High fault tolerant image processing system based on stochastic computing," *Proc. Intl. Conf. Computer Science & Service System*, pp. 1587-1590, 2012.

[14] A. Moini, *Vision Chips*, Kluwer, 2000.

[15] W. Mokwa, "Retinal implants to restore vision in blind people," *Proc. Transducers*, pp. 2825-2830, Beijing, 2011.

[16] A. Naderi *et al.*, "Delayed stochastic decoding of LDPC codes," *IEEE Tran. Signal Proc.*, **59**, pp. 5617-5626, 2011.

[17] N.L. Opie *et al.*, "Heating of the eye by a retinal prosthesis: modeling, cadaver and *in vivo* studies," *IEEE Trans. Biomed. Engin.*, **59**, pp. 339-345, 2012.

[18] W. Qian *et al.*, "An architecture for fault-tolerant computation with stochastic logic," *IEEE Trans. Comp.*, **60**, pp.93-105, 2011.

[19] E.M. Sentovich, *et al.*, "SIS: A system for sequential circuit synthesis," Univ. of California, Berkeley, Tech. Report UCB/ERL M92/41, Electronics Research Lab, 1992.

[20] F. Taherian & D. Asemani, "Design and implementation of digital image processing techniques in pulse-domain," *Proc. Asia Pacific Conf. Ccts. & Sys.* (APCCAS), pp. 895-898, 2010.

[21] H. Yamashita & C.G. Sodini, "A CMOS imager with a pro-grammable bit-serial column-parallel SIMD/MIMD processor," *IEEE Trans. Electron. Dev.*, **56**, pp. 2534-2545, 2009.

An Event-Driven Simulation Methodology for Integrated Switching Power Supplies in SystemVerilog

Ji Eun Jang, Myeongjae Park, and Jaeha Kim

Department of Electrical Engineering and Computer Science, Inter-university Semiconductor Research Center
Seoul National University, Seoul, Korea

jieunjang@mics.snu.ac.kr, mjpark@mics.snu.ac.kr, jaeha@snu.ac.kr

ABSTRACT

Emerging power-supply-on-chip applications such as on-chip DC-DC conversion, energy harvesting, and LED drivers use switching regulator ICs integrated with digital controllers. Although the resulting mixed-signal systems call for efficient system-level behavioral simulation, this remains difficult due to the fast switching and slow transients of the regulator and the high complexity of the controller. This paper presents a truly event-driven approach for modeling and simulating such integrated power systems entirely in SystemVerilog. By modeling various switching regulator topologies as switched linear networks whose responses can be expressed as a sum of complex exponentials, $ct^{m-1}e^{-at}u(t)$, the accurate voltage/current waveforms can be captured by updating the coefficients, c, at each input or switching event. The model is applied to two examples, a power factor corrector and switched-capacitor DC-DC converter, and the results demonstrate that the proposed simulator can achieve $20{\sim}100\times$ improvements in speed while maintaining SPICE-level accuracy in evaluating power efficiency, steady-state ripples, and power factor.

Categories and Subject Descriptors

EDA6.3 [Analog Design and Simulation] Analog, mixed-signal, and RF simulation.

General Terms

Algorithms, Design, Verification.

Keywords

Event-driven simulation, Behavioral modeling, Switching-mode power supplies, SystemVerilog.

1. INTRODUCTION

The design and verification of state-of-the-art switching-mode power supplies demands advanced methodologies for mixed-signal IC systems. This is because a large portion of these switching regulators is now being realized in ICs to minimize their power, cost, and size when used in various emerging applications such as energy harvesting systems [1], LED drivers [2], and on-chip power management systems [3]. These so-called

Permission to make digital or hard copies of part or all of this work for personal or classroom use is granted without fee provided that copies are not made or distributed for profit or commercial advantage and that copies bear this notice and the full citation on the first page. Copyrights for components of this work owned by others than ACM must be honored. Abstracting with credit is permitted. To copy otherwise, to republish, to post on servers or to redistribute to lists, requires prior specific permission and/or a fee.

DAC '13, May 29 - June 07 2013, Austin, TX, USA

power-supplies-on-chip [4] have brought two notable changes to the previous discrete-component-based switching regulator designs: first, complex converter topologies such as time-interleaved architectures are frequently used because the cost is no longer set by the component count, and second, many controllers are digitally implemented to support multiple operating modes and maintain high efficiency in different conditions. Hence, the resulting switching-mode power supplies typically comprise a complex mixed-signal system with an analog power converting stage (e.g. buck or boost) and a digital controller. This paper presents an efficient, event-driven behavioral simulator suited for these integrated switching power supplies.

For accurate performance estimation and complete functional verification of switching-mode power supplies, a behavioral simulator should be fast at handling large digital circuits while accurately simulating analog signals. However, the existing simulation tools do not meet these requirements. For instance, SPICE, a versatile circuit simulator capable of nonlinear transient simulation, is not the best solution for simulating integrated switching power supplies, as it cannot achieve the required speed and accuracy [5]. The problem is that switching supplies are characterized by high-frequency switching activities that demand fine-grained simulation for accuracy, but also by slow transients that require long simulation times. The large transistor counts in the digital components further aggravate the situation.

To address this problem, several approximate models, such as average models and sampled-data models, have been proposed [6,7]. The average models capture only the slow-moving transients of the system dynamics and neglect the fast-moving transients due to switching. Similarly, the sampled-data models capture only the cycle-to-cycle behavior of the system by dealing only with the samples at a specific time instant in each switching period. A device-level average model is another alternative [8], with which it is possible to compose an average model of an arbitrary converter topology and simulate it directly in SPICE. The pulse-width-modulated switch model is such an example [9]. While these approximate models can significantly improve the simulation time by simplifying the system to a non-switching, time-invariant one, a critical shortcoming of these models is that they cannot model the behavior or estimate the performance related to the high-frequency switching activity of the regulator, such as the conversion efficiency and power factor. For instance, although the magnitude of the output voltage ripples is one of the key factors influencing the power efficiency, it is not well modeled by the average or sampled-data models.

This paper presents a fast, event-driven simulation methodology to accurately simulate the high-frequency switching behavior of integrated switching power supplies. Event-driven simulation is a preferred solution because the simulation cost only increases with the system activities and not with the system size. However, until recently, there were no good methods for simulating analog signals in an event-driven way. For instance, an event-driven oscillator model [10] and an analog filter model based on a look-up table [11] have been reported, but these techniques share the common limitation that they can be applied only to specific types of systems or inputs.

The event-driven approach presented in this paper builds on the method described in [12], whose key idea is to express an analog waveform as a linear combination of basis functions. The approach was demonstrated in a system containing linear time-invariant (LTI) analog components, whose output signals can always be expressed as a series of complex exponentials. With this signal representation, it was shown that the coefficients in the output series need to be updated only when those in the input series change, thus enabling a truly event-driven simulation. In addition, a series of complex exponentials is very expressive in representing various analog signals encountered in circuit applications, including steps, ramps, sinusoids, exponentials, and all of their combinations. Nonetheless, the approach in [12] cannot be directly applied to the modeling of switching-mode power supplies as it assumes that each analog block has a fixed linear transfer function over time.

This paper extends the event-driven simulation method described in [12] to switched linear systems, which include the majority of the switching-mode power supply topologies. This is achieved by allowing the system to change its input-to-output transfer function, while preserving previously-stored state values such as the currents through inductors or voltages across capacitors. To enable this, the states of the energy-storing elements are made explicit in the system formulation so that their initial values can be supplied as additional inputs to the system. As in [12], the proposed simulation method is implemented on a single logic simulation platform of SystemVerilog so that both the analog and digital components can be simulated in an event-driven fashion.

The rest of this paper is organized as follows. Section 2 shows that a general switching-mode power supply can be modeled as a switched linear system, using a boost converter as a working example. Section 3 describes the proposed event-driven simulation method for switched linear systems and Section 4 discusses its implementation in SystemVerilog. Finally, in Section 5, the proposed method is demonstrated with two power converter examples, a power factor correction (PFC) boost converter [13] and a time-interleaved switched-capacitor DC-DC converter [14].

2. SWITCHED LINEAR SYSTEM MODEL FOR SWITCHING POWER SUPPLIES

A common characteristic of switching regulators is that the circuit configuration changes depending on the switching phases. For example, a boost converter has two operation phases alternated by a switch connection as shown in Fig. 1(a). With the switch in position 1 (phase 1), the right-hand side of the inductor is connected to the ground, resulting in the network shown in Fig. 1(b). With the switch in position 2 (phase 2), the inductor is connected to the output, leading to the circuit shown in Fig. 1(c).

In this work, we model such switching regulators as a switched linear system rather than as an average linear system or sampled-data system. Even though the input/output relationship changes at each switching instant, the circuit can be modeled as a linear time-invariant system within each operation phase. For example, the relationship between the input $v_{IN}(t)$ and output $v_{OUT}(t)$ of the

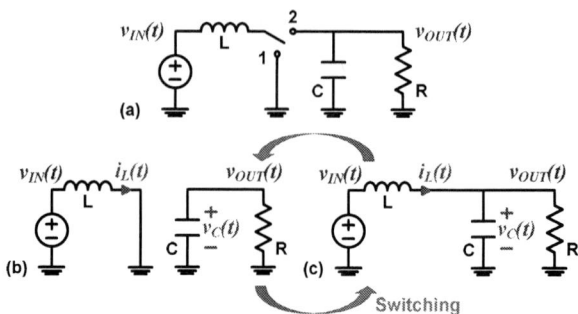

Figure 1. (a) A boost converter circuit and its linear system model in (b) switching phase 1 and (c) switching phase 2.

boost converter in phases 1 and 2 can be modeled as a set of differential equations listed below:

For phase 1:

$$\begin{cases} \frac{dv_C(t)}{dt} & = -\frac{1}{RC}v_C(t) \\ \frac{di_L(t)}{dt} & = \frac{1}{L}v_{IN}(t) \\ v_{out}(t) & = v_C(t) \end{cases}$$

For phase 2:

$$\begin{cases} \frac{dv_C(t)}{dt} & = \frac{1}{C}i_L(t) - \frac{1}{RC}v_C(t) \\ \frac{di_L(t)}{dt} & = \frac{1}{L}v_{IN}(t) - \frac{1}{L}v_C(t). \\ v_{OUT}(t) & = v_C(t) \end{cases} \tag{1}$$

This set of differential equations can be converted to a Laplace s-domain equivalent, using the Laplace transformation formula for a function derivative in Eq. (2), resulting in Eq. (3). This Laplace transformation is used to apply the event-driven simulation method in [12], which is explained in more detail in the next section. Note that the initial conditions of the internal state variables v_C and i_L are made explicit in the s-domain equations. In Eq. (3), the capital letters denote s-domain signals while the italic letters denote their initial conditions in the time domain.

$$\mathcal{L}\{f^n(t)\} = s^n \mathcal{L}\{f(t)\} - \sum_{k=1}^{n} s^{k-1} f^{n-k}(0) \tag{2}$$

For phase 1:

$$\begin{cases} V_C(s) & = \frac{RC}{sRC+1}v_C(0) \\ I_L(s) & = \frac{1}{sL}V_{OUT}(s) + \frac{1}{s}i_L(0) \\ V_{OUT}(s) & = V_C(s) \end{cases}$$

For phase 2:

$$\begin{cases} V_C(s) & = \frac{V_{IN}(s)+sLC \cdot v_C(0)+L \cdot i_L(0)}{s^2LC+sL/C+1} \\ I_L(s) & = \frac{(sC+1/R)V_{IN}(s)-C \cdot v_C(0)+(sLC+L/R)i_L(0)}{s^2LC+sL/C+1} \\ V_{OUT}(s) & = V_C(s) \end{cases} \tag{3}$$

As this simple boost converter example shows, the main difference between this switched linear system model and a linear time-invariant system model is that the former requires additional terms related to the initial conditions of the energy-storing elements in the circuit. In Eq. (3), for instance, $v_C(0)$ and $i_L(0)$ represent the initial voltage across the capacitor and the initial current through the inductor, respectively. These terms in Eq. (3) model how the stored energy in the circuit affects the output responses after $t = 0$. Each time a switch occurs, the final state (of

952

phase 1 or phase 2) becomes the initial state of the next phase (phase 2 or phase 1, respectively), thus preserving the energy stored within the system.

3. TRUE EVENT-DRIVEN SIMULATION OF SWITCHED LINEAR SYSTEMS

This section describes how the above switched linear system model can be simulated in a truly event-driven fashion. The main difficulty in performing event-driven simulations for analog circuits stems from the fact that their output response can continuously change even when there is no change in the input. If the output signal is represented by a set of data points, a one-time update (equivalent to one data point) cannot capture this continuous waveform. The method proposed in [12] addressed this problem by representing an analog signal with a linear combination of basis functions, $c \cdot t^{m-1} e^{-at} \cdot u(t)$, and updating the coefficients, c, when the input changes. For instance, considering a first-order filter with a step input, $c_{in}u(t)$, its output is known to be in the analytical form of $c_{out}u(t) - c_{out}e^{-\omega_p t}u(t)$, where ω_p is its pole frequency (see Fig. 2). When the input coefficient, c_{in}, changes, the corresponding change in the output only affects the output coefficient, c_{out}. Therefore, it is sufficient to update the output only once at the time of the input change.

Note that this method also computes the output waveform analytically in the s-domain without involving time integration, provided that both the input signal and the system transfer function are expressed in the s-domain representation after the Laplace transformation. Again, for the first-order filter example, the output signal of the filter in the s-domain is nothing but a product of a step function, $1/s$, and a filter transfer function, $\omega_p/(\omega_p+s)$, which can be decomposed to the sum of a step and an exponential functions. The two main advantages of this method are that fast simulation is possible because the output is explicitly calculated without solving an ordinary differential equation, and its accuracy is virtually infinite because the signal is represented as a function, not as a data point.

This work extends the method described in [12] to switched linear systems by adopting a signal representation with a series of complex exponentials and the s-domain output computation method. There are two main differences between our method and that in [12]. First, in addition to the input events, the switching event can also trigger an output event. Second, while the output of a time-invariant system can be determined only by an input and its input-to-output transfer function, a switching system additionally requires the information on the initial states and its relationship (transfer function) to the system dynamics, as explained in Section 2.

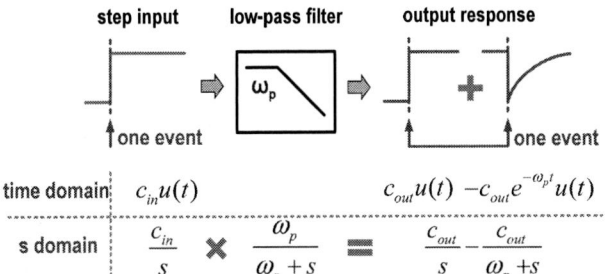

Figure 2. A simple analog filter example to demonstrate the event-driven simulation method for linear time-invariant (LTI) systems proposed in [12].

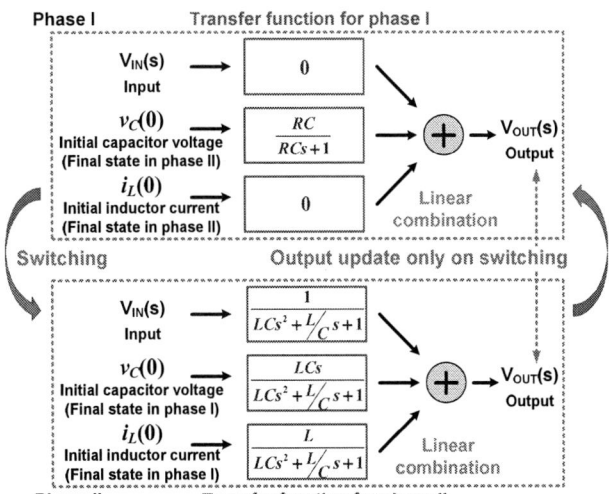

Figure 3. The proposed method applied to the boost converter example in Section 2.

Based on these differences, our proposed event-driven simulation methodology can be summarized as follows: 1) model a switched linear system with multiple transfer functions, each of which describes the circuit network during each switching phase, 2) when the switching occurs, sample the initial states of the reactive elements, and 3) compute the output by summing the effects from both the input signal and the initial states of the reactive elements. Fig. 3 illustrates the presented method applied to the boost converter example described in Section 2. For each phase, the three transfer functions define the relationship between the input (V_{IN}), the initial capacitor voltage ($v_C(0)$), the initial inductor current ($i_L(0)$), and the output (V_{OUT}). Every time the circuit switches between phases, the initial capacitor voltage and the inductor current are sampled and their transfer functions are redefined according to Eq. (3). Then, just as in the LTI system described in Fig. 2, the output voltage induced by each of the input and initial conditions can be evaluated simply by multiplying each one by its corresponding transfer function. The zero transfer gains for V_{IN} and $i_L(0)$ in phase 1 imply that the output is not related to the input and the inductor current, as they are disconnected from the output during this phase. When the circuit switches to the next phase, however, V_{IN} and the energy stored in the inductor start to increase V_{OUT} again. Finally, the output is updated as a linear combination of these three components.

4. IMPLEMENTATION IN SYSTEMVERILOG

This section outlines the implementation of the proposed event-driven simulation methodology in SystemVerilog. SystemVerilog was chosen as our simulation platform as it can serve as a true event-driven simulation engine that seamlessly integrates analog and digital models. A further advantage of SystemVerilog over other HDL standards is that it offers a composite data type, *struct*, and hence the set of multiple parameters can be exchanged between the block modules as though they are a single bundled signal. We adopt the digital signal data type defined in [15] ("XBIT") and the analog signal data type defined in [12] ("XREAL"). The digital data type named XBIT has two member variables: *value*, and *t_offset*. *Value* is a digital data bit and *t_offset* is a real-valued variable that indicates the actual time instant of the event. The analog data type named XREAL comprises three variables: *param_set*, *t_offset*, and *flag*.

```
module boost_converter(
    input XBIT switching,
    input XREAL in,
    output XREAL out);

    // transfer functions in phase 1
    chandle TF_in_out_ph1;
    chandle TF_vc0_out_ph1;
    chandle TF_il0_out_ph1;
    // transfer functions in phase 2
    chandle TF_in_out_ph2;
    chandle TF_vc0_out_ph2;
    chandle TF_il0_out_ph2;

    always @(switching) begin
        vc0=sample(vc);   // sampling initial states
        il0=sample(il);

        if (phase1)        // evaluating outputs
          out.param_set = multiply(in,TF_in_out_pout_ph1)
                        + multiply(vc0,TF_vc0_out_ph1)
                        + multiply(il0,TF_il0_out_ph1);
        If (phase2)
            out.param_set= multiply(in,TF_in_out_ph2)
                        + multiply(vc0,TF_vc0_out_ph2)
                        + multiply(il0,TF_il0_out_ph2);

        out.t_offset = switching.t_offset;
        -> out.flag;       // output triggering
    end
endmodule
```

Figure 4. Pseudo-code in SystemVerilog for the boost converter example in Section 2.

Param_set is a set of coefficients for expressing the analog waveforms, and *flag* is an event variable that indicates whether the event change has happened for the signal in the current time step, which is generally used within *always* statements.

An outline of a boost converter model in SystemVerilog is given in Fig. 4. The input and output signals are defined as XREAL, while the switching input signal is XBIT. Six transfer functions, three for each of the two switching phases, are given as a C-pointer handle pointing to a linked list in C that stores information on the poles, zeros, and gain of the system. The *always* statement within the module is triggered when the circuit switches between phases. The initial states of the capacitors and inductors are then sampled and the *param_set* of output XREAL signals is updated according to the current input and sampled initial conditions. The *multiply()* function is a DPI function written in C that performs s-domain multiplications of XREAL signals and transfer functions. As the output update is aligned in time with a switching event, the *t_offset* of the output has the same value as the switching. Once the *param_set* and *t_offset* outputs are updated, the event variable *out.flag* is triggered, thus notifying subsequent blocks of the change event.

5. EXPERIMENTAL RESULTS

5.1 POWER FACTOR CORRECTION (PFC) BOOST CONVERTER

The speed and accuracy of the event-driven simulation method are demonstrated using the example of a power factor correction circuit (PFC) composed of a bridge-diode rectifier and a boost converter (Fig. 5). The power factor is one of the key performance metrics of AC-DC power converters required by many regulatory standards. It is defined as in Eq. (4), which expresses the ratio of

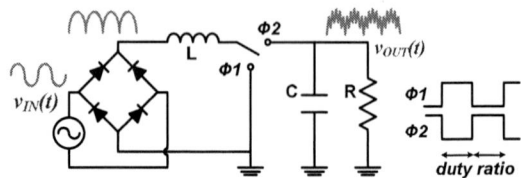

Figure 5. A power factor correction boost converter.

the real power flowing to the load and the apparent power in the circuit:

$$power\ factor = \frac{average\ power}{(rms\ voltage)(rms\ current)}. \qquad (4)$$

For a high power factor, the circuit should basically behave as a pure resistive load. A boost converter is a widely used topology for power factor correction circuits because the switched inductor at the input conducts a current that is proportional to the input voltage with very low harmonics [13].

One difficulty in simulating such an AC-DC power converter is that there is a big gap between the input AC frequency and the switching frequency of the boost converter. For instance, in most applications, the input source is 50~60-Hz 110~220V AC power, while the switching frequency is typically in the 100kHz~1MHz range. The average behavior analysis mentioned in the introduction is not suitable here because accurate measurement of the power factor requires detailed information on the input voltage/current waveforms, such as ripples. The required simulation time is long, typically several tens of milliseconds, to simulate a few cycles of the 60-Hz AC input.

Fig. 6 illustrates the accuracy of the waveforms simulated by the proposed event-driven simulation method. Fig. 6(a) is the simulated output voltage of the boost converter, $v_{OUT}(t)$, for one 60-Hz input cycle. The zoom-in waveforms in Figs. 6(b) and (c), simulated with the proposed method and HSPICE, respectively, demonstrate that they are well matched, and illustrate the switching ripples of the converter. It is noteworthy that HSPICE requires many data points to express the switching ripples (marked by the blue dots in Fig. 6(c)) while our event-driven method generates only two events per switching cycle, as indicated by the arrows in Fig. 6(b). The power factor can be measured from the simulated input current waveform, which has a similar level of accuracy to SPICE (not shown). The comparison between the simulated power factors as a function of switching frequency and duty cycle in Fig. 7 confirms that the proposed method indeed achieves the same level of accuracy as SPICE.

The proposed event-driven simulator demonstrates significant improvements in speed compared with HSPICE, yet retains the equivalent accuracy. On a Linux machine with an AMD Phenom II X4 945 processor, the total execution time to simulate a 0.1-second period with a 100-ns time step is 8.2 seconds. Under the same conditions, the HSPICE simulation takes 920.5 seconds, which is 110× slower.

The execution time of the proposed method varies weakly with the time step. For instance, when the time step is reduced from 100-ns to 10-ps (1/10,000X reduction), the execution time of the proposed method increases by only 15% (from 8.2 to 9.4 seconds), while that of HSPICE increases by 15000%. The reason the execution time hardly varies is that the number of switching events within the 0.1-second period remains the same regardless of the time step, which confirms that the proposed simulation

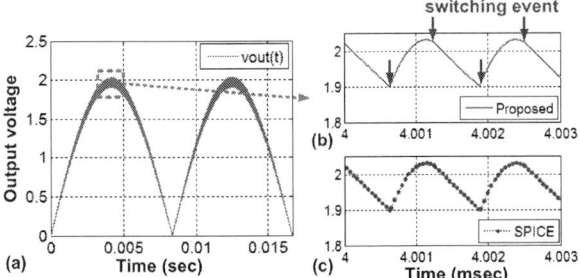

Figure 6. (a) The output voltage waveform $v_{out}(t)$ simulated for one 60-Hz input cycle, (b) 5000× zoom-in view of $v_{out}(t)$, (c) $v_{out}(t)$ simulated by HSPICE.

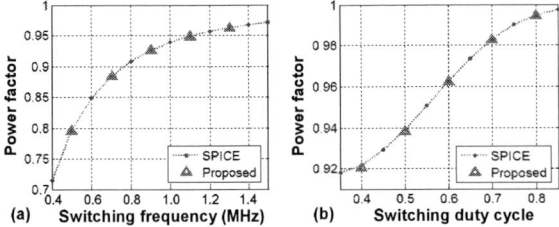

Figure 7. The comparison of the simulated power factors vs. (a) frequency and (b) switching duty cycle.

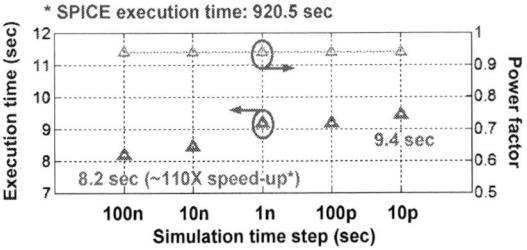

Figure 8. Execution time and simulated power factor vs. time step.

indeed operates in a purely event-driven fashion. This remarkable speed-up does not incur any penalty in accuracy, as the power factor measured using the proposed method is virtually constant, independent of the time step (Fig. 8).

5.2 TIME-INTERLEAVED SWITCHED-CAPACITOR DC-DC CONVERTER

For the second example, we choose the multi-phase time-interleaved switched-capacitor (TI-SC) DC-DC converter described and analyzed in [14]. The switched-capacitor DC-DC converter topology is becoming a common choice for power-supplies on chips, as the IC technology is more amenable to integrating high-density capacitors than low-loss inductors. One difficulty in simulating a TI-SC DC-DC converter is that the number of switching activities increases with the number of time-interleaving phases.

Our TI-SC converter example is composed of N interleaved 2:1 step-down converter units, as shown in Fig. 9(a). The total capacitance is divided into a set of small units and the switching is controlled by a set of N equally spaced clocks ($\Phi 1$, ..., ΦN). Fig. 9(b) illustrates the basic operation of a 4-phase TI-SC converter with the output waveforms. The output voltage ripple is inherent in a switched-capacitor converter, and generally decreases as the switching frequency increases. Time-interleaving is an alternate way of reducing the ripples without increasing the switching frequency.

The TI-SC converter in Fig. 9 can be modeled as in Fig. 10, where the unit capacitors, C_{fly}, switch their configurations between a series and parallel connections depending on the controlling clock phase (Φ). The model includes the on-resistance of the switches, R_{sw}, and parasitic top- and bottom-plate capacitances, C_{par}, to account for the conduction and switching losses, respectively. To simplify the model, the top- and bottom-plate capacitances are combined into a single capacitor because they experience approximately the same voltage swings in steady states [14]. The s-domain transfer function of each phase can be derived from Eq. (5), where $v_{cap}[i]$ denotes the voltage across the i-th unit capacitor.

$$V_{out}(s) = \frac{\frac{s}{2R_{sw}C_{par}}V_{in}(s) + \frac{N_S}{4R_{sw}^2 N_S C}v_{out}(0) + \frac{s}{2R_{sw}C_{par}N_S}v_{cap}(0)}{\left\{ s^2 + s\left(\frac{R_L N_S C_{par} + 2R_{sw}C + NCR_L}{2R_{sw}CR_L N_S C_{par}}\right) + \frac{1}{2R_{sw}CR_L N_S C_{par}}\right\}}$$

where $v_{cap}(0) = \sum_{parallel} v_{cap}[i](0) - \sum_{series} v_{cap}[i](0)$

$$V_{cap}[i](s) = \begin{cases} \text{for series capacitor:} \\ \frac{1}{2R_{sw}Cs+1}V_{in}(s) + \frac{1}{2R_{sw}Cs+1}V_{out}(s) + \frac{2R_{sw}C}{2R_{sw}Cs+1}v_{cap}[i](0) \\ \text{for parallel capcitor:} \\ -\frac{1}{2R_{sw}Cs+1}V_{out}(s) + \frac{2R_{sw}C}{2R_{sw}Cs+1}v_{cap}[i](0) \end{cases}$$

(5)

Fig. 11(a) plots the power efficiency of the TI-SC converter as a function of the number of time-interleaving phases, N, and compares the results produced by HSPICE and the proposed method. When the total amount of charge delivered to the load is fixed with N, the output voltage ripple initially decreases by a factor of N, as the amount of charge delivered per clock transition

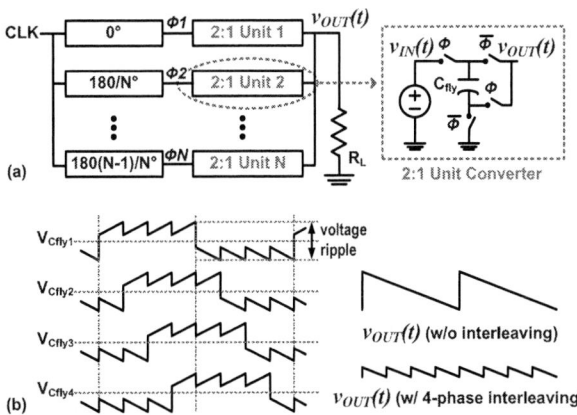

Figure 9. (a) N time-interleaved 2:1 step-down switched-capacitor DC-DC converter, (b) the waveforms of its internal capacitor voltages and final output voltage when N=4 [14].

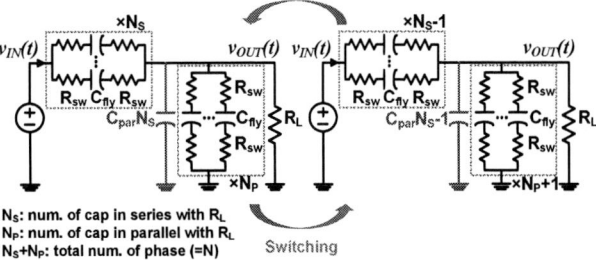

N_S: num. of cap in series with R_L
N_P: num. of cap in parallel with R_L
N_S+N_P: total num. of phase (=N)

Figure 10. Switched linear circuit model of an N time-interleaved, 2:1 step-down TI-SC converter.

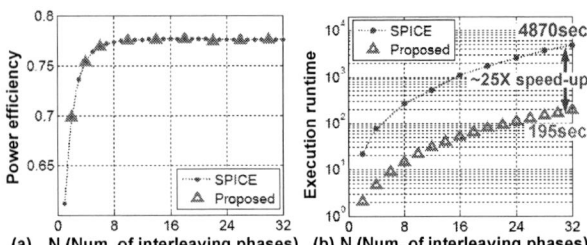

Figure 11. (a) The simulated power efficiency of the TI-SC converter and (b) execution time vs. the number of time-interleaving phases (N).

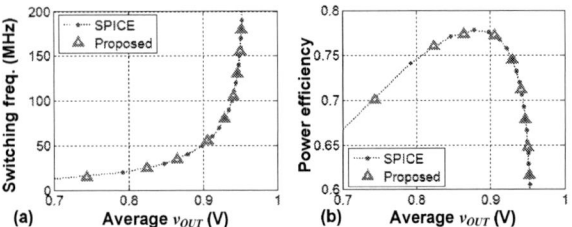

Figure 12. The simulated (a) switching frequency and (b) power efficiency of the TI-SC converter vs. the average output voltage.

is smaller. Therefore, the better power efficiency can be achieved with a higher N. However, increasing N above a certain value produces diminishing returns because the other losses, such as the conduction loss of the switches and the switching loss of the parasitic capacitors, increase. Fig. 11 illustrates this tendency, and the simulation results of the proposed method match well with the SPICE simulation results.

The improvement in speed with the proposed method is moderate compared with the boost converter case, as SPICE is better at simulating switched capacitors than inductors. When simulated on a Linux machine with an AMD Phenom II X4 945 processor, the proposed method shows a ~20X overall speed improvement compared with the HSPICE simulation (see Fig. 11(b)).

The proposed method predicts the well-known dependencies of the switching frequency and power efficiency for the output voltage v_{OUT}, described in [14]. Fig. 12 verifies the accuracy of the obtained results against those of HSPICE. Fig. 12(a) plots the switching frequency vs. the average v_{OUT} and Fig. 12(b) plots the power efficiency vs. the average v_{OUT} when v_{IN} is 2V and N=16. The v_{OUT} dependency on the switching frequency is similar to the IR-drop phenomenon in linear regulators, in which the output voltage drops when the load current is higher than the current that the TI-SC converter can nominally supply. As a result, slower switching leads to a lower average v_{OUT} and lower power efficiencies (Fig. 12). Nonetheless, an excessively high switching frequency is also undesirable as the power efficiency can be degraded due to the loss in the switching capacitors.

6. CONCLUSIONS

This paper presents an event-driven simulation methodology for switching regulators. A common characteristic of various switching regulators such as buck, boost, and switched-capacitor converters is that they can all be modeled as switched linear systems. By extending the previously published event-driven

methodology for LTI systems, the proposed method realizes a truly event-driven simulation of these switched linear systems, in which the output response is updated only once with each input change or switching event. Two power converter examples simulated in SystemVerilog demonstrate that the proposed method achieves SPICE-level accuracy with 20~100× faster simulation speeds. The experimental results also show that the simulation accuracy and execution time are not influenced by the simulation time step, thus confirming that the proposed method achieves a truly event-driven simulation.

7. ACKNOWLEDGMENTS

This work was supported by the National Research Foundation of Korea Grant funded by the Korean Government (2012-0003320). The CAD tool licenses are supported by the IC Design Education Center (IDEC) in Korea.

8. REFERENCES

[1] G. K. Ottman, et al., "Adaptive piezoelectric energy harvesting circuit for wireless remote power supply," *IEEE Trans. on Power Electronics*, pp.669-676, May 2002.

[2] H. Broeck, et al., "Power driver topologies and control schemes for LEDs," *IEEE Applied Power Electronics Conf. (APEC)*, pp.1319-1325, Mar. 2007.

[3] G. Patounakis, et al., "A fully integrated on-chip DC-DC conversion and power management system," *IEEE J. of Solid-State Circuits*, pp.443-451, Mar. 2004.

[4] S. S. Kudva, et al., "Fully integrated on-chip DC-DC converter with a 450x output range," *IEEE J. of Solid-State Circuits*, pp.1940-1951, Aug. 2011.

[5] T. G. Wilson, Jr., "Life after the schematic: the impact of circuit operation on the physical realization of electronic power supplies," *Proc. of the IEEE*, pp.325-334, Apr.1988.

[6] D. Maksimovic, et al., "Modeling and simulation of power electronic converters," *Proc. of the IEEE*, pp.898-912, Jun. 2001.

[7] H. Jin, "Behavior-mode simulation of power electronic circuits," *IEEE Trans. on Power Electronics*, pp.443-452, Mar. 1997.

[8] G.W. Wester, et al., "Low-frequency characterization of switched dc-dc converters," *IEEE Trans. on Aerospace and Electronic Systems*, pp.376-385, Mar. 1973.

[9] E. Dijk, et al., "PWM-switch modeling of DC-DC converters," *IEEE Trans. on Power Electronics*, pp.659-665, Jun. 1995.

[10] R. B. Staszewski, et al., "Event-driven simulation and modeling of phase noise of an RF oscillator," *IEEE Trans. on Circuits and Systems I*, pp.723-733, Apr. 2005.

[11] M. Ierssel, et al., "Event-driven modeling of CDR jitter induced by power-supply noise, finite decision-circuit bandwidth, and channel ISI," *IEEE Trans. on Circuits and Systems I*, pp.1306-1315, Jun. 2008.

[12] J.-E. Jang, et al., "True event-driven simulation of analog/mixed-signal behaviors in SystemVerilog: a decision-feedback equalizing (DFE) receiver example," *IEEE Custom Integrated Circuits Conf. (CICC)*, pp.1–4, Sep. 2012.

[13] B. Singh, et al., "A review of single-phase improved power quality AC-DC converters," *IEEE Trans. on Industrial Electronics*, pp.962-981, May 2003.

[14] H.-P. Le, et al., "Design techniques for fully integrated switched-capacitor DC-DC converters," *IEEE J. of Solid-State Circuits*, pp.2120-2131, Sep. 2011.

[15] M.-J. Park, et al., "Fast and accurate event-driven simulation of mixed-signal systems with data supplementation," *IEEE Custom Integrated Circuits Conf. (CICC)*, pp.1-4, Sep. 2011.

SUPPLEMENTAL MATERIAL

This supplemental section provides additional experimental results to further demonstrate the efficiency and accuracy of the proposed method.

S.1. DISCONTINUOUS CURRENT MODE (DCM) BUCK CONVERTER EXAMPLE

The discontinuous current mode (DCM) buck converter illustrated in Fig. 13 is chosen as an additional power converter example. Under low load conditions, the power converter needs to operate in a DCM to achieve high efficiency. In this case, the average current flow may be low compared with the current ripple magnitude, and a reverse current flow may exist during phase 2, as shown in Fig. 14(a). Because this reverse current lowers the converter's power efficiency, a DCM converter disconnects switch 2 ($\Phi2$) to block the reverse current (phase 3 in Fig. 14(b)). However, as it is difficult to determine a precise period in phase 2, any discrepancy from the ideal turn-on period will result in performance degradation. For instance, if the switch is turned off later than the zero-crossing instant of the inductor current, the converter will suffer from reverse current. If the switch is disconnected too early, then a large negative voltage spike will appear at the v_X node and turn on the parasitic diode of the MOS pull-down switch (source/drain-to-body pn-junction), resulting in a loss of power due to the diode voltage drop and conduction loss. Most of the DCM power converters use digital control loops to determine the precise time for turning off this switch, and therefore an accurate co-simulation with digital circuitry is key to simulating its performance.

Fig. 15 demonstrates the aforementioned operating phases of a DCM buck converter. The converter can be modeled as a switched linear system with four phases depending on the configuration of the two switches. During the first phase (Fig. 15(a)), the load is connected to the input and then switches to ground during the second phase (Fig. 15(b)). When the circuit switches into the third phase, the parasitic diode either kicks in or is disconnected, depending on the v_X node voltage. If the v_X node voltage is lower than the ground due to a diode voltage turning on at the switching instant, then the parasitic diode turns on and charges the v_X node (Fig. 15(c)). In this phase, the diode is modeled as an ideal voltage source and a current-limit resistor.

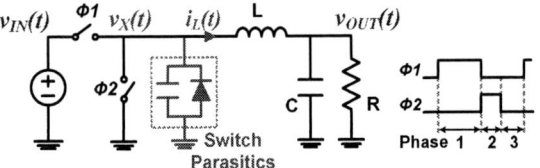

Figure 13. A discontinuous current mode buck converter and its switching signals.

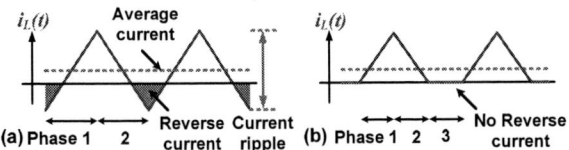

Figure 14. (a) A continuous current mode (CCM), and (b) a discontinuous current mode (DCM) for a low-power condition.

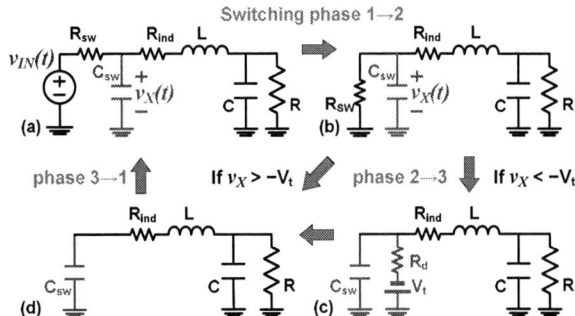

Figure 15. A DCM buck converter: (a) phase 1, (b) phase 2, (c) phase 3 with parasitic diode on, and (d) phase 3 with parasitic diode off.

Once the voltage across the diode becomes smaller than the turn-on voltage, the diode is disconnected, leaving only the parasitic capacitor connected to the inductor (Fig. 15(d)).

Fig. 16 shows the simulated inductor current waveforms for three cases in which switch 2 is disconnected 1) too early, 2) precisely, and 3) too late. The proposed method produces the same waveforms as HSPICE, but again this signal waveform is expressed with far fewer events, which are indicated by arrows in Fig. 16. Based on the inductor current and output voltage, the power efficiency is calculated as shown in Fig. 17. When the period in phase 2 is too short (case 1 in Fig. 16), the power efficiency drops due to the conduction loss of the parasitic diode. With a longer period in phase 2 (case 3 in Fig. 16), the converter efficiency is degraded by the inductor reverse current. The simulation execution time is 82.4 sec for 1-msec simulation time, while SPICE takes 3053.7 sec (~37× slower) under the same conditions.

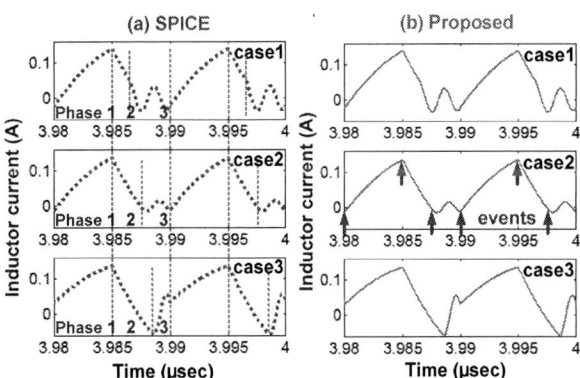

Figure 16. Inductor current waveforms simulated with (a) HSPICE and (b) the proposed method (case1: switch-off too early, case2: switch-off precisely, case3: switch-off too late).

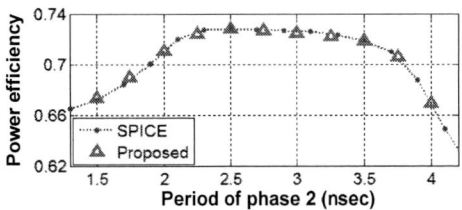

Figure 17. Simulated power efficiency vs. period of phase 2.

A New Time-Stepping Method for Circuit Simulation

G. Peter Fang

Texas Instruments, Inc., Dallas, Texas

g-fang1@ti.com

ABSTRACT

Adaptive time-stepping is crucially important for the efficiency of a circuit simulator. Existing time-stepping methods rely on information at prior time point(s) to select step sizes, which can be problematic when the circuit is undergoing a fast transition. In this work, we propose a new time-stepping method that solves the circuit equations together with the condition for local truncation error (LTE) as *one* nonlinear system. Circuit solution and step size are obtained *simultaneously* for the current time point. It allows designers to have direct control of LTE so the errors can be distributed more evenly along non-uniformed time grid. Experiments show the new method generates significantly less time points and is faster for the same accuracy settings. It is also more accurate for the simulation of non-dissipative circuits.

Categories and Subject Descriptors

B.7.2 [**Integrated Circuits**]: Design Aids - *simulation*

General Terms

Algorithms, Design, Performance

Keywords

Circuit Simulation, Differential Equations, Stepsize Control

1. INTRODUCTION

Modern analog and mixed signal circuits are typically stiff systems characterized by a wide range of time constants. During a transient simulation, it is desirable to vary the step size dynamically by taking small steps when the circuit is under going fast transitions in order to preserve accuracy and large steps when there is little activity. Adaptive time-stepping is essential for the efficiency of the simulator. A good time-stepping algorithm needs to monitor the activity within the system of differential equations to determine the maximum allowable step size while satisfying conditions of stability and local truncation error.

The step size distribution of transient simulation of modest sized class-D audio amplifier is shown in Figure 1. The step size varies nearly 6 orders of the magnitude during the portion of the transient run.

In 1972, Gear [1] introduced an adaptive stepping method for solving ordinary differential equations or ODE, which is still widely used among circuit simulators. A number of improvements [2-5] have been developed over past four decades. However, all those methods are essentially trial and error, predicting a step size based on prior information and hoping it is good enough. We will

Permission to make digital or hard copies of all or part of this work for personal or classroom use is granted without fee provided that copies are not made or distributed for profit or commercial advantage and that copies bear this notice and the full citation on the first page. To copy otherwise, or republish, to post on servers or to redistribute to lists, requires prior specific permission and/or a fee.

show in Section 2 that those methods become inefficient in many situations. A new time-stepping method will be proposed in Section 3. And experimental results will be discussed in Section 4, followed by a conclusion.

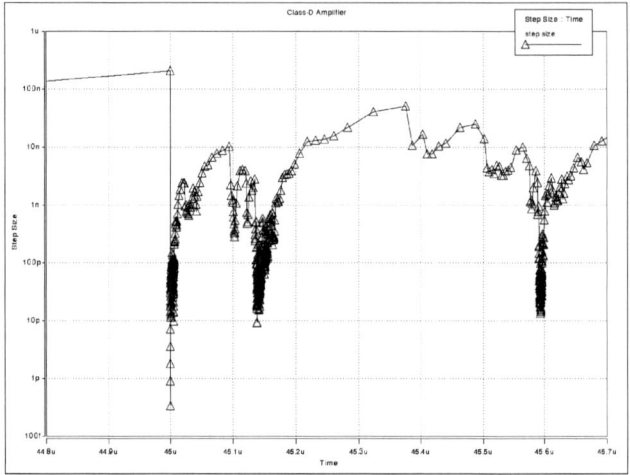

Figure 1. Step size distribution of a transient run

2. BACKGROUND

A Spice-like general purpose circuit simulator solves a system of nonlinear differential algebraic equations (DAE) in form of,

$$\bar{f}_{ckt}\left(\bar{v}(t)\right) = \frac{d}{dt}\bar{q}\left(\bar{v}(t)\right) + \bar{i}\left(\bar{v}(t)\right) + \bar{u}(t) = 0, \qquad (1)$$

with initial conditions,

$$\bar{v}(t_0) = \bar{v}_0, \ t > t_0,$$

in which $\bar{f} \in R^N$ is the vector function, $\bar{v} \in R^N$ is the solution vector that consists of nodal voltages and branch currents, $\bar{i}\left(\bar{v}(t)\right), \bar{q}\left(\bar{v}(t)\right) \in R^N$ are vectors of resistive node currents and node charges or fluxes.

2.1 Time Integration

Implicit methods are commonly used for the numerical time integration of the differential terms $\dot{\bar{q}} = d\bar{q}/dt$ in form of

$$\dot{\bar{q}}_m = \frac{1}{h_m}\sum_{i=0}^{n}\alpha_i\bar{q}_{m-i} + \sum_{i=1}^{n}\beta_i\dot{\bar{q}}_{m-i}, \qquad (2)$$

where m is the time index of the mth time point, h_m is the step size between $(m-1)^{th}$ and m^{th} time points, $\bar{q}_m = \bar{q}\left(\bar{v}(t_m)\right)$ is the charge vector at the mth time point, and the coefficients α_i and β_i are specific for each integration method. For trapezoidal rule, $n=1$, $\alpha_0=1$, $\alpha_1=-2$, and $\beta_1=-1$.

958

$$\dot{\bar{q}}_m = \frac{\bar{q}_m - \bar{q}_{m-1}}{h_m/2} - \dot{\bar{q}}_{m-1}. \tag{3}$$

For Gear methods or backward differentiation formula (BDF), $\beta_i = 0, i = 1, \cdots, n$,

$$\dot{\bar{q}}_m = \sum_{i=0}^{n} \alpha_i(h_m)\bar{q}_{m-i}. \tag{4}$$

Almost all the circuit simulators only use low-order ($n \le 2$) Gear methods which are guaranteed A-stable for uniform step sizes. With the time differential term discretized, the system of differential equations in (1) becomes a system of nonlinear equations in form of,

$$\bar{f}_{ckt}(\bar{v}_m) = 0. \tag{5}$$

2.2 Truncation Error

Truncation error occurs when the differential term is replaced by a discrete-time approximation. Local truncation error or LTE is the error made in a single step assuming all previous steps are accurate. To obtain LTE for the current time step, a typical industrial circuit simulator calculates the difference between the computed solution and the polynomial extrapolation from previous time steps, and considers the maximum value a good estimate for the local truncation error, which is given by,

$$\varepsilon_m = \left| v_i(t_m) - v_{i,extrapolated} \right|, \tag{6}$$

where i is the index of the variable with the maximum difference. In most of cases the variable is a node voltage. The corresponding node is often referred to as controlling LTE node, which may vary from time point to time point. Since the accumulated local truncation error or global truncation error is difficult to estimate, almost all the simulators use LTE to control time step. A typical LTE condition for accepting a time step is,

$$\varepsilon_m < \gamma \cdot \tau_m, \tag{7}$$

where τ_m is the tolerance for the LTE and γ is a coefficient greater than 1.

2.3 Adaptive Time-Stepping

In 1972, Gear [1] proposed the original elementary step size selection algorithm based on LTE in the form of,

$$h_{m+1} = \left(\frac{\tau_m}{\varepsilon_m}\right)^{\frac{1}{n+1}} h_m, \tag{8}$$

where n is the order of time integration. This algorithm has been widely used among adaptive solvers. A number of improved approaches have been developed over the past four decades, most notably the step size controller based on digital filter theory [3,4], which is given by,

$$h_{m+1} = \left(\frac{\tau_m}{\varepsilon_m}\right)^{\frac{1}{4(n+1)}} \left(\frac{\tau_{m-1}}{\varepsilon_{m-1}}\right)^{\frac{1}{4(n+1)}} \left(\frac{h_m}{h_{m-1}}\right)^{-\frac{1}{4}} h_m. \tag{9}$$

The digital filter step size controller has been implemented in DASPK [5] and is more suitable for problems where the smoothness of step sizes is desired.

A simplified flow for a typical transient adaptive solver is shown in Figure 2. The adaptive solver starts with selecting a step size for the current time point based information at previous time point(s), e.g., equation (8) or (9). It then solves nonlinear circuit equations to obtain a solution. If the condition for local truncation error, e.g., equation (7), is not satisfied, the step size will be reduced (backup) and the circuit equations will be solved again. This process will be repeated until the LTE condition is satisfied.

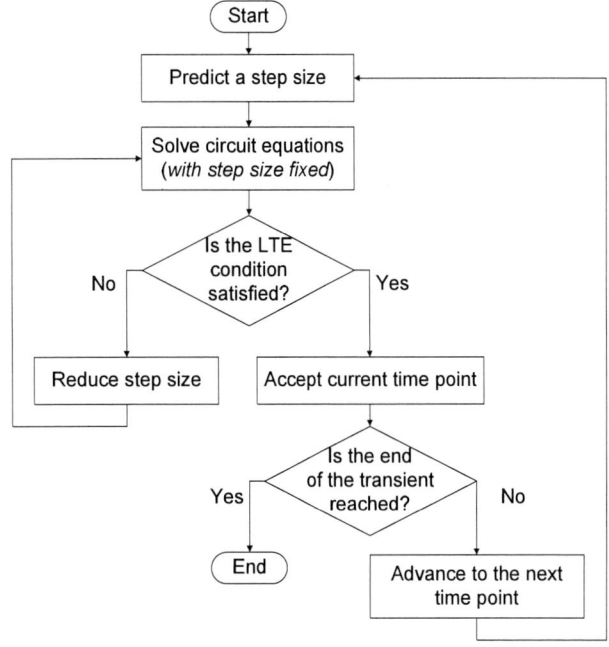

Figure 2. Simplified flow for a transient adaptive solver

The existing trial and error time-stepping methods predict or select the step size for the current time point based on information such as step sizes and local errors at previous time point(s). Ideally the LTE of the current time point will be close to the tolerance. However, for stiff circuits with strong nonlinearity, the dynamics of the DAE system can change dramatically from time point to time point. Moreover, the existing methods were derived for initial condition problems, assuming there is no external stimulus. But in practical cases, circuits are often driven by time-varying independent sources for transient simulation. As a result, the prediction or selection of step size based on prior information can be too optimistic, causing time steps being rejected (backup); or too pessimistic, keeping step sizes unnecessarily small. Either way, the efficiency of the adaptive solver suffers.

3. NEW TIME-STEPPING METHOD

To overcome the drawback of the existing methods, we propose a new time-stepping method that solves the circuit equations together with the condition for local truncation error as *one* nonlinear system. The solution vector and step size are computed *simultaneously* for the current time point. There is no backup due to LTE. Newton method is used to solve the coupled nonlinear system and the predicted step size is used as an initial guess. The resulting step size may be either larger or smaller than the predicted step size. The top-level flow for the new time-stepping method is shown in Figure 3.

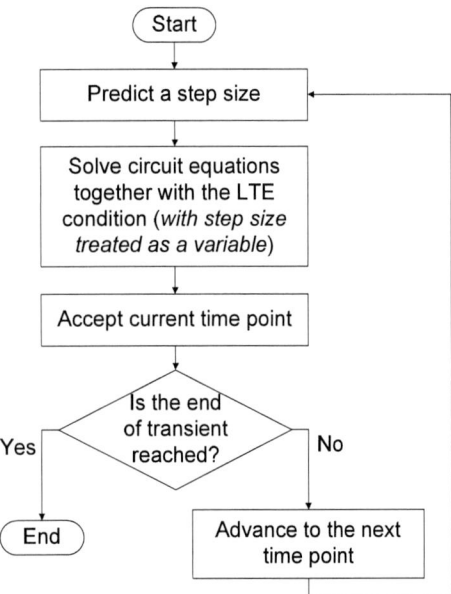

Figure 3. Top-level flow for the new time-stepping method

3.1 Coupled Nonlinear System

From equations (6-9), we can create an LTE equation in a general form, which is an explicit function of solution variables and step size,

$$f_{lte}\left(\bar{v}_m, h_m\right) = \varepsilon_m\left(\bar{v}_m, h_m\right) - \tau_m = 0. \tag{10}$$

In order to compute the proper time step size and the circuit solution for the time step simultaneously, we treat the time step h_m as an independent variable or an unknown, and solve the LTE equation together with circuit equations (2) as one coupled system of nonlinear equations,

$$\begin{cases} \bar{f}_{ckt}\left(\bar{v}_m, h_m\right) = 0 \\ f_{lte}\left(\bar{v}_m, h_m\right) = 0 \end{cases} \tag{11}$$

or

$$\overline{F}_{coupled}\left(\bar{v}_m, h_m\right) = 0$$

where $\overline{F}_{coupled} \in R^{N+1}$, i.e., we now solve $N+1$ nonlinear equations for $N+1$ unknowns.

Standard Newton flow is modified to solve the coupled nonlinear system as shown in Figure 4. Since the controlling LTE node can change from iteration to iteration, we separate Newton iteration into two stages. In the first stage, the regular circuit equations (N system) are solved to obtain an updated solution. The controlling LTE node can be identified for the iteration based on the updated solution. If the LTE condition is not satisfied, the solution will be discarded and a LTE equation will be formed in the second stage. The coupled system ($N+1$ system) is then solved to obtain a new solution and a new step size simultaneously.

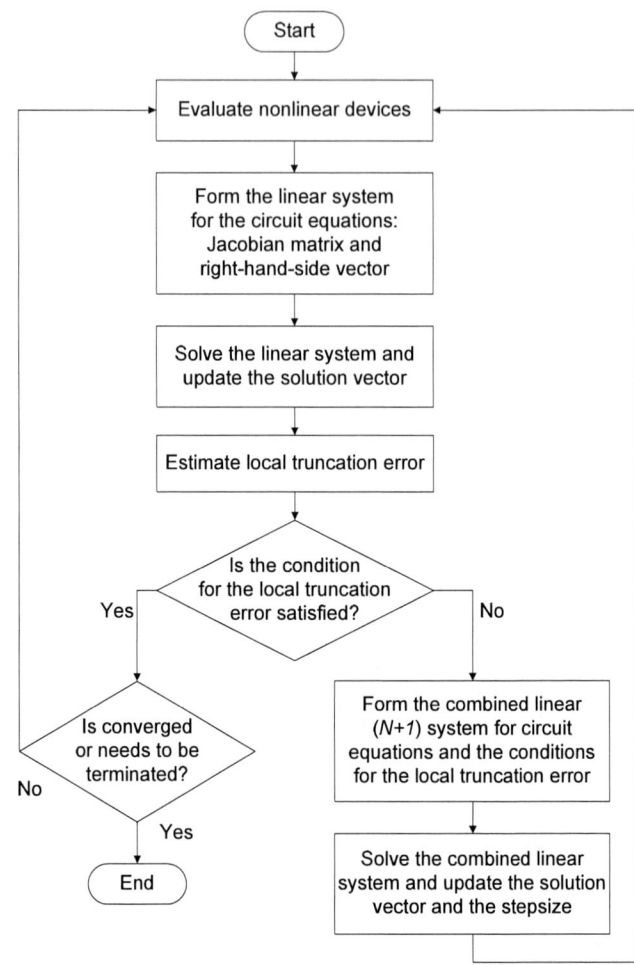

Figure 4. Newton flow for the new time-stepping method

In [6], a similar coupled Newton method was proposed for solving threshold-crossing events. Since the additional cross equation is not a function of time step, the resulting linear system can be solved in a decoupled way. The linear system for this problem is closely coupled and has to be solved differently.

3.2 Solving $N+1$ Linear System

The $N+1$ linear system in the second stage of the Newton iteration is given by,

$$\begin{pmatrix} \overline{\overline{J}} & \bar{p} \\ \bar{q}^T & d \end{pmatrix} \begin{pmatrix} \Delta \bar{v}_m^{k+1} \\ \Delta h_m^{k+1} \end{pmatrix} = \begin{pmatrix} -\bar{f}_{ckt}\left(\bar{v}_m^k, h_m^k\right) \\ -f_{lte}\left(\bar{v}_m^k, h_m^k\right) \end{pmatrix}, \tag{12}$$

where k is the Newton iteration index, $\bar{p} = \dfrac{\partial \bar{f}_{ckt}}{\partial h_m} \in R^{N \times 1}$,

$\bar{q}^T = \dfrac{\partial f_{lte}}{\partial \bar{v}_m} \in R^{1 \times N}$, $d = \dfrac{\partial f_{lte}}{\partial h_m} \in R^{1 \times 1}$, and $\overline{\overline{J}} = \dfrac{\partial \bar{f}_{ckt}}{\partial \bar{v}_m} \in R^{N \times N}$

is the original circuit Jacobian matrix. $\bar{p}, \bar{q}^T$, and d can be computed explicitly with negligible computational costs [6].

If a direct matrix solver is used in the simulator, the Jacobian matrix is already decomposed into LU factors in the first stage. It is desirable to reuse those factors to avoid costly full factorization

960

in the second stage. This can be achieved using partial LU factorization for the matrix solvers based on Doolittle method [7], in which matrix elements are processed in the order of rows. In the first stage, top N rows of the $(N+1)\times(N+1)$ matrix are factorized, with the last row for the LTE equation undetermined. We can still obtain solution to identify the LTE node using the fully factorized $N \times N$ sub-matrix. If LTE condition is not satisfied, the last row will be loaded and factorized to obtain a new solution and a new step size in the second stage.

Alternatively we can reduce the $N+1$ linear system in equation (12) to the following N linear system,

$$\left(\overline{\overline{J}} - \frac{1}{d}\,\overline{p}\,\overline{q}^{\,T}\right)\Delta\overline{v}_m^{k+1} = -\overline{f}_{ckt} + \frac{1}{d}\,f_{lte}\,\overline{p}\,, \qquad (13)$$

and

$$\Delta h_m^{k+1} = -\frac{1}{d}\left(f_{lte} + \overline{q}^{\,T}\Delta\overline{v}_m^{k+1}\right). \qquad (14)$$

Rank-one update technique [8] can be readily used to update LU factors in the second stage.

For simulator employing iterative matrix solvers [9,10], the matrix vector product part needs to be modified based on equation (13) to solve the coupled linear system. The pre-conditioner for the first stage can be reused in the second stage.

3.3 Convergence Criteria

Other than the original criteria for circuit equations, two additional convergence criteria for the LTE equation have to be satisfied before the Newton iteration is considered converged and, therefore, terminated. The first criterion is the new LTE condition, which specifies the lower and upper bounds for local truncation error,

$$\gamma_{\min}\tau_m \le \varepsilon_m^k\!\left(\overline{v}_m^k, h_m^k\right) \le \gamma_{\max}\tau_m\,, \qquad (15)$$

where $\gamma_{\max}$ is a coefficient greater than 1, and $\gamma_{\min}$ is a coefficient between 0 and 1. The introduction of the lower bound $\gamma_{\min}$ prevents step sizes from being unnecessarily small. By adjusting $\gamma_{\max}$ and $\gamma_{\min}$, we can precisely control the LTE allowed at each step. The second criterion assures that the change of step size between two iterations is small enough,

$$\left|\Delta h_m^{k+1}\right| \le \eta \cdot h_m^k\,, \qquad (16)$$

where η is relative tolerance for step size.

3.4 Approximate Newton Method

The coupled nonlinear system sometimes is very sensitive to the change of step size, especially during a fast transition. Sometimes step size change needs to be limited or damped to avoid convergence problems. It is unnecessary and computationally expensive to obtain high accuracy for the step size. A typical value for relative tolerance η is 15%. Based this observation, we developed an approximate Newton method. In the second stage of the Newton iteration, when step size needs to be changed, we use Gear's formula (8) to determine new step size based on the step

size of previous iteration, h_m^k and the LTE computed in the first stage, $\varepsilon_m^{k+\frac{1}{2}}$,

$$h_m^{k+1} = \left(\frac{\tau_m}{\varepsilon_m^{k+\frac{1}{2}}}\right)^{\frac{1}{n+1}} h_m^k. \qquad (17)$$

A new solution vector can be obtained by correcting the solution computed in the first stage,

$$\Delta\overline{v}_m^{k+1} = \Delta\overline{v}_m^{k+\frac{1}{2}} - \overline{\overline{J}}^{\,-1}\,\overline{p}\!\left(h_m^{k+1} - h_m^k\right). \qquad (18)$$

Though polynomial extrapolation or interpolation is sometimes preferred for large step size changes.

Without the need to modify the matrix solver of the simulator, the approximate Newton method is straightforward to implement and carries very little overhead.

4. RESULTS

The new time-stepping method has been implemented in our in-house circuit simulator, TISpice. We introduced two new user options, Itemax and Itemin, which define the upper and lower bounds for the local truncation error at each time point. In this section we show the effectiveness of this new stepping method in transient simulation of industry analog and mixed-signal circuits, including a Class-D amplifier, a suite of performance benchmarks, and a Colpitts oscillator. We compare the new approach with an improved version of Gear's time-stepping method, which we refer to as standard time-stepping method. The standard method predicts current step size based on equation (8). If the normalized local truncation error is greater than a given value (we used 4.63), the current time point will be recomputed with a smaller time step. Results shown in this section were generated using 2nd-order Gear method, the default time integration method for our simulator; though the new time-stepping method works well for other commonly used time integration formula such as backward Euler and trapezoidal rule.

4.1 Error Distribution

In the first example, we simulate the Class-D audio amplifier mentioned previously in Section 1. With local truncation errors bounded between 0.7 (Itemin) and 3.0 (Itemax), the new method generates 39% less time points than the standard method, and the total wall clock runtime improved 17%. 35% of the time steps have been altered and about half of them are larger than the predicted ones. It can be seen in Figure 5 that the local truncation errors are more evenly distributed along the non-uniform time grid for the new method. On the other hand, the standard method sometimes generates unnecessary small steps that are overly conservative, especially during fast transitions (e.g., around 3.9us). The maximum allowable LTE for the new method, 3.0 (Itemax) is actually smaller than that for the standard method (4.63), but the average LTE for the new method could be larger. The stability condition, output, and some behavioral models can also limit the step size besides local truncation error [11]. It is possible and acceptable for LTE to go below the lower bound at certain time points.

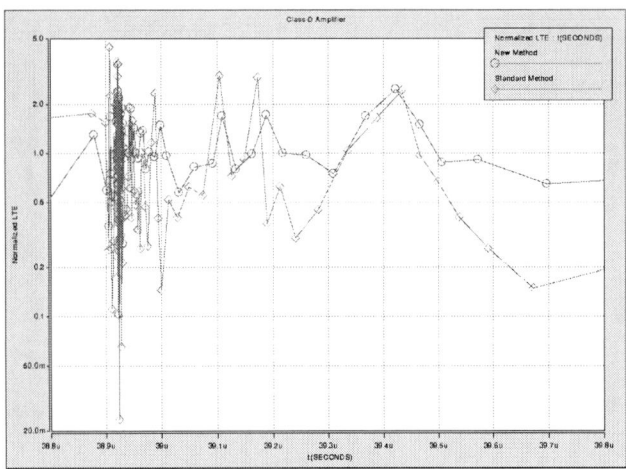

Figure 5. LTE distributions of two time-stepping methods

4.2 Performance Benchmarks

Next, we simulate a performance benchmark suite of 21 industry analog and mixed-signal circuits as shown in Table 1. A wider or more aggressive range of local truncation errors is chosen for the new time-stepping method with |temin=0.7 and |temax=4.0. Comparing with the standard method, we achieved speedup and time point reduction for all the benchmarks. In the best case scenario, 48% runtime speedup is obtained for LDO 2 and the number of time points is reduced 48.5% for the power stage circuit. The average runtime speedup and time point reduction are 20.3% and 40.3% respectively. The average number of Newton iterations increases from 3.52 to 4.35, while the mean percentage of rejected time point decreased from 10.5% to 2.5%. Although the new time-stepping method eliminates the need of backup, a time step still could be rejected due to a Newton failure or a request of a behavioral model.

Table 1. Performance comparison of two stepping methods

Benchmark circuits	Size	Speedup	Time Point Reduction
Boost regulator	420	17.5%	37.9%
Synchronous rectifier	5,761	5.6%	34.7%
Power management	57,730	22.1%	41.4%
DC-DC converter 1	27,065	30.4%	40.4%
DC-DC converter 2	3,192	14.3%	35.1%
Controller	15,552	25.3%	40.5%
Gate driver 1	4,228	23.7%	39.1%
LED driver	3,890	30.9%	46.9%
Power interface	7,510	14.1%	40.0%
LDO 1	6,689	18.3%	38.9%
Power stage	4,415	31.2%	48.5%
LDO 2	6,759	48.0%	44.0%
Motor driver converter	82,951	13.1%	44.4%
Preamp 1	46,062	16.9%	34.3%
Large mixed-signal	59,173	15.0%	43.6%
Preamp 2	53,003	21.2%	47.4%
Small mixed-signal	2,869	5.4%	32.7%
PLL	11,696	20.5%	43.2%
Digital PLL	2,873	23.9%	43.2%
Class-D amplifier	4,649	19.3%	41.1%
Gate driver 2	2,619	8.8%	28.8%
Average	--	**20.3%**	**40.3%**

The percentage of time point reduction is relatively consistent across the board, while runtime speedup fluctuates considerably from circuit to circuit. We believe the fluctuation is caused by strong nonlinear behaviors in some of the circuits. The sparser time grid generally leads to poorer initial guesses for the Newton method and the simulator has a hard time to capture those strong nonlinear behaviors. This also slightly increases the average number of iterations per time point but the total numbers of iterations have been greatly reduced. This benchmark suite was originally built for testing solver performance and the runtimes are dominated by model evaluation and matrix solving. The new time-stepping method achieves better performance for reliability simulation and full chip power analysis, for which transient runtimes are dominated by output/post-processing and, therefore, proportional to the number of computed time points.

4.3 Non-Dissipative Circuits

The new time-stepping method needs significantly less time points to satisfy the same or slightly tighter LTE criterion. However, there is not an effective way to estimate global truncation errors for general circuit simulation. We plotted a number of signals and most waveforms appear to match well or overlay. But this type of visual checking is rather subjective.

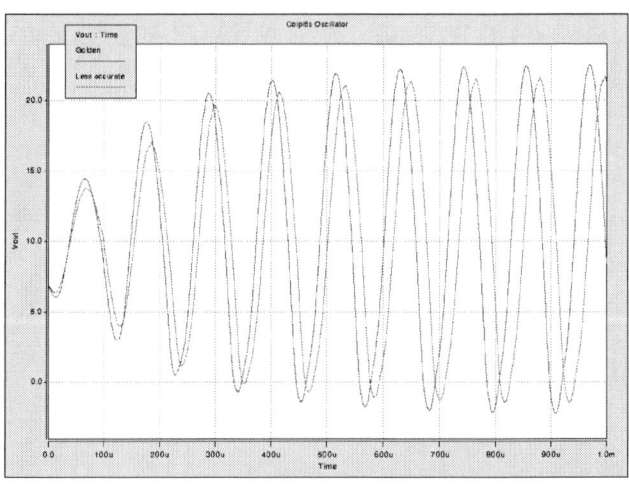

Figure 6. Phase shift between golden waveform and a less accurate waveform

In the last example, we study the accuracy of the new method in the simulation of a Colpitts oscillator. The Colpitts oscillator outputs near perfect sine waves and has been widely used in low-frequency signal generators. It is a typical non-dissipative circuit. The local truncation errors tend to accumulate during a transient simulation and cause the phase of the oscillator to shift as shown in Figure 6. So the phase shift becomes a good measure of accumulated truncation error or global truncation error. For this experiment, a narrow range of LTE is chosen for the new time-stepping method with |temin=0.8 and |temax=1.25. We ran a number of transient simulations with different tolerance settings and measured the phase shifts (against the golden waveform) of the last cycle before 2.0ms. The results have been summarized in Table 2.

Table 2. Comparison of phase shifts for two methods

Standard Stepping Method			New Stepping Method		
Relative tol	#time points	Phase shift	Relative tol	#time points	Phase shift
6E-04	Failed	Failed	Golden	4,026	0.00
2E-04	702	69.76	6E-04	424	71.01
1E-04	895	51.08	1E-04	750	51.50
6E-05	1,006	38.56	5E-05	980	36.41
1E-05	1,531	16.47	1E-05	1,443	16.60
1E-06	2,568	3.31	1E-06	2,494	3.50
1E-07	3,501	0.61	1E-07	3,482	0.61

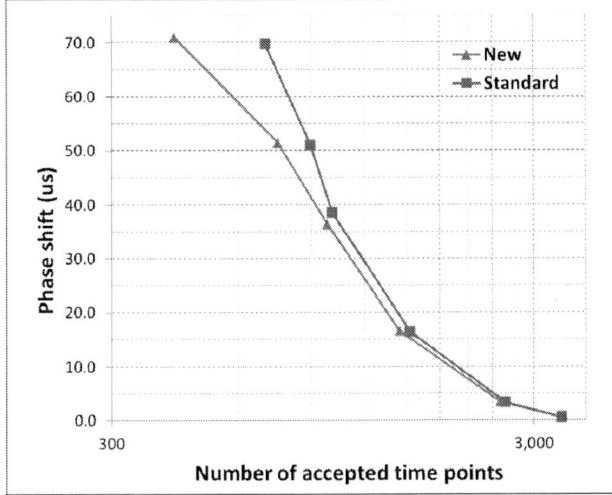

Figure 7. Phase shifts for two time-stepping methods

It is clear from Figure 7 that the new time-stepping method is more accurate when the tolerances are looser and less time points are computed. The evenly distributed local truncation errors lead to a smaller global truncation error. For one relatively liberal tolerance setting, the Colpitts oscillator fails to start when the standard time-stepping is used. As we tighten the tolerances, the accuracy advantage of the new method starts to diminish. The two methods eventually perform the same with very tight tolerances.

5. CONCLUSION

In this paper, we present a new method for adaptive time-stepping. LTE condition is solved together with circuit equations as a coupled nonlinear system, and time step is treated as an independent variable. As a result, a user can directly control LTE at each time step. Experiments show the new method generates significantly less time points and is faster for the same accuracy settings. It is also more accurate for the simulation of non-dissipative circuits.

Although this new time-stepping method was developed for solving circuit simulation problem, it should also benefit a large number of engineering and scientific problems which are modeled by various nonlinear differential equations.

6. REFERENCES

[1] C.W. Gear, Numerical Initial Value Problems in Ordinary Differential Equations, Prentice Hall, 1971

[2] K. Gustafsson, "Control-theoretic techniques for stepsize selection in implicit Runge-Kutta methods", ACM TOMS, December 1994

[3] G. Söderlind, "Digital filters in adaptive time-stepping", ACM TOMS, March 2003

[4] G. Söderlind, L. Wang, "Adaptive Time-Stepping and Computational Stability". Journal of Computational and Applied Mathematics, January 2006

[5] K. Meeker, C. Homescu, L. Petzold, H. El-samad, M. Khammash, G. Söderlind, "Digital Filter Stepsize Control in DASPK and its Effect on Control Optimization Performance", Proc. Sandia Real-Time Optimization PDE Optimization Conf. 2005

[6] G.P. Fang, "An efficient method to simulate threshold-crossing events", BMAS 2008

[7] W.J. McCalla, Fundamentals of Computer-Aided Circuit Simulation, Kluwer Academic, 1988

[8] P.E. Gill, G.H. Golub, W. Murray, M.A. Saunders, "Method for modifying matrix factorizations", Math. Comp., 1974

[9] Z. Li, C.-J. R. Shi, "A quasi-Newton preconditioned Newton-Krylov method for robust and efficient time-domain simulation of integrated circuits with strong parasitic couplings", IEEE TCAD, December 2006

[10] H.K. Thornquist, E.R. Keiter, R.J. Hoekstra, D.M. Day, E.G. Boman, "A parallel preconditioning strategy for efficient transistor-level circuit simulation", ICCAD 2009

[11] C.W. Gear, "Efficient step size control for output and discontinuities", Technical report, Dept. of CS, University of Illinois at Urbana-Champaign, 1982

Supplemental Material

This supplemental material includes more detailed figures and data for the numerical experimental results discussed in Section 4 of the research manuscript.

S1. CLASS-D AMPLIFIER

Time step and local truncation error distributions for a transient run of the class-D audio amplifier are shown in larger figures for better viewing.

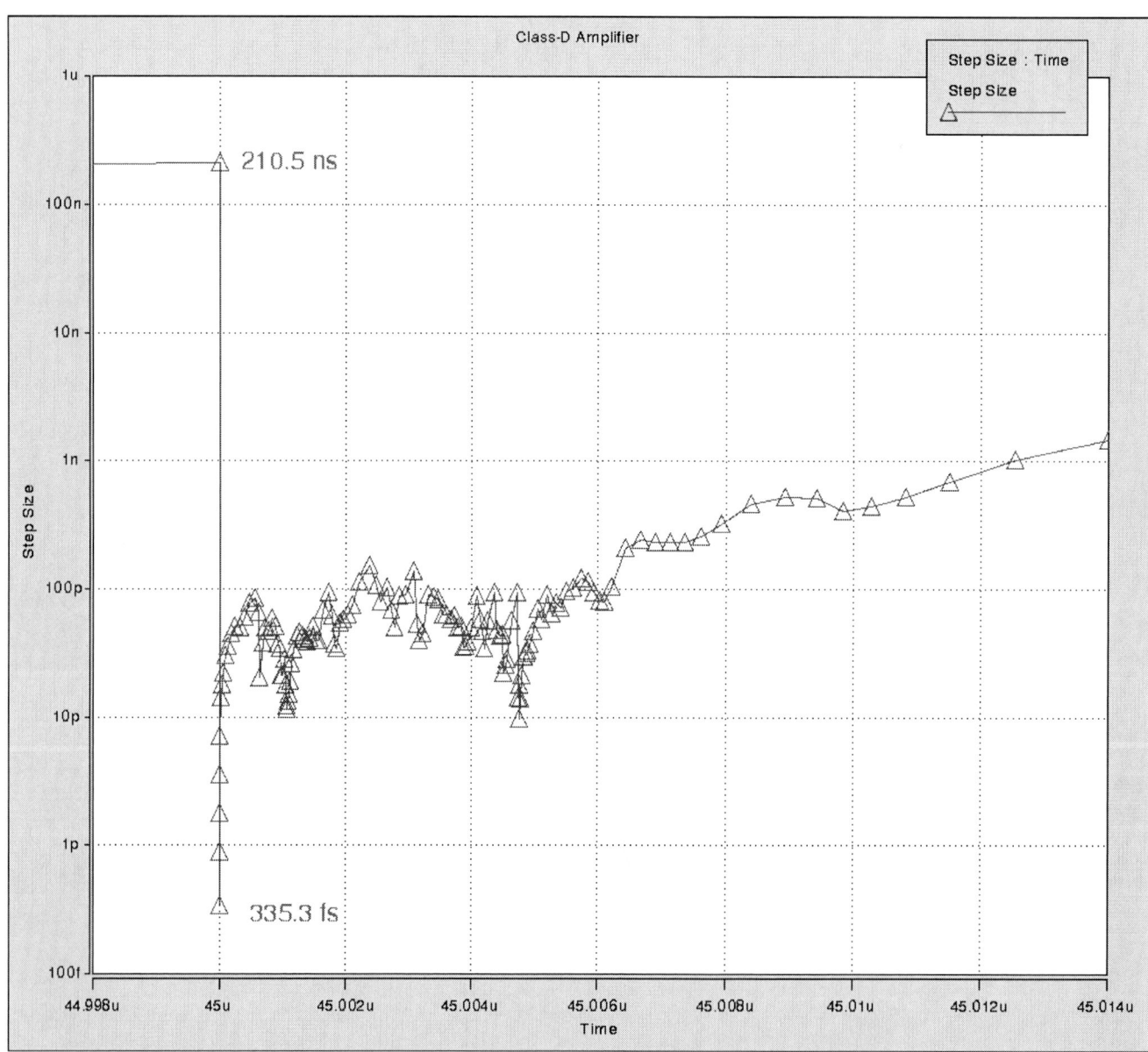

Figure 8. Zoomed-in plot of Figure 1: Modern analog and mixed signal circuits are typically stiff systems characterized by a wide range of time constants. Adaptive time-stepping is essential for the efficiency of a simulation. Time step distribution between 44.998u and 45.014u is shown for the transient simulation of a class-D audio amplifier. Note that the step size drops from 210.5ns to 335.3fs, nearly 6 orders of the magnitude, around t=45us.

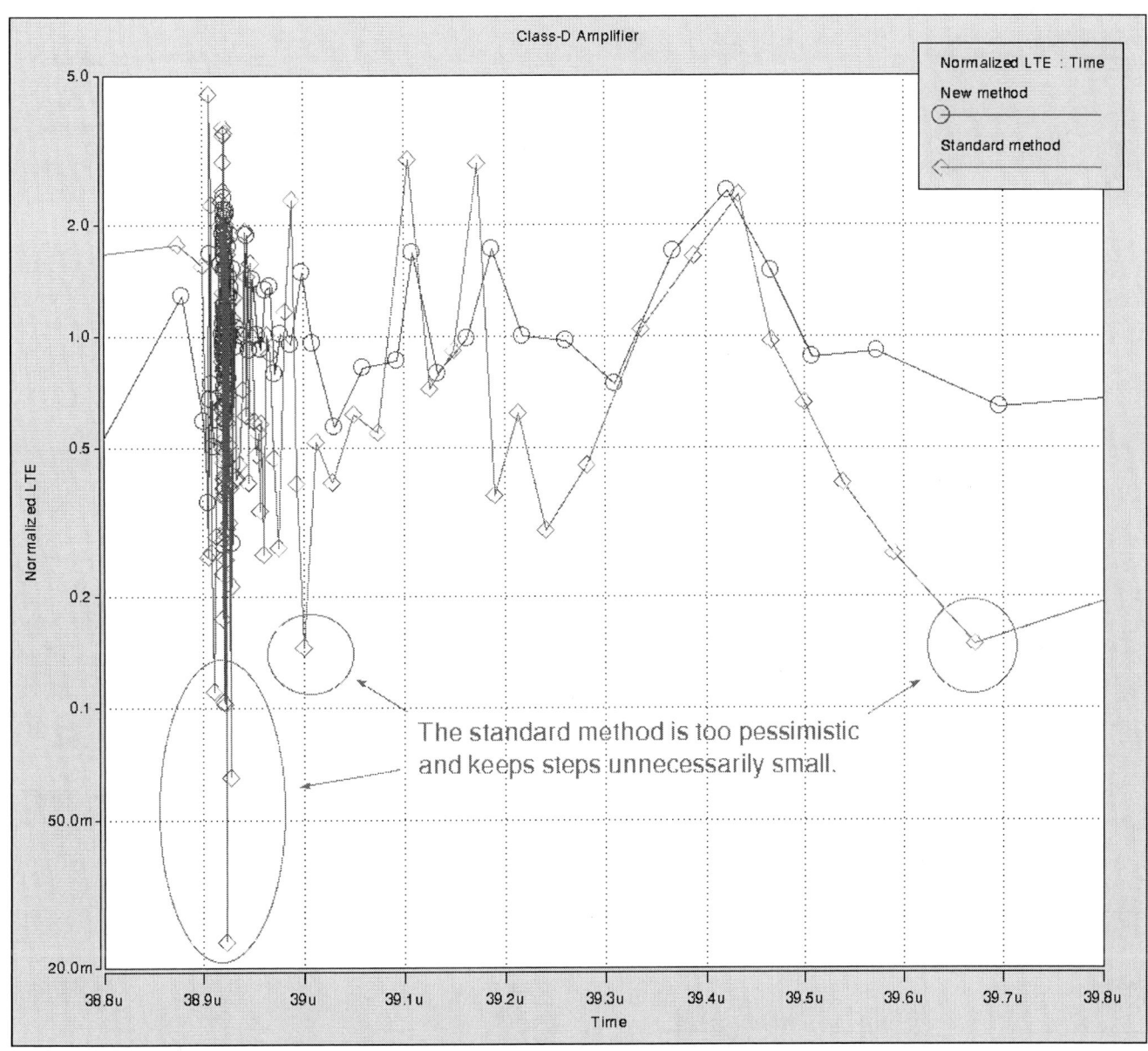

Figure 9. Enlarged view of Figure 5: Normalized local truncation errors at accepted time points between 38.8us and 39.8us are shown for both the new and the standard time-stepping methods. The standard method is too pessimistic and keeps steps unnecessarily small, while the errors for the new method are more evenly distributed and mostly within the specified range between 0.7 and 3.0.

S2. PERFORMANCE BENCHMARKS

Two time-stepping methods are tested on a suite of 21 industry analog and mixed-signal circuits. Detailed results have been summarized in Table 3 and Figure 10.

Table 3. More detailed performance comparison of two stepping methods: The best case and worst case for each column have been highlighted. The new method generates less time points, runs faster, and has less rejected time points across the board. The average iterations per time point increases slightly for the new method but the total numbers of iterations have been greatly reduced.

Benchmark circuits	Size	Standard method		New method		Speedup	Time Point Reduction
		Rejection ratio	#iterations per step	Rejection ratio	#iterations per step		
Boost regulator	420	9.75%	3.24	1.91%	4.14	17.5%	37.9%
Synchronous rectifier	5,761	9.94%	3.08	3.35%	3.96	5.6%	34.7%
Power management unit	57,730	12.41%	3.61	3.36%	4.45	22.1%	41.4%
DC-DC converter 1	27,065	10.83%	3.96	1.30%	4.42	30.4%	40.4%
DC-DC converter 2	3,192	12.95%	3.95	5.89%	4.59	14.3%	35.1%
Controller	15,552	10.30%	3.59	1.07%	4.43	25.3%	40.5%
Gate driver 1	4,228	9.66%	3.17	1.67%	4.03	23.7%	39.1%
LED driver	3,890	9.89%	3.18	3.40%	4.46	30.9%	46.9%
Power interface	7,510	15.82%	3.33	6.27%	4.27	14.1%	40.0%
LDO 1	6,689	13.98%	3.73	1.26%	4.42	18.3%	38.9%
Power stage	4,415	6.01%	2.88	1.89%	4.19	31.2%	48.5%
LDO 2	6,759	8.96%	3.09	1.67%	4.03	48.0%	44.0%
Motor driver converter	82,951	8.21%	2.90	0.03%	3.80	13.1%	44.4%
Preamp 1	46,062	13.78%	5.61	3.65%	5.99	16.9%	34.3%
Large mixed-signal	59,173	9.67%	3.71	2.65%	5.24	15.0%	43.6%
Preamp 2	53,003	7.63%	5.17	1.55%	5.66	21.2%	47.4%
Small mixed-signal	2,869	11.90%	3.58	1.89%	4.06	5.4%	32.7%
PLL	11,696	9.26%	3.01	0.09%	3.81	20.5%	43.2%
Digital PLL	2,873	7.69%	3.01	0.01%	3.92	23.9%	43.2%
Class-D amplifier	4,649	10.34%	3.03	0.76%	3.85	19.3%	41.1%
Gate driver 2	2,619	11.51%	3.02	8.89%	3.68	8.8%	28.8%
Average	--	**10.50%**	**3.52**	**2.50%**	**4.35**	**20.3%**	**40.3%**

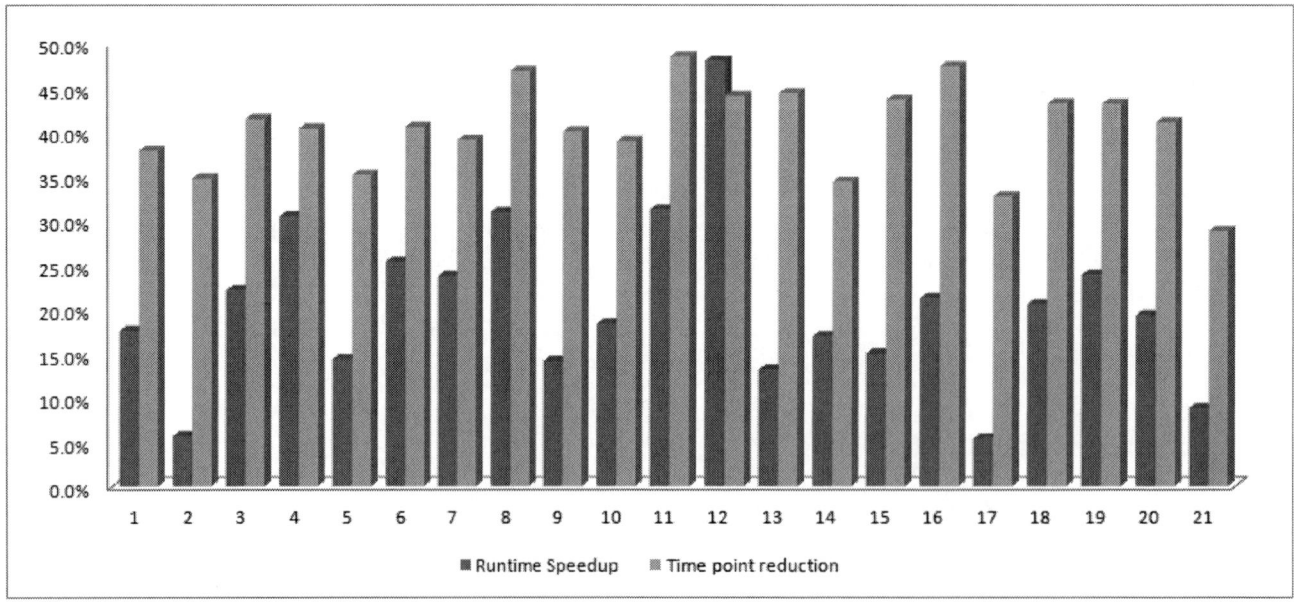

Figure 10. Speedup and time point reduction for the new method (last two columns in Table 3)

S3. COLPITTS OSCILLATOR

The Colpitts oscillator example shown in Figure 11 was used to test the accuracy of two time-stepping methods. Figure 12 shows, for loose tolerances, the oscillator may fail to start due to the large accumulated truncation error when the standard time-stepping method is used.

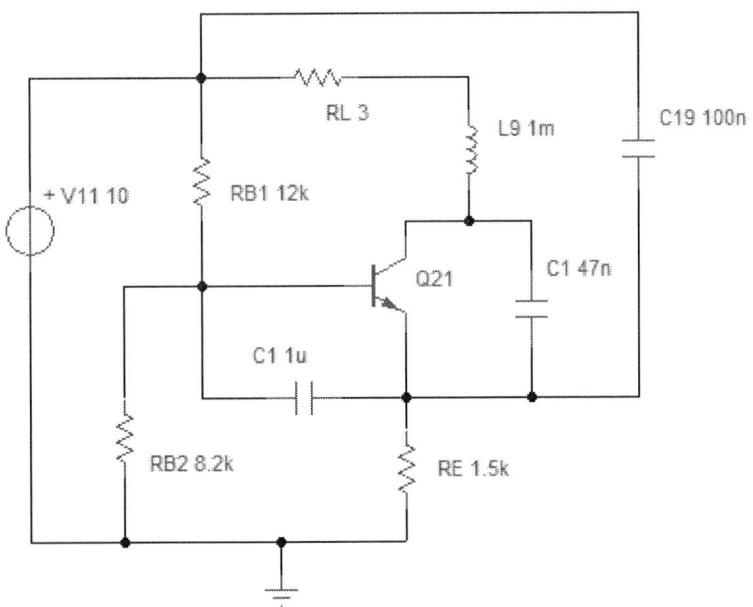

Figure 11. A practical Colpitts oscillator circuit

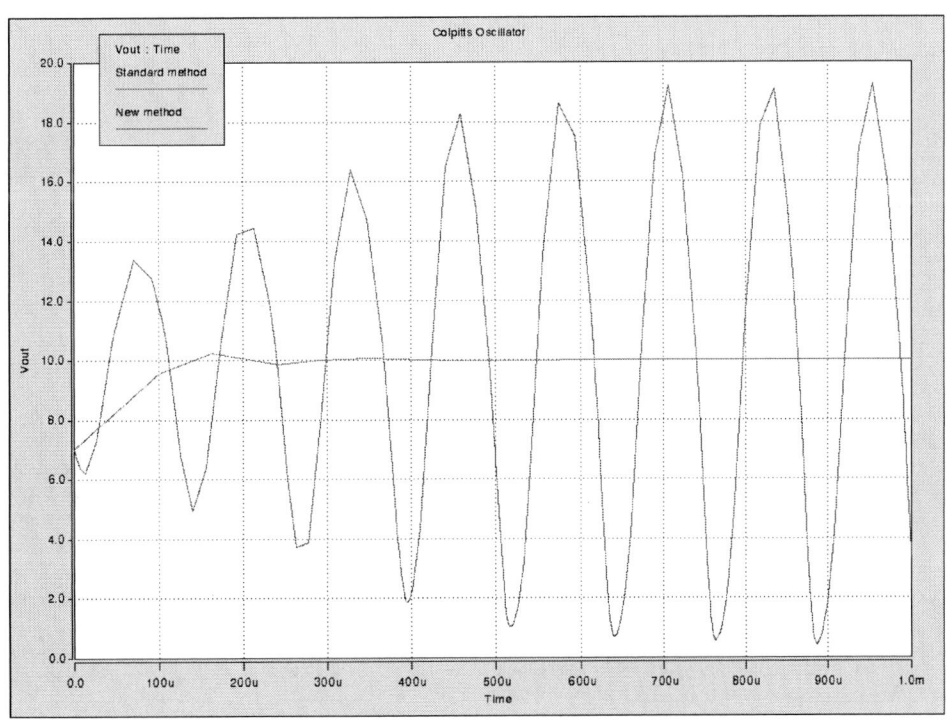

Figure 12. The oscillator fails to start when the standard method is used.

Time-Domain Segmentation based Massively Parallel Simulation for ADCs

Zuochang Ye [†] Bichen Wu [†], Song Han [‡], Yang Li [*]
[†] Tsinghua University, [‡] Stanford University, [*] University of Texas at Austin
zuochang@tsinghua.edu.cn

ABSTRACT

The great availability of massively parallel computing platforms gives rise a question to the EDA industry–how can this be really helping the productivity of circuit designs. Scalability of traditional parallel methods have shown to be limited as the computational resources keep increasing. In this paper we propose a time-domain segmentation method for massively parallel transistor-level simulation for short-memory circuits. SNDR simulation for ADCs is selected as the application as ADCs are typical short-memory circuits and the SNDR simulation is very time consuming. Experiments with realistic Flash and SAR ADCs demonstrate 64x-78x speed-ups with 100 CPU cores. With minor, yet important modifications, the proposed method can even be applied to simulation of Σ-Δ modulator, which does not satisfy the short-memory condition due to the presence of integrator, and 52x speed-up is observed with 100 CPU cores. The implementation of the proposed method is extremely simple and no modification to simulator is needed.

1. INTRODUCTION

With technology advances, the availability of computational power has kept increasing for many years. Nowadays, a small cluster with more than 100 CPU cores, e.g. 10 nodes, 12 cores per node, is easily affordable for a small design house, and computing platforms with more than 1000 CPU cores are also available via HPC or cloud computing service vendors.

This availability of computational power gives rise opportunities as well as challenges for circuit simulation. The ever-increasing circuit complexity calls for the advent of efficient parallel simulation technique. This is especially true for a certain types of applications which require very long simulation time. Signal-to-noise and distortion ratio (SNDR) simulation for ADCs, for example, is a typical simulation problem which is very time-consuming, as the ratio between the total simulation time and the smallest time step is generally very large.

Accelerating circuit simulation with parallelization has been studied for more than two decades. The most traditional methods are to decomposed the circuit either at topology level or matrix level [1–3]. It has been shown that such kind of methods are still being actively studied [4] because of the generality. The performance of

Permission to make digital or hard copies of all or part of this work for personal or classroom use is granted without fee provided that copies are not made or distributed for profit or commercial advantage and that copies bear this notice and the full citation on the first page. To copy otherwise, to republish, to post on servers or to redistribute to lists, requires prior specific permission and/or a fee.

such methods usually depends on the size of the circuit. For very large problems, especially for those from post-layout simulations, such methods may have 10x~20x speed-up. However, for smaller circuits the gain from parallelization is limited.

Parallel relaxation [5–7], parallel numerical integration [8], and multi-algorithm parallelization [9, 10] methods try to perform the parallelization in different dimensions other than just the circuit topology. The scalability of such methods are still limited. In addition, implementing such algorithms in existing simulator is nontrivial.

In this paper, instead of looking to parallelize simulation for general circuits, we narrow the focus on simulation problems for a subset of applications which are much easier to parallelize, i.e. those circuits with short memory. For such problemss, we propose a simple yet effective time-domain segmentation based parallel simulation method. The basic idea of this method is to partition the transient simulation task to segments with shorter period of time and perform simulations on each segment individually. It is obvious that due to the dependence of solution on initial-condition such method will not work for general circuits. However, for specific applications which are usually of most concern in circuit simulation, such simple method can be applied directly, or via some minor, yet important modifications. Experiments with SNDR simulation for Flash, SAR, and Σ-Δ ADCs show that this method provides 52x-78x speed-ups with 100 CPU cores compared with sequential simulations.

This paper is organized as follows. Section II provides background knowledge for transient circuit simulation. Section III explains the basic principle of this method. Section IV gives application on SNDR simulations for Flash, SAR and Σ-Δ ADCs. Finally, conclusions are drawn in Section V.

2. BACKGROUND

In general, transient circuit simulation is to solve the initial condition problem of the following differential equation

$$f(x(t)) + \frac{d}{dt}q(x(t)) + u(t) = 0, \qquad (1)$$

where $x(t)$ is the vector of nodal voltages and/or branch currents. $u(t)$ denotes the input, and $f(\cdot)$ and $q(\cdot)$ denotes static and dynamic nonlinear functions. Numerical integration methods such as Backward Euler method can be applied to convert the equation into a sequence of non-linear algebra equations. Such non-linear equations can be further converted to linear equations through, e.g., Newton's method.

In essence, parallelization is to decompose the computational task and assign each sub-task to different computing units, e.g. CPU cores, threads, or even GPU processing elements. Task decomposition can be applied at different level, and the efficiency of parallelization depends tightly on the coupling between sub-tasks.

For example, device evaluation is embarrassingly parallel, as there is not coupling between sub-tasks. In contrast, solution of linear equations usually subject to coupling which generally limits the efficiency of parallelization.

3. METHOD DESCRIPTION

3.1 Motivative Example

As a motivating example, Fig.1 illustrates the typical output of an ADC driven by a sinusoid input. Slightly different from traditional simulation, here the starting time of the simulation is altered. Green curve is the output when simulation is done for $t \in [t_1^{\text{start}}, t_1^{\text{end}}]$, and red curve is the output when simulation is done for $t \in [t_2^{\text{start}}, t_2^{\text{end}}]$. After a transient time which is shorter than $t_1^{\text{end}} - t_2^{\text{start}}$, the two curves converge with indistinguishable deviation.

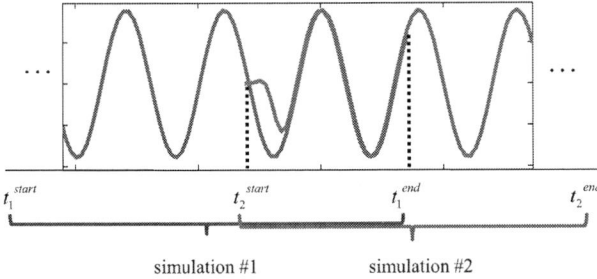

Figure 1: Output of an ADC driven by a sinusoid input. Green: output waveform when simulation is done for t_1^{start} and t_1^{end}. Red: output waveform when simulation is done for t_2^{start} and t_2^{end}. After a short transient time period, red curve converges to the green curve.

This illustrative example give hints to potential parallelization based on task decomposition in time domain. The two simulations can be done concurrently and there is no communication between the two computational tasks. The complete waveform can be generated by conjoining the two waveforms and discarding the starting portion, i.e. $t \in (t_2^{\text{start}}, t_1^{\text{end}})$, of the second waveform.

3.2 Time Domain Segmentation (TDS) Method

This procedure can be generalized and the resulting flow is described in Algorithm 1.

Algorithm 1 Parallel Simulation with Time-Domain Segmentation
---
Step 1: Given t_{end}, the period of transient simulation.
Step 2: Partition the total time $[0, t_{\text{end}}]$ as n equal-length segments $[t_k^{\text{start}}, t_k^{\text{end}}]$, for $k = 1 \cdots n$ with overlap time T_{ov} between adjacent time segments, i.e. $t_{k-1}^{\text{end}} - t_k^{\text{start}} = T_{ov}$.
Step 3: Perform transient simulations for all time segments concurrently with multiple CPU cores. The initial states for each segment is set, e.g. with DC operating points, as in regular transient simulation.
Step 4: Assemble the result by conjoining the waveforms, and discarding the beginning T_{ov} waveforms in each time segment except for the first segment.

Notice that in practice no modification to the simulator is needed. The parallel simulation can be implemented by running multiple simulations for the same circuit, and altering the starting and stoping time.

It is obvious that the proposed flow does not work for general circuits, as in general the solution of DAE equations depends on the initial condition, and the time segmentation essentially neglects this inter-segment dependence and thus may lead to incorrect results. However, in practice, a lot of circuits of interest satisfies the so-called "**short-memory**" condition, i.e. the state of the circuit at a certain time point is correlated only to a limited history of the input. Most of the time, this feature is intensionally designed. For example, consider the case in amplifiers, combinational logic circuits, and more of interest, ADCs.

Suppose the total transient simulation time is T_{total}, the overlap time is T_{ov}, the number of CPU cores is n, which is the same as the number of segments. The total simulation time decreases with increasing CPU cores. Suppose the sequential simulation time is $T_{\text{sequential}}$, and assume that the simulation time is proportional to the length of time segments. The time for TDS parallel simulation will be

$$T_{\text{parallel}} = T_{\text{sequential}}(1/N + \alpha) \qquad (2)$$

where $\alpha = T_{ov}/T_{\text{sequential}}$.

Fig.2 plots the total simulation time with respect to the number of CPU cores for different α. As the number of cores increases, the

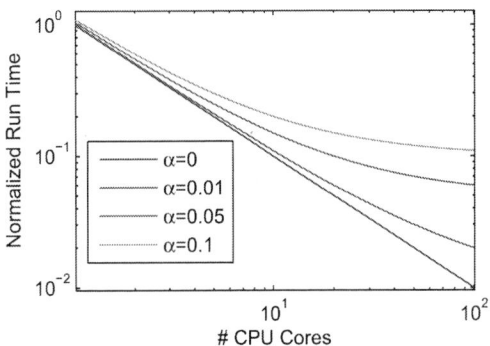

Figure 2: Normalized run time with respect to number of CPU cores.

length of each segment decreases and T_{ov} takes larger portion of each core's simulation, and the run time behaves sub-linear with respect to CPU cores. When $\alpha \ll 1$, the run time reduction approaches linear. In many practical simulations, such as in SNDR simulation for ADCs, T_{total} is very large and thus the ratio α is very small, and near linear speed-up can be observed even for $n = 100$.

3.3 Adaptive Setting of Overlap Time T_{ov}

Obviously proper T_{ov} is circuit-dependent. Large T_{ov} ensures the adjacent waveforms converge with each other. However, overly conservative T_{ov} may also limit the efficiency of parallelization.

In practice, T_{ov} can be set adaptively by the following procedure.

1. Set T_{ov} to 0 or any estimated value.

2. Perform the TDS algorithm and verify the convergence by checking each pair of adjacent waveforms.

3. If not converged, extend T_{ov} by a small step and continue the simulation, otherwise stops the simulation and return. Repeat until converged.

3.4 Limitations and Scope of Application

Obviously the proposed TDS method is too simple to handle general situations in circuit simulation. The key assumption is that the circuit to be simulated should be with short-memory, i.e., it will converge to its nominal state determined only by the current input or a short input history. In general there is no general "rules" that determine whether or not a certain circuit can be applied with the proposed method. Instead, designer's insight is required. For example, circuits with registers in general is not suitable for the proposed method. However, if the data bits stored in the register is known a prior to be with "short memory", such as the SAR ADC illustrated in the next section, then the proposed method is applicable.

Notice that in some cases, the circuit may behave abnormally for some special input history. It is not guaranteed that such kind of abnormal behavior can be detected by the proposed method. The scenarios which best fit the proposed method should be those in which functional verification has been done and one wants to obtain the performance of the circuit, as in the case of SNDR simulation for ADCs.

4. APPLICATIONS ON ADC SIMULATION

ADCs are circuits that perform analog-to-digital conversion. The most important performance metric which designers want to obtain through simulation is the Signal to Noise and Distortion Ratio (S-NDR). SNDR simulation can only be done with transient simulation, in which a sinusoid signal is fed into the ADC and the output of ADC is recorded. The output is then converted to frequency domain, and SNDR is computed by dividing the output signal power by the summation of all the noise and distortion power. The step size of simulation is determined by the fastest switching signal in the whole circuit, which in general is the clock. The total simulation time is determined by the frequency precision which is usually a fraction of bandwidth. In practice, the ratio between total simulation time and the time step is very large, making SNDR simulation of ADC extremely time consuming.

It is obvious that except for very special cases which are out of the scope of this paper, ADCs are by-design short-memory circuits. Ideally, the output of any ADC should follow the input immediately subject to quantization error determined by the bit-width of ADCs. Even with taking the electrical noise and non-linearity into account, the output should be a function of the current input, and irrelevant to history. In realistic circuits, the output of ADCs depends on the history of the input due to dynamic behavior of the ADCs caused by transistor delay, RC time constant, etc. However, this history dependence is limited to a very short period of time, especially compared with the total simulation time. Thus ADCs are perfect examples of short memory circuit.

In the following experiments, all the simulations are done by Cadence *spectre*. Parallelization is realized by splitting the time segments and running *spectre* simulation individually. Distributed computing is performed on an 8-node cluster with 12 cores per node plus a single-node server with 8 cores, enabling a maximally 104x parallelization.

4.1 Flash and SAR ADC

Flash ADC is also known as direct conversion ADC. It mainly consists of voltage dividing resistor array, comparator array, encoding logic and an output register. The input signal is compared with evenly-distributed voltage levels simultaneously and generate binary output representing the approximation to input signal. Despite the difference in architecture, Flash ADC spends constant number of clock cycles to sample, compare, encode and etc. to complete the conversion. When the total simulation time is much larger than the sum of these times, Flash ADCs are short memory circuits.

In our experiment, a 5-bit Flash ADC targeted at high-speed communication is studied. The topology is similar to [11], and

there are totally 7011 MOSFET transistors. The sampling clock frequency is 2 GHz. In simulation, the input signal is set to 111 MHz. The overlap time is set to 6 clock cycles to ensure that the circuit settles completely. The total simulation time is 2 us, which corresponds to 1000 clock cycles.

The Flash ADC is simulated with 1x, 2x, 4x, 8x and 100x parallelization. The result from serial simulation (1x) is taken as the reference. The frequency-domain magnitude of the outputs in dB for 1x, 2x, and 100x parallelization are given in Fig. 3, which shows perfect agreement between sequential simulation and parallel simulations. Table 1 gives quantitative comparison of run times and computed SNDR values. 64x speed-up is obtained with 100x parallelization. The sub-linear speed-up is mainly due to both memory bandwidth limitation and unequal run times for equal-length time segments. The run time of parallelized simulation is computed as the maximum run time of all sub-tasks.

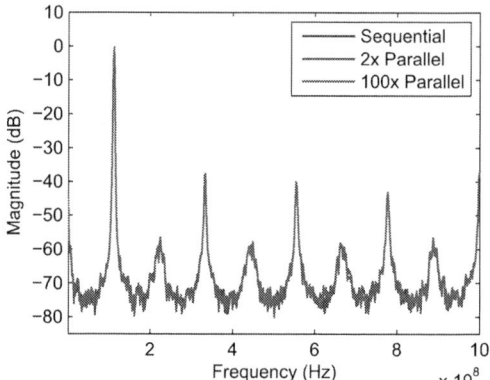

Figure 3: Output of Flash ADC

Table 1: Simulation Results for FLASH ADC

#CPU Cores	SNDR (dB)	Run Time (hours)	Speed-ups
1	30.94	38	1x
2	30.94	20	1.9x
4	30.94	10	3.7x
8	30.94	5.3	7.1x
100	30.93	0.6	64x

SAR ADC, or successive approximation register ADC is based on the binary search algorithm. A typical SAR ADC consists of a comparator which compares the input signal with the voltage stored digitally in the registers and converted to analog via the DAC. Although registers presence in SAR ADCs, the data bits stored in the registers updates with the input signal and the life-time of data compared with the total simulation time is still very short. Thus the TDS method should still applicable for SAR ADCs.

In our experiment, a 6-bit SAR ADC also targeted at high-speed communication is studied. The topology is similar to [12], and there are totally 2673 transistors. The sampling clock frequency is 1 GHz. In simulation, the input signal is set to 400 MHz. The overlap time is set to 6 clock cycles to ensure that the circuit settles completely. Total simulation time is 8 us, which corresponds to 8000 clock cycles.

The SAR ADC is also simulated with 1x, 2x, 4x, 8x and 100x parallelization. The frequency-domain magnitude of the outputs for 1x, 2x, and 100x parallelization are given in Fig. 4, which also shows perfect agreement between sequential simulation and

970

parallel simulations. Table 2 gives quantitative comparison of run times and computed SNDR values. 78x speed-up is obtained with 100x parallelization.

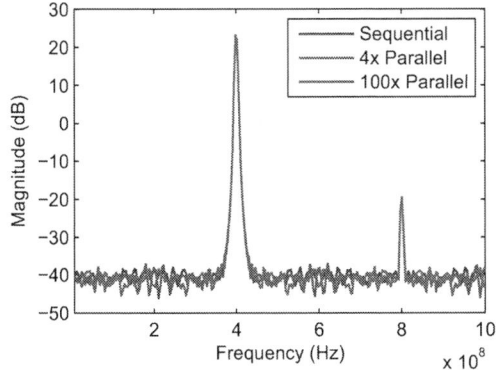

Figure 4: Output of SAR ADC

Table 2: Simulation Results for SAR ADC

#CPU Cores	SNDR (dB)	Run Time (hours)	Speed-ups
1	37.19	41	1x
2	37.19	21	1.9x
4	37.19	11	3.7x
8	37.19	5.7	7.2x
100	37.18	0.5	78x

4.2 Σ-Δ ADC

Illustrated in Fig.5, a typical Σ-Δ ADC [13] consists of an analog Σ-Δ modulator, and a digital low-pass filter followed by a digital decimation filter. In Σ-Δ ADC design, circuit simulations are only performed on the modulator, as bringing the digital filters into simulation is neither practical nor necessary.

The input to the Σ-Δ modulator is a normal analog signal, and the output of the modulator is a bitstream valued either 0 or 1. The percentage of 1s in the bitstream reflects the magnitude of input voltage. Digital low-pass filter recovers the magnitude from the bitstream and the decimation filter down-samples the filtered result to generate the output with given data rate. Typical structure of Σ-Δ modulator is illustrated in Fig. 6. It consists of a subtractor, an integrator, a comparator and a D/A convertor. Practical Σ-Δ modulators may have multi-stage (high-order) or even multi-loop structures.

In our experiment, a typical single-loop 3rd-order Σ-Δ modulator targeted for audio application is studied. There are totally 464 transistors. The sampling clock frequency is 15 MHz, the input signal frequency is 25 kHz. The total simulation time is 2.68 ms, which corresponds to 40000 clock cycles.

At a first glance, the presence of integrator causes long-term history dependence, and thus violates the short-memory condition which is a necessary condition for the proposed TDS method to be applicable. Fig 7 shows the output of Σ-Δ modulator resulting from simulations for two adjacent time segments. It shows that after many clock cycles the two adjacent curves do not exhibit any trend of convergence.

However, considering the fact that the information encoded in the bitstream, i.e. the percentage of 1s, still depends only on the current input, it could be possible that the TDS scheme still works

if applied with some special treatments. First we will look into the details of TDS results.

Fig. 8 shows the frequency-domain magnitude of the output signal before and after digital filter when the waveforms from 4x TDS simulation are concatenated. The noise floor is erroneously shifted from around -120dB to around -90dB even when the overlap time is long enough (2000 clock cycles in this case). To show this discrepancy in the time domain, the deviations between digital-filtered output waveforms from serial and 4x parallel simulations is generated as illustrated in Fig 9. Fig. 10 shows the deviations corresponding to different overlap times. It clearly shows that the deviations occurs only at the conjunctions of adjacent segments. Although the magnitudes change for different overlap times, the widths of deviations are constant. Converting the deviation to frequency domain shows that such impulse-like deviations are the root cause of the artificial white noise in the output which leads to the shifted noise floor in Fig. 8.

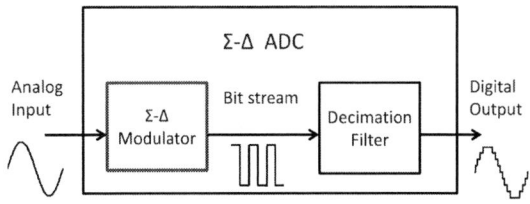

Figure 5: Block diagram of Sigma-Delta ADC. Usually only the modulator is included in circuit simulation.

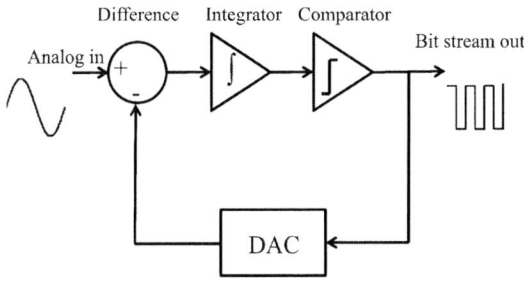

Figure 6: Block diagram of Sigma-Delta modulator. The presence of the integrator violates the short-memory condition.

Figure 7: Output of Σ-Δ modulator resulting from simulations for two adjacent time segments. No sign of convergence is observed with many clock cycles.

It would be interesting where these deviations are coming from and why they are all impulse-like. Following we will give an explanation and after which a solution of fixing this noise-floor-shifting problem will be given.

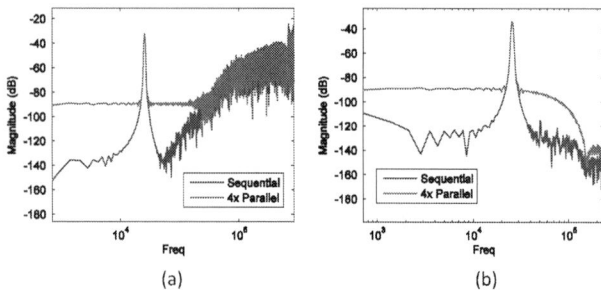

(a)   (b)

Figure 8: Spectrum of output signal (a) Before digital filter. (b) After behavior-level digital filter output.

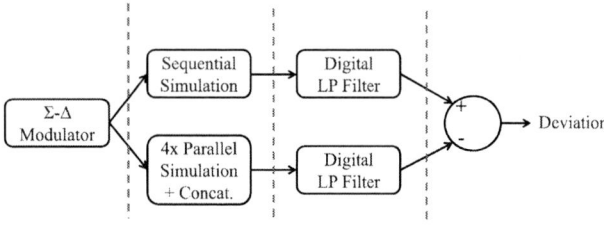

Figure 9: Flow to generate output deviation between digitally filtered output signal by sequential and parallel simulation.

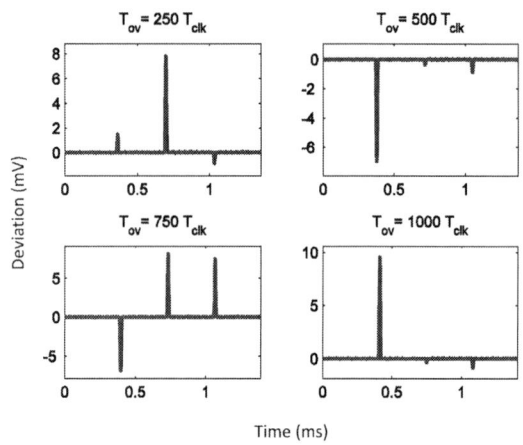

Figure 10: Waveform deviation produced with the flow described in Fig 9. Deviations occurs at the conjunction of adjacent segments.

Consider a simple case, suppose the reference voltage of ADC is 1V, and the input to the Σ-Δ ADC is constantly 0.5V. And consider the filtering window of digital low-pass filter is 4 bits and decimation rate is 2. Depending on the initial condition the output of Σ-Δ modulator has two possibilities showing in Fig. 11 (a) and (b). Both bitstreams represent a value 2/4=0.5 as shown in the "filtered and decemated" result above the bitstreams. Suppose the two bitstreams are resulting from two independent simulation, and one concatenate the bitstreams by taking the first half of (a) and the second half of (b), the result will be as shown in Fig. 11 (c), in which when the filtering window crosses the two halves of the

bitstream, error codes, i.e. the two '3/4's, will be generated. It is clear that the width of deviation, i.e. 2 bits, is determined by the filtering window, i.e. 4 bits, divided by decimation rate, which is 2. This has been examined in the true Σ-Δ ADC simulation, in which the digital filter consists of three stages of identical low-pass filter with timing window 32 plus decimation filter with decimation factor 4. Thus in theory the width of the deviations should be 8. Fig. 12 shows the zoom-in view of the deviation in realistic simulation, which confirms this prediction.

Algorithm 2 TDS for Σ-Δ Modulator

Step 1: Perform Step 1 to Step 3 of Algorithm 1.

Step 2: Perform behavior-level low-pass and decimation digital filtering on the outputs for all time segments.

Step 3: Perform Step 4 of Algorithm 1 on the filtered result.

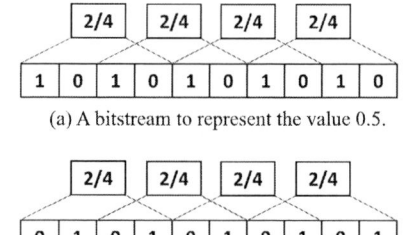

(a) A bitstream to represent the value 0.5.

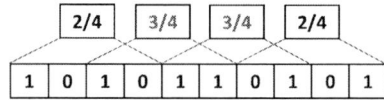

(b) Another bitstream that also represents the value 0.5.

(c) Take the first half of (a) and the second half of (b).

Figure 11: Illustration of concatenation error caused by digital filter. Deviation only occurs when the window of digital filter crosses two segments.

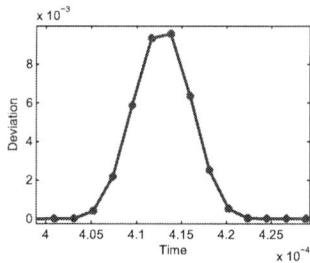

Figure 12: Zoom-in view of the waveform deviation in Fig. 10. The width of deviation, i.e. 8, agrees with theoretical estimation.

Now it is clear that the error is caused by the digital filtering during the filtering window across two adjacent segments. One way to avoid this unwanted phenomenon is to perform the digital filtering before the concatenation. This way, instead of concatenating the waveform of the Σ-Δ modulator, we are actually concatenating the waveform of a virtual Σ-Δ ADC, and the latter, by design, satisfies the short-memory condition for TDS method to be applicable.

Algorithm 2 describes the TDS simulation for Σ-Δ modulator. Notice that the result produced by this algorithm is the digitally filtered waveform instead of the bitstream. As long as the digital filter is not the performance limiter, the SNDR computed from the filtered output will be the same as that computed from the bitstream directly. Notice that the digital filter can be easily implemented in Matlab, and it is not necessary to be exactly the same as the filter that will be used in the actual Σ-Δ ADC.

Algorithm 2 is applied for 8x and 100x parallelization with $T_{\text{ov}} = 400T_{\text{clk}}$. Fig. 13 shows the comparison of results with sequential and 100x parallel simulation. The computed spectrums agree with each other, validating the accuracy of TDS parallel simulation. Table 3 gives quantitative results for 1x, 8x, and 100x parallel simulation. The difference of computed SNDR in dB is less than 1%, and about 52x speed-up is observed with 100x parallel simulation compared with sequential simulation.

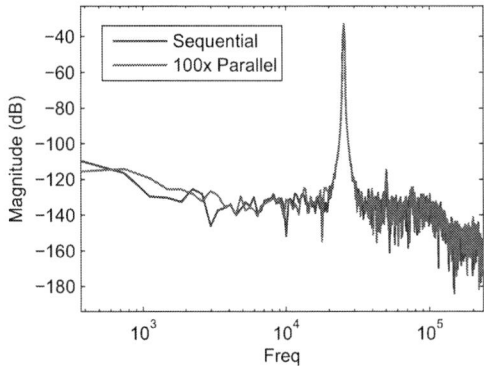

Figure 13: Output signal produced by filtering first and concatenating last. This avoids the problem with concatenating first, and produces correct result.

Table 3: Simulation Results for Sigma-Delta ADC

#CPU Cores	SNDR (dB)	Run Time (hours)	Speed-ups
1	73.7	71	1x
8	74.3	11	6.4x
100	73.7	1.4	52x

5. CONCLUSION

In this paper, parallel circuit simulation for ADCs is studied. Short memory nature of ADCs are exploited and the simulation problem becomes embarrassingly parallel after time domain segmentation. Simulations for realistic Flash, SAR ADCs have been done and 64X and 78X speed-ups have been observed with 100 CPU cores in a distributed computing environment. Even for Σ-Δ modulators, which are not short memory circuits due to the presence of integrator, the TDS approach can still be applied after minor modification, and about 52 speed-ups is observed with 100 CPU cores. Examples also show that the accuracy is not scarified with the proposed algorithm. The proposed approach provides orthogonal speed-ups with other parallelization techniques, and no modification to simulator is needed, thus implementation is extremely simple.

6. REFERENCES

[1] P. Cox, R. Burch, D. Hocevar, P. Yang, and B. Epler, "Direct circuit simulation algorithms for parallel processing [vlsi]," *Computer-Aided Design of Integrated Circuits and Systems, IEEE Transactions on*, vol. 10, no. 6, pp. 714 –725, jun 1991.

[2] M.-C. Chang and I. Hajj, "ipride: a parallel integrated circuit simulator using direct method," in *Computer-Aided Design, 1988. ICCAD-88. Digest of Technical Papers., IEEE International Conference on*, nov 1988, pp. 304 –307.

[3] J. Demmel, J. Gilbert, and X. Li, "An asynchronous parallel supernodal algorithm for sparse gaussian elimination," *SIAM Journal on Matrix Analysis and Applications*, vol. 20, no. 4, pp. 915–952, 1999.

[4] X. Chen, Y. Wang, and H. Yang, "An adaptive lu factorization algorithm for parallel circuit simulation," in *Design Automation Conference (ASP-DAC), 2012 17th Asia and South Pacific*, 30 2012-feb. 2 2012, pp. 359 –364.

[5] Y.-C. Wen, K. Gallivan, and R. Saleh, "Improving parallel circuit simulation using high-level waveforms," in *Circuits and Systems, 1995. ISCAS '95., 1995 IEEE International Symposium on*, vol. 1, apr-3 may 1995, pp. 728 –731 vol.1.

[6] E. Lelarasmee, A. Ruehli, and A. Sangiovanni-Vincentelli, "The waveform relaxation method for time-domain analysis of large scale integrated circuits," *Computer-Aided Design of Integrated Circuits and Systems, IEEE Transactions on*, vol. 1, no. 3, pp. 131 – 145, july 1982.

[7] M. Al-Khaleel, A. Ruchli, and M. Gander, "Optimized waveform relaxation methods for longitudinal partitioning of transmission lines," *Circuits and Systems I: Regular Papers, IEEE Transactions on*, vol. 56, no. 8, pp. 1732 –1743, aug. 2009.

[8] W. Dong, P. Li, and X. Ye, "Wavepipe: Parallel transient simulation of analog and digital circuits on multi-core shared-memory machines," in *Design Automation Conference, 2008. DAC 2008. 45th ACM/IEEE*, june 2008, pp. 238 –243.

[9] X. Ye, W. Dong, P. Li, and S. Nassif, "Maps: Multi-algorithm parallel circuit simulation," in *Computer-Aided Design, 2008. ICCAD 2008. IEEE/ACM International Conference on*, nov. 2008, pp. 73 –78.

[10] ——, "Hierarchical multialgorithm parallel circuit simulation," *Computer-Aided Design of Integrated Circuits and Systems, IEEE Transactions on*, vol. 30, no. 1, pp. 45 –58, jan. 2011.

[11] B. Verbruggen, P. Wambacq, M. Kuijk, and G. Van der Plas, "A 7.6 mw 1.75 gs/s 5 bit flash a/d converter in 90 nm digital cmos," in *VLSI Circuits, 2008 IEEE Symposium on*, june 2008, pp. 14 –15.

[12] J. Yang, T. Naing, and R. Brodersen, "A 1 gs/s 6 bit 6.7 mw successive approximation adc using asynchronous processing," *Solid-State Circuits, IEEE Journal of*, vol. 45, no. 8, pp. 1469 –1478, aug. 2010.

[13] R. Schreier, S. R. Norsworthy, and G. C. Temes, *Delta-Sigma Data Converters: Theory, Design, and Simulation*. New York: John Wiley and Sons, Inc, 1996.

A Direct Finite Element Solver of Linear Complexity for Large-Scale 3-D Circuit Extraction in Multiple Dielectrics

Bangda Zhou, Haixin Liu, and Dan Jiao
School of Electrical and Computer Engineering, Purdue University, West Lafayette, IN 47907

ABSTRACT

We develop a direct finite-element solver of linear (optimal) complexity to extract broadband circuit parameters such as S-parameters of arbitrarily shaped 3-D interconnects in inhomogeneous dielectrics. Numerical experiments demonstrate a clear advantage of the proposed solver as compared with existing finite-element solvers that employ state-of-the-art direct sparse matrix solutions. A linear complexity in both CPU time and memory consumption is achieved with prescribed accuracy satisfied. A finite-element matrix from the analysis of a large-scale 3-D circuit in multiple dielectrics having 5.643 million unknowns is directly factorized in less than 2 hours on a single core running at 2.8 GHz.

Categories and Subject Descriptors

B.7.2 [**Integrating Circuits**]: Design Aids - simulation, verification

General Terms

Algorithms

Keywords

Circuit extraction, interconnect, finite element methods, direct solvers

1. INTRODUCTION

Multicore and many-core computing have become a new form of equivalent scaling to accompany the continuation of Moore's Law. As information needs to be transferred in and out of the processors at a throughput proportional to the computational performance, a concurrent rapid scaling of processor input/output (I/O) bandwidth is required. Thus, the design of high-bandwidth I/O has become a new challenge in current and future integrated circuit and system design. To meet such a challenge, accurate and efficient

modeling of non-quasi-static effects, substrate noise, high-frequency noise, and parasitic coupling is called for. Modeling of effects that have a more global influence such as cross talk, substrate return path, substrate coupling, and electromagnetic radiation is also demanded. The advent of new technologies such as 3D integration by the use of TSVs and wireless signaling with passive devices further challenges the traditional quasi-statics and/or 2D based circuit modeling approaches.

The fullwave based methods for 3-D circuit modeling can be categorized into two broad classes: integral equation (IE) based methods and partial differential equation based methods. In the latter, a representative method is the finite element method (FEM). A surface IE-based method reduces a 3-D volumetric problem to a surface problem. However, its formulation becomes cumbersome when the circuit to be extracted involves complicated materials. Little work has been reported in a surface IE-based fullwave extraction of 3-D lossy circuits in nonuniform dielectrics. A volume IE-based method is capable of handling arbitrary nonuniform materials with great ease, however, the resulting linear system of equations is not only dense but also large involving volume unknowns in the entire 3-D circuit. Therefore, when the problem of interest involves complicated materials, the FEM method has been a popular method for choice because of its great capability in handling both irregular geometries and inhomogeneous materials.

As a partial differential equation based method, the FEM produces a sparse system matrix for solving Maxwell's equations. Although the system matrix is sparse, its inverse as well as LU factors are generally dense. As a result, solving it can be a computational challenge when the matrix size is large. A traditional direct solution is computationally expensive. It is shown in [1] that the optimal operation count of the direct solution of an FEM matrix in exact arithmetic is $O(N^{1.5})$ for 2-D problems, and $O(N^2)$ for 3-D problems, where N is the matrix dimension. Although there have been successes in speeding up the direct finite element solution by a multifrontal based algorithm [2] or by an $\mathcal{H}$-matrix based mathematical framework [3] for 3D fullwave analysis, as yet, no $O(N)$ complexity, i.e. optimal complexity, has been accomplished for the FEM-based direct solution of general 3-D circuit problems.

State-of-the-art fast FEM-based solvers rely on iterative approaches to solve large-scale circuit problems. The computational complexity of an iterative solver is $O(N_{it}N_{rhs}N)$ at best, where N_{it} is the number of iterations, and N_{rhs} the number of right hand sides. When N_{it} and N_{rhs} are large,

Permission to make digital or hard copies of all or part of this work for personal or classroom use is granted without fee provided that copies are not made or distributed for profit or commercial advantage and that copies bear this notice and the full citation on the first page. To copy otherwise, to republish, to post on servers or to redistribute to lists, requires prior specific permission and/or a fee.
DAC '13, May 29 - June 07 2013, Austin, TX, USA.

state-of-the-art iterative solutions become inefficient since the entire iteration procedure has to be repeated for each right hand side. To give an example, assuming the CPU cost for each port is 30 minutes, to assess crosstalk between 64 ports in a high-speed I/O, one has to wait for 1.3 days. Furthermore, the complexity of an iterative solver is problem dependent since the iteration number N_{it} is, in general, problem dependent.

The contribution of this paper is a direct FEM solver of linear complexity for extracting fullwave circuit parameters from arbitrarily shaped 3-D lossy conductors in inhomogeneous materials. Both theoretical analysis and numerical experiments have demonstrated its linear complexity in both CPU time and memory consumption with prescribed accuracy satisfied. The proposed direct solver successfully factorizes an FEM matrix having 5,643,240 unknowns resulting from the analysis of a large-scale 3-D lossy circuit in multiple dielectrics in less than 2 hours on a single core running at 2.8 GHz. Comparisons with state-of-the-art direct FEM solvers that employ the most advanced direct sparse matrix solutions as well as a commercial iterative FEM solver have shown a clear advantage of the proposed direct solver.

2. VECTOR FINITE-ELEMENT ANALYSIS

Consider a general 3-D circuit with arbitrarily shaped lossy conductors and inhomogeneous dielectrics that can be lossless, lossy, and dispersive. The physical phenomena in such a circuit are governed by the second-order vector wave equation

$$\nabla \times (\frac{1}{\mu_r} \times \mathbf{E}) + jk_0\eta_0\sigma\mathbf{E} - k_0^2\varepsilon_r\mathbf{E} = -jk_0Z_0\mathbf{J}, \quad (1)$$

where $\mathbf{E}$ is electric field intensity, μ_r is relative permeability, ε_r is relative permittivity, σ is conductivity, k_0 is free-space wave number, Z_0 is free-space wave impedance, and $\mathbf{J}$ is current density. A finite element based solution of (1) results in the following linear system of equation:

$$\mathbf{Y}\{u\} = \{b\}, \quad (2)$$

in which b denotes a vector of current sources, $\mathbf{Y}$ is a sparse matrix, which can be written as

$$\mathbf{Y} = -k_0^2\mathbf{T} + jk_0\mathbf{G} + \mathbf{S} \quad (3)$$

where $\mathbf{T}$, $\mathbf{G}$, and $\mathbf{S}$ are assembled from their elemental contributions as the following

$$\begin{aligned}
\mathbf{T}_{ij}^e &= \iiint_V \varepsilon_r\mathbf{N}_i \cdot \mathbf{N}_j\mathrm{d}V \\
\mathbf{S}_{ij}^e &= \iiint_V \frac{1}{\mu_r}(\nabla \times \mathbf{N}_i) \cdot (\nabla \times \mathbf{N}_j)\mathrm{d}V \quad (4) \\
\mathbf{G}_{ij}^e &= \iiint_V \eta_0\sigma\mathbf{N}_i \cdot \mathbf{N}_j\mathrm{d}V + \iint_S (\hat{n} \times \mathbf{N}_i) \cdot (\hat{n} \times \mathbf{N}_j)\mathrm{d}S,
\end{aligned}$$

where V denotes the volume of each element e, S is the outermost boundary of the entire circuit, $\hat{n}$ is a unit normal vector, and $\mathbf{N}$ is the vector basis function used to expand unknown $\mathbf{E}$ in each element.

3. MATHEMATICAL BACKGROUND

In state-of-the-art multifrontal based direct sparse solvers [2, 5], the overall factorization of a sparse matrix is organized into a sequence of partial factorizations of smaller dense frontal matrices. Various ordering techniques have been adopted to reduce the number of fill-ins introduced during the direct matrix solution process. The computational cost of a multifrontal based solver depends on the number of nonzero elements in the $\mathbf{L}$ and $\mathbf{U}$ factors of the sparse matrix. This solver is based on exact arithmetic, and hence its optimal complexity is higher than linear complexity [1].

Recently, it is proved in [3] that the sparse matrix resulting from a finite-element based analysis of electromagnetic problems can be represented by an $\mathcal{H}$ matrix [4] without any approximation, and the inverse as well as $\mathbf{L}$ and $\mathbf{U}$ of this sparse matrix has a data-sparse $\mathcal{H}$-matrix approximation with a controlled error. It is shown that an $\mathcal{H}$-matrix based direct FEM solver has a complexity of $O(NlogN)$ in storage and a complexity of $O(Nlog^2N)$ in CPU time for solving general 3-D circuit problems.

In an $\mathcal{H}$ matrix [4], the matrix blocks are partitioned into multilevel admissible blocks and inadmissible blocks. The admissible blocks appear in the off-diagonal blocks. Consider a matrix block $\mathbf{Y}^{t,s}$ formed by unknown sets t and s, if it satisfies the following admissibility condition

$$\min\{diam(\Omega_t), diam(\Omega_s)\} < \eta dist(\Omega_t, \Omega_s), \quad (5)$$

then it is admissible. In the above, Ω_t denotes the support of all basis functions belonging to set t, Ω_s denotes the support of all basis functions belonging to set s, $diam(\cdot)$ is the Euclidean diameter, $dist(\cdot, \cdot)$ is the Euclidean distance between two sets, and η is a positive parameter that can be used to control the accuracy of the admissibility condition. A matrix block that does not satisfy the above admissibility condition is called inadmissible. In an $\mathcal{H}$ matrix, an inadmissible block keeps its original full matrix form, while an admissible block is represented by a low-rank matrix shown as the following:

$$\tilde{\mathbf{Y}}^{t \times s} = \mathbf{A}_{\#t \times k} \cdot \mathbf{B}_{\#s \times k}^T, \quad (6)$$

where k is the rank which is bounded by a constant for circuit problems [3], and $\#$ denotes the cardinality of a set. The error of a rank-k approximation can be evaluated as [4]

$$\|\mathbf{Y}^{t \times s} - \tilde{\mathbf{Y}}^{t \times s}\|_2 = \sigma_{k+1}, \quad (7)$$

in which σ_{k+1} is the maximum singular value among truncated singular values. By applying (7), the error of an $\mathcal{H}$-matrix representation can be quantitatively controlled. With a rank-k representation, an $\mathcal{H}$ matrix significantly accelerates matrix-related operations and reduces storage cost for a prescribed accuracy as compared to a full-matrix based representation.

4. PROPOSED LINEAR-COMPLEXITY DIRECT FINITE ELEMENT SOLVER

In the proposed method, we fully take advantage of the zeros in the original FEM matrix, and also the zeros in the $\mathbf{L}$ and $\mathbf{U}$ factors. Instead of treating $\mathbf{L}/\mathbf{U}$ as a whole $\mathcal{H}$-matrix like what is done in [3], we only store the nonzeros in $\mathbf{L}$ and $\mathbf{U}$ with a compact error-controlled $\mathcal{H}$-matrix representation, compute these nonzeros with fast $\mathcal{H}$-matrix based arithmetic, while completely removing all the zeros in $\mathbf{L}$ and $\mathbf{U}$ from storage and computation. Since the geometry of a 3-D circuit is known, we maximize the zeros in $\mathbf{L}$ and $\mathbf{U}$ by nested dissection ordering [1]. We build an elimination tree based on the nested dissection ordering, and precisely

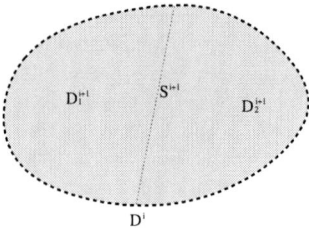

Figure 1: An example of nested dissection on domain cluster D^i

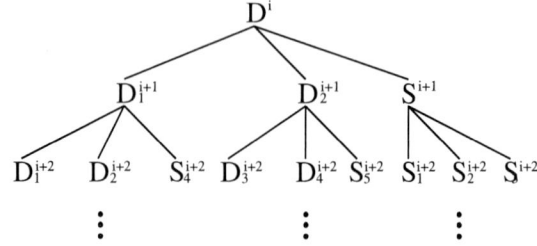

Figure 2: An example of cluster tree $T_{\mathcal{I}}^i$ of domain cluster D^i

identify the nonzeros in $\mathbf{L}$ and $\mathbf{U}$ by finding a tight boundary of each node in the elimination tree. The boundary of each node is the union of the unknowns that have a crosstalk with the unknowns contained in the node, in $\mathbf{L}$ and $\mathbf{U}$. Since each nonleaf node in the elimination tree is a 2-D separator, the boundary of each node is essentially the union of the unknowns residing on the bounding box of the 2-D separator, thus the boundary size is proportional to the node size, which is 2-D. Since the computation associated with each node in the elimination tree is essentially a dense matrix operation of size (node+boundary), we reduce the factorization of the original large 3-D FEM matrix to a sequence of factorizations of 2-D dense matrices. We then develop efficient $\mathcal{H}$-matrix based algorithms to accelerate the computation of all the intermediate 2-D dense matrices. After adding the cost associated with each node in the elimination tree, the computational complexity of the proposed direct finite element solver can be theoretically proved to be $O(N)$, which will be detailed in Section 5. The overall algorithm of the proposed direct solver has six major steps:

1. Build cluster tree $T_{\mathcal{I}}$ based on nested dissection

2. Build elimination tree $E_{\mathcal{I}}$ from $T_{\mathcal{I}}$

3. Do symbolic factorization guided by $E_{\mathcal{I}}$ to obtain the boundary for each node $N^{\{j\}}$ in $E_{\mathcal{I}}$

4. Generate $\mathcal{H}$-matrix structure $\mathbf{F}_{N^{(j)}}^{\mathcal{H}}$ for each node $N^{\{j\}}$ in $E_{\mathcal{I}}$

5. Do numerical factorization guided by $E_{\mathcal{I}}$ by $\mathcal{H}$-matrix-based algorithms

6. Solve for a right hand side based on $E_{\mathcal{I}}$. $\qquad$ (8)

4.1 Build cluster tree $T_{\mathcal{I}}$ from nested dissection

Nested dissection [1] is an ordering scheme based on geometrical information. Different ordering techniques can result in a different number of fill-ins and the nested dissection ordering has been proved to be optimal for a finite element matrix in the sense that it minimizes the number of fill-ins. With nested dissection, a computational domain is recursively divided into two separated subdomains and one interface. An example is shown in Figure 1, where at level i, a computational domain D^i is divided into three parts D_1^{i+1}, D_2^{i+1}, and S^{i+1}, in which D denotes a domain and S denotes an interface. Subdomain D_1^{i+1} and D_2^{i+1} are separated by interface S^{i+1}. Let $\mathbf{L}_{D_2^i \times D_1^i}$ be a matrix block in $\mathbf{L}$ in the D_2^i-row and D_1^i-column, and $\mathbf{U}_{D_1^i \times D_2^i}$ be a matrix block in $\mathbf{U}$ in the D_1^i-row and D_2^i-column. We can order the

unknowns in domain D_i as $\{D_1^i, D_2^i, S^i\}$, then the $\mathbf{L}_{D_2^i \times D_1^i}$ and $\mathbf{U}_{D_1^i \times D_2^i}$ will become zero, and hence reducing fill-ins. Moreover, if the interface S^{i+1} is also further divided, i.e., S^{i+1} is divided and ordered as S_1^{i+2}, S_2^{i+2} and S_3^{i+2}, then S_1^{i+2} and S_3^{i+2} are also completely separated by S_2^{i+2}. The recursive procedure of the nested dissection results in a tree structure shown in Figure 2. Each node of the tree is called a cluster in an $\mathcal{H}$-matrix-based representation, and the entire reversed tree that is rooted at the whole unknown set $\mathcal{I}$ and ended at the leaf unknown sets is called a cluster tree. Figure 2 shows a sub-tree rooted at the D^i cluster. The nodes at the leaf level are called leaf clusters, and those at the non-leaf levels are called non-leaf clusters. The partition is recursively done until the unknown number in each cluster is no greater than $leafsize$, a predetermined constant.

4.2 Build elimination tree $E_{\mathcal{I}}$ from $T_{\mathcal{I}}$

The elimination tree of a matrix [2] is defined as the structure with n nodes $\{N^{(1)}, \ldots, N^{(n)}\}$ such that node $N^{(p)}$ is the parent of $N^{(j)}$ if and only if the following is satisfied,

$$p = \min\{i > j | \mathbf{L}_{N^{(i)} \times N^{(j)}} \neq 0\} \qquad (9)$$

in which $\mathbf{L}_{N^{(i)} \times N^{(j)}}$ is one matrix block in factor $\mathbf{L}$ with node $N^{(i)}$ being the row and node $N^{(j)}$ being the column. Because the sparse matrix $\mathbf{Y}$ obtained from a finite-element method is irreducible, the data structure generated from (9) is a tree.

In the proposed method, a domain cluster D^i in the cluster tree is decomposed and ordered by nested dissection as $\{D_1^{i+1}, D_2^{i+1}, S^{i+1}\}$. If we treat each cluster as a node in the elimination tree, it is easy to verify that node S^{i+1} is the parent of node D_1^{i+1} and D_2^{i+1} based on (9). Hence, the elimination tree generated from a nested-dissection based cluster tree is a tree with the interface cluster being the parent of subdomain clusters. Such a rule can be applied to each domain cluster recursively. The final elimination tree constructed for the entire unknown set has nonleaf nodes being the interface clusters at different levels and leaf nodes being the domain clusters of leaf size. As an example, the elimination tree corresponding to the cluster tree shown in Figure 2 is given in Figure 3. Note that each node $N^{(j)}$ in elimination tree has its own cluster tree structure. The elimination tree $E_{\mathcal{I}}$ is built upon selected nodes in $T_{\mathcal{I}}$. Combining all nodes in $E_{\mathcal{I}}$ will result in $\mathcal{I}$, the entire unknown set.

In an LU factorization procedure, from the definition of an elimination tree we can find that factorizing one node $N^{(j)}$ in the elimination tree, i.e. $\mathbf{Y}_{N^{(j)} \times N^{(j)}}$, will modify the matrix content belonging to all its ancestors. In other words,

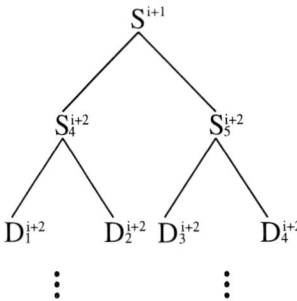

Figure 3: Illustration of an elimination tree.

$\mathbf{U}_{N^{(j)} \times ans(N^{(j)})}$ and $\mathbf{L}_{ans(N^{(j)}) \times N^{(j)}}$ will be updated if we use $ans(N^{(j)})$ to denote the combined set of all ancestors of node $N^{(j)}$. The LU factorization of matrix $\mathbf{Y}$ is a bottom-up or a post-order traversal of the elimination tree $E_{\mathcal{I}}$.

4.3 Symbolic factorization

As mentioned in previous section, factorizing one node in elimination tree would only affect its ancestors. However, the combined set of all its ancestors is larger than what the factorization of one node actually modifies. We therefore need to find out the minimal set of nodes affected by the factorization of one node to save time and storage. In this paper, this minimal set is termed the boundary of each node, denoted by $bnd(N^{(j)})$ for each node $N^{(j)}$ in the elimination tree. As a simple example, a 3-level nested dissection on a domain cluster D^i is shown in Figure 4, where the interface cluster S^{i+1} at level $i+1$ is decomposed as $\{S_1^{i+1}, S_2^{i+1}, S_3^{i+1}\}$, and the S^{i+2} at level $i+2$ is decomposed as $\{S_1^{i+2}, S_2^{i+2}, S_3^{i+2}\}$. The boundary of D^{i+3}, denoted by $bnd(D^{i+3})$, can be identified geometrically as

$$bnd(D^{i+3}) = \{S_1^{i+1}, S_2^{i+1}, S_2^{i+2}, S_3^{i+2}, S^{i+3}\}. \tag{10}$$

Mathematically, the $\mathbf{L}_{D^{i+3} \times bnd(D^{i+3})}$ is filled due to the assembly of an FEM matrix or the factorization of lower-level nodes because the $bnd(D^{i+3})$ is in contact with node D^{i+3}, while we have

$$\mathbf{L}_{D^{i+3} \times \{ans(D^{i+3}) - bnd(D^{i+3})\}} = 0. \tag{11}$$

Hence, if the above block is removed from memory as well as the factorization process, then the time and storage can be significantly reduced.

Identifying $bnd(N^{(j)})$ for each node $N^{(j)}$ in elimination tree based on geometrical information is straightforward. However, due to the complication of mesh and the irregularity in clusters having a small size, a geometrical way may not generate $bnd(N^{(j)})$ precisely. We hence develop the following symbolic factorization procedure to identify the boundary of each node in the elimination tree. Notice that a post-ordering or bottom-up tree traversal scheme is

adopted to go through all nodes in the elimination tree.

Initialize $bnd(N^{(i)})$ from $\mathbf{Y}_{\mathcal{I} \times \mathcal{I}}$

For each node $N^{(i)}$ in elimination tree $E_{\mathcal{I}}$

    For each cluster C_j (cluster index) in $bnd(N^{(i)})$

        idx = node index of cluster C_j in $E_{\mathcal{I}}$

        For each cluster $C_k \neq C_j$ in $bnd(N^{(i)})$

            Put C_k into $bnd(N^{(idx)})$

        end

    end

end $\tag{12}$

4.4 Generate the $\mathcal{H}$-matrix representation for each node in $E_{\mathcal{I}}$

After obtaining the boundary for each node in the elimination tree, the structure of the factorized matrix is known. This permits the construction of an $\mathcal{H}$-matrix representation $\mathbf{F}_j^{\mathcal{H}}$ for each node as the following

$$\mathbf{F}_j^{\mathcal{H}} = \begin{pmatrix} \mathbf{Y}_{N^{(j)} \times N^{(j)}}^{\mathcal{H}} & \mathbf{Y}_{N^{(j)} \times bnd(N^{(j)})}^{\mathcal{H}} \\ \mathbf{Y}_{bnd(N^{(j)}) \times N^{(j)}}^{\mathcal{H}} & \hat{\mathbf{Y}}_{bnd(N^{(j)}) \times bnd(N^{(j)})}^{\mathcal{H}} \end{pmatrix} \tag{13}$$

where $\mathbf{F}_j^{\mathcal{H}}$ is the $\mathcal{H}$-matrix representation of frontal matrix, the $\mathbf{Y}_{N^{(j)} \times N^{(j)}}^{\mathcal{H}}$, $\mathbf{Y}_{N^{(j)} \times bnd(N^{(j)})}^{\mathcal{H}}$, and $\mathbf{Y}_{bnd(N^{(j)}) \times N^{(j)}}^{\mathcal{H}}$ are unique matrix blocks associated with node $N^{(j)}$. The $\hat{\mathbf{Y}}_{bnd(N^{(j)}) \times bnd(N^{(j)})}^{\mathcal{H}}$ is composed of multiple pointers that link different parts to corresponding parts in the $\mathcal{H}$-matrix representation of the nodes that $bnd(N^{(j)})$ belongs to. Therefore, we build the $\mathcal{H}$-matrix representation for three matrix blocks and form the 4th one with the link to other matrices. The $\mathcal{H}$-matrix representation of $\mathbf{F}_j^{\mathcal{H}}$ is constructed based on the tree described in Section 4.1. Usually, the size of $bnd(N^{(j)})$ is larger than that of $N^{(j)}$, therefore, an adaptive division scheme is introduced to prevent skewed $\mathcal{H}$-matrix representations, namely a 2×1 or 1×2 division will be adopted if the size of row cluster and the size of column cluster are very different so that the row and column dimension can be made similar in an admissible block.

4.5 Numerical factorization and Solution

Numerical factorization is done as the following:

For each node $N^{(j)}$ in $E_{\mathcal{I}}$

    collect $\mathbf{F}_j^{\mathcal{H}}$, make it ready for factorization

    compute $\mathbf{L}_{N^{(j)} \times N^{(j)}}^{\mathcal{H}}$, $\mathbf{U}_{N^{(j)} \times N^{(j)}}^{\mathcal{H}}$ by $\mathcal{H}$-LU

        $\mathbf{Y}_{N^{(j)} \times N^{(j)}}^{\mathcal{H}} = \mathbf{L}_{N^{(j)} \times N^{(j)}}^{\mathcal{H}} \cdot \mathbf{U}_{N^{(j)} \times N^{(j)}}^{\mathcal{H}}$

    compute $\mathbf{U}_{N^{(j)} \times bnd(N^{(j)})}^{\mathcal{H}}$ by solving

        $\mathbf{L}_{N^{(j)} \times N^{(j)}}^{\mathcal{H}} \cdot \mathbf{U}_{N^{(j)} \times bnd(N^{(j)})}^{\mathcal{H}} = \mathbf{Y}_{N^{(j)} \times bnd(N^{(j)})}^{\mathcal{H}}$

    compute $\mathbf{L}_{bnd(N^{(j)}) \times N^{(j)}}^{\mathcal{H}}$ by solving

        $\mathbf{L}_{bnd(N^{(j)}) \times N^{(j)}}^{\mathcal{H}} \cdot \mathbf{U}_{N^{(j)} \times N^{(j)}}^{\mathcal{H}} = \mathbf{Y}_{bnd(N^{(j)}) \times N^{(j)}}^{\mathcal{H}}$

    update $\hat{\mathbf{Y}}_{bnd(N^{(j)}) \times bnd(N^{(j)})}^{\mathcal{H}}$ by

        $\hat{\mathbf{Y}}_{bnd(N^{(j)}) \times bnd(N^{(j)})}^{\mathcal{H}} = \hat{\mathbf{Y}}_{bnd(N^{(j)}) \times bnd(N^{(j)})}^{\mathcal{H}}$

        $- \mathbf{L}_{bnd(N^{(j)}) \times N^{(j)}}^{\mathcal{H}} \cdot \mathbf{U}_{N^{(j)} \times bnd(N^{(j)})}^{\mathcal{H}}$

end $\tag{14}$

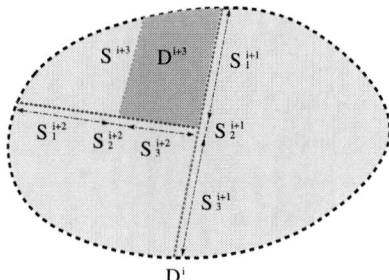

Figure 4: Illustration of symbolic factorization on multi-level domain cluster D^i

where the first three operations can be done based on $\mathcal{H}$-based arithmetic [4], and the last one is performed by developing an efficient algorithm to update the local $\mathcal{H}$-matrices associated with the nodes affected the node being factorized. With $\mathbf{L}/\mathbf{U}$ known in their $\mathcal{H}$-format, the solution can also be efficiently obtained for any right hand side.

5. COMPLEXITY AND ACCURACY

Consider a computational domain with the number of unknowns along each dimension being n. It is split into 8 smaller domains using nested dissection recursively. One intermediate stage is shown in Figure 5, where three shaded surface separators are combined to form one interface cluster S. The total number of unknowns is $N = n^3$, and the depth of elimination tree is $L = \log_2 n$. For a node $N^{(j)}$, the dimension of its frontal matrix $\mathbf{F}_j^{\mathcal{H}}$ satisfies

$$dim(\mathbf{F}_j^{\mathcal{H}}) = \#\{N^{(j)}\} + \#\{bnd(N^{(j)})\}. \quad (15)$$

As shown in Figure 5, the number of unknowns of the boundary for a cube domain with side length a is proportional to the surface area, $6a^2$, and the number of unknowns contained in the interface node is $3a^2$. Therefore, the following satisfies:

$$\#\{bnd(N^{(j)})\} = \alpha \#\{N^{(j)}\} \quad (16)$$

where α is a constant for an arbitraily-shaped domain, and equal to 2 for a cube domain. Assuming $\mathbf{F}_j$ is at level l with $l = L$ being the root level, its dimension can be approximated as $dim(\mathbf{F}_j) = 3\#\{N^{(j)}\} = 9 \times (2^l \times 2^l)$. It is proven that the complexity of an $\mathcal{H}$-matrix LU factorization is bounded by the complexity of an $\mathcal{H}$-matrix multiplication [4], which is $O(m \log^2 m)$ for a matrix of dimension m. For each node $N^{(j)}$ in the elimination tree $E_{\mathcal{I}}$, the operation associated with this node is hence

$$P_{lu}(\mathbf{F}_j^{\mathcal{H}}) = P(\mathbf{F}_j^{\mathcal{H}} \otimes \mathbf{F}_j^{\mathcal{H}})$$
$$= O\big(2^l \times 2^l \log^2(2^l \times 2^l)\big) = O\big(2^{2l}(2l)^2\big). \quad (17)$$

The total operation complexity is the summation of the operation complexity of each node in the elimination tree, thus

$$P_{lu}(\mathbf{Y}^{\mathcal{H}}) = \sum_{j, N^{(j)} \in E_{\mathcal{I}}} P_{lu}(\mathbf{F}_j^{\mathcal{H}})$$
$$\sim \sum_{l=0}^{L} 8^{(L-l)} \times (2^l)^2 (2l)^2 = N \sum_{l=0}^{L} (\frac{4l^2}{2^l}) = O(N), \quad (18)$$

which is linear. Similarly, the total storage complexity of

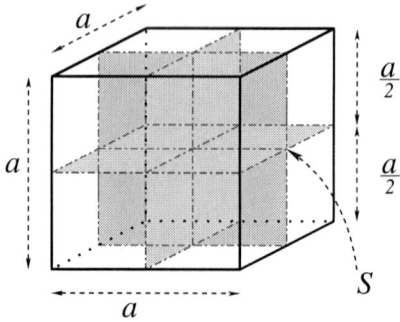

Figure 5: One step in nested dissection

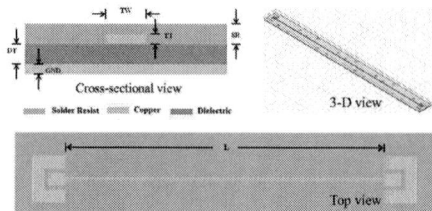

Figure 6: Illustration of a package interconnect.

the proposed solver can be evaluated as

$$Storage_{lu}(\mathbf{Y}^{\mathcal{H}}) \sim \sum_{l=0}^{L} 8^{(L-l)} \times (2^l)^2 (2l)$$
$$= N \sum_{l=0}^{L} (\frac{2l}{2^l}) = O(N). \quad (19)$$

In the proposed solver, we represent the update and frontal matrices associated with each node in the elimination tree by an $\mathcal{H}$ matrix. Such an $\mathcal{H}$-matrix representation exists because the update and frontal matrices are related to the Schur complement. Since the original FEM matrix has an exact $\mathcal{H}$-matrix representation and its inverse has an error bounded $\mathcal{H}$-matrix representation [3], it can be proven that the intermediate Schur complements obtained during the factorization process and their inverses both can be represented by $\mathcal{H}$-matrices with error controlled.

6. NUMERICAL RESULTS

6.1 Package interconnect with measured data

We first validated the accuracy of the proposed method on a package interconnect provided by a company. Figure 6 illustrates its cross sectional view, top review, and 3-D view. Due to a nondisclosure agreement, detailed geometrical data are not shown. In Figure 7, we plot the S-parameters extracted by the proposed method in comparison with the measured data and the results generated from a commercial FEM-based tool. Excellent accuracy of the proposed solver can be observed in the entire frequency band.

6.2 Large-scale inductor array

We then simulated a large-scale 3-D inductor array [3] to examine the performance of the proposed direct solver. The computer used is a unix server with an AMD CPU running at 2.8 GHz. The frequency simulated is 10 GHz. The num-

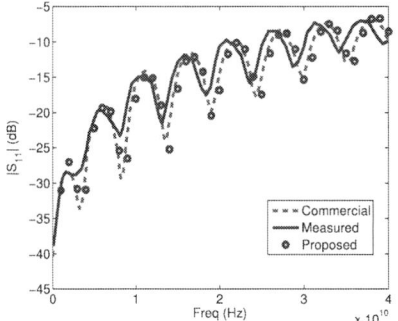 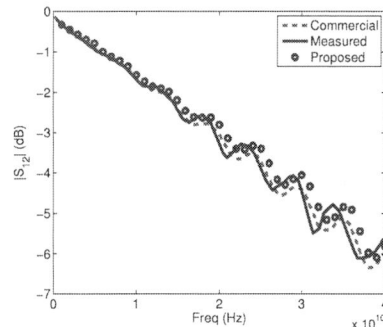

Figure 7: Simulation of a package interconnect. (a)$|S_{11}|$. (b)$|S_{12}|$.

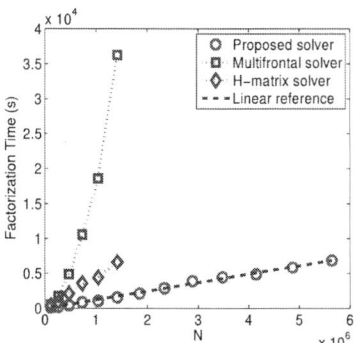 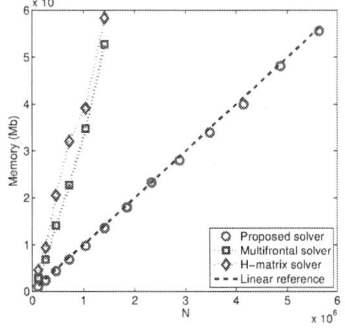 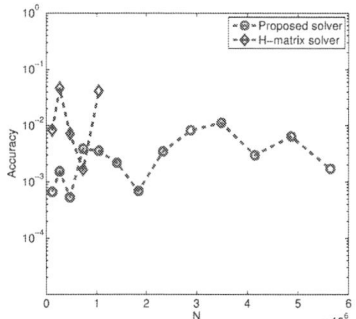

Figure 8: Simulation of a large-scale 3-D inductor array from 117,287 to 5,643,240 unknowns at 10 GHz. (a) LU factorization time. (b) Memory. (c) Accuracy.

ber of inductors varies from 2×2 to 14×14. The number of unknowns ranges from 117,287 to 5,643,240. The simulation parameters used are $leafsize=8$ and $\eta=3$. The relative error for (7) is set as $6e-5$. In Figure 8(a), (b), and (c), we plot the factorization time, the memory cost, and the accuracy of the proposed solver respectively with respect to N in comparison with both a state-of-the-art multifrontal based solver [5] and an $\mathcal{H}$-matrix based direct FEM solver. The accuracy is measured by relative residual. As can be seen from Figure 8, first, the proposed direct solver demonstrates a clear linear complexity with good accuracy achieved in the entire unknown range; second, the proposed method outperforms both the multifrontal based solver and the $\mathcal{H}$-matrix based solver. Notice that the multifrontal solver is based on exact arithmetic instead of prescribed accuracy, thus its accuracy is not shown in Fig. 8 (c). We also compared the performance of the proposed direct solver with a state-of-the-art iterative FEM solver available in the market. The computer used is a PC having an Intel CPU running at 2.4 GHz. Due to limited memory available on this PC, a 5×5 inductor array is simulated. To extract the S-parameters at 50 ports of the inductor array, the proposed direct solver only costs 558.39 s, whereas the commerical iterative FEM solver costs 2208 s.

7. CONCLUSION

A linear-complexity direct sparse matrix solution is developed for FEM-based large-scale 3-D circuit extraction with arbitrarily shaped lossy conductors in inhomoegenous materials. A comparison with both state-of-the-art direct FEM

solvers and an iterative FEM solver has demonstrated its superior performance.

8. ACKNOWLEDGMENT

This work was supported by a grant from SRC (Task 1292.073). The authors would also like to thank Dr. Henning Braunisch at Intel for providing the interconnect structure and measured data.

9. REFERENCES

[1] A. George. Nested dissection of a regular finite element mesh. *SIAM J. on Numerical Analysis*, 10(2):345–363, April 1973.

[2] J. W. H. Liu. The multifrontal method for sparse matrix solution: Theory and practice. *SIAM Review*, 34(1):82–109, March 1992.

[3] H. Liu and D. Jiao. Existence of $\mathcal{H}$-matrix representations of the inverse finite-element matrix of electrodynamic problems and $\mathcal{H}$-based fast direct finite-element solvers. *IEEE Trans. MTT*, 58(12):3697–3709, December 2010.

[4] S. Borm, L. Grasedyck, and W. Hackbusch. *Hierarchical matrices*. Lecture note 21 of the Max Planck Institute for Mathematics, 2003.

[5] UMFPACK5.0. [on line] http://www.cise.ufl.edu/research/sparse/umfpack/.

FPGA Code Accelerators - The Compiler Perspective

Walid Najjar
University of California Riverside
Computer Science & Engineering
najjar@cs.ucr.edu

Jason Villarreal
Jacquard Computing Inc.
Riverside, CA
jason@jacquardcomputing.com

ABSTRACT

FPGA-based accelerators have repeatedly demonstrated superior speed-ups on an ever-widening spectrum of applications. However, their use remains beyond the reach of traditionally trained applications code developers because of the complexity of their programming tool-chain. Compilers for high-level languages targeting FPGAs have to bridge a huge abstraction gap between two divergent computational models: a temporal, sequentially consistent, control driven execution in the stored program model versus a spatial, parallel, data-flow driven execution in the spatial hardware model. In this paper we discuss these challenges to the compiler designer and report on our experience with the ROCCC toolset.

Categories and Subject Descriptors

D.3.4 [Programming Languages]]: Processors—Retargetable compilers; optimizations; B.5.2 [Register-Transfer-Level Implementation]: Design Aids

General Terms

Algorithms, Design, Performance.

Keywords

FPGAs. Compiler. Hardware Accelerators.

1. INTRODUCTION

In recent years we have witnessed a dramatic widening of the scope of use of FPGAs as computing devices. It is driven by a variety of factors including their larger size enabling a very large degree of parallelism, a richer set of embedded functionalities (RAM, DSP etc.), high efficiency, as compared to software, coupled with re-programmability, lower energy per task, high I/O bandwidth that eliminates the need for memory off-loading of data, etc. A general-purpose use of FPGAs as accelerators was already described a few years after the introduction of the first device [2]. However, the main obstacle facing a wider use of FPGAs as code accelerators is their poor programmability using high-level programming languages (HLL). The challenge lies in the translation of a stored-program machine, the high-level language, to a spatial and parallel computing structure with no instruction set architecture and no pre-determined control structures.

In this paper we revisit the earliest documented use of FPGAs as

Permission to make digital or hard copies of all or part of this work for personal or classroom use is granted without fee provided that copies are not made or distributed for profit or commercial advantage and that copies bear this notice and the full citation on the first page. To copy otherwise, to republish, to post on servers or to redistribute to lists, requires prior specific permission and/or a fee.

code accelerators and discuss various aspects of the challenge of compiling HLLs to accelerators mapped onto FPGAs (Section 2). Section 3 describes the difficulties inherent in bridging the abstraction gap between high-level languages and hardware circuits. In Section 4 we describe the ROCCC (Riverside Optimizing Compiler for Configurable Computing) [3] approach to compiling code accelerators for FPGAs. The programming model and code optimizing features of ROCCC are described in Section 5 with an emphasis on its use for high-level design space exploration.

2. THE HISTORICAL PERSPECTIVE

The first documented use of FPGAs as code accelerators appeared, to our knowledge, just four years after the introduction of the first SRAM-based FPGA device (Xilinx, 1985). The PAM (Programmable Active Memory) [2], built at the DEC Paris Research Lab, is described as "universal hardware co-processor closely coupled to a standard host computer." Ten benchmark codes were implemented and evaluated on PAM [3], including: long multiplication, RSA cryptography, Ziv-Lempel compression, edit distance calculations, heat and Laplace equations, N-body calculations, binary 2D convolution, Boltzman machine model, 3D graphics (including translation, rotation, clipping and perspective projection) and discrete cosine transform. The authors' conclusions were that PAM delivered a performance comparable to that of ASIC chips or supercomputers, of the time, and was one to two orders of magnitude faster than software. They also state that because of the PAM's large off-chip I/O bandwidth (6.4 Gb/s) it was ideally suited for "... on-the-fly data acquisition and filtering, ..."

What has changed in the nearly 25 years since the first PAM? FPGAs are much larger and faster; the application domains have grown in scope following the growth in size and speed of the devices. However, the main challenge to FPGAs as code accelerators, namely the abstraction gap between application development and FPGA programming, not only remains un-changed but has probably gotten worse due to increase in complexity of the applications enabled by the larger device sizes.

3. THE ABSTRACTION GAP

In this section we discuss two issues that define the complexity of compiling HLLs to hardware circuits: (1) the semantic gap between the sequential stored-program execution model implicit in these languages and (2) the effects of virtualization, or lack thereof, on the complexity of the compiler.

3.1 Semantics of the Execution Model

CPUs and GPUs are inherently stored-program machines (or von Neumann machines) and so are the programming languages used on these, essentially most of the languages in use today. As such they are bound by the sequential consistency of that model, both at the language and machine levels. While CPU and GPU architectures exploit various forms of parallelism, instruction, data and thread-level, they do so circumventing the sequential consistency implied in the source code internally (branch prediction, out-of-order execution, SIMD parallelism, etc.), while preserving the appearance of a sequentially consistent execution externally (reorder buffers, precise interrupts etc.). Von Neumann, or imperative, languages are even more constrained by sequential consistency: sequential execution, pre-determined control flow structures, etc. The compiling of a HLL code to a CPU or GPU is therefore the translation from one stored program machine model, the HLL, to another, the machine's ISA. A digital circuit, on the other hand, is inherently parallel, spatial, with distributed storage, timed behavior etc. the abstraction and semantic gap between the hardware and software computing models is summarized in Table 1. Translating a HLL to a circuit requires a transformation of the sequential to a spatial/parallel, with the creation of custom sequencing, timed synchronizations, distributed storage, pipelining, etc.

Table 1. Features of hardware and software computing models

	Stored Program	Spatial Computing
Storage & data access	Central, large, virtualized. Memory resident, multi-level caches	Distributed, small, physical. Streaming, limited memory and caching. No virtualization
Parallelism	Dynamic - separate ILP, DLP, TLP	Static - combined ILP, DLP, TLP
Sequencing	Central, static, appearance of SC	Data-flow, asynchronous
Data-Path	Pre-designed, one size fits all. Dynamic data dependencies	Customized, very deep pipelines. No dynamic data dependencies

Raising the abstraction level of FPGA programming to that of CPU or GPU programming is a daunting task that is yet to be fully completed. It is of critical importance in the programming of accelerators as opposed to the high-level design of arbitrary digital circuit, which is the focus of high- level synthesis.

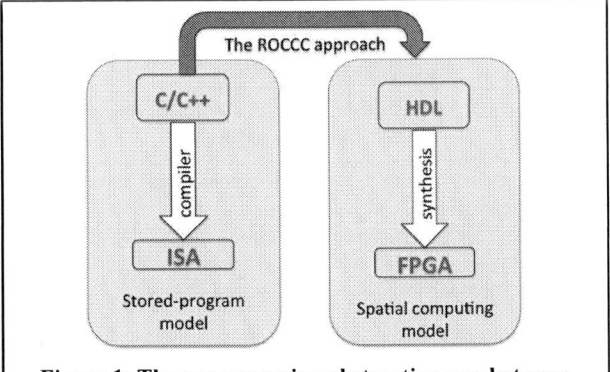

Figure 1: The programming abstraction gap between HLLs and FPGAs. The ROCCC approach focuses on a subset of C to generate code accelerators

Code accelerators differ from general purpose logic design in one important way: the starting point of logic design is a device whose behavior is specified by a hardware description code implemented in a hardware description language (HDL) such as VHDL, Verilog, SystemC, SystemVerilog, or Bluespec. The starting point

of a code accelerator is an existing software application a subset of which, being frequently executed, is translated into hardware. That subset is, quasi by definition, a loop nest. Hopefully that loop nest is parallelizable and can therefore exploit the FPGA resources. By focusing on loop nests, the task of compiling HLLs to FPGAs is simplified and opportunities for loop transformations and optimizations abound. This is the approach taken by the ROCCC compiler (Figure 1) and is described in the rest of this paper.

3.2 Virtualization

Virtualization is probably one of the greatest achievements of modern computer system design: when a CPU issues a *load* instruction to an address in memory it is not aware of its actual physical location: the loaded data may not be in its cache or physical memory, it may not even be in the same time zone as the CPU! Thanks to multiple layers of hardware and software support, the world of the CPU is a single dimensional memory as large as its address space[1]. Obviously, this storage model is a perfect fit to the one implicit in all commonly used HLLs: one large flat array of bytes.

FPGA-based accelerators do not enjoy, yet, such sophisticated levels of virtualization. Rather, the compiler must be aware of all data placements, on and off-chip, and actively manage the interfaces to one or more memory modules as well as data streaming interfaces (e.g. PCI, USB, Ethernet etc.). Furthermore, most HLLs do not support streams as programming constructs or indications of physical data locations. The compiler must therefore manage the data location, both off and on-chip, with no support in the HLL, through pragmas or GUI-based indications from the user.

Each of these interfaces, to physical memories or streams, implies a preset data width, addressing modality (bursts or singletons) and mapping to a single or multiple data channels on the circuit. None of these parameters is supported in the HLL let alone in the intermediate representation (IR) the compiler uses to generate the code. On a CPU, or GPU, all data values entering the data-path come from the L1 cache with a pre-determined timing pattern. Consider a loop body that accesses four arrays, or streams, from two separate memory modules and two streams. Its data-path on an FPGA requires four physical data interfaces each with its own timing patterns that could raise very significant timing and synchronization issues.

4. THE ROCCC APPROACH

As mentioned above, the objective of ROCCC [1] is the generation of efficient customized hardware accelerators for frequently executing code segments, namely loop nests. Its target audience is application code developers with hardly any training as hardware designers. The objective being to make FPGA-based code accelerators accessible to a wider spectrum of users. As such, ROCCC is not a general-purpose high-level logic design tool, rather its focus is on generating hardware accelerators from existing C codes with minimal modifications to the source code. The same code can be compiled for software execution or translated to hardware.

The ROCCC compiler supports an extensive set of loop optimizations and transformations. One of the driving

---

[1] This has not always been the case; there was a time when users, or compilers, had to actively manage the memory allocation of data and code.

philosophies of ROCCC is that there should be one source code description of an accelerator that could be compiled, using different transformations, into multiple hardware implementations. All transformations are therefore done in the GUI by the user, and not by re-implementing the source code. Users are given the flexibility to choose which optimizations and transformations to perform, on each individual loop and exactly how to apply them to different parts of their application. Control of optimizations is given to the user as options in the GUI and each can have a dramatic effect on the generated hardware [7][5][4]. It would have been possible, in some limited instances, to let the compiler automatically decide which transformations should be applied. This would imply having, in the compiler, knowledge of all possible FPGA platforms, i.e. system and board architectures, current and future.

The objectives of ROCCC in generating code accelerators is maximizing throughput through (1) parallelism, (2) minimizing the area occupied by the circuit, (3) the reuse of data fetched off-chip [6] and (4) pipelining to reduce clock cycle time. ROCCC favors throughput over space, so, under user control, it could generate as much hardware as necessary to maximize parallelism. The data-path generated is purely data driven with no FSM created for resource sharing: data is pushed onto the top and flows through without any control. There is minimal control logic generated to keep track of which pipeline stages are active so the hardware can output values at the correct clock cycle.

4.1 The ROCCC Programming Model

ROCCC code is a subset of C. All ROCCC code can be compiled and run with a normal software compiler such as *gcc* and will generate the same output as the ROCCC-generated hardware from the same source. The limitations of ROCCC, compared to C, are (1) no recursion, (2) no arbitrary use of pointers that the compiler cannot un-alias statically. The use of dynamic pointers inside loop bodies would result in multiple memory de-referencing accesses being serialized, for consistency reasons, and would eliminate the parallelism.

4.1.1 Bottom-up design and code reuse

Just as in software construction, designs for hardware accelerators can benefit from opportunities for code reuse and raising the abstraction level. ROCCC is designed to support a modular approach to hardware accelerator design, enabling reuse of components and ease of design space exploration [8][9].

C code compiled by ROCCC falls under one of two categories: modules or systems. Both modules and systems are represented in C as a function call and can be compiled with *gcc* to perform the same operations in software as in hard- ware.

Module code describes components, which perform a computation on scalar inputs and generate a set of scalar outputs. They are translated into pipelined hardware structures that can take a set of inputs every clock cycle and generate a set of outputs every clock cycle. Each module is itself a complete hardware implementation and may be used by itself as a complete design, or may be included as a component in larger modules or systems.

Each module may have different optimizations performed in order to best suit the user's specific needs with regard to clock speed or area. Modules included in larger designs are treated as black boxes by the compiler so as not to affect any implementation decisions made at the lower level. Treating module instantiations as black boxes could obscure some optimization opportunities, so inlining is given as an option if the user wants to take advantage

of coding at a higher level but has no fixed requirement for the low level components. All modules and systems are stored in a database, supported in the GUI, from where the user can drag and drop them in other projects. An example module that sorts two values is shown in Figure 2. In C, this code takes two integers and returns two integers in sorted order. When compiled with ROCCC, this generates a pipelined component that can take two integers every clock cycle and generated two sorted integers every clock cycle. The generated hardware is purely computational and consists of a pipeline that performs a comparison and two multiplexors.

```
void BitonicSort2(int a, int b, int& o1, int& o2)
{ if (a < b) { o1 = a; o2 = b;}
  else {o1 = b; o2 = a;} }
```
Figure 2: Bitonic sort module for two numbers

System code describes computational kernels that may apply large amounts of computation on input streams of data and generate output streams of data. Streams connections in hardware are inferred from array accesses in C. Figure 3 shows an example system that performs the Median filter operation on a 3x3 window of an NxN stream. The call to BitonicSort9 is a function call in C, but is translated into an instantiation of the BitonicSort9 module and placed in a pipeline when converted to hardware. The BitonicSort9 module is not shown, but is constructed by

```
#include "roccc-library.h"
void MedianFilter(int** A, int N, int** Out) {
  int i, j ;
  int s1, s2, s3, s4, s5, s6, s7, s8, s9 ;
  for (i = 0 ; i < N ; ++i) {
    for (j = 0 ; j < N ; ++j) {
      BitonicSort9(A[i][j], A[i][j+1],
              A[i][j+2], A[i+1][j], A[i+1][j+1],
              A[i+1][j+2], A[i+2],[j], A[i+2][j+1],
              A[i+2][j+2], s1, s2, s3, s4, s5,
              s6, s7, s8, s9) ;
      Out[i][j] = s4; } } }
```
Figure 3: Median filter on a 3X3 window using the bitonic sort module

instantiating many copies of the BitonicSort2 module in a butterfly network. An input stream and an output stream are inferred from the parameters A and Out respectively, and result in hardware that communicates with memory in order to feed the pipeline elements from A and stores output to Out.

The generated data-path with no optimizations specified requires nine elements from A each clock cycle in order to generate one output each clock cycle. The first iteration, all of these values must be fetched from memory, but subsequent iterations only need fetch three new elements from memory and can reuse six elements.

The parameter N to the function in Figure 3 is translated into an input scalar. A connection is made in the generated hardware to a register that is read once at the beginning of execution and then kept constant.

4.1.2 Data types

In addition to the standard C data types (*char, int, long, float, double*) ROCCC supports variable bit width data types both integer and fixed-point.

ROCCC does not assume a fixed target data-path width so operations such as addition and multiplication do not need to be truncated after every step. The user can elect to maximize precision or adopt a C-like truncation model. For example, the addition of two eight-bit numbers in software will result in an eight-bit value, but in the generated hardware the result can be stored and used as a nine-bit value. Floating-point operations are assumed to be present in software, but require hardware components. Different FPGA platforms may have varying levels of support for floating point operations and since the hardware generated is not specific to a certain platform, there can be no assumptions made about the target platform's resources. As a solution, ROCCC gives the user the ability to manage a library of intrinsics, which include floating point operations and integer division. These libraries are reflections of hardware libraries such as cores generated by Xilinx Core Generator [12] and generate connections to include platform specific cores to handle the floating-point operations. Changing the libraries can affect the performance of the hardware, but is purely done through the GUI and has no effect on the source code.

ROCCC supports user-defined tables to be accessed by the data-path in some instances. These can be read-only or random access. Some operations are more efficient when implemented as a look-up table rather than an actual circuit; division is used as an example later in this paper. These tables are implemented using BRAMs when available. Random access tables may be written once per loop iteration but may be read as many times as necessary in each loop iteration.

4.1.3 Importing external modules

In many cases the development of hardware accelerators requires IP that was created outside the scope of the project and must be integrated into a larger design. Just as ROCCC is designed to integrate modules into larger designs, external IP can be imported and instantiated. Importing external IP requires the user provide a description of the inputs, outputs, and latency of the core through the GUI. A wrapper with default parameters such as stall and done signals connects the external IP to the generated data-paths. Calling external IP is identical to a module instantiation and appears as a function call in C. External IP calls, as well as module instantiations, can be inserted into application code directly through the GUI.

4.2 Transformations and Optimizations

A major goal of ROCCC is to enable the exploration of large design spaces through the tuning of optimizations on unchanging source code. Two types of transformations are exposed to the user in the GUI that facilitates design space exploration: High-level transformations control the overall structure of the generated hardware and can be used to create different memory configurations. These include inlining of modules, redundancy, loop optimizations, temporal common subexpression elimination, and systolic array generation. Low-level transformations control the utilization of the underlying hardware. These include pipelining control and fan-out tree generation.

Loop unrolling typically increases parallelism but also increases the necessary bandwidth to sustain a high throughput as well as the area used. For example, if the loop in Figure 3 is unrolled once so that two loop bodies are performed each iteration, the throughput is doubled as two values are generated every clock cycle. However, the resulting hardware requires two new data elements from the A input stream every clock cycle in order to maintain this. ROCCC provides fine-grained control over loop unrolling and stream connections to a degree not normally seen. Individual loops can be unrolled different amounts independently of one another, creating memory access requirements specific to individual streams.

By default, each input and output stream has one channel to memory through which all values must go. If there are multiple values generated in one clock cycle but only one output stream channel, the data must be serialized. The number of channels to memory may be configured on a stream by stream basis for each input and output stream. Each stream may be configured have the number of memory channels specified to support the highest possible throughput. Conversely, the streams can be tuned to read fewer elements per clock cycle on hardware platforms that cannot support the ideal bandwidth. For multidimensional streams support, the memory channels are further split up into address channels and data channels. Loop unrolling in multiple dimensions has different consequences on the resulting hardware depending on which loop is unrolled. Unrolling the outer loop results in more rows being fetched every clock cycle, which can be processed by increasing the number of data channels. Unrolling the inner loop, results in an increase to the size of each burst that is fetched but not the number of channels available. Unrolling either loop has the potential to increase parallelism.

Temporal common subexpression elimination [9] identifies computations that are identical across consecutive iterations of a loop and replaces those computations with a register. This can drastically reduce the area requirements by eliminating large blocks of hardware. A consequence of this optimization is that some memory fetches might be determined to be unnecessary, changing the stream interface.

Systolic array generation [5] completely transforms two-dimensional computation into a one-dimensional computation with much less area and high throughput. The memory connections of the generated hardware are changed by this optimization.

Different hardware platforms have different characteristics, such as number of inputs per LUT, which can have an effect on the relative cost of individual operations. When generating a hardware pipeline, the decision of how many basic operations to put into each level of the pipe is dependent on this information. As the compiler has no knowledge of the underlying restrictions, this control is again passed to the user.

The GUI provides both a basic slider to control the pipeline construction and the advanced capability to specify the relative cost of each basic operation on the underlying platform. Without changing the source code, many different pipelines can be created exploring the tradeoff of clock speed versus latency and area.

Another characteristic that differs from device to device is the amount of routing resources. While high fan-out is to be avoided in general, the specific limit on the amount of fan-out per element is platform specific. Again, this control is given to the user in order to control potential routing issues at the high level without rewriting the application.

5. DESIGN SPACE EXPLORATION

In this section we examine the effect of the high and low level transformations on clock speed and area on a concrete hardware platform. The implementations were synthesized and placed and routed for a Xilinx Virtex 6 LX760 FPGA.

Median Filter – Loop Unrolling and Throughput.

Shown in Figure 3, the median filter works on a 3x3-sliding window of 8-bit data over a large 2D array. The 8-bit data is meant to be representative and not restrictive, similar results can be achieved for other bit widths. It uses the bitonic sort module (Figure 2) and has 50 cycles latency. The application is synthesized, placed and routed on the Xilinx Virtex 6 LX760, with a generic wrapper consisting of two sets of dual clock BRAMs connected to the I/O pins and acts as input and output to the ROCCC generated code.

Results for Median Filter are shown in Table 2. Each row shows the effect on area, clock speed, throughput, and throughput per unit area resulting from unrolling the outer loop and adjusting the input and output memory channels appropriately. Throughput per unit area is reported in MB/s/slice and represents the gain in throughput with respect to the amount of area added with each transformation.

Table 2: Impact of loop unrolling on Median Filter

In/Out Channels	Clock (MHz)	Area (slices)	Throughput (MB/s)	Through put / area
1/1	225	735	75	0.102
3/1	225	766	225	0.294
4/2	225	1215	450	0.370
8/6	200	3160	1200	0.380

The first row of Table 2 represents the base configuration, where no transformations have taken place and the code was compiled with the default options. In this case ROCCC generates hardware that has only one input channel and one output channel. Before any input can be processed, the hardware has to read three elements from the one input channel, which takes three clock cycles, effectively cutting the throughput into one third of its potential.

The second row shows the effect of specifying three input memory channels with no other transformations. This allows all the necessary data to be read in one clock cycle, allowing the output to be generated every clock cycle resulting in a tripling of throughput. The area is slightly larger as the hardware has to deal with multiple connections, but some internal hardware components that serialized the incoming data are actually simplified in this implementation leading to a small increase in area.

The third and fourth row show the effect of unrolling the outer loop once and six times, corresponding to connecting to an interface of 32-bits and 64-bits respectively. Each unrolling allows the number of input and output channels to increase and still produce all output every clock cycle, resulting in a large increase in throughput and maximizing the throughput per unity area for this experiment.

Average Filter – Lookup Tables and Arithmetic Cores.

Average Filter computes the average of each 3x3-sliding window in the input array. We compare two versions where the division is either implemented as a look-up table or as an instantiation of an IP core generated by Xilinx Core Generator. Results are shown in Table 3.

For all transformations the achievable clock speed was 225 MHz. Again, the first row shows the default configuration with one input and one output, the second row shows three input channels but no transformations, and the third and fourth row show the configuration of unrolling the outer loop once and six times to interface with a 32-bit and 64-bit interface. In addition to loop transformations, the Average Filter example was synthesized using both a ROCCC compiled look up table (as reported in the column labeled Area Table) and an integer division core generated (as reported in the column labeled Area Divider).

Table 3: Average Filter implementations using table lookup or integer division core (clock in both cases is 225 MHz)

In/Out Channels	Area Table (slices)	Area Divider (slices)	Throughput (MB/s)	Through put / area
1/1	218	283	75	0.344
3/1	225	275	225	1.00
4/2	253	351	450	1.78
8/6	498	826	1350	2.71

The results of these transformations provide similar throughput while the Table implementation takes less area, even when unrolling causes duplication of the table to support multiple reads per clock cycle. The throughput per unit area reported in Table 3 is reported for the Table implementation, which is the more space efficient design. Using lookup tables and unrolling the loop provides nearly 8X improvement in terms of throughput per unit area over the default case.

Max Filter – Temporal Common Sub-expression Elimination.

Max Filter computes the maximum value in a sliding 3x3 window on a 2D array (image) of *height x width* as shown in Figure 4. We use it to show the impact of temporal common sub-expression elimination (TCSE), when combined with loop unrolling, in area and throughput.

The results are shown in Table 4. The original implementation, with no optimizations, is in the first row and has three input channels and generates one output element every clock cycle. It consists of four Max modules taking up 311 slices. When TCSE is applied, two of these components are removed and only one new data element is needed each cycle resulting in a lower area for the same throughput.

The third row of Table 4 shows the results when the outer loop is unrolled five times, taking in seven elements each clock cycle and generating five outputs. Applying TCSE (fourth row) results in smaller area, increased in clock speed and two variables being

```
void MaxFilterSystem(int** A, int N, int** Out) {
  int i, j ;
  int maxCol1, maxCol2, maxCol3, winMax ;
  for (i = 0 ; i < N ; ++i) {
    for (j = 0 ; j < N ; ++j) {
      MaxFilter(A[i][j], A[i][j+1], A[i][j+2], maxCol1);
      MaxFilter(A[i+1][j],A[i+1][j+1],A[i+1][j+2], maxCol2);
      MaxFilter(A[i+2][j],A[i+2][j+1],A[i+2][j+2], maxCol3);
      MaxFilter(maxCol1, maxCol2, maxCol3, winMax);
      Out[i][j] = winMax ; } } }
```

Figure 4: Max filter on a 3X3 window

reused across iterations requiring only five input elements every clock cycle. Assuming the necessary memory bandwidth is available, this exploration shows that a 48% increase in area results in 5X higher throughput and a 3.38X higher throughput per unit area.

Table 4: Impact of TCSE on Max Filter with loop unrolling

In/Out Channels	Clock (MHz)	Area (slices)	Through put (MB/s)	Through put / area
3/1	225	311	225	0.723
1/1 with TCSE	225	266	225	0.846
7/5	220	526	1100	2.092
5/5 with TCSE	225	460	1125	2.446

6. CONCLUSION

The automatic translation of programs written in HLLs to FPGA-based hardware accelerators is a daunting task. These tools have to (1) overcome a large semantic gap between temporal, sequential and control driven programs and spatial, parallel and data/event driven circuits; and (2) without any of the virtualizations commonly available with CPUs and GPUs. In this paper we describe the ROCCC C to VHDL compilation tool, one of over 40 similar tools developed in academia and industry. The focus of ROCCC is on compiling a subset of C into hardware accelerators while providing an extensive set of compiles time transformations and optimizations under user control via a GUI-based console. We report the experimental evaluation of the impacts of some of these transformations on the circuit costs (area) and performance (throughput).

7. ACKNOWLEDGMENTS

This work was supported in part by NSF Awards CCF-1219180 and IIS-1161997 and by AFRL Contract FA945309C0173.

8. REFERENCES

[1] ROCCC 2.0 - www.jacquardcomputing.com, 2013.

[2] Bertin, P., Roncin, D, and Vuillemin. J. *Introduction to programmable active memories*, pages 300–309. Prentice Hall, 1989.

[3] Bertin, P., Roncin, D, and Vuillemin. J. Programmable active memories: a performance assessment. In *Parallel*

Architectures and Their Efficient Use. Paderborn, Germany, Nov. 1992, Lecture Notes in Computer Science, pages 119–130. Springer Verlag, 1992.

[4] Buyukkurt, B., Cortes, J., Villarreal, J. and Najjar, W. A. Impact of high-level transformations within the ROCCC framework. *ACM Trans. Architecture and Code Optimizations (TACO)*, 7(4):17:1–17:36, Dec. 2010.

[5] Buyukkurt, B. and Najjar, W. A. Compiler Generated Systolic Arrays for Wavefront Algorithm Acceleration on FPGAs. In *Int. Conference on Field Programmable Logic and Applications (FPL)*, September 2008.

[6] Buyukkurt, B., Guo, Z., and Najjar, W. A. Impact of loop unrolling on area, throughput and clock frequency in ROCCC: C to VHDL compiler for FPGAs. *In Proc. Int. Workshop On Applied Reconfigurable Computing (ARC)*, March 2006.

[7] Guo, Z., Buyukkurt, B. and Najjar, W. A. Input data reuse in compiling window operations onto reconfigurable hardware. In *ACM SIGPLAN/SIGBED Conference on Languages, Compilers and Tools for Embedded Systems (LCTES)*, pages 249–256, New York, NY, USA, June 2004. ACM Press.

[8] Guo, Z., Najjar, W. and Buyukkurt, B. Efficient hardware code generation for FPGAs. *ACM Trans. on Architecture and Code Optimizations (TACO)*, 5(1):26, May 2008.

[9] Hammes, J., Bohm, A.P.W., Ross, C., Chawathe, M., Draper, B., Rinker, R., and Najjar, W. Loop Fusion and Temporal Common Sub-expression Elimination in Window-based Loops. In *Reconfigurable Architecture Workshop*, April 2001.

[10] Villarreal, J., Park, A., Najjar, W. and Halstead, R. Designing modular hardware accelerators in C with ROCCC 2.0. In *18th IEEE Ann. Int. Symp. on Field-Programmable Custom Computing Machines (FCCM)*, 2010, pages 127 – 134, May 2010.

[11] Villarreal. J. *Compiled acceleration of C programs on FPGAs*. Ph.D. Thesis, U. California Riverside, USA, 2008. AAI3332643.

[12] Xilinx Core Generator System www.xilinx.com/tools/coregen.htm

Can CAD Cure Cancer?

Smita Krishnaswamy[1], Bernd Bodenmiller[2], Dana Pe'er[1]
[1]Columbia University, Biological Sciences, New York, NY 10027
[2]University of Zurich, Institute of Molecular Life Sciences, Zurich, Switzerland CH-8057

ABSTRACT

Eukaryotic cells have complex regulatory systems that sense adversity (e.g. DNA damage, heat shock, external death-induction signals) and respond by invoking programmed cell-death or apoptosis. Cancer cells have evolved the ability to thwart such sensory information and associated regulation. As biologists begin to understand cells in circuit-like terms, we can also begin to derive and simulate druggable targets of cellular networks that cause a cancerous cell to kill its self. Towards this goal, we have performed preliminary experiments that test the impact of siRNA (RNA silencing) on diverse cellular signaling pathways, at unprecedented single-cell resolution. We propose ways in which a CAD system can process such data to automatically derive drug targets.

1. INTRODUCTION

Cancer is a unique disease in which cells have acquired enough mutations to be able to grow and divide by thwarting internal and external mechanisms that stop and limit growth. Cell growth and division are normally carefully controlled processes. For instance, normal mature cells do not grow in the presence of DNA damage or absence of growth factor signals. Commonly mutated genes in cancer, also known as oncogenes include 1) p53 which is responsible for DNA damage response, 2) pRB which controls the progression of a cell from the G phase to the synthesis phase of the cell cycle and 3) Src, a kinase, that resides downstream of receptors and regulates growth. Note that only one cell has to "figure out" how to be cancerous, because then it can found a colony of other cells that grow similarly. These cells are often referred to as "cancer stem cells" in literature.

However, once cells become cancerous there is often genomic instability that allows cells to acquire other spurious mutations. Therefore, researchers have searched for "driving mutations", i.e., mutations that are truly driving the development of the cancer rather than passenger mutation acquired later [6]. By contrast, in our work, we are searching directly for drugabble target proteins that have the desired effect on cellular signaling.

Permission to make digital or hard copies of all or part of this work for personal or classroom use is granted without fee provided that copies are not made or distributed for profit or commercial advantage and that copies bear this notice and the full citation on the first page. To copy otherwise, or republish, to post on servers or to redistribute to lists, requires prior specific permission and/or a fee.

Cancer is generally treated by inhibiting or knocking out key parts of the cellular machinery that allows for growth and proliferation. For instance, one of the first chemotherapy drugs 'Taxol' is a mitotic inhibitor that prevents microtubules in the cell from breaking down. This successfully inhibits cell division because microtubule breakdown is a key step in cell division. Another commonly used chemo therapy drug "Dasatinib" is an inhibitor of kinases that are associated with cell growth. However, since chemotherapy drugs have very specific targets, they are not effective for all kinds of cancers and all people.

Most researchers believe that the future of cancer treatment is in personalized medicine where the mutations and signaling in each individual is tested, detected and treated specifically, possibly with RNA based gene silencing that is delivered to cancerous cells. Patient-specific information is becoming plausible with the advent of new high-throughput, single-cell genomic and proteomic technologies that allow for the measurement of several proteins and genes simultaneously.

2. IMPACT OF GENE-SILENCING ON SIGNALING

In our experiment, we measure the abundance of 30 key signaling proteins from various parts of the cellular machinery under the knockdown of thousands of other proteins in ovarian cancer cells. This allows us to assess the functionality of the proteins with respect to key parts of the cellular machinery and determine their viability as targets for cancer therapy.

Figure 1 shows a new technology, known as mass cytometry, that has been developed to measure protein abundance. Here, antibodies with chemically-chelated metal ions are injected in cells to target specific proteins. Then these cells are essentially vaporized in a plasma chamber and sent through a time-of-flight mass spectrometer to measure the metal abundance. The metal abundance is proportional to the respective protein abundance. This technology allows us to measure up to 100 proteins simultaneously at a single-cell resolution.

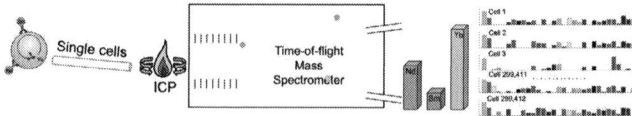

Fig. 1 Schematic of a mass cytometer measuring cell proteins.

Recently siRNAs , short for "small interfering RNA" has been used in large-scale screens to assess the effect of silencing particular proteins [5]. Small interfering RNA work by "covering up" mRNA transcripts before proteins are synthesized, thereby stopping the synthesis of the targeted protein. RNA screens generally try knocking out one protein at a time and the average level of other proteins is measured as a result.

In our experiment, we combine mass cytometry with siRNA screening for the first time in order to assess the impact of thousands of protein-knockdowns on a wide panel of measured proteins in ovarian cancer cells. Since proteins are interconnected in a complex and dynamic network inside cells, the knockdown of a particular protein has affects on its neighbors in the protein interaction network. In order to assess the impact of knockdowns we have measured proteins representing a wide variety of functions and pathways in the cell including cell growth, apoptosis, damage response, cell cycle, mitosis and stress response.

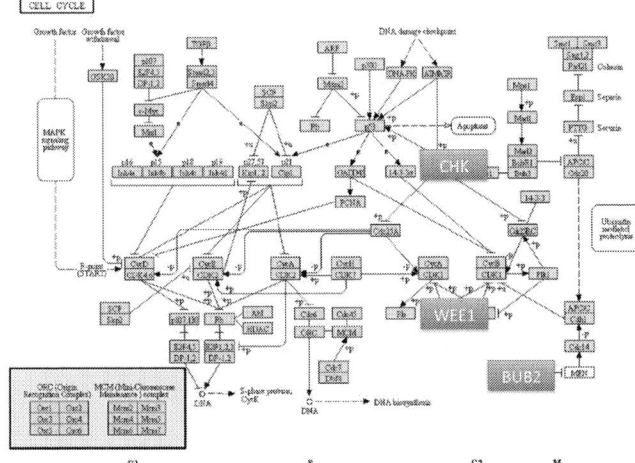

Fig 2. KEGG diagram of mammalian cell-cycle pathways.

Figure 3 shows the impact of WEE1 (cell cycle protein) knockdown as a 2-d image, derived by compressing 30-dimensional signaling-protein-abundance information with a non-linear dimensionality reduction method [4]. The result is the increase in cells undergoing apoptosis (programmed self-death). The apoptotic cells cluster together in the image due to the highly elevated levels of the measured cPARP, a key member of the apoptosis pathway.

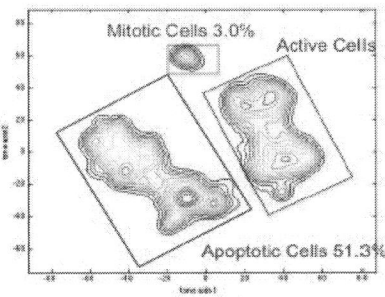

Fig 3. Cells under WEE1 knockdown form a large apoptotic population. Dimensionality reduced [4] rendering is shown.

3. CAD FOR CANCER

Often, cancer is thought of as faulty or compromised machinery. However, key to our viewpoint is the notion o cancer

as an "evolved organism" that has perfected its machinery to allow for growth and proliferation. If we think of the cellular machinery of cancer as our design, the key CAD-problem is to induce a minimal set of faults that destroy cancer cells.

In principle, it is possible for CAD-methods to handle biological network analysis and simulation because cells contain around 2000 proteins, and a few thousand regulatory pieces of miRNA--- far fewer than the number of components in modern VLSI systems. However, unlike traditional CAD, the system design is unspecified and contains a large number of unknowns. With the advent of new technologies, this problem is getting mitigated and researchers are uncovering "pathways" and mechanisms of interaction. For example, there are now large databases such as the KEGG database (http://www.genome.jp/kegg/pathway.html) that contain pathway maps for all kinds of cells and processes. Figure 2 shows the pathways involved in cell cycle regulation.

Biological CAD systems of the future must infer missing parts of the network and potential mechanisms of interaction in order to test if a set of drug targets have the desired effect upon perturbation. Given a signaling network with thousands of components, it seems computationally tractable to simulate a handful of mechanisms of interaction in cases where mechanisms are largely unknown. In addition, recent research has shown that network and pathway constraints can further reduce problem complexity. For instance, Vandin et. al [6] use the following two combinatorial constraints in deriving sets of cancer driver genes: 1) *mutual exclusivity of mutations*, as in there is generally one driver mutation per biological pathway, 2) *coverage* as in the set of driver mutations covers a variety of cancer pathways [6]. Similarly druggable targets may also need to cover multiple pathways downstream of the mutation.

In summary, CAD systems for deriving sets of druggable targets for cancer must involve: 1) the integration of biological pathway database information, 2) wide-pathway coverage in assessment/measurement of candidate drug targets, 3) optimization under combinatorial constraints such as pathway coverage and exclusivity, 4) the ability to validate or simulate the perturbation of multiple genes while accounting for pathway crosstalk.

This brings us to a vision for a future hospital. A patient arrives for screening of tumor cells, cell signaling is measured in real-time and then a CAD system automatically derives her personal treatment cocktail. This is perhaps a future that will occur sooner than many of us expect.

4. REFERENCES

[1] Bendall, S., et al., "Single-cell Mass Cytometry of Differential Immune and Drug Responses across a Human Hematopoietic Continuum," Science, vol. 332(6030), 6 May 2011, pp. 687-696.

[2] D. Hanahan, R.A. Weinberg, "Hallmarks of cancer: the next generation," *Cell,* vol. 144(5) March 4 2011, pp.646-674.

[3] KEGG Database http://www.genome.jp/kegg/pathway.html)

[4] L.J.P. van der Maaten, G.E. Hinton, "Visualizing High-Dimensional Data Using t-SNE," *Journal of Machine Learning Research* vol. 9, Nov. 2008, pp. 2579-2605.

[5] J. Moffat, D.M. Sabatini, "Building Mammalian Signalling Pathways with RNAi Screens," *Nature Reviews Molecular Cell Biology*, vol. 7, March 2006, pp. 177-187.

[6] F. Vandin, J. Upfal, B.J. Raphael, De Novo discovery of mutated driver pathways in Cancer, *Genome Research*, Feb 22(2), 2012.

Let's put the Car in your Phone!

Martin Geier, Martin Becker, Daniel Yunge, Benedikt Dietrich,
Reinhard Schneider, Dip Goswami, Samarjit Chakraborty
Institute for Real-Time Computer Systems, TU Munich, Germany
lastname@rcs.ei.tum.de

ABSTRACT

Today high-end cars have extremely complex E/E architectures – with 50–100 electronic control units (ECUs), connected by communication buses like CAN, FlexRay and Ethernet. They are used to run several (control) applications with many million lines of code. We propose a radically new architecture where all these applications are instead run on a mobile phone being carried by the driver. The car now has a considerably simpler architecture with few or no ECUs, using RF links to connect sensors and actuators to the mobile phone with a powerful multicore processor. We discuss the advantages and challenges and describe a small prototype implementation with an adaptive cruise control application.

Categories and Subject Descriptors

C.3 [**Special-purpose and application-based systems**]:
Real-time and embedded systems

General Terms

Design, Experimentation, Performance, Reliability

Keywords

Automotive, Mobile Phone, Consolidation

1. INTRODUCTION

Over the past one decade automotive electrical and electronic (E/E) architectures have become extremely complex and distributed, to the extent that this growth in complexity is no longer sustainable. High-end cars now contain 50–100 electronic control units (ECUs) or processors. They are connected by several different communication buses like CAN, FlexRay and Ethernet, with their wiring running into several kilometers in length. Such an architecture is then connected to various sensors and actuators and is used to run several control applications related to safety-critical, driver assistance and comfort functionalities. In total, all applications sum up to many million lines of code. However, the hardware infrastructure including the cabling weighs several tens of kilograms and has an impact on fuel or energy consumption, which is especially critical for electric vehicles. Such an architecture and the growth in the number of ECUs can be partially attributed to the need for more computational power; e.g., cars now have a rich set of driver assistance functions that process inputs from multiple cameras, radars and LIDAR sensors. However, this growth may also be attributed to the business models followed by the automotive industry, where a network of Tier 1 suppliers provide ECUs with specific functionalities that the OEM then integrates.

But increasingly it is being realized that this practice cannot continue for long – the large number of ECUs and complex networks are difficult to maintain, verify and debug,

they lead to compatibility problems during architecture extensions and they increase the weight and the cost of the car. This has led to a lot of recent work to enable a move from such *federated* to *integrated* architectures [7], where a number of functionalities are integrated into a single, possibly multicore ECU, thereby reducing the number of ECUs. The use of software platforms like AUTOSAR [1] and virtualization [6] to provide isolation between applications is being explored to realize such integrated architectures. Another major drawback of current automotive architectures is that its hardware gets outdated within 1–2 years, given the fast pace at which technology and, thus, processors are currently developing, while a typical car has a lifetime of 10–15 years. This is problematic since now most innovation in modern cars is in their electronics and software.

Phone in the loop: As a possible solution to these problems, we propose a radically new automotive E/E architecture where all applications (including control) are instead run on a smartphone being carried by the driver. The car will now have a considerably simpler architecture with few or no ECUs, and all its sensors and actuators will instead communicate with the mobile phone via RF links. While currently available smartphones have quadcore processors, 100-core processors in such phones are not unforeseeable in the future, which can then replace the 100 ECUs currently in the car. Further, smartphones today already have multiple communication interfaces like Bluetooth and WiFi and this is likely to grow, too. As the driver enters the car, the phone will connect the available sensors and actuators in the car and run the computations that currently run on the ECUs. While a number of challenges need to be overcome in order to realize this, some of the advantages are obvious. The phone can be replaced easily, as we already know today. Software upgrades will also be simplified. The setup will be significantly more cost-effective, since the compute power in the ECUs currently is not used when the car is not being driven. With these ECUs being "buried" in the phone, they can be used for other purposes when the car is not in use. Further, reliability of ECUs is a major concern today, since they are exposed to harsh environments and extreme temperatures. The phone on the other hand is always in the passenger cabin. The phone may also carry driver preferences/profiles such as seat adjustments, display layouts and even emails, entertainment-related information and daily schedule. Many high-end cars today allow a smartphone to be connected to the car for accessing emails, music, etc. These functionalities along with personalization information can now be more seamlessly integrated with the car [9]. A video illustrating the concept and our prototype is available online [10].

2. TECHNICAL CHALLENGES

A number of technical challenges need to be overcome to realize our proposed architecture, some of which even require new technological developments. First, current mobile phones cannot support the computational bandwidth required to replace all the ECUs. Thus, the path towards our proposal can also have a number of intermediate stops, i.e., few ECUs and the phone might be more realistic now. Furthermore, the phone could serve as a gateway to a cloud-based compute infrastructure, which we discuss in Section 4. Virtualization techniques that are already being investigated

Permission to make digital or hard copies of all or part of this work for personal or classroom use is granted without fee provided that copies are not made or distributed for profit or commercial advantage and that copies bear this notice and the full citation on the first page. To copy otherwise, to republish, to post on servers or to redistribute to lists, requires prior specific permission and/or a fee.
DAC '13, May 29 - June 07 2013, Austin, TX, USA.

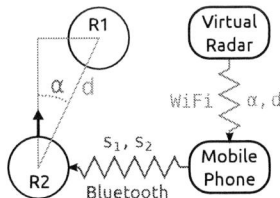

Figure 1: Prototype setup

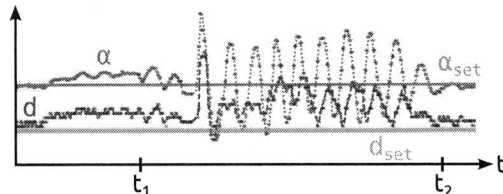

Figure 2: Oscillation due to RF interference

in the automotive context may be implemented inside the phone in order to isolate safety-critical applications from others. We believe that this can already be realized to a large extent and is a good research problem. The next obvious challenge is that of communication between the mobile phone and the sensors and actuators in the car. While Bluetooth is currently used for in-vehicle communication, it is restricted to the infotainment domain. Almost all current in-vehicle communication is performed over wired networks [8]. New reliable, energy-efficient, high-bandwidth, short range communication technologies will be needed to realize our proposal and a significant amount of research effort should go in this direction. While security issues are largely ignored in current automotive architectures and are only being studied recently, e.g., for firmware updates [5], they will become much more important in the proposed architecture. Finally, software and control systems development will require attention, too. Since sensor and control messages will now be transmitted over wireless networks, they will suffer variable delays and probably also loss. Therefore, controller design techniques will have to be more aware of computation and communication resource constraints [4] than they are today.

3. PROTOTYPE IMPLEMENTATION

As a starting point for investigating some of the challenges and potential solutions associated with the proposed architecture, we have implemented an *adaptive cruise control* (ACC) application on a robot. The setup (depicted in Figure 1) is that of a *Robotic Road Train* recently reported in [3], where a semiautonomous vehicle follows another vehicle in front, while maintaining a safe distance to it.

The first robot (R1) follows an arbitrary trajectory. The second robot (R2) receives two wheel speed values (s_1, s_2) from the mobile phone via a Bluetooth link. A virtual radar sensor (implemented using a Vision Tracking System) determines distance d and angle α between the robots. Using WiFi, this information is sent to the ACC application running on the mobile phone which finally closes the control loop and enables the second robot to follow the first.

This system is susceptible to delay, jitter and loss from various sources, such as RF interference and scheduling on the mobile phone. To show the effects of RF interference between sensor and controller, we set up a second WiFi link on the same channel as the sensor connection. By adding load we increase the collision probability of the two WiFi links until packet loss occurs which leads to control instability. Figure 2 shows set-points and actual values of d and α as recorded by the mobile phone during system operation. The second WiFi link was loaded for a short interval centered around t_1. It can be seen that for $t_1 < t < t_2$ both d and α oscillate around their respective set points which can be observed as snaking movement as shown in our video [10].

The problem of network failures shown in the above small-scale example will become even more prominent in our proposed architecture. To overcome this and the aforementioned problems, advanced design techniques will be required in multiple domains. Increased reliability and fault tolerance of the wireless communication network may be achieved using *link diversity schemes* and *integrated control strategies*.

The former uses multiple diverse wireless communication paths to avoid network failures. Modern mobile phones support several RF interfaces, each of them having different transmission characteristics (link rate, latency, resilience) resulting from parameters such as frequency, bandwidth and transmission power. Based on availability of individual communication paths and on current control requirements, a *communication management unit* may dynamically modify the information routing during runtime.

The latter takes the underlying communication network into account to better deal with its delay, jitter and loss. These factors lead to multiple *switching* control subsystems and create a major challenge in designing *platform-aware* control algorithms to stabilize the resulting switched system.

4. OUTLOOK

Moving a bulk of the computation away from automotive-specific ECUs and onto a smartphone carried by the driver is a paradigm shift in automotive architectures. But it has a number of advantages and certainly can be realized given the current rate of progress in processor architectures and wireless communication technologies. Eventually, the phone can serve as a gateway to a cloud that will support computations from many cars. This will also result in new business models where along with the car, one needs to additionally buy a service, i.e., computation time on the cloud. Thus, the latest hardware/software technologies will be available to a car and largely free it from maintenance requirements. This will be especially useful for electric cars where most of the functionalities will anyway be implemented in software and the need to cut down weight and energy consumption is severe. However, such a scheme will also need several developments in cloud computing, such as support for real-time computing and low-delay communication. The design of control algorithms will also be influenced, with cloud-based implementations of compute-intensive control algorithms (e.g., visual servoing) already being studied [11]. Such a setup will also open up the possibility of new advanced driver assistance functionalities and autonomous cars [2] that require too much computation (and hence energy and cooling infrastructure) to be feasibly implemented *inside* a car.

5. REFERENCES

[1] Automotive open system architecture, "AUTOSAR Specification", release 4.0, 2012.

[2] D. Bernstein, N. Vidovic, and S. Modi. A cloud PAAS for high scale, function, and velocity mobile applications - with reference application as the fully connected car. In *Int. Conf. on Systems and Networks Comm. (ICSNC)*, 2010.

[3] E. Coelingh and S. Solyom. All aboard the robotic road train. In *IEEE Spectrum*, November 2012.

[4] D. Goswami, R. Schneider, and S. Chakraborty. Co-design of cyber-physical systems via controllers with flexible delay constraints. In *ASP-DAC*, 2011.

[5] M. S. Idrees, H. Schweppe, Y. Roudier, M. Wolf, D. Scheuermann, and O. Henniger. Secure automotive on-board protocols: A case of over-the-air firmware updates. In *Nets4Cars/Nets4Trains*, 2011.

[6] A. Masrur, S. Drössler, T. Pfeuffer, and S. Chakraborty. VM-based real-time services for automotive control applications. In *RTCSA*, 2010.

[7] M. D. Natale and A. L. Sangiovanni-Vincentelli. Moving from federated to integrated architectures in automotive: The role of standards, methods and tools. *Proceedings of the IEEE*, 98(4):603 – 620, 2010.

[8] N. Navet, Y. Song, F. Simonot-Lion, and C. Wilwert. Trends in automotive communication systems. *Proceedings of the IEEE*, 93(6):1204 – 1223, 2005.

[9] D. Siewiorek. Generation smartphone. In *IEEE Spectrum*, September 2012.

[10] DAC2013 WACI Video: Let's put the Car in your Phone!, December 2012. http://www.rcs.ei.tum.de/en/research/byop/.

[11] H. Wu, L. Lou, C.-C. Chen, S. Hirche, and K. Kühnlenz. Cloud-based networked visual servo control. *IEEE Tran. Industrial Electronics*, 60(2):554 – 566, 2013.

The Undetectable and Unprovable Hardware Trojan Horse

Sheng Wei   Miodrag Potkonjak
Computer Science Department
University of California, Los Angeles (UCLA)
Los Angeles, CA 90095
{shengwei, miodrag}@cs.ucla.edu

ABSTRACT

We have developed an approach for automatic embedding of customizable hardware Trojan horses (HTHs) into an arbitrary finite state machine. The HTH can be used to facilitate a variety of security attacks and does not require any additional gates, because it is morphed into the specified design. Even after the HTH induces provable damage, one is not capable of proving that any malicious circuitry is embedded into the design. The main ramification of the developed HTH is that hardware and system techniques should move from HTH detection toward synthesis for trusted systems.

1. INTRODUCTION

Hardware Trojans horses (HTHs) [1] are malicious modifications to integrated circuits (ICs) during design (e.g., by untrusted CAD tools) or manufacture (e.g., by untrusted foundries). Recently, HTH has posed great concerns with regard to the security and integrity of ICs with the rapid growth of IC outsourcing. The detection of HTHs has triggered a great deal of attention in the IC community, due to the fact that the consequences of HTHs can be extremely severe for security-sensitive systems.

The existing HTH research [1][2][3][4] targeted only on Trojans that are physically present and thus observable on the target IC, either in the form of additional malicious components or modifications toward the target circuit. Although these types of HTHs can be well hidden under the target circuit, the difficulty level for detection is limited due to the following two reasons. (1) the embedded HTH would result in at least one type of variation in the observable properties of the IC, including but not limited to physical structures (e.g., layout and wiring) and side channels (e.g., delay and power); and (2) the HTHs under consideration are identical on different chips of the same design due to the high cost of customizing the design and manufacturing for HTH insertion. As a result, once one chip compromised by HTHs is detected during the IC test, all other chips under attack can be easily identified in a straightforward way.

Permission to make digital or hard copies of all or part of this work for personal or classroom use is granted without fee provided that copies are not made or distributed for profit or commercial advantage and that copies bear this notice and the full citation on the first page. To copy otherwise, to republish, to post on servers or to redistribute to lists, requires prior specific permission and/or a fee.
DAC'13, May 29 - June 07 2013, Austin, TX, USA.

We argue that it is completely feasible to create challenging HTHs and bypass the existing detection schemes that are subject to the above limitations. It is rather important to investigate on these HTH attacks and motivate security primitives from a completely new angle. Based on these thoughts, we develop a zero-overhead, customizable HTH model that an attacker could leverage to create untrusted CAD tools and trigger undetectable security attacks. Our undetectable HTHs have the following features:

Zero-overhead. Our HTH model leverages the redundant states (called HTH states) in the finite state machine (FSM) of the target circuit for security attacks, which does not require any additional hardware to trigger the HTH during normal IC operations and, therefore, exposes no observable variations in the IC properties.

Customizable. The proposed HTH model induces different and customizable security attacks on different ICs without introducing additional manufacturing costs. Consequently, even if one instance of the HTH is detected, it is extremely difficult for the detection procedures to prove the presence of HTHs, nor can they generalize the found instance to other chips under test. We achieve this goal by employing post-silicon device aging caused by the negative bias temperature instability (NBTI) effect [5]. The attacker could intentionally age the target ICs after manufacture in such a way that unpredictable delay faults occur at runtime to transition the IC from normal states to the HTH states.

Therefore, the main ramification of our proposed HTH model is that it forces the detection mechanisms to move from traditional detection to sequential synthesis. Not only the cost for detection is significantly increased, but also the fundamental paradigms in the existing detection approaches have to be revisited and reconsidered in order to achieve reliable HTH detection schemes.

2. MOTIVATIONAL EXAMPLE

Figure 1 shows a motivational example of our proposed undetectable HTH model. Figure 1(a) is the finite state machine of a mod-3 up/down counter, including two inputs (x_1, x_0) that control the counter to stop, count up, and count down; and 4 states that can be implemented by 2 flip-flops. Among all 4 states, only 3 of them are valid states of the counter, i.e., representing the count number 0, 1, and 2. The shaded state S_3 is a redundant state (or don't-care state) that cannot be reached from any other states using any inputs. An attacker could leverage S_3 to trigger a variety of attacks, such as leaking confidential information or consuming higher energy.

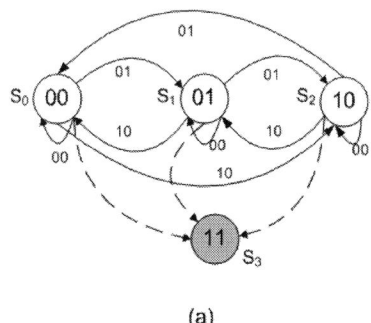 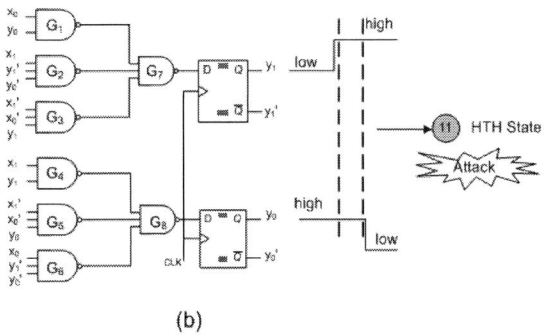

(a)   (b)

Figure 1: Motivational example of the undetectable hardware Trojan horses: (a) finite state machine of a mod-3 up/down counter, which includes 3 normal states (i.e., S_0, S_1, and S_2) and 1 redundant state (i.e., S_3); and (b) demonstration of the HTH state transition using device aging.

Figure 1(b) demonstrates the design of the sequential circuit based on the FSM, which shows the transition from normal states (i.e., states S_0, S_1, and S_2) to the don't-care state (i.e., state S_3), so that the desired HTH attack can be triggered at runtime. Our approach is to intentionally age (i.e., stress the corresponding transistors) a certain set of gates and trigger delay faults at the circuit output. For example, the attacker could age gate G_8 so that the signal transmitted to F_2 is delayed. It is possible that the delayed signal for F_2 causes delay fault, e.g., both F_1 and F_2 stay at signal 1, which transitions the circuit into the HTH state.

The trigger of delay fault and thus the state transition is fully customizable, in the sense that the attacker can selectively age different components for different chips post-silicon, which transitions the target circuit to different HTH states from different normal states. Even in small designs, there are exponentially many combinations of transitions that can be leveraged by the attacker to complicate and obfuscate the attacks.

3. FEASIBILITY STUDY AND VALIDATION

The feasibility of the proposed HTH attack is based on the assumption that there are large numbers of redundant states available in the target circuit. We argue that the assumption holds for the following two reasons. Firstly, the design of modern sequential ICs often results in large numbers of redundant states for the consideration of performance and ease of integration. Secondly, even if the original design specification does not indicate enough don't-care states, the attacker could easily minimize the FSM [6] to create equivalent designs that include many redundant states.

4. HTH DETECTION REVISITED

As a consequence of the undetectable HTH model, the traditional HTH detection mechanisms have to be revisited to accommodate the elevated difficulty level for ensuring a trusted IC system. In order to achieve this goal, we argue that the current HTH detection approaches [1], which rely on the monitoring of the end system in the post-silicon stage, have to be moved to sequential synthesis at the design time. In other words, the detection process must examine the redundant states generated by the untrusted CAD tools and exclude the possibility of HTH attacks early at the design time, which is an extremely difficult task.

Our idea to address the problem is to enforce a specified system at design time, where all or a part of the don't-care states are either explicitly removed or incorporated as a well defined state. In this way, we can limit the freedom of manipulating the FSM that is exposed to the untrusted tools. The downside of this solution is that it may compromise the performance gains obtained from the don't-care states. Therefore, a careful design is required to balance the tradeoff between performance and security of the system.

5. CONCLUSION

We have developed a zero-overhead and customizable hardware Trojan horse that cannot be detected by existing HTH detection approaches. The attack model leverages redundant states in the finite state machine and triggers the malicious state transition using device aging. We show that the HTH detection schemes must move from post-silicon monitoring to complex design-time synthesis in order to capture the new attack. By presenting the new HTH model, we aim to motivate new HTH detection research in the community to ensure the security and integrity of the ICs.

6. ACKNOWLEDGEMENTS

This work was supported in part by the NSF under Award CNS-0958369, Award CNS-1059435, and Award CCF-0926127, and in part by the Air Force Award FA8750-12-2-0014.

7. REFERENCES

[1] M. Tehranipoor, F. Koushanfar, A Survey of Hardware Trojan Taxonomy and Detection, IEEE Design and Test of Computers, Vol. 27, No. 1, 2010, pp. 10-25.

[2] S. Wei, S. Meguerdichian, M. Potkonjak, Gate-level Characterization: Foundations and Hardware Security Applications, DAC 2010, pp. 222-227.

[3] S. Wei, K. Li, F. Koushanfar, M. Potkonjak, Hardware Trojan Horse Benchmark via Optimal Creation and Placement of Malicious Circuitry, DAC 2012, pp. 90-95.

[4] S. Wei, L. Kai, F. Koushanfar, M. Potkonjak, Provably Complete Hardware Trojan Detection Using Test Point Insertion, ICCAD 2012, pp. 569-576.

[5] H. Baba, S. Mitra, Testing for Transistor Aging, VTS 2009, pp. 215-220.

[6] L. Yuan, G. Qu, Information Hiding in Finite State Machine, IH 2004, pp. 340-354.

Path to a TeraByte of On-Chip Memory for Petabit per Second Bandwidth with < 5Watts of Power

Swaroop Ghosh
Computer Science and engineering, University of South Florida, Tampa, Florida-33647
sghosh@cse.usf.edu

ABSTRACT

We propose a path to achieve an ambitious target that has never been tried before: a terabyte of on-chip memory for petabit/second of bandwidth with < 5W of power. Conventional methodology of on-chip memory design is bottom up where the choice of bitcell topology and associated peripherals are predetermined. The resulting memory is sub-optimal and often suffers from high power and poor bandwidth. We approach this problem from top down where the capacity, bandwidth and power specifications guide the choice of bitcell. Our evaluation shows that domain wall memory (DWM) can be a potential technology that can meet TB capacity and Pb/s bandwidth with shoestring power budget.

Categories and Subject Descriptors
B.7.1Types & Design Styles – Advanced Technologies

General Terms
Performance, Design

Keywords
High density, low-power memory, Domain wall memory.

1. INTRODUCTION

Need for bandwidth (BW): The need for BW is rooted in two basic theories: (a)Von-Neumann architecture (Fig. 1(a)) which prescribes processing unit interacting with memory for computation and, (b) Moore' law that dictates transistor scaling. Faster pipelines (due to speedy scaled transistors) need data quicker to keep itself busy. The advent of multicore designs and integration of graphics accentuate this issue further. Assuming frame size of 1920x1080 pixels, 24 bit color depth and 30 fps speed, the BW requirement of a video application would be 1.4GB/s. With modern graphics capabilities the BW requirement will easy shoot upto TB/s. There is already a ~2X BW gap between CPU and external memory today giving rise to a "Memory Wall" [1] (Fig.1(a-b)). In absence of proactive BW boosting steps, the gap will continue to worsen in future.

Need for capacity: The need for on-chip memory capacity is linked with the miss rate. From [2], number of misses (M) is related to number of cores (P), cache size (S) and an empirical constant (a where $0.3 < a < 0.7$) by $M = P*M_0S^{(-a)}$. Therefore, if the number of cores and cache size changes from P_1 to P_2 and S_1 to S_2 respectively (due to tech scaling), the new miss rate is given by, $M_2 = (P_2/P_1)(S_2/S_1)^{-a}M_1$. To keep miss rate fixed, new cache size will be $S_1(P_2/P_1)^2$ meaning thereby that the cache size will grow quadratically. Cache size projections using this formulation (assuming 32MB cache [3]) with tech scaling indicate 52X increase in cache size for 2.5X increase in number of cores (Fig. 1(c)). Obviously an on-chip cache size of > 1TB (32MB*52*52) will be common if the number of cores is grows by 5X.

Permission to make digital or hard copies of all or part of this work for personal or classroom use is granted without fee provided that copies are not made or distributed for profit or commercial advantage and that copies bear this notice and the full citation on the first page. To copy otherwise, or republish, to post on servers or to redistribute to lists, requires prior specific permission and/or a fee.
DAC '13, May 29 - June 07 2013, Austin, TX, USA

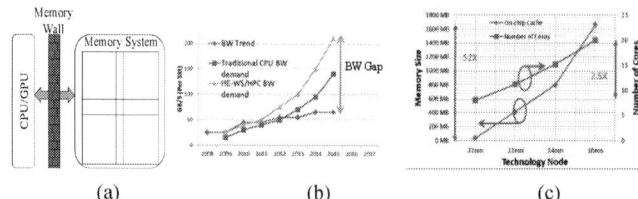

Fig. 1 (a) Memory wall, (b) Need for BW, (c) Need for memory capacity in multicore designs

Need for low power: Power plays crucial role in determining the battery lifetime. Active power is important but leakage also starts to add up with growing memory capacity. Considering the leakage of SRAM subarray[4], a TB size memory would consume 42kW of leakage power (equivalent to battery life of 5 sec!). *Therefore, containing the leakage power of a TB size memory is crucial.*

Array footprint: The die footprint of today's state-of-the-art processor is ~160mm^2 [5]. Considering 0.092um^2 SRAM size [6] the cache footprint for 1TB would be ~1350000mm^2 which is clearly prohibitive. In order to fit 1TB memory into 160mm^2 footprint, the bitcell must be 8000X smaller than SRAM or 8000 layers of 3D integration [7] would be required. Emerging high density embedded memories such as eDRAM and STTRAM [8] are 4X denser than SRAM however they are still 2000X away from the target. Stacking memories [7,9] is a possible solution however the power dissipation of the peripherals would still dominate.

Meeting tremendous BW requirement while conforming to power and array footprint poses serious challenges in future memory system design. Our study reveals that Domain Wall Memory (DWM) [10] could be a promising candidate for 1TB capacity, Pb/s BW with <5W of power consumption.

2. RELATED WORK

Terascale memory design is not new for the hard disk industry. Seagate demonstrated storage density of 1TB/in^2 [11] which is 1000X better than the state-of-the-art embedded memory density (~1.38GB/in^2 [3]). Obviously, both external disks and HDDs are tailored for density however they suffer from poor performances (~10^{6}X slower than SRAM). Therefore these memories are not suited for embedded application. On the other hand, embedded memory is tailored for quick access and not for density. SRAM in 22nm node allows access every fraction of a nanosec[6]. IBM Power7[3] describes 32MB eDRAM architecture with nanosec access time. Hybrid Memory Cube (HMC)[12] addresses the tremendous BW and capacity challenges in exascale computing by 3D integration of DRAM. Although promising, HMC is expected to be overshadowed by cost. Another initiative for TB/s bandwidth is taken by Rambus[13] using ZRAM technology.

3. EVALUATION OF MEMORIES

Table-1 shows evaluation of various memory technologies for a TB capacity in 22nm tech. The leakage power/subarray is estimated by assuming 0.1nA/um leakage [14] and typical sizes of peripherals. The total leakage of 1024x256 subarray is found to be ~20uW. It is evident from Table-1 that although non-volatile technologies such as STTRAM and RRAM[8] provide smaller footprint due to MLC

Table 1 Comparison of various technologies for 1TB of on-chip memory

Memory	Cell Size	Area(F=22nm) in nm²	MLC	RowxCol/subarry	Area (mm²)	Standby Power (W)	3D Layer
SRAM	50F²	0.0242	1	512x512	304,093	209715	1,901
eDRAM	6F²	0.00484	1	1024x256	60,818	4505 (refr+ leak)	380
STTRAM	6F²	0.002904	4	1024x256	9,122	168	57
RRAM	5F²	0.000484	4	1024x256	7602	168	48
DWM	16F²	0.007744	288	1024x256	338	2.33	2

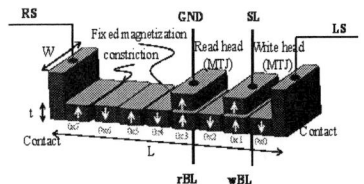
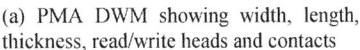

(a) PMA DWM showing width, length, thickness, read/write heads and contacts

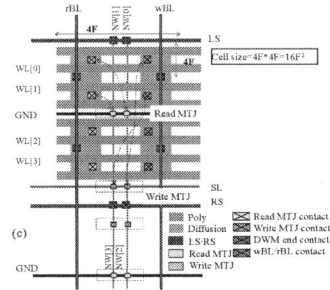

(b) DWM layout showing WL,rBL,wBL, SL,LS&RS connections. MTJ connections to access transistors are shown by dashed lines

capability, their standby power is still dominated by the peripheral leakage. In order to reduce the number of peripherals the MLC capability needs to be boosted by two orders of magnitude which is extremely challenging. Furthermore, significant number of 3D integration layers would be required to fit the 1TB array within 160mm² using latest MLC capabilities of STTRAM/RRAM (4th and 8th column, Table-1). DWM indicates potential to meet the power and footprint challenge due to its superior MLC capability. IBM has demonstrated MLC=6 in 90nm tech with *in-plane* magnetic anisotropy nanowire (NW) [15] which can be projected to reach 288 with 22nm *perpendicular magnetic anisotropy (PMA)* NW (3X due to in-plane→PMA transition and 16X due to device scaling) with a bitcell footprint of 16F² (Fig. 2(b)).The standby power of DWM is > 1W but it can be brought down by (a) employing deeper sleep of peripherals and (b) by sharing the peripherals (e.g., 4096X256 array to share WL drivers with extra senseamps). *Features of desired memory tech:* The discussion above indicates following desired features for TB capacity: (a) high MLC capability, (b) small cell size and, (c) low idle power.

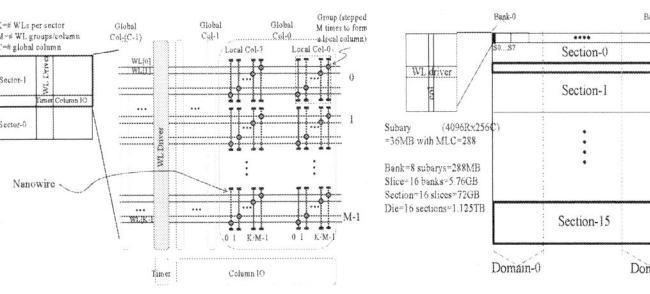

(c) DWM array organization showing local and global columns

(d) 1TB memory array architecture. Subarray, bank, slice and section is shown

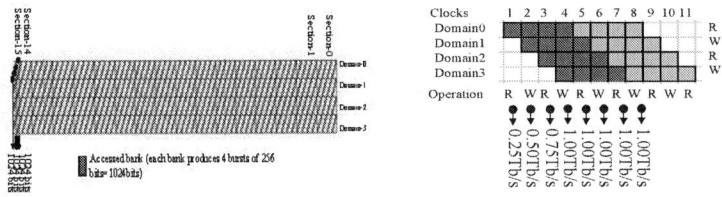

(e) Concurrent access of 16 sections and bank interleaving for 1Pb/s BW

(f) Interleaved access of domains to amortize the busy time of each bank

Fig. 2

4. ARCHITECTURE FOR BANDWIDTH

DWM Basics: DWM is partially serial in nature (Fig. 2(a)). The access starts with injection of spin polarized current (through LS and RS) to shift the bits in lockstep fashion in order to bring the desired bit under read or write head. Read is performed by sensing the resistance of MTJ formed by DW under the read head. Write is performed by injecting new domains through the write head.

Subarray design: The proposed array organization for one sector is shown (Fig. 2(c)). K WLs per sector is divided into M groups and each group has (K/M-1) WLs that connects to corresponding NWs to form a local column. The local column is stepped to create a global column. The layout of a single column with WL[3:0] connecting with NW[3:0] is shown (Fig.2(b)). The connection of access transistors with MTJs are shown by dashed lines. The size of the bitcell/NW is 16F².

System Organization: Fig. 2 (d) shows the organization of 1TB memory. The memory sub-system is divided into 16 sections (72GB each), each section is composed of 16 slices of 5.76GB, each slice contains 16 banks of 288MB and each bank has 8 subarrays of 36MB. Furthermore, the die is divided into 4 independent domains for interleaved access. Each bank has access latency of 4 cycles (equivalent to 1ns at 4GHz to meet read/write time [16] at scaled values in 22nm tech) and produces 1024 data bits (4 bursts, each 256 bits). One bank from each slice is accessed in parallel for getting 256Mb (i.e., 0.25Pb/s) of data. To boost the BW to 1Pb/s, three more banks from remaining domains are accessed in consecutive cycles (Fig. 2(e-f)).

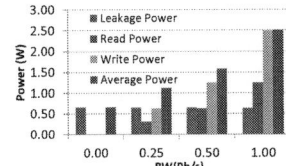

Fig. 3 Power vs BW in 22nm tech. 1Pb/s BW can be sustained with <5W.

Power consumption wrt BW at 4GHz is shown (Fig. 3). At zero BW, the leakage power dominates. A BW of 0.25Tb/s is achieved

when 256 banks within domain-0 are accessed (where active power starts dominating). The active power is calculated by scaling read/write currents in 22nm tech (19uA/bit and 38uA/bit respectively [17]). Interestingly, we achieve ~200Tb/s BW with ~1W which is 500X better than previously reported values [18]. A BW of 1Pb/s is sustained with less than 2.5W of power.

5. SUMMARY

We provided a solution path to achieve a TB of memory and Pb/s BW with sub-5W power. Our findings indicate that DWM can be the potential memory technology for future petascale computing. Although attractive, power consumption of DWM heavily depends on scaling of read/write powers with technology. Furthermore, datapath power could be critical and would need to be contained.

6. REFERENCES

[1] Asanovic, Krste, et al,Technical Report, 2006. [2] Rogers, B M., et al, SIGARCH, 2009. [3] Kalla, R et al, Micro, 2010. [4] Wang,Y. et al, JSSC, 2010. [5] http://www.anandtech.com/show/5771/the-intel-ivy-bridge-core-i7-3770k-review/3. [6] E. Karl et al., ISSCC 2012. [7] Faye A Briggs, ICAF, 2011. [8] M. H. Kryder et al, 2009. [9] Wordeman, M., et al, ISSCC, 2012. [10] Parkin, S. et al, Science, 2008. [11] http://news.cnet.com/8301-21546_3-57400009-10253464/seagate-reaches-1tb-per-square-inch-hard-drive-to-reach-60tb-capacity/. [12]http://hybridmemorycube.org/files/SiteDownloads/20120710_HPCWire_HMCAnglesforExascale.pdf. [13] http://www.realworldtech.com/terabyte-intiative/1/. [14] http://www.asu.edu/~ptm. [15] A. J. Annunziata et al, IEDM 2011. [16] R. Venkatesan et al, ISLPED, 2012. [17] Thomas, L., et al, IEDM, 2011. [18] Sekiguchi, T., et al, JSSC 2011.

993

Reconciling real-time guarantees and energy efficiency through unlocked-cache prefetching

Emilio Wuerges, Romulo S. de Oliveira
Dept. of Automation and Systems Engineering
Federal University of Santa Catarina, UFSC
Florianopolis, Brazil
emilio@inf.ufsc.br, romulo@das.ufsc.br

Luiz C. V. dos Santos
Department of Computer Sciences
Federal University of Santa Catarina, UFSC
Florianopolis, Brazil
santos@inf.ufsc.br

ABSTRACT

For real-time tasks, cache behavior must be constrained via cache locking or predicted by WCET analysis. Since the former gives up energy efficiency for predictability, this paper proposes a novel code optimization that reduces the miss rate of unlocked instruction caches and, provenly, does not increase the WCET. We optimized the 37 programs from the Mälardalen WCET benchmark for 36 cache configurations and two technologies. By exploiting software prefetching on top of on-demand fetching, we reduced the memory's contribution to the energy consumption (by 11.2%), to the average case execution time (by 10.2%), and to the WCET (by 17.4%).

1. INTRODUCTION

With the rise of smartphones and tablets, Mobile Computing requires increasing energy efficiency to execute programs whose complexity keeps raising. Mobile devices are essentially a combination of two subsystems – a "PC" and a "radio". The former runs the end-user interface and application programs on a multithreading environment supported by a conventional operating system, whereas the latter implements baseband, protocol-stack, and security processing by relying on a multi-tasking environment built on top of an RTOS. This scenario asks for techniques that do not jeopardize predictability when improving energy efficiency. Since an instruction cache may consume around 40% of an embedded processor's energy [3] and it impacts predictability and throughput, it becomes a relevant optimization target.

Cache controllers exploit locality of reference through *on-demand fetching*. When it is fully exploited, further miss rate reductions can only be obtained by fetching in advance the items that will not be in cache before they are referenced. To keep the processor from stalling, such *prefetching*

This work was partially supported by the Brazilian Council for Science and Technological Development (CNPq) through grants 141732/2010-5 (PNM), 303748/2009-5 (PQ), and 306654/2009-1 (PQ).

Permission to make digital or hard copies of all or part of this work for personal or classroom use is granted without fee provided that copies are not made or distributed for profit or commercial advantage and that copies bear this notice and the full citation on the first page. To copy otherwise, to republish, to post on servers or to redistribute to lists, requires prior specific permission and/or a fee.
DAC '13, May 29 - June 07 2013, Austin, TX, USA

mechanism relies on a non-blocking cache port or prefetch buffer. A smaller miss rate not only decreases the dynamic consumption, but it also shrinks the static energy consumption as it shortens the *average-case execution time* (ACET).

Despite its impact on consumption and *worst-case execution time* (WCET), cache prefetching is underexploited in Real-Time Systems, although a solid basis for accurately predicting cache behavior [8, 21] has been laid. One work extended cache abstract semantics to take prefetching into account [22], another exploited it for optimizing the WCET [5], and a recent one combined it with cache locking [2], but their impact on energy efficiency was not evaluated.

This paper presents a novel technique that inserts prefetch instructions for improving the energy efficiency of instruction caches. In contrast with most real-time optimization techniques [5, 4, 14], which target the minimization of the WCET as a single objective, our algorithm relies on the results of preliminary WCET analysis to identify the most profitable prefetches and to determine their insertion points in the execution flow. We claim that our non-conventional use of static WCET analysis drives code optimization towards energy-efficient binaries for real-time applications. We provide theoretical guarantees that the new algorithm does not increase the memory's contribution to the WCET. Our experiments show that, as compared to standard fetching alone, the technique can provide energy reductions up to 21% with cache capacities from 2 to 4 times smaller, while sustaining the same or superior performance.

The next section reviews conventional prefetching and its tailoring to low-power, real-time systems. Section 3 shows the models adopted for static analysis and formulates the target problem. Section 4 describes the proposed solution by means of examples. Section 5 discusses experimental results and Section 6 draws the overall conclusions. Supplement S.1 formally describes our algorithms and their complexity analysis. Supplement S.2, provides proofs for our theoretical claim. Supplements S.3 to S.5 present an extra example, the detailed set up, and extra experimental results.

2. RELATED WORK AND CONTRIBUTION

Sequential prefetching [18] assumes that the line contiguous to the one containing the current instruction is likely to be referenced and deserves to be loaded to the cache in advance depending on some criterion (*next-line always*, *next-line on miss*, or *next-line tagged*). It can be extended to multiple lines (*next-N-line prefetching*). However, it does not handle branches efficiently, since a target instruction typically does not lie in a line contiguous to the one containing the branch instruction. This led to more sophisticated techniques. *Target prefetching* [19] keeps a reference prediction table (RPT). When a branch is taken, its target address is stored in some RPT entry, which is tagged with the instruction's own address. When a branch is executed anew, the matching of a tag at some RPT entry induces the prefetch of the block corresponding to the entry's target address. Note that this implicitly assumes the branch as always taken. To exploit prefetching when the branch is untaken, *wrong-path prefetching* [13] stores two addresses (target and fall-through) for each branch in the RPT. Although it can be profitable regardless of the taken path, the number of ineffective prefetches may be increased.

As opposed to the techniques discussed above, whose mechanisms are hardwired, *software prefetching* relies on a special instruction to load a memory block into a cache line. It allows the preclusion of unnecessary prefetches, which pollute the cache and reduce its effective capacity. The use of dominance trees in control flow graphs was proposed as a way of exploiting static program analysis for prefetch placement [12]. By moving prefetch instructions earlier enough in the control flow, their latency is hidden and their potential of migrating out of loop bodies is raised.

Hardware mechanisms often guess the required prefetches, but they do not issue them early enough so as to produce the desired effect. To reduce cache pollution [9], *cooperative prefetching* [12] was proposed. Hardware control is limited to sequential prefetching while non-sequential flows are handled by software prefetching.

2.1 Prefetching for energy efficiency

Instead of wasting energy in hardware-controlled prefetch, the performance gain obtained by software prefetching can be directly translated into an increase of energy efficiency when software prefetching is combined with dynamic voltage scaling [1]. A recent work [20] confirms the energy inefficiency of hardware prefetching for old technologies, but indicates a distinct scenario for newer ones. Since hardware prefetching contributes to shortening the average execution time, the resulting static energy profit can be larger than the energy cost of hardware prefetching. Therefore, to completely rule out the need for hardware prefetching, a software prefetching technique should not increase the ACET.

2.2 Prefetching under real-time constraints

There are two conflicting views on how to handle caches under real-time constraints. Those who prescribe cache locking [4, 14] (to trade-off performance for predictability) argue that cache-aware WCET analysis [8, 21] often neglects the interference between tasks [15]. They prescribe a combination of instruction prefetching and cache locking [16] [2]. Such works, however, target the minimization of WCET as a single objective and do not report the impact on energy efficiency. On the other hand, those who prescribe the accurate prediction [8, 21] of cache behavior (during WCET analysis) argue that cache locking may unnecessarily give up performance [2]. Under such assumption, the original cache abstract semantics proposed in [8] was extended in [22] to incorporate the effect of *next-N-line prefetching*. Based on such extension, a later work [5] exploited software prefetching for minimizing the WCET. Unfortunately, since it inserts a prefetch at the beginning of the basic block where the prefetched instruction belongs, the distance between them might be insufficient to hide the latency of the former.

To the best of our knowledge, none of the works advocating the use of prefetching has reported the impact on WCET, ACET, and energy consumption all together.

2.3 The intended reconciliation

Any reduction in performance is unwelcome, since it may impair energy efficiency through static consumption, especially for nanometer caches [17]. Cache locking tends to become less energy efficient as CMOS technology scales down, since it saves dynamic energy by increasing the ACET, i.e. it lenghtens the interval during which static power is drained. Thus, as technology evolves and applications ask for higher throughputs under low power budgets, cache-aware WCET analysis seems more adequate for preserving real-time guarantees under increasing energy efficiency requirements.

To design a proper prefetching algorithm for unlocked caches, all the following conditions should hold for each prefetch: *Condition 1*: It does not increase the WCET (to preserve real-time guarantees); *Condition 2*: It reduces the miss rate (to save dynamic energy); *Condition 3*: It does not increase the ACET (to avoid wasting static energy).

To check Condition 1, we reuse *classical* static analysis [11, 8, 21]. To verify Condition 2, we propose a *novel* static analysis technique that visits instructions in reverse execution order and employs extended abstract semantics to identify prefetching points. As trace-based ACET estimates would be inefficient to check Condition 3, we propose an *alternative* approach relying on recent evidence that the ACET is often reduced when the WCET is decreased [6, 7]. This indicates that, despite their different magnitudes, the *derivatives* of the WCET and the ACET with respect to target technology and cache configuration tend to be *correlated*. Therefore, if Condition 1 holds, then Condition 3 is likely to hold as far as the optimization mechanism is based on *iterative improvement* (where the derivative is the optimization driver). That is why we propose an algorithm that transforms a program iteratively as far as an improvement can be observed. Indeed, the expected correlation between Conditions 1 and 3 is confirmed by the results in Section 5.

Although prefetching was already used to optimize the WCET [5], a criterion to select the most energy-efficient prefetches that preserve real-time guarantees was *not* proposed so far. Our new criterion is described in Section 4.3.

3. MODELING AND FORMULATION

3.1 Cache behavior

For self-containement, we review the main concepts from [8, 21]. The main storage and the cache are divided in *blocks* of equal capacity. A program *item* (instruction or data) always resides in a memory block and may also lie in a cache block. A memory block may contain one or more items. A group of a cache blocks is organized as a cache *line* (or set), where a is the cache's associativity. A cache is represented by a set of *lines* $L = \{l_1, \cdots, l_n\}$ and the main storage by a set of *blocks* $S = \{s_1, \cdots, s_m\} \cup \{I\}$, where I represents an *invalid* block. A *concrete cache state* is a function $c : L \to S$. The expression $c(l_i) = s_j$ means that block s_j is in cache line l_i. C_c denotes the set of all concrete cache states.

DEFINITION 1. *An update function* $\mathcal{U} : C_c \times S \to C_c$ *defines the new cache state from the state immediatly before a reference to a memory block.*

To represent the distinct concrete cache states leading to the WCET scenario, the notion of abstract state is used:

DEFINITION 2. *An abstract cache state is defined by* $\hat{c} : L \to 2^S$. $\hat{C}$ *is the set of all possible abstract cache states. A state where all blocks are invalid is denoted as* $\hat{c}_I$.

An *abstract update function* $\hat{\mathcal{U}} : \hat{C} \times S \to \hat{C}$ handles abstract states. The abstract update functions used in this work are described in [8].

During the concrete execution of a program, when a path branches off, only one of the divergent paths is executed. In abstract interpretation, however, all paths are taken into account. That is why, a *join function* has to be defined to merge the abstract cache states prior to the convergence point into a single abstract state after it. The join functions used in this work for WCET analysis are described in [8].

Although we rely on such classical functions for preliminary WCET analysis, we propose novel update and join functions to drive code optimization in Sections 4 and S.1.

3.2 Conditional execution

We assume a conventional representation as starting point:

DEFINITION 3. *Given a program, its control flow graph is a directed graph $CFG = (B, F)$ where $bb_i \in B$ represents a basic block and $(bb_i, bb_j) \in F$ represents the precedence between bb_i and bb_j in a concrete execution of that program.*

The *Implicit Path Enumeration Technique* (IPET) [11] casts the properties of execution paths into an integer linear programming (ILP) formulation, providing efficient static analysis [21] and accurate WCET bounds [8]. It encodes the conservation of execution flow on entry to and on exit from every basic block, instead of explicitly encoding execution paths. For instance, assume that a basic block bb_1 reaches two mutually exclusive basic blocks bb_2 and bb_3 and let n_{bb} be the number of executions of a basic block. The corresponding ILP constraint is $n_{bb_1} = n_{bb_2} + n_{bb_3}$. This implicitly encodes the fact that bb_2 and bb_3 cannot be executed simultaneously, i.e. if the WCET scenario corresponds to the execution through (bb_1, bb_2), then $n_{bb_3} = 0$ in such scenario.

3.3 Determination of the WCET scenario

Given a program p and a referenced memory item r, let $t_w^p(r)$ denote the time spent, in the WCET scenario, when accessing that item. Given a basic block bb, let $t_w^p(bb) = \sum_r t_w^p(r)$ be the time spent, in the WCET scenario, when accessing all the memory items referenced in one execution of that basic block. The overall contribution to the WCET induced by all memory items referenced by bb is $t_w^p(bb) \times n_{bb}$. The objective function for the ILP problem is:

$$maximize : \sum_{bb \in B} t_w^p(bb) \times n_{bb}, \tag{1}$$

whose solution leads to the *number of executions* of each basic block bb in the *WCET scenario*, written n_{bb}^w. Note that $n_{bb}^w = 0$ for every bb not belonging to the WCET path. The overall contribution of an item r to the WCET is:

$$\tau_w^p(r) = t_w^p(r) \times n_{B(r)}^w, \tag{2}$$

where $B(r)$ represents the basic block to which r belongs.

Given a program p, the overall contribution of the memory system to the WCET is:

$$\tau_w^p = \sum_{bb \in B} t_w^p(bb) \times n_{bb}^w \tag{3}$$

3.4 Problem formulation

DEFINITION 4. *The latency of a prefetch instruction, written Λ, is the time it takes to place a block in cache.*

DEFINITION 5. *Programs p and p' are prefetch-equivalent, written $p \equiv p'$, iff they are indistinguishable, except for their prefetch instructions.*

PROBLEM 1. *Given a program p, find a prefetch-equivalent program p' such that $\tau_w^{p'} \le \tau_w^p$ and it minimizes the number of cache misses, for a given prefetch latency Λ, a given cache configuration, and a given process technology.*

4. THE PROPOSED TECHNIQUE

To solve Problem 1, we deliberately adopted *iterative improvement* so as to increase the chances that program p' leads to higher energy efficiency than program p (as explained in Section 2.3). A joint improvement criterion was designed to evaluate the impact of each prefetch on *both* miss rate *and* WCET. From the original program, prefetch-equivalent programs are iteratively generated one after another as far as the joint improvement criterion is satisfied.

4.1 Abstract program representation

As we target the memory subsystem, our program representation abstracts the references to memory items from the concrete instructions of the actual program. It assumes that loops were virtually unrolled beforehand, by applying the transformation (VIVU) proposed in [8], leading to an implicit loop representation where back edges are broken:

DEFINITION 6. *Given a program, its abstract control flow graph is a polar, directed acyclic graph $ACFG = (R, E)$ where each vertex $r_i \in R$ is a reference to a memory item and each edge $(r_i, r_j) \in E$ represents the order of precedence between the references r_i and r_j in a concrete execution of that program. The poles are the source ($\bullet$) and the sink ($\odot$).*

To denote that r_i reaches r_j through a path in the ACFG, we write $r_i \rightsquigarrow r_j$. Each edge defines a *program point* between successive references. Given two references belonging to convergent execution paths, to stress the precedence between each of them and a third post-dominating reference, we include special join vertices.

DEFINITION 7. *Given an $ACFG = (R, E)$, its reverse abstract control flow graph is a directed acyclic graph $ACFG^* = (R, E^*)$ such that there exists an edge $(r_j, r_i) \in E^*$ for every edge $(r_i, r_j) \in E$ and vice-versa.*

We define the predecessors and the successors of a given vertex r in $ACFG^*$ as $PRED^*(r) = \{r' \in R \mid (r', r) \in E^*\}$ and $SUCC^*(r) = \{r' \in R \mid (r, r') \in E^*\}$, respectively. A given predecessor of r is denoted as $pred^*(r)$.

DEFINITION 8. *Given an item r_i, we write $\mathcal{S}(r_i)$ to denote the memory block where r_i is stored. Conversely, given a memory block s, we write $\mathcal{R}(s)$ to denote the reference to the item in s with the smallest address (i.e. the first item).*

Let us now link the proposed representation with the classical model of cache behavior reviewed in Section 3.1.

DEFINITION 9. *The set of blocks in cache at a given state $\hat{c}$, written $\mathcal{B}(\hat{c})$, is $\bigcup_{i=1}^{|L|} \{\hat{c}(l_i)\}$.*

Let $\hat{c}(r_i, r_j)$ be the cache state at program point (r_i, r_j). Let $\mathcal{B}(r_i, r_j)$ be a shorthand notation for $\mathcal{B}(\hat{c}(r_i, r_j))$. Given three successive references r_{i-1}, r_i, and r_{i+1}, the following properties hold:

PROPERTY 1. *When $\mathcal{B}(r_i, r_{i+1}) - \mathcal{B}(r_{i-1}, r_i) = \emptyset$, the access to item r_i resulted in a hit.*

PROPERTY 2. *When $\mathcal{B}(r_i, r_{i+1}) - \mathcal{B}(r_{i-1}, r_i) = \{s\}$, the access to item r_i resulted in a miss and r_i is stored in memory block s.*

PROPERTY 3. *When $\mathcal{B}(r_{i-1}, r_i) - \mathcal{B}(r_i, r_{i+1}) = \{s'\}$, the access to item r_i replaced the memory block s'.*

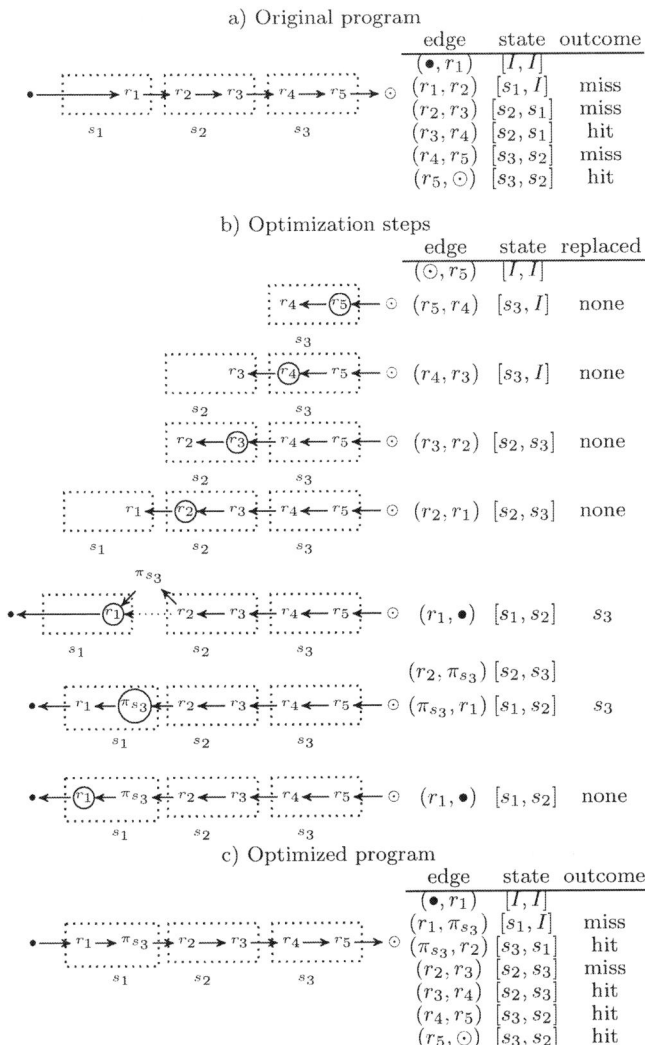

Figure 1: The technique applied to a straight-line program

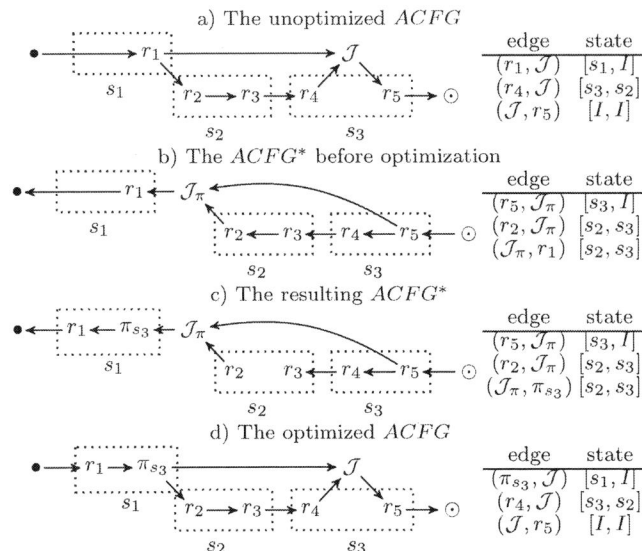

Figure 2: How conditional control flows are handled

cached item s_3 is replaced. Therefore, a prefetch for the replaced item, denoted as π_{s_3}, is inserted at the program point (r_2, r_1). This is done by removing the edge (r_2, r_1) and adding the edges (r_2, π_{s_3}) and (π_{s_3}, r_1). When π_{s_3} is visited, the effect of the prefetch to the cache is merely recalculated (despite the detection of s_3 as a replaced block, which was already treated by π_{s_3} itself). Then the vertex r_1, which is the successor of the inserted prefetch, is revisited. The analysis ends when the node $\bullet$ is reached.

As shown in Figure 1c, the optimized program is obtained by simply reversing the edges of the resulting $ACFG^*$. Note that, although the references to r_1 and r_2 induce cache misses, the accesses to r_3, r_4, and r_5 do not.

A second example handles conditional constructs with the help of join functions. When a reference r is reached from distinct paths, the state at its leaving edge depends on the taken path. To derive a single output state from multiple input states, *join vertices* are added to the $ACFG$ ($ACFG^*$) and their behaviors are modeled by *join functions*, as illustrated in Figure 2. Figure 2a shows that a vertex $\mathcal{J}$ performing a conventional join function determines the state at its leaving edge as the intersection of the states of its entering edges [8]. Figure 2b shows that a vertex $\mathcal{J}_\pi$ performing a join function tailored to prefetching simply propagates to its leaving edge the state of the entering edge that belongs to the WCET path. Figures 2c and 2d directly show the resulting $ACFG^*$ and $ACFG$.

4.2 Illustrative examples

We show how our technique works by means of examples. For simplicity, we assume that all the references map to the same line of a 2-way LRU cache with 2 items per block.

In Figure 1, our technique is applied to a simple straight-line program. From the $ACFG$ of the original program (1a), it shows the $ACFG^*$s representing the intermediate optimization steps for each visited vertex (1b) until the $ACFG$ of the optimized program is obtained (1c). Each memory block is represented as a dotted box. The cache states at each program point are displayed at the right-hand side. They help track either the number of misses in program order (1a, 1c) or the replaced blocks in reverse order (1b). The blocks of a cache line are denoted as $[MRU, LRU]$, to indicate the most and the least recently used blocks.

Figure 1a shows the states at each edge of the ACFG. By applying Properties 1 and 2 to every successive pair of edges, we obtain whether the outcome was a hit or a miss.

Figure 1b presents our reverse analysis step-by-step from sink to source. Initially, the edge $(\odot, r_5)$ is assigned a state where all blocks are invalid. By applying Property 3 to each successive pair of edges, a replaced block can be identified. When r_5, r_4, r_3, and r_2 are visited, since no cache item is replaced, no action is taken but visiting the next vertex. However, when r_1 is visited, the technique detects that the

4.3 The joint improvement criterion

Let us denote the contribution to the WCET of all items referenced on a path starting at r_i and ending at r_j as:

$$\tau_w^p(r_i, r_j) = \sum_{r \in \{x \mid x \in R \wedge r_i \rightsquigarrow x \rightsquigarrow r_j\}} t_w^p(r) \times n_{B(r)}^w \quad (4)$$

Let p_{n-1} and p_n denote programs containing $n-1$ and n prefetches, respectively, such that $p_{n-1} \equiv p_n$. Let r_j be a reference to an item stored in some memory block s'. We denote an instruction that prefetches the block s' into cache as $\pi_{s'}$. Finally, let (r_i, r_{i+1}) denote some program point such that $r_i \rightsquigarrow r_j$. To check if the insertion of $\pi_{s'}$ at program point (r_i, r_{i+1}) precludes the miss on access to r_j without increasing the WCET, five notions are required.

The first notion tracks prefetch effectiveness, i.e. the guarantee that the prefetched block is in cache before it is referenced, despite the prefetch latency (Definition 4). Given a program p_{n-1}, the time spent, in the WCET scenario, to perform all memory access in between r_i and r_j is:

$$t_w^{p_{n-1}}(r_{i+1}, r_{j-1}) = \sum_{r \in \{x | x \in R \wedge r_{i+1} \leadsto x \leadsto r_{j-1}\}} t_w^{p_{n-1}}(r) \quad (5)$$

DEFINITION 10. *A prefetch instruction inserted at some program point (r_i, r_{i+1}) is effective iff $\Lambda \leq t_w^{p_{n-1}}(r_{i+1}, r_{j-1})$.*

The second notion tracks the contribution to the WCET of a reference r_j to an item *missing* in cache at a given point, say (r_{j-1}, r_j), of a program p_{n-1}:

$$mcost(r_j) = \tau_w^{p_{n-1}}(r_j) \quad (6)$$

The third notion tracks the contribution to the WCET of a reference r_j to an item *hitting* in cache as a result of an effective prefetch instruction $\pi_{s'}$ inserted at point (r_i, r_{i+1}) in program p_{n-1}, leading to a new program p_n:

$$pcost(r_i) = \tau_w^{p_n}(\pi_{s'}) + \tau_w^{p_n}(r_j) \quad (7)$$

The fourth notion tracks the contribution to the WCET resulting from the *relocation* of all references preceding r_i in the address space, as a result of the insertion of a prefetch instruction $\pi_{s'}$ at point (r_i, r_{i+1}) in program p_{n-1}, turning it into a new program p_n:

$$rcost(r_i) = \sum_{r \in \{x | x \in R \wedge x \leadsto r_i\}} \tau_w^{p_n}(r) - \sum_{r \in \{x | x \in R \wedge x \leadsto r_i\}} \tau_w^{p_{n-1}}(r) \quad (8)$$

The fifth notion combines all the previous concepts to define the profitability of a prefetch. Given two references r_i and r_j such $r_i \leadsto r_j$ in the $ACFG$, the *profit* of inserting an instruction, at the program point (r_i, r_{i+1}), to prefetch the memory block storing the item r_j is:

$$profit(r_i, r_j) = \begin{cases} 0 & \text{if } r_j \text{ is a prefetch} \\ 0 & \Lambda > t_w^{p_{n-1}}(r_{i+1}, r_{j-1}) \\ 0 & rcost(r_i) > 0 \\ mcost(r_j) - pcost(r_i) & \text{otherwise} \end{cases} \quad (9)$$

A prefetch is profitable if and only if it is effective, the induced relocation does not increase the WCET, and the gain of suppressing a miss (induced by program p_{n-1}) is higher than the cost of inserting a prefetch to supress that miss (in a program p_n).

4.4 The novel optimization algorithm

As a precondition, our algorithm assumes that traditional WCET analysis (to determine $t_w^p(r)$ for each $r \in R$ and n_{bb}^w for every $bb \in B$) and classical VIVU analysis (to transform a cyclic CFG into an acyclic $ACFG$) [8] were performed beforehand. Our algorithm, which is formally described in Supplement S.1, runs a *non-conventional* static analysis in *reverse execution order* to find the profitable prefetches that do not increase the WCET. The optimization algorithm relies on the novel update function informally described in Figure 1 and the novel join function illustrated in Figure 2. Their formal descriptions are available in Supplement S.1.

When the program order is preserved for the *memory* operations at execution time, our optimization algorithm provenly does not increase the contribution of the memory system to the WCET (see Theorem 1 in Supplement S.2).

5. EXPERIMENTAL EVALUATION

We ran the proposed optimization on all 37 programs of the Mälardalen WCET benchmark [10], each one under 36 cache configurations and two technologies (45nm and 32nm), leading to 2664 use cases. For reproducibility, Supplement S.4 provides the detailed set up. We assume that each program fully owns the instruction cache. Since in practice

many programs compete for the cache, the adopted sizes should be interpreted as *effective* cache capacities allocated to *individual* programs (not as the overall capacity). We selected them so that the average miss rate lies in a large span from 1% to 10% before the proposed optimization is applied. A 128MB DRAM was employed as level-two memory.

Figure 3 shows average improvements for distinct cache sizes. The overall average improvement was 10.2% for the ACET and 11.2% for energy consumption. Indeed, energy savings were obtained for all use cases without increasing the memory's contribution to the ACET. To reach such energy efficiency, the maximal increase in the number of executed instructions was 1.32% (see Supplement S.5). The non-increasing ACET has two consequences: the memory's static energy and the average number of cycles per instruction are not increased (if time anomalies are second-order effects). As the amount of inserted instructions is negligible, our optimization of the *memory* subsystem may only marginally increase the static consumption of the *rest* of the system.

Figure 3 also plots the average improvement in the WCET. It confirms the correlation between WCET and ACET variations, which was a premise for the design of our algorithm. Note that, to preserve real-time guarantees, our technique employs the WCET as a constraint and therefore does not try to optimize it. However, an average improvement of 17.4% was observed (see Supplement S.5 for an illustration).

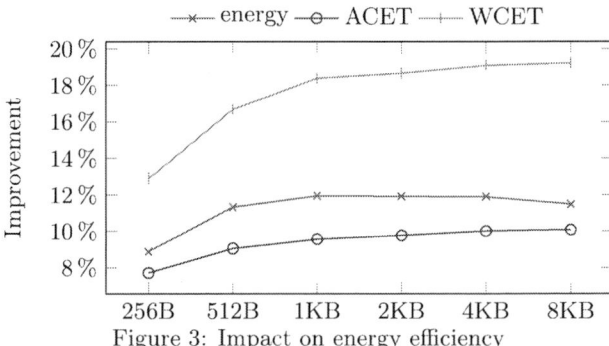

Figure 3: Impact on energy efficiency

Figure 4 shows the impact on miss rate. The overall average miss rate reduction was 28%. As the arrow shows, the optimized programs require less cache capacity than the unoptimized ones to sustain the same miss rate.

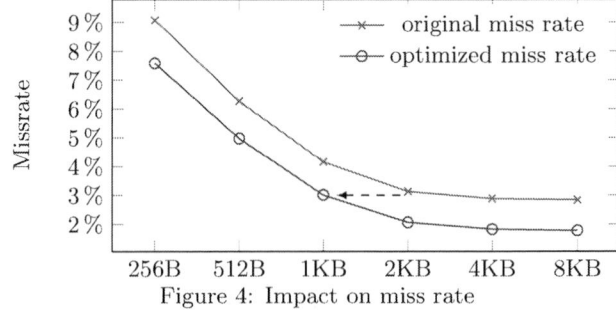

Figure 4: Impact on miss rate

As our technique enables the use of smaller caches, it can exploit the resulting reduction in static and dynamic consumption for further improving the energy efficiency, as follows. Figure 5 plots average reductions, but the cache size used to run the optimized programs was set to 1/4 and 1/2 of the cache size used to run the original programs. Note that, within the shaded areas, the optimized programs sustained ACETs less or equal to the unoptimized ones with

1/2 and 1/4 of the original cache size. Although our technique provides WCET guarantees when the original and the optimized program run on the same cache configuration, it cannot keep such guarantee when comparing their behaviors on configurations with arbitrarily selected sizes. However, as Figure 5 indicates, the WCET did not grow for any use case when cache sizes were reduced. This is an evidence that our technique, by enabling the use of smaller caches, can lead to energy reductions up to 21% while sustaining the same or superior performance and preserving real-time guarantees.

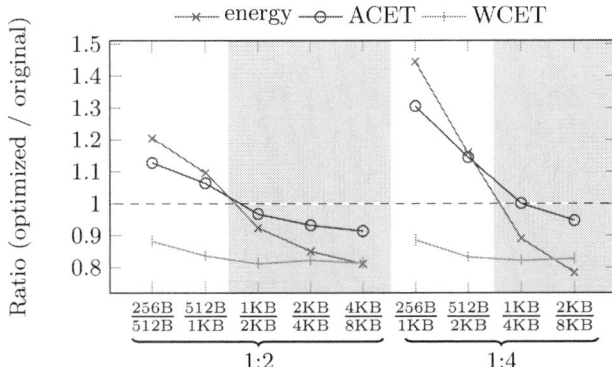

Figure 5: Higher energy efficiency with smaller caches

For the program that took the longest (adpcm), our first prototype took on average 0.43 seconds to run each WCET analysis and 8 minutes to perform the proposed optimization on a workstation i7-2600K running at 3.4 GHz.

6. CONCLUSIONS AND PERSPECTIVES

We showed that, since conventional WCET analysis should be run anyway to provide for real-time guarantees, a polynomial algorithm can exploit the analysis' outcome to increase the energy efficiency of a program for a given cache configuration and process technology, while preserving the WCET.

Being independent from locality, prefetching can reduce the required cache size to reach the same performance level as obtained through on-demand fetching on larger caches. This diminishes *static* and *dynamic* consumption at *level-1* caches. Besides, prefetching seems to harmonize with future hierarchies, where the sensitivity of power consumption to associativity is likely to be reduced [17] to the point of enabling level-2 caches with higher associativities than the current ones. This allows for smaller level-1 caches and raises the potential of prefetching for energy efficiency.

As the impact of cache locking on energy efficiency has not been reported (to the best of our knowledge), we intend to bridge such gap by implementing techniques like [16] and [2] within our experimental environment. We also intend to generalize our algorithms for handling unlocked *data* caches.

7. REFERENCES

[1] D. Agarwal, S. Pamnani, G. Qu, and D. Yeung. Transferring performance gain from software prefetching to energy reduction. In *Proc. of Int. Symp. on Circuits and Systems*, pages 241–244, 2004.

[2] L. C. Aparicio, J. Segarra, C. Rodriguez, and V. Vinals. Combining prefetch with instruction cache locking in multitasking real-time systems. In *Proc. of IEEE Int. Conf. Embedded and Real-Time Computing Systems and Applications*, pages 319–328, 2010.

[3] W. Dally, J. Balfour, D. Black-Shaffer, J. Chen, R. Harting, V. Parikh, J. Park, and D. Sheffield. Efficient embedded computing. *IEEE Computer Magazine*, 41(7):27–32, 2008.

[4] H. Ding, Y. Liang, and T. Mitra. WCET-centric partial instruction cache locking. In *Proc. of*

IEEE/ACM Design Automation Conference, pages 412–420, 2012.

[5] Y. Ding, J. Yan, and W. Zhang. Optimizing instruction prefetching to improve worst-case performance for real-time applications. *Journal of Computing Science and Engineering*, 3(1):59–71, 2009.

[6] H. Falk and J. C. Kleinsorge. Optimal static WCET-aware scratchpad allocation of program code. In *Proc. of IEEE/ACM Design Automation Conference*, pages 732–737, 2009.

[7] H. Falk, N. Schmitz, and F. Schmoll. WCET-aware register allocation based on integer-linear programming. In *Proc. of Euromicro Conference on Real-Time Systems*, pages 13–22, 2011.

[8] C. Ferdinand, F. Martin, R. Wilhelm, and M. Alt. Cache behavior prediction by abstract interpretation. *Science of Computer Programming*, 35(2-3):163–189, 1999.

[9] R. Gupta and C.-H. Chi. Improving instruction cache behavior by reducing cache pollution. In *Proc. of ACM/IEEE Conference on Supercomputing*, pages 82–91, 1990.

[10] J. Gustafsson, A. Betts, A. Ermedahl, and B. Lisper. The mälardalen wcet benchmarks–past, present and future. In *Proc. of Int. Workshop on Worst-Case Execution Time Analysis*, pages 137–147, 2010.

[11] Y.-T. S. Li and S. Malik. Performance analysis of embedded software using implicit path enumeration. In *Proc. of IEEE/ACM Design Automation Conference*, pages 456–461, 1995.

[12] C.-K. Luk and T. C. Mowry. Architectural and compiler support for effective instruction prefetching: a cooperative approach. *ACM Transactions on Computer Systems*, 19(1):71–109, 2001.

[13] J. Pierce and T. Mudge. Wrong-path instruction prefetching. In *Proc. of ACM/IEEE Int. Symposium on Microarchitecture*, pages 165–175, 1996.

[14] S. Plazar, J. C. Kleinsorge, P. Marwedel, and H. Falk. WCET-aware static locking of instruction caches. In *Proc. of ACM Int. Symposium on Code Generation and Optimization*, pages 44–52, 2012.

[15] I. Puaut. Cache analysis vs static cache locking for schedulability analysis in multitasking real-time systems. In *Proc. of Int. Workshop on Worst-Case Execution Time Analysis*, 2002.

[16] I. Puaut. WCET-centric software-controlled instruction caches for hard real-time systems. In *Proc. of Euromicro Conf. on Real-Time Systems*, pages 217–226, 2006.

[17] S. Rodriguez and B. Jacob. Energy/power breakdown of pipelined nanometer caches. In *Proc. of ACM Int. Symp. on Low Power Electronics and Design*, pages 25–30, 2006.

[18] A. J. Smith. Sequential program prefetching in memory hierarchies. *IEEE Computer Magazine*, 11(12):7–21, 1978.

[19] J. E. Smith and W.-C. Hsu. Prefetching in supercomputer instruction caches. In *Proc. of ACM/IEEE Conference on Supercomputing*, pages 588–597, 1992.

[20] J. Tang, S. Liu, Z. Gu, C. Liu, and J.-L. Gaudiot. Prefetching in embedded mobile systems can be energy-efficient. *IEEE Computer Architecture Letters*, 10(1):8–11, 2011.

[21] H. Theiling, C. Ferdinand, and R. Wilhelm. Fast and precise wcet prediction by separated cache and path analyses. *Real-Time Systems*, 18(2/3):157–179, 2000.

[22] J. Yan and W. Zhang. WCET analysis of instruction caches with prefetching. In *Proc. of ACM SIGPLAN/SIGBED Conf. on Languages, Compilers, and Tools for Embedded Systems*, pages 175–184, 2007.

SUPPLEMENTAL MATERIAL

S.1. ALGORITHMS AND COMPLEXITY

We propose the *prefetching update function* $\hat{\mathcal{U}}_\pi : \hat{C} \times R \rightarrow \hat{C}$ defined in Algorithm 1. It detects the need for a prefetch (line 2) and checks if it is profitable (line 4). If so, it inserts the prefetch in the $ACFG^*$ (lines 5-7) and relocates all memory items affected by such insertion (new block boundaries up to the source vertex). Then it is applied recursively to the inserted prefetch (line 9). If it detects no need for prefetching or an unprofitable prefetch, the conventional update function is applied and the resulting state is returned (line 10).

We also propose the *prefetching join function* $\mathcal{J}_\pi : \hat{C} \times \hat{C} \rightarrow \hat{C}$ defined in Algorithm 2. Essentially, it propagates, to the edge leaving a join, the cache state from the entering edge that belongs to the WCET path.

Algorithm 1 The proposed update function $\hat{\mathcal{U}}_\pi(\hat{c}, r_i)$

```
1   s = S(r_i)
2   if ∃s' ∈ S | B(ĉ) − B(Û(ĉ,s)) = {s'} ≠ {I}:
3       r_j = R(s')
4       if profit(r_i, r_j) > 0:
5           R  := R ∪ {π_s'}
6           E* := E* ∪ {(pred*(r_i), π_s'), (π_s', r_i)}
7           E* := E* − {(pred*(r_i), r_i)}
8           relocate_upwards(r_i)
9           return Û_π(ĉ, π_s')
10  return Û(ĉ, s)
```

Algorithm 2 The proposed join function $\mathcal{J}_\pi(\hat{c}_1, \hat{c}_2)$

```
1   let (r_x, J) ∈ E* | ĉ(r_x, J) = ĉ_1
2   let (r_y, J) ∈ E* | ĉ(r_y, J) = ĉ_2
3   if mcost(r_x) < mcost(r_y)
4       return ĉ(r_y, J)
5   else
6       return ĉ(r_x, J)
```

Algorithm 3 builds the $ACFG^*$ (line 1) and finds a topological ordering $\prec_T$ of its vertices (line 2). Then it visits vertices in that order from *sink* (line 5) to *source* (line 6). If it visits a join, the proposed join function is invoked (line 10); otherwise, the proposed update function is called (line 13). Finally, the optimized $ACFG$ is built from the $ACFG^*$ that was modified by the proposed update and join functions (line 15).

Lines 1, 3, and 5-7 of Algorithm 1 take $O(1)$. Lines 2 and 10 also take $O(1)$ when cache states are precomputed during the preliminary WCET analysis and stored in a hash table. At line 4, the evaluation of Equations 6 and 7 takes $O(1)$. Although the second summation of Equation 8 can benefit from precalculated values (and, therefore, takes $O(1)$), the first summation takes $O(|R|)$. The relocation at line 8 also takes $O(|R|)$. Therefore, Algorithm 1 takes $O(|R|)$.

Algorithm 1 is called at most $|R|$ times from the line 13 of Algorithm 3 and, recursively, at line 9, as many times as the number of inserted prefetches, which is at most $|R|$. Therefore, the line 13 of Algorithm 3 contributes $O(|R|^2)$ to the overall complexity. All lines of Algorithm 2 take $O(1)$ due to the hash table and it is invoked at most $|R|$ times.

As a result, lines 6-14 of Algorithm 3 contribute $O(|R|^2)$ to the overall complexity, whereas lines 1, 2 and 15 take $O(|R| + |E|)$. Thus, the overall worst case complexity of Algorithm 3 is $O(|R|^2)$.

Besides, when generating the $ACFG = (R, E)$ from the $CFG = (B, F)$, we bound the set R by virtually unrolling each loop at most once when applying the VIVU transformation [8].

Algorithm 3 The proposed prefetching optimization

```
1   build ACFG* = (R, E*) from program p
2   ≺_T = {(u,v) ∈ R × R | (u,v) ∈ E* ∨ (v,u) ∉ E*}
3   let succ_{≺_T}(r) be the successor of r in ≺_T
4   c(⊙, succ*(⊙)) := ĉ_I
5   r := ⊙
6   while (succ_{≺_T}(r) != •)
7       {r_z} := SUCC*(r)
8       if r is a join vertex:
9           {r_x, r_y} := PRED*(r)
10          ĉ(r_z, r)  := J_π(ĉ(r_x, r), ĉ(r_y, r))
11      else
12          {r_x} := PRED*(r)
13          ĉ(r_z, r)  := Û_π(ĉ(r_x, r), r)
14      r := succ_{≺_T}(r)
15  build ACFG = (R, E) for program p'
```

S.2. FORMAL GUARANTEES

To improve readability, this section adopts $\sum_{r \rightsquigarrow r_i} \tau_w^{p_n}(r)$ as a shorthand notation for $\sum_{r \in \{x | x \in R \wedge x \rightsquigarrow r_i\}} \tau_w^{p_n}(r)$.

LEMMA 1. *Given the ACFG representing a program p_{n-1} and a path $(r_{i+1}, \cdots, r_j)$, if Algorithm 1 inserts, at program point (r_i, r_{i+1}), a prefetch $\pi_{s'}$ for a block $s' = S(r_j)$, thereby generating a program p_n, the overall contribution to the WCET of all memory items referenced on the new path $(\pi_{s'}, r_{i+1}, \cdots, r_j)$ is smaller than on path $(r_{i+1}, \cdots, r_j)$, i.e. $\tau_w^{p_n}(\pi_{s'}, r_j) < \tau_w^{p_{n-1}}(r_{i+1}, r_j)$.*

PROOF. Line 4 of Algorithm 1 guarantees, via Equation 9, that a prefetch $\pi_{s'}$ is inserted only if $pcost(r_i) < mcost(r_j)$, which from Equations 6 and 7 leads to $\tau_w^{p_n}(\pi_{s'}) + \tau_w^{p_n}(r_j) < \tau_w^{p_{n-1}}(r_j)$ (I). Since $\pi_{s'}$ is inserted immediatly before r_{i+1} and every vertex r such that $r_{i+1} \rightsquigarrow r \rightsquigarrow r_{j-1}$ is untouched by Algorithm 1, we can write $\tau_w^{p_n}(r_{i+1}, r_{j-1}) = \tau_w^{p_{n-1}}(r_{i+1}, r_{j-1})$ (II). Thus, from (I) and (II) we conclude that $\tau_w^{p_n}(\pi_{s'}) + \tau_w^{p_n}(r_{i+1}, r_{j-1}) + \tau_w^{p_n}(r_j) < \tau_w^{p_{n-1}}(r_{i+1}, r_{j-1}) + \tau_w^{p_{n-1}}(r_j)$. Therefore, from Equations 2 and 4, we can write $\tau_w^{p_n}(\pi_{s'}, r_j) < \tau_w^{p_{n-1}}(r_{i+1}, r_j)$. □

LEMMA 2. *Given the ACFG representing a program p_{n-1} and a path $(r_{i+1}, \cdots, r_j)$, if Algorithm 1 inserts, at program point (r_i, r_{i+1}), a prefetch $\pi_{s'}$ for a block $s' = S(r_j)$, thereby generating a program p_n, the overall contribution to the WCET of all memory items reaching r_i is not increased, i.e. $\sum_{r \rightsquigarrow r_i} \tau_w^{p_n}(r) \leq \sum_{r \rightsquigarrow r_i} \tau_w^{p_{n-1}}(r)$.*

PROOF. Line 4 of Algorithm 1 guarantees, via Equation 9, that a prefetch $\pi_{s'}$ is inserted only if $rcost \leq 0$, which from Equation 8 leads to $\sum_{r \rightsquigarrow r_i} \tau_w^{p_n}(r) - \sum_{r \rightsquigarrow r_i} \tau_w^{p_{n-1}}(r) \leq 0$, i.e. $\sum_{r \rightsquigarrow r_i} \tau_w^{p_n}(r) \leq \sum_{r \rightsquigarrow r_i} \tau_w^{p_{n-1}}(r)$. □

THEOREM 1. *Given a program p, Algorithm 3 produces a program p' such that $p' \equiv p$ and $\tau_w^{p'} \leq \tau_w^p$ if all memory operations are kept in program order at execution time.*

1000

PROOF. Let p_{n-1} denote the program generated by Algorithm 3 after inserting $n-1$ prefetches prior to some invocation of Algorithm 1 in which the condition in line 4 holds. This means that a prefetch $\pi_{s'}$ for a block $s' = \mathcal{S}(r_j)$ will be inserted at point r_i, r_{i+1} of program p_{n-1}, thereby generating a program p_n with n prefetches. From Equations 2 and 3, we can write:

$$\tau_w^{p_{n-1}} = \sum_{bb \in B} t_w^{p_{n-1}}(bb) \times n_{bb}^w = \sum_{r \in R} \tau_w^{p_{n-1}}(r),$$

$$\tau_w^{p_n} = \sum_{bb \in B} t_w^{p_n}(bb) \times n_{bb}^w = \sum_{r \in R} \tau_w^{p_n}(r),$$

which can be rewritten, with the help of Equation 4, as follows:

$$\tau_w^{p_{n-1}} = \sum_{r \rightsquigarrow r_i} \tau_w^{p_{n-1}}(r) + \tau_w^{p_{n-1}}(r_{i+1}, r_j) + \sum_{r_{j+1} \rightsquigarrow r} \tau_w^{p_{n-1}}(r)$$

$$\tau_w^{p_n} = \sum_{r \rightsquigarrow r_i} \tau_w^{p_n}(r) + \tau_w^{p_n}(\pi_{s'}, r_j) + \sum_{r_{j+1} \rightsquigarrow r} \tau_w^{p_n}(r)$$

When p_{n-1} is turned into p_n, all paths starting at r_{j+1} are untouched by Algorithm 1 and Lemma 2 holds. Therefore, we can write:

$$\sum_{r \rightsquigarrow r_i} \tau_w^{p_n}(r) + \sum_{r_{j+1} \rightsquigarrow r} \tau_w^{p_n}(r) \le \sum_{r \rightsquigarrow r_i} \tau_w^{p_{n-1}}(r) + \sum_{r_{j+1} \rightsquigarrow r} \tau_w^{p_{n-1}}(r)$$

Therefore, we conclude that:

$$\tau_w^{p_n} - \tau_w^{p_n}(\pi_{s'}, r_j) \le \tau_w^{p_{n-1}} - \tau_w^{p_{n-1}}(r_{i+1}, r_j) \Leftrightarrow \tau_w^{p_n} \le \tau_w^{p_{n-1}} - K,$$

where $K = \tau_w^{p_{n-1}}(r_{i+1}, r_j) - \tau_w^{p_n}(\pi_{s'}, r_j)$.

Since p_n and p_{n-1} are indistinguishable except for $\pi_{s'}$ and we know from Lemma 1 that $K > 0$, we conclude that $p_n \equiv p_{n-1}$ and $\tau_w^{p_n} \le \tau_w^{p_{n-1}}$ hold for any integer $n > 1$, i.e. $p' \equiv p$ and $\tau_w^{p'} \le \tau_w^p$ hold for any program $p' \ne p$ produced by Algorithm 3. If, however, Algorithm 3 does not insert any prefetches ($n = 0$), i.e. $p' = p = p_0$, we obviously have $p \equiv p'$ and $\tau_w^{p'} = \tau_w^p$. Thus, $p' \equiv p$ and $\tau_w^{p'} \le \tau_w^p$ hold for any program p' produced by Algorithm 3 from p. $\square$

S.3. HOW LOOPS ARE HANDLED

This supplemental example illustrates that, to handle loops, our technique relies on the VIVU transformation [8] (which is often employed by conventional WCET analysis) to derive an acyclic $ACFG$ from a cyclic CFG. Figure 6a shows a CFG prior to the VIVU transformation, where a back edge closes a loop. Figure 6b shows the transformation's effect: the back edge is broken and the loop body is instanciated twice, leading to an $ACFG$ where the effect of loop iteration is implicitly encoded in the conditional control flow. In that figure, r_2^f denotes the reference to an item r_2 in the *first* loop iteration and r_2^o denotes the reference to the same item in *other* loop iterations. From the $ACFG$ in Figure 6b, our technique obtains the $ACFG^*$ in Figure 6c, according to the mechanisms already illustrated in the previous examples. Figure 6d shows the resulting CFG for the optimized program.

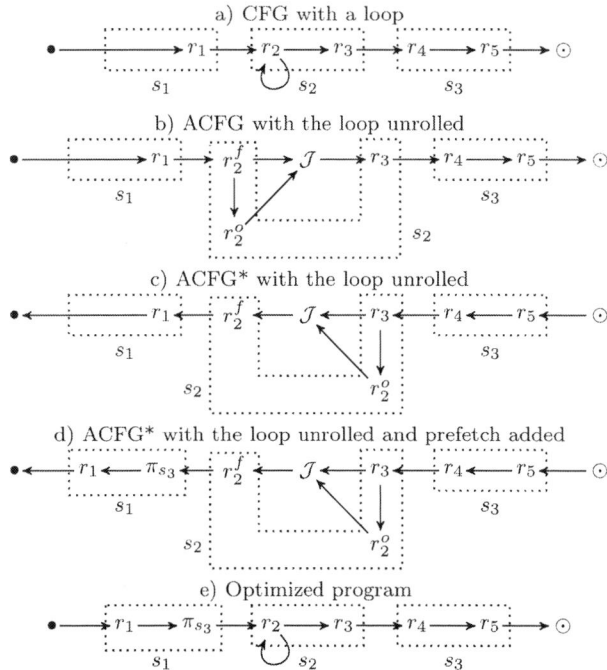

Figure 6: How the technique handles loops

S.4. DETAILED EXPERIMENTAL SETUP

Table 1 labels all the adopted benchmark programs.

The cache configurations employed in our experiments are denoted as $k = (a, b, c)$ in Table 2, where a is the associativity, b is the block size (in bytes), and c is the cache capacity (in bytes).

Our technique was integrated into the GNU compiler (version 4.6.1). We used the '-O2' optimization level and targeted ARMv7. First, the source code of each program was compiled by disabling the proposed technique, leading to a single non-optimized executable file for each program p, denoted as e_p^r, to be used as a *reference*. Then each source file was compiled by applying our technique to every cache configuration k and each process technology t, resulting in a set of optimized executable files for each program p, denoted as $\{e_{pkt}^o\}$.

For each executable file e, we estimated the contributions of the memory system to the WCET and to the ACET, written $\tau_w(e)$ and $\tau_a(e)$, respectively, and the energy consumption of the memory system in the ACET scenario, written $\epsilon_a(e)$. The first estimate was obtained with conventional WCET analysis, whereas the last two were obtained through a traditional trace-based approach. For trace generation, we employed an instruction-set simulator available within the GEM5 simulation environment. Since our optimization heavily relies on WCET analysis, we implemented our own WCET analyzer based on [8, 21] and integrated its components into the optimizing tool prototype.

We employed the CACTI 6.5 power/energy model to obtain energy and access times for the primary cache and the level-two memory. Since we did not model the processor's micro-architecture, we did not estimate the impact of the instruction overhead on the processor's energy consumption and execution time. However, since the measured increase in the number of executed instructions was negligible, the impact is likely to be marginal.

Table 1: Program identification

program	ID	program	ID	program	ID
adpcm	p1	bs	p2	bsort100	p3
cnt	p4	compress	p5	cover	p6
crc	p7	duff	p8	edn	p9
expint	p10	fac	p11	fdct	p12
fft1	p13	fibcall	p14	fir	p15
insertsort	p16	janne_complex	p17	jfdctint	p18
lcdnum	p19	lms	p20	loop3	p21
ludcmp	p22	matmult	p23	minmax	p24
minver	p25	ndes	p26	ns	p27
nsichneu	p28	prime	p29	qsort-exam	p30
qurt	p31	recursion	p32	select	p33
sqrt	p34	statemate	p35	st	p36
ud	p37				

Table 2: Cache configurations

(a, b, c)	ID	(a, b, c)	ID	(a, b, c)	ID
(1, 16, 256)	k1	(2, 16, 256)	k2	(4, 16, 256)	k3
(1, 32, 256)	k4	(2, 32, 256)	k5	(4, 32, 256)	k6
(1, 16, 512)	k7	(2, 16, 512)	k8	(4, 16, 512)	k9
(1, 32, 512)	k10	(2, 32, 512)	k11	(4, 32, 512)	k12
(1, 16, 1024)	k13	(2, 16, 1024)	k14	(4, 16, 1024)	k15
(1, 32, 1024)	k16	(2, 32, 1024)	k17	(4, 32, 1024)	k18
(1, 16, 2048)	k19	(2, 16, 2048)	k20	(4, 16, 2048)	k21
(1, 32, 2048)	k22	(2, 32, 2048)	k23	(4, 32, 2048)	k24
(1, 16, 4096)	k25	(2, 16, 4096)	k26	(4, 16, 4096)	k27
(1, 32, 4096)	k28	(2, 32, 4096)	k29	(4, 32, 4096)	k30
(1, 16, 8192)	k31	(2, 16, 8192)	k32	(4, 16, 8192)	k33
(1, 32, 8192)	k34	(2, 32, 8192)	k35	(4, 32, 8192)	k36

To evaluate the improvement in energy efficiency (ACET and energy consumption), we used the inequations:

$$\frac{1}{|\{e^o_{pkt}\}|} \sum_{p,k,t} \frac{\epsilon_a(e^o_{pkt})}{\epsilon_a(e^r_p)} < 1 \qquad (10)$$

$$\frac{1}{|\{e^o_{pkt}\}|} \sum_{p,k,t} \frac{\tau_a(e^o_{pkt})}{\tau_a(e^r_p)} \leq 1 \qquad (11)$$

To evaluate the reduction in the WCET, we used the inequation:

$$\forall(p,k,t) : \frac{\tau_w(e^o_{pkt})}{\tau_w(e^r_p)} < 1 \qquad (12)$$

First, such inequations were used to check all use cases individually. Then, by restricting them to a given cache size and taking the average, we obtained the values plotted in the Figures 3 and 5. Inequation 12 was also used to plot Figure 7.

S.5. EXTRA RESULTS

Figure 7 illustrates the WCET reduction for every use-case scenario, according to Inequation 12, when targeting 32nm.

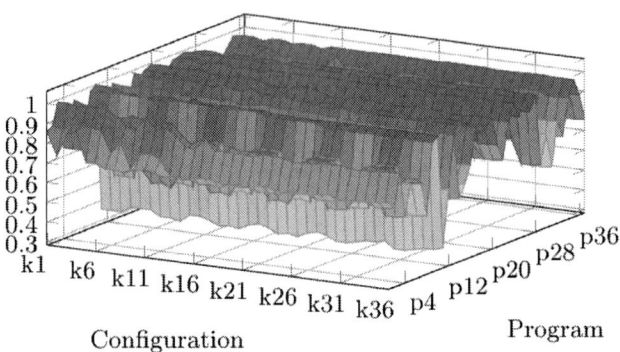

Figure 7: Reduction in the WCET for 32nm

Figure 8 plots the average ratio between the number of executed instructions of the optimized program as compared to the original one. It shows a maximal increase of 1.32%.

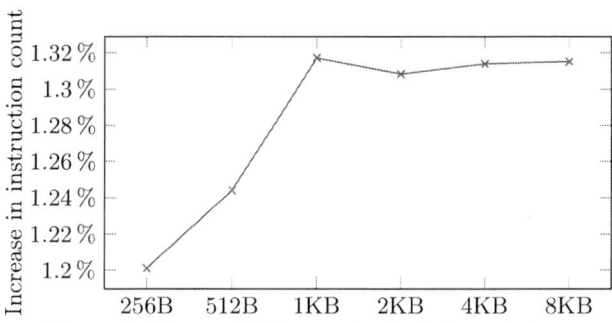

Figure 8: The negligible overhead of the technique

Integrated Instruction Cache Analysis and Locking in Multitasking Real-time Systems

Huping Ding[2], Yun Liang[1], and Tulika Mitra[2]

[1]Center for Energy-efficient Computing and Applications, School of EECS, Peking University
[2]School of Computing, National University of Singapore
{d-huping,tulika}@comp.nus.edu.sg, ericlyun@pku.edu.cn (contact author)

ABSTRACT

Cache locking improves timing predictability at the cost of performance. We explore a novel approach that opportunistically employs both cache analysis and locking to enhance schedulability in preemptive multi-tasking real-time systems. The cache is spatially shared among the tasks by statically locking a portion of the cache per task. To overcome the issue of limited cache space per task, we keep a portion of the cache unlocked and let all the tasks use it through time-multiplexing. Compared to locking the entire cache for each task during execution, our approach obviates the cost of reloading locked blocks at preemption. But we require static cache analysis for WCET estimation and cache related preemption delay (CRPD) analysis of the unlocked cache space. We design an algorithm to make appropriate locking decisions through accurate cost-benefit analysis. Experimental results show that our integrated approach leads to substantially improved schedulability results compared to cache analysis and cache locking employed individually.

Categories and Subject Descriptors

C.3 [**Special-purpose and Application-based Systems**]: [Real-time and embedded systems]

General Terms

Algorithm, Design, Performance.

Keywords

Real-Time, Multi-tasking, WCET, CRPD, Cache Locking.

1. INTRODUCTION

Multi-tasking real-time systems demand predictable timing behavior from the underlying architectural mechanisms employed to execute the tasks. Modern micro-architectures, however, perform aggressive dynamic optimizations to improve performance at the cost of timing predictability. The instruction cache is one such architectural feature that is ubiquitous in real-time embedded systems. But the instruction cache introduces two challenging problems in multi-tasking real-time systems. First, in the presence of

an instruction cache, it is difficult to estimate the *Worst-case Execution Time (WCET)* [28] of a task through program analysis as it is not known statically whether a memory access will hit or miss in the cache. Second, in a preemptive multi-tasking system, when a task T gets preempted by another task T', the memory blocks of T are replaced by those of T' in the cache. Once T resumes execution, it needs to bring in the replaced memory blocks again into the cache. This cost is known as *Cache Related Preemption Delay (CRPD)* [13] and adds a variable delay to the fixed context switching cost. The delay depends on the number of replaced memory blocks that are used again in task T. The schedulability analysis of a task set requires both the WCET and the CRPD values.

At the other end of the spectrum, we have cache locking [8, 9, 19] that offers completely predictable cache behavior and avoids the complexity of cache modeling altogether. The cache is locked with selected memory blocks from a task before execution. The memory blocks locked in the cache are guaranteed to be cache hits, while the remaining blocks are guaranteed to be cache misses, obviating the need for cache modeling in WCET analysis.

In multi-tasking systems, there exist two different locking approaches. Puaut and Decotigny [23] propose space sharing of the entire cache by locking a portion of the cache per task. We call it *PD-Locking* approach following the last names of the authors. The advantage of this approach is that the cache content remains unchanged throughout the execution of the tasks. The downside is that each task has access to only a fraction of the cache. Aparicio et al. [4] observe this limitation and introduces a time-multiplexed sharing of the cache through locking, called *ASRV-Locking* approach in the rest of the paper. In this approach, a task has exclusive access to the entire cache during execution and locks the cache with its own memory blocks leading to improved WCET per task compared to *PD-Locking* approach. However, when a task resumes execution after preemption, it has to reload the locked cache blocks leading to significantly higher (but fixed) preemption cost. Note that both approaches can bypass CRPD analysis as *PD-Locking* does not require cache reloading at preemption, while *ASRV-Locking* has fixed cache re-loading/locking cost at preemption.

We propose a non-traditional approach that judiciously combines cache locking with cache modeling in preemptive multi-tasking real-time systems and overcomes the space limitation of *PD-Locking* and cache reloading cost at preemption for *ASRV-Locking*. Similar to *PD-Locking*, we adopt space sharing by statically locking a portion of the cache per task but with a crucial difference. We leave a portion of the cache unlocked and let the tasks take advantage of this unlocked portion during execution through normal cache replacement policy. That is, the locked portion of the cache is statically shared, while the unlocked portion is time-multiplexed among the tasks. This relaxes the space constraint for each task and elimi-

---

Permission to make digital or hard copies of all or part of this work for personal or classroom use is granted without fee provided that copies are not made or distributed for profit or commercial advantage and that copies bear this notice and the full citation on the first page. To copy otherwise, to republish, to post on servers or to redistribute to lists, requires prior specific permission and/or a fee.
DAC'13, May 29-June 07 2013, Austin, TX, USA

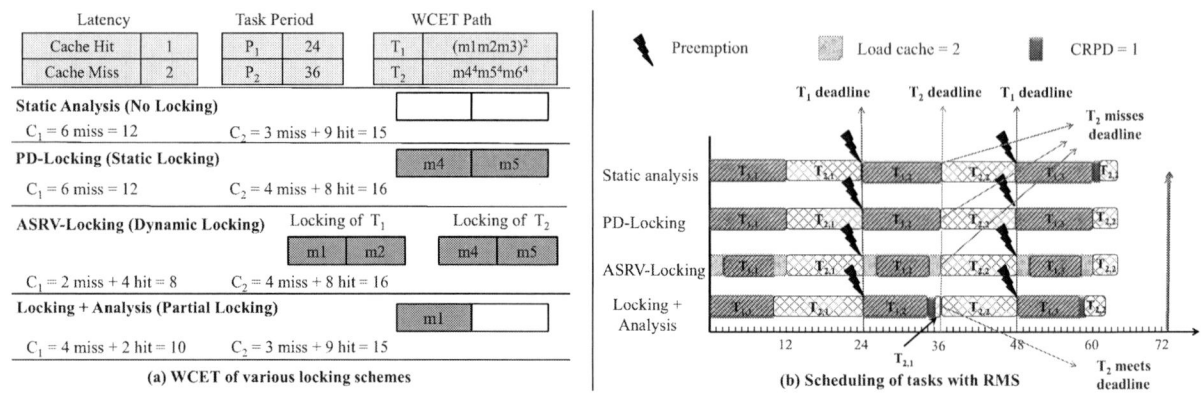

Figure 1: Motivating example.

nates cache re-loading at preemption. However, WCET and CRPD analysis are required for the unlocked portion of the cache.

Why do we want to give up the predictability offered through full cache locking, but not embrace static cache analysis all the way? We believe cache locking comes to rescue when cache analysis is unable to conclusively classify a memory access as hit or miss. At the same time, advancement in both WCET and CRPD analysis ensures that we can analyze the cache behavior quite precisely for most of the memory blocks. These memory blocks with predictable access patterns can reside in the unlocked portion of the cache providing improved performance. Indeed recent work [8] shows that, for a single task, a partially locked cache provides optimal WCET compared to both static analysis and full cache locking.

While Ding et al. [8] develops partial cache locking solution for a single task, it is challenging to make a locking decision for multi-tasking systems. First, the change in execution time of a task impacts the schedulability of the other tasks. Thus, we need to adopt a global optimization approach rather than a local per task approach. Second, the unlocked portion of the cache requires CRPD analysis. We propose an algorithm that employs accurate cost-benefit analysis to capture the impact of locking a memory block on the WCET, CRPD, and schedulability of all the tasks and thereby makes an informed decision to choose the appropriate memory blocks for locking. We perform detailed experimental evaluation to validate the improved schedulability results of our approach.

Motivating Example. We illustrate the benefit of our integrated cache analysis and locking using the example in Figure 1.

WCET comparison of various locking schemes. Figure 1(a) compares the WCET under various locking schemes. We assume two tasks T_1 and T_2 with periods 24 and 36. Cache hit latency is 1 cycle and miss latency is 2 cycles. The WCET paths of T_1 and T_2 are $(m_1 m_2 m_3)^2$ and $(m_4^4 m_5^4, m_6^4)$ where m_i represent a memory block. We assume all the memory blocks are mapped to the same cache set in a 2-way set associative cache. To simplify discussion, we also assume that timing is solely determined by instruction cache effects and the WCET paths do not change after locking.

Static Analysis estimates the WCET via cache modeling with no locking [24]. For T_1, all the accesses are miss because 3 memory blocks compete for 2 cache blocks resulting in 6 cache miss and WCET $C_1 = 12$. For T_2, it has 3 cold miss while remaining accesses are hit resulting in $C_2 = 15$. *PD-Locking* statically locks the entire cache with memory blocks from T_2 by selecting memory blocks with highest *access frequency/period* [23]. Thus, accesses of T_1 miss, while for T_2 it depends on which blocks are locked. *ASRV-Locking* employs dynamic locking where each task has exclusive

access to the entire cache. WCET of T_1 is reduced; but not the WCET of T_2. Finally, our *Locking + Analysis* judiciously chooses to lock only one block of T_1 and leaves the other cache block unlocked so that T_2's accesses are still hits after the cold misses.

Scheduling results of RMS. Let us assume that the reloading overhead for *ASRV-Locking* is 2 and the CPRD for our *Locking + Analysis* is 1. The execution of the tasks over the hyper-period is shown in Figure 1(b). The tasks are scheduled with Rate Monotonic Scheduling (RMS) policy where the task with the shortest period (T_1) has the highest priority. $T_{X,Y}$ represents the Yth instance of task T_X. For *Static Analysis* and *PD-Locking*, they fail to meet the deadline due to high WCET. The CRPD for *Static Analysis* is not shown as the task already misses the deadline. For *ASRV-Locking*, each task locks the entire cache and thus have lower WCET. However, every time a new task instance starts execution or a preempted task resumes execution, we need to reload and lock the cache with 2 cycle penalty. The additional cache reloading overhead makes T_2 miss its first deadline. Due to time-multiplexing of the unlocked cache space among all the tasks and lower preemption cost, our *Locking + Analysis* solution has lower WCET compared to *Static Analysis* and *ASRV-Locking*. This enables both T_1 and T_2 to meet their deadlines. A comparison with real task sets will be presented later in the experimental evaluation section.

2. RELATED WORK

Caches are problematic for WCET estimation due to their timing unpredictability. Static analysis has been widely used to bound the WCET [15, 24] of a single task. In multi-tasking preemptive real-time systems, a number of techniques have been proposed to accurately model the CRPD [13, 25, 21, 12].

Modern architectures often feature cache locking for better timing predictability. By carefully selecting the memory blocks for locking, WCET can be improved. There exist two locking mechanisms, static cache locking [9, 19, 22, 16] and dynamic cache locking [5, 26]. However, all of the above techniques lock the entire cache. Recently, Ding et al. [8] demonstrated that by partially locking the cache, WCET can be improved significantly.

In the context of multi-tasking real-time systems, Puaut and Decotigny [23] propose two static low complexity locking algorithms. Aparicio et al. [4] propose a dynamic locking solution based on Integer Linear Programming. Campoy et al. [7] develop static locking solutions using generic algorithms. Liu et al. [20] also propose dynamic locking solution within a task. We do not consider dynamic locking within a task as it requires code modification by inserting cache reloading instructions as shown in [20]. Again, these techniques lock the entire cache. Verma et al. [27] propose a hybrid approach for scratchpad memory allocation in multiprocess systems. Each process is allocated a disjoint region while the rest portion is shared by all processes. Their approach aims to minimize the energy consumption. Scratchpad memory does not have any address mapping constraint as in cache. Moreover, there is no timing unpredictability issue in scratchpad memory, making the problem somewhat simpler. Meanwhile, real-time scheduling is not considered in their work.

3. SYSTEM MODEL

In this section, we present the basic models of caches and tasks.

Cache Model. Cache design involves a few parameters: line (block) size L, which defines the unit of data or instruction transfer between the cache and main memory; number of cache sets K that the cache is divided into; associativity A, which determines the number of cache lines in a set. Then, the capacity of the cache is $L \times K \times A$. We assume LRU (Least Recently Used) cache replacement policy.

Given a memory block m, it can be mapped to only one cache set (m *modulo* K). To simplify the following discussion, we assume there is only one cache set as the different cache sets do not interfere with each other. However, our locking algorithm works with multiple cache sets. We use M to represent the set of memory blocks mapped to a cache set. We also use $\perp$ to indicate the absence of any memory block in a cache line.

DEFINITION 1 (**Concrete Cache State**). *A concrete cache state c is a vector $\langle c[0], ..., c[A-1] \rangle$ of length A where $c[j] \in M \cup \{\perp\}$. If $c[j] = m$, then m is the jth most recently used memory block in the cache set. We also define a special concrete cache state $c_{\perp} = \langle \perp, ..., \perp \rangle$ called the empty concrete cache state.*

DEFINITION 2 (**Abstract Cache State**). *An abstract cache state a is a vector $\langle a[0], ..., a[A-1] \rangle$ of length A where $a[j] \in 2^M$.*

An abstract cache state maps a cache block to a set of memory blocks. At a program point, the abstract cache state is a safe approximation of the concrete cache states along all the incoming program paths, and hence is a more compact representation. Both concrete and abstract cache states have been widely used in real-time systems for timing analysis [17, 24].

Task Model. We assume a preemptive multi-tasking real-time system running on uni-processor with a set of N independent periodic tasks $\mathscr{T} = \{T_1, ..., T_N\}$. For each task T_i, we use P_i to represent its period and C_i to represent its WCET. We assume the deadline $D_i = P_i$. The C_i value is obtained by performing intra-task WCET analysis for T_i. In other words, the WCET analysis is performed in isolation per task. In processors with caches, we also need to account for the delay due to preemption: the CRPD and the context switching cost. For each task T_i, we use Δ_i to denote the delay due to preemption. Let U be the total processor utilization for the task set. A necessary condition for feasible scheduling of the task set is

$$U = \sum_{i=1}^{N} \frac{C_i + \Delta_i}{P_i} \leq 1 \qquad (1)$$

The delay due to preemption for task T_i is defined as follows.

$$\Delta_i = \sum_{T_j \in pt(T_i)} (CRPD(T_i, T_j) + CSC) \times n(T_i, T_j) \qquad (2)$$

where $pt(T_i)$ is the set of tasks that may preempt T_i, $CRPD(T_i, T_j)$ is the CRPD of T_i imposed by T_j in one preemption, $n(T_i, T_j)$ is the bound for the number of preemption of T_i imposed by T_j, and CSC represents the context switching cost.

EDF Scheduling. Earliest Deadline First (EDF) is a dynamic priority based scheduling policy. The priority of a task is determined by its deadline. At any time instance, EDF chooses the ready task with the closest deadline for execution. For EDF, Equation 1 ($U \leq 1$) is both sufficient and necessary condition for feasible schedule. The task set that may preempt T consists of all the tasks that may have earlier deadline than T [11].

RMS Scheduling. Rate Monotonic Scheduler (RMS) is a static priority based scheduling policy. The priority of a task is determined statically by its period. Task T_i has higher priority than task T_j if $P_i < P_j$. Therefore, the set of tasks that may preempt T is the set of tasks with higher priority. Unlike EDF, $U \leq 1$ is not a sufficient condition for feasible schedule with RMS. There exists no polynomial time schedulability test for RMS. An iterative method is employed to estimate the response time of each task and compare it against the deadline.

$$S_i^{n+1} = C_i + \sum_{T_j \in hp(T_i)} \lceil \frac{S_i^n}{P_j} \rceil (C_j + CRPD(T_i, T_j) + CSC) \qquad (3)$$

where S_i^n is the response time of T_i in the nth iteration, and $hp(T_i)$ represents the set of tasks that have higher priority than T_i.

4. FRAMEWORK OVERVIEW

We first provide an overview of our integrated analysis and locking approach. We propose to statically lock a part of the cache per task, while a part is left unlocked to be used by all the tasks. The locked cache space is spatially shared among the tasks, while the unlocked cache space is temporally shared by all the tasks.

According to Equation 1, the execution time of a task depends on both intra-task WCET and inter-task CRPD. For the locked memory blocks, they do not incur any CRPD as they can not be evicted from the cache and their impact on the WCET can be easily determined. However, for the unlocked memory blocks, we still need to perform static analysis for both intra-task WCET and inter-task CRPD analysis as they use the remaining unlocked cache space.

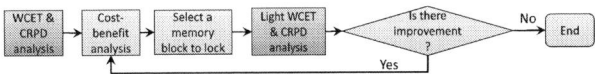

Figure 2: Framework for *Locking + Analysis* approach.

Figure 2 illustrates the flow of our *Locking + Analysis* approach. We first perform intra-task WCET analysis with abstract interpretation [24]. Meanwhile, we also perform inter-task CRPD analysis [12]. Then, for each memory block in the task set, a cost-benefit analysis on WCET and CRPD is carried out for cache locking. This cost-benefit analysis captures the impact of locking a memory block on the WCET, CRPD, and schedulability of all the tasks. Based on this cost-benefit analysis, we choose the most profitable memory block to lock. We perform intra-task WCET analysis and inter-task analysis again after locking this memory block. We call it light WCET & CRPD analysis, because it avoids some unnecessary cache analysis compared to the full-fledged WCET & CRPD analysis. If either the schedulability or the utilization improves, we continue to lock other memory blocks. Otherwise, the iterative process stops and we obtain the final solution.

5. WCET AND CRPD ANALYSIS

In the following, we present a brief description of the static analysis techniques for intra-task WCET and inter-task CRPD estimation (see Figure 3). This analysis ignores cache locking. But this

1005

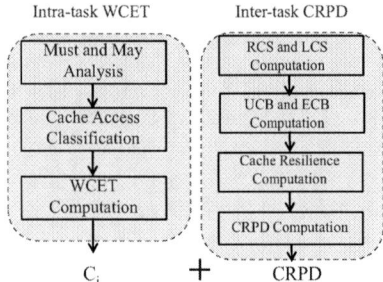

Figure 3: WCET and CRPD Analysis.

background material is required to appreciate the WCET and CRPD estimation in the presence of cache locking for our cost-benefit analysis presented in the next section.

5.1 Intra-Task WCET

As shown in Figure 3, intra-task WCET analysis involves three steps. First, it performs abstract interpretation — must and may analysis — based on abstract cache states [24]. Must analysis determines the set of memory blocks that are guaranteed to be present in the cache at a program point. May analysis captures the set of memory blocks that may be present in the cache at a program point. Next, the memory blocks are classified based on the must and may analysis. The memory blocks present in the abstract cache states of must analysis are classified as always hits; the memory blocks not present in the abstract cache states of may analysis are classified as always misses; the remaining memory blocks belong to non-classified category, i.e., they are assumed to be cache misses during WCET estimation. Finally, the WCET is derived based on the memory accesses classification.

5.2 Inter-Task CRPD

A preempted task T incurs CRPD as some of its useful memory blocks are evicted from the cache by the preempting task T'. The "useful" memory blocks are those memory blocks that have been loaded into the cache before preemption and may be accessed again after preemption. Thus, the key to CRPD analysis is to determine the useful cache blocks of the preempted task and verify whether they could survive after the preemption.

Recently, Kleinsorge et al. [12] proposed a CRPD estimation approach for set associative caches that combines techniques for direct mapped caches [21] with resilience analysis for set associative caches [3]. Concrete cache states are fundamental to inter-task CRPD analysis [12, 21]. Given a program point, it may be reached via multiple program paths, which leads to multiple concrete cache states. Thus, in general, it is infeasible to maintain all the possible concrete cache states for large programs with complex control flow. Inter-task CRPD analysis aims to identify the cache states with the largest number of useful memory blocks and thus higher preemption delay. The subsumed cache states (with lower number of useful memory blocks) can be safely removed. Hence, it is feasible to use concrete cache states for inter-task CRPD analysis as shown in [12, 21]. We adopt the technique in [12] to estimate the CRPD for a single preemption. The approach in [12] depends on the computation of UCB (useful cache blocks) and ECB (evicting cache blocks). The resilience of a UCB defines the maximum number of allowed memory accesses from the preempting task before it can be evicted. Figure 3 shows the four steps of CRPD analysis in [12] that are detailed in Appendix A.

During the execution of task T, it is possible that higher priority task T' preempts lower priority task T multiple times. The preemption bound imposed by T' on T is denoted as $n(T, T')$ in Equation 2. $n(T, T')$ depends on the scheduling policy. For EDF scheduling policy, we use the approach in [11] to bound $n(T, T')$; for RMS scheduling, $n(T, T')$ is a by-product of the response time computation as shown in [18] and Equation 3.

6. LOCKING ALGORITHM

As we have mentioned, existing locking techniques for multi-tasking systems allocate the entire cache for locking [4, 23]. Such locking techniques eliminate the CRPD analysis at the expense of poor performance. Thus, it is relatively easy to compute the memory blocks to be locked. While in our approach, there is a complex interplay between cache locking and its impact on schedulability analysis. When a memory block of task T is locked in the cache, T generally benefits from the locking. However, it also takes away valuable cache space from the remaining tasks and also changes their CRPD. Any exact locking algorithm for our approach will have exponential complexity. Thus we design an efficient heuristic to decide on the memory blocks to be locked.

As noted earlier, we first perform intra-task WCET analysis and inter-task CRPD for each task in the task set (see Figure 2) when no memory block is locked. Then we compute the processor utilization and response time for each task. As a by-product of intra-task WCET analysis, we have the abstract cache states (must and may) at each program point. We also collect the memory blocks along the WCET path and their execution frequencies for each task. Similarly, we record the worst-case preemption point and the corresponding UCB and ECB for each task during CRPD analysis. We design an iterative solution to select the memory blocks for locking. In each iteration, we choose the most beneficial memory block for locking. The benefit of locking a memory block is defined differently for different scheduling policy (see section 6.3). We stop this process when there is no benefit due to locking and the remaining cache space is left unlocked.

The cost and benefit of locking is based on the following observations. Given a memory block $m \in T$, if m is locked, then all the accesses to m are cache hits. But as cache size is reduced, it might have negative impact on the other memory blocks mapped to the same cache set for all the tasks including T. For task T, its intra-task WCET might be improved if the benefit of locking m is greater than the cost on other memory blocks. However, for other tasks, there is no positive effect on their intra-task WCET. Finally, the CRPDs for all the tasks are usually reduced as the effective cache size is reduced after locking m. In the following, we show how to estimate the cost and benefit of locking memory block m. We assume $m \in T$ and m is mapped to cache set s.

6.1 Cost-benefit analysis within a task

We only consider the memory blocks of task T along the WCET path for locking as locking the other memory blocks has no benefit. Let m be a memory block along the WCET path of T and f_m be the execution frequency of m along the WCET path of T. We use lat_m to denote the access latency of memory block m. lat_m is determined by the classification (cache hit or cache miss) of memory block m in must/may analysis. We use lat_{hit} and lat_{miss} to represent the cache hit and miss latency, respecitively. Then, the benefit of locking m on the WCET of T is

$$wcet\_benefit_m^T = (lat_m - lat_{hit}) \times f_m$$

However, locking m may also have negative impact on the other memory blocks of T mapped to the same cache set s as the number of cache blocks in set s is now reduced by one. Let C be the abstract cache state for must analysis of set s. If $m \in C$, m is classified as cache hit before cache locking, thus locking m does not evict any other memory block from the cache and $wcet\_cost_m^T = 0$. However, if $m \notin C$, locking m will evict out the memory block m' with age $A - 1$ in C from the cache, which results in cache miss for the accesses of m'. In this case, the cost of locking m is

$$wcet\_cost_m^T = \sum_{(m' \in M_s) \wedge (age_{m'}^C = A - 1)} (lat_{miss} - lat_{hit}) \times f_{m'}$$

where M_s is the set of memory blocks mapped to set s in T, and $f_{m'}$ indicates the execution frequency of m' along the worst-case path. Therefore the WCET gain of T by locking m is

$$wcet\_gain_m^T = wcet\_benefit_m^T - wcet\_cost_m^T$$

Apart from the influence on the intra-task WCET of T, locking m may also affect the CRPD of T. We assume T is preempted by another task T'. Obviously, locking m will not generate any new useful cache block because the cache size is reduced. As mentioned before, we record the UCB, ECB and the preemption point that lead to the worst-case CRPD for this preemption. Suppose M_s^u is the set of useful cache blocks in T mapped to set s at this point and $M_s^{u'}(M_s^{u'} \subset M_s^u)$ is the set of blocks that contribute to the CRPD before locking m. To model the effect of locking m, we update ECB of set s and the resilience of any block in M_s^u. With the new ECB and updated resilience, we can obtain $M_s^{u''} \subset M_s^{u'}$, the new set of blocks that contribute to the CRPD after locking m. So the CRPD gain by locking m due to one preemption by T' is

$$crpd\_gain_m^{TT'} = (|M_s^{u'}| - |M_s^{u''}|) \times (lat_{miss} - lat_{hit})$$

Therefore, the total CRPD gain of T by locking m is

$$crpd\_gain_m^T = \sum_{T' \in pt(T)} crpd\_gain_m^{TT'} \times n(T, T')$$

where $pt(T)$ is the set of tasks that may preempt T and $n(T, T')$ is the bound on the number of preemptions of T imposed by T'. Finally the overall execution time gain of T by locking m is

$$time\_gain_m^T = wcet\_gain_m^T + crpd\_gain_m^T$$

6.2 Cost-benefit analysis of other tasks

Let $T' \neq T$ and $m' \in T'$ be a memory block along the WCET path of T'. We assume m' and m are mapped to the same cache set s and m' is in the abstract cache state of must analysis C. If the age of m' is $A - 1$, then locking m will evict m' out of cache. Thus, locking m has negative impact on the WCET of other tasks. We define the WCET cost on T' by locking m as follows

$$wcet\_cost_m^{T'} = \sum_{(m' \in M_s') \wedge (age_{m'}^C = A - 1)} (lat_{miss} - lat_{hit}) \times f_{m'}$$

where M_s' is the set of memory blocks mapped to set s in T' and $f_{m'}$ is the execution frequency of m' along the WCET path.

The CRPD gain of T' by locking m, $crpd\_gain_m^{T'}$, can be obtained via the same approach as in section 6.1. Thus, the overall execution time gain of task T' by locking m is

$$time\_gain_m^{T'} = crpd\_gain_m^{T'} - wcet\_cost_m^{T'}$$

6.3 Memory block selection strategy

We design different memory block selection strategies for EDF and RMS scheduling policies.

EDF Scheduling. Equation 1 is a sufficient and necessary condition for feasible schedule. Thus we select the memory blocks based on their impact on total processor utilization as follows

$$util\_gain_m = \frac{time\_gain_m^T}{P} + \sum_{T' \in \mathscr{T} \setminus \{T\}} \frac{time\_gain_m^{T'}}{P'}$$

where P is the period of task T, P' is the period of task T' and $\mathscr{T}$ is the task set. The utilization gain of locking a memory block is used as a metric to select the memory blocks for locking. In each iteration, we select the memory block with maximum utilization gain over all memory blocks in the task set.

RMS Scheduling. Utilization (Equation 1) is not a sufficient condition for feasible schedule in RMS. Thus, for RMS, we first need to ensure the schedulability of the task set. For each task, its response time can be computed using the iterative method provided by Equation 3. We focus on the tasks with response time greater than their deadline, and among them try to optimize the response time of the task with highest priority first. Based on Equation 3, in order to improve the response time of a task T, we can either reduce the execution time of T, or improve the execution time of the tasks with higher priority than T. So, when we try to lock a memory block $m \in T$, the corresponding response time gain of T is

$$rsp\_gain_m^T = wcet\_gain_m^T + \sum_{T' \in hp(T)} (crpd\_gain_m^{TT'} - wcet\_cost_m^{T'}) \times n(T, T')$$

where $hp(T)$ is the set of tasks with higher priority than T. When we try to lock a memory block $m' \in T'$ with higher priority than T, the corresponding response time gain of T is

$$rsp\_gain_{m'}^T = (crpd\_gain_{m'}^{TT'} + wcet\_gain_{m'}^{T'}) \times n(T, T') - wcet\_cost_{m'}^T + \sum_{T'' \in hp(T) \setminus \{T'\}} (crpd\_gain_{m'}^{TT''} - wcet\_cost_{m'}^{T''}) \times n(T, T'')$$

where T'' is a task with higher priority than T and $T'' \neq T'$, and $n(T, T'')$ represents the number of preemption bound imposed on T by T''. The WCET and CRPD gain are different for T' and T'' through locking of m'. But both of them contribute to the response time gain of T. Thus, $rsp\_gain_{m'}^T$ includes both of them. Because $m' \in T'$, $wcet\_gain_{m'}^{T'}$ is obtained via the approach in section 6.1. Meanwhile, $m' \notin T$ and $m' \notin T''$, thus, $wcet\_cost_{m'}^T$ and $wcet\_cost_{m'}^{T''}$ are computed similarly via the approach in section 6.2. $crpd\_gain_{m'}^{TT'}$ and $crpd\_gain_{m'}^{TT''}$ can be obtained via the same approach as in section 6.1.

We select the memory block with the maximum response time gain to lock, while at the same time we ensure the utilization gain of this block to be non-negative. After all the tasks are schedulable, we apply the same method used for EDF scheduling to further minimize the utilization. For both scheduling policies, after each iteration, we recompute the abstract cache states of set s where the selected memory block m is mapped to, and then recompute the WCET. Similarly, the cache states for CRPD computation at each program point are also updated. We then recompute the UCB and ECB for each task in task set and obtain the new CRPD. Based on the new WCET and CRPD, we derive the metric value. If there is improvement, we continue to lock. The iterative approach stops only when all the memory blocks are locked or there is no improvement after locking any memory block. The pseudo-code of the algorithms appear in Appendix C.

7. EXPERIMENTAL EVALUATION

In this section, we quantitatively compare our approach with static analysis [24, 12], *ASRV-Locking* [4], and *PD-Locking* [23].

Experiments Setup. We use similar task sets used in [23, 4]. The task sets are shown in Table 1. They contain one small and one medium task set. All the tasks are from MRTC benchmark suite [10].

We assume the deadline of a task is equal to its period. Our framework is built on top of the open-source WCET analysis tool Chronos [14]. All the tasks are compiled with gcc cross-compiler for an ARM-like instruction set [6].

We assume there is only one level of instruction cache. Instruction hit latency is 1 cycle, while the cache miss latency is 30 cycles. The locking routine is stored in non-cacheable memory and it uses five instructions to load and lock a memory block [2, 1]. Thus, the cost of locking a memory block is 150 cycles. The cache is 4-way set-associative with block size of 32 bytes. We also assume each context switch takes 1,000 cycles per preemption for all the approaches. For a fair comparison, we assume there is no line buffer for Aparicio et al.'s approach [4].

Table 1: Characteristics of task sets

Task set	Task	Code size (bytes)	Period
small	jfdctint	5,512	1,500,000
	crc	2,032	2,000,000
	fir	1,144	4,200,000
	matmult	1,632	3,900,000
medium	minver	6,256	720,000
	qurt	2,048	44,000
	jfdctint	5,512	680,000
	fdct	5,176	370,000

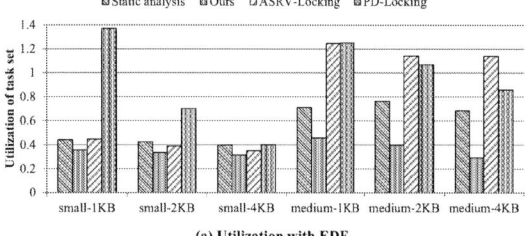

(a) Utilization with EDF

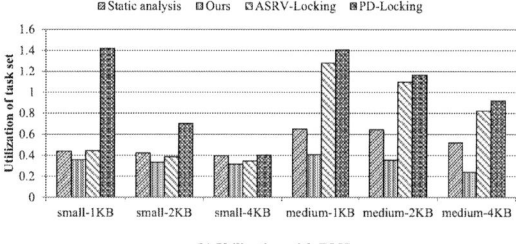

(b) Utilization with RMS

Figure 4: Utilization comparison of different approaches.

Utilization Comparison. Figure 4 (a) and (b) present the utilization comparison of different approaches under EDF and RMS scheduling. *small-X KB* (*medium-X KB*) denotes small (medium) task set with cache size of X KB. As shown, our integrated cache analysis and locking substantially improves the utilization irrespective of task set size, scheduling policy, and cache size. *PD-Locking* has high utilization when the cache size is small. For *PD-Locking*, the locked memory blocks for each task are very limited and most of the memory accesses are serviced from main memory instead of cache. As a result, the WCET of the tasks and utilization of the task set are high. In the medium task set, the utilization of *ASRV-Locking* is also high. First, the code size for tasks in medium task set is large. Thus, there are still many unlocked memory blocks. Second, the period of task *qurt* is much smaller than the other tasks, and these tasks suffer many preemptions from *qurt*. Thus, the re-locking cost also contributes a lot to the utilization.

More results are detailed in Appendix B.

8. CONCLUSION

In this paper, we present an approach that integrate instruction cache analysis and locking in multitasking preemptive real-time systems. A portion of the cache is locked by the tasks in the task set, while the remaining portion is used by all the tasks. We propose an algorithm based on accurate cost-benefit analysis to select the appropriate memory contents to lock. Experimental results show that our approach outperforms previous techniques that either time-multiplexes the cache among all the tasks or statically shares and fully locks the cache.

9. ACKNOWLEDGMENTS

This work was partially supported by Singapore Ministry of Education Academic Research Fund Tier 2, MOE2009-T2-1-033.

10. REFERENCES

[1] ARM 940T (rev 2) technical reference manual.
[2] Intel XScale core developer manual.
[3] S. Altmeyer, C. Maiza, and J. Reineke. Resilience analysis: tightening the crpd bound for set-associative caches. In *LCTES*, 2010.
[4] L. C. Aparicio et al. Improving the WCET computation in the presence of a lockable instruction cache in multitasking real-time systems. *J. Syst. Archit.*, 57(7), 2011.
[5] A. Arnaud and I. Puaut. Dynamic instruction cache locking in hard real-time systems. In *RTNS*, 2006.
[6] D. Burger and T. M. Austin. The simplescalar tool set, version 2.0. *SIGARCH Comput. Archit. News*, 25(3), 1997.
[7] A. M. Campoy et al. Cache contents selection for statically-locked instruction caches: An algorithm comparison. In *ECRTS*, 2005.
[8] H. Ding, Y. Liang, and T. Mitra. WCET-centric partial instruction cache locking. In *DAC*, 2012.
[9] H. Falk, S. Plazar, and H. Theiling. Compile-time decided instruction cache locking using worst-case execution paths. In *CODES+ISSS*, 2007.
[10] J. Gustafsson et al. The mälardalen WCET benchmarks - past, present and future. In *WCET*, 2010.
[11] L. Ju, S. Chakraborty, and A. Roychoudhury. Accounting for cache-related preemption delay in dynamic priority schedulability analysis. In *DATE*, 2007.
[12] J. C. Kleinsorge, H. Falk, and P. Marwedel. A synergetic approach to accurate analysis of cache-related preemption delay. In *EMSOFT*, 2011.
[13] C.-G. Lee et al. Analysis of cache-related preemption delay in fixed-priority preemptive scheduling. *IEEE Trans. Comput.*, 47(6), 1998.
[14] X. Li, Y. Liang, T. Mitra, and A. Roychoudhury. Chronos: A timing analyzer for embedded software. *Sci. Comput. Program.*, 69(1-3):56–67, 2007.
[15] Y.-T. S. Li, S. Malik, and A. Wolfe. Cache modeling for real-time software: beyond direct mapped instruction caches. In *RTSS*, 1996.
[16] Y. Liang, H. Ding, T. Mitra, A. Roychoudhury, Y. Li, and V. Suhendra. Timing analysis of concurrent programs running on shared cache multi-cores. *Real-Time Syst.*, 48(6):638–680, 2012.
[17] Y. Liang and T. Mitra. Cache modeling in probabilistic execution time analysis. In *DAC*, 2008.
[18] C. L. Liu and J. Layland. Scheduling algorithms for multiprogramming in a hard realtime enviroment. *Journal of the ACM*, 20(1), 1973.
[19] T. Liu, M. Li, and C. J. Xue. Minimizing wcet for real-time embedded systems via static instruction cache locking. In *RTAS*, 2009.
[20] T. Liu, M. Li, and C. J. Xue. Instruction cache locking for multi-task real-time embedded systems. *Real-Time Syst.*, 48(2), 2012.
[21] H. S. Negi, T. Mitra, and A. Roychoudhury. Accurate estimation of cache-related preemption delay. In *CODES+ISSS*, 2003.
[22] S. Plazar et al. WCET-aware static locking of instruction caches. In *CGO*, 2012.
[23] I. Puaut and D. Decotigny. Low-complexity algorithms for static cache locking in multitasking hard real-time systems. In *RTSS*, 2002.
[24] H. Theiling, C. Ferdinand, and R. Wilhelm. Fast and precise WCET prediction by separated cache andpath analyses. *Real-Time Syst.*, 18(2/3), 2000.
[25] H. Tomiyama and N. D. Dutt. Program path analysis to bound cache-related preemption delay in preemptive real-time systems. In *CODES*, 2000.
[26] X. Vera, B. Lisper, and J. Xue. Data cache locking for tight timing calculations. *ACM Trans. Embed. Comput. Syst.*, 7(1), 2007.
[27] M. Verma et al. Scratchpad sharing strategies for multiprocess embedded systems: a first approach. In *ESTIMedia*, 2005.
[28] R. Wilhelm et al. The worst-case execution-time problem – overview of methods and survey of tools. *ACM Trans. Embed. Comput. Syst.*, 7(3), 2008.

APPENDIX

A. INTER-TASK CRPD ANALYSIS

In this section, we detail the steps to compute the CRPD by Kleinsorge et al. [12].

RCS and LCS Computation. The useful memory blocks are computed using two different types of cache states: the reaching cache states (RCS) and live cache states (LCS). At a program point p, RCS_p is the set of possible cache states when p is reached via any incoming program path. Conversely, at a program point p, LCS_p represents the set of possible cache states via any outgoing program path from p. The cache states RCS and LCS can be computed via forward/backward fix-point data flow analysis [12, 21].

UCB and ECB Computation. We use UCB_p to denote the set of useful memory blocks at program point p. UCB_p is computed as

$$UCB_p = \{c \cap c' | c \in RCS_p, c' \in LCS_p\}$$

where $c \cap c'$ is defined as

$$c \cap c' = \{b | \exists\, 0 \le j < K \ s.t. \ c[j] = b, \exists\, 0 \le k < K \ s.t. \ c'[k] = b\}$$

Evicting cache block (ECB) captures the memory blocks that may be accessed during the execution of the preempting task. Thus

$$ECB = RCS_{exit}$$

where *exit* is the exit point of the preempting task.

Cache Block Resilience. Given a useful memory block m at a program point p, its survival upon preemption depends on its resilience to the preempting task. We define its resilience res_p^m as the maximum number of allowed memory accesses from the preempting task before m can be evicted and is computed as follows. We define the distance of useful memory block m at program point p

$$distance_p^m = \begin{cases} A - 1 & if\ age_m^\downarrow + age_m^\uparrow \ge A \\ age_m^\downarrow + age_m^\uparrow & otherwise. \end{cases}$$

where $age_m^\downarrow$ and $age_m^\uparrow$ denote the maximum age of m in RCS_p and LCS_p, respectively. Then, the resilience is defined as

$$res_p^m = (A - 1) - distance_p^m$$

CRPD Computation. We can now bound the CRPD based on the UCB of preempted task T and ECB of preempting task T'. Let UCB_p be the set of useful cache blocks at a program p of the preempted task T and ECB be the set of evicting cache blocks of the preempting task T'. For any $u \in UCB_p$ and $e \in ECB$

$$CRPD_{(u,e)}^p = |u \setminus \{m | res_p^m \ge |e|\}| \times CRT$$

where CRT is the reloading overhead of one memory block. Then, the CRPD at this program point p is the maximum among all the possible combinations of UCB of T and ECB of T'

$$CRPD^p = \max_{u \in UCB_p, e \in ECB} CRPD_{(u,e)}^p$$

The CRPD for this preemption is the maximum CRPD over all the program points. That is

$$CRPD(T, T') = \max_{p \in PP} CRPD^p$$

where PP is the set of program points of the preempted task T.

B. EXTRA EXPERIMENTAL RESULTS

In this section, we show more experimental results with our approach, including response time speedup, utilization breakdown, unlocked cache space, and runtime of our approach. The details are shown in the following subsections.

B.1 Response Time Speed-up

We compare the different approaches using response-time speedup metric proposed in [4] for RMS policy. It is defined as follows.

$$speedup = \frac{period}{response\ time}$$

It is calculated for the lowest priority task and indicates the slack available in the schedule. Thus, a speedup greater than or equal to 1 implies that the task set is schedulable. Figure 5 shows the response time speed-up for the task sets with varying cache size. Clearly, with our approach, the tasks with lowest priority are always schedulable. However, the lowest priority task in medium task set with *ASRV-Locking* and *PD-Locking* are not schedulable in most of the cases.

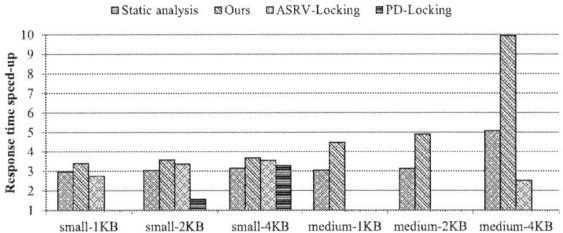

Figure 5: Response time speed-up.

B.2 Utilization Breakdown

Figure 6 details the contribution to the utilization by WCET, CRPD and re-locking overhead, respectively for the medium task set with 2KB cache size under RMS scheduling policy. Compared to static analysis, our approach either significantly reduces the WCET (*qurt*) or nearly eliminates the CRPD (*minver*, *jfdctint* and *fdct*). While for *ASRV-Locking*, we observe a great contribution to utilization due to re-locking overhead (*jfdctint* and *fdct*). Finally, the WCET using *ASRV-Locking* and *PD-Locking* are usually large, because the unlocked memory blocks are all serviced by main memory instead of cache.

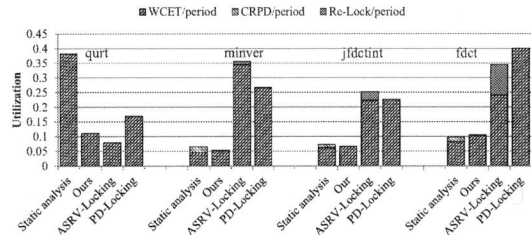

Figure 6: Utilization breakdown for medium-2KB.

B.3 Unlocked Cache Space

Figure 7 (a) and (b) show the percentage of the unlocked cache lines of our approach under EDF and RMS, respectively. The percentage of the unlocked cache space depends on the cache size and the scheduling policy. As shown, with our approach, there is a portion of cache space left unlocked for all the settings. The unlocked cache space can be used by all the tasks in the task set. We also notice that the percentage of unlocked cache lines of 2KB cache is smaller than that of 1KB and 4KB cache. When the cache is small, our approach decides to lock only a small portion of the cache. It is because locking more memory blocks may have significant negative impact on the WCET of the tasks. On the other hand, when the cache is big, more memory blocks can be classified as cache hits and locking those memory has no benefit. Thus, our approach decides to lock only a small portion of a big cache.

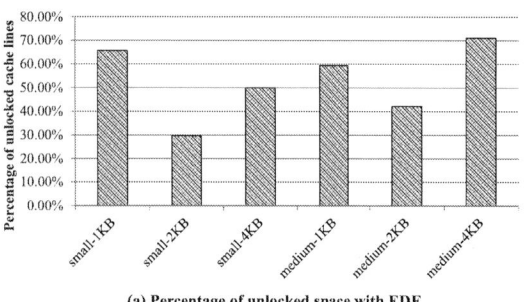

(a) Percentage of unlocked space with EDF

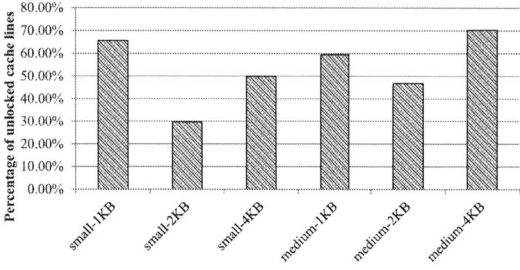

(b) Percentage of unlocked space with RMS

Figure 7: Percentage of unlocked cache lines with our approach.

B.4 Runtime of Our Approach

Table 2 presents the runtime of our approach under both EDF and RMS scheduling policy with different cache sizes. We perform all the experiments on 2.53GHz Intel Xeon CPU with 24GB memory. Clearly, our locking algorithm runs efficiently. The overall runtime depends on the number of locked memory blocks, WCET analysis and CRPD analysis. We notice that for the small task set, the runtime of 2KB cache is higher than that of 4KB cache. In small task set, the code size of crc is about 2KB. When the cache size is set to 2KB, crc has complicated RCS and LCS analysis that leads to long CRPD computation time.

Table 2: Runtime of our approach

Task set	Cache size	EDF (seconds)	RMS (seconds)
small	1KB	4.65	4.68
	2KB	22.72	22.82
	4KB	4.32	4.29
medium	1KB	1.67	1.67
	2KB	33.94	37.19
	4KB	133.44	137.05

C. INTEGRATED LOCKING + ANALYSIS ALGORITHMS

In this section, we present the detailed locking + analysis Algorithms used in our approach, including cost-benefit analysis algorithm, utilization optimization algorithm and schedulability improvement algorithm. The details are shown as follows.

C.1 Cost-benefit Analysis Algorithm

Algorithm 1 presents the detailed cost-benefit analysis by locking a memory block m. For each task T_i in the task set $\mathscr{T}$, if m belongs to T_i, then there may be WCET benefit for T_i by locking m as all the accesses to m are cache hits after locking (line 5). On the other hand, locking m also impacts the other memory blocks mapped to the same cache set in T_i as the effective cache size is reduced after locking m, which may leads to more cache misses. Therefore we compute the cost by locking m for T_i (line 6). If m does not belong to T_i, obviously, there is no benefit for T_i by locking m. Thus we only calculate the cost for T_i by locking m (line 8-9). The WCET gain for T_i should consider both WCET benefit and WCET cost (line 10). Locking memory block m also impacts the CRPD as the effective cache size is reduced after locking. Thus, we also compute the corresponding cost and benefit of CRPD for task T_i by locking m (line 11). Finally, The overall execution time gain for task T_i includes both the WCET gain and CRPD gain (line 12).

Algorithm 1: Cost-benefit analysis on WCET and CRPD

Input: Task set $\mathscr{T} = \{T_1, T_2 ... T_N\}$, cache configuration $config$ and candidate memory block m

Output: WCET gain $wcet\_gain_m^{T_i}$ and CRPD gain $crpd\_gain_m^{T_i}$ by locking memory block m for each task $T_i \in \mathscr{T}$

1 **begin**
2   **foreach** $T_i \in \mathscr{T}$ **do**
3   Suppose M_i is the set of memory blocks of T_i;
4   **if** $m \in M_i$ **then**
5       $wcet\_benefit_m^{T_i}$ = wcet_benefit_self();
6       $wcet\_cost_m^{T_i}$ = wcet_cost_self();
7   **else**
8       $wcet\_benefit_m^{T_i} = 0$;
9       $wcet\_cost_m^{T_i}$ = wcet_cost_others();
10     $wcet\_gain_m^{T_i} = wcet\_benefit_m^{T_i} - wcet\_cost_m^{T_i}$;
11     $crpd\_gain_m^{T_i}$ = crpd_cost_benefit_analysis();
12     $time\_gain_m^{T_i} = wcet\_gain_m^{T_i} + crpd\_gain_m^{T_i}$;
13
14 **end**

C.2 Utilization Optimization Algorithm

Algorithm 2 shows the details of utilization optimization. We first perform one round of WCET and CRPD analysis for each task $T_i \in \mathscr{T}$ (line 3-9). For each task T_i, we perform abstract cache state analysis and compute the WCET (line 4-5). We also perform RCS and LCS analysis for each task T_i (line 6). Based on the RCS and LCS analysis results, we calculate the UCB and ECB, as well as the resilience for each useful cache block (line 7-8). Then we do the CRPD analysis for the task set (line 9). With the CRPD and WCET, the initial utilization of the task set is then carried out (line 10). Later, we iteratively select memory blocks with the maximum utilization gain to lock. For each candidate memory block m, we first check whether it is locked or not (line 16). Meanwhile, we check whether the corresponding cache set is fully locked or not (line 16). If m has been locked or there is no free space in the cor-

1010

responding cache set that m mapped to, we skip m and try other candidates. When we find a memory block m that can be locked, we first perform the cost-benefit analysis on WCET and CRPD by using Algorithm 1 (line 17). Then, we calculate the utilization gain for each task in the task set (line 18). The total utilization gain for the entire task set by locking m is the summation of utilization for all the tasks in the task set (line 19). We compare m with the candidate memory block $mblk$ that currently has the most utilization gain (line 20). If m has more utilization gain than $mblk$, we update $mblk$ with m (line 21-22). We continue to do this until all candidate memory blocks are considered. If we find no memory block with positive utilization gain, this algorithm will terminate (line 42-43). Otherwise, we will end up with a memory block $mblk$ that has the maximum utilization gain. We lock $mblk$ into the cache (line 27). For each task, we update the abstract cache states in the cache set that $mblk$ mapped to, and recompute the WCET (line 29-30). We also update the RCS and LCS for this particular cache set, and recompute the UCB, ECB and resilience (line 31-33). After the resilience for each useful cache block is updated, we perform the CRPD analysis again to get the new CRPD (line 34). Based on the new WCET and CRPD, we obtain the new utilization of the task set (line 35). If there is improvement on utilization of the task set, we update the utilization and add $mblk$ to the set of locked memory blocks, and continue to lock other memory blocks (line 37-18). Otherwise we stop locking and obtain the final results (line 40).

C.3 Schedulability Improvement Algorithm

Algorithm 3 presents the detailed approach to improve schedulability for RMS. For a task set $\mathscr{T}$ in RMS, since Equation 1 is not a sufficient condition for feasible schedule in RMS, we should first check the schedulability of $\mathscr{T}$ based on the response time of each task. Therefore, We also need to perform one round of WCET and CRPD analysis first for the task set as we did in Algorithm 2 (line 3-9). Then, apart from computing the utilization for the task set (line 10), we also need to calculate the corresponding response time for each task in the task set (line 11). We check the schedulability by comparing response time of each task with its deadline. If all tasks meet their deadline, we set the boolean variable is_sch to $true$ (line 13). In this case, we stop locking for improving schedulability, and continue to optimize utilization with Algorithm 2 (line 14-15). Otherwise, we choose the highest priority task T among the tasks that do not meet their deadline, and try to improve its response time first (line 16). Based on Equation 3, the response time of T is mainly determined by T and the tasks that can preempt T. Thus, we only consider locking memory blocks belong to T or tasks that can preempt T (line 19-20). For such a memory block m belongs to T or tasks that can preempt T, if it is not locked and there is free space in the corresponding cache set, we carry out its cost and benefit analysis (line 23-24). After that, we perform utilization gain analysis by locking m as we did in Algorithm 2 (line 25-26). Apart from the utilization gain, we also compute the response time gain by locking m (line 27). We compare the response time gain between m and $mblk$ that currently has the most response time gain. If m has higher response time gain than $mblk$ and the utilization gain of m is not negative, we update $mblk$ with m (line 28-30). We continue to do this until all candidate memory blocks are considered. If there is no suitable memory block to lock, we stop locking (line 51). Otherwise, we select the memory block $mblk$ with the maximum response gain on T to lock (line 35). Then, we recompute the new utilization as we do in Algorithm 2, as well as the new response time (line 36-44). If utilization of the task set does not become worse and there is response time improvement on T, we add $mblk$ to the set of locked memory blocks and continue to check the

schedulability for the task set (line 46) Otherwise, we stop locking (line 51).

Algorithm 2: Utilization Optimization for EDF and RMS

Input: Task set $\mathscr{T} = \{T_1, T_2 \ldots T_N\}$ and cache configuration $config$
Output: Set of locked memory blocks $lock\_set$ and utilization after locking $util$

1 **begin**
2       $stop\_locking := false$; $lock\_set := null$;
3   **foreach** $T_i \in \mathscr{T}$ **do**
4   abstract_cache_states_analysis(T_i, $config$);
5   wcet_analysis();
6   rcs_lcs_analysis(T_i, $config$);
7   ucb_ecb_computation();
8   resilience_computation();
9   crpd_analysis($\mathscr{T}$);
10      $util = $ utilization_computation($\mathscr{T}$);
11   **while** (!$stop\_locking$) **do**
12          $mblk := null$; $util\_gain_{mblk} := 0$;
13   **foreach** $T_i \in \mathscr{T}$ **do**
14   **foreach** $m \in M_i$ **do**
15   Suppose m is mapped to cache set s;
16   **if** $m \notin lock\_set \wedge !is\_fully\_locked(s)$ **then**
17   cost_benefit_analysis($\mathscr{T}$, $config$, m);
                     foreach $T_i \in \mathscr{T}$ **do**
18                          $util\_gain_m^{T_i} = time\_gain_m^{T_i}/P_i$;
19                      $util\_gain_m = \sum_{T_i \in \mathscr{T}} util\_gain_m^{T_i}$;
20   **if** $util\_gain_m > util\_gain_{mblk}$ **then**
21                        $util\_gain_{mblk} = util\_gain_m$;
22                        $mblk = m$;
23
24
25
26   **if** $mblk \neq null$ **then**
27   lock_to_cache($mblk$);
28   **foreach** $T_i \in \mathscr{T}$ **do**
29   update_abstract_cache_state(T_i, $mblk$, $config$);
30   wcet_analysis();
31   update_rcs_lcs(T_i, $mblk$, $config$);
32   ucb_ecb_computation();
33   resilience_computation();
34   crpd_analysis($\mathscr{T}$);
35              $new_util = $ utilization_computation($\mathscr{T}$);
36   **if** $new\_util < util$ **then**
37                  $util = new\_util$;
38                  $lock\_set := lock\_set \cup \{mblk\}$;
39   **else**
40                  $stop\_locking := true$;
41
42   **else**
43              $stop\_locking := true$;
44
45
46 **end**

Algorithm 3: Schedulability Improvement for RMS

Input: Task set $\mathscr{T} = \{T_1, T_2 ... T_N\}$ and cache configuration $config$

Output: Set of locked memory blocks $lock\_set$ and utilization after locking $util$

1 **begin**
2      $stop\_locking := false$; $lock\_set := null$;
3   **foreach** $T_i \in \mathscr{T}$ **do**
4   abstract_cache_states_analysis(T_i, $config$);
5   wcet_analysis();
6   rcs_lcs_analysis(T_i, $config$);
7   ucb_ecb_computation();
8   resilience_computation();
9   crpd_analysis($\mathscr{T}$);
10      $util$ = utilization_computation($\mathscr{T}$);
11   response_time_computation($\mathscr{T}$);
12   **while** (*!stop_locking*) **do**
13   is_sch = check_schedulability($\mathscr{T}$);
14   **if** $is\_sch == true$ **then**
15   break;
16   Suppose T is the task with highest priority that cannot be scheduled;
17          $rsp\_gain^T_{mblk} := 0$;
18          $mblk := null$;
19   suppose $hp(T)$ is the set of task with higher priority than T;
20   **foreach** $T_i \in T \cup hp(T)$ **do**
21   **foreach** $m \in M_i$ **do**
22   Suppose m is mapped to cache set s;
23   **if** $m \notin lock\_set \wedge !is\_fully\_locked(s)$ **then**
24   cost_benefit_analysis($\mathscr{T}$, $config$, m);
                     foreach $T_i \in \mathscr{T}$ **do**
25                          $util\_gain^{T_i}_m = time\_gain^{T_i}_m / P_i$;
26                      $util\_gain_m = \sum_{T_i \in \mathscr{T}} util\_gain^{T_i}_m$;
27                      $rsp\_gain^T_m$ = response_time_gain();
28   **if** $rsp\_gain^T_m > rsp\_gain^T_{mblk} \wedge util\_gain_m >= 0$ **then**
29                          $rsp\_gain^T_{mblk} = rsp\_gain^T_m$;
30                          $mblk = m$;
31
32
33
34   **if** $mblk \neq null$ **then**
35   lock_to_cache($mblk$);
36   **foreach** $T_i \in \mathscr{T}$ **do**
37   update_abstract_cache_state(T_i, $mblk$, $config$);
38   wcet_analysis();
39   update_rcs_lcs(T_i, $mblk$, $config$);
40   ucb_ecb_computation();
41   resilience_computation();
42   crpd_analysis($\mathscr{T}$);
43   new_util = utilization_computation($\mathscr{T}$);
44   response_time_computation($\mathscr{T}$);
45   **if** $new\_rsp_T > rsp_T \wedge new\_util <= util$ **then**
46                  $lock\_set := lock\_set \cup \{mblk\}$;
47   **else**
48                  $stop\_locking := true$;
49
50   **else**
51              $stop\_locking := true$;
52
53
54 **end**

Precise Timing Analysis for Direct-Mapped Caches

Sidharta Andalam
TUM CREATE, Singapore
sidharta.andalam@tum-create.edu.sg

Roopak Sinha, Partha Roop
University of Auckland, New Zealand
{r.sinha,p.roop}@auckland.ac.nz

Alain Girault
INRIA - Grenoble, France
alain.girault@inria.fr

Jan Reineke
Saarland University- Saarbrücken, Germany
reineke@cs.uni-saarland.de

ABSTRACT

Safety-critical systems require guarantees on their worst-case execution times. This requires modelling of speculative hardware features such as caches that are tailored to improve the average-case performance, while ignoring the worst case, which complicates the Worst Case Execution Time (WCET) analysis problem. Existing approaches that precisely compute WCET suffer from state-space explosion. In this paper, we present a novel cache analysis technique for direct-mapped instruction caches with the same precision as the most precise techniques, while improving analysis time by up to 240 times. This improvement is achieved by analysing individual control points separately, and carrying out optimisations that are not possible with existing techniques.

Categories and Subject Descriptors: B.3.3 [Performance Analysis and Design Aids]: Worst-case analysis

General Terms: Verification, Algorithms

Keywords: Instruction, Direct-Mapped, Cache Analysis.

1. INTRODUCTION

Hard real-time systems require accurate guarantees on the functionality as well as the timing characteristics of programs. Traditional speculative architectural features such as multi-level caches and deep pipelines render the worst-case execution. Two types of memory architectures are used in real-time systems: specialized compiler assisted caches, called *scratchpads* [2], and (widely available) conventional caches [3, 9, 11]. This article focuses on the static analysis [12] for predictable direct-mapped instruction caches [9, 11], where locations in main memory are mapped to unique cache lines.

Cache analysis involves computing the number of cache misses that can happen in the instruction cache at specific control points in a program. The program is usually translated to a control flow graph (CFG) [9], which contains control points as its nodes. Analytically, the cache analysis problem boils down to statically determining all possible cache states (and therefore the number of misses in the worst case) at each node using a suitable fixed point computation.

While a number of cache analysis approaches exist [11, 9], some are not scalable (but more precise) while others overestimate cache misses (but are more scalable). In *concrete* techniques like [9], all possible cache states are enumerated explicitly at each control point, and very precise results can

be obtained. However, these techniques suffer from state-explosion, and do not scale for large programs. In [8], a probabilistic approach for modelling cache behaviour is presented, which is used for design-space exploration to reduce overall analysis time by exploiting the structural similarities among related cache configurations. In [7], the idea of cache conflict graphs is introduced where cache lines are analysed one at a time. This approach is more scalable than concrete approaches, but loses precision as any relations between cache lines are abstracted out. On the other hand, *abstract* techniques like [10, 11] collapse multiple possible cache states at every control point into a single abstract cache state. This abstraction allows the algorithm to reach a fixed point much faster, and hence larger programs can be analysed, but at the cost of sacrificing precision.

In this article, we present a novel cache analysis technique for direct-mapped caches that maintains the same precision as the concrete techniques while significantly improving scalability (for large benchmarks, analysis time is less than 2 minutes, instead of 12 hours as in [9]). This improvement is achieved by analysing individual control points at once, capturing and aggregating instructions in a *relative* manner, and carrying out certain optimisations that are not possible in other techniques. The key contribution of this paper is a new algorithm for static analysis that offers the same precision as the most precise algorithms while being extremely scalable in comparison. It also compares very favourably with abstraction-based analysis techniques with respect to analysis time.

This paper is organised as follows. The cache analysis problem is formalized in Sec. 2. In Sec. 3 we present the proposed approach. Qualitative and quantitative comparison with the concrete and abstract approaches is presented in Sections 4 and 5 respectively. Finally, conclusions are presented in Sec. 6. The Appendix further illustrates key aspects of the proposed algorithm with additional experimental results to demonstrate its advantages. Also, in Appendix E, we discuss how the proposed approach for direct-mapped caches can be extended to set-associative caches, which are also widely used in predictable systems [3].

2. THE CACHE ANALYSIS PROBLEM

Our analytical *cache model* is based on the model of [9], and is defined below. To improve readability, we represent the terms "basic blocks" and "memory blocks" as "blocks" and "instructions", respectively.

DEFINITION 1 (CACHE MODEL). *The cache model for a given program is defined as a tuple $CM = \langle I, C, CI, G, BI \rangle$, where I is a finite set of instructions in the program, $C = \{c_0, c_1, \ldots, c_{N-1}\}$ is an ordered set of cache lines with $N = |C|$ as the total number of cache lines, and $CI : C \to 2^I$ is the direct-mapping function.*

Also, G is a directed graph $G = \langle B, b_{init}, E \rangle$ where B is a finite set of basic blocks with $b_{init} \in B$ as the initial block, and $E \subseteq B \times B$ is the set of edges.

Finally, $BI : B \to (I \cup \{\mp\})^N$ is the block to instruction mapping function, where $\mp$ represents no instruction.

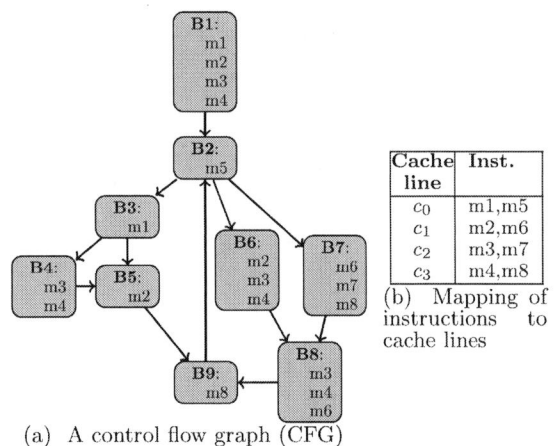

(a) A control flow graph (CFG)

Cache line	Inst.
c_0	m1,m5
c_1	m2,m6
c_2	m3,m7
c_3	m4,m8

(b) Mapping of instructions to cache lines

Figure 1: An example cache model

Figure 1 shows the cache model of a sample program. It presents a CFG in Figure 1(a), which contains 9 blocks ($B = \{B1, \ldots, B9\}$) with $B1$ as the initial block. Each block executes one or more instructions. The CFG has 8 instructions ($I = \{m1, \ldots, m8\}$) that are mapped to 4 cache lines ($C = \{c_0, c_1, c_2, c_3\}$), as presented in Figure 1(b). Each cache line has two unique instructions mapped to it (e.g., $CI(c_0) = \{m1, m5\}$). An instruction can be mapped only to a unique cache line, representing a direct-mapped cache. A cache line, at any time, can contain only one of the instructions that are mapped to it (or is empty).

The edges describe program flow (sequential and branching) between the control points of the given program. We use the short-hand $B1 \to B2$ to describe a CFG edges such as $(B1, B2) \in E$.. Each block also contains instructions (from I) that it executes using the mapping function BI. For example, $B1$ contains four instructions $m1$, $m2$, $m3$ and $m4$. We restrict the model such that blocks can contain at most one instruction mapped to any cache line (as per CI), which allows us to represent the block to instruction mapping using a vector indexed by C. E.g., $BI(B1) = [m1, m2, m3, m4]$, with $BI(B1)[0] = m1$ describing the instruction mapped to cache line c_0. If a block b does not contain an instruction mapped to a cache line c_i, then $BI(b)[i] = \mp$ (e.g., $BI(B2)[1] = \mp$).

The contents of the cache at any time during the program execution is called a *cache state*, and is represented as a vector $cs = [inst_0, \ldots, inst_{N-1}]$ where each $inst_i$ represents the instruction contained in cache line $c_i \in C$. When execution begins, we assume that there is no instruction (represented by $\top$) in each cache line. The cache is assumed to be empty. For the example presented of Fig. 1, the empty cache state is represented by $cs_\top = [\top, \top, \top, \top]$. During execution, or traversal of the CFG, instructions are loaded into the cache as the basic blocks are executed (starting from the initial block). E.g., after executing the instructions in $B1$ the cache state is $cs_1 = [m1, m2, m3, m4]$. In this example, $cs_\top$ is a *reaching cache state* of $B1$, while cs_1 is a *leaving cache state* of $B1$. Since all 4 instructions needed by $B1$ were not present in $cs_\top$, we say that there were 4 cache *misses*.

Given any reaching cache state cs of a block b, we can compute the number of cache misses by comparing the instructions in cs and $BI(b)$. E.g., given a reaching cache state $cs = [m1, m2, m3, m4]$ of block $B2$ (with $BI(B2) = [m5, \mp, \mp, \mp]$), there is a single miss on cache line c_0 because the instruction $m5$ needed by the block (as per $BI(b)[0] = m5$) is not present in the cache state ($cs[0] = m1$).

It is generally possible for a block to have multiple reaching cache state. Here, we compute the *worst-case* miss count wmc_b as the maximum number of cache misses possible, as defined below.

DEFINITION 2 (THE CACHE ANALYSIS PROBLEM).
Given a cache model CM, the cache analysis problem is the computation of the number of worst case cache misses (wmc_b), for all basic blocks $b \in B$, along all possible executions.

3. PROPOSED APPROACH

Our approach for the static analysis of direct-mapped caches is based on the intuition that analysing a single basic block b_{ref} of the CFG at a time allows us to (a) reduce the number of blocks needed to compute the worst and best case miss counts for b_{ref}, and (more importantly), (b) we can abstract cache states computed during the fixed point algorithm w.r.t. the instructions executed by b_{ref}. This may significantly reduce the number of possible cache states and consequently reduce analysis time. We use Fig. 1 to illustrate these benefits.

First principle: During the analysis of block $b_{ref} = B8$, since $B8$ does not execute any instruction on cache line c_0 ($BI(B8)[0] = \mp$), we can ignore block $B2$ as the execution of $B2$ can only affect the cache line c_0. We call $B2$ a *vacuous* block because it does not interfere with any of the cache lines used by $B8$, and it can be removed when analysing $B8$.

Second principle: During the analysis of block $b_{ref} = B8$, the instructions contained in another block, say $B1$, can be abstracted such that they only refer to their *effect* on the analysis of $B8$. Given that $BI(B8) = [\mp, m6, m3, m4]$ and $BI(B1) = [m1, m2, m3, m4]$, the instructions in $B1$ can be abstracted as the vector $[\times, 1, 0, 0]$. Here, the first element '$\times$' means that the instruction is *not of interest* as the reference block does not use this cache line ($BI(B8)[0] = \mp$). The second element '1' means that for cache line c_1, the instruction in $B1$ is *different* from the instruction in $B8$ ($BI(B1)[1] = m2 \neq m6 = BI(B8)[1]$). Finally, for the third and the fourth elements '0' means that for cache line c_2, the instruction in $B1$ is the *same* as the instruction in $B8$ ($BI(B8)[2] = m3 = BI(B1)[2]$). Also, when there is *no instruction* on cache line c_i in a block the abstract representation is '$\mp$'. The ability of reducing the number of instructions ($|I|$) in the CFG to just four *relative instructions* ($\times, \mp, 0, 1$) significantly reduces the memory foot print and analysis time, without sacrificing precision. This is a key optimization that enables us to propose a scalable analysis technique without sacrificing precision.

Alg. 1 presents an overview of our approach. Each block in the CFG is analysed individually (described using the for-loop on lines 1-5). For each reference block b_{ref} in B, on line 2, we first reduce the CFG by removing the vacuous blocks and then compute the relative instructions w.r.t. b_{ref}. Next, on line 3, a fixed point algorithm is used to compute all possible cache states of the reference block. Finally, on line 4, the number of cache misses in the worst case are computed. We now present the details of each of these steps.

Algorithm 1 Overview of the proposed approach

Input: A cache model $CM = \langle I, C, CI, G, BI \rangle$.
Output: Compute the worst miss count (wmc) for all basic blocks.

1: **for each** $b_{ref} \in B$ **do**
2:    {Reduced CFG, compute relative instructions (Sec. 3.1)}
    $(G^r, BI^r) = Reduce(CM, b_{ref})$
3:    {Compute reaching relative cache states (Section 3.2.3)}
    $RCS^r_{b_{ref}} = FP(CM, G^r, BI^r)$
4:    {Compute cache misses in the worst-case (Section 3.3)}
    $wmc_{b_{ref}} = MAXmc(RCS^r_{b_{ref}})$
5: **end for**
6: **return** wmc_b for all blocks b in B {Solution for the cache analysis problem (Definition 2)}

3.1 CFG Reduction

Alg. 2 presents the pseudocode of the *Reduce* algorithm that returns a reduced graph G^r from a given CFG G. Given G and a reference block b_{ref}, we first represent the instructions in each block w.r.t. the reference block, and then remove any vacuous blocks. Line 1 initializes G^r as a copy of G. Then, for each block b in G^r, and each cache line c_i, a relative block to instructions mapping BI^r is created (lines 2–7). Depending on whether the instruction originally contained in b (in G) is of no-interest to the reference block (line 3), different (line 4) or identical (line 5) to the instruction contained in the reference blocks, or is equal to $\mp$ (line 6).

Algorithm 2 *Reduce*: Reduce the CFG and abstract inst.

Input: Cache model $CM = \langle I, C, CI, G, BI \rangle$ and a reference block $b_{ref} \in B$.
Output: $G^r = \langle B^r, b_{init}, E^r \rangle$ and $BI^r : B^r \rightarrow (\{\times, \mp, 1, 0\})^N$
1: Initialize G^r as a copy of G with $BI^r(b) = \emptyset$ for all $b \in B^r$.
2: **for each** $b \in B^r$ and for each $c_i \in C$ **do**
3:    $BI^r(b)[i] = \times$, if $(BI(b_{ref})[i] = \mp)${not of interest}
4:    $BI^r(b)[i] = 1$, if $(BI(b_{ref})[i] \neq BI(b)[i])$ {different}
5:    $BI^r(b)[i] = 0$, if $(BI(b_{ref})[i] = BI(b)[i])$ {identical}
6:    $BI^r(b)[i] = \mp$, if $(BI^r(b)[i] \neq \times \land BI(b)[i] = \mp)${no inst.}
7: **end for**
8: **for each** $b \in B^r$ **do**
9:    **if** $(BI(b) \in (\{\times, \mp\})^N) \land (b \neq b_{init})$ {check for all vacuous blocks, excluding initial block} **then**
10:      Remove b from B^r, and adjust E^r
11:    **end if**
12: **end for**

Next, on lines 8–11, blocks for which the mapping function BI^r returns $\times$ or $\mp$ for every element of the vector $BI^r(b)$ are declared as vacuous and are removed from the graph. The removal of a vacuous block b_r involves adjusting the edges E^r of the graph G^r such that each predecessor of b_r now has a direct edge to each of the successors of b_r. More details and the full *Reduce* Algorithm appear in Appendix A.

Fig. 2 shows the reduced CFG returned by the algorithm *Reduce* when $b_{ref} = B8$. Note that the vacuous block $B2$ is removed from G^r. Also, every block now contains the relative instructions.

3.2 Fixed point Computation

3.2.1 Relative cache states

Since block to instructions mapping is described using relative instructions, we also compute cache states in a relative fashion. A *relative cache state* is defined as follows.

DEFINITION 3 (RELATIVE CACHE STATE). *A relative cache state cs^r is a vector $[inst^r_0, \ldots, inst^r_{N-1}]$, where each element $inst^r_i \in \{1, 0, \top, \bot, \times\}$. The set of all possible relative cache states is denoted as CS^r.*

Each relative instruction $inst^r_i$ of a relative cache state $cs^r = [inst^r_0, \ldots, inst^r_{N-1}]$ is described w.r.t. the instruction ($BI(b_{ref})$

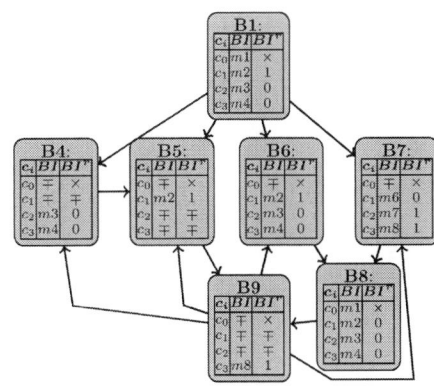

Figure 2: The reduced graph G^r obtained from the *Reduce* algorithm, with the reference block $b_{ref} = B8$.

$[i]$, $i \in [0, |C|])$ in the reference block b_{ref} for the cache line $c_i \in C$. $inst^r_i = 1$ or $inst^r_i = 0$ means the instruction in the cache is *different* or *identical* respectively to the instruction executed in the reference block b_{ref}. E.g., given $BI(b_{ref})[i] = m1$, $inst^r_i = 1$ or $inst^r_i = 0$ means the instruction on cache c_i is not $m1$ or $m1$ respectively. $inst^r_i = \top$ means that cache line c_i is *empty*, whereas $inst^r_i = \bot$ means that the cache has an *unknown* instruction. Finally, $inst^r_i = \times$ means that the instruction on this cache line is *not of interest* during the analysis of b_{ref}. Also, a relative cache state before executing block b is know as a *reaching relative cache state* of b. Similarly, a relative cache state after executing block b is know as a *leaving relative cache state* of b. More details about relative cache states are presented in Appendix B.

3.2.2 The Transfer Function

An important operation in the fixed point computation is the transformation of a reaching relative cache state into a leaving relative cache state. This transformation, called the *transfer* function $T : CS^r \times B \rightarrow CS^r$, is illustrated as follows.

$$
\begin{array}{ll}
cs^r_1 & = [\times, 1, 0, 1] \\
BI^r(b) & = [\times, 0, \mp, 1] \\
\hline
T(cs^r_1, b) = cs^r_2 & = [\times, 0, 0, 1]
\end{array}
$$

For any cache line c_i, the instruction $cs^r_2[i]$ in the leaving relative cache state is equal to $BI^r(b)[i]$ only if there is an instruction in block b ($BI^r(b)[i] = 1$ or $BI^r(b)[i] = 0$). Otherwise, the instruction $cs^r_2[i]$ is the same as the instruction $cs^r_1[i]$ in the reaching cache state. For the above example, given $BI^r(b) = [\times, 0, \mp, 1]$ and $cs^r_1 = [\times, 1, 0, 1]$, after execution of block b, $cs^r_2 = [\times, 0, 0, 1]$. For cache lines c_1 and c_3, block b executes instructions that relate to the reference block, and for cache lines c_0 and c_2, its does not execute any relevant instructions. Therefore, the contents of the leaving cache states are updated to be the same as the instructions executed by the block on cache lines c_1 and c_3, and remain the same as the reaching cache state for cache lines c_0 and c_2.

3.2.3 Fixed point computation

Alg. 3 shows the fixed point algorithm FP used to compute all possible reaching relative cache states for the referenced block in the reduced graph G^r. We illustrate the fixed point computation using Tab. 1, which shows the possible reaching and leaving relative cache states of each block in the reduced graph (Fig. 2) in every iteration of FP.

During initialization, the reaching cache state of the initial block is set to $cs^r_\top$ (line 2) because the cache is considered empty initially. For every other block, on line 2, the initial reaching cache state is unknown ($cs_\perp^r$). As shown in Tab. 1, The initial reaching relative cache state of the initial block $B1$ is set to $cs_\top^r = \{[\times, \top, \top, \top]\}$, while for every other block, the initial reaching relative cache state is $cs_\perp^r = \{[\times, \perp, \perp, \perp]\}$ (iteration 1, column 3, Tab. 1).

Algorithm 3 FP: Fixed point computation

Input: A cache model $CM = \langle I, C, CI, G, BI \rangle$, reduced graph $G^r = \langle B^r, b_{init}, E^r \rangle$ and $BI^r : B^r \to (\{\times, \mp, 1, 0\})^N$.
Output: Reaching relative cache states of block b_{ref} ($RCS_{b_{ref}}^r$).

1: Create initial cache state $cs_\top^r$ where for every $c_i \in C$, $cs_\top^r[i] = \times$ if $BI^r(b_{ref}) = \times$, or $cs_\top^r[i] = \top$ otherwise.
2: $RCS_{b_{init}}^{r^1} = \{cs_\top^r\}$, and $RCS_b^{r^1} = \{cs_\perp^r\}$ for all other $b \in B^r$.
3: $i = 1$
4: **repeat**
5:    **for each** $b \in B^r$ **do**
6:       $LCS_b^{r^i} = T(RCS_b^{r^i}, b)$
7:    **end for**
8:    $i = i + 1$; {Next iteration}
9:    **for each** $b \in B^r$ **do**
10:      **if** $b = b_{init}$ **then**
11:         $RCS_{b_{init}}^{r^i} = \{cs_\top^r\} \cup (\bigcup LCS_{b'}^{r^i} \mid (b', b) \in E^r)$
12:      **else**
13:         $RCS_b^{r^i} = \bigcup LCS_{b'}^{r^i} \mid (b', b) \in E^r$
14:      **end if**
15:    **end for**
16: **until** $\forall b \in B^r$, $RCS_b^{r^i} = RCS_b^{r^{i-1}}$ {Termination condition}
17: **return** $RCS_{b_{ref}}^r$

Table 1: Computing all possible reaching relative cache states of the reference block b_{ref} (B8).

Itr. (i)	Block (b)	Reaching Relative Cache States (RCS_b^i)	Leaving Relative Cache States (LCS_b^i)
1	B1	$\{[\times, \top, \top, \top]\}$	$\{[\times, 1, 0, 0]\}$
	B4	$\{[\times, \perp, \perp, \perp]\}$	$\{[\times, \perp, 0, 0]\}$
	B5	$\{[\times, \perp, \perp, \perp]\}$	$\{[\times, 1, \perp, \perp]\}$
	B6	$\{[\times, \perp, \perp, \perp]\}$	$\{[\times, 1, 0, 0]\}$
	B7	$\{[\times, \perp, \perp, \perp]\}$	$\{[\times, 0, 1, 1]\}$
	B8	$\{[\times, \perp, \perp, \perp]\}$	$\{[\times, 0, 0, 0]\}$
	B9	$\{[\times, \perp, \perp, \perp]\}$	$\{[\times, \perp, \perp, 1]\}$
2	B1	$\{[\times, \top, \top, \top]\}$	$\{[\times, 1, 0, 0]\}$
	B4	$\{[\times, 1, 0, 0], [\times, \perp, \perp, 1]\}$	$\{[\times, 1, 0, 0], [\times, \perp, 0, 0]\}$
	B5	$\{[\times, 1, 0, 0], [\times, \perp, \perp, 1]\}$	$\{[\times, 1, 0, 0], [\times, 1, \perp, 1]\}$
	B6	$\{[\times, 1, 0, 0], [\times, \perp, \perp, 1]\}$	$\{[\times, 1, 0, 0]\}$
	B7	$\{[\times, 1, 0, 0], [\times, \perp, \perp, 0]\}$	$\{[\times, 0, 1, 1]\}$
	B8	$\{[\times, 1, 0, 0], [\times, 0, 1, 1]\}$	$\{[\times, 0, 0, 0]\}$
	B9	$\{[\times, 1, \perp, 1], [\times, 0, 0, 0]\}$	$\{[\times, 1, \perp, 1], [\times, 0, 0, 1]\}$
3	...	...	...
4	...	...	...
5	B1	$\{[\times, \top, \top, \top]\}$	
	B4	$\{[\times, 1, 0, 0], [\times, 1, 0, 1],$ $[\times, 0, 0, 1]\}$	
	B5	$\{[\times, 1, 0, 0], [\times, 1, 0, 1],$ $[\times, 0, 0, 1], [\times, 0, 0, 0]\}$	
	B6	$\{[\times, 1, 0, 0], [\times, 1, 0, 1],$ $[\times, 0, 0, 1]\}$	
	B7	$\{[\times, 1, 0, 0], [\times, 1, 0, 1],$ $[\times, 0, 0, 1]\}$	
	B8	$\{[\times, 1, 0, 0], [\times, 0, 1, 1]\}$	
	B9	$\{[\times, 1, 0, 0], [\times, 0, 0, 0],$ $[\times, 1, 0, 1]\}$	

The repeat-until loop (lines 4–16) is the fixed point iteration. In each iteration i, each reaching relative cache state contained in the set $RCS_b^{r^i}$ for every block b is transformed into a leaving relative cache state (in set $LCS_b^{r^i}$) by applying the transfer function (lines 5–7). E.g., for block $B1$ in iteration 1, given the only reaching relative cache state $[\times, \top, \top, \top]$, the corresponding leaving relative cache state is $T([\times, \top, \top, \top], B1) = \{[\times, 1, 0, 0]\}$ (see Tab. 1).

Then we compute, the reaching relative cache states of each block for the next iteration. For each block, the set of reaching relative cache states is the union of the sets of the leaving relative cache states of all of its predecessors (line 13). For b_{init}, the additional reaching cache state $cs_\top^r$ is also added to this set (line 11). E.g., the predecessors of $B4$ are $B1$ and $B9$ (see Fig. 2). Hence, their sets of leaving cache states (resp. $\{[\times, 1, 0, 0]\}$ and $\{[\times, \perp, \perp, 1]\}$) for iteration 1 are aggregated

together to form the reaching cache state of $B4$ in iteration 2.

The iterations continue until the fixed point is reached, i.e., when two consecutive iterations yield the same sets of reaching relative cache states for all blocks (line 35). For the reduced CFG shown in Fig. 2, the fixed point is reached in the 5th iteration. Also during the fixed point, as an optimisation, a relative cache state cs_k^r is dropped if there exists cs_j^r such that, if for all cache lines (c_i), when the relative instruction in cs_j^r is the same as the instruction in cs_k^r ($cs_j^r[i] = cs_k^r[i]$) or, the relative instruction in cs_j^r captures a cache miss ($cs_j^r[i] = 1$). E.g., given four possible reaching cache states $\{[0, 0, 0, 1], [0, 1, 0, 1], [0, 1, 1, 1], [1, 0, 0, 1]\}$, we can safely ignore the first two cache states, because the third state captures the worst case behaviour. However, we must still carry the last state, resulting in the reduced set $\{[0, 1, 1, 1], [1, 0, 0, 1]\}$.

3.3 Computing the number of cache misses

The final step in the cache analysis algorithm is the computation of the number of cache misses in the worst case. This is done by analysing the relative cache states of the reference block b_{ref} as computed by the fixed point algorithm. For each reaching cache state $cs^r = [inst_0^r, \ldots, inst_{N-1}^r]$, and for every cache line $c_i \in C$, $inst_i = 1$ represents a cache miss on c_i. The total number of misses when the reaching cache state cs^r is the number of 1's contained in cs^r. The reaching cache states with the highest number of 1's correspond to the worst-case miss counts of b_{ref} respectively. E.g., as shown in Tab. 1, the two reaching cache states of the reference block $B8$ as computed by the fixed point algorithm FB are $\{[\times, 1, 0, 0]$ and $[\times, 0, 1, 1]\}$. The first reaching cache state has one occurrence of '1' while the second one has two occurrences of '1'. Thus, the maximum miss count for $B8$ is 2, i.e., $wmc_{B8} = 2$.

4. QUALITATIVE COMPARISON

Tab. 2 provides a qualitative comparison between the concrete [9], abstract [4], and the proposed approaches. Figure 3 illustrates the basic block $B8$ of the CFG shown in Fig. 2 with its two predecessors $B6$ and $B7$. We use this example to show the differences in the ways the three approaches represent and aggregate cache states.

Table 2: Qualitative comparison of the three approaches.

App.	Fixed point	Precision	Time	Optimisation	Max no. of cs at each program point
Conc.	all blocks	high	slow	none	$(I/N)^N$
Abs.	all blocks	low	fast	merge cache states	constant
Pro.	one block at a time	high	med.	(1) reduce graph (2) reduce cache lines (3) relative instructions	$(3)^N$

In the **concrete** approach [9], a cache state (cs) is described as a vector $[inst_0, \ldots, inst_{N-1}]$ where each $inst_i$ represents a single instruction contained in cache line $c_i \in C$ or is empty ($\top$), i.e., $inst_i \in CI(c_i) \cup \{\top\}$. E.g., the sets of leaving cache states for $B6$ and $B7$ are $\{m5, m2, m3, m4\}$ and $\{m5, m6, m7, m8\}$ respectively (see Fig. 3(a)). The set of reaching cache states of $B8$ is the *union* of these sets, as shown in Fig. 3(a). This set represents the fact there are only two states in which the cache can be before $B8$ is executed.

In the **abstract** approach [4], a cache state is described as a vector $[set_0, \ldots, set_{N-1}]$ where each set_i represents a set of instructions that must be contained in cache line $c_i \in C$. That is, $set_i \subseteq CI(c_i) \cup \{\top\}$.. E.g., the abstract leaving cache states for $B6$ and $B7$ are $[\{m5\}, \{m2\}, \{m3\}, \{m4\}]$ and $[\{m5\}, \{m6\}, \{m7\}, \{m8\}]$ respectively (see Fig. 3(b)). For

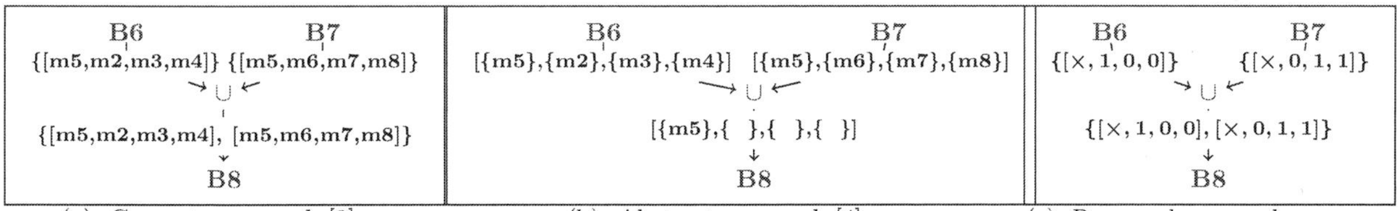

B6	B7
{[m5,m2,m3,m4]}	{[m5,m6,m7,m8]}

$$\searrow \cup \swarrow$$

{[m5,m2,m3,m4], [m5,m6,m7,m8]}

B8

(a) Concrete approach [9]

B6	B7
[{m5},{m2},{m3},{m4}]	[{m5},{m6},{m7},{m8}]

$$\searrow \cup \swarrow$$

[{m5},{ },{ },{ }]

B8

(b) Abstract approach [4]

B6	B7
$\{[\times, 1, 0, 0]\}$	$\{[\times, 0, 1, 1]\}$

$$\searrow \cup \swarrow$$

$\{[\times, 1, 0, 0], [\times, 0, 1, 1]\}$

B8

(c) Proposed approach

Figure 3: Illustration of the combining cache states.

each of these abstract cache states, each cache line contains precisely one instruction. Hence, these two abstract cache states are equivalent to the concrete cache states $[m5, m2, m3, m4]$ and $[m5, m6, m7, m8]$ shown in Fig. 3(a). Next, the abstract reaching cache state of $B8$ is the *pair-wise intersection* over the vector elements of the above abstract cache states, giving $[\{m5\}, \{ \}, \{ \}, \{ \}]$ (see Fig. 3(b)). A cache line c_i, with an empty set represents *any instruction* (that is mapped to c_i, $CI(c_i)$) or *no instruction*. E.g., the empty set on cache line c_1, represents three possible states $\{m2, m6, \top\}$. Similarly, $[\{m5\}, \{ \}, \{ \}, \{ \}]$ represents $[\{m5\}, \{m2, m6, \top\}, \{m3, m7, \top\}, \{m4, m8, \top\}]$. This abstract reaching cache state represents 27 possible concrete cache states, which are computed using cross product as:

$$\{m5\} \times \{m2, m6, \top\} \times \{m3, m7, \top\} \times \{m4, m8, \top\}$$
$$=\{[m5, m2, m3, m4], [m5, m2, m3, m8], [m5, m2, m7, m4],$$
$$[m5, m2, m7, m8], ...\}$$

While two of these possible concrete states $[m5, m2, m3, m4]$ and $[m5, m6, m7, m8]$ are valid (as discussed above), many others, such as $[m5, m6, m3, m4]$, are not reachable because instructions $m6$ and $m3$ cannot be loaded into the cache by any one of the predecessor blocks of $B8$. As the above example shows, the **abstract** approach may lose precision due to joins, which may introduce non-reachable cache states. Yet, the fixed point algorithm converges faster for the same reason.

Finally, in the **proposed approach**, cache states are relative to the instructions of the reference block. As shown in Fig. 3(c), the reaching relative cache states of $B8$ are computed by taking a *union* of the predecessor's leaving relative cache states, $\{[\times, 1, 0, 0]\}, \{[\times, 0, 1, 1]\}\}$, as shown in Fig. 3(c). The relative cache state representation allows us to optimize performance without sacrificing precision. E.g., as shown in Fig. 3(c), one of the reaching relative cache state of $B8$ is $\{[\times, 1, 0, 0]\}$. It represents 6 possible concrete cache states, which are computed using cross product as:

$$\{[\times, 1, 0, 0]\} = \{\top, m1, m5\} \times \{\top, m2\} \times \{m3\} \times \{m4\}$$
$$= \{[\top, \top, m3, m4], [\top, m2, m3, m4], [m1, \top, m3, m4],$$
$$[m1, m2, m3, m4], [m5, \top, m3, m4], [m5, m2, m3, m4]\}$$

Thus, the translation results in 6 concrete cache states. However, the extra states, unlike in the abstract approach, do not effect the precision of the analysis.

Table 3: Comparing the cache states and WCET estimates between the three approaches as we analyse block $B8$.

App.	cache state	wmc_{B8} (Worst)
Concrete	$\{[m5, m2, m3, m4], [m5, m6, m3, m4]\}$	2
Abstract	$[\{m5\}, \{m2, m6\}, \{m3, m7\}, \{m4, m8\}]$	3
Proposed	$\{[\times 100], [\times 011]\}$	2

Comparison of the precision: By analysing the reaching cache states of $B8$, we can compute wmc_{B8} and compare the precision among the three approaches. We present this comparison using Table 3. For block $B8$ ($BI(B8) = [\mp, m6, m3, m4]$), given the set of possible reaching cache states for each approach (in column 2), the WCET estimate (wmc) is presented in column 3. We note that the **concrete** approach has higher precision (smaller WCET), compared the

abstract approach. Also, the **proposed** approach maintains this high precision. The optimised use of relative cache states is only possible in the **proposed** approach because it analyses one block at once. Both the concrete and abstract approaches analyse all blocks in the CFG together.

Comparison of the complexity: The time complexity of each of the three approaches depends on the number of blocks in the CFG, the complexity of the merging and equivalence operations, and the maximum number of cache states that can be created (or the height of the lattice). The number of blocks of the CFG is fixed at $|B|$, while we assume that the merging and equivalence operations between cache states can be performed in $O(N)$ times (N is the number of cache lines). Therefore, the complexities are of the form $O(|B| \times N \times HeightOfLattice)$. To compute the height of the lattice, we assume that each cache line has an equal share $\lceil |I|/N \rceil$ (I is the set of instructions) of instructions mapped to it, and that $|I|$ and N are sufficiently large so that we can ignore the presence of $\bot$ and $\top$ in the computed cache states.

For the **concrete** approach, there are $(\lceil |I|/N \rceil)^N$ possible reaching cache states. Hence the complexity of the approach is $O(|B| \times N \times (\lceil |I|/N \rceil)^N)$. For the **abstract** approach, because of the point-wise intersection operation used to merge cache states for every cache line, the height of the lattice is constant at 1 for a single cache line, or N for all cache lines. The complexity of this approach is therefore $O(|B| \times N^2)$. Finally, for the **proposed** approach, each cache line has precisely 5 relative instructions $\{\times, 0, 1, \bot, \top\}$ mapped to it. However, recall that if a cache line has $\times$ assigned to it, it means the cache line is not used (and hence cannot contain any other instruction during the analysis). Similarly, $\top$ is used only for the initial node. The number of cache states possible (over N cache lines) in our approach is therefore 3^N. The complexity of our approach is therefore $O(|B| \times N \times 3^N)$ for *one* call to the fixed point algorithm. Given that we repeat this process $|B|$ times (once for every block), and that the CFG reduction operation is linear to the size of the CFG (again, $|B|$), the complexity of our approach is therefore $O(|B|^3 \times N \times 3^N)$.

As can be seen above, the complexity of the abstract approach is significantly lower than the concrete approach as well as our approach. Also, the complexity of the proposed approach does not depend on the number of instructions in the program, which helps us achieve much better performance than the concrete approach, as discussed in the next section.

5. BENCHMARKING AND RESULTS

We compare the precision and analysis time of the concrete, abstract, and proposed approaches over a set of benchmarks consisting of five control applications from [6]. In addition, we created two examples: a BubbleSort program and a Synthetic example. The benchmarking results are presented in Tab. 4. The first column presents the examples followed by their description (column 2). The number of C lines of the program and its memory footprint is presented in columns 3 and 4 respectively. The largest examples are CruiseController

1017

and RailRoadCrossing, with more than 4000 lines of code.

Each program is compiled to execute on the MicroBlaze (MB) processor [1]. We choose MB due to the availability of timing analysis tools [5]. The CFG for each example is extracted automatically from its compiled binary, and each loop of the CFG is unrolled once for more precise cache analysis [4]. For MicroBlaze with 64 MB main memory, the size of the cache can be configured from 128 bytes to 64 KB [1]. Using these proportions, in our experiments we explore cache sizes between 0.1% and 1% of the program's size.

Table 4: Benchmark programs and their characteristics.

Example		Description	LOC	Size
BubbleSort	(BS)	Bubble sort algorithm	128	2KB
Synthetic	(SY)	Branching and loops	180	4KB
Flasher	(FL)	Distributed lights	384	9KB
DrillStation	(DS)	Drilling station	1800	62KB
ConvBelt	(CB)	Airport conveyor belt	1280	44KB
RailRoadCros.	(RR)	Rail road cnt.	4613	163KB
CruiseCntroler	(CC)	Cruise control model	4194	146KB

For the first two examples (BS, SY), the WCET estimates from the proposed approach was identical to that of the concrete approach (see Table 5 in Appendix D). However, for the rest of the examples, the concrete approach failed to terminate (analysis time is more than 12 hours, represented using "T.O" in Table 5). Thus, we only focus on comparing between the proposed and the abstract approaches.

For a cache size of 1% of the program size, normalised WCET estimates w.r.t. the results from the abstract approach are presented in Figure 4(a). Across all the benchmarks, we observe that the WCET computed by the proposed approach is always less than or equal to the estimates from the abstract approach. On average, the WCET estimate from the proposed approach is 15% smaller than the abstract approach. For a 0.1% relative cache size, the WCET analysis results are presented in Figure 4(b). Here, the proposed approach does not gain extra precision (w.r.t. the abstract approach), because the cache size is too small.

In terms of analysis time, the proposed approach always takes less than 3 minutes for each example, which is significantly faster than the concrete approach. However, the abstract approach is even much faster (always takes less than 4 seconds) than the proposed approach. For the largest example (RR), the proposed approach takes 142 seconds compared to the 4 seconds taken by the abstract approach, but the WCET estimate is tighter by 19% $(1 - 250725/308505)$.

Finally, Fig. 5 shows the WCET vs analysis time for the control applications (last 5 benchmarks). Since the concrete approach failed to terminate, its WCET is represented by the WCET of the proposed approach (since both yield identical results). On average, compared to the abstract approach, the WCET from the proposed approach is 16% tighter. For the analysis time, the proposed approach always completes in less than 3 minutes, compared to the timeout after 12 hours for the concrete approach.

In Appendix D, we explore eight other cache sizes between 0.2% and 0.9%. For the RailRoadCrossing example, on average across the eight cache sizes, the WCET of proposed approach gives 16 % much tighter results on average, and up to 19 % tighter result than the abstract approach.

6. CONCLUSIONS

We proposed a new cache analysis approach for precise analysis of instructions in direct-mapped caches. The proposed approach presents a new abstraction, and compared to the concrete approach it significantly reduces the analysis time without sacrificing the precision. This is unlike the concrete

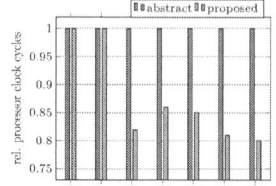 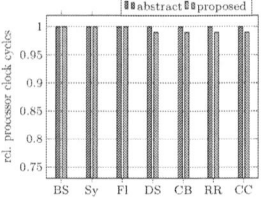

(a) 1% relative cache size   (b) 0.1% relative cache size

Figure 4: Comparing the WCET estimates (the smaller the better) of the abstract and proposed approaches

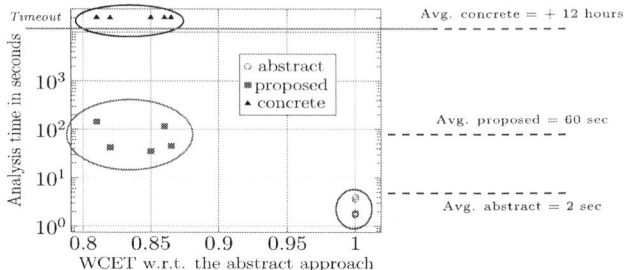

Figure 5: Comparing WCET and analysis time for the last five examples (1% relative cache size)

or the abstract approaches, where scalability or precision must be sacrificed. Overall, the proposed approach enables precise and efficient WCET analysis even for larger programs. In the future, we will be extending our approach for analysing set associative caches and design space exploration of caches.

7. REFERENCES

[1] *MicroBlaze Processor Reference Guide.* www.xilinx.com (Last accessed: 01/10/2012).

[2] D. Bui, E. Lee, I. Liu, H. Patel, and J. Reineke. Temporal isolation on multiprocessing architectures. In *Proc. of DAC'11*, pages 274–279, San Diego, California, June 2011.

[3] H. Ding, Y. Liang, and T. Mitra. WCET-centric partial instruction cache locking. In *Proc. of DAC'12*, pages 412–420, San Francisco, California, June 2012.

[4] C. Ferdinand, F. Martin, R. Wilhelm, and M. Alt. Cache Behavior Prediction by Abstract Interpretation. *Science of Computer Programming*, 35:163–189, November 1999.

[5] M. Kuo, R. Sinha, and P. S. Roop. Efficient WCRT Analysis of Synchronous Programs using Reachability. In *Proc. of DAC'11*, San Diego, USA, June 2011.

[6] M. Kuo, L. H. Yoong, S. Andalam, and P. Roop. Determining the worst-case reaction time of IEC 61499 function blocks. In *Proc. INDIN'10*, pages 1104 –1109, July 2010.

[7] Y.-T. S. Li, S. Malik, and A. Wolfe. Performance estimation of embedded software with instruction cache modeling. *ACM Trans. Des. Autom. Electron. Syst.*, 4(3):257–279, July 1999.

[8] Y. Liang and T. Mitra. Static analysis for fast and accurate design space exploration of caches. In *Proc. of CODES+ISSS'08*, pages 103–108, Atlanta, GA, USA, 2008.

[9] H. S. Negi, T. Mitra, and A. Roychoudhury. Accurate Estimation of Cache-Related Preemption Delay. In *Proc. of CODES+ISSS'03*, pages 201–206, CA, USA, 2003.

[10] J. Reineke, D. Grund, C. Berg, and R. Wilhelm. Timing predictability of cache replacement policies. *Real-Time Systems*, 37(2):99–122, November 2007.

[11] H. Theiling, C. Ferdinand, and R. Wilhelm. Fast and Precise WCET Prediction by Separated Cache and Path Analyses. *Real-Time Systems*, 18:157–179, 1999.

[12] R. Wilhelm et al. The Worst-Case Execution-Time Problem—Overview of Methods and Survey of Tools. *Transactions on Embedded Computing Systems*, 7(3):1–53, 2008.

APPENDIX

A. REDUCING THE CFG AND COMPUTING RELATIVE INSTRUCTIONS

Given a cache model $CM = \langle I, C, CI, G, BI \rangle$ and the reference block $b_{ref} \in B$, the objective of the algorithm is: (1) to compute a new reduced graph $G^r = \langle B^r, b_{init}, E^r \rangle$ which contains only these blocks that are relevant for the analysis of block b_{ref} (described earlier in Example 1) and, (2) to compute the function BI^r which describes the relative instructions executed by the blocks in B^r w.r.t. b_{ref}, referred as the *relative instruction mapping* function (described earlier in Example 2). The algorithm contains the following three steps.

Step 1: Initialise (lines 2 to 4) On line 2, we initialise G^r to have the same content as G, i.e., same set of blocks ($B^r = B$), initial block (b_{init}), edges ($E^r = E$). On lines 3 to 4, we initialise the function BI^r such that it does not contain any relative instructions for any block. For the example CFG (presented in Figure 1(a)), the graph G^r is presented in Figure 6(a).

Algorithm 4 *Reduce*: Reduce the CFG and compute relative instructions

---

Input: Cache model $CM = \langle I, C, CI, G, BI \rangle$ and a reference block $b_{ref} \in B$.
Output: $G^r = \langle B^r, b_{init}, E^r \rangle$ and $BI^r : B^r \to (\{\times, \mp, 1, 0\})^N$
1: {Step 1: Initialise}
2: $B^r = B$, $E^r = E$, $G^r = \langle B^r, b_{init}, E^r \rangle$ {Copy all blocks and edges}
3: **for each** $b_1 \in B^r$ **do**
4:     $BI^r(b_1) = \emptyset$ {Initialise the vector $BI^r(b_1)$}
5: **end for**

6: {Step 2: Relative instruction mapping}
7: **for each** $b_1 \in B^r$ **do**
8:     **for each** $c_i \in C$ **do**
9:         **if** $(BI(b_{ref})[i] = \mp)$ {not of interest} **then**
10:             $BI^r(b_1)[i] = \times$
11:         **end if**
12:         **if** $(BI(b_{ref})[i] \neq BI(b_1)[i])$ {different instruction} **then**
13:             $BI^r(b_1)[i] = 1$
14:         **end if**
15:         **if** $(BI(b_{ref})[i] = BI(b_1)[i])$ {same instruction} **then**
16:             $BI^r(b_1)[i] = 0$
17:         **end if**
18:         **if** $(BI(b_1)[i] = \mp)$ {no instruction} **then**
19:             $BI^r(b_1)[i] = \mp$
20:         **end if**
21:     **end for**
22: **end for**

23: {Step 3: Remove vacuous blocks and update edges}
24: **for each** $b_1 \in B^r$ **do**
25:     **if** $(BI(b_1) \in (\{\times, \mp\})^N) \wedge (b_1 \neq b_{init})$ {check for all vacuous blocks, excluding initial block} **then**
26:         {Compute predecessors and successors of b_1}
27:         $Preds = \{b_2 | b_2 \to b_1 \in E^r\}$
28:         $Succ = \{b_2 | b_1 \to b_2 \in E^r\}$
29:         {Remove incoming and outgoing edges of b_1}
30:         **for each** $b_2 \in Preds$ **do**
31:             $E^r = E^r \setminus \{b_2 \to b_1'\}$
32:         **end for**
33:         **for each** $b_2 \in Succ$ **do**
34:             $E^r = E^r \setminus \{b_2 \to b_1'\}$
35:         **end for**
36:         {Add new edge from each predecessor to each successor }
37:         **for each** $b_p' \in Preds$ **do**
38:             **for each** $b_s' \in Succ$ **do**
39:                 $E^r = E^r \cup \{b_p' \to b_s'\}$
40:             **end for**
41:         **end for**
42:         $B^r = B^r \setminus \{b_1\}$ {Remove block b_1}
43:     **end if**
44: **end for**

Step 2: Relative instruction mapping (lines 7 to 22)

Given a reference block $b_{ref} \in B$ and a cache line c_i, the instruction of any blocks $b_1 \in B^r$ can be expressed as *different* (1) when $BI(b_1)[i] \neq BI(b_{ref})[i]$ (checked on line 12), or *same* (0) when $BI(b_1)[i] = BI(b_{ref})[i]$ (checked on line 15), or *no instruction* ($\mp$) when there is no instruction in b_1 on cache line c_i, $BI(b_1)[i] = \mp$ (checked on line 18), or *not of interest* ($\times$) when there is no instruction in b_{ref} on cache line c_i, $BI(b_{ref})[i] = \mp$ (checked on line 9). This relation is captured using the relative instruction mapping $BI^r(b_1)$. For illustration, refer to Example 2 in Section 3.

For the graph G^r in Figure 6(a), the relative mapping (w.r.t. B8) for every block is shown in Figure 6(b). Note that for all blocks b in B^r, if the reference block (b_{ref}) does not have an instruction on cache line c_i ($BI(b_{ref})[i] = \mp$), then for cache line c_i, the relative instruction for all blocks is $\times$ ($BI^r(b)[i] = \times$). This shows that during the analysis of b_{ref}, the cache line c_i is not considered. For example, in Figure 6(b), the reference block B8 does not have an instruction on cache line c_0 ($BI(B8)[0] = \mp$). Thus, for cache line c_0, the relative instruction for all blocks is $\times$ ($BI^r(b)[i] = \times$).

Step 3: Remove vacuous blocks and update edges (lines 24 to 12). As described in Example 1, we can remove vacuous blocks which do not affect the precision of the reference block b_{ref}. We define a block $b_1 \in B^r$ to be vacuous, if $BI^r(b_1) \in (\{\times, \mp\})^N$. This check is done on line 25. It is possible for the initial block (b_{init}) to be vacuous. However, if the initial block has more than one successors, removing the initial block may result in multiple initial blocks. Thus, to simplify the analysis, we do not remove the initial block even when it is vacuous (line 25). If a block b_1 is vacuous, we first compute the predecessors (line 27) and the successors (line 28) of b_1. Secondly, we remove the incoming edges to b_1 from each predecessor block (on lines 30 to 32) and, we remove the outgoing edges from b_1 to each successor block (on lines 33 to 35). Thirdly, we create a transition from each predecessor to each successor of b_1. This is achieved using the nested loop on lines 37 to 41. Finally, on line 42, the vacuous block b_1 is removed. For the graph shown in Figure 6(b), blocks B2 and B3 do not contain any relative instructions ($BI^r(B2) = [\times, \mp, \mp, \mp]$ and $BI^r(B3) == [\times, \mp, \mp, \mp]$, thus, they are removed for the graph and the updated graph G^r is presented in Figure 6(c). This is the *reduced graph* G^r (w.r.t. to $b_{ref} = B8$).

B. RELATIVE CACHE STATES

We describe the contents of a cache using the notion of *relative cache states*. It is described as a vector $[inst_0^r, inst_1^r, \ldots, inst_{N-1}^r]$, where each element $inst_i^r$ is described w.r.t. the instruction ($BI(b_{ref})[i]$) in the block b_{ref}.

DEFINITION 4 (RELATIVE CACHE STATE). *Given a reference block b_{ref}, a relative cache state $cs^r \in (\{1, 0, \top, \bot, \times\})^N$, is a vector $[inst_0^r, inst_1^r, \ldots, inst_{N-1}^r]$, where each element $inst_i^r \in \{1, 0, \top, \bot, \times\}$. Also, the set of all possible relative cache states (w.r.t b_{ref}) is denoted as CS^r.*

Before we illustrate relative cache states, we introduce two key terms essential to cache analysis. For any basic block b, the *reaching relative cache states* represent the set of relative cache states prior to the execution of a basic block and the *relative leaving cache states* represent the set of cache states

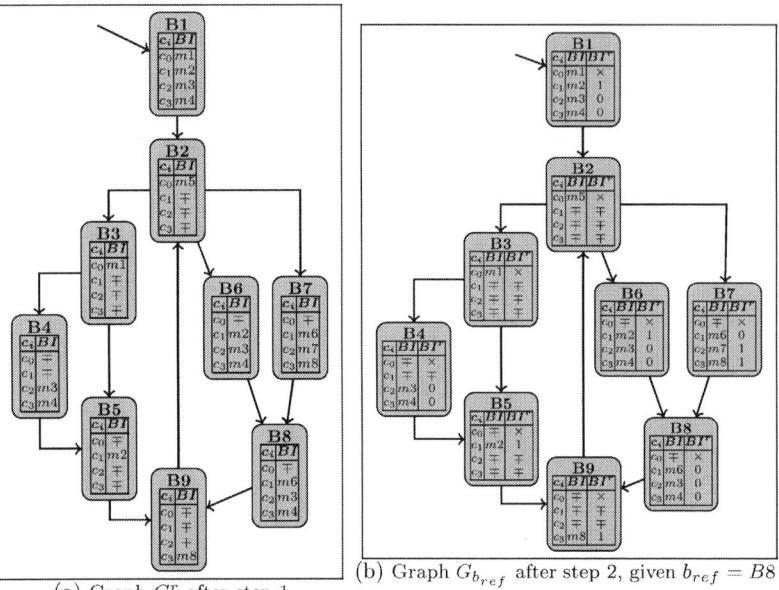

(a) Graph G^r after step 1

(b) Graph $G_{b_{ref}}$ after step 2, given $b_{ref} = B8$

(c) Graph $G_{b_{ref}}$ after step 3, given $b_{ref} = B8$ (reduced graph).

Figure 6: Illustration of Algorithm 4 with $b_{ref} = B8$.

after the execution of the basic block. We denote the reaching relative cache states of a basic block b as RCS^r_b and the relative leaving cache states of a basic block b as LCS^r_b.

B.1 Illustration

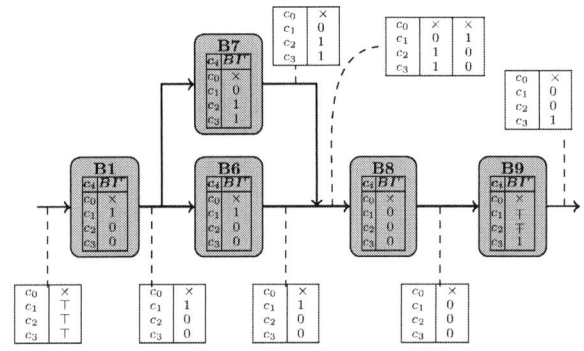

Figure 7: Illustration of the relative cache states.

A fragment of the reduced graph (Figure 6(c)) is presented in Figure 7. Using this graph we illustrate relative reaching and leaving cache states of a block. Given the reference block $b_{ref} = B8$ with $BI(b_{ref}) = [\mp, m6, m3, m4]$ and the relative instruction mapping of each block (described by the function BI^r), we illustrate the relative cache state as we start executing the initial block $B1$. Initially, the cache is empty and the reference block does not have an instruction on cache line c_0 ($BI^r(b_{ref})[0] = \mp$). Thus the instruction on cache line c_0 is *not of interest* ($\times$) resulting in the initial cache state of $[\times, \top, \top, \top]$, denoted as $cs^r_\top$. Note this initial cache state is unlike the concrete approach, where the initial state is represented using the vector $[\top, \top, \top, \top]$.

In our approach, given the reference block b_{ref} (in this case $b_{ref} = B8$), we are only interested in the cache line c_1, c_2, c_3. Thus, we ignore the cache state w.r.t cache line c_0, as it does not affect the precision of the reference block. Also, by ignoring c_0, allows for a memory efficient implementation.

In this case, the relative cache state can be represented using a vector with only three elements, instead of four elements, saving 25% of memory.

After execution of block $B1$, where the relative instructions are described by $BI^r(B1) = [\times, 1, 0, 0]$, cache lines c_1, c_2, c_3 will contain relative instructions 1, 0, 0. Hence after $B1$ executes, the relative cache state $cs^r = [\times, 1, 0, 0]$. Here, $cs^r[0] = \times$ represents the instruction on cache line c_0 is *not of interest*. $cs^r[1] = 1$ represents that the instruction in the cache is not $m6$, because $BI(b_{ref})[1] = m6$. $cs^r[2] = 0$ represents that the instruction in the cache is $m3$, because $BI(b_{ref})[2] = m3$. Similarly, $cs^r[3] = 0$ represents the instruction in the cache is $m4$ ($BI(b_{ref})[3] = m4$).

The relative cache state $cs^r_\top$ is the state of the cache prior to the execution of the basic block $B1$. Thus, $RCS^r_{B1} = \{[\times, \top, \top, \top]\}$ is the set of reaching relative cache states of block $B1$. Similarly, $LCS^r_{B1} = \{[\times, 1, 0, 0]\}$ is the set of relative leaving cache states of block $B1$.

Now, the control reaches a branch due to which, it is possible to execute either block $B6$ or block $B7$. In this case, $RCS^r_{B6} = LCS^r_{B1} = RCS^r_{B7}$. After executing blocks $B6$ (or $B7$), the state of the cache is $[\times, 1, 0, 0]$ (or $[\times, 0, 1, 1]$).

Block $B8$ has two incoming edges: from blocks $B6$ and $B7$. To compute RCS^r_{B8}, we need to *join* LCS^r_{B6} and LCS^r_{B7}. In this case, the join function is a union over the set of relative cache states. Thus, $RCS^r_{B8} = LCS^r_{B6} \cup LCS^r_{B7}$, resulting in $RCS^r_{B8} = \{[\times, 0, 1, 1], [\times, 1, 0, 0]\}$.

C. COMPUTING ALL POSSIBLE REACHING RELATIVE CACHE STATES OF THE REFERENCE BLOCK

The first step for cache analysis involves the computation of all possible reaching relative cache states of b_{ref} ($RCS^r_{b_{ref}}$), using the fixed point computation algorithm presented in Algorithm 5.

As illustrated in Figure 7, the initial state of the cache is empty ($cs^r_\top$), and is based on the instructions of the reference block. Similarly, like the concrete and the abstract ap-

1020

proaches, we must also introduce the unknown cache state $(cs_\perp^r)$, which is explained later during this algorithm. In general, the empty/unknown cache state is different for each $b_{ref} \in B$. Thus, for a given reference block b_{ref}, we first compute the empty cache state $cs_\top^r$, and unknown cache state $cs_\perp^r$ on lines 2 to 8.

For each cache line c_i, if the reference block b_{ref} does not have an instruction (in this case, $BI^r(b_{ref}) = \times$), then the cache state on c_i is *not of interest* during the analysis of the reference block b_{ref}. Thus, the relative instruction on cache line c_i of $cs_\top^r$ and $cs_\perp^r$ is set to $\times$ (line 4). Otherwise $(BI^r(b_{ref}) \neq \times)$, on line 6, for cache line c_i, the empty cache state is set to be $\top$ ($cs_\top^r[i] = \top$), and the unknown cache state is set to be $\perp$ ($cs_\perp^r[i] = \perp$).

Using relative cache states $cs_\top^r$ and $cs_\perp^r$, we initialise the reaching relative cache states for all blocks (lines 11 to 17). Since we assume that initially the state of the cache is empty, on line 13 for the initial block b_{init} we set its reaching as $RCS_{b_{init}}^{r1} = \{cs_\top^r\}$. Here, the notation RCS_b^{ri} represents the reaching relative cache states of block b in iteration i. E.g., $RCS_{b_{init}}^{r1}$ represents the reaching relative cache states of block b_{init} for iteration 1. For rest of the blocks, the initial state of the cache is unknown. Thus, on line 15, we set their reaching cache states as $\{cs_\perp^r\}$. After initialisation, we compute the relative leaving cache states of each block, on lines 19 to 22. We apply the transfer function(T) to every block and its corresponding reaching relative cache states.

The iteration index (i) is incremented (line 23) to signal the start of the next iteration. Next, on lines 25 to 34, the reaching relative cache states of each block are computed. For the initial block b_{init}, we know that the reaching relative cache state is always empty. Thus, on line 27, we always set its reaching as $RCS_{b_{init}}^{ri} = \{cs_\top\}$. For rest of the blocks, we first initialise the reaching relative cache state as empty set (line 29), and on lines 30 to 32, the reaching cache states are computed by looking at the relative leaving cache states of the predecessors (b') of the block b and using the union operation.

The iterative process, repeat-until loop on lines 18 to 35, is repeated until a fixed point is reached, i.e., if two consecutive iterations have the same sets of reaching relative cache states for all blocks (line 35).

D. RESULTS

Example	abstract		proposed		concrete		Gain
	WCET (clks)	AT (sec)	WCET (clks)	AT (sec)	WCET (clks)	AT (sec)	(col5/ col2)
BubbleSort	2571	0.7	2571	36.6	2571	44	1
Synthetic	14134	1.4	14134	8.2	14134	311	1
Flasher	117508	1.8	95908	41.9	T.O	T.O	0.82
DrillStation	31881	1.9	27453	45.1	T.O	T.O	0.86
ConvBeltModel	21344	1.7	18104	35.1	T.O	T.O	0.85
RailRoadCrossing	308505	4.0	250725	142.0	T.O	T.O	0.81
CruiseController	357206	3.6	288374	113.7	T.O	T.O	0.80

Table 5: Quantitative comparison between the abstract, proposed and concrete approaches

The WCET and the analysis time for abstract is presented in columns 2 and 3 respectively of Table 5. Similarly, the following columns present results for the proposed and concrete approaches. Final column presents the WCET estimate of the proposed approach w.r.t. the abstract approach. The proposed approach is tighter by 13% on average, and up to 19% tighter than the abstract approach.

Using the cache size between 0.2% and 0.9% of the program size, for each example, the WCET analysis results are presented in Figures 8(a)–8(h). Across all the benchmarks,

---

Algorithm 5 FP: Fixed point computation for the proposed approach

Input: A cache model $CM = \langle I, C, CI, G, BI \rangle$, reduced graph $G^r = \langle B^r, b_{init}, E^r \rangle$ and $BI^r : B^r \to (\{\times, \mp, 1, 0\})^N$.
Output: Reaching relative cache states of block b_{ref} ($RCS_{b_{ref}}^r$).

1: {Initialise $cs_\top^r$ and $cs_\perp^r$ }
2: **for each** $c_i \in C$ **do**
3:     **if** $BI^r(b_{ref})[i] = \times$ **then**
4:         $cs_\top^r[i] = \times$, $cs_\perp^r[i] = \times$ {When b_{ref} has no instruction (in this case, $BI^r(b_{ref}) = \times$) on cache line c_i, initialise empty/unknown cache states as *not of interest*.}
5:     **else**
6:         $cs_\top^r[i] = \top$, $cs_\perp^r[i] = \perp$ {When b_{ref} has an instruction on cache line c_i, initialise empty/unknown cache states as $\top/\perp$.}
7:     **end if**
8: **end for**
9: $i = 1$ {iteration counter}
10: {Initialise RCS^r for all blocks }
11: **for each** $b \in B$ **do**
12:     **if** $b = b_{init}$ **then**
13:         $RCS_{b_{init}}^{r1} = \{cs_\top^r\}$ {for the initial block, the initial state of the cache is always empty}
14:     **else**
15:         $RCS_b^{r1} = \{cs_\perp^r\}$ {for rest of the blocks, the initial state of the cache is unknown}
16:     **end if**
17: **end for**

18: **repeat**
19:     {Compute LCS^r for all blocks}
20:     **for each** $b \in B^r$ **do**
21:         $LCS_b^{ri} = T(RCS_b^{ri}, b)$
22:     **end for**

23:     $i = i + 1$; {Next iteration}
24:     {Compute RCS^r for the next iteration $i + 1$}
25:     **for each** $b \in B^r$ **do**
26:         **if** $b = b_{init}$ **then**
27:             $RCS_{b_{init}}^{ri} = \{cs_\top^r\}$
28:         **else**
29:             $RCS_b^{ri} = \emptyset$
30:             **for each** $LCS_{b'}^{ri}$, where $(b', b) \in E^r$ **do**
31:                 $RCS_b^{ri} = RCS_b^{ri} \cup LCS_{b'}^{ri}$
32:             **end for**
33:         **end if**
34:     **end for**

35: **until** $\forall b \in B^r$, $RCS_b^{ri} = RCS_b^{ri-1}$ {Termination condition}
36: **return** $RCS_{b_{ref}}^r$

---

we observe that the WCET estimates from the proposed approach is always less than or equal to the estimates from the abstract approach. With 1% relative cache size, on average, the WCET of proposed approach gives 13 % much tighter results and upto 19% tighter result than the abstract approach.

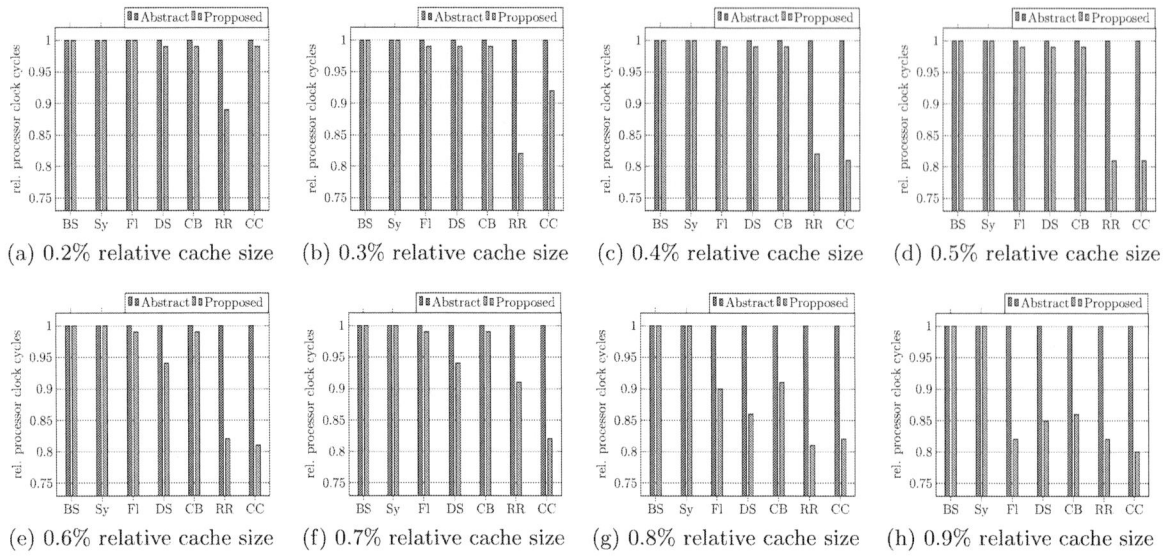

(a) 0.2% relative cache size   (b) 0.3% relative cache size   (c) 0.4% relative cache size   (d) 0.5% relative cache size

(e) 0.6% relative cache size   (f) 0.7% relative cache size   (g) 0.8% relative cache size   (h) 0.9% relative cache size

Figure 8: Comparing the WCET of the abstract and proposed approaches between 0.2% and 0.9% relative cache sizes

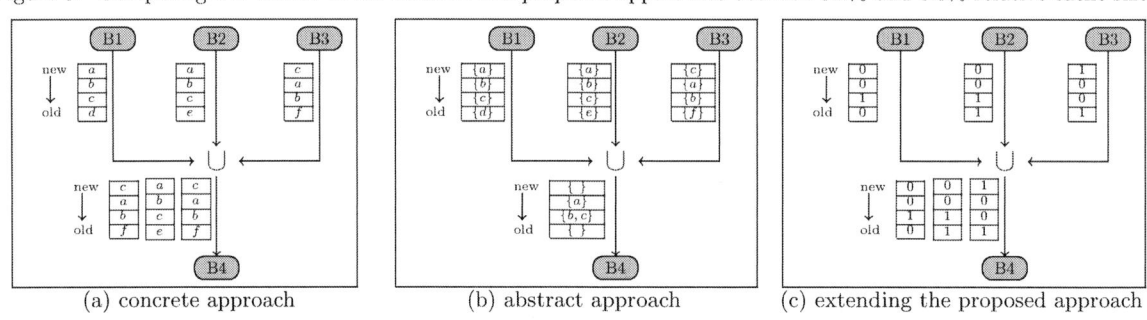

(a) concrete approach   (b) abstract approach   (c) extending the proposed approach

Figure 9: Extending and comparing the proposed approach to set associative caches

E. EXTENDING OUR APPROACH TO SET ASSOCIATIVE CACHES

The proposed approach can easily be extended for timing analysis of set associative caches. For a 4 way set associative cache with the least recently used replacement (LRU) policy [3], a comparison of the *join operation* between the concrete and abstract approaches is presented in Figure 2 of [3]. Using this exact example, we reproduce Figures 9(a) and 9(b). Here, block $B4$ has three predecessor blocks (B1, B2 and B3). The stack(s) next to the transitions represent the state of the cache, for each cache line. The order represents the history, the recent instruction is on top of the stack, and the least recent instruction is at the bottom of the stack.

For the concrete approach (Fig. 9(a)), the reaching cache states of $B4$ are computed as the union of all the leaving cache states of the predecessor blocks, resulting in three cache states. This is very precise, but will not scale for large programs. In the case of the abstract approach (Fig. 9(b)), the instructions are combined based on the *upper bound of its age*. E.g., for the leaving cache states of $B1$, $B2$ and $B3$, instruction 'a' resides on *top* (first), *top* (first) and *second* positions, respectively. Since, the upper age bound (oldest) is the *second* position, the join function abstracts the age of instruction 'a' as *second* position. Similarly, instruction 'b' and 'c' are computed as

third position. In contract, instruction 'd', only exists in the leaving cache state of $B1$, and not in $B2$ and $B3$. In this case, we cannot guarantee its presence, so it is removed from the stack. Similarly, instructions 'e' and 'f' are removed during the join operation. This abstraction, improves scalability, but lacks precision.

For the proposed approach (Fig. 9(c)), the instructions in the cache are presented w.r.t. the instructions of the reference block, 0 if *identical* , or 1 otherwise. E.g., let us assume that instructions a, b, d are abstracted as 1, and c, e, f are abstracted as 0. This abstraction is similar to the idea of *relative cache states* presented earlier in Sec. 3.2.1. Once again, the join operation performs the union over the reaching relative cache states, maintaining both precision and scalability.

Since the join operations of each approach is similar to their counter parts in direct-mapped analysis, we believe the complexity of the three approaches may be similar to Tab. 2. Also, the WCET and the analysis time may reflect the trends seen in Figures 4 and 5.

SSDM: Smart Stack Data Management for Software Managed Multicores (SMMs)

Jing Lu, Ke Bai[*] and Aviral Shrivastava
Compiler Microarchitecture Laboratory
Arizona State University, Tempe, Arizona 85287, USA
{Jing_Lu, Ke.Bai, Aviral.Shrivastava}@asu.edu

ABSTRACT

Software Managed Multicore (SMM) architectures have been proposed as a solution for scaling the memory architecture. In an SMM architecture, there are no caches, and each core has only a local scratchpad memory. If all the code and data of the task to be executed on an SMM core cannot fit on the local memory, then data must be managed explicitly in the program through DMA instructions. While all code and data need to be managed, an efficient technique to manage stack data is of utmost importance since an average of 64% of all accesses may be to stack variables [16]. In this paper, we formulate the problem of stack data management optimization on an SMM core. We then develop both an ILP and a heuristic - SSDM (Smart Stack Data Management) to find out where to insert stack data management calls in the program. Experimental results demonstrate SSDM can reduce the overhead by 13X over the state-of-the-art stack data management technique [10].

Categories and Subject Descriptors

D.3.4 [**Software**]: Processors—*Code generation, Compilers, Optimization*

General Terms

Algorithm, Design, Experimentation, Performance

Keywords

Stack data, local memory, scratchpad memory, SPM, embedded systems, multi-core processor

1. INTRODUCTION

As we scale the number of cores in a processor, scaling the memory hierarchy is a major challenge. Several computer architects believe that completely cache coherent architectures will not scale when there are hundreds and thousands of cores. Recently, Intel manufactured a

---

[*]This author contributed equally to this work.

Permission to make digital or hard copies of all or part of this work for personal or classroom use is granted without fee provided that copies are not made or distributed for profit or commercial advantage and that copies bear this notice and the full citation on the first page. To copy otherwise, to republish, to post on servers or to redistribute to lists, requires prior specific permission and/or a fee.
DAC '13, May 29 - June 07 2013, Austin, TX, USA.

48-core non-cache-coherent architecture, called Single-chip Cloud Computer or SCC [3]. However, caches still consume large amounts of power and die area. A promising option for a more power-efficient and scalable memory hierarchy is to have only scratchpad memory (SPM) in the cores. Since scratchpads consume 30% less area and power than a direct mapped cache of the same effective capacity [11], Software Managed Multicore (SMM) architectures can be extremely power-efficient. A very good example of SMM memory architecture is the Cell processor that is used in the Sony Playstation 3. Its power efficiency is around 5 GFlops per watt [14], while the power efficiency of Intel i7 4-core Bloomfield 965 XE is only 0.5 GFlops per watt [1,2].

Software Managed Multicore (SMM) architecture is a truly "distributed memory architecture on-a-chip." Therefore, applications on it require programmers to write several interacting tasks. The tasks are then mapped to the cores of the SMM architecture. Conventionally, *main task* executes on main core and creates *execution tasks*, which are then distributed and executed on execution cores. Main core has a large global or main memory, but execution cores have only a small local memory (the scratchpad memory). The execution cores can directly access only their local memory. To access other memories, including the global memory, explicit DMA instructions are needed in the application. In such architectures, the local memory is shared among code, and all data (stack, global and heap) of the task executing on the core. If the task can fit into the local memory, then extremely power-efficient execution can be achieved – and this is indeed the promise of SMM architectures.

However, for the general case, when all the code and data of the task do not fit in the local memory, explicit data management must be done to enable its execution. The programmer can do this, by bringing in the data/code before they need it, and evicting it back to the global memory after it is no longer needed. This is very difficult, since the programmer must now not only be aware of the local memory available in the architecture, but also be cognizant of the memory requirement of the task at every point in the execution of the program. Estimating the memory requirement is difficult for C/C++ programs, since stack and heap sizes may be variable and input data dependent. This difficulty of programming these SMM architectures has been the biggest roadblock in the success of extremely power-efficient SMM architectures.

To enable execution on the core of an SMM architecture, all code and data must be managed on the local scratchpad. We have started to develop techniques to manage code [18],

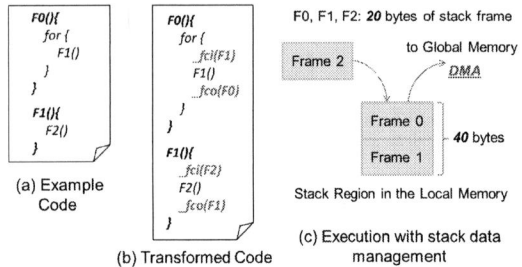

Figure 1: Function-level Stack Management - *(a) an example code, (b) the same code with function stubs _fci and _fco inserted before and after each function call. (c) when the program executes, _fci() may evict existing function frames to the global memory to make space for the incoming function frame, and _fco() may bring back the calling function.*

stack data [10, 29] and heap data ([6, 8, 9] for its form in C, [7] for its form in C++) on the cores with only scratchpad memories. Of these techniques, developing efficient approaches to manage stack data is especially important, since an average of 64% of all accesses in embedded applications may be to stack variables [16].

While the state-of-the-art stack data management scheme [10] enables managing stack data of any task on any SPM size (as long as the SPM size is larger than the size of the largest stack frame), there is a lot of room for improving the efficiency of stack data management. The opportunities lie in i) increasing the granularity of management, ii) not performing management when not absolutely needed, iii) performing minimal work each time management is performed, i.e., low instruction overhead of management library. To perform these optimizations, this paper makes two contributions:

- **Problem Formulation:** We formulate the optimization problem of where to insert the management functions so as to minimize the management overhead. We show that the function placement problem can be described as that of finding an optimal cutting of a weighted call graph (WCG). We believe problem definition is very important, and think that lack of formal problem definition is the reason behind high overheads of previous approaches to stack data management.
- **Efficient Heuristic:** Insights from the problem formulation enable us to design an effective heuristic, which we name SSDM. SSDM takes the WCG of the program, and then generates an efficient function placement of data management functions that satisfies the memory constraint on the local memory, while minimizing the management overhead.

Experimental results on several benchmarks from MiBench demonstrate SSDM can reduce the overhead by 13X over the current state-of-the-art stack management technique [10].

2. BACKGROUND AND STATE-OF-THE-ART

Scratchpad memories have been used in embedded systems for a long time, since they may be faster, and lower-power than caches [11]. However, unlike caches (in which the data management is in hardware and software is completely oblivious of it), the data management must be done explicitly in the software in order to use them. As a result, techniques have been developed to manage code [5, 13, 17, 30],

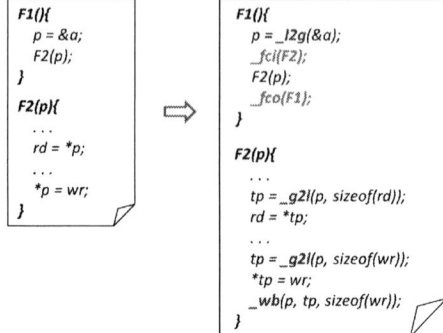

Figure 2: Pointer Management - *Function F2 accesses the pointer p, which points to a local variable 'a' of function F1. Since 'a' is a local variable on the stack of F1, it has a local address. When F2 is called, if F1 is evicted from the local memory, then the pointer p will point to a wrong value. This is fixed by assigning a global address to the pointer when it is created (through _l2g), and then when needed, it is accessed through _g2l. Finally it is written back using _wb.*

global variables [19, 20, 26, 30], stack data [15, 22, 23, 25, 27, 28, 30] and heap data [12, 24, 28] on scratchpad memories. However, these solutions are not applicable for SMM cores because of the difference in memory hierarchy of SMM cores and the traditional embedded cores. In typical embedded cores, the scratchpad memory is in addition to the regular cache hierarchy. This implies that applications can execute on embedded cores without using the scratchpad. However, frequently needed data can be mapped to the scratchpad memory to improve performance and power. On the other hand, the scratchpad is the only memory in the core of SMM architecture. Therefore everything must be accessed through the scratchpad, the only question is how to perform the management correctly and efficiently.

This paper focuses on stack data management, since an average of 64% of all accesses in embedded applications may be to stack variables [16]. Previous stack data management techniques (both [10, 29]) propose to manage stack data at function level granularity. This is done through code transformations shown in Figure 1. Figure 1 (a) shows an example original code, and (b) shows the transformed code. The _fci() and _fco() calls are inserted before and after each function call. The function stub _fci() makes space for the about-to-be-called function (by removing previous function frames). The function stub _fco() brings back the frame of the calling function, in case it was evicted. The execution of the transformed program is depicted in (c), which shows that if the space for stack was 40 bytes, and each function frame was 20 bytes, then when function F2 is called, there is no more space for it. The _fci() will evict the frame of F0 out of the local memory to make space for the stack frame of F2. The _fco() at return from function F1, will bring the function frame of F0 back in the local memory.

If a function accesses stack variables of another (ancestor) function through pointers (that may be passed to it as function parameters, or in other data structures), then there may be a problem. The problem, as shown in Figure 2 is that the pointer to a stack variable will be to a local address, since the stack is created in the scratchpad. However, when the pointer to a stack variable of an ancestor func-

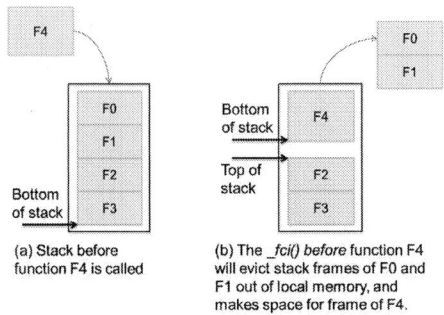

Figure 3: Circular Stack Management

tion is accessed, that function stack frame may have been evicted by the stack data management. Then the pointer will point to a wrong value. Bai et al. [10] extend the stack management approach to handle pointers correctly. To resolve pointers, they convert the local addresses of the pointers to their global addresses at the time of their definition (through the use of _l2g function stub), and at the time of pointer access, the data pointed to is brought into the local memory (through the use of _g2l function stub), and after the program is done accessing, it is finally written back to the global memory (through the use of _wb function stub). In this paper, we adopted the stack pointer management scheme in [10].

3. MOTIVATION

The state-of-the-art stack data management scheme [10] enables managing stack data of any task on any amount of space on the scratchpad and manages all stack pointers correctly. However, the management overhead is high, and the management is not optimized. The objective of this paper is to optimize stack data management, and reduce its overhead. Optimization opportunities lie in:

Opt1 - Increasing the granularity of management: Not only in SMM architectures, but in all multicore architectures, as the number of cores increases, the memory latency of a task will be very strongly dependent on the number of memory requests. This is because memory pipelines are becoming longer, and a large part of latency is the waiting time to get the chance to access memory. Therefore, it will be better to make small number of large requests, than large number of small memory requests. So the question is: how to increase the granularity of stack data management, even beyond function stack frames.

Opt2 - Not performing management when not absolutely needed: In existing approaches, the function _fci() and _fco() are inserted before and after each function call. Many times, these functions will not result in any data movement. For example, if there is space for the stack frame of the to-be-called function, then no DMA is required, only some book keeping happens. Much of the overhead is due to calling these functions, even though they are not needed. So, the question is: how do we not insert _fci() and _fco() functions when not needed.

Opt3 - Performing minimal work each time management is performed: In the existing approach, circular stack management, the older function frames are evicted from the top, and new frames can be instantiated as soon as enough space is available. Figure 3 shows that although this results in a judicious usage of local memory space for

stack management, it makes the book-keeping of the space extremely complicated. As different functions may have different stack frame sizes, the stack space will get fragmented after some time. To be able to track the status of stack space, a data structure is required. It needs to reserve stack size of each function, where the frame is stored in the global memory, what the starting address and the end address of the free slots in the scratchpad memory are, etc. In the library, we need to check these variables and update them accordingly, which therefore slows down the application.

4. OVERVIEW OF OUR APPROACH

To optimize the stack data management, we propose to perform stack data management (i.e., transfer stack data between scratchpad and global memory) at the whole stack space granularity. In other words, we keep on instantiating stack frames in the local memory until the management point. At the time of management, the whole stack space is written out to the global memory. When returning from the last frame in the local memory, the whole stack state is copied from the memory to the scratchpad. Since this is no longer at function level, we rename the management functions to _sstore, and _sload. This approach of performing management at stack space level granularity has several advantages: First is that the granularity of stack data management is much coarser (than function level), and therefore there will be fewer DMA calls (Opt1). Second is that the management library (_sstore and _sload) becomes simpler, since now the scratchpad is managed as a linear queue, rather than circular queue (Opt3). Table 1 shows our runtime stack management functions and their functionalities.

A problem that can happen in this scheme is that of thrashing. This happens when the stack space is full just before entering a loop with high execution count in which another function is called. Then every time the function is called, the stack state will be written back to the global memory, and reloaded on return. However, this can be avoided by carefully placing the scratchpad functions _sstore, and _sload in the program. In the next section we formulate the problem of optimal placement of these stack data management functions. We show that the management function placement problem can be described as that of finding an optimal cutting of a weighted call graph (WCG). We formulate an Integer Linear Program solution to the problem (explained in the Appendix, section A), and then propose a heuristic (SSDM) to solve this problem efficiently.

5. PROBLEM FORMULATION

A *weighted call graph* (V, E, W, T) contains a function node set V and a directed edge set E. Each node represents a

Table 1: Library on stack data and stack pointers

Library	Functionality
_sstore()	uses DMA to evict all stack frame(s) from local memory to global memory
_sload()	uses DMA to get all stack frame(s) in previous stack state back to local memory
_g2l(ga,size)	converts global address to a local address; gets the value from global mem. if misses
_l2g(la)	converts local address to a global address
_wb(ga,la,size)	updates data to ancestor frame

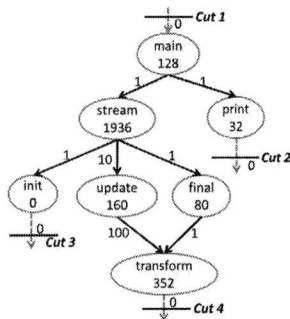

Figure 4: *WCG with cuts of benchmark SHA. The edge with dashed red arrow represents an artificial edge for root node or leaf node.*

function, and each directed edge pointing from the caller to the callee represents the calling relationship between two functions. Weight set $W = \{w_{f_1}, w_{f_2}, ...\}$ represents *stack sizes* of function nodes. Value on each edge e_{ij} ($e_{ij} \in E$) from the value set $T = \{t_1, t_2, ...\}$ corresponds to the number of times function node v_i calls v_j. Figure 4 shows the Weighted Call Graph (WCG) of the benchmark SHA.

A *root node* is the node with no in-coming edges. There is only one root node in the weighted call graph, which is usually the "main" function in a program. A *leaf node* is the node that has no out-going edges. Those are functions that do not call any other functions. However, for the convenience of our problem formulation, we add an *artificial* in-coming edge to the root node with value 0, and an *artificial* out-going edge to the leaf node with value 0. A *root-leaf path* is a sequence of nodes and edges from the root to any leaf node. For example, *main-stream-init* is a *root-leaf path* in Figure 4.

A *cutting of the graph* is defined as a set of cuts on graph edges. A *cut* on an edge e_{ij} ($e_{ij} \in E$) corresponds to a pair of function _sstore and _sload inserted respectively before and after function v_i calls function v_j. As shown in Figure 4, a set of cuts have been added on *artificial edges* in advance.

We use a list to represent the collection of nodes on a root-leaf path between two cuts. We call such a list of nodes as a *segment*. In Figure 4, the segment between cut 1 and cut 2 is $<main, print>$. A node can belong to multiple segments, e.g., node *stream* can be in both segment $<main, stream, init>$ and $<main, stream, update, transform>$. As the total function frame sizes in the local scratchpad memory cannot exceed the size limit of stack space, a positive weight (the size of stack space) constraint $\mathbb{W}$ is imposed on each segment so that the total weight (stack sizes) of functions in a segment will not exceed $\mathbb{W}$. Therefore, given a segment $s = \{f_1, f_2, ...\}$ with function weights $\{w_{f_1}, w_{f_2}, ...\}$, the total weight must satisfy the weight constraint

$$\sum_{f_i \in s} w_{f_i} \leq \mathbb{W} \qquad (1)$$

The cost of stack data management for each segment s comprises of two components: i) the running time spent on extra instructions caused by _sstore and _sload function calls, and ii) the time spent on data movement between the global memory and the local scratchpad memory. Let us assume a segment $s = \{f_1, f_2, ...\}$ is formed with two cuts on edges e_{start} and e_{end}, the functions in this segment have weights $\{w_{f_1}, w_{f_2}, ...\}$, and the two edges have values t_{start} and t_{end} (the number of function calls), the first part of the cost can

be represented as

$$cost_1 = t_{end} \times \tau_0 \qquad (2)$$

where τ_0 is a constant which represents the average execution time for extra instructions in run-time library (in both _sstore and _sload function). The time spent on data movement is linearly correlated to the size of DMA, which equals to the total function stack sizes in a segment. As a result, the second cost can be represented as

$$cost_2 = t_{end} \times 2(\tau_{base} + \tau_{slope} \times \sum_{f_i \in s} w_{f_i}) \qquad (3)$$

where τ_{base} is the base latency for any DMA transfer, τ_{slope} is the additional latency increasing rate with data size, and 2 shows the consideration for DMA data transfer *in* and *out*. Therefore, the total cost for each segment s is

$$cost_s = cost_1 + cost_2 \qquad (4)$$

For a set of cuts on a Weighted Call Graph (WCG) that forms a set of segments $S = \{s_1, s_2, ...\}$, the total cost can be represented as

$$cost_{WCG} = \sum_{s_i \in S} cost_{s_i} \qquad (5)$$

It should be noted that we treat each recursive function as a single segment and always assign a cut to it to ensure a pair of _sstore and _sload is placed right before and after recursive function calls. The detailed handling could be found in both ILP (Appendix, section A) and SSDM heuristic (Appendix, section B).

DEFINITION 1. **(*Optimal Cutting of a Weighted Call Graph*)** *An optimal cutting of a weighted call graph G contains a set of cuts that forms a set of segments, where each segment satisfies the weight constraint and the total cost of the segments is minimal.*

6. OUR HEURISTIC: SSDM

SSDM initially cuts all edges, and then checks all edges to see whether there is a cut on the edge. When a cut is found, our algorithm searches upward and downward through each *root-leaf path* to get its nearest neighboring cuts. Next we form all segments related to this cut by extracting all function nodes between the cut and its neighboring cuts. Thereafter, the total cost of those segments is calculated with Equation 2-5. Now we can assume this cut is removed, and construct new segments by combing upward segment and downward segment in the same *root-leaf path*. If none of these new segments violates the memory constraint of stack space, we can again calculate the new total cost. Otherwise, this cut could not be removed. By subtracting the newer one from the older one, we can get the *removing benefit* of this cut. We can calculate the *removing benefit* of other cuts through the same method. When all calculations are done, SSDM picks the largest one and indeed removes the cut associated with it. It keeps removing the cuts on WCG until no more cuts can be eliminated. The complete algorithm is presented in the Appendix, section B.

7. EXPERIMENTAL RESULTS

In this section we evaluate the efficiency of our SSDM technique by comparing it against the ILP (details are presented in Appendix, section A) and previous CSM heuristic approaches [10]. We have implemented our heuristic in

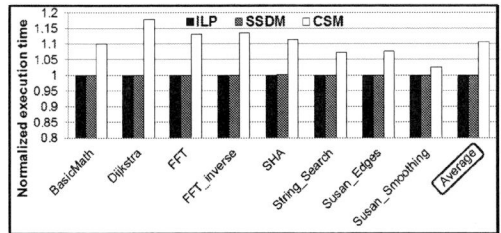

(a) SSDM against ILP and CSM.

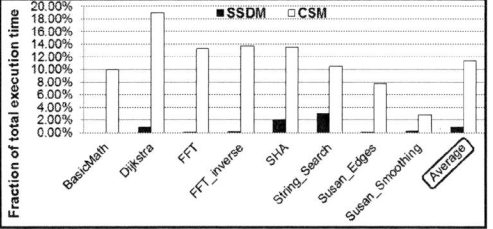

(b) Overhead comparison between SSDM and CSM.

Figure 5: *SSDM reduces the data management overhead and improves performance.*

the GCC 4.1.1 cross compiler for the Cell SPE (Synergistic Processing Element). We consider eight applications from MiBench suite [16]. The other applications in MiBench suite cannot be executed on SPEs because, to some extent, they lack standard library support, or they have large application code size. The eight applications are modified to be multi-threaded by keeping all I/O functionality of the benchmark in the main thread on Power Processor Element (PPE) and the core functionality is executed on the Synergistic Processing Elements (SPEs) [14]. τ_{base} and τ_{slope} used in Equation 3 are $2.1\mu s$ and $0.075\mu s/KB$ respectively [21]. Table 2 shows detailed information of all benchmarks.

We first utilized PPE and 1 SPE available in the IBM Cell BE and compared our SSDM performance against the results from ILP and CSM [10]. The number of function calls used in Weighted Call Graph (WCG) is estimated from profile information. As observed from Figure 5(a), our SSDM shows very similar performance to ILP approach. This means our heuristic approaches the optimal solution when the benchmark has a small call graph. Compared the CSM scheme, our SSDM demonstrates up to 19% and average 11% performance improvement. The overhead of the management comprises of i) time for data transfer, ii) execution of the instructions in the management library functions. Figure 5(b) compares the execution time overhead of CSM and the proposed SSDM. Results show that when using CSM, an average 11.3% of the execution time was spent on stack data management. With our new approach SSDM, the overhead is reduced to a mere 0.8% – a reduction of 13X. Next we break down the overhead and explain the effect of our techniques on its different components:

Opt1 - Increase in the granularity of management: Due to our stack space level granularity of management, the number of DMA calls have been reduced. Table 3 shows the number of stack data management DMAs executed when we use CSM, vs. the new technique SSDM. Note that there are no DMAs required for *Basicmath*. This is because the whole stack fits into the stack space allowed for this benchmark. Our technique performs well for all benchmarks, except for *Disjkstra*. This is because of the recursive function

print_path in *Dijkstra*. CSM will perform a DMA only when the stack space is full of recursive function instantiations, while we have to evict recursive functions every time with unused stack space. As a result, our technique does not perform very well on recursive programs. However, since many embedded programs are non-recursive, we have left the problem of optimizing for recursive functions as a future work.

Opt2 - Not performing management when not absolutely needed: Our SSDM scheme reduces the number of library function calls because of our compile-time analysis. In Table 4, we compare the number of _sstore and _sload function calls when using SSDM, vs. _fci and _fco calls when using CSM. We can observe that our scheme has much less number of library function calls. The main reason is that our SSDM considers the thrashing effect discussed in Section 4. Our approach tries to avoid (if possible) placing _sstore and _sload around a function call that executes many times, for example, within a loop. However, CSM always inserts management functions at all function call sites.

Opt3 - Performing minimal work each time management is performed: Our management library is simpler, since we only need to maintain a linear queue, as compared to a circular queue in CSM. Table 5 shows the amount of local memory required by SSDM and CSM, where we can find our runtime library has much less footprint than CSM does. It is very important for improving the performance, since stack frames will obtain less space in the local memory if the library occupies more space. The reason for larger footprint of CSM is that it needs to handle memory fragmentation, while our SSDM doesn't have this circumstance.

Table 6 shows the cost of extra instructions per library function call. We ran all benchmarks with both schemes and approximately calculated the average additional instructions incurred by each library call. As demonstrated in Table 6, our SSDM performs much better than CSM. There is no cost in SSDM when the stack region is sufficient to hold the incoming frames. However, CSM still needs extra instructions, since it checks the status of the stack region at runtime. *hit* for _g2l and _wb means the accessing stack data is residing in the local memory when the function is called, while *miss* denotes stack data is not in the local memory.

Table 2: Benchmarks, the number of nodes and edges in their WCG, their stack sizes, and the scratchpad space we manage them on.

Benchmark	Nodes	Edges	Stack Size (B)	Scratchpad Size (B)
BasicMath	7	6	400	512
Dijkstra	11	12	1712	1024
FFT	22	21	656	512
FFT_inverse	22	21	656	512
SHA	13	12	2512	2048
String_Search	11	10	992	768
Susan_Edges	8	7	832	768
Susan_Smoothing	7	6	448	256

Table 3: Comparison of number of DMAs

Benchmark	CSM	SSDM
BasicMath	0	0
Dijkstra	108	364
FFT	26	14
FFT_inverse	26	14
SHA	10	4
String_Search	380	342
Susan_Edges	8	2
Susan_Smoothing	12	4

1027

Table 4: Number of _sstore/_fci and _sload/_fco calls

Benchmark	_sstore/_fci		_sload/_fco	
	CSM	SSDM	CSM	SSDM
BasicMath	40012	0	40012	0
Dijkstra	60365	202	60365	202
FFT	7190	8	7190	8
FFT_inverse	7190	8	7190	8
SHA	57	2	57	2
String_Search	503	143	503	143
Susan_Edges	776	1	776	1
Susan_Smoothing	112	2	112	2

Table 5: Code size of stack manager (in bytes)

	_sstore/_fci	_sload/_fco	_l2g	_g2l	_wb
CSM	2404	1900	96	1024	1112
SSDM	184	176	24	120	80

In CSM approach, more instructions are needed for the *hit* case than the *miss* case in the function _wb. It is because the library directly writes back the data to the global memory when *miss*, but looking up the management table is required to translate the address. More importantly, as the table itself occupies space and therefore needs to be managed, CSM may need additional instructions to transfer table entries.

Besides comparing results between SSDM and CSM, we also examined the impact of stack space size and the scalability of our heuristic. We found that i) performance improves as we increase the space for stack data (Appendix, section C), ii) our SSDM scales well with different number of cores (Appendix, section D).

8. SUMMARY AND FUTURE WORK

This paper focuses on managing stack data, since the majority of the accesses in embedded applications may be to stack variables. We formulated the problem of efficiently placing library functions at the call sites. In addition, we proposed a heuristic algorithm called SSDM to generate the efficient function placement. Our experimental results show that SSDM generates function placement which leads to significant performance improvement compared to CSM.

Our optimization works under the assumption that Weighted Call Graph (WCG) could be constructed. However, future work could be devising a scheme to handle function pointers in the construction of WCG. In addition, the number of function calls are profile-based. A static estimation method should be proposed to get those values. Finally, previous scheme for pointers to stack data is directly adopted, but a proper scheme might be developed to further reduce the stack pointer management cost.

9. ACKNOWLEDGMENT

This research was partially funded by grants from National Science Foundation CCF-0916652, IIP-0856090, and NSF I/UCRC for Embedded Systems.

10. REFERENCES

[1] Intel Core i7 Processor Extreme Edition and Intel Core i7 Processor Datasheet, Volume 1. In *White paper*. Intel.

[2] Raw Performance: SiSoftware Sandra 2010 Pro (GFLOPS).

[3] The SCC Programmer's Guide. Technical report.

[4] A. V. Aho, M. S. Lam, R. Sethi, J. D. Ullman. *Compilers: Principles, Techniques, and Tools*. Addison Wesley, 1986.

[5] F. Angiolini et al. A Post-Compiler Approach to Scratchpad Mapping of Code. In *Proc. of CASES*, pages 259–267, 2004.

[6] K. Bai, and A. Shrivastava. A Software-Only Scheme for Managing Heap Data on Limited Local Memory (LLM) Multi-core Processors. *ACM TECS*, 2013.

Table 6: Dynamic instructions per function

	_sstore/_fci		_sload/_fco		_l2g	_g2l		_wb	
	F	NF	F	NF		H	M	H	M
CSM	180	100	148	95	24	45	76	60	34
SSDM	46	0	44	0	6	11	30	4	20

* F: stack region is full when function is called; NF: stack region is enough for the incoming function frame; H: hit of stack data; M: miss of stack data.

[7] K. Bai, D. Lu, and A. Shrivastava. Vector Class on Limited Local Memory (LLM) Multi-core Processors. In *Proc. of CASES*, 2011.

[8] K. Bai and A. Shrivastava. Heap Data Management for Limited Local Memory (LLM) Multi-core Processors. In *Proc. of CODES+ISSS*, 2010.

[9] K. Bai and A. Shrivastava. Automatic and Efficient Heap Data Management for Limited Local Memory Multicore Architectures. In *Proc. of DATE*, 2013.

[10] K. Bai, A. Shrivastava, and S. Kudchadker. Stack Data Management for Limited Local Memory (LLM) Multi-core Processors. In *Proc. of ASAP*, pages 231–234, 2011.

[11] R. Banakar et al. Scratchpad Memory: Design Alternative for Cache on-chip Memory in Embedded Systems. In *Proc. of CODES+ISSS*, pages 73–78, 2002.

[12] A. Dominguez, S. Udayakumaran, and R. Barua. Heap Data Allocation to Scratch-pad Memory in Embedded Systems. *J. Embedded Comput.*, 1(4):521–540, 2005.

[13] B. Egger et al. A Dynamic Code Placement Technique for Scratchpad Memory Using Postpass Optimization. In *Proc. of CASES*, pages 223–233, 2006.

[14] B. Flachs et al. The Microarchitecture of the Synergistic Processor for A Cell Processor. *IEEE Solid-state circuits*, 41(1):63–70, 2006.

[15] L. Gauthier and T. Ishihara. Implementation of Stack Data Placement and Run Time Management Using a Scratch-Pad Memory for Energy Consumption Reduction of Embedded Applications. *IEICE*, 94-A(12):2597–2608, 2011.

[16] M. R. Guthaus et al. Mibench: A Free, Commercially Representative Embedded Benchmark Suite. *Proc. of Workload Characterization*, pages 3–14, 2001.

[17] A. Janapsatya et al. A Novel Instruction Scratchpad Memory Optimization Method Based on Concomitance Metric. In *Proc. of ASP-DAC*, pages 612–617, 2006.

[18] S. C. Jung, A. Shrivastava, and K. Bai. Dynamic Code Mapping for Limited Local Memory Systems. In *Proc. of ASAP*, pages 13–20, 2010.

[19] M. Kandemir and A. Choudhary. Compiler-directed Scratch pad Memory Hierarchy Design and Management. In *Proc. of DAC*, pages 628–633, 2002.

[20] M. Kandemir et al. Dynamic Management of Scratch-pad Memory Space. In *Proc. of DAC*, pages 690–695, 2001.

[21] M. Kistler et al. Cell Multiprocessor Communication Network: Built for Speed. *IEEE Micro*, 26(3):10–23, May 2006.

[22] L. Li, L. Gao, and J. Xue. Memory Coloring: A Compiler Approach for Scratchpad Memory Management. In *Proc. of PACT*, pages 329–338, 2005.

[23] M. Mamidipaka and N. Dutt. On-chip Stack Based Memory Organization for Low Power Embedded Architectures. In *Proc. of DATE*, pages 1082–1087, 2003.

[24] R. McIlroy et al. Efficient Dynamic Heap Allocation of Scratch-pad Memory. In *ISMM*, pages 31–40, 2008.

[25] N. Nguyen, A. Dominguez, and R. Barua. Memory Allocation for Embedded Systems with A Compile-time-unknown Scratch-pad Size. In *Proc. of CASES*, pages 115–125, 2005.

[26] P. Panda et al. On-chip vs. Off-chip Memory: the Data Partitioning Problem in Embedded Processor-based Systems. In *ACM TODAES*, pages 682–704, 2000.

[27] S. Park et al. A Novel Technique to Use Scratch-pad Memory for Stack Management. In *Proc. of DATE*, pages 1478–1483, 2007.

[28] F. Poletti et al. An Integrated Hardware/Software Approach for Run-time Scratchpad Management. In *Proc. of DAC*, pages 238–243, 2004.

[29] A. Shrivastava et al. A Software-only Solution to Use Scratch Pads for Stack Data. *IEEE TCAD*, 28(11):1719–1728, 2009.

[30] S. Udayakumaran, A. Dominguez, and R. Barua. Dynamic Allocation for Scratch-pad Memory Using Compile-time Decisions. *ACM TECS*, 5(2):472–511, 2006.

APPENDIX

A. INTEGER LINEAR PROGRAMMING FORMULATION

In this section, we present our Integer Linear Programming (ILP) formulation for placing _sstore and _sload functions. For a given segment, the cost and total weight can be calculated with Equation 1-5. Given a graph G, all the possible segments can be found out in advance by randomly picking two edges from the graph and putting two cuts on them respectively. Therefore, the optimal _sstore and _sload placement problem can be transformed as to pick out a set of segments from all the possible segments whose total cost is minimal, and they also satisfy the following two conditions: i) the set of segments can make up the complete weighted call graph G, and ii) each segment satisfies the weight constraint.

The weight constraint can be checked with Equation 1, while checking the first constraint is more complicated. For a graph, we can cut each edge of the graph, and define a smallest segment as an *element*, which contains exactly one node and two edges. In the example shown in Figure 6, the graph is composed of five elements, namely, $<e_0\text{-}F_0\text{-}e_{01}>$, $<e_{01}\text{-}F_1\text{-}e_{13}>$, $<e_{13}\text{-}F_3\text{-}e_3>$, $<e_0\text{-}F_0\text{-}e_{02}>$ and $<e_{02}\text{-}F_2\text{-}e_2>$. Similarly, any segment S in a graph can be represented as a set of elements $S = \{el_1, el_2, ...\}$. In the previous example, the segment formed by the cuts on e_0 and e_{13} contains two elements, which are $<e_0\text{-}F_0\text{-}e_{01}>$ and $<e_{01}\text{-}F_1\text{-}e_{13}>$. For a segment S and a root-leaf path P, if all nodes in elements that belong to S are also contained in P, we say $S \subseteq P$, and we define the segment S as a *subset-segment* of P. For example, in Figure 6, the segment $<F_0, F_1>$ is a *subset-segment* of path $F_0\text{-}F_1\text{-}F_3$. Apparently, each segment must be a *subset-segment* to at least one root-leaf path. Now we can check if a set of picked segments can make up the complete weight call graph G. If each element in the path P_i is contained in one and only one subset-segment of P_i, then we can claim that the picked segments can cover path P_i. If the picked segments can cover all paths in G, then we can claim that the picked segments $\mathcal{S}$ can make up the complete graph G.

Eventually, the problem can be presented as follows:

Input:

- $\mathbb{W}$: total weight constraint, it is the size of local scratchpad memory

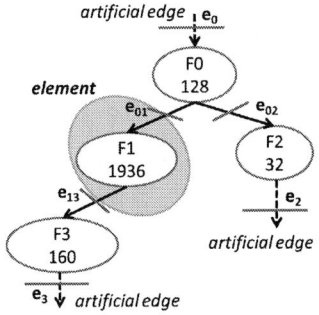

Figure 6: *WCG has many elements, which is composed of 1 node and 2 edges between cuts.*

- E: a set of elements
- S: a set of segments
- P: a set of root-leaf paths
- $cost_s$: cost of each segment s, where $s \in S$
- $weight(s)$: total weight of each segment s, where $s \in S$
- $In(e, s)$: binary value. For any segment s and element e, it is one if $e \in s$, zero if otherwise.
- $subset(s, p)$: binary value. For any segment s and root-leaf path p ($p \in P$), it is one if $s \subseteq p$, zero if otherwise.
- $E(p)=\{e_1, e_2, ...\}$: a set of elements such that $e_i \in p$, $p \in P$.

Variable:

$$x_s = \begin{cases} 1 & \text{if segment } s \text{ is picked} \\ 0 & \text{otherwise} \end{cases}$$

Objective Function:

$$\text{minimize} \quad \sum_{s \in S} cost_s \times x_s$$

Constraints:

$$weight(s) \times x_s \leq \mathbb{W}, \ for \ s \in S$$

$$\sum_{s \in S} subset(s, p) \times In(e, s) \times x_s = 1, \ \forall \ p \in P, \ and \ \forall \ e \in E(p)$$

The first constraint is the weight constraint, and the second constraint guarantees that the picked segments can make up the complete graph. It should be noted that we must treat each recursive function as a single segment, and add one more constraint for each as follows:

$$x_s = 1, \forall s \ that \ indicates \ a \ recursive \ function$$

It ensures a pair of _sstore and _sload is placed right before and after recursive function calls.

B. SSDM HEURISTIC

In this section, we present the complete SSDM heuristic for placing _sstore and _sload library functions. As observed from Algorithm 1, Line 1 preprocesses all recursive edges by placing a cut on them. Since _sstore and _sload are statically placed at compile time and recursive function calls itself, we must put a cut on the recursive edge to eliminate the nondeterminacy of recursive functions. In line 8-10, we first find out the segments that are associated with each cut x_{ij} on edge e_{ij} ($e_{ij} \in E$). To do this, we need to find out all root-leaf path P_i, where $e_{ij} \in P_i$. Then we search upward through each P_i, until we meet a cut x_{up}. Similarly, we search downward through each root-leaf path P_i, until we meet a cut x_{down}. The segment between x_{ij} and x_{up} or x_{down} is defined as associated with x_{ij}. For example, in Figure 6, the segments that are associated with cut on e_{02} is the segment $<F_0>$ and the segment $<F_2>$. Then we calculate the cost of each segment with Equation 2-5, and the total cost by summing up the cost of all the associated segments. In Line 11-19, we assume the cut is removed, and we can get a new set of associated segments. Those segments are formed by merging the segment between x_{ij} and x_{up} with the segment between x_{ij} and x_{down} on each root-leaf path P_i. As an edge might belong to several root-leaf paths, there might be many x_{up} and x_{down} accordingly. In Figure 6, after removing the cut on e_{02}, the two associated segments are merged into one segment, which is $<F_0, F_2>$. Similarly, we can calculate the cost of each new segment

1029

Algorithm 1: SSDM(WCG(V,E))

1 Place cuts on recursive edges, if there are recursive functions.
2 Define vector $\mathcal{C}$, in which x_{ij} indicates if a cut should be placed on edge e_{ij} ($e_{ij} \in E \setminus E_{recursive}$). set all $x_{ij}=1$.
3 **while** *true* **do**
4   Define vector $\mathcal{B}$ to store *removing benefit* of each cut.
5   **foreach** $x_{ij} == 1$ **do**
6   Set boolean *violate* to *false*, it shows if removing this cut would violate the weight constraint.
7   Define total cost $Cost_{before} = 0$.
8   **foreach** *segment* s_old_i *that are associated with* x_{ij} **do**
9   Calculate cost $cost\_old_i$ with Equation 2-5.
10 $Cost_{before}+ = cost\_old_i$
11   Assume the cut of x_{ij} is removed, and get a new set of associated segments.
12   Define total cost $Cost_{after} = 0$.
13   **foreach** *new associated segment* s_new_i **do**
14   Check weight constraint with Equation 1.
15   **if** *weight constraint is violated* **then**
16 $violate = true$
17   break
18   Calculate cost $cost\_new_i$ with Equation 2-5.
19 $Cost_{after}+ = cost\_new_i$
20   **if** *violate* **then**
21   continue
22   Calculate the benefit of removing the cut as $B_{ij} = Cost_{before} - Cost_{after}$.
23   **if** $B_{ij} > 0$ **then**
24   Store B_{ij} into vector $\mathcal{B}$.
25   **if** $\mathcal{B}$ *contains no element* **then**
26   break
27   Find out the largest benefit value B_{max} from $\mathcal{B}$, and set the corresponding cut $x_{max} = 0$.
28 **foreach** $x_{ij}==1$ **do**
29   Place a cut on edge e_{ij}, i.e., the compiler places _sstore and _sload right before and after the call instruction respectively.

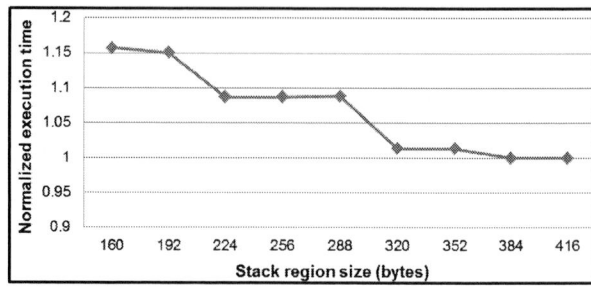

Figure 7: *Performance - different stack region sizes.*

we constructed another set of experiments that evaluates our SSDM technique under tight size constraints. The benchmark *Dijkstra* contains many nested function calls within loop structures, making it a good candidate for showing the impact of different stack region sizes. We expanded the region size from 160 bytes to 416 bytes with the step size of 32 bytes. The resulted performances are demonstrated in Figure 7, where the execution time with different stack region sizes were normalized to the smallest one. The execution time decreases when we increase stack region size. When the size reaches 384 bytes, the performance hardly improves. The primary reason is that we conservatively manage the recursive function by always placing a pair of library function around all its call sites. Therefore, although the region size is large enough, no more benefit can be obtained as only the insertion for recursive function *print_path* is left.

D. SCALABILITY OF SSDM

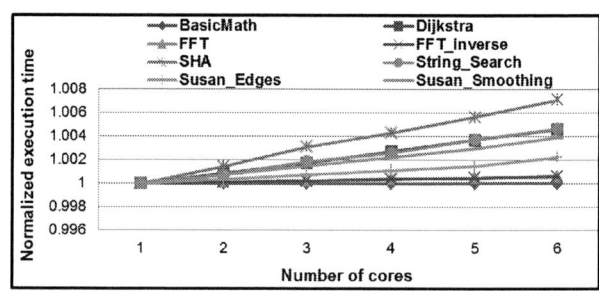

Figure 8: *Performance - different number of cores.*

Figure 8 shows the results we examined the scalability of our SSDM heuristic. We normalized the execution time of each benchmark with number of SPEs to its execution time with only one SPE, and show them on y-axis. In this experiment, we executed the same application on different number of cores. This is very aggressive, since DMA transfers occur almost at the same time when stack frames need to be moved between the global memory and the local memory. This will lead to the competition of DMA requests. As shown in Figure 8, the execution time increases gradually as we scale the number of cores, but no more than 1%. Benchmark *SHA* increases most steeply, as there are many stack pointer accesses in this program. Because of this, more data transfers are conducted for objects pointed by those stack pointers.

with Equation 2-5, and the total cost of all associated segments after removing the cut. Line 14-17 check if weight constraint is satisfied by removing this cut. If the constraint is violated, this cut will not be considered to be removed (line 20-21). Line 27 removes the cut with largest positive benefit among all the cuts whose removal will not violate the weight constraint. Line 25-26 is the exit condition of the WHILE loop. The procedure stops until no more cut can be removed from the graph. At this point of time, the rest cuts either have negative removing benefit, or cannot be removed due to weight constraint. The last two lines in the algorithm shows the operations that need to be made in our modified compiler.

C. IMPACT OF STACK SPACE

The experiment for each application in Section 7 was conducted under the scratchpad size specified in Table 2. Next

Taming the Complexity of Coordinated Place and Route

Jin Hu, Myung-Chul Kim and Igor L. Markov

{jinhu, mckima}@us.ibm.com, imarkov@eecs.umich.edu

ABSTRACT

IC performance, power dissipation, size, and signal integrity are now dominated by interconnects. However, with ever-shrinking standard cells, blind minimization of interconnect during placement causes routing failures. Hence, we develop Coordinated Place-and-Route (CoPR) with (i) a Lightweight Incremental Routing Estimation (LIRE) frequently invoked during placement, (ii) placement techniques that address three types of routing congestion, and (iii) an interface to congestion estimation that supports new types of incrementality. LIRE comprehends routing obstacles and non-uniform routing capacities, and relies on a cache-friendly, fully-incremental routing algorithm. Our implementation extends and improves our winning entry at the ICCAD 2012 Contest.

1. INTRODUCTION

The nature of global routing has changed since the 1980s as interconnect stacks grew from three metal layers to 9-12 layers with non-uniform pitches [19,20]. Router runtimes have increased, and so has the impact of routing on design quality. Modern *global routing* cannot be viewed as a standalone optimization because of signal integrity concerns and the impact of coupling capacitance on interconnect delays. *Placement* is also no longer standalone, as it interacts with numerous other optimization steps to control interconnect lengths and delays. In the last 15 years, global placement has often been guided by routability estimation [16] in commercial EDA tools and academic contests [18–20]. The development of such *integrative optimizations* requires understanding the strengths and weaknesses of *dedicated optimizations*, as well as invoking the right primitive at the right time. Indeed, *complexity* — both the number of steps executed at runtime and the number of lines of code — is the main gating factor for what can be achieved by EDA tools in the foreseeable future. Moreover, new optimization primitives must be justified by their context and intended use.

In this work, we develop a streamlined system for Coordinated Place-and-Route (CoPR) that (i) uses cache-friendly routing primitives to quickly and accurately estimate routing congestion (LIRE), (ii) leverages incrementality in routing and congestion updates in new ways, and (iii) offers a new categorization of congestion and new congestion-relief techniques during placement. CoPR achieves unprecedented trade-offs between speed and placement quality on large industry netlists, as we illustrate using ICCAD 2012 contest benchmarks from IBM Research [20].

The remainder of this paper is structured as follows. Section 2 presents our fast and accurate routing estimation technique. Section 3 introduces our placement techniques that proactively alleviate routing congestion. Section 4 describes the interactions between the placer and the routing estimator. Section 5 compares our techniques to currently-known approaches. Section 6 empirically validates the scalability of our techniques. Section 7 concludes our discussion. Supplemental material is provided in the Appendices.

2. LIRE: ROUTING ESTIMATION

We develop a Lightweight Incremental Routing Estimator (LIRE) that quickly produces congestion maps as accurate as those by a global router (Figure 5). Empirically, we target 75K nets per second,[1] but also facilitate a tradeoff between quality and runtime. In contrast, modern routers [4,12] complete 6K nets per second.[1]

Notation. We consider an $X \times Y$ routing grid $G(V, E)$ with (i) a set V of *GCells* (nodes) where each GCell $v \in V$ has integer coordinates (x_v, y_v), and (ii) and a set E of directed edges $e = (v_1, v_2)$, where the weight w_e of edge e encapsulates routing congestion and history costs (Lagrangian multipliers). Each node $v \in V$ is adjacent to its four cardinal neighbors: NORTH $(x_v, y_v + 1)$, SOUTH $(x_v, y_v - 1)$, EAST $(x_v + 1, y_v)$ and WEST $(x_v - 1, y_v)$. Consider a point-to-point connection π between two distinct GCells $S, T \in V$. When $x_S \leq x_T$ and $y_S \leq y_T$, a *forward* edge for π is an edge (v_1, v_2) such that $x_{v_1} < x_{v_2}$ or $y_{v_1} < y_{v_2}$, i.e., EAST or NORTH, and a *backward* edge for π is an edge (v_1, v_2) such that $x_{v_1} > x_{v_2}$ or $y_{v_1} > y_{v_2}$, i.e., WEST or SOUTH. Definitions for the three other orientations of π are symmetrical.

Key definitions. A *route segment* is a directed path in the routing grid. A *flat* route segment is a set of directed edges that are all NORTH, SOUTH, EAST or WEST. A *monotonic* segment is a connected set of flat segments such that each flat segment is either: (i) NORTH or EAST, (ii) NORTH or WEST, (iii) SOUTH or EAST, or (iv) SOUTH or WEST. Each monotonic segment is classified as NORTH-EAST, NORTH-WEST, SOUTH-EAST, or SOUTH-WEST. A *route* r_π is a collection of routing segments linking S and T.

2.1 Faster Routing

Global routing spends a large fraction of runtime finding weighted shortest paths in highly-congested regions [20]. Such shortest-path (maze) routing is necessary in congested regions for both *global routing* and *accurate congestion estimation* because, unlike pattern routing, it adequately captures detours. Detours are shaped by edge weights, which include congestion and history costs [4]. These weights must be maintained with sufficient accuracy and can be neither binned nor rounded without adverse impact on resulting routes. Therefore, shortest-path routing in congested regions is performed by A*-search. However, (i) the priority queue in A*-search is responsible for an extra $O(\log V)$ term in the overall complexity of the algorithm, (ii) priority queues, even when implemented using Fibonacci heaps, are too slow [10] and their pointer-based algorithms can experience costly cache misses, (iii) typical A* admissible functions based on straight-line distance become ineffective when history-based costs become large, and (iv) A*-search cannot leverage incrementality, i.e., given a candidate path, it cannot check optimality or perform an incremental improvement.

Permission to make digital or hard copies of all or part of this work for personal or classroom use is granted without fee provided that copies are not made or distributed for profit or commercial advantage and that copies bear this notice and the full citation on the first page. To copy otherwise, to republish, to post on servers,requires prior specific permission and/or a fee.
DAC'13, May 29 - June 07 2013, Austin, TX, USA.

---

[1]Median single-thread router performance on placements by the top three ICCAD 2012 contestants (Intel Xeon 3.4GHz CPU).

Algorithm 1 Bellman-Ford Algorithm with Non-negative Weights

Input: Point-to-point connection π, Search Space (V', E')
Output: route r_π

1: **for** i from $1 \rightarrow |V'|$ **do**
2:      $cost[v_i] = \infty$;
3: **end for**
4: $cost[S] = 0$;
5: **for** i from $1 \rightarrow |V'|$ **do**
6:      **for** $j = 1$ from $1 \rightarrow |E'|$ **do**
7:          $e_j = (v_1, v_2)$;
8:          **if** $cost[v_2] < cost[v_1] +$ COST(e_j) **then** $\triangleright$ relaxation
9:              $cost[v_2] = cost[v_1] +$ COST(e_j);
10:             $parent[v_2] = v_1$;
11:         **end if**
12:     **end for**
13: **end for**
14: $r_\pi = $ TRACE_PATH(π);

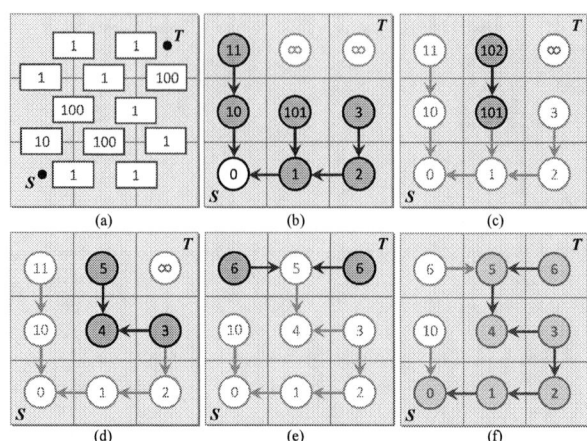

Figure 1: **Applying one BF pass with duplex-edge relaxation and echo-relaxation to a point-to-point connection $S \rightarrow T$ without via-cost modeling. Arrows point to the previous node in the path. (a) The routing grid and edge costs (congestion). Let S have coordinate $(0,0)$. (b) The partial costs of the first row and the center-left node have been populated. (c) Relaxing the** NORTH $(1,1) \rightarrow (1,2)$ **and** SOUTH $(1,2) \rightarrow (1,1)$ **edges at node with coordinate $(1,1)$. (d) Relaxing the** EAST $(1,1) \rightarrow (2,1)$ **and** WEST $(2,1) \rightarrow (1,1)$ **edges at node with coordinate $(1,1)$. The cost at $(1,1)$ has been updated by the** WEST **edge and is propagated to $(1,2)$. (e) The remaining nodes are considered, and partial costs are populated through T. (f) An optimal path with three monotonic segments is found in a single BF pass.**

Linear-time cache-friendly routing. Given that A*-search is derived from Dijkstra's algorithm [1, Section 24.3], we hope to avoid these priority-queue-based approaches. Of the classic weighted shortest-path algorithms, the *Bellman-Ford (BF) algorithm* [1, Section 24.1] is array-based and moreover preserves memory locality. However, it may require V linear-time passes, taking $O(EV)$ time.

Notably, the *worst-case complexity* of Bellman-Ford (BF) can be avoided in global routing. Recall that each BF pass performs $E \times O(1)$ *relaxation* steps. When no relaxations in a pass result in improvement, no further improvement is achieved in later passes. Thus, BF can be terminated early without the loss of optimality.

During global routing, we consider one point-to-point connection $(S \rightarrow T)$ at a time. Routing is limited to a subgrid $G'(V', E') \subseteq G$ enclosed in an isothetic (coaxial) bounding box that contains S and T. To generate a route, we visit the nodes of G' in a specified ordering $v_0, v_1, \ldots, v_{|V'|-1}$.[2] While the Bellman-Ford algorithm supports any node visitation ordering, we specify an ordering that not only affords us the highest memory locality, but also caters to the common case of monotonic paths. Starting from S, the nodes are traversed in the row containing S, and at each node v, relaxation is performed (lines 8-9 in Algorithm 1) along the four v-incident edges pointing toward T. The nodes in the next row closer to T are traversed, and so on until the row that contains T. When the node traversal follows the in-memory array layout (by rows or by columns), this method maintains the locality of memory access.

We propose to optimize BF passes with *duplex-edge relaxation*. At each edge considered by this technique, relaxations are attempted in both directions, but only *forward-looking edges* are considered at each vertex. While the same number of edges is considered per pass, cache utilization and memory locality are improved because for each adjacent vertex (and edge cost) loaded from memory, two relaxations can be attempted rather than just one. Furthermore, if the first relaxation succeeds, the second one cannot occur — this saves an extra comparison. For example, at node v with coordinate (x, y), we relax either the outgoing NORTH edge $(x, y) \rightarrow (x, y+1)$ *or* the incoming SOUTH edge $(x, y+1) \rightarrow (x, y)$. A similar duplex relaxation is performed in the EAST and WEST directions. By explicitly modeling via costs within these traversals [4, Section 3.4], BF will prefer fewer-bend routes.

Monotonic routing with one linear-time BF pass. As a special case, an optimal monotonic route can be found by (i) considering only forward edges (e.g., NORTH and EAST), and (ii) fixing the

---
[2]Edges are traversed in the increasing order of adjacent vertices.

considered space to the bounding box b with dimensions $w \times h$, $w = x_T - x_S + 1$ and $h = y_T - y_S + 1$, that minimally contains S and T. Let t be the $w \times h$ matrix where $t[x][y]$ stores the partial cost from S with coordinates $(0,0)$ to a node $v = (x, y)$. By construction, the cost at (x, y) depends solely on the costs at $(x-1, y)$ and $(x, y-1)$. Therefore, by visiting the nodes in row order (or column order) from S toward T, we visit every node in b exactly once. Since b has $w \times h$ nodes, the runtime complexity is $O(wh)$.

Non-monotonic routing with one linear-time BF pass. Recall that BF supports any node (and edge) ordering. Some optimal non-monotonic routes can be found in linear time within the bounding box b that minimally contains π by (i) employing duplex-edge relaxation and (ii) *echo-relaxation* if the relaxation succeeded in the direction opposite to node ordering (from a greater-numbered node to a smaller-numbered node). That is, in the forward-going node ordering, if a backward edge at node $v(x, y)$ results in a smaller-cost route, we forward-propagate the smaller cost through all recently-relaxed edges incident to v. Figure 1 illustrates finding an optimal route with three distinct monotonic segments in one BF pass. This improvement is effective in detouring short nets, and a majority of nets are short in practice. A more powerful variant of echo-relaxation would propagate costs through *all* incident edges, and allows one BF pass to find longer detours (not used in this work).

Non-monotonic routing with BF and Yen's improvement. J. Y. Yen [23] suggested that *reversing* the node ordering between BF passes reduces the number of passes required to find an optimal path. We refer to the Bellman-Ford algorithm with early termination and Yen's improvement as BFY. Two and three BFY passes can quickly find long detours, as illustrated in Figure 2. This is especially applicable for large nets.

THEOREM 1. *Let π be a point-to-point connection. Finding a minimal-cost route r_π^{min} with m (distinct) monotonic segments requires at most m BFY passes.*

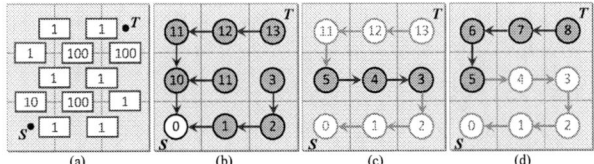

Figure 2: Applying BFY to a point-to-point connection $S \rightarrow T$ without via-cost modeling. (a) The routing grid and edge costs (congestion). (b) The first forward pass finds the optimal monotonic path of cost 13. (c) The backward pass finds a detour. (d) The second forward pass finds the optimal path of cost 8.

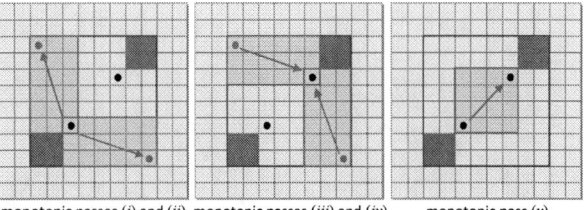

monotonic passes (*i*) and (*ii*)   monotonic passes (*iii*) and (*iv*)   monotonic pass (*v*)

Figure 3: Non-monotonic routing using the Bellman-Ford Algorithm with an expanded bounding box. The red arrows represent monotonic passes.

Theorem 1 (proved in Appendix A) is significant in practice because many connections are routed with very few monotonic segments. In particular, most connections have very few bends [14], and the number of monotonic segments is upper-bounded by the number of bends. Furthermore, a route with many bends can still be monotonic (Figure 6b). In this context, Theorem 1 suggests that BFY typically finds shortest-path routes in $O(1)$ passes, and explains why BFY finds optimal paths faster than A*-search for most nets in our experiments. In a $w \times h$ bounding box, m BFY passes take $O(mwh)$ runtime. Limiting the number of passes further reduces BFY runtime. This may lead to a small loss of optimality in standalone BFY, but our main focus is on incremental calls to BFY during routing estimation (see the "LIRE in CoPR" columns in Table 1).

Incremental routing with BFY can use any existing route, including those previously found by A*-search. Instead of propagating the costs in an ∞-initialized table, we record the partial costs along an existing route (this is significantly faster than populating the entire table). Subsequent BFY passes find an optimal route, but require less runtime when a near-optimal initial route is available (Figure 6). Multiple such initial routes can be recorded in the BFY table before the first pass.[3] *This type of incrementality speeds up not only rip-up-and-reroute and negotiated-congestion methods, but also repeated invocations of LIRE during placement* (Section 4 also outlines other types of incrementality supported by LIRE).

Coarse-grid routing is based on the observation that large nets often admit near-optimal routes with long flat segments. Therefore, we reduce the search space by only considering every i^{th} row and j^{th} column. This allows us to find a reasonable-cost route quickly, and then incrementally relax it on a finer subgrid (multilevel BFY).

2.2 Fast and Accurate Estimation

Unlike true global routing, constructive congestion estimation needs not optimize routes in congestion-free regions. Finding routes that avoid congested GCells is sufficient. Therefore, existing methods first evaluate several pattern routes (L, C, Z) and invoke more sophisticated algorithms only when needed. For similar reasons, eligible GCells are initially limited to the bounding box of the connection, which is gradually expanded until an acceptable route is found. LIRE too estimates congestion by catering to the common case first. For each point-to-point connection $\pi = S \rightarrow T$, LIRE initially limits the search space to the bounding box b with size $w \times h$ that minimally contains π. It considers the two L routes, with preference for congestion-free routes. If both routes are congested, BFY finds a route with $O(1)$ monotonic segments. If this route is congested, LIRE expands b to to $W \times H$, $w < W \leq X, h < H \leq Y$, which may be based on congestion [13].

---

[3]A speed-up common in A*-search (for large nets) limits search to GCells in narrow corridors around known routes. This speed-up is equally applicable to BFY, but not used in our implementation.

Within this expanded rectangular bounding box (Figure 3), we only consider a partial rectilinear bounding box for two reasons. First, we reduce the overall runtime by limiting the bounding box. Second, we observe that the space omitted only contributes to routes that require multiple detours. Since congestion estimation needs not generate heavily-detoured routes, we omit this search space to reduce detours and runtime. Within the rectilinear bounding box B, consider the upper-left (UL) and bottom-right (BR) corners. We perform five monotonic-routing passes: (*i*) $S \rightarrow UL$, (*ii*) $S \rightarrow BR$, (*iii*) $UL \rightarrow T$, (*iv*) $BR \rightarrow T$, and (*v*) $S \rightarrow T$. For nets with extreme aspect ratios, we found that this is more effective than repeating many monotonic passes. Our implementation expands bounding boxes up to twice the original size.

3. CONGESTION RELIEF

The main precept of routability-driven placement is to increase the *porosity* of placement regions with high routing congestion. Regardless of how congestion is estimated, porosity has traditionally been increased in two ways: (*i*) *after global placement*, by shifting cell locations [17, 24] and using congestion-driven detailed placement [3, 5, 8], and (*ii*) *during global placement*, by inflating cells based on early congestion estimates and pin density [3, 5, 8].

While studying the impact of these techniques on challenging IC layouts, we observed their insufficiency. Modifications performed after the global-placement phase must preserve the structure of resulting placement or risk unbearable deterioration of interconnect length. Cell inflation performed during global placement is more flexible and powerful. However, when inflated cells move outside the congested region, new cells must be inflated, and this process may consume all available whitespace without addressing the root cause of congestion in a given region (this phenomenon was confirmed to us by several industrial colleagues and academic colleagues). Further analysis revealed two previously unreported types of routing congestion, which we include below as types 2 and 3.

1) *cell-based* congestion caused by cell-to-cell proximity,
2) *local layout-based* congestion caused by static design properties, such as blockages and reduced routing capacities,
3) *remotely-induced layout-based* congestion attributed to non-local factors, e.g., long nets.

These congestion types are illustrated in Figure 4. The distinctions among them can be blurred by inaccurate congestion maps and also during congestion reduction *after* global placement [17, 24] when cells do not drastically move across the layout. However, these distinctions become sharper when guiding global-placement iterations by accurate congestion maps. Conceptually, type-2 congestion requires whitespace injection into relevant regions *in such a way that whitespace remains in these regions even when cells relocate*.

Cell-based congestion. As the placer spreads cells, it often implicitly keeps cells close together to decrease HPWL. However, this

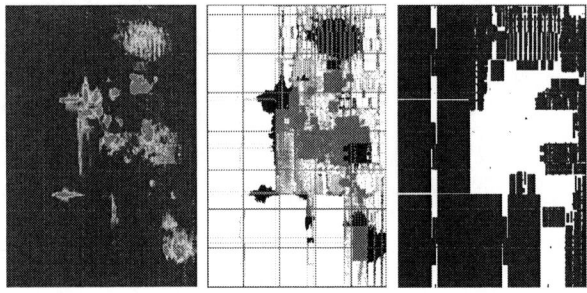

Figure 4: Congestion map produced after one BFG-R [4] iteration (left), placement map of cell locations (center), and blockages (right) for SUPERBLUE2 **[19]. In the center, blue indicates movable cells, and black indicates congested GCells over blockages. Congestion is present around blockages (layout-based) and blockage-free regions (cell-based).**

"clumping" creates difficult-to-route regions, as there may be too few tracks to accommodate all incident nets. This type of congestion is easily mitigated through cell inflation. However, inflating too many cells *or* inflating some cells by too much can exhaust whitespace too soon, inhibit convergence and undermine quality. To ensure steady improvement, we inflate cells in the top 5% most congested GCells. Details are given in Appendix B.

Layout-based congestion. During HPWL-driven placement, the target density is often high, as this facilitates low-HPWL placement solutions. However, if the placer is not congestion-aware, it may pack cells in regions of high congestion. To this end, we seek to locally increase whitespace to encourage cells to spread elsewhere. However, analytic placement frameworks are not always amenable to techniques that change (local) target density. Instead, we enforce non-uniform target densities in localized regions. We distinguish layout-based congestion as either *local*, which is caused primarily by static constraints such as custom routing-edge capacity reductions, or *remotely-induced*, where congested GCells contain no standard cells but have few routing tracks traversed by long nets. While the former can be addressed through locally injecting whitespace, the latter cannot, as there are no cells to move out of the congested region. In the remainder of this section, we discuss our method of enforcing non-uniform target density by (i) creating a *packing peanut* (fixed cell) at the center of every GCell, and (ii) modifying its size based on congestion.

Implementation. To address *local layout-based congestion*, we modify the size of packing peanuts during the initial HPWL optimization stage based on *pin density*, and during the global placement stage based on routing congestion. During initial placement, we coarsely estimate routing congestion of the design based on available routing capacity and cell pin density. We first divide the layout into 8×8 GCell regions and compute the number of pins in each region. We then (pessimistically) estimate that each pin in the region will occupy two routing tracks, and increase the packing peanuts' size based on the ratio of estimated usage and routing capacity. This approach of coarsely dividing the layout gives the placer a high-level outlook and encourages cells spreading to regions of lower pin density. We define two parameters: (i) the maximum expandable area $PA(g)_{max}$, which is based on the surrounding non-overlapping GCell areas, and (ii) the minimum area $PA(g)_{min}$, which is based on GCell pin density. Let $C(g)^k$ be the congestion of GCell g at routing iteration k. Then the packing peanut area $PA(g)$ at g is initially $PA(g)_{min}$ and

$$PA(g) + 0.15 \cdot \left(PA(g)_{max} - PA(g)\right) \qquad (1)$$

if $C(g)^k > C(g)^{k-1}$ and $C(g)^k > 1$. If the congestion is reduced but not removed, i.e., $C(g)^{k-1} > C(g)^k > 1$, then the packing peanut size remains the same. Otherwise, if congestion is removed, the size is $\max\{PA(g)_{min}, PA(g) \times 40\%\}$.

To address *remotely-induced layout-based congestion*, we increase the packing peanuts in GCells *closest to the congested region* and their neighboring GCells. Such modifications often occur around placement blockages. Across placement iterations, the packing peanuts increase placement porosity by reducing the demand in regions without blockages as well as customizing the resource distribution around blockages. Unlike rectangular macro expansion [5], our approach affords the placer a high degree of flexibility to where long nets can be shifted (Figure 7). It complements cell-inflation techniques by preventing allocated whitespace from moving away.

4. COORDINATED PLACE AND ROUTE

The integration of routing estimation within placement allows us to leverage the existing infrastructure and avoids task redundancy. Giving LIRE up-to-date access to cell locations simplifies the construction of new congestion maps when placement changes.

Incremental placement updates. After its first invocation, LIRE maintains the overall congestion map and keeps track of the GCells traversed by each point-to-point connection π. At subsequent LIRE invocations, if the endpoints of π remain in the same GCells (despite changes in their continuous-valued locations), π's route and its contribution to the congestion map are left unchanged. While this type of incrementality has limited use in early placement iterations, its effect is more pronounced in later iterations and during detailed placement, when the locations have stabilized.

Incremental route updates. When invoked for the first time, LIRE generates routes from scratch. Subsequently, it tries to reuse existing routes where possible. Nets whose terminals relocated to different GCells are rerouted using the original net ordering, as outlined in Section 2. For remaining nets, we check if their routes are congested. Congestion is mitigated by single incremental BFY passes. This helps replicate the accuracy of a maze router and improves runtime by reusing full and partial routes.

Placement-routing interface for coordinated place-and-route:

- LIRE::Initialize() reads in the benchmark information, sets up the routing environment, and computes the *static* routing-edge capacities (e.g., due to blockages or custom capacity reductions). Dynamic capacity adjustments such as pin blockages in Section 6, are accounted for by LIRE::updatePlacement().
- LIRE::updatePlacement() restructures the nets based on any placement changes, and maintains lists of nets that require full modification, as well as those that can be reused. Dynamic routing capacities are adjusted due to cell-location updates.
- LIRE::route() generates routes on an as-needed (lazy) basis. It decomposes each multi-pin net into two-pin subnets based on its MST, and follows the protocol outlined in Section 2.
- LIRE::genCongMap() translates edge capacities and usages to a GCell-centric congestion map as in [8], where a GCell is congested if any surrounding edge is congested.

The handling of design hierarchy is entrusted to the placer and does not add complexity to the place-and-route interface (Figure 8). In summary, we advocate a *coordinated* integration style of physical optimizations, where each component uses algorithms that are independently-meaningful and independently-efficient, but also are capable of taking external suggestions. Unlike *simultaneous place-and-route* advocated in [8], this type of integration limits software complexity, allows for component replacement and unit testing. It eases the integration of timing analysis and other components necessary for effective timing closure of modern SoC designs [10].

5. COMPARISONS TO PRIOR ART

Comparing our techniques to prior art, we consider (*i*) point-to-point routing algorithms, (*ii*) using global routes versus probabilistic congestion maps, (*iii*) incremental routing techniques, and (*iv*) handling congestion around blockages.

Fast routing. The closest publication to our material in Section 2 is [13]. It advocates replacing A*-search with fast linear-time routing algorithms that exploit a different notion of monotonic routes (our work was completed before [13] was published or available to us). It uses multiple passes to find non-monotonic routes and does not claim optimality. It does not consider CPU cache effects and the connection we establish to the Bellman-Ford algorithm with Yen's improvement. Empirically, the RCE estimator [13] is not used to drive a competitive global placer, whereas we report successful results for coordinated place-and-route using LIRE. We believe that congestion-driven bounding-box expansion pioneered in RCE can be valuable, but have not had the time to implement and evaluate it.

The only modern description of an industry router that we could find is in [10]. It concedes that Dijkstra's algorithm [1, Section 24.3] (from which A*-search is derived) is "much too slow" for large modern netlists, even with Fibonacci heaps. However, rather than replace Dijkstra with linear-time algorithms as we do, the authors speed it up with sophisticated data structures (interval-based route-cost representations) and algorithms (sharper admissible functions for A*-search based on landmarks). Direct comparisons would be difficult to make, even if we had access to their source code, because advanced data structures use more memory and require significant up-front set-up, along with maintenance. However, a single-threaded version of LIRE takes only <15% of runtime in our entire place-and-route flow, despite frequent (>10) invocations by the placer (Table 1). Speeding it up further would have limited impact. Most importantly, we have advanced the goal of our research — to tame the complexity of place-and-route — by entirely avoiding sophisticated routing algorithms and data structures.

Congestion estimation must accurately identify hotspots and guide the placer to relieve congestion. While *probabilistic congestion maps* are easy to implement, they can be slower per net than constructive routing, as shown in [22]. They are also highly inaccurate, as recently articulated by IBM researchers [11]. Nevertheless, most routability-driven placers [3, 5] still use probabilistic methods. As illustrated in Figure 5, congestion maps built using *LZ*-routing [14] significantly differ from router-based maps. Table 2 compares total overflow (TOF) between *L*-routing, *LZ*-routing, LIRE, and maze routing [4]. On average, LIRE overestimates TOF by 4% with no significant outliers.

Incremental routing techniques. All modern routability-driven placers [3, 5, 8] use built-in congestion estimation to construct new estimates from scratch on every invocation. This process is unnecessarily time-consuming, especially when the placement has not changed significantly. While some prior techniques rip-up and reroute some congested nets [25], they assume a static routing (and placement) instance. In contrast, our incremental techniques account for dynamic placement (and routing) instances, take advantage of previous (partial) routes, and update routes on an as-needed basis. These techniques are especially applicable to congestion estimators based on constructive global routing, but also should be helpful in full-fledged routers. Empirically, we matched the accuracy of a full global router with limited runtime overhead.

Placement and routing blockages, e.g., macro blocks, often lead to congestion around their borders. Previous work [5] proactively reserves resources by expanding macros. However, (rectangular) macro inflation is rather crude in controlling whitespace — it either allows all cells or prevents all cells in a given rectangular region. Our non-uniform target density, as implemented with packing peanuts, provides much more flexible control of whitespace, as shown in Figure 7. By increasing the packing peanut sizes in areas of congestion *and* in selected neighboring GCells, we allow cells to move into congestion-free regions around macro borders, whatever shape those regions may assume.

6. EMPIRICAL VALIDATION

Our algorithms are implemented in a tool called CoPR (pronounced "copper") in C++ using the OpenMP library [2] and compiled with g++ 4.7.0. Our global placer is derived from SimPL [9], which was the case for three out of the top four teams at the ICCAD 2012 Contest [20]. Thus, the choice of the global placement algorithm is not a significant factor in relative performance.

Empirical results are reported on the ICCAD 2012 benchmark suite [20] derived by IBM researchers from industry designs. Some of these benchmarks were released only after the results of the ICCAD 2012 Contest were announced. The overall figure of merit combines quality metrics (interconnect length, routing congestion evaluated by a router, and pin blockage) and runtime. Table 1 compares CoPR to official contest results [20] for the top three contestants. With quality metrics based on the NCTUgr router (without runtime), CoPR outperforms NTUplace4h by 7% and SimPLR by 2%, while matching the overall quality of Ripple, which is $5.7\times$ slower. CoPR runtime intentionally matches SimPLR (the fastest top-3 contestant, which trails CoPR in quality) so that officially-reported runtime ratios between SimPLR and other contestants also apply to CoPR. The last two columns in Table 1 show that LIRE is called by CoPR 14-22 times per run, amounting to <15% of CoPR's runtime. With quality metrics based on the BFG-R router, CoPR outperforms NTUplace4h by 3%, SimPL by 6% and Ripple by 2%, respectively (Table 3). With respect to scoring formulas used at the ICCAD 2012 contest, CoPR outperforms the winner SimPLR (from which CoPR was derived).

7. CONCLUSIONS

Our work deals with an alarming trend in the design or digital random-logic blocks, where interconnect's dominance in area, volume, delay, power and signal-integrity is increasing with every new technology node [7]. If unchecked, this trend is threatening to render Moore's law irrelevant — packing more devices on a chip is useless if they cannot be effectively connected. The most direct and effective remedy known today is to reduce interconnect demand, which can be done by optimizing standard-cell locations and wire routes. As articulated recently by IBM researchers, design flows with separate placement and routing steps have become ineffective for modern ICs [19], but combining the two brings tangible and significant benefits in IC cost [17]. However most of physical-design research continues focusing on standalone optimizations, partly due to the complexities involved in place-and-route integration. These complexities include sophisticated data structures and elaborate multistep optimizations used by state-of-the-art algorithms [10], unmaintainable source-code bases that are unnecessarily entangled, large sets of tuning parameters that may need to be adjusted to individual benchmarks, and of course significant runtime. In this work, we develop an algorithmic framework for coordinated place-and-route (CoPR) that combines independently-meaningful components and systematically reduces the complexities of place-and-route. Our contributions fall into four categories: (*i*) dramatic acceleration of constructive routing estimation through linear-time cache-friendly algorithms that do not require sophisticated data structures, (*ii*) significant reductions in the amount of work through pervasive incrementality at the interface between

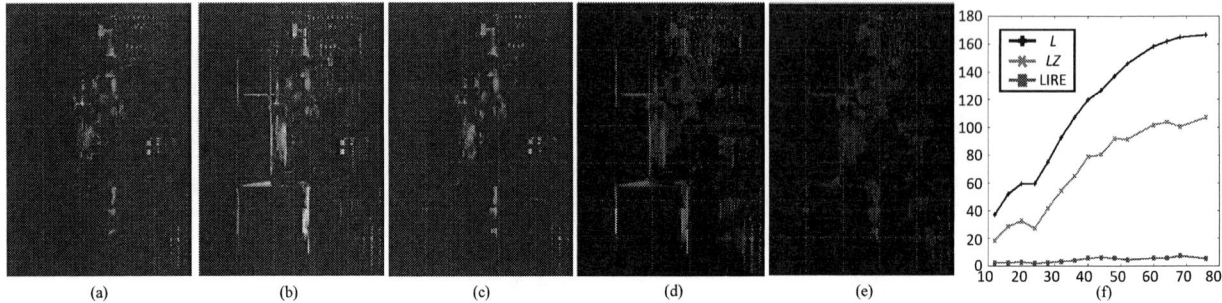

Figure 5: Comparison of different routing estimation techniques on the SUPERBLUE2 benchmark [19]. The congestion map in (a) is produced by BFG-R [4], in (b) — by LZ-routing, and in (c) — by LIRE. Images in (d) and (e) show how well (b) and (c) match (a) — ratios of congestion values are plotted. Orange areas indicate large differences and black areas — no difference. (f) plots the error percentage of total overflow for L-routing, LZ-routing, and LIRE relative to (a) over the placement iterations of CoPR. While all techniques overestimate congestion, LZ-routing and L-routing produce many false positives, whereas LIRE does not.

| Benchmark | Nodes | Nets | Quality metrics using NCTUgr [12] (e8) | | | | Runtime ratio | LIRE in CoPR | |
			SimPLR (1)	Ripple (2)	NTUplace4 (3)	CoPR	CoPR/SimPLR	calls	LIRE %
SUPERBLUE1	847K	822K	2.789	2.889	2.850	2.860	1.058	14	12.4%
SUPERBLUE3	920K	898K	3.439	3.604	4.477	3.457	0.962	15	14.8%
SUPERBLUE4	600K	567K	2.434	2.269	2.360	2.366	1.445	22	13.9%
SUPERBLUE5	772K	787K	3.603	3.486	4.217	3.510	1.089	14	11.8%
SUPERBLUE7	1.36M	1.34M	4.313	4.291	4.137	4.360	1.099	15	11.9%
SUPERBLUE10	1.20M	1.15M	6.909	6.111	7.190	6.505	1.054	22	16.4%
SUPERBLUE16	699K	697K	2.857	2.840	2.833	2.797	0.687	14	15.3%
SUPERBLUE18	1.27M	469K	1.823	1.791	1.709	1.676	0.816	17	19.0%
Ratios of averages ($\times$)			1.02$\times$	1.00$\times$	1.07$\times$	1.00$\times$	1.01$\times$		14.3%

Table 1: Quality metrics (based on NCTUgr [12]) *without runtime* for the top three contestants as reported at the ICCAD 2012 Routability-driven Placement Contest [20]. CoPR runtimes are compared to those of the fastest top-3 contestant SimPLR by running both tools on the same server. The last two columns show the runtime of LIRE as a percent of total CoPR runtime, and the number of LIRE invocations on each benchmark. Full results for SimPLR, RippleCUHK and NTUplace4h are available at [20]. Using BFG-R [4] (rather than NCTUgr) to evaluate results results in greater advantage for CoPR as shown in Table 3.

placement and routing, (iii) identification of two new types of routing congestion, as well as mechanisms by which a global placer can diagnose them and respond effectively, and (iv) strong empirical results on the most recent benchmarks from IBM Research.

Our place-and-route improvements should (a) lead to more compact (less costly) IC layouts in a way illustrated in [17], and (b) reduce back-end turn-around-time so that IC designers can evaluate a greater number of micro-architectural configurations.

8. REFERENCES

[1] T. H. Cormen, C. E. Leiserson, R. L. Rivest and C. Stein, *Introduction to Algorithms*, Second Edition, MIT Press and McGraw-Hill, 2001.

[2] L. Dagum and R. Menon, "OpenMP: An Industry Standard API for Shared-memory Programming," *Computational Science and Engineering* 1998, pp. 46-55.

[3] X. He, T. Huang, L. Xiao, H. Tian, G. Cui and E. F. Young, "Ripple: An Effective Routability-driven Placer by Iterative Cell Movement", *ICCAD* 2011, pp. 74-79.

[4] J. Hu, J. A. Roy and I. L. Markov, "Completing High-quality Global Routes", *ISPD* 2010, pp. 35-41.

[5] M.-K. Hsu, S. Chou, T.-H. Lin and Y.-W. Chang, "Routability-driven Analytical Placement for Mixed-size Circuit Designs", *ICCAD* 2011, pp. 80-84.

[6] L. Hsu, R. Iyer, S. Makineni, S. Reinhardt and D. Newell, "Exploring the Cache Design Space for Large Scale CMPs," *Computer Architecture News* 2005, pp. 24-33.

[7] International Technology Roadmap for Semiconductors (ITRS).

[8] M.-C. Kim, J.Hu, D.-J. Lee and I. L. Markov, "A SimPLR Method for Routability-driven Placement", *ICCAD* 2011, pp. 67-73.

[9] M.-C. Kim, D.-J. Lee and I. L. Markov, "SimPL: An Effective Placement Algorithm", *TCAD* 31(1) (2012), pp. 50-60.

[10] B. Korte, D. Rautenbach and J. Vygen, "BonnTools: Mathematical Innovation for Layout and Timing Closure of Systems on a Chip", *Proc. IEEE* 95(3) (2007), pp. 555-572.

[11] Z. Li, C. J. Alpert, G.-J. Nam, C. Sze, N. Viswanathan and N. Y. Zhou,

"Guiding a Physical Design Closure System to Produce Easier-to-route Designs with More Predictable Timing", *DAC* 2012, pp. 465-470.

[12] W.-H. Liu, W.-C. Kao, Y.-L. Li, K.-Y. Chao, "Multi-threaded Collision-aware Global Routing with Bounded-length Maze Routing", *DAC* 2010, pp.200-205.

[13] W.-H. Liu, Y.-L. Li, C.-K. Kok,"A Fast Maze-free Routing Congestion Estimator With Hybrid Unilateral Monotonic Routing",*ICCAD* 2012, pp.713-719.

[14] M. Pan, Y. Xu, Y. Zhang and C. Chu, "FastRoute: An Efficient and High-quality Global Router", *VLSI Design* 2012, 18 pages.

[15] S. K. Raman, V. Pentkovski and J. Keshava, "Implementing Streaming SIMD Extensions on the Pentium III Processor", *Micro* 20(4)(2000), pp. 47-57.

[16] P. N. Parakh, R. B. Brown and K. A. Sakallah, "Congestion Driven Quadratic Placement, *DAC* 1998, pp. 275-278.

[17] J. A. Roy, N. Viswanathan, G.-J. Nam, C. J. Alpert and I. L. Markov, "CRISP: Congestion Reduction by Iterated Spreading during Placement", *ICCAD* 2009, pp. 357-362.

[18] N. Viswanathan, C. J. Alpert, C. Sze, Z. Li, G.-J. Nam and J. A. Roy, "The ISPD-2011 Routability-driven Placement Contest and Benchmark Suite", *ISPD* 2011, pp. 141-146.

[19] N. Viswanathan, C. J. Alpert, C. Sze, Z. Li, Y. Wei, "The DAC 2012 Routability-driven Placement Contest and Benchmark Suite", *DAC* 2012, pp. 774-782.

[20] N. Viswanathan, C. J. Alpert, C. Sze, Z. Li and Y. Wei, "ICCAD-2012 CAD Contest in Design Hierarchy Aware Routability-driven Placement and Benchmark Suite", *ICCAD* 2012, pp. 345-348. cad_contest.cs.nctu. edu.tw/CAD-contest-at-ICCAD2012/problems/p2/p2.html

[21] Y. Wei, C. Sze, N. Viswanathan, Z. Li, C. J. Alpert, L. N. Reddy, A. D. Huber, G. E. Terez, D. Keller and S. S. Sapatnekar, "GLARE: Global and Local Wiring Aware Routability Evaluation", *DAC* 2012, pp. 768-773.

[22] J. Westra and P. Groeneveld, "Is Probabilistic Congestion Estimation Worthwhile?" *SLIP* 2005, pp. 99-106.

[23] J. Y. Yen, "An Algorithm for Finding Shortest Routes From All Source Nodes to a Given Destination in General Networks", *Proc. Quarterly of Applied Mathematics* 27 (1970), pp.526-530.

[24] Y. Zhang and C. Chu, "CROP: Fast and Effective Congestion Refinement of Placement", *ICCAD* 2009, pp. 344-350.

[25] Y. Zhang and C. Chu, "GDRouter: Interleaved Global Routing and Detailed Routing for Ultimate Routability", *DAC* 2012, pp. 597-602.

Appendix A. Proof of Theorem 1

Let t be the $w \times h$ matrix, where $t[x][y]$ stores the cost of the optimal path from its cardinal neighbors. Consider the first pass where partial costs are not yet propagated. By construction, $t[x][y]$ only depends on $t[x-1][y]$ and $t[x][y-1]$, and requires one BF(Y) pass. Therefore, an optimal route r_π with $m = 1$ monotonic segments is found after $m = 1$ passes. Consider the general case where r_π has k distinct monotonic segments. By assumption, r_π is formed using k BFY passes. By the early termination criterion, if BFY changes no costs in t, then $r_\pi = r_\pi^{min}$. If relaxation is successful during the backward pass, then r_π is allowed to *detour* through some intermediate node v' such that the route cost from S to v is reduced by going through v'. If such v' exists, then there exists a new path from S to v through v' such that the new path has an additional monotonic segment. During the forward pass, the full path of S to T through v. If going through v reduces the cost, then there is an additional monotonic segment $v \to T$. Therefore, for two additional BFY passes for an r_π with k distinct monotonic segments, we will generate a new path with $k + 2$ monotonic segments. Because we consider all intermediate nodes v as detours, the best-cost path will be stored. Therefore, if r_π^{min} has $m = k + 2$ monotonic segments, it will require $k + 2$ BFY passes. $\square$

Appendix B. Cell Inflation

We inflate each cell in the top 5% most congested GCells by computing its new width as follows.

$$\max\{\text{width}(cell) + 1, 1 + \theta(G) \cdot \Lambda(cell) \cdot \deg(cell)\} \quad (2)$$

Here, $cell$ is a movable cell in a congested GCell, $\text{width}(cell)$ and $\deg(cell)$ are the width and connectivity of $cell$, respectively. $\theta(G)$ is an adaptive function (described below) of the routing grid G, and $\Lambda(cell)$ is the number of times $cell$ has been in a congested GCell. We define θ similarly to [8, Equation 12], except that we upperbound θ to limit how much a cell can be inflated.

$$\theta = \min\{0.5, \max\{0, \alpha \cdot \eta(G) \cdot \xi(G) + \beta\}\} \quad (3)$$

Here, $\eta(G)$ and $\xi(G)$ represent the respective *difficulty* and *routability* of the design, where $\eta(G)$ is the sum of every GCell congestion in G, and $\xi(G)$ is the ratio of the total GCell congestion in G. α and β are constants based on linear regression. Unlike previous cell-inflation approaches [8], our formula *does not include* the GCell's congestion. By excluding the numeric congestion value, we only rely on the routing estimator's accuracy for congestion *locations*, and less on the reported congestion value. This prevents excessive inflation, and facilitates a smooth placement transition.

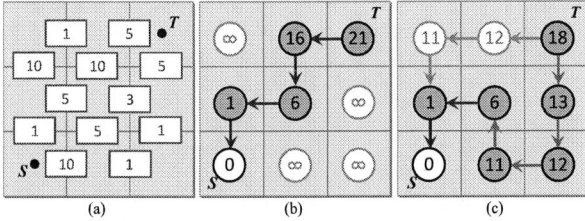

Figure 6: Applying BFY to an initial route for a point-to-point connection $S \to T$. (a) The routing grid and edge costs (congestion). (b) The initial route with cost 21. (c) Through relaxation, BFY can preserve part of the route, and find a better partial segment, resulting in a new route with cost 18.

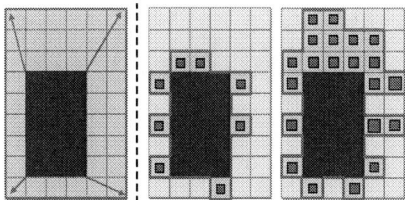

Figure 7: Congestion-driven rectangular macro expansion [5] (left) versus our technique (right).

Iter. #	Total overflow (e5)				Comparison vs. maze		
	maze	L	LZ	LIRE	L	LZ	LIRE
12	31.04	42.59	36.78	31.80	1.372	1.185	1.024
16	20.41	31.00	26.30	20.91	1.519	1.289	1.024
20	16.00	25.49	21.22	16.45	1.594	1.327	1.039
24	15.13	24.13	19.31	15.13	1.595	1.276	1.020
28	11.68	20.44	16.58	11.96	1.749	1.420	1.024
32	7.880	15.17	12.16	8.149	1.925	1.544	1.034
36	6.424	13.29	10.59	6.684	2.069	1.649	1.041
40	5.452	11.99	9.745	5.755	2.199	1.787	1.056
44	5.051	11.44	9.108	5.359	2.266	1.803	1.061
48	4.636	10.98	8.895	4.898	2.369	1.919	1.057
52	4.375	10.75	8.382	4.575	2.458	1.916	1.046
60	3.825	9.876	7.721	4.043	2.582	2.019	1.057
64	3.718	9.736	7.572	3.931	2.618	2.036	1.057
68	3.697	9.796	7.410	3.964	2.650	2.004	1.072
76	3.503	9.337	7.254	3.684	2.665	2.071	1.052
Avg					2.06×	1.65×	1.04×

Table 2: Total overflow estimation comparisons of L-routing, LZ-routing, the initial (maze) routing of BFG-R [4], and LIRE inside CoPR for the SUPERBLUE2 benchmark [19] (Figure 5f).

Benchmark	Quality metrics using BFG-R [4] (e8)			
	SimPLR	Ripple	NTUplace4	CoPR
SUPERBLUE1	3.023	3.341	2.962	3.084
SUPERBLUE3	3.803	3.906	4.609	3.757
SUPERBLUE4	2.865	2.659	2.773	2.530
SUPERBLUE5	3.980	3.654	3.919	3.646
SUPERBLUE7	4.479	4.502	4.283	4.439
SUPERBLUE10	8.114	7.080	7.810	7.378
SUPERBLUE16	3.117	2.929	3.032	2.989
SUPERBLUE18	2.461	2.207	1.838	2.163
Ratios (×)	1.06×	1.02×	1.03×	1.00×

Table 3: Quality metrics (based on BFG-R [4]) *without runtime* for the top three contestants as reported at the ICCAD 2012 Routability-driven Placement Contest [20] and CoPR.

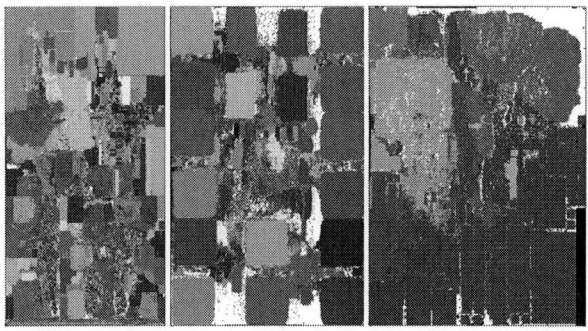

Figure 8: CoPR placements of the SUPERBLUE7 (left), SUPERBLUE10 (center), and SUPERBLUE18 (right) testcases [20].

1037

Routability-Driven Placement for Hierarchical Mixed-Size Circuit Designs *

Meng-Kai Hsu[1], Yi-Fang Chen[1], Chau-Chin Huang[1], Tung-Chieh Chen[3], and Yao-Wen Chang[1,2]

[1]Graduate Institute of Electronics Engineering, National Taiwan University, Taipei, Taiwan
[2]Department of Electrical Engineering, National Taiwan University, Taipei, Taiwan
[3]Synopsys Inc., Hsinchu, Taiwan

{kaie, yifang, wlkb83}@eda.ee.ntu.edu.tw, donnie.chen@synopsys.com, ywchang@cc.ee.ntu.edu.tw

ABSTRACT

A wirelength-driven placer without considering routability could introduce irresolvable routing-congested placements. Therefore, it is desirable to develop an effective routability-driven placer for modern mixed-size designs employing hierarchical methodologies for faster turnaround time. This paper presents a novel two-stage technique to effectively identify design hierarchies and guide placement for better wirelength and routability. To optimize wirelength and routability simultaneously during placement, a new analytical net-congestion-optimization technique is also proposed. Compared with the participating teams for the 2012 ICCAD Design Hierarchy Aware Routability-driven Placement Contest, our placer can achieve the best quality (both the average overflow and wirelength) and the best overall score (by additionally considering running time).

Categories and Subject Descriptors

B.7.2 [**Integrated Circuits**]: Design Aids [Placement and Routing]

General Terms

Algorithms, Performance

Keywords

Physical Design, Placement, Routability

1. INTRODUCTION

Due to the advancement of process technology, a modern circuit design could contain more than billions of transistors in a single chip. Although the placement problem has been studied for several decades, modern placement still remains very tough due to the fast-growing design complexity and advanced process technologies. Therefore, many new placement algorithms were developed recently, and most of these algorithms try to optimize total wirelength for better circuit performance. However, because existing wirelength models cannot accurately model the congestion objective in placement, a wirelength-driven placer that does not consider routability could introduce irresolvable routing-congested

---

*This work was partially supported by IBM, SpringSoft, TSMC, Academia Sinica, and NSC of Taiwan under Grant No's. NSC 101-2221-E-002-191-MY3, NSC100-2221-E-002-088-MY3, NSC 99-2221-E-002-207-MY3 and NSC 99-2221-E-002-210-MY3.

Permission to make digital or hard copies of all or part of this work for personal or classroom use is granted without fee provided that copies are not made or distributed for profit or commercial advantage and that copies bear this notice and the full citation on the first page. To copy otherwise, to republish, to post on servers or to redistribute to lists, requires prior specific permission and/or a fee.
DAC'13, May 29 - June 07 2013, Austin, TX, USA

placements. Furthermore, the big difference of large macros and small standard cells brings new challenges to modern mixed-size placement. Because large macros usually occupy several metal layers (called macro porosity), routing resources for a modern mixed-size circuit design are more constrained than traditional circuit designs.

In recent years, hierarchical design methodologies are often employed in modern circuit designs for faster turnaround time [7]. While placing circuit elements which belong to a particular hierarchy closely is expected to have better routability and circuit performance [7, 22], how to honor the design hierarchy of a circuit during placement incurs big challenges to modern circuit designs. Therefore, it is desirable to develop an effective routability-driven placer for modern hierarchical mixed-size designs.

1.1 Previous Work

The routability-driven placement problem has attracted increasing attention in recent literature. Congestion-driven Dragon [23] first distributes whitespace into rows and then into bins within a row. As it may increase the row imbalance, a lower bound and an upper bound on the amount of whitespace available in a row are imposed. Li *et al.* [16] distributes whitespace by adjusting the cut-lines of hierarchically sliced placement and re-placing cells to avoid congested regions. ROOSTER [18] uses rectilinear spanning minimal trees as its routing congested model during global placement and applies whitespace allocation during detailed placement. BonnPlace [2], CRISP [19], Ripple [9], and SimPLR [15] implicitly allocate whitespace by inflating cell area according to estimated routing congestions. Although allocating whitespace is pervasive in fixed-die placement, inappropriate whitespace distribution might induce longer wirelength and hurt circuit routability. Instead of allocating whitespace during placement, RUDY [20] estimates routing congestion and modifies the density term to consider both the routing density and module density. Jiang *et al.* [14] and Chuang *et al.* [7] reduce congestion by removing overlaps between net bounding boxes. In addition to reducing overlaps between net bounding boxes, Chuang *et al.* [7] utilize design-heirarchy information to further improve circuit routability. Tsota *et al.* [21] integrates the wire density term into an analytical placement framework. Hsu *et al.* [12] optimizes routability by introducing a sigmoid function based overflow refinement method during analytical placement. Incorporating routability models in the optimization objective during placement could effectively optimize routability. However, how to accurately estimate the exact effect of routing incurs big challenges.

1.2 Our Contributions

This paper presents a new routability-driven analytical placement algorithm for hierarchical mixed-size circuit designs. As preserving design hierarchies physically is significantly important for modern circuit designs, we propose a novel two-stage technique to effectively identify design hierarchies and guide placement for better wirelength and routability. Furthermore, unlike most existing works which usually optimize routability by implicitly or explicitly reallocating whitespace, we propose a new analytical net-congestion-optimization technique to optimize routability and wirelength simultaneously during placement. We summarize the main contributions of our proposed algorithm as follows.

- A new routability-driven analytical placement algorithm is proposed. Different from most existing works which optimize routability by explicitly or implicitly reallocating whitespace, our algorithm optimizes routability from three major aspects: (1) design hierarchy identification, (2) narrow channel handling, and (3) net congestion optimization.

- We propose a smoothed congestion estimation method to model the global routing behavior and try to give more accurate congestion estimation during placement.

- A novel two-stage technique is proposed to effectively identify design hierarchies and guide placement for better wirelength and routability.

- To consider the constrained routing resource incurred by big macros, a new narrow-channel-handling technique is proposed.

- A novel net-congestion-optimization technique is proposed to optimize routability and wirelength simultaneously during placement.

- Experimental results show that our proposed algorithm is effective. Compared with the participating teams for the 2012 ICCAD Design Hierarchy Aware Routability-driven Placement Contest [1], our placer can achieve the best quality (both the average overflow and wirelength) and the best overall score (by additionally considering running time).

The remainder of this paper is organized as follows. Section 2 formulates the routability-driven placement problem and introduces our analytical placement framework. Section 3 introduces our proposed routability-driven placement algorithm. Experimental results are shown in Section 4, and Section 5 concludes this paper.

2. PRELIMINARIES

In this section, we introduce the routability-driven placement problem and our analytical placement framework.

2.1 Problem Formulation

The circuit placement problem can be formulated as a hypergraph $H = (V, E)$ placement problem. Let vertices $V = \{v_1, v_2, ..., v_n\}$ represent blocks and hyperedges $E = \{e_1, e_2, ..., e_m\}$ represent nets. Let x_i and y_i be the x and y coordinates of the center of block v_i, respectively. The blocks to be placed can be categorized into two types: pre-placed blocks and movable blocks. Pre-placed blocks have fixed x and y coordinates and cannot be moved.

Global routing is often used to evaluate the routability of a placement. After dividing the routing region into uniform and non-overlapping grids, a routing edge is the edge between two adjacent routing grids, and the capacity of each routing edge is defined by the available routing resources on the edge. Then, the circuit routability can be evaluated by the total overflow of all routing edges. It should be noted that, the capacities of routing edges which overlap with big macros would decrease due to the macro porosity issue. Moreover, to model local congestions during global routing, pin density in each routing grid is usually considered. In [1], to model local congestions more precisely, pin densities are formulated as routing blockages, and thus will also decrease the capacities of routing edges. The routability-driven placement can then be defined as follows: Given a netlist and global routing resources, we intend to determine the desired positions of movable blocks so that the total wirelength and total overflow are minimized.

2.2 Our Analytical Placement Framework

The placement problem is usually solved in three steps: (1) global placement, (2) legalization, and (3) detailed placement. Global placement evenly distributes the blocks and finds the best position for each block while minimizing the target cost (e.g., wirelength). Then legalization removes all overlaps. Finally, detailed placement refines the placement solution. Generally, global placement has the most crucial impact on the overall placement quality, and is thus considered as the most critical step [4].

After the placement region is divided into uniform non-overlapping grids, the global placement problem can be formulated as a constrained minimization problem as follows:

$$
\begin{aligned}
\min \quad & W(\mathbf{x}, \mathbf{y}) \\
\text{s.t.} \quad & D_b(\mathbf{x}, \mathbf{y}) \leq M_b, \quad \text{for each bin } b,
\end{aligned}
\tag{1}
$$

where $W(\mathbf{x}, \mathbf{y})$ is the wirelength function, $D_b(\mathbf{x}, \mathbf{y})$ is the potential function that is the total area of movable blocks in bin b, and M_b is the maximum allowable area of movable blocks in bin b. M_b can be computed by $M_b = t_{density}(w_b h_b - P_b)$, where $t_{density}$ is a user-specified target density value for each bin, w_b (h_b) is the width (height) of bin b, P_b is the base potential equal to the pre-placed block area in bin b, and M_b is a fixed value as long as all pre-placed block positions are given and the bin size is determined.

During placement, the wirelength $W(\mathbf{x}, \mathbf{y})$ is usually defined as the total half-perimeter wirelength (HPWL),

$$
\begin{aligned}
W(\mathbf{x}, \mathbf{y}) &= \sum_{e \in E} (\max_{v_i, v_j \in e} |x_i - x_j| + \max_{v_i, v_j \in e} |y_i - y_j|) \\
&= \sum_{e \in E} (\max_{v_i \in e} x_i - \min_{v_i \in e} x_i + \max_{v_i \in e} y_i - \min_{vi \in e} y_i).
\end{aligned}
\tag{2}
$$

Since $W(\mathbf{x}, \mathbf{y})$ in Equation (2) is not smooth, it is hard to minimize it directly. As a result, several smooth wirelength approximation functions have been proposed. We use the weighted-average (WA) wirelength model,

$$
\sum_{e \in E} \left(\begin{array}{c} \dfrac{\sum_{v_i \in e} x_i \exp\left(\frac{x_i}{\gamma}\right)}{\sum_{v_i \in e} \exp\left(\frac{x_i}{\gamma}\right)} - \dfrac{\sum_{v_i \in e} x_i \exp\left(\frac{-x_i}{\gamma}\right)}{\sum_{v_i \in e} \exp\left(\frac{-x_i}{\gamma}\right)} + \\ \dfrac{\sum_{v_i \in e} y_i \exp\left(\frac{y_i}{\gamma}\right)}{\sum_{v_i \in e} \exp\left(\frac{y_i}{\gamma}\right)} - \dfrac{\sum_{v_i \in e} y_i \exp\left(\frac{-y_i}{\gamma}\right)}{\sum_{v_i \in e} \exp\left(\frac{-y_i}{\gamma}\right)} \end{array} \right),
\tag{3}
$$

proposed in [11], to approximate the HPWL in Equation (2). The WA wirelength model converges to the HPWL in Equation (2), as γ converges to 0.

The potential function is defined as

$$
D_b(\mathbf{x}, \mathbf{y}) = \sum_{v \in V} P_x(b, v) P_y(b, v),
\tag{4}
$$

where P_x and P_y are the overlap functions of bin b and block v along the x and y directions. Since $D_b(\mathbf{x}, \mathbf{y})$ is neither smooth nor differentiable, the sigmoid potential model [6] can be used to smooth the density for each block.

Equation (1) can be solved by the quadratic penalty method, implying that we solve a sequence of unconstrained minimization problems of the form

$$
\min \quad \hat{W}(\mathbf{x}, \mathbf{y}) + \lambda \sum_b (\hat{D}_b(\mathbf{x}, \mathbf{y}) - M_b)^2,
\tag{5}
$$

with increasing λ's, where $\hat{W}(\mathbf{x}, \mathbf{y})$ and $\hat{D}_b(\mathbf{x}, \mathbf{y})$ are the smoothed wirelength and density functions. The solution of the previous problem is used as the initial solution for the next one. We solve the unconstrained problem in Equation (5) by the nonlinear conjugate gradient (CG) method.

3. THE PROPOSED ALGORITHM

In this paper, we propose a new routability-driven analytical placement algorithm. During global placement, we optimize the routability from three major aspects: (1) Design hierarchy identification: we propose a new two-stage technique to effectively identify design hierarchies and guide placement for better wirelength and routability; (2) Narrow channel handling: our algorithm adopts a new narrow-channel-handling technique to overcome the macro porosity issue; (3) Net congestion optimization: Our algorithm optimizes net congestions while minimizing wirelength during analytical placement. Moreover, we propose a smoothed congestion estimation method to model the global routing behavior and try to give more accurate congestion estimation during placement. Besides, due to the increasing complexity of the placement problem, we use the multilevel framework [3, 5] for global placement to improve the scalability. There are two main stages for the multilevel framework: (1) the coarsening stage and (2) the uncoarsening stage. During the coarsening stage, blocks are clustered iteratively to reduce the problem size. Then, during the uncoarsening stage, the placement problem is solved from the coarsest level to the finest

1039

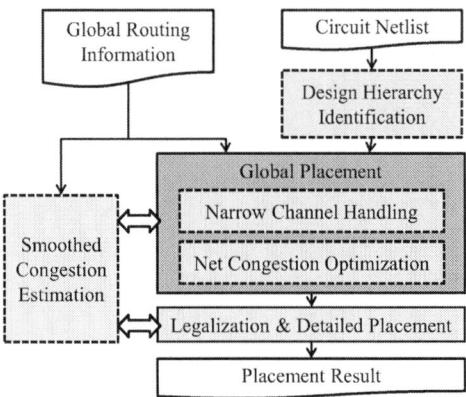

Figure 1: An Overview of our proposed algorithm.

level, where the placement for the current level provides the initial placement for the next level.

After global placement, we adopt the Tetris legalization technique [10] to legalize block positions with minimum displacement, and detailed placement techniques, such as cell matching and cell swapping [5] are used to further improve the placement results. It should be noted that, during legalization and detailed placement, instead of optimizing wirelength only, we use our proposed smoothed congestion estrinization method to evaluate routing overflows cost, and try to place each block with minimum resulting total overflows. Figure 1 summarizes the overall flow of the proposed algorithm. We detail each technique in the following subsections.

3.1 Smoothed Congestion Estimation

To estimate the routing congestion, the L-shaped probabilistic model is one of the most popular models for placement [20]. However, the L-shaped probabilistic model might not accurately capture global routing behavior, such as routing detours and rip-up and re-route during global routing. To model the global routing behavior more precisely, we apply Gaussian smoothing [8] to smooth the L-shaped approximation model. In the two-dimensional case, a circularly symmetric Gaussian has the form

$$G(x,y) = \frac{1}{2\pi\sigma^2} \exp\left(-\frac{x^2 + y^2}{2\sigma^2}\right), \qquad (6)$$

where σ is the standard deviation of the Gaussian distribution. Figure 2 shows an example of our smoothed congestion estimation. Figure 2(a) shows an original congestion estimation, and Figure 2(b) illustrates the congestion estimation after Gaussian smoothing.

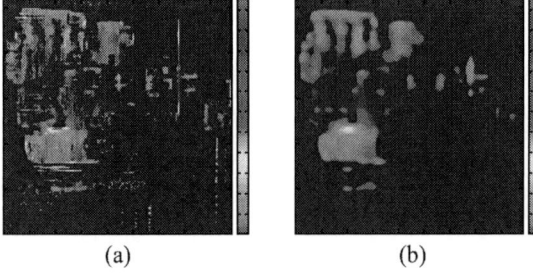

Figure 2: Illustration of smoothed congestion estimation. (a) Original congestion estimation. (b) Congestion estimation after Gaussian smoothing.

3.2 Design Hierarchy Identification

Preserving design hierarchies physically during placement is expected to guide the placement to minimize wirelength as well as

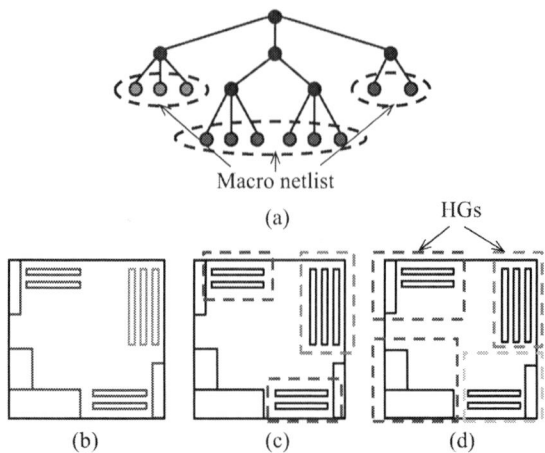

Figure 3: Illustration of macro-based hierarchy grouping. (a) A macro hierarchical structure tree (MHST) of a set of macros. (b) The corresponding placement of these macros. (c) The grouping of neighboring macros with the same dimension. (d) Four HGs constructed according to the MHST. Note that the macros on the top-left corner and bottom-right corners should not be identified as the same HG because the distance of them is too large.

routing congestion [7, 22]. However, inappropriate handling of design hierarchies might hurt the wirelength and routability substantially. As mentioned before, our algorithm adopts the multilevel framework to improve the scalability. While the initial block positions are determined after the coarsening stage, clustering blocks with deeper hierarchical relations could help to preserve design hierarchies during the uncoarsening stage and thus guide the placement for better wirelength and routability. Therefore, in this section, we propose a two-stage technique which includes (1) macro-based hierarchy grouping and (2) hierarchy group (HG)-based clustering to identify design hierarchies and guide placement for better wirelength and routability. In the macro-based hierarchy grouping stage, we construct a set of HGs based on the design hierarchy of macros, where an HG is a group of blocks which are candidates for clustering. Then, we cluster blocks based on these HGs. By using HGs, blocks could be clustered efficiently while considering the design hierarchy for better placement results. We detail each stage in the following subsections.

3.2.1 Macro-based Hierarchy Grouping

From designers' perspective, macros and the standard cells which have strong connectivity with those macros would be seen as in the same hierarchy. In modern mixed-size circuit designs [22], most macros are pre-placed and fixed, and thus clustering blocks only based on design hierarchy information [7] and without considering the physical information of macros might not be sufficient. Therefore, in our design hierarchy identification method, we determine a set of HGs based on the hierarchy and positions of macros for better clustering quality.

In our macro-based hierarchy grouping method, we first construct macro hierarchical structure trees (MHSTs) according to the design hierarchy of macros. Figure 3(a) shows an example of an MHST for a set of macros, and the corresponding placement of these macros is shown in Figure 3(b). After the construction of an MHST, macros in the MHST with the same dimension and placed close to each other are grouped first, as shown in Figure 3(c), and other macros are then grouped in a bottom-up manner according to the MHST, as shown in Figure 3(d). As mentioned before, the positions of pre-placed macros should be considered carefully during clustering. As shown in Figure 3(d), the macros on the top-left and bottom-right corners belong to the same hierarchy, but they should not be identified as the same HG because their distance is too large. Hence, during constructing the HGs, we should consider not only the design hierarchy information, but also the distance between two macros. To finish the construction of HGs, for each macro, we cluster the standard cells with the same hierarchy as that of the macros into the same group. For standard cells which do not have the same hierarchy as any macro, they are grouped according to the original design hierarchy information [7].

3.2.2 HG-based Clustering

To simultaneously consider the design hierarchy and net connectivities during the coarsening stage, we propose an HG-based clustering to evaluate the affinity between two blocks which denotes the relation between the two blocks. We attempt to cluster blocks with higher affinities earlier. Unlike the score function of clustering in [7], the HGs are introduced to emphasize the effect of the design hierarchy of macros, and the affinity value Φ is computed as follows:

$$\Phi(v_i, v_j) = \frac{\Gamma_{i,j}}{|e_{i,j}| \cdot |a_i + a_j|}, \tag{7}$$

where $|e_{i,j}|$ is the number of terminals of a net connecting the two blocks v_i and v_j, and a_i and a_j denote the respective areas of blocks v_i and v_j. The hierarchy factor $\Gamma_{i,j}$ is defined as

$$\Gamma_{i,j} = \begin{cases} \alpha \cdot k_{i,j}, & \text{if } v_i, v_j \text{ are in the same HG,} \\ k_{i,j}, & \text{others,} \end{cases} \tag{8}$$

where $k_{i,j}$ is the common naming factor which is defined as the number of common hierarchy parts between the two blocks v_i and v_j. In a hierarchical design, the name of a block usually consists of several hierarchy parts to represent its hierarchy [22], such as A/B/C/D, and the common hierarchy parts of two blocks means the common part of these two blocks' name. A large $k_{i,j}$ implies that these two blocks have a high hierarchy relation. The user-defined parameter α denotes that two blocks have a larger $\Gamma_{i,j}$ if they are in the same HG. In our implementation, we set α equal to 10.

3.3 Narrow Channel Handling

As mentioned in Section 1, the limited routing resource induced by big macros make the routability-driven placement problem more difficult. Due to the macro porosity issue, many routing overflows occur along macro boundaries, especially for narrow channels between macros. Narrow channels are formed between two or more close fixed macros, and placing too many blocks in these narrow channels would incur significant routing overflows.

To reduce the congestion caused by narrow channels, one intuitive method is to block these channels during placement. However, this method might lack flexibility and overkill placement solutions because either it blocks all the narrow channels or it does not block them at all. There is also an existing technique which applies Gaussian smoothing to handle preplaced macros [5]. However, directly smoothing the density of big macros could still induce deep ravines between neighboring macros which trap movable blocks during placement and limit the final placement quality. To solve this problem, we propose a narrow-channel-handling technique which adaptively modifies the density constraints around narrow channels to reduce routing congestions.

In our narrow-channel-handling technique, we first detect the locations of narrow channels, and then modify the maximum allowable area in each bin to prevent blocks from getting stuck in the narrow channels during global placement. As mentioned in Section 2.2, the maximum allowable area in each bin b is defined as $M_b = t_{density}(w_b h_b - P_b)$, where $t_{density}$ is a user-specified target density value for each bin, w_b (h_b) is the width (height) of bin b and P_b is the base potential equal to the pre-placed block area in bin b. We define P_b' as the modified base potential in bin b after considering narrow channels:

$$P_b' = \begin{cases} P_b \left(\beta + (1 - \beta) \left(1 - \left(\frac{\frac{P_b}{w_b h_b} - \beta}{1 - \beta} \right)^2 \right) \right), & \text{if } \frac{P_b}{w_b h_b} \geq \beta, \\ P_b, & \text{others,} \end{cases} \tag{9}$$

where β is a base potential threshold such that if $P_b/w_b h_b$ is greater than β, then bin b is treated as in a narrow channel. We set β

equal to 0.6 in our implementation. We then smooth the base potential by Gaussian smoothing [8] to make the density model differentiable. The detailed formulation for the Gaussian function has been introduced in Section 3.1. With first increasing the base potential at narrow channels by using Equation (9), our proposed narrow-channel-handling technique could prevent blocks from being trapped in narrow channels and achieve better placement quality.

3.4 Net Congestion Optimization

Because minimizing routed wirelength is always an important objective for practical applications, we propose a new net congestion optimization technique considering wirelength. Different from RUDY [20] and Tsota et al. [21] which estimate routing congestion and modify the density term to consider both the routing density and module density, we optimize net congestion directly based on a nonlinear formulation. In particular, instead of using the quadratic [20] or CHKS [21] wirelength approximations, we adopt the weighted-average wirelength approximation [11] to smooth the horizontal and vertical overlap lengths between nets and placement bins. In the following, we detail our net congestion optimization technique.

We optimize the net congestion by solving a constrained minimization problem of the form

$$\begin{aligned} \min \quad & W(\mathbf{x}, \mathbf{y}) \\ \text{s.t.} \quad & D_b(\mathbf{x}, \mathbf{y}) \leq M_b, \text{ for each bin } b, \\ & C_b(\mathbf{x}, \mathbf{y}) \leq S_b, \text{ for each bin } b, \end{aligned} \tag{10}$$

where $C_b(\mathbf{x}, \mathbf{y})$ denotes the net congestion of bin b, and S_b is defined as the total allowable routing area in bin b. Note that $C_b(\mathbf{x}, \mathbf{y})$ is a function of block positions, and S_b would decrease due to the macro porosity issue.

To evaluate the net congestion of bin b, we first define the weighted perimeter-per-area (p/a) ratio of a net e as follows:

$$d_e(\mathbf{x}, \mathbf{y}) = \frac{L_{x,e}(\mathbf{x}, \mathbf{y}) \cdot \sigma_v + L_{y,e}(\mathbf{x}, \mathbf{y}) \cdot \sigma_h}{L_{x,e}(\mathbf{x}, \mathbf{y}) \cdot L_{y,e}(\mathbf{x}, \mathbf{y})}, \tag{11}$$

where $L_{x,e}$ and $L_{y,e}$ are the respective horizontal and vertical bonding box lengths of net e, and σ_v and σ_h are the respective average vertical and horizontal wire pitches. The average vertical (horizontal) pitch could be obtained by dividing the summation of vertical (horizontal) pitches in every routing layers by the total number of routing layers. As shown in Equation (11), while minimizing $d_e(\mathbf{x}, \mathbf{y})$, a net with smaller HPWL, i.e., $L_{x,e} + L_{y,e}$, and square bounding box is preferred.

The overlaps between a net e and a bin b can be computed by

$$\phi_{b,e}(\mathbf{x}, \mathbf{y}) = o_h(b, e) \cdot o_v(b, e), \tag{12}$$

where $o_h(b, e)$ and $o_v(b, e)$ are the horizontal and vertical overlaps between net e and bin b. With the net density defined in Equation (11), the net congestion of a bin b is then defined as follows:

$$C_b(\mathbf{x}, \mathbf{y}) = \sum_e \left(d_e(\mathbf{x}, \mathbf{y}) \cdot \phi_{b,e}(\mathbf{x}, \mathbf{y}) \right). \tag{13}$$

The net congestion $C_b(\mathbf{x}, \mathbf{y})$ consists of two major parts: (1) overlaps between nets and bins and (2) the weighted p/a ratio of each net. So, if we minimize $C_b(\mathbf{x}, \mathbf{y})$, both the net lengths and overlaps between nets and bins are expected to be reduced.

We then use the quadratic penalty method to solve Equation (10) by transforming the constrained optimization problem into an unconstrained problem as follows:

$$\min \; \lambda_1 \hat{W}(\mathbf{x}, \mathbf{y}) + \lambda_2 \sum_b (\hat{D}_b(\mathbf{x}, \mathbf{y}) - M_b)^2 + \lambda_3 \sum_e (\hat{C}_b(\mathbf{x}, \mathbf{y}) - S_b)^2 \tag{14}$$

where λ_1, λ_2, and λ_3 are weights for wirelength, density, and net congestion, respectively, and $\hat{C}_b(\mathbf{x}, \mathbf{y})$ is the smoothed net congestion of a bin b. Since both the net lengths ($L_{x,e}$ and $L_{y,e}$) and the overlap lengths ($o_h(b, e)$ and $o_v(b, e)$) can be formulated as maximum and minimum operations, smoothing techniques for HPWL can be directly used to smooth the net lengths and overlap lengths. Since the weighted-average wirelength model gives the best approximation to HPWL [11], we adopt the weighted-average wirelength model to smooth $L_{x,e}$, $L_{y,e}$, $o_h(b,e)$, and $o_v(b,e)$ in our analytical formulation. Furthermore, in our implementation, we set $\lambda_1 = \frac{||\nabla \hat{C}(\mathbf{x},\mathbf{y})||}{||\nabla \hat{W}(\mathbf{x},\mathbf{y})||}$, $\lambda_2 = \frac{||\nabla \hat{C}(\mathbf{x},\mathbf{y})||}{||\nabla \hat{D}(\mathbf{x},\mathbf{y})||}$, and $\lambda_3 = 1$. During the execution of our algorithm, λ_1 and λ_3 are fixed, and λ_2 is gradually doubled until blocks are spread sufficiently.

4. EXPERIMENTAL RESULTS

To evaluate our proposed algorithm, we conducted several experiments based on the ICCAD'12 placement contest benchmarks [22]. Please refer to [22] for detailed statistics of the benchmark. We implemented our algorithm in the C++ programming language. For routing congestion analysis, NCTU-GR 2.0 [17] and BFG-R 2.0 [13] were used. We compared our algorithm with the top placers in the ICCAD'12 Design Hierarchy Aware Routability-Driven Placement Contest [1] based on the ICCAD'12 contest evaluation metric. Our results were obtained on a Linux workstation with Intel Xeon E5-2620 2.0GHz CPU with 16GB maximum allowed memory. The placement results in the contest [1] were derived from a Linux workstation with Intel Xeon X7560 2.27GHz CPU with 16GB maximum allowed memory, which is faster than our environment.

Table 1 shows the comparisons of HPWL and placement runtime among our algorithm (NTUplace4h) and the top placers in the ICCAD'12 contest, where Columns 2 to 7 give the results of SimPLR, Ripple, and VDAPlace, respectively, and Columns 8 and 9 give the results of our algorithm. For fair comparisons, we scaled our machine performance to match with the contest environment. As shown in Table 1, without considering the cost of global-routing-based congestion, our proposed algorithm can achieve 2% to 12% shorter HPWL compared with SimPLR, Ripple, and VDAPlace.

In Tables 2 and 3, we compared the scaled HPWL which includes the global-routing-based congestion, and the total routed wirelength evaluated by NCTU-GR 2.0 [17] and BFG-R 2.0 [13], respectively. As shown in Tables 2 and 3, our algorithm can achieve the best global-routing-based congestion and scaled HPWL for both routers. The results show the robustness of our proposed algorithm. Figure 4 shows the placement result and routing congestion map for superblue7.

Table 4 summarizes the overall evaluation, where normalized costs which include HPWL, global-routing-based congestion, and placement runtime are compared. Columns 2 to 7 give the results of Ripple, SimPLR, and VDAPlace, and Columns 8 and 9 give the results of our algorithm. Based on NCTU-GR 2.0 and BFG-R 2.0, our algorithm can achieve the best overall score, with 1%, 8%, and 26% smaller normalized cost compared to SimPLR, Ripple, and VDAPlace, respectively.

5. CONCLUSIONS

This paper has proposed a new routability-driven analytical placement algorithm for hierarchical circuit designs. The three techniques have been adopted to optimize the routability: (1) design hierarchy identification, (2) narrow channel handling, and (3) net congestion optimization. Moreover, our placer uses a smoothed congestion estimation method to model global routing behavior for better congestion estimation during placement. Experimental results have shown that our placer can achieve the best quality (both the average overflow and wirelength) among all participating teams for the 2012 ICCAD Placement Contest.

6. REFERENCES

[1] ICCAD 2012 Placement Contest. http://cad_contest.cs.nctu.edu.tw/CAD-contest-at-ICCAD2012/.

[2] U. Brenner, M. Struzyna, and J. Vygen. BonnPlace: Placement of leading-edge chips by advanced combinatorial algorithms. IEEE Tran. on CAD, 27(9):1607–1620, September 2008.

[3] T. F. Chan, J. Cong, T. Kong, and J. R. Shinnerl. Multilevel optimization for large-scale circuit placement. In Proc. of ICCAD, pages 171–176, 2000.

[4] Y.-W. Chang, Z.-W. Jiang, and T.-C. Chen. Essential issues in analytical placement algorithms. IPSJ Tran. on SLDM, 2:145–166, August 2009.

[5] T.-C. Chen, Z.-W. Jiang, T.-C. Hsu, H.-C. Chen, and Y.-W. Chang. NTUplace3: An analytical placer for large-scale mixed-size designs with preplaced blocks and density constraints. IEEE Tran. on CAD, 27(7):1228–1240, July 2008.

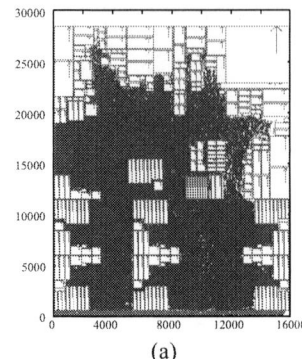 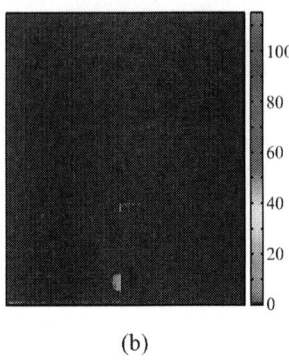

(a)   (b)

Figure 4: (a) The placement result and (b) routing congestion map routed by NCTU-GR 2.0 for superblue7. Our algorithm achieves 9%, 8%, and 26% shorter scaled HPWL compared with SimPLR, Ripple, and VDAPlace, respectively.

[6] S. Chou, M.-K. Hsu, and Y.-W. Chang. Structure-aware placement for datapath-intensive circuit designs. In Proc. of DAC, pages 762–767, 2012.

[7] Y.-L. Chuang, G.-J. Nam, C. J. Alpert, Y.-W. Chang, J. Roy, and N. Viswanathan. Design-hierarchy aware mixed-size placement for routability optimization. In Proc. of ICCAD, pages 663–668, 2010.

[8] R. C. Gonzales and R. E. Woods. Digital Image Processing. Addison-Wesley, 1992.

[9] X. He, T. Huang, L. Xiao, H. Tian, G. Cui, and E. F. Y. Young. Ripple: An effective routability-driven placer by iterative cell movement. In Proc. of ICCAD, pages 74–79, 2011.

[10] D. Hill. US patent 6,370,673: Method and system for high speed detailed placement of cells within an integrated circuit design. 2002.

[11] M.-K. Hsu, Y.-W. Chang, and V. Balabanov. TSV-aware analytical placement for 3D IC designs. In Proc. of DAC, pages 664–669, 2011.

[12] M.-K. Hsu, S. Chou, T.-H. Lin, and Y.-W. Chang. Routability-driven analytical placement for mixed-size circuit designs. In Proc. of ICCAD, pages 80–84, 2011.

[13] J. Hu, J. A. Roy, and I. L. Markov. Completing high-quality global routes. In Proc. of ISPD, pages 35–41, 2010.

[14] Z.-W. Jiang, B.-Y. Su, and Y.-W. Chang. Routability-driven analytical placement by net overlapping removal for large-scale mixed-size designs. In Proc. of DAC, pages 167–172, 2008.

[15] M.-C. Kim, J. Hu, D.-J. Lee, and I. L. Markov. A SimPLR method for routability-driven placement. In Proc. of ICCAD, pages 67–73, 2011.

[16] C. Li, M. Xie, C.-K. Koh, J. Cong, and P. H. Madden. Routability-driven placement and white space allocation. In Proc. of ICCAD, pages 394–401, 2004.

[17] W.-H. Liu, W.-C. Kao, Y.-L. Li, and K.-Y. Chao. Multi-threaded collision-aware global routing with bounded-length maze routing. In Proc. of DAC, pages 200–205, 2010.

[18] J. A. Roy, J. F. Lu, and I. L. Markov. Seeing the forest and the trees: Steiner wirelength optimization in placement. IEEE Tran. on CAD, 26(4):632–644, April 2007.

[19] J. A. Roy, N. Viswanathan, G.-J. Nam, C. J. Alpert, and I. L. Markov. CRISP: Congestion reduction by iterated spreading during placement. In Proc. of ICCAD, pages 357–362, 2009.

[20] P. Spindler and F. M. Johannes. Fast and accurate routing demand estimation for efficient routability-driven placement. In Proc. of DATE, pages 1226–1231, 2007.

[21] K. Tsota, C. Koh, and V. Balakrishnan. Guiding global placement with wire density. In Proc. of ICCAD, pages 212–217, 2008.

[22] N. Viswanathan, C. Alpert, C. Sze, Z. Li, and Y. Wei. ICCAD-2012 CAD contest in design hierarchy aware routability-driven placement and benchmark suite. In Proc. of ICCAD, pages 345–348, 2012.

[23] X. Yang, B.-K. Choi, and M. Sarrafzadeh. Routability-driven white space allocation for fixed-die standard-cell placement. In Proc. of ISPD, pages 410–419, 2002.

Table 1: Resulting HPWLs and CPU times for placement of our algorithm and the top placers in the ICCAD'12 Design Hierarchy Aware Routability-Driven Placement Contest [1] on the ICCAD'12 placement contest benchmarks [22].

Circuit Name	SimPLR		Ripple		VDAPlace		NTUplace4h	
	HPWL ($\times 10^7$)	CPU (min)	HPWL ($\times 10^7$)	CPU (min)	HPWL ($\times 10^7$)	CPU (min)	HPWL ($\times 10^7$)	CPU (min)
superblue1	27.12	38.65	27.86	170.22	28.20	14.75	26.73	91.19
superblue3	33.83	45.10	33.31	251.90	41.26	19.55	32.94	108.90
superblue4	23.00	21.12	21.83	142.92	24.57	10.07	21.82	74.27
superblue5	34.89	35.90	34.47	180.55	35.78	14.57	35.27	80.09
superblue7	42.23	54.15	41.79	383.62	45.74	28.75	39.16	161.76
superblue10	60.77	80.62	57.80	438.53	58.87	21.93	58.39	153.13
superblue16	27.30	29.95	26.70	158.23	27.37	11.52	26.08	61.06
superblue18	16.41	27.42	16.37	183.15	19.16	10.97	15.08	69.70
Normalized	1.04	0.41	1.02	2.35	1.12	0.17	1.00	1.00

Table 2: Resulting global-routing-based congestions (RC) evaluated by NCTU-GR 2.0 [17] and scaled HPWLs (sHPWL) of our algorithm and the top placers in the ICCAD'12 Design Hierarchy Aware Routability-Driven Placement Contest [1] on the ICCAD'12 placement contest benchmarks [22].

Circuit Name	SimPLR		Ripple		VDAPlace		NTUplace4h	
	RC	sHPWL ($\times 10^7$)	RC	sHPWL ($\times 10^7$)	RC	sHPWL ($\times 10^7$)	RC	sHPWL ($\times 10^7$)
superblue1	100.94	27.89	101.23	28.89	110.39	36.98	101.48	27.90
superblue3	100.56	34.39	102.73	36.04	117.69	63.16	103.81	36.71
superblue4	101.94	24.34	101.32	22.69	104.96	28.23	102.01	23.14
superblue5	101.09	36.03	100.38	34.86	119.72	56.96	100.57	35.87
superblue7	100.71	43.13	100.89	42.91	103.11	50.01	100.39	39.62
superblue10	104.56	69.09	101.91	61.11	125.33	103.60	101.48	61.67
superblue16	101.55	28.57	102.13	28.40	105.91	32.23	102.19	27.80
superblue18	103.72	18.23	103.15	17.91	117.31	29.11	102.84	16.36
Normalized	1.00	1.04	1.00	1.02	1.11	1.47	1.00	1.00

Table 3: Resulting global-routing-based congestions (RC) evaluated by BFG-R 2.0 [13] and scaled HPWLs (sHPWL) of our algorithm and the top placers in the ICCAD'12 Design Hierarchy Aware Routability-Driven Placement Contest [1] on the ICCAD'12 placement contest benchmarks [22].

Circuit Name	SimPLR		Ripple		VDAPlace		NTUplace4h	
	RC	sHPWL ($\times 10^7$)	RC	sHPWL ($\times 10^7$)	RC	sHPWL ($\times 10^7$)	RC	sHPWL ($\times 10^7$)
superblue1	103.89	30.28	106.64	33.41	109.27	36.04	103.49	29.58
superblue3	104.74	38.64	105.75	39.06	114.55	59.28	106.67	39.53
superblue4	108.05	28.56	107.27	26.59	103.88	27.43	108.54	27.42
superblue5	104.20	39.29	102.00	36.54	107.31	43.63	102.55	37.97
superblue7	102.39	45.25	102.57	45.02	105.37	53.10	101.59	41.03
superblue10	110.87	80.58	107.50	70.80	123.44	100.26	107.52	72.65
superblue16	104.72	31.17	103.23	29.29	110.87	36.30	106.01	30.78
superblue18	116.66	24.61	111.62	22.07	131.16	37.07	104.32	17.03
Normalized	1.02	1.09	1.01	1.05	1.08	1.36	1.00	1.00

Table 4: Resulting normalized costs which include HPWL, global-routing-based congestion and CPU times for placement of our algorithm and the top placers in the ICCAD'12 Design Hierarchy Aware Routability-Driven Placement Contest [1] on the ICCAD'12 placement contest benchmarks [22]. Two routers, NCTU-GR 2.0 [17] and BFG-R 2.0 [13], are applied separately for evaluating global-routing-based congestions.

Circuit Name	SimPLR		Ripple		VDAPlace		NTUplace4h	
	NCTU-GR	BFG-R	NCTU-GR	BFG-R	NCTU-GR	BFG-R	NCTU-GR	BFG-R
superblue1	0.94	0.97	1.06	1.16	1.18	1.08	1.00	1.00
superblue3	0.88	0.92	1.02	1.03	1.54	1.34	1.00	1.00
superblue4	0.97	0.96	1.01	1.00	1.07	0.88	1.00	1.00
superblue5	0.95	0.98	1.01	1.00	1.42	1.03	1.00	1.00
superblue7	1.01	1.03	1.13	1.14	1.13	1.16	1.00	1.00
superblue10	1.07	1.06	1.04	1.03	1.48	1.21	1.00	1.00
superblue16	0.98	0.96	1.07	1.00	1.04	1.06	1.00	1.00
superblue18	1.07	1.39	1.17	1.39	1.61	1.97	1.00	1.00
Average	1.01		1.08		1.26		1.00	

1043

Ripple 2.0: High Quality Routability-Driven Placement via Global Router Integration

Xu He, Tao Huang, Wing-Kai Chow, Jian Kuang, Ka-Chun Lam, Wenzan Cai and
Evangeline F.Y. Young
Department of Computer Science and Engineering
The Chinese University of Hong Kong, Shatin, N.T., Hong Kong
email: {xhe, thuang, wkchow, jkuang, lamkc, wzcai, fyyoung}@cse.cuhk.edu.hk

ABSTRACT

Due to a significant mismatch between the objectives of wirelength and routing congestion, the routability issue is becoming more and more important in VLSI design. In this paper, we present a high quality placer Ripple 2.0 to solve the routability-driven placement problem. We will study how to make use of the routing path information in cell spreading and relieve congestion with tangled logic in detail. Several techniques are proposed, including (1) lookahead routing analysis with pin density consideration, (2) routing path-based cell inflation and spreading and (3) robust optimization on congested cluster. With the official evaluation protocol, Ripple 2.0 outperforms the top contestants on the ICCAD 2012 Contest benchmark suite.

1. INTRODUCTION

Today, the routability problem has become a very important issue in VLSI physical design. The existence of macro blockages and the large problem size with millions of standard cells and nets make the routability problem increasingly challenging. Common placement metrics may not capture the key aspects of solution quality [1][2]. The major objective of traditional placements is to minimize wirelength which is often estimated by the half-perimeter wirelength (H-PWL) model. However, over-optimizing of wirelength may lead to routing failures, since the wires may be distributed unevenly. Therefore, we have to trade off between wirelength and routing congestion during the placement process.

Routability evaluation is itself a problem that needs to be studied [1]. In the ISPD Contest 2011 [3], only the total overflow provided by a global router [4] is considered. However, the metric of total overflow usually fails to provide a clear picture of the design routability [5]. Therefore, the work [5] introduces average congestion of G-Cell edges (ACE) to give a more accurate congestion measure.

Routability-driven optimizations could be performed during 1) global placement, 2) detailed placement and 3) the post-placement process. During global placement, many techniques [6][7][8][9][10] model the congestion information in the objective function. Besides, several works [11][12][13][14][15][16][17] allocate whitespace in the congested regions explicitly or implicitly to distribute the routing demand evenly. Due to the improvement in quality and runtime of re-

cent routers, a placement layout can be evaluated by a global router with a short time-out. IPR [18] and SimPLR [12] integrate global router FastRoute [19] and BFG-R [20] respectively in their placements and make proper adjustments on the intermediate placement solutions. In detailed placement, the most common approach is to change the objective of cell swapping to alleviate routing congestion [11][12][21][18][22]. In post-placement process, there are many previous works addressing routability, such as Crop [22] and CRISP [16]. After obtaining the congestion map, Crop [22] adjusts the boundary of each G-Cell and spreads cells within G-Cell. CRISP [16] applies cell inflation and spreading to allocate cells more sparsely in the congested regions.

In Ripple 2.0, we propose several techniques to further improve routability. **First**, instead of using probabilistic methods, we simplify the global router FastRoute [19] with pin density consideration to do routing analysis. Comparing with probabilistic congestion estimations, this lookahead routing estimation can not only produce more accurate congestion map, but also can provide a routing solution for each net. The runtime of our Simplified FastRoute is fast enough to be called many times during the global placement. **Second**, we propose a routing path-based method for cell inflation and spreading. By considering the routing paths, we spread cells whose connections are passing through congested regions, even if the cells themselves are not in those regions. Our strategy of cell inflation and spreading is more effective to alleviate congestion than the method that only considers spreading cells in congested regions. **Third**, we detect and reduce the overflow from groups of tangled logic. A group of tangled logic invariably have a higher degree of inter-connectivity [23]. Placement may naturally want to pull these cells tightly together, resulting in routing hotspot. We detect those cells in tangled nets and spread them properly to trade off between congestion and wirelength.

Our work has the following contributions:

1. Present an accurate routing analysis method by simplifying FastRoute and considering pin density. This Simplified FastRoute is efficient enough to be called hundreds of times during placement.

2. Propose a routing path-based method for cell inflation and spreading. Different from previous methods that only spread cells in congested regions, we also consider cells whose routing path passes through congested regions. Our strategy can thus alleviate both local and global congestion effectively.

3. Present a method of handling tangled logic to trade off between congestion and wirelength. We detect the congestion from tangled logic, and spread the cells there more sparsely without sacrificing HPWL unnecessarily.

The rest of this paper is organized as follows. In Section 2, we review related preliminaries. Section 3 gives an overview of our placer.

Permission to make digital or hard copies of all or part of this work for personal or classroom use is granted without fee provided that copies are not made or distributed for profit or commercial advantage and that copies bear this notice and the full citation on the first page. To copy otherwise, to republish, to post on servers or to redistribute to lists, requires prior specific permission and/or a fee.

DAC 13, May 29 - June 07 2013, Austin, TX, USA

Section 4 describes the techniques used in our global placement in detail. Section 5 explains the methods used after global placement. Experimental results are presented in Section 6, followed by a conclusion in Section 7.

2. PRELIMINARIES

Given a netlist $N = (E, V)$ with nets E and nodes V, a routability-driven placer is required to compute the positions of the movable nodes (cells) to minimize the routing congestion and wirelength. The placement layout is evaluated by a given router within a limited amount of time. The major goal is to trade off between the congestion measure obtained by the router and the HPWL.

HPWL is an objective used by many traditional placers. The H-PWL function is usually approximated by a differentiable function, such as the quadratic objective shown in equation (1).

$$f(\vec{x}, \vec{y}) = \sum_{i,j \in V} w_{i,j}((x_i - x_j)^2 + (y_i - y_j)^2) \quad (1)$$

where $\vec{x}$ and $\vec{y}$ are coordinate vectors of the cell locations, and $w_{i,j}$ is the connectivity weight between cells i and j. In our implementation, B2B net model [24] is applied to calculate the connectivity weight.

The lower-upper bound framework [25] is used in our placer. In the lower bound computation, the quadratic objective shown in equation (1) is applied to minimizing HPWL. In the upper bound computation, a fast lookahead legalization is used to roughly remove overlaps between cells. The two bound computations are invoked alternatively until the computations of these two bounds converge. Ripple [11] uses many techniques to improve routability. After routing congestion estimation, cells are spread sparsely in the congested regions by a modified lookahead legalization. A modified FastPlace-DP [26] with congestion consideration is applied in legalization and detailed placement.

3. OVERVIEW

Ripple 2.0 is a flat placer with a lower-upper-bound framework [25] as shown in Figure 1. In the lower bound computation, we focus on minimizing HPWL by solving the quadratic objective shown in equation (1). In the upper bound computation, we spread cells by considering routing congestion.

The process of upper bound computation is shown in Figure 2. We apply Simplified FastRoute with pin density consideration to analyze congestion of intermediate layouts. This global router not only give a more accurate congestion map comparing with previous probabilistic estimation, but also provide routing solutions for nets, which enables us to use the strategy of routing path-based cell inflation and spreading to alleviate congestion. It is known that the congestion of a region can be caused by three types of nets: (1) local nets connecting cells within the region, (2) semi-global nets coming into/out of the region, and (3) global nets passing through the region without connecting any cell inside the region. Different from many previous works that only spread cells in congested regions, the cells whose connections passing through congested regions will be inflated as well. Therefore, the congestion due to various types of nets can be alleviated effectively. A global placement solution is generated when the congestion and H-PWL obtained by the lower and upper bound computations converge.

After a number of iterations (> 30 iterations in our implementation), the congestion and HPWL will nearly converge. However, several congested clusters caused by tangled logic may appear. The cells of tangled logic may not be spread sparsely enough, even if we have already used the technique of cell inflation and spreading. Therefore, those cells in the tangled logic should be identified and made to spread more sparsely. Section 4.3 gives further discussion about this issue.

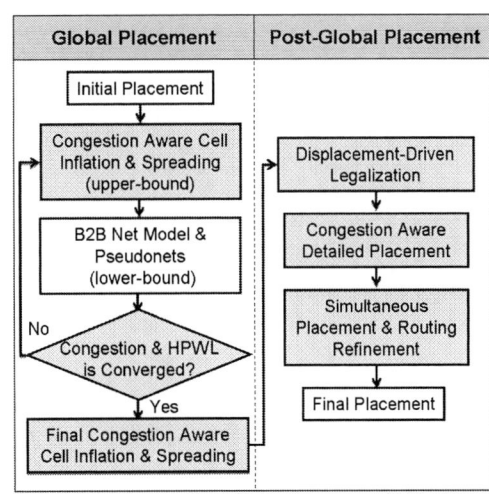

Figure 1: The framework of our routability-driven placer.

A displacement-driven legalization is performed after global placement. We use FastPlace-DP [26] as our detailed placer to further improve HPWL. FastPlace-DP uses many techniques: (1) cell clustering, (2) cell swapping and (3) local reordering. In order to avoid worsening congestion during these processes, we need to determine when to terminate each step during detailed placement. Finally, a post-processing step SRP [27] is applied to relocate cells directly for congestion refinement. The details will be discussed in section 5.

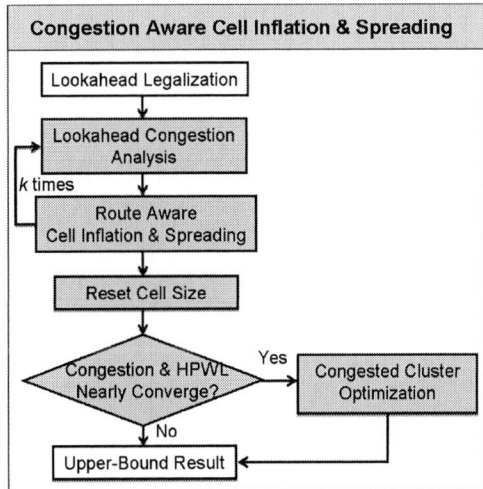

Figure 2: The flow of congestion-aware cell inflation and spreading.

4. GLOBAL PLACEMENT

4.1 Lookahead Routing Analysis

In order to optimize routability while preserving wirelength, a lookahead routing is invoked during global placement. Unlike previous congestion estimations that only report congestion map, our lookahead routing analysis reports both routing congestion and interconnection paths. Besides, our lookahead routing accounts for pin density as well.

The chip is divided into uniform G-Cells whose width and height are $TileSizeX$ and $TileSizeY$ respectively. Each pair of adjacent G-Cells have an edge connecting them. In our lookahead routing analysis, the routing supply and demand are computed in the horizontal and vertical directions separately. We will discuss below the computation of the horizontal supply and demand, while the vertical supply and demand can be obtained similarly.

4.1.1 Routing Supply Calculation

The routing supply of a G-Cell edge is equal to the total number of tracks supplied in all metal layers. Since different metal layers have different wire widths and wire spacings, we need to sum up the number of tracks of each metal layer. The default supply of a horizontal G-Cell edge e is obtained by equation (2).

$$DefaultSupplyH_e = \sum_{l=1}^{NumLayer} \frac{TileSizeY}{MinWidth_l + MinSpace_l}, \quad (2)$$

where $MinWidth_l$ and $MinSpace_l$ are the minimum wire width and spacing on metal layer l, and $NumLayer$ is the number of metal layers.

Since there are many fixed routing blockages occupying the routing resources on the metal layers, the supply of G-Cell edges need to exclude those blocked routing resource. In order to calculate the number of blocked tracks, we first obtain the union of all the blockages on a G-Cell edge. Then, we obtain the number of blocked tracks of each blocked region separately. As shown in equation (3), the supply of a horizontal G-Cell edge e is equal to $DefaultSupplyH_e$ minus the number of blocked tracks of e.

$$SupplyH_e = DefaultSupplyH_e - BlockTrackH_e$$
$$BlockTrackH_e = \sum_{l=1}^{NumLayer} \frac{BlockLength_{e,l}}{MinWidth_l + MinSpace_l} \times (1-p),$$
$$(3)$$

where $BlockTrackH_e$ is the number of blocked tracks of edge e, $BlockLength_{e,l}$ is the length of the blocked region in the G-Cells connected by e on metal layer l and p is the porosity of the routing blockage (zero implies complete blockage).

The calculation of the routing supply is demonstrated in Figure 3. In this example, $TileSizeY = 40$. Let both the minimum wire width and spacing be 2, and porosity p is 0. The $DefaultSupplyH_{e_1}$ and $DefaultSupplyH_{e_2}$ are both 40/(2+2)=10. Since two blockages overlap with G-Cells connected by e_1 and the total blocked length is 32, $BlockTrackH_{e_1}$ is 32/(2+2)=8 and $SupplyH_{e_1}$ is 10-8=2. Similarly, $BlockTrackH_{e_2}$ is 16/(2+2)=4 and $SupplyH_{e_2}$ is 6.

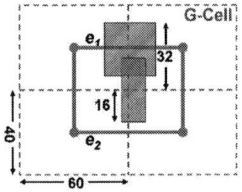

Figure 3: The G-Cells connected by horizontal edge e_1 and e_2 are overlapped with routing blockages. The supply of e_1 and e_2 have to exclude the blocked routing resources.

4.1.2 Routing Demand Calculation

We use FastRoute [19] to do routing congestion analysis. In order to check the overflow fast, we simplify FastRoute by removing the steps of monotonic routing and maze routing. Since global routing captures the wires that pass between G-Cell edges, internal congestion of G-Cells is ignored [16]. Therefore, we need to consider pin density within G-Cells to account for local congestion when computing the routing demand before invoking Simplified FastRoute.

In this Simplified FastRoute, multi-pin nets are decomposed into two-pin subnets by a congestion-driven method [19]. Then, a rip-up and reroute process will be invoked by using L/Z shape pattern routing. According to the current congestion map, if a net passes through congested regions, it will be rerouted. Since local peaks of pin density within G-Cell often cause routing congestion [16], pin density is added when constructing the initial congestion map. The demand of pin density on G-Cell edge e is obtained by equation (4).

$$PinDemandH_e = (PinNumL + PinNumR) \times \beta, \quad (4)$$

where $PinNumL$ and $PinNumR$ are the number of pins in the left and right G-Cells connected by e and β is the pin density factor. In our implementation, $\beta = 0.025$ at the beginning iterations and $\beta = 0.05$ when the congestion and HPWL nearly converge and further optimization step (section 4.3) is to be invoked.

After finishing Simplified FastRoute, we obtain the routing demand of each G-Cell edge e. The horizontal demand of e is shown in equation (5).

$$DemandH_e = PinDemandH_e + TrackH_e, \quad (5)$$

where $TrackH_e$ is the number of tracks passing through e which is obtained from Simplified FastRoute.

4.2 Routing Path-based Cell Inflation & Spreading

After routing congestion analysis, we can identify some congested regions where the routing demand is larger than the supply. In the next step, we make use of cell inflation and spreading to reduce the congestion problem. Different from many previous cell inflation approach in which only the cells within the congested regions will be inflated, we propose a routing path-based cell inflation and spreading technique to deal with the congestion problem. This new technique is based on the following two reasons.

1. Empirical study shows that a large part of congestion is caused by global nets [28], i.e., the nets with all its connecting cells located outside the congested region. This part of routing congestion cannot be solved by inflating only the cells within the congested region. This problem is more critical when congestion appears on top of macro blockages, which is often the case in modern designs. In such situation, there would be no cell in those congested regions and traditional cell inflation will fail to deal with the problem.

2. Since we make use of a router to perform routing congestion analysis, the routing path of each net in the congestion map is known. We know that, among various options, the router will try to route a net by choosing the least congested path. This kind of information should be utilized in dealing with the congestion problem but was often overlooked.

Consider the horizontal direction in the following example. The key idea of the cell inflation and spreading step can be illustrated in Figure 4. For each decomposed two-pin net, we trace its routing path for a certain distance (at most 10 G-Cell edges in our implementation) and find the most congested horizontal G-Cell edge e. If the supply $SupplyH_e$ is less than the demand $DemandH_e$, we will inflate the height of the corresponding cells of this two-pin net by equation (6)

1046

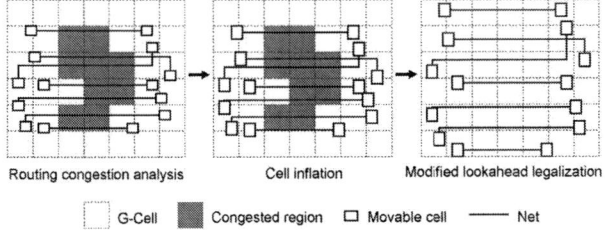

Routing congestion analysis   Cell inflation   Modified lookahead legalization

□ G-Cell   ▨ Congested region   ▢ Movable cell   —— Net

Figure 4: An example of routing path-based cell inflation and spreading.

$$InflateH_i = InflateH_i \times \frac{DemandH_e}{SupplyH_e}, \qquad (6)$$

where $InflateH_i$ is the inflated height of cell i. $InflateH_i$ is equal to the original height H_i of cell i before inflation. For a cell associated with several nets, we will inflate it according to the maximum inflation ratio.

After cell inflation, some regions may contain more cells than it can contain. Then, the modified lookahead legalization [11] is performed to roughly legalize the layout. Cell inflation combined with the modified lookahead legalization will spread the cells in the vertical direction. In this case, more routing resources will be provided to the problematic nets that pass through congested regions. This can effectively reduce the congestion problem without a significant change of the layout. For vertical congestion, we inflate the cell width and perform horizontal spreading in the same way. As we can see, this path-based cell inflation technique can relieve the congestion caused by the local, semi-global and more importantly the global nets.

After the horizontal and vertical inflation and spreading, we will perform routing congestion analysis again and this process will be repeated a number of times k ($k = 3$ in our implementation). The rationale behind is that we want to spread the cells in a more careful way to avoid over-spreading. Each time, we spread the cells slightly and perform analysis to see the impact, and the resulting congestion map will be used to guide the next spreading. The inflated height $InflateH_i$ and width $InflateW_i$ will be reset to H_i and W_i before the next iteration starts.

4.3 Congested Cluster Optimization

When congestion and HPWL nearly converge after a number of iterations, we find that there may still be some congested clusters in the layout. Many of these congested clusters have cels with a lot of interconnections [23]. The cells in these clusters need to be spread more sparsely. We show the placement layout and its routing result by NCTUgr [29] of superblue3 in Figure 5(a) and (b) respectively. Figure 5(b) shows the routing hotspots in dark red or violet color. We use white circles to highlight these routing hotspots. Many congested clusters appearing are occupied by a lot of cells (by comparing Figure 5(a) and (b)). If the cells are spread more sparsely, the routing hotspots will be eliminated.

Identification of congested clusters is performed by applying NC-TUgr [29] directly. NCTUgr uses the techniques of pattern routing, monotonic routing, maze routing and post routing to relieve congestion. After using these techniques, the remaining congested regions usually contain cells with too many interconnections with each other and the congestion problem cannot be resolved yet.

Adjustments of cell sizes and pseudo-net weight are used after identification of congested clusters. If cell i is in a congested region, we choose to adjust either its height or width with a larger inflation ratio. We budget the adjustment in height by equation (7), and the

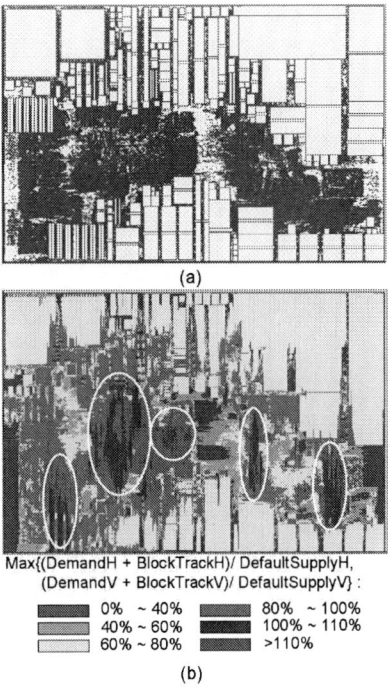

(a)

Max{(DemandH + BlockTrackH)/ DefaultSupplyH, (DemandV + BlockTrackV)/ DefaultSupplyV} :

▨ 0% ~ 40%   ▨ 80% ~ 100%
▨ 40% ~ 60%   ▨ 100% ~ 110%
▨ 60% ~ 80%   ▨ >110%

(b)

Figure 5: The placement layout (a) and its routing result by NC-TUgr [29] (b) of superblue3.

adjustment in width can be similarly computed.

$$H_i = H_i \times \frac{DemandH_{e_1} + DemandH_{e_2}}{SupplyH_{e_1} + SupplyH_{e_2}}, \qquad (7)$$

where e_1 and e_2 are the edges connecting the G-Cell containing cell i on the left and right side respectively. The adjusted size will be remembered and used in the following iterations of the global placement. It should be mentioned that the cells will be reset to their original sizes after global placement.

Artificial two-pin pseudo-nets are introduced in the lower bound computation to consider the target positions of cells obtained in the upper bound computation. The pseudo-net weight is computed by equation (8).

$$pseudo_i = \alpha/(|lx_i - ux_i| + |ly_i - uy_i|) \qquad (8)$$

where (lx_i, lx_j) and (ux_i, uy_i) are the coordinates of cell i obtained by the lower and upper bound computation of the previous iteration and $\alpha = 0.01 \times (1+ \text{iteration number})$.

After changing cell sizes, HPWL will be increased. In order to avoid worsening HPWL, we perform some additional iterations to reduce HPWL. During these additional iterations, the impact of the upper bound computation is reduced by decreasing the pseudo-net weight in the lower bound computation. After these additional iterations, we will identify congested clusters and adjust cell sizes again. This process of congested cluster optimization will be repeated several times until the congestion is improved obviously. The value of α in the pseudo-net weight (shown in equation (8)) is illustrated in Figure 6. At the beginning, α is increased linearly. When we perform the process of congested cluster optimization on the intermediate layouts, α will be adjusted to a lower value and increased linearly again in the subsequence iterations. Since the process of congested cluster optimization is repeated several times, α will be adjusted repeatedly as well.

1047

Table 1: Evaluation of routing path-based cell inflation and spreading (RPB) and congested cluster improvement (CCI)

Benchmark	Method	ACE (%)				HPWL	ScaledWL	Runtime (seconds)
		0.50	1.00	2.00	5.00			
superblue1	Base	111.78	110.43	107.57	103.03	272906304	340071852	1989
	RPB	118.59	115.79	112.48	106.53	276418525	387109145	1636
	RPB+CCI	102.63	101.31	100.66	100.26	278613308	288763344	5462
superblue3	Base	129.61	122.59	117.29	110.89	307528119	492931490	2646
	RPB	111.26	108.96	106.19	102.47	343317959	417703300	2584
	RPB+CCI	105.29	103.19	101.59	100.64	333315621	360085825	8487
superblue4	Base	106.06	103.03	101.51	100.61	218230511	236570795	1423
	RPB	105.18	102.59	101.3	100.52	219891835	235700081	1127
	RPB+CCI	102.75	101.37	100.69	100.27	218321746	226639292	4504
superblue5	Base	118.04	114.54	109.83	103.93	335332413	451886534	2178
	RPB	104.87	102.43	101.22	100.49	345713557	369066484	1543
	RPB+CCI	100.79	100.4	100.2	100.08	344935609	348733540	5728
superblue7	Base	111.04	108.81	106.75	103.11	395288349	483366222	4274
	RPB	102.97	101.48	100.74	100.3	425010166	442509990	3928
	RPB+CCI	101.91	100.96	100.48	100.19	417956482	429041785	11387
superblue10	Base	117.63	114.65	110.66	105.72	565020331	771229468	5868
	RPB	105.19	102.95	101.47	100.59	582988405	627566932	4393
	RPB+CCI	101.79	100.9	100.45	100.18	583014725	597515266	14383
superblue16	Base	115.36	114.35	113.18	109.46	249202445	347035229	1470
	RPB	108.03	104.23	102.11	100.85	263035858	293059642	1267
	RPB+CCI	104.48	102.24	101.12	100.45	267097907	283689195	4942
superblue18	Base	115.95	112.97	110.2	105.2	171609483	228658479	2061
	RPB	112.19	111.09	108.84	104.41	170358833	217035361	1945
	RPB+CCI	107.18	103.88	101.94	100.78	166702202	183918099	6267
Avg.	Base	1.00	1.00	1.00	1.00	1.00	1.00	1.00
	RPB	-6.18%	-5.75%	-4.86%	-3.06%	+4.44%	-10.80%	-15.91%
	RPB+CCI	-10.66%	-9.67%	-7.97%	-4.64%	+3.77%	-18.90%	+179.15%

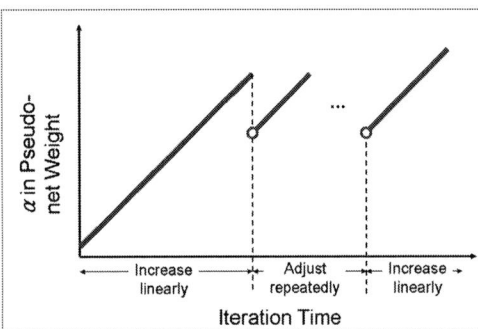

Figure 6: The weight of pseudo-net in each iteration of global placement.

Figure 7 shows the routing result after applying the congestion cluster optimization. Comparing with Figure 5 (b), the area of the congested regions has been reduced a lot.

5. POST GLOBAL PLACEMENT

We use FastPlace-DP [26] for our legalization and detailed placement. However, FastPlace-DP is HPWL-driven and that may increase congestion and reduce routability. Therefore, the HPWL-driven approach need to be modified to trade off between congestion and HPWL. Besides, a refinement method of simultaneous routing and placement SRP [27] is applied to relocate cells for routability.

Legalization has three major steps: (1) move cells out of macro blocks, (2) spread cells to makes sure that the total cell area in a segment does not exceed the segment capacity and (3) legalize cells within segment. The segments are strips in each row divided by fixed nodes. The cells will be placed in their segments without overlapping. During legalization, cells may have large displacements after

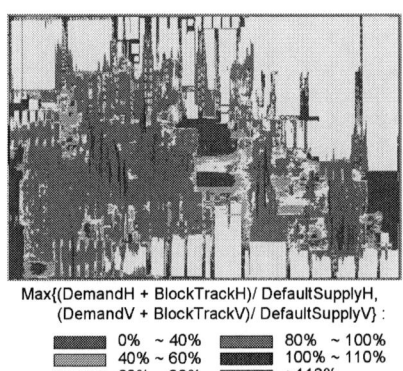

Max{(DemandH + BlockTrackH)/ DefaultSupplyH, (DemandV + BlockTrackV)/ DefaultSupplyV} :

- 0% ~ 40%
- 40% ~ 60%
- 60% ~ 80%
- 80% ~ 100%
- 100% ~ 110%
- >110%

Figure 7: After congested cluster optimization, the area of the routing congested regions of superblue3 is much less than the result shown in Figure 5 (b).

spreading within long segments. Therefore, we limit the maximum length of a segment. In our implementation, the maximum length of a segment is set to $\max\{3 \times maxCellWidth, 0.02 \times chipWidth\}$, where $maxCellWidth$ is the maximum width of a movable node and $chipWidth$ is the width of the chip.

Detailed placement is based on a greedy method. It has several steps: (1) cell clustering, (2) cell swapping and (3) local reordering. Each of these three steps contains many iterations and they are performed one by one. For example, in the cell clustering step of FastPlace-DP [26], clustering is called recursively for many iterations until the HPWL is not improved obviously. In order to avoid worsening congestion while minimizing HPWL, we will check the overflow at each iteration using Simplified FastRoute. If the overflow of the current iteration is larger than a threshold (e.g., the overflow after legalization), the placement result will be reset to the result of

Table 2: Comparison with the top results of the ICCAD 2012 contest [30]
(The runtime (s) of VDAPlacer, SimPLR, NTUPlacer4h are obtained from the ICCAD 2012 contest.)

	VDAPlace		SimPLR		NTUPlace4h		Ripple 2.0	
Benchmark	ScaledWL	Runtime	ScaledWL	Runtime	ScaledWL	Runtime	ScaledWL	Runtime
superblue1	369837790	885	278880496	2319	284990372	8769	288763344	5462
superblue3	631616365	1173	343918131	2706	447690064	7193	360085825	8487
superblue4	282282824	604	243416746	1267	236022004	4866	226639292	4504
superblue5	569553239	874	360305914	2154	421690925	7322	348733540	5728
superblue7	500120583	1725	431325680	3249	413677502	15005	429041785	11387
superblue10	1036030901	1316	690852663	4837	718978960	12352	597515266	14383
superblue16	322305959	691	285718339	1797	283297667	6024	283689195	4942
superblue18	291093797	658	182345027	1645	170935389	4622	183918099	6267
Avg.	+47.25%	N/A	+3.62%	N/A	+9.52%	N/A	1.00	N/A

the previous iteration and the current step will stop.

Refinement SRP [27] relocates cells directly by considering routing and placement simultaneously. Given a placement layout and a global routing result by NCTUgr [29], three major steps are used to do congestion refinement: (1) identify and rip-up problematic cells (a cell is identified as problematic if its routing paths pass through congested regions), (2) search for a new location for a problematic cell by considering routing paths of its associated nets and (3) move the problematic cell and reroute it from its new location to its associated nets. If the overflow by rerouting is not improved, the cell will be moved back to its original location and the rerouting result will be reset.

6. EXPERIMENTS

Our implementation is written in C++ and compiled with g++ 4.1.2. All the benchmarks were run on an Intel Xeon Linux workstation with 3.40GHz and 32 GB RAM, using one CPU core. In our placer, the target density is set to 0.95 for all the benchmarks. The benchmarks of the ICCAD 2012 contest [30] are used in this section, and our current version does not consider the hierarchy information.

Routability evaluation is performed by scaled wirelength (ScaledWL) with the same configuration as in the ICCAD contest 2012 [30]. Both HPWL and congestion are considered in ScaledWL. After obtaining the routing result by NCTUgr [29], the ACE metric [5], which is based on the distribution of the G-Cell congestion, is used to evaluate congestion where $ACE(x), x \in (0.5, 1, 2, 5)$, is the average congestion of the top $x\%$ congested G-Cell edges.

Routing path-based cell inflation and spreading (RPB) and **congested cluster improvement (CCI)** are evaluated by comparing with the placer using probabilistic congestion estimation [11] without these techniques (Base). We report the (1) $ACE(x)$ where $x \in (0.5, 1, 2, 5)$, (2) HPWL, (3) ScaledWL and (4) Runtime in Table 1. We can see that most of the runtime is spend on the technique of CCI. Both RPB and CCI can improve congestion and scaled wirelength obviously.

Routability comparisons with the top results of the ICCAD 2012 Contest is shown in Table 2. In terms of scaled wirelength, Ripple 2.0 outperforms the other top results on average. Since different machines were used and we did not use multi-threading, the runtimes reported in Table 2 are just for reference.

7. CONCLUSIONS

In this paper, we present a high quality placer to address the routability problem with several techniques: (1) lookahead routing analysis with pin density consideration, (2) routing path-based cell inflation and spreading, and (3) robust congested cluster optimization. Results show that these techniques are very effective in alleviating routing congestion.

8. REFERENCES

[1] C. Alpert, Z. Li, M. Moffitt, G. Nam, J. Roy, and G. Tellez, "What makes a design difficult to route," in *ISPD*, pp. 7–12, ACM, 2010.

[2] J. Roy and I. Markov, "Seeing the forest and the trees: Steiner wirelength optimization in placement," *TCAD*, vol. 26, no. 4, pp. 632–644, 2007.

[3] N. Viswanathan and et al, "The ispd-2011 routability-driven placement contest and benchmark suite," in *ISPD*, pp. 141–146, ACM, 2011.

[4] H. Shojaei, A. Davoodi, and J. Linderoth, "Congestion analysis for global routing via integer programming," in *ICCAD*, pp. 256–262, IEEE, 2010.

[5] Y. Wei and et al, "Glare: Global and local wiring aware routability evaluation," in *DAC*, pp. 768–773, ACM, 2012.

[6] K. Tsota, C. Koh, and V. Balakrishnan, "Guiding global placement with wire density," in *ICCAD*, pp. 212–217, IEEE, 2008.

[7] A. Kahng and Q. Wang, "Implementation and extensibility of an analytic placer," *TCAD*, vol. 24, no. 5, pp. 734–747, 2005.

[8] Z. Jiang and et al, "Routability-driven analytical placement by net overlapping removal for large-scale mixed-size designs," in *DAC*, pp. 167–172, 2008.

[9] Y. Chuang and et al, "Design-hierarchy aware mixed-size placement for routability optimization," in *ICCAD*, pp. 663–668, IEEE, 2010.

[10] P. Spindler and F. Johannes, "Fast and accurate routing demand estimation for efficient routability-driven placement," in *DATE*, pp. 1–6, IEEE, 2007.

[11] X. He, T. Huang, L. Xiao, H. Tian, G. Cui, and E. Young, "Ripple: An effective routability-driven placer by iterative cell movement," in *ICCAD*, pp. 74–79, 2011.

[12] M. Kim, J. Hu, D. Lee, and I. Markov, "A simplr method for routability-driven placement," in *ICCAD*, pp. 67–73, IEEE Press, 2011.

[13] X. Yang, B. Choi, and M. Sarrafzadeh, "Routability-driven white space allocation for fixed-die standard-cell placement," *TCAD*, vol. 22, no. 4, pp. 410–419, 2003.

[14] C. Li, M. Xie, C. Koh, J. Cong, and P. Madden, "Routability-driven placement and white space allocation," *TCAD*, vol. 26, no. 5, pp. 858–871, 2007.

[15] U. Brenner and A. Rohe, "An effective congestion-driven placement framework," *TCAD*, vol. 22, no. 4, pp. 387–394, 2003.

[16] J. Roy and et al, "CRISP: congestion reduction by iterated spreading during placement," in *ICCAD*, pp. 357–362, ACM, 2009.

[17] W. Hou and et al, "A new congestion-driven placement algorithm based on cell inflation," in *ASP-DAC*, pp. 605–608, IEEE, 2001.

[18] M. Pan and C. Chu, "Ipr: an integrated placement and routing algorithm," in *DAC*, pp. 59–62, 2007.

[19] M. Pan and C. Chu, "Fastroute: A step to integrate global routing into placement," in *ICCAD*, pp. 464–471, ACM, 2006.

[20] J. Hu and et al, "Completing high-quality global routes," in *ISPD*, pp. 35–41, 2010.

[21] M. Hsu, S. Chou, T. Lin, and Y. Chang, "Routability-driven analytical placement for mixed-size circuit designs," in *ICCAD*, pp. 80–84, IEEE, 2011.

[22] Y. Zhang and C. Chu, "Crop: Fast and effective congestion refinement of placement," in *ICCAD*, pp. 344–350, IEEE, 2009.

[23] T. Jindal and et al, "Detecting tangled logic structures in vlsi netlists," in *Design Automation Conference (DAC), 2010 47th ACM/IEEE*, pp. 603–608, IEEE, 2010.

[24] P. Spindler and et al, "Kraftwerk2: A fast force-directed quadratic placement approach using an accurate net model," *TCAD*, 27(8), pp. 1398–1411, 2008.

[25] M. Kim, D. Lee, and I. Markov, "simpl: an effective placement algorithm," *TCAD*, vol. 31, no. 1, pp. 50–60, 2012.

[26] M. Pan, N. Viswanathan, and C. Chu, "An efficient and effective detailed placement algorithm," in *ICCAD*, pp. 48–55, IEEE Computer Society, 2005.

[27] X. He, W.-K. Chow, and E. F. Young, "Srp: Simultaneous routing and placement for congestion refinement," in *ISDP*, ACM, March, 2013.

[28] G. Nam and J. Cong, *Modern circuit placement: best practices and results*. Springer Publishing Company, Incorporated, 2007.

[29] W. Liu, W. Kao, Y. Li, and K. Chao, "Multi-threaded collision-aware global routing with bounded-length maze routing," in *DAC*, pp. 200–205, ACM, 2010.

[30] N. Viswanathan and et al, "Iccad-2012 cad contest in design hierarchy aware routability-driven placement and benchmark suite," in *ICCAD*, pp. 345–348, 2012.

Optimization of Placement Solutions for Routability

Wen-Hao Liu[1,2,3], Cheng-Kok Koh[2], and Yih-Lang Li[1]

[1]Department of Computer Science, National Chiao-Tung University, Hsin-Chu, Taiwan
[2]School of Electrical and Computer Engineering, Purdue University, West Lafayette, IN, USA
[3]Department of Computer Science, National Tsing-Hua University, Hsin-Chu, Taiwan
dnoldnol@gmail.com, chengkok@purdue.edu, ylli@cs.nctu.edu.tw

Abstract – **Routability has become a critical issue in VLSI design flow. To avoid producing an unroutable design, many placers [4-7] invoke global routers to get a congestion map and then move cells to reduce congestion based on this map. However, as cells move, the accuracy of the congestion map degrades, thereby affecting the effectiveness of the placer in minimizing congestions. Moreover, most global routers [8-13] ignore local congestion. If placers are guided by these routers, it may produce hard-to-route placement solutions in terms of detailed routing. This work develops a routability optimizer, called Ropt, to reduce both global and local routing congestion levels of a given placement. Based on a local-routability-aware routing model, Ropt builds a global routing instance to obtain global and local congestion information for guiding global re-placement. In addition, this work presents a new legalization scheme to preserve the global routing instance after legalization. Finally, local detailed placement further minimizes the local congestion and wirelength. For the evaluation of Ropt, we use an academic global router and a commercial router to obtain both global and detailed routing results, respectively. Experimental results reveal that Ropt can improve the routing quality (in terms of congestion, wirelength, and violation) and routing runtime of a given placement solution.**

Categories and Subject Descriptors

B.7.2 [**Integrated Circuits**]: Design Aids - Placement and Routing.

General Terms

Algorithms, Design.

Keywords

Placement, global routing, detailed routing, routability optimization.

1. INTRODUCTION

Routability is of primary concern in nanometer-scale design. Many placers have considered the routability issue in the placement stage in order to avoid producing an unroutable design. Some placers [1-3] adopt probabilistic methods to estimate the routing congestion. However, this method typically fails to capture the actual routing behavior, and thus has low estimation accuracy. To obtain routing-based congestion information, routability-driven placers in [4-7] invoke global routers to obtain a routing congestion map, and then move cells to reduce the congestion based on the map. However, as cells move, the accuracy of the congestion map degrades, thereby affecting the effectiveness of the placer in minimizing the congestion.

To maintain the accuracy of the congestion map, IPR in [5] adopts FastRoute, developed in [8], to obtain a congestion map at the beginning of the detailed placement stage, and then incrementally updates the congestion map as cells move. For each cell c_i, IPR identifies the optimal region of c_i that minimizes the half-perimeter wirelength (HPWL) associated with c_i and swaps c_i with a cell in the optimal region. The cell chosen for swapping is the one that minimizes the congestion the most. After that, IPR reroutes the nets connected to the swapped cells. This method always maintains a global routing instance based on the current placement to offer accurate congestion information. However, IPR decides the cell locations according to its HPWL optimal region. If the optimal region is in a congested area, the congestion is hardly reduced.

To further compound the problem, most global routers [8-13] ignore the local congestion. As a consequence, placers that are guided by these routers may produce hard-to-route placement solutions in terms of detailed routing. In most global routing models, the placement is typically partitioned into an array of uniform bins, and global routers identify the bin-to-bin routes for each global net, a net whose terminals reside in different bins. However, the effects of the local nets, each of which has terminals residing within a bin, are typically ignored. To minimize local congestion, an estimation metric for local-routability was used by the detailed placer in [14] to guide the iterative movements of cells from high-cost bins to low-cost bins. A bin with a lower cost means that the routing within this bin is easier. However, the experiment in [14] reveals that this method may increase global congestion. To simultaneously consider global and local congestions, the authors of [15] took local pin density into account during global routing. If a bin has high pin density, the routing capacities from the bin to its neighboring bins are reduced, in order to avoid routing too many global paths through these bins. The remaining routing resources in the bin can be used for local routing.

This work presents a routability optimizer **Ropt** that takes a placement solution and optimizes it to improve the routing congestion, wirelength, and runtime. It combines both placement and global routing, and always maintains a global routing instance based on the current placement solution. The global routing instance is generated based on a local-routability-aware model. Therefore, the global routing instance provides both global and local congestion information to guide the placement algorithms. In contrast to CRISP [6] and CROP [16], which locally spread and shift cells, Ropt globally re-places cells to significantly improve the routability. This work has following contributions:

(a) This work proposes new capacity and demand estimations in the global routing model to account for both global and local congestion levels simultaneously, and generates a global routing instance based on the proposed model. Similar to [15], the proposed model also considers pin density. However, in contrast to [15], the proposed model also places emphasis on the effect of blockages on routability. The case study in [24] indicates that the effect of blockages for routability is often underestimated in the placement and global routing stages, which may cause routing violations in the detailed routing results.

(b) Rather than minimizing HPWL, the proposed routing-cost-driven global re-placement directly improves routability by minimizing the routing cost of nets, as the routing cost is defined in terms of congestion. The estimation of routing cost is based on the global routing instance, which is updated after every cell movement.

(c) This work presents a legalization scheme to preserve the global routing instance after legalization. After that, the proposed local detailed placement further minimizes the local congestion and wirelength without increasing global congestion.

(d) In addition to using a global router to evaluate the routability of the placement solutions, this work also uses commercial router Wroute [23] (version 3.0.61) to obtain detailed routing results of the optimized placement solutions for the evaluation of routability.

2. PROBLEM DESCRIPTION

A circuit can be defined as a set of nets, where each net comprises a set of pins, each of which can be located on a cell, a macro, or a pad. The placement problem deals with the determination of the positions of cells and macros, with the goal of minimizing the (routed) wirelength used to connects all the nets, subject to the constraints that there are no overlaps between cells and macros. The placement area is partitioned into rows, the cells and macros must be aligned with the rows.

For the ISPD11 and DAC12 benchmark circuits [17, 18] used in this work, the positions of macros are fixed. Therefore, the objective of this work is to re-place cells to improve routability, subject to the constraints that the cells are aligned with the rows and that there are no overlaps in the final placement solutions. This work uses NCTU-GR 2.0 [10] and Wroute to evaluate the routability of the placement solutions produced by the proposed optimizer Ropt on the circuits in [17, 18]. NCTU-GR 2.0 is an academic global router that has been used to evaluate the routability of placement solutions in the DAC12 placement contest. However, NCTU-GR 2.0 does not route local nets. Therefore, the impact of local congestion on the real routability cannot be evaluated properly. Indeed, a good global routing solution does not imply that a feasible detailed routing solution exists. For a thorough evaluation of the real routability of the placement solutions produced by Ropt, we also feed the placement solutions to Wroute to obtain the detailed routing results.

3. PROPOSED ROUTABLITY OPTIMIZER

Figure 1 shows the design flow of Ropt. Given a set of nets and an initial placement, Ropt first constructs a global routing model with local routability taken into consideration. Next, each net is decomposed into two-pin nets based on the topology of a rectilinear minimum spanning tree (RMST), and then an initial global routing instance is generated by the initial routing stage that is formed by pattern routing and monotonic routing. The details of the initial routing stage are introduced in paper [10]. Thereafter, routing-cost-driven global re-placement stage re-places each cell and immediately updates the global routing instance after each cell moves. Legalization stage aligns the cells on the row and removes overlaps; the goal of this stage is to preserve the global routing instance after legalization. Finally, local detailed placement further minimizes the local congestion and wirelength. Ropt iteratively refines the placement until a user defined terminating condition is met. In our implementation, we iterate the process up to two times (see Table 1 in Section 4 on the experimental setup).

3.1. Local-Routability-Aware Global Routing Model

Two global routing approaches are generally used in multi-layer global routing problem. The first one performs global routing on a three-dimensional (3D) grid graph to get a 3D routing result. The second approach first projects a 3D grid graph into a two-dimensional (2D) grid graph. Then, a 2D global router is used to obtain a 2D routing result. Finally, a layer assignment step transforms the 2D routing result to a 3D routing result. The former has a larger solution space than the latter has, but the latter is much faster than the former. In order to quickly extract the congestion

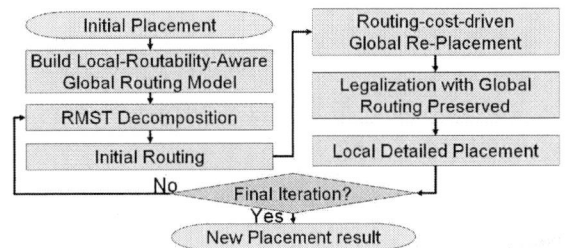

Fig. 1. Design flow of the proposed routability optimizer

information to guide the placement process, this work builds the global routing instance based on a 2D routing approach.

In the 2D global routing approach, the given placement is typically partitioned into an 2D array of uniform bins modeled in the form of a 2D grid graph $G(V, E)$, where V denotes the set of bins, and E refers to the set of grid edges. Each grid edge connects two adjacent bins. The capacity $c(e)$ of grid edge e is the number of routing tracks that can legally cross the abutting boundary, and the demand $d(e)$ of e is the number of global routing path passing through e. If $d(e)$ exceeds $c(e)$, too many global routing paths are passing through e, which implies global congestion. Thus, global congestion can be measured in terms of $d(e)$ and $c(e)$.

This work accounts for local-routability in the estimations of capacities and demands of grid edges. Therefore, when the global router in the proposed flow finds the least-cost routing paths whose cost formulation is based on the new capacity and demand estimations, the global and local congestion can be simultaneously considered and reduced. The following equation has been used in many 2D global routers to estimate grid edge capacity [8-13]:

$$c(e) = \sum_{l \in L} (T_l(b_1, b_2) - BT_l(b_1, b_2)), \qquad (1)$$

where b_1 and b_2 are the adjacent bins of grid edge e, L is the set of layers, and $T_l(b_1, b_2)$ and $BT_l(b_1, b_2)$ represent the number of routing tracks and the number of routing tracks blocked by big macros at layer l between b_1 and b_2, respectively. This capacity estimation disregards the vias and local routes in the adjacent bins. However, the presence of vias and local routes can make it hard for a design to be detailed routed. We observed that the vias and local routes in b_1 and b_2 increase as the number of blocked tracks between b_1 and b_2 increases. The increased vias and local routes would further block the tracks between b_1 and b_2, and more blocked tracks necessitate more detour and vias for a path to cross the boundary between these two bins. As the number of blocked tracks increases, the edge capacity decreases more than linearly. To capture this effect, this work estimates the capacity of a grid edge with the following equation:

$$c'(e) = \left\lceil (1 - \frac{\sum_{l \in L} BT_l(b_1, b_2)}{\sum_{l \in L} T_l(b_1, b_2)})^{\alpha} \times \sum_{l \in L} T_l(b_1, b_2) \right\rceil, \qquad (2)$$

where α is a user defined constant; we set α to be 2.5 in this work. (In this work, there are many parameters that can be tuned. We select these parameters based on empirical results. However, once set, these parameters stay the same throughout the experiment. We do not change the parameters according to the benchmark circuits, although it may be wise to do so to obtain better results.) Moreover, it has been demonstrated in [15] that the pin densities in b_1 and b_2 also affect the routability between b_1 and b_2. Thus, we estimate the demand placed on e as follows:

$$d'(e) = d(e) + \left\lceil \beta \times (\frac{p(b_1)}{\sum_{l \in L} A_l(b_1)} + \frac{p(b_2)}{\sum_{l \in L} A_l(b_2)}) \right\rceil, \qquad (3)$$

where $d'(e)$ denotes the number of routing tracks used by global and local paths between b_1 and b_2, in which $d(e)$ is the number of routing tracks used by global paths, and the second term denotes the number

Algorithm Solving FOPB problem
Input: a movable cell c_i, global routing instant O
1. $R \leftarrow$ identifyPlacingRegion(c_i)
2. Rip_up(c_i, $N(c_i)$, O)
3. $Tc \leftarrow$ getBinDesityCost(R)
4. **foreach** two-pin net $n(p_i, p_j) \in N(c_i)$
5. $Tr \leftarrow$ getRoutingCost($b(p_j)$, R, O)
6. $Tc = Tc + Tr$
7. **end foreach**
8. Place c_i into the bin with the minimal placing cost
9. Monotonic_Routing($N(c_i)$, O)

Fig. 2. Pseudo code of the algorithm for FOPB problem.

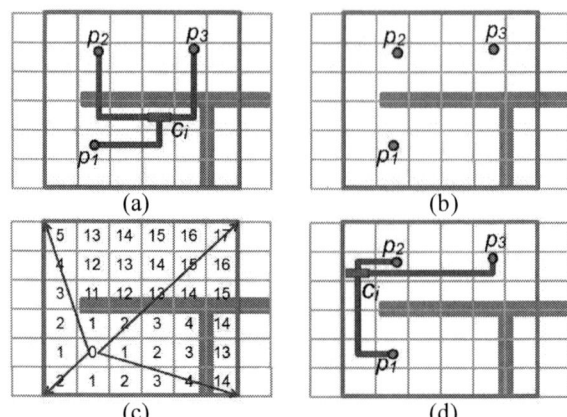

(a)   (b)

(c)   (d)

Fig. 3. (a) the bins in the red box are the placing candidates for c_i; (b) rip-up c_i and the two-pin nets in $N(c_i)$; (c) the number in each bin denotes the least-cost of monotonic path from $b(p_1)$ to each bin; (d) place c_i into the bin with minimal placing cost and reroute the two-pin nets in $N(c_i)$.

of estimated routing tracks used by local paths. Since detailed routing is yet to be performed to realize local paths, we simply use pin density to estimate the number of local paths, the concept is similar to [15]. In Eq. (3), $p(b_1)$ and $p(b_2)$ denote the number of pins in b_1 and b_2, respectively. $A_l(b_1)$ and $A_l(b_2)$ denote the areas which are not covered by macros in b_1 and b_2 at layer l, respectively. β is a user defined constant, and we set it to be 1500.

Many global routing papers [8, 9, 11, 13] discuss how to formulate the routing cost in terms of the global congestion based on $c(e)$ and $d(e)$ (and possibly other relevant metrics) and then find the least-cost routing path for each net to minimize the global congestion. This work adopts the routing cost formulation presented in [9], but replaces $c(e)$ and $d(e)$ by $c'(e)$ and $d'(e)$, respectively, in order to consider local-routability. This local-routability–aware routing cost function is used in the initial routing and global re-placement stages.

3.2. Routing-cost-driven Global Re-Placement

As minimizing the total routing cost can directly improve the congestion, we formulate such a routing cost minimization problem as that of finding the optimal placement bin b_m for a movable cell c_i in the global re-placement stage to minimize the routing cost. Also, to avoid placing too many cells into a bin, each bin has to satisfy a bin density constraint. The problem of Finding Optimal Placement Bin (FOPB) for c_i is concisely formulated by the following equations (for ease of explanation, if c_i is placed into b_m, we assume all pins of c_i are inside b_m):

$$\min_{b_m \in R} \sum_{n(p_i, p_j) \in N(c_i)} \sum_{e \in r(b_m, b(p_j))} routC(e) \qquad (4)$$

$$routC(e) = C_1 + C_2 \big/ (1 + C_3^{\,C_4 \times (c'(e) - d'(e))}) \qquad (5)$$

$$\text{s.t.} \qquad D(b_m) \le D_t \qquad (6)$$

Equation (4) is the objective equation of FOPB problem, R is the set of candidate bins for placing c_i, $N(c_i)$ denotes the set of two-pin nets connecting to c_i, and $n(p_i, p_j)$ represents a two-pin net whose terminals are pins p_i and p_j. Pin p_i is on c_i and p_j is on the other cell, macro or pad. When c_i is placed into b_m, p_i is in b_m; $b(p_j)$ denotes the bin containing p_j. $r(b_m, b(p_j))$ represents the least-cost global routing path from b_m to $b(p_j)$, e is a grid edge passed by $r(b_m, b(p_j))$, and $routC(e)$ denotes the routing cost of e. The formulation of $routC(e)$ in Eq. (5) is inspired by [9], in which C_1, C_2, C_3 and C_4 are user defined constants; we set them to 25, 20, 2.72 and 0.3, respectively. In Eq. (5), when $d'(e)$ is close to $c'(e)$ and increases, $routC(e)$ increases significantly. The larger $routC(e)$ means that grid edge e is congested, so minimizing $routC(e)$ by Eq. (4) can reduce congestion. Moreover, Eq. (6) is the bin density constraint; $D(b_m)$ denotes the bin density of b_m, which is the total area of cells in b_m divided by the area of b_m. If $D(b_m)$ exceeds D_t, it is called bin overflow. In this work, we set D_t to 0.9.

The proposed routing-cost-driven global re-placement stage uses a heuristic method to solve the FOPB problem. This method first identifies a set of candidate bins for placing c_i and then evaluates the cost of placing c_i in each candidate bin. Finally, c_i is placed into the

bin with minimal placing cost. The cost of placing c_i in b_m is calculated as follows:

$$p(b_m, c_i) = desC(b_m) \times h(b_m) + \sum_{n(p_i, p_j) \in N(c_i)} \sum_{e \in mr(b_m, b(p_j))} routC(e), \quad (7)$$

$$desC(b_m) = \begin{cases} \kappa \times D(b_m) & \text{if } D(b_m) \le D_t \\ \kappa \times D_t + \mu \times (D(b_m) - D_t) & \text{otherwise} \end{cases} \quad (8)$$

where $p(b_m, c_i)$ denotes the cost of placing c_i in b_m, $desC(b_m)$ denotes the penalty because of the bin density of b_m, and $mr(b_m, b(p_j))$ denotes the least-cost monotonic routing path from b_m to $b(p_j)$. Moreover, $h(b_m)$ is the history cost of b_m; it has an initial cost of 1 and it increases by 1 when the placement of a cell in b_m causes bin overflow. Incrementally increasing history cost can gradually decrease bin overflows. In Eq. (8), κ and μ are user defined constants. In this work, we set κ and μ to 1 and 1000, respectively. Accordingly, if $D(b_m)$ exceeds D_t, the bin density cost becomes very large.

Figure 2 shows the algorithm to solve the FOPB problem. Line 1 first identifies a minimum bounding box enclosing c_i and the pins with connection to c_i, and then extends the boundaries of the box by γ units of bins. The bins within the box are regarded as the candidate bins for placing c_i. For example, in Figure 3(a), the bins in the red box are the candidate bins for placing c_i when γ is one. A larger γ offers a bigger solution space but longer runtime; γ is set to 5 in this work. In line 2, c_i is ripped-up from the placement and the two-pin nets in $N(c_i)$ are also ripped-up from the global routing instance O (Figure 3(b)).

Line 3 computes the bin density and the associated penalty (via Eq. (8)) of each bin in R and stores the penalties in matrix Tc. In Line 5, monotonic routings from $b(p_j)$ to four corner bins in R are performed. Monotonic routing can obtain a least-cost monotonic path from the source to each node in the minimum bounding box enclosing the source and target. The monotonic routings from $b(p_j)$ to four corner bins in R can obtain the least-cost of the monotonic path from $b(p_j)$ to every bin in R. After that, the routing costs are stored in matrix Tr. In Figure 3(c), the gray rectangles represent congested regions, the arrows depict the directions of monotonic routings, and the number in each bin denotes the least-cost of the monotonic path from $b(p_1)$ to each bin. For simplicity, we assume in this example that $routC(e)$ is 10 if e crosses a congested region, and 1 otherwise. When the iterations from lines 4 to 7 terminate, Tc contains for each bin in R, the cost of placing c_i in that bin. Finally, line 8 places c_i into the bin with minimal cost and line 9 routes the monotonic paths of every two-pin net in $N(c_i)$ and updates O (Figure

1052

Algorithm Local Detailed Placement
Input: Grid Graph $G(V, E)$
1.   **for** it =1 to u
2.   **foreach** bin $b_m \in V$
3.   **foreach** cell $c_i \in C(b_m)$
4.   Cell_Swapping($c_i, C(b_m)$)
5.   **foreach** segment s_j within b_m
6.   Sliding_Window($s_j, C(b_m)$)
7.   **foreach** cell $c_i \in C(b_m)$
8.   Moving_to_Empty_Spot($c_i, C(b_m), b_m$)
9.   **end foreach**
10. **end for**

Fig. 4. Pseudo code of local detailed placement

3(d)). The time complexity of this algorithm is $O(|R|*|N(c_i)|)$ where $|R|$ and $|N(c_i)|$ denote the number of bins in R and the number of two-pin nets connected to c_i, respectively.

The global re-placement stage performs the algorithm in Figure 2 for every cell in each round. The placing order of cells influences the solution quality, but we simply place cells in the increasing order according to their id. In our implementation, this stage runs for four rounds. Notably, this stage simply places cells at the center of the assigned bins. The definite locations of cells are decided in the following stages: legalization and local detailed placement.

3.3. Legalization with Global Routing Preserved

After global re-placement, many cells have overlaps and are not aligned with the rows. The duty of legalization is to align cells on the rows and remove overlaps. Most legalizers [19, 20] minimize the total displacement of cells as the main objective to ensure consistency between the global placement solution and the legalized solution. Instead, the objective of our legalizer is to preserve the global routing instance after legalization. In order to minimize the change to the global routing instance, if the global re-placement stage places a cell in bin b_m, our legalizer attempts to keep this cell in b_m.

Our legalizer is similar to Abacus [19], but with a different objective. First, every cell is sorted in increasing ordering according to its x-coordinate. Next, for each cell c_i, our legalizer tentatively moves c_i to its neighboring rows r_k and calculates the cost of moving c_i to r_k. Then, c_i is moved to the best row with the lowest cost. In Abacus, the cost formulation is based on the displacement of c_i. Instead, this work calculates the cost as follows,

$$mc(c_i, b_m, r_k) = \mu \times ao(c_i, b_m, r_k) + \sigma \times \sum_{p \in P(c_i)} d(b_m, b(p_k)) \quad (9)$$
$$+ m(c_i, r_k) + cu(r_k)$$

where c_i is placed into bin b_m in the global re-placement stage, $mc(c_i, b_m, r_k)$ denotes the cost of moving c_i to row r_k, $ao(c_i, b_m, r_k)$ denotes the area of c_i out of b_m when c_i is moved to r_k, $P(c_i)$ denotes the set of pins belonged to c_i, $b(p_k)$ denotes the bin containing p when c_i is moved to r_k, and $d(b_m, b(p_k))$ denotes the index distance between $b(p_k)$ and b_m. For example, if the indexes of b_m and $b(p_k)$ in the grid graph are (x, y) and (x', y'), respectively, $d(b_m, b(p_k))$ is $|x-x'|+|y-y'|$. Moreover, $m(c_i, r_k)$ is the displacement of c_i when c_i is moved to r_k, $cu(r_k)$ is the capacity utilization of r_k which is between zero and one. μ and σ are user defined constants. In our implementation, μ and σ are set to 1000 and 100, respectively. Since the global routing instance would change if c_i moves out of b_m, Eq. (9) gives the high penalty of $mc(c_i, b_m, r_k)$ if c_i moves out of b_m.

3.4. Local Detailed Placement

This stage addresses the problem of minimizing the local wirelength in each bin to reduce local congestion. To avoid degrading the global routability, this stage does not move a cell from its original bin to another bin so as to ensure that the global routing instance remains the same. Performing detailed routing in each bin

TABLE 1 ROPT WITH DIFFERENT FEATURES

Versions	Features				
	LGM	RGP	LRP	LDP	iteration
Ropt$_1$		✓			1
Ropt$_2$		✓	✓		1
Ropt$_3$		✓	✓		2
Ropt$_4$	✓	✓	✓		2
Ropt$_5$		✓	✓	✓	2
Ropt$_6$	✓	✓	✓	✓	2

can get accurate local wirelength and congestion information but is time-consuming. Thus, this stage simply uses Manhattan distance to measure the local wirelength in each bin. If the terminals of a two-pin net n are both in bin b_m, the local wirelength of n is measured by the Manhattan distance between its two terminals. If a terminal of n is outside b_m, the terminal is projected on the boundary of b_m and the local wirelength of n in b_m is measured by the Manhattan distance between the projected terminal and inside terminal.

Figure 4 shows the pseudo code of local detailed placement, in which $C(b_m)$ denotes the set of cells that are entirely in b_m, i.e. the cells in $C(b_m)$ do not cross the boundaries of b_m; s_j denotes a segment of rows within b_m, u is a user defined terminating condition, we set it to be 5. Local detailed placement adopts cell swapping, sliding window and moving cells to empty spots methods to greedy improve local wirelength. In Figure 4, line 4 tentatively swaps c_i with other cells in $C(b_m)$. If there exists at least a legal swap that can reduce the local wirelength, line 4 picks the legal tentative swap with the best improvement to perform an actual swap. A legal swap means that the swapped cells are entirely in b_m and they do not cause cell overlap. Line 6 adopts the sliding windows method [21] to re-order the cells in s_j for minimizing the local wirelength. We use a window size of 5 cells in this work. Line 8 tentatively moves c_i to an empty spot in b_m. If there exists at least a legal move that can reduce the local wirelength, line 8 picks the legal tentative move with the best improvement to perform a real move. Again, a legal move is one that does not cause cell overlap.

4.   EXPERIMENTAL RESULTS

The proposed algorithms are implemented in C/C++ on a quad-core 2.4 GHz Xeon-based linux server with a 50GB memory (only a single core is used). This work uses the placement solutions of NTUplace to be the input placement for Ropt. NTUplace won the DAC12 routability-driven placement contest [18], and the placement solutions of NTUplace in the contest are downloaded from [22].

This work proposes four features: local-routability-aware global routing model (LGM) with new resource and demand estimates, routing-cost-driven global re-placement (RGP), legalization with global routing preserved (LRP) and local detailed placement (LDP). To show the effectiveness of each feature, several versions of Ropt are built to evaluate the effects of these features. Table 1 lists the features of each version, in which the column "iteration" is the terminal condition in Figure 1. Notably, the versions without LGM adopt $c(e)$ and $d(e)$ rather than $c'(e)$ and $d'(e)$ to formulate the routing cost in Eq. (5), and the Ropt$_1$, which does not use LRP, uses Abacus [19] instead of the proposed legalizer to perform legalization.

4.1.   Global Routability: Evaluation by NCTU-GR 2.0

For the evaluation of the effect of the proposed algorithm on global routability, we use Ropt$_1$, Ropt$_2$ and Ropt$_3$ to optimize the placement solutions of NTUplace, and then use NCTU-GR 2.0 to generate the global routing result of each placement solution. Because LGM and LDP address the issues of local wirelength and congestion, the effectiveness of Ropt$_4$, Ropt$_5$ and Ropt$_6$ cannot be evaluated by global routers. Therefore, Ropt$_4$, Ropt$_5$ and Ropt$_6$ are evaluated by Wroute in Section 4.2. Notably, NCTU-GR 2.0 is a public global routing tool with several tunable parameters. We set

TABLE 2 GLOBAL ROUTING RESULT COMPARISON BETWEEN NTUPLACE, ROPT₁, ROPT₂ AND ROPT₃

	NTUplace			NTUplace+Ropt$_1$				NTUplace+Ropt$_2$				NTUplace+Ropt$_3$			
	WL	Via	R_{cpu}	WL	Via	R_{cpu}	P_{cpu}	WL	Via	R_{cpu}	P_{cpu}	WL	Via	R_{cpu}	P_{cpu}
s2	178.43	55.37	471.62	179.51	62.81	325.89	611.98	173.75	49.00	236.08	594.14	173.07	48.22	234.00	1143.15
s3	109.15	50.75	270.28	109.99	57.45	283.69	483.58	104.44	46.37	204.97	479.23	103.61	45.62	184.03	908.00
s6	104.37	51.64	169.73	107.09	61.26	175.29	445.01	100.67	47.11	120.18	445.15	100.09	46.34	115.86	829.15
s7	127.57	75.00	128.12	129.39	85.02	118.74	746.92	122.02	69.24	100.28	739.33	121.11	68.27	94.44	1409.29
s9	77.08	41.15	88.69	78.99	49.06	78.65	638.75	73.89	37.59	62.85	643.26	73.37	36.97	60.62	1152.00
s11	103.38	45.77	109.75	105.69	53.35	102.85	322.33	100.38	41.86	90.86	322.34	99.82	41.33	83.11	615.59
s12	109.52	70.80	111.74	113.16	83.85	108.24	1002.11	104.22	64.02	82.20	1048.66	103.20	62.97	81.92	1821.92
s14	69.19	33.95	87.11	70.85	38.72	90.25	307.41	67.63	31.20	72.07	320.11	67.31	30.82	66.02	576.04
s16	79.13	33.06	154.94	81.84	41.34	250.24	261.45	76.77	29.69	91.18	263.02	76.42	29.12	81.75	500.49
s19	46.79	25.19	41.38	48.02	30.40	39.03	608.24	44.68	22.46	32.30	612.08	44.42	22.11	31.33	1101.72
Ratio$_{ind}$	1	1	1	1.022	1.173	1.009		0.964	0.908	0.722		0.958	0.894	0.681	
Ratio$_{sum}$	1	1	1	1.020	1.167	0.963		0.964	0.909	0.669		0.958	0.895	0.632	

*The units of R_{cpu} and P_{cpu} are second.

the parameters of via cost, wirelength optimization level, pattern routing iteration, monotonic routing iteration and post routing iteration in NCTU-GR 2.0 to 1, 50, 2, 2 and 1, respectively. The rip-up and rerouting stage in NCTU-GR 2.0 iterates until either an overflow-free result is obtained or the iteration number is more than 25. In addition, NCTU-GR 2.0 is a parallel router, but it only uses one thread in the following experiment.

Table 2 shows the routing results of each placement solution, in which WL, Via, R_{cpu} and P_{cpu} denote the wirelength (10^5), via count (10^5), NCTU-GR 2.0's runtime (sec) and Ropt's runtime (sec), respectively. The routing results of NTUplace are treated as the baseline; Ratio$_{ind}$ is the average of the ratio of individual entries in the same column, while Ratio$_{sum}$ denotes the ratio of the sum of each column. The routing overflows are not listed in Table 2 since every routing result is overflow-free. Table 2 reveals that Ropt$_1$ has worse WL, Via and R_{cpu} than NTUplace. This implies that even though the global re-placement may optimize a global routing instance, the legalization stage can worsen the global routing result when the legalizer is oblivious to the global routing instance. On the other hand, Ropt$_2$, which uses the proposed legalizer to preserve the global routing instance, can on the average improve WL, Via and R_{cpu} by 3.6%, 9.2% and 27.8%, respectively, when compared to NTUplace. In addition, the overflow issue in the Ropt$_2$'s solutions can be easily resolved by the pattern and monotonic routing stages in NCTU-GR 2.0. Consequently, NCTU-GR 2.0 does not have to invoke the more time-consuming maze routing stage as frequently. Therefore, the runtime of NCTU-GR 2.0 decreases. For example, in case s16, after the monotonic routing stage in NCTU-GR 2.0, 105430 and 14363 overflows remain in the placement solutions of NTUplace and Ropt$_2$, respectively. Also, Ropt$_2$ allows many nets to have simple routing solution, thereby reducing via count. Moreover, Ropt$_3$ can on the average reduce WL, Via and R_{cpu}, respectively, by 4.2%, 10.6% and 31.9%, when compared to NTUplace. Table 2 shows that the placement solutions can be further improved as the Ropt runs more iterations, but the improvement gradually diminishes.

4.2. Effective Routability: Evaluation by Wroute

Because global routers ignore the local nets within bins, using global routers to evaluate placement solutions may encourage placers to push many nets into a bin to improve global routing results. However, the local congestion would make it harder to route such designs in the detailed routing stage. To examine that Ropt can really improve routability, this work evaluate the placement solutions by their detailed routing results obtained by Wroute. Because of the format issue and the lack of design rules in the benchmark circuits in [17, 18], we used a recently developed translator [24] to feed the placement solutions of [17, 18] to Wroute (version 3.0.61). The setup of Wroute in this work refers to [24]. Table 3 shows the detailed routing results of the placement solutions obtained by

TABLE 3 DETAILED ROUTING RESULTS OF NTUPLACE

Bench marks	NTUplace				
	NUN	Vio	WL (10^7)	Via (10^6)	R_{cpu}
s2	2169	725813	67.80	12.28	48:33:08
s3	1450	988	39.99	11.05	13:06:50
s6	921	637	39.17	11.49	11:02:01
s7	419	168	48.31	16.93	13:17:31
s9	1112	89	29.19	9.62	08:49:18
s11	313	697	38.63	10.19	11:52:41
s12	120	94	42.73	17.01	11:26:41
s14	2352	18446	26.81	7.46	21:57:51
s16	78	24	29.19	7.72	06:59:38
s19	414	15114	18.11	5.99	16:42:50

NTUplace, in which NUN is an indicator in Wroute to estimate the global routability (lower NUN means better global routability), Vio denotes the routing violations which are caused by opens, shorts or spacing errors, and R_{cpu} (hh:mm:ss) denotes the runtime of Wroute. Note that, the units of WL and Via in Table 2 and 3 are different, because Table 2 shows global routing results while Table 3 shows detailed routing results.

Table 4 compares the detailed routing results of Ropt$_3$, Ropt$_4$, Ropt$_5$ and Ropt$_6$ with NTUplace. Because Ropt$_3$ improves the global routing results, NUN is reduced by 35.2% on the average. However, Ropt$_3$ does not consider the local wirelength and congestion, resulting in increases in violations, wirelength and vias. Furthermore, Ropt$_4$ uses LGM to consider global and local congestions simultaneously. Therefore, Ropt$_4$ yields fewer NUN and violations than Ropt$_3$. Because Ropt$_4$ does not minimize the local wirelength, local routing can still cause violations, see s14 for example. To minimize the local wirelength and congestion, Ropt$_5$ uses LDP to get fewer violations, shorter wirelength, and fewer vias than Ropt$_3$. Notably, since the only difference between Ropt$_5$ and Ropt$_3$ is LDP, the global routing results of Ropt$_5$ and Ropt$_3$ are similar. Thus, Ropt$_5$ and Ropt$_3$ yield similar NUN. Finally, Ropt$_6$ involves all features proposed in this work, it can minimize NUN, violations, wirelength and runtime. In particular, compared to the results for NTUPlace in Table 3, the runtime for s2 is reduced from 48 hours to 13 hours; the number of violations in s19 is reduced from 15114 to 265.

By comparing Table 2 and Table 4, a big gap between global and detailed routing results can be found. Ropt$_3$ seems to get better results than NTUplace in Table 2, but it increases violations, wirelength and vias in its detailed routing results. In addition, the via improvement of Ropt$_3$ in Table 2 and Table 4 have a big mismatch because global routing model does not consider the vias generated by local routes. However, the vias generated by local routes are considerable. These imply that optimizing a placement for

1054

improving quality of its global routing result may not help in improving its effective routability in the detailed routing stage.

5. CONCLUSIONS

This work develops a routability optimizer Ropt that takes a placement solution and then optimizes its routability for both global routing and detailed routing. Ropt combines both placement and global routing. A global routing instance is built to provide the congestion information for placement algorithms. This work presents local-routability-aware global routing model, routing-cost-driven global re-placement, legalization with global routing preserved, and local detailed placement to optimize the routability. Finally, NCTU-GR 2.0 and Wroute are both adopted to evaluate the routability of Ropt's placement solutions. The experiment results obtained by both routers reveal that Ropt can improve routing congestion, wirelength and runtime of a given placement.

REFERENCES

[1] X. He et al, "Ripple: an effective routability-driven placer by iterative cell movement", in Proc. ICCAD, pages 74–79, 2011.

[2] M.-K. Hsu et al, "Routability-driven analytical placement for mixed-size circuit designs", in Proc. ICCAD, pages 80–84, 2011.

[3] K. Tsota et al, "Guiding global placement with wire density", in Proc. ICCAD, pages 212–217, 2008.

[4] M.-C. Kim et al, "A SimPLR method for routability-driven placement", in Proc. ICCAD, pages 80–84, 2011.

[5] M. Pan and C. Chu, "IPR: An integrated placement and routing algorithm", in Proc. DAC, pages 67–73, 2007.

[6] J. Roy et al, "CRISP: Congestion reduction by iterated spreading during placement", in Proc. ICCAD, pages 357–362, 2009.

[7] K.-R. Dai et al, "GRPlacer: Improving routability and wire-length of global routing with circuit replacement", in Proc. ICCAD, 2009.

[8] M. Pan and C. Chu, "FastRoute 2.0: A high-quality and efficient global router", In Proc. ASP-DAC, pages 250-255, 2007.

[9] K.-R. Dai et al, "NCTU-GR: Efficient Simulated Evolution-Based Rerouting and Congestion-Relaxed Layer Assignment on 3-D Global Routing", *IEEE TVLSI*, vol. 20(3), pp. 459-472, 2012.

[10] W.-H. Liu et al, "Multi-Threaded Collision-Aware Global Routing with Bounded-Length Maze Routing" in Proc. DAC, pages 200-205, 2010.

[11] J. Hu et al, "Completing high-quality global routes", in Proc. ISPD, pages 35-41, 2010.

[12] W.-H. Liu et al, "A Fast Maze-Free Routing Congestion Estimator With Hybrid Unilateral Monotonic Routing", in Proc. ICCAD, 2012.

[13] Y.-J. Chang et al, "NTHU-Route 2.0: a fast and stable global router," in Proc. ICCAD, pages 338-343, 2008.

[14] T. Taghavi et al, "New Placement Prediction and Mitigation Techniques for Local Routing Congestion", in Proc. ICCAD, 2010.

[15] Y. Wei et al, "GLARE: Global and Local Wiring Aware Routability Evaluation", in Proc. DAC, 2012.

[16] Y. Zhang and C. Chu, "Fast and Effective Congestion Refinement of Placement", in Proc. ICCAD, 2009.

[17] N. Viswanathan et al, "The ISPD-2011 Routability-driven Placement Contest and Benchmark Suite", in Proc. ISPD, 2011.

[18] N. Viswanathan et al, "The DAC 2012 Routability-driven Placement contest and benchmark suite", in Proc. DAC, 2012.

[19] P. Spindler et al, "Abacus: Fast Legalization of Standard Cell Circuits with Minimal Movement", in Proc. ISPD, 2008.

[20] Y.-M. Lee et al, "A Hierarchical Bin-Based Legalizer for Standard-Cell Designs with Minimal Disturbance", in Proc. ASP-DAC, 2010.

[21] A. Agnihotri et al, "Fractional cut: Improved recusive bisecction placement", in Proc. ICCAD, 2003.

[22] http://archive.sigda.org/dac2012/contest/dac2012_contest.html

[23] http://www.cadence.com/products/di/soc_encounter/

[24] W.-H. Liu et al, "Case Study for Placement Solutions in ISPD11 and DAC12 Routability-Driven Placement Contests", in Proc. ISPD, 2013.

[25] N. Viswanathan et al, "ICCAD-2012 CAD Contest in Design Hierarchy Aware Routability-Driven Placement and Benchmark Suite", in Pro. ICCAD, 2012.

TABLE 4 DETAILED ROUTING RESULT COMPARISON BETWEEN NTUPLACE, ROPT₃, ROPT₄, ROPT₅ AND ROPT₆

	NTUplace+Ropt₃						NTUplace+Ropt₄					
	NUN	Vio	WL(10^7)	Via(10^6)	R_{cpu}	P_{cpu}	NUN	Vio	WL(10^7)	Via(10^6)	R_{cpu}	P_{cpu}
s2	1003	800	67.25	12.30	14:03:40	00:18:45	552	693	67.27	12.27	13:55:41	00:24:13
s3	728	369	39.62	11.40	11:34:13	00:14:58	270	205	39.57	11.37	10:31:41	00:17:28
s6	243	267	39.46	12.03	11:27:09	00:14:14	113	217	39.36	12.00	10:38:46	00:16:27
s7	192	152	48.11	17.34	12:44:58	00:24:46	110	132	48.06	17.33	12:05:44	00:27:17
s9	394	875	29.44	9.97	10:11:47	00:20:38	51	37	29.42	9.95	07:24:35	00:22:59
s11	258	464	38.80	10.50	11:30:38	00:11:17	119	421	38.85	10.50	10:22:15	00:12:52
s12	119	1226	43.20	18.17	16:14:58	00:33:23	65	431	42.94	18.07	14:41:42	00:35:32
s14	2007	15736	27.18	7.78	23:19:04	00:10:40	1482	22656	27.09	7.72	19:27:56	00:11:46
s16	65	26	29.70	8.22	06:45:03	00:08:40	0	22	29.66	8.23	06:38:53	00:10:13
s19	386	6814	18.30	6.31	16:10:48	00:19:23	287	2411	18.20	6.29	11:49:00	00:21:31
Ratio$_{ind}$	0.648	2.763	1.005	1.040	0.971		0.312	0.924	1.003	1.037	0.851	
Ratio$_{sum}$	0.577	0.035	1.003	1.039	0.818		0.326	0.036	1.001	1.036	0.718	
	NTUplace+Ropt₅						NTUplace+Ropt₆					
	NUN	Vio	WL(10^7)	Via(10^6)	R_{cpu}	P_{cpu}	NUN	Vio	WL(10^7)	Via(10^6)	R_{cpu}	P_{cpu}
s2	989	693	66.55	11.96	14:26:49	00:23:58	538	664	66.59	11.94	12:49:11	00:27:01
s3	771	312	39.00	11.00	10:41:40	00:19:37	279	201	38.95	10.95	09:10:16	00:20:01
s6	267	283	38.79	11.60	09:59:04	00:18:55	108	218	38.71	11.56	09:38:43	00:19:37
s7	191	174	47.42	16.94	11:46:12	00:29:23	113	136	47.38	16.92	12:20:41	00:31:32
s9	357	69	28.91	9.67	07:40:39	00:25:57	58	24	28.91	9.65	07:48:57	00:26:16
s11	234	441	38.27	10.23	10:35:36	00:14:30	147	422	38.32	10.22	10:53:57	00:15:07
s12	103	259	42.11	17.41	12:11:41	00:42:11	72	96	41.83	17.32	10:54:10	00:42:57
s14	1998	14913	26.82	7.54	22:40:07	00:12:35	1457	10374	26.74	7.48	17:22:44	00:13:53
s16	62	26	29.15	7.84	06:15:12	00:11:25	0	24	29.13	7.86	06:10:00	00:13:03
s19	367	2745	17.93	6.07	10:01:30	00:23:06	268	265	17.83	6.05	06:40:36	00:23:51
Ratio$_{ind}$	0.619	0.803	0.988	1.005	0.826		0.322	0.483	0.987	1.002	0.759	
Ratio$_{sum}$	0.571	0.026	0.987	1.005	0.710		0.325	0.016	0.985	1.002	0.634	

*The units of R_{cpu} and P_{cpu} are hh:mm:ss.

6. SUPPLEMENT

6.1. Legalization: Global Placement Preserved versus Global Routing Preserved

Table 5 shows the detailed comparison between Abacus [19] used in $Ropt_1$ and our legalizer used in $Ropt_2$, in which D_{max}, D_{avg}, BD_{max} and BD_{avg} denote the maximum displacement, average displacement, maximum bin displacement and average bin displacement of cells after legalization, respectively. Notably, if the center of a cell before and after legalization is respectively located at bins b_i and b_j, the bin displacement of this cell is the index distance between b_i and b_j. Table 5 reveals that our legalizer obtains longer D_{avg} but shorter BD_{avg} than Abacus.

Traditional legalizers usually focus on minimizing D_{avg} to ensure consistency between the global placement solution and the legalized solution. Instead, our legalizer attempts to keep cells in the bins assigned by the global re-placement stage in order to preserve the global routing instance. Therefore, the average bin displacement is reduced. Table 2 and Table 5 reveal that a legalizer minimizing average bin displacement may identify better placement solutions than minimizing average displacement in the routability-driven placement problem.

6.2. Optimizing the Placement Solutions in ISPD11 and DAC12 contests

In last two years, three routability-driven placement contests are respectively held at ISPD11 [17], DAC12 [18] and ICCAD12 [25], which motivate many researchers to develop the new-generation placers to address the routability issue. To further validate the effectiveness of this work, we perform $Ropt_6$ to optimize the placement solutions of the leading placers in ISPD11 and DAC12 contests, and then use Wroute to evaluate the routability of the placement solutions. However, the benchmark circuits used in ICCAD12 contest does not provide the information of blockages' height, thereby the placement solutions in ICCAD12 contest cannot be evaluated by Wroute.

The top four placers in both ISPD11 and DAC12 placement contests, in alphabetical order, are mPL, NTUplace, Ripple, and SimPLR. The optimized placement solutions of NTUplace in DAC12 contest have been shown in Table 2 and Table 4. Moreover, Tables 6, 7 and 8 show that $Ropt_6$ can reduce NUN, violations, total wirelength, via count and routing runtime in most placement solutions of mPL, Ripple, and SimPLR in DAC12 contest. Finally, Table 9 treats the detailed routing results of NTUplace as the baseline to compare the detailed routing results of mPL, NTUplace, Ripple, SimPLR and $Ropt_6$.

The benchmark circuits used in ISPD11 placement contest are different to that used in DAC12 contest. Ripple won the ISPD11 placement contest, so Table 10 shows the effectiveness of using $Ropt_6$ to optimize Ripple's placement solutions in ISPD11 contest. Moreover, Table 11 treats the detailed routing results of Ripple as the baseline, and then shows the improvement of detailed routing results of using $Ropt_6$ to optimize the placement solutions of the top four placers in ISPD11 placement contest. Note that, the version of NTUplace for ISPD11 contest was named RADIANT during the contest.

TABLE 5   COMPARISON BETWEEN ABACUS AND OUR LEGALIZER

	Abacus [19] used in $Ropt_1$					Our legalizer used in $Ropt_2$				
	D_{max}	D_{avg}	BD_{max}	BD_{avg}	Time(sec)	D_{max}	D_{avg}	BD_{max}	BD_{avg}	Time(sec)
s2	306	11.025	10	0.742	2.98	333	17.738	11	0.138	4.08
s3	945	12.867	24	0.728	2.76	913	19.612	27	0.174	3.82
s6	840	13.830	24	0.749	2.82	836	20.649	24	0.168	4.14
s7	1391	13.944	43	0.721	3.6	1391	20.363	43	0.207	5.04
s9	691	12.680	22	0.754	2.21	691	19.604	22	0.190	2.87
s11	3089	13.594	83	0.771	2.09	3083	20.170	83	0.199	2.99
s12	595	12.285	17	0.791	4.5	595	19.935	17	0.159	7.53
s14	1292	14.770	38	0.718	1.43	1305	21.122	37	0.198	33
s16	872	13.421	22	0.738	1.86	872	20.210	22	0.139	2.86
s19	165	12.582	4	0.746	1.1	142	19.443	4	0.151	1.8
Ratio$_{ind}$	**1**	**1**	**1**	**1**	**1**	**1.088**	**1.609**	**1.100**	**0.185**	**1.369**
Ratio$_{sum}$	**1**	**1**	**1**	**1**	**1**	**0.998**	**1.518**	**1.010**	**0.231**	**2.688**

TABLE 6   DETAILED ROUTING RESULT COMPARISON BETWEEN MPL AND MPL+$ROPT_6$

	mPL					mPL+$Ropt_6$					
	NUN	Vio	WL(10^7)	Via(10^6)	R_{cpu}	NUN	Vio	WL(10^7)	Via(10^6)	R_{cpu}	P_{cpu}
s2	2068	113991	80.08	13.40	40:44:26	520	3033	76.79	12.65	18:00:13	00:25:15
s3	649	231	44.93	11.56	10:32:16	83	175	42.85	11.25	09:13:55	00:22:15
s6	892	231	45.42	12.22	10:13:03	125	243	43.92	12.00	09:36:14	00:20:49
s7	1525	170	55.55	18.44	14:13:09	153	144	53.13	17.93	12:28:56	00:29:16
s9	573	68	34.22	10.32	08:34:57	42	55	33.08	10.14	07:19:49	00:27:09
s11	531	11009	47.48	11.31	22:16:53	218	527	45.93	10.97	11:17:07	00:14:50
s12	994	343637	47.58	19.18	48:49:04	537	23309	44.69	18.71	20:16:41	00:39:59
s14	1717	271799	31.10	7.96	36:07:43	709	75677	30.17	7.72	15:47:54	00:13:43
s16	127	36	31.90	8.00	07:11:05	1	36	31.27	7.88	06:25:10	00:10:49
s19	811	276	20.74	6.43	05:57:55	164	358	19.89	6.27	05:51:20	00:25:31
Ratio$_{ind}$	**1**	**1**	**1**	**1**	**1**	**0.227**	**0.618**	**0.962**	**0.973**	**0.722**	
Ratio$_{sum}$	**1**	**1**	**1**	**1**	**1**	**0.258**	**0.140**	**0.961**	**0.972**	**0.568**	

TABLE 7   DETAILED ROUTING RESULT COMPARISON BETWEEN RIPPLE AND RIPPLE+ROPT$_6$

	Ripple					Ripple+Ropt$_6$					
	NUN	Vio	WL(10^7)	Via(10^6)	R$_{cpu}$	NUN	Vio	WL(10^7)	Via(10^6)	R$_{cpu}$	P$_{cpu}$
s2	1743	81227	72.77	12.67	32:26:26	802	5928	70.18	12.26	23:01:26	00:25:32
s3	566	243	43.55	11.58	10:47:19	323	180	41.53	11.45	09:50:43	00:19:45
s6	267	232	40.96	11.77	10:35:16	146	235	40.07	11.83	09:34:46	00:18:26
s7	703	300	53.83	17.77	13:28:38	383	128	51.90	17.58	13:23:08	00:28:40
s9	125	8136	32.46	10.03	09:21:58	31	32	31.49	10.01	07:52:09	00:27:18
s11	115	433	40.22	10.48	11:14:55	89	428	38.70	10.42	10:50:42	00:13:37
s12	167	155	47.13	17.91	12:05:03	56	113	44.87	17.96	11:23:14	00:39:18
s14	1220	19086	27.79	7.59	11:16:24	961	10559	27.41	7.64	09:49:46	00:13:05
s16	129	38	29.17	7.92	06:13:12	41	50	29.11	8.04	06:16:49	00:10:51
s19	518	110	19.35	6.21	05:37:44	76	111	18.83	6.20	05:02:08	00:22:00
Ratio$_{ind}$	**1**	**1**	**1**	**1**	**1**	**0.473**	**0.685**	**0.970**	**0.996**	**0.904**	
Ratio$_{sum}$	**1**	**1**	**1**	**1**	**1**	**0.524**	**0.162**	**0.968**	**0.995**	**0.870**	

TABLE 8   DETAILED ROUTING RESULT COMPARISON BETWEEN SIMPLR AND SIMPLR+ROPT$_6$

	SimPLR					SimPLR+Ropt$_6$					
	NUN	Vio	WL(10^7)	Via(10^6)	R$_{cpu}$	NUN	Vio	WL(10^7)	Via(10^6)	R$_{cpu}$	P$_{cpu}$
s2	553	876	69.67	12.47	14:37:14	170	686	67.65	11.99	13:37:42	00:23:23
s3	487	194	45.59	11.71	10:26:24	226	206	43.21	11.21	09:30:40	00:20:03
s6	443	361	41.54	11.80	10:26:33	71	236	40.56	11.63	09:26:07	00:17:41
s7	518	5402	55.44	18.15	15:40:00	302	170	50.91	17.46	12:08:27	00:29:15
s9	786	22	31.09	9.94	08:53:54	51	32	30.35	9.81	08:04:55	00:24:18
s11	979	1840	39.27	10.49	15:41:12	504	913	38.70	10.40	11:56:58	00:13:31
s12	715	241	46.98	17.80	12:04:46	277	117	44.89	17.76	11:06:29	00:37:47
s14	1459	224239	28.65	7.71	24:24:14	823	50725	28.05	7.53	12:51:33	00:12:23
s16	434	135	30.31	7.97	06:42:16	126	39	29.95	7.99	06:03:54	00:10:35
s19	510	72777	18.72	6.12	17:55:16	226	388	18.38	6.08	05:56:17	00:21:53
Ratio$_{ind}$	**1**	**1**	**1**	**1**	**1**	**0.378**	**0.549**	**0.968**	**0.982**	**0.787**	
Ratio$_{sum}$	**1**	**1**	**1**	**1**	**1**	**0.403**	**0.175**	**0.964**	**0.980**	**0.736**	

TABLE 9   COMPARISON BETWEEN THE DETAILED ROUTING RESULTS OF THE PLACEMENT SOLUTIONS IN DAC12 CONTEST.

	Ratio$_{ind}$					Ratio$_{sum}$				
	NUN	Vio	WL	Via	R$_{cpu}$	NUN	Vio	WL	Via	R$_{cpu}$
NTUplace	1	1	1	1	1	1	1	1	1	1
NTUplace+Ropt$_6$	0.322	0.483	0.987	1.002	0.759	0.325	0.016	0.985	1.002	0.634
Ripple	0.846	9.882	1.068	1.036	0.827	0.594	0.144	1.072	1.038	0.752
Ripple+Ropt$_6$	0.356	0.616	1.036	1.032	0.755	0.311	0.023	1.037	1.033	0.654
SimPLR	1.952	6.097	1.070	1.037	0.975	0.736	0.402	1.072	1.040	0.836
SimPLR+Ropt$_6$	0.751	0.891	1.035	1.018	0.748	0.297	0.070	1.033	1.019	0.615
mPL	2.082	369.029	1.153	1.078	1.378	1.058	0.973	1.155	1.083	1.249
mPL+Ropt$_6$	0.672	25.639	1.109	1.048	0.842	0.273	0.136	1.110	1.053	0.710

TABLE 10   DETAILED ROUTING RESULT COMPARISON BETWEEN RIPPLE AND RIPPLE+ROPT$_6$ USING ISPD11 BENCHMARKS

	Ripple in ISPD11 contest					Ripple in ISPD11 contest + Ropt$_6$					
	NUN	Vio	WL(10^7)	Via(10^6)	R$_{cpu}$	NUN	Vio	WL(10^7)	Via(10^6)	R$_{cpu}$	P$_{cpu}$
s1	2070	99	33.70	10.23	11:31:24	876	93	33.13	10.31	08:35:42	00:20:04
s2	1086	820	78.83	12.75	16:57:30	479	699	71.89	12.15	13:07:19	00:27:37
s4	632	242	27.50	6.87	07:04:18	84	274	26.93	6.87	05:33:11	00:09:47
s5	1761	805	41.97	9.23	13:09:14	1025	679	40.61	9.11	09:36:24	00:15:58
s10	657	837	67.88	13.72	17:50:19	396	639	66.40	13.57	14:32:26	00:23:50
s12	501	179	55.25	18.24	13:09:41	165	110	50.06	17.88	11:10:18	00:43:09
s15	985	118	44.71	13.70	10:37:41	478	56	42.91	13.43	08:03:19	00:18:53
s18	2515	64478	32.03	7.99	23:54:27	1823	962	28.48	7.32	08:04:56	00:22:10
Ratio$_{ind}$	**1**	**1**	**1**	**1**	**1**	**0.428**	**0.803**	**0.955**	**0.985**	**0.780**	
Ratio$_{sum}$	**1**	**1**	**1**	**1**	**1**	**0.522**	**0.052**	**0.944**	**0.977**	**0.689**	

TABLE 11   COMPARISON BETWEEN THE DETAILED ROUTING RESULTS OF THE PLACEMENT SOLUTIONS IN ISPD11 CONTEST.

	Ratio$_{ind}$					Ratio$_{sum}$				
	NUN	Vio	WL(10^7)	Via(10^6)	R$_{cpu}$	NUN	Vio	WL(10^7)	Via(10^6)	R$_{cpu}$
NTUplace	2.656	10332.015	0.953	1.085	2.750	1.924	227.559	0.962	1.102	2.353
NTUplace+Ropt$_6$	1.642	6860.152	0.903	1.020	1.066	0.988	148.777	0.909	1.032	0.846
Ripple	1	1	1	1	1	1	1	1	1	1
Ripple+Ropt$_6$	0.428	0.803	0.955	0.985	0.780	0.522	0.052	0.944	0.977	0.689
SimPLR	1.200	541.762	0.992	1.020	1.448	1.053	16.276	0.987	1.020	1.330
SimPLR+Ropt$_6$	0.471	20.016	0.922	0.973	0.469	0.459	1.887	0.922	0.979	0.450
mPL	2.379	5058.336	0.957	1.073	1.936	1.875	112.558	0.965	1.085	1.881
mPL+Ropt$_6$	0.748	3069.438	0.914	1.022	0.715	0.571	65.542	0.922	1.035	0.655

Exploration with Upgradeable Models Using Statistical Methods for Physical Model Emulation

Bailey Miller
Dept. of Computer Science
and Engineering
University of California, Riverside
bmiller@cs.ucr.edu

Frank Vahid
Dept. of Computer Science
and Engineering
University of California, Riverside
Also with CECS, UC Irvine
vahid@cs.ucr.edu

Tony Givargis
Center for Embedded Computer
Systems (CECS)
University of California, Irvine
givargis@uci.edu

ABSTRACT

Physical models capture environmental phenomena such as biochemical reactions, a beating heart, or neuron synapses, using mathematical equations. Previous work has shown that physical models can execute orders of magnitude faster on FPGAs (Field-Programmable Gate Arrays) compared to desktop PCs. Different models of the same physical phenomenon may vary, with "upgraded" models being more accurate but using more FPGA area and having slower performance. We propose that design space exploration considering upgradable models can dramatically increase the useful design space. We present an analysis of the solution space for utilizing networks of processing-elements (PEs) on FPGAs to emulate physical models, implement a web-based frontend to a compiler and cycle-accurate simulator of PE networks to estimate solution metrics, and utilize design-of-experiments (DOE) statistical methods to identify Pareto points. By considering upgradeable models during the design space exploration of a human lung physical model, the solution space of possible speedup, area, and accuracy is increased by 6X, 7.3X, and 1.5X, respectively, compared to evaluating a single model.

Categories and Subject Descriptors

B.7.0 [**Integrated Circuits**]: General

General Terms

Design, Performance

Keywords

Design space exploration, FPGA, cyber-physical systems

1. INTRODUCTION

Fast physical model emulation is important in various domains for research and testing purposes. Iterative step-solvers can be used to compute the state of the physical model at real-time or faster than real-time speeds. Fast physical model emulation is especially important in the cyber-physical domain for testing purposes, since physical models can be used as a replacement for environments that are dangerous, expensive, or difficult to recreate.

Previous work has shown that emulating physical models on networks of processing-elements on FPGAs can provide orders of magnitude speedup over desktop processors and graphical

Permission to make digital or hard copies of all or part of this work for personal or classroom use is granted without fee provided that copies are not made or distributed for profit or commercial advantage and that copies bear this notice and the full citation on the first page. To copy otherwise, to republish, to post on servers or to redistribute to lists, requires prior specific permission and/or a fee.
DAC '13, May 29 – June 07 2013, Austin, TX, USA.

processing units [2][3], due largely to parallel execution on tens/hundreds of processing elements. Such parallel execution is enabled by physical models consisting of independent and locally communicating equations. Previous work applied traditional design space exploration, partitioning equations among different types and numbers of processing elements to achieve area and performance tradeoffs.

However, physical systems provide a rather unique additional solution option. The same physical system can be modeled with different equations. Each model may have tradeoffs in terms of the number of equations, ease of computation, and accuracy. We denote sets of models that are functionally similar as *upgradeable* models, since a designer may 'upgrade' to a more accurate model at the expense of area and performance. For example, Figure 1(a) shows three models that capture the behavior of the same physical system of lung airway mechanics. A simple RC model can coarsely capture the behavior using a single ordinary differential equation (ODE). For higher accuracy, a binary-tree shaped Weibel model [13] with variable levels of complexity can be used at the expense of higher computational costs. Accuracy also depends on the step size and type of equation solver. A smaller step size yields higher accuracy but slower performance. Likewise, more accurate solvers yield slower performance. Figure 1(b) illustrates the accuracy of each model, where dashed lines represent some deviations in accuracy due to different step solvers or step sizes.

Upgradeable models substantially increase the solution space that must be explored, not only via expanding area and performance ranges, but also by adding the design metric of accuracy. Figure 2 shows various tradeoffs in terms of speedup, area, and accuracy for a set of upgradeable models.

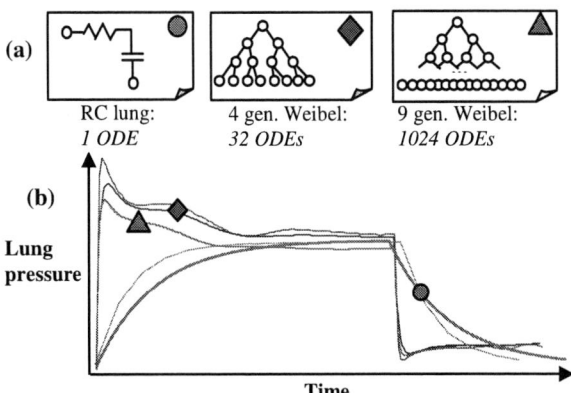

Figure 1: (a) A set of upgradeable models that implement similar lung airway mechanics behavior. (b) Relative accuracy of each model. Dashed lines show variations in accuracy when different solvers or step sizes are used.

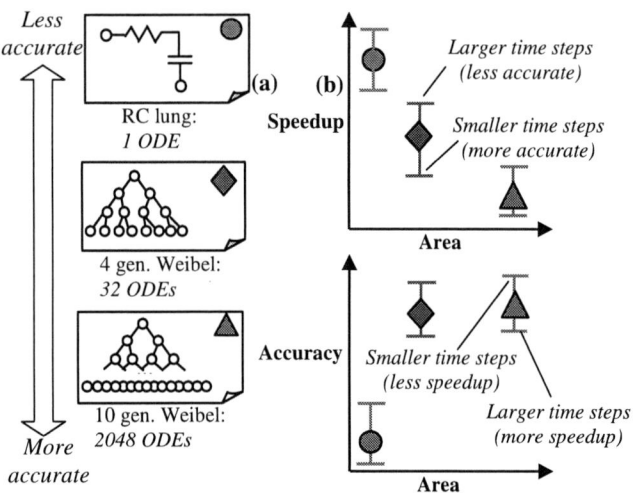

Figure 2: Impacts of various physical models on metrics. (a) Set of upgradeable models, and (b) comparisons of each model's speedup, area, and accuracy. The vertical scales show how solver time step affects speedup or accuracy.

Upgradable models introduce numerous additional parameters that influence and tremendously increase the design space. The influence on design metrics of those parameters can be complex and interdependent. To deal with these new parameters, we apply the statistical method known as design-of-experiments (DOE), which efficiently determines the impacts and dependencies of parameters, to enable efficient search of the design space. We introduce an approach that expands design space exploration to consider upgradeable models. To search the large design space, we utilize DOE statistical method to generate the Pareto points of the design. By generating the Pareto points, the design space can be pruned to enable a feasible exploration of solutions. We present a web-based tool that uses a processing-element (PE) network compiler and cycle-accurate simulator to automatically generate a PE network from an MML-language [7] based input model specification and evaluate the relevant size, performance, and accuracy metrics of PE network implementations. The web-based tool also supports automatic exploration of the design space using DOE techniques to aid in finding and appropriate model to use from an upgradeable set of models and an appropriate underlying PE network implementation that meets given constraints.

2. RELATED WORK

Previous research has demonstrated a compiler for translating MML-language based input specifications of physical models to VHDL descriptions of networks of PEs, which provides orders of magnitude speedup over standard desktop PCs [2]. Model equations are partitioned to PEs using various heuristics, scheduled to exploit parallelism, and connected via custom point-to-point networks. Various types of PE networks have been investigated, including homogeneous networks of programmable ALU-based general PEs, homogeneous networks of PEs with custom datapaths to solve specific equations, and heterogeneous networks consisting of both general and custom PEs [3]. Implementations of physical systems models on FPGAs exist elsewhere in literature [9], however all of these solutions are ad-hoc application specific solutions, whereas the PE network approach is a general CAD flow for any physical system model.

Considering how to select the appropriate physical model to emulate, to the best of our knowledge, has not been investigated

elsewhere in literature. The problem however has analogs to other domains in the CAD literature. The work on algorithm selection is a close corollary [10], in which a suitable algorithm from a set of functionally equivalent algorithms must be selected in order to optimize a given goal such as throughput, cost, or power.

Design space exploration is a well-known and often addressed issue [1][11]. The generation of Pareto points or Pareto curves has long been of interest as a method for reducing the amount of system configurations that need to be considered. Givargis introduced Platune [1], a tool for automatically exploring the design space of System-on-Chip (SoC) architectures. Platune implements an automated algorithm to explore the design space of a given SoC, and provides estimations of the resulting solution. Sheldon improved on the automated exploration by using statistical methods to calculate platform independent parameter interdependencies [12]. Our work is similar to Platune, except that we specifically target PE network implementations.

3. UPGRADEABLE MODELS

Upgradeable models in the context of physical systems emulation refers to having multiple underlying sets of equations that are each able to emulate the same physical model, with tradeoffs among accuracy versus area and performance. In this section, we define how to determine if relative models are part of the same upgradeable set, and discuss size-scalable upgradeable models.

3.1 Functional equivalence

We consider different models to be a part of the same upgradeable set if they meet the following requirements:

1. The models contain the base input/output interface required to support the physical system behavior.
2. The models are functionally similar, i.e. they produce similar output for all possible inputs.

The first requirement ensures that all models can operate on the same inputs and can provide the same outputs. Physical model emulations are usually a part of a larger design, often for testing purposes, thus ensuring that all models in an upgradeable set have a similar interface ensures smooth transitions and reduces the potential to introduce new errors. Some small differences in interfaces may be acceptable, as long as a correct transformation is available. For example, a lung model may require either an air flow input or an air pressure input. Flow can be easily converted into air pressure, and vice versa, thus we may still consider the models to be functionally equivalent. Models may also provide a supplemental input/output interface, in addition to the required base interface. For example, the Weibel lung model of Figure 2(a) provides output pressures at each of the leaf nodes, whereas the RC lung model provides only a single output pressure node below the capacitor. The supplemental interface is not required, but may improve model accuracy or provide additional information about the internal model state to the designer.

The second requirement demands that all models in an upgradable set produce similar outputs for given inputs. This requirement ensures that the physical model being emulated is similar in functionality, despite any differences in the underlying equations that are computed. Similarity can be determined by both qualitative and quantitative methods. Figure 1(b) shows the output of three various lung models; the models are considered interchangeable because they all produce an output of the same physical system, yet they have different quantitative and qualitative measures of accuracy. A designer can either determine that models are close enough to be functionally interchangeable, or a distance measure could be automatically calculated.

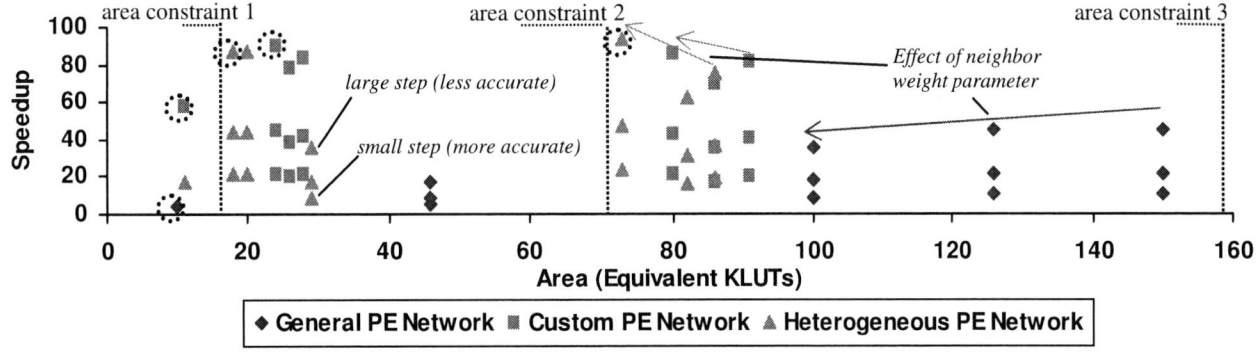

Figure 3: Possible PE networks for lung model. The three vertical lines indicate area constraints. Each solution is shown with a step size of 1e-2, 0.5e-2, and 0.25e-2 milliseconds. Arrows show the effect of the compiler neighbor weight parameter. Dashed circles show possible area/performance metric Pareto points.

3.2 Scaling model size

Many physical models have a common, repeating pattern or structure. The previously introduced Weibel model has a binary tree structure, which resembles the 23 bifurcating branches of a human lung. Neuron or cell models can consist of hundreds or thousands of individual elements that are connected to neighboring elements in mesh or grid structures. Previous work has shown that physical model structure can even be utilized to aid placement of PE networks on FPGA fabrics [6]. Physical models with regular structures can be considered upgradeable if the physical model can be scaled in size by adding new elements into the structure.

Scaling a physical model may or may not affect the accuracy of the model, but certainly impacts the resulting area and performance of the implementation. The Weibel model can be scaled in size to have more or less tree generations - having more generations implies a higher level of accuracy because the number of branches is closer to actual lung physiology. However, doubling the number of cells in a cell tissue model does not necessarily imply that the equations of each individual cell are more accurate. Even so, a designer may want to know how many cells can be included, given some area or performance constraints.

4. PE NETWORK PARAMETERS AND METRICS

For a given physical model of sufficient size and complexity, the solution space for a PE network that emulates the model is extremely large. This is due mostly to the parameters available during PE network synthesis, and partly to the non-deterministic heuristics used during equation partitioning. The considered key parameters are the model specification itself, PE network type, equation partitioning neighbor function weightings, the given resource constraints, step size, and solver type. The key solution metrics are FPGA area (LUTs, memory, and DSP usage), performance (speedup over real-time), and accuracy (closeness to exact solution).

Figure 3 depicts a chart of possible PE network solutions for a neuron model with 300 equations. Three sections are shown which depict solutions yielded by using different area constraints. For each area constraint, the PE network type, neighbor weight, and step size parameters are varied. Area/speedup metric Pareto points are circled; the accuracy metric is not measured explicitly in the figure, though points towards the bottom typically use smaller time steps and thus would be more accurate.

The model itself is an important parameter. The PE network solutions that are generated depend highly on how many equations are in the model, the complexity of each equation, and the data dependencies between the equations. The user also must specify a coefficient that quantitatively captures the quality of the model compared to the others in the set. For example, the 9 generation Weibel lung model may be considered to be the most accurate, and have a coefficient of 1. The RC model may be considered to have a coefficient of 0.4, because it only coarsely captures realistic behavior. The coefficients are used when comparing relative accuracies of different models.

PE type is a critical parameter that determines the type of PE network that is generated to emulate the model. There are three options: homogeneous general PE network, homogeneous custom PE network, and heterogeneous PE network. A general PE is a flexible, programmable, ALU-based processor that can solve any equation. A custom PE uses a pipelined datapath to solve a single specific equation much faster than a general PE, but may use more FPGA resources and incur higher routing congestion cost. Heterogeneous PE networks combine general and custom PEs to create a network with balanced performance and area metrics.

Neighbor weight refers to an option within the PE network compiler that controls whether the equation partitioning favors size or performance. This option is a sliding scale that can be set from 1 (favor size) to 10 (favor performance).

Area constraints detail the available LUTs, DSPs, and BRAMs on the target platform. The PE compiler will not allocate more than the available resources. The area constraint may have a very large area or performance impact on sufficiently complex models. An area constraint which is too small for the given model will not allow enough PEs to be allocated and reduce the possible speedup. Small models are not affected by the area constraint.

Solver type selects the iterative step solver to use. The currently supported solvers are Euler and Runge-Kutta4. Euler is the most simple solver, but can be inaccurate or diverge with medium to large time steps. The Runge-Kutta solver type is much more accurate, but may require up to 4X more computation time than the Euler solver.

Step size determines the amount of time between iterative solutions of the model equations. Decreasing the step size requires more computations per second, which reduces the performance of the model, but allows the solvers to be more accurate.

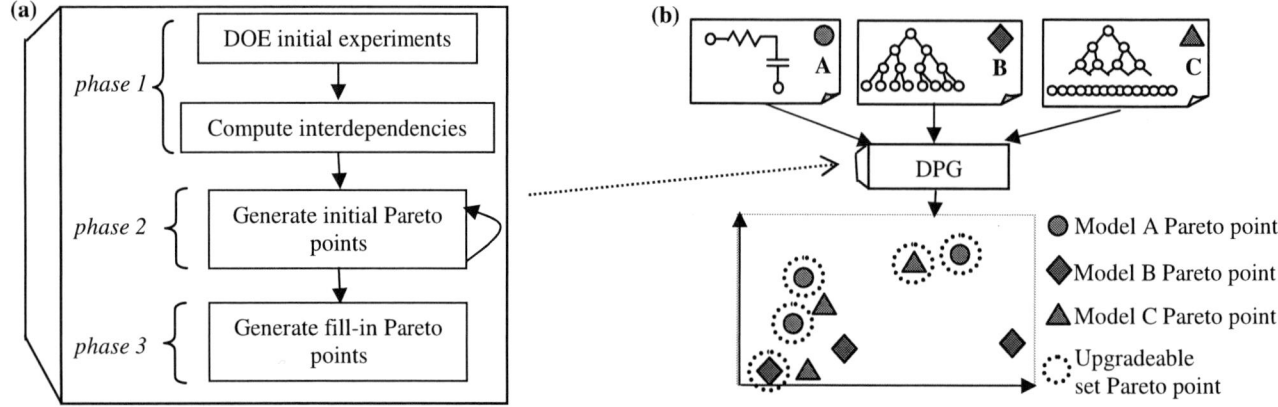

Figure 4: (a) The DPG algorithm flow. (b) Finding the Pareto points for a set of upgradeable models by applying DPG.

5. METHODOLOGY AND FRAMEWORK

To explore the space of PE network solutions for a set of upgradeable physical models, we have developed a visual web-based frontend coupled with a design-of-experiments statistical approach to identifying Pareto points that span an upgradeable set of physical models. In the following section we describe briefly what DOE is, and how our tool utilizes DOE.

5.1 DOE-based exploration of PE networks

Design-of-experiments is a statistical technique that identifies a minimal set of experiments that provide maximal cover of the possible solution space. Originally, DOE was developed for use in agriculture, but has since been developed into a powerful statistical technique used in many fields. DOE automatically identifies each parameter's magnitude of influence on the solution, since the differences between physical models (complexity, connectivity of equations, etc.) can impact how much a specific parameter like PE network type or neighbor weight matters. For example, a model with few equations will not be sensitive to area constraints.

5.1.1 *The DPG algorithm*

The DOE-based Pareto-point Generation (DPG) algorithm [12] can be used to apply DOE to and identify PE network solution Pareto points. By applying DPG to the upgradeable models, such that the models themselves are a parameter to the algorithm, the Pareto points that span across the set can be easily located. A

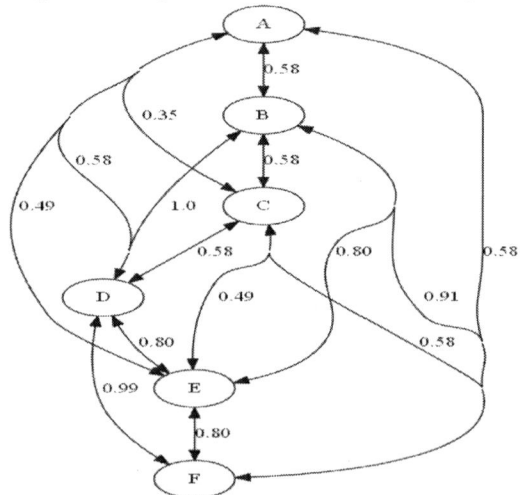

Figure 5: (a) Weighted parameter interdependency graph generated by DPG for a lung model.

basic flow chart of DPG is given in Figure 4. DPG consists of three phases: running initial experiments to identify parameter interdependencies, generating initial Pareto points, and filling in gaps in the Pareto curve. DOE uses either two or three-level parameters. Since PE networks have some continuous parameters, such as step size, we always select the minimum and maximum and midpoint values for continuous parameters to ensure we cover the space well enough. Phase 3 fills of DPG fills in the gaps of the solution space left by this discretization.

Phase 1 of DPG runs an initial Plackett-Burman [8] set of experiments to automatically generate a *weighted parameter interdependency graph*. This graph details the relationship between parameters for each metric in a single description. DPG generates the graph by first estimating the solution metrics for every pair of parameters in the system, and then running the experiment. The amount of error between the estimated and actual value suggests the amount of interdependency between the parameters. Figure 5 shows the interdependency graph for the speedup metric for a RC lung, 6 gen. Weibel lung, and 9 gen. Weibel set of upgradeable models. Each node represents a parameter: A is the model from the set, B is the area constraint, C is the step size, D is the PE network type, E is the solver, and F is the neighbor weights. Higher edge weights represent higher levels of interdependency; for example, the 0.99 weight between D and F indicates that the effects of the PE type and neighbor weight options on the solution depend on one another, which is observable in by examining the changes in speedup due to different neighbor weights.

Phase 2 of DPG generates initial Pareto points from the parameter interdependency graph. The algorithm starts by evaluating the edge with the highest error weighting, and exhaustively searching the possible ranges of the two associated parameters. DOE uses either two or three level parameter values, so there is a maximum of nine possible configurations to run. The solutions of the search are pruned to only the local Pareto points, and the two parameter nodes of the graph are merged. This continues until only one node remains which contains a set of Pareto points for the entire design.

Phase 3 of DPG identifies regions which were not explored, due to the reduction of continuous parameters into a discrete three-level parameter. Parameters which are constant around the region are locked, and a local search within the region takes place. New Pareto points are added to the set identified in phase two.

5.2 Tool

The tool to explore the PE network solution space consists of a web page frontend and server DPG backend, implemented in

1062

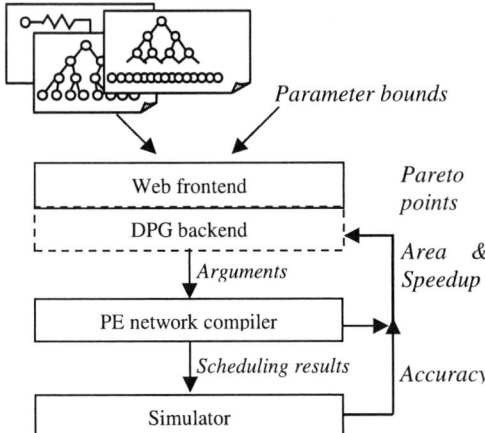

Figure 6: Architecture of the PE network exploration tool.

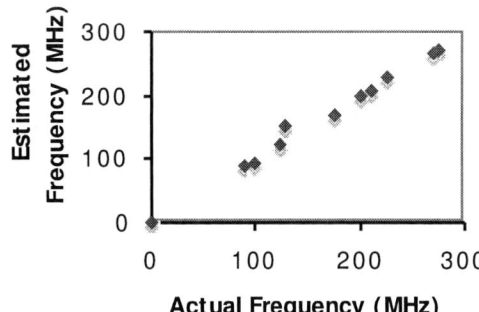

Figure 7: Regression model for estimating circuit frequency.

ASP.NET 4.0. A PE network compiler and simulator are implemented as .NET WCF web services. Figure 6 shows the architecture of the tool. The set of upgradeable models and parameter bounds are entered by the user. Parameter bounds include minimum and maximum area constraints, step sizes, etc. The DPG backend selects a set of experiments to run, and iteratively runs the compiler and simulator to generate area, speedup, and accuracy metrics. Pareto points are selected from the results by DPG and plotted visually on the web page.

5.2.1 PE network compiler and simulator

The PE network compiler accepts arguments generated by the DPG algorithm to partition the equations of a specific model across PEs. Once the equations are partitioned, each PE is scheduled. The partitioning and schedule information is enough to generate the area and speedup metrics. Area is reported by the compiler in terms of the number of LUTs, DSPs, and BRAM components used. The compiler automatically calculates the number of these components based on the type and frequency of each PE type. General PEs use one DSP and one BRAM each, while custom PEs may use arbitrary numbers of DSPs and typically a single BRAM. The final area metric is *equivalent LUTs* [5], which is a method for comparing resource usage for designs with various usage of logic cells and hard macros like DSPs. For a Xilinx Virtex6-240T, we use the following equation to calculate equivalent LUTs, where L_{EQ} is the equivalent LUTs, L is the number of LUTs, K_{DSP} is the equivalent LUTs per DSP (250), D is the number of DSPs, K_{BRAM} is the equivalent LUTs per BRAM (360), and B is the number of BRAMs:

$$L_{EQ} = L + K_{DSP}D + K_{BRAM}B$$

To calculate the speedup metric, the frequency of the resulting circuit must be estimated. The maximum frequency for a single PE is approximately 300 MHz when targeting a Virtex6, thus the maximum frequency that a larger PE network could achieve is also 300 MHz. As the number of PEs and connections between PEs grows larger, the place-and-route tools (Xilinx ISE 14.2) can not maintain the same timing due to congestion. We have created a regression model to estimate the frequency based on FPGA resource usage and the number of connections in the design:

$$Freq = K_0 - K_1W - K_2R_{DSP} + K_3R_{BRAM} - K_4R_{LUT}$$

Freq is the estimated frequency of the design, K_0, K_1, K_2, and K_3 are regression coefficients based on experimental data from PE networks targeting a Virtex6. W is the number of wires in the design (PE-to-PE connections), and R_{DSP}, R_{BRAM}, and R_{LUT} are resource usage ratios. This model is able to estimate frequencies to within 5% of their actual values, as shown in Figure 7.

Once frequency has been estimated, total speedup is calculated:

$$Speedup = \frac{1}{\frac{1}{Freq} * C * S}$$

C is the number of cycles required to compute a iteration of the model. S is iterations per second, derived from the step size parameter. The factor $1/Freq*C$ yields the amount of time to compute one iteration; multiplying by S yields the time to simulate 1 second. The inverse of the equation yields the speedup.

Accuracy is determined by simulating the PE network. A cycle-accurate simulator executes an iteration worth of PE instructions. The simulation is performed twice: once using the given solver and step size parameters, and once using a 'golden' set of parameters that consists of the most accurate configuration. For the golden parameters, we use RK4 solver and 0.01 ms step size.

After the simulations are complete, the time-series traces of each variable are compared. The simulator finds the variable in the user-defined simulation that is of a maximal distance from the golden standard simulation trace and returns the error. The error is then multiplied by the coefficient describing the model's relative accuracy in the upgradeable set, as described in section 4.

6. EXPERIMENT

We present an exploration of a set of upgradeable RC and Weibel models. We target a Xilinx Virtex6-240T FPGA, which consists of 150K LUTs, 716 DSPs, and 417 BRAMs. Table 1 enumerates the parameters and bounds that are input into the DPG algorithm for each set of models, since DOE uses 2 or 3-level parameters.

6.1 Lung models

The set of upgradeable lung models includes the RC, 6-generation and 9-generation Weibel models previously described. We use coefficients of 0.4, 0.9, and 1.0 to describe relative model accuracy, respectively. Figure 8 shows three plots comparing the area, speedup, and accuracy metrics of the 57 Pareto points found by the DPG algorithm. We consider the total design space size to be over 7,200 configurations, if the area constraint is discretized into just ten levels. Thus, the DPG algorithm prunes more than 99% of the design space, making exploration of the solutions more feasible. Filled circles indicate where groupings of Pareto points originate from the same model. For example, the left-most plot showing speedup vs. area has a group of Pareto points in the top-left corner that all are related to the RC model. Since the RC model is relatively simple, it has high speedup and low area

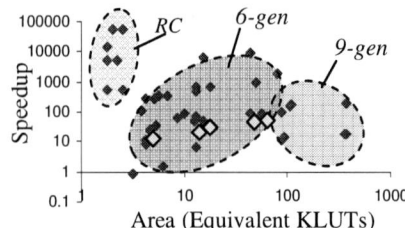 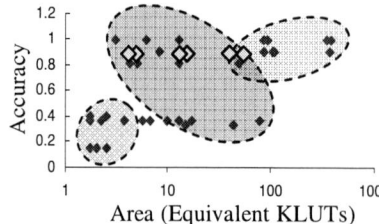 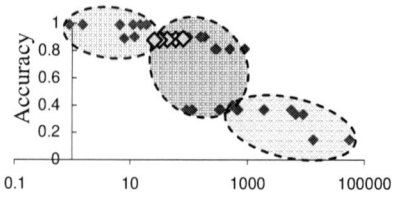

Figure 8: *3-dimensional Pareto plots projected onto 2-dimensions.* **Yellow points show where accuracy parameters (model, step size, and solver type) are constant, emphasizing design space exploration of area and speedup without the accuracy metric. Considering upgradeable models during exploration tremendously expands the solution space.**

Parameter	Low	Mid	High
LUTs	10K	50K	150K
DSPs	20	200	716
BRAMs	20	200	417
PE Type	General	Custom	Hybrid
Neighbor weight	1	5	10
Solver type	Euler	-	RK4
Step size (ms)	0.01	0.1	1.0

requirements; however, the other plots that include accuracy indicate the low fidelity of the solution.

The lighter points of Figure 8 illustrate Pareto points that correspond to configurations using an RK4 solver, 0.01 ms step size, and the 9-generation model. The points represent the 'normal' design space exploration of a PE network that has configurable type, number of PEs, etc. Considering only the middle case of the 6-generation Weibel model yields a solution space with speedups between 8X and 9200X, area between 4KLUTs and 55 KLUTs, and accuracy between 0.33 and 0.89. By considering the RC and 9-generation Weibel models during exploration, the solution space expands to speedups between 0.86X and 55000X, area between 1.7KLUTs and 376KLUTs, and accuracy between 0.15 and 0.99. Overall, the solution space that can be considered has increased in size by 6X in terms of speedup, 7.3X in terms of area, and 1.5X in terms of accuracy.

7. CONCLUSION

Physical models implemented on FPGAs provide large speedups over other implementation methods. Physical models introduce the feature of upgradeable models into design space exploration. We demonstrated how to include upgradeable models into a search approach and demonstrated improvements in the solution space of 5X on average of the area, speedup, and accuracy metrics. We utilized a design-of-experiments approach to enable rapid finding of Pareto points. Upgradeable models are not limited to physical models, but in fact may also apply to domains like signal processing and video processing where different algorithms can be considered that tradeoff quality with size and performance.

8. ACKNOWLEDGEMENTS

This work was supported in part by the National Science Foundation (CNS1016792, CPS1136146), the Semiconductor Research Corporation (GRC 2143.001), and a U.S. Department of Education GAANN fellowship. Special thanks also to David Sheldon for his help with the DPG algorithm and data analysis.

9. REFERENCES

[1] Givargis, T., and Vahid, F. 2002. Platune: a tuning framework for system-on-a-chip platforms. Computer-Aided Design of Integrated Circuits and Systems, IEEE Transactions on, 21.11, 1317-1327.

[2] Huang, C., Vahid, F., and Givargis, T. A Custom FPGA Processor for Physical Model Ordinary Differential Equation Solving. *IEEE Embed. Syst. Lett.* 3, 4, Dec. 2011, 113-116.

[3] Huang, C., Miller, B., Vahid, F., and Givargis, T. 2012. Synthesis of custom networks of heterogeneous processing elements for complex physical system emulation. In Proceedings of the eighth IEEE/ACM/IFIP international conference on Hardware/software codesign and system synthesis (CODES+ISSS '12). ACM, 215-224.

[4] Kahng, A.B., Li, B., Peh, L., Samadi, K. 2009. ORION 2.0: a fast and accurate NoC power and area model for early-stage design space exploration. Proceedings of the Conference on Design, Automation and Test in Europe (DATE'09).423-428.

[5] Meyer, J., Kocan, F. 2007. Sharing of SRAM Tables Among NPN-Equivalent LUTs in SRAM-Based FPGAs, Very Large Scale Integration (VLSI) Systems, IEEE Transactions on, vol.15, no.2, pp.182-195, Feb. 2007.

[6] Miller, B., Vahid, F., and Givargis, T. 2013. Embedding-based placement of processing element networks on FPGAs for physical model simulation. In Proceedings of the ACM/SIGDA international symposium on Field programmable gate arrays (FPGA '13). ACM, 181-190.

[7] Miller, J. A., Nair, R. S., Zhang, Z., Zhao, H. 1997. JSIM: A JAVA-based simulation and animation environment. Simulation Symposium,Proceedings. 30th Annual, pp. 31-42.

[8] Petersen, R. 1985. Design and Analysis of Experiments. Mercel Dekker Inc. New York, New York, 1985.

[9] de Pimentel, J. C. G., Y. G., Tirat-Gefen. "Hardware Acceleration for Real Time Simulation of Physiological Systems". Engineering in Medicine and Biology Society, 2006. EMBS'06. 28th Annual International Conference of the IEEE (pp. 218-223). IEEE.

[10] Potkonjak, M., Rabaey, J. "Algorithm Selection: A Quantitative Computation-intensive Optimization Approach". Computer-Aided Design, 1994., IEEE/ACM International Conference on , vol., no., pp.90-95, 6-10 1994.

[11] J. M. Rabaey, C. Chu, P. Hoang, and M. Potkonjak. 1991. Fast Prototyping of Datapath-Intensive Architectures. *IEEE Des. Test* 8, 2 (April 1991), 40-51.

[12] Sheldon, D., Vahid, F. "Making good points: application-specific pareto-point generation for design space exploration using statistical methods. International Symposium on Field Programmable Gate Arrays, 2009, pp. 123-132. ACM.

[13] Weibel, E. R. "Morphometry of the human lung". Anesthesiology vol. 26, no., pp., 1965.

Modular System-Level Architecture for Concurrent Cell Balancing

Matthias Kauer, Swaminathan Naranayaswami, Sebastian Steinhorst, Martin Lukasiewycz
TUM CREATE, Singapore
matthias.kauer@tum-create.edu.sg

Samarjit Chakraborty
TU Munich, Germany
samarjit@tum.de

Lars Hedrich
University of Frankfurt/Main, Germany
hedrich@em.cs.uni-frankfurt.de

ABSTRACT

This paper proposes a novel modular architecture for Electrical Energy Storages (EESs), consisting of multiple series-connected cells. In contrast to state-of-the-art architectures, the presented approach significantly improves the energy utilization, safety, and availability of EESs. For this purpose, each cell is equipped with a circuit that enables an individual control within a homogeneous architecture. One major advantage of our approach is a direct and concurrent charge transfer between each cell of the EES using inductors. To enable a system-level modeling and performance analysis of the architecture, a detailed investigation of the components and their interaction with the Pulse Width Modulation (PWM) control was performed at transistor-level. At system-level, we propose a control algorithm for the charge transfer that aims at minimizing the energy loss and balancing time. The results give evidence of the significant advantages of our architecture over existing passive and active balancing methods in terms of energy efficiency and charge equalization time.

Categories and Subject Descriptors: B.7.1 [Integrated Circuits]: Types and Design Styles
General Terms: Algorithms, Design
Keywords: Active Cell Balancing, Charge Equalization, Battery Management, Modeling, Simulation

1 Introduction

Electrical Energy Storages (EESs) are widely used in many applications such as mobile devices, electric vehicles, or smart grids. To cope with high power and energy demands, series-connected EES topologies of Lithium-Ion (Li-Ion) cells are used in many domains. For instance, electric vehicles require a voltage of about $300V$ to $400V$ to drive the electric motor, resulting in about 100 Li-Ion series-connected cell modules with an individual voltage of about $3.7V$.

State-of-the-art EES architectures are strictly static and do not enable a charge transfer between cells. The drawback of these common architectures is a severe lack of flexibility, requiring a passive cell balancing such that the weakest cell determines the capacity of the entire EES. As a result, the lifetime, availability, and efficiency of these systems are dete-

---

This work was financially supported in part by the Singapore National Research Foundation under its Campus for Research Excellence And Technological Enterprise (CREATE) programme.

Permission to make digital or hard copies of all or part of this work for personal or classroom use is granted without fee provided that copies are not made or distributed for profit or commercial advantage and that copies bear this notice and the full citation on the first page. To copy otherwise, to republish, to post on servers or to redistribute to lists, requires prior specific permission and/or a fee.
DAC '13, May 29–June 7 2013, Austin, TX, USA.

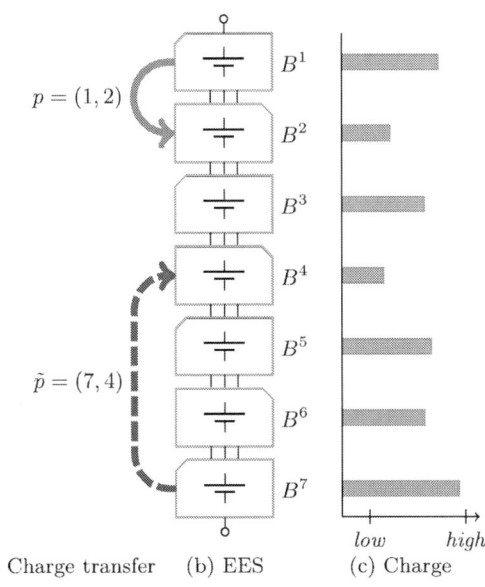

(a) Charge transfer   (b) EES   (c) Charge

Figure 1: Illustration of the concurrent charge transfer (a) between adjacent cells B^1 and B^2 and non-adjacent cells B^7 and B^4 using the proposed EES architecture (b) to equalize the charge of the cells (c).

riorated. To cope with these drawbacks, major efforts have been made within the recent years to reduce variations of individual Li-Ion cells in the manufacturing process. However, this had even further negative effects since it leads to very high production costs and prevents innovations in cell chemistries that often come along with significant variations.

As a remedy, improved architectures were proposed that use balancing circuits, allowing a charge transfer between cells. These active cell balancing methods can cope with variances in the energy capacity and discharge behavior of different cells, improving the total capacity and lifetime of EESs. However, known approaches are still in an early development and require detailed investigations at system-level and transistor-level. Moreover, these approaches only allow balancing between adjacent cells, reducing the efficiency significantly. Therefore, we propose a novel modular architecture that enables bi-directional and non-adjacent active cell balancing, considering the control at system-level and the interaction with the transistor-level.

Contributions of the paper. In this paper, we propose a novel EES architecture and control as illustrated in Fig. 1. A major advantage of the proposed architecture over other approaches is the bi-directional and non-adjacent active cell balancing that is performed concurrently at runtime. This significantly reduces energy loss during balancing, cell fatigue, and also the time that is required to perform the cell balancing. As a result, the lifetime, availability, and efficiency of this EES is significantly improved. Moreover, by enabling the usage of low-cost cells with high variations, the additional costs of the balancing circuits might be compensated.

The contributions of this paper comprise (1) a novel homogeneous architecture based on basic building block circuits, (2) a switching scheme that enables the charge transfer between non-adjacent cells, (3) an analytical nonlinear closed-form model for the charge transfer behavior, and (4) a system-level control algorithm for charge transfer that takes advantage of the proposed architecture:

1. We propose an EES architecture that consists of homogeneous modular blocks, combining a cell with six power Metal-Oxide-Semiconductor Field-Effect Transistors (MOSFETs) and an inductor. This enables the exchange of charge between individual cells while it also makes the the overall cell integration more flexible.

2. Our proposed switching scheme enables a concurrent charge transfer between non-adjacent cells in both directions. This significantly reduces the energy loss and improves the charge transfer time compared to other known active cell balancing architectures.

3. We propose an analytical nonlinear closed-form model for the charge transfer behavior that not only applies to our circuit, but that also enables a detailed analysis of related work that until now relied on time-consuming numerical simulation. A speed-up of three orders of magnitude can be achieved at similar accuracy. This enables the development of control algorithms and system-level optimization techniques for charge transfer beyond the scope of this paper.

4. At system-level, we propose a control algorithm that performs the charge transfer for the proposed architecture. The case study gives evidence of the practicability of the proposed balancing architecture and control by significantly reducing overall energy loss and charge equalization time compared to passive balancing and other active balancing methods.

Organization of the paper. The remainder of the paper is organized as follows: Section 2 discusses related work in the domain of architectures and control for EESs. Our novel balancing architecture is introduced in Section 3, comprising the architecture and switching scheme. In Section 4, an analytical model is proposed that enables a system-level analysis and a development of a control algorithm for the charge balancing. Section 5 presents a model validation and a detailed case study, comparing the proposed architecture and model to results from previous work. Finally, Section 6 makes concluding remarks.

2 Related Work

With a growing amount of electronics and control in EESs, the design of complex EESs is becoming increasingly relevant in the embedded systems domain. Already common system-level battery management systems as discussed in [1] require a significant amount of embedded control. In [2] and [3], hybrid EESs and appropriate control strategies are discussed, proposing an optimization for cycle efficiency and charge management algorithms, respectively. Integrating more intelligence at the cell-level of a modular battery and thus decentralizing its management is discussed in [4].

In all architectures, cell balancing is a crucial part of the EES control and a system-level analysis and optimization is becoming increasingly important. A comprehensive overview of cell balancing methods is presented in [5]. In [6], a passive cell balancing is presented where energy dissipation of cells with higher charge levels is realized using switched resistors. Although this approach is easy to implement and very common, it is wasting energy during balancing in form of heat. An approach to isolate cells from the series-connected battery is introduced in [7] with the goal to reach an equal charge level during a charging process. However, an active balancing with this approach is not possible.

Active cell balancing approaches aim at equalizing the charge of cells in an EES. Charge equalization using an inductor as an active charge transfer element is proposed in [8] while another DC-DC converter based balancing technique is presented in [9]. However, all these approaches have the drawback that the charge transfer is only possible between adjacent cells or with further circuit complexity between close cells [10]. Therefore, significant energy losses occur when transferring charge between distant cells in a series-connected EES.

In this paper, we propose an architecture that enables active cell balancing between non-adjacent cells, improving the efficiency of the EES significantly compared to all existing approaches. The proposed architecture consists of modular building blocks that make a flexible and extensible EES design possible. Furthermore, a system-level model has been developed that allows a fast and accurate analysis of our architecture in a high-level EES framework to propose and simulate charge transfer control algorithms.

3 EES Architecture

In this section, the proposed EES architecture is introduced. First, the system-level concept for charge transfer is presented before the basic blocks are explained. A switching scheme to enable the concurrent charge transfer between non-adjacent cells is proposed and explained based on the example from Fig. 1.

System-level charge transfer. The charge transfer at system-level is controlled by the set of pairs $\mathcal{P}$ that is determined at each time step. The set $\mathcal{P}$ consists of elements $p = (\sigma, \delta)$ where σ denotes the source cell and δ denotes the destination cell of a charge transfer. An example of this charge transfer at system-level is illustrated in Fig. 1 where $\mathcal{P} = \{(1, 2), (7, 4)\}$. Note that $\sigma, \delta \in \mathbb{N}$ are cell indexes denoting a cell's position in the series-connected EES.

The set $\mathcal{P}$ needs to fulfill the following requirements to be considered feasible for charge transfer. Source and destination of a transfer cannot be identical:

$$\sigma \neq \delta \quad \forall p = (\sigma, \delta) \in \mathcal{P} \tag{1}$$

Additionally, transfers have to be performed on disjoint sets of cells with at least one cell between pairs from $\mathcal{P}$. This avoids merging currents of concurrent balancing processes. Formally, it has to hold for any $n \in \mathbb{N}$ and any two non-identical pairs $p = (\sigma, \delta), \tilde{p} = (\tilde{\sigma}, \tilde{\delta}) \in \mathcal{P}$:

$$\big(n \in [\min(\sigma, \delta) - 1, \max(\sigma, \delta) + 1]\big)$$
$$\Rightarrow \big(n \notin [\min(\tilde{\sigma}, \tilde{\delta}), \max(\tilde{\sigma}, \tilde{\delta})]\big) \tag{2}$$

Module-level charge transfer building block. Our proposed architecture consists of homogeneous modular charge transfer blocks, comprising the individual cells as illustrated in Fig. 2. Note that each cell might be a parallel composition of single cells, increasing the energy capacity of the EESs. The blocks are asymmetrically (observe the location of L^i) connected in series as illustrated in Fig. 1 while Fig. 3 illustrates the identical architecture at module-level.

One basic building block consists of six power MOSFETs used as switches as well as one inductor. Each MOSFET might be *closed* or *open*, corresponding to the N-MOS tran-

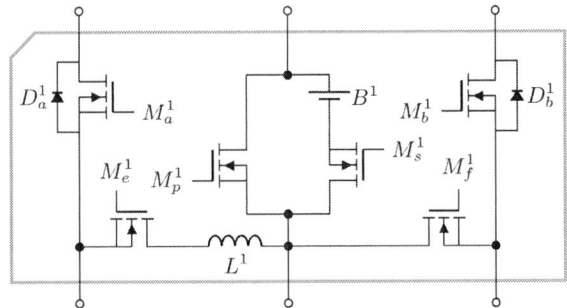

Figure 2: The basic building block of the modular charge transfer circuit attached to a cell (B^1), consisting of six MOSFETs and an inductor.

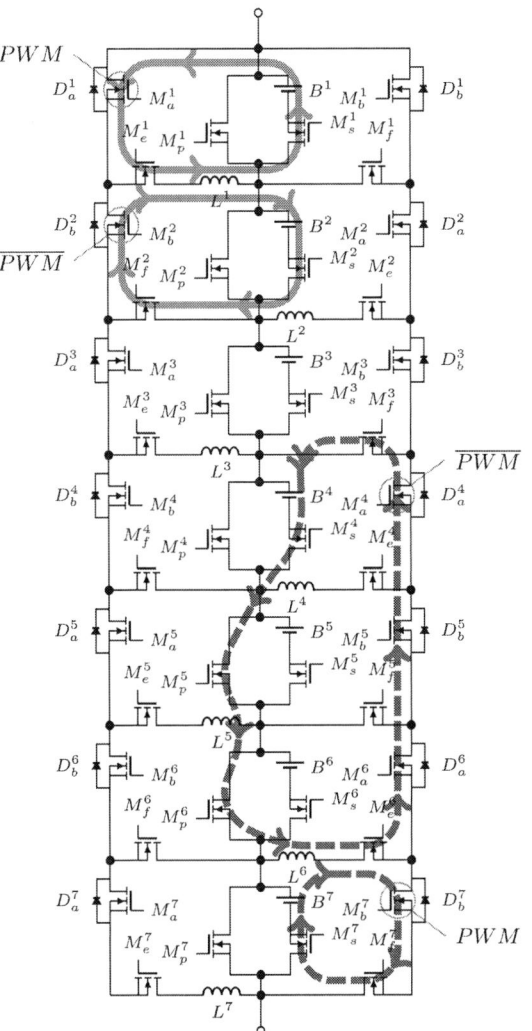

sistor conducting in state ON (closed switch, logical 1) and not conducting in state OFF (open switch, logical 0), respectively. The M_s^i is in series with the cell while the parallel M_p^i adds the capability of bypassing or isolating cells which is used to increase the safety and reliability of the EES in case of failure of one cell. During normal operation of the EES where all cells are series-connected, all M_s^i are closed while the M_p^i remain open. For balancing purposes, M_a^i and M_b^i can be controlled via PWM signals, enabling the active charge transfer process via the inductors. Diodes D_a^i and D_b^i are protection diodes of the power MOSFETs, preventing destructive voltage spikes during switching of the PWM when the inductor current cannot flow across a transistor. M_e^i and M_f^i are routing switches that are required to avoid horizontal currents. Most of the MOSFETs can remain constant during one particular balancing operation.

While the MOSFETs M_s^i and M_p^i have to be capable of coping with high voltages and currents, the remaining MOSFETs are only exposed to significantly lower voltages and currents that occur during the cell balancing. This ensures that the charge transfer block remains a cost-efficient solution.

Switching scheme. To ensure the correct charge transfer for a given $\mathcal{P}$, the MOSFETs have to be switched correctly. For this purpose, we define the function

$$x_{M_j^i} : 2^{\mathcal{P}} \to \{0, 1, PWM, \overline{PWM}\} \qquad (3)$$

that determines whether the corresponding switch is open (0), closed (1), or controlled by a PWM signal (PWM or $\overline{PWM}$). Here, M_j^i refers to MOSFET j in the i-th cell of the string. More precisely, superscript $i \in \{1, \dots, 7\}$ in Fig. 3 and subscript $j \in \{a, b, p, s, e, f\}$. The switching rules are defined in Section S1.

Note that accurately controlled signals PWM and $\overline{PWM}$ minimize usage of the protection diodes of the MOSFETs, resulting in a significantly improved efficiency over previous approaches that used diodes instead of a non-overlapping PWM control. For this purpose, the discharge time of the inductor determining T_{OFF} needs to be calculated very precisely such that the cell cannot discharge itself through the inductor.

Switching scheme example. For a better understanding of the circuit and the switching scheme, the example depicted in Fig. 3 is considered. The resulting switching is summarized in Table 1 and explained in the following.

First, the transfer of charge between adjacent cells shall be considered which is the case for the transfer from B^1 to B^2. We thus transfer the excess charge via inductor L^1. M_s^1 and M_s^2 are closed and M_p^1 and M_p^2 open in order to discharge and charge the cells, respectively. Note that M_p^3 and M_s^3 are open to isolate the current charge transfer process. M_a^1

Figure 3: Illustration of the charge transfer in the proposed EES architecture for a determined MOSFET switching. Charge is transfered concurrently between neighbor cells with $p = (1, 2)$ (⟶) and non-adjacent cells with $\tilde{p} = (7, 4)$ (⟹).

and M_b^1 are activated by an alternating PWM signal. During the time the signal PWM closes M_a^1 – a period that we will refer to as T_{ON} – inductor L^1 is charged from cell B^1 via M_a^1. Afterwards, M_a^1 is opened for a time of T_{OFF} and the inductor is discharged through M_b^2 into cell B^2. M_b^2 is closed by the corresponding non-overlapping signal $\overline{PWM}$, see Fig. 4. Non-overlapping signals are required in order to avoid energy loss due to shortened current paths by introducing a dead time. Note that charge transfer would even be possible if the charge level of B^2 was higher than that of B^1 due to the DC-DC converter behavior of the circuit.

In addition to charge transfer between adjacent cells, the proposed architecture can directly transfer charge between non-adjacent cells. This is illustrated in Fig. 3 where cell B^7 transfers charge to cell B^4. For this purpose, M_s^i is closed for $i = 4, 7$ while M_p^i is opened for $i = 5, 6$. This isolates cells B^5 and B^6 from the electric circuit and allows the current flow to bypass them. Switch M_a^7 is activated by signal PWM

1067

i	M_a^i	M_b^i	M_e^i	M_f^i	M_p^i	M_s^i
1	PWM	1	1	0	0	1
2	1	$\overline{PWM}$	0	1	0	1
3	1	1	0	1	0	0
4	$\overline{PWM}$	1	0	0	0	1
5	1	1	0	0	1	0
6	1	1	1	0	1	0
7	1	PWM	0	1	0	1

Table 1: MOSFET switch states for concurrent charge transfer from B^1 to B^2 and from B^7 to B^4. 0 denotes OFF (open) and 1 denotes ON (closed). Note that inductors L^i of adjacent cells are on opposite sides.

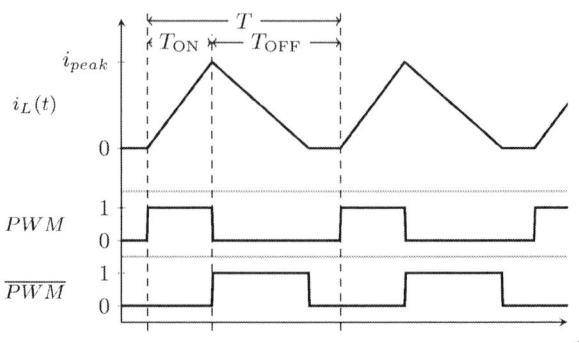

Figure 4: Inductor current $i_L(t)$ during T_{ON} and T_{OFF} with corresponding signals PWM and $\overline{PWM}$.

and M_a^4 by its non-overlapping corresponding $\overline{PWM}$. During T_{ON}, cell B^7 charges inductor L^6 which – during T_{OFF} – discharges into cell B^4 using the path $M_e^6 \rightarrow M_a^6 \rightarrow M_b^5 \rightarrow M_a^4 \rightarrow M_f^3 \rightarrow B^4 \rightarrow M_s^4 \rightarrow M_p^5 \rightarrow M_p^6$. To prevent horizontal currents during T_{OFF}, switches M_e^4 and M_f^5 remain open. Switches M_b^5 and M_a^6 remain closed to avoid the voltage drop incurred from routing over the diodes.

4 System-level Model

In this section, the development of a nonlinear analytical closed-form model for the behavior of the proposed circuit is presented that enables system-level simulation, optimization, and control algorithm engineering.

Analytical system-level model for transistor-level abstraction. A verification of the qualitative behavior of the proposed balancing circuit from Section 3 is possible, using analog circuit simulators such as LTspice IV [11] where cells are modeled by a capacitance-based model. This is, however, only feasible for a small number of our building blocks as the numerical simulation is not well scalable. The numerical solver of the simulator uses very short time steps for the iterative solutions of the system that contains continuous analog behavior in conjunction with discrete PWM signals, resulting in very long computation times. Furthermore, analog circuit simulators run into numerical problems for complex systems like full scale EESs. Hence, a transistor-level analysis of the proposed architecture with numerical approaches becomes infeasible and a system-level model has to be developed for the analysis of balancing algorithms for the EES.

It is therefore necessary to develop faster, scalable, and more abstract simulation models that retain the accuracy of the transistor-level analysis at system-level. The goal is to capture the quantitative behavior of 100 and more circuit building blocks within a real-world EES for an accurate analysis, enabling the development of system-level charge routing algorithms.

For an accurate analysis of the circuit behavior, the system configurations during T_{ON} and T_{OFF} need to be considered separately. Fig. 4 illustrates the behavior of the inductor current $i_L(t)$ of one basic building block controlled by PWM and $\overline{PWM}$, respectively. For each of the phases T_{ON} and T_{OFF}, an equivalent circuit can be identified for the path of the current flow through the inductor, cell, and transistors. For these components in the current flow, according to KIRCHHOFF's Voltage Law, an equation can be set up to describe the circuit behavior. Consider, for instance, the loop of the current flow through cell B^1 in Fig. 3. Applying KIRCHHOFF's Voltage Law to this part of the circuit results in the following equation:

$$L \cdot \frac{\mathrm{d}}{\mathrm{d}t}i + R_\sigma \cdot i + \frac{1}{C}\int_0^{T_{\text{ON}}} i(\tau)\,\mathrm{d}\tau - V_1 = 0 \qquad (4)$$

The series resistance R_σ models the inductor series resistance R_L, the cell series resistance R_C, and the ON-resistance R_M of the MOSFETs in the current path. All other current flow equations within the system can be modeled correspondingly as described in detail in Section S2 by adapting the series resistance R_σ as described in Table 3. Once R_σ is obtained as well as suitable initial conditions, we can treat the various cases using methods from [12] to determine a very accurate *nonlinear* closed-form solution for the charge transfer. This proposed nonlinear equation system is detailed in Section S2. To simplify and implicitly speed-up the nonlinear model with a reasonable loss in accuracy (see Section 5.1), a *linear* model is obtained by assuming a linear inductor current as presented in Section S2.6.

The individual PWM signals between charge transfer pairs in $\mathcal{P}$ do not need to be synchronized mutually. Instead for each $p = (\sigma, \delta) \in \mathcal{P}$, the source σ and destination δ need to synchronize their respective PWM and $\overline{PWM}$ signals. In the following, we consider the peak current i_{peak} as in Fig. 4 as input from the control level. We can then calculate T_{ON} and T_{OFF} using the formulas from Section S2.3 to S2.6.

System-level cell balancing control algorithm. In order to enable the cell balancing for the presented architecture from a system-level perspective, we propose a control algorithm $A_{K,r}$ that is outlined in Algorithm 1. The main parameters that can be used to adjust the behavior are K that is the number of maximal concurrent transfers and r that is the maximal allowed distance for charge transfer. The algorithm balances the charge levels Q of N cells until a normalized variance of the charge level falls below a predefined threshold (line 1). The charge equalization is performed in time steps T_M by defining the charge transfer pairs $\mathcal{P}$. Initially, the set $\mathcal{P}$ is empty while $\mathcal{V}$ is the set of all available cells that can be used for the charge transfer (line 2). In each time step, the algorithm determines pairs for charge transfer iteratively until there are no available cells or the maximum of concurrent transfers K is reached (line 3). The control scheme selects as sender σ the available cell with the highest charge (line 4) and transfers from there into the direction of the lower mean. For this purpose, the average charge levels of cells preceding and succeeding the sender σ, $\bar{Q}_{\text{prec}}$ and $\bar{Q}_{\text{succ}}$, respectively, are calculated and compared to determine the direction 1 or -1, respectively (line 5-7). Once the transfer direction is determined, the algorithm checks for the maximal possible distance ν that allows a transfer via available cells (line 8). The destination cell δ is then chosen as the cell with the least amount of charge among those between σ and ν in the determined direction (line 9). If the difference in charge levels of σ and δ is above a certain threshold (line 10),

1068

Algorithm 1 System-level cell balancing control algorithm $A_{K,r}$ where K is the number of maximal concurrent transfers and r is the maximal distance for charge transfer.

IN: Unbalanced charge array Q, macro step size T_M
OUT: Balanced Charge Array Q

1: **while** $\mathrm{Var}(Q)/\mathrm{avg}(Q) > 0.01$ **do**
2:     $\mathcal{P} = \{\}, \mathcal{V} = \{1, \ldots, N\}$
3:     **while** $\mathcal{V} \neq \{\} \wedge |\mathcal{P}| < K$ **do**
4:         $\sigma = \arg\max_{j \in \mathcal{V}} Q^{(j)}$
5:         $\bar{Q}_{\mathrm{prec}} = \frac{1}{\sigma-1} \sum_{j=1}^{\sigma-1} Q^{(j)}$
6:         $\bar{Q}_{\mathrm{succ}} = \frac{1}{N-\sigma-1} \sum_{j=\sigma+1}^{N} Q^{(j)}$
7:         $\mathrm{dir} = \mathrm{signum}(\bar{Q}_{\mathrm{prec}} - \bar{Q}_{\mathrm{succ}})$
8:         $\nu = \max\left\{k \in \mathbb{N}_{[0,r]} \mid (\sigma + \mathrm{dir} \cdot l) \in \mathcal{V} \quad \forall l \in \mathbb{N}_{[0,k]}\right\}$
9:         $\delta = \underset{j \in \{\sigma, \ldots, \sigma + \mathrm{dir} \cdot \nu\}}{\arg\min} Q^{(j)}$
10:        **if** $\left| Q^{(\sigma)} - Q^{(\delta)} \right| > 0.001$ **then**
11:            $\mathcal{P} = \mathcal{P} \cup \{(\sigma, \delta)\}$
12:            $\mathcal{V} = \mathcal{V} \setminus \{\min(\sigma, \delta) - 1, \ldots, \max(\sigma, \delta) + 1\}$
13:        **else**
14:            $\mathcal{V} = \mathcal{V} \setminus \{\sigma\}$
15:        **end if**
16:     **end while**
17:     Perform transfers in $\mathcal{P}$ for a duration of T_M
18:     Adjust Q according to transfers
19: **end while**

	T_{ON} [ms]	T_{OFF} [ms]	q_σ [As]	q_δ [As]
SPICE	0.12539	0.16582	3.14e-4	4.14e-4
linear	0.125	0.16442	3.1447e-4	4.1102e-4
rel. error	0.3%	1%	0.1%	0.7%
nonlinear	0.12546	0.16552	3.1371e-4	4.1288e-4
rel. error	0.05%	0.2%	0.1%	0.2%

Table 2: Comparison of the results of the switching time and transferred charge of the linear and nonlinear analytical model to the SPICE simulation. The relative error remains very small, validating the proposed system-level models.

(σ, δ) is added to $\mathcal{P}$ (line 11) while all cells in between and the outer neighbor cells are removed from the set of available cells (line 12). Otherwise, only the current source cell is removed from the set of available cells (line 14). After the set $\mathcal{P}$ is determined, the circuit then transfers charge according to $\mathcal{P}$ for the selected macro time T_M (line 17), before adjusting the Q values and continuing in the next time step (line 18). This is repeated until distortion measure $\mathrm{Var}(Q)/\mathrm{avg}(Q)$ is sufficiently small, i.e., until the cells are considered balanced (line 1).

Note that K is the most important adaptation parameter. If it is small, energy conservation is valued higher, if it is large, balancing time is given more attention. On the other hand, r is constrained by the circuit which is $r = 1$ for active cell balancing with adjacent charge transfer only and for the non-adjacent charge transfer circuit in the proposed approach it can be chosen more flexibly with $r \leq N$.

5 Experimental Results

This section presents experimental results on the validation of our analytical system-level model as well as a case study to illustrate the benefits of our EES architecture and the proposed control algorithm. All experiments were carried out on an Intel i5 @ 2.50 Ghz with 4GB RAM.

5.1 Model Validation

In this section, we compare the analytical nonlinear model and its linear approximation described in Section 4 and detailed in Section S2 to the LT SPICE IV [11] numerical simulation for validation purposes.

Accuracy. To determine the accuracy of the proposed analytical approach, we compared a simulation run in SPICE with the results from our system-level model. Both the nonlinear and linear model stay within a very close range of the SPICE simulation. The linear model for $i(t)$ does not exceed a relative error of 1% to the SPICE reference solution while the nonlinear model remains within even tighter bounds, not exceeding 0.1% relative error.

Additionally, we compared the different approaches for calculating the PWM periods T_{ON} and T_{OFF} as well as the transferred charge amounts q_σ and q_δ. T_{ON} and T_{OFF} are obtained from the SPICE simulation as the points in time where $i(t)$ crosses the corresponding threshold values $i(t) = 0$ and $i(t) = i_{\mathrm{peak}}$, respectively. The values q_σ and q_δ are the changes in charge that correspond to T_{ON} and T_{OFF} in SPICE. Section S2 details how the nonlinear and the linear models handle these computations. Table 2 summarizes the results of a representative comparison with $i_{\mathrm{peak}} = 5.0A$ (see Fig. 4). Again, the relative error of the linear model remains below 1% while the nonlinear model achieves almost negligible relative errors of less than 0.3%.

Speed-up. To estimate the speed-up of the analytical solutions over the numerical SPICE simulation, we ran various simulations over a length of 200 PWM cycles. The SPICE simulator averaged a runtime of 130s over different scenarios with optimized solver settings which was compared to our prototypical Python Scipy [13] implementation. Note that every PWM cycle was evaluated individually. Further speed-up might be obtained by combining PWM cycles and extending the evaluation of the first cycle to all of them. The nonlinear model ran for 0.1277s on average (min: 0.1207s, max: 0.1627s) while the average runtime of the linear model is further reduced to 0.0221s (min: 0.0201s, max: 0.0366s). This corresponds to a speed-up of more than 1000 for the nonlinear model and more than 5000 for the linear model, enabling a fast and accurate system-level model for the charge transfer control algorithm.

5.2 Case Study

In this section, we present the results of a case study, analyzing the cell balancing of an EES using the system-level model of our architecture, controlled by Algorithm 1 in comparison to state-of-the-art approaches.

Setup. To compare the capabilities of the proposed circuit, we considered the charge balancing of an EES of $N = 100$ cells. We randomly initialized the cell voltages according to

$$V^{(i)} \sim \mathcal{N}(3.6, 0.05^2)[V] \qquad (5)$$

and calculated the corresponding initial charge vector using $Q = C \cdot V$. Other parameters were:

$$R_M = 1m\Omega \qquad R_L = 1m\Omega \qquad R_C = 1m\Omega$$
$$L = 100\mu H \qquad C = 10kF \qquad V_d = 0.8V \qquad (6)$$

Additionally, we set $i_{\mathrm{peak}} = 5A$ (see Fig. 4) – a balance between speed and energy dissipation in the switches – and assumed a charging efficiency of the cells of $\eta = 0.97$, meaning that 3% of energy is dissipated if a cell is charged.

Balancing strategies. In our case study, we compared four different strategies. *Passive* balancing simply dissipates all excess energy over a small resistor. We assumed a dissipation

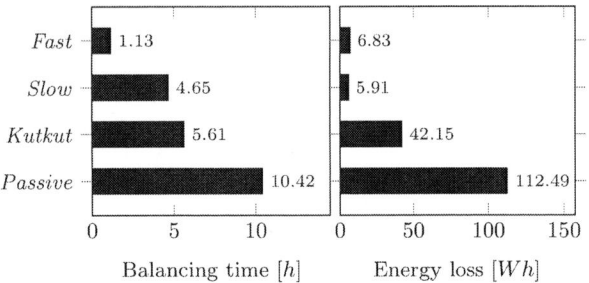

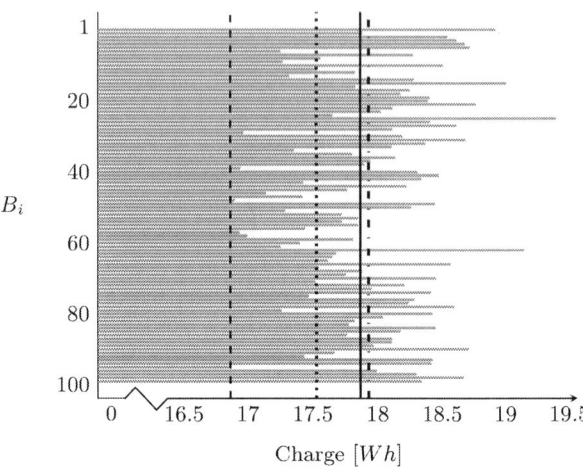

Figure 5: Results of the case study comparing balancing time and energy loss between different approaches.

rate of $10.8W$ as it would be achieved by nine MAX11068 [14] battery management micro-controllers, each handling up to twelve cells. *Kutkut* is a fully concurrent but neighbor-only balancing approach which is carried out on the circuit from [8]. Since [8] only proposes the architecture but no control algorithm for charge transfer, we assume a strategy corresponding to Algorithm 1 with $K = N$ and $r = 1$ that can be simulated using our analytical modeling approach by taking the lower MOSFET count and the diode in the current path of this circuit into account as denoted in Table 3. *Slow* and *Fast* are strategies for our proposed architecture, fully employing its long-range charge transfer capabilities. *Slow* (Algorithm 1 with $K = 1, r = N$) is a non-concurrent approach and therefore only does the energy-optimal transfers. *Fast* (Algorithm 1 with $K = N, r = N$) is fully concurrent and therefore compromises on energy efficiency to reduce balancing time.

Results. Fig. 6 shows the initial distribution of the randomly initialized cell charge. The four vertical lines indicate the minimal charge levels in the pack after cell balancing of each respective approach is finished. The passive cell balancing corresponds to the lowest initial charge of a cell to which all other cells are discharged for equalization. The *Kutkut* strategy is significantly better than passive balancing, but is still clearly outperformed by the *Fast* strategy, using our proposed architecture. Moreover, the *Slow* strategy using our proposed architecture has the lowest energy loss.

A comparison of the various strategies performed with respect to balancing time and dissipated energy is given in Fig. 5. As expected, the *Passive* approach is both the slowest and dissipates by far the most energy. Among the remaining strategies, *Fast* is both 80% faster than *Kutkut* and dissipates 85% less energy. *Slow* further reduces the energy loss by another 15%, but needs four times as long – which is still almost 20% faster than *Kutkut*.

The results show that our proposed architecture controlled by Algorithm 1 is by far outperforming previous approaches. They also give evidence that Algorithm 1 can be adapted by parameter K with respect to the achieved balancing time and energy loss where *Fast* with $K = N$ has fastest equalization speed and *Slow* with $K = 1$ provides the best energy conservation. A supplemental case study is presented in Section S3.

6 Concluding Remarks

This paper introduces and thoroughly analyzes a novel efficient architecture and control for active cell balancing of EES. In contrast to state-of-the-art approaches, the underlying proposed circuit allows to transfer charges between nonadjacent cells in a concurrent fashion. An analytical nonlinear model has been developed for the circuit and further approximated to a linear model, enabling fast and accurate simulations of a complete charge equalization process of an EES at system-level. A speed-up of three orders of magnitude over

Figure 6: Illustration of the initial charge distribution and the lowest final charge of the case study with 100 cells. Most charge is preserved with the *Slow* strategy (‒ · ‒) followed by *Fast* (——) still clearly outperforming *Kutkut* (......) while the *Passive* (‒ ‒ ‒) balancing reduces the charge to the level of the lowest cell.

SPICE-level analysis has been achieved. Finally, a detailed case study shows that our approach with the proposed charge transfer control algorithm significantly outperforms existing approaches in terms of energy efficiency and balancing time.

7 References

[1] M. Brandl et al. Batteries and Battery Management Systems for Electric Vehicles. In *Proc. of DATE*, 2012.

[2] Y. Kim, S. Park, Y. Wang, Q. Xie, N. Chang, M. Poncino, and M. Pedram. Balanced Reconfiguration of Storage Banks in a Hybrid Electrical Energy Storage System. In *Proc. of ICCAD*, 2010.

[3] Q. Xie, X. Lin, Y. Wang, M. Pedram, D. Shin, and N. Chang. State of Health Aware Charge Management in Hybrid Electrical Energy Storage Systems. In *Proc. of DATE*, 2012.

[4] S.K. Mandal, P.S. Bhojwani, S.P. Mohanty, and R.N. Mahapatra. IntellBatt: Towards Smarter Battery Design. In *Proc. of DAC*, 2008.

[5] Jian Cao, N. Schofield, and A. Emadi. Battery Balancing Methods: A Comprehensive Review. In *Proc. of VPPC*, 2008.

[6] M.J. Isaacson, R.P. Hollandsworth, P.J. Giampaoli, F.A. Linkowsky, A. Salim, and V.L. Teofilo. Advanced Lithium Ion Battery Charger. In *Proc. of BCAA*, 2000.

[7] H. Shibata, S. Taniguchi, K. Adachi, K. Yamasaki, G. Ariyoshi, K. Kawata, K. Nishijima, and K. Harada. Management of Serially-connected Battery System Using Multiple Switches. In *Proc. of PEDS*, 2001.

[8] N. H. Kutkut. A Modular Nondissipative Current Diverter for EV Battery Charge Equalization. In *Proc. of APEC*, 1998.

[9] Xi Lu, Wei Qian, and Fang Zheng Peng. Modularized Buck-Boost + Cuk Converter for High Voltage Series Connected Battery Cells. In *Proc. of APEC*, 2012.

[10] A.C. Baughman and M. Ferdowsi. Double-Tiered Switched-Capacitor Battery Charge Equalization Technique. *IEEE Transactions on Industrial Electronics*, 55(6):2277–2285, 2008.

[11] Linear Technology. Design Simulation and Device Models – LTspice IV, 2012.

[12] Katsuhiko Ogata. *Modern Control Engineering*. Prentice Hall, 1997.

[13] Eric Jones, Travis Oliphant, Pearu Peterson, et al. SciPy: Open source scientific tools for Python.

[14] Maxim Integrated. MAX11068: 12-Channel, High-Voltage Sensor, Smart Data-Acquisition Interface.

APPENDIX

S1 Switching Scheme

In the following, the switching rules for the proposed balancing architecture are proposed. The switching rules are defined as follows:

$$x_{M_s^i}(\mathcal{P}) = \begin{cases} 1 & \text{if } \exists(\sigma,\delta) \in \mathcal{P} : \sigma = i \vee \delta = i \\ 0 & \text{otherwise} \end{cases} \quad (7)$$

$$x_{M_p^i}(\mathcal{P}) = \begin{cases} 1 & \text{if } \exists(\sigma,\delta) \in \mathcal{P} : \\ & \quad i \in [\min(\sigma,\delta) + 1, \max(\sigma,\delta) - 1] \\ 0 & \text{otherwise} \end{cases} \quad (8)$$

$$x_{M_a^i}(\mathcal{P}) = \begin{cases} PWM & \text{if } \exists p \in \mathcal{P} : \sigma = i \wedge \sigma < \delta \\ \overline{PWM} & \text{if } \exists p \in \mathcal{P} : \delta = i \wedge (\sigma > \delta \oplus |\delta - \sigma|\%2 = 0) \\ 1 & \text{otherwise} \end{cases} \quad (9)$$

$$x_{M_b^i}(\mathcal{P}) = \begin{cases} PWM & \text{if } \exists p \in \mathcal{P} : s = i \wedge \sigma > \delta \\ \overline{PWM} & \text{if } \exists p \in P : \delta = i \wedge (\sigma > \delta \oplus |\delta - \sigma|\%2 = 1) \\ 1 & \text{otherwise} \end{cases} \quad (10)$$

$$x_{M_e^i}(\mathcal{P}) = \begin{cases} 1 & \text{if } \exists(\sigma,\delta) \in \mathcal{P} : (\sigma < \delta \wedge \sigma = i) \vee \\ & \quad (\sigma < \delta \wedge \delta = i \wedge |\delta - \delta|\%2 = 0) \vee \\ & \quad (\sigma > \delta \wedge \sigma - 1 = i) \vee \\ & \quad (\sigma > \delta \wedge \delta - 1 = i \wedge |\delta - \sigma|\%2 = 0) \\ 0 & \text{otherwise} \end{cases} \quad (11)$$

$$x_{M_f^i}(\mathcal{P}) = \begin{cases} 1 & \text{if } \exists(\sigma,\delta) \in \mathcal{P} : (\sigma > \delta \wedge \sigma = i) \vee \\ & \quad (\sigma < \delta \wedge \delta = i \wedge |\delta - \sigma|\%2 = 1) \vee \\ & \quad (\sigma < \delta \wedge \sigma - 1 = i) \vee \\ & \quad (\sigma > \delta \wedge \delta - 1 = i \wedge |\delta - \sigma|\%2 = 1) \\ 0 & \text{otherwise} \end{cases} \quad (12)$$

Eq. (7) controls M_s^i and closes it in case the respective cell is a source or destination of charge, otherwise it remains open. Eq. (8) controls M_p^i and closes it when charge is transferred between a preceding and succeeding cell, otherwise it remains open. Eq. (9) and (10), respectively, are used to generate the PWM signals. If charge is transfered forward, M_a^i is used for the PWM signal at the block of the source cell. Correspondingly, for transfer of charge backwards, M_b^i is used for the PWM signal. The $\overline{PWM}$ signal is used for controlling the charging of the destination cell that depends on the direction of charge transfer and the absolute distance of the cells. Eq. (11) and (12), respectively, are used to close the electric circuits and prevent horizontal currents.

S2 Model Derivation

The contribution of this section is the detailed development of an analytical model for the proposed circuit. Starting from a circuit analysis for phases T_{ON} and T_{OFF}, the analytical solution of the resulting second-order nonlinear Ordinary Differential Equation (ODE) system is presented. The relevant distinction between different damping cases is then introduced. Finally, a linearization of the inductor current behavior leads to a simplified linear model.

S2.1 From Circuit to Equation

Numerical circuit simulation can be slow for large-scale systems as discussed in Section 4. Hence it is desirable to obtain analytical, closed-form solutions that describe the system behavior as accurately as possible and at the same time execute orders of magnitude faster than the numerical simulation.

We consider the system configurations during T_{ON} and T_{OFF} separately since the behavior significantly changes af-

ter the current routing is altered by the PWM signal. Fig. 7 shows the equivalent circuit for the relevant part of the architecture during T_{ON} when the inductor is being charged.

The structure of the equivalent circuit does not depend upon whether transfer occurs only between neighbors as in [8] or more flexibly as in the proposed architecture. The only difference is the amount of series resistance R_σ that we have to introduce to model the inductor series resistance R_L, the cell series resistance R_C, and the ON-resistance of the MOSFETs R_M. The contributions to R_σ are summarized in Table 3.

Circuit behavior during T_{ON}. Applying KIRCHHOFF's Voltage Law to the circuit in Fig. 7 results in

$$L \cdot \frac{\mathrm{d}}{\mathrm{d}t} i + R \cdot i + \frac{1}{C} \int_0^{T_{ON}} i(\tau) \, \mathrm{d}\tau - V_1 = 0 \quad (13)$$

with $R = R_\sigma$. Differentiating with respect to t, we obtain the following second-order ODE:

$$L \cdot \frac{\mathrm{d}^2}{\mathrm{d}t^2} i + R \cdot \frac{\mathrm{d}}{\mathrm{d}t} i + \frac{1}{C} i = 0 \quad (14)$$

For such a second-order ODE system to have a unique solution, two initial conditions need to be provided. We assume the inductor is fully discharged initially, $i(0) = i_0 := 0$, which gives the first condition. Applying this, we can deduce the second condition from Eq. (13) with

$$L \cdot \frac{\mathrm{d}}{\mathrm{d}t} i(0) + 0 + 0 - V_1 = 0 \quad (15)$$

and therefore

$$\Rightarrow \frac{\mathrm{d}}{\mathrm{d}t} i(0) = \mathrm{d}i_0 := \frac{V_1}{L}. \quad (16)$$

Circuit behavior during T_{OFF}. The equivalent circuit of the architecture segment that is relevant during T_{OFF} is shown in Fig. 8. R_δ summarizes series resistances similarly to R_σ. Its calculation is also detailed in Table 3. Applying KIRCHHOFF's Voltage Law yields

$$L \cdot \frac{\mathrm{d}}{\mathrm{d}t} i + R \cdot i + \int_0^{T_{ON}} i(\tau) \, \mathrm{d}\tau + V_d + V_2 = 0 \quad (17)$$

with $R = R_\delta$. We can differentiate with respect to t and obtain Eq. (14) again. Concerning the initial values, we can assume the current to start where it ended during T_{ON}, meaning $i(0) = i_0 := i_{\text{peak}}$. With this in mind, transforming Eq. (17) yields the second condition.

$$0 = L \cdot \frac{\mathrm{d}}{\mathrm{d}t} i(0) + R \cdot i_{\text{peak}} + 0 + V_d + V_2$$

$$\Rightarrow \frac{\mathrm{d}}{\mathrm{d}t} i(0) = \mathrm{d}i_0 := -\frac{V_2 + V_d + Ri_{\text{peak}}}{L} \quad (18)$$

Circ.	Symb.	$\lvert \sigma - \delta \rvert \bmod 2 = 1$	$\lvert \sigma - \delta \rvert \bmod 2 = 0$
K	R_σ	$R_C + R_L + R_M$	-
	R_δ	$R_C + R_L$	-
	V_d	V_d	-
N	R_σ	$R_C + R_L + 3R_M$	$R_C + R_L + 3R_M$
	R_δ	$R_\sigma + (2d-1)R_M$	$R_\sigma + R_L + (2d-1)R_M$
	V_d	0	0

Table 3: Lookup table for the sum of the series resistances and the diode voltage drop V_d. $d = \lvert \sigma - \delta \rvert$ is the distance between source and destination cell. K: Circuit from [8], N: Proposed circuit employing non-overlapping PWM signal.

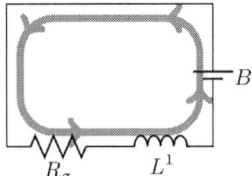

Figure 7: Equivalent circuit of the architecture segment relevant during T_{ON}.

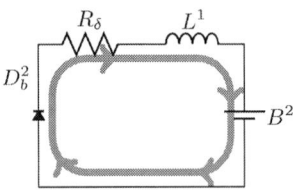

Figure 8: Equivalent circuit of the architecture segment relevant during T_{OFF}.

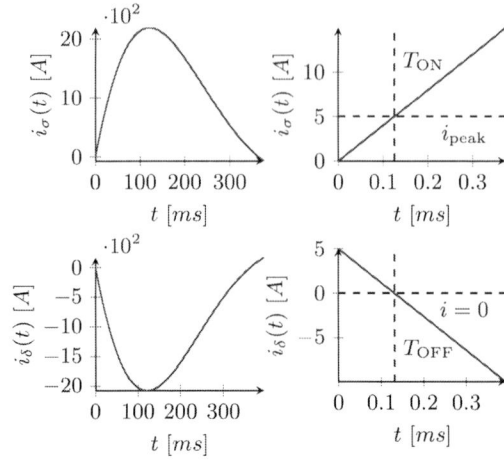

Figure 9: Current plots for the under-damped case; upper row: charging current i_σ; lower row: discharging current i_δ.

S2.2 Ordinary Differential Equation (ODE) Solution

This section presented a solution approach for the ODE in Eq. (14) with initial conditions in Eq. (16) and Eq. (18) for T_{ON} and T_{OFF}, respectively. Combined, they form the following system:

$$L \cdot \frac{\mathrm{d}^2}{\mathrm{d}t^2} i + R \cdot \frac{\mathrm{d}}{\mathrm{d}t} i + \frac{1}{C} i = 0$$

$$i(0) = i_0 \qquad \frac{\mathrm{d}}{\mathrm{d}t} i(0) = \mathrm{d}i_0 \qquad (19)$$

The solution process of such a second-order system is detailed in control engineering literature such as [12]. Following these solution processes, we rewrite Eq. (19) to

$$0 = \frac{\mathrm{d}^2}{\mathrm{d}t^2} i + \frac{R}{L} \cdot \frac{\mathrm{d}}{\mathrm{d}t} i + \frac{1}{LC} i =: \frac{\mathrm{d}^2}{\mathrm{d}t^2} i + 2\xi\omega_n \cdot \frac{\mathrm{d}}{\mathrm{d}t} i + \omega_n^2 i. \quad (20)$$

The behavior of the system largely depends upon its natural frequency ω_n and its damping ratio ξ. In case of the proposed circuit, ω_n and ξ are given by the following equations:

$$\omega_n = \frac{1}{\sqrt{LC}} \qquad \xi = \frac{1}{2} \frac{R}{L} \sqrt{LC} \qquad (21)$$

The characteristic equation of the ODE whose roots lead to the general solution of the system can be obtained by introducing variable s and modeling Eq. (20) as:

$$s^2 + 2\xi\omega_n \cdot s + \omega_n^2 \qquad (22)$$

The roots of the characteristic Eq. (22) of the ODE results in

$$s_{1/2} = -\xi\omega_n \pm \omega_n \sqrt{\xi^2 - 1} \qquad (23)$$

and the general time domain solution of ODE system in Eq. (19) is therefore given by:

$$i(t) = \gamma_1 e^{s_1 t} + \gamma_2 e^{s_2 t} \qquad (24)$$

Solving for the constants γ_1 and γ_2, using the initial values from Eq. (19) yields the following relation:

$$i(t) = \frac{\mathrm{d}i_0 - i_0 s_2}{s_1 - s_2} e^{s_1 t} - \frac{\mathrm{d}i_0 - i_0 s_1}{s_1 - s_2} e^{s_2 t} \qquad (25)$$

Using Eq. (23), this can be further reformulated to:

$$i(t) = e^{-\xi\omega_n t} \cdot \left\{ \right.$$
$$\frac{\mathrm{d}i_0 - i_0(-\xi\omega_n - \omega_n \sqrt{\xi^2 - 1})}{2\omega_n \sqrt{\xi^2 - 1}} e^{(\omega_n \sqrt{xi^2 - 1})t}$$
$$\left. - \frac{\mathrm{d}i_0 - i_0(-\xi\omega_n + \omega_n \sqrt{\xi^2 - 1})}{2\omega_n \sqrt{\xi^2 - 1}} e^{-(\omega_n \sqrt{xi^2 - 1})t} \right\} \quad (26)$$

From this point, we need to differentiate between three cases, $\xi < 1$ (under-damping), $\xi = 1$ (critical damping) and $\xi > 1$ (over-damping) because they represent entirely different system behaviors. Under-damping leads to a resonating signal that is slowly damped away. Over-damping on the other hand is not resonating, but creeps very slowly to its equilibrium. Critical damping is an interim situation where the system signal resonates exactly once and is then damped away. This is the fastest way to reach the equilibrium and critical damping is therefore used as a design methodology in certain situations. Since our approach ends the system signals prematurely, all three cases can be handled and do in fact barely differ on the relevant time scale as we will see in the following sections.

S2.3 Under-Damping($\xi < 1$)

If $\xi < 1$, the roots of the characteristic equation are not real, but complex:

$$s_{1/2} = -\xi\omega_n \pm j\omega_n \sqrt{1 - \xi^2} \qquad (27)$$

Using this, we can transform Eq. (26):

$$i(t) = e^{-\xi\omega_n t} \cdot \left\{ \frac{\mathrm{d}i_0 - i_0(-\xi\omega_n - j\omega_n \sqrt{1 - \xi^2})}{2j\omega_n \sqrt{1 - \xi^2}} \right.$$
$$\cdot (\cos(\omega_n \sqrt{xi^2 - 1})t) + j\sin(\omega_n \sqrt{xi^2 - 1}t)$$
$$- \frac{\mathrm{d}i_0 - i_0(-\xi\omega_n + j\omega_n \sqrt{1 - \xi^2})}{2j\omega_n \sqrt{1 - \xi^2}}$$
$$\left. \cdot (\cos(\omega_n \sqrt{xi^2 - 1})t) - j\sin(\omega_n \sqrt{xi^2 - 1}t) \right\}$$

1072

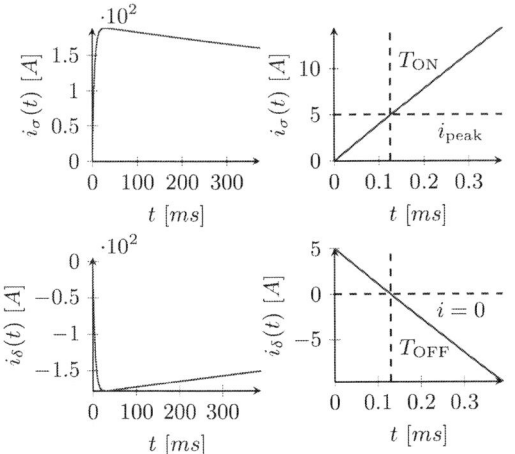

Figure 10: Current plots for the over-damped case; upper row: charging current i_σ; lower row: discharging current i_δ.

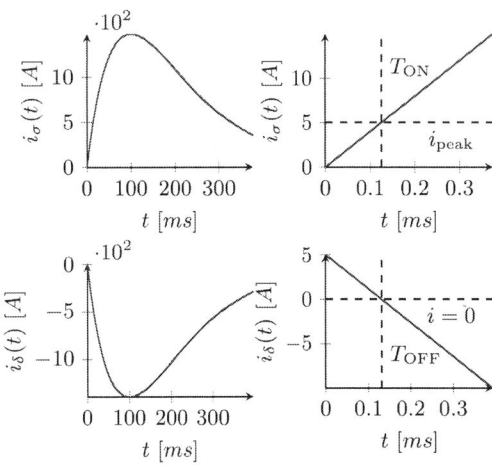

Figure 11: Current plots for the critically damped case; upper row: charging current i_σ; lower row: discharging current i_δ.

$$i(t) = e^{-\xi\omega_n t} \cdot \Big\{ i_0 \cos(\omega_n \sqrt{1-\xi^2}t)$$
$$+ \frac{di_0 + i_0\xi\omega_n}{\omega_n\sqrt{1-\xi^2}} \sin(\omega_n\sqrt{1-\xi^2}t) \Big\} \quad (28)$$

Abstracting $i(t) = e^{-ct}(A\cos(at) + B\sin(at))$, we can integrate Eq. (28) to obtain the transferred charge q.

$$q(T) = \frac{-A}{a^2+c^2}\Big[c(e^{-Tc}\cos(Ta)-1) - ae^{-Tc}\sin(Ta)\Big]$$
$$- \frac{B}{a^2+c^2}\Big[a(e^{-Tc}\cos(Ta)-1) + ce^{-Tc}\sin(Ta)\Big] \quad (29)$$

The remaining task is to calculate T. For the charging phase, we can use $T = T_{\text{ON}}$, but for $T = T_{\text{OFF}}$, we need to calculate when the inductor is actually empty, i.e., we solve the following:

$$0 = A\cos(at) + B\sin(at)$$
$$\Leftrightarrow -\frac{A}{B} = \frac{\sin(at)}{\cos(at)}$$
$$\Leftrightarrow at = \arctan(-\frac{A}{B}) \quad (30)$$

Since there is no closed-form for T_{ON} in the nonlinear model, a short binary search using the nonlinear model can be used to improve the accuracy of the linear estimation. Fig. 9 gives an impression on how i_σ and i_δ behave and how small T_{ON}, T_{OFF} are relatively to the time constants of the sine waves.

S2.4 Over-Damping ($\xi > 1$)

If $s_{1/2} \in \mathbb{R}$, we can directly abstract $i(t) = e^{-ct}[Ae^{at} - Be^{-at}]$ from Eq. (26) and integrate it to obtain:

$$q(T) = \frac{B}{a+c}\Big[e^{-Ta-Tc}-1\Big] + \frac{A}{a-c}\Big[e^{Ta-Tc}-1\Big] \quad (31)$$

To calculate T_{OFF}, we need to solve the following:

$$0 = Ae^{at} - Be^{-at}$$
$$\Leftrightarrow \log(e^{at}) = \log(\frac{B}{A}e^{-at}) = \log(\frac{B}{A}) + -at$$
$$\Leftrightarrow 2at = \log(\frac{B}{A}) \quad (32)$$

As illustrated in Fig. 10, it becomes obvious that the damping actually has only little influence during the time frame we are interested in.

S2.5 Critical Damping ($\xi = 1$)

If $\xi = 1$, the characteristic equation has co-located roots and the solution therefore becomes

$$i(t) = (A + Bt)e^{-ct} \quad (33)$$

with $A = i_0$, $B = di_0 + \omega_n i_0$, $c = \omega_n$. Fig. 11 shows the corresponding plots. Again, the behavior does not differ much from the other cases as far as very short intervals are concerned. Integrating $i(t)$ yields the transferred charge as in the other cases:

$$q(T) = \frac{B}{c^2}\big(1 - (Tc+1)e^{-Tc}\big) - \frac{A}{c}\big(e^{-Tc}-1\big) \quad (34)$$

The calculation of T_{OFF} is done with Eq. (33). We have to solve the following:

$$0 = A + Bt \qquad \Rightarrow t = \frac{-A}{B} \quad (35)$$

S2.6 Linear Analysis

We developed a nonlinear model in Section S2.2. It results in many different cases to consider and it might become tedious to implement certain cases. Optimization and control scenarios in particular need the system model to be as simple as possible. If we assume the current to be linear – a reasonable assumption for the small time steps we are treating as seen in Figs. 9, 10 and 11 – we can model i_σ and i_δ as

$$i(t) = i_0 + di_0 \cdot t \quad (36)$$

with i_0, di_0 as in Eq. (16) or Eq. (18) for T_{ON} and T_{OFF}, respectively. In practice, this results in:

$$i_\sigma(t) = \frac{V_1}{L} \cdot t$$
$$i_\delta(t) = i_{\text{peak}} - \frac{V_2 + V_d + R_\delta i_{\text{peak}}}{L} \cdot t \quad (37)$$

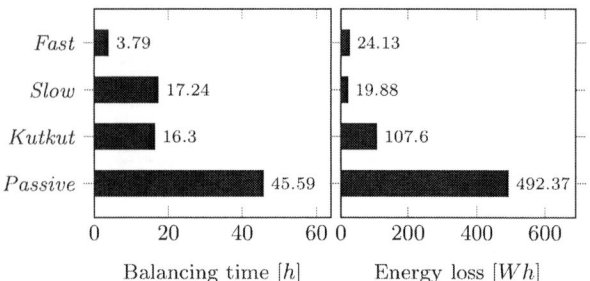

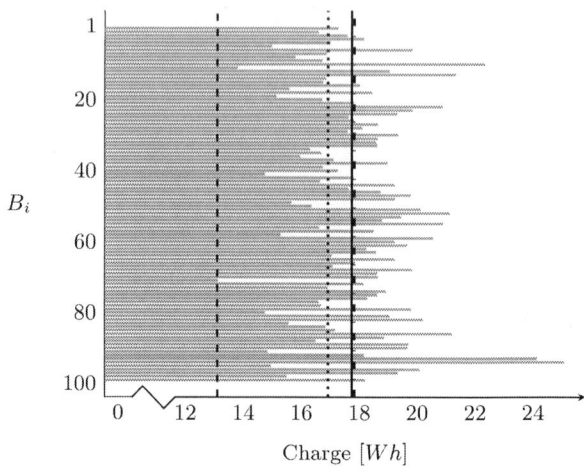

Figure 12: Results of the supplemental case study with a higher variance in charge levels, comparing balancing time and energy loss between different approaches. *Fast* and *Slow* are two proposed approaches for the presented architecture while *Kutkut* and *Passive* constitute state-of-the-art approaches.

Figure 13: Illustration of the initial and final charge distribution of the supplemental case study with a higher variance in charge levels. Most charge is preserved with the *Slow* strategy (‒ · ‒) followed by *Fast* (——) still clearly outperforming *Kutkut* (......) while the *Passive* (- - -) balancing reduces the charge to the level of the lowest cell.

From there, we can again integrate to obtain the transferred charge $q(T)$ and obtain the following:

$$q_\sigma(T) = \int_0^T i_\sigma(\tau)\,d\tau = \int_0^T \frac{V_1}{L}\cdot\tau\,d\tau$$

$$= \frac{V_1}{L}\cdot\frac{T^2}{2} \tag{38}$$

$$q_\delta(T) = \int_0^T \left[i_{\text{peak}} - \frac{V_2 + V_d + R_\delta i_{\text{peak}}}{L}\cdot\tau\right]d\tau$$

$$= i_{\text{peak}}T - \frac{V_2 + V_d + R_\delta i_{\text{peak}}}{L}\cdot\frac{T^2}{2} \tag{39}$$

Again, we need to provide values for T. In the linearized case, we can calculate both T_{ON} and T_{OFF} directly from Eq. (37). We obtain the following:

$$T_{\text{ON}} = i_{\text{peak}}\frac{L}{V_1}$$

$$T_{\text{OFF}} = i_{\text{peak}}\frac{L}{V_2 + V_d + R_\delta i_{\text{peak}}} \tag{40}$$

With two system models available that are both suitable for system-level analysis, the user is left with a choice. Both methods are significantly faster than transistor-level simulations and both provide sufficient accuracy as shown in our experimental results in Section 5.1.

S3 Supplemental Case-study with Higher Variance

To validate the results from section 5.2, we performed a supplemental case-study with significantly higher initial distortion of the charge levels. This poses a greater challenge to all balancing methodologies and it can be expected that their individual strengths and weaknesses become even more apparent than in Section 5.2. We initialized the cell voltages according to

$$V^{(i)} \sim \mathcal{N}(3.6, 0.2^2) \tag{41}$$

and left the other parameters from Eq. (6) unchanged. The strategies under consideration are still *Passive* balancing, the *Kutkut* approach, as well as our *Fast* and *Slow* methodology as defined in Section 5.2.

Figs. 12 and 13 give an overview of the performed simulation. As expected with higher initial variance among the cells, the final charge values in Fig. 13 are lower than those from Fig. 6. A similar observation can be made when com-

paring Fig. 12 to Fig. 5. The balancing time as well as the unrecoverable energy is increased among all the approaches.

Between the different strategies, we observe that active balancing now performs even better than before with respect to the *Passive* approach. For instance, *Fast* improved from 85% time savings and almost 95% less energy dissipation to 92% and over 95%, respectively. With respect to the other active approaches, *Fast* keeps its lead in balancing speed of over 75%. At the same time, it requires only 25% more energy than *Slow*, making it a very good choice for most scenarios. Concerning balancing speed, *Slow* is now only on par with *Kutkut*. On the other hand, it firmly establishes itself as the first choice with respect to energy dissipation outperforming *Kutkut* by over 80% and *Fast* by 18%.

S4 Nomenclature

Charge Transfer

σ	Source cell
δ	Destination cell
p	Cell pair (σ, δ), indicating a transfer from σ to δ
$\mathcal{P}$	Set of concurrent transfers $\mathcal{P} = \{p_1, \ldots, p_n\}$

Circuit Elements

R	Resistance
R_s	Series resistance during T_{ON} (see Table 3)
R_d	Series resistance during T_{OFF} (see Table 3)
L	Inductance
C	Capacitance
B	Battery cell
M	MOSFET or switch for current routing

Time-Dependent Functions

$i(t)$	Current at time t
$q(T)$	Charge transferred until time T
V	Voltage

Pulse Width Modulation (PWM)

PWM	PWM controlled signal during T_{ON}
$\overline{PWM}$	PWM controlled signal during T_{OFF}
T_{ON}	Duration of PWM ON interval
T_{OFF}	Duration of PWM OFF interval

A Method to Abstract RTL IP Blocks into C++ Code and Enable High-Level Synthesis

Nicola Bombieri[1], Hung-Yi Liu[3], Franco Fummi[1,2], Luca Carloni[3]

Dip. Informatica - Università di Verona, Verona - Italy[1]
EDALab s.r.l., Verona - Italy[2]
Dept. Computer Science - Columbia University, NY - USA[3]
nicola.bombieri@univr.it[1], franco.fummi@{univr.it, edalab.it}[1,2], {hungyi, luca}@cs.columbia.edu[3]

ABSTRACT

We present a method to automatically generate a synthesizable C++ specification from the given RTL design of an IP block, by abstracting away most of its micro-architectural characteristics while preserving its functionality. The goal is twofold: recover the IP block specification for system-level design, and enable the derivation of more optimized implementations through high-level synthesis. The C++ specification can be generated with different interfaces thus allowing the IP model to be reused across different system platforms. Experimental results show that the proposed approach not only enhances the reusability of the recovered IP block but also unveils a richer design space to explore.

Categories and Subject Descriptors

B.5 [**REGISTER-TRANSFER-LEVEL IMPLEMEN-TATION**]: Design Aids - Optimization

General Terms

Design, Performance

Keywords

RTL IP reuse, System-level Design

1. INTRODUCTION

Reuse of existing and already verified RTL IP components is a key strategy to cope with the complexity of designing modern SoCs under ever stringent time-to-market requirements. To achieve a 10x gain in design productivity by the year 2020 is expected to require that a complex SoC will consist of 90% reused components [9]. The reusability of an RTL IP component is not always guaranteed since it depends on the designers' ability to implement it independently from a specific integration context.

While design reuse methodologies have been proposed for almost a decade, often the main priority of RTL designers is to optimize a given component for a particular SoC product. Furthermore, the level of abstraction of RTL is inherently limited in its capability of expressing efficiently many alternative micro-architectural choices and interface configurations for a given IP: typically a Verilog or VHDL description that is aimed at deriving an efficient logic synthesis implementation only specifies one I/O interface protocol (e.g. how many input data are sampled at each clock cycle) and one internal micro-architecture (e.g., the depth of the internal pipeline of a datapath).

Sustained by the increasing need to start the design and validation of multi-core SoCs at higher levels of abstraction (system-level design) [5], the use of high-level synthesis (HLS) is gaining consensus [11]: indeed, there are now several commercial HLS tools which are capable to take a single

Permission to make digital or hard copies of all or part of this work for personal or classroom use is granted without fee provided that copies are not made or distributed for profit or commercial advantage and that copies bear this notice and the full citation on the first page. To copy otherwise, to republish, to post on servers or to redistribute to lists, requires prior specific permission and/or a fee.
DAC '13, May 29 - June 07 2013, Austin, TX, USA.

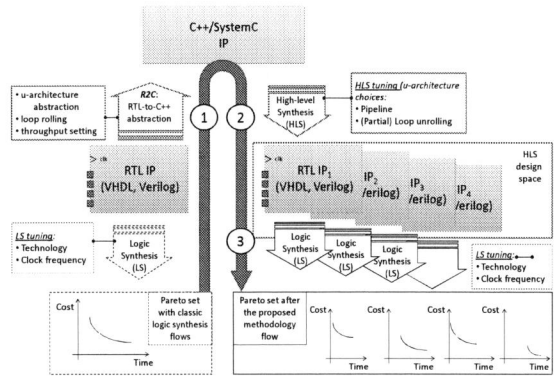

Figure 1: The proposed methodology for RTL IP block recovery and reuse.

C/C++ or SystemC specification of an IP component and generate many alternative RTL implementations, each optimized for a given SoC. Starting from a specification given in one of these languages, designers can configure the rich set of knobs offered by HLS tools to explore a design space that is much richer than the one offered by the combination of an RTL specification and a logic synthesis tool.

Still, while new SoC designs are increasingly started at the system level, there is a large body of RTL IP blocks which have been designed in VHDL or Verilog by both industry designers and third-party vendors over the years. The motivation for our work is precisely the observation that it would be nice to *recover* the core functionality of these designs, make them suitable for HLS and system-level design, and, ultimately, enhance their reusability.

Our main contribution is a method to automatically generate a C++ code specification optimized for HLS from the RTL design of a given component, by abstracting away most of its micro-architectural characteristics while preserving its functionality.

We implemented this method in a new tool, called *R2C*, which is used in Step 1 of our proposed methodology for RTL IP block recovery and reuse enhancement (as illustrated in Fig. 1). The C++ code generated by *R2C* can be used for efficient design-space exploration using a commercial HLS tool (Step 2) and, after the application of traditional logic synthesis, ultimately to obtain a final implementation that is optimized for a given SoC design context (Step 3.) The enhanced reusability is the combined result of: (a) enabling design-space exploration at the system-level where simulations of the whole SoC can be done with more complex user case scenarios, (b) leveraging HLS to evaluate alternative micro-architectures, and (c) generating the C++ specification with different I/O interfaces, which is a feature of *R2C*.

The application of the proposed methodology to two case studies not only confirms this reusability enhancement but it also typically leads to final design implementations that are better than the original RTL design in terms of either performance or area occupation and, sometimes, in both cases.

Related Work. Several methods for translating RTL VHDL and Verilog models into C/C++ descriptions tar-

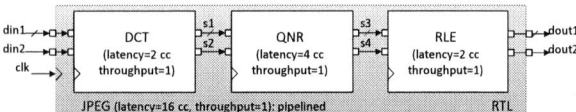

Figure 2: A pipelined RTL JPEG model

geting verification of hardware models via simulation have been proposed in the literature or implemented in commercial tools [7, 8, 14, 15, 4, 3]. In [7], a VHDL to C++ converter transforms VHDL test-benches to C++ source. During the conversion, the C++ source is compiled into a small simulation kernel that runs the whole simulation with the interconnected hardware board. In [8, 14], various translation tools allow designers to use C++ executable files in place of VHDL models for decreasing simulation time compared to the typical acceleration process with hardware description language simulators. In [15], a methodology, which was then implemented in the tool VTOC, is proposed to convert synthesizable Verilog into C++. VTOC tries to reduces the number of delta cycles by topological sorting all processes and by applying process merging. Since the goal of these methods is the verification of the given RTL design, all the implementation details included in the RTL descriptions and strictly related to the hardware modeling (e.g., clock accuracy, bit accuracy) are maintained during the translation to the software domain.

Carbon Design System [4] provides commercial products that convert Verilog or VHDL RTL models into cycle-accurate and register-accurate SystemC models. Carbon's tools aim at creating complete virtual platforms in order to gain both a fast and accurate system validation. In [3], the C++ generation from RTL IPs aims at abstracting many architectural details for fast simulation. The process synchronization relies on a dynamic scheduling, which is embedded into the generated C++ code.

Differently from all the techniques presented in literature that target C++ code generation for simulation and verification, the proposed method is aimed at the generation of C++ for HLS. As discussed in the following sections, the different goal leads to important differences in the approach of generating C++ code.

2. METHODOLOGY

The main technical problem we address in this paper is the following: Given a synthesizable RTL IP implemented in hardware description language, generate a sequential C++ code that preserves the RTL IP semantics and that can be synthesized by HLS, with the aim of:

- Minimizing the implementation cost of the RTL designs that can be synthesized by the C++ code through HLS; and
- Maximizing the design space of the synthesized RTL designs.

The proposed method relies on three key concepts:

1. During the generation of C++ code, the scheduling of RTL statements is resolved statically at compile time. Motivated by area optimization purposes, this approach is different from the related works mentioned above, which preserve dynamic scheduling kernels. In fact, even though dynamic scheduling allows great simulation performance, it leads to area overhead costs once synthesized, as explained in Section 2.1.

2. Static variables in the C++ code are generally synthesized into registers at gate level. As explained in 2.2, *R2C* tries to reduce the number of static variables during the generation of C++ code, again to minimize area overhead.

3. *R2C* performs loop-rolling transformations on both internal logic and the I/O interface. The goal is to generate loops in the C++ code. Loops are strategic in helping HLS tools to produce good quality synthesis results as well as a richer design space to explore, as explained in Section 2.3.

Consider for example the JPEG of Fig. 2, a synthesizable RTL IP model implemented in Verilog, and the generated C++ implementation of Fig. 3(a). The JPEG consists of three components: A discrete cosine transform (DCT) module, a quantization (QNR) module, and an entropy coding

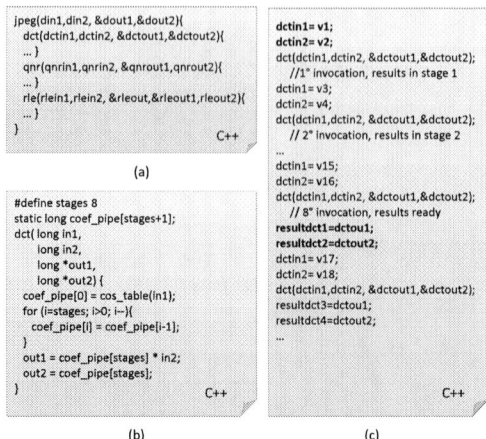

Figure 3: (a) The main structure of the JPEG example when translated in C++, (b) the C++ code generated from the pipelined RTL DCT, and (c) the sequence of *dct*() invocations due to the pipeline nature of the RTL IP component.

that runs the run-length coding (RLE). *R2C* implements each RTL IP component through a C++ function. The main function implementing the JPEG is obtained by the sequential call of the three functions, ordered according to the RTL IP dataflow.

Fig. 3(b) shows the C++ code generated from the pipelined RTL code of the DCT component. The RTL signals implementing the pipeline stages are translated into C++ static variables (*coef_pipe*[ ] in the example). As a consequence, for returning the first results, the C++ function implementing the DCT module must be invoked as many time as the number of pipeline stages (see Fig. 3(c)). Then, after such a kind of *functional latency*, each function invocation returns one result.

2.1 The static scheduling in the C++ code

The RTL-to-C++ abstraction consists of mapping each synchronous process ps of the RTL model M_{RTL} into a C++ function fs of the C++ model M_{C++}, where each sequential statement of the process is mapped into a C++ statement. The translation of RTL statements into C++ statements is merely syntactic and the order of statements in the RTL process is preserved in the C++ function.

In the same way, each asynchronous process or global statement (i.e., statement outside synchronous processes) of M_{RTL} is mapped into a C++ function fa of M_{C++}.

Then, RTL signals are mapped into C++ variables, by preserving the corresponding data width and type. In certain specific cases, as explained in Section 2.2, signals are mapped into *static* variables to preserve the IP semantics. For the sake of clarity and without loss of generality, we do not dwell on other hardware description language declarations (e.g., variables, constants, subprograms, etc.) for which the RTL semantics is easily mappable into the C++ semantics.

Data types used at RTL, which are bit-accurate and implement the multi-value logic, are maintained in the generated C++ implementation. Different libraries of such bit-accurate data types and corresponding operators are available (e.g., Accellera Systems Initiative [2], Mentor [12], etc.). The proposed approach is independent of the data type library and any of those already existing can be adopted.

To preserve the RTL IP semantics, the C++ functions generated from the RTL processes have to be executed in the same *partial* order as the corresponding RTL processes are executed in the RTL model. That is, concurrent processes can execute in a non-deterministic order, while the order between non concurrent processes must be preserved.

To do that, *R2C* implements a *static scheduling* of functions. At compile time, *R2C* resolves the order of function calls according to the order of RTL processes in the process communication graph. Consider, for example, the

1076

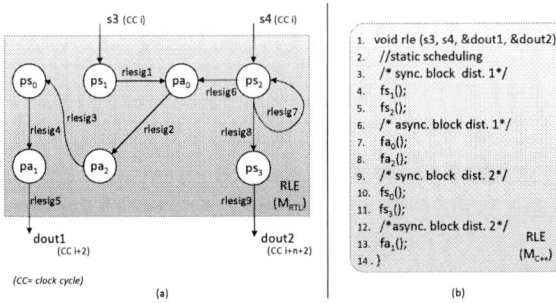

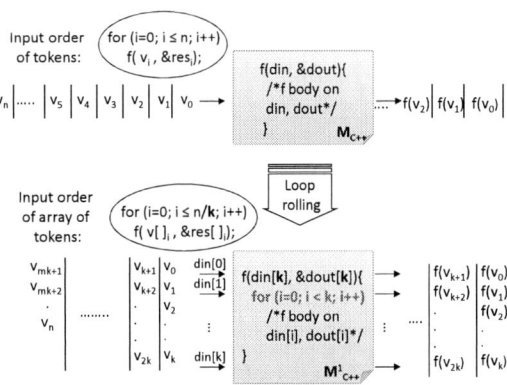

Figure 4: (a) Process communication graph (PCM) of *RLE*, (b) the corresponding static scheduling of functions in the C++ code.

Figure 5: Loop rolling on I/O

RLE component of the *JPEG* RTL IP. Fig. 4(a) represents such a component through a graph, each process being a vertex and each signal being an oriented edge. The graph, which is automatically extracted from the RTL description, represents the synchronization and communication net among processes. The *RLE* component consists of four synchronous processes (ps_0-ps_3), three asynchronous processes (pa_0-pa_2), two input ports ($s3$, $s4$), two output ports ($dout1$, $dout2$), and seven internal signals ($rlesig1$-$rlesig9$).

The generated functions are firstly grouped into synchronous and asynchronous blocks, according to their distance (in clock cycles of latency) from the input ports. The idea is that the functions are invoked by alternating a group of synchronous to a group of asynchronous functions, according to the block distance from the inputs. Fig. 4(b) shows, for example, an overview of the generated C++ code (and in particular the order of the function calls) generated from the RTL model of Fig. 4(a). Notice that the outputs of the RTL model have different latencies, since the PCM has a n-iteration loop on the data flow. This implies *rle*() to be called multiple times for generating the first result on *dout2* and a static variable to implement the memory element for *rlesig7*, as explained in the next section.

R2C generates the C++ model with static scheduling since it targets better synthesis results rather than simulation performance. In contrast, the techniques presented in literature [3, 4] implement a *dynamic scheduling* of functions, targeting C++ code generation for fast simulation and verification. Even though dynamic scheduling outperforms static scheduling in traditional simulation environments, it takes extra logic to be implemented. Such an extra logic would imply more area in the synthesized circuit with no benefits to delay. A more detailed comparison between static vs. dynamic scheduling is given in Section A.1 of the Appendix.

2.2 Static variable minimization

When producing C++ code from the given RTL IP code, *R2C* generates static variables from RTL signals as part of two possible scenarios::

- From RTL communication signals between synchronous processes. If there are asynchronous processes in the path between them, only the signals outcoming the synchronous processes are mapped into *static* variables (e.g., $rlesig1$, $rlesig4$, $rlesig6$, $rlesig8$, and $rlesig9$ in Fig. 4(a));

- From RTL signals that form dataflow loops (e.g., $rlesig7$ in Fig. 4(a) that implements a pipeline barrier).

Since *static* variables in C++ code always lead to the synthesis of registers, *R2C* aims at area optimization by generating C++ code with a reduced number of static variables. Beside area optimization, static variable minimization in the C++ code is a key aspect as it allows the C++ *functional latency* to be reset (set to the minimal value one):

DEFINITION 1. *Considering the generated M_{C++}, we define* functional latency *as the number of invocations of the C++ main function (e.g.,* jpeg() *) for reading the inputs and producing the output results. The latency of the original IP is reset when the generated M_{C++} has functional latency equal to one invocation.*

If the functional latency is not reset, the latency of any M_{RTL_i} synthesized from the C++ code is:

$$Latency_{M_{RTL_i}} \quad Latency_{HLS} \times Functional\_latency_{M_{C++}}$$

where $Latency_{HLS}$ is the latency inferred to the RTL model by HLS (e.g., to break critical path, etc.).

R2C abstracts the static variables generated from the RTL communication signals between processes, into *non-static* variables. Consider, for example, the RTL RLE dataflow $(s3, s4) \rightarrow dout1$ shown in the leftmost side of Fig. 4(a). It traverses at least two synchronous processes thus involving a latency of two clock cycles on *dout1*. The corresponding C++ implementation consists of the function-call sequence $< fs_1(), fs_2(), fa_0(), fa_2(), fs_0(), fa_1() >$. In this case, the main C++ function *rle*() reads inputs $s3$, $s4$ through $fs_1()$, $fs_2()$ and, after the function-call sequence, generates the result *dout1* with a functional latency equal to one invocation. Signals $rlesig1$-$rlesig6$ are all mapped into (non-static) variables.

On the contrary, static variables generated from RTL signals that form loops cannot be directly abstracted. Consider for example static variables implementing pipeline barriers. The data transition towards the output over the pipeline stages requires one input to be read for each stage. This behavior, in M_{C++}, can be implemented only through multiple invocations of the main C++ function. A more detailed analysis of the static variable minimization and latency reset is given in Section A.2 of the Appendix.

In the example of Fig. 4(a), the dataflow $s4 \rightarrow dout2$ contains a loop on ps_2. The corresponding C++ implementation consists of the function call sequence $< fs_2(), fs_3() >$, which must be invoked n times for generating the first result. Then, since $fs_2()$ and $fs_3()$ are subfunctions of *rle*(), the whole *rle*() must be invoked n times (functional latency equal to n) for generating the first result on *dout2*.

For resetting the functional latency during abstraction of RTL components with loops in the dataflow, *R2C* performs *loop rolling on I/O*, as explained in the following section.

2.3 The loop rolling

To maximize the design space of the recovered RTL models, *R2C* performs *loop rolling* transformations on the internal logic and on I/O interfaces.

When multiple instances of the same block (i.e., module in Verilog, entity in VHDL) are explicitly instantiated in the RTL code, *R2C* rolls up the instances into a loop and resolves the binding. The loop rolling on internal logic applies over different hierarchy levels of M_{RTL}, by generating, as a result, nested C++ loops in M_{C++}. A more detailed analysis of loop rolling on internal logic is given in Section A.3 of the Appendix.

Loop rolling is also applied to abstract the interface from single input values to arrays of values. Consider a M_{C++} model, which implements functionality f, reads an input value (called *input token* hereafter), elaborates, and returns the output result (*output token*), as shown in the upper side of Fig. 5. Consider that the sequence of n input values $(v_0, .., v_n)$ is generated from the environment in which the model is inserted (e.g., a component upstream of M_{C++}). The computation of f over the ordered sequence of input

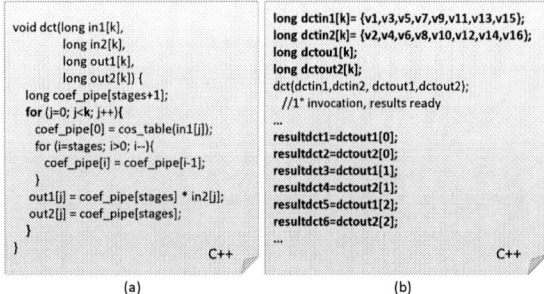

Figure 6: (a) The C++ code generated from the pipelined RTL implementation with loop rolling, (b) and the single function invocation due to the latency reset (considering k equal to the RTL DCT latency).

tokens is represented by a `for` loop, in which f is called over one new token at each iteration, by producing an ordered sequence of output tokens $(f(v_0), f(v_1), f(v_2)$, etc.).

The loop rolling on the model interface consists of augmenting the model interface from single tokens to arrays of tokens (see, for example, the *din* and *dout* I/Os of Fig. 5). Then, function f is enriched with a `for` loop of the main body over the input and output arrays, in order to preserve the ordered sequence of read inputs, elaborations, and writes of the results.

The model M^1_{C++} obtained through loop rolling preserves the semantics of M_{C++}. After accounting for the difference between the input and output token cardinality, the iterations of function calls over the data tokens do not change. In particular, the main *for* loop that is used to call f is split into two nested loops, one of them moved inside function f. In this way, the C++ function can iteratively read and elaborate multiple inputs (i.e., through a loop added to the code) during one single invocation.

Loop rolling is strategic since it enriches the C++ code with loops, which are fundamental in HLS for exploring and enhancing the design space. In addition, loop rolling can reduce static variables even in RTL models with cyclic dataflow (e.g., pipelined architectures), by resetting the *functional latency* of the C++ code. For example, Fig. 6 shows how the C++ function implementing the DCT with loop rolling is invoked once over arrays of input values for returning arrays of results.

On the other hand, loop rolling also involves an increase of area. This is due to the fact that the data read as input becomes an array of input values, whose size corresponds to the number of loop iterations (i.e., the k value in Fig. 6(a)).

With loop rolling, the interface of the generated C++ and of the RTL models synthesized from such a C++ code differ from the interface of the starting RTL IP. This is an advantage of our approach since it can generate both the implementation that preserves the original interface as well as many different ones.

In contrast, if the starting interface is a strict requirement, the new RTL models must be extended with parallel/serial wrappers, which also reset the latency to the value of the starting RTL IP. In this case, however, the area introduced by these wrappers is negligible w.r.t. the area saved by the proposed flow.

Since M^1_{C++} works on arrays of data tokens, the model throughput is augmented, while the functional latency is reduced. In particular, the new functional latency involved by loop rolling is the following:

$$Functional\_latency_{M^1_{C++}} \quad \lceil \frac{Functional\_latency_{M_{C++}}}{k} \rceil$$

where k is the parameter that sets the array size and the loop iterations (see Fig. 5).

Parameter k plays a key role in the C++ model generation and its HLS. Considering that the functional latency of M_{C++} is automatically extracted during the M_{C++} generation, if k is set equal to that value, the functional latency of M^1_{C++} is reset. To set k with a value greater than the functional latency would imply an increasing of the model throughput.

Table 1: Dynamic vs. Static Scheduling on JPEG

Component	DCT	QNR	RLE
Min Area (um^2) with dynamic sched.	88,681	20,758	4,151
Max Area (um^2) with static sched.	80,039	9,661	1,321
Dynamic_sched./Static_sched.	1.1X	2.1X	3.1X

Table 2: Static variable reduction on JPEG

Component	Dynamic → Static sched.		Static sched. + Loop rolling	
	# of var	size of var	# of var	size of var
DCT	11	102 bits	64	768 bits
QNR	12	89 bits	46	402 bits
RLE	19	75 bits	0	0 bit

However, in general, we cannot state that setting k to reset the functional latency or to increase throughput is always the best solution in terms of HLS quality of results. In fact, any increase of k involves an increase of the interface size, with a consequent impact on the area. In particular, the parameters of the generated C++ function (e.g., *dctin1*, *dctin2*, *dctout1*, *dctout2* in Fig. 3 and Fig. 6) are synthesized by an HLS tool into interface pins or memory elements at gate level, depending on the way they are passed (i.e., by value or by reference). Thus, the value of k should be evaluated to find the best tradeoff between reducing or resetting the latency (i.e., $k = latency$) and thus saving area from static variables, versus an increasing of size of the new interface.

3. EXPERIMENTAL RESULTS

We applied *R2C* to two RTL IP designs: *(i)* a **Reed-Solomon** decoder composed of five subcomponents (*BM_lamba*, *Lambda_roots*, *Omega_Phy*, *Error_correction*, *Out_stage*), and *(ii)* a **JPEG** decoder composed of three pipelined components (*DCT*, *QNR*, *RLE*) from [1]. Note that although our designs represent moderate RTL IPs, we treat them as *system* designs in our study, whereas their components as *IP* components w.r.t. the systems. Thus, our experimental results can demonstrate how effectively *R2C* can recover the individual component designs as well as enrich the system-design space. For the RTL-to-C++ part in our flow, (Step 1 in Fig. 1) we implemented our translation tool in C++. For the C++-to-RTL part, (Steps 2 and 3 in Fig. 1) we leveraged a recent HLS design-space-exploration tool [10]. For our experiments, we used commercial logic-synthesis and HLS tools with an industrial 45nm technology.

Dynamic vs. Static Scheduling. To evaluate the area-saving advantage by adopting static scheduling, we also implemented a dynamic-scheduling engine and tested it on JPEG as follows. For each JPEG component, given a same clock period, we searched for the component's *minimum* implementation area with dynamic scheduling and its *maximum* area with static scheduling. The results are reported in Table 1. Even under the biased condition in favor of dynamic scheduling, we can still see a 1.1X–3.1X area saving using our static-scheduling approach.

Static-Variable minimization. Table 2 shows the number (n) and size (s) of static variables that can be reduced by moving from dynamic to static scheduling and the contribution of the loop rolling. By solely replacing dynamic scheduling with static scheduling, *R2C* can on average decrease n by 14 and s by 89 bits over the three components. In addition, with the aid of loop rolling on I/O interfaces, *R2C* can further reduce n by 55 and s by 585 bits over DCT and QNR. Note that for RLE, all the static variables are reset after we apply static scheduling. Therefore, the combination of static scheduling and loop rolling can indeed effectively minimize the static variables, which also implies fewer implementation costs. We will see in the following experiments, how much area saving *R2C* can actually achieve when these techniques are combined.

Component-Level Exploration of Reed-Solomon. In this set of experiments, we applied *R2C* with loop-rolling on internal logic only, i.e., not on I/O interfaces. Under these constraints, we still see interesting results from exploring the component design spaces with our *R2C* flow. In Fig. 7(a) to 7(e), we plotted the *area vs. effective IO latency* exploration results, where **effective IO latency** is defined as the product of the number of clock cycles and the length of the clock period. For the original RTL designs (represented by red cross points in the figures and all the figures

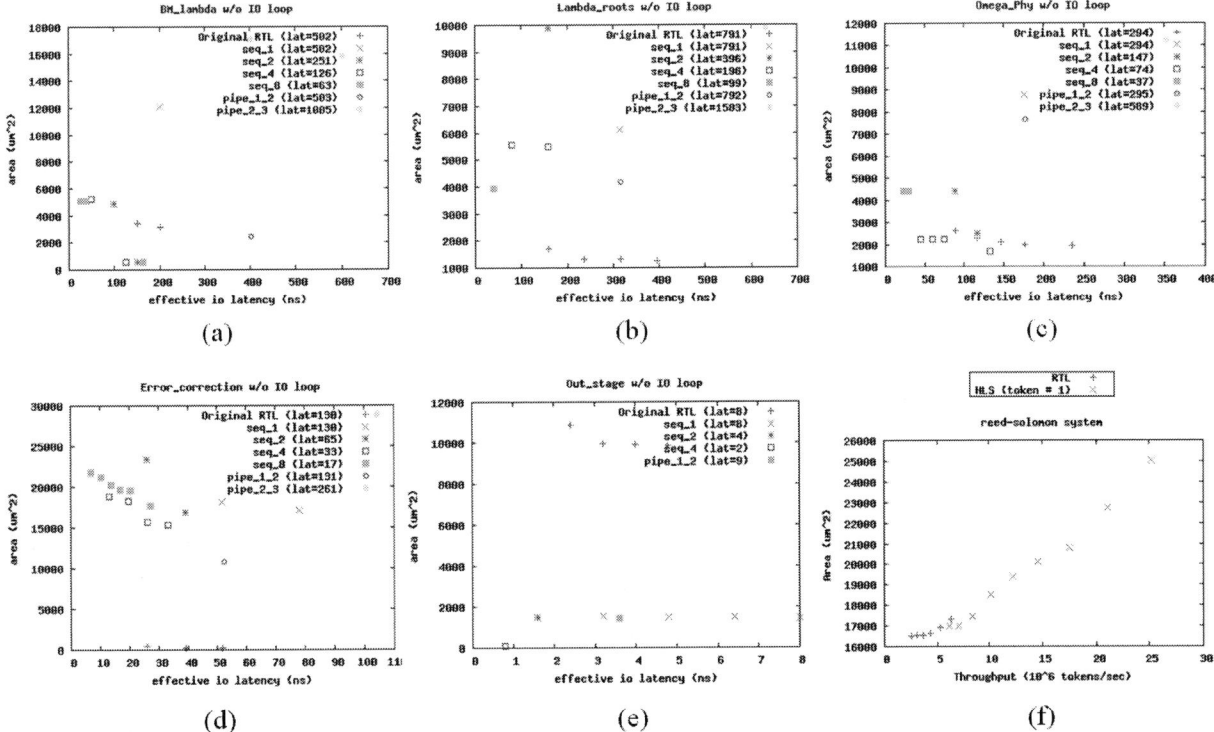

(a)   (b)   (c)

(d)   (e)   (f)

Figure 7: Design space exploration of Reed-Solomon: (a) to (e) show the component-level exploration results of the five components, while (f) shows the system-level Pareto curve of Reed-Solomon.

hereafter), we explored the design space by adjusting the clock period used for logic synthesis. In contrast, our approach could derive various RTLs using different HLS knobs, which changes both clock cycles and clock periods. In particular, for four out of the five Reed-Solomon components (see Fig. 7(a) to 7(d)), the derived RTLs can outperform the original RTLs in terms of effective IO latency, at the cost of increased area. Consider, for instance, BM_lambda (see Fig. 7(a)): the left-most derived RTL has an effective IO latency that is 6.0X shorter than the left-most original RTL, while having only 1.5X more area. Even better, for Out_stage (see Fig. 7(e)), one *R2C*-derived RTL dominates all the original RTLs in both objectives. However, we also observe that for Error_correction (see Fig. 7(d)), R2C cannot derive less area-consuming RTLs by any means. The Reed-Solomon is recovered from sequential-cell-rich RTLs. Therefore, the recovered Error_correction C++ code can still include many state-preserving variables, which require registers and/or memory for implementation. We believe the Error_correction result can be improved by recently proposed memory optimization techniques for HLS, e.g. [13, 6], but this topic is beyond the scope of this paper.

Component-Level Exploration of JPEG. In this set of experiments, we applied *R2C* with loop-rolling on internal logic as well as I/O interfaces, and static-variable reduction on component-communication signals as well as pipeline-stage signals. Since the original RTLs of JPEG are pipelined implementations, we plotted their *area vs. throughput* exploration results in Fig. 8. In particular, Fig. 8(a) to 8(c) show the results *without* loop rolling on I/O interfaces (i.e., the token size per input/output equals one), while Fig. 8(d) to 8(f) show the results *with* loop rolling on the interfaces with a token size equal to 16. In the case *without* I/O loop rolling, for the components DCT and RLE (see Fig. 8(a) and 8(c), respectively), compared with the original RTLs, *R2C* successfully derives area-economic implementations with comparable throughputs. The maximum area savings are around 2.8X–3.0X. For the component QNR (see Fig. 8(b)), *R2C* is limited by few applicable HLS knobs for exploring the design space, but can still recover the RTL with comparable quality in both objectives. Fortunately, this limitation can be

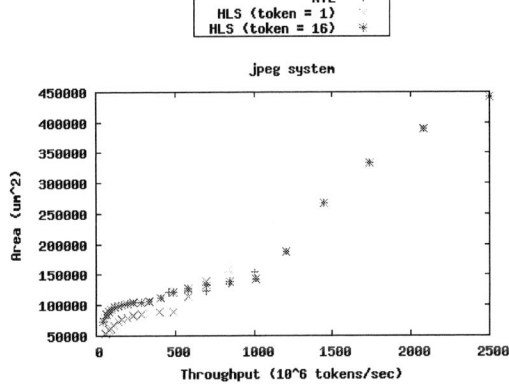

Figure 9: System design space exploration of JPEG.

relaxed in the case *with* I/O loop rolling, i.e., we can have more flexible options for loop transformation during HLS. Therefore, in Fig. 8(e), we observe that *R2C* can then derive QNR with higher throughput by up to 2.9X, at the cost of at most 3.3X area. This capability also applies to the components DCT and RLE (see Fig. 8(d) and 8(f), respectively). The maximum throughput improvements are around 2.5X–5.7X, while the area overheads are around 2.7X–4.2X. In contrast to the Reed-Solomon case, on the JPEG components, we see the full power of *R2C*, which can indeed enhance component reusability by reducing the area cost or increasing the throughput performance.

System-Level Exploration of Reed-Solomon & JPEG. We now examine how those *R2C*-derived RTL components can lead to better system composition. In our cases, this corresponds to better Reed-Solomon and JPEG IPs. We use the system-level design-space-exploration tool [10] to select the optimal RTL components for composing Pareto-optimal systems. The composition results of Reed-Solomon and JPEG are shown in Fig. 7(f) and 9, respectively. For Reed-Solomon (see Fig. 7(f)), *R2C* can improve the system throughput by

1079

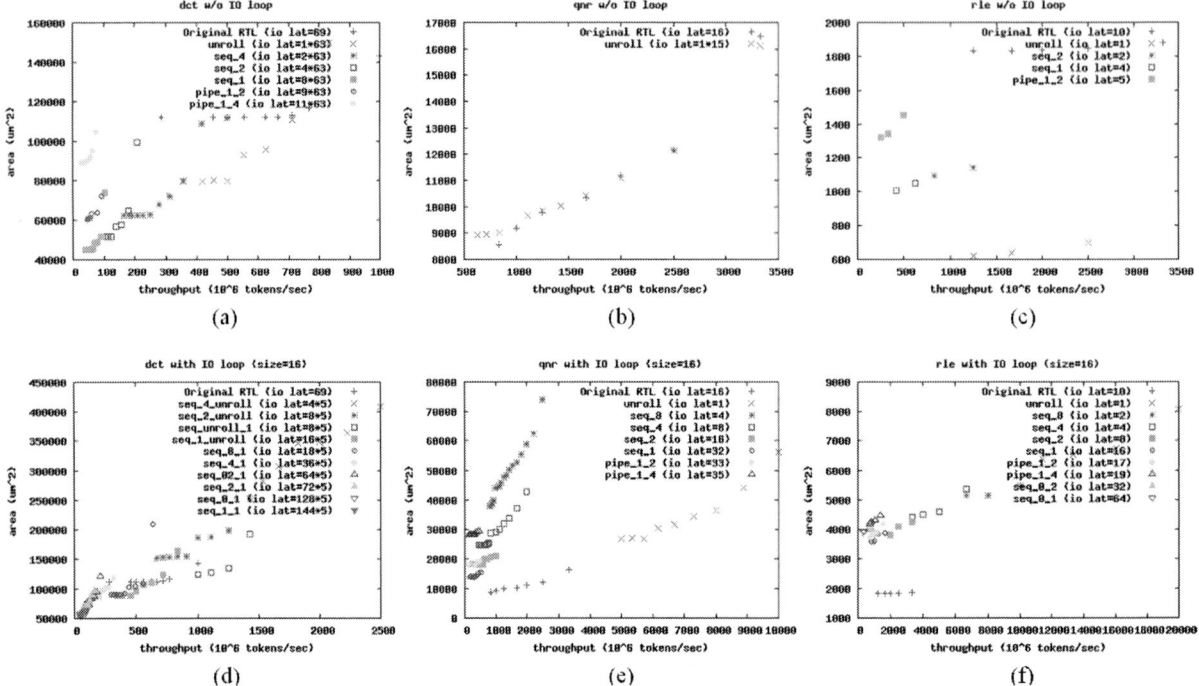

Figure 8: Component design space exploration of JPEG: (a) to (c) without loop rolling on I/O interfaces; (d) to (f) with the loop rolling.

up tp 4.0X, but cannot reduce much the system area, due to the bottleneck component `Error_correction`, for the reasons we have discussed, which happens to dominate the total area of Reed-Solomon. However, for the system throughput of roughly 6-million tokens per second, the *R2C*-derived Reed-Solomon requires less area than the original design. As a result, *R2C* not only enhances the reusability of Reed-Solomon as an IP core, but also re-optimizes the IP design. For JPEG (see Fig. 9), we observe effective improvements on both area (by up to 3.0X) and throughput (by up to 2.5X) over the original RTLs. Since I/O loop-rolling also raises the bandwith of the JPEG I/O interfaces, the JPEG can thus be reused as an IP core in a broader system design context. **Main limitations**. The main limitations of the proposed method, as confirmed by the experimental results, are related to the RTL IP architecture rather than its size. The more complex and with multi-cycle data flow the RTL model, the richer of static variables the corresponding C++ code, with a proportional limitation in area optimization. Even though loop rolling on I/O can reduce the number of such variables, it also involves an increase of the interface size. Thus, RTL IPs with no (or very short) clock cycle latency, with complex cyclic data flows, and with single instances of simple sub-components have less opportunities to find benefits from the proposed method.

4. CONCLUSIONS

We proposed a method to recover the RTL design of IP components and make it suitable for system-level design and high-level synthesis (HLS). Our main contribution is an automatic tool that can generate a synthesizable C++ specification by abstracting away the micro-architectural characteristics of the original design while preserving its functionality. The generated C++ specification is not only a better starting point to explore a richer design space to implement the component in different SoC contexts, but, after completing high-level and logic synthesis, it also typically leads to a more optimized implementation w.r.t. the original one.

While the main motivation of our work is enabling IP recovery and reuse, our tool can also be used for another practical goal: as the interest for HLS is growing, often designers who are used to work at the RTL level are eager to compare the results that can be obtained using the combination of HLS and logic synthesis tools against the implementation of their designs. The problem is that their design were orig-

inally completed at the RTL using Verilog or VHDL. By using *R2C* these designers would be able to jump start this process. In turn, this could speed-up the adoption of high-level synthesis.

5. ACKNOWLEDGMENTS

This work is partially supported by the EU large-scale integrating project SMAC (SMArt systems Co-design, FP7-ICT-2011-7-288827), an ONR Young Investigator Award and the National Science Foundation (Awards: 1018236 and 1219001).

6. REFERENCES

[1] Opencores. www.opencores.org.
[2] Accellera Systems Initiative. http://www.accellera.org.
[3] N. Bombieri, F. Fummi, and G. Pravadelli. Abstraction of RTL IPs into embedded software. In *Proc of ACM IEEE DAC*, pages 24–29, 2010.
[4] Carbon Model Studio. *http car on esi ns stems com* .
[5] W. Cesário et al. Component-based design approach for multicore SoCs. In *Proc of ACM IEEE DAC*, pages 789–794, 2002.
[6] J. Cong, P. Zhang, and Y. Zou. Optimizing memory hierarchy allocation with loop transformations for high-level synthesis. In *Desi n A tomation Conference DAC , 2012 49th ACM EDAC IEEE*, pages 1229–1234, june 2012.
[7] DVM. http://aldec.com.
[8] FreeHDL-V2CC. http://linux.die.net/man/1/freehdl-v2cc.
[9] http://www.itrs.net/Links/2011ITRS/2011Chapters/2011SysDrivers.pdf. *International Technolo Roa map for Semicon ctors - 2011*, 2011.
[10] H.-Y. Liu, M. Petracca, and L. P. Carloni. Compositional system-level design exploration with planning of high-level synthesis. In *Proc of ACM IEEE DATE*, pages 641–646, 2012.
[11] G. Martin and G. Smith. High-level synthesis: Past, present, and future. *IEEE Desi n Test of Comp ters*, 26(4):18–25, Aug. 2009.
[12] Mentor Graphics: Algorithmic C Datatypes. http://www.mentor.com/esl/catapult/algorithmic.
[13] C. Pilato, F. Ferrandi, and D. Sciuto. A design methodology to implement memory accesses in high-level synthesis. In *ACM IEEE CODES+ISSS*, pages 49–58, oct. 2011.
[14] W. Snyder, P. Wasson, and D. Galbi. Verilator - Convert Verilog code to C++/SystemC. *http www veripool or wiki verilator*.
[15] W. Stoye, D. Greaves, N. Richards, and J. Green. Using RTL-to-C++ translation for large SoC concurrent engineering: A case study. *IEEE Electronics S stems an Software*, 1(1):20Ü–25, 2003.

APPENDIX

A. R2C: A DIFFERENT APPROACH FOR GENERATING C++ CODE TARGETING HLS

Differently from all the techniques presented in literature that target C++ code generation for simulation and verification, the proposed method is aimed at the generation of C++ for HLS. The different goal leads to important differences in the approach of generating C++ code. The following sections underline such differences and deepen some C++ generation characteristics of *R2C*. In particular, Section A.1 extends the concepts presented in Section 2.1 and explains why *R2C* generates C++ code that, differently from the related works that preserve *dynamic scheduling* kernels, it resolves the RTL statement scheduling *statically* at compile time to guarantee area optimization. Section A.2 extends Section 2.1 and gives some additional details on the static variable minimization and latency reset. Finally, Section A.3 extends Section 2.2 by detailing the loop rolling process on the internal logic of the model.

A.1 The function scheduling and the function statement order

To preserve the RTL IP semantics, the C++ functions generated from the RTL processes have to be executed in the same *partial* order as the corresponding RTL processes are executed in the RTL model. That is, concurrent processes can execute in a non-deterministic order, while the order between non concurrent processes must be preserved.

To do that, the C++ IP can be implemented in two different ways: by preserving the event-driven model of computation (i.e., *dynamic scheduling* of functions) or by resolving the scheduling (i.e., *static scheduling* of functions).

In dynamic scheduling, the process execution order is known at run time. Processes are woke up if and only if there has been an event to which they are sensitive. Fig. 10(a) recalls, for example, the process communication graph of the RLE component, while Fig. 10(b) shows the corresponding process execution order resolved at run time.

In the C++ model, the functions generated from the RTL processes are thus run as many times and in the same partial order as the RTL processes are run by the hardware description language simulator. To do that, the dynamic scheduling activity is implemented through an event queue, a runnable processes queue and extra scheduling functions (see Fig. 10(c)), which are additional control logic with regard to the IP functionality.

In traditional single core simulation environment, dynamic scheduling outperforms static scheduling. Nevertheless, it takes extra logic to be implemented (compare for example, Fig. 10(c) and Fig. 10(d)) and it would imply more area in the synthesized circuit with no benefits to delay. Since the goal of the proposed method is to obtain better synthesis results rather than simulation performance, *R2C* generates the C++ model with static scheduling.

Finally, for each RTL process, *R2C* maps each sequential statement into a C++ statement of the function. The translation of RTL statements into C++ statements is merely syntactic and the order of statements in the RTL process is preserved in the C++ function.

In general, optimizations on translation from RTL to C++ sequential statements that could provide benefits for simulation performance (e.g., removing temporary variables, merging many RTL statements into single C++ statements, etc.) do not necessarily involve benefits on synthesis results. Rather, these optimizations may provide different results depending on the adopted HLS tool.

A.2 Static variable minimization and latency reset

The *R2C* capability of minimizing static variables and reducing the functional latency during the generation of the C++ code relies on the process organization in the starting M_{RTL}.

Fig. 11 shows how RTL processes and communication be-

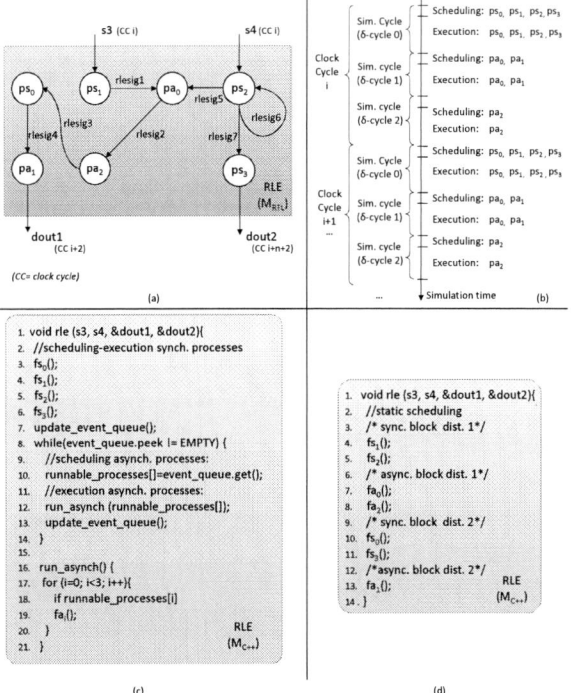

Figure 10: (a) Process communication graph of *RLE*, (b) the corresponding process execution order, (c) the dynamic scheduling of functions in the C++ code, and (d) the static scheduling of functions in the C++ code.

tween them influence both the number of static variables and the functional latency, and how *R2C* minimizes them during the RTL-to-C++ abstraction. Consider, in a M_{RTL} model, two communicating processes implementing functionality f and g, a sequence of values $< v0, v1, v2, .. >$ that are read from an input signal S_{in}, an internal signal C, and an output signal S_{out}.

If the two processes are asynchronous and they implement functionality f and g by means of pure combinational logic (Fig. 11(a)), they do not imply any clock cycle to the latency of M_{RTL} (latency equal to 0 cc). *R2C* generates such a functionality as $g(f(v_i))$ in the abstracted M_{C++}, with no need of static variables for implementing C (functional latency equal to one invocation).

In the case of Fig. 11(b), the synchronous process implies one clock cycle to the M_{RTL} latency (τ represents one clock cycle of latency). That is, C involves a register at gate level once synthesized. In the case of Fig. 11(c1), the two synchronous processes implementing f and g infer two clock cycles to the M_{RTL} latency. In both cases, *R2C* generates the functionality as $g(f(v_i))$ in the abstracted M_{C++}, with no need of static variables for implementing C, since there are no loops in the data flow. The functional latency is thus reset to one invocation.

It is important to note that the functionality implemented by the two processes could be pipelined (Fig. 11(c2)). In this case, the register barrier of the pipeline stage is intrinsically represented by C. Also in this case, *R2C* generates such a functionality as $g(f(v_i))$ in the abstracted M_{C++}. Since there are no feedback loops and C is translated into a non-static variable (which does not infer on the functional latency), the pipelined behavior of the model is abstracted as well.

In general, a path of n synchronous processes infers a latency of n clock cycles. If there are no loops in the data path, the functionality implemented in M_{RTL} by the sequence of processes $ps_1, .., ps_n$ is implemented, in M_{C++}, by one procedural function $fs_n(fs_{n-1}(..fs_1(v_i)))$, which reads the input value, elaborates, and writes the results in one

1081

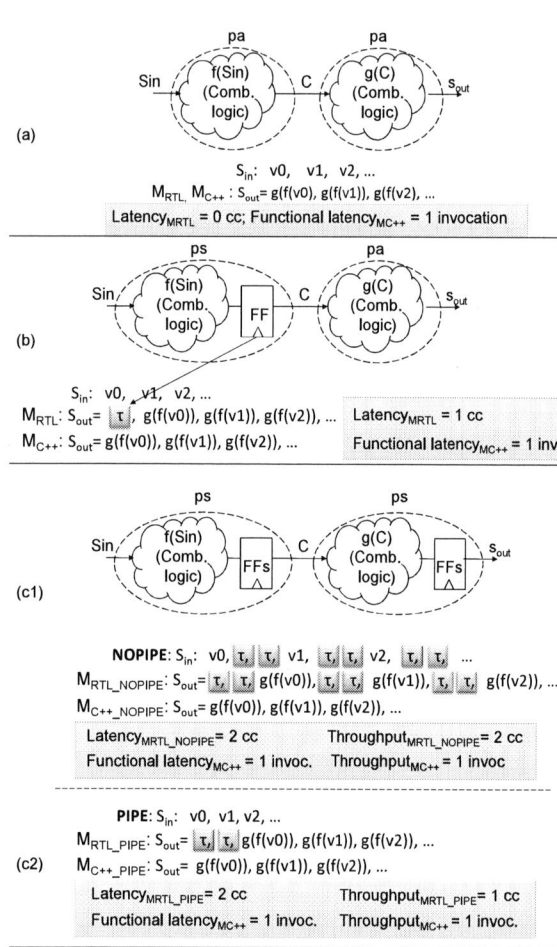

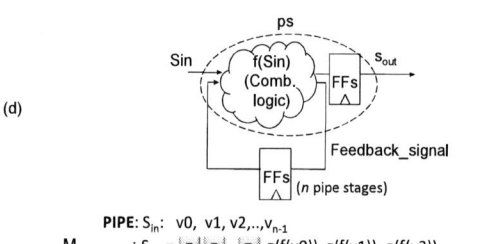

Figure 11: Latency inferred by internal signals for hardware description language process communication and latency reset through RTL-to-C++ abstraction: the latency is reset in (a), (b), and (c) (τ represents one clock cycle of latency). The latency cannot be reset in presence of feedback loops in the datapath (d). In this case, the feedback signal involves static variables in the C++ model.

invocation.

Fig. 11(d) shows an example of a single synchronous process that implements a n stage pipeline. In this case, the register inferred by the signal outcoming the synchronous process is abstracted, while the registers implementing the pipeline stages (register barriers) cannot. This is due to the feedback signal, i.e., the transition over the pipeline stages for computing the result requires one input to be read for each stage. This behavior, in M_{C++}, can be implemented

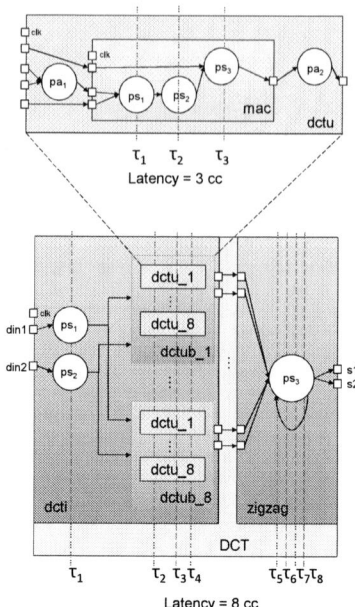

Figure 12: The example of latency minimization of a FDCT IP. The system level FDCT RTL IP has a latency of eight clock cycles. Given the pipelined nature of process ps_3 (in the zig-zag subcomponent), the functional latency of the generated M_{C++} will be ~~five~~ four cycles.

through multiple invocations of the procedural function. As a consequence, $R2C$ preserves the latency of the pipelined M_{RTL} in the generated *pipelined* M_{C++}.

In general, the total latency of a M_{RTL} is given by the number of synchronous processes traversed by the data path. The latency inferred by the internal signals between synchronous processes, as in the examples of Fig. 11(a,b,c) can be directly reset during abstraction. In contrast, the latency is maintained in the case of Fig. 11(d).

Consider, for example, the *pipelined DCT* component of Fig. 12. It is composed of different subcomponents instantiated over four levels of hierarchy. The total latency of the model is eight clock cycles. The four clock cycles of latency inferred by the *mac* and *dcti* subcomponents is reset during abstraction (see Fig. 11(c2)). In contrast, for the characteristic of the pipelined *zigzag* subcomponent (see Fig. 11(d)), the latency of four clock cycles is mapped into a functional latency of four invocations in the generated M_{C++}.

A.3 The loop rolling on internal logic

To maximize the design space of the recovered RTL models, $R2C$ performs *loop rolling* transformations on the internal logic and on I/O interfaces.

Given an RTL component X_{RTL}, which consists of n instances of subcomponent Y_{RTL}[1]:

```
X : for i in 0 to n generate
  Y : Y_component
  -generic map ()
  -port map ();
end generate X;
```

Considering that the syntax of the hardware description language *loop generate* statement is similar to the syntax of the C++ *for loop* statement and that component Y_{RTL} (a

---

[1] For the sake of clarity, we consider subcomponents as loop body. The VHDL and Verilog syntax allows designers to include a set of statements as loop body instance. The proposed mechanism applies in any case, even though the result of loop unrolling for few statements does not lead to remarkable results.

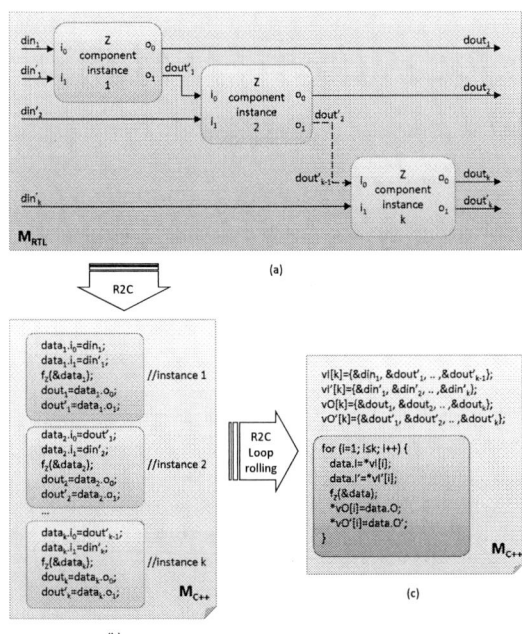

Figure 13: Loop rolling on internal logic

module in Verilog, an entity in VHDL) is translated into a C++ function f_Y, component X_{RTL} is translated into function f_X as follows:

```
f_X (&dataX) {
  for (int i=0; i ≤ n; i++) {
    //function parameter binding
    f_Y (&dataY);
    //function parameter binding
  };
};
```

In contrast, when multiple instances of the same block (i.e., module in Verilog, entity in VHDL) are explicitly instantiated in the RTL code, $R2C$ rolls up the instances into a loop and resolves the binding, as shown in Fig. 13.

Given the M_{RTL} model, each block instance (Z) is firstly translated into a C++ function $f_Z()$ (Fig. 13a, b). Then, considering k instances, a vector of k input/output variables (i.e., $vI_i[k]$, $vO_j[k]$) is generated for each block interface.

A for loop is finally generated by relying on such vectors, which, thanks to the C++ pointers, guarantee the correct I/O variables resolutions for each loop iteration without incurring in extra area during synthesis. The loop rolling on internal logic applies over different hierarchy levels of M_{RTL}, by generating, as a result, nested C++ loops in M_{C++}.

DMR3D: Dynamic Memory Relocation in 3D Multicore Systems

Dean Michael Ancajas   Koushik Chakraborty   Sanghamitra Roy

USU BRIDGE LAB, Electrical and Computer Engineering, Utah State University
dbancajas@gmail.com, {koushik.chakraborty, sanghamitra.roy}@usu.edu

ABSTRACT

Three-dimensional Multicore Systems present unique opportunities for proximity driven data placement in the memory banks. Coupled with distributed memory controllers, a design trend seen in recent systems, we propose a Dynamic Memory Relocator for 3D Multicores (DMR3D) to dynamically migrate physical pages among different memory controllers. Our proposed technique avoids long interconnect delays, and increases the use of vertical interconnect, thereby substantially reducing memory access latency and communication energy. Our techniques show 30% and 25% average performance and communication energy improvement on real world applications.

1. INTRODUCTION

A key factor limiting the system performance stems from the increasing gap between the processor and memory speed. Recent fabrication techniques such as 3D die stacking seek to solve this problem by bringing the memory physically closer to the processor. 3D die-stacking is used to bond multiple wafers in a vertical manner using low-latency and high bandwidth vertical interconnects [5,7], allowing faster communication between the processor and the memory.

Several previous works evaluate the potential benefits of 3D-stacking in system performance. Loh studied aggressive memory configurations that take advantage of a 3D processor-memory setup [12]. Other studies evaluated the benefit of placing memory directly on top of the processor and have reported huge performance speedups [11, 13]. However, many of these studies considered systems with a centralized memory controller (MC). In a traditional 2D system, this setup was suitable because of the limited pin count and the slow off-chip communication interface, which dominates the memory latency. On the contrary, such design limitations are not present in 3D systems because of the abundance of low latency Through Silicon Via (TSV) interconnects. Moreover, with the increasing core count in multicore systems, we can expect MCs to grow in number to accommodate the bandwidth needed. This idea has been recently incorporated in several traditional 2D systems [2,21].

Permission to make digital or hard copies of all or part of this work for personal or classroom use is granted without fee provided that copies are not made or distributed for profit or commercial advantage and that copies bear this notice and the full citation on the first page. To copy otherwise, to republish, to post on servers or to redistribute to lists, requires prior specific permission and/or a fee.
DAC '13, May 29 - June 07 2013, Austin, TX, USA.

In this work, we identify a new problem that arises in 3D systems with multiple MCs: the problem of suboptimal data placement that precludes the performance brought about by 3D die stacking. In a 3D system with distributed MCs, data located on a memory bank directly above the requesting processor will experience a smaller delay as compared to data that needs to traverse an interconnect system. A typical modern on-die interconnect such as HyperTransport [9] still requires 100+ ns in order to send 8 bytes of data. Meanwhile, vertical interconnect delays using TSVs and 3D dram access times are just a small fraction of this at 12ps [13] and 7ns [19], respectively. Therefore, *we can achieve 10x access latency improvement if data is directly accessed through TSVs rather than being transferred using the interconnect.*

In a 3D Multicore, the placement of data in a memory bank and the location of the processing core determines if that data can be accessed through the TSV or if it must traverse the interconnect. However, current system designs are agnostic about this critical consideration, thereby substantially undermining the potential performance and energy efficiency in a 3D Multicore. In this work, we study hardware mechanisms that alleviate the interconnect problem by remapping data to an optimal location in a 3D Multicore System. Our contributions are as follows:

- We show that the data mapping policy of a modern OS can hurt performance of 3D multicore systems with multiple MCs. Our analysis with a detailed full system simulation for real world applications shows 117–280% (average 170%) performance improvement if all non-local accesses are transformed into local accesses as a result of efficient data placement (Section 3).

- We propose DMR3D, a hardware technique that minimizes non-local accesses by identifying critical memory pages and migrating them to a closer memory bank. We present two algorithms for DMR3D: a *Global Scheme* that dynamically allocates migration slots to different threads and a *Thread On-Demand Scheme* that statically allocates equal migration slots to individual threads (Section 4).

- We perform a thorough evaluation of our two DMR3D schemes using a state-of-the-art full system simulator (Section 5). Our best performing scheme increases performance by 7–72% (ave: 30%), increases local accesses by 9–95% (ave: 50%) and improves communication energy by up to 48% (ave: 25%) (Section 6) compared to the baseline. We also compare our schemes with a representative scheme for NUCA caches (Victim Replication [22]), and observe an average of 33% improved

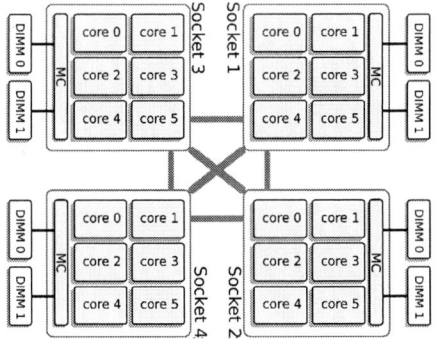

Figure 1: Logical organization of HP G7 series composed of AMD Istanbul processors. The processors communicate through point-to-point Direct Connect Architecture. DIMMs attached to each socket are local memory.

performance.

2. BACKGROUND

In this section, we first describe a modern Non-Uniform Memory Access (NUMA) system. We then introduce our 3D NUMA processor-memory system based on recent work on 3D integration.

Modern server machines typically distribute work to several chip multiprocessors. Because each CMP is composed of multiple aggressive cores, recent designs [2,21] have included a memory controller (MC) on-chip in order to satisfy increasing bandwidth demands. In this setup, the entire memory address space is equally divided among all MCs such that independent requests to different memory sections can be serviced simultaneously. Figure 1 shows an example of this setup. In the figure, processors accessing memory locations residing in a different socket will experience a longer delay compared to the ones which reside in the same socket (also known as local access). Previous studies estimate the non-local access to be 33%-100% longer than the local access [15, 20].

2.1 Baseline System Organization

In our simulation setup, we model a 64-bit processor system with 8GB of total addressable main memory. As depicted in Figure 2, we have a 3D system with the first layer containing all 8 processors along with four memory controllers connected by a ring network interconnect. The second and third layers contain the DRAM cells. Each node in our NUMA system is composed of two processors, one MC and two memory ranks in the topmost two layers.

We recognize three different access types based on source-destination proximity: *local*, *neighbor* and *remote*. Figure 3 shows an example of how different nodes interact with each other. The most basic type of access is when a processor requests data that belongs to the same node (Figure 3a). In Figure 3b, if a processor in node 1 requests data which resides in local memory of node 2, it requires a single hop in the interconnect network. We refer to this request as *neighbor*. Consequently, a processor in node 1 requesting data situated in the local memory of node 3 would now require 2 hops (Figure 3c). We refer to this type of access as *remote*.

2.2 DRAM organization

The 8GB main memory is composed of eight 1 GB ranks. Data bus width is 64-bits. Each rank is constructed using

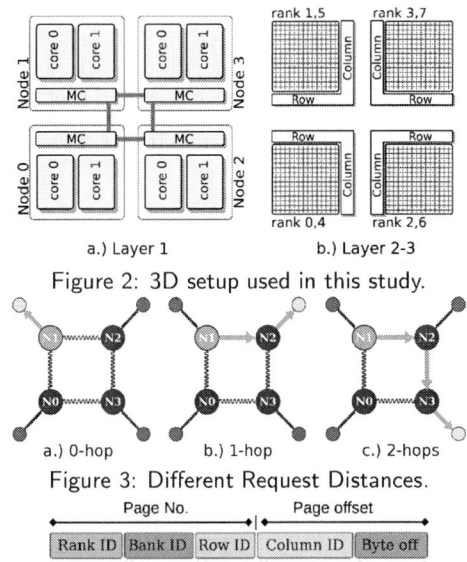

a.) Layer 1   b.) Layer 2-3

Figure 2: 3D setup used in this study.

a.) 0-hop   b.) 1-hop   c.) 2-hops

Figure 3: Different Request Distances.

Figure 4: DRAM Address Mapping Policy.

4 128 Mbit, x16 DDR3 devices. We based our DRAM device model on Micron MT41J128M16 [1]. Timing details are indicated in Table 2 of Section 5. The address mapping policy that we used is shown in Figure 4 and is similar to Intel 845G Memory Controller. With this mapping, all data from a memory page being worked by a particular thread is always located in the same *rank*, *bank* and *row* which allows easier migration of data from one bank to another.

3. MOTIVATION

In this section, we first discuss why current technological trends demand a more efficient mapping of data. Then, we analyze how data from different programs are mapped across the memory banks. Along with the results of our motivational study, we will argue why such intelligent mapping policies are desired.

3.1 Effects of 3D integration on NUMA ratio

In current multicore systems, the NUMA ratio[1] is around 1.5 [15]. With 3D integration, this ratio will further increase because on-chip 3D DRAMs will be naturally faster while interconnect latency in the processor layer will remain the same. To explain this point clearly, we gathered values of different NUMA system parameters from three machines manufactured in the last decade (2004, 2006, 2011). We combine data from the work of [3], [16] and Tezzaron 3D DRAMs [19] to estimate NUMA ratios of future 3D systems. Both [3] and [16] use the same benchmarking methodology so we were able to combine their data with little effort.

To get the main memory latency values, both studies ran a program that strides through an array, the stride size is then increased until all accesses effectively miss the last level of cache. The interconnect latency is then the difference between the remote and local access latency. We consolidate all this data and show it in Figure 5(a). The first two data points from the graph (V1280 and Opteron) are taken from [3]'s work, while the third one (MagnyCours) is taken from [16]'s study. Our estimates for these parameters in a 3D

---

[1]NUMA ratio is the ratio between the remote and local memory latency

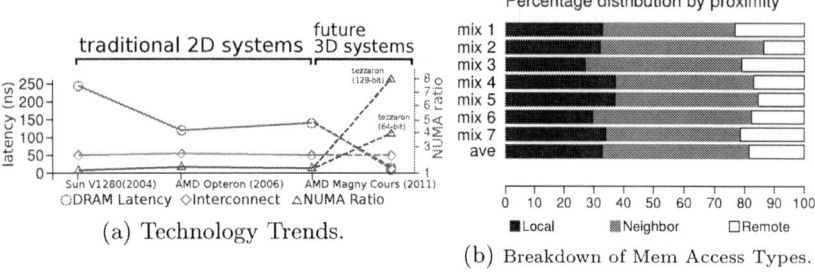
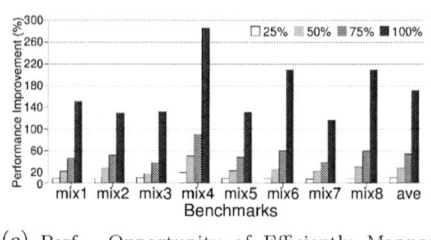

(a) Technology Trends.

(b) Breakdown of Mem Access Types.

(c) Perf. Opportunity of Efficiently Mapped Data.

Figure 5: Technology Trends, Access Profile and Performance Opportunity in 3D Systems.

multicore system are sketched in broken lines.

3.1.1 NUMA ratio

The NUMA ratio can be calculated using two parameters: interconnect and DRAM access latency. Local memory requests only experience DRAM latency while non-local requests include interconnect latency. The increase in NUMA ratio in 3D systems is largely driven by the inability of interconnects to scale well with 3D DRAM latency. We show dramatic leaps in the value of this parameter in Figure 5(a). Going from a 2D system to a 3D, the penalty of a non-local access increases by 700%, or to a ratio of 8. Next, we discuss how trends in interconnects and DRAMs drive this parameter to enormous values.

3.1.2 Interconnect

One notable difference from the interconnect used by the three machines in Fig. 5(a) is that both Opteron and V1280 use **off-chip** communication while MagnyCours uses **on-chip** communication. However, looking at the interconnect delay in Figure 5(a), the values are almost the same. This poor scaling of interconnect delay primarily stems from the long RC delays within a given chip, which dominates most of the communication through the interconnect fabric.

3.1.3 DRAM Latency

3D DRAM modules use TSVs to reduce latency and increase bandwidth. As such, the latency of 3D DRAM modules are substantially smaller compared to that of its 2D counterparts. Prototypes of 3D DRAMs from Tezzaron have access times of 7ns [19]. We used data sheets from Tezzaron to calculate latency to get a single cache line, both for 64 and 128-bit versions of DRAM. These values are shown as two estimations in the DRAM latency curve in Figure 5(a).

3.2 Workload Access Patterns

In the light of the trends shown above, we perform workload characterization of modern benchmarks on our setup (discussed in Section 5) to determine the amount of non-local accesses in each benchmark. Figure 5(b) shows the distribution of memory accesses of programs composed of high memory bandwidth SPEC 2006 programs. From the figure, workload 3 (*mix3*), about 25% of its accesses are local while 75% of its memory accesses incur large interconnect latencies. On an average, for all workloads, about 68% of the memory accesses are to non-local nodes.

3.3 Performance Opportunity in 3D Systems

From these results, we also conducted another experiment to see the performance improvement if fractions of the non-

local requests were transformed to local requests as a result of a more efficient data mapping. For this experiment, we assume the NUMA ratio to be 8. Figure 5(c) shows the result of this study. The four bars represent the performance improvement if fractions of non-local memory requests were routed locally. For instance, if all non-local accesses of *mix4* were routed to the local memory as a result of migration, its performance would improve by 280%. As a whole, there is an average improvement of 10%, 27%, 54% and 171% if we can transform 25%, 50%, 75% and 100% of non-local accesses into local accesses, respectively.

The results of our motivational study show a tremendous opportunity for performance improvement. We now discuss our proposed techniques to realize this opportunity in a system design.

4. DESIGN OVERVIEW

In this section, we discuss the design of *Dynamic Memory Relocator for 3D Multicores* (DMR3D). We first give a brief overview of our design, discuss two schemes of implementing DMR3D, and then explain hardware implementation overheads.

The key principle of DMR3D is identifying and managing data that needs to be swapped to increase the relative percentage of local requests in the whole system. To this end, DMR3D creates an online access profile of data from different memory banks to determine which data needs to be migrated. At certain time intervals (hereafter referred to as *epoch*), a DMA copy is issued to swap data between memory banks. We propose two DMR3D schemes that differ in the way the profiling and migration of data is done. The first scheme (*Global*) relies on a global epoch to synchronize profiling and migration of data, while the second (*Thread On-Demand*) uses a per-thread epoch.

We now discuss the basic working principle of our proposed DMR3D (Section 4.1), two proposed schemes (Sections 4.2 and 4.3), and the corresponding overheads (Section 4.5).

4.1 Hardware Structures in DMR3D

DMR3D is composed of two major hardware structures: the *Profile Table* (PT) and the *Address Remapping Table* (ART). Both of these tables are added to each memory controller. The PT is used to identify access patterns of threads while the ART is used as a layer of indirection for memory pages that were remapped.

To explain DMR3D in detail, we provide an example of the sequence of events that happen during a memory request in Figure 6. In this example, it is assumed that the system

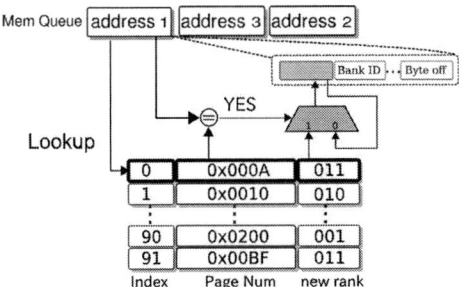

Figure 6: Mechanism of Address Remapping Table.

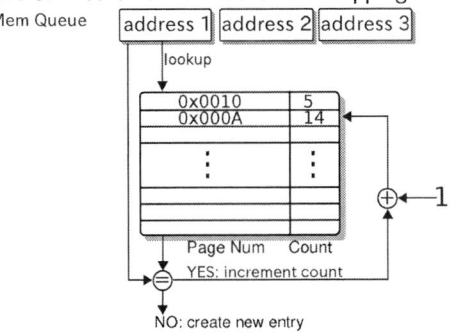

Figure 7: Profile Table Mechanism.

Parameter	Value
MP Size and Freq	8-core, 2Ghz
Re-order Buffer	64 entries
Fetch/Dispatch	4/cycle
Exec/Commit	4/cycle
L1 I-cache	32 KB/4-way, private, 2-cycle
L1 D-cache	48 KB/4-way, private, 2-cycle
L2 cache	256 KB/8-way, shared, 16-cycle.
Cache Line	64 Bytes

Table 1: NUMA Parameters

has already swapped some data in different memory banks. When a memory request is added in the memory queue, its page number is scanned in the ART. If there is a hit, the address in the memory queue is modified to reflect the local location of the data. Otherwise, the request is immediately redirected to a non-local MC (not shown in figure) or stays in the queue if it is local to the node.

There are also cases when an originally local address is now mapped to a non-local bank because of a data migration initiated by a non-local MC. For correctness issue, we handle these cases by storing such mappings in ARTs of MCs involved in the swap.

DMR3D reshuffles data between different nodes by changing certain sections of the memory address. Since the memory *ranks* are distributed uniformly across all nodes, DMR3D relies on this information to construct the new address of memory pages. The exact mechanisms for address translation and migration are discussed in Section S5.

The next key structure in DMR3D is the PT that is used to create an online access profile of memory addresses requested by the running threads. As the whole execution is divided into several *epochs*, DMR3D uses the profile of a current epoch to anticipate memory accesses in the next *epoch* by exploiting the temporal and spatial locality property of programs. Shown in Figure 7 is the mechanism behind the use of the PT. For each request received in the memory controller, the address is analyzed using the PT to get insight on which memory pages are accessed the most by the running programs. The PT contains the page numbers and its access count that are used to guide data migration at the end of every *epoch*.

While data migration is being done, accesses to these remote data will still be serviced by the non-local MC. Once migration finishes, the ART table is then populated with new address mappings which are then used by their respective memory controllers.

4.2 Proposal 1: Global Scheme (GS)

In the Global scheme, the migration of data is done at constant time intervals. The main motivation of this scheme is the fact that different programs running on a multicore present different demands to the memory system. This scheme gives more priority to threads that access the memory more and less to threads that do not, particularly CPU-bound threads. GS improves the performance by devoting more migration opportunities to memory-intensive threads that are severely crippled by the interconnect latency.

At the start of each *epoch*, GS will migrate the top 200 pages servicing non-local requests and migrate them at appropriate locations. In this work, we chose a 1 million cycle *epoch* size and 200-page size migration in each epoch. We also vary these two parameters to see their effects on DMR3D (See Section S6 for our design space exploration).

4.3 Proposal 2: Thread on-demand Scheme (TODS)

While GS focuses on improving the overall system locality by giving more migration opportunities to memory-intensive threads, TODS allocates equal migration opportunities for each running thread in the system. In TODS, the threads strive to maintain a specific ratio (threshold) between their local accesses and non-local accesses. Once this threshold has been crossed, the MC then triggers migration in order to satisfy the threshold requirement. We discuss the motivation behind TODS in more details in Section S2.

4.4 Cache Coherency Issues

During page relocation, coherent data could be moved around and might be corrupted if not synchronized properly. As such, any pending writes to a memory location being migrated are stalled in the remote MC until the data has been completely transferred. The write requests, along with succeeding queued requests are then forwarded to the new MC. Read requests on a memory location being transferred are serviced by the original MC.

4.5 Migration Overhead

The performance increase resulting from the use of DMR3D is based on transferring memory pages to a physically closer bank when they are likely to be referenced again soon, thus avoiding the repeated cost of interconnects. We assume the presence of an overlaid DMA channel responsible for moving data around DRAM banks, similar to [6] and [18].

Other pertinent overheads such as the sizing of PT, ART and the latencies are discussed in Section S4.

5. METHODOLOGY

Our detailed full system simulator is built upon the Windriver Simics [14] platform. Important parameters of our system are shown in Table 1. We modeled a modern multicore system composed of 8 superscalar processors with a

Parameter	Value
Device	Micron MT41J128M16 [1] DDR3
Configuration	1 rank/DIMM, 64-bit channel, 4 devices/DIMM
Clock	2 Ghz
Timings	$t_{CCD} = t_{RCD} = 20$ns
DRAM Capacity	8 GB
NUMA ratio	4 (350 cpu cycles)

Table 2: Memory System

Name	Threads
Mix 1	bwaves-cactus-mcf-gems-lbm-zeusmp-sjeng-leslie
Mix 2	bwaves-bwaves-cactus-cactus-mcf-mcf-gems-gems
Mix 3	leslie-leslie3d-sjeng-sjeng-zeusmp-zeusmp-lbm-lbm
Mix 4	namd-gromacs-leslie-bwaves-cactus-sjeng-mcf-namd
Mix 5	milc-zeusmp-sjeng-sjeng-leslie-leslie-namd-namd
Mix 6	milc-milc-namd-namd-lbm-lbm-mcf-mcf
Mix 7	zeus-zeus-milc-gems-gems-mcf-gromacs-bwaves
Mix 8	namd-namd-gromacs-gromacs-milc-milc-sjeng-sjeng

Table 3: Workload Mix

detailed memory system, DRAMSim2 [17]. All aspects of DRAM device operation are accurately modeled using state machines. Since the main memory is now placed on chip, we run it at a higher clock speed but all timing parameters remain the same. The parameters of our memory system are shown in Table 2.

Our two schemes are evaluated using full system simulation of modern benchmark programs. We combined individual programs from SPEC 2006 suite to construct memory-intensive multiprogram workloads. The threads were chosen according to the profile of [8]. We ran all benchmarks to approximately 1 billion instructions before checkpointing. Each workload has eight threads running on the *reference* input set. The thread composition of the workloads is shown in Table 3.

Before doing detailed simulation for 100 million instructions, our simulator fast forwards the simulation by 25 million instructions to warm-up the caches. Doing so averts bias in our results as streaming accesses to fill-in cold caches could disrupt DMR3D algorithms. Additionally, we scale down cache sizes to optimize simulation time, as done in [18].

For performance evaluation we used the Fair-Speedup (FS) [4] metric as it provides a better measure of the overall system improvement. FS is essentially the harmonic mean of speedups seen in individual threads. More details about the FS metric are discussed in Section S1. We also evaluate communication energy improvement using the McPAT [10] framework.

6. RESULTS

In this section, we present the results from our experiments. We show the increase in local accesses, performance and communication energy improvement of DMR3D against a baseline configuration.

6.1 DMR3D Schemes

We present results for four different schemes: baseline, GS, TODS, and Perfect. Perfect is the same as GS except that there are no restrictions on the number of page migrations per *epoch* and migrations incur no overhead. This is equivalent to an OS that can almost perfectly predict forthcoming memory accesses and bring those data closer to the processor. Our results show that this is enough to establish an upper-bound performance on our schemes. We use an epoch length of 1 million cycles and migrate 200 pages per epoch. We also compare our technique with Victim Replication (VR) [22], which attempts to improve cache performance using a data placement technique similar to DMR3D.

Since the goal of DMR3D is to increase the overall system locality, we first present results on the improvement of the number of local accesses achieved by GS and TODS. Figure 8(a) shows that both GS and TODS can increase memory access locality significantly across a range of benchmarks. Except for *mix 2* and *mix 7*, our schemes achieve at least 60% of the locality compared with the Perfect scheme. *Mix 8* has the most locality increase at 97%, while *mix 6* has the lowest at 14%. *Mix 6* has poor temporal and spatial locality as even an ideal scheme would only obtain a 20% increase. We omit VR in Fig. 8(a) as it does not affect page placement.

Figure 8(b) shows the collective performance improvement at the application level resulting from our proposed techniques. We estimate the performance of these workloads using Fair Speedup (see Section 5). Except on *mix 3* and *mix 1*, GS consistently outperforms TODS. GS performs better because migration slots are appropriately allocated based on global demand of the thread, while in TODS, slots can go unused if a thread does not need a non-local memory page. The best performance for GS and TODS are 71% in *mix 8* and 29% in *mix 4*, respectively. It is worth noting that *mix 8* and *mix 4* are among the four workloads with the most increase in locality. On an average, GS and TODS show 29% and 16% performance improvement in all benchmarks, respectively.

VR's performance improvement ranges from -6% to 5%. Performance degradation happens when some threads are favored at the expense of others. In some benchmarks, the memory access pattern allows VR to speed up some threads. These threads in turn can starve other slow-running threads in the benchmark. The primary difference between VR (and similar schemes) and DMR3D is that VR works at the cache latency level, and is unable to optimize speedy access of memory pages through the vertical interconnect.

6.2 Communication Energy

We show in Figure 8(c) the percentage improvement in communication energy when using the DMR3D schemes. Except for *mix 7*, all schemes show good improvements. The maximum improvement is achieved by GS on *mix 5* at 48%. Benchmark *mix 7* shows a degradation on GS and TODS. This degradation can be attributed to the program behavior changing its access patterns because the Perfect scheme shows a good improvement, increasing its total energy. On an average, we see an improvement of 25%, 17% and 45% for GS, TODS and Perfect, respectively.

Figure 9 shows the breakdown of the energy spent on memory access. The energy consumption was obtained using both the DRAMSim2 and McPat tools. The bars in each cluster represent the schemes evaluated. From left to right, they are the baseline, GS, TODS and Perfect schemes. The interconnect energy, which is used to send data across the network, accounts for more than 50% of the total energy. Note that since we are showing the breakdown in percentages, our schemes show smaller reduction compared with the baseline even though we have lowered the overall energy consumption. Benchmarks *mix 6* and *mix 7* have higher or

1088

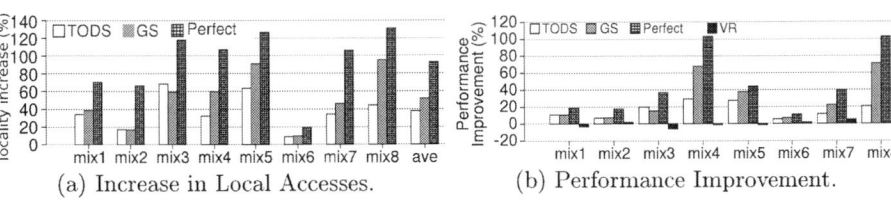
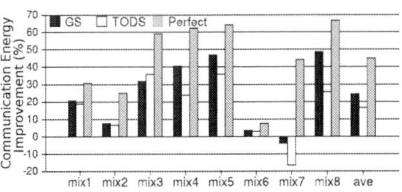

(a) Increase in Local Accesses.

(b) Performance Improvement.

(c) Comm. Energy Improvement.

Figure 8: Locality, Performance and Communication Energy Improvements (Higher is better).

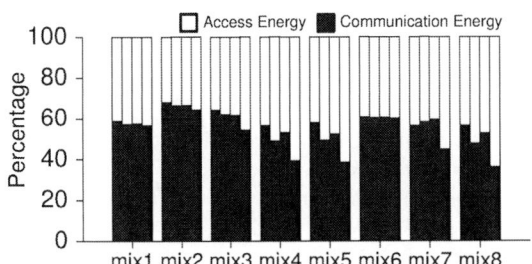

Figure 9: Breakdown of Energy Consumption.

almost equal interconnect energy compared to the baseline. This is consistent with our observation in Figure 8(c), where the unpredictable access patterns from programs may occasionally hurt the performance when using DMR3D schemes.

7. RELATED WORK

3D processor-memory integration has received a lot focus in industry and academia recently. Loi et. al showed that vertically integrating the processor and memory can achieve an impressive 65% performance speedup [13]. Liu's [11] work obtained a 90+% speedup but focused mostly on different cache-memory configurations for 3D architectures. However, they do not consider accesses to the main memory, which can be the source of a lot of traffic. In this context, Loh's work analyzed the performance improvement of aggressive configurations of 3D DRAM systems [12]. None of these works fully explore the context of a 3D NUMA multicore system, where memory latency is significantly smaller than the interconnect latency. We leveraged data from recent literature [3, 16, 19] to accurately model NUMA parameters for a 3D multicore system. The results of our work show that there is a significant performance difference between a system that is aware of data placement and one that is not.

8. CONCLUSION

In this work, we investigate the unique opportunities of exploiting the short vertical interconnect delay instead of long horizontal interconnect delays in a 3D Multicore System. We propose two schemes based on a hardware mechanism to dynamically migrate pages between distributed memory controllers: GS and TODS. GS allocates migration slots based on global demand of threads, while TODS insures fairness by equally distributing migration slots. Our techniques show 30% and 25% average performance and communication energy improvements on real world applications.

Acknowledgments

This work was supported in part by National Science Foundation grants CNS-1117425 and CAREER-1253024, and donation from the Micron Foundation.

9. REFERENCES

[1] Micron Technology Inc. Micron DDR3 SDRAM Part MT41J128M16, 2006.

[2] ABTS, D. AND OTHERS Achieving predictable performance through better memory controller placement in many-core CMPs. In *Proc. of ISCA* (2009), pp. 451–461.

[3] ANTONY, J. AND OTHERS Exploring Thread and Memory Placement on NUMA Architectures: Solaris and Linux, UltraSPARC-FirePlane and Opteron-HyperTransport. In *HIPC* (2006), pp. 338–352.

[4] CHANG, J., AND SOHI, G. S. Cooperative cache partitioning for chip multiprocessors. In *ICS* (2007), pp. 242–252.

[5] DAS, S. AND OTHERS Technology, performance, and computer-aided design of three-dimensional integrated circuits. In *Proc. of ISPD* (2004), pp. 108–115.

[6] DONG, X. AND OTHERS Simple but Effective Heterogeneous Main Memory with On-Chip Memory Controller Support. In *Proceedings of HPCNSA* (2010), SC '10, pp. 1–11.

[7] GUPTA, S. AND OTHERS Techniques for Producing 3D ICs with High-Density Interconnect. VMIC '04, pp. 1–5.

[8] HENNING, J. L. SPEC CPU2006 benchmark descriptions. *SIGARCH Comput. Archit. News 34*, 4 (2006), 1–17.

[9] HOLDEN, B. Latency Comparison Between HyperTransport and PCI-Express In Communications Systems. White paper, HyperTransport Consortium Technical Group, November 2006. Available online (11 pages).

[10] LI, S. AND OTHERS McPAT: An integrated power, area, and timing modeling framework for multicore and manycore architectures. In *Proc. of MICRO* (2009), pp. 469 –480.

[11] LIU, C. AND OTHERS Bridging the processor-memory performance gap with 3D IC technology. *Design Test of Computers, IEEE 22*, 6 (nov.-dec. 2005), 556 – 564.

[12] LOH, G. H. 3D-Stacked Memory Architectures for Multi-core Processors. In *Proc. of ISCA* (2008), pp. 453–464.

[13] LOI, G. L. AND OTHERS A thermally-aware performance analysis of vertically integrated (3-D) processor-memory hierarchy. In *Proc. of DAC* (2006), pp. 991–996.

[14] MAGNUSSON, P. S. AND OTHERS Simics: A Full System Simulation Platform. *IEEE Computer 35*, 2 (Feb 2002), 50–58.

[15] MARATHE, J., AND MUELLER, F. Hardware profile-guided automatic page placement for ccNUMA systems. In *PPOPP* (2006), pp. 90–99.

[16] MCCORMICK, P. AND OTHERS Empirical Memory-Access Cost Models in Multicore NUMA Architectures. In *ICPP* (September 2011).

[17] ROSENFELD, P. AND OTHERS DRAMSim2: A Cycle Accurate Memory System Simulator. *Computer Architecture Letters* (jan. 2011), 16–19.

[18] SUDAN, K. AND OTHERS Micro-pages: increasing DRAM efficiency with locality-aware data placement. *SIGPLAN Not. 45* (March 2010), 219–230.

[19] TEZZARON. FaStackÂŏ Creates 3D Integrated Circuits @ONLINE, May 2011.

[20] TIKIR, M. M., AND HOLLINGSWORTH, J. K. Using Hardware Counters to Automatically Improve Memory Performance. In *SC* (2004), IEEE Computer Society, p. 46.

[21] WENTZLAFF, D. AND OTHERS On-Chip Interconnection Architecture of the Tile Processor. *Proc. of MICRO* (2007), 15–31.

[22] ZHANG, M., AND ASANOVIC, K. Victim Replication: Maximizing Capacity while Hiding Wire Delay in Tiled Chip Multiprocessors. In *Proc. of ISCA* (2005), pp. 336–345.

Supplemental Materials

S1. FAIR SPEEDUP

To measure the performance in our multiprogrammed work-load, we use a metric called *Fair Speedup* (FS) [SR1]. FS is defined as the harmonic mean of per-thread Speedups. We use FS as it follows the Pareto efficiency principle where improvements on the system should increase the performance of all running threads.

$$FS(scheme) = \frac{n}{\sum \frac{IPC_i(base)}{IPC_i(scheme)}}$$

Additionally, FS is a fair metric because the harmonic mean of speedup rewards uniform improvements across all threads while at the same time penalizing slow downs. Such fairness is not exhibited by other metrics such as speedup of aggregate IPCs (Agg_{ipc}).

We give an example below to illustrate this point clearly. In this example, there are 4 threads running with IPC values indicated in Table 4. Suppose we introduce a scheme that degrades the performance of the first three threads by 10% while increasing the last one by 100%.

Setup	IPC1	IPC2	IPC3	IPC4	Agg_{ipc}	FS
Baseline	1	2	3	4	-	-
NewScheme	0.9	1.8	2.7	8	**1.17**	**1.04**

Table 4: Sample IPC values

Both performance improvements from Agg_{ipc} and FS are shown in the table (last two columns). Agg_{ipc} reports this as an impressive 17% performance increase while FS will report this at a modest 4% improvement. We believe FS gives a more accurate picture of the actual improvement because outlier values cannot heavily influence the final speedup as compared to other metrics based on arithmetic averages.

S2. PROFILING RESULTS

We present results for profiling of pages in Figure 10. We show both the average and the maximum unique pages accessed in each *epoch*, where the trend in the maximum value is 10-15$\times$ larger than the average value. Although the Profile Table needs to accommodate the maximum number of pages, on an average, it only uses a small fraction of its entire capacity. This translates to a low overhead hardware logic required to search for an entry in the table.

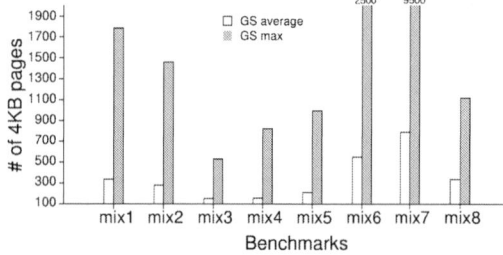

Figure 10: Page reach of benchmarks using the Global scheme. Both the average and maximum unique pages per epoch are shown.

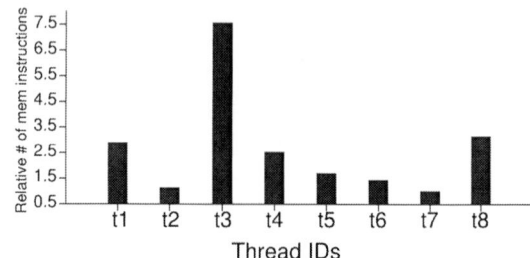

Figure 11: Breakdown of # of mem instructions executed for w1. Values relative to thread 7.

S2.1 TODS Rationale

TODS is largely motivated by the natural imbalance of the thread execution speed present in different applications. We show an example of this execution speed asymmetry in Figure 11, where thread 3 has executed 7.5$\times$ more memory instructions than thread 7. Other threads also exhibit varying number of memory instructions executed. With such diverse demands on the memory, thread fairness can suffer if less memory-demanding threads are not given migration slots by DMR3D.

S3. SIZING PT AND ART

To estimate the practical size of PT, we obtained profiles on how many unique pages are accessed by different threads in each *epoch*. We show these data in Figure 10, where on an average, most threads access less than 700+ unique pages in a 1 million cycle *epoch* size. There are also outlier workloads such as *w7*, which access at most 9500 unique pages in 1 epoch. To accommodate all workloads, we size the PT to be 10000 entries.

For the sizing of the ART, there are two factors to be considered. First, the table must be as minimum as possible to get less latency. Second, the size chosen must be enough to hold address mappings of frequently accessed data, in order to justify the cost of migration. We have experimented many parameters and found that for most benchmarks, only 200 pages need to be migrated in each new epoch to saturate the performance of DMR3D. We explore the design space of DMR3D to arrive at this choice (see Section S6).

S4. LATENCIES OF PT AND ART

Since the PT and ART are accessed for every memory request, the latency to query these tables must not add significant overhead to the overall memory access time. Using CACTI 6.0 [SR2] at 45nm, we find that associative lookups of these tables take 0.52ns and 0.30ns for PT and ART, respectively. These overheads are negligible considering that the average memory access times can reach 100ns.

S4.1 Memory Thrashing

In the context of DMR3D, thrashing can occur if data are constantly being swapped between two same memory banks such that the performance improvement is nullified by the overhead of constantly migrating data. This is usually caused by threads that share memory pages. However, modern parallel programs are designed to share data minimally in order to harness multicore performance. Since thrashing can degrade performance, we profile our benchmarks and find that sharing of pages occurs less than 0.45% of the time.

S5. PAGE REMAPPING TECHNIQUE

In a system with multiple MCs, the whole address space is divided into smaller contiguous address spaces such that each memory controller services independent requests to different memory sections simultaneously. Figure 12 shows the distribution of the memory address space across four MCs. Memory rank 0 is managed by MC0, rank 1 by MC1 and so forth. Note that we ignore other fields (Bank, Column Id etc.) in the address to simplify our discussion. The two MSBs in the address determine which MC manages its data. As such, for DMR3D to migrate and move around memory pages across memory controllers, it only needs to keep track of the previous and new rank locations of the data.

We next show the sequence of events during the migration of memory pages. The first step in migrating is to determine the source MC where the page belongs. This is taken as the two MSB in the address. The second step is to obtain the new page number by replacing the rank field with the one from the destination MC. All this information is then added to the ART of the destination MC after data migration has been done.

Once a page is listed for migration (i.e. *alien* page), the original data (i.e. *victim* page) residing at the new location must be evicted and moved to the previous address of the *alien* page, effectively swapping the two pages. The *victim* page undergoes the same process as the *alien* page but with opposite source-destination locations. Requests for the *victim* page arriving at the original MC will now be detected by ART and then routed to the remote MC. Accordingly, request for *victim* page from the remote MC will be detected by its own ART and treated as a local request.

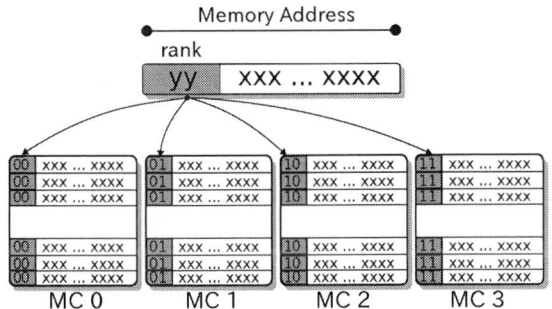

Figure 12: Memory Mapping across different MCs.

S6. DESIGN SPACE EXPLORATION

We performed two experiments: varying the number of pages migrated per epoch and changing the epoch length. To simplify our discussion, we use the term *Migration Slot Size* (MSS) to refer to the total number of pages migrated per epoch. The MSS values are chosen as 200, 400 and 800. These are taken from the range of values in our page reach profile experiment (Fig. 10). The *epoch* lengths are then chosen such that the migration overhead for the page is around 10% of the whole *epoch*.

S6.1 Increasing Epoch Lengths

Figures 14 and 15 show the results for the design exploration of GS and TODS. For an MSS of 200 and 400, increasing the epoch length to 2M and 4M slightly decreases the

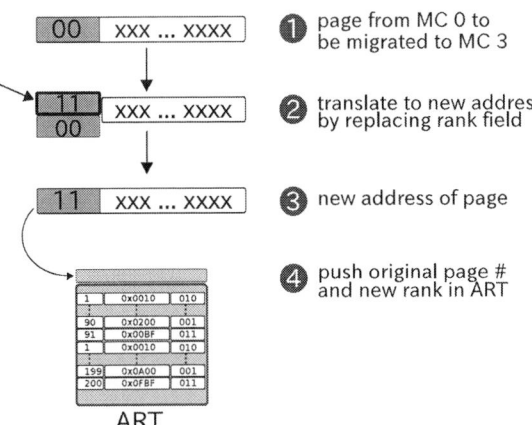

Figure 13: Sequence of events for Address Translation.

performance by 0.8% and 2.5%, respectively. This degradation trend is also exhibited on TODS but with smaller values (0.2 and 0.5%) on all MSS. The decrease in performance can be caused by two things. First, the increase in *epoch* length could increase the page reach of the program and can cause interference on the profiles generated by the PT. This can do more harm than good. Second, due to the dynamic nature of programs, the profile used to predict accesses to locations might only be useful at a fraction of the epoch length. As such, when the MSS of GS is 800, there is now an average of 4% improvement because it can handle the large page reach of threads. Hence, increasing epoch lengths does not automatically mean improved performance as it has to be matched with a much larger MSS to accommodate the corresponding large page reach of running programs.

S6.2 Increasing Migration Slot Size

The same set of figures (Figures 14 and 15) show the results for increasing the MSS. We first discuss its effect on GS and then on TODS. For *epoch* lengths of 1M and 2M, using an MSS of 400 instead of 200, increases the performance by 4.25% on an average. For a 4M *epoch* length, the performance improvement drops to 3%. However, further increasing the MSS to 800 pages hurts the performance because of the migration overhead. The performance drops for 1M and 2M *epoch* lengths are 6% and 0.35%, respectively. The former has more degradation because a shorter *epoch* length translates to less chances of using migrated data before another *epoch* is reached. To overcome the migration overhead, we can use a 4M *epoch* length that increases performance by 1.2%.

While GS experiences degradation on some configurations, TODS consistently increases its performance with an increase in MSS. On an average, it improves by 3% going from an MSS of 200 to 400-pages and 400 to 800-pages. This improvement is the same for all *epoch* lengths. Although the MSS for TODS is increased, each thread will use only the migration slots if its ratio of local to non-local accesses fall below the threshold. As such, the migration overhead is reduced to threads that need to migrate and performance is realized.

S6.3 DMR3D Design Choice

From the results of our design space exploration, we chose MSS to be 200-pages and epoch length to be 1 Million cy-

1091

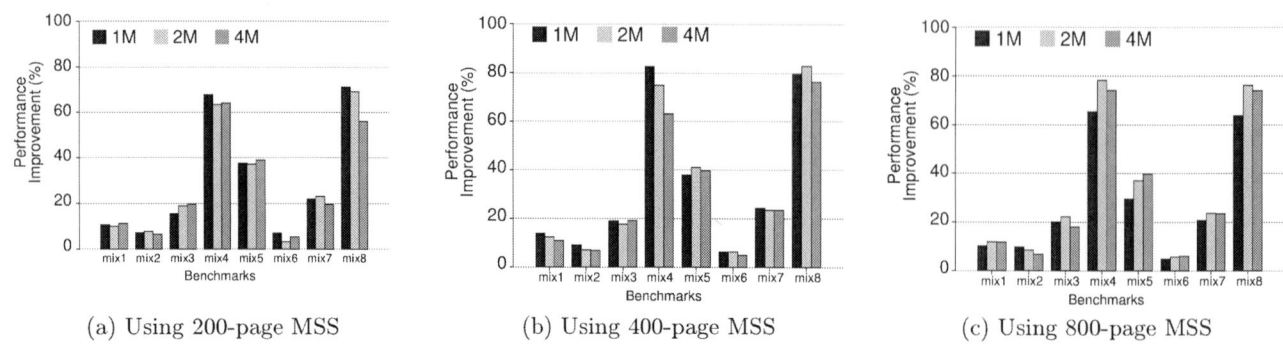

(a) Using 200-page MSS

(b) Using 400-page MSS

(c) Using 800-page MSS

Figure 14: Varying Epoch Sizes for Global Scheme under different Migration Slot Sizes

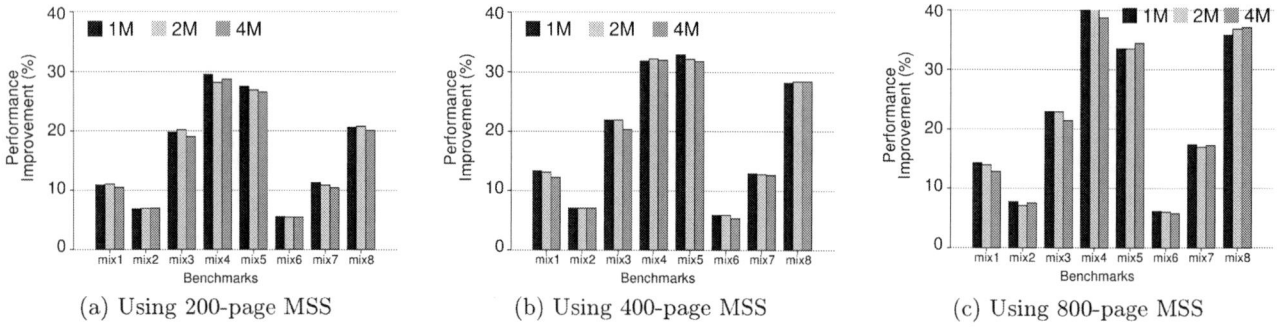

(a) Using 200-page MSS

(b) Using 400-page MSS

(c) Using 800-page MSS

Figure 15: Varying Epoch Sizes for TODS under different Migration Slot Sizes

cles. Although there is at most 5% average performance improvement using larger MSS, we cannot justify its hardware overhead on the ART size. For instance, using an MSS of 800 can improve the GS by 5% but needs an ART that is 300% larger. Furthermore, using a larger ART could lead to longer latency in table lookup. Thus, the best configuration for DMR3D is one that has fewer MSS/ART size.

S7. REFERENCES

[SR1] CHANG, J., AND SOHI, G. S. Cooperative cache partitioning for chip multiprocessors. In *International Conference on Supercomputing* (2007).

[SR2] MURALIMANOHAR AND OTHERS CACTI 6.0: A tool to model large caches. University of Utah, School of Computing, Technical Report (2007).

Power Gating Applied to MP-SoCs for Standby-Mode Power Management

David Flynn
Research & Development
ARM Ltd
Cambridge, UK
+44 1223 400438
david.flynn@arm.com

ABSTRACT

Complex SoCs from servers to intelligent sensors are increasingly built up from heterogeneous IP cores and subsystems. Accelerator blocks or additional processor cores support both general purpose and graphics optimized processing in mobile SoCs, but the number of cores that may be simultaneously active is typically restricted for both battery life and thermal package limits. Power gating is the primary approach to cutting the leakage power for inactive blocks, while state retention and standby voltage scaling can be valuable enhancements for improving energy and latency costs for such leakage mitigation schemes. This paper describes work on techniques that look promising to build on current multi-voltage EDA tools and power intent, without the costs of resorting to full-custom design techniques.

Categories and Subject Descriptors

B.5.1 [**Register Transfer Level Implementation**] – *design*.
B.6.1 [**Logic Design**]: Design Styles – *sequential circuits*.
B.7.1 [**Hardware**]: Integrated Circuits – *types and design styles*.
C.3 [**Special-purpose and Application-based systems**].

General Terms

Design, Performance, Experimentation.

Keywords

Central Processor Unit (CPU), Dynamic Voltage and Frequency Scaling (DVFS), Electronic Design Automation (EDA), energy-efficiency, Implementation IP (IIP), Intellectual Property (IP), IP-deployment, Logical IP (LIP), low-power, Multi-Threshold CMOS (MTCMOS), Multi-Voltage (MV), Physical IP (PIP), Power-Gating (PG), power intent, standard-cell, State-Retention (SR), System-on-Chip (SoC).

1. INTRODUCTION

As an IP provider, ARM faces the challenge of ensuring licensees and customers can address low power product design and implementation in consumer markets where responsive performance must be traded off against battery life constraints for active and standby power [10], while using standard EDA tools and design-flows. In the market areas where ARM specializes, the

Permission to make digital or hard copies of all or part of this work for personal or classroom use is granted without fee provided that copies are not made or distributed for profit or commercial advantage and that copies bear this notice and the full citation on the first page. To copy otherwise, to republish, to post on servers or to redistribute to lists, requires prior specific permission and/or a fee. DAC '13, May 29 - June 07 2013, Austin, TX, USA.

success metrics for end products are qualified in terms of time-to-market, volume manufacturability, and end-user active and standby experience on a variety of advanced process technologies. The power management required increasingly involves the intervention of device drivers and the operating system software layers that support the user interface and applications software [6].

Academic research on mitigating dynamic and static power, such as Dynamic Voltage and Frequency Scaling and Power-Gating respectively is well understood and exploited in full-custom and 'expert' design flows for standard products and integrated circuits [9]. However, enabling such techniques to be deployed successfully by SoC design teams that need to rely on standard-cell design flows and industry standard test and sign-off tools has required significant work engaging with EDA companies and standards committees [8].

Addressing high performance requirements has driven industry to move from single-processor designs, with straightforward programming and operating systems models, to more complex multi-core and multi-processor system designs that can meet the power density limits of SoC technologies and the thermal limits of low-cost consumer applications in low-cost packages. However technology scaling challenges are now limiting the multi-core approaches and the number of cores that can be active at a time, leading to the advent of 'dark-silicon' [1][2][3]. This paper addresses standby power modes and on-chip sub-system power switching required to optimize the battery life of multi-core SoC designs.

2. POWER GATING TODAY

Standard design flows for SoC design now include Multi-Voltage enhancements to conventional RTL and timing constraints.

The two industry standards for describing 'Power Intent' annotation of synthesizable RTL designs, Common Power Format (CPF) and Unified Power Format (UPF) are converging under the work of the IEEE P1801 standardization body [12].

Power Intent covers four main areas in relation to power gating:

- Associating voltage domains with the specific parts of the design hierarchy

- Associating power switches with switched domains and inferring control ports

- Associating isolation strategies at boundaries of switched domains for defined interface signaling, and inferring clamp control ports

- Defining legal power states for the hierarchical design

The UPF and CPF convergence work is driving better levels of abstraction of such power intent to support building up the power intent constraints by successive refinement. This will address separating out the technology independent constraints and descriptions of the multiple IP components and the design-specific technologies and power management control systems required at integration.

The designer and integrator is responsible for the design of the power management state machines and controllers that must connect up and drive the inferred power gating and isolation controls through the defined legal power states, and the UPF/CPF aware multi-voltage simulation and formal verification checks to validate the legal power state transitions and coverage. And they are also required to address the Design-for-Testability complexities introduced by the power gating and clamp circuitry inferred.

The functional inference of power gating in the MV extensions to EDA tools should be well suited to multi-core and fine-grain subsystem power gating on chip. However the implementation and verification challenges are however still tough and a long way from being as transparent to the designer as clock gating has become with current synthesis, verification and test tools.

The power gates - PMOS 'Header' or NMOS 'Footer' switches are technology dependent and are not perfect switches. The on-resistance is not zero, the off-resistance is not infinite so the "I_{ON}:I_{OFF}" ratio is the primary metric to optimize. The resultant IR drop in series with a high-performance core in a multi-core SoC causes some degradation in the F_{MAX} operating frequency, addressed by adding more and more switches in parallel, but the off-leakage currents grow in proportion to the number of power gates; over designing for 'On' resistance impacts the 'Off' leakage.

The SoC integrator is also responsible for managing the surge of inrush currents when turning power-gated blocks on. Typically a dual- or multi-stage power gating approach is used where a small subset of power gates is turned on to provided resistive charging of the bulk capacitance of a power-gated region. This limits the dI/dt impact - which otherwise results in ground or supply rail 'bounce' which will impact the timing of other functional sub-systems sharing the same power grid. Sizing the weak switch networks is often a complex challenge given the Process/Voltage/Temperature extreme corners.

The analysis of dynamic IR drop across power-gated subsystems is well supported with EDA tools, but does require the additional analysis of the power-grids before and after the switches. Optimizing decoupling capacitance insertion is still a challenge in addition to the standard challenge of developing representative dynamic activity profiles for the analysis.

The implementer typically chooses one of two topologies for the power gating. *Distributed coarse-grain* power gates are pre-placed in columns, rows, or checker-board styles, that locally switch the standard cell power rails. The transistor wells need to remain powered to ensure that the power gating control buffering and isolation networks are biased correctly. The advantage of distributed power gating is that the impedance of the switches, plus the via stacks to connect the power gate transistors to the top-level thick metal low resistance power and ground grids is minimized, but a local power switched power grid is implemented in mid-layer metal to provide current sharing across the switches. The power gates can appear as single- or multi-height standard cells but that drive the standard cell power or ground rail from

switches that have grid power delivered to intermediate cell tracks.

The alternative approach is that of *ring* or *periphery* switches. In this case the switches can be co-located in strips around the core to be power gated, so different gate oxide transistors or custom height/width switch cells can be optimized for the technology independent of the standard cell track height. In this case the wells can be power gated with the power gated block, but the downside is that the switched-rail IR drop can be harder to close than in the distributed switch case. As well as the top-level thick metal power grids that typically are shared across the SoC, another high-current switched grid must be routed to provide low-impedance supply between the switches at the periphery and delivered to the conventionally routed standard cell region.

3. STATE RETENTION TODAY

It is power gating that cuts the leakage power when a core or subsystem is in standby or stopped mode, but State Retention with Power gating, SRPG, addresses the wake up latency and energy cost of restarting after power gating.

Simplistically one can in theory provide hardware schemes to reuse manufacturing scan chains to shift state out to an area of memory and back, but in practice this is a significant implementation challenge and the simultaneous switching power has to be addressed by throttling the clock so this is not an easy approach to integrate at system level.

Baseline power gating cuts the supply to combinatorial logic and registers alike. On re-powering the state values are unknown and reset must be asserted by the power control state sequencer. For some simple DSP or data-flow engines this may be fine - no persistent data needs to be held. Where CPUs are involved there may be cases where the architectural state may have to be saved in software before power gating and restored after re-powering (and resetting); GPUs may have no need to maintain transitory frame data but typically contain certain configuration state that is expensive to re-program or re-load repeatedly.

EDA tools support hardware state retention inference through the UPF and CPF power intent standards where this is mapped onto a special form of register. The retention register is augmented with control signal port(s) that can control saving and restoring state to some internal form of low leakage "balloon" latch. This latch and control must be independently supplied with backup power and isolated from the rest of the flip flop and clock, reset and data ports when the main logic is power gated. There are many forms of state retention register standard cell designs; the tools simply expect (non-power-gated) save and restore control nets and save restore conditions as well as a retention latch supply.

Retention registers incur complexities in relation to clock sensitivity - master/slave designs have to be designed defensively to allow the register state to be saved independent of whether the clock is stopped in the high or low phase. More complications arise with the latches in integrated clock gates; if the clock in a rising-edge design is stopped in the high phase the latches inferred by the clock gating tools have to be mapped to retention latches in order to maintain clock enable terms while power gated.

The cost functions for implementing with state retention in terms of additional area per retention register or latch and the associated performance degradation mean that such SRPG approaches are conventionally only applied to small micro-controllers that benefit from near-instant wake-up for minimizing interrupt response latencies. Application to large IP cores is usually prohibitive in terms of overhead, over and above power gating.

Power Intent specific to state retention power gating supports:

- Associating retention domains with the specific parts of the design hierarchy (or even named sequential processes)

- Associating backup supplies with retention domains and inferring retention control ports

The design challenge to IP providers is to validate selective or partial retention schemes; designing blocks to come out or reset cleanly without having the overhead of resetting every register in the design is often enough of a challenge. There are no verification tools currently that help reason about proving that the state-space explosion or arbitrary retained state with other re-initialized state will never break clock enable/gating terms etc. when implementation tools treat the entire RTL hierarchy as a whole when synthesizing and optimizing designs [7].

So the simplest verifiable approach is often to treat IP subsystems or hierarchies as full-state retention, with the associated area and performance penalty.

4. ENHANCED POWER GATING

One promising approach to exploiting the current power gating inference and tool chains is to add more advanced techniques to the power gating. Providing the same abstraction is presented to the multi-voltage tools it is possible to take gate-bias approaches that are well understood in full-custom chip design and add MVCMOS [11] techniques such as Super Cut-off to the baseline power gates. In the case of PMOS header switches this exploits the over-driving of the gate of the power-switch transistors by a technology-dependent voltage - typically 50 to 100 mV above the primary VDD power rail that is being gated. The Gate Bias supply requires minimal current capacity but enables a stronger ratio of turn-off leakage to on-current. This can mean that more aggressive IR-reduction strategies can be adopted while maintaining the same level of off current. Alternatively the super cut-off mode can be exploited to achieve better I_{OFF} behavior than the baseline PG implementation.

Figure 1 illustrates an example of such a SCCMOS header switch. A very low cost, limited range level shifting function is integrated in the power gate control buffering such that the MV tools 'think' they are working with standard logic-level-drive power gates but with the simple provision of a weak VDDGB supply the gate bias can be abstracted away inside the cell. Depending on whether such power gates are deployed as periphery ring switches or distributed stand cell switching the device wells have to be handled carefully - as 'hot' wells within standard cell designs incur design rules with significant spacing rules that impact area.

The case of header-switch super cutoff is shown for ease of deployment - typically there is a higher I/O VDD supply voltage rail that can be chopped down efficiently to provide the low current voltage offset above the core VDD rail to be power gated.

Conventional tools expect to target one switched rail within a region - typically rectilinear for floor planning of voltage domains.

However it becomes possible to work with a hybrid approach where a distributed power gated primary domain can be augmented with a secondary power gated domain controlled by ring switches around the main power-gated region.

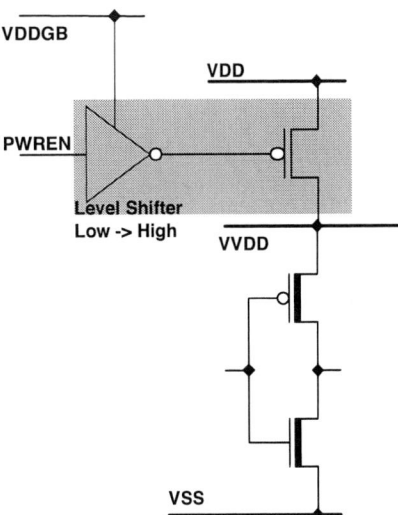

Figure 1: SCCMOS Power Gating

5. ENHANCED STATE RETENTION

A promising experimental approach adopted has been to amortize the cost of state retention across multiple registers by splitting the power rails for high performance flip-flops at near-zero area cost. The retention control cost is amortized by managing the clamping of clocks and resets efficiently in the SoC implementation flow such that the speed and area impacts are minimized over and above the cost of Power Gating that designers well understand.

Figure 2 illustrates how the retention power domain is distributed to manage "live-slave" state retention between clock-gates and registers. For registers with asynchronous set or reset functionality, such controls must also be explicitly clamped similarly [5].

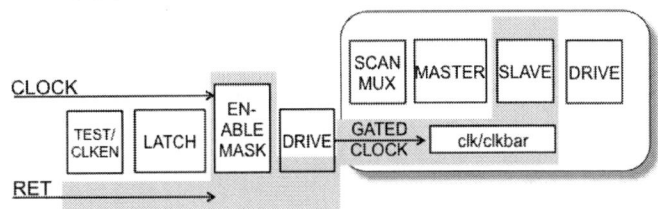

Figure 2: Enhanced SRPG control partitioning

For short-term SRPG retention the slave latches and associated clamping domains must be kept powered. For deep sleep this domain (shown with gray overlay) is power-gated off as well (state lost PG, typically requiring software state save and restore).

Voltage scaling of the state retention rail is attractive to provide an extended SRPG mode of operation, but simple techniques such as adding a V_{th}-drop that was safe at higher-voltage process nodes do not provide sufficiently safe state-integrity margin for latch structures on sub 90nm technologies with higher inter-device variation on latch feedback structures. Figure 3 shows the addition of a Boosted-Gate "drowsy" retention to the buffered SCCMOS power-gate of Figure 1 where the raised-voltage gate bias supply provides additional headroom to the scaled retention voltage.

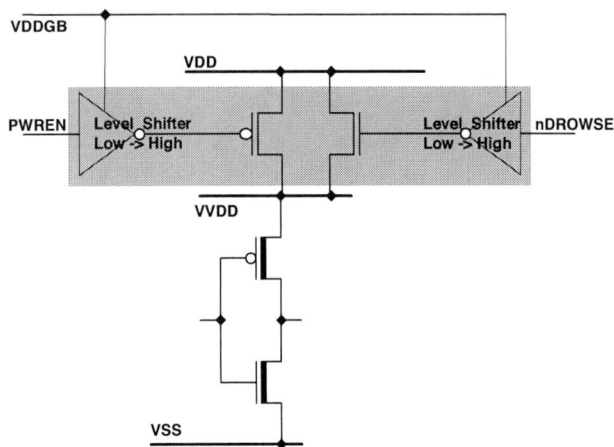

Figure 3: SCCMOS PG with Drowsy SRPG

6. Technology Demonstrators

Bringing together the experimental IP with standard EDA tools and power intent is important to validate the model abstractions and methodology, and enable evaluation of measured versus predicted data. A dual-core ARM® Cortex™-A5 cached microprocessor design was used to implement the enhanced PG and SRPG approaches using prototype power gates dual-rail live-slave registers and integrated clock gating cells in addition to standard ARM Artisan™ standard-cell libraries and compiled memories, building on standard UPF power intent and flows.

Figure 4 illustrates the floor-plan for the one of the CPU cores. The cache memories are implemented with ring-switching – shown top and bottom of the lower region of the clustered cache-RAM instances (with the clamp cells clustered at the boundary to protect the memories when in dormant modes when the standard cell area is power gated). The upper standard cell region has enhanced power gating implemented as distributed gridded SCCMOS power gates, and additional ring switches implement the power-gating and drowsy retention for the lower current retention voltage grid that supplies the 'live-slave' latches and the clamping ICGs. All isolation clamps are inferred with standard power intent.

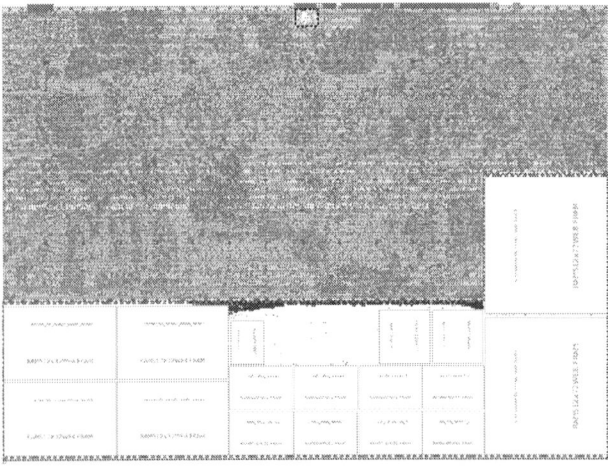

Figure 4: Example ARM Cortex-A5 SRPG floor-plan

Implementation flows were developed on a mature 65nm technology and then ported to 28nm to showcase the techniques on a performance process node with more challenging leakage power.

Figure 5 shows the layout of test silicon implemented in 2012 to validate the physical IP cell abstractions and EDA flow compatibility. Fabrication of rapid prototypes was done in collaboration with the University of Southampton under the Europractice program [4] on a tiny (2 x 2mm!) die-size. An ARM Cortex-M0 CPU is the control processor for the power management control plus local SRAM in the lower-left corner, the dual Cortex-A4 cores are implemented as pre-hardened macros and the multi-core coherence support and a bank of Level-2 RAM are implemented in the rest of the SoC. Local adaptive charge pumps are implemented for the two Cortex-A5 macros to support independent gate-bias and controlled drowsy retention voltages.

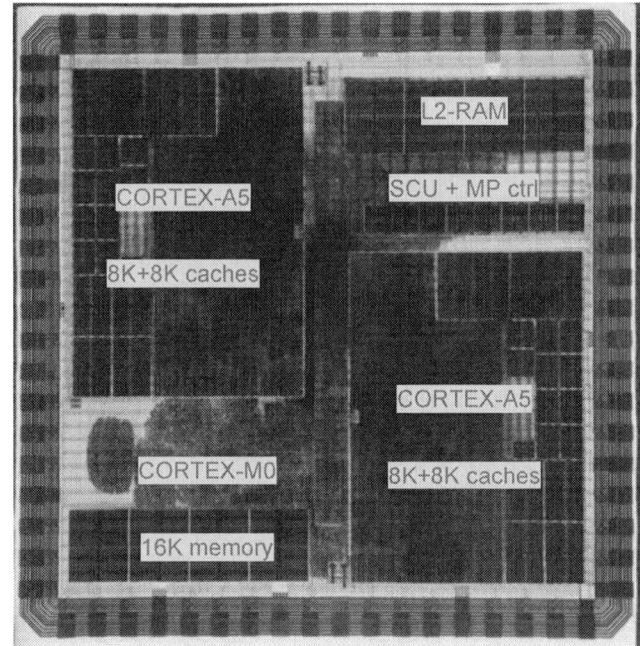

Figure 5: 65nm enhanced SRPG technology demonstrator

Packaged silicon was delivered in early 2013 and actively under characterization to confirm the predicted versus measured performance for the extended low-power standby modes implemented over and above the baseline power gating.

7. ACKNOWLEDGMENTS

Thanks are expressed to colleagues in ARM R&D, especially James Myers for the technology demonstrator integration and Anand Savanth for the adaptive charge pumps developed for the SCCMOS gate bias and drowsy-voltage retention; to Europractice for the MiniASIC test silicon fabrication team, post-graduate researchers and staff at the University of Southampton, UK and to Synopsys Inc for the award of the Charles Babbage prize in 2008 to the university that provided access to the full suite of multi-voltage tools that enabled the implementation and signoff.

8. REFERENCES

[1] Bose, P. 2013. Is dark silicon real?: technical perspective. *Commun. ACM* 56, 2 (February 2013), 92-92. DOI=10.1145/2408776.2408796 http://doi.acm.org/10.1145/2408776.2408796

[2] Esmaeilzadeh, H., Blem, E, St. Amant, R., Sankaralingam, K., and Burger, D. 2011. Dark silicon and the end of multicore scaling. In *Proceedings of the 38th annual international symposium on Computer architecture* (ISCA '11). ACM, New York, NY, USA, 365-376. DOI=10.1145/2000064.2000108 http://doi.acm.org/10.1145/2000064.2000108

[3] Esmaeilzadeh, H., Blem, E, St. Amant, R., Sankaralingam, K., and Burger, D. 2013. Power challenges may end the multicore era. *Commun. ACM* 56, 2 (February 2013), 93-102. DOI=10.1145/2408776.2408797 http://doi.acm.org/10.1145/2408776.2408797

[4] EUROPRACTICE mini@sic programme: http://www.europractice-ic.com/prototyping_minisic.php

[5] Flynn, D. 2012. High Performance State Retention with Power Gating applied to CPU subsystems – design approaches and silicon evaluation. Poster in *Hot Chips 24* archives http://www.hotchips.org/wp-content/uploads/hc_archives/hc24/HC24-Posters/HC24.30.p10-State-Retention-Gating-Flynn-ARM.pdf

[6] Flynn, D. 2012. An ARM perspective on addressing low-power energy-efficient SoC designs. In *Proceedings of the 2012 ACM/IEEE international symposium on Low power electronics and design* (ISLPED '12). ACM, New York, NY, USA, 73-78. DOI=10.1145/2333660.2333680 http://doi.acm.org/10.1145/2333660.2333680

[7] Flynn. D., Gibbons, A. 2008. Design for State Retention: Strategies and Case Studies. *Synopsys User Group* SNUG San Jose 2008, Track TA2

[8] Keating, M., Flynn, D., et al. 2007. Low Power Methodology Manual - for System-on-Chip Design, *Springer* ISBN: 978-0-387-71818-7 http://www.lpmm-book.org/

[9] Mutoh S. et al. "A 1v multi-threshold voltage CMOS DSP with an efficient power management technique for mobile phone applications" ISSCC1996, pages 168–169, 1996.

[10] Mudge, T. 2001. Power: A First-Class Architectural Design Constraint. *Computer* 34, 4 (April 2001), 52-58. DOI=10.1109/2.917539 http://dx.doi.org/10.1109/2.917539

[11] Stan, M.. 1998. Low threshold CMOS circuits with low standby current. In *Proceedings of the 1998 international symposium on Low power electronics and design* (ISLPED '98). ACM, New York, NY, USA, 97-99. DOI=10.1145/280756.280807 http://doi.acm.org/10.1145/280756.280807

[12] 1801-2009 IEEE Standard for Design and Verification of Low Power Integrated Circuits http://standards.ieee.org/develop/wg/UPF.html http://standards.ieee.org/findstds/standard/1801-2009.html

Power Management and Delivery for High-Performance Microprocessors

Tanay Karnik
Intel® Corporation
Hillsboro, OR, USA
tanay.karnik@intel.com

Mondira (Mandy) Pant
Intel® Corporation
Hudson, MA, USA
mondira.pant@intel.com

Shekhar Borkar
Intel® Corporation
Hillsboro, OR, USA
shekhar.y.borkar@intel.com

ABSTRACT

This paper provides an introduction to advanced power management and delivery techniques that have been employed in leading microprocessor designs. The techniques need multiple voltage rails supplied by independent voltage regulators. We provide justification for near-load regulators and explain the practical challenges associated with the regulator integration.

Categories and Subject Descriptors

B.7.0 [Hardware]: Integrated Circuits General

General Terms

Performance, Design

Keywords

Voltage regulators, power management, microprocessors, power delivery.

1. INTRODUCTION

Aggressive technology scaling has enabled very high transistor integration capacity. Complex functions are integrated into hardware with multiple heterogeneous cores and caches. Managing total power consumption has emerged as the most challenging task in today's highly complex microprocessor systems. Independent per-core dynamic voltage/frequency scaling (DVFS) is proven to be an effective way to minimize power consumption. It needs multiple voltage rails supplied by independent voltage regulators on the platform.

We provide an introduction to power management techniques that have been employed in leading designs in Section 2. It is followed by justification for near-load voltage regulators (VR) as a possible solution. Section 3 attempts to explain the practical design considerations and challenges associated with regulator integration in microprocessor systems.

1.1 Power Trends

Aggressive technology scaling has enabled integration of billions of transistors on a single die to enable high-performance multi-core servers, heterogeneous multi-core client processors and small form factor handheld devices. The ever-increasing transistor density has led to decreasing wire widths, increased current density, higher supply currents, larger transients, and large die sizes while lowering both the supply V_{MIN} and V_{MAX} of operation. Off-die dimensions, such as die-package interface, are not

Permission to make digital or hard copies of all or part of this work for personal or classroom use is granted without fee provided that copies are not made or distributed for profit or commercial advantage and that copies bear this notice and the full citation on the first page. To copy otherwise, to republish, to post on servers or to redistribute to lists, requires prior specific permission and/or a fee.
DAC '13, May 29 - June 07 2013, Austin, TX, USA.

shrinking at the same rate. This has led to high power densities, high power consumption, strict supply impedance targets and expensive metal stacks.

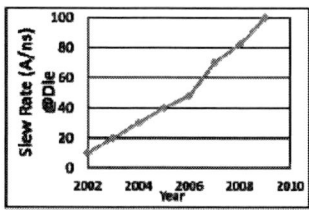

(a) Slew Rate

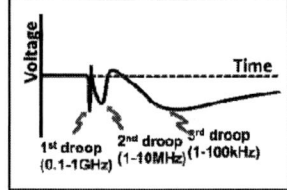

(b) Die Voltage Droops

Figure 1: Power Delivery Requirements and Effects

Figure 1(a) includes the slew rate trends on a high-performance server processor. It is exceeding 100A/nS and causing three distinct voltage droops on microprocessor die, as shown in Figure 1(b). The largest of those, known as the first droop is due to on-die capacitance and on-package inductance in the power delivery network.

2. MICROPROCESSOR POWER MANAGEMENT

DVFS is a well-known power reduction technique employed in current multi-core products offered by major microprocessor vendors.

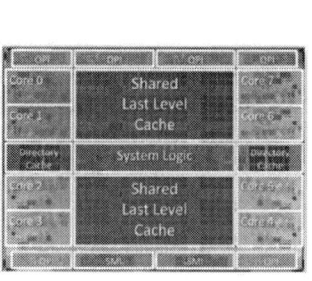

(c) Itanium™ Processor

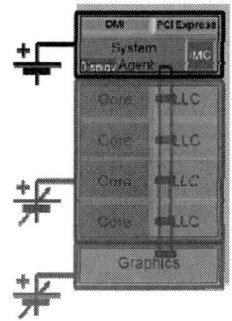

(d) Core™ Processor

Figure 2: Power Domains on 32nm Microprocsssors

Figure 2 shows multiple supply domains on two recent 32nm Intel® microprocessors. The latest Itanium™ processor requires 10+ independent voltage rails [1] while the second generation Core™ processor includes 3 independent rails, two of which were scaled dynamically [2]. Off-chip voltage regulator modules (VRM) supply these voltage rails from motherboard to socket to package to the die with various forms of bulky and expensive decoupling capacitors and power routing planes in the system.

2.1 Recent Techniques

The obvious solution to the platform level power delivery problem is to have an efficient VR on the motherboard with switching FETs, inductors for conversion, capacitors to control droops, and a low resistance path to socket with minimal communication with load die. Top metal layers on the die are dedicated to power distribution and hence made very thick compared to the rest. Dedicated power and ground tracks are added on every metal layer with opportunistic on-die device decap placement.

To mitigate non-ideal power delivery requirements, a concept of load line was established for VRMs. It follows a linear V-I supply characteristic to satisfy triple operating voltage constraints for I_{MAX}, I_{MIN} and thermal design power (I_{TDP}) conditions. VRMs also employ active voltage positioning (AVP) to maintain the supply at the low voltage point of the 3rd droop to expose the system only to "excess" 1st droop beyond the 3rd droop.

First generation Core[TM] processors included adaptive clocking, integrated per-core power gates to minimize standby leakage and a Power Control Unit (PCU) to optimize power across various states of operation. The PCU was a proprietary 32-bit microcontroller. Temperature, current and power sensors on the die feed information to PCU which in return decided on a power gating algorithm. PCU also communicated to external regulator control and accepted OS inputs [3].

3. NEAR-LOAD VOLTAGE REGULATORS

For effective DVFS, microprocessors require multiple independent variable voltages to the die that can be changed by mV increments. e.g. Equation (1) explains a sample 6.25mV V_{STEP} for a7-bit voltage identification (VID) code that is sent from a processor to a motherboard VRM.

$$\frac{(V_{MAX} - V_{MIN})}{VID} = \frac{(1.3V - 0.5V)}{2^7} = 6.25 mV \quad (1)$$

In addition to VID requests, the VRM is also supposed to compensate the voltage droops. In case of the 1st and 2nd droops the VRM is too far to respond effectively. Hence voltage regulators should be placed as close to the die as possible.

3.1 Conversion Benefits

If a second stage VR is inserted between a motherboard VRM and a microprocessor die, the VRM can provide low supply current at high output voltage. High output voltage enables higher allowance of output variation, thereby relaxing the VRM to 2nd stage VR impedance requirement significantly. Figure 3 shows different voltage conversion ratios benefits for an assumed conversion efficiency of 85% [4]. VRM current, off-chip decoupling requirement and resistive losses decrease significantly with the voltage ratio.

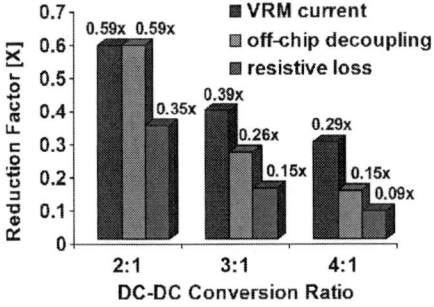

Figure 3: Voltage Conversion Benefits

If the VR is integrated on the package or in a microprocessor die, it will improve load response; voltage control by as much as 50% and response time up to 10x. Small parasitic capacitances in the VR to load power delivery network enable switching frequencies of 100MHz+ [7]. Due to miniaturization of the overall solution, including power FETs and passive components, complex multi-phase designs are easily implemented for droop mitigation.

3.2 Regulator Topologies

The VRM on the motherboard converts a battery output or AC-DC brick output to die voltages. The voltage conversion is one aspect of the design and regulating the die voltage to 6.25mV resolution is another aspect. Three most common topologies for near-load VRs are linear [5], switched capacitor [6] and inductor-based buck regulators [7]. The description of the individual topologies is out of scope for this paper and the reader is referred to [3] for detailed explanation. Linear regulation achieves fine resolution by manipulating device R_{OUT} while buck regulators do this by adjusting the duty cycle of power FET inputs. Switched capacitors are easy to integrate for high to low voltage conversion, but switch capacitor based regulators lack the fine resolution ability. It can be achieved at the expense of loss of efficiency or complex topologies.

4. INTEGRATION CHALLENGES

Near-load VRs seem to be the panacea for all of the power management and power delivery problems, but they require practical design considerations, and pose engineering and financial challenges for integration in high volume products. We describe some of the issues in this section.

System-Level Requirements: Integrated VRs must conform to all the VID, adaptation, and droop response requirements. They have to interact with power control signals, satisfy DVFS and thermal throttling requests.

Testability: The silicon-integrated VRs should be testable across the entire voltage range. They may require power noise injectors.

Across-Load Efficiency: VRs are designed to provide high efficiencies at higher current loads. They suffer from light load inefficiency. Load-adaptive phase or driver shedding is required for reasonable light load efficiency.

Power FETs: If the incoming voltage is higher than $2*V_{MAX}$, high-voltage tolerant FETs may be necessary.

Incoming Power Delivery: The on-die routing from the bumps to integrated VRs is as important as the VR output power routing. It is challenging to route high currents in high device density areas.

Passives: If high-Q inductors or capacitors are integrated, they should be area efficient. If inductors are processed with on die magnetics [8], additional processing is involved.

Thermal considerations: Integrated VRs will typically occupy 10% of the load die area. The current delivery system at the output of VR should not cause thermal hotspots.

Total cost of integration: Near-load VRs will require changes in devices, passives, package, and also in HW-SW interface. The incurred cost by all of these changes should reduce system-level energy consumption to reduce the total system level cost.

Tools: EDA supported tools are required to be able to effectively and efficiently simulate the behavior of the near-load VRs in conjunction with die activity to project benefits and any potential issues.

It should be noted that the increase in number of heterogeneous components on die operating on different isolated voltage domains with varying current and load requirements will result in a proliferation of the near-load VRs which will exacerbate the integration complexities highlighted above.

5. CONCLUSIONS

Advanced power management techniques are employed in high power microprocessor systems and low energy handheld products. Near-load voltage regulators are key enablers for future products, however, they pose some design challenges and offer tool opportunities. Voltage regulator integration is the wave of the future, but it has to be done in the right way with the support of effective tools.

6. ACKNOWLEDGMENTS

Our thanks to Gerhard Schrom, Fabrice Paillet, Vivek De, Shamala Chickamenahalli, Kaladhar Radhakrishnan and Dave Ayers of Intel Corporation.

7. REFERENCES

[1] Reidlinger, R.J., et al. 2011. A 32nm 3.1 billion transistor 12-wide-issue Itanium® processor for mission-critical servers. *IEEE ISSCC.* (Feb. 2011), 84-86. DOI= 10.1109/ISSCC.2011.5746230.

[2] Yuffe, M., et al. 2012. A Fully Integrated Multi-CPU, Processor Graphics, and Memory Controller 32-nm Processor. *IEEE JSSC.* 47, 1. (Jan. 2012), 194-205. DOI= 10.1109/JSSC.2011.2167814.

[3] Karnik, T., 2012. *Power Management using Integrated Voltage Regulators.* Tutorial T6. *IEEE ISSCC.* (Feb. 2012).

[4] Schrom, G., et al. 2004. Feasibility of monolithic and 3D-stacked DC-DC converters for microprocessors in 90nm technology generation. *IEEE ISLPED.* (Aug. 2004), 263-268. DOI= 10.1109/LPE.2004.1349348.

[5] Hazucha, P., et al. 2005. Area-efficient linear regulator with ultra-fast load regulation. *IEEE JSSC.* 40, 4. (April. 2005), 933-940. DOI= 10.1109/JSSC.2004.842831.

[6] Le, H.P., et al. 2013. A Sub-ns Response Fully Integrated Battery-Connected Switched-Capacitor Voltage Regulator Delivering 0.19W/mm2 at 73% Efficiency. *IEEE ISSCC* (Feb. 2013), 371-372.

[7] Schrom, G., et al. 2004. A 480-MHz, multi-phase interleaved buck DC-DC converter with hysteretic control. *IEEE PESC.* 35. (Jun. 2004), 4702-4707.

[8] Gardner, D.S., et al. 2009. Review of On-Chip Inductor Structures With Magnetic Films. *IEEE TMAG.* 45, 10. (Oct. 2009), 4760-4766.

Flexible On-Chip Power Delivery for Energy Efficient Heterogeneous Systems

Benton H. Calhoun and Kyle Craig
University of Virginia, Electrical and Computer Engineering
351 McCormick Rd. PO Box 400743, Charlottesville, VA, 22904
(434) 243-2076
<bcalhoun, kylecraig>@virginia.edu

ABSTRACT

Heterogeneous systems-on-chip pose a challenge for power delivery given the variety of needs for different components. In this paper, we describe recent work that leverages power switches and conventional EDA toolflows to implement a set of power delivery schemes that provide a flexible, adaptable range of options for power management of SoCs for which energy efficiency is important. We first present an enhanced dynamic voltage scaling (DVS) scheme that uses power switches to provide rapid changes in the energy-speed operating point to match workloads at a component level. To demonstrate this approach, we describe a data flow processor chip in 90nm CMOS that supports flexible operation from 0.25V with super high energy efficiency up to GHz speeds at 1.2V. This chip shows that our low overhead method to scale energy consumption with the performance requirement supports both high performance and ultra low energy (>10X reduction in energy per operation) in the same circuit. We discuss power switch design for this scheme and investigate strategies for optimizing power switches for different operating modes. Finally, we show how segmented power switches offer several advantages for flexibly managing leakage and for modulating local voltages with low overhead.

Categories and Subject Descriptors

B. Hardware

General Terms

Management, Performance, Design

Keywords

Dynamic Voltage Scaling, PDVS, Variable Weighted Headers, Low Power Design, Leakage

1. INTRODUCTION

Modern integrated circuits across a broad design space ranging from ultra low power (ULP) sensor systems-on-chip (SoCs) to high performance multicore processors are moving toward the integration of heterogeneous blocks onto a chip. At the same time, energy efficiency has clearly become a driving force for all of

Permission to make digital or hard copies of all or part of this work for personal or classroom use is granted without fee provided that copies are not made or distributed for profit or commercial advantage and that copies bear this notice and the full citation on the first page. To copy otherwise, to republish, to post on servers or to redistribute to lists, requires prior specific permission and/or a fee.

these systems. At the low power, lower performance end of the design space, embedded SoCs seek energy efficient operation to extend battery lifetimes or to enable operation from harvested power. For wall-powered multicore processors, energy efficiency improves performance, reduces hot spots and cooling costs, and helps avoid the dark silicon problem. In part, the demand for energy efficiency helps to motivate heterogeneous solutions, since optimizing blocks for specific tasks improves energy efficiency. A heterogeneous SoC under a tight energy budget presents a significant challenge for power delivery and management. Since different blocks will exhibit different load characteristics, activity factors, operating patterns, and workloads, a flexible power delivery system seems critical for energy efficient operation.

Dynamic voltage scaling (DVS) is the conventional solution for adjusting energy consumption to performance requirements, and some form of DVS makes a lot of sense for heterogeneous SoCs. DVS adjusts the supply voltage (V_{DD}) and frequency to match a circuit's workload, providing quadratic energy savings at lower processing rates when timing slack exists. Recent DVS implementations limit the spatial granularity (the ability to assign each component to different voltages at any given time) with which V_{DD} can vary to the microprocessor core level or above [1][2][3]. Existing techniques also limit the temporal granularity (the ability to adjust a component's voltage quickly) by relying on DC-DC converters to adjust V_{DD}, which take tens to hundreds of μsecs for an output voltage transition [4]. The coarse granularity of conventional DVS means that blocks can only save energy for spatially broad and temporally slow changes in workload.

Further, these core-wide or chip-wide DVS blocks are typically provided a voltage that is modulated directly by a DC-DC converter. Since each converter costs on-chip area and, in many cases, requires off chip passives, there is a cost-driven limit on how many blocks can be reasonably supplied with separate voltages. This constraint opposes the emerging need to support DVS across a broad voltage range to maximize energy efficiency across various operating modes. Even high performance multicore processors are predicted to move toward a dim silicon regime in which cores are intentionally slowed (with lower V_{DD}) to allow more active cores under a chip-wide power constraint, e.g. [5].

This paper reports on recent work to address these problems using a flexible power delivery approach called Panoptic (all encompassing) DVS (PDVS). PDVS combines readily available and EDA-supported features (multiple V_{DD}s and power switches) to offer wide range DVS at low overhead cost. We describe this scheme in Section 2 and report its use in two different types of chips. Section 3 discusses design tradeoffs for implementing the scheme. Section 4 describes several additional uses for the header switches that might provide additional benefits for the scheme for some scenarios, and Section 5 concludes the paper.

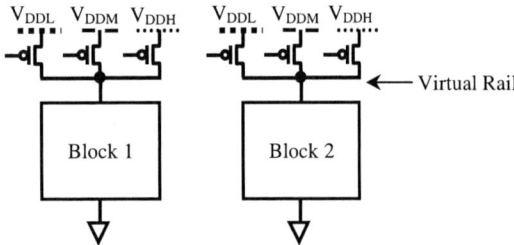

Figure 1. PDVS architecture. Power switches let blocks connect to any of several shared V_{DD}s.

2. PANOPTIC DVS

Panoptic DVS (PDVS) extends DVS to finer granularity in space and time, allowing for much more flexible and energy efficient designs. Figure 1 shows the PDVS architecture, which uses a small set (2-4) of V_{DD} rails throughout the chip. Each component uses multiple header switches to select its own local V_{DD}. As number of components wanting an independent supply increases, the incremental overhead of adding new headers is minor (a few % of the block size) compared to the cost of adding a new DC-DC converter for each block in the conventional case.

The fine spatial granularity provided by PDVS results in higher energy efficiency by allowing more non-critical components to work at the lowest voltage while still meeting performance needs on the critical path. In contrast, chip- or core-wide DVS only allows for V_{DD} changes when the workload of the entire chip or core decreases. Locally controlling voltage for smaller blocks also allows for fast switching from low to high V_{DD} due to the lower virtual rail capacitance of smaller voltage islands. We have measured this V_{DD} virtual rail switching delay to be on the order of a few nanoseconds, which is orders of magnitude faster than the resettling of a DC-DC converter.

The PDVS scheme is very well suited for use in heterogeneous SoCs. Since different blocks on the SoC will necessarily exhibit different average and instantaneous workloads, there is a substantial opportunity to save energy by matching the effective V_{DD} of each block to its workload and corresponding performance requirement. PDVS provides several mechanisms to do this. Clearly, a block can simply hop to a voltage that is nearest the desired V_{DD}, but PDVS also supports dithering (dividing operations between a higher and a lower voltage to provide an effective performance-energy point in between the endpoints) to allow blocks to approximate any effective performance across a broad range. With a straightforward implementation and limited overhead, PDVS can provide near-optimal energy efficiency across a broad range.

2.1 PDVS in a Digital Signal Processor

To demonstrate PDVS in a system, we implemented a data flow processor, capable of executing arbitrary data flow graphs (DFGs) to support a wide range of signal processing functions, in a commercial 90nm bulk CMOS technology. The two million transistor chip (4.3mm x 3.3mm) has the architecture shown in Figure 2 [6]. To implement a DFG, the processor pulls operands each cycle from the register file and feeds them to the correct arithmetic units through a crossbar switch. Each arithmetic component (adders and multipliers) in the PDVS data path has its own set of three header switches to connect it to one of the three V_{DD}s on the chip. Rather than scheduling operations to be

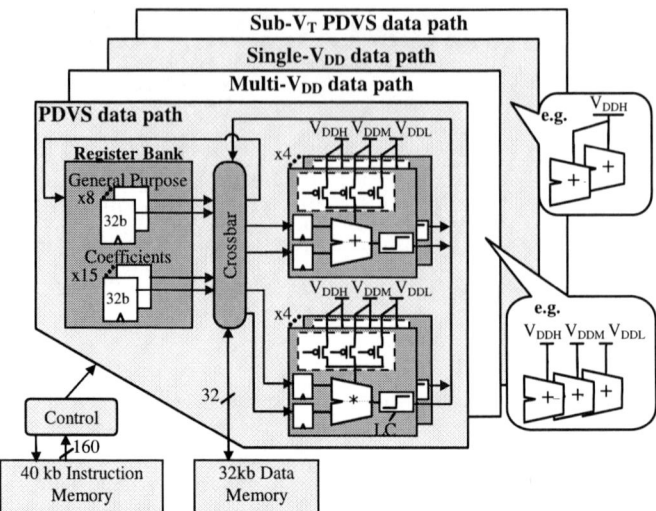

Figure 2. Block diagram of the PDVS data flow processor [6]. Four data paths allow direct comparison of PDVS with SV_{DD} & MV_{DD}.

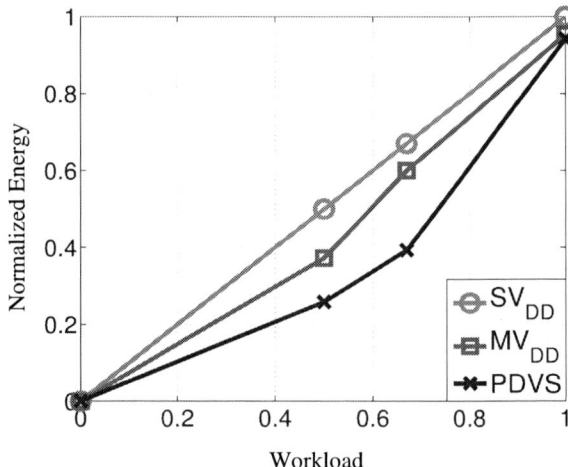

Figure 3. Average measured energy (w/ overheads) vs. workload across 4 different data flow graphs [6].

performed on a specific block (as is the case for Multi-V_{DD} implementations), we assign a voltage to each operation based on how much time is available for that operation to occur. The operation then can be scheduled, along with the correct voltage, to occur on any free arithmetic block based on whichever scheduling algorithm is preferable. By selecting V_{DD}s that cause a slow down of components by an integer number of clock cycles at the fastest rate, we can simplify retiming by simply scheduling the regfile to receive data the correct number of cycles after an operation is launched to a component operating at lower voltage.

The PDVS chip has a measured operational range from 0.25V to 1.2V and was to our knowledge the first to demonstrate single clock cycle V_{DD} switching at the component level, integrated V_{DD} dithering for near optimal energy scalability, and the capability to switch efficiently and rapidly between high performance DVS and sub-threshold modes. We added two different conventional architectures (Single-V_{DD} (SV_{DD}) and Multi-V_{DD} (MV_{DD})) for comparison on the same chip, and Figure 3 shows the measured energy vs. workload. The PDVS design allows reduced energy at lower required performance, supports dithering to approximate

1102

operation between the specific supplied V_{DD} values, and closely approximates an ideal DVS energy profile.

2.2 PDVS in a Body Sensor Node SoC

As a second example of PDVS, we incorporated the scheme into the digital section of a body sensor node (BSN) SoC [7]. The SoC includes a four channel ECG/EMG/EEG (electrocardiogram/ electromyogram/electroencephalogram) front end, analog to digital converter (ADC), sub-threshold general purpose processor, hardware accelerated DSP, SRAMs, MICS band radio, on-chip power management, and power harvesting. The whole chip can run exclusively off of harvested energy from body heat using a thermoelectric generator (TEG) without any battery at all. It consumes only 19 µW while running on body heat and measuring ECG, extracting heart rate, and sending heart rate information over the radio [7].

To meet the clearly stringent power constraints of this system, all of the digital blocks have two header switches that allow them to connect either to a 0.5V sub-threshold supply or to a variable voltage supply, whose value is set by the chip controller. Regulators for both supplies are on-chip. The digital components include a programmable 8bit PIC processor and hardware accelerators for programmable FIR filtering, heart-rate extraction, atrial fibrillation detection, envelope detection, and a packetizer for the radio. Incorporating all of these blocks into a PDVS implementation allows each block to spend most of its time in a low power mode in sub-threshold at 0.5V. For example, the microcontroller is just 474 nW at 200 kHz and 0.5 V. However, if the need for either higher speed or lower power arises, the blocks use PDVS to switch to an alternative voltage to meet that need. This illustrates again the flexibility of the PDVS scheme for supporting energy efficient processing across a broad range of requirements. The digital subsystem on this SoC allows for the extraction of medical information locally to reduce the required load on the radio, dramatically lowering chip power, and the PDVS scheme allows this to happen while still retaining the option to support high speed operation when necessary.

3. POWER SWITCH DESIGN FOR DVS

The design of the header switches that connect PDVS blocks to the multiple power supplies clearly will affect the details of the scheme. PMOS headers have already entered mainstream use as power gating devices to reduce leakage in idle blocks. Numerous commercial designs, especially in the low power space, use power gating with PMOS (and/or NMOS) header (footer) switches. These designs are tending to include more numerous power regions with finer grained gating switches in the search for lower standby power (e.g. [8]). Due to the popularity of this approach, many design issues related to using headers are well-understood. For example, layout and design to reduce noise [9][10], sizing to achieve bounded active delay [11], layout with standard cells [12][13], and decap placement tradeoffs [13]. We can readily adopt this knowledge for PDVS headers, but we need to include some additional analysis of newly relevant metrics.

3.1 Header Overhead and Sizing

Two types of timing overhead result from the switches. First, the on resistance of the header switch causes a drop in the voltage at the virtual power node relative to the selected V_{DD}. This small drop slows down the attached circuit, since it sees a lower supply voltage. This effect is well-known for leakage reduction circuits that use power switches, and sizing of the header device is used to

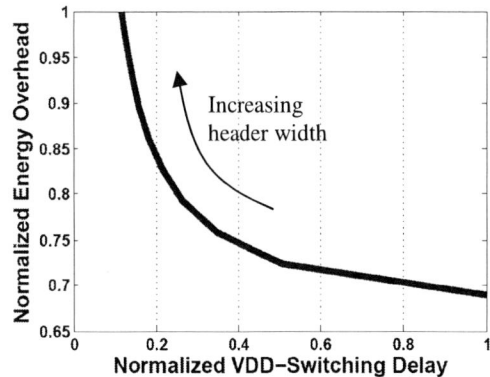

Figure 4. Variation of V_{DD}-switching delay and energy vs. header size.

prevent the delay penalty from increasing above a desired level (e.g. [11]). Second, switching between two V_{DD}s for a block cannot occur instantaneously. Instead, drivers must switch the gate voltages of the header transistors, and the virtual power rail must settle to its new value. In some cases, such as during multi-mode changes or dithered DVS, the PDVS methodology requires a change in the V_{DD} applied to a block that should occur as quickly as possible. Fast changes between V_{DD}s will encourage more frequent and more valuable transitions to the lower V_{DD} to save power. In addition to timing overhead, the headers create energy overhead. A transition to a lower V_{DD} should only occur if the energy consumed by operating at V_{DDL} plus the overhead of transitioning there and back is less than the energy of performing the same operations at the higher V_{DD}. This imposes a breakeven time. If the transition to V_{DDL} will be for less than this time, then the switch is not worthwhile. For the chip in [6], the measured break even time was less than 4 cycles for an addition and less than 1 cycle for a multiplication.

In order to understand these overheads, we must characterize the capacitance of the virtual rail. The intrinsic capacitance of the headers is relatively easy to determine based strictly on the header size. The capacitance of the wires attached to the virtual power rail depends on the layout of the local block, but is readily estimated from process parameters and layout. The effective capacitance looking into the power terminal of the logic attached to the virtual rail depends on the state of the circuit, but this component becomes roughly independent of inputs (due to averaging) for blocks having more than a few gates, except in pathological cases. It becomes straightforward to characterize the virtual rail capacitance of a block using just a few simulations. Once the block capacitance is known, a simple header size sweep can reveal the tradeoff between V_{DD}-switching delay and V_{DD}-switching energy. For example, Figure 4 shows the tradeoff between these metrics for a representative block as a function of the PMOS header size.

3.2 Headers for wide range DVS

Header design becomes somewhat more complicated for designs that seek to support a wide range of potential voltages. For example, the best choice of the body connection for a PMOS header will vary depending on the desired performance. Further, using a conventional PMOS header does not always produce the best result. Figure 5 shows three options for header topologies. These are a basic PMOS device, an NMOS device with a swing at its gate of 0V to V_{DDH}, and a transmission gate (TX) consisting of a parallel NMOS and PMOS.

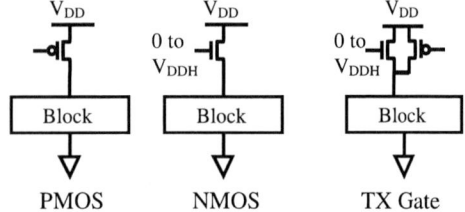

Figure 5. Header switch options for wide range DVS.

For a rail voltage near the nominal voltage for the technology (e.g. near V_{DDH}), the PMOS device provides the best active delay in the block and the best virtual rail switching speed since it supplies the most current. As the rail voltage drops, the choice of PMOS header body voltage introduces a tradeoff. Connecting the body to V_{DDH} lowers virtual rail capacitance at a cost of decreasing the header current. Generally, for a rail lower than V_{DDH}, connecting both the header body and the bodies of the PMOS devices in the block to the virtual rail provides the best choice.

For a rail voltage that drops low enough to approach or enter the sub-threshold region, the situation changes [14]. Figure 6 shows the max frequency of a digital block as a function of header switch width for both NMOS and PMOS at a rail voltage of 0.3V. Clearly, the PMOS switch must be much wider than the NMOS switch (with its gate at V_{DDH}) to support the same frequency. This is because the extra overdrive on the NMOS substantially increases its current relative to the PMOS biased into the sub-threshold region. This result suggests a strategy for header selection. For a rail voltage well above the threshold voltage, a PMOS header is the best choice. For a purely sub-threshold rail, an NMOS header will be much smaller than a PMOS for the same target block speed. For a rail that might change across this entire range (e.g. the variable voltage rail in the BSN SoC [7]), the best choice is a wide PMOS sized to provide the desired speed at high voltage in parallel with a near minimum sized NMOS for supplying current in the sub-threshold and near-threshold regions [14].

4. HEADER EXTRAS

In the previous sections, we have described how header switches provide the basic functionality for PDVS, and we have discussed how to size those headers for various design situations. In this section, we describe two areas of further work that use headers for further optimization of a PDVS system.

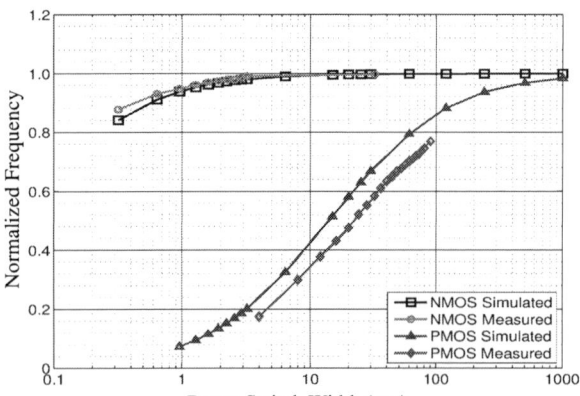

Figure 6. Simulated and measured frequency at 0.3V with an activity factor of 1.0 [14] showing header trade offs at low V_{DD}.

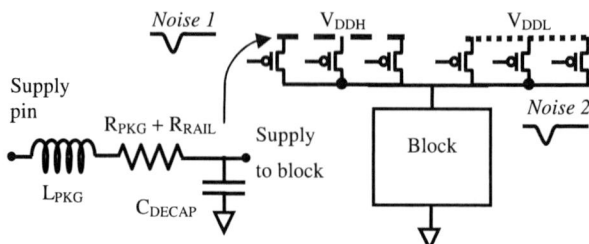

Figure 7. Circuit block with headers and supply rail RLC. While switching from V_{DDL} to V_{DDH}, noise is generated on V_{DDH}.

4.1 Managing V_{DD} Noise

Power supply noise during mode changes is a primary concern in designs employing DVS and power gating. Power rail resistance and pad, pin, and package resistance and inductance cause the voltage rail to droop or spike whenever there is a sudden, large current draw. A block coming out of power gating or switching from a low voltage (V_{DDL}) to a higher voltage (V_{DDH}) are situations that exhibit significant noise on V_{DDH} because the virtual rail of the block needs to get to the final V_{DD} in a short amount of time. This impacts other blocks operating on the V_{DDH} rail at the same time by exposing them to rail droop. One mitigation technique is to apply a slow ramp to the header gate such that the header turns on slowly, which creates a tradeoff between power supply noise and the time to charge the virtual rail to the nominal voltage. The most appropriate tradeoff point may be different for different programming of a system, for different blocks within a system, and for different applications being executed on a system.

In this section, we explore the idea of using split headers to mitigate supply droop during header transitions. This idea has been proposed before, e.g. in [16], which divides the header into 48 equally sized segments with individual control. Furthermore, the delay between the turn-on of consecutive headers is configurable. The design in [16] sets all intermediate delays equally and uses all header partitions of equal sizes as well.

Here we extend the work in [16] by exploring header sizes and intermediate turn-on delays that would minimize supply noise given a total header size. Figure 7 shows the elements involved in creating and determining noise on the voltage supply rail and the virtual voltage rail, *Noise 1* and *Noise 2*, respectively. We assume a total header size that is fixed by considerations of active operation of the block at a fixed V_{DD} through traditional header sizing methodologies as mentioned previously. We fix the number of parts of the header to three, but the methodology can be applied to an arbitrary number. We also assume that the data inputs to the block remain constant and the block is idle. We need to find values for four variables as shown in Figure 8. W1 and W2 are sizes of the first two header parts, and D1 and D2 are the intermediate delays between turn-on of the three legs. W3 is determined based on the size of the first two legs (W1 and W2) and the total header size. The header sizes themselves remain fixed but can be turned on in different combinations to create effective header sizes of various combinations.

Increasing both D1 and D2 monotonically decreases the worst case voltage droop on the supply rail. However, as D1 and D2

1104

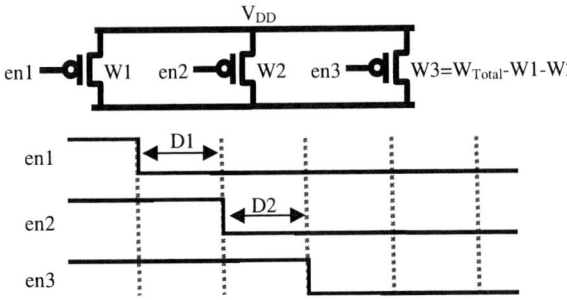

Figure 8. Schematic and timing diagram of header partitioned into three parts.

increase, the time to charge the virtual rail using smaller headers increases. This translates into lower noise, but also a higher charging time, with a monotonic trend. Increasing D1 and D2 shows diminishing returns for noise reduction. For example, if we keep increasing D1, W1 will end up charging the virtual rail most of the way to the final V_{DD}. This would result in W2 and D2 being no longer relevant. We observe that D1 or D2 greater than the original V_{DD} switching delay (with all header parts on at the same time) results in little further noise reduction.

The behavior of noise as a function of W1 and W2 is non-monotonic. Finding optimal values of W1 and W2 only makes sense for larger values of D1 and D2, or else the three parallel headers would effectively act as a single header. Also, D1 should not be too large or else W2 would not matter. Similarly, if D2 is too large, W3 becomes irrelevant. Both scenarios would call for the monolithic header to be broken into fewer pieces.

There are two observations that emerge from our analysis of header sizing for noise mitigation. First, the headers that turn on earlier have a greater impact on noise (as the earlier parts get a chance to charge up the rail starting from the initial V_{DD}, which in the case of power gating is near 0V). Second, no header part should be too large or too small. This is because all header parts generate noise when they turn on. Thus, the optimal scenario is when all of the headers generate the same amount of noise. This does not mean that we size all header segments equally, because the first observation states that the first header has a greater impact (or weight) on noise than the second header. If we combine the two observations, we can conclude that the optimal relationship would be W1<W2<W3, with no large jumps in sizing.

Figure 9 shows the impact of W1 and W2 on noise for a given total header width of 60 μm. The plot shows that there is an optimal point for both W1 and W2 to minimize noise. As predicted, the optimal W1 is less than the optimal W2. Thus, linear division of a header into parts is not the best solution to minimize noise.

Figure 10 shows the V_{DDH} noise reduction benefit of the optimally configured weighted headers compared to a same-size monolithic header. The total header size is 60 μm with the ratio W1:W2:W3 = 1:1.25:3.25, and the total turn on delay for the weighted headers (D1+D2) is 6 ns. This approach reduced the maximum voltage

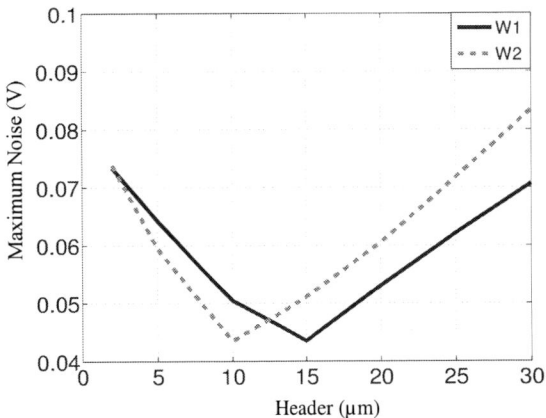

Figure 9. Maximum voltage supply noise with sweeping W1 and W2, while holding the other constant at 10 μm and 15 μm, respectively. The total header width is 60 μm for D1=D2= 3 ns.

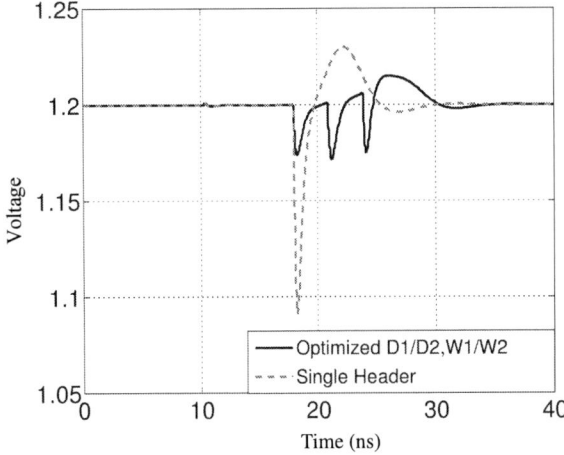

Figure 10. Noise on the voltage supply rail induced by switching with the optimally sized weighted headers and with the same-sized monolithic header.

supply noise from 140 mV to 40 mV in our test case – a 3.5X reduction. As mentioned above, this sizing makes the noise contribution of each weighted header almost the same, leading to optimal noise on V_{DDH} given some modest additional delay for charging the virtual rail up to the nominal voltage. Applying this segmentation approach to PDVS allows us to limit V_{DD} noise below an acceptable threshold without unreasonable complexity.

4.2 Modulating Header Resistance

There are additional potential benefits to segmenting the individual PDVS headers into sets of parallel headers that can be separately controlled. Variable width headers can provide multiple effective header sizes to dynamically select a V_{DD} switching energy-delay trade off based on the specific requirements of a system, each component within a system, and each application and application phase being executed on a system. Figure 11 shows the switching energy-delay of a 32b Kogge Stone adder as a result of sweeping header size using a variable weighted header scheme. There is a slight energy and delay overhead due to the increased parasitic capacitances on the virtual rail from headers that are off, but through simulations, we found the overhead to be less than 4% for this Kogge Stone adder. This figure shows that the headers can be controlled at run time to

1105

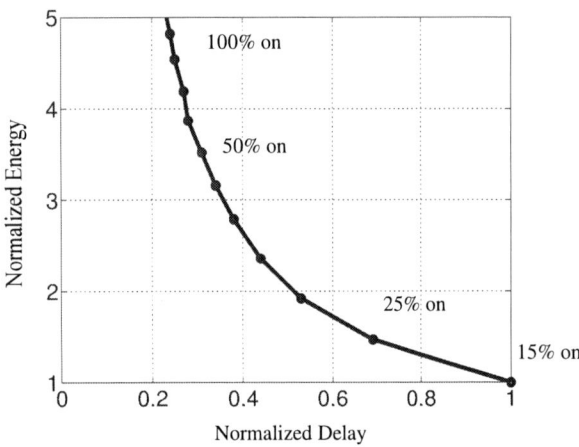

Figure 11. V_{DD} switching energy-delay curve of a 32b Kogge Stone Adder with various sized headers as enabled by variable weighted headers.

modulate the trade-off between V_{DD}-switching energy and delay. The wider header configuration of variable weighted headers can meet timing constraints requiring a quick V_{DD} transition at the expense of energy. The smaller header configuration can switch the rail more energy efficiently but over a longer period of time, which is not problematic when timing slack permits. As header size is reduced, the gate capacitance is reduced, thus saving energy. The gate capacitance (including parasitic capacitances within layout) scales linearly with the width and results in a linear savings. Energy savings are more significant for larger header sizes, due to the large gate capacitances. The V_{DD} switching energy for rail capacitance does not change.

Finally, these same variable width headers can be used to create crude linear regulators for fine tuning the voltage supplied to local blocks [15]. By turning on a fraction of the width of an individual header, we can increase the header resistance. The load current through this header will reduce the voltage seen at the virtual rail in proportion with the current, generating a linear reduction in energy with the lower voltage. This scheme can save up to 30% extra active energy in local blocks and was demonstrated on a Bulldozer processor core [15]. In combination with PDVS, it adds to the flexibility of the overall power delivery strategy.

5. CONCLUSION

Heterogeneous SoCs present a wide range of performance and energy operating points, and a flexible power delivery scheme would be ideal. PDVS uses headers and fixed voltage rails to provide near optimal DVS energy efficiency from sub-threshold to high performance with minimal overhead and maximum flexibility. The headers further provide a tuning mechanism for limiting voltage noise and for fine tuning local block voltages.

6. REFERENCES

[1] J. Howard et al., "A 48-Core IA-32 Message-Passing Processor with DVFS in 45nm CMOS," *ISSCC* , pp. 22-33, 2010.

[2] D. Truong et al, "A 167-processor 65 nm Computational Platform with Per-Processor Dynamic Supply Voltage and Dynamic Clock Frequency Scaling," *Symposium on VLSI Circuits,* pp.22-23, 2008.

[3] B. Nam et al., "A 52.4mW 3D Graphics Processor with 141Mvertices/s Vertex Shader and 3 Power Domains of Dynamic Voltage and Frequency Scaling," *ISSCC*, pp.278-603, 2007.

[4] C. Zheng and D. Ma, "A 10MHz 92.1%-Efficiency Green-Mode Automatic Reconfigurable Switching Converter with Adaptively Compensated Single-Bound Hysteresis Control," *ISSCC*, pp.204-205, 2010.

[5] W. Huang et al., "Scaling with Design Constraints: Predicting the Future of Big Chips," *IEEE Micro*, 2011.

[6] Y. Shakhsheer, S. Khanna, K. Craig, S. Arrabi, J. Lach, and B. H. Calhoun, "A 90nm Data Flow Processor Demonstrating Fine Grained DVS for Energy Efficient Operation from 0.25V to 1.2V," *CICC*, 2011.

[7] F. Zhang, Y. Zhang, J. Silver, Y. Shakhsheer, M. Nagaraju, A. Klinefelter, J. Pandey, J. Boley, E. Carlson, A. Shrivastava, B. Otis, and B. H. Calhoun, "A Battery-less 19µW MICS/ISM-Band Energy Harvesting Body Area Sensor Node SoC," *ISSCC*, 2012.

[8] Y. Kanno, H. Mizumo, Y. Yasu, et al., "Hierarchical Power Distribution with Power Tree in Dozens of Power Domains for 90-nm Low-Power Multi-CPU SoCs," IEEE Journal of Solid-State Circuits, Vol. 42, No. 1, pp. 74-83, January 2007.

[9] S. Kim, S. V. Kosonocky, D. R. Knebel, K. Stawiasz, D. Heidel, and M. Immediato, "Minimizing inductive noise in system-on-a-chip with multiple power gating structures," *ESSCIRC*, pp. 635 – 638, September 2003.

[10] S. Kim, S. V. Kosonocky, and D. R. Knebel, "Understanding and minimizing ground bounce during mode transition of power gating structures," IEEE *ISLPED*, pp. 22-25, August 2003.

[11] M. Anis, S. Areibi, and M. Elmasry, "Design and optimization of multithreshold CMOS (MTCMOS) circuits," IEEE Transactions on Computer-Aided Design of Integrated Circuits and Systems, Vol. 22, No. 10, pp. 1324 – 1342, October 2003.

[12] P. Babighian, L. Benini, A. Macii, and E. Macii, "Post-layout leakage power minimization based on distributed sleep transistor insertion," IEEE *ISLPED*, pp. 138-143, August 2004.

[13] J. Tschanz, S. Narendra, Y. Ye, B. Bloechel, S. Borkar, and V. De, "Dynamic sleep transistor and body bias for active leakage power control of microprocessors," IEEE Journal of Solid-State Circuits, pp. 1838 - 1845, November 2003.

[14] K. Craig, Y. Shakhsheer, and B. H. Calhoun, "Optimal Power Switch Design for Dynamic Voltage Scaling from High Performance to Subthreshold Operation," *ISLPED*, 2012.

[15] K. Craig, Y. Shakhsheer, S. Khanna, S. Arrabi, J. Lach, B. H. Calhoun, and S. Kosonocky, "A Programmable Resistive Power Grid for Post-Fabrication Flexibility and Energy Tradeoffs," *ISLPED*, 2012.

[16] D.N. Truong, et al., "A 167-processor computational platform in 65 nm CMOS," IEEE Journal of Solid-State Circuits, vol. 44, no. 4, pp. 1130-1144, April 2009

Power and Signal Integrity Challenges in 3D Systems

Miguel Miranda Corbalan, Anup Keval, Thomas Toms, Durodami Lisk, Riko Radojcic, Matt Nowak

Qualcomm Technologies, Inc., 5775 Morehouse Dr, San Diego, CA

ABSTRACT

Power/signal delivering network for 2D systems comprising a package and an Integrated Circuit (IC) are design tasks that can be concurrently handled today. Design iterations can be locally carried out in each subsystem part without the need to modify the other one's decisions. This is unfortunately not the case in 2.5D/3D stacked systems. Finer system integration technology, either via Through Silicon Stack (TSS) and/or Through Silicon Interposer (TSI), involves tighter evaluation of the coupling effects in the system-wide PDN impedance and Signal Integrity (SI) characteristics. If these interactions are not properly accounted early in the design cycle, undesired design loop iterations, affecting design productivity is possible. Therefore, new tools and flows incorporating abstracted physical information of the PDN and signal interconnect stack architecture are needed for early design exploration. This paper elaborates on the problems, tool flows and methods necessary to address these challenges for 2.5D/3D stacked systems.

Categories and Subject Descriptors

B.4.3 **[Interconnections (Subsystems)]**: Interfaces, Physical Structures; B.4.4 **[Performance Analysis & Design Aids]** Simulation, Verification

General Terms

Performance, Design, Standardization, Verification.

Keywords

Power and Signal Integrity, Power Delivering Networks, Through Silicon Stack (TSS) and Through Silicon Interposer (TSI) design.

1. INTRODUCTION

3D heterogeneous stacking integration offers several advantages over monolithic SoC. On the one hand, it allows a unique mix & match of functionality and specifications to technologies. This important feature potentially enables lower cost system implementations and facilitates silicon IP reuse. In addition it may potentially improve design productivity by bypassing technology scaling threats such as for global interconnect; and/or by preserving modularity in design and implementation.

However, it requires a considerable learning curve due to the new process integration and technologies needed. There is a need to re-design interfaces and specifications, both in terms of I/O and Electrostatic Discharge (ESD) protection circuitry. There are additional constraints to consider in the already time consuming physical design and timing closure phases. Thermal awareness is a must and last but not least, PI/SI becomes very challenging. This paper focuses on the PI/SI challenges of 2.5D/3D integration.

Permission to make digital or hard copies of all or part of this work for personal or classroom use is granted without fee provided that copies are not made or distributed for profit or commercial advantage and that copies bear this notice and the full citation on the first page. To copy otherwise, to republish, to post on servers or to redistribute to lists, requires prior specific permission and/or a fee.
DAC'13, May 29 – June 07 2013, Austin, TX, USA.

With respect to power integrity, the role of the Power Delivery Network (PDN) is to provide a very stable supply voltage to the standard cells of the IC via a very low impedance power/ground network. The current consumed by the cells follow the cell's switching activity profile which depends on the use case and the actual power management technique. These results on a changing temporal profile of the current draw, hence I/R drop. Current fluctuations reflect on voltage fluctuations in the supply rail.

If the PDN would be ideal, its impedance would be zero and the voltage noise in the power/ground rail would be zero as well. However the power/ground metal lines have non-zero impedance that unfortunately leads to noise impact. 2.5/3D integration with limited PDN resources makes this problem even worse.

It is important to have very low voltage noise in the PDN because noise fluctuations are handled by assuming worst case supply corners. An overly pessimistic supply voltage would force designers to consider overly pessimistic worse case corners. Timing cell characterization would lead to overly pessimistic performance, thus indirectly limiting achievable clock frequency. In addition, power noise couples to other noise sensitive parts of system via substrate, thus creating negative impact on RF/mixed signal circuit performance.

For signal integrity (SI), the goal is to transmit signal from I/O driver to I/O receiver over interconnect within acceptable timing margins. In case of bus interface, the group of signals is required to propagate over interconnect channels and reach within target latency and acceptable skew margin with respect to each other. The quality of the signal transmission is dependent on multiple factors including: I/O driver strength, physical interconnect and receiver load. In case of 2.5D/3D stacked systems, interconnect is comprised of through silicon via (TSV) and small feature size of wires that can yield significant insertion loss which can compromise the maximum operating frequency of the propagating signal.

2. POWER INTEGRITY CHALLENGES FOR 2.5D/3D SYSTEMS

From a PDN viewpoint a classical battery operated system comprises some of the following blocks:

- A Power Management System (PMS): it provides multiple internal voltages and regulators.
- A Printed Circuit Board (PCB), together with the package (PKG): it mechanically supports and electrically connects the different electronic components (passives and ICs).
- A SoC implemented as a single IC.

Each system part delivers power to the other part. System parasitic contributes to different impedance profile regions at frequency ranges where impedance profile does not overlap. This interaction can be modeled by using the concept of resonant models that allow back of the envelope estimations of the overall impedance response [1]. The goal is to have low impedance across the frequency of interest.

Figure 1 illustrates the concept for a simplistic system example consisting on an IC and a PKG and their respective parasitic impedance contributions (capacitance and inductance).

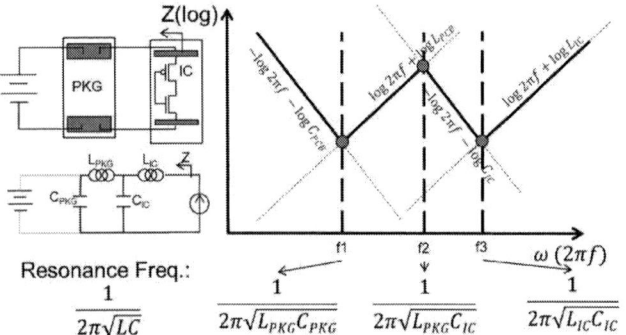

Figure 1 Simplified Resonator model for a typical 2D system

In real-life cases these models get more complex than depicted in Figure 1. Yet, they all share the capability to identify which PDN subsystem parasitic dominates at which frequency range. In a 2D system this allows a divide and conquers approach for the impedance analysis in frequency domain.

Unfortunately, with 2.5D/3D integration, the opportunities for a divide and conquer design approach are much less obvious. In these systems PI challenges comprise capturing both the spatial and temporal distribution of current demand of the IC so as to ensure low impedance at the multitude of operating frequencies they handle.

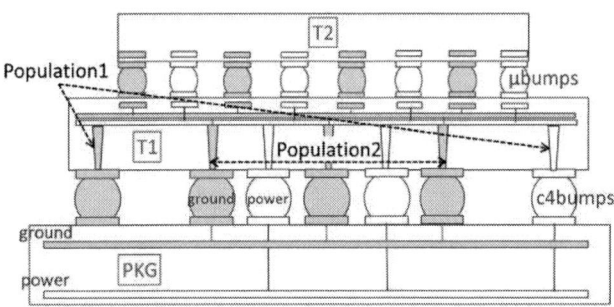

Figure 2 PDN schematic of a Through Silicon Stack (TSS) system comprising two tiers showing two TSV populations.

Figure 2 depicts a PDN integration scheme for a simple TSS system comprising of two tiers (T1 and T2) with an IC in each of them. There are new PDN design problems in these systems:

1. In a 2D system, the SoC power/ground interconnect mesh is directly fed by the package connections and the number of package connections can be very large However in 3D systems such direct connection is only possible for tier T1 while Tier T2 has much more restricted access to package resources via a more limited

number of TSV connections through Tier T1. In a 2.5D system, the situation is not much better. The SoC has no direct access to the package. A limited number of power/ground TSVs, placed between the Silicon Interposer and the package, service the SoC PDN. This considerably limits power/supply resources and challenges the impedance requirements for I/R drop since they require current summing for large areas from the high order tiers and create DC hot spots.

2. New parasitics get introduced, both of capacitive but more importantly also inductive nature. Depending on the operating frequency, these parasitics may couple the impedance of the dies they connect through via TSVs. For instance there may be additional resonance between Tier N and N+1 as there are two caps separated by a low resistance inductive path.

Figure 3 shows the impedance frequency profile corresponding to a system representative of that one shown in Figure 2. The plot shows the frequency domain response of the impedance *seen* looking into each of the ubumps from T2. The plot shows the package to 3D-Si impedance resonance frequency between 10 and 100MHz along with two populations of TSV resonance. Note that the IC resonance frequency is not visible. Our observation point is at the TSV PDN, thus excluding the IC from the observation path.

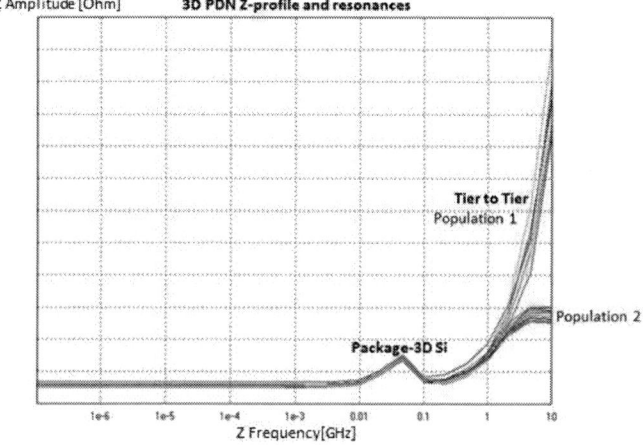

Figure 3 PDN frequency profile of a system comprising of a package and two TSV stacked dies

While not shown in this diagram, the package to 3D-Si PDN impedance is altered by the addition of the T2 stack due to the increased Si capacitance per bump in a 3D-Si stack. Thus the temporal response of T1 clearly changes by considering the impact of T2 PDN. One could expect a lower impedance resonance frequency and amplitude by considering the 3D-Si stack vs. the tiers in isolation.

Moreover Tier 2 shows in Figure 3 a change in spatial response with two populations of TSV responses (see Figure 2), Population 1 is associated with PDN connections that are isolated from the rest of T2T PDN. They are either placed in corners and/or have large gaps between connections. Therefore, Population 1 presents both larger loop inductance (due to less parallel inductive TSV parasitic) and slightly more Si Capacitance (due to increased

1108

fringe impact). Note the resonance peak of the two populations may not overlap. Population 1 may present a resonance peak far from Population 2, not observable at the maximum frequency (10GHz) considered by the analysis.

3. SIGNAL INTEGRITY CHALLENGES FOR 2.5D/3D SYSTEMS

For SI in 2.5D/3D stacked system, the impact of TSV would need to be accounted for in the overall interconnect model. Signal transmission through TSV is frequency dependent with significant insertion loss seen at high-frequency [2]. Since TSV is embedded in silicon substrate, the signal loss through TSV is attributed to substrate leakage and coupling as shown in Figure 4.

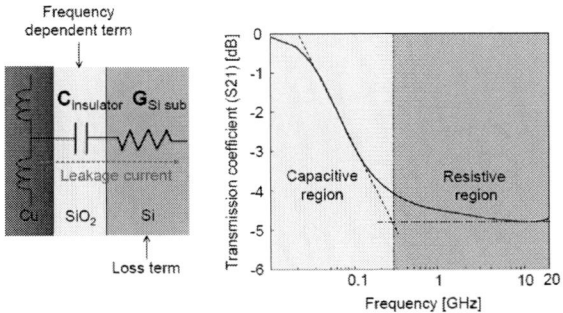

Figure 4: Frequency-dependent loss of TSV (Courtesy of [2])

In the case of 2.5D system where ICs are placed side-by-side on Silicon Interposer substrate, the need for dense die-to-die (D2D) interconnect routes would lead to significant insertion and return loss [4] due to following reasons:

- Fine routing segments would be inherently resistive and therefore signal would attenuate.
- Long routing segments would yield loss due to noise coupling (capacitive and potentially inductive).
- Routing segments exposed to short height from Si substrate (lossy material) and small inter-metal dielectric (IMD) thickness.

Additionally, trade-offs for I/O driver strength vs. size would need to be explored to meet SI requirements. Large I/O driver would occupy more die area and therefore may not be advantageous should dense interconnect routing pitch be necessary.

4. TOOL FLOWS FOR 2.5/3D POWER & SIGNAL INTEGRITY ANALYSIS

Today IC power & signal integrity analysis flows (either for time domain or frequency domain analysis) require full GDS or netlist level information [4][5]. These flows are intended for detailed design verification and to obtain voltage/current/power maps and reports, frequency domain response of the IC via e.g., S,Y,Z parameters at selected observation points. PKG information can be brought via extensions of these flows that include a 3D model of the whole system structure (i.e., PKG, board, etc) via full wave simulations [6], or simplified resonant models [1].

The disadvantage is however verification flows require a very low abstraction level entry point. For 3D verification, S-domain analysis will need all of the Tiers PDN design data to be available

previous to any step and the time domain analysis will need additional library/technology information (ie, lef, def, lib).

Under this scenario, much detailed design information only available at the end of the design process is required. This is not feasible for early design exploration. Moreover, small changes in one of the system parts PDN would require a time consuming re-evaluation of the system wide response.

This situation is clearly not advisable, much less desirable. If a change in e.g., planning of TSVs in one of the dies turns out to lead to a system impedance increase that is incompatible with the signaling or operating frequency specs, it may require to reconsider the die partitioning decision made early in the design cycle. This will have a negative impact in design productivity.

In addition 2.5/3D PDN design flows would require supporting both PathFinding as well as design authoring mode. From an EDA tool perspective the difference between these modes boils down to the tech file and the difficulty or ease of setting those up. In PathFinding mode the designer needs to vary the parameters that are contained in the tech file - often locked by the foundry. In design authoring mode the designer prefers to have a single tech file that represents the already selected process. Supporting these two modes in a single environment is not trivial.

5. SYSTEM-WIDE POWER & SIGNAL INTEGRITY FOR 2.5D/3D SYSTEMS

For early design, 2.5D/3D power and signal integrity architects need quick feedback on the system-wide feasibility of architectural choices. This is in terms of I/R drop analyses, with a need to understand spatial distribution characteristics of current demands. In addition there is also a need for estimating the impact of the PDN impedance/frequency profile in space and with particular focus on insertion losses in case of signal distribution.

For that purpose, we need a higher abstraction level entry for impedance models of the various subsystem parts. It is simply not feasible to fix all PDN design parameters (i.e., GDS, lib, lef and def, files) while exploring the 3D stacked system architecture. Many architecture decisions need to be considered while ensuring they do not present roadblocks for PI/SI.

Given the regular structure of the power delivering and signal interconnect networks; much of their physical properties could be captured via compact architectural parameters [7]. Indeed some high-level information of the power/signal interconnect is available. For instance, metal hierarchy, thickness, pitches of the metal lines, dimensions and electrical properties of the metal planes in case of a PKG/interposer subsystem, and pitch/diameter and routing density information in case of TSVs. Of course not all possible implementation scenarios can be captured with such high level information, but they may be also not necessary at this phase. Finer implementation decisions having less impact on the overall system-wide response could be captured during the more elaborated bottom-up verification flow (see Section 0.)

We need tools able to perform frequency-domain impedance analysis, featuring multi-scale/multi-physics for PI/SI modeling, which are fast enough to handle multiple hierarchies of TSV/bump interconnections. We need these tools to be capable to extract the scattering/impedance matrices (S/Z-parameters) of the various PDN system parts. And we need to be able to come up with a linked S(Z) parameter model representation of the system-

1109

wide impedance [8], capable of handling interconnection realizations in the order of to 10K-100K integration points.

Moreover, we also need design flows capable of the ease of use inherent to the simplistic resonance models for early design, but overcoming the modeling limitations of these approaches. They should account for the new parasitic interactions that 2.5/3D integration brings and accurately predict the sensitivity of the system-wide response to high-level design parameters (e.g., TSV planning decisions, TSV density, wire properties, pitches etc).

6. CONCLUSION

Exploring systems architectures for 2.5/3D systems involves using new integration technologies with limited interconnect resources. Many system design trade-offs need to be considered in this early phase for a variety of system architectures. Moreover, use case dependent PI/SI issues may lead to non-implementable systems. An early design flow providing quick feedback of the PI/SI characteristic of the considered system during exploration is therefore required.

7. REFERENCES

[1] L. Smith, et al., "Chip–Package Resonance in Core Power Supply Structures for a High Power Microprocessor" *In Proc. IPACK 2001.*

[2] Joungho Kim et al, "Signal Integrity Design of TSV-Based 3D IC." Georgia Tech IPC 3D System Integration Workshop. 14 Jun 2010. http://www.ipc.gatech.edu/workshop/2010/5.pdf

[3] Mandy Jin et al. "3D Si Interposer Design and Electrical Performance Study." DesignCon 2013. 30 Jan 2013.

[4] http://www.apache-da.com/flows/chip-package-system

[5] http://www.sigrity.com/products/xcitepi/xcitepi.htm

[6] http://www.cst.com/Content/Documents/Articles/article673/CST_Application_note_RichardSjiariel.pdf]

[7] R. Chen et al, "Analytical Model for the Rectangular Power-Ground Structure Including Radiation Loss", *IEEE Trans. Electromagn. Compat.,* 47(1). pp. 10-16, Feb. 2005

[8] T. Okoshi, et. al, "The segmentation method—An approach to the analysis of microwave planar circuits," *IEEE Trans. Microwave Theory Tech.* 24(10), pp. 662–668, Oct. 1976

Underpowering NAND Flash: Profits and Perils

Hung-Wei Tseng   Laura M. Grupp   Steven Swanson
The Department of Computer Science and Engineering
University of California, San Diego
{h1tseng,lgrupp,swanson}@cs.ucsd.edu

Abstract

MLC Flash memory is getting more popular in computer systems ranging from sensor networks and embedded systems to large-scale server systems. However, MLC flash has many reliability concerns, including the potential for corruption due to supply voltage fluctuations. This paper characterizes MLC flash when the chip is underpowered (i.e., power fading and voltage droops). We demonstrate that underpowering flash can cause serious errors, but also help saving up to 45% of operation energy without incurring failure.

1. INTRODUCTION

In recent years, computer systems ranging from sensor networks to portable devices to data centers have embraced flash-based solid-state drives (SSDs) to provide low power, light weight and shock resistant storage. These applications can expose flash memories to a range of stresses including unstable or unreliable power supplies that can underpower flash memories.

Flash-based SSDs have several characteristics that make unstable or unreliable power supplies particularly dangerous. For instance, an inopportune droop in supply voltage could corrupt flash translation layer (FTL) metadata and render the whole storage device unusable. If designers hope to implement reliable storage systems, they must understand the behavior of flash devices under these conditions.

This paper characterizes the behavior of flash memory devices when supply voltage droops unexpectedly during an operation. We show that supply voltage droops can incur errors for all operations. We also demonstrate that, for some operations, significantly reducing the supply voltage does not impact reliability. In these cases, we can exploit the chips' tolerance for drooping voltages to save energy.

To characterize the behavior of flash memory chips, we developed a testing platform to repeatedly change the supply voltage to a raw flash device during read, program and erase operations. We examined 5 popular MLC chips from different vendors. Our data show several interesting behaviors of underpowering flash memory chips. First, underpowering flash memory during an operation does cause data corruption in some cases, but not always lead to incomplete operations. Second, the minimum voltage each operation requires to complete an operation without increasing the error rate is different. Third, underpowering flash chips can negatively impact the latency of operations but positively improve the energy consumption. We utilize our characterization results to design a dy-

namic voltage scaling mechanism that adjusts the supply voltages to flash chips according to the type of operations. We can achieve up to 45% energy saving for the flash storage device without the assistance of any special data encoding scheme.

The rest of this paper is organized as follows: Section 2 describes the aspects of flash memory pertinent to this study. Section 3 describes our experimental platform and methodology for characterizing flash memory's behavior when supply voltage droops. Section 4 presents our characterization results. Section 5 discusses the potential energy saving using our characterization result. Section 6 provides a summary of related work to put this project in context, and Section 7 concludes the paper.

2. FLASH MEMORY

This section presents a brief introduction of flash's characteristics that are most relevant to understanding its behavior in the face of unstable power supplies.

A cell in the flash array stores data by trapping electrons using a floating gate transistor. The electrical charge of the floating gate affects the transistor's voltage level, and the chip compares this level voltage and with threshold voltages to read the data that the cell currently stores.

According to the number of voltage levels of each cell, flash cells can divide into single-level cell (SLC) and multi-level cell (MLC). SLC devices store one bit per cell, while MLC devices store two or more. MLC chips obtain higher densities by representing n bits using 2^n threshold voltage levels. SLC chips provides better and more consistent performance and lifetime than MLC chips. The empirical measurements in [3] show that MLC chips need 30 μs – 50 μs to perform a read operation, 300 μs – 2 ms to perform a program operation, and 2 ms – 4 ms to perform an erase operation. In this paper, we focus on 2-bit MLC cells, since they are most prevalent in current systems.

The flash chip divides cell arrays into blocks (between 32 and 256 pages) and blocks into pages (between 2 and 8 KB). The erase operation sets all the bits in a block to '1'. The erase operation operates at the granularity of one block. The program operation converts 1s to 0s and performs operates at page granularity. Each block in flash memory has limited number of erase cycles (lifetime). Due to the limited lifetime and the difference in granularity between programs and erases, SSDs apply complex flash translation layers (FTLs) to perform out-of-place update and remapping operations to improve performance and lifetime. FTLs need to store metadata in the flash storage array along with the user data.

The erase operations is iterative. The chip removes the electrical charges from cells within the block and checks if all the cells reaches the erased state. The chip will continue to remove electrons until the threshold voltage levels of cells reach the erased state.

The program operation injects electrons into floating gates to change the threshold voltage for the cells and then perform a read-verify operation to check if the cells have reached the target threshold voltage. If any of the cells in a page failed to pass the check, the chip will repeat the program and the read-verify process [13, 7].

Permission to make digital or hard copies of all or part of this work for personal or classroom use is granted without fee provided that copies are not made or distributed for profit or commercial advantage and that copies bear this notice and the full citation on the first page. To copy otherwise, to republish, to post on servers or to redistribute to lists, requires prior specific permission and/or a fee.
DAC'13, May 29 - June 07, 2013, Austin, TX, USA.

Voltage Levels	Logic Bits			
	Gray coding		2's complement coding	
	1st page bit	2nd page bit	1st page bit	2nd page bit
Lowest	1	1	1	1
	1	0	1	0
	0	0	0	1
Highest	0	1	0	0

Table 1: The mapping of voltage level and logic bits in 2-bit MLC chips using Gray coding and 2's complement coding.

For 2-bit MLC chips, each cell stores one bit from two different, logical pages for two different pages. Manufactures require that pages within a block be programmed in order, so to differentiate between the two pages in a cell, we refer to them as "first page" and "second page". Programming a first page only sets the threshold voltages of cells to contain one bit data. Programming a second page further divide the existing 1s and 0s into four different states to present 2-bit data. Table 1 shows the mappings between threshold voltages and logic bits of a 2-bit MLC cell using gray coding and 2's complement coding after the chip programmed a second page of a cell. Programming a second page takes longer time than programming a first page, because programming the second page requires a more complex process.

3. METHODOLOGY

To study behavior flash memory under an unstable power supply, we designed and implemented a test system that can issue commands to raw flash chips and change the supply voltage to the chip precisely. This section presents our test system, experimental methodology, and the flash chips we examined in this study.

3.1 Experimental hardware

Our test system contains a Xilinx XUP board, a custom flash testing board, and the voltage control circuit. The Xilinx XUP FPGA board implements a custom flash controller. The PowerPC core on the FPGA board allows the system to host a full-fledged Linux distribution. The system also provides libraries that allow benchmark applications to directly access the flash device on the flash testing board through our custom controller and control the supply voltage. The flash testing board provides sockets for installing flash devices and facilities for measuring power and performance. The flash testing board uses 3.3 V as the default supply voltage. The voltage control circuit consists of high-speed transistors and accepts signals from the FPGA board to change the supply voltage to the flash chips. We use a high-resolution oscilloscope to measure the energy consumption of operations. To minimize switching time, we remove all capacitors on the flash testing board. The oscilloscope shows that the resulting system can change the chip's supply voltage within 10 ns and the time takes to reach a stable voltage level is slightly less than 2 μs. We also examined the behavior of flash memory devices between 2.7 V – 3.3 V with the capacitors. Removing the capacitors does not result in significant differences in bit error rates and latencies.

3.2 Test procedure

To examine the behavior of underpowering flash memory chips during read, program and erase operations, we change supply voltage of flash memory chips from the suggested operating voltage (2.7 V – 3.3 V) to a voltage below the chip's specified minimum operating voltage (1.8 V – 2.6 V) during the operations. We define the *voltage change interval* as the time between issuing a command to a flash device and when the FPGA board triggers the change of supply voltage. The voltage change interval starts after the chip receives the last byte of the command. The high-resolution oscilloscope shows that the chip starts executing the command within 10 μs .

Abbrev.	Manufacturer	Cell Type	Cap. (GBit)	Tech. Node (nm)	Page Size (B)	Pgs/ Blk
A-MLC16	A	MLC	16		4096	128
B-MLC32	B	MLC	32	34	4096	256
B-MLC128	B	MLC	128	34	4096	256
E-MLC8	E	MLC	8		4096	128
F-MLC16	F	MLC	16	41	4096	128

Table 2: Parameters for flash devices we studied in this work

For read operations, we use voltage change intervals varying from 0.4 μs to 60 μs at increments of 0.4 μs. For program tests, we use voltage change intervals varying from 0.4 μs to 2.4 ms at increments of 0.4 μs. For erase, we use voltage change intervals varying from 2 μs to 4.8 ms at increments of 2 μs.

3.3 Flash devices

Flash memory chips from different manufacturers demonstrate various behaviors because of their architectural differences within the devices and manufacturing technologies. In this work, we selected 5 MLC chips that cover a variety of technologies and capacities to better understand the variation of flash devices when voltage droop occurs. According to the data-sheets, all these chips have suggested operation voltage range between 2.7 V and 3.6 V. Each chip that we used in this project guarantees 5000 erase cycles for their blocks. We perform all our experiments under the suggested lifetime of blocks.

Table 2 illustrates the flash memory chips that we studied in this work. These devices come from five different manufacturers with process technologies ranging from 72 nm to 34 nm. Their capacities range from 8 GBits to 128 GBits. We obtain values that are not publicly available from the manufacturer from [3].

4. EXPERIMENTAL RESULTS

We found interesting behavior when the supply voltage droops during operations. It appears that the flash memory chips can tolerate the changing of supply voltage within a certain range and allow operations to complete without incurring additional errors. The voltage ranges that each chip can tolerate are different for different type of operations. The supply voltage also affects the latencies and energy of flash operations. We describe the details of our results in this section.

4.1 Program and power fade

To understand the impact of underpowering flash chips during program operations, we begin by programming each page in a block with random data and changing the supply voltage from the 3.3 V to a lower voltage level (1.8 V – 2.6 V) at different voltage change intervals. Then, we measure the resulting bit error rate and latency.

We found that all chip behave similarly when the supply voltage droops during program operations. First, lowering the supply voltage during a program operation does not always incur more errors. Second, lowering the supply voltage will increase the latency to complete a program operation. Finally, we find that lowering the voltage during the program operation can potentially corrupt data already present in flash memory.

4.1.1 Programming pages

To illustrate our experimental result, we use E-MLC8 as an example. Figure 1 shows the average bit error rates when the supply voltage droops from 3.3 V to 1.8 V – 2.6 V at different voltage change intervals. When the target supply voltage is above 2.1 V, the bit error rates do not show significant difference comparing with performing the whole operation at 3.3 V. When the voltage droops to 2.0 V, the bit error rate curve shows some spikes indicating that

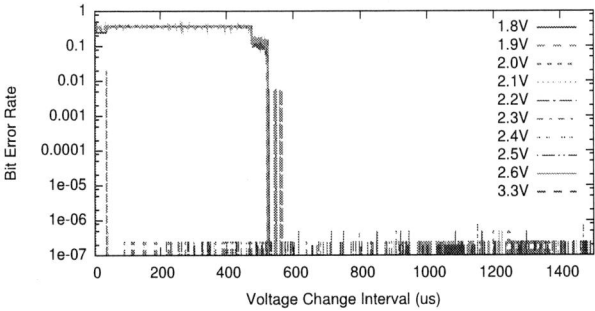

Figure 1: The average bit error rate for E-MLC8 when the supply voltage droops to 1.8 V – 2.6 V at different voltage change intervals.

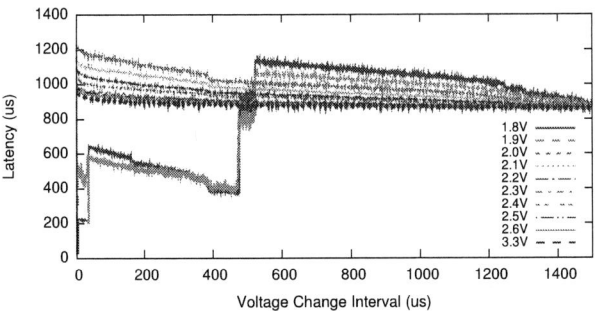

Figure 2: The average program latency for E-MLC8 when the supply voltages droops to 1.8 V – 2.6 V at different voltage change intervals

the bit error rate can suddenly increase at a certain voltage change intervals because the insufficient power supply corrupts the process of changing threshold voltages in some cells. Underpowering the chip to 2.0 V during programming is dangerous. If we change the supply voltage to 1.9 V or 1.8 V within 4 μs of the operation starting, the bit error rate reaches 50%, which means the program operation completely failed. For voltage change interval between 4 μs and 480 μs , some cells fail to finish programming their first page bits and all cells fail to program the second page bits resulting in an 37.5% bit error rate during this period. At 480 μs , the chip applies a different threshold voltage to distinguish the second page data, and the bit error rate drops significantly. However, the bit error rates of the second page bits remain high until the voltage change interval reaches 556 μs .

We also found that lowering the supply voltage will increase the program latency for both first and second pages. Figure 2 shows the average latency of E-MLC8 when the supply voltage droops from 3.3 V to 1.8 V – 2.6 V. The latencies for voltages 1.8 V and 1.9 V shows very low latency when the voltage change interval is smaller than 556 μs . The chip aborts the program and read-verify process because the chip cannot reprogram and correct these high bit error rates using the lower power supply. When the supply voltage droops to 2.1 V – 2.6 V during the operations, the program latency increases. In general, lower voltages result in longer latencies. For 2.1 V, which is the minimum supply voltage that allows the program operation to complete without incurring significant program error rates, the latency increases by 20% if we change the voltage immediately after the operation begins.

Programming a flash page with lower supply voltage increases the latency but can help reduce the energy consumption. For example, in E-MLC8, programming at 2.1 V can reduce the energy consumption by 32% for a first page, and 34% for a second page.

4.1.2 Retroactive data corruption

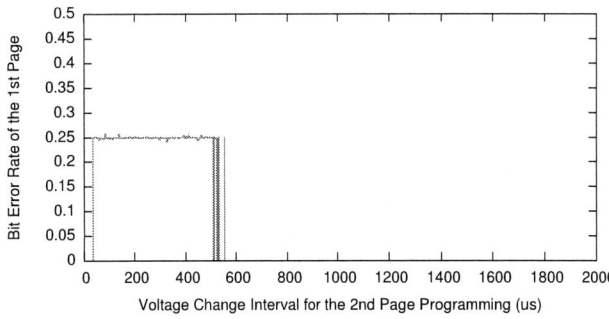

Figure 3: The retroactive data corruption problem occurs when we change the supply voltage for E-MLC8 during a program operation

The retroactive data corruption [14] is a phenomenon that a later incomplete program operation can corrupt data that the chip already successfully programmed. The prior work demonstrates this phenomenon can corrupt 50% of first page bits if power failure occurs during second page program operations. We also examined this problem when the supply voltage droops during a program operation.

Figure 3 illustrates this effect for E-MLC8. In this graph, we program random data into the first page bits of E-MLC8 without changing the supply voltage. However, when we program the second page, we change the supply voltage to 1.8 V during the operations at different voltage change intervals. The x-axis shows the voltage change intervals for the second page program operations, and the y-axis shows the bit error rates of the first pages after we programmed the second page.

For E-MLC8, the bit error rate reaches 25% if the voltage droop happens between 36 μs and 556 μs during the second page program operation. This result indicates that SSDs can potentially lose data that are already present in the chip if power supply is unstable.

4.1.3 Read disturb sensitivity

Another data integrity issue that we examined is the relationship of read disturb and power fade. Read disturb happens because the read operation of flash array applies weak programming voltages to cells in pages not involved in the read operation. It typically takes several million read operations to cause significant errors due to read disturb [3]. In this work, we examine the read disturb sensitivity if we underpowering flash chips during program operations.

To examine the read disturb problem for programming flash pages with low voltages, we program flash pages within the chip using the lowest voltage that does not incur more errors (e.g. 2.1 V for E-MLC8) than programming at 3.3 V for any voltage change interval.

Our result shows that for programming operation finished with the selected voltage level, the read disturb sensitivity does not show any difference comparing with programming at normal voltage.

4.1.4 Data retention

The previous experiments demonstrate that program operation can potentially complete with lower supply voltage. In this section, we also examine the long-term stability of the programming data under lower supply voltages. We use a laboratory oven to bake the chips for 9.33 hours at 125°C to simulate the aging effect of one year [5], and repeat the same process until each chip reaches the aging of 10 years. For each chip, we programmed 5 blocks for two conditions: (1) Programming at minimum program voltage: We program flash pages using the lowest voltage that does not incur more errors (e.g. 2.1 V for E-MLC8) than programming at 3.3 V for the whole operation. (2) Programming at normal voltage: the

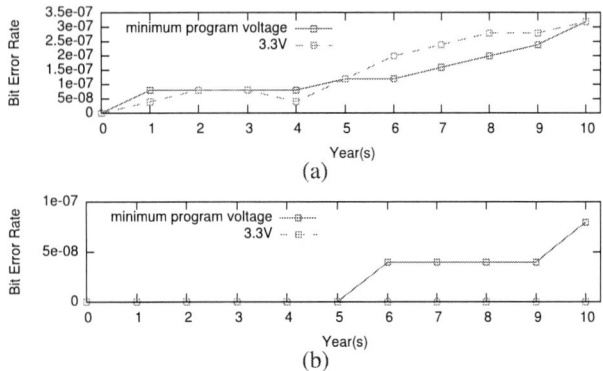

Figure 4: Accelerating aging experiments reveal that program flash chips using lower supply voltage can affect the long-term reliability.

program operation completes using 3.3 V.

The experimental result falls into two categories. We use two chips, F-MLC16 (Figure 4(a)) and A-MLC16 (Figure 4(b)), as examples. The x-axes of Figure 4 are accelerated age in years. The y-axes show the average bit error rates in a block after aging for each accelerated year. For F-MLC16, the data retention of programming at minimum program voltage does not exhibit any difference from programming at normal voltage. For A-MLC16, programming at lowest program voltage results in 1 to 3 bits errors in each page with an average error rate of 7.95×10^{-8} rather than 0 for programming at normal voltage. Though lowering the programming voltage can potentially hurt data retention, the error rates are still manageable with usual error correction codes.

4.2 Read and power fade

To examine the behavior of flash memory chips when the supply voltage droops during read operations, we first program each page in a block with random data at the normal voltage level (3.3 V). Then, we read back the data but change the supply voltage during the operations and measure the bit error rate and the latencies.

The behavior for read operation shows some diversity across chips. For B-MLC32 and B-MLC128, these two chips can never safely read beyond the recommend operating voltages. Take Figure 5(a) as an example, the chip hits 25% error rate when the voltage droops to 1.9 V. If the target voltage is between 2.1 V and 2.6 V, the chip seems to work properly when the voltage change intervals is less than 10 μs . But the read operation, again, shows significant errors if the voltage change intervals are more than 10 μs . Reading these flash memory chips at any voltage below the suggested range is dangerous. There is a cliff at 42 μs because the read operation completes before the voltage changes.

For the rest of the chips we examined, the behaviors are similar to E-MLC8 in Figure 5(b). For E-MLC8, if the voltage droops to below 2.2 V, the chip will show significant errors no matter when we switch the supply voltage. However, if the voltage only droops to 2.5 V, which is still below the suggested range, the bit error rate is as low as reading this chip at 3.3 V. Comparing with program operations, the read operations for all chips that we examined require higher supply voltages. The minimum voltages that allow read operations to complete without incurring significant errors are also very close to the lower bound of the suggested operation voltage.

Lowering the supply voltage also increases the latency across all chips we tested. Figure 6 shows the read latencies of E-MLC8. For E-MLC8, dropping the supply voltage to 2.5 V during a read operation increases the latency by 1%.

As in program operations, lowering the operating voltage for read operations can also help reducing the energy consumption

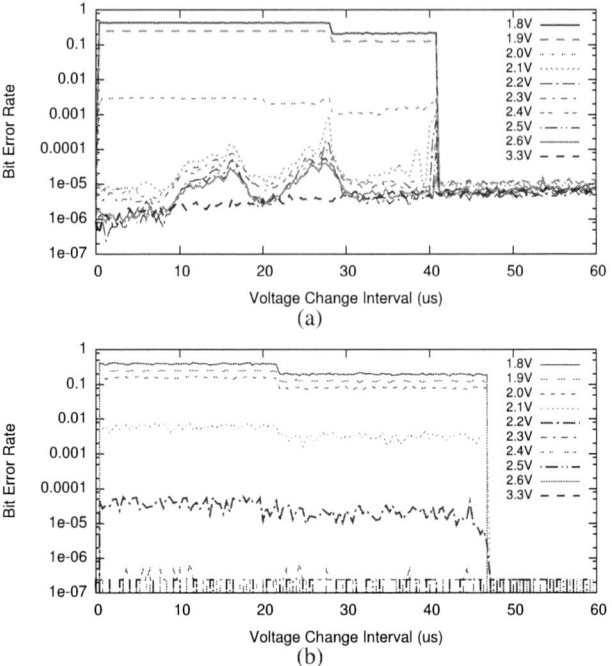

Figure 5: The bit error rate of reading random data from (a) B-MLC128 (b) E-MLC8 when supply voltage droops to 1.8 V – 2.6 V at different voltage change intervals

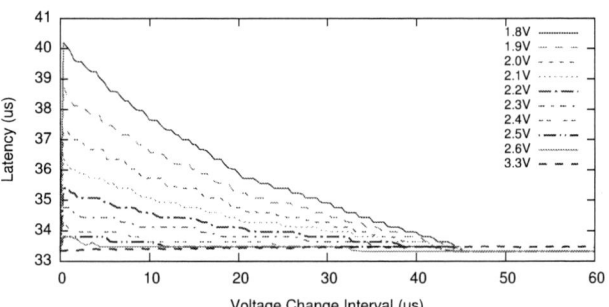

Figure 6: The latency of reading random data from E-MLC8 when supply voltage droops to 1.8 V – 2.6V at different voltage change intervals

even though the latency increases. For E-MLC8, we can reduce 36% read energy by operating the chip at 2.5 V.

4.3 Erase and power fade

To examine the effect of power fade for erase operations, we first program a block with random data, and then change the supply voltage while erasing the block. We use the number of zero bits as the bit error rate for erase operations since a block should contain 1s after a successful erase.

The behaviors of erase operations under power fade are similar across all the chips we examined. In general, if the voltage is lower than a certain level, the chip cannot erase all the bits within a block. However, for the supply voltage above a certain level, the chip seems to complete the erase operation.

Figure 7 shows the result of E-MLC8. For 1.8 V, if the voltage change interval is less than 137 μs , the erasing block have about 50% bit error rates, which implies that the erase operation fails completely. For 1.9 V, if the voltage changes at 4.8 μs , the error rate still reaches 50%. However, when the voltage only drops

1114

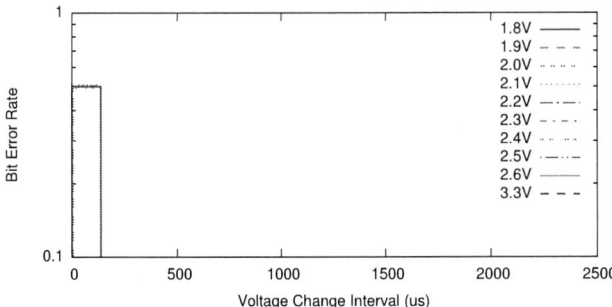

Figure 7: The bit error rate when erasing E-MLC8 with various voltage change intervals

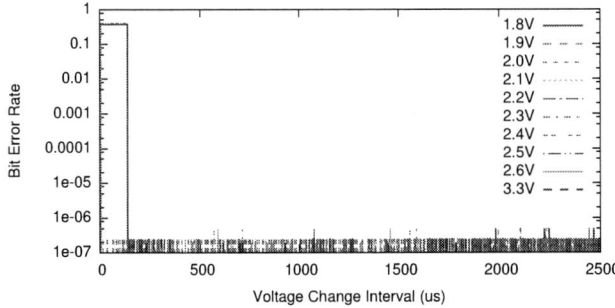

Figure 8: The program error rates of E-MLC8 of a block when supply voltage droops to 1.8 V – 2.6 V during the previous erase operation

to 2.0 V during the erase operation, the bit error rate reaches 0% (0.1 V lower than voltage required for programming), the operation succeeds.

In [14], the authors show that erasing a block with power failure can increase the error rate of later program operations even if all bits in the block seems to be in the erased state. To examine this problem, we first erase a block of random data with voltage droops. Then, we program random data into the block and measure the program error rate. However, we did not find the same result as power failure. The program error rate shows no difference if the block seems to be erased under a lower voltage level.

Figure 8 uses E-MLC8 to illustrate the result. The x-axis is the voltage change interval for the previous erase operation, and the y-axis is the bit error rate of the later program operations on the same block. The program error rate only becomes high under erasing voltage at 1.9 V and 2.0 V, where erasing the block will have high error rates. When we program a block erased under 2.0 V in E-MLC8, the error rate shows no difference comparing with programming a block erased with 3.3 V.

Figure 9 shows the latency of erase operations of E-MLC8 when we change the supply voltage during an operation. The trend is consistent with other operations. Lowering the supply voltage increases latency. In terms of energy consumption, erasing a block in E-MLC8 at 2.1 V can help reduce the energy consumption by 38%.

5. DYNAMIC VOLTAGE SCALING FOR FLASH MEMORY

As flash memory makes inroads into the mainstream storage devices for embedded systems and sensor networks, the energy consumption of flash memory becomes a design issue in these energy-limited devices. In this section, we will present a dynamic voltage scaling scheme that helps saving energy based on our characterization results.

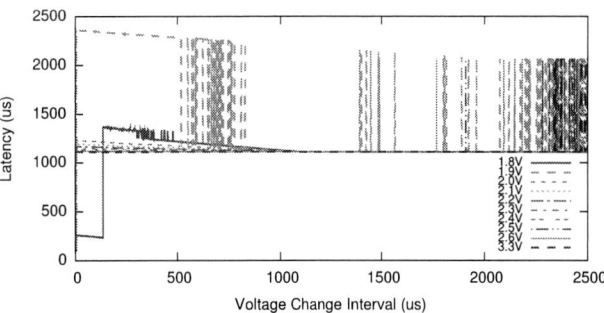

Figure 9: The latency of erasing E-MLC8 when supply voltage droops to 1.8 V – 2.6 V at different voltage change intervals

Name	Program	Read	Erase
A-MLC16	2.3V	2.5V	2.1V
B-MLC32	2.2V	2.7V	2.0V
B-MLC128	2.1V	2.7V	1.7V
E-MLC8	2.1V	2.5V	2.0V
F-MLC16	2.4V	2.4V	2.2V

Table 3: The minimum operating voltage for each chip we applied in our dynamic voltage scaling scheme

Our experiments in the previous section indicates that lowering the supply voltage during an operation does not always result in higher error rates. High-resolution power measurement also shows that operating flash memory chips at lower voltages can reduce energy consumption. We define the lowest voltage that allows a chip to complete an operation without increasing the error rate using any voltage change interval as the *minimum operating voltage* for the corresponding operation. Table 3 lists the minimum operating voltage for each chip we examined in our work.

To achieve energy saving, our scheme switches the supply voltage to the minimum operating voltage immediately after the chip starts an operation using high-speed transistors. Since the error rates do not increase when the chip is operating at its minimum operating voltage, we do not need to modify the existing error coding scheme. After the operation completes, we immediately switch back to the maximum of minimum operating voltages across all operations for that chip. For example, in E-MLC8, we will switch back to 2.5 V. For B-MLC32 and B-MLC128, because reading at lower than suggested voltage range cannot guarantee the data correctness, we only switch the supply voltage to 2.7 V during read operations. We validated the error rates of all the chips that we examined in this work to make sure these chips function correctly at the maximum of minimum operating voltages.

To understand the potential energy saving and performance

Name	Reads	Description
Berkeley-DB Btree	34%	Transactional updates to a B+tree key/value store
Build	94%	Compilation of the Linux 2.6 kernel
Financial	15%	Live OLTP trace for financial transactions [1].
Patch	83%	Applies patches to the Linux kernel from version 2.6.0 to 2.6.29
OLTP	80%	Real-time processing of SQL transactions
Postmark	97%	Models an email server
ptfdb	97%	Palomar Transient Factory database realtime transient sky survey queries
run14	45%	24 hour trace of a software development work station.
WebSearch3	99%	I/O traces from a popular search engine [1].

Table 4: We use traces from nine benchmarks and workloads to evaluate the dynamic voltage scaling for flash memory chips.

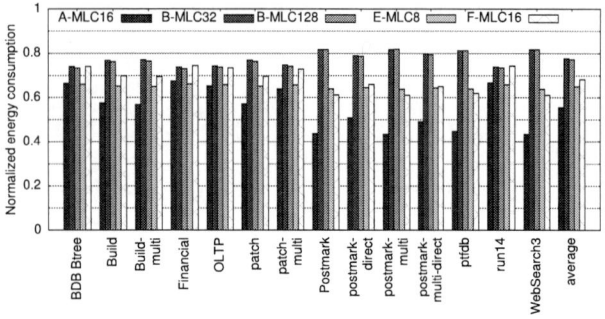

Figure 10: The relative active energy consumption after using the dynamic voltage scaling scheme

impact, we implemented the proposed dynamic voltage scaling scheme into a trace-based simulator. We use the timing and power that we measured in Section 4 for simulation. We also include the overhead of switching among different supply voltages that we described in Section 3. Table 4 summarizes the traces we used in our experiments. These traces cover a variety of applications including Internet services and databases. We ran each trace on our simulator with the latency and energy data that we measured in the previous experiments.

Figure 10 shows the normalized energy consumption (not including idle power) of our dynamic voltage scaling scheme. We normalize the energy value with respect to the energy consumption of the system that does not use any energy saving scheme. On average, we can save 22% to 45% energy using the proposed scheme. For B-MLC32 and B-MLC128, because the supply voltages we use to read data are relatively higher than other chips, the energy saving is smaller, especially in traces consist of intensive reads, such as postmark, WebSearch3, and ptfdb.

In terms of performance, the dynamic voltage scaling mechanism only hurts the performance for less than 0.1% in most benchmarks. The only exception is the run14 trace, where we suffer 2% to 19% performance degradation depending on the chip we used.

6. RELATED WORK

Flash manufactures release very little information about the behavior of flash memory chips under adverse operating conditions. Our work compliments previous efforts[14, 3] to quantify chip-level integrity properties that are otherwise unavailable. Boboila and Desnoyers [2] characterize the timing, endurance, and FTL designs for SSDs. However, none of the above works focuses on data integrity issues for power fade.

Though flash-based storage devices consumes relatively low power than conventional storage technologies, the energy efficiency of flash-based storage system is still an issue for devices that require ultra low power storage systems. The flash-based storage system still contribute 14.5% of a sensor node after using a highly optimized file system [9].

Previous efforts to optimize energy and power consumption in SSDs have tried to utilize heterogeneous storage technologies to maximize the energy efficiency for flash-based storage system [8, 10]. Lee and Chang [8] presents a memory allocation strategy that maximizes the energy efficiency for handheld devices. Mathur, Desnoyers, Ganesan, and Shenoy [10] proposes using a surface-mount NAND flash modules to replace the power hungry on-board NOR flash memory. Song, Choi, Cha, and Ha [12] saves energy for multimedia intensive embedded system by improving the flash file system to avoid redundant data compression for write and by-pass page caching for reads. Our paper is orthogonal to the above works and can help further reduce the energy consumption of ex-

isting systems.

The programming data can affect energy consumption for program operation. Joo, Cho, Shin, and Chang [6] proposed an energy-aware data compression scheme that minimizes the flash programming energy rather than data size. Salajegheh, Wang, Fu, Jiang, and Learned-Miller [11] presented software-only coding schemes to allow NOR flash chips to program at lower voltage and save energy. However, our work focus not only on the program operation, but also read and erase operations. We also demonstrate the potential of save flash energy without using special coding schemes for any flash operations.

Revising the chip design helps reducing the energy consumption for flash operations. Ishida et al. [4] uses 3D die stacking to allow multiple flash dies to share a common charge pump. Our work only requires the change in micro-controller but still saves energy.

7. CONCLUSION

This paper examines the data integrity issue when unstable or unreliable power supplies underpower flash memory chips. Underpowering flash memory chips can corrupt both programming data and data already stored in the flash memory. Underpowering flash chips can also affect the correctness of reading data and result in incomplete block erase. However, underpowering flash memory chips is not always harmful. We found that flash memory chips can tolerate voltage droops, and the tolerable range is different from operations. We utilize the result to dynamically adjust supply voltages to flash chips and achieve up to 45% energy saving.

8. REFERENCES

[1] Umass trace repository. http://traces.cs.umass.edu/index.php/Storage/Storage.

[2] S. Boboila and P. Desnoyers. Write endurance in flash drives: measurements and analysis. In *FAST '10: Proceedings of the 8th USENIX conference on File and storage technologies*, pages 9–9, Berkeley, CA, USA, 2010. USENIX Association.

[3] L. Grupp, A. Caulfield, J. Coburn, S. Swanson, E. Yaakobi, P. Siegel, and J. Wolf. Characterizing flash memory: Anomalies, observations, and applications. In *MICRO-42: 42nd Annual IEEE/ACM International Symposium on Microarchitecture*, pages 24 –33, 12 2009.

[4] K. Ishida, T. Yasufuku, S. Miyamoto, H. Nakai, M. Takamiya, T. Sakurai, and K. Takeuchi. A 1.8v 30nj adaptive program-voltage (20v) generator for 3d-integrated nand flash ssd. In *Solid-State Circuits Conference - Digest of Technical Papers, 2009. ISSCC 2009. IEEE International*, pages 238 –239,239a, feb. 2009.

[5] JEDEC. Preconditioning of Plastic Surface Mount Devices Prior to Reliability Testing. http://www.jedec.org/sites/default/files/docs/22a113F.pdf.

[6] Y. Joo, Y. Cho, D. Shin, and N. Chang. Energy-aware data compression for multi-level cell (mlc) flash memory. In *Proceedings of the 44th annual Design Automation Conference*, DAC '07, pages 716–719, New York, NY, USA, 2007. ACM.

[7] T.-S. Jung, Y.-J. Choi, K.-D. Suh, B.-H. Suh, J.-K. Kim, Y.-H. Lim, Y.-N. Koh, J.-W. Park, K.-J. Lee, J.-H. Park, K.-T. Park, J.-R. Kim, J.-H. Yi, and H.-K. Lim. A 117-mm2 3.3-v only 128-mb multilevel nand flash memory for mass storage applications. *IEEE Journal of Solid-State Circuits*, 31(11):1575 –1583, Nov. 1996.

[8] H. G. Lee and N. Chang. Low-energy heterogeneous non-volatile memory systems for mobile systems. *Journal of Low Power Electronics*, 1:52–62, 2005.

[9] G. Mathur, P. Desnoyers, D. Ganesan, and P. Shenoy. Capsule: an energy-optimized object storage system for memory-constrained sensor devices. In *Proceedings of the 4th international conference on Embedded networked sensor systems*, SenSys '06, pages 195–208, New York, NY, USA, 2006. ACM.

[10] G. Mathur, P. Desnoyers, D. Ganesan, and P. Shenoy. Ultra-low power data storage for sensor networks. In *Information Processing in Sensor Networks, 2006. IPSN 2006. The Fifth International Conference on*, pages 374 –381, 0-0 2006.

[11] M. Salajegheh, Y. Wang, K. Fu, A. Jiang, and E. Learned-Miller. Exploiting half-wits: smarter storage for low-power devices. In *Proceedings of the 9th USENIX conference on File and stroage technologies*, FAST'11, pages 4–4, Berkeley, CA, USA, 2011. USENIX Association.

[12] H. Song, S. Choi, H. Cha, and R. Ha. Improving energy efficiency for flash memory based embedded applications. *Journal of System Architecture*, 55:15–24, January 2009.

[13] K. Takeuchi, T. Tanaka, and T. Tanzawa. A multipage cell architecture for high-speed programming multilevel NAND flash memories. *IEEE Journal of Solid-State Circuits*, 33(8):1228 –1238, Aug. 1998.

[14] H.-W. Tseng, L. M. Grupp, and S. Swanson. Understanding power loss behavior on flash memory. In *DAC 2011: Proceedings of 48th Design Automation Conference*.

New ERA: New Efficient Reliability-Aware Wear Leveling for Endurance Enhancement of Flash Storage Devices

Ming-Chang Yang, Yuan-Hao Chang, Che-Wei Tsao, and Po-Chun Huang
Institute of Information Science
Academia Sinica, Taipei 115
Taiwan, R.O.C.
{mcyang, johnson, bearman.sky, aufbu}@iis.sinica.edu.tw

ABSTRACT

As the program/erase (P/E) cycles of flash memory keep decreasing, improving the lifetime/endurance of flash memory has become a fundamental issue in the design of flash devices. This work is motivated by the observation that flash blocks endured the same P/E cycles usually have different bit error rates. In contrast to the existing wear-leveling techniques that try to distribute erases to flash blocks as evenly as possible, we propose an efficient reliability-aware wear-leveling scheme to distribute block erases based on the bit error rates of blocks so as to even out the error rate among flash blocks, to maximize the number of good blocks, and thus to ultimately prolong the lifetime of flash storage devices. The experiments were conducted based on representative realistic workloads to evaluate the efficacy of the proposed scheme, for which the results are very encouraging.

Categories and Subject Descriptors

D.4.2 [**Operating Systems**]: Storage Management—*Secondary storage*

General Terms

Design, Management, Measurement, Performance

Keywords

flash memory, wear leveling, endurance, reliability

1. INTRODUCTION

Due to the advances of manufacturing technologies and the pressure of cost reduction, the low-price, high-density multi-level-cell (MLC) flash memory has gained their market share in the design of flash storage devices in recent years, as compared to its counterpart single-level-cell (SLC) flash memory. Such a development trend makes the endurance of flash storage devices keep decreasing. In order to resolve the endurance issue, wear leveling is an effective way to improve the lifetime/endurance of flash memory by distributing erases to flash blocks evenly, so as to prevent any flash block from enduring too many erases, i.e., program/erase (P/E) cycles. Using P/E cycle as the design metric, a number of excellent wear-leveling designs have been proposed and proved to be very effective to improve the endurance of

flash storage devices [1, 2, 5, 6, 7, 17]. However, due to the manufacturing process variation [8], blocks of the same flash chip usually have different bit error rates when they have endured the same number of P/E cycles; given the same ECC capability to correct bit errors and to reduce bit error rates, some blocks could endure (much) more P/E cycles than other blocks [9, 19]. As a consequence, using P/E cycle as the metric to design wear leveling might underestimate or overestimate the endurance of flash blocks; thus, it might not be able to ultimately prolong the lifetime and minimize the error rate of flash storage devices when the same ECC hardware is supported. In contrast, the hardware-dependent information such as the bit error rate and the number of error bits in each flash block might be a better metric for the wear-leveling design. Such an observation motivates us to explore the possibility on utilizing the hardware-dependent information (such as bit error rates and error types) in the wear-leveling design so as to estimate the endurance of each flash block more precisely and to ultimately prolong the lifetime of flash storage devices.

A flash storage device usually adopts multiple flash chips, and each chip usually consists of multiple planes. Each plane is composed of many blocks, each of which is made up of a fixed number of pages, where a block is the basic unit for erase operations and could endure a limited number of erases. A page is the smallest access unit for read/write operations; it consists of a data area and a spare area. The data area is for the data storage and the spare area is used to store the house-keeping information such as error correction codes, flags, and logical block addresses. Due to the *write-once property*, a page could not be overwritten unless its residing block is erased, so that *out-place update* is usually adopted to improve the write performance by writing the updated data to free pages. Thus, to access data, address translation is needed to translate the logical addresses from the host to the physical address in flash memory. Thus, multiple versions of data could coexist in flash memory. Pages containing the up-to-date (resp. out-of-date) data version are called live pages (resp. dead pages). When there are not enough free blocks for data writes, "garbage collector" should be activated to reclaim space of dead pages by erasing their residing blocks, where *live-page copying* is needed to copy live pages from the to-be-erased block to other locations before the block is erased. In the past decade, many different address translation designs (i.e., FTL designs) for flash memory have been proposed to improve the access performance and/or to reduce the RAM space usage on maintaining the address translation information. Well-known examples are DFTL and FAST [10, 11], where DFTL adopts a page-level address translation mechanism and FAST adopts a hybrid (page/block) address translation mechanism. In addition, in order to extend the lifetime of flash storage devices, some proposed different wear leveling designs to prevent wearing out some blocks excessively, based on the P/E cycles of blocks. As for wear leveling designs, many excellent designs based on dynamic wear leveling (DWL) [1, 2, 6] have been proposed to achieve wear leveling by recycling blocks of hot (i.e., frequently updated) data; in contrast, more recent designs were based on static wear leveling (SWL) [5, 7, 17], which pro-actively moves static or cold (i.e., infrequently updated) data to other locations so as to prevent any cold data from staying at any block for a long period of time.

In contrast to existing wear leveling designs, we are interested in the wear leveling designs with considering the hardware-dependent information, such as the bit error rate, error type, and number of error bits, so as to estimate the reliability level/status of each block more precisely. In this paper, we propose a new efficient reliability-aware wear leveling scheme to improve the lifetime of flash storage devices. By utilizing the hardware information, the proposed scheme could separate different types of errors to better estimate the reliability level/status of each block, so that it could even out the error rates among blocks and maximize the number of good blocks as long as possible in order to ultimately prolong the lifetime/endurance of flash storage devices with limited extra overheads on live-page copyings and block erases. In particular, transient errors in blocks could be detected and eliminated to prevent from underestimating the endurance of any block. A series of experiments based on realistic workloads was conducted to evaluate the capability of the proposed scheme. We show that the proposed scheme could outperform both DWL and SWL by 2.5 times.

The rest of this paper is organized as follows: Section 2 introduces the typical system architecture and the motivation of this work. In Section 3, a new efficient reliability-aware wear leveling scheme is proposed. Section 4 summarizes the experiment results. Section 5 is the conclusion.

2. BACKGROUND AND MOTIVATION

NAND flash memory could be classified into two types, single-level-cell (SLC) and multiple-level-cell (MLC) flash memory, according to the number of bits that could be contained in one flash cell. Each cell of an SLC, $MLC_{\times2}$, and $MLC_{\times3}$ (also referred to as triple-level-cell (TLC)) flash memory could store data of one, two, and three bits, respectively. As shown in Table 1 [14, 18], an SLC (resp. MLC) block is usually composed of 128 (resp. 256 or 384) pages, each of which usually could store 4KB (resp. 8KB or 16KB) user data and 128B (resp. 218B or 976B) spare information. Consequently, MLC flash memory usually has lower cost but has worse endurance and access performance than SLC flash memory does. Besides, MLC flash memory has two constraints on writing pages: (1) The content of a page could not be partially modified/programmed, and (2) pages of a block must be written sequentially from the first page to the last one.

Cell type	SLC	$MLC_{\times2}$	$MLC_{\times3}$(TLC)
Price (USD/32Gbs)	25.1	4.08	3.87
Page size (KBs)	4	8	8 or 16
Block size (pages)	128	256	384
Page read time (μsec.)	35	75	100
Page write time (msec.)	0.3	1.3	2.5
Block erase time (msec.)	0.7	3.8	3
Endurance (P/E cycles)	≥ 10000	≈ 3000	≤ 1000

Table 1: Comparisons of flash chips [14, 18].

As shown in Figure 1, a typical flash storage device usually equips with two firmware layers, i.e., the flash translation layer (FTL) driver and the memory technology device (MTD) driver. The MTD driver provides an unified

interface for the FTL driver to control flash chips through primitive access operations such as read, write, and erase. The FTL driver is named after its main functionality on translating any logical address to its corresponding physical address, and usually consists of three components: *allocator*, *garbage collector*, and *wear-leveler*. The allocator handles address translation from logical block addresses (LBAs) to the physical addresses of their corresponding valid data because multiple data versions of an LBA could coexist in flash memory, where each LBA represents the logical address of a sector. The garbage collector is to reclaim the space of dead pages by erasing their residing blocks when there is not enough free space in flash memory, where "live-page copying" should be performed to copy live pages to free space before their residing block is erased. The wear-leveler is an optional component to extend the lifetime/endurance of flash memory by evenly distributing erases over blocks.

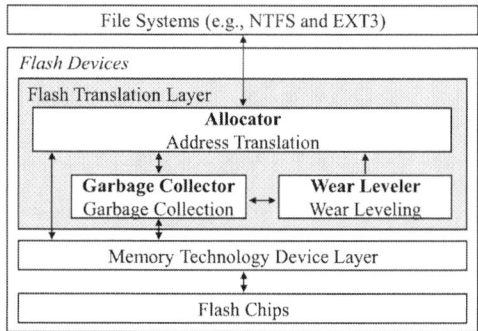

Figure 1: A typical flash system architecture.

Bit errors in a flash page mainly come from two types of errors: *write error* and *transient error*. Write errors are the errors caused by over-programming to flash cells or fast-erasing cells when a page is programmed [4]; and the average number of write error bits in a page is increased as its residing block has endured more P/E cycles [3, 15]. Transient errors could be classified into *disturb errors* and *retention errors*; disturb errors of a page mainly come from the disturbs caused by reading and writing data in the same block of the page, and retention errors of a page are caused by the charge losses over time since the page was programmed [12]. Thus, on average, write error bits in a page are increased when the number of erases to its residing block is increased; and transient error bits in a page are reset after the page's residing block is erased although they are accumulated (1) when the time passes after the page was programmed and (2) when the number of reads/writes to the page's residing block is increased. However, *write error bits are hard to be distinguished from transient error bits because only the total number of error bits could be detected when the error correction code (ECC) is applied to correct error bits*[1]. On the other hand, due to the manufacturing process variation [8], blocks of the same flash chip have different bit error rates when they have endured the same number of erases (or P/E cycles); among them, some even have very different bit error rates, and some could be worn out prematurely [9, 19]. Thus, given the same ECC capability to correct bit errors and reduce bit error rates, some blocks could endure (much) more P/E cycles than other blocks. That is, the endurance of some blocks is higher than that of other blocks and might be higher than that defined in the specification of flash chips.

---

[1]ECCs are usually used to correct error bits in each page with the ECC redundancy stored in the spare area of the corresponding page.

As a result, using P/E cycle as the metric to design wear leveling might not be able to ultimately prolong the lifetime of flash memory because the endurance of some blocks could be underestimated.

This work is inspired by the observation that the bit error rate or the number of error bits could be a better metric to design wear leveling schemes than the P/E cycle (or erase cycle) does, even though the P/E cycle is widely used in existing wear leveling designs [5, 7, 17]. Our goal is to propose an error-rate-aware and error-type-aware wear-leveling module to ultimately improve the endurance/lifetime of flash memory with the considerations of the serious challenges in restricted resource in embedded systems and embedded memory cards, where the wear-leveling module should be easily integrated into existing FTL designs. The technical problem is how to effectively distinguish write errors from transient errors and how to even out the write errors in each block/page so as to prevent turning any block into a bad one prematurely because transient errors could be reset or eliminated after the erases of blocks; nonetheless, it is also a challenging issue on how to prevent transient errors from failing the supported ECC on correcting page errors without excessive block erases or management overheads.

3. NEW EFFICIENT RELIABILITY-AWARE WEAR LEVELING

3.1 Overview

In this section, a new efficient reliability-aware wear leveling scheme (referred to as *New ERA*) is proposed to perform wear leveling based on the bit error rate and error type (i.e., write error and transient error), instead of the P/E cycle. The objective of the proposed scheme is to separate write errors from transient errors, to minimize the write error differences among blocks, and to minimize the probability of accumulating too many error bit in pages to be corrected by the supported ECC, so that the lifetime and the reliability of flash storage devices could be maximized. In this work, we consider a modular design for the reliability-aware wear leveling (ERA-WL) so that it can be integrated into many existing FTL designs with limited overheads/efforts.

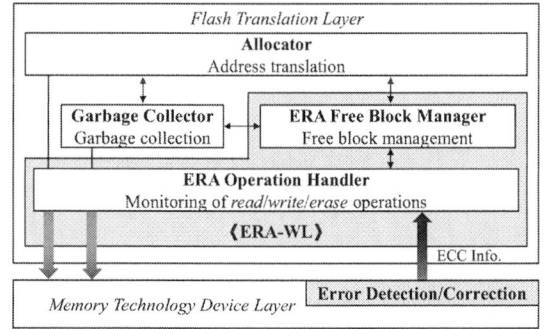

Figure 2: New ERA module.

As shown in Figure 2, the proposed scheme is designed as a modular component, called efficient reliability-aware wear leveler *ERA-WL*, in the FTL driver. The ERA-WL consists of two major components: *ERA operation handler* and *ERA free block manager* (see Section 3.3). The ERA operation handler intercepts read, write, and erase operations from the allocator and the garbage collector, and redirects them to the MTD driver, so that it can monitor and maintain the reliability information of blocks and the usage states of blocks; that is to maintain the write errors in the *write*

error table in order to distinguish write errors from transient errors, and to remember the states, e.g. *bad, free,* and *allocated,* of each block in the *block state table* (see Section 3.2). On the receiving of an operation, the ERA free block manager would get the ECC information from the underlying MTD driver (see Figure 2) to check the number of error bits in the accessed page, and perform reliability-aware wear leveling when the accessed page contains too many error bits. On the other hand, the ERA free block manager manages free blocks in order of their reliability level. When the garbage collector erases a block, the erased block is put in the free block list managed by the ERA free block manager, and when the allocator needs a new free block, it simply invokes the ERA free block manager to select and return a free block to it. Note that in order to prevent wearing out blocks, the ERA free block manager would perform reliability-aware wear leveling when all of the free blocks in the free block manager contains too many write error bits.

3.2 Write Error Table and Block State Table

The *write error table (WET)*, as shown in Figure 3, is a counter array and each counter is used to keep track of the write error status of each block so as to distinguish write errors from transient errors for each block, because the write error status could reflect the reliability level of a block. However, ECC is usually used to correct error bits in the unit of one flash page. Thus, each counter of the WET is only used to remember the largest *write error bit count*, i.e., the largest number of write error bits, among the pages in the corresponded block, so that the write error status of each block will not be underestimated with a minimized table size. In order to estimate the largest write error bit count for each block more accurately, we apply the *read-after-write* validation to check the number of error bits right after a page is programmed[2]. That is, after a page is written, the page data are read back to check the number of error bits in this page immediately because there are minimal number of transient error bits in this page at this moment. If this number is larger than the write error bit count of the page's residing block, it becomes the new largest write error bit count of this block; otherwise, it is simply ignored. Note that counters in the WET are initialized to 0, but as the time passes, the write error bit count maintained in each counter will approach the real largest write error bit count of each block.

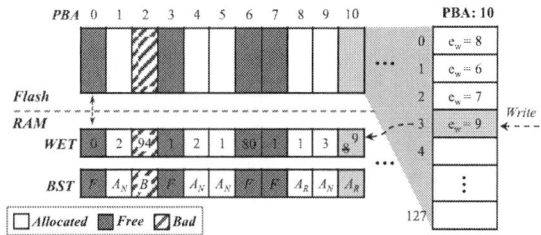

Figure 3: WET and BST.

On the other hand, the *block state table (BST)* is to remember the state of each block, and each block could be in one of the four states: bad, free, recently-allocated, and non-recently-allocated states (as shown in Figure 3). If the largest write error count of a block exceeds the pre-determined threshold, it is marked as a bad one and enters the bad state (denoted as 0 or "B" state); it will no longer be allocated to

---

[2]The read-after-write validation is adopted when flash storage devices use the less reliable MLC/TLC chips; it could be implemented in hardware and done by the hardware automatically [16].

store data. Meanwhile, when a block is erased to become a free block, it enters the free state (denoted as 1 or "F" state) and is put into the free block list for the future free block allocation. Once a free block is allocated to store data, it becomes a recently-allocated block and enters the recently-allocated state (denoted as 2 or "A_R" state), and will be advanced to the non-recently-allocated state (denoted as 3 or "A_N" state) after a while; thus, a block in the A_N state usually contains more transient errors than it stays in the A_R state.

3.3 ERA-WL

The ERA-WL is composed of three procedures, ERA-OpHandler, FreeBlk-Alloc, and FreeBlk-Ins, to handle the reliability-aware wear leveling, where ERA-OpHandler is implemented in the *ERA operation handler* and both FreeBlk-Ins and FreeBlk-Alloc are implemented in the *ERA free block manager* (see Figure 2). ERA-OpHandler intercepts read, write, and erase operations and redirects them to the MTD driver so as to monitor and maintain the reliability levels (i.e., write errors) of blocks and the states of blocks in WET and BST respectively, and to perform reliability-aware wear leveling when the accessed page in some allocated block contains too many error bits; thus, it could significantly reduce the ECC's failure rate on correcting error bits. Besides, FreeBlk-Ins is usually invoked by the garbage collector after the garbage collector erases a block and wants to insert the erased block into the free block list. FreeBlk-Alloc is usually invoked by the allocator to allocate a free block from the free block list when there are not enough free blocks.

In order to make the reliability level of blocks more evenly (or to even out the write errors among blocks), an efficient reliability-aware free block list (referred to as *ERA free block list*) is maintained in the ERA-WL so as to maintain free blocks and return free blocks with low write error bit counts the allocator efficiently. As shown in Figure 4, the ERA free block list is a multi-list indexed in order of the (largest) write error bit count, and free blocks are linked to one list according to the value in their corresponding WET counters. In the ERA-WL, two thresholds T_L and T_H are defined to classify blocks into different types of blocks according to the reliability level, i.e., the largest write error bit count, of blocks, where T_L and T_H could be defined according to the supported ECC capability, the specification of flash chips, and the reliability-level of the flash storage devices. Blocks with their largest write error bit counts smaller than T_L are considered *strong blocks*; blocks with those between T_L and T_H are weak blocks; and to simply the discussion, blocks with those between T_H and the supported ECC capability are simply marked as bad blocks by setting their corresponding entries in the BST to the B state, so as to avoid allocating these blocks for data storage unless there is no strong block in the system.

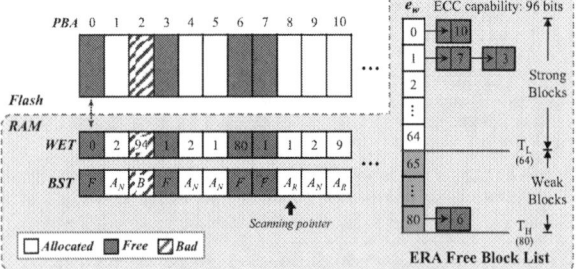

Figure 4: ERA free block list.

Algorithm 1 shows the algorithm of ERA-OpHandler,

where *op* is the received operation, *data* is the page buffer for the read/write operation, and *ppa* and *pba* are the address of the accessed page and the address of this page's residing block respectively. ERA-OpHandler is invoked to intercept operations when the allocator or the garbage collector issues a read, write, or erase operation to the underlying MTD driver. If a read operation is received (Lines 1-5), ERA-OpHandler invokes the standard read operation provided by the MTD driver to read the page data from flash memory and gets the number err_{cnt} of error bits in this page from the MTD driver or from the ECC hardware (Lines 2-3). If ecc_{cnt} is larger than T_H, block *pba* is replaced with the strongest free block in the ERA free block list, because block *pba* contains too many error bits, especially transient error bits, in this page and needs to be erased to prevent failing the ECC on correcting error bits and to reset the transient errors (Lines 4-5); note that the strongest free block is the free block whose largest write error bit count is the smallest among the free blocks in the ERA free block list.

Algorithm 1: ERA-OpHandler

Input: *op*, *ppa*, *pba*, *data*, *WET*, and *BST*
Output: NULL

```
1  if op = READ then                              /* READ */
2  |   data ← READ(ppa);
3  |   err_cnt ← ECC-CHECK();
4  |   if err_cnt > T_H then
5  |   └   BLOCK-REPLACE(pba);

6  else if op = Write then                        /* WRITE */
7  |   WRITE(ppa, data);
8  |   err_cnt ← READ-ECC-CHECK(ppa);
9  |   if err_cnt > WET[pba] then      // Update write error
10 |   └   WET[pba] ← err_cnt;
11 |   if err_cnt > T_H then           // Check bad block
12 |   |   BST[pba] ← B;
13 |   └   BLOCK-REPLACE(pba);
14 |   else if err_cnt > T_L and WET[BestFreeBlk] < T_L then
15 |   └   BLOCK-REPLACE(pba);

16 else if op = ERASE then                        /* ERASE */
17 └   ERASE(block_pba);
```

If a write operation is received (Lines 6-15), the data are written to the designated page *ppa* (Line 7) and then read back to verify the write error bit count, i.e., the number err_{cnt} of error bits, in this page (Line 8). If err_{cnt} is larger than the largest write error bit count of the page's residing block *pba*, the block's corresponding WET counter is updated with err_{cnt} (Lines 9-10). Then, if err_{cnt} is larger than T_H, the page's residing block *pba* is marked as a bad one and replaced with the strongest free block in the ERA free block list, because there are too many write error bits (Lines 11-13); otherwise, err_{cnt} is compared with T_L; if $err_{cnt} > T_L$ and the strongest free block in the ERA free block list is good enough (i.e., smaller than T_L), block *pba* is replaced with the strongest free block because this block *bpa* has already got too many write errors and is not strong enough; *thus, such a block replacement could prevent wearing out weak blocks excessively so as to ultimately prolong the lifetime of weak blocks; as a result, the number of bad blocks could be minimized to ultimately prolong the lifetime of the flash storage device.* (Lines 14-15). At last, if the received operation is an erase one, ERA-OpHandler simply invokes the standard erase function provided by the MTD driver to erase the designated block pba (Lines 16-17).

Algorithm 2: FREEBLK-ALLOC

Input: p, N, WET, and BST
Output: a free block b

1   **if** IS-STRONG-FREEBLK() $= TRUE$ **then**
2 $b \leftarrow$ the strongest free block;
3   **else if** IS-WEAK-FREEBLK() $= TRUE$ **then**
4   **for** $n = 1$ **to** N **do**
5   **if** $BST[p] = A_R$ **then**
6 $BST[p] = A_N$;
7   **else if** $BST[p] = A_N$ *and* $WET[p] < T_L$ **then**
8   Exchange block p with the weakest free block;
9   Erase(p);
10 $b \leftarrow p$;
11   Move p to next block circularly;
12   **return** b;
13   Move p to next block circularly;
14 $b \leftarrow$ the strongest free block;
15   **else**
16 $b \leftarrow null$;
17   **return** b;

FREEBLK-ALLOC, as shown in Algorithm 2, is invoked by the allocator when the allocator needs a new free block, where p is an index that points to flash blocks for performing circular scan, and N is the maximum number of blocks scanned in each invocation of the procedure. It would perform the reliability-aware wear leveling by switching weak free blocks with the strong blocks that are in the non-recently-allocated state when the ERA free block list only contains weak free blocks, so as to prevent wearing out any block (in terms of write errors) prematurely and to even out the write errors among blocks pro-actively. As shown in Algorithm 2, when FREEBLK-ALLOC is invoked, it simply selects and returns the strongest free block in the ERA free block list when there are strong free blocks in the list (Lines 1-2). If there are only weak free blocks in the list (Line 3), the reliability-aware wear leveling is activated. When wear leveling is activated, FREEBLK-ALLOC checks the states of the consecutive N blocks from the block pointed by p circularly with the next-fit algorithm until a strong block in the non-recently-allocated state is found (Lines 4-13). If a block in the recently-allocated state is found, the block's state is advanced to the non-recently-allocated state because this block has just been allocated to store new data, which are hard to be identified as hot data or cold data at the moment (Lines 5-6). However, if this block is checked next time and it is still in the non-recently used state, it is very likely that the data stored in this block are cold data. Thus, if a strong block in the non-recently-allocated state is found, it is swapped with the weakest free block in the ERA free block list by exchanging their roles; then this strong block is erased to become a free block and is returned to the allocator (Lines 7-12). Here, Figure 5 shows an example on releasing a strong block (i.e., Block 9) in the non-recently-allocated state. Nonetheless, if no strong block in the non-recently-allocated state is found after searching for N consecutive blocks, the currently strongest free block in the ERA free block list is allocated for the allocator (Line 14). Note that N should be much smaller than the number of blocks in the storage device to reduce the time delay on free block allocation; meanwhile, a small N can reduce the frequency on mistakenly switching free weak blocks with allocated strong blocks that contains

hot data, because hot data tend to be invalided in the near future to let their residing blocks have higher probability to be erased by the garbage collector due to the consideration of the live-page copying overheads. Finally, if the ERA free block list has no free blocks, no free block is returned to the allocator (Line 16). When the allocator fails to get a free block from FREEBLK-ALLOC, it usually invokes the garbage collector to reclaim space by erasing blocks of invalid space. After that, FREEBLK-INS is invoked to insert the erased block in to the ERA free block list and mark the block as free, so that the allocator could obtain free blocks in the next invocation of FREEBLK-ALLOC.

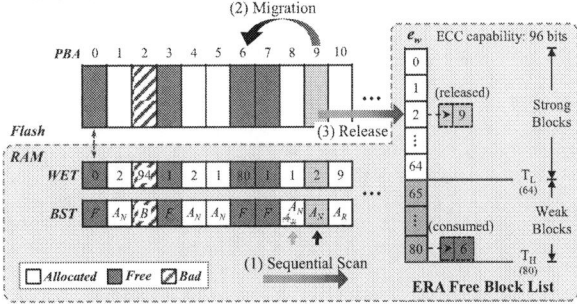

Figure 5: Search strong blocks and release them.

4. PERFORMANCE EVALUATION

4.1 Evaluation Metrics and Experiment Setup

In this section, we evaluate the efficacy of the proposed efficient reliability-aware wear leveling scheme (referred to as ERA-WL), in terms of lifetime and performance. The lifetime is evaluated based on the first failure time (referred to as FFT), i.e., the first time that an error could not be corrected by the supported ECC, and the performance is based on the average response time (referred to as ART) per read/write request. The proposed ERA-WL is evaluated when it works with the well-known FTL designs: $DFTL$ and $FAST$. Meanwhile, the efficacy of the proposed ERA-WL is also compared with some existing well-know wear leveling approaches, DWL and SWL, that are designed to distribute the P/E cycles of blocks evenly [7]. The first approach is a dynamic wear leveling design (referred to as DWL) that recycles blocks of dynamic data in a greedy fashion according to the live-page-copying overhead on reclaiming a block, and the second approach adopts a static wear leveling design (referred to as SWL) that proactively reclaims blocks of cold/static data so as to prevent any data from staying in the same block for a long time. In order to evaluate the efficacy of SWL when there are some extra blocks to replace bad blocks, we also investigate SWL with extra space that is 20% (referred to as *enhanced SWL*) of the storage capacity for bad block replacement.

# of chips	4	Random Read	$60\mu s$
Chip size	16GB	Serial Read	$52.8\mu s$/page
Block size	1MB(256 pages)	Write	$800\mu s$
P/E cycles	1000	Erase	$1.5ms$
ECC ability	256 bits/page	BER variation	10^{-12}–10^{-3}

Table 2: The evaluated MLC flash storage [14, 18]

As shown in Table 2, a 64GB four-chipped MLC NAND flash storage device was under investigation. The range of the bit error rate (BER), including write error and transient error, of a block was set between 10^{-12} to 10^{-3} when the ECC is not supported, and the BER is increased in consistent with the number of P/E cycles [15], where the distri-

1121

bution of BERs of blocks is simulated with a normal distribution. In this experiment, the supported ECC capability is 256 bits per 4KB page, and the thresholds T_L and T_H are set as 192 and 224 (bits) respectively. Three representative traces collected from enterprise servers by Microsoft Research Cambridge, acquired from the SINA IOTTA repository [13], were used to evaluate the efficacy of the investigated wear leveling approaches.

4.2 Experimental Results

Figure 6 shows the lifetime achieved by the investigated approaches, where the x-axis denotes the trace type and the y-axis denotes the first failure time normalized to that of DWL. It is observed that the lifetime of the proposed ERA-WL was at least multiple times of that of both DWL and SWL under all of the evaluated traces when DFTL or FAST was adopted, because the proposed ERA-WL performs wear leveling based on the write errors in blocks and could precisely identify the time to reset transient errors by erasing blocks. In contrast, DWL and SWL only consider the P/E cycles of blocks, but neither erase blocks to prevent unrecoverable errors caused by transient errors in time nor distribute more eases to the blocks having fewer write error bits. Thus, the ERA-WL could achieve at least 3.5 and 4.06 times of the lifetime of enhanced SWL under DFTL and FAST respectively. In addition, since DWL only recycles blocks of dynamic data in a greedy fashion, it achieves poorer lifetime as the access locality is higher. For example, the File Server trace is a trace with very high locality, so that both SWL and ERA-WL could outperform DWL for more times under the File Server trace but fewer times under the Media Server trace. In addition, the proposed ERA-WL could perform better under read-intensive traces because it could proactively erase blocks to reset the transient errors caused by read operations, but DWL and SWL are not aware of the transient errors caused by read disturbances. Figure 7 shows the performance overheads incurred by the investigated approaches. We observe that the proposed ERA-WL introduces more performance overheads than DWL and SWL since it requires to read data back to verify error bits after each write operation. However, it is interesting to see that the proposed ERA-WL introduces more performance overheads under DFTL than under FAST. This is because FAST itself introduces more writes than DFTL does.

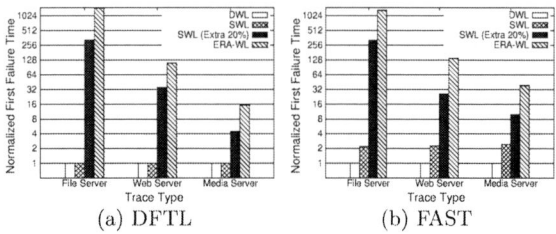

(a) DFTL   (b) FAST

Figure 6: Normalized FFT over different FTLs.

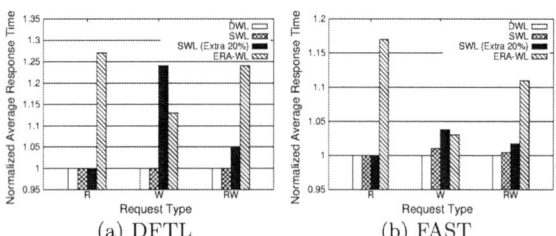

(a) DFTL   (b) FAST

Figure 7: Normalized ART (Web Server).

5. CONCLUSIONS

This paper proposes a reliability-aware wear leveling scheme to improve the endurance of flash storage devices with the considerations of the error rates and error types. Through the read-after-write validation, the write errors could be separated from transient errors, so that the reliability level of blocks could be better estimated. Thus, the error rates among blocks could be even out to maximize the number of good blocks as long as possible, so as to significantly improve the lifetime of flash storage devices with limited extra overheads on live-page copyings and block erases. Meanwhile, transient errors could be reset/eliminated to reduce the ECC's failure rate on correcting page errors and to prevent underestimating or overestimating the endurance of blocks. We conducted extensive experiments to evaluate the lifetime improvement and overhead of the investigated approaches based on representative real-world traces. The experimental results show that the proposed scheme could outperform both DWL and SWL by 2.5 times. In the future research, we shall investigate the capability of the proposed scheme with more different hardware configurations and exploit more hardware information to improve the wear-leveling designs and flash management designs.

6. REFERENCES

[1] Increasing Flash Solid State Disk Reliability. Technical report, SiliconSystems, Apr 2005.

[2] A. Ban. Wear Leveling of Static Areas in Flash Memory. US Patent 6,732,221. *M-systems*, 2004.

[3] Y. Cai, E. F. Haratsch, O. Mutlu, and K. Mai. Error Patterns in MLC NAND Flash Memory: Measurement, Characterization, and Analysis. In *DATE*, 2012.

[4] P. Cappelletti, C. Golla, P. Olivo, and E. Zanoni. *Flash Memories*. Kluwer Academic Publishers, 1999.

[5] L.-P. Chang. On Efficient Wear Leveling for Large-scale Flash-memory Storage Systems. In *the 2007 ACM symposium on Applied computing (SAC)*, 2007.

[6] L.-P. Chang and T.-W. Kuo. An Adaptive Striping Architecture for Flash Memory Storage Systems of Embedded Systems. In *the IEEE Real-Time and Embedded Technology and Applications Symposium (RTAS)*, pages 187–196, 2002.

[7] Y.-H. Chang, J.-W. Hsieh, and T.-W. Kuo. Endurance Enhancement of Flash-Memory Storage Systems: An Efficient Static Wear Leveling Design. In *the 44th ACM/IEEE Design Automation Conference (DAC)*, June 2007.

[8] M. Dietrich and J. H. (Eds.). *Process Variations and Probabilistic Integrated Circuit Design*. Springer, 2012.

[9] G. Dong, N. Xie, and T. Zhang. Techniques for embracing intra-cell unbalanced bit error characteristics in mlc nand flash memory. In *IEEE GLOBECOM Workshops*, 2010.

[10] A. Gupta, Y. Kim, and B. Urgaonkar. DFTL: a flash translation layer employing demand-based selective caching of page-level address mappings. In *ASPLOS*, 2009.

[11] S.-W. Lee, D.-J. Park, T.-S. Chung, D.-H. Lee, S. Park, and H.-J. Song. A Log Buffer-Based Flash Translation Layer Using Fully-Associative Sector Translation. *ACM Transactions on Embedded Computing Systems*, 6(3), July 2007.

[12] R.-S. Liu, C.-L. Yang, and W. Wu̧. Optimizing NAND Flash-Based SSDs via Retention Relaxation. In *USENIX FAST*, 2012.

[13] SNIA. IOTTA repository. http://iotta.snia.org/tracetypes/3.

[14] Micron. *SLC NAND Flash Memory Features MT29F8G08ABACA*, 2010.

[15] N. Mielke, T. Marquart, N. Wu, J. Kessenich, H. Belgal, E. Schares, F. Trivedi, E. Goodness, and L. R. Nevill. Bit Error Rate in NAND Flash Memories. In *IRPS*, 2008.

[16] S. Moon and A. L. N. Reddy. Write Amplification due to ECC on Flash Memory or Leave those Bit Errors Alone. In *IEEE MSST*, 2012.

[17] M. Murugan and David.H.C.Du. Rejuvenator: A Static Wear Leveling Algorithm for NAND Flash Memory with Minimized Overhead. In *IEEE MSST*, 2011.

[18] SpecTek. *64Gib TLC NAND Flash Features FNNB74A*, 2011.

[19] G. Tressler, D. Vanstee, and T. Griffin. Enterprise MLC NAND Industry Comparison. In *Flash Memory Summit*, 2011.

SAW: System-Assisted Wear Leveling on the Write Endurance of NAND Flash Devices

Chundong Wang and Weng-Fai Wong
School of Computing
National University of Singapore
Email: {wangc.nus@gmail.com, wongwf@comp.nus.edu.sg}

ABSTRACT

The write endurance of NAND flash memory adversely impacts the lifetime of flash devices. A flash cell is likely to wear out after undergoing excessive *program/erase (P/E) flips*. Wear leveling is hence employed to spread erase operations as evenly as possible. It is traditionally conducted by the *flash translation layer* (FTL), a management firmware residing in flash devices. In this paper, we shall propose a novel wear leveling algorithm involving the operating system (OS). We will show that our *operating System-Assisted Wear leveling* (SAW) algorithm can significantly improve the wear evenness. SAW takes advantage of OS's knowledge about files at a higher level of abstraction, and provides useful hints to the lower-level FTL to accommodate data. A prototype based on a file system and an FTL has been developed to verify the effectiveness of SAW. Experiments show that wear evenness can be improved by as much as 85.0% compared to the state-of-the-art FTL wear leveling schemes.

1. INTRODUCTION

Today it is economically feasible to utilize NAND flash devices for secondary storage in both embedded systems and general-purpose computing systems. The issue of *write endurance*, however, hinders the further use of NAND flash as the lifespan of a flash device is inherently limited.

Write endurance refers to the limited cycles of *program/erase (P/E) flips* on a flash *page*. A page is the unit for write and read operations in NAND flash. It consists of thousands of cells. A page cannot be directly reprogrammed with updates due to the physical characteristics of flash cells. It has to be erased first. Programming a page is to selectively set some bits to '0'. All bits are reset to '1' by an erase operation. Erase operations must be performed in the unit of a *block* that consists of multiple pages. Excessive P/E flips will wear out a page, and damage its ability to retain data. The limitation of P/E flips for single-level cell (SLC) NAND flash which stores one bit in a cell, is 100,000 cycles. For the denser multi-level cell (MLC) flash, it is about

10,000. A flash block that has a worn-out pages, i.e., a worn-out block, cannot be used any further. Too many worn-out blocks will lead to the complete failure of a flash device [19].

Wear leveling [20, 10, 2, 14, 19] has been devised to target the write endurance of NAND flash. Generally, wear leveling attempts to intelligently put data in suitable blocks to avoid the skewness of erase operations. To do so, it first needs to classify data and blocks, respectively. *Hot* data are ones that are frequently updated, while data that are never or seldom rewritten are deemed to be *cold*. On the other side, the erase count of each flash block is maintained in a *block aging table* [14]. A block that has a smaller erase count is *younger* than one that has a larger erase count. Wear leveling tries to put hot data into younger blocks, and cold data into elder blocks. It is expected that updates of hot data can result in many erasures, while blocks occupied by cold data can avoid being frequently erased. However, the correct identification of hot or cold data, and how to site them in a block of a suitable age are two challenging issues.

Wear leveling, and other modules of flash management, are included in a firmware of flash devices called the *flash translation layer* (FTL) [8, 12, 20, 10, 7]. The FTL is designed to be self-contained. The host operating system (OS) communicates with the flash device through interfaces like USB or SATA, and is generally oblivious to the management of flash memory. The OS sends requests to the FTL, and waits for replies in a client-server way, treating the flash device as a black box. File system-aware FTLs [9, 8, 12] have been proposed, but they focus only on deleted data, metadata or critical data. Our paper, however, will exploit OS's knowledge of files for wear leveling in a collaborative way.

Data from the OS have to be assessed for the purpose of wear leveling. Traditionally, the FTL estimates data by itself, and allocates pages and blocks accordingly [20, 1]. The estimation is achieved either through recording data's update frequencies [5, 17], or with the aid of address translation scheme in use [20]. Nonetheless, the FTL's classification of data within a flash device is arduous and costly on time and space. Worse, at that level, the FTL has no idea of the intended use of data. On the other hand, the OS is aware of the file type data belong to, and what applications are using them. The scheme proposed in this paper, namely *operating System-Assisted Wear leveling* (SAW), will leverage on the OS knowledge of the files being accessed so that the FTL can do a better job in wear leveling. The key ideas of SAW, also the main contributions of this paper, are as follows:

- Data of files are *quantitatively* analyzed and classified using a measure of the *temperature*. The temperatures

---

Permission to make digital or hard copies of all or part of this work for personal or classroom use is granted without fee provided that copies are not made or distributed for profit or commercial advantage and that copies bear this notice and the full citation on the first page. To copy otherwise, to republish, to post on servers or to redistribute to lists, requires prior specific permission and/or a fee.
DAC '13, May 29 - June 07 2013, Austin, TX, USA.

of files of various types are periodically updated by the OS, which is based on a succinct and reasonable mathematical model. The computed temperature is sent along with data to the FTL.

- Flash blocks that are available for use, i.e., *free blocks*, are organized into groups using an *exponential division*. The FTL interprets the temperature of data, and performs allocation from a relevant group accordingly.

- A prototype based on the Linux virtual file system (VFS), an open-source file system and its special FTL, i.e., UBIFS and UBI [13], has been developed to verify the effectiveness of SAW.

UBIFS and UBI are specifically designed for raw flash that is found in embedded systems like smartphones. With the above prototype, experiments show that the wear evenness can be significantly improved by as much as 85.0% compared to the state-of-the-art FTL-based wear leveling schemes.

The rest of this paper is organized as follows. Section 2 shows background and related works. Section 3 presents the details of SAW. Section 4 describes the experimental evaluation. Section 5 is our acknowledgement and Section 6 will conclude the paper.

2. BACKGROUND AND RELATED WORKS

The FTL is a firmware that autonomously manages flash device. It processes access requests from upper-level file system by translating logical address to physical address into the form of flash block and page. It also conducts wear leveling, as well as garbage collection for resource reclamation. The latter is used to reclaim pages and blocks that are occupied by obsolete data owing to out-of-place updating.

Wear leveling is an important module of the FTL. It may either be *static* or *dynamic* [10, 14]. Dynamic wear leveling generally selects the youngest free block for new data, while static wear leveling may vacate the block currently occupied by cold data for use. The latest FTL wear leveling schemes include dual-pool algorithm [1], BET [10], lazy wear leveling [2] and OWL [20]. We shall focus on BET and OWL as they represent two different strategies. BET takes the perspective of the flash block. It has a *Block Erasing Table* (BET) to maintain the erase status of each flash block in every fixed interval. If the count of erasures over the number of erased blocks exceeds a predefined threshold, BET will repeatedly pick un-erased blocks of the last interval, and perform data transfers, after which it will erase them until the skewness is smoothed out. OWL, however, emphasizes on data. OWL also has a table, namely the *Block Access Table* (BAT), in which a block is a logical block instead of a flash block. The BAT stores access frequencies of logical blocks that have been recently rewritten. OWL ranks data of logical blocks with the BAT, and allocates flash blocks accordingly. The ranking is used to predict a logical block's access frequency compared to other logical blocks in the near future. Doing so OWL attempts to put data into suitable blocks in a proactive way to avoid wear unevenness.

Note that both BET and BAT are data structures maintained by the FTL inside flash device. There are non-trivial spatial and temporal overheads in doing so.

There are FTLs that were devised to take file system into account. MFTL [8] interposes a filter between file system and the FTL to separate metadata and real data of files.

It specifically manages metadata that are small and frequently updated. MFTL was implemented for ext2 and ext3 file systems, and performance improvement was reported. FSAF [9] focuses only on deleted data in FAT32. It is similar to the TRIM command of modern OS [3]. FSAF detects the deletion by utilizing its knowledge about the format of FAT32 in flash devices. Meta-Cure [12] is similar to MFTL. It adds a filter between file system and FTL to enhance the reliability of "critical data" to avoid being damaged. Meta-Cure does not change the file system; it is transparent to the FTL. In all these works, either the OS is unaware of FTL's workings, or vice-versa.

3. SAW

In this section we will present details of how the OS collaborates with the FTL for wear leveling in SAW. The OS manages files for applications. It segments or assembles data of files to satisfy applications' access requests. Thus, the OS knows which file a data segment belongs to, and which application is requesting it. Such information is invisible to the lower-level FTL. Traditionally, data are either coarsely identified to be hot or cold, or arduously classified by the FTL within the limited computation resource of a flash device. It is where SAW is to make a difference. In SAW, the OS is responsible for *quantitatively* classifying files based on a mathematical model. For a file type, the OS detects its files' updates, and generates a *temperature* that is sent along with each data segment to the FTL. When a data segment arrives in the flash device, the FTL extracts the temperature information, and processes the allocation request accordingly.

3.1 File Type Temperatures

Files have attributes, such as filename, extension, access mode, and last modified time. Mesnier et al. [6] revealed that files' properties, like access pattern, can be predicted based on their attributes. Take a text file for example. It is likely to be rewritten more often than a video file. Mesnier et. al. [6] did an offline mining over collected files. Online exploration of the files' attributes, however, is not straightforward. We need a succinct and reasonable mathematical model for doing so.

For simplification, SAW only considers two attributes of a file, namely its filename extension and access mode. Read-only files are hardly rewritten, and will be specially dealt with. A rewritable file will be assigned a temperature degree according to it *type*. Here a file's type refers to its filename extension, although it is conceivable that other attributes can be used too. Files without any extension will be treated separately. Previous qualitative ways to identify data to be hot or cold are somewhat lacking. With the assistance of the OS, we will perform the classification in a quantitative way. The temperature of a file type, as will be derived below, depends on files' update *frequency* and *recency*.

3.1.1 Update Frequency of A File Type

Measuring the update frequency of a file type is the key issue of SAW. FTLs can record the number of writes to a logical block or a logical page [20]. For files, however, it is not so simple. The OS manages a large number of files of the same type. It is neither reasonable nor scalable to keep access information for each file. Moreover, two files with the same type may have completely different update frequencies. Hence, we need an approximation to represent

1124

the access frequency for a **type** of files. Since not all files are accessed at runtime, we will not consider *dormant* files but focus only on *active* ones. This simplifies the online analysis, and also reduces the overhead of resuming SAW at boot-up.

SAW maintains several variables for a file type. Given a file type t, $S_{(t)}$ records the total number of active files of type t. $\varsigma_{(t)}$ is the number of accessed files of type t. This includes the files of type t that have been opened (and possibly then closed) after the current system boot, as well as newly created files. $\delta_{(t)}$ is the number of files of type t that have been deleted (since the last boot). $\omega_{(t)}$ counts the *rewrites* to all t files. $\varsigma_{(t)}$, $\delta_{(t)}$ and $\omega_{(t)}$ are used to compute the update frequencies of file type t. Note that we are interested in rewrites detected in the kernel module of file system, not writes, because the latter is not a good estimate for update frequency. For example, a video file triggers a vast number of writes during its creation. Afterwards its contents are hardly rewritten again. So a video file's update frequency is low. When a text file is reopened, however, it may be inserted, appended or replaced with new data. Thus, its update frequency is much higher due to many rewrites.

Let $\varphi_{(t)}$, the *update frequency* of type t, be defined as

$$\varphi_{(t)} = \frac{\omega_{(t)}}{S_{(t)}}. \tag{1}$$

At the first sight, $\varphi_{(t)}$ seems to be the average rewrite of active files of type t. Nonetheless, as is mentioned, files of the same type may differ significantly in update behavior. Moreover, files are being created and deleted at runtime. So Equation (1) is imprecise. But it is infeasible to keep too much information for each file. We shall place more constraints to enhance the accuracy of Equation (1).

First, the OS will collect the values of ς, δ and ω **periodically**. The interval is defined as I. The total number of active files of type t after the nth I is to be $S_{(t)}^n$. The base case, i.e., at boot-up, is defined as $S_{(t)}^0$, and initialized to be zero. In the nth I interval, $\varsigma_{(t)}^n$ files were newly accessed or created, and $\delta_{(t)}^n$ files were removed. So the number of type t files before the start of the $(n+1)$th I is

$$S_{(t)}^{n+1} = S_{(t)}^n + (\varsigma_{(t)}^n - \delta_{(t)}^n). \tag{2}$$

Hence, the absolute increment of type t files is

$$S_{(t)}^{n+1} - S_{(t)}^n = \varsigma_{(t)}^n - \delta_{(t)}^n. \tag{3}$$

The rate of increase of type t files, $s_{(t)}^n$, of the nth I, is

$$s_{(t)}^n = \frac{\varsigma_{(t)}^n - \delta_{(t)}^n}{S_{(t)}^n}, \tag{4}$$

where $n \geq 1$ because at boot-up $S_{(t)}^0 = 0$, and it is in the first I that files are accessed or created.

$s_{(t)}^n$ could be positive or negative, as the number of type t files may increase or decrease. β is a bound such that

$$-\beta \leq s_{(t)}^n \leq \beta, \tag{5}$$

or put in another way,

$$S_{(t)}^{n+1} = S_{(t)}^n \cdot (1 \pm \beta), \tag{6}$$

and Equation (1) is hence valid for the calculation of temperature. In this paper, β is set to be 10%. We do not expect the number of active files of type t changes sharply. If $|s_{(t)}^{n+1}| > \beta$, we will identify t to be an *outlier*. An outlier deserves special attention since many t files are likely to be created or removed in a short period of time.

3.1.2 Update Recency

After the nth I, Equation (1) can be rewritten as

$$\varphi_{(t)}^n = \frac{\omega_{(t)}^n}{S_{(t)}^n}. \tag{7}$$

Equation (7) gives the rewrite frequency on a file type t, and it estimates the update behavior of type t files in the $(n+1)$th interval. However, $S_{(t)}^n$ accumulates the number of active files during past n intervals. As time goes by, the updates of type t may change a lot due to the context switch of applications. Hence a value from a long time ago may mislead the estimation. Generally, the most recent intervals are more relevant to the coming interval, and this *recency* should be factored into Equation (7).

We introduce another variable to improve Equation (7), $f_{(t)}^n$, which is defined to be the predicted value for $\varphi_{(t)}^n$ of the nth interval. Next, we define an *exponentially average* value of $f_{(t)}^{n+1}$ for the $(n+1)$th I as

$$f_{(t)}^{n+1} = \alpha \cdot \varphi_{(t)}^n + (1 - \alpha) \cdot f_{(t)}^n, \tag{8}$$

in which $0 \leq \alpha \leq 1$. When $\alpha = 0$, the recent interval will have no effect. With $\alpha = 1$, the past history is assumed to have no influence. Given an α that $0 < \alpha < 1$, we have

$$
\begin{aligned}
f_{(t)}^{n+1} &= \alpha \cdot \varphi_{(t)}^n + (1 - \alpha) \cdot f_{(t)}^n \\
&= \alpha \cdot \varphi_{(t)}^n + (1 - \alpha) \cdot \left[\alpha \cdot \varphi_{(t)}^{n-1} + (1 - \alpha) \cdot f_{(t)}^{n-1} \right] \\
&= \ldots \\
&= \alpha \cdot \varphi_{(t)}^n + (1 - \alpha) \cdot \alpha \cdot \varphi_{(t)}^{n-1} + \ldots + (1 - \alpha)^i \cdot \alpha \cdot \varphi_{(t)}^{n-i} \\
&\quad + (1 - \alpha)^{(i+1)} \cdot \alpha \cdot \varphi_{(t)}^{n-(i+1)} + \ldots + (1 - \alpha)^{n+1} \cdot f_{(t)}^0.
\end{aligned}
$$

Because

$$\alpha > (1 - \alpha) \cdot \alpha > \ldots > (1 - \alpha)^i \cdot \alpha > \ldots > (1 - \alpha)^n \cdot \alpha, \tag{9}$$

we can conclude for $f_{(t)}^{n+1}$, the farther an interval is, the less the effect it has ($f_{(t)}^0 = 0$ and $\varphi_{(t)}^n = 0$, so the last $(1-\alpha)^{n+1}$ is ignorant). In other words, $f_{(t)}^{n+1}$ depends the most on $\varphi_{(t)}^n$, and also takes the past history into consideration when $0 < \alpha < 1$. Now we can use $f_{(t)}^{n+1}$ to predict the future update behavior to files of type t in the $(n+1)$th interval.

3.1.3 Temperature of File Types

Now that we have f^n for all file types, we can compute their temperature before each interval. The temperature used in this paper is from 0 to T. T is a predefined constant. A file with the zero degree is very cold, effectively like a read-only file. If a file's temperature is near to T, it is very hot. Given a set of file types, each one with an f^{n+1} for the $(n+1)$th interval, we sort them by their f values in an ascending order. The type t then has a *position number* in the sequence, $P_{(t)}^{n+1}$, where $0 \leq P_{(t)}^{n+1} \leq \Theta - 1$. Θ is the number of active file types. For example, there are five file types (i.e., $\Theta = 5$), and the sorting sequence is

$$f_{(t_0)}^{n+1} \leq f_{(t_1)}^{n+1} \leq f_{(t_2)}^{n+1} \leq f_{(t_3)}^{n+1} \leq f_{(t_4)}^{n+1}.$$

So $P_{(t_0)}^{n+1} = 0$ and $P_{(t_3)}^{n+1} = 3$. Then we can calculate the temperature for type t, $C_{(t)}^{n+1}$, using

$$C_{(t)}^{n+1} = \frac{P_{(t)}^{n+1}}{\Theta} \cdot T. \tag{10}$$

If T is set to be 5, $C^{n+1}_{(t_3)} = 3$ for type t_3. Note that Equation (10) is valid when $n \geq 1$. The temperature of each type t for the first interval, i.e., $C^1_{(t)}$, is initialized to be zero.

It may seem tedious to have to perform a sort over Θ file types after each interval. However, since the access behaviors of the majority of file types are stable, a complete re-sorting is not yet necessary. Instead, SAW scans the previous sequence with updated f values, performing the necessary reordering. This is fairly inexpensive.

According to Equation (10), it is not possible for a file to have a temperature of T, as T is reserved for outlier files.

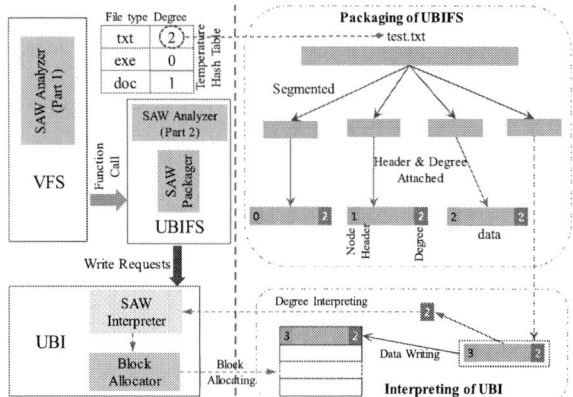

Figure 1: A Sketch of SAW Prototype

3.2 Wear Leveling with Temperature

3.2.1 Exponential Division of Flash Blocks

We use the temperature of a file to allocate blocks and pages to its data. First, we need a hash table to maintain the temperatures for file types. The hash key is a file type t that is hashed to its $f^n_{(t)}$ for the nth interval. This table is managed by the OS in main memory, not in flash devices.

The basic idea of wear leveling is to allocate young blocks to hot data, and old blocks to cold data. To make use of the temperature, free blocks in flash device should be well organized. SAW sorts them in an ascending order by their erase counts. As we have T degrees, all free blocks are divided into T groups. The division is not equal but in an *exponential* way. Assuming there are Γ sorted free blocks, the first group has $\Gamma/2$ blocks that have the smallest erase counts. The second group has $\Gamma/2^2$ blocks. By analogy, the gth group has $\Gamma/2^g$ $(0 \leq g \leq T-1)$ blocks. The Tth group, however, is an exceptional one that keeps $\Gamma/2^{(T-1)}$ blocks that are the most worn at that time. This is because

$$\Gamma = (\frac{\Gamma}{2^1} + \frac{\Gamma}{2^2} + \frac{\Gamma}{2^3} + ... + \frac{\Gamma}{2^g} + ... + \frac{\Gamma}{2^{(T-1)}}) + \frac{\Gamma}{2^{(T-1)}}. \quad (11)$$

In SAW, an allocation request with a temperature of d is satisfied by the $(T-d)$th block group. Whether to allocate a page or a block depends on the FTL's allocation policy. As is mentioned, SAW specially treats read-only and outlier files. The former corresponds to the Tth group, and the latter will be handled with pages and blocks from the first group. There are usually not that many read-only files, so $\Gamma/2^{T-1}$ blocks should be sufficient. Outlier files are quite active. They are accommodated into the youngest $\Gamma/2$ blocks.

The exponential division is due to our intention to make the best use of young blocks that are the least worn. SAW maintains more blocks to the higher temperatures. Ones with the smallest erase counts are given more chance to be utilized, while elder block can avoid being frequently picked.

3.2.2 Temperature Adjustment

The temperature is re-calculated in every interval. Hence, cold data would lag behind with outdated temperature since they are infrequently updated. Their temperatures should be adjusted. To look for such cold data is not easy. SAW will not do it by itself. As is mentioned, there is a module called garbage collection in flash management to clean up obsolete data that are generated due to out-of-place updating. Cold data are left with them. SAW works alongside when garbage collection are being conducted. At this time, SAW checks data to be moved and changes their temperature. They are written back by garbage collection with updated temperature then. In this way, the overhead is minimized.

3.3 A Prototype of SAW

We have developed a prototype of SAW based on UBIFS and UBI [13]. Generally, there are two types of flash device. One is found in solid-state drives (SSDs), SD cards and USB thumb drives. On equipment such as smartphones, raw flash may be used. UBIFS is designed for the latter, and UBI can be viewed as its special FTL. We chose UBIFS and UBI to implement SAW because they are open-source.

UBIFS is a log-structure file system. UBI serves UBIFS to access data and performs functionalities of flash management. Several features of UBI and UBIFS facilitate the implementation of SAW. First, UBIFS roughly classifies data to be LONGTERM, SHORTTERM and UNKNOWN. For example, all files' data are hot, i.e., SHORTTERM. Second, data are encapsulated by UBIFS in a *node* with information like the inode number that they belong to [13]. Note that the coarse identification of data is not embedded into nodes. Though, the node structure makes it possible to add our temperature degree into each node. Third, their original wear leveling and garbage collection are not complicated and can be easily replaced or enhanced.

The prototype of SAW has three components, as is shown in Figure 1. The *SAW analyzer* is implemented in the Linux VFS and UBIFS. It maintains the hash table and performs SAW calculations. The *SAW packager* is in UBIFS. It packages data along with relevant temperature into a node. The *SAW interpreter* of UBI supports block allocations using the temperature. Figure 1 also gives a sample on text file "test.txt". The temperature degree of "txt" is 2. The file is segmented into four parts, each packed with the temperature. When a node arrives in UBI, SAW interpreter will suggest to the allocator what would be a suitable age for the block to be allocated. The temperate would be written to flash along with data. Note that in real implementation the temperate is in the header. Here we separate it out for ease of discussion. For the same reason, the writing sequence of the nodes does not adhere strictly to their header numbers.

4. EXPERIMENTAL EVALUATION

The evaluation of SAW was done in two ways. The first is within the above prototype. We compiled the Linux kernel 3.1.6 in Ubuntu 12.04.1. A flash device of 1GB was simulated using the nandsim simulator of Linux kernel. BET was implemented for comparison. The second way we evaluated SAW was with the FlashSim simulator [11], in which we implemented OWL, BET, lazy wear leveling and SAW. The simulated flash was also 1GB. We went on further to enhance BET and lazy wear leveling with the basic idea of SAW on block allocation.

The reason why we did experiments in two ways is that OWL and lazy wear leveling work within hybrid address mapping [20, 2], so they cannot be implemented in UBI. BET does not have such a limitation [10, 20]. The NAND flash in the simulation was configured according to a recent datasheet [15]. The wear evenness is measured using the average erase count and its standard deviation over all flash blocks [20, 19]. For similar average erase counts, the smaller the standard deviation is, the better the wear evenness is.

We did not find any file system benchmarks that target the write endurance of flash memory. What we want are ones that operate on a large number of files and generate sufficient write requests. We examined the analysis of Traeger et al. [4] on various benchmarks, and selected two macro-benchmarks: postmark [16] and filebench [18]. Postmark is single-thread, while filebench can be multi-thread. However, they both name file in sequential numbers without any extension. We modified them in order to append a suffix to each file in the form of ".ϵ". ϵ is a lower-case English letter from 'a' to 'z' randomly picked for a file.

The parameters of SAW are set as follows. $T = 10$ and $\alpha = 0.5$. I is relatively measured in terms of write requests. Its default length is 10,000 write requests.

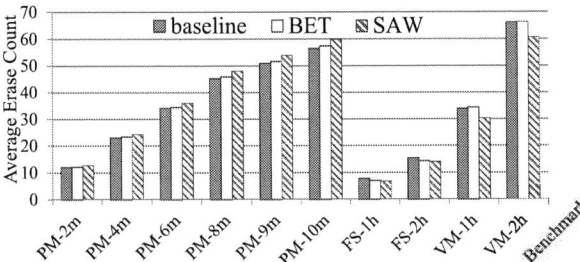

Figure 2: Average Erase Count with Prototype

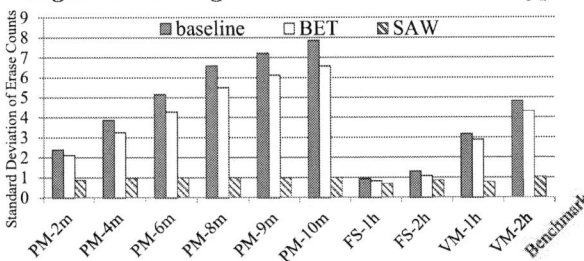

Figure 3: Standard Deviation of Erase Counts with Prototype

4.1 The Effectiveness of SAW

Figure 2 and 3 show the average erase count and standard deviation of baseline, BET and SAW with the prototype. baseline has the original wear leveling of UBIFS and UBI. BET and SAW refer to implementations of BET and SAW, respectively. We ran postmark with ten settings, from 1 million to 10 million transactions. The number of simultaneous files was 50,000. Because of space limitation, we present the results of 2, 4, 6, 8, 9 and 10 million transactions, and they are referred to as PM-2m, PM-4m and so on. We ran filebench with two public workloads: fileserver and varmail. For each workload we ran for an hour and two hours, respectively. They are referred to as FS-1h, FS-2h, VM-1h and VM-2h. The number of files was also set to be 50,000. As both postmark and filebench have random behaviors at runtime [16, 4], we ran our experiments with each

setting thrice, and the results shown in Figure 2 and 3 are the mean values. Full results are presented in the Appendix.

The effectiveness of SAW is evident. From Figure 2 we can see that in each case SAW performed a similar number of erasures compared to baseline and BET. However, in Figure 3, SAW's standard deviation of erase counts significantly decreases compared to BET, as much as 85.0% with PM-10m. Even with FS-1h and FS-2h that are read-dominant workloads, the reductions can reach 17.3% and 22.8% compared to BET, respectively. Hence we conclude that SAW effectively avoids wear skewness with the cooperation of the OS.

Measuring the performance overheads is not straightforward with the involvement of the OS. Moreover, the changing behaviors of postmark and filebench during each run make direct comparison difficult. We have recorded counts of write, read and erase operations for each case as indicators of the performance. They can be found in the Appendix.

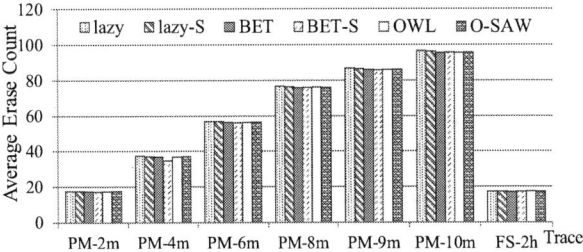

Figure 4: Average Erase Count with FlashSim

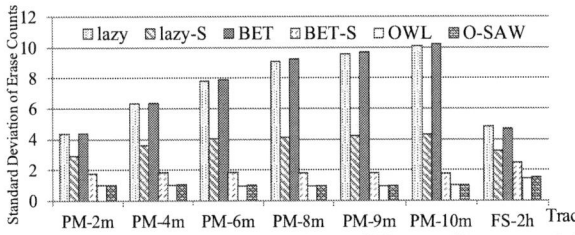

Figure 5: Standard Deviation of Erase Counts with FlashSim

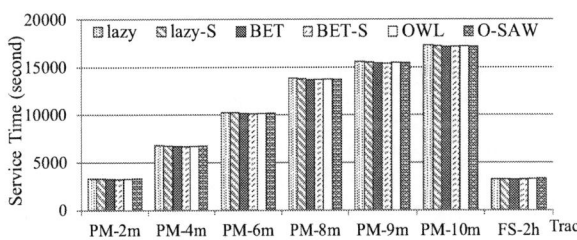

Figure 6: Service time with FlashSim

OWL, lazy wear leveling, and BET were implemented in FlashSim. The latter two were enhanced with SAW's idea in their block allocation. Their implementation are referred to as OWL, lazy, BET, lazy-S and BET-S, respectively. OWL has already considered block allocation. We did not enhance it. Instead, we replaced OWL's block allocation with SAW's. This implementation is referred to as O-SAW.

Note that FlashSim is a trace-driven simulator. Previous experiments on FlashSim utilized traces collected from various machines. However, since write requests of those traces have no temperature information, they are not suitable. Instead, we recorded access request in UBI. There, each request does have a temperature. These traces were then fed to FlashSim. Experimental results are partially shown in Figure 4, 5 and 6 due to the space limitation.

Figure 4 and 5 show that the average erase counts on a trace for each scheme is similar, but the standard deviation has decreased significantly for `lazy-S` and `BET-S`, by as much as 55.9% and 82.6%, respectively. Thus, wear evenness was highly improved in the presence of SAW. On the other side, `O-SAW` has comparable wear evenness to `OWL`, and the standard deviation of the former is at most 7.0% more. But `OWL` allocates blocks according to its own calculation utilizing the lower computation capability of a flash device, while `O-SAW` just needs to use the temperature of each incoming request. Hence, `O-SAW` has a much lower computation and resource overhead, while achieving a similar level of wear evenness.

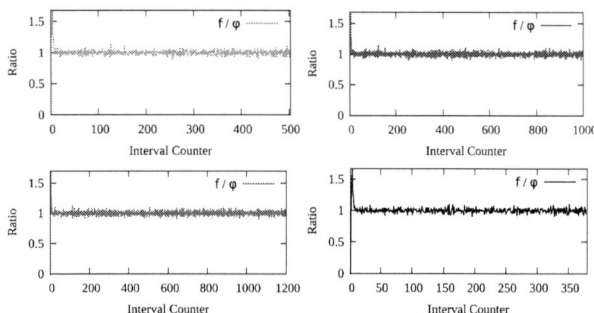

Figure 7: Fluctuation of f/φ (Clockwise: PM-5m, PM-10m, FS-2h, VM-2h)

The performance overhead can be measured using trace driven simulation because it is entirely deterministic. The time needed to service all requests of a trace is a good indicator of the performance overhead incurred by wear leveling [20, 19]. The more the service time, the greater the performance degradation. Figure 6 shows the service time for traces with each scheme. It is obvious that the addition of SAW has little performance impact.

4.2 The Accuracy of f for φ

f is used to predict φ for the next interval using Equation (8), which is the basis of the temperature calculation. We ran experiments to verify the accuracy of f for φ. Without loss of generality, we selected the file type whose filename extension is ".c". We collected f and φ in every interval with PM-5m, PM-10m, FS-2h and VM-2h, and calculated f/φ, as is shown in Figure 7. We can see after system boot-up, f/φ fluctuates within tight bounds around 1.0. Hence, we conclude that the prediction of f for φ is accurate.

4.3 Impact of Interval Length

I is an important parameter of SAW. Its default length is 10,000 write requests. We also experimented with lengths of 5,000, 15,000, 20,000 and 25,000. They are referred to as `5k`, `10k`, `15k`, `20k` and `25k`, respectively. Because of space limitation, we could only show their standard deviation in each case in Table 1. There are the mean values over five intervals, as well as the absolute mean differences between the value of each I and the mean. From Table 1 we can see the fluctuation caused by changes of I is insignificant.

5. ACKNOWLEDGEMENT

This paper is supported by the Ministry of Education of Singapore under the grant MOE2010-T2-1-075. We also thank Sudipta Chattopadhyay of NUS for his valuable help.

6. CONCLUSION

In this paper, we revisit the write endurance issue of flash device, and propose a novel scheme named *operating System-Assisted Wear leveling* (SAW). In SAW, the OS participates in the process of wear leveling by exploiting its higher level view of files. Using our proposed model, the OS quantitatively estimates the temperature of files, and sends this information to the lower-level FTL that in turn uses it for block allocation. We have developed a prototype based on an open-source flash file system and FTL. Experiments show that the collaboration between the OS and the FTL improves the wear evenness by as much as 85.0% compared to the latest FTL-based wear leveling algorithms.

Table 1: Mean Difference of Standard Deviation with Five Intervals (I)

Benchmark	Mean	Absolute Mean Difference				
		5k	10k	15k	20k	25k
PM_2m	0.865	0.009	0.001	0.001	0.002	0.013
PM_4m	0.960	0.018	0.004	0.014	0.007	0.007
PM_6m	0.981	0.019	0.002	0.004	0.017	0.004
PM_8m	0.986	0.007	0.025	0.002	0.002	0.018
PM_9m	0.982	0.015	0.012	0.002	0.003	0.022
PM_10m	0.985	0.007	0.011	0.023	0.014	0.009
FS-1h	0.692	0.017	0.001	0.024	0.004	0.007
FS-2h	0.861	0.047	0.009	0.016	0.029	0.007
VM-1h	0.789	0.012	0.004	0.011	0.002	0.003
VM-2h	1.036	0.004	0.021	0.017	0.026	0.026

7. REFERENCES

[1] L.-P. Chang. On efficient wear leveling for large-scale flash-memory storage systems. In *SAC '07*.

[2] L.-P. Chang and L.-C. Huang. A low-cost wear-leveling algorithm for block-mapping solid-state disks. In *LCTES '11*.

[3] Intel Corporation. What are the advantages of TRIM and how can I use it with my SSD? http://www.intel.com/support/ssdc/hpssd/sb/CS-031846.htm.

[4] A. Traeger et al. A nine year study of file system and storage benchmarking. *Trans. Storage*, 4(2):5:1–5:56, May 2008.

[5] J.-W. Hsieh et al. Efficient identification of hot data for flash memory storage systems. *Trans. Storage*, 2(1):22–40, Feb. 2006.

[6] M. Mesnier et al. File classification in self-* storage systems. In *ICAC '04*.

[7] P.-C. Huang et al. Joint management of RAM and flash memory with access pattern considerations. In *DAC '12*.

[8] P.-L. Wu et al. A file-system-aware FTL design for flash-memory storage systems. In *DATE '09*.

[9] S. K. Mylavarapu et al. FSAF: file system aware flash translation layer for NAND flash memories. In *DATE '09*.

[10] Y.-H. Chang et al. Endurance enhancement of flash-memory storage systems: an efficient static wear leveling design. In *DAC '07*.

[11] Y. Kim et al. FlashSim: A simulator for NAND flash-based solid-state drives. In *SIMUL '09*.

[12] Y. Wang et al. Meta-Cure: a reliability enhancement strategy for metadata in NAND flash memory storage systems. In *DAC '12*.

[13] A. Hunter. A brief introduction to the design of UBIFS, 2008.

[14] Micron Technology Inc. TN-26-61: Wear-leveling in Micron® NAND flash memory. Technical report, Oct 2011.

[15] Micron Technology, Inc. NAND flash memory datasheet (MT29F16G08AJADAWP), Feburary 2012.

[16] J. Katcher. Postmark: A new file system benchmark. Technical Report TR3022, Oct. 1997.

[17] D. Park and D. H.-C. Du. Hot data identification for flash-based storage systems using multiple bloom filters. In *MSST '11*.

[18] File system and Storage Lab. Filebench benchmark, 2011. http://sourceforge.net/projects/filebench/.

[19] C. Wang and W.-F. Wong. Extending the lifetime of NAND flash memory by salvaging bad blocks. In *DATE '12*.

[20] C. Wang and W.-F. Wong. Observational wear leveling: an efficient algorithm for flash memory management. In *DAC '12*.

APPENDIX

A. MORE EXPERIMENTAL RESULTS

A.1 The Impact of β

It was mentioned in Section 3.1 that we use β as a bound for identifying whether a file type is an outlier or not. Without loss of generality, we collected the s value in every interval for the file type ".c" using the experimental configurations PM-5m, PM-10m, FS-2h and VM-2h. The results are presented in Figure A-1. At boot-up, many files are accessed, so s is somewhat large. After the system has warmed up, s fluctuates marginally around 0. Note that β was set to be 10% by default. In summary, experiments show that s is typically much less than the selected threshold.

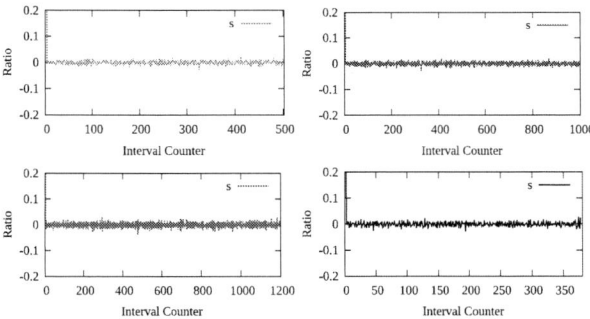

Figure A-1: s and β at runtime (Clockwise: PM-5m, PM-10m, FS-2h, VM-2h)

A.2 Experimental Results with the Prototype

As mentioned in the paper, due to the essentially non-deterministic nature of the operating system, there can be differences between each run of the experiments. Here, we present the full experimental results of baseline, BET and SAW in terms of the average erase count, standard deviation, the counts of write and read operations in Table A-1, A-2 and A-3. Table A-1, A-2 and A-3 show results recorded at each time, respectively. The count of erase operations is not separately listed because it can be computed using the average erase count in each table.

The detailed experimental results of five settings on the interval I are presented in Table A-4, including the average erase count and standard deviation.

A.3 Experimental Results with FlashSim

The average erase count, standard deviation and service time of lazy, lazy-S, BET and BET-S, OWL and O-SAW are shown in three tables for readability. They are Table A-5, A-6 and A-7. The traces of VM-1h and VM-2h were not fed to FlashSim because they are too short. Note that the computation time of OWL's sorts is not included in service time due to FlashSim's limitation. Instead, we recorded the count of sorting for each trace. It is also shown in Table A-7.

The result of the prototype and that of the trace-driven simulation under the same setting may be different, or even vary significantly. Take the average erase count for example. The value of PM-5m using BET in Table A-1 is 28.855, but in Table A-6 the average erase count is 46.355. We ascribe this to the different simulators (nandsim vs Flashsim) used. nandsim is within a full-system simulator while Flashsim performs standalone trace-driven simulation.

Table A-1: Average Erase Count, Standard Deviation, the Counts of Write and Read Operations of baseline, BET and SAW (1st Time)

Benchmark		Average Erase Count			Standard Deviation			Write Operations			Read Operations		
		baseline	BET	SAW	baseline	BET	SAW	baseline	BET	SAW	baseline	BET	SAW
Postmark	PM-1m	6.572	6.624	6.842	1.471	1.350	0.791	2,972,904	3,000,917	3,119,305	5,949,162	5,634,356	5,049,137
	PM-2m	12.043	12.196	12.643	2.387	2.115	0.860	5,877,722	5,950,703	6,196,399	8,443,477	9,615,839	9,516,198
	PM-3m	17.512	17.771	18.464	3.167	2.731	0.912	8,779,117	8,896,513	9,281,541	12,810,709	13,366,750	14,236,254
	PM-4m	22.985	23.430	24.276	3.864	3.271	0.957	11,681,921	11,836,146	12,363,103	16,897,790	17,397,386	20,548,770
	PM-5m	28.446	28.855	30.075	4.497	3.798	0.952	14,578,317	14,767,427	15,437,939	21,119,642	22,706,330	23,939,011
	PM-6m	33.956	34.461	35.929	5.186	4.319	0.992	17,500,788	17,737,843	18,544,059	26,133,797	25,403,395	32,996,678
	PM-7m	39.590	40.135	41.868	5.828	4.894	0.976	20,485,814	20,746,099	21,692,520	30,624,022	29,920,760	49,397,386
	PM-8m	45.249	45.916	47.881	6.606	5.509	0.975	23,484,623	23,796,587	24,881,391	35,741,196	34,436,772	55,940,048
	PM-9m	50.798	51.627	53.847	7.207	6.098	0.979	26,429,906	26,812,825	28,042,582	40,370,526	40,111,468	47,798,743
	PM-10m	56.377	57.412	59.750	7.817	6.525	0.991	29,384,145	29,829,525	31,170,566	45,729,523	47,253,426	48,540,694
Filebench	FS-1h	7.628	7.224	6.655	0.915	0.825	0.678	4,035,325	3,788,685	3,518,173	253,787,028	233,511,118	213,650,267
	FS-2h	15.463	14.195	12.767	1.285	1.068	0.817	8,192,460	7,440,788	6,761,944	520,335,165	465,269,296	417,412,195
	VM-1h	34.619	34.574	27.985	3.219	2.892	0.738	18,354,325	18,180,748	14,829,527	161,193,400	164,616,315	135,871,795
	VM-2h	65.281	68.581	55.862	4.747	4.346	0.976	34,619,953	36,225,506	29,611,400	364,102,368	326,448,894	270,803,909

Table A-2: Average Erase Count, Standard Deviation, the Counts of Write and Read Operations of baseline, BET and SAW (2nd Time)

Benchmark		Average Erase Count			Standard Deviation			Write Operations			Read Operations		
		baseline	BET	SAW	baseline	BET	SAW	baseline	BET	SAW	baseline	BET	SAW
Postmark	PM-1m	6.572	6.621	6.839	1.472	1.328	0.777	2,977,457	2,996,229	3,117,289	5,603,129	4,346,563	5,790,084
	PM-2m	12.060	12.173	12.667	2.398	2.094	0.864	5,886,879	5,935,576	6,208,200	8,700,539	8,570,968	9,520,877
	PM-3m	17.476	17.743	18.469	3.126	2.693	0.917	8,791,218	8,880,335	9,284,858	12,817,376	13,057,314	14,646,823
	PM-4m	23.035	23.379	24.270	3.877	3.255	0.942	11,708,459	11,834,954	12,361,087	17,244,826	19,895,452	19,788,998
	PM-5m	28.509	28.853	30.085	4.561	3.752	0.958	14,610,057	14,766,016	15,442,771	21,478,833	26,090,482	23,782,072
	PM-6m	34.014	34.431	35.895	5.097	4.242	1.004	17,528,201	17,721,035	18,525,806	25,659,251	28,238,393	33,947,237
	PM-7m	39.626	40.155	41.861	5.818	4.811	0.961	20,503,432	20,752,174	21,688,147	30,191,413	32,778,028	45,098,977
	PM-8m	45.270	45.907	47.865	6.602	5.470	0.987	23,498,782	23,801,510	24,869,822	34,692,895	36,817,108	59,203,647
	PM-9m	50.900	51.613	53.799	7.272	6.098	0.966	26,478,332	26,802,807	28,014,754	39,780,798	41,270,430	71,700,908
	PM-10m	56.457	57.172	59.724	7.861	6.566	0.989	29,427,218	29,770,773	31,155,659	44,148,945	44,942,244	47,564,490
Filebench	FS-1h	7.731	6.754	6.695	0.941	0.816	0.684	4,091,164	3,553,758	3,538,685	256,187,246	218,886,483	215,527,014
	FS-2h	15.284	14.553	14.135	1.296	1.086	0.837	8,097,963	7,673,349	7,486,036	508,522,787	481,981,880	462,880,636
	VM-1h	33.276	34.123	29.758	3.130	2.833	0.761	17,642,494	18,003,351	15,769,747	156,901,333	160,309,605	147,363,881
	VM-2h	66.426	63.870	61.740	4.873	4.222	1.003	35,227,359	33,746,903	32,729,150	316,867,995	385,058,182	277,678,522

Table A-3: Average Erase Count, Standard Deviation, the Counts of Write and Read Operations of baseline, BET and SAW (3rd Time)

Benchmark		Average Erase Count			Standard Deviation			Write Operations			Read Operations		
		baseline	BET	SAW	baseline	BET	SAW	baseline	BET	SAW	baseline	BET	SAW
Postmark	PM-1m	6.568	6.635	6.828	1.466	1.342	0.782	2,975,544	3,005,347	3,112,347	5,269,322	5,423,570	7,082,480
	PM-2m	12.031	12.201	12.644	2.382	2.115	0.866	5,871,655	5,951,349	6,194,713	8,804,954	10,184,471	9,911,950
	PM-3m	17.504	17.769	18.445	3.165	2.718	0.918	8,773,932	8,890,051	9,272,150	13,180,203	16,262,391	14,982,829
	PM-4m	23.012	23.407	24.278	3.837	3.245	0.964	11,692,882	11,826,157	12,363,732	17,205,703	17,782,500	19,032,648
	PM-5m	28.446	28.845	30.067	4.475	3.776	0.961	14,577,788	14,760,804	15,433,961	21,460,584	21,652,666	23,525,050
	PM-6m	33.976	34.449	35.902	5.163	4.259	0.979	17,509,942	17,729,383	18,528,649	26,179,472	27,685,902	37,342,566
	PM-7m	39.569	40.159	41.872	5.806	4.831	0.975	20,471,142	20,758,460	21,692,750	30,716,852	32,483,543	50,465,286
	PM-8m	45.233	45.946	47.891	6.581	5.484	1.012	23,476,279	23,806,545	24,878,595	36,004,233	40,954,853	61,744,099
	PM-9m	50.820	51.693	53.808	7.152	6.176	0.969	26,437,793	26,828,372	28,014,460	39,524,900	48,308,253	68,480,672
	PM-10m	56.366	57.392	59.686	7.849	6.633	0.974	29,380,836	29,811,485	31,137,180	44,703,237	44,835,663	47,361,632
Filebench	FS-1h	7.935	7.251	6.893	0.942	0.842	0.691	4,200,841	3,825,111	3,645,013	261,152,625	238,413,134	221,023,278
	FS-2h	15.579	14.295	14.859	1.334	1.093	0.852	8,253,135	7,544,699	7,870,924	526,400,710	475,705,393	484,400,972
	VM-1h	33.592	34.077	32.638	3.154	2.920	0.793	17,810,247	17,984,757	17,296,913	159,574,537	160,924,500	141,305,567
	VM-2h	66.478	66.029	63.885	4.794	4.295	1.057	35,254,799	34,859,154	33,866,482	327,086,848	322,980,052	296,867,320

Table A-4: Average Erase Count and Standard Deviation of 5k, 10k, 15k, 20k and 25k

Benchmark		Average Erase Count					Standard Deviation				
		5k	10k	15k	20k	25k	5k	10k	15k	20k	25k
Postmark	PM-1m	6.822	6.828	6.831	6.828	6.832	0.782	0.782	0.772	0.792	0.786
	PM-2m	12.642	12.644	12.644	12.656	12.631	0.873	0.866	0.866	0.867	0.852
	PM-3m	18.448	18.445	18.433	18.442	18.452	0.913	0.918	0.913	0.900	0.921
	PM-4m	24.249	24.278	24.239	24.260	24.237	0.942	0.964	0.973	0.967	0.952
	PM-5m	30.037	30.067	30.058	30.046	30.041	0.989	0.961	0.966	0.963	0.953
	PM-6m	35.877	35.902	35.888	35.891	35.884	1.000	0.979	0.992	0.964	0.978
	PM-7m	41.848	41.872	41.843	41.865	41.823	0.966	0.975	0.992	0.959	0.970
	PM-8m	47.877	47.891	47.884	47.896	47.858	0.980	1.011	0.985	0.988	0.968
	PM-9m	53.797	53.808	53.832	53.825	53.806	0.966	0.969	0.984	0.985	1.004
	PM-10m	59.722	59.686	59.702	59.635	59.687	0.978	0.974	1.009	0.971	0.995
Filebench	FS-1h	7.664	6.893	7.016	7.152	7.378	0.708	0.691	0.668	0.692	0.698
	FS-2h	15.172	14.859	14.173	14.234	14.749	0.908	0.852	0.845	0.832	0.868
	VM-1h	33.422	32.638	32.741	31.086	32.857	0.800	0.793	0.778	0.786	0.785
	VM-2h	63.635	63.885	62.587	63.971	63.940	1.032	1.057	1.019	1.010	1.062

Table A-5: Average Erase Count, Standard Deviation and Service Time of lazy and lazy-S

Trace	Average Erase Count		Standard Deviation		Service Time (second)	
	lazy	lazy-S	lazy	lazy-S	lazy	lazy-S
PM-1m	7.887	7.876	2.952	2.184	1,556.628	1,554.672
PM-2m	17.670	17.622	4.352	2.888	3,294.608	3,286.003
PM-3m	27.436	27.395	4.481	3.364	5,029.318	5,022.122
PM-4m	37.263	37.067	6.310	3.627	6,775.204	6,740.659
PM-5m	47.056	46.739	7.159	3.484	8,513.718	8,457.832
PM-6m	56.904	56.667	7.833	4.027	10,263.168	10,221.351
PM-7m	66.756	66.518	8.458	4.127	12,014.693	11,972.629
PM-8m	76.791	76.521	9.103	4.126	13,795.465	13,747.889
PM-9m	86.744	86.437	9.586	4.229	15,564.870	15,510.796
PM-10m	96.480	96.183	10.089	4.329	17,295.008	17,242.518
FS-1h	8.25	8.236	3.256	2.431	1,619.643	1,616.433
FS-2h	17.415	17.312	4.833	3.258	3,246.716	3,228.453

Table A-6: Average Erase Count, Standard Deviation and Service Time of BET and BET-S

Trace	Average Erase Count		Standard Deviation		Service Time (second)	
	BET	BET-S	BET	BET-S	BET	BET-S
PM-1m	7.759	7.759	2.932	1.519	1,534.515	1,534.153
PM-2m	17.400	17.400	4.379	1.785	3,246.938	3,246.937
PM-3m	27.048	27.048	5.488	13833	4,960.660	4,960.621
PM-4m	36.732	34.732	6.360	1.875	6,679.744	6,679.732
PM-5m	46.355	46.355	7.255	1.873	8,389.552	8,389.545
PM-6m	56.096	56.096	7.901	1.867	10,119.220	10,119.206
PM-7m	65.948	65.948	8.436	1.839	11,870.850	11,870.873
PM-8m	75.953	75.953	9.256	1.825	13,646.346	13,646.327
PM-9m	85.873	85.873	9.672	1.815	15,409.453	15,458.531
PM-10m	95.618	95.618	10.201	1.772	17,141.411	17,141.443
FS-1h	8.139	8.139	3.214	1.908	1,597.788	1,597.665
FS-2h	17.190	17.190	4.687	2.484	3,202.400	3,202.490

Table A-7: Average Erase Count, Standard Deviation and Service Time of OWL and O-SAW

Trace	Average Erase Count		Standard Deviation		Service Time (second)		The Count of OWL's Sorts
	OWL	O-SAW	OWL	O-SAW	OWL	O-SAW	
PM-1m	7.926	7.929	1.017	0.977	1,584.611	1,585.733	34,547
PM-2m	17.569	17.569	0.987	1.026	3,297.739	3,297.769	76,558
PM-3m	27.211	27.217	1.000	1.054	5,010.163	5,012.342	118,522
PM-4m	36.883	36.888	1.014	1.045	6,730.632	6,732.437	160,669
PM-5m	46.511	46.513	1.002	1.053	8,438.900	8,439.666	202,563
PM-6m	56.248	56.252	0.972	1.011	10,168.723	10,169.981	244,955
PM-7m	66.098	66.102	0.942	0.987	11,919.311	11,920.698	287,988
PM-8m	76.103	16.104	0.934	0.980	13,696.460	13,697.167	331,575
PM-9m	86.021	86.022	0.944	0.970	15,458.531	15,458.991	374,842
PM-10m	95.763	95.762	0.993	1.028	17,189.121	17,188.892	417,425
FS-1h	8.331	8.338	1.049	1.148	1,665.058	1,667.493	38,522
FS-2h	17.381	17.390	1.440	1.541	3,276.431	3,279.699	80,686

Performance Enhancement of Garbage Collection for Flash Storage Devices: An Efficient Victim Block Selection Design

Che-Wei Tsao
Institute of Information
Science
Academia Sinica, Taipei 115,
Taiwan, R.O.C.
bearman.sky@-
iis.sinica.edu.tw

Yuan-Hao Chang
Institute of Information
Science
Academia Sinica, Taipei 115,
Taiwan, R.O.C.
johnson@iis.sinica.edu.tw

Ming-Chang Yang
Institute of Information
Science
Academia Sinica, Taipei 115,
Taiwan, R.O.C.
mcyang@iis.sinica.edu.tw

ABSTRACT

Motivated by the needs to enhance the performance of garbage collection in low-cost flash storage devices, we propose a victim block selection design to efficiently identify the blocks for erases and reclaim the space of invalid data without extensively scanning flash memory for the status of data stored in the storage, so as to achieve improved performance of garbage collection on reclaiming space of invalid data. At the same time, this design could also easily identify and reclaim the space released by file systems. A series of experiments based on benchmark traces demonstrates the significantly improved performance of garbage collection with limited system overheads.

Categories and Subject Descriptors

D.4.2 [**Operating Systems**]: Storage Management—*Secondary storage*

General Terms

Design, Management, Performance, Reliability

Keywords

Flash memory, performance, reliability

1. INTRODUCTION

Flash storage devices have fast growing market size in recent years because of the booming markets in mobile devices such as laptops, tablets, and smartphones. Well known examples are solid-state drives (SSDs) and eMMCs. A flash storage device usually contains one or more NAND flash-memory chips (referred to as *flash chips* for short). Each flash chip consists of one or more subchips (also referred to

as *dies*) that can operate independently. A subchip might contain more than one plane. Each plane is partitioned into blocks and each block is divided into a fixed number of pages. A block is the smallest unit for erase operations, and a page is the basic unit for read and write operations. A page contains a user area and a spare area. The user area is for data storage, and the spare area is for storing out-of-band (OOB) data. Due to the "write-once property", a page that is written can not be overwritten unless its residing block is erased. Thus, "out-place update" is usually adopted to write the updated data to free pages to avoid many overheads on block erases and (live) page copies. Pages with the latest copy of data (called *valid data*) are regarded as *live or valid pages*, and those containing old data (called *invalid data*) versions are considered *dead or invalid pages*. As a result, "address translation" is needed to map the LBAs of data to their physical block addresses (PBAs), and "garbage collection" is needed to reclaim space of dead pages by selecting blocks with dead pages as *victim blocks* and erasing them when the free space is insufficient. Before the victim block is erased, *live-page copyings* must be performed to copy live pages from the victim blocks to free pages.

As for the address translation, the traditional FTL designs require large RAM space to maintain the fine-grained address mapping information, so that some hybrid address mapping designs (*e.g.*, AFTL, BAST, and FAST) were proposed to manage frequently-accessed addresses with a fine-grained mapping table and infrequently-accessed addresses with a coarse-grained mapping table adaptively [4,6,10]. As for the garbage collection, the simplest baseline approach is the greedy policy (*e.g.*, cost-benefit policy (CB) [3,7], the cost-age-time policy (CAT) [1], and the cost-age-time policy with age sort (CATA) [2]) to maximize the amount of space reclaimed by each block erase and to incorporate the concepts of the age of data (*i.e.*, the time since a block was allocated or since any page of a block was invalidated) with a proper compromise between the space reclamation efficiency and the garbage collection overheads. Although these garbage collection policies are efficient on space reclamation, there is still a large room to improve their performance. This is because they do not consider to reclaim the space released by file systems, and might need to spend a lot of time (or overheads) on identifying the page status of blocks during the process on selecting victim blocks and copying live pages

Permission to make digital or hard copies of all or part of this work for personal or classroom use is granted without fee provided that copies are not made or distributed for profit or commercial advantage and that copies bear this notice and the full citation on the first page. To copy otherwise, to republish, to post on servers or to redistribute to lists, requires prior specific permission and/or a fee.
DAC '13, May 29 - June 07 2013, Austin, TX, USA

before erasing the selected (victim) blocks.

This work is motivated by the needs to enhance the performance of garbage collection in low-cost flash storage devices with the support of crash recovery. In this paper, we propose an efficient victim block selection (module) design to improve the performance of the existing garbage collection policies by minimizing the time on (1) selecting the victim blocks with maximal invalid pages, (2) identifying the status of each flash page, and (3) copying live pages from the selected victim blocks. In particular, a victim block selection strategy is designed to recommend the best victim blocks (*i.e.*, the blocks with the maximal number of invalid pages) and page status of the selected victim blocks for the existing garbage collection policies with limited accesses to the flash memory, such that the victim block selection overheads and the live-page copying overheads could be minimized even when the storage capacity grows. A series of experiments based on realistic workloads was conducted to evaluate the efficacy of the proposed design. The results are encouraging.

The rest of this paper is organized as follows. Section 2 presents the system architecture and the motivation of this work. In Section 3, we present the proposed the victim block selection design. The experimental results are reported in Section 4. Section 5 is the conclusion.

2. BACKGROUND AND MOTIVATION

NAND flash memory could be classified into single-level-cell (SLC) flash memory and multiple-level-cell (MLC) flash memory. Each cell of SLC flash memory can store one-bit data and that of high-density $MLC_{\times n}$ flash memory could contain n-bit information. Compared to SLC flash memory, although MLC flash memory has lower unit cost, it has lower access performance, lower endurance, and higher bit error rate (as shown in Table 1). In addition, MLC flash memory imposes two new write constraints. That is, partial page programming is not allowed, and pages in the same block should be programmed in sequential order.

Type & Part Number	SLC	$MLC_{\times 2}$	$MLC_{\times 3}$(TLC)
Price (USD/GB, 2012)	5.76	0.80	0.49
Serial Access	$20ns$	$20ns$	$25ns$
Random Read	$35\mu s$	$75\mu s$	$150\mu s$
Write/Program	$350\mu s$	$1300\mu s$	$2300\mu s$
Erase	$1.5ms$	$3.8ms$	$3ms$
Erase Cycles	60,000	3,000	500
Bit Error Rate	1×10^{-9}	1×10^{-6}	$\gg 1\times 10^{-6}$

Table 1: The Characteristics of Flash Memory

Figure 1 shows a typical architecture of flash-memory storage systems. In the device side, flash memory chips are controlled by the memory technology device layer that provides primitive functions such as read, write, and erase to the Flash Translation Layer. Thus, the flash translation layer could control various flash memory chips with the unified interface. Most of the existing flash management designs are implemented in this flash translation layer to support address translation, garbage collection, and wear leveling. The address translation is to manage flash-memory space and translate any given LBA to its corresponding PBA, where LBAs are the addresses of sectors accessed by the operating systems and PBAs are usually the addresses of flash pages. The garbage collection is to reclaim the space of dead pages

by erasing their residing blocks when the flash memory has not enough free space. In other words, the garbage collection selects some blocks (that contain dead pages) as *victim blocks*, copies live pages from the victim blocks to free pages (referred to as *live-page copyings*). The wear leveling is an optional feature that evenly distributes block erases to extend the lifetime of flash memory.

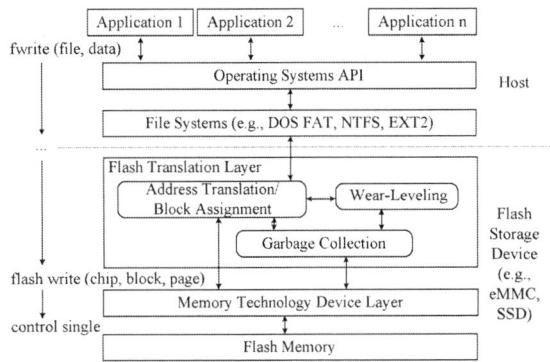

Figure 1: System Architecture

The flash translation layer usually emulates the underlying flash memory chips as a block device. Thus, file systems in the host side could access the flash storage device transparently. File systems usually consist of two parts: *userdata* and *metadata*. Userdata contains the user data or file content, and metadata contain the general information of file systems and the attributes of files/directories. On deleting a file, the clusters (or data units) allocated to the file can be "freed" (or "released" interchangeably) by simply updating their corresponding bit flags in the bitmap table through sending write requests to the storage device. The flash translation layer only receives write requests and does not know which clusters or LBAs are freed. It still maintains the mapping information for the freed LBAs such that the physical space allocated for the freed LBAs could not be released. As a result, the available physical space (*i.e.*, free pages and blocks) in the flash storage device keeps decreasing even after the allocated LBAs or clusters are freed. With less available physical space, the garbage collection would be activated more frequently and introduce more overheads such as block erases and live-page copyings on reclaiming space of dead pages or invalid data.

On the other hand, existing flash management designs are implemented in the flash translation layer and usually equip with a garbage collection policy to reclaim space of invalid data by erasing victim blocks. Some garbage collection policies propose to find a block with a comparatively small number of live pages as the victim block by scanning the spare area of flash pages, so that they could minimize the garbage collection overheads such as block erases and live-page copyings [1,3]. However, the process to identify the status of flash pages by scanning the spare area of each page in each block is very time-consuming and could become the performance bottleneck of garbage collection. Thus, some garbage collection policies proposed a method to maintain information (*e.g.*, counters to count the number of dead pages in each block) in RAM so as to perform the victim block selection without scanning the spare area of flash pages and without

1133

identifying the status of each page in each block [5]. This method needs to read each page of the victim block and also needs to look up the mapping information to see whether data read from each page are valid or not before the victim block could be erased. This is because a page of MLC flash memory could not be marked as a dead one by updating its status flag in the spare area due to the prohibition of partial page programming. Thus, this method could be very inefficient when most pages in the victim block are dead ones. Based on the above observations, it is not easy to select and reclaim a proper victim block efficiently with limited RAM space usage, but both are critical to the performance of garbage collection.

This research is motivated by the needs to enhance the performance of garbage collection in low-cost flash storage devices with considering the space released by the file system and the time on victim block selection. We are interested in a modular design to enhance the garbage collection performance of existing garbage collection policies. This modular design should be easily integrated into the design of many existing garbage collection policies. The technical problem is how to identify and reclaim physical space allocated to the logical space that has been released by the file system with and without the support of the TRIM command, and how to select and reclaim victim blocks efficiently with minimized overheads on copying valid data from victim blocks and with minimal accesses to dead pages. At the same time, the proposed design should also consider the limited RAM space and limited computing power in low-cost MLC-based flash storage devices.

3. VICTIM BLOCK SELECTION

3.1 Overview

In this section, an efficient victim block selection (VBS) design is proposed to enhance the performance of garbage collection with limited modifications to the existing flash management designs and with the consideration of the limited main-memory (or RAM) space in flash storage devices, where the characteristics of file systems are considered. In a file system, it usually maintains the *space allocation information* to indicate the usage status of clusters for the storage of user data. In this work, we consider a modular design for the victim block selection (VBS) so that it can be easily integrated into many existing implementations. As shown in Figure 2, the victim block selection module is inserted into the flash translation layer. It consists of three major components, *i.e.*, *request filter*, *file system identifier*, and *garbage manager*. The request filter is to intercept I/O requests and redirect them to the garbage manager or the underlying address translator according to the request's type. The file system identifier is to identify the file-system's type and the location of the file-system's *space allocation information* in the flash storage device by analyzing the partition table and the file-system's formation information during the system's start-up; it passes the retrieved information to the garbage manager such that the garbage manager could distinguish whether a (write) request is to update the file-system's space allocation information or not. The garbage manager is to manage the information of invalid space (*i.e.*, invalid pages), called *garbage* here, so as to help the underlying existing garbage collector to perform garbage collection efficiently.

When the request filter receives an I/O request that is a

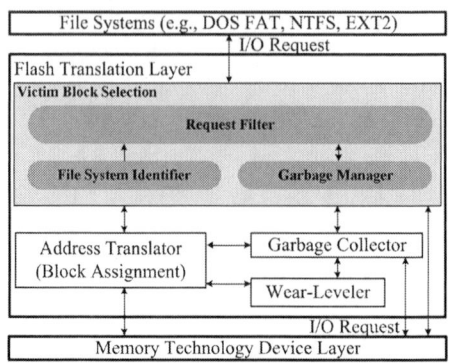

Figure 2: Architecture with Victim Block Selection

read request, it bypasses the received request to the underlying address translator as a traditional I/O request; otherwise, it redirects the request to the garbage manager. When the garbage manager receives an I/O request from the request filter, it analyzes the I/O request and then issues necessary requests to the underlying address translator, and the address translator should report which pages are invalidated. The cases of the I/O requests that the garbage manager should handle are threefold: (1) the request redirected to the garbage manager is a TRIM command, (2) the request redirected from the request filter is a write request to the location other than that of the space allocation information, and (3) the received write request is to update the space allocation information. The above three handling cases to interact with the address translator make the garbage manager able to collect and manage the invalid space information such as the number of invalid pages in a block and the status (*e.g.*, live and dead) of each page in a block. When the garbage collector is activated to reclaim invalid space (*i.e.*, space of invalid data), it can simply invokes the garbage manager to recommend the victim blocks and to provide the status of the pages inside the victim blocks without the time-consuming process on scanning pages of blocks.

3.2 Invalid Space Management

3.2.1 Page Status Management

In this work, the address translator would inform the garbage manager which pages are invalidated. Thus, we propose to manage the *invalid space information* in the garbage manager. In order to enhance the garbage collection performance with limited extra RAM space requirement, the garbage manager maintains the invalid space information such as the *invalid page table (IPT)* and *invalid page count (IPC)* of blocks in flash memory and only temporarily keeps the currently used IPTs and IPCs in main memory (or RAM). An IPT is a bit array to indicate the status of each page in a block and an IPC is an integer counter to keep the number of invalid pages in a block. The IPCs are used to help the garbage collector on selecting the best victim block to reclaim so as to minimize the number of live pages that need to be copied out, and the IPTs are used to minimize the time on detecting the status of each page in the victim block (in terms of the number of pages being checked/scanned) as well as the time on live-page copies. In order to reduce the overheads and RAM space requirement on managing the IPTs

1134

and IPCs, blocks of flash memory are partitioned into a *reserved region* and some *block regions*, as shown in Figure 3; blocks in the reserved region are managed by the garbage manager to store the IPTs and IPCs of the blocks in block regions, and block regions, each of which consists of a fixed number of blocks, are managed by the existing FTL designs (or flash management designs) implemented as the address translator and garbage collector. The IPTs and IPCs of the blocks in the same block region are managed together and saved in the same *block* (referred to as *recycle bin*) of the reserved region.

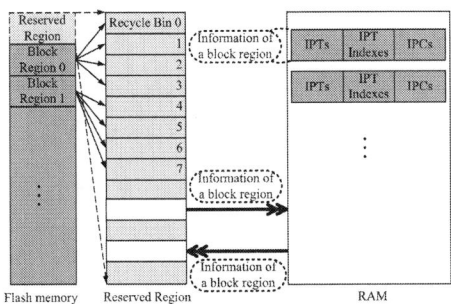

Figure 3: Flash Layout and RAM Usage

The IPCs of blocks of a block region are loaded from recycle bins of the reserved region together and are temporarily kept/cached in RAM for the future updates when pages of the blocks in the block region are recently invalidated. In order to reduce the RAM space usage, only the IPTs of the blocks whose residing pages are recently invalidated are kept in RAM; thus, we propose a least-recently-used based (referred to as *LRU-based*) search tree structure, referred to as *LRU-AVL tree*, to maintain the recently updated IPTs for each block region with reasonable performance on searching the desired IPTs in RAM. Note that the AVL-tree structure could be any efficient (binary) search tree structure because search tree usually has good search performance when the tree is stored in byte-addressable RAM. As shown in Figure 4(a), each node of the tree is used to maintain the *IPT* of a block with *block address* of its corresponding block as its index key and with a *dirty* flag to indicate whether its corresponding IPT is dirty (*i.e.*, modified and has not been written back to flash memory); each node is also associated with four pointers, two to its two child nodes in the AVL tree and the other two to its previous and next nodes in the LRU list[1]. For each LRU-AVL tree, a counter c_d is maintained to indicate the number of dirty IPTs in the tree so as to efficiently identify the best time moment to write the cached IPTs to flash memory. For example, in Figure 4(b), the tree consists of five nodes, and four of them have dirty IPTs, while the least-recently-updated sequence of their IPTs is logically depicted in Figure 4(c).

Due to the consideration of the write-back atomicity and crash recovery, the number of dirty IPTs (or nodes) maintained in an LRU-AVL tree is at most equal to the maximal number N_{max} of IPTs that could be stored in the same flash page. Once the number of dirty IPTs in the LRU-AVL

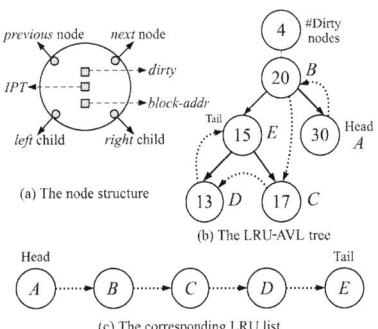

Figure 4: The LRU-AVL tree

tree equals to N_{max}, the dirty IPTs and their corresponding block addresses in the tree are written back to the their corresponding recycle bin in order of their corresponding block address; meanwhile, all of the dirty flags in the tree node are cleared with the tree's corresponding counter c_d being reset to 0 because all the cached IPTs in the tree are clean now.

The dirty IPTs written back to the same page of their corresponding recycle bin are usually the IPTs of some blocks in the same block region because the IPTs of the blocks in the same block region are usually too large to fit in a flash page. Thus, the latest IPT of each block in the same block region might be spread in different pages of different recycle bins. In order to efficiently identify the location of the latest IPT of each block in the same block region, we propose to maintain an *IPT index array* for each block region, and each element in the IPT index array is a pointer (referred to as an *IPT index*) that points to the page storing the latest version of its corresponding IPT. When the IPCs of a block region is loaded to RAM, the IPT index array (*i.e.*, the IPT indexes) of the same block region is also loaded to RAM as well. Whenever the cached dirty IPTs are written to a page of their corresponding recycle bin, their corresponding IPT indexes are updated to indicate the new page storing their latest version. Due to the considerations of the data consistency and system crash recovery, the IPCs and IPT indexes of the same block region should be written to the following page(s) of the same recycle bin right after the dirt IPTs are written to the recycle bin, so as to "commit" the page status changes in the block region. In the following section, we shall discuss how the IPTs, IPT indexes, and IPCs are managed in flash memory.

3.2.2 Recycle Bin Management

In order to reduce the RAM consumption, the *invalid space information* (*i.e.*, IPTs, IPT indexes, and IPCs) of the blocks in the same block region are stored in the blocks, each of which is called a recycle bin, of the reserved region. To reduce the management complexity, each block region is statically mapped to a fixed number of recycle bins and the mapping information between each block region and its corresponding recycle bins could be statically stored in the system block[2]. The example in Figure 3 is to map each block region to four recycle bins, and the recycle bins of each block

---

[1]The LRU-AVL tree is space-efficient when realized by a link list constructed by a two-dimensional array such that each of the four pointers in a tree node takes only 1 byte as the number of nodes in each tree is no more than 256.

[2]The system block is usually the physical block 0 of each flash chip and is usually guaranteed being valid for a certain number (*e.g.*, 1000) of erase cycles; it is usually used to store system data.

region are used in a round-robin fashion.

The layout of the IPTs, IPT indexes, and IPCs in recycle bins could be depicted in Figure 5 as an example, because the dirty IPTs of a block region are usually written to a page of their corresponding recycle bin, followed by the IPT indexes and IPCs of the same block region in the next/following page(s) of the same recycle bin to commit the page status changes. Suppose that the IPT indexes and IPCs are stored in two different pages, the newly flushed IPTs are written in one page called the *IPT area*, and the IPT indexes and IPCs are written in the next two pages called *IPC index area* and *IPC area* respectively, as shown in Figure 5. The IPT area consists of IPT records, each of which indicates the page status of one block and consists of two fields, ba and ipt, to indicate the page status (stored in ipt) of the block, where ba is a block address and ipt is a bit array; each bit in ipt indicates the status of one page: a bit "1" indicates that its corresponding page is invalid, but a bit "0" means that the status of its corresponding page is unknown (instead of valid or free) because the system might crash or lose power at any time. Note that each bit in ipt is initialized to zero because pages are not invalid initially. The IPT index area consists of all the IPT indexes of blocks in the block region; each IPT index has two fields, irb and ip, to indicate that the IPT of its corresponding block is stored at Page ip of the irb_{th} recycle bin of the block region. The IPC area consists of all the IPCs of the blocks in the block region and each IPC is a one-byte counter cnt to indicate the number of invalid pages in its corresponding block; in other words, each cnt denotes the number of 1's in the IPT of its corresponding block.

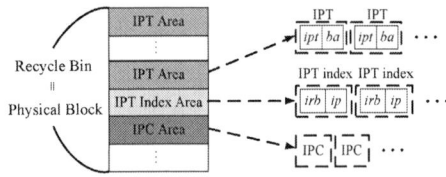

Figure 5: Recycle Bin and Data Structure

3.3 Counter Arrays and Global Counter Array

In this work, the garbage collector would request the garbage manager to recommend victim blocks for space reclamation, and inform the garbage manager which blocks are erased. On each invocation, the garbage manager would return the recommended (or selected) victim blocks and their corresponding page status (*i.e.*, their corresponding IPCs and IPTs)[3].

In order to reduce the overheads on selecting the best victim blocks for the garbage collector, the garbage manager maintains a *counter array* for each block region, and the i-th entry of the counter array is a counter to keep the number of blocks having i invalid pages in the block region. In other words, the i-th entry of the counter array is to keep the number of IPCs whose value is i in the corresponding block region. Meanwhile, the garbage manager maintains a *global*

---

[3]The proposed garbage manager could work with many of the existing garbage collection strategies, especially the greedy ones, because they usually need to find victim blocks that have fewer valid pages.

counter array, and the i-th entry of the array is a counter to keep the total number of blocks having i invalid pages in all of the block regions. That is, the i-th entry in the global counter array is the sum of the i-th counter in each counter array. Figure 6 shows an example when each block is made of 128 pages.

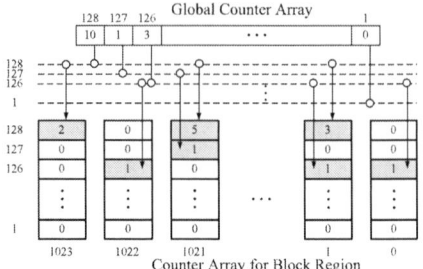

Figure 6: Counter Arrays for Victim Block Selection

To reduce the overheads on block erases, the garbage manager always recommends the blocks with the largest number of invalid pages in every block regions as the victim blocks. Thus, when the garbage manager is invoked by the garbage collector for the victim block selection/recommendation, the garbage manager searches the global counter array in decreasing order of the entry index in the first step. Once the largest non-zero i-th counter is identified, the IPCs of the block region are searched for value i and the blocks with their corresponding IPCs equal to i are selected as the victim blocks. Then, the selected victim blocks and their corresponding IPTs are passed to the garbage collector.

4. PERFORMANCE EVALUATION

4.1 Experimental Setup

In this section, the performance of the proposed VBS strategy will be compared to that of three existing strategies, namely Round-Robin (RR), cost-benefit (CB) and the optimal strategies. For simplicity, we assume that page-level FTL is used throughout all the experiments. As there is no sufficient free space for further write requests, the garbage collection process is initiated, and a victim block will be selected for cleaning by either of the four strategies until sufficient free flash space is available for the write request. With the RR policy, the next block of the last-cleaned block with any invalid pages will be selected. With the CB policy, the block with the *cost-benefit factor* $(c - b)$ larger than 0 is selected as the victim block, where c is the number of invalid pages and b is the number of valid pages of the block. Meanwhile, the optimal policy maintains a RAM-resident multi-list, where each list in the multi-list is a *least-recently invalidated (LRI)* list with the blocks that have the same number of invalid pages. When the garbage collection process is triggered, the least-recently invalidated block among the blocks with the largest number of invalid pages will be selected as the victim block. Although the optimal policy performs well in capturing the data usage pattern to reduce live data copying overheads, the major problem of the optimal policy is on its huge RAM usage.

For the trace-driven simulation, we implemented a filter driver to collect the I/O workload of two well-known benchmarks, namely IOZone [9] and Bonnie++ [8], running on a

1136

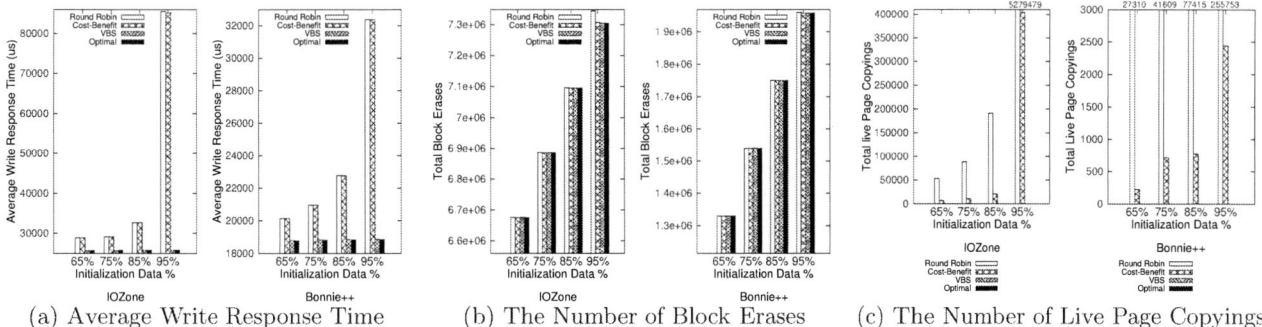

(a) Average Write Response Time   (b) The Number of Block Erases   (c) The Number of Live Page Copyings

Figure 7: Average Write Response Time and Garbage Collection Overheads

1TB hard disk (file system: FAT32) with Linux operating system (distribution: Debian 6.0). Throughout all the experiments, we simulated a 1TB MLC flash storage device whose statistics are shown in Table 2.

Number of Chips	1
Chip Size	2,097,152 Blocks (1TB)
Block Size	128 Pages (512KB)
Page Size	4KB + 128B
Write/Program	$800us.$
Erase	$1.5ms.$
Random Read (Set-up Time)	$60us.$
Serial Access	$25ns./Byte$

Table 2: The Evaluated Flash Storage Device

4.2 Experimental Results

Figure 7(a) shows that the average write performance of the VBS and the optimal strategies significantly outperform the Round Robin and cost-benefit strategies. Specifically, the VBS strategy outperforms Round Robin & cost-benefit strategies by up to 69.75% and 41.69% for the IOZone and Bonnie++ traces, respectively. In the meantime, *the difference between the average write response time of the VBS and the optimal strategies in all configurations of initial space utilization is less than 1%.* As astute readers may point out, the average write response time of Round Robin and cost-benefit policies deteriorate seriously as the initial space utilization grows higher, because the Round Robin strategy might clean a block with many live pages while the cost-benefit strategy might need to scan many pages to locate a victim block with the cost-benefit factor.

The number of block erases and copied live pages of the four strategies with respect to different benchmarks are summarized in Figures 7(b) and 7(c), respectively. Similar to the average write response time, the number of block erases and copied live pages increase significantly as the space utilization grows. While all four strategies have a similar number of block erases, inappropriate selection of the victim blocks of the Round Robin strategy have resulted in excessive live page copying overheads (more than 10 times against the other strategies in all configurations), as observed from Figure 7(c).

5. CONCLUSION

In this paper, we proposed an efficient victim block selection strategy to enhance the performance of garbage col-

lection in flash storage devices. The aim of the proposed strategy is to reduce the write response time, which is often the performance bottleneck of flash storage devices. In contrast to the existing approaches, the proposed strategy goes a step further by incorporating the considerations of the management overheads on the selection of proper victim blocks for garbage collection. Meanwhile, an efficient identification method of invalid pages is proposed to further alleviate the performance impact of garbage collection. Experimental results show that the proposed strategy achieves an average write response time that outperforms the Round Robin and Cost-Benefit strategies by up to 69.89%, and is comparable to the optimal strategy (< 1%).

6. REFERENCES

[1] M.-L. Chiang and R.-C. Chang. Cleaning Policies in Mobile Computers Using Flash Memory. *Journal of System Software*, 48(3):213–231, 1999.

[2] L. Han, Y. Ryu, and K. Yim. CATA: A Garbage Collection Scheme for Flash Memory File Systems. In *Ubiquitous Intelligence and Computing*, volume 4159, pages 103–112, 2006.

[3] A. Kawaguchi, S. Nishioka, and H. Motoda. A Flash-memory based File System. In *Proc. of the USENIX Technical Conference*, 1995.

[4] S.-W. Lee, D.-J. Park, T.-S. Chung, D.-H. Lee, S. Park, and H.-J. Song. A log buffer-based flash translation layer using fully-associative sector translation. *ACM TECS*, 6(3), July 2007.

[5] W.-H. Lin and L.-P. Chang. Dual Greedy: Adaptive Garbage Collection for Page-Mapping Solid-State Disks. In *ACM/IEEE DATE*, 2012.

[6] D. Ma, J. Feng, and G. Li. LazyFTL: A Page-level Flash Translation Layer Optimized for NAND Flash Memory. In *the ACM SIGMOD Conference (SIGMOD)*, 2011.

[7] M. Rosenblum and J. K. Ousterhout. The Design and Implementation of A Log-structured File System. *ACM TOCS*, 10(1):26–52, 1992.

[8] Russell Coker. Bonnie++ benchmark suite. http://www.coker.com.au/bonnie++/.

[9] William D. Norcott, Don Capps. Iozone filesystem benchmark. http://http://www.iozone.org/.

[10] C.-H. Wu and T.-W. Kuo. An Adaptive Two-level Mnagement for the Flash Translation Layer in Embedded Systems. In *ICCAD*, 2006.

DuraCache: A Durable SSD Cache Using MLC NAND Flash

Ren-Shuo Liu[1], Chia-Lin Yang[1,2], Cheng-Hsuan Li[1], Geng-You Chen[1]

[1] Department of Computer Science and Information Engineering,National Taiwan University, Taipei, Taiwan

[2] Graduate Institute of Networking and Multimedia, National Taiwan University, Taipei, Taiwan

renshuo@ntu.edu.tw; yangc@csie.ntu.edu.tw; r01922029@ntu.edu.tw; r01922070@ntu.edu.tw

ABSTRACT

Adopting SSDs as caches for HDD arrays has gained popularity in datacenters because SSDs are superior in handling random reads that HDDs cannot efficiently deal with. Two types of flash memory cells are available for building SSD caches, single-level cells (SLC) and multi-level cells (MLC). MLC is more appealing than SLC because it can achieve higher cache capacity at the same cost. However, we see a critical issue for SSD caches to adopt MLC NAND flash: the endurance of modern MLC NAND flash is too low to sustain datacenter workloads. In this paper, we propose DuraCache that addresses the durability issue of SSD caches. DuraCache exploits the fact that SSD caches are write-through caches in datacenters. Therefore, uncorrectable errors in SSD caches can be handled like cache misses which bring in correct data from HDD arrays. In addition, DuraCache gradually allocates more ECC parities associated with data when NAND flash reaches wearout thresholds. This allows SSD caches to continue operating by sacrificing available capacity. We conduct empirical experiments and demonstrate that DuraCache enables MLC SSD caches to achieve 4.1 years of service life assuming a TPC-C workload.

Categories and Subject Descriptors

B.8.1 [**Performance and Reliability**]: Reliability, Testing, and Fault-Tolerance; D.4.2 [**Operating Systems**]: Storage Management—*Storage hierarchies*

General Terms

Design, Management, Reliability

Keywords

NAND flash, SSD, Cache, Endurance, MLC

1. Introduction

NAND flash-based solid-state drives (SSDs) have gained popularity in datacenter storage. Although whether SSDs could ultimately replace hard disk drives (HDDs) is controversial, it has been widely accepted that utilizing SSDs

---

Permission to make digital or hard copies of all or part of this work for personal or classroom use is granted without fee provided that copies are not made or distributed for profit or commercial advantage and that copies bear this notice and the full citation on the first page. To copy otherwise, to republish, to post on servers or to redistribute to lists, requires prior specific permission and/or a fee.

DAC '13 May 29 - June 07 2013, Austin, TX, USA.

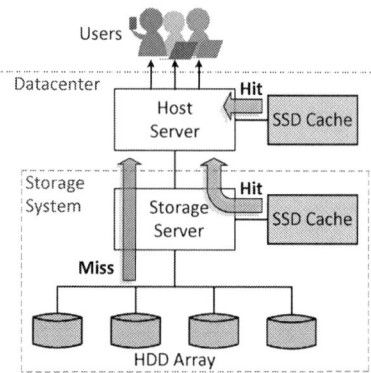

Figure 1: SSD caches in a datacenter

as caches is a cost-effective solution for optimizing storage performance. Major storage vendors, such as EMC, Fusion-io, and NetApp, all have announced their SSD cache solutions [2, 3, 4].

SSD caches can be deployed in both host servers and storage servers. One benefit of equipping servers with SSD caches is accelerating applications that demand high storage performance such as online transaction processing. Employing SSD caches has another important benefit that the storage performance can be decoupled from the number of HDD spindles. Traditional design strategies increase the number of HDD spindles to boost the performance of a storage system. However, with single-disk capacity growing much faster than the performance nowadays, adding HDDs for performance becomes inappropriate because additional spindles mean more power, more space, more cooling, and under-utilized capacity. By intelligently caching data blocks in SSD caches, storage systems can deliver high performance without the drawbacks of HDD over-deployment.

Figure 1 shows a scenario that SSD caches supplement a HDD array in a datacenter. The storage server is responsible for communicating to the attached HDDs and serving I/O requests from the host server. Both the storage server and the host server are equipped with an SSD cache. Reads that hit the SSD caches can be directly served by the SSD caches. Since many requests are filtered out by SSD caches, less pressure is posed to the HDDs whose performance is limited by the speed of head movement and platter rotation. One thing worth noting is that in a production environment, SSD caches are **write-through** caches. Data that are written to SSD caches have a valid copy in HDD arrays as well. Adopting a write-through policy is for guaranteeing that writes persist to the back-end storage array to offer high availability, data reliability, end-to-end data integrity, and disaster recovery [1, 7, 12].

There are two types of flash[1] memory cells, single-level cells (SLC) and multi-level cells (MLC). SLC stores one bit of information per cell while MLC stores multiple bits by dividing the threshold voltage window of a cell into multiple levels. MLC is appealing because MLC-based SSD caches can have up to ten times more capacity than SLC ones at the same cost [15]. This is not only because MLC stores more bits per cell than SLC, but more importantly because MLC flash accounts for the majority of worldwide flash consumption and enjoys the economies of scale in its production.

However, due to the low-endurance issue of MLC, most SSD cache solutions have to adopt SLC flash [6]. Let's take a typical datacenter workload, TPC-C, for example. TPC-C can cause an SSD cache to be programmed and erased 39 cycles per day on average [6]. Modern 25-nm two-bit MLC flash can only sustain 3k program and erase cycles (P/E cycles) before wearout. That means, the lifetime of SSD caches using MLC flash is just 77 days ($\frac{3000}{39} = 77$). Therefore, to enable the adoption of MLC flash in SSD caches, it is critical to resolve the endurance issue.

In this paper, we take a fresh look at designing a reliable SSD cache. The fact that SSDs serve as storage caches instead of storage leads to different design philosophies to tackle the endurance issue. First, unlike storage where data errors must be corrected via the ECC mechanism built in SSDs, erroneous data in SSD caches can be recovered through accessing HDD arrays. Second, in a storage system, capacity is a key system specification. A shortage or change in capacity is unacceptable to customers [5] and may cause unexpected errors in some applications and OSes. Different from common SSDs, SSD caches are actually transparent to applications and end users. Therefore, part of the capacity can be utilized to increase the ECC strength when flash reaches wearout thresholds. Based on these observations, in this paper, we propose *DuraCache* that is able to allow SSD caches to be used beyond wearout based on the design principles of fault tolerance and graceful degradation.

DuraCache features two mechanisms, Error Transformation and Dynamic-Rate SSD (DR-SSD). Error Transformation handles uncorrectable errors by converting them into cache misses that bring in valid data from HDD arrays. Evaluation shows that with Error Transformation, the lifetime of an MLC SSD cache can be improved by 1.7× at the cost of less than 1% of cache hit loss. DR-SSD allocates more ECC parities associated with data when flash reaches wearout thresholds. This allows SSD caches to continue operating by reducing available capacity. With DR-SSD, an MLC SSD cache can increase its lifetime by 3.5× at the expense of 11% capacity reduction. The combination of Error Transformation and Dynamic-Rate SSD enables an MLC SSD cache to sustain 4.1 years of service life assuming a write-intensive workload like TPC-C.

The rest of this paper is organized as follows: Section 2 describes background and related works. Section 3 describes the proposed mechanisms. Section 4 presents experiments evaluating the effectiveness of the proposed mechanisms. Section 5 concludes the paper.

2. Background and Related Work

Flash stores data using floating gate transistors. Through injecting different amount of charge on the floating gate, a transistor presents different threshold voltages to represent different data. Flash is prone to data errors because the amount of charge on a floating gate can be different from intended due to noises, coupling, and leakage over time. The

---

[1]"Flash" and "NAND flash" are used interchangeably in this paper

probability that a data bit in flash is erroneous (i.e., bit error rate, or BER for short) increases as flash transistors gradually wear out due to programming and erasing cycles.

To handle bit errors in flash, common practices employ error correction codes (ECCs) and limit the available endurance of flash. ECCs add parities to user data to form codewords. With the redundant parity information, a codeword with a certain number of bits corrupted can be recovered. The available endurance of flash varies dramatically with different process technologies and different numbers of bits stored per cell. SLC flash usually can have endurance of 100k P/E cycles, but modern two-bit MLC flash is 3k cycles only.

Several techniques have been proposed to improve the lifetime of flash-based storage given limited endurance. First, wear leveling aims at evenly spreading writes to the entire flash device to prevent the early wearout of blocks due to concentrated writes. Wear leveling can effectively ensure a reasonable lifetime for flash devices with moderate write intensiveness such as memory cards, USB drives, and SSDs in PCs. However, wear leveling is less effective for SSD caches to sustain a long enough lifetime given the high P/E rate of datacenter workloads.

The second lifetime-prolonging technique exploits the property that if the retention requirement of flash is lowered, the errors due to leakage over time are reduced. Therefore, flash memory can keep working until higher P/E cycles [9, 15, 16]. We refer to such techniques as Retention Relaxation. Retention Relaxation is well suited to SSD caches in datacenters for two reasons. First, since datacenters are operated continuously, data in SSD caches can be periodically refreshed to reduce their retention requirement. For example, previous works propose that SSDs can be refreshed on a daily basis [9, 16]. Second, since a datacenter workload can program and erase SSD caches up to 39 cycles per day [6], the data lifetime in SSD caches is likely to be shorter than a day. That means, the refreshing overhead is small because many data are updated and thus have their retention capability renewed before a refresh is required [13].

The third technique aims at increasing ECC strength under a constant-rate constraint, which means that the ratio between parities and user data has to be kept constant. Common practices, such as Dynamic Codeword Transition, enlarge ECC codewords for strengthening ECCs [9, 17]. With increased ECC strength, flash can have higher endurance.

Compared with the aforementioned schemes, DuraCache exploits properties unique to SSD caches and features two techniques, *Error Transformation* and *Dynamic-Rate SSD*. The former technique exploits the fact that SSD caches are write-through caches, so data errors in SSDs can also be handled through accessing underlying disk arrays. The latter exploits the observation that SSD caches are transparent to applications and users, so more parities can be allocated with user data in SSDs for lifetime prolonging purpose. The two proposed techniques are orthogonal to wear leveling and Retention Relaxation and are demonstrated to be able to outperform Dynamic Codeword Transition.

3. Design of DuraCache

3.1 Fault Tolerance: Error Transformation

The first technique of DuraCache is fault tolerance through *Error Transformation*. Error Transformation exploits the observation that unlike common SSDs where data errors must be corrected via the ECC mechanism built in the SSDs, if data errors are uncorrectable in SSD caches, valid data can always be retrieved from the back-end HDD array like a cache miss. That is, data errors are "transformed" into cache misses.

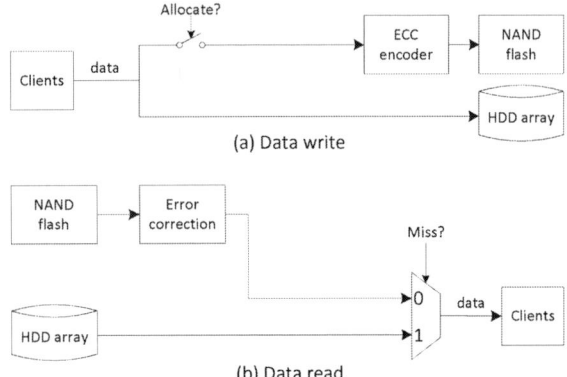

(a) Data write

(b) Data read

Figure 2: Baseline architecture and data flow

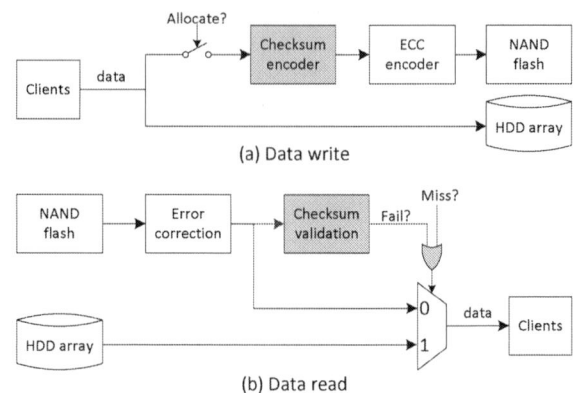

(a) Data write

(b) Data read

Figure 3: DuraCache with Error Transformation

Figure 2 and Figure 3 compare the baseline storage and DuraCache with the proposed Error Transformation technique. Figure 2(a) and (b) depict the data path for writes and reads in baseline storage. On writing data into HDD arrays, data are selectively cached into flash. On a read request, data are read from either flash or HDD arrays according to whether the read is a hit or a miss in flash. In industrial standards, data read from storage need to guarantee $\leq 10^{-16}$ bit error rate (BER). Due to wearout, the BER of data read from flash increases with more P/E cycles. Flash reaches its end of lifetime once its error rate exceeds 10^{-16} after error correction.

Figure 3(a) and (b) illustrate the data path for writes and reads with the proposed Error Transformation mechanism. The key idea behind Error Transformation is to detect errors that cannot be corrected by ECC and bring in valid data from HDD arrays accordingly. Therefore, as shown in Figure 3, error detection via checksums is employed. Checksums are added to data (before ECC encoding) when flash is written, and the checksums are validated (after error correction) when the data are read out of the flash. With checksum validation, uncorrectable errors can be detected and transformed into misses: if wearout incurs too many errors in raw data such that the ECC cannot handle, checksum validation fails, and valid data are retrieved from HDD arrays like a cache miss.

Checksum Selection

Error Transformation relies on checksum validation to detect potential uncorrected errors and trigger cache misses that bring in correct data from HDD arrays. Therefore, checksum selection can affect the resulting data reliability. Below we discuss the coding scheme and the size of the checksum required by DuraCache.

DuraCache adopts cyclic redundancy check (CRC) as the checksum, which can be implemented with simple hardware called *linear feedback shift registers*. The required size of CRC to guarantee sufficient data reliability is actually small as analyzed below. Given an r-bit CRC, its error detection coverage, λ, is as follows:

$$\lambda = 1 - 2^{-r} \qquad (1)$$

That means, the rate that a CRC successfully detects data that contain uncorrectable errors is exponentially close to one in terms of the CRC size. Therefore, for those data that pass CRC validation, their BER (BER_{pass}) is as follows:

$$BER_{pass} = BER_{ECC} \times (1 - \lambda) \qquad (2)$$

Here BER_{ECC} is the BER of data after error correction but before CRC validation. Since $BER_{ECC} \leq 1$ (i.e., all bits are erroneous in the worst case), based on Eq. (1) and Eq. (2) we have:

$$BER_{pass} \leq (1 - \lambda) = 2^{-r} \qquad (3)$$

That means, the worst-case BER_{pass} is an exponentially decreasing function of the CRC size. To guarantee $BER_{pass} \leq 10^{-16}$ like conventional storage, based on Eq. (3), we have:

$$2^{-r} \leq 10^{-16} \Rightarrow r \geq 54 \; (bits) \qquad (4)$$

That means, through including 54 bits (about seven bytes) of CRC in data, the BER of the data can be guaranteed to be no worse than conventional storage. Modern flash stores 1024-byte data and 42-byte ECC parities in 1080-byte space. Therefore, a seven-byte CRC can be stored in the remaining fourteen bytes of space ($1080 - (1024 + 42) = 14$).

In addition to CRC, there are other advanced checksum schemes such as SHA-1 and MD5. Error Transformation does not require such advanced schemes. The error detection coverage of SHA-1 and MD5 is similar to that of CRC (i.e., Eq. (1)). The advantage of SHA-1 and MD5 over CRC is their high security. That is, it is relatively difficult to deliberately produce two distinct data that share the same SHA-1 or MD5 checksum. SHA1 and MD5 are required for storage that has to guard collision attacks conducted by malicious user [10, 11]. For SSD caches, attackers have little chance to observe and manipulate the raw data in the flash. Therefore, CRC is appropriate for the proposed Error Transformation scheme.

Hit Count Impact

Error Transformation prolongs SSD lifetime at the expense that some read hits turn out to be misses due to errors. Let $R_{effective}$ stand for the ratio of the effective hit count to the original hit count. To prevent noticeable performance loss, Error Transformation should only prolong SSD lifetime to a degree that $R_{effective}$ does not significantly deviate from 100%. $R_{effective}$ is a function of two factors, BER_{ECC} and the size of read request (n_{read}):

$$R_{effective} = (1 - BER_{ECC})^{n_{read}} \qquad (5)$$

Here $(1 - BER_{ECC})$ is the probability that a bit is correct, and its n_{read}-th power is thus the probability that all bits read are correct.

Figure 4 plots $R_{effective}$ against the two factors, BER_{ECC} and n_{read}. We set BER_{ECC} to 10^{-16}, 10^{-8}, and 10^{-7}, to represent flash with different P/E cycles. Higher BER_{ECC}

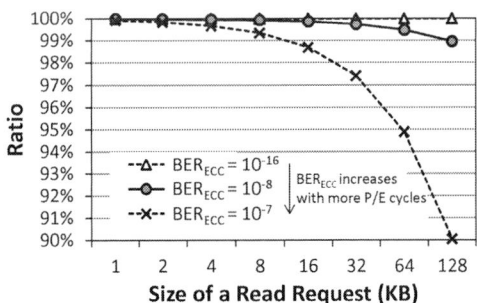

Figure 4: Effective ratio of read hits

means higher P/E cycles, and conventionally, flash has to retire when BER_{ECC} reaches 10^{-16}. We set read size to be 128KB at most because the main purpose of SSD caches is to handle small and random reads that HDDs are hard to deal with. SSD caches typically do not intercept requests larger than 128KB [7]. We can see that $R_{effective}$ is above 99% when BER_{ECC} is within 10^{-8}. The hit count impact becomes noticeable when BER_{ECC} reaches 10^{-7}. That means, with Error Transformation, the endurance cycles of flash can be prolonged from P/E cycles that BER_{ECC} equals to 10^{-16} to P/E cycles that BER_{ECC} equals to 10^{-8} without a noticeable hit count impact. The exact value of the prolonged endurance cycles is estimated based on the strength of ECC and the characterization results of flash. Quantitative analysis on lifetime improvement is presented in Section 4.2.

3.2 Graceful Degradation: Dynamic-Rate SSD

An SSD's lifetime is determined by the strength of the employed ECC. A stronger ECC can prolong SSD lifetime at the cost of higher ECC encoding and decoding complexity. Therefore, dynamically increasing ECC strength according to the wearout of flash is an appealing scheme that can prolong SSD lifetime without paying high ECC cost upfront.

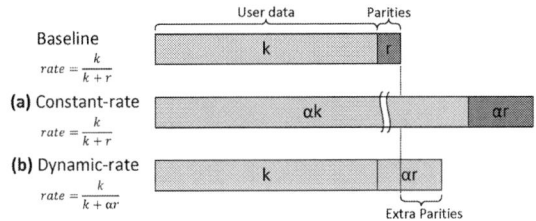

Figure 5: Comparison of different codeword compositions

To dynamically increase ECC strength in an SSD, common practices have to resort to an approach we called the "constant-rate" approach [9, 17]. Conventionally, SSDs have to expose constant capacities to users and applications. A shortage or a change in disk capacity is unacceptable to customers [5] and may cause unexpected errors in some applications and OSes. The constant-rate approach enlarges codeword size without changing the ratio between user data and parities and thus is suitable for common SSDs. Figure 5(a) shows the constant-rate approach using an α times larger codeword assuming that a baseline codeword contains k-bit user data and r-bit parities.

We observe that unlike common SSDs, SSD caches serve as accelerators in datacenters, so their capacities are transparent to applications and end users. SSD caches possess a unique opportunity to gradually sacrifice capacity for pro-

longing their service life. Therefore, the proposed DuraCache architecture features a graceful-degradation technique named *Dynamic-Rate SSD* (DR-SSD) that employs a variable number of extra parities associated with the same number of user data during run time. Figure 5(b) illustrates such an approach with α times more parities for k-bit user data. Generally speaking, the dynamic-rate approach is more effective in increasing ECC strength than the constant-rate one at the cost of disk capacity overhead.

Deciding where to store extra parities is the main challenge in applying the dynamic-rate approach to MLC SSDs. If extra parities are stored with user data in the same flash page, the effective page size which is the basic address-translation granularity in SSDs varies, and address translation becomes difficult[2]. Storing extra parities and user data into separate pages can keep the effective page size unchanged, but it poses the following design issues. First, MLC flash has a restriction that a page can be written only once after the page is erased. Therefore, once extra parities are stored into a page, the left space in this page becomes unusable until an erasure. This unacceptably degrades the space utilization of SSD. Second, to maintain the page mappings between user data and their extra parities, an extra address-translation table is required. Last, if user data and their extra parities are in different blocks, the wearout (P/E cycles) differences between the two blocks need also be handled carefully.

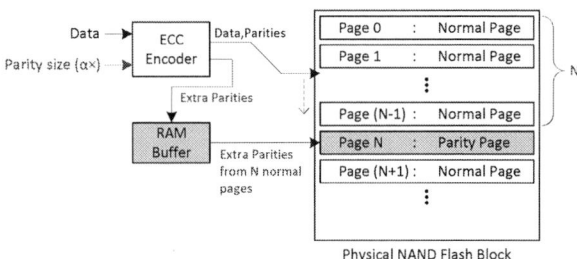

Figure 6: Architecture and data flow of DR-SSD

The architecture of DR-SSD is shown in Figure 6. We take the approach of storing extra parities into dedicated parity pages. In a flash block, for every N successive normal pages, a parity page is allocated to store the extra parities of the N normal pages. Here N is determined based on the ratio between the size of a flash page and the size of extra parities associated with a normal page. By doing so, DR-SSD does not require additional address mapping tables. When a normal page is read, the associated parity page can be directly found using the following equation:

$$p_{parity} = min(\lceil \frac{(p_{read}+1)}{(N+1)}\rceil(N+1) - 1, p_{last}) \qquad (6)$$

where p_{read}, p_{parity}, and p_{last} are the page addresses of the read page, the associated parity page, and the last page of a block, respectively. DR-SSD does not incur additional wearout differences. Normal pages and associated parity pages are placed in the same block, and their P/E cycles (wearout degree) are thus the same.

To resolve the challenge that an MLC flash page can only be written once after it is erased, DR-SSD employs a RAM buffer to allow the extra parities of N normal pages being written into the associated parity page at a time. One may worry about loss of user data if extra parities in the RAM

---

[2]DuraCache adopts a page-level address mapping scheme similar to [8]

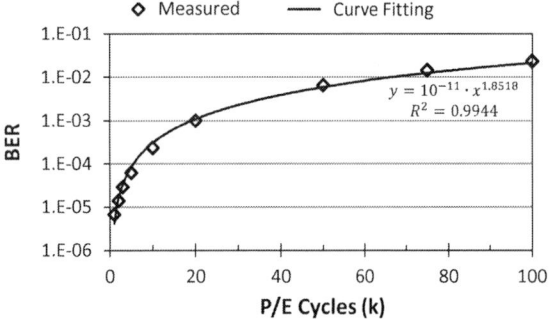

Figure 7: BER in the tested NAND flash vs. P/E cycles

is lost due to power failure. SSD caches do not have such concern because valid data are always available from HDD arrays.

Handling additional parity pages incurs performance overhead in addition to capacity decreases. As mentioned earlier, all these costs are not paid upfront. An SSD cache just behaves as usual when P/E cycles do not reach the baseline wearout threshold. For the graceful-degradation design principle, it is acceptable to sacrifice performance and capacity to keep SSD caches alive beyond wearout thresholds.

4. Evaluation

In this section we evaluate the effectiveness of DuraCache. We target at a modern 25-nm two-bit MLC NAND flash memory. The rated endurance of this flash is 3k P/E cycles. Detailed specification of the tested flash is shown in Table 1. We first perform characterization on the flash to understand its BER versus P/E cycles. Then we estimate the lifetime improvement of DuraCache based on the characterization results.

Table 1: Specification of the tested NAND flash

Technology	25-nm, 2-bit per cell
Density	8GB per die
Page size	8640 bytes
Block size	256 pages
Baseline ECC requirement	24-bit error correction per 1080 bytes
Rated endurance	3000 program/erase cycles

Table 2 lists the six scenarios we evaluate. All the scenarios support wear leveling and the baseline ECC strength (24-bit error correction per 1080 bytes). Baseline-2 further reduces the retention requirement of flash to one day to increase flash endurance [9, 15, 16]. Baseline-3 further adopts Dynamic Codeword Transition (DCT) [17] and can increase the ECC strength through enlarging codeword size by two times or four times. For the proposed DuraCache architecture, we evaluate three configurations. DuraCache-1 adopts Error Transformation, DuraCache-2 adopts DR-SSD, and DuraCache-3 adopts both Error Transformation and DR-SSD. The target BER after ECC decoding (i.e., BER_{ECC}) is set to 10^{-8} for Error Transformation. DR-SSD can employ two times or four times more ECC parities to increase the ECC strength. The error correction capability of DCT and DR-SSD with different codeword sizes or different parity sizes is listed in Table 3 assuming that BCH codes are used.

When DR-SSD employs two times and four times more parities, one parity page are allocated for every 25 ($\lfloor \frac{8640}{336} \rfloor = 25$) and eight ($\lfloor \frac{8640}{336 \times 3} \rfloor = 8$) normal pages, respectively. The effective size of a flash block thus changes from 256 pages to 246 pages (i.e., 96%) and 227 pages (i.e., 89%), respectively.

Table 2: Scenarios compared in the experiments

Scenario	Wear Leveling	Retention Relaxation	Error Transformation	ECC Strength Adjustment
Baseline-1	✓			
Baseline-2	✓	✓		
Baseline-3	✓	✓		DCT (2x or 4x)
DuraCache-1	✓	✓	✓	
DuraCache-2	✓	✓		DR-SSD (2x or 4x)
DuraCache-3	✓	✓	✓	DR-SSD (2x or 4x)

Table 3: Error correction capability of different schemes

		User Data Field (Bytes)	Parity Field (Bytes)	Number of Correctable Bit Errors
Baseline		1038	42	24
DCT	(2x)	2076	84	44
	(4x)	4152	168	84
DR-SSD	(2x)	1038	84	48
	(4x)	1038	168	96

4.1 NAND Flash Characterization

We design a platform based on a Xilinx Virtex-5 FPGA board to perform real-chip characterization on NAND flash memory. The platform can generate pseudo-random data, access flash, and locate bit errors in data retrieved from flash. A software running on the platform repeatedly programs flash blocks with pseudo-random data and erases the blocks until 100k P/E cycles. At 1k, 2k, 5k, 10k, 20k, 50k, 75k, and 100k P/E cycles, written data are kept in the flash for one day and are read back for measuring BER. The measured BER thus includes errors presenting right after programming the blocks and errors emerging within one day. The entire measurement sequence takes about 12 days.

Figure 7 plots the measured BER in the tested flash. We can see that the BER increases from 7×10^{-6} to 2×10^{-2} with more P/E cycles. We can also see that the BER trend follows a power-law model (i.e., $y = a \cdot x^b$) well.

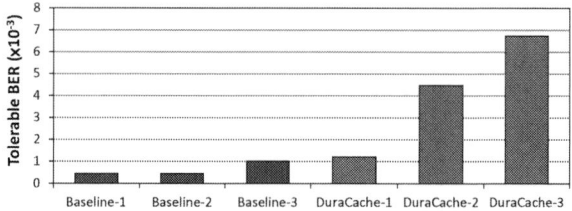

Figure 8: Tolerable BER in flash

4.2 Lifetime Improvement

Figure 8 compares the tolerable BER for each scenario based on formulas in [14]. Baseline-1 and Baseline-2 both can handle a BER of 4.5×10^{-4}. Baseline-3 adopts DCT with up to four times larger ECC codewords than the baseline and can handle a BER of 1×10^{-3}. In comparison,

1142

DuraCache-1, which utilizes the baseline ECC and the proposed Error Transformation scheme, can handle a BER of 1.2×10^{-3}. That means, Error Transformation can be more effective than DCT without incurring additional ECC cost. DuraCache-2 that can handle a BER of 4.5×10^{-3} with up to four times more parities is also more effective than DCT with similar ECC complexity. Combining Error Transformation and DR-SSD, DuraCache-3 can handle a BER up to 6.7×10^{-3}. That means, DuraCache-3 can deal with a 6.7 times higher BER than Baseline-3 with similar ECC complexity.

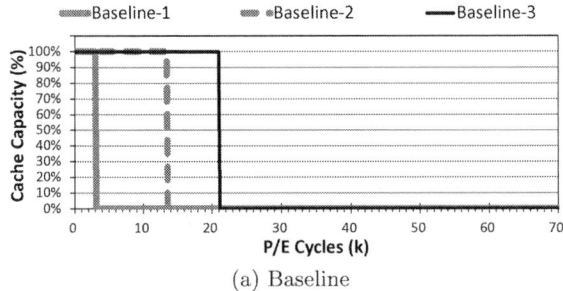

(a) Baseline

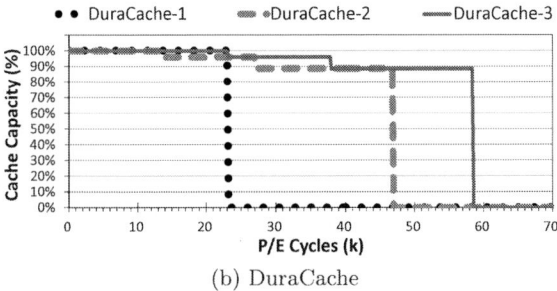

(b) DuraCache

Figure 9: Available cache capacity vs. achievable P/E cycles

Based on the BER characterization on flash (i.e., Figure 7) and the tolerable BER of various scenarios (i.e., Figure 8), Figure 9 shows the available cache capacity versus achievable P/E cycles for different scenarios. We can see that Baseline-1 sustains 3k P/E cycles as suggested in the datasheet of the tested flash. Baseline-2 sustains 14k cycles because it periodically refreshes flash and requires only one-day retention capability for flash. Baseline-3 achieves 21k cycles with DCT. In comparison, DuraCache can achieve higher P/E cycles. DuraCache-1 can sustain 23k P/E cycles with Error Transformation. DuraCache-2 can sustain 47k cycles with DR-SSD. The capacity of DuraCache-2 is reduced by 4% at 14k P/E cycles because two times more parities are employed and is reduced by 11% at 27k P/E cycles because four times more parities are employed. We can see that both Error Transformation and DR-SSD are shown more effective than DCT in terms of lifetime. Combining the two mechanisms, DuraCache-3 can sustain 59k P/E cycles. The capacity of DuraCache-3 is reduced by 4% and 11% at 23k and 38k P/E cycles, respectively.

As mentioned earlier, datacenter workloads such as TPC-C can cause SSD caches to be program and erased 39 cycles per day, and this prevents SSD caches from adopting modern MLC NAND flash memory [6]. Our evaluation demonstrates that with DuraCache, SSD caches using 25-nm two-bit MLC NAND flash can sustain such a write-intensive workload for 4.1 years ($\frac{59 \times 10^3}{39 \times 365} = 4.1$).

5. Conclusions and Future Work

In this paper, we present DuraCache that features the proposed Error Transformation and Dynamic-Rate SSD (DR-SSD) schemes to enable architecting durable SSD caches using modern MLC NAND flash. Error Transformation exploits the fact that SSD caches are write-through caches. Therefore, uncorrectable errors in SSD caches can be transformed into cache misses that bring in correct data from HDD arrays. DR-SSD dynamically employs more ECC parities to handle the increasing BER with more P/E cycles. We conduct real-chip characterization on a modern 25-nm two-bit MLC NAND flash memory. Evaluation results show that with DuraCache, SSD caches can have 4.1-year service life assuming a write-intensive datacenter workload like TPC-C. This paper is the first work pointing out that the lifetime of MLC SSD caches can be effectively prolonged through adopting different design philosophies from that of conventional SSDs. We leave quantitative analysis on the performance impact of DuraCache and further optimization as future works.

Acknowledgments

We would like to thank the anonymous reviewers for their insightful comments and constructive suggestions. This work is supported in part by research grants from ROC National Science Council NSC-101-2220-E-002-013-, NSC-101-2220-E-002-017-, NSC-100-2221-E-002-248-MY3; Excellent Research Projects of National Taiwan University 102R890822; Macronix International Co., Ltd. 101-S-C54; and Etron Technology Inc. 102R7150.

6. References

[1] Datasheet: NetApp Flash Cache. NetApp. DS-2811-1211.
[2] Flash Cache™. NetApp.
[3] ioCache™. Fusion-io.
[4] VFCache™. EMC Corporation.
[5] Western Digital settles hard-drive capacity lawsuit. FOX News, June 2006.
[6] Considerations for choosing SLC versus MLC flash, REV A01. Technical Report 300-013-740, EMC Corporation, 2012.
[7] Wite paper: Introduction to EMC VFCache, 2012. H10502.2.
[8] A. Birrell, et al. A design for high-performance flash disks. *SIGOPS Oper. Syst. Rev.*, 41(2):88–93, 2007.
[9] Y. Cai, et al. Flash correct-and-refresh: Retention-aware error management for increased flash memory lifetime. In *Proc. ICCD '12.*
[10] F. Chen, et al. CAFTL: a content-aware flash translation layer enhancing the lifespan of flash memory based solid state drives. In *Proc. FAST '11.*
[11] A. Gupta, et al. Leveraging value locality in optimizing NAND flash-based SSDs. In *Proc. FAST '11.*
[12] R. Koller, et al. Write policies for host-side flash caches. In *Proc. FAST '13.*
[13] R.-S. Liu, et al. Optimizing NAND flash-based SSDs via retention relaxation. In *Proc. FAST '12.*
[14] N. Mielke, et al. Bit error rate in NAND flash memories. In *Proc. IRPS '08.*
[15] V. Mohan, et al. reFresh SSDs: Enabling high endurance, low cost flash in datacenters. Technical Report CS-2012-05, University of Virginia, 2012. Presented at FMS '12.
[16] Y. Pan, et al. Quasi-nonvolatile SSD: Trading flash memory nonvolatility to improve storage system performance for enterprise applications. In *Proc. HPCA '12.*
[17] S. Tanakamaru, et al. Post-manufacturing, 17-times acceptable raw bit error rate enhancement, dynamic codeword transition ECC scheme for highly reliable solid-state drives, SSDs. In *Proc. IMW '10.*

Distributed Stable States for Process Networks – Algorithm, Analysis, and Experiments on Intel SCC

Devendra Rai, Lars Schor, Nikolay Stoimenov and Lothar Thiele
Computer Engineering and Networks Laboratory, ETH Zurich, 8092 Zurich, Switzerland
firstname.lastname@tik.ee.ethz.ch

ABSTRACT

Technology scaling is a common trend in current embedded systems. It has promoted the use of multi-core, multiprocessor, and distributed platforms. Such systems usually require run-time migration of distributed applications between the different nodes of the platform in order to balance the workload or to tolerate faults. Before an application can be migrated, it needs to be brought to a *stable state* such that restarting the application after migration does not violate its functional correctness. An application in a *stable state* does not change its context any further, and therefore, *stabilization* is a prerequisite for any application migration. Process networks are a common model of computation for specifying distributed applications. However, most results on the migration of process networks do not provide an algorithm to put a general process network into a *stable state*, suitable for migration. This paper proposes a technique which efficiently and correctly brings a process network executing on a distributed system to a known *stable state*. The correctness of the technique is independent of the temporal characteristics of the system and the topology of the process network. The required modifications of a process network are lightweight and preserve its original functionality. A model characterizing the timing properties of the technique is provided. The feasibility and efficiency of the proposed approach and the respective model are validated with experimental results on Intel's SCC platform.

1. INTRODUCTION

Nowadays, increasing computational demands in the embedded systems domain have required the use of distributed many-core platforms. One example is the automotive industry where a contemporary car has many driver assistant systems with tens of cameras, each of them supplying a video stream that needs to be processed in real-time.

However, the increased performance from such platforms comes at the price of increased power consumption per unit area. Such systems may experience high chip temperatures which may require that applications are migrated at runtime between different processing nodes in order to cool down

parts of the chip. Moreover, load-balancing also requires runtime migration in order to optimize the performance.

Migration requires that upon detecting an event, an application can be brought to a *stable state* where all processes involved in migration have stopped their execution, collected all data packets sent to them, do not send any new data, and all local variables have been saved, including the program counters. Only when such a consistent state is reached, contexts can be saved correctly, applications (or parts of them) can be migrated and safely restarted from the point where they have been interrupted.

Bringing an application (or parts of it) to a stable state is not trivial when the application is distributed, composed of many asynchronously executing processes which do not share clocks or memory, with possibly asynchronous communication, and no prior knowledge of the amount of data being produced (or consumed) by a process in any given interval of time. Such is the case for applications specified as Kahn Process Networks (KPNs) [12]. The model is quite often used for specification and design of control and signal-processing applications which are ubiquitous today.

The paper focuses on the stabilization problem. Given a process network executing on a distributed memory system, upon the detection of an event, the process network needs to be brought to a stable state which is suitable for the migration of any of its processes. The technique should be lightweight, safe, correct, and work independently of the timing properties of the system or the topology of the process network.

The contributions of the paper are summarized as follows: 1) A technique is proposed that brings a process network executing on a distributed memory system to a stable state; 2) Timing analysis for the technique is provided; 3) Experiments are performed on a state-of-the-art multiprocessor system (Intel SCC [11]). They validate the efficiency and applicability of the technique, and the correctness of the provided timing model.

2. RELATED WORK

Lots of research results have been published on process migration techniques, see e.g. [1, 2, 4, 13], however, they usually target shared memory systems or do not provide any details on how to bring a general process network to a stable state where contexts can be saved correctly. Moreover, timing models are rarely provided or discussed. Similarly, this is the case for load-balancing literature [15, 18, 20].

A process migration technique for Polyhedral Process Networks (PPNs) has been proposed in [5]. PPNs are a restricted form of KPNs since all loop bounds, array indices, and index expressions must be affine expressions and a pro-

---

Permission to make digital or hard copies of all or part of this work for personal or classroom use is granted without fee provided that copies are not made or distributed for profit or commercial advantage and that copies bear this notice and the full citation on the first page. To copy otherwise, to republish, to post on servers or to redistribute to lists, requires prior specific permission and/or a fee.
DAC '13, May 29 - June 07 2013, Austin, TX, USA.

cess cannot change these parameters at run-time. Therefore, the technique proposed in [5] is not applicable to general KPNs. In the proposed technique, a process execution can be stopped at any time. However, this may require re-execution of the same code after migration which is in contrast to our solution which does not require re-executions. Moreover, the approach in [5] relies on a complex middleware system that continues to run on the processing node even if the application is migrated which makes the technique unsuitable in cases the reason for migration is high temperature. In contrast, in our approach, the affected core can completely stop, after a known time, the *stabilization time*. Furthermore, the authors do not discuss memory requirements, timing properties, or the correctness of their stabilization technique.

Kernel-based approaches to do process migration usually require the usage of specific features of an operating system (OS) or modifications to the OS kernel making them non-portable, e.g. [6,16]. In this paper, we focus on solutions that work in user-space so that they do not depend on any specific OS features, guaranteeing the portability of the solution.

Checkpointing provides a means to manage the context of a migrating process, e.g., the Berkeley Labs Checkpointing and Restore (BLCR) algorithm [19]. However, checkpointing requires a fairly complex bookkeeping process where all processes must log all incoming tokens, all calculations, and all output tokens between each checkpoint. Thus, checkpointing can easily overwhelm the computational capabilities of a typical embedded system [21]. In this paper, we focus on a *lightweight* approach which avoids rolling back, but is able to bring the process network to a stable state which is ready for migration.

Chandy and Lamport [7] have proposed an approach to taking snapshots of a process network on a distributed system which is similar to the stabilization problem that we handle. However, they restrict themselves to (theoretical) systems with infinite FIFO channels and rule out the possibility that a process may block when attempting to write on a full output channel. In contrast, our technique is applicable to practical systems with bounded FIFO channels. Furthermore, we provide an implementation and a timing model which are validated with experiments.

3. MOTIVATIONAL EXAMPLES

In this section, we illustrate the challenges involved in the stabilization of a process network executing on a distributed memory system by using two simple examples.

Example 1: Decentralized Stabilization. Consider the process network shown in Fig. 1, with three processes v_1, v_2, and v_3 and the possibility that v_1 and v_2 must be stabilized. Assume that v_1 exits successfully, but v_3 is blocked when it attempts to read from the input FIFO $\mathcal{F}(v_3, 1)$ due to insufficient number of tokens. On the other hand, v_2 blocks when attempting to write to the full FIFO $\mathcal{F}(v_3, 2)$. This creates a deadlock and v_2 can never proceed to collect its context. In principle, process v_3 can be unblocked when process v_1 resumes normal operation after migration, allow-

ing v_2 to also stabilize, and then migrate. However, such an approach serializes the stabilization of the process network, which may be unacceptable. The example shows that it should be possible to unblock processes that are performing possibly blocking operations (read/writes). Also some form of coordination is needed between processes v_1, v_2, and v_3 in order to lead them to stable states.

Example 2: Coordinated Stabilization. For the given discussion, assume that a separate master process is connected to all processes via event channels and it initiates the stabilization of the processes by sending a `stop` event. Consider the producer (v_1) and consumer (v_2) process network shown in Fig. 2. Once the affected processes receive the `stop` event, they can start the stabilization procedure.

Suppose that the master sends the `stop` event to both v_1 and v_2. Let v_2 receive the event before v_1. Stabilization requires that v_2 must not receive anymore data tokens from its parents, it must stop any further computation, and it must not transmit any tokens to its children. However, v_1 is not aware that v_2 is entering stabilization, and hence, v_1 keeps transmitting data tokens until it receives the `stop` event or its output channel becomes full. Process v_2 cannot go into a stable state because it does not know if v_1 has already received the `stop` event and which data token will be the last data token transmitted by v_1. If v_2 ignores any data tokens which continuously arrive over its input channels, then the execution context for v_2 cannot be calculated correctly and tokens may be lost.

The example shows that simply having a master which tries to bring all processes to a stable state is not sufficient. Due to different network latencies, processes may go into stable states at different times. Therefore, the correctness of the coordination technique needs to be time insensitive. Further, a mechanism is needed which indicates when all processes have stopped and special tokens have to be sent to child processes to notify that the last data token has arrived.

4. MODEL AND DEFINITIONS

A process network $\mathcal{N}$ is defined as the tuple $\mathcal{N} = (V, C, I, O, F, i, o, c, f)$, where V is a set of processes, C is a set of data channels, I is a set of input ports, O is a set of output ports, and F is a set of bounded first-in first-out (FIFO) buffers. The function $i : V \to \mathcal{P}(I)$ maps a process to a set of input ports, where $\mathcal{P}(S)$ denotes the power set of a set S. The function $o : V \to \mathcal{P}(O)$ maps a process to a set of output ports. Processes cannot share input or output ports, i.e., $i(v_k) \cap i(v_j) = \emptyset$ and $o(v_k) \cap o(v_j) = \emptyset$ for all $v_k, v_j \in V, v_k \neq v_j$. A process $v \in V$ reads data tokens from its input ports $i(v)$ and writes data tokens to its output ports $o(v)$. The function $c : U \to C$, where $U = \{(a, b) : a \in o(v_k), b \in i(v_j), v_k, v_j \in V, v_k \neq v_j\}$, maps pairs of output and input ports, belonging to different processes, to a data channel.

The function $par : V \to \mathcal{P}(V)$ returns the set of parent processes for a given process, and the function $ch : V \to \mathcal{P}(V)$ returns the set of child processes for a given process.

The function $f : V \times I \to F$ provides a FIFO buffer $\mathcal{F}(v, m)$ to an input port m of a process v. The buffer has a finite size denoted as $|\mathcal{F}(v, m)|$. A process attempting to write to an output port connected to an input port with a full FIFO will block until there is sufficient space available. Similarly, a process attempting to read data tokens from an input port with an empty FIFO will block until there are sufficient tokens available. Conventionally, KPNs [12] assume unbounded FIFOs, however, practical systems with finite

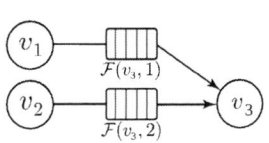

Figure 1: Example 1.

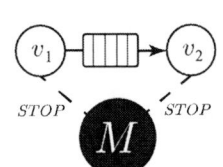

Figure 2: Example 2.

resources impose maximum sizes on the FIFOs [9, 10, 17]. Therefore, the notation adheres more closely to the typical implementation of a process network.

Bringing a process network $\mathcal{N}$ into a state that is ready for migration requires that each process $v \in V$ reaches a stable state which is defined as follows:

DEFINITION 1. (STABLE STATE) *A process $v \in V$ enters a stable state if it does not perform any more computations, parents $par(v)$ do not send any new data tokens, v does not send any new data tokens to its children $ch(v)$, and it has received all data tokens already sent by its parents $par(v)$.*

Once such a stable state has been reached, the context of each process can be saved and migrated. The context of a process includes all unread tokens (all not yet processed input tokens), all produced but unsent tokens, and the program state. Given this context, a process can be safely restarted.

Having the ability to bring each process to a stable state may require that the original processes are slightly modified. Such modifications should preserve the correct functionality of the network. Thus, it must be ensured that the original process network $\mathcal{N}$ is functionally equivalent to the modified process network $\mathcal{N}'$. In other words, the solution must comply with the notion of *Correctness* defined as:

DEFINITION 2. (CORRECTNESS) *Given two process networks $\mathcal{N}$ and $\mathcal{N}'$, where $\mathcal{N}'$ is a modified version of $\mathcal{N}$ such that it has mechanisms to be brought to a stable state. We say that $\mathcal{N}'$ is correct, if for any process $v' \in V'$ of $\mathcal{N}'$ which corresponds to process $v \in V$ of $\mathcal{N}$, for any vector of input sequences of data tokens In, the following relationship holds:*

$$In \xrightarrow[(v \in V)]{} Out \implies In \xrightarrow[(v' \in V')]{} Out \qquad (1)$$

where $In \xrightarrow[(v \in V)]{} Out$ means that process v produces the vector of output sequences of data tokens Out, when given with the vector of input sequences of data tokens In.

Thus, the overall problem of this paper can be summarized as: Extend the process network $\mathcal{N}$ to $\mathcal{N}'$, such that:

1. $\mathcal{N}'$ is functionally equivalent to $\mathcal{N}$;

2. $\mathcal{N}'$ can be brought into a stable state independent of computation or communication delays.

5. PROPOSED TECHNIQUE

Interrupting the normal execution of a process network is initialized and coordinated by a central authority, which can be either an external process or a process of the existing network. Without loss of generality, we assume that the central authority is an external process called the "master".

We start by defining a set of coordination events which will be used by the master for the communication with all processes. The set of coordination events $\mathcal{E}$ contains the stop event: a process receiving this event will *eventually* stop any computations and data transmissions to children, and must then acknowledge the reception of the event; and the proceed event: a process receiving this event must proceed to collect its context. The master sends the proceed event only when it has received all acknowledgments for the stop events.

In this paper we focus on stabilizing the entire process network, therefore, the master process always broadcasts the stop event to all processes in the network. However, the

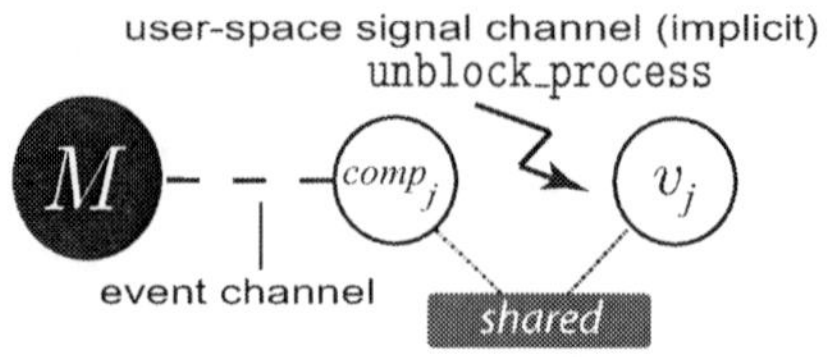

Figure 3: Process, companion process, master, event channel, and signal.

algorithm can be easily extended to stop only specific processes in the network, see Appendix B. Furthermore, a prototype implementation of the proposed stabilization technique is discussed in Appendix A.

The master uses a (bidirectional) event channel $e_j \in E$ to communicate with process v_j. If the process is blocked because of reading from an empty FIFO or writing to a full FIFO, it will not be able to detect (and process) events from the master, therefore the event channel $e_j \in E$ is not directly connected to process v_j, but to a companion process $comp_j$, as shown in Fig. 3. The companion process $comp_j$ is very lightweight. It only receives and processes events from the master and makes them available to its process via a shared variable *shared*. When $comp_j$ receives the stop event, it sets the shared variable *shared* to stop and sends a signal unblock_process to v_j that cancels any blocking read or write of process v_j.

Process v_j checks the variable *shared* at the beginning of each communication primitive (a read or write statement) and exits normal execution if *shared* is set to stop. If process v_j is blocked and receives the signal unblock_process, it also exits normal execution, otherwise the signal has no effect. Once the process exits execution, it sets the shared variable *shared* to done, and then, the companion process can send an acknowledgment back to the master.

The above mechanism is only a conceptual description of our technique. The actual implementation of a separate companion process or unblocking signals will depend on the underlying platform.

5.1 Collecting Data Tokens

When a process exits normal execution, it executes a special function called *Wrapup*. Here no more data transmissions or computations are performed. First, the function waits until the process has received the proceed event from the master, i.e., meaning that all processes have suspended their normal computational activity. Then, it collects the process context and performs any other housekeeping steps such as returning any allocated memory.

During the collection of the context, a process must collect all tokens that are sent by its parents. If these tokens are not collected, the data tokens are "lost", which leads to incorrect behavior. This is further complicated by the fact that there might be a number of tokens arriving late due to late arrival of stop events in parent processes. Therefore, it must be ensured that there are no data tokens which are "in flight", i.e., written by a parent process but not yet received in the local FIFO of the child process. Otherwise, the technique would not be delay-independent.

In order to remedy this problem, in the *Wrapup* function, each process, after receiving the proceed event, sends an end-of-stream (EOS) token to all its children. Thus, a process must continue to collect late arriving tokens from its input channels until the EOS marker has been received on each channel.

5.2 Bounding the Size of Contexts

For the purpose of discussion, the FIFO $\mathcal{F}_v(m)$ for an input port m of process v is divided into two FIFOs: $\mathcal{M}_v(m)$ and $\mathcal{L}_v(m)$. The size of $\mathcal{F}_v(m)$ is the sum of sizes of $\mathcal{M}_v(m)$ and $\mathcal{L}_v(m)$. Tokens move from $\mathcal{M}_v(m)$ to $\mathcal{L}_v(m)$ when there is sufficient space in $\mathcal{L}_v(m)$. The separation is made in order to reflect more closely real implementations of process networks, where $\mathcal{M}_v(m)$ refers to buffers of the interprocess communication layers and the capacity of communication links, while $\mathcal{L}_v(m)$ refers to the FIFO local to a process. The process has only the knowledge of the current status of $\mathcal{L}$ but not of $\mathcal{M}$.

We assume that the number of late-arriving tokens to each process can be bounded. This is the case for any NoC-based communication where the network capacity can be statically calculated (or at least upper bounded) by analyzing its topology, and it is the case for many communication libraries where the buffer sizes of data links are finite.

The memory space to absorb all late-arriving tokens is provided by a statically allocated set of "backup FIFOs" $\mathcal{B}$, in particular, one for each input port FIFO. The maximum number of late arriving tokens that a process v must absorb on each input port m and store in a backup FIFO is upper bounded by: $|\mathcal{B}_v(m)| = |\mathcal{M}_v(m)| + |\text{EOS}|$ where $|\mathcal{M}_v(m)|$ denotes the size of FIFO $\mathcal{M}_v(m)$, and $|\text{EOS}|$ denotes the size of the EOS marker. The bound is correct even if the local FIFO $\mathcal{L}_v(m)$ is full, since tokens in the input FIFOs are not considered as late arriving tokens and therefore not saved in the backup FIFOs. The backup FIFOs $\mathcal{B}$ are not available during the normal course of operation. Upon reception of the **proceed** event and before sending the EOS tokens, a process v_j swaps all regular FIFOs $\mathcal{L}_{v_j}(m)$ with the corresponding backup FIFOs $\mathcal{B}_{v_j}(m)$.

Consequently, an upper bound on the size of the context of process v is:

$$D_v^* = \sum_{\forall j} \left\{ |\mathcal{L}_v(j)| + |\mathcal{B}_v(j)| \right\} + |\text{LN}| + |\text{LV}| \tag{2}$$

where $|\text{LN}|$ is the memory space required to store the line number of the program when v exited the normal execution, and $|\text{LV}|$ is the memory space required to store all local variables (loop indexes, unsent tokens, etc.).

Note that many existing solutions for migration do not rely on backup FIFOs but simply use a constantly running middle-ware system that will re-direct any late arriving data, no matter how late it is. However, such solutions are not always feasible if, for example, a processing node is close to reaching peak temperature and any processing activity on it needs to be stopped after a certain time.

5.3 Timing Analysis

The correct behavior of the proposed algorithm to stabilize a process network is delay-independent. However, in case that the maximum time to transmit a token between two processing nodes and the maximum time that a process is executing without calling a communication primitive are known, an upper bound on the overall stabilization time for a process network can be calculated. Such timing parameters can be obtained either with formal analysis and then the computed bounds would be hard real-time ones, or by measurements (or simulations), and then the bounds would be soft real-time ones.

In order to analyze the timing, we consider two phases of the algorithm. In the first one (denoted as $phase1$), the master (denoted as M) broadcasts the **stop** token to all processes and waits for all acknowledgments. In the second one (denoted as $phase2$), it broadcasts the **proceed** token and then each process waits until it receives an EOS marker on its input ports.

The maximum time between the instance when the master broadcasts the **stop** token and the instance it receives the acknowledgment from process v is composed of four time periods: (a) the maximum time $t_{M \to v}^*$ for the **stop** token to travel from the processing node of the master to the one of process v, (b) the maximum time $t_{read,v|write,v}^*$ that process v requires to perform a single read or write of a data packet of maximum size, (c) the maximum time $t_{c,v}^*$ that process v is executing without calling a communication primitive, and (d) the maximum time $t_{v \to M}^*$ for the **ack** token to travel from the processing node of process v to the one of the master. In other words, the master receives the acknowledgment from process v no later than after the following time period:

$$t_{phase1,v}^* = t_{M \to v}^* + t_{read,v|write,v}^* + t_{c,v}^* + t_{v \to M}^* \tag{3}$$

and can broadcast the **proceed** token no later than after the following time period:

$$t_{phase1,\mathcal{N}}^* = \max_{v \in V} \left\{ t_{phase1,v}^* \right\}. \tag{4}$$

Afterwards, each process waits until it receives the **proceed** token, swaps all regular FIFOs with the backup FIFOs, and waits until it receives an EOS marker on each of its input ports. The time between the instance the master broadcasts the **proceed** token and the instance process v can start to collect its context is upper bounded either by the sequence that v receives the **proceed** token, swaps all regular FIFOs, and waits until it receives an EOS marker, or by the sequence that a parent of v receives the **proceed** token and then process v receives an EOS. Thus, $phase2$ takes no longer than the following time period:

$$t_{phase2,v}^* = \max \left\{ t_{M \to v}^* + \max_{u \in par(v)} \left\{ t_{u \to v}^* \right\}, \\ \max_{u \in par(v)} \left\{ t_{M \to u}^* + t_{u \to v}^* \right\} \right\} \tag{5}$$

where $t_{u \to v}^*$ is the maximum time for the EOS marker to travel from the processing node of process u to the one of process v.

Finally, the stabilization time of process v and process network $\mathcal{N}$ is upper bounded by:

$$t_{stab,v}^* = t_{phase1,\mathcal{N}}^* + t_{phase2,v}^* \tag{6}$$

$$t_{stab,\mathcal{N}}^* = t_{phase1,\mathcal{N}}^* + \max_{v \in V} \left\{ t_{phase2,v}^* \right\}. \tag{7}$$

The property of Correctness. The proposed technique requires three modifications of the process network, namely the addition of a companion process, a *Wrapup* function, and a conditional check before proceeding with a blocking read or write. The addition of the companion process $comp_j$ to process v_j does not change the order of any tokens in any channel for any process. It only retains the information that an event $\{\text{stop, proceed}\} \in \mathcal{E}$ was dispatched from the master. The *Wrapup* function simply stores late-arriving tokens in backup FIFOs, maintaining the relative order of arrival of tokens. Finally, the conditional check before proceeding with a blocking read or write does not interfere with computations or the tokens that are already read or need to be written. Thus, all three modifications preserve the original functionality of the process network and the correctness property is satisfied.

Table 1: Measured stabilization time $t_{stab,v}$ vs. its upper bound $t^*_{stab,v}$ for each process of the Demosaicing application. In addition, the maximum context size D_v is compared to its upper bound D^*_v.

process	$t_{phase1,v}$ avg	max	$t^*_{phase1,v}$	$t_{phase2,v}$ avg	max	$t^*_{phase2,v}$	$t_{stab,v}$ avg	max	$t^*_{stab,v}$	D_v max	D^*_v
load image	0.024 s	0.13 s	0.13 s	19.46 μs	21.39 μs	29.33 μs	0.77 s	3.26 s	4.20 s	52 B	52 B
pre processing	0.042 s	0.16 s	0.16 s	40.80 μs	44.39 μs	48.12 μs	0.77 s	3.26 s	4.20 s	632472 B	632472 B
pre demosaicing	0.32 ms	0.55 ms	0.55 ms	77.07 μs	78.03 μs	79.74 μs	0.77 s	3.26 s	4.20 s	36 B	632484 B
demosaicing_0	0.41 s	1.04 s	1.05 s	60.58 μs	61.54 μs	66.91 μs	0.77 s	3.26 s	4.20 s	157704 B	163370 B
demosaicing_1	0.53 s	1.05 s	1.07 s	60.30 μs	62.63 μs	66.91 μs	0.77 s	3.26 s	4.20 s	160296 B	163370 B
demosaicing_2	0.61 s	1.04 s	1.07 s	58.70 μs	60.98 μs	66.91 μs	0.77 s	3.26 s	4.20 s	160296 B	163370 B
demosaicing_3	0.60 s	1.05 s	1.05 s	58.86 μs	61.32 μs	66.91 μs	0.77 s	3.26 s	4.20 s	157704 B	163370 B
post demosaicing	0.70 s	2.28 s	2.40 s	149.09 μs	159.04 μs	179.65 μs	0.77 s	3.26 s	4.20 s	338964 B	2538900 B
post processing	0.77 s	3.26 s	4.20 s	73.63 μs	76.26 μs	80.43 μs	0.77 s	3.26 s	4.20 s	620750 B	1255992 B
write result	0.36 ms	0.53 ms	0.57 ms	37.02 μs	37.32 μs	56.37 μs	0.77 s	3.26 s	4.20 s	368 B	623460 B
process network		3.26 s	4.20 s		159.04 μs	179.65 μs		3.26 s	4.20 s	2176.4 KB	6188.3 KB

6. EXPERIMENTS

The feasibility and efficiency of the proposed stabilization technique are validated using two representative multiprocessing benchmarks: Demosaicing and a distributed Motion-JPEG (MJPEG) decoder algorithm, detailed in Appendix C. We aim to measure the time to bring the benchmark applications into a stable state, and to compare the time with the (theoretical) upper bound described in Section 5.3. The experiments were performed on Intel's SCC platform [11], a 48-cores (24-tiles) experimental prototype of future on-chip many-core platforms detailed in Appendix C.1.

Experimental Setup. Both benchmarks are running bare-metal to avoid timing jitter due to the operating system. Cache-related timing variations are reduced by hosting one process per tile. Since the SCC implements a deterministic X-Y routing, timing variations due to router contention are reduced by carefully binding the processes onto the tiles. Inter-process communication is implemented using the iRCCE library [8]. For the timing measurements, all tiles establish a common time reference when they boot using the barrier operation available in the communication library. L2 caches and interrupts are disabled on all tiles. Data messages are at most of size 3 KB each (longer ones are split) and control tokens are of size 16 B. The master process was placed on a separate tile so that it does not interfere with the application.

In order to achieve our goal, i.e., to compare the observed stabilization time with its upper bound, we proceed in three steps: 1) Calibration experiments are performed to derive a communication model of the target platform and to obtain the characteristics of the benchmark applications. As a result of this step, we calculate the upper bounds $t^*_{phase1,v}$, $t^*_{phase2,v}$, D^*_v for each process v, and $t^*_{stab,\mathcal{N}}$ for the network, see Section 5. 2) Stabilization experiments of the benchmarks are executed to observe the actual time taken by a process v to complete phase1 and phase2, the time $t_{stab,v}$ to stabilize, and the context size D_v. 3) The observed values are compared with the bounds calculated in Step 1.

Calibration. The communication model was derived by observing the time taken to deliver a packet with size ranging from 4 B through 3 KB over hop distances ranging from one through eight. A total of 585 observations were made. The communication latency under high cross-traffic between any two processes u and v (including the master) mapped onto different tiles, was observed to be upper bounded by:

$$t^*_{u \to v} = 5.182|P| + 9935 \quad \text{(cl.cycles)} \qquad (8)$$

where $|P|$ is the size of the payload in bytes. Because of the high cross-traffic, a dependency on the number of hops be-

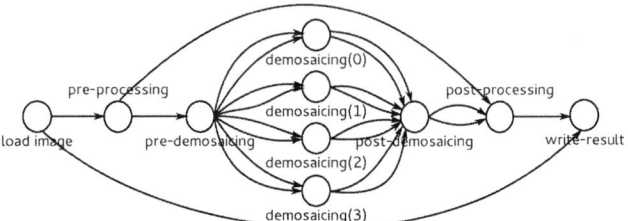

Figure 4: The Demosaicing application.

tween the processing nodes of the communicating processes is not observed.

Another set of calibration experiments was performed in order to obtain the maximum computation time $t^*_{c,v}$ for each process. The Demosaicing application was executed using five RAW images of different sizes and for the MJPEG decoder, the execution time of each process was measured over each frame. The detailed results of all calibration experiments are reported in Appendix C.

Demosaicing. Demosaicing [14] is both a compute and data intensive application consisting of 10 processes, see Fig. 4. To measure the stabilization times, the experiment was repeated 20 times with different inputs and randomly varying the instants at which the master starts a stop token broadcast. Both the average and maximum values of the 20 runs are reported.

Table 1 compares the measured stabilization time $t_{stab,v}$ with the calculated upper bound $t^*_{stab,v}$. It can be seen that all processes did indeed stabilize before the expected time bounds. For some of the processes, the observed measurements are very close to the expected upper bounds. This means that the estimated bounds can be very accurate. The gaps between observed values and bounds are explained by the fact that $t^*_{stab,v}$ is mainly composed of the time the master waits until it receives all acknowledgments, and considers that a process can be in its longest computation section $t^*_{c,v}$. As shown in Table 1, $t^*_{c,v}$ is particularly large for the post processing process.

In addition, the maximum measured size of the context D_v is compared in Table 1 with its upper bound D^*_v calculated using Eq. (2). For some of the processes, the observed measurements are equal to the estimated upper bounds which means that equation (2) is tight. The total size of the upper bound is about three times larger than the measured maximum size. The former assumes that all FIFO channels are full when the context is calculated, but in practice, some of the channels are only partly filled.

MJPEG Decoder. The second example is a parallelized version of the MJPEG decoder application taken from the benchmark suite of the Artist Network of Excellence [3].

Table 2: Measured stabilization time $t_{stab,v}$ vs. its upper bound $t^*_{stab,v}$ for each process of the MJPEG decoder.

process	$t_{phase1,v}$		$t^*_{phase1,v}$	$t_{phase2,v}$		$t^*_{phase2,v}$	$t_{stab,v}$		$t^*_{stab,v}$
	avg	max		avg	max		avg	max	
trigger	$55.5\,\mu s$	$129\,\mu s$	$155\,\mu s$	$20.1\,\mu s$	$23.7\,\mu s$	$24.1\,\mu s$	$401\,\mu s$	$671\,\mu s$	$899\,\mu s$
splitstream	$102\,\mu s$	$157\,\mu s$	$167\,\mu s$	$31.1\,\mu s$	$44.8\,\mu s$	$47.7\,\mu s$	$412\,\mu s$	$698\,\mu s$	$923\,\mu s$
splitframe	$48.8\,\mu s$	$96.6\,\mu s$	$98.9\,\mu s$	$49.4\,\mu s$	$76.4\,\mu s$	$77.5\,\mu s$	$430\,\mu s$	$699\,\mu s$	$952\,\mu s$
iqzigzagidct	$380\,\mu s$	$653\,\mu s$	$875\,\mu s$	$47.5\,\mu s$	$74.2\,\mu s$	$78.4\,\mu s$	$428\,\mu s$	$716\,\mu s$	$953\,\mu s$
mergeframe	$71.1\,\mu s$	$116\,\mu s$	$116\,\mu s$	$43.1\,\mu s$	$65.1\,\mu s$	$66.9\,\mu s$	$424\,\mu s$	$683\,\mu s$	$942\,\mu s$
mergestream	$53.8\,\mu s$	$107\,\mu s$	$114\,\mu s$	$24.7\,\mu s$	$37.2\,\mu s$	$37.6\,\mu s$	$405\,\mu s$	$687\,\mu s$	$913\,\mu s$
process network		$653\,\mu s$	$875\,\mu s$		$76.4\,\mu s$	$78.4\,\mu s$		$699\,\mu s$	$953\,\mu s$

Table 3: Overhead in terms of execution time and binary code size for adding the ability to stabilize compared to the original implementation.

application	memory overhead	time overhead
Demosaicing	8624 B	$43.01\,\mu s\,(< 0.05\%)$
MJPEG decoder	7104 B	$43.01\,\mu s\,(< 0.05\%)$

The application consists of six processes and its structure is outlined in Fig. 5.

Similar to the first benchmark example, we compare the measured stabilization time $t_{stab,v}$ of each process v with its upper bound $t^*_{stab,v}$, see Table 2. The experiment was repeated 20 times and the average and maximum results are reported in Table 2. The results confirm the trend observed with the Demosaicing application. In particular, all processes stabilized before the expected time bounds. In many cases, the bounds are actually very accurate. For the MJPEG decoder application, the upper bound is mainly composed of the maximum execution time $t^*_{c,v}$ of the iqzigzagidct process.

Time and Memory Overheads. The time and memory overhead generated by the additional code required to accomplish process network stabilization is presented in Table 3. The time overhead is mainly due to the additional logic to check for the **stop** token from the master and related housekeeping activities. In particular, an individual checking for the **stop** token has taken on average $43.01\,\mu s$.

Summary. Using realistic applications, it has been shown that the proposed technique can bring a process network to a stable state. Performance metrics such as upper bounds on the maximum stabilization time and maximum context sizes are also presented. The maximum stabilization time is dominated by the maximum time a process can execute without calling any communication primitive, i.e., *process compute time*. It may be possible to further reduce the stabilization time by inserting additional checks for events in the process' compute segments. Finally, detailed results from experiments on the Intel SCC baremetal platform were presented, validating the ideas presented in this paper.

7. CONCLUSION

The paper presented a technique to bring a process network executing on a distributed system into a stable state, suitable for migration. The proposed technique has been shown to be lightweight, and preserves the original functionality of the network. The correctness of the technique has been shown to be independent of the temporal characteristics of the system and the topology. We have shown that if the token communication time and *process compute time*

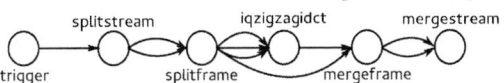

Figure 5: The MJPEG application.

are upper bounded, then an upper bound on the overall stabilization time can be calculated. Finally, we validated the feasibility and efficiency of the proposed approach and the respective timing models with representative experiments on Intel's SCC platform.

8. REFERENCES

[1] A. Acquaviva et al. Assessing Task Migration Impact on Embedded Soft Real-Time Streaming Multimedia Applications. *EURASIP J. Embedded Syst.*, pages 9:1–9:15, 2008.

[2] G. M. Almeida et al. An Adaptive Message Passing MPSoC Framework. *Int'l J. of Reconfigurable Computing*, 2009.

[3] Artist. Benchmarks. http://www.artist-embedded.org/artist/Benchmarks.html, 2008.

[4] S. Bertozzi et al. Supporting Task Migration in Multi-Processor Systems-on-Chip: A Feasibility Study. In *Proc. DATE*, pages 15–20, 2006.

[5] E. Cannella et al. Adaptivity Support for MPSoCs Based on Process Migration in Polyhedral Process Networks. *VLSI Design*, pages 2:2–2:17, 2012.

[6] S. Chakravorty et al. Proactive Fault Tolerance in MPI Applications via Task Migration. In *Proc. HPC*, pages 485–496, 2006.

[7] K. M. Chandy. Distributed Snapshots: Determining Global States of Distributed Systems. *ACM Trans. on Computer Systems*, 3:63–75, 1985.

[8] C. Clauss et al. iRCCE: A Non-Blocking Communication Extension to the RCCE Communication Library for the Intel Single-chip Cloud Computer. Technical report, RWTH Aachen, 2011.

[9] M. Geilen and T. Basten. Kahn Process Networks and a Reactive Extension. In *Handbook of Signal Processing Systems*, pages 967–1006. Springer, 2010.

[10] W. Haid et al. Efficient Execution of Kahn Process Networks on Multi-Processor Systems Using Protothreads and Windowed FIFOs. In *Proc. ESTIMedia*, pages 35–44, 2009.

[11] J. Howard et al. A 48-Core IA-32 Message-Passing Processor with DVFS in 45nm CMOS. In *Proc. ISSCC*, pages 108–109, 2010.

[12] G. Kahn. The Semantics of a Simple Language for Parallel Programming. In *Proc. IFIP Congress*, pages 471–475, 1974.

[13] C. Lee, H. Kim, H.-W. Park, S. Kim, H. Oh, and S. Ha. A Task Remapping Technique for Reliable Multi-Core Embedded Systems. In *Proc. CODES/ISSS*, pages 307–316, 2010.

[14] X. Li. Demosaicing by successive approximation. *Trans. Img. Proc.*, 14(3):370–379, 2005.

[15] C. Lu and S.-M. Lau. A Performance Study on Load Balancing Algorithms with Task Migration. In *Proc. TENCON*, pages 357–364, 1994.

[16] D. F. Mark Claypool. Transparent Process Migration for Distributed Applications in a Beowulf Cluster. In *Proc. INC*, pages 459–466, 2002.

[17] H. Nikolov et al. Systematic and Automated Multiprocessor System Design, Programming, and Implementation. *IEEE Trans. Comput. Aided Design*, 27(3):542–555, 2008.

[18] J.-C. Ryou and J. Wong. A Task Migration Algorithm for Load Balancing in a Distributed System. In *Proc. System Sciences*, pages 1041–1048, 1989.

[19] S. Sankaran et al. The LAM/MPI Checkpoint/Restart Framework: System-Initiated Checkpointing. *Journal of High Performance Computing Applications*, 19(4):479–493, 2005.

[20] T. Suen and J. Wong. Efficient Task Migration Algorithm for Distributed Systems. *IEEE Trans. on Parallel and Distributed Systems*, 3:488–499, 1992.

[21] C. Wang et al. Hybrid Checkpointing for MPI Jobs in HPC Environments. In *Proc. ICPADS*, pages 524–533, 2010.

APPENDIX

A. PROTOTYPE IMPLEMENTATION

In this section, we illustrate a prototype implementation of the proposed stabilization technique.

A.1 Process Network Specification

We start with illustrating a high-level API for specifying process networks. A process $v \in V$ starts executing by first initializing itself at Line 2 in Algorithm 1, and then repeatedly invoking the *Fire* function at Line 4. The function can consist of any number and order of data-token read/write steps, and compute steps (e.g. branches, loops, assignments, etc.) depending on the actual functionality of the process.

Algorithm. 1: Basic structure of a process $v \in V$.

```
1: process v
2:   INIT();                                         ▷ Initialization
3:   while true do                            ▷ Call Fire repeatedly
4:     FIRE();              ▷ Communication and computation
5:   end while
6: end process
```

A.2 Integrating the Stabilization Mechanism

Next, we will illustrate how the original process network specification from Algorithm 1 can be extended to integrate the stabilization mechanism. The structure of a process network should be minimally modified so that each process obtains the ability to go into the stable state and collect its context. The modifications discussed in Section 5 are incorporated into the pseudo-code shown in Algorithm 2. First, notice that Lines 10-19 are thread-safe. Line 10 checks for the **stop** event *before* starting a potentially blocking data token read or write step. If no event has been posted, the process starts the data-token read or write step. If the process is blocking on a read or write step while the **stop** event from the master is received, the signal **unblock_process** will unblock the process.

Algorithm. 2: New Structure of process v_j.

```
1: process v_j
2:   INIT();                                         ▷ Initialization
3:   while !cancelled do          ▷ Until the stop event cancels further
                                                        execution
4:     cancelled ← FIRE();     ▷ Communication and computation
5:   end while
6:   WRAPUP();    ▷ Call the Wrapup function to finish stabilization
7:   function FIRE
8:     . . .
9:            ▷ Must not start read or write step if shared is stop
10:    if shared = stop then
11:      shared ← done
12:      return (cancelled ← true);
13:    else
14:      Start a blocking R/W step
15:      if unblocked by signal then
16:        shared ← done
17:        return (cancelled ← true)
18:      end if
19:    end if
20:    . . .
21:  end function
22: end process
```

Unblocking upon reception of an event is easily accomplished by using user-space signals from threading libraries such as the POSIX library. Thus, the reception of the **stop** event by a process effectively cancels the currently blocked token-write or token-read operation.

A.3 The Wrapup Function

The *Wrapup* function is introduced in Section 5.1. Pseudo-code illustrating the function is given in Algorithm 3. First, in Line 2, it waits until it receives the **proceed** event from the master. It swaps all regular FIFOs $\mathcal{L}(v_j, m)$ with the corresponding backup FIFOs $\mathcal{B}(v_j, m)$ in Line 3. Afterwards, it sends an end-of-stream (EOS) token to all its children and waits in Line 5 until it receives an EOS marker from all its parents. Finally, some cleanup operations are performed to return the memory to the system.

Notice that swapping of the local FIFO $\mathcal{L}(v_j, m)$ with the backup FIFO $\mathcal{B}(v_j, m)$ preserves the correctness of the process network. This is because:

- A backup FIFO is brought online *only* after reception of the **proceed** signal. Note that a companion process transmits the acknowledgement only when the process changes the shared variable to **done**. Thereafter, the algorithm guarantees that the process will not transmit any more tokens. Therefore, in the worst case, the backup FIFO must be able to accommodate the tokens which are still in flight, which are upper bounded to $|\mathcal{M}(m)| + |\text{EOS}|$.

- The only token that a process transmits post-reception of the **proceed** event is the $|\text{EOS}|$ marker, which is accommodated in the backup FIFO.

Therefore, it can be seen that (assuming that the communication network is lossless), none of the data tokens are lost in the process of stabilization. Further, the FIFO data structure maintains the relative ordering on the tokens on each channel.

Algorithm. 3: Basic structure of the *Wrapup* function.

```
1: function WRAPUP
2:   WAIT-TO-PROCEED();    ▷ Wait for proceed event from master
3:   Switch L(v_j, k) with B(v_j, k);
4:   FORWARD-EOS();       ▷ Forward EOS token to all children
5:   COLLECT-TOKENS();          ▷ Collect "late-arriving" tokens
6:   CLEANUP();           ▷ Return memory to the system, etc.
7: end function
```

A.4 The Companion Process

The companion process $comp_j$ of process v_j is responsible for the communication of process v_j with the master process. Algorithm 4 describes the companion process $comp_j$, which is executed independently of process v_j. The assumption is that **stop** events from the master cannot overlap. In particular, the companion process waits for an event of the master. If it finds a **stop** event, it updates the *shared* variable and waits until the variable is set to **done**. Afterwards, it sends an acknowledgement to the master and waits until it receives the **proceed** event.

It is possible to optimize this structure such that not every process in the process network has a companion process, but instead a group of processes residing on one processing element share a companion process. However, such optimization is beyond the scope of the paper.

As a process v_j might be blocked because of reading from an empty FIFO or writing to a full FIFO channel, the companion process $comp_j$ has to be implemented as an additional object that is running in parallel to process v_j and just shares a single variable with process v_j. In case the platform supports multi-threading, the companion process

Algorithm. 4: Pseudo-code illustrating the functionality of the companion process $comp_j$.

```
 1: process comp_j          ▷ Runs as a separate concurrent process
 2: while true do
 3:     Read event from event channel
 4:     if found stop then
 5:         shared ← stop
 6:     end if
 7:     Sleep on the shared variable until it is changed to done
 8:     Send acknowledgment to master
 9:     Read event from event channel
10:      if found proceed then
11:          shared ← proceed;
12:      end if
13: end while
14: end process
```

$comp_j$ can be implemented as an additional thread. Otherwise, one can use stack-less threads as described in section A.6. As the companion process is in a known state when the stabilization is completed, the companion process is not part of the context of process v_j, but can be re-initialized after migration.

A.5 Additional Note on Stabilization Time

It can be seen from (7) that the overall stabilization time depends on the time it takes for *phase 1* and *phase 2* to complete. The length of phase 1 is dominated by the maximum *process compute time*. The length of phase 2 is dominated by network speed, with time taken to swap FIFO being negligible. Notice from Algorithm 2, lines 14 - 17 that the a write by a process can be canceled by a signal, and therefore, the stabilization time is (largely) independent of the amount of data written by a process. The network speed accounts for a small part in the overall duration of the stabilization time, since a maximum of $|\mathcal{M}_v(m)|$ tokens need to be collected in phase 2 before the stabilization is complete.

A.5.1 |EOS| *Send and Receive Times*

Equation (5) does not consider the |EOS| send and receive times by a process individually. This is because the time it takes for a process v to receive the |EOS| token from its parent already covers the time it takes for v's parent to send the |EOS| tokens (plus a small communication time), and thus the use of $t_{u \to v}^*$.

A.6 Implementing Unblocking

Finally, we present an overview of how the unblocking functionality is implemented in the prototype implementation used in Section 6. There are various mechanisms that provide the ability to cancel a blocked read or write operation. For example, one could separately schedule each block of code between two communication calls using a kernel-space thread. However, this might lead to a large scheduling overhead due to the full preemption and memory protection, both undesired when scheduling such code blocks. Another option is to use stack-less threads as, for example, protothreads [1]. Protothreads have already been successfully applied to KPNs to provide lightweight scheduling [2]. The functionality of protothreads is implemented as a set of macros that enclose the communication calls. The embedding of a KPN process into a protothread process can also be automated at the software synthesis step.

In protothread, a control structure is used to store the local data of a process together with a variable that represents the line number of the process. Whenever the process exits the *Fire* function, it updates this variable either to the current line number or, if the process has reached the end of the *Fire* function, to the beginning of the *Fire* function. On the other hand, at the beginning of the *Fire* function, the line number variable of the control structure is read and the program counter jumps to this line. In order to extend the protothread library with the unblocking functionality, we change the process structure in two ways: First, we extend the `PT_WAIT_UNTIL` macro of protothreads with the abilities to check for the *shared* variable and to be unblocked by the `unblock_process` signal, as outlined in Algorithm 2. Originally, the `PT_WAIT_UNTIL` macro just blocked a process until the read or write is successful. Second, we enclose each communication call with the extended `PT_WAIT_UNTIL` macro to obtain the functionality described in Algorithm 2.

A.7 Summary of Distributed Process Network Stabilization Approach

A short summary of the approach is presented here:

1. The algorithm stabilizes the distributed process network *correctly*, i.e., no tokens are lost, and relative ordering amongst tokens in each channel is maintained. All tokens not consumed by a process v are stored in the process context in the correct order.

2. POSIX signals are used to unblock a process blocked on a full output FIFO or an empty input FIFO. The Intel SCC implementation uses a custom communication layer with the following properties:

   - Blocks a process on an empty input FIFO or a full output FIFO;
   - Unblocks the process upon receiving the *signal* from the *master*. All messages are broken into *chunks*. The maximum chunk-size is carefully selected so that the each core has *guaranteed* space to receive the messages from the master. The communication layer intersperses normal message read and write steps with checking for control messages from the master.

3. The stabilization procedure is composed of two phases, *phase 1* and *phase 2*.

   - Phase 1 brings all processes in the process network into a *known* state. The completion of phase 1 guarantees that no process in the network performs any further compute, write, or read steps. The length of phase 1 is dominated by *process compute times*.
   - Phase 2 required each process to swap local FIFO with backup FIFOs. The backup FIFOs are sized appropriately in order to accommodate all possible in-flight tokens from each of process' parents, plus the |EOS| token. Phase 2 is determined by the communication times, and hence dependent upon the network characteristics.

4. The algorithm is independent of the size of local FIFOs, and is the same as in the original process network. The size of backup FIFOs are statically determined.

B. STABILIZING INDIVIDUAL PROCESSES

Our main section introduced an approach in which the entire process network was stabilized. However, the same principles can be applied to stabilize a part of the process network. Consider a process $v \in V$, which must stabilized.

Therefore, all parents of v are informed of the stabilization of v, and as a result, process v receives the EOS from all its parents. The process v must also send EOS to all its children, so that v's children do not continue to expect tokens from the previous location of v. Once process v stabilizes, it can migrate. Subsequently, new channels must be established between the parents of v and v's children. Since the network structure is statically known, recreating channels is straightforward.

C. ADDITIONAL EXPERIMENTAL RE-SULTS

In addition to the results presented in Section 6, we summarize the Intel SCC platform and other experimental results in this section.

C.1 Intel Single-Chip Cloud Computer

The Intel Single-chip Cloud Computer (SCC) is a prototype of future embedded on-chip many-core platforms [3]. The processor consists of 24 tiles that are organized into a 4×6 grid and linked by a 2D mesh on-chip network. A tile contains a pair of P54C processor cores, a router, and a 16 KB block of SRAM. Each core runs at 533 MHz and each router runs at 800 MHz. The on-tile SRAM block is also called "message passing buffer" (MPB) as it enables the exchange of information between cores in the form of messages. Figure 6 schematically outlines the SCC processor.

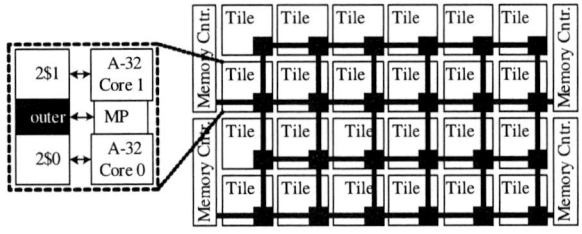

Figure 6: Schematic representation of Intel's SCC processor [3].

C.2 Additional Calibration Results

In Section 6, we have shown that the communication latency under high cross-traffic between any two processes u and v can be upper bounded. In particular, we have seen that the latency is independent of the number of hops between the processing nodes of the communicating processes. In addition, we observed the communication latency under low data traffic conditions, which is bounded by:

$$t_{u \to v}^{\text{low traffic}} = 5.182|P| + 307.4H + 531 \quad \text{(cl.cycles)} \quad (9)$$

Here, the latency depends on the number of hops H between the nodes of the communicating processes. However, this model cannot be used as an upper bound as it considers only low data traffic in the network.

C.3 The Demosaicing Benchmark: Additional Results

Table 4 summarizes the characteristics of the Demosaicing application obtained when measuring the computation and read/write times of the individual processes. The binding reports the identification number of the SCC core, on which the process is executed. The maximum execution

Table 4: Characteristics of the Demosaicing application.

process	bind-ing	max. exec-ution time	port	max. data / iteration
load image	26	0.13 s	0	618 KB
pre processing	12	0.13 s	0	618 KB
pre demosaicing	14	0.4 ms	1 2 3 4 5 6 7 8	160 KB 2 B 160 KB 2 B 160 KB 2 B 160 KB 2 B
demosaicing_0	28	1.05 s	2 3	160 KB 471 KB
demosaicing_1	18	1.07 s	2 3	160 KB 471 KB
demosaicing_2	02	1.07 s	2 3	160 KB 471 KB
demosaicing_3	40	1.05 s	2 3	160 KB 471 KB
post demosaicing	30	2.4 s	8 9	618 KB 618 KB
post processing	20	4.2 s	2	618 KB
write result	22	0.5 ms		

time corresponds to the maximum time that a process is executing without calling a communication primitive and is calculated when the Demosaicing application was executed under five RAW images of different sizes. Finally, the maximum amount of data that is transmitted per outgoing channel and iteration is reported. The values given in Table 4 have been used in Table 1 to calculate the upper bounds on the stabilization times.

Context Sizes: More Details.

The overall context sizes reported in Table 1 is the sum of:

- The space required to store the line number (in order to restore the context), local variables (except those which store unsent output data tokens, and unread input data tokens), denoted as S_1.

- The space required for storing unsent output data tokens, and unread input data tokens, denoted as S_2.

Therefore, S_2 can be considered as the application-dependent context storage requirement, while S_1 is largely independent of the application. The contribution due to S_1 in the overall context sizes reported in Table 1 is presented in Table 5. It is clear from Table 5 that the total size of the context is dominated by the nature of the application itself.

C.4 The MJPEG Benchmark: Additional Results

Next, the characteristics of the MJPEG decoder application are summarized in Table 6. The following values are shown: The identification number of the SCC core, on which the process is executed; the maximum time that a process is executing without calling a communication primitive; and the maximum amount of data that is transmitted per outgoing channel and iteration. The time of the longest compute segment was measured over each frame of an example video with resolution 320×240 pixels.

Table 5: Context size *exclusively* due to S_1.

process	context size for S_1
load image	48
pre processing	24
pre demosaicing	36
demosaicing_0	72
demosaicing_1	72
demosaicing_2	72
demosaicing_3	72
post demosaicing	84
post processing	152
write result	68

Table 6: Characteristics of the MJPEG decoder application.

process	binding	max. execution time	port	max. data / iteration
trigger	02	$58\,\mu s$	0	4 B
splitstream	04	$167\,\mu s$	0	4 B
			1	10 KB
splitframe	06	$98\,\mu s$	2	307.2 KB
			3	64 B
			4	4 B
			5	8 B
iqzigzagidct	18	$875\,\mu s$	3	76.8 KB
mergeframe	20	$116\,\mu s$	2	7.68 KB
			3	8 B
mergestream	22	$114\,\mu s$		

D. REFERENCES

[1] A. Dunkels, O. Schmidt, T. Voigt, and M. Ali. Protothreads: Simplifying Event-Driven Programming of Memory-Constrained Embedded Systems. In *Proc. SenSys*, pages 29–42, 2006.

[2] W. Haid et al. Efficient Execution of Kahn Process Networks on Multi-Processor Systems Using Protothreads and Windowed FIFOs. In *Proc. ESTIMedia*, pages 35–44, 2009.

[3] J. Howard et al. A 48-Core IA-32 Message-Passing Processor with DVFS in 45nm CMOS. In *Proc. ISSCC*, pages 108–109, 2010.

Distributed run-time resource management for malleable applications on many-core platforms

Iraklis Anagnostopoulos, Vasileios Tsoutsouras, Alexandros Bartzas, Dimitrios Soudris
School of Electrical and Computer Engineering, National Tech. Univ. of Athens, Greece

ABSTRACT

Todays prevalent solutions for modern embedded systems and general computing employ many processing units connected by an on-chip network leaving behind complex superscalar architectures In this paper, we couple the concept of distributed computing with parallel applications and present a workload-aware distributed run-time framework for malleable applications on many-core platforms. The presented framework is responsible for serving in a distributed way and at run-time, the needs of malleable applications, maximizing resource utilization avoiding dominating effects and taking into account the type of processors supporting platform heterogeneity, while having a small overhead in overall inter-core communication. Our framework has been implemented as part of a C simulator and additionally as a run-time service on the Single-Chip Cloud Computer (SCC), an experimental processor created by Intel Labs, and we compared it against a state-of-art run-time resource manager. Experimental results showed that our framework has on average 70% less messages, 64% smaller message size and 20% application speed-up gain.

1. INTRODUCTION

The current trend in computing and embedded architectures is to replace complex superscalar architectures with many processing units connected by an on-chip network. Modern embedded, server, graphics and network processors already include tens to hundred of cores on a single die. Intel has already created platforms with 80, 48 and 50 processing cores [10, 16, 18, 11], while Tilera currently features up to 100 cores per chip. Also, Networks-on-Chip (NoC) capable of supporting such large number of cores, are already supported by the industry (such as the Æthereal NoC [9] from NXP and the STNoC [17] from STMicroelectronics). Moreover, the development of such multi-core platforms is driven by the increase of highly parallel/multi-threading and demanding applications. Thus, the challenge focuses on finding an efficient way to deal with this varying amount of re-

This work is partially supported by the E.C. funded FP7-248716 2PARMA Project, www.2parma.eu and ENIAC-2010-1 TOISE project, www.toise.eu

Permission to make digital or hard copies of all or part of this work for personal or classroom use is granted without fee provided that copies are not made or distributed for profit or commercial advantage and that copies bear this notice and the full citation on the first page. To copy otherwise, to republish, to post on servers or to redistribute to lists, requires prior specific permission and/or a fee.
DAC'13, May 29 - June 07 2013, Austin, TX, USA.

sources at run-time leading to performance improvements.

The run-time resource management paradigm has been revealed as a key challenge to modern multi-core systems and it has become prominent due to the run-time dynamicity of modern parallel applications and platforms. In modern execution environments run-time resource availability may vary due to system dynamism as resources can be added or removed from such environments at any time. As described in [12] malleability is used for autonomous application reconfiguration in response to dynamic changes in platform's resource availability, thus allowing applications to optimize the use of platform's features (e.g., number of processors). In other words, malleable applications use varying amounts of platform resources during their execution and they may specify the minimum and maximum number of processors they require. As minimum can be considered the minimum number of processors a malleable job needs while maximum describes the maximum one. Any allocation of cores more than the maximum number is a waste or platform resources.

As applications and computing systems are continuously getting more and more complex, it becomes harder to manage them from one central point. Traditionally, a central core periodically analyzes applications' malleability and platform resources and tries to find the best match between them. However, such centralized approaches [15] limit scalability due to bottlenecks appeared from processing and communication functions, especially in environments that require frequent configuration changes. Also, the large number of cores in modern systems, increase the failure rate of single processors resulting in system erros when parallel applications are executing [3]. Last, centralized run-time managers lack the concept of self-adaptation and self-organization, actions that trend to be a solution to modern platforms [13].

In this work, we couple the concept of distributed computing with parallel applications and we present a workload-aware distributed run-time framework for malleable applications running on many-core platforms. The proposed framework is based on the idea of local controllers and managers while an on-chip intercommunication scheme ensures decision distribution. The presented framework is responsible (i) for serving, at run-time, the needs of malleable applications, in terms or processing cores; (ii) makes sure that the application will get the optimum number of cores avoiding dominating effects; (iii) it takes into account the type of processors best utilizing any platform's heterogeneity; and (iv) it has a small overhead in overall core intercommunication.

The rest of the paper is organized as follows. An overview of previous works is presented in Section 2, while the proposed methodology framework is presented in Section 3. The evaluation results are presented in Section 4, and finally conclusions are drawn in Section 5.

2. RELATED WORK

The run-time resource management on many-core platforms can be either centralized or distributed. Nollet et al. [14] present a centralized runtime resource management scheme that is able to efficiently manage an NoC containing fine grain reconfigurable hardware tiles and two task migration algorithms. The resource management heuristic consists of a basic algorithm completed with reconfigurable add-ons. Al Faruque et al. [1] present a distributed cluster oriented framework for homogeneous platforms which is based on agents for the task-to-cluster mapping. Anagnostopoulos et al. [2] present a divide and conquer based distributed run-time mapping framework for both homogeneous and heterogeneous platforms with the introduction of a matching factor. In order to reduce the on-chip node intercommunication Cui et al. [4] present a decentralized cluster-based scheme for task mapping, designed for reduction of the communication traffic between agents. *However, even though these approaches handle application mapping at run-time in a good way, they are designed for fixed-size applications without any malleability aspect.* Thus, no application reconfiguration is performed in response to any dynamic changes of platform's available resources and no resizing or remapping of the applications is allowed.

On the field of self-organized and dynamic systems and from the aspect of malleable or parallel applications in general, Sabin et al. [15] present a greedy centralized scheduling strategy and demonstrate that the importance of efficiency varies with respect to the characteristics of the workload a scheduler encounters. Desell et al. [6] show that the application malleability provides up to a 15% speedup over component migration alone on a dynamic cluster environment. Kobe et al. [12] present a distributed agent-based task mapping for malleable applications supporting also self-organization. The agent is assigned at run-time to a random core when a new application arrives having the disadvantage of a possible communication bottleneck when a randomly selected already occupied core serves the new request.

In this work we present a distributed run-time resource management framework for malleable applications. The novelty of the paper is threefold: (i) it takes into account platform's heterogeneity by best utilizing any cores' types thus offering flexibility; (ii) it has a small overhead in overall core intercommunication; and (iii) it has been integrated as a run-time service on a real many-core platform.

3. METHODOLOGY FRAMEWORK

The goal of the proposed methodology framework is to perform run-time resource management on many-core platforms, both homogeneous and heterogeneous, for parallel applications in a distributed way. According to the parallel job classification scheme presented in [8] we can separate parallel applications into two categories based on their capabilities. *Moldable applications* can be stopped at any point but the number of processors that occupy cannot be changed during run-time. *Malleable applications* can be stopped at any point of execution at run-time and they have the flexibility to change the number of assigned processors at run-time.

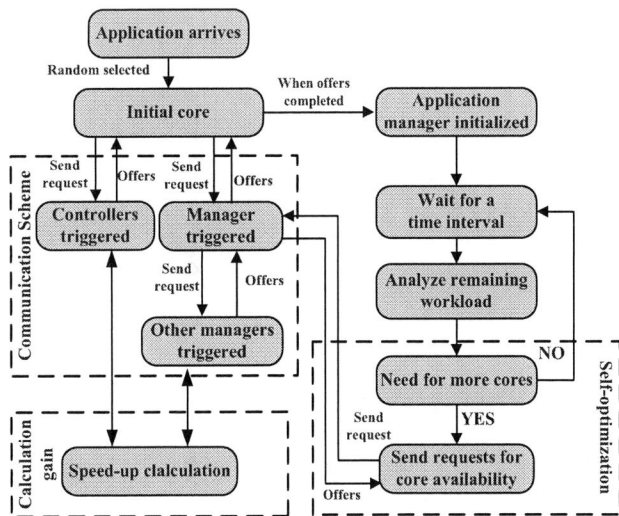

Figure 1: Overall flow of the proposed methodology.

The proposed framework is designed for malleable applications. Even though it can support both moldable and migratable ones, it best utilizes the platform available resources for malleable applications. In this work each core can have one of the following roles concerning the resource management of the platform: *(i) initial core; (ii) controller core;* and *(iii) manager core.* The *initial core* is randomly chosen and triggered when a new application arrives in the system being responsible for determining the cores on which the application will start running. The *controller core* is responsible for handling all the unoccupied cores of a certain region of the platform. It is defined at the initialization of the platform and cannot be changed in run-time. It also maintains a list of all manager cores that possess a core in its region. This information is provided to initial or manager cores if and when it is asked for. Last, the *manager core* manages an application searching for new cores and instructing the resizing of the application whenever it has more or less cores to run on. This core does not execute any part of the application. If it does not possess any other cores for the actual application to run on, a self optimization is necessary in order to for the manager to acquire at least one working node. Although an initial core can be a manager one and vice versa, a controller core cannot change its functionality.

An overview of our methodology framework in terms of node intercommunication is presented in Figure 1. Once a new application arrives on a random core (this is the initial core), this core sends messages to controller cores found near it and asks for an available core to serve as the actual application manager. Then the controller cores search into their area for any unoccupied cores and also send requests to any managers into this area. According to the application's characteristics, cores with the appropriate type respond, provided that the speed-up gain of the requesting application is greater than the speed-up loss of the offering manager. After that, the initial core receives all offers and determines the new manager which is initialized by a signal. Then, the new manager distributes evenly the workload to

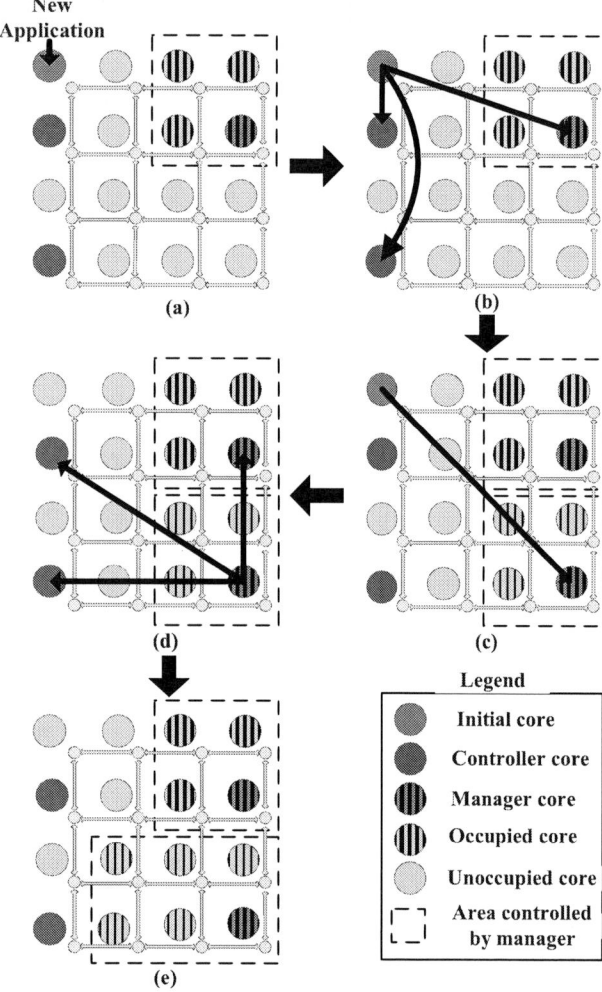

Figure 2: Example of the communication scheme.

the working node he manages. After a predefined time interval the manager, if there is a need and the application is not close to its finish, requests for more cores in order to increase application's performance giving to the framework a self-optimization aspect.

3.1 Definitions

The application and the speed-up model used in this work is the common malleable application model described in [7, 5] and used by other distributed approaches [12]. Each application is described by four parameters W, var, A and Q, where W is the workload, var is the parallelism variance, A is the average parallelism and Q is most preferred Processing Element (PE) type that the application is supposed to be executed on. A many-core platform topology and its communication infrastructure can be uniquely described by a strongly connected directed graph $P(I, N)$. The set of vertices N is composed of two mutually exclusive subsets N_{PE} and N_C containing the available platform's PEs and the platform's on-chip interconnection elements (C), such as routers in an NoC technology. Each platform's PE can be of a specific

type and differ from the other platform types (supporting heterogeneous platforms) or all PEs can be of the same type and thus have the same functionality (homogeneous platform). In our framework $T_{pe_i} \forall pe_i \in N_{PE}$ specifies the type of the PE pe_i. In an heterogeneous platform, the T_{pe_i} varies for each PE while in an homogeneous platform T_{pe_i} is the same for all PEs. In order to comply with the application's requirements in terms of required PE classes and have the best $Q \to T_{pe_i}$ utilization, we define $Util[pe_i] \in [0, 1]$ that implies how good the $app(W, var, A, Q)$ is served by the pe_i with T_{pe_i} type. $Util[pe_i]$ can also be considered as a priority factor when an application is asking for additional cores, meaning that when $Util[pe_i] = 1$ we want the best match while when $Util[pe_i] = 0$ the application is not requesting any specific T_{pe_i}. Last, we define the sets F and $offers[]$, which describe all the nodes that can appropriately serve an application based on their type and all the cores offered to the application respectively.

3.2 Communication Scheme

In order to be consistent with the rest of the document and with Algorithms 1 and 2 we declare that the index dst specifies the manager core that is requesting more resources while src is the manager core that is offering them. Also, we define the set R which contains, for each controller core, the PEs in a manhattan distance specified by the equation 1 where $size$ is the size of the platform and $num\_controllers$ is the number of controllers cores on it on the X dimension.

$$distance = size/num\_controllers_X \qquad (1)$$

Algorithm 1 Communication scheme algorithm

```
    // Initial core actions
1:  analyze(W, var, A, Q)
2:  req_send(control[], core_id, app, R)
3:  start_timer()
4:      offers[] = receive_offers
5:  end_timer()
6:  sel_pe = best{offers[]}
7:  initialize_manager(sel_pe, offer, R)

    // Controller core actions
8:  analyze(W, var, A, Q, R)
9:  for each (N_PE ∈ F̄ && N_PE ∈ R)
10:     If calculate(gain(app)) > 0 // Algorithm 2
11:        offers[] = offers[] + new_offer
12: send_offers(offers, core_id)

    // Manager core actions
13: // Actions for offering cores
14: analyze(W, var, A, Q, R)
15: If calculate(gain(app)) > 0 // Algorithm 2
16:     offers[] = offers[] + new_offer
17:     send_offers(offers, core_id)
18: // Actions for self-optimization
19: while (app(W, var, A, Q)! = finished) {
20:     analyze(W, var, A, Q, offers)
21:     timer()
22:     If ((app.threshold = max) || (app.left_time < timer()))
23:        continue
24:     else
25:        req_send(control[], R)
26:        start_timer()
27:           offers[] = offers[] + receive_offers
28:        end_timer()
29: end while
```

As aforementioned, when a new application arrives on a core, the initial core task is executed and the communication inside the platform takes place in order to establish a manager core for the application. The communication between initial, controller and manager cores is described in Algorithm 1.

Initial core (lines 1-7): When a new application arrives on an initial core (Figure 2a), this core analyzes the application's characteristics, sends a message to the controllers and managers (Figure 2b) that are inside its region R and fires a timer in order to check for their responses. After the end of the timer, the initial core selects the best offer and sends a signal, according to the offer, and starts the initialization of the manager that will handle the application (Figure 2c).

Controller core (lines 8-12): A controller cores has a variety of responsibilities. Besides maintaining regional information about the managers existing on its region, it has to provide this information to any cores requesting it. It also informs these cores about the position of controller cores in other regions. When the controller receives the signal form the initial core, it analyzes the application and starts to find cores to offer. The controller core can offer any unoccupied core he owns, inside the region R of the initial core, provided that it serves the application's characteristics.

Manager core (lines 13-29): The manager core has two tasks. During the first one, the manager checks if it can offer a core to the new application without loosing more in terms of application speed-up than the gain that the new application will have with the new addition. The second task has to do with the self-optimization process of the already running application. Specifically, there is a time threshold in which the manager checks if the application has taken all the necessary resources it needs or it is near to its completion. If the application has maximized its speed-up [7] there is no need to search for more cores. The same happens if the application is almost finished. In other words, if the remaining time of the application is less than the time interval there is no need to search for more cores. Otherwise, the application enters a self-optimization phase and the manager follows the same communication scheme and sends a message to the controllers (Figure 2d) that are inside its region R and fires a timer in order to check for their responses. After the end of the timer, the manager core checks the offers form the controller cores or from any other manager cores and starts the resize of the application (Figure 2e). Both the controller and manager cores use a function (lines 10 and 15 respectively) in order to calculate the gain or the loss to the application speed-up when offering a core (Algorithm 2).

3.3 Gain calculation

Algorithm 2 describes the required steps that both the controller and manager cores take in order to decide which cores should be offered when an application starts its self-optimization process asking for cores. It has three discrete parts: (i) the actions regarding the calculation of the speedup of the destination node requesting application (lines 4-8); (ii) the actions regarding the calculation of the speedup of the source node offering application (lines 9-13); (iii) and the calculation of the final total gain of this core trade (lines 14-21). During the actions regarding the destination node, for each $N_{PE} \in ((PE_{src} \cap R \cap F) - offers[])$ we calculate the speed-up of the application taking into account the core utilization ($Util[N_{PE}]$) in order to offer to the application the best choices in terms of PE type. Once the speed-up is calculated we check whether the gain of adding the new core to our working set results to an overall gain for the appli-

cation. On the other side, the actions regarding the source node check whether the loss of a core results in a bigger performance degradation on an already running application. In order to verify it, we calculate the loss speed-up both in terms of performance power and in terms of the $Util[N_{PE}]$ that are occupied and are needed by the application. Since both destination and source actions are greedy, the source offers cores to the destination only when the gain of the destination is bigger than the loss of the source.

Algorithm 2 Gain calculation algorithm

1: $offers[] = \emptyset$
2: while ($gain > 0$) {
3:    for each $N_{PE} \in ((PE_{src} \cap R \cap F) - offers[])$ {
   // Actions regarding the destination node
4:       $PE_{dst} = PE_{dst} \cup N_{PE} \cup offers[]$
5:       $ord\_PE_{dst} = order\{PE_{dst}\}$
6:       for $pos = 1$ to $ord\_PE_{dst}.length()$ {
7:         $SP_{dst} = Util\{ord\_PE_{dst}[pos]\} * (SP[pos] - SP[pos - 1])$
8:         $gain_{dst} = SP_{dst} - previous\_SP_{dst}$
    // Actions regarding the source node
9:       $PE_{src} = PE_{src} - offers[] - N_{PE}$
10:     $ord\_PE_{src} = order\{PE_{src}\}$
11:     for $pos = 1$ to $ord\_PE_{src}.length()$ {
12:       $SP_{src} = Util\{ord\_PE_{src}[pos]\} * (SP[pos] - SP[pos - 1])$
13:       $loss_{src} = previous\_SP_{src} - SP_{src}$
    // Calculate total gain
14:     $gain\_temp = gain_{dst} - loss_{src}$
15:     if $((gain\_temp > gain) \; || \; ((gain\_temp = gain) \; \&\& \; D(manger, N_{PE}) < D(manger, prev\_N_{PE})))$
16:       $gain = gain\_temp$
17:       $prev\_N_{PE} = N_{PE}$
18:    end for
19:    if ($gain > 0$)
20:     $offers[] = offers[] \cup N_{PE}$
21: end While

---

4. EXPERIMENTAL RESULTS

In order to validate our framework we have performed extensive simulation experiments in two steps: (i) using a c-based simulator (Section 4.1) and (ii) integrating the framework on the Intel Single Cloud Chip (SCC) many-core platform [10] (Section 4.2). In both cases, we compared the performance of the presented framework to the state-of-art distributed run-time resource manager DistRM [12]. As malleable applications input we have used the benchmarks provided by the parallel workload archive [8] and produced a file representing scenarios of applications arriving to our system. Each scenario consists of the time the application arrives on the system, its parameters as defined in section 3.1 and a random workload.

4.1 Evaluation on c simulator

Firstly, as aforementioned, we have developed a c-based simulator capable of simulating the behavior of malleable applications and all the necessary actions required by the on-chip communication scheme. For more accurate simulation, every node was represented by a different process. The inter-node communication is implemented using traditional Linux signals, while messages are passed using a pipe between the sender and the receiver. For synchronization semaphores are used. The goal of this step was to have a quick and abstract view of our methodology and estimate the cost of the developed distributed on-chip communication scheme. Also, the simulation method offers the capability of functional error correction and fast debugging in temps of possible communication deadlocks. The simulator sup-

Table 1: Comparison of the proposed technique to the DistRM [12] in the C simulator.

Plat. sizes	Msg. cnt.		Msg. size		Avg. sp.		Comp. eff.	
	Number of applications							
	32	64	32	64	32	64	32	64
6x6	72.3	71.4	73.2	72.4	3.8	13.8	86.8	86.6
8x8	64.3	63.4	64.8	63.6	4.5	9.3	83.3	83.0
12x12	49.1	52.3	44.8	48.7	-1.4	1.5	66.0	68.3
16x16	42.6	45.8	40.0	41.4	0.5	3.1	61.3	64.9
20x20	32.6	36.1	27.7	30.5	3.1	2.7	53.4	57.8
24x24	25.1	27.2	18.5	19.2	2.0	2.4	50.1	51.7
28x28	20.6	22.3	13.5	14.1	1.5	2.8	42.7	48.7
32x32	17.9	14.6	9.9	2.5	1.6	2.8	41.9	42.4
Average	41%		37%		3%		62%	

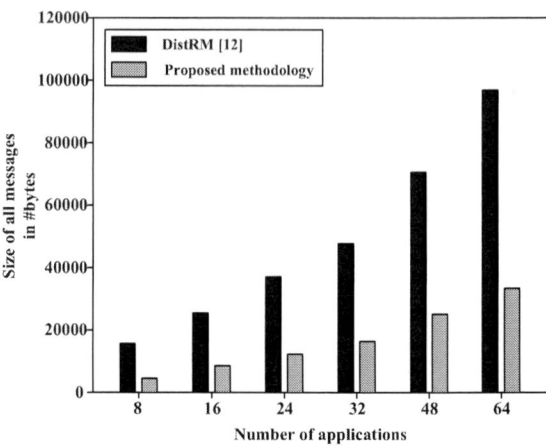

Figure 4: Total size of sent messages in *bytes*.

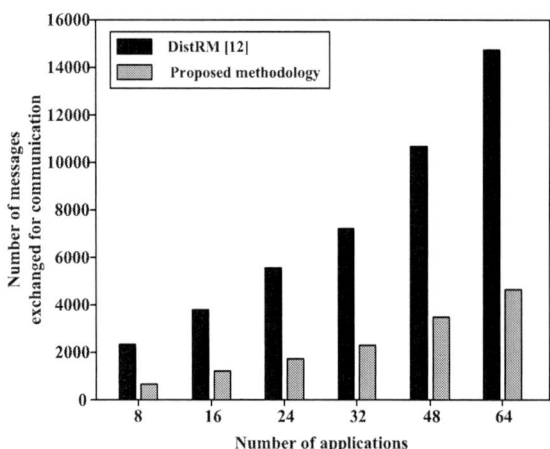

Figure 3: Total number of messages sent for inter-communication by all nodes for various applications.

ports big topologies (up to 32×32 core system), numerous application inputs.

Table 1 presents the results of the comparison of the proposed technique against the DistRM [12] distributed run-time manager. We evaluated the two run-time managers for various platform platform sizes, from 6×6 up to 32×32, and for 32 and 64 applications. The comparison metrics are: (i) message count (*Msg. cnt.*), which is the total number of messages sent by all nodes during the whole duration of the simulation both for resource management and application execution; (ii) message size (*Msg. size*), which is the total size of sent messages; (iii) application average speed-up (*Avg. sp.*); and (iv) computational effort (*Comp. eff.*), which is the total number of speed-up function calls. Table 1 presents the *percentage gains* of the presented framework in comparison to DistRM. Concerning the message count, simulation results showed an average gain of 41% for the presented methodology due to the fact that the core request messages are sent only inside the area R while DistRM sends messages in many areas, smaller than R and probably overlapping, thus increasing the number of messages used for sending requests and receiving answers. Also, the total size of these messages, measured in *bytes*, for our framework is on average 37% smaller that then size needed by DistRM.

Thus, the network burden of our framework is on average 38% smaller. In terms of average application speed-up, the proposed framework achieves on average 3% better results than DistRM. The speed-up function used for this metric is the application speed-up function presented in [12]. Last, the gain of our methodology regarding the total computational effort is on average 62% compared to the DistRM.

4.2 Evaluation on a many-core platform

After the first evaluation of our framework in the C simulator, we integrated the whole framework as a run-time service on the many-core Intel SCC platform [10] in order to test our framework on a real platform and not only at the simulation level. The Single-Chip Cloud Computer (SCC) experimental processor [10] is a 48-core "concept vehicle" created by Intel Labs as a platform for many-core software research. Since the platform size is fixed (48 cores) we compared the performance of our distributed run-time manager to the DistRM [12] for various number of applications running on the platform (from 8 to 64).

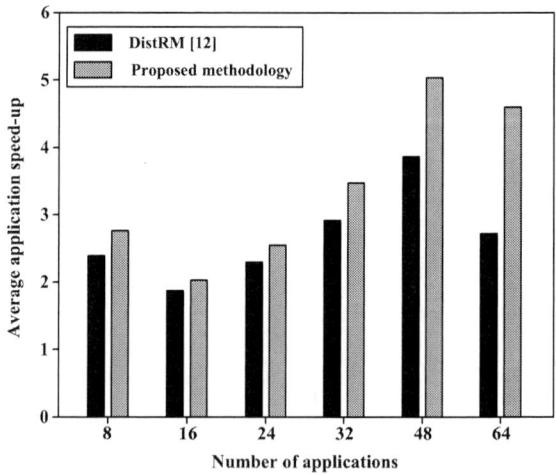

Figure 5: Average application speed-up using the speed-up function presented in [12].

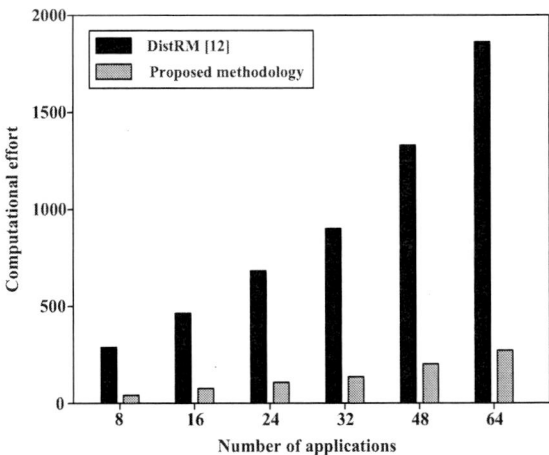

Figure 6: Computational effort comparison.

Figures 3 and 4 present the total number of messages sent by all nodes during the whole duration of the simulation and the total size of these messages in *bytes* respectively. The messages are used for application initialization, self-optimization and application resizing by manager cores and application execution. The proposed framework sends on average 70% less messages in order to perform all the necessary actions since the messages are sent only inside the area R. Whereas, DistRM searches for available cores in more areas, smaller than R, thus increasing the number of messages used for sending requests and receiving answers. Another reason that the presented algorithm has less messages than DistRM, is the fact that the framework performs application self-optimization under very specific criteria, only when the application has not maximized its speed-up or it is not near to its completion. Also, the size of the messages sent are on average 64% less than the ones used by the DistRM. The size is not proportional to the number of the messages since each message varies in size. For example, an offer message about four cores can have up to 20 *bytes* while the answer to this offer is 1 *byte*. Last, the proposed framework burdens network resources 66% less than DistRM.

Figure 5 presents the average application speed-up using the speed-up function presented in [12]. Speed-up is defined as the ratio of the total number of turnarounds performed for all applications divided by the total workload. The presented framework achieves on average 20% better application speed-up than DistRM. This can be explained by the fact that in the presented framework cores are not disturbed so often by messages during their application execution and thus completing the applications faster. On the other hand, due to the large number of messages sent by the application agents in DistRM, cores stop their functionality more frequently in order to answer to these messages, thus delaying the execution.

Figure 6 shows the comparison of the computational effort between the presented framework and DistRM. Computational effort is defined as the total number of speed-up function calls during the whole simulation. The presented framework has on average 85% less speed-up function calls than DistRM. This can be explained by the fact that, as aforementioned, the presented framework performs less application self-optimizations due to specific criteria. So the man-

ager cores calculate the speed-up function less frequently.

5. CONCLUSIONS

In this paper we presented a distributed run-time manager for malleable applications. The presented framework has a small communication overhead, takes into account platform's heterogeneity and makes sure that the application will maximize its speed-up function. Our framework has been implemented as part of a C simulator and additionally as a run-time service on a real many-core platform and we compared it against the DistRM [12] state-of-art run-time resource manager. Experimental results showed that our framework has on average 70% less messages, 64% smaller message size and 20% application speed-up gain.

6. REFERENCES

[1] M. A. Al Faruque et al. Adam: run-time agent-based distributed application mapping for on-chip communication. In *Proc. of DAC*, pages 760–765. ACM, 2008.

[2] I. Anagnostopoulos et al. A divide and conquer based distributed run-time mapping methodology for many-core platforms. In *Proc. of DATE*, pages 111–116, 2012.

[3] A. Beguelin et al. Application level fault tolerance in heterogeneous networks of workstations. *J. Parallel Distrib. Comput.*, 43(2):147–155, June 1997.

[4] Y. Cui et al. Decentralized agent based re-clustering for task mapping of tera-scale network-on-chip system. In *Proc. of ISCAS*, pages 2437–2440. IEEE, 2012.

[5] D. Feitelson. Parallel Workloads Archive, http://www.cs.huji.ac.il/labs/parallel/workload.

[6] T. Desell et al. Malleable applications for scalable high performance computing. *Cluster Computing*, 10(3):323–337, 2007.

[7] A. B. Downey. A model for speedup of parallel programs. Technical report, 1997.

[8] D. G. Feitelson and L. Rudolph. Toward convergence in job schedulers for parallel supercomputers. In *Proc. of JSSPP*, pages 1–26. Springer-Verlag, 1996.

[9] K. Goossens et al. Æthereal network on chip: Concepts, architectures, and implementations. *IEEE Des. Test*, 22(5):414–421, 2005.

[10] J. Howard et al. A 48-core ia-32 message-passing processor with dvfs in 45nm cmos. In *Proc. of ISSCC*, pages 108–109, feb. 2010.

[11] Intel. The Intel® Xeon Phi™ Coprocessor, http://www.intel.com/content/www/us/en/high-performance-computing/high-performance-xeon-phi-coprocessor-brief.html.

[12] S. Kobbe et al. DistRM: distributed resource management for on-chip many-core systems. In *Proc. of CODES+ISSS*, pages 119–128. ACM, 2011.

[13] S. Kounev et al. Towards self-aware performance and resource management in modern service-oriented systems. In *Proc. of SCC*, pages 621–624. IEEE CS, 2010.

[14] V. Nollet et al. Centralized run-time resource management in a network-on-chip containing reconfigurable hardware tiles. In *Proc. of DATE*, pages 234–239. IEEE CS, 2005.

[15] G. Sabin et al. Moldable parallel job scheduling using job efficiency: an iterative approach. In *Proc. of JSSPP*, pages 94–114. Springer-Verlag, 2007.

[16] L. Seiler et al. Larrabee: a many-core x86 architecture for visual computing. *ACM Trans. Graph.*, 27:18:1–18:15, August 2008.

[17] STMicroelectronics. STNoC: Building a new system-on-chip paradigm. White Paper, 2005.

[18] S. Vangal et al. An 80-Tile 1.28 TFLOPS Network-on-Chip in 65nm CMOS. In *Proc. of ISSCC*, pages 98–589. IEEE, 2007.

netShip: A Networked Virtual Platform for Large-Scale Heterogeneous Distributed Embedded Systems

YoungHoon Jung
Dept. of Computer
Science
Columbia University
New York, NY 10027
jung@cs.columbia.edu

Jinhyung Park
The Fancy
Thing Daemon, Inc.
New York, NY 10014
jp@thefancy.com

Michele Petracca
Cadence Design
Systems, Inc.
San Jose, CA 95134
petracca@cadence.com

Luca P. Carloni
Dept. of Computer
Science
Columbia University
New York, NY 10027
luca@cs.columbia.edu

ABSTRACT

From a single SoC to a network of embedded devices communicating with a backend cloud-computing server, emerging classes of embedded systems feature an increasing number of heterogeneous components that operate concurrently in a distributed environment. As the scale and complexity of these systems continues to grow, there is a critical need for scalable and efficient simulators. We propose a networked virtual platform as a scalable environment for modeling and simulation. The goal is to support the development and optimization of embedded computing applications by handling heterogeneity at the chip, node, and network level. To illustrate the properties of our approach, we present two very different case studies: the design of an Open MPI scheduler for a heterogeneous distributed embedded system and the development of an application for crowd estimation through the analysis of pictures uploaded from mobile phones.

Categories and Subject Descriptors

D.4.7 [**Organization and Design**]: distributed systems, realtime systems and embedded systems

General Terms

Distributed Embedded Systems Design

Keywords

Android, Embedded Systems, MPI, OpenCV, OVP, QEMU, Simulation, System Design, Virtual Platform

1. INTRODUCTION

Computing systems are becoming increasingly more concurrent, heterogeneous, and interconnected. This trend happens at all scales: from multi-core systems-on-chip (SoC), which host a variety of processor core and specialized accelerators, to large-scale data-center systems, which feature racks of blades with general purpose processors, graphics-processor units (GPUs) and even accelerator boards based on FPGA technology. Furthermore, nowadays many embedded devices operate while being connected to one or more networks: e.g., modern video-game consoles rely on the Ethernet protocol [30], millions of TVs and set-top boxes are connected through DOCSIS networks [12], and most smartphones can access a variety of networks including 3G, 4G, LTE, and WLAN [17, 14, 32].

Permission to make digital or hard copies of all or part of this work for personal or classroom use is granted without fee provided that copies are not made or distributed for profit or commercial advantage and that copies bear this notice and the full citation on the first page. To copy otherwise, to republish, to post on servers or to redistribute to lists, requires prior specific permission and/or a fee.
DAC '13, May 29 - June 07 2013, Austin, TX, USA.

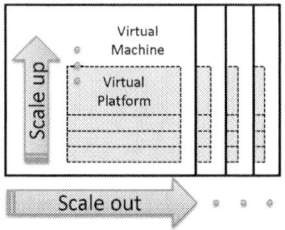

Fig. 1: The two orthogonal scalabilities of NETSHIP.

As a consequence, a growing number of software applications involve computations that run concurrently on embedded devices and backend servers, which communicate through heterogeneous wireless and/or wired networks. For example, *mobile visual search* is a class of applications which leverages both the powerful computation capabilities of smart phones as well as their access to broadband wireless networks to connect to cloud-computing systems [11, 31].

We argue that the design and programming of these systems offer many new unique opportunities for the electronic design automation (EDA) community. For instance, system and sub-system architects need tools to model, simulate, and optimize the interaction of many heterogeneous devices; hardware designers need tools to characterize the applications, software and network stack that they must support; and software developers need early high-level modeling environments of the underlying hardware architecture, often much before all its components are finalized.

As a step in this direction, we present NETSHIP, a networked virtual platform to develop simulatable models of large-scale heterogeneous systems and support the programming of embedded applications running on them. Users of NETSHIP can model their target systems by combining multiple different virtual platforms with the help of an infrastructure that facilities their interconnection, synchronization, and management across different virtual machines.

Given a target system, NETSHIP can be used to set up a simulation environment where each VP works as single-device simulator running a real software stack, e.g. the Linux operating system, with drivers and applications. Thus, it makes it possible to run real applications over the entire distributed system, without actually deploying the devices. This allows users both to jump start the functional verification process of the software and to drive the design optimization process of the hardware and the network.

While in certain areas the terms *virtual platform (VP)* and *virtual machine (VM)* are often used without a clear distinction, in this paper it is particularly important to distinguish them. A VP is a simulatable model of a system that includes processors and peripherals and uses binary translation to simulate the target binary code on top of a host instruction-set architecture (ISA). VPs enable system-level co-simulation of the hardware and the software parts of a given system before the actual hardware implementation is

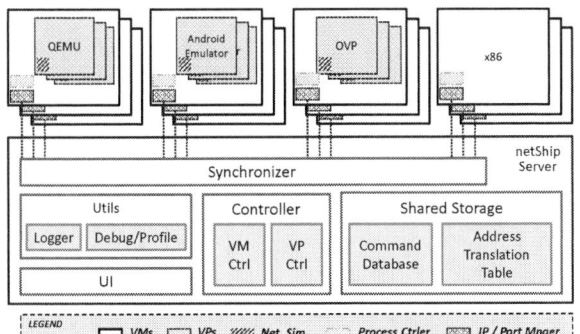

Fig. 2: The architecture of NETSHIP.

finalized. Instead, a VM is the management and provisioning of physical resources in order to create a virtualized environment. The resources are mostly provided by one or more server computers and the management is performed by a *hypervisor*. Examples of VPs include OVP, VSP, and QEMU, while KVM, VMware, and the instances enabled by the Xen hypervisor are examples of VMs. [1]

Thanks to its novel *VP-on-VM model*, the NETSHIP infrastructure simplifies the difficult process of modeling a system with multiple different VPs. In fact, the ability to support multiple VPs interconnected through a network makes NET-SHIP free from the limitation of one specific VP while providing access to the superset of their features. For example, users who are interested in modeling an application running in part on certain ARM-based mobile phones and in part on MIPS-based servers can use NETSHIP to build a network of Android emulators [1] and OVP nodes.

The VP-on-VM model makes NETSHIP scalable both horizontally and vertically, as illustrated in Fig. 1. The users can scale the system out by adding more VM instances to the network (horizontal scalability) and scale the system up by assigning to each VM instance more CPU cores on which more VP instances can run (vertical scalability).

Another pivotal advantage the VP-on-VM model adds to NETSHIP is access to the features of VMs, i.e. pausing, resuming the VM instances, duplicating instanced preconfigured for specific VP types, or migrating them across physical machines.

Contributions. The main goal of this research work is to understand how to build and use a *Networked Virtual Platform* for the analysis of distributed heterogeneous embedded systems. To do so, we built NETSHIP as a prototype based on the VP-over-VM model with the main objectives of supporting heterogeneity and scalability. To the best of our knowledge, this is the first paper that presents this type of CAD tool. To evaluate NETSHIP we have completed a series of experiments including two complete case studies. The first case study shows how a networked virtual platform can be used to better utilize the computational resources that are available in the target system while guaranteeing certain performance metrics. The second case study shows how a networked virtual platform can be used to develop a software application running on a heterogeneous distributed system that consists of many personal mobile devices and multiple computer servers while, at the same time, obtaining an estimation of the resource utilization of the entire system.

2. NETWORKED VIRTUAL PLATFORMS

A heterogeneous distributed embedded system can consists of a network connecting a variety of different components. In our approach, we consider three main types of

heterogeneity: first, we are interested in modeling systems that combine computing nodes based on different types of processor cores supporting different ISAs (*core-level heterogeneity*); second, nodes that are based on the same processor core may differ for the configuration of hardware accelerators, specialized coprocessors like GPUs, and other peripherals (*node-level heterogeneity*); third, the network itself can be heterogeneous, e.g. some nodes may communicate via a particular wireless standard, like GSM or Wi-Fi, while others may communicate through Ethernet (*network-level heterogeneity*.)

NETSHIP provides the infrastructure to connect multiple VPs in order to create a networked VP that can be used to model one particular system architecture having one or more of these heterogeneity levels. For example, Fig. 2 shows one particular instance of NETSHIP which is obtained by connecting multiple instances of the QEMU machine emulator [6], the Android mobile-device emulator [1], and the Open Virtual Platform (OVP) [3].

Each VP instance runs an operating system, e.g. Linux, with all the required device drivers for the available peripherals and accelerators. The application software is executed on top of the operating system. Each VP typically supports the modeling of a different subset of peripherals: e.g., OVP supports various methods to model the hardware accelerators of an SoC: users can write models in SystemC TLM 2.0 or take advantage of the BHM (Behavioral Hardware Modeling) and PPM (Peripheral Programming Model), which are C-compatible Application Programming Interfaces (APIs) that can be compiled using the OVP-supplied PSE toolchain[2].

In addition to the features supported by each particular VP, we equipped NETSHIP with all the necessary instrumentation to: (1) enable multiple instance executions; (2) configure port forwarding; and (3) measure the internal simulation time. Furthermore, any node in the network of VPs could potentially be a real platform, instead of being a virtual one: e.g. in Fig. 2, each of the x86 processors runs native binary code and still behaves as a node of the network.

One of the main novelty aspects of NETSHIP is the *VP-on-VM model* which is critical for the scalability of modeling and simulations. We designed NETSHIP so that multiple VP instances (e.g., 2 to 8) can be hosted by the same VM. By adding more VMs, the number of VPs in the system can be increased with a small performance penalty, as discussed in Section 3. Notice that the simple action of cloning a VM image that includes several VPs often represents a convenient way to scale out the model of the target system.

Next, we describe the main building blocks of NETSHIP.

Synchronizer. VPs vary in the degree of accuracy of the timing models for the CPU performance that they support. Some VPs do not have any timing model and simply execute the binary code as fast as possible. This is often desirable, particularly when a VP runs in isolation. In NETSHIP, however, we are running multiple VPs on the same VM and, therefore, we must prevent a VP from taking too much CPU resources and starving other VPs. QEMU provides a crude way to keep simulation time within a few seconds of realtime. OVP, instead, controls the execution speed so that the simulated time never surpasses the wall clock time. Multiple OVP instances, however, still show different time developments which require a synchronization method across the VPs in the network.

We equipped NETSHIP with a *synchronizer* module to support synchronization across the heterogeneous set of VPs in the networked platform, as shown in Fig. 2. The synchronizer is a single process that runs on just one particular VM and is designed in a way similar to the fixed-time step

---

[1] Recent efforts to run VMs on embedded cores [7, 19] remain within the VM definition as they do not adopt binary translation.

[2] PSE is Imperas Peripheral Simulation Engine [3].

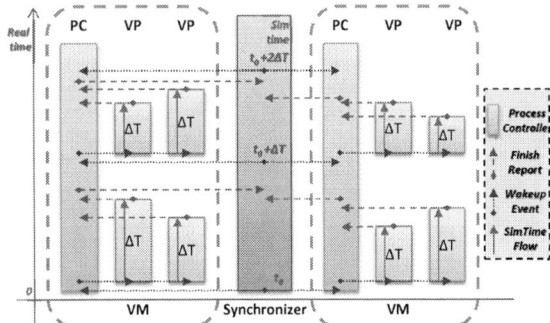

Fig. 3: Synchronization process example.

synchronization method presented in [8]: at each iteration, a central node increases the base timestamp and the client nodes stop after reaching the given timestamp. However, we considered two aspects in our synchronizer:

- we must synchronize VPs that might be scattered over several physically-separated machines;
- we must preserve the scalability provided by the VP-on-VM model.

NETSHIP targets large-scale systems which involve deployments across physically- separated machines where millisecond-level network packet travelling is actually required to synchronize. Hence, NETSHIP supports the modeling of applications that have running times ranging from a few seconds to multiple hours or days, rather than simulations at nanosecond-level.

To support synchronization over the VP-on-VM model, we designed a *Process Controller* (PC) that allows us to manage the VPs in a hierarchical manner. Each VM hosts one PC, which controls all the VPs on that VM. In particular, all messages sent by a VP to the synchronizer pass through the PC. The PC supports also running programs on a host machine: e.g. in the case of Fig. 2, the PCs manage the synchronization of the processes running on a x86 through the two POSIX signals SIGSTOP and SIGCONT, in the same way as the UNIX command cpulimit limits the CPU usage of a process.

Fig. 3 illustrates an example of the synchronization process with two VMs, each hosting two VP instances. The following steps happen at each given iteration i:

1. the synchronizer issues a future simulation time $t_i = t_{i-1} + \Delta T$ to the VPs and wakes them up;
2. the VPs run until they reach the appointed time t_i and report to their PC;
3. As soon as a PC receives reports from all the connected VPs, it reports to the synchronizer;
4. After the synchronizer has received the reports from all the PCs, it loops back to Step 1.

The users can configure the time step ΔT to adjust the trade-off between the accuracy and the simulation speed. We briefly discuss the complexity comparison of this hierarchical method in Appendix A.2.

Command Database. NETSHIP was designed to support the modeling of systems with a large scale of target networked VPs. In these cases, to manually manage many VP instances becomes a demanding effort involving many tasks, including: add/remove new VP instances to/from a system, start the execution of applications in every instance, and modify configuration files in the local storage of each instance. In order to simplify the management of the networked VP as a whole, we developed the *Command Database* that stores the script programs used by the different NET-SHIP modules. For example, the network simulation module and IP/Port forwarding module load the corresponding scripts from the database and execute them. Table 8 contains a detailed list of the commands in the database.

VM and VP Management. Whereas the commands in the Command Database are dedicated to VP configuration, we developed specialized modules to manage the VPs and the VMs (for the latter we integrated tools provided by the VM vendor). These modules manage the disk images of the VMs and VPs, for creating, copying, and deleting their instances. Since many VPs are still in the early stages of development and are frequently updated by the vendors, the VP management module checks the availability of new updates for all the installed VPs.

Network Simulation. The VP models of NETSHIP are provided with their own models of the network interface card (NIC). These models, however, are purely behavioral and do not capture any network performance property, such as bandwidth or latency [8]. Consequently, we developed a *Network Simulation* module that enables the specification of bandwidth, latency, and error rates, thus supporting the modeling of network-level heterogeneity in any system modeled with NETSHIP. As shown in Fig. 2, a Network Simulation module resides in each particular VP and uses the traffic-shaping features based on the *tc* command, which manipulates the traffic control settings of the Linux kernel.

Address Translation Table. In NETSHIP there are two points where packet forwarding plays a critical role:

1. To allow incoming connections to the VPs through their emulated NIC model, most VPs provide a way to redirect a port of the host to a port of the VP, so that packets that arrive to that VM port are redirected to the corresponding VP port. We leverage this redirection mechanism so that the applications running on the VPs can open ports to receive packets from other VPs, even if those are located on a physically separated VM. [3] More details are described in the Appendix A.3.

2. Since certain applications required that each VP must be accessible through a unique IP address and generally there is only one physical IP address per VM, we must map each VP to a virtual IP address. Each VP must know such mapping for all other VPs in the system. Hence, we used the UNIX command iptables to create a table of assignments within the kernel of each VP. NETSHIP stores the translation information in the *Address Translation Table*, which is loaded through the network commands stored in the Command Database.

3. SCALABILITY EVALUATION

In this section, we experiment NETSHIP from the synchronization, scalability, performance, and network-fairness perspectives. Functional validations of NETSHIP will be covered in each case study, in Section 4 and 5.

Simulated Time and Synchronization. Eight OVP instances and eight QEMU instances are running in this simulation setup. The three figures in Fig. 4 show the *simulated time* in each. The red solid line represents the time graph of an ideal VP, with $y = x$, where y is the *wall-clock time* and x is the simulated time. While there were multiple instances running together, in the figure we show only the fastest and slowest instances for each VP family, in order to summarize the range of variations within each VP family and to better compare the VP families.

Fig. 4(a) measures the simulated time of unloaded VPs. Each VP advances its simulated time linearly, but differently from each other. In particular, the range of simulated time among QEMU instances is wide: from 4% slower up to 25% faster than the wall-clock time. Instead, the OVP instances show almost the same simulation speed (0.3% variation), which is 8% slower than the slowest QEMU instance. This

---

[3]While certain VPs provides a network bridge feature that allows more generic network functionalities, we use port redirection because it is commonly supported by every VP family.

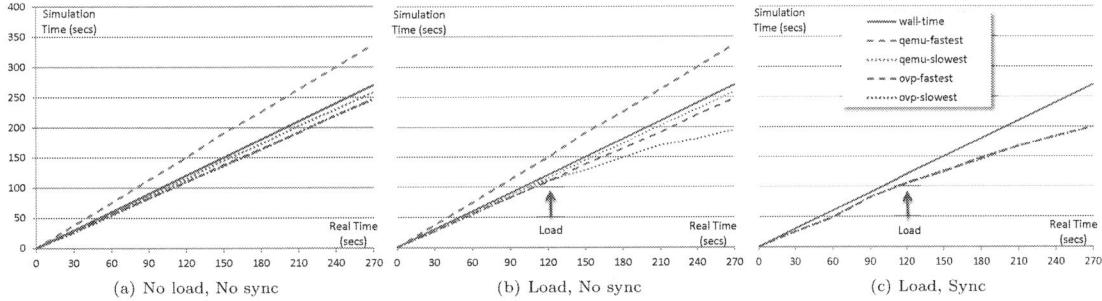

(a) No load, No sync   (b) Load, No sync   (c) Load, Sync

Fig. 4: Simulation time measurements.

VP Type	Core Model	CPU use	Preferred #VPs
OVP	Accelerator	~ 24%	4
OVP	MIPS	~ 6%	16
QEMU	PowerPC	~ 12%	8
VMWare	x86	~ 5%	20

Table 1: Host CPU use of each VP.

reflects the fact that OVP has a better method to control the simulation speed.

Fig. 4(b) shows the case when a VP is subject to a heavy workload. In particular, at simulated time $x = 120s$ one OVP instance starts using a high-performance accelerator. From that point on, the OVP instance gets slower than every other instances, as shown by the deviation among the OVP lines in the figure. This is natural when the peripherals are modeled at a very high level of abstraction. In a fair host VM, all VPs are granted the same amount of CPU time to be executed. Simulating the use of a hardware accelerator on a VP typically requires the VP process on the VM to executes a non-negligible computation. In the other words, running the functional model of the accelerator uses the VM's CPU resources and requires a certain amount of wall-clock time. From the viewpoint of simulated time, however, this computation happens in a short period of time (due to the accelerator's timing model); therefore the given VP instance becomes slower than the others. The misalignment of the simulated time among VP instances is a concern when simulating distributed systems, because it might cause the simulated behaviors to be not representative of reality.

To address this problem, we implemented the synchronization mechanism explained in Section 2. Fig. 4(c) shows the behavior of all VP instances under the same conditions but with the synchronization mechanism turned on (with a synchronization cycle of 300ms). The simulated time of all VPs becomes the same as the slowest instance. The synchronization cycle can be decided by the users. Our experiments show that it should not be too small ($\geq$ 1ms) because: i) a synchronization that is much more frequent than the OS scheduling time slice[4] may disturb the timely execution of the VPs, and ii) the synchronization is an overhead and slows down the overall simulation.

Vertical Scalability. By *vertical scalability* we mean the behavior of the networked VP as more VPs are added to a single VM. As discussed above, although the synchronizer preserves the simultaneity of the simulation among VPs, it makes them all run at the speed of the slowest instance , i.e. even one slow VP instance is enough to degrade the simulation performance of the whole system. Therefore, an excessive number of VP instances on the same VM will likely cause a simulation slowdown.

Table 1 shows the amount of CPU of the host VM that is used by a VP instance. For example, when an OVP instance fully utilizes one accelerator, it takes up to 24% of the

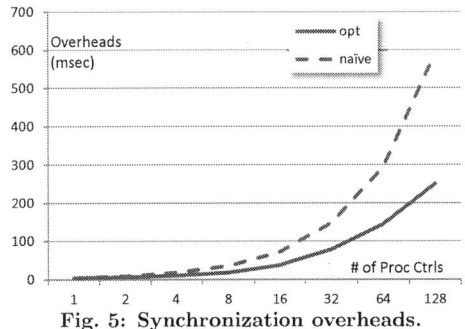

Fig. 5: Synchronization overheads.

host CPU resource in the hosting VM. This means that 4 is the optimal number of OVP instances, equipped with that accelerator, which can co-exist on the same VM without performance penalty. Likewise, the CPU of a QEMU PowerPC that is fully busy, i.e. a simulated 100% utilization, uses up to 12% of the host VM's CPU resources: hosting up to 8 QEMU PowerPC instances in the same VM is performance optimal.

Note that even if the number of VP instances goes over the optimal value, the synchronizer still preserves the simultaneous simulation of all nodes. However, balancing out the number of VP instances hosted across the VPs, or alternatively increasing the computational resources available to the VM, helps to increase the overall simulation performance.[5]

Synchronization Overheads and Horizontal Scalability. *Horizontal scalability* describes the behavior of the simulated VP as we scale the number of VMs. The synchronizer is the entity in the networked VP that communicates with all VMs in order to keep all VPs aligned. Fig. 5 shows the overhead increase as the number of VMs grows. We measured the overhead as the time elapsed from when the slowest VP instance reports to have terminated the execution step to the time the same instance starts the new one. We experimented with ten VP instances insisting on each PC. In the figure we compare a naïve implementation with an optimized implementation of the synchronizer. For both version the principle of the synchronization is the same; however, in the optimized version we used more advanced techniques to reduce the communication latency and overhead, such as multicasting wakeup, local reporting by atomic operations on a shared memory, and POSIX signals, as described in Appendix B.1. The overhead for 128 PCs is approximately $250ms$, which slows down the networked VP by about 25% if the simulation step is set to $1s$.

Although the optimized implementation significantly reduces the overhead, both slopes increase linearly with re-

---

[4]Linux O(1) scheduler dynamically determines the time slice, ranging from milliseconds to a few hundreds milliseconds.

[5]The CPU resources of the VM might not be the only bottleneck. For a more generic approach, an analysis of disks, network congestion, memory bandwidth, bus capacity, and cache interference are required. In our experiments, however, the constraint due to the VM's processing power was the most dominating factor that decides Vertical Scalability.

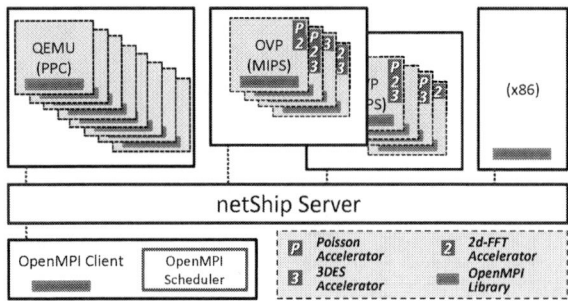

Fig. 6: The system architecture for Case Study I.

Algorithm	Operations per Hour		Speedup
	CPU	Accelerator	
Poisson	349	1183	3.39
2d-FFT	314	517	1.65
3DES	632	1339	2.12

Table 2: Case Study I: performance comparisons.

VP Type	Network Type	Bandwidth	Latency
OVP MIPS	DOCSIS 2.0	30.72Mbps	30ms
QEMU PowerPC	Evolved EDGE	1.00Mbps	80ms
Host x86	IEEE 802.11g	54.00Mbps	45ms

Table 3: Case Study I: configured bandwidth & latency.

spect to the number of PCs in the network (notice that the x-axis is logarithmic). This is because synchronization involves all PCs, each of which is located in separate machine, and all reports require a packet transmission across the network and linear-time computation to parse the reports.

In summary, the synchronization across VMs limits the horizontal scalability, in the sense that the simulation step after which all VPs are synchronized, must be (much) bigger than the time it takes to actually perform the synchronization, which strongly depends on the characteristics of the hosting VMs and how they are connected.

4. CASE STUDY I - MPI SCHEDULER

We modeled a distributed embedded system as a networked VP, which runs Open MPI (Message Passing Interface) applications. This system features all the three kinds of heterogeneities: it has three different models of CPUs, it has arbitrarily scattered accelerators over the VPs, and we also vary the types of network the devices are connected to. The goal of this case study is to use a networked VP for designing a static scheduler that optimizes the execution time by better distributing the work across the system in Fig. 6.

Open MPI is an open source MPI-2 implementation, which is a standardized and portable message passing system for various parallel computers [2]. In this case study, we used Open MPI to establish a computation and communication model over NETSHIP. Since the mainstream implementation of Open MPI does not support MIPS or ARM architectures (because it misses the implementation of atomic operation backends) we wrote and applied patches for Open MPI to run on these ISAs.

We simultaneously run three MPI applications over the distributed system: Poisson [22], 2d-FFT [29], and Triple DES [5]. Each application is a standalone executable program and is configured to process a small amount of data so that they act as embarrassingly parallel. Every application is designed to either use the hardware accelerator, whenever available on the VP, or run purely in software, otherwise. Accelerators are modeled to run basically the same algorithm as the applications. We modeled one iteration of the algorithm to take a few milliseconds.

According to our timing model for the accelerators and to the native timing model of the CPUs, accelerators show $1.65\times \sim 3.39\times$ speedup over CPUs, as summarized in Table 2. Note that the speedup introduced by hardware acceleration with respect to pure software execution is not the main point here. Instead, the comparative analysis is a demonstration of the type of analysis that a designer can carry by using the networked VP paradigm. In fact, the speedup mentioned above is actually conservative with respect to the literature in order to keep the design exploration of our case study interesting [28, 27].

Based on such time model, Fig. 7 shows the performance profile for different applications on a few VPs: the OVP instances are always equipped with an accelerator, while the

other VPs are not. We also consider models for the network, whose bandwidth and latency parameters are summarized in Table 3. Note that, since NETSHIP allows designers to use time models to better simulate the characteristics of the network, those models are inputs to the networked VP and their derivation goes beyond the scope of this paper.

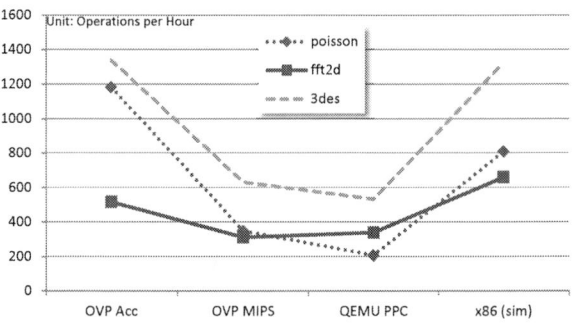

Fig. 7: Case Study I: performance of different cores.

In order to improve the application performance by taking advantage of the known properties of the system, i.e. performance profile of the nodes and network characteristics, we designed an *OpenMPI Scheduler* and we used the networked VP to evaluate its effectiveness. [6] As shown in Table 4, the scheduler delivers a speedup ranging from $1.3\times$ to over $4\times$, depending on the user request. Such achievement is very encouraging since it is obtained without using any additional resource, but only by re-assigning tasks to the nodes that are better equipped for each of them. Note that the design, the verification, and also an initial assessment on the effectiveness of the scheduler have been carried out on the networked VP, without having to deploy the real system.

5. CASE STUDY II - CROWD ESTIMATION

Crowd estimation, or crowd counting, is the problem of predicting how many people are passing by or are already in a given area [21]. A number of researches have focused on the crowd estimation based on image processing of pictures [9, 20]. The crowd estimation application we developed in this section is based on user-taken pictures, from mobile phones, targeting relatively wide areas, e.g. a city.

We built a networked VP (Fig. 8) that is representative of the typical distributed platform required to host this kind of application. The networked VP features *Android Emulators* to model the phones and a cluster of *MIPS-based servers* based on the multiple OVP instances (on the right-hand side of the figure). The Android Emulators emulate mobile phones that take pictures through the integrated camera and upload them to the cloud. The pictures are stored

---

[6]Details on the scheduler design are available in Section C of the Appendix.

# of operations			Time in ms		
Poisson	2d FFT	Triple DES	without Sched	with Sched	Speed Up
60	30	800	199,642	48,924	4.08
60	30	1800	344,422	117,380	2.93
50	20	40	102,293	45,210	2.26
60	30	150	103,700	46,693	2.22
250	80	140	198,927	113,383	1.75
20	100	10	161,462	122,527	1.32

Table 4: Case Study I: scheduler performance.

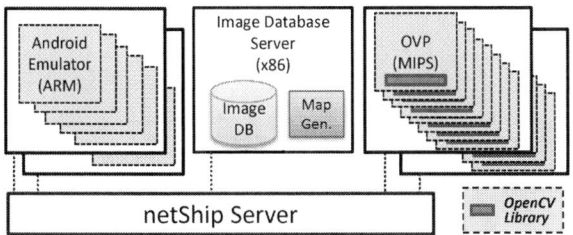

Fig. 8: The system architecture for Case Study II.

# of Emulator in the model	# of Pic Upload / Hour	Incoming Traffic (KB/s)		
		Max	Min	Avg.
1	13333	N/A	N/A	N/A
2	6666	380.4	372.5	379.3
4	3333	384.1	376.4	381.2
8	1666	377.8	361.8	374.2
16	833	389.0	367.5	381.5

Table 5: Case Study II: impact of varying number of Android-emulator instances.

Image Size (KB)	8	32	128	512	74(Avg)
Process Time (s)	3.48	13.42	49.7	247.1	31.5
Throughput (KB/s)	2.30	2.38	2.57	2.07	2.34

Table 6: Case Study II: image processing (human recognition) performance.

on an *Image Database Server* (IDS), to which both phones and servers have access. The servers emulate the cloud, and run image processing algorithms on the pictures. Specifically, we developed a *Human Recognition* application based on OpenCV [4] to count the people in each picture and store the result on the IDS. Then, a *Map Generator* process running on the IDS reads the people counting from the IDS and plots it on a map.

Given the application requirements, we used the networked VP to gain insights on the amount of resources required to process pictures in real-time. Note that our main concern is the opportunity to build and study the networked VP, and to use it to analyze the properties of the application that runs on it. In other words, we used this application primarily as a case study to test the capabilities of NETSHIP, while the optimization of the quality of the crowd estimation was only a secondary concern.

Android Emulator Scalability. We used several Android Emulators to model millions of mobile phones that sporadically take pictures (instead of using millions of emulators). To validate whether the emulators realistically reflect the actual devices' behavior with respect to network utilization, we performed multiple tests after making the following practical assumptions:

1. There are 3,000,000 mobile phone users in Manhattan and 2% of them upload 2 pictures a day.
2. The uploading of the pictures is evenly spread over the daytime (09:00~ 18:00).
3. The average image file size is 74KB, as the image size we have in the DB.

Given the assumptions above, we summarize in Table 5 the number of pictures an emulator must upload in an hour and the actually measured incoming traffic of the DB server, with respect to the number of available emulators in the networked VP and accordingly configured the number of pictures uploaded by each emulator per hour. For example, if the networked VP has only one Android emulator (first row), we can achieve the desired load for the cluster when this emulator uploads $3,000,000 * 0.02 * 2/9 \approx 13333$ pictures per hour. Since one single emulator fails to upload 13333 pictures per hour, because of insufficient emulator performance, we must increase the number of emulators to at least 2. The measured incoming traffic is rather consistent independently of how many emulators we use to split the job. This implies that we can deploy less emulators than the number of nodes

we would have in reality, i.e., 4 vs. 3 million, as long as those emulators generate more traffic than they would in reality, i.e. 3333/hour vs. 2/day, after verifying that they also simulate fast enough to sustain the traffic generation. We leave the modeling of more complicated traffic patterns than *Assumption 1*, e.g. bursty traffic, as future work.

Bottleneck Analysis. The data in Table 6 show the average time required by one MIPS server to run the Human Recognition application on a given picture. Based on this data and on the characterization of the traffic load, the application designer can attain a number of meaningful design considerations.

1. The designer can measure the number of required MIPS servers that support the required volume of image processing, given the input and output data rates. For example, when images are fed to the DB at $380KB/s$, then, based on the throughput for the average image size in Table 6, the cluster must have more than $380KB/2.34KB/s = 162.4$ MIPS servers to guarantee real-time performance. Note that $2.34KB/s$ is the throughput of the average image size in Table 6.

2. On the other side, if the number of available servers cannot be changed, the designer can reason on the appropriate image size. If we assume to have 80 MIPS servers in the cluster, then they can process only up to $80 \times 2.34KB/s = 187.2KB/s$. In that case, the average image size must be less than $74KB \times (187.2/380) = 36.5KB$ for the application to work in real-time.

3. The network traffic through the DB server includes picture uploading from mobile phones, picture downloading by the MIPS clusters, updating and reading of geolocation information and people count. Based on an analysis of the network traffic and how it scales as the system grows, the designer can evaluate the best database architecture, e.g. distributed rather than centralized.

6. RELATED WORK

A number of studies have previously focused on helping system architects to better design distributed embedded systems by providing ways to optimize the process scheduling and the communication protocols [15, 23], tools to ease design space explorations [26, 16, 13], estimation models [34], and network behavior simulations [10], or methodologies [18, 13]. Nonetheless, these tools or systems only generate quantitative guidelines that must be then applied to the physical devices, thereby precluding their usability without already having the physical devices in place and the application deployed on them. There are also contributions obtained through the use of VPs [8, 33]; however, none of these works consider the three levels of heterogeneity that characterize more and more distributed embedded systems (Section 2).

Synchronization between the VP instances, one of the key features of our networked VP, has been inspired by [25, 24]. However, they do not consider node-level or network-level heterogeneity.

7. CONCLUSIONS

We have designed and implemented NETSHIP, a framework for building networked VPs that model heterogeneous distributed embedded systems. Networked VPs can be utilized for various purposes, including: i) simulation of distributed applications, ii) systems, power, and performance analysis, and iii) costs modeling and analysis of embedded networks' characteristics.

We also designed hardware accelerators for specific algorithms. We analyzed that accelerators might require more resources of the CPUs that host the simulation. We quantified how this phenomenon partially limits the scalability of the entire networked VP, and provided guidelines on how to distribute the VPs in order to counter balance this loss of simulation performance.

Finally, we used NETSHIP to develop two networked VPs. We used one VP to design a scheduler based on MPI and to verify through simulation how the scheduler is able to optimize the execution of many MPI jobs over a network of heterogeneous machines, by simply distributing the jobs among the available machines on the basis of their performance-per-application profile. We used the other VP to design and validate an application distributed among portable devices and a cloud of servers, and also to derive potential insight about the number of servers and the image size that guarantee the entire application to run in real-time.

Acknowledgements

This project is partially supported by the National Science Foundation under Awards #644202 and #1147406 and by an ONR Young Investigator Award. We gratefully acknowledge Yosinori Watanabe for useful discussions.

8. REFERENCES

[1] Android (developer.android.com).
[2] Open MPI (www.open-mpi.org).
[3] Open Virtual Platforms (www.ovpworld.org).
[4] OpenCV (opencv.org).
[5] W. C. Barker and E. B. Barker. SP 800-67 Rev. 1. Recommendation for the Triple Data Encryption Algorithm (TDEA) Block Cipher. Technical report, Jan. 2012.
[6] F. Bellard. QEMU, a fast and portable dynamic translator. In USENIX Annual Technical Conference, pages 41–46, Feb. 2005.
[7] C. Dall and J. Nieh. KVM for ARM. In Proc. of the Linux Symp., pages 45–56, July 2010.
[8] M. D.Angelo et al. A simulator based on QEMU and SystemC for robustness testing of a networked Linux-based fire detection and alarm system. In Proc. of the Conf. on ERTS², pages 1–9, Feb. 2012.
[9] T. Fei, L. SunDong, and G. Sen. A novel method of crowd estimation in public locations. In Int. Conf. on FBIE, pages 339–342, Dec. 2009.
[10] E. Giordano et al. MoViT: the mobile network virtualized testbed. In Proc. of the Int Workshop on VANET, pages 3–12, June 2012.
[11] B. Girod et al. Mobile visual search. IEEE Signal Processing Magazine, 28(4):61–76, July 2011.
[12] S. Howard and J. Martin. DOCSIS performance evaluation: piggybacking versus concatenation. In Proc. of Southeast Regional Conf., volume 2, pages 43–48, Mar. 2005.
[13] Z.-M. Hsu, J.-C. Yeh, and I.-Y. Chuang. An accurate system architecture refinement methodology with mixed abstraction-level virtual platform. In Proc. of DATE, pages 568–573, Mar. 2010.

[14] M. Ismail. WiMAX: a competing or complementary technology to 3G? In Proc. of the Conf. on Integrated Circuits and Syst. Design, pages 3–3, Sept. 2007.
[15] V. Izosimov et al. Design optimization of time-and cost-constrained fault-tolerant distributed embedded systems. In Proc. of the conf. on DATE, pages 864–869, Mar. 2005.
[16] D.-I. Kang et al. A software synthesis tool for distributed embedded system design. In Proc. of the Workshop on Languages, Compilers, and Tools for Embed. Syst., pages 87–95, May 1999.
[17] D. Koutsonikolas and Y. C. Hu. On the feasibility of bandwidth estimation in wireless access networks. Wireless Net., 17(6):1561–1580, Aug. 2011.
[18] J. Kruse et al. Introducing flexible quantity contracts into distributed SoC and embedded system design processes. In Proc. of DATE, pages 938–943, Mar. 2005.
[19] S.-M. Lee et al. Fine-grained I/O access control of the mobile devices based on the xen architecture. In Proc. of the Int. Conf. on Mobile Comp. and Net., pages 273–284, Sept. 2009.
[20] W. Li et al. Crowd density estimation: An improved approach. In Int. Conf. on Signal Processing, pages 1213–1216, Oct. 2010.
[21] A. Marana et al. Estimating crowd density with Minkowski fractal dimension. In Proc. of ICASSP, volume 6, pages 3521–3524, Mar. 1999.
[22] N. Ng, N. Yoshida, and K. Honda. Safe parallel programming with message optimisation. In Proc. of Int. Conf. on Objects, Models, Components, Patterns, volume 7304, pages 202–218, May 2012.
[23] T. Pop, P. Eles, and Z. Peng. Design optimization of mixed time/event-triggered distributed embedded systems. In Proc. of CODE+ISSS, pages 83–89, Sept. 2003.
[24] D. Quaglia et al. Timing aspects in QEMU/SystemC synchronization. Proc. of the Int. QEMU Users' Forum, pages 11–14, Mar. 2011.
[25] H. Raj et al. Enabling semantic communications for virtual machines via iConnect. In Proc. of the Int. Workshop on VTDC, pages 1:1–1:8, Nov. 2007.
[26] D. E. Setliff, J. K. Strosnider, and J. A. Madriz. Towards a design assistant for distributed embedded systems. In Proc. of the Int. Conf. on ASE, pages 311–312, Nov. 1997.
[27] M. Shand, P. Bertin, and J. Vuillemin. Hardware speedups in long integer multiplication. SIGARCH Comp. Arch. News, 19(1):106–113, Mar. 1991.
[28] T. Shimokawabe et al. An 80-fold speedup, 15.0 TFlops full GPU acceleration of non-hydrostatic weather model ASUCA production code. In Proc. of the Int. Conf. for High Perf. Comp., Net., Storage and Anal., pages 1–11, Nov. 2010.
[29] R. C. Singleton. On computing the fast Fourier transform. Comm. of the ACM, 10(10):647–654, Oct. 1967.
[30] D. Taylor. Need for speed: PS3 Linux! Linux Journal, (156):5–6, Apr. 2007.
[31] S. S. Tsai, D. Chen, J. P. Singh, and B. Girod. Rate-efficient, real-time cd cover recognition on a camera-phone. In Proc. of the 16th ACM Intl. Conf. on Multimedia, pages 1023–1024, Oct. 2008.
[32] C. H. K. van Berkel. Multi-core for mobile phones. In Proc. of DATE, pages 1260–1265, Mar. 2009.
[33] C.-C. Wang et al. NetVP: A system-level network virtual platform for network accelerator development. In IEEE Int. Symp. on Circuits and Systems, pages 249–252, May 2012.
[34] Y. Xiangzhan et al. Research on performance estimation model of distributed network simulation based on PDNS conservative synchronization mechanism in complex environment. In Proc. of the ICCIT, pages 2576–2580, Dec. 2010.

APPENDIX

A. ARCHITECTURE

A.1 Scalability and Detailed Configurations

In general, Horizontal Scalability is the ability to have more VP instances running in NETSHIP by adding more VM instances. Vertical Scalability is the ability to have more VP instances running in NETSHIP by adding more CPU cores to a VM. For example, the preferred number of OVP instances with accelerators that can run in one VM (with one CPU core) is four, as shown in Table 1. If we add one more CPU core to the VM, we can run up to eight OVP instances with accelerators in that VM.

The configuration of Case Study I, shown in Fig. 6, includes: one VM that runs eight QEMUs, two VMs that run four OVPs each, and one VM that supports x86. This is an optimal configuration for the purpose of this case study. However, we also tested Horizontal Scalability (e.g. by adding one VM that runs eight other QEMUs and another VM that runs four other OVPs) and Vertical Scalability (e.g. by adding one CPU core to the VM running eight QEMUs so that it can sustain up to 16 QEMUs.)

For Case Study II we varied the number of Android Emulator in "Android Emulator Scalability" and the number of OVP instances in "Bottleneck Analysis " in Section 5. When we run 1 ~ 4 Android Emulator, we used one VM, for 8 two VMs (5 + 3), for 16 four VMs (5 + 5 + 5 + 1, because the preferred VP number of Android Emulator is 5). For VMs running OVP instances, we have used two CPU cores for each VM, hosting eight OVPs on each instance, for a total of four VMs for 32 OVP instances.

A.2 Synchronization Complexity Comparison

In the synchronization algorithm in [8], if the number of VP is $|VP|$ the synchronization process should receive and count $|VP|$ reports to make sure that all the VPs have reached to the appointed simulation time. This results in a $\Theta(|VP|)$ algorithm complexity in *Synchronizer*, whereas in NETSHIP it is $\Theta(\sqrt{|VP|})$ because *Synchronizer* manages $\sqrt{|VP|}$ PCs, each of which controls $\sqrt{|VP|}$ VPs.[7]

A.3 Port Forwarding

Port forwarding is the technique of redirecting the traffic incoming on one network port of the OS running on the host VM towards a specific port of the OS running on the hosted VP. For example, when a packet arrives to Port 10020 of the VM's OS, the VP to which Port 10020 is assigned intercepts the packet and forwards it to Port 22 of the VP's OS. Hence, when users connects through SSH to the host's IP and Port 10020 they are forwarded to Port 22 of the VP. This is configured in the behavioral model of the VP and performed through the NIC model.

Unlike SSH, some libraries require a random port to be accessed by clients; for instance Open MPI communicates through random ports ranging from 1025 to 65535 [A4]. However, most libraries also provide a way to change or reduce the required port range as shown in Table A.3. We reduced the range and mapped it to the same port range on the virtual addresses, 200.0.0.x. One of these addresses is allocated to each of VP instances using *iptables* through the Port Management module in Fig. 2.

---

[7]It may be enough for *Synchronizer* only to count the number of reports from PC to know that every VP instance is ready and advance the simulation time. However, this method is unreliable in the sense that there is no way for *Synchronizer* to tolerate a PC malfunctioning. If a hash table, for example, is used to map a PC's IP to the data structure for checking that the PC is reporting more than once in a cycle, the average complexity of the algorithm in [8] is $O(|VP| + |VP|^2/k)$ and for our algorithm it is $O(\sqrt{|VP|} + \frac{|VP|}{k})$, where k is the number of buckets in the hash table and searching n times in a hash table takes $n * O(1 + n/k)$.

Library	Option	Default Value
SSH (fixed)	Port	22
Hadoop (fixed)	dfs.http.address	50070
	dfs.datanode.http.address	50075
	mapred.job.tracker.http.address	50030
	mapred.task.tracker.http.addres	50060
Open MPI (random)	oob_tcp_port_min_v4	0
	oob_tcp_port_range_v4	65535
	btl_tcp_port_min_v4	1025
	btl_tcp_port_range_v4	65525

Table 7: Example of library port uses.

A.4 Command Database

Name	Behavior
vp_ctrl_pwr	turns the VP on/off
net_set_bw	sets the VP's network bandwidth to simulate
net_set_delay	sets the VP's network delay to simulate
net_set_error	sets the VP's network error rate to simulate
net_load_rt	loads the address/port settings to use
cmd_execute	executes a command in all the VPs
acc_gen	loads driver modules and creates a device node for the specified accelerator
report_local	reports the local time in the VP
report_cpu	reports the cpu time in the VP

Table 8: List of commands in the command database.

B. EXPERIMENTS

B.1 Optimizing Synchronization

Multicasting-based Wakeup. In order to reduce the serial latency of the wakeup packets delivered from the synchronizer to the PCs, we used multicast UDP.

Atomic Operations in Shared Memory for In-Machine Reporting. The PC must check that VPs correctly report the end of the current simulation cycle. This is done by having each VP increase a shared counter through an atomic operation. This is possible because all VPs are on the same machine.

Disabling Nagle's Algorithm. Unlike the waking-up message of the synchronizer to the PCs (1-to-N), multicast UDP cannot be used to carry reports from the PCs to the synchronizer (N-to-1). In the Linux kernel, TCP sockets typically use by default an optimization technique, Nagle's algorithm, which combines a number of small outgoing packets and sends them all in one single message [A5]. This method, however, increases the latencies of these small packets (up to 30ms in our experiments), which is a critical issue in our synchronization design, since latency is way more important than throughput. We then disabled the Nagle's algorithm by turning on the socket option TCP_NODELAY for each TCP socket.

Using POSIX Signals to Sleep and Wake up. In order to stop and wake-up a VP instance our PC uses two signals: *SIGSTOP* and *SIGCONT*. The use of standard Linux signals provides several advantages. First, the PC can be easily implemented in a separate user space program, without the knowledge of the internals of the VPs. Second, once implemented, the PC is portable across the VPs, requiring no modifications. Third, the PC can stop all threads in the process, while sleeping works only for the thread of the current context. Most importantly, this also enables a synchronized execution with processes that run natively on a host VM, e.g. x86 server, outside of any VP.[8]

B.2 Network Fairness Depending on Deployment

---

[8]In NETSHIP x86 binaries are executed on a VM, not a VP. Through the stop and continue signals PC synchronizes the process without modifying the binary executables.

1167

Target	PNI[9]	RTT (ms)
self VP (the loopback interface)	no	0.15
local VM (a VM where the testing VP runs)	no	0.17
local VP (another VP on the local VM)	no	0.19
remote VM (a VM, except the local VM)	yes	0.17
remote VP (a VP on the remote VM)	yes	0.19

Table 9: Ping test from a VP.

Based on the measurement of the latency of packet responses, through tools such as *ping* or *traceroute*, it is possible to determine whether two VM instances are deployed on the same physical machine or not [A6]. Likewise, VP instances may experience variations in the network latency, depending on how they are deployed.

However, our experimental results in Table 9 show that, given a VP instance, the difference in latency to reach a *local VP* on the same VM, versus a *remote VP* on another VM, is $\leq 0.05ms$. In particular, this value is small enough to be effectively hidden by the latencies set by the designer to model the network-level heterogeneity configuration, as shown in Table 3.

C. CASE STUDY I

Scheduler Design. We designed the scheduler to run on a client machine and to follow these steps:

1. It receives the *user request*, which includes the number of times each MPI application must run, e.g. 300 Poissons, 200 2d-FFTs, and 500 3DESs.

2. It loads the *performance profile* of each VP in executing the given applications, e.g. VP_0 takes 3.182s to execute Poisson, while VP_1 takes 10.427s , and so on, as shown in Fig. 7.

3. It derives the objective function to be minimized.

4. It converts the constraints (user request and performance profile) and the objective function into a matrix, and solves it by running a linear programming algorithm.

5. It distributes the workload according to the solution.

Examples of the constraints and the objective function to 1) minimize the total execution time and 2) minimize the power dissipation are illustrated in Appendix C.1.

C.1 Linear Programming Examples

For the sake of simplicity, the following two examples assume that there are two devices, d_1 and d_2, and we have two applications, x_1 and x_2. In both examples, variables T_{ij} denotes the amount of time that device d_i spent on an application x_j. For instance, a variable T_{12} is the time period that device d_1 spent running application x_2. After executing a linear programming algorithm, the solution includes the best (the minimun or the maximun) possible value of the objective function along with the values of the variable used for the best value of the objective function.

Minimizing the Execution Time. Let's assume we have the following profile data:

1. application x_1 runs on device d_1 110 times per unit time.

2. application x_2 runs on device d_1 250 times per unit time.

3. application x_1 runs on device d_2 150 times per unit time.

4. application x_2 runs on device d_2 100 times per unit time.

The user's request may be like the following:

1. Execute application x_1 400 times and application x_2 320 times.

Then we have two inequations from them:

---
[9]Physical Network Involved.

1. $110T_{11} + 150T_{21} >= 400$
2. $250T_{12} + 100T_{22} >= 320$

To bring the total execution time in the calculation, we introduce a new variable t.

1. $T_{11} + T_{12} <= t$
2. $T_{21} + T_{22} <= t$

Finally, the objective function p to be minimized will be equal to t. The resulted matrix is given in Table 10.

$T11$	T_{12}	T_{21}	T_{22}	t	
110	0	150	0	0	400
0	250	0	100	0	320
-1	-1	0	0	1	0
0	0	-1	-1	1	0
0	0	0	0	1	0

Table 10: Linear programming matrix for minimizing execution time.

The optimal solution for this example is $p = 52/25$; $T_{11} = 4/5$, $T_{22} = 32/25$, $T_{21} = 52/25$, $T_{22} = 0$, $t = 52/25$. This means that the devices can finish the user request in $52/25$ time units when the system follows this solution. The solutions for each variable, as the definition, stand for the execution time, e.g. T_{11} requires device d_1 to run application x_1 for $4/5$ unit time or 88 times.

Minimizing the Power Dissipation. In this example, we assume the same user request and the profile data used in the execution time minimization example. In addition to these conditions, the power dissipation profile data is required:

1. device d_1 dissipates 30W per unit time when executing application x_1.

2. device d_1 dissipates 50W per unit time when executing application x_2.

3. device d_2 dissipates 20W per unit time when executing application x_1.

4. device d_2 dissipates 70W per unit time when executing application x_2.

Then the objective function derived is $p = 30T_{11} + 50T_{12} + 20T_{21} + 70T_{22}$. The resulted matrix to minimize this objective function subject to the constraints is given in the Table 11.

T_{11}	T_{12}	T_{21}	T_{22}	
110	0	150	0	400
0	250	0	100	320
30	50	20	70	0

Table 11: Linear programming matrix for minimizing power dissipation.

The execution of a linear programming algorithm for this matrix gives the solution $p = 352/3$; $T_{11} = 0$, $T_{12} = 32/25$, $T_{21} = 8/3$, $T_{22} = 0$. This means that the user requests can be executed with consuming $352/3$ power units.

D. CASE STUDY II
D.1 Application Design

The application iterates the following work flow:

1. The mobile phone users take pictures and upload then to the Image DB along with their geolocation.

2. The cluster of MIPS servers fetches one image at the time from the DB and counts the people in it, by means of a human recognition algorithm.

3. The number of people in each image is stored back into the DB.

4. The Map Generator creates a plotted image as the result.

1168

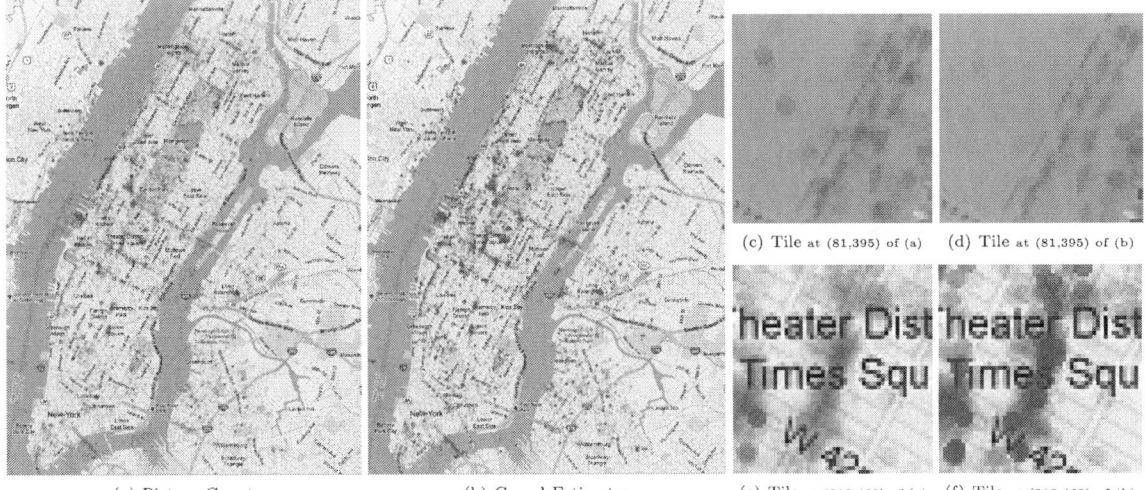

(a) Picture Count   (b) Crowd Estimates   (c) Tile at (81,395) of (a)   (d) Tile at (81,395) of (b)

(e) Tile at (210,460) of (a)   (f) Tile at (210,460) of (b)

Fig. 9: Case Study II: visualization of (a) picture count and (b) estimated crowd based on the pictures.

Each iteration is done in parallel, in the sense that the multiple Android Emulators upload images and the MIPS servers process the images concurrently.

The application consists of the following modules.

Android Camera App. One instance of the Android Camera App runs on each Android Emulator. To simulate the smart phones that users use to take pictures, we took images publicly available on the Picasa and Flickr's image databases [A1,A2], and we distributed them across the local storage of the Android Emulators, before starting the simulation. For the experiment, we considered 55,831 images with the geolocation information of Manhattan[10], assuming the users are taking pictures in this area. We modeled the act of a phone user taking a picture with the App loading a picture from the local storage.

Image Database. Every time the App takes a picture, it immediately uploads it, together with its metadata, i.e. latitude and longitude, to the Image DB. Also, the MIPS cluster fetches images and metadata from this database to process them, and stores back the results.

Human Recognition. The human recognition program is based on OpenCV. It is stored in the NETSHIP Server's storage, and runs on the MIPS cluster. To detect human bodies in a given picture, we used a head-and-shoulder detecting Haar model [A7] and an upper-and-lower-body detecting Haar model [A3]. It is, however, difficult to grasp human bodies from multiple directions, in particular from a side view [A8].

Map Generator. The Map Generator program reads the people counting from the Image DB and plots it on the map[11] with a resolution 717×944, translating latitude and longitude to the pixel position.

D.2 Experiments

Although the quality of the developed application's result is not the primary concern of this work, we present the resulted maps from two possible alternative variations of the crowd estimation application: one based on counting only *the number of pictures* taken at a particular location, and the other based on counting *the number of people* showing in those pictures. The results of Fig. 9 are interesting but they can be substantially improved by using NETSHIP to analyze various possible optimizations of the application.

Fig. 9(a) shows how many pictures are taken by users

---

[10] For the geolocation of Manhattan we used longitude -74.015 ~ -73.928 and latitude 40.700 ~ 40.816.

[11] The map is extracted from the Google Maps service.

and Fig. 9(b) shows the estimated crowds based on such pictures. One red circle on the map corresponds to an area of approximately $2500m^2$ or $2990yd^2$. The density of the crowds is presented with the opacity, where the transparent area indicates no people and an opaque circle indicates more than 80 people in that area.

Fig. 9(c) is a tile taken from Fig. 9(a) at the pixel position <81,395> and Fig. 9(d) is a tile taken from Fig. 9(b) at the same position. Likewise, Fig. 9(e) is a tile taken from Fig. 9(a) at the pixel position <210,460> and Fig. 9(f) is from the same pixel position of Fig. 9(b).

Both Fig. 9(a) and Fig. 9(b) give the idea about which areas are more crowded than others, based on the opacity of the red circles on the map. However, the comparison of these two figures shows that our crowd estimation algorithm, based on human recognition, gives a more accurate outcome than simply counting the total number of taken pictures. For example, as shown in the comparison of the pair of Fig. 9(c) and Fig. 9(d), the estimated crowds on the river was decreased by the human recognition algorithm because the pictures taken over the river are mostly Manhattan skyline photos taken on a boat, a helicopter, or an airplane. On the other hand, in the case of Fig. 9(e) and Fig. 9(f), the actual crowds on the ground might be greater than the number of pictures taken on the same spot.

E. REFERENCES

[A1] The Flickr API (www.flickr.com/services/api).

[A2] The Picasa Web Albums Data API (developers.google.com/picasa-web).

[A3] H. Kruppa, M. Castrillon-Santana, and B. Schiele. Fast and robust face finding via local context. In *Proc. of the Int. Workshop on VS-PETS*, pages 157-164, Oct. 2003.

[A4] R. L. Graham, T. S. Woodall, and J. M. Squyres. Open MPI: a flexible high performance MPI. In *Proc. of the Int. Conf. on PPAM*, pages 228-239, Feb. 2006.

[A5] J. Nagle. Congestion control in IP/TCP internetworks. *SIGCOMM Comp. Comm. Rev.*, 14(4):11-17, Oct. 1984.

[A6] T. Ristenpart et al. Hey, you, get off my cloud: exploring information leakage in third-party compute clouds. In *Proc. of the ACM Conf. on Comp. and Comm. Sec.*, pages 199-212, Nov. 2009.

[A7] M. C. Santana et al. Face and facial feature detection evaluation. In *Proc. of VISAPP*, pages 167-172, Apr. 2008.

[A8] P. Viola, M. Jones, and D. Snow. Detecting pedestrians using patterns of motion and appearance. *In Proc. of ICCV*, volume 2, pages 734-741, Oct. 2003.

Exploiting Just-enough Parallelism when Mapping Streaming Applications in Hard Real-time Systems

Jiali Teddy Zhai
tzhai@liacs.nl

Mohamed A. Bamakhrama
mohamed@liacs.nl

Todor Stefanov
stefanov@liacs.nl

Leiden Institute of Advanced Computer Science
Leiden University, Leiden, The Netherlands

ABSTRACT

Embedded streaming applications specified using parallel *Models of Computation* (MoC) often contain ample amount of parallelism which can be exploited using Multi-Processor System-on-Chip (MPSoC) platforms. It has been shown that the various forms of parallelism in an application should be explored to achieve the maximum system performance. However, if more parallelism is revealed than needed, it will overload the underlying MPSoC platform. At the same time, the revealed parallelism should be sufficient such that the MPSoC platform is fully utilized. Therefore, the amount of revealed and exploited parallelism has to be just-enough with respect to the platform constraints. In this paper, we study the problem of exploiting just-enough parallelism by application task unfolding, when mapping streaming applications modeled using the Synchronous Data Flow (SDF) MoC onto MPSoC platforms in hard real-time systems. We show that our problem of simultaneously unfolding and allocating tasks under hard real-time scheduling has a bounded solution space and derive its upper bounds. Subsequently, we devise an efficient algorithm to solve the problem, while the obtained solution meets a pre-specified quality. The experiments on a set of real-life streaming applications demonstrate that our algorithm results, within reasonable amount of time, in a system specification with large performance gain. Finally, we show that our proposed algorithm is on average 100 times faster than one of the state-of-the-art meta-heuristics, i.e., NSGA-II genetic algorithm, while achieving the same quality of solutions.

1. INTRODUCTION

Streaming applications are widely used in embedded systems in several application domains, such as image processing, video/audio processing, and digital signal processing. The ample amount of parallelism in streaming applications matches perfectly the processing power of Multi-Processor System-on-Chip (MPSoC) platforms, which contain an increasing number of Processing Elements (PEs). Having many PEs available has imposed huge challenges on both application developers and design tools to identify and exploit the right amount of parallelism which can utilize the PEs efficiently. To tackle the challenges, *Models-of-Computation* (MoC) have been adopted as the parallel application specification in most of the design approaches and tools [7]. For example, Kahn Process Networks [11] and Synchronous Data Flow (SDF) [14] are two prominent parallel MoCs. In such MoCs, a streaming application is modeled as a directed graph, where the graph nodes represent the application tasks and the graph edges represent the data communication FIFO channels. The tasks execute concurrently and communicate data explicitly via the FIFOs. In this case, the task-level parallelism is naturally exposed.

However, in most cases, an initial graph given by application developers to specify application behavior is not the most suitable for a given MPSoC platform. This is because application developers mainly focus on realizing certain application behavior, including the identification of the functionality of tasks and the synchronization/communication between these tasks. The computational capacity of the MPSoC platform is often not fully taken into account.

The authors in [12] showed that, for a set of representative streaming benchmarks, the maximum achievable speedup of mapping the initial graphs can only reach up to a limited number. To better utilize the underlying MPSoC platform, the initial graph of an application should be transformed to an alternative one that exposes more parallelism while preserving the same application behavior. To this end, task unfolding is an effective technique to generate such alternative graphs. Basically, task unfolding replicates the functionality of a task by a certain number of times, referred as *unfolding factor*. Then, replicas of tasks concurrently process different data, thereby exploring also data-level parallelism next to the task-level parallelism.

Unfolding individual tasks in an initial graph by different unfolding factors results in a large number of possible alternative graphs. To transform the initial graph to an alternative one by unfolding, the main problem is to determine a proper unfolding factor for each task. This problem is challenging because platform constraints must be considered during unfolding. The platform constraints can be the number of available PEs and temporal scheduling of tasks on the PEs. On the one hand in [22, 5, 20], the authors determine an unfolding factor for each task such that the obtained alternative graph exposes the maximum data-level parallelism, without considering the platform constraints. However, unfolding a task too many times reveals more parallelism than needed. The overwhelming parallelism leads to an inefficient mapping of replicas of tasks. That is, the excessive number of replicas cannot be efficiently allocated and temporally scheduled on the available PEs. Moreover, the excessive number of replicas introduces significant memory overhead for both code and data. On the other hand in [9, 12, 17], the authors assume that the unfolding factor of a task cannot exceed the number of available PEs. This assumption, however, restricts the amount of revealed parallelism because a proper unfolding factor is not necessarily less than or equal to the number of available PEs. As a consequence, the aforementioned assumption might lead to under-utilized PEs. From the discussion above, we can see that exploiting excessive or insufficient parallelism may result in sub-optimal system utilization and performance. Therefore, in this paper, we address the problem of determining a proper unfolding factor of each task in a given initial graph, such that the obtained alternative graph exposes *just-enough* parallelism to fully utilize the available PEs. This is achieved by considering the platform constraints when determining the unfolding factors.

We solve the problem explained above when a streaming application is modeled using the SDF MoC and mapped onto MPSoC platforms with hard real-time constraints. The SDF MoC has been successfully adopted in both industrial and academic tools. We consider the problem in the context of hard real-time systems, because many streaming applications nowadays require hard real-time execution. For instance, collision avoidance algorithms used in the avionics or automotive industry require very strict timing guarantees. At the same time, it has been reported in [1] that these algorithms require approximately 170 million calculations for each frame update, with the expectation of being executed on up to 64 PEs.

1.1 Paper Contributions

We propose an efficient approach to exploit just-enough parallelism in streaming applications modeled using the SDF MoC in hard real-time systems, in order to increase the performance that can be guaranteed on an MPSoC platform. More specifically, our problem is to determine simultaneously which actors (i.e., tasks) to unfold by what factor, and the allocation of unfolded actors onto PEs. We show that the solution space of our problem is bounded and derive its upper bounds. We then propose an efficient

---

Permission to make digital or hard copies of all or part of this work for personal or classroom use is granted without fee provided that copies are not made or distributed for profit or commercial advantage and that copies bear this notice and the full citation on the first page. To copy otherwise, to republish, to post on servers or to redistribute to lists, requires prior specific permission and/or a fee.
DAC'13, May 29 - June 07 2013, Austin, TX, USA.

algorithm to find a solution to the problem, while the obtained solution meets a pre-specified quality. In addition, we evaluate the efficiency and time complexity of the proposed algorithm on 11 real-life applications. Finally, we show that our algorithm is, on average, 100 times faster than a state-of-the-art meta-heuristic, i.e., NSGA-II genetic algorithm [4], while achieving the same quality of the solution.

1.2 Scope of Work

In this paper, we assume that a given SDF graph is acyclic. Such assumption covers a large set of applications as it has been empirically shown in [18] that around 90% of streaming applications can be modeled as acyclic SDF graphs. Once a cycle exists in an SDF graph, one can always fuse all actors in the cycle into a single stateful actor. A stateful actor is the one whose next execution depends on the current execution. As a consequence, our approach does not unfold stateful actors. Furthermore, the data source and sink actors, which are connected to the external environment, are not unfolded. The target platform assumed in this work is a homogeneous programmable MPSoC with distributed memory. The interconnection structure between PEs must provide guaranteed communication latency, e.g., Æthereal network-on-chip [8].

2. RELATED WORK

The approach in [17] is closely related to our work, although the considered problem is relaxed, i.e., without considering timing constraints, compared to our problem. A genetic algorithm based heuristic is proposed to determine the unfolding factor of an actor and allocation of all replicas. The unfolding factor of an actor cannot exceed the number of PEs, which might result in sub-optimal solutions as we show later in Sec. 5. Moreover, we show in the experiments that our approach outperforms significantly the genetic algorithm based heuristic in terms of running time.

In [12], an Integer Linear Programming (ILP) formulation gives exact solutions to minimize makespan on any PE while simultaneously unfolding actors in an SDF graph and allocating them on PEs. In the ILP formulation, an unfolding factor of an actor cannot exceed the number of available PEs. This assumption might lead to sub-optimal system performance as discussed previously. Moreover, it has been shown in [5] that the ILP formulation is even intractable for benchmarks with medium graph size. For instance, it takes around 70 hours to solve the ILP formulation for the FFT benchmark with 26 actors on 4 PEs (see Table 2 in [5]). In practice, real-life applications have been shown to contain up to 2868 actors [18]. Therefore, it is clear that the ILP-based approach suffers from severe scalability issues. In contrast, our proposed algorithm solves the combined problem within a reasonable amount of time as demonstrated later in Sec. 7.

To address the scalability issue of [12], the authors in [5] propose to decompose the combined problem into two problems and solve them separately. The separation of the two problems often leads to inferior performance, as both problems are strongly related. In contrast, our proposed algorithm is capable of solving the combined problem simultaneously. Moreover, our algorithm takes into account timing constraints, while the work in [5] does not.

In the context of synthesizing an SDF graph using dedicated hardware, the authors in [10] also determine which actors to unfold and by what factor. The addressed problem is easier than ours because there is no need to consider allocation of actors after unfolding in case of hardware synthesis.

In [13], a synchronous programming model is used for the application specification under hard real-time scheduling. The term "synchronous" in this context refers to the fact that a master thread can *fork* a job into several parallel execution segments and they *join* upon completion. These parallel execution segments are, to some extent, similar to unfolded actors in our case. There is also no need to consider allocation of parallel segments at compile-time because migration at run-time is allowed targeting MPSoC platforms with shared memory. In contrast, we solve the problem of allocating actors at compile-time. Recall from Sec. 1.2 that we consider the MPSoC platforms with distributed memory. On such platforms, migration of actors at run-time introduces non-negligible overhead.

3. BACKGROUND

In Sec. 3.1, we first introduce the application specification, i.e., the SDF MoC, and the unfolding operation on SDF graphs. After that in Sec. 3.2, hard real-time scheduling of the SDF MoC is reviewed. These are essential for better understanding our problems formally defined in Sec. 4 and contributions presented in Sec. 5 and 6.

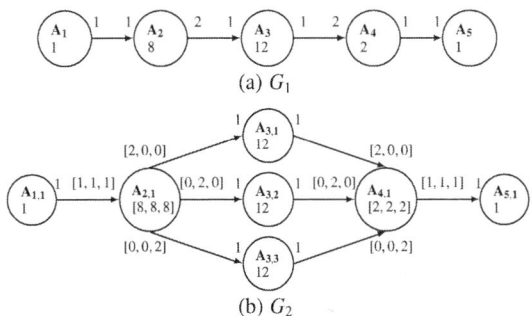

Figure 1: (a) An example of an SDF graph and (b) its equivalent CSDF graph by unfolding actor A_3 by factor 3.

3.1 Unfolding of SDF Graphs

The SDF MoC is defined as a directed graph $G = (\mathcal{A}, \mathcal{E})$, where $\mathcal{A}$ is a set of n actors and $\mathcal{E}$ is a set of communication edges. An actor $A_i \in \mathcal{A}$ executes by first consuming data tokens from all its incoming edges, performing certain computation, and subsequently producing data tokens to all its outgoing edges. The number of tokens consumed from an edge or produced to an edge is known a-priori and given as a constant integer. It has been shown in [14] that, to have a valid periodic schedule, an SDF graph has to be *consistent* with a non-trivial *repetition vector* $\vec{q} \in \mathbb{N}^n$. An entry $r(A_i) \in \vec{q}$ denotes how many times an actor $A_i \in \mathcal{A}$ has to be executed in every graph iteration of G. Additionally, for each actor A_i, we associate a Worst-Case Execution Time (WCET) C_i and its code size S_i. For a summary of all the notations used in the paper, please refer to Appendix A.

An SDF graph G_1 is shown in Figure 1(a). The actors A_1 and A_5 are the data source and sink actors, respectively. G_1 has five actors and a repetition vector $\vec{q} = [1, 1, 2, 1, 1]^T$. The WCET of each actor is shown below its name, e.g., $C_3 = 12$ for actor A_3.

The unfolding operation on an SDF graph used in this paper is conceptually similar to the one used in [5, 9, 10, 12], in which two special constructs *splitter* and *joiner* are employed for the unfolded actors. Given a vector $\vec{f} \in \mathbb{N}^n$ of unfolding factors, where f_i denotes the unfolding factor for actor A_i, the unfolding operation replaces A_i by f_i replicas of itself. Then, instead of inserting a splitter and joiner before and after the f_i replicas of A_i, we transform the initial SDF graph to a functionally equivalent Cyclo-Static Data Flow (CSDF) [3] graph. The CSDF MoC generalizes the SDF MoC in the sense that each CSDF actor may produce or consume a variable but predefined number of data tokens in consecutive executions, called *production/consumption sequence*. Similar to the SDF MoC, the necessary condition for the existence of a valid periodic schedule for a given CSDF graph is to have a non-trivial repetition vector $\vec{q}'$. To ensure the functional equivalence, the production and consumption rates of an SDF actor are modified accordingly to the production and consumption sequences in the resulting CSDF graph. This modification results in a different repetition vector of the obtained CSDF graph to ensure its consistency.

The algorithm for performing the unfolding of actors in SDF graphs is given in Algorithm 2 in Appendix C. The algorithm accepts as inputs an SDF graph G and a vector $\vec{f}$ of unfolding factors. The algorithm produces as an output a CSDF graph G', where $A_{i,f}$ denotes the fth replica of A_i with repetition $r(A_{i,f})$ given by:

$$r(A_{i,f}) = \frac{r(A_i) \cdot \text{lcm}(\vec{f})}{f_i}, \qquad (1)$$

where $r(A_i)$ is the repetition of actor A_i in the initial SDF graph and $\text{lcm}(\vec{f})$ denotes the least common multiple of $f_i \in \vec{f}$. It follows that the repetition vector of G', denoted by $\vec{q}' \in \mathbb{N}^{n'}$ where $n' = \sum_{A_i \in \mathcal{A}} f_i$, is given by $\vec{q}' = [r(A_{1,1}), \cdots, r(A_{1,f_1}), \cdots, r(A_{n,f_n})]^T$. After obtaining $\vec{q}'$ using Eq. 1, production/consumption sequences of each CSDF actor are generated accordingly.

Suppose that a vector of unfolding factors is given as $\vec{f} = [1, 1, 3, 1, 1]$ for G_1 in Figure 1(a). Algorithm 2 outputs a CSDF graph G_2 shown in Figure 1(b) with three replicas $A_{3,1}$, $A_{3,2}$ and $A_{3,3}$ for actor A_3 in G_1. The unfolding results in a repetition vector of G_2 as $\vec{q}' = [r(A_{1,1}), r(A_{2,1}), r(A_{3,1}), r(A_{3,2}), r(A_{3,3}), r(A_{4,1}), r(A_{5,1})]^T = [3, 3, 2, 2, 2, 3, 3]^T$. For example, SDF actor A_4 executes only once ($\mathsf{r}(A_4) = 1$) in G_1 per graph iteration, while executing three times ($\mathsf{r}(A_{4,1}) = 3$) in G_2 per graph iteration. Three consumption sequences of actor $A_{4,1}$ in G_2 behave similar to a joiner, with which $A_{4,1}$ collects data tokens from the three replicas $A_{3,1}$, $A_{3,2}$ and $A_{3,3}$. Analogous to a splitter, actor $A_{2,1}$ with three production sequences distributes tokens to the three replicas.

3.2 Hard Real-time Scheduling of (C)SDF

The authors in [2] showed that the actors in acyclic CSDF graphs can be executed in a strictly periodic fashion. Note that this result applies also to acyclic SDF graphs, since the SDF MoC is a special case of the CSDF MoC. As a result, a variety of hard-real-time scheduling algorithms, such as Earliest Deadline First (EDF, [15]), can be applied to temporally schedule the actors allocated on a PE.

To execute the actors in an acyclic (C)SDF graph in a strictly periodic fashion, the period of each actor needs to be computed first. To do so, we introduce the following definition:

Definition 1. The *workload* of a (C)SDF actor A_i per graph iteration, denoted by W_i, is given by $W_i = \mathsf{r}(A_i)C_i$, where $\mathsf{r}(A_i)$ is the repetition of A_i and C_i is the WCET of A_i.

Accordingly, we define the maximum workload per graph iteration, denoted by $\hat{W}_G$, as $\hat{W}_G = \max_{A_i \in \mathcal{A}}(\mathsf{r}(A_i)C_i)$. The period of an actor A_i, denoted by T_i where $T_i \in \mathbb{N}$, has a lower bound, denoted by $\check{T}_i$. It is given in [2] as follows:

$$\check{T}_i = \frac{\mathrm{lcm}(\vec{q})}{\mathsf{r}(A_i)}\left\lceil \frac{\hat{W}_G}{\mathrm{lcm}(\vec{q})} \right\rceil, \qquad (2)$$

where $\mathrm{lcm}(\vec{q})$ is the least common multiple of all repetition entries in $\vec{q}$. The actual period T_i of an actor A_i is given by $T_i = s\check{T}_i$, where $s \in \mathbb{N}$ is the period *scaling factor* of the graph. For a given (C)SDF graph G, s is a constant that is used to scale the periods of all its actors. The deadline of each actor is the start of its next period, i.e., the deadline is equal to the period (often called *implicit deadline*). Once a period T_i of each actor is derived, we can compute the *utilization* $u_i \in (0, 1]$ of actor A_i as $u_i = C_i/T_i$. Based on this, we can also compute the utilization of a CSDF graph G, denoted by U_G, as $U_G = \sum_{A_i \in \mathcal{A}} C_i/T_i$. The throughput of a graph G when its actors are scheduled as strictly periodic actors is determined by the period of the sink actor. In the rest of the paper, we denote the lower bound on the period of the sink actor by $\check{T}_{\mathrm{snk}}$, and the actual period of the sink actor by T_{snk}. Accordingly, the throughput of the graph is $1/T_{\mathrm{snk}}$.

In this work, we consider that the schedule on each PE is built according to the EDF scheduling algorithm, which is known to be optimal on a uniprocessor system [15]. A set of actors allocated on a PE is schedulable using EDF if and only if their total utilization does not exceed 1. The schedule itself can be built either off-line for efficiency, or on-line for flexibility according to the system requirements. The problem of allocating actors onto PEs is similar to the bin-packing problem and can be solved using either *exact* or *approximate* allocation algorithms. An example of an exact allocation algorithm is proposed in [16], which returns an optimal allocation of actors. One disadvantage of using an exact algorithm is its high computational complexity. Therefore, to have a trade-off between optimality of the allocation and computational complexity, we choose in this paper an approximate allocation algorithm, namely the First-Fit Decreasing (FFD) algorithm, which has a proven worst-case approximation ratio $R_{\mathrm{FFD}} = \frac{11}{9}$ [21].

For instance, consider the CSDF graph G_2 in Figure 1(b) with repetition vector $\vec{q'} = [3, 3, 2, 2, 2, 3, 3]^T$ as computed in Sec. 3.1. The WCET $C_{i,f}$ of each actor $A_{i,f}$ is shown below its name. The throughput of G_2 is determined by the period of the sink actor $A_{5,1}$. We first compute $\hat{W}_{G_2} = \mathsf{r}(A_{3,1}) \cdot C_{3,1} = 24$ and $\mathrm{lcm}(\vec{q'}) = 6$. By solving Eq. 2, we obtain that the minimum period of the sink actor $A_{5,1}$ is $\check{T}_{\mathrm{snk}} = \frac{6}{3} \cdot \lceil \frac{24}{6} \rceil = 8$. This means that the maximum throughput of G_2 is $\frac{1}{8}$. The strictly periodic execution of all actors in G_2 is visualized in Figure 2. The bars indicate the execution of the actors in their periods and the up arrows denote the earliest starting time of each actor. It can be seen that actor $A_{5,1}$ executes for 1 time unit every 8 time units. The total utilization of G_2 is $U_{G_2} = \frac{1}{8} + \frac{8}{8} + \frac{12}{12} + \frac{12}{12} + \frac{12}{12} + \frac{2}{8} + \frac{1}{8} = 4.5$. To achieve the maximum throughput $\frac{1}{8}$ assuming the EDF scheduling algorithm and the FFD allocation algorithm, 5 PEs are required.

Using the scaling factor s defined earlier, we can have a trade-off between processing resources and guaranteed performance as shown in the following lemma:

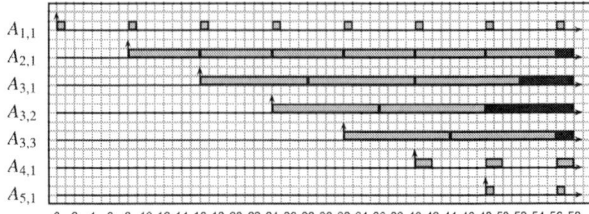

Figure 2: A schedule showing the strictly periodic execution of the actors in G_2. The x-axis represent the time. The last execution of $A_{2,1}$, $A_{3,1}$, $A_{3,2}$, and $A_{3,3}$ in the figure is truncated due to space limits.

LEMMA 1. *Let G be a CSDF graph that is schedulable using a scheduling algorithm SA and an allocation algorithm AA on $\breve{m}$ PEs, when the period of each actor A_i is equal to $\check{T}_i$. G is schedulable using the same SA and AA on $\lceil \frac{\breve{m}}{s} \rceil$ PEs, when the period of each actor A_i is equal to $s\check{T}_i$.*

The proof of Lemma 1 can be found in Appendix B. Considering G_2 in Figure 1(b), it can be scheduled on $\lceil \frac{5}{2} \rceil = 3$ PEs achieving a period $T_{\mathrm{snk}} = 2 \times \check{T}_{\mathrm{snk}} = 16$, i.e., throughput $\frac{1}{16}$ by scaling all minimum periods by $s = 2$.

Now, suppose that AA is an approximate allocation algorithm with an approximation ratio R_{AA}. Then, we can have the following corollary of Lemma 1:

COROLLARY 1. *Let G be a CSDF graph that is schedulable using a scheduling algorithm SA and any exact allocation algorithm on $\breve{m}$ PEs, when the period of each actor A_i is equal to $s\check{T}_i$. G is schedulable using SA and any approximate allocation algorithm AA, with approximation ratio R_{AA}, on $\breve{m}$ PEs, when the period of each actor A_i is equal to $sR_{AA}\check{T}_i$.*

4. PROBLEM FORMULATION

Now, we formally define our problem introduced in Sec. 1 as follows:

Problem 1. Given an SDF graph G, where the actors are scheduled as strictly periodic actors, and m available PEs. Suppose that each actor A_i in G is to be unfolded by an unfolding factor f_i. Find, for each actor A_i, the minimum value of f_i and the allocation of each replica $A_{i,f}$, where $1 \le f \le f_i$, such that the period of the sink actor T_{snk} in the unfolded graph is minimized.

If Problem 1 is considered as *primal*, we can have its equivalent *dual* problem defined as follows:

Problem 2. Given an SDF graph G, where the actors are scheduled as strictly periodic tasks, and m available PEs. Suppose that each actor A_i in G is to be unfolded by an unfolding factor f_i. Find, for each actor A_i, the minimum value of f_i and the allocation of each replica $A_{i,f}$, where $1 \le f \le f_i$, such that the total utilization $\sum_{A_{i,f} \in \mathcal{A}'} C_{i,f}/T_{i,f}$ of the unfolded graph on m PEs is maximized.

It can be seen that Problems 1 and 2 are not trivial. In general, for a given SDF graph, the number of possible alternative graphs that can be generated using unfolding grows exponentially as the number of actors increases. Furthermore, for each alternative graph, we have to perform allocation of unfolded actors which is by itself an NP-hard problem.

5. BOUNDING THE SOLUTION SPACE

In order to solve Problems 1 and 2 defined in Sec. 4, we need first to bound the solution space, i.e., to bound the values of the unfolding factors f_i. Bounding the solution space ensures that the algorithm devised in Sec. 6 terminates. We define the *upper bound* on unfolding factors as follows:

Definition 2. Let G be an SDF graph, where the actors in G are scheduled as strictly periodic actors, and assume that the number of PEs is unlimited. Suppose that every actor A_i in G is to be unfolded by a factor f_i resulting in a CSDF graph G', where $\check{T}_{i,f}$ is the minimum period of each replica $A_{i,f}$ and $C_{i,f} = C_i$ is its WCET. The *upper bound* on f_i, denoted by $\hat{f}_i$, is the minimum value which results in utilization $C_{i,f}/\check{T}_{i,f} = 1.0$ for each replica $A_{i,f}$ in G'.

In other words, unfolding an SDF graph G by a vector of unfolding factors $\hat{\vec{f}} = [\hat{f}_1, \cdots, \hat{f}_n]$ results in a graph G' with utilization $U_{G'} = n'$, where n' is the number of actors in the unfolded graph. Hence, unfolding any actor A_i by an unfolding

1172

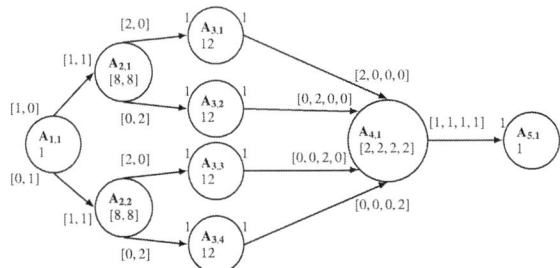

Figure 3: G_3: **Optimal alternative graph of** G_1 **in Figure 1(a) with unfolding factors** $f_2 = 2, f_3 = 4$ **when scheduled on 2 PEs.**

factor $f_i^* > \hat{f}_i$ cannot result in any increase in the total utilization of the unfolded graph. Moreover, the unfolded graph achieves the maximum achievable throughput since the sink actor fully utilizes the PE on which it executes. Therefore, $\vec{f}$ defines the solution space that has an impact on the total utilization of the unfolded graph.

Determining the upper bound $\vec{f}$ is not trivial. One common assumption, e.g., in [9] and [12], is to set $\vec{f} = [m, m, \cdots, m]$, where m is the number of PEs. In this section, we show, using an example, that this assumption sometimes limits the solution space. As a consequence, the limited solution space might not contain the optimal solution to Problems 1 and 2.

Let us consider G_1 in Figure 1(a) and suppose that 2 PEs are available. The optimal alternative graph of G_1 is G_3, shown in Figure 3, when the vector of unfolding factors is $\vec{f} = [1, 2, 4, 1, 1]$. The repetition vector of G_3 can be computed according to Eq. 1 as $\vec{q'} = [r(A_{1,1}), r(A_{2,1}), r(A_{2,2}), r(A_{3,1}), r(A_{3,2}), r(A_{3,3}), r(A_{3,4}), r(A_{4,1}), r(A_{5,1})]^T = [4, 2, 2, 2, 2, 2, 2, 4, 4]^T$. It follows that $\hat{W}_{G_3} = 2 \times 12$ and $\text{lcm}(\vec{q'}) = 4$. Solving Eq. 2 yields the minimum period of the sink actor $A_{5,1}$ as $\check{T}_{\text{snk}} = \frac{4}{4} \cdot \lceil \frac{24}{4} \rceil = 6$. To achieve $\check{T}_{\text{snk}} = 6$, 6 PEs are required. Then, we can scale all periods of the actors in G_3 by $s = 3$, which yields a period $T_{\text{snk}} = 3\check{T}_{\text{snk}} = 18$. According to Lemma 1, the graph G_3 is schedulable on $\lceil \frac{6}{3} \rceil = 2$ PEs. After scaling the periods of all actors, the total utilization U_{G_3} of G_3 on 2 PEs is 2.0, thereby no shorter period can be achieved. Thus, G_3 is the optimal alternative graph of G_1 for 2 PEs with an unfolding factor $f_3 = 4$, which is greater than the number of PEs available. Therefore, this example shows that the optimal solution is beyond $\vec{f} = [2, 2, 2, 2, 2]$, which defines the solution space if we set $\vec{f} = [m, m, \cdots, m]$. Hence, we conclude that the upper bound on an unfolding factor is not necessarily equal to the number of PEs.

Now, we derive the upper bound on the unfolding factor for each actor in the initial SDF graph by stating the following theorem:

THEOREM 1. *Given an SDF graph G, where the actors are scheduled as strictly periodic actors. Suppose that each actor A_i is to be unfolded by a factor f_i. The upper bound on f_i according to Definition 2 can be computed as follows:*

$$\hat{f}_i = \text{lcm}(x_1, x_2, \cdots, x_n)/x_i, \tag{3}$$

where $x_i = \text{lcm}\{W_1, W_2, \cdots, W_n\}/W_i$ (W_i is the workload of actor A_i given by Definition 1).

The proof of Theorem 1 is given in Appendix B.

Now, we give an example on how to compute $\vec{f}$. For G_1 in Figure 1(a), $\vec{x}$ containing the values of x_i is given by $\vec{x} = [24, 3, 1, 12, 24]$. Then, we obtain $\text{lcm}(\vec{x}) = 24$, and $\vec{f} = [1, 8, 24, 2, 1]$.

6. THE ALGORITHM

Considering the upper bounds on unfolding factors $\vec{f}$ derived in Sec. 5, we devise, in this section, an efficient algorithm to solve Problems 1 and 2 as defined in Sec. 4.

The algorithm accepts as an input the following: 1) the initial SDF graph G; 2) the number of available PEs m; 3) the vector containing the upper bounds on the unfolding factors $\vec{f}$ computed using Eq. 3; and 4) a pre-specified quality factor $\rho \in (0, 1]$, which is used to terminate the algorithm. The outputs of the algorithm are: 1) a vector of unfolding factors that is the solution to Problems 1 and 2; 2) the allocation of the unfolded SDF graph on m PEs; 3) the minimum achievable period of the sink actor in the unfolded

SDF graph on m PEs which is the objective of Problem 1; and 4) the maximum utilization of the unfolded SDF graph on m PEs which is the objective of Problem 2.

6.1 Algorithm Description

The algorithm builds, incrementally during its execution, a list of nodes in which each node represents a possible vector of unfolding factors $\vec{f}$. Initially, the list contains only a single node which corresponds to the given initial SDF graph with a vector of unfolding factors $\vec{f} = \vec{1}$. Then, we compute the minimum period of the sink actor T_{snk} in the initial SDF graph G, when G is allocated on m PEs, and its total utilization U_G. Both values initialize a tuple $(T_{\text{best}}, U_{\text{best}})$ which holds the period and total utilization of the current best solution. During the execution of the algorithm, new nodes are created and added to the list, where a node represents an alternative CSDF graph G' of the initial graph G with a vector $\vec{f}$ of unfolding factors. Each entry $f_i \in \vec{f}$ ranges from 1 up to $\hat{f}_i$ derived in Eq. 3.

A newly created node inherits from its previous node a copy of the unfolding factors vector $\vec{f}_{\text{prev}}$ used by the previous node to generate the unfolded graph G'_{prev}. After that, we search in G'_{prev} for the bottleneck actor, denoted by $A_{b,f}$, which is the one with the maximum workload $\hat{W}_G$ as defined in Sec. 3.2. If multiple actors have the same maximum workload, then the one with the smallest code size is selected. Next, we increment by one the entry f_b in the inherited unfolding factors vector $\vec{f}_{\text{prev}}$, thereby, obtaining $\vec{f}_{\text{curr}}$. Then, we unfold the initial graph G by the factors in $\vec{f}_{\text{curr}}$ which results in a CSDF graph G'_{curr}. The next step is to evaluate the unfolded graph G'_{curr} when it is allocated on m PEs. The procedure for evaluating G'_{curr} is explained in details in Sec. 6.2. The result of the evaluation procedure is the minimum period of the sink actor T_{snk} in G'_{curr}, when G'_{curr} is allocated on m PEs, and the total utilization of the graph U_{curr}. If the obtained U_{curr} is higher than U_{best} corresponding to the current best solution (i.e., T_{snk} smaller than T_{best}), then T_{best} and U_{best} are updated with T_{snk} and U_{curr}, respectively. Otherwise, T_{best} and U_{best} remain unchanged.

The creation of new nodes is terminated when one of the following conditions is met:

1. The total utilization $U_{G'}$ of the CSDF graph at the current node satisfies $U_{G'} \geq \rho m$, where $\rho \in (0, 1]$ is the quality factor given as an input to the algorithm. If $\rho = 1$, then this means that each PE is fully utilized, which means that no shorter period can be obtained.

2. The unfolding factor f_i of an actor A_i exceeds either its upper bound $\hat{f}_i$ if A_i is stateless, or 1 if A_i is stateful or a data source/sink actor. Recall that stateful actors together with the data source and sink actors cannot be unfolded.

After the creation of new nodes is terminated, we select the first node in the list that has a minimum sink period and a total graph utilization equal to T_{best} and U_{best}, respectively. The selected node contains the solution to Problems 1 and 2.

6.2 Evaluating the Unfolded Graphs

As explained in Sec. 6.1, at each node, the initial SDF graph G is unfolded to produce a CSDF graph $G' = \{\mathcal{A}', \mathcal{E}'\}$. Then, we compute two values for G': 1) the minimum sink actor period T_{snk} when G' is allocated on m PEs; and 2) its total utilization $U_{G'}$. In this section, we explain in details how these two values are computed. Recall from Sec. 3.2 that T_{snk} is given by $T_{\text{snk}} = s\check{T}_{\text{snk}}$, and $U_{G'}$ can be computed as follows:

$$U_{G'} = \sum_{A_{i,f} \in \mathcal{A}'} \frac{C_{i,f}}{s\check{T}_{i,f}}. \tag{4}$$

Recall also that the objective of Problem 2 is to maximize the utilization. Therefore, we need to find a value of scaling factor s, such that all actors in G' are schedulable on m PEs and $U_{G'}$ is maximized. To do so, we first bound the search range for s by deriving its lower and upper bounds. Using any allocation algorithm, we have from Lemma 1 a lower bound on s, denoted by $\check{s}$, as follows:

$$\check{s} = \left\lceil \frac{1}{m} \sum_{A_{i,f} \in \mathcal{A}'} \frac{C_{i,f}}{\check{T}_{i,f}} \right\rceil. \tag{5}$$

That is, for any AA, the scaling factor s cannot be smaller than $\check{s}$. From Corollary 1 in Sec. 3.2, we compute, using the approximation

Algorithm 1: The procedure for evaluating an unfolded graph.

Input: A CSDF graph G', number of available PEs m, and the period and total utilization corresponding to the current best solution T_{best} and U_{best}.

Result: *alloc* which is an m-partition describing the allocation of the actors in G' onto m PEs

1   Compute $\check{s}$ using Eq. 5 and $\hat{s}$ using Eq. 6 ;
2   **for** $s = \check{s}$ **to** $\hat{s}$ **do**
3   Compute the period $T_{i,f}$ of each actor $A_{i,f}$ as $T_{i,f} = s\check{T}_{i,f}$;
4   **if** $T_{snk} \geq T_{best}$ **then**
5   **return** $\emptyset$;
6   Compute the utilization $U_{G'}$ using Eq. 4;
7   Find an m'-partition of the actors in G', denoted by *alloc*, using the FFD algorithm and assuming the EDF scheduling algorithm;
8   **if** $m' \leq m$ **then**
9       $U_{best} = U_{G'}$, $T_{best} = T_{snk}$;
10   **return** *alloc* ;

ratio of the FFD allocation algorithm $R_{FFD} = 11/9$ given in Sec. 3.2, the upper bound on the scaling factor s, denoted by $\hat{s}$, as follows:

$$\hat{s} = \left\lceil \frac{11}{9m} \sum_{A_{i,f} \in \mathcal{A}'} \frac{C_{i,f}}{\check{T}_{i,f}} \right\rceil + 1. \quad (6)$$

Once the lower and upper bounds of s are found using Eq. (5) and Eq. (6), respectively, we perform a linear search to seek the smallest s, such that a CSDF graph G' is schedulable on m PEs. Specifically, we check if an m-partition of all actors in G' exists, assuming the EDF scheduling algorithm and the FFD allocation algorithm explained in Sec. 3.2. The complete procedure for evaluating the unfolded graphs is depicted in Algorithm 1. If the period resulting from a given scaling factor s is greater than T_{best}, then Algorithm 1 terminates immediately to speed-up the search (see line 4 in Algorithm 1).

6.3   Example

Now, we illustrate our algorithm using graph G_1 in Figure 1(a) and schedule it on 2 PEs (i.e., $m = 2$). Suppose that $\rho = 0.95$, i.e., the algorithm terminates when $U_{G'} \geq 0.95 \times 2 = 1.9$. The whole list produced by the algorithm is illustrated in Figure 4. The numbers inside the nodes correspond to the sequence in which the nodes are created. The algorithm starts with the initial G_1 in node 0 and computes the scaling factors $\check{s}$ and $\hat{s}$ which result in $U_{G_1} = 1.5$ and period $T_{snk} = 24$. At this point, U_{best} is initialized to 1.5 and T_{best} to 24. Node 1 inherits from node 0 a vector of unfolding factors equal to $[1, 1, 1, 1, 1]$. After that, we search in $G'_{prev} = G_1$ for the bottleneck actor which is A_3. Next, we increment f_3 in the inherited vector of unfolding factors at node 1 resulting in $\vec{f} = [1, 1, 2, 1, 1]$. Then, G' is generated and Algorithm 1 is invoked. Since U_{best} cannot be improved (see line 4 in Algorithm 1), the algorithm continues by creating node 2. At node 2, a new bottleneck actor $A_{2,1}$ is introduced. Therefore, at node 3, the unfolding factor f_2 is incremented by 1. Then, the algorithm continues to node 4, at which one termination criterion is met, namely $U_{G'} \geq 1.9$. As a result, $\vec{f} = [1, 2, 4, 1, 1]$ is the solution with $T_{best} = 18$ and $U_{best} = 2.0$.

7.   EVALUATION

In this section, we present the results of evaluating our algorithm using a set of real-life streaming applications. We evaluate the algorithm by performing two experiments. In the first experiment, we run our algorithm on the applications and report the following: 1) the performance gain resulting from mapping the SDF graph unfolded using the unfolding factors obtained from our algorithm, compared to mapping the initial SDF graph without unfolding; and 2) the total time needed to execute our algorithm.

In the second experiment, we compare our proposed algorithm with one of the state-of-art search meta-heuristics, since problems 1 and 2 in general can be readily formulated and solved by these meta-heuristics, such as genetic algorithms, simulated annealing, etc. However, meta-heuristics normally require parameter tuning to achieve a good solution. In this work, we select a particular meta-heuristic, namely Genetic Algorithms (GA) for two reasons: 1) they have been applied by several researchers to solve similar problems (e.g., [17]), and 2) several researchers have reported the

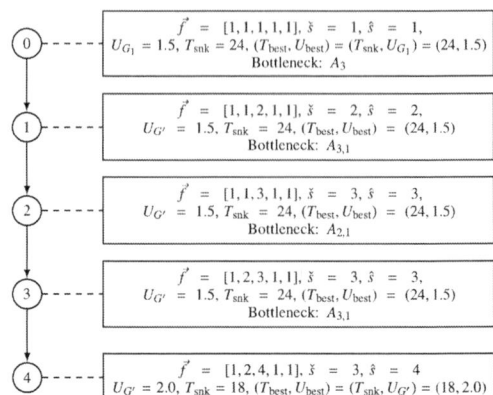

Figure 4: The list produced by the algorithm for G_1 in Figure 1(a) on 2 PEs.

optimal parameter settings for GA in the context of our problem (e.g., [19]). In particular, we compare our proposed algorithm with the NSGA-II genetic algorithm [4]. Specifically, we compare two metrics: 1) the total execution time needed by each algorithm to find a solution; and 2) the total code size of the returned solution.

We conducted all experiments on 11 real-life streaming applications from the StreamIt benchmarks suite [9]. The exact characteristics of the benchmarks are outlined in Table 2 in Appendix D. The experiments were performed on an Intel Core 2 Duo T9600 CPU running at 2.80 GHz with Linux Kubuntu 10.4.

7.1   Evaluating the Proposed Algorithm

First, we present the performance gain resulting from mapping the unfolded SDF graph, compared to mapping the initial SDF graph without unfolding. We do so by running the algorithm on the benchmarks and mapping each application on a number of PEs that varies from 2 up to 128 PEs. We evaluate the trade-off between the performance gain and total execution time by setting different quality factors $\rho \in \{0.8, 0.85, 0.9, 0.95\}$. To measure the performance gain, we compute, for each benchmark, the ratio between the sink actor period resulting from mapping the unfolded SDF graph, and the period resulting from mapping the initial SDF. This ratio is denoted by Ω and is given by $\Omega = (T_{snk}$ of $G')/(T_{snk}$ of $G)$, where G' is the unfolded graph, and G is the initial SDF graph. A lower value of Ω indicates a shorter sink actor period in the unfolded graph, and therefore, a higher throughput. In Figure 5(a), each vertical line shows the variations in Ω for all the benchmarks. The marker at the middle of each vertical line represents the Geometric Mean (GM) of Ω, while the upper and lower ends of the line represent the maximum and minimum values of Ω, respectively. It can be seen that mapping the unfolded SDF graphs of the benchmarks achieves significant performance improvement compared to mapping the initial SDF graphs of the benchmarks. As the number of PEs increases, the unfolded SDF graphs utilize the PEs much better than the initial SDF graphs. For example, on 64 and 128 PEs, mapping the unfolded SDF graphs with quality factor $\rho = 0.95$ achieves a GM of Ω equal to 0.2 and 0.1, respectively. The DCT benchmark benefits significantly from the algorithm and achieves a GM of Ω equal to 0.021 and 0.042 on 128 and 64 PEs, respectively. Even when a small number of PEs is available, the unfolded SDF graphs still achieve, with quality factor $\rho = 0.95$, a GM of Ω equal to 0.92 and 0.85 on 2 and 4 PEs, respectively.

During the experiment, we also find that the unfolding factor of an actor, obtained using our algorithm, is not necessarily equal to the number of PEs. For example, the obtained unfolded SDF graph of the Vocoder benchmark, when mapped onto 8 PEs, requires the RectangularToPolar actor in the initial SDF graph to be unfolded by a factor of 20. This confirms our statement in Sec. 5.

We also evaluate the total execution time of our algorithm, denoted by t_{ours}, when it is invoked on the benchmarks. Figure 5(b) shows the total execution time of our algorithm in seconds for all the benchmarks. For all benchmarks, our algorithm takes a GM of 6.07 seconds for 128 PEs with utilization ratio $\rho = 0.95$. The Serpent benchmark (the largest graph size with 120 actors) takes the longest running time (78.90 seconds), while the DCT benchmarks takes the shortest running time (1.09 seconds). As the quality factor ρ is decreased from 0.95 to 0.9, the GM of the running time drops to 2.49 seconds for 128 PEs. These results show clearly that our

1174

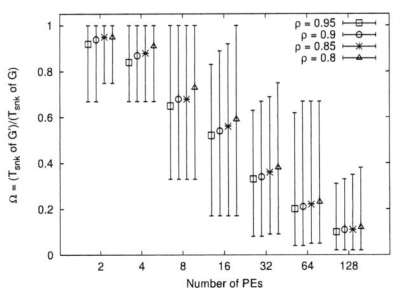

(a) Period ratio (lower is better)

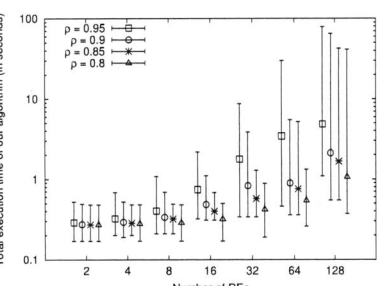

(b) Running time of our algorithm

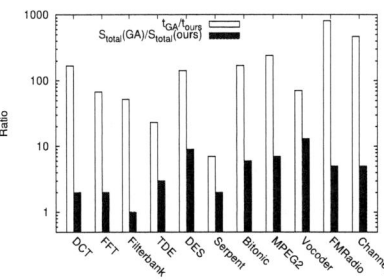

(c) The ratios of total execution time and total code size for the GA and our algorithm

Figure 5: (a and b) Results of evaluating our proposed algorithm and (c) comparing our algorithm vs. GA.

algorithm results, within a reasonable amount of time, in a large performance gain.

7.2 Comparison with Genetic Algorithm

To compare our algorithm with the GA-based heuristic, we perform the following steps. First, we run the GA to map each benchmark onto 64 PEs. It outputs an achievable period T and total utilization U_{GA}. Then, we run our algorithm to map the same benchmark onto 64 PEs with a termination criterion $U_{G'} \geq U_{GA}$. This criterion ensures a fair comparison since our algorithm runs till it finds the same or better solution in terms of the sink actor period and total utilization compared to the best solution found by the GA-based heuristic. Then, we compare two metrics: 1) the total execution time of each algorithm; and 2) the total code size resulting from the unfolding factors returned by each algorithm. The total code size is computed as $\sum_{A_{i,f} \in \mathcal{A}'} S_{i,f}$, where $S_{i,f}$ is the code size for actor $A_{i,f}$.

In this work, we use the NSGA-II implementation from the DEAP framework [6]. For the GA-based heuristic, each individual (also known as a chromosome) encodes a particular unfolding vector $\vec{f}$ of the initial SDF graph and the allocation of the replicas on m PEs. The structure of an individual is visualized in Figure 6. Basically, in an individual, each SDF actor A_i in the initial graph has $\hat{f}_i$ cells as derived in Eq. 3, indicating that A_i may have up to $\hat{f}_i$ replicas. Each cell may have a value varying from 0 up to m. A value of 0 denotes that the replica does not exist, while a value of 1 up to m denotes the PE on which the replica is allocated. Then, we formulate Problem 1 as a multi-objective optimization problem with two objectives. The first objective is to minimize the sink actor period, and the second one is to minimize the total code size of the unfolded graph. During the search, we use the evaluation function shown in Algorithm 3 in Appendix C. The GA outputs a set of Pareto points, for which we select the one with the shortest achievable period. In order to control the GA, we use the parameters reported in [19], because the target application domain and used platforms are similar to ours. The values of these parameters are given in Table 3 in Appendix D.

Figure 5(c) shows two ratios. The first ratio (shown in white bars) is the total execution time ratio given by t_{GA}/t_{ours}, where t_{GA} is the total time needed by the GA, and t_{ours} is the total time needed by our algorithm. The second ratio in Figure 5(c) (shown in black bars) is the total code size ratio given by $S_{total}(GA)/S_{total}(ours)$, where $S_{total}(GA)$ is the total code size of the solution obtained using the GA, and $S_{total}(ours)$ is the total code size of the solution obtained using our algorithm. Our algorithm is on average 104 times faster than the GA-based heuristic. For example, to unfold and map the FMRadio benchmark onto 64 PEs, our algorithm takes only 3 seconds, while the GA-based heuristic takes 2439 seconds. This means that our algorithm, for the FMRadio benchmark, is 813 times faster. We also see from Figure 5(c) that our algorithm results in less total code size compared to the GA-based heuristic. These results show clearly that our algorithm outperforms the GA-based heuristic in terms of: 1) the time needed to obtain the solution; and 2) the total code size of the obtained solution.

8. CONCLUSIONS

In this paper, we addressed the problem of exploiting just-enough parallelism when mapping a streaming application modeled using the SDF MoC in hard real-time systems. Exploiting just-enough parallelism is achieved by simultaneously unfolding and allocating

the SDF actors onto an MPSoC platform, while considering the number of available PEs and hard real-time scheduling of actors on the PEs. We showed that the solution space to our problem is bounded and subsequently derived its upper bound. We devised an efficient algorithm to solve the problem and evaluated the algorithm on a set of real-life applications. The experiments showed that our algorithm results a system specification with large performance gain. We also compared our algorithm with one of the state-of-the-art meta-heuristics, i.e., NSGA-II genetic algorithm, and showed that our algorithm is on average 100 times faster than the GA, while achieving the same quality of the solution.

9. ACKNOWLEDGMENT

This work is supported by the Dutch STW NEtherlands STreaming (NEST) project and CATRENE/MEDEA+ TSAR project.

10. REFERENCES

[1] EU FP-7 parMERASA project. http://www.parmerasa.eu/.
[2] M. Bamakhrama and T. Stefanov. Hard-real-time scheduling of data-dependent tasks in embedded streaming applications. In *Proc. EMSOFT*, 2011.
[3] G. Bilsen et al. Cyclo-static data flow. *IEEE Trans. Signal Process.*, 44:397–408, 1996.
[4] K. Deb et al. A fast and elitist multiobjective genetic algorithm: NSGA-II. *IEEE Trans. Evol. Comput.*, 6(2):182–197, 2002.
[5] S. M. Farhad et al. Orchestration by approximation: mapping stream programs onto multicore architectures. In *Proc. ASPLOS*, 2011.
[6] F.-A. Fortin et al. DEAP: Evolutionary algorithms made easy. *J. Mach. Learn. Res.*, 2171–2175(13), 2012.
[7] A. Gerstlauer et al. Electronic system-level synthesis methodologies. *IEEE Trans. Comput.-Aided Design Integr. Circuits Syst.*, 28(10):1517–1530, 2009.
[8] K. Goossens et al. Æthereal network on chip: concepts, architectures, and implementations. *IEEE Des. Test. Comput.*, 22(5):414 – 421, 2005.
[9] M. I. Gordon, W. Thies, and S. Amarasinghe. Exploiting coarse-grained task, data, and pipeline parallelism in stream programs. In *Proc. ASPLOS*, 2006.
[10] A. Hagiescu et al. A computing origami: folding streams in FPGAs. In *Proc. DAC*, 2009.
[11] G. Kahn. The semantics of a simple language for parallel programming. In *Proc. of IFIP Congress*. 1974.
[12] M. Kudlur and S. Mahlke. Orchestrating the execution of stream programs on multicore platforms. In *Proc. PLDI*, 2008.
[13] K. Lakshmanan, S. Kato, and R. Rajkumar. Scheduling parallel real-time tasks on multi-core processors. In *Proc. RTSS*, 2010.
[14] E. A. Lee and D. G. Messerschmitt. Static scheduling of synchronous data flow programs for digital signal processing. *IEEE Trans. Comput.*, 36:24–35, 1987.
[15] C. Liu and J. Layland. Scheduling algorithms for multiprogramming in a hard-real-time environment. *J. ACM*, 20(1):46–61, 1973.
[16] S. Martello and P. Toth. *Knapsack Problems: Algorithms and Computer Implementations*. John Wiley & Sons, 1 edition, 1990.
[17] A. Stulova et al. Throughput driven transformations of synchronous data flows for mapping to heterogeneous MPSoCs. In *ICSAMOS*, 2012.
[18] W. Thies and S. Amarasinghe. An empirical characterization of stream programs and its implications for language and compiler design. In *PACT*, 2010.
[19] M. Thompson. *Tools and techniques for efficient system-level design space exploration*. PhD thesis, University of Amsterdam, 2012.
[20] H. Yang and S. Ha. Pipelined data parallel task mapping/scheduling technique for MPSoC. In *Proc. DATE*, 2009.
[21] M. Yue. A simple proof of the inequality FFD (L) ≤11/9 OPT (L) + 1, ∀L for the FFD bin-packing algorithm. *Acta Mathematicae Applicatae Sinica*, 1991.
[22] J. T. Zhai, H. Nikolov, and T. Stefanov. Mapping of streaming applications considering alternative application specifications. *ACM Trans. Embed. Comput. Syst.*, 12:34:1–34:21, 2013.

APPENDIX

A. NOTATIONS

Table 1: Notations used in the paper.

$\mathbb{N}$	the set of natural numbers excluding zero
$\check{x}$	lower bound (minimum) of a value x
$\hat{x}$	upper bound (maximum) of a value x
lcm	least common multiple
$\lceil x \rceil$	smallest integer that is greater or equal to x
A_i	ith actor, where $1 \le i \le n$
C_i	worst-case execution time of the actor A_i (equivalent to μ_i in [2])
f_i	unfolding factor for actor A_i
G	a (C)SDF graph, $G = \{\mathcal{A}, \mathcal{E}\}$
m	number of PEs
n	number of actors in a (C)SDF graph
$r(A_i)$	repetition of actor A_i in one graph iteration
ρ	quality factor $\rho \in (0, 1]$
s	scaling factor for periods of all actors in a (C)SDF graph
S_i	code size of actor A_i
T_i	period of actor A_i (equivalent to λ_i in [2])
u_i	utilization factor of actor A_i, $u_i = \frac{C_i}{T_i}$
U_G	total utilization of (C)SDF graph G, $U_G = \sum_{A_i \in \mathcal{A}} C_i / T_i$
W_i	workload of actor A_i per graph iteration, $W_i = r(A_i) \cdot C_i$

B. PROOFS

PROOF. (of Lemma 1) Let U_{SA} be the utilization bound of a scheduling algorithm SA. If G is schedulable on $\check{m}$ PEs using SA and any AA, then this means that the total utilization of the actors on each PE j, where $1 \le j \le \check{m}$, is $U_{PE_j} \in (0, U_{SA})$. If we scale the periods of the actors in G by s, then this means that $U_{PE_j} \in (0, \frac{U_{SA}}{s}]$. Therefore, it is possible to combine the actors in every s PEs into 1 PE. Hence, the number of PEs needed after scaling the periods is $\lceil \frac{\check{m}}{s} \rceil$. □

PROOF. (of Theorem 1) Suppose that G' is the CSDF graph obtained by unfolding each actor A_i in the initial SDF graph G by $\hat{f}_i$. From Definition 2, it follows that every replica $A_{i,f}$ in G' has $\check{T}_{i,f} = C_{i,f} = C_i$. Therefore, we can re-write Eq. 2 as:

$$C_i = \frac{\text{lcm}(\vec{q'})}{r(A_{i,f})} \left\lceil \frac{\hat{W}_{G'}}{\text{lcm}(\vec{q'})} \right\rceil \tag{7}$$

where $r(A_{i,f})$ is the repetition of $A_{i,f}$ in G'. Eq. 7 can be re-written as:

$$r(A_{i,f})C_i = \text{lcm}(\vec{q'}) \left\lceil \frac{\hat{W}_{G'}}{\text{lcm}(\vec{q'})} \right\rceil \tag{8}$$

Since $\text{lcm}(\vec{q'})\lceil \hat{W}_{G'} / \text{lcm}(\vec{q'}) \rceil$ is constant, then we re-write Eq. 8 as:

$$r(A_{1,1})C_1 = r(A_{1,2})C_1 = ... = r(A_{1,f_1})C_1 = ... = r(A_{n,f_n})C_n \tag{9}$$

Now, we can write $r(A_{i,f}) = x_i \cdot r(A_i)$, where $r(A_i)$ is the repetition of A_i in the initial SDF graph and x_i is an integer factor. That is:

$$x_1 r(A_1)C_1 = x_2 r(A_2)C_2 = \cdots = x_n r(A_n)C_n \tag{10}$$

Eq. 10 can be re-written as:

$$x_1 W_1 = x_2 W_2 = \cdots = x_n W_n \tag{11}$$

where W_i is the workload of actor A_i according to Definition 1. The minimum solution to Eq. 11 is:

$$x_i = \text{lcm}\{W_1, W_2, \cdots, W_n\}/W_i \tag{12}$$

Since $r(A_{i,f}) = x_i r(A_i)$ and the graph is unfolded by $\vec{f}$, we can substitute this in Eq. 1 to get:

$$x_i r(A_i) = \frac{r(A_i)\,\text{lcm}(\vec{f})}{\hat{f}_i} \tag{13}$$

which can be re-written as:

$$x_i \hat{f}_i = \text{lcm}(\vec{f}) \tag{14}$$

Since $\text{lcm}(\vec{f})$ is constant, then Eq. 14 can be re-written as:

$$x_1 \hat{f}_1 = x_2 \hat{f}_2 = \cdots = x_n \hat{f}_n \tag{15}$$

The minimum solution to Eq. 15 is:

$$\hat{f}_i = \frac{\text{lcm}\{x_1, x_2, \cdots, x_n\}}{x_i} \tag{16}$$

□

C. ALGORITHMS

Algorithm 2: Unfolding an SDF graph.

Input: An SDF graph $G = \{\mathcal{A}, \mathcal{E}\}$ with a vector $\vec{f}$ of unfolding factors.
Result: The equivalent CSDF graph $G' = \{\mathcal{A}', \mathcal{E}'\}$

1   $\mathcal{A}' = \emptyset, \mathcal{E}' = \emptyset$;
2   **foreach** $A_i \in \mathcal{A}$ **do**
3   Add $f_i \in \vec{f}$ replicas of A_i to $\mathcal{A}'$;
4   Set repetition entry $r(A_{i,ii}) = \frac{r(A_i)\cdot\text{lcm}(\vec{f})}{f_i}, \forall ii \in [1, f_i]$;
5   **foreach** $E \in \mathcal{E}$ **do**
6   Get source actor A_i and sink actor A_j of edge E ;
7   Get production rate $prd(E)$ and consumption rate $cns(E)$;
8     $lcm\_pc = \text{lcm}(prd(E), cns(E))$;
9   **if** f_j is dividable by f_i **then** $OP = f_j/f_i$; $IP = 1$;
10   **else if** f_i is dividable by f_j **then** $IP = f_i/f_j$; $OP = 1$;
11   **else** $IP = f_i/f_j$; $OP = 1$;
12   **for** $ii = 1$ **to** f_i **do**
13   Add OP output ports to $A_{i,ii}$;
14   **for** $k = 1$ **to** OP **do**
15   Initialize a production sequence $\mathcal{P}_{i,ii}$ of length $r(A_{i,ii})$ to 0;
16       $\mathcal{P}_{i,ii}[p] = prd(E), \forall p \in [(k-1)\frac{lcm\_pc}{prd(E)} + 1, k\frac{lcm\_pc}{prd(E)}]$;
17   **if** f_j is dividable by f_i **then** $jj = (ii-1)OP + k$;
18   **else if** f_i is dividable by f_j **then** $jj = ii/IP$;
19   **else** $jj = k$;
20   Initialize a consumption sequence $C_{j,jj}$ of length $r(A_{j,jj})$ to 0;
21       $C_{j,jj}[c] = cns(E), \forall c \in [(ii-1)\frac{lcm\_pc}{cns(E)} + 1, ii\frac{lcm\_pc}{cns(E)}]$;
22   Create a new channel E' connecting replica $A_{i,ii}$ to replica $A_{j,jj}$;
23   Add channel E' to $\mathcal{E}'$;

24   Compact the production and consumption sequences of each actor in $\mathcal{A}'$;

Algorithm 3: Evaluation function in the GA-based meta-heuristic

Input: An individual to be evaluated
Result: An achievable period and total code size.

1   Check if the given individual is valid ;
2   **if** *the individual is invalid* **then return** $(-1, -1)$;
3   Build the vector of unfolding factors $\vec{f}$ from the individual ;
4   Generate the CSDF graph G' by unfolding G with $\vec{f}$ using Algorithm 2;
5   Compute the minimum achievable period $\check{T}_{i,f}$ of each actor $A_{i,f}$ using to Eq. 2 ;
6   Compute $\check{s}$ according to Eq. 5 ;
7   $s = \check{s}$;
8   **while** *true* **do**
9   Compute the period $T_{i,f}$ of each actor $A_{i,f}$ as $T_{i,f} = s\check{T}_{i,f}$;
10   **if** G' is schedulable on m PEs **then**
11   Compute total code size $S_{total} = \sum_{A_{i,f} \in \mathcal{A}'} S_{i,f}$;
12   Get the period T_{snk} of the sink actor in G' ;
13   **return** (T_{snk}, S_{total}) ;
14   **else**
15      $s = s + 1$;

D. EXPERIMENTS

Table 2: Benchmark characteristics.

Benchmark	Num. of Actors	Num. of Edges	Has Stateful Actors?
DCT	8	7	No
FFT	17	16	No
Filterbank	85	99	No
TDE	29	28	No
DES	53	60	No
Serpent	120	128	No
Bitonic	40	46	No
MPEG2	23	26	Yes
Vocoder	114	147	Yes
FMRadio	43	53	No
Channel	55	70	No

$A_{1,1}$	...	$A_{1,\hat{f}_1}$	...	$A_{n,1}$	...	$A_{n,\hat{f}_n}$
j	...	0	...	1	...	2

Figure 6: An example of an individual. The first replica of A_1 is allocated on the jth PE and the $\hat{f}_1$th replica of A_1 does not exist.

Table 3: Parameters for the genetic algorithm.

Parameter	Recommended value in [19]
Population size	80
Number of generations	300
Crossover rate	0.9
Mutation rate	0.05
Mating rate	0.1

On Robust Task-Accurate Performance Estimation

Yang Xu   Bo Wang   Ralph Hasholzner
Intel Mobile Communications
Munich, Germany
{yang.a.xu, bo1.wang,
ralph.hasholzner}@intel.com

Rafael Rosales   Jürgen Teich
University of Erlangen-Nuremberg
Erlangen, Germany
{rafael.rosales,
teich}@informatik.uni-erlangen.de

ABSTRACT

Task-accurate performance estimation methods are widely applied in early design phases to explore different architecture options. These methods rely on accurate annotations generated by software profiling or real measurements to guarantee accurate results. However, in practice, such accurate annotations are not available in early design phases due to lack of source code and hardware platform. Instead, estimated mean or worst-case annotations are usually used, which makes the final result inaccurate because of the errors induced by the estimations, especially for designs with tight time constraints. In this paper, we propose a novel methodology that combines Distributionally Robust Monte Carlo Simulation with task-accurate performance estimation method to guarantee robust system performance estimation in early design phases, i.e., determining the lower bound of the confidence level of fulfilling a specific time constraint. Instead of using accurate annotations, our method only uses estimated annotations in the form of intervals and it does not make any assumptions of the distribution types of these intervals.

Categories and Subject Descriptors: B.8.2 [Performance and Reliability]: Performance Analysis and Design Aids

General Terms: Performance

Keywords: task-accurate, robust performance estimation, distributionally robust Monte Carlo simulation

1. INTRODUCTION

In early System-on-Chip (SoC) design phases, to evaluate different system architectures, high-level performance estimation methods are usually applied during the exploration of the design space. Correct design decisions made in early design phases are very important for avoiding significant modification efforts and cost in later phases. Therefore, robust early performance estimations are required in order to take correct early design decisions. Additionally, owing to the increasing complexity of modern SoCs, the design space becomes very huge. Thus, fast performance estimation methods are mandatory to allow for an efficient design space exploration.

To achieve such a high efficiency during design space exploration, task-accurate performance estimation methods have been proposed [5, 13, 15]. In these methods, basic system operations, such as functions or communication transactions, are modeled as tasks. The execution time of each task is back-annotated into the task model for fast performance estimation through either simulation or model-based analysis. Since a task is the finest granularity in such kind of method, it is called

Permission to make digital or hard copies of all or part of this work for personal or classroom use is granted without fee provided that copies are not made or distributed for profit or commercial advantage and that copies bear this notice and the full citation on the first page. To copy otherwise, to republish, to post on servers or to redistribute to lists, requires prior specific permission and/or a fee.

Task-Accurate Performance Estimation (TAPE). In [5], the execution time of each task is modeled by annotating the delay budgets to the communication events. Then, a virtual processing unit is introduced to simulate the timing behavior of executing the tasks on the corresponding system. Thanks to its XML-based performance model, the design space exploration is significantly accelerated. Similarly, in [15], an action-accurate virtual processing component approach is proposed. It differentiates itself from [5] by applying an actor-oriented modeling approach [9] [10]. By strictly separating data flow from control flow within each actor, timing may be annotated individually to the actions instead of a full task, given a finer timing granularity and thus precision during timing analysis. A hybrid method is proposed in [13] where Worst-Case Execution Time (WCET) analysis is first performed and the resulting WCET values are back-annotated into task models to simulate system performance. Dynamic timing variations, such as caused by data dependencies, are also incorporated by proposing dynamic correction.

All these methods can simulate system performance very efficiently. However, none of them can guarantee a robust early performance estimation because in early design phases the annotation values these methods rely on are usually inaccurate. With the term robust estimation, we designate the estimation methods that enable a determination of the error and error bounds of a performance estimation, e.g., using confidence level. In early design phases where software and hardware might not yet be available for generating accurate annotation values, estimated timing information must be used to estimate system performance. This may make the final estimation result inaccurate and unreliable. Even though in early design phases relatively accurate results can still help the designers differentiate different architecture options, robust performance estimation is still mandatory when critical design decisions or tight time constraints are involved. For example, given two baseband processor architectures $arch1$ and $arch2$, the TAPE methods can only confirm that $arch1$ is faster than $arch2$ in processing Long Term Evolution (LTE) packets. But because of the inaccuracy of the performance estimation they cannot confirm whether these two architectures can finish the packet processing within 1 ms (real-time constraint of LTE) especially when the estimated time is too close to the real-time constraint. Therefore, it is highly desired to have a robust performance estimation method providing at least some confidence information, i.e., the probability of a architecture meeting a specific performance constraint, given errors existed in the estimation results.

A few low-level methods have been proposed to estimate performance with confidence information, e.g., [2, 4]. In [4], the software behavior is modeled by a sequence of virtual instructions with each one representing a type of real instructions. Then linear regression methods are used to determine a predictor equation, which estimates the resulting performance along with confidence levels. The drawback of this method is that the statistical predictor equation is only trained for a specific application population; whenever a new type of application is under evaluation, it needs a new dedicated equation. To overcome this limitation, the work in [2] proposes a trace-based estimation method. Instead of training a statistical equation for each application, for each virtual instruction it models the inaccuracy of its execution time as a statistical model, which is consequently accumulated to estimate the performance and the corresponding confidence intervals of a pre-generated instruction trace. It simply assumes that the statistical model follows Gaussian distribution. However, in practice, the knowledge of the exact distribution types of inaccuracy is not available, especially in early design phases. How to estimate the system performance with confidence levels without any exact statistical information of the inaccuracy still remains as a challenge.

In this paper, we propose a novel methodology that combines Distributionally Robust Monte Carlo Simulation (DRMCS) [1, 6–8] with task-accurate performance estimation method to guarantee robust system performance estimation in early design phases. The contributions of this paper are:

1. We model the TAPE result as a function of statistical variables whose exact distribution types may be unknown and we prove that DRMCS can be applied on TAPE to extract robust system performance estimation, i.e., performance estimation with worst-case confidence levels of fulfilling specific time constraints.

2. Instead of accurate timing annotations, which are usually unavailable in early design phases, our method only requires estimated annotations, which are specified by intervals;

3. Our method complements traditional TAPE methods by adding the capability to provide worst-case confidence information on meeting a performance constraint. This information may therefore guide designers to explore architecture options with respect to real-time constraints.

The rest of this paper is organized as follows: in Section 2 we introduce the motivation of our work and some basics of DRMCS. In Section 3 we present our robust task-accurate performance estimation method in details. Thereafter, experimental results are given in Section 4 and the paper is concluded in Section 5.

2. MOTIVATION & PRELIMINARIES

In this section, we will first present the motivation of our work. Then, we will introduce some preliminary knowledge of DRMCS, a known statistical method that will be used in our methodology.

2.1 Motivation

As mentioned previously, in order to predict the performance of a new design in early SoC design phases, estimated execution time of tasks must be used to execute TAPE simulations. These estimated values may be generated by extrapolation, taking measurements from previous products as a baseline and projecting the new execution time based on the differences between two generations and sometimes even based on the experience of the designers. Such estimated values are usually in the form of intervals, e.g., the decoding time of 1 byte of MP3 data is specified as $220\ us \pm 20\ us$ on a 312 MHz processor. The exact distribution type of such intervals are usually unknown. Since traditional TAPE typically relies on constant annotation values, mean values of such intervals, e.g., 220 us in this case, are used to simulate average system performance, which does not cover all the possible cases. Therefore, such kind of results cannot be used to guide designs with tight absolute time constraints. Additionally, corner-based worst-case execution time analysis is usually too pessimistic and will lead to over-engineering.

To solve this practical problem, we propose a statistical method that can make good use of the estimated intervals to generate robust system performance estimation. In this method, the execution time of each task is modeled as a mean value + a random variable representing the estimated intervals whose exact distribution types may be unknown. The expected results of this method are performance estimations with worst-case confidence levels of fulfilling specific performance constraints.

2.2 DRMCS

In statistics, such kind of problem has been tackled by proposing Distributionally Robust Monte Carlo Simulation (DRMCS) [1, 6–8]. DRMCS first evaluates whether the set of so-called good values of the random variables is of specific shapes, e.g., convex. In our context, an example of such good values would be those resulting in the fulfillment of a time constraint. If positive, then it can efficiently estimate the worst-case probabilistic performance metric such as the confidence level of meeting a time constraint, given the intervals and the class of distributions of the variables. The difference between DRMCS and the traditional Monte Carlo (MC) simulation is that instead of requiring the exact distribution types of the variables, DRMCS only needs the class of distributions of the variables and can still find the worst-case value of the probabilistic performance metric.

2.2.1 Notations
Here we first introduce some basic notations related to DRMCS.

1. Uncertain parameter space: considering a system with uncertain parameters $\mathbf{Q} = [q_1, q_2, ..., q_L]^T \in \mathbf{R}^L$, radius $\mathbf{r} = [r_1, r_2, ..., r_L]^T$, we define the uncertain parameter space as

$$Q_r \doteq \{\mathbf{Q} : |q_i| \leq r_i, i = 1, 2, ...L\} \qquad (1)$$

given bounds $|q_i| \leq r_i$ for $i = 1, 2, ..., L$. In our context, q_i correspond to the random variables representing the estimated intervals and r_i are the ranges of the intervals.

2. Probabilistic performance metric: we define the probabilistic performance metric as the probability of fulfilling a performance constraint γ

$$\Phi(f) \doteq Prob\{\phi(\mathbf{Q}^f) \leq \gamma\} \qquad (2)$$

where $\phi(\mathbf{Q})$ is the deterministic performance metric and $\mathbf{Q}^f$ means the joint probability function of $\mathbf{Q}$ is f. For example, the probability that the execution time is smaller than γ assuming the random variable vector $\mathbf{Q}$ has the statistical distribution f.

3. Good value space: we define the good value space as

$$Q_g \doteq \{\mathbf{Q} \in Q_r : \phi(\mathbf{Q}^f) \leq \gamma\} \qquad (3)$$

One example of Q_g in our context would be all timing annotation value samples making the performance constraint γ fulfilled.

4. Worst-case probabilistic performance metric is defined as

$$\hat{\Phi}(\hat{f}) \doteq \min_{\forall f \in \mathcal{F}} Prob\{\phi(\mathbf{Q}^f) \leq \gamma\} = \min_{\forall f \in \mathcal{F}} Prob\{\mathbf{Q}^f \in Q_g\}$$
$$(4)$$

where $\mathcal{F}$ is the class of distribution of the random variables and $\hat{f} \in \mathcal{F}$ is the distribution type that minimizes $\Phi(f)$. In our case, $\hat{\Phi}(\hat{f})$ is a worst-case confidence level of fulfilling a specific performance constraint.

2.2.2 Principles of DRMCS
The key point of DRMCS is that with respect to certain classes of distributions, the worst-case probabilistic performance metric $\hat{\Phi}(\hat{f})$ is obtained when $\hat{f}$ is the uniform distribution [1, 6–8]. Additionally, it turns out that different classes of distributions and the shape of Q_g, e.g., convex or unirectangular, also have a large impact on achieving $\hat{\Phi}(\hat{f})$. In the following, we will introduce the DRMCS principles for the most interesting class of distributions.

Class of Independent Symmetric Distributions $\mathcal{F}_r^{SI}$ requires 1) the probability density function (pdf) f of each variable must be symmetric and nonincreasing with $|q_i|$; 2) the uncertain parameters $q_i, i = 1, 2, ..., L$ must be independent, i.e., $f(\mathbf{Q}) = \prod_{i=1}^{L} f(q_i)$. Many distributions can fall into this class,

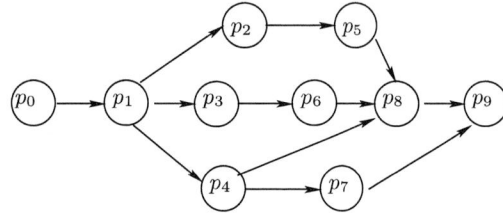

Figure 1: An example of task graph

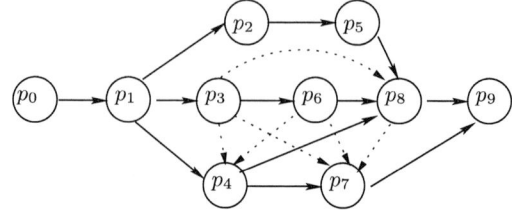

Figure 2: One example of serialized task graph

e.g., independent Gaussian, student-t and Triangular distributions.

For class $\mathcal{F}_r^{SI}$ DRMCS has two principles for robust estimation of $\hat{\Phi}(\hat{f})$. The first one is the *Truncation principle* [1]:

$$\inf_{f \in \mathcal{F}_r^{SI}} Prob\{\mathbf{Q}^f \in Q_g\} = \inf_{u^t \in U_r^T} Prob\{\mathbf{Q}^{u^t} \in Q_g\} \quad (5)$$

where $\mathcal{F}_r^{SI}$ represents the set of symmetric and independent distributions bounded by $\mathbf{r}$ and U_r^T is a set of truncated uniform distributions with a maximum bound $\mathbf{r}$. U_r^T is defined as

$$u_i^{t_i}(q) = \begin{cases} 1/2t_i & \text{if } q \in [-t_i, t_i] \\ 0 & \text{otherwise} \end{cases}$$

where $0 \le t_i \le r_i$. If $t_i \to 0$, then $u_i^{t_i}$ is the delta function. This principle indicates that the robust estimation of $\hat{\Phi}(\hat{f})$ for all $f \in \mathcal{F}_r^{SI}$ can be obtained by only evaluating the set of truncated uniform distributions U_r^T. In other words, we are guaranteed to obtain a robust estimation by executing many MC simulations on the set of truncated uniform distributions with maximum radius $\mathbf{r}$ and computing the worst-case value of $\Phi(f)$.

Running many MC simulations is time-consuming though. The *Uniformity principle* enables efficient robust estimation by constraining the shape of Q_g: if Q_g is convex or unirectangular, we have

$$\inf_{f \in \mathcal{F}_r^{SI}} Prob\{\mathbf{Q}^f \in Q_g\} = Prob\{\mathbf{Q}^{u^r} \in Q_g\} \quad (6)$$

This principle suggests that if the shape of Q_g is convex or unirectangular, a single MC simulation on the uniform distribution with the original bounds $\mathbf{r}$ ($t_i = r_i$) is sufficient for robust probabilistic performance metric estimation; the MC simulations with the bounds $t_i < r_i$ are no longer needed.

The other two classes of distributions are class of independent asymmetric distributions $\mathcal{F}_r^{AI}$ and class of dependent symmetric distributions $\mathcal{F}_r^{SD}$. The DRMCS principles for these two classes are out of scope of this paper. Refer to [1,6–8] for more details.

3. ROBUST TASK-ACCURATE PERFORMANCE ESTIMATION

In oder to use DRMCS to extract robust TAPE results, we need first to prove that the DRMCS principles are applicable to TAPE problems. In [17], the authors prove that the DRMCS method can be applied to estimate robust timing yield with partial statistical information on process variations by showing approaches for the three classes of distributions. In this paper, we will show that DRMCS can also be used to estimate robust TAPE results with estimated annotations, and hence provide robust performance estimation at the system level.

3.1 Task Graph Expression of TAPE Simulation

To prove the TAPE problem meets the conditions of DRMCS, we first transform the TAPE simulation into a task graph expression [16] [3] and then formulate the simulation result in a mathematical form, which can be directly evaluated against the DRMCS conditions.

A task graph is a directed acyclic graph with nodes representing tasks and edges representing data dependencies between

tasks, as shown in Fig. 1. Each task is an atomic operation in the system. It only starts its execution when all its input data are available and releases its outputs after the execution terminates. This level of abstraction is a perfect match for the specification of TAPE simulation, which is equivalent to a repeated execution of the functionality described by the task graph. The raw execution time of each task is determined by the hardware component it is mapped to. In TAPE, it usually exists in the form of annotation values and it is independent of the token data passed between tasks.

The TAPE result (e.g., total execution time) is equivalent to the latency of the *Longest Execution Path* (LEP) that crosses the first started task and the last terminated task. However, the LEP cannot be directly obtained from the task graph, since the task graph does not contain any resource sharing dependency information, which, in addition to data dependencies, also influences the LEP. For example, if task p_3 and task p_4 in Fig. 1 are mapped to the same hardware resource, e.g., a processor, even though there is no data dependency between them, they cannot be executed in parallel due to the resource sharing. This means in order to obtain the LEP, such kind of tasks must be serialized in the task graph. Therefore, we define a *Serialize operation* to incorporate the resource sharing dependency.

Serialize operation. *If task p_i and task p_j, $i \neq j$, are mapped to the same resource and there is no data dependency between them, they are serialized by adding a resource dependency edge (dot edge in Fig. 2) between them. The direction of the edge is determined by the scheduling algorithm assigned to the resource.*

In Fig. 2, we assume p_3, p_4, p_6, p_7, p_8 are mapped to the same resource. By applying the *serialize operation*, the LEP can be easily obtained from the serialized task graph, namely, $p_0 \to p_1 \to p_3 \to p_6 \to p_4 \to p_8 \to p_7 \to p_9$ assuming all the tasks have the same execution time. Note that some of the resource dependency edges are redundant and can be removed, e.g., the one from $p_3 \to p_8$.

After the task graph transformation, we can start to formulate the TAPE simulation result in a mathematical form. We model the execution time of each task as a mean value + a random variable, e.g., for task p_i we have

$$D_i = D_{i0} + a_i q_i, \quad i = 1, 2, ..., N \quad (7)$$

where D_{i0} is the mean value of the execution time of p_i, q_i represents the Errors Induced by Estimation (EIE), N is number of tasks in the system and a_i are constant coefficients.

For an arbitrary path k in the serialized task graph, the path delay is

$$D^k = \sum_{i \in path(k)} D_i = \sum_{i \in path(k)} D_{i0} + \sum_{i \in path(k)} a_i q_i = D_0^k + A^k \mathbf{Q} \quad (8)$$

where $\mathbf{Q} = [q_1, q_2, ..., q_M]^T$, $A^k = [a_1, a_2, ..., a_M]$, M is the number of tasks on path k.

Now, the delay of the LEP can be computed as

$$D_{LEP} = \max(D^1, D^2, ..., D^P) \quad (9)$$

where P is the number of paths in the serialized task graph. For a given time constraint D_t, the confidence level of fulfilling this constraint is defined as

$$CL(D_t) = Prob(D_{LEP} \le D_t) \quad (10)$$

1180

If the exact pdf of the EIE q_i is known, we can calculate the exact $CL(D_t)$ for a specific time constraint. However, in early design phases, we only have partial statistical information, such as the intervals of q_i. This is a similar problem as in DRMCS in Section 2. By applying the DRMCS method, we can obtain the worst-case confidence level $\hat{CL}(D_t)$ if the pdf of q_i belongs to a class of distributions $\mathcal{F}$

$$\hat{CL}(D_t) = \min_{\forall f \in \mathcal{F}} CL(D_t) \qquad (11)$$

where $\mathcal{F}$ is the class of distributions such as $\mathcal{F}_r^{SI}$, $\mathcal{F}_r^{AI}$ and $\mathcal{F}_r^{SD}$, as mentioned in Section 2. Note that the mathematical formulae are only used to evaluate against the DRMCS conditions. The performance is still estimated per simulation. In the following we will match the pdf of EIE to one of these classes and derive the principle for robust TAPE.

3.2 Robust Confidence Level for TAPE with EIE

We assume the random variables q_i representing the EIE are independent, symmetric and nonincreasing with $|q_i|$. This assumption is reasonable because 1) in early design phases, the execution time of each task is estimated independently so that the errors induced by estimation, the EIE, are also independent of each other; 2) the symmetry assumption suggests that the positive and negative deviations from the mean execution time are equally likely; 3) the small estimation errors are more likely than large errors. Therefore, they are nonincreasing with $|q_i|$. Based on this assumption, we can category the pdf of EIE as the $\mathcal{F}_r^{SI}$ class in Section 2.

For class $\mathcal{F}_r^{SI}$ the *Truncation principle* can be applied without any restrictions. To apply the *Uniformity principle* under class $\mathcal{F}_r^{SI}$ for efficient robust estimations, we need to prove that the Q_g of TAPE is convex or unirectangular (6).

Lemma 1. *The shape of Q_g of TAPE is convex with respect to EIE $\mathbf{Q}$ when the path delay is given by (8) and the LEP delay is given by (9).*

Proof. According to the definition of Q_g, the Q_g of TAPE is defined as $Q_g^{TAPE} = \{\mathbf{Q} \in Q_r : D_{LEP} \leq D_t\}$. In order to prove the shape of Q_g^{TAPE} is convex, we first prove that the shape of the Q_g of a single path is convex. Based on the definition of convexity [11], we only need to show that for any two EIE vector Q_i, Q_j, $i \neq j$, if these two vectors result in a fulfillment of the time constraint, the new vector $\alpha Q_i + (1-\alpha)Q_j$, $\alpha \in [0,1]$, also results in meeting the constraint.

According to (8), for any two vectors $Q_i \in Q_g$, $Q_j \in Q_g$, $i \neq j$, by definition of Q_g we have

$$\begin{aligned} D_0^k + A^k Q_i &\leq D_t \\ D_0^k + A^k Q_j &\leq D_t \quad \forall k = 1, 2, \dots P \end{aligned} \qquad (12)$$

where $Q_i = [0, 0, \dots q_i, \dots 0]^T$, $Q_j = [0, 0, \dots q_j, \dots 0]^T$, $i \neq j$; P is the number of paths in the system. This is equivalent to

$$\begin{aligned} A^k Q_i \leq D_t - D_0^k &\Leftrightarrow A^k \alpha Q_i \leq \alpha (D_t - D_0^k) \\ A^k Q_j \leq D_t - D_0^k &\Leftrightarrow A^k (1-\alpha) Q_j \leq (1-\alpha)(D_t - D_0^k) \end{aligned} \qquad (13)$$

When we sum up the two equations above, we obtain

$$\begin{aligned} A^k[\alpha Q_i + (1-\alpha)Q_j] &\leq (D_t - D_0^k) \\ D_0^k + A^k[\alpha Q_i + (1-\alpha)Q_j] &\leq D_t \quad \forall k = 1, 2, \dots P \end{aligned} \qquad (14)$$

which means $\alpha Q_i + (1-\alpha)Q_j \in Q_g$ and this proves that the shape of Q_g of a single path is convex. Because the delay of the LEP is the maximum delay of all the paths and the max operation preserves convexity [11], we can finally conclude that the shape of Q_g^{TAPE} is also convex. This completes the proof. $\square$

Now, we can apply the *Uniformity principle* for efficient robust TAPE with worst-case confidence levels, which requires

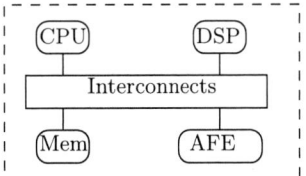

Figure 3: Simplified platform architecture

only a single MC simulation [1] with each random variable q_i uniformly distributed in $[-r_i, r_i]$. The confidence level of fulfilling a specific time constraint is obtained by calculating the probability of the total execution time less than the constraint. The DRMCS method guarantees that this confidence level is the worst-case one. Within each iteration of MC simulation, one traditional TAPE simulation with sampled annotation values is executed. Since a single TAPE simulation is very efficient and the number of tasks are usually quite small, usually less than 100, the whole simulation can be finished very fast, as shown in the experimental results. Additionally, Latin Hypercube Sampling (LHS) [12] can be used to further accelerate the simulation. Note that our method is not restricted to simulation-based performance estimation approach; it can be easily applied to analytic approach by substituting the TAPE simulation inside each MC iteration with model-based analysis.

3.3 Frequency Scaling Induced Dependency

Dynamic Voltage & Frequency Scaling (DVFS) is widely applied on modern SoC chips for dynamic power management. This technique is also modeled in some TAPE framework [18] to estimate power and performance jointly. Modeling DVFS in TAPE makes the robust TAPE even more complex, since it induces the frequency scaling dependency. For example, a task p is mapped to a CPU that supports two frequency settings, a fast mode (1 GHz) and a slow mode (500 MHz). In TAPE, two annotations may therefore be assigned to the same task p in the two speed modes, p_{fast} and p_{slow}, since it has different execution time in these modes. Apparently, the EIE for p_{fast} and p_{slow} are dependent on each other, e.g., if p_{fast} has a positive error, the p_{slow} should have a positive error, too. This frequency scaling induced dependency therefore violates our previous assumption of independent variables, which makes the *Uniformity principle* inapplicable. Fortunately, the EIE for p_{fast} and p_{slow} are fully correlated, i.e., if the execution time of p_{fast} is 1 ms, the counterpart of p_{slow} must be 2 ms due to the frequency scaling factor. Therefore, when we formulate the delay of a path in a mathematical form, the random variables for the same task in different frequency modes can be merged together [2] and the merged variable is still independent of the variables of other tasks. This makes sure that the *Uniformity principle* can still be used.

In our implementation, this dependency is incorporated by a *linked sampling* method. This means we only sample for the annotations of tasks in the fastest mode and take the sample values multiplied by a frequency scaling factor as the samples for a slower mode.

4. EXPERIMENTAL RESULTS

We applied the proposed methodology to a real world scenario, namely, MP3 decoding time estimation, to evaluate its effectiveness. We choose an actor-oriented TAPE framework [18] as the base of our robust TAPE simulation. To further accelerate the simulation, we applied LHS and parallelized MC simulation at the same time. The simulations were executed on a Linux machine with 4 GB RAM and a 3.4 GHz CPU, which has 4 cores and each core supports 2 threads executing in parallel.

---

[1] A single MC simulation means one MC simulation with a series of iterations.

[2] It is guaranteed that they are on the same path because they share the same resource.

Table 1: Annotations

Task	I			II					
	Mapping	T_{fast} (us)	s	Mapping		T_{fast} (us)		s	
				cpu-dsp	all-cpu	cpu-dsp	all-cpu	cpu-dsp	all-cpu
Src	-	0	-	-	-	0	0	-	-
PrePro	CPU	[86.4, 105.6]	2	CPU	CPU	[432.0, 528.0]	[432.0, 528.0]	2	2
Dec	CPU	[199.1, 243.3]	2	CPU	CPU	[995.4, 1216.6]	[995.4, 1216.6]	2	2
PostPro	DSP	[582.1, 711.4]	2.33	DSP	CPU	[2587.2, 3880.8]	[1509.2, 2263.8]	2.33	2
Play	AFE	20	-	AFE	AFE	20	20	-	-

4.1 Robust TAPE vs Traditional TAPE

We first compare our method with the traditional TAPE with worst-case annotations to show its advantages during design space exploration. We modeled a simplified embedded platform which consists of one CPU, one DSP, memory subsystem, interconnections and a dedicated hardware, as illustrated in Fig. 3. The AFE (Audio Front End) is an analog hardware responsible for decoded data playback. The functionality of MP3 decoding is modeled by a data flow graph as shown in Fig. 4, which is composed of five actors, namely, source (Src), pre-processing (PrePro), decoding (Dec), post-processing (PostPro) and Playback (Play). The MP3 decoding time is defined as the time between the data streaming into the PrePro stage and leaving the PostPro stage. Both CPU and DSP support two frequency settings, fast and slow. Whenever their computing load exceeds a specific threshold, they are switched to the fast mode automatically. The actor mappings, execution time intervals and frequency scaling factor $s = f_{fast}/f_{slow}$ are specified in TABLE 1, part I. We assume the uncertain radius $|r_i|$ is 10% of the mean execution time for all tasks. Since the Src models the data source, it has no real corresponding processing unit and its execution time is annotated as 0 us. Note that the communication latencies can be easily modeled by adding communication tasks between any pair of tasks and considering their execution time, too.

We first use our robust method to estimate the time of decoding 2 KB MP3 data. Then we run the traditional TAPE simulation with annotation values being the upper bound of the intervals (worst case). Fig. 5 shows us the Cumulative Distribution Function (CDF) of the decoding time estimated by our robust method (labeled as Robust), from which we can see that the estimation reaches 100% confidence level at around 28.2 ms. However, the execution time estimated by the traditional TAPE with worst-case annotations is 28.53 ms, which is apparently too pessimistic. This pessimism will become even worse if the $|r_i|$ are larger. Therefore, we can conclude that in spite of the unavoidable EIE of annotation values, our method can still provide more confident and realistic estimation than traditional mean and worst-case based TAPE.

In order to evaluate the robustness, we compare the CDF generated by our method with the ones generated by normal MC simulations with the distribution types of the intervals being Gaussian and Triangular. The comparison is also illustrated in Fig. 5, which confirms that both Gaussian and Triangular predict more optimistic confidence levels; our method guarantees the worst-case confidence level.

We also found that the confident decoding time cannot be simply obtained by adding a margin to the mean decoding time even though all the tasks have the same margin (10% in this case), because different tasks contribute differently to the total execution time, thus they have different EIE sensitivities. We analyze the sensitivities of the tasks by computing the Standardized Regression Coefficients for each task and found that PostPro has the highest sensitivity while PrePro has the lowest. To verify this, each time we artificially improve the EIE of one task by 5% and rerun the simulation. Then we compare the confidence level improvement of each task, as shown in Fig. 6.

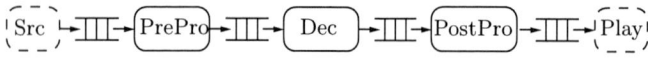

Figure 4: Actor-based MP3 decoding model

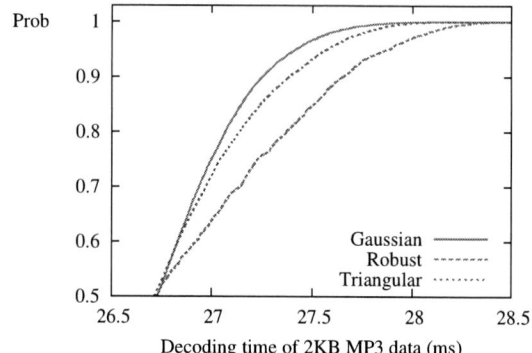

Figure 5: CDF of MP3 decoding time

The CDF marked with PrePro is nearly the same as the old CDF, which indicates the confidence level remains the same although the EIE of PrePro has been improved; while the CDF marked with PostPro is improved significantly. This suggests that in early design phases we can use our framework to analyze the sensitivity of tasks with respect to timing guarantees and improve the annotations of the tasks with high EIE sensitivity to effectively improve the confidence level of the estimations.

4.2 Architecture Exploration with Respect to Time Constraints

In the following we will describe a case study to demonstrate the effectiveness of our method to explore different architectures with respect to real-time constraints. In this case study, we would like to reduce the cost of the platform in Fig. 3. One architecture option would be to replace the processors with slower ones (5 times slower), noted as cpu-dsp. And another more aggressive option is to even get rid of the DSP by mapping all actors to the CPU, noted as all-cpu. Before making these design decisions, we need to evaluate whether these new architectures will meet a given latency, respectively throughput constraint for processing a chunk of 2 KB MP3 data.

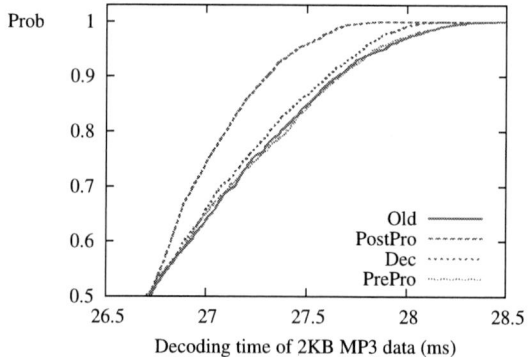

Figure 6: Confidence level refinement

Given a sampling frequency of 44.1 KHz and a compression

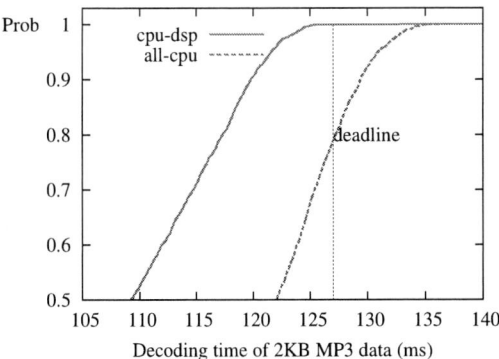

Figure 7: CDF of all-cpu and dsp-cpu

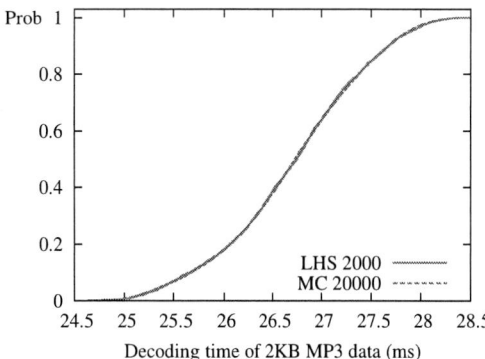

Figure 8: CDFs of LHS 2000 and MC 20000

ration of 1 : 11 for 2 KB MP3 data, the decoding must be finished within $(2048 \times 11)/4/44.1 = 127$ ms. The mappings, execution time intervals and frequency scaling factor s of these new architectures are specified in TABLE 1, part II.

Fig. 7 shows the CDFs of the decoding time for both cpu-dsp and all-cpu architectures. According to the mean decoding time estimated by the traditional TAPE (110.1 ms, 121.5 ms), both architectures meet the time constraint (127 ms). However, the confidence levels tell us that only the cpu-dsp architecture can meet the real-time constraint with nearly 100% confidence while the all-cpu architecture can meet the deadline with only about 80% confidence. Therefore, we can confirm that the cpu-dsp architecture can be applied to lower the cost and it would be very risky to use the all-cpu architecture (20% chance to miss the deadline). From this case study, we can conclude that applying traditional TAPE to guide designs with tight time constraints may lead to a constraint violation; in contrast our robust method is able to provide the worst-case confidence levels of fulfilling time constraints, which can be safely used to guide time constrained designs despite of the existence of EIE.

4.3 Simulation Time

It took a quad-core CPU 29.67 s to finish one robust estimation. Since there are no general rules for the sample size of LHS, we choose a conservative sample size, i.e., 2000 samples, for our simulation, which makes the ratio of sample size and variables (2000/3) much larger than the ones used in literature [14] (10-20). This conservative sample size guarantees convergence of the results, as shown in Fig. 8 where the CDF of LHS with 2000 samples matches very well with the CDF of MC simulation with 20000 samples. It also indicates that the simulation speed of our method still has a lot of improvement potential.

5. CONCLUSIONS

In this paper, we propose a novel robust task-accurate performance estimation method that only needs estimated annotation values to generate robust worst-case timing confidence levels, which can be used to guide the search of designs satisfying tight time constraints. The DRMCS method is applied within this method to guarantee an estimation with worst-case confidence level of fulfilling a specific time constraint. Experimental results confirm that our method can provide more reliable and realistic estimations than the traditional mean value based TAPE methods; in early design phases, our method can be used to win confidence in the fulfillment of time constraints during exploration of time-critical architecture and mapping decisions.

Acknowledgments

This work was supported in part by the Project PowerEval (funded by Bayerisches Wirtschafsministerium, support code IUK314/001).

6. REFERENCES

[1] B. Barmish and C. Lagoa. The Uniform Distribution: A Rigorous Justification for Its Use in Robustness Analysis. *Mathematics of Control, Si nals, an S stems MCSS* , 10(3):203–222, 1997.

[2] P. Bjuréus and A. Jantsch. Performance Analysis with Confidence Intervals for Embedded Software Processes. In *Procee in s of the 14th international s mposi m on S stems s nthesis*, pages 45–50, 2001.

[3] P. Eles, K. Kuchcinski, Z. Peng, A. Doboli, and P. Pop. Process Scheduling for Performance Estimation and Synthesis of Hardware/Software Systems. In *E romicro Conference, 1998 Procee in s 24th*, volume 1, pages 168–175. IEEE, 1998.

[4] P. Giusto, G. Martin, and E. Harcourt. Reliable Estimation of Execution Time of Embedded Software. In *DATE*, pages 580–589, 2001.

[5] T. Kempf, M. Doerper, R. Leupers, G. Ascheid, H. Meyr, T. Kogel, and B. Vanthournout. A Modular Simulation Framework for Spatial and Temporal Task Mapping onto Multi-Processor SoC Platforms. In *DATE*, pages 876–881, 2005.

[6] C. Lagoa. *Contri tions to The Theor of Pro a ilistic Ro stness*. PhD thesis, University of Wisconsin–Madison, 1998.

[7] C. Lagoa. Probabilistic Enhancement of Classical Robustness Margins: A Class of Nonsymmetric Distributions. *A tomatic Control, IEEE Transactions on*, 48(11):1990–1994, 2003.

[8] C. Lagoa and B. Barmish. Distributionally Robust Monte Carlo Simulation: A Tutorial Survey. In *Procee in s of the I AC Worl Con ress*, pages 1–12, 2002.

[9] E. Lee, S. Neuendorffer, and M. Wirthlin. Actor-Oriented Design of Embedded Hardware and Software Systems. *o rnal of Circ its S stems an Comp ters*, 12(3):231–260, 2003.

[10] E. A. Lee and S. Neuendorffer. Actor-Oriented Models for Codesign: Balancing Re-Use and Performance. *ormal Metho s an Mo els for S stem Desi n*, pages 33–56, 2004.

[11] R. Rockafellar. *Conve Anal sis*, volume 28. Princeton University Press, 1996.

[12] T. Santner, B. Williams, and W. Notz. *The Desi n an Anal sis of Comp ter E periments*. Springer, 2003.

[13] J. Schnerr, O. Bringmann, A. Viehl, and W. Rosenstiel. High-Performance Timing Simulation of Embedded Software. In *DAC*, pages 290–295, 2008.

[14] M. Stein. Large Sample Properties of Simulations Using Latin Hypercube Sampling. *Technometrics*, 29(2):143–151, 1987.

[15] M. Streubühr, J. Gladigau, C. Haubelt, and J. Teich. Efficient Approximately-Timed Performance Modeling for Architectural Exploration of MPSoCs. In *A vances in Desi n Metho s from Mo elin an a es for Em e e S stems an SoC's*, volume 63, pages 59–72. 2010.

[16] W. Wolf. An Architectural Co-Synthesis Algorithm for Distributed, Embedded Computing Systems. *TV SI*, 5(2):218–229, 1997.

[17] L. Xie and A. Davoodi. Robust Estimation of Timing Yield with Partial Statistical Information on Process Variations. *TCAD*, 27(12):2264–2276, 2008.

[18] Y. Xu, R. Rosales, B. Wang, M. Streubühr, R. Hasholzner, C. Haubelt, and J. Teich. A Very Fast and Quasi-Accurate Power-State-Based System-Level Power Modeling Methodology. In *ARCS'12*, pages 37–49. Springer-Verlag.

Stochastic Response-Time Guarantee for Non-Preemptive, Fixed-Priority Scheduling Under Errors

Philip Axer, Rolf Ernst
Institut für Datentechnik
TU Braunschweig
Germany
{axer, ernst}@tu-bs.de

ABSTRACT

Error recovery mechanisms, such as automatic repeat request (ARQ) for e.g. the CAN protocol, are a crucial part of safety critical embedded systems. These can have a strong impact on the timing behavior of the system and an unpropitious combination of error events may cause a real-time application to miss deadlines with potentially hazardous consequences. Therefore, formal analysis of the worst-case timing including errors is indispensable for certification. We present a new convolution-based stochastic analysis in which we model errors as additional execution time to bound the probability for an activation to exceed a response-time value in the worst-case.

Categories and Subject Descriptors

C.4 [**Performance of Systems**]: fault tolerance; C.3 [**Special-purpose and application-based systems**]: Real-time and embedded systems

General Terms

Real-Time, Fault Tolerance, Embedded Systems

1. INTRODUCTION

Embedded systems as used in the automotive and avionics industry are to a great extent time critical systems. This means that the correctness of the system behavior depends not only on the logical result, but also on the time at which the result is produced [11]. To ensure correct timing, the used scheduling policy must be able to handle such real-time applications. Due to its simplicity and proven field-application, fixed-priority scheduling is commonly used. It is used in the OSEK operating system, Controller Area Network (CAN) [15] and others.

Many real-time applications are used in a safety-critical context (e.g. active steering in a car) and for any certifiable, safety-critical system it is crucial to follow safety standards.

Permission to make digital or hard copies of all or part of this work for personal or classroom use is granted without fee provided that copies are not made or distributed for profit or commercial advantage and that copies bear this notice and the full citation on the first page. To copy otherwise, to republish, to post on servers or to redistribute to lists, requires prior specific permission and/or a fee.
DAC '13 May 29 - June 07 2013, Austin, TX, USA

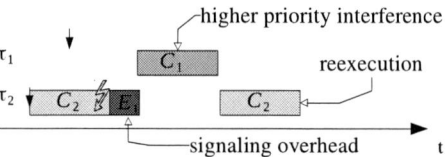

Figure 1: Response time for task τ_2 is increased caused by error handling and additional higher priority load.

Safety standards such as the industrial-oriented IEC-61508 [9] or the automotive domain specific ISO-26262 [10] pose very specific requirements on the reliability of the used infrastructure. In this context, the consideration of errors and their effects is key for any certifiable safety-critical system. Most systems which use fault-tolerance mechanisms exhibit additional overhead in case of errors, such as signaling overhead and recovery mechanisms (e.g. an automatic repeat request (ARQ) mechanism that triggers a retransmission).

The major concern in the context of real-time systems is that latency is severely affected by recovery actions caused by error-events (cf. Figure 1) compared to an error-free case. Thus, for hard real-time systems which operate in an environment under heavy electromagnetic interference (EMI), such as electric cars, the transmission latency of CAN, which is predicted by traditional formal approaches assuming the absence of errors, may not be valid.

The key problem is that the interference caused by errors (i.e. the total recovery overhead) cannot be bounded conservatively because there is no upper bound for the amount of errors due to their non-deterministic nature. It is possible but unlikely to have an arbitrary large amount of errors in a very small interval of time, leading to infeasible real-time constraints.

In this paper we present an algorithm to obtain probabilistic guarantees on worst-case response times under errors. We derive an upper bound on the probability that a response-time is exceeded under worst-cast conditions. Hence, the probability that a given activation in the mission time exceeds the response-time is lower than our bound.

The approach is applicable to any non-preemptive fixed priority scheduling policy under the effects of random error events. Contrary to comparable approaches [3] the probabilistic bound that our approach yields is by orders of magnitude tighter.

The rest of the paper is structured as follows: After summarizing related work in section 2, the system model is presented in section 3. An in-depth discussion on the stochastic schedulability analysis is given in section 4 followed by an experiment in section 5 where we apply the algorithm to the CAN protocol. Finally, we conclude the paper in section 6. In the supplemental material we provide pseudo-code for the algorithm, further proofs and an additional experiment.

2. RELATED WORK

Classical response-time analyses are available from real-time research for a large variety of different scheduling policies. They can be directly applied to fault-free analysis. For example, when computing the worst-case response time of a task, which is the largest time from its release until completion, one can rely on the *busy window* technique [12, 19].

In order to include effects of errors (e.g. recovery overhead) the busy window approach was extended in [13], to include error overhead which is modeled as higher priority load with a known minimum inter-arrival time T_F. A reliability value was derived in [4] by calculating the probability that the distance between two error events is never smaller than T_F over a given mission time.

In [2], a tree-based approach is presented where different error scenarios are evaluated iteratively. Here, an error scenario is a certain pattern of errors in the busy window. In a second step, these scenarios are translated to probabilities. The tree-based algorithm was superseded by a simpler, more accurate approach [3]. This algorithm was extended to arbitrary deadlines and arbitrary event models in [1].

By construction, the approaches [1, 3] induce additional pessimism because recovery overhead is always assumed to be the worst case recovery overhead among all higher priority tasks.

Instead of considering only the worst-case, more instances of a task can be considered, leading to tighter results [17], however, this only works for synchronized, periodic tasks.

The problem of obtaining the response time under the probabilistic effect of errors is very similar to effects known from variable execution time research where execution times are modeled by probability mass functions (pmf). In [7] an algorithm was presented to derive stochastic response times for individual activations of a given trace (e.g. periodic). Based on this methodology, an average response-time distribution can be derived [6]. As we will show in the following sections, we can adopt the underlying idea.

3. SYSTEM AND SCHEDULING MODEL

We consider systems consisting of a set of tasks Γ mapped to a single resource such as a communication bus or a processor. Each task $\tau = (C, p, \delta)$ is characterized by its worst-case execution time (or transmission time) C, a priority p and an event-model δ describing the activation pattern of τ. Fixed priority scheduling is based on the priority of a task, if multiple tasks are ready to execute, the one with the highest priority is admitted. Once a task is running, it cannot be preempted by a higher priority task. In case of an error event, the error-signaling mechanism causes additional execution time overhead and eventually triggers further recovery operations. We call the worst-case error-signaling overhead E. Throughout the paper, we assume re-execution recovery for erroneous executions. That is a re-execution of a task takes at most C time (cf. Figure 1). It is assumed that after the error-signaling a re-arbitration process takes place. That means previously released higher priority load is admitted before the recovery takes place. Note that an error can occur at any time, e.g. an error event occurring during a re-execution causes the re-execution to fail and leads to

another error-signaling and a re-execution.

Throughout the paper, we denote τ_i as the i-th task and $\tau_{i,j}$ as the j-th instance of the i-th task.

3.1 Event Models

The dataflow into a task is modeled with the help of *event models*. Event models abstract the activation of tasks from an actual trace by representing only the worst-case (best-case) behavior.

Following this concept $\eta^+(\Delta t)$ ($\eta^-(\Delta t)$) describe the maximum (minimum) number of events which can arrive during any given time window Δt at the input and be queued for processing. Thus, a given event model comprised of η^+ and η^- is always a conservative approximation for any actual event trace that is smaller (larger) than η^+, (η^-).

An alternative representation is the notion of a minimum (maximum) distance between n subsequent events $\delta^-(n)$ ($\delta^+(n)$). As shown in [16], both representations η and δ are pseudoinverse and can be converted to each other. Standard event models [14] such as periodic with jitter, sporadic and others can be expressed in this way. The δ^- function for a bursty input with a given period $\mathcal{P}$, jitter $\mathcal{J}$ and minimal distance d^{min} between any two events is defined as:

$$\delta^-(n) = max((n-1)d^{min}, (n-1)\mathcal{P} - \mathcal{J}) \qquad (1)$$

3.2 Fault and Error Model

Similar to related work (e.g. [3]), we use the single-bit-error model for the following analysis. That is, the probability that a bit is affected by an error is constant and independent of previous errors and time. The occurrence of error events is modeled using a Poisson process with a given error rate λ. According to [8], typical bit error-rates for CAN in harsh environments are in the order of 10^{-5}.

For the Poisson model, the following equations give the probability that exactly m error events occur (i.e. bit-flips), the probability that no error at all occurs and the converse probability that at least one error occurs during a given time interval Δt.

$$P(m, \Delta t) = \frac{e^{-\lambda \Delta t}(\lambda \Delta t)^m}{m!} \qquad (2)$$

$$P^{ok}(\Delta t) = e^{-\lambda_i \Delta t} \qquad (3)$$

$$P^{errors}(\Delta t) = 1 - e^{-\lambda_i \Delta t} \qquad (4)$$

4. STOCHASTIC ANALYSIS

4.1 Problem Statement

The goal of the following steps is to derive the worst-case exceedance function for a task τ_i. This function gives an upper bound on the probability that an activation $\tau_{i,j}$ exceeds a response-time value t.

$$X_i^+(t) \geq \max_{\forall j} P[R_{i,j} > t] \qquad (5)$$

The response time $R_{i,j}$ is the time interval from the activation of a job $\tau_{i,j}$ until the it has fully been processed. Practically, if we pick an activation by random, $X_i^+(t)$ gives an upper bound on the probability that it's response time will exceed a threshold t. The worst-case exceedance function is a conservative bound for each activation in isolation which is seen over the mission time. Note that this work significantly differs from the work in [6] where the average-case distributions of all activations is derived. However, we apply a similar methodology.

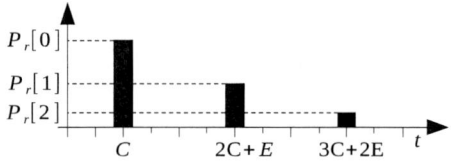

Figure 2: Worst-Case execution time pmf $e(t)$.

The following stochastic error analysis consists of two steps. First, the task and error model are transformed into an equivalent variable execution time problem by using a probability mass function (pmf) instead of a single worst case execution time. In a second step, similar to [7], we iteratively apply *splitting, convolution* and *merging* of the pmfs to calculate conservative stochastic response-time bounds.

4.2 From Errors to Variable Execution Times

We model the variable execution time overhead caused by error events by using execution time pmfs. This approach assesses the overhead in a fine grained fashion, rather than using the worst-case error overhead among all higher priority tasks. This is possible because the Poisson error model is memoryless, hence it allows us to examine the execution time behavior of tasks in isolation. For the following steps it is necessary to distinguish between error-events (i.e. the bit flip) and the manifestation (i.e. recovery operation consisting of error signaling and re-transmission).

LEMMA 1. *Considering the execution of one job $\tau_{i,j}$, a lower bound on the probability for exactly $k \geq 0$ recovery operations is given by:*

$$P_r[k] = \begin{cases} P^{ok}(C) & \text{if } k=0 \\ P^{errors}(C) \cdot (P^{errors}(C+E))^{k-1} \cdot P^{ok}(C+E) & \text{if } k>0 \end{cases}$$

$$(6)$$

We can obtain a lower bound on the probability that we observe less or equal than k recoveries by summation $\sum_{i=0}^{k} P_r[i]$. In that sense, a lower bound for $P_r[k]$ is conservative because this implies that the likelihood $(1 - P_r[k])$ for more than k recoveries is higher, which cause even more recovery overhead.

Now we need to bound the time which is spend by $\tau_{i,j}$ for handling k recoveries. By using the worst-case execution time C and the worst-case error signaling overhead E, we can bound the execution time for error situations with exactly k recoveries by $C + k(C + E)$. By combining the overhead for k recoveries with the probability, we retrieve an execution-time probability mass function (pmf).

DEFINITION 1 (WORST-CASE EXECUTION TIME PMF). *The worst-case execution time pmf $e(t)$ is given by the probability that the overall execution time of a task including errors and recoveries is of length t.*

LEMMA 2. *The worst-case execution time pmf $e(t)$ including error-signaling and recovery operations is given by:*

$$e(t) = \begin{cases} P_r[k] & \text{if } t = C + k(C+E), k \geq 0 \\ 0 & \text{otherwise} \end{cases}$$

$$(7)$$

PROOF. The proof is straightforward. An upper bound for the execution time with exactly k recoveries is given by $C + k(C + E)$. The probability that a k-recovery situation will occur is bounded by $P_r[k]$ (Equation 6). □

Based on the lower-bound notion observed for $P_r[k]$, we can conclude that $\sum_{x=0}^{t} e(x)$ is a lower bound that the observed execution time (including error handling and recovery) of job $\tau_{i,j}$ is less or equal than t. An example for the worst-case execution time pmf is shown in Figure 2. Here the probability for the task executing for time C is given by $e(C)$ which corresponds to the error-free case, i.e. the case where no recoveries are observed.

4.3 Stochastic Busy Window

To determine the worst-case response-time we use the concept of the busy window. In fixed-priority scheduling, a level-i busy window of a task τ_i is a time interval during which the resource is busy processing instances of task τ_i as well as tasks of higher priority tasks. In the non-stochastic case it was shown that the *critical instant* assumption - all activations arrive as early as possible - leads to the worst-case busy window [12]. The worst-case response time can then be found among all activations in the worst-case busy-window.

As for the non-stochastic busy window approach, the critical instant assumption remains valid also for the case where the execution time is a random variable. The workload is maximized if all arrivals arrive as early as possible (independent of their execution time). The remaining question is, what is the likelihood that a worst-case busy window of length t will occur under errors.

DEFINITION 2 (LEVEL-i BUSY-WINDOW PMF). *The level-i busy-window pmf $\omega_i(t)$ is the probability that the critical instant initiates a level-i busy-window of length t. That is, time t after the critical instant there exists no ready workload from any task with priority higher or equal than τ_i.*

In the following steps we explain, how to determine the worst-case level-i busy-window pmf.

The idea can be sketched informally as follows: We iterate over the activations released after the critical instant in the order of their earliest release, pick up their execution time and "add" them up. After a sufficient amount of activations has been considered, we know the stochastic busy window. A sufficient amount has been considered, if the probability of new activations falling in the busy window is very small.

We assume job and release-time indices are ordered in increasing release times from the start of the critical instant. That is, $\Delta_i(n)$ denotes the release time of the n-th job among all jobs of higher and same priority than τ_i. In case release times are equal, activations are ordered according to their priority and their event ordering (activations of the same task cannot pass each other). Figure 3 shows an example. τ_0 has a high priority and a periodic with jitter event model $P=11$, $J=4$. Task τ_1 has a low priority and a periodic event model $P=15$. Arrows denote the earliest release times of the associated event, j gives the index w.r.t. to its release time and priority

$$\Delta_i(n) = \inf_{\Delta t \geq 0} \{\Delta t \mid \sum_{\forall j \in hp(i)} \eta_j^+(\Delta t) \geq n\} \quad (8)$$

Accordingly, for the rest of the paper $e_n(t)$ is the execution time pmf of the n-th activation.

In the following algorithm, we use the fact that the pmf of the sum of execution time variables C_0 and C_1 can be calculated by convolution, assuming they are independent.

$$P[C_0 + C_1 = t] = (e_0 * e_1)(t) \quad (9)$$

1186

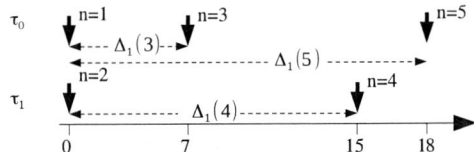

Figure 3: Critical-instant scenario of task τ_0 and τ_1. All activations are numbered according to their release time (i.e. $n=5$ is the 5-the activation released at time $\Delta_1(5)$.)

where

$$(e_0 * e_1)(t) = \sum_{\tau=-\infty}^{\infty} e_0(t-\tau)e_1(\tau) \tag{10}$$

Independence is a valid assumption for the consideration of error effects under the Poisson model.

Before we formalize the approach, we sketch the algorithm by using the illustrative Gantt-chart from Figure 3 and derive the stochastic busy window for task τ_1. This process is shown step by step in Figure 4. By using eq. 7, the execution time pmfs of both tasks can be calculated. For τ_0 and τ_1 they are shown in Figure 4 a) on the left. The busy-window pmf $\omega_1(t)$ is then retrieved incrementally by iterating over all activations of both tasks in order of their releases.

Like for the non-stochastic analysis, we start with a busy window of length zero. Hence, the initial busy-window pmf $\omega_1^0(t)$ has a peak at $t=0$ as shown in Figure 4 a) on the right. The first activation $n=1$ arrives at $t=0$ earliest which will increase the busy-window pmf by "adding" its execution time pmf to $\omega_1^0(t)$. Activation $n=1$ is associated with task τ_0 (cf. Figure 3). The pmf of the sum of the busy-window plus execution time $e_0(t)$ of the new activation can be calculated by using eq. 9.

The resulting busy-window pmf $\omega_1^1(t)$ is shown in Figure 4 b). The second activation $n=2$ arrives at $t=0$ earliest. We can see in Figure 3 that activation $n=2$ is associated with task τ_1, and adds the execution-time pmf of task τ_1 to the busy-window pmf, thus we apply eq. 9 a second time to get $\omega_1^2(t)$. The busy-window pmf now includes the first two activations.

The third activation $n=3$ at $t=7$ resembles the most interesting case. It depends on the previously seen workload if the activation falls in the busy window:

1. If the busy window is smaller than $\Delta_1(3)$ (i.e. <7), the third activation does not contribute to the busy window at all, since it arrives after the window ended.

2. If the busy window is larger than $\Delta_1(3)$ (i.e. ≥ 7), the third activation will further increase the length of the window.

The first case implies that due to causality, activation 3 cannot alter the probability that a busy window is smaller than $\Delta_1(3)$. Thus, we call the intervall $[0, \Delta_1(3)]$ of $\omega_1^2(t)$ *stable* (cf. bottom left in Figure 4).

In the other cases where the busy window is larger than $\Delta_1(3)$, the execution time of activation 3 must be added to the busy window. Practically, this is achieved by applying eq. 9 only to the tail of $\omega_1^2(t)$ as shown in the bottom part of Figure 4 d). Thus, only busy windows which are larger than $\Delta_1(3)$ are increased by the execution time of $n=3$. The bottom of Figure 4 d) shows how the unstable part of the busy-window is convolved with the execution time pmf of $n=3$ and forms the new tail. After the new tail is

merged with the stable part, we have computed $\omega_1^3(t)$. The probability that the 4-th event at time $t=15$ falls into a busy window is sufficiently small, so we stop the iterative process.

To formalize the process, we assume we knew $\omega_i^n(t)$ which includes the first n events. We will now show how to derive $\omega_i^{n+1}(t)$ which includes the first $n+1$ events. The $n+1$-th event will obviously add additional workload to $\omega_i^n(t)$, but in what way does it alter the busy-window? Therefore, we first introduce the notion of backlog.

DEFINITION 3 (RELATIVE BACKLOG PMF). *We define the relative backlog pmf $b_i^n(t)$ as the probability that the accumulated workload of activations of higher or the same priority than τ_i released before activation n is time t.*

LEMMA 3. *The relative backlog pmf of activation $n+1$ can be calculated using the stochastic busy-window, since it already includes all activations released prior to $n+1$. The relative backlog pmf $b^{n+1}(t)$ of activation $n+1$ is given by the tail of $\omega_i^n(t)$.*

$$b_i^{n+1}(t) = \begin{cases} \omega_i^n(t + \Delta_i(n+1)) & \text{if } t \geq 0 \\ 0 & \text{otherwise} \end{cases} \tag{11}$$

PROOF. The proof is given in [6]. □

Note that the relative backlog pmf does not necessarily sum up to one. This is because the probability that a new activation $n+1$ falls in a previously busy window can be smaller than one (i.e. $n+1$ only falls in the busy window if a certain minimum of recoveries was observed).

This operation is called *splitting*. It divides the busy window pmf in two parts, a *stable* and an *unstable* part. Due to causality, the new activation $n+1$ cannot alter the busy window pmf in the stable interval $[0, \Delta_i(n+1)]$. Vice-versa only the unstable part of the busy window pmf is altered by activation $n+1$.

In the next step, the relative backlog of the busy-window pmf and the workload originating from the new event $n+1$ have to be added.

The resulting backlog which includes the old backlog plus additional execution time $e_{n+1}(t)$ is called $\tilde{b}_i(n+1)(t)$. It is calculated by applying eq. 9:

$$\tilde{b}_i^{n+1}(t) = \left(b_i^{n+1} * e_{n+1}\right)(t) \tag{12}$$

The busy-window function ω_i^{n+1} which includes the first $n+1$ events is generated by *merging* the stable busy-window interval and the new busy-window tail:

$$\omega_i^{n+1}(t) = \begin{cases} \omega_i^n(t) & \text{if } t < \Delta_i(n+1) \\ \tilde{b}_i^{n+1}(t) & \text{otherwise} \end{cases} \tag{13}$$

The procedure for the next event $n+2$ consists of the same three steps: *split* (eq. 11), *convolution* (eq. 12) and *merge* (eq. 13). The algorithm terminates if the probability that a new event falls in the busy window is below a threshold. This threshold is then a bound of the probability that a larger busy window is seen and can be used as a confidence value for the analysis.

In the previous steps we have elaborated how to derive $\omega_i^{n+1}(t)$ from $\omega_i^n(t)$. For the starting value of the iterative process, we assume that the busy-window size is the length of the lower priority blocker as described in [5]. This allows us to start the algorithm with the following initial busy window:

$$\omega_i^0(t) = \begin{cases} 1 & \text{if } t = B_i \\ 0 & \text{otherwise} \end{cases} \tag{14}$$

1187

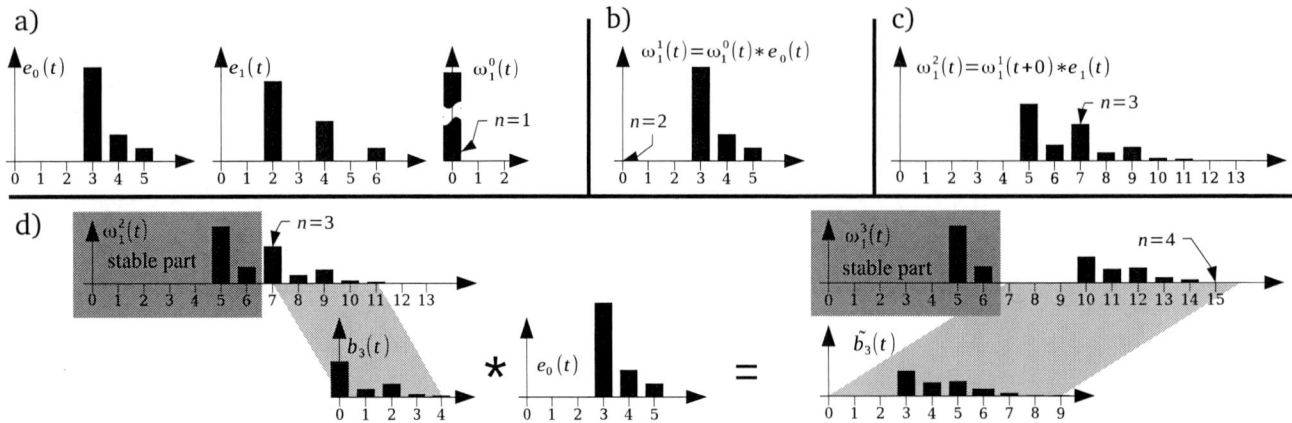

Figure 4: Illustrative example for constructing the busy-window pmf ω_1.

with

$$B_i = \max_{\forall j \in lp(i)} C_j \qquad (15)$$

As explained previously, the iterative process is stopped if the probability that a new event falls in the busy window is sufficiently small or the stable part of the busy-window pmf is sufficiently long.

4.4 Stochastic Queuing Delay and Response Time

Similar to the error-free case, the response time can be obtained by checking all activations in the busy-window and find the one with the largest response-time. For the stochastic case this means we need to evaluate the response-time distribution for the first q^+ events and find a worst-case among them. We will later elaborate on q^+.

Analogous to the non-stochastic/error-free case, a task τ_i can start executing once all previously released activations of τ_i and other higher priority tasks have executed. Once τ_i starts executing, it cannot be interrupted. Additionally, in the error-case we must also wait until all higher and same priority execution attempts and their recoveries have finished.

The total time the q-th activation is queued is called the *queuing delay*. Similarly to the busy-window we introduce a probabilistic version.

DEFINITION 4 (QUEUING DELAY PMF). *The queuing delay pmf $\omega_{q,i}(t)$ is defined as the probability that the q-th activation is queued for time t and fully transmits without errors right after.*

The definition implies that all erroneous transmission and recovery attempts of the q-th activation itself have finished before time t, so that the response time pmf of the q-th activation can be calculated as the queuing delay pmf shifted by the execution time.

The computation of the queuing delay pmf associated with the q-th activation is very similar to the computation of the busy window with minor modifications: Again, we iterate over the activations in order of their release. However, for the computation of $\omega_{q,i}(t)$ only the first q activations of τ_i must be included as well as activations of the tasks of higher priority.

Naturally, the stochastic queuing delay includes execution and recovery time spend for the first $q-1$ transmissions of task τ_i. Only for the q-th activation itself the execution time *must not* be included, hence, we subtract C_i from its

execution time pmf of the q-th activation of τ_i:

$$e_n^q(t) = e_n^q(t + C_i) \qquad (16)$$

Note that all other tasks have an execution time pmf according to eq. 7.

To determine the actual queuing delay pmf we use the same algorithm as for the busy-window pmf. That is, we apply eq. 11, eq. 12 and eq. 13 to $\omega_{q,i}(t)$ rather than $\omega_i(t)$. The starting condition obtained by eq. 14 can also be used for $\omega_{q,i}^0$. For the sake of completeness we will provide the slightly modified equations. Split, to compute the relative backlog pmf (this resembles eq. 11):

$$b_i^{n+1}(t) = \begin{cases} \omega_{q,i}^n(t + \Delta_i(j+1)) & \text{if } t \geq 0 \\ 0 & \text{otherwise} \end{cases} \qquad (17)$$

Convolution, to include the execution time of activation $n+1$. Note that this step is slightly different from eq. 12: If the q-th activation is encountered, the modified execution time 16 is used.

$$\tilde{b}_{q,i}^{n+1}(t) = \begin{cases} \left(b_i^{n+1} * e_{n+1}^q\right)(t) & \text{if } j+1 = \tau_{i,q} \\ \left(b_i^{n+1} * e_{n+1}\right)(t) & \text{otherwise} \end{cases} \qquad (18)$$

Merging of the new backlog in the queuing delay pmf (this resembles eq. 13:

$$\omega_{q,i}^{n+1}(t) = \begin{cases} \omega_{q,i}^n(t) & t < \Delta_i(n+1) \\ \tilde{b}_{q,i}(t) & \text{otherwise} \end{cases} \qquad (19)$$

Based on the queuing delay $\omega_{q,i}$, we can compute the probabilistic finishing time of the q-th event by adding (right-shifting) the execution time of the q-th event of τ_i:

$$F_{q,i}(t) = \omega_{q,i}(t - C_i) \qquad (20)$$

Accordingly, the response time is calculated by subtracting (left-shifting) the earliest arrival time.

$$R_{q,i}(t) = F_{q,i}(t + \delta_i^-(q)) \qquad (21)$$

At that point we need to evaluate $R_{q,i}(t)$ for some q, to determine the worst-case. Theoretically, an infinite amount of activations must be considered, since the derived stochastic busy window is infinitely long. In practice, the evaluation can be stopped when the probability that the q-th event falls in the busy-window is smaller than a given threshold.

$$q^+ = \max_{q>0, q \in \mathbb{N}} \left\{ q \left| \left(1 - \sum_{t=0}^{\delta_i^-(q)} \omega_i(t) \right) < \epsilon \right. \right\} \qquad (22)$$

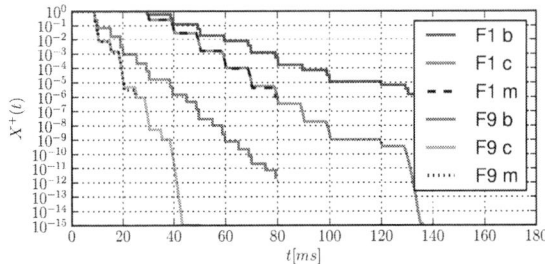

Figure 5: Exceedance functions according to modified Broster (b), convolution based approach (c) and Monte Carlo simulation (m) with 10^6 samples. λ=0.00024 per bit $\cong$30 per sec @ 125kbit/s.

The probability that the q-th event exceeds some response-time t is given as:

$$X_{q,i}(t) = 1 - \sum_{x=0}^{t} R_{q,i}(x) \qquad (23)$$

An upper bound to the probability that any event in any possible busy window exceeds a response time t due to interference and errors is:

$$X_i^+(t) = \max_{1 \le q \le q^+} X_{q,i}(t) \qquad (24)$$

5. EXPERIMENT

Now we apply the presented approach to the Controller Area Network (CAN). A typical CAN network consists of a set of ECUs connected to a CAN bus. Applications, mapped to the ECUs communicate by transmitting CAN frames over the bus.

In case multiple controller start transmitting at the same time, the protocol uses the CAN Identifier of the messages to determine the transmission order. Once a frame is in-flight it cannot be interrupted by other frames.

As already mentioned, CAN implements a set of mechanisms to detect erroneous transmissions. Framing consistency and correctness of payload data is monitored by all bus controllers. Hence, bit-errors for instance caused by electromagnetic interference (EMI) are detected. After an error is detected, the controller sends an error-frame (cf. signaling-overhead in Figure 1), assuring that all other bus-nodes remain in a consistent state. After an error-frame a new arbitration phase begins. Exhaustive information on the protocol and the underlying mechanisms can be found in [15].

Each CAN frame modeled as a task τ with a worst-case execution time C and a fixed priority p according to the payload and the CAN identifier. This model transformation is described in-depth in [3,5], in this paper further details are omitted.

To evaluate the presented algorithm we use the 17-messages SAE benchmark as presented in [18]. The bus speed is 125kbit/s which leads to a bus utilization of roughly 82%. Throughout all experiments, we use exactly the same parameters as in [3], this makes the results more transparent and comparable to related work.

To show the accuracy of our approach we compared the computed exceedance function $X^+(t)$ with a Monte Carlo based approach as well as Broster's approach [3]. For the Monte Carlo analysis, we simulated the critical instant 10^6 times under the presented error-model and recorded the worst-case response times. The exceedance function was computed via the empirical CDF. Figure 5 shows the results for Frames F1 and F9. Note that the Monte Carlo re-

sults are only statistically significant down to 10^{-5} due to the sample size. Unsurprisingly, our approach matches the Monte Carlo results, since we do not introduce additional pessimism during the computation. Thus, according to the model our approach is significantly tighter than Broster's.

6. CONCLUSION

Adequate, formal reliability analysis plays an important role in the design of dependable systems. In this paper, we have presented a novel convolution based reliability analysis, which allows more accurate calculation of the reliability of non-preemptive fixed priority scheduling under the effect of errors over existing methods by considering not only the worst case overhead condition, but rather the effects of each task individually. We showed that the implied pessimism of related work could be entirely removed.

7. REFERENCES

[1] P. Axer, M. Sebastian, and R. Ernst. Probabilistic response time bound for can messages with arbitrary deadlines. In *Proc. of DATE*, pages 1114–1117, march 2012.

[2] I. Broster, A. Burns, and G. Rodríguez-Navas. Probabilistic analysis of CAN with faults. In *Proc. of Real-Time Systems Symposium*, pages 269–278. IEEE, 2002.

[3] I. Broster, A. Burns, and G. Rodriguez-Navas. Comparing real-time communication under electromagnetic interference. In *Proc. 16th ECRTS*, pages 45–52, 2004.

[4] A. Burns, S. Punnekkat, L. Strigini, and D. R. Wright. Probabilistic scheduling guarantees for fault-tolerant real-time systems. In *Proc. of Dependable Computing for Critical Applications*, pages 361–378, 1999.

[5] R. Davis, A. Burns, R. Bril, and J. Lukkien. Controller area network (can) schedulability analysis: Refuted, revisited and revised. *Real-Time Systems*, 35(3):239–272, 2007.

[6] J. L. Diaz, D. F. Garcia, K. Kim, C.-G. Lee, L. Lo Bello, J. M. Lopez, S. L. Min, and O. Mirabella. Stochastic analysis of periodic real-time systems. In *Proc. 23rd IEEE Real-Time Systems Symp. RTSS 2002*, pages 289–300, 2002.

[7] J. L. Diaz, J. M. Lopez, and D. F. Garcia. Probabilistic analysis of the response time in a real time system. In *Proc. of the 1st CARTS Workshop on Advanced Real-Time Technologies,*, 2002.

[8] J. Ferreira, A. Oliveira, P. Fonseca, and J. Fonseca. An experiment to assess bit error rate in can. In *Proc. of RTN*, 2004.

[9] International Electrotechnical Commission (IEC). Functional safety of electrical / electronic / programmable electronic safety-related systems ed2.0, 2010.

[10] International Organization for Standardization (ISO). Iso/fdis 26262: Road vehicles – functional safety, 2000.

[11] H. Kopetz. *Real-Time Systems: Design Principles for Distributed Embedded Applications*. Kluwer Academic Publishers, Norwell, MA, USA, 1997.

[12] J. Lehoczky. Fixed Priority Scheduling of Periodic Task Sets with Arbitrary Deadlines. *Proc. 11th RTSS*, pages 201–209, Dec 1990.

[13] S. Punnekkat and A. Burns. Analysis of checkpointing for schedulability of real-time systems. In *Proc. of Int. Workshop Real-Time Computing Systems and Applications*, pages 198–205, 1997.

[14] K. Richter. *Compositional scheduling analysis using standard event models*. PhD thesis, TU Braunschweig, 2005.

[15] Robert Bosch GmbH, Postfach 30 02 40, D-70442 Stuttgart. CAN Specification version 2.0, 1991.

[16] S. Schliecker, J. Rox, M. Ivers, and R. Ernst. Providing accurate event models for the analysis of heterogeneous multiprocessor systems. In *Proc. of 6th CODES-ISSS*, pages 185–190, Atlanta, GA, USA, October 2008. ACM New York.

[17] M. Sebastian and R. Ernst. Reliability Analysis of Single Bus Communication with Real-Time Requirements. In *Proc. of PRDC*, pages 3–10, 2009.

[18] K. Tindell and A. Burns. Guaranteeing message latencies on control area network (can). In *Proceedings of the 1st International CAN Conference*. Citeseer, 1994.

[19] K. W. Tindell, A. Burns, and A. J. Wellings. An Extendible Approach for Analyzing Fixed Priority Hard Real-Time Tasks. *Real-Time Systems*, 6(2):133–151, 1994.

Table 1: SAE CAN Benchmark

Prio	DLC	C [1] (bits)	E [2] (bits)	T (ms)	D (ms)	R (ms)
17	1	62	13	1000	5	1.416
16	2	72	13	5	5	2.016
15	1	62	13	5	5	2.536
14	2	72	13	5	5	3.136
13	1	62	13	5	5	3.656
12	2	72	13	5	5	4.256
11	6	112	13	10	10	5.016
10	1	62	13	10	10	8.376
9	2	72	13	10	10	8.976
8	2	72	13	10	10	9.576
7	1	62	13	100	100	10.096
6	4	92	13	100	100	19.096
5	1	62	13	100	100	19.616
4	1	62	13	100	100	20.136
3	3	82	13	1000	1000	28.976
2	1	62	13	1000	1000	29.496
1	1	62	13	1000	1000	29.52

[1] for standard ID, without intermission space
[2] active error frame minus end of frame

APPENDIX

S1.   ADDITIONAL PROOFS

Proof for Lemma 6

PROOF. The first case is straightforward: to see no recovery there must be no errors during the execution. For the second case, the argument is as follows: the first execution attempt of execution time C must have been erroneous otherwise there would be no recovery operation in the first place. This is followed by $k-1$ erroneous error-signaling/execution attempts until the final k-th attempt succeeds (otherwise there may be more recoveries). Since all time intervals are non-overlapping, repeated application of eq. 3 and 4 directly leads to eq. 6.

□

S2.   ADDITIONAL EXPERIMENT

In the second experiment our approach is applied to a more realistic setup with a much lower error rate. The results are shown in Figure 6 for an bit error rate of 10^{-5}. By construction, our approach is always better than Broster's. The improvement of the convolution-based approach is especially noticeable if there is some slack available which can accommodate multiple error recoveries. This is because our approach evaluates a broad variation of recovery scenarios might fit in the slack time by evaluating all possibilities. In Broster's approach, only the largest task of higher priority is considered for the recovery overhead which imposes pessimism. Obviously, this effect adds up if the slack can accommodate multiple recoveries.

For realistic error rates the improvement for F9 is in the order of one magnitude (at the deadline $t=10$ms) and up to 10^7 at $t=50$. For the extreme error rate (Figure 5) it can be as high as 10^9.

S3.   BUSY WINDOW PSEUDO CODE

Algorithm 1 gives a pseudo code implementation of the stochastic busy-window computation. As described, the individual steps are composed of splitting which separates the unstable from the stable part of the busy-window pmf, convolution, which integrates the arrived workload and the backlog, as well as merging which integrates the new backlog in the busy window pmf.

It is assumed that $\omega_i(t)$ is implemented as a vector and the operation $[t:]$ slices the vector at the t-th index and returns

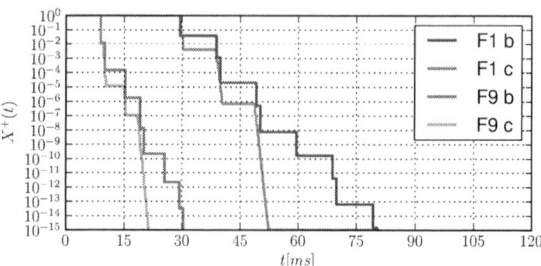

Figure 6: Exceedance functions according to modified Broster (b) and the convolution based approach (c). $\lambda=10^{-5}$ per bit.

Algorithm 1 Discrete Computation of the stochastic busy period

function BUSY WINDOW(τ_i, Δ, w_i^0, set of pmfs $e(t)$)
   $n \Leftarrow 0$
   $w \Leftarrow w_i^0$
   while $1 - \sum_{t=0}^{\Delta_i(n)-1} w(t) > \epsilon$ **do**
     $t \Leftarrow \Delta_i(n)$
     $u = w[t:]$ ▷ split, unstable part
     $s = w[:t-1]$ ▷ split, stable part
     $c = u * e_n$ ▷ convolution
     $w \Leftarrow [s, c]$ ▷ merge
     $n \Leftarrow n + 1$
   end while
   return w
end function

---

the right slice. Analogous, $[:t]$ slices the vector at the t-th index and returns the left slice. Similarly, $[f, c]$ concatenate vectors f and c.

HaDeS: Architectural Synthesis for Heterogeneous Dark Silicon Chip Multi-processors

Yatish Turakhia[1], Bharathwaj Raghunathan[2], Siddharth Garg[2,*] and Diana Marculescu[3]

[1]Department of Electrical Engineering, Indian Institute of Technology Bombay, Mumbai, India
[2]Department of Electrical and Computer Engineering, University of Waterloo, Waterloo, ON
[3]Department of Electrical and Computer Engineering, Carnegie Mellon University, Pittsburgh, PA
[*]Corresponding author contact: s6garg@ecemail.uwaterloo.ca

Abstract—**In this paper, we propose an efficient iterative optimization based approach for architectural synthesis of dark silicon heterogeneous chip multi-processors (CMPs). The goal is to determine the optimal number of cores of each type to provision the CMP with, such that the area and power budgets are met and the application performance is maximized. We consider general-purpose multi-threaded applications with a varying degree of parallelism (DOP) that can be set at run-time, and propose an accurate analytical model to predict the execution time of such applications on heterogeneous CMPs. Our experimental results illustrate that the synthesized heterogeneous dark silicon CMPs provide between 19% to 60% performance improvements over conventional homogeneous designs for variable and fixed DOP scenarios, respectively.**

I. INTRODUCTION

Technology scaling has enabled increasing on chip integration to the extent that, in the near future, a chip will have more transistors than can be simultaneously powered on within the peak power and temperature budgets. This has been referred to as the dark silicon era [6] where, at any given point in time, only a percentage of transistors on the die are operational. Dark silicon chips are expected to be heterogeneous in nature, consisting of, for example, a multitude of dedicated hardware accelerators to assist the on-chip cores [7] or heterogeneous CMPs consisting of different types of general-purpose cores. In this paper, we address the optimal synthesis of heterogeneous dark silicon CMPs.

The problem of architectural synthesis of heterogeneous dark silicon CMPs can be defined as follows: given (i) a library of general-purpose cores, (ii) a set of multi-threaded benchmark applications, (iii) a peak power budget, and (iv) an area budget; determine the optimal number of cores of each type to provision the heterogeneous dark silicon CMP with, such that the average performance over all benchmarks applications is maximized. Compared to prior work on architectural synthesis for application-specific multi-processor systems [8], [3],

architectural synthesis for dark silicon CMPs introduces a number of new challenges and metrics of interest. *First*, accurate analytical models for the execution time of multi-threaded benchmark applications running on heterogeneous cores are not readily available, although these are essential for optimization.

This is in contrast to the application specific-domain where the performance models are well defined, often using formal models of computation such as data-flow graphs [8]. *Second*, typical multi-threaded applications, for example the applications from the SPLASH-2 and PARSEC benchmark suites, can be executed with a variable degree of parallelism (DOP), *i.e.*, a varying number of parallel threads, at run-time. The architectural synthesis algorithm must be aware of the optimal DOP and optimal run-time scheduling of threads to cores for each application. In fact, we show that assuming a fixed DOP for the benchmark applications can result in suboptimal architectures. *Finally*, from an empirical perspective, an important metric of interest is the performance benefit of heterogeneous versus homogeneous CMP architectures with increasing dark silicon area. With increasing dark silicon area, a greater number of cores specialized for each application can be included. Therefore, the performance benefits of heterogeneous CMPs over homogeneous CMPs should increase with increasing dark silicon – empirical validation of this trend is of immense interest to system designers.

In this paper, we propose Hades – a framework for architectural synthesis of heterogeneous dark-silicon CMPs. The Hades framework consists of the following novel features:

- A new analytical performance model for multi-threaded applications from the SPLASH-2 and PARSEC-2.1 benchmark suites. The proposed performance model is shown to be accurate for applications scheduled on both homogeneous and heterogeneous CMPs, and across a wide range of DOP values for each application.
- An efficient, iterative algorithm to determine the optimal number of cores of each type to provision the heterogeneous dark silicon CMP with. The algorithm takes into account run-time optimization of the DOP and mapping of application threads to cores.
- Comprehensive validation and evaluation of the proposed

Permission to make digital or hard copies of all or part of this work for personal or classroom use is granted without fee provided that copies are not made or distributed for profit or commercial advantage and that copies bear this notice and the full citation on the first page. To copy otherwise, to republish, to post on servers or to redistribute to lists, requires prior specific permission and/or a fee. DAC '13, May 29 - June 07 2013, Austin, TX, USA.

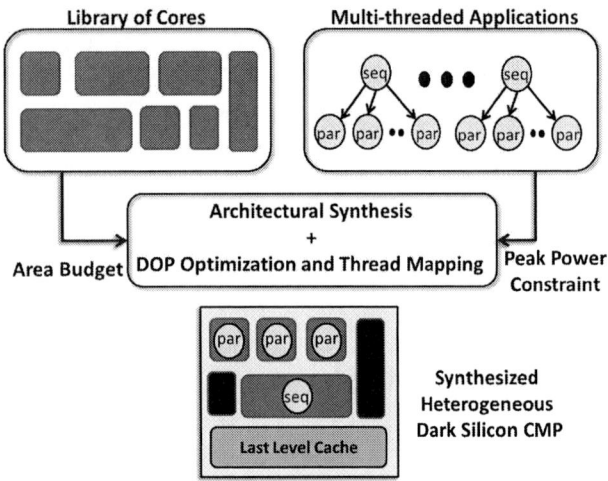

Fig. 1. Overview of the Hades framework. In the synthesized heterogeneous dark silicon CMP, a subset of cores are dark.

models and optimization technique on Sniper [5], a detailed software simulator for multi-core systems. Our results show that heterogeneous dark silicon CMPs provide up to 60% and 19% performance gains over homogeneous dark silicon CMPs with fixed and optimized DOPs, respectively.

To the best of our knowledge, Hades is the first architectural synthesis framework targeted at optimal synthesis of heterogeneous, dark silicon CMP architectures. Figure 1 provides an overview of the Hades framework.

II. PRIOR WORK

General-purpose, heterogeneous CMPs were first proposed by Kumar et al. [9] in the context of single-threaded applications running on CMPs in a multi-programmed fashion. In [10], the authors propose hill-climbing based solutions to for heterogeneous CMP design, but only for multi-programmed worklaods and do not model or evaluate the impact of dark silicon. Lee et al. [11] have also performed similar studies.

The notion of dark silicon was recently introduced by Goulding et al. [7] and Esmailzadeh et al. [6]. Goulding et al. focus on provisioning the dark silicon area with application-specific accelerators while Esmailzadeh et al. focus on using only general-purpose cores. In their work, Esmailzadeh et al. use a highly simplistic performance model based on Amdahl's Law to select cores. In addition, a different architecture is synthesized for each application. In contrast, Hades (i) efficiently explores the entire design space of heterogeneous CMP architectures using a novel iterative optimization strategy, (ii) uses a more realistic performance model that is validated against detailed simulations on both homogeneous and heterogeneous CMPs, and (iii) optimizes over a set of benchmark applications. More recent work on architectural synthesis for dark silicon chips has focused on synthesizing and provisioning application specific accelerators [14] and general-purpose cores [2]. In the latter work, the authors focus

only on multi-programmed workloads and use a simulated annealing based heuristic.

Finally, architectural synthesis has been widely studied in the application-specific domain for multi-processor systems-on-chip (MPSoC). Wolf et al. [15] provide a comprehensive account of work in this area. As mentioned before, the architectural synthesis problem for MPSoCs is significantly different because accurate performance models are readily available, the DOP of the application is typically predetermined, and the optimization objective is typically to meet hard or soft timing deadlines. In addition, this problem has not been explored in the dark silicon context.

III. PRELIMINARIES AND ASSUMPTIONS

We begin by discussing the assumptions and mathematical notation relevant to our work.

1) Applications: We assume that we are given a set of N representative benchmark applications. Each application is multi-threaded and consists of both sequential and parallel phases. A single sequential thread executes in the sequential phase, while multiple parallel threads execute in the parallel phase.

The DOP of an application, *i.e.*, the number of threads in the parallel phase can be determined at compile time and is represented as D_i. Without loss of generality, we assume that $D_i \in \mathbb{N}$ and $1 \leq D_i \leq D_{max}$. Note that, in practice, the DOP for some applications can be restricted to a certain subset of values, values that are powers of two, for example.

In this paper, we assume that each multi-threaded application runs independently on the CMP. The run-time scheduler determines the optimal DOP and the optimal mapping of threads to cores to minimize the execution time within a peak power budget, P_{budget}. Each core executes only one application thread.

2) Core Library: We assume a library of M different, general-purpose cores — in other words, each core is capable of executing each application. Each core consists of the micro-architectural pipeline and private instruction and data caches. In the experimental results section, we discuss the micro-architectural parameters, *e.g.*, issue width, cache size *etc.*, that we vary to generate a library of cores.

The peak power dissipation of core j ($j \in [1, M]$) executing one of the parallel threads of application i ($i \in [1, N]$) is denoted by P_{ij}, and includes both the peak dynamic and leakage power components. When a core is idle, *i.e.*, no thread is mapped to it, its leakage power dissipation is denoted by P_j^{idle}. The area of core of type j is A_j, including the pipeline and private caches. Since our goal is to maximize performance, we assume that each core runs at at its highest voltage and frequency level.

3) Uncore Components: In this paper, we focus only on optimizing the number of cores (including their private caches) of each type. We assume that all cores share a single, global last-level cache (LLC) with a uniform access penalty. For fairness, the LLC size is kept constant over all experiments. By the same token, the configuration of other uncore components, for example, the number and bandwidth of the off-chip DRAM memory controllers, is fixed.

IV. PROPOSED HADES FRAMEWORK

Hades enables the efficient exploration of the vast design space of heterogeneous dark silicon chip multi-processor architectures to pick the optimal design that maximizes application performance within area and power budgets. In addition, the optimization is inherently aware of and accounts for run-time decisions including the optimal DOP for each application and mapping of threads to cores — *i.e.*, Hades performs joint design-time and run-time optimization.

We now discuss the two components of the Hades framework: we first discuss the proposed application performance model, and then the iterative optimization used to determine the optimal heterogeneous dark silicon CMP architecture.

A. Application Performance Model

We focus on multi-threaded applications from the scientific computing domain, such as those found in the SPLASH-2 and PARSEC benchmark suites. These applications consist of two phases of execution — a sequential phase, which consists of a single thread of execution; and a parallel phase in which multiple threads process data in parallel.

1) Homogeneous CMPs: As the DOP of the application is increased, the execution latency of the parallel phase decreases and is ideally inversely dependent on the DOP. However, due to increased contention for shared hardware and software resources, the benefits of increasing the DOP begin to saturate and, for some applications, the execution time of the parallel phase might actually *increase* beyond a certain DOP. This can be observed in Figure 2(a).

Based on this observation, we propose the following performance model for multi-threaded applications executing on homogeneous cores, and then generalize the model for heterogeneous CMPs. The execution time E_{ij} of an application i ($i \in [1, N]$) running on a homogeneous CMP with cores of type j ($j \in [1, M]$) is expressed as:

$$E_{ij} = t_{ij}^s + \frac{t_{ij}^p}{D_i} + D_i K_{ij} \quad i \in [1, N], \; j \in [1, M] \quad (1)$$

In this equation, t_{ij}^s is the execution time of the sequential phase, t_{ij}^p is the execution time of the parallelizable part of the parallel phase, and K_{ij} represents the increase in execution latency in the parallel phase because of resource contention.

The model parameters of Equation 1 are learned by executing each application with different DOP values on homogeneous CMPs and using standard regression techniques to minimize the error between the execution times obtained from simulation and from analytical prediction. Therefore, at most $N \times M \times D_{max}$ simulations are required to learn the model parameters. Figure 2(a) shows that the proposed model is, in fact, able to provide very accurate estimates of the actual execution time for varying DOPs.

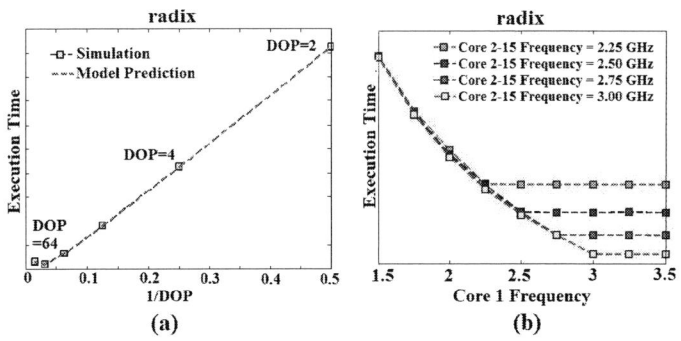

Fig. 2. (a) Execution time for the *radix* benchmark on a homogeneous 64 core CMP with varying DOP. (b) Execution time for the *radix* benchmark with DOP=16 on a heterogeneous CMP. In the experiment, heterogeneity is introduced by varying the core frequency.

2) Heterogeneous CMPs: In a heterogeneous CMP, the execution time of each parallel thread is different. In the parallel phase of execution, threads typically synchronize on a barrier, *i.e.*, all threads must finish execution before the application can proceed to the next phase. Therefore, the latency of a parallel phase is determined by the worst case execution latency across all parallel threads.

This observation is empirically validated in Figure 2(b), again using the *radix* benchmark. Although the application is executed on a homogeneous CMP, heterogeneity is introduced by varying the frequency of one of the parallel cores. Observe that the slowest core does indeed determine the application execution time.

Based on this observation and Equation 1, we propose the following model for the execution time of an application executing on a heterogeneous CMP:

$$E_i = t_{im^s}^s + \max_{v \in [1, D_i]} \left(\frac{t_{im^p(v)}^p}{D_i} + D_i K_{im^p(v)} \right) \quad (2)$$

In this equation, m^s represents the type of core that the sequential thread executes on, while $m^p(v)$ represents the type of core that the vth parallel thread executes on.

B. Architectural Synthesis

We now formulate the architectural synthesis problem as an integer program. As we will show, the direct formulation that utilizes Equation 2 results in a non-linear, integer program that can be converted, in polynomial time, to an integer linear program (ILP) without loss of optimality. Although powerful commercial-off-the-shelf ILP solvers exist, we observe empirically that for even reasonable problem instances, the computational cost of using an ILP solver is significant. To address this concern, we then propose an iterative optimization procedure that provides orders-of-magnitude speed-up without significant loss in optimality.

1) Non-linear Integer Programming Formulation: We begin by discussing the performance maximization objective function and then incorporate the constraints.

Objective Function: The goal is to minimize the weighted sum of execution time for each benchmark application,

$$\min\left(\sum_{i=1}^{N}\rho_i(l_i^s + l_i^p)\right) \quad (3)$$

where l_i^s and l_i^p are the sequential and parallel execution times for each benchmark. The weight ρ_i is a designer specified constant.

Sequential Execution Time: Let $s_{ij} \in \{0,1\}$ be an indicator variable that is 1 if the sequential thread of application i executes on a core of type j. The sequential execution time of application i can be written as:

$$l_i^s = \sum_{j=1}^{M} s_{ij} t_{ij}^s \quad \forall i \in [1, N] \quad (4)$$

In addition, each application is allowed to use exactly one sequential core. Therefore:

$$\sum_{j=1}^{M} s_{ij} = 1 \quad \forall i \in [1, N] \quad (5)$$

Parallel Execution Time: Let $b_{ij} \in \{0,1\}$ be an indicator variable that is 1 if at least one parallel thread of application i executes on a core of type j. The execution time of the parallel phase is determined by the slowest parallel thread. Therefore:

$$l_i^p \geq b_{ij}\left(\frac{t_{ij}^p}{D_i} + K_{ij}D_i\right) \quad \forall i \in [1, N], j \in [1, M] \quad (6)$$

Note that since the DOP values, D_i, are also variables in the formulation, Equation 6 is a *non-linear* constraint.

Let $r_{ij} \in \mathbb{N}$ be the number of cores of type j used by parallel threads of application i. The DOP of the application must be equal to the total number of parallel cores utilized:

$$\sum_{j=1}^{M} r_{ij} = D_i \quad \forall i \in [1, N] \quad (7)$$

and must be less than the maximum value:

$$1 \leq D_i \leq D_{max} \quad \forall i \in [1, N] \quad (8)$$

Finally, r_{ij} should be zero when b_{ij} is zero and at most D_{max} when $b_{ij} = 1$:

$$r_{ij} - D_{max}b_{ij} \leq 0 \quad \forall i \in [1, N], j \in [1, M] \quad (9)$$

Number of Cores of Each Type: The design vector $Q \in \mathbb{Z}^M$ represents the number of cores of each type: $Q = \{Q_1, Q_2, \ldots, Q_M\}$, and is the ultimate objective of the architectural synthesis problem.

The number of cores of type j should be at least larger than the number of cores of that type used by the sequential and parallel threads of any application. Therefore:

$$Q_j \geq s_{ij} + r_{ij} \quad \forall i \in [1, N] \quad (10)$$

Peak Power Constraint: The peak power dissipation of the dark silicon heterogeneous CMP must be below the peak power budget for each application:

$$\sum_{j=1}^{M} r_{ij}P_{ij} + (Q_j - r_{ij})P_j^{idle} \leq P_{budget} \quad \forall i \in [1, N] \quad (11)$$

Dark Silicon Area Constraint: All cores must fit in the prescribed area budget.

$$\sum_{j=1}^{M} Q_j A_j \leq A_{budget} \quad \forall i \in [1, N] \quad (12)$$

The objective function in Equation 3 and the constraints in Equations 4 to 12 represent a non-linear integer optimization problem which we refer to as **NILP-OPT**. Solving the problem yields the optimal number of cores of each type, *i.e.*, the vector Q, *and* the optimal DOP for each application, $D_i, \forall i \in [1, N]$.

2) ILP Formulation: We now show that the non-linear constraint in NILP-OPT, *i.e.*, Equation 6, can be readily linearized to result in a standard ILP problem.

We introduce an indicator variable, $y_{iw} \in \{0,1\}$ ($i \in [1, N], w \in [1, D_{max}]$) that is 1 if and only if application i has an optimal DOP of w. Equation 6 can be re-written as:

$$l_i^p \geq \sum_{w=1}^{D_{max}} b_{ij}y_{iw}\left(\frac{t_{ij}^p}{w} + K_{ij}w\right) = \sum_{w=1}^{D_{max}} m_{ijw}\left(\frac{t_{ij}^p}{w} + K_{ij}w\right) \quad (13)$$

where $m_{ijw} = b_{ij}y_{iw} = \min(b_{ij}, y_{iw})$ and $m_{ijw} \in \{0,1\}$. The following three linear equations express the relationship between m_{ijw}, b_{ij} and y_{iw}:

$$m_{ijw} \leq b_{ij} \quad \forall i \in [1, N], j \in [1, M], w \in [1, D_{max}] \quad (14)$$

$$m_{ijw} \leq y_{iw} \quad \forall i \in [1, N], j \in [1, M], w \in [1, D_{max}] \quad (15)$$

$$m_{ijw} \geq b_{ij} + y_{iw} - 1 \quad \forall i \in [1, N], j \in [1, M], w \in [1, D_{max}] \quad (16)$$

Equations 13 to 16 are the linearized versions of Equation 6. This completes the ILP formulation. We refer to this formulation as **ILP-OPT**.

3) Iterative Optimization: Although ILP-OPT guarantees optimality, we find that empirically the computational time of running the ILP optimization to completion can be significant. To address this issue, we propose an iterative optimization approach, **ITER-OPT**, that separates the architectural optimization from DOP optimization.

ITER-OPT operates as follows: it starts with an initial guess for the optimal architectural design vector Q^*, determines the optimal DOP for each application for this heterogeneous design, re-synthesizes the optimal heterogeneous architecture for this *fixed* DOP, and iterates till convergence, *i.e.*, till no further improvement in performance is observed. The solution to which ITER-OPT converges represents a *local* optima in the design space. Figure 3 as an illustration of how the ITER-OPT algorithm works. We now discuss the two primary components of the ITER-OPT algorithm: (i) the architecture

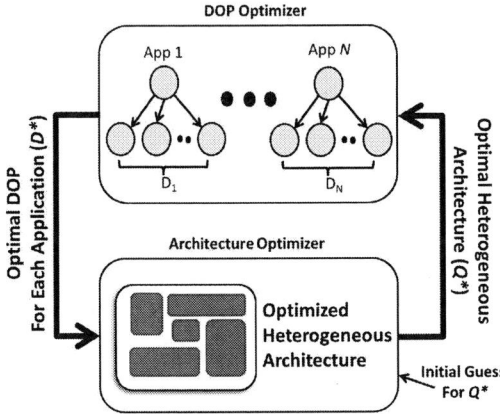

Fig. 3. Illustration of the **ITER-OPT** algorithm.

optimizer and (ii) the DOP optimizer.

Architecture Optimizer: Note that if the DOP of each application is fixed, then Equation 6 is, in fact, linear. Thus NILP-OPT is converted to an ILP problem with only $\mathcal{O}(MN)$ variables, and can therefore be solved significantly faster than ILP-OPT. The architecture optimizer takes a fixed DOP, D^*, vector as input and outputs the optimal heterogeneous architecture Q^* for these input DOPs.

DOP Optimizer: The DOP optimizer (described in Algorithm 1 of Appendix) determines, in polynomial time, the optimal DOP for each application, given a heterogeneous architecture, Q^*, *i.e.*, the number of cores of each type. Intuitively, Algorithm 1 is based on the fact that the execution time of any scheduling of threads to cores depends only on the performance of two cores — the sequential core and the slowest parallel core. Thus, it is sufficient to search over all possible pairs of cores types in the heterogeneous architecture, which is polynomial in the number of core types [13].

V. EXPERIMENTAL METHODOLOGY

All our experiments are run on the Sniper [5] multi-core simulator that provides the ability to model heterogeneous core configurations and detailed models for the memory hierarchy.

A library of 108 cores was generated by varying a number of key micro-architectural parameters as shown in TABLE I, yielding a rich design space of area, power, and performance values. These are obtained from the McPAT tool [12] for a 22 nm technology node and a 1.0V nominal supply voltage.

The last-level L3 cache size is set to 32 MB and has a uniform access latency. We model a 1 GB DRAM main memory that is accessed through a DRAM memory controller with aggregate bandwidth of 7.6 GBps.

We experiment with five multi-threaded applications from the SPLASH-2 [16] and PARSEC [4] benchmarks suites — *blackscholes*, *fft*, *fluidanimate*, *radix*, and *swaptions*. The maximum DOP is set to 32 for each application. Simulating these applications on the entire core library reveals that only 15 cores

TABLE I
MICRO-ARCHITECTURAL PARAMETERS OF CORE LIBRARY

Core Parameter	Values
Dispatch Width	{1,2,4}
ROB Window Size	{8,16,32,64}
L1-I/D Cache size	{64, 128, 256} KB
L2 cache (private)	256 KB
Frequency	{2.5, 3.5, 4.5} GHz

are Pareto optimal in terms of area, power or performance for at least one application. These 15 cores are retained for further experimentation. We have used the Gurobi ILP tool-box [1] to implement both ILP-OPT and ITER-OPT.

VI. EXPERIMENTAL RESULTS

We conduct our first set of experimental results with an area budget $A_{budget} = 180mm^2$ and for different peak power budgets: $P_{budget} = \{40W, 60W, 80W\}$.

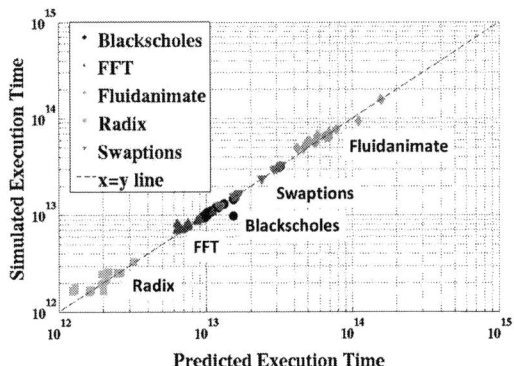

Fig. 4. Validation of proposed analytical performance model against Sniper simulations for both homogeneous and heterogeneous dark silicon CMPs.

A. ITER-OPT vs. ILP-OPT

We begin by noting that the proposed iterative optimization scheme (ITER-OPT) compares favorably with the optimal ILP (ILP-OPT) solution. For $A_{budget} = 180mm^2$ and $P_{budget} = 60W$, ITER-OPT took only 17 seconds to provide a solution within 2.5% of the solution provided by ILP-OPT in 2 hours. This represents a $430\times$ reduction in run-time with only marginal loss in optimality.

B. Performance Model Validation

Figure 4 shows the scatter plot of predicted performance using the proposed model and the simulated performance for a variety of homogeneous and heterogeneous dark silicon CMP architectures that we experimented with. As it can be seen, the performance model agrees very well with simulated values and provides 5.2% error on average over 180 experiments.

C. Heterogeneous Vs. Homogeneous

We synthesized heterogeneous dark silicon CMPs using the Hades framework for the three power budgets mentioned above. These are compared with: (i) a homogeneous dark

1195

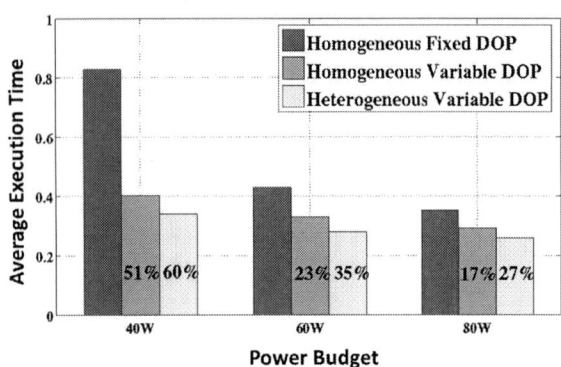

Fig. 5. Average execution time over all benchmarks for different architectural configurations and power budgets.

silicon CMP where every application has a fixed DOP=16 and (ii) a homogeneous dark silicon CMP where the DOP for every application was optimized for the homogeneous architecture. The heterogeneous CMP always uses DOPs optimized for its architecture.

For power budgets of 40W, 60W and 80W, the heterogeneous CMP has 60%, 35% and 27% higher performance than a homogeneous CMP with fixed DOPs, respectively. Assuming a homogeneous CMP with variable DOPs, the heterogeneous CMP still has 19%, 16% and 12% higher performance. Low power budgets correspond to more dark silicon, and we observe that the benefits of heterogeneity are, as expected, greater with increasing dark silicon.

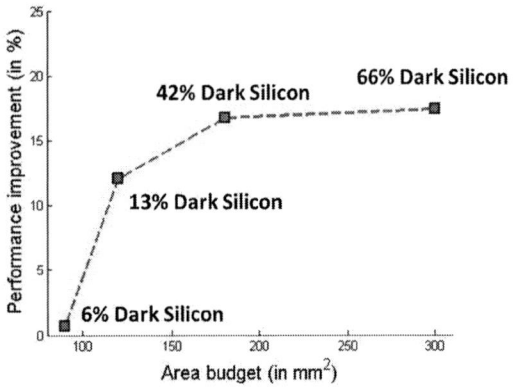

Fig. 6. Performance improvement of a heterogeneous dark silicon CMP with increasing area budgets and increasing proportion of dark silicon transistors.

D. Performance Benefits with Increasing Dark Silicon

We plot in Figure 6 the performance improvement of heterogeneous versus homogeneous for increasing area budgets, from $90mm^2$ to $300mm^2$. Also shown is the % dark silicon for each area budget. It is evident that the benefits of heterogeneity saturate after a certain point.

VII. CONCLUSION

In this paper we have proposed Hades, a framework for optimal architectural synthesis of heterogeneous dark silicon

CMPs. As part of Hades, we propose a new, analytical performance model for general-purpose multi-threaded applications running on heterogeneous platforms, and an iterative optimization algorithm, ITER-OPT, that determines the optimal number of cores of each type to provision the heterogeneous dark silicon CMP with. ITER-OPT takes into account the optimal DOP of each application while synthesizing the heterogeneous architecture. As future work, we plan to incorporate application specific accelerators in the Hades framework.

REFERENCES

[1] Gurobi optimizer (www.gurobi.com).

[2] J. Allred, S. Roy, and K. Chakraborty. Designing for dark silicon: a methodological perspective on energy efficient systems. In *Proceedings of the 2012 ACM/IEEE ISLPED*, 2012.

[3] F. Angiolini, J. Ceng, R. Leupers, F. Ferrari, C. Ferri, and L. Benini. An integrated open framework for heterogeneous mpsoc design space exploration. In *DATE'06. Proceedings*, 2006.

[4] C. Bienia, S. Kumar, J.P. Singh, and K. Li. The parsec benchmark suite: Characterization and architectural implications. In *Proceedings of the 17th international conference on Parallel architectures and compilation techniques*, 2008.

[5] T.E. Carlson, W. Heirmant, and L. Eeckhout. Sniper: exploring the level of abstraction for scalable and accurate parallel multi-core simulation. In *High Performance Computing, Networking, Storage and Analysis (SC), 2011 International Conference for*, pages 1–12. IEEE, 2011.

[6] H. Esmaeilzadeh, E. Blem, R.S. Amant, K. Sankaralingam, and D. Burger. Dark silicon and the end of multicore scaling. In *Computer Architecture (ISCA), 2011 38th Annual International Symposium on*, pages 365–376. IEEE, 2011.

[7] N. Goulding, J. Sampson, G. Venkatesh, S. Garcia, J. Auricchio, J. Babb, M.B. Taylor, and S. Swanson. Greendroid: A mobile application processor for a future of dark silicon. In *Hot Chips*, 2010.

[8] A. Kumar, B. Mesman, B. Theelen, H. Corporaal, and Y. Ha. Analyzing composability of applications on mpsoc platforms. *Journal of Systems Architecture*, pages 369–383, 2008.

[9] R. Kumar, K.I. Farkas, N.P. Jouppi, P. Ranganathan, and D.M. Tullsen. Single-isa heterogeneous multi-core architectures: The potential for processor power reduction. In *Microarchitecture, 2003. MICRO-36. Proceedings*. IEEE, 2003.

[10] R. Kumar, D.M. Tullsen, and N.P. Jouppi. Core architecture optimization for heterogeneous chip multiprocessors. In *Proceedings of the 15th international conference on Parallel architectures and compilation techniques*, pages 23–32. ACM, 2006.

[11] B.C. Lee and D.M. Brooks. Illustrative design space studies with microarchitectural regression models. In *High Performance Computer Architecture, 2007. HPCA 2007. IEEE 13th International Symposium on*, pages 340–351. IEEE, 2007.

[12] S. Li, J.H. Ahn, R.D. Strong, J.B. Brockman, D.M. Tullsen, and N.P. Jouppi. Mcpat: an integrated power, area, and timing modeling framework for multicore and manycore architectures. In *Microarchitecture, 2009*. IEEE, 2009.

[13] B. Raghunathan, Y. Turakhia, S. Garg, and D. Marculescu. Cherry-picking: Exploiting process variations in dark-silicon homogeneous chip multi-processors. In *Proceedings of the Design Automation and Test in Europe Conference*. EDAA, 2013.

[14] G. Venkatesh, J. Sampson, N. Goulding-Hotta, S.K. Venkata, M.B. Taylor, and S. Swanson. Qscores: trading dark silicon for scalable energy efficiency with quasi-specific cores. In *Proceedings of the 44th Annual IEEE/ACM International Symposium on Microarchitecture*, pages 163–174. ACM, 2011.

[15] W. Wolf, A.A. Jerraya, and G. Martin. Multiprocessor system-on-chip (mpsoc) technology. *Computer-Aided Design of Integrated Circuits and Systems, IEEE Transactions on*, pages 1701–1713, 2008.

[16] S.C. Woo, M. Ohara, E. Torrie, J.P. Singh, and A. Gupta. The splash-2 programs: Characterization and methodological considerations. In *ACM SIGARCH Computer Architecture News*, pages 24–36, 1995.

APPENDIX

We note that ITER-OPT is guaranteed to converge since: (a) the objective function, *i.e.*, the application execution time, always decreases in each iteration; and (b) the globally optimal execution time is bounded. As an initial guess for Q^*, the available area budget is divided equally between all core types.

Algorithm 1 Optimal DOP for an application $i \in [1, N]$ given heterogeneous architecture Q^*

1: $L \leftarrow$ List of cores in ascending order of power dissipation
2: $t^* \leftarrow \infty$ $D_i^* \leftarrow 1$
3: **for** $d \in [1, D_{max}]$ **do**
4:    **for** $j \in [1, M]$ **do**
5:       $Q^{used} \leftarrow Q^*$
6:       **if** $Q_j^{used} \geq 1$ **then**
7:          $t^{seq} \leftarrow t_{ij}^s$
8:          $Q_j^{used} \leftarrow Q_j^{used} - 1$
9:       **else**
10:          $t^{seq} \leftarrow \infty$
11:       **end if**
12:       **for** $k \in [1, M]$ **do**
13:          **if** $Q_k^{used} \geq 1$ **then**
14:             $t^{par} \leftarrow \frac{t_{ik}^p}{d} + K_{ik}d$
15:             $Q_k^{used} \leftarrow Q_k^{used} - 1$
16:             $P_{used} \leftarrow P_{used} + P_{ik}$
17:          **else**
18:             $t^{par} \leftarrow \infty$
19:          **end if**
20:          $req \leftarrow d - 1$
21:          **for** $l \in [1, M]$ **do**
22:             $c \leftarrow L_l$ /* l^{th} element of list L */
23:             **if** $\frac{t_{il}^p}{d} + K_{il}d \leq \frac{t_{ik}^p}{d} + K_{ik}d$ **then**
24:                $add \leftarrow \min\left(Q_c^{used}, req\right)$
25:                $req \leftarrow req - add$
26:                $P_{used} \leftarrow P_{used} + add \times P_c^{ik}$
27:                **if** $P_{used} > P_{budget}$ **or** $req = 0$ **then**
28:                   Go to line 37
29:                **end if**
30:             **end if**
31:          **end for**
32:          **if** $req = 0$ **then**
33:             $t_{jk} \leftarrow t^{seq} + t^{par}$
34:          **end if**
35:       **end for**
36:    **end for**
37:    **if** $\min_{jk}(t_{jk}) \leq t^*$ **then**
38:       $t^* \leftarrow \min_{jk}(t_{jk})$ $D_i^* \leftarrow d$
39:    **end if**
40: **end for**
41: **return** D_i^*

Hierarchical Power Management for Asymmetric Multi-Core in Dark Silicon Era

Thannirmalai Somu Muthukaruppan[1], Mihai Pricopi[1], Vanchinathan Venkataramani[1],
Tulika Mitra[1] and Sanjay Vishin[2]
[1]School of Computing, National University of Singapore
[2]Cambridge Silicon Radio
{tsomu,mihai,vvanchi,tulika}@comp.nus.edu.sg, Sanjay.Vishin@csr.com

ABSTRACT

Asymmetric multi-core architectures integrating cores with diverse power-performance characteristics is emerging as a promising alternative in the dark silicon era where only a fraction of the cores on chip can be powered on due to thermal limits. We introduce a hierarchical power management framework for asymmetric multi-cores that builds on control theory and coordinates multiple controllers in a synergistic manner to achieve optimal power-performance efficiency while respecting the thermal design power budget. We integrate our framework within Linux and implement/evaluate it on real ARM big.LITTLE asymmetric multi-core platform.

Categories and Subject Descriptors

C.1.4 [**PROCESSOR ARCHITECTURES**]: Parallel Architectures

General Terms

Algorithms, Design, Performance

Keywords

Asymmetric Multi-core, Power Management, Feedback controller.

1. INTRODUCTION

Computing systems have made an irreversible transition towards parallel architectures with multi-cores and many cores. However, power and thermal limits are rapidly bringing the computing community to another crossroad where a chip can have many cores but a significant fraction of them are left un-powered, or dark, at any point in time [7]. This phenomenon, known as dark silicon, is immediately visible in the embedded computing space where the restricted form factor rules out elaborate cooling mechanisms.

The dark silicon era is driving the emergence of asymmetric multi-cores where the cores share the same ISA (instruction- set architecture) but their micro-architectures offer diverse power/ performance characteristics. This heterogeneity enables better match between application demand and computation capabilities leading to substantially improved energy-efficiency [18, 9]. Indeed, ARM

Permission to make digital or hard copies of all or part of this work for personal or classroom use is granted without fee provided that copies are not made or distributed for profit or commercial advantage and that copies bear this notice and the full citation on the first page. To copy otherwise, to republish, to post on servers or to redistribute to lists, requires prior specific permission and/or a fee.
DAC '13 May 29 - June 07 2013, Austin, TX, USA.

has recently announced big.LITTLE [2] processing for mobile platforms where high performance, out-of-order Cortex A15 cluster is coupled with energy-efficient in-order Cortex A7 cluster in the same chip as shown in Figure 1.

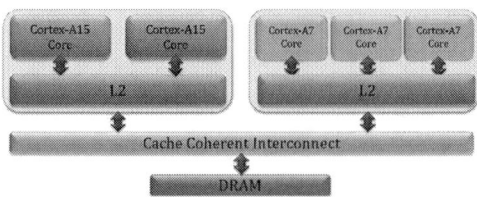

Figure 1: ARM big.LITTLE asymmetric multi-core.

We present a comprehensive power management framework for asymmetric multi-cores — in particular ARM big.LITTLE architecture in the context of mobile embedded platforms — that can provide satisfactory user experience while minimizing energy consumption within the Thermal Design Power (TDP) constraint. Compared to homogeneous multi-cores, power management is challenging on asymmetric multi-cores under limited TDP budget. We set out to design our framework with the following objectives:

- The dramatically different power-performance behavior of the cores implies that we need to identify the right core for the right task at runtime and migrate the tasks accordingly.

- The power hungry complex cores should be employed sparingly and only when absolutely necessary.

- Dynamic Voltage and Frequency Scaling (DVFS) as a control knob is available per cluster rather than per core within a cluster necessitating appropriate load balancing strategies. A cluster should run at the minimum frequency level required for adequate user experience so as to conserve energy.

- The restricted TDP budget precludes certain combination of frequencies for the different clusters. For example, it may be necessary to power down A7 cluster when A15 cluster is running at maximum frequency, a canonical example that illustrates the impact of the dark silicon era. Thus power budget has to be allocated opportunistically among clusters.

- If a system exceeds the power budget, the quality-of-service (QoS) of the tasks should degrade gracefully.

- The framework should be integrated in a commodity operating system without altering any of its desirable properties.

While, there exists solutions in the literature focusing on at least a subset of the objectives mentioned earlier, each of these solution have been generally designed to operate independently. It should be clear that deploying them together requires a carefully coordinated approach that is aware of the complex interplay among the individual solutions. For example, once the system exceeds the TDP of the entire chip, the power budgets for the clusters have to be reduced, which implies scaling down the voltage and frequency levels of the clusters, and consequently degrading the QoS of the tasks that triggered the thermal emergency in the first place. However, once the system load decreases (e.g., some tasks leave the system), this process has to be reversed and the QoS of the tasks should be restored back to the original level. This requires synergistic interaction among the different solutions so as to ensure *safety* (operate under power budget) and *efficiency* (optimal trade-off between power and performance), while maintaining *stability*, i.e., avoiding oscillation between different operating points.

We design a *hierarchical power management framework* that is based on the solid foundation of *control theory* and integrates multiple controllers to collectively achieve the goal of optimal energy-performance tradeoff under restricted power budget in asymmetric multi-core architectures. Moreover, we build our framework as an extension of *Linux completely-fair scheduler* while preserving all of its desirable properties such as fairness, non- starvation etc. We take advantage of Heart Rate Monitor [8] infrastructure in Linux to set the performance goal for a task and to monitor its execution progress as a measure of QoS. Finally, our Linux-based hierarchical power management framework is implemented on *real ARM big.LITTLE platform* exploiting all the control knobs provided on the platform, namely, per cluster DVFS, cluster power down, and task migration within and across clusters.

To the best of our knowledge, ours is the first work to *provide a comprehensive power management approach for asymmetric multi-cores under limited power budget* and definitely the first one to *integrate the solution in a commodity operating system (Linux) running on real platform (ARM big.LITTLE)*. Our key contributions are

- Our power management framework successfully achieves all the objectives enumerated earlier.
- Our solution builds on a formal control-theoretic approach that provides guarantees for safety, efficiency, and stability.
- Our hierarchical framework carefully coordinates the controllers to avoid inter-controller interference.
- We integrate our framework within the confines of Linux and implement it on a test version of the ARM Big.Little asymmetric multi-core architecture and report power, performance results from this real chip (as opposed to simulation).
- We experimentally evaluate and establish the superiority of our approach compared to the state-of-the-art.

2. RELATED WORK

The asymmetric multi-cores, due to the power/performance tradeoff [10, 6] introduce additional complexity to the scheduler. [11] proposed a scheduling algorithm for asymmetric cores that are simply symmetric cores using different frequency levels. [9] identified the key metrics for mapping a task to the appropriate core to optimize performance. [17] proposed an asymmetric- aware scheduler, where ILP and TLP threads are scheduled in fast and small cores, respectively. Operating system support for heterogeneous architecture with non-identical ISA was proposed in [12]. However, none of these techniques consider power management.

There exist plethora of works [5, 13, 14, 20] focusing on power management of homogeneous multi-core systems based on the control theory. [5] adapts the number of cores and frequency using an offline regression technique to keep the power below threshold. [14] allocates the chip power budget to each of the power islands based on absolute performance metric. [13] provided power

capping technique for many core system consisting of both single thread and multi-threaded applications. [15] present a hierarchical control system for power management in server farms. In contrast, our power management framework operates on an asymmetric architecture and selectively penalizes the cores and the tasks under thermal emergency. [6] developed energy-aware scheduling for a single task on Intel QuickIA heterogeneous platform with two cores. In contrast, we provide a general framework that can handle any number of tasks and cores, satisfy QoS and thermal constraints, minimize energy, and is implemented in Linux on a real platform.

3. ARM BIG.LITTLE ARCHITECTURE

The big.LITTLE architecture is a system-on-chip comprising high performance Cortex-A15 cluster and power efficient Cortex-A7 cluster (see Figure 1). The test chip we use in this work contains three A7 cores and two A15 cores. All the cores implement ARM v7A ISA. Table 2 in Appendix details the architecture configurations. The architecture provides DVFS feature per cluster. Note that all the cores within a cluster should run at the same frequency level. Moreover an idle cluster can be powered down if necessary. We now provide a detailed power- performance characterization of the architecture. The detailed experimental setup used for this study appears in Appendix 8.1.

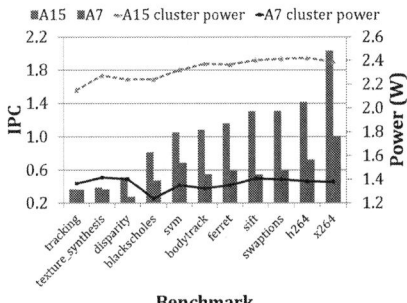

Benchmark

Figure 2: IPC & Power of A7 and A15 at 1GHz

Power-Performance Tradeoff. Figure 2 plots the Instructions Per Cycle (IPC) and the average power (Watt) for each benchmark on A7 and A15 cluster, respectively. For this experiment we set the frequency level of both clusters at 1GHz. Note that we can only measure the power at cluster level rather than individual core level. So the power reported in this figure corresponds to the power in a cluster even though only one core is running the benchmark application, while other cores are idle. Clearly, A15 has significantly better IPC compared to A7 (average speedup of 1.86) but far worse power behavior (1.7 times more power than A7 on an average).

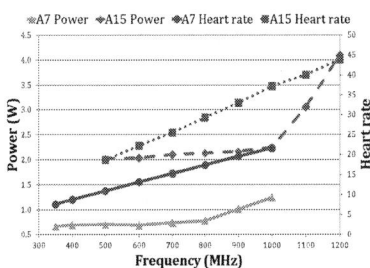

Figure 3: Power and heart rate with varying frequency.

Impact of DVFS. As mentioned earlier, our objective is to provide satisfactory user experience or QoS at minimal energy. We employ Heart Rate Monitor [8] infrastructure to set the performance goal and measure the QoS of a task. Heart rate is defined as the throughput of the critical kernel of a task, for example, number of frames per second for a video encoder. Figure 3 plots the heart rate and

1199

power for `blacksholes` from PARSEC benchmark suite on A7 and A15 at difference frequency levels. We observe that the heart rate increases linearly with increasing frequency on a core. Also as the IPC of A15 is better than A7, the heart rate can be improved by migrating a task from A7 to A15 at the same frequency level (but at higher power cost). Finally, the power generally increases linearly with increasingly frequency on a core; but there is a sudden jump at 800MHz for A7 and 1GHz for A15 due to change in voltage level.

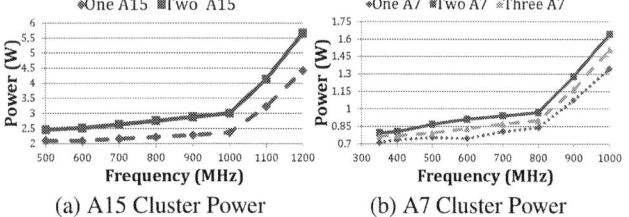

(a) A15 Cluster Power   (b) A7 Cluster Power

Figure 4: Impact of number of active cores on cluster power.

Impact of active cores on cluster power. As noted earlier, we can set frequency and measure power only at cluster level. Also we can only power down a cluster, but not individual cores. Thus, even if a core is idle, it still consumes power. Here we evaluate the impact of active cores on power consumption of the cluster. For this experiment, we run the same benchmark application on one, two, and three cores in A7 cluster as well as one and two cores on A15 cluster at different frequency levels. It is interesting to observe(Figure 4) that the A7 and A15 cluster have completely different power behavior with respect to the number of active cores.

In A7 cluster, even at the highest frequency level (1GHz), there is only 0.3 Watt difference between one active core and three active cores. In the A15 cluster, on the other hand, there is roughly 1.5 Watt difference in power between one active core and two active cores. For both clusters, it is important to perform load balancing and run all the cores at the lowest possible frequency level.

Migration Cost. Task migration across clusters is important to exploit the unique advantage of asymmetric multi-cores. We perform a set of experiments to quantize the migration cost within and across clusters (refer Appendix). We observed that the migration cost across clusters is higher than the cost within the cluster. Thus, task migration for load balancing within a cluster can be performed more frequently, whereas migration decisions across clusters should be taken at longer time intervals.

4. POWER MANAGEMENT FRAMEWORK

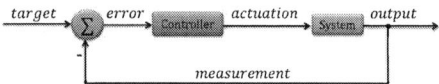

Figure 5: Feedback based Controller.

An overview of our hierarchical power management framework is presented in Figure 6. We incorporate several feedback based controllers in our framework. A controller measures the output metric and compares it with the reference or target metric as shown in Figure 5. The error is minimized by manipulating the actuators of the target system. The actuation policy is determined by the model of the target system being designed. We employ PID (Proportional-Integral-Derivative) controllers $z(t) = K_p e(t) + K_i \int e(t)dt + K_d \frac{de(t)}{dt}$, where $z(t)$, $e(t)$, K_p, K_i and K_d are the output of the controller, error, proportional gain, integral gain, and derivative gain, respectively.

We have two types of tasks in our system; *QoS* and *non-QoS* tasks. A *QoS* task is one that demands certain user-defined throughput (e.g., video encoder, music player), while the *non-QoS* tasks do not specify any QoS requirement. As noted in Section 3, we specify the QoS of a task in terms of its heart rate.

The framework consists of three different types of controllers: per-task resource share controller, per-cluster DVFS controller, and per-task QoS controller. Each QoS task in the system is assigned a resource share controller and a QoS controller. The resource share controller (bottom left in Figure 6) of a QoS task Q_i manipulates the CPU share available to Q_i so that it can meet the target heart rate $hr_{ref}(Q_i)$. The per-task QoS controller (top in Figure 6) is inactive when the entire system is lightly loaded. However, when the total power of the chip exceeds the TDP, the QoS controller slowly throttles the target heart rate $hr_{ref}(Q_i)$ so that the workload in the system decreases to a sustainable level and brings it back to the user- defined level when the thermal emergency is over.

We have two cluster controllers corresponding to A7 and A15 clusters. The objective of the controller for cluster Cl_m (bottom right in Figure 6) is to apply DVFS such that the utilization remains close to the target utilization $u_{ref}(Cl_m)$. The utilization of a cluster is determined by the maximum utilization of its cores. Thus, we periodically invoke a load balancer to ensure even utilization among the cores within a cluster. We also invoke a migrator periodically (at a much longer interval compared to the load balancer) to migrate the tasks between the clusters if necessary. Finally, we have a chip-level power allocator (extreme right in Figure 6) that throttles the frequency of the clusters and forces QoS controller to degrade target heart rates when the total power exceeds the TDP.

The key challenge here is to coordinate the various controllers, load balancer, migrator, and chip- level power allocator. We achieve a synergistic coordination with two mechanisms. First, the different components in our framework are invoked at different timescales. The per-task resource share controller and load balancer are invoked most frequently, followed by per-cluster DVFS controller and per-task QoS controller, then the migrator, and finally the chip-level power allocator. This ensures that a task attempts to reach its QoS target by first manipulating its share in a core or through migration within a cluster. If this fails, then it tries to change the frequency of the cluster. As a last resort, the task is migrated to another cluster. The thermal emergency takes quite a long time to develop; hence the power allocator is invoked least frequently.

Second, the controllers communicate with each other through designated channels. For example, the resource shares of the tasks within a core (both QoS and non-QoS) determines its utilization, which is provided as input to the cluster controller. More interestingly, when the power exceeds TDP, the power allocator increases the target utilization levels of the clusters $u_{ref}(Cl_m)$. This indirectly achieves the goal of decreasing power as the cluster controller is forced to lower its frequency in order to meet the increased target utilization. In parallel, the power allocator also sends a heart rate throttling factor ($hr_{throttle}(Q_i)$) to each QoS controller which makes them slowly degrade their target heart rate. This reduced heart rate is communicated to the resource share controller, who in turn, reduces the CPU share of the QoS tasks and hence the processor utilization to a more sustainable level. Overall, the system stabilizes to a level where the total power is just below the TDP.

Per-Task Resource Share Controller. We employ one resource share controller per QoS task. The target heart rate of a task Q_i is defined as a range $hr_{ref}(Q_i) = [hr_{ref}^{min}(Q_i), hr_{ref}^{max}(Q_i)]$ and is set by the QoS controller. The objective of the resource share controller is to keep the measured heart rate $hr(Q_i)$ in the target heart rate range. This is achieved by regulating the slice $s(Q_i)$ of time provided to the task Q_i in the scheduler. For example, a task that does not meet the minimum heart rate would demand more resource, which translates to more slices of time. The manipulation

1200

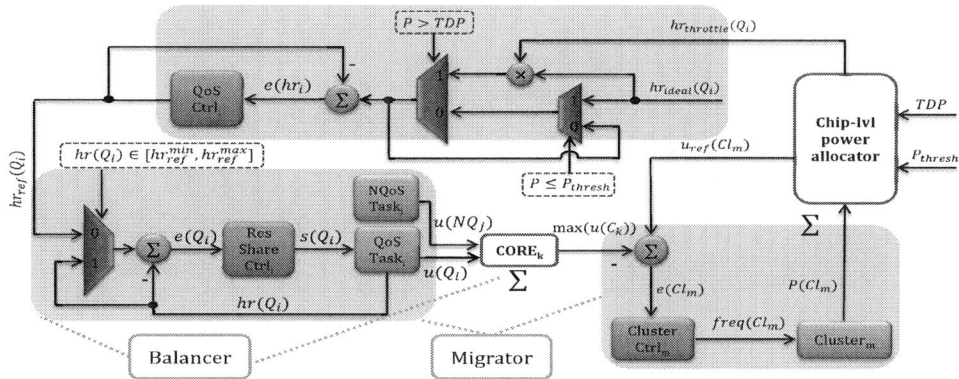

Figure 6: Overview of the hierarchical power management system coordinating multiple controllers.

of the slice value of a task within Linux completely fair scheduler is explained in detail in Appendix 8.1. If the measured heart rate is within the reference range, then the controller does not need any action and hence the target heart rate $hr_{ref}(Q_i)$ is set to the measured heart rate $hr(Q_i)$ so that error is zero in the controller.

Per-Cluster DVFS Controller. Let a core C_k consist of N QoS and P non-QoS tasks. Then its current utilization is $u(C_k) = \sum_{i=0}^{N} u(Q_i) + \sum_{j=0}^{P} u(NQ_j)$ where $u(Q_i)$ and $u(NQ_j)$ are the utilizations of the QoS task Q_i and non-QoS tasks NQ_j, respectively. The *core* component in Figure 6 is responsible for measuring the utilization of each individual core. As the frequency can be controlled only at cluster level, the utilization of cluster Cl_m defined as $u(Cl_m)$ is set to the maximum utilization $max(u(C_k))$ across all the cores within the cluster.

The DVFS controller attempts to achieve the target utilization $u_{ref}(Cl_m) = max(u_{ideal}, u_{target}(Cl_m))$ where u_{ideal} is a constant specifying the ideal target utilization and $u_{target}(Cl_m)$ is the target utilization set by the power allocator under thermal emergency. Using $max(u(C_k))$ as the measured metric and $u_{ref}(Cl_m)$ as the reference metric, the cluster-level PID controller actuates the frequency of the cluster.

Chip-Level Power Allocator. When the total power of the chip exceeds the TDP, the power allocator needs to throttle the frequency of the clusters and the QoS of the tasks. Let P_m be the current power measured for cluster Cl_m. The target power $\overline{P_m}$ for cluster Cl_m is calculated using the following equation

$$\overline{P_m} = P_m - \left((P - TDP) \times \frac{(T^{qos} - T_m^{qos})}{T^{qos}} \right) \quad (1)$$

where P is the total power of the chip given by $P = \sum_{m=0}^{M} P_m$, T^{qos} is the total number of QoS tasks in the system and T_m^{qos} is the total number of QoS tasks in the cluster Cl_m. From Equation 1, it is evident that the reference power allocated to the cluster is proportional to the number of QoS tasks in the cluster. From the reference power budget allocated to each cluster, the power allocator computes $u_{target}(Cl_m)$ using the following equation,

$$u_{target}(Cl_m) = u_{ideal} + u_{ideal} \times \frac{P_m - \overline{P_m}}{P_m} \quad (2)$$

In the event of TDP violation, the power allocator increases the target utilization $u_{ref}(Cl_m)$ of the cluster, which in turn causes cluster-level DVFS controller to set a lower frequency for the cluster. As our controllers are reactive in nature, the power may exceed the TDP for a short time interval. The gain factors within the DVFS controller are set appropriately so that it stabilities the power below the TDP within the specified time interval (typically few seconds [16]) as demonstrated in Section 5.

When TDP is violated, the power allocator also sets a throttle factor $hr_{throttle}(Q_i)$ for each QoS task Q_i in a hierarchical manner. The throttle factor $hr_{throttle}(Cl_m)$ for a cluster is proportional to its penalty factor as defined via higher than ideal utilization.

$$hr_{throttle}(Cl_m) = 1 - \frac{(u_{target}(Cl_m) - u_{ideal})}{u_{target}(Cl_m)} \quad (3)$$

The cluster throttle factor is further divided among the cores

$$hr_{throttle}(C_k) = hr_{throttle}(Cl_m) \times \frac{u(C_k)}{u_{avg}(Cl_m)} \quad (4)$$

where $u_{avg}(Cl_m)$ is the average utilization in cluster Cl_m across all the cores. Finally, the throttle factor of a QoS task in a core ensures that the penalty of a task is proportional to its utilization.

$$hr_{throttle}(Q_i) = hr_{throttle}(C_k) \times \frac{\sum_{i=0}^{N} u(Q_i)}{u(C_k)} \quad (5)$$

Once the system escapes from the thermal emergency, the power allocator needs to set back $hr_{throttle}(Q_i) = 1$. During the thermal emergency, the clusters reduce their frequency and the QoS tasks reduce their workload, the power decreases just below the TDP. However, the QoS of the tasks cannot be brought back to their ideal QoS level as the system will again oscillate back to thermal emergency. The QoS of the tasks can be restored only when the workload decreases because (a) one or more tasks leave the system and/or b) the tasks exhibit phase behavior. This is reflected in the drop in power consumption of the system. Thus, we chose an empirically determined power threshold P_{thresh} below which the $hr_{throttle}$ is set to one (as shown in Figure 6).

Per-Task QoS Controller. The QoS controller provides the graceful degradation of the QoS measure in case of thermal emergency by manipulating the target heart rate $hr_{ref}(Q_i)$. The input to this controller is the user-defined ideal heart rate range $hr_{ideal}(Q_i) = [hr_{ideal}^{min}(Q_i), hr_{ideal}^{max}(Q_i)]$. When the power is below the TDP, the power allocator sets $hr_{throttle}(Q_i) = 1$ and this controller sets $hr_{ref}(Q_i) = hr_{ideal}(Q_i)$. In case of thermal emergency, the controller strives to set the reference heart rate $hr_{ref}(Q_i) = hr_{throttle}(Q_i) \times hr_{ideal}(Q_i)$.

Load Balancer and Migrator. In our framework, the *Balancer* ensures that the cores within the cluster are evenly load balanced in terms of the utilization. The *Migrator* migrates the set of tasks that do not achieve their target heart rate at maximum frequency in the A7 cluster to the A15 cluster. Similarly, a task is migrated from A15 cluster to A7 cluster when the measured heart rate $hr(Q_i)$ is above the maximum target heart rate $h_{ref}^{min}(Q_i)$ at the minimum frequency in the A15 cluster.

1201

5. EXPERIMENTAL EVALUATION

We implement our hierarchical power management framework, called *HPM*, on Versatile Express Development Platform [2] that includes ARM big.LITTLE chip. We use Ubuntu 12.10 Linaro release for Versatile Express [3] and Linux kernel release 3.6.1. The chip is equipped with sensors to measure frequency, voltage, power, and energy consumption of each cluster. Detailed description of the prototype implementation appears in Appendix 8.1.

We deploy PID controllers for per-task resource share controllers, per-task QoS controllers, and per-cluster DVFS controller. The gain parameters used for these controllers, the intervals for invocation of different components in our framework, and additional parameters are all presented in Table 6 in Appendix.

We use applications from PARSEC [4], Vision [19] benchmark suites and *h264* from SPEC benchmark [1]. We use the sequential version of the PARSEC benchmarks as QoS tasks. We specify and track the heart rates for the QoS tasks using Heart Rate Monitor infrastructure [8] integrated with our Linux kernel. We use the applications from Vision benchmark suite as non-QoS tasks. The details of these benchmarks appear in Table 3 in Appendix. Note that some of the benchmarks are computationally demanding (e.g., *x264*) and requires hardware accelerators for execution. As we run software-only versions of these benchmark, they achieve low heart rate even on A15 core at highest frequency.

The evaluations are designed to demonstrate that HPM achieves the following objectives: (1) HPM can exploit asymmetry to provide significant energy savings compared to symmetric multi-cores, (2) HPM performs better than the Linaro scheduler, (3) HPM can respond to thermal emergency in a graceful manner, and (4) HPM does not interfere with the desired properties of Linux CFS, namely, fairness and non-starvation of the non-QoS tasks (see Appendix).

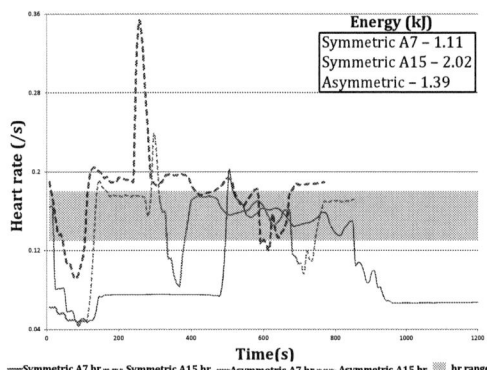

Figure 7: x264: Heart rate on symmetric & asymmetric multi-core.

Asymmetric versus symmetric multi-core. We use *x264* benchmark that exhibits phases with varying performance requirements during execution. The symmetric architectures are emulated using only A7 cluster or A15 cluster. We run *x264* benchmark on each of these configuration. All the configurations use HPM framework; but inter-cluster migration is disabled for symmetric architectures. Figure 7 plots the heart rate on the asymmetric and symmetric configurations. The heart rate line type specifies the cluster on which the task is running: continuous line corresponds to A7 cluster and dashed line corresponds to A15 cluster. The gray shaded area shows the specified heart rate range.

On symmetric configurations, the measured heart rate is below the minimum heart rate most of the time when executing on A7 cluster, while the heart reate mostly exceeds the maximum heart rate when running on A15 cluster. As expected, the energy consumption is very low (1.11kJ) in A7 cluster and quite high in A15 cluster (2.02kJ). The asymmetric multi-core provides the best of both worlds. On the asymmetric architecture, we can see that the application migrates to A15 cluster for the demanding phases and moves back to A7 cluster as the computational demand decreases. The HPM manages to maintain the heart rate within the reference range with a very low energy consumption (1.39kJ), which is 68% less than the energy consumption on A15 cluster alone.

HPM versus Linaro scheduler. We compare HPM scheduler with Linaro scheduler kernel release 3.6.1, where we activate the power conservative governor. The Linaro scheduler is aware of the different performance capabilities of the asymmetric cores, but it does not react to different performance requirements of the QoS tasks. Once the task load (defined as time spent on the runqueue of the processor) increases above a predefined threshold, the Linaro scheduler moves the task to the more powerful core. However, it never migrates the task back to the weaker core when workload reduces. We launch three QoS tasks, *x264*, *bodytrack*, *h264*, on three A7 cores. The results are shown in Figure 8. In all the subgraphs the X-axis shows the time in seconds. The Y-axis in the first three subgraphs shows the measured heart rate of the QoS tasks under HPM and Linaro. Additionally, the figure shows the specified heart rate range for the tasks as grey shared area. The last subgraph shows power comparison between the two approaches.

bodytrack and *h264* meet their specified heart rate on A7 cluster. As *x264* does not meet its heart rate on A7 all the time, it is migrated to A15 cluster by HPM when necessary. All the while, HPM keeps the heart rate of all the applications within the specified range. The Linaro scheduler, on the other hand, migrates all the tasks to the A15 cluster based on task load. As a result, the tasks complete execution much earlier compared to HPM; but exceeds the heart rate by a large margin consuming significantly more energy. On an average, the system consumes 2.27W using our scheduler compared with 5.83W consumed under Linaro scheduler.

Table 1 quantitatively shows the average power consumption and heart rate miss percentage (i.e., how much time a QoS task spends below its minimum specified heart rate) for HPM and Linaro scheduler using identical experimental setup but five different combination of QoS benchmarks. In general, a small loss in performance of the QoS tasks in our framework is heavily compensated by the average power reduction. The Linaro scheduler performs quite badly even in terms of performance in the two highlighted experiments. This is because the benchmarks are very demanding. Linaro scheduler moves them all to the A15 cluster, where they suffer from lack of resources, even at the highest frequency level. HPM uses the resources more efficiently and miss rate is reduced along with considerable reduction in power consumption. The results clearly demonstrate that HPM exploits the asymmetric architecture much more efficiently than current Linux scheduler.

Benchmarks	HPM scheduler		Linaro scheduler	
	Avg Power(W)	hr miss %	Avg Power(W)	hr miss %
swap_h264_x264	3.35	8.27	6.18	5.95
swap_h264_body	3.88	13.39	6.06	9.80
h264_body_black	**4.19**	**15.65**	**6.00**	**33.99**
black_x264_h264	**4.21**	**19.93**	**6.19**	**29.76**
x264_body_h264	2.27	9.61	5.83	7.41

Table 1: Quantitative comparison of HPM with Linaro scheduler.

Response under TDP constraint. This experiment evaluates the efficiency of HPM in managing the chip power below the TDP through DVFS and graceful degradation of the QoS of the tasks if necessary. For fair comparison, we add a feature to the Linaro

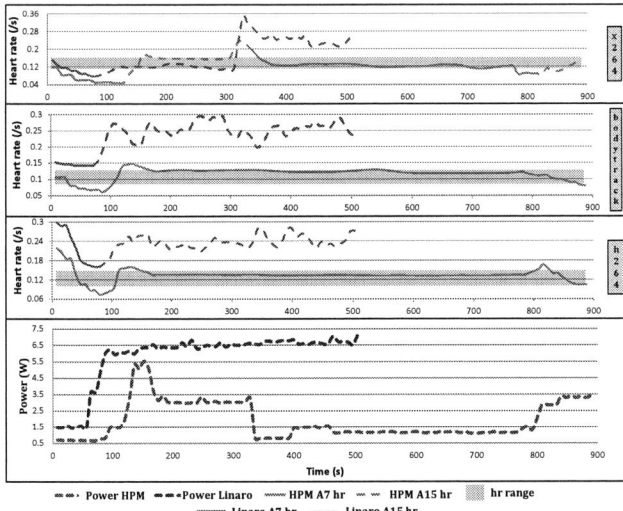

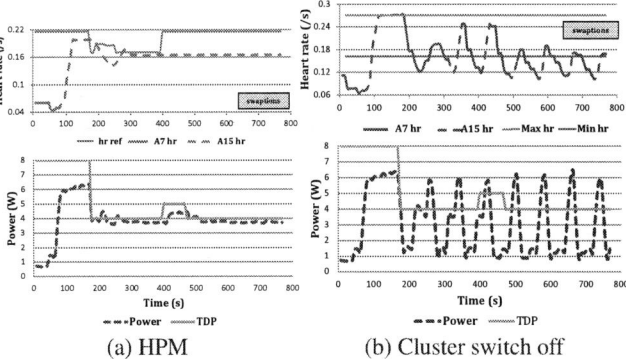

(a) HPM   (b) Cluster switch off

Figure 9: Comparison of HPM and Linaro extended with cluster switch-off policy under TDP constraint.

Figure 8: HPM versus stock Linaro scheduler equipped with DVFS governor and inter-cluster migration.

scheduler that switches off the A15 cluster once the power exceeds the TDP threshold. We use *bodytrack*, *swaptions*, and *h264* where the first two benchmarks have high workload and are migrated to A15 cluster. *swaptions* is the most demanding one and sets the frequency of the A15 cluster to the highest value. As we cannot control the frequency of individual cores, the core with *bodytrack* is forced to run at a higher frequency than required and hence its heart rate exceeds the target. Figure 9a shows the heart rate of *swaptions* (the application with maximum impact on power) together with the median value of the target heart rate range. The subgraph at the bottom of Figure 9a shows the chip power and the specified TDP cap. In this experiment, we dynamically change the TDP cap between 4-8W to demonstrate how the scheduler adapts to TDP budget. Once the chip power exceeds the TDP, the power allocator immediately increases the target utilization value of the clusters, which forces the DVFS controllers to decrease the frequency, and thereby reduce total chip power. Meanwhile, the power allocator also sets the heart rate throttle values, which in turn makes the QoS controllers reduce the target heart rates correspondingly bringing down the workload to a more sustainable level. HPM always maintains the heart rate of *swaptions* at the target value. Note that the target heart rate is decreased by the QoS controller when the power is above the TDP, thereby degrading the performance of the tasks. Once TDP is increased, the target heart rate switches back to the user-specified ideal value.

In case of the modified Linaro scheduler (Figure 9b) the A15 cluster is switched on and off frequently in response to increase in chip power beyond TDP. This oscillation happens because the workload is not throttled when the A15 cluster is switched off. As soon as A15 cluster is switched off, the power decreases much below the TDP, the tasks again migrate back to A15, the power increases above TDP, and the cycle continues. This frequent powering down of clusters and consequent migration makes *bodytrack* and *swaptions* run below their target heart rate most of the time under modified Linaro scheduler. This experiment confirms that a holistic approach is required to maintain the chip power below TDP; our approach not only decreases the frequency of the clusters but also solves the root cause of increased power by slowly degrading the QoS of the tasks. As a result, our approach reaches a stable and sustainable level both w.r.t. the heart rate and the chip power.

6. CONCLUSION

We present a power management framework for asymmetric multi-cores that carefully coordinates multiple controllers. It is integrated with Linux on ARM big.LITTLE platform. It exploits asymmetry among the cores through selective migration and employs DVFS to minimize power consumption while satisfying QoS constraints. Our technique combines graceful QoS degradation along with DVFS to reach stable and sustainable execution under TDP.

ACKNOWLEDGMENTS: This work was supported by CSR research funding, and Singapore Ministry of Education Academic Research Fund Tier 2 MOE2009-T2-1-033, MOE2012-T2-1-115.

7. REFERENCES

[1] SPEC CPU Benchmarks. http://www.spec.org/benchmarks.html.
[2] ARM Ltd., 2011. http://www.arm.com/products/tools/development-boards/versatile-express/index.php.
[3] Linaro Ubuntu release for Vexpress, November 2012. http://releases.linaro.org/12.10/ubuntu/vexpress/.
[4] Bienia et al. The PARSEC benchmark suite: characterization and architectural implications. PACT, 2008.
[5] Cochran et al. Pack & Cap: Adaptive DVFS and thread packing under power caps. In *MICRO*, 2011.
[6] Cong et al. Energy-efficient scheduling on heterogeneous multi-core architectures. In *ISLPED*, 2012.
[7] Esmaeilzadeh et al. Dark silicon and the end of multicore scaling. In *ISCA*, 2011.
[8] Hoffmann et al. Application heartbeats: A generic interface for specifying program performance and goals in autonomous computing environments. In *ICAC*, 2010.
[9] Koufaty et al. Bias scheduling in heterogeneous multi-core architectures. In *EuroSys*, 2010.
[10] Kumar et al. Single-ISA heterogeneous multi-core architectures: The potential for processor power reduction. In *MICRO*, 2003.
[11] Li et al. Efficient operating system scheduling for performance-asymmetric multi-core architectures. In *ACM/IEEE conference on Supercomputing*, 2007.
[12] Li et al. Operating system support for overlapping-ISA heterogeneous multi-core architectures. In *HPCA*, 2010.
[13] Ma et al. Scalable power control for many-core architectures running multi-threaded applications. *ACM SIGARCH*, 2011.
[14] Mishra et al. CPM in CMPs: Coordinated power management in chip-multiprocessors. In *High Performance Computing, Networking, Storage and Analysis (SC), 2010 International Conference for*.
[15] Raghavendra et al. No power struggles: Coordinated multi-level power management for the data center. In *ACM SIGOPS*, 2008.
[16] Rotem et al. Power-management architecture of the intel microarchitecture code-named sandy bridge. *MICRO*, 2012.
[17] Saez et al. A comprehensive scheduler for asymmetric multicore systems. In *EuroSys*. ACM, 2010.
[18] Van Craeynest et al. Scheduling heterogeneous multi-cores through performance impact estimation (pie). In *ISCA*, 2012.
[19] Venkata et al. Sd-vbs: The San Diego Vision Benchmark suite. In *IISWC*, 2009.
[20] Wang et al. Adaptive power control with online model estimation for chip multiprocessors. *Parallel and Distributed Systems, IEEE Transactions on*, 2011.

8. APPENDIX

8.1 big.LITTLE Platform with Linux

In this section, we explain in detail our target platform and prototype implementation of our HPM technique. In our evaluation, we used the real Versatile Express development platform [2] as shown in Figure 10. It is a flexible, configurable and modular developing platform that allows quick prototyping of hardware and software projects. The system comprises a motherboard on which modular daughter boards can be plugged. The big.LITTLE processor is part of the daughter board (TC2) pointed in the Figure 10. Table 2 describes the architecture of the A15 (big cores) and A7 (LITTLE cores). The motherboard handles the interconnection between the daughter board and the peripherals by using a FPGA bus interconnection network.

Figure 10: Picture of the Vexpress board.

Processor	Cortex-A7	Cortex-A15
Issue Width	2	3
Pipeline Stages	8-10	15-24
L1$	32kB 4-way	512kB 2-way
L2$	512kB 8-way	1MB 16-way
Frequency Levels	8	8
Frequency Range(MHz)	350-1000	500-1200
Voltage Range(mV)	900 - 1050	900 - 1050

Table 2: Cortex-A7 and Cortex-A15 specifications.

The board boots an Ubuntu 12.10 Linaro release for Versatile Express [3]. The platform firmware runs on an ARM controller (MCC) embedded on the motherboard and handles the load of the Linux kernel while booting. The Linux file system is installed on the Secure Digital (SD) card where all our benchmarks are saved. The TC2 daughter board is also equipped with sensors for measuring the frequency, voltage, current, power and energy consumption per cluster. The board also supports the change of voltage and frequency per cluster.

The migration cost among cores within A15 cluster is 54 μsec – 105 μsec depending on the frequency level, while the cost within A7 cluster is 71 μsec – 167 μsec. However, the migration costs between clusters are somewhat high: 1.88ms – 2.16ms for moving from A7 to A15 cluster at different frequency levels, and 3.54ms – 3.83ms for a move from A15 to A7 cluster.

Heart Rate Monitor. We use the Application Heartbeats framework proposed in [8] as a mechanism to measure the performance of an application. The API provided in this framework provides a QoS metric in terms of *heartbeats* which are periodic signals sent by an application to track its progress. The QoS metric provided by the framework is called *heartrate* (i.e, the number of heartbeats per second). For example, in video encoding applications the heartbeats can be registered every frame. Thus, the heart rate measured would be the number of frames per second. The interested reader is referred to [8] for more information on Heartbeats Framework.

Table 3 describes the benchmarks used in our experiments together with the inputs. Table 4 summarizes heartbeat insertions in the benchmarks [8].

Benchmark	Heartbeat Location
swaptions	Every "swaption"
h264	Every frame
bodytrack	Every frame
x264	Every frame
blackscholes	Every 25000 options

Table 4: Heartbeats in QoS benchmarks.

Profiling Section. The Linux version [3] provides hardware monitor (*hwmon*) interface to communicate with the sensors located in the test chip. We use the perf tool provided with the kernel [3] to obtain performance related metrics like instructions per cycle (IPC). Powering off the cluster and adjusting the clock frequency were made possible by accessing the oscillator related drivers provided in the kernel. The legal voltage and frequency ranges for the clusters are shown in Table 2.

Managing task slice and Migrator. Linux scheduler uses the notion of time slicing for allocating the resources to the running tasks in the system. At every system tick (10ms in our experiments), the kernel computes the time slice that the next tasks should receive. By default, the CFS scheduler fairly divides a relatively fixed period of time (6ms in our experiments) and allocates the slice to the task. The slice dictates the duration for which the task can consume the core. Our Resource Share Controller manipulates the computed slice for the QoS task by the original Linux mechanism by gradually increasing the time slice when a higher utilization is required or reducing the slice when less utilization is required. For non-QoS tasks, the CFS scheduler will try to fairly share the remaining time period among them. Linux kernel uses cpumask to decide the affinity of the tasks. Migrator component in our HPM alters the cpumask associated with each task to attain the desired scheduling decision.

In Table 5 we show the minor modifications that we did in the Linux kernel in order to implement our HPM scheduler.

8.2 Controller Features

Table 6 summarizes our HPM framework, describes the terminologies and provides the gain factor values associated with each of the controllers employed in our experiments. The invocation period of RSC is a user-defined value. For example, for video encoding it can typically be 30 frames per second, which translates to RSC being evoked every 30 frames. The invoke period was chosen in such a way that the per-task resource share controller and load balancer are invoked most frequently followed by DVFS controller, per-task QoS controller, migrator and finally the chip-level power allocator.

Benchmark	Benchmarks suite	Description	Inputs
swaptions	PARSEC	QoS — Monte Carlo (MC) simulation to compute the prices.	sim_native
bodytrack	PARSEC	QoS — Tracks a human body with multiple through a series of image sequence.	sim_native
x264	PARSEC	QoS — Video encoder.	sim_native
blackscholes	PARSEC	QoS — Solves partial differential equation to calculate the prices for a portfolio.	sim_native
h264	SPEC 2006	QoS - Video encoder.	foreman
disparity	Vision	non-QoS — Motion, tracking and stereo vision	fullhd
sift	Vision	non-QoS — Image Analysis	fullhd
tracking	Vision	non-QoS — Motion, tracking and stereo vision	fullhd

Table 3: Benchmarks description.

Function	Description	# lines
scheduler_tick()	Fire controllers based on system tick.	30
load_balance()	HPM Balancer within the cluster.	12
run_rebalance_domains()	HPM Migrator across the clusters.	47
sched_slice()	Manipulate the QoS time slices.	5

Table 5: Linux kernel modifications.

Controller name	Metrics	Symbol	Value
Resource Share Controller (RSC)	target heart rate	hr_{ref}	tuned by QoSC
	measured heart rate	hr	measured by the task
	slice	s	actuator tuned by RSC
	proportional gain	K_p^{RSC}	0.8512
	integral gain	K_i^{RSC}	0.01241
	derivative gain	K_d^{RSC}	0.00941
	invoked period	$T^{RSC} = \beta \times \frac{1}{hr_{ideal}}$	determined by the hr_{ideal}
	heart rate measurement frequency factor	β	user-defined
CORE component	core utilization	u_k	measured by each core
	invoked period	$T^c = 4 \times max(T^{RSC}(Q_i))$	determined by the task with max hr_{ideal}
DVFS Controller (DVFSC)	target cluster utilization	u_{ref}	estimated by CHIP component
	measured cluster utilization	$max(u(C_k))$	measured by CORE component
	cluster frequency	$freq$	tuned by DVFSC
	proportional gain	K_p^{DVFSC}	0.9533
	integral gain	K_i^{DVFSC}	0.2572
	derivative gain	K_d^{DVFSC}	0.0014
	invoked period	$T^{DVFSC} = 5 * T^c$	slower than the CORE component
QoS Controller (QoSC)	ideal hr	hr_{ideal}	user-defined
	throttle factor	$hr_{throttle}$	estimated by CHIP component
	target reference hr	$hr_{ideal} \times hr_{throttle}$	product of ideal hr and throttle factor
	measured reference hr	hr_{ref}	measured by the task
	proportional gain	K_p^{QoSC}	0.74175
	integral gain	K_i^{QoSC}	0.0214
	derivative gain	K_d^{QoSC}	0.0045
	invoked period	$T^{QoSC} = 30 * T^{RSC}$	much slower than RSC
CHIP level power allocator	thermal design power	TDP	user-defined
	threshold power	P_{thresh}	user-defined
	throttle factor	$hr_{throttle}$	estimated by CHIP component
	invoked frequency	$T^{ch} = 2 \times T^{QoSC}$	slower than DVFSC
Balancer	invoked period	$T^b = 2 \times T^{RSC}$	faster than DVFSC
Migrator	invoked period	$T^m = 4 \times T^{QoSC}$	slower than QoSC

Table 6: Controller Parameters.

8.3 Additional Experiment Results

We now present some additional experimental results.

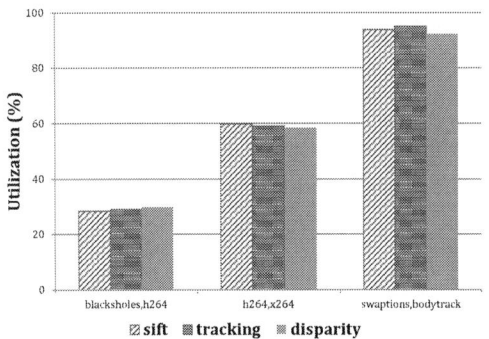

Figure 11: Fairness of non-QoS tasks.

Fairness of Non-QoS tasks. Our HPM framework is built on top of the existing Linux kernel scheduler. This set of experiments validate that we do not interfere with the scheduling of the non-QoS tasks handled by the Completely Fair Scheduler (CFS), which guarantees equal (fair) share of processor utilization among the tasks.

We run three experiments each with three non-QoS tasks (*sift*, *tracking*, *disparity*) and two QoS tasks. The behavior of the QoS tasks dictates the amount of A7 cluster utilization that CFS can provide to the non-QoS tasks. The first experiment uses *blackscholes* and *h264* as QoS tasks that satisfy the target heart rate on A7 cluster with close to maximum target utilization. Thus the CFS scheduler clusters together non-QoS tasks on the third available core. Figure 11 shows that the non-QoS tasks have equal share of utilization (33%).

The second experiment involves *h264* and *x264*; *x264* has a demanding execution phase where HPM migrates it to A15 cluster. Mostly the three non-QoS tasks run on their cores receiving 60% utilization, while *h264* runs on A7 cluster.

The final experiment uses *swaptions* and *bodytrack*, both of which migrate to A15 cluster and non-QoS tasks receive almost 100% of the A7 cluster utilization.

Peak Power Reduction and Workload Balancing by Space-Time Multiplexing based Demand-Supply Matching for 3D Thousand-core Microprocessor

Sai Manoj P. D., Kanwen Wang, and Hao Yu
School of Electrical and Electronic Engineering
Nanyang Technological University, Singapore 639798

ABSTRACT

Space-time multiplexing is utilized for demand-supply matching between many-core microprocessors and power converters. Adaptive clustering is developed to classify cores by similar power level in space and similar power behavior in time. In each power management cycle, minimum number of power converters are allocated for space-time multiplexed matching, which is physically enabled by 3D through-silicon-vias. Moreover, demand-response based task adjustment is applied to reduce peak power and to balance workload. The proposed power management system is verified by system models with physical design parameters and benched power traces, which show 38.10% peak power reduction and 2.60x balanced workload.

Categories and Subject Descriptors: B.7.2 [Design Aids]

Keywords: Demand-supply matching, Peak power reduction, Workload balancing, 3D thousand-core

1. INTRODUCTION

Exa-scale cloud computing for big-data applications requires integration of many-core microprocessors on a single chip [1, 2] at thousand-core scale. Though 3D integration is one promising solution [3] to increase integration density and communication bandwidth, the provision of many-core power supply voltages with maintenance of low power density has become an unresolved issue to address [4, 5, 6, 7, 8]. Supplying same voltage-level to all cores will result in high power density because the demand of each core can be different at different time instant. As such, a demand-supply matched dynamic voltage and frequency scaling (DVFS) scenario needs to be employed during power management for both peak power reduction and workload balancing.

From physical hardware perspective, an optimal demand-supply matching requires on-chip power converters [5, 6, 7, 8], which can provide prompt DVFS management with efficient power delivery. However, one power converter for one core has large area overhead in presence of non-scalable buck inductor. The design of single-inductor-multiple-output (SIMO) power converters [6] utilizes one common single buck inductor to provide different voltage-levels at different time slots in a time-multiplexed manner. The capability of SIMO is, however, still limited for many-core microprocessors at thousand-core scale. Moreover, considering hundreds of cores to be integrated

on one chip, the remaining area is quite limited to consider on-chip power converter with buck inductor. The 3D integration introduces additional room for on-chip power converters. The recent work in [8] has demonstrated the possibility to design power converter on one die and 64-tile network-on-chip on the other die, which are integrated by through-silicon-via (TSV).

From cyber management perspective, the power management for many-core power-supply system will no longer be the same as the one for the traditional single-core. For big-data applications, there may exist various power patterns deployed on many-cores with multi-time-scale demands for power supply. Moreover, there are many microprocessor cores but limited power converters. A number of power management works for many-core microprocessor system have been explored before [5, 6, 7, 8] but with not fully resolved challenge that requires to not only match various demands from microprocessors with limited number of power converters, but also to reduce peak power and to balance workload on a power converter. As such, the smart power management of many-core microprocessor has similarity as smart-grid though at different time-scale with different workload behaviors. Thereby, the study of workload behavior with classification and also the demand-response can be leveraged from smart-grid management [9] to deal with the on-chip demand-supply matching problem.

In this paper, a space-time multiplexing (STM) based DVFS power management is utilized for demand-supply matching between many-core microprocessors and power converters. In each power management cycle an adaptive clustering is developed such that the minimum number of power converters are allocated for different groups classified by power-magnitudes, called *space multiplexing*. In one group, power converters are further reused in different time slots for different subgroups classified by power-phases, called *time multiplexing*. Such a space-time multiplexed matching is physically enabled by designing a reconfigurable power switch network with the use of 3D through-silicon-vias (TSVs). Moreover, demand-response based task adjustment is applied to reduce peak power and to balance workload. The proposed power management system is verified by system models in SystemC-AMS. The physical design parameters are based on 130nm CMOS process with TSV models. Experiment results show that the proposed power management can achieve 38.10% peak power reduction and 2.60x balanced workload.

The rest of this paper is organized as follows. In Section 2, we present the 3D many-core microprocessor system architecture with space-time multiplexing (STM) problem formulation towards demand-supply matching. In Section 3, we show the solution by STM-based resource allocation of power converters with use of adaptive clustering, which is based on singular-value-decomposition (SVD) analysis of workload correlation. We further show the demand-response based task scheduling to utilize demand slacks and to adjust tasks for both peak power reduction and workload balancing. The experiment results are included in Section 5 with conclusion in Section 6.

Permission to make digital or hard copies of all or part of this work for personal or classroom use is granted without fee provided that copies are not made or distributed for profit or commercial advantage and that copies bear this notice and the full citation on the first page. To copy otherwise, to republish, to post on servers or to redistribute to lists, requires prior specific permission and/or a fee.

DAC '13, May 29-June 07 2013, Austin, TX, USA.

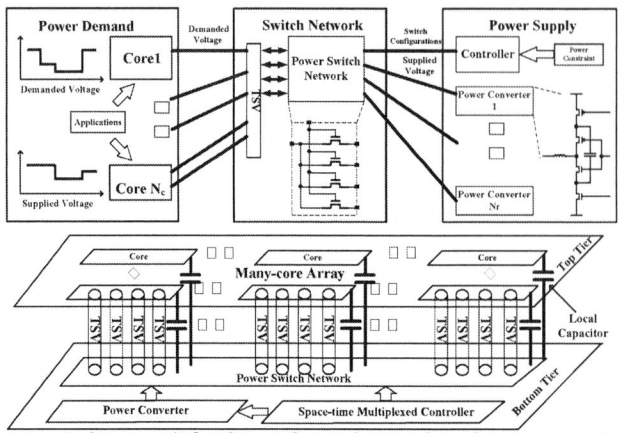

Figure 1: 3D reconfigurable power switch network for demand-supply matching between on-chip multi-output power converters and many-core microprocessors

Table 1: Notations and definitions

Notations	Definitions
$V = \{v_1, \ldots, v_{N_v}\}$	Set of voltage levels
$I = \{i_1, \ldots, i_{N_v}\}$	Set of core current loads
$R = \{r_1, \ldots, r_{N_r}\}$	Set of power converters
$C = \{c_1, \ldots, c_{N_c}\}$	Set of cores
$P = \{p_1, \ldots, p_{N_c}\}$	Set of power trace patterns
$S = \{s_1, \ldots, s_{N_s}\}$	Set of switch boxes
$G = \{g_1, \ldots, g_{N_v}\}$	Set of groups
$K = \{k_1, \ldots, k_{N_k}\}$	Set of subgroups
$A = \{a_1, \ldots, a_{N_k}\}$	Set of slacks
$L = \{l_1, \ldots, l_w\}$	Set of workloads
$B = \{b_1, \ldots, b_r\}$	Set of priorities
$v_d(c_i) \in V$	Demanded voltage-level of core c_i
$v_a(c_i) \in V$	Supplied voltage-level to core c_i
$v(r_i) \in V$	Output voltage-level of converter r_i
$d(r_i) \in V$	Output driving ability of converter r_i
I_L	Maximum converter inductance current
ΔV	Maximum core supply-voltage drop
H	Time slot for time-multiplexing
P_{th}	Peak power threshold

2. 3D SYSTEM ARCHITECTURE WITH SPACE-TIME MULTIPLEXING

In this section, 3D many-core microprocessor system architecture with a reconfigurable power switch network is reviewed with a space-time multiplexing (STM) problem formulated for power management. Table. 1 summarizes necessary notations used in this paper.

2.1 3D System Architecture

As shown in Fig. 1, the 3D many-core microprocessor system architecture is basically composed of two tiers. The bottom tier is for power management, including arrays of power converters and power switches. Each power converter is SIMO type, capable of supplying multiple voltage-levels by one buck inductor. The top tier includes array of many-core microprocessors. In between these two array-structured tiers, there are through-silicon-vias (TSVs), controlled by power switches, to connect power converters and cores. Moreover, there is one local super-capacitor for each core, working as local power storage to supply voltage during the multiplexing when power converter is not available.

The proposed 3D system architecture can be described by a demand-supply system model composed of the following three components:

- *Power Demand*: a set of cores C with demanded voltage-levels with set-size N_c. Each core c_i has a demanded voltage-level $v_d(c_i)$ to meet the deadline of its running workload. In addition, $v_a(c_i)$ is the allocated voltage-level to c_i after power management.

- *Power Supply*: a set of power converters R with set-size N_r. Each power converter outputs the voltage-level $v(r_i) \in V$ to supply the cores, where V is the set of available voltage-levels before power management;

- *Power Switch Network*: a set of reconfigurable switch-boxes S with set-size N_s to connect between R and C for demand-supply matching.

2.2 Space-Time Multiplexing Problem

As aforementioned in the introduction, the primary challenge in 3D thousand-core system to support exa-scale computing is to solve a large-scale demand-supply matching problem. Though there are various big-data applications with different power patterns, most of their power profiles can be still classified by magnitudes and phases. As such, if one can perform a detailed power profile characterization by clustering cores with similar power behaviors, the complexity in matching may be accordingly reduced. With the further consideration for implementation with the minimum cost of power converters, it is still feasible to formulate a resource (power converter) allocation problem with constraints of demand and supply matching. As such, one can formulate the first subproblem as follows.

Subproblem 1: Resource Allocation Problem is to decide the minimum number of power converters such that demand-supply matching can be satisfied.

What is more, due to spatial and temporal variation of power profiles, there may exist lots of power slacks to be utilized for a demand-response based workload scheduling. Without violating the workload execution priority or deadline, one can delay over-loaded workloads in one time-slot to the other time-slot with under-loaded workloads. As such, the peak power can be reduced as well as the workload can be balanced at power converters, which can be formulated as the second subproblem below.

Subproblem 2: Workload Scheduling Problem is to delay over-loaded workloads to under-loaded time-slots based on availability of slack and without violation of priority.

In this paper, we show that based on the aforementioned 3D system architecture, a space-time multiplexing (STM) based power management can be developed to solve the two subproblems in sections 3 and 4, respectively.

3. ADAPTIVE CLUSTERING BASED RESOURCE ALLOCATION

This section deals with resource allocation by adaptive clustering, resulting in the use of the minimum number of power converters for matched demand-supply. To deal with a large-scale demand-supply matching problem, we start with classification of cores into clusters by studying their power

1208

profile characteristic within one power management control-cycle T_c.

3.1 Grouping by Power Magnitude for Space Multiplexing

Grouping is the process of clustering different cores, which have similar power magnitudes and hence will demand the similar voltage-level.

Note that z-th group g_z, $g_z \in G$, can be formed by the following criteria

$$g_z = \{c_i; v_d(c_i) = v_d(c_j) = v_z, \forall i, j = 1, ...N_c, z \leq N_v\}. \quad (1)$$

Here, v_z, $v_z \in V$ represents the z-th voltage-level and c_i, $c_i \in C$ and $v_d(c_i) \in V$.

Based on the power magnitude levels, different groups are formed. Each group may contain different number of cores, which can have similar power magnitudes but maybe different power phases. The group formulation can change at different control-cycle. Based on the partitioned groups, power converters can be also partitioned in space to provide the specified voltage-levels for groups. This grouping process has less complexity because it involves just numerical comparisons.

3.2 Subgrouping by Power Phase for Time Multiplexing

Subgrouping is the process of clustering different cores, which have similar power phase (or pattern) and are within the same group.

Subgroup k_s, $k_s \in K$, can be formed by the following criteria

$$k_s = \{c_i; (v_d(c_i) = v_d(c_j) = v_z) \& (p_i \sim p_j), \forall i, j = 1, ...N_c\}. \quad (2)$$

Here, p_i, $p_i \in P$, represents the phase or pattern of one power trace of the core c_i, $c_i \in C$. $v_d(c_j)$ represents the demanded voltage-level of core c_i and v_z represents the z-th voltage-level, v_z, $v_d(c_j) \in V$. However, the subgrouping by phase is more difficult than grouping by magnitude and may consume bit more time in clustering. In the next subsection, we show a solution by means of spectral clustering to perform subgrouping of power profiles, which can be easily deployed to make power management faster compared to the one without subgrouping. Moreover, all the computations can be pre-stored in a look-up-table for implementing a real-time control.

3.3 Spectral Clustering for Subgrouping

Spectral clustering algorithm is discussed below. To find similarity between two power profiles p_i and p_j, $p_i, p_j \in P$, with N samples in one control-cycle, correlation in term of covariance matrix can be evaluated by

$$X = \frac{1}{N} \sum_{i,j=1}^{N} (p_i - \overline{P})(p_j - \overline{P})^T \quad (3)$$

where $\overline{P}$ is the mean of all power profiles ($\frac{1}{N} \sum_{i=1}^{N} (p_i)$).

Based on the order of covariance matrix, number of clusters, K can be analyzed by the singular-value-decomposition (SVD) of covariance matrix

$$X = U \times S \times V^{-1}. \quad (4)$$

Matrices U and V are orthogonal matrices with S as the diagonal matrix. Based on the rank analysis of S, the number of clusters K can be decided. A new matrix can be formed with K independent vectors, extracted from either of orthogonal matrices. Let the newly formed matrix be V_K, assuming it is extracted from V. The product of V_K with the covariance matrix X

$$X_K = X \times V_K \quad (5)$$

will result in a reduced matrix X_K, which becomes the basis of spectral clustering for subgrouping. For example, one core will be allocated to i-th subgroup if the value of $X_K(j, i)$ is the maximum in jth-row. The procedure for subgrouping is described in Algorithm 1.

Algorithm 1 Subgrouping by correlation extraction and spectral clustering

---

INPUT: Power trace matrix P with p_i power trace vectors after grouping
1. Compute covariance matrix $R \in R^{p_i \times p_i}$
2. Perform SVD: $R = U \times S \times V^{-1}$
3. Determine number of clusters: $K = rank(S)$
4. Compute the first K singular-value vectors $v_1, ...v_K$ of V
5. Let $V_K = [v_1, ..., v_K] \in R^{N \times K}$ and $R_K = R \times V_K$
6. Add *ith* core to *jth* cluster if $R_K(i, j)$ is maximum in the *ith* row
7. Form P_K matrices by finding corresponding indices in power trace matrix P
OUTPUT: New clustered subgroup matrices $P_K, (k = 1, ..., K)$

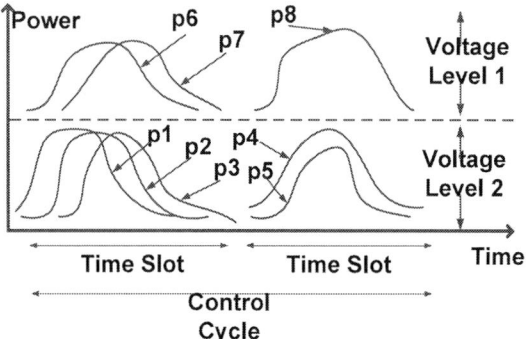

Figure 2: Grouping and subgrouping based on power levels and power phases

The formulation of groups and subgroups are illustrated in Fig. 2. At one control-cycle, power traces p_1, p_2, p_3, p_4 and p_5 are operating at one power magnitude level and other cores are working at a different power magnitude level. As such, one can form two groups with two voltage-levels v1 and v2. Inside the group supplied by voltage-level v1, one can observe that p_1, p_2, p_3, have a similar power phase compared to p_4 and p_5; so p_1, p_2 and p_3 further form a subgroup and p_4 along with p_5 forms another subgroup. The formed groups and subgroups can change at the next control-cycle.

In the following, we show that with the help of adaptive clustering, one can find the minimum number of power converters to satisfy the demand-supply matching. Moreover, by clustering, the complexity from the demand (power profiles) can be significantly reduced. As such, the large-scale demand-supply matching can be efficiently solved by the proposed two-step clustering in every control-cycle.

3.4 Solution to Subproblem 1

Once subgroups are formed, the maximum workloads of one subgroup can be determined. As such, the minimum number of power converters can be also determined to supply that subgroup. This results in one feasible solution to solve the Subproblem 1 in Section 2 as rephrased below.

$$
\begin{aligned}
\text{min:} \quad & \sum_{i=1}^{N_v} r_i \\
\text{s.t.:} \quad & \text{(i)} \; v_a(c_j) \geq v_d(c_j), \forall c_j \in C. \\
& \text{(ii)} \; d(r_i) \leq N_{max}, \forall r_i \in R.
\end{aligned}
\quad (6)
$$

If one can determine the minimum number of power converters r_i for each group, the total number of power converters can be correspondingly minimized. Note that constraint (i) guarantees that the supplied voltage-level $v_a(c_j)$ from power converter will

1209

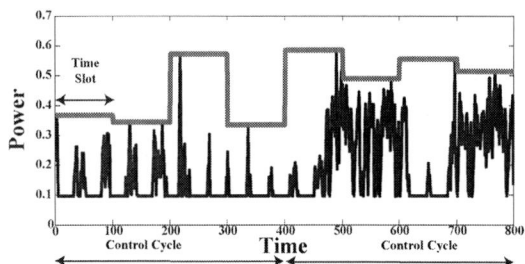

Figure 3: Peak power envelope extracted in each time-slot in one control cycle

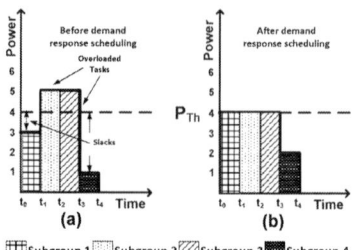

Figure 4: (a) Load before demand response scheduling (b) Peak reduction by demand response scheduling

satisfy the demanded voltage-level $v_d(c_j)$ from core c_j. Moreover, constraint (ii) imposes the driving ability $d(r_i)$ of each power converter is N_{max}, i.e., the maximum number of cores to drive. The driving ability can vary with the voltage-level: the higher the voltage-level is, the lower the number of cores that one power converter could drive.

Next, we show that the minimization of total number of power converters can be solved by grouping and subgrouping. By performing grouping, power converters are shared in space among N_v number of groups and subgrouping makes sharing of power converters inside a group in time. Based on the driving capability d_i^j of i-th power converter in group g_j, $g_j \in G$, having k subgroups, and the maximum number of cores among different subgroups, $max(c_i)$, $c_i \in C$, the minimum number of power converters for group g_j can be determined as

$$r_{g_j} = max(c_i)/d_i^j.$$

As such, for the whole system, the total number of power converters needed will be $\sum(r_{g_j})$, which is the minimum number to satisfy the demand-supply matching.

4. DEMAND RESPONSE BASED WORKLOAD SCHEDULING

This section deals with peak reduction and load balancing after the minimum number of power converters are allocated. A demand-response based workload scheduling will be developed towards uniform distribution of workload with reduction in peaks at one power converter.

4.1 Peak Power Envelope Extraction

To deal with peak reduction and load balancing, we first discuss the extraction of *peak power envelope* in one control-cycle, because it is impractical to perform power management in continuous form. Based on the extracted peak power envelope, one can build workload behavior model for each subgroup to be used in scheduling.

Assume that in one control-cycle T^i for the ith-group, g_i, $g_i \in G$ with N_k number of subgroups, each core is assigned with one workload. One can have *time slot* T_j^i, which is is the amount of time to finish all workloads in a subgroup, k_j, $k_j \in K$. Relation between T^i and T_j^i is

$$T^i = \sum_{j=1}^{N_j} T_j^i. \tag{7}$$

As such, in one time-slot T_j^i, peak power envelope Pe is extracted for workloads $p(t)$ of one subgroup by

$$Pe(T_j^i) = max(p(t)). \tag{8}$$

This is repeated for whole control cycle T^i. Thus peaks are extracted and one envelope is formed. Peak extraction by forming one envelope is shown in Fig. 3. The control-cycle T^i is 400ns with time-slot T_j^i of 100ns. At each time slot, the power envelope is formed on the peak value.

4.2 Peak Reduction and Load Balancing

When the peak envelope of subgroup k_i, $k_i \in K$ is compared with one threshold power P_{th}^i of group g_i, the *slack* can be calculated by

$$a_j^i = Pe(T_j^i) - P_{th}^i. \tag{9}$$

If the value of slack is positive, then the allocated power converter, r_j, $r_j \in R$, is overloaded and not capable of handling extra load at that time-slot; otherwise, the power converter r_j is underloaded and can be allocated with additional workloads. After calculating the amount of slack, the workload of the power converter r_j can be rescheduled such that priority is not violated.

We call such a scheduling as *demand-response* based workload scheduling. The procedure for scheduling is described in Algorithm 2. It is deployed after clustering to decide the time slot. The first step in scheduling is to calculate the threshold and slack. Line 2-4 of Algorithm 2 explains the scheduling of task from a power converter that is overloaded and reduction of corresponding load. Similarly Line 6-8 describes adding of workloads on an underloaded power converter. In short, it can be viewed as re-clustering or refinement. The overhead includes the time to perform the calculation and movement, which is negligible in the whole control cycle.

Algorithm 2 Demand-response based workload scheduling

1: **INPUT:** Initial set Workload L, Slack A
2: **if** $a_j^i > 0$ **then**
3:     Decrease workload on r_j
4:     $l(r_j) - -$;
5: **else**
6:     **while** $a_j^i < 0$ **do**
7:         Increase workload on r_j
8:         $l(r_j) + +$;
9:         $a_j^i + +$;
10:     **end while**
11: **end if**

Example in Fig. 4 shows the peaks of four subgroups. Before performing demand-response based workload scheduling, subgroup 2 and subgroup 3 are overloaded and subgroups 1 and 4 have slacks for scheduling. The peak value in subgroup 2 and 3 is 5, which means there are 5 peaks in those two subgroups. The peak power reduction is then achieved with the comparison of the highest value in subgroups before and after the demand-response scheduling. After the demand-response scheduling, the peak value will be reduced to 4. So, a 20% peak power reduction will be achieved.

4.3 Solution to Subproblem 2

The aforementioned demand-response based workload scheduling can be deployed to solve the Subproblem 2 addressed in Section 2, which is reformulated as

$$\text{min:} \quad \sum_{j=1} |\sum_{i=1} s_j^i| \tag{10}$$

$$\text{s.t.:} \quad Pe(T_j^i) < P_{th}^i$$

Table 3: Clustering result for 64 cores

	Cluster 1	Cluster 2	Cluster 3	Cluster 4
Group 1	31, 37, 52 58, 59, 63	12, 49, 54	33, 43	7, 8, 14
Group 2	27, 29	17, 40, 41, 50 51, 56, 62	22, 42	N/A
Group 3	6, 21, 32 36, 39, 46 47, 64	9, 15, 16 20, 26 ,28 35, 53, 55	1, 5 11, 18 19, 38	N/A
Group 4	2, 3, 23, 25 34, 44, 45, 48 57, 60, 61	10, 13	N/A	4, 24, 30

Table 4: Comparison of number of allocated power converters under different PM schemes

		STM	SM	TM	STM/SM	STM/TM
32-core	Group 1	1	2	3	-50.00%	-66.67%
	Group 2	1	2	2	-50.00%	-50.00%
	Group 3	3	7	5	-57.14%	-40.00%
	Group 4	4	9	4	-55.56%	0.00%
	Total	9	20	14	-55.00%	-35.71%
64-core	Group 1	2	4	6	-50.00%	-66.67%
	Group 2	3	4	7	-25.00%	-57.14%
	Group 3	5	12	9	-58.33%	-44.44%
	Group 4	11	16	11	-31.25%	0.00%
	Total	21	36	33	-41.67%	-36.36%

Table 5: Comparison of peak power reduction and workload balancing by demand-response scheduling

	Peak Reduction	Balance before	Balance after
Group 1	33.33%	1.00	0.58 (1.72X)
Group 2	42.86%	1.00	0.50 (2X)
Group 3	33.33%	0.91	0.50 (1.82X)
Group 4	42.86%	0.63	0.13 (4.85X)
Average	38.10%	0.89	0.43 (2.60X)

Solution to this problem is to minimize the overall sum of slacks. This can be achieved by rescheduling workloads that demand power more than the threshold. So, initially peak reduction has to be performed followed by load balancing. Based on the value of slack for a subgroup k_j, if the slack is positive, then the workload on that subgroup needs to be delayed or advanced to other time-slot. As such, the workloads are allocated to subgroups with highly negative slack, and the differences in slack is reduced. As a result, peak reduction and load balancing can be achieved eventually.

5. SIMULATION RESULTS

5.1 System Modeling and Settings

The proposed system is validated by Matlab and system-level models built from SystemC-AMS. Table 2 summarizes the system design specifications. All units are scaled or modeled at CMOS 130nm CMOS process. The specification of low-power MIPS microprocessor core [10] is taken as the core model. Each core has the nominal frequency of 250MHz with the maximal power consumption of 0.4W. Benchmarks from SPEC2000 [11] are simulated by Wattch [12] to generate power profiles. The extracted power profiles are used as workload models, which are distributed to different cores randomly. The typical control cycles for power management is 400ns.

A 2-phase multi-output power converter [13] is designed to generate 4 different voltage-levels. As driving ability of power converter depends on supply voltage-level, driving abilities are set as 4, 3, 2, 1 for voltage-levels of 0.6V, 0.8V, 1.0V and 1.2V respectively. Moreover, the inductance value in power converter is set as $1nH$ per phase to support the maximum current on the buck inductor. Such an inductor requires an area of $0.25mm^2$, occupying 30% area of the power converter. The local super-capacitor for each core is set as $1\mu F$ to support time-multiplexing scheme between clusters. The design of on-chip power converter thereby needs to consider the limitation of inductor and capacitor area, which are both placed in 3D fashion and hence has the minimum area overhead to cores all on the other tier.

In addition, the vertical TSV [14] with size of $500\mu m^2$ works as connections between cores and power converters. According to the model in [15], it has a dc-resistance of $20m\Omega$. Considering

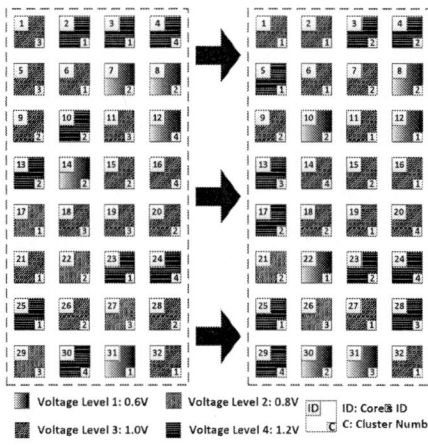

Voltage Level 1: 0.6V   Voltage Level 2: 0.8V   ID: Core ID
Voltage Level 3: 1.0V   Voltage Level 4: 1.2V   C: Cluster Number

Figure 5: Results of adaptive clustering at two continuous control-cycles

the maximum current of $330mA$, the IR-drop of is around $7mV$, which is quite small. Note that the capacitor of TSV is in fF-scale and hence does not influence the load capacitance. What is more, for each TSV channel, one switch box is assigned with Nr power switches to support the core-converter connection. The switch box offers a compact reconfigurable unit driven by the controller. The power switch inside each switch box occupies $520\mu m^2$ and is able to deliver the maximum core current with switching time of 300ns. As such, the TSV coupling is also quite small to consider under such a slow power switching.

5.2 Results and Comparisons

Firstly, we take 32-core and 64-core microprocessors as two examples to show results under adaptive clustering. The input power traces are first grouped into 4 based on the power magnitudes, then in each group subgroups are formed based on their power phases.

Fig. 5 illustrates the adaptive clustering result of 32-core between two consecutive control cycles. Different filling-shapes represent different groups or voltage-levels. Different clustering numbers on the downright-corner of cores represent different subgroups. For example, in the first control cycle, the 30th core will be assigned to subgroup 4 with voltage-level 4 (group 4). And in the next control cycle, it will be assigned to subgroup 2 with voltage-level 1 (group 1). For 64-core case, Table. 3 summarized the clustering results with the value in the table to represent the core ID. One can also observe that the runtime of clustering is small at the scale of 200ms.

Next, we use the space-time multiplexing (STM) scheme to perform the demand-supply matching. The first step is for resource allocation and adaptive clustering is deployed. After clustering, we extract simplified workload models to represent the peak power in one control cycle; and also determine the minimum number of power converters for each group. When comparing to two schemes, namely space-multiplexing (SM) and time-multiplexing (TM) with the same driving ability and time slot, the STM-based approach takes the advantage of both space and time to minimize the number of power converters. Table. 4 shows the comparison for 32-core and 64-core cases with the three schemes. One can observe that 55.00% (SM) and 35.71% (TM) number of power converters can be reduced for the case of 32-core, while 41.67% (SM) and 36.36% (TM) number of power converters can be reduced for the case of 64-core. Therefore, STM based adaptive clustering can satisfy the demand-supply matching with the minimum number of power converters to reduce the area overhead and also on-chip implementation cost.

Lastly, we perform demand-response based workload scheduling for time-multiplexing of power converters inside one

1211

Table 2: System settings of 3D many-core microprocessors, on-chip power converters, TSVs and power switches

Item	Description	Symbol	Value	Size
Microprocessor	Performance	N.A.	410 DMIPS	1.5mm²
	Frequency	f_c	250MHz	
	Power Consumption	P_c	0.4W	
Power Converter	Input Voltage	V_{in}	2.4V	1.6mm²
	Output Voltage	V_{out}	0.6V, 0.8V, 1.0V, 1.2V	
	Load Current	I_L	120mA, 150mA, 220mA, 350mA	
	Number of Phases	N.A.	2	
	Inductor per Phase	L	1nH	
	Switching Frequency	f_s	50-200MHz	
	Peak Efficiency	N.A.	77%	
TSV	Length	l	25μm	500μm^2
	Diameter	W	5μm	
	Isolation Film	r	120nm	
	Resistance	R_{TSV}	20mΩ	
	Capacitance	C_{TSV}	37fF	
Power Switch	Width	w_s	4mm	520μm^2
	Length	l_s	130nm	
	Switching Time	N.A.	300ns	

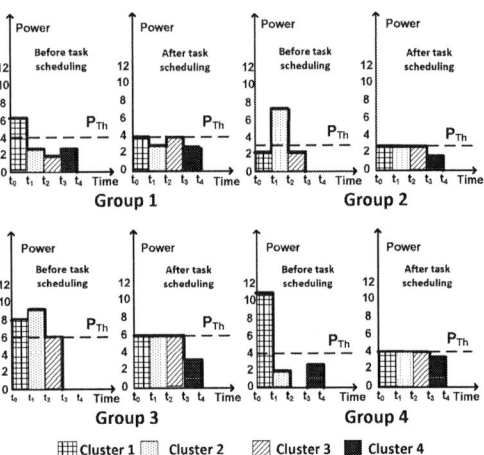

Figure 6: Peak power reductions for 4 subgroups of 64-core case

group. The peak power reduction is defined as the difference of peak power value before and after the scheduling. The workload balancing is defined as the number of cores which one power converter drives over control cycles. We compare the peak power reduction by averaging the reduction in each group; and compare workload balancing by averaging the standard-deviation (SD) of workload on each power converter. For a 64-core microprocessor results shown in Fig. 6, in Group 3, the peak power value has been reduced from 9 to 6 with 33.33% peak power reduction. The average standard deviation of workload on each power converter before and after scheduling are 0.91 and 0.50 respectively, with a standard deviation improvement by 1.82x. Table. 5 shows the summarized results for peak reduction and workload balancing by demand-response scheduling. One can observe an average of 38.10% peak power reduction and 2.60x workload balancing.

6. CONCLUSION

A space-time multiplexed power management is developed for large-scale demand-supply matching between on-chip power converters and many-core microprocessors. The power switch network is configured to perform space-time multiplexing between power converters and cores by vertical TSVs in 3D. Based on adaptive clustering of cores classified by both power magnitudes and power phases, the minimum number of power converters are allocated to supply the demanded voltage-levels from cores. What is more, demand-response based workload scheduling is deployed by utilizing the power slacks, such that

peak power can be reduced as well as workload can be balanced. As verified by system-level behavior models implemented in SystemC and SystemAMS, and also physical-level models with design parameters, experiment results show that the space-time multiplexing can reduce peak power by 38.10% and improve load balancing by 2.60x improvement on average with the minimum number of allocated power converters.

7. ACKNOWLEDGMENTS

This work is sponsored by Singapore MOE TIER-2 fund MOE2010-T2-2-037 (ARC 5/11) and A*STAR SERC-PSF fund 11201202015. Please address comments to haoyu@ntu.edu.sg.

8. REFERENCES

[1] S. Vangal and et.al., "An 80-Tile 1.28TFLOPS network-on-chip in 65 nm CMOS," in *IEEE ISSCC*, 2007.

[2] S. Bell and et.al., "TILE64^{TM} processor: a 64-core SoC with mesh interconnect," in *IEEE ISSCC*, 2008.

[3] M. Healy and et.al., "Design and analysis of 3D-MAPS: a many-core 3D processor with stacked memory," in *IEEE CICC*, 2010.

[4] H. Yu, J. Ho, and L. He, "Allocating power ground vias in 3d ics for simultaneous power and thermal integrity," *ACM TODAES*, vol. 14, no. 3, 2011.

[5] W. Kim and et.al., "System level analysis of fast, per-core DVFS using on-chip switching regulators," in *IEEE PCA*, 2008.

[6] R. Bondade and D. Ma, "Hardware-software codesign of an embedded multiple-supply power management unit for multicore SoCs using an adaptive global/local power allocation and processing scheme," *ACM TODAES*, vol. 16, no. 3, 2012.

[7] J. Howard and et. al, "A 48-core ia-32 processor in 45 nm cmos using on-die message-passing and dvfs for performance and power scaling," *IEEE SSC*, vol. 46, pp. 173–183, January 2011.

[8] N. Sturcken and et.al., "A 2.5D integrated voltage regulator using coupled-magnetic-core inductors on silicon interposer delivering 10.8A/mm²," in *IEEE ISSCC*, 2012.

[9] R. H. Katz and et. al, "An information-centric energy infrastructure: The berkley view," *S staina le Comp tin Informatics an S stems*, no. 1, pp. 7–22, March 2011.

[10] "MIPS processor cores," http://www.mips.com/products/processor-cores/.

[11] "SPEC 2000 CPU benchmark suits," http://www.spec.org/cpu/.

[12] "Wattch version 1.02," http://www.eecs.harvard.edu/~dbrooks/wattch-form.html.

[13] W. Kim and et.al., "A fully-integrated 3-level DC/DC converter for nanosecond-scale DVS with fast shunt regulation," in *IEEE ISSCC*, 2011.

[14] V. der Plas and et.al., "Design issues and considerations for low-cost 3D TSV IC technology," in *IEEE ISSCC*, 2010.

[15] G. Katti and et.al., "Electrical modeling and characterization of through silicon via for three-dimensional ICs," *IEEE Trans on Electron Devices*, vol. 57, no. 1, pp. 256–262, 2010.

Techniques for Energy-Efficient Power Budgeting in Data Centers

Xin Zhan
School of Engineering
Brown University
Providence, RI 02912
xin_zhan@brown.edu

Sherief Reda
School of Engineering
Brown University
Providence, RI 02912
sherief_reda@brown.edu

ABSTRACT

We propose techniques for power budgeting in data centers, where a large power budget is allocated among the servers and the cooling units such that the aggregate performance of the entire center is maximized. Maximizing the performance for a given power budget automatically maximizes the energy efficiency. We first propose a method to partition the total power budget among the cooling and computing units in a self-consistent way, where the cooling power is sufficient to extract the heat of the computing power. Given the computing power budget, we devise an optimal computing budgeting technique based on knapsack-solving algorithms to determine the power caps for the individual servers. The optimal computing budgeting technique leverages a proposed on-line throughput predictor based on performance counter measurements to estimate the change in throughput of heterogeneous workloads as a function of allocated server power caps. We set up a simulation environment for a data center, where we simulate the air flow and heat transfer within the center using computational fluid dynamic simulations to derive accurate cooling estimates. The power estimates for the servers are derived from measurements on a real server executing heterogeneous workload sets. Our budgeting method delivers good improvements over previous power budgeting techniques.

ACM Categories & Subject Descriptors C.5.5 [Computer System Implementation]: Servers.
General Terms: Management, Performance, Algorithms.
Keywords: Power, Budgeting, Management, Data Centers.

1. INTRODUCTION

Data center and computing clusters with hundreds or thousands of servers consume excessive amounts of power, with large facilities consuming up to 20 MW for a total cost of $12 million per year [14, 3]. As a result, the total cost of ownership of data centers is dominated by power consumption, which constrains total performance and scalability [7, 14, 11]. In many cases, the power consumption of a facility at any moment of time must be capped below a maximum limit that is specified by the electric grid operators and the electrical current carrying capacity of its power cables [7, 8].

Permission to make digital or hard copies of all or part of this work for personal or classroom use is granted without fee provided that copies are not made or distributed for profit or commercial advantage and that copies bear this notice and the full citation on the first page. To copy otherwise, to republish, to post on servers or to redistribute to lists, requires prior specific permission and/or a fee.
DAC '13, May 29 - June 07 2013, Austin, TX, USA.

One of the challenges in server power management is that different workloads trigger different power consumption patterns, and thus the power management settings that work for one set of workloads do not necessarily work for another set of workloads [9, 5]. As a result, one needs to find settings for each server that lead to a global optimal for the entire computing facility. Another challenge is that the total power consumption of a data center is the sum of the power consumption of its computing servers and the Computer Room Air Conditioning (CRAC) units, where the power consumption of the CRAC units depends on the power consumption of servers and the hot spots in the layout of the center [1]. The goal of this paper is to devise new power budgeting technique, where the total power budget is allocated among the servers and cooling equipment to maximize the total throughput, or equivalently minimize the average runtime. We summarize our contributions as follows.

1. We propose a novel method to partition the total power budget between the computing servers and cooling units in a *self-consistent* way, where the cooling power meets the heat removal requirements for the computing power, which is allocated using an optimal power budgeting technique.

2. We propose a novel *throughput predictor* for servers with heterogeneous workload sets, where the measurements from the performance counters are used to estimate the change in the throughput as a function of the server power cap.

3. Leveraging the throughput predictor, we propose an *optimal computing power budgeting* technique that is inspired by methods for solving the well-known knapsack problem. The budgeting technique identifies the optimal power caps for the servers, such that the total server power meets the computing budget and the total throughput is maximized.

4. We setup a realistic simulation environment for a data center with a large number of servers, where the power estimates for the servers are derived from real measurements on a server executing heterogeneous workload sets. We use Computational Fluid Dynamics (CFD) simulations to ensure accurate modeling of air flow and heat transfer within the center, and use the CFD results to estimate the cooling power. We experimentally demonstrate the advantages of our power budgeting method compared to previous approaches.

The organization of this paper is as follows. In Section 2, we describe previous related techniques in the literature. We formulate the power budgeting problem and describe our proposed framework in Section 3. Our experimental results are presented in Section 4, and Section 5 provides the conclusions of this work and directions for future work.

2. RELATED WORK

A number of models have been proposed in the literature to capture the relationship between the throughput and power of a single server. For example, Rajamani *et al.* proposed linear models [17] and Gandhi *et al.* proposed linear and cubic models [8]. The coefficients of these models are functions of the server configuration and the workload characteristics. In these previous works, fixed values for these coefficients were assumed irrespective of the workload characteristics. These values were obtained through prior characterization of standard benchmarks. As a result, these models are likely to show prediction errors for throughput and power in case heterogeneous applications with wide range of characteristics are executed on a cluster.

To enforce a required power cap, a number of previous approaches have proposed equipping each server with a feedback controller that computes the observed difference between the measured power and the power cap, and accordingly adjusts the p-state of the server using dynamic frequency and voltage scaling (DVFS) [10, 16]. If the difference is positive then DVFS is decreased, and if the difference is negative then DVFS is increased. To determine the power cap of each server, a number of power budgeting methods have been proposed [13, 8]. Ghandi *et al.* proposed power budgeting methods for servers that execute the same workload. This situation can be useful for data centers that execute transactional workloads of the same nature; however, they are not relevant for computing facilities that execute high-performance computing (HPC) applications. These later facilities typically have high utilizations where most of the servers are fully utilized executing a large range of workloads with heterogeneous characteristics. Nathuji *et al.* consider the case of power budgeting for heterogeneous workloads and servers [13]. The main proposed approach is a greedy method, where the throughput per Watt for the servers are calculated, and then servers with higher throughput per Watt are allocated more power during budgeting.

A related problem to power budgeting in data centers is the problem of power allocation in multi-core processors [9, 18]. Power budgeting for data centers is different in a number of ways: (1) unlike independent servers, multi-core processors do not offer power cap controllers for the individual cores; (2) workloads on a multi-core processor are likely to show memory interference issues, whereas workloads servers are relatively independent unless they explicitly communicate using message passing; (3) data centers feature air conditioning units that have to be considered during power budgeting; and (4) the interactions between computing and cooling power in data center are highly complex in nature.

3. PROPOSED APPROACH

We assume that a data center or a computing cluster is composed of n servers with identical hardware configuration and m CRAC units. We make the general assumption of heterogeneous workload sets, where different servers and different cores within the same server can be executing different workloads, and that the set points of the CRAC units can be controlled independently. We assume a closed-loop queueing model where all servers are fully utilized. As a result, maximizing the total throughput is equivalent to minimizing the response time [8].

Problem Formulation: Given n fully utilized servers with heterogeneous workloads, m CRAC units, and a total B power budget, the objective is to distribute the total power among the n servers and m CRACs, such that the total throughput is maximized or equivalently the average response time is minimized. That is, if $\tau_i(p_i)$

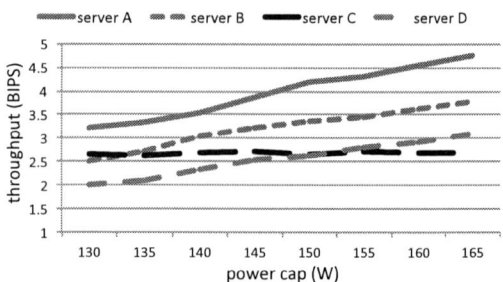

Figure 1: Impact of power cap on the throughput of a server.

and p_i denote the throughput and allocated power for server i respectively and c_k denote the cooling power of CRAC k, then the goal is to maximize $\sum_{i=1}^{n} \tau_i(p_i)$ such that $B_s + B_{CRAC} \le B$, where $B_s = \sum_{i=1}^{n} p_i$ is the total server computing power, and $B_{CRAC} = \sum_{k=1}^{m} c_k$ is the total cooling power. The budgeting has to be done in a *self-consistent* way, where cooling power is able to extract the heat generated from the servers.

Motivation. If the workloads on all servers are identical then the budgeting problem is trivial since the total power can be divided uniformly among all servers. To get a better understanding of the relationship between throughput and the allocated power cap in case of heterogeneous workloads, we equip our experimental server with a power capping controller. The capping controller executes once every second and adjusts the p-state of the server using DVFS based on the difference between the allocated power cap and actual power consumption [10]. We report the average throughput as a function of the power cap for four identical servers with different workload sets in Figure 1, where each server is executing a heterogeneous workload mix from the SPEC CPU06 benchmarks. The plot leads to a number of observations.

1. The observed throughput is highly dependent on the workload characteristics. Servers A and B show large improvement in throughput with increased power allocations. Server D shows modest improvements, while server C shows little or no improvements. Thus, some servers will not be able to leverage their allocated power caps to improve throughput.

2. The plot shows that the slope of an individual server plot changes as a function of the operating power cap. For instance, Server A shows a larger slope in the range of 140-150 W compared to other regions of operation. Thus, accurate modeling requires considering the impact of the operational power cap of the server on its slope.

3. The plots of servers C and D show that greedy allocation methods based on the throughput/Watt alone (e.g., [13]) will not give optimal results. For example, if the current power allocations for servers C and D are at the lowest cap, then Figure 1 shows that server C has higher throughout than server D, which can lead to the wrong conclusion that it is better to allocate more power to Server C. However, the plots of the two servers eventually cross over, where server D attains large throughput than server C at higher power caps.

Our power budgeting approach consists of two components: (i) a total power budgeting method that partitions the total power budget, B, into the computing budget, B_s, and the cooling budget, B_{CRAC}, in a self-consistent way; and (ii) an *optimal computing power budgeting* method that identifies the power cap for each server, such that the total computing power budget, B_s is met and the total throughput is maximized. Our optimal computing power budgeter makes use of a novel *throughput predictor* that takes as inputs the measurements, e.g., throughput, power and performance counters, of servers at the current power cap, and uses them to predict the change in throughput of each server for every possible power cap. We describe each of these components in the next subsections.

3.1 Total power budgeting

Our goal is to apply a total power budget for both computing power and cooling power in a self-consistent way, where the cooling power, $B_{CRAC} = \sum_{k=1}^{m} c_k$, extracts the heat generated from the computing power. The cooling power is a function of many factors, including the layout of the data center, the spatial allocation of the computing power, the air flow rate, and the efficiency of the CRAC units. The power consumption, c_k, of a CRAC unit k is equal to

$$c_k = \frac{\sum_i p_i}{CoP}, \tag{1}$$

where $\sum_i p_i$ is the power consumption of servers with their heat flow directed towards the CRAC unit, and CoP is the *coefficient of performance* that gives the performance of the CRAC units [12]. For example, based on physical measurements, an empirical model for CoP of a commercial water-chilled CRAC unit is equal to

$$CoP(t) = 0.0068t^2 + 0.0008t + 0.458, \tag{2}$$

where t is the supply air temperature of the CRAC unit in degrees Celsius [12]. To find the minimum sufficient cooling power for an allocation of a certain computing power, it is necessary to maximize the supply air temperature t, while ensuring that the inlet temperatures of all the servers will not exceed the manufacturer's redline temperature T_{red}. Identifying the inlet temperature for the servers requires accurate CFD models for the air flow and the heat transfer dynamics inside the data center. If the results of the CFD simulation show that the inlet temperature of any server violates T_{red}, then t should be lowered to bring the inlet temperature back under T_{red}, and if the inlet temperature has not reached T_{red}, then t should be increased without causing an inlet temperature of racks increase beyond T_{red}. Given a spatial layout of computing power consumption inside the data center, Equation (1) and Equation (2) and the associated CFD simulations enable us to compute the required cooling power $B_{CRAC} = \sum_k c_k$.

To ensure that the sum of the computing power, B_s, and the cooling power, B_{CRAC}, meets the total power budget B, we propose an algorithm, given in Figure 2, to identify a self-consistent partitioning of the total power budget. The main loop of the iterative algorithm first calculates the computing power budget B_s in step 3,

---

Procedure: Self-Consistent power budgeting algorithm
Input: Total power budget B; data center configuration; T_{red}.
Output: Computing power B_s and cooling power B_{CRAC}.

---

1. initialize B_{CRAC} based on initial CFD simulation.

2. repeat:

3.   let $B_s = B - B_{CRAC}$.

4.   budget B_s using the multi-choice knapsack algorithm.

5.   run CFD simulations to get minimum B_{CRAC} given T_{red}.

6. until B_{CRAC} is equal to $B - B_s$.

Figure 2: Algorithm for self-consistent total power budgeting.

and then in Step 4, the computing budget, B_s, is allocated optimally among the servers using the multi-choice knapsack algorithm described in the next subsection. Given power allocation and the data center configuration, Step 5 estimates the minimum required cooling power, B_{CRAC}, as described in the previous paragraph. If it happens that $B_{CRAC} + B_s = B$ (step 6), then the algorithm has converged to a solution; otherwise, it continues iterating. The proposed algorithm is guaranteed to converge; proof is available in the supplemental material.

3.2 Computing Power Budgeting

Our goal is to maximize the total throughput under a total computing power budget B_s. We consider a discrete set of individual server power caps with a fixed increment (e.g., 130 W, 135 W, ..., 165 W). The choice of a discrete number of power caps is natural given that p-states are discrete and changing them does not lead to a continuous power range. Thus, the power cap of a server can be described as

$$p_i = p_0 + \sum_{j=1}^{r} w_j x_{ij}, \tag{3}$$

where p_0 is the least possible power cap, r is the number of individual server power caps, w_j is the increment power for each cap over the least possible cap, and $x_{ij} \in \{0, 1\}$, where x_{ij} is only equal to 1 when server i is assigned a power cap equal to $p_0 + w_j$. For the case of power caps: 130 W, 135 W, ..., 165 W, we have $r = 8$, $p_0 = 130W$ and $w_1 = 0, w_2 = 5, w_3 = 10, ..., w_8 = 35$.

A challenging aspect is that we need to estimate the impact of a change in power cap on the throughput of a server. We propose the following *throughout predictor*. Suppose that $\hat{p}_i$ denotes current allocated power cap to server i, and that the attained throughput for the server from using the power cap controller is equal to $\tau_i(\hat{p}_i)$. Given the measurements at the current power cap, the *objective* of the *throughput predictor* is to estimate the throughput of the server resulting from allocating a new power cap p_i to the server. We propose a piecewise linear model, where the predicted throughput is equal to,

$$\tau_i(p_i) = \tau_i(\hat{p}_i) + s_i(\hat{p}_i)(p_i - \hat{p}_i), \tag{4}$$

where $s_i(\hat{p}_i)$ is the *slope* of the throughput-power plot of server i at the server's current power cap. To predict the throughput, we need to identify the slope from the observations at the current operating point. To get an insight into the factors that determine the slope of the throughput-power characteristics, we analyzed a large number of performance counters from off-line characterization data. We have found that the Last Level Cache (LLC) misses is one of the most reliable predictor of the slope. Figure 3 illustrates the relationship between the slope and LLC using a large volume of characterization data collected from the SPEC CPU 2006 benchmarks. The results show a trend where workloads with larger LLC suffer throughput degradation. This trend is plausible as LLC misses show memory boundedness [2, 6], and as a result allocating more power caps to memory bound workloads give little improvements to throughput. In addition to the LLC misses, we have found that the ratio, i.e., $\tau_i(\hat{p}_i)/\hat{p}_i$, is a good predictor of the slope at the setting. Figure 4 illustrates the relationship between the two using our off-line characterization data. The results show that servers with higher throughput per Watt usually have higher slopes.

Our slope estimator makes uses of both $\tau_i(\hat{p}_i)/\hat{p}_i$ and $L\hat{L}C_i$. We experimented with a number of models for the slope, and we found the following model to give the best results:

$$s_i = \alpha_1 + \alpha_2 \frac{\tau_i(\hat{p}_i)}{\hat{p}_i} + \alpha_3 e^{\alpha_4 \cdot L\hat{L}C_i}, \tag{5}$$

1215

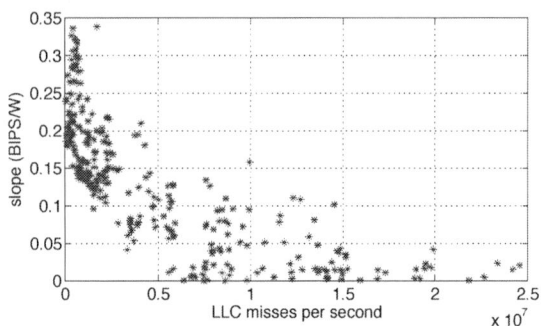

Figure 3: Relationship between LLC and the slope for a large number of heterogeneous workload sets.

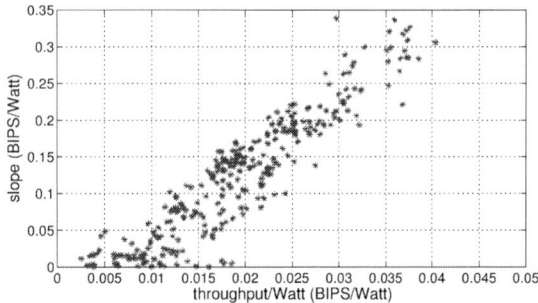

Figure 4: Relationship between current throughput/Watt and the slope for a large number of heterogeneous workload sets.

where $\alpha_1, \ldots, \alpha_4$ are the model coefficients for the current power cap. The coefficients can be easily found through off-line training on subset of workload characterization data.

Using Equations (3) and (4), and given the current $\tau_i(\hat{p}_i)$ and $\hat{p}_i$, it can be shown (see supplemental material) that the throughput objective can be recast as follows:

$$\max \sum_{i=1}^{n} \tau_i(p_i) \quad \Leftrightarrow \quad \max \sum_{i=1}^{n} \sum_{j=1}^{r} v_{ij} x_{ij}, \qquad (6)$$

where $v_{ij} = s_i w_j$. Thus, the entire optimization formulation is given by

$$
\begin{aligned}
\text{maximize} \quad & \sum_{i=1}^{n} \sum_{j=1}^{r} v_{ij} x_{ij} \\
\text{subject to} \quad & \sum_{i=1}^{n} \sum_{j=1}^{r} w_j x_{ij} \le B_s - n p_0, \\
& \sum_{j=1}^{r} x_{ij} = 1 \quad \forall i = \{1 \ldots n\}, \\
& x_{ij} \in \{0, 1\}.
\end{aligned}
$$

We observe the similarity between power budgeting formulation and the *multiple-choice knapsack* problem [15]. In the multiple-choice knapsack problem, there are a number of classes, where each class has a few items, each with its own value and weight, and we have to select one item from each class to maximize the total value for the given total weight of the knapsack. Our problem naturally leads to a multiple-choice formulation, where the each server corresponds to a class, and the items within the class correspond to the power cap settings that can be applied to the server, each with its own throughput value ($v_{ij} x_{ij}$) and weight (power cap $w_j x_{ij}$). The multiple-choice knapsack problem is readily solved using dynamic programming. The supplemental material provides the details of the dynamic programming algorithm, which has a complexity of $O(nrB_s)$. In a computing cluster with hundreds or thousands of servers, it is easy to envision a server dedicated to carrying out the computations necessary for power budgeting.

4. EXPERIMENTAL RESULTS

In our data center configuration, we assume 320 servers forming 8 U40 racks with 40 servers per rack. To simulate the heat flow and air flow in the data center, we use TileFlow [20], which is a CFD software tool for simulating cooling characteristics of data centers. We provide the center's layout details in Experiment 3. The throughput and power estimates for the servers are derived from measurements on a real server executing heterogeneous workload sets. The Linux-based server has a quad-core Intel Core i7 processor and 8 GB of memory. To measure power consumption, the 120 V AC power lines to the server are intercepted and the electric current is measuring using an Agilent 34410A digital multimeter. The total power measurements are read back to the server over USB using the SCPI interface and provided as inputs to the power cap controller. The engagement period of the feedback power cap controller is 1 second.

We use the experimental server to construct a database of 320 execution traces of workload sets selected from the SPEC CPU06 [19] and PARSEC benchmarks [4]. For the SPEC CPU06 benchmarks, each workload set consists of four randomly chosen benchmarks, so that all the cores of our server are fully utilized. For the PARSEC benchmarks, all workloads are executed with four-thread configuration. We measured the number of retired instructions per second and LLC misses using the `pfmon` tool library interface. To train our predictor, we collected a large volume of characterization, where the throughput and LLC are measured for different workloads under different power caps. The database enables us to simulate the impact of different power budgets on a large number of servers in an extremely fast way that preserves the accuracy of results. In particular, each time a new power budget is applied, the power and performance outcomes are computed by reassembling the proper sections of the workload set traces of different servers from the database.

Exp 1. Throughput Predictor Accuracy. In the first experiment, we evaluate the accuracy of our throughput predictor compared to the actual throughput results. We compare our predictor in three versions: (i) `predictor` which uses the measurements of throughput, power and LLC as described in Subsection 3.2; (ii) `predictor-LLC` that just uses LLC measurements, and (iii) `predictor-TP` just uses the throughput and power. We also compare against the linear (`previous-linear`) model [17, 8] and cubic (`previous-cubic`) model proposed in previous works [8]. The average absolute error of the predictors are reported in Table 1. The results show that our predictor leads to better throughput prediction, and that combining LLC measurements together with throughput and power leads to more accurate results. Both linear and cubic models previously proposed in the literature trail our models in accuracy.

Exp 2. Computing Power Budgets. In the second experiment, we evaluate the effectiveness of the proposed knapsack-based optimal

prediction method	throughput prediction error
`predictor`	3.57%
`predictor-LLC`	7.83%
`predictor-TP`	4.89%
`previous-linear` [17, 8]	15.94%
`previous-cubic` [8]	8.16%

Table 1: Error in throughput prediction for various models.

1216

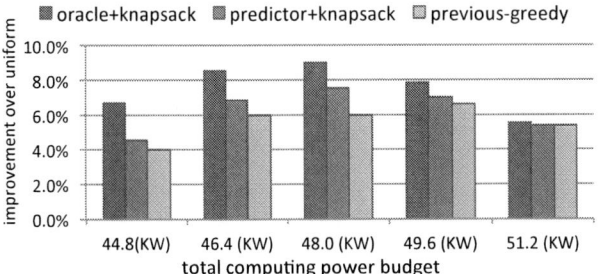

Figure 5: Throughput improvement over baseline uniform power allocation for heterogeneous workloads across servers, homogenous within server.

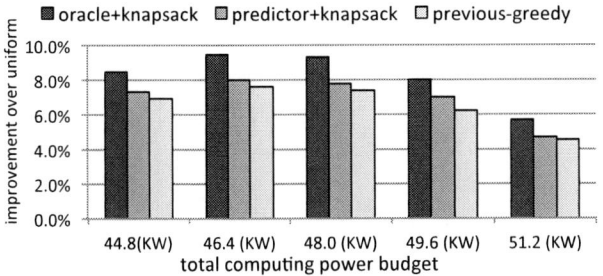

Figure 6: Throughput improvement over baseline uniform power allocation for heterogeneous workloads across servers, heterogeneous within server.

power budgeting method given a total computing power budget. We do not consider the cooling power consumption in this experiment. We refer to our technique by predictor+knapsack. We report the improvement in throughput over a baseline method (uniform), where the budget is allocated uniformly among the servers. We also compare against a previously proposed approach (previous-greedy) [17], which utilizes a greedy approach for power budgeting, where servers with higher throughput per Watt at the moment of re-calculating the power budget are allocated more power. Finally, we compute an upper bound on the attainable throughput by using the optimal knapsack algorithm on the true throughput and power for each server at the power cap, which are not known during runtime, but can be computed in our simulation environment. We refer to this method by oracle+knapsack. We consider two cases:

a) Heterogeneous across servers, homogenous within server: In the first case, the servers execute different workload sets, but the workload set assigned for each server is homogenous, e.g., a PARSEC workload with four threads or four instances of the same SPEC CPU06 benchmark. This is the most common case in modern clusters as administrators prefer to eliminate the interference between workloads arising from execution on the same server. The results are given in Figure 5 for a number of total computing power budgets. The results demonstrate that our predictor+knapsack method consistently outperforms other methods. For the case of 48 KW, we increase the throughput by 7.56% compared to uniform allocation case, whereas the previous-greedy method only increases the throughput by 5.98%. The results from our predictive method are close to the results from the oracle case.

b) Heterogeneous across servers, heterogeneous within server: In the second case, the servers execute different workload sets, and each workload set on a server consists of different benchmarks (e.g., four instances of different SPEC CPU06 applications). The

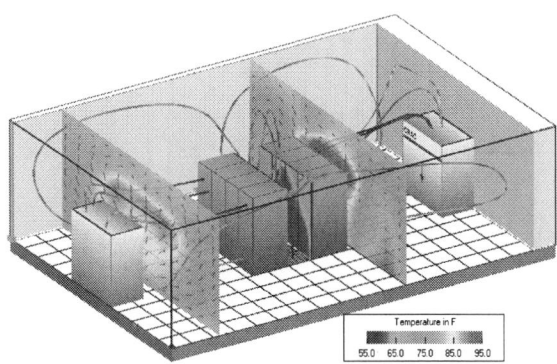

Figure 7: The inlet temperatures of racks and the spatial temperature maps in Fahrenheit.

results are given in Figure 6 for five total computing power budgets. The results show that our proposed method consistently outperforms other methods. For example, for total budget of 49.6 KW, our method increases throughput by 7.02% over uniform, whereas the previously proposed greedy method increases throughput by 6.12%. The relative improvements in this case are less than the first case, which is expected given the heterogeneity of workloads within the server. This heterogeneity causes averaging in characteristics, which leads to less differentiation among the ensemble of servers. Furthermore, the interactions between the workloads within the servers reduce the accuracy of the throughput predictor. Therefore, there might be further room for improvement through better throughput predictors.

In both cases, it is natural to expect that the relative advantages among the methods would disappear when the total power budget is too high or too low. If the total budget is too high, then all servers can afford to run at the highest power cap and throughout irrespective of the method, and similarly when the total budget is too low, then all servers will be forced to the lowest power state.

Exp 3. Total Power (Computing+Cooling) Budgeting. Our first three experiments assume that the power budget is entirely applied to the computing servers. In the fourth experiment, we include cooling power into our method and calculate the optimal partition between cooling power and computing power of a given total power budget. In our data center configuration, the 8 racks are arranged into two symmetric rows at the center of the room as illustrated in Figure 7. Two down flow CRAC units are located at two sides of the center. Cold air comes from under floor through perforated tiles between the two front side rack rows. The fans integrated with the racks draw the cold air through servers, which removes the heat generated by the operation of servers. The air heated by servers leaves the racks from the back side and is sucked into the CRAC units at the sides. The CRAC units extract the heat from the hot air and push cold air back into data center from perforated tiles on the floor at user specified temperature. We assume a redline inlet temperature of racks is $24\,^{\circ}C$.

We consider five total power budgets 62 KW, 66 KW, 70 KW, 74 KW and 78KW. We execute the self-consistent budgeting algorithm of Figure 2 to find the optimal partition of total power budget between computing power and cooling power under several total power budgets. After each simulation, TileFlow returns a report about the estimated maximum inlet temperature of racks. We can check the temperature at any specific point by the temperature mapping tool in Figure 7. The partitioning of the total power into its computing and cooling components is given in Figure 8. From Fig-

1217

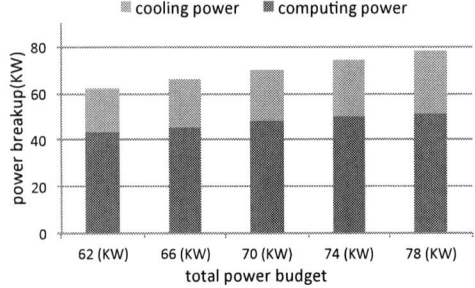

Figure 8: The breakup of cooling power and computing power under different total power budgets.

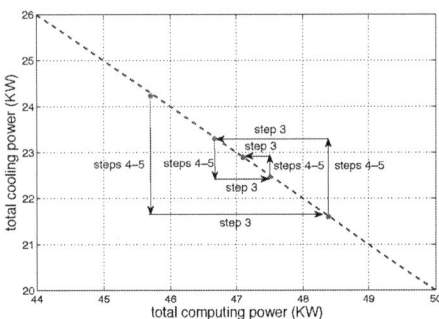

Figure 9: Illustration of the self-consistent power budgeting of the algorithm in Figure 2 for the case of 70 KW.

ure 8, we can observe that the cooling power consumption typically takes $30\% - 35\%$ of total power consumption. Another interesting observation from the results in Figure 8 is that the proportion of cooling power increases with the increase in total power budget, and that the rate of this increment also increases. Figure 9 illustrates the application of the self-consistent budgeting algorithm of Figure 2 to the case of 70 KW total power budget. The dashed blue line gives the power partitions that sum to 70 KW, and the red points give the intermediate partitions before convergence. The algorithm requires only 5 iterations to converge to a self-consistent solution.

5. CONCLUSIONS & FUTURE WORK

In this work we considered the problem of optimal power budgeting for servers with heterogeneous workloads. It is well-known that workloads exhibit different power and performance characteristics depending on their memory or processor boundedness. We leveraged this observation to devise a power budgeting method that allocates power to servers that can efficiently translate their power allocation to increases in throughput. During runtime, a power budgeting system has only one snapshot of the servers' status based on their current measurements. Thus, we devised a throughput prediction method that estimates the changes in throughput as functions of potential changes to allocated power caps. We have demonstrated that our throughput predictor is capable of providing accurate predictions under different power cap and workload characteristics. We have devised an optimal computing power budgeting method based on the multiple-choice knapsack formulation to identify the optimal power allocations for each server such that the total throughput is maximized. Furthermore, we proposed a self-consistent method to partition the total power budget between the computing and the cooling component of the data center. Our results show good improvements over previous methods.

Future work. Our future work will consider the possibility of under-utilized servers and use of other Quality of Service (QoS) metrics besides the total throughput and average response time [3].

6. REFERENCES

[1] F. Ahmad and T. Vijaykumar, "Joint Optimization of Idle and Cooling Power in Data Centers while Maintaining Response Time," in *Proceedings of Architectural Support for Programming Languages and Operating Systems*, 2010, pp. 243–256.

[2] H. Amur, K. Schwan, and M. Prvulovic, "Towards Optimal Power Management: Estimation of Performance Degradation due to DVFS on Modern Processors," Georgia Tech, Tech. Rep. GIT-CERCS-10-02, 2010.

[3] L. A. Barroso and U. Holzle, *The Datacenter as a Computer*. Morgan and Claypool Publishers, 2009.

[4] C. Bienia and K. Li, "PARSEC 2.0: A New Benchmark Suite for Chipmultiprocessors," in *In Proceedings of the Annual Workshop on Modeling, Benchmarking and Simulation*, 2009.

[5] R. Cochran, C. Hankendi, A. Coskun, and S. Reda, "Identifying the Optimal Energy-Efficient Operating Points of Parallel Workloads," in *ACM/IEEE International Conference on Computer-Aided Design*, 2011, pp. 608–615.

[6] S. Eyerman and L. Eeckhout, "A Counter Architecture for Online DVFS Profitability Estimation," *IEEE Transactions on Computers*, vol. 59, no. 11, pp. 1576–1583, 2010.

[7] X. Fan, W. Weber, and L. Barroso, "Power Provisioning for a Warehouse-sized Computer," *International Symposium on Computer Architecture*, pp. 13–23, 2007.

[8] A. Gandhi, M. Harchol-Balter, and R. Das, "Optimal Power Allocation in Server Farms," in *International Conference on Measurement and Modeling of Computer Systems*, 2009, pp. 157–168.

[9] C. Isci, A. Buyuktosunoglu, C.-Y. Cher, P. Bose, and M. Martonosi, "An Analysis of Efficient Multi-Core Global Power Management Policies: Maximizing Performance for a Given Power Budget," in *International Symposium on Microarchitecture*, 2006, pp. 347 –358.

[10] C. Lefurgy, X. Wang, and M. Ware, "Power Capping: A Prelude to Power Shifting," *Cluster Computing*, vol. 11, pp. 183–105, 2008.

[11] J. Leverich and C. Kozyrakis, "On the Energy (In)efficiency of Hadoop Clusters," *ACM SIGOPS Operating Systems Review*, vol. 44, no. 1, pp. 61–65, 2010.

[12] J. Moore, J. Chase, P. Ranganathan, and R. Sharma, "Making Scheduling "Cool": Temperature-Aware Workload Placement in Data Centers," in *Proceedings of USENIX Annual Technical Conference*, 2005, pp. 61–75.

[13] R. Nathuji, C. Isci, E. Gorbatov, and K. Schwan, "Providing Platform Heterogeneity-Awareness for Data Center Power Management," *Cluster Computing*, vol. 11, pp. 159–271, 2008.

[14] C. Patel and A. Shah, "Cost Model for Planning, Development and Operation of a Data Center," *Hewlett-Packard Laboratories Technical Report*, 2005.

[15] D. Pisinger, "Algorithms for Knapsack Problems," Ph.D. dissertation, University of Copenhagen, 1995.

[16] R. Raghavendra, P. Ranganathan, V. Talwar, Z. Wang, and X. Zhu, "No "Power" Struggles: Coordinated Multi-Level Power Management for the Data Center," in *Architectural Support for Programming Languages and Operating Systems*, 2008, pp. 48–59.

[17] K. Rajamani, H. Hanson, J. Rubio, S. Ghiasi, and F. Rawson, "Application-Aware Power Management," in *International Workshop on Workload Characterization*, 2006, pp. 39–48.

[18] J. Sartori and R. Kumar, "Three Scalable Approaches to Improving Many-core Throughput for a Given Peak Power Budget," in *International Conference on High-Performance Computing*, 2009, pp. 89–98.

[19] C. D. Spradling, "SPEC 2006 Benchmark Tools," *SIGARCH Computer Architecture News,*, vol. 35, no. 1, pp. 13–134, 2007.

[20] TileFlow, "http://inres.com/products/tileflow."

7. SUPPLEMENTAL MATERIAL

7.1 Dynamic Programming Algorithm for Computing Power Budgeting

Using Equations (3) and (4), and using the fact that $\tau_i(\hat{p}_i)$ and $\hat{p}_i$ are given inputs, we can re-cast the throughput objective as follows:

$$\max \sum_{i=1}^{n} \tau_i(p_i) \;\Leftrightarrow\; \max \sum_{i=1}^{n} (\tau_i(\hat{p}_i) + s_i(p_i - \hat{p}_i))$$

$$\Leftrightarrow\; \max \sum_{i=1}^{n} s_i \sum_{j=1}^{r} w_j x_{ij}$$

$$\Leftrightarrow\; \max \sum_{i=1}^{n} \sum_{j=1}^{r} v_{ij} x_{ij}, \qquad (7)$$

where $v_{ij} = s_i w_j$. Thus, the entire optimization formulation is given by

maximize $\quad \sum_{i=1}^{n} \sum_{j=1}^{r} v_{ij} x_{ij}$

subject to $\quad \sum_{i=1}^{n} \sum_{j=1}^{r} w_j x_{ij} \leq B_s - np_0,$

$\qquad\qquad \sum_{j=1}^{r} x_{ij} = 1 \quad \forall i = \{1 \ldots n\},$

$\qquad\qquad x_{ij} \in \{0, 1\}.$

We observe the similarity between power budgeting formulation and the *multiple-choice knapsack* problem [15]. In the multiple-choice knapsack problem, there are a number of classes, where each class has a few items, each with its own value and weight, and we have to select one item from each class to maximize the total value with the knapsack total weight. Our problem naturally leads to a multiple-choice formulation, where the each server corresponds to a class, and the items within the class correspond to the power cap settings that can be applied to the server, each with its own throughput value ($v_{ij} x_{ij}$) and weight (power cap $w_j x_{ij}$). The standard dynamic programming given in Figure 10 can be used to solve the problem optimally. It has a complexity of $O(nrB_s)$.

Procedure: Optimal computing power budgeting algorithm.
Input: values and weights for the servers, n, and B_s.
Output: Power allocated for every server.

Let V be a vector that holds the total knapsack's value for each
    possible budget. V is initialized to all zero.

for $i := 1 : n$

  for $k := B_s : -1 : 1$

    for $j := 1 : r$

    $p_i := p_0 + w_j$

    if $k \geq p_i$ and $V(k) \leq v_{ij} + V(k - p_i)$

      let $V(k) = v_{ij} + V(k - p_i)$

      let $x_{ij} = 1$ and let $x_{il} = 0$ for all $l \neq j$

Figure 10: Algorithm for optimal power budgeting.

7.2 Proof of Convergence of the Self-Consistent Power Budgeting Algorithm

In this subsection, we will prove the convergence of the self-consistent power budgeting algorithm given in Figure 2 of Subsection 3.1. Let (B_s^*, B_{CRAC}^*) denote the self-consistent solution of a total power budget $B_s^* + B_{CRAC}^* = B$ at a maximum CRAC supply temperature of t^*. Let the computing power at iteration k of the algorithm is denoted by $B_s(k)$, and the minimum cooling power required for heat extraction is $B_{CRAC}(k)$ at a maximum CRAC supply temperature of t_k. Define $\delta_p(k) = |B_s(k) - B_s^*|$ and $\delta_c(k) = |B_{CRAC}(k) - B_{CRAC}^*|$.

When $B_s(k) > B_s^*$, the CRAC unit's supply temperature, t_k, is higher than t^*. According to the CRAC model of Equation (2), $CoP(t_k) > CoP(t^*)$, which will also hold true for any monotonically increasing CRAC model as a function of temperature. We can derive the following relationship between $\delta_p(k)$ and $\delta_c(k)$:

$$\delta_p(k) \;=\; |B_s(k) - B_s^*|$$

$$\frac{\delta_p(k)}{CoP(t_k)} \;=\; \left| \frac{B_s(k)}{CoP(t_k)} - \frac{B_s^*}{CoP(t_k)} \right|$$

$$> \left| \frac{B_s(k)}{CoP(t_k)} - \frac{B_s^*}{CoP(t^*)} \right|$$

$$> |B_{CRAC}(k) - B_{CRAC}^*|$$

$$> \delta_c(k).$$

With a CoP with a numerical value greater than 1, as expected from Equation (2), we conclude that

$$\delta_p(k) > \delta_c(k). \qquad (8)$$

A similar argument can be made for the case of $B_s(k) < B_s^*$. For iteration $k + 1$, our method will update the computing power as:

$$B_s(k+1) \;=\; B - B_{CRAC}(k)$$

$$=\; B_s^* + B_{CRAC}^* - B_{CRAC}(k),$$

which can be re-arranged to

$$B_s(k+1) - B_s^* \;=\; B_{CRAC}^* - B_{CRAC}(k)$$

$$|B_s(k+1) - B_s^*| \;=\; |B_{CRAC}^* - B_{CRAC}(k)|$$

$$\delta_p(k+1) \;=\; \delta_c(k)$$

$$\delta_p(k+1) \;<\; \delta_p(k).$$

Thus, the distance, $\delta_p(k+1)$, between $B_s(k+1)$ and B_s^* is less than the distance, $\delta_p(k)$, between $B_s(k)$ and B_s^*. Therefore, the computing power approaches B_s^* with every iteration and finally converges to B_s^*.

Temperature Aware Thread Block Scheduling in GPGPUs

Rajib Nath
University of California, San Diego
rknath@ucsd.edu

Raid Ayoub
Strategic CAD Labs, Intel Corporation
raid.ayoub@intel.com

Tajana Simunic Rosing
University of California, San Diego
tajana@ucsd.edu

ABSTRACT

In this paper, we present a first general purpose GPU thermal management design that consists of both hardware architecture and OS scheduler changes. Our techniques schedule thread blocks from multiple computational kernels in spatial, temporal, and spatio-temporal ways depending on the thermal state of the system. We can reduce the computation slowdown by 60% on average relative to the state of the art techniques while meeting the thermal constraints. We also extend our work to multi GPGPU cards and show improvements of 44% on average relative to existing technique.

1. INTRODUCTION

General purpose graphics processor unit ($GPGPU$) provides an energy efficient computing platform for a wide range of parallel applications. The rising trend in the number of cores per $GPGPU$ chip, in addition to technology scaling, results in high power densities. High power dissipation causes thermal hot spots that may have a significant effect on reliability, performance, and leakage power [2]. Meanwhile the duty cycle time of $GPGPUs$ is getting shorter because of the advancements in the $PCIe$ bus design, $GPGPU's$ ability to hide the data transfer with overlapping computation, and $GPGPU$ multiuser mode. As a result, it is becoming more challenging to dissipate the heat using existing cooling mechanisms without sacrificing performance.

In $NVIDIA$ $GPGPUs$, the submitted jobs, usually referred to as kernels, wait in a queue called $KQueue$. Each kernel has a massive number of threads, which are divided into disjoint groups called thread blocks (TB). A kernel has hundreds to thousands of TBs, which have a very short lifetime (μs to ms). Threads inside each TB may synchronize, though TBs are completely independent of each other. TBs are scheduled to the available streaming multiprocessors (SM) by a hardware TB scheduler. Kernels from $KQueue$ are processed one at a time unless there are unused SMs to launch more TB from the next kernel. Since

Permission to make digital or hard copies of all or part of this work for personal or classroom use is granted without fee provided that copies are not made or distributed for profit or commercial advantage and that copies bear this notice and the full citation on the first page. To copy otherwise, to republish, to post on servers or to redistribute to lists, requires prior specific permission and/or a fee.
DAC '13, May 29 - June 07 2013, Austin, TX, USA

all concurrently running TBs in modern $GPGPUs$ are usually from a single kernel, their temperature is dominated by that kernel. A highly compute intensive kernel occupying all the cores can increase the temperature very quickly, causing performance degradation due to dynamic thermal management (DTM) techniques like throttling or dynamic voltage frequency scaling ($DVFS$).

In this work, we propose temperature aware TB scheduling (T^ABS), the first ever $GPGPU$ specific thermal management technique. T^ABS exploits the data parallel computation pattern of $GPGPUs$, the short life time of TBs, the abundance of TBs, and the thermal heterogeneity in $GPGPU$ workloads to reduce the thermal hotspots. T^ABS intelligently intermixes kernels without doing any thread migration to eliminate the overhead of context switching whereas most of the thermal aware workload scheduling in CPU require thread migrations. We present three classes of intermixing algorithms for T^ABS: (a) temporal (alternate), (b) spatial (mixed), (c) spatio-temporal (mixed-alternate). T^ABS has great opportunities in new $GPGPUs$ like Kepler where kernels from multiple applications submitted by more than one user can run concurrently in a virtualized $GPGPU$ environment. We provide the necessary architectural and software changes which make it possible to implement T^ABS in $GPGPUs$. We also explore the use of T^ABS in multi-$GPGPU$ graphics cards. We present a thorough evaluation of T^ABS which reduces performance cost due to thermal hotspots by 60% on average, while meeting thermal constraints.

2. RELATED WORK

Thermal management for general purpose processors has been an active research area in recent years. There are two classes of core level thermal management techniques: reactive and proactive. Popular reactive dynamic thermal management (DTM) [5] techniques remove the excess heat aggressively by slowing down the cores by using pipeline throttling or $DVFS$ at associated performance cost, which our proposed technique reduces dramatically. Activity migration has also been proposed to manage the excess temperature by rescheduling computation across redundant units [6, 9, 8]. This technique does not fit well in the context of $GPGPU$ programming model due to the thread structure, the computation pattern, and the migration cost in short life time of the threads. Moreover, migrations of a kernel across $GPGPUs$ may involve lots of data transfer thereby increasing the performance and energy overhead. In order to address the performance overhead and non uniform thermal distribution in reactive techniques, a set of proactive thermal management techniques [7, 16] have been introduced that leverage temperature predictors. $T^A BS$ can be complemented farther by such temperature predictors if future $GPGPU$ kernels show heterogeneous thermal phases in a single run. Sheaffer, et al. [13] has explored multiple thermal management techniques such as global clock gating, fetch gating, dynamic voltage scaling (DVS), and DVS in multiple clock domains for graphics workloads in $GPUs$. However, all these techniques have performance slowdown due to the induced throttling. Despite all the existing work in $CPUs$ and recent comparison of DTM techniques for graphics workloads in $GPUs$, the thermal management of $GPGPUs$ has not been explored. Our work exploits the thermal heterogeneity in $GPGPU$ kernels and intermixes thread blocks from multiple kernels to have a better thermal distribution over time and space. Although spatial scheduling has been recently proposed for $GPGPUs$ to maximize the resource usage that ultimately leads to increased performance [1], it does not take temperature into account; as a result it may co-schedule two hot kernels, thus experiencing performance overhead due to thermal hotspot.

3. T^A BS DESIGN

We motivate our work by running $GPGPU$ workloads on $NVIDIA$'s $GTX280$ [12] graphics card which has a maximum total power consumption of $236W$ (more in Section 4.1). We measure temperature difference as high as $27°C$ when running different kernels, with the hottest kernel running at the maximum allowed temperature with the $GPGPU$ fan running at the maximum operating speed. Interleaving the execution of the hottest and coldest kernels lowers the measured temperature by $15°C$ below maximum. To investigate the performance overhead caused by thermal hotspots, we develop a thermal simulator for $GTX280$ (details in Section 4) and run a micro kernel that consumes $120W$ power in the SMs. In the execution period of $800s$, we have found that all the SMs occasionally reach the critical temperature and thus experiences 22% performance overhead as a result of the $GPGPU$'s default throttling mechanism. We observe no throttling when running a low power kernel that consumes a total of $60W$ SM power. Interleaving these two kernels with $100ms$ time interval leads to 6% overhead, a large reduction. Since throttling rate relates to the SM temperature, the kernel alternation pattern eventually lowers the temperature. Time multiplexing is beneficial from the thermal and performance points of view when there is no context switching overhead. While such overhead can be very large in general purpose processor, we show that this is negligible and avoidable in $GPGPUs$. We observe similar thermal benefits by distributing available SMs among hot and cold kernels (SM multiplexing). These results have motivated us to intermix TBs from thermally heterogeneous kernels and to modify the existing HW TB and OS schedulers.

3.1 T^A BS System Architecture

Our proposed system architecture for temperature aware thread block scheduler ($T^A BS$), shown in Figure 1, has two components: HW TB scheduler and OS scheduler. They coordinate with each other to execute temperature aware TB scheduling algorithms in order to reduce thermal hotspots in $GPGPUs$. HW TB scheduler maintains power and lifetime

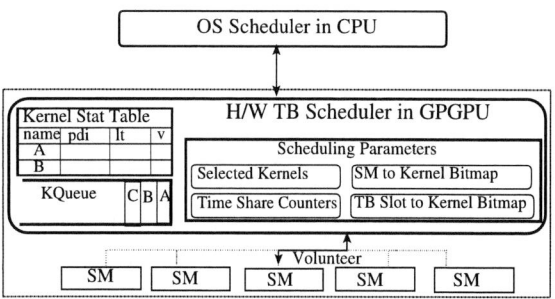

Figure 1: T^A BS Architecture

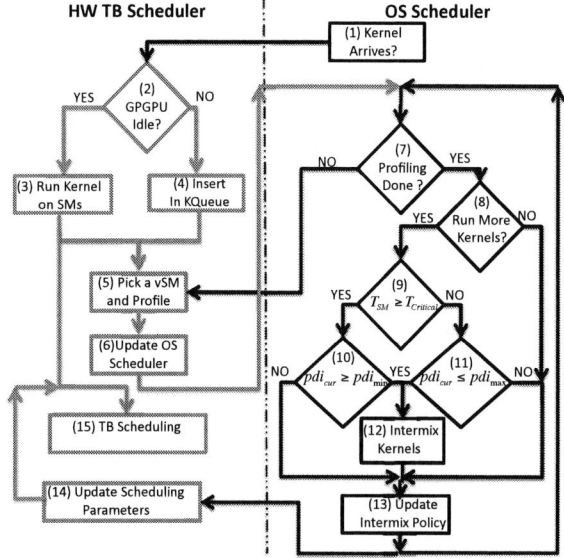

Figure 2: Interactions in $T^A BS$

statistics of TBs of each kernel in a kernel stat table. The key to determining which kernel should run next is comparing the currently running kernel with kernels in $KQueue$ in terms of power density, and therefore the chance to reduce the temperature. OS $scheduler$ uses the kernel stat table along with the instantaneous temperatures of SMs to set the TB scheduling policy for the next OS scheduling tick. OS $scheduler$ periodically updates HW TB scheduler about the set of intermixed kernels and their relative share of $GPGPU$ resources (e.g., time, SMs), which define and parametrize various intermixing policies for $T^A BS$.

Figure 2 shows the interaction between HW TB scheduler and OS $scheduler$. When a $GPGPU$ job arrives in our system (step 1), the OS $scheduler$ sends it to $GPGPU$. In step 2 when the $GPGPU$ is idle, the arriving kernel starts execution immediately. Otherwise the kernel waits in the $KQueue$ (step 4), which is usually served in the first come first served ($FCFS$) manner. As soon as a kernel arrives in $KQueue$, the TB scheduler selects a volunteer SM (vSM) to profile the new kernel (step 5). The coldest SM in the $GPGPU$ is selected as vSM. There can be multiple $vSMs$ simultaneously profiling multiple kernels if several kernels arrive in $KQueue$ in a short duration of time. The number of $vSMs$ cannot be more than the number of SMs in the $GPGPU$. The TB scheduler waits for the running TBs in a vSM to finish before scheduling a batch of TBs from that kernel.

The profile of each kernel, provided by vSM, is stored in a kernel stat table. Each entry in the kernel stat table has four fields: (a) name, (b) power density index, (c) average TB lifetime, (d) valid bit. Power density index (pdi) is an estimate of the average dynamic power consumption of one SM while running TBs from that kernel. We use the model in Equation 1, derived from [10], to estimate the pdi based on activity rates in integer unit (INT), floating point unit (FP), special function unit (SFU), cache ($CACHE$), register file (REG), shared memory (SHM), arithmetic logic unit (ALU), and fetch decode unit (FDS)

$$P_{SM} = \sum_{i \in Units} P_i = P_{INT} + P_{FP} + P_{SFU} + P_{ALU}$$
$$+ P_{FDS} + P_{REG} + P_{CACHE} + P_{SHM}$$
(1)

where $P_i = MaxPower_i \times AccessRate_i$ for each unit i. This model has the sufficient accuracy to compare thermal behavior of two kernels based on power consumption [10]. The performance monitoring unit (PMU) in the vSM monitors the performance counters in Equation 1 until one of the TB from the batch retires. This way the monitoring is performed with a constant thread level parallelism (TLP) in the vSM which is necessary for an accurate estimate of the pdi in $GPGPUs$. Once the activity data and average lifetime (lt) of TBs are available from the vSM, the TB scheduler computes the pdi, updates the corresponding kernel entry in the kernel stat table and marks it as profiled by setting the valid bit(v). One time sampling of pdi is adequate to classify the power profile of the kernels because the dynamic power consumption of $GPGPU$ applications is mostly constant as shown in many publications [10]. $OS\ scheduler$ uses pdi to assess whether a particular kernel will run hotter or colder on average than a competitor kernel. Since pdi is unbiased with respect to the thermal history, the current temperature of the SMs, and the activity in the neighboring SMs, it is a more reliable metric than direct temperature sensor reading.

Once the TB scheduler informs the $OS\ scheduler$ that the kernels in the $KQueue$ are profiled (steps 6 & 7), the $OS\ scheduler$ looks for intermixing opportunities in every OS scheduling tick ($50ms$) (steps 8 through 12). When the $GPGPU$ has space for running more parallel kernels (step 8), the $OS\ scheduler$ explores $KQueue$ for the best option while checking the thermal state of the $GPGPU$ (step 9). During a thermal emergency situation when the SM temperature reaches emergency threshold (step 10), the $OS\ scheduler$ intermixes the pending kernels with the lowest pdi with the running kernel if possible (step 10) and assigns an initial small percentage (e.g., 5%) of total $GPGPU$ resources (e.g., time, SMs) to it (step 13). The $OS\ scheduler$ takes proactive action in the absence of temperature problem by intermixing the maximum pdi pending kernel with the running kernel whenever possible (step 11). Such decision is beneficial because it is ideal to spread the heat uniformly over time. When more than one kernel is running in a thermal emergency, the $OS\ scheduler$ updates the policy with a new distribution of resources if needed (step 13).

Each time the SM temperature reaches a critical value, the $OS\ scheduler$ progressively takes away resources from the kernel with higher pdi and assigns those to the kernel with a lower pdi. Although taking away resources from kernels may affect the order at which kernels finish, this helps to increase the overall $GPGPU$ throughput in the presence of thermal emergencies. As the thermal problem disappears, the $OS\ scheduler$ does the reverse to ensure that the victim kernel with a high pdi is not falling behind in execution. The minimum resource assigned to the victim kernel that arrived first in the $KQueue$ among two concurrently running kernels does not go below 50% of the total $GPGPU$ resources. This way, the $OS\ scheduler$ prevent starvation of the kernels with high pdi. In order to ensure fairness, the $OS\ scheduler$ maintains a non-overlapping window of N kernels that can be intermixed. The window size represents a trade-off between fairness and intermixing opportunity. For example, $T^A BS$ will act as a first come first serve ($FCFS$) scheduler that performs no intermixing when $N = 1$. While we pick $N = 32$ to match $KQueue$ size, any smart fairness algorithm can be implemented with this framework.

The $OS\ scheduler$ updates the $HW\ TB$ scheduler with new scheduling parameters (step 14). The scheduling parameters, describing the distribution of resources (time and SM) among intermixed kernels, defines different classes of intermixing policies: (a) alternate, (b) mixed, and (c) mixed alternate. Alternate (A) intermixing policy time shares the $GPGPU$ between multiple heterogeneous kernels in an interleaved fashion. The TB scheduler executes the alternation of kernels inside each $GPGPU$ while the time sharing interval of each kernel is dynamically determined by the $OS\ scheduler$ at each OS scheduling tick. The time sharing interval of each kernel should be an integer multiple of the lifetime of its TB. In this way, the $T^A BS$ does not interrupt execution to maximize performance while keeping the same time share ratio. With mixed intermixing policy, the TB scheduler schedules multiple heterogeneous kernels with different power densities on a single $GPGPU$ simultaneously. Mixed intermixing is further divided into two subtypes depending on the heterogeneity of TBs in a single SM. In mixed uniform (MU), each SM is allowed to host TBs from a single kernel. The number and the topological location of SMs that run TBs from each kernel is an optimization parameter and is dynamically determined by the $OS\ scheduler$ depending on the pdi of individual kernels and the thermal state of the system. $T^A BS$ usually schedules TBs with higher pdi in the corner SMs. The $OS\ scheduler$ updates the $HW\ TB$ scheduler with a SM to kernel bitmap which describes the SM distribution. Mixed nonuniform (MNU) allows TBs from multiple kernels to coincide together in a single SM. The $OS\ scheduler$ distributes available TB slots in each SM among the concurrently running heterogeneous kernels and updates the TB slot to kernel bitmap. Since SMs will have heterogeneous threads, MNU is analogous to alternate at finer granularity. Mixed alternate (MA) is a hybrid technique that uses mixed and alternate approaches at the same time. Initially MA allocates x SMs to a hot kernel and $(M - x)$ SMs to a cold kernel where M is the total number of SMs in a GPGPU. In the subsequent alternating intervals, each SM alternates between hot and cold kernels.

For each intermixing policy, TB scheduler allows the running TBs to finish before adopting to new scheduling parameters provided by the $OS\ scheduler$ (step 15). Since TBs have very short lifetime (μs to few ms) and the thermal time constraint of $GPGPU$ is usually in the range of seconds [14], delaying the update is not an issue. The short lifetime of TBs, the abundance of TBs, and TB scheduler's ability to schedule new TBs with near zero context switch overhead have allowed us to use $T^A BS$ as an efficient thermal management technique for modern $GPGPUs$. This unique property of $GPGPU$ has given us the freedom to manage the thermal problems of $GPGPU$ without doing any context switching or migration of threads which are normally the key techniques of temperature management in $CPUs$. The same techniques cannot be applied to $CPUs$ because CPU threads have a much longer lifetime that often exceeds the thermal time constant. A scheduler in the cluster level or supercomputer level can reuse the power profiles of each kernel to maintain a heterogeneous mixture of kernels in the intermixing window of each compute node. In the absence of heterogeneous kernels in the KQueue, $T^A BS$ applies $DVFS$ to handle thermal emergency situations if required.

3.2 T^BS Across Multiple GPGPUs

In a multi-$GPGPU$ graphics card (e.g., $NVIDIA$ GTX 690), a single fan is shared by two $GPGPUs$, so the fan speed rises with the temperature of the hottest $GPGPU$ chip. Instead of executing two heterogeneous kernels (*hot* and *cold*) in two separate $GPGPUs$, TBs from each kernel can be distributed intelligently among them. This also saves cooling energy since the cooling cost is cubically related to the fan speed. OS $schedulers$ contacts TB scheduler in all the $GPGPUs$ to find a heterogeneous mix of kernels. Similar to a single $GPGPU$ case, the percentage of TBs from concurrently running kernels in each $GPGPU$ is an optimization parameter that the OS $scheduler$ sets dynamically.

3.3 T^BS Overhead

In this section, we discuss both area and performance overhead of our techniques. Area overhead is dominated by the kernel stat table whose size is bound by the size of $KQueue$ (e.g., 32). The overhead of maintaining kernel statistics is 832 bits where each entry needs 26 bits: 5 bits for kernel name, 4 bits for pdi, 16 bits for lt and 1 bit for v. To record the resource sharing info for different intermixing policies, TB scheduler also needs two counters to keep track of the time quantum progress of intermixed kernels, one bit map (bounded by number of SM in the chip) to map SM to kernels, and one common bit map for all SMs (bounded by 8) to map TB slot per SM to kernels. These additional hardware requirements (2 counters + 30 bits + 8 bits) also represent a very low overhead.

Since the lifetime of each TB is typically in the range of μs to few ms and each kernel is profiled once, the kernel profiling overhead is very small. TB scheduler launches a system kernel (one TB with 8 threads) which computes pdi from the PMU activity data using using Equation 1. TB scheduler also uses two other system kernels (one TB with 32 threads) to find the kernels with minimum and maximum pdi from the kernel stat table. OS $scheduler$ reads these values instead of reading the whole kernel stat table. We use system kernel to save additional HW requirement. Even though the run time of these system kernels is $500ns$ while running alone, the overhead is negligible if executed with other TBs because these are $GPGPU$ reduction (e.g., sum, max, min) kernels with less than 100 instructions. Moreover, they run once in each OS scheduling tick.

$T^A BS$ in multi-$GPGPU$ graphics card has to pay the penalty of accessing memory from the other $GPGPU$ chip on the board. Such overhead is amortized when two kernels with very diverse pdi are intermixed.

4. RESULTS
4.1 Methodology

While our solution is applicable to $GPGPUs$ from different vendors, we have selected $NVIDIA$'s $GTX280$ for this study since it is a representative of modern $GPGPUs$ and its power data, power model and floor plan are available from published work [10]. $GTX280$ has 30 streaming multiprocessors (SM), which share a $L2$ cache and an off chip memory. Three SMs share a common L1 cache. Each SM is equipped with 8 $CUDA$ cores, big register bank, and shared local memory. Device memory is interleaved into 8 memory modules providing bandwidth of $140GB/s$. The theoretical peak performance of $GTX280$ is $933GFlops/s$ in single precision at $1.3GHz$ clock speed. The maximum total power of this graphics card is $236W$ while the idle power is $80W$ [10].

Since our contribution requires both architecture and hardware scheduler changes, it is not possible to implement this in today's $GPGPU$ cards. We extend Hotspot [14] simulator with $GPGPU$ thermal simulation based on the floor plan and package characteristics of $GTX280$. The heatsink dimension and the case to ambient thermal resistance (K/W) are used as $0.06m$ and 0.25 respectively based on the data in [15]. Each simulation starts from an initial temperature of $45°C$ with a warm-up period of $200s$. We keep the $GPGPU$ fan running at a fixed speed throughout the simulation to maintain a constant case temperature. There is one temperature sensor per SM. The local ambient temperature within PC and the critical temperature threshold on chip are set at $45°C$ and $90°C$ respectively. Since we are interested in understanding the effectiveness of TB scheduling for thermal management, we focus on SMs only and do not do thermal management of memory or handle cooling. We leverage the power model from [10] to generate the dynamic power traces for our benchmarks. We also account for thermally dependent leakage power based on the model from [3]. Like in [4], we include a baseline power of the $GPGPU$ chip in the simulator.

Bench	SM Power(W)	Bench	SM Power(W)
SVM	87	Bino	67
Sepia	44	Conv	74
Bs	50	Cmem	48
Dotp	60	Madd	61
Dmadd	69	Mmul	64

Table 1: Benchmarks on a GTX 280

We use ten benchmarks [10]. A subset (*SVM, Bino, Sepia, Conv, Bs*) is from merge benchmark suite [11] which represents financial and image processing sectors. Others are memory bound (*Dotp, Madd, Dmadd, Mmul*) and compute bound (*Cmem*) benchmarks, which we find in many scientific computing and graphics applications. The dynamic power consumption of SMs while running each benchmark is reported in Table 1. The dynamic range of the SM power consumption is $43W$. We form 10 workloads using these 10 benchmarks to keep a representative mix of hot and cold kernels shown in Table 2. Each benchmark in the workload has computation load of $400s$.

In our simulations, the baseline (BL) DTM policy throttles the $GPGPU$ when the SM temperature reaches the critical threshold and clock gates the $GPGPU$ until the temperature falls below $85°C$. We have also implemented $DVFS$ at five different settings: $(1.18V, 1.33GHz), (1.14V,$

$1.28GHz$), ($1.13V$, $1.26GHz$) , ($1.12V$, $1.20GHz$), and ($1.12V$, $1.15GHz$). We compare T^ABS, the first ever thermal management techniques for $GPGPUs$, against throttling (BL) and $DVFS$, two well known DTM techniques. We show the performance of T^ABS for four different intermixing policies discussed in Section 3.1. T^ABS adopts $DVFS$ if the OS *scheduler* fails to find heterogeneous kernels during a thermal emergency.

Workload	Mix	Workload	Mix
WL1	SVM + Cmem	WL6	Conv+ Bs
WL2	SVM + Bs	WL7	SVM + SVM
WL3	SVM + Dotp	WL8	SVM + Conv
WL4	Conv+ Dmadd	WL9	Conv + Conv
WL5	Conv+ Mmul	WL10	Sepia + Bs

Table 2: Workload Description

4.2 Measured Overhead Due to Intermixing

Running two concurrent kernels using different intermixing policies may cause performance overhead due to sharing of resources, e.g., bandwidth, L1 cache, L2 cache, and SMs. Due to the alternation of context, the contents of the caches might get flushed. While the impact of cache thrashing can be severe in general purpose $CPUs$, this effect is small in $GPGPUs$ because of the underlying computation patterns. $GPGPU$ kernels perform data parallel computation where multiple threads execute identical code on separate data sets. Instead of depending on high cache hit rates and efficient branch prediction, most of the highly optimized $GPGPU$ kernels depend on the large bandwidth and the high TLP to hide the memory latency.

Intermixing	Performance Improvements (%)		
	Mean	Stdev	Minimum
Alternate	1.26	3.45	-1
Mixed Uniform	5.43	7.35	-1
Mixed Non Uniform	2.83	7.76	-1
Mixed Alternate	2.87	7.01	-1

Table 3: Performance Improvements by Intermixing

Even though new $GPGPUs$ have limited support for parallel kernel execution, they do not intermix TBs. The only way to evaluate the performance overhead due to intermixing TBs from multiple kernels is by manually editing the source code to merge them. In a merged kernel, we keep the functionality of the original kernels unchanged. Inside each merged kernel, threads execute different function based on the TB ID and host SM IDs. For example, in case of mixed uniform intermixing, TB executes function of kernel 1 when SM ID is less than 8 and vice versa. The alternate intermixing is simulated by switching between different kernel functions every 100 TBs. Since TB in $GPGPUs$ are scheduled in monotonically increasing order of their TB IDs, we could emulate our intermixing policies by implementing these tricks. We form 23 different intermixed kernel combinations using benchmarks from Table 1 and run them on actual $GPGPU$. Table 3 shows the mean and standard deviation of performance improvements of all the 23 merged kernels for different intermixing policies. The results show that on average performance improved by 3%, and the worst case overhead is only 1%. Clearly intermixing is beneficial to both performance and on-chip temperature in a large fraction of the cases.

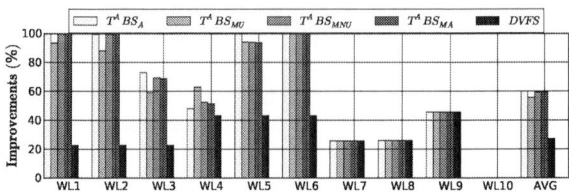

Figure 3: Improvements over BL in Single GPGPU

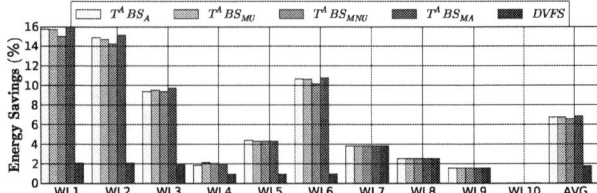

Figure 4: Energy Savings in Single GPGPU

4.3 Simulation Results

Single GPGPU: Our baseline policy (BL), which clock gates the $GPGPU$ until the temperature falls below $85°C$, experiences computation slowdown ranging between 2.06% and 17.79% with an average of 6.75% for the workloads in Table 2. As expected, the performance cost due to throttling is higher for workload with *hot* benchmarks, e.g., $WL7$. Meanwhile, workloads with only *cold* benchmarks (e.g., $WL10$) do not experience any thermal hotspot. Figure 3 shows improvements of T^ABS and $DVFS$ over BL. The improvement is the reduction in computation slowdown due to thermal throttling over the baseline policy. $DVFS$ works better than BL by 27% on average. Our proposed technique T^ABS periodically checks the $KQueue$ and intermix TBs from heterogeneous kernel proactively that results in reduction in number of thermal hotspots and improvement in performance. For different TB intermixing policies: alternate (A), mixed uniform (MU), mixed non uniform (MNU), and mixed alternate (MA) described in Section 3.1, the average reduction in computation slowdown by T^ABS compared to BL ($DVFS$) is from 57%(40%) to 60%(45%). However, for the subset of workloads representing our target cases, T^ABS improves over BL ($DVFS$) by 82%(74%) to 89%(86%). Among the four proposed intermixing policies, alternate works best since it does a good job spreading heat over time. Mixed non-uniform and mixed alternate work better than mixed uniform since mixed uniform runs hot TBs on the same subset of SM while the other two do some kind of alternation of hot and cold kernels.

Figure 4 shows the percentage of energy savings of T^ABS and $DVFS$ relative to BL. Interestingly, all of the intermixing policies of T^ABS save 6.75% on average compared to 1.8% with $DVFS$. The savings of energy come from two different sources. The reduction in computation throttling helps the jobs finish faster. In addition, T^ABS saves leakage power by reducing the average temperature of the SMs. For the heterogeneous workloads representing our target cases ($WL1$, $WL2$, $WL3$, $WL4$, $WL5$, and $WL6$), T^ABS saves 9.48% and 8.12% energy on average relative to BL and $DVFS$ respectively. Energy savings are maximized when we mix the hottest and coldest kernels. For example, the SM power consumption of $Cmem$, Bs, $Dotp$, and $Conv$ are $48W$, $50W$, $60W$ and $74W$ respectively. When we

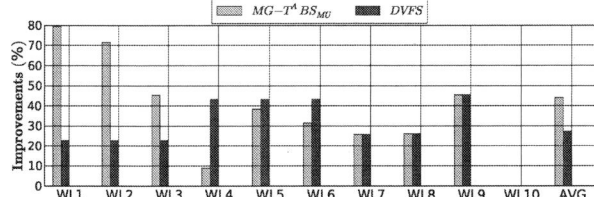

Figure 5: Improvements over BL in Multi GPGPUs

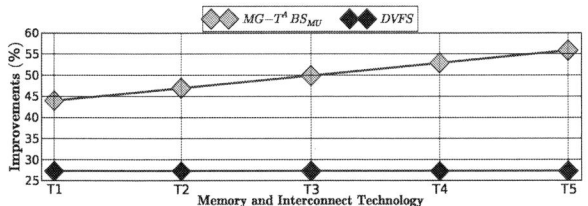

Figure 6: Effects of Memory and Interconnect Technology in Multi $GPGPU$ TB Scheduling Policy

intermix them with SVM, the energy savings are 15.75%, 14.88%, 9.39%, and 1.55% for workload $WL1$, $WL2$, $WL3$, and $WL8$ respectively. In the absence of heterogeneous kernels, the energy savings of our technique is the same as the $DVFS$ ($WL7$ and $WL9$).

Multi GPGPUs: We extend our simulator for graphics card with multiple $GPGPUs$. We use the benchmark data from [10] to estimate memory overhead in our simulator. Figure 5 shows the improvements of $DVFS$, and multi-$GPGPU$ T^ABS ($MG-T^ABS$) with MU intermixing policy over default technique BL. BL and $DVFS$ schedule each kernel to a separate $GPGPU$ and experiences the same amount of throttling that we have seen in single $GPGPU$ case. T^ABS improves over BL by 44% for mixed uniform intermixing policy. Even though the performance of T^ABS is worse than their respective single $GPGPU$ case due to memory overhead, the benefit is still substantial in our target cases, e.g., for $WL1$ T^ABSA improves over BL and $DVFS$ by 80% and 75% respectively.

Results in Figure 5 suggests that scheduling TBs across $GPGPUs$ is beneficial only when the performance gained through T^ABS is greater than the memory overhead. When the thermal profiles of two kernels are very diverse, we get the maximum gain, e.g., $WL1$. For workloads $WL4$, $WL5$ and $WL6$, T^ABS performs worse than $DVFS$ because the benefit gained through reduction in thermal hotspots could not amortize the memory overhead. This observation also implies that as the memory & interconnect technologies in multi-$GPGPU$ cards improve, the gain through T^ABS grows. Figure 6 shows the reduction in computation slowdown for T^ABS and $DVFS$ comparing to BL for different memory & interconnect technologies (from $T1$ to $T5$ memory & interconnect technology gets better). The result suggests that while T^ABS will benefit with memory and interconnect improvements, $DVFS$ will not perform any better than today.

5. ACKNOWLEDGEMENT

This work has been funded by NSF SHF grant 0916127, NSF CCF 1218666, NSF grant 1029783, NSF ERC CIAN EEC-0812072 NSF Variability, CNS, Oracle, Google, Microsoft, MuSyC, and SRC grant P11816.

6. CONCLUSION

The nature of data parallel computation in $GPGPUs$ provides us a unique opportunity to manage the thermal problems by intermixing thread blocks from multiple thermally heterogeneous kernels. In this work, we have proposed T^ABS, the first ever thermal management technique for modern $GPGPUs$, which exploit the opportunity to spread the heat in time and/or space. We have provided the required architectural and software changes to incorporate T^ABS in modern $GPGPUs$. Our results show that T^ABS performs 60% better than state of the art thermal management techniques with energy savings of 6.75% on average. T^ABS's prospect in multi-$GPGPU$ graphics card is also very promising, 44% improvements over well known techniques.

7. REFERENCES

[1] J. T. Adriaens, K. Compton, N. S. Kim, and M. J. Schulte. The case for gpgpu spatial multitasking. *HPCA*, 2012.

[2] A. Ajami, K. Banerjee, and M. Pedram. Modeling and analysis of nonuniform substrate temperature effects on global interconnects. *IEEE Trans. on CAD*, 2005.

[3] R. Ayoub, K. Indukuri, and T. Rosing. Temperature aware dynamic workload scheduling in multisocket cpu servers. *TCAD*, 2011.

[4] R. Ayoub, R. Nath, and T. Rosing. Jetc: Joint energy thermal and cooling management for memory and cpu subsystems in servers. *HPCA*, 2012.

[5] D. Brooks and M. Martonosi. Dynamic thermal management for high-performance microprocessors. *HPCA*, 2001.

[6] J. Choi, C. Cher, H. Franke, H. Hamann, A. Weger, and P. Bose. Thermal-aware task scheduling at the system software level. *ISLPED*, 2007.

[7] A. Coskun, T. Rosing, and K. Gross. Proactive temperature management in mpsocs. *ISLPED*, 2008.

[8] J. Donald and M. Martonosi. Techniques for multicore thermal management: Classification and new exploration. *ISCA*, 2006.

[9] S. Heo, K. Barr, and K. Asanovic. Reducing power density through activity migration. *ISLPED*, 2003.

[10] S. Hong and H. Kim. An integrated gpu power and performance model. *ISCA*, 2010.

[11] M. D. Linderman, J. D. Collins, H. Wang, and T. H. Meng. Merge: a programming model for heterogeneous multi-core systems. *ASPLOS*, 2008.

[12] NVIDIA. Gtx280 http://www.geforce.com.

[13] J. W. Sheaffer, K. Skadron, and D. P. Luebke. Studying thermal management for graphics-processor architectures. *ISPASS*, 2005.

[14] K. Skadron, M. Stan, K. Sankaranarayanan, W. Huang, S. Velusamy, and D. Tarjan. Temperature-aware microarchitecture: Modeling and implementation. *TACO*, 2004.

[15] J. Wang and W. Chen. Vapor chamber in high-end vga card. *IMPACT*, 2010.

[16] I. Yeo, C. Liu, and E. Kim. Predictive dynamic thermal management for multicore systems. *DAC*, 2008.

VAWOM: Temperature and Process Variation Aware WearOut Management in 3D Multicore Architecture

Hossein Tajik[1], Houman Homayoun[2], Nikil Dutt[1]
[1]Center for Embedded Computer Systems, University of California Irvine
[2]Department of Electrical and Computer Engineering, George Mason University
{tajikh, dutt}@uci.edu, hhomayou@gmu.edu

ABSTRACT

Three dimensional (3D) integration attempts to address challenges and limitations of new technologies such as interconnect delay and power consumption. However, high power density and increased temperature in 3D architectures accelerate wearout failure mechanisms such as Negative Bias Temperature Instability (NBTI). In this paper we present VAWOM (Variation Aware WearOut Management), an approach that reduces the NBTI effect by exploiting temperature and process variation in 3D architectures. We demonstrate the efficacy of VAWOM on a two-layer 3D architecture with 4x4 cores on the first layer and 4x4 last level caches on the second layer, and show that VAWOM reduces NBTI induced threshold voltage degradation by 30% with only a small degradation in performance.

Categories and Subject Descriptors

B.3.2 [**Design Style**]: Cache Memories; B.3.m [**Miscellaneous**]

General Terms

Algorithms, Management, Design, Reliability.

Keywords

Wearout, NBTI, 3D Integration, Variation.

1. INTRODUCTION

Advances in technology and reduction in feature sizes have made transistors faster and also allow larger die sizes that permit the designer to exploit many computational cores for parallel computing. Due to the increased complexity in newer technologies, interconnection power consumption and delay are becoming critical challenges that require new methodologies. 3D integration is a new manufacturing technology that reduces power consumption and increases the speed of interconnections, while improving packaging density [1]. In a 3D architecture, more than one layer of electronic circuits are integrated in a single chip. Vertically stacked dies are connected by TSVs (Through Silicon Vias) which leads to shorter and faster connections.

However, 3D integration does not come free. In fact higher performance and considerable reduction in energy consumption enabled by this technology, comes with the cost of higher power density and increased heat conduction paths [2][36]. As a result, the chip's temperature is much higher in 3D architecture compared to 2D architectures. Increased cooling cost, higher probability of timing errors, physical damages, and lifetime reduction are just a few of many consequences due to this higher power density.

Such operational conditions in 3D processors put more stress

Permission to make digital or hard copies of all or part of this work for personal or classroom use is granted without fee provided that copies are not made or distributed for profit or commercial advantage and that copies bear this notice and the full citation on the first page. To copy otherwise, or republish, to post on servers or to redistribute to lists, requires prior specific permission and/or a fee.
DAC'13 May 29 – June 7 2013, Austin, TX, USA.

on the circuit and activate failure mechanisms such as Negative Bias Temperature Instability (NBTI) [3]. NBTI is one of the most important wearout mechanisms in PMOS transistors [4] [5] which gradually adds delay to transistors and leads to timing errors. Studies have shown that NBTI is highly dependent on the operating temperature [6].

The average temperature in a 3D architecture is much higher than for 2D architectures and temperature of vertically adjacent components in different layers are highly correlated [7][28][37]. Long conduction paths for layers farther from the heat-sink make them hotter (from this point we assume that upper layers are farther from the heat-sink). Placing cache banks in upper layers helps to reduce thermal hotspots, due to low power density in cache banks. The higher temperature of a 3D architecture, specifically in the layer most distant from the heatsink, has a significant effect in accelerating NBTI degradation.

The common architecture of a 3D processor, which stacks cache vertically on top of the logic, makes cache temperature dependent on the workload running in the logic underneath. Therefore, contrary to a 2D architecture, spatial temperature differences in cache banks could be very high, depending on the underlying logic power density.

Moreover, new technologies create process and operational variations that impose major challenges to processor design [8] [9]. Threshold voltage and other types of variation lead to large performance variation between different device components. Frequency and cache access time variation exist in new technologies which make some cores or cache banks more prone to wearout failure [13].

In this paper, we present VAWOM (Variation Aware WearOut Management), an approach that reduces wear-out induced by NBTI in cache banks and cores of a 3D architecture, using temperature and process variation aware schemes. VAWOM sacrifices some banks in the cache layer to proactively perform wearout recovery. During recovery, the SRAM cache and the associated router is deactivated. By exploiting task migration in the core layer, VAWOM balances the temperature in cores and cache memories to mitigate NBTI effects in the cores and the cache banks.

To the best of our knowledge VAWOM is the first work that highlights the large threshold voltage variation over time due to high power density in 3D architecture and investigates activity migration in the core layer and wearout recovery mechanism in cache layer to mitigate this variation. VAWOM makes three major contributions:

- Demonstrates the large thermal and threshold voltage variation in 3D cache, unlike 2D cache architectures.
- Presents novel activity migration and proactive recovery mechanisms that exploit thermal variation in 3D architecture to mitigate wearout effects in multiple stacked layers in 3D architecture. VAWOM balances wearout in different parts of a 3D architecture.

- Reduces wearout in 3D architectures by exploiting frequency variation in cores and access time variability in the cache banks. VAWOM balances wearout across multiple layers by imposing less stress on the slower cores and cache banks.

2. Related Work

For prolonging the lifetime of digital circuits in the presence of aging mechanisms such as NBTI, several techniques have been proposed at different layers of abstraction. At the circuit level, the most effective method is using delay guard-bands [10], where a delay margin is added to the nominal design. Adding a large guard band negatively affects the performance of the circuit. At the system level, DVS has been used to speed up the circuit or reduce the stress condition [39]. Some works propose to use redundancy in functional units, caches, or cores for mitigation of aging in both reactive and proactive manner [14] [15] [16]. Reducing duty cycle (e.g., with bit flipping) has also been proposed to decrease the effects of NBTI [4] [17]. [11] uses frequency variation to hide aging in processors by a careful scheduling of hot jobs to fast cores. This technique negatively affects cache banks in the 3D architecture (as we'll discuss later in Section 4).

At the architecture level, in [18] the authors proposed to balance workloads among active cores. NBTI aware Dynamic Instruction Scheduling was proposed by [19] which distributes tasks over functional units to balance aging. Using BIST, [20] tries to monitor NBTI degradation in SRAM cells and reconfigures cache arrays.

These methods do not consider cache banks and large process variation in circuits. Unlike VAWOM, none of these works consider temperature variation between cache banks in 3D architecture. Moreover, the cache access time variation and interplay of variation-aware techniques in different layers are not considered in these previous efforts.

3. Preliminaries (NBTI)

Negative Bias Temperature Instability (NBTI) is one of the important wearout mechanisms that mostly affects PMOS transistors. Based on transistor bias voltage, NBTI has two modes: stress and recovery. When gate-source voltage is negative, the transistor is under stress; this increases the threshold voltage and leads to performance degradation (device aging). A cumulative performance degradation results in timing errors after a sufficiently long stress period. In the recovery mode (PMOS biased with positive voltage), some threshold voltage increase can be compensated [12]. We use the well-known Reaction-Diffusion (R-D) model for measuring threshold voltage shift when applying VAWOM. In this model, Si-H bonds are dissociated under negative gate-source bias by positive holes from channel (stress mode) and make interface traps and H atoms. Traps formed at the interface of gate oxide, cause increase in threshold voltage of PMOS transistor [12]:

$$\Delta V_{th} = q\, N_{IT}/C_{ox} \tag{1}$$

N_{IT} is number of interface traps, q is basic charge and C_{ox} is oxide capacitance.

When gate-source voltage becomes positive, some of the hydrogen atoms diffuse back to the gate oxide interface and anneal some of the broken bonds. This phenomenon reduces the number of interface traps and compensates threshold voltage shift partially (recovery mode).

3.1 Stress

Negative voltage on the gate-source of PMOS transistor creates some interface traps. Equation (2) is a differential expression that describes formation of H, H2 and interface traps in the R-D model [5]:

$$\frac{N_{IT}}{t} - \frac{\delta k_H \left(k_f N_0 - \frac{N_{IT}}{t}\right)^2}{k_r^2 N_{IT}^2} + \frac{\delta k_{H2} N_{IT}}{\sqrt{6 D_{H2} t}} = 0, \tag{2}$$

where t is time, k_f is Si–H bond-breaking rate, k_r is Si–H bond-annealing rate, and N_0 is the initial bond density available before stress. k_H, k_{H2}, D_{H2}, and δ are generation rate of H_2, dissociation rate of H_2, diffusion rate of H_2, and interfacial thickness (~1−2 Å) respectively.

In different time slots, Equation (2) can be reduced to different analytical solutions [5]. A considerable amount of degradation occurs in the primary stages of stress mode. In the long stress time, threshold voltage degradation rate decreases due to the small time exponent factor. Initial bond density (N_0) is affected by operating voltage and initial threshold voltage. Reducing gate-source voltage or increasing threshold voltage slows down the degradation [22].

3.2 Recovery

A positive voltage on the gate-source of PMOS transistor recovers some interface traps. From [23], if at the beginning of the recovery phase (time t_0), N_{IT} is equal to $N_{IT}^{(0)}$ and $\xi = 1/2$, the number of interface traps at the end of the recovery phase (time t) equals:

$$N_{IT} = N_{IT}^{(0)}[1 - (\xi t/t_0)^{1/2} / (1 + t/t_0)^{1/2}] \tag{3}$$

A large portion of recovery occurs in the first few seconds after beginning of the recovery.

3.3 Temperature Dependence of NBTI

The operating temperature changes NBTI sub-process rates such as diffusion rate, conversion rate, and reaction rates during stress time. In general, higher temperature exacerbates degradation due to NBTI. NBTI thermal dependence and direct effect of temperature on sub-process rates have been investigated and formulated in several recent works [6] [24].

4. VAWOM

4.1 A. Baseline Configuration

To illustrate VAWOM, we assume an exemplar baseline 3D floorplan[1], where a single cache layer stacks on top of a core layer (Figure 1). Our studied 3D architecture has 16 dual issue cores all in one layer. To avoid thermal build up in the core layer as a result of 3D stacking, we place the core layer close to the heatsink while the cache layer is placed farther from the heatsink. In the bottom layer (Layer 1), 16 core routers are connected through a mesh topology network. In the upper layer (Layer 2), 16 shared last level cache banks are connected via 16 routers connected in a mesh topology. The baseline configuration in Figure 1 does not show inter-layer interconnections (between routers).

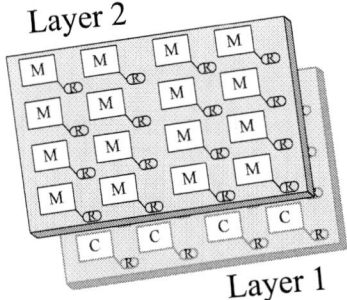

Figure 1. Baseline Configuration (C= Core, M = cache Memory)

---

[1] Of course our method could be applied to any 3D architecture.

4.2 VAWOM Overview

VAWOM deploys a proactive recovery method to manage temperature and process variation induced wearout. VAWOM uses several temperature and variation aware policies for activity migration and proactive recovery. These policies take into account device degradation (aging) in different layers, including cache bank access time degradation and core frequency degradation.

VAWOM runs periodically and in each run, two types of system change are performed. First, *activity migration in the core layer* with respect to NBTI thermal dependence improves the lifetime reliability of the cache banks and the cores. Second, *proactive recovery in the cache layer* improves cache lifetime, since aging is a more intense problem in the layers farther from the heatsink. In each run, VAWOM decides which cache bank should be placed in the recovery mode (called "recovery bank"), moves data from the recovery bank to the previously disabled/recovered cache bank, updates routing information, and disables new recovery bank. We first describe the experimental framework and the assumptions before describing VAWOM policies.

Unlike 2D architecture, temperature variation between different cache banks is noticeable in the 3D architecture. To better understand this characteristic, we study thermal variation in 2D and 3D architectures running with the same benchmark. We used Hotspot 5.0 for thermal estimation [25]. We observe large temperature variations of 25°C across various cache banks in 3D. This is noticeably higher that the 4°C thermal variation in the 2D design. Furthermore, cache temperature in the 3D architecture is about 24°C higher than the 2D architecture. Thus, SRAM cache in the 3D architecture suffers from NBTI degradation significantly more than the 2D architectures.

In the baseline configuration without any wearout management technique, all of the cache banks are active and they are constantly under stress. This steady stress increases threshold voltage constantly over time. For compensating this degradation, an effective solution is to put cache blocks into the recovery mode. [15] proposed a circuit technique to put SRAM arrays of a cache bank in the recovery mode. VAWOM deploys a proactive recovery method to fight back the wearout in cache banks and puts whole cache bank in the recovery mode.

Furthermore, studies show that temperature has a significant effect on NBTI. Data in [24] indicates that in just 1000 seconds, threshold voltage increases by 10 mV as operating temperature increases from 75 °C to 125 °C. As the total power consumed in 3D architecture is significantly dominated by the core power, dynamic thermal management techniques to reduce overall temperature is more effective if applied to the core layer rather than the cache layer. In all of these thermal management techniques, contrary to VAWOM, wearout is not considered and the only goal is to reduce the temperature.

NBTI is dependent on many different parameters such as supply voltage, switching activity, temperature, process variation etc. We assume a fixed supply voltage for each part of the architecture.

Variation in threshold voltage and channel length of transistors leads to gate delay variation and therefore critical path delay variation between different cores. Therefore, different cores have different frequencies. To estimate this frequency variation we used the method described in [8].
It's determined empirically that gate delay is nearly linear with respect to the threshold voltage.

There are many SRAM cells in each cache bank. Variation in transistor delay also exists in cache banks and therefore different SRAM cells have different access times. The probability distribution of critical path delay for different SRAM cells is computed. The maximum value of SRAM cell access times in one bank is considered as the initial access time for the entire cache bank. NBTI-induced v_{th} shift increases the access time for each cache bank and we use the model introduced in [8] to compute the access time change in the cache banks.

VAWOM assumes a constant temperature for each component between two consecutive runs (because of the long period), even though right after the activity migration, cores and cache banks have a transient temperature.
From the perspective of NBTI recovery, most of the recovery happens in a short period of time: a few seconds after the start of the recovery [6]. Therefore, keeping a cache bank in the recovery mode for more than few seconds is not very beneficial.
VAWOM assumes all transistors in a single core or cache bank will have the same degradation and each core is equipped with a thermal sensor to obtain its temperature.

4.3 VAWOM's Temperature-Aware Scheme

Several algorithms for activity migration in a multicore architecture have been proposed [26]. VAWOM's temperature-aware scheme keeps the design and performance overhead of the activity migration low, by applying it very infrequently. VAWOM's temperature aware scheme examines two different algorithms for activity migration:

1) *Temperature-Balancing algorithm (T-balancing)*: in this algorithm, tasks are migrated due to their current temperature. VAWOM sorts cores based on their temperature and swap tasks as follows: swap hottest and coldest cores' tasks, swap second hottest and the second coldest cores' tasks and so on. This algorithm intuitively will help to balance the temperature (and the degradation) between different cores and caches.
2) *Aging-Balancing algorithm (A-Balancing)*: temperature-balancing algorithm and other activity migration algorithms do not consider chip degradation (aging) in the cores and the cache banks. Due to the idiosyncrasy of NBTI stress and recovery, it is important to consider degradation in the activity migration policy. In the aging-balancing algorithm, VAWOM assigns estimated coldest jobs to the most degraded core and estimated hottest job to the least degraded core. Frequency degradation is VAWOM's metric for aging in the core layer.
VAWOM's temperature-aware scheme uses one of the following policies for proactive recovery in the cache layer: 1) *Round-Robin (RR)*: where banks are put into recovery mode cyclically, or 2) *Most Degraded (Max)*, where the bank with the highest V_{th} degradation is put into the recovery mode. Appendix 9.3 presents more details about VAWOM's temperature-aware scheme.

4.4 VAWOM's Process Variation-Aware Scheme

The move to deep submicron process technologies have caused process variations to become a major challenge that must be dealt with during design of chip multiprocessors. In particular, device parameters such as threshold voltage, resistance per unit length, gate length and width suffer large variation. According to ITRS, threshold voltage variation is projected to become more and more severe with scaling due to stochastic dopant variation [38]. [29] reported that in 35nm technology, the standard deviation of threshold voltage is 30.28 mV. Process variation could be manifested as frequency variation between different processors or access time latency variation between different cache banks. In this work we exploit these variations to reduce wearout in memories and processors.

Process variation in processors can be used to extend system lifetime [30] [11]. [11] uses the frequency variation between

1228

different cores and attempts to schedule tasks such that hotter tasks run on cores with higher frequency. This method results in temperature variation between cores. The cores with lower operating frequency will be colder and have smaller degradation, which is desirable in a 2D architecture. However, in a 3D architecture, higher temperature in core layer makes cache layer hotter and thus imposes more degradation in the upper levels. In contrast with 2D architecture, the temperature variation aggravates the NBTI effects in the upper layer. Therefore, using the scheme proposed in [11] aggravates lifetime in the upper layer and is not effective for 3D architectures.

VAWOM's variation-aware scheme considers two types of variation: frequency variation between different cores [11] and cache access time variation between different cache banks [31]. Similar to our temperature-aware scheme, VAWOM's process variation-aware scheme uses activity migration in the core layer and proactive recovery in the cache layer for improving chip lifetime. This is done by imposing less threshold voltage degradation on slower cores (with lower operating clock frequency) and slower caches (higher access latency). For activity migration, VAWOM's process variation-aware scheme follows two policies: migration of hotter tasks to 1) faster cores or 2) cores whose adjacent cache bank in the upper layer is faster (Figure 7 in Appendix 9.4 details these two policies). The first policy (core based policy) has a negative effect on the performance degradation (wearout) of the cache layer. In case a slow cache bank is vertically adjacent to a faster core, this policy will impact the slow cache bank temperature which could potentially result in access delay degradation and therefore makes it even slower. The second policy (cache based policy) may have a negative effect on the core layer because slow cores may be vertically adjacent to fast cache banks. Therefore it is important to study the tradeoff between these two policies and develop a policy that considers all the layers simultaneously. We call this policy "Interleaved Policy".

It's likely that a 3D architecture will have different components in different layers. For example (in our case), we have cores in one layer and cache banks in the other layer. NBTI reduces the speed of a circuit until occurrence of a timing failure. We should specify a threshold for each component to be operational. For example, if the access time of a cache bank exceeds acc_{th} we consider it as a failure and that cache bank can't be used anymore. Similarly, for a core, if the critical path delay exceeds cp_{th} we have a failure in the core and it's not operational anymore. To exploit the process and temperature variation in different components, we need a unified criterion to compare different components. Therefore we use the term Distance to Failure (DF) which computes the distance of current state to the failure state:

$$DF_{core}(i) = \frac{cp_{th} - \mathrm{cp}(i)}{cp_{th}} \quad (4)$$

$$DF_{cache}(i) = \frac{acc_{th} - \mathrm{acc}(i)}{acc_{th}} \quad (5)$$

$cp(i)$ is the current critical path delay of core i, cp_{th} is the threshold critical path delay, $acc(i)$ is the current access time of cache bank i, and acc_{th} is the threshold access time. By using DF (which is between 0 and 1), we have a unified metric for different components. As temperatures in vertically adjacent components are highly correlated, we just consider core layer temperature. $Temp(i)$ is the temperature of $Task(i)$ (the task running on the core i). Figure 2 outlines the VAWOM which is run periodically and in each run tries to solve this problem:
Inputs:
Access time for caches, frequency for cores, cores temperature
Outputs:

1) New task mapping (task migration): for all tasks and cores
$Core(i) \leftarrow Task(j)$
2) New disabled cache bank.
   Objectives:
1) Maximize Minimum Distance to Failure of components after running VAWOM for certain time: This objective tries to postpone the first failure of the components as much as possible.
2) Reduce the number of Hotspots.

It's noteworthy that in this work we want to keep components operational as much as possible. Any other objective can be selected based on the context.

The optimal output would be obtained by brute-force search (examining every combination of task assignment and disabling cache bank). But this method needs examining $n! \times n$ different combinations which is infeasible for even small values of n. Consequently, we introduce a heuristic for running VAWOM.

```
1:  Inputs: Critical Path Delay: cp[1..n];  Cache access time: acc[1..n];
    CoreTemp: T[1..n];  TaskHash: task_hash[T]→{1,...,n}
2:  For i = 1 to n:   ---<Compute Distance to failure>
3:        DF_core [i] = (cp_th−cp[i])/cp_th
4:        DF_cache [i] = (acc_th−acc(i))/acc_th
5:  A = Sort_ascending(DF_cache) ---<find the cache for disabling>
6:  Disable (A[0])
7:  Remove (DF_cache, A[0])
8:  For i = 1 to n:   ---<Group Vertically Adjacent Components >
9:        If method == min:
10:            DF[i] =min {DF_core [i], DF_cache[i]}
11:       If method == mean:
12:            DF[i] =mean {DF_core [i], DF_cache[i]}
13: Sort_ascending(DF)
14: Sort_ascending(T)
15: For i = 1 to n:   ---<task assignment><avoiding hotspots>
16:       For j =1  to window_size :
17:           Neighbors[j] = Number_of_assigned_neighbors(DF[j])
18:       max_i = index(max(Neighbors))
19:       Task[task_hash(T[i])] → Core[DF_hash(DF[max_i])]
20:       Remove(DF, DF[max_i])
21: Run (benchmark)
22: T ← new Temperature
23: cp ←new_cp(core_model, T)   ---<update degradation>
24: acc ←new_acc(cache_model, T)
```

Figure 2. Interleaved+Max policy, VAWOM's process variation-aware scheme.

We call the policy introduced in Figure 2 "Interleaved + Max" (which disables max degraded cache bank). In case of using Round Robin scheme for recovery in cache banks, we call the policy "Interleaved + RR". In case of not using the proactive recovery in the cache layer (not disabling any cache bank), we just use the proposed activity migration and call it "Interleaved".

Figure 2 outlines the interleaved heuristic as described below (numbers in parenthesis are referring to the related lines in the figure). First, we compute the Distance to Failure (DF) for different parts (2-4). Minimum DF in the cache banks is the most degraded cache and should be disabled in the next period (5-6). We remove this cache from the cache DF list (7). In the next step, the criteria for task migration should be applied (8-20). For doing task migration, the order of the task assignment is from the hottest task to the coldest task. We group all vertically adjacent components as a unit entity, and examine two different methods (8-12): first, we consider the minimum $\{DF_{core}[i], DF_{cache}[i]\}$ as the representative of DF[i]; and second, we consider the average $\{DF_{core}[i], DF_{cache}[i]\}$ as the representative of DF[i]. Our results showed that the first one outperforms the second one. In the result section we used the first method (minimum).

We sort the DF list and the Temperature list (*13-14*). If we simply assign the hottest task to the entity with the maximum DF, we may have some hot parts in one layer adjacent to each other which can increase the risk of creating hotspot. For dealing with this problem, instead of picking the maximum DF, we look at a window of entities with the maximum DFs (for example, 3 least degraded entities). Then for each of them, we count the number of neighbors in the architecture which have already been assigned a task. Since we start assigning the tasks from the hottest to the coldest, a larger number of already assigned neighbors means that the entity might be hotter in the future (*16-18*). Therefore, we assign the hottest task to the entity within the window, with the minimum number of already-assigned-task neighbors (*19*). Then we remove the hottest task and the selected entity from the lists (*20*) and do this operation for the remaining tasks. After that, with new configuration, we run the benchmarks (*21*), get the new temperatures (*22*), and compute the new frequency and access time for the cores and cache banks (*23-24*).

4.5 Performance Overhead

In all policies and mechanisms for wearout management, VAWOM maintains negligible performance overhead. This is mainly due to the fact that VAWOM is applied very infrequently. In all simulations, the largest performance loss is found to be approximately 3% (detailed results are in our Technical Report [40]). This small performance impact comes from the three following sources:
1- Disabling only a small part of the cache; one out of 16 banks, for proactive recovery.
2- Infrequent transferring of data from a cache bank to previously disabled cache bank.
3- Infrequent task migration in the core layer.

Observing these performance losses, we can state that our policies have negligible performance penalty and we have a lightweight wearout management scheme.

5. Experiments
5.1 Experimental Setup

The platform introduced in Figure 1 is implemented for evaluating VAWOM. In this platform, a cache layer with 16 shared banks is stacked on top of 16 cores in a 3D architecture. Cores and cache banks use routers to form a mesh topology for interconnection.

SMTSIM [32] is used for processor simulation. 64 multi-program workloads consisting of 16 threads each are used for simulation. The applications are selected from the SPEC2006 benchmark suites and results are averaged among these 64 workloads. For network simulation HORNET1.0 [33] is integrated into SMTSIM simulation. For core power estimation, we used McPAT [34]. Using the HORNET network simulator, power consumption of routers and number of accesses to cache banks are obtained. Cache bank power consumption is computed using Cacti [35]. Hotspot 5.01 [25] is used to compute the temperature of each cache bank and core. After obtaining the temperature of the cores and cache banks, threshold voltage degradation is computed using Equations (1)-(3) and the equations and parameters described in [5] [24].

We assumed an initial threshold voltage for different transistors with normal distribution ($\mu = 340mV$ and $\sigma = 3.2mV$) and a 6σ variance.

To determine the efficacy of VAWOM, the baseline configuration and different schemes of VAWOM are implemented and compared in the following section. First, we show the efficacy of temperature-aware schemes in reducing maximum

degradation in the cache and core layer. Second we show the similarity of temperature-aware schemes in reducing average degradation. Finally, effectiveness of variation-aware scheme in the cache and core layer is investigated.

5.2 Experimental Results

Figure 3 shows the maximum access time increase among cache banks w.r.t nominal access time for different temperature-aware schemes introduced in Section 4.3. Moreover, Figure 3 shows the results for the maximum critical path delay (frequency) increase (decrease) in the core layer w.r.t nominal frequency. All results are normalized to the baseline configuration with no recovery support. Degradation is computed after 100000s of system work where the time step between each run of VAWOM is 10 seconds.

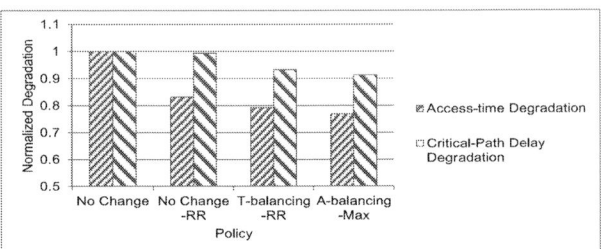

Figure 3. Maximum normalized access-time increase in caches and critical-path delay increase in cores after 100000s for temperature-aware schemes.

Because we use proactive recovery in the cache layer, degradation in this layer is much smaller than the degradation in the core layer. Using task migration in the core layer and proactive recovery in the cache layer (A-balancing - Max), we can achieve 23% improvement in the access time degradation in the cache layer. Also 8% improvement in the frequency degradation is obtained in the core layer. The A-balancing - Max scheme achieves the lowest degradation compared to the other schemes, as it applies the migration to the hottest core and applies the recovery to the most degraded cache bank. Our results show that change in the cache disabling policy (round-robin and max-degraded) affects the results at most 2%.

Figure 4 shows the average access time degradation in the cache layer w.r.t to nominal access time. VAWOM policies balance aging between different components. As a result, all VAWOM policies are almost uniformly effective in reducing average degradation effect.

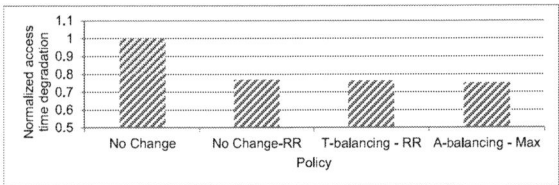

Figure 4. Average access time degradation in different cache banks after 100000s for temperature-aware scheme.

Figure 5 shows degradation improvement in variation-aware scheme introduced in Section 4.4 w.r.t nominal frequency and nominal access time. Note that in the four left bars no proactive recovery has been deployed. As shown in these figures, the interleaved schemes are more effective in lowering the degradation.

The core-based policy (Column 2) increases degradation in cache banks. Meanwhile, the cache-based policy (Column 3) aggravates degradation in cores (normalized degradation is bigger than 1 in both cases). Using the interleaved policy (fourth Column), we are able to achieve the benefit of both policies and avoid their drawbacks (compare Columns 2, 3, and 4).

In the two right bars, effects of using proactive recovery methods are shown. In both studied methods (round-robin and max-degradation), more than 10% improvement is achieved. Our results indicate that degradation improvement is not very sensitive to cache disabling policy, as long as there is uniform distribution of the disabled cache.

Our results confirm that by considering both core and cache layers simultaneously for activity migration and using proactive recovery in the most degraded cache bank, we see improvement in the access time degradation by 31 % in the cache memories and improvement in frequency degradation by 16% in the core layer w.r.t to the nominal frequency. Furthermore, using the interleaved method, the maximum temperature could be reduced up to 4 °C.

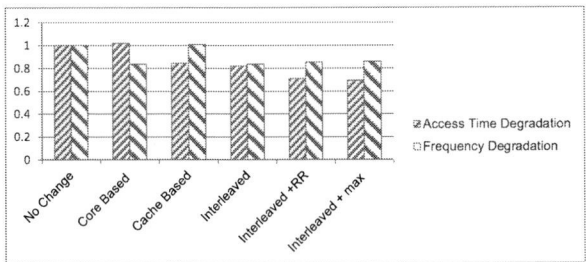

Figure 5. Normalized access-time degradation in the cache banks and frequency degradation in the core layer for variation-aware schemes.

6. CONCLUSION

In this paper, we proposed VAWOM, which takes into account the temperature and the process variation in 3D architecture and exploits activity migration mechanisms with respect to NBTI thermal dependence to improve the lifetime reliability of the system. We proposed and evaluated several temperature and process variation aware policies. We observed that by using very low overhead process variation-aware policies, threshold voltage degradation due to NBTI effect, could be reduced by 31% while maintaining performance at almost the same level. We believe VAWOM is the first effort to investigate simultaneous temperature and process variation aware wearout management for 3D multicore architecture. Future work will consider more than one recovery block and more sophisticated policies to mitigate the effects of wearout.

7. ACKNOWLEDGMENTS

This research was partially supported by NSF Variability Expedition Grant Number CCF-1029783.

8. REFERENCES

[1] J. Burns et al. Design, CAD and technology challenges for future processors: 3D perspectives. In *DAC*, 2011.

[2] A. K. Coskun et al. Dynamic thermal management in 3D multicore architectures. In *DATE*, 2009.

[3] J. Srinivasan et al. The Impact of Technology Scaling on Lifetime Reliability. In *DSN*, 2004.

[4] J. Abella et al. Penelope: The NBTI-Aware Processor. In *MICRO*, 2007.

[5] A. E. Islam et al. Recent Issues in Negative-Bias Temperature Instability: Initial Degradation, Field Dependence of Interface Trap Generation, Hole Trapping Effects, and Relaxation. *IEEE Trans. Electron Devices*, 2007.

[6] W. Wang et al. The Impact of NBTI Effect on Combinational Circuit: Modeling, Simulation, and Analysis. *IEEE Trans. Very Large Scale Integr. (VLSI) Syst*, 2010.

[7] X. Zhou et al. Thermal Management for 3D Processors via Task Scheduling. In *ICPP*, 2008.

[8] S. R. Sarangi et al. VARIUS: A Model of Process Variation and Resulting Timing Errors for Microarchitects. *IEEE Trans. Semicond. Manuf*, 2008.

[9] S. R. Nassif. Modeling and analysis of manufacturing variations. In *CICC*, 2001.

[10] S. V. Kumar et al. NBTI-Aware Synthesis of Digital Circuits. In *DAC*, 2007.

[11] A. Tiwari et al. Facelift: Hiding and slowing down aging in multicores. In *MICRO*, 2008.

[12] R. Vattikonda et al. Modeling and Minimization of PMOS NBTI Effect for Robust Nanometer Design. In *DAC*, 2006.

[13] A. Makhzan et al. Process Variation Aware Cache for Aggressive Voltage-Frequency Scaling. In *DATE*, 2009.

[14] J. Srinivasan et al. Exploiting structural duplication for lifetime reliability enhancement. In *ISCA*, 2005.

[15] J. Shin et al. A Proactive Wearout Recovery Approach for Exploiting Microarchitectural Redundancy to Extend Cache SRAM Lifetime. In *ISCA*, 2008.

[16] L. Huang et al. Characterizing the lifetime reliability of manycore processors with core-level redundancy. In *ICCAD*, 2010.

[17] E. Gunadi et al. Combating Aging with the Colt Duty Cycle Equalizer. In *MICRO*, 2010.

[18] J. Sun et al. NBTI Aware Workload Balancing in Multi-core Systems. In *ISQED*, 2009.

[19] T. Siddiqua et al. A Multi-Level Approach to Reduce the Impact of NBTI on Processor Functional Units. In *GLSVLSI*, 2010.

[20] F. Ahmed et al. Reliable cache design with on-chip monitoring of NBTI degradation in SRAM cells using BIST. In *VLSI Test Symposium*, 2010.

[21] A. Ricketts et al. Investigating the impact of NBTI on different power saving cache strategies. In *DATE*, 2010.

[22] T.-B. Chan et al. On the efficacy of NBTI mitigation techniques. In *DATE*, 2011.

[23] M. A. Alam et al. A comprehensive model of PMOS NBTI degradation. *Microelectronics Reliability*, 2005.

[24] Seyab et al. NBTI modeling in the framework of temperature variation. In *DATE*, 2010.

[25] W. Huang et al. HotSpot: Thermal Modeling for CMOS VLSI Systems. *IEEE Trans. Compon. Packag. Technol.*, 2005.

[26] C. Zhu et al. Three-Dimensional Chip-Multiprocessor Run-Time Thermal Management. *IEEE Trans. Computer Aided Design*, 2008.

[27] J. Wu. A Fault-Tolerant and Deadlock-Free Routing Protocol in 2D Meshes Based on Odd-Even Turn Model. *IEEE Transactions on Computers*, 2003.

[28] L. Tran et al. Heterogeneous Memory Management for 3D-DRAM and External DRAM with QoS. In *ASP-DAC*, 2013.

[29] D. Reid et al. Analysis of Threshold Voltage Distribution Due to Random Dopants: A 100□000-Sample 3-D Simulation Study. *IEEE Trans. Electron Devices*, 2009.

[30] X. Fu et al. NBTI tolerant microarchitecture design in the presence of process variation. In *MICRO*, 2008.

[31] B. Zhao et al. Variation-tolerant non-uniform 3D cache management in die stacked multicore processor. In *MICRO*, 2009.

[32] D. M. Tullsen. Simulation and Modeling of a Simultaneous Multithreading Processor. 1996.

[33] M. Lis et al. Scalable, accurate multicore simulation in the 1000-core era. In *ISPASS*, 2011.

[34] S. Li et al. McPAT: an integrated power, area, and timing modeling framework for multicore and manycore architectures. In *MICRO*, 2009.

[35] N. Muralimanohar et al. CACTI 6.5. HP Laboratories. 2009.

[36] H. Homayoun et al. Dynamically heterogeneous cores through 3D resource pooling. In *HPCA*, 2012.

[37] D. Zhao et al. Temperature Aware Thread Migration in 3D Architecture with Stacked DRAM. In *ISQED*, 2013.

[38] ITRS, 2009. [Online]. Available: http://www.itrs.net.

[39] L. Zhang et al. Scheduled Voltage Scaling for Increasing Lifetime in the Presence of NBTI. In *ASP-DAC*, 2009.

[40] H. Tajik et al. Policies for Variation Aware Wearout Management in 3D Multicore Architecture using VAWOM, CECS Technical Report 13-01, UC Irvine, 2013.

9. Appendix

9.1 Incorporating Temperature and mode change in VAWOM

In VAWOM, two key parameters determine the NBTI effect (Vth degradation): the temperature and the operation mode (stress or recovery). VAWOM runs periodically and because of the long period between two consecutive runs, we assume constant temperature between each run and ignore the transient temperature after running VAWOM.

The number of interface traps is calculated at the end of each period. In the next period, with the initial N_{IT}, new temperature, new mode and equations described in [5] [24] we can find the number of interface traps at the end of the period.

Duty cycle is the percentage of time during which a PMOS transistor is under stress. Most papers assume the duty cycle to be 0.5. However, usually the most degraded parts of the circuit lead to timing error and some parts of the circuit may have a duty cycle close to one. Therefore, in this work we assume the duty cycle is equal to one. However, our work is still applicable when using other methods that change the duty cycle of transistors (e.g., [21]).

9.2 Variation Model

Dependence of gate delay to threshold voltage is described in (4) [8]:

$$T_g = \frac{V(1 + v_{th}/v_{th}^0)}{(v - v_{th})^\alpha} \qquad (4)$$

Where T_g is the delay of the gate, V is the supply voltage, v_{th} is the threshold voltage, v_{th}^0 is the nominal value of v_{th}, and α is typically 1/3.

Each modern core has thousands of critical paths. For a core:

max(T)=probability distribution of longest critical path delay
f = 1/max (T)

Where T is the critical path delay and f is the frequency. We use this model to measure the frequency of different cores and explore the effects of NBTI induced threshold voltage shift on the frequency of the cores.

9.3 VAWOM's Temperature-Aware scheme

Figure 6 shows the activity migration policy for VAWOM's *aging-balancing* algorithm. Array T contains core temperatures and array D contains current degradation of each core. c_h is a hash table with current degradation as its key and core number as its value. t_h is a hash table with temperature as its key and task number as its value.

```
Inputs:   CoreTemp:   T[1..n];   CoreDegradation:   D[1..n];
CoreHash:      c_h[degradation]→{1,...,n};      TempHash:
t_h[Temp]→{1,...,n}
Sort_ascending(T)
Sort_descending(D)
For i = 1 to n:  ---<new task assignment>
    Task[ t_h(T[i])] → Core[c_h(D[i])]
Run (benchmark)
D ←D + new_degradation   ---<update degradation>
```

Figure 6. Aging-balancing activity migration algorithm in VAWOM's temperature-aware scheme.

We list several variations of VAWOM's temperature-aware policies, that combine the core layer migration policies, and the cache layer proactive recovery policies described earlier. In the paper, we evaluate the efficacy of these policies for extending the lifetime reliability of the chip compared to our baseline architecture (*No Change* policy in Table 1) where there is no recovery support.

Table 1- VAWOM temperature-aware policies.

VAWOM temperature-aware policies	Policy in the Core Layer	Policy in the Cache Layer
No Change (baseline)	no change	no change
No Change - RR	no change	recovery in round robin manner
T-Balancing - RR	temperature-balancing algorithm	recovery in round robin manner
A-balancing -Max	aging-balancing algorithm	recover max degraded cache bank

9.4 VAWOM's Variation-Aware scheme

Figure 7 describes two basic policies for VAWOM's process variation-aware scheme. *T, co_init, co_d, ca_init,* and *ca_d* are 5 arrays containing the temperature of cores, core initial threshold voltage, core degradation, cache initial threshold voltage, and cache degradation respectively. *ca_hash* and *co_hash* map total threshold voltage to caches and cores respectively. *task_hash* maps temperatures to the tasks. In the core layer based policy, after sorting the temperature and sorting the current speed of the cores, colder tasks will be assigned to the slowest cores. In the cache layer based policy, after sorting the core temperatures and sorting the current speed of the cache banks, colder tasks will be assigned to the cores adjacent to the slower cache banks.

```
Inputs:   CoreTemp:   T[1..n];   CoreInitThresh:   co_init[1..n];
CoreDegrade:    co_d[1..n];    CacheInitThresh:    ca_init[1..n];
CacheDegrade:   ca_d[1..n];   Cachehash:   ca_hash[ca_init +
ca_d]→{1,...,n};  CoreHash: co_hash[co_init + co_d]→{1,...,n};
TaskHash: task_hash[T]→{1,...,n}
1- Core Layer Based:
For i = 1 to n:   ---<calculate core speed>
    A[i] = co_init[i] + co_d[i]
Sort_ascending(T)
Sort_descending(A)
For i = 1 to n:   ---<task assignment>
    Task[task_hash(T[i])] → Core[co_hash(A[i])]
Run (benchmark)
T ← new Temperature
co_d ←co_d + new_degradation   ---<update degradation>
2- Cache Layer Based:
For i = 1 to n:   ---<calculate cache speed>
    A[i] = ca_init[i] +ca_d[i]
Sort_ascending(T)
Sort_descending(A)
For i = 1 to n:   ---<task assignment>
    Task[task_hash(T[i])] → Core[ca_hash(A[i])]
Run (benchmark)
T ← new Temperature
ca_d ←ca_d + new_degradation   ---<update degradation>
```

Figure 7. Migration policies for VAWOM's process variation-aware scheme.

Similar to the temperature-aware scheme, VAWOM's process variation-aware scheme uses one of the *round-robin* or *most degraded* policies for cache layer proactive recovery (Table 2). In the paper, we examine a combination of various migration and proactive recovery policies for this process variation-aware scheme. Maximum threshold voltage shift with respect to nominal voltage is our metric to measure the efficacy of each of these policies.

Table 2- VAWOM process variation-aware policies.

VAWOM process variation-aware policies	Policy in the Core Layer	Policy in the Cache Layer
No Change (baseline)	no change	no change
Core Based	migration of hotter tasks to faster cores	no change
Cache Based	migration of hotter tasks to cores with adjacent faster bank	no change
Interleaved	Periodically select between core based and cache based	no change
Interleaved+RR	Periodically select between core based and cache based	recovery in round robin manner
Interleaved+Max	Periodically select between core based and cache based	recover max degraded cache bank

9.5 Experimental Setup and Parameters

In each VAWOM scheme, a number of cache memory banks and their associated routers need to be disabled. By disabling routers, change of routing is inevitable. We used x-y routing as the base routing algorithm. For deadlock-free routing in presence of disabled routers we used the method proposed in [27].

We examined different periods for running VAWOM, from 5 to 10 seconds and the results didn't show a significant difference. We picked 10 seconds to be the period of VAWOM. This number keeps the overheads low. In addition, larger period provides enough time to reach the stable temperature which is more consistent with the constant temperature assumption between two consecutive runs. Based on these results, any number between 5 and 10 seconds can be used as the period of VAWOM. This frequency of activity migration may not reduce the peak temperature, but will reduce the average temperature of cache banks over time.

Table 3 summarizes the configuration parameters used in our study. Each application in the workloads is fast-forwarded by 200 million instructions to skip part of the initialization phase and then simulated until each thread executes 200 million instructions.

Table 3- System Configuration

Processor Cores	16 dual issue cores (out-of-order), 2 GHz
L1 Instruction/Data Cache	8KB, 4-way (LRU), 64 B Blocks, 2 cycle
L2 cache (shared)	8 MB NUCA, 16 banks, 8-way (LRU), 64 B Blocks, 20 cycle latency (nominal per bank)
Memory Latency	250 cycles
Interconnect	NoC of mesh (4x4 banks and cores) 32-byte links (2 flits per memory access), 1-cycle link latency, 2-cycle router, XY routing

On the Potential of 3D Integration of Inductive DC-DC Converter for High-Performance Power Delivery

Sergio Carlo, Wen Yueh, and Saibal Mukhopadhyay

School of ECE, Georgia Institute of Technology, Atlanta, GA-30332

sergio.carlo@gatech.edu, wyueh3@gatech.edu, and saibal@ece.gatech.edu

ABSTRACT

This paper studies the potential and challenges of integrating an inductor based DC-DC converter based voltage regulator module (VRM) as a separate die with processor for high-performance power delivery network (PDN). The frequency domain analysis of PDN considering the converter shows 3D integration of VRM improves PDN impedance but the effectiveness depends on the converter design and whether the LC filter is integrated on-board, on-package, or on-die with the die-stack. The methodologies to co-design the converter with PDN and packaging scenarios are discussed and implications on PDN impedance and power losses are studied to maximally exploit the advantage of 3D stacking.

Categories and Subject Descriptors

B.0. [Hardware General]: Power Delivery, 3D Integration

General Terms

Design, Performance

Keywords

Power Delivery, DC-DC Converter, 3D Integration

1. INTRODUCTION

The higher current demand and decreasing operating voltage make power delivery and voltage regulation in high-performance digital systems an increasingly difficult challenge in successive technology generation [1]. Due to package-die and package-board resonances, it is difficult to ensure power integrity over a wide frequency range making reliable power saving techniques a challenging task [2]. Increased transient noise also implies the need for higher cost and complexity of on-die decoupling. The addition of voltage margin to tolerate worst-case IR and Ldi/dt noise may improve robustness but increases the average power dissipation. On-chip integration of the linear regulators (LDOs) may improve transient noise because of their bandwidth but at the expense of a reduced efficiency. Fully integrated switched capacitor converters have also been reported to improve performance and noise margin [3]-[5]. As the efficiency/performance of the inductive converters tends to be better, the inductive converters for higher efficiency [6],[7], fast response [8], and advanced efficiency control over wide load range [1],[9],[10] are being studied.

This paper explores the potential of Through-Silicon-Vias

Permission to make digital or hard copies of all or part of this work for personal or classroom use is granted without fee provided that copies are not made or distributed for profit or commercial advantage and that copies bear this notice and the full citation on the first page. To copy otherwise, or republish, to post on servers or to redistribute to lists, requires prior specific permission and/or a fee.

DAC'13, May 29–June 7, 2013, Austin, TX, USA.

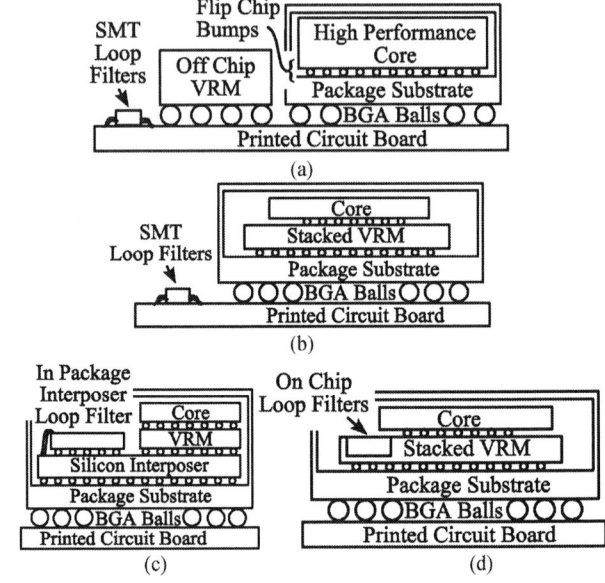

Figure 1: The different integration options for VRM, loop filters, and processor: (a) off-chip VRM with off-chip filter (baseline system); (b) 3D VRM with off-chip filter; (c) 3D VRM with on-package filter; and (d) 3D VRM with on-die filters.

(TSV) based 3D stacking for high-performance power delivery. 3D technology allows heterogeneous technologies to be integrated through distributed TSVs. We explore the potential of using different integration schemes and distributed TSVs to improve power integrity. This is achieved by integrating the voltage regulator module (VRM) (i.e. the DC-DC power converter and regulator) as a separate die in 3D with the processor. The rationale behind our approach is that the integration of the VRM within the 3D stack eliminates the resistance/inductance (R/L) associated with the off-chip power lines including the printed circuit board (PCB) routing and bond-wire connections. The resistance and inductance of TSVs added between the VRM and the processor is comparatively much smaller; it has also been suggested to use TSV capacitance to reduce de-coupling capacitors ("decap") size [11]. Moreover, compared to the solder bumps, the TSVs can be distributed at a much finer grain between the tiers providing larger number of equidistant vertical connections (parallel paths) between the VRM and processor cores. Hence, the PDN impedance can be significantly reduced resulting in much lower IR drop, Ldi/dt noise, and reduced off-chip/on-chip decoupling capacitors.

Although the promise of 3D integration of VRM has been mentioned in few recent literatures (see section 2), a detailed analysis of the potentials and challenges has not been performed yet. This paper evaluates the effects of 3D integration of VRM on the PDN impedance as well as the design of the power converter. A detailed frequency domain analysis of the PDN impedance is

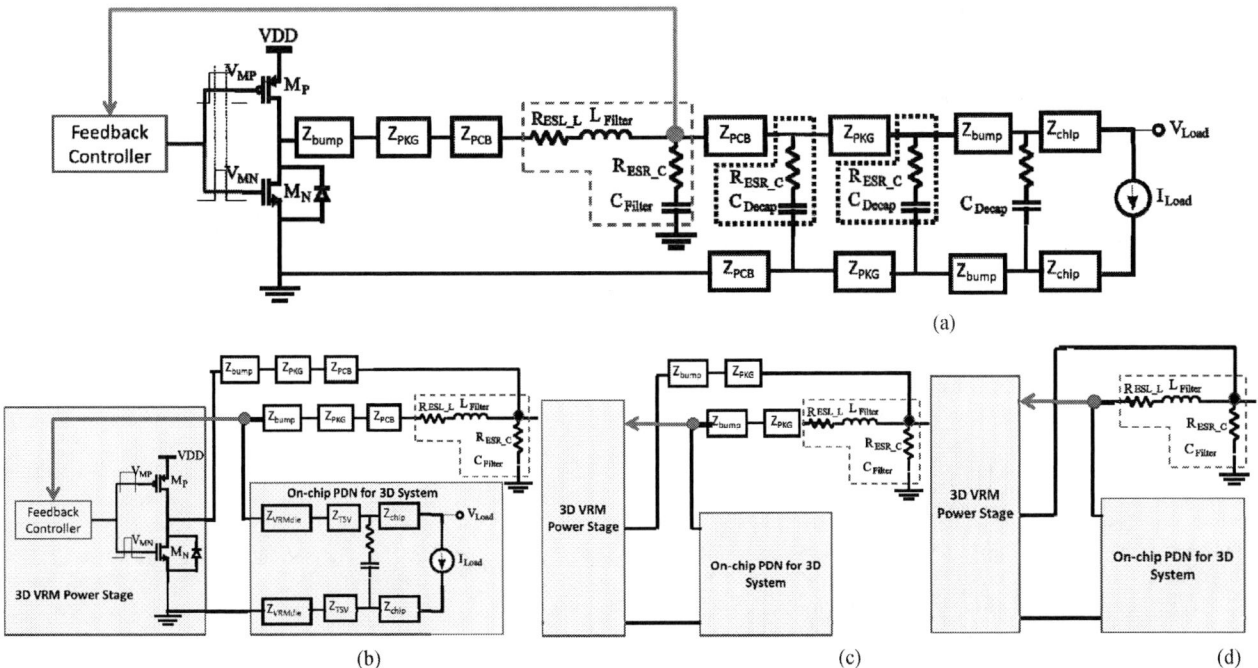

Figure 2: The modeling of the PDN considering the buck converter: The simplified electrical model of the system for (a) off-chip converter, (b) 3D VRM with on-board (off-chip) loop filter, (c) 3D VRM with on-package loop filter and (c) 3D VRM with loop filter on the VRM die.

performed for the 3D VRM-Processor stack considering the inductive DC-DC converter and integration options (on-board, on-package, or on-chip) of the LC loop filter (Figure 1). Our analysis shows that 3D integration of VRM improves PDN impedance but the effectiveness depends on the design of the converter and integration scenarios for the LC filter. The methodologies to co-design the converter with the PDN and packaging scenarios are discussed and implications on PDN impedance, performance, and power losses are studied. The implications of the potentially higher operating temperature for VRM due to the die-to-die thermal coupling between VRM and processor dies are studied.

The rest of the paper is organized as follows: Section 2 discusses the related work; Section 3 presents the frequency domain analysis approach; Section 4 presents the results; and Section 5 summarizes the paper.

2. RELATED WORK

This paper builds on related prior work of two areas: first, power converters in 3D and second, on-package/on-chip passives. Earlier work on 3D power converters focused on chip-stacking technologies. The chip-stacked 3D buck regulators are discussed in [12]. Schrom et. al. proposed the use of flip-chip and through-hole packaging technology to vertically integrate a converter with a processor [13]. More recently, Sun. et. al has explored the potential of using inter-wafer-via (like TSVs) based 3D integration of the buck converter [14]. The authors designed a prototype buck converter in 2D BiCMOS technology and discussed the potential for the 3D integration of the converter. The 3D converter design operates interleaved converters at a very high switching frequency to address the small area and poor quality on-chip inductance/capacitance. The peak efficiencies are quoted at 62% to 64%, which is considerably low compared to off-chip converters. While the design of the converter itself is discussed in detail, the challenges of integrating it on 3D are briefly mentioned. Using interposed high quality inductors helps improve the efficiency on 3D stacked converters because the equivalent

series resistance (ESR) can be very low. Hence, on-chip/in-package high quality inductors are also under intense investigation to match the demand from on-chip/in-package regulators. The interposer based inductor in [15] opens the possibility of high quality in-package integration. Advanced on chip inductor material from [5][16] also made fully integrated regulators possible. However, the advanced on-chip inductors' quality is still not on par with the discrete or in-package components, thus limiting the overall system efficiency.

This work, instead of circuit design of DC-DC converter, focuses on the system-level characterization of the overall PDN impedance considering 3D VRM module and the LC loop filter. Using a unified frequency domain simulation method considering distributed RLC based PDN model (including TSVs) and power converters, we study reliability, performance, and losses associated with different VRM-loop-filter-processor integration. The co-analysis results and co-design studies can provide guidelines to chip and system designers.

3. PDN MODELING CONSIDERING CONVERTER

A full-chip PDN model for one die is developed considering the impedances due to on-chip grids (distributed RLC), and off-chip RLC ladders including decaps [17]. The model is extended for the PDN network of 3D die-stacks where PDN is composed of two planar power meshes connected through vertical P/G TSVs. In the 3D VRM case, one planar mesh represents the output of the VRM module and provides power to the bottom mesh through the P/G TSVs. The power consuming processor is connected to the bottom planar mesh. Section A.1 describes the PDN models and the associated parameter values. The framework is used to model the following PDN configurations including the VRM module. First, the conventional 2D scenario (Figure 1(a)) is modeled where the converter is off-chip and connected through PCB routing, flip-chip bumps and bond-wires. For the 3D VRM-Processor stack the converter is brought in as a stacked die. For the 3D cases, we first consider the LC loop filter components of the VRM remains off-chip Figure 1(b)). Next we consider the case where the LC components are within the package but outside the die (Figure 1(c)). This can be achieved using system-in-package integration and/or interposers. Figure 1(d) shows the case where the loop filter is implemented on-chip on the VRM die. The PDN analysis is often performed under assuming an ideal power converter source, which overestimates the performance of the PDN. Using an ideal VRM also neglects the pole-zero interactions between the network and the VRM. Therefore, we perform a detail converter-aware PDN analysis. The full system PDN model for different integration scenarios are shown in Figure 2. Figure 2a shows the structure considering off-chip PDN while Figure 2b, 2c, and 2d shows the structure considering 3D VRM but off-chip, on-package, and on-die loop filters respectively.

3.1 Converter Structure

Our goal is to evaluate the possibility of using a high-efficiency converter and evaluate the integration schemes in order to exploit the advantages of lower parasitics and achieve a similar performance as the linear regulator. This work will use a pulse-width modulation (PWM) based buck converter as the VRM [18]. It was decided to use a closed-loop converter because of their common use. The closed loop converter provides tolerance to process, voltage, and temperature. The use of a feedback sampling point closer to the load provides better line and load regulation. Since 3D integration provides the opportunity to sample very close to the load, there is an improvement in the regulation to the on-chip PDN since the converter now compensates for the path losses due to PCB, packaging and bumps (see Figure 2).

The introduction of feedback and better control has a price; it creates the possibility of instability, especially because at this point a complex conjugate dual pole pair appears due to the filter. Therefore, it is important to know how the loop of the whole signal chain behaves to compensate and ensure stability while exploiting the opportunity of performance. To ensure stability we can either use dominant pole compensation (which sacrifices the gain at higher frequencies to ensure stability) or use a controller which shapes the frequency response of the loop filter to effectively cancel out the undesired parts of the overall loop response. As we will show in Section 4.2, a simplistic converter design cannot take full advantage of the 3D integration schemes; furthermore, the design of the VRM will finally determine the performance gain of the 3D integration scheme.

3.2 Design of the Loop filter

The closed loop transfer function of the converter can be represented as:

$$\frac{V_o(s)}{V_i(s)} = \frac{G_C(s) \cdot H(s)}{[1 + \beta \cdot G_C(s) \cdot H(s)]} \quad (1)$$

where $G_C(s)$ represents the controller transfer function, β is the feedback ratio, the controller gain was modeled as:

$$G_C(s) = A_{DC} \frac{[(s + z_1)(s + z_2)]}{[(s + p_1)(s + p_2)(s + p_3)]} \quad (2)$$

where A_{DC} the DC gain of the controller, p_1, p_2 and p_3 are left hand plane poles and z_1 and z_2 are left hand plane zeros. A larger number of poles than zeros were introduced in order to implement a controller which is physically realizable. Since the filter comes inside the loop equation, we must also model its transfer function:

$$H(s) = \frac{\omega_o^2}{s^2 + 2\xi\omega_o s + \omega_o^2} \quad (3)$$

The damping factor (ξ) and the resonance frequency (ω_o) are related to the circuit parameters through:

$$\xi = \frac{1}{2R_{load}}\sqrt{\frac{L_{filter}}{C_{filter}}} \text{ and } \omega_o = \frac{1}{\sqrt{L_{filter}C_{filter}}} \quad (4)$$

Since we want a stable loop gain (phase<180° @ zero dB crossing), we can use the two zeroes to cancel the resonant filter poles. The dominant pole p_1 is placed at very low frequency to act as an integrator and the second (p_2) and third poles (p_3) are used to reduce the gain at very high frequencies since there are gain peaks due to the parasitics such as packaging, PCB and bumps.

We must be aware that the LC filter has non-idealities associated with them. Parasitic zeros in the transfer function appear due to the resistive components of the inductor and capacitance. The locations of these zeros are:

$$\omega_{z,ESR,C} = \frac{1}{R_{ESR,C}C_{filter}} \text{ and } \omega_{z,ESR,L} = \frac{R_{ESR,L}}{L_{filter}} \quad (5)$$

where the ESR stands for the equivalent series resistance. The two zeroes are fundamentally different: the inductor ESR zero appears at low frequencies and the impedance is boosted as frequency increases. The capacitor ESR zero appears at very high frequencies, at which the capacitor is an equivalent short and the output impedance flattens out to a minimum of $R_{ESR,C}$. Although these zeros boost the loop gain at high frequencies, they have losses associated with them that impact the efficiency. Because of this we will try and keep these two zeros to higher frequencies and have then outside the loop filter bandwidth. Section 4.4 deals with the losses of a switching regulator and both of these effects are explained there.

If the loop gain is large enough, the closed-loop transfer function across the converters bandwidth can be approximated as:

$$V_o = \lim_{A_{ol}\to\infty}\left(\frac{A_{ol}(s)}{1 + A_{ol}(s)\beta}V_i\right) = \frac{V_i}{\beta} \quad (6)$$

Then, assuming we're using a scaled down version of the supply $V_i = DV_{dd}$ to feed into our loop filter and that the feedback network $\beta = 1$, then: $V_o = DV_{dd}$.

3.3 Modeling of VRM Output Impedance

The output impedance of the VRM plays a very important role in the impedance profile for both the 2D and 3D VRM integration scenarios. The open-loop output impedance of the converter can be calculated in the s-domain as:

$$Z_{OL} = R_{ESR,C}\frac{\left(s + \frac{1}{R_{ESR,C}C}\right)\left(s + \frac{R_{ESR,L}}{L}\right)}{\left[s^2 + \left(\frac{R_{ESR,C}+R_{ESR,C}}{L}\right)s + \frac{1}{LC}\right]} \quad (7)$$

We can see that the DC value of the output impedance is $R_{ESR,L}$ and the high-frequency open loop output impedance will saturate to $R_{ESR,C}$, this neglects the series inductance with the capacitance $L_{ESL,C}$ which kicks in at much higher frequencies and adds up to the $R_{ESR,C}$. The controller loop of choice uses shunt sampling feedback to obtain the benefit of impedance attenuation; it is known that the closed-loop output impedance can be estimated as:

$$Z_{CL} = \frac{Z_{OL}}{1 + T(s)} \quad (8)$$

where $T(s)$ is the frequency-dependent loop gain of the converter. The net effect of shunt feedback is that $R_{ESR,L}$ is attenuated by the loop gain factor at low frequencies. At high frequencies, however, $R_{ESR,C}$ cannot be attenuated since the loop gain of the converter eventually dies out and $Z_{CL} = Z_{OL} = R_{ESR,C}$. We then see a limitation to how much improvement we can get by placing the converter in 3D.

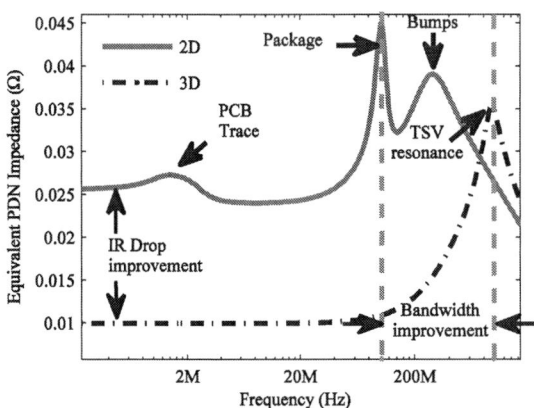

Figure 3: The ideal impedance profile for 3D VRM neglecting the effect of the converter.

4. CO-ANALYSIS AND CO-DESIGN OF VRM AND PDN

The on-chip PDN is modeled as a distributed RLC network considering a die-dimension of 12mm×12mm. The RLC network uses an equivalent distributed power mesh derived from the lumped impedance model of a Pentium 4 processor [16]. The converter is designed in 130nm CMOS technology.

4.1 Ideal PDN Impedance with 3D VRM

Figure 3 shows the impedance plot of the PDN using an ideal voltage source termination for a 2D PDN and a 3D PDN to demonstrate the potential of 3D VRM. The most noticeable effects are a DC impedance drop and a shift in the resonant peaks (i.e. bandwidth extension). The resonant peak due to the packaging and the bond-wires are removed and the only peak that remains is due to the TSV inductance. A higher TSV density reduces the TSV non-idealities, resulting in a lower DC drop and Ldi/dt droop shift to higher frequencies. The above analysis shows the potential of an ideal 3D VRM module. Next, we discuss the impact of the converter characteristics in the above analysis.

4.2 Co-analysis and Design of VRM and PDN with 3D Integration

A co-simulation the loop gain of the VRM plus filter plus PDN for different integration scenarios. Figure 4 shows the effect

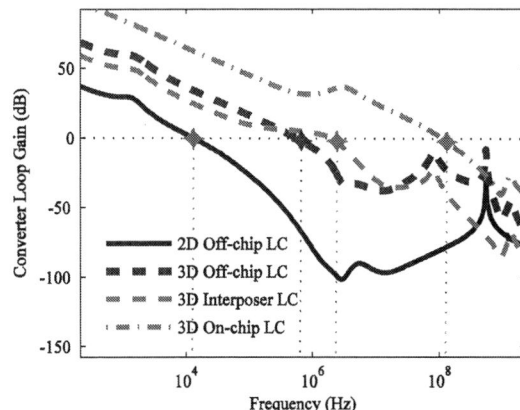

Figure 5: Loop gain of the converters co-designed with the PDN.

of the VRM on the PDN, in these plots the same converter was used for all integration scenarios. As expected, using a 2D off-chip converter adds a large amount of routing and package peaks which limit the achievable bandwidth (without using a non-specialized controller). The DC regulation at the load is also worse due to the IR drop across the PCB and package which do not fall inside the control loop. We also observe that adding the VRM introduces a low frequency impedance peak. This peak occurs at the unity crossover frequency, where the converter loses the ability to self-regulate its output impedance. This peaks corresponds to the open loop impedance of the converter, it disappears at higher frequencies due to the impedance reduction of the capacitor and settles to $R_{ESR,C}$.

Next, we consider the characteristics of the 3D VRM for different integration scenarios. Initially, the same converter as the 2D off-chip integration is used. The PDN impedance in Figure 4 shows that, although we see a reduction in the PDN impedance at all frequencies, we still observe peaks at lower frequencies compared to the ideal case these are due to the low bandwidth of the converter used for the 2D case. The 2D VRM has to have low bandwidth since parasitics appear at low frequencies.

The controller loop gain is then modified to optimize the integration scheme; the poles/zeroes are shifted to take advantage of the low-impedance flatness in 3D integrated structures. Figure 5 shows the loop gain response of the 2D off-chip converters and the optimized 3D VRM converters, where the PDN is included as part of the loop gain. In order to compensate the networks the

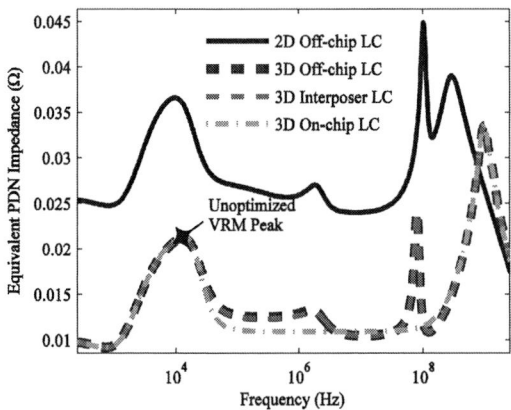

Figure 4: PDN impedance with same 2D converter for all designs.

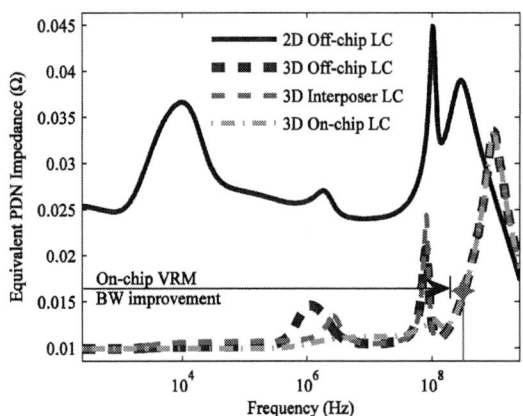

Figure 6: Impedance with compensated converters

1237

controller function described in section 3.2 was used.

The parasitic reduction obtained from 3D integration can be used to extend the VRM bandwidth; which can, in turn, be used to maximize the flatness of the impedance profile. Figure 6 shows the impedance response after co-designing the converter. We now discuss our observations:

<u>3D VRM with off-chip LC:</u> This approach improves the low-frequency impedance as well as the regulation into the PDN. The converter can now compensate losses in the packaging and PCB since the converter can sample very close to the load. However, the bumps, package, and PCB parasitics between the VRM module and the on-board loop filters result in low to medium frequency peaks in the loop transfer function. Thus the routing parasitics limit the bandwidth of the converter (hence, response time). The upside of having higher quality off-chip components is offset by the PDN parasitics.

<u>3D VRM with on-package LC:</u> This approach removes low-frequency resonant peaks due to PCB which allows an improvement in the VRM bandwidth. Higher bandwidth helps in better response time. However, we still observe the high frequency peak due to the package resonance, which is the main limiting factor for the bandwidth.

<u>3D VRM with on-die LC</u>: Using an on-die LC filter allows us to extend the bandwidth and exploit the advantage of 3D integration. The impedance profile is very close to the ideal 3D behavior, this is because of two things. First, removing the package and PCB components enables us to improve the bandwidth without sacrificing stability so the loop gain is now only limited by the resonance of the TSV impedance. Secondly, using an on-chip LC requires us to lower the LC component values such that the bandwidth can be extended without the need for frequency shaping using the controller poles and zeroes. However, more than a design choice, a lower LC is constraint as typical on-chip inductances range up to nH and on-chip capacitors are in the nF range. The most critical constraint of an on-chip inductor is the low quality factor.

4.3 Transient Performance Analysis

Figures 7(a) and (b) show the converter startup and its step response to a transient 0-20 Amp current step that was applied at a node in the middle of the core PDN mesh. When the current step is applied, two things happen. First, the local decap for the node is discharged to provide quick charge to the load and the voltage droops, when the converter senses the droop it begins injecting charge back and the voltage recovers from the droop. After this happens, the voltage settles to the desired target voltage minus the IR drop in the PDN. Figure 7(b) shows that the 3D VRM can significantly improve the response to load transient and reduce the IR drop. As expected the best performance is obtained for the 3D VRM with on-die loop filter but 3D VRM with on-package filter is close in performance. The low frequency voltage droop seen in the 2D case can be taken reduced using larger decaps. However, decaps consume area and have been shown to become leaky as process dimensions shrink. The startup response of the converter clearly indicates that 3D VRM with on-die filters has much faster start-up time compared to the other 3D VRM options.

4.4 Analysis of Power Loss

The losses in the power delivery are can be separated into the losses in the converter and the losses in the PDN. The DC loss can be obtained by considering the PDN impedance profile at low frequency and the load current. Extensive studies have also been

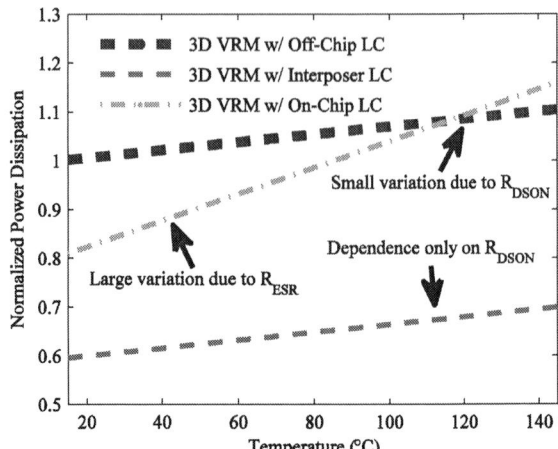

Figure 8: Normalized power dissipation as a function of temperature.

performed in the literature evaluating the losses associated with synchronous buck converters [3].

In order to evaluate the advantage of 3D VRM power loss is estimated using CMOS 130nm process. An important factor that requires careful analysis here is the effect of **<u>thermal coupling in 3D VRM</u>**. Since the 3D VRM die is in the 3D stack with the processor die, the heat generated in the processor will strongly impact the temperature of the 3D VRM die. Assuming the processor is close to the heat sink, the on-chip temperature of the 3D VRM die will be higher than that of the processor. Hence, we expect a higher operating temperature for the VRM when integrated in 3D. In the case of off-chip VRM, the operating temperature is much lower since it does not experience the thermal coupling from the processor core. While the feedback loop can be designed to ensure robust operation of the converter across temperature, a higher temperature will increase the R_{RDS_ON} losses in the transistor. The on-chip PDN losses may also increase due on the increased metal resistance.

Figure 8 shows the normalized power losses of the 3D VRMs with respect to losses in a 2D VRM considering effect of higher temperature in 3D VRM. The model considers each integration scheme's switching frequency, filter design and thermal dependent losses. We first observe that all of 3D VRM cases provide lower losses than the 2D off-chip integration scenario mainly due to elimination of the PCB and packaging losses. The higher switching losses for the 3D VRM are overshadowed by the reduced resistive losses in the PDN impedance. Among the different 3D VRM solutions, the maximum loss is observed for the off-chip LC case due to presence of parasitics in the filter path. The on-chip LC filter has a higher loss compared to the on-package integration due to higher ESR of the on-die inductors. The temperature sensitivity is observed to be strongest for the 3D VRM with on-die filter. This is explained by the fact that an increase in processor temperature would increase the PDN, inductor ESR and MOSFET $R_{RDS,ON}$ losses. In the 3D VRM with interposer filter case, the loss model considers the thermal dependency of the copper lines and the power transistors but it neglects the ESR temperature increase in the inductor since it is not directly below the processor. Thus it has lower ESR than the on-chip inductor counter-part and has low temperature sensitivity which is mainly due increasing R_{DSON} with temperature.

5. CONCLUSIONS

We have analyzed the potential of 3D integration of VRM with a switched inductor DC-DC VRM. The analysis of the PDN behavior considering the effect of the converter shows that the 3D VRM can provide appreciable advantage but methods for integration of the loop filter of the converter scheme has a strong impact on the stability and the achievable PDN+VRM impedance. The analyses indicate a reduction in the IR losses obtained from the 3D PDN and an opportunity to improve VRM bandwidth. The results also show that a better regulation can be achieved through the close proximity of the converter. Finally, the losses associated with the implementation of these schemes are evaluated considering die-to-die thermal coupling in a 3D stack. It is observed that although the 3D VRM has higher temperature sensitivity, the overall losses are lower. In particular, the 3D VRM with the interposer filter configuration has great potential due to low losses as well as lower temperature sensitivity.

This paper establishes the potential of the 3D VRM in reducing losses and improving performance of power delivery system. 3D VRM reduces both low-frequency and high-frequency PDN impedance resulting in potentially much lower IR and Ldi/dt in the digital circuits. As technologies scale down, digital circuits are becoming highly sensitive to voltage noise, and thus lower supply noise leads to better robustness. The reduced IR and Ldi/dt noise also provide opportunities for more aggressive voltage scaling in processor hence, lower operating power. The power efficiency is further improved due to lower losses in the 3D PDN.

New challenges to VRM design are also introduced by 3D stacking. The effect of higher temperature due to proximity is observed in our analysis. Considering the fact that increasing temperature will increase power losses, it is important to evaluate the thermal stability of the processor-VRM system. The implications of TSV-device interactions on the design of the power converters also require further investigation. Finally, the cost associated with the 3D integration needs to be considered and whether it can be amortized by lower processor power needs to be evaluated. A separate 3D VRM die also provides opportunities to place other mixed-signal components in multi-tier systems.

6. ACKNOWLEDGMENTS

This work is based on materials supported in part by National Science Foundation (CNS-1218745), Semiconductor Research Corporation (1836.110), and Sandia National Lab.

7. REFERENCES

[1] W.J Lambert, et. al, "Fast Load Transient Regulation of Low-Voltage Converters with the Low-Voltage Transient Processor," IEEE Transactions on Power Electronics, , pp.1839-1854, July 2009

[2] G. Huang et. al., "Power Delivery for 3D Chip Stacks: Physical Modeling and Design Implication," *Electrical Performance of Electronic Packaging, 2007* pp.205-208, 29-31 Oct. 2007

[3] M. Gildersleeve, et. al, "A comprehensive power analysis and a highly efficient, mode-hopping DC-DC converter," *ASIC, 2002. Proceedings. 2002 IEEE Asia-Pacific Conference on* , vol., no., pp. 153- 156, 2002

[4] G. Sizikov et. al., "Frequency dependent efficiency model of on-chip DC-DC buck converters," *Electrical and Electronics Engineers in Israel (IEEEI), 2010 IEEE 26th Convention of* , vol., no., pp.000651-000654, 17-20 Nov. 2010

[5] N. Sturcken, et. al, "Design of Coupled Power Inductors with Crossed Anisotropy Magnetic Core for Integrated Power Conversion" Applied Power Electronics Conference (APEC), 2012.

[6] X. Zhang et. al, "Monolithic/Modularized Voltage Regulator Channel," Power Electronics, IEEE Transactions on , vol.22, no.4, pp.1162-1176, July 2007

[7] Qun Zhao; Stojcic, G.; , "Characterization of Cdv/dt Induced Power Loss in Synchronous Buck DC–DC Converters," Power Electronics, IEEE Transactions on , vol.22, no.4, pp.1508-1513, July 2007

[8] Pengfei Li; Bhatia, D.; Lin Xue; Bashirullah, R.; , "A 90–240 MHz Hysteretic Controlled DC-DC Buck Converter With Digital Phase Locked Loop Synchronization," Solid-State Circuits, IEEE Journal of , vol.46, no.9, pp.2108-2119, Sept. 2011

[9] Mengmeng Du; Hoi Lee; , "A 5-MHz 91% peak-power-efficiency buck regulator with auto-selectable peak- and valley-current control," Custom Integrated Circuits Conference (CICC), 2010 IEEE , vol., no., pp.1-4, 19-22 Sept. 2010

[10] Hong-Wei Huang; Ke-Horng Chen; Sy-Yen Kuo; , "Dithering Skip Modulation, Width and Dead Time Controllers in Highly Efficient DC-DC Converters for System-On-Chip Applications," Solid-State Circuits, IEEE Journal of , vol.42, no.11, pp.2451-2465, Nov. 2007.

[11] A. Trivedi et. al; "Impact of Through-Silicon-Via capacitance on high frequency supply noise in 3D-stacks," *EPEPS, 2011* .

[12] K. Onizuka et. al., "Stacked-Chip Implementation of On-Chip Buck Converter for Distributed Power Supply System in SiPs," *Solid-State Circuits, IEEE Journal of* , vol.42, no.11, pp.2404-2410, Nov. 2007

[13] Schrom, et al., "Feasibility of monolithic and 3D-stacked dc-dc converters for microprocessor in 90 nm technology generation," in Proceedings of ISLPED'04, pp. 263-268, 2004.

[14] J. Sun et. al., "Fully Monolithic Cellular Buck Converter Design for 3-D Power Delivery," *Very Large Scale Integration (VLSI) Systems, IEEE Transactions on* , vol.17, no.3, pp.447-451, March 2009

[15] N. Sturcken, et. al. "A 2.5D Integrated Voltage Regulator Using Coupled- Magnetic-Core Inductors on Silicon Interposer Delivering 10.8A/mm2" ISSCC, 2012.

[16] N. Wang, et. al., "Integrated on-chip inductors with electroplated magnetic yokes", J. Appl. Phys. 111, 07E732 (2012),

[17] M.S. Gupta et al., "Understanding voltage variations in chip multiprocessors using a distributed power-delivery network," in *DATE*, Apr. 2007, pp.1-6.

[18] M. Lu et. al. "A sub-1V voltage-mode DC-DC buck converter using PWM control technique," *EDSSC*, vol., no., pp.1-4, 15-17 Dec. 2010

[19] Buck Converter Design Issues, MS Thesis, Muhammad Saad Rahman;Thesis No: LiTH-ISY-EX--06/3854—SEG.

8. APPENDIX

8.1 PDN Modeling

The on-chip PDN is modeled as a distributed RLC network. The RLC network uses an equivalent distributed power mesh derived from the lumped impedance model of a Pentium 4 processor [2]. The grid has 12×12 grid nodes for VDD and corresponding 12x12 grid nodes for ground. The die dimension is 12 mm × 12 mm and forms unit cell dimension of 250 μm × 250 μm. The off-chip impedances are modeled with RLC ladders as well to capture the low frequency noise. The first segment of the ladder models the board level lump impedance and the second segment of the ladder modes the package impedance. The package ladder is evenly distributed to points on the on-die grid with partially-lumped controlled collapse chip connection (C4) bump impedances. The values of the PDN parameters used for simulations are shown in the Figure A.1. Note both the VDD and GND grids are modeled as distributed RLC to accurately account for the effect of return path.

The PDN network for the 3D stacked VRM is composed of two planar power meshes connected through vertical P/G TSVs. The P/G TSVs are modeled as lumped R, L, and C. Depending on the TSV density and grid size, the number of TSV per grid are determined assuming uniform TSV placement. Based on the estimated number of TSVs, the equivalent TSV impedance is attached to each grid in the planar meshes. The top planar mesh represents the VRM die. The output of the power converter module is connected to one point of the top planar mesh. The power consuming processor is connected to the bottom planar mesh. The power is provided to the processor from the output of the VRM module; and through the on-die RLC grid of VRM die, distributed TSVs, and the on-die RLC grid of the bottom die. Therefore, a complete distributed RLC model is considered for the 3D VRM configuration as well. The power dissipation of the processor is assumed to be at the center grid of the bottom plain. The one point power delivery and power consumption model helps account for the distributed RCL effect in both plains and TSVs. The PDN models are developed considering distributed decap i.e. a fixed decap is attached per grid node during simulation.

Figure 2 shows the PDN diagram for different 3D VRM configurations. The PDN between the power stage of the VRM and power dissipation in the processor remains the same. However, the parasitic component between power stage output and the loop filter are different for the different cases as shown in the Figure 2. The off-chip PDN models, shown in Figure A.1 are used to model these parasitic components. No package/board level decap is considered for the 3D VRM analysis.

8.2 Loop Filter Design

The loop filter design was a very important aspect of this work. We used a controller transfer function in the forward path to improve the stability and stability of the system. Implementations that can track circuit non-idealities and compensate losses dynamically are not easy to implement in real-life systems. However, we did choose to keep our design moderately complex (2 zeros & 3 poles) to maintain high performance but in the realizable domain, i.e. a finite bandwidth and with a reasonable amount of zero-pole pairs. Higher performance can be achieved through pole-zero cancelation of higher frequency non-idealities but proper implementation and characterization and PVT compensation is extremely difficult to achieve. [19] Offers various implementations of different realistic buck converter compensators, the one chosen for this work is called a type III

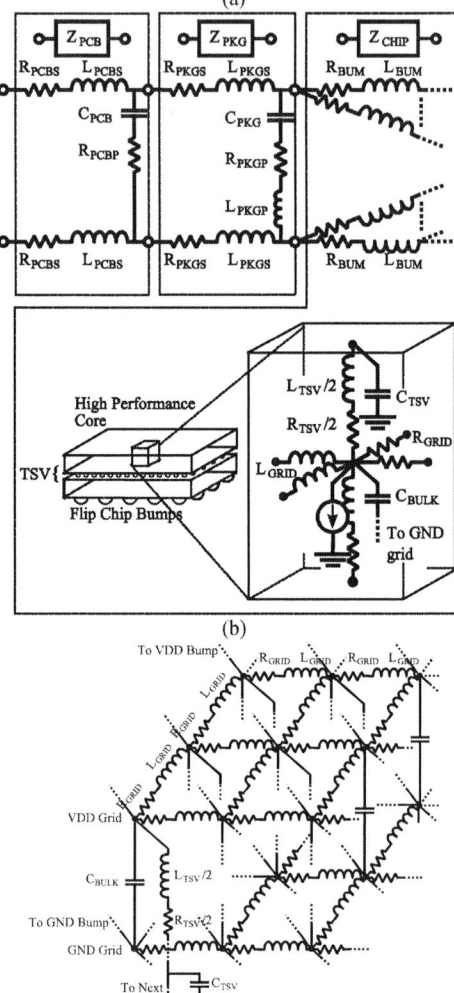

Parameters	R (ohm)	L (H)	C (F)
PCB	94 μ (s)	21 p	240 μ
	166.6 μ (p)		
PKG	1000 μ (s)	120 p	26 μ
	541.5 μ (p)		
BUM	40 m	72 p	
GRID	28.1 m	3.1 f	93.8 μ
TSV	7.735 μ	5.710 p	313.2 f

(a)

(b)

(c)

Figure A.1: The basic modeling components for the power delivery network. (a) the parameter values; (b) the off-chip ladder for PDN modeling; and (c) the on-chip distributed RLC based PDN models. Note the TSVs are modeled only for 3D VRM analysis . For 2D VRM analysis, the C4 bumps are directly connected to the bottom grid in part (c) which represents the processor grid.

compensator. If designed properly, this implementation maximizes the performance of the loop while maintaining stability.

8.3 Power Loss Analysis

We can separate the power losses through two categories, the DC (average) losses and the AC (switching) losses:

$$P_{DC,HS} = D R_{ds,on,HS} I_{load}^2$$
$$P_{DC,LS} = (1 - D) R_{ds,on,HS} I_{load}^2$$
$$P_{PDN,Total} = (R_{PDN} + R_{ESR_L}) I_{load}^2$$

Table A.2. Summary of Observations for Different VRM integration.			
2D Off-Chip VRM	**3D VRM Off-Chip Filter**	**3D VRM Interposer Filter**	**3D VRM with On-Chip Filter**
Ease of Integration for both VRM and loop filters	Integration of VRM requires 3D process but Loop filters are easy to integrate	Integration of VRM requires 3D process. Integration of loop filters on package have been demonstrated but challenging	Integration of VRM requires 3D process. Integration of loop filters on diehave been demonstrated but challenging
Poor regulation due to long distance feedback point	Good regulation due to closer to processor feedback	Better Regulation due to closer to processor feedback	Best Regulation due to closer to processor feedback and ellimination of all paraistiocs in the feedback path
High PDN IR Drops	Less PDN IR Drops	Lesser PDN IR Drops	Least PDN IR Drops
Higher Ldi/dt noise	Less Ldi/dt noise	Lesser Ldi/dt noise	Least Ldi/dt noise
Higher impedance at at all frequencies	Small low frequency impedance. High frequency impedance is limited by package and PCB paraistics in the loop filter path	Small low impedance. High frequency impedance is limited by package paraistics in the loop filter path	Low PDN Impedance over wide band. High frequency impedance is limited by the TSV and on-chip parasitcs.
Low ESR inductors are easily available	Low ESR inductors are easily available	On-package inductors with moderate ESR are challenging but available.	On-die inductors have higher ESR (~10s to 100s of mΩ)
Die-to-die thermal coupling is not critical	Moderate sensitivity to die-to-die thermal coupling (only VRM)	Moderae sensitivity to die-to-die thermal coupling (only VRM)	Higher sensitivity to die-to-die thermal coupling (VRM + Filter ESR)
High power loss in PDN + VRM	High power loss in PDN + VRM	Minimum power loss in the system assuming previously reported ESR of on-chip inductors	Lesser power loss in PDN + VRM (Depends on the ESR of on-chip inductor)
Conventional system majority of the application	Suitable for high-power processor with need good regulation and for easier integration of passive elements	Suitable for high-power processor with tight regulation and loss constraints but availability of interposer technology	Suitable for low-to-moderate power processor with tight performance constraints and availavilibity of on-chip passives technology
Conventional converter design	High bandwidth converter can be used at the expense for complexity.	Higher bandwidth converter can be used at the expense for complexity.	Very high bandwidth converter can be used resulting in excellent performance but require complex design/control

The losses associated with the AC ripple are:

$$P_{HS} = D \, I_{Ripple}^2 R_{RDS,ON_{HS}}$$

$$P_{LS} = (1 - D)I_{Ripple}^2 R_{RDS,ON_{LS}}$$

$$P_{PDN,Total} = (R_{ESR_L} + R_{ESR_C} + R_{PDN})I_{Ripple}^2$$

where, R_{RDSON} represents the on-resistance of the transistors and R_{PDN} is the DC resistance in the path of the current all the way through to the load.

Switching the transistors involve injecting charge in the gate capacitor so they can turn on, the switching loss is estimated by:

$$P_{driver} = f_s C_{total,avg} V_i^2$$

The circuit simulations in 130nm CMOS for transistor parameters coupled with the PDN analysis performed in section 4.3 are used for power loss analysis. The effect of temperature on the properties of transistors and PDN components are considered for loss analysis.

8.4 Performance Summary

As explained in section 4, the VRM for different integration scenarios is optimized to ensure maximum robustness. Figure A.2 shows the loop gain and phase response of the all the converters optimized for corresponding PDN configurations. The associated PDN impedance response, performance, and loss characteristics are discussed in section 4. Here we summarize the overall observation in Table A.2 to help designers.

Full-Chip Multiple TSV-to-TSV Coupling Extraction and Optimization in 3D ICs

Taigon Song[1], Chang Liu[2], Yarui Peng[1], and Sung Kyu Lim[1]
[1]School of ECE, Georgia Institute of Technology, Atlanta, GA
[2]Broadcom Corporation, Irvine, CA
taigon.song@gatech.edu, limsk@ece.gatech.edu

ABSTRACT

TSV-to-TSV coupling is a new parasitic element in 3D ICs and can become a significant source of signal integrity problem. Existing studies on its extraction, however, becomes highly inaccurate when handling more than two TSVs on full-chip scale. In this paper we investigate the multiple TSV-to-TSV coupling issue and propose an accurate model that can be efficiently used for full-chip extraction. Unlike the common belief that only the closest neighboring TSVs affect the victim, our study shows that non-neighboring aggressors also cause non-negligible impact. Based on this observation, we propose an effective method of reducing the overall coupling level in multiple TSV cases.

Categories and Subject Descriptors

B.8.2 [**Performance and Reliability**]: Performance Analysis and Design Aids

General Terms

Design

Keywords

3D IC, TSV, Coupling, TSV-to-TSV Coupling

1. INTRODUCTION

Through-silicon-via (TSV) and three-dimensional integrated circuits (3D ICs) are expected to be the key technology trend in high performance and low power systems. Industries are already designing 3D DRAMs using TSVs [6], and academia are reporting the impact of TSVs on 3D ICs in many studies [4].

One of the essential signal integrity (SI) characteristics in studying TSVs is coupling. Recognizing that the impact of the coupling of TSVs is non-negligible, many studies have reported methodologies for reducing TSV-to-TSV coupling [3] [5] [2]. However, these studies have mostly focused on the impact of TSV-to-TSV coupling on only single TSV-pair cases and not on multiple TSV-pair cases [5]. Studies have also focused on the analysis of multiple

Permission to make digital or hard copies of all or part of this work for personal or classroom use is granted without fee provided that copies are not made or distributed for profit or commercial advantage and that copies bear this notice and the full citation on the first page. To copy otherwise, to republish, to post on servers or to redistribute to lists, requires prior specific permission and/or a fee.
DAC'13, May 29 - June 07 2013, Austin, TX, USA.

TSV coupling, but they only analyzed TSV arrays that were not on a full-chip level [2]. A study proposed a methodology that performs full-chip TSV-to-TSV coupling analysis [3], but the analysis may not have been accurate because the analytical model they used overestimated the coupling capacitance.

Therefore, in this paper, we study the multiple TSV-to-TSV coupling effect inside 3D ICs on a full-chip level. We describe the true phenomena that take place inside the ICs and propose a compact model that captures the coupling effect between multiple TSVs. Then, we propose a methodology that performs an analysis of multiple TSV coupling on a full-chip level. The main contributions of this work include the following: (1) A physical limit of the coupling capacitance: We prove that TSV-to-TSV coupling has a maximum capacitance limit. (2) Non-neighboring aggressor impact on TSV-to-TSV coupling: Unlike wire coupling, we show that TSV coupling is affected not only by the closest neighbor, but also by the non-neighboring aggressors. (3) A compact multiple TSV-to-TSV coupling model and extraction algorithm: To the best of the authors' knowledge, we propose the first compact multiple TSV-to-TSV coupling model and extraction algorithm that can be applied on a full-chip level. (4) TSV coupling optimization methodology: We propose a design optimization methodology that reduces TSV-to-TSV coupling in large-scale full-chip 3D IC designs.

2. MOTIVATION

In this section, we describe the motivation of our work and show our findings. We also show why [3] is inaccurate. In this paper, we use the TSVs of a radius of $2\mu m$, a height of $60\mu m$, a SiO_2 liner of 0.5 μm, and a minimum pitch of $10\mu m$.

2.1 Maximum Coupling Capacitance

In [3], the authors assumed that silicon substrate capacitance depends only on the distance between two TSVs. We describe why this assumption is inaccurate. When a victim TSV is surrounded by more than one aggressor, the total coupling capacitance of the silicon substrate has a maximum limit and does not increase linearly.

Many TSV modeling papers [3] [5] claim that the silicon substrate capacitance follows Eq. 1, which is the capacitance between two parallel, circular conducting wires,

$$C_{\text{si}} = \frac{\pi \epsilon_0 \epsilon_{\text{si}} L}{\ln[(P/2r) + \sqrt{(P/2r)^2 - 1}]} \quad (1)$$

in which, ϵ_{si}, L, P, and r are the permittivity of the silicon substrate, the height of the TSVs, the pitch between the TSVs, and the radius of the TSVs, respectively. By this equation, when the coupling capacitance between an aggressor and a victim in a certain pitch is 1x, the victim will see 8x coupling capacitance when there are eight aggressors in every direction.

However, Eq. 1 is correct only when there are no other neighbors near the two TSVs. When TSV aggressors are close to another aggressor, the total substrate capacitance that a victim sees will increase but not linearly. Fig. 1 illustrates this concept when the radius is $2\mu m$ and the pitch between TSVs is $10\mu m$. We simulated the total coupling capacitance using Synopsys Raphael when different number of aggressors are near a victim TSV. Fig. 1 shows that although more TSVs are near the victim, the increase in total coupling capacitance is minor. For example, (d) has two more aggressors than (c), but the total capacitance increase is only 0.51x. For (e), four more aggressors are added than (d), but the capacitance increase is only 0.05x. From this study, we prove that Eq. 1 cannot be used for multiple TSV coupling analysis. We also point that even when there are same number of aggressors, TSV coupling capacitance changes when aggressors are in different locations. For example, Fig. 1 (b) and (c) have same number of aggressors but the total capacitance is different by 0.1x. This is because the E-field that forms capacitance changes due to different locations of the TSVs. Thus, we conclude that the coupling capacitance is a function of location, as well as a function of distance.

We show that a maximum substrate capacitance limit exists for a TSV victim when the radius (r) and the minimum pitch (P) are given. Even when an infinite number of aggressors are near a victim, the maximum substrate capacitance cannot be larger than that of a coaxial TSV, whose inner conductor radius is r, and the outer conductor, whose inner radius is P. We show this formula of a coaxial TSV in Eq. 2 [1].

$$C_{\text{si,max}} \quad \frac{2\pi\epsilon_0\epsilon_{\text{si}}L}{\ln{(P/r)}} \tag{2}$$

Regardless of how many aggressors surround a victim TSV, the total sum of TSV coupling capacitance will be smaller than Eq. 2. In other words, no matter how many aggressors surround a victim (Fig. 1 (f)), the E-field between the victim and the aggressors cannot be formed as strongly as a coaxial TSV (Fig. 1 (g)). Although the values of the maximum coupling capacitance will vary on different TSV radii and pitches, when the radius is $2\mu m$ and the minimum pitch between TSVs is $10\mu m$, the maximum capacitance will be around 2.26x. We conclude that the capacitance sum between a victim and the aggressors has a physical limit. Therefore, it cannot be larger than Eq. 2.

2.2 Neighbor Effect on TSV Coupling

Unlike the common belief that only the nearest aggressors impact TSV coupling, TSV coupling occurs even between the non-neighbor aggressors. In this section, we prove this and also prove that neighbor TSVs can reduce the capacitance of other aggressor TSVs.

First, we show that TSV coupling occurs between the far aggressor and the victim. Assume a simple layout where a victim TSV is neighboring two aggressor TSVs in a straight line (see Fig. 2 (a)). We performed modeling using the proposed model in Section 3.1 and the model was validated using Ansys HFSS. We intuitively think that the far aggressor will not affect coupling because a closer neighbor is near by. However, Fig. 3 shows that the far aggressor affects as much coupling voltage (139.6mV) as the close aggressor (184.6mV) when 1GHz signal is applied in 45nm Nangate technology. This is because the far aggressor also has a significant amount of capacitance between the victim (close aggressor: 9.46fF, far aggressor: 4.14fF, see Fig. 4 Case 3). Though the close aggressor shields the E-field between the victim and the far aggressor, it cannot be perfect. A strong E-field detours the first aggressor and forms capacitance between the far aggressor and the victim (see

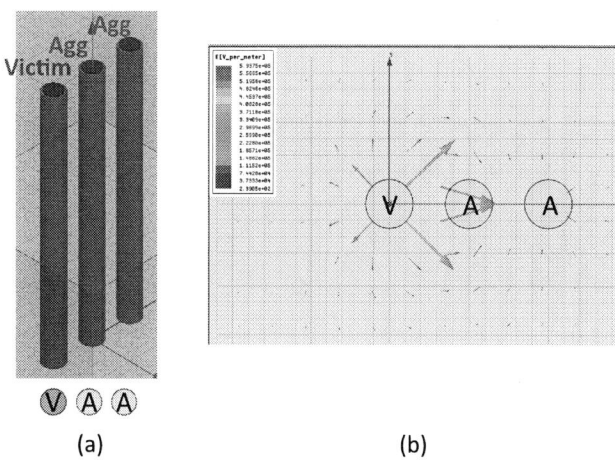

Figure 2: Neighbor Effect. (a) Two aggressor model in HFSS, (b) the E-field distribution between the TSVs.

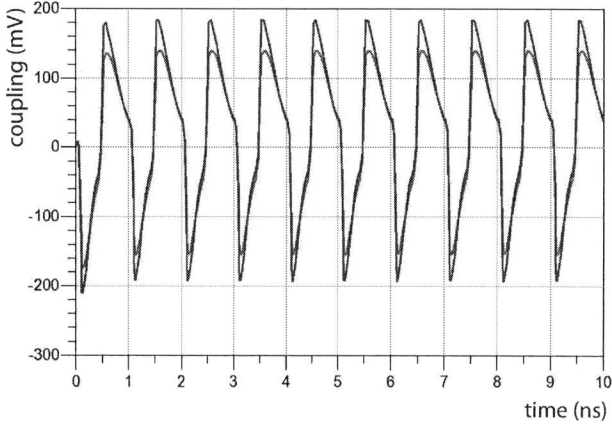

Figure 3: Coupling voltage of the near (blue) and far (red) aggressors shown in Figure 2.

Fig. 2 (b), field distribution simulated using Ansys Q3D). Despite the far aggressor has less than 50% capacitance of the close aggressor, V_{far} reduces by only 40mV. This is because of the complicated coupling network that TSVs compose, explained in [7].

Second, we show that neighbor TSVs can reduce the capacitance of other aggressor TSVs. Fig. 4 describes the far aggressor impact in terms of capacitance. Assume there are only two TSVs as Case 1 and Case 2. Each capacitance is 12.4fF (near aggressor) and 8.5fF (far aggressor). However, in a layout where two aggressors are together (Case3), the coupling capacitance of both aggressors decreases to 9.4fF and 4.1fF. This is because the TSVs in the layout correlate each other and create a new E-field distribution. We call this "Neighbor Effect". Using the Neighbor Effect, if we want to reduce the coupling capacitance between an aggressor and a victim, adding another TSV near the original aggressor will reduce the capacitance of both the original aggressor and the new TSV. Described in Section 2.1 Eq. 2, since there is a physical limit to the total coupling capacitance, no matter how many TSV neighbors are added, the total capacitance will be smaller than a certain value. Therefore, we conclude that the coupling capacitance is a function of distance, location, and also a function of neighbors.

3. MULTI-TSV COUPLING EXTRACTION

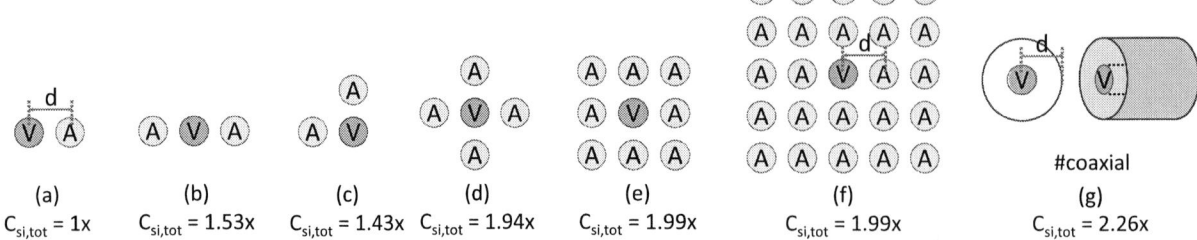

Figure 1: Illustration showing non-linear capacitance increase when the number of aggressors increase, and (g) the maximum limit of coupling capacitance of a TSV.

(a) $C_{si,tot} = 1x$ (b) $C_{si,tot} = 1.53x$ (c) $C_{si,tot} = 1.43x$ (d) $C_{si,tot} = 1.94x$ (e) $C_{si,tot} = 1.99x$ (f) $C_{si,tot} = 1.99x$ (g) $C_{si,tot} = 2.26x$

(case1) $C_1 = 12.4fF$

(case2) $C_2 = 8.5fF$

(case3) $C_1 = 9.4fF$ $C_2 = 4.1fF$

Figure 4: Neighbor Effect case study on how neighbor TSVs affect other aggressors.

In this section, we propose a compact multiple TSV-to-TSV coupling model and an extraction algorithm for full-chip analysis.

3.1 Compact Multi-TSV Coupling Model

[8] proposed an analytical multiple TSV model that can be used for performing coupling analysis. However, this model consists of many RLC components even when there are only few TSVs. Thus, we propose a compact multi-TSV-to-TSV coupling model that can be easily used on full-chip analysis. Fig. 11 in the supplement (a) shows the original model, and (b) shows our proposed model.

To describe the formulas of our model, we explain the concepts used in [8]. Assume three aggressors (N 3) are near a victim. An $N + 1$ system can be considered as N-conductor transmission line. We assume the victim TSV (#0) as the ground and use the multi-conductor transmission line theory. Thus, the victim TSV does not have inductance and only have resistance. A TSV can be expressed as a resistor (R_{TSV}) and an inductor (L_{TSV}) in series. A SiO$_2$ liner surrounds the TSV for isolation, and is expressed as a capacitor (C_{ox}). Silicon substrate can be expressed as a resistor ($R_{si,ij}$) and a capacitor ($C_{si,ij}$) in parallel, of which is the resistance and the capacitance between aggressor i and aggressor j. When i j, it is the resistance and the capacitance of the substrate between victim and the aggressor.

For $R_{si,ij}$ and $C_{si,ij}$, we calculate $L_{si,ij}$, which is the substrate inductance between two TSVs. L_{si} is expressed in matrix ($[L_{si}]$), and consists of self-loop inductance and mutual-loop inductance. The following equations describe how to calculate these values,

$$L_{si,11} \quad \frac{\mu L}{\pi} \ln \left[\frac{P_{i0}}{r + t_{ox}} \right] \tag{3}$$

$$L_{si,ij} \quad \frac{\mu L}{2\pi} \ln \left[\frac{P_{i0} P_{j0}}{P_{i0}(r + t_{ox})} \right] \tag{4}$$

where P_{i0} is the pitch between the victim TSV (#0), and the aggressor TSV(#i) and P_{ij} is the pitch between two aggressor TSVs (#i, and #j). By the relation between the inductance matrix and the capacitance matrix in a homogeneous medium [10], we calculate $[C_{si}]$,

$$[C_{si}] \quad \mu_0 \epsilon_0 \epsilon_{si} L^2 [L_{si}]^{-1} \tag{5}$$

where C_{si} can be defined as Eq. 6.

$$[C_{si}] \begin{bmatrix} \sum_{k=1}^{N} C_{1k} & -C_{12} & \dots & -C_{1N} \\ -C_{21} & \sum_{k=1}^{N} C_{2k} & \dots & -C_{2N} \\ \vdots & \vdots & \ddots & \vdots \\ -C_{N1} & C_{N2} & \dots & \sum_{k=1}^{N} C_{Nk} \end{bmatrix} \tag{6}$$

The conductance matrix $[G_{si}]$ can be defined as

$$[G_{si}] \quad \frac{\sigma}{\epsilon_0 \epsilon_{si}} [C_{si}] \tag{7}$$

In our compact model, we only use $C_{si,ii}$ and $G_{si,ii}$. The other RLC components can be reduced. This is reasonable because we only consider the impact between a victim and an aggressor, not the impact between two different aggressors. Using our model, we can reduce the RLC count around 60% when $N=3$. The RLC count reduces more as N increases. We ignore self inductance and mutual inductance in our model for two reasons. First, the TSV inductance impacts in a very high frequency (> 10 GHz). Second, a coupling path (C_{ox}) exists before the TSV inductance can impact coupling.

To validate our model, we first place aggressor TSVs around the victim TSV randomly in a fixed space. Then, we perform modeling using 3D EM solver HFSS, and also generate a SPICE netlist based on our compact model. We generate 10 layouts for each sample cases, and we compare the S-parameter of these two and report the maximum error of insertion loss. Fig. 5 shows the S-parameter comparison when N=3, and Table 1 shows the validation result. We show that our model is very accurate, even in a multiple TSV structure, by reporting the maximum difference in insertion loss less than 0.02dB.

3.2 Extraction Algorithm

In our previous discussions (Section 2.1 and 2.2), we showed that TSV coupling capacitance is a function of distance, location, and neighbor aggressors. To extract TSV-to-TSV coupling capacitance accurately, an approach considering only the closest neighbor or limiting the maximum target distance to calculate coupling capacitance cannot be used. Therefore, we propose an algorithm that considers distance, direction, and Neighbor Effect all in a holistic manner when extracting the coupling capacitance for all nets in the layout for full-chip analysis. Algorithm 1 describes how this works.

From a given layout, we first extract the (x,y) coordinate of each TSVs. Then, for a victim TSV, we sort all neighbor aggressor TSVs

1244

Table 1: Model validation on general layouts

Radius	Min. pitch	Height	# TSVs	Average error (dB)	Max. error (dB)
2		3		.	. 1
				. 11	. 1
			1	.	. 1
			12	. 11	. 1
				.	. 1
				. 11	. 1
			1	. 11	. 1
			12	.	. 1
1		3		.	. 1
				.	. 1
			1	. 11	. 1
			12	. 11	. 1
				.	. 1
				.	. 1
			1	. 1	. 1
			12	.	. 1

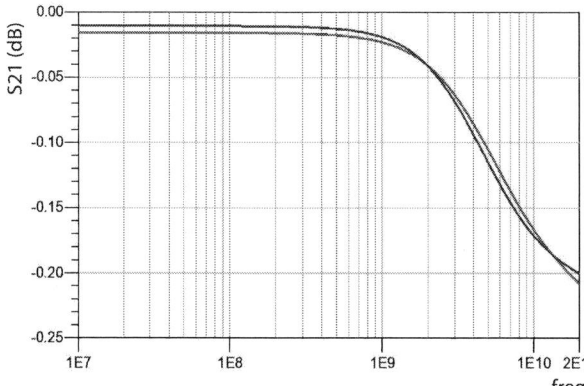

Figure 5: S-parameter comparison between our model and HFSS (red: HFSS, blue: our model)

by the closest Euclidean distance to the victim. We choose N neighbor aggressor TSVs (N: a significantly large number) from the sorted result that are closest from the victim and calculate the capacitance between the victim and the chosen aggressors. Once we calculate the capacitance of the aggressors, we create a coupling network between the victim and the aggressor that the capacitance is higher than a certain value (e.g., $C > 0.01$fF).

We choose a significant number of aggressors (N: more than 100) after sorting to guarantee that we do not neglect any meaningful, physically far but does not have any closer neighbors in the pathway, aggressors. Fig. 6 illustrates this idea. Unless we choose a certain number of aggressors for analysis, we can accidentally miss the valid aggressors that must be considered for extraction. For example, when N=10, the aggressor circled in blue is ignored. This can be considered only when N is bigger than 114. Therefore, N has to be a big number that can consider all the effective neighbors in a layout. By performing this extraction on every victim TSVs, we can extract the coupling capacitance on full-chip scale.

The advantage of this algorithm is that it is fast and considers all effective aggressors that affect the victim. In a layout, it is not the distance, but the location and the neighbors that is important. Since our algorithm calculates the coupling capacitance from a very large number of aggressors, not by distance, it does not neglect any aggressors that must be considered.

Algorithm 1: Multiple TSV-to-TSV capacitance extraction

1 Algorithm: Multiple TSV-to-TSV capacitance extraction
2 Locate all TSVs by its coordinate (x,y);
3 while *For a victim TSV* **do**
4   **while** *For all neighbor aggressor TSVs* **do**
5   Calculate the Euclidean distance of the aggressor TSVs to the victim TSV
6   **end**
7   Sort the neighbor aggressor TSVs by the closest Euclidean distance to the victim TSV;
8   Choose N aggressors that is closest to the victim;
9   Calculate the coupling capacitance of the N aggressors using the formula in Section 3.1;
10   **if** *The calculated TSV capacitance is higher than C* **then**
11   Generate a coupling network between the aggressor and the victim;
12   **else**
13   Assign the TSV coupling capacitance to be zero;
14   **end**
15 end

Table 2: TSV coupling impact on crosstalk and timing.

	W/O coupling	W/ coupling [3]	W/ coupling (Our results)
Footprint (μm)	970×823	970×823	970×823
Total coupling noise (V)	590.77	732.75	815.01
Longest path delay (ns)	2.734	3.165	2.852
Total negative slack (ns)	-61.65	-115.07	-75.24

4. FULL-CHIP ANALYSIS

Using our extraction flow, we perform full-chip SI analysis in this section and compare our results to [3].

4.1 Full Chip 3D SI Analysis Flow

Since existing SI analysis tools cannot analyze 3D circuits accurately, we modified the 3D SI analysis flow in [3] to implement our results. First, we extract the SPEF file for each dies using RC extraction tool. Then, we run our script that implements the algorithm developed in Section 3.2 to create the SPEF file of TSV parasitics that can be plugged in to our flow. Then, we create a top-level verilog file. Once these files are prepared, we use Synopsys Prime-Time to read the verilog file, and create a top-level stitched SPEF file that contains RC information of all dies and the TSVs. Then, we analyze the stitched SPEF file and generate a SPICE netlist for each individual net for performing coupling noise simulation. The SPICE netlist has all the coupling information including wire-coupling, TSV coupling network by the extraction algorithm, and the aggressor signal and the victim driver models. We run HSPICE on each nets one by one, and report the peak noise at each port.

4.2 Design and Analysis Results

We designed FFT 256-8, which is a 256 point with 8 bit precision, real and imaginary, FFT as a test circuit. The circuit has 140K gates and 211 TSVs. The design is a 2-tier 3D IC based on Nangate 45nm technology. Our TSVs are $2\mu m$ in radius, $60\mu m$ in height, $0.5\mu m$ SiO_2 in liner, and $10\mu m$ on minimum pitch. Landing pad is $5 \times 5\mu m$, and each TSVs have a $0.5\mu m$ keep-out zone that no standard cells can be placed inside. The designs were based on our Cadence Encounter design flow to generate 3D layouts [9].

In Fig. 7 and Table 2, coupling analysis results of top-hierarchy

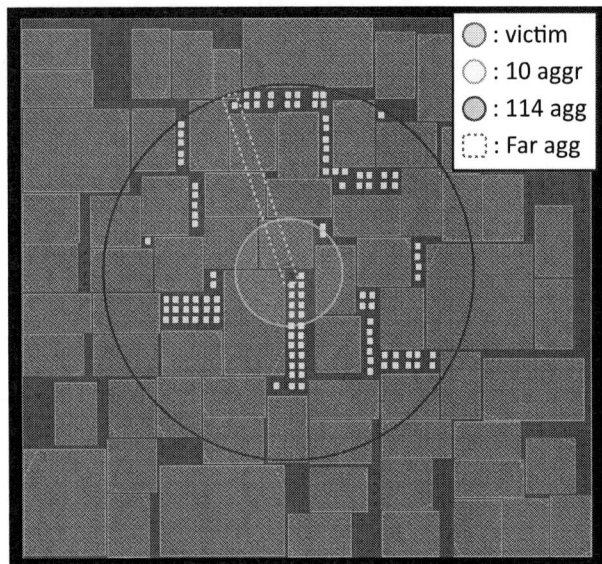

Figure 6: Comparison between a small N (10 aggressors) and a large N (114 aggressors) in the proposed algorithm.

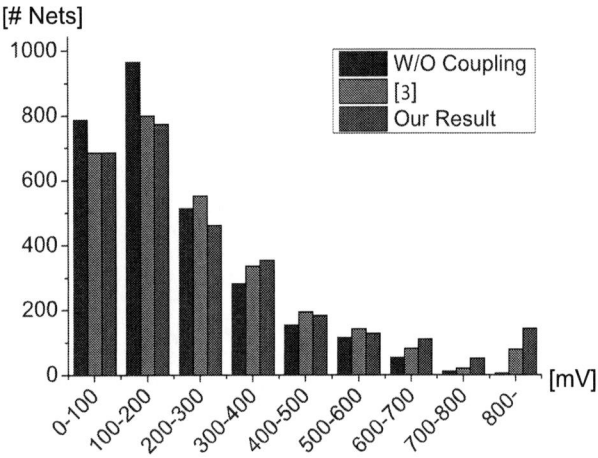

Figure 7: Coupling analysis result.

nets are shown and compared with [3]. Based on the results, we observe the following impacts: First, both approaches calculate higher coupling noise than w/o TSV coupling (590V). Second, [3] is missing a significant amount of TSV coupling impact that must be considered because it considers only the closest neighbors. Third, despite [3] is overestimating the coupling capacitance by linear superposition, our results show higher total noise voltage. The total coupling noise is 732V using the flow in [3] and 815V in our result. This is because our model considers more neighbor aggressors than [3] that must be considered. Therefore, we conclude that we cannot ignore Neighbor Effect. Second, in timing analysis, because [3] overestimates the total coupling capacitance, it also overestimates the timing degradation by TSVs as well. Since the maximum substrate capacitance is limited by Eq.2, by using the correct TSV model, we can save a significant amount of timing margin. In summary, we save more than 83V coupling voltage, more than 300ps in longest path delay, and more than 40ns total negative slack compared to [3].

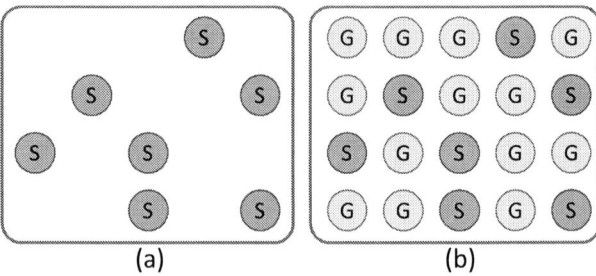

Figure 8: TSV Path Blocking in a layout: (a) Before TSV Path Blocking, (b) after TSV Path Blocking.

Table 3: Impact of TSV Path Blocking - block level design

	W/O Path Blocking	W/ Path Blocking
Footprint (μm)	970×823	970×823
Total coupling noise (V)	815.01	745.02
Longest path delay (ns)	2.852	2.811
Total negative slack (ns)	-75.24	-79.62
3D coupling noise (V)	224.24	154.25

5. TSV-TO-TSV COUPLING REDUCTION

Based on our findings, we propose a TSV-to-TSV coupling reduction method. We validate our methodology in block-level and wide-I/O design.

5.1 TSV Path Blocking

We propose a design optimization method based on that TSV coupling capacitance is a function of distance, direction, and neighbor aggressors. For a layout that has an aggressor and a victim, when an additional TSV is included in the design, the capacitance of the aggressor and the additional TSV both decrease (Section 2.2). Thus, whenever a space between an aggressor and a victim exists, we add GND TSVs between them. We name our coupling reduction method "TSV Path Blocking". By adding GND TSVs between an aggressor and a victim, we block the E-field path between the aggressor and the victim, and thus reduce the coupling capacitance. When applying this method in the layout, we assign as many GND TSVs as possible in empty spaces as Fig. 8. We may think that by adding GND TSVs, the total capacitance of the victim will increase. However, in a layout, a TSV is surrounded by many neighbors that the total coupling capacitance will saturate in a range around 2x (when $C_{\text{one victim}-\text{one aggressor}} = 1$x) . Thus, adding GND TSVs near the neighbor does not have a big impact on increasing the total coupling capacitance (Section 2.1) of a victim.

Table 3 shows the impact of our method. By adding TSVs inside the empty space, the total coupling noise reduces from 815V to 745V. Considering 3D noise only, we reduce the 3D coupling noise by 32% from 224V to 154V. We report that TSV Path Blocking has a minor impact on timing. When GND TSVs are added, the total capacitance will increase slightly since more TSVs are placed near the victim. By the increased capacitive load, the total negative slack increases, but the impact is minor since the total capacitance has a maximum limit, and it is shared by the aggressor and the GND TSVs. We conclude that TSV Path Blocking is an effective way in reducing TSV-to-TSV coupling that has minor impact on timing performance. In a situation where we do not have enough space to insert GND TSVs, we can increase the area occupied by the TSVs and apply our technique. We show the impact of increasing the area occupied by the TSVs in wide-IO design.

Table 4: Impact of TSV Path Blocking - wide I/O design

	Original array	Spread array	W/ Path Blocking
Area by TSV (μm)	160×140	320×140	320×140
Total coupling noise (V)	824.26	797.9	742.37
Longest path delay (ns)	2.907	2.963	2.925
Total negative slack (ns)	-77.26	-74.51	-82.04
3D coupling noise (V)	193.99	157.41	105.81

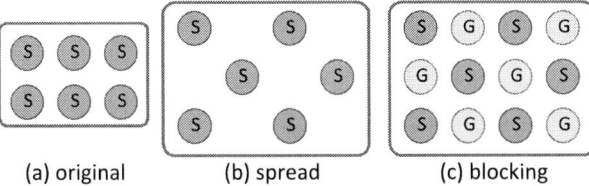

(a) original   (b) spread   (c) blocking

Figure 9: (a) Initial wide-I/O design (b) wide I/O design with spread TSVs (c) wide-I/O design with TSV Path Blocking

5.2 Optimization for Wide-I/O Design

We show the impact of TSV Path Blocking in wide-IO design. TSV Path Blocking can be an effective way to reduce coupling with the cost of increased TSV area. We designed three wide-I/O layouts and compare the SI impact. Fig. 9 (a) is our initial wide I/O design (original), (b) is the wide-I/O design with increased area (spread), and (c) is the wide-I/O design with our technique applied (blocking). Fig. 10 shows an actual layout applying our technique. For fair comparison, we did not modify the placement of the blocks and only increased the area used by TSVs. If the total die size changes due to increased TSV area, then the whole design will change. Therefore, the die size is the same for all cases.

By our technique, we see that the TSV occupied area doubles, but the total coupling noise reduces from 824V to 742V. Considering 3D noise only, we reduce the 3D coupling noise by 45% from 193V to 105V. Note that just by spreading the wide I/O array like Fig. 10 (d), the total coupling noise reduces too. However, if we include GND TSVs as (b), we observe more TSV coupling reduction. Wide I/O with spread TSV shows less total negative slack because the capacitance that a victim sees reduces due to the increased distance. When TSV Path Blocking is applied, we observe more coupling reduction in cost of a minor increase in total negative slack due to increased capacitance. In conclusion, in wide I/O design, we obtain a significant amount of coupling reduction by our technique in the cost of TSV area and small timing performance.

6. CONCLUSIONS

In this paper, we proposed a compact multiple TSV-to-TSV coupling model and an extraction algorithm for 3D ICs. Based on our model and simulations, we demonstrated that TSV-to-TSV coupling has a maximum capacitance limit, and the effect of non-neighboring aggressors is also critical to the total coupling capacitance. We developed a compact multiple TSV-to-TSV coupling model and an algorithm that can accurately consider the impact of far-neighbors on full-chip 3D signal integrity analysis. Using this model, we demonstrated that the far-neighbor aggressors have a significant impact on TSV-to-TSV coupling. To reduce the TSV-to-TSV coupling noise, we proposed a coupling reduction technique: TSV Path Blocking. We applied our technique on block level and wide-I/O design, and experimental results show that by TSV path blocking, 45% 3D coupling reduction can be obtained.

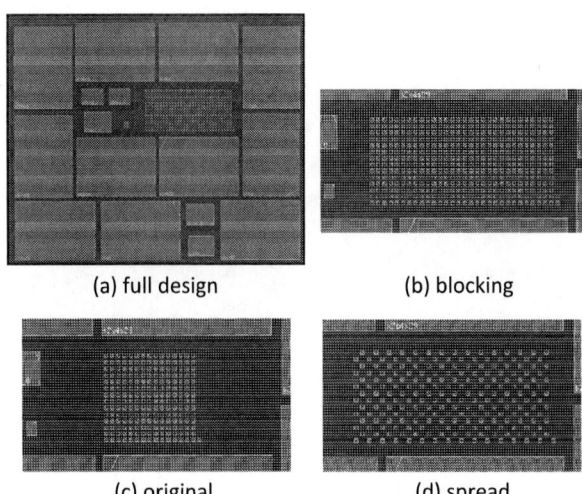

(a) full design   (b) blocking

(c) original   (d) spread

Figure 10: (a) TSV Path Blocking in Wide-I/O layout, (b) zoom-in photo of (a), (c) initial wide I/O design, (d) wide-I/O with spread TSVs

7. ACKNOWLEDGMENTS

This material is based upon the work supported by the NSF (CCF-1018216, CCF-0917000), SRC (1836.075, 2239.001), and the CISS funded by the MEST Global Frontier Project of the South Korean Government (CISS-2-3).

8. REFERENCES

[1] D. K. Cheng. *Field and Wave Eletromagnetics*. Addison Wesley, Boston, MA, second edition, 1992.

[2] B. X. et al. Coupling analysis of through-silicon via (tsv) arrays in silicon interposers for 3d systems. In *Electromagnetic Compatibility (EMC), 2011 IEEE International Symposium on*, pages 16 –21, aug. 2011.

[3] C. L. et al. Full-chip tsv-to-tsv coupling analysis and optimization in 3d ic. In *Design Automation Conference (DAC), 2011 48th ACM/EDAC/IEEE*, pages 783 –788, june 2011.

[4] D. H. K. et al. A study of through-silicon-via impact on the 3d stacked ic layout. In *Computer-Aided Design - Digest of Technical Papers, 2009. ICCAD 2009. IEEE/ACM International Conference on*, pages 674 –680, nov. 2009.

[5] J. K. et al. High-frequency scalable electrical model and analysis of a through silicon via (tsv). *Components, Packaging and Manufacturing Technology, IEEE Transactions on*, 1(2):181 –195, feb. 2011.

[6] J.-S. K. et al. A 1.2v 12.8gb/s 2gb mobile wide-i/o dram with 4x128 i/os using tsv-based stacking. In *Solid-State Circuits Conference Digest of Technical Papers (ISSCC), 2011 IEEE International*, pages 496 –498, feb. 2011.

[7] T. S. et al. Analysis of tsv-to-tsv coupling with high-impedance termination in 3d ics. In *Quality Electronic Design (ISQED), 2011 12th International Symposium on*, pages 1 –7, march 2011.

[8] Y.-J. C. et al. Novel crosstalk modeling for multiple through-silicon-vias (tsv) on 3-d ic: Experimental validation and application to faraday cage design. In *Electrical Performance of Electronic Packaging and Systems, 2012. EPEPS '12. IEEE 21th Conference on*, pages 232 –235, October 2012.

[9] Y.-J. Lee and S. K. Lim. Timing analysis and optimization for 3d stacked multi-core microprocessors. In *3D Systems Integration Conference (3DIC), 2010 IEEE International*, pages 1 –7, nov. 2010.

[10] C. R. Paul. *Analysis of multiconductor transmission lines*. John Wiley and Sons, Lexington, KY, 1994.

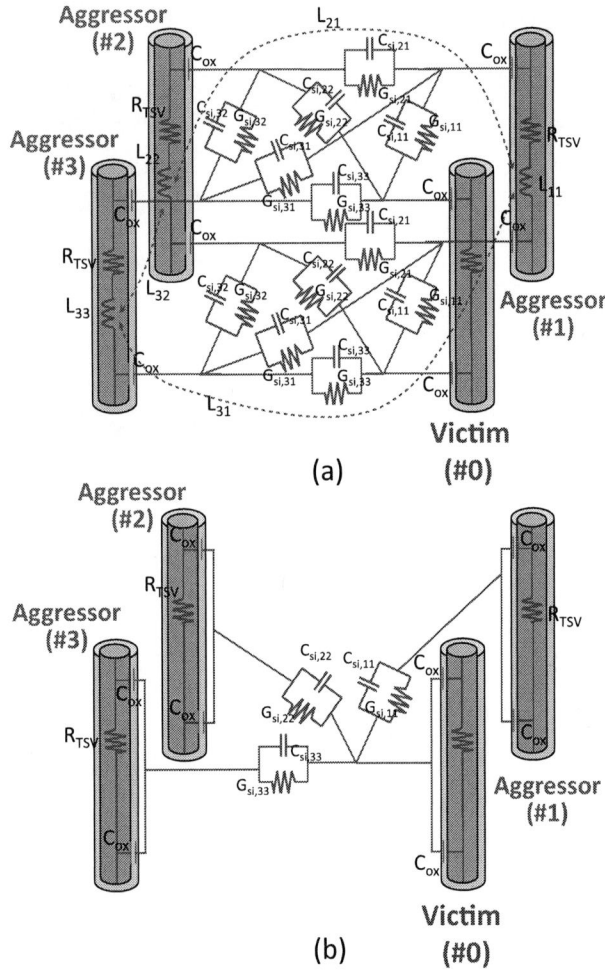

Figure 11: (a) Original model proposed in [8], (b) proposed compact TSV model for full-chip analysis.

SUPPLEMENT
S1 Multiple TSV-to-TSV Model

In this supplement, we provide an illustration and compare the multiple TSV-to-TSV coupling model that was proposed in [8] to our model (see Fig. 11). The model proposed in [8] (a) models the interaction between all TSVs. [8] takes into account for not only the interaction between the victim and the aggressors, but also the interaction between the different aggressors as well. Therefore, [8] may be feasible to model the interaction between multiple TSVs for a small number, but it may not be a feasible model for full chip analysis when the TSV count increases to a high number (such as more than 100) due to the high total RLC count. However, our model (b) reduces the RLC count significantly by considering the interaction between the victim and the each aggressor only. Therefore, our model gives us a reasonable amount of RLC count that enables full-chip analysis.

Note that even when there are only 3 aggressor TSVs, [8] uses 42 RLC components (20 capacitors, 16 resistors, three inductors, and three mutual inductors). However, our model uses only 18 components (11 capacitors and 7 resistors). This is about 60% reduction in the total component count. The RLC count reduction will be more significant when the number of aggressor TSVs in the layout increases.

An Accurate Semi-Analytical Framework for Full-Chip TSV-induced Stress Modeling

Yang Li   David Z. Pan

Department of Electrical and Computer Engineering
The University of Texas at Austin
jerryyangli@gmail.com   dpan@ece.utexas.edu

Abstract

TSV-induced stress is an important issue in 3D IC design since it leads to serious reliability problems and influences device performance. Existing finite element method can provide accurate analysis for the stress of simple TSV placement, but is not scalable to larger designs due to its expensive memory consumption and high run time. On the contrary, linear superposition method is efficient to analyze stress in full-chip scale, but sometimes it fails to provide an accurate estimation since it neglects the stress induced by interactions between TSVs. In this paper we propose an accurate two-stage semi-analytical framework for full-chip TSV-induced stress modeling. In addition to the linear superposition, we characterize the stress induced by interactions between TSVs to provide more accurate full-chip modeling. Experimental results demonstrate that the proposed framework can significantly improve the accuracy of linear superposition method with reasonable overhead in run time.

Categories and Subject Descriptors

B.7.2 [*Hardware, Integrated Circuits*]: Design Aids

General Terms

Design

Keywords

3D IC, TSV, stress, analytical model

1. Introduction

Through-silicon-via (TSV) induces thermo-mechanical stress during the fabrication process and operation phase of 3D IC due to the mismatch of coefficients of thermal expansion (CTE) between the materials of TSV and substrate. The thermo-mechanical stress further induces mobility variation and influences the device performance [1, 2], and may cause serious reliability issues like crack growth in the chip [3, 4]. Hence, it is important to accurately analyze the TSV-induced stress in the design phase in order to predict circuit performance and avoid reliability issues. Most previous works on thermo-mechanical stress employ finite element method (FEM) to characterize the stress [5–7]. While FEM can accurately analyze the stress of simple TSV placement, it encounters enormous difficulties when tackling larger design due to its expensive memory consumption and high run time. To overcome the difficulties of FEM, some analytical methods have been proposed [3, 8]. Although these analytical methods are generally efficient, they usually assume

an over-simplified TSV structure and sometimes lead to results with unacceptable error [9].

Besides FEM and analytical methods, linear superposition method has been proposed in [9–11] to both consider realistic TSV structure and complete the simulation within reasonable run time and memory usage. The fundamental idea of linear superposition method is to calculate the stress contribution of each TSV separately, and then superpose them. However, when calculating the stress contribution of a certain TSV, linear superposition method assumes the TSV is a single one in isolation and neglects the influence of nearby TSVs. This assumption neglects the fact that nearby TSVs have different mechanical properties compared with substrate and will behave differently, and therefore they have considerable influences on stress contribution of the TSV. We term the stress induced by the interactions between TSVs *interactive stress*. Since linear superposition method fails to characterize interactive stress, its accuracy will suffer with high TSV integration density.

To overcome the limitations of previous works, we propose in this paper an accurate semi-analytical framework for full-chip TSV-induced stress modeling. The proposed framework uses complex variable method in elasticity to characterize the interactive stress between TSVs, and achieves an analytical solution to it. Based on the analytical solution, the proposed framework calculates the interactive stress for each simulation point on chip, and uses it to adjust the analysis result of linear superposition method. Experimental results demonstrate the proposed framework significantly benefits the accuracy of linear superposition method under short additional run time. For example, in a placement consisting of two baseline TSVs with 8um pitch, the proposed framework reduces the average error rate of linear superposition from 36.8% to 14.3% in the region surrounding TSVs with only 1% additional run time.

The main contributions of this paper include the following: (1) For the first time, the paper proposes the concept of interactive stress, analyzes its mechanism and points out its importance to stress modeling. (2) The paper shows that the existing linear superposition method will induce more error when the pitch between TSVs decreases due to the neglect of interactive stress. (3) The paper applies the complex variable method in elasticity to interactive stress, and proposes an analytical solution. (4) The paper proposes a semi-analytical framework which can accurately characterize the TSV-induced stress in full-chip scale within reasonable run time and memory usage.

The rest of paper is organized as follows. In Section 2, we will first introduce the baseline TSV structure and stress simulation methodology used in the paper. Then we will define and analyze the interactive stress between TSVs, and point out its importance for stress modeling. In Section 3, we will first propose a 2D analytical stress model for a single TSV which considers the influence of liner on stress field. And then we apply the complex variable method in elasticity to characterize the interactive stress. In Section 4, based on the analytical solution to interactive stress, we will propose an accurate semi-analytical framework for full-chip TSV-induced stress modeling. In Section 5, we will validate our proposed framework on several placements, and justify its accuracy and scalability. Conclusions are drawn in Section 6. More details of the theory and experimental results can be found in Appendix.

Permission to make digital or hard copies of all or part of this work for personal or classroom use is granted without fee provided that copies are not made or distributed for profit or commercial advantage and that copies bear this notice and the full citation on the first page. To copy otherwise, to republish, to post on servers or to redistribute to lists, requires prior specific permission and/or a fee.

DAC '13, May 29 - June 07 2013, Austin, TX, USA

2. Preliminary

2.1 Baseline TSV Structure and Stress Simulation

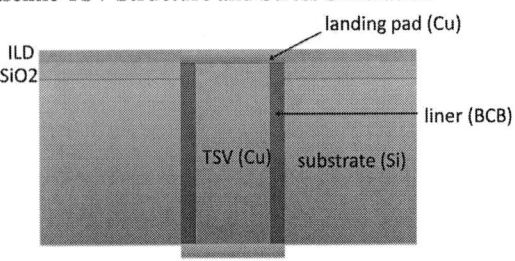

Figure 1. A typical TSV structure.

Various TSV structures have been reported in previous literature [9, 12, 13]. Our baseline TSV structure shown in Figure 1 is based on a typical one [9]. In the structure, TSV consists of copper TSV body and a benzocyclobutene (BCB) liner. We validate our proposed framework on the baseline TSV structure. However, we also use SiO2 as an alternative liner material and test the proposed framework on it (Appendix A.2).

We use commercial FEM tool to obtain the golden result of the stress simulation. In our FEM simulation, the entire structure is assumed to be linear elastic and isotropic with constant material properties. We also assume it is stress free at the annealing temperature, and bears a -250K thermal load during the annealing process.

2.2 Mechanism of Interactive Stress

Interactive stress is defined as the stress induced by the interactions between TSVs due to the mismatch of mechanical properties (Young's modulus and Poisson's ratio) of TSV body, liner and substrate. Here we use a placement consisting of two TSVs to illustrate how interactive stress originates (Figure 2). To simplify the problem, we consider one TSV as the aggressive TSV, and the other as the victim, and then vice versa. In the stress field induced by the aggressive TSV, the victim TSV just acts as a part of medium with different mechanical material properties from substrate.

If we assume victim TSV has the same mechanical property as substrate, the stress field of entire placement is the same as that of a single TSV in isolation. Under this circumstance, we term the stress field of entire placement *ideal stress distribution*, term the stress load along the boundary Γ_1 between the victim and substrate *ideal stress load* and term the deformation of the victim and surrounding substrate *ideal deformation*. However, in reality, the stress load along the boundary Γ_1 must be different from ideal stress load. Otherwise surrounding substrate will deform as ideal deformation, but the victim will deform more significantly than ideal deformation since Young's modulus of TSV is smaller than that of substrate, hence under the same stress load, victim TSV will deform more significantly. In that sense, if the stress load along the boundary Γ_1 does not change, there will be some overlap or discrepancy between victim TSV and substrate, which definitely cannot happen in reality. Interactive stress just originates from the change of stress load along the boundary Γ_1. And then, interactive stress will propagate into the substrate and TSV, and influence the stress distribution in these regions.

Linear superposition method [9] neglects the existence of victim TSV when considering the stress contribution of aggressive TSV, and thus fails to consider interactive stress induced by the interaction between TSVs. For the TSV structure with SiO2 liner, interactive stress is not very severe due to a smaller Young's modulus difference between liner and substrate materials. And hence the linear superposition method can achieve acceptable accuracy though our proposed framework can still significantly improve the accuracy of stress analysis. However, the interactive stress is considerable for the TSV structure with BCB liner due to a much bigger Young's modulus difference between liner and substrate materials. Figure 3 shows the device layer distribution of stress component σ_{xx} calculated by FEM and linear superposition method

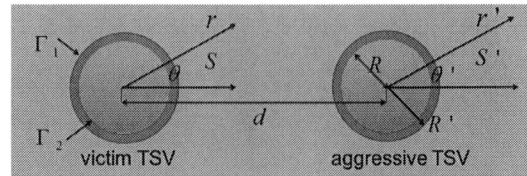

Figure 2. Illustrations for the characterization of interactive stress.

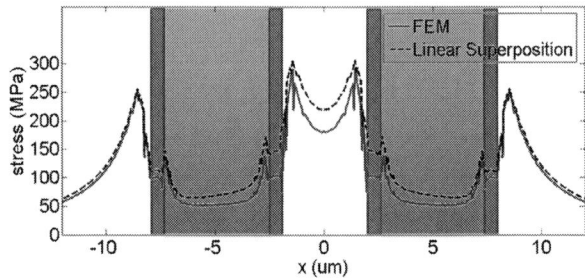

Figure 3. Comparison between FEM and linear superposition method on σ_{xx} distribution along the line through the centers of two TSVs.

along the line through the centers of two baseline TSVs. If we assume FEM simulation contains negligible errors, the discrepancy of results is totally caused by interactive stress. In this figure, we can see linear superposition method overestimates stress level generally, and leads the analysis to an error rate as high as 50% in some regions of interests. Therefore, it is very necessary to consider and characterize interactive stress.

The key points to characterize interactive stress lay in two aspects. One is how to decouple the interaction between aggressive and victim TSV; the other one is how to achieve equilibrium state between the deformation of TSV body, liner and substrate. Firstly, interactive stress originates from the victim TSV under the stress field of the aggressive, but reversely influences the deformation of the aggressive. The influenced deformation might in turn change the stress field of the aggressive, and lead to the change of interactive stress. Therefore, it is important to decouple the interaction between TSVs. Secondly, we need to investigate what is the smart stress load of victim TSV that ensures TSV body, liner and substrate deform harmonically and prevents the occurrence of overlap or discrepancy between these regions.

3. Analysis of Interactive Stress

In this section, we analyze and characterize the interactive stress. We first introduce the basics of elasticity theory which serves as the fundamental of derivation. Then we propose a 2D analytical stress model for a single TSV which has considered the influence of liner on stress field. This model can be used as the ideal stress distribution induced by the aggressive TSV. Based on this model, we employ the complex variable method in elasticity to characterize the interactive stress.

3.1 Basics of Elasticity Theory

Elasticity theory generally uses a second-order tensor to represent the stress in solids. For example, in a 3D problem, a second-order stress tensor is shown as (1).

$$\sigma = \begin{bmatrix} \sigma_{11} & \sigma_{12} & \sigma_{13} \\ \sigma_{21} & \sigma_{22} & \sigma_{23} \\ \sigma_{31} & \sigma_{32} & \sigma_{33} \end{bmatrix} \quad (1)$$

In this tensor, the first subscript of each component represents the normal direction of plane the stress component aims at. The second subscript represents the direction of component. In a 3D Cartesian coordinate system, index 1,2,3 represent index x,y,z respectively; while in a cylindrical coordinate system, they represent index r,θ,and z. The stress tensor is generally symmetric.

Since stress tensor and other tensors (like strain tensor) can be represented in various coordinate systems, it is necessary to introduce how to transform tensors between different coordinate systems. If the

1250

original and new coordinate system are respectively cylindrical and Cartesian system, the transform matrix and transform formula is shown as (2).

$$Q = \begin{bmatrix} \cos\theta & -\sin\theta & 0 \\ \sin\theta & \cos\theta & 0 \\ 0 & 0 & 1 \end{bmatrix}, \qquad \sigma_{xyz} = Q \cdot \sigma_{r\theta z} \cdot Q^T \qquad (2)$$

where $\sigma_{r\theta z}$ and σ_{xyz} are stress tensors respectively in the cylindrical and Cartesian coordinate system; θ is the angle between x-axis of the Cartesian system and r-axis of the cylindrical system.

When we focus on the stress on the surface of solids, we can simplify the problem into a 2D problem. In a 2D problem, we only consider the stress component σ_{rr}, $\sigma_{r\theta}$ and $\sigma_{\theta\theta}$. Other components are either null or combinations of these components. Since the key issue of this paper is to characterize the interactive stress in device layer, it is reasonable to analyze the interactive stress using the complex variable method in elasticity for 2D problems [14]. In the complex variable method, the pursuit of stress tensor is transformed to the pursuit of two analytical functions ϕ and ψ (in complex analysis meanings) which satisfy the boundary condition. These two analytical functions are termed complex potential. If complex potentials are successfully obtained, stress tensor and displacement vector can be achieved through (3 - 5).

$$\sigma_{rr} + \sigma_{\theta\theta} = 2[\phi'(z) + \overline{\phi'(z)}] \qquad (3)$$

$$\sigma_{\theta\theta} - \sigma_{rr} + 2i\sigma_{r\theta} = 2\exp(2i\theta)[\bar{z}\phi''(z) + \psi'(z)] \qquad (4)$$

$$u_r + iv_\theta = \frac{1+v}{E}\exp(-i\theta)[\frac{3-v}{1+v}\phi(z) - z\overline{\phi'(z)} - \overline{\psi(z)}] \qquad (5)$$

where z represents the position of the point (r, θ) in the complex plane, $z = r\exp(i\theta)$; u_r and v_θ are displacement of the point (r, θ) in r-direction and θ-direction; $\phi(z)$ and $\psi(z)$ are complex potentials; E and v are Young's modulus and Poisson's ratio of corresponding materials.

3.2 Stress Field induced by the Aggressive TSV

In this section, we propose a 2D analytical stress model for a single TSV. The model can be used to represent the stress field induced by the aggressive TSV. As what is shown in [9, 15], liner has considerable influences on the stress field of TSV. Hence, different from previous 2D analytical models, the proposed model incorporates the influence of liner into its derivation. It can be applicable to liner with any material and thickness. In this model, as we focus on the stress of device layer, plane stress condition is assumed. Although it is a 2D model, it provides a simple while accurate enough basis to characterize interactive stress since interactive stress is kind of a second order effect of the original stress field. Due to the page limit, we omit the derivation of model, and directly give the formula for the stress field in the substrate shown in (6). We adopt the cylindrical coordinate system S' (Figure 2).

$$\sigma_{rr}|_{ideal} = \frac{K}{r'^2} \qquad \sigma_{\theta\theta}|_{ideal} = -\frac{K}{r'^2} \qquad \sigma_{r\theta}|_{ideal} = 0, \quad r' > R' \qquad (6)$$

where R' is the radius of TSV; r' is the distance between the simulation point and the center of TSV; K is a constant which can be directly calculated from the material property and geometry specification of TSV structure (Appendix A.4); $\sigma_{rr}|_{ideal}$, $\sigma_{\theta\theta}|_{ideal}$ and $\sigma_{r\theta}|_{ideal}$ are stress components. The formula shows stress components decrease with r'^{-2} in the substrate.

Since we will take advantage of the boundary condition along Γ_1 (Figure 2) to derive the interactive stress, we replace the current coordinate system with the cylindrical coordinate system S (Figure 2), and factorize the following stress combination along Γ_1 based on (6). We omit the derivation and give the result in (7). The corresponding displacement combination factorized along Γ_1 is shown in (8).

$$(\sigma_{rr} - i\sigma_{r\theta})|_{\Gamma_1, ideal} = \sum_{m=2}^{\infty} \frac{K(m-1)}{R'^2}(\frac{R}{d})^m \exp(im\theta) \qquad (7)$$

$$(u_r + iv_\theta)|_{\Gamma_1, ideal} = \sum_{m=-\infty}^{\infty} \frac{K}{R'}\frac{1+v_s}{E_s}(\frac{d}{R'})^m \exp(im\theta) \qquad (8)$$

In (7) and (8), u_r and v_θ represents the displacement in r-direction and θ-direction; d is the pitch between TSVs; E_s and v_s are the Young's modulus and Poisson's ratio of silicon. (7) and (8) are the ideal stress distribution and displacement field factorized along Γ_1. In next section, we will investigate how they change in reality, and use them to derive the interactive stress.

3.3 Characterization of Interactive Stress

To characterize the interactive stress, we need to decouple interactions between TSVs, and establish an equilibrium state between the deformation of TSV body, liner and substrate. Similar to previous analysis, for a TSV pair, we consider one as the aggressive and the other one as the victim in one round, and then vice versa in another round. In each round, we observe the phenomenon that the interactive stress between the aggressive and victim TSV has little influence on the stress field around aggressive TSV (Details can be referred to Appendix A.1). It infers that the boundary condition around the aggressive can be ignored. Specifically, we can just consider the stress induced by the reaction of victim under the stress field of aggressive, and ignore the influence of such reaction on the deformation of aggressive. In this way, we avoid iterations of interaction between the stress field of aggressive and victim TSV, and thus decouple the interactions between them.

We use the complex variable method in elasticity to achieve the equilibrium state between the deformation of victim TSV and substrate. The general flow is as follows. First, we adopt the method of undetermined coefficients, and assume a general form of complex potentials in TSV body, liner and substrate respectively. We also assume a stress load along the boundary of victim TSV. Second, we investigate how victim TSV and substrate deform under this stress load, and establish the relations of assumed complex potentials to the assumed stress load. Finally, based on the boundary condition between victim TSV and substrate, we solve the assumed stress load and obtain the analytical solution to interactive stress. Section 3.3.1 and 3.3.2 respectively investigate how substrate and victim TSV deforms under the assumed stress load. Section 3.3.3 solves the assumed stress load and obtains the solution to interactive stress.

3.3.1 Elasticity Analysis of Substrate

In this section we investigate how substrate deforms under a certain stress load. The stress load that the substrate bears comes from two parts. One is from the boundary of aggressive TSV; the other one is from the boundary of victim TSV. The stress load along the boundary of the aggressive is almost known. But the stress load along the boundary of the victim is unknown. Therefore, we need to find a general relation of substrate deformation to arbitrary stress load along the boundary of the victim. Additionally, the non-uniform distribution of stress load along the boundary of the victim also increases the difficulty of solution.

To tackle the problem, we decompose the stress load along the boundary Γ_1 into two parts $\hat{f}_0$ and $\hat{f}_1$, where $\hat{f}_0$ is the ideal stress load defined in Section 3.2, $\hat{f}_1$ is the change of ideal stress load along Γ_1 in reality. In section 3.2, we have obtained $\hat{f}_0$ and corresponding displacement of substrate under the stress field of aggressive TSV and $\hat{f}_0$, and factorize them along the boundary Γ_1 shown in (7) and (8). Hence, we only need to further investigate how the substrate deforms under $\hat{f}_1$, and then superpose them to obtain the deformation of substrate in reality. Since $\hat{f}_1$ is unknown, we represent it as a series factorized along Γ_1 in (9).

$$\sigma_{rr} - i\sigma_{r\theta}|_{\Gamma_1, change} = \sum_{m=-\infty}^{\infty} f_m \exp(im\theta) \qquad (9)$$

where f_m are undermined coefficients. We will continue to see how the substrate deforms under this assumed stress load.

To investigate the deformation of substrate under $\hat{f}_1$, we assume the corresponding complex potentials in the region of substrate. Based on the fact that the interactive stress has little influence on the stress field around the aggressive TSV (Appendix A.1), we ignore the boundary condition of aggressive TSV, and consider the substrate as pure silicon. Therefore, according to the theory of complex analysis, complex potentials $\phi_s(z)$ and $\psi_s(z)$, which are analytical functions in an infinite region containing a circular hole, can be further represented as a series in the complex plane as (10).

$$\phi_s'(z) = \sum_{m=-\infty}^{0} A_{s,m} z^m \qquad \psi_s'(z) = \sum_{m=-\infty}^{0} B_{s,m} z^m \qquad (10)$$

1251

where $\phi'_s(z)$ and $\psi'_s(z)$ are derivatives of $\phi_s(z)$ and $\psi_s(z)$; $A_{S,m}$ and $B_{s,m}$ are undetermined coefficients; z represents the position of simulation point in the complex plane. We can further adapt these complex potentials to the stress load in (9) to determine the unknown coefficients of complex potentials, and then achieve the displacement of substrate under the assumed stress load.

Based on (3 - 5) and (9 - 10), we can achieve the displacement field $(u_r + iv_\theta)|_{change}$ under the stress load $\hat{f}_1$. Then we superpose it with the displacement field $(u_r + iv_\theta)|_{ideal}$ under ideal stress load $\hat{f}_0$ and the stress load of aggressive TSV, and obtain the displacement field of substrate $(u_r + iv_\theta)|_{substrate}$. Specifically, the displacement along the boundary Γ_1 will be

$$(u_r + iv_\theta)|_{\Gamma_1, substrate} = (u_r + iv_\theta)|_{\Gamma_1, ideal} + (u_r + iv_\theta)|_{\Gamma_1, change} \quad (11)$$

We will further use the displacement field of substrate obtained from (11) to fit the displacement field of TSV to determine the assumed stress load along the boundary Γ_1.

3.3.2 Elasticity Analysis of TSV

In this section, we investigate how TSV body and liner deforms under the assumed stress load $\hat{f}_0 + \hat{f}_1$. This problem encounters even more complexity than the previous one since the mechanical properties of TSV body and liner are generally different, and we need to preserve an equilibrium state between their deformation under arbitrary stress load.

To tackle this problem, we assume different complex potentials for TSV body and liner, and take advantage of the boundary condition between TSV body and liner to establish their relationship. After that we try to find their relation to the assumed stress load. Specifically, since the region of TSV body is circular and the region of liner is ring, according to the theory of complex analysis, the complex potentials of TSV body $\phi_c(z)$ and $\psi_c(z)$, and the complex potentials of liner $\phi_l(z)$ and $\psi_l(z)$ can be assumed to be

$$\phi'_c(z) = \sum_{m=0}^{\infty} A_{c,m} z^m \qquad \psi'_c(z) = \sum_{m=0}^{\infty} B_{c,m} z^m \quad (12)$$

$$\phi'_l(z) = \sum_{m=-\infty}^{\infty} A_{l,m} z^m \qquad \psi'_l(z) = \sum_{m=-\infty}^{\infty} B_{l,m} z^m \quad (13)$$

where $\phi'_c(z)$, $\psi'_c(z)$, $\phi'_l(z)$ and $\psi'_l(z)$ are the derivatives of $\phi_c(z)$, $\psi_c(z)$, $\phi_l(z)$ and $\psi_l(z)$; $A_{c,m}$, $B_{c,m}$, $A_{l,m}$ and $B_{l,m}$ are undetermined coefficients. We adapt these assumed complex potentials to the boundary condition along Γ_1 and Γ_2, and obtain the corresponding displacement field within TSV.

Specifically, these complex potentials need to satisfy the boundary condition along Γ_1 and Γ_2. In details, they should satisfy the continuous condition of stress component combination along Γ_1 and Γ_2 shown as (14) and (15), and satisfy the continuous condition of displacement along Γ_2 shown as (16).

$$\sigma_{rr} - i\sigma_{r\theta}|_{\Gamma_1, liner} = \sum_{m=2}^{\infty} \frac{K(m-1)}{R'^2} \left(\frac{R'}{d}\right)^m \exp(im\theta) + \sum_{m=-\infty}^{\infty} f_m \exp(im\theta) \quad (14)$$

$$\sigma_{rr} - i\sigma_{r\theta}|_{\Gamma_2, liner} = \sigma_{rr} - i\sigma_{r\theta}|_{\Gamma_2, copper} \quad (15)$$

$$(u_r + iv_\theta)|_{\Gamma_2, liner} = (u_r + iv_\theta)|_{\Gamma_2, copper} \quad (16)$$

Based on (3 - 5) and (12 - 16), we obtain the displacement field in TSV body and liner $(u_r + iv_\theta)|_{copper}$ and $(u_r + iv_\theta)|_{liner}$. We will establish an equilibrium state between the displacement field of liner $(u_r + iv_\theta)|_{liner}$ and that of substrate $(u_r + iv_\theta)|_{\Gamma_1, substrate}$ to determine the stress load along Γ_1 in next section.

3.3.3 Analytical Solution to Interactive Stress

After finishing the elasticity analysis of substrate and TSV, we obtain the displacement field within TSV and silicon. We further take advantage of the continuous condition of displacement along the boundary Γ_1 and obtain (17).

$$(u_r + iv_\theta)|_{\Gamma_1, substrate} = (u_r + iv_\theta)|_{\Gamma_1, liner} \quad (17)$$

Algorithm 1 Full-chip TSV-induced stress modeling framework.

Input: TSV list, stress library
Output: stress map

Stage I Linear Superposition Method (Table Look-up Method)
for each simulation point p
    find nearby TSVs;
    for each nearby TSV of p
        calculate stress contribution;
        add it to p.linear;
    end
end

Stage II Interactive Stress Calculation (Analytical Method)
for each simulation point p
    find nearby TSV pairs;
    for each nearby TSV pair of p
        calculate interactive stress contribution using (18);
        convert the interactive stress contribution using (2);
        add it to p.interactive;
    end
end

Final:
for each simulation point p
    stress tensor p.stress ← p.linear + p.interactive (tensor addition);
end

Up to now, all the necessary equations have been established. Based on (3 - 5) and (7 - 17), we can solve the problem, and achieve the ultimate analytical solution to the interactive stress represented in the coordinate system S (Figure 2) shown in (18).

$$\sigma_{rr} = \frac{K}{R'^2} \sum_{m=2}^{\infty} \cos(m\theta) \left[\left(\frac{r}{d}\right)^m \left(h_{i1}(m) - \frac{R'^2}{r^2} h_{i2}(m) \right) \right.$$
$$\left. + \left(\frac{R'^2}{rd}\right)^m \left(h_{i3}(m) - \frac{R'^2}{r^2} h_{i4}(m) \right) \right]$$

$$\sigma_{\theta\theta} = \frac{K}{R'^2} \sum_{m=2}^{\infty} \cos(m\theta) \left[\left(\frac{r}{d}\right)^m \left(h_{i5}(m) + \frac{R'^2}{r^2} h_{i2}(m) \right) \right.$$
$$\left. + \left(\frac{R'^2}{rd}\right)^m \left(h_{i6}(m) + \frac{R'^2}{r^2} h_{i4}(m) \right) \right] \quad (18)$$

$$\sigma_{r\theta} = \frac{K}{R'^2} \sum_{m=2}^{\infty} \sin(m\theta) \left[\left(\frac{r}{d}\right)^m \left(h_{i7}(m) + \frac{R'^2}{r^2} h_{i2}(m) \right) \right.$$
$$\left. + \left(\frac{R'^2}{rd}\right)^m \left(h_{i8}(m) - \frac{R'^2}{r^2} h_{i4}(m) \right) \right]$$

where r, θ are the position of simulation point in the coordinate system S; $h_{ij}(m)$, $i = 1, 2, 3$, $j = 1, 2, \ldots, 8$, are functions which only depend on material properties and geometry specification of TSV, and are irrelevant to TSV placement (Appendix A.4). In (18), if $r < R$ (TSV body), $i = 1$, $h_{1j} = 0$, $j = 3, 4, 6, 8$; if $R < r < R'$ (liner), $i = 2$; if $r > R'$ (substrate), $i = 3$, $h_{ij} = 0$, $j = 1, 2, 5, 7$.

In the above analytical solution, it is not hard to see each term of the solution generally decreases with a speed no slower than (R'/d) as m increases. Since $R'/d < 0.5$, the series will converge with a reasonable amount of terms. We employ 9 terms in practice ($m_{max} = 10$).

4. Full-Chip Stress Modeling Framework

In this section, we propose an interactive stress aware two-stage semi-analytical framework for full-chip TSV-induced stress modeling as shown in Algorithm 1. The first stage performs linear superposition [9] to obtain a rough stress estimation. Since stress decays to a negligible intensity after a certain distance from TSV, the framework only considers the stress induced by nearby TSVs, and ignores other TSVs' contribution in order to improve the efficiency of algorithm. Hence, in Stage I, we firstly determine what are the nearby TSVs for each simulation point. Here, we consider a TSV with distance less than a certain value (e.g. 25um) from the simulation point as nearby TSV. After that, we employ a table look-up method to find the stress contribution of nearby TSVs and superpose them for each simulation point. The complexity of first stage is $O(n)$, where n is the amount of simulation points.

The second stage calculates interactive stress for each simulation point. The general idea is to first calculate the interactive stress contribution of each TSV pair, and then superpose them together. Since a certain TSV may interact with multiple TSVs and each induces interactive stress, during the calculation, a TSV may form multiple pairs with other TSVs. For each TSV pair, although there may exist other TSVs nearby the pair and make the substrate not as pure silicon, we can still use the analytical formula derived in Section 3.3. This is because of the fact that the interactive stress induced by the interaction of a TSV pair is nearly irrelevant to the existence of other TSVs nearby the pair (Details can be referred to Appendix A.1).

Similar to the first stage, the second stage only considers the interactive stress contribution of nearby TSV pairs, and firstly determines what are the nearby pairs for each simulation point. Here, we consider any two TSVs as a nearby pair for a certain point if the pair satisfies 1) the pitch of pair is within a certain distance (e.g. 25um) and 2) the victim TSV is located within a certain distance (e.g. 25um) from the point. For a TSV pair, if the pitch is large, the interactive stress will become too small to be considered. For a simulation point, since interactive stress decreases no slower than r^{-2} (Section 3.3), r is the distance from the victim TSV to the point, if the distance is too large, we can ignore the interactive stress contribution of the pair in one round to the point. Please note in a TSV pair, a TSV can both become aggressive and victim in two rounds. Therefore, although a TSV pair may not become nearby pair for a simulation point in one round due to the large distance from the victim to the point, when the roles of TSVs exchange, the pair may become a nearby pair for the same point. After determining the nearby pairs, we use analytical formula to calculate and superpose the interactive stress contribution of each nearby pair, and obtain the interactive stress for each simulation point. Since we only consider nearby TSV pair's contribution, the complexity of second stage is nearly irrelevant to the amount of TSVs for large designs but relevant to TSV integration density. However, since TSV integration density faces a upper bound in real applications, its influence on the run time of second stage also faces a limit. Therefore, the complexity of second stage mainly depends on the amount of simulation point and is linear in terms of it. Finally, we superpose the stress contribution calculated in the first and second stage, and obtain the stress analysis result.

5. Experimental Results

In this section, we validate the performance of proposed framework. We implement the linear superposition method and proposed framework in MATLAB. The golden result of stress analysis is generated by FEM simulation tool COMSOL [16]. The main material properties used in our modeling are as follows: Young's modulus (GPa) for copper = 110, BCB = 3, SiO2 = 71, silicon = 188; CTE (ppm/K) for copper = 17, BCB = 40, SiO2 = 0.5, silicon = 2.3. The radius of TSV body is 2.5um. The thickness of liner is 0.5um. And the dimension of landing pad is 6um.

5.1 Validation: A Placement of Two TSVs

We first validate the proposed framework in a placement containing two TSVs. In this placement, we vary the pitch between TSVs from 8um to 30um. Just as claimed by [17], the minimal pitch in the current process technology is 10um. However, since future TSV-based design will require higher TSV integration density to fully take advantage of the benefit of vertical integration, we validate the proposed framework in a broader range of pitch. Figure 4 compares the error of stress component σ_{xx} obtained by the linear superposition method and proposed framework under a 10um pitch. Due to symmetry, only right half analysis result is provided. Figure 4(a) shows linear superposition method leads to an error as much as 70MPa in some regions. Figure 4(b) shows the proposed framework can significantly alleviate the error of linear superposition method, and estimate the stress level around TSV with an error less than 25MPa generally.

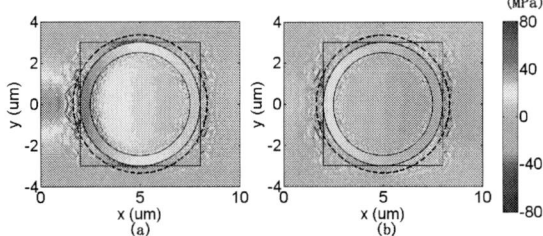

Figure 4. Comparison on the error of stress component σ_{xx} obtained by linear superposition and proposed framework for the placement of two TSVs (right half shot). (a) linear superposition method. (b) proposed framework.

	d (um)	Avg. Error (MPa)	Threshold 10MPa		Threshold 50MPa		Threshold 50MPa Critical Region	
			Avg. Error (MPa)	Avg. Error Rate(%)	Avg. Error (MPa)	Avg. Error Rate(%)	Avg. Error (MPa)	Avg. Error Rate(%)
LS	8	3.24	6.42	13.5	20.5	20.7	35.3	36.8
	9	2.63	5.35	11.3	16.0	16.1	27.6	28.9
	10	2.22	4.67	9.62	13.1	13.1	22.7	23.7
	11	1.90	4.16	7.99	10.9	10.9	19.4	19.9 .
	12	1.65	3.63	6.76	9.27	9.14	17.0	16.9
	18	0.92	1.88	3.20	4.82	4.84	10.4	8.58
	30	0.54	0.95	1.39	3.57	2.96	7.83	5.14
PF	8	1.96	4.01	8.94	11.7	11.0	16.0	14.3
	9	1.47	3.05	6.43	8.81	8.25	13.2	11.8
	10	1.19	2.58	4.85	7.35	6.78	11.8	10.4
	11	1.01	2.30	3.86	6.29	5.77	10.8	9.33
	12	0.89	2.07	3.28	5.57	5.04	10.1	8.48
	18	0.61	1.31	1.87	3.55	3.16	8.12	5.81
	30	0.47	0.87	1.12	3.44	2.77	7.64	4.97

Table 1. Comparison on the error of stress component σ_{xx} obtained by linear superposition and proposed framework for the placement of two TSVs. LS: linear superposition; PF: proposed framework.

Table 1 presents a quantitative comparison on the error of linear superposition method and proposed framework. Before we talk about this comparison, we need to define two concepts which facilitate the comparison. The first concept "monitored region" refers to the region which is influenced by TSV-induced stress to a certain extent. In this placement, we use a rectangular region as the monitored region. The center of monitored region is on the midpoint of the segment through the centers of two TSVs. Along x and y dimension shown in Figure 4, the dimensions of monitored region are respectively 60um and 30um. We use it as a baseline region of concern, and do the comparison listed in Table 1 in the monitored region except for the last two columns. The second concept "critical region" refers to the region influenced considerably by stress. We define it as the region within 3.3um to the center of each TSV, which is shown as the region within the dashed line around each of TSVs in Figure 4. The last two columns of Table 1 are the comparison for the critical region. We compare the linear superposition method and proposed framework in monitored region to give a general comparison on the stress estimation capability of two methods, while still compare them in the critical region which shows the difference on estimating stress in the region of more reliability concern. When we perform the comparison, we set a threshold on the stress intensity of simulation point, and only consider the point with intensity exceed the threshold. It is logical since only point with large stress intensity matters the reliability and mobility variation most, and thus should be emphasized.

Table 1 shows that the proposed framework considerably improves the accuracy of linear superposition method. For example, under the 8um pitch, the proposed framework can achieve a 14.3% average error rate for the stress component σ_{xx} in the critical region with 50MPa as threshold while the linear superposition method only achieve a 36.8% average error rate under the same condition. Since the interactive stress

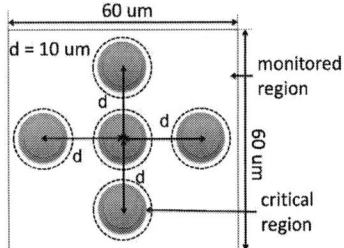

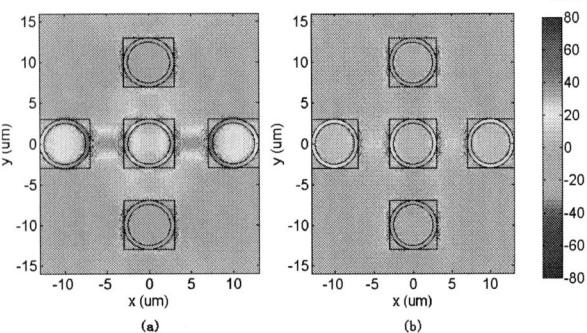

Figure 5. A placement of five TSVs.

Figure 6. Comparison on the error of stress component σ_{xx} obtained by the linear superposition method and proposed framework for the placement of five TSVs. (a) linear superposition method. (b) proposed framework.

	Stress Type	Avg. Error (MPa)	Threshold 10MPa		Threshold 50MPa		Threshold 50MPa Critical Region	
			Avg. Error (MPa)	Avg. Error Rate(%)	Avg. Error (MPa)	Avg. Error Rate(%)	Avg. Error (MPa)	Avg. Error Rate(%)
LS	σ_{xx}	2.84	5.45	11.6	13.9	13.1	22.8	23.0
	Von Mises	3.11	3.41	5.63	8.39	6.39	20.1	15.1
PF	σ_{xx}	1.60	3.05	5.74	8.19	7.19	13.8	12.6
	Von Mises	1.75	1.94	2.66	5.14	3.00	14.3	8.22

Table 2. Comparison on the stress error obtained by the linear superposition method and proposed framework for the placement consisting of five TSVs. LS: linear superposition; PF: proposed framework.

framework for full-chip TSV-induced stress modeling. Experimental results have demonstrated that the proposed framework can significantly reduce the error of linear superposition method with short additional run time. The remaining error of proposed framework is caused by the 2D nature of the analytical interactive stress model. Due to the generality of proposed framework, it can be applicable to TSV structures with various kinds of materials and geometry specifications. Therefore, we expect the proposed framework can be widely used in full-chip 3D IC reliability analysis and layout optimization.

7. Acknowledgment

This work is supported by NSF under Award No. 1018750 and SRC under Task No. 2238.001. The authors would like to thank Prof. Sung Kyu Lim from Georgia Tech and Dongjie Jiang from UT Austin for their helpful discussion.

References

[1] K. Athikulwongse, A. Chakraborty, J.-S. Yang, D. Z. Pan, and S. K. Lim, "Stress-driven 3D-IC placement with TSV keep-out zone and regularity study," in *Proc. ICCAD*, 2010.

[2] J. Yang, K. Athikulwongse, Y. Lee, S. K. Lim, and D. Z. Pan, "TSV stress aware timing analysis with applications to 3D IC layout optimization," in *Proc. DAC*, 2010.

[3] S.-K. Ryu, K.-H. Lu, X. Zhang, J.-H. Im, P. Ho, and R. Huang, "Impact of near-surface thermal stresses on interfacial reliability of through-silicon vias for 3-D interconnects," *IEEE Trans. Device and Materials Reliability*, vol. 11, no. 1, pp. 35–43, 2011.

[4] M. Jung, X. Liu, S. Sitaraman, D. Z. Pan, and S. K. Lim, "Full-chip through-silicon-via interfacial crack analysis and optimization for 3D IC," in *Proc. ICCAD*, 2011.

[5] J. Zhang, M. Bloomfield, J.-Q. Lu, R. Gutmann, and T. Cale, "Modeling thermal stresses in 3-D IC interwafer interconnects," *IEEE Trans. Semiconductor Manufacturing*, vol. 19, no. 4, pp. 437–448, 2006.

[6] C. Khong et al., "Numerical modeling of through silicon via (TSV) stacked module with micro bump interconnect for biomedical device," in *Proc. EPTC*, 2010.

[7] C. Zhang and L. Li, "Characterization and design of through-silicon via arrays in three-dimensional ICs based on thermomechanical modeling," *IEEE Trans. Electron Devices*, vol. 58, no. 2, pp. 279–287, 2011.

[8] K. Lu, X. Zhang, S.-K. Ryu, J. Im, R. Huang, and P. Ho, "Thermo-mechanical reliability of 3-D ICs containing through silicon vias," in *Proc. ECTC*, 2009.

[9] M. Jung, J. Mitra, D. Z. Pan, and S. K. Lim, "TSV stress-aware full-chip mechanical reliability analysis and optimization for 3D IC," in *Proc. DAC*, 2011.

[10] M. Jung, D. Z. Pan, and S. K. Lim, "Chip/package co-analysis of thermo-mechanical stress and reliability in TSV-based 3D ICs," in *Proc. DAC*, 2012.

[11] J. Mitra, M. Jung, S.-K. Ryu, R. Huang, S. K. Lim, and D. Z. Pan, "A fast simulation framework for full-chip thermo-mechanical stress and reliability analysis of through-silicon-via based 3D ICs," in *Proc. ECTC*, 2011.

[12] C. Ko et al., "Structural design, process, and reliability of a wafer-level 3D integration scheme with Cu TSVs based on micro-bump/adhesive hybrid wafer bonding," in *Proc. ECTC*, 2012.

[13] Y. Chen, E. Kursun, D. Motschman, C. Johnson, and Y. Xie, "Analysis and mitigation of lateral thermal blockage effect of through-silicon-via in 3D IC designs," in *Proc. ISLPED*, 2011.

[14] N. Muskhelishvili, *Some Basic Problems of the Mathematical Theory of Elasticity*. Moscow: Nauka, 1966.

[15] S. Marella, S. Kumar, and S. Sapatnekar, "A holistic analysis of circuit timing variations in 3D-ICs with thermal and TSV-induced stress considerations," in *Proc. ICCAD*, 2012.

[16] COMSOL in *www.comsol.com*.

[17] G. Van der Plas et al., "Design issues and considerations for low-cost 3D TSV IC technology," in *Proc. ISSCC*, 2010.

[18] L. Yu, W.-Y. Chang, K. Zuo, J. Wang, D. Yu, and D. Boning, "Methodology for analysis of TSV stress induced transistor variation and circuit performance," in *Proc. ISQED*, 2012.

diminishes as the TSV pitch increases, linear superposition method and proposed framework achieve similar accuracy under the pitch of 30um. We also compare the run time of two methods, and find the proposed framework only takes around 1% additional run time compared to the linear superposition method. More results including the comparison on von Mises stress and comparison on the TSV structure with SiO2 liner are provided in Appendix A.2.

5.2 Validation: A Placement of Five TSVs

We also validate the proposed framework in a placement containing five TSVs shown as Figure 5. The minimal pitch is 10um in this placement. In this experiment, we also define monitored region and critical region. Since the placement has been changed, we slightly change the definition of monitored region in Section 5.1, and use a square region with dimension of 60um as the monitored region. We follow the previous definition of critical region, and the radius of each critical region is 3.3um. Figure 6 shows the comparison on stress component σ_{xx} obtained by linear superposition method and proposed framework. Figure 6(a) shows that linear superposition method leads to a stress error as high as 60MPa. While Figure 6(b) shows that the proposed framework generally ensures stress error within 25MPa. Table 5 shows the quantitative comparison on stress error obtained by the linear superposition method and proposed framework for this placement. We can see that the proposed framework significantly improves the accuracy of linear superposition method. For example, under the 50MPa threshold, the proposed framework achieves an average error rate of 12.6% and 8.22% respectively for the stress component σ_{xx} and von Mises stress in the critical region, while the linear superposition only achieves an error rate of 23.0% and 15.1% under the same circumstance. Though we use von Mises stress as a reliability metric here, since our proposed framework can achieve accurate stress tensor, it is also applicable to other reliability metric like maximum tensile stress. In the experiment, the proposed framework takes 4% additional run time compared to the linear superposition method.

6. Conclusions

In this paper, we have proposed the concept of interactive stress, analyzed its mechanism and characterized this stress using the complex variable method in elasticity. Based on the analytical solution to interactive stress, we have further proposed an accurate semi-analytical

1254

A. Appendix

A.1 Characteristics of Interactive Stress

1. The interactive stress between the aggressive and victim TSV has little influence on the stress distribution around aggressive TSV.

It is because interactive stress originates from the boundary of the victim, and has a similar stress level as the ideal stress distribution in that region. Suppose the stress level around the aggressive is N, according to [9], the level of ideal stress distribution around the victim will roughly be $N(R'/d)^2$, d is the pitch between TSVs, R' is the radius of TSV, $R'/d \leq 0.5$. Hence, the interactive stress has an intensity of $\alpha N (R'/d)^2$ around the victim. Since the interactive stress also decays with similar speed in the substrate, it will roughly decay to an negligible intensity of $\alpha N (R'/d)^4$ around the aggressive and thus have little influence on the stress distribution around the aggressive.

2. The interactive stress induced by the interaction of a TSV pair is nearly irrelevant to the existence of other TSVs nearby the pair.

Similar to the previous analysis, suppose the stress level around the aggressive is N, the interactive stress around the victim will be $\alpha N (R'/d)^2$. Suppose the distance between the victim TSV to the third TSV is also d, then the level of interactive stress around the third TSV will be $\alpha N (R'/d)^4$. Due to the mechanical property mismatch between the third TSV and substrate, the interactive stress will be changed around the third TSV, but the changes will still be in the similar level of $\alpha N (R'/d)^4$ and thus is negligible. Please note the interaction between victim TSV and the third TSV also induces interactive stress, but that is not a portion of interactive stress induced by the interaction between the original TSV pair.

A.2 Validation: More Results for the Case of Two TSVs

We employ von Mises yield criterion as a metric to access the reliability of 3D IC. Von Mises yield criterion determines whether a material starts to yield under a certain stress load. It can be calculated via

$$\sigma_v = \sqrt{\begin{array}{c}\frac{1}{2}(\sigma_{xx}-\sigma_{yy})^2 + \frac{1}{2}(\sigma_{yy}-\sigma_{zz})^2 + \frac{1}{2}(\sigma_{zz}-\sigma_{xx})^2 \\ +3(\sigma_{xy}^2 + \sigma_{yz}^2 + \sigma_{zx}^2)\end{array}}$$

Where σ_{xx}, σ_{xy}, σ_{xz}, σ_{yy}, σ_{yz} and σ_{zz} are stress components defined in Section 3.1. Please note that although von Mises yield criterion is used here, since our proposed framework can accurately calculate stress tensor, it is also applicable to other reliability metric like maximum tensile stress which is the maximum eigenvalue of stress tensor.

The comparison of von Mises stress obtained by linear superposition method and the proposed framework for the placement of two TSVs (Section 5.1) is given in Table 3. In this table, the definition of monitored region (column 3 - 7), critical region (column 8 - 9) and threshold is the same as those in Section 5.1. Table 3 shows that our proposed framework improves the accuracy of linear superposition method. For example, it improves the average error rate from 24.3% to 10.4% for the pitch of 8um case in the critical region under the threshold 50MPa.

We have demonstrated the performance of the proposed framework for TSV structure with BCB liner. As for SiO2 liner, we also compare the performance of linear superposition method and proposed framework for the placement of two TSVs, and list the comparison on stress component σ_{xx} and von Mises stress respectively in Table 4 and 5. The definition of monitored region (column 3 - 7), critical region (column 8 - 9) and threshold in these tables also follow those defined in Section 5.1. These tables show that linear superposition method can achieve acceptable accuracy. However, the proposed framework still improves the accuracy of linear superposition method. For example, it improves the average error of von Mises stress within critical region from 22.4MPa to 14.1MPa under the threshold of 50MPa in the case of 8um pitch.

	d (um)	Avg. Error (MPa)	Threshold 10MPa		Threshold 50MPa		Threshold 50MPa Critical Region	
			Avg. Error (MPa)	Avg. Error Rate(%)	Avg. Error (MPa)	Avg. Error Rate(%)	Avg. Error (MPa)	Avg. Error Rate(%)
LS	8	3.43	4.54	7.80	9.26	8.98	28.2	24.3
	9	2.62	3.41	5.95	7.15	6.33	20.6	15.6
	10	2.14	2.74	4.73	6.20	5.05	16.5	10.9
	11	1.81	2.29	3.87	5.42	4.21	14.3	8.25
	12	1.60	2.00	3.26	4.87	3.62	13.3	6.64
	18	0.98	1.14	1.64	2.85	1.83	10.8	3.70
	30	0.70	0.76	0.93	2.08	1.05	10.6	3.43
PF	8	2.22	2.91	3.98	6.73	5.59	17.3	10.4
	9	1.65	2.13	2.82	5.22	4.06	14.1	7.28
	10	1.34	1.73	2.16	4.56	3.22	12.5	5.69
	11	1.16	1.48	1.74	4.05	2.70	11.8	4.87
	12	1.06	1.34	1.46	3.76	2.36	11.6	4.42
	18	0.75	0.89	0.89	2.51	1.44	10.6	3.39
	30	0.64	0.70	0.77	2.00	0.95	10.6	3.42

Table 3. Comparison on the error of von Mises stress obtained by linear superposition and proposed framework for the placement of two TSVs. LS: linear superposition; PF: proposed framework.

	d (um)	Avg. Error (MPa)	Threshold 10MPa		Threshold 50MPa		Threshold 50MPa Critical Region	
			Avg. Error (MPa)	Avg. Error Rate(%)	Avg. Error (MPa)	Avg. Error Rate(%)	Avg. Error (MPa)	Avg. Error Rate(%)
LS	8	2.15	3.10	7.27	8.69	7.07	27.3	21.6
	9	1.75	2.55	5.91	7.12	5.63	21.7	17.0
	10	1.47	2.14	4.89	5.97	4.61	18.0	13.9
	11	1.26	1.84	4.13	5.12	3.88	15.6	11.7
	12	1.10	1.60	3.54	4.45	3.33	13.7	9.97
	18	0.62	0.97	1.57	2.58	1.97	8.66	5.68
	30	0.49	0.70	1.18	2.07	1.47	7.76	5.51
PF	8	1.99	3.15	6.77	8.80	7.32	14.6	10.1
	9	1.50	2.34	5.16	6.56	5.23	11.8	7.94
	10	1.17	1.81	4.05	5.07	3.94	10.0	6.64
	11	0.97	1.47	3.26	4.15	3.13	9.37	5.96
	12	0.81	1.23	2.68	3.43	2.56	8.57	5.40
	18	0.48	0.75	1.25	1.97	1.39	7.37	4.38
	30	0.47	0.63	1.11	1.88	1.27	7.20	4.94

Table 4. SiO2 liner: comparison on the error of stress component σ_{xx} obtained by linear superposition and proposed framework for the placement of two TSVs. LS: linear superposition; PF: proposed framework.

	d (um)	Avg. Error (MPa)	Threshold 10MPa		Threshold 50MPa		Threshold 50MPa Critical Region	
			Avg. Error (MPa)	Avg. Error Rate(%)	Avg. Error (MPa)	Avg. Error Rate(%)	Avg. Error (MPa)	Avg. Error Rate(%)
LS	8	2.19	2.31	3.03	4.51	3.29	22.4	12.5
	9	1.70	1.78	2.38	3.47	2.48	16.7	9.43
	10	1.38	1.43	1.92	2.84	1.93	13.1	7.16
	11	1.18	1.21	1.58	2.49	1.57	11.0	5.53
	12	1.03	1.06	1.33	2.30	1.36	9.48	4.34
	18	0.70	0.71	0.71	1.56	0.73	6.74	2.02
	30	0.66	0.68	0.69	1.25	0.63	6.40	1.82
PF	8	1.91	2.02	2.54	3.91	2.33	14.1	5.07
	9	1.51	1.58	1.93	3.15	1.87	11.4	3.94
	10	1.25	1.30	1.53	2.67	1.55	9.45	3.16
	11	1.09	1.12	1.25	2.42	1.35	8.47	2.66
	12	0.96	0.99	1.05	2.24	1.19	7.60	2.27
	18	0.68	0.68	0.57	1.58	0.75	6.49	1.86
	30	0.64	0.65	0.65	1.21	0.57	6.38	1.81

Table 5. SiO2 liner: comparison on the error of von Mises stress obtained by linear superposition and proposed framework for the placement of two TSVs. LS: linear superposition; PF: proposed framework.

Case #	1	2	3	4	5	6	7
TSV #	100	500	1000	100	100	100	100
TSV Density ($\times 10^{-2} \cdot \mu m^{-2}$)	1	1	1	0.69	0.25	1	1
Simulation Point #	0.5M	0.5M	0.5M	0.5M	0.5M	1M	2M
AR (%)	12	13	14	7.9	3.9	13	13

Table 6. Run time of the proposed framework. AR = additional run time of proposed framework / run time of linear superposition.

A.3 Run Time of the Proposed Framework

In order to examine the scalability of proposed framework, we run it on several placements which contain large amount of TSVs, and list the run time in Table 6. Case1, 2 and 3 show that AR (ratio of additional run time of proposed framework to the run time of linear superposition) nearly remains constant when the amount of TSVs increase. It accords with the theoretical analysis that both the run time of proposed framework and linear superposition method is irrelevant to the amount of TSVs. Case 1, 4 and 5 show that AR increases when TSV integration density increases. However, since the TSV integration density faces a upper bound in real application, the increase of run time due to the increase of TSV integration density also faces a upper bound. For example, in a very dense square TSV array with 10um pitch [18], the TSV integration density is only $1.0 \times 10^{-2} um^{-2}$. Hence, the TSV integration density in Case 1 approximates the upper bound but only makes AR as 12%. Case 1, 6 and 7 show that AR nearly remains constant when the amount of simulation points increases. It also accords with the previous theoretical analysis that both the run time of proposed framework and linear superposition method is linear with the amount of simulation points.

A.4 Constants & Functions

E_c, E_l and E_s, v_c, v_l and v_s, α_c, α_l and α_s are respectively Young's modulus, Poisson's ratio and CTE of corresponding materials, where subscript c, l and s respectively represent copper, BCB and silicon; T is the thermal load; R is the radius of TSV body; R' is the radius of TSV (including liner); $k = R/R'$.

$$K = -TR'^2 \left[\left(\frac{1-v_c}{E_c} + \frac{1+v_l}{E_l} \right)(\alpha_l - \alpha_s) + \left(\frac{1-v_c}{E_c} + \frac{1+v_l}{E_l} \right)(\alpha_c - \alpha_l) k^2 \right.$$
$$\left. - \left(\frac{1-v_c}{E_c} - \frac{1-v_l}{E_l} \right)(\alpha_c - \alpha_s) k^2 \right] / \left[\left(\frac{1-v_c}{E_c} + \frac{1+v_l}{E_l} \right) \right.$$
$$\left. \left(\frac{1+v_s}{E_s} + \frac{1-v_l}{E_l} \right) - \left(\frac{1-v_c}{E_c} - \frac{1-v_l}{E_l} \right) \left(\frac{1+v_s}{E_s} - \frac{1+v_l}{E_l} \right) k^2 \right]$$

$$a_1 = \left(1 + \frac{E_c}{E_l} \frac{3-v_l}{1+v_c} \right) / \left(1 - \frac{E_c}{E_l} \frac{1+v_l}{1+v_c} \right)$$

$$a_2 = \left(1 - \frac{E_c}{E_l} \frac{3-v_l}{3-v_c} \right) / \left(1 + \frac{E_c}{E_l} \frac{1+v_l}{3-v_c} \right)$$

$$G_1(m) = \frac{16 \left(k^2-1\right)^2}{E_l^2} + \left\{ \frac{4a_1 k^{2m+2}-4}{E_l} + \left(\frac{1+v_l}{E_l} - \frac{1+v_s}{E_s} \right) \right.$$
$$\left. \left[a_1 a_2 k^4 - a_1 k^{2m+2} - a_2 k^{2-2m} + \left(1-k^2\right)\left(m^2-1\right) + 1 \right] \right\}$$
$$\left\{ \frac{4a_2 k^{2-2m}-4}{E_l'} + \left(\frac{1+v_l}{E_l} + \frac{3-v_s}{E_s} \right) \left[a_1 a_2 k^4 - a_1 k^{2m+2} \right. \right.$$
$$\left. \left. - a_2 k^{2-2m} + \left(1-k^2\right)^2\left(m^2-1\right) + 1 \right] \right\} / \left(m^2-1\right)$$

$$G_2(m) = \frac{16}{E_l E_s} \left(1-k^2\right) \left[a_1 a_2 k^4 - a_1 k^{2m+2} - a_2 k^{2-2m} + 1 + \right.$$
$$\left. + (1-k^2)^2(m^2-1) \right]$$

$$G_3(m) = \frac{16 \left(k^2-1\right)^2}{E_l^2} + \left\{ \frac{4a_1 k^{2-2m}-4}{E_l} + \left(\frac{1+v_l}{E_l} - \frac{1+v_s}{E_s} \right) \right.$$
$$\left. \left[a_1 a_2 k^4 - a_1 k^{2-2m} - a_2 k^{2m+2} + \left(1-k^2\right)^2\left(m^2-1\right) + 1 \right] \right\}$$
$$\left\{ \frac{4a_2 k^{2m+2}-4}{E_l} + \left(\frac{1+v_l}{E_l} - \frac{1+v_s}{E_s} \right) \left[a_1 a_2 k^4 - a_1 k^{2-2m} \right. \right.$$
$$\left. \left. - a_2 k^{2m+2} + \left(1-k^2\right)^2\left(m^2-1\right) + 1 \right] \right\} / \left(m^2-1\right)$$

$$F(m) = \begin{cases} G_2(m)/G_1(m), & \text{if } m \le -2 \\ G_3(m)/G_1(-m), & \text{if } m \ge 2 \end{cases}$$

$$F_1(m) = a_1 a_2 k^4 - a_1 k^{2m+2} - a_2 k^{2-2m} + 1 + \left(1-k^2\right)^2\left(m^2-1\right)$$

$$F_2(m) = \left(1-k^2\right)(m+1)F(m) + \left(a_2 k^{2-2m}-1\right)(F(-m)+m+1)$$

$$F_3(m) = \left(1-k^2\right)(m+1)(F(m)-m+1) + \left(a_1 k^{2-2m}-1\right)F(-m)$$

$$H(m) = \begin{cases} F_2(m)/F_1(m), & \text{if } m \le -2 \\ F_3(m)/F_1(-m), & \text{if } m \ge 2 \end{cases}$$

$$h_{11}(m) = (1-a_2)(2-m)H(m)$$

$$h_{12}(m) = (m-1) + (a_1-1)k^{2-2m}H(-m) + (a_2-1)k^2(m-1)H(m)$$

$$h_{15}(m) = (1-a_2)(2+m)H(m)$$

$$h_{17}(m) = (1-a_2)mH(m)$$

$$h_{13}(m) = h_{14}(m) = h_{16}(m) = h_{18}(m) = 0$$

$$h_{21}(m) = (2-m)H(m)$$

$$h_{22}(m) = (m-1) + (1-m)k^2 H(m) + a_1 k^{2-2m}H(-m)$$

$$h_{23}(m) = (2+m)H(-m)$$

$$h_{24}(m) = (m+1)k^2 H(-m) + a_2 k^{2m+2}H(m)$$

$$h_{25}(m) = mH(m)$$

$$h_{26}(m) = mH(-m)$$

$$h_{27}(m) = (2+m)H(m)$$

$$h_{28}(m) = (2-m)H(-m)$$

$$h_{33}(m) = -(2+m)F(m)$$

$$h_{34}(m) = F(-m) - (m+1)F(m)$$

$$h_{36}(m) = (m-2)F(m)$$

$$h_{38}(m) = -mF(m)$$

$$h_{31}(m) = h_{32}(m) = h_{35}(m) = h_{37}(m) = 0$$

Speeding up Computation of the max/min of a set of Gaussians for Statistical Timing Analysis and Optimization

Vimitha Kuruvilla[1], Debjit Sinha, Jeff Piaget, Chandu Visweswariah and Nitin Chandrachoodan[2]

[1]IBM Systems and Technology Group, Bangalore, India

IBM Systems and Technology Group, Hopewell Junction, USA

[2]EE, Indian Institute of Technology, Chennai, India

vimitha.k@in.ibm.com, {dsinha, piaget, chandu}@us.ibm.com, nitin@ee.iitm.ac.in

ABSTRACT

Statistical static timing analysis (SSTA) involves computation of maximum (max) and minimum (min) of Gaussian random variables. Typically, the max or min of a set of Gaussians is performed iteratively in a pair-wise fashion, wherein the result of each pair-wise max or min operation is approximated to a Gaussian by matching moments of the true result obtained using Clark's approach [1]. The approximation error in the final result is thus a function of the order in which the pair-wise operations are performed.

In this paper, we analyze known "run-time expensive" ordering techniques that attempt to reduce this error in the context of SSTA and SSTA driven optimization. We propose new techniques to speeding up the computation of the max/min of a set of Gaussians by special handling of prevalent "zero error" cases. Two new methods are presented using these techniques that provide more than 60% run-time savings (3X speed-up) in max/min operations. This translates to an overall run-time improvement of 2-17% for a single SSTA run and an improvement of up to 8 hours (55%) in an SSTA driven optimization run.

Categories and Subject Descriptors

B.8.2 [**Hardware**]: Performance and Reliability—
Performance Analysis and Design Aids

General Terms

Algorithms, Performance

Keywords

Statistical timing, variability

1. INTRODUCTION

Advances in technology leading to reduced transistor sizes and increased logic density on a chip has brought focus on process variability and its impact on robustness of a design. Statistical Static Timing Analysis (SSTA) has emerged as a key approach to account for such variability [2–4]. SSTA models timing quantities (like delay, arrival time/AT, required arrival time/RAT, slew) as Gaussian distributions that are functions of sources of variability [5]. An important step in block based statistical timing analysis is the propagation of these distributions along the timing graph model of a given circuit. Timing propagation involves fundamental mathematical operations including addition, subtraction, maximum, and minimum. Unlike *add/sub* operations, finding the *max/min* of a set of Gaussians is non-linear and non-trivial. The *max* of a set of Gaussians is not necessarily Gaussian.[1] However, a Gaussian result is desired for timing propagation. Analytical approaches to computing the moments of the *max* of two Gaussians have been presented by Clark [1] and Cain [6]. Subsequently, using moment matching, the result of the *max* of two Gaussians can be approximated to a Gaussian. Multiple flavors of this technique have been suggested for SSTA [5, 7]. This approach is performed iteratively as pair-wise *max* operations while computing the *max* of more than two Gaussians. Mathematically,

$$max(X_1, \ldots, X_n) \approx max(X_1, max(X_2, \ldots, max(X_{n-1}, X_n))).$$

This operation can be represented as a binary tree with the n Gaussians $X_1, \ldots, X_n$ denoted by the leaf nodes of a tree, referred to as a Max Binary Tree (MBT) in [8]. Each internal node represents the result of a *max* operation on its children and hence the root node is the final result. The error in approximating this final resultant to a Gaussian distribution depends on how the MBT was constructed, that is, the order of each pair-wise *max* operation.

In [8], Sinha *et al.* suggest a method to quantify the error introduced in the moment matching based approximation of the result of the *max* of two Gaussians to a Gaussian. Using this metric, they propose algorithms to compute the *max* of n Gaussians such that the error in the final result is minimized. These methods use heuristics and are generic to all applications. Our work focuses on how the *max* operation can be improved further, particularly in the context of SSTA. Although a simpler variant of the greedy MBT method suggested in [8] was employed in an industrial tools framework for yield optimization through statistical timing [9], the run-time and memory impact of the approach was not documented. The complexity of this method is $\Theta(n^2 lg\ n)$ indicating significant overhead for large n, where n is the number of Gaussian operands for the *max* operation. Design parameters like number of fan-ins, fan-outs, and clocks indirectly impact the number of operands (n) encountered for various *max/min* operations during statistical timing analysis of the design. Apart from the *max/min* operations on ATs and RATs for propagat-

---

Permission to make digital or hard copies of all or part of this work for personal or classroom use is granted without fee provided that copies are not made or distributed for profit or commercial advantage and that copies bear this notice and the full citation on the first page. To copy otherwise, to republish, to post on servers or to redistribute to lists, requires prior specific permission and/or a fee.

DAC'13, May 29 - June 07 2013, Austin, TX, USA.

---

[1]The *min* operation is similar to *max* in all respects and for the rest of this paper all discussions on *max* operations would implicitly imply the same for *min* operations as well.

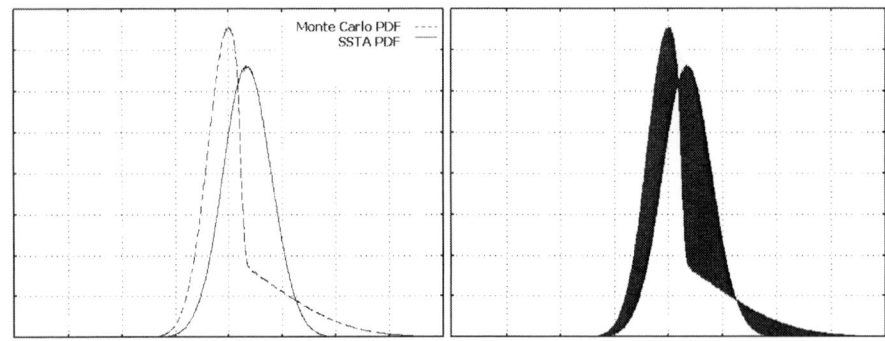

Figure 1: Error in approximating Z as Z_G is represented by the area of the shaded region

ing these quantities along the timing graph, there are several other instances where these operations are used in SSTA. Calculation of the worst chip slack is one such operation where *min* of Gaussian slacks corresponding to various timing tests in the design is computed. This typically involves a *min* operation involving thousands of Gaussians and the operation itself is performed iteratively at various stages of timing optimization. While an intelligent pair-wise *max/min* ordering is desired for accuracy, the run-time overheads of Greedy MBT like approaches are found to be significant in some cases based on experimental results on industrial designs.

The objectives of our work are:

- To understand the general pattern of Gaussian distributions typically encountered in SSTA and optimize the order of pair-wise *max* operations based on the nature of such workloads.

- To improve the run-time and memory overheads of existing *max* computation approaches with no loss of accuracy of the result.

Although techniques presented are in the context of SSTA, they are expected to provide similar benefits when applied in other domains where the Gaussians exhibit similar behavior.

2. APPROXIMATION ERROR ANALYSIS

Consider a pair of random variables X and Y with Gaussian distributions $N(\mu_X, \sigma_X^2)$ and $N(\mu_Y, \sigma_Y^2)$, respectively. μ_X, μ_Y denote their respective means and σ_X, σ_Y denote their respective standard deviations. ρ denotes their correlation coefficient. The operation $Z \triangleq max(X, Y)$ results in a non-Gaussian distribution (except for special cases listed later). The mean μ_Z and standard deviation σ_Z of Z can be computed as [1]:

$$\mu_Z = \mu_X \Phi(\alpha) + \mu_Y \Phi(-\alpha) + a\phi(\alpha) \tag{1}$$

$$\sigma_Z^2 = (\sigma_X^2 + \mu_X^2)\Phi(\alpha) + (\sigma_Y^2 + \mu_Y^2)\Phi(-\alpha) + (\mu_X + \mu_Y)a\phi(\alpha) - \mu_Z^2 \tag{2}$$

where $\phi(x)$, $\Phi(y)$, θ and α are defined as follows:

$$\phi(x) \triangleq \frac{1}{\sqrt{2\pi}} exp(-\frac{x^2}{2}) \tag{3}$$

$$\Phi(y) \triangleq \int_{-\infty}^{y} \phi(x)dx \tag{4}$$

$$\theta \triangleq (\sigma_X^2 + \sigma_Y^2 - 2\rho\sigma_X\sigma_Y)^{1/2} \tag{5}$$

$$\alpha \triangleq \frac{\mu_X - \mu_Y}{\theta}. \tag{6}$$

Using the method of moment matching described in [6], Z can be approximated to a Gaussian $Z_G = N(\mu_Z, \sigma_Z^2)$ by matching the first and second order moments and ignoring the higher order moments. The error in approximating Z as Z_G can be pre-computed and expressed as a function of the bounded parameters α, ρ and σ_Y' ($\triangleq \frac{\sigma_Y}{\sigma_X}$) [8]. Given any two Gaussian distributions, the error introduced by approximation can thus be obtained by a simple look-up operation of a table that stores these pre-computed values. The error represents the non-overlapping region between the probability density functions (pdfs) of Z and Z_G. It can thus take any value in the range [0, 2]. The shaded region in Figure 1 illustrates this error.

We analyze the conditions for which this error becomes zero. A zero error implies that the result of the *max* operation is a true Gaussian which needs no approximation. The zero error conditions are expressed as a function of α and θ. Based on the findings in [8], we categorize the conditions leading to zero error into two cases:

- **Case I:** $|\alpha| > k$. α is a measure of the dominance of Gaussian X over Y. A large α indicates domination of X over Y and the result of the *max* operation in this case is the dominant input X, with little to no error. The error is bound by $\Phi(k)$. One can choose an appropriate value for k based on the error tolerance. For our experiments we choose $k = 4$ which bounds the error to 0.006% and assumes zero error in computing the *max/min* of Gaussians pairs with $|\alpha| > 4$.

Consider the following known results:

$$\phi(\alpha) \approx 0 \quad for \; |\alpha| > 4 \tag{7}$$

$$\Phi(\alpha) \approx 0 \quad for \; \alpha < -4 \tag{8}$$

$$\Phi(\alpha) \approx 1 \quad for \; \alpha > 4 \tag{9}$$

Applying (7)-(9) to (1)-(2), we get $\mu_Z = \mu_X$ or μ_Y and $\sigma_Z = \sigma_X$ or σ_Y. Thus the result of the *max* operation is the dominant input Gaussian for sufficiently large k.

Such cases of dominance are common in SSTA. Consider the case of arrival times (ATs) of two paths merging at a point. Dominance of AT on one path means that one AT is consistently larger than the other AT even considering variability.

- **Case II:** $\theta = 0$. Analyzing (5) there are two cases that lead to this condition.

  1. $\sigma_X = 0$ and $\sigma_Y = 0$: This is when X and Y are deterministic numbers, in which case the result is

the output of deterministic *max* operation on X and Y.

2. $\sigma_X = \sigma_Y$ and $\rho=1$: This is when X and Y are either equal Gaussians or their means differ by a constant. The independently random part in such distributions will be zero since any non-zero random part automatically makes ρ not equal to 1. This scenario is possible in the case of very symmetric circuits (like arithmetic or bit slice circuits). The result of the *max* operation in such cases is the input Gaussian with the larger mean. It can be found using a trivial deterministic *max* operation on the mean of the inputs.

From the above analysis, we conclude that the result of the *max* operation on a pair with zero error can be approximated by the larger of the input Gaussians, determined by a deterministic *max* operation on the mean of the two input Gaussians. Thus a relatively more expensive statistical *max* operation of two Gaussians is avoided under these conditions. We make use of this result in the new approaches suggested in the next section.

To understand the typical value of errors for Gaussian pairs encountered in SSTA, we choose a set of industrial designs in 45 and 65 nm technology. As suggested in [8] we use a precomputed 3-way look-up table that stores the approximation error, with α, ρ and σ_Y' as the indices. We measure the total number of pair-wise *max/min* Gaussian operations in a single statistical timing run, and categorize the errors into three buckets from the possible range of 0 to 2. The findings are shown in Table 1 as a percentage of the total *max/min* operations. In all the designs, more than 85% of the Gaussian pairs had zero error. Based on this finding, we propose ways to improve the *max* operation on a set of Gaussians by special handling of the zero error pairs.

Table 1: Error distribution for 5 designs *A-E*

Err	A	B	C	D	E
0	87.9%	92.1%	90.2%	91.2%	85.3%
(0,0.1]	11.0%	6.9%	8.8%	7.7%	13.7%
(0.1,2]	1.1%	1.0%	1.0%	1.1%	1.0%

Further in Table 2, we show the split-up of zero error among the two cases discussed earlier. The dominant Gaussian case of $|\alpha| > 4$ is seen to be the cause of zero errors in over 95% of the total zero errors. We use this finding as a motivation to propose the Dominant MBT approach for computing *max* in the next section.

Table 2: Case wise break-up of zero-error pairs

Case	A	B	C	D	E
$|\alpha| > 4$	99.8%	99.7%	99.8%	99.8%	95.1%
$\theta = 0$	0.2%	0.3%	0.2%	0.2%	4.9%

3. TECHNIQUES FOR MAX AND MIN OPERATION

We propose two methods to efficiently perform the *max* operation on a set of Gaussians: $S = \{X_1, \ldots, X_n\}$, based on observations from the previous section.

3.1 Smart Greedy MBT

This method is an improvement of the Greedy MBT method suggested in [8], with special handling of zero error Gaussian pairs. In Greedy MBT, each possible Gaussian pair (X_i, X_j) that can be formed among the Gaussians in the set S is picked, its approximation error $\Xi_{X_i X_j}$ found, and the tuple $\langle i, j \rangle$ with the error information is added to a heap H. H is sorted based on the error and the pair at the top of the heap (the one with minimum error) is worked on at each stage of the *max* operation. The resultant Gaussian X_{index} from every intermediate *max* operation is added to the set S of remaining Gaussians. The process of error computation is repeated for this new entry with all other elements in the set and the new tuples with error information are added to the heap. This process repeats till the final result is computed. The size of the heap thus formed for a set of n Gaussians is $\Theta(n^2)$, which causes the overall complexity of the approach to be $\Theta(n^2 lg\, n)$.

In Smart Greedy MBT, we perform an additional check each time an error is computed for a pair, to determine if it is zero. If the error $\Xi_{X_i X_j}$ is zero, Gaussians X_i, X_j are not further used for error pair computation since they cannot yield another pair with smaller error (zero being the minimum error by definition). X_i, X_j are removed from a set P (initialized with the original input set S) which is used to track Gaussians that still need to be used for error pair computation. Each element in set P indicates that a zero error pair has not yet been found for that element with any other element in set S.

Algorithm: Smart Greedy MBT
Input: Set of n Gaussians $S = \{X_1, \ldots, X_n\}$
Output: Gaussian $X_G^{Smart} \approx \max(X_1, \ldots, X_n)$

```
1    num ← index ← n
2    P ← S
3    for i in (1, ..., n − 1)
4        for j in (i + 1, ..., n)
5            if (Xi ∈ P ∧ Xj ∈ P)
6                Lookup error Ξ_XiXj
7                Min-Heap-Insert(H, ⟨i, j⟩)
                    (using Ξ_XiXj as comparison key)
8                if (Ξ_XiXj = 0)
9                    P ← P − {Xi, Xj}
10   while (num ≠ 1)
11       ⟨i, j⟩ ← Min-Heap-Extract(H)
12       if (Xi ∈ S ∧ Xj ∈ S)
13           num ← num − 1
14           index ← index + 1
15           X_index ← max(Xi, Xj)
16           S ← S − {Xi, Xj}
17           S ← S ∪ {X_index}
18           if (Xi ∈ P)
19               P ← P − {Xi, Xj}
20               P ← P ∪ {X_index}
21               foreach Xk ∈ P : k ≠ index
22                   Lookup error Ξ_X_index Xk
23                   Min-Heap-Insert(H, ⟨index, k⟩)
                        (using Ξ_X_index Xk as comparison key)
24                   if (Ξ_X_index Xk = 0)
25                       P ← P − {X_index, Xk}
26                       break from foreach loop (step 21)
27   return X_G^Smart ← only element in S
```

Figure 2: Algorithm for Smart Greedy MBT

Figure 2 shows the pseudo-code for this approach. In steps *3-9*, a heap H is initialized. For any tuple $\langle i, j \rangle$ in the heap with zero error, it is guaranteed that no other tuple $\langle i, k \rangle : k \neq j$ or $\langle k, j \rangle : k \neq i$ exists in the heap with zero error. The current minimum error tuple is then extracted iteratively from H in steps *10-11* until the final result is obtained in step *27*. If a *max* operation has not yet been performed on the corresponding Gaussians (indicated by their presence in set S as shown in step *12*), a *max* is performed and the intermediate result is added to the set S after removing the operands from the same set (steps *14-17*). If the operands exist in set P, they are removed from that set as well. It is sufficient to check the presence of only one operand in P (step *18*) since either both would exist (case of non zero error) or none would exist (zero error case) in P. Iteratively, new tuples of the calculated *max* with remaining elements in P are added to H in steps *21-26*. Similar to the heap initialization steps, once a zero error pair is found for X_{index} with any other element in P, no additional tuples are inserted in H since the minimum (zero) error pair for X_{index} has already been found.

Given that the majority of computed errors are zero (based on results from Table 1), the heap size can be drastically reduced using the proposed approach in comparison to the Greedy MBT approach [8], resulting in both run-time and memory improvements. In the best case when all Gaussians can form at least one zero error pair, the heap size reduces to $\Theta(\frac{n}{2})$. The new method retains the accuracy of the original Greedy MBT algorithm since only pairs that have already resulted in the minimum (zero) error are removed from the set.

3.2 Dominant MBT

In this method, we target further reduction in heap size of the previous approach, primarily through two improvements:

- We make use of the finding that the *max* of a Gaussian pair with zero error is the dominant of the two input Gaussians. Rather than adding this pair to the heap and then later retrieving it for *max* computation, the pair is operated on immediately. The operation involves removing the non-dominant Gaussian from the set S of Gaussians whose *max* needs to be found. This further reduces the size of the heap by the number of zero-error pairs formed from the set.

- The key to achieving drastic reduction in heap size is to find the zero-error pairs early in the iteration. To expedite this process, whenever a zero error is found, the dominant Gaussian is used in the next iteration of error computation from among the set P of Gaussians. The dominant Gaussian is more likely to yield further zero errors and thus, this out-of-order iterative method can hasten the error computation process.

A pseudo-code of this algorithm is shown in Figure 3.

Algorithm: Dominant MBT
Input: Set of n Gaussians $S = \{X_1, \ldots, X_n\}$
Output: Gaussian $X_G^{Dominant} \approx \max(X_1, \ldots, X_n)$

1   Heap $H \leftarrow$ Find-All-Error-Pairs(S)
2   $S \leftarrow$ Find-Max-Using-Error-Pairs(H, S, n)
3   return $X_G^{Dominant} \leftarrow$ only element in S

Figure 3: Algorithm for Dominant MBT

Algorithm: Find-All-Error-Pairs
Input: Set of n Gaussians $S = \{X_1, \ldots, X_n\}$
Output: Heap H containing Gaussian pair tuples

1   $P \leftarrow S$
2   $nxtIdSet \leftarrow false$
3   $start_{id} \leftarrow id \leftarrow nxtId \leftarrow 1$
4   while $(P \neq \emptyset)$
5   for i in $start_{id}, \ldots, n$
6   if $(i \neq id \wedge X_i \in P)$
7   Lookup $\Xi_{X_i X_{id}}$
8   if $(\Xi_{X_i X_{id}} \neq 0)$
9   Min-Heap-Insert$(H, \langle i, id \rangle)$
       (using $\Xi_{X_i X_{id}}$ as comparison key)
10   else
11       $\langle p, q \rangle \leftarrow$ Get-Dominant(X_i, X_{id})
       // Remove non-dominant
12       $S \leftarrow S - X_q$
13       $P \leftarrow P - X_q$
14   if $(p \neq id)$
15   // Use dominant X_i as next X_{id}
16        $nxtIdSet \leftarrow true$
17        $nxtId \leftarrow i$
18   break from for loop (step 5)
19    $id \leftarrow$ GetNxtId$(P, start_{id}, nxtIdSet, nxtId)$
20   return H

Figure 4: Heap initialization in Dominant MBT

Algorithm: Get-Dominant
Input: Gaussian pair (X_i, X_j) with zero error
Output: Id pair $\langle p, q \rangle$ indicating
       $\langle$Dominant$_{id}$, Non-dominant$_{id}\rangle$

1   if $(X_i.\text{mean}() > X_j.\text{mean}())$
2    $p \leftarrow i; q \leftarrow j$
3   else
4    $p \leftarrow j; q \leftarrow i$
5   return $\langle p, q \rangle$

Figure 5: Finding dominant Gaussian

The algorithm to initialize the heap is shown in Figure 4. This algorithm uses a set P (initialized with set S) to keep track of which Gaussians (tuples) have either been added to the heap or are dominated by other Gaussians in the set. It iteratively looks up the error for Gaussian pairs $\langle i, id \rangle$ (steps *5-7*) where id is an index indicating a potentially dominating Gaussian (initialized to 1). Once a tuple is found with zero error (step *10*), the Gaussian tuple is not inserted in the heap. Instead, the dominated Gaussian is removed from the sets S and P (steps *12-13*), and the index of the dominating Gaussian is preferred for subsequent iterations of error lookup (steps *14-18*). Once the heap is initialized, the *max* of the set S of Gaussians is computed using the algorithm in Figure 7. In this algorithm, minimum error tuples are iteratively extracted from the heap (step *3*). It is expected that the error for any extracted tuple is non zero since zero error pairs are never stored in the heap thereby significantly reducing the heap size. Once the *max* is computed for an extracted tuple (step *6*), the error of the result with other elements in the set are looked up (steps *9-10*). Zero error tuples are handled right away by eliminating the dominated Gaussian and only non zero error tuples are inserted in the heap (steps *11-17*). The result is returned once all tuples are extracted from the heap in step *18*.

1260

Algorithm: GETNXTID
Input: Set P of pending Gaussians for finding error
Index $start_{id}$ - Minimum id in the set P
Flag $nxtIdSet$ indicating if $nxtId$ to be used
Preset index $nxtId$ to use for next iteration
Output: Index id of next Gaussian to compute errors

1	if $(nxtIdSet)$
	// Set id to preset dominant index
2	$id \leftarrow nxtId$
3	$nxtIdSet \leftarrow false$
4	else
5	$P \leftarrow P - \{X_{id}\}$
6	while $(X_{start_{id}} \notin P \wedge start_{id} < n)$
7	$start_{id} \leftarrow start_{id} + 1$
8	$id \leftarrow start_{id}$
9	return id

Figure 6: Finding next id for error pair computation

This algorithm can be further optimized for early identification of the dominant Gaussian. A complete sort of the Gaussian set or even a simple scan operation to pick the minimum and maximum based on the mean of the Gaussians and using these as the starting pair for error computation can potentially hasten the early finding of the dominant Gaussian. We have not currently applied this technique in our method since there were sufficiently large number of dominant pairs in our experiments such that a dominant Gaussian could be found early in the operation.

Algorithm: FIND-MAX-USING-ERROR-PAIRS
Input: Heap H with Gaussian pair tuples
Set S of pending Gaussians
n: Number of Gaussians in original set S
Output: Set S reduced to a single Gaussian

1	$index \leftarrow n$
2	while $(H \neq \emptyset)$
3	$\langle i, j \rangle \leftarrow$ MIN-HEAP-EXTRACT(H)
4	if $(X_i \in S \wedge X_j \in S)$
5	$index \leftarrow index + 1$
6	$X_{index} \leftarrow max(X_i, X_j)$
7	$S \leftarrow S - \{X_i, X_j\}$
8	$S \leftarrow S \bigcup \{X_{index}\}$
9	foreach $X_k \in S : k \neq index$
10	Lookup $\Xi_{X_{index}X_k}$
11	if $(\Xi_{X_{index}X_k} = 0)$
12	$\langle p, q \rangle \leftarrow$ GET-DOMINANT(X_{index}, X_k)
13	$S \leftarrow S - \{X_q\}$
14	if $(p \neq index)$
15	break from foreach loop (step 9)
16	else
17	MIN-HEAP-INSERT$(H, \langle index, k \rangle)$
	(using $\Xi_{X_{index}X_k}$ as comparison key)
18	return S

Figure 7: Max computation loop in Dominant MBT

4. RESULTS

The Smart Greedy MBT and Dominant MBT methods have been implemented and incorporated into an industrial SSTA tools framework. For comparisons, a Greedy MBT approach [8] is used as reference (referred as "Ref. [8]" in the rest of this section). Results are presented for the speed-ups seen in

Table 3: Design parameters

Design	Num. gates	Max n	$\dfrac{\text{Max } n}{\text{Num. gates}} \times 10^3$
A	61K	91	1.5
B	258K	792	**3.1**
C	402K	388	1.0
D	3600K	1223	0.3

the max/min computation functions as well as its impact on overall timing runs. The run-time impact is also measured for iterative timing optimization runs. We choose a set of industrial designs (A-D) from 65nm and 45nm technology for evaluation, as illustrated in Table 3.

Table 4 compares the total time spent in min and max operations during a single SSTA run, with different ordering techniques. It is observed that the proposed Smart Greedy MBT and Dominant MBT techniques achieve significant run-time improvements: more than 60% (3X speed-up) over Ref. [8] for all designs. The maximum improvement obtained is 87% (7.5X speed-up) for the max operation in design B using Dominant MBT. The average improvement is about 70% for Smart Greedy MBT and 75% for Dominant MBT.

Table 5 compares the total time taken for a single SSTA run. The average run-time improvement is observed to be 5% while the best run-time improvement being 17%. The variance in improvement can be attributed to the variance in the time spent in min and max operations as a fraction of the total timing run, shown in the "%Total run-time in max/min" column of Table 5. To explain this variance, we observe two key parameters of the designs as shown in Table 3.

It is evident that the time spent in a max/min operation depends on the size of the the Gaussian set on which the operation is performed, and the benefits of the proposed methods are attractive for cases where the set is large. The size of the largest Gaussian set encountered in min /max operations during an SSTA run is captured as a design parameter "Max n". This value depends on various parameters like number of fanins, fan-outs and clocks in the design. The expected run-time improvements from the proposed methods are also a function of the size of the design since the time spent in a few max/min operations with large n may be amortized over a large number of base timing calculations (delay/slew calculations). Consequently, the expected overall run-time improvement is large when the ratio of Max n to a design size metric is large. As observed from Table 3, Design B with this ratio of 3.1 shows the maximum improvement of 17%, whereas design D with a ratio of 0.3 shows 2% improvement. The speed-up obtained using the Dominant MBT technique is generally expected to be larger than that obtained using the Smart Greedy MBT technique. However, the speed-ups obtained for design D exhibit a small anomaly in Table 5. This may be due to uneven load balancing in the machine while evaluating the run-times for the large design.

Placement and routing aware SSTA driven timing optimization of a circuit is a run-time expensive step in the design closure flow. Timing runs are made incrementally for several iterations in this step, before the design can find an appropriate placement and routing that meets the timing requirement. Table 6 compares the time taken for this step using the various ordering approaches. The proposed methods provide more than 6% improvement on all designs. The improvement is over 8 hours (55%) for design B as shown in Figure 8. The over 2X speed-up obtained on this design for an original run

Table 4: Run-time results and gain for *max/min* operation

Design	Min					Max				
	Run-time (secs)			Speedup(%)		Run-time (secs)			Speedup(%)	
	Ref. [8]	Smart	Dominant	Smart	Dominant	Ref. [8]	Smart	Dominant	Smart	Dominant
A	14.4	5.8	3.9	60	72	17.1	6.4	4.5	63	74
B	234.1	34.3	31.5	**85**	**87**	238.2	35.1	32.3	**85**	**87**
C	321.3	87.4	81.2	73	75	280.3	92.4	75.2	67	73
D	1704.2	468.8	370.9	72	78	1714.2	427.0	370.9	75	78

Table 5: Run-time results and gain for a single SSTA run

Design	SSTA Run-time (hh:mm:ss)			% Total run-time in max/min	Speedup			
					(hh:mm:ss)		(%)	
	Ref. [8]	Smart	Dominant		Smart	Dominant	Smart	Dominant
A	00:08:10	00:07:49	00:07:45	6	00:00:21	00:00:25	4	5
B	00:58:20	00:49:54	00:48:10	14	00:08:26	00:10:10	**14**	**17**
C	04:55:13	04:50:12	04:49:30	3	00:05:01	00:05:43	2	2
D	17:04:02	16:28:18	16:43:30	5	00:35:44	00:20:32	3	2

Table 6: Run-time results and gain for a timing optimization run

Design	Optimization Run-time (hh:mm:ss)			% Total run-time in max/min	Speedup			
					(hh:mm:ss)		(%)	
	Ref. [8]	Smart	Dominant		Smart	Dominant	Smart	Dominant
A	05:00:00	04:42:38	04:39:15	7	00:17:22	00:20:45	6	7
B	15:06:51	06:40:37	06:29:47	61	**08:26:14**	**08:37:04**	**56**	**57**
C	03:51:05	03:26:17	03:23:48	14	00:24:48	00:27:17	11	12

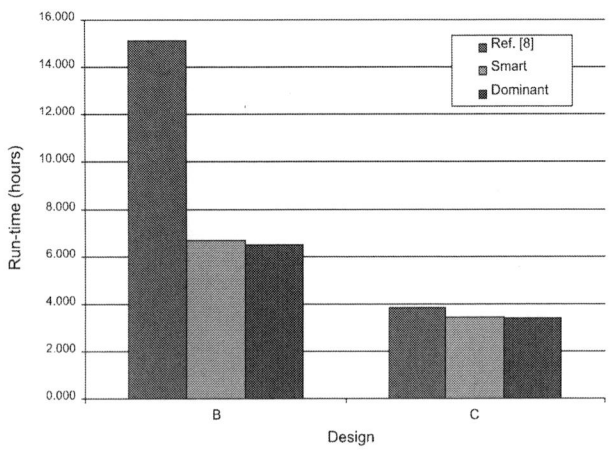

Figure 8: SSTA driven optimization run-time results and gains

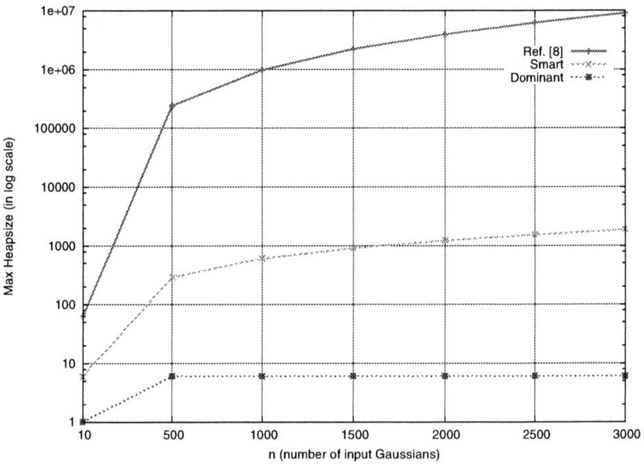

Figure 9: Max Heap-size variation with n

taking over 15 hours is a significant productivity improvement. The variance in improvement across designs for this step is dependent on design parameters as mentioned in Table 5. However, in addition to these design parameters, there are other factors that influence optimization runs given the iterative and incremental nature of timing analysis as compared to a single full timing analysis run. In this case, the proportion of time spent in *max/min* operations can be more than delay and sensitivity computation operations, which are restricted to only parts of the design undergoing changes during optimization. Design B demonstrates this behavior and has a high "%Total run-time in max/min" of 61% in optimization runs compared to 14% in single timing run, resulting in the 57% run-time improvement.

The proposed methods, by construction, match the accuracy of our reference method [8].

Using the two proposed methods, we compute the worst chip slack for each design, which involves a *min* operation on all the Gaussian slack (corresponding to various timing tests) distributions in the design. The resultant slack show no loss of accuracy as compared to [8]. We use this operation to measure the memory improvement with the proposed methods, by measuring the maximum size of the heap in the three methods. Figure 9 shows the max heap-size for different number of Gaussians (n) in the input set. Heap-size reductions ranging from 150X to 4500X are seen with Smart Greedy MBT. The Dominant MBT has a near constant heap size with varying n and provides up to 300X reduction over the Smart Greedy MBT. Analyzing the input Gaussian set for this experiment shows the presence of a large number of dominant Gaussians, resulting in almost all input Gaussians forming at least one zero error producing pair. The drastic reduction in heap size is attributed to this property of the input set which is not uncommon in operations like worst chip slack computation.

5. CONCLUSIONS AND FUTURE WORK

Using the Greedy MBT method [8] for finding the *max/min* of a set of Gaussians can have significant run-time impact when the number of operands in the set is large, as indicated by our results. We propose techniques to improve this method by special handling of zero error pairs with no loss of accuracy. Experimentally, we show that a majority of the operand pairs encountered in *max/min* operations in SSTA have zero error. We additionally prove that for an operand pair with zero approximation error, the resultant is the dominant input operand, and it can be found using a deterministic *max/min* operation on the mean of the input Gaussian random variables. Based on these observations we suggest two new methods: Smart Greedy MBT and Dominant MBT.

These methods are implemented in an industrial SSTA tool and in comparison to Greedy MBT show run-time improvements of over 60% (3X speed-up). The impact of this improvement on the overall run-time of single and iterative SSTA runs are evaluated. The improvements are more significant for designs where the number of input operands to the *max/min* operations are larger. The largest net improvement observed is over 8 hours (2X run-time speed-up).

Apart from the core timing propagation step, which is the primary focus of our study, there are other key operations in an SSTA flow that use these *max/min* operations heavily. An example for such a step is the *min* operation on all statistical slacks in a design to find the worst chip slack (or chip yield). We expect the improvement from these techniques to be more pronounced and consistent across designs for such steps, since the Gaussian set would be larger, independent of the design topology. Early evaluation of our techniques in chip slack computation show heap-size reduction of over 150X. As future work, we propose evaluating and employing our techniques for such steps in the SSTA flow.

6. REFERENCES

[1] C. E. Clark, "The greatest of a finite set of random variables," in *Operations Research, Vol. 9, No. 2 (Mar - Apr)*, 1961, pp. 145–162.

[2] C. Visweswariah, "Death, taxes and failing chips," in *Proc. of the Design Automation Conf.*, 2003, pp. 343–347.

[3] A. Devgan and C. Kashyap, "Block-based static timing analysis with uncertainty," in *Proc. Intl. Conf. on Computer-Aided Design*, 2003, pp. 607–614.

[4] D. Blaauw, K. Chopra, A. Srivastava, and L. Scheffer, "Statistical timing analysis: From basic principles to state-of-the-art," in *IEEE Transactions on Computer-Aided Design, 27(4) April 2008*, pp. 589–607.

[5] C. Visweswariah, K. Ravindran, K. Kalafala, S. G. Walker, and S. Narayan, "First-order incremental block-based statistical timing analysis," in *Proc. of the Design Automation Conf.*, 2004, pp. 331–336.

[6] M. Cain, "The moment-generating function of the minimum of bivariate normal random variables," *The American Statistician*, vol. 48, no. 2, pp. 124–125, May 1994.

[7] H. Chang and S. S. Sapatnekar, "Statistical timing analysis considering spatial correlations using a single PERT-like traversal," in *Proc. Intl. Conf. on Computer-Aided Design*, 2003, pp. 621–625.

[8] D. Sinha, H. Zhou, and N. V. Shenoy, "Advances in computation of the maximum of a set of Gaussian random variables," in *IEEE Transactions on Computer-Aided Design, 26(8) August 2007*, pp. 1522–1533.

[9] D. Sinha, N. V. Shenoy, and H. Zhou, "Statistical timing yield optimization by gate sizing," in *IEEE Transactions on Very Large Scale Integration (VLSI) Systems, 14(10) October 2006*, pp. 1140–1146.

InTimeFix: A Low-Cost and Scalable Technique for In-Situ Timing Error Masking in Logic Circuits

Feng Yuan and Qiang Xu
CUhk REliable Computing Laboratory (CURE)
Department of Computer Science & Engineering
The Chinese University of Hong Kong, Shatin, N.T., Hong Kong
Email: {fyuan,qxu}@cse.cuhk.edu.hk

ABSTRACT

With technology scaling, integrated circuits (ICs) suffer from increasing process, voltage, and temperature (PVT) variations and adverse aging effects. In most cases, these reliability threats manifest themselves as timing errors on critical speed-paths of the circuit, if a large design guard band is not reserved. This work presents a novel in-situ timing error masking technique, namely InTimeFix, by introducing fine-grained redundant approximation circuit into the design to provide more timing slack for speed-paths. The synthesis of the redundant circuit relies on simple structural analysis of the original circuit, which is easily scalable to large IC designs. Experimental results show that InTimeFix significantly increases circuit timing slack with low area/power cost.

1. INTRODUCTION

With technology scaling, integrated circuits (ICs) suffer from increasing process, voltage, and temparature (PVT) variations and aging effects [1]. In most cases, these reliability threats manifest themselves as timing errors on circuit speed-paths (i.e., critical paths that determine circuit speed) [2, 3]. Conventional design methodology requires designers to embed a large design guardband to prevent any timing failure. This conservative design methodology, however, inevitably diminishes the benefit of technology scaling [4]. Consequently, there is a growing research interest to achieve online timing error resilience.

1.1 Related Work

In order to achieve timing error resilience, we can either predict the error occurrence and take proactive actions to avoid them or detect and correct timing errors (or their effects) when they occur. Generally speaking, timing error prediction techniques (e.g., [5]) are applicable to detect gradual increase of circuit delay resulting from aging effects only. In this work, we focus on the more general timing error detection and correction techniques presented in the literature.

Timing Error Recovery: Most existing solutions for timing error resilience try to restore the state of the system to a known-good pre-error state. For example, *RAZOR* [6] implemented such a recovery scheme with microarchitectural support. In this technique, those flip-flops that are driven by speed-paths (denoted as *suspicious FFs* in this paper), are replaced with the so-called *RAZOR-FFs*, which contains a main flip-flop, an additional shadow latch and some control logic. The main flip flop latches the output signal at the clock edge with possible timing error, while the shadow latch, controlled by a delayed clock signal, latches the signal a fraction of a cycle later, which guarantees to re-

ceive the correct value. Consequently, when the shadow and the main FF values do not agree, indicated by the comparator, the timing error is detected. By flushing the pipeline and replaying failed instructions at lower clock frequency, the processor operates correctly with little performance penalty. Bowman *et al.* [7] implemented two novel metastability-immune timing error detectors (namely *error-detection sequential* in their work) and the corresponding error-recovery circuits in an Intel test chip. By removing design guard band used to guarantee "always correct" operations, the above techniques enable *better than worst-case designs* that are more energy-efficient [4], and has inspired a large amount of later research work (e.g., [8–12]).

Timing Error Masking: While *RAZOR*-like techniques are very effective for timing error correction (TEC) in microprocessor datapath with the help of instruction replay, they are very difficult, if not impossible, to be applied to general logic circuits, due to the high cost to checkpoint error-free states in them. It is therefore imperative to develop *in-situ* timing error correction techniques that are able to mask errors without any rollback. There are a few such techniques presented in the literature and they can be classified into two categories: *temporal error masking* and *logic error masking*.

Temporal error masking techniques replace suspicious FFs with sequential circuit elements having time-borrowing capability (e.g., soft-edge flip-flops) and correct timing errors by delaying the arrival time of the correct data to the next logic level (e.g., [13–15]). While effective in many cases, such time-borrowing techniques have the inherent weakness of error effect propagation. That is, even if a suspicious FF can borrow some time from its successive logic level, the timing slack of this level is reduced and hence some initially non-suspicious flip-flops in this level may become suspicious ones and need to be replaced by sequential elements with time-borrowing capability again. Due to this timing error propagation effect, the hardware cost for such temporal error masking techniques can be quite high.

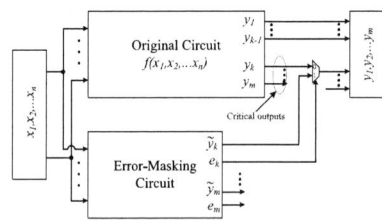

Figure 1: Timing Error Masking Scheme in [16].

In [16], Choudhury and Mohanram proposed to add a redundant logic block to predict the outputs of the circuit upon application of inputs that sensitize speed-paths. With this exact sensitization constraint, the *error-masking circuit* tends to have more timing slack when compared to the original circuit, and hence is immune to timing errors. As shown in Fig. 1, targeting those timing-critical outputs $y_k, ..., y_m$, error masking circuit generates two outputs for each of them, e.g., $\tilde{y}_k$ and e_k for y_k with potential timing error. To be specific, when a speed-path driving y_k is sensitized, e_k is set correspondingly and the original circuit's output y_k is substituted with the fast predicted value $\tilde{y}_k$ to achieve timing error resilience.

Permission to make digital or hard copies of all or part of this work for personal or classroom use is granted without fee provided that copies are not made or distributed for profit or commercial advantage and that copies bear this notice and the full citation on the first page. To copy otherwise, to republish, to post on servers or to redistribute to lists, requires prior specific permission and/or a fee.

1264

1.2 Motivation and Summary of Contributions

The idea of adding redundant logic into the original circuit to mask timing errors on speed-paths as shown in [16] is quite interesting. The main limitation of this work lies in the difficulty of error-masking circuit generation. To be specific, [16] requires to synthesize the so-called *speed-path characteristic function* that represents the set of *all* speed-path activation patterns, which is only practical for small circuit blocks. The associated area/power overhead is also quite significant, as demonstrated in their experimental results.

In this paper, we propose a novel *in-situ* timing error correction technique, namely *InTimeFix*. Unlike [16] that adds *coarse-grained* redundancy by synthesizing the Boolean function that represents the activation of all speed-paths for timing error correction, the redundant TEC circuit in *InTimeFix* is generated at a much more *fine-grained* manner based on the concept of *approximation circuit*, which can be obtained by simple structural analysis of the original circuit. The proposed solution is therefore of low cost and is easily scalable to large IC designs. The main contributions of this paper include:

- we present a novel technique to add redundant approximation circuit into the original design to create a logically-equivalent yet timing-improved circuit, and prove its correctness;

- we propose a low-cost and scalable technique to synthesize timing error masking logic based on simple structural analysis, without necessarily acquiring characteristic function for speed-paths;

From another perspective, *InTimeFix* can be also regarded as a circuit timing optimization technique, since it facilitates to improve circuit timing slack with low hardware cost, as demonstrated in our experimental results. It is also important to emphasize that, as a redundancy scheme, *InTimeFix* is compatible with other timing/power optimization techniques such as gate sizing [19] and dual V_{th} allocation [20], and in fact, these techniques can be combined to further improve circuit performance under variation.

The remainder of this paper is organized as follows. In Section 2 and Section 3, we detail the proposed *InTimeFix* technique for *in-situ* correction of timing errors on speed-paths. Experimental results on various benchmark circuits are then presented in Section 4. We discuss the limitations of *InTimeFix* in Section 5. Finally, Section 6 concludes this paper.

2. IN-SITU TIMING ERROR MASKING WITH APPROXIMATE LOGIC

The concept of *approximation circuit* was first presented in [17], which tries to increase a microprocessor's clock frequency by replacing a complete logic function with a simplified circuit that mimics the function and uses rough calculations to speculate results. In [18], the authors used approximation circuit for concurrent error detection (for random error detection istead of timing error detection) and proposed a methodology to tradeoff error detection capability and area overhead.

In [18], the authors defined *approximate logic* as: Given two Boolean functions F and G, $G0$ is a *0-approximate logic* of F if $G0 = 0 \Rightarrow F = 0$. Similarly, $G1$ is a *1-approximate logic* of F if $G1 = 1 \Rightarrow F = 1$. According to counter-positive law, we can further obtain two statements that are $F = 1 \Rightarrow G0 = 1$ and $F = 0 \Rightarrow G1 = 0$. Consider a Boolean function $F = a + b + \bar{a}cd$, there are 13 on-set minterms[1] and 3 off-set minterms[2] in its truth table. Similarly, a 0-approximate logic function of F, $G0 = a + b + c$, covers 2 out of 3 off-set minterms of F. Generally speaking, since the approximate logic circuit is much simpler when compared to the original circuit, its computational latency is smaller. The basic idea of the proposed *InTimeFix* technique is to generate approximate logic for the original logic function of suspicious FFs in such manner that it covers all the logic minterms that sensitize speed-paths.

---

[1] The on-set minterm is the minterm where the function outputs logic '1'.

[2] The off-set minterm is the minterm where the function outputs logic '0'.

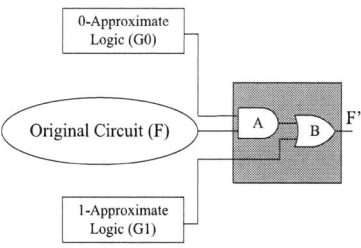

Figure 2: Equivalent Circuit with Approximate Logic.

2.1 Equivalent Circuit Construction with Approximate Logic

Given a logic circuit whose Boolean function is F, suppose $G0$ is a 0-approximate logic for F and $G1$ is a 1-approximate logic for F. we define: P is all the minterms in F's truth table, $P0$ is the off-set minterms of F covered by $G0$ and $P1$ is the on-set minterms of F covered by $G1$. Now, let us construct a circuit $F' = F \cdot G0 + G1$ as shown in Fig. 2. We define $d_{G0}(P0)$, $d_{G1}(P0)$ and $d_F(P0)$ as the worst-case delay among all minterms in P0 through circuit $G0$, $G1$ and F, respectively. Similarly, $d_{G0}(P1)$, $d_{G1}(P1)$, $d_F(P1)$ and $d_F(P - P0 - P1)$ are defined. d_{AB} is the total propagation delay of *AND* gate A and *OR* gate B. Assuming approximate logic $G0$ and $G1$ are implemented with simpler logic structures when compared to original circuit F and hence has less computation latency (e.g., $d_{G0}(P0) < d_F(P0)$, $d_{G1}(P0) < d_F(P0)$ and $d_{G1}(P1) < d_F(P1)$), we have the following theorem:

THEOREM 1. *The circuit shown in Fig. 2, $F' = F \cdot G0 + G1$, is logically-equivalent to the original circuit F, and its worst-case timing delay is $\max\{d_F(P - P0 - P1), d_{G0}(P0), d_{G1}(P0), d_{G1}(P1)\} + d_{AB}$.*

PROOF. When the original circuit F outputs 1, by applying counter-positive law, its 0-approximate logic must also output 1 (i.e., $G0 = 1$), and hence $F' = 1$. Similarly, when the original circuit F outputs 0, by applying counter-positive law, its 1-approximate logic must also output 0 (e.g., $G1 = 0$), and hence $F' = 0$. Consequently, F and F' are logically-equivalent.

To obtain the worst-case delay for this equivalent circuit F', let us consider the circuit delay before the shaded logic block in Fig. 2 for the following three cases, corresponding to the application of inputs belonging to different set of minterms of the truth table of F/F'.

- When the inputs applied to the circuit belong to $P0$, $G0$ outputs controlling value 0 for *AND* gate A after $d_{G0}(P0)$ and dominates the path through the original circuit F with longer delays. The worst-case delay would be $\max\{d_{G0}(P0), d_{G1}(P0)\}$. Note that $d_{G1}(P0)$ is the time spent to settle $G1$ to be non-controlling value 0 for *OR* gate B.

- When the inputs applied to the circuit belong to $P1$, $G1$ outputs controlling value 1 for *OR* gate B after $d_{G1}(P1)$ and it dominates the path through the original circuit F and $G0$. The worst-case delay in this case is therefore simply $d_{G1}(P1)$.

- When the inputs applied to the circuit belong to $P - P0 - P1$, we have to wait for the original circuit F to settle down, and hence the worst-case delay would be $d_F(P - P0 - P1)$.

The worst-case timing delay for circuit F' is therefore $\max\{d_F(P - P0 - P1), d_{G0}(P0), d_{G1}(P0), d_{G1}(P1)\} + d_{AB}$, after considering the time spent on gates A and B. ∎

2.2 Timing Error Masking with Approximate Logic

Generally speaking, a suspicious FF is driven by multiple paths, and timing errors may occur only when speed-paths are sensitized. In other words, timing errors may be activated by only a few minterms of the truth table for a suspicious FF, denoted as *critical minterms*. Motivated by this observation, according to *Theorem 1*, if all the critical minterms are covered with approximate logic, we can achieve large timing slack and mask potential timing errors. *The question now becomes how to efficiently construct redundant approximate logic for speed-paths?*

1265

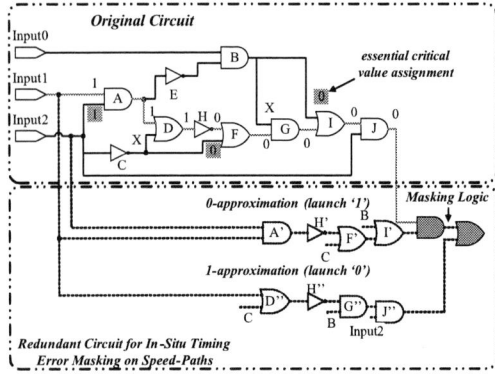

Figure 3: Speed-Path Approximation.

Let us consider an example circuit shown in Fig. 3 for explanation, wherein path P {$Input1, A, D, H, F, G, I, J$} is a speed-path. When logic '1' is applied at $Input1$ and propagates along this path to generate logic '0' at the receiving end, we have to assign logic '1' at the side-input[3] of gate A. This is because, logic '1' is a non-controlling value of AND gate, and the output of A will be dominated by the side-input if it is assigned with controlling value. Similarly, side-inputs of gate F and gate I have to be assigned as non-controlling values (see Fig. 3).

Let us define such side-inputs on the path that need to have deterministic non-controlling values (marked in shade) as *essential side-inputs*. Based on the path sensitization theory in [23], we have the following lemma to functionally activate a speed-path:

LEMMA 2. *To cover all the critical minterms that sensitize a particular speed-path is equivalent to approximate its essential side-inputs.*

With the above, we can construct redundant approximate logic for each speed-path by simple structural analysis. Again, take path P in Fig. 3 as an example. Suppose we would like to construct 0-approximate logic for this path, we first duplicate the entire path and then gradually remove those gates without essential side-inputs. To be specific, the removing process is conducted structurally by analyzing the targeted speed-path P reversely from the ending gate (i.e., gate G) to the sending gate (i.e., gate A). Consider gate J, since it is dominated by its on-input with controlling value 0, we can remove it from the approximate logic. Similarly, gate G and gate D are not needed. While the outputs of gates F and A are determined by both on-input and side-input signals, two gates need to be duplicated in the 0-approximate logic, and the side-inputs are connected to the same net as path P in the original circuit. As shown in Fig. 3, the 0-approximate logic constructed as above will output logic '0' if and only if the speed-path P in the original circuit is sensitized with launching value logic '1'. The 1-approximate logic for speed-path P can be constructed similarly (see Fig. 3). Since the approximate logic is with much simpler logic structure and the delay of the masking logic is usually insignificant (without necessarily sizing it up), we can achieve large timing slack and mask potential timing errors on speed-paths.

Note that, not all kinds of logic cells have controlling values for their inputs, e.g., XOR/XNOR gate. If a speed-path contains such kind of logic cells, their side-inputs will be treated as essential side-inputs to have deterministic values to approximate and the proposed methodology is applicable to such designs.

3. INTIMEFIX SYNTHESIS

Adding redundant approximate logic facilitates to achieve timing error resilience on speed-paths. As the construction of the approximate logic needs to take side-input signals from the original circuit, however, two potential problems arise: (i) the latencies for side-inputs may become a concern for the propagation delay of the approximate logic and such side-inputs are denoted as *critical side-inputs* (formally defined later); (ii) the increased loading capacitance on side-inputs can

---
[3]Given a path P = {$G_0, G_1,...G_m$}, for a specific gate G_i, G_{i-1} is the *on-input* signal of G_i, while other input signals of G_i are its *side-inputs*.

prolong the delay of those paths going through them. The above observations motivate us to propose a cost-efficient and scalable synthesis framework for *InTimeFix*, as illustrated in Fig. 4.

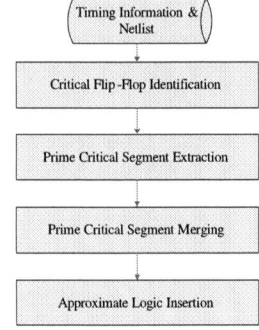

Figure 4: *InTimeFix* Synthesis Framework.

3.1 Overall Flow

Fig. 4 describes the synthesis flow of InTimeFix. With the optimized circuit netlist and the corresponding timing information in standard delay format (SDF), we firstly identify those suspicious FFs that are driven by speed-paths and thus need to be considered. Speed-paths are defined as those paths whose propagation delays exceed a threshold value, e.g., 80% of the maximum path delay. Note that, although static timing analysis is not able to output accurate timing values, its accuracy is sufficient for suspicious FF identification because we only need a comparative relationship among paths and we can always tune the threshold to tradeoff between the hardware cost and protection strength.

Next, a set of so-called *prime critical segments* is extracted from speed-paths, defined as the segments of speed-paths that do not include any critical side-inputs, with which approximate logic can be safely generated with more timing slack. Then, a heuristic method is used to merge prime critical segments to minimize the hardware cost of the approximate logic and the extra loading capacitance of side-inputs. Finally, approximate logic is generated for the merged prime critical segments and inserted into the original circuit for timing error correction.

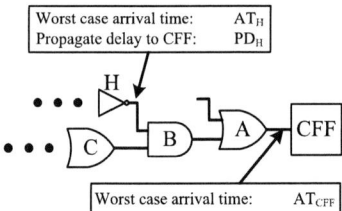

Figure 5: Example to Illustrate the Critical Side-Input.

3.2 Prime Critical Segment Extraction

Before introducing the details of our proposed algorithm, let us first formally define the criteria used to identify critical side-input, as depicted in Fig. 5. Taking side-input H as an example, when considering the critical flip-flop CFF, we have the worst case arrival time AT_H on side-input H, the propagation delay PD_H from H to CFF, and the worst case arrival time AT_{CFF} on CFF. Suppose RD is a predefined reduced delay (i.e., extra slack) that we want to achieve and MD is the delay of the masking logic, then H is a critical side-input if $AT_{CFF} - (AT_H + PD_H + MD) > RD$. The basic idea behind this definition is that, if this criteria is not satisfied, there is sufficient timing slack on this side-input and it does not affect the timing of the approximate logic at all. Based on the above definition, we denote a gate on speed-path to be prime critical gate if all its side-inputs are non-critical. Furthermore, a segment of a speed-path is a prime critical segment if it only consists of prime critical gates.

1266

Our prime critical segment extraction algorithm (denoted as *Ex-PriSeg*) is shown in Algorithm 1, where, *Gate* is one gate on a speed-path and *Segment* is the parameter to store the targeted segment; *SegmentSet* denotes the set of extracted prime critical segments; while *Ogate* and *Cside* represent the on-input and critical side-inputs of *Gate*.

Algorithm 1: Extract Prime Critical Segment(ExPriSeg)

```
   input: Gate,Segment
1  begin
2  |  if PI or FF is reached then
3  |  |  if Segment is not illegal then
4  |  |  |  return FailToExtract;
5  |  |  else
6  |  |  |  add Segment into SegmentSet;
7  |  |  |  return SucceedToExtract;
8  |  if Gate is prime critical gate then
9  |  |  add Gate into Segment;
10 |  |  find the on-input Ogate;
11 |  |  return ExPriSeg (Ogate,Segment);
12 |  else
13 |  |  if Segment is legal then
14 |  |  |  add Segment into SegmentSet;
15 |  |  |  return SucceedToExtract;
16 |  |  else
17 |  |  |  foreach Cside do
18 |  |  |  |  new Segment;
19 |  |  |  |  if ExPriSeg (Cside,Segment) ==FailToExtract then
20 |  |  |  |  |  return FailToExtract;
21 |  |  |  return SucceedToExtract;
```

To relax the timing slack for a critical flip-flop by *RD*, we need to reduce the delay of all the speed-paths connected to it. Here we regard an extracted prime critical segment as a *legal* one if the approximate logic for it is able to reduce the delay of the targeted speed-path for at least *RD*. Starting from a specific critical flip-flop, our algorithm extracts the prime critical segments by recursively tracing its fan-in cone in a depth-first manner. Initially, *Segment* is empty and *Gate* is the input gate of the targeted critical flip-flop. During the tracing procedure, once a primary input or a flip-flop, denoted as *PI* or *FF*, is reached (Line 2), function returns *FailToExtract* if there is no legal prime critical segment found, otherwise we keep the *Segment* and return *SucceedToExtract*. Suppose *Gate* is detected to be a prime critical gate, we add *Gate* into *Segment* and keep on tracing its on-input gate. On the other hand, if *Gate* is not a prime critical gate (Line 12), we first check whether the current *Segment* is legal or not. The searching process is stopped by storing *Segment* and return *SucceedToExtract* if *Segment* is legal prime critical segment, otherwise, we empty the current *Segment* and start to trace each critical side-input separately. Clearly, the extracted prime critical segments can cover all the speed-paths ending at the targeted flip-flop if the final returned value is *SucceedToExtract*. Suppose it returns *FailToextract*, we will try to reduce the value of *RD* by a pre-defined ratio, and conduct the same search again. One thing to note is that our method tends to extract legal prime critical segments that are close to the targeted flip-flops, and they are able to cover more speed-paths, if any.

3.3   Prime Critical Segment Merging

The prime critical segments extracted from different critical flip-flops are likely to be merged to further reduce hardware cost. As the example shown in Fig. 6, two prime critical segments $\{E,D,...C,A\}$ and $\{E,D,...C,B\}$ can share a merged prime critical segment $\{E,D,...C\}$. Furthermore, we notice that our extracted prime critical segments are not in the most compact format. Taking segment $\{E,D,...C,A\}$ as an example, it is not necessary to approximate the entire segment, instead, only by approximating $\{E,D,...C\}$ is enough to achieve *RD* delay reduction. We denote the length (i.e., the number of logic gates) of the shortest legal subpart of prime critical segment as the *essential length*, and represent the length of the remaining part as the *redundant length*. For the sake of simplicity, we regard a subpart of a prime critical segment still legal if the length of the subpart is larger than the essential

length. Therefore, two prime critical segments can be merged if the length of the shared part is larger than both of their essential lengths.

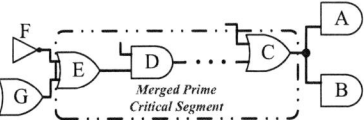

Figure 6: Prime Critical Segment Merging: An Example.

The main flow of our proposed prime critical segment merging algorithm is shown in Fig. 7, comprising the following steps:

1. Starting from the extracted prime critical segment set, we first sort them in non-decreasing order in terms of their redundant length. The basic idea behind this step is that we need to first fix those prime critical segment with less flexibility. Then, all the segments are merged by iteratively applying the following steps.

2. We always select the top un-processed segment and employ a heuristic method to find the optimal subpart. First of all, the closest-to-output gate on the selected segment shared by most remaining un-processed segments is identified. Taking the circuit shown in Fig. 6 as example, gate C is picked first if the selected segment is $\{E,D,C,A\}$ since it has higher probability to replace the most remaining un-processed segments by backwardly tracing the selected segment from this gate, say, segment $\{E,D,C\}$. Suppose that the length of $\{E,D,C\}$ is less than the essential length, we extend this sub-segment to $\{E,D,C,A\}$ and such extension is conducted until the length requirement is satisfied. Suppose the length of $\{E,D,C\}$ is larger than the essential length, we enumerate all the possible legal subparts to identify the one that includes minimal number of side-inputs in the hope that increased loading capacitance is minimized when inserting approximate logic. Finally, the optimal segment is fixed.

3. The remaining un-processed segments are checked to determine whether they can be replaced by newly fixed segment. The replaceable segments are labeled as processed.

4. The procedure terminates if all the segments have been processed, otherwise it goes back to step 2.

4. EXPERIMENTAL RESULTS

4.1   Experimental Setup

To evaluate the effectiveness of our proposed *InTimeFix* technique, we conduct experiments on two large ISCAS'89 benchmark circuits, *s38417* and *s38584*, as well as three large IWLS benchmark circuits, *wb_conmax*, *ethernet* and *des_perf*.

In the experimental flow, we first synthesize the benchmark circuits with Synopsys Design Compiler to obtain the optimized circuit netlist and its SDF timing information. They are then fed to *InTimeFix* to generate redundant approximate logic to mask potential timing errors, targeting speed-paths within 20% of the longest path delay. Finally, Synopsys PrimeTime is applied to evaluate the solution, and the quality of the solution is demonstrated by the extra timing slack achieved with the new circuit when compared with the original one.

4.2   Experimental Results

When comparing the worst case delay (WCD) between the original circuit and the one equipped with redundant approximate logic (see Table 1, Columns 7-10), it can be observed that the proposed solution is able to achieve 11.15% timing slack relaxation on average. On the other hand, as shown in Columns 5-6, the hardware cost (the unit of cost is the area of smallest 2 input *AND* gates) introduced in the proposed *InTimeFix* technique to achieve the above timing slack is extremely low, less than 0.89% on average. As can be seen from Column 11, the runtime to process the largest benchmark circuit *ethernet* takes less than one second. Consequently, we believe the proposed methodology can be easily scalable to large industrial designs.

1267

Benchmark	Circuit Size (# of gates)	FF (#)	Critical FF (#)	Cost (# of gates)	Increased Ratio (%)	Ori. WCD (ns)	Our WCD (ns)	Relaxed Slack (ns)	Improved Ratio (%)	Runtime (s)
s38417	24370	1636	78	570	2.34	35.34	31.93	3.41	9.64	0.09
s38584	21066	1426	12	98	0.47	20.50	18.49	2.01	9.80	0.1
des_perf	154323	9105	89	592	0.38	7.80	6.68	1.12	14.36	0.95
wb_conmax	75352	3316	277	1160	1.54	8.47	7.74	0.73	8.59	0.4
ethernet	157841	10752	28	386	0.24	8.21	7.11	1.10	13.38	1.083
Ave.					0.89				11.15	

Table 1: Experimental Results on Improved Timing Slack and Hardware Cost.

Benchmark	Gate Sizing with Area Constraint vs. *InTimeFix*				*InTimeFix* on top of Gate Sizing				
	Area Constraint (# of gates)	Gate Sizing WCD (ns)	*InTimeFix* WCD (ns)	Improvement (%)	Gate Sizing WCD (ns)	Area Cost (# of gates)	*InTimeFix* WCD (ns)	Area Cost (# of gates)	Further Improvement (%)
s38417	570	34.70	31.93	7.99	31.93	1050	27.90	608	12.63
s38584	98	19.96	18.49	7.36	15.18	1217	13.85	171	8.81
des_perf	592	6.95	6.68	3.88	6.43	5937	6.06	1309	5.85
wb_conmax	1160	7.47	7.74	-3.61	5.38	4008	4.88	1561	9.42
ethernet	386	7.23	7.11	1.73	6.15	6257	5.25	614	14.61
Ave.				3.47					10.26

Table 2: Comparison on Timing Slack Improvement: Gate Sizing vs. *InTimeFix*.

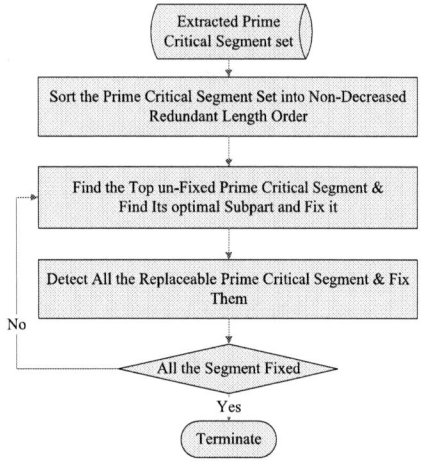

Figure 7: Flowchart of Prime Critical Segment Merging.

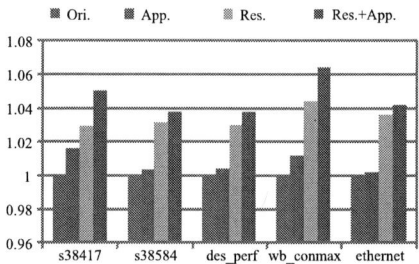

Figure 8: Power Comparison Results.

A close examination of the experimental results show that benchmark circuits *s38417* and *wb_conmax* consume the largest percentage of hardware overhead. The reason is that the speed-paths in these two benchmarks are quite evenly distributed and the prime critical segments identified from different critical flip-flops are more likely to be independent to each other. Therefore, the hardware cost is higher than other benchmark circuits. To have a better design tradeoff, we try to reduce the hardware cost for these two benchmarks by gradually removing the approximation logic on the shortest speed-paths. Even when the hardware cost is reduced to less than 1%, we can still achieve 7.1% and 11.6% slack relaxation for *wb_conmax* and *s38417*, respectively.

With the above experimental results, we can observe clear advantages of the proposed *InTimeFix* technique over [16], without performing direct comparisons. In [16], the authors conducted experiments on a number of benchmark circuits whose sizes are less than 2000 gates. Their experimental results show that, on average 18% area overhead is required to mask timing errors on speed-paths within 10% of the longest path delay (the minimum overhead is 4%).

While not targeting timing error resilience, gate sizing is an effective technique to improve circuit timing. In our next experiment, we compare our *InTimeFix* solution against a greedy gate sizing technique [19]. Firstly, when we constrain the area overhead for gate sizing solution to be the same as the one with *InTimeFix* in earlier experiment, it can be seen that *InTimeFix* outperforms gate sizing in most cases (except *wb_conmax*), and the average improvement is 3.47%, which proves the cost-efficiency of the proposed solution. Next, since *InTimeFix* is compatible with gate sizing technique, we combine the two solutions in such manner that we first employ gate sizing to improve circuit timing until no further benefits can be achieved and then apply *InTimeFix* on top of it. We can observe that, without area constraints, gate sizing can significantly improve circuit performance, but at con-siderable area and power cost. *InTimeFix* is able to provide additional 10.26% timing slack on average, and the area cost is still quite small.

Fig. 8 compares the power consumption of the following circuit netlists: (i) original circuits denoted as *"Ori."*; (ii) original circuits equipped with *InTimeFix* logic, denoted as *"App."*; (iii) circuits after gate sizing, denoted as *"Res."*; (iv) resized circuits equipped with *InTimeFix* logic, denoted as *"Res.+App."*. The power values are obtained with Synopsys PrimeTIme PX tool and normalized with respect to that of original circuit. From this figure, we can observe the power impact of applying *InTimeFix* is not significant and vary with different circuits, on average about 1-2%. Another observation is that the power impact of applying *InTimeFix* with a resized circuit is usually higher than applying it with the original circuit for the same circuit. This is because critical paths are more balanced after gate sizing and we need to approximate more circuit speed-paths.

Finally, we evaluate the impact of process variation on the proposed *InTimeFix* architecture, using Monte Carlo simulation. According to [21], we assume there is 10% variation on each standard cell. The results are depicted in Fig. 9, where we plot and compare the WCD distributions of two sets of circuits (i.e. the black pile represents set of processed circuits and the gray one denotes the original set of circuits) for the three large IWLS circuits. As can be seen from the figure, even for circuit with *InTimeFix* approximate logic under the worst case process variation corner, it has smaller or similar WCD when comparing with that of the original circuit under the best case scenario. Moreover, it can be observed that the number of the closer-to-mean chips increases and the standard deviation of WCD distribution shrinks with *InTimeFix*. In particular, as can be observed in Fig. 9(a), benchmark circuit *des_perf* with *InTimeFix* has only one third of the distribution width when compared to that of the original circuit.

The above phenomenon demonstrate that the proposed *InTimeFix* technique facilitates to tolerate process variation effects. The reason is that our proposed method effectively reduces the number of logic elements on the critical paths and also shrinks the variation, behind which the mathematic principle can be explained by the example given in [22]: assuming inverters have independent gaussian delay distribution (μ, σ), the delay of a path including n inverters obey the gaussian distribution $(n\mu, \sqrt{n}\sigma)$. Clearly, less n leads to smaller deviation.

1268

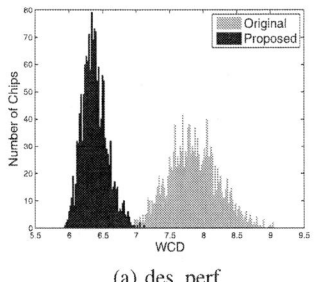
(a) des_perf

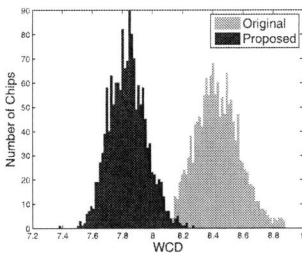
(b) wb_conmax

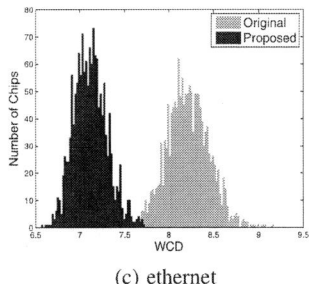
(c) ethernet

Figure 9: Circuit Timing under Process Variation

5. DISCUSSIONS

5.1 Limitations of InTimeFix

The effectiveness of *InTimeFix* is highly related to the original circuit structure with its current synthesis flow. In this section, we discuss certain circuit structure that are difficult to be optimized with *InTimeFix*. As shown in *Lemma 2*, for a particular speed-path, we need to approximate its essential side-inputs when generating the redundant approximation circuit. Consequently, if the number of essential side-inputs is large, the timing slack provided by the approximation circuit delay would be small. In the extreme case (e.g., to generate 1-approximation for a speed-path that contains nothing but a series of *AND* gates), the redundant circuit may simply be a duplication of the speed-path, rendering *InTimeFix* not effective at all.

To mitigate the above problems, it would be beneficial to manipulate the original circuit structure for *InTimeFix*. That is, we could synthesize the circuit in a *InTimeFix-friendly* manner so that speed-paths are easy to approximate, whenever possible. Alternatively, we could resort to circuit re-synthesis techniques such as retiming and/or rewiring to make the circuit more *InTimeFix-friendly*. We are currently investigating the above techniques to mitigate the limitations of *InTimeFix*.

5.2 Testing Circuits with InTimeFix

From manufacturing test perspective, adding redundancy into the circuit may introduce untestable faults, and there is no exception for our design. For example, for the circuit shown in Fig. 3, the stuck-at-1 fault at the output of $G0$ is untestable, because to activate this fault, we need to set it as logic '0', but when $G0 = 0$, the original circuit F will also output logic '0', preventing the propagation of its faulty effect to outputs. Similarly, the stuck-at-0 fault at the output of $G1$ is untestable. It is important to note that, the presence of such faults would not affect the functional correctness of the circuit. At the same time, they do render the corresponding timing error masking logic to be ineffective, but the likelihood for such faults is quite low due to their small sizes.

One way to make these faults testable is to add some design-for-testability (DfT) circuit, e.g., adding additional flip-flop to be driven by G0/G1 directly. This, however, increases the hardware cost and also prolongs the propagation delay on approximate logic paths due to extra load. Alternatively, we can rely on delay testing to guarantee the timing correctness of the corresponding speed-paths. In other words, as long as the path can pass at-speed test, its timing correctness is guaranteed and we do not need to care whether these untestable faults exit or not.

6. CONCLUSION

The timing behavior of integrated circuits has become increasingly uncertain with technology scaling. In this paper, we propose the so-called *InTimeFix* technique to achieve low-cost and scalable timing error resilience in logic circuits, by introducing fine-grained redundant *approximate logic* for speed-paths in the circuit. As the proposed synthesis methodology for the redundant circuit only relies on simple structural analysis of the original circuit, it can be easily scalable to large IC designs. Experimental results demonstrate that the proposed solution can effectively increase circuit timing slack with very low cost. In addition, as a redundancy scheme, *InTimeFix* can be combined with other timing optimization techniques (e.g., gate sizing) to further improve circuit performance under variation.

7. ACKNOWLEDGEMENT

This work was supported in part by the Hong Kong SAR Research Grants Council under General Research Fund No. CUHK418111 and No. CUHK418812.

8. REFERENCES

[1] S. Borkar. Designing reliable systems from unreliable components: the challenges of transistor variability and degradation. *IEEE Micro*, 25(6):10–16, 2005.

[2] K. Bowman, et al. Circuit techniques for dynamic variation tolerance. In *Proc. DAC*, pp. 4–7, 2009.

[3] D. Frank, R. Puri, and D. Toma. Design and CAD Challenges in 45nm CMOS and beyond. In *Proc. ICCAD*, pp. 329–333, 2006.

[4] T. Austin and V. Bertacco. Deployment of better than worst-case design: solutions and needs. In *Proc. ICCD*, pp. 550–555, 2005.

[5] M. Agarwal, B. C. Paul, M. Zhang, and S. Mitra. Circuit Failure Prediction and Its Application to Transistor Aging. In *Proc. VTS*, pp. 277–286, 2007.

[6] D. Ernst, et al. Razor: a low-power pipeline based on circuit-level timing speculation. In *Proc. MICRO*, pp. 7–18, 2003.

[7] K. Bowman, et al. Energy-efficient and metastability-immune resilient circuits for dynamic variation tolerance. *IEEE JSSC*, 44(1):49–63, 2009.

[8] S. Das, et al. RazorII: In situ error detection and correction for PVT and SER tolerance. *IEEE JSSC*, 44(1):32–48, 2009.

[9] B. Greskamp, et al. Blueshift: Designing processors for timing speculation from the ground up. *Proc. HPCA*, pp. 213-224, 2009.

[10] R. Ye, F. Yuan and Q. Xu. Online clock skew tuning for timing speculation. *Proc. ICCAD*, pp. 442–447, 2011.

[11] Y. Liu, et al. On logic synthesis for timing speculation. *Proc. ICCAD*, pp. 591–596, 2012.

[12] R. Ye, et al. Post-placement voltage island generation for timing-speculative circuits. *Proc. DAC*, 2013.

[13] M. R. Choudhury and K. Mohanram. TIMBER: Time borrowing and error relaying for online timing error resilience. In *Proc. DATE*, pp. 1554–1559, 2010.

[14] M. Kurimoto, et al. Phase-adjustable error detection flip-flops with 2-stage hold driven optimization and slack based grouping scheme for dynamic voltage scaling. In *Proc. DAC*, pp. 884–889, 2008.

[15] K. Hirose, et al. Delay-compensation flip-flop with in-situ error monitoring for low-power and timing-error-tolerant circuit design. *Japan Journal of Applied Physics*, 47(4):2779–2787, Apr. 2008.

[16] M. R. Choudhury and K. Mohanram. Masking timing errors on speed-paths in logic circuits. In *Proc. DATE*, pp. 87–92, 2009.

[17] S.-L. Lu. Speeding up processing with approximation circuits. *Computer*, 37(3):67–73, Mar. 2004.

[18] M. R. Choudhury and K. Mohanram. Approximate logic circuits for low overhead, non-intrusive concurrent error detection. In *DATE*, pp. 903–908, 2008.

[19] O. Coudert, *et al.*, New Algorithm for Gate Sizing: A Comparative Study. In *Proc. DAC*, pp. 734-739, 1996.

[20] M. Mani, *et al.*, An Efficient Algorithm for Statistical Minimization of Total Power under Timing Yield Constraints. In *Proc. DAC*, pp. 309-314, 2005.

[21] G. Yu, et al. Statistical Static Timing Analysis Considering Process Variation Model Uncertainty. In *Proc. IEEE TCAD*, pp. 1880-1890, 2008.

[22] J. L. Tsai, et al. A Yield Improvement Methodology Using Pre- and Post-Silicon Statistical Clock Scheduling. In *Proc. ICCAD*, pp. 611-618, 2004.

[23] K.-T. Cheng and H.-C. Chen, Classification and Identification of Nonrobust Untestable Path Delay Faults. In *IEEE TCAD*, 15(8):845-853, Aug. 1996.

Improving PUF Security with Regression-based Distiller

Chi-En Yin and Gang Qu

Department of Electrical and Computer Engineering & Institute for Systems Research
University of Maryland, College Park, USA
{chienyin,gangqu}@umd.edu

Abstract—Silicon physical unclonable functions (PUF) utilize fabrication variation to extract information that will be unique for each chip. However, fabrication variation has a very strong spatial correlation and thus the PUF information will not be statistically random, which causes security threats to silicon PUF. We propose to decouple the unwanted systematic variation from the desired random variation through a regression-based distiller. In our experiments, we show that information generated by existing PUF schemes fail to pass NIST randomness test. However, our proposed method can provide statistically random PUF information and thus bolster the security characteristics of existing PUF schemes.

Index Terms—ring oscillator (RO), physical unclonable functions (PUFs), linear regression, variation decomposition

I. INTRODUCTION

In order to provide a secure storage for cryptographic keys, contemporary tamper-resistant devices such as smart cards arm themselves with a number of countermeasures to defeat various kinds of physical attacks. Nevertheless, it is still possible for attackers to read, and possibly write, the secret bits in the non-volatile memory through the electron beam of a scanning electron microscope once the surface of the chip is exposed by, for instance, focused ion beam. Physical unclonable functions (PUFs), in contrast, are 'inseparable' because the underlying nano-scale structural disorder will most likely be damaged during the course of physical tampering of the device, so will the keys. Since the first introduction of PUFs, many types of circuitry have been proposed to realize the notion. Most notable are Arbiter PUF [1], RO PUF [2], SRAM PUF [3] and Bistable Ring PUF [4]. Many methodologies have been proposed to advance the art in terms of reliability, security, and hardware efficiency. The typical workflow of a RO PUF involves the following steps.

- **Fabrication Variation Extraction** The very first task of PUFs is to measure the unique characteristics endowed from the uncontrollable fabrication process. The analog-to-digital transformation is part of the physical entropy source subject to tests; in our case, this step corresponds to a full frequency characterization of a RO array [6].

- **Secret Selection** This step selects secure and reliable secrecy out of the variation profile measured in the previous step. Existing approaches include the classic 1-out-of-8 coding [2], the index-based syndrome (IBS) coding [8] and the chain-like neighbor coding [9], [6], temperature-aware cooperative coding [7] and group-based coding [5], [16].

- **Error Correction** To further enhance reliability, error correcting code (ECC) may be applied. Codes have been used for RO PUFs include Hamming and BCH codes [2].

- **Tests for Security and Reliability**
  Randomness test ensures the security of the PUF secrecy. Reliability test, on the other hand, verifies the PUF secrecy can be regenerated even under environmental fluctuations such as in temperature and supply voltage.

Many cryptography applications such as key generation require random numbers. NIST has established several standards for cryptographically secure pseudo-random number generator (PRNG) as well as a statistical test suite for random and pseudo-random number generators [10]. If the numbers produced by a PRNG fail to pass the NIST test, it is considered vulnerable against cryptanalysis. Therefore, it is critical to verify that the secrecy generated by PUF is random and can pass the NIST test.

We consider the public available RO PUF data obtained from frequency characterization on 125 FPGA devices [6]. To our surprise, none of their random sequence can pass all the NIST tests that are applicable to their sequence length. Table I shows the detailed testing results. Column 1 lists the 9 statistical random tests we find applicable to the length of our test sequences.[1] Take Frequency Test for example, it examines whether the number of 1's and 0's in a sequence are approximately the same as would be expected for a truly random sequence, for which the number of 1's and 0's in a sequence should be about the same. If a sequence has a very disproportional 1's to 0's such that its *P-value*, the probability for events that at least as extreme as this instance to occur, is smaller than a significant level α, 1% in our case, the event is regarded significant. If it turns out that more than 4% of the

---

Permission to make digital or hard copies of all or part of this work for personal or classroom use is granted without fee provided that copies are not made or distributed for profit or commercial advantage and that copies bear this notice and the full citation on the first page. To copy otherwise, to republish, to post on servers or to redistribute to lists, requires prior specific permission and/or a fee.
DAC 13, May 29 - June 07 2013, Austin, TX, USA.

---

[1] 192 bits for the 1-out-of-8 coding. 480 bits for the chain-like neighbor coding.

STATISTICAL TEST	1-out-of-8		chain-like neighbor		decoupled neighbor	
	P-VALUE	PROPORTION	P-VALUE	PROPORTION	P-VALUE	PROPORTION
Frequency	0.013689	122/125	0.000072 *	125/125	0.000003 *	115/125 *
BlockFrequency	0.166594	125/125	0.000000 *	125/125	0.050764	120/125
CumulativeSums (m-2)	0.231636	121/125	0.000000 *	125/125	0.000000 *	119/125 *
CumulativeSums (m-3)	0.059743	122/125	0.000000 *	125/125	0.000000 *	118/125 *
Runs	0.002320	117/125 *	0.000000 *	0/125 *	0.302788	120/125
LongestRun	0.000603	123/125	0.000000 *	62/125 *	0.000062 *	124/125
ApproximateEntropy	0.000001 *	117/125 *	0.000000 *	0/125 *	0.000001 *	119/125 *
Serial (forward)	0.004904	124/125	0.000000 *	1/125 *	0.070160	116/125 *
Serial (backward)	0.552185	125/125	0.000000 *	117/125 *	0.192277	123/125

TABLE I

NIST TEST RESULTS WITH RESPECT TO THE RANDOM SEQUENCES GENERATED BY **1-OUT-OF-8** CODING, **CHAIN-LIKE NEIGHBOR** CODING AND **DECOUPLED NEIGHBOR** CODING, WHERE '*' MARKS A FAILURE.

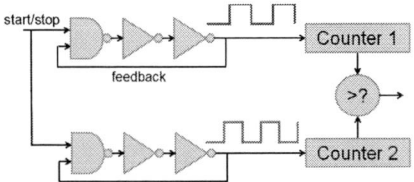

Fig. 1.   The physical structure of a RO PUF [2]

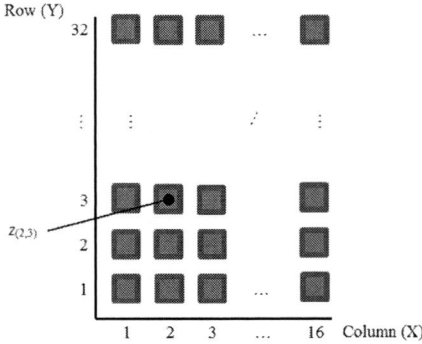

Fig. 2.   The placement of 512 ROs as a 16 (columns) by 32 (rows) array; for site $RO_{x,y}$, its running frequency is denoted as $z_{x,y}$

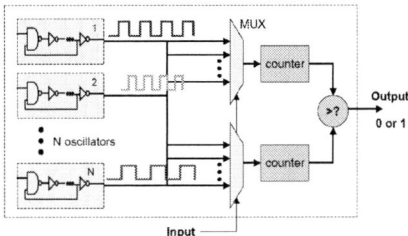

Fig. 3.   The hardware structure of the 1-out-of-8 RO PUF [2]

total test sequences are significant, the 'PROPORTION' of a test fails; otherwise, it passes. Furthermore, the *P-values* of all the test sequences are expected to be uniformly distributed. To examine this, each of the 9 tests also calculates the *P-value* of the *P-values* using the χ^2 statistic. The 'P-VALUE (OF P-VALUES)' of a test fails if the *P-value* of χ^2 is smaller than 0.0001; otherwise, it passes.

Table I clearly indicate that none of the 125 PUF sequences can be deemed 'ideally random' and therefore cannot be used for critical cryptography applications. In this paper, we study the architectures of the current RO PUFs and the aforementioned 4-step secret generation workflow. We propose to add one more step before the secret selection, which we refer to as *entropy distillation*, such that we can decouple the unwanted systematic variation from the desired stochastic variation. This will enable us to generate PUF secret from the true random variation and pass the NIST randomness tests. Although RO PUF is used as the example, our approach can be applied to other PUFs to enhance their security (of course, for those that are more resistant to systematic variation, it will be less effective). As for the implementation of the proposed method, one can either implement the distiller with hardware or rely on a secure ALU for the data process.

II. PRELIMINARY

A. Basics on RO PUF

A RO PUF extracts fabrication variations through comparing the frequencies of ring oscillators that are identically designed. As depicted in Figure 1, the basic RO PUF consists of two ring oscillators, followed by two counters and one comparator at the end. When the start/stop control signal is asserted, the two ROs start to oscillate until the control signal is negated. The result of the race between the two ROs is determined by fabrication variations. During the course of the race, the two counters count the number of logic cycles run by

the respective RO. At the end of race, the comparator outputs a binary result based on the two counter values, say,

$$x_{i,j} = \begin{cases} 1 & \text{if } RO_i < RO_j \\ 0 & \text{otherwise.} \end{cases} \quad (1)$$

To generate a secret in greater length, a RO PUF typically implements hundreds of ROs and placed them in a 2-dimensional array. As illustrated in Figure 2, the dataset we use implements ROs in 16 (columns) by 32 (rows) on each their 125 FPGAs [6].

B. 1-out-of-8 Coding

The result of the race between the same two ROs may differ when the environmental conditions change. For example, a RO running faster than its peer at low temperature can actually be slower than the same peer at high temperature [2]. To prevent this, the 1-out-of-8 coding scheme uses a multiplexer to select the pair with the largest frequency difference out of 8 RO pairs as depicted in Figure 3. For the two dimensional RO array illustrated in Figure 2, we may generate one random bit for each row j from its 16 ROs $RO_{1,j} \ldots RO_{16,j}$. However, in order to generate more random bits to better serve the statistical test purpose, we made a variation by forming two blocks $(RO_{1,j} \ldots RO_{8,j})$ and $(RO_{9,j} \ldots RO_{16,j})$ for each row. The 8 ROs in the same block are referenced by a 3-bit index: $000,001,010 \ldots 111$, and the index to the fastest RO is output as the random bits. This way we generate 6 random bits for each row.

C. Chain-like neighbor coding

Another well-known pairing strategy is the chain-like neighbor coding, which consists of two design principles: 1) place ROs as close as possible and 2) pair ROs located adjacent to each other. In the two dimensional setting of Figure 2, we may derive 15 random bits for each row j by pairing $(RO_{1,j}, RO_{2,j})$, $(RO_{2,j}, RO_{3,j})$, $(RO_{3,j}, RO_{4,j}) \ldots (RO_{15,j}, RO_{16,j})$.

III. SECURITY ANALYSIS

A. Failure Cause 1: Chain Dependency

The high failure rate of the chain-like neighbor coding can be attributed to the non-independent comparison chain. Take 3 ROs RO_A, RO_B and RO_C for example, two random bits are generated by comparing RO_A with RO_B and RO_B with RO_C. As we know, to pass NIST test for randomness, the random sequence is expected to demonstrate no significant deviation from the probability mass function (p.m.f) of tossing a fair coin twice, i.e., the 4 possible outcomes '00', '01', '10' and '11' are expected to occur equally with probability 1/4 is not the case for the two bits we generate from the 3 ROs. Let RO_i also denote the running frequency of RO_i. For three ROs, their running frequencies can have six different orders: $RO_A < RO_B < RO_C$, $RO_A < RO_C < RO_B$, $RO_B < RO_A < RO_C$, $RO_B < RO_C < RO_A$, $RO_C < RO_A < RO_B$, $RO_C < RO_B < RO_A$, where each order happens with probability 1/6 when the running frequencies are random and identical and independent distributed (i.i.d.). According to Eqn. (1), both bits x_{AB} and x_{BC} will be equally likely to be '0' or '1'. However, the 2-bit data $x_{AB}x_{BC}$ will be '00', '01', '10', and '11' with probabilities 1/6, 1/3, 1/3, and 1/6 respectively. This means that '01' or '10' occur twice as frequent as '00' or '11', clearly not the p.m.f. of the ideal random sequences.

A simple solution to fix this problem caused by the chain dependency is to break the chain such that each RO will only be paired with its neighbor once as follows: $(RO_1, RO_2), (RO_3, RO_4) \ldots (RO_{2i-1}, RO_{2i}) \ldots$. We refer to this as *decoupled neighbor coding*. Apparently this is less efficient than the original chain-like neighbor coding. For example, when there are n ROs, the chain-like neighbor coding will generate $n-1$ bits, but the decoupled neighbor coding can only generate $\frac{n}{2}$ bits. However, even with this hardware cost, we are unable to produce true random sequence. As the last two columns in Table I show, the decoupled neighbor coding scheme helps the original chain-like neighbor coding

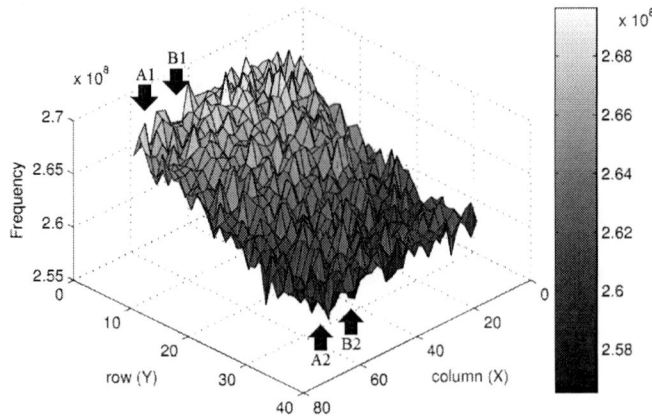

Fig. 4. The across-die frequency topology of a RO array. The roughness of the surface represents the random variation while the slope represents the systematic [11]

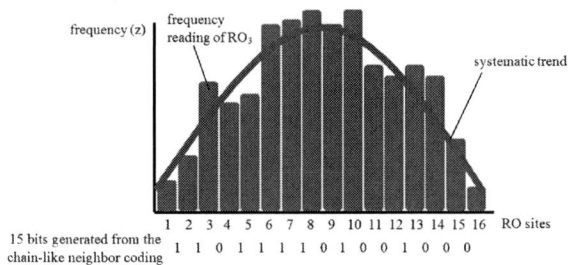

Fig. 5. Illustration of the impact from systematic variation even after decoupling.

in passing four out of the nine 'P-VALUE (OF P-VALUES)' tests, and improves the 'PROPORTION' tests too. But it still fails half of the NIST statistical randomness tests.

B. Failure Cause 2: Spatial Correlation

Chain dependency does not exist in the 1-out-of-8 coding so to investigate the cause of the failures of the 1-out-of-8 coding, we investigate the raw data from the physical measurement of fabrication variation. Figure 4 shows how the fabrication variation of the semiconductor process portrays: the roughness of the surface (random variation) is superimposed upon a spatial trend (systematic variation). With the existence of systematic variation, the 'random' bits generated by RO PUF arrays may have very low min-entropy[2], which means that they may not be secure for cryptographic purpose. For example, as we can see in Figure 4, along the row (Y), the RO's frequency tends to increase as the Y index increases. Therefore, for the two bits x_{A1A2} and x_{B1B2} generated by two pairs of ROs (A1, A2) and (B1, B2), they are more likely to be '0' at the same time. Similarly, on a different die where the frequency tends to increase along the row (Y), these two bits are more likely to be '1' at the same time. This means that the systematic variation (the spatial correlation for two ROs on the same row (Y) in this case) will render the probability

---

[2]the min-entropy of a discrete random event x with possible outcomes $1 \ldots n$ and corresponding probabilities $p_1 \ldots p_n$ is $H_\infty(X) = \min_{i=1}^n (-\log p_i)$.

of $x_{A1A2} = x_{B1B2}$ much higher than 0.5, making them not as random nor independent.

While spatial correlation may explain the reason why none of the coding strategies in Table I passes all tests, it is interesting to note that in fact this threat has been reported in the chain-like neighbor coding, where they attempt to let the systematic effect cancel out with each other, extracting secrecy out of the random effect [9]. Similar principles have been used in [12]. However, the results we have in Table I indicate that such treatment is not sufficient to pass the NIST randomness tests. We postulate that a small remnant of the systematic variation can still be captured by the tests, causing the above failures. To illustrate this, we consider a hypothetical frequency characterization of 16 ROs as shown in Figure 5. Based on the chain-like neighbor coding, these 16 ROs will generate the following 15-bit sequence: 1101,1110,1001,000. The first bit is a '1' because $RO_1 < RO_2$ and the third bit is a '0' because $RO_3 > RO_4$, and so on. If our proposed decoupled neighbor coding is used, we will only have 8-bit data: 1011,1000. Although in both cases we have about the same number of 0's as the number of 1's, there is a clear trend that 1's are more likely to be in the first half of the sequence and the 0's in the second half. When we fit the frequencies into the curve in Figure 5, we see clearly the systematic trend of 'going up slope' first and then 'going down slope', which causes 0's and 1's not distributed uniformly in the sequences. Finally, we mention that this systematic trend can stay undiscovered when one tallies the total number of 0's and 1's or calculates the inter-die uniqueness [2].

IV. SYSTEMATIC VARIATION ELIMINATION

A. The Causes of Process Variation

The main causes of the systematic variation can be attributed to equipment and process non-uniformity such as the focus shift of photolithography, the gradient of thermal annealing, dissimilar interactions between circuit layout and the chemical mechanical polishing process [13], [14]. The random component, on the other hand, accounts for the difference between the model estimates and the observed data; its constituents include atomic-level stochastic phenomena such as random dopant profiles, measurement errors and any unidentified patterns [15], [14].

It is important to clarify that our goal is not to build a new variation model. Indeed, the proposed distiller does not require the accuracy of the variation model to be as high as those for power or performance driven applications. In the rest of this section, we will illustrate how the distiller can improve PUF data's randomness using the simple polynomial regression model.

B. Regression-based Distiller

We believe that one of the main causes of the failures in randomness tests for the RO PUF generated sequences is the systematic process variation. We propose to model such variation and thus remove them to build RO PUF sequences based on the true random part of the process variation. This

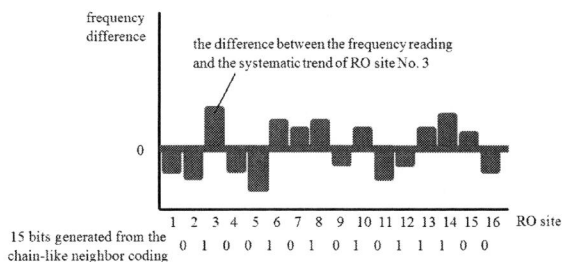

Fig. 6. The distilled random fabrication variation after the systematic trend is removed.

idea can be illustrated by the example in Figure 6. It reports the frequency information of the same 16 RO PUF as in Figure 5 except that the Y-axis now shows the difference between each RO's frequency and the systematic trend (the bell-shape curve in Figure 5). When we use the same chain-like neighbor coding, the 15-bit sequence becomes 0100,1010,1011,100. Compared to the original sequence 1101,1110,1001,000 in Figure 5, we don't see the 'unrandomness' in having most 1's in the first half and most 0's in the second half of the sequence.

When we apply polynomial regression models to capture the systematic variation trend, higher order models have better accuracy and generate smaller residual terms. While they may lead to sequences that are more random and secure, they incur more computational cost and more importantly, the small magnitude of the residual terms can cause difficulties in error correction phase and damage the efficiency of RO PUF. For example, Figure 7 shows the histograms of the random variation of a data set (see the result section for detailed description of the data) after regression models with polynomials of degrees 0 to 6 are applied. Clearly we see as the order increases, the number of ROs whose frequencies are far from the center decreases quickly, yielding a smaller variance. Nevertheless, they all appear normal distribution and it is difficult to judge which model is the best choice without running the standard randomness tests. Therefore, our goal is to find the polynomial regression model in minimal order that can successfully distill the ideal random variation. We propose to conduct this distillation procedure after the 'fabrication variation extraction' phase. The remaining question is in the next 'secret selection' phase that how to build the RO PUF sequence based on the residual terms, or the true random variations.

Suppose we have a 2-dimensional array of ROs placed in r rows and c columns (see Figure 2), there are many ways to define RO PUF bits from the distilled RO frequency information. For example, in our implementation of the 1-out-of-8 coding, each row generates $\frac{c}{8} \times 3$ bits; in the chain-like neighbor coding, we have $c - 1$ bits from each row; in the decoupled neighbor coding, this number reduces to $\lfloor \frac{c}{2} \rfloor$. Of course, instead of focusing on each row, we can define RO PUF bits by comparing the ROs in the same column. In addition to these three coding schemes, we study the following two generic sequences, S and T, to gauge if there is still any

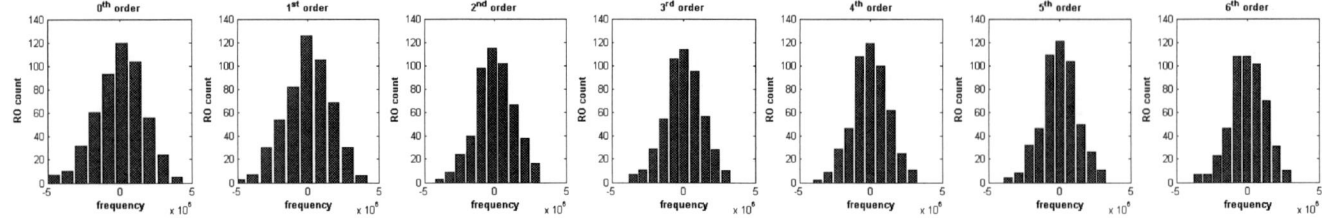

Fig. 7. The histogram of the distilled random variation after applying 0th through 6th-order polynomial regression to the dataset of Chip No. 1.

trace of spatial correlation in the distilled random component:

$$S = X_1, \ldots, X_{l_X}, \ldots, X_{L_X} \text{ where } X_{l_X}$$

$$= \begin{cases} 0 & \text{if } z_{u_X, v_X} \leq z_{u_X + \lfloor \frac{c}{2} \rfloor, v_X} \\ 1 & \text{otherwise} \end{cases} \quad (2)$$

$$T = Y_1, \ldots, Y_{l_Y}, \ldots, Y_{L_Y} \text{ where } Y_{l_Y}$$

$$= \begin{cases} 0 & \text{if } z_{u_Y, v_Y} \leq z_{u_Y, v_Y + \lfloor \frac{r}{2} \rfloor} \\ 1 & \text{otherwise} \end{cases} \quad (3)$$

where in (2), $u_X = ((l_X - 1) \mod \lfloor \frac{c}{2} \rfloor) + 1, v_X = \lfloor (l_X - 1)/\lfloor \frac{c}{2} \rfloor \rfloor + 1, 1 \leq l_X \leq L_X = r \times \lfloor \frac{c}{2} \rfloor$; similarly in (3), $u_Y = \lfloor (l_Y - 1)/\lfloor \frac{r}{2} \rfloor \rfloor + 1, v_Y = ((l_Y - 1) \mod \lfloor \frac{r}{2} \rfloor) + 1, 1 \leq l_Y \leq L_Y = c \times \lfloor \frac{r}{2} \rfloor$.

Intuitively, S and T are formulated by cutting each row (or column in T) in the RO array into two equal halves, pairing up ROs in the two halves, and comparing their residual variation terms. Recall that the principle in neighbor coding is to pair up ROs that are next to each other in order to reduce the systematic variation. In S and T, we have purposely done the opposite to pair up ROs that are far from each other to amplify the effect of systematic variation in order to test the effectiveness of the proposed regression-based entropy distiller. In the next section, we will report our detailed findings on such randomness tests.

V. RESULTS AND ANALYSIS ON RANDOMNESS TESTS

In this section, we conduct standard NIST randomness tests to validate that the proposed regression-based entropy distiller will improve the randomness of the RO PUF sequences. We use the test bench in the public domain which consists of the frequency characterization of 125 RO PUFs implemented on 125 Xilinx Spartan-3 FPGAs, where 512 ROs were placed on each FPGA as shown Figure 2 [6].

A. The Polynomial Regression Models

We first report the systematic variation distillation procedure and then results. For each chip, we apply regression models of different orders to its 512 averages of frequency readings. In the 0th order, the systematic variation is the average of the 512 averages. In the 1st order linear model, we observe that the ROs have higher frequency as their Y coordinates (in Figure 2) decrease. As we use higher order polynomials, it starts to show trend similar to Figure 4.

Figure 8 show the random variation after distillation. We see the radial pattern close to the center for the 0th and 1st models, which is known as the 'bull's eye' and a clear indication of

'non-randomness'. However, it vanishes in the cases of 2nd model and beyond. This suggests us that polynomials of 2nd degree or higher should be used.

B. NIST Randomness Tests

There are nine randomness tests in the NIST statistical test suite applicable to the length of our test sequences: Frequency Test, Block Frequency Test, Cumulative Sums Test (with block size $m = 2$ and $m = 3$), Runs Test, Longest Run Test, Serial Test (both forward and backward) and Approximate Entropy Test. According to [10], empirical results have to be interpreted in two forms of analysis: First, the proportion of sequences passing a test shall be above a minimum rate, 0.96 in our case, i.e., to pass 120 sequences out of a sample size of 125 sequences at significance level $\alpha = 0.01$. Secondly, the P-values of all the random sequences shall be uniformly distributed. Based on χ^2 Goodness-of-Fit Test, the underlying distribution is deemed uniform if the P-value of the P-values is equal or greater than 0.0001 for a population of 125 sequences. Whenever either of these two approaches fails, further tests based on a different sample space will help clarify whether the failure is a statistical anomaly or a clear non-randomness.

We now report the detailed test results on the generic S-sequence, T-sequence, and the sequences generated by the coding schemes of 1-out-of-8, chain-like neighbor, and decoupled neighbor.

1) S-sequence and T-sequence: The 512 ROs will generate a 256-bit S-sequence and a 256-bit T-sequence. The S-sequence and T-sequence for NIST randomness test are 32000 bits long obtained by concatenating the 125 such 256-bit sequence from the 125 chips.

As the 0th-order section shows, random sequences generated without entropy distillation fail miserably for both forms of analysis 'PROP. (PROPORTION)' and 'P-VAL. (P-VALUE OF P-VALUES)'. This strongly suggests the existence of systematic variation in the raw data. The failure rate decreases sharply when applied with 1st-, 2nd- or 3rd-order distiller in the case of S and with 2nd- or 3rd-order distiller in the case of T.

Unfortunately, there is at least one failure with respect to S, though the failure is only slightly below the cutting value. In such a boarder case where a weak existence of systematic variation is inferred, further investigation with different dataset is necessary to conclude the entropy source, i.e., RO PUF plus the distillation model, 'good' or 'bad'. If simply taking the sum of failure rates with respect to S and T, either 2nd- or 3rd-order distillers can be considered optimal.

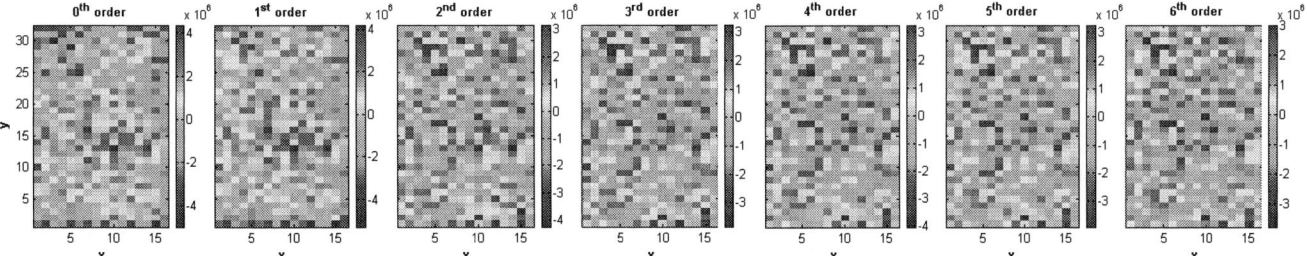

Fig. 8. The distilled random variation after applying 0th through 6th-order polynomial regression to the dataset of Chip No. 1.

Finally, we mention that the pass rate of the 'P-VALUE OF P-VALUES' analysis drops when applied with a model in 4th order or beyond. A further investigation reveals that this is caused by model over-fitting. This also indicates that the 2nd or 3rd order model should suffice for our purpose.

2) 1-out-of-8 Coding: For our implementation of the 1-out-of-8 coding, a 3-bit index '000', '001'... or '111' is generated by pointing to the fastest RO out of 8 consecutive ROs on the same row, i.e., 192 bits per chip or $125 \times 192 = 24000$ bits for the test sequence.

The 1-out-of-8 coding does very well even without the entropy distillation with only one clear 'P-VALUE' failure and two marginal 'PROPORTION' failures. The best linear model can fix all these three failures, but introduced a different 'P-VALUE' failure which is marginal. Also, distillers of 4th order and higher pass all the tests and can be deemed 'good'. However, the 2nd and 3rd order models fail about half of the tests. We suspect that this is caused by the fact that we are collecting three bits at a time from the 3-bit index of the eight ROs.

3) Neighbor Coding: In the case of the chain-like neighbor coding, 15 bits are generated per row by pairing up with row neighbors, which yields 480 bits per chip. Thus, the length of the test sequence is $125 \times 480 = 60000$ bits.

None of the polynomial distiller makes meaningful improvement in the randomness of the test sequence. This phenomenon aligns with our expectation that the failures are caused by the intrinsic chain dependencies of the pairing strategy rather than spatial correlation. Moreover, consider our treatment to this problem, the decoupled neighbor coding, the 1st order linear model is capable of helping it to pass all the randomness tests.

VI. CONCLUSION

The systematic component of fabrication variation has long posted a security threat to RO PUFs. This work provides experimental data to demonstrate that none of the current coding schemes can pass all the NIST randomness tests. To address the issue, we propose a family of entropy distillers based on polynomial regression. We affirm their effectiveness in improving the randomness of the PUF output.

Acknowledgement: This work is supported in part by the Air Force Office of Scientific Research under grant FA95501010140 and by the National Science Foundation of China under grant 61228204.

REFERENCES

[1] M. van Dijk, B. Gassend, D. Clarke and S. Devadas, "Silicon physical random functions," *Proceedings of 9th ACM Computer and Communications Security Conference (CCS)*, Nov. 2002.

[2] G. E. Suh and S. Devadas, "Physical unclonable functions for device authentication and secret key generation," *Proceedings of 44th ACM/IEEE Design Automation Conference (DAC) pp. 9–14*, Jun. 2007.

[3] W. B. D. Holcomb and K. Fu, "Initial sram state as a fingerprint and source of true random numbers for rfid tags," *Proceedings of the Conference on RFID Security 07*, Jul. 2007.

[4] Q. Chen, G. Csaba, P. Lugli, U. Schlichtmann, and U. Ruhrmair, "The bistable ring puf: A new architecture for strong physical unclonable functions," *Proceedings of 4th IEEE International Workshop on Hardware Oriented Security and Trust (HOST)*, Jun. 2011.

[5] C.E. Yin and G. Qu, ,"LISA: Maximizing RO PUFs Secret Extraction," *Proceedings of 3rd IEEE International Workshop on Hardware Oriented Security and Trust (HOST)*, Jun. 2010.

[6] A. Maiti and P. Schaumont, "A large scale characterization of ro-puf," *Proceedings of 3rd IEEE International Workshop on Hardware Oriented Security and Trust (HOST)*, Jun. 2010.

[7] C.E. Yin and G. Qu, "Temperature-Aware Cooperative Ring Oscillator PUF," *Proceedings of 2nd IEEE International Workshop on Hardware Oriented Security and Trust (HOST)*, Jun. 2009.

[8] M.-D. Yu and S. Devadas, "Secure and robust error correction for physical unclonable functions," *IEEE Journal of Design & Test Computers, Vol. 27, Issue 1*, Jan. 2010.

[9] A. Maiti and P. Schaumont, "Improving the quality of a physical unclonable function using configurable ring oscillators," *Proceedings of 19th IEEE International Conference on Field Programmable Logic and Applications (FPLA)*, Sep. 2009.

[10] A. Rukhin, J. Soto, J. Nechvatal, and et al, "A statistical test suite for random and pseudorandom number generators for cryptographic applications," *NIST Special Publication 800-22 Revision 1a*, Apr. 2010.

[11] P. Sedcole and P. Y. K. Cheung, "Within-die delay variability in 90nm fpgas and beyond," *Proceedings of 16th IEEE International Conference on Field Programmable Technology (FPT) pp. 97–104*, Dec. 2006.

[12] X. Wang and M. Tehranipoor, "Novel physical unclonable function with process and environmental variations," *Design, Automation & Test in Europe (DATE)*, Mar. 2010.

[13] B. E. Stine, D. S. Boning, and J. E. Chung, "Analysis and decomposition of spatial variation in integrated circuit processes and devices," *IEEE Transactions on Semiconductor Manufacturing, Vol. 10, Issue 1, pp. 24–91*, Feb. 1997.

[14] K. Bernstein, D. J. Frank, A. E. Gattiker, W. Haensch, B. L. Ji, S. R. Nassif, E. J. Nowak, D. J. Pearson, and N. J. Rohrer, "High-performance cmos variability in the 65-nm regime and beyond," *IBM Journal of Research and Development*, vol. 50, no. 4.5, pp. 433–449, 2006.

[15] B. E. Stine, T. Maung, R. Divecha, and et al, "Using a statistical metrology framework to identify systematic and random sources of die- and wafer-level ild thickness variation in cmp processes," *International Electron Devices Meeting*, pp. 499–502, Dec. 1995.

[16] C.E. Yin, G. Qu, and Q. Zhou, "Design and Implementation of a Group-based RO PUF, Design, Automation and Test in Europe (DATE13), March 2013.

On the Convergence of Mainstream and Mission-Critical Markets

Sylvain Girbal[†], Miquel Moretó[*,⊙,‡], Arnaud Grasset[†], Jaume Abella[‡], Eduardo Quiñones[‡],
Francisco J. Cazorla[‡,Φ], Sami Yehia[Ψ,1]

[†]Thales Research & Technology, France
[⊙]International Computer Science Institute, Berkeley, USA
[Φ]Spanish National Research Council (IIIA-CSIC), Spain

[*]Universitat Politecnica de Catalunya, Spain
[‡]Barcelona Supercomputing Center, Spain
[Ψ]Intel Corporation, USA

ABSTRACT

The computing market has been dominated during the last two decades by the well-known convergence of the high-performance computing market and the mobile market. In this paper we witness a new type of convergence between the mission-critical market (such as avionic or automotive) and the mainstream consumer electronics market. Such convergence is fuelled by the common needs of both markets for more reliability, support for mission-critical functionalities and the challenge of harnessing the unsustainable increases in safety margins to guarantee either correctness or timing. In this position paper, we present a description of this new convergence, as well as the main challenges and opportunities that it brings to computing industry.

Categories and Subject Descriptors

B.8 [**Performance and Reliability**]: Miscellaneous; D.2.8 [**Metrics**]: Performance measures

General Terms

Measurement, Performance, Reliability

Keywords

Mission Critical, High Performance, Quality of Service

1. INTRODUCTION

The computing market has been dominated during the last two decades by the convergence between high-performance (HP) computing market pursuing for power efficiency and the embedded/mobile market pursuing for more performance and functionalities. This evolution has lead to: 1) several open standards to control the balance between low power and high performance in current processors, 2) several APIs

---

[1]This work was done while Sami Yehia was a Research Engineer at Thales Research and Technology.

Permission to make digital or hard copies of all or part of this work for personal or classroom use is granted without fee provided that copies are not made or distributed for profit or commercial advantage and that copies bear this notice and the full citation on the first page. To copy otherwise, to republish, to post on servers or to redistribute to lists, requires prior specific permission and/or a fee.
DAC '13, May 29 - June 07 2013, Austin, TX, USA.

to control the low-power hardware features from the software layers (e.g. Operating System, OS) and 3) processor core designs that can be easily retargeted to provide different performance-power design points depending on the target market.

In this position paper we present our witnessed new type of convergence between the mission critical market (such as the avionic, automotive, healthcare and robotic) and the mainstream consumer electronics market. Such convergence is fuelled by the increasing requirements of the mission-critical market for performance and functionalities and the growing needs of the consumer electronics market (such as mobile phones) to embed more critical functionalities and interact with other critical systems (such as cars and health monitoring systems). At the same time, we are confronted with the challenge of harnessing the unsustainable increases in safety margins to guarantee either correctness or timing (worst-case execution time margins). We also show that this new convergence brings opportunities and challenges in the way hardware and software have to be designed to deal with mission-critical requirements.

In Section 2 we show the main motivation behind the new convergence after revisiting the past convergence between low power and high-performance markets. Section 3 shows challenges and opportunities brought by this new convergence, while Section 4 presents novel timing analysis initiatives. We conclude in Section 5.

2. CONVERGENCE OF MAINSTREAM AND MISSION-CRITICAL MARKETS

We witness a new type of convergence between the the mainstream and the mission-critical (MS-MC) markets resulting from the previous convergence between the mobile and high-performance markets, as shown in Figure 1.

2.1 The Past Convergence: Low Power and High Performance Markets

In the 1990s, the mobile market was a niche market guided only by low-power constraints with very low performance and functionality requirements. Three factors motivated the convergence between mobile market and the mainstream market, see Figure 1. First, the increase in performance and functionality requirements of the mobile market made low-power processors including high-performance features. Second, in the 90's and 00's processors increased their performance at the expense of a rapid increase in power dissipation. For instance, Intel processors increased power from

Early 90's	Now	Future

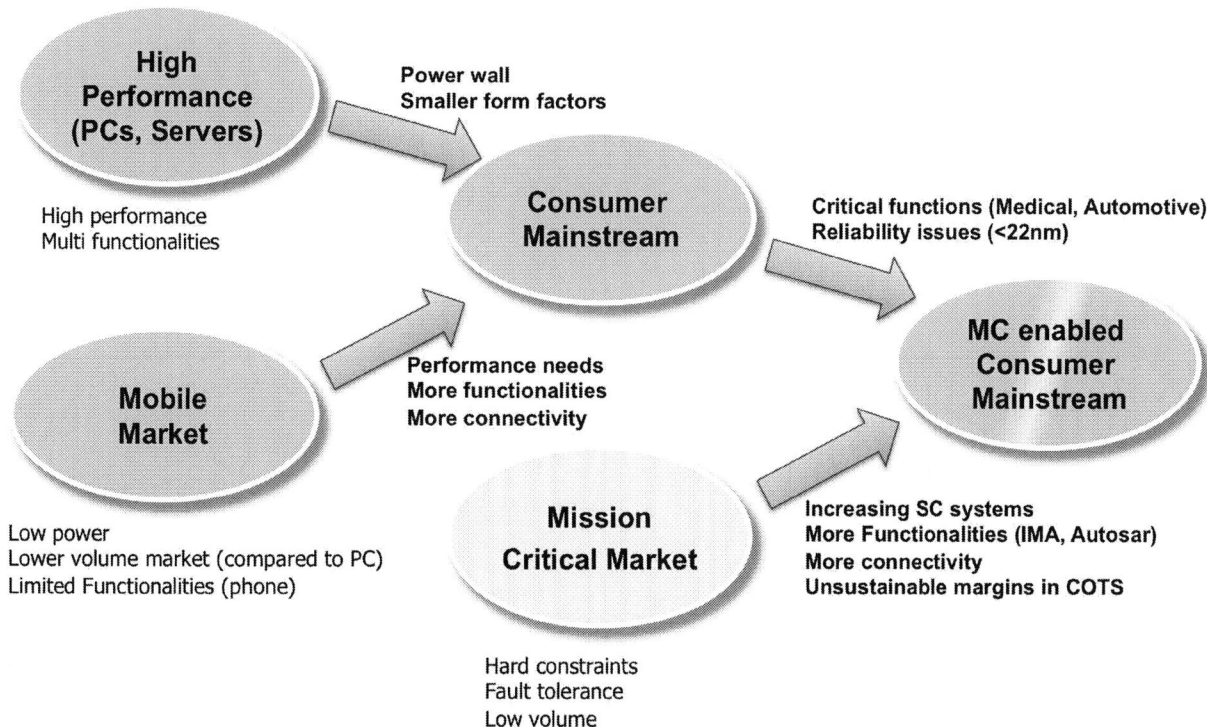

Figure 1: The past convergence between mobile and high-performance markets as a reference to explain our devised new convergence between mainstream and mission-critical markets.

15W (Pentium [10]) to 115W (Pentium4 [11]) in less than 10 years. Limitations in heat dissipation as well as the increase in the energy cost led to a U-turn in high-performance processor design. Low-power features were increasingly incorporated to provide the highest performance under a given power envelope, e.g. Intel released Pentium M, right after Pentium 4, that dissipated up to 27W. And third, processors need several years and astronomical costs to be designed, verified and validated. Hence, reusing designs across domains drastically decreases costs.

After several years of convergence, several standards have been defined to balance power and performance in all high-performance processors. For instance, the Advanced Configuration and Power Interface (ACPI) [15] defines several low-power states as well as means to allow the OS to control the current state, so that different tradeoffs between performance and power can be used dynamically. This enables the same processor design being used in mobile and HP markets despite their different constraints.

Recently, market convergence has extended also towards the lowest performance/power range of the mobile segment. Processors initially designed for handheld devices such as smartphones are widely used in high-performance segments such as tablet and netbook segments. For instance, some servers have been released soon based on, for instance, the ARM11 and Intel Atom processors [12] given that they can provide higher performance/density (e.g., within the volume and power envelope of a particular server).

2.2 Mission-Critical Application Market

The mission-critical system designers have traditionally relied on low performance components to address the diverse, but limited in number, service requirements of mission-critical applications. In recent years this design approach is challenged, and found to be limited and cost-ineffective, by the growing need for more functionalities and performance, while still ensuring stringent reliability and safety requirements as well as tight hard real-time constraints. We observe an upcoming increase - unprecedented in diversity and level - in performance requirements in the avionic, automotive and medical domains, between others. For instance:

- In the avionic domain future generation of navigation systems (four-dimension trajectory and N-Fly zones management systems [7]) in aircrafts, 5th generation cockpit (with the possible introduction of single-crew capable airliners [5]), and collision avoidance are all functionalities with increasing performance requirements. This can be inferred from Figure 2, where code size is used as a proxy for the demand for computational power [6].

- In the automotive domain, future driver assistance systems for vehicles requires a supercomputer (*high-performance computing*) level of performance for processing data from cameras, radar, LIDAR and other sensors to detect and decide about warnings and in critical situations autonomous breaking or steering. Both

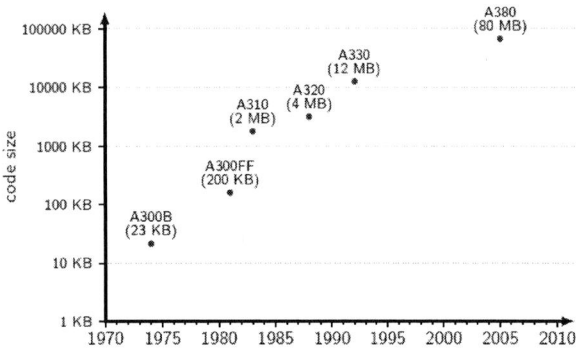

Figure 2: Growing of embedded software size in avionics

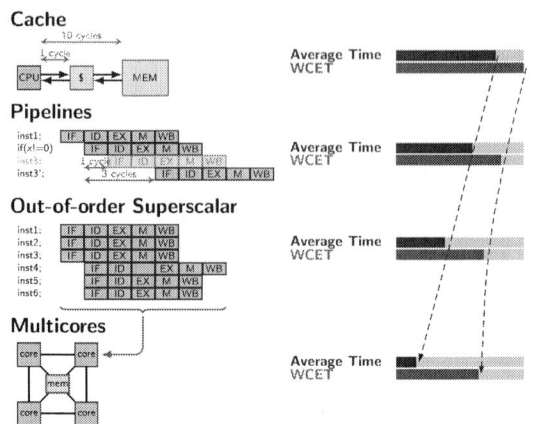

Figure 3: Evolution of average execution time and WCET under different processor setups with increasing high-performance features

domains are overarched by stringent requirements of lower fuel consumption, CO_2 emissions, with high-speed real time processing and fault-tolerance.

- In the medical market, the microcontrollers used in the implantable devices, pacemakers, infusion pumps and digital hearing aids have also significant fault-tolerance, low-power and high-performance requirements.

These required levels of performance can only realistically be realised by deploying high-performance hardware acceleration features such as caches or multiple cores per chip and smaller technology nodes (i.e. smaller transistor sizes). These technology advances enable (i) reducing energy consumption and temperature, and (ii) integrating more hardware functionalities per chip, which in turn enables running more system functionalities per chip, thus reducing overall SWaP (system size, weight and power consumption) costs.

However, because of the increasing complexity of those features, analysis of the worst-case scenarios for mission-critical applications is extremely difficult, especially with the emergence of multicores in the embedded market. One typical example in the time domain, is the calculation of worst-case execution time (WCET) [16]. The extra complexity of emerging hardware complicates the analysis timing behavior of mission-critical systems in two ways.

First, there is an increasing gap between the actual WCET and the average best effort average performance, as depicted in Figure 3. The use of complex hardware such as caches, pipelining, superscalar execution and multicores, reduces much more average performance than WCET. For instance, in the case of a multicore with a hardware shared resource, the execution time of a task is much smaller when no other cores use that shared resource. If the tasks running on the other cores, introduce high load in the shared resource, each task is going to suffer a significant slowdown due to these inter-task interferences. While on average this effect may not be significant, the actual WCET of the task is much more affected.

Second, the WCET estimates that can be provided by current timing analysis tools are becoming increasingly pessimistic with respect to the actual WCET. This is so because current analysis approaches, mainly static timing analysis, require acquiring sufficient knowledge of all factors of influence, at hardware and software level, on the timing behavior of the program. The addition of high-performance features, increases the complexity of acquiring the knowleged required to carry out the analysis. In order to appreciate this phenomenon, consider a cache model with a Least Recently Used (LRU) replacement policy. The accuracy in predicting the hit/miss outcome of a memory access depends on knowing the full sequence and addresses of the previous memory accesses made by the program up to the point of interest. This information is needed to build a complete and correct representation of the cache state so as to determine the hit/miss outcome of memory accesses with sufficient accuracy. Any reduction in the available knowledge, e.g. when the addresses of some memory accesses are unknown, leads to a rapid degradation of the tightness of the WCET estimation. In fact, partial knowledge can lead to results as inaccurate (i.e. inordinately pessimistic) as those obtained with no information at all. As a result, many of the analysis techniques in the literature rely on over-provisioning the architecture. At software levels all the hardware resources that cannot be analyzed, such as the cache are disabled. Aggressive architectural techniques such as branch prediction and data path forwarding are also disabled to ease computing WCET estimates [16]. Also, some manufacturers just turn off all but one core in a multicore platform if highly-critical system components exist.

It is also the case that the relatively low volume of the mission-critical market makes the development of specific processors meeting the specific requirements of mission-critical applications extremely costly because of the high Non-Recurring Engineering (NRE) costs.

2.3 Mainstream Consumer Electronics

The microprocessor systems in mainstream or consumer electronics market are also facing challenges. At the hardware level, semiconductor technology poses serious challenges in the design of future microprocessor systems. First the shrinking technology nodes (14nm in 2013-2014 according to Intel [3]) will make the processing elements subject to variability constraints, wear-out and soft errors such that the reliability of the basic computing elements will no longer be guaranteed. Thus, reliability will be a first-order design constraint similar to power and performance. Faults at the transistor and circuit levels will be more common than ever such that microprocessor and system designers will have to provide additional fault tolerance features at several design layers.

At the application level, we foresee an increasing demand in autonomy, decision-making and artificial intelligence. This demand will very likely be overarched by stringent requirements in term of safety, quality of service (QoS) and hard real time constraints. Examples of such applications are future domestic robots, healthcare applications [4], intelligent cars, augmented reality, human++ applications [14]. Also the increasing demands in connectivity will naturally lead to a situation where mobile devices directly connect with more mission-critical systems such as cars and medical devices. Hence, providing mobiles with mission-critical capabilities will enable a new type of applications for end users [9].

Our view is that most future applications will require some form of hard real-time behavior for at least part of their operation. For domestic robots, cars, planes, telesurgery, and Human++ implants, it is clearly necessary to impose limitations on the delay between sensing and giving the appropriate response. For parallel applications, it is important that all processes running in parallel have balanced execution time in order to maximally exploit the parallel resources of the platform, and limit the synchronization overhead. Especially on heterogeneous multicores, being able to accurately estimate execution times is crucial for performance optimization.

2.4 The Next Convergence

Overall, we observe a second challenging convergence trend, akin to the convergence between the computing PC market and mobile market during the last two decades (See Figure 1), between mainstream consumer electronics and mission-critical markets. On the one hand, the performance requirements of current and future mission-critical applications, as well as their connectivity with the less critical mobile electronic devices, make the use of low performance or over-provisioned architectures not a viable solution anymore. On the other hand, applications such as health care monitoring will start to hit the mainstream market and the increasing safety margins for reliability issues start to make the available transistors and available performance more and more difficult to exploit. One key common aspect in mission-critical and mainstream systems is that they are both relying on over-provisioning some of their resources in order to make some guarantees.

3. CHALLENGES OF THE MC AND MS MARKETS CONVERGENCE

Next, we attempt to pave the way toward achieving convergence between the mission-critical and the consumer electronic markets so as to reduce the high NRE cost of the mission-critical market and provide efficient mission-critical capabilities to the general purpose market.

To achieve such a convergence we need to enrich existing mainstream processing architectures with capabilities that provide non-functional guarantees to address the mission critical needs, without over-provisioning the processing architecture or diminishing its efficiency. In other words, we want to provide the user a (Quality of Service) QoS according to her or his functional and non-functional requirements.

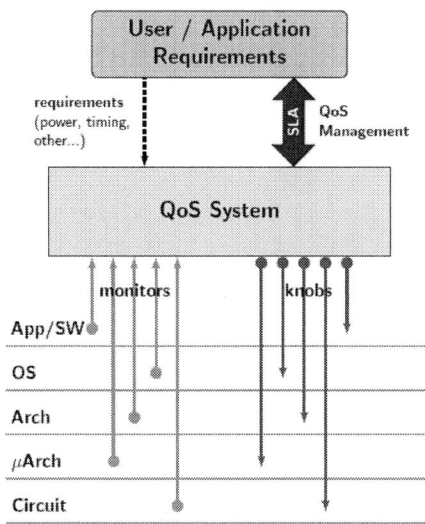

Figure 4: Integral QoS Approach

3.1 User Requirements

Because QoS objectives are often application specific, the envisioned QoS approach provides an efficient and general interface that can satisfy QoS objectives over a range of applications. It is required to define the necessary API and language support to express the requirements. A common factor for all type of applications is that the application requirements provided to the QoS system have to be architecture independent. This removes the need of the application programmer to provide a different set of metrics for every target architecture, guaranteeing performance and requirements portability. In particular our envisioned QoS system combines the requirements coming from mission-critical-like applications and general-purpose applications, possibly running them concurrently (mixed criticality systems). In the case of mission critical applications, we need much stronger guarantees and a pure best effort approach is not sufficient. Several approaches have previously advocated for the introduction of non-functional properties in programming languages, especially domain specific ones [8, 13]. Those languages consider timing characteristics, which are of paramount importance in the development of the application as well as the architecture.

Historically QoS standards have been widely used in networked systems in order to ensure high-quality performance of data-flow, especially for real-time multimedia applications, without reactively expanding or over-provisioning the networks. The advent of multicore, reliability issues and the demands of mission-critical applications make the adoption of QoS and resource reservation mechanisms in multiprocessors systems the only way to efficiently use these systems and meet these mission critical demands.

3.2 QoS Service-Level Agreements

Establishing a Service-Level Agreement (SLA) between the application and the OS is the key to offering a guaranteed QoS to the user, especially for mission critical systems. For example, a user may want to guarantee the completion of a task within a specified deadline. In such a situation, the application will attempt to establish an SLA with the

OS before starting the task. The OS will accordingly allocate and block the necessary resources needed to guarantee the completion of the task within the deadline. In the case where the OS is unable to allocate the necessary resources to offer such guarantee, the application will be notified before starting the task and can trigger a backup mechanism. The advantage of such a scheme is that the designer can analyze its application based on a reasonable assumption of available resources, environmental conditions and performance degradation (due to wear-out effects) and not on the assumption of the worst-case scenario where no resources are available for the task or the assumption of an over-provisioned system that will guarantee the availabilities of resources.

3.3 Example of non-Functional Requirement: Specifying Timing QoS Requirements

In mission-critical systems, temporal aspects of their behavior are part of their specification. That is, their correctness depends on the correctness of their functional behavior and also its timing behavior: the time at which the results are produced. This is so because mission-critical systems interact with their environment, which also changes with time.

The software component of real-time functions are usually implemented as an endless loop that makes some data sensing, processing and actuation over an actuator. In each iteration of the loop, a new instance of the program/task is generated. These instances are called jobs. Jobs have a deadline prior to which its execution must end.

Timing properties of mission-critical systems are usually specified by means of the concept of *deadline misses*. In particular, the percentage of deadline misses stands for the percentage of instances that end after their deadline. Depending on the particular system, the system *value function* may be sharp or progressive. Sharp functions are those where the success is an "all or nothing" function. Conversely, progressive functions have a decreasing value after the deadline.

We follow the approach presented by Bernat et al. [2] for specifying the QoS time requirements of systems that can tolerate occasional losses of deadlines, such as mission-critical. QoS time requirements cannot be adequately specified with a single parameter, for example, with the percentage of deadline misses. This is so because the percentage of missed deadlines is an average measure and does not allow to determine the temporal behavior of the missed deadlines. For instance, for a task with 10% deadline misses, we lack the information about whether the task misses one deadline every ten, ten consecutive deadlines every 100, etc. This temporal frequency of misses is critical, as some real-time systems are more sensitive to consecutive deadline misses.

Therefore, it is necessary to provide metrics to express the QoS timing requirements of a system exact distribution of missed deadlines. This could be in the form of hit/miss patterns, e.g. (1 1 1 0 1 1 1 0) where '1' represents hit and '0' miss; or by specifying the minimum number of consecutive hit deadlines between two missed deadlines [2]. This can be further combined with progressive value functions.

3.4 Challenges and Opportunities at SW level

At software level, a promising research area are QoS API, Middleware and architecture designs, that allow applications or users to be provided a guaranteed level of service by effi-

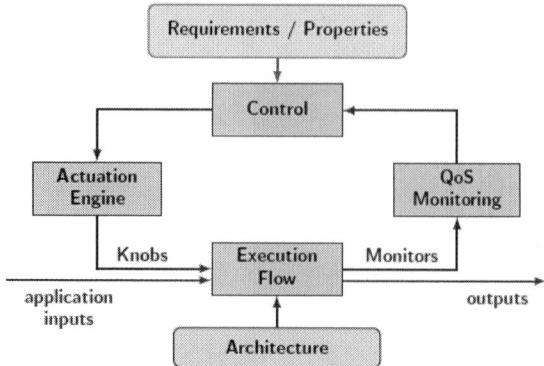

Figure 5: Detailed view of the Integral QoS Approach

ciently using the available resources, according to the above-mentioned SLA, and without over-provisioning the architecture to achieve the required level of service. This represents a major departure and different philosophy from previous design approaches that in general address QoS with narrow scope that is neither cost-effective nor scalable. We advocate for holistic approaches across computing layers in which QoS becomes part of the design approach at each computing layer. Similarly to Instruction Set Architectures (ISA) that provide the abstract interface between low level programming and the hardware implementation, interfaces at each layer must be able to express QoS requirements. QoS compliant designs will need to provide means for bidirectional communication for monitors to observe and knobs to control the system, as shown in Figure 4. Some primitive notions of QoS principles are present in existing systems (such as load balancing, dynamic voltage scaling, etc. [1]) but are not pervasive enough and not meant to provide QoS. In other words we advocate for defining the necessary services required for meeting non-functional requirements (or properties), the necessary micro-architectural, architectural, middleware and system support to efficiently implement these services, how these services are expressed in the application and how they are deployed over all layers down to the hardware.

3.5 Challenges and Opportunities at HW level

Monitors are used to collect the different parameters and state of the system periodically and verify that the tasks progress according to the requirements, as illustrated in Figure 5. The concept of monitors generalizes the standard concept of Performance Monitoring Counters (PMCs) in several aspects. Monitors provide much richer information than PMCs (e.g. information about the interaction between tasks), cover several metrics (not just performance) and provide feedback that allows deducing part of the future behavior of the running applications.

Depending on the behavior of the system, the knobs regulate the different tasks and system behavior to ensure a guaranteed QoS. Monitors and knobs can exist at different layers of the system. For example hardware counters at the architecture levels can be used to monitor cache misses, job queues at the OS level can be used to monitor the system load, and temperature sensors at the circuit level can be used to monitor the temperature. Similarly knobs can consist of migrating tasks (OS level), shutting down some cores for power saving (architecture level) or performing some DVFS actions at the circuit level.

4. TIMING ANALYSIS INITIATIVES

As mentioned in Section 2, there is an increasing performance demand in the mission-critical market, such as the avionic, automotive or the medical domain. However, the benefits that the additional computational power of advanced hardware/software features can potentially provide cannot be fully realized unless accompanied by analysis techniques that enable their use. In particular, in this section we focus on providing some approaches to analyze the timing behavior of future mission-critical/mainstream systems.

Several studies work in that direction by proposing different techniques to analysis the time behavior of applications running in a Commercial Off-The-Shelf (COTS) multicore processor having a varying degree of hardware shared resources. Annex I describes several of the main software-only solutions proposed in that line.

In addition to changing timing analysis techniques, new advanced hardware/software can be delivered with a twofold objective: (i) Improving performance and (ii) Enabling simple means to provide QoS guarantees. In that direction, Probabilistic Timing Analysis (PTA) has emerged as an attractive solution to respond to the demand for trustworthy timing analysis in the critical real-time embedded system domain in the face of more performance aggressive processors. Compared to conventional static timing analysis, PTA requires much less knowledge about the program and the processor, which allows users to obtain high performance and timing guarantees with little effort. This is a significant advantage considering that (1) market competition drives programs to become increasingly complex and (2) processor manufacturers increasingly use black-box IP components. In Annex II, we provide more details about PTA techniques and their advantages for the MS-MC convergence.

5. CONCLUSIONS

We envision a new type of convergence between the mission-critical market and the mainstream consumer electronics market: both markets have common needs for increased support for mission-critical functionalities together with an increase in safety margins to guarantee either correctness or timing. To reach these goals future computer designs should implement an integral QoS approach in which QoS becomes part of the design approach at each computing layer. The hardware has to provide features that enable software analyzing and controlling processor internal resource allocation through intelligent knobs and sensors. These hardware features could be enabled from the processor, depending on the criticality and performance needs of the target market, so that the same processor design can be used for different mission-critical requirements. The software stack should be able to provide interfaces that allow user QoS requirements to reach the appropriate layer as well as the proper knobs to provide non-functional guarantees to address the mission-critical needs.

We have shown how QoS requirements should be specified and measured using metrics beyond percentage of deadline misses to better assess the non-functional requirements of a MS-MC system.

Finally, providing guarantees on the timing behavior, as a representative of the non-functional requirements of the system, will also introduce changes aimed at controlling the interaction between software components. The probabilistic approach offers an interesting alternative path to simplify the analysis of the timing behavior of complex MS-MC systems.

Acknowledgements

M. Moretó, J. Abella, F. J. Cazorla and E. Quiñones have been partially supported by the PROARTIS FP7 European Project under grant agreement number 249100, the Spanish Ministry of Science and Innovation under grant TIN2012-34557 and the HiPEAC Network of Excellence. M. Moretó is supported by a Fulbright/MEC Fellowship. E. Quiñones has also been supported by the Spanish Ministry of Science and Innovation under the grant Juan de la Cierva JCI-2009-05455. The authors thank all the anonymous reviewers for their constructive comments and suggestions.

6. REFERENCES

[1] A. Bartolini et al. A virtual platform environment for exploring power, thermal and reliability management control strategies in high-performance multicores. In *GLSVLSI*, 2010.

[2] G. Bernat, A. Burns, and A. Llamoso. Weakly hard real-time systems. *IEEE Trans. Comput.*, 2001.

[3] M. Bohr. Silicon technology leadership for the mobility era, September 2012. Intel Developer Forum.

[4] S. Cantrill. Computers in patient care: the promise and the challenge. *Commun. ACM*, 53(9):42–47, 2010.

[5] A. Doyle. Thales outlines thinking on single-crew cockpits. *Flight-globale*, July 2010.

[6] G. Edelin. Embedded Systems at THALES: the Artemis challenges for an industrial group. In *ARTIST Summer School in Europe*, 2009.

[7] European Organisation For The Safety of Air Navigation. Study report on avionics systems for the time frame 2007, 2011 and 2020. *Eurocontrol*, 2004.

[8] T. Henzinger, B. Horowitz, and C. Kirsch. Giotto: A time-triggered language for embedded programming. *Proceedings of the IEEE*, 91:84–99, 2003.

[9] M. Holzbock et al. Evolution of aeronautical communications for personal and multimedia services. *IEEE Communications*, 2003.

[10] Intel Corporation. Datasheet: Intel pentium processor 75/90/100/120/133/150/166/200, 1997.

[11] Intel Corporation. Datasheet: Intel pentium 4 processors 570/571, 560/561, 550/551, 540/541, 530/531 and 520/521 supporting hyper-threading technology, 2005.

[12] Intel Newsroom. Chip shot: Intel Atom processors codenamed "Centerton" to power first HP's extreme low-energy production servers, 2012.

[13] E. A. Lee. Computing needs time. *Commun. ACM*, 52(5):70–79, 2009.

[14] J. Penders et al. Human++: Emerging technology for body area networks. *VLSI-SoC: Research Trends in VLSI and Systems on Chip*, pages 377–397, 2008.

[15] White Paper. Intel® Turbo Boost Technology in Intel® Core™ Microarchitecture (Nehalem)Based Processors, 2008.

[16] R. Wilhelm et al. The worst-case execution-time problem—overview of methods and survey of tools. *ACM Transactions on Embedded Computing Systems*, 7(3), 2008.

APPENDIX

A. TIMING ANALYSIS OF COTS MULTI-CORE PROCESSORS

The mission-critical domain industries have been using Commercial Off-The-Shelf (COTS) architectures to reduce the non-recurring engineering costs (NRE) and time-to-market (TTM) [1], while trying to deal with the variability in runtime to satisfy real-time constraints.

The recent shift to multicore in the embedded COTS market should effectively allow the industry to cope with the exponential increase in performance needs. However this shift worsens the runtime variability problem as contentions on shared hardware resources bring new variability sources.

A.1 Timing Analysis Techniques

A common practice to guarantee the deadlines of a safety-critical application is to determine the application Worst-Case Execution Time (WCET). This WCET computation usually relies on analysis tools based on static program analysis tools based on extensive program analysis [17, 14], detailed hardware model, as well as measurement techniques through execution or simulation [6]. However, those WCET analysis tools are not currently able to determine the exact WCET: they are providing an upper bound, introducing some safety margins as depicted in Figure 6.

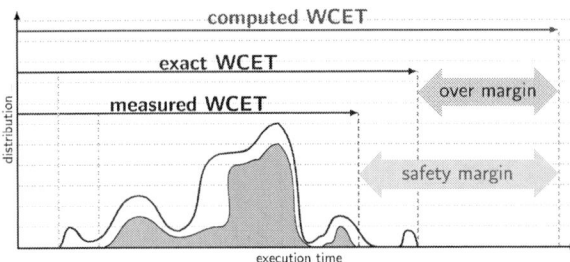

Figure 6: WCET estimation and the over-estimation problem

A measurement-based approach for single-threaded architectures computes the WCET estimate by multiplying the longest observed execution time by a safety margin, usually provided by an expert with understanding of both the target hardware architecture and the reference applications [11]. This method has been successfully used in the past to determine WCET estimate of applications running on single-threaded processors with a moderate difference between measured and computed WCET.

A.2 Analysis Technique Limitations

A direct extension to the measurement-based analysis presented in Figure 6 to multicore COTS processors would consist of running several reference applications simultaneously on the same processor, and monitoring the execution time for each application in the workload.

During the last decades, despite all the improvements in the WCET estimation domain [8, 3], the over-estimation remained mostly constant as the predictability of the architecture decreased [17], making the use of WCET analysis tools difficult for real industrial programs running on multicore COTS architectures [7, 11] for the following reasons: (1) WCET analysis of real industrial programs with a vast number of possible execution paths is a challenging task. (2) The implementation of accurate hardware models for new architectures requires a significant effort and a detailed description of the hardware, which is not always available. (3) Possible interference on shared hardware resources among co-running tasks significantly increases the complexity of timing analysis, forcing it to have a full knowledge of co-running tasks at software level, and detailed resource contention models at hardware level.

A.3 Quantifying Variability on Multicores

Before proposing alternative timing analysis techniques coping with multicore COTS, it is important to quantify the runtime variability for such systems.

To perform this study, we used the Freescale P4080DS platform detailed in Table 1, a complex 8-core architecture, with private L1 instruction & data caches and a private L2 unified cache. The eight cores are connected through a proprietary CoreNet Fabric to two shared 1MB L3 caches, each connected to a DDR memory controller.

Core	8 Power Architecture e500mc at 1.5GHz
Pipeline	7-stage pipeline, superscalar, out-of-order
Distributed L1 caches	32kb, 8-way associative, 64-byte line, PLRU
Distributed L2 cache	128kb, 8-way associative, 64-byte line, PLRU
Shared L3 cache (x2)	1MB, 32-way associative, 64-byte line, PLRU
Memory controller	Two DDR memory controllers
Interconnect	CoreNet Coherency Fabric

Table 1: Freescale P4080 specifications

Figure 7 illustrates the runtime variation for some applications of the MiBench benchmark suite [4] during 1000 successive execution iterations on the P4080 platform. Each application runs standalone on the barebone platform, the other cores being idle to minimize variability sources.

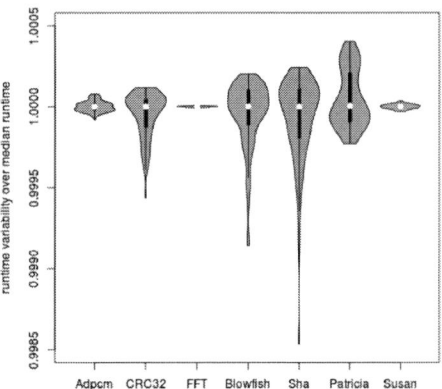

Figure 7: Variability of mibench applications running standalone on the 8-core P4080 platform

Each violin plot from Figure 7 is showing the runtime distribution around the median runtime for this benchmark. The width of a violin plot for a specific runtime is proportional to the population ratio with this particular runtime. If Figure 7 shows quite different variability schemes, the variation remains very small (below 1%).

Figure 8 quantifies the impact of co-running benchmarks. Each mibench application is now run concurrently with 2 benchmarks dedicated at stressing the interconnect resource.

Each violin plot of Figure 8 illustrates the runtime variation of each application around the previously computed

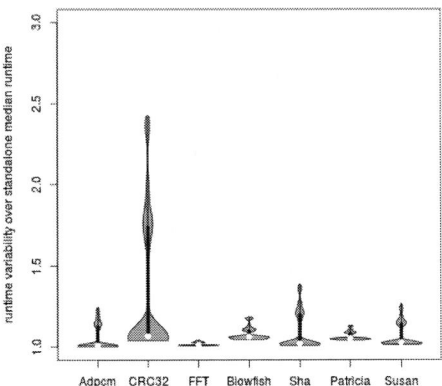

Figure 8: Variability of mibench applications running with 2 co-running benchmarks

standalone median. If the average variation remains quite low, the impact on the worse case is much more significant with an average variation of x1.34 (+35%) and a maximum variation of x2.42 (+140%) to be compared with the previous variation of 1%.

A similar study [12] from EADS shows on the same architecture that using actual WCET analysis techniques forces the industry to multiply the WCET value by a value close to the number of cores being used. This is of the same order of magnitude than the overall potential performance gain, and will provide no performance benefits over a single-core deployment. Therefore, particular care should be taken for WCET analysis techniques to scale for multicore systems.

A.4 Hardware Alternatives

Above mentioned limitations have motivated studies that analyze if changes in hardware can facilitate the effective timing analysis for multicore architectures. Several hardware proposals [13, 5, 16, 10, 9] aim at easing the computation of composable WCET bounds of tasks running on multithreaded architectures. However as COTS architecture are dominated by the consumer electronic market which does not actually share such predictability concerns, it is very unlikely that such costly mechanism to be integrated by general chip providers.

A.5 Modeling Alternatives

In order to provide timing guarantees for future mission-critical systems running on multicore COTS, new timing analysis mechanisms have to be developed. Recently a new set of approaches based on resource stressing benchmarks [15, 2] have been proposed, where the target application is tested against the resource stressing benchmarks to 1) identify which hardware resources are effectively shared, and to 2) identify the shared hardware resources each target application is sensitive to.

By combining this information for co-running benchmarks, we will be able to identify the benchmarks that will run smoothly together avoiding worst-case contentions. Such a scheme will allow us to better predict and control the variability of execution time on multicore COTS architecture by selecting interference friendly mappings of co-runners.

A.6 Additional Appendix Authors

Petar Radojković, Barcelona Supercomputing Center, Spain.

Jingyi Bin, Thales Research & Technology, France and Fundamental Electronic Institute, France.

A.7 References

[1] T. G. Baker. Lessons learned integrating COTS into systems. In *Proceedings of the First International Conference on COTS-Based Software Systems*, ICCBSS '02, pages 21–30, 2002.

[2] J. Bin, A. Grasset, D. G. Perez, S. Girbal, P. Bonnot, and A. Merigot. Controlling execution time variability using cots in for safety critical systems. *ACACES*, page 191, 2012.

[3] C. Ferdinand, F. Martin, C. Cullmann, M. Schlickling, I. Stein, S. Thesing, and R. Heckmann. Program analysis and compilation, theory and practice. chapter New developments in WCET analysis, pages 12–52. 2007.

[4] M. Guthaus, J. Ringenberg, D. Ernst, T. Austin, T. Mudge, and R. Brown. Mibench: A free, commercially representative embedded benchmark suite. In *Proceedings of the Workload Characterization, 2001*, WWC '01, pages 3–14, 2001.

[5] A. Hansson, K. Goossens, M. Bekooij, and J. Huisken. Compsoc: A template for composable and predictable multi-processor system on chips. *ACM Trans. Des. Autom. Electron. Syst.*, 14(1):2:1–2:24, Jan. 2009.

[6] R. Heckmann and C. Ferdinand. Verifying safety-critical timing and memory-usage properties of embedded software by abstract interpretation. In *Proceedings of the conference on Design, Automation and Test in Europe*, DATE'05, pages 618–619, 2005.

[7] R. Kirner and P. Puschner. Obstacles in worst-case execution time analysis. In *Proceedings of the 2008 11th IEEE Symposium on Object Oriented Real-Time Distributed Computing*, ISORC '08, pages 333–339, 2008.

[8] R. Kirner, I. Wenzel, B. Rieder, and P. Puschner. Using measurements as a complement to static worst-case execution time analysis. In *Intelligent Systems at the Service of Mankind*, volume 2. Dec. 2005.

[9] H. Kopetz and G. Bauer. The time-triggered architecture. *Proceedings of the IEEE*, page 91, 2003.

[10] B. Lickly, I. Liu, S. Kim, H. D. Patel, S. A. Edwards, and E. A. Lee. Predictable programming on a precision timed architecture. In *Proceedings of the 2008 international conference on Compilers, architectures and synthesis for embedded systems*, CASES '08, pages 137–146, 2008.

[11] E. Mezzetti and T. Vardanega. On the industrial fitness of wcet analysis. In *Proceedings of the 11th International Workshop on Worst Case Execution Time Analysis (WCET2011)*. 2011.

[12] J. Nowotsch and M. Paulitsch. Leveraging multi-core computing architectures in avionics. *European Dependable Computing Conference*, pages 132–143, 2012.

[13] R. Obermaisser, H. Kopetz, and C. Paukovits. A cross-domain multi-processor system-on-a-chip for embedded real-time systems. *IEEE Trans. Industrial Informatics*, 6(4):548–567, 2010.

[14] P. Puschner and A. Burns. Guest editorial: A review of worst-case execution-time analysis. *Real-Time Systems*, 18(2/3):115–128, 2000.

[15] P. Radojkovic, S. Girbal, A. Grasset, E. Quiñones, S. Yehia, and F. J. Cazorla. On the evaluation of the impact of shared resources in multithreaded cots processors in time-critical environments. *TACO*, 8(4):34, 2012.

[16] T. Ungerer, F. Cazorla, P. Sainrat, G. Bernat, Z. Petrov, C. Rochange, E. Quinones, M. Gerdes, M. Paolieri, J. Wolf, H. Casse, S. Uhrig, I. Guliashvili, M. Houston, F. Kluge, S. Metzlaff, and J. Mische. Merasa: Multicore execution of hard real-time applications supporting analyzability. *IEEE Micro*, 30(5):66–75, 2010.

[17] R. Wilhelm, J. Engblom, A. Ermedahl, N. Holsti, T. Mitra, S. Thesing, D. Whalley, G. Bernat, C. Ferdinand, I. Puaut, R. Heckmann, F. Mueller, P. Puschner, J. Staschulat, and P. Stenström. The worst case execution time problem, overview of methods and survey of tools. *ACM Trans. Embed. Comput. Syst.*, pages 36–53, May 2008.

B. PROBABILISTIC TIME-ANALYSABLE COMPUTING SYSTEMS

Aggressive hardware acceleration features like deep memory hierarchies and many-cores; and software like complex motor control algorithms that require sophisticated mathematics [2], are needed to respond to the increasing demand for more software functionality due to the ever-rising proportion of system value that is delivered in software.

More powerful hardware/software also offers the opportunity to schedule a larger number of applications, while the use of many-cores allows co-hosting several critical and non-critical applications on a common powerful platform, providing a better performance/Watt ratio than a single core solution with similar performance. Such a reduction in the number of processors will also offer tremendous scope for cost reduction.

B.1 Existing Timing Analysis Techniques

The benefits that the additional computational power of these advanced hardware/software features can potentially provide cannot be realized in the mission critical market unless accompanied by analysis techniques that enables their use. In particular, there must be guarantees on the achievable performance. However, the introduction of complex hardware and software (1) challenges current practice in the timing analysis of real-time embedded systems; and (2) produces abrupt performance variations due to rather small variations in the program or the execution environment.

Timing analysis is a complex process, mainly due to the fact that the variations in execution time of programs is caused by the characteristics of the software itself, as well as by the hardware platform upon which the program is to run and the complex interaction between both. As a result, all characteristics of software and hardware must be thoroughly understood in order to provide meaningful time guarantees.

Static timing analysis (STA) techniques rely on the construction of a cycle-accurate model of the system and a mathematical representation of the application code which makes it possible to determine the timing behavior on that model. STA approaches have the important limitation that they are expensive to carry out requiring exhaustive knowledge of all factors, both hardware and software, that determine the execution history of the program under analysis. For instance, to determine whether an access to a cache hits/misses in cache, the addresses of previous accesses have to be known. The introduction of multicore architectures and the software running on top of them will dramatically increase this cost. It is also the case that multicore architectures may be subject to intellectual property restrictions or have huge documentation, likely without the level of detail required to make a cycle-accurate model, making it altogether impossible to use STA.

Alternatively, measurement-based timing analysis (MBTA) techniques have been devised to sort out the limitations of STA. MBTA relies on extensive testing performed on the real system under analysis using stressful, high-coverage input data, recording the the longest observed execution time; and adding to it an engineering margin to make safety allowances for the unknown. However, the engineering margin is extremely difficult – if at all possible – to determine, especially when the system may exhibit discontinuous changes in timing due to inter-task interferences in shared resources in many-core processors or pathological cache access pat-

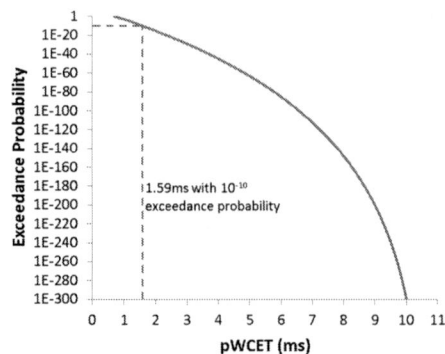

Figure 9: Synthetic example of a pWCET function

terns. Moreover, uncontrolled interactions of applications with mixed criticality levels further challenge the WCET estimates obtained with MBTA.

B.2 Probabilistic Timing Analysis (PTA)

New approaches have recently emerged to provide an alternative to those based on partitioning the resources. One of such approaches is the probabilistic approach [4][5][3][1]. PTA techniques enable probabilistic guarantees of timing correctness to be derived. For example, if the requirements placed on the reliability of a sub-system indicate that the probability of a timing failure must be less than 10^{-9} per hour of operation, then the PTA techniques aim to translate this reliability requirement into a probabilistic WCET (pWCET) for the sub-system. PTA effectively provides a continuum of WCETs for different confidence levels. Thus a sub-system may have a probability of 10^{-8} per hour of exceeding an execution time of 1.43ms, and probabilities of 10^{-9}, and 10^{-10} per hour of exceeding 1.51ms and 1.59ms, respectively (see Figure 9). The absolute WCET may be 10ms and can occur with a probability of 10^{-300} per hour.

The main idea of PTA techniques is that for future real-time systems, such probabilistic guarantees offer significant advantages over deterministic approaches which attempt to make absolute guarantees, severely limiting the use of advanced hardware features and inevitably offering significantly lower performance guarantees. PTA allows filling the gap between the execution times observed and the WCET attaching trustworthy probabilities to the execution times in between, thus enabling the (safe) use of lower execution time bounds. For instance, in the example before the effective utilization of the system is increased by a factor above 5x if the pWCET used for scheduling is as reliable as the hardware on which the program runs. Overall, using the absolute WCET that can occur with a probability of 10^{-300} per hour makes no sense if our system (e.g., an aircraft) is not resilient against higher probability catastrophic events such as an asteroid impact, which can occur with a probability higher than 10^{-30} per hour.

Under the probabilistic approach to time analysis, the timing behavior of certain hardware resources must be randomized, such as the cache placement and replacement [6]. Randomizing the timing behavior of hardware enables the use of PTA to analyze the timing behavior of the system. Randomization breaks many of the dependences between hardware and software components, simplifying the analysis. Techniques like *Extreme Value Theory* can be used to predict the timing behavior of applications [3].

PTA does not require randomizing the timing of all resources, but the timing behavior of resources can be characterized with a random variable, which is expressed as an *execution time profile* (ETP). In other words, each potential latency that a resource may take must have an associated probability. Those resources with constant latency, l, have a probability of 1 to have that latency and hence are also amenable to PTA. Consequently, PTA needs that any hardware resource has either (i) constant response time or (ii) different response times with an associated (true) probability each.

In order to treat a resource with PTA, the probability assigned to each latency in the timing vector of the ETP for that resource must be a *true* probability. The probability for a given latency is different from its frequency. This is best shown by an example. Consider a resource R_1 with $\overrightarrow{ETP_1} = \{\{t_1, t_2\}, \{0.5, 0.5\}\}$: each latency in the timing vector (t_1 and t_2) would have a true probability of occurrence of 0.5 if – in the implementation of that resource – on every request we flipped a coin and the request had latency t_1 if we saw heads and t_2 otherwise. In contrast, if for a deterministic stateful resource R_2 with the same potential latencies (t_1 and t_2) we *observed* that, for a given program, 50% of the requests take t_1 and 50% t_2, we would have a 50% observed frequency for each possible latency, but not a true 0.5 probability. This is so because for events that are strictly dependent on the history of execution, cumulative information on past events *cannot* be used to provide guarantees about the appearance of future events.

The example of the cache. Exploiting the execution history of the program, is one of the most common principles of processor design. A typifying example of this strategy is the cache. The use of caches is widespread in general-purpose processors because they dramatically improve average performance by exploiting locality (either temporal or spatial) in memory access patterns, reducing access times by several orders of magnitude. However, this strategy has an important downside: the execution time of programs, and their WCET, heavily depend on execution history, which challenges timing analysis. In particular, the combination of deterministic cache placement and replacement polices such as modulo placement and Least Recently Used (LRU) replacement make cache behavior to depend on the particular location of data and instructions in memory. A minor modification in such location may produce abrupt effects in performance by varying completely which cache lines compete for the space in each cache set.

Cache behavior imposes severe constraints on STA since all addresses must be known to provide tight WCET bounds. STA keeps track of the cache state on each access. However, dynamic memory, stack placement and operating system interaction among others make this process costly – if at all feasible. Any lack of information implies that accesses must be assumed to be missed and, even worse, assume that unknown accesses can occur in any cache set, thus evicting many cache lines from the guaranteed cache state tracked by STA. Alternatively, MBTA can run experiments to record the highest execution times. Unfortunately, there is no guarantee on whether those memory layouts leading to the WCET are included in the tests. This is a serious issue given that performance variation can be huge and those undesirable memory layouts may occur systematically once the system is deployed.

Conversely, PTA requires cache designs with random placement and replacement so that the timing behavior of caches does not depend on data and instructions location in memory. This has been proven to be feasible recently by means of either (i) hardware means such as parametric random placement [6] or (ii) software means such as stack and function placement randomization [7]. Those techniques make placement of objects in cache random so that each placement has a true probability of occurrence and hence, if measurement-based PTA (MBPTA) is used [3], observed execution times are relevant of those that will be observed in the system once deployed. Further, random placement and replacement have been proven to avoid abrupt performance variations, which facilitates providing QoS guarantees.

B.3 PTA benefits

The benefits of PTA can be summarized as follows:
System reliability is expressed in terms of probabilities for hardware failures, memory failures, software faults and for the system as a whole. PTA techniques fit and extend this probabilistic approach to timing correctness so that probabilistic guarantees can be provided holistically for the different system components. In that sense, PTA is in match with the aging and reliability behavior of the hardware itself, which common wisdom accepts may fail with a given (though very low) probability. The first steps towards adapting PTA to work in the presence of faults at hardware level have been presented in [8].

PTA allows obtaining trustworthy and tight WCET estimates at low cost, which is attractive for the mission critical market, especially for the mainstream one, where cost constraints for QoS features are more severe.

Based on the fact that PTA provides probabilistic WCET estimates for any target exceedance probability, it can be used to obtain WCET estimates for applications with different criticality levels at low cost. In other words, a single (low cost) approach can be used to provide the timing guarantees needed across mixed criticalities.

Overall, we regard probabilistic timing analysis as a technically viable and economically attractive solution to respond to the demand for providing guaranteed timing analysis in future mission-critical/mainstream systems in the face of more performance aggressive software and hardware.

B.4 References

[1] F. Cazorla et al. PROARTIS: Probabilistically analysable real-time systems. Technical Report to appear in ACM TECS, 2013.

[2] M. Copeland. Implementing Complex Motor Control Algorithms with a Standard ARM® Processor. 2012.

[3] L. Cucu-Grosjean et al. Measurement-based probabilistic timing analysis for multi-path programs. In *ECRTS*, 2012.

[4] S. Edgar and A. Burns. Statistical analysis of WCET for scheduling. In *RTSS*, 2001.

[5] D. Griffin and A. Burns. Realism in Statistical Analysis of Worst Case Execution Times. In *WCET Workshop*, 2010.

[6] L. Kosmidis, J. Abella, E. Quinones, and F. Cazorla. A cache design for probabilistically analysable real-time systems. In *DATE*, 2013.

[7] L. Kosmidis, C. Curtsinger, E. Quinones, J. Abella, E. Berger, and F. Cazorla. Probabilistic timing analysis on conventional cache designs. In *DATE*, 2013.

[8] M. Slijepcevic, L. Kosmidis, J. Abella, E. Quinones, and F. J. Cazorla. Degraded test mode for fault-aware probabilistic timing analysis. In *ECRTS*, 2013.

AUTHOR INDEX

Abadir, M. 847
Abella, J. 590, 1276
Abousamra, A. 237
Abraham, J. 711
Aceituno, P. 110
Acharyya, D. 410
Adeniyi-Jones, C. 583
Agosta, G. 570
Agrawal, P. 913
Akesson, B. 161
Alaghi, A. 945
Alexander, C. 219
Alle, M. 355
Alpert, C. 645
Amaru, L. 328, 868
Ambrose, J. 437, 931
Anagnostopoulos, I. 1154
Ancajas, D. 275, 721, 1084
Andalam, S. 665, 1013
Angione, C. 298
Asadi, H. 705
Asanovic, K. 528
Ascheid, G. 404
Asenov, A. 219
Atienza, D. 365
Auras, D. 404
Axer, P. 1184
Ayoub, R. 808, 1220
Badaroglu, M. 169
Baert, R. 169
Bai, K. 1023
Ballal, B. 169
Bamakhrama, M. 1170
Banerjee, D. 389
Banerjee, P. 62
Bank, J. 859
Barenghi, A. 570
Bario, P. 169
Bartolini, A. 11
Bartolini, D. 528, 538
Bartzas, A. 1154
Basu, A. 137
Batterywala, S. 181
Bauer, L. 695
Becker, M. 988
Benazouz, M. 17
Benini, L. 11, 104
Beretta, I. 365
Bernstein, G. 756
Bhadra, J. 847
Bhattacharya, S. 181
Bhunia, S. 420
Bird, S. 528
Bobba, S. 868
Bodenmiller, B. 986

Bodin, B. 17
Bombieri, N. 1075
Borkar, S. 1098
Brisk, P. 319
Bruschi, F. 365
Burleson, W. 86
Busseuil, R. 583
Cai, W. 1044
Calhoun, B. 1101
Calimera, A. 781
Carapezza, G. 298
Carlo, S. 775, 1234
Carloni, L. 348, 1075, 1160
Cattaneo, R. 538
Catthoor, F. 913
Cazorla, F. 1276
Chakrabarty, K. 307, 516
Chakraborty, K. 275, 721, 1084
Chakraborty, R. 410
Chakraborty, S. 665, 671, 686, 988
Chakradhar, S. 799
Chandrachoodan, N. 1257
Chandrasekar, K. 161
Chandrikakutty, H. 564
Chang, N. 680
Chang, S. 478
Chang, W. 665
Chang, Y. 24, 30, 36, 62, 175, 187, 1038, 1117, 1132
Charbon, E. 882
Chatterjee, A. 389
Chattopadhyay, A. 921
Chattopadhyay, S. 213
Chava, B. 169
Chen, D. 68
Chen, F. 878
Chen, G. 1138
Chen, H. 872
Chen, I. 878
Chen, J. 478, 899
Chen, T. 24, 1038
Chen, W. 847
Chen, X. 404, 808
Chen, Y. 42, 1038
Cher, C. 711
Chiang, C. 472
Chien, H. 24, 30
Chippa, V. 799
Chiprout, E. 459
Cho, H. 711
Choi, J. 889
Chou, D. 528
Choudhury, M. 334
Chow, W. 1044
Colmenares, J. 528
Cong, J. 54, 68, 78

AUTHOR INDEX

Cong, K. 201
Corbalan, M. 1107
Costanza, J. 298
Courant, Y. 503
Cozzens, B. 721
Craig, K. 1101
Cristal, A. 377
Croes, K. 169
Csaba, G. 756
Dally, W. 659
Daneshtalab, M. 269
Das, A. 815
Das, C. 247
Debole, M. 939
Derrien, S. 355
Dev, K. 510
Devarakond, S. 389
Dieny, B. 886
Dietrich, B. 988
Ding, H. 1003
Dousti, M. 291
Drechsler, R. 822
Du, Y. 653
Durelli, G. 538
Dutt, N. 695, 1226
Eads, G. 528
Ebrahimi, M. 705
Eklow, B. 516
Enz, C. 137
Ernst, R. 1184
Faeder, J. 48
Fahmy, S. 665
Fan, D. 763
Fan, J. 98
Fang, G. 958
Fang, S. 175
Fariborzi, H. 878
Fattah, M. 269
Faubet, P. 503
Feng, Z. 604, 624
Flynn, D. 1093
Franke, B. 145
Fu, B. 263
Fu, X. 576
Fummi, F. 1075
Gaillardon, P. 328, 868
Gao, J. 490
Garg, S. 1191
Garibotti, R. 583
Ge, Y. 396
Geier, M. 988
Georakos, G. 686
Ghosh, S. 992
Gielen, G. 872
Girault, A. 1013

Girbal, S. 1276
Givargis, T. 1059
Gluzman, B. 528
Goncalves, O. 886
Goossens, K. 161
Goswami, D. 671, 988
Gould, M. 145
Graeb, H. 225
Grasset, A. 1276
Gratz, P. 808
Grissom, D. 319
Gross, W. 383
Große, D. 822
Grossman, J. 859
Grupp, L. 1111
Gu, C. 453, 459
Gupta, P. 695
Gupta, R. 104
Ha, H. 889
Ha, S. 889
Hamzeh, M. 119
Han, L. 624
Han, S. 968
Han, Y. 263
Hao, K. 828
Haralampos-G., S. 503
Hartman, M. 913
Hashemian, M. 420
Hasholzner, R. 1178
Hassoun, S. 314
Hayes, J. 945
He, X. 1044
Henkel, J. 1, 110, 437, 695, 899
Herdt, V. 822
Hills, G. 746, 872
Ho, K. 36
Ho, T. 307
Ho, Y. 187
Hofmeyr, S. 528
Hornayoun, H. 1226
Hsu, M. 1038
Hsu, S. 193
Hsueh, C. 598
Hu, J. 808, 1031
Hu, M. 42
Hu, X. 756
Hu, Y. 576
Huang, C. 834, 840, 1038
Huang, P. 1117
Huang, T. 42, 1044
Hujsa, T. 17
Hutin, L. 878
Hwu, W. 68
Irwin, M. 939
Jahn, J. 899

AUTHOR INDEX

Jang, J. 951
Javaid, H. 151
Jha, N. 92
Jiang, I. 472
Jiang, J. 342
Jiang, L. 516
Jiao, D. 974
Jone, W. 207, 793
Jones, A. 237
Joo, D. 632
Jouppi, N. 769
Juels, A. 86
Jung, Y. 1160
Kahng, A. 231, 638
Kang, S. 638
Kapur, R. 213
Karakostas, V. 377
Karnik, T. 1098
Karri, R. 557
Karthik, A. 444
Kauer, M. 665, 671, 1065
Keckler, S. 659
Kestur, S. 939
Keval, A. 1107
Khalid, A. 921
Kim, H. 808
Kim, J. 632, 889, 951
Kim, M. 1031
Kim, T. 632
Kim, Y. 680
Kinsman, A. 853
Kishinevsky, M. 808
Kistler, M. 522
Kleeberger, V. 225
Ko, H. 853
Kobbe, S. 899
Koh, C. 1050
Kourouma, M. 583
Koushanfar, F. 86
Krishnawarny, S. 986
Kuan, Y. 24
Kuang, J. 484, 1044
Kubiatowicz, J. 528
Kumar, A. 1, 815
Kundu, S. 213
Kuruvilla, V. 1257
Lai, C. 834
Lam, K. 1044
Larnech, C. 410
Le, H. 822
Leblebici, Y. 868
Lee, H. 638
Lee, K. 314
Lee, R. 878
Lee, S. 522

Lee, Y. 598, 736
Lei, L. 201
Leupers, R. 404
Li, C. 945, 1138
Li, H. 42, 371
Li, J. 193, 478
Li, P. 78, 466
Li, T. 437
Li, X. 263, 453, 459, 497
Li, Y. 645, 968, 1050, 1249
Li, Z. 645
Liang, Y. 68, 1003
Liao, K. 193
Liljeberg, P. 269
Lim, S. 736, 1242
Limbrick, D. 736
Lin, G. 472
Lin, H. 466
Lin, S. 478
Lin, T. 62
Lingamneni, A. 137
Lio, P. 298
Lisk, D. 1107
Liu, B. 42
Liu, C. 1242
Liu, D. 129
Liu, H. 348, 974, 1075
Liu, I. 175
Liu, L. 129
Liu, R. 1138
Liu, S. 756
Liu, T. 169, 878
Liu, W. 645, 1050
Liu, Y. 207
Lu, H. 263
Lu, J. 1023
Lukasiewycz, M. 665, 671, 1065
Luk-Pat, G. 653
Luo, Y. 307
Ma, Q. 653
Macii, E. 781
Mackin, C. 746
Maggi, M. 570
Maggio, M. 538
Malachowsky, C. 659
Mallik, A. 169
Manoj, S. 1207
Mao, Z. 42
Marchi, M. 868
Marculescu, D. 48, 1191
Maric, B. 590
Markov, I. 1031
Masrur, A. 671
Melham, R. 237
Mercati, P. 11

AUTHOR INDEX

Mercha, A. 169
Micheli, G. 328, 868
Miller, B. 1059
Miloslavsky, A. 653
Min, S. 151
Mirkhani, S. 711
Mishra, A. 247
Mishra, V. 618
Miskov-Zivanov, N. 48
Mitra, S. 711, 746, 872
Mitra, T. 1003, 1198
Mohamed, F. 503
Mor, N. 528
Moreto, M. 528, 1276
Morvan, A. 355
Mukhopadhyay, S. 775, 1234
Mundhenk, P. 665
Munier-Kordon, A. 17
Muralimanohar, N. 769
Muthukaruppan, T. 1198
Mutlu, O. 247
Myers, C. 466
Nacci, A. 365
Najar, W. 980
Naranayaswami, S. 1065
Narasimhan, S. 548
Narayanan, V. 257, 939
Nassif, S. 695
Nath, R. 1220
Nathanael, R. 878
Nemirovsky, M. 377
Nickerson, J. 275
Nicolici, N. 853
Nicosia, G. 298
Niemier, M. 756
Niu, D. 769
Nowak, M. 1107
Ogras, U. 808
Oh, H. 889
Oliveira, R. 994
Onizawa, N. 383
Orshansky, M. 314
Ost, L. 583
Ou, H. 24, 30, 36
Ouyang, J. 257
Pagani, S. 899
Palem, K. 137
Pan, D. 334, 490, 1249
Pant, M. 1098
Papakonstantinou, A. 68
Parameswaran, S. 151, 437, 931
Parandhaman, A. 787
Park, J. 1160
Park, M. 939, 951
Park, S. 680

Paterna, F. 11
Paul, G. 921
Pedram, M. 285, 291
Pe'Er, D. 986
Pelosi, G. 570
Pendina, G. 886
Peng, L. 576
Peng, Y. 1242
Perre, L. 913
Petracca, M. 1160
Piaget, J. 1257
Piguet, C. 137
Pimentel, A. 907
Plosila, J. 269
Plusquellic, J. 410
Poncino, M. 781
Porod, W. 756
Potkonjak, M. 990
Prenat, G. 886
Pricopi, M. 1198
Puri, R. 334
Qiu, Q. 396
Qu, G. 1270
Quan, W. 907
Quinones, E. 1276
Radojcic, R. 1107
Raghavan, P. 913
Raghunathan, A. 92, 787, 799
Raghunathan, B. 1191
Rahimi, A. 104
Rai, D. 1144
Rajagopalan, S. 181
Rakossy, Z. 921
Ramasubramanian, S. 787
Ramesh, S. 671
Rana, V. 365
Ray, S. 828
Reda, S. 510, 1213
Regazzoni, F. 882
Rehman, S. 110, 437
Reineke, J. 1013
Rellermeyer, J. 522
Reparaz, O. 98
Rethy, J. 872
Riddet, C. 219
Robert, M. 583
Roman, E. 528
Roop, P. 1013
Rosales, R. 1178
Rosing, T. 11, 1220
Rostami, M. 86
Roy, D. 548
Roy, K. 763, 799
Roy, S. 275, 334, 721, 1084
Roychowdhury, J. 444

AUTHOR INDEX

Rozic, V. 98
Rutenbar, R. 497
Ryckaert, J. 169
Sacchetto, D. 868
Saeedi, M. 285
Sagstetter, F. 665
Sakurai, T. 875
Salodkar, N. 181
Santambrogio, M. 538
Santos, L. 994
Sapatnekar, S. 618
Sassatelli, G. 583
Saxena, S. 497
Schlichtmann, U. 225, 686
Schneider, R. 686, 988
Schor, L. 1144
Schurmans, S. 404
Sciuto, D. 365
Sekitani, T. 875
Sen, S. 389
Sengupta, I. 213
Shafaei, A. 285
Shafique, M. 1, 110, 437, 695
Shahzad, K. 921
Shanker, S. 665
Sharad, M. 763
Sharma, N. 913
Shaw, D. 859
Shiely, J. 653
Shrivastava, A. 119, 1023
Shulaker, M. 746, 872
Singh, A. 1, 815
Sinha, D. 1257
Sinha, R. 1013
Sironi, F. 538
Someya, T. 875
Song, H. 653
Song, T. 1242
Soudris, D. 1154
Stefanov, T. 1170
Steinhorst, S. 665, 671, 1065
Stoimenov, N. 257, 1144
Stojanovic, V. 878
Strojwas, A. 497
Sun, S. 453
Sun, Z. 371, 429, 793
Swanson, S. 1111
Sze, C. 645
Tahoori, M. 695, 705
Tajik, H. 1226
Takamiya, M. 875
Teich, J. 1178
Tessier, R. 564
Thiele, L. 257, 1144
Tiang, L. 931

Tomic, S. 377
Toms, T. 1107
Topham, N. 145
Towie, E. 219
Towles, B. 859
Trivedi, A. 775
Tsao, C. 1117, 1132
Tsao, H. 36
Tseng, H. 1111
Tsoutouras, V. 1154
Tu, K. 342
Turakhia, Y. 1191
Unnikrishnan, D. 564
Unsal, O. 377
Vahid, F. 1059
Valero, M. 590
Varga, E. 756
Venkataramani, S. 787
Venkataramani, V. 1198
Verbauwhede, I. 98
Verkest, D. 169
Villarreal, J. 980
Vishin, S. 1198
Viswanathan, N. 645
Visweswariah, C. 1257
Vrudhula, S. 119
Wagstaff, H. 145
Wang, B. 1178
Wang, C. 1123
Wang, F. 453
Wang, J. 612
Wang, K. 1207
Wang, L. 404, 847
Wang, X. 548, 557
Wang, Y. 78
Wang, Z. 729
Waszecki, P. 665
Wehn, N. 161, 695
Wei, H. 746
Wei, L. 429
Wei, S. 129, 990
Wei, Y. 645
Weis, C. 161
Wen, W. 478
Wolf, T. 564
Wong, H. 746, 872
Wong, M. 653
Wong, W. 1123
Woods, G. 510
Wu, B. 840, 968
Wu, C. 834
Wu, W. 371
Wuerges, E. 994
Xiao, B. 54
Xie, F. 201, 828

AUTHOR INDEX

Xie, Y. 257, 769
Xiong, X. 612
Xu, C. 769
Xu, Q. 207, 429, 516, 793, 1264
Xu, Y. 1178
Xu, Z. 808
Yan, G. 263
Yang, C. 1138
Yang, M. 1117, 1132
Yang, Z. 828
Ye, F. 516
Ye, R. 793
Ye, Z. 968
Yehia, S. 1276
Yin, C. 1270
Yin, S. 129
Yokota, T. 875
Young, E. 484, 1044
Yousofshahi, M. 314
Yu, B. 490
Yu, H. 1207
Yu, Y. 472
Yuan, F. 207, 429, 793, 1264
Yuan, K. 490
Yueh, W. 548, 1234
Yunge, D. 988
Zhai, J. 1170
Zhan, J. 257
Zhan, X. 1213
Zhang, C. 78, 939
Zhang, D. 404
Zhang, J. 429, 746
Zhang, M. 92
Zhang, P. 78
Zhang, W. 42, 453, 497
Zhang, Y. 396, 576
Zhao, X. 624
Zheng, Y. 420, 548
Zhou, B. 974
Zuber, P. 169